CHILTON®

ASIAN SERVICE MANUAL 2008 EDITION VOLUME IV

Mazda
Mitsubishi
Subaru
Suzuki

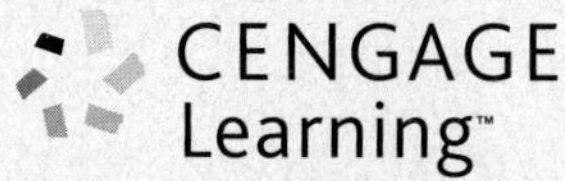

Australia • Brazil • Japan • Korea • Mexico • Singapore • Spain • United Kingdom • United States

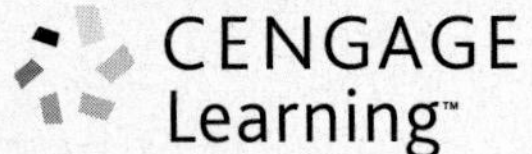

CHILTON®
Asian Service Manual
2008 Edition
Volume IV
Mazda, Mitsubishi, Subaru, Suzuki

Vice President, Technology & Trades Professional Business Unit:
Gregory L. Clayton

Publisher, Technology & Trades Professional Business Unit:
David Koontz

Director of Marketing:
Beth A. Lutz

Marketing Manager:
Jennifer Stall

Marketing Assistant:
Rachel Conover

Production Director:
Carolyn Miller

Editorial Assistant:
Jason Yager

Production Manager:
Andrew Crouth

Publishing Coordinator:
Paula Baillie

Sr. Content Project Manager:
Elizabeth C. Hough

Managing Editor:
Terry L. Blomquist

Editors:
Tim Crain
Eugene F. Hannon, Jr.
Marla Himes
Doug Lee
James R. Marotta
David G. Olson
Christine Sheeky

Graphical Designer:
Melinda Possinger

For more information contact:
Cengage Learning
Executive Woods
5 Maxwell Drive, PO Box 8007,
Clifton Park, NY 12065-8007
Visit us at **www.chilton.cengage.com**
For more learning solutions, visit **www.cengage.com**
For permission to use material from
the text or product, contact us by
Tel. (800) 730-2214
Fax (800) 730-2215
www.cengage.com/permissions

Cengage Learning products are represented in Canada by Nelson Education, Ltd.

ISBN10: 1-4283-2218-3
ISSN 13: 978-283-2218-9
ISSN: 1939-621X

NOTICE TO THE READER

Publisher does not warrant or guarantee any of the products described herein or perform any independent analysis in connection with any of the product information contained herein. Publisher does not assume, and expressly disclaims, any obligation to obtain and include information other than that provided to it by the manufacturer.

The reader is expressly warned to consider and adopt all safety precautions that might be indicated by the activities herein and to avoid all potential hazards. By following the instructions contained herein, the reader willingly assumes all risks in connection with such instructions.

The publisher makes no representation or warranties of any kind, including but not limited to, the warranties of fitness for particular purpose or merchantability, nor are any such representations implied with respect to the material set forth herein, and the publisher takes no responsibility with respect to such material. The publisher shall not be liable for any special, consequential, or exemplary damages resulting, in whole or part, from the readers' use of, or reliance upon, this material.

Printed in the United States of America
1 2 3 4 5 xx13 12 11 10 09 08

Table of Contents

Sections

Model Index

USING THIS INFORMATION

Organization

To find where a particular model section or procedure is located, look in the Table of Contents. Main topics are listed with the page number on which they may be found. Following the main topics is an alphabetical listing of all of the procedures within the section and their page numbers.

Manufacturer and Model Coverage

This product covers 2007–2008 Asian models that are produced in sufficient quantities to warrant coverage, and which have technical content available from the vehicle manufacturers before our publication date. Although this information is as complete as possible at the time of publication, some manufacturers may make changes which cannot be included here. While striving for total accuracy, the publisher cannot assume responsibility for any errors, changes, or omissions that may occur in the compilation of this data.

Part Numbers & Special Tools

Part numbers and special tools are recommended by the publisher and vehicle manufacturer to perform specific jobs. Before substituting any part or tool for the one recommended, you must be completely satisfied that neither your personal safety, nor the performance of the vehicle will be endangered.

ACKNOWLEDGEMENT

The publisher would like to express appreciation to the following vehicle manufacturers for their assistance in producing this manual. No further reproduction or distribution of the material in this manual is allowed without the expressed written permission of the vehicle manufacturers and the publisher. Mazda Motor Corporation, Mitsubishi Motors North America, Inc., Fuji Heavy Industries Ltd., including Subaru Motors Ltd, Suzuki Motor Corporation.

PRECAUTIONS

Before servicing any vehicle, please be sure to read all of the following precautions, which deal with personal safety, prevention of component damage, and important points to take into consideration when servicing a motor vehicle:

- Always wear safety glasses or goggles when drilling, cutting, grinding or prying.
- Steel-toed work shoes should be worn when working with heavy parts. Pockets should not be used for carrying tools. A slip or fall can drive a screwdriver into your body.
- Work surfaces, including tools and the floor should be kept clean of grease, oil or other slippery material.
- When working around moving parts, don't wear loose clothing. Long hair should be tied back under a hat or cap, or in a hair net.
- Always use tools only for the purpose for which they were designed. Never pry with a screwdriver.
- Keep a fire extinguisher and first aid kit handy.
- Always properly support the vehicle with approved stands or lift.
- Always have adequate ventilation when working with chemicals or hazardous material.
- Carbon monoxide is colorless, odorless and dangerous. If it is necessary to operate the engine with vehicle in a closed area such as a garage, always use an exhaust collector to vent the exhaust gases outside the closed area.
- When draining coolant, keep in mind that small children and some pets are attracted by ethylene glycol antifreeze, and are quite likely to drink any left in an open container, or in puddles on the ground. This will prove fatal in sufficient quantity. Always drain the coolant into a sealable container.
- To avoid personal injury, do not remove the coolant pressure relief cap while the engine is operating or hot. The cooling system is under pressure; steam and hot liquid can come out forcefully when the cap is loosened slightly. Failure to follow these instructions may result in personal injury. The coolant must be recovered in a suitable, clean container for reuse. If the coolant is contaminated it must be recycled or disposed of correctly.
- When carrying out maintenance on the starting system be aware that heavy gauge leads are connected directly to the battery. Make sure the protective caps are in place when maintenance is completed. Failure to follow these instructions may result in personal injury.
- Do not remove any part of the engine emission control system. Operating the engine without the engine emission control system will reduce fuel economy and engine ventilation. This will weaken engine performance and shorten engine life. It is also a violation of Federal law.
- Due to environmental concerns, when the air conditioning system is drained, the refrigerant must be collected using refrigerant recovery/recycling equipment. Federal law requires that refrigerant be recovered into appropriate recovery equipment and the process be conducted by qualified technicians who have been certified by an approved organization, such as MACS, ASI, etc. Use of a recovery machine dedicated to the appropriate refrigerant is necessary to reduce the possibility of oil and refrigerant incompatibility concerns. Refer to the instructions provided by the equipment manufacturer when removing refrigerant from or charging the air conditioning system.
- Always disconnect the battery ground when working on or around the electrical system.
- Batteries contain sulfuric acid. Avoid contact with skin, eyes, or clothing. Also, shield your eyes when working near batteries to protect against possible splashing of the acid solution. In case of acid contact with skin or eyes, flush immediately with water for a minimum of 15 minutes and get prompt medical attention. If acid is swallowed, call a physician immediately. Failure to follow these instructions may result in personal injury.
- Batteries normally produce explosive gases. Therefore, do not allow flames, sparks or lighted substances to come near the battery. When charging or working near a battery, always shield your face and protect your eyes. Always provide ventilation. Failure to follow these instructions may result in personal injury.
- When lifting a battery, excessive pressure on the end walls could cause acid to spew through the vent caps, resulting in personal injury, damage to the vehicle or battery. Lift with a battery carrier or with your hands on opposite corners. Failure to follow these instructions may result in personal injury.
- Observe all applicable safety precautions when working around fuel. Whenever

servicing the fuel system, always work in a well-ventilated area. Do not allow fuel spray or vapors to come in contact with a spark, open flame, or excessive heat (a hot drop light, for example). Keep a dry chemical fire extinguisher near the work area. Always keep fuel in a container specifically designed for fuel storage; also, always properly seal fuel containers to avoid the possibility of fire or explosion. Do not smoke or carry lighted tobacco or open flame of any type when working on or near any fuel-related components.

• Fuel injection systems often remain pressurized, even after the engine has been turned OFF. The fuel system pressure must be relieved before disconnecting any fuel lines. Failure to do so may result in fire and/or personal injury.

• The evaporative emissions system contains fuel vapor and condensed fuel vapor. Although not present in large quantities, it still presents the danger of explosion or fire. Disconnect the battery ground cable from the battery to minimize the possibility of an electrical spark occurring, possibly causing a fire or explosion if fuel vapor or liquid fuel is present in the area. Failure to follow these instructions can result in personal injury.

• The EPA warns that prolonged contact with used engine oil may cause a number of skin disorders, including cancer! You should make every effort to minimize your exposure to used engine oil. Protective gloves should be worn when changing oil. Wash your hands and any other exposed skin areas as soon as possible after exposure to used engine oil. Soap and water, or waterless hand cleaner should be used.

• Some vehicles are equipped with an air bag system, often referred to as a Supplemental Restraint System (SRS) or Supplemental Inflatable Restraint (SIR) system. The system must be disabled before performing service on or around system components, steering column, instrument panel components, wiring and sensors. Failure to follow safety and disabling procedures could result in accidental air bag deployment, possible personal injury and unnecessary system repairs.

• Always wear safety goggles when working with, or around, the air bag system. When carrying a non-deployed air bag, be sure the bag and trim cover are pointed away from your body. When placing a non-deployed air bag on a work surface, always face the bag and trim cover upward, away from the surface. This will reduce the motion of the module if it is accidentally deployed.

• Electronic modules are sensitive to electrical charges. The ABS module can be damaged if exposed to these charges.

• Brake pads and shoes may contain asbestos, which has been determined to be a cancer-causing agent. Never clean brake surfaces with compressed air. Avoid inhaling brake dust. Clean all brake surfaces with a commercially available brake cleaning fluid.

• When replacing brake pads, shoes, discs or drums, replace them as complete axle sets.

• When servicing drum brakes, disassemble and assemble one side at a time, leaving the remaining side intact for reference.

• Brake fluid often contains polyglycol ethers and polyglycols. Avoid contact with the eyes and wash your hands thoroughly after handling brake fluid. If you do get brake fluid in your eyes, flush your eyes with clean, running water for 15 minutes. If eye irritation persists, or if you have taken brake fluid internally, immediately seek medical assistance.

• Clean, high quality brake fluid from a sealed container is essential to the safe and proper operation of the brake system. You should always buy the correct type of brake fluid for your vehicle. If the brake fluid becomes contaminated, completely flush the system with new fluid. Never reuse any brake fluid. Any brake fluid that is removed from the system should be discarded. Also, do not allow any brake fluid to come in contact with a painted or plastic surface; it will damage the paint.

• Never operate the engine without the proper amount and type of engine oil; doing so will result in severe engine damage.

• Timing belt maintenance is extremely important! Many models utilize an interference-type, non-freewheeling engine. If the timing belt breaks, the valves in the cylinder head may strike the pistons, causing potentially serious (also time-consuming and expensive) engine damage.

• Disconnecting the negative battery cable on some vehicles may interfere with the functions of the on-board computer system (s) and may require the computer to undergo a relearning process once the negative battery cable is reconnected.

• Steering and suspension fasteners are critical parts because they affect performance of vital components and systems and their failure can result in major service expense. They must be replaced with the same grade or part number or an equivalent part if replacement is necessary. Do not use a replacement part of lesser quality or substitute design. Torque values must be used as specified during reassembly to ensure proper retention of these parts.

MAZDA

Mazda3 • Mazdaspeed3 • Mazda6 • Mazdaspeed6 • MX-5 Miata

1

SPECIFICATIONS AND MAINTENANCE CHARTS

ENGINE AND VEHICLE IDENTIFICATION

Engine						
Code ①	Liters (cc)	Cu. In.	Cyl.	Fuel Sys.	Engine Type	Eng. Mfg.
LF ③	2.0 (1999)	121.9	4	MPFI	DOHC	Mazda
LF ④	2.0 (1999)	121.9	4	MPFI	DOHC	Mazda
L3 ⑤	2.3 (2261)	137.9	4	MPFI	DOHC	Mazda
L3 ⑥	2.3 (2261)	137.9	4	MPFI	DOHC	Mazda
AJ	3.0 (2999)	181.1	6	MPFI	DOHC	Mazda

Model Year	
Code ②	Year
7	2007
8	2008

MPFI: Multi-Point Fuel Injection

DOHC: Double Over Head Cam

① Located above the starter

② 10th digit of the Vehicle Identification Number (VIN)

③ Mazda3

④ MX-5 Miata

⑤ Mazda6

⑥ Mazdaspeed6 and Mazdaspeed3 are a 2.3L engine with a turbocharger

22140_MAZD_C0001

GENERAL ENGINE SPECIFICATIONS

Year	Model	Engine Displacement Liters (VIN)	Net Horsepower @ rpm	Net Torque @ rpm (ft. lbs.)	Bore x Stroke (in.)	Com-pression Ratio	Oil Pressure @ rpm
2007	Mazda 3 (I)	2.0 (LF)	①	②	3.44x3.27	10.0:1	49.0-85.8@3000
	Mazda 3 (S)	2.3 (L3)	160@6500	150@4500	3.44x3.70	9.7:1	49.0-85.8@3000
	Mazdaspeed3	2.3 (L3 Turbo)	263@5500	280@3000	3.44x3.70	9.5:0	43.0-79.0@3000
	Mazda 6 (I)	2.3 (L3)	160@6000	155@4000	3.44x3.70	9.7:1	33.9-75.5@3000
	Mazda 6 (S)	3.0 (AJ)	220@6300	192@5000	3.50x3.13	10.0:1	20-45@1500
	Mazdaspeed6	2.3 (L3 Turbo)	274@5500	280@3000	3.44x3.70	NA	43-79@3000
	MX5-Miata	2.0 (LF)	170@6700	140@5000	3.44x3.27	10.0:1	49-85@3000
2008	Mazda 3 (I)	2.0 (LF)	①	②	3.44x3.27	10.0:1	49.0-85.8@3000
	Mazda 3 (S)	2.3 (L3)	160@6500	150@4500	3.44x3.70	9.7:1	49.0-85.8@3000
	Mazdaspeed3	2.3 (L3 Turbo)	263@5500	280@3000	3.44x3.70	9.5:0	43.0-79.0@3000
	Mazda 6 (I)	2.3 (L3)	160@6000	155@4000	3.44x3.70	9.7:1	33.9-75.5@3000
	Mazda 6 (S)	3.0 (AJ)	220@6300	192@5000	3.50x3.13	10.0:1	20-45@1500
	Mazdaspeed6	2.3 (L3 Turbo)	274@5500	280@3000	3.44x3.70	NA	43-79@3000
	MX5-Miata	2.0 (LF)	170@6700	140@5000	3.44x3.27	10.0:1	49-85@3000

NA: Not Available

EFI: Electronic Fuel Injection

① Partial Zero Emission Vehicle (PZEV): 144@6500
Except PZEV states: 148@6500

② Partial Zero Emission Vehicle (PZEV): 132@4500
Except PZEV states: 135@4500

22140_MAZD_C0002

ENGINE TUNE-UP SPECIFICATIONS

Year	Engine Displacement Liters (VIN)	Spark Plug Gap (in.)	Ignition Timing (deg.) MT	Ignition Timing (deg.) AT	Fuel Pump (psi)	Idle Speed (rpm) MT	Idle Speed (rpm) AT	Valve Clearance In.	Valve Clearance Ex.
2007	2.0 (LF) ①	0.049-0.053	8B	8B	51-59	600-700	650-750	0.008-0.0011	0.010-0.012
	2.0 (LF) ②	0.050-0.053	8B	8B	51-59	700-800	700-800	0.008-0.0011	0.010-0.012
	2.3 (L3) ③	0.049-0.053	8B	8B	55-65	650-750	650-750	0.008-0.0011	0.010-0.012
	2.3 (L3 Turbo) ④	28-31	10B	10B	60-71	650-750	700-800	0.008-0.0011	0.010-0.012
	3.0 (AJ)	0.049-0.053	10B	10B	63-73	700-800	700-800	NA	NA
2008	2.0 (LF) ①	0.049-0.053	8B	8B	51-59	600-700	650-750	0.008-0.0011	0.010-0.012
	2.0 (LF) ②	0.050-0.053	8B	8B	51-59	700-800	700-800	0.008-0.0011	0.010-0.012
	2.3 (L3) ③	0.049-0.053	8B	8B	55-65	650-750	650-750	0.008-0.0011	0.010-0.012
	2.3 (L3 Turbo) ④	28-31	10B	10B	60-71	650-750	700-800	0.008-0.0011	0.010-0.012
	3.0 (AJ)	0.049-0.053	10B	10B	63-73	700-800	700-800	NA	NA

NOTE: The Vehicle Emission Control Information label often reflects specification changes made during production. The label figures must be used if they differ fror

NA: Not Available

B: Before top dead center

HYD: Hydraulic

① Mazda3

② MX-5 Miata

③ Mazda6

④ Mazdaspeed6 and Mazdaspeed3

22140_MAZD_C0003

CAPACITIES

Year	Model	Engine Displacement Liters (VIN)	Engine Oil with Filter (qts.)	Transmission (pts.) Man	Transmission (pts.) Auto.	Drive Axle Front (pts.)	Drive Axle Rear (pts.)	Fuel Tank (gal.)	Cooling System (qts.)
2007	Mazda 3 (I)	2.0 (LF)	4.5	6.06	15.2	①	—	14.5	7.9
	Mazda 3 (S)	2.3 (L3)	4.5	6.06	15.2	①	—	14.5	7.9
	Mazdaspeed3 (S)	2.3 (L3 Turbo)	6.0	5.38	—	—	2.3	15.9	8.5
	Mazda 6 (I)	2.3 (L3)	4.5	6.06	15.2	①	2.3	14.5	7.9
	Mazdaspeed6 (S)	2.3 (L3 Turbo)	6.0	5.38	—	—	2.3	15.9	8.5
	Mazda 6 (S)	3.0 (AJ)	6.0	4.8	19.4	①	—	14.5	7.9
	MX5-Miata	2.0 (LF)	4.5	4.4	15.6	—	1.7	12.7	7.9
2008	Mazda 3 (I)	2.0 (LF)	4.5	6.06	15.2	①	—	14.5	7.9
	Mazda 3 (S)	2.3 (L3)	4.5	6.06	15.2	①	—	14.5	7.9
	Mazdaspeed3 (S)	2.3 (L3 Turbo)	6.0	5.38	—	—	2.3	15.9	8.5
	Mazda 6 (I)	2.3 (L3)	4.5	6.06	15.2	①	2.3	14.5	7.9
	Mazdaspeed6 (S)	2.3 (L3 Turbo)	6.0	5.38	—	—	2.3	15.9	8.5
	Mazda 6 (S)	3.0 (AJ)	6.0	4.8	19.4	①	—	14.5	7.9
	MX5-Miata	2.0 (LF)	4.5	4.4	15.6	—	1.7	12.7	7.9

NOTE: All capacities are approximate. Add fluid gradually and check to be sure a proper fluid level is obtained.

① Included in transaxle

22140_MAZD_C0005

FLUID SPECIFICATIONS

Year	Model	Engine Displacement Liters	Engine ID/VIN	Engine Oil	Auto. Trans.	Manual Trans.	Power Drive Axle	Brake Master Cylinder	Engine Coolant ② ③
2007	Mazda 3 (I)	2.0 (LF)	F	5W-20	①	75W-90	75W-90	DOT 3	Ethylene-glycol
	Mazda 3 (S)	2.3 (L3)	C	5W-20	①	75W-90	75W-90	DOT 3	Ethylene-glycol
	Mazdaspeed3 (S)	2.3 (L3 Turbo)	L	5W-20	①	75W-90	75W-90	DOT 3	Ethylene-glycol
	Mazda 6 (I)	3.0 (AJ)	D	5W-20	①	75W-90	75W-90	DOT 3	Ethylene-glycol
	Mazdaspeed6 (S)	2.3 (L3 Turbo)	L	5W-20	①	75W-90	75W-90	DOT 3	Ethylene-glycol
	Mazda 6 (S)	2.3 (L3)	C	5W-20	①	75W-90	75W-90	DOT 3	Ethylene-glycol
	MX5-Miata	2.0 (LF)	F	5W-20	①	75W-90	75W-90	DOT 3	Ethylene-glycol
2008	Mazda 3 (I)	2.0 (LF)	F	5W-20	①	75W-90	75W-90	DOT 3	Ethylene-glycol
	Mazda 3 (S)	2.3 (L3)	C	5W-20	①	75W-90	75W-90	DOT 3	Ethylene-glycol
	Mazdaspeed3 (S)	2.3 (L3 Turbo)	L	5W-20	①	75W-90	75W-90	DOT 3	Ethylene-glycol
	Mazda 6 (I)	3.0 (AJ)	D	5W-20	①	75W-90	75W-90	DOT 3	Ethylene-glycol
	Mazdaspeed6 (S)	2.3 (L3 Turbo)	L	5W-20	①	75W-90	75W-90	DOT 3	Ethylene-glycol
	Mazda 6 (S)	2.3 (L3)	C	5W-20	①	75W-90	75W-90	DOT 3	Ethylene-glycol
	MX5-Miata	2.0 (LF)	F	5W-20	①	75W-90	75W-90	DOT 3	Ethylene-glycol

DOT: Department Of Transpotation

① The Asian Warner 6-speed transaxle uses JWS3309 a special fluid produced by Exxon/Mobil

② Do not use coolants containing Alcohol, Methanol, Borate or Silicate. These coolants could damage the cooling system.

③ If the "FL22" mark is shown on or near the cooling system cap, use FL22 type engine coolant.

22140_MAZD_C0004

VALVE SPECIFICATIONS

Year	Engine Displacement Liters (VIN)	Seat Angle (deg.)	Face Angle (deg.)	Maximum out of Square (in.)	Spring Free Length (in.)	Stem-to-Guide Clearance (in.)		Stem Diameter (in.)	
						Intake	Exhaust	Intake	Exhaust
2007	2.0 (F)	NA	NA	NA	NA	NA	NA	NA	NA
	2.3 (C)	NA	NA	NA	NA	NA	NA	NA	NA
	2.3 (L)	NA	NA	NA	NA	NA	NA	NA	NA
	3.0 (D)	NA	NA	NA	NA	NA	NA	NA	NA
2008	2.0 (F)	NA	NA	NA	NA	NA	NA	NA	NA
	2.3 (C)	NA	NA	NA	NA	NA	NA	NA	NA
	2.3 (L)	NA	NA	NA	NA	NA	NA	NA	NA
	3.0 (D)	NA	NA	NA	NA	NA	NA	NA	NA

NA: Not Available

22140_MAZD_C0006

CRANKSHAFT AND CONNECTING ROD SPECIFICATIONS

All measurements are given in inches.

	Engine	Crankshaft				Connecting Rod		
	Displacement	Main Brg.	Main Brg. Oil	Shaft	Thrust	Journal	Oil	Side
Year	Liters (VIN)	Journal Dia.	Clearance	End-play	on No.	Diameter	Clearance	Clearance
2007	2.0 (F)	NA	NA	NA	NA	NA	NA	NA
	2.3 (C)	NA	NA	NA	NA	NA	NA	NA
	2.3 (L)	NA	NA	NA	NA	NA	NA	NA
	3.0 (D)	NA	NA	NA	NA	NA	NA	NA
2008	2.0 (F)	NA	NA	NA	NA	NA	NA	NA
	2.3 (C)	NA	NA	NA	NA	NA	NA	NA
	2.3 (L)	NA	NA	NA	NA	NA	NA	NA
	3.0 (D)	NA	NA	NA	NA	NA	NA	NA

NA: Not Avilable

22140_MAZD_C0007

PISTON AND RING SPECIFICATIONS

All measurements are given in inches.

	Engine		Ring Gap			Ring Side Clearance		
	Displacement	Piston	Top	Bottom	Oil	Top	Bottom	Oil
Year	Liters (VIN)	Clearance	Compression	Compression	Control	Compression	Compression	Control
2007	2.0 (F)	NA	NA	NA	NA	NA	NA	NA
	2.3 (C)	NA	NA	NA	NA	NA	NA	NA
	2.3 (L)	NA	NA	NA	NA	NA	NA	NA
	3.0 (D)	NA	NA	NA	NA	NA	NA	NA
2008	2.0 (F)	NA	NA	NA	NA	NA	NA	NA
	2.3 (C)	NA	NA	NA	NA	NA	NA	NA
	2.3 (L)	NA	NA	NA	NA	NA	NA	NA
	3.0 (D)	NA	NA	NA	NA	NA	NA	NA

NA: Not Available

22140_MAZD_C0008

TORQUE SPECIFICATIONS

All readings in ft. lbs.

Year	Engine Displacement Liters (VIN)	Cylinder Head Bolts	Main Bearing Bolts	Rod Bearing Bolts	Crankshaft Damper Bolts	Flywheel Bolts	Manifold Intake	Manifold Exhaust	Spark Plugs	Oil Pan Drain Plug
2007	2.0 (LF)	①	NA	NA	②	③	④	32-47	8-10	22-30
	2.3 (L3)	①	NA	NA	②	③	④	32-47	8-10	19-22
	3.0 (AJ)	⑤	NA	NA	⑥	⑦	⑧	14-18	⑨	19-22
2008	2.0 (LF)	①	NA	NA	②	③	④	32-47	8-10	22-30
	2.3 (L3)	①	NA	NA	②	③	④	32-47	8-10	19-22
	3.0 (AJ)	⑤	NA	NA	⑥	⑦	⑧	14-18	⑨	19-22

NA: Not Available

① Step 1: 97 inch lbs.
Step 2: Tighten12.5 ft. lbs
Step 3: Tighten 34.6 ft. lbs.
Step 4: Tighten 88 degrees
Step 5: Tighten 88 degrees

② 71-77 ft. lbs. plus 87-93 degrees

③ Manual transmission: 80-85 ft. lbs.
Automatic transmission: 71-76 ft. lbs.

④ Mazda 3 with the 2.0L (LF) and 2.3L (L3) engines: 142-177 ft. lbs.
Mazda 6 with the 2.3L (L3) engine: 71-101 inch lbs.

⑤ Step 1: 17-19 ft. lbs.
Step 2: Tighten 85-95 degees
Step 3: Repeat step 2

⑥ Step 1: 88 ft. lbs.
Step 2: loosen one full turn
Step 3: Tighten 35-39 ft. lbs.
Step 4: Tighten 85-95 degrees

⑦ G35M-R A26MX-R manual or FN4A-EL automatic transmission: 80-85 ft. lbs.
A65M-R manual or JA5A-EL automatic transmission: 54-64 ft. lbs.

⑧ Lower intake: 72-106 inch lbs.
Upper intake plenum: 72-106 inch lbs.

⑨ 79-177 inch. lbs.

22140_MAZD_C0009

WHEEL ALIGNMENT

Year	Model		Caster Range (+/-Deg.)	Caster Preferred Setting (Deg.)	Camber Range (+/-Deg.)	Camber Preferred Setting (Deg.)	Toe-in (in.)
2007	Mazda 3	F	1.00	+2.58	1.00	-0.40	0.04 +/- 0.12
		R	—	—	1.00	0	0.08 +/- 0.16
	Mazdaspeed3	F	1.00	+2.58	1.00	-0 17' +/- 1.00	0 +/- 0 22'
		R	—	—	1.00	-0 09' +/- 1.00	0 +/- 0 22'
	Mazda 6	F	1.00	①	1.00	②	0.12 +/- 0.16
		R	—	—	1.00	③	0.08 +/- 0.16
	Mazdaspeed6	F	1.00	+2.58	1.00	-0 17' +/- 1.00	0 +/- 0 22'
		R	—	—	1.00	-0 09' +/- 1.00	0 +/- 0 22'
	MX5-Miata	F	1.00	④	1.00	⑤	⑥
		R	—	—	1.00	⑦	⑧
2008	Mazda 3	F	1.00	+2.58	1.00	-0.40	0.04 +/- 0.12
		R	—	—	1.00	0	0.08 +/- 0.16
	Mazdaspeed3	F	1.00	+2.58	1.00	-0 17' +/- 1.00	0 +/- 0 22'
		R	—	—	1.00	-0 09' +/- 1.00	0 +/- 0 22'
	Mazda 6	F	1.00	①	1.00	②	0.12 +/- 0.16
		R	—	—	1.00	③	0.08 +/- 0.16
	Mazdaspeed6	F	1.00	+2.58	1.00	-0 17' +/- 1.00	0 +/- 0 22'
		R	—	—	1.00	-0 09' +/- 1.00	0 +/- 0 22'
	MX5-Miata	F	1.00	④	1.00	⑤	⑥
		R	—	—	1.00	⑦	⑧

① Except wagon: 0 degrees 11', plus or minus 0 degrees 22'
Wagon: 3 degrees 33' plus or minus 1 degree

② Except wagon:-0 degrees 16', plus or minus 1 degree
Wagon: -0 degrees 16', plus or minus 1 degree

③ Except wagon:-1 degree 09', plus or minus 1 degree
Wagon: -1 degree 02', plus or minus 1 degree

④ 16 inch wheel: 5 degrees 59', plus or minus 1 degree
17 inch wheel: 6 degrees 06', plus or minus 1 degree

⑤ 16 inch wheel:-0 degrees 06', plus or minus 1 degree
17 inch wheel: -0 degrees 14', plus or minus 1 degree

⑥ 16 inch wheel: 0.05 +/- 0.09
17 inch wheel: 0.06 +/- 0.11

⑦ 16 inch wheel: -1 degrees 04', plus or minus 1 degree
17 inch wheel: -1 degrees 11', plus or minus 1 degree

⑧ 16 inch wheel: 0.071 +/- 0.094
17 inch wheel: 0.083 +/- 0.11

22140_MAZD_C0010

TIRE, WHEEL AND BALL JOINT SPECIFICATIONS

Year	Model	OEM Tires		Tire Pressures (psi)		Wheel	Ball Joint	Lug
		Standard	Optional	Front	Rear	Size	Inspection	Nut
2007	Mazda 3	①	①	32	32	②	③	65-87
	Mazda 6	P205/60R16	P215/50R17	32	32	④	⑤	65-87
	Mazdaspeed6	P205/50R16	P215/50R17 ⑧	32	32	⑨	⑤	65-87
	MX5-Miata	P205/50R16	P205/45R17	29	29	⑥	⑦	65-87
2008	Mazda 3	①	①	32	32	②	③	65-87
	Mazda 6	P205/60R16	P215/50R17	32	32	④	⑤	65-87
	Mazdaspeed6	P205/50R16	P215/50R17 ⑧	32	32	⑨	⑤	65-87
	MX5-Miata	P205/50R16	P205/45R17	29	29	⑥	⑦	65-87

OEM: Original Equipment Manufacturer
PSI: Pounds Per Square Inch
STD: Standard
OPT: Optional

① Standard Steel rim: P195/65R15
Aluminum alloy: P205/55R16
② Standard Steel rim: 15 x 6 J
Aluminum alloy: 16 x 6 1/2 J or 17 x 6 1/2 J
③ Lower arm ball rotation torque, front: 3-10 ft. lbs.
④ Standard Steel rim: 16 x 6 1/2JJ
Aluminum alloy: 16 x 7 JJ or 17 x 7 JJ
⑤ Lower arm ball rotation torque, front: 10.5-19.7 inch lbs.
Lower arm ball rotation torque, rear: 8.86-19.6 inch lbs.
Upper arm ball rotation torque: 13.2 inch lbs. (max)
⑥ Standard rim: 16 x 6 1/2 J
Optional rim:17 x 7 J
⑦ Lower arm ball rotation torque, front: 4-25 inch lbs.
Upper arm ball rotation torque: 3-19 inch lbs. (max)
⑧ Also avaialable with 215/45/R18
⑨ Standard Steel rim: 16 x 6 1/2JJ
Aluminum alloy: 17 x 7 JJ or 18 x 7

22140_MAZD_C0011

BRAKE SPECIFICATIONS

All measurements in inches unless noted

Year	Model		Brake Disc			Brake Drum			Minimum	Brake Caliper	
			Original Thickness	Minimum Thickness	Maximum Runout	Original Inside Diameter	Max. Wear Limit	Maximum Machine Diameter	Lining Thickness	Bracket Bolts (ft. lbs.)	Mounting Bolts (ft. lbs.)
2007	Mazda 3	F	NA	0.910	0.002	—	—	—	0.079	75-87	19-22
		R	NA	0.350	0.002	—	—	—	0.079	43-56	19-22
	Mazdaspeed 3	F	NA	0.910	0.002	—	—	—	0.079	75-87	19-22
		R	NA	0.350	0.002	—	—	—	0.079	43-56	19-22
	Mazda 6	F	NA	0.910	0.002	—	—	—	0.079	58-75	36-39
		R	NA	0.310	0.002	—	—	—	0.079	37-49	27-36
	Mazdaspeed 6	F	NA	0.910	0.002	—	—	—	0.059	58-75	16-23
		R	NA	0.310	0.002	—	—	—	0.039	37-49	16-23
	MX5-Miata	F	NA	0.790	0.002	—	—	—	0.079	58-75	15-23
		R	NA	0.310	0.002	—	—	—	0.079	36-50	14-18
2008	Mazda 3	F	NA	0.910	0.002	—	—	—	0.079	75-87	19-22
		R	NA	0.350	0.002	—	—	—	0.079	43-56	19-22
	Mazdaspeed 3	F	NA	0.910	0.002	—	—	—	0.079	75-87	19-22
		R	NA	0.350	0.002	—	—	—	0.079	43-56	19-22
	Mazda 6	F	NA	0.910	0.002	—	—	—	0.079	58-75	36-39
		R	NA	0.310	0.002	—	—	—	0.079	37-49	27-36
	Mazdaspeed 6	F	NA	0.910	0.002	—	—	—	0.059	58-75	16-23
		R	NA	0.310	0.002	—	—	—	0.039	37-49	16-23
	MX5-Miata	F	NA	0.790	0.002	—	—	—	0.079	58-75	15-23
		R	NA	0.310	0.002	—	—	—	0.079	36-50	14-18

NA: Not Avilable
F: Front
R: Rear

22140_MAZD_C0012

SCHEDULED MAINTENANCE INTERVALS

2007-08 Mazda Car—Mazda 3, Mazdaspeed3, Mazda 6, Mazdaspeed 6 and MX5-Miata

TO BE SERVICED	TYPE OF SERVICE	VEHICLE MILEAGE INTERVAL (x1000)												
		7.5	15	22.5	30	37.5	45	52.5	60	67.5	75	82.5	90	97.5
Engine oil & filter	R	✓	✓	✓	✓	✓	✓	✓	✓	✓	✓	✓	✓	✓
Air cleaner element	R					✓					✓			
Engine coolant ① ② ③	R													
Spark plugs	R										✓			
Bolts & nuts on chassis & body	S/I				✓				✓				✓	
Brake lines, hoses & connections	S/I				✓				✓				✓	
Cooling system	S/I				✓				✓				✓	
Cabin filter	R			✓										
Disc brakes	S/I				✓				✓				✓	
Drive belts ④	S/I				✓				✓				✓	
Drive shaft dust boots	S/I				✓				✓				✓	
Exhaust system heat shield	S/I				✓				✓				✓	
Front & rear suspension ball joints	S/I				✓				✓				✓	
Fuel lines & hoses	S/I				✓				✓				✓	
Steering operation & linkages	S/I				✓				✓				✓	
Fuel filter	R								✓					
Valve clearance ⑤	I								✓					
Hose & tube for emission	S/I								✓					

R: Replace S/I: Service or Inspect

① Mazda 6: replace initially at 105,000 miles or 6 months and every 30, 000 miles or 24 months thereafter

② Mazda 3: replace initially at 60,000 miles or 4 years and every 24 months thereafter

③ Models with FL22 long-life coolant can be identified by the FL22 marking on the cooling system cap label. It is recommended that FL22 coolant continue to be used for models originally filled with FL22 coolant from the factory.

④ 2.0L LF engine and 2.3L L3 engine: inspect every 37, 500 miles. 3.0L AJ engine: inspect every 30,000 miles

⑤ 2.0L LF and 2.3L L3 engine: Audible inspect every 75, 000 miles and if noisy adjust.

FREQUENT OPERATION MAINTENANCE (SEVERE SERVICE)

If a vehicle is operated under any of the following conditions it is considered severe service

- Extremely dusty areas.

- 50% or more of the vehicle operation is in 32°C (90°F) or higher temperatures, or constant operation in temperatures below 0°C (32°F).

- Prolonged idling (vehicle operation in stop and go traffic).

- Frequent short running periods (engine does not warm to normal operating temperatures).

- Police, taxi, delivery usage or trailer towing usage.

Oil & oil filter: change every 5000 miles.

Oil & oil filter (Puerto Rico): change every 3000 miles.

Air cleaner element: service or inspect every 15,000 miles

Automatic transaxle fluid: service or inspect every 15,000 miles.

Bolts & nuts on chassis & body: tighten every 15,000 miles.

Disc brakes: service or inspect every 15,000 miles.

22140_MAZD_C0013

PRECAUTIONS

Before servicing any vehicle, please be sure to read all of the following precautions, which deal with personal safety, prevention of component damage, and important points to take into consideration when servicing a motor vehicle:

- Never open, service or drain the radiator or cooling system when the engine is hot; serious burns can occur from the steam and hot coolant.
- Observe all applicable safety precautions when working around fuel. Whenever servicing the fuel system, always work in a well-ventilated area. Do not allow fuel spray or vapors to come in contact with a spark, open flame, or excessive heat (a hot drop light, for example). Keep a dry chemical fire extinguisher near the work area. Always keep fuel in a container specifically designed for fuel storage; also, always properly seal fuel containers to avoid the possibility of fire or explosion. Refer to the additional fuel system precautions later in this section.
- Fuel injection systems often remain pressurized, even after the engine has been turned OFF. The fuel system pressure must be relieved before disconnecting any fuel lines. Failure to do so may result in fire and/or personal injury.
- Brake fluid often contains polyglycol ethers and polyglycols. Avoid contact with the eyes and wash your hands thoroughly after handling brake fluid. If you do get brake fluid in your eyes, flush your eyes with clean, running water for 15 minutes. If eye irritation persists, or if you have taken brake fluid internally, IMMEDIATELY seek medical assistance.
- The EPA warns that prolonged contact with used engine oil may cause a number of skin disorders, including cancer. You should make every effort to minimize your exposure to used engine oil. Protective gloves should be worn when changing oil. Wash your hands and any other exposed skin areas as soon as possible after exposure to used engine oil. Soap and water, or waterless hand cleaner should be used.
- All new vehicles are now equipped with an air bag system, often referred to as a Supplemental Restraint System (SRS) or Supplemental Inflatable Restraint (SIR) system. The system must be disabled before performing service on or around system components, steering column, instrument panel components, wiring and sensors. Failure to follow safety and disabling procedures could result in accidental air bag deployment, possible personal injury and unnecessary system repairs.
- Always wear safety goggles when working with, or around, the air bag system. When carrying a non-deployed air bag, be sure the bag and trim cover are pointed away from your body. When placing a non-deployed air bag on a work surface, always face the bag and trim cover upward, away from the surface. This will reduce the motion of the module if it is accidentally deployed. Refer to the additional air bag system precautions later in this section.
- Clean, high quality brake fluid from a sealed container is essential to the safe and proper operation of the brake system. You should always buy the correct type of brake fluid for your vehicle. If the brake fluid becomes contaminated, completely flush the system with new fluid. Never reuse any brake fluid. Any brake fluid that is removed from the system should be discarded. Also, do not allow any brake fluid to come in contact with a painted surface; it will damage the paint.
- Never operate the engine without the proper amount and type of engine oil; doing so WILL result in severe engine damage.
- Timing belt maintenance is extremely important. Many models utilize an interference-type, non-freewheeling engine. If the timing belt breaks, the valves in the cylinder head may strike the pistons, causing potentially serious (also time-consuming and expensive) engine damage. Refer to the maintenance interval charts for the recommended replacement interval for the timing belt, and to the timing belt section for belt replacement and inspection.
- Disconnecting the negative battery cable on some vehicles may interfere with the functions of the on-board computer system(s) and may require the computer to undergo a relearning process once the negative battery cable is reconnected.
- When servicing drum brakes, only disassemble and assemble one side at a time, leaving the remaining side intact for reference.
- Only an MVAC-trained, EPA-certified automotive technician should service the air conditioning system or its components.

BRAKES

ANTI-LOCK BRAKE SYSTEM (ABS)

GENERAL INFORMATION

When wheel slip is detected during a brake application, the ABS enters antilock mode. During antilock braking, hydraulic pressure in the individual wheel circuits is controlled to prevent any wheel from slipping. A separate hydraulic line and specific solenoid valves are provided for each wheel. The ABS can decrease, hold, or increase hydraulic pressure to each wheel brake. The ABS cannot, however, increase hydraulic pressure above the amount which is transmitted by the master cylinder during braking.

During antilock braking, a series of rapid pulsations is felt in the brake pedal. These pulsations are caused by the rapid changes in position of the individual solenoid valves as the hydraulic control module (HCU) responds to wheel speed sensor inputs and attempts to prevent wheel slip. These pedal pulsations are present only during antilock braking and stop when normal braking is resumed or when the vehicle comes to a stop. A ticking or popping noise may also be heard as the solenoid valves cycle rapidly. During antilock braking on dry pavement, intermittent chirping noises may be heard as the tires approach slipping. These noises and pedal pulsations are considered normal during antilock operation.

Vehicles equipped with ABS may be stopped by applying normal force to the brake pedal. Brake pedal operation during normal braking is no different than that of previous non-ABS systems. Maintaining a constant force on the brake pedal provides the shortest stopping distance while maintaining vehicle stability.

The instrument panel cluster illuminates the ABS indicator when the following occurs:

The electronic brake control module (HCU) detects a malfunction with the antilock brake system.

The IPC performs the displays test at the start of each ignition cycle. The indicator illuminates for approximately 3 seconds.

WHEEL SPEED SENSORS

REMOVAL & INSTALLATION

Front

Mazda3 and Mazdaspeed3

See Figure 1.

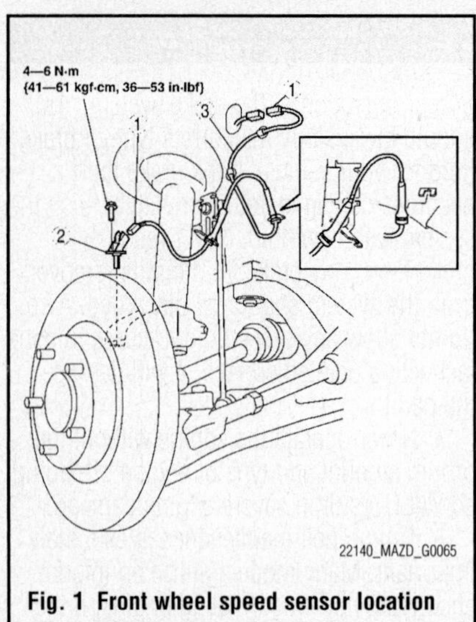

Fig. 1 Front wheel speed sensor location

1. Before servicing the vehicle, refer to the precautions section.
2. Remove the front wheel.
3. Remove the mud guard.
4. Disconnect the wheel speed sensor electrical connector.
5. Remove the fasteners attaching the wheel speed sensor harness.
6. Remove the wheel speed sensor retaining bolts and sensor.
7. Installation is the reverse of removal.
8. Tighten the mounting bolt to 53 inch. lbs. (6 Nm)

MX-5 Miata

See Figure 2.

1. Before servicing the vehicle, refer to the precautions section.
2. Remove or disconnect the following:
   - Wheel
   - Mudguard
   - Connector
   - Bolt
   - Front ABS speed sensor

To install:

3. Install or connect the following:
   - Front ABS speed sensor
   - Bolt
   - Connector
   - Mudguard
   - Wheel

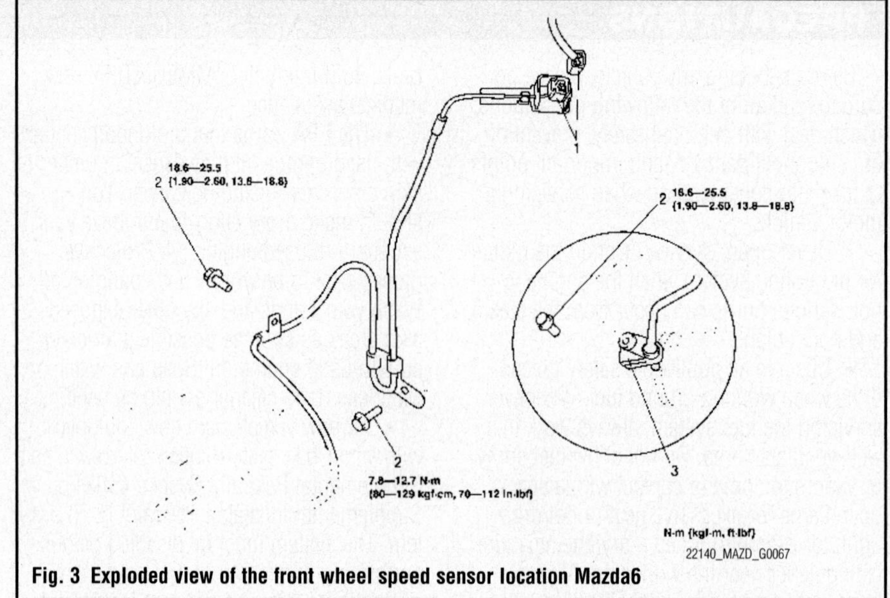

Fig. 3 Exploded view of the front wheel speed sensor location Mazda6

Mazda6 and Mazdaspeed6

See Figure 3.

1. Before servicing the vehicle, refer to the precautions section.
2. Remove or disconnect the following:
   - Wheel
   - Mudguard
   - Connector
   - Bolt
   - Front ABS speed sensor

To install:

3. Install or connect the following:
   - Front ABS speed sensor
   - Bolt
   - Connector
   - Mudguard
   - Wheel

Rear

Mazda3 and Mazdaspeed3

See Figure 4.

1. Before servicing the vehicle, refer to the precautions section.
2. Remove the rear undercover.
3. Disconnect the wheel speed sensor electrical connector.
4. Remove the rear ABS wheel-speed sensor.
5. Press the tab of the ABS hole cover to separate the ABS hole cover from the body and remove the hole cover. Unplug the connector and remove the sensor wiring harness.
6. Installation is the reverse of removal, tighten the sensor fasteners to 36–53 in. lbs. (4–6 Nm).
7. Pass the rear wheel speed sensor wiring harness outside the rear parking brake cable.
8. Install the ABS hole cover into the body so that the arrow on it is facing toward the outer side of the vehicle.

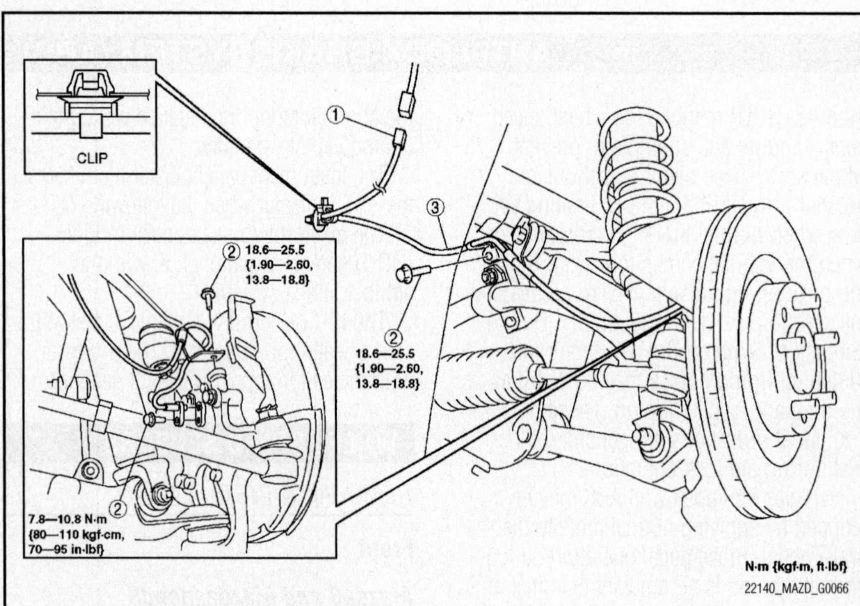

Fig. 2 MX-5 Miata front wheel speed sensor location

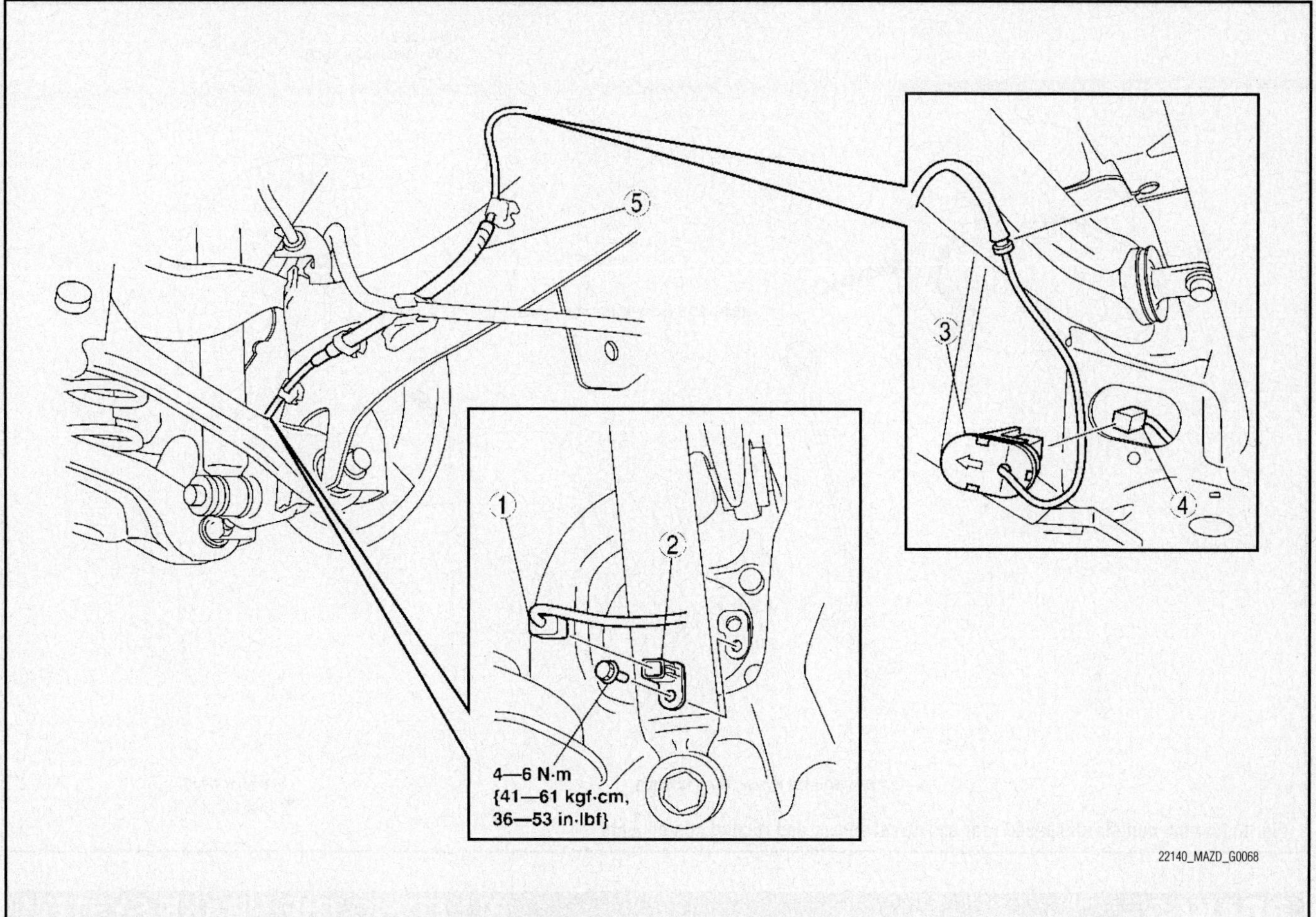

5
3
4
1
2
4—6 N·m
{41—61 kgf·cm,
36—53 in·lbf}
22140_MAZD_G0068

Fig. 4 Rear abs wheel sensor and related components

MX-5 Miata

See Figure 5.

1. Remove the trunk end trim.
2. Remove the partition board.
3. Remove the trunk side trim.
4. Remove the fuel-filler pipe protector.
5. Disconnect the wheel speed sensor electrical connector.
6. Remove the rear ABS wheel-speed sensor.
7. Installation is the reverse of removal.

Mazda6 and Mazdaspeed6

See Figure 6.

1. Before servicing the vehicle, refer to the precautions section.
2. Remove the tire house trim.
3. Remove the ABS speed sensor connector.
4. Remove the ABS speed sensor bolt.
5. Remove the rear ABS speed sensor.
6. Installation is the reverse of removal.

CLIP
1
3
2 18.6—25.5 N·m
{1.9—2.6 kgf·m, 13.8—18.8 ft·lbf}
2 7.8—10.8 N·m
{80—110 kgf·cm, 70—95 in·lbf}
22140_MAZD_G0069

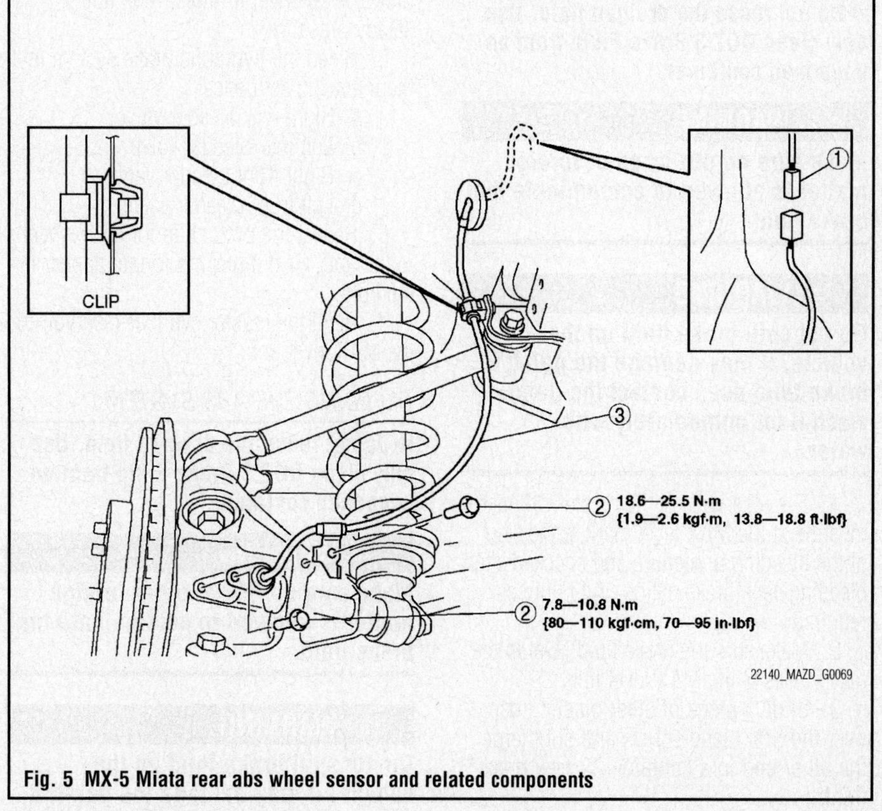

Fig. 5 MX-5 Miata rear abs wheel sensor and related components

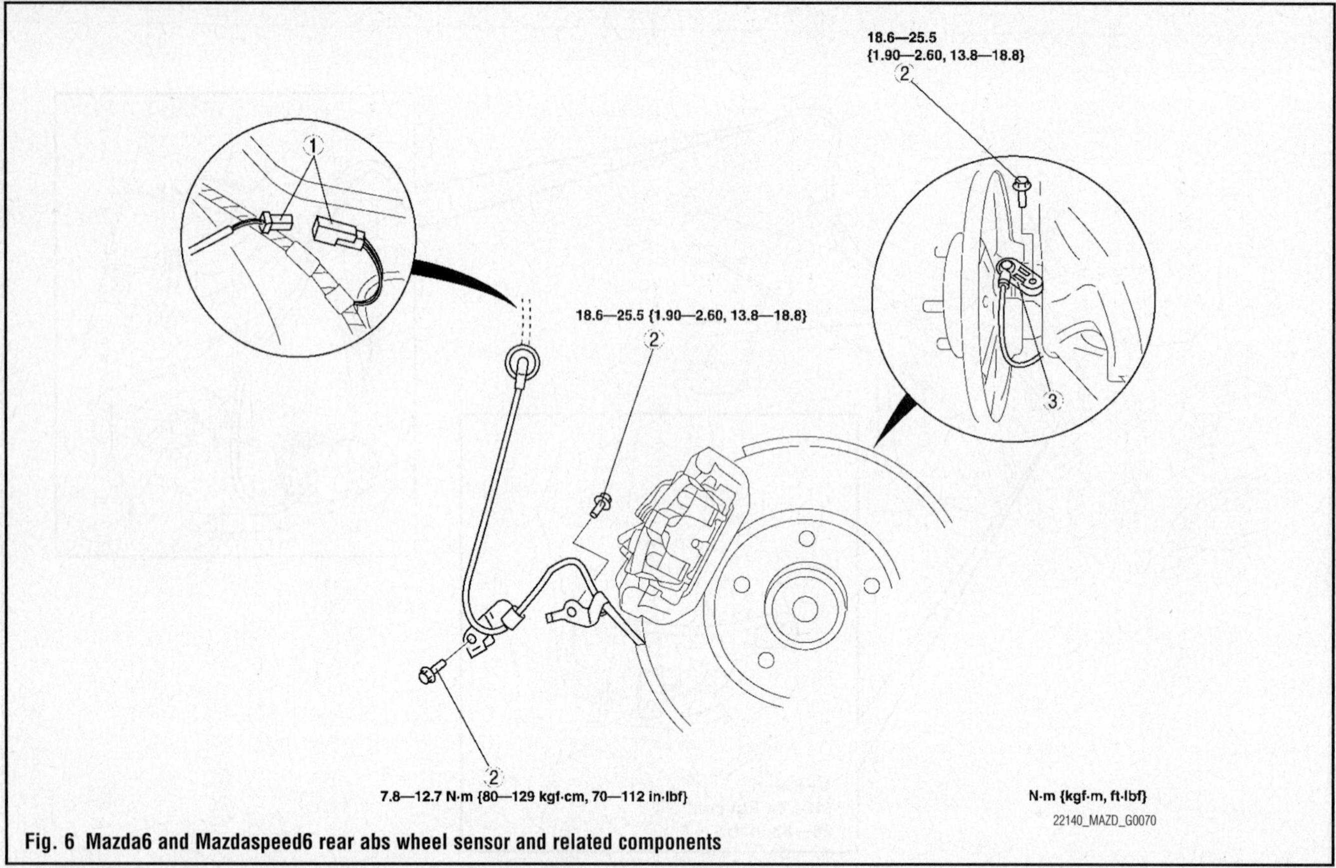

Fig. 6 Mazda6 and Mazdaspeed6 rear abs wheel sensor and related components

BRAKES — BLEEDING THE BRAKE SYSTEM

BLEEDING PROCEDURE

➡**Do not reuse the drained fluid. Use only clean DOT 3 Brake Fluid from an unopened container.**

✲✲ WARNING

Make sure no dirt or other foreign matter is allowed to contaminate the brake fluid.

✲✲ WARNING

Do not spill brake fluid on the vehicle, it may damage the paint; if brake fluid does contact the paint, wash it off immediately with water.

1. The reservoir on the master cylinder must be at the MAX level mark at the start of the bleeding procedure and checked after bleeding each brake caliper. Add fluid as required.
2. Make sure the brake fluid level in the reservoir is at the MAX level line.
3. Slide a piece of clear plastic hose over the first bleed screw, and submerge the other end in a container of new brake fluid.
4. Have someone slowly pump the brake pedal several times, then apply steady pressure.
5. Bleed the hydraulic brake system in the following sequence:
    a. Right rear bleeder valve
    b. Left rear bleeder valve
    c. Right front bleeder valve
    d. Left front bleeder valve
6. Repeat the procedure for each wheel in the until air bubbles no longer appear in the fluid.
7. Refill the master cylinder reservoir to the MAX level line.

BLEEDING THE ABS SYSTEM

➡**Do not reuse the drained fluid. Use only clean DOT 3 Brake Fluid from an unopened container.**

✲✲ WARNING

Make sure no dirt or other foreign matter is allowed to contaminate the brake fluid.

✲✲ WARNING

Do not spill brake fluid on the vehicle, it may damage the paint; if brake fluid does contact the paint, wash it off immediately with water.

1. The reservoir on the master cylinder must be at the MAX level mark at the start of the bleeding procedure and checked after bleeding each brake caliper. Add fluid as required.
2. Make sure the brake fluid level in the reservoir is at the MAX level line.
3. Slide a piece of clear plastic hose over the first bleed screw, and submerge the other end in a container of new brake fluid.
4. Have someone slowly pump the brake pedal several times, then apply steady pressure.
5. Bleed the hydraulic brake system in the following sequence:
    a. Right rear bleeder valve
    b. Left rear bleeder valve
    c. Right front bleeder valve
    d. Left front bleeder valve
6. Repeat the procedure for each wheel in the until air bubbles no longer appear in the fluid.
7. Refill the master cylinder reservoir to the MAX level line.

BRAKES

FRONT DISC BRAKES

CAUTION

Dust and dirt accumulating on brake parts during normal use may contain asbestos fibers from production or aftermarket brake linings. Breathing excessive concentrations of asbestos fibers can cause serious bodily harm. Exercise care when servicing brake parts. Do not sand or grind brake lining unless equipment used is designed to contain the dust residue. Do not clean brake parts with compressed air or by dry brushing. Cleaning should be done by dampening the brake components with a fine mist of water, then wiping the brake components clean with a dampened cloth. Dispose of cloth and all residue containing asbestos fibers in an impermeable container with the appropriate label. Follow practices prescribed by the Occupational Safety and Health Administration (OSHA) and the Environmental Protection Agency (EPA) for the handling, processing, and disposing of dust or debris that may contain asbestos fibers.

BRAKE CALIPER

REMOVAL & INSTALLATION

MX5-Miata

See Figures 7 and 8.

1. Before servicing the vehicle, refer to the precautions section.

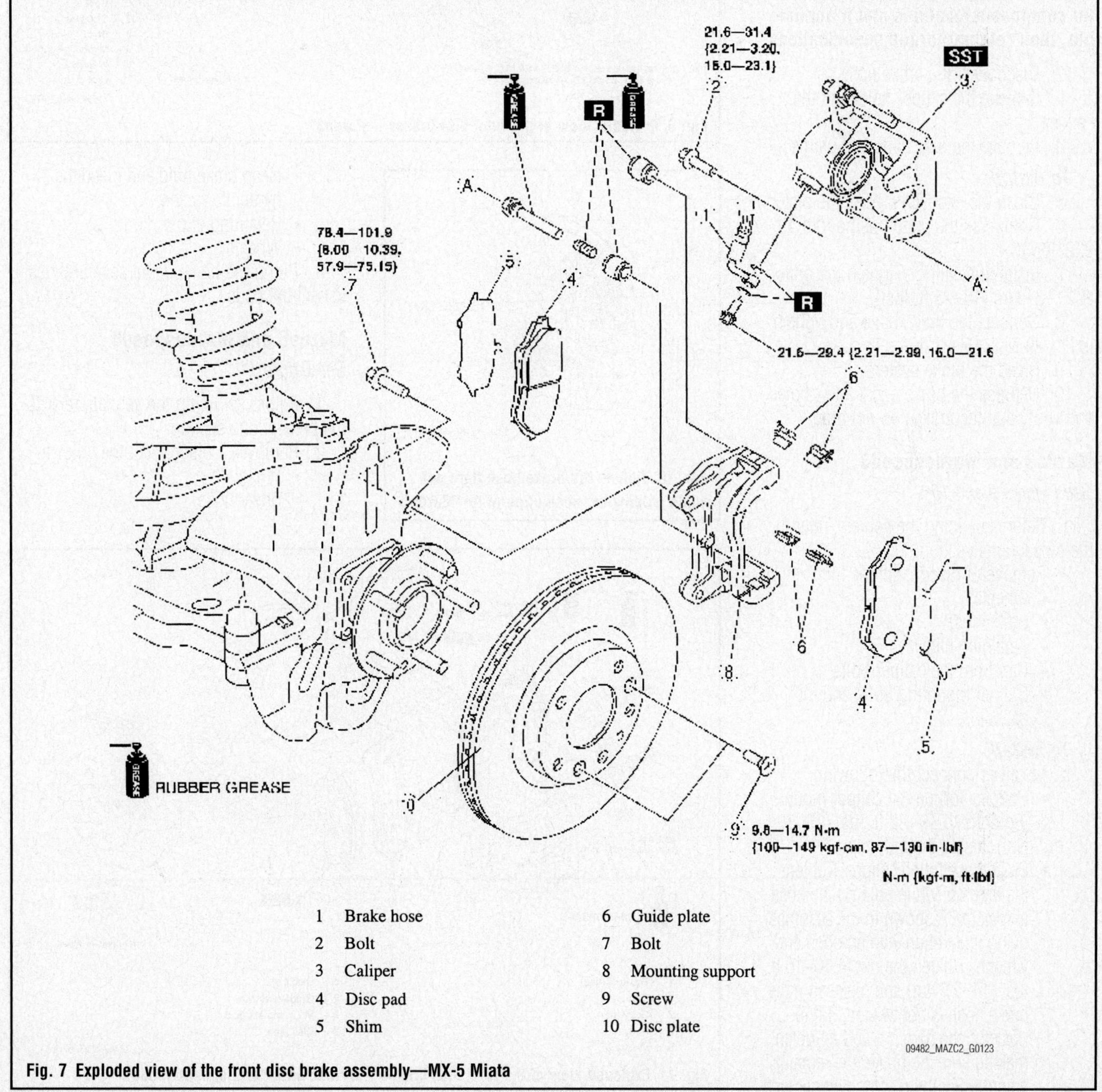

Fig. 7 Exploded view of the front disc brake assembly—MX-5 Miata

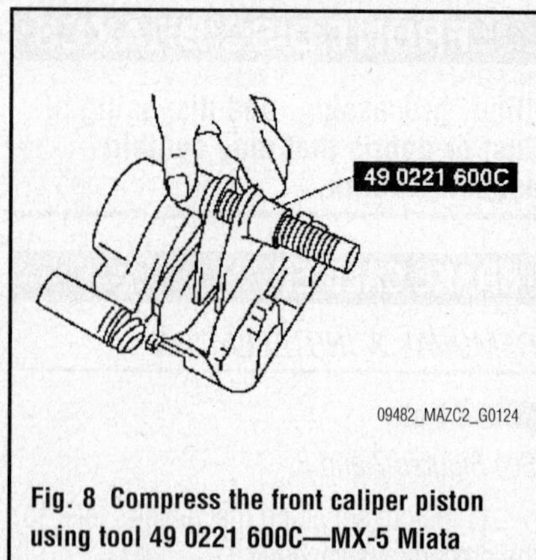

Fig. 8 Compress the front caliper piston using tool 49 0221 600C—MX-5 Miata

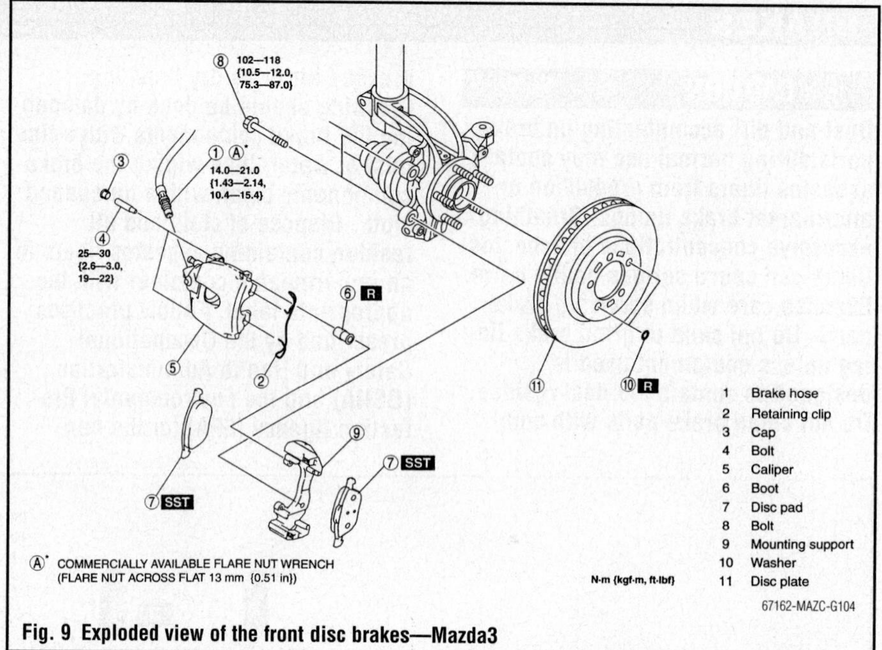

Fig. 9 Exploded view of the front disc brakes—Mazda3

➡Refer the exploded view illustration for component locations and if applicable, their retainer torque specifications.

2. Disconnect the brake hose.
3. Remove the caliper bolts and the caliper.
4. Remove the disc pads and shims.

To install:

5. Clean the exposed area of the piston.
6. Compress the piston using tool 49 0221 600C
7. Install the caliper. Tighten the bolts to 21–31 Nm (15–23 ft. lbs.).
8. Connect the brake hose and tighten to 21–29 Nm (15–21 ft. lbs.).
9. Bleed the brake system.
10. Depress the brake pedal a few times and verify that the brakes do not drag

Mazda3 and Mazdaspeed3

See Figures 9 and 10.

1. Before servicing the vehicle, refer to the precautions section.
2. Remove or disconnect the following:
   - Wheels
   - Brake hose
   - Retaining clip
   - Cap from the caliper bolts
   - Caliper mounting bolts and the caliper

To install:

3. Install or connect the following:
   - Caliper. Torque the caliper mounting bolts to 19–22 ft. lbs. (25–30 Nm). Install the bolt caps.
   - Brake hose to the caliper. Tighten the flare nut while holding the hose at location A shown in the accompanying illustration with an open end wrench. Tighten the nut to 10–15 ft. lbs. (14–21 Nm) and make sure the brake hose is not twisted, if it is unfasten the flare nut and retighten making sure the brake line remains straight. Fill the master cylinder with clean brake fluid and bleed the hydraulic system.
   - Retaining clip
   - Wheels
4. Pump the brake pedal several times to seat the pads.

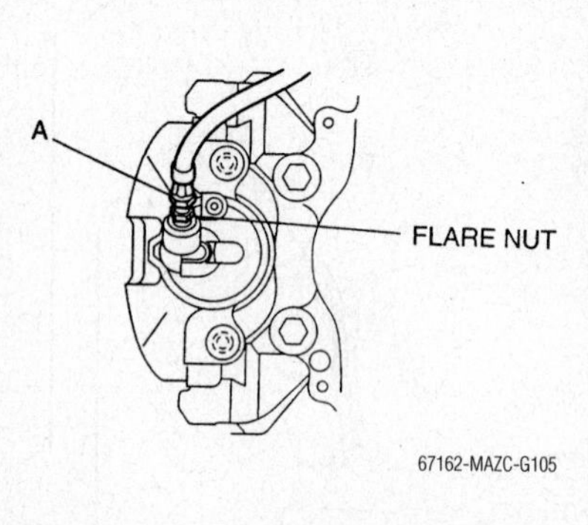

Fig. 10 Tighten the brake hose flare nut while holding the hose at point A—Mazda3

Mazda6 and Mazdaspeed6

See Figure 11.

1. Before servicing the vehicle, refer to the precautions section.
2. Remove or disconnect the following:
   - Wheels
   - Brake hose

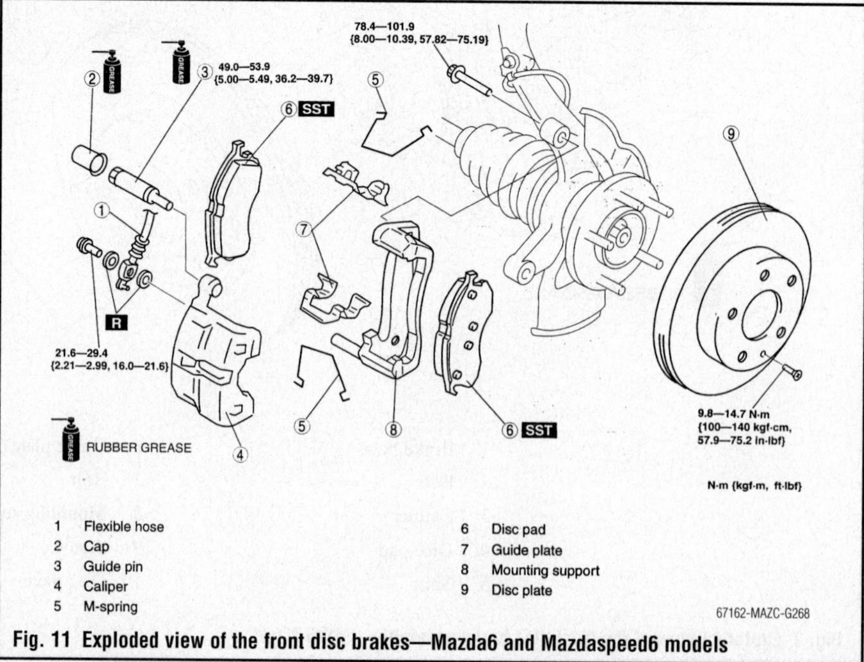

Fig. 11 Exploded view of the front disc brakes—Mazda6 and Mazdaspeed6 models

- Cap from the caliper bolts
- Caliper mounting bolts and the caliper

To install:

3. Install or connect the following:
   - Caliper. Torque the caliper mounting bolts to 36—39 ft. lbs. (49–54 Nm). Install the bolt caps.
   - Brake hose to the caliper. Tighten the bolt to 16–21 ft. lbs. (21–29 Nm) and make sure the brake hose is not twisted. Fill the master cylinder with clean brake fluid and bleed the hydraulic system.
   - Wheels
4. Pump the brake pedal several times to seat the pads.

DISC BRAKE PADS

REMOVAL & INSTALLATION

MX5-Miata

See Figure 12.

1. Before servicing the vehicle, refer to the precautions section.

➡Refer the exploded view illustration for component locations and if applicable, their retainer torque specifications.

2. Remove the lower caliper bolt and pivot the caliper up.
3. Remove the pads and shims.
4. Remove the guide plate.

To install:

5. Compress the piston into the bore.
6. Installation is the reverse of removal. Tighten the lower caliper bolt to 15–23 ft. lbs. (21–31 Nm.

Mazda3 and Mazdaspeed3

See Figures 13 through 15.

1. Before servicing the vehicle, refer to the precautions section.
2. Remove or disconnect the following:
   - Wheels
   - Pad retaining clip
   - Cap from the bottom caliper bolt
   - Lower caliper mounting bolt and swing the caliper up
   - Brake pads

To install:

3. Press the caliper piston back into the cylinder using tool 49 0221 600C.

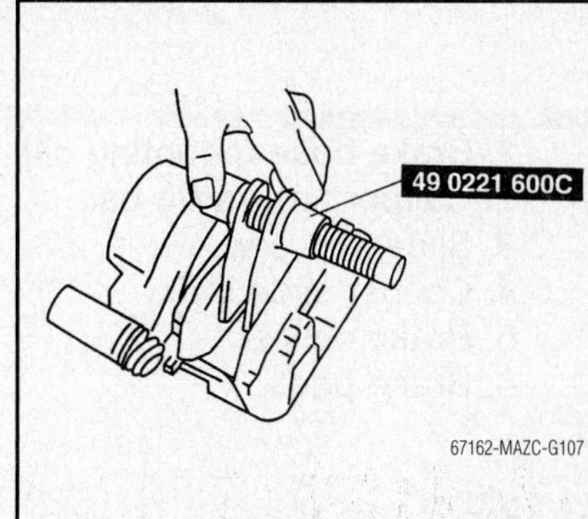

Fig. 13 Press the front caliper piston back into the cylinder—Mazda3

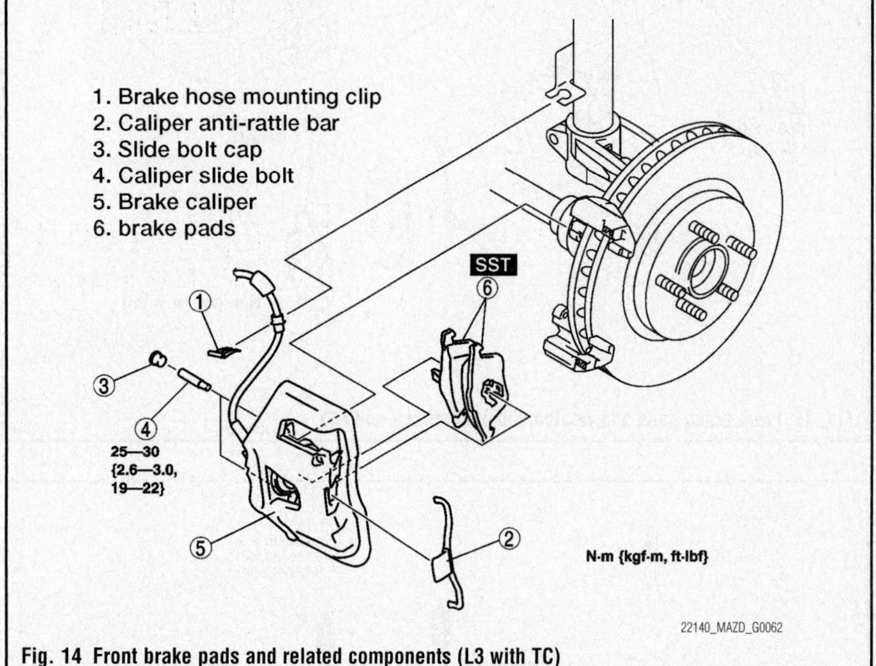

Fig. 14 Front brake pads and related components (L3 with TC)

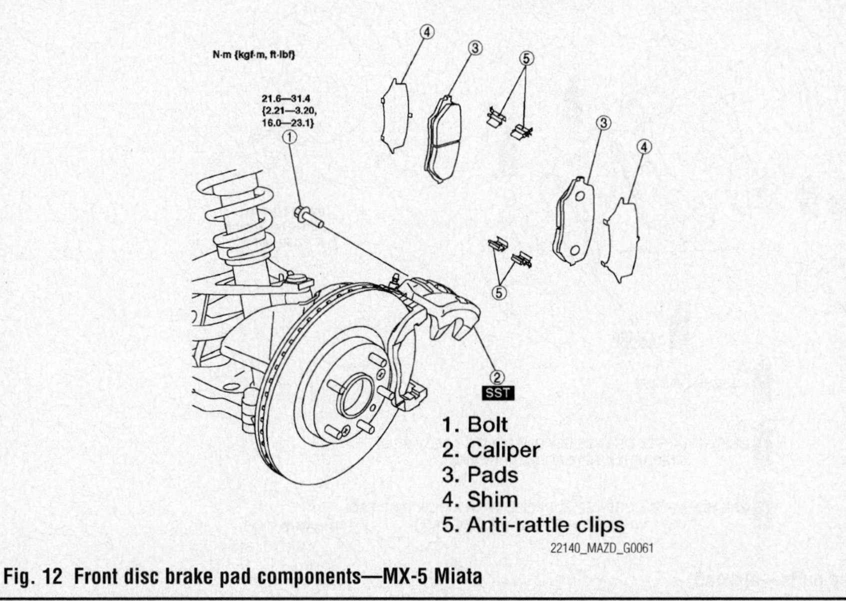

Fig. 12 Front disc brake pad components—MX-5 Miata

4. Install or connect the following:
   - Outer pad to the mounting support and inner pad to the caliper
   - Caliper. Torque the caliper lower mounting bolt to 19–22 ft. lbs. (25–30 Nm). Install the bolt cap.
   - Retaining clip
   - Wheels
5. Pump the brake pedal several times to seat the pads.

Mazda6 and Mazdaspeed6

See Figures 16 and 17.

1. Before servicing the vehicle, refer to the precautions section.
2. Remove or disconnect the following:
   - Wheels
   - Caliper bolts
   - Remove and support caliper
   - Springs

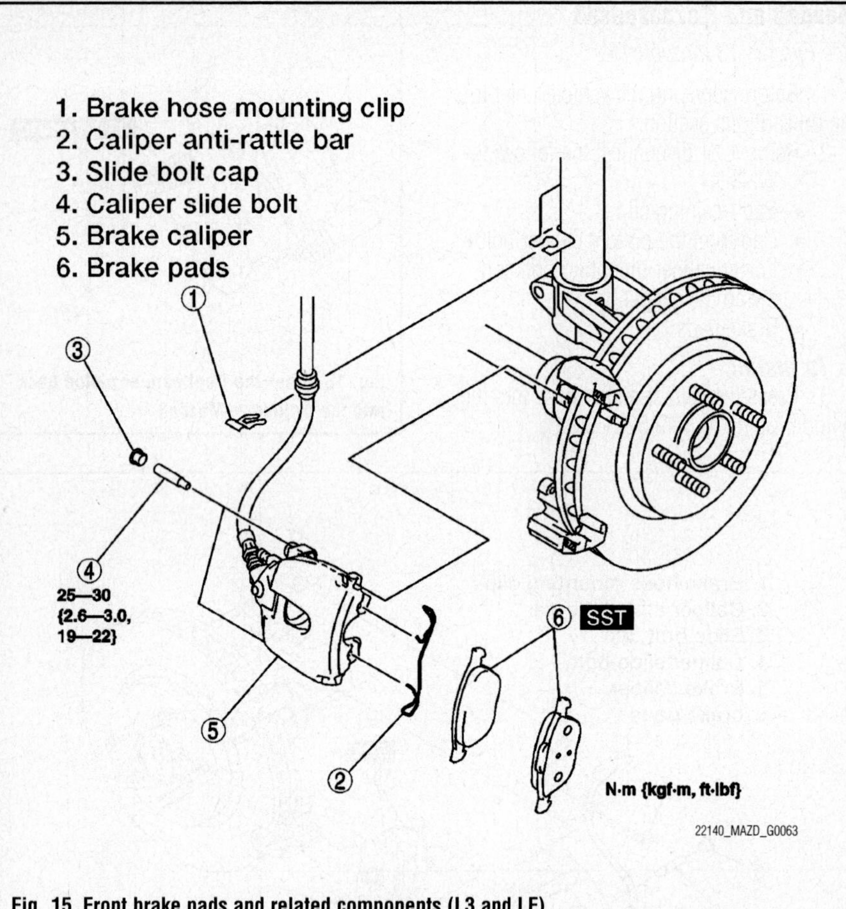

Fig. 15 Front brake pads and related components (L3 and LF)

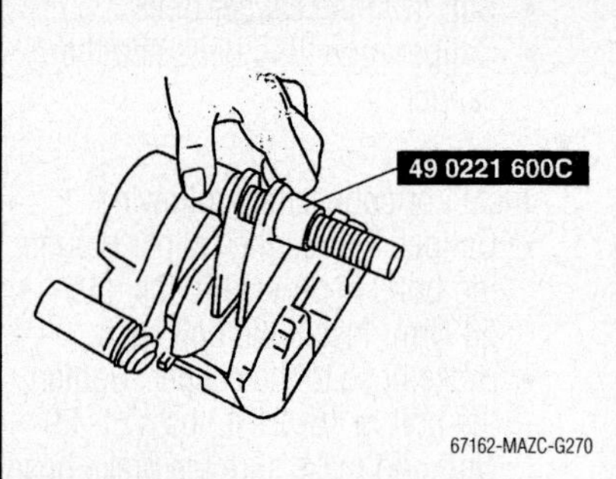

Fig. 17 Press the front caliper pistons back into their bores using tool 49 0221 600C—Mazda6 and Mazdaspeed6 models

- Disc pads
- Shims
- Guide plates

To install:

3. Press the caliper pistons back into their cylinders using tool 49 0221 600C
4. Install or connect the following:
   - Guide plates
   - Shims
   - Wheels
   - Disc pads
   - Springs
   - Caliper
5. Install the caliper mounting bolts and tighten to 23 ft. lbs. (31 Nm).
6. Install the wheels

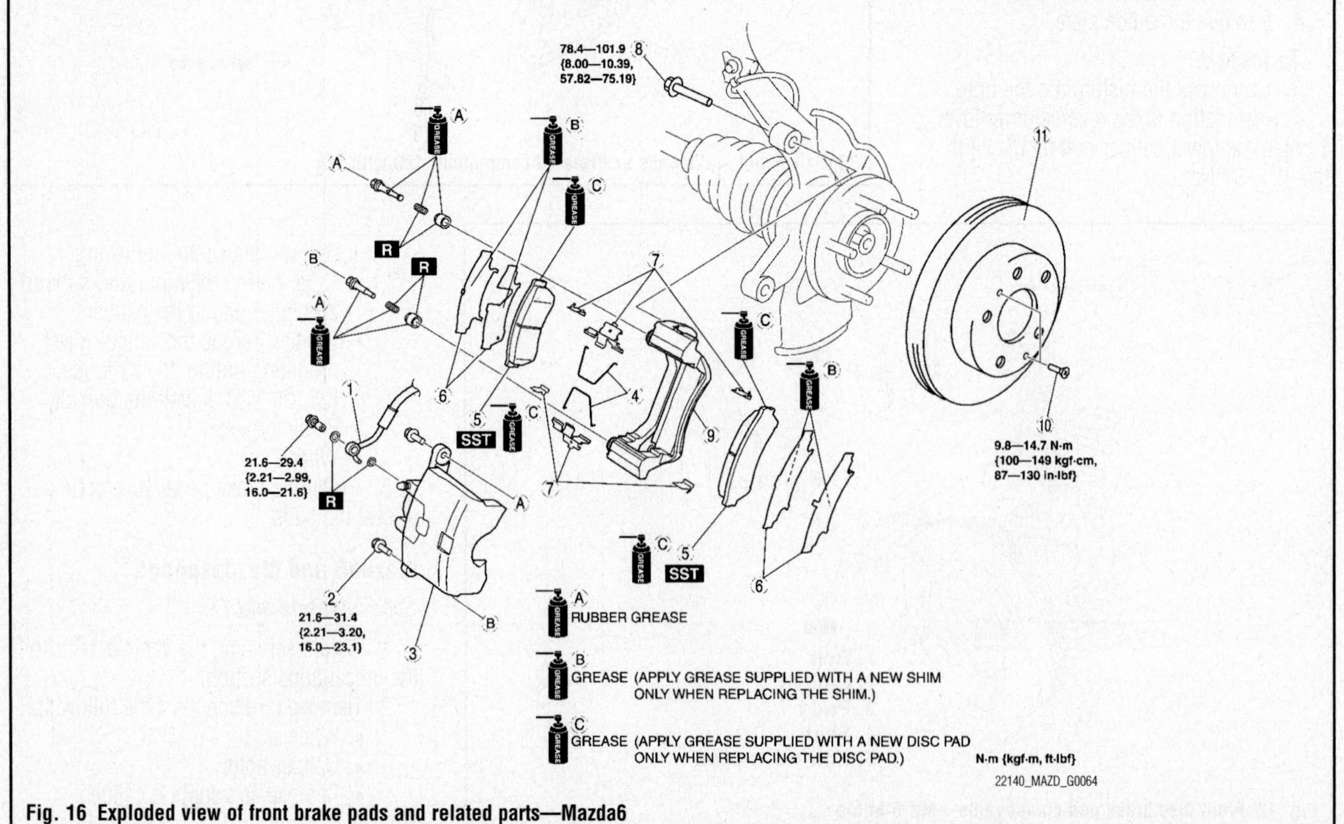

Fig. 16 Exploded view of front brake pads and related parts—Mazda6

CAUTION

Dust and dirt accumulating on brake parts during normal use may contain asbestos fibers from production or aftermarket brake linings. Breathing excessive concentrations of asbestos fibers can cause serious bodily harm. Exercise care when servicing brake parts. Do not sand or grind brake lining unless equipment used is designed to contain the dust residue. Do not clean brake parts with compressed air or by dry brushing. Cleaning should be done by dampening the brake components with a fine mist of water, then wiping the brake components clean with a dampened cloth. Dispose of cloth and all residue containing asbestos fibers in an impermeable container with the appropriate label. Follow practices prescribed by the Occupational Safety and Health Administration (OSHA) and the Environmental Protection Agency (EPA) for the handling, processing, and disposing of dust or debris that may contain asbestos fibers.

BRAKE CALIPER

REMOVAL & INSTALLATION

MX5-Miata

See Figures 18 through 20.

1. Before servicing the vehicle, refer to the precautions section.

➡Refer the exploded view illustration for component locations and if applicable, their retainer torque specifications.

2. Disconnect the parking brake cable.
3. Disconnect the brake hose.
4. Remove the caliper bolts and the caliper.
5. Remove the disc pads and shims.

To install:

6. Clean the exposed area of the piston.
7. Rotate the piston clockwise slowly using the tool shown to push the piston completely until the piston grooves are in the position shown in the figure.
8. Install the caliper. Tighten the bolts to 20–25 Nm (14–18 ft. lbs.).
9. Connect the brake hose and tighten to 21–29 Nm (15–21 ft. lbs.).
10. Bleed the brake system.

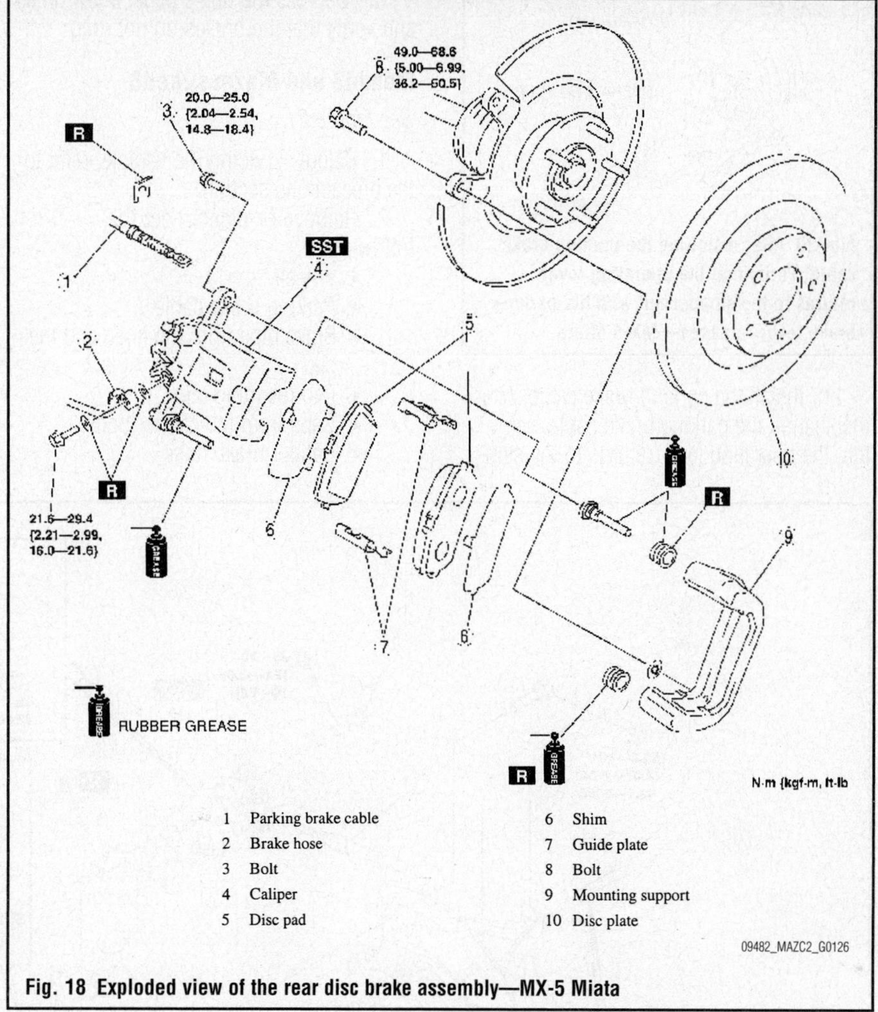

Fig. 18 Exploded view of the rear disc brake assembly—MX-5 Miata

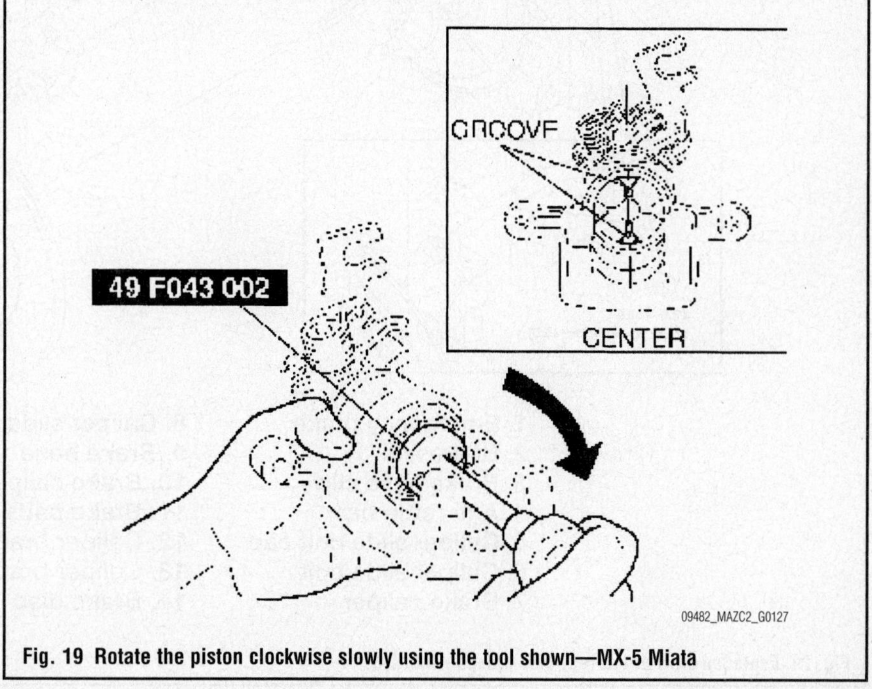

Fig. 19 Rotate the piston clockwise slowly using the tool shown—MX-5 Miata

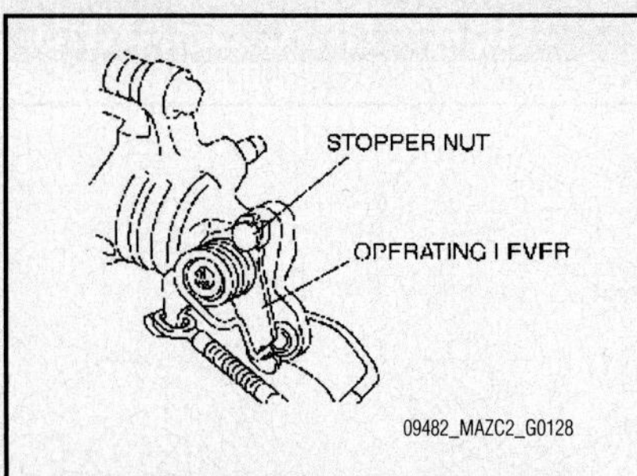

Fig. 20 After installing the parking brake cable, verify that the operating lever returns to the stopper nut with the parking brake lever released—MX-5 Miata

11. Install the parking brake cable. After installation the parking brake cable, verify that the operating lever returns to the stopper nut with the parking brake lever released.

12. Depress the brake pedal a few times and verify that the brakes do not drag

Mazda3 and Mazdaspeed3

See Figure 21.

1. Before servicing the vehicle, refer to the precautions section.
2. Remove or disconnect the following:
   - Wheels
   - Parking brake cable
   - Brake pipe from the hose and the clip
   - Pad retaining clip
   - Caps from the caliper bolts
   - Caliper brake hose
   - Caliper mounting bolts and the caliper

To install:

3. Install or connect the following:
   - Caliper. Torque the caliper mounting bolts to 19–22 ft. lbs. (25–30 Nm). Install the bolt caps.
   - Brake hose and pipe and tighten to 14–16 ft. lbs. (19–23 Nm). Reinstall a new clip and fill the master cylinder with clean brake fluid and bleed the hydraulic system.
   - Pad retaining clip and parking brake cable
   - Wheels
4. Pump the brake pedal several times to seat the pads. Inspect the parking brake

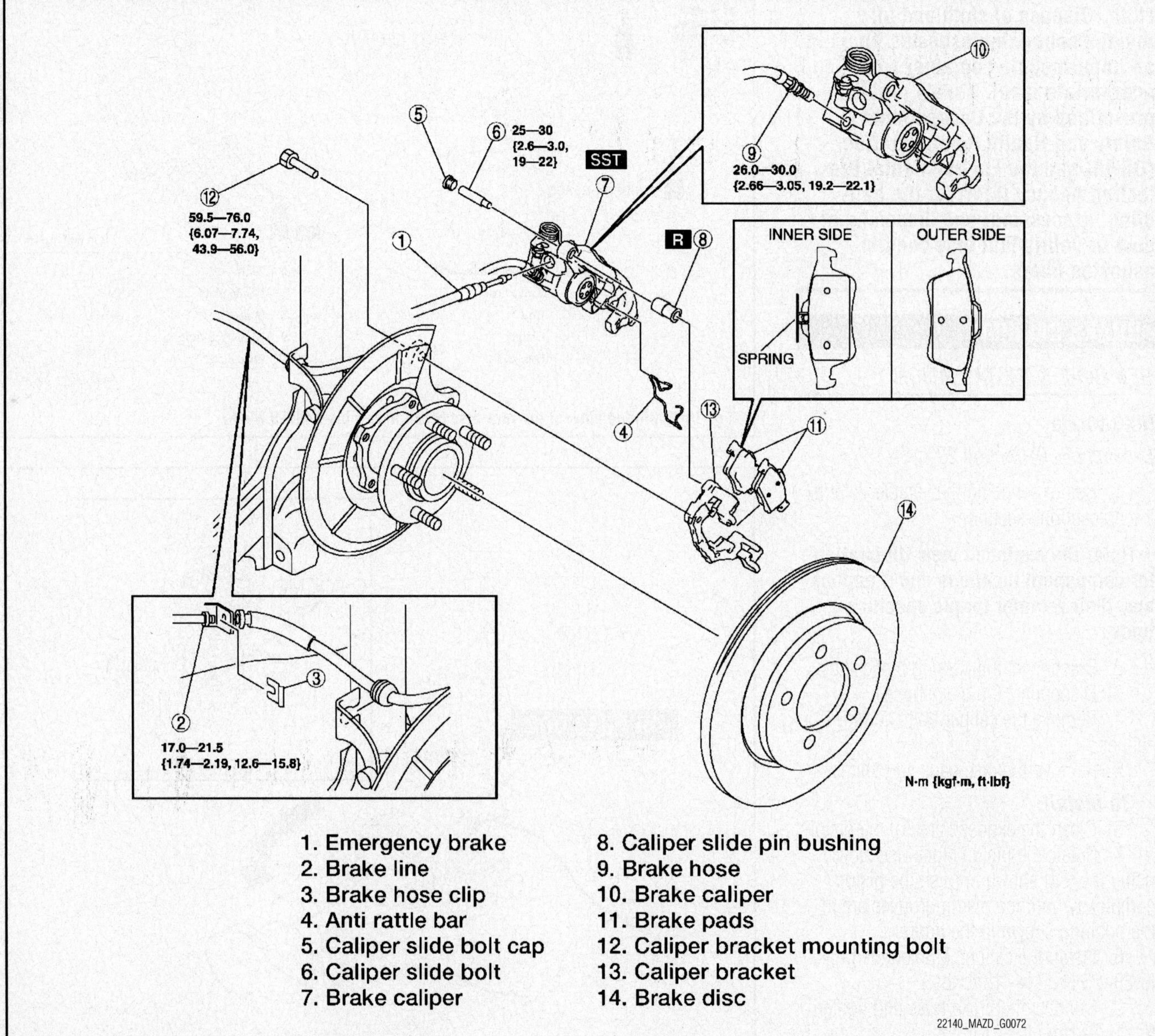

1. Emergency brake
2. Brake line
3. Brake hose clip
4. Anti rattle bar
5. Caliper slide bolt cap
6. Caliper slide bolt
7. Brake caliper
8. Caliper slide pin bushing
9. Brake hose
10. Brake caliper
11. Brake pads
12. Caliper bracket mounting bolt
13. Caliper bracket
14. Brake disc

22140_MAZD_G0072

Fig. 21 Exploded view of the rear disc brakes—Mazda3

lever stroke and brake drag and adjust as necessary.

Mazda6 and Mazdaspeed6

See Figure 22.

1. Before servicing the vehicle, refer to the precautions section.
2. Remove or disconnect the following:
   - Wheels
   - Parking brake cable clip
   - Caliper brake hose bolt
   - Caliper brake hose
   - Caliper mounting bolts and the caliper

To install:

3. Install or connect the following:
   - Caliper. Torque the caliper mounting bolts to 27–36 ft. lbs. (37–49 Nm) on Mazda6models, or 16–23 ft. lbs. (21–31 Nm) on Mazdaspeed6 models.
   - Brake hose and pipe and tighten to 16–21 ft. lbs. (21–29 Nm).
   - Parking brake cable clip and fill the master cylinder with clean brake fluid and bleed the hydraulic system.
   - Wheels
4. Pump the brake pedal several times to seat the pads. Inspect the parking brake lever stroke and brake drag and adjust as necessary.

DISC BRAKE PADS

REMOVAL & INSTALLATION

MX5-Miata

See Figure 23.

1. Before servicing the vehicle, refer to the precautions section.

➡Refer the exploded view illustration for component locations and if applicable, their retainer torque specifications.

2. Disconnect the parking brake cable.
3. Remove the upper caliper bolt and pivot the caliper downwards.
4. Remove the disc pads and shims.

To install:

5. Clean the exposed area of the piston.
6. Rotate the piston clockwise slowly using the tool shown to push the piston completely until the piston grooves are in the position shown in the figure.
7. Install the components in the reverse order of removal. Tighten the caliper bolts to 20–25 Nm (14–18 ft. lbs.).
8. Install the parking brake cable. After installation the parking brake cable, verify that the operating lever returns to the stopper nut with the parking brake lever released.
9. Depress the brake pedal a few times and verify that the brakes do not drag.

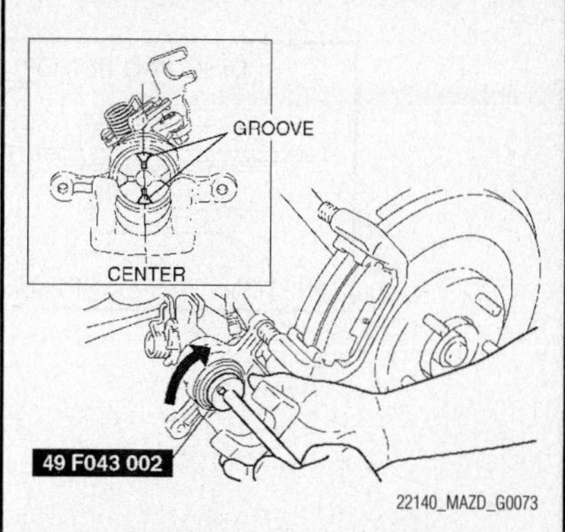

Fig. 23 Rotate the piston clockwise slowly using the tool shown

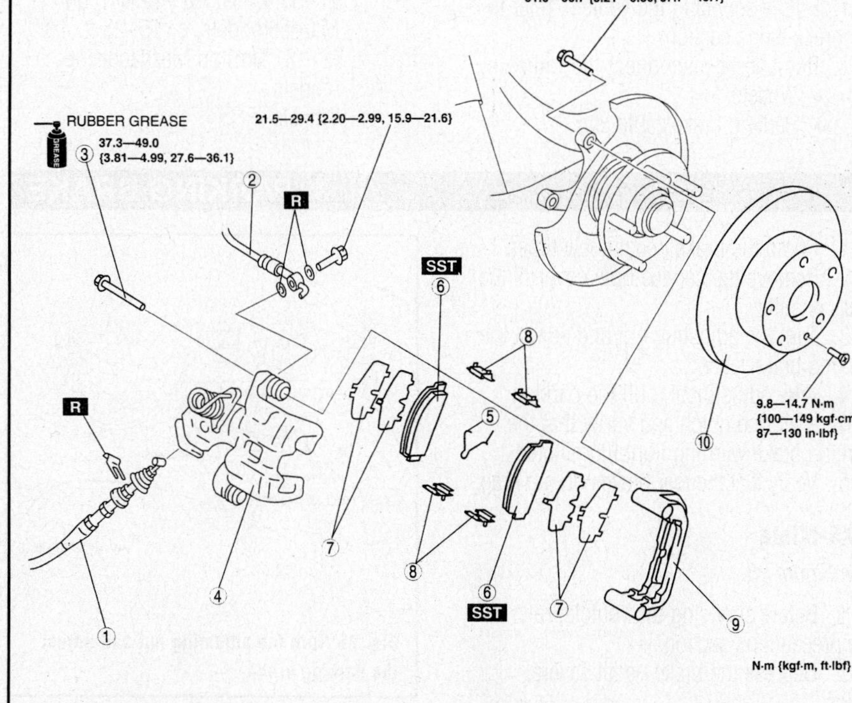

Fig. 22 Exploded view of the rear disc brakes—Mazda6, shown, Mazdaspeed6 models similar

Mazda3 and Mazdaspeed3

See Figures 24 and 25.

1. Before servicing the vehicle, refer to the precautions section.
2. Remove or disconnect the following:
   - Wheels
   - Parking brake cable
   - Pad retaining clip
   - Caps from the caliper bolts
   - Caliper mounting bolts and the caliper
   - Outer brake pad from the mount support and pull the inner pad from the caliper

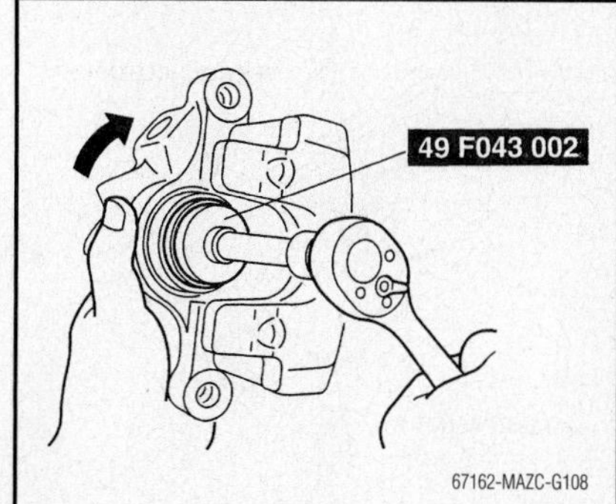

Fig. 24 Turn the rear caliper piston clockwise slowly until the piston is fully seated in its bore—Mazda3

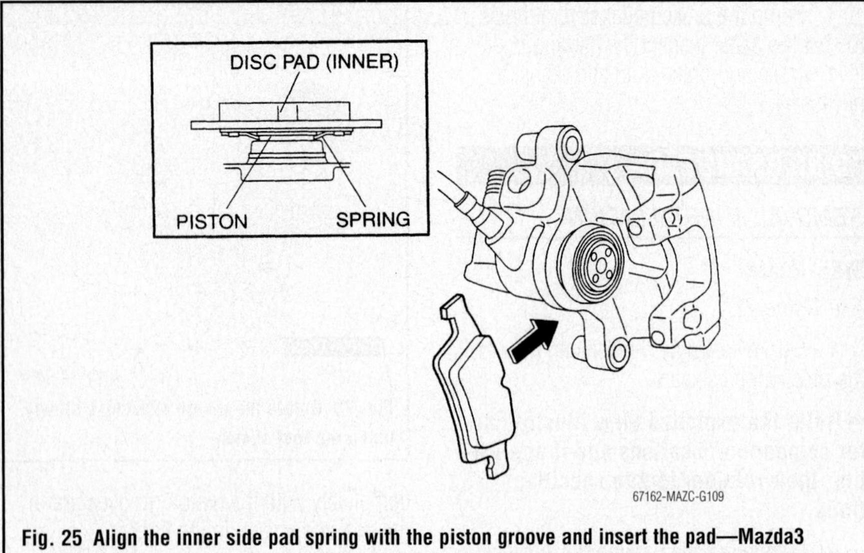

Fig. 25 Align the inner side pad spring with the piston groove and insert the pad—Mazda3

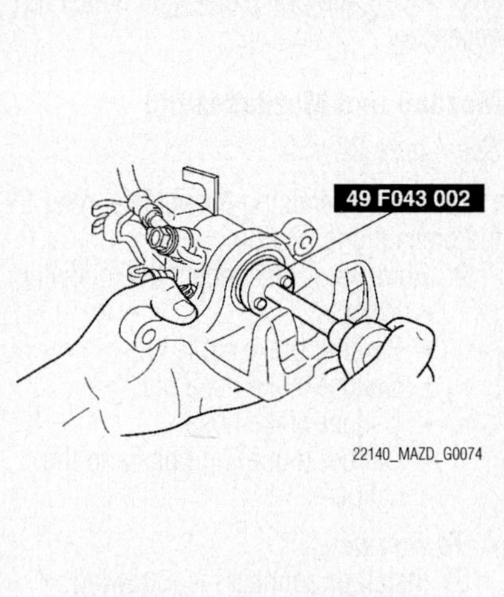

Fig. 26 Press the rear caliper pistons back into their bores using tool 49 FD43 002

To install:

3. Install the out pad to the mounting support and clean the piston area.
4. Using tool 49 F043 002, turn the piston clockwise slowly until the piston is fully seated in its bore.
5. Align the inner side pad spring with the piston groove and insert the pad. Refer to the illustration for spring location and inner pad installation arrow.
6. Install or connect the following:
   - Caliper. Torque the caliper mounting bolts to 19–22 ft. lbs. (25–30 Nm). Install the bolt caps.
   - Pad retaining clip and parking brake cable
   - Wheels
7. Pump the brake pedal several times to seat the pads. Inspect the parking brake lever stroke and brake drag and adjust as necessary.

Mazda6 and Mazdaspeed6

See Figure 26.

1. Before servicing the vehicle, refer to the precautions section.
2. Remove or disconnect the following:
   - Wheels
   - Parking brake cable clip
   - Upper caliper bolt and rotate the caliper downwards
   - V-springs, pads and the shims from the pads

To install:

3. Press the rear caliper pistons back into their bores using tool 49 FD43 002.
4. Install or connect the following:
   - Pads, shims, and springs
   - Caliper and torque the bolt to 27–36 ft. lbs. (37–49 Nm) on Mazda6models, or 16–23 ft. lbs. (21–31 Nm) on Mazdaspeed6 models
   - Wheels

BRAKES

PARKING BRAKE

PARKING BRAKE CABLES

ADJUSTMENT

Mazda3 and Mazdaspeed3

See Figure 27.

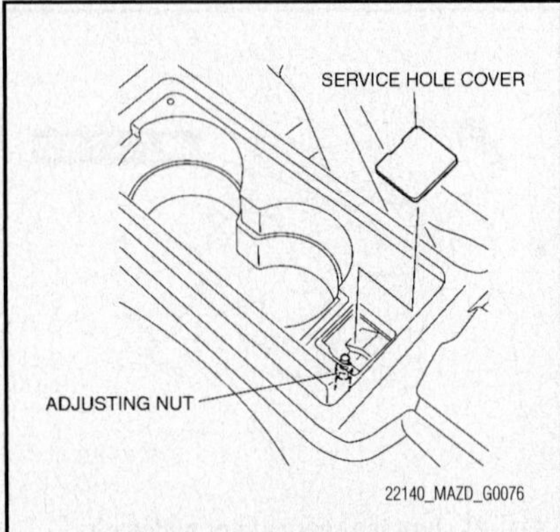

Fig. 27 Remove the service hole cover and adjust the parking brake

1. Pump the brake pedal a few times.
2. Remove the service hole cover of the rear console.
3. Turn the adjusting nut and adjust the parking brake lever.
4. After adjustment, pull the parking brake lever one notch and verify that the parking brake warning light illuminates.
5. Verify that the rear brakes do not drag.

MX5-Miata

See Figure 28.

1. Before servicing the vehicle, refer to the precautions section.
2. Depress the brake pedal several times.
3. Remove the parking brake lever boot
4. Turn the adjusting nut and adjust the parking brake lever.
5. After adjustment, pull the parking brake lever one notch and verify that the parking brake warning light illuminates.
6. Verify that the rear brakes do not drag.

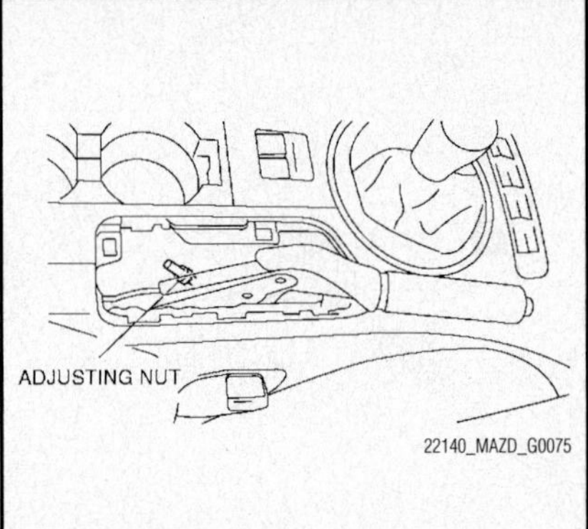

Fig. 28 Turn the adjusting nut and adjust the parking brake

Mazda6 and Mazdaspeed6

See Figure 29.

1. Before servicing the vehicle, refer to the precautions section.
2. Start the engine and depress the brake pedal several times.

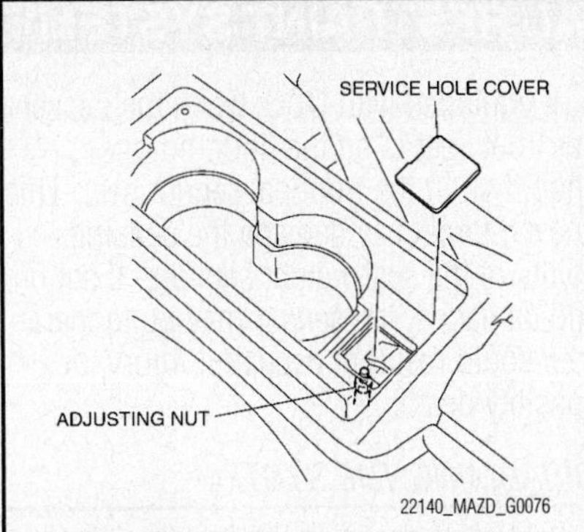

Fig. 29 Verify the position of the adjusting nut while looking in from above the console (A) as shown, and set the socket on the nut

3. Stop the engine.
4. Pull the underside of the cover out and remove the clips from the console.
5. Disengage the hooks from the console and remove the cover.
6. Verify the position of the adjusting nut while looking in from above the console, and set the socket on the nut.
7. Turn the adjusting nut to adjust the parking brake lever.
8. After adjustment, inspect the following points:
9. Turn the ignition switch to ON, pull the parking brake lever one notch, and verify that the parking brake warning light illuminates.
10. Verify that the rear brakes do not drag.

PARKING BRAKE SHOES

REMOVAL & INSTALLATION

Mazda3 and Mazdaspeed3

See Figure 30.

The parking brake system uses the rear brake pads as the parking brake. Refer to rear brake shoe removal and installation in this section. The parking brake cable actuates the rear caliper to apply and hold the brake pads.

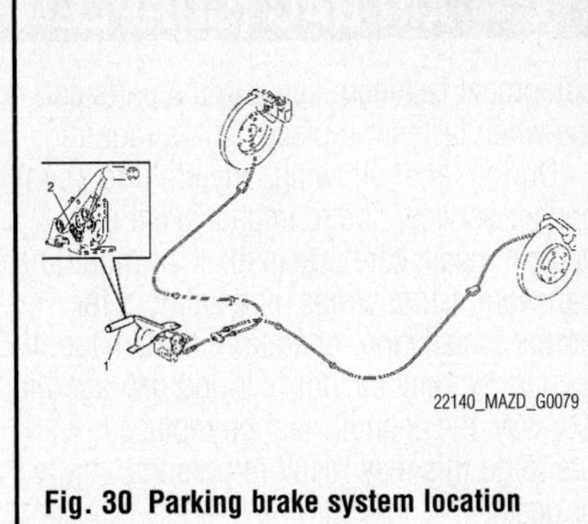

Fig. 30 Parking brake system location index

1. Before servicing the vehicle, refer to the precautions section.
2. Remove or disconnect the following:
   - Wheels
   - Parking brake cable at the lever and caliper

To install:

3. Installation is the reverse of removal.

MX5-Miata

See Figure 31.

1. Before servicing the vehicle, refer to the precautions section.

The parking brake system uses the rear brake pads as the parking brake. Refer to rear brake shoe removal and installation in this section. The parking brake cable actuates the rear caliper to apply and hold the brake pads.

2. Before servicing the vehicle, refer to the precautions section.
3. Remove or disconnect the following:
   - Wheels
   - Parking brake cable at the lever and caliper

To install:

4. Installation is the reverse of removal.

Mazda6 and Mazdaspeed6

See Figure 32.

1. Before servicing the vehicle, refer to the precautions section.

The parking brake system uses the rear brake pads as the parking brake. Refer to rear brake shoe removal and installation in this section. The parking brake cable actuates the rear caliper to apply and hold the brake pads.

2. Before servicing the vehicle, refer to the precautions section.
3. Remove or disconnect the following:
   - Wheels
   - Parking brake cable at the lever and caliper

To install:

4. Installation is the reverse of removal.

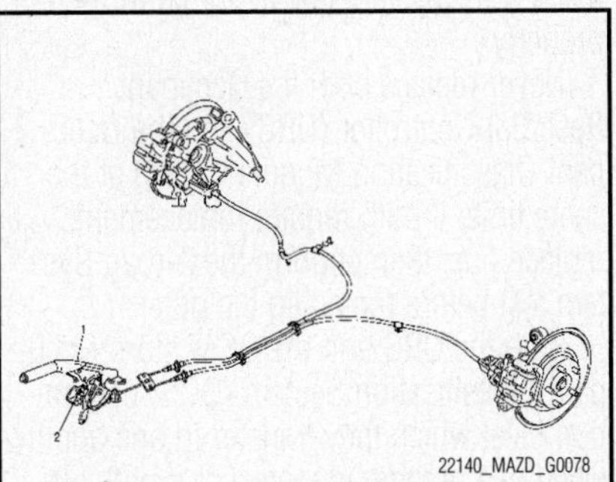
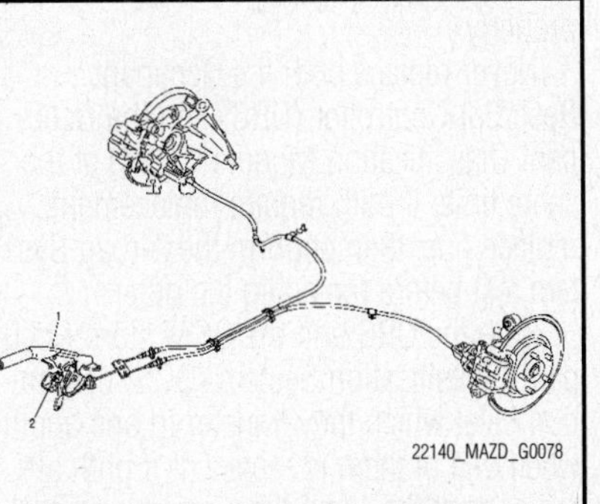

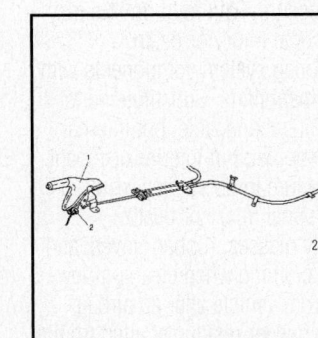

Fig. 31 Parking brake system location index

22140_MAZD_G0077

Fig. 32 Parking brake system location index

GENERAL INFORMATION

Mazda vehicles are equipped with an air bag system. The system must be disarmed before performing service on, or around, system components, the steering column, instrument panel components, wiring and sensors. Failure to follow the safety precautions and the disarming procedure could result in accidental air bag deployment, possible injury and unnecessary system repairs.

SERVICE PRECAUTIONS

Disconnect and isolate the battery negative cable before beginning any airbag system component diagnosis, testing, removal, or installation procedures. Allow system capacitor to discharge for two minutes before beginning any component service. This will disable the airbag system. Failure to disable the airbag system may result in accidental airbag deployment, personal injury, or death.

Do not place an intact undeployed airbag face down on a solid surface. The airbag will propel into the air if accidentally deployed and may result in personal injury or death.

When carrying or handling an undeployed airbag, the trim side (face) of the airbag should be pointing towards the body to minimize possibility of injury if accidental deployment occurs. Failure to do this may result in personal injury or death.

Replace airbag system components with OEM replacement parts. Substitute parts may appear interchangeable, but internal differences may result in inferior occupant protection. Failure to do so may result in occupant personal injury or death.

Wear safety glasses, rubber gloves, and long sleeved clothing when cleaning powder residue from vehicle after an airbag deployment. Powder residue emitted from a deployed airbag can cause skin irritation. Flush affected area with cool water if irritation is experienced. If nasal or throat irritation is experienced, exit the vehicle for fresh air until the irritation ceases. If irritation continues, see a physician.

Do not use a replacement airbag that is not in the original packaging. This may result in improper deployment, personal injury, or death.

The factory installed fasteners, screws and bolts used to fasten airbag components have a special coating and are specifically designed for the airbag system. Do not use substitute fasteners. Use only original equipment fasteners listed in the parts catalog when fastener replacement is required.

During, and following, any child restraint anchor service, due to impact event or vehicle repair, carefully inspect all mounting hardware, tether straps, and anchors for proper installation, operation, or damage. If a child restraint anchor is found damaged in any way, the anchor must be replaced. Failure to do this may result in personal injury or death.

Deployed and non-deployed airbags may or may not have live pyrotechnic material within the airbag inflator.

Do not dispose of driver/passenger/curtain airbags or seat belt tensioners unless you are sure of complete deployment. Refer to the Hazardous Substance Control System for proper disposal.

Dispose of deployed airbags and tensioners consistent with state, provincial, local, and federal regulations.

After any airbag component testing or service, do not connect the battery negative cable. Personal injury or death may result if the system test is not performed first.

If the vehicle is equipped with the Occupant Classification System (OCS), do not connect the battery negative cable before performing the OCS Verification Test using the scan tool and the appropriate diagnostic information. Personal injury or death may result if the system test is not performed properly.

Never replace both the Occupant Restraint Controller (ORC) and the Occupant Classification Module (OCM) at the same time. If both require replacement, replace one, then perform the Airbag System test before replacing the other.

Both the ORC and the OCM store Occupant Classification System (OCS) calibration data, which they transfer to one another when one of them is replaced. If both are replaced at the same time, an irreversible fault will be set in both modules and the OCS may malfunction and cause personal injury or death.

If equipped with OCS, the Seat Weight Sensor is a sensitive, calibrated unit and must be handled carefully. Do not drop or handle roughly. If dropped or damaged, replace with another sensor. Failure to do so may result in occupant injury or death.

If equipped with OCS, the front passenger seat must be handled carefully as well. When removing the seat, be careful when setting on floor not to drop. If dropped, the sensor may be inoperative, could result in occupant injury, or possibly death.

If equipped with OCS, when the passenger front seat is on the floor, no one should sit in the front passenger seat. This uneven force may damage the sensing ability of the seat weight sensors. If sat on and damaged, the sensor may be inoperative, could result in occupant injury, or possibly death.

DISARMING THE SYSTEM

1. Before servicing the vehicle, refer to the precautions section.
2. If equipped, deactivate the audio anti-theft system.
3. Turn the ignition switch to LOCK.
4. Disconnect and isolate the negative battery cable and wait for more than 1 minute to allow the backup power supply to deplete its stored power.

ARMING THE SYSTEM

1. Before servicing the vehicle, refer to the precautions section.
2. Connect the negative battery cable, turn the ignition switchONand verify the air bag warning light cones on for 6 seconds. If the light does not illuminate there are problems with the system.
3. If equipped, activate the audio anti-theft system.

CLOCKSPRING CENTERING

Mazda3 and Mazdaspeed3

See Figures 33 through 37.

**** WARNING**

Handling the air bag module improperly can accidentally operate (deploy) the air bag module, which may seriously injure you.

1. Before servicing the vehicle, refer to the precautions section.
   a. Remove the air bag cover.
   b. Remove the bolt.
   c. Using a flathead screwdriver, pry out the connector stopper plate and unplug the connector.
   d. Remove the driver's side air bag module.
   e. Remove the lockbolt.

**** WARNING**

Handling the air bag module improperly can accidentally deploy the air bag module, which may seriously injure you.

   f. Remove the locknut.

**** CAUTION**

Do not try to remove the steering wheel by hitting the shaft with a hammer. The column will collapse.

g. Set the vehicle in the straight-ahead position.

h. Remove the steering wheel using a suitable puller.

i. Remove the column cover.

j. Remove the clock spring connector.

k. Remove the steering angle sensor connector, if equipped with a steering angle sensor.

l. Remove the screw.

m. Remove the clock spring.

n. Remove the steering angle sensor by detaching the tabs at the four locations shown in the figure and remove the steering angle sensor.

To install:

2. Install the clock spring.
3. Adjust the clock spring as follows:

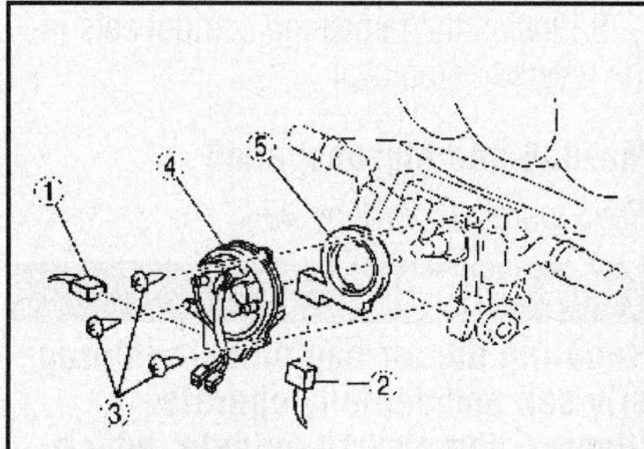

Fig. 33 Exploded view of clock spring assembly—Mazda3

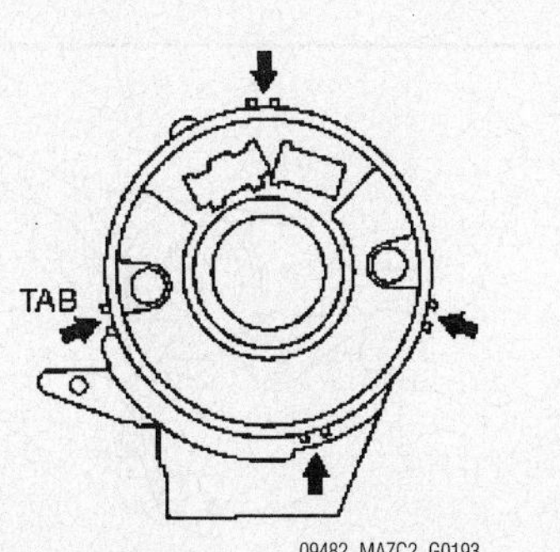

Fig. 34 Remove the steering angle sensor by detaching the tabs at the four locations shown—Mazda3

**** CAUTION**

If the clock spring is not adjusted, the spring wire in the clock spring could over-wind and break when the steering wheel is turned. Always adjust the clock spring after installing it.

- Set the front tires straight-ahead

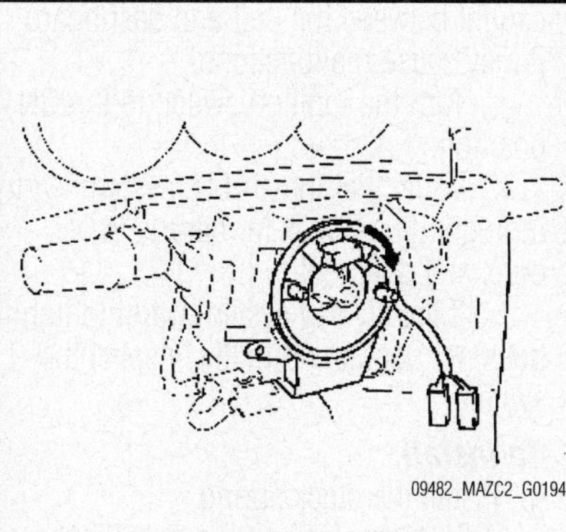

Fig. 35 Turn the clock spring clockwise until it stops—Mazda3

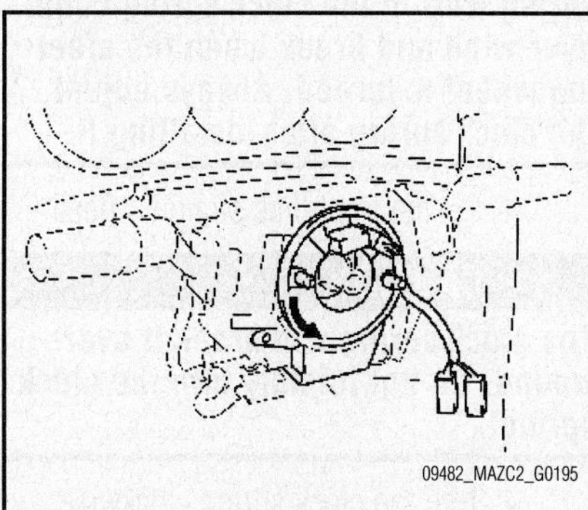

Fig. 36 From the stopped position, turn the clock spring counterclockwise 2¾ turns—Mazda3

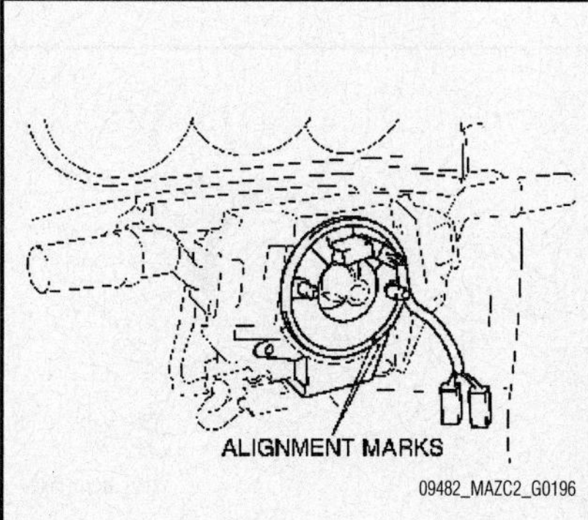

Fig. 37 Align the marks on the clock spring—Mazda3

**** CAUTION**

The clock spring will break if over-wound. Do not forcibly turn the clock spring.

- Turn the clock spring clockwise until it stops
- From the stopped position, turn the clock spring counterclockwise 2 ¾ turns
- Align the marks.

4. Install the remaining components in the reverse of removal.

MX5-Miata

See Figures 38 through 40.

**** WARNING**

Handling the air bag module improperly can accidentally deploy the air bag module, which may seriously injure you.

1. Before servicing the vehicle, refer to the precautions section.
2. Remove the steering wheel as follows:

a. Remove the locknut.

**** CAUTION**

Do not try to remove the steering wheel by hitting the shaft with a hammer. The column will collapse.

b. Set the vehicle in the straight-ahead position.

c. Remove the steering wheel using a suitable puller.

d. Remove the column cover.

e. Remove the clock spring screw, connector and clock spring.

f. Make sure the wheels in the straight-ahead position, before installing the steering wheel.

g. After installing the combination switch, perform the steering angle sensor initialization procedure

**** WARNING**

Unless the initialization procedure of the steering angle sensor is completed, the DSC will not operate, causing an unexpected accident. Therefore, always perform the initialization procedure to ensure DSC operation if the power supply to the steering angle sensor has been cut off due to disconnection of the steering angle sensor connector or negative battery cable, or any other cause.

➡The initialization value of the steering angle sensor is stored using the battery power supply. Therefore, the battery power supply of the steering angle sensor is cut and the stored initialization value is cleared when any of the following items are performed.

- Negative battery cable disconnection
- Steering angle sensor connector disconnection
- Fuse (ROOM 15A) removal
- Wiring harness disconnection between battery and steering angle sensor connector
- Inspect the wheel alignment, inflation pressure, and the installation condition of the steering wheel
- If there is any malfunction, adjust the applicable part
- Connect the negative battery cable
- Turn the ignition switch to the ON position
- Confirm that the DSC indicator light illuminates and that the DSC OFF light flashes
- Turn the steering wheel to full right lock, then turn it to full left lock
- Confirm that the DSC OFF light goes out
- Turn the ignition switch off
- Turn the ignition switch to the ON position again, and confirm that the DSC indicator light goes out
- If the DSC indicator light does not go out, disconnect the negative battery cable, and perform the procedure again
- Drive the vehicle for about 10 minutes and confirm that the ABS warning and DSC indicator lights do not illuminate

h. If the initialization procedure of the steering angle signal is not completed, the DSC will not operate properly and may cause an accident. Therefore, always perform initialization of the DSC HU/CM steering angle signal to ensure proper DSC operation when any of the following items are performed.

- Steering angle sensor replacement DSC HU/CM replacement
- Inspect the wheel alignment and inflation pressure.
- Park the vehicle on level ground
- Turn the ignition switch off
- Connect the WDS or equivalent to the DLC-2
- Access the active command mode, select the following commands, and then follow the indication on the monitor
- Drive the vehicle forward
- After 5 min of driving, verify that the DSC system is normal

i. When installing the center panel unit, make sure that the wiring harness and antenna feeder are not caught between the unit and dashboard. If the wiring harness or the antenna feeder is caught between the unit and dashboard, it may cause malfunctions.

j. Turn the ignition switch to the ON position.

k. Verify that the air bag system warning light illuminates for approx. 6 seconds and goes out.

l. If the air bag system warning light does not operate normally, inspect the system.

To install:

3. Install the clock spring.
4. Adjust the clock spring as follows:

✻✻ CAUTION

If the clock spring is not adjusted, the spring wire in the clock spring could over-wind and break when the steering wheel is turned. Always adjust the clock spring after installing it.

- Set the front tires straight-ahead

✻✻ CAUTION

The clock spring will break if overwound. Do not forcibly turn the clock spring.

- Turn the clock spring clockwise until it stops
- From the stopped position, turn the clock spring counterclockwise 2¾ turns
- Align the marks.

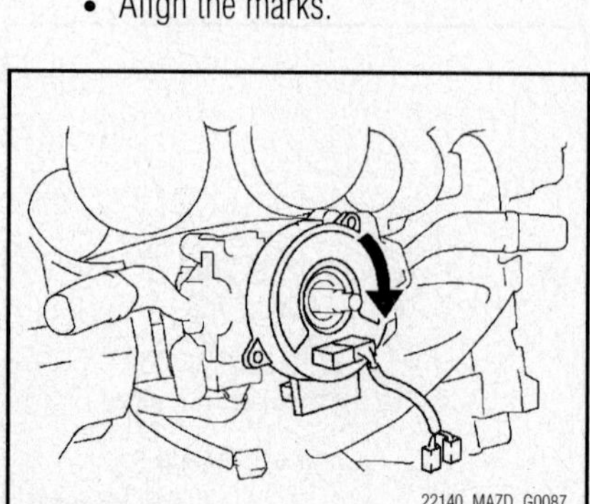

Fig. 38 Turn the clock spring clockwise until it stops

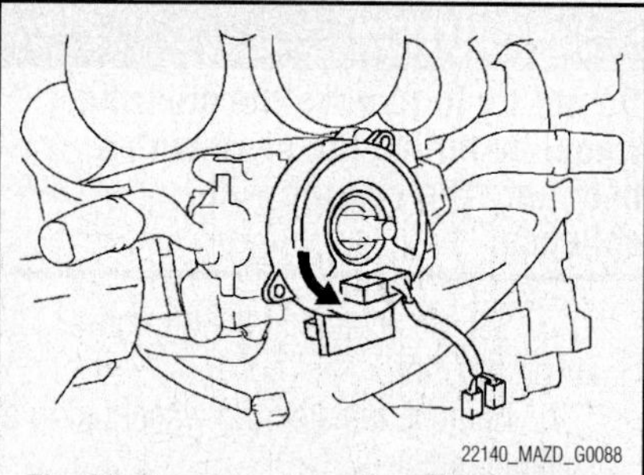

Fig. 39 From the stopped position, turn the clock spring counterclockwise 2¾ turns

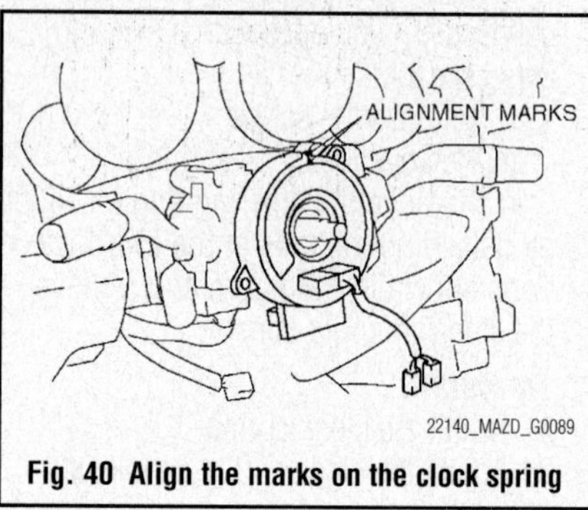

Fig. 40 Align the marks on the clock spring

5. Install the remaining components in the reverse of removal.

Mazda6 and Mazdaspeed6

See Figures 41 through 43.

✻✻ WARNING

Handling the air bag module improperly can accidentally operate (deploy) the air bag module, which may seriously injure you.

1. Before servicing the vehicle, refer to the precautions section.
   a. Remove the air bag cover.
   b. Remove the bolt.
   c. Using a flathead screwdriver, pry out the connector stopper plate and unplug the connector.

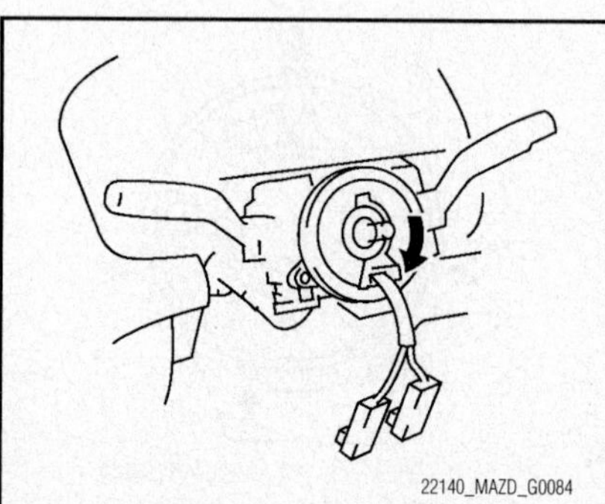

Fig. 41 Turn the clock spring clockwise until it stops—Mazda6 and Mazdaspeed6

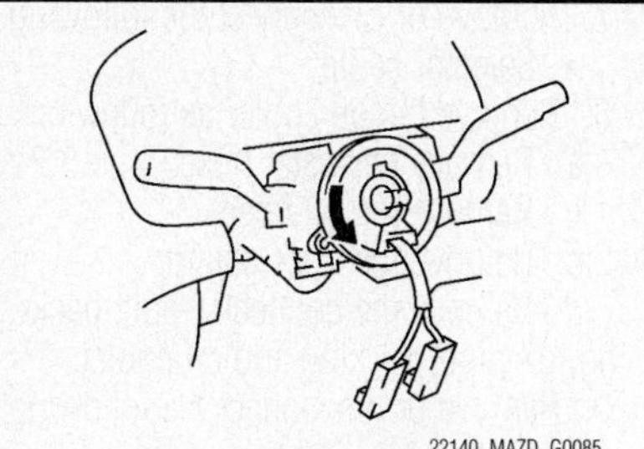

Fig. 42 From the stopped position, turn the clock spring counterclockwise 2 ¾ turns—Mazda6 and Mazdaspeed6

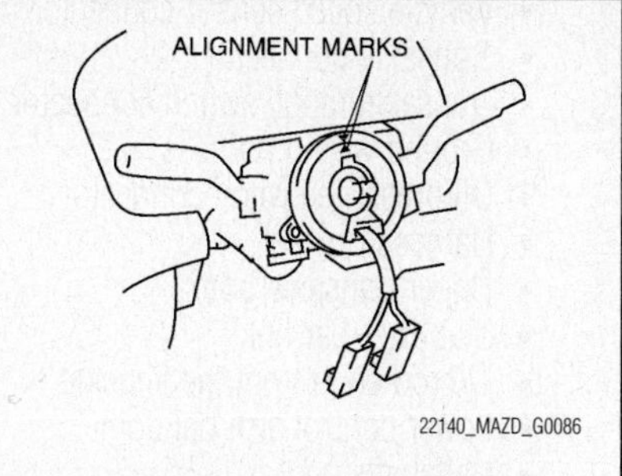

Fig. 43 Align the marks on the clock spring—Mazda6 and Mazdaspeed6

d. Remove the driver's side air bag module.

e. Remove the lockbolt.

⁂ WARNING

Handling the air bag module improperly can accidentally deploy the air bag module, which may seriously injure you.

f. Remove the locknut.

⁂ CAUTION

Do not try to remove the steering wheel by hitting the shaft with a hammer. The column will collapse.

g. Set the vehicle in the straight-ahead position.

h. Remove the steering wheel using a suitable puller.

i. Remove the column cover.

j. Remove the clock spring connector.

k. Remove the screw.

l. Remove the clock spring.

m. Adjust the clock spring as follows:

⁂ CAUTION

If the clock spring is not adjusted, the spring wire in the clock spring could over-wind and break when the steering wheel is turned. Always adjust the clock spring after installing it.

- Set the front tires straight-ahead

⁂ CAUTION

The clock spring will break if overwound. Do not forcibly turn the clock spring.

- Turn the clock spring clockwise until it stops
- From the stopped position, turn the clock spring counterclockwise 2¾ turns
- Align the marks

DRIVETRAIN

AUTOMATIC TRANSMISSION ASSEMBLY

REMOVAL & INSTALLATION

MX-5 Miata (SJ6J-EL)

See Figure 44.

1. Before servicing the vehicle, refer to the precautions section.
2. Remove the battery cover.
3. Disconnect the negative battery cable.
4. Drain the ATF.
5. Loosen the starter installation bolts only enough that the starter is loose, but not removed.
6. Remove or disconnect the following:
   - Tunnel member component
   - Middle pipe
   - Manual shaft lever component
   - Transverse member
   - Undercover
   - Torque converter installation nuts
   - Oil pipe, oil hose
   - Insulator
   - TR switch connector
   - Solenoid valve connector
   - VSS connector
   - Turbine sensor connector
   - Oil pressure switch connector (for oil filter)

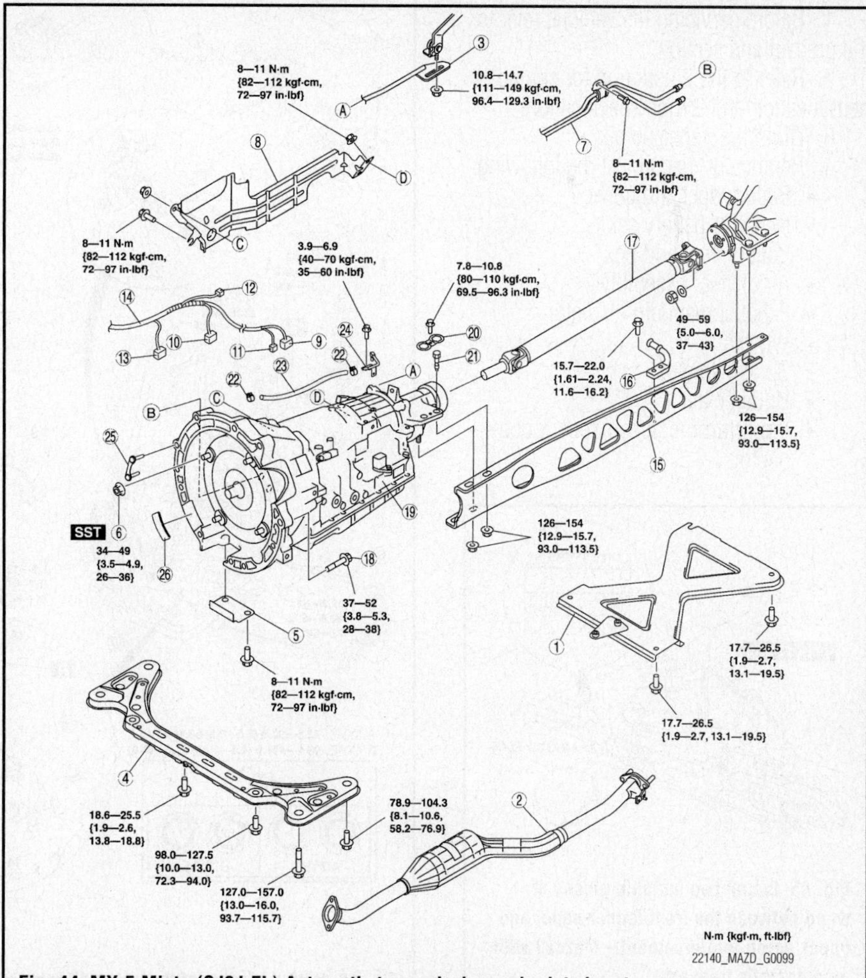

Fig. 44 MX-5 Miata (SJ6J-EL) Automatic transmission and related parts

- Wiring harness
- Power plant frame
- Hanger bracket
- Transmission installation bolt and nut

7. To prevent the torque converter and transmission from separating, remove the transmission without tilting it toward the torque converter.
8. Remove the transmission assembly.

To install:

9. Install in the reverse order of removal.
10. Add ATF and, with the engine idling, inspect the ATF level and inspect for leakage.
11. Inspect selector lever operation.
12. Inspect for leakage of ATF from all connecting points.
13. Perform the road test.

AUTOMATIC TRANSAXLE ASSEMBLY

REMOVAL & INSTALLATION

Mazda3 and Mazdaspeed3

See Figures 45 through 51.

1. Before servicing the vehicle, refer to the precautions section.
2. Refer to the illustration for component location and torque specifications.
3. Drain the transaxle oil.
4. Remove or disconnect the following:
   - Battery duct and cover
   - Negative battery cable
   - Battery, tray and box
   - Air cleaner assembly
   - Exhaust manifold insulator
   - Wheels
   - Splash shields
   - Undercover
   - Input/turbine speed sensor connector
   - Vehicle speed sensor connector
   - Transaxle connector
   - Transaxle range switch connector
   - Ground wiring harness
   - Oil pressure switch connector
   - Harness bracket
   - Upper transaxle bolts
   - Stabilizer bar link
   - Tie rod ends from the knuckle
   - Lower control arm ball joint
   - Halfshafts
5. Remove the joint shaft as follows:
   a. Disconnect the right halfshaft from the joint shaft by tapping the transaxle side outer ring with a brass bar and a hammer. Disconnect the joint shaft bracket from the block and remove the joint shaft.
   b. Install tool 49 G030 455 to hold the side gears after removal
6. Support the engine assembly with a engine support assembly such as 49 E017 5A0.
7. Remove or disconnect the following:
   - Selector cable
8. Remove the oil cooler as follows:
   a. Remove the water hose.
   b. Remove the oil hose.
   c. Remove the hose clamp.
   d. Remove the connector bolt, packing, oil pipe, packing and oil cooler.
9. Remove or disconnect the following:
   - Starter
   - End plate cover
   - Torque converter bolts, through the starter opening
10. Install an engine support device such as 49 E017 5A0.
    - Number 1 engine mount rubber
    - Battery tray bracket
11. Remove the number 4 engine mount and joint bracket as follows:
    a. Install two suitable pieces of wood between the front fender panel and upper apron reinforcement as illustrated.

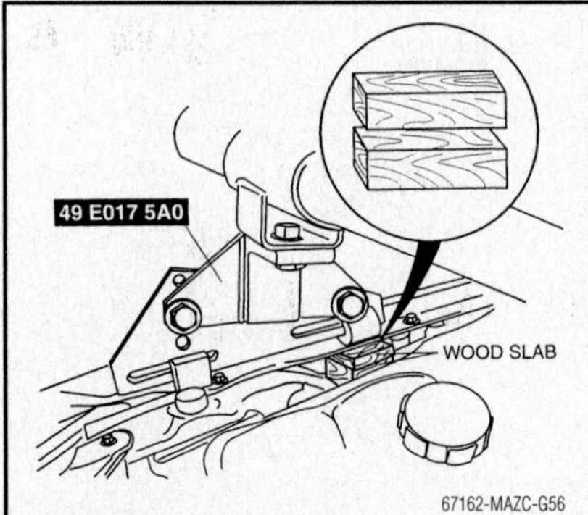

Fig. 45 Install two suitable pieces of wood between the front fender panel and upper apron reinforcement—Mazda3 with an automatic transaxle

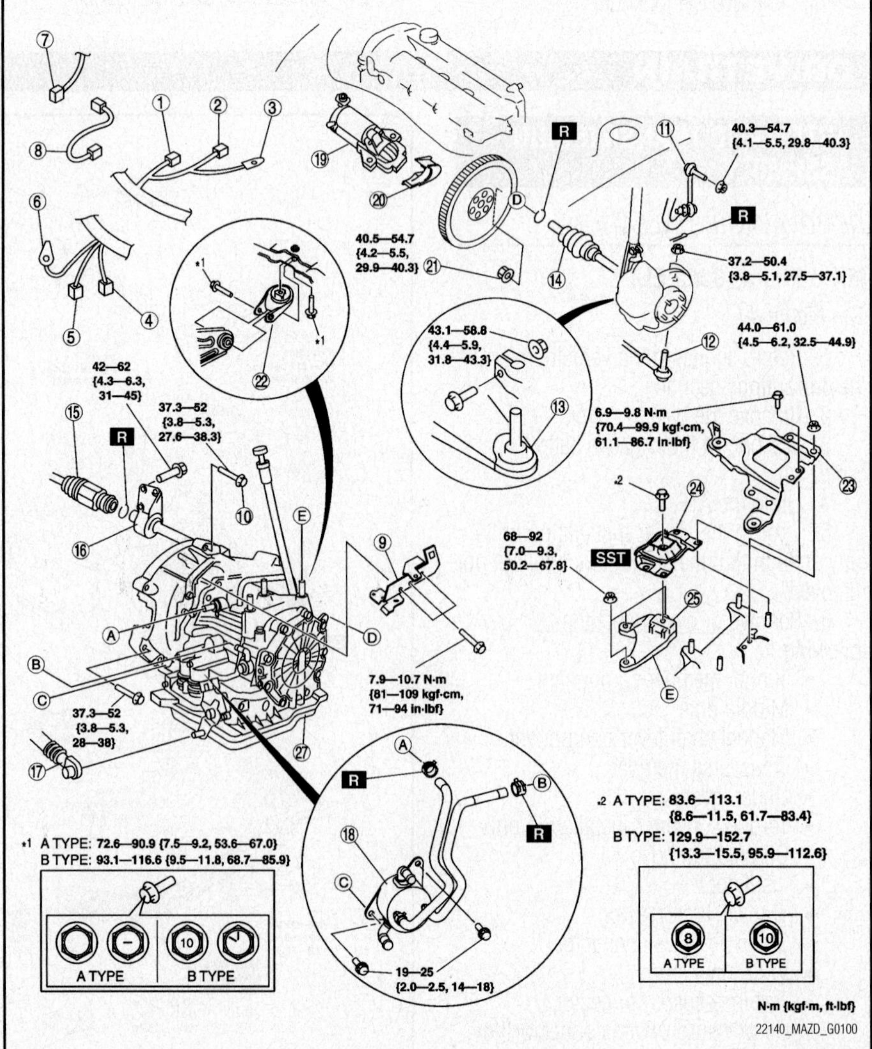

Fig. 46 Exploded view of the automatic transaxle assembly and related parts Mazda3 models

The wood size should be approximately 1.38 inch (35mm) on 4 door models or 2.36 inch (60mm) on 5 door models.

12. Loosen the engine support assembly and lean the engine towards the transaxle.

13. Support the transaxle with a jack, remove the transaxle bolts and the transaxle.

To install:

14. Place the transaxle onto a jack and raise into position.

15. Install the transaxle bolts and tighten the bolts to 28–38 ft. lbs. (37–52 Nm). Refer to the exploded view illustration for bolt locations.

16. Install or connect the following:

- Number 4 engine mount bracket to the transaxle case and tighten the bolts to 49–68 ft. lbs. (66–93 Nm)
- Number 1 engine mount rubber to the crossmember and hand tighten the bolts
- Number 4 engine mount rubber with the body stud passing through the holes and tighten the bolts to 61–87 ft. lbs. (83–113 Nm)
- Battery tray bracket on the number 4 engine mount rubber with the body stud bolts passing through the holes. Tighten retainers identified in the illustration as 2 to , 32–45 ft. lbs. (44–61 Nm) and the retainers identified as 1 in the same illustration to 61–86 inch lbs. (7–10 Nm).

17. Remove the engine support device and tighten the number 1 mount rubber bolts to 68–85 ft. lbs. (116 Nm).

- Torque converter bolts and tighten to 25–44 ft. lbs. (34–60 Nm)
- End plate cover
- Starter

18. Install the oil cooler in the reverse order of removal keeping in mind the following steps and referring to the oil cooler exploded illustration for torque specifications:

a. Apply compressed air to the cooler side opening to clear any debris. Apply the air for less than a minute only.

b. If reusing the same oil hose, install a new clamp right on the mark left by the old clamp and apply force in the direction of the arrow to secure the clamp. Align the marks and slide the oil hose onto the pipe until seated as illustrated. Make sure the hose clamp does not interfere with any other component.

- Selector cable

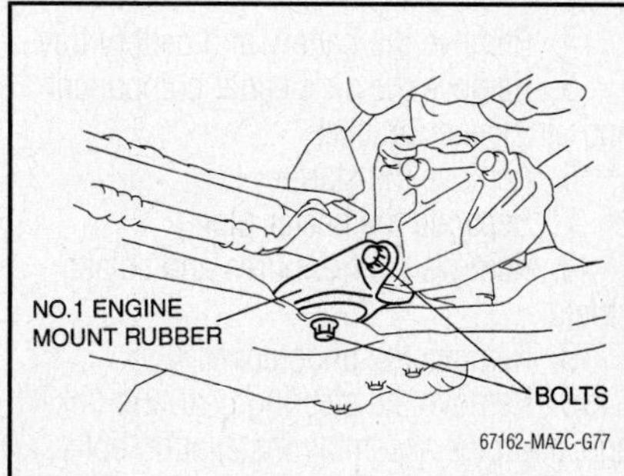

Fig. 47 Number 1 engine mount rubber bolt locations—Mazda3 models with an automatic transaxle

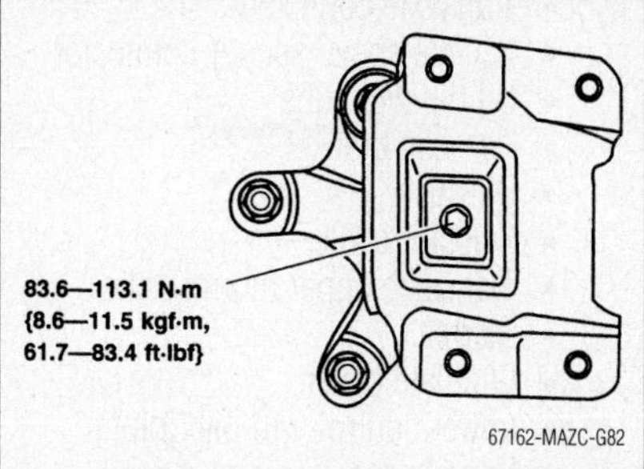

Fig. 48 Install the number 4 engine mount rubber with the body stud passing through the holes—Mazda3 models with an automatic transaxle

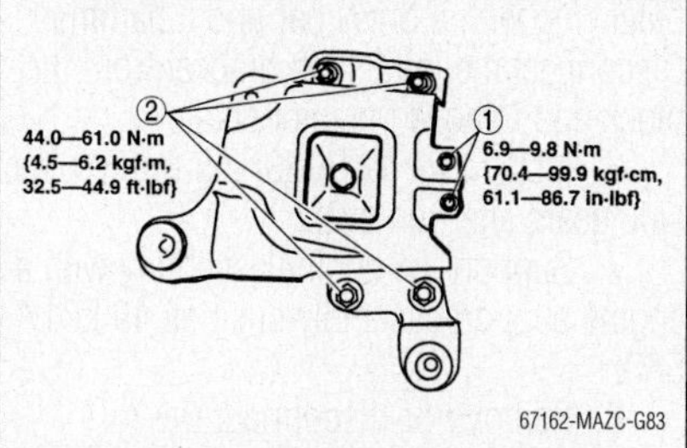

Fig. 49 Install the battery tray bracket on the number 4 engine mount rubber with the body stud bolts passing through the holes—Mazda3 models with an automatic transaxle

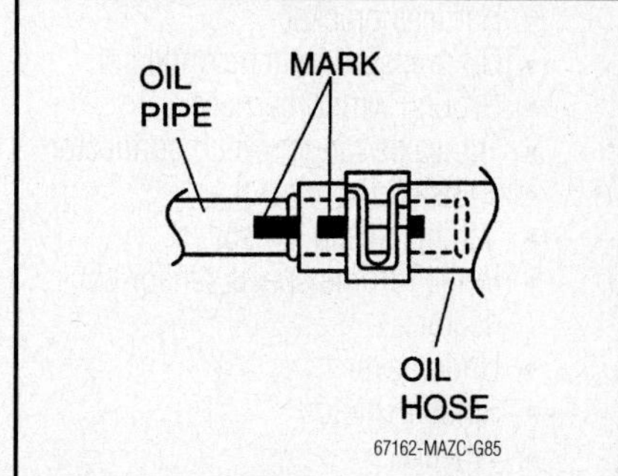

Fig. 51 Align the marks and slide the oil hose onto the pipe until seated—Mazda3 models with an automatic transaxle

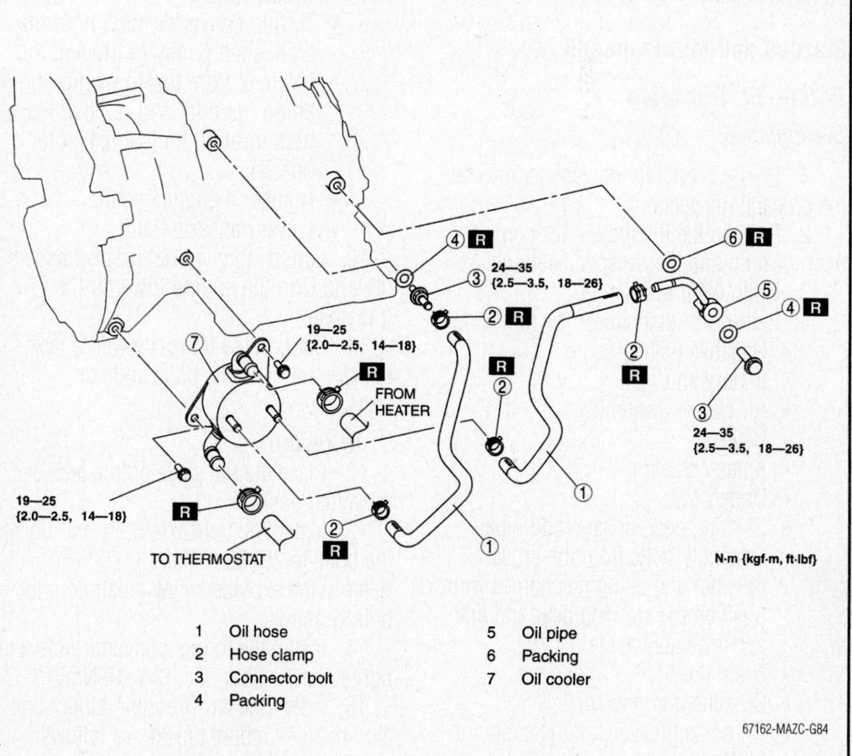

Fig. 50 Exploded view of the oil cooler assembly and related components—Mazda3 models with an automatic transaxle

- Joint shaft. Install a new circlip with the opening facing up. Hand tighten the bolts, then tighten the bolts to 31–45 ft. lbs. (42–62 Nm). Refer to the illustration for bolt location.
- Halfshafts
- Lower control arm ball joint. Tighten the retainers to 31–43 ft. lbs. (43–59 Nm).
- Tie rod ends to the knuckle. Tighten the retainers to 27–37 ft. lbs. (37–50 Nm).
- Stabilizer bar link. Tighten the retainers to 30–40 ft. lbs. (40–54 Nm).
- Harness bracket
- Oil pressure switch connector
- Ground wiring harness
- Transaxle range switch connector
- Transaxle connector
- Vehicle speed sensor
- Input/Turbine speed sensor connector
- Undercover
- Splash shields
- Wheels
- Exhaust manifold insulator
- Air cleaner assembly
- Battery tray and battery

19. Fill the transaxle with fluid. Road test the vehicle and check for leaks. Top off all fluids as needed.

Mazda6 and Mazdaspeed6

FS5A—EL Transaxle

See Figure 52.

1. Before servicing the vehicle, refer to the precautions section.
2. Refer to the illustration for component location and torque specifications.
3. Drain the transaxle oil.
4. Remove or disconnect the following:
   - Negative battery cable
   - Battery and battery tray
   - Air cleaner assembly
   - Wheels
   - Splash shields
   - Undercover
   - Steering gear, linkage and pipe assembly bolts from the crossmember and using mechanics wire, position the steering gear and linkage assembly aside.
   - Drain the ATF
   - O2 sensor connector
   - Oil pressure switch connectors for (ATX)
   - Input/turbine speed connector
   - Transmission range switch connector
   - Transaxle connector
   - Vehicle speed sensor connector
   - Oil dipstick tube
   - Harness bracket
   - Oil hose
   - Selector cable
   - Transaxle upper mount bolts
   - Starter
   - Endplate cover
   - Lower control arm ball joints
   - Damper fork
   - Tie rod ends from the knuckle
   - Stabilizer bar link
   - Halfshafts
5. Disconnect the right halfshaft from the joint shaft by tapping the transaxle side outer ring with a brass bar and a hammer. Disconnect the joint shaft bracket from the block and remove the joint shaft.
6. Install tool 49 G030 455 to hold the side gears after removal.
7. Support the engine assembly with a engine support assembly such as 49 E017 5A0.
8. Remove or disconnect the following:
   - Number 1 engine mount
   - Lower front shock absorber bolt
   - No. 1 engine mount center bolt
9. Support the crossmember with a jack and remove the nuts and the crossmember bracket.
   - Crossmember
   - Torque converter nuts, hold the crankshaft pulley to prevent the flywheel from turning while loosening the nuts and remove the nuts through the starter motor opening
   - Number 4 engine mount
   - Lower transaxle bolts
10. Loosen the engine support assembly and lean the engine towards the transaxle.
11. Support the transaxle with a jack, remove the transaxle bolts and the transaxle.

To install:

12. Place the transaxle onto a jack and raise into position.
13. Install the transaxle bolts and tighten the bolts to 28–38 ft. lbs. (37–52 Nm). Refer to the exploded view illustration for bolt locations.
14. Install the torque converter nuts and tighten to 25–32 ft. lbs. (34–44 Nm).
15. Install the No. 1 engine mount and No. 4 engine mount bracket as follows:
16. Install the No. 1 engine mount and No. 4 engine mount bracket as follows:
    a. Make sure the mount is installed as illustrated.

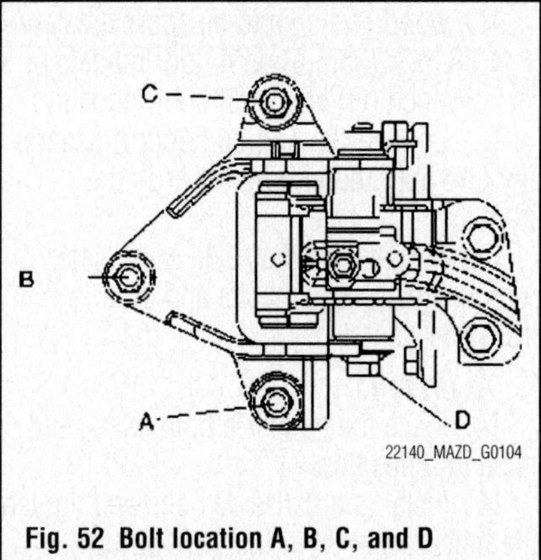

Fig. 52 Bolt location A, B, C, and D

    b. Align the bolt holes with the stud bolts, install the No. 4 mount bracket to the transaxle.
    c. Align the holes with the stud bolts and install the No. 1 engine mount to the transaxle.
    d. Align the hole on the No. 4 engine mount bracket with the No. 4 mount rubber on the car and hand tighten bolt D.
17. Refer to the illustration for bolt locations.
18. Hand tighten bolts B, C and A.
    a. Tighten bolts B and C then bolt A to 49–68 ft. lbs. (66–93 Nm).
    b. Tighten bolt D to 63–86 ft. lbs. (85–116 Nm).
    c. Tighten bolt E on the No. 1 mount to 63–86 ft. lbs. (85–116 Nm).
19. Remove the engine support device.
20. Fill the transaxle with fluid. Road test the vehicle and check for leaks. Top off all fluids as needed.

AW6A–EL Transaxle

See Figure 53.

1. Before servicing the vehicle, refer to the precautions section.
2. Refer to the illustration for component location and torque specifications.
3. Disconnect the negative battery cable.
4. Remove the battery and battery tray.
5. Remove the air cleaner component and air cleaner bracket.
6. Remove the starter.
7. Separate the heater pipe.
8. Remove the front tires and splash shield.
9. Remove the undercover.
10. Remove the steering gear and linkage, and pipe assembly installation bolts from the front crossmember, and then suspend the steering gear and linkage with a cable.
11. Drain the ATF.

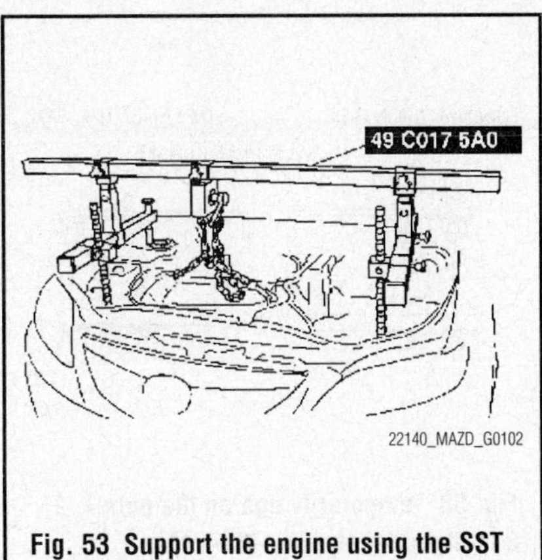

Fig. 53 Support the engine using the SST

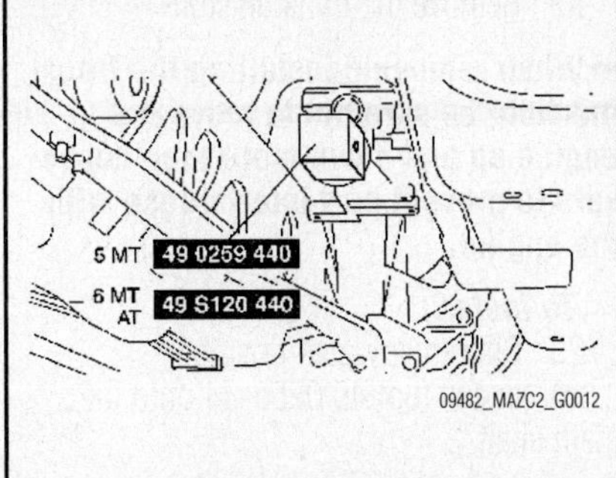

Fig. 54 Install the tools shown depending on transmission type, onto the to the extension housing—MX-5 Miata

12. Remove or disconnect the following:
    - Clips
    - Selector cable
    - Cable bracket
    - GND harness
    - TCM connector
    - Wiring harness component
    - Wiring harness bracket
    - Wiring harness component, wiring harness bracket
    - Transaxle mounting bolt (Upper side)
    - Oil hose
    - Lower arm (front, rear) ball joint
    - Damper fork (right side)
    - Tie-rod end ball joint
    - Stabilizer control link
    - Drive shaft
    - Drive shaft, joint shaft
    - No.1 engine mount
    - No.4 engine mount
    - Crossmember bracket
    - Crossmember
    - Endplate cover
    - Torque converter installation nuts
    - Transaxle mounting bolt (lower side)
    - Transaxle
    - Oil filter tube, Dipstick, breather hose

To install:

13. Install in the reverse order of removal.
14. Remove the engine support device.
15. Fill the transaxle with fluid. Road test the vehicle and check for leaks. Top off all fluids as needed.

MANUAL TRANSMISSION ASSEMBLY

REMOVAL & INSTALLATION

MX-5 Miata

See Figures 54 through 61.

1. Before servicing the vehicle, refer to the precautions section.

➡Refer the exploded view illustration for component locations and if applicable, their retainer torque specifications.

2. Remove the battery cover.
3. Disconnect the negative battery cable.
4. Loosen the starter installation bolts only enough that the starter is loose, but not removed.
5. Remove the shift lever knob.
6. Remove the console.

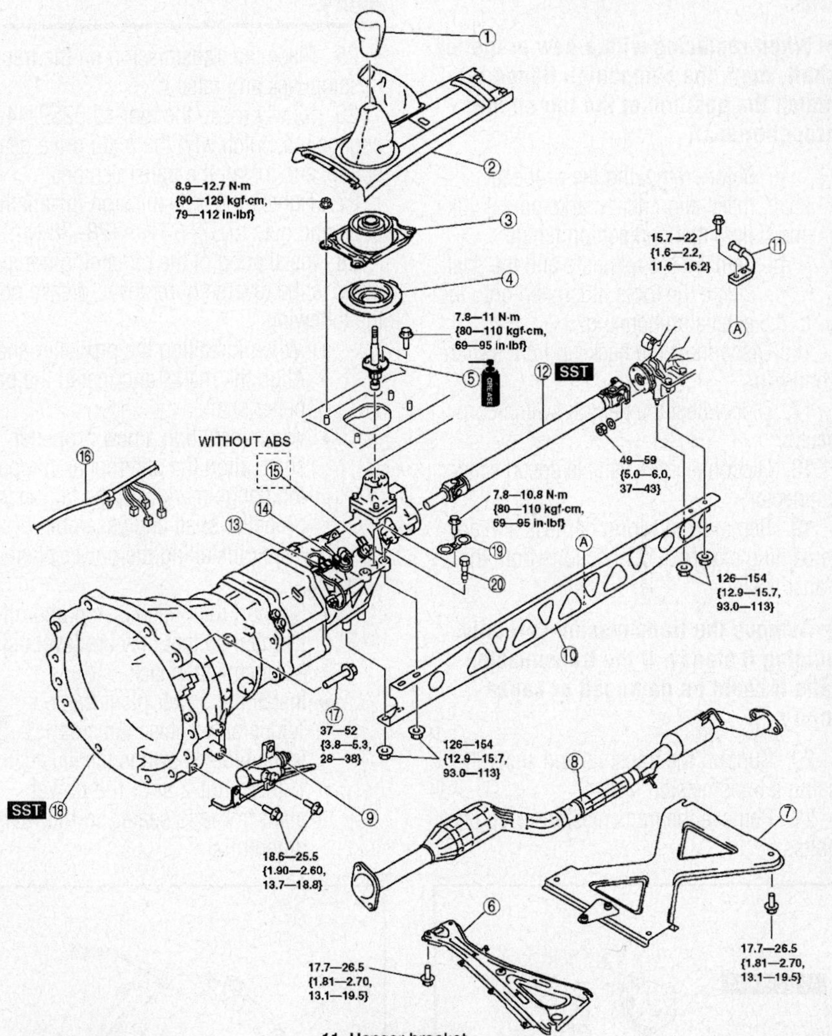

1. Shift lever knob
2. Console
3. Shift insulator component (outer)
4. Shift insulator component (inner)
5. Shift lever component
6. Member bracket
7. Tunnel member
8. Catalytic converter, middle pipe
9. Clutch release cylinder
10. Power plant frame
11. Hanger bracket
12. Propeller shaft
13. Back-up light switch connector
14. Neutral switch connector
15. Vehicle speed sensor connector
16. Wire
17. Transmission installation bolt
18. Transmission
19. Stopper
20. Bolt

N·m {kgf·m, ft·lbf}

22140_MAZD_G0117

Fig. 55 Exploded view of the manual transmission assembly and related components—MX-5 Miata

7. Remove the outer, then inner shift insulator components.
8. Remove the shift lever component.
9. Remove the member bracket
10. Remove the tunnel member.
11. Remove the catalytic converter and middle pipe.
12. Remove the clutch release pipe and cylinder.
13. Remove the power plant frame as follows:
a. Support the transmission using a transmission jack.
b. Remove the power plant frame.
14. Remove the hanger bracket.
15. Remove the propeller shaft as follows:

➡When replacing with a new propeller shaft, mark the companion flange to match the position of the tag on the propeller shaft.

a. Before removing the propeller shaft, make alignment marks on the yoke and differential companion flange.
b. Remove the retainers and the shaft.
c. Install the tools illustrated onto the to the extension housing.
16. Disconnect the back-up light switch connector.
17. Disconnect the neutral switch connector.
18. Disconnect the vehicle speed sensor connector
19. Remove the wiring harness and any remaining electrical connections from the transmission.

➡Remove the transmission carefully, holding it steady. If the transmission falls it could be damaged or cause injury.

20. Support the transmission securely using a transmission jack.
21. Remove the transmission-to-engine bolts.

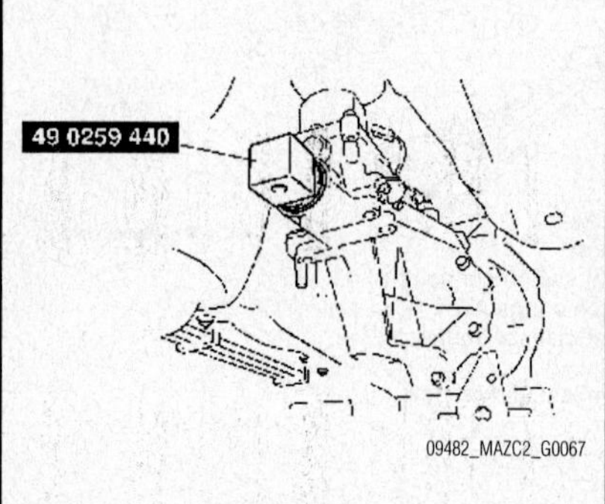

Fig. 56 Install tool 49 0259-44 onto the main shaft—MX-5 Miata

22. Remove the transmission.

➡When removing/installing the transmission, be sure not to move the engine up and down more than necessary to prevent part interference with the engine.

To install:

23. Shift to any gear position.
24. Install tool 49 0259-44 onto the main shaft.

⁂ WARNING

Install the transmission carefully, holding it steady. If the transmission falls it could be damaged or cause injury.

25. Place the transmission on the transmission jack and raise it.
26. Slowly rotate the tool 49 0259-44 to engage the clutch with the main drive gear spline, and install the transmission.
27. Tighten the transmission installation bolts and nuts to 37–52 Nm (28–38 Nm).
28. Installation of the remaining components is the reverse of removal, please note the following:
a. When installing the propeller shaft:
- Align the marks and install the propeller shaft.
- When installing a new propeller shaft, align the differential companion flange mark with the tag on the propeller shaft and assemble.

b. When installing the power plant frame:
- Support the transmission and differential so that they are level using a transmission jack.
- Install the power plant frame.
- Temporarily tighten the nuts 1, 2, 3 in the order shown in the illustration.
- Tighten nut 2 until the power plant frame is seated in the rear differential.

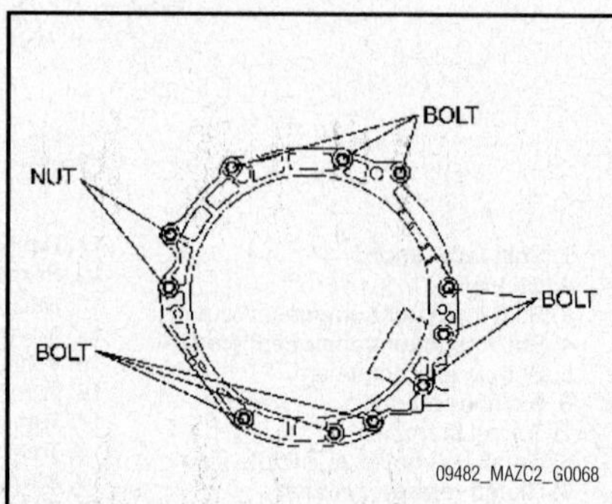

Fig. 57 Location of the manual transmission bolts and nuts—MX-5 Miata

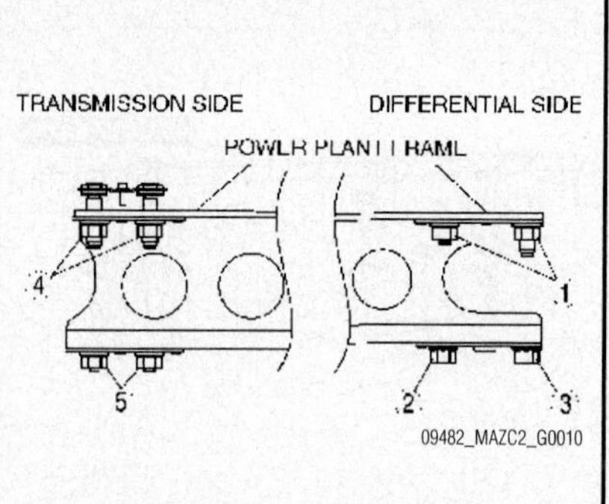

Fig. 58 Temporarily tighten the nuts 1, 2, 3 in the order shown—MX-5 Miata

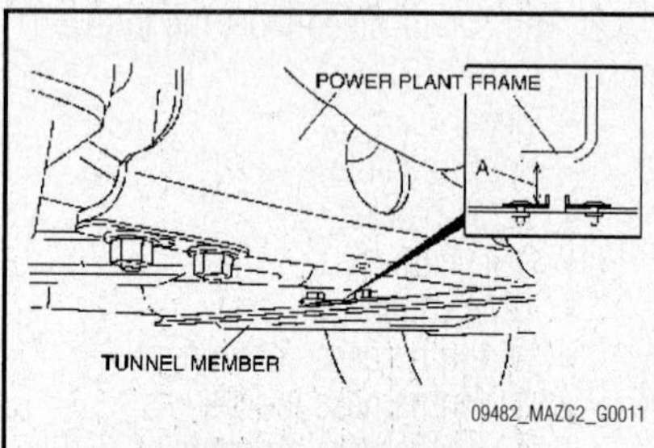

Fig. 59 Raise the front end of the power plant frame (transmission side) or the transmission with the transmission jack, and adjust dimension A to 26.7–34.7 mm (1.06–1.36 in—MX-5 Miata

- Temporarily tighten the nuts 4, 5 in order shown in the illustration.
- Install the middle pipe and tunnel member.
- Raise the front end of the power plant frame (transmission side) or the transmission with the transmission jack, and adjust dimension A to 26.7–34.7 mm (1.06–1.36 in) (lower surface of power plant frame-upper surface of the tunnel member) as shown in the illustration.
- Tighten the power plant frame installation nuts to 126–154 Nm (93–113.5 ft. lbs)
- Verify that dimension A is within the specification with the transmission jack and the adjustment bolt removed.
- If not within the specification, adjust dimension A again.

c. To position the plug hole plate, grasp rubber 1 and 2, as shown in the illustration, with your hands and press them in.
d. Apply grease to the areas of the shift lever component as shown in the illustration.

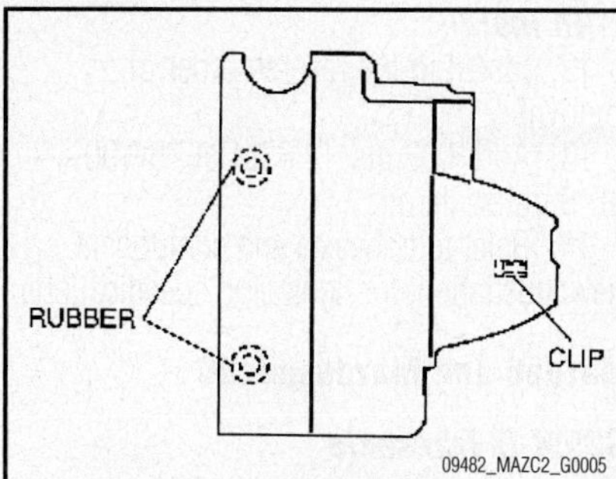

Fig. 60 Install the plug hole plate by grasping rubber and pressing them in—MX-5 Miata

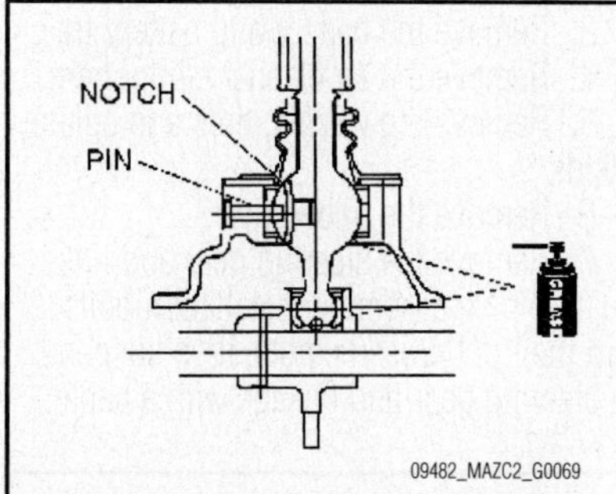

Fig. 61 Apply grease to the areas of the shift lever component indicated—MX-5 Miata

e. Align the shift lever component notch with the shift control case pin and install the shift lever component.

MANUAL TRANSAXLE ASSEMBLY

REMOVAL & INSTALLATION

Mazda3 Mazdaspeed3

A26M-R Manual Transaxle

See Figures 62 and 63.

1. Before servicing the vehicle, refer to the precautions section.

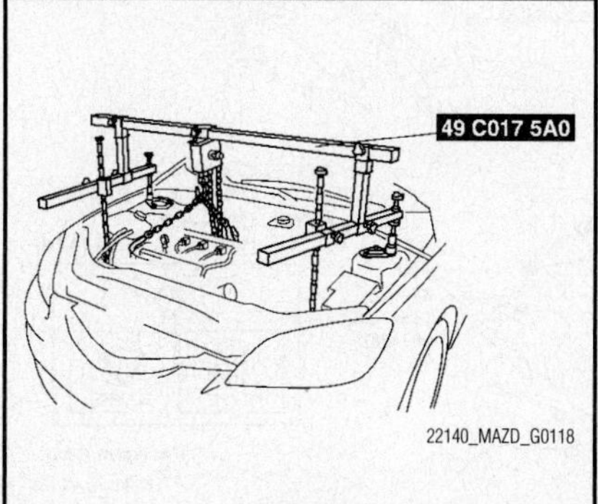

Fig. 62 Engine support system

2. Remove the battery and battery tray.
3. Remove the air cleaner component.
4. Remove the charge air cooler.
5. Remove in the fuel pump resistor.
6. Remove the front wheels.
7. Remove the undercover and splash shields.
8. Remove the mudguard. (LF side)
9. Remove the starter.
10. Drain the transaxle oil into a suitable container.
11. Support the engine using the SST.
12. Remove or disconnect the following:
   - Selector cable
   - Shift cable
   - Wiring harness bracket
   - Cable bracket
   - Back-up light switch connector
   - Neutral switch connector
   - Wiring harness bracket
   - Clutch release cylinder
   - Transaxle mounting bolt (upper side)
   - Tie-rod end ball joint
   - Stabilizer control link
   - Lower arm ball joint
   - Drive shaft
   - Joint shaft
   - No.1 engine mount bracket
   - No.1 engine mount rubber

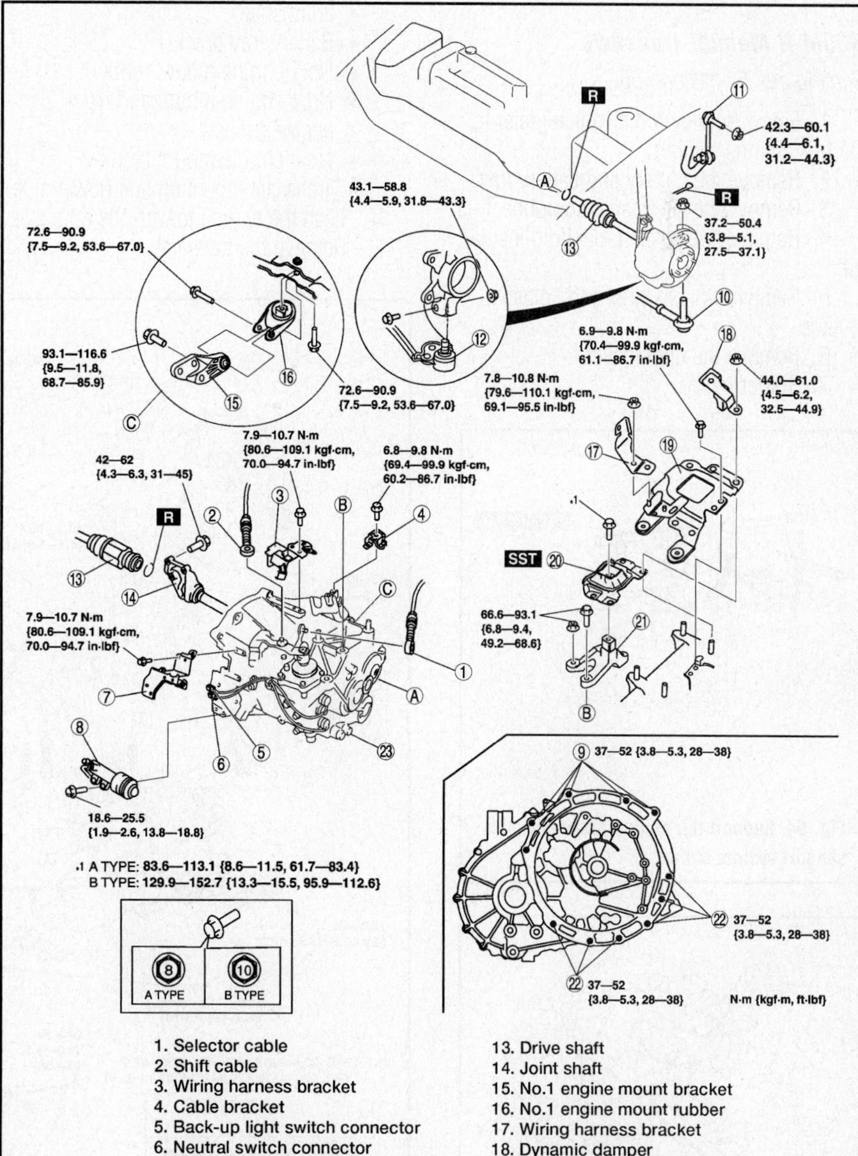

Fig. 63 A26M-R manual transaxle view and components

- Wiring harness bracket
- Dynamic damper
- Battery tray bracket
- No.4 engine mount rubber
- No.4 engine mount bracket
- Transaxle mounting bolt (lower side)
- Manual transaxle

To install:

13. Install in the reverse order of removal.

14. Add 2.6 qts. (2.5L) of 75W90—GL-5 transaxle oil.

15. Refer to transaxle and component view illustration, for tightening specifications.

G35M-R Manual Transaxle

See Figures 64 through 66.

1. Before servicing the vehicle, refer to the precautions section.
2. Remove the battery and battery tray.
3. Remove the air cleaner component.
4. Remove the exhaust manifold insulator.
5. Remove wheels, tires and splash shields.
6. Remove the undercover.
7. Starter motor.
8. Drain the transaxle oil into a suitable container.
9. Remove or disconnect the following:
   - Back-up light switch connector
   - Neutral switch connector
   - Vehicle speed sensor connector (without ABS)
   - Harness bracket
   - Clutch release cylinder
   - Transaxle mounting bolt (upper side)
   - Tie-rod end ball joint
   - Front crossmember
   - Stabilizer control link
   - Lower arm ball joint
   - Drive shaft
   - Joint shaft
   - Battery tray bracket
   - No.1 engine mount rubber
   - No.1 engine mount and No.4 engine mount
   - No.4 engine mount bracket
   - Transaxle mounting bolt (lower side)
10. Lean the engine toward the transaxle.
11. Remove the transaxle.

To install:

12. Install in the reverse order of removal.

13. Add 3.03 qts. (2.87L) of 75W90—GL-5 transaxle oil.

14. Refer to transaxle and component view illustration, for tightening specifications.

Mazda6 and Mazdaspeed6

G35M-R Transaxle

See Figures 67 through 69.

1. Before servicing the vehicle, refer to the precautions section.
2. Disconnect the negative battery cable.
3. Remove the battery and battery tray.
4. Remove the air cleaner component.
5. Remove the wheels, tires and splash shields.
6. Remove the undercover.
7. Remove the steering gear and linkage, and pipe assembly installation bolts from the front crossmember, then suspend the steering gear and linkage with a cable.

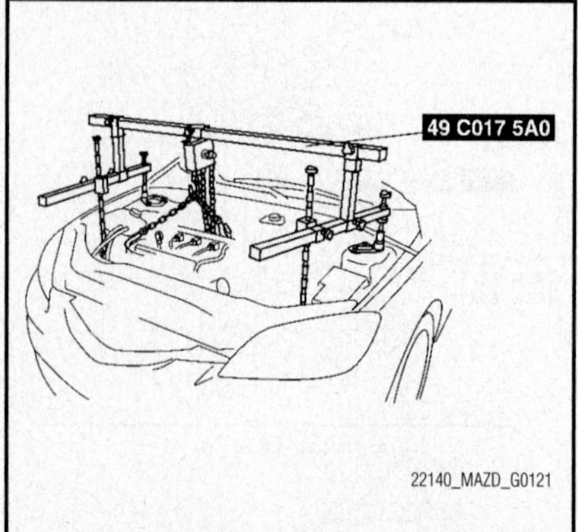

Fig. 64 Support the engine assembly with support system shown

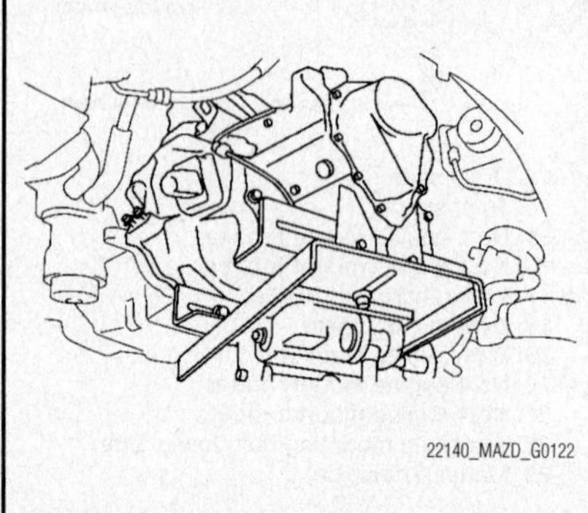

Fig. 65 Support and remove the transaxle assembly

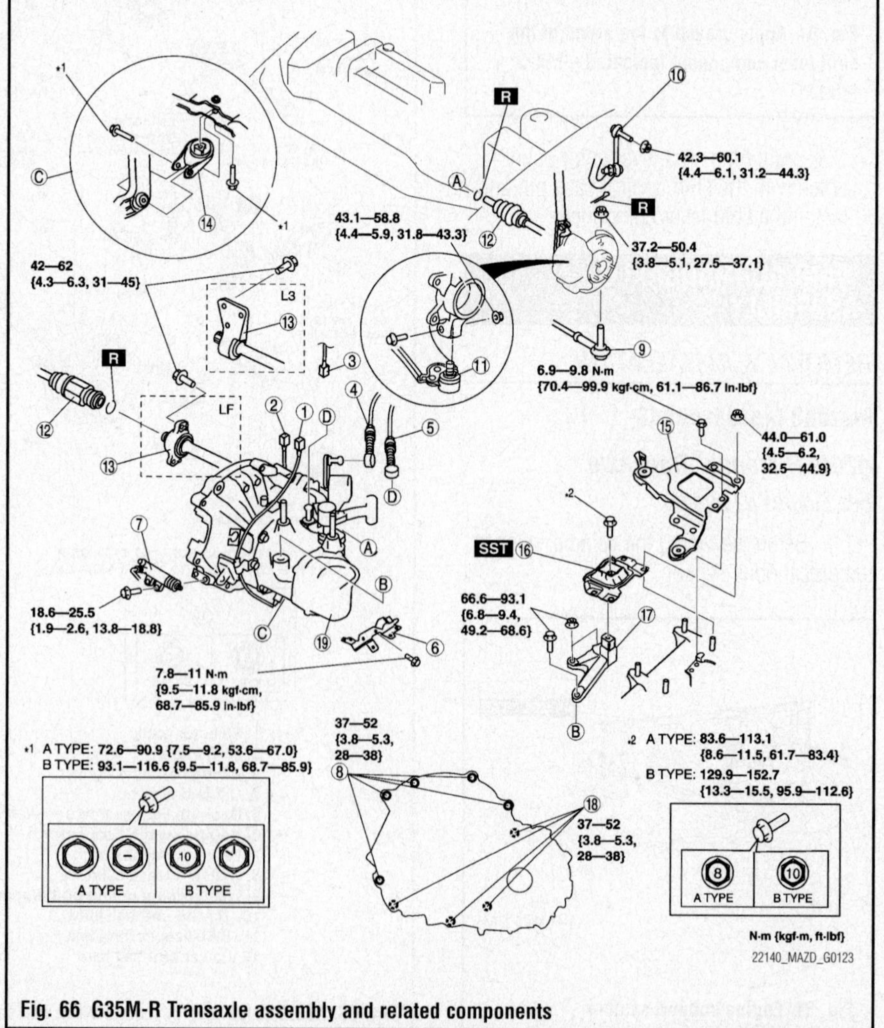

Fig. 66 G35M-R Transaxle assembly and related components

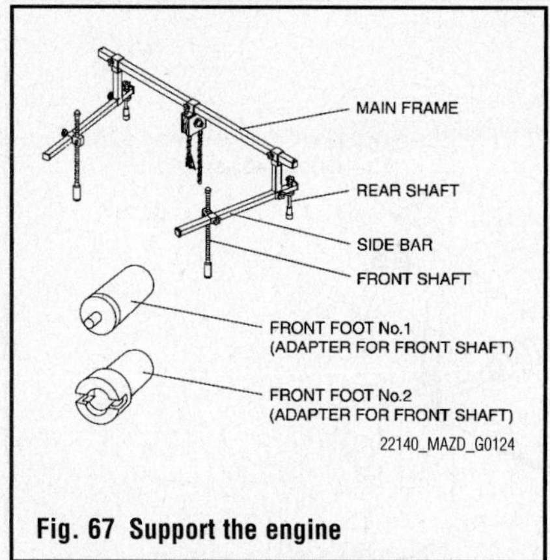

Fig. 67 Support the engine

8. Drain the transaxle oil into a suitable container.
9. Remove or disconnect the following:
- HO2S connector
- GND wiring harness
- Back-up light switch connector
- Neutral switch connector
- Harness bracket
- Vehicle speed sensor connector (Without ABS)
- Select cable
- Shift cable
- Cable bracket
- Clutch release cylinder
- Starter
- Endplate cover
- Transaxle mounting bolt (upper side)
- Lower arm (front, rear) ball joint
- Damper fork
- Tie-rod end ball joint
- Stabilizer control link
- Drive shaft
- Joint shaft
- No.1 engine mount
- Crossmember bracket
- Crossmember assembly
- No.4 engine mount
- Dynamic damper

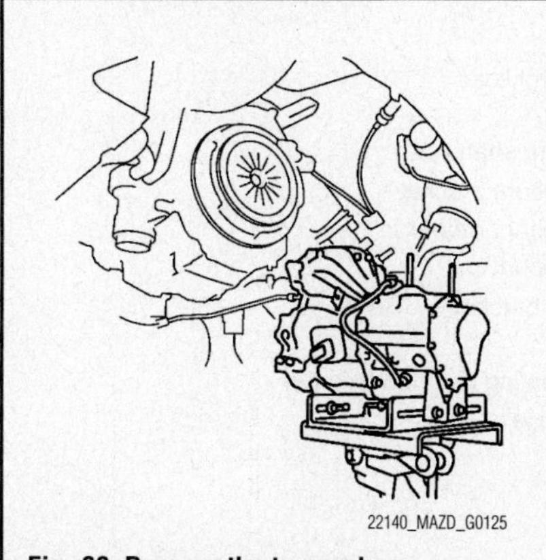

Fig. 68 Remove the transaxle

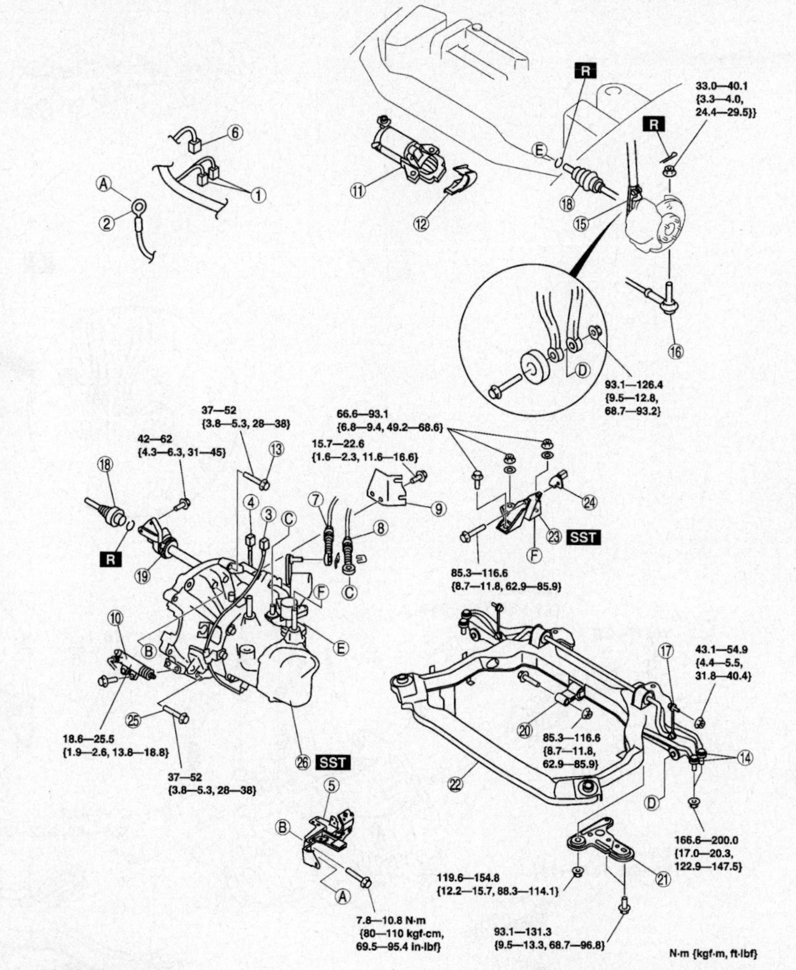

Fig. 69 Exploded view of transaxle and related components

- Transaxle mounting bolt (lower side)
- Manual transaxle

To install:

10. Install in the reverse order of removal.
11. Add 3.03 qts. (2.87L) of 75W90—GL-5 transaxle oil.
12. Refer to transaxle and component view illustration, for tightening specifications.

A65M-R Transaxle

See Figures 70 through 75.

1. Before servicing the vehicle, refer to the precautions section.
2. Refer to the illustration for component location and torque specifications.
3. Drain the transaxle oil.
4. Remove or disconnect the following:
- Battery and battery tray
- Air cleaner assembly
- Starter
- Wheels
- Splash shields
- Undercover
- Steering gear, linkage and pipe assembly bolts from the cross-

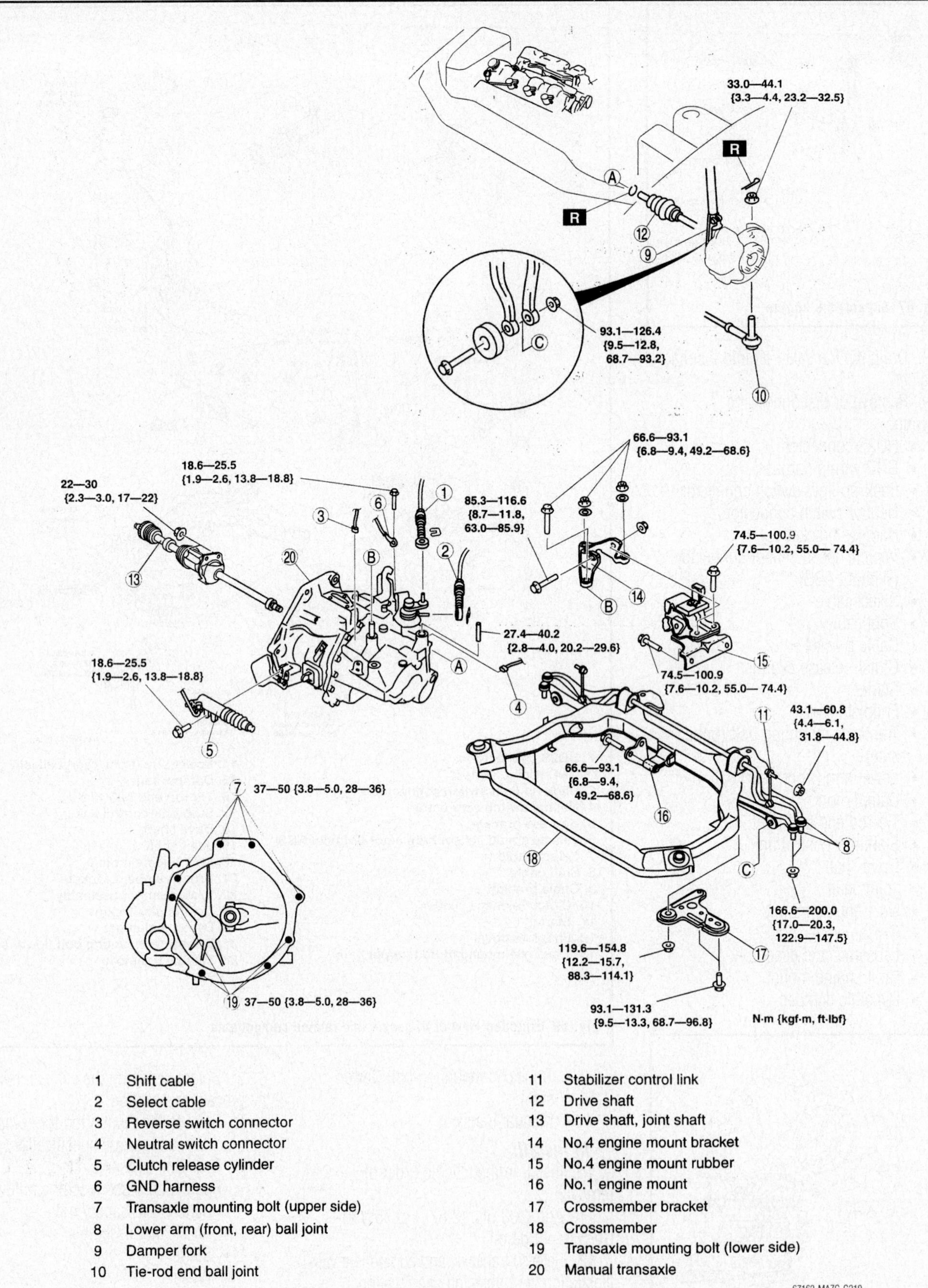

1 Shift cable
2 Select cable
3 Reverse switch connector
4 Neutral switch connector
5 Clutch release cylinder
6 GND harness
7 Transaxle mounting bolt (upper side)
8 Lower arm (front, rear) ball joint
9 Damper fork
10 Tie-rod end ball joint
11 Stabilizer control link
12 Drive shaft
13 Drive shaft, joint shaft
14 No.4 engine mount bracket
15 No.4 engine mount rubber
16 No.1 engine mount
17 Crossmember bracket
18 Crossmember
19 Transaxle mounting bolt (lower side)
20 Manual transaxle

67162-MAZC-G219

Fig. 70 Exploded view of the A65M-R manual transaxle assembly mounting—Mazda6 models

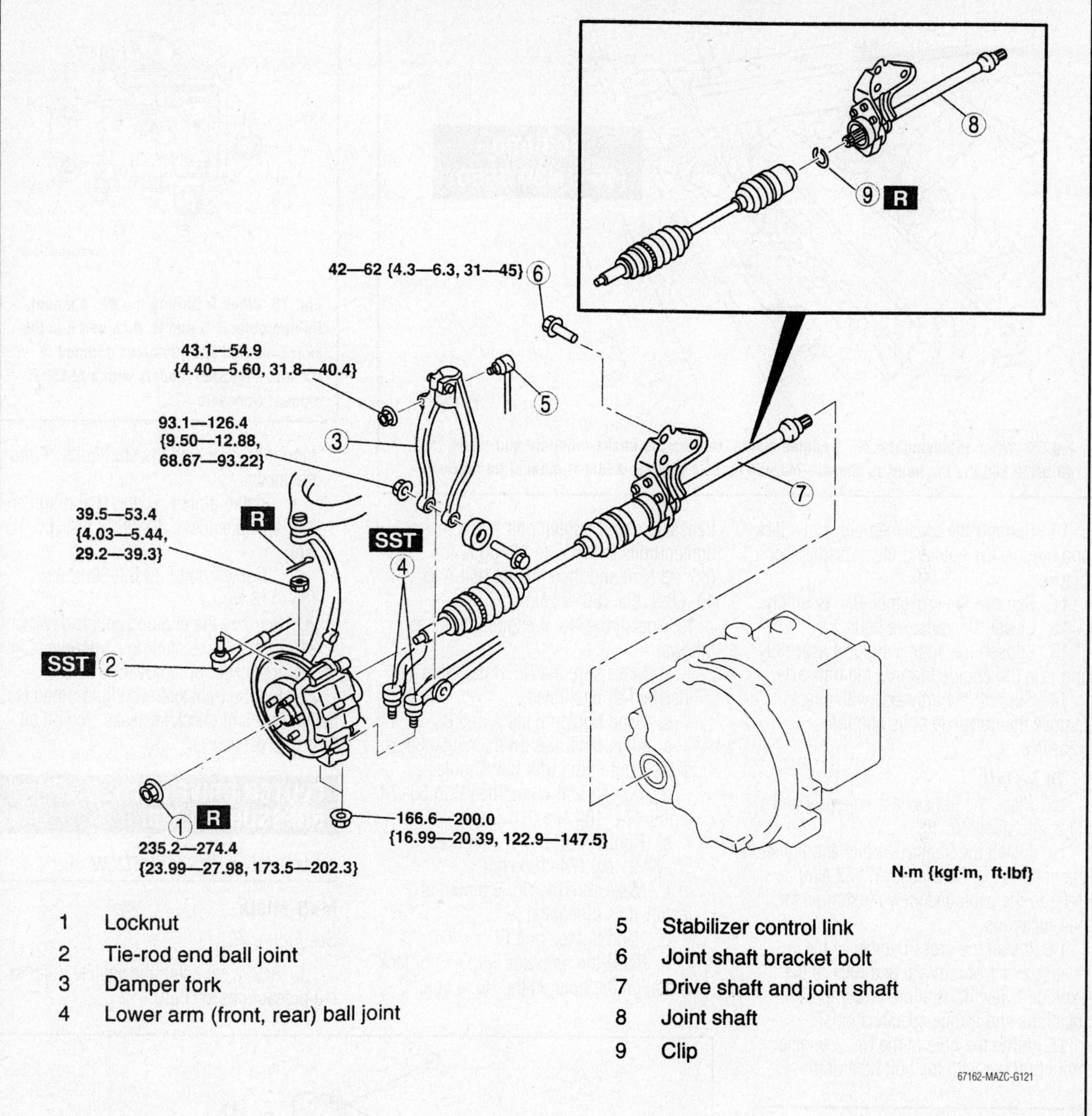

Fig. 71 Exploded view of the joint shaft assembly—Mazda6 models with the 3.0L (AJ) engine

member and using mechanics wire, position the steering gear and linkage assembly aside.

- Shift cable and select cable
- Back up light switch connector
- Neutral safety switch connector
- Clutch release cylinder
- Ground harness
- Transaxle upper mount bolts
- Lower control arm ball joints
- Damper fork
- Tie rod ends from the knuckle
- Stabilizer bar link
- Halfshafts

5. Disconnect the left halfshaft from the joint shaft by inserting a pry bar between the transaxle and the halfshaft outer ring. Disconnect the joint shaft bracket from the block and remove the joint shaft.

6. Install tool 49 G030 455 to hold the side gears after removal.

7. Remove the No. 4 engine mount bracket and rubber.

8. Support the engine assembly with a engine support assembly such as 49 E017 5A0.

9. Remove or disconnect the following:

- Number 1 engine mount bracket
- Lower front shock absorber bolt
- No. 1 engine mount. Remove the intake manifold and attach tool 49 UN30 3050 to the head as shown and remove the mount.

Fig. 72 When removing the No. 1 engine mount, remove the intake manifold and attach tool 49 UN30 3050 to the head as shown—Mazda6 models with the A65M-R manual transaxle

10. Support the crossmember with a jack and remove the nuts and the crossmember bracket.
11. Remove the crossmember assembly.
12. Lower the transaxle bolts.
13. Loosen the engine support assembly and lean the engine towards the transaxle.
14. Support the transaxle with a jack, remove the transaxle bolts and the transaxle.

To install:

15. Place the transaxle onto a jack and raise into position.
16. Install the transaxle bolts and tighten the bolts to 28–38 ft. lbs. (37–52 Nm). Refer to the exploded view illustration for bolt locations.
17. Install the crossmember in the reverse order of removal and refer to the exploded view illustration for component locations and torque specifications.
18. Align the hole of the No. 1 engine mount rubber with the bolt hole of the transaxle. Hand tighten bolt A and then tighten bolts B and C to 49–68 ft. lbs. (66–93 Nm) and then tighten bolt A to 49–68 ft. lbs. (66–93 Nm).
19. Install the No. 4 engine mount as follows:
    a. Make sure the No. 4 mount is installed as illustrated.
    b. Hand tighten bolts A and B.
    c. Align the holes on the contact area of the front frame with bolt C hole.
    d. Tighten bolt A and then B to 55–74 ft. lbs. (74–100 Nm).
    e. Tighten bolt C and then D to 55–74 ft. lbs. (74–100 Nm).
    f. Make sure the No. 4 mount is installed as illustrated.
    g. Hand tighten bolt F.
    h. Raise the transaxle with a floor jack and align the hole on the No. 4 engine mount bracket with the stud bolts on the transaxle.
    i. Install bolts F, G and H and tighten in that sequence to 49–68 ft. lbs. (66–93 Nm).
    j. Tighten bolt E to 63–86 ft. lbs. (85–116 Nm).
20. Remove the engine support device.
21. Install the remaining components in the reverse order of removal.
22. Fill the transaxle with fluid. Road test the vehicle and check for leaks. Top off all fluids as needed.

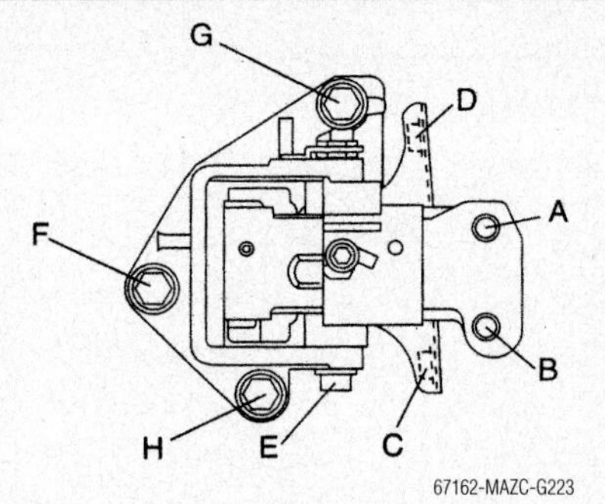

Fig. 75 When installing the No. 4 mount, tighten bolts F, G and H, then bolt E in the sequence and specifications outlined in the text—Mazda6 models with a A65M-R manual transaxle

CLUTCH DRIVEN DISC & PRESSURE PLATE

REMOVAL & INSTALLATION

MX5-Miata

See Figure 76.

1. Before servicing the vehicle, refer to the precautions section.

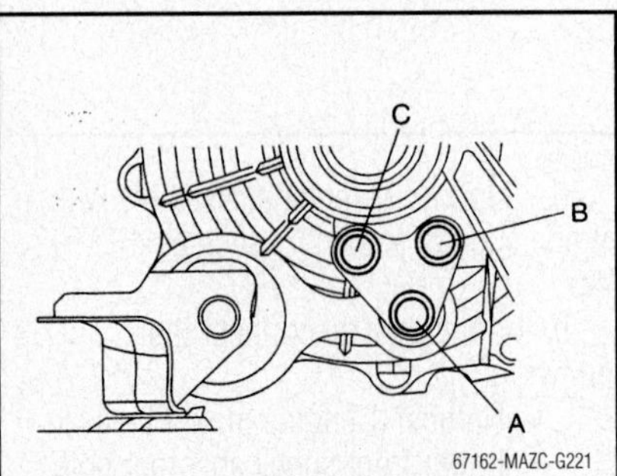

Fig. 73 Tighten the No. 1 mount in the sequence and specifications outlined in the text—Mazda6 models with a A65M-R manual transaxle

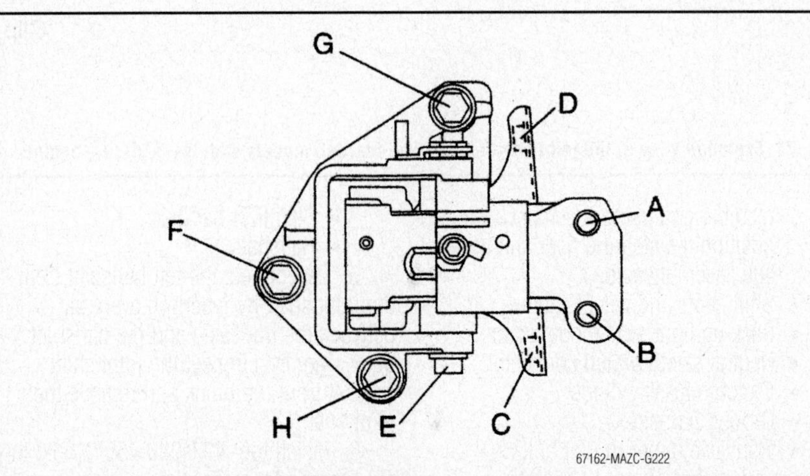

Fig. 74 Install the No. 4 mount and tighten bolts A thru D in the sequence and specifications outlined in the text—Mazda6 models with a A65M-R manual transaxle

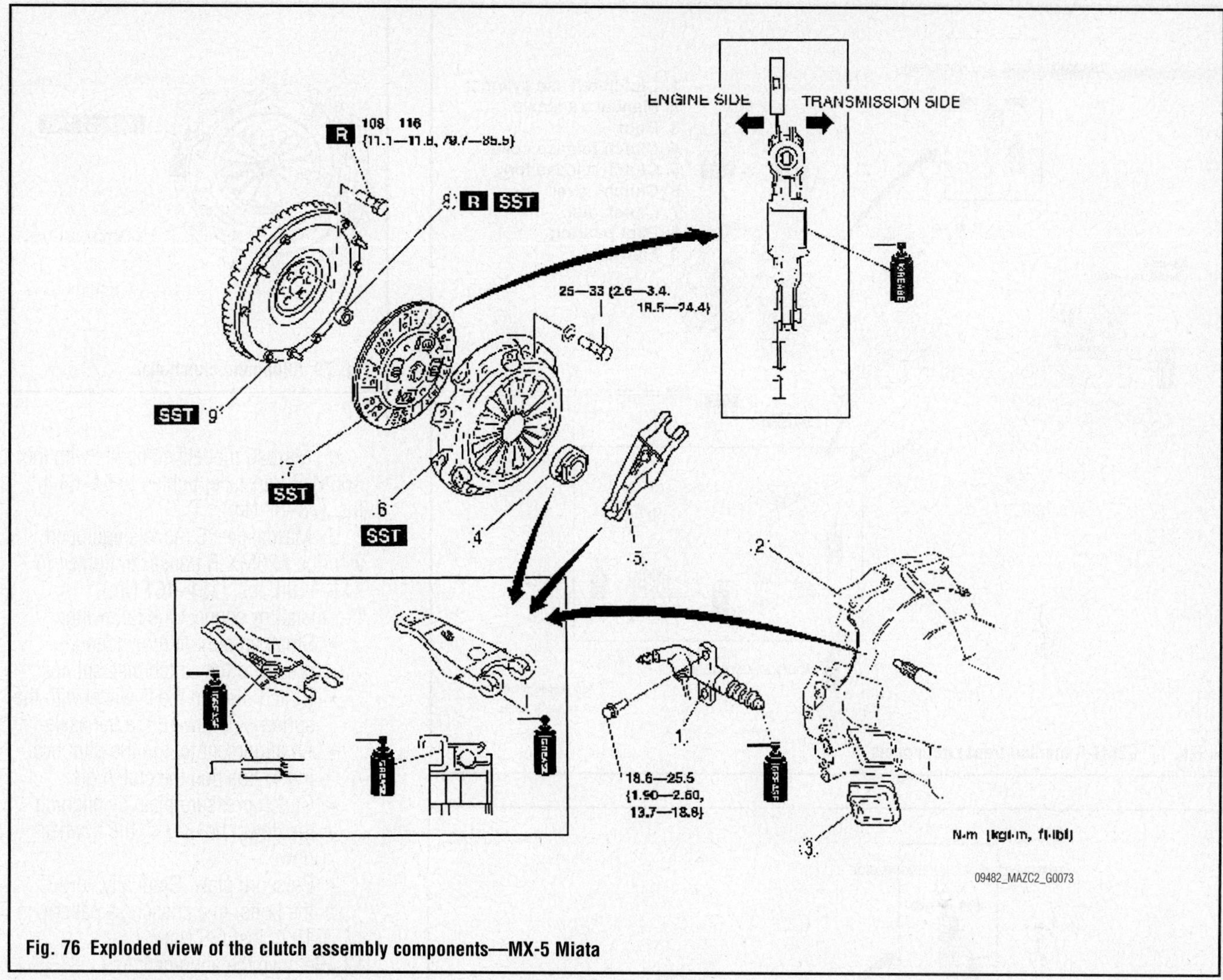

Fig. 76 Exploded view of the clutch assembly components—MX-5 Miata

➡Refer the exploded view illustration for component locations and if applicable, their retainer torque specifications.

2. Clutch release cylinder.
3. Manual transmission.
4. Clutch release collar.
5. Clutch release fork.
6. Clutch cover bolts in a criss-cross pattern using several steps.
7. Clutch disc.

To install:

8. Installation is the reverse of removal. Tighten the clutch cover bolts to 25–33 Nm (18–24 ft. lbs.).

Mazda3, Mazdaspeed3, Mazda6 and Mazdaspeed6 Models

See Figures 77 through 79.

1. Before servicing the vehicle, refer to the precautions section.
2. Remove or disconnect the following:
   - Negative battery cable
   - Clutch release cylinder
   - Transaxle
   - Rubber boot
   - Clutch release collar
   - Clutch release fork
   - Pressure plate loosening the bolts one turn each in a criss—cross pattern
   - Clutch disc
3. Inspect the pilot bearing. If it is worn or damaged and does not turn easily by hand, remove it using a puller/slide hammer.
4. Check the flywheel surface for scoring, cracks or burning and machine or replace, as necessary.
5. Install Holder tool 49 E011 1A0 to keep the flywheel from turning. Loosen the flywheel bolts evenly and gradually in a crisscross pattern. Remove the flywheel.
6. Inspect the clutch release bearing for wear. Replace it if it sticks or does not turn easily.
7. Inspect the release fork for wear or damage and replace as necessary.

To install:

8. Lubricate the release fork fingers and pivot with molybdenum grease and install in the release fork boot.
9. Install or connect the following:
   - Clutch release bearing on the release fork
   - New pilot bearing in the flywheel, if removed, using an installation tool
10. Be sure the flywheel mounting surface and the crankshaft or eccentric shaft mounting surfaces are clean. Remove any old sealant from the flywheel bolt hole threads and the flywheel bolts.
    - Flywheel
    - Sealant to the flywheel bolt threads and install them hand tight
    - Flywheel holding tool. Tighten the bolts, in a crisscross pattern as follows:

    a. Mazda3 models, tighten to 79–85 ft. lbs. (108–116 Nm).

    b. Mazda6 models equipped with the G35M-R transaxle, tighten to 79–84 ft. lbs. (108–115 Nm).

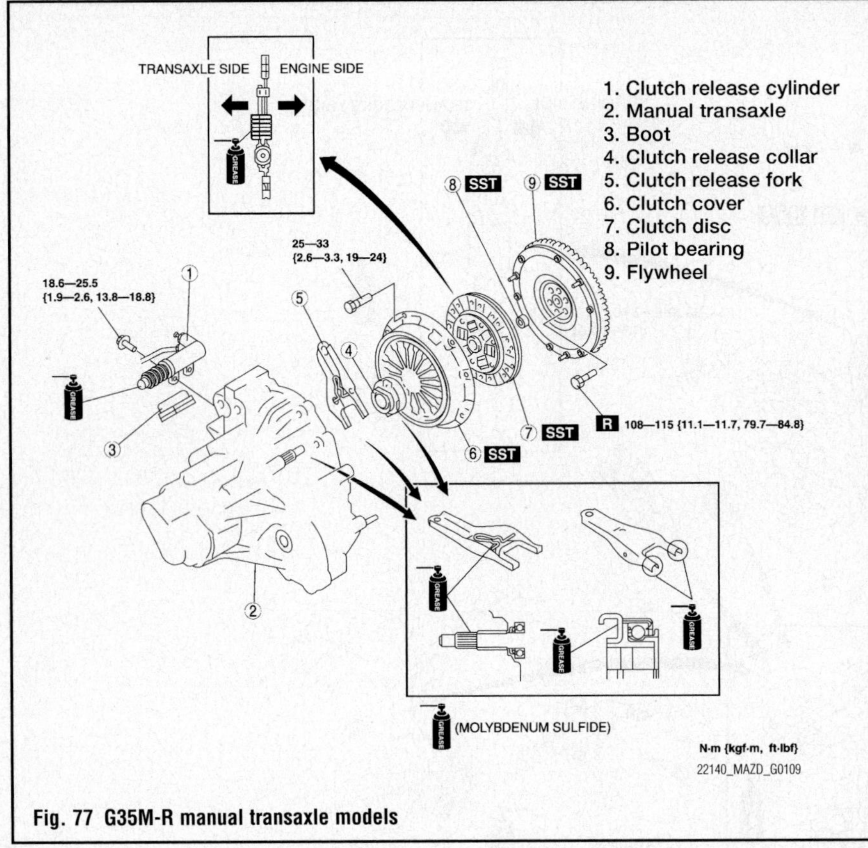

Fig. 77 G35M-R manual transaxle models

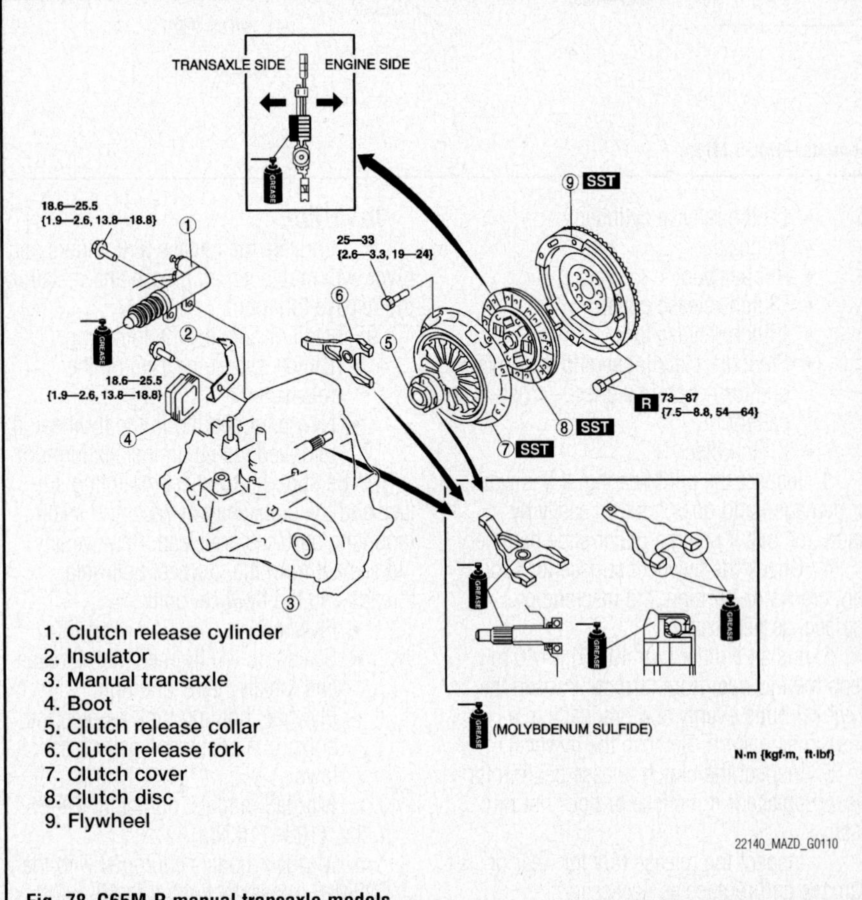

Fig. 78 G65M-R manual transaxle models

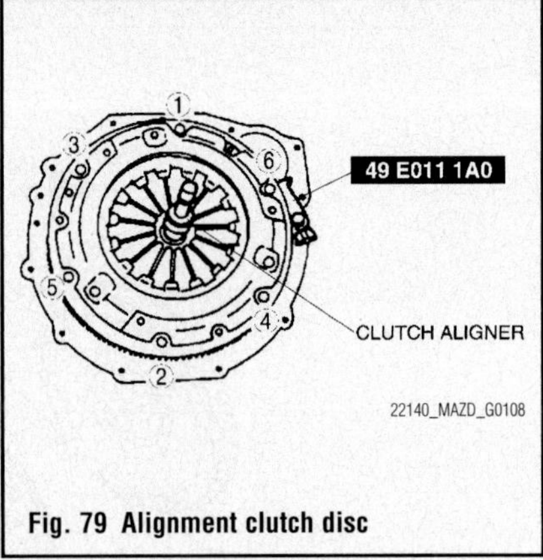

Fig. 79 Alignment clutch disc

c. Mazda6 models equipped with the A65M-R transaxle, tighten to 54–64 ft. lbs. (73–87 Nm).

d. Mazdaspeed6 models equipped with the A26MX-R transaxle, tighten to 111–118ft. lbs. (151–161 Nm).

11. Install or connect the following:
- Small amount of molybdenum grease to the clutch disc splines
- Clutch disc on the flywheel with the spring side toward the transaxle
- An alignment tool in the pilot bearing to position the clutch disc
- Clutch pressure plate by aligning the dowel holes with the flywheel dowels
- Pressure plate. Gradually, torque the bolts, in a crisscross pattern, to 19 ft. lbs. (26 Nm).

12. Remove the alignment tool.
- Clutch release fork
- Clutch release collar
- Rubber boot
- Transaxle
- Clutch release cylinder
- Negative battery cable

ADJUSTMENTS

The hydraulic clutch system does not require any adjustment.

CLUTCH MASTER CYLINDER

REMOVAL & INSTALLATION

Mazda3 and Mazdaspeed3

See Figure 80.

1. Remove the battery and battery tray.
2. Separate the steering shaft.
3. Remove the reserve hose from the master cylinder while pressing on the coupling.
4. Pull the clutch master cylinder clip down and pull out the clutch pipe connector straight to detach it

5. Plug the clutch pipe after removing it to avoid leakage.
6. Unplug the clutch pedal position switch connector and starter interlock switch connector.
7. Remove the starter interlock switch.
8. Remove the nuts and the clutch pedal assembly
9. Rotate the clutch master cylinder in the counterclockwise direction and remove.
10. Press on the tabs on the left and right sides of the push rod using a flathead screwdriver and remove the rod.

To install:

11. Push the push rod in until the tabs lock. On the master cylinder.
12. Rotate the clutch master cylinder in a clockwise direction until it stops.
13. Install the clutch pedal assembly and tighten the nuts to 13–18 ft. lbs. (18–25 Nm).
14. Install the starter interlock switch.
15. Connect the clutch pedal position switch connector and starter interlock switch connector.
16. Install the clutch pipe and engage the retaining clip.

➡Verify that there is no chipping or damage to the seal ring of the clutch pipe connector.

17. Insert the clutch pipe connector straight.
18. Pull the clutch pipe to verify that it does not come off, and reinsert it completely.
19. Insert the reserve hose connector straight until a click is heard.
20. Pull the reserve hose to verify that it does not come off, and reinsert it completely.
21. Fill and bleed the clutch hydraulic system.

✲✲ WARNING

When performing the following procedure, make sure that the area around the vehicle is free of people or objects as the vehicle might move suddenly.

22. Apply the parking brake and fix the front and rear of the wheels with the wheel chocks.
23. Start the engine at idle.
24. Without depressing the clutch pedal, move the shift lever slowly to the reverse position.
25. Hold the lever when the gear noise is heard.
26. Depress the clutch pedal slowly.

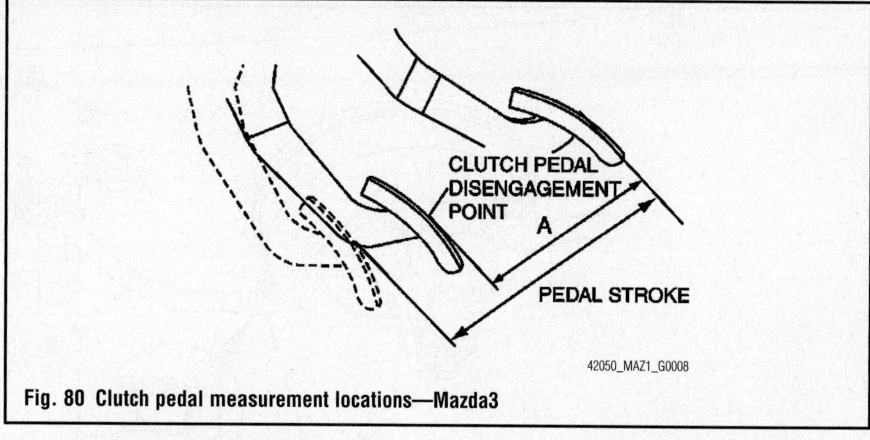

Fig. 80 Clutch pedal measurement locations—Mazda3

27. Hold the pedal when the gear noise stops (clutch pedal disengagement point).
28. Measure distance A, verify that they are within the following specifications .Refer to the illustration for further detail.
   a. The clutch pedal disengagement stroke should be 3.55–4.33 in. (90–110 mm)
   b. The pedal stroke should be 5.31 in. (135 mm).

MX5-Miata

See Figure 81.

1. Disconnect the hose clip.
2. Disconnect the clutch pipe.
3. Remove the nut, master cylinder and packing.

To install:

4. Installation is the reverse of removal. Tighten the nuts to 13–18 ft. lbs. (18–25 Nm).
5. Fill and bleed the clutch hydraulic system.

Mazda6 and Mazdaspeed6

See Figure 82.

1. Remove the battery and tray.
2. Disconnect the hose clip.
3. Disconnect the clutch pipe.

➡Plug the clutch pipe after removing it to avoid leakage.

4. Remove the nut.
5. Remove the master cylinder.
6. Remove the gasket.

To install:

7. Installation is the reverse of removal. Tighten the nuts to 13–18 ft. lbs. (18–25 Nm).
8. Fill and bleed the clutch hydraulic system.

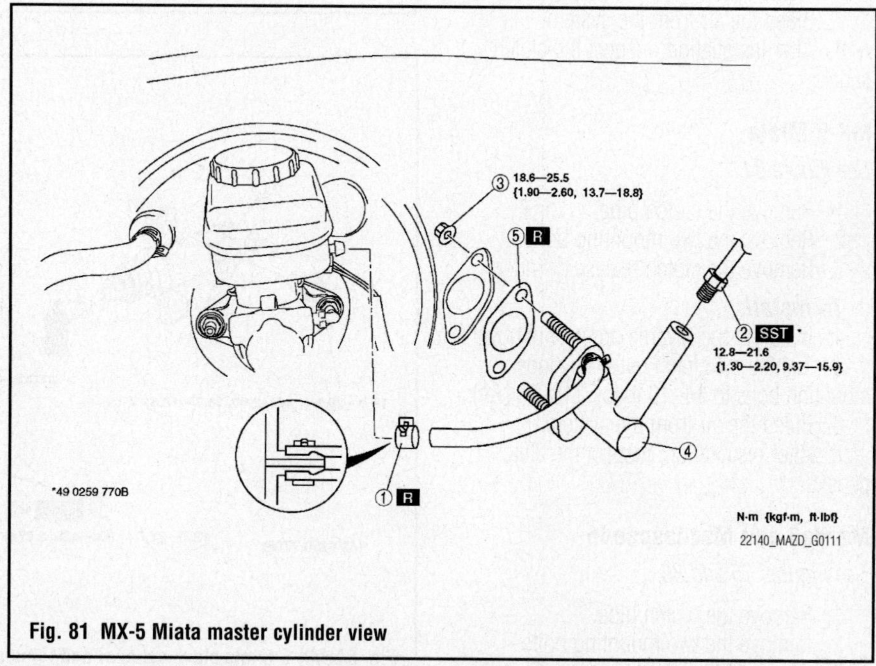

Fig. 81 MX-5 Miata master cylinder view

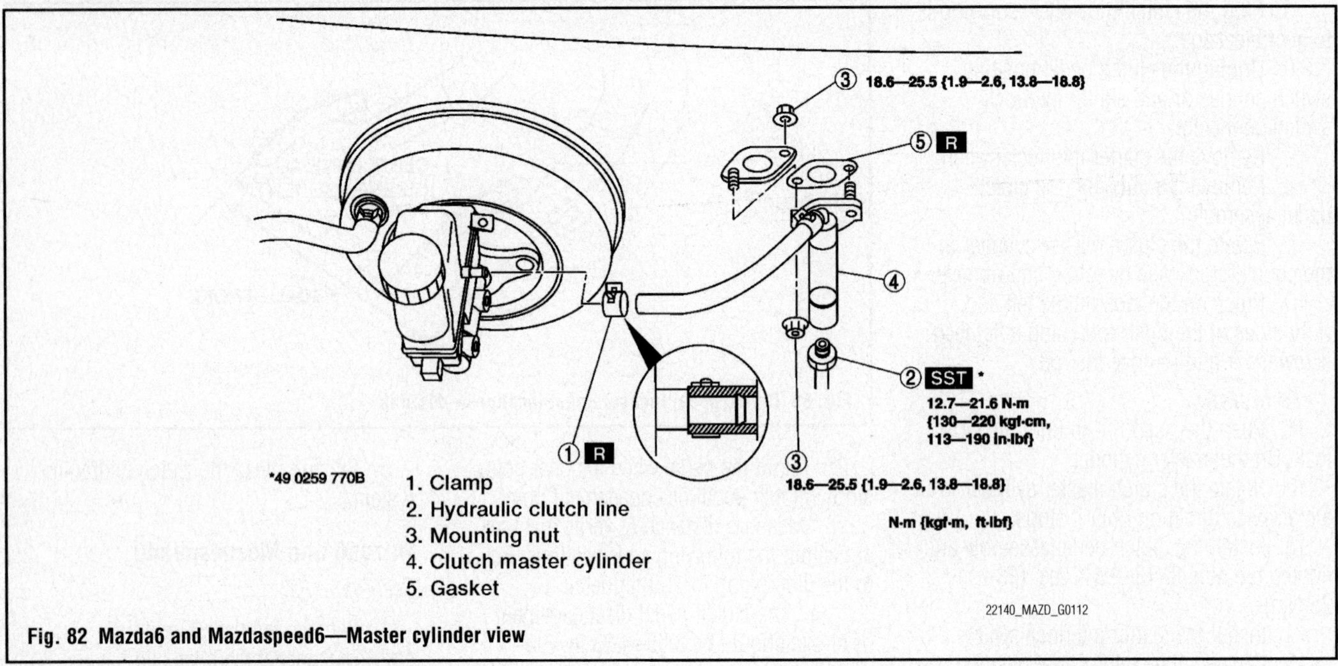

Fig. 82 Mazda6 and Mazdaspeed6—Master cylinder view

CLUTCH SLAVE CYLINDER

REMOVAL & INSTALLATION

Mazda3 and Mazdaspeed3

See Figure 83.

1. Remove the undercover.
2. Remove the clutch pipe.
3. Remove the two mounting bolts.
4. Remove the clutch release cylinder.

To install:

5. Install in the reverse order of removal.
6. Tighten the clutch release cylinder mounting bolts to 14–19 ft. lbs. (18–25 Nm).
7. Bleed the air from the system.
8. After installation, inspect the clutch pedal.

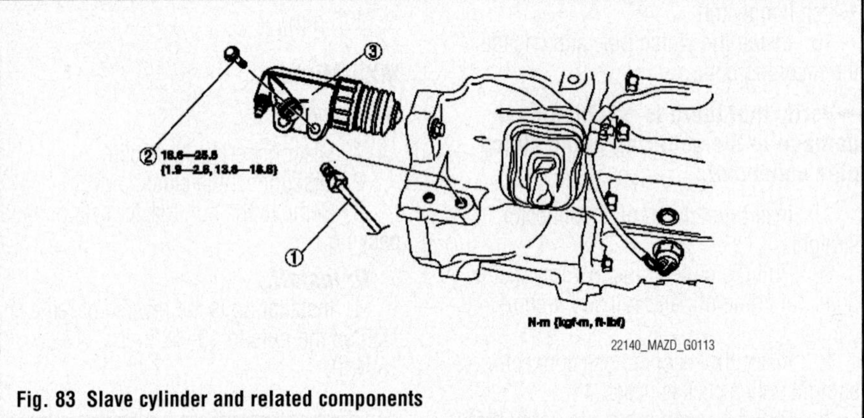

Fig. 83 Slave cylinder and related components

MX-5 Miata

See Figure 84.

1. Remove the clutch pipe.
2. Remove the two mounting bolts.
3. Remove the clutch release cylinder.

To install:

4. Install in the reverse order of removal.
5. Tighten the clutch release cylinder mounting bolts to 14–19 ft. lbs. (18–25 Nm).
6. Bleed the air from the system.
7. After installation, inspect the clutch pedal.

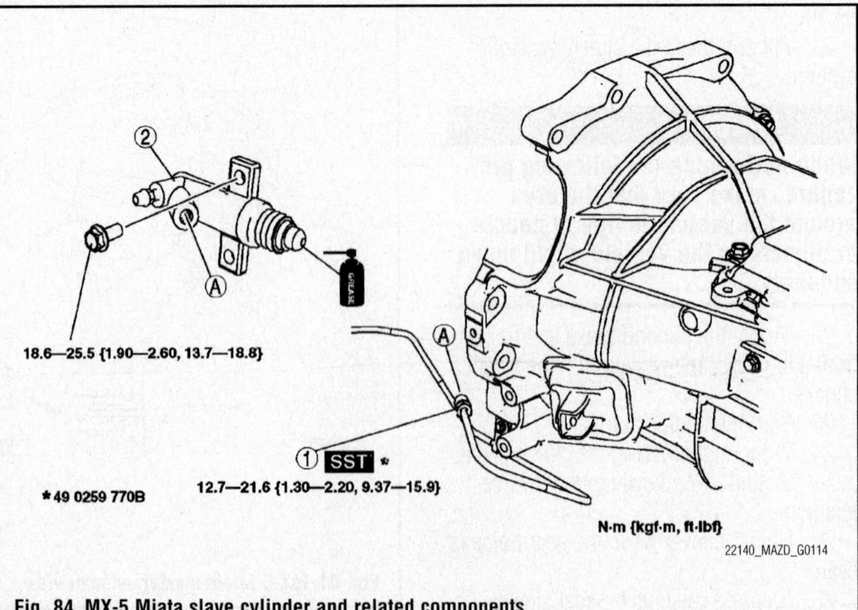

Fig. 84 MX-5 Miata slave cylinder and related components

Mazda6 and Mazdaspeed6

See Figures 85 and 86.

1. Remove the clutch pipe.
2. Remove the two mounting bolts.
3. Remove the clutch release cylinder.

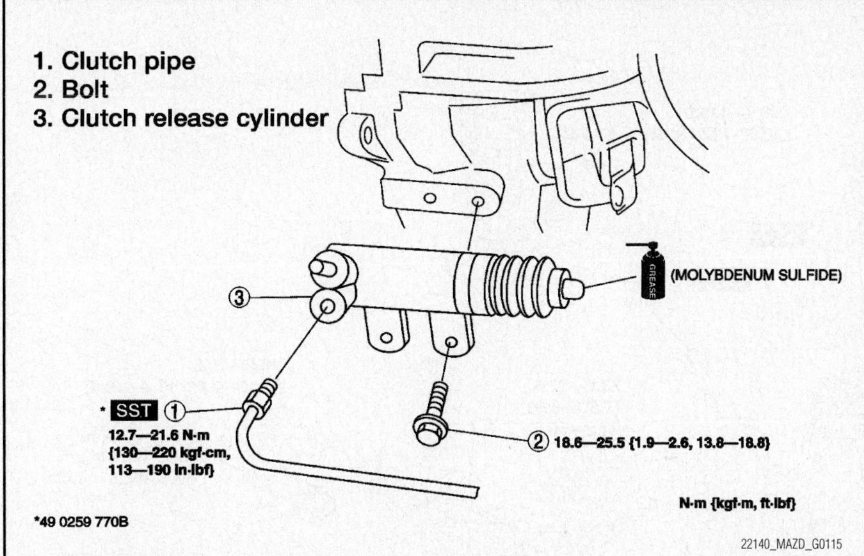

Fig. 85 Mazda6 and Mazdaspeed6 slave cylinder and related components—G35M-R transmission

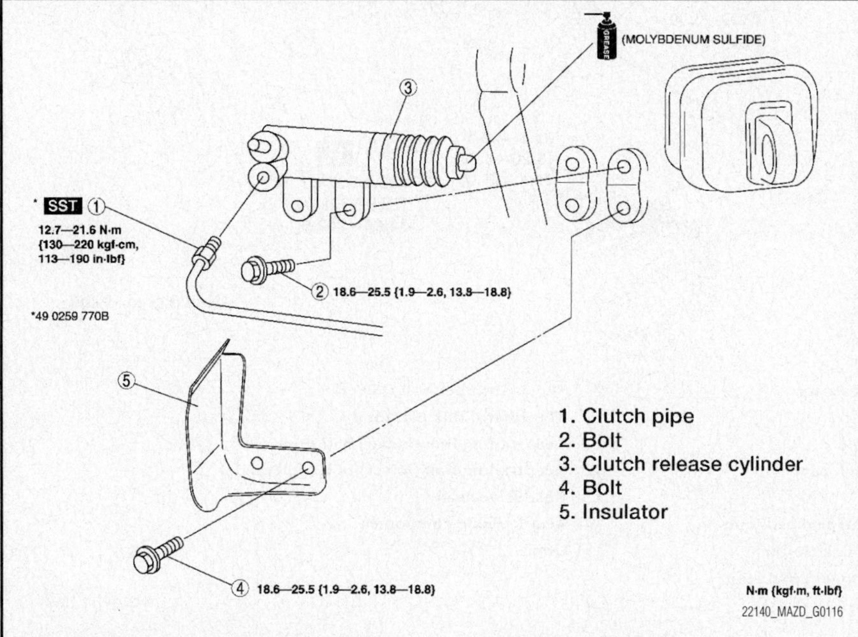

Fig. 86 Mazda6 and Mazdaspeed6slave cylinder and related components— G65M-R transmission

To install:

4. Install in the reverse order of removal.
5. Tighten the clutch release cylinder mounting bolts to 14–19 ft. lbs. (18–25 Nm).
6. Bleed the air from the system.
7. After installation, inspect the clutch pedal.

CLUTCH HYDRAULIC SYSTEM BLEEDING

1. Before servicing the vehicle, refer to the precautions section.
2. Remove the rubber cap from the bleeder screw on the release cylinder.
3. Place a bleeder tube over the end of the bleeder screw.
4. Submerge the other end of the tube in a jar half filled with hydraulic brake fluid.
5. Slowly pump the clutch pedal fully and allow it to return slowly, several times.
6. While pressing the clutch pedal to the floor, loosen the bleeder screw until the fluid starts to run out. Then, close the bleeder screw. Keep repeating this Step, while watching the hydraulic fluid in the jar. As soon as the air bubbles disappear, close the bleeder screw.
7. During the bleeding procedure the reservoir must be kept at leastC\vfull.

FRONT HALFSHAFTS

REMOVAL & INSTALLATION

MX5-Miata

See Figures 87 through 89.

1. Before servicing the vehicle, refer to the precautions section.

➡Refer the exploded view illustration for component locations and if applicable, their retainer torque specifications.

⁂ CAUTION

Performing the following procedures without first removing the ABS wheel-speed sensor may possibly cause an open circuit in the wiring harness if it is pulled by mistake. Before performing the following procedures, remove the ABS wheel-speed sensor (axle side) and fix it to an appropriate place where the sensor will not be pulled by mistake while servicing the vehicle.

2. Disconnect the ABS wheel-speed sensor.
3. Lock the disc plate by applying the brakes.
4. Knock the crimped portion of the locknut outward using a chisel and a hammer.
5. Remove the locknut.
6. Disconnect the parking brake cable.
7. Disconnect the brake caliper and position aside with the hose still attached. Secure the caliper so that no strain is put on the hose.
8. Remove the rotor.
9. Remove the rear upper lateral link ball joint.
10. Remove the lower stabilizer control link.
11. Remove the rear lower lateral link ball joint.
12. Remove the shock absorber lower bolt.
13. Remove the toe control link ball joint.
14. Remove the rear trailing link upper ball joint.
15. Remove the rear trailing link lower bolt on the outer side.
16. Temporarily install a spare nut to the end of the rear drive shaft.
17. Strike the nut with copper hammer lightly and remove the rear drive shaft from the wheel hub.

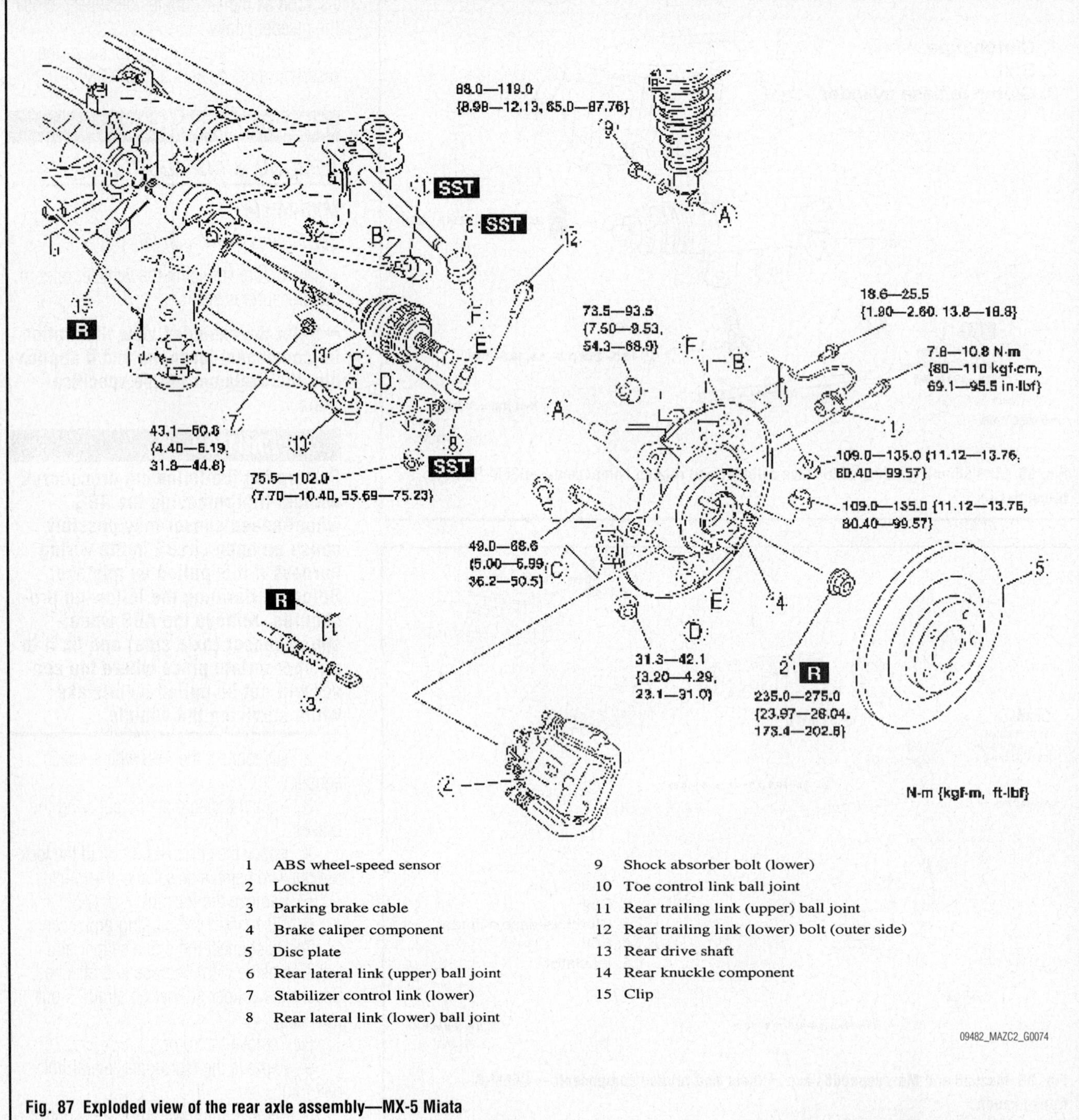

Fig. 87 Exploded view of the rear axle assembly—MX-5 Miata

18. Separate the rear drive shaft from the wheel hub.
19. Insert a pry bar between the rear differential and differential side outer ring, and then remove the rear drive shaft.

CAUTION

The drive shaft edges are sharp and can damage the oil seal. Be careful not to damage the oil seal when removing the drive shaft from the differential.

20. Pull the rear drive shaft to the outer side of the vehicle and disconnect it from the rear differential.
21. Secure the rear knuckle component by installing the upper rear lateral link to the rear knuckle temporarily after disconnecting the rear drive shaft.

To install:

22. Apply differential oil to the differential oil seal lip.

CAUTION

The drive shaft edges are sharp and can damage the oil seal. Be careful not to damage the oil seal when installing the rear drive shaft to the rear differential.

23. Insert the rear drive shaft into the rear differential with the clip opening facing upward. Point the opening of the new drive shaft clip upward, install it to the clip groove at the end of the rear drive shaft with

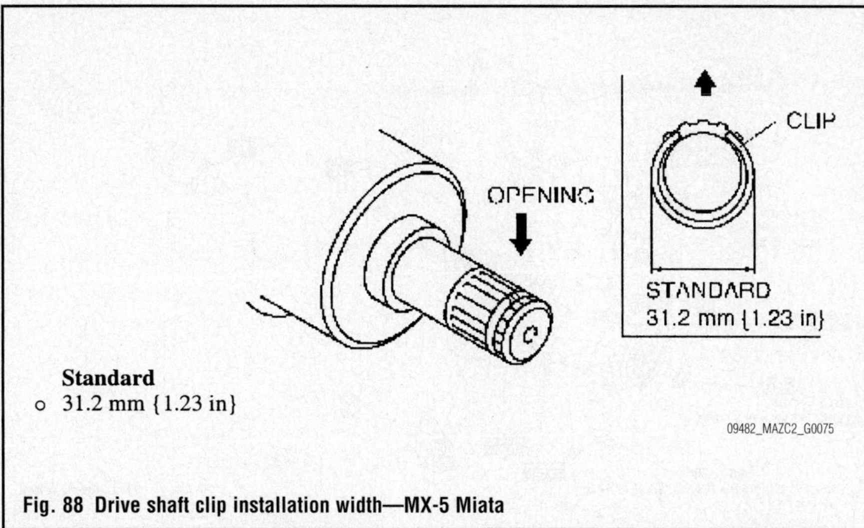

Fig. 88 Drive shaft clip installation width—MX-5 Miata

the installation width within the specification shown in the illustration. After installing the clip, measure the outer diameter. If it exceeds the specification, reinstall the new clip.

24. After installation, verify that the rear drive shaft is securely held by the clip by pulling the outer ring on the differential side towards the axle.
25. Install the rear trailing link lower bolt.
26. Install the rear trailing link upper ball joint.
27. Install the toe control link ball joint.
28. Install the shock absorber lower bolt.
29. Install the rear lower lateral link ball joint.
30. Install the lower stabilizer control link.
31. Install the rear upper lateral link ball joint.
32. Install the rotor.
33. Install the brake caliper.
34. Connect the parking brake cable.
35. Install the locknut and tighten to 235–275 Nm (173–202 ft. lbs.).

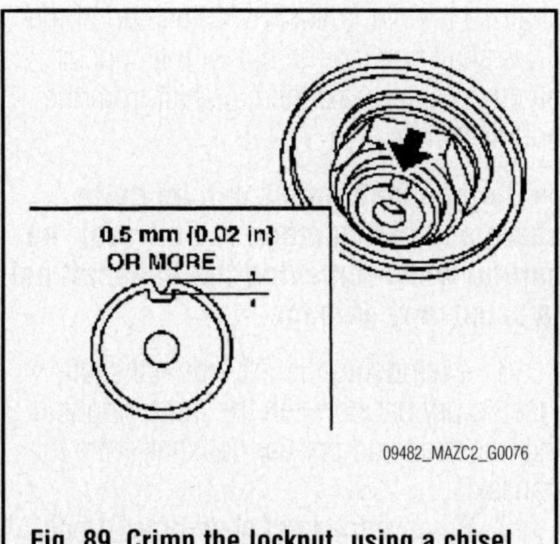

Fig. 89 Crimp the locknut, using a chisel and hammer—MX-5 Miata

36. Crimp the locknut, using a chisel and hammer.
37. Connect the parking brake cable.
38. Install the remaining components in the reverse order of removal.

Mazda3 and Mazdaspeed3

See Figure 90.

1. Before servicing the vehicle, refer to the precautions section.
2. Drain the transaxle oil.
3. Remove or disconnect the following:
   - Wheels
   - ABS sensor
   - Stabilizer control link upper nut
   - Front lower arm ball joint
   - Joint shaft
   - Clip

To install:

4. Install a new circlip on the end of the shaft, if removed, with the end gap facing upward. Measure the outer diameter of the clip, if it exceeds 1.19–1.23 inch (30–31mm) replace the clip.
5. Install the left side shaft as follows:
   a. Insert the shaft into the hub.
   b. Apply clean engine oil to the oil seal lip.
   c. Push the shaft into the transaxle and once the shaft clicks into place, pull on the outer ring to make sure the clip is engaged.
6. Install the right side shaft as follows:
   a. Insert the shaft into the hub.
   b. Insert the halfshaft into the joint shaft and once the shaft clicks into place, pull on the outer ring to make sure the clip is engaged.
7. Install the remaining components in the reverse order of removal. Refer to the exploded view of the halfshaft assembly and related components for torque specifications.

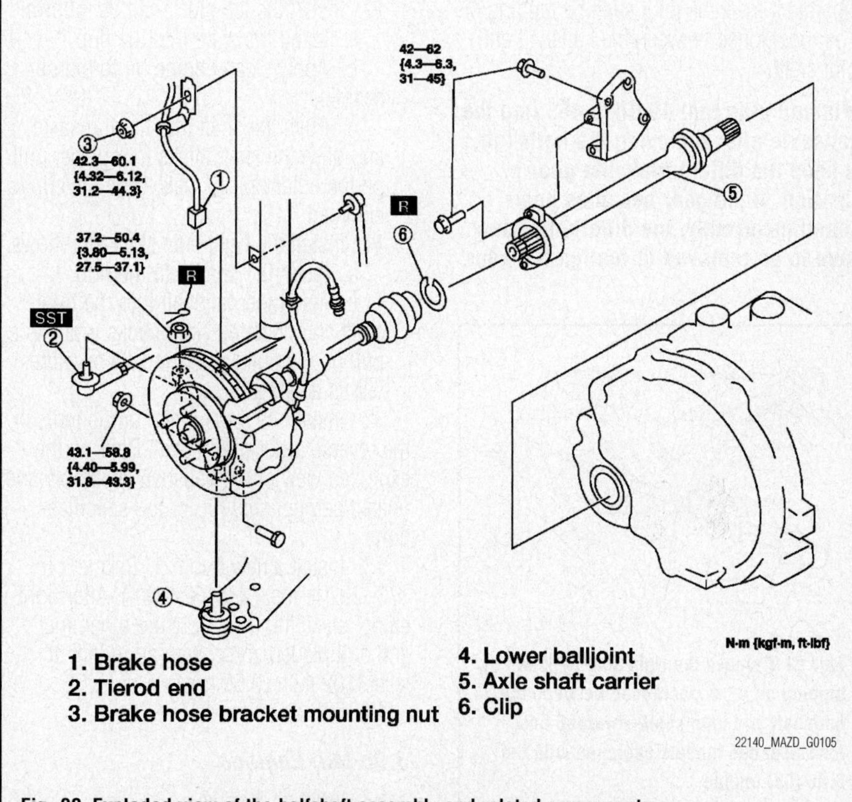

Fig. 90 Exploded view of the halfshaft assembly and related components

Mazda6 and Mazdaspeed6

2.3L (L3) Engines

See Figures 91 and 92.

1. Before servicing the vehicle, refer to the precautions section.
2. Drain the transaxle oil.
3. Raise the staked portion of the hub locknut with a hammer and chisel.
4. Lock the hub by applying the brakes and remove the nut.
5. Remove or disconnect the following:
   - Wheels
   - Tie rod end from the knuckle
   - Stabilizer bar link from the damper fork
   - Bolt attaching the fork to the knuckle
   - Front and rear lower arm ball joints
6. Install a spare nut onto the drive shaft so the nut is flush with the end of the drive shaft and tap the nut with a copper hammer to separate the halfshaft from the hub assembly.

➡The halfshaft edges can be quite sharp and can damage the oil seal. be careful when removing the halfshaft not to cause any damage.

7. If removing the left side halfshaft, insert a pry bar between the outer ring and the transaxle and pry the halfshaft from the transaxle.
8. If removing the right side halfshaft, separate it from the joint shaft by tapping on a bar positioned between the halfshaft and joint shaft.

➡Install plug tool 49 G030 455 into the transaxle after removing the halfshaft, to keep the differential side gear in position. If the gear becomes positioned incorrectly, the differential may have to be removed to realign the gear.

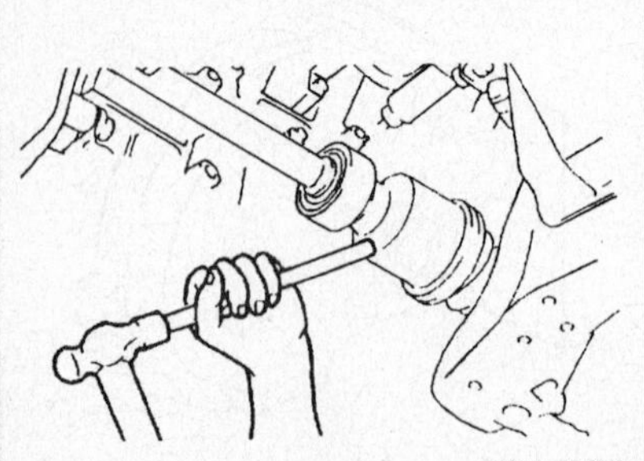

Fig. 91 Remove the right side halfshaft by tapping on a bar positioned between the halfshaft and joint shaft—Mazda6 and Mazdaspeed6 models equipped with the 2.3L (L3) engine

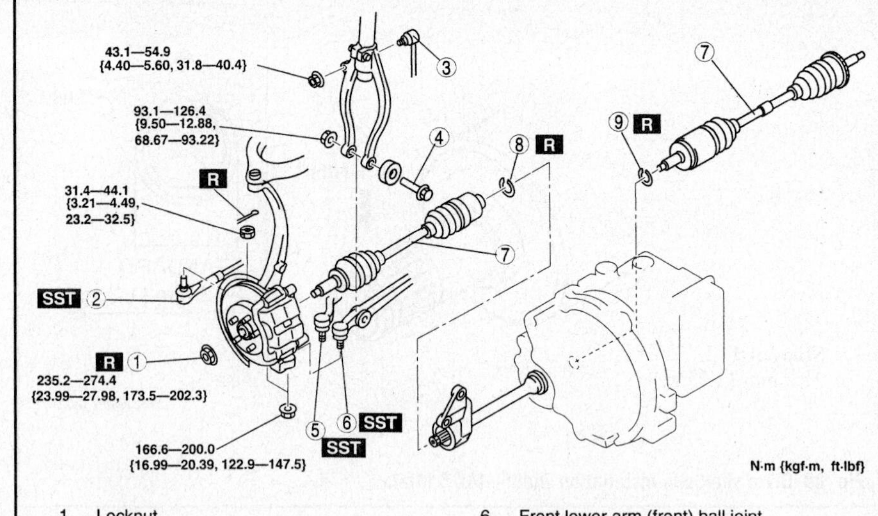

Fig. 92 Exploded view of the halfshaft assembly and related components—Mazda6 and Mazdaspeed6 models equipped with the 2.3L (L3) engine

To install:

9. Install a new circlip on the end of the shaft, if removed, with the end gap facing upward. Measure the outer diameter of the clip, if it exceeds 1.19–1.23 inch (30–31mm) replace the clip.
10. Install the left side shaft as follows:
    a. Insert the shaft into the hub.
    b. Apply clean engine oil to the oil seal lip.
    c. Push the shaft into the transaxle and once the shaft clicks into place, pull on the outer ring to make sure the clip is engaged.
11. Install the right side shaft as follows:
    a. Insert the shaft into the hub.
    b. Insert the halfshaft into the joint shaft and once the shaft clicks into place, pull on the outer ring to make sure the clip is engaged.
12. Install the remaining components in the reverse order of removal. Refer to the exploded view of the halfshaft assembly and related components for torque specifications.
13. Install a new hub nut. Torque it to 173–202 ft. lbs. (235–274 Nm). After tightening, stake the locknut using a hammer and dull bladed chisel with an indent at least 0.02 inch (0.5mm).
14. Fill the transaxle.

3.0L (AJ) Engines

See Figures 93 and 94.

1. Before servicing the vehicle, refer to the precautions section.
2. Drain the transaxle oil.
3. Remove the knuckle and hub assembly.
4. Raise the staked portion of the hub locknut with a hammer and chisel.
5. Lock the hub by applying the brakes and remove the nut.
6. Remove or disconnect the following:
   - Wheels
   - Tie rod end from the knuckle
   - Damper fork
   - Front and rear lower arm ball joints
   - Stabilizer bar link from the damper fork
7. Install a spare nut onto the drive shaft so the nut is flush with the end of the drive shaft and tap the nut with a copper hammer to separate the halfshaft from the hub assembly.

➡The halfshaft edges can be quite sharp and can damage the oil seal. be careful when removing the halfshaft not to cause any damage.

8. If removing the left side halfshaft, insert a pry bar between the outer ring and the transaxle and pry the halfshaft from the transaxle.
9. Remove the joint shaft bracket bolt and remove the right side halfshaft. Secure

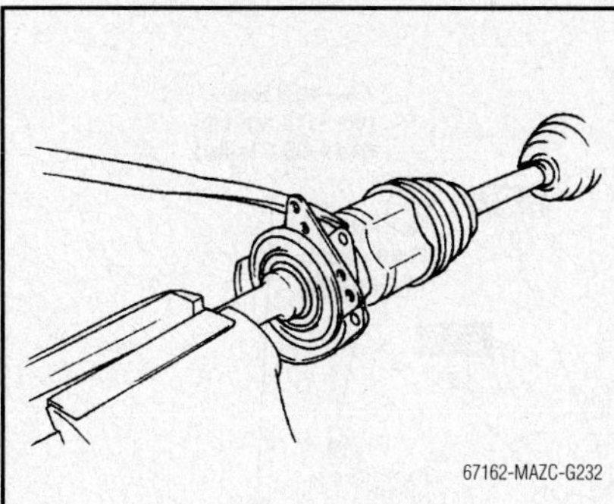

Fig. 93 Remove the right side halfshaft by securing the shaft in a vise, inserting a pry bar between the halfshaft and joint shaft and separating them—Mazda6 models equipped with the 3.0L (AJ) engine

the shaft in a vise and insert a pry bar between the halfshaft and joint shaft and separate them.

➡Install plug tool 49 G030 455 into the transaxle after removing the halfshaft, to keep the differential side gear in position. If the gear becomes positioned incorrectly, the differential may have to be removed to realign the gear.

To install:

10. Install a new circlip on the end of the shaft, if removed, with the end gap facing upward. Measure the outer diameter of the clip, if it exceeds 1.26–1.30 inch (32–33mm) replace the clip.
11. Install the left side shaft as follows:
    a. Insert the shaft into the hub.
    b. Apply clean engine oil to the oil seal lip.
    c. Push the shaft into the transaxle and once the shaft clicks into place, pull on the outer ring to make sure the clip is engaged.
12. Install the right side shaft as follows:
    a. Place the halfshaft and joint shaft assembly in a vise.
    b. Insert the halfshaft into the joint shaft and use a plastic hammer to attach the joint shaft as illustrated.
    c. Tighten the joint shaft bracket bolt to 31–45 ft. lbs. (42–62 Nm).

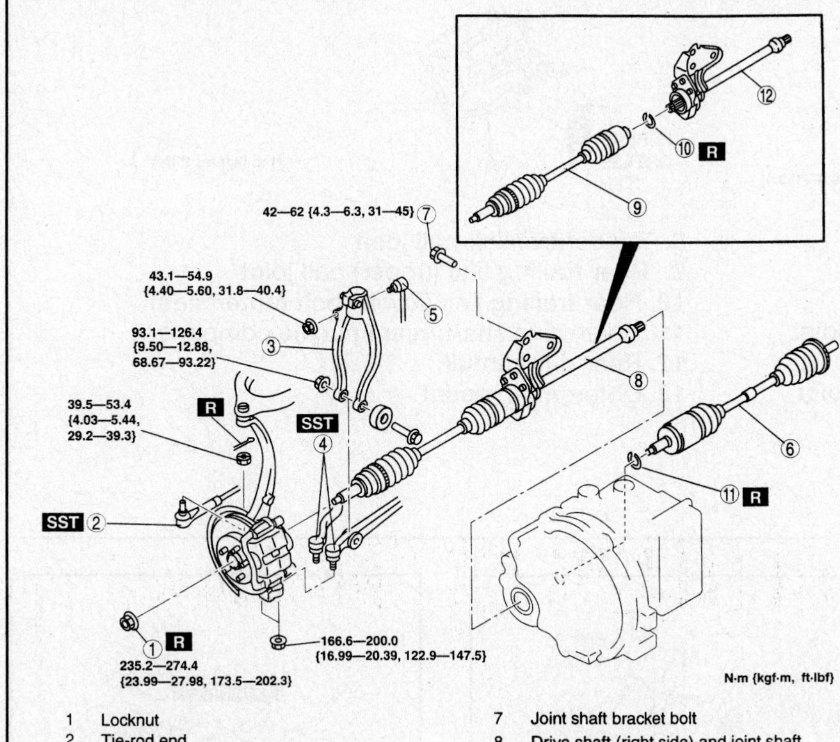

Fig. 94 Exploded view of the halfshaft assembly and related components—Mazda6 models equipped with the 3.0L (AJ) engine

13. Install the remaining components in the reverse order of removal. Refer to the exploded view of the halfshaft assembly and related components for torque specifications.
    - New hub nut. Torque it to 173–202 ft. lbs. (235–274 Nm). After tightening, stake the locknut using a hammer and dull bladed chisel with an indent at least 0.02 inch (0.5mm).
14. Fill the transaxle.

REAR AXLE HOUSING

REMOVAL & INSTALLATION

MX-5 Miata

See Figure 95.

⁂ WARNING

Performing the following procedures without first removing the ABS wheel-speed sensor may possibly cause an open circuit in the wiring harness if it is pulled by mistake. Before performing the following procedures, remove the ABS wheel-speed sensor (axle side) and fix it to an appropriate place where the sensor will not be pulled by mistake while servicing the vehicle.

1. Before servicing the vehicle, refer to the precautions section.
2. Drain the rear differential oil.
3. Remove the middle exhaust pipe.
4. Remove the propeller shaft.
5. Remove the power plant frame.
6. Remove or disconnect the following:
    - ABS wheel-speed sensor
    - Parking brake cable
    - Brake caliper component
    - Rear lateral link (upper) ball joint
    - Stabilizer control link (lower)
    - Rear lateral link (lower) ball joint
    - Shock absorber bolt (lower)
    - Toe control link ball joint
    - Rear trailing link (upper) ball joint
    - Rear trailing link (lower) bolt (outer side)
    - Rear drive shaft, rear knuckle component
    - Rear differential
    - Differential mount

To install:

7. Refer to illustration for tightening specifications
8. Install in the reverse order of removal.
9. Add differential oil: API service GL-5, SAE 75W-90

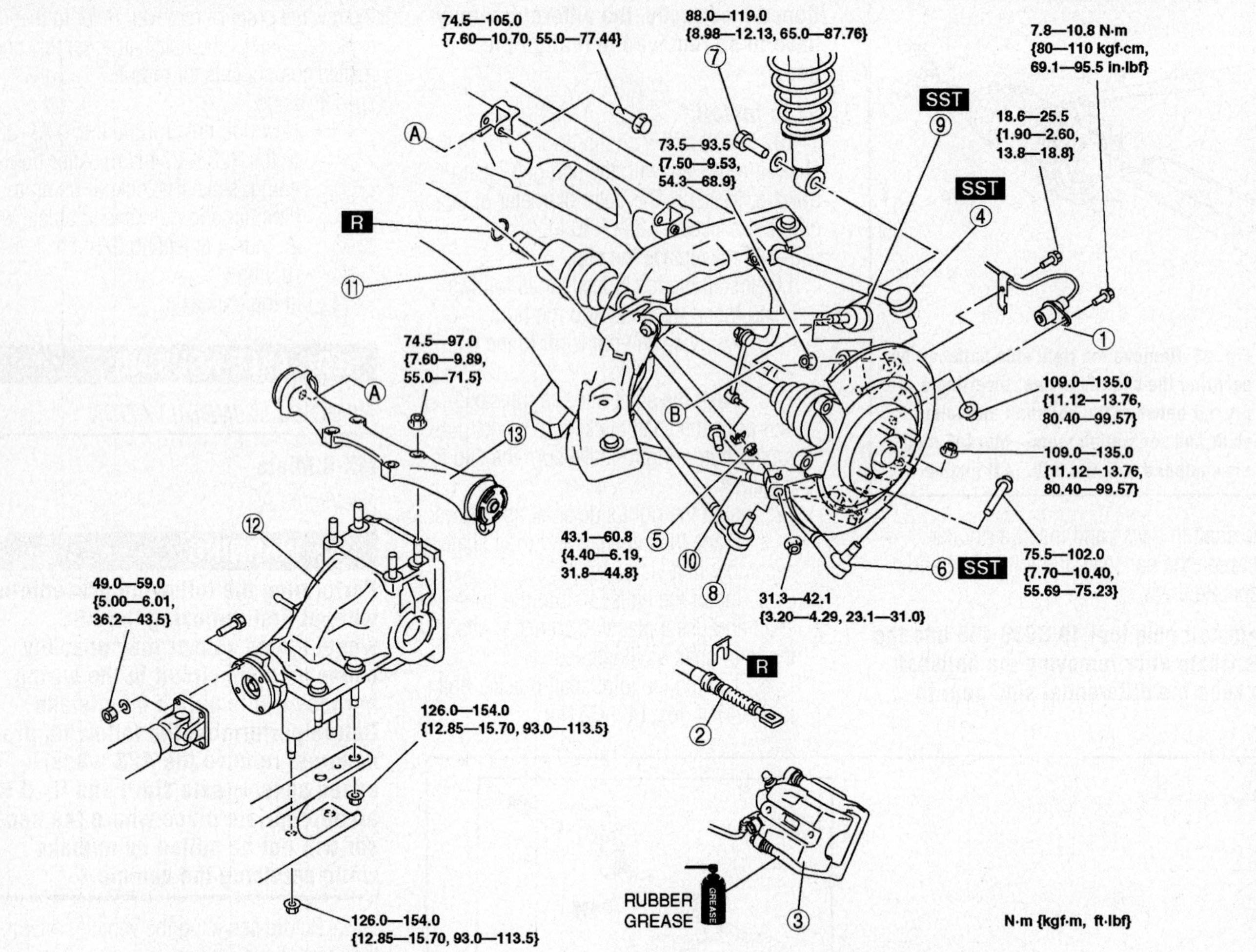

74.5—105.0
{7.60—10.70, 55.0—77.44}
88.0—119.0
{8.98—12.13, 65.0—87.76}
7.8—10.8 N·m
{80—110 kgf·cm,
69.1—95.5 in·lbf}
SST
18.6—25.5
{1.90—2.60,
13.8—18.8}
73.5—93.5
{7.50—9.53,
54.3—68.9}
SST
R
74.5—97.0
{7.60—9.89,
55.0—71.5}
109.0—135.0
{11.12—13.76,
80.40—99.57}
109.0—135.0
{11.12—13.76,
80.40—99.57}
43.1—60.8
{4.40—6.19,
31.8—44.8}
SST
75.5—102.0
{7.70—10.40,
55.69—75.23}
49.0—59.0
{5.00—6.01,
36.2—43.5}
31.3—42.1
{3.20—4.29, 23.1—31.0}
R
126.0—154.0
{12.85—15.70, 93.0—113.5}
RUBBER
GREASE
GREASE
126.0—154.0
{12.85—15.70, 93.0—113.5}
N·m {kgf·m, ft·lbf}

1. ABS wheel-speed sensor
2. Parking brake cable
3. Brake caliper component
4. Rear lateral link (upper) ball joint
5. Stabilizer control link (lower)
6. Rear lateral link (lower) ball joint
7. Shock absorber bolt (lower)
8. Toe control link ball joint
9. Rear trailing link (upper) ball joint
10. Rear trailing link (lower) bolt (outer side)
11. Rear drive shaft, rear knuckle component
12. Rear differential
13. Differential mount

22140_MAZD_G0131

Fig. 95 MX-5 Miata

REAR AXLE SEAL

REMOVAL & INSTALLATION

MX-5 Miata

See Figures 96 and 97.

1. Before servicing the vehicle, refer to the precautions section.
2. Drain the rear differential oil.
3. Remove the rear halfshaft.
4. Remove the oil seal from the differential carrier using a flathead screwdriver.

To install:

5. Clean seal installation surface.

22140_MAZD_G0132

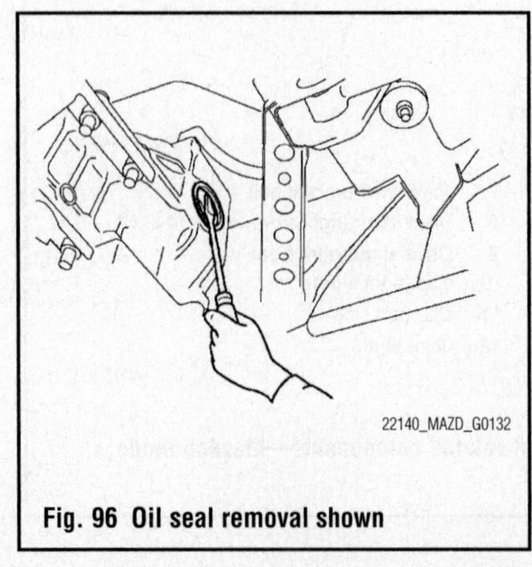

Fig. 96 Oil seal removal shown

49 G030 797
49 U027 003

22140_MAZD_G0133

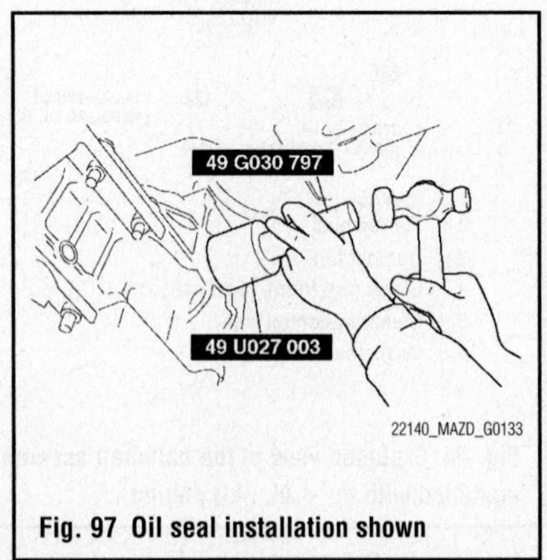

Fig. 97 Oil seal installation shown

6. Install differential oil on the seal surface.
7. Tap in the new oil seal until it reaches the differential carrier using the SSTs.
8. Install the halfshaft.
9. Add differential oil: API service GL-5, SAE 75W-90

REAR PINION SEAL

REMOVAL & INSTALLATION

MX-5 Miata

See Figures 98 through 100.

1. Before servicing the vehicle, refer to the precautions section.
2. Drain the rear differential oil.
3. Mark and remove the propeller shaft.
4. Remove the locknut while fixing the companion flange using the SST.
5. Remove the companion flange using the SST.
6. Carefully remove the pinion flange
7. Use seal remover and remove the pinion seal.

To install:

8. Apply differential oil to the lip of a new oil seal.
9. Assemble the oil seal using the SST.
10. Tighten a new locknut with the torque recorded at the drive pinion preload adjustment using the SST.
11. Verify that the drive pinion preload is within the specification.
12. Drive pinion preload: 12–15 inch. lbs. (1.3–1.8 Nm).
13. Install the propeller shaft.
14. Add differential oil: API service GL-5, SAE 75W-90
15. Check seal for leaks.

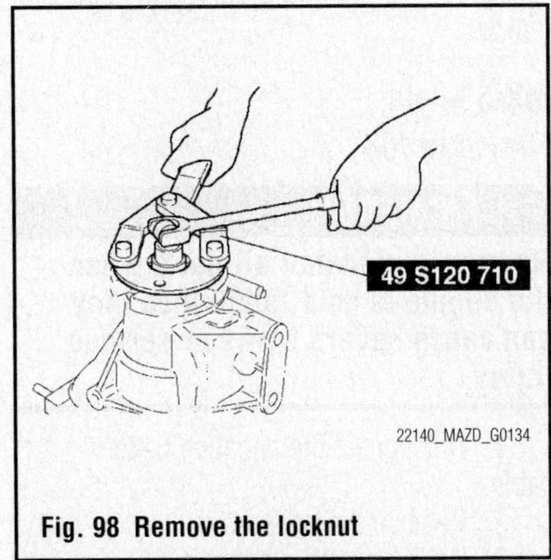

Fig. 98 Remove the locknut

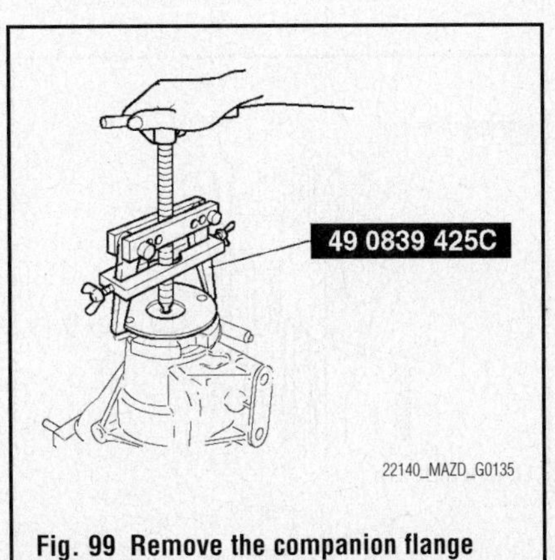

Fig. 99 Remove the companion flange

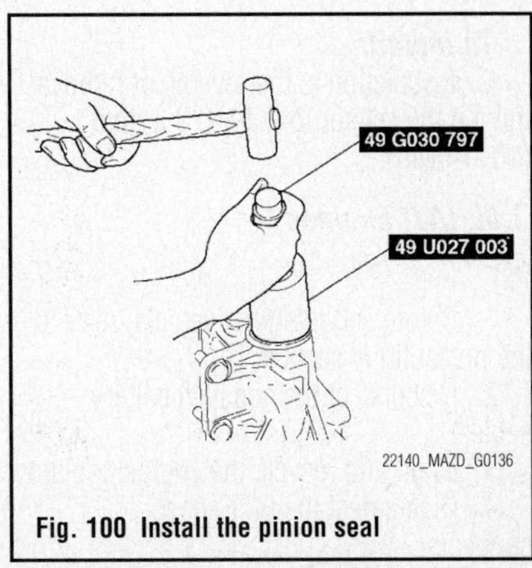

Fig. 100 Install the pinion seal

ENGINE COOLING

COOLANT TEMPERATURE SENSOR

REMOVAL & INSTALLATION

Mazda3 and Mazdaspeed3

See Figure 101.

1. Remove the battery cover.
2. Disconnect the negative battery cable.
3. Remove the plug hole cover.
4. Drain and recycle the engine coolant.
5. Disconnect the ECT sensor connector.
6. Remove the ECT sensor.

To install:

7. Installation is the reverse of removal, tighten the sensor to 89–123 in. lbs. (10–14 Nm).

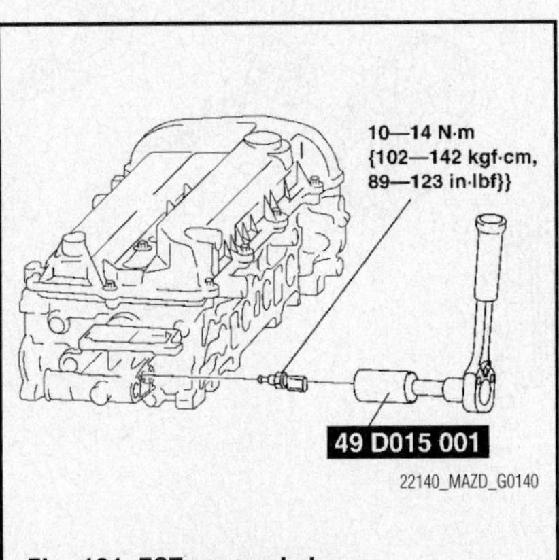

Fig. 101 ECT removal shown

MX-5 Miata

See Figure 102.

1. Before servicing the vehicle, refer to the precautions section.
2. Remove the battery cover.
3. Disconnect the negative battery cable.
4. Drain and recycle the engine coolant.
5. Remove the service hole cover.
6. Remove the suspension tower bar (joint), (right side) and (left side).
7. Remove the wiper arm.
8. Remove the cowl grille.
9. Remove the side cowl grille.
10. Move the cooler pipe No.3 and heater pipe slightly out of the way.
11. Remove the service hole cover.
12. Disconnect the heater hose and move the heater pipe slightly out of the way.
13. Disconnect the ECT sensor connector.
14. Remove the ECT sensor.
15. Installation is the reverse of removal, tighten the sensor to 7.4–10.3 ft. lbs. (10–14 Nm).

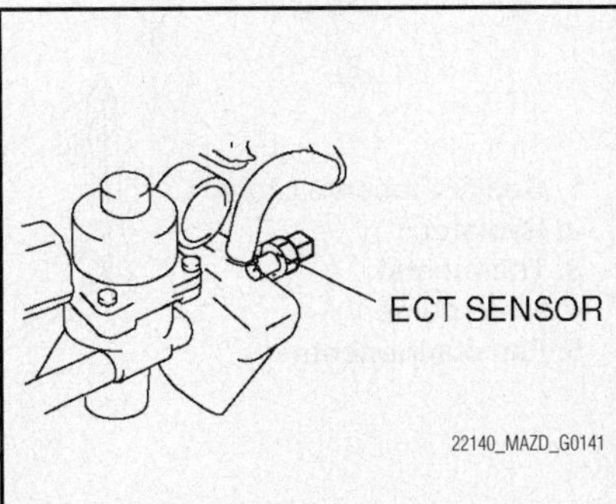

Fig. 102 MX-5 Miata—ECT sensor location

Mazda6 and Mazdaspeed6

2.3L (L3) Engines

See Figure 103.

1. Before servicing the vehicle, refer to the precautions section.
2. Disconnect the negative battery cable.
3. Drain and recycle the engine coolant.
4. Disconnect the ECT sensor connector.
5. Remove the ECT sensor.

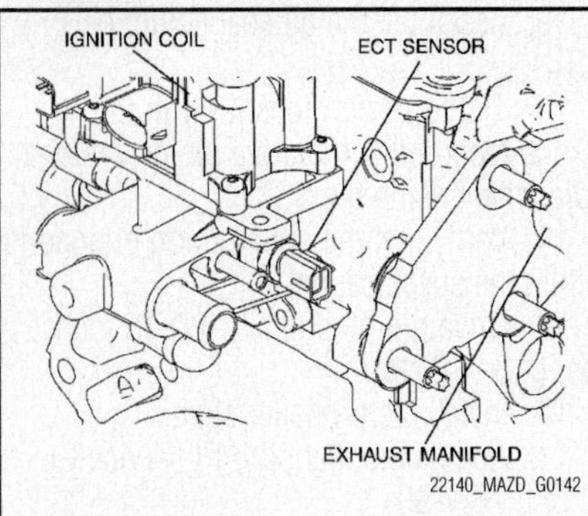

Fig. 103 Mazda6 and Mazdaspeed6—ECT sensor location

To install:

6. Installation is the reverse of removal, tighten the sensor to 7.4–10.3 ft. lbs. (10–14 Nm).

3.0L (AJ) Engines

See Figure 104.

1. Before servicing the vehicle, refer to the precautions section.
2. Disconnect the negative battery cable.
3. Drain and recycle the engine coolant.
4. Disconnect the ECT sensor connector.
5. Remove the ECT sensor.
6. Replace the gasket.

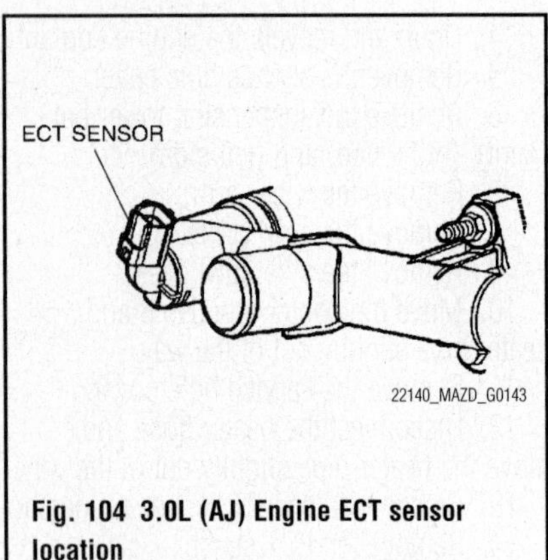

Fig. 104 3.0L (AJ) Engine ECT sensor location

To install:

7. Install in the reverse order of removal.
8. Refill the engine coolant.
9. Check for coolant leaks.

ENGINE FAN

REMOVAL & INSTALLATION

Mazda3 and Mazdaspeed3

See Figures 105 and 106.

✲✲ CAUTION

Remove and install all parts when the engine is cold, otherwise they can cause severe burns or serious injury.

1. Before servicing the vehicle, refer to the precautions section.
2. Remove the battery cover.
3. Disconnect the negative battery cable.
4. Remove the undercover.
5. Drain and recycle the engine coolant.

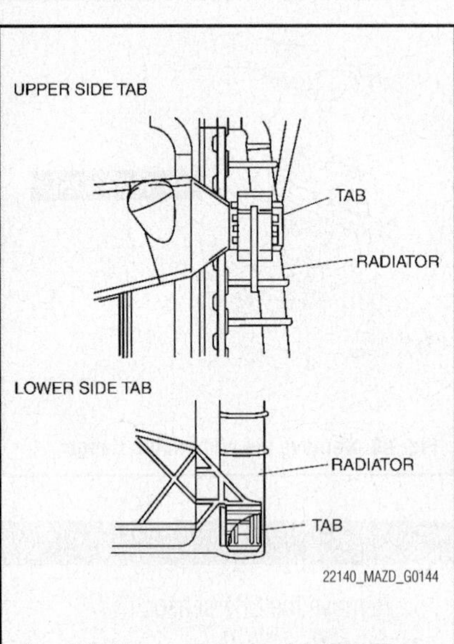

Fig. 105 Upper and lower side tabs shown

6. Disconnect the coolant reservoir hose from the radiator.
7. Position the front wiring harnesses out of the way.
8. Remove the fan control module connector.
9. Unlock the cooling fan component lower side tabs a at two points by pressing the cooling fan side tabs, remove upper side tabs, then remove the cooling fan component from the radiator

To install:

10. Install the components in the reverse of removal.
11. Fill the cooling system and check for leaks.

MX-5 Miata

See Figure 107.

✲✲ CAUTION

Remove and install all parts when the engine is cold, otherwise they can cause severe burns or serious injury.

1. Disconnect the negative battery cable.
2. Remove the undercover.
3. Drain and recycle the engine coolant.
4. Remove the battery, battery tray, and battery duct.
5. Remove the air cleaner.
6. Remove the PCM and air cleaner insulator.
7. Remove the coolant reserve tank.

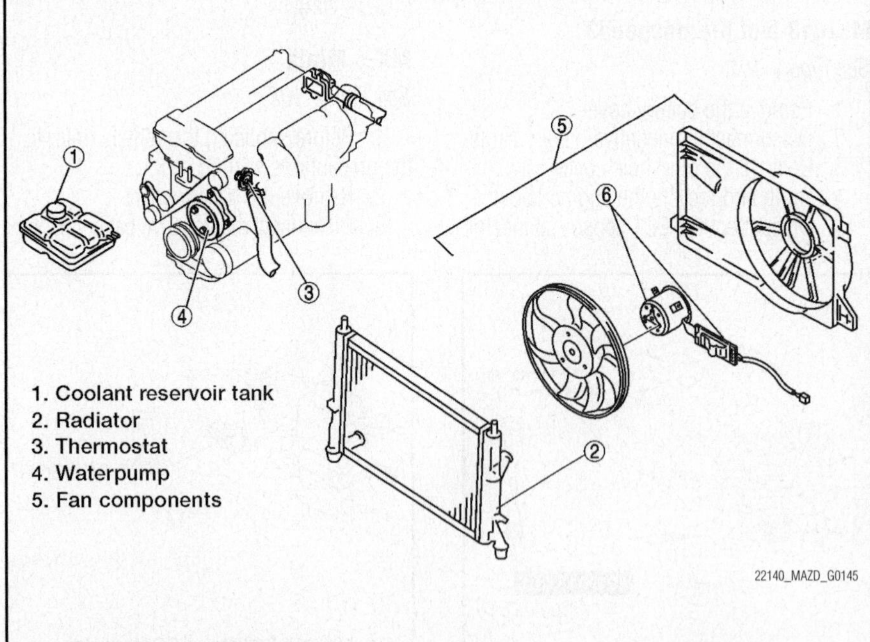

Fig. 106 Cooling system index view

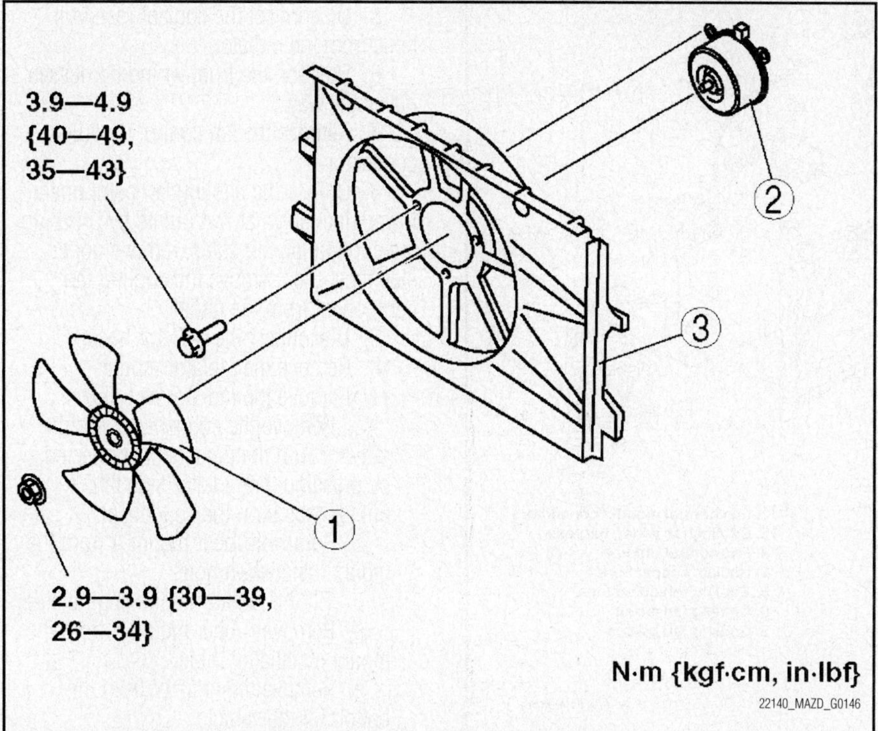

Fig. 107 Exploded view of the cooling fan components—MX-5 Miata

6. Remove the fan motor connector and wiring harness.
7. Remove the coolant reservoir hose.
8. Remove the cooling fan assembly.
9. Remove the fan motor and blades if needed. Refer to the illustration for components location and torque values.

To install:

10. Install the components in the reverse of removal.
11. Fill the cooling system and check for leaks.
12. Check the transaxle fluid level.

3.0L (AJ) Engines

See Figure 109.

1. Before servicing the vehicle, refer to the precautions section.
2. Disconnect the negative battery cable.
3. Remove the shroud panel.
4. Remove the cooler pipe stay.
5. Remove the oil hose stay on models with an automatic transaxle.
6. Remove the coolant reservoir hose.
7. Unplug the fan control module connectors.
8. Disconnect the ATF oil cooler hose on models with an automatic transaxle.
9. Remove the cooling fan connector and the PCM duct.
10. Disconnect the radiator and reserve tank hose.
11. Remove the P/S cooling pipe with the hoses still connected.
12. Remove the condenser with the cooler pipes still connected
13. Remove the rubber mount brackets.
14. Remove the radiator and fan as an assembly.
15. Remove the cooling fan.

To install:

16. Install the components in the reverse of removal.
17. Fill the cooling system and check for leaks.
18. Check the transaxle fluid level.

Mazda6 and Mazdaspeed6

2.3L (L3) Engines

See Figure 108.

1. Before servicing the vehicle, refer to the precautions section.
2. Disconnect the negative battery cable.
3. Remove the shroud panel.
4. Remove the cooler pipe stay.
5. Remove the oil hose stay on models with an automatic transaxle.

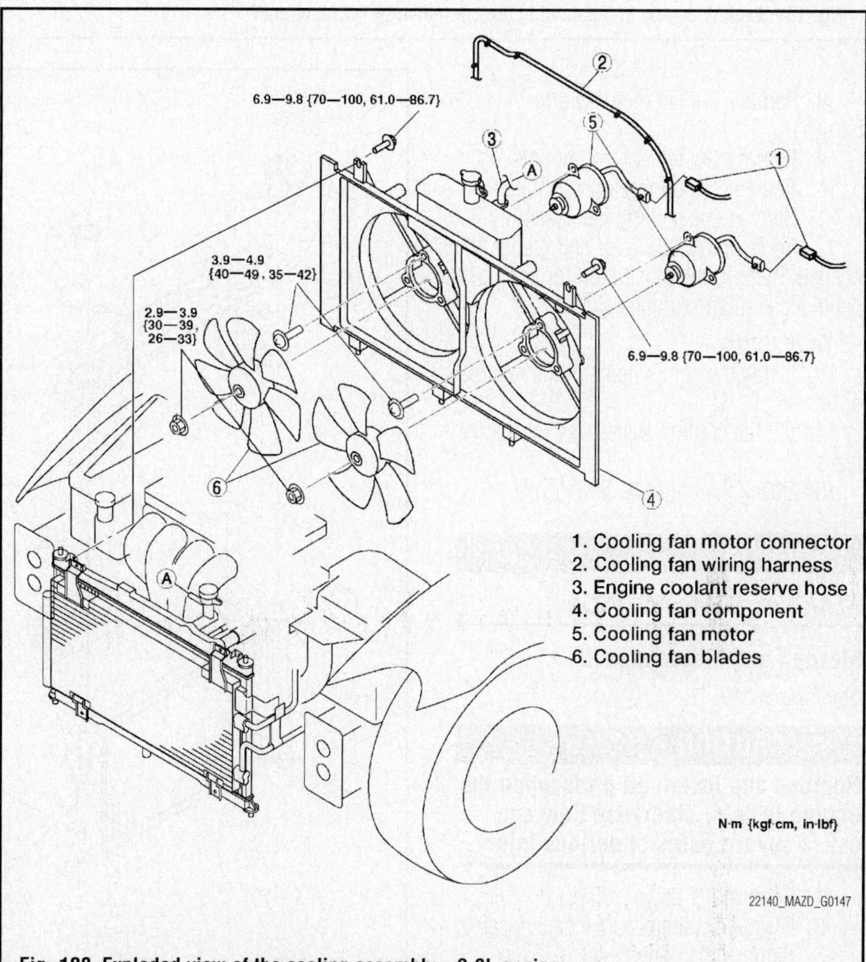

Fig. 108 Exploded view of the cooling assembly—2.3L engine

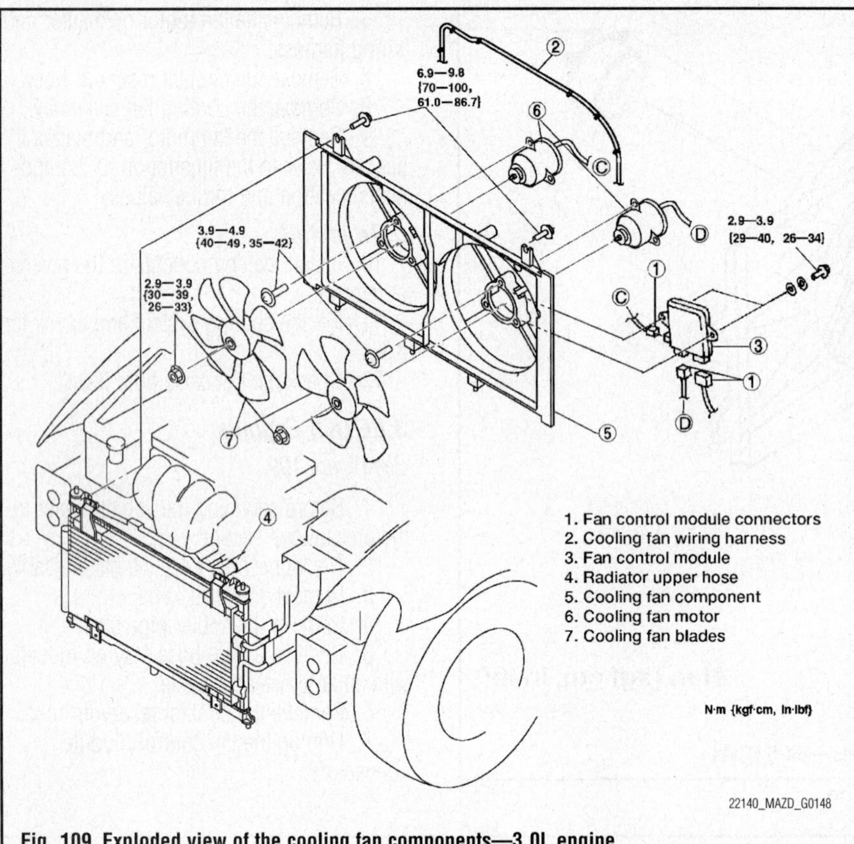

Fig. 109 Exploded view of the cooling fan components—3.0L engine

8. Remove the fan motor wiring harness.
9. Remove the fan control module.
10. Remove the upper radiator hose
11. Remove the cooling fan assembly.
12. Remove the fan motor and blades if needed. Refer to the illustration for components location and torque values.

To install:

13. Install the components in the reverse of removal.
14. Fill the cooling system and check for leaks.
15. Check the transaxle fluid level.

RADIATOR

REMOVAL & INSTALLATION

Mazda3 and Mazdaspeed3

See Figure 110.

✲✲ CAUTION

Remove and install all parts when the engine is cold, otherwise they can cause severe burns or serious injury.

1. Remove the battery cover.
2. Disconnect the negative battery cable.
3. Remove the undercover.
4. Drain and recycle the engine coolant.
5. Disconnect the coolant reservoir hose from the radiator.
6. Position the front wiring harnesses out of the way.
7. Remove the fan control module connector.
8. Unlock the cooling fan component lower side tabs a at two points by pressing the cooling fan side tabs, remove upper side tabs, then remove the cooling fan component from the radiator
9. Disconnect the radiator hoses.
10. Remove the radiator mount.
11. Remove the radiator as follows:
    a. Remove the condenser from the radiator with the pipes still connected, by pressing the radiator side tabs to unlock tabs b on the condenser.
    b. Remove rubber mount A from the mount installation hole.
    c. Tilt the radiator to the engine side.
    d. Remove rubber mount B from the mount installation hole.
    e. Remove the radiator from the bottom of the vehicle.

To install:

12. Install the radiator, mount and hoses.
13. Install the condenser to the radiator by aligning lower side tabs with the radiator

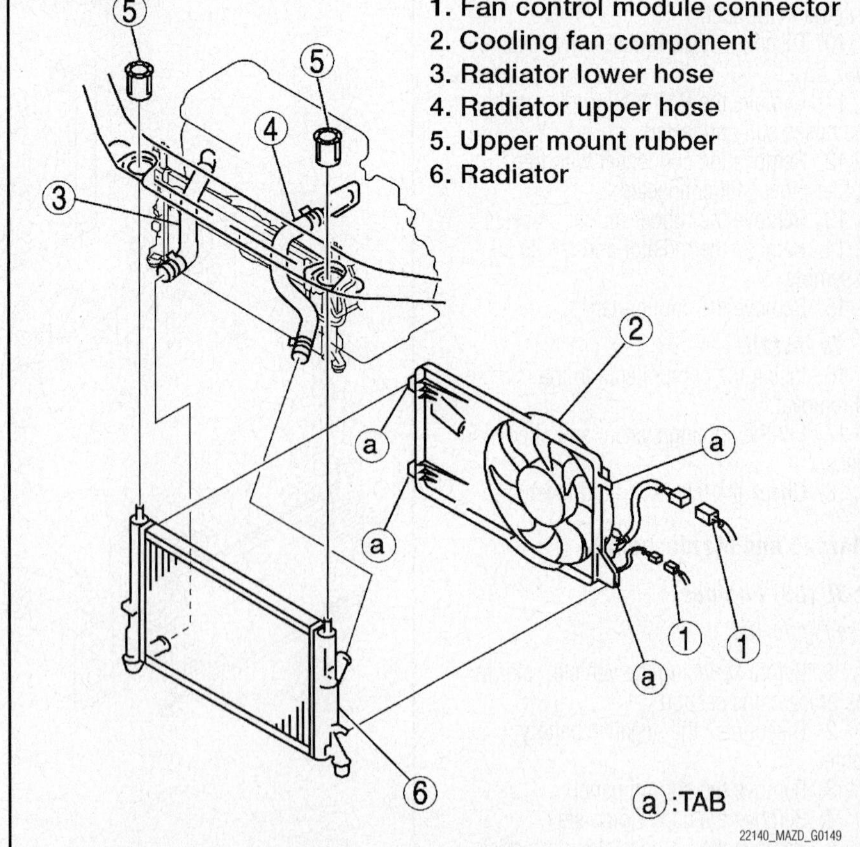

Fig. 110 2.3L Engines—radiator and related components

side tabs, install upper tabs b, then install lower side tabs.

14. Insert the cooling fan component tabs to the radiator to install the cooling fan component.

15. Install the remaining components in the reverse of removal.

16. Fill the cooling system and check for leaks.

MX-5 Miata

See Figure 111.

⁂ CAUTION

Remove and install all parts when the engine is cold, otherwise they can cause severe burns or serious injury.

1. Disconnect the negative battery cable.
2. Remove the undercover.
3. Drain and recycle the engine coolant.
4. Remove the battery, battery tray, and battery duct.
5. Remove the air cleaner.
6. Remove the PCM and air cleaner insulator.
7. Remove the coolant reserve tank.
8. Disconnect the ATF oil cooler hose on models with an automatic transaxle.
9. Remove the cooling fan connector and the PCM duct.
10. Disconnect the radiator and reserve tank hose.
11. Remove the P/S cooling pipe with the hoses still connected.
12. Remove the condenser with the cooler pipes still connected
13. Remove the rubber mount brackets.
14. Remove the radiator and fan as an assembly.
15. Remove the cooling fan.
16. Remove the radiator.

To install:

17. Install the components in the reverse of removal.
18. Fill the cooling system and check for leaks.
19. Check the transaxle fluid level.

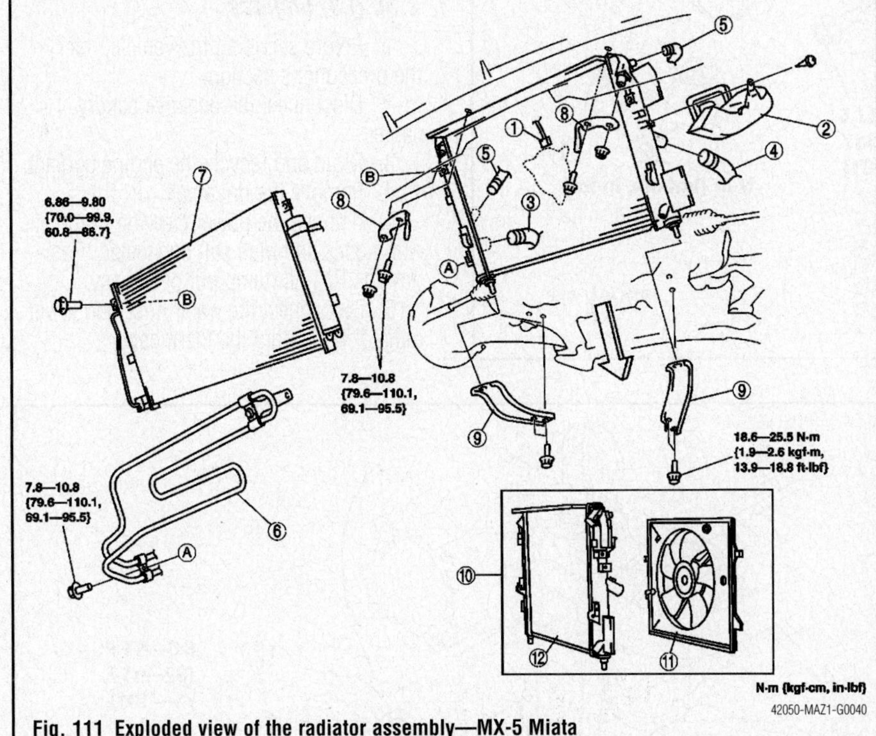

Fig. 111 Exploded view of the radiator assembly—MX-5 Miata

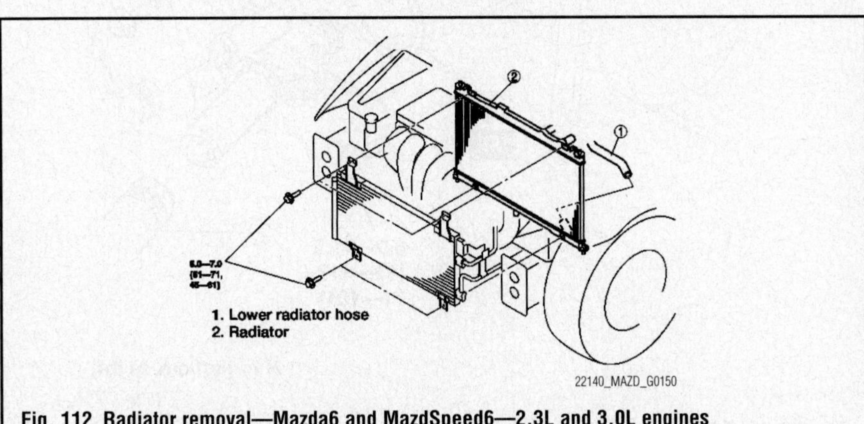

Fig. 112 Radiator removal—Mazda6 and MazdSpeed6—2.3L and 3.0L engines

Mazda6 and Mazdaspeed6

2.3L and 3.0L Engines

See Figure 112.

1. Before servicing the vehicle, refer to the precautions section.
2. Disconnect the negative battery cable.
3. Drain and recycle the engine coolant.
4. Remove the air cleaner.
5. Remove the shroud panel.
6. Remove the cooling fan.
7. Disconnect the upper radiator hose from the radiator, on the 2.3L engine.
8. Disconnect the oil hose from the radiator if equipped with an automatic transaxle.
9. Remove the engine coolant reserve hose on the 3.0L engine.
10. Remove the radiator hoses, radiator bolts and the radiator.

To install:

11. Install the components in the reverse of removal. Tighten the radiator bolts to 45–61 inch lbs. (5.0–7.0 Nm).
12. Fill the cooling system and check for leaks.
13. Check the transaxle fluid level.

THERMOSTAT

REMOVAL & INSTALLATION

Mazda3 and Mazdaspeed3

See Figure 113.

⁂ CAUTION

Remove and install all parts when the engine is cold, otherwise they can cause severe burns or serious injury.

1. Remove the battery cover.
2. Disconnect the negative battery cable.
3. Remove the undercover and splash shield as a single unit.
4. Drain and recycle the engine coolant.

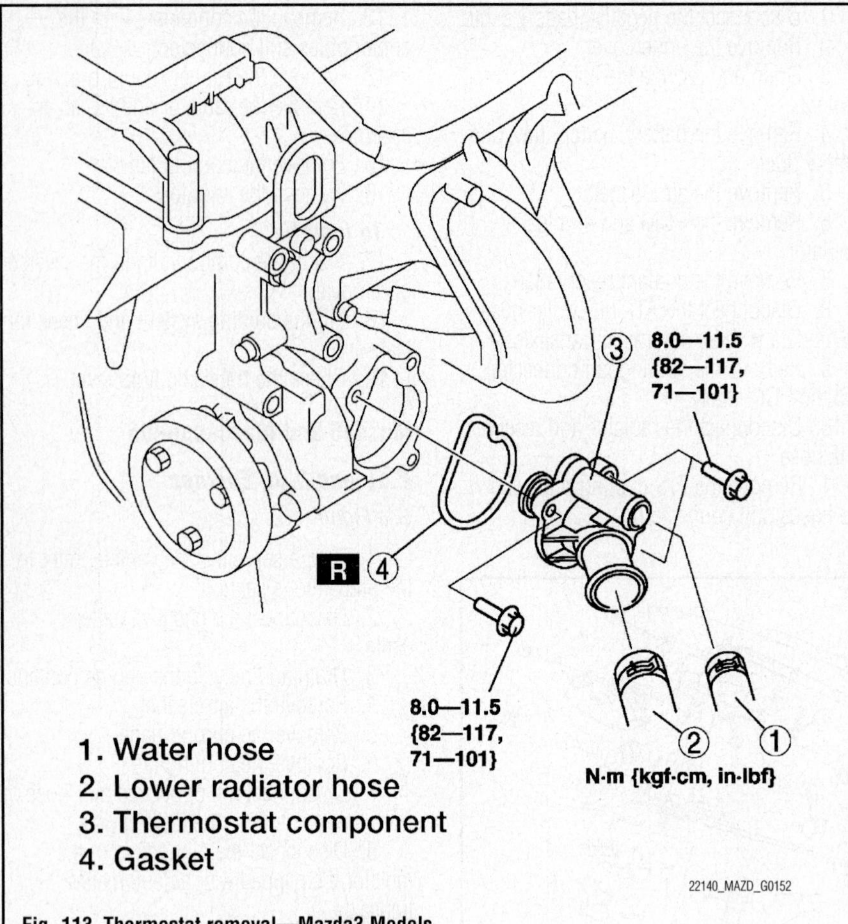

Fig. 113 Thermostat removal—Mazda3 Models

5. Position the coolant reserve tank out of the way.

6. Remove the plug hole plate by lifting off and removing the plug hole plate from the areas shown in the accompanying illustration.

7. Position the drive belt out of the way.

8. Remove the drive belt tensioner.

9. Disconnect the water hose and lower radiator hose from the thermostat.

10. Remove the thermostat retainers, thermostat housing, thermostat and gasket.

To install:

11. Installation is the reverse of removal. Use a new thermostat housing gasket and tighten the thermostat housing retainers to 71–101 in. lbs. (8–11.5 Nm).

12. Fill the cooling system and check for leaks.

MX-5 Miata

See Figure 114.

**** CAUTION**

Remove and install all parts when the engine is cold, otherwise they can cause severe burns or serious injury.

1. Remove the battery cover.

2. Disconnect the negative battery cable.

3. Drain and recycle the engine coolant.

4. Remove the throttle body.

5. Disconnect the bypass hose and lower radiator hose from the thermostat.

6. Remove the thermostat retainers, thermostat housing, thermostat and gasket.

To install:

7. Installation is the reverse of removal. Use a new thermostat housing gasket and tighten the thermostat housing retainers to 71–101 in. lbs. (8–11.5 Nm).

8. Fill the cooling system and check for leaks.

Mazda6 and Mazdaspeed6

2.3L (L3) Engines

1. Before servicing the vehicle, refer to the precautions section.

2. Disconnect the negative battery cable.

3. Drain and recycle the engine coolant.

4. Remove the drive belt.

5. Remove the power steering pump with hoses and pipe still connected. Position the P/S oil pump out of the way.

6. Disconnect the water hose and lower radiator hose from the thermostat.

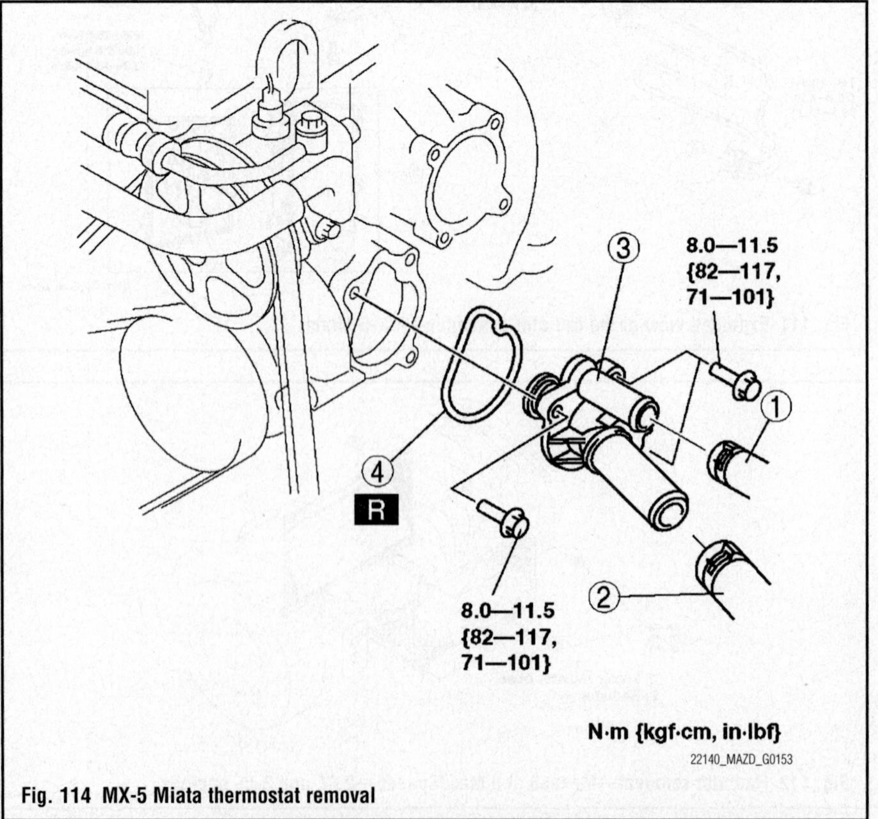

Fig. 114 MX-5 Miata thermostat removal

7. Remove the thermostat retainers, thermostat housing, thermostat and gasket.

To install:

8. Installation is the reverse of removal. Use a new thermostat housing gasket and tighten the thermostat housing retainers to 71–101 in. lbs. (8–11.5 Nm).
9. Fill the cooling system and check for leaks.

3.0L (AJ) Engines

See Figure 115.

1. Before servicing the vehicle, refer to the precautions section.
2. Disconnect the negative battery cable.
3. Drain and recycle the engine coolant.
4. Remove the drive belt.
5. Disconnect upper and lower radiator hoses.
6. Remove the thermostat retainers, thermostat housing, thermostat and gasket.

To install:

7. Installation is the reverse of removal. Make sure to install the thermostat with the jiggle valve at the top. Use a new thermostat housing gasket and tighten the thermostat housing retainers to 71–101 in. lbs. (8–11.5 Nm).
8. Fill the cooling system and check for leaks.

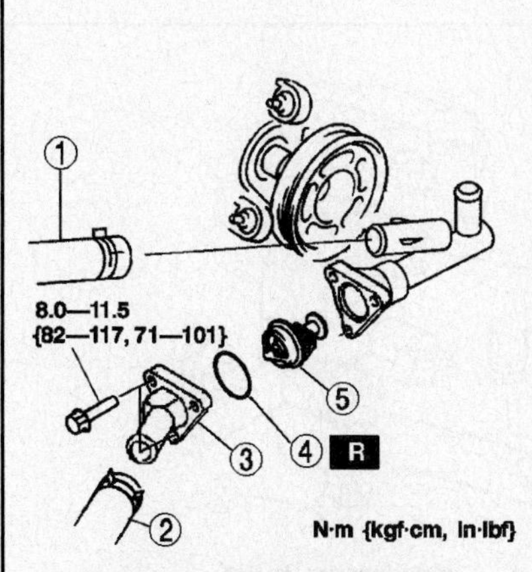

Fig. 115 Make sure to install the thermostat with the jiggle valve at the top—Mazda6 with 3.0L engine

WATER PUMP

REMOVAL & INSTALLATION

MX-5 Miata

See Figure 116.

1. Before servicing the vehicle, refer to the precautions section.

⁂ CAUTION

Never remove the cooling system cap or loosen the radiator drain plug while the engine is running, or when the engine and radiator are hot. Scalding engine coolant and steam may shoot out and cause serious injury. It may also damage the engine and cooling system.

2. Remove the battery cover.
3. Disconnect the negative battery cable.
4. Drain the engine coolant.
5. Remove the air cleaner.
6. Loosen the water pump pulley bolt and remove the drive belt.
7. Remove the water pump pulley, pump and the O–ring
8. Installation is the reverse of removal, refer to the illustration for component location and torque specifications.
9. Refill the engine coolant, start the vehicle and inspect for leaks.

Mazda3 and Mazda6 Models

2.3L (L3) Engines

See Figure 117.

1. Before servicing the vehicle, refer to the precautions section.
2. Remove the battery cover and disconnect the negative battery cable.
3. Remove the undercover and splash shield as an assembly.
4. Drain the cooling system.
5. Position the coolant reservoir tank aside with the hose still attached.

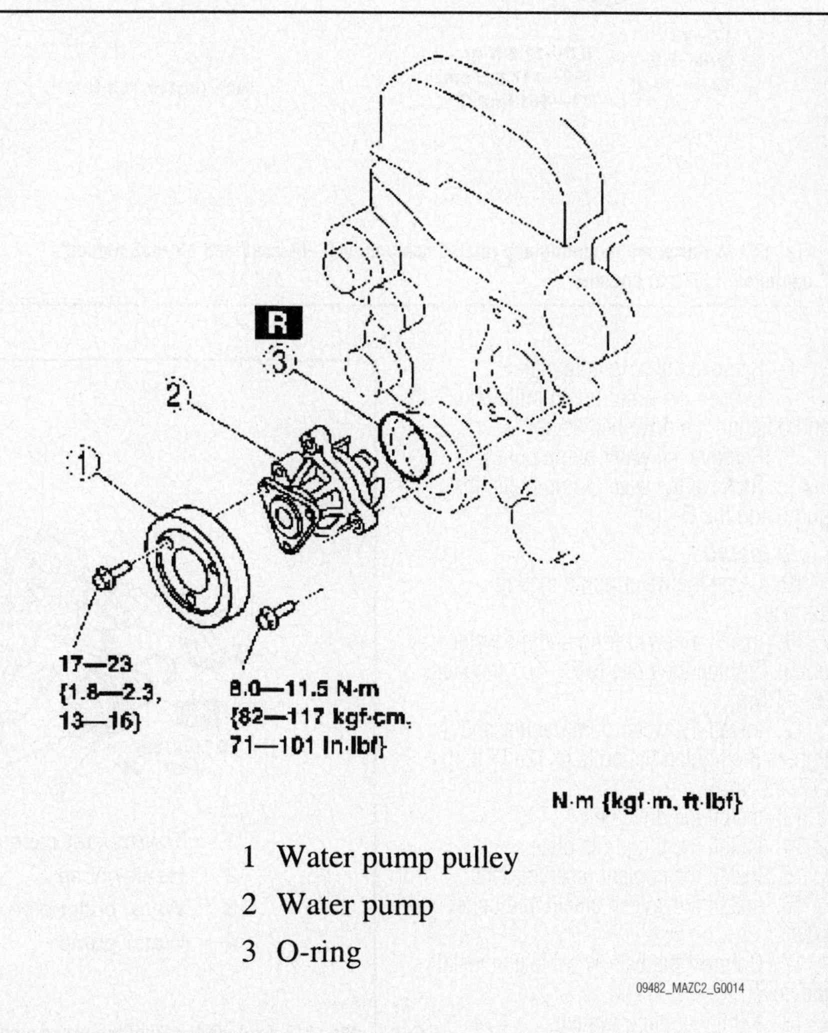

Fig. 116 Exploded view of the water pump assembly—MX-5 Miata

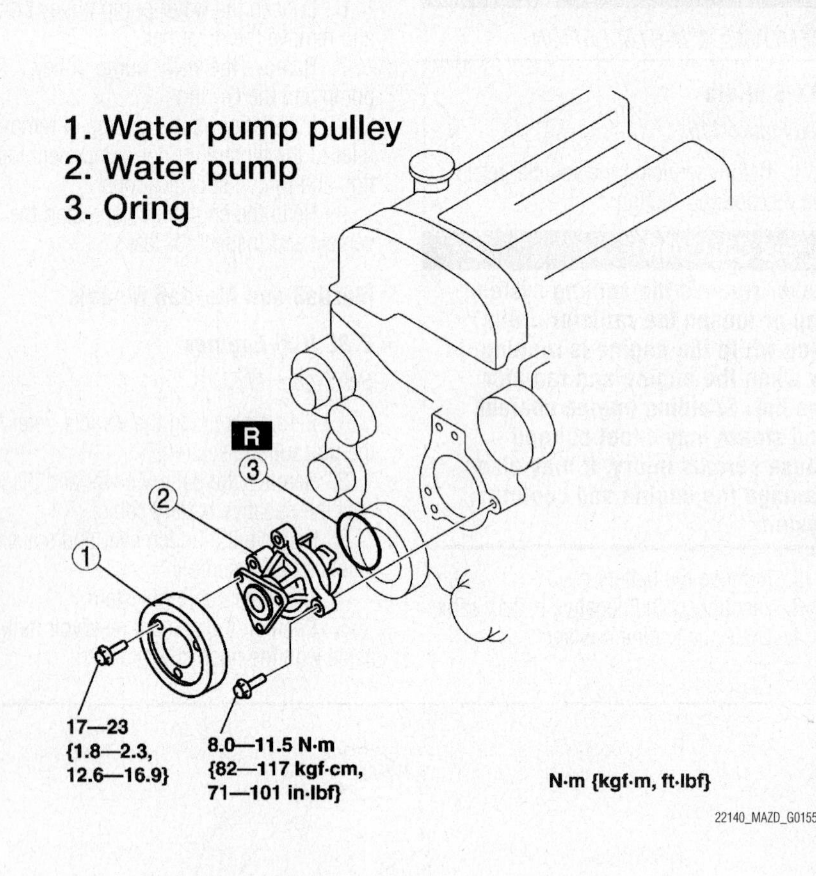

Fig. 117 Water pump mounting and related components—Mazda3 and Mazda6 models equipped 2.3L (L3) engines

6. Remove the plug hole plate.
7. Loosen the water pump pulley bolt and position the drive belt aside.
8. Remove the water pump pulley.
9. Remove the water pump bolts, the pump and the O–ring.

To install:

10. Clean the water pump mating surfaces.
11. Install a new O–ring and the water pump. Tighten the bolts to 71–101 inch lbs. (8–11 Nm).
12. Install the water pump pulley and tighten the tighten the bolts to 12–17 ft. lbs. (17–23 Nm).
13. Install the drive belt.
14. Install the plug hole plate
15. Install the coolant reservoir tank.
16. Install the splash shield and undercover.
17. Connect the battery cable and install the cover.
18. Fill the cooling system.
19. Start the engine, check for leaks and repair if necessary.

3.0L (AJ) Engines

See Figures 118 through 120.

1. Before servicing the vehicle, refer to the precautions section.
2. Disconnect the negative battery cable.
3. Remove the undercover.
4. Drain the cooling system.
5. Remove the air cleaner.
6. Remove the water pump drive belt pulley.
   a. Replace part of tool 49 UN30 3009 with tool 49 UN30 3457.
   b. Install the tools as illustrated and remove the water pump drive pulley.
7. Remove the thermostat case, heater hose and the water outlet pipe.
8. Remove the water pump bolts and the pump.

To install:

9. Install the water pump and tighten the bolts 89 inch lbs. (10 Nm), plus an additional 85–95 degrees.
10. Install the water outlet pipe, heater hose and thermostat case.
11. Install a new water pump drive belt pulley using tool 49 UN21 1185.
12. Install the air cleaner.
13. Fill the cooling system and connect the negative battery cable.
14. Start the vehicle and check for leaks, repair if necessary.
15. Install the undercover if no leaks are found.

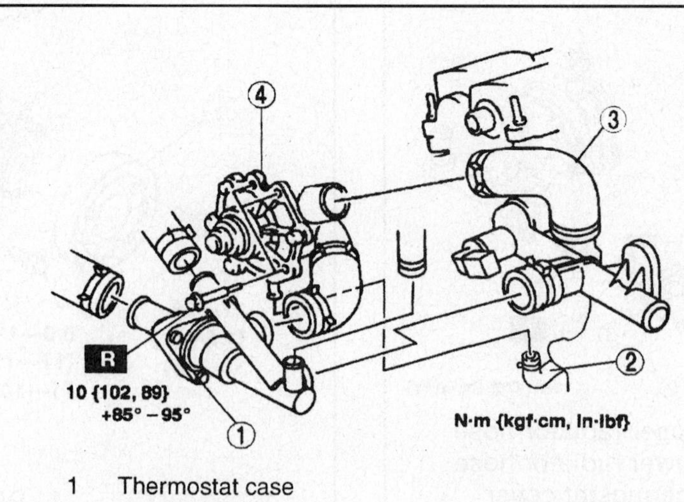

Fig. 118 Exploded view of the water pump and related components—Mazda6 models with the 3.0L (AJ) engine

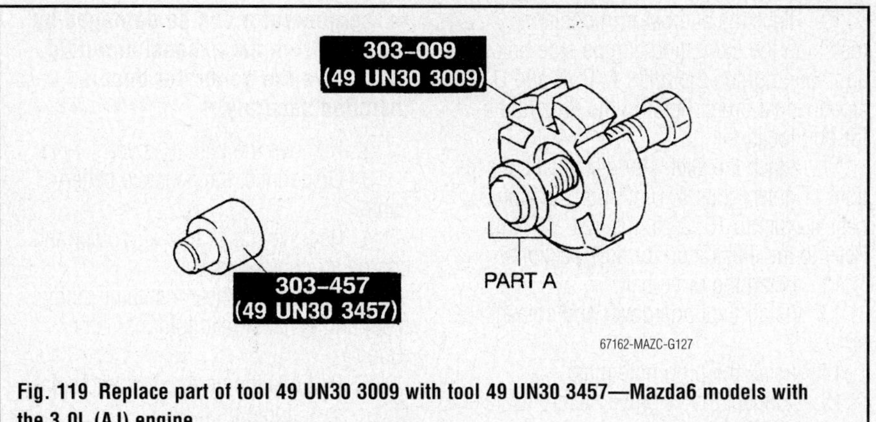

Fig. 119 Replace part of tool 49 UN30 3009 with tool 49 UN30 3457—Mazda6 models with the 3.0L (AJ) engine

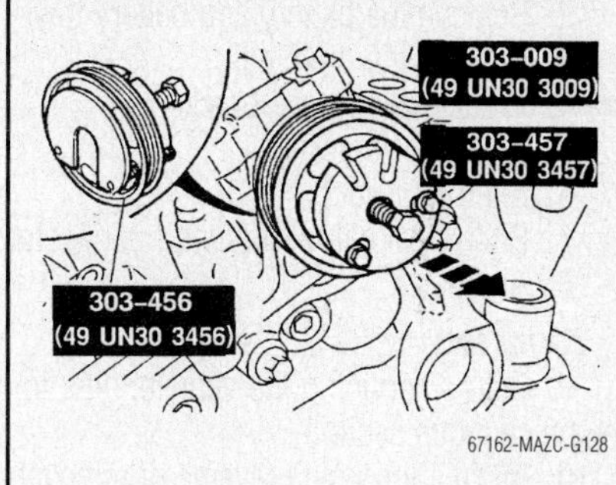

Fig. 120 Install the tools as illustrated and remove the water pump drive pulley—Mazda6 models with the 3.0L (AJ) engine

ENGINE ELECTRICAL

CHARGING SYSTEM

ALTERNATOR

REMOVAL & INSTALLATION

MX-5 Miata

See Figure 121.

1. Before servicing the vehicle, refer to the precautions section.

CAUTION

Remove and install all parts when the engine is cold, otherwise they can cause severe burns or serious injury.

CAUTION

The alternator can be damaged by the heat from the exhaust manifold. Make sure the alternator duct is installed securely.

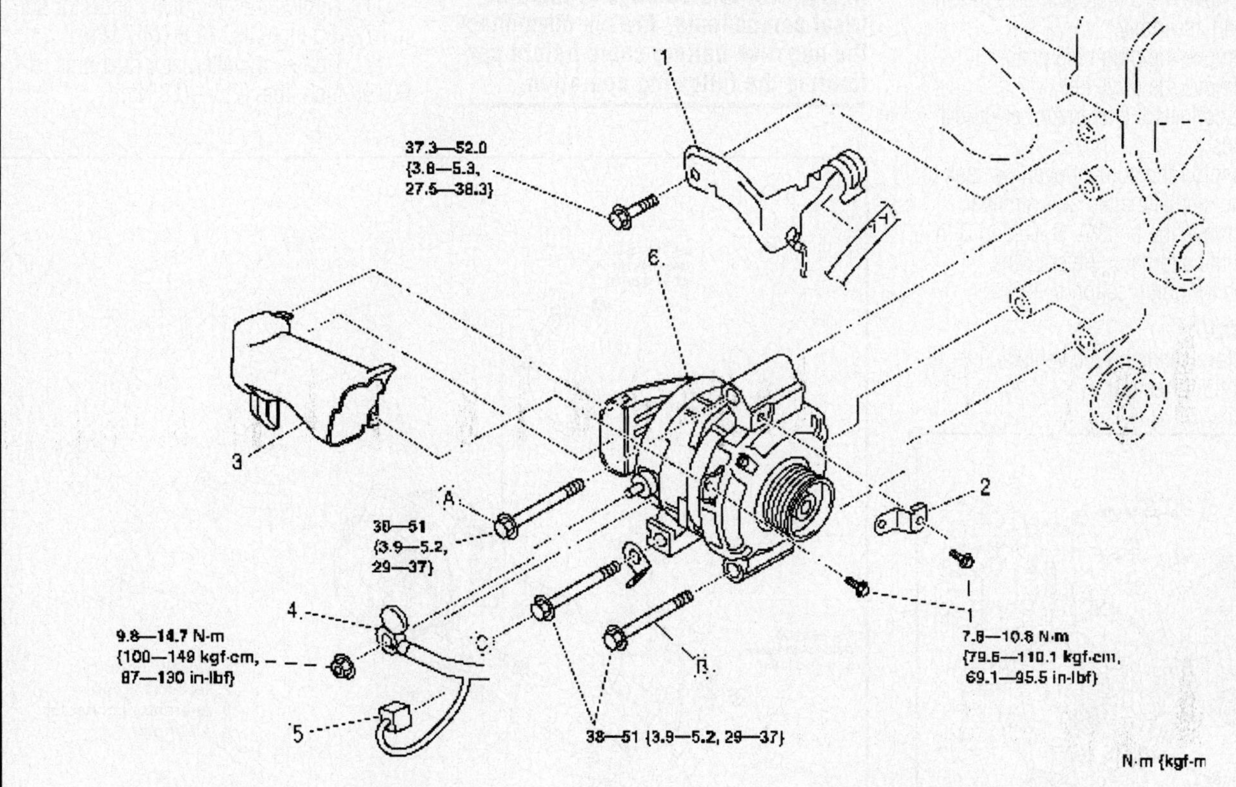

Fig. 121 Alternator mounting—MX-5 Miata

2. Remove the battery and battery tray.
3. Remove the drive belt.
4. P/S pressure hose bracket
5. Alternator bracket
6. Alternator duct
7. B terminal cable Alternator connector
8. Alternator

To install:

9. Before servicing the vehicle, refer to the precautions section.
10. Installation is the reverse of removal. Please note the following:
    a. Tighten bolt A temporarily.
    b. Tighten bolt B, C to 29–37 ft. lbs. (38–51 Nm).
    c. Tighten bolt A to 29–37 ft. lbs. (38–51 Nm).

Mazda3

2.3L (LF and L3) Engines

See Figure 122.

1. Before servicing the vehicle, refer to the precautions section.
2. Remove the battery cover and disconnect the negative battery cable.
3. Remove the undercover and splash shield as an assembly.
4. Remove the plug hole plate.
5. Remove the drive belt.
6. Disconnect the alternator electrical connections.
7. Position the coolant overflow tank to one side to facilitate alternator removal.
8. Remove the bolts A, B, C, and D in order and the alternator. Refer to the illustration for bolt location.

To install:

9. Before servicing the vehicle, refer to the precautions section.
10. Place the alternator in position matching the fixing and engine side holes and hand tighten the bolts A, B, C and D in alphabetical order. Refer to the illustration for bolt locations.
11. Attach the alternator electrical connections. Tighten bolts A, B, C and D in alphabetical order to 16–22 ft. lbs. (21–30 Nm). Refer to the illustration for bolt locations.
12. Install the drive belt.
13. Install the undercover and splash shield.
14. Install the plug hole plate.
15. Connect the negative battery cable and install the battery cover.

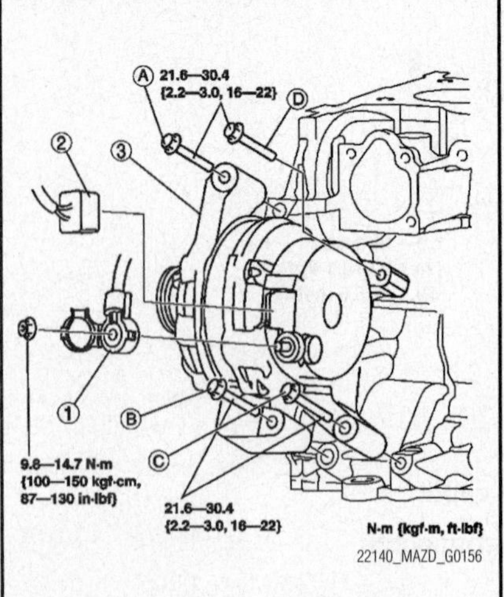

Fig. 122 Location of alternator mounting bolts A, B, C, and D—Mazda3 models

2.3L (L3 with TC) Engines

See Figure 123.

1. Before servicing the vehicle, refer to the precautions section.

✻✻ CAUTION

When the battery cables are connected, touching the vehicle body with generator terminal B will generate sparks. This can cause personal injury, fire, and damage to the electrical components. Always disconnect the negative battery cable before performing the following operation.

➡The generator can be damaged by the heat from the exhaust manifold. Make sure the generator duct is installed securely.

2. Remove the battery cover.
3. Disconnect the negative battery cable.
4. Remove the charge air cooler and charge air cooler bracket.
5. Remove the heat insulator (body side) and exhaust manifold insulator (upper).
6. Remove the undercover and splash shield as a single unit.
7. Remove the drive belt.
8. Remove or disconnect the following:
   - Generator duct
   - Terminal B cable
   - Alternator connector
   - Alternator

To install:

9. Install in the reverse order of removal.
10. Tighten the alternator mounting bolts to 29–37 ft. lbs. (38–51 Nm).
11. Tighten the electrical connector nut to 87–130 inch. lbs. (9.8–14.7 Nm).
12. Tighten the alternator duct nuts to 60–96 inch. lbs. (7.8–10.8 Nm).

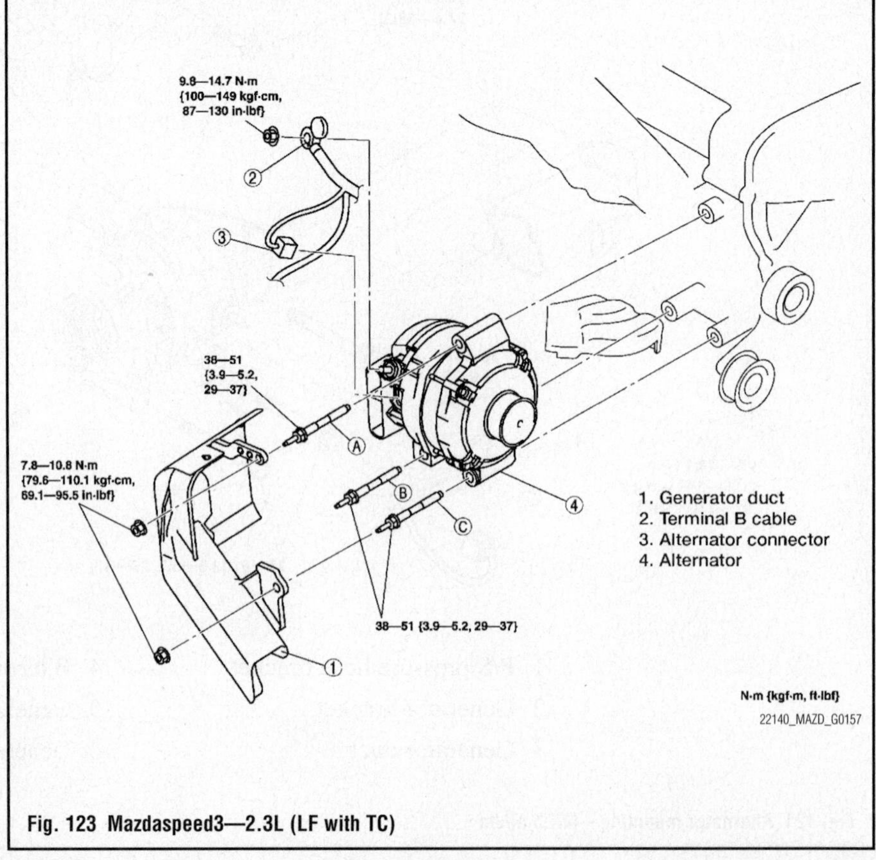

Fig. 123 Mazdaspeed3—2.3L (LF with TC)

Mazda6

2.3L (L3) Engines

See Figure 124.

1. Before servicing the vehicle, refer to the precautions section.
2. Disconnect the negative battery cable.
3. Remove the undercover.
4. Remove the drive belt.
5. Remove the alternator duct and heat insulator.
6. Disconnect the alternator electrical connections.
7. Remove the bolts A, B and C. Refer to the illustration for bolt location.

To install:

8. Before servicing the vehicle, refer to the precautions section.
9. Install the alternator, tighten bolt A temporarily, then tighten the bolts in the following order; B, A and C to 29–41 ft. lbs. (40–55 Nm).
10. Connect the alternator electrical connections.
11. Install the heat insulator and tighten the bolt to 18–47 inch lbs. (2–5 Nm).
12. Install the alternator duct
13. Install the undercover.
14. Install the drive belt.
15. Connect the negative battery cable.

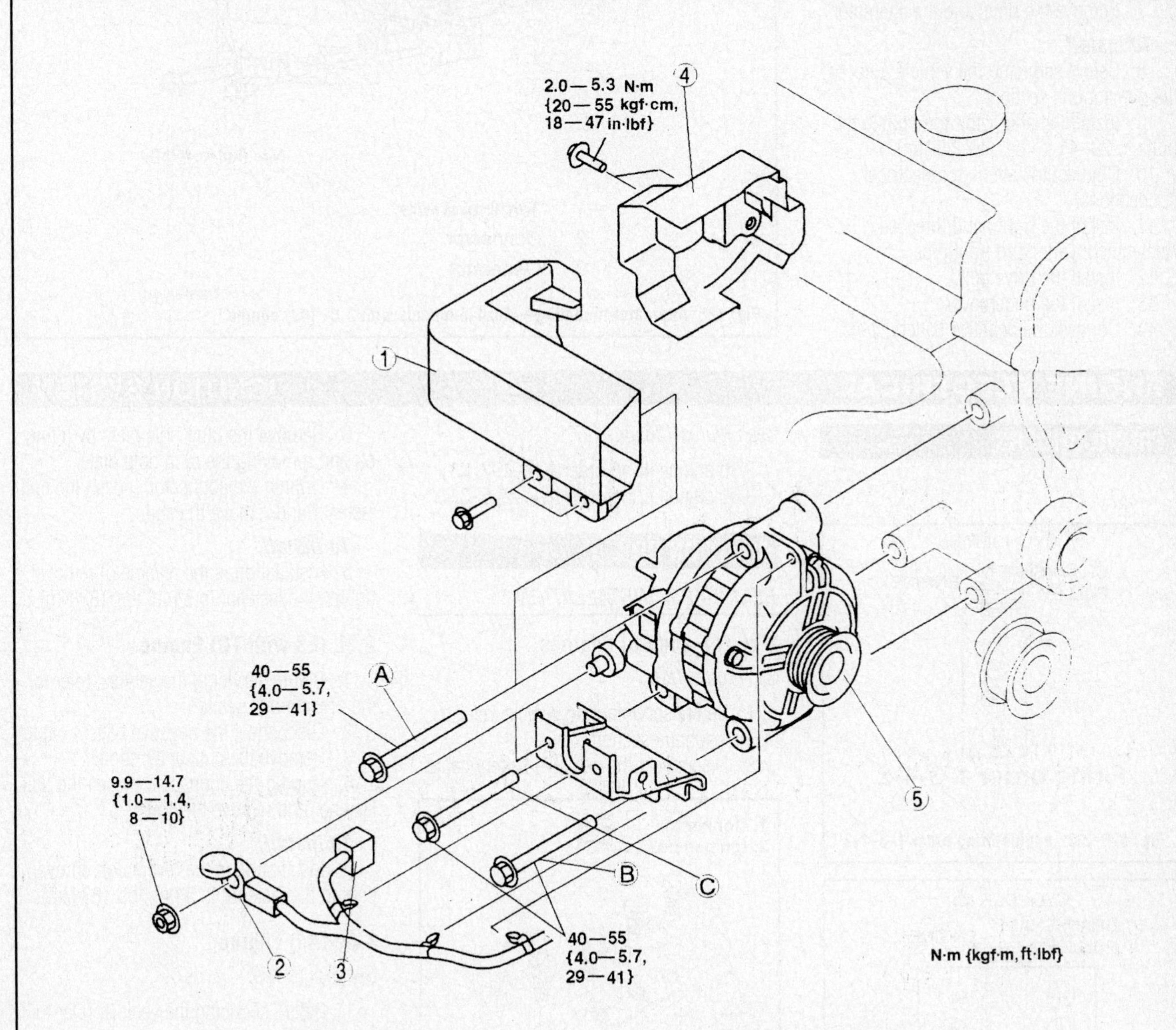

Fig. 124 Location of alternator mounting bolts A, B and C—Mazda6 and Mazdaspeed6 models with 2.3L (L3) engine

3.0L (AJ) Engines

See Figure 125.

1. Before servicing the vehicle, refer to the precautions section.
2. Disconnect the negative battery cable.
3. Remove the undercover.
4. Remove the drive belt.
5. Remove the right hand three way catalytic converter and front pipe.
6. Disconnect the alternator electrical connections.
7. Remove the bolts and the alternator.

To install:

8. Before servicing the vehicle, refer to the precautions section.
9. Install the alternator and tighten the bolts to 29–41 ft. lbs. (40–55 Nm).
10. Connect the alternator electrical connections.
11. Install the right hand three way catalytic converter and front pipe.
12. Install the drive belt.
13. Install the undercover.
14. Connect the negative battery cable.

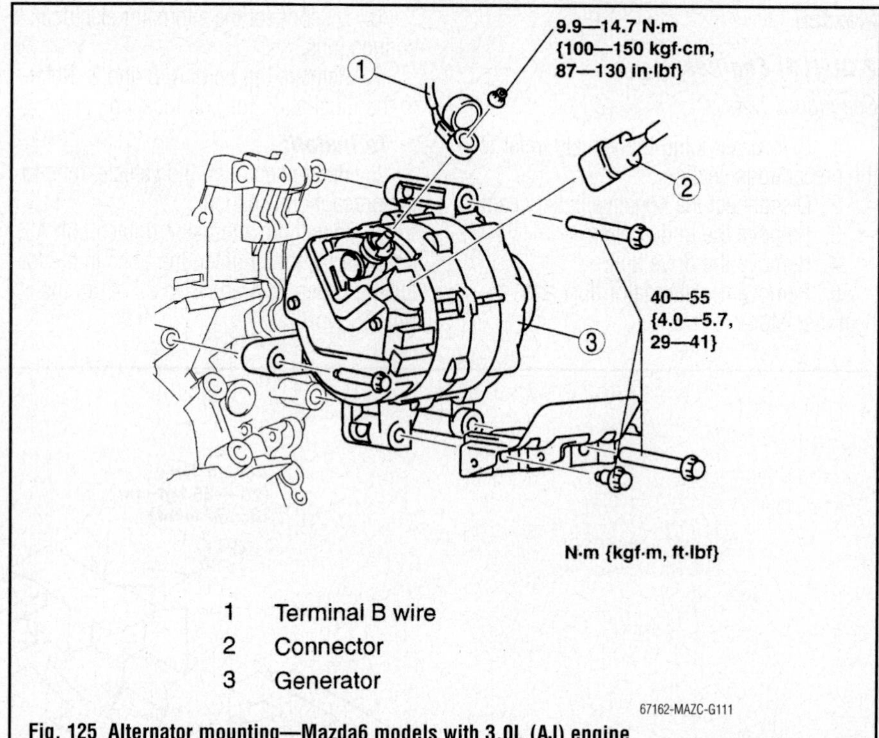

Fig. 125 Alternator mounting—Mazda6 models with 3.0L (AJ) engine

ENGINE ELECTRICAL

IGNITION SYSTEM

FIRING ORDERS

See Figures 126 and 127.

2.3L engine firing order 1–3–4–2
3.0L engine firing order 1–4–2–5–3–6

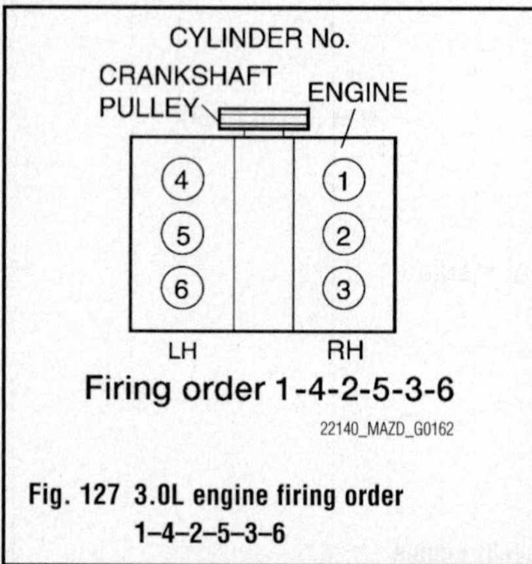

Fig. 126 2.3L engine firing order 1–3–4–2

Fig. 127 3.0L engine firing order 1–4–2–5–3–6

IGNITION COIL

REMOVAL & INSTALLATION

2.3L (LF and L3) Engines

See Figure 128.

1. Before servicing the vehicle, refer to the precautions section.
2. Disconnect the negative battery cable.

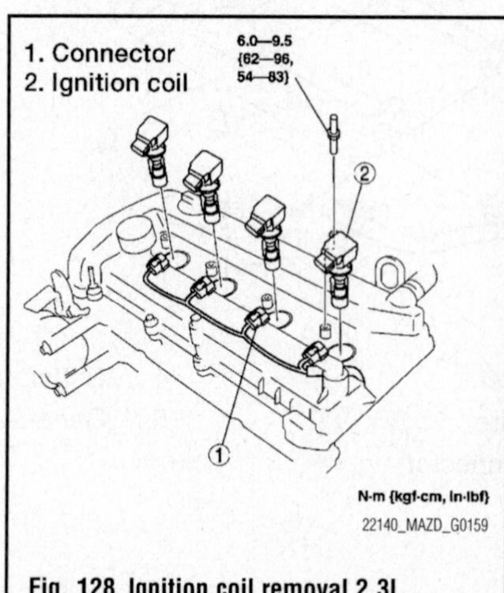

Fig. 128 Ignition coil removal 2.3L (LF and L3) Engines

3. Remove the plug hole plate by lifting off and removing the plug hole plate.
4. Unplug the connector, loosen the coil fastener and remove the coil.

To install:

5. Installation is the reverse of removal, tighten the fastener to 54 in. lbs. (83 Nm).

2.3L (L3 with TC) Engine

1. Before servicing the vehicle, refer to the precautions section.
2. Disconnect the negative battery cable.
3. Remove the charge air cooler.
4. Unplug the connector, loosen the coil fastener and remove the coil.

To install:

5. Installation is the reverse of removal, tighten the fastener to 54 in. lbs. (83 Nm).

3.0L (AJ) Engine

See Figure 129.

1. Before servicing the vehicle, refer to the precautions section.
2. Disconnect the negative battery cable.
3. Remove the plug hole plate LH.
4. Remove the dynamic chamber RH.
5. Unplug the connector, loosen the coil fastener and remove the coil.

To install:

6. Installation is the reverse of removal, tighten the fastener to 54 in. lbs. (83 Nm).

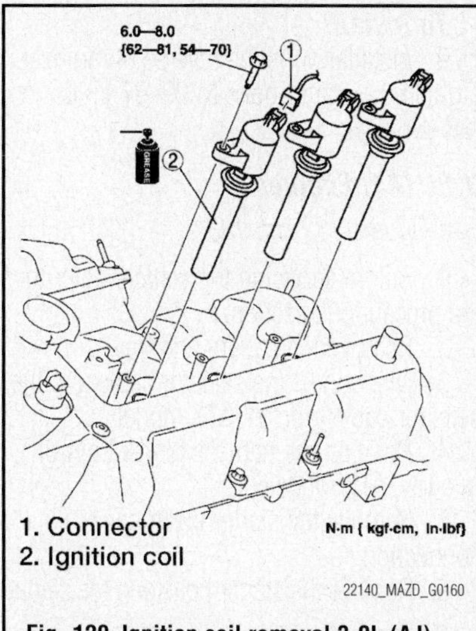

Fig. 129 Ignition coil removal 3.0L (AJ) Engine

IGNITION TIMING

ADJUSTMENT

1. Before servicing the vehicle, refer to the precautions section.

➡If the information given in the following procedures differs from that on the emission information label located in the engine compartment, follow the directions given on the label. The label often reflects production changes made during the model year.

The timing is controlled by the computer. Ignition timing adjustment is not possible or necessary.

2. If the timing is still not within specification. the following components may be defective:
   - Camshaft position (CMP) sensor
   - Crankshaft Position (CKP) sensor
   - Throttle Position (TP) sensor
   - Engine Coolant Temperature (ECT) sensor
   - Neutral switch if equipped with a manual transaxle
   - Clutch switch if equipped with a manual transaxle
   - Transaxle range switch if equipped with an automatic transaxle
3. If the above components are normal, replace the Powertrain Control Module (PCM).

SPARK PLUGS

REMOVAL & INSTALLATION

2.3L (LF and L3) Engines

1. Before servicing the vehicle, refer to the precautions section.
2. Remove the battery cover.
3. Disconnect the negative battery cable.
4. Remove the plug hole plate.
5. Remove the ignition coils.
6. Remove the spark plugs.

To install:

7. Installation is the reverse of removal, tighten the plugs to 8–10 ft. lbs. (10–14 Nm).
8. Spark plug gap: 0.028–0.031
9. Mazda6 (L3) models: 0.050–0.053

2.3L (L3 with TC) Engine

1. Before servicing the vehicle, refer to the precautions section.
2. Remove the battery cover.
3. Disconnect the negative battery cable.
4. Remove the charge air cooler.
5. Remove the plug hole plate.
6. Remove the ignition coils.
7. Remove the spark plugs.

To install:

8. Installation is the reverse of removal, tighten the plugs to 8–10 ft. lbs. (10–14 Nm).
9. Spark plug gap: 0.028–0.031

3.0L (AJ) Engine

1. Before servicing the vehicle, refer to the precautions section.
2. Remove the battery cover.
3. Disconnect the negative battery cable.
4. Remove the plug hole plate.

➡Apply small amount of the specified dielectric grease to the inside of the ignition coils, then install the ignition coils to the spark plugs.

5. Remove the ignition coils.
6. Remove the spark plugs.
7. Spark plug gap: 0.051–0.057

To install:

8. Installation is the reverse of removal, tighten the plugs to 79–177 inch. lbs. (9–20 Nm).

ENGINE ELECTRICAL — STARTING SYSTEM

STARTER

REMOVAL & INSTALLATION

MX5-Miata

See Figure 130.

1. Before servicing the vehicle, refer to the precautions section.

⁂ CAUTION

Remove and install all parts when the engine is cold, otherwise they can cause severe burns or serious injury.

⁂ CAUTION

When the battery cables are connected, touching the vehicle body with starter terminal B will cause sparks. This can cause personal

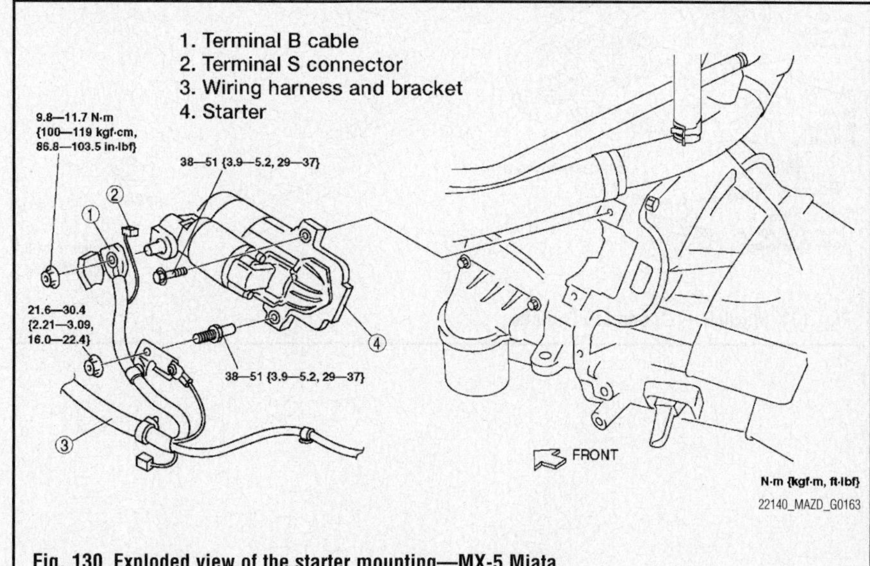

Fig. 130 Exploded view of the starter mounting—MX-5 Miata

injury, fire, and damage to the electrical components. Make sure to always disconnect the negative battery cable before removing any electrical component.

2. Remove the battery cover.
3. Disconnect the negative battery cable.
4. Remove the left hand side cover.
5. Remove the undercover on models equipped with a manual transmission.
6. Remove the oil filter. (Vehicles with oil cooler)
7. On models equipped with a manual transmission, remove the clutch release cylinder with the pipes still connected. Position the clutch release cylinder so that it is out of the way.
8. Remove the wiring from the starter, wiring harness bracket, starter bolts and the starter.

To install:

9. Installation is the reverse of removal.
10. Tighten the starter bolts to 38–51 Nm (29–37 ft. lbs.)

Mazda3—(L3 and LF) Engines

See Figure 131.

1. Before servicing the vehicle, refer to the precautions section.
2. Remove the battery cover and disconnect the negative battery cable.
3. Remove the undercover.
4. If equipped with a manual transaxle, remove the clutch release cylinder with the line still attached.
5. Remove the starter electrical connections and wiring harness bracket.
6. Remove the starter bolts and the starter.

To install:

7. Installation is the reverse of removal. Tighten the wiring harness bracket retainers to 70–104 inch lbs. (8–11 Nm) and the starter bolts to 29–37 ft. lbs. (38–51 Nm).

Mazda6

2.3L (L3) Engines

1. Before servicing the vehicle, refer to the precautions section.
2. Remove the battery cover and disconnect the negative battery cable.
3. Remove the air cleaner.
4. Remove the undercover.
5. Remove the clutch release cylinder with the pipes still connected. MTX models.
6. Remove the water hose bracket.
7. Remove the starter electrical connections.
8. Remove the starter bolts and the starter.

To install:

9. Installation is the reverse of removal. Tighten the starter bolts to 29–37 ft. lbs. (38–51 Nm).

3.0L (AJ) Engines

See Figures 132 and 133.

1. Before servicing the vehicle, refer to the precautions section.
2. Remove the battery and tray.
3. Disconnect the selector cable and the selector cable bracket. ATX models.
4. Position the harness bracket out of the way. ATX models.
5. Remove the starter electrical connection.
6. Remove the starter bolts and the starter.

To install:

7. Installation is the reverse of removal. Tighten the starter bolts to 28–38 ft. lbs. (37.2–52 Nm).

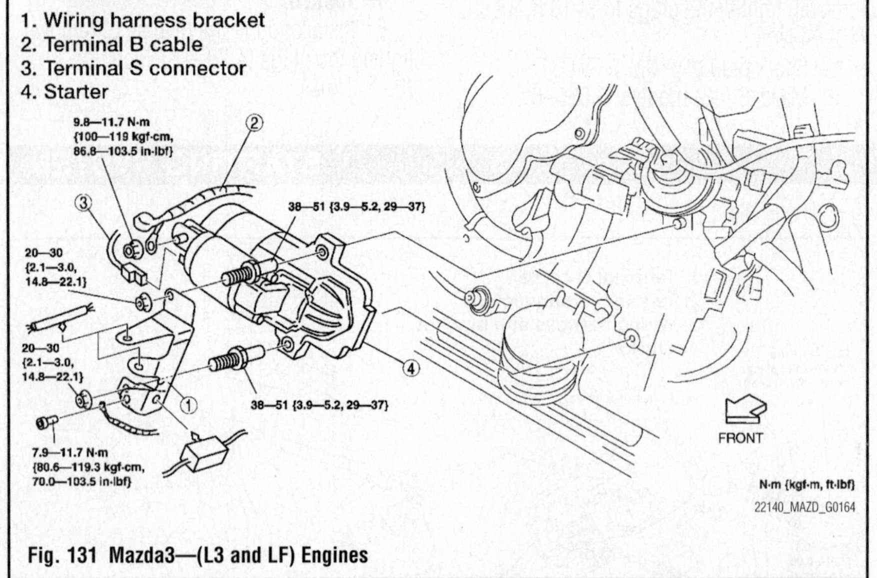

Fig. 131 Mazda3—(L3 and LF) Engines

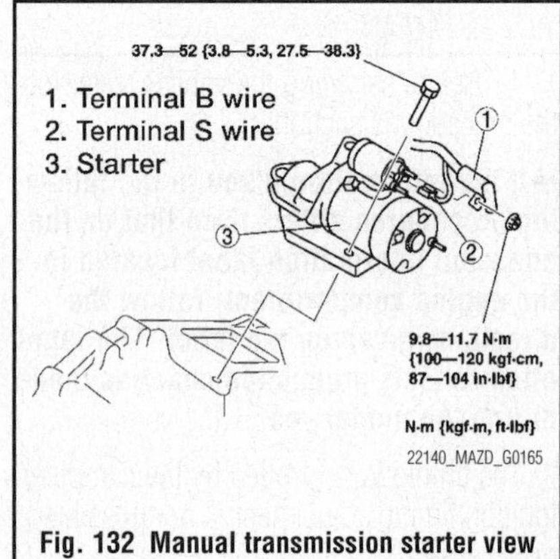

Fig. 132 Manual transmission starter view

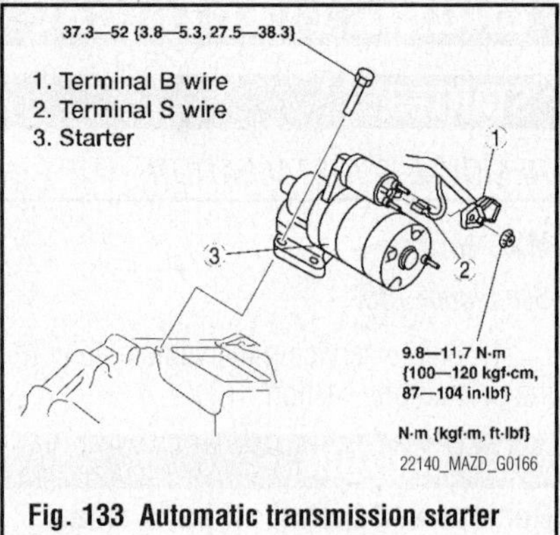

Fig. 133 Automatic transmission starter view

ENGINE MECHANICAL

ACCESSORY DRIVE BELTS

ACCESSORY BELT ROUTING

See Figures 134 and 135.

Refer to the accompanying illustrations for drive belt routing.

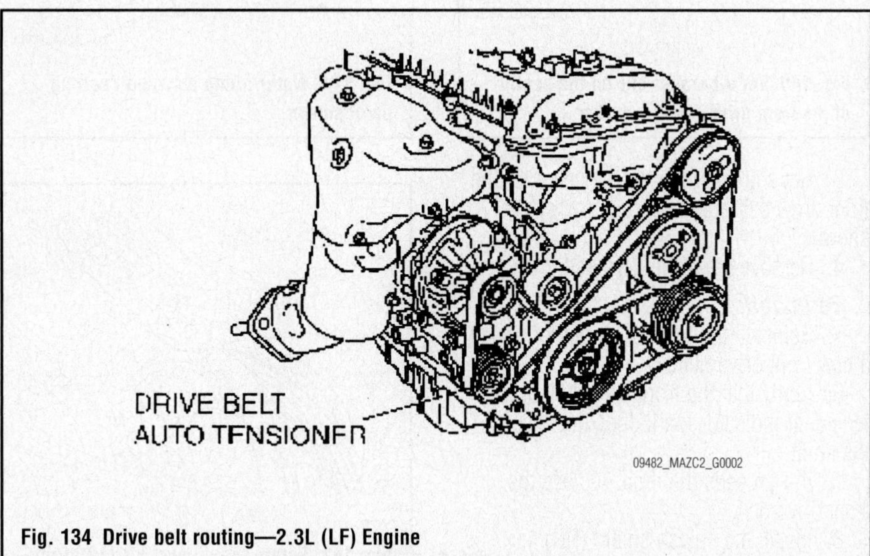

Fig. 134 Drive belt routing—2.3L (LF) Engine

INSPECTION

Inspect the drive belt for signs of glazing or cracking. A glazed belt will be perfectly smooth from slippage, while a good belt will have a slight texture of fabric visible. Cracks will usually start at the inner edge of the belt and run outward. All worn or damaged drive belts should be replaced immediately

ADJUSTMENT

The belt adjustment is maintained by an auto tensioner. Inspect the tensioner as outlined below.

1. Remove the drive belt.
2. Verify that the alternator drive belt auto tensioner moves smoothly in a clockwise and counterclockwise direction.
3. If it does not move smoothly, replace the drive belt auto tensioner.
4. Turn the drive belt auto tensioner pulley by hand and verify that it rotates smoothly.
5. If it does not move smoothly, replace the drive belt auto tensioner.

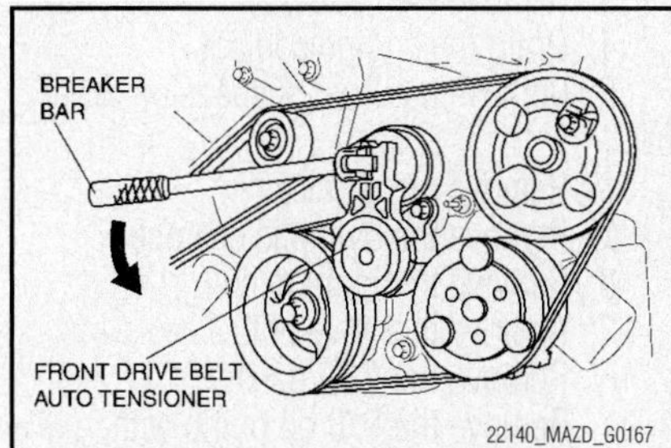

Fig. 135 Drive belt routing—3.0L (AJ) Engine

REMOVAL & INSTALLATION

Mazda3

See Figures 136 and 137.

1. Before servicing the vehicle, refer to the precautions section.
2. Remove/Install the A/C belt as follows:
   a. Remove the right side engine undercover and splash shield.
   b. Cut the A/C drive belt using scissors.
3. Install a jig which comes with the new belt to the crankshaft pulley as shown in the illustration.

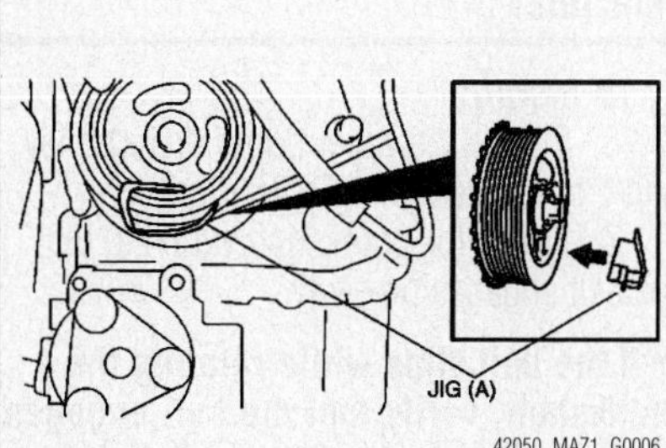

Fig. 136 Install jig (A) which comes with the new A/C belt to the crankshaft pulley as shown—Mazda 3

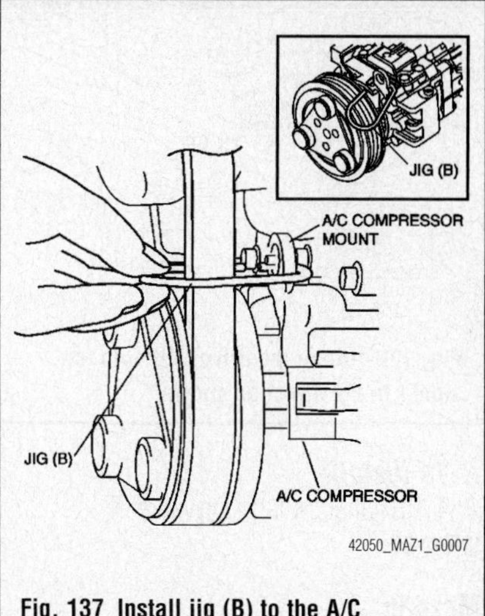

Fig. 137 Install jig (B) to the A/C compressor mount as shown in the illustration—Mazda 3

   a. Install a new A/C drive belt to the A/C compressor pulley, move jig (A) upward, and then install the A/C drive belt to the crankshaft pulley.
   b. Install jig (B) to the A/C compressor mount as shown in the illustration.

➡The A/C drive belt cannot be reused. The jig is prepackaged with a new A/C drive belt. Do not pass jig (B) through the A/C compressor mount hole.

   c. Rotate the crankshaft pulley to the right using a wrench and install the A/C drive belt.
   d. Install the undercover and splash shield (RH).
4. Remove/Install the alternator belt as follows:
   a. Remove the plug hole plate.
   b. Remove the A/C drive belt.
   c. Turn the center of the auto tensioner pulley counterclockwise to release tension to the drive belt tension.
   d. Remove the alternator drive belt.
   e. Install a new alternator drive belt.
   f. Install the A/C drive belt.
   g. Install the plug hole plate.

MX-5 Miata

See Figure 138.

1. Before servicing the vehicle, refer to the precautions section.
2. Remove the battery and battery tray.
3. Rotate the drive belt auto tensioner in the direction shown in the figure and remove the drive belt.

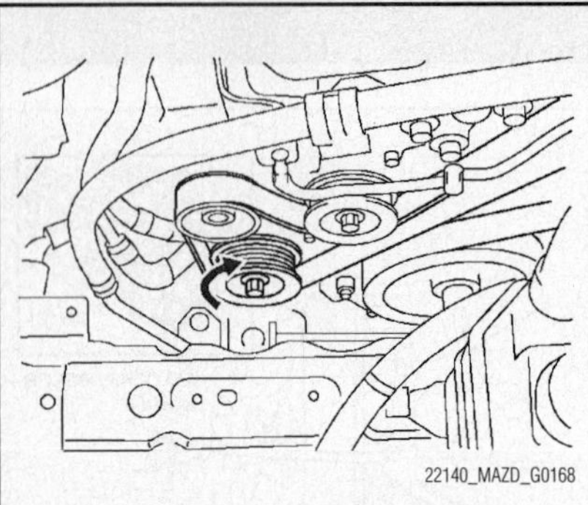

Fig. 138 Rotate the drive belt auto tensioner in the direction shown

4. Installation is the reverse of removal.

Mazda6

See Figure 139.

1. Before servicing the vehicle, refer to the precautions section.
2. Remove the splash shield.
3. Rotate the drive belt auto tensioner and remove the drive belt.

To install:

4. Reinstall the drive belt or install a new drive belt.
5. Verify that the drive belt auto tensioner indicator mark does not exceed the limit.
6. If it exceeds the limit, replace the drive belt.
7. Install the splash shield (RH).

Mazdaspeed6

Front Drive Belt—3.0L (AJ) Engine

See Figure 140.

1. Before servicing the vehicle, refer to the precautions section.
2. Remove the undercover and splash shield (RH).

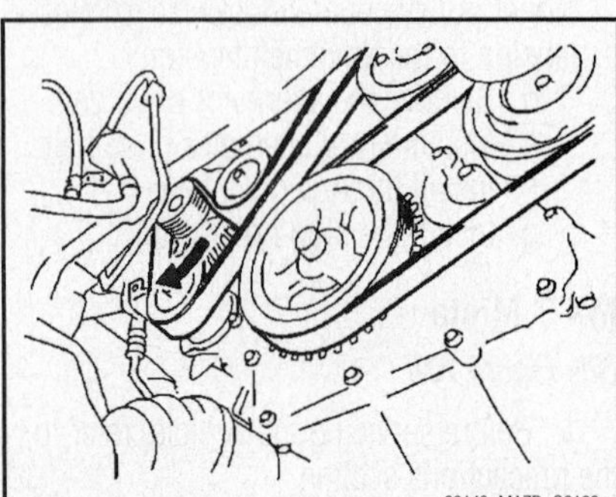

Fig. 139 Rotate the drive belt auto tensioner on the (L3) engine as shown

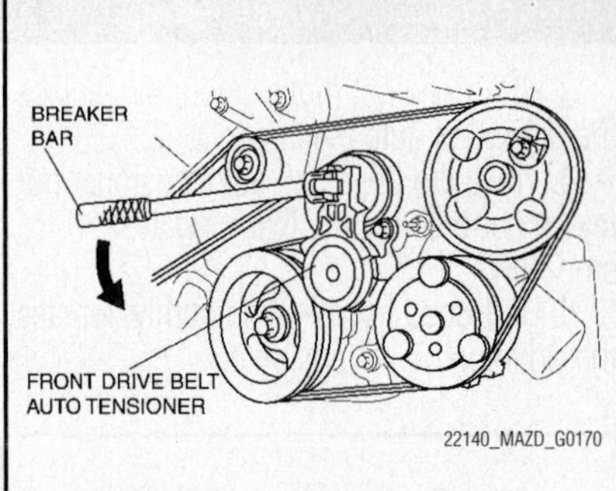

Fig. 140 Set a breaker bar on the center of the front drive belt as shown

3. Set a breaker bar on the center of the front drive belt auto tensioner pulley as shown.
4. Remove the front drive belt.

To install:

5. Reinstall the front drive belt or install a new front drive belt.
6. Verify that the front drive belt auto tensioner indicator mark does not exceed the limit.
7. If it exceeds the limit, replace the front drive belt.
8. Install the splash shield (RH).

Water Pump Drive Belt—3.0L (AJ) Engine

See Figures 141 and 142.

1. Before servicing the vehicle, refer to the precautions section.
2. Remove the plug hole plate.
3. Remove the front wheel and tire (RH).
4. Remove the splash shield (RH).
5. Hold down the belt while cutting as shown in the figure.

⁂ CAUTION

To prevent the belt from snapping back, always hold down the belt while cutting it. Also, keep your face away from the belt. The belt is not reusable.

To install:

6. Secure the water pump drive pulley and the new belt using a tie wrap.
7. Rotate the crankshaft verifying that the belt engages properly.

➡If the belt slips while rotating the crankshaft, verify that the belt engages properly by rotating the crankshaft further.

8. Remove the tie wrap.
9. To prevent damage to the belt, cut the tie wrap at the pulley side using the nipper or scissors when removing the tie wrap by cutting it.

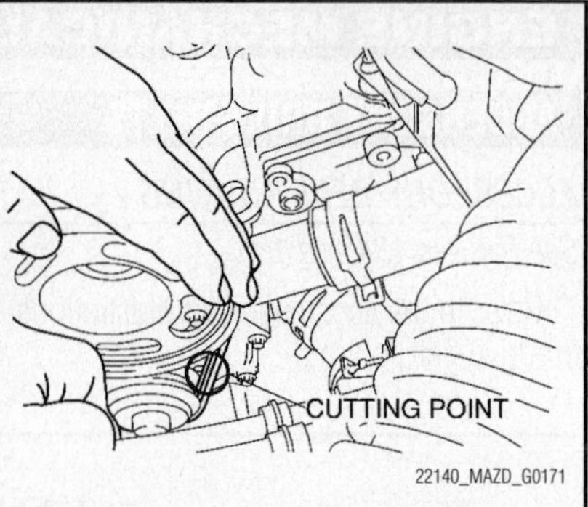

Fig. 141 Water pump drive belt cutting point shown

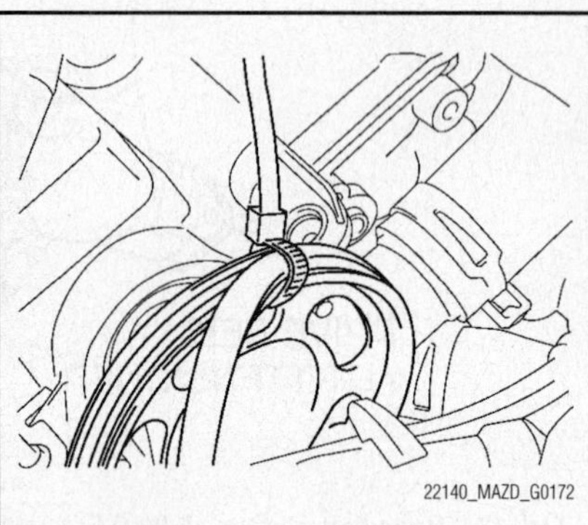

Fig. 142 Rotate the crankshaft 110°verifying that the belt engages properly

10. Install the splash shield (RH)
11. Install the front wheel and tire (RH).
12. Install the plug hole plate.

CAMSHAFT AND VALVE LIFTERS

REMOVAL & INSTALLATION

MX-5 Miata

See Figures 143 and 144.

1. Before servicing the vehicle, refer to the precautions section.
2. Properly relieve the fuel system pressure.
3. Remove the battery and battery tray.
4. Drain the engine coolant.
5. Remove the front suspension tower bar.
6. Remove the air cleaner.
7. Remove the dynamic chamber.
8. Remove the ignition coil.
9. Remove the drive belt.
10. Remove the CKP sensor.
11. Remove the P/S oil pump with the oil hose still connected and position the P/S oil pump so that it is out of the way.

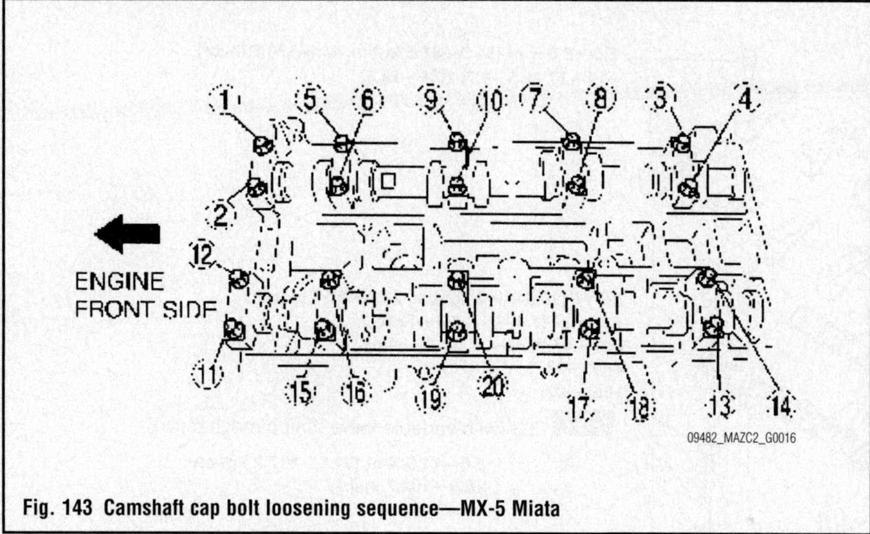

Fig. 143 Camshaft cap bolt loosening sequence—MX-5 Miata

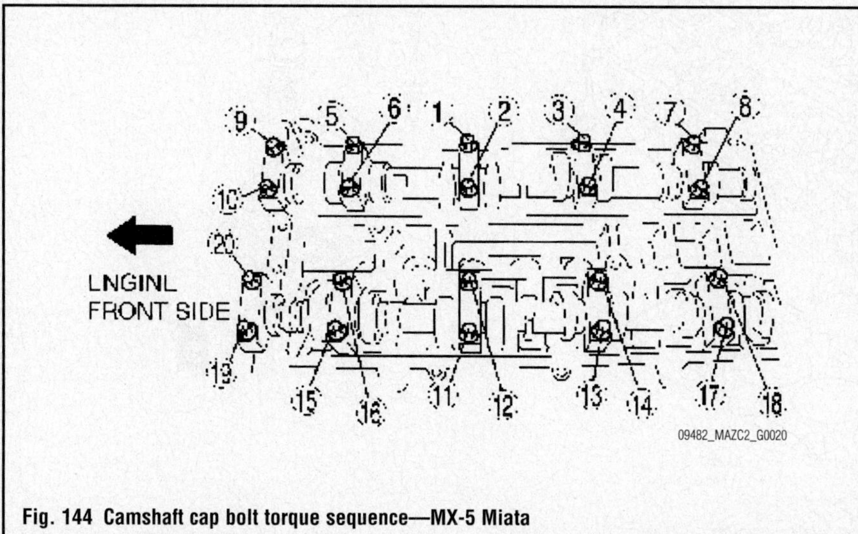

Fig. 144 Camshaft cap bolt torque sequence—MX-5 Miata

12. Remove the timing chain.
13. Remove the wiper arm.
14. Remove the cowl grille.
15. Remove the side cowl grille.
16. Remove the service hole cover.
17. Disconnect the alternator, but do not remove it from the vehicle.
18. Remove the exhaust manifold.
19. Remove the cylinder head cover.
20. Remove the OCV sensor, if equipped with variable valve timing.
21. Mark the camshaft cap locations so they may be reinstalled in their original positions.
22. Remove camshafts.

To install:

23. Set the cam position of the number 1 cylinder to top dead center, install the camshafts, making sure to install the caps in their original locations.
24. Tighten the camshaft cap bolts using two to three steps in the sequence illustrated to 5–9 Nm (44.3–79.6 in. lbs.) and then to 14–17 Nm (10.4–12.5 ft. lbs.)
25. Install the remaining components in the reverse order of removal
26. Start the engine and inspect for leaks.
27. Check the ignition timing, idle speed and idle mixture.

Mazda3 Model

See Figures 145 through 148.

1. Before servicing the vehicle, refer to the precautions section.
2. Disconnect the negative battery cable.
3. Remove the timing chain.
4. Remove the intake manifold.
5. Disconnect the following components:
   - Warm Up–Three Way Converter (WU–TWC)
   - Upper radiator hose

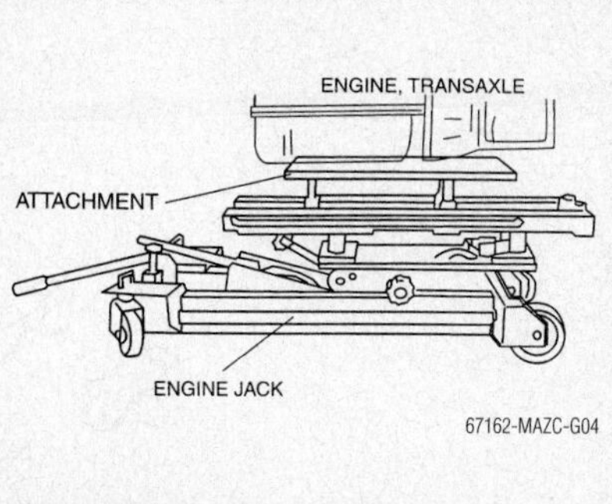

Fig. 145 Support the engine using an engine jack and attachment as shown—Mazda3 models

   - Water and heater hose
   - Wiring harness
6. Support the engine using an engine jack and attachment as shown in the illustration.
7. Remove the camshafts.

➡The cylinder head and camshaft caps are numbered and must be reassembled in their original locations. When removing a when removed, keep the caps with the cylinder head they were removed from, do not switch the caps.

8. Loosen the camshaft cap bolts using 2–3 passes in the sequence illustrated.
9. Once all the bolts have been removed, remove the caps, keeping them in the original positions and remove the camshafts.

To install:

10. Set the cam position of the number 1 cylinder at Top Dead center (TDC) and install the camshaft.
11. Temporarily install the camshaft caps in their original positions.
12. Tighten the camshaft cap bolt in two steps:
    a. Step 1:44–71 inch lbs. (5–8 Nm).
    b. Step 2:10.4–12.5 ft. lbs. (14–17 Nm).
13. Remove the engine jack and attachment.
14. Connect the following components:
    - Wiring harness
    - Water and heater hose
    - Upper radiator hose
    - WU–TWC converter
15. Install the intake manifold.
16. Install the timing chain.
17. Connect the negative battery cable.
18. Fill and bleed the cooling system.
19. Change the oil and filter.
20. Run the engine and check for proper operation.

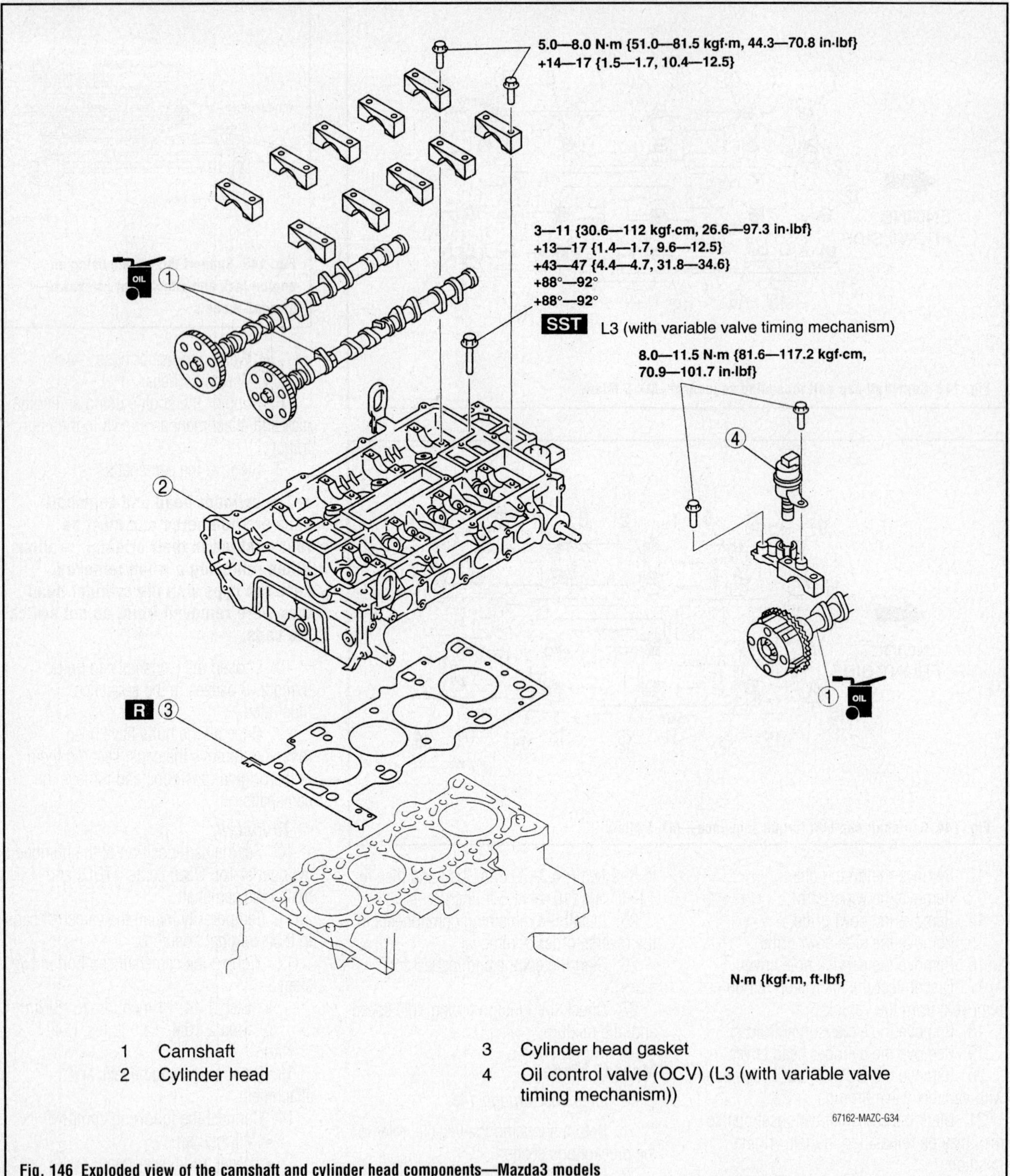

Fig. 146 Exploded view of the camshaft and cylinder head components—Mazda3 models

Mazda6 and Mazdaspeed6

2.3L (L3) Engines

See Figures 145 and 149 through 151.

1. Before servicing the vehicle, refer to the precautions section.
2. Disconnect the negative battery cable.
3. Remove the timing chain.
4. Remove the ignition coil and spark plug wires.
5. Disconnect the alternator and move it aside.
6. Remove the front exhaust pipe.
7. Remove the intake manifold.
8. Remove the upper radiator hose, bypass and heater hoses.
9. Support the engine using an engine jack and attachment as shown in the illustration.
10. Remove the oil control valve.
11. Remove the camshafts.

67162-MAZC-G35

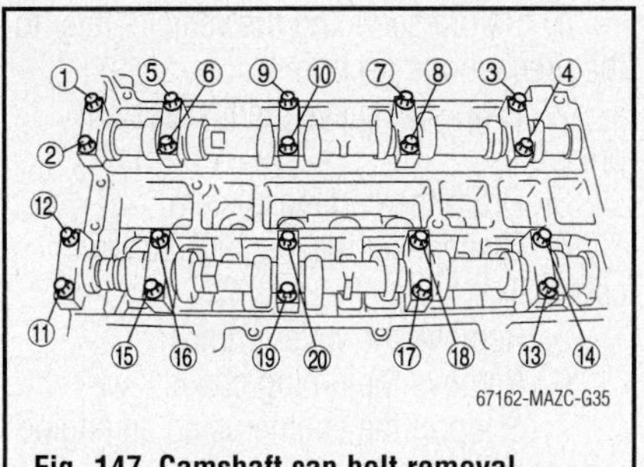

Fig. 147 Camshaft cap bolt removal sequence—Mazda3 models

67162-MAZC-G36

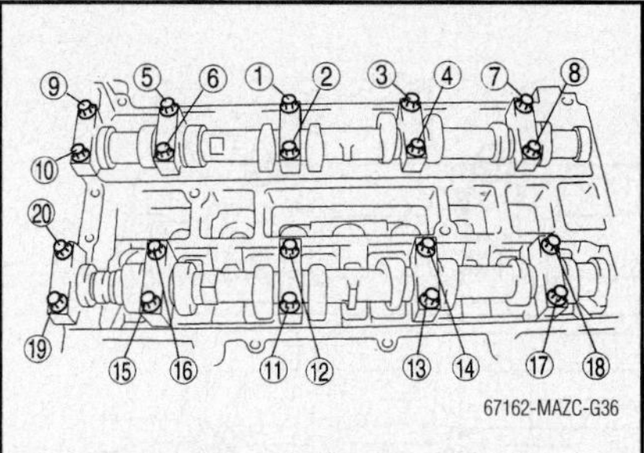

Fig. 148 Camshaft cap bolt tightening sequence—Mazda3 models

➡The cylinder head and camshaft caps are numbered and must be reassembled in their original locations. When removing a when removed, keep the caps with the cylinder head they were removed from, do not switch the caps.

12. Loosen the camshaft cap bolts using 2–3 passes in the sequence illustrated.

13. Once all the bolts have been removed, remove the caps, keeping them in the original positions and remove the camshafts.

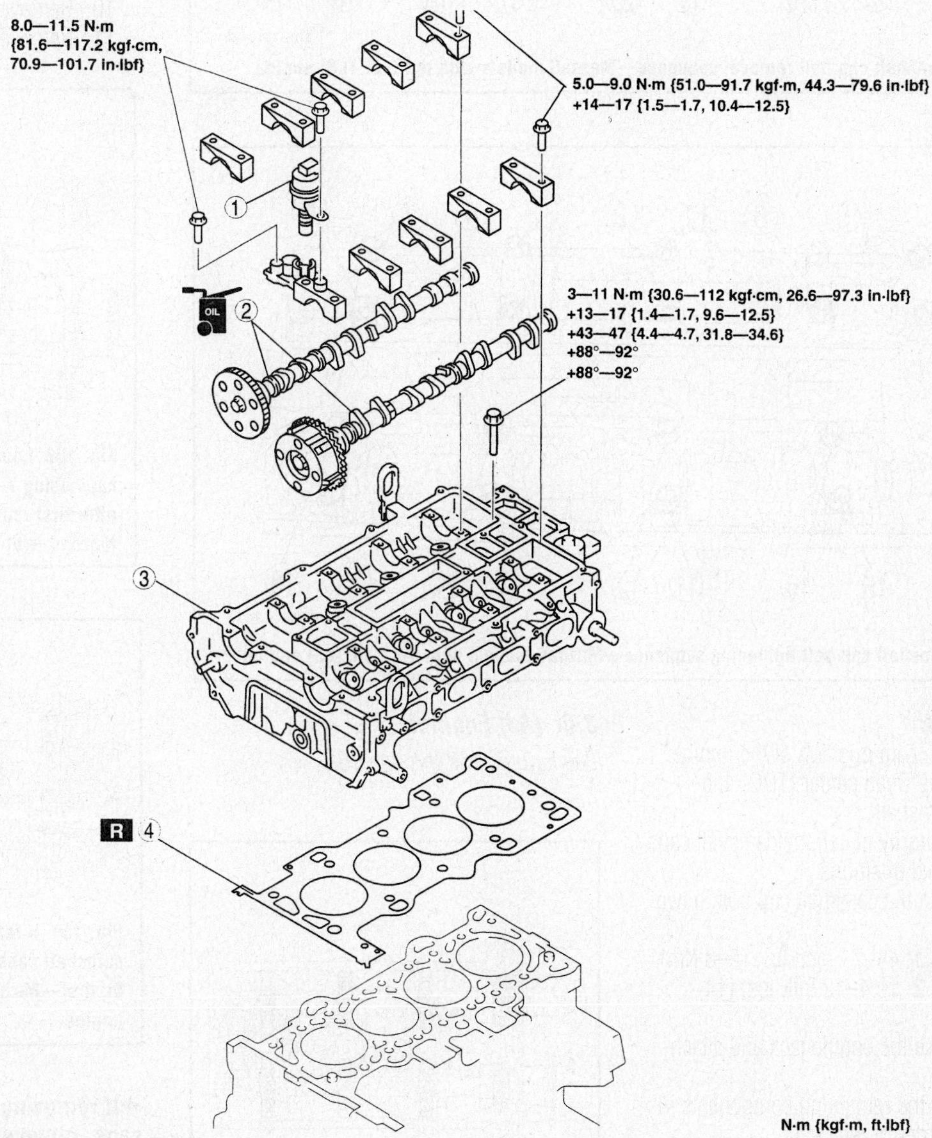

8.0—11.5 N·m
{81.6—117.2 kgf·cm,
70.9—101.7 in·lbf}
5.0—9.0 N·m {51.0—91.7 kgf·m, 44.3—79.6 in·lbf}
+14—17 {1.5—1.7, 10.4—12.5}
3—11 N·m {30.6—112 kgf·cm, 26.6—97.3 in·lbf}
+13—17 {1.4—1.7, 9.6—12.5}
+43—47 {4.4—4.7, 31.8—34.6}
+88°—92°
+88°—92°
OIL
R
N·m {kgf·m, ft·lbf}

1	Oil control valve (OCV)	3	Cylinder head
2	Camshaft	4	Cylinder head gasket

67162-MAZC-G168

Fig. 149 Exploded view of the camshaft and cylinder head components—Mazda6 models with the 2.3L (L3) engine

1. Before servicing the vehicle, refer to the precautions section.
2. Properly relieve the fuel system pressure.
3. Drain the cooling system.
4. Disconnect the negative battery cable.
5. Remove the water pump.
6. Remove the timing chain.
7. Support the engine using an engine jack and attachment.
8. Remove the water bypass tube stud bolt and bolt and disconnect the tube from the cylinder head.
9. Remove the camshaft(s).
10. Remove the rocker arm(s).
11. Remove the camshafts.

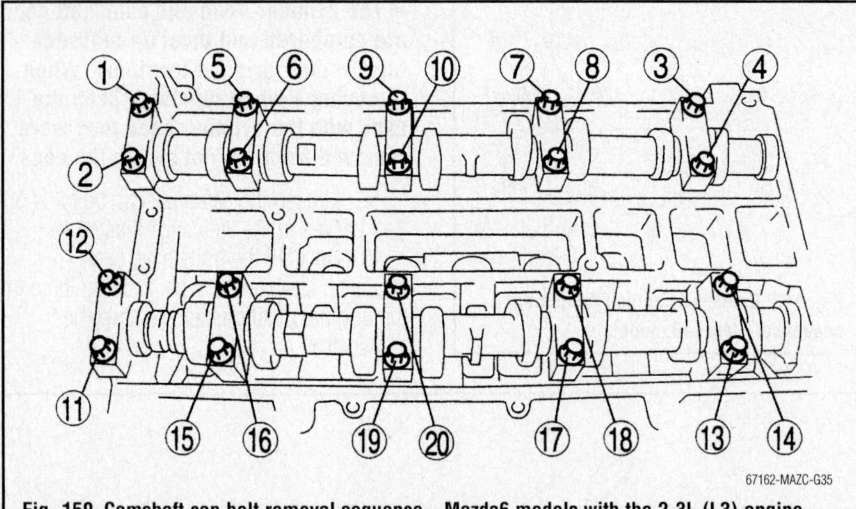

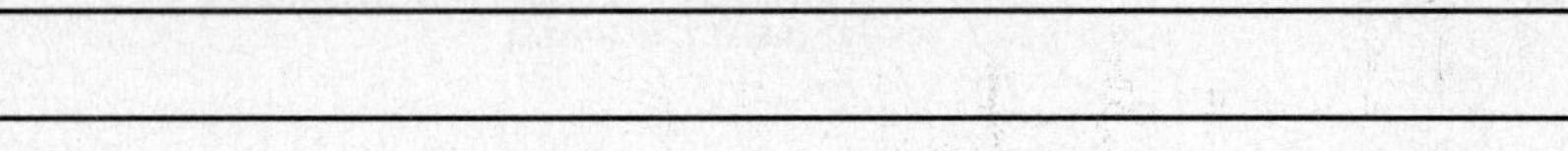
Fig. 150 Camshaft cap bolt removal sequence—Mazda6 models with the 2.3L (L3) engine

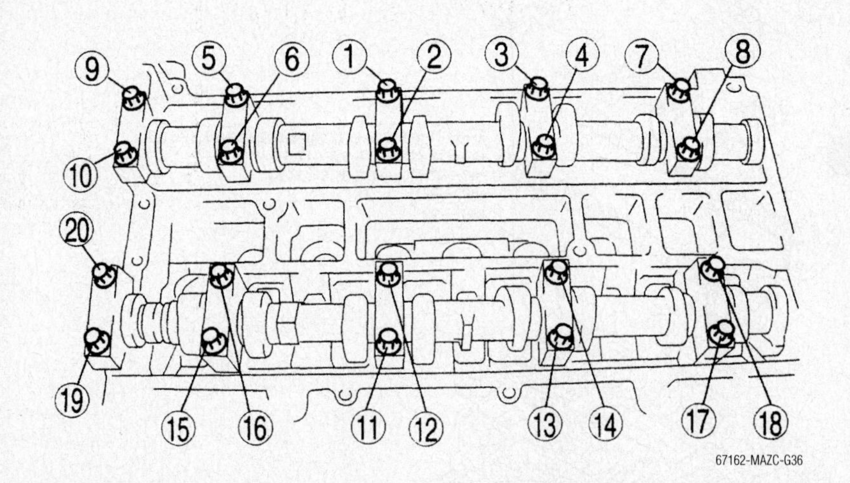

Fig. 151 Camshaft cap bolt tightening sequence—Mazda6 models with the 2.3L (L3) engine

To install:

14. Set the cam position of the number 1 cylinder at Top Dead center (TDC) and install the camshaft.
15. Temporarily install the camshaft caps in their original positions.
16. Tighten the camshaft cap bolt in two steps:
    a. Step 1: 44–71 inch lbs. (5–8 Nm).
    b. Step 2: 10.4–12.5 ft. lbs. (14–17 Nm).
17. Remove the engine jack and attachment.
18. Install the remaining components in the reverse order of removal.
19. Connect the negative battery cable.
20. Fill and bleed the cooling system.
21. Change the oil and filter.
22. Run the engine and check for proper operation.

3.0L (AJ) Engines

See Figures 152 through 161.

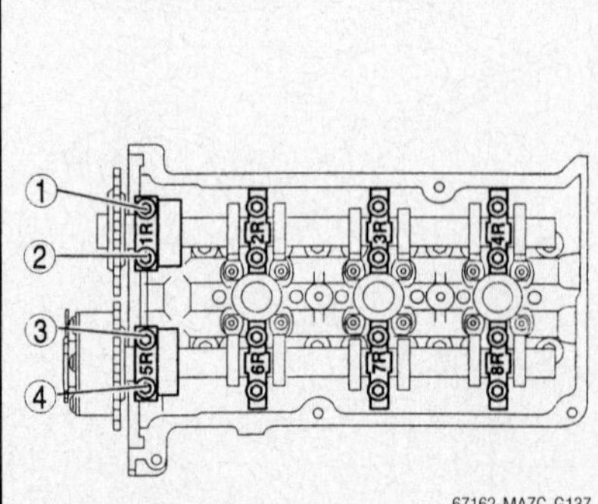

Fig. 152 If removing the right hand camshaft caps, remove thrust caps 1R and 5R first—Mazda6 with the 3.0L (AJ) engine

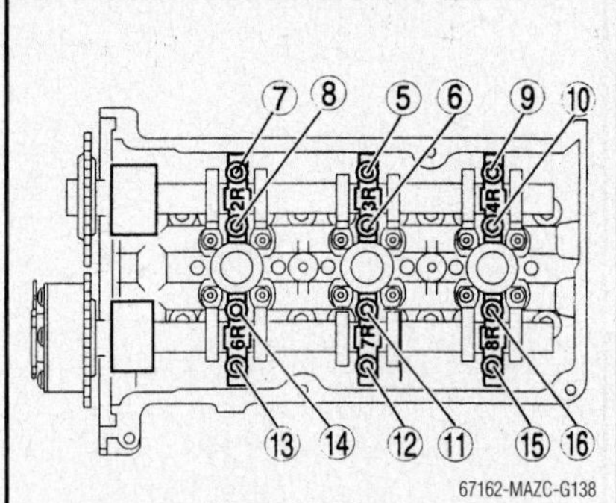

Fig. 153 Loosen the right hand camshaft caps using 7–8 passes in this sequence after first removing the thrust caps—Mazda6 with the 3.0L (AJ) engine

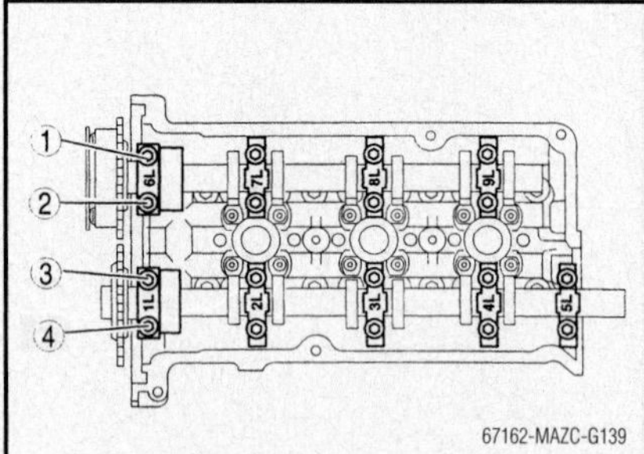

Fig. 154 If removing the left hand camshaft caps, remove thrust caps 1L and 6L first—Mazda6 with the 3.0L (AJ) engine

➡If removing the right hand camshaft caps, remove thrust caps 1R and 5R first. Do not loosen any of the other cap bolts until the thrust caps are removed or you could damage the thrust caps. If removing the left hand camshaft caps, remove thrust caps 1L and 6L first. Do not loosen any of the other cap

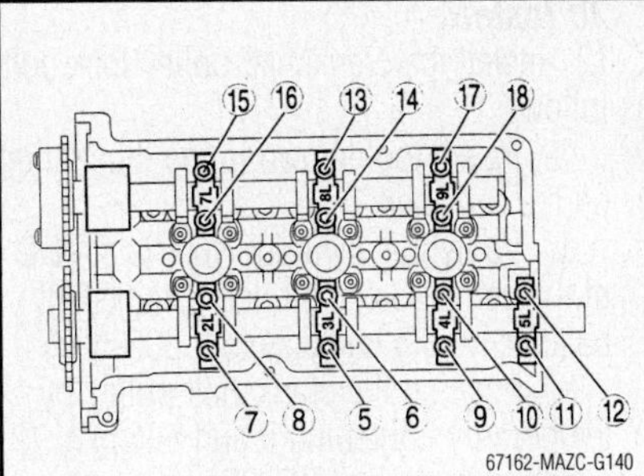

Fig. 155 Loosen the left hand camshaft caps using 7–8 passes in this sequence after first removing the thrust caps—Mazda6 with the 3.0L (AJ) engine

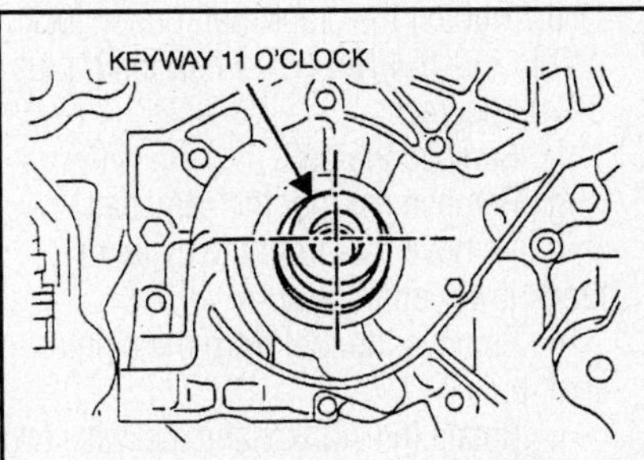

Fig. 156 When installing the left hand camshaft place the crankshaft keyway at the 11 o'clock position—Mazda6 with the 3.0L (AJ) engine

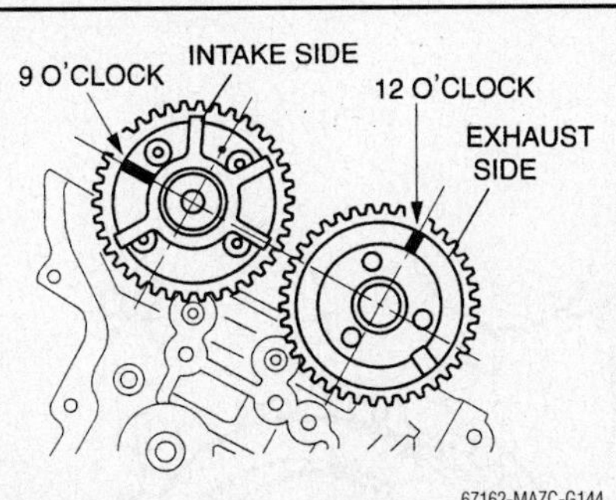

Fig. 157 Install the left hand camshaft and position the mark on the intake camshaft to the 9 o'clock and the exhaust camshaft to the 12 o'clock position—Mazda6 with the 3.0L (AJ) engine

bolts until the thrust caps are removed or you could damage the thrust caps.

12. Loosen the camshaft caps using 7–8 passes in sequence after first removing the thrust caps to allow the camshafts to be slowly raised. Keep the caps in the order they were removed so they are re-installed in their original positions.

To install:

13. Install the camshafts in their original positions.

14. Install the left hand camshaft as follows:
 a. Place the crankshaft keyway at the 11 o'clock position by rotating the crankshaft clockwise.
 b. Install the camshaft and the caps in their original positions and hand tighten the bolts.

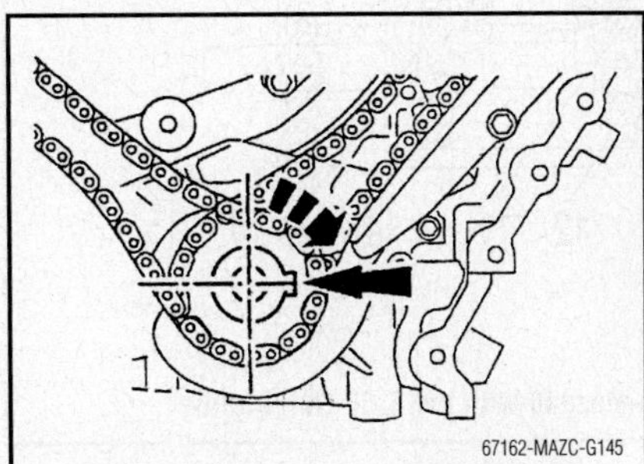

Fig. 158 When installing the right hand camshaft place the crankshaft keyway at the 3 o'clock position—Mazda6 with the 3.0L (AJ) engine

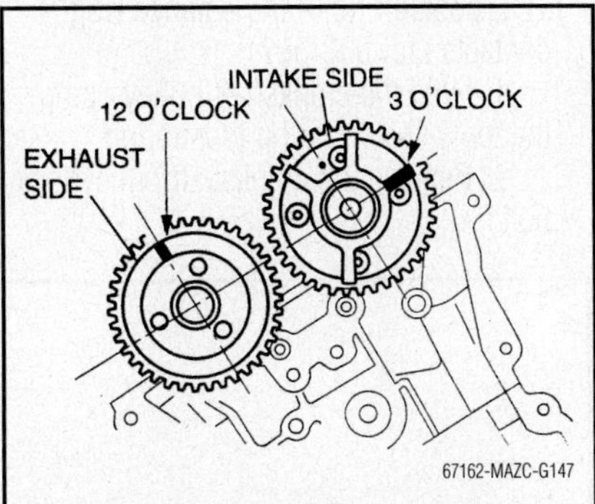

Fig. 159 Install the right hand camshaft and position the mark on the intake camshaft to the 3 o'clock and the exhaust camshaft to the 12 o'clock position—Mazda6 with the 3.0L (AJ) engine

 c. Position the mark on the intake camshaft to the 9 o'clock position as illustrated.
 d. Position the mark on the exhaust camshaft to the 12 o'clock position.
 e. Install the timing chain.

➡Tighten the camshaft journal thrust caps last to avoid damaging the thrust caps.

 f. Tighten the camshaft caps using several passes to 71–106 inch lbs. (8–12 Nm). After adjust the camshaft endplay using the thrust caps 1L and 6L, tighten the remaining caps.

15. Install the right hand camshaft as follows:
 a. Install the camshaft and the caps in their original positions and hand tighten the bolts.
 b. Place the crankshaft keyway at the 3 o'clock position.
 c. Position the mark on the exhaust camshaft to the 12 o'clock position by rotating the crankshaft clockwise.
 d. Position the mark on the intake camshaft to the 3 o'clock position as illustrated.
 e. Install the timing chain.

➡Tighten the camshaft journal thrust caps last to avoid damaging the thrust caps.

 f. Tighten the camshaft caps using several passes to 71–106 inch lbs. (8–12 Nm). After adjust the camshaft endplay using the thrust caps 1R and 5R, tighten the remaining caps.

16. Install the water bypass tube.
17. Remove the engine jack and attachment.
18. Install the water pump.
19. Fill the cooling system.

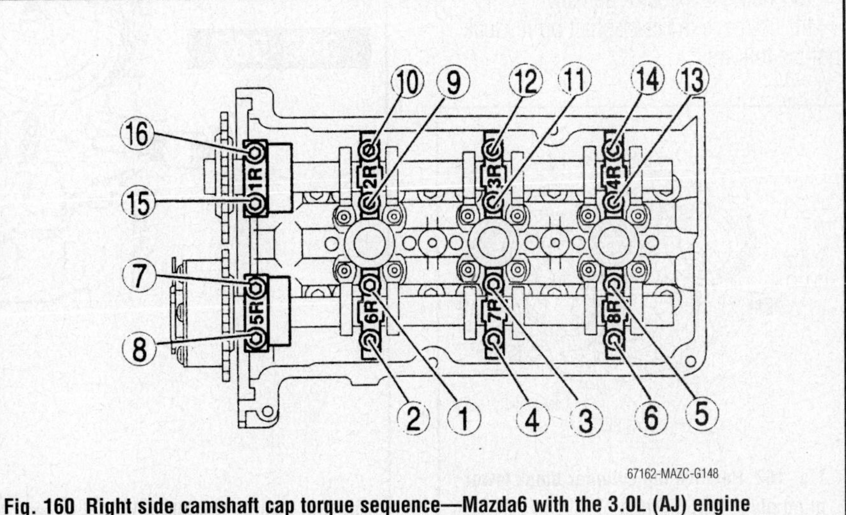

Fig. 160 Right side camshaft cap torque sequence—Mazda6 with the 3.0L (AJ) engine

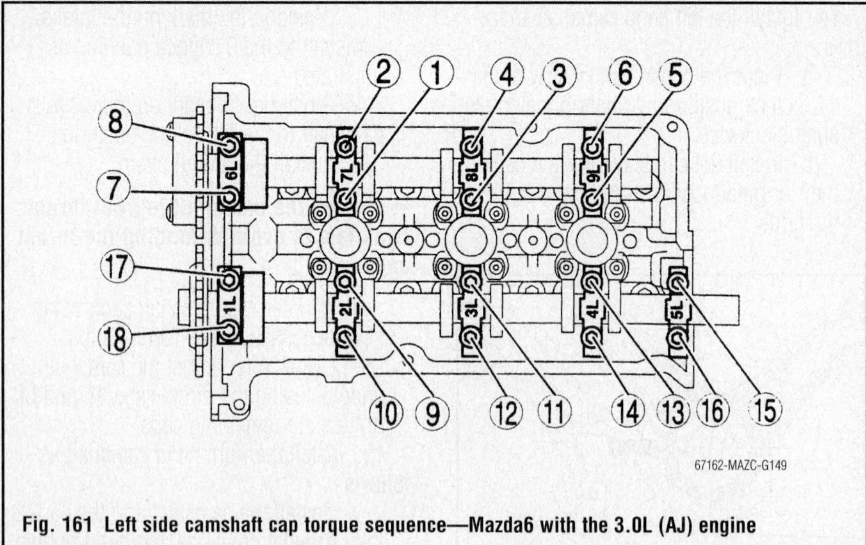

Fig. 161 Left side camshaft cap torque sequence—Mazda6 with the 3.0L (AJ) engine

20. Fill and bleed the cooling system.
21. Change the oil and filter.
22. Run the engine and check for proper operation and leaks.

CRANKSHAFT DAMPER

REMOVAL & INSTALLATION

MX-5 Miata

See Figures 162 through 166.

1. Before servicing the vehicle, refer to the precautions section.
2. Remove the battery and battery tray.
3. Remove the air cleaner.
4. Remove the drive belt.
5. Remove the undercover.
6. Remove the front suspension tower bar.
7. Remove the ignition coil.
8. Remove the OCV connector.
9. Remove the cylinder head cover.
10. Remove the CKP sensor.
11. Remove the crankshaft pulley lock bolt as follows:

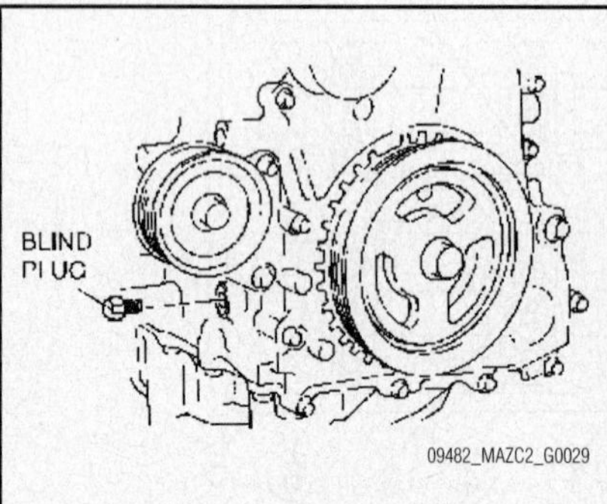

Fig. 162 Remove the cylinder block lower blind plug—MX-5 Miata

a. Remove the cylinder block lower blind plug.
b. Install tool 303-507.
c. Turn the crankshaft clockwise until the crankshaft is in the No.1 cylinder TDC position (until the balance weight contacts tool 303-507).
d. Hold the crankshaft pulley using the tools shown in the illustration.
e. Remove the crankshaft pulley lock bolt.

To install:

12. Install the crankshaft pulley lock bolt as follows:

a. Install tool shown in the illustration on the camshaft.
b. Verify that cylinder No.1 is at TDC of the compression stroke. (Crankshaft balance weight contacts tool 303-507)
c. Position the crankshaft pulley by temporarily tightening it and, using a suitable length bolt 25–35mm(0.99–1.37 in.), fix the crankshaft pulley to the engine front cover.
d. Install the tools illustrated to the crankshaft pulley, lock the crankshaft against rotation.
e. Tighten the crankshaft pulley lock bolt to 96–104 Nm (71–77 ft. lbs.), plus 87–93 degrees.
f. Remove the bolt.
g. Remove the tool the camshaft.
h. Remove tool from the cylinder block lower blind plug.
i. Remove the tool from the crankshaft pulley.
j. Rotate the crankshaft clockwise two turns until the TDC position.
k. If not aligned, loosen the crankshaft pulley lock bolt and repeat from the first step.

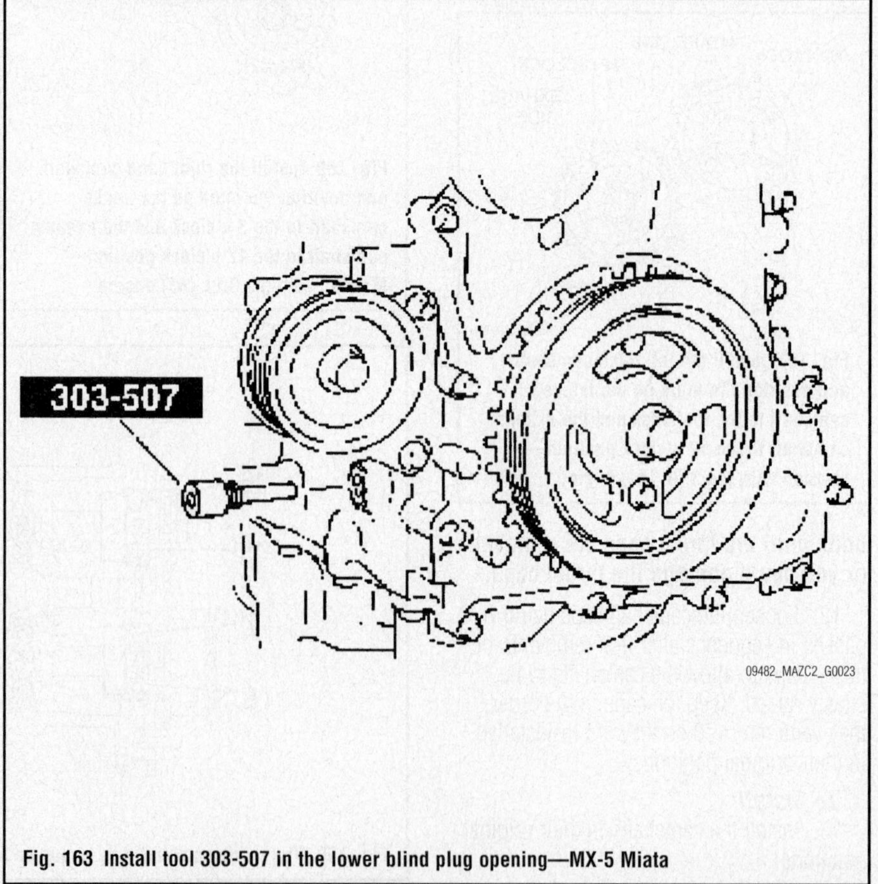

Fig. 163 Install tool 303-507 in the lower blind plug opening—MX-5 Miata

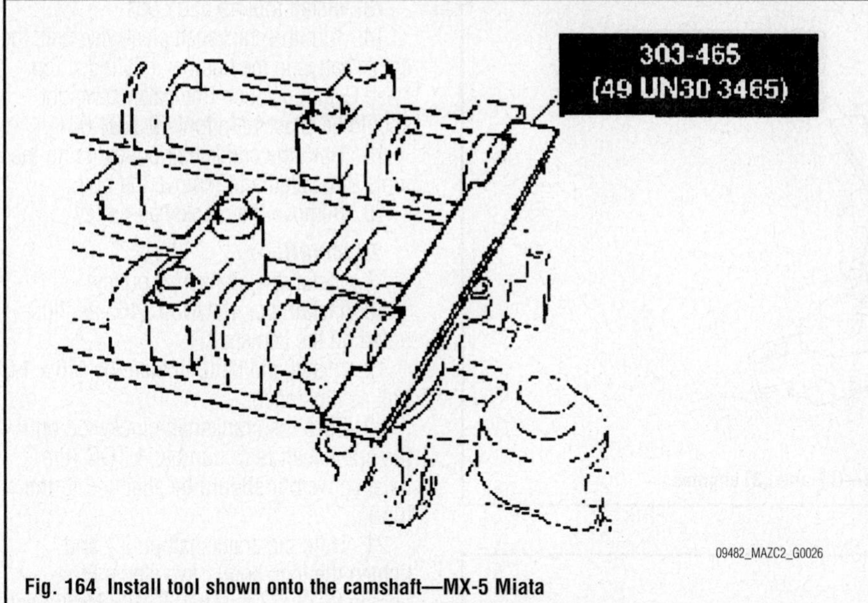

Fig. 164 Install tool shown onto the camshaft—MX-5 Miata

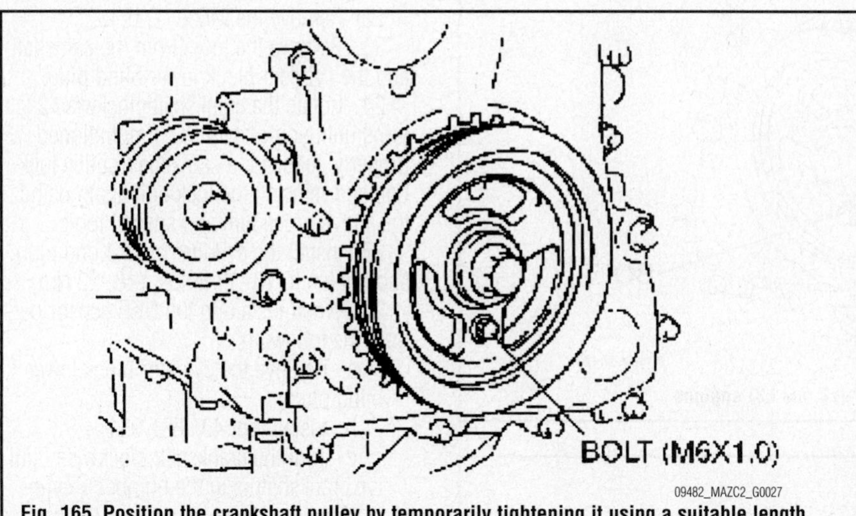

Fig. 165 Position the crankshaft pulley by temporarily tightening it using a suitable length bolt—MX-5 Miata

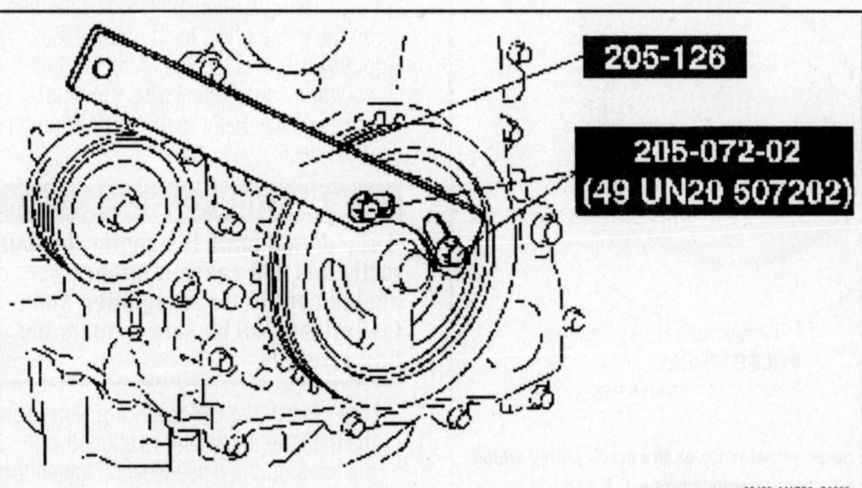

Fig. 166 Install the tools shown onto the crankshaft pulley to lock the crankshaft against rotation—MX-5 Miata

l. Install the cylinder block lower blind plug and tighten to 18–22 Nm (13–16 ft. lbs.)

13. Install the remaining components in the reverse order of removal.

Mazda3 and Mazda6 Models

2.3L (L3) and 2.0L (LF) Engines

See Figures 167 through 171.

1. Before servicing the vehicle, refer to the precautions section.
2. Remove the plug hole plate and bracket.
3. Remove the battery cover.
4. Disconnect the negative battery cable.
5. Disconnect the wiring harness.
6. Remove the ignition coils, spark plugs, valve cover and accessory drive belt.
7. Remove the front wheels.
8. Remove the undercover and splash shield, if equipped.
9. On Mazda6 models, remove the right hand halfshaft from the joint shaft.
10. Remove the CKP sensor.
11. Remove the crankshaft pulley bolt and pulley.
12. Remove the cylinder block lower blind plug.

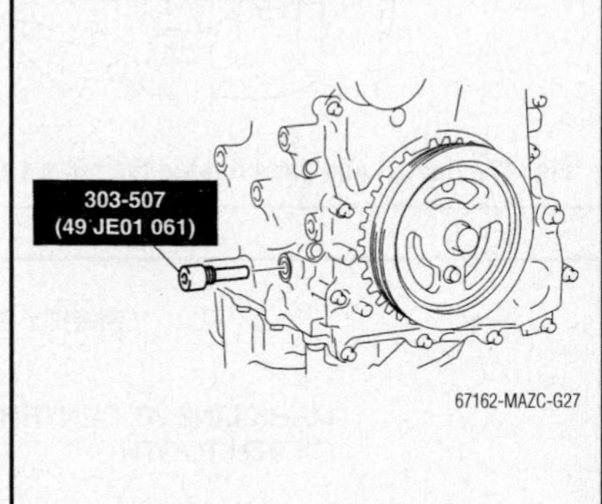

Fig. 167 Install tool 49 JE01 061—(LF and L3) engines

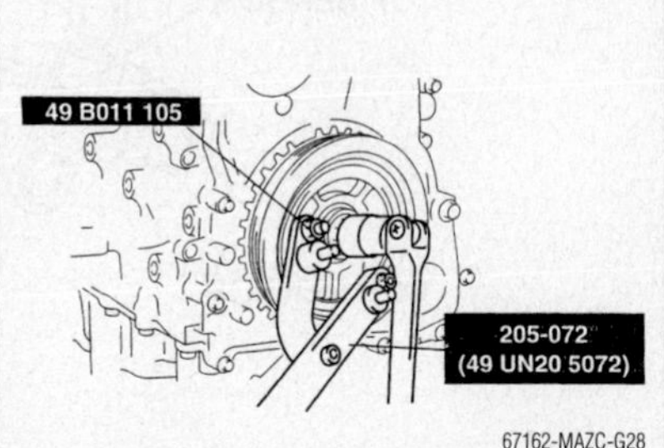

Fig. 168 Install tool 49 JE01 061 Hold the crankshaft pulley using the tools illustrated and remove the bolt—(LF and L3) 2.3L engines

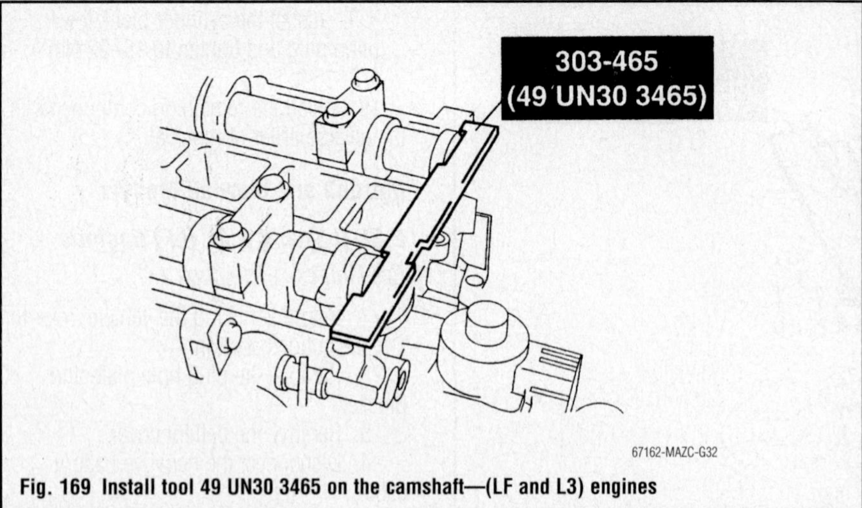

Fig. 169 Install tool 49 UN30 3465 on the camshaft—(LF and L3) engines

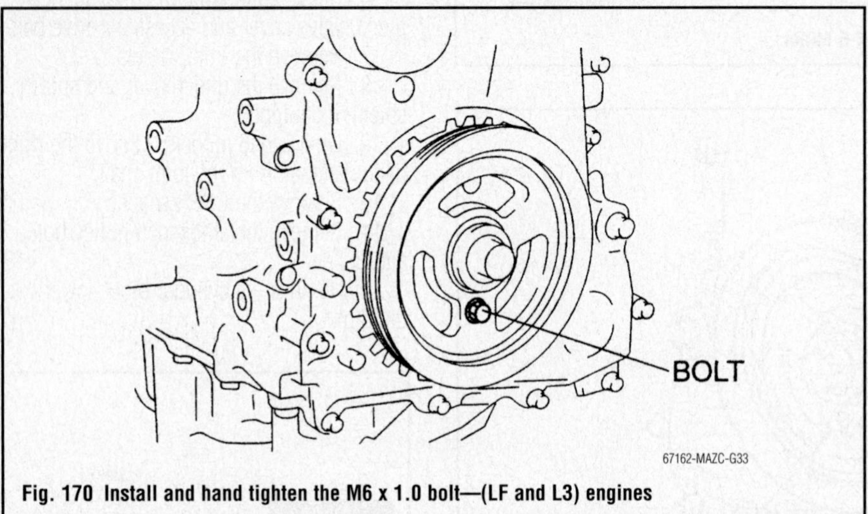

Fig. 170 Install and hand tighten the M6 x 1.0 bolt—(LF and L3) engines

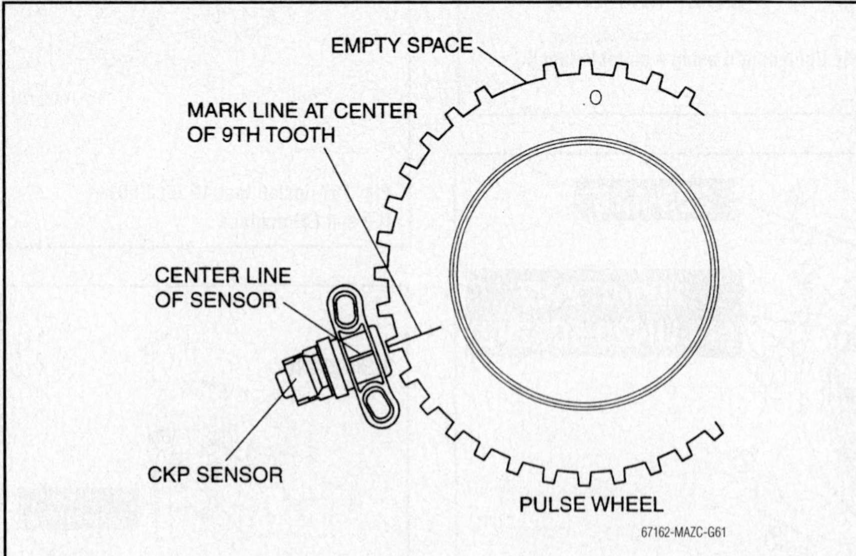

Fig. 171 Using a ruler, mark the center line on the pulse wheel teeth on the crank pulley which is located at the 9th tooth counting counterclockwise from the empty space—(LF and L3) engines

13. Install tool 49 JE01 061.
14. Turn the crankshaft clockwise until the crankshaft is in the number 1 cylinder Top Dead Center position (the balance weight should be attached to tool 49 JE01 061).
15. Hold the crankshaft pulley using the tools illustrated and remove the bolt.
16. Remove the crankshaft pulley.

To install:

17. Install the crankshaft pulley.
18. Install tool 49 UN30 3465 as illustrated on the camshaft.
19. Install and hand tighten the M6 x 1.0 bolt as illustrated.
20. Turn the crankshaft clockwise until the crankshaft is in number 1 TDC (the balance weight should be attached to the tool).
21. Hold the crankshaft pulley and tighten the lock bolt in two steps. First tighten to 71–77 ft. lbs. (96–104 Nm). Then final tighten an additional 87–93 degrees.
22. Remove the M6 x 1.0 bolt.
23. Remove the tools from the camshaft and the cylinder block lower blind plug.
24. Rotate the crankshaft clockwise 2 turns until you reach TDC. If not aligned properly, loosen the crankshaft pulley lock bolt and reinstall the lock bolt again using the above procedure and special tools.
25. Install the cylinder block blind plug and tighten to 13–16 ft. lbs. (18–22 Nm).
26. When installing the CKP sensor perform the following:
    a. Remove the cylinder block lower blind plug.
    b. Install tool 49 JE01 061.
    c. Turn the crankshaft clockwise until the crankshaft is in the number 1 cylinder Top Dead Center position (the balance weight should be attached to tool 49 JE01 061).
    d. Using a ruler, mark the center line on the pulse wheel teeth on the crank pulley which is located at the 9th tooth counting counterclockwise from the empty space. Refer to the illustration for more detail.

✲✲ CAUTION

If you do not mark the center line correctly this will cause improper engine control for the ignition and fuel system, so be sure to mark the line carefully.

    e. Install the CKP sensor making sure the mark on the sensor is aligned with the mark on the pulse wheel. Tighten the sensor bolt to 49–66 inch lbs. (5.5–7.5 Nm)and attach the electrical connector.

f. Remove the tool from the cylinder block lower blind plug.

g. Rotate the crankshaft clockwise 2 turns until you reach TDC. If not aligned properly, loosen the crankshaft pulley lock bolt and reinstall the lock bolt.

h. Install the cylinder block blind plug and tighten to 13–16 ft. lbs. (18–22 Nm).

27. Install the remaining components in the reverse order of removal.

28. Start the vehicle and check for leaks, repair if necessary.

3.0L (AJ) Engines

See Figures 172 through 174.

1. Before servicing the vehicle, refer to the precautions section.
2. Disconnect the negative battery cable.
3. Remove the accessory drive belt.
4. Hold the crankshaft pulley using the tools illustrated and remove the pulley bolt.
5. Remove the A/C compressor and set aside with the lines still attached.
6. Remove the crankshaft pulley using a suitable puller such as the one illustrated.

To install:

7. Install the pulley using the tools illustrated.
8. While holding the pulley, tighten the crankshaft pulley bolt as follows:
   a. Tighten to 88 ft. lbs. (120 Nm).
   b. Loosen one full turn.
   c. Tighten to 35–39 ft. lbs. (47–53 Nm).
   d. Tighten 85–95 degrees.
9. Install the A/C compressor.
10. Install the accessory drive belt.
11. Connect the negative battery cable.
12. Start the vehicle and check for leaks, repair if necessary.

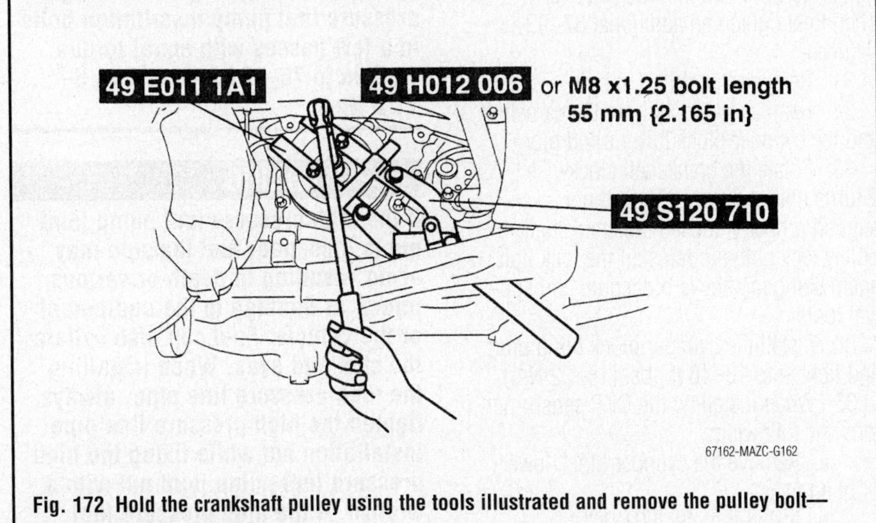

Fig. 172 Hold the crankshaft pulley using the tools illustrated and remove the pulley bolt—Mazda6 with the 3.0L (AJ) engine

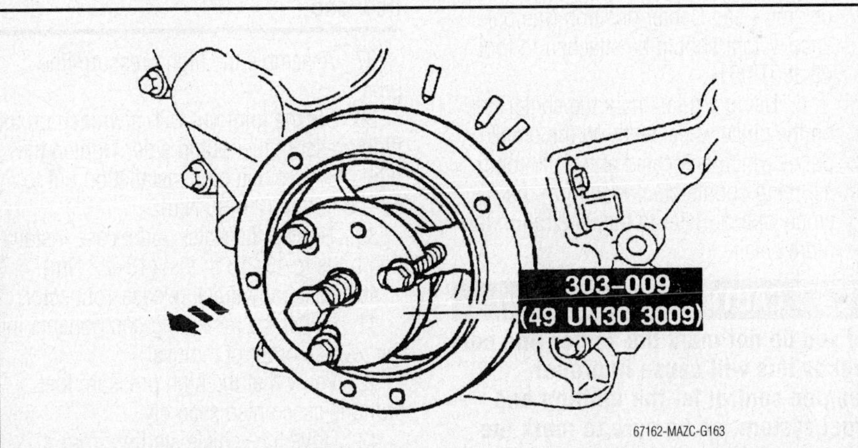

Fig. 173 Remove the crankshaft pulley using a suitable puller such as the one shown—Mazda6 with the 3.0L (AJ) engine

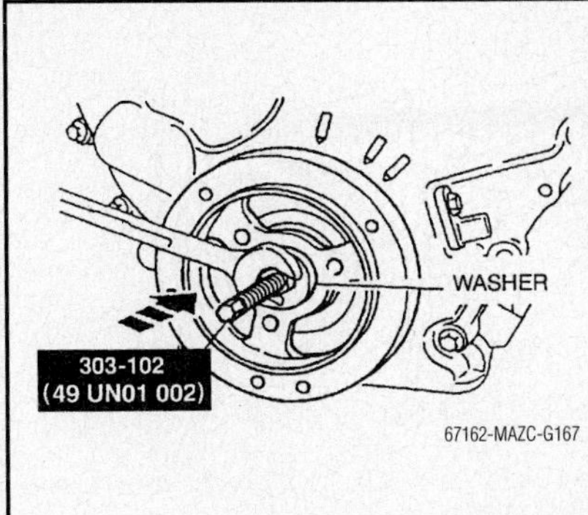

Fig. 174 Install the pulley using the tools shown—Mazda6 with the 3.0L (AJ) engine

2.3L (L3 with TC) Engines

See Figures 175 through 177.

1. Before servicing the vehicle, refer to the precautions section.
2. Disconnect the negative battery cable.
3. Remove the right side front tire.
4. Remove the undercover.
5. Remove the right side splash shield.

➡If the high pressure fuel pump is removed, replace the O-ring with a new one.

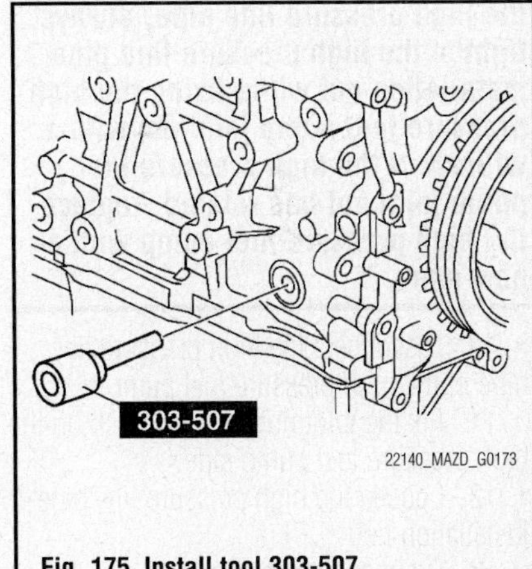

Fig. 175 Install tool 303-507

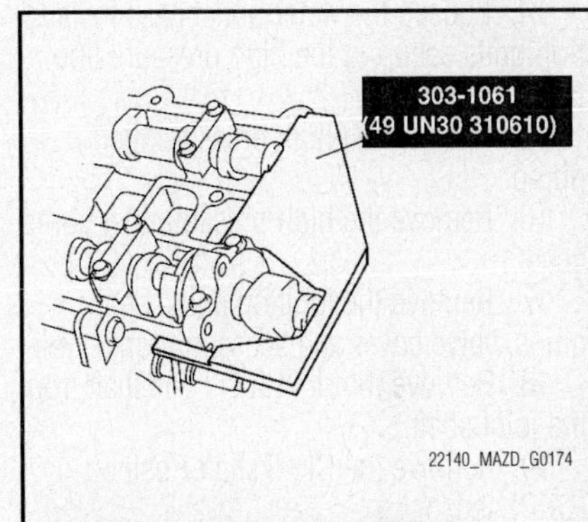

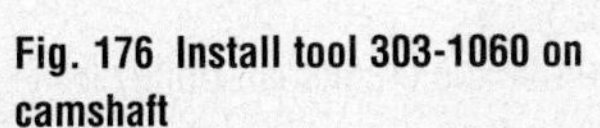

Fig. 176 Install tool 303-1060 on camshaft

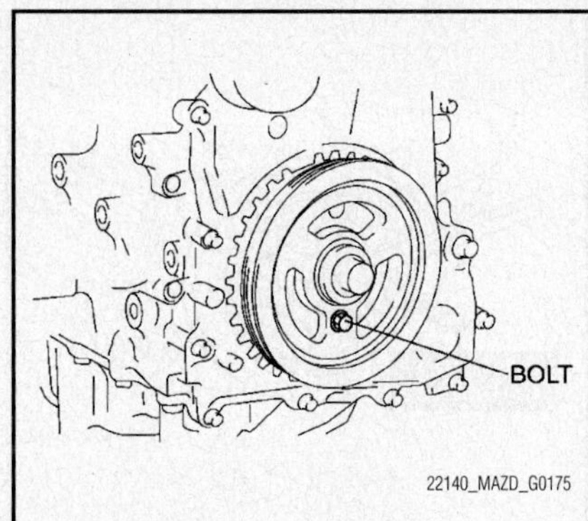

Fig. 177 Install and hand tighten the M6 x 1.0 bolt

6. Remove the charge air cooler cover.
7. Disconnect the spill valve control solenoid valve connector.
8. Disconnect the quick release connector on the high pressure fuel pump.
9. Remove the battery and battery tray.

⁂ WARNING

If the high pressure fuel pump joint nut is loosened, fuel leakage may occur resulting in death or serious injury, or damage to the equipment or the vehicle. Fuel can also irritate the skin and eyes. When removing the high pressure line pipe, always tighten the high pressure line pipe installation nut while fixing the high pressure fuel pump joint nut with a wrench. If the high pressure fuel pump joint nut has rotated, replace the high pressure fuel pump with a new one.

10. Disconnect the high pressure line pipe of the high pressure fuel pump.
11. Fix the joint nut with a wrench on the high pressure fuel pump side.
12. Loosen the high pressure line pipe installation nut.
13. Drain and recycle the engine coolant.
14. Loosen the water outlet case installation bolts securing the high pressure line pipe.
15. Remove the high pressure fuel pump.
16. Remove the high pressure fuel pump cover.
17. Remove the ignition coils, spark plugs, valve cover and accessory drive belt.
18. Remove the right hand halfshaft from the joint shaft.
19. Remove the Crankshaft Position (CKP) sensor.
20. Remove the crankshaft pulley bolt and pulley.
21. Remove the cylinder block lower blind plug.
22. Install tool 303-507.
23. Turn the crankshaft clockwise until the crankshaft is in the number 1 cylinder Top Dead Center position (the balance weight should be attached to tool 49 JE01 061).
24. Hold the crankshaft pulley using the tools illustrated and remove the bolt.
25. Remove the crankshaft pulley.

To install:

26. Install the crankshaft pulley.
27. Install tool 303-1060 as illustrated on the camshaft.
28. Install and hand tighten the M6 x 1.0 bolt as illustrated.
29. Turn the crankshaft clockwise until the crankshaft is in number 1 TDC (the balance weight should be attached to the tool).
30. Hold the crankshaft pulley and tighten the lock bolt in two steps. First tighten to 71–77 ft. lbs. (96–104 Nm). Then final tighten an additional 87–93 degrees.
31. Remove the M6 x 1.0 bolt.
32. Remove the tools from the camshaft and the cylinder block lower blind plug.
33. Rotate the crankshaft clockwise 2 turns until you reach TDC. If not aligned properly, loosen the crankshaft pulley lock bolt and reinstall the lock bolt again using the above procedure and special tools.
34. Install the cylinder block blind plug and tighten to 13–16 ft. lbs. (18–22 Nm).
35. When installing the CKP sensor perform the following:
    a. Remove the cylinder block lower blind plug.
    b. Install tool 49 JE01 061.
    c. Turn the crankshaft clockwise until the crankshaft is in the number 1 cylinder Top Dead Center position (the balance weight should be attached to tool 49 JE01 061).
    d. Using a ruler, mark the center line on the pulse wheel teeth on the crank pulley which is located at the 9th tooth counting counterclockwise from the empty space. Refer to the illustration for more detail.

⁂ CAUTION

If you do not mark the center line correctly this will cause improper engine control for the ignition and fuel system, so be sure to mark the line carefully.

    e. Install the CKP sensor making sure the mark on the sensor is aligned with the mark on the pulse wheel. Tighten the sensor bolt to 49–66 inch lbs. (5.5–7.5 Nm)and attach the electrical connector.
    f. Remove the tool from the cylinder block lower blind plug.
    g. Rotate the crankshaft clockwise 2 turns until you reach TDC. If not aligned properly, loosen the crankshaft pulley lock bolt and reinstall the lock bolt.
    h. Install the cylinder block blind plug and tighten to 13–16 ft. lbs. (18–22 Nm).
36. Install the high pressure fuel pump cover and tighten to 13–16 ft. lbs. (18–22 Nm).

⁂ CAUTION

If the high pressure fuel pump installation bolts are tightened with the high pressure fuel pump tilted, the high pressure fuel pump may not operate correctly. Tighten the high pressure fuel pump installation bolts in a few passes with equal torque. Tighten to 76–101 inch lbs. (8.5–11.5 Nm).

⁂ WARNING

If the high pressure fuel pump joint nut is loosened, fuel leakage may occur resulting in death or serious injury, or damage to the equipment or the vehicle. Fuel can also irritate the skin and eyes. When installing the high pressure line pipe, always tighten the high pressure line pipe installation nut while fixing the high pressure fuel pump joint nut with a wrench. If the high pressure fuel pump joint nut has rotated, replace the high pressure fuel pump with a new one.

37. Assemble the high pressure line pipe.
38. Fix the joint nut with a wrench on the high pressure fuel pump side. Tighten the high pressure line pipe installation nut to 17–26 ft. lbs. (23–35 Nm).
39. Tighten the water outlet case installation bolts to 13–16 ft. lbs. (18–22 Nm).
40. Install the quick release connector.
41. Install the remaining components in the reverse order of removal.
42. Verify that the high pressure fuel pump is assembled securely.
43. Drive the vehicle starting from a standstill and brake suddenly five to six times at a low speed.

44. Stop the vehicle and verify from outside the vehicle that there is no fuel leakage around the high pressure fuel pump.

CRANKSHAFT FRONT SEAL

REMOVAL & INSTALLATION

MX-5 Miata

See Figures 163 through 165 and 178 through 180.

1. Before servicing the vehicle, refer to the precautions section.
2. Remove the battery and battery tray.
3. Remove the air cleaner.
4. Remove the drive belt.
5. Remove the undercover.
6. Remove the front suspension tower bar.
7. Remove the ignition coil.
8. Remove the OCV connector.
9. Remove the cylinder head cover.
10. Remove the CKP sensor.
11. Remove the crankshaft pulley lock bolt as follows:
    a. Remove the cylinder block lower blind plug.
    b. Install tool 303-507.
    c. Turn the crankshaft clockwise until the crankshaft is in the No.1 cylinder TDC position (until the balance weight contacts tool 303-507).
    d. Hold the crankshaft pulley using the tools shown in the illustration.
    e. Remove the crankshaft pulley lock bolt.
12. Remove the front oil seal as follows:
    a. Cut the oil seal lip using a razor knife.
    b. Remove the oil seal using a screwdriver wrapped with a rag.

To install:

13. Install the oil seal as follows:
    a. Apply clean engine oil to a new oil seal.
    b. Push the front oil seal in the engine front cover by hand.
    c. Tap the oil seal in evenly using tool shown and a hammer to 0–10 mm (0–0.039 in.).
14. Install the crankshaft pulley lock bolt as follows:
    a. Install tool shown in the illustration on the camshaft.
    b. Verify that cylinder No.1 is at TDC of the compression stroke. (Crankshaft balance weight contacts tool 303-507)

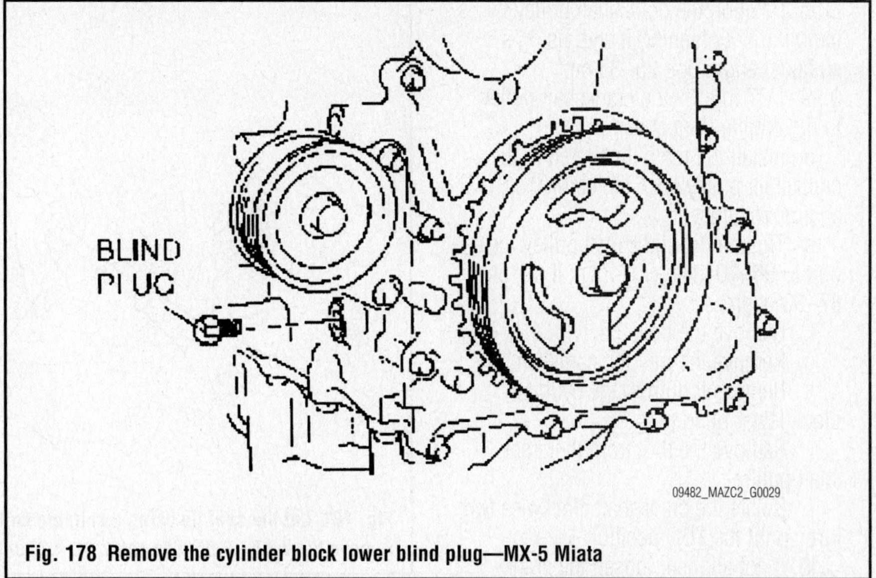

Fig. 178 Remove the cylinder block lower blind plug—MX-5 Miata

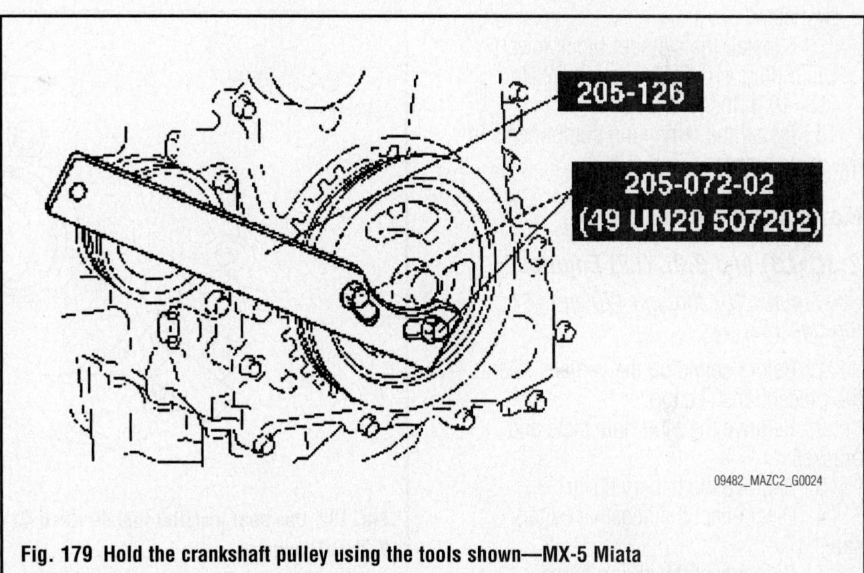

Fig. 179 Hold the crankshaft pulley using the tools shown—MX-5 Miata

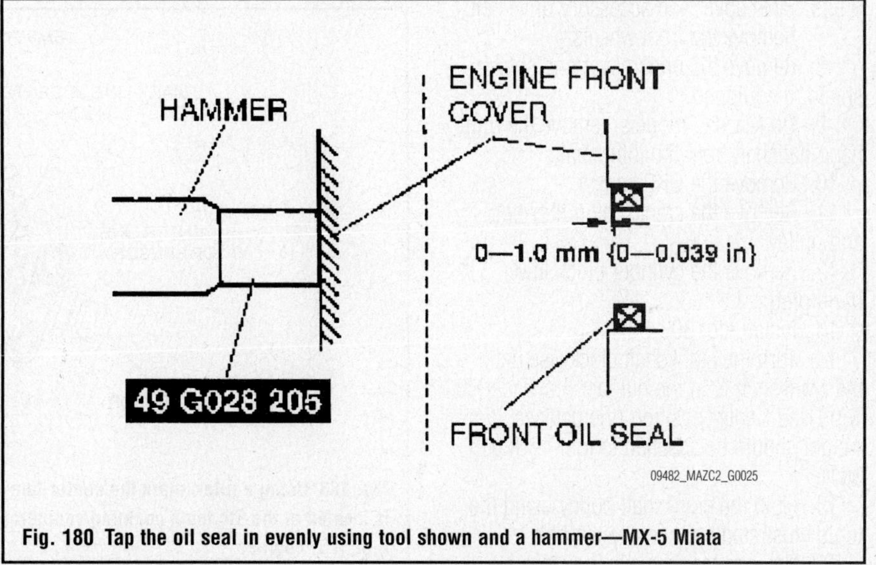

Fig. 180 Tap the oil seal in evenly using tool shown and a hammer—MX-5 Miata

c. Position the crankshaft pulley by temporarily tightening it and, using a suitable length bolt 25–35mm(0.99–1.37 in.), fix the crankshaft pulley to the engine front cover.

d. Install the tools illustrated to the crankshaft pulley, lock the crankshaft against rotation.

e. Tighten the crankshaft pulley lock bolt to 96–104 Nm (71–77 ft. lbs.), plus 87–93 degrees.

f. Remove the bolt.

g. Remove the tool the camshaft.

h. Remove tool from the cylinder block lower blind plug.

i. Remove the tool from the crankshaft pulley.

j. Rotate the crankshaft clockwise two turns until the TDC position.

k. If not aligned, loosen the crankshaft pulley lock bolt and repeat from the first step.

l. Install the cylinder block lower blind plug and tighten to 18–22 Nm (13–16 ft. lbs.)

15. Install the remaining components in the reverse order of removal.

Mazda3 and 6 Models

2.3L (L3) and 2.0L (LF) Engines

See Figures 167 through 170 and 181 through 183.

1. Before servicing the vehicle, refer to the precautions section.
2. Remove the plug hole plate and bracket.
3. Remove the battery cover.
4. Disconnect the negative battery cable.
5. Disconnect the wiring harness.
6. Remove the ignition coils, spark plugs, valve cover and accessory drive belt.
7. Remove the front wheels.
8. Remove the undercover and splash shield, if equipped.
9. On Mazda6 models, remove the right hand halfshaft from the joint shaft.
10. Remove the CKP sensor.
11. Remove the crankshaft pulley bolt and pulley.
12. Remove the cylinder block lower blind plug.
13. Install tool 303-507.
14. Turn the crankshaft clockwise until the crankshaft is in the number 1 cylinder Top Dead Center position (the balance weight should be attached to tool 49 JE01 061).
15. Hold the crankshaft pulley using the tools illustrated and remove the bolt.
16. Remove the crankshaft pulley.

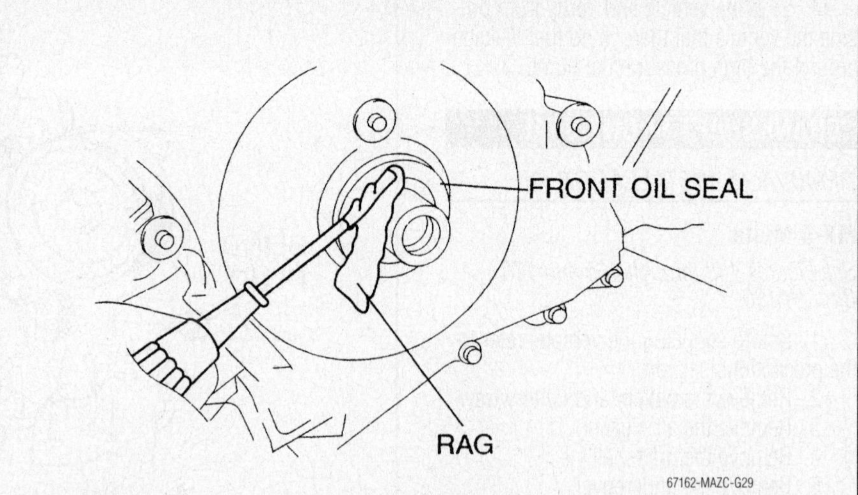

Fig. 181 Cut the seal lip using a suitable knife and using a suitable pry tool wrapped in a rag, remove the seal—2.0L (LF) and 2.3L (L3) engines

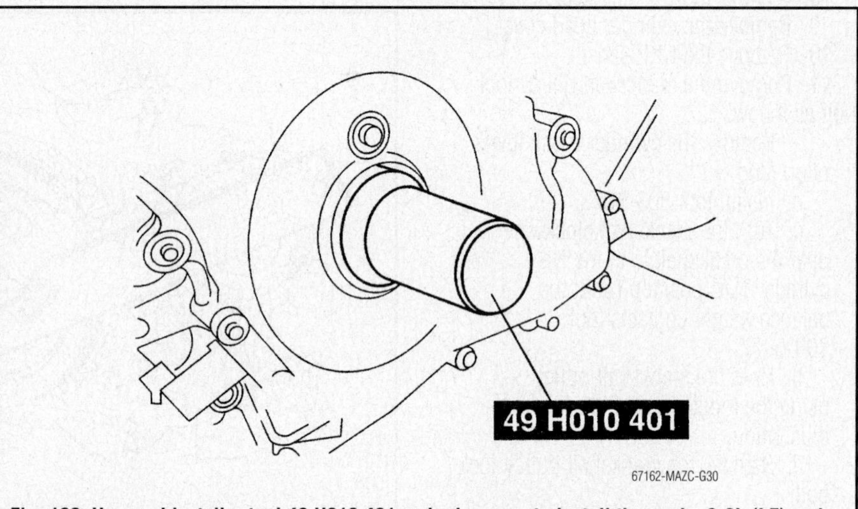

Fig. 182 Use seal installer tool 49 H010 401 and a hammer to install the seal—2.0L (LF) and 2.3L (L3) engines

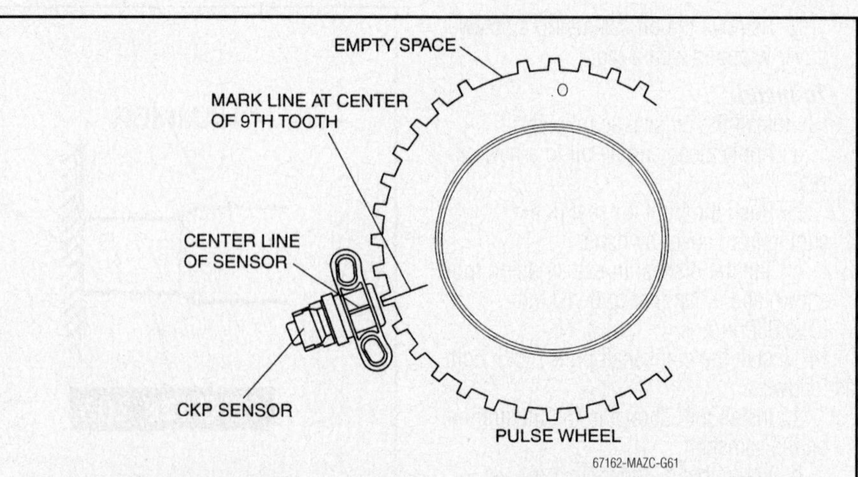

Fig. 183 Using a ruler, mark the center line on the pulse wheel teeth on the crank pulley which is located at the 9th tooth counting counterclockwise from the empty space—2.0L (LF) and 2.3L (L3) engines

17. Cut the seal lip using a suitable knife and using a suitable pry tool wrapped in a rag, remove the seal.

To install:

18. Coat the new seal with clean engine oil.
19. Push the seal in by hand to get it started.
20. Use seal installer tool 49 H010 401 and a hammer to install the seal so that it is recessed 0–0.019 inch (0–0.5mm) as shown in the accompanying illustration.
21. Install the crankshaft pulley.
22. Install tool 49 UN30 3465 as illustrated on the camshaft.
23. Install and hand tighten the M6 x 1.0 bolt as illustrated.
24. Turn the crankshaft clockwise until the crankshaft is in number 1 TDC (the balance weight should be attached to the tool).
25. Hold the crankshaft pulley and tighten the lock bolt in two steps. First tighten to 71–77 ft. lbs. (96–104 Nm). Then final tighten an additional 87–93 degrees.
26. Remove the M6 x 1.0 bolt.
27. Remove the tools from the camshaft and the cylinder block lower blind plug.
28. Rotate the crankshaft clockwise 2 turns until you reach TDC. If not aligned properly, loosen the crankshaft pulley lock bolt and reinstall the lock bolt again using the above procedure and special tools.
29. Install the cylinder block blind plug and tighten to 13–16 ft. lbs. (18–22 Nm).
30. When installing the CKP sensor perform the following:
    a. Remove the cylinder block lower blind plug.
    b. Install tool 49 JE01 061.
    c. Turn the crankshaft clockwise until the crankshaft is in the number 1 cylinder Top Dead Center position (the balance weight should be attached to tool 49 JE01 061).
    d. Using a ruler, mark the center line on the pulse wheel teeth on the crank pulley which is located at the 9th tooth counting counterclockwise from the empty space. Refer to the illustration for more detail.

**** CAUTION**

If you do not mark the center line correctly this will cause improper engine control for the ignition and fuel system, so be sure to mark the line carefully.

    e. Install the CKP sensor making sure the mark on the sensor is aligned with the mark on the pulse wheel. Tighten the sensor bolt to 49–66 inch lbs. (5.5–7.5 Nm)and attach the electrical connector.
    f. Remove the tool from the cylinder block lower blind plug.
    g. Rotate the crankshaft clockwise 2 turns until you reach TDC. If not aligned properly, loosen the crankshaft pulley lock bolt and reinstall the lock bolt.
    h. Install the cylinder block blind plug and tighten to 13–16 ft. lbs. (18–22 Nm).

31. Install the remaining components in the reverse order of removal.
32. Start the vehicle and check for leaks, repair if necessary.

3.0L (AJ) Engines

See Figures 184 through 189.

1. Before servicing the vehicle, refer to the precautions section.
2. Disconnect the negative battery cable.
3. Remove the accessory drive belt.
4. Hold the crankshaft pulley using the tools illustrated and remove the pulley bolt.
5. Remove the A/C compressor and set aside with the lines still attached.
6. Remove the crankshaft pulley using a suitable puller such as the one illustrated.
7. Remove the front oil seal using a pry tool as illustrated.

To install:

8. Assemble the seal using part (A) of tool 49 UN01 002 and tool 49 UN01 002 as illustrated.
9. Apply clean oil to the seal and push the seal in by hand.
10. Install the seal using the installation tools until the seal is recessed 0–0.039 inch (0–1mm).

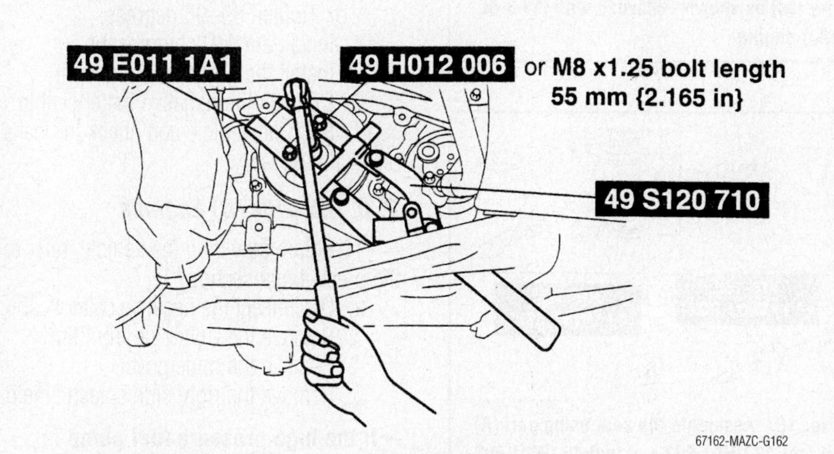

Fig. 184 Hold the crankshaft pulley using the tools illustrated and remove the pulley bolt—Mazda6 with the 3.0L (AJ) engine

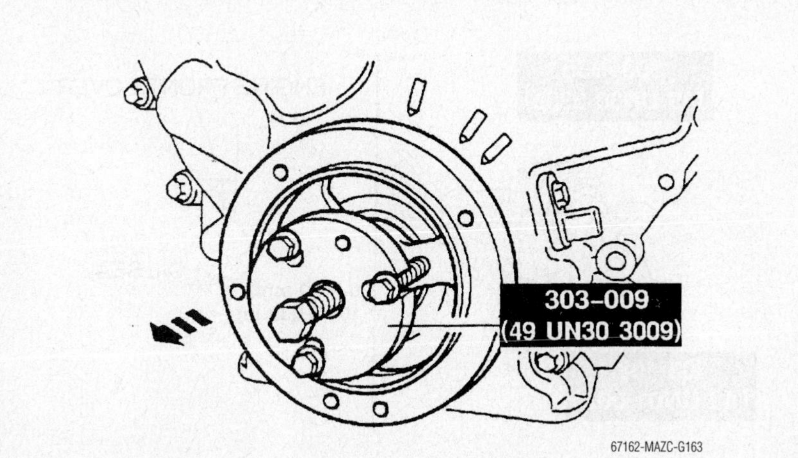

Fig. 185 Remove the crankshaft pulley using a suitable puller such as the one shown—Mazda6 with the 3.0L (AJ) engine

11. Seal the crankshaft pulley using a silicone sealant.
12. Install the pulley using the tools illustrated.
13. While holding the pulley, tighten the crankshaft pulley bolt as follows:
    a. Tighten to 88 ft. lbs. (120 Nm).
    b. Loosen one full turn.

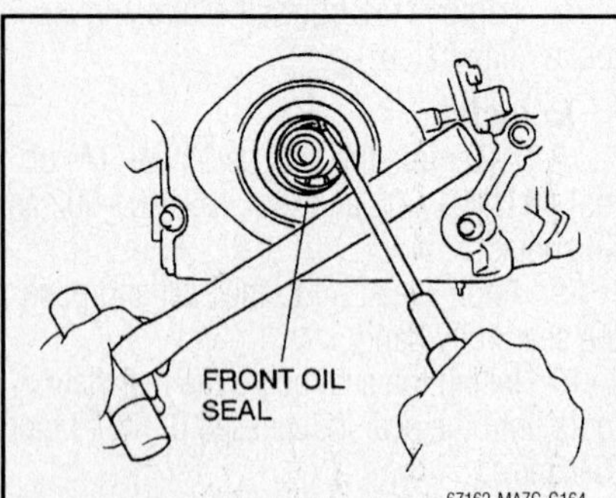

Fig. 186 Remove the front oil seal using a pry tool as shown—Mazda6 with the 3.0L (AJ) engine

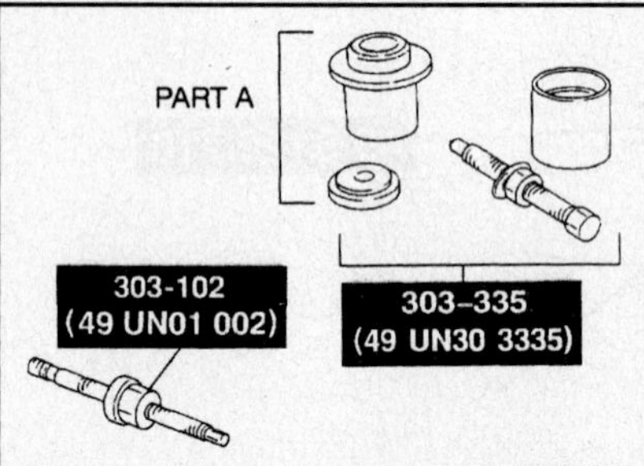

Fig. 187 Assemble the seal using part (A) of tool 49 UN01 002 and tool 49 UN01 002 as shown—Mazda6 with the 3.0L (AJ) engine

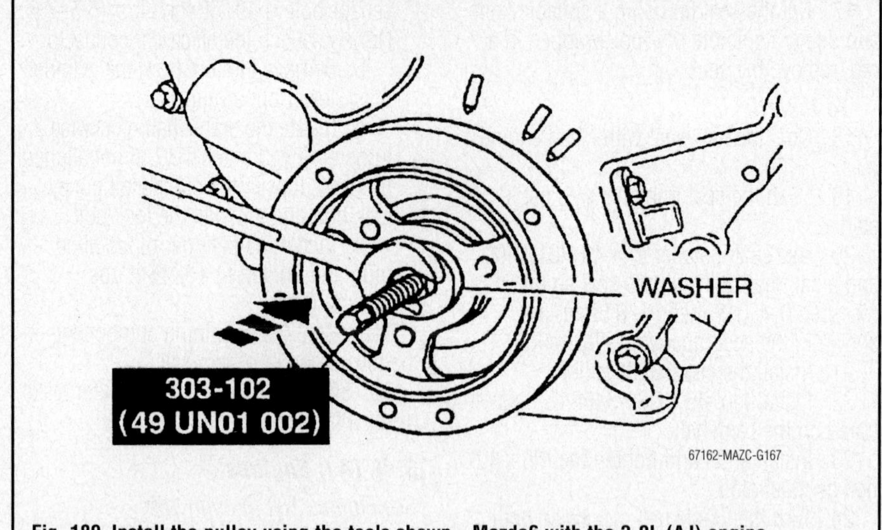

Fig. 189 Install the pulley using the tools shown—Mazda6 with the 3.0L (AJ) engine

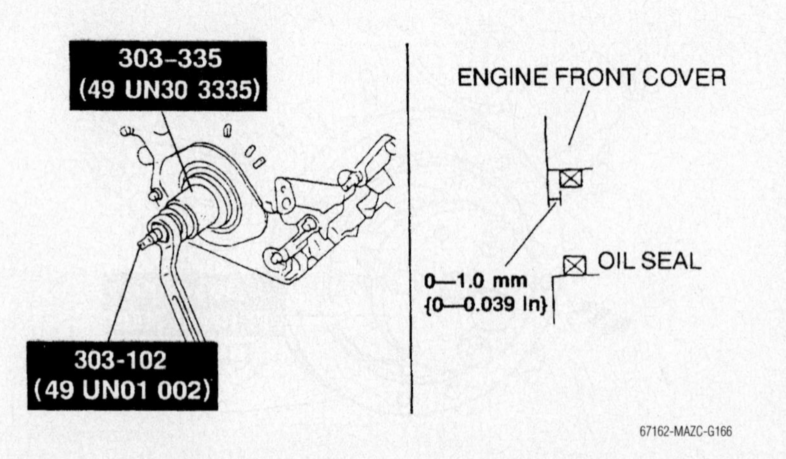

Fig. 188 Install the seal using the installation tools until the seal is recessed 0–0.039 inch (0–1mm)—Mazda6 with the 3.0L (AJ) engine

    c. Tighten to 35–39 ft. lbs. (47–53 Nm).
    d. Tighten 85–95 degrees.
14. Install the A/C compressor.
15. Install the accessory drive belt.
16. Connect the negative battery cable.
17. Start the vehicle and check for leaks, repair if necessary.

2.3L (L3 with TC) Engines

1. Before servicing the vehicle, refer to the precautions section.
2. Disconnect the negative battery cable.
3. Remove the right side front tire.
4. Remove the undercover.
5. Remove the right side splash shield.

➡If the high pressure fuel pump is removed, replace the O-ring with a new one.

6. Remove the charge air cooler cover.
7. Disconnect the spill valve control solenoid valve connector.
8. Disconnect the quick release connector on the high pressure fuel pump.
9. Remove the battery and battery tray.

⁂ WARNING

If the high pressure fuel pump joint nut is loosened, fuel leakage may occur resulting in death or serious injury, or damage to the equipment or the vehicle. Fuel can also irritate the skin and eyes. When removing the high pressure line pipe, always tighten the high pressure line pipe installation nut while fixing the high pressure fuel pump joint nut with a wrench. If the high pressure fuel pump joint nut has rotated, replace the high pressure fuel pump with a new one.

10. Disconnect the high pressure line pipe of the high pressure fuel pump.
11. Fix the joint nut with a wrench on the high pressure fuel pump side.
12. Loosen the high pressure line pipe installation nut.
13. Drain and recycle the engine coolant.
14. Loosen the water outlet case installation bolts securing the high pressure line pipe.
15. Remove the high pressure fuel pump.
16. Remove the high pressure fuel pump cover.
17. Remove the ignition coils, spark plugs, valve cover and accessory drive belt.
18. Remove the right hand halfshaft from the joint shaft.

19. Remove the Crankshaft Position (CKP) sensor.
20. Remove the crankshaft pulley bolt and pulley.
21. Remove the cylinder block lower blind plug.
22. Install tool 49 JE01 061.
23. Turn the crankshaft clockwise until the crankshaft is in the number 1 cylinder Top Dead Center position (the balance weight should be attached to tool 49 JE01 061).
24. Hold the crankshaft pulley using the tools illustrated and remove the bolt.
25. Remove the crankshaft pulley.
26. Cut the seal lip using a suitable knife and using a suitable pry tool wrapped in a rag, remove the seal.

To install:

27. Coat the new seal with clean engine oil.
28. Push the seal in by hand to get it started.
29. Use seal installer tool 49 H010 401 and a hammer to install the seal so that it is recessed 0–0.019 inch (0–0.5mm) as shown in the accompanying illustration.
30. Install the crankshaft pulley.
31. Install tool 49 UN30 3465 as illustrated on the camshaft.
32. Install and hand tighten the M6 x 1.0 bolt as illustrated.
33. Turn the crankshaft clockwise until the crankshaft is in number 1 TDC (the balance weight should be attached to the tool).
34. Hold the crankshaft pulley and tighten the lock bolt in two steps. First tighten to 71–77 ft. lbs. (96–104 Nm). Then final tighten an additional 87–93 degrees.
35. Remove the M6 x 1.0 bolt.
36. Remove the tools from the camshaft and the cylinder block lower blind plug.
37. Rotate the crankshaft clockwise 2 turns until you reach TDC. If not aligned properly, loosen the crankshaft pulley lock bolt and reinstall the lock bolt again using the above procedure and special tools.
38. Install the cylinder block blind plug and tighten to 13–16 ft. lbs. (18–22 Nm).
39. When installing the CKP sensor perform the following:
 a. Remove the cylinder block lower blind plug.
 b. Install tool 49 JE01 061.
 c. Turn the crankshaft clockwise until the crankshaft is in the number 1 cylinder Top Dead Center position (the balance weight should be attached to tool 49 JE01 061).
 d. Using a ruler, mark the center line on the pulse wheel teeth on the crank pulley which is located at the 9th tooth counting counterclockwise from the empty space. Refer to the illustration for more detail.

✲✲ CAUTION

If you do not mark the center line correctly this will cause improper engine control for the ignition and fuel system, so be sure to mark the line carefully.

 e. Install the CKP sensor making sure the mark on the sensor is aligned with the mark on the pulse wheel. Tighten the sensor bolt to 49–66 inch lbs. (5.5–7.5 Nm)and attach the electrical connector.
 f. Remove the tool from the cylinder block lower blind plug.
 g. Rotate the crankshaft clockwise 2 turns until you reach TDC. If not aligned properly, loosen the crankshaft pulley lock bolt and reinstall the lock bolt.
 h. Install the cylinder block blind plug and tighten to 13–16 ft. lbs. (18–22 Nm).

40. Install the high pressure fuel pump cover and tighten to 13–16 ft. lbs. (18–22 Nm).

✲✲ CAUTION

If the high pressure fuel pump installation bolts are tightened with the high pressure fuel pump tilted, the high pressure fuel pump may not operate correctly. Tighten the high pressure fuel pump installation bolts in a few passes with equal torque. Tighten to 76–101 inch lbs. (8.5–11.5 Nm).

✲✲ WARNING

If the high pressure fuel pump joint nut is loosened, fuel leakage may occur resulting in death or serious injury, or damage to the equipment or the vehicle. Fuel can also irritate the skin and eyes. When installing the high pressure line pipe, always tighten the high pressure line pipe installation nut while fixing the high pressure fuel pump joint nut with a wrench. If the high pressure fuel pump joint nut has rotated, replace the high pressure fuel pump with a new one.

41. Assemble the high pressure line pipe.
42. Fix the joint nut with a wrench on the high pressure fuel pump side. Tighten the high pressure line pipe installation nut to 17–26 ft. lbs. (23–35 Nm).
43. Tighten the water outlet case installation bolts to 13–16 ft. lbs. (18–22 Nm).
44. Install the quick release connector.
45. Install the remaining components in the reverse order of removal.
46. Verify that the high pressure fuel pump is assembled securely.
47. Drive the vehicle starting from a standstill and brake suddenly five to six times at a low speed.
48. Stop the vehicle and verify from outside the vehicle that there is no fuel leakage around the high pressure fuel pump.

CYLINDER HEAD

REMOVAL & INSTALLATION

MX-5 Miata

See Figures 190 through 195.

1. Before servicing the vehicle, refer to the precautions section.
2. Properly relieve the fuel system pressure.
3. Remove the battery and battery tray.
4. Drain the engine coolant.
5. Remove the front suspension tower bar.
6. Remove the air cleaner.
7. Remove the dynamic chamber.
8. Remove the ignition coil.
9. Remove the drive belt.
10. Remove the CKP sensor.
11. Remove the P/S oil pump with the oil hose still connected and position the P/S oil pump so that it is out of the way.
12. Remove the timing chain.
13. Remove the wiper arm.
14. Remove the cowl grille.
15. Remove the side cowl grille.
16. Remove the service hole cover.
17. Disconnect the alternator, but do not remove it from the vehicle.
18. Remove the exhaust manifold.
19. Remove the OCV sensor, if equipped with variable valve timing.
20. Mark the camshaft cap locations so they may be reinstalled in their original positions.
21. Remove camshafts.
22. Mark the cylinder head bolt locations so they may be reinstalled in their original positions.

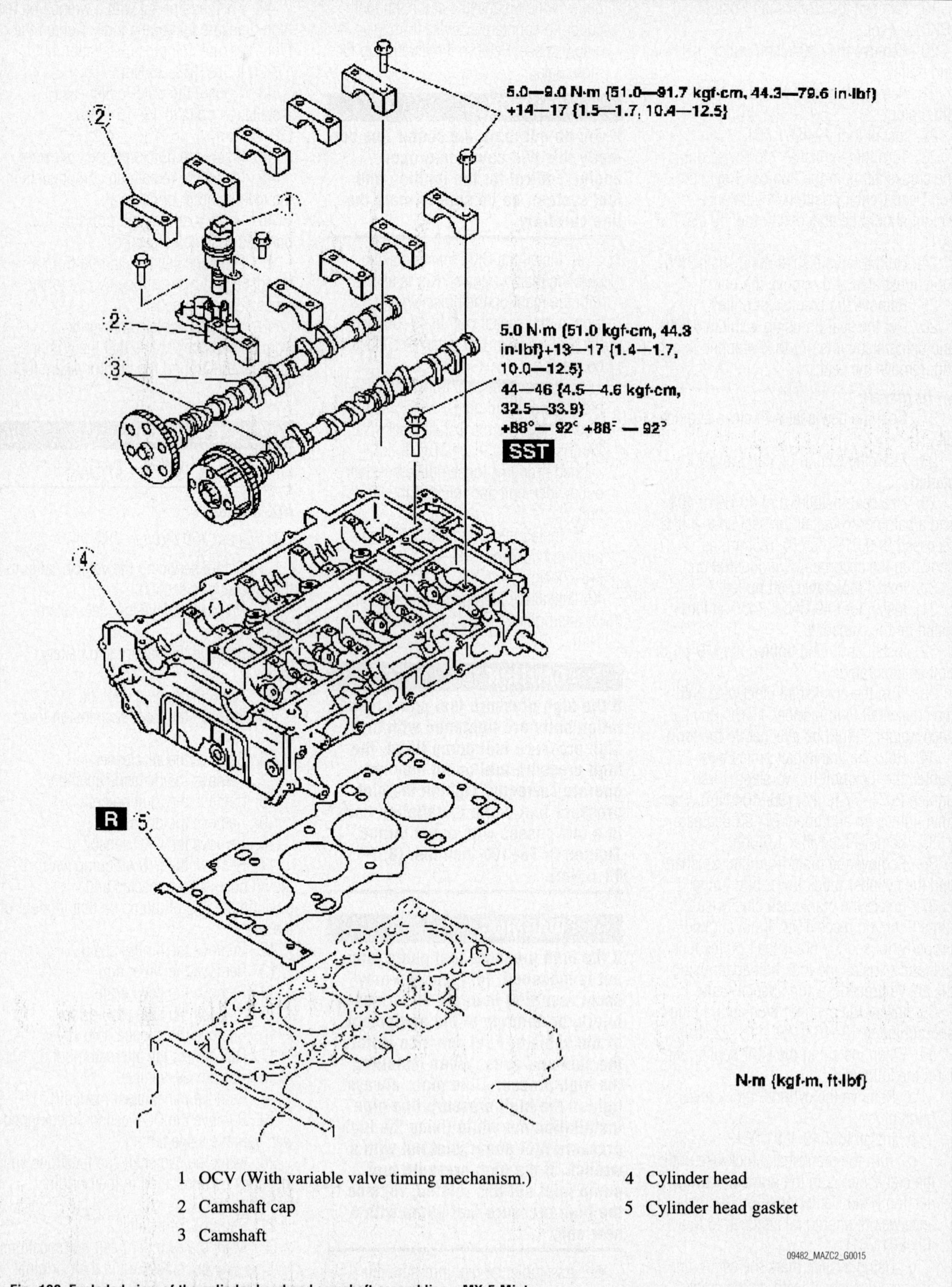

Fig. 190 Exploded view of the cylinder head and camshaft assemblies—MX-5 Miata

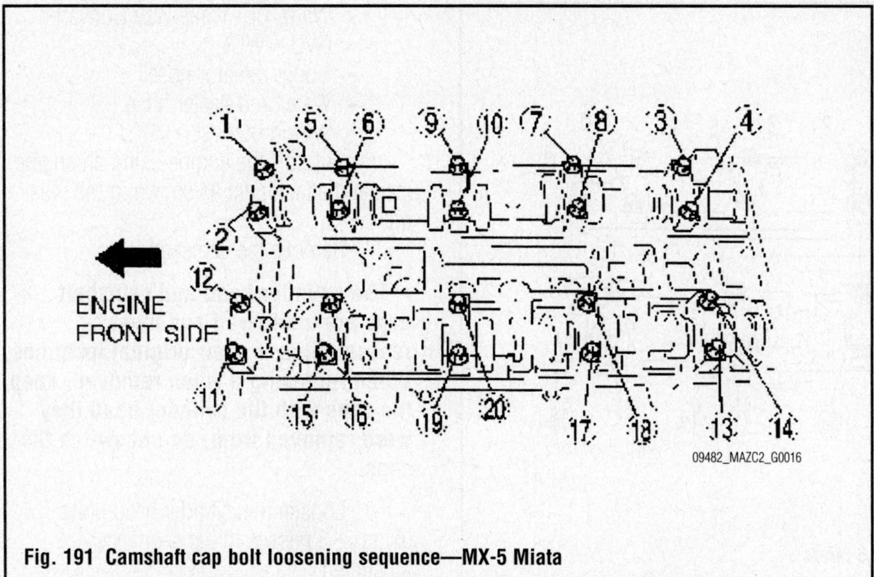

Fig. 191 Camshaft cap bolt loosening sequence—MX-5 Miata

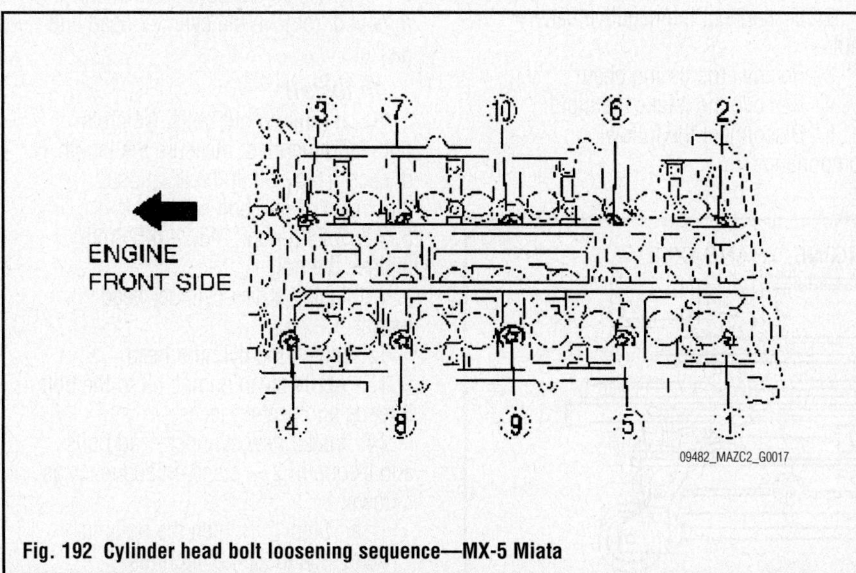

Fig. 192 Cylinder head bolt loosening sequence—MX-5 Miata

23. Loosen the cylinder head bolts in the sequence illustrated using two or three passes.
24. Remove the cylinder head.
25. Measure the length of the cylinder head bolts and replace any bolts that exceed the maximum length specification of 146. mm (5.77 in.).

To install:

26. Install the cylinder heads.
27. Install the cylinder head bolts in their original locations.
28. Tighten the cylinder head bolts in the sequence illustrated using 5 passes as follows:
    a. Step 1: Tighten to 5 Nm (44.3 in. lbs.)
    b. Step 2: Tighten 13–17 Nm (10–12.5 ft. lbs.)
    c. Step 3: Tighten 44–46 Nm (32.5–33.9 ft. lbs.)
    d. Step 4: Tighten and additional 88–92 degrees.
    e. Step 5: Tighten and additional 88–92 degrees.
29. Set the cam position of the number 1 cylinder to top dead center, install the camshafts, making sure to install the caps in their original locations.
30. Tighten the camshaft cap bolts using two to three steps in the sequence illustrated to 5–9 Nm (44.3–79.6 in. lbs.) and then to 14–17 Nm (10.4–12.5 ft. lbs.)
31. Install the remaining components in the reverse order of removal
32. Start the engine and inspect for leaks.
33. Check the ignition timing, idle speed and idle mixture.

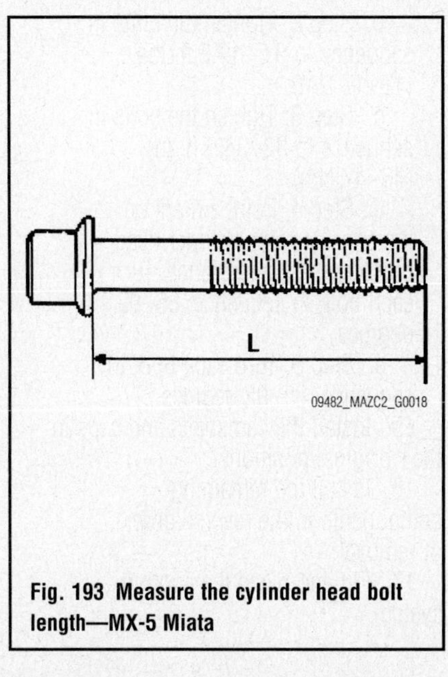

Fig. 193 Measure the cylinder head bolt length—MX-5 Miata

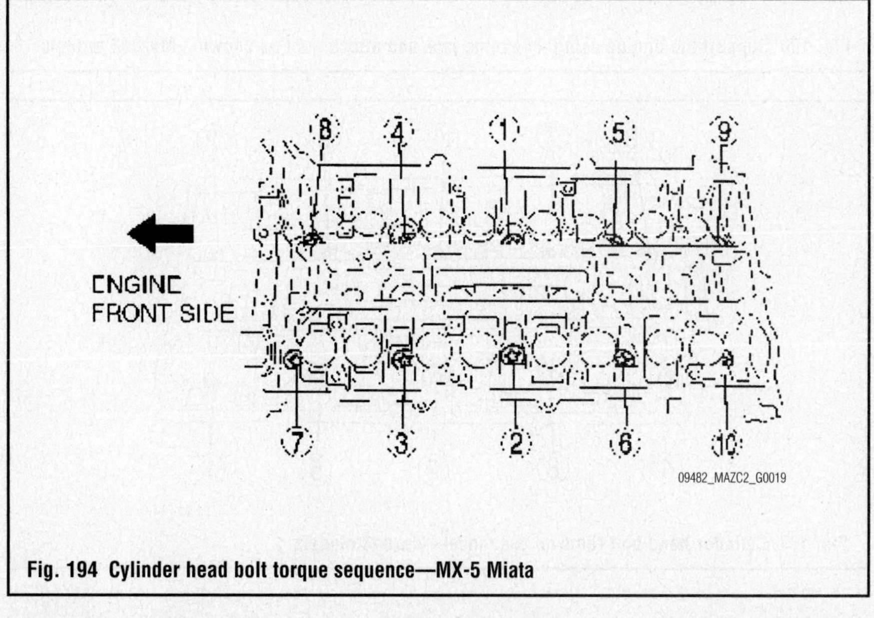

Fig. 194 Cylinder head bolt torque sequence—MX-5 Miata

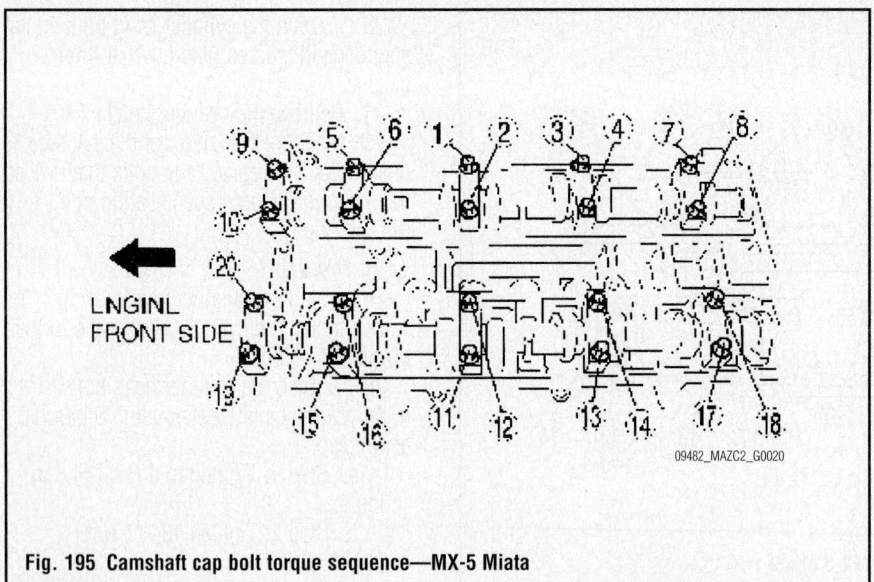

Fig. 195 Camshaft cap bolt torque sequence—MX-5 Miata

Mazda3

2.0L (LF) and 2.3L (L3) Engines

See Figures 196 through 199.

1. Before servicing the vehicle, refer to the precautions section.
2. Disconnect the negative battery cable.
3. Remove the timing chain.
4. Remove the intake manifold.
5. Disconnect the following components:

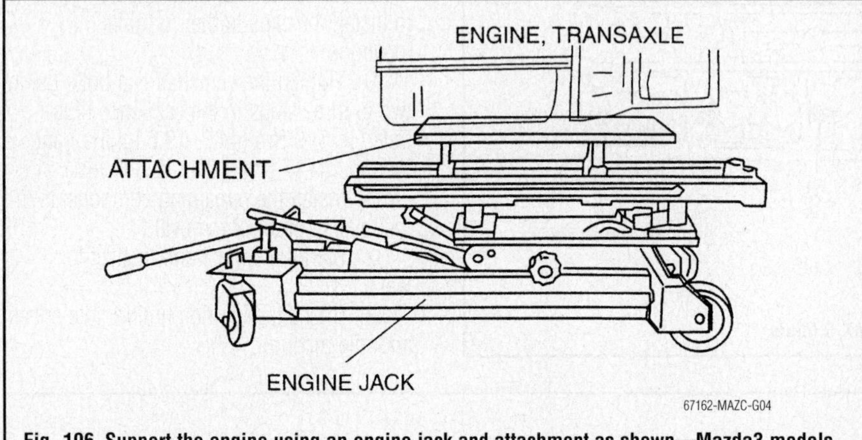

Fig. 196 Support the engine using an engine jack and attachment as shown—Mazda3 models

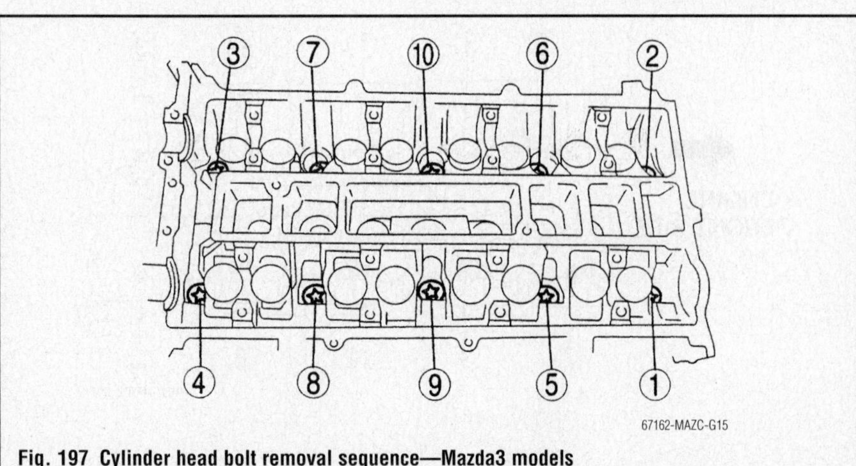

Fig. 197 Cylinder head bolt removal sequence—Mazda3 models

- Warm Up–Three Way Converter (WU–TWC)
- Upper radiator hose
- Water and heater hose
- Wiring harness

6. Support the engine using an engine jack and attachment as shown in the illustration.
7. Remove the camshafts.

➡The cylinder head and camshaft caps are numbered and must be reassembled in their original locations. When removing a when removed, keep the caps with the cylinder head they were removed from, do not switch the caps.

8. Loosen the cylinder head bolts using 2–3 passes in the sequence illustrated.
9. Once all the bolts have been removed, remove the cylinder head and gasket.

To install:

10. If reusing old bolts, which is not recommended; measure the length of each head bolt at the locations illustrated. If any bolt exceeds 5.717–5.740 inch (145.2–145.8mm) replace the bolt.
11. Install a new cylinder head gasket.
12. Install the cylinder head.
13. Apply clean engine oil to the bolt threads and seating faces.
14. Install new cylinder head bolts and torque in 2–3 steps, in sequence as follows:
    a. Step 1: Tighten the bolts in sequence to 27–97 inch lbs. (3–11 Nm).
    b. Step 2: Tighten the bolts in sequence to 9.5–12.5 ft. lbs. (13–17 Nm).
    c. Step 3: Tighten the bolts in sequence to 32–34.5 ft. lbs. (43–47 Nm).
    d. Step 4: Paint a mark on the edge of each cylinder head bolt to use as a reference. Turn each bolt, in sequence, 85–92 degrees.
    e. Step 5: Turn each bolt, in sequence, 85–92 degrees.
15. Install the camshafts and caps in their original positions.
16. Install the remaining components in the reverse order of removal.
17. Fill and bleed the cooling system.

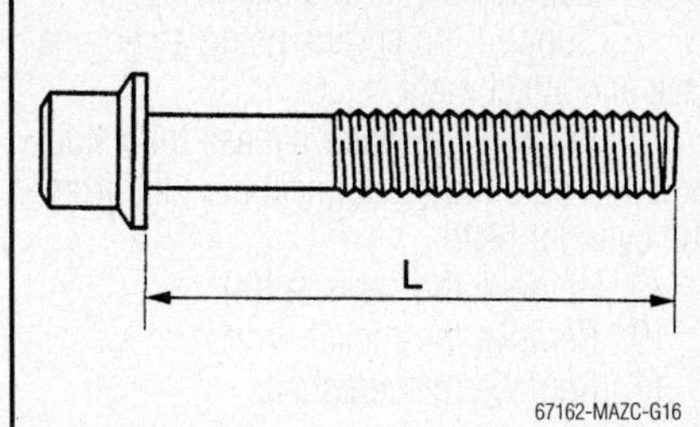

Fig. 198 Measure the length of each head bolt at the locations illustrated, if reusing the old bolts. If any bolt exceeds 5.717–5.740 inch (145.2–145.8mm) replace the bolt—Mazda3 models

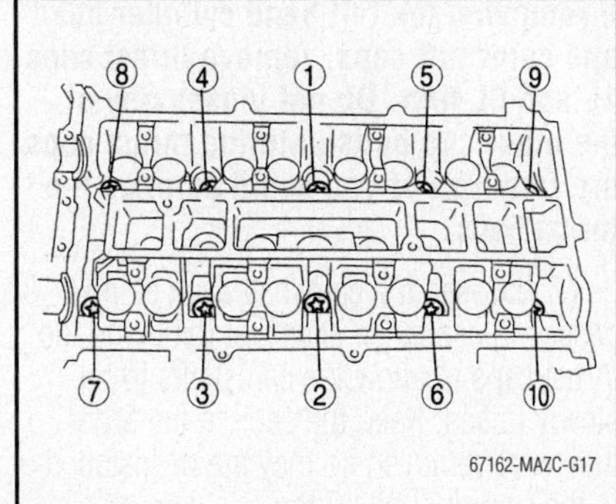

Fig. 199 Cylinder head bolt tightening sequence—Mazda3 models

18. Change the oil and filter.
19. Run the engine and check for proper operation.

Mazda6

2.3L (L3) Engines

See Figures 196 and 200 through 202.

1. Before servicing the vehicle, refer to the precautions section.
2. Disconnect the negative battery cable.
3. Remove the timing chain.
4. Remove the ignition coil.
5. Unbolt the alternator and move it aside with the connectors still attached.
6. Remove the front exhaust pipe.
7. Remove the intake manifold.
8. Disconnect the following components:
   - Upper radiator hose
   - Water and heater hose
9. Support the engine using an engine jack and attachment as shown in the illustration.
10. Remove the oil control valve.
11. Remove the camshafts.

➡The cylinder head and camshaft caps are numbered and must be reassembled in their original locations. When removing a when removed, keep the caps with the cylinder head they were removed from, do not switch the caps.

12. Loosen the cylinder head bolts using 2–3 passes in the sequence illustrated.
13. Once all the bolts have been removed, remove the cylinder head and gasket.

To install:

14. If reusing old bolts, which is not recommended; measure the length of each head bolt at the locations illustrated. If any bolt exceeds 5.717–5.740 inch (145.2–145.8mm) replace the bolt.
15. Install a new cylinder head gasket.
16. Install the cylinder head.
17. Apply clean engine oil to the bolt threads and seating faces.
18. Install new cylinder head bolts and torque in 2–3 steps, in sequence as follows:
    a. Step 1: Tighten the bolts in sequence to 27–97 inch lbs. (3–11 Nm).

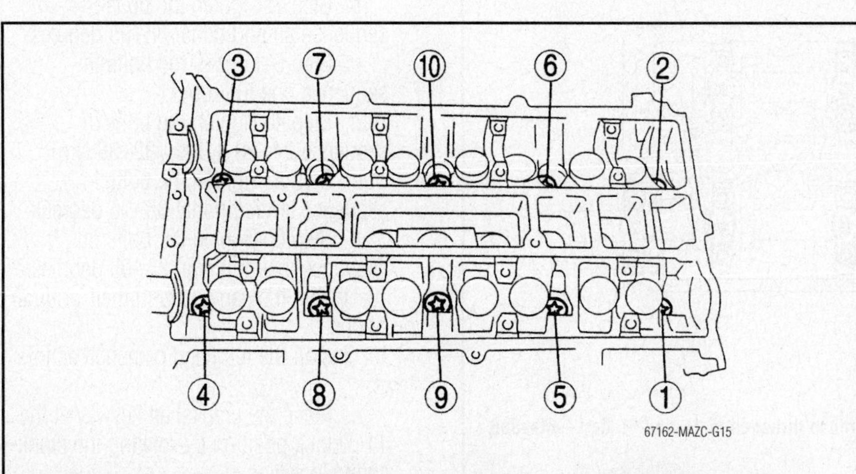

Fig. 200 Cylinder head bolt removal sequence—Mazda6 and Mazdaspeed6 models

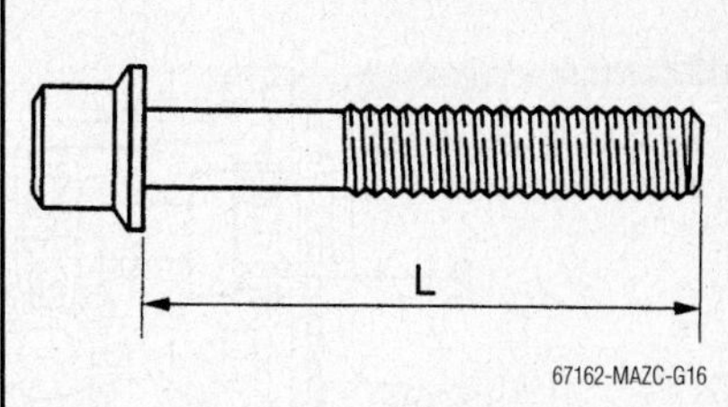

Fig. 201 Measure the length of each head bolt at the locations illustrated, if reusing the old bolts. If any bolt exceeds 5.717–5.740 inch (145.2–145.8mm) replace the bolt—Mazda6 models

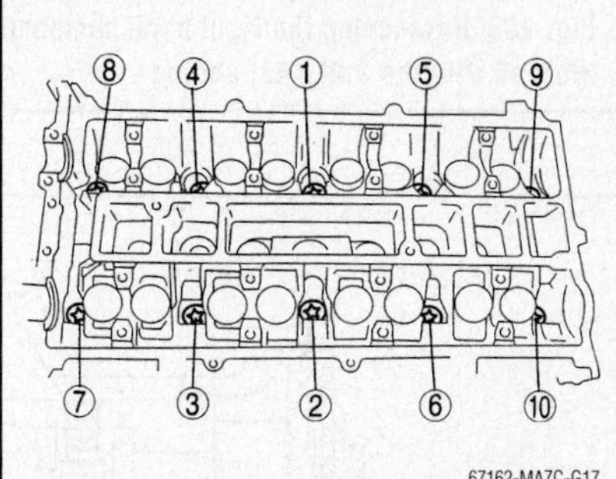

Fig. 202 Cylinder head bolt tightening sequence—Mazda6 and Mazdaspeed6 with the 2.3L (L3) engine

    b. Step 2: Tighten the bolts in sequence to 9.5–12.5 ft. lbs. (13–17 Nm).
    c. Step 3: Tighten the bolts in sequence to 32–34.5 ft. lbs. (43–47 Nm).
    d. Step 4: Paint a mark on the edge of each cylinder head bolt to use as a reference. Turn each bolt, in sequence, 85–92 degrees.
    e. Step 5: Turn each bolt, in sequence, 85–92 degrees.
19. Install the camshafts and caps in their original positions.
20. Install the remaining components in the reverse order of removal.
21. Fill and bleed the cooling system.
22. Change the oil and filter.
23. Run the engine and check for proper operation.

3.0L (AJ) Engines

See Figures 203 through 214.

1. Before servicing the vehicle, refer to the precautions section.
2. Properly relieve the fuel system pressure.
3. Drain the cooling system.
4. Disconnect the negative battery cable.
5. Remove the water pump.

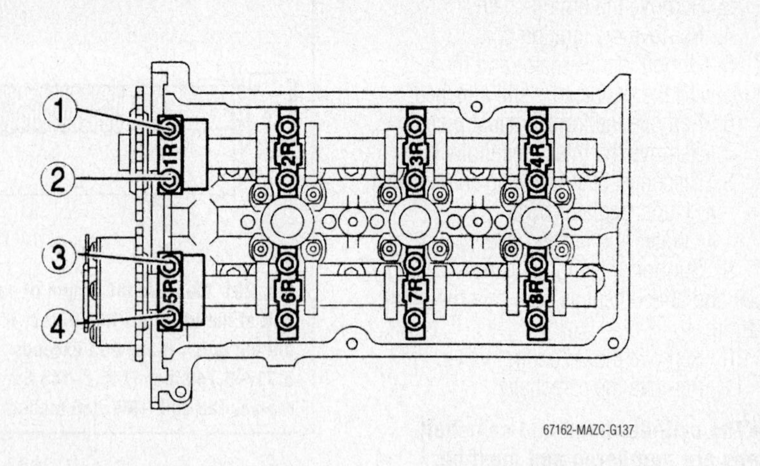

Fig. 203 If removing the right hand camshaft caps, remove thrust caps 1R and 5R first—Mazda6 with the 3.0L (AJ) engine

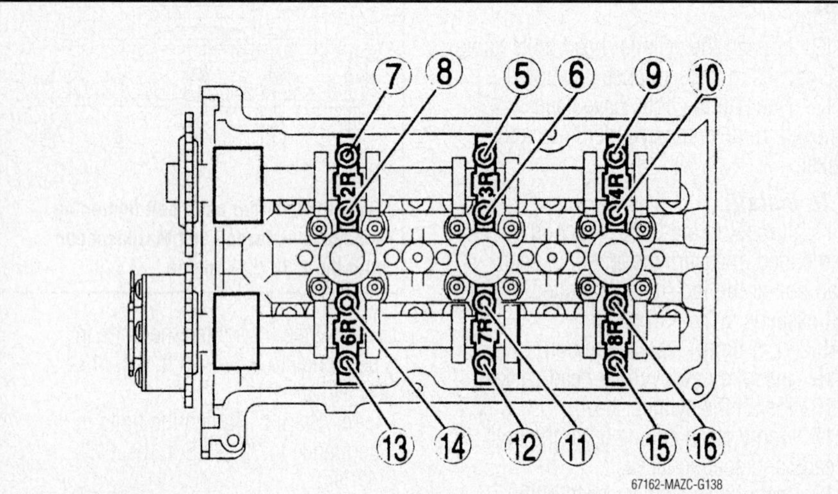

Fig. 204 Loosen the right hand camshaft caps using 7–8 passes in this sequence after first removing the thrust caps—Mazda6 with the 3.0L (AJ) engine

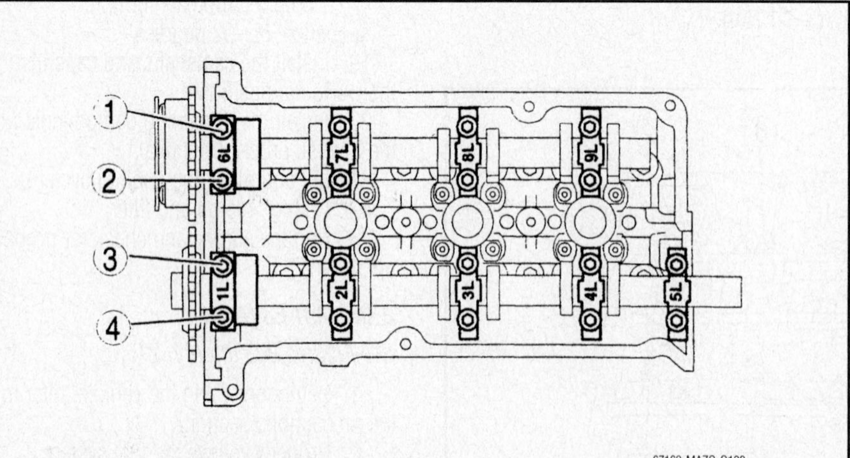

Fig. 205 If removing the left hand camshaft caps, remove thrust caps 1L and 6L first—Mazda6 with the 3.0L (AJ) engine

6. Remove the timing chain.
7. Support the engine using an engine jack and attachment.
8. Remove the water bypass tube stud bolt and bolt and disconnect the tube from the cylinder head.
9. Remove the camshaft(s).
10. Remove the rocker arm(s).
11. Remove the camshafts.

➡If removing the right hand cylinder head and camshaft caps, remove thrust caps 1R and 5R first. Do not loosen any of the other cap bolts until the thrust caps are removed or you could damage the thrust caps.

If removing the left hand cylinder head and camshaft caps, remove thrust caps 1L and 6L first. Do not loosen any of the other cap bolts until the thrust caps are removed or you could damage the thrust caps.

12. Loosen the camshaft caps using 7–8 passes in sequence after first removing the thrust caps to allow the camshafts to be slowly raised. Keep the caps in the order they were removed so they are re-installed in their original positions.
13. Loosen the cylinder head bolts using 2–3 passes in the sequence illustrated.
14. Once all the bolts have been removed, remove the cylinder head and gasket.

To install:

15. Install a new head gasket and the cylinder head.
16. Lubricate the cylinder head bolt threads.
17. Torque the cylinder head bolts in the proper sequence as follows:
    a. Step 1: Tighten the bolts in sequence 24–28 ft. lbs. (32–38 Nm).
    b. Step 2: Tighten the bolts in sequence an additional 85–95 degrees.
    c. Step 3: Loosen the bolts in sequence one full turn.
    d. Step 4: Tighten the bolts in sequence 24–28 ft. lbs. (32–38 Nm).
    e. Step 5: Tighten the bolts in sequence an additional 85–95 degrees.
    f. Step 6: Tighten the bolts in sequence an additional 85–95 degrees.
18. Install the camshafts in their original positions.
19. Install the left hand camshaft as follows:
    a. Place the crankshaft keyway at the 11 o'clock position by rotating the crankshaft clockwise.

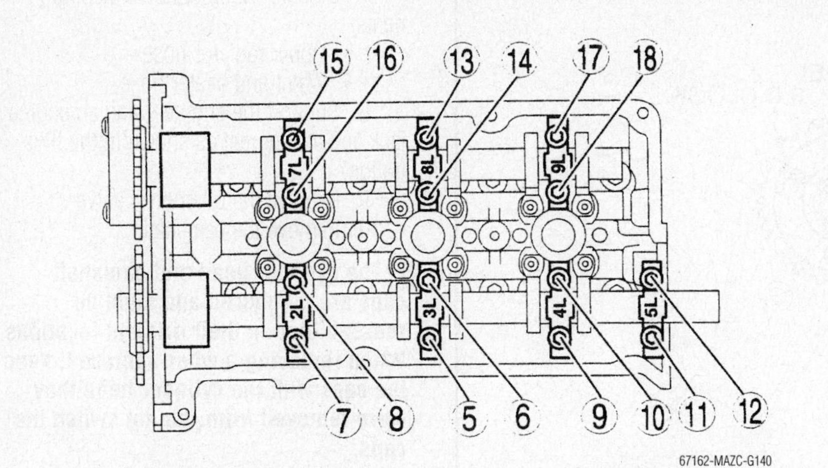

Fig. 206 Loosen the left hand camshaft caps using 7–8 passes in this sequence after first removing the thrust caps—Mazda6 with the 3.0L (AJ) engine

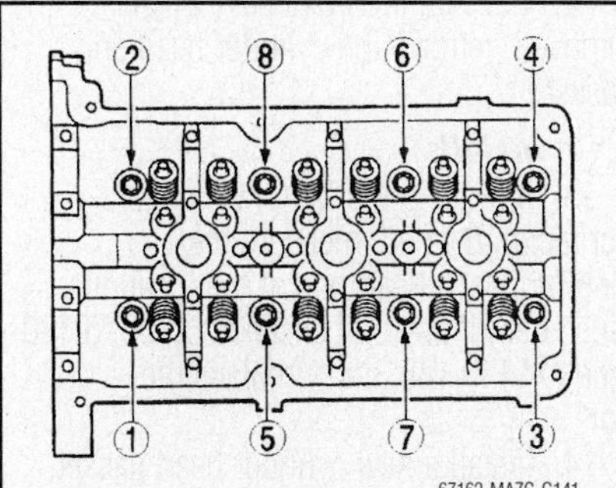

Fig. 207 Cylinder head bolt loosening sequence—Mazda6 with the 3.0L (AJ) engine

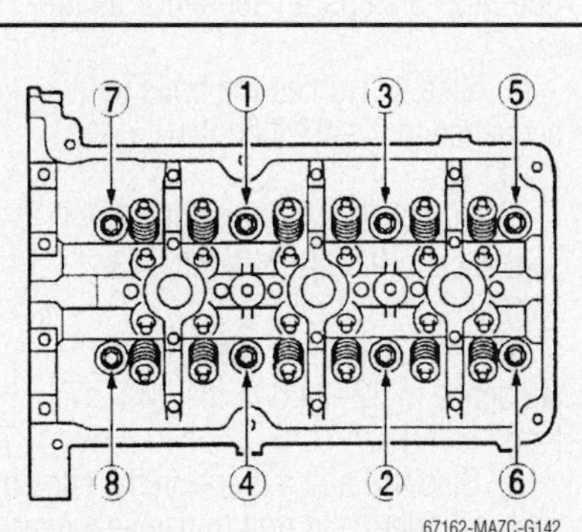

Fig. 208 Cylinder head bolt torque sequence—Mazda6 with the 3.0L (AJ) engine

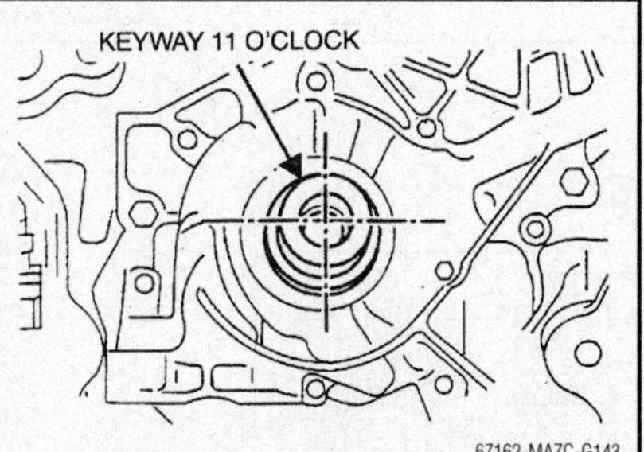

Fig. 209 When installing the left hand camshaft place the crankshaft keyway at the 11 o'clock position—Mazda6 with the 3.0L (AJ) engine

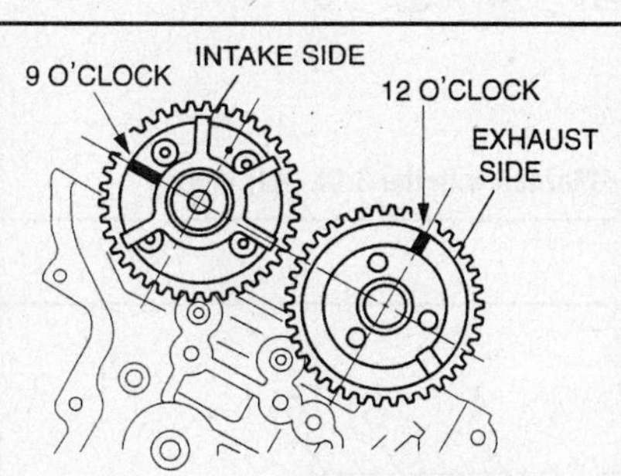

Fig. 210 Install the left hand camshaft and position the mark on the intake camshaft to the 9 o'clock and the exhaust camshaft to the 12 o'clock position—Mazda6 with the 3.0L (AJ) engine

b. Install the camshaft and the caps in their original positions and hand tighten the bolts.

c. Position the mark on the intake camshaft to the 9 o'clock position as illustrated.

d. Position the mark on the exhaust camshaft to the 12 o'clock position.

e. Install the timing chain.

➡Tighten the camshaft journal thrust caps last to avoid damaging the thrust caps.

f. Tighten the camshaft caps using several passes to 71–106 inch lbs. (8–12 Nm). After adjust the camshaft endplay using the thrust caps 1L and 6L, tighten the remaining caps.

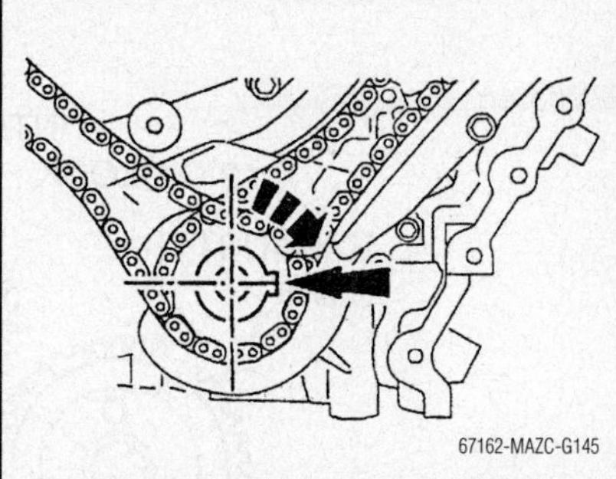

Fig. 211 When installing the right hand camshaft place the crankshaft keyway at the 3 o'clock position—Mazda6 with the 3.0L (AJ) engine

20. Install the right hand camshaft as follows:

a. Install the camshaft and the caps in their original positions and hand tighten the bolts.

b. Place the crankshaft keyway at the 3 o'clock position.

c. Position the mark on the exhaust camshaft to the 12 o'clock position by rotating the crankshaft clockwise.

d. Position the mark on the intake camshaft to the 3 o'clock position as illustrated.

e. Install the timing chain.

➡Tighten the camshaft journal thrust caps last to avoid damaging the thrust caps.

f. Tighten the camshaft caps using several passes to 71–106 inch lbs. (8–12 Nm). After adjust the camshaft endplay using the thrust caps 1R and 5R, tighten the remaining caps.

21. Install the water bypass tube.
22. Remove the engine jack and attachment.
23. Install the water pump.
24. Fill the cooling system.
25. Fill and bleed the cooling system.
26. Change the oil and filter.
27. Run the engine and check for proper operation.

2.3L (L3 with TC) Engines

1. Before servicing the vehicle, refer to the precautions section.
2. Disconnect the negative battery cable.
3. Remove the timing chain.
4. Unbolt the alternator and move it aside with the connectors still attached.
5. Remove the front exhaust pipe.
6. Remove the intake manifold.

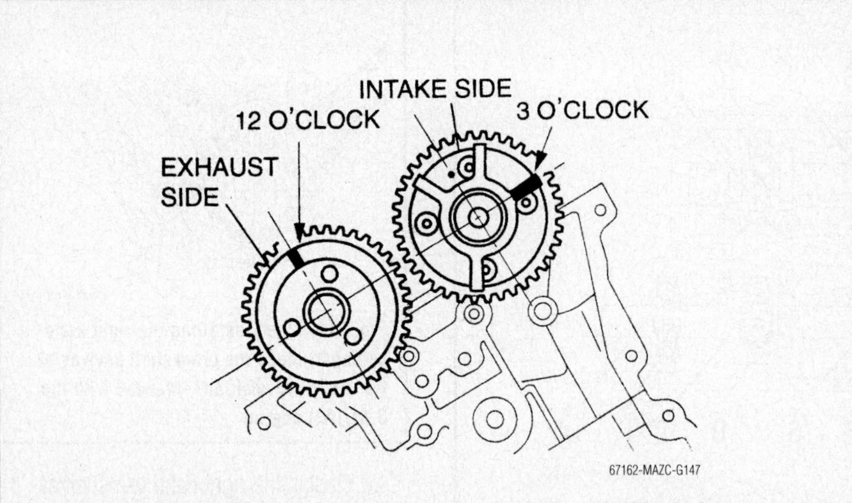

Fig. 212 Install the right hand camshaft and position the mark on the intake camshaft to the 3 o'clock and the exhaust camshaft to the 12 o'clock position—Mazda6 with the 3.0L (AJ) engine

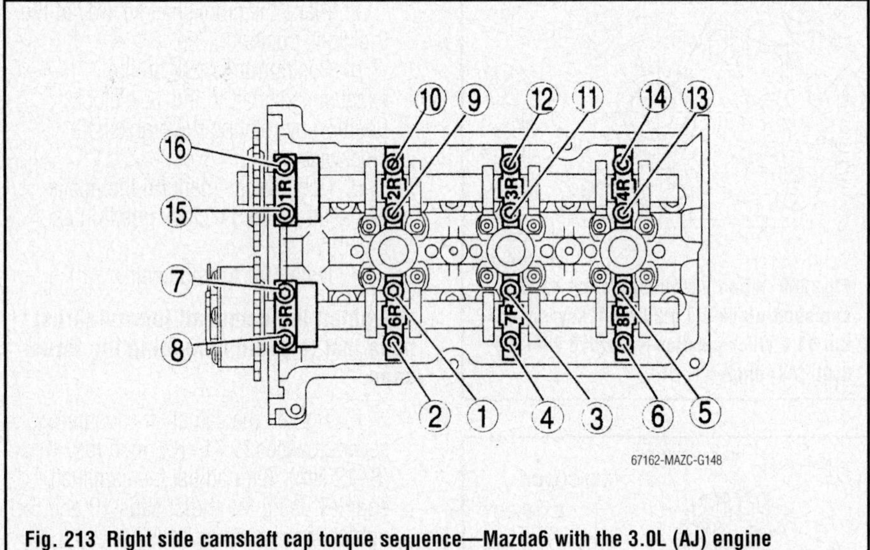

Fig. 213 Right side camshaft cap torque sequence—Mazda6 with the 3.0L (AJ) engine

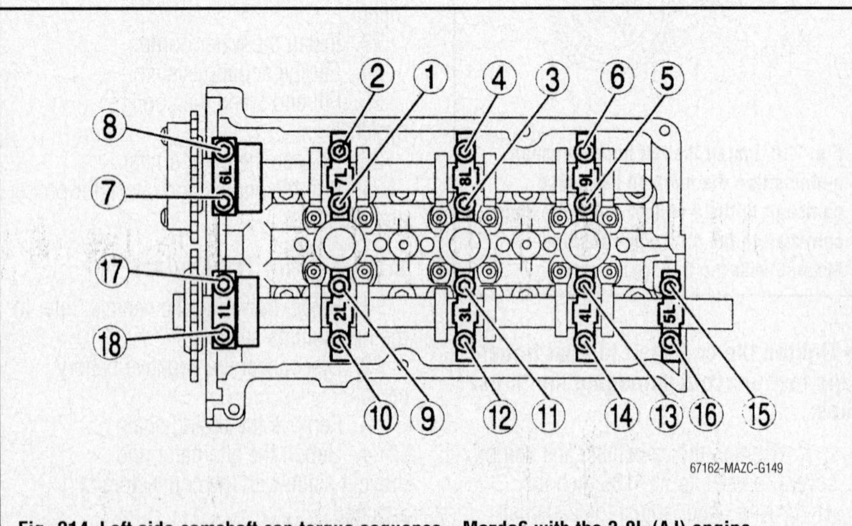

Fig. 214 Left side camshaft cap torque sequence—Mazda6 with the 3.0L (AJ) engine

7. Disconnect the following components:
   - Upper radiator hose
   - Water and heater hose
8. Support the engine using an engine jack and attachment as shown in the illustration.
9. Remove the oil control valve.
10. Remove the camshafts.

➡The cylinder head and camshaft caps are numbered and must be reassembled in their original locations. When removing a when removed, keep the caps with the cylinder head they were removed from, do not switch the caps.

11. Loosen the cylinder head bolts using 2–3 passes in the sequence illustrated.
12. Once all the bolts have been removed, remove the cylinder head and gasket.

To install:

13. If reusing old bolts, which is not recommended; measure the length of each head bolt at the locations illustrated. If any bolt exceeds 5.700–5.720 inch (144.7–145.3mm) replace the bolt.
14. Install a new cylinder head gasket.
15. Install the cylinder head.
16. Apply clean engine oil to the bolt threads and seating faces.
17. Install new cylinder head bolts and torque in 2–3 steps, in sequence as follows:
   a. Step 1: Tighten the bolts in sequence to 27–97 inch lbs. (3–11 Nm).
   b. Step 2: Tighten the bolts in sequence to 9.6–12.5 ft. lbs. (13–17 Nm).
   c. Step 3: Tighten the bolts in sequence to 32–34.6 ft. lbs. (43–47 Nm).
   d. Step 4: Paint a mark on the edge of each cylinder head bolt to use as a reference. Turn each bolt, in sequence, 88–92 degrees.
   e. Step 5: Turn each bolt, in sequence, 88–92 degrees.
18. Install the camshafts and caps in their original positions.
19. Install the remaining components in the reverse order of removal.
20. Fill and bleed the cooling system.
21. Change the oil and filter.
22. Run the engine and check for proper operation.

ENGINE ASSEMBLY

REMOVAL & INSTALLATION

MX-5 Miata

See Figures 215 through 225.

➡The engine, transmission, and cross-member assembly is removed as a unit from under the vehicle.

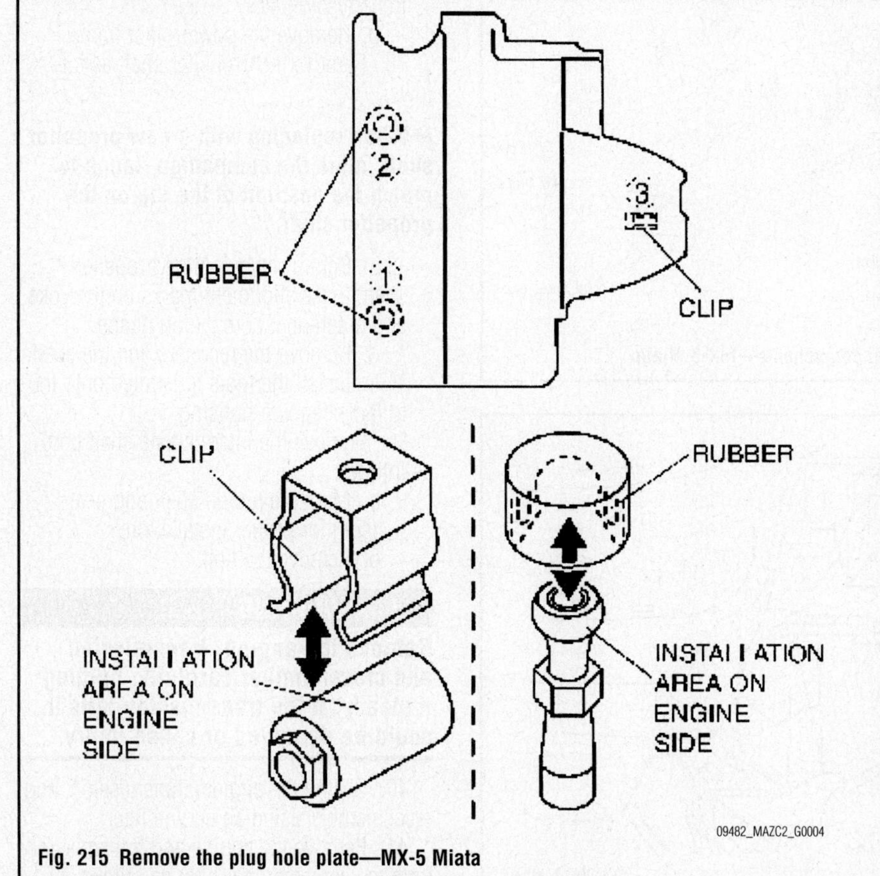

Fig. 215 Remove the plug hole plate—MX-5 Miata

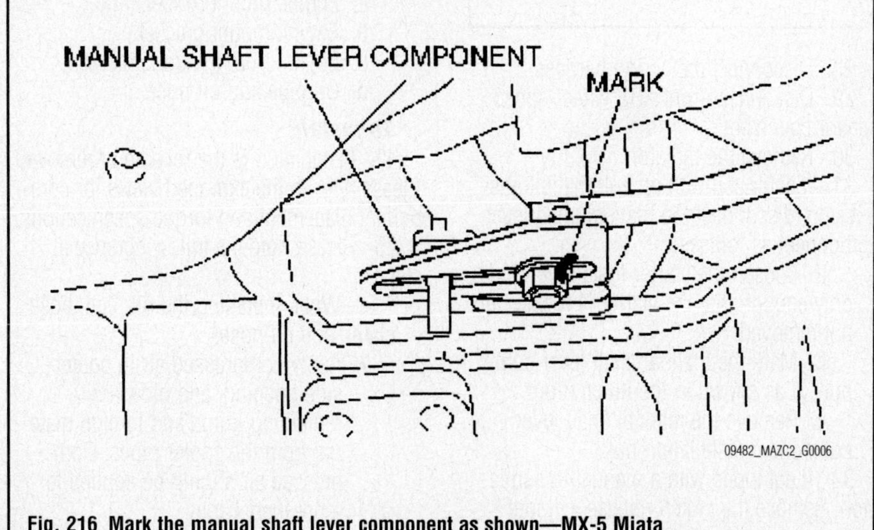

Fig. 216 Mark the manual shaft lever component as shown—MX-5 Miata

1. Before servicing the vehicle, refer to the precautions section.
2. Remove the fuel-filler cap to release the pressure inside the fuel tank.
3. Remove the fuel pump relay.
4. Start the engine.
5. After the engine stalls, crank the engine several times .
6. Turn the ignition switch to the LOCK position.
7. Install the fuel pump relay
8. Drain the engine coolant.
9. Remove front tires.
10. Remove the plug hole plate by lifting off and removing the plug hole plate from the installation areas (rubber and clips) as shown in the illustration.
11. Remove the battery cover, battery, battery box, battery tray and battery duct
12. Remove the air cleaner.
13. Remove the throttle body.
14. Remove the PCM, PCM duct and air cleaner insulator, please perform the following to remove the PCM:
    a. Connect the WDS or equivalent to the DLC-2.
    b. Set up the WDS or equivalent (including the vehicle recognition).
    c. Select "Module Programming".
    d. Select "Programmable Module Installation".
    e. Select "PCM" and perform procedures according to directions on the WDS or equivalent screen.

➡If the PCM is replaced with a new one, the PCM stores DTC P0602 and illuminates the MIL even though no malfunction is detected. This means the PCM has not been configured yet.

    f. Retrieve DTC's by the WDS or equivalent, then verify that there is no DTC present.
    g. If DTC is present, perform an applicable DTC inspection.
    h. Move the water hose from the PCM cover slightly out of the way.
    i. Remove the PCM cover, connector and PCM.
15. Remove the coolant reserve tank.
16. Remove the center console by removing the components as follows:
    a. Selector lever knob if equipped with an automatic transmission or the shift lever knob if equipped with a manual transmission.
    b. Front cover.
    c. Rear cover.
    d. Hole cover.
    e. Parking brake lever boot.
    f. Parking brake lever boot plate.
    g. Console.
    h. Power window main switch connector.
    i. Indicator compartment connector if equipped with an automatic transmission.
17. Disconnect the P/S oil pump hoses and drain the P/S fluid reservoir.
18. Remove the splash shield, undercover and mud guards.

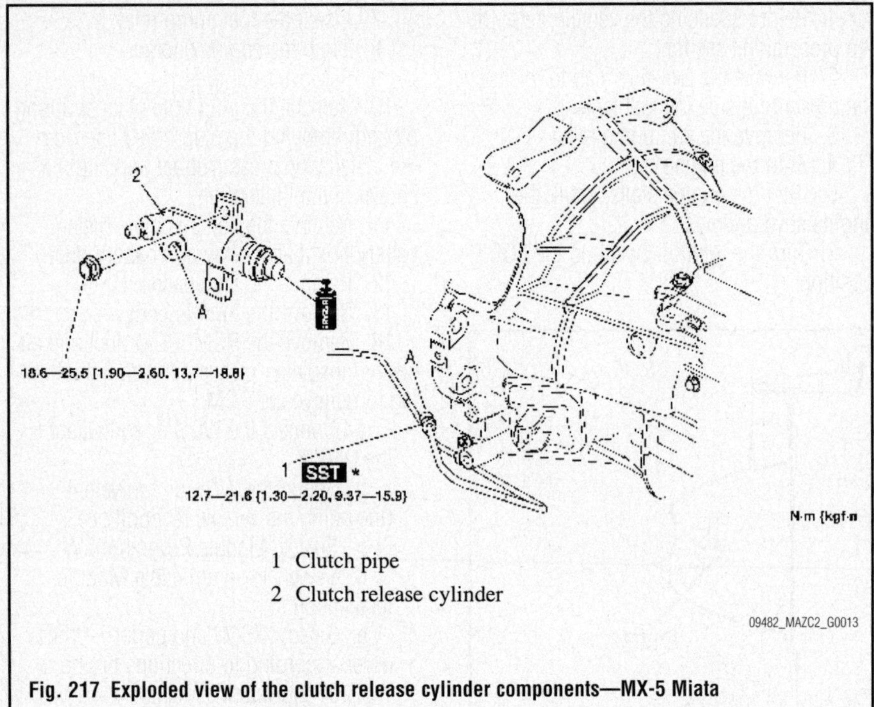

Fig. 217 Exploded view of the clutch release cylinder components—MX-5 Miata

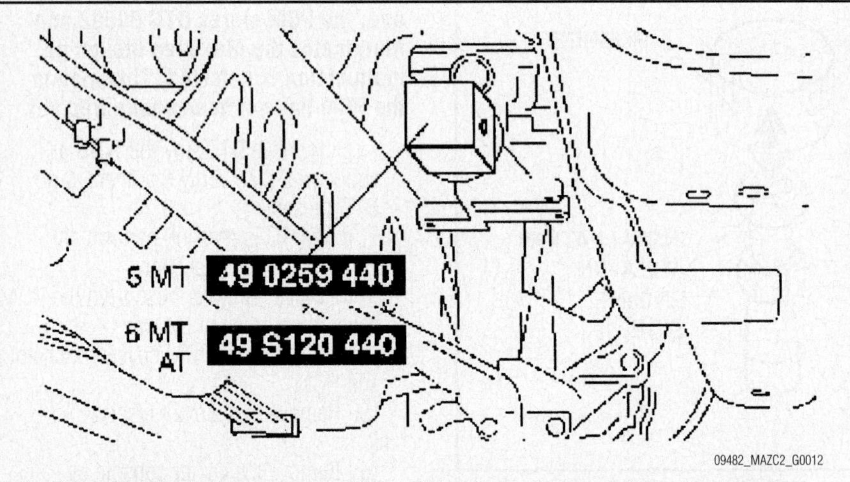

Fig. 218 Install the tools shown depending on transmission type, onto the to the extension housing—MX-5 Miata

19. Remove the alternator duct.
20. Drain the transmission fluid.
21. Disconnect the brake vacuum hose.
22. Disconnect the quick release connector from the dynamic chamber.
23. Disconnect the quick release connector from the fuel distributor.
24. Remove the drive belt.
25. Remove the A/C compressor with the pipes connected and attach the A/C compressor using wire so that it is out of the way.
26. Disconnect the water hose and heater hose.
27. Secure the front caliper using wire so that it is out of the way.
28. Disconnect the wiring harness.
29. Disconnect front ABS wheel-speed sensor connector.
30. Remove the radiator.
31. If equipped with an automatic transmission, disconnect the manual shaft lever component as follows:
    a. Loosen the starter installation bolts only enough that the starter is loose, but not removed.
    b. Mark the manual shaft lever component as shown in the illustration.
    c. Remove the manual shaft lever component installation nut.
32. If equipped with a manual transmission, Remove the clutch release cylinder with the pipes connected and secure the clutch release cylinder using wire or rope so that it is out of the way. Remove the shift lever knob.
33. Remove the tunnel member.
34. Remove the member bracket.
35. Remove the transverse member.
36. Remove the exhaust middle pipe.
37. Remove the power plant frame as follows:
    a. Support the transmission using a transmission jack.
    b. Remove the power plant frame.
38. Remove the propeller shaft as follows:

➡When replacing with a new propeller shaft, mark the companion flange to match the position of the tag on the propeller shaft.

    c. Before removing the propeller shaft, make alignment marks on the yoke and differential companion flange.
    d. Remove the retainers and the shaft.
    e. Install the tools illustrated onto the to the extension housing.
39. Remove the intermediate shaft bolt as follows:
    a. Mark the pinion shaft and gear housing for proper installation.
    b. Remove the bolt.

⁂ WARNING

Remove the engine, transmission and crossmember carefully, holding it steady. If the transmission falls it could be damaged or cause injury.

40. Secure the engine, transmission, and crossmember using an engine lifter
41. Remove the engine and transmission from the crossmember lifter as follows by suspending them with a crane:
    a. Engine mount rubber.
    b. Engine mount bracket.
    c. Engine from the transmission.
    d. Oil pipe and oil hose.

To install:

42. Installation is the reverse of removal. Please refer to the exploded views for component placement and torque specifications.
43. Please note the following special steps:
    a. When installing the oil pipe, hose clamp and oil hose:
        - Apply compressed air to cooler-side opening, and blow any remaining grime and foreign material from the cooler pipes. Compressed air should be applied for more than 1 min.

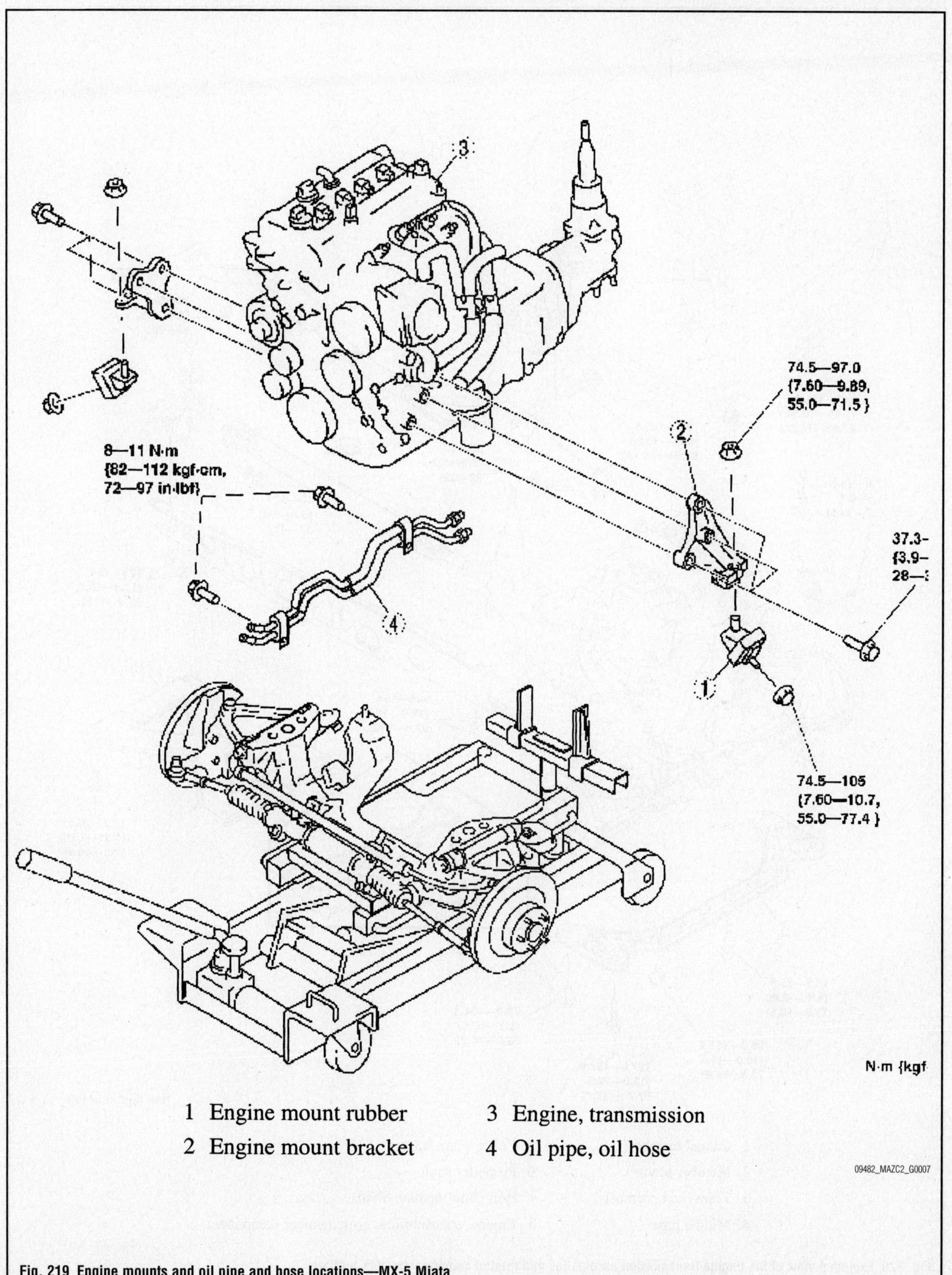

Fig. 219 Engine mounts and oil pipe and hose locations—MX-5 Miata

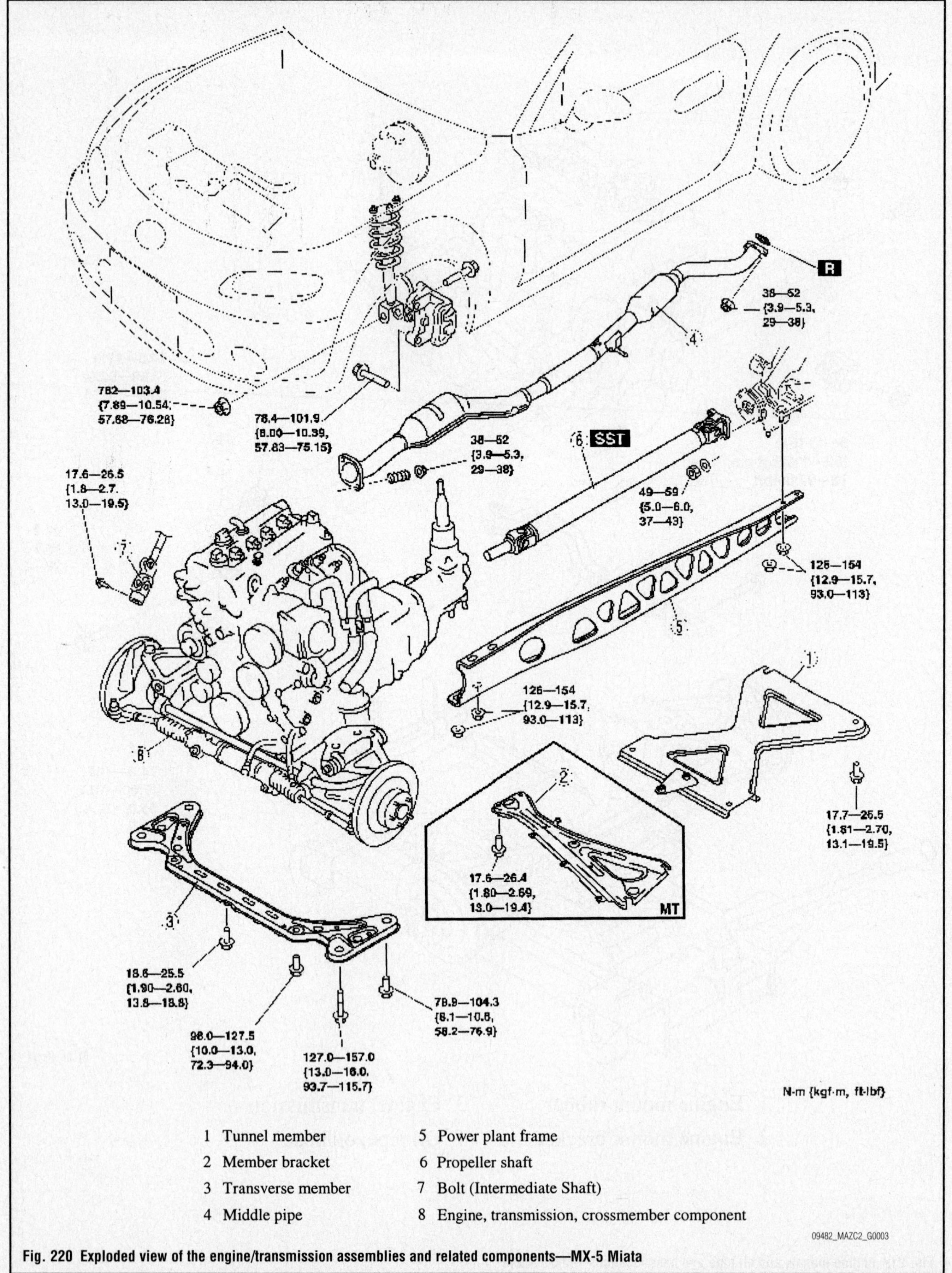

Fig. 220 Exploded view of the engine/transmission assemblies and related components—MX-5 Miata

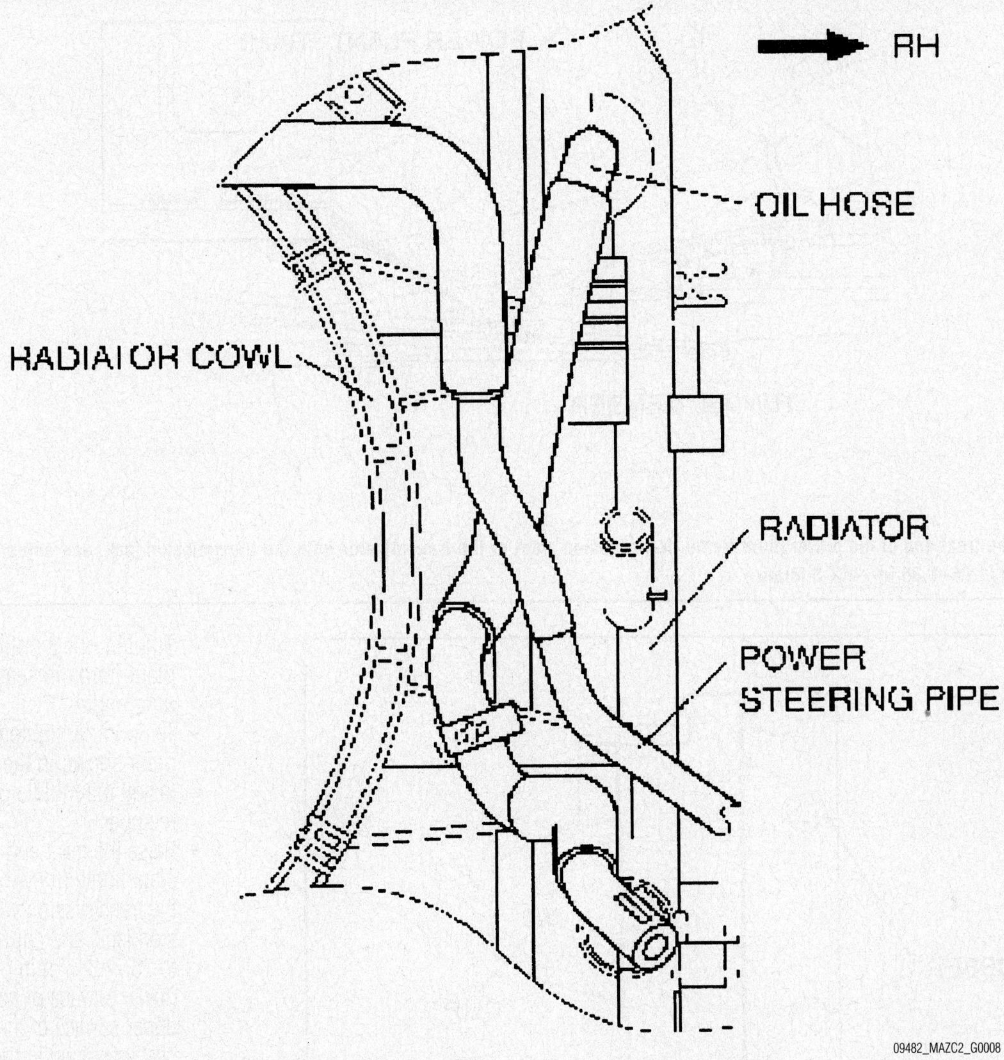

09482_MAZC2_G0008

Fig. 221 Be sure to install the oil hose between the power steering pipe and the radiator cowl—MX-5 Miata

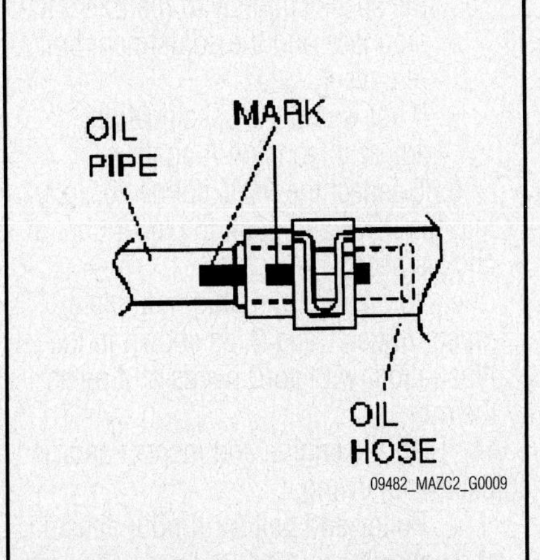

Fig. 222 Align the marks, and slide the oil hose onto the oil pipe until it is fully seated—MX-5 Miata

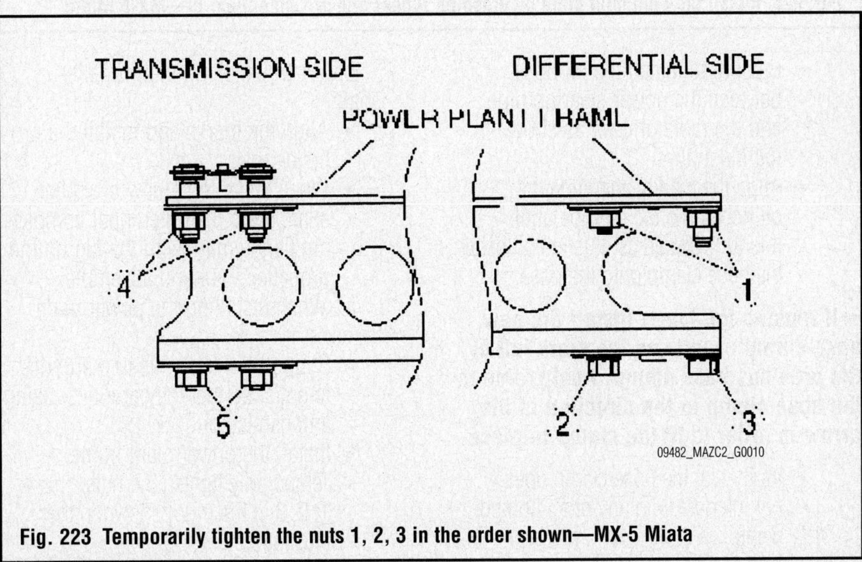

Fig. 223 Temporarily tighten the nuts 1, 2, 3 in the order shown—MX-5 Miata

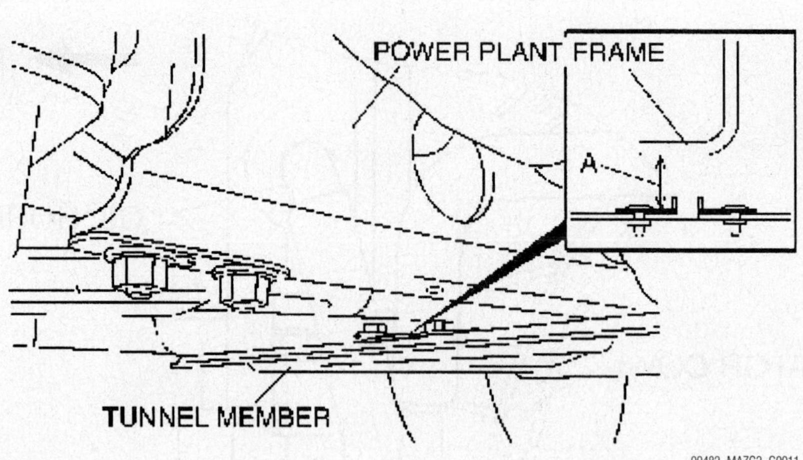

Fig. 224 Raise the front end of the power plant frame (transmission side) or the transmission with the transmission jack, and adjust dimension A to 26.7–34.7 mm (1.06–1.36 in—MX-5 Miata

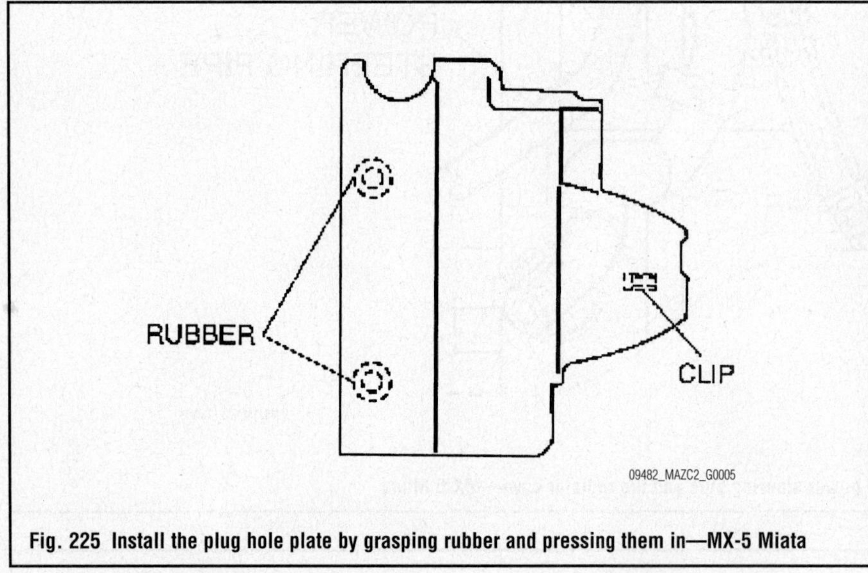

Fig. 225 Install the plug hole plate by grasping rubber and pressing them in—MX-5 Miata

- Be sure to install the oil hose between the power steering pipe and the radiator cowl as shown in the illustration.
- Align the marks, and slide the oil hose onto the oil pipe until it is fully seated as illustrated. Install the hose clamp onto the hose.

➡If reusing the hose, install the new hose clamp exactly on the mark left by the previous hose clamp. Apply force to the hose clamp in the direction of the arrow in order to fit the clamp in place.

- Verify that the hose clamp does not interfere with any other components.

b. When installing the propeller shaft:

- Align the marks and install the propeller shaft.
- When installing a new propeller shaft, align the differential companion flange mark with the tag on the propeller shaft and assemble.

c. When installing the power plant frame:

- Support the transmission and differential so that they are level using a transmission jack.
- Install the power plant frame.
- Temporarily tighten the nuts 1, 2, 3 in the order shown in the illustration.
- Tighten nut 2 until the power plant frame is seated in the rear differential.
- Temporarily tighten the nuts 4, 5 in order shown in the illustration.
- Install the middle pipe and tunnel member.
- Raise the front end of the power plant frame (transmission side) or the transmission with the transmission jack, and adjust dimension A to 26.7–34.7 mm (1.06–1.36 in) (lower surface of power plant frame-upper surface of the tunnel member) as shown in the illustration.
- Tighten the power plant frame installation nuts to 126–154 Nm (93–113.5 ft. lbs)
- Verify that dimension A is within the specification with the transmission jack and the adjustment bolt removed.
- If not within the specification, adjust dimension A again.

d. Connect the PCM connector fully into the PCM and push the lever until a click is heard.

e. To position the plug hole plate, grasp rubber 1 and 2, as shown in the illustration, with your hands and press them in.

44. Start the engine and inspect and/or adjust the following:

a. Pulley and belt for runout, tension, and contact

b. Leakage of engine oil, coolant, ATF or MT fluid, and check for fuel leaks

c. Ignition timing, idle speed, and idle.

d. Front wheel alignment.

e. Perform a road test and verify that there is no vibration or noise.

Mazda3 Models

See Figures 196 and 226 through 233.

1. Before servicing the vehicle, refer to the precautions section.

2. Remove the plug hole plate by lifting off and removing the plug hole plate from the areas shown in the accompanying illustration.

3. Remove the air hose and air cleaner assembly.

4. Remove the battery cover, duct, clamp, battery and battery tray.

5. Relieve the fuel system pressure and disconnect the fuel hoses.

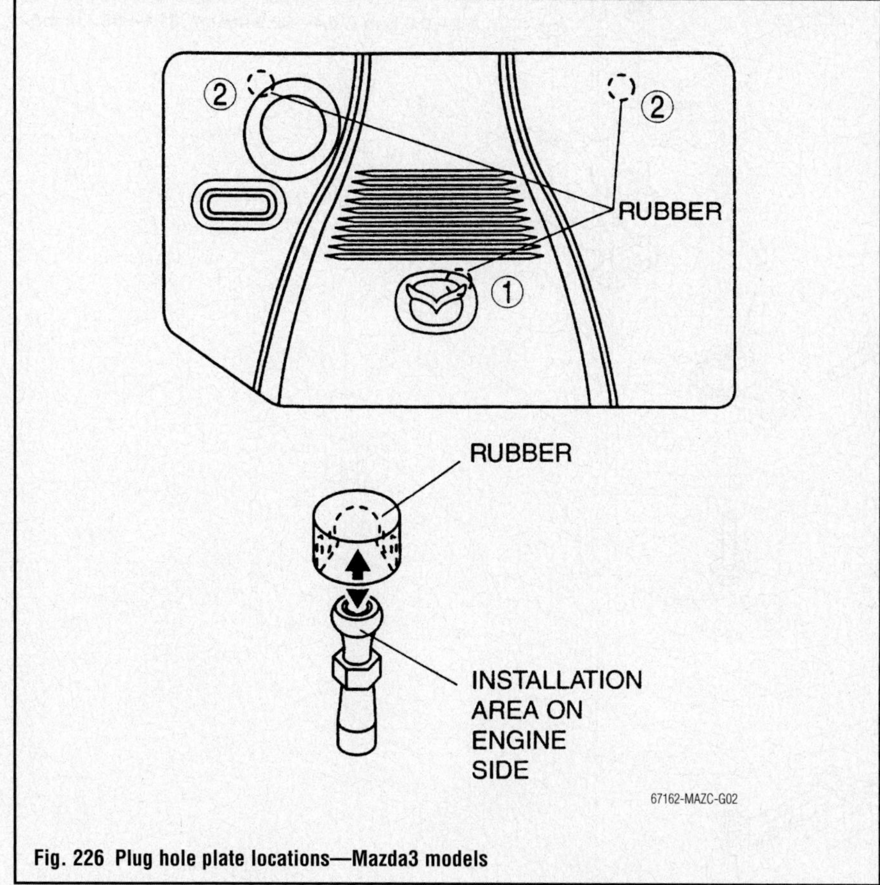

Fig. 226 Plug hole plate locations—Mazda3 models

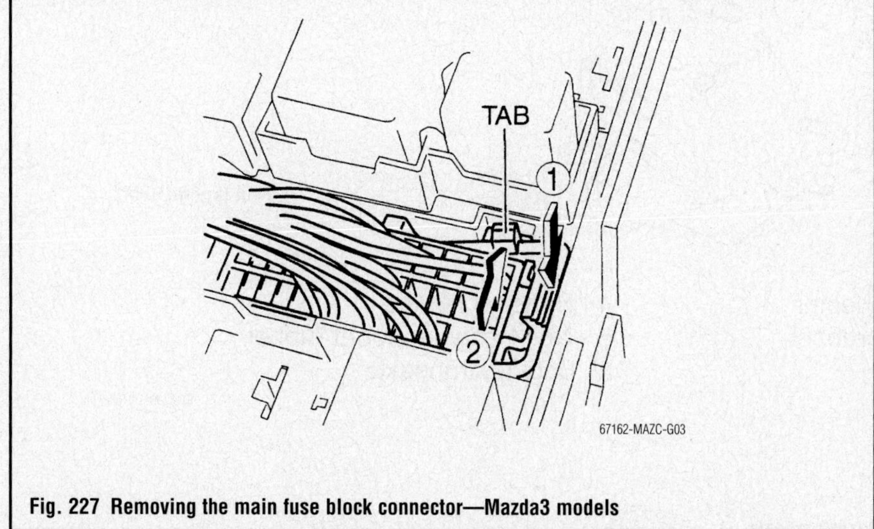

Fig. 227 Removing the main fuse block connector—Mazda3 models

6. Remove the accelerator cable and bracket.

7. Remove the front wheels, undercover and splash shields.

8. Remove the A/C drive belt.

9. Remove the A/C compressor with the lines still attached and wire the compressor out of the way.

10. Drain the transaxle fluid.

11. Drain the coolant.

12. Disconnect the brake booster vacuum hose.

13. Remove the exhaust system member.

14. Remove the front crossmember, front stabilizer bar, lower control arm, steering gear and the No. 1 engine mount.

15. Remove the drive shafts.

16. Remove the coolant over flow tank with the hose attached and wire it to one side.

17. Remove the cooling fan assembly.

18. If equipped with an automatic transaxle, disconnect the transaxle fluid lines, selector cable and wiring harness.

19. If equipped with a manual transaxle, remove the shift cable and the clutch release cylinder with the line still attached.

20. Disconnect the heater hoses.

21. Remove the radiator hoses.

22. Remove the exhaust system main silencer.

23. Remove the main fuse block connector by releasing the tab in the order shown in the accompanying illustration and pulling the lock lever up and remove the connector.

24. Secure the engine and transaxle using an engine jack and attachment as shown in the accompanying illustration and remove the No. 3 engine mount.

25. Remove the battery bracket.

26. Remove the No. 4 engine mount.

27. Remove the engine and transaxle assembly from the vehicle as an assembly.

To install:

28. Installation is the reverse of removal, please keep in mind the following:

29. When installing the No. 4 engine mount, secure the engine and transaxle using an engine jack and attachment as shown in the accompanying illustration. Install the No. 1 and No. 4 engine mounts, do not tighten the retainers at this time.

30. Use a new No. 4 mount bolt and tighten it to 61–87 ft. lbs. (83–113 Nm).

31. Tighten the No. 4 engine mount and battery bracket nuts and bolts in the sequence shown in the accompanying illustration. Tighten the retainers (1) to 32–45 ft. lbs. (44–61 Nm) and the retainers (2) to 61–86 inch lbs. (7–10 Nm).

32. Tighten the No. 3 mount bracket stud bolts to 62–115 inch lbs. (7–13 Nm).

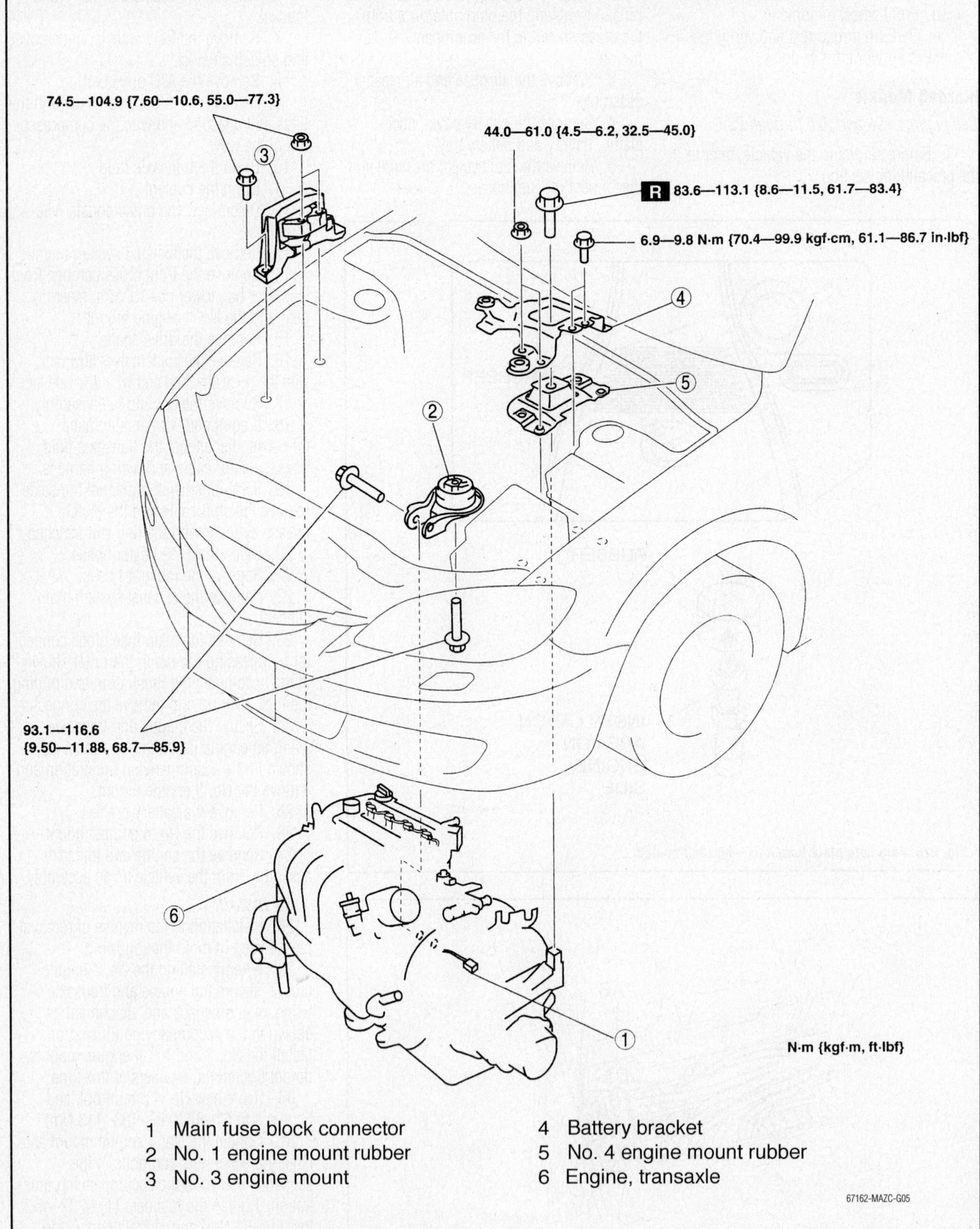

Fig. 228 Exploded view of the engine mounting—Mazda3 models

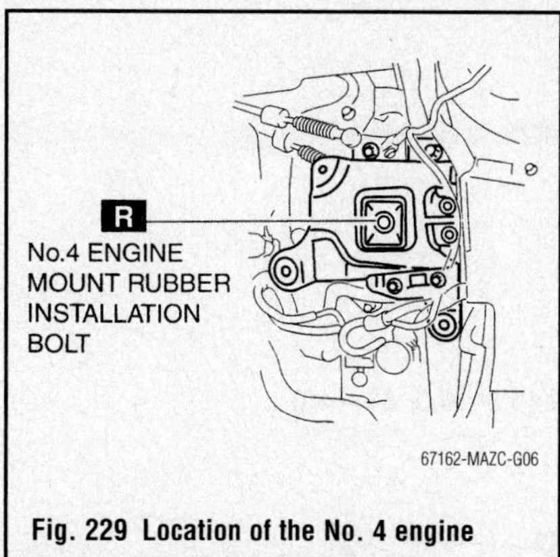

Fig. 229 Location of the No. 4 engine mount rubber bolt—Mazda3 models

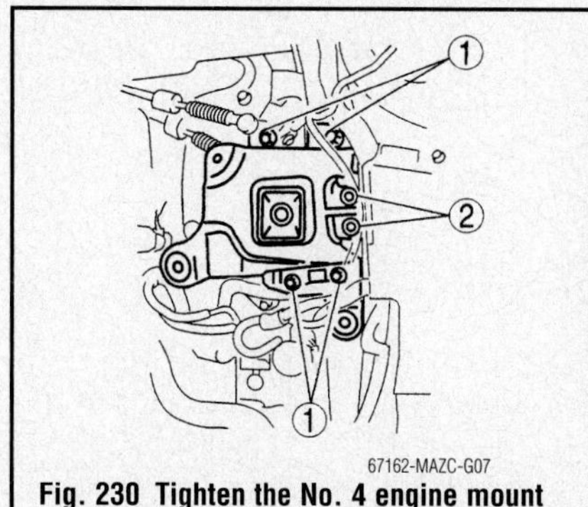

Fig. 230 Tighten the No. 4 engine mount and battery bracket nuts and bolts in this sequence—Mazda3 models

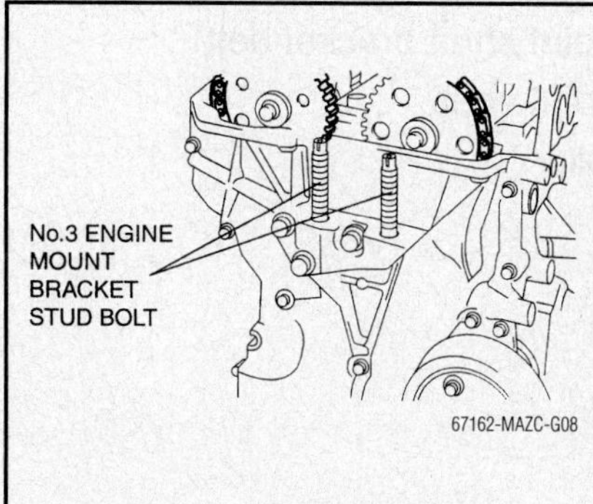

Fig. 231 Location of the No. 3 engine mount bracket stud bolts—Mazda3 models

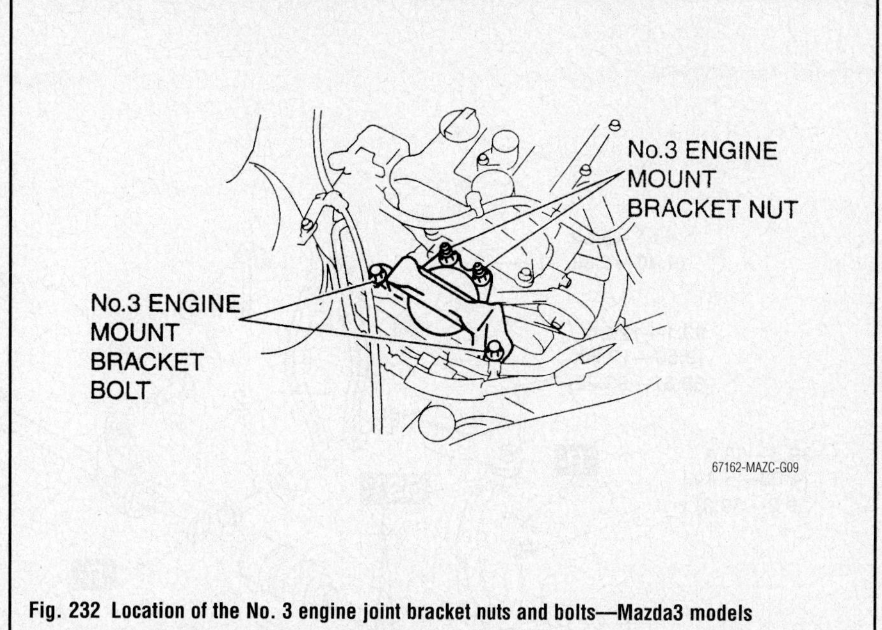

Fig. 232 Location of the No. 3 engine joint bracket nuts and bolts—Mazda3 models

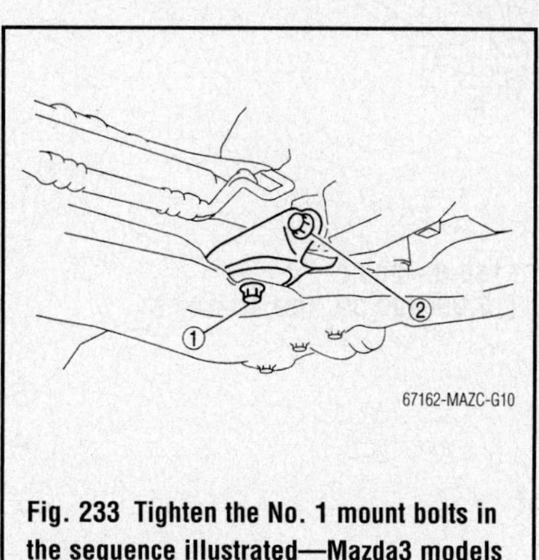

Fig. 233 Tighten the No. 1 mount bolts in the sequence illustrated—Mazda3 models

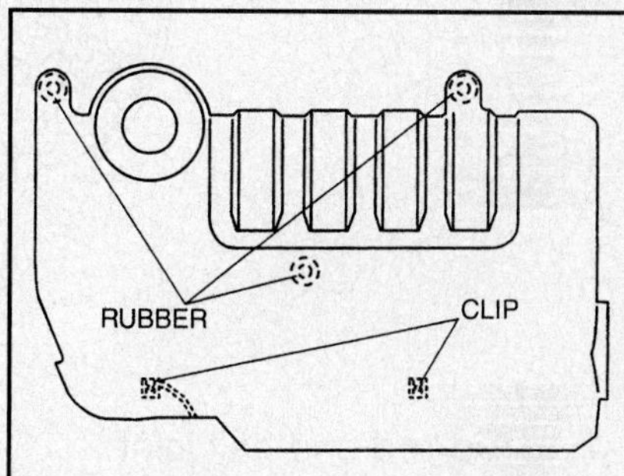

Fig. 234 Plug hole plate assembly—Mazda6 shown, Mazdaspeed6 similar with 2.3L (L3) engine

33. Tighten the No. 3 joint bracket nuts and bolts to 55–77 ft. lbs. (74–105 Nm).
34. Remove the jack and the attachment and tighten the No. 1 mount bolts in the sequence illustrated to 69–86 ft. lbs. (93–116 Nm).
35. Fill the engine and the transaxle with the proper type and amount of fluids. Fill the cooling system.
36. Connect the negative battery cable, start the engine and check for leaks.
37. Check the ignition timing and the idle speed.
38. Check all fluid levels.
39. Check the vehicle alignment.

Mazda6 Models

2.3L (L3) Engines

See Figures 234 through 242.

1. Before servicing the vehicle, refer to the precautions section.
2. Remove the battery and battery tray.
3. Remove the shroud panel.
4. Remove the radiator.
5. Drain the transaxle fluid.
6. Remove the plug hole plate.
7. Remove the power steering pump with the lines attached and set aside.
8. Remove the A/C compressor with the lines still attached and wire the compressor out of the way.
9. Remove the joint shaft as follows:
    a. Remove the ABS sensor.
    b. Separate the tie rod ends from the knuckle.
    c. Remove the damper fork bolt.
    d. Separate the lower control arm front and rear ball joints.
    e. Remove the stabilizer bar link nut and the link from the damper fork.
    f. Remove the joint shaft bracket bolt.
    g. Remove the halfshafts.
    h. Disconnect the right halfshaft from the joint shaft by tapping the transaxle side outer ring with a brass bar and a hammer. Disconnect the joint shaft bracket from the block and remove the joint shaft.
    i. Install tool 49 G030 455 to hold the side gears after removal.
10. Remove the air cleaner assembly, air intake duct, bracket and vacuum hoses.
11. On models with an automatic transaxle, disconnect the transmission fluid hose and the selector cable.
12. Remove the vacuum and heater hoses
13. If equipped with a manual transaxle, remove the control cable and the clutch release cylinder with the line still attached.

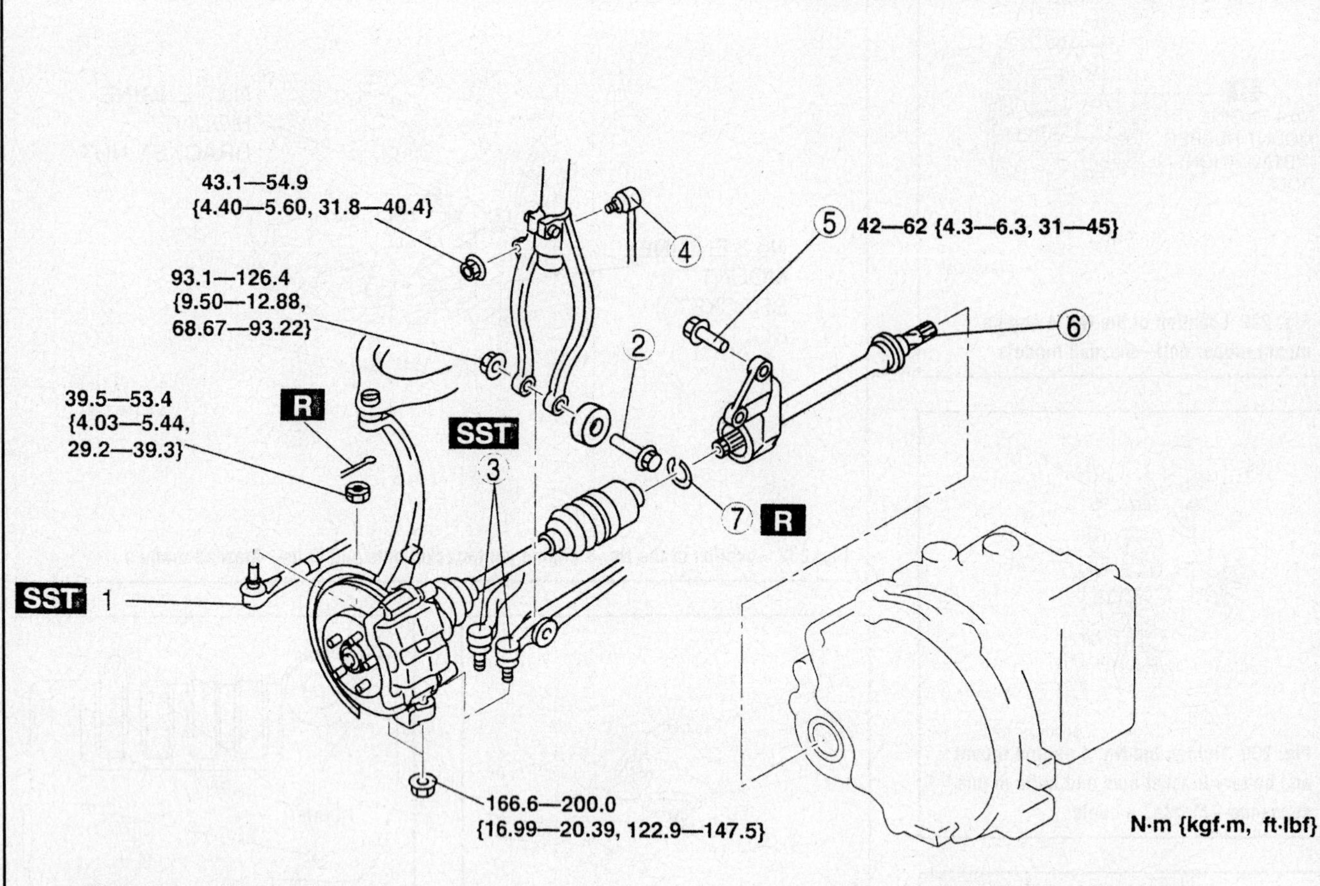

Fig. 235 Exploded view of the joint shaft assembly— Mazda6 shown, Mazdaspeed6 similar with the 2.3L (L3) engine

14. Relieve the fuel system pressure and disconnect the fuel hoses.
15. Disconnect the wiring harness at the engine side.
16. Remove the front exhaust pipe.
17. Remove the No. 1 mount rubber bolt A on the engine mount bracket side, loosen bolt B on the crossmember side until about 3 pitches are showing. DO NOT remove the NO, 1 mount rubber from the vehicle.
18. Remove the No. 1 engine mount bracket.
19. Secure the engine and transaxle using a hoist.
20. Remove the No. 4 mount bracket and rubber as a unit.
21. Remove the engine ground.
22. Remove the No. 3 engine joint bracket.
23. Remove the engine and transaxle from the vehicle as an assembly.
24. Remove the engine-to-transaxle bolts and separate the engine from the transaxle.

To install:

25. Installation is the reverse of removal, please keep in mind the following:
26. On models with an automatic transaxle, tighten the engine-to-transaxle bolts to 28–38 ft. lbs. (37–52 Nm)
27. On models with a manual transaxle, tighten the engine-to-transaxle bolts to 28–36 ft. lbs. (37–50 Nm)
28. When installing the No. 3 engine joint bracket, tighten the mount bracket stud bolt to 62–115 inch lbs. (7–13 Nm) and the joint bracket bolt and nut in the order illustrated to 55–76 ft. lbs. (74–105 Nm).
29. Install the No. 4 engine mount bracket and rubber. Tighten the No.4 mount bracket and rubber bolts and nuts in the order illustrated as follows:
   a. Bolts numbered 1, 2 and 3: 43–58 ft. lbs. (58–80 Nm).

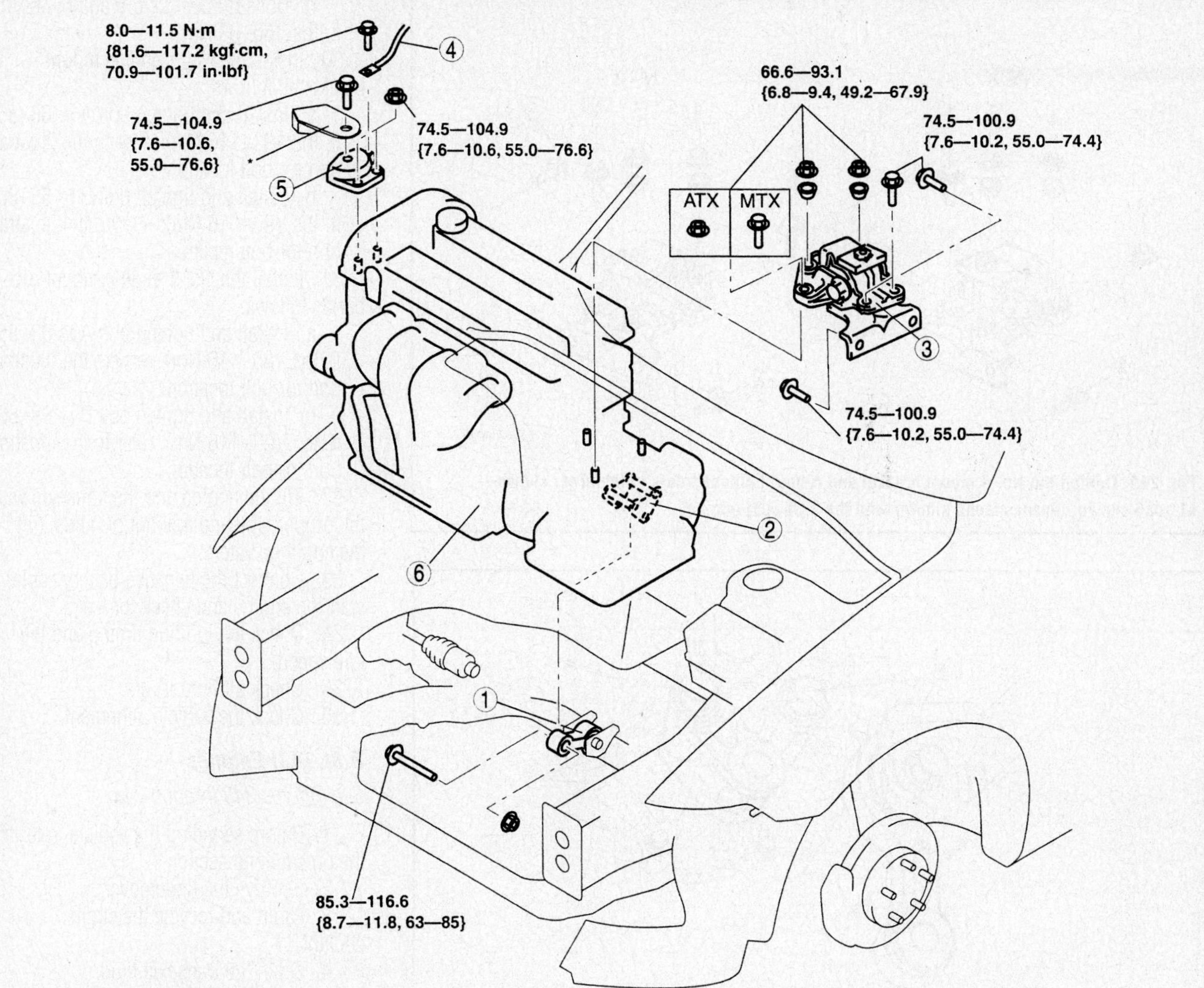

8.0—11.5 N·m
{81.6—117.2 kgf·cm,
70.9—101.7 in·lbf}
74.5—104.9
{7.6—10.6,
55.0—76.6}
74.5—104.9
{7.6—10.6, 55.0—76.6}
66.6—93.1
{6.8—9.4, 49.2—67.9}
74.5—100.9
{7.6—10.2, 55.0—74.4}
ATX
MTX
74.5—100.9
{7.6—10.2, 55.0—74.4}
85.3—116.6
{8.7—11.8, 63—85}

*: Only MTX

1 No.1 Engine mount rubber
2 No.1 Engine mount bracket
3 No.4 Engine mount bracket and No.4 Engine mount rubber
4 Engine ground
5 No.3 Engine joint bracket
6 Engine, transaxle

67162-MAZC-G113

Fig. 236 Exploded view of the engine mounting—Mazda6 shown, Mazdaspeed6 similar with the 2.3L (L3) engine

B
A

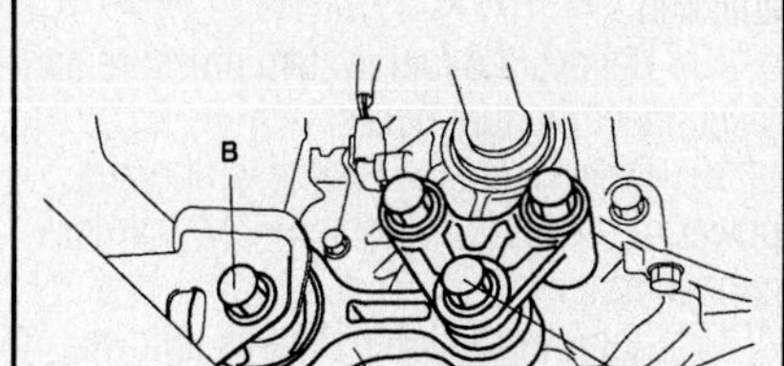

67162-MAZC-G114

Fig. 237 Locations of bolts A and B on the No. 1 mount rubber—Mazda6 shown, Mazdaspeed6 similar with the 2.3L (L3) engine

No.3
ENGINE
MOUNT
BRACKET
STUD BOLT

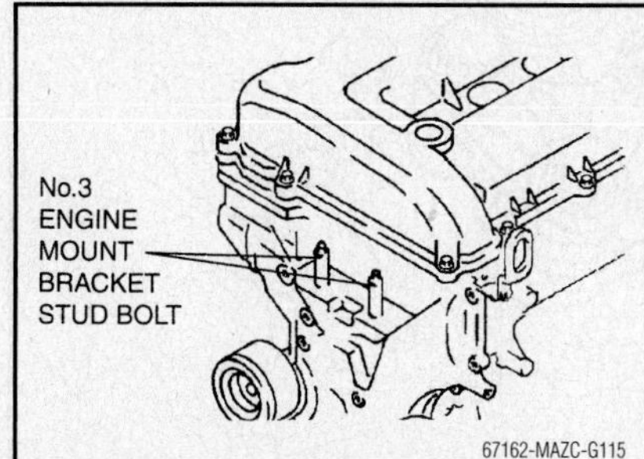

67162-MAZC-G115

Fig. 238 No. 3 engine mount bracket stud bolt locations—Mazda6 shown, Mazdaspeed6 similar with the 2.3L (L3) engine

ATX
MTX

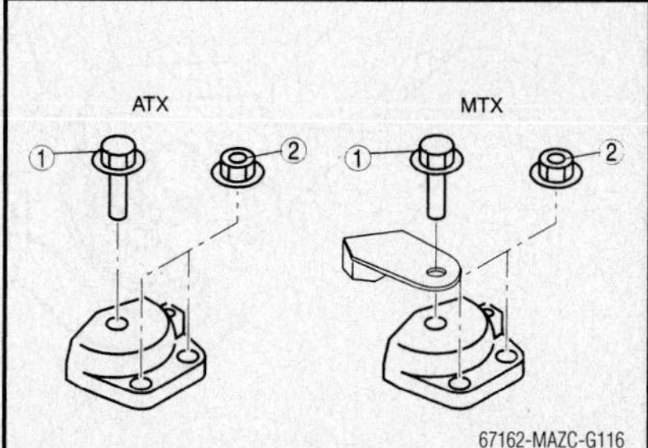

67162-MAZC-G116

Fig. 239 Tighten the No. 3 joint bracket bolt and nut in the order shown—Mazda6 shown, Mazdaspeed6 similar with the 2.3L (L3) engine

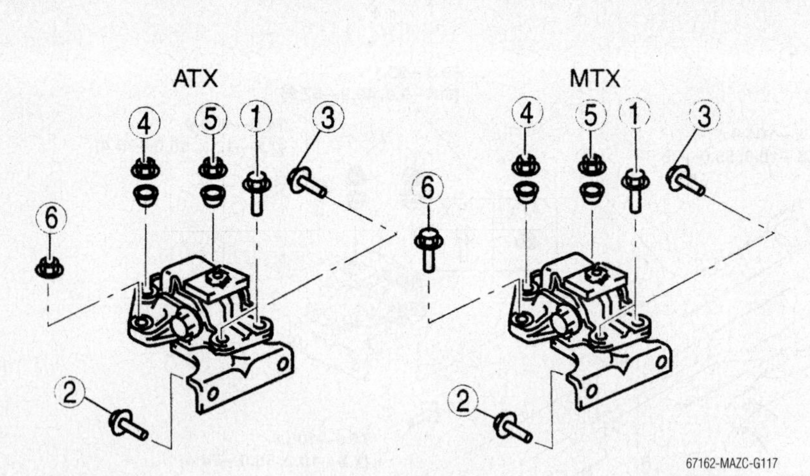

Fig. 240 Tighten the No. 4 mount bracket and rubber bolts and nuts in the order shown—Mazda6 shown, Mazdaspeed6 similar with the 2.3L (L3) engine

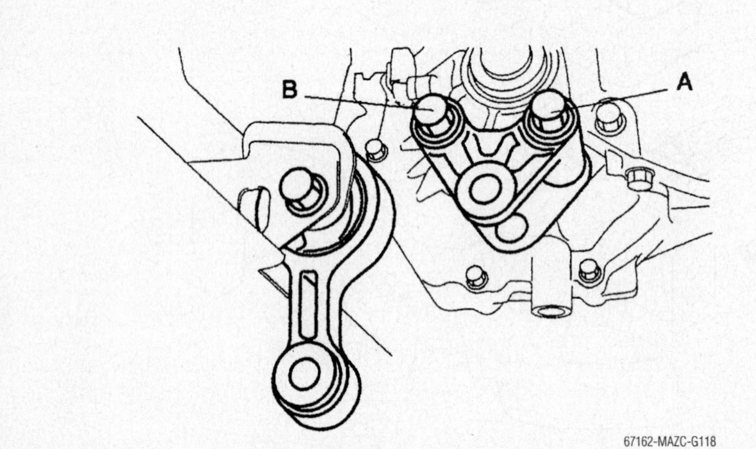

Fig. 241 Tighten the No. 1 mount bracket bolts in the order shown—Mazda6 shown, Mazdaspeed6 similar with the 2.3L (L3) engine

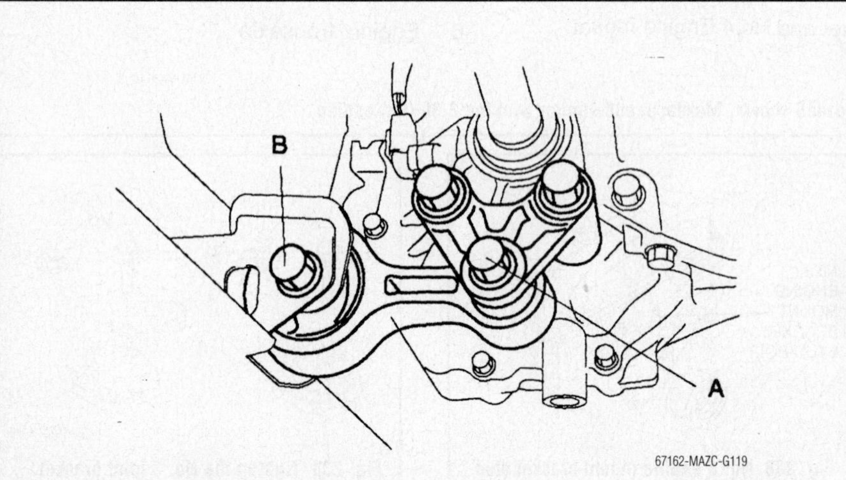

Fig. 242 Tighten the No. 1 mount rubber bolts in the order shown—Mazda6 shown, Mazdaspeed6 similar with the 2.3L (L3) engine

b. Bolts numbered 4, 5 and 6: 49–68 ft. lbs. (66–93 Nm).

30. Install the No. 1 engine mount bracket as follows:

a. Install and tighten bolt A to 68–86 ft. lbs. (93–116 Nm). refer to the illustration for bolt location.

b. Install and tighten bolt B to 68–86 ft. lbs. (93–116 Nm). refer to the illustration for bolt location.

31. Install the No. 1 engine mount rubber as follows:

a. Install and tighten bolt A to 63–85 ft. lbs. (85–116 Nm). refer to the illustration for bolt location.

b. Install and tighten bolt B to 68–86 ft. lbs. (93–116 Nm). refer to the illustration for bolt location.

32. Fill the engine and the transaxle with the proper type and amount of fluids. Fill the cooling system.

33. Connect the negative battery cable, start the engine and check for leaks.

34. Check the ignition timing and the idle speed.

35. Check all fluid levels.

36. Check the vehicle alignment.

3.0L (AJ) Engines

See Figures 243 through 248.

1. Before servicing the vehicle, refer to the precautions section.
2. Remove the undercover.
3. Drain and recycle the engine coolant.
4. Drain the transaxle fluid.
5. Remove the battery and battery tray.
6. Remove the air cleaner assembly, air intake duct, bracket and vacuum hoses.
7. On models with an automatic transaxle, disconnect the transmission fluid hose and the selector cable.
8. If equipped with a manual transaxle, remove the shift and selector cables and the clutch release cylinder with the line still attached.
9. Relieve the fuel system pressure and disconnect the fuel hoses.
10. Disconnect the engine wiring and powertrain Control Module (PCM) harness connections.
11. Remove the joint shaft as follows:

a. Remove the ABS sensor.

b. Remove the halfshaft lock nut.

c. Separate the tie rod ends from the knuckle.

d. Remove the damper fork.

e. Separate the lower control arm front and rear ball joints.

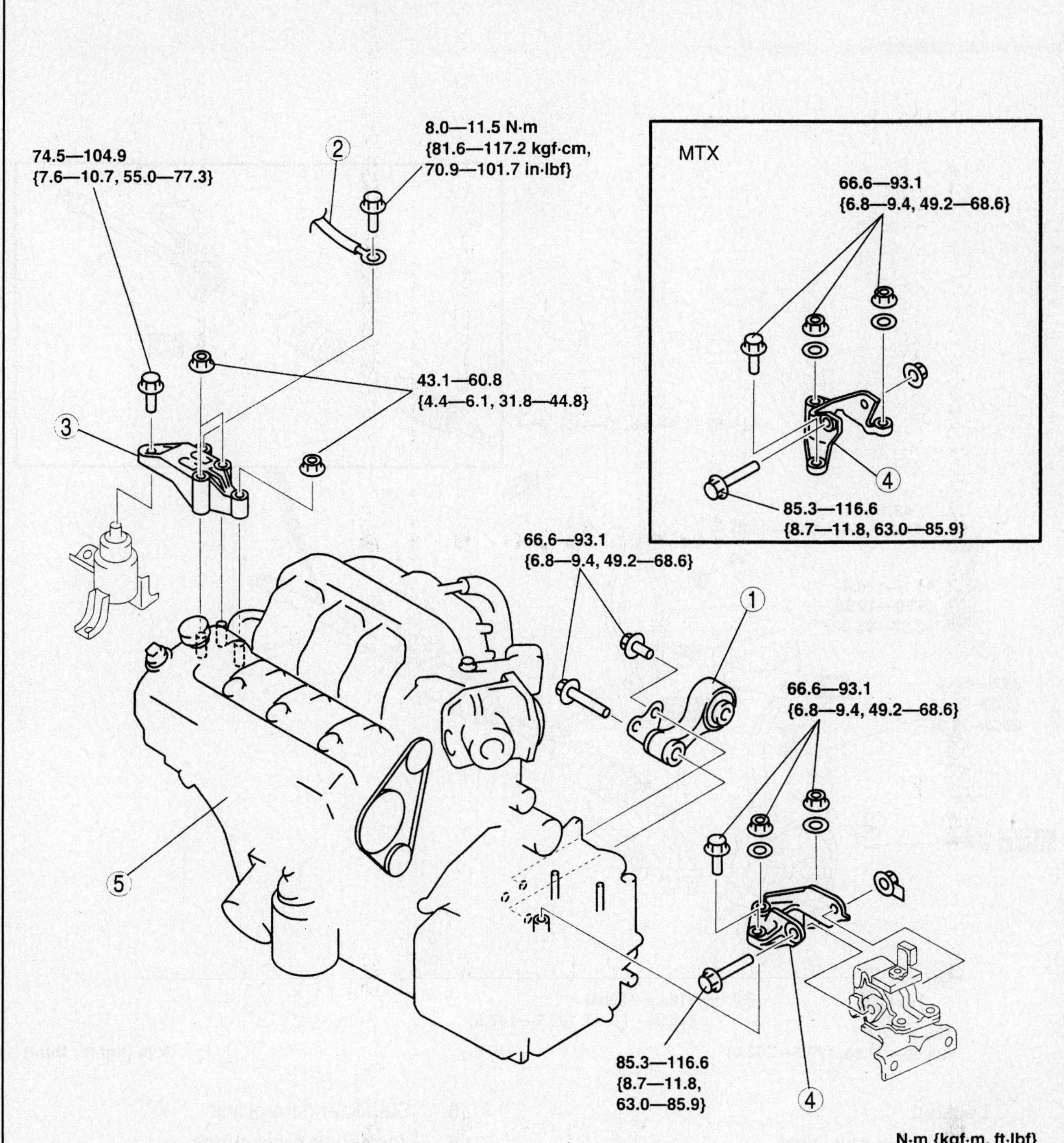

Fig. 243 Exploded view of the engine mounting—Mazda6 models with the 3.0L (AJ) engine

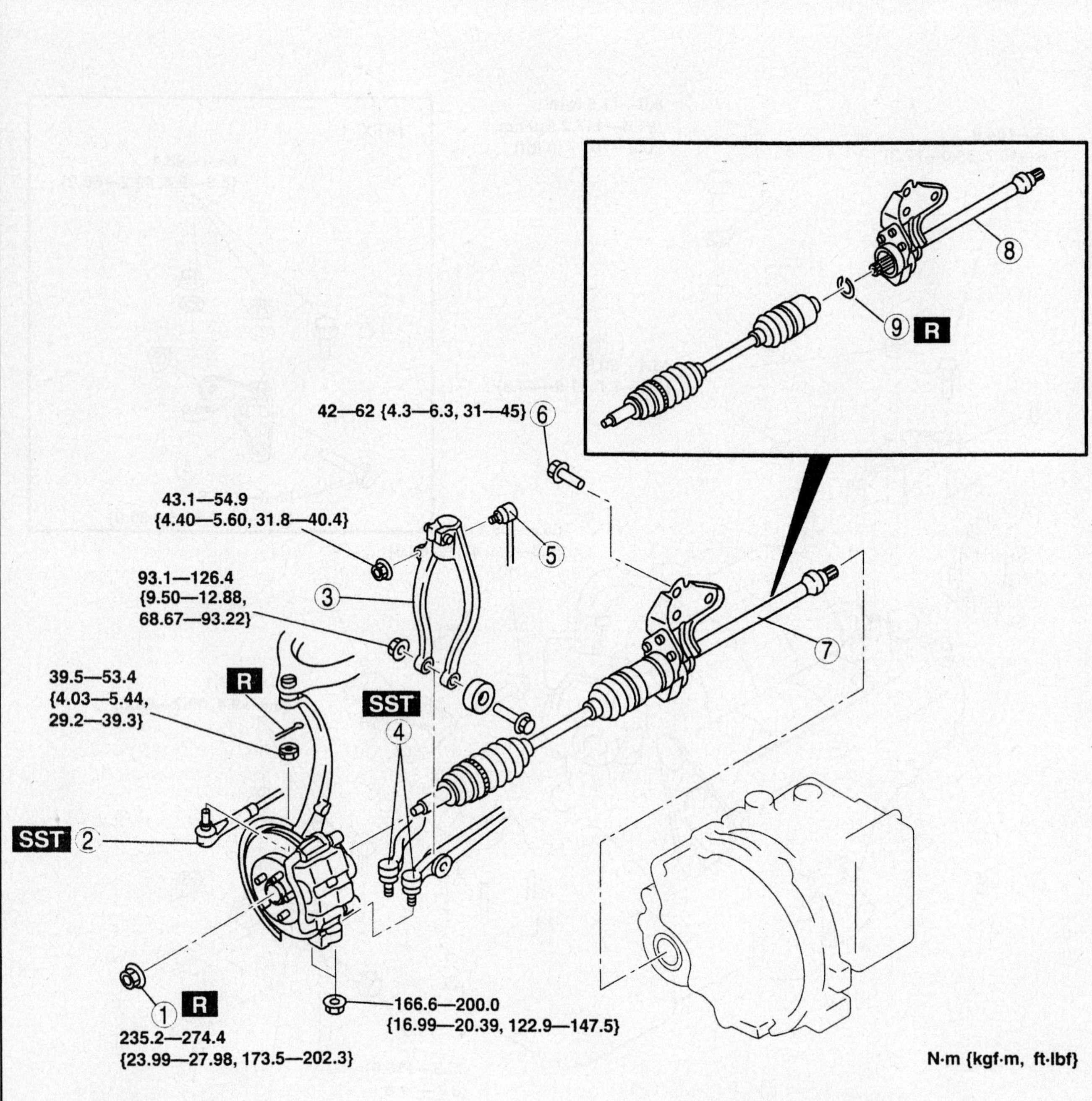

Fig. 244 Exploded view of the joint shaft assembly—Mazda6 models with the 3.0L (AJ) engine

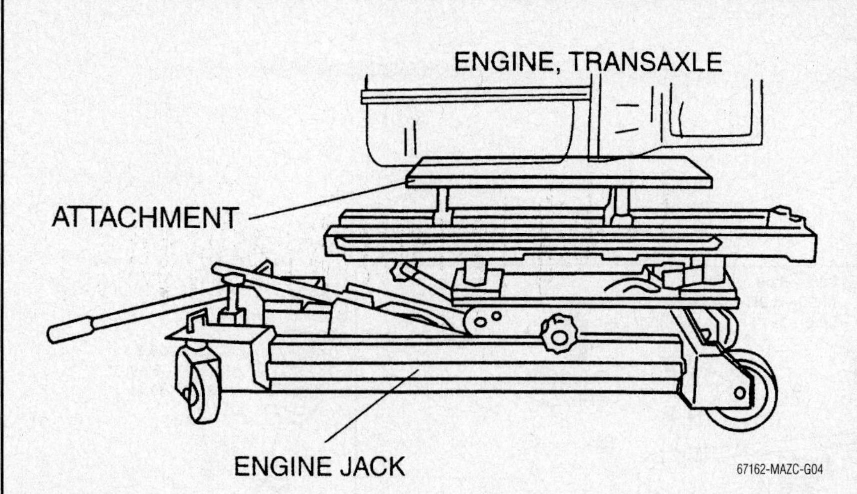

Fig. 245 Support the engine using an engine jack and attachment as shown—Mazda6 models with the 3.0L (AJ) engine

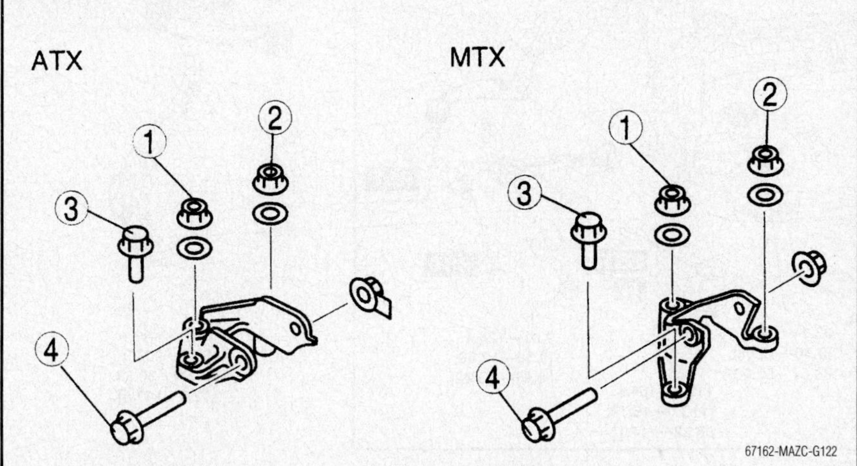

Fig. 246 Tighten the No. 4 engine joint bracket as shown—Mazda6 models with the 3.0L (AJ) engine

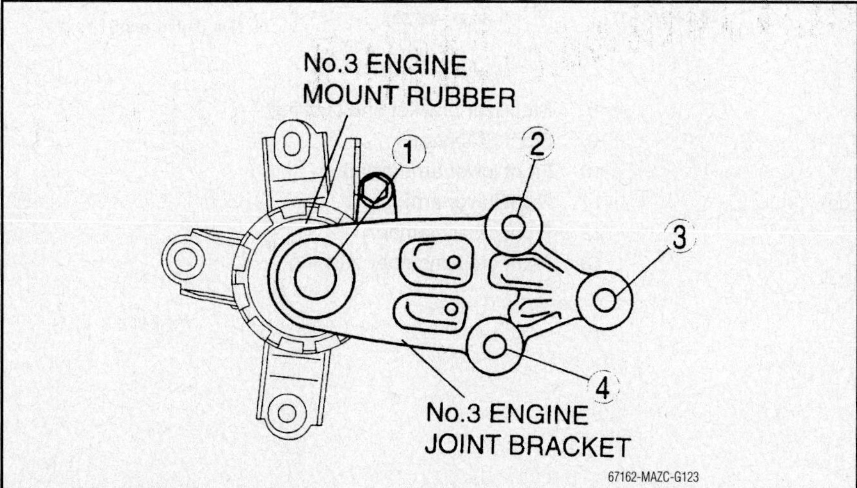

Fig. 247 Tighten the No. 3 engine joint bracket as shown—Mazda6 models with the 3.0L (AJ) engine

f. Remove the stabilizer bar link nut and the link from the damper fork.

g. Remove the joint shaft bracket bolt.

h. Remove the halfshafts.

i. Disconnect the left halfshaft from the joint shaft by inserting a pry bar between the transaxle and the halfshaft outer ring. Disconnect the joint shaft bracket from the block and remove the joint shaft.

j. Install tool 49 G030 455 to hold the side gears after removal

12. Remove the transverse member.
13. Remove the steering gear and linkage assembly from the front crossmember and use mechanics wire to support the gear and linkage away from the crossmember.
14. Lower front shock absorber bolt.
15. No. 1 engine mount center bolt.
16. Support the crossmember with a jack and remove the nuts and the crossmember bracket.
17. Remove the crossmember.
18. Remove the drive belt and the dipstick tube.
19. Disconnect the power steering hoses and drain the power steering reservoir.
20. Properly evacuate the A/C system using approved equipment.
21. Remove the A/c compressor.
22. Remove the left hand three way catalytic converter.
23. Remove the front exhaust pipe.
24. Remove the radiator and heater hoses.
25. Remove the No. 1 engine mount rubber.
26. Disconnect the engine ground.
27. Support the engine using an engine jack and attachment as shown in the illustration.
28. Remove the No. 3 joint bracket and the No. 4 engine mount bracket.
29. Remove the No. 4 engine mount bracket.
30. Remove the engine and transaxle from the vehicle as an assembly.
31. Remove the engine-to-transaxle bolts and separate the engine from the transaxle.

To install:

32. Installation is the reverse of removal, please keep in mind the following:
33. Tighten the engine-to-transaxle bolts to 28–38 ft. lbs. (37–52 Nm)
34. Install the No. 4 engine mount bracket. Tighten the No.4 mount bracket bolts and nuts in the order illustrated as follows:

a. Bolts numbered 1, 2 and 3: 49–68 ft. lbs. (66–93 Nm).

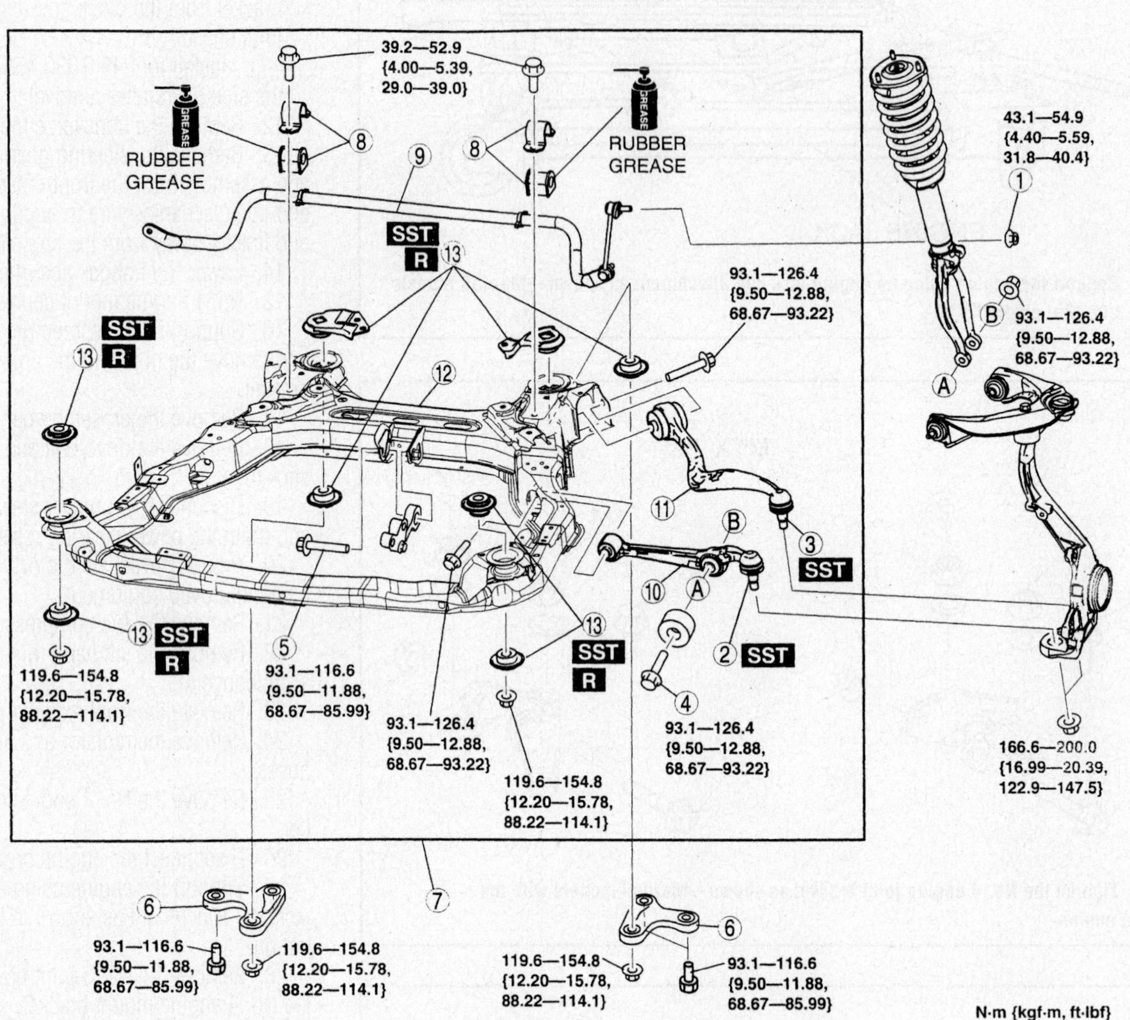

1 Nut (stabilizer control link)
2 Front lower arm (front) ball joint
3 Front lower arm (rear) ball joint
4 Bolt (front shock absorber lower side)
5 No.1 engine mount center bolt
6 Crossmember bracket
7 Crossmember component
8 Stabilizer bracket and bushing
9 Front Stabilizer
10 Front lower arm (front)
11 Front lower arm (rear)
12 Front crossmember
13 Front crossmember bushing

67162-MAZC-G124

Fig. 248 Exploded view of the front crossmember, related components and their torque specifications—Mazda6 models with the 3.0L (AJ) engine

b. Bolt 4: 63–86 ft. lbs. (85–116 Nm).

35. When installing the No. 3 engine joint bracket, tighten the No.3 mount bracket bolts and nuts in the order illustrated as follows:

a. Bolt 1: 55–77 ft. lbs. (74–105 Nm).

b. Bolts 2, 3 and 4: 31–44 ft. lbs. (43–60 Nm).

36. Install the No. 1 engine mount bracket and tighten the bolts to 49–68 ft. lbs. (66–93 Nm).

37. Install all remaining components in the reverse of removal.

38. Fill the engine and the transaxle with the proper type and amount of fluids. Fill the cooling system.

39. Recharge the A/C system.

40. Connect the negative battery cable, start the engine and check for leaks.

41. Check the ignition timing and the idle speed.

42. Check all fluid levels.

43. Check the vehicle alignment.

2.3L (L3 with TC) Engines

1. Before servicing the vehicle, refer to the precautions section.

2. Remove the battery, battery tray and bracket.

3. Remove the charge air cooler cover.

4. Remove the air cleaner.

5. Remove the left side front mudguard, then the resonance chamber.

6. Disconnect the MAF/IAT sensor.

7. Remove the charge air cooler.

8. Remove the charge air cooler bracket

9. Remove the shroud panel.

10. Remove the radiator.

11. Drain the transaxle fluid.

12. Remove the power steering pump with the lines attached and set aside.

13. Remove the A/C compressor with the lines still attached and wire the compressor out of the way.

14. Disconnect the right hand front drive shaft from the joint shaft side.

15. Disconnect the left hand front drive shaft from the transaxle side.

16. Remove the propeller shaft.

17. Remove the air cleaner assembly, air intake duct, bracket and vacuum hoses.

18. Remove the control cable and the clutch release cylinder with the line still attached.

19. Remove the vacuum and heater hoses.

20. Disconnect the wiring harness at the engine side.

21. Remove the three way catalytic converter.

22. Remove the front exhaust pipe.

23. Relieve the fuel system pressure and disconnect the fuel hoses.

24. Remove the No. 1 mount rubber bolt A on the engine mount bracket side, loosen bolt B on the crossmember side until about 3 pitches are showing. DO NOT remove the NO, 1 mount rubber from the vehicle.

25. Remove the No. 1 engine mount bracket.

26. Secure the engine and transaxle using a hoist.

27. Remove the No. 4 mount bracket and rubber as a unit.

28. Remove the engine ground.

29. Remove the No. 3 engine joint bracket.

30. Remove the engine and transaxle from the vehicle as an assembly.

31. Remove the engine–to–transaxle bolts and separate the engine from the transaxle.

To install:

32. Installation is the reverse of removal, please keep in mind the following:

33. Tighten the engine–to–transaxle bolts to 28–36 ft. lbs. (37–50 Nm)

34. When installing the No. 3 engine joint bracket, tighten the mount bracket stud bolt to 62–115 inch lbs. (7–13 Nm) and the joint bracket bolt and nut in the order illustrated to 55–76 ft. lbs. (74–105 Nm).

35. Install the No. 4 engine mount bracket and rubber. Tighten the No.4 mount bracket and rubber bolts and nuts in the order illustrated as follows:

a. Bolts numbered 1, 2 and 3: 43–58 ft. lbs. (58–80 Nm).

b. Bolts numbered 4, 5 and 6: 49–68 ft. lbs. (66–93 Nm).

36. Install the No. 1 engine mount bracket as follows:

a. Install and tighten bolt A to 68–86 ft. lbs. (93–116 Nm). refer to the illustration for bolt location.

b. Install and tighten bolt B to 68–86 ft. lbs. (93–116 Nm). refer to the illustration for bolt location.

37. Install the No. 1 engine mount rubber as follows:

a. Install and tighten bolt A to 63–85 ft. lbs. (85–116 Nm). refer to the illustration for bolt location.

b. Install and tighten bolt B to 68–86 ft. lbs. (93–116 Nm). refer to the illustration for bolt location.

38. Fill the engine and the transaxle with the proper type and amount of fluids. Fill the cooling system.

39. Connect the negative battery cable, start the engine and check for leaks.

40. Check the ignition timing and the idle speed.

41. Check all fluid levels.

42. Check the vehicle alignment.

EXHAUST MANIFOLD

REMOVAL & INSTALLATION

MX-5 Miata

See Figure 249.

1. Before servicing the vehicle, refer to the precautions section.

2. Disconnect the front and oxygen sensor connectors.

3. Move the water pipe and heater hose slightly out of the way.

4. Disconnect the ventilation hose from the cylinder head

5. Remove the upper exhaust manifold insulator as follows:

a. Remove the battery and battery tray.

b. Remove the drive belt.

c. Remove the undercover.

d. Remove the suspension tower bar (joint) and (right side).

e. Remove the alternator.

f. Remove the upper insulator.

6. Remove the lower exhaust manifold insulator.

7. Remove the exhaust manifold bracket.

8. Remove the exhaust manifold.

To install:

9. Installation is the reverse of removal, please note the following:

a. Install the manifold and tighten the retainers to 43–64 Nm (32–47 ft. lbs.)

b. Tighten the exhaust manifold bracket to 38–52 Nm (29–38 ft. lbs.).

c. Tighten the upper and lower insulators to 8–11 Nm (70–95 in. lbs.).

d. If removed, tighten the front and oxygen sensors to 29–39 Nm (22–36 ft. lbs.)

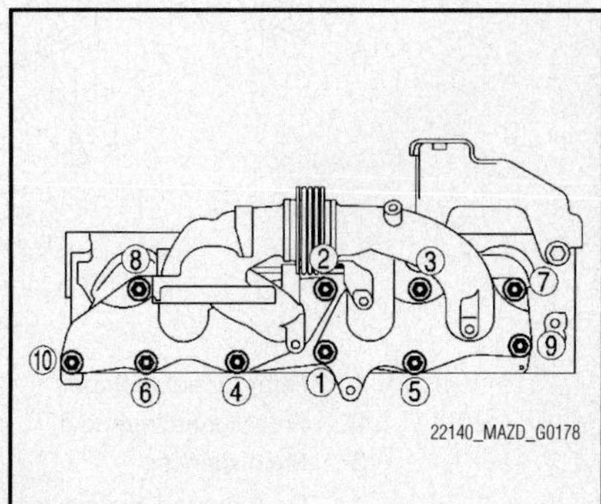

Fig. 249 MX-5 Exhaust manifold and related components

Mazda3

2.0L (LF) and 2.3L (L3) Engines

See Figures 250 through 253.

1. Before servicing the vehicle, refer to the precautions section.
2. Remove the plug hole plate.
3. Remove the battery cover and duct.
4. Remove the undercover.
5. Remove the rear and front tunnel members.
6. If necessary, remove the main silencer as follows:

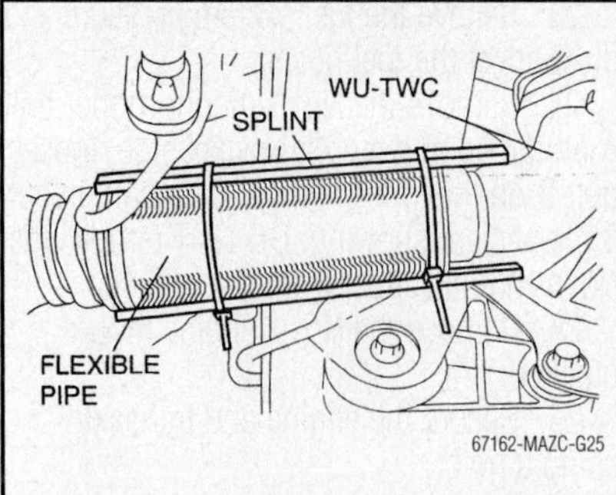

Fig. 250 Support the flexible pipe with a support wrap or splint—Mazda3 models

a. Disconnect the ABS sensor wiring harness connector.

b. Disengage the brake pipe mount from the bracket and remove the lower shock absorber bolts.

c. Loosen the rear crossmember bolts and lower the crossmember approximately 2.8 inches (70mm) and remove the main silencer.

7. Unplug both Oxygen Sensor (O_2S) connectors.
8. Remove the exhaust manifold bracket, heat shield and clip.

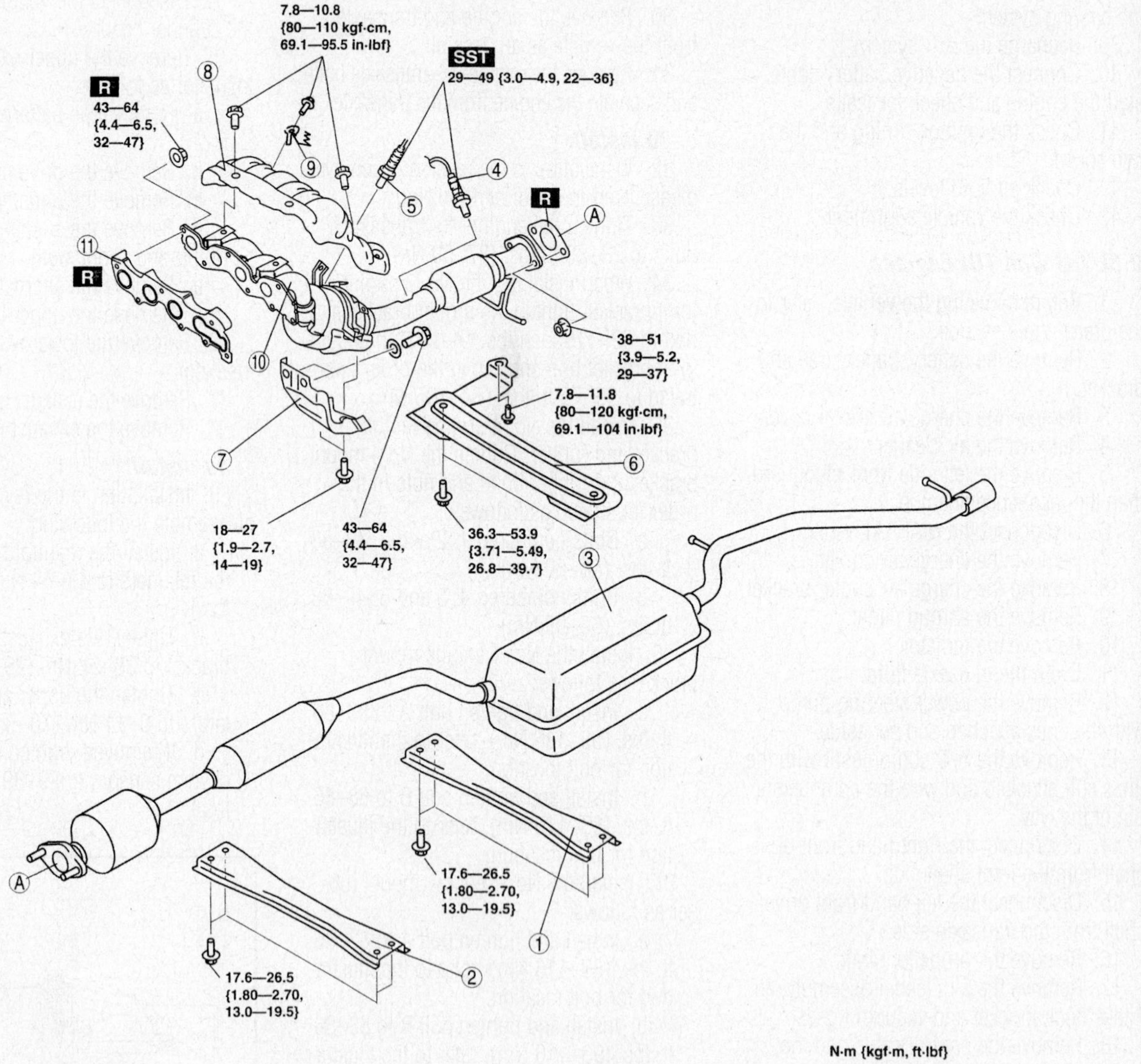

N·m {kgf·m, ft·lbf}

1	Rear tunnel member	6	Member
2	Front tunnel member	7	Exhaust manifold bracket
3	Main silencer	8	Exhaust manifold insulator
4	Rear heated oxygen sensor	9	Clip
5	Front heated oxygen sensor	10	Exhaust manifold
		11	Exhaust manifold gasket

67162-MAZC-G23

Fig. 251 Exploded view of the exhaust manifold assembly and related components–Non California emissions models—Mazda3 models

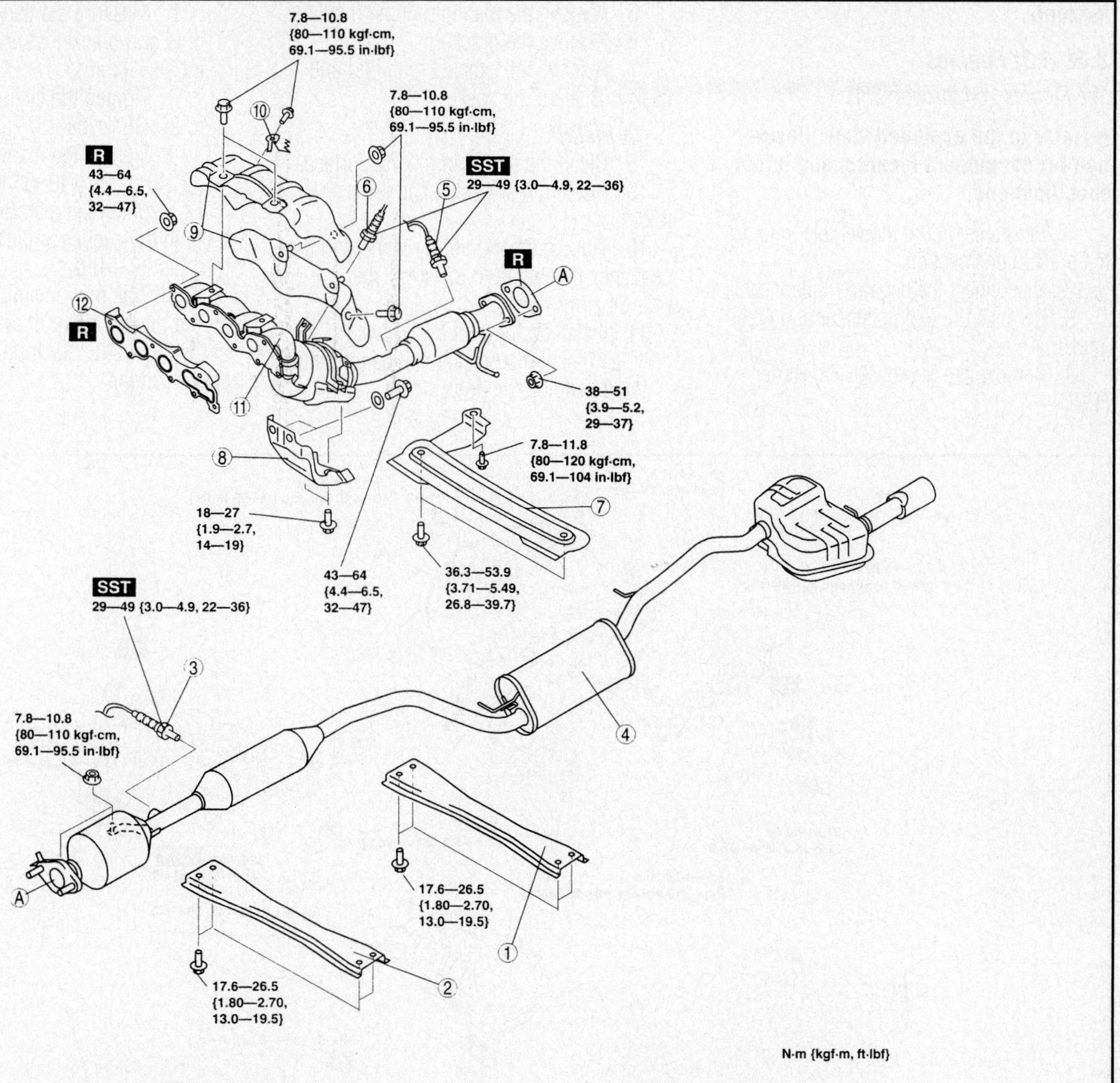

N·m {kgf·m, ft·lbf}

1 Rear tunnel member
2 Front tunnel member
3 Rear heated oxygen sensor
4 Main silencer
5 Middle heated oxygen sensor
6 Front heated oxygen sensor
7 Member
8 Exhaust manifold bracket
9 Exhaust manifold insulator
10 Clip
11 Exhaust manifold
12 Exhaust manifold gasket

67162-MAZC-G24

Fig. 252 Exploded view of the exhaust manifold assembly and related components–California emissions models—Mazda3 models

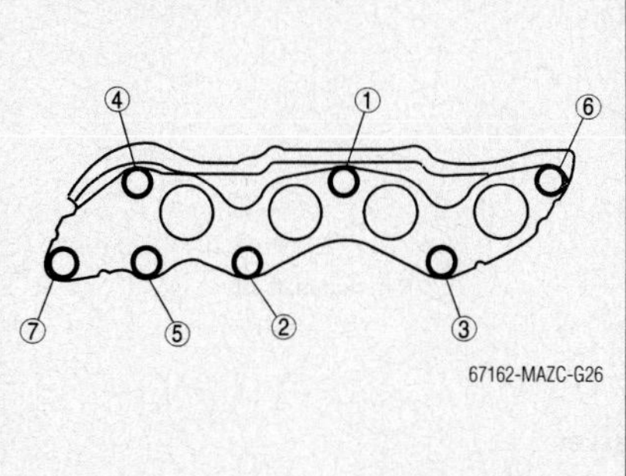

Fig. 253 Tighten the manifold retainers in this sequence—Mazda3 models

9. Remove the front wheels.
10. Disconnect the steering shaft from the steering gear and linkage side.
11. Support the engine and remove the No. 1 engine mount.
12. Loosen the exhaust manifold bolts.
13. Remove the front stabilizer and stabilizer control link bolts.
14. Loosen the front crossmember bolts and lower the crossmember approximately 3.94 inches (100mm).
15. Support the flexible pipe with a support wrap or splint as illustrated.
16. Remove the exhaust manifold and gasket.

To install:

17. Installation a new exhaust manifold gasket and the manifold . Tighten the manifold retainers in the sequence illustrated to 32–47 ft. lbs. (43–64 Nm).
18. Install the remaining components in the reverse order of removal, using the exploded view of the exhaust system for component location and torque specifications.

Mazda6

2.3L (L3) Engines

See Figures 254 through 257.

➡**Refer to the exploded view illustration for component location and torque specifications.**

1. Before servicing the vehicle, refer to the precautions section.
2. Disconnect the negative battery cable.
3. Remove the main silencer and presilencer.
4. Remove the Three Way Converter (TWC).
5. Remove the manifold upper insulator.
6. Remove the bracket.
7. Remove the exhaust manifold and gasket. Discard the gasket.

To install:

8. Clean the manifold mating surfaces.
9. Install a new gasket and the manifold.
10. Tighten the manifold retainers in the sequence illustrated to 32–47 ft. lbs. (43–64 Nm).
11. Install the bracket as follows :
    a. Install the bracket and hand tighten the manifold side bolts.
    b. Measure the gap between the manifold and bracket it should be 0.08–0.18 inch (2–4mm).
    c. Tighten the cylinder block side bolt to 27–38 ft. lbs. (37–52 Nm).
    d. Tighten the manifold side bolts to 29–40 ft. lbs. (40–55 Nm).
12. Tighten the manifold upper insulator bolts in the sequence illustrated to 70–95 inch lbs. (7–10 Nm).
13. Install the remaining components on the reverse order of removal, referring to the illustration for torque specifications.

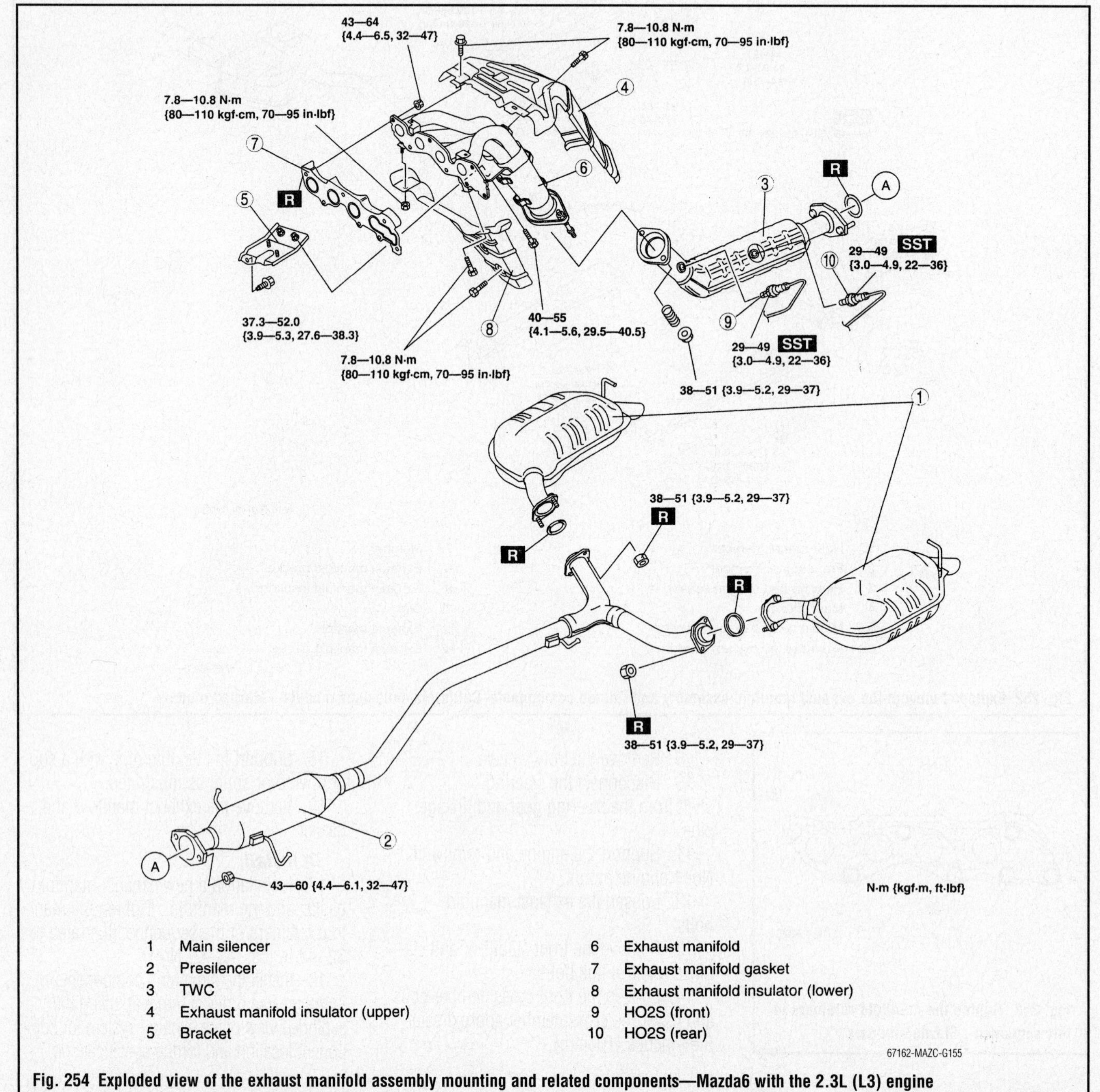

Fig. 254 Exploded view of the exhaust manifold assembly mounting and related components—Mazda6 with the 2.3L (L3) engine

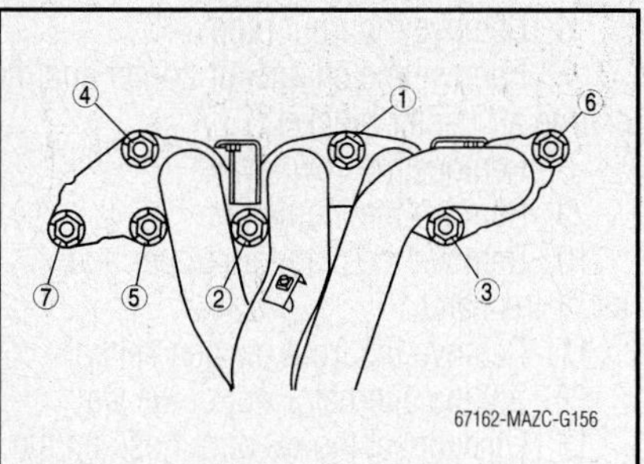

Fig. 255 Exhaust manifold torque sequence—Mazda6 with the 2.3L (L3) engine

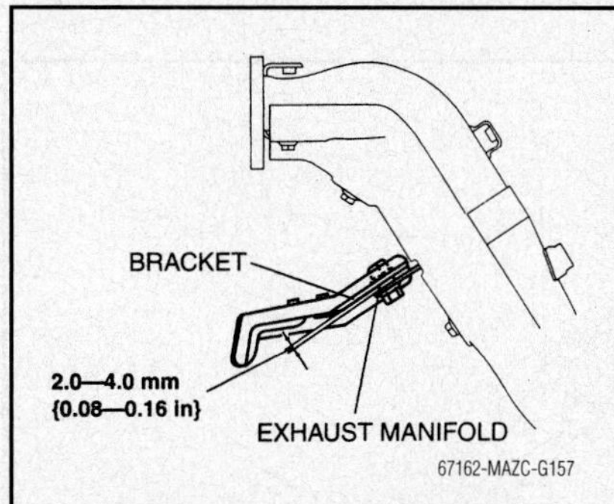

Fig. 256 Measure the gap between the manifold and bracket—Mazda6 with the 2.3L (L3) engine

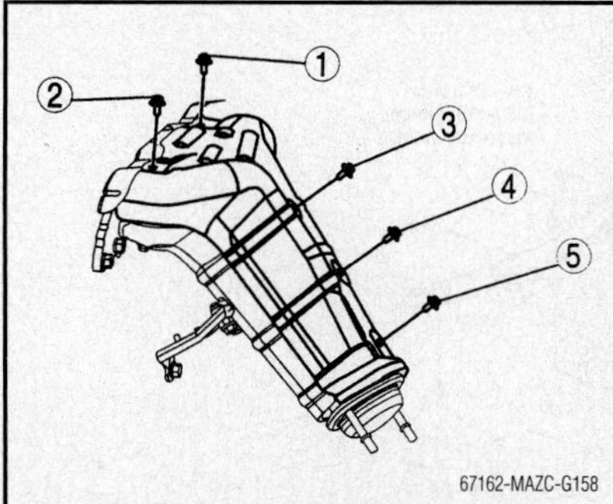

Fig. 257 Tighten the manifold upper insulator bolts in this sequence—Mazda6 with the 2.3L (L3) engine

3.0L (AJ) Engines

See Figures 258 through 260.

➡Refer to the exploded view illustration for component location and torque specifications.

1. Before servicing the vehicle, refer to the precautions section.
2. Disconnect the negative battery cable.
3. Remove the main silencer and presilencer.
4. Remove the Three Way Converter (TWC) from the left or right side depending on which manifold is being replaced.
5. Remove the stud bolt.
6. Remove the front pipe.
7. Unplug both Heated Oxygen Sensor (HO_2S) connectors.
8. remove the Exhaust Gas Recirculation (EGR) pipe.
9. Remove the exhaust manifold and gasket, from the left or right side depending on which manifold is being replaced. If removing the right hand side manifold, remove the alternator bracket first.
10. Discard the manifold gasket.

To install:

11. Clean the manifold mating surfaces.
12. Install a new gasket and the manifold.
13. Tighten the manifold retainers in the sequence illustrated to 14–19 ft. lbs. (19–26 Nm).

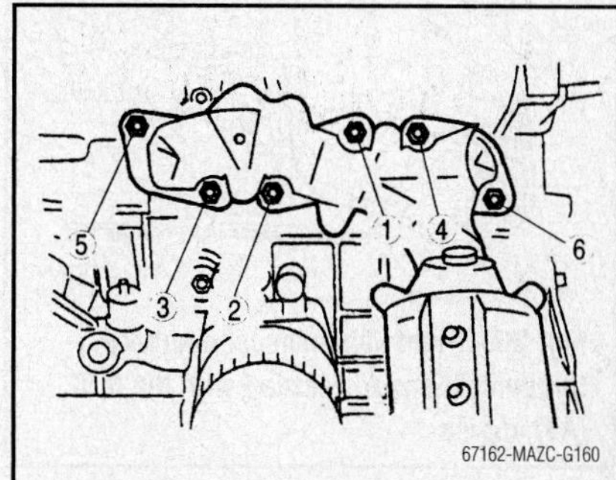

Fig. 259 Left side exhaust manifold torque sequence—Mazda6 with the 3.0L (AJ) engine

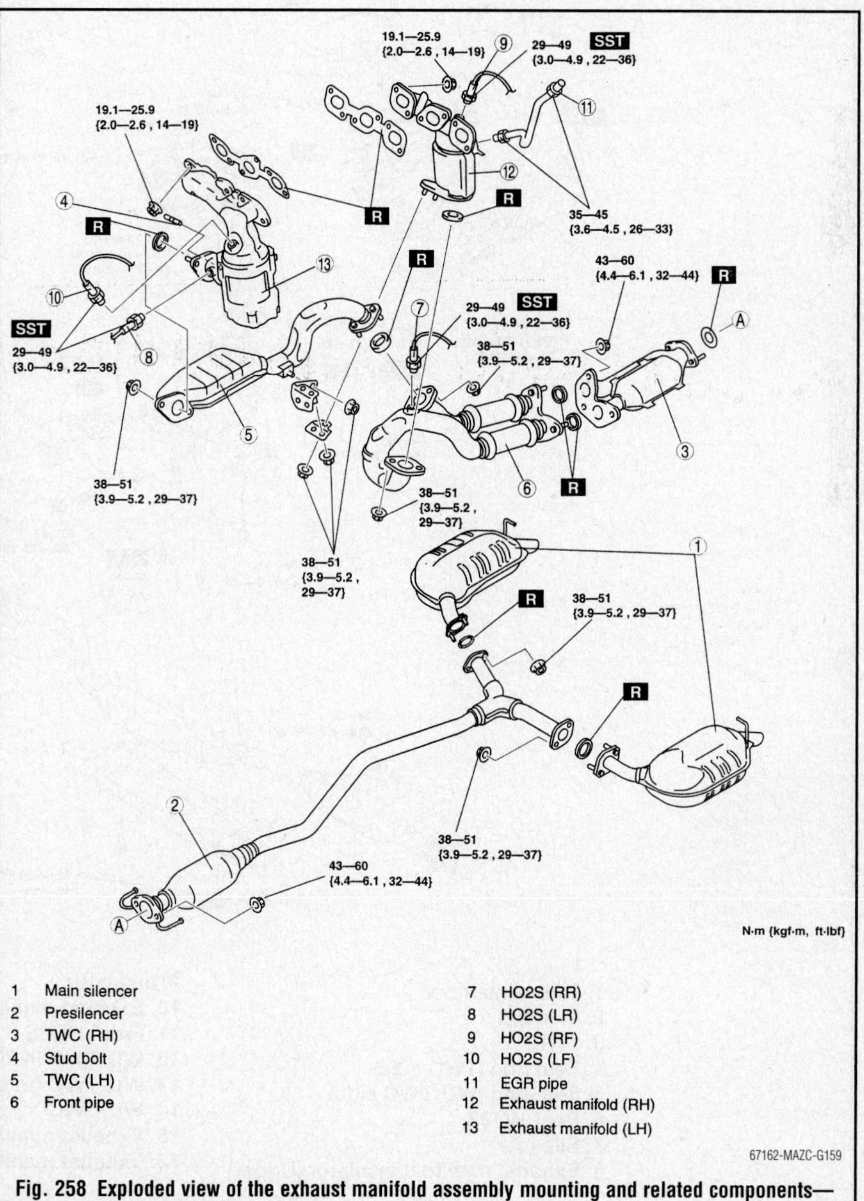

Fig. 258 Exploded view of the exhaust manifold assembly mounting and related components—Mazda6 with the 3.0L (AJ) engine

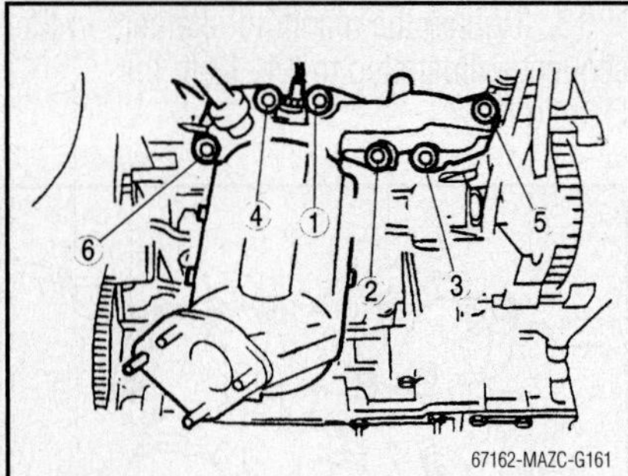

Fig. 260 Right side exhaust manifold torque sequence—Mazda6 with the 3.0L (AJ) engine

14. Install the remaining components on the reverse order of removal, referring to the illustration for torque specifications.

2.3L (L3 with TC) Engines

See Figures 261 and 262.

1. Before servicing the vehicle, refer to the precautions section.
2. Disconnect the negative battery cable.
3. Remove the main silencer.
4. Remove the Three Way Converter (TWC).
5. Remove the rear oxygen sensor.
6. Remove the front pipe.
7. Remove the charge air cooler and the charge air cooler bracket.
8. Remove the cowl grille.
9. Remove the insulators
10. Remove the manifold upper and lower insulator.
11. Remove the front oxygen sensor.
12. Set the alternator out of the way.
13. Disconnect the vacuum hose on the master cylinder side.
14. Remove the warm up three way converter.
15. Remove the exhaust manifold and gasket. Discard the gasket.

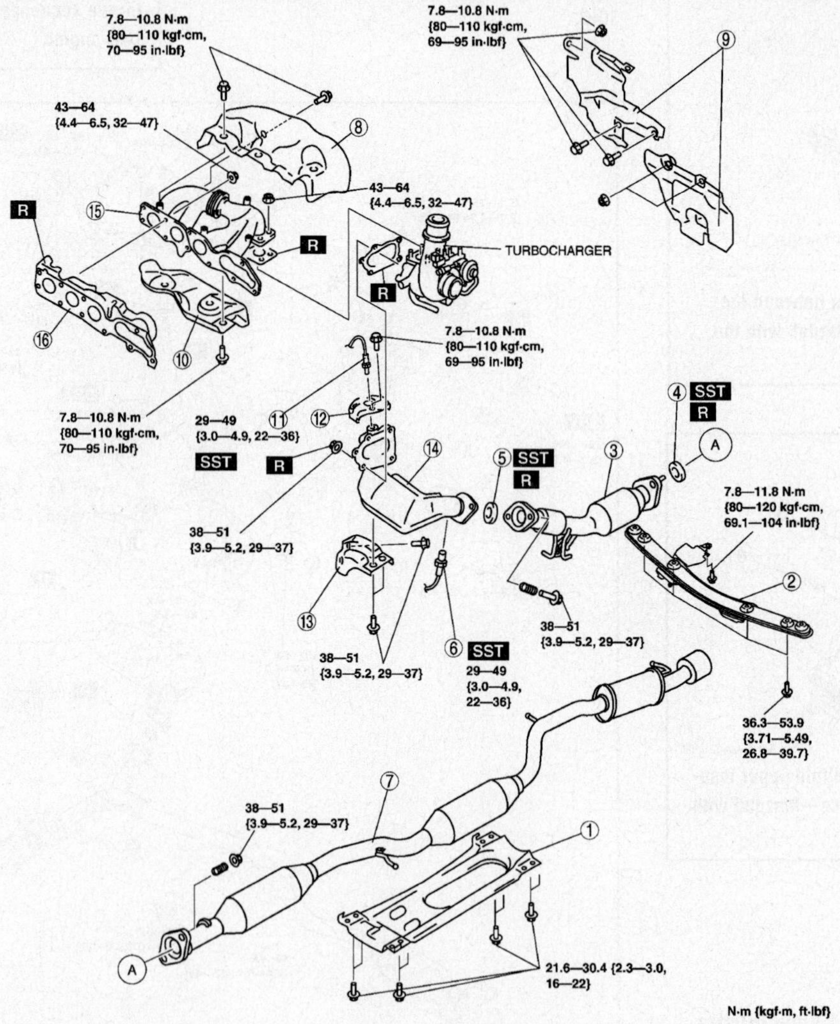

1. Tunnel member
2. Member
3. TWC
4. Seal ring (TWC side)
5. Seal ring (WU-TWC side)
6. Rear HO2S
7. Silencer
8. Exhaust manifold insulator (Upper)
9. Insulator
10. Exhaust manifold insulator (Lower)
11. Front HO2S
12. WU-TWC insulator
13. WU-TWC bracket
14. WU-TWC
15. Exhaust manifold
16. Exhaust manifold gasket

22140_MAZD_G0177

Fig. 261 Exhaust manifold and related components 2.3L (L3 with TC) engine

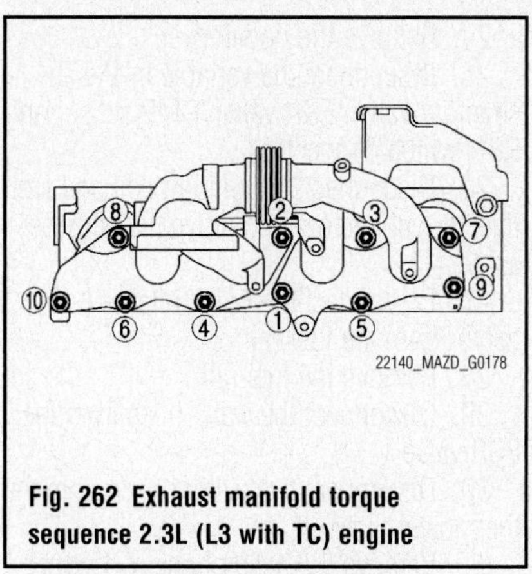

Fig. 262 Exhaust manifold torque sequence 2.3L (L3 with TC) engine

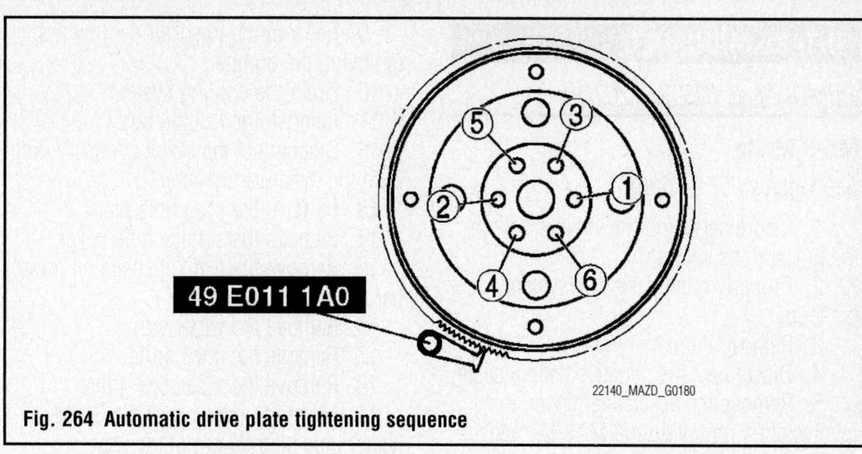

Fig. 264 Automatic drive plate tightening sequence

To install:

16. Installation is the reverse of removal, note the following:
    a. Clean the manifold mating surfaces.
    b. Install a new gasket and the manifold.
    c. Tighten the manifold retainers in the sequence illustrated to 32–47 ft. lbs. (43–64 Nm).
    d. Tighten the installation nuts between the catalytic converter and middle pipe. Tighten to 32–47 ft. lbs. (43–64 Nm).
    e. Tighten the installation nuts between main silencer and middle pipe in the order of left side, then right side. Tighten to 29–37 ft. lbs. (38–51 Nm).

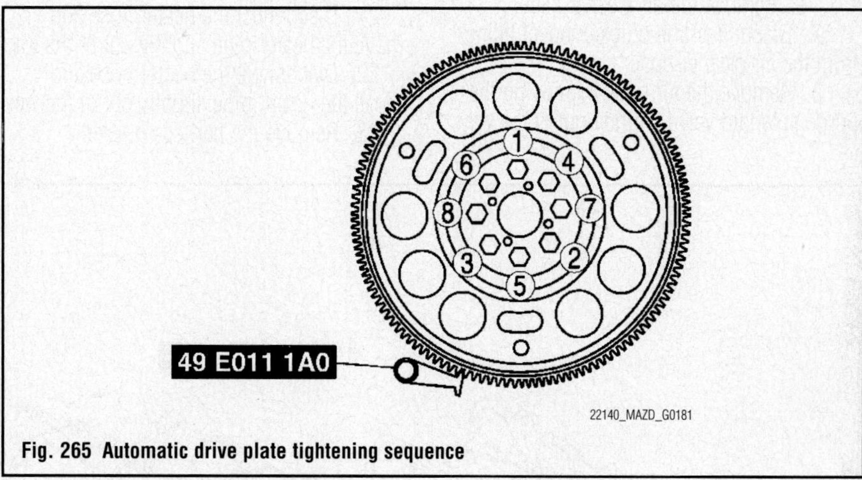

Fig. 265 Automatic drive plate tightening sequence

FLYWHEEL

REMOVAL & INSTALLATION

All 4 Cylinder Engines

See Figures 263 through 265.

1. Before servicing the vehicle, refer to the precautions section.
2. Remove the transmission assembly.
3. Remove the flywheel on manual transmissions, or the drive plate on automatic transmissions.

To install:

4. Install the flywheel or drive plate.
5. Tighten the drive plate and flywheel mounting bolts in two or three steps to 80–85.5 ft. lbs. (108–116 Nm). In the order as shown.
6. Install the transmission assembly.

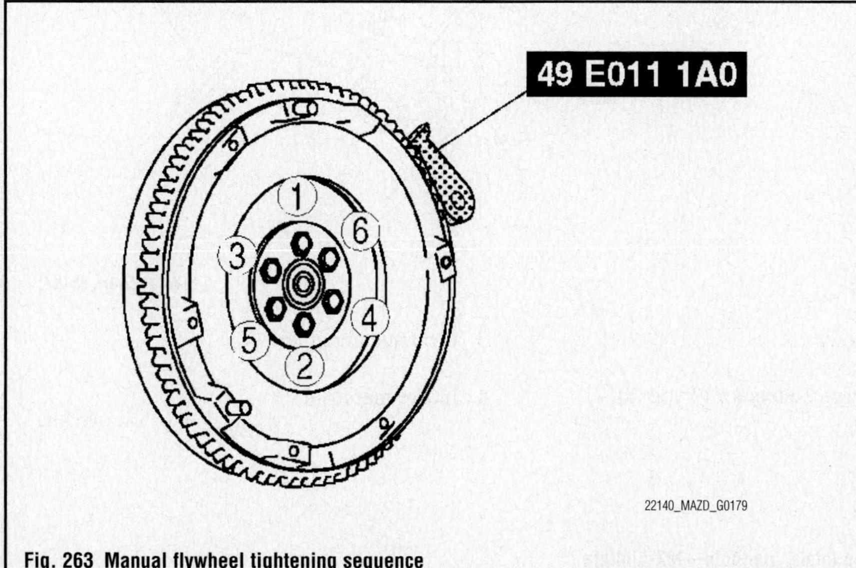

Fig. 263 Manual flywheel tightening sequence

3.0L (AJ) Engine

See Figure 265.

1. Before servicing the vehicle, refer to the precautions section.
2. Remove the transaxle.
3. Remove the drive plate mounting bolts.
4. Remove the drive plate.

To install:

5. Clean the crankshaft threads before tightening the bolts. The threads may be damaged if the bolts are tightened with any old sealant remaining.
6. Clean the crankshaft thread holes.
7. Install the drive plate.
8. Tighten the drive plate and flywheel mounting bolts in two or three steps to 54–64 ft. lbs. (37–87 Nm). In the order as shown.

INTAKE MANIFOLD

REMOVAL & INSTALLATION

MX-5 Miata

See Figures 267 and 268.

1. Before servicing the vehicle, refer to the precautions section.
2. Properly relieve the fuel system pressure.
3. Remove the battery cover.
4. Disconnect the negative battery cable.
5. Remove the air cleaner cover, this involves first removing the MAF/IAT sensor.
6. Remove the air cleaner element.
   a. Remove the air cleaner case.
7. Disconnect the quick connect fitting from the air cleaner hose.
8. Remove the air hose by moving the purge solenoid valve slightly out of the way.
9. Remove the fresh-air duct by first removing the bumper.
10. Drain the cooling system.
11. Remove the throttle body.
12. Disconnect the quick connect fitting from the dynamic chamber.
13. Remove the plug hole plate.
14. Remove the service hole cover.
15. Remove the front suspension tower bar.
16. Remove the wiper arm.
17. Remove the cowl grille.
18. Remove the side cowl grille.
19. Move the cooler pipe No.3 and heater pipe slightly out of the way.
20. Remove the service hole cover.
21. Disconnect the heater hose and move the heater pipe slightly out of the way.
22. Disconnect the heater hose and move the heater pipe slightly out of the way.
23. Remove the harness bracket.
24. Remove the undercover.
25. Disconnect the variable intake air solenoid valve, EGR valve, CMP sensor and PSP switch connectors.
26. Disconnect the ignition coil and fuel injector connectors and move the harness aside.
27. Disconnect the quick release connector from the fuel rail.
28. Remove the fuel rail.
29. Disconnect the water hose from the EGR valve.
30. Disconnect two water hoses from the thermostat.
31. Remove the heater hose and heater pipe from the dynamic chamber.
32. Remove the variable intake air solenoid valve.
33. Remove the dynamic chamber installation bolts.
34. Remove the EGR pipe.

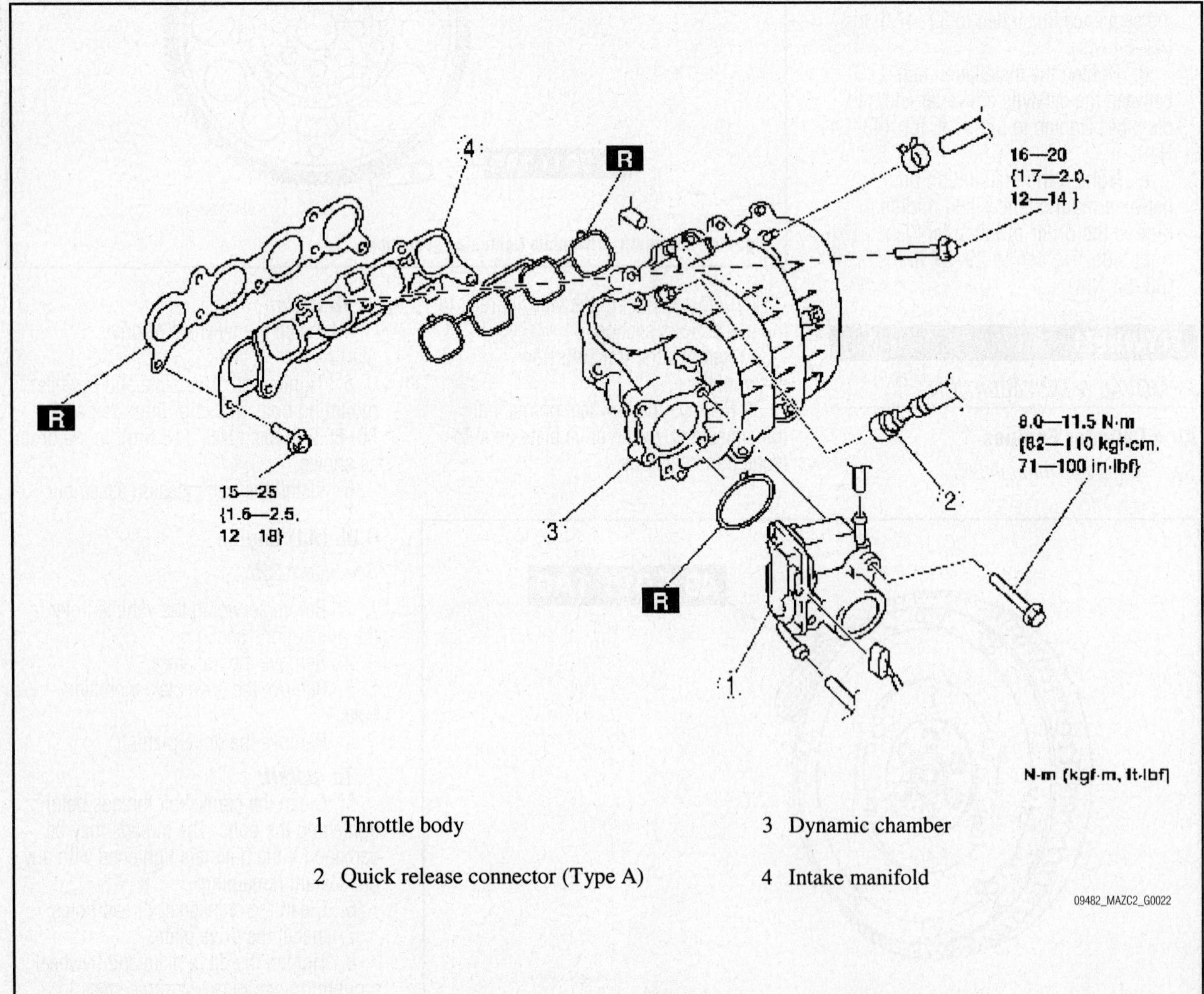

Fig. 267 Exploded view of the dynamic chamber and intake manifold—MX-5 Miata

3
1
2
4
09482_MAZC2_G0021

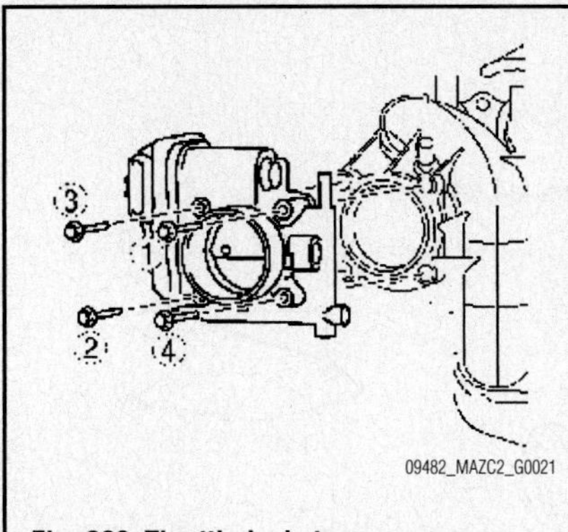

Fig. 268 Throttle body torque sequence—MX-5 Miata

35. Disconnect the connector from the A/C compressor.

36. Disconnect the knock sensor connector.

37. Move the vacuum hose between the purge solenoid valve and the charcoal canister aside.

38. Move the clutch release cylinder aside.

39. Disconnect the evaporative hose with the dynamic chamber raised.

40. Remove the dynamic chamber.

41. Remove the intake manifold.

To install:

42. Install the intake manifold and tighten the bolts to 15–25 Nm (12–18 ft. lbs.).

43. Install the dynamic chamber and tighten the bolts to 16–20 Nm (12–14 ft. lbs.)

44. Installation is the reverse order of removal, make sure to tighten the throttle body bolts in the sequence shown to 8–11 Nm (71–100 in. lbs).

45. Check and adjust all fluid levels, start the vehicle and check for proper operation.

Mazda3

See Figures 269 through 273.

1. Before servicing the vehicle, refer to the precautions section.

2. Drain the cooling system.

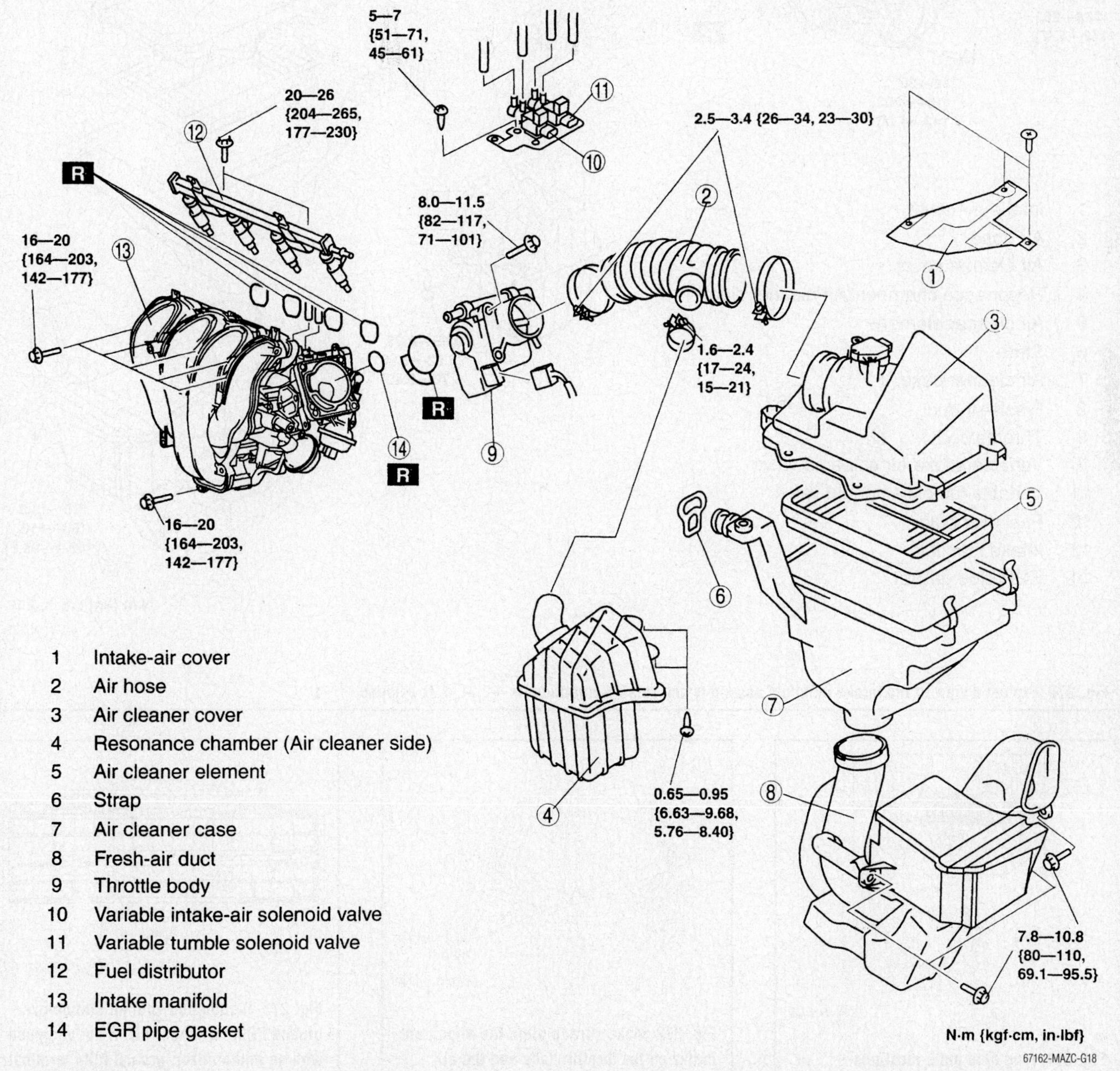

5—7 {51—71, 45—61}
20—26 {204—265, 177—230}
2.5—3.4 {26—34, 23—30}
8.0—11.5 {82—117, 71—101}
16—20 {164—203, 142—177}
1.6—2.4 {17—24, 15—21}
16—20 {164—203, 142—177}
0.65—0.95 {6.63—9.68, 5.76—8.40}
7.8—10.8 {80—110, 69.1—95.5}
R
R
R
1 Intake-air cover
2 Air hose
3 Air cleaner cover
4 Resonance chamber (Air cleaner side)
5 Air cleaner element
6 Strap
7 Air cleaner case
8 Fresh-air duct
9 Throttle body
10 Variable intake-air solenoid valve
11 Variable tumble solenoid valve
12 Fuel distributor
13 Intake manifold
14 EGR pipe gasket
N·m {kgf·cm, in·lbf}
67162-MAZC-G18

Fig. 269 Exploded view of the intake manifold assembly and related components—2.0L (LF) engines

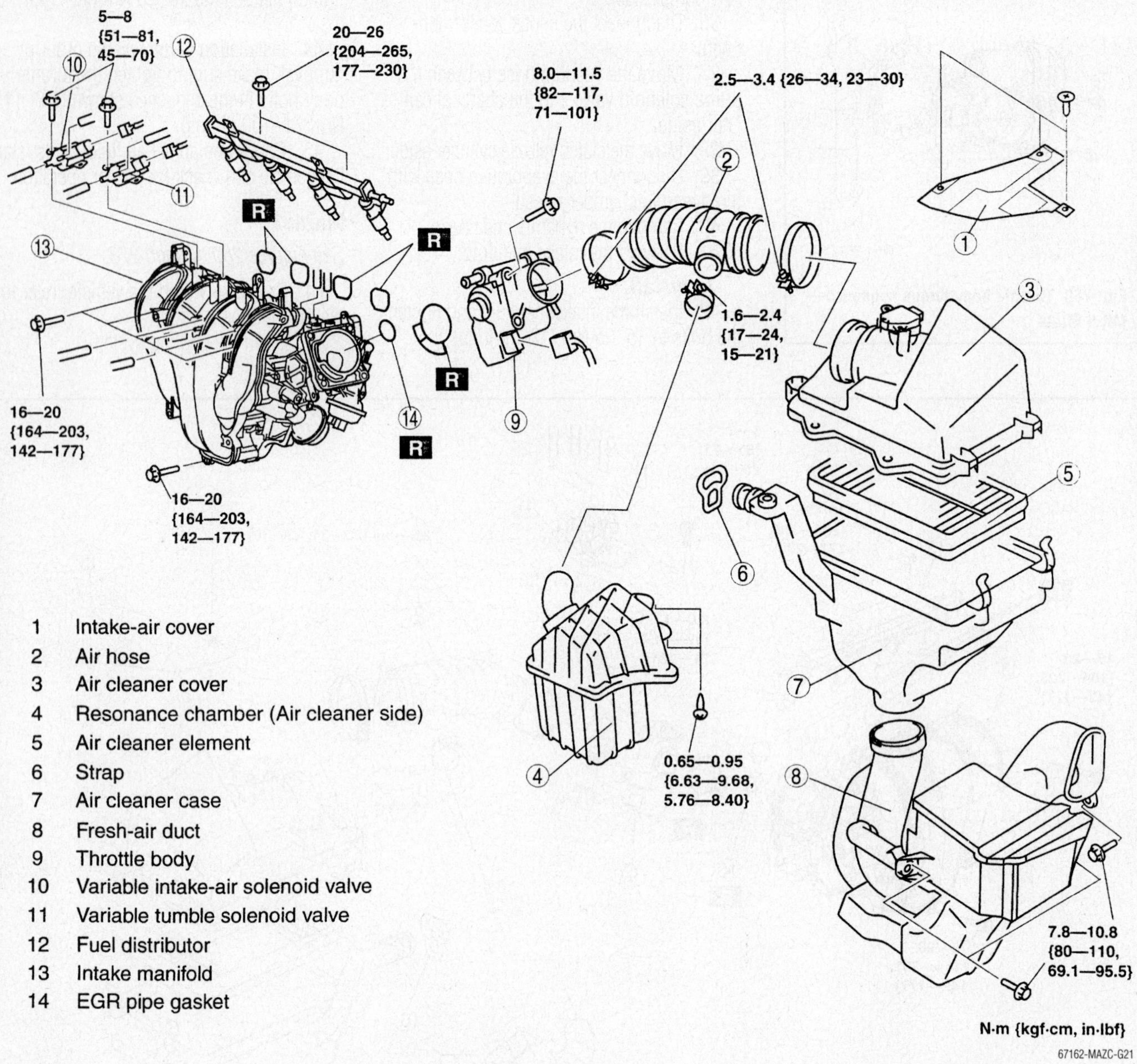

5—8
{51—81,
45—70}
20—26
{204—265,
177—230}
8.0—11.5
{82—117,
71—101}
2.5—3.4 {26—34, 23—30}
1.6—2.4
{17—24,
15—21}
16—20
{164—203,
142—177}
16—20
{164—203,
142—177}
0.65—0.95
{6.63—9.68,
5.76—8.40}
7.8—10.8
{80—110,
69.1—95.5}
R
1 Intake-air cover
2 Air hose
3 Air cleaner cover
4 Resonance chamber (Air cleaner side)
5 Air cleaner element
6 Strap
7 Air cleaner case
8 Fresh-air duct
9 Throttle body
10 Variable intake-air solenoid valve
11 Variable tumble solenoid valve
12 Fuel distributor
13 Intake manifold
14 EGR pipe gasket
N·m {kgf·cm, in·lbf}
67162-MAZC-G21

Fig. 270 Exploded view of the intake manifold assembly and related components—2.3L (L3) engines

RUBBER
RUBBER
INSTALLATION
AREA ON
ENGINE
SIDE
67162-MAZC-G02

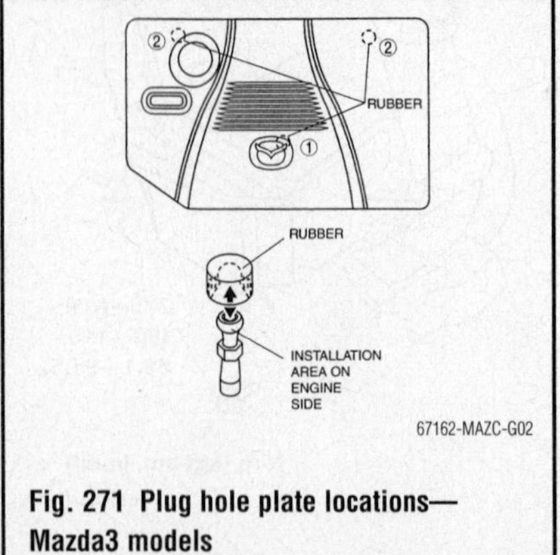

Fig. 271 Plug hole plate locations—Mazda3 models

ALIGNMENT
MARK
67162-MAZC-G20

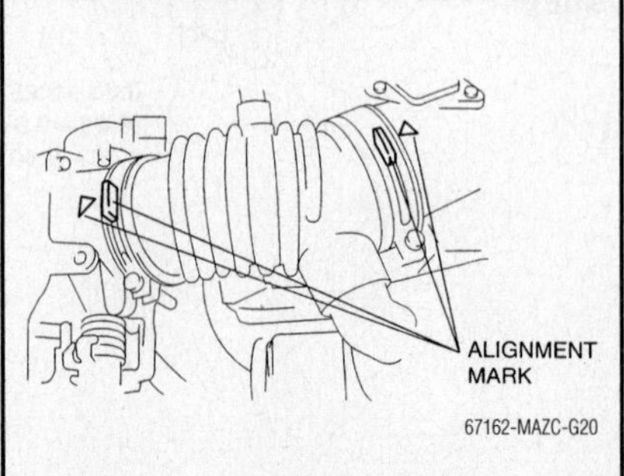

Fig. 272 Make sure to align the alignment marks on the throttle body and the air hose—Mazda3 models

PCM
WIRING HARNESS-SIDE CONNECTOR
1AQ
1AR
67162-MAZC-G22

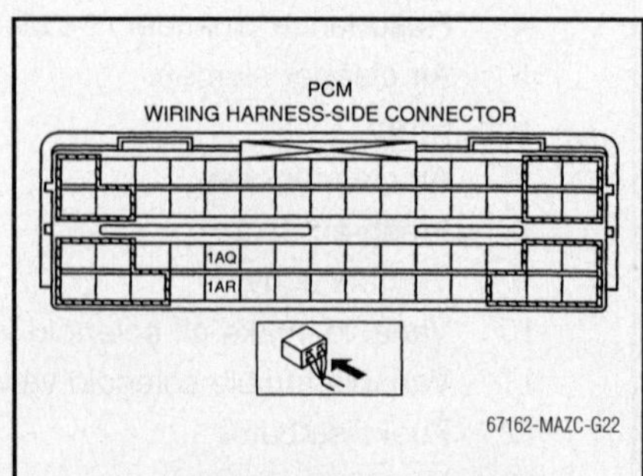

Fig. 273 If equipped with an immobilizer, ground PCM terminal 1AR, if not equipped with an immobilizer, ground PCM terminal 1AQ—Mazda3 models

3. Relieve the fuel system pressure.

4. Remove the plug hole plate by lifting off and removing the plug hole plate from the areas shown in the accompanying illustration.

5. Remove the battery cover and battery duct.

6. Remove the undercover and disconnect the negative battery cable.

7. Remove the intake air cover, the air hose and the air cleaner element.

8. Remove the resonance chamber and air filter element.

9. Remove the strap and the air cleaner case.

10. Remove the fresh air duct.

11. Remove the throttle body.

12. Remove the variable intake air solenoid valve and the variable tumble solenoid valve.

13. Remove the fuel rail and injectors as an assembly.

14. Disconnect the vacuum hose from the intake manifold.

15. Remove the engine oil dipstick tube.

16. Remove the intake manifold bolts, the manifold and gasket.

17. Discard the gasket.

18. Clean the mating surfaces of any gasket material.

19. If necessary, remove the Exhaust Gas Recirculation (EGR) pipe gasket.

To install:

20. If removed, install the EGR pipe gasket.

21. Install a new gasket and the manifold. Tighten the manifold bolts to 142–177 inch lbs. (16–20 Nm).

22. Install the engine oil dipstick tube.

23. Connect the vacuum hose to the intake manifold.

24. Install the fuel rail and injectors.

25. Install the variable intake air solenoid valve and the variable tumble solenoid valve.

26. Install the throttle body and tighten the bolts to 71–101 inch lbs. (8–11.5 Nm).

27. Install the fresh air duct.

28. Verify the rubber mounts on the battery support bracket are still in place. Insert the air cleaner case into the rubber mounts, using soapy water if necessary to ease installation.

29. Use the strap to secure the shroud panel and the air cleaner case as shown in the accompanying illustration.

30. Install the air cleaner element and the resonance chamber.

31. Install the air cleaner cover.

32. Install the air hose, make sure to align the alignment marks on the throttle body and the air hose.

33. Install the intake air cover.

34. Install the undercover.

35. Connect the negative battery cable.

36. Install the battery duct and cover.

37. Install the plug hole plate.

38. Fill the cooling system.

39. Start the vehicle and check for leaks as follows:

a. Using a jumper wire, ground the Powertrain Control Module (PCM) terminals. If equipped with an immobilizer, ground terminal 1AR, if not equipped with an immobilizer, ground terminal 1AQ. Refer to the illustration for terminal location.

b. Turn the ignition switch the ON position to activate the fuel pump.

c. Check the hoses, clips and other fuel system components for leaks.

d. If there are any leaks, replace the fuel hoses and clips. If the is damage to the seal on the fuel pipe side, replace the pipe.

e. The system must be leak free for five minutes with the terminal grounded. If any component is replaced because of a system, leak, turn the ignition key OFF; remove the jumper wire from the terminal. reapply the jumper wire, turn the ignition On and check for leaks.

Mazda6

2.3L (L3) Engines

See Figures 274 and 275.

1. Before servicing the vehicle, refer to the precautions section.

2. Drain the cooling system.

3. Relieve the fuel system pressure.

4. Disconnect the negative battery cable.

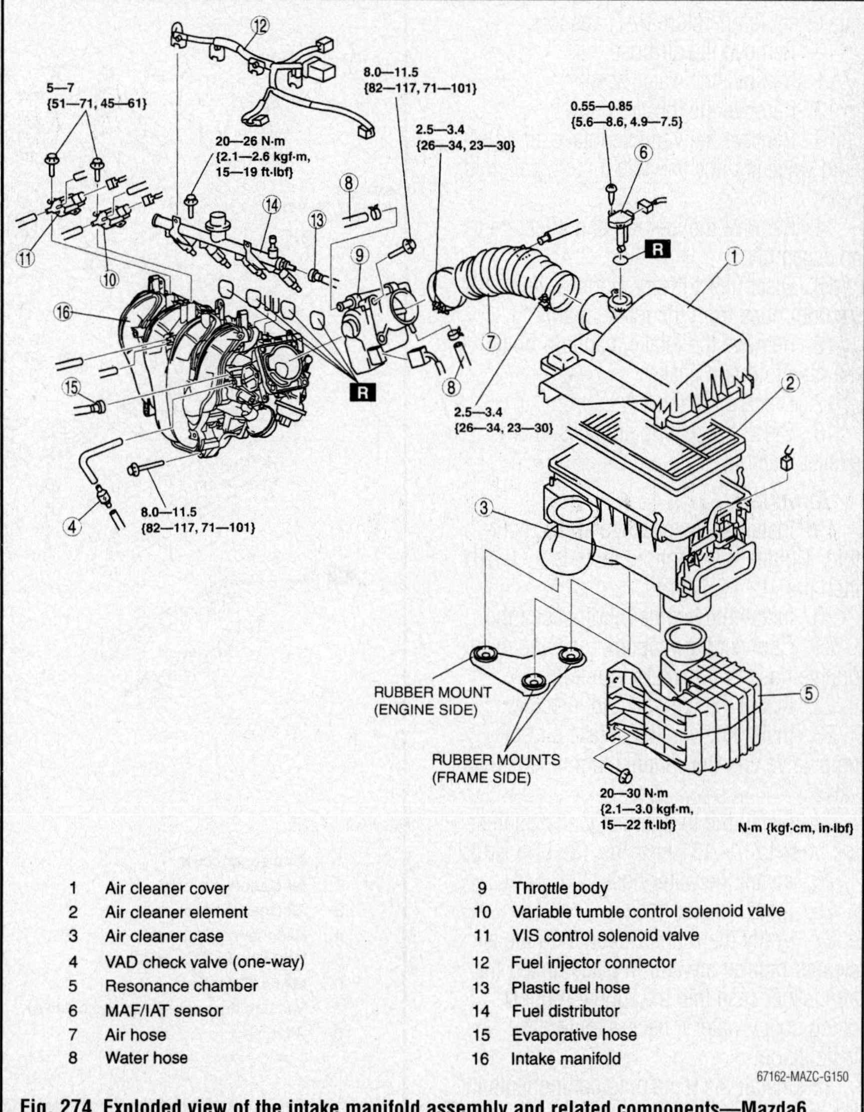

Fig. 274 Exploded view of the intake manifold assembly and related components—Mazda6 with the 2.3L (L3) engine

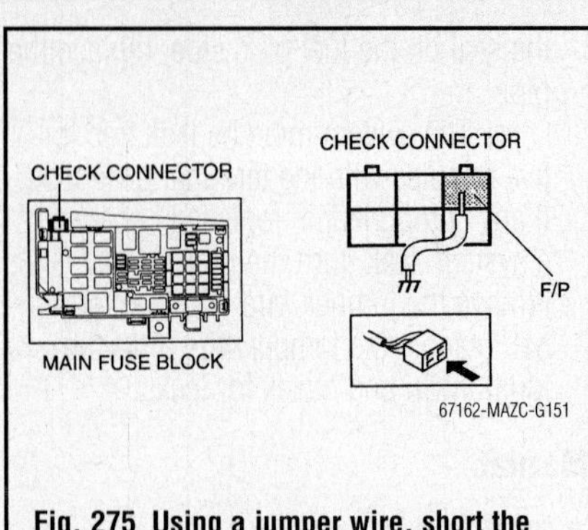

Fig. 275 Using a jumper wire, short the check connector to terminal F/P to a body ground—Mazda6 models

5. Remove the intake air cover and element.
6. Remove the air cleaner case.
7. Remove the VAD check valve.
8. Remove the left front mud guard and resonance chamber.
9. Remove the Mass Airflow (MAF) and Intake Air Temperature (IAT) sensors.
10. Remove the air hose.
11. Remove the water hose.
12. Remove the throttle body.
13. Remove the variable intake air solenoid valve and the variable tumble solenoid valve.
14. Remove the fuel rail and injectors as an assembly.
15. Disconnect the evaporative hose and vacuum hose from the intake manifold.
16. Remove the intake manifold bolts, the manifold and gasket.
17. Discard the gasket.
18. Clean the mating surfaces of any gasket material.

To install:

19. Install a new gasket and the manifold. Tighten the manifold bolts to 171–101 inch lbs. (8–11.5 Nm).
20. Install the engine oil dipstick tube.
21. Connect the vacuum hose and evaporative hose to the intake manifold.
22. Install the fuel rail and injectors.
23. Install the variable intake air solenoid valve and the variable tumble solenoid valve.
24. Install the throttle body and tighten the bolts to 71–101 inch lbs. (8–11.5 Nm).
25. Install the water hose.
26. Install the air hose.
27. Verify the rubber mounts on the air cleaner bracket are still in place. Insert the air cleaner case into the rubber mounts, using soapy water if necessary to ease installation.
28. Install the remaining components in the reverse order of removal.
29. Fill the cooling system.
30. Start the vehicle and check for leaks as follows:
  a. Using a jumper wire, short the check connector to terminal F/P to a body ground.
  b. Turn the ignition switch the ON position to activate the fuel pump.
  c. Check the hoses, clips and other fuel system components for leaks.
  d. If there are any leaks, replace the fuel hoses and clips. If the is damage to the seal on the fuel pipe side, replace the pipe.
  e. The system must be leak free for five minutes with the terminal grounded. If any component is replaced because of a system, leak, turn the ignition key OFF; remove the jumper wire from the terminal. reapply the jumper wire, turn the ignition On and check for leaks.

3.0L (AJ) Engines

See Figures 276 through 278.

1. Before servicing the vehicle, refer to the precautions section.
2. Remove the left front mud guard and resonance chamber.
3. Remove the Mass Airflow (MAF) and Intake Air Temperature (IAT) sensors.
4. Remove the purge solenoid valve vacuum hose.
5. Remove the air hose.
6. Remove the water hose.
7. Remove the throttle body.
8. Remove the upper radiator hose, disconnect the Exhaust Gas Recirculation (EGR) pipe, EGR electrical connector and valve.
9. Remove the dynamic chamber bolts, chamber and gasket and discard the gasket.

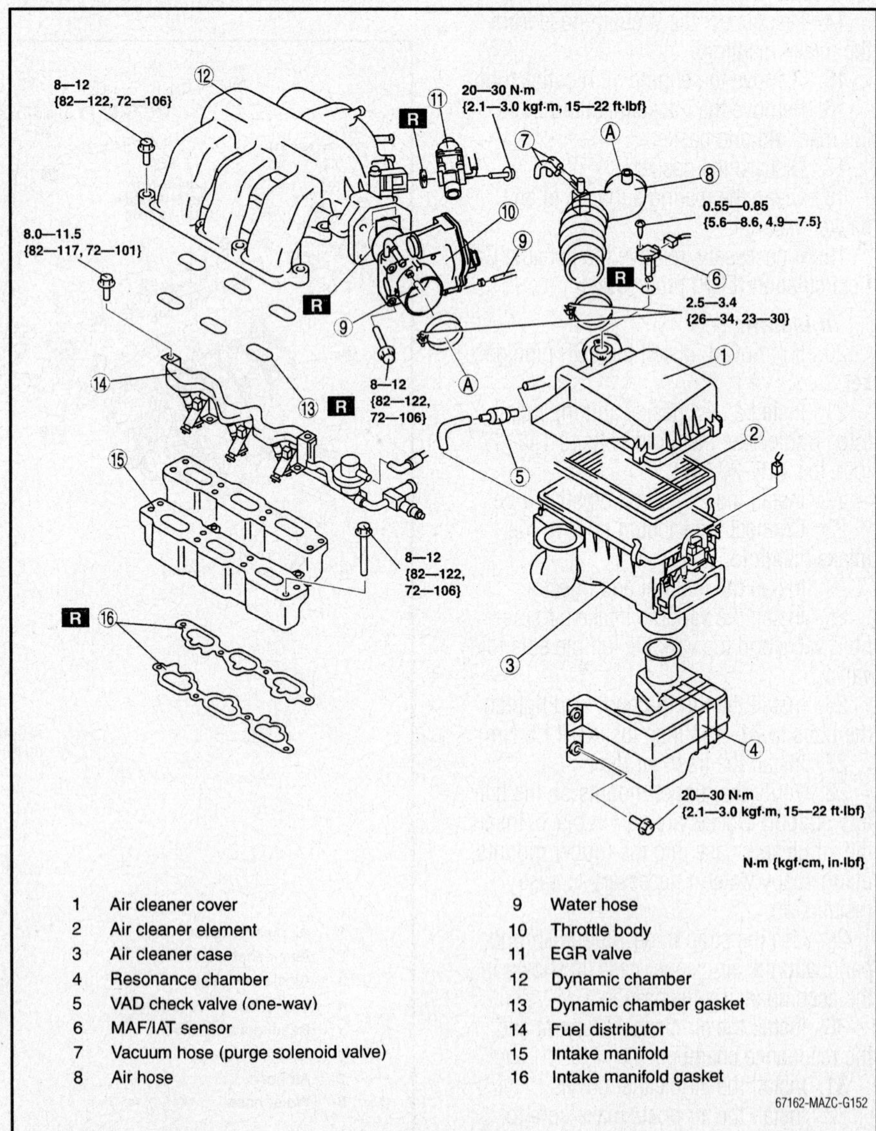

Fig. 276 Exploded view of the intake manifold assembly and related components—Mazda6 with the 3.0L (AJ) engine

10. Remove the fuel rail and injectors as an assembly.
11. Remove the intake manifold bolts, the manifold and gasket.
12. Discard the gasket.
13. Clean the mating surfaces of any gasket material.

To install:

14. Installation is the reverse of removal, please note the following:
    a. Install a new gasket and the manifold. Tighten the manifold bolts in the sequence illustrated to 72–106 inch lbs. (8–12 Nm).
    b. Install a new dynamic chamber gasket and the chamber. Tighten the bolts in the sequence illustrated to 72–106 inch lbs. (8–12 Nm).
    c. Install the EGR valve and tighten the retainers to the specifications shown in the accompanying illustration.
    d. Connect the EGR connector, and pipe.
15. Fill the cooling system.
16. Start the vehicle and check for leaks as follows:
    a. Using a jumper wire, short the check connector to terminal F/P to a body ground.
    b. Turn the ignition switch the ON position to activate the fuel pump.
    c. Check the hoses, clips and other fuel system components for leaks.
    d. If there are any leaks, replace the fuel hoses and clips. If the is damage to the seal on the fuel pipe side, replace the pipe.
    e. The system must be leak free for five minutes with the terminal grounded. If any component is replaced because of a system, leak, turn the ignition key OFF; remove the jumper wire from the terminal. reapply the jumper wire, turn the ignition On and check for leaks.

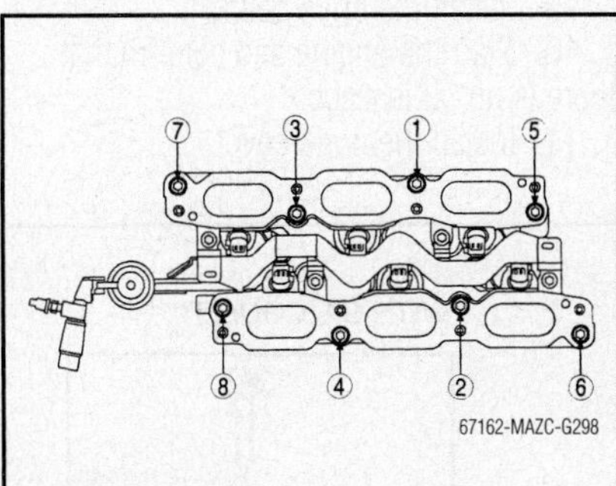

Fig. 277 Tighten the lower intake manifold bolts in this sequence—Mazda6 with the 3.0L (AJ) engine

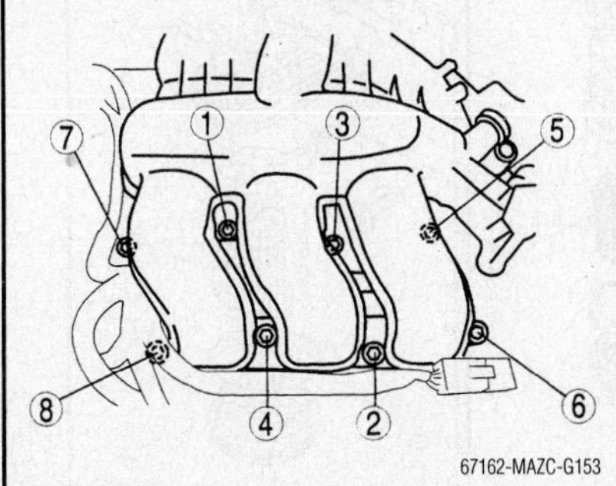

Fig. 278 Tighten the dynamic chamber bolts in this sequence—Mazda6 with the 3.0L (AJ) engine

Mazdaspeed3 and Mazdaspeed6—2.3L (L3 with TC) Engines

See Figure 279.

1. Disconnect the negative battery cable.
2. Before servicing the vehicle, refer to the precautions section.
3. Remove the charge air cooler cover.
4. Remove the battery and tray.
5. Remove the air cleaner.
6. Remove the resonance chamber, remove the left hand front mudguard before removing the resonance chamber.
7. Disconnect the MAF/IAT sensor.
8. Remove the charge air cooler.
9. Remove the air bypass valve
10. Remove the throttle body.
11. Remove the air hose.
12. Remove the intake manifold.
13. Remove the air bypass valve.
14. Drain and recycle the engine coolant.
15. Remove the throttle body.

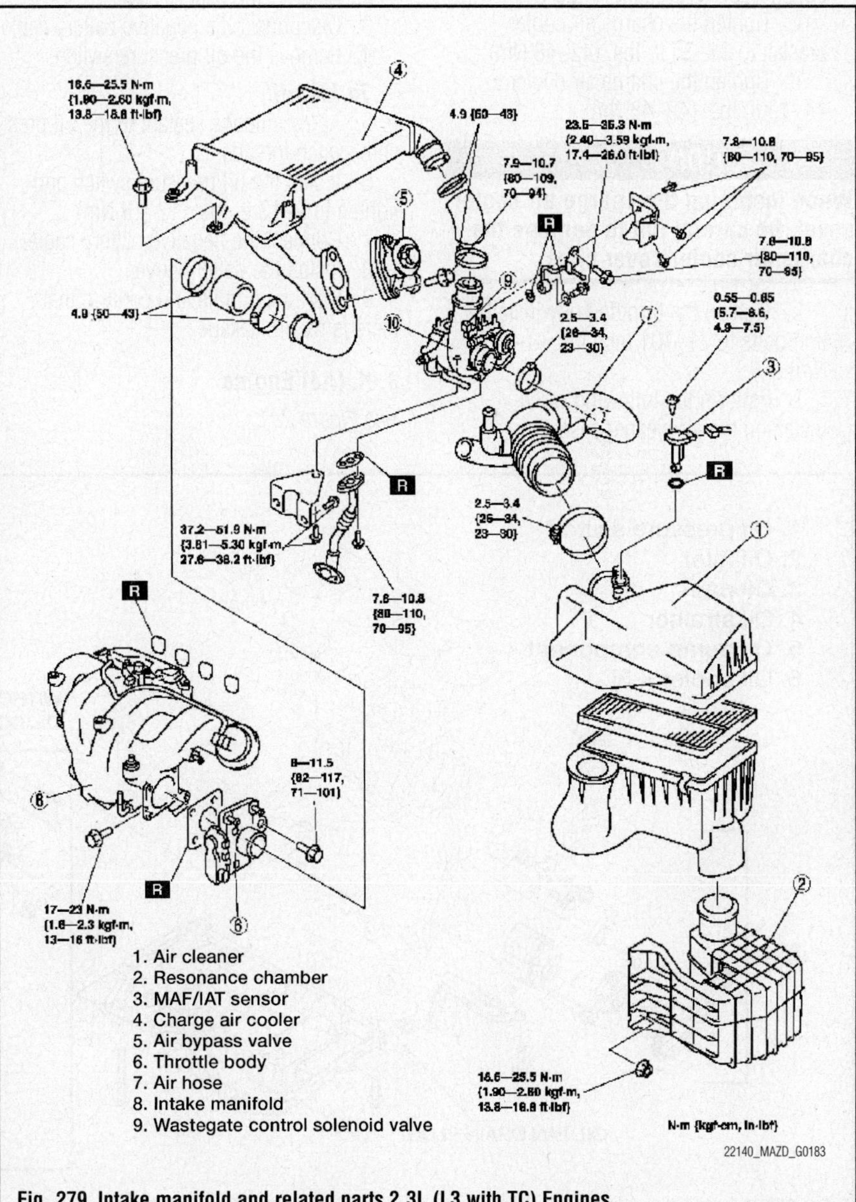

Fig. 279 Intake manifold and related parts 2.3L (L3 with TC) Engines

16. Disconnect the EGR valve connector.
17. Remove the air hose.
18. Remove the high pressure fuel pump bracket and the EGR pipe bracket.
19. Remove the fuel delivery pipe cover.
20. Remove the oil level gauge pipe.
21. Remove the drive belt.
22. Remove the power steering pump out of the way with the lines still attached.
23. Remove the vacuum hose.
24. Remove the intake manifold.

To install:

25. Installation is the reverse of removal, note the following:
   a. Tighten the intake manifold bolts to 13–16 ft. lbs. (17–23 Nm).
   b. Tighten the fuel delivery pipe cover bolts to 13–16 ft. lbs. (17–23 Nm).
   c. Tighten the charge air cooler bracket to 31–35 ft. lbs. (42–48 Nm).
   d. Tighten the charge air cooler to 14–19 ft. lbs. (42–48 Nm).

CAUTION

When installing the charge air cooler cover, be careful not to damage the charge air cooler cover clips.

   e. Tighten the throttle body installation bolts to 71–101 inch lbs. (8–11.5 Nm).
   f. Perform the following when installing the air cleaner case:
   - Verify that the rubber mounts are set in the air cleaner bracket
   - Install the projections on the frame side.
   - Verify that the projections on the frame side are installed securely.
   - Install the projection on the engine side.
   - Verify that the projection on the engine side installed securely.

OIL PRESSURE SENSOR

REMOVAL & INSTALLATION

All 4 Cylinder Engines

See Figure 280.

1. Before servicing the vehicle, refer to the precautions section.
2. Remove the battery cover.
3. Disconnect the negative battery cable.
4. Remove the oil pressure switch.

To install:

5. Apply silicone sealant to the oil pressure switch threads.
6. Install the oil pressure switch and tighten to 9–13 ft. lbs. (12–18 Nm).
7. Connect the negative battery cable.
8. Install the battery cover.
9. Start the engine and confirm that there is no oil leakage.

3.0L (AJ) Engine

See Figure 281.

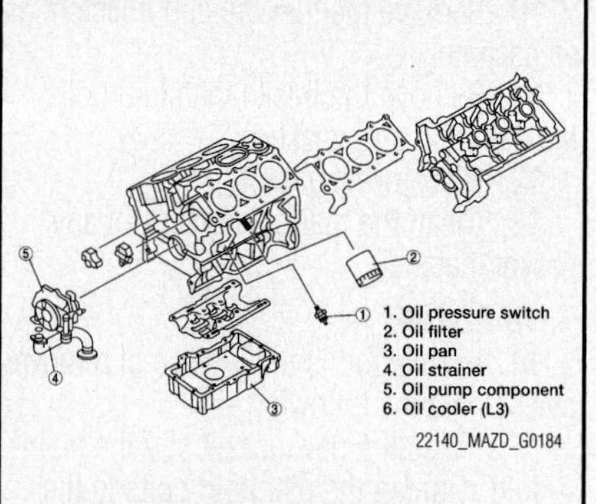

Fig. 281 3.0L (AJ) engine oil pressure switch and components

1. Before servicing the vehicle, refer to the precautions section.
2. Remove the undercover.
3. Remove the drive belt.
4. Remove the A/C compressor with the pipe still connected. Position the A/C compressor so that it is out of the way.
5. Remove the oil pressure switch.

To install:

6. Apply silicone sealant to the oil pressure switch threads as shown in the figure.
7. Install the oil pressure switch.
8. Install the A/C compressor.
9. Install the drive belt.
10. Start the engine and confirm that there is no oil leakage.
11. Install the undercover.

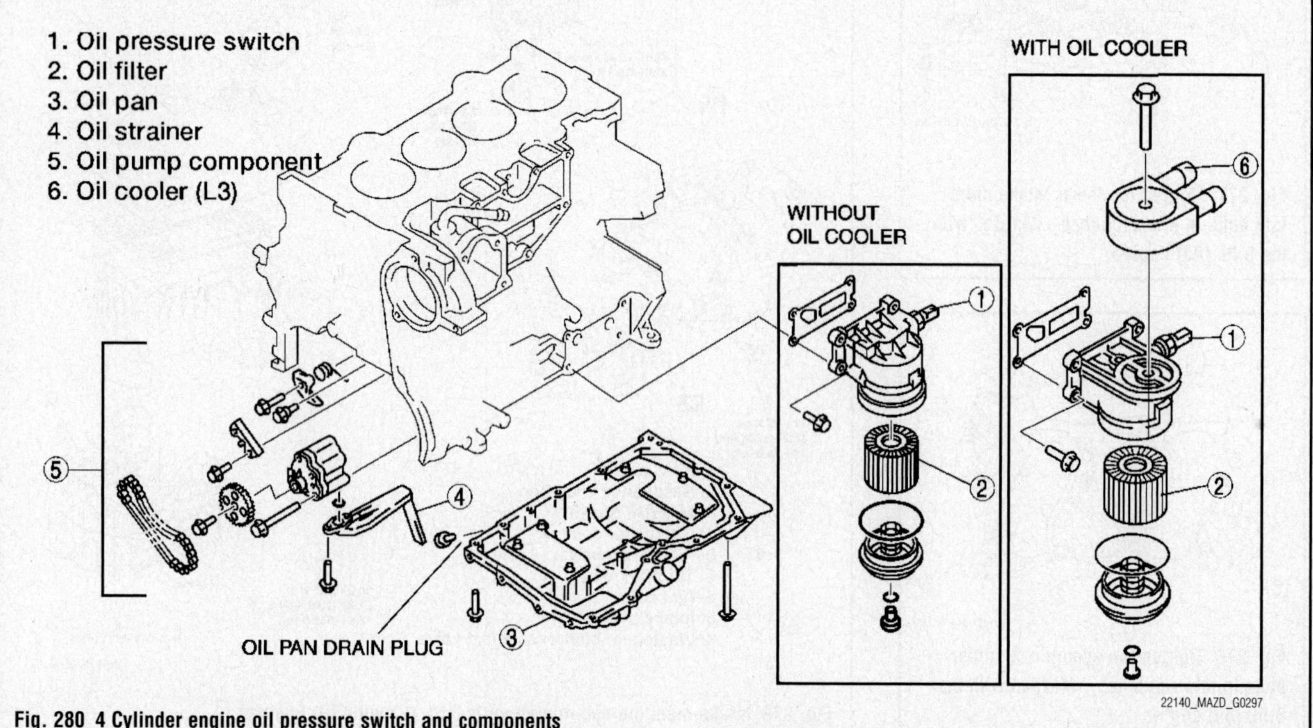

Fig. 280 4 Cylinder engine oil pressure switch and components

OIL PAN

REMOVAL & INSTALLATION

MX5-Miata

See Figures 282 through 289.

1. Before servicing the vehicle, refer to the precautions section.

CAUTION

Remove and install all parts when the engine is cold, otherwise they can cause severe burns or serious injury.

2. Remove the battery and battery tray.
3. Remove the air cleaner.
4. Drain the engine oil.
5. Loosen the water pump pulley bolt and remove the drive belt.
6. Remove the front suspension tower bar from the joint, right side and left side.
7. Remove the plug hole plate.
8. Remove the ignition coils.
9. Remove the P/S oil pump with the hose and pipe sill connected. Position the P/S oil pump out of the way.
10. Remove the Crankshaft Position (CKP) sensor.
11. Remove the engine front cover.
12. Remove the transverse member.
13. Remove the member bracket if equipped with a manual transmission.
14. Remove the windshield wiper arm.
15. Remove the cowl grille.

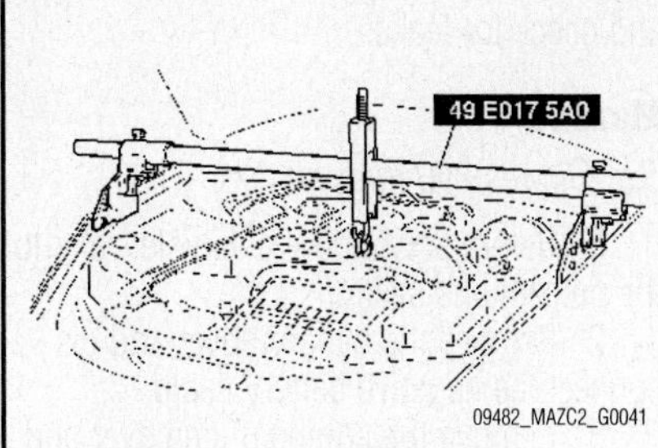

Fig. 283 Support the engine using the an engine support tool such as 49 ED17 5AO—MX-5 Miata

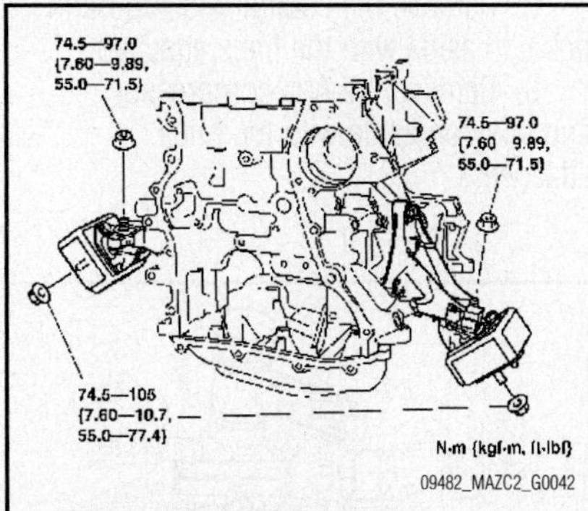

Fig. 284 Remove the engine mount rubber installation nuts—MX-5 Miata

16. Remove the side cowl grille.
17. Remove the engine compartment service hole cover.
18. Remove the front tires.

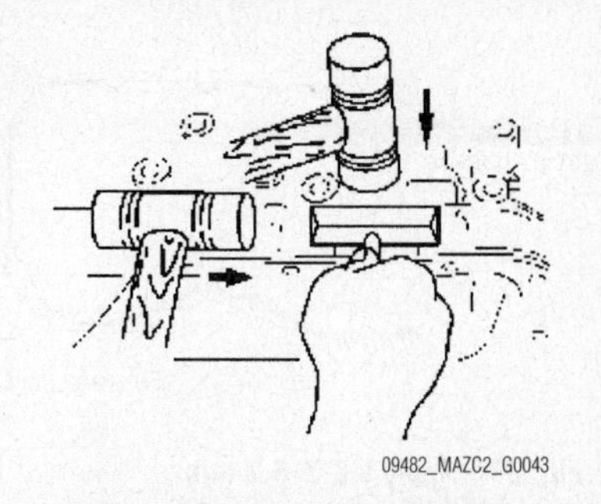

Fig. 285 Remove the oil pan using a separator tool to break the seal between the gasket and the pan/block—MX-5 Miata

19. Support the engine using the an engine support tool such as 49 ED17 5AO.
20. Remove the engine mount rubber installation nuts
21. Lift up the engine approx. 25 mm (0.98 in.) to assure clearance for the oil pan removal, then remove the oil pan bolts.
22. Remove the oil pan using a separator tool to break the seal between the gasket and the pan/block.

To install:

➡Apply the silicon sealant in a single, unbroken line around the whole perimeter.

➡Using bolts with the old seal adhering could cause cracks in the housing.

23. Completely clean and remove any oil, dirt, sealant or other foreign material that may be adhering to the housing and oil pan.
24. When reusing the oil pan installation bolts, clean any old sealant from the bolts.
25. Use a square ruler to align the oil pan and the cylinder block junction side on the engine front cover side.

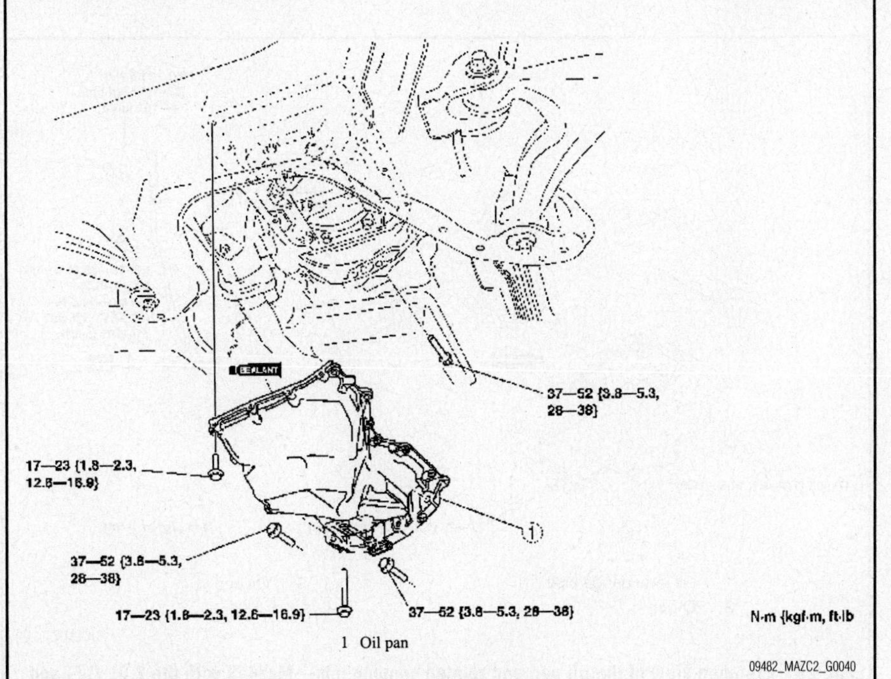

Fig. 282 Exploded view of the oil pan mounting—MX-5 Miata

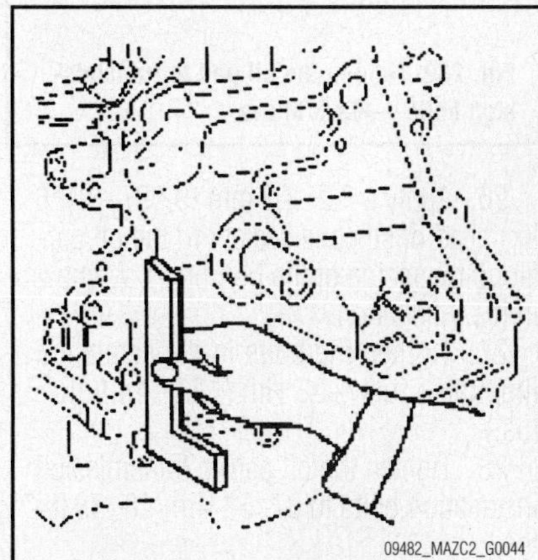

Fig. 286 Use a square ruler to align the oil pan and the cylinder block junction side on the engine front cover side—MX-5 Miata

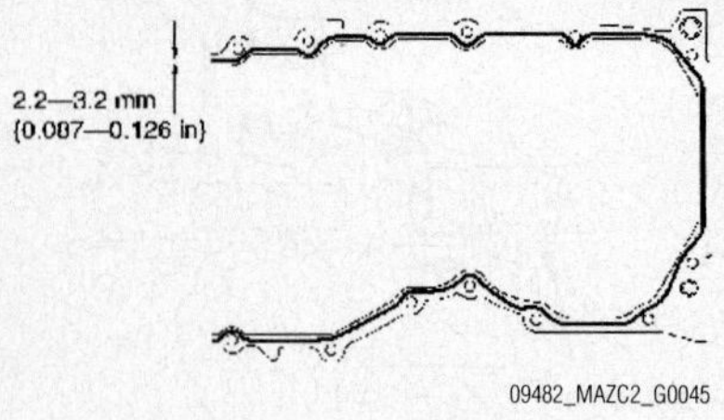

Fig. 287 Apply a 2.2–3.2 mm (0.087–0.126 in.) bead of silicone sealant to the oil pan along the inside of the bolt holes—MX-5 Miata

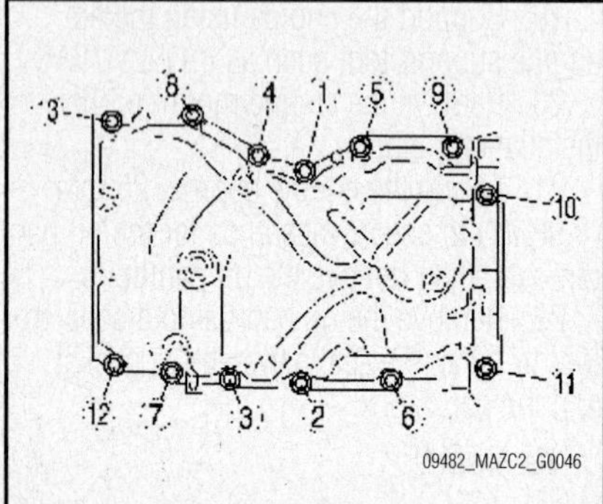

Fig. 288 Tighten the bolts in this sequence—MX-5 Miata

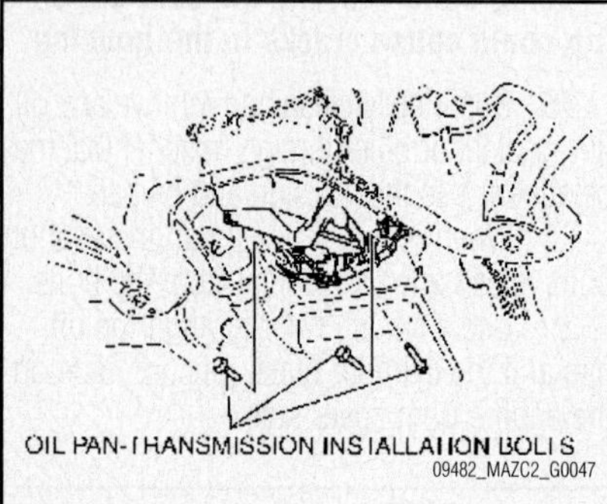

Fig. 289 Tighten the oil pan to transmission bolts —MX-5 Miata

26. Apply a 2.2–3.2 mm (0.087–0.126 in.) bead of silicone sealant to the oil pan along the inside of the bolt holes as shown in the illustration.
27. Tighten the bolts in the sequence illustrated to 17–23 Nm (12.6–16.9 ft. lbs.).
28. Tighten the oil pan to transmission installation bolts to 37–52 Nm (28–38 ft. lbs.).
29. Install the remaining components in the reverse order of removal, refer to the illustrations accompanying this procedure for component locations and related torque specifications.
30. Refill the engine oil, start the engine and check for leaks.

Mazda3

See Figures 290 through 295.

1. Before servicing the vehicle, refer to the precautions section.
2. Remove the battery cover and disconnect the negative battery cable.
3. Remove the engine undercover and splash shield as an assembly.
4. Remove the right front wheel.
5. Remove the plug hole plate.
6. Drain the engine oil.
7. Remove the drive belt.
8. Remove the coolant reservoir and position aside with the lines attached.
9. Remove the A/C compressor and position aside with the lines attached.
10. Remove the ignition coil and position the accelerator cable bracket aside.
11. Remove the Crankshaft Position (CKP) sensor.
12. Remove the engine front cover. Refer to the timing chain removal procedure in this section.
13. Remove the dipstick tube pipe and O–ring.
14. Remove the oil pan bolts and use a separator tool such as the one illustrated to separate the oil pan from the block.
15. Remove the oil pan.

To install:

16. Clean the oil pan mating surfaces.
17. Apply a 0.087–0.126 inch (2.2–3.2mm) bead of sealant around the perimeter of the oil pan as illustrated.
18. Use a square ruler to align the oil pan and the block junction side on the engine front cover side as illustrated.
19. Install the oil pan within 5 minutes of applying the sealant.
20. Install the lower oil pan bolts and tighten in sequence to 12.6–16.9 ft. lbs. (17–23 Nm) and the oil pan–to–transaxle bolts to 23–38 ft. lbs. (32–52 Nm).
21. When installing the CKP sensor perform the following:
   a. Remove the cylinder block lower blind plug.
   b. Install tool 49 JE01 061.
   c. Turn the crankshaft clockwise until the crankshaft is in the number 1 cylinder Top Dead Center position (the balance weight should be attached to tool 49 JE01 061).

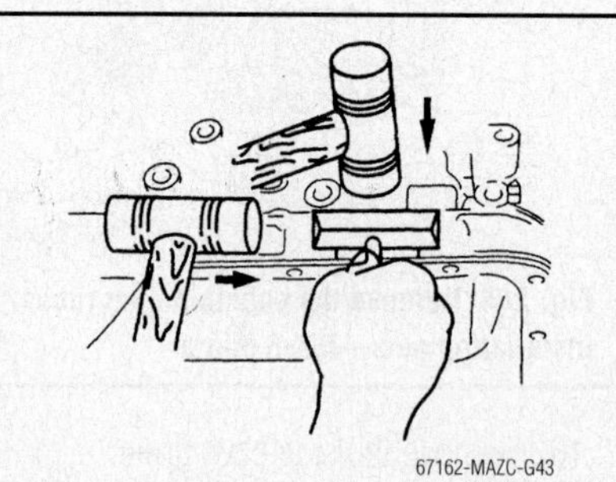

Fig. 290 Use a separator tool such as the one illustrated to separate the oil pan from the block—2.0L (LF) and 2.3L (L3) engines

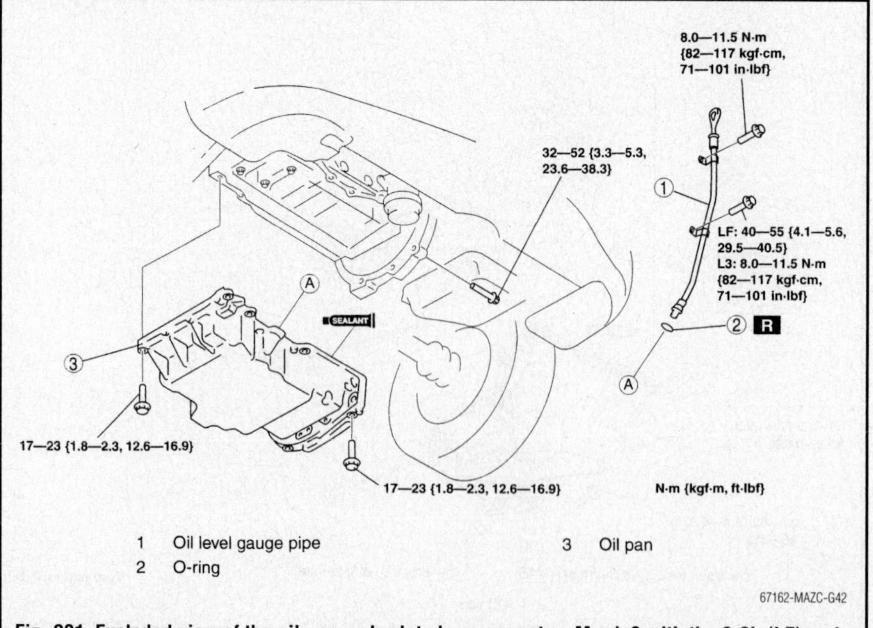

Fig. 291 Exploded view of the oil pan and related components—Mazda3 with the 2.0L (LF) and 2.3L (L3) engines

d. Using a ruler, mark the center line on the pulse wheel teeth on the crank pulley which is located at the 9th tooth counting counterclockwise from the empty space. Refer to the illustration for more detail.

**** CAUTION**

If you do not mark the center line correctly this will cause improper engine control for the ignition and fuel system, so be sure to mark the line carefully.

e. Install the CKP sensor making sure the mark on the sensor is aligned with the mark on the pulse wheel. Tighten the sensor bolt to 49–66 inch lbs. (5.5–7.5 Nm)and attach the electrical connector.

f. Remove the tool from the cylinder block lower blind plug.

g. Rotate the crankshaft clockwise 2 turns until you reach TDC. If not aligned properly, loosen the crankshaft pulley lock bolt and reinstall the lock bolt.

h. Install the cylinder block blind plug and tighten to 13–16 ft. lbs. (18–22 Nm).

22. Install the remaining components in the reverse order of removal.
23. Fill the engine with clean oil.
24. Start the vehicle, check for leaks and repair if necessary.

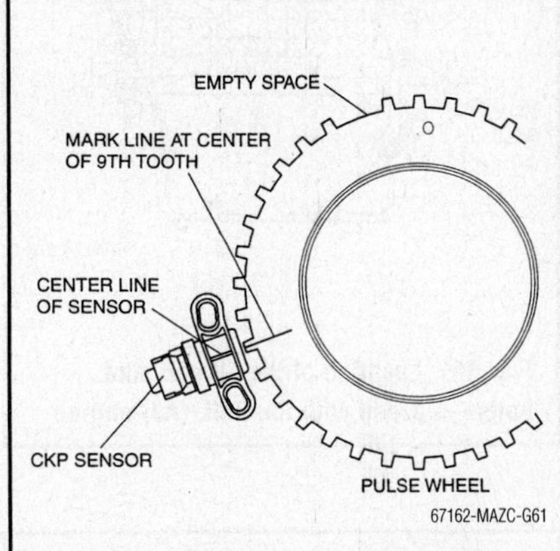

Fig. 292 Using a ruler, mark the center line on the pulse wheel teeth on the crank pulley which is located at the 9th tooth counting counterclockwise from the empty space—2.0L (LF) and 2.3L (L3) engines

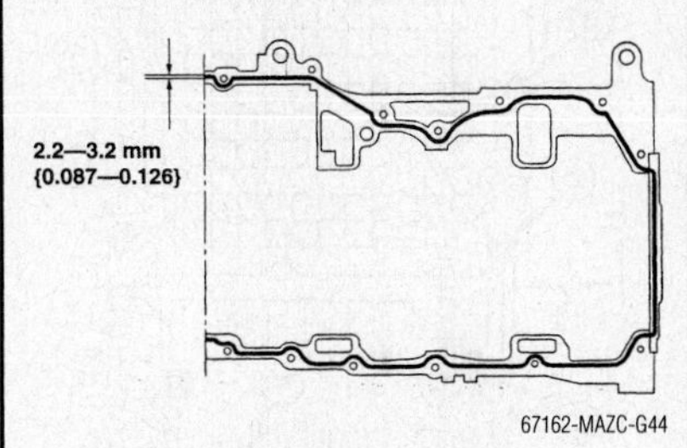

Fig. 293 Apply a 0.087–0.126 inch (2.2–3.2mm) bead of sealant around the perimeter of the oil pan—Mazda3 with the 2.0L (LF) and 2.3L (L3) engines

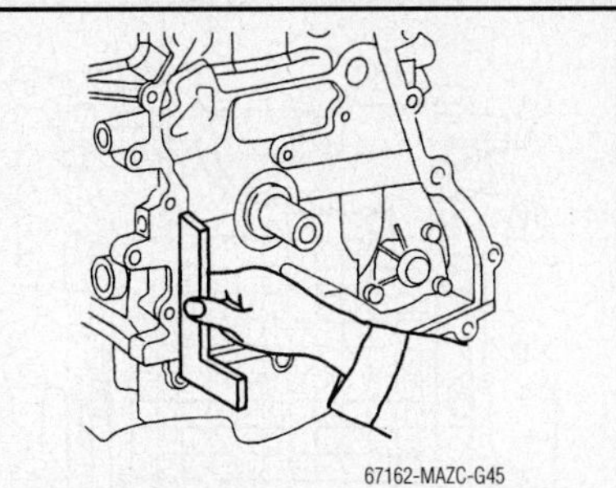

Fig. 294 Use a square ruler to align the oil pan and the block junction side on the engine front cover side—2.0L (LF) and 2.3L (L3) engines

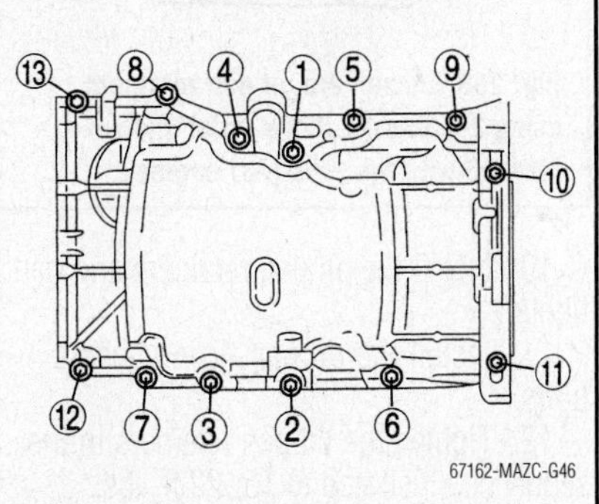

Fig. 295 Oil pan bolt torque sequence—2.0L (LF) and 2.3L (L3) engines

Mazda6

2.3L (L3) Engines

See Figures 296 and 297.

1. Before servicing the vehicle, refer to the precautions section.
2. Disconnect the negative battery cable.
3. Remove the engine undercover.
4. Remove the right front wheel.
5. Drain the engine oil.
6. Remove the engine front cover. Refer to the timing chain removal procedure in this section.
7. Remove the dipstick tube pipe and O–ring.
8. Remove the oil pan bolts and use a separator tool to separate the oil pan from the block.
9. Remove the oil pan.

To install:

10. Clean the oil pan mating surfaces.
11. Apply a 0.098 inch (2.5mm) bead of sealant around the perimeter of the oil pan as illustrated.

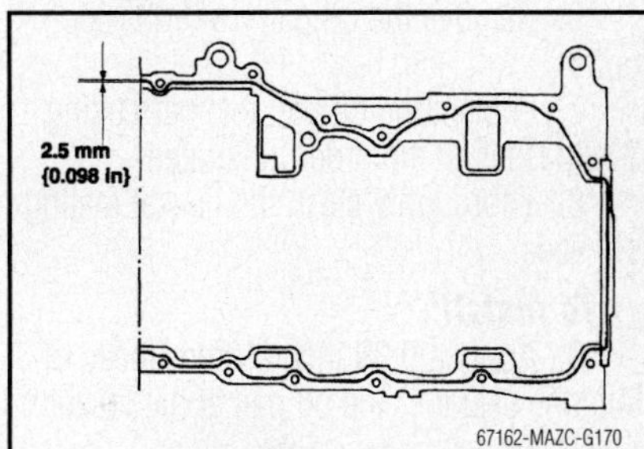

Fig. 297 Apply a 0.098 inch (2.5mm) bead of sealant around the perimeter of the oil pan—Mazda6 with the 2.3L (L3) engine

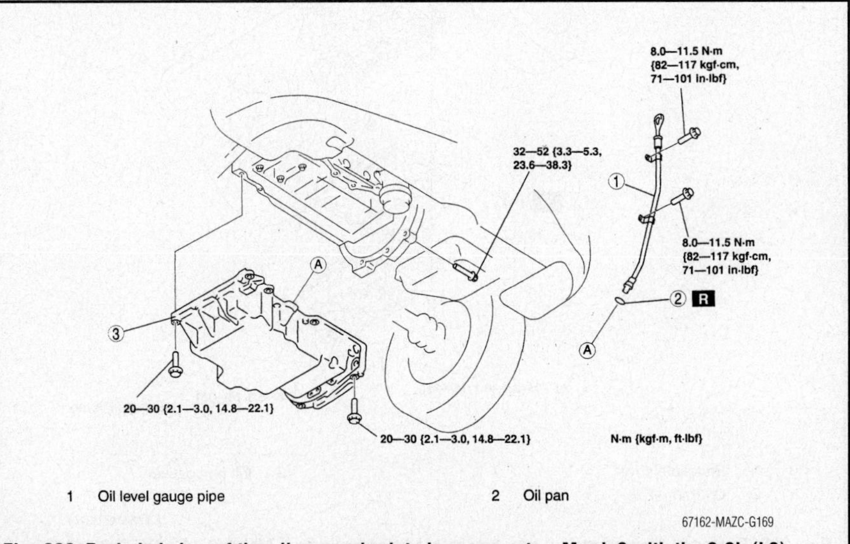

Fig. 296 Exploded view of the oil pan and related components—Mazda6 with the 2.3L (L3) engine

12. Use a square ruler to align the oil pan and the block junction side on the engine front cover side.
13. Install the oil pan within 5 minutes of applying the sealant.
14. Install the lower oil pan bolts and tighten in sequence to 15–22 ft. lbs. (20–30 Nm) and the oil pan–to–transaxle bolts to 23–38 ft. lbs. (32–52 Nm).
15. Install the remaining components in the reverse order of removal.
16. Fill the engine with clean oil.
17. Start the vehicle, check for leaks and repair if necessary.

3.0L (AJ) Engine

See Figures 298 through 302.

1. Before servicing the vehicle, refer to the precautions section.
2. Drain the engine oil.
3. Disconnect the negative battery cable.
4. Remove the right hand three way converter.
5. Remove the end plate cover.
6. Remove the oil pan–to–transaxle bolts.
7. Loosen the oil pan retainers using 2–3 passes in the order illustrated.
8. Thoroughly clean the gasket mating surfaces.

To install:

9. Apply a 0.39 inch (10mm) bead of silicone sealer to the oil pan at the locations illustrated.

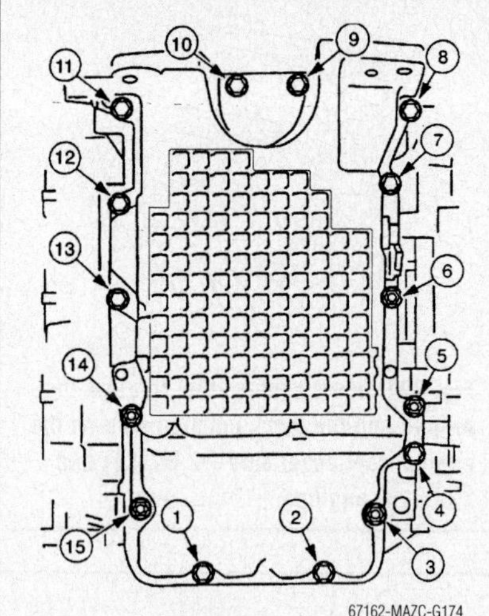

Fig. 298 Loosen the oil pan retainers using 2–3 passes in the order shown—Mazda6 with the 3.0L (AJ) engine

10. Install the oil pan retainers and hand tighten.
11. Install the oil pan–to–transaxle bolts.
12. Tighten the oil pan retainers in the sequence illustrated to 15–22 ft. lbs. (20–30 Nm).
13. Tighten the oil pan–to–transaxle bolts to 28–38 ft. lbs. (38–51 Nm)
14. Install the remaining components in the reverse order of removal.
15. Fill the engine with clean oil.
16. Start the vehicle, check for leaks and repair if necessary.

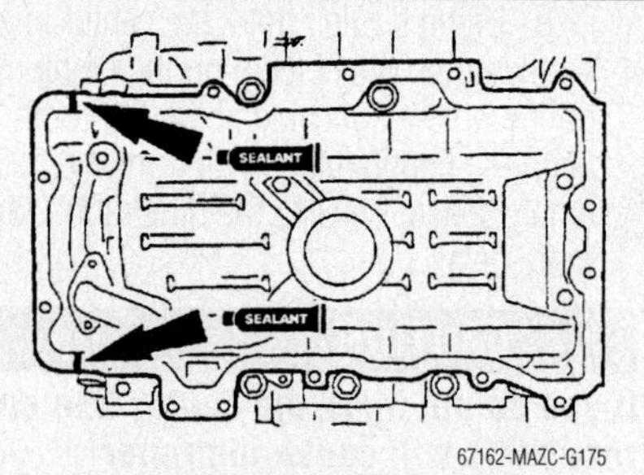

Fig. 300 Apply a 0.39 inch (10mm) bead of silicone sealer to the oil pan at the locations shown—Mazda6 with the 3.0L (AJ) engine

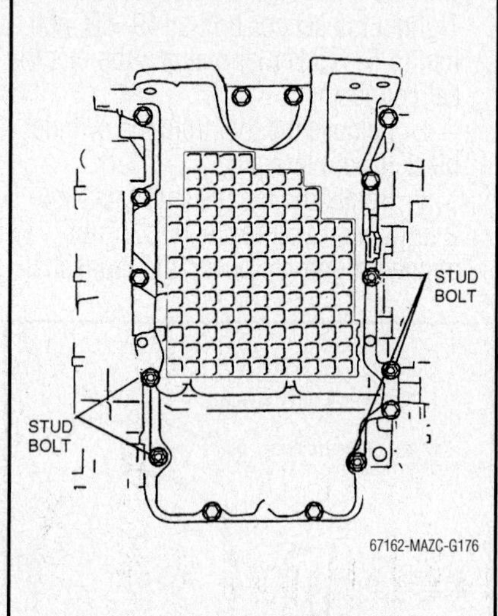

Fig. 301 Location of the oil pan stud bolts—Mazda6 with the 3.0L (AJ) engine

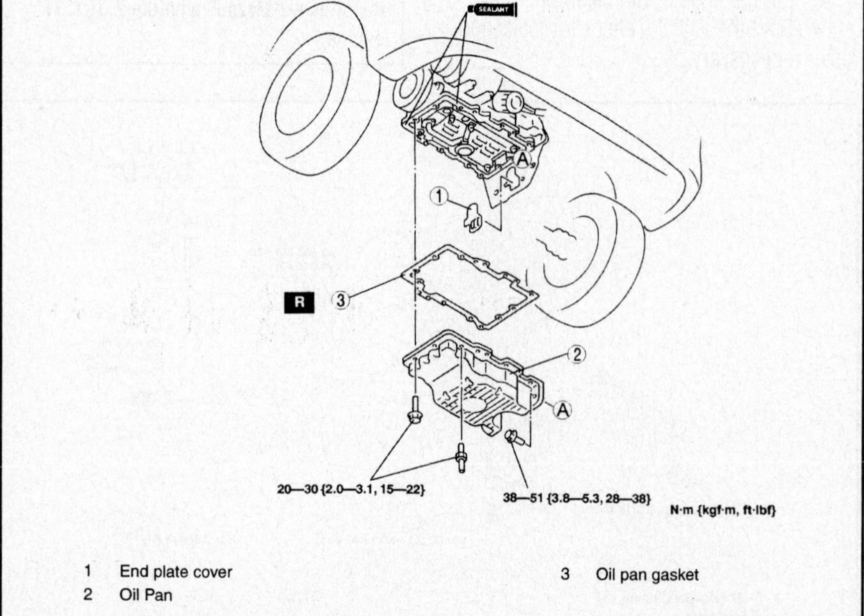

Fig. 299 Exploded view of the oil pan and related components—Mazda6 with the 3.0L (AJ) engine

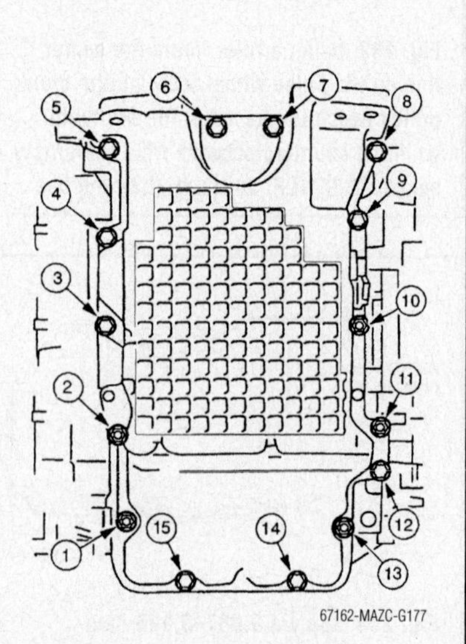

Fig. 302 Oil pan torque sequence—Mazda6 with the 3.0L (AJ) engine

2.3L (L3 with TC) Engine

See Figure 303.

1. Before servicing the vehicle, refer to the precautions section.
2. Remove the battery, battery tray and PCM component.
3. Remove the front tire RH.
4. Remove the undercover and splash shield as a single unit.
5. Drain the engine oil.
6. Drain the engine coolant.
7. Remove the air cleaner, charge air cooler, and air hose.

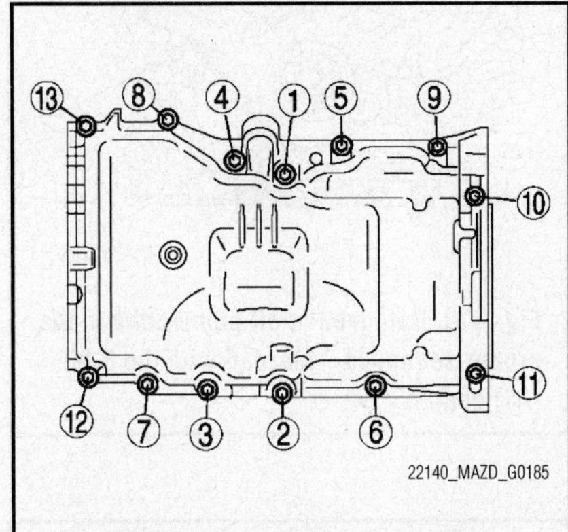

Fig. 303 Oil pan tightening sequence

8. Disconnect the quick release connector on the high pressure fuel pump.
9. Remove the high pressure fuel pump.
10. Remove the ignition coils.
11. Loosen the water pump pulley bolts before removing the drive belt.
12. Remove the drive belt.
13. Remove the P/S oil pump with hose and pipe still connected. Position the P/S oil pump out of the way.
14. Remove the crankshaft position sensor.
15. Remove the engine front cover.
16. Remove oil pan.

To install:

17. Install in the reverse order of removal.
18. Install and tighten the oil pan bolts to 13–17 ft. lbs. (17–23 Nm).
19. Refill with the specified type and amount of the engine oil.
20. Refill the engine coolant.
21. Start the engine and confirm that there is no oil leakage.
22. Inspect the oil level.
23. Inspect for engine coolant leakage.
24. Inspect for the ignition timing and idle speed.

OIL PUMP

REMOVAL & INSTALLATION

MX5-Miata

See Figures 304 through 306.

1. Before servicing the vehicle, refer to the precautions section.

CAUTION

Remove and install all parts when the engine is cold, otherwise they can cause severe burns or serious injury.

2. Remove the oil pan.
3. Remove the oil strainer.
4. Remove the oil pump chain guide.
5. Remove the oil pump chain tensioner.
6. Remove the oil pump chain.
7. Install the tool 49 G032 354 onto the oil pump sprocket to stop the oil pump from rotating.
8. Remove the oil pump sprocket.
9. Remove the oil pump bolts and pump.

To install:

10. Install the oil pump, tighten the bolts in the sequence illustrated to 8–12 Nm (71–105 inch lbs.), then final tighten to 17–23 Nm (12.6–17 ft. lbs.).
11. Install the pump sprocket. Tighten the bolt to 20–30 Nm (14.8–22 ft. lbs.).
12. Install the pump chain and tensioner and chain guide. Tighten the tensioner and guide bolts to 8–11.5 Nm (71–101 inch lbs.).
13. Install the oil strainer. Tighten the bolts to 8–11.5 Nm (71–101 inch lbs.).
14. Install the oil pan.

Mazda3 and 6 Models

2.0L (LF) and 2.3L (L3) Engines

See Figures 307 through 309.

1. Before servicing the vehicle, refer to the precautions section.
2. Remove the oil pan.

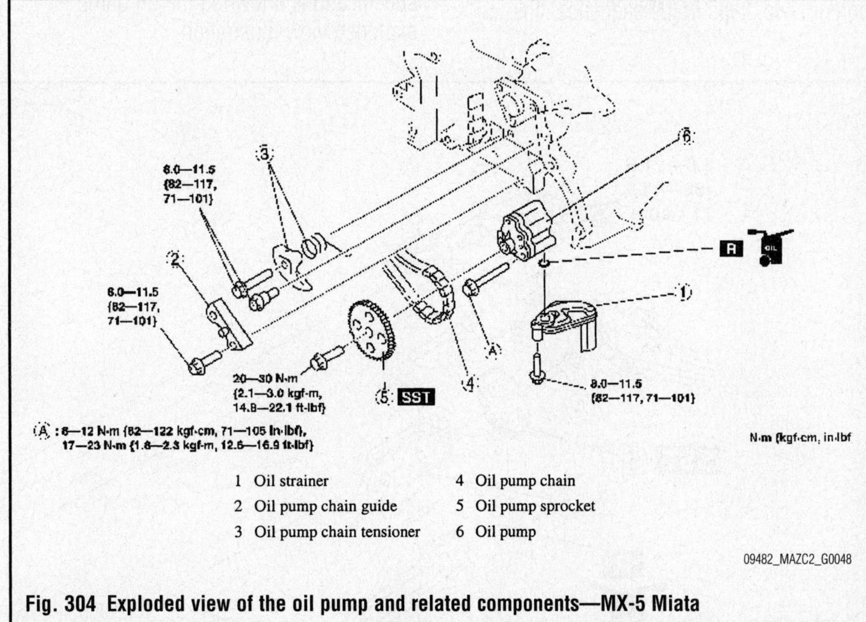

Fig. 304 Exploded view of the oil pump and related components—MX-5 Miata

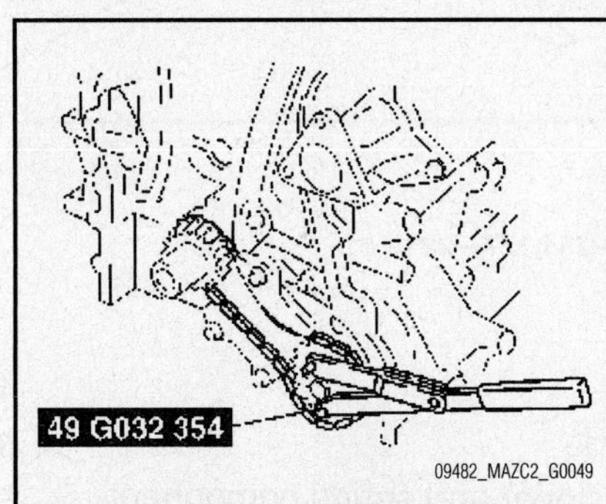

Fig. 305 Install the tool 49 G032 354 onto the oil pump sprocket to stop the oil pump from rotating—MX-5 Miata

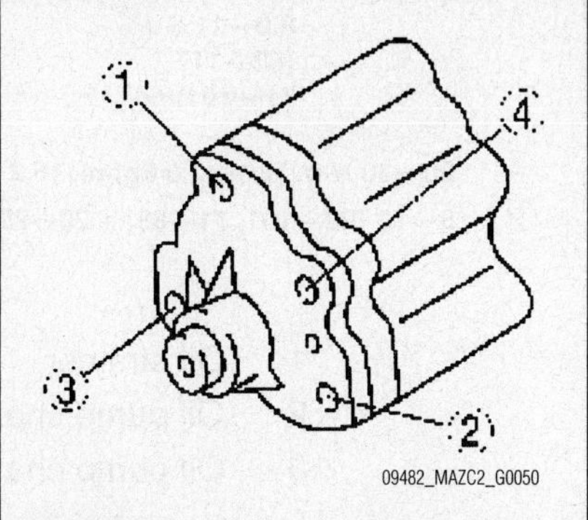

Fig. 306 Oil pump bolt torque sequence—MX-5 Miata

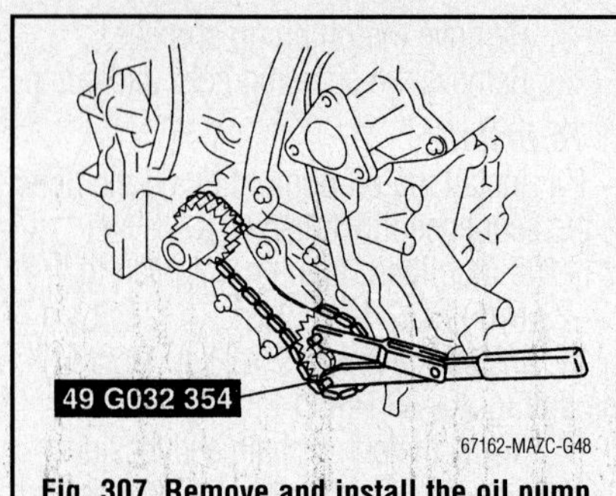

Fig. 307 Remove and install the oil pump sprocket using tool 49 G032 354 to stop the pump rotating—2.0L (LF) and 2.3L (L3) engines

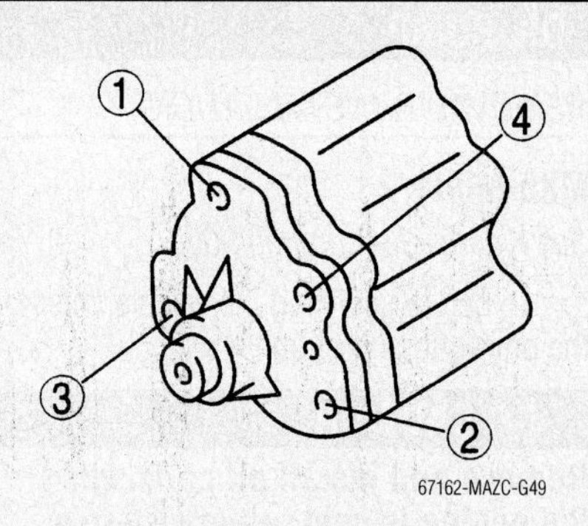

Fig. 309 Oil pump torque sequence—2.0L (LF) and 2.3L (L3) engines

3. Remove the oil strainer.
4. Remove the oil pump chain guide, tensioner, spring and chain.
5. Remove the oil pump sprocket using tool 49 G032 354 to stop the pump rotating.
6. Remove the oil pump bolts and the pump.

To install:

7. Install the oil pump and tighten the bolts in two steps in the sequence illustrated. Tighten the bolts in sequence to 71–88 inch lbs. (8–10 Nm) and final tighten to 15.2–18.4 ft. lbs. (20–27 Nm).
8. Install the oil pump sprocket using tool 49 G032 354 to stop the pump rotating. Tighten the bolts to the specifications shown in the oil pump exploded view illustration.
9. Install the oil pump chain, spring, tensioner and guide. Tighten the bolts to the specifications shown in the oil pump exploded view illustration.
10. Install the oil strainer. Tighten the bolts to the specifications shown in the oil pump exploded view illustration.
11. Install the oil pan.
12. Fill the engine with clean oil.
13. Start the vehicle, check for leaks and repair if necessary.

3.0L (AJ) Engines

See Figures 310 through 312.

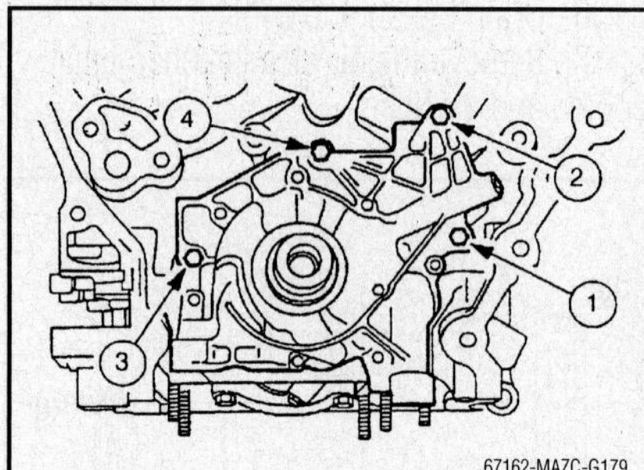

Fig. 310 Remove the oil pump bolts in the proper sequence—Mazda6 with the 3.0L (AJ) engine

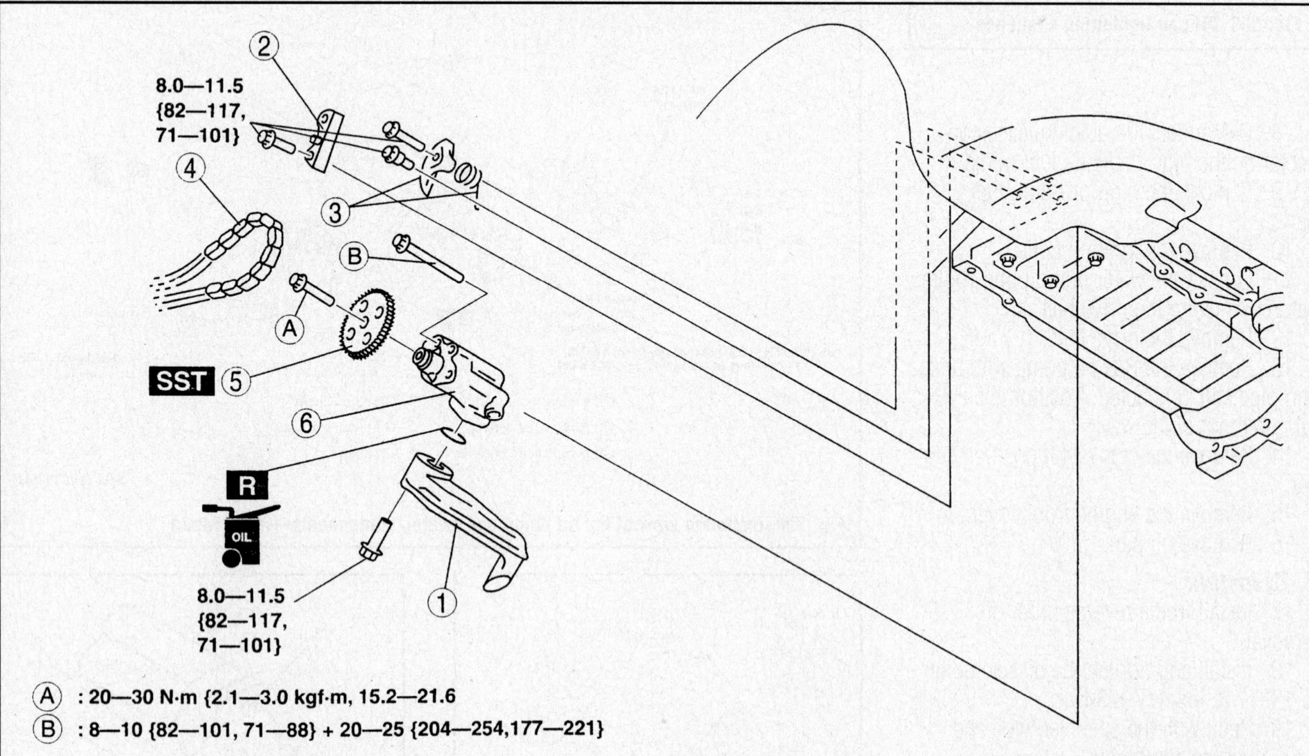

Fig. 308 Exploded view of the oil pump and related components—2.0L (LF) and 2.3L (L3) engines

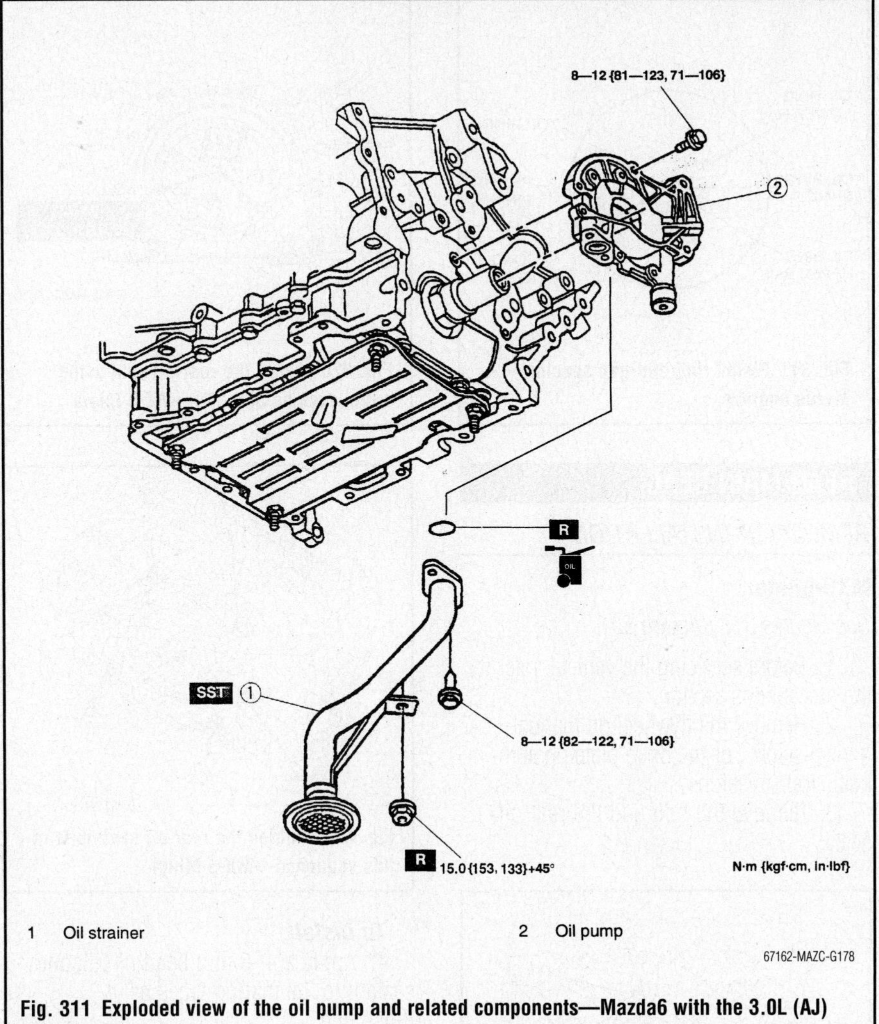

Fig. 311 Exploded view of the oil pump and related components—Mazda6 with the 3.0L (AJ) engine

1. Before servicing the vehicle, refer to the precautions section.
2. Drain the engine oil.
3. Disconnect the negative battery cable.
4. Remove the oil pan.
5. Remove the timing chain.
6. Remove the oil strainer.
7. Remove the oil pump bolts in the sequence illustrated.

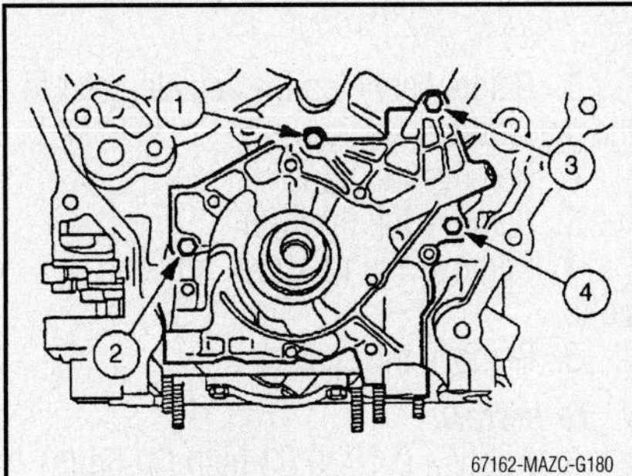

Fig. 312 Tighten the oil pump bolts in the proper sequence—Mazda6 with the 3.0L (AJ) engine

8. Thoroughly clean the gasket mating surfaces.

To install:

9. Install the oil pump and tighten the bolts in the proper sequence. Torque the bolts to 71–106 inch lbs. (8–12 Nm).
10. Install the oil strainer. Tighten the strainer bolts to 71–106 inch lbs. (8–12 Nm) and the oil strainer stay nut to 133 inch lbs. (15 Nm) plus an additional 45 degree turn.
11. Install the remaining components in the reverse order of removal.
12. Refill the engine with clean oil.
13. Start the engine and check for leaks; repair if necessary.

2.3L (L3 with TC) Engines

See Figure 313.

1. Before servicing the vehicle, refer to the precautions section.
2. Remove the battery, battery tray and PCM component.
3. Remove the front tire RH.
4. Remove the undercover and splash shield as a single unit.
5. Drain the engine oil.
6. Drain the engine coolant.
7. Remove the air cleaner, charge air cooler, and air hose.
8. Disconnect the quick release connector on the high pressure fuel pump.
9. Remove the high pressure fuel pump.
10. Remove the ignition coils.
11. Loosen the water pump pulley bolts before removing the drive belt.
12. Remove the drive belt.
13. Remove the P/S oil pump with hose and pipe still connected. Position the P/S oil pump out of the way.
14. Remove the crankshaft position sensor.
15. Remove the engine front cover.
16. Remove the oil pan.
17. Remove the oil pump as follows:
    - Oil strainer
    - Oil pump chain guide
    - Oil pump chain tensioner
    - Oil pump chain
    - Oil pump sprocket
    - Oil pump

To install:

18. Install the oil pump.
19. Tighten the oil pump bolts in two steps in the order shown in the figure.
    - Step 1: 71–105 inch. lbs. (8–12 Nm).
    - Step 2: 13–17 inch. lbs. (17–23 Nm).
20. Finish installation in the order of removal.
21. Refill with the specified type and amount of the engine oil.
22. Refill the engine coolant.
23. Start the engine and confirm that there is no oil leakage.
24. Inspect the oil level.
25. Inspect for engine coolant leakage.

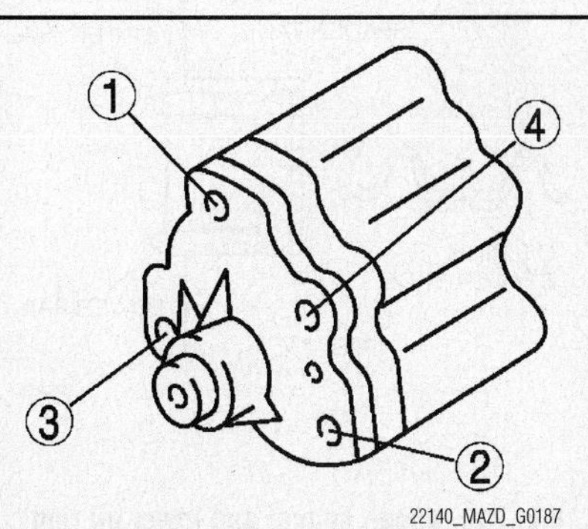

Fig. 313 Oil pump tightening sequence 2.3L (L3 with TC)

26. Inspect for the ignition timing and idle speed.
27. Inspect the oil pressure.

PISTON AND RING

POSITIONING

See Figures 314 through 317.

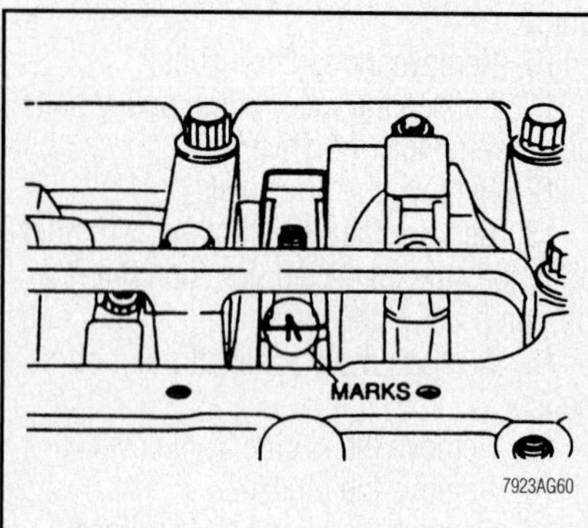

Fig. 314 Before removing the caps from the connecting rods, be sure to matchmark them—Mazda engines

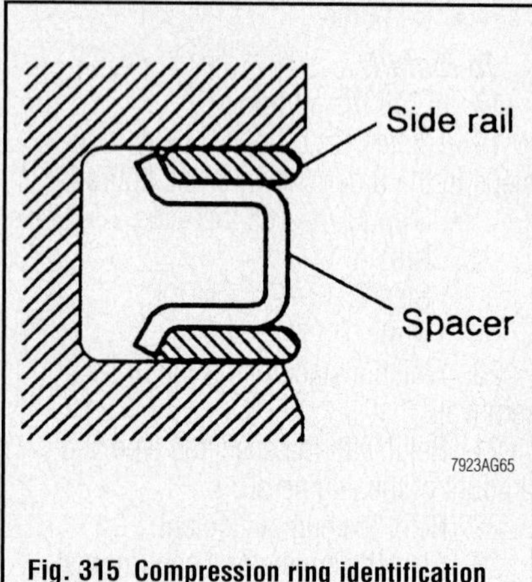

Fig. 315 Compression ring identification and positioning—Mazda engines

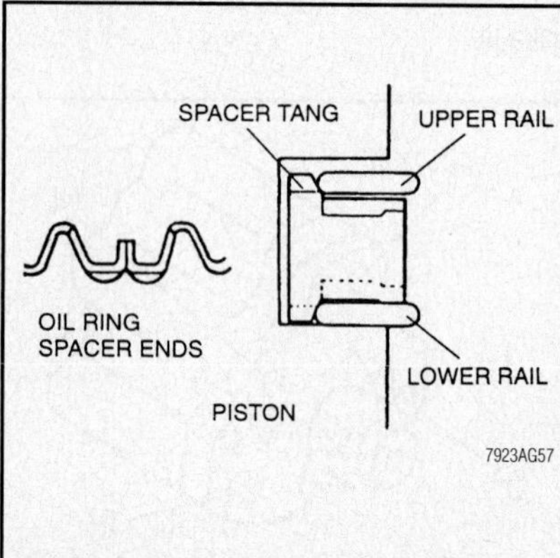

Fig. 316 Upper, spacer and lower oil ring identification and positioning—Mazda engines

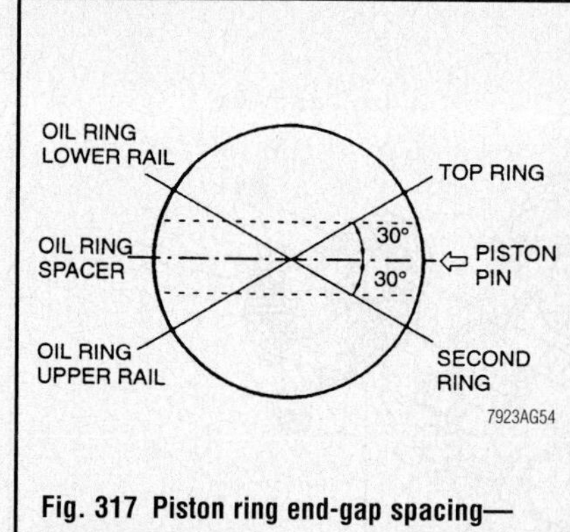

Fig. 317 Piston ring end-gap spacing—Mazda engines

REAR MAIN SEAL

REMOVAL & INSTALLATION

MX5-Miata

See Figures 318 through 321.

1. Before servicing the vehicle, refer to the precautions section.
2. Remove the flywheel on manual transmissions, or the drive plate on automatic transmissions.
3. Remove the bolt and the rear oil seal.

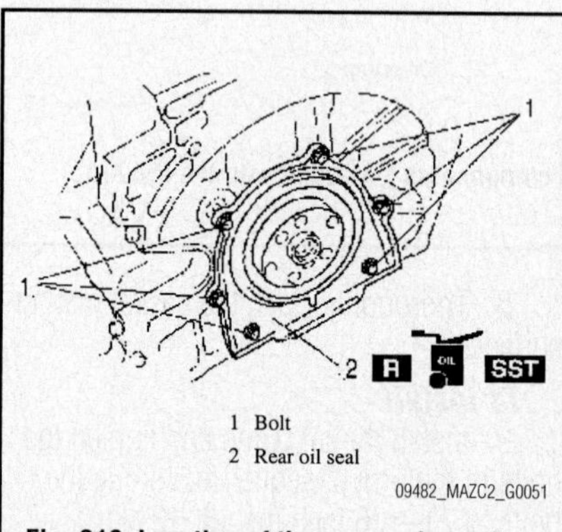

Fig. 318 Location of the rear main seal bolts—MX-5 Miata

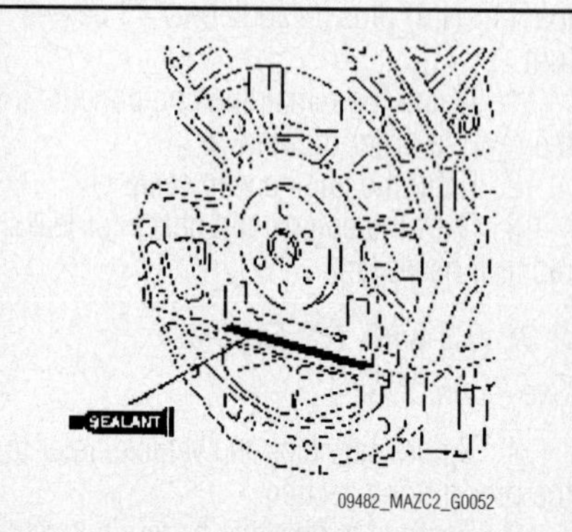

Fig. 319 Apply a 4–6 mm bead of silicone sealant to the mating faces—MX-5 Miata

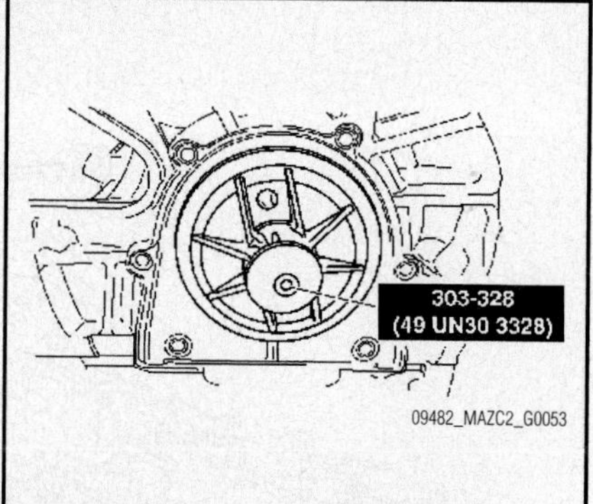

Fig. 320 Install the rear oil seal using a tool such as 303-328—MX-5 Miata

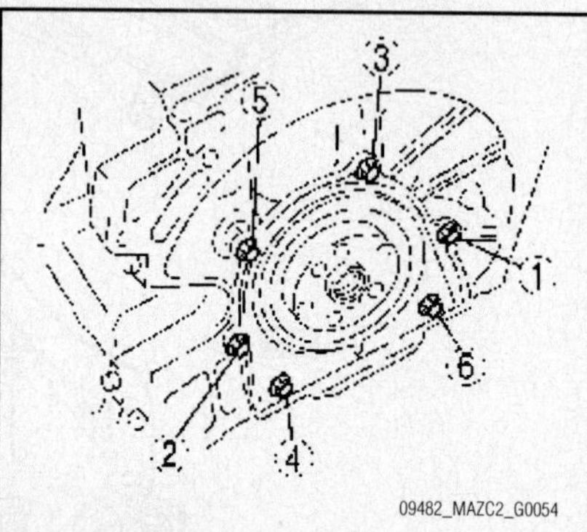

Fig. 321 Tighten the rear oil seal bolts in this sequence—MX-5 Miata

To install:

4. Apply a 4–6 mm bead of silicone sealant to the mating faces as illustrated.
5. Apply clean engine oil to the new oil seal lip.
6. Install the rear oil seal using a tool such as 303-328.
7. Tighten the rear oil seal bolts in the sequence illustrated to 8–11.5 Nm (71–101 inch lbs.).
8. Install the flywheel or drive plate.

Mazda3, Mazda6 Models

2.0L (LF) and 2.3L (L3) Engines

See Figures 322 through 327.

1. Before servicing the vehicle, refer to the precautions section.
2. Remove the transmission.
3. Remove the flywheel.
4. Remove the rear seal housing bolts.
5. Remove the seal.

To install:

6. Apply a 0.16–0.23 inch (4–6mm) bead of silicone sealant to the seal mating surfaces as illustrated. Install the seal within 10 minutes of applying the sealant.

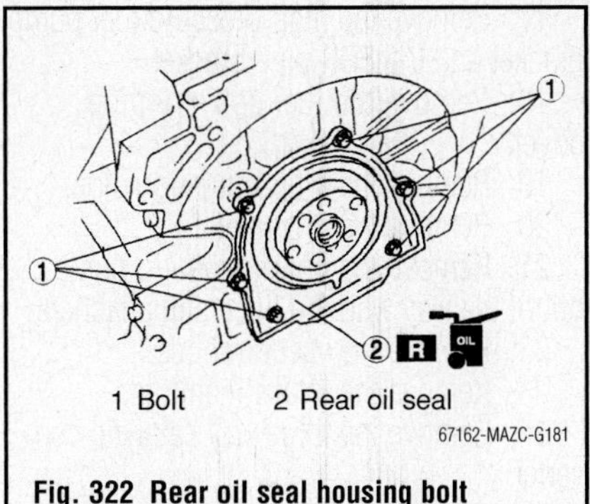

Fig. 322 Rear oil seal housing bolt removal sequence—2.0L (LF) and 2.3L (L3) engines

7. Apply a coat of clean engine oil to the new seal lip.
8. Using a suitable installer tool, install the seal as illustrated.
9. Tighten the seal housing bolts in sequence to 71–101 inch lbs. (8–11.5 Nm).
10. Install the flywheel, use tool 49 E011 1A0 to prevent the unit from turning. If reusing the old bolts coat them with a thread locking compound. If install new bolts no locking compound is needed.
11. On Mazda3 models, tighten bolts in sequence 80–85 ft. lbs. (108–116 Nm) on models equipped with a manual transaxle.

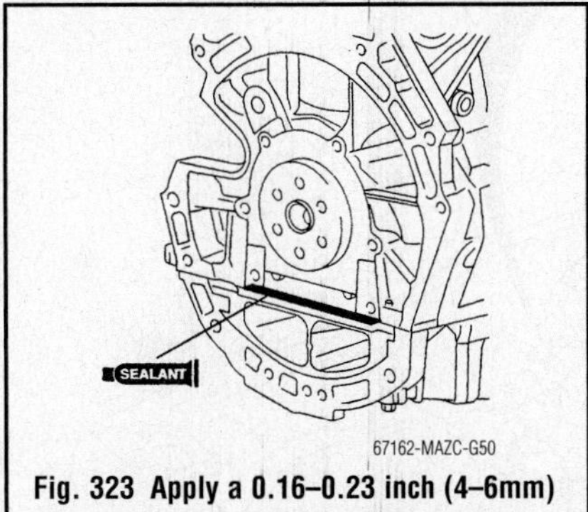

Fig. 323 Apply a 0.16–0.23 inch (4–6mm) bead of silicone sealant to the seal mating surfaces—2.0L (LF) and 2.3L (L3) engines

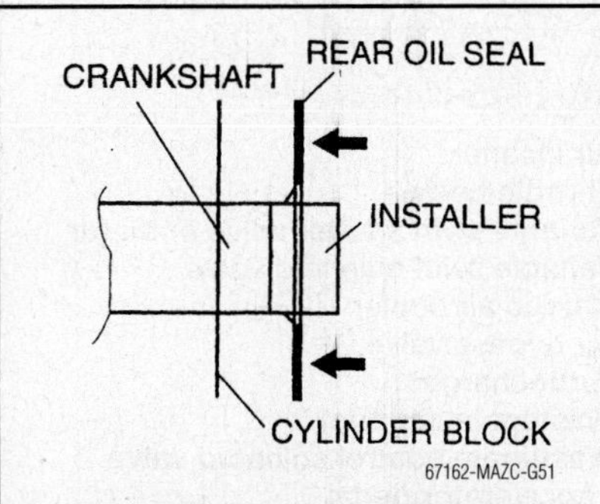

Fig. 324 Using a suitable installer tool, install the seal as shown—2.0L (LF) and 2.3L (L3) engines

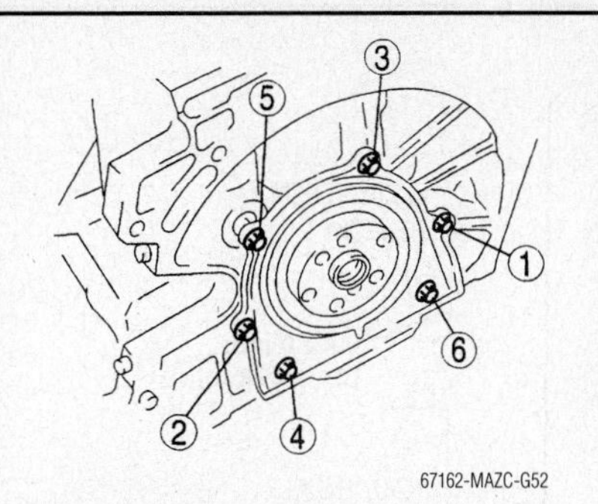

Fig. 325 Rear oil seal housing bolt torque sequence—2.0L (LF) and 2.3L (L3) engines

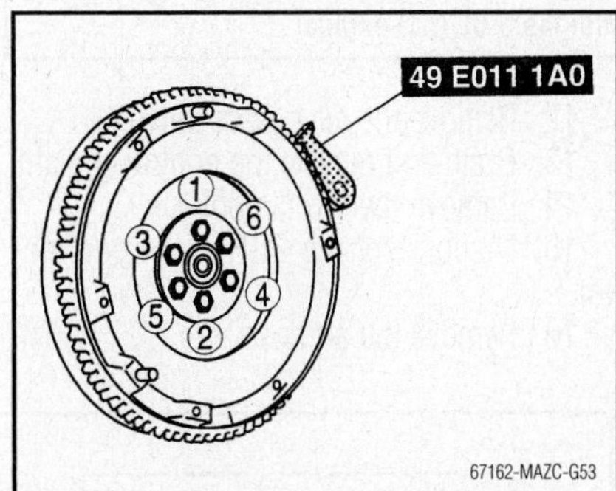

Fig. 326 Flywheel bolt torque sequence—2.0L (LF) and 2.3L (L3) engines with a manual transaxle

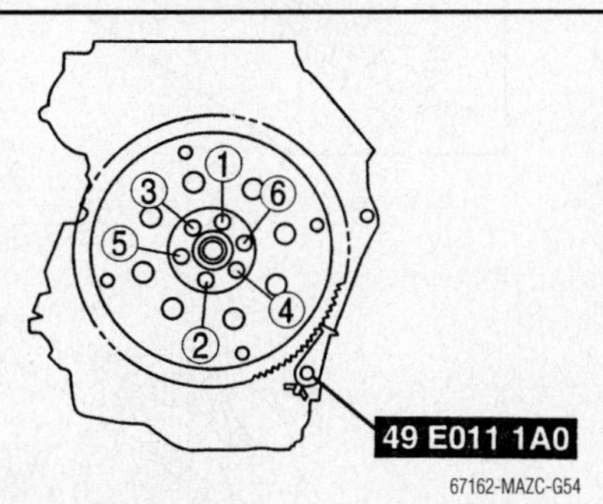

Fig. 327 Flywheel bolt torque sequence—2.0L (LF) and 2.3L (L3) engines with an automatic transaxle

On models equipped with an automatic transaxle, tighten the bolts to 80–86 ft. lbs. (96–103 Nm).

12. On Mazda6 models, if equipped with a G35M-R manual transaxle, tighten the bolts to 79–84 ft. lbs. (108–115 Nm). If equipped with a A65M-R manual transaxle, tighten the bolts to 54–64 ft. lbs. (73–87 Nm). On models equipped with an FN4A-EL automatic transaxle, tighten the bolts to 80–86 ft. lbs. (96–103 Nm). On models equipped with an JA5A-EL automatic transaxle, tighten the bolts to 54–64 ft. lbs. (73–87 Nm).
13. Install the transaxle.

3.0L (AJ) Engine

See Figures 328 through 331.

1. Before servicing the vehicle, refer to the precautions section.
2. Remove the transmission.
3. Remove the flywheel.
4. Cut the rear seal with a razor and remove the seal using a suitable pry tool.

To install:

5. Assemble the new oil seal with part (A) of tool 49 UN01 070 and tool 49 UN30 3384.

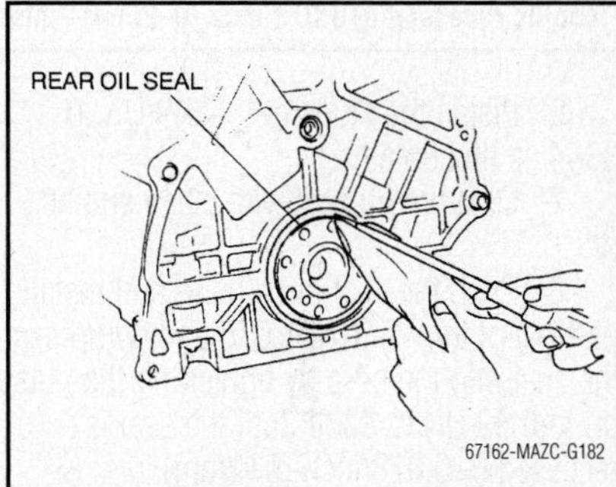

Fig. 328 Remove the seal using a suitable pry tool—Mazda6 with the 3.0L (AJ) engine

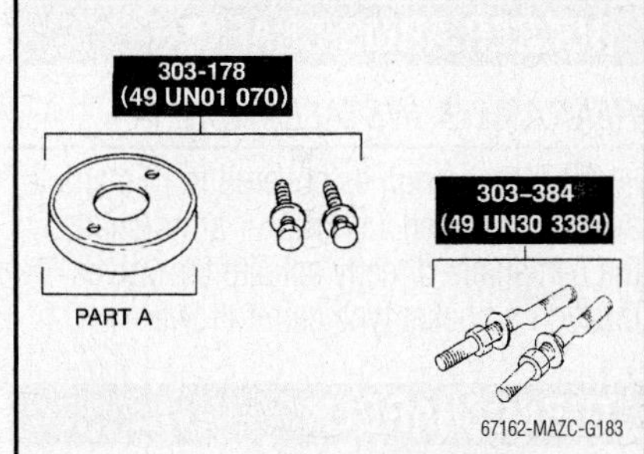

Fig. 329 Assemble the new oil seal with part (A) of tool 49 UN01 070 and tool 49 UN30 3384—Mazda6 with the 3.0L (AJ) engine

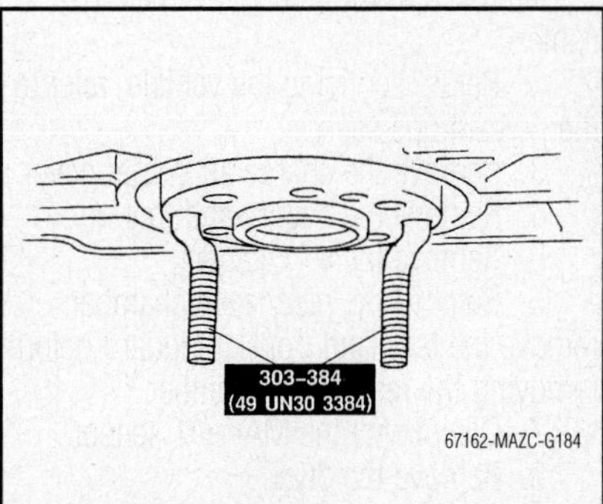

Fig. 330 Install the studs of tool 49 UN30 3384—Mazda6 with the 3.0L (AJ) engine

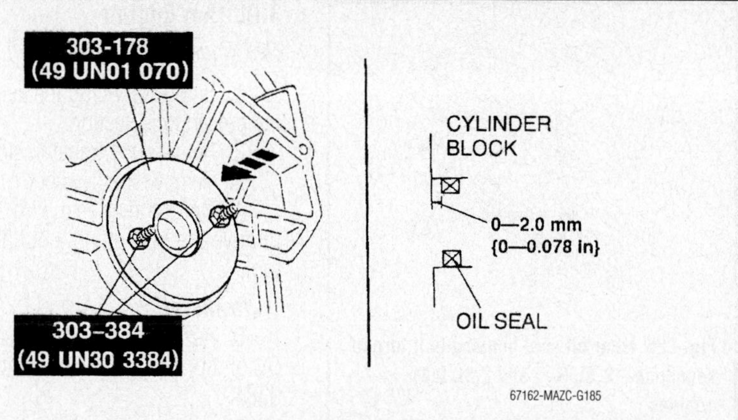

Fig. 331 Compress the seal into the bore by tightening the nuts on tool 49 UN30 3384 until the seal is recessed 0–0.078 inch (0–2mm)—Mazda6 with the 3.0L (AJ) engine

6. Install the studs of tool 49 UN30 3384 as illustrated.
7. Coat the oil seal with clean engine oil.
8. Push the seal in by hand and install part (A) of tool 49 UN01 070 and compress the seal into the bore by tightening the nuts on tool 49 UN30 3384 until the seal is recessed 0–0.078 inch (0–2mm).
9. Remove the tools.
10. Install the flywheel and tighten bolts in sequence 80–85 ft. lbs. (108–116 Nm).
11. Install the transaxle.

ROCKER ARMS/SHAFTS

REMOVAL & INSTALLATION

All Mazda engines covered in this article are not equipped with rocker arms/shafts, the camshafts directly actuate the valves through a bucket type cam follower.

TURBOCHARGER

REMOVAL & INSTALLATION

2.3L (L3 with TC) Engine

See Figure 332.

1. Disconnect the negative battery cable.
2. Before servicing the vehicle, refer to the precautions section.
3. Remove the charge air cooler cover.
4. Remove the battery and tray.
5. Remove the air cleaner.
6. Remove the resonance chamber, remove the left hand front mudguard before removing the resonance chamber.
7. Disconnect the MAF/IAT sensor.
8. Remove the duct.
9. Remove the charge air cooler cover.
10. Remove the charge air cooler.
11. Remove the charge air cooler bracket.
12. Remove the air bypass valve.
13. Drain and recycle the engine coolant.
14. Remove the throttle body.
15. Disconnect the EGR valve connector.
16. Remove the air hose.
17. Remove the high pressure fuel pump bracket and the EGR pipe bracket.
18. Remove the fuel delivery pipe cover.
19. Remove the oil level gauge pipe.
20. Remove the drive belt.
21. Remove the power steering pump out of the way with the lines still attached.
22. Remove the vacuum hose.
23. Remove the intake manifold.
24. Remove the three way catalytic converter.
25. Disconnect the rear oxygen sensor connector.
26. Remove the front exhaust pipe.

CAUTION

When removing the cowl, a part or tool may hit the edge of the windshield and could damage it. Protect the windshield by covering it with a clean rag to prevent damage to the windshield.

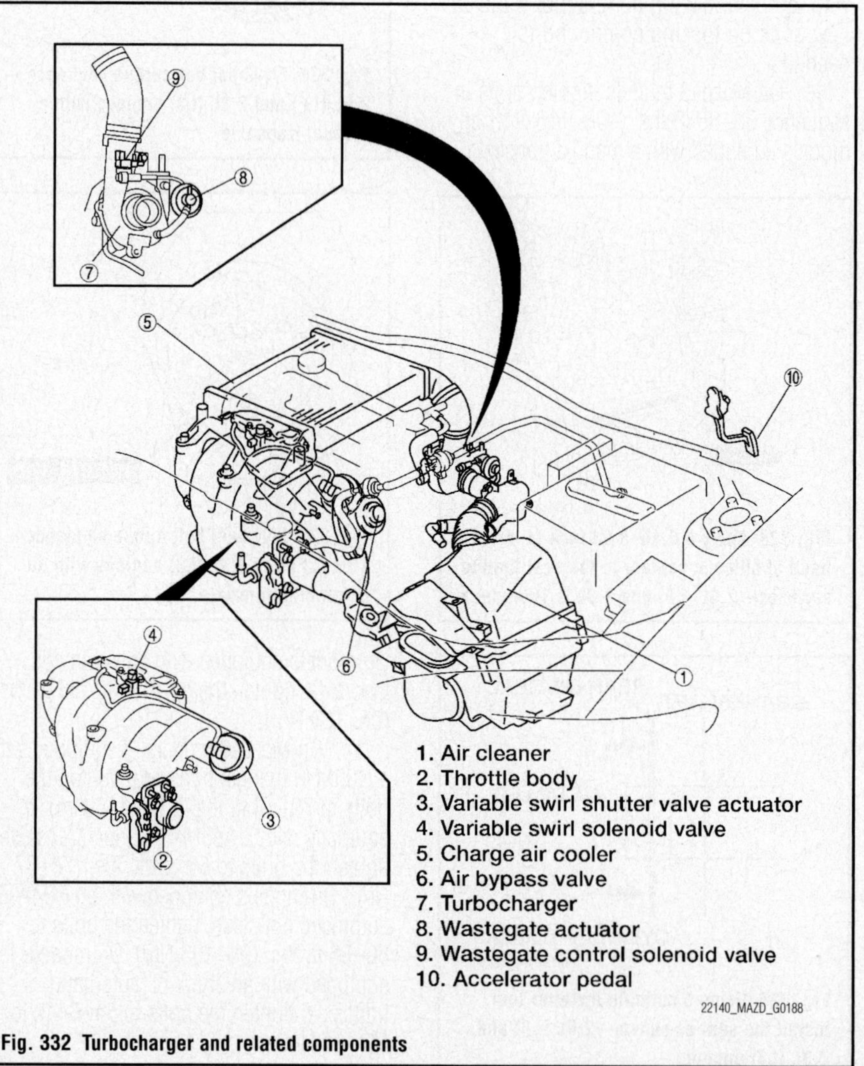

Fig. 332 Turbocharger and related components

27. Remove the cowl and the insulator under the cowl.
28. Remove the exhaust manifold upper side insulator.
29. Remove the alternator.
30. Remove the exhaust manifold lower side insulator.
31. Disconnect the front oxygen sensor connector.
32. Remove the warm up three-way catalytic converter top side insulator.
33. Remove the brake master back side vacuum hose.
34. Remove the oil pipes.
35. Remove the front oxygen sensor.
36. Remove the warm up three-way catalytic converter.
37. Remove the turbocharger.

To install:

38. Installation is the reverse of removal, note the following:
   a. Tighten the turbocharger bolts to 29–37 ft. lbs. (38–51 Nm).
   b. Tighten the oil pipe installation bolt while the stopper of the oil pipe is faced to the turbocharger.
   c. Tighten the waste gate Control Solenoid Valve bolts to 70–95 inch lbs. (8–11 Nm).
   d. Tighten the intake manifold bolts to 13–16 ft. lbs. (17–23 Nm).
   e. Tighten the fuel delivery pipe cover bolts to 13–16 ft. lbs. (17–23 Nm).
   f. Tighten the charge air cooler bracket to 31–35 ft. lbs. (42–48 Nm).
   g. Tighten the charge air cooler to 14–19 ft. lbs. (42–48 Nm).

⁂ CAUTION

When installing the charge air cooler cover, be careful not to damage the charge air cooler cover clips.

   h. Tighten the throttle body installation bolts to 71–101 inch lbs. (8–11.5 Nm).
   i. Perform the following when installing the air cleaner case:
   - Verify that the rubber mounts are set in the air cleaner bracket
   - Install the projections on the frame side.
   - Verify that the projections on the frame side are installed securely.
   - Install the projection on the engine side.
   - Verify that the projection on the engine side installed securely.

TIMING CHAIN AND SPROCKETS

REMOVAL & INSTALLATION

MX5-Miata

See Figures 333 through 345.

1. Before servicing the vehicle, refer to the precautions section.
2. Remove the battery and battery tray.
3. Remove the air cleaner.
4. Disconnect the ventilation hose.
5. Loosen the water pump pulley bolt and removal the drive belt.
6. Remove the front suspension tower bar.
7. Remove the camshaft position sensor.
8. Disconnect the OCV connector.
9. Remove the ignition coils.
10. Remove the drive belt.
11. Remove the undercover.
12. Remove the crankshaft position sensor.
13. Remove the P/S oil pump with the oil hose still connected and position the P/S oil pump so that it is out of the way.
14. Move the cooler pipe No.3 and heater pipe slightly out of the way.
15. Remove the dipstick.
16. Remove the cylinder head cover.
17. Remove the crankshaft pulley lock bolt as follows:
   a. Remove the cylinder block lower blind plug.
   b. Install tool 303-507.
   c. Turn the crankshaft clockwise until the crankshaft is in the No.1 cylinder TDC position (until the balance weight contacts tool 303-507).
   d. Hold the crankshaft pulley using the tools shown in the illustration.
   e. Remove the crankshaft pulley lock bolt.
18. Remove the crankshaft pulley.
19. Remove the front oil seal as follows:
   a. Cut the oil seal lip using a razor knife.
   b. Remove the oil seal using a screwdriver wrapped with a rag.
20. Remove the water pump pulley.
21. Remove the drive belt idler pulley.
22. Remove the engine front cover.
23. Remove the oil seal using a screwdriver.
24. Remove the chain tensioner as follows:
   a. Using a thin screwdriver, hold the chain tensioner ratchet lock mechanism away from the ratchet stem.

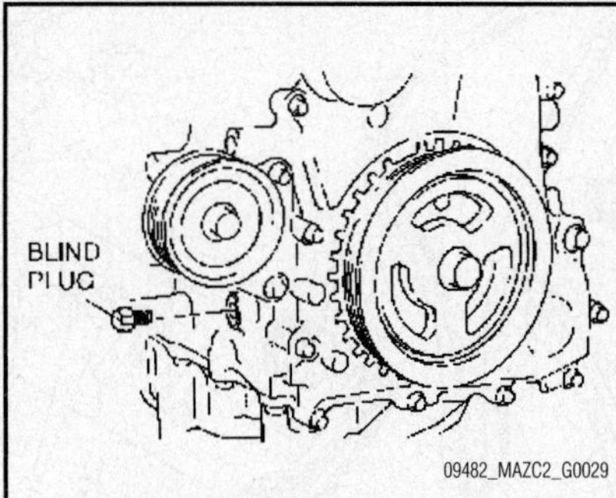

Fig. 333 Remove the cylinder block lower blind plug—MX-5 Miata

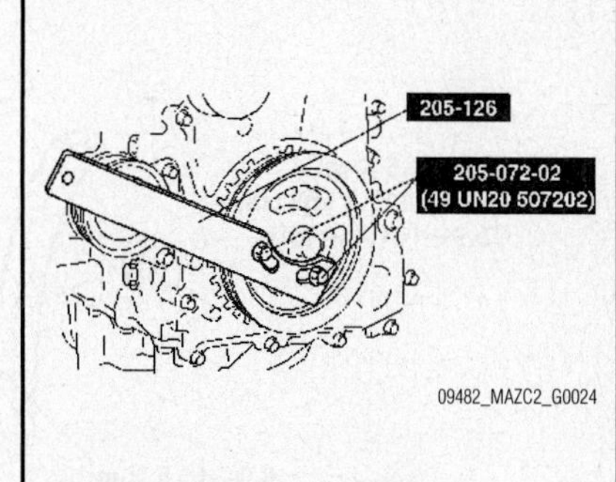

Fig. 335 Hold the crankshaft pulley using the tools shown—MX-5 Miata

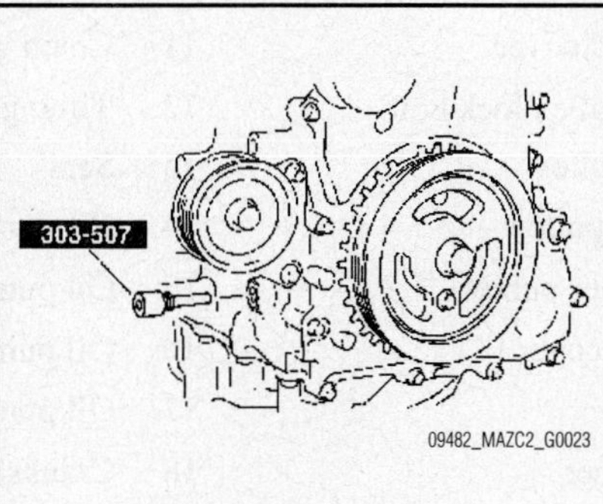

Fig. 334 Install tool 303-507 in the lower blind plug opening—MX-5 Miata

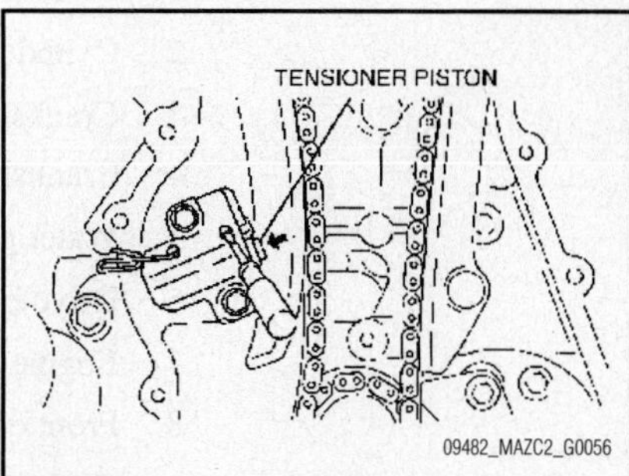

Fig. 336 Hold the tensioner piston using a 1.5 mm (0.059 in.) wire or paper clip—MX-5 Miata

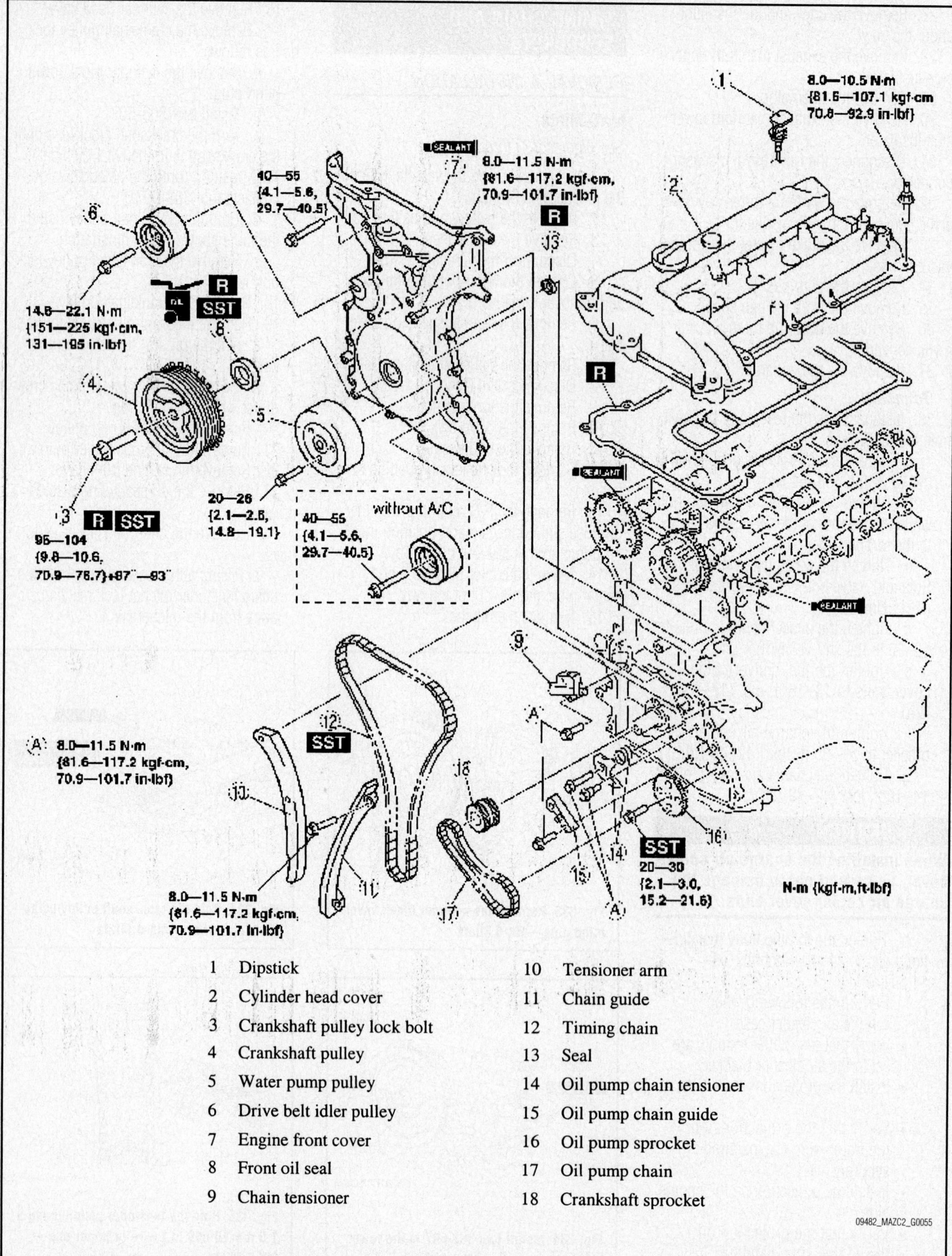

Fig. 337 Exploded view of the timing chain and related components—MX-5 Miata

b. Slowly compress the tensioner piston.

c. Hold the tensioner piston using a 1.5 mm (0.059 in.) wire or paper clip.

25. Remove the tensioner arm.
26. Remove the chain guide.
27. Remove the timing chain

To install:

28. Install tool 303-465 onto the camshaft as illustrated.
29. Install the timing chain.
30. Install the chain guide. Tighten the bolt to 8–11.5 Nm (71–101 inch lbs.).
31. Install the tensioner arm. Tighten the bolt to 8–11.5 Nm (71–101 inch lbs.).
32. Remove the retaining wire or paper clip from the chain tensioner to apply tension to the timing chain.
33. Install the oil seal.
34. Install the engine front cover as follows:

a. Install the front oil seal before installing the cover.

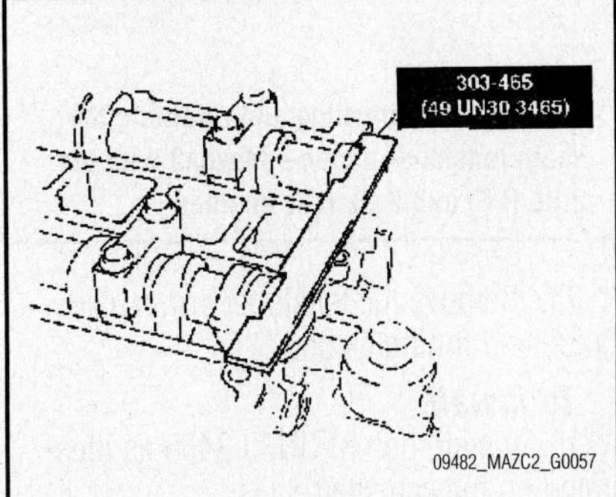

Fig. 338 Install tool 303-465 onto the camshaft—MX-5 Miata

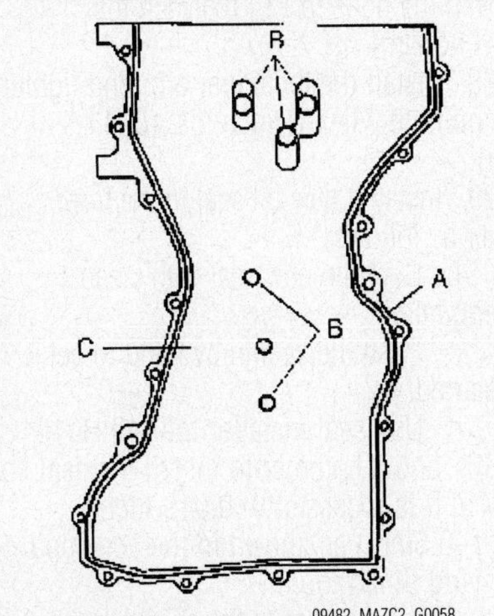

Fig. 339 Apply sealant at the locations shown in the illustration. Refer to the illustration for sealant thickness—MX-5 Miata

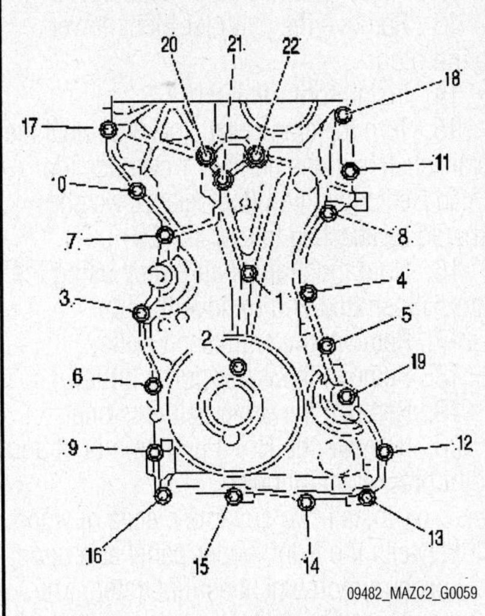

Fig. 340 Front cover bolt locations—MX-5 Miata

b. Apply silicone sealant to the engine front cover as illustrated.

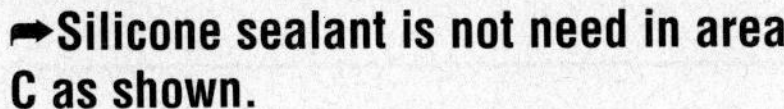

CAUTION

Install the engine front cover within 10 minutes of applying the silicone sealant.

➡Silicone sealant is not need in area C as shown.

35. Apply silicone sealant at location **A**. The sealant thickness should be 2–3mm (0.079–0.118 inch. At location **B** the sealant thickness should be 1.5–2.5mm (0.059–0.098 inch). Refer to the illustration for locations.
36. Install the front cover. On bolts 1 through 18 tighten to 8–11.5 Nm (71–101 in. lbs). Bolts 19 through 22, tighten to 20–55 Nm (30–40 ft. lbs.). Refer to the illustration for locations.
37. Install the crankshaft pulley lock bolt as follows:

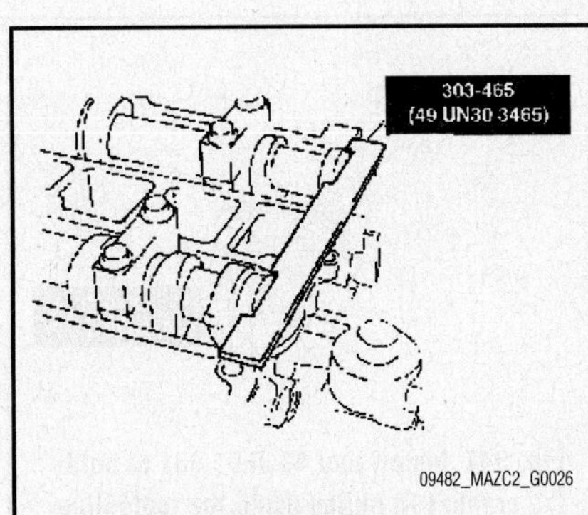

Fig. 341 Install tool shown onto the camshaft—MX-5 Miata

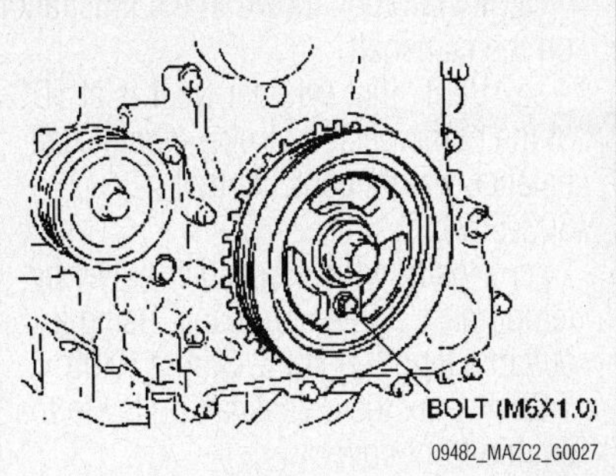

Fig. 342 Position the crankshaft pulley by temporarily tightening it using a suitable length bolt—MX-5 Miata

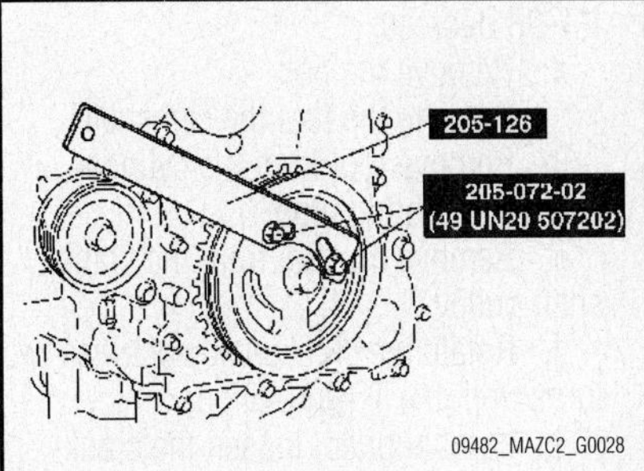

Fig. 343 Install the tools shown onto the crankshaft pulley to lock the crankshaft against rotation—MX-5 Miata

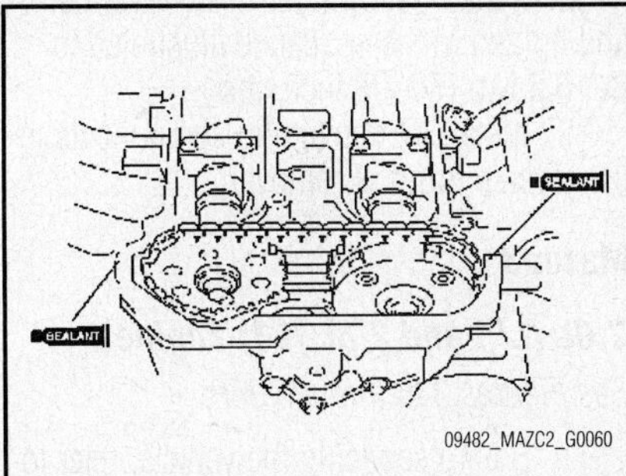

Fig. 344 Prior to install the cylinder head cover, apply a 4–6mm (0.16–0.23 in) bead of sealant at these locations—MX-5 Miata

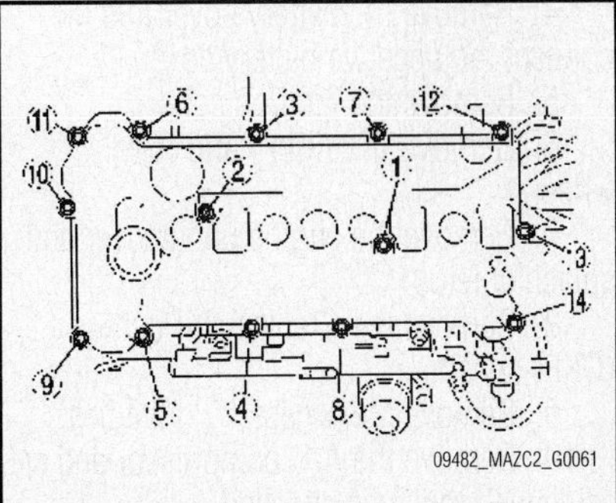

Fig. 345 Cylinder head cover torque sequence—MX-5 Miata

a. Install tool shown in the illustration on the camshaft.

b. Verify that cylinder No.1 is at TDC of the compression stroke. (Crankshaft balance weight contacts tool 303-507)

c. Position the crankshaft pulley by temporarily tightening it and, using a suitable length bolt 25–35mm (0.99–1.37 in.), fix the crankshaft pulley to the engine front cover.

d. Install the tools illustrated to the crankshaft pulley, lock the crankshaft against rotation.

e. Tighten the crankshaft pulley lock bolt to 96–104 Nm (71–77 ft. lbs.), plus 87–93 degrees.

f. Remove the bolt.

g. Remove the tool the camshaft.

h. Remove tool from the cylinder block lower blind plug.

i. Remove the tool from the crankshaft pulley.

j. Rotate the crankshaft clockwise two turns until the TDC position.

k. If not aligned, loosen the crankshaft pulley lock bolt and repeat from the first step.

l. Install the cylinder block lower blind plug and tighten to 18–22 Nm (13–16 ft. lbs.).

38. Install the cylinder head cover bolts and tighten in the sequence illustrated to 8–10.5 Nm (70–93 inch. lbs.).

39. Install the remaining components in the reverse order of removal.

Mazda3

2.0L (LF) and 2.3L (L3) Engines

See Figures 346 through 360.

1. Before servicing the vehicle, refer to the precautions section.
2. Remove the plug hole plate and bracket.
3. Remove the accelerator cable and bracket.
4. Remove the battery cover and disconnect the negative battery cover.
5. Remove the ignition coils.
6. Remove the right hand front wheel.
7. Remove the engine undercover and splash shields.
8. Remove the Crankshaft Position (CKP) sensor.
9. Remove the accessory drive belt.
10. Remove the A/C compressor and set aside with the lines attached.
11. Remove the coolant reservoir tank and set aside with the lines still attached.
12. Remove the cylinder head cover.
13. Remove the cylinder block lower blind plug.
14. Install tool 49 JE01 061.
15. Turn the crankshaft clockwise until the crankshaft is in the number 1 cylinder Top Dead Center position (the balance weight should be attached to tool 49 JE01 061).
16. Hold the crankshaft pulley using the tools illustrated and remove the bolt.
17. Remove the crankshaft pulley.
18. Remove the water pump pulley.
19. Remove the drive belt tensioner.
20. Remove the No. 3 engine mount and joint bracket as follows:
    a. Install two suitable pieces of wood between the front fender panel and upper apron reinforcement as illustrated. The wood size should be approximately 1.38 inch (35mm) on 4 door models or 2.36 inch (60mm) on 5 door models.
21. Install an engine support device such as 49 E017 5A0.
22. Remove the engine front cover and oil seal.
23. Unlock the chain tensioner using a suitable tool to slowly compress the tensioner piston. Insert a 0.059 inch (1.5mm) wire or a paper clip to hold the piston in its compressed position.

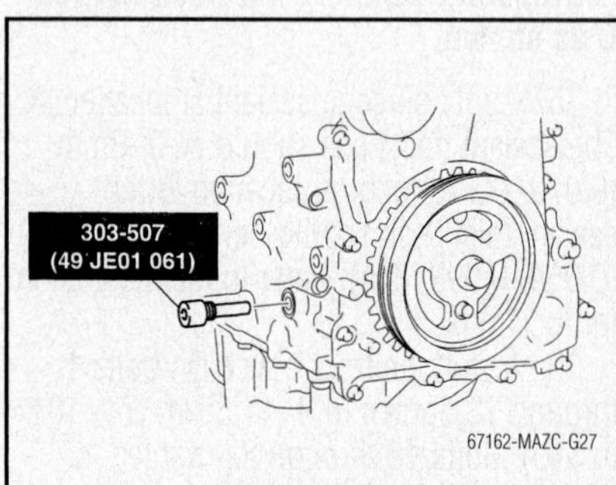

Fig. 346 Install tool 49 JE01 061—Mazda3 with the 2.0L (LF) and 2.3L (L3) engines

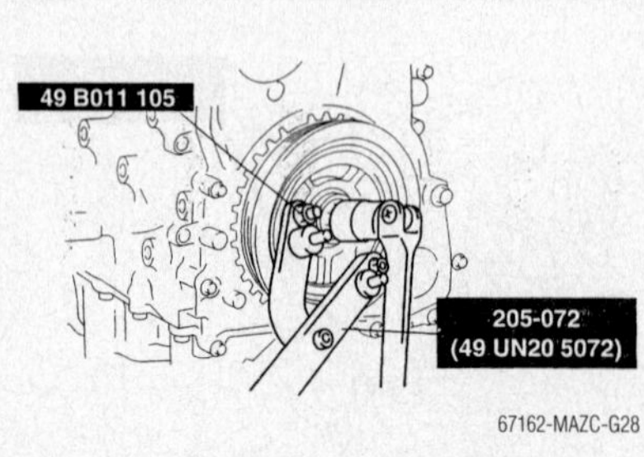

Fig. 347 Install tool 49 JE01 061 to hold the crankshaft pulley using the tools illustrated and remove the bolt—Mazda3 with the 2.0L (LF) and 2.3L (L3) engines

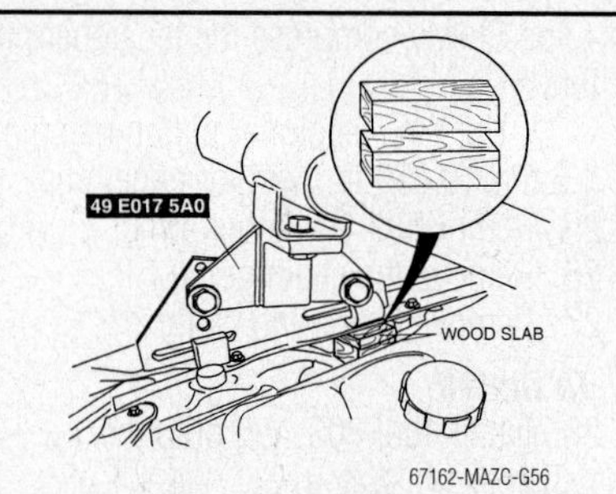

Fig. 348 Install two suitable pieces of wood between the front fender panel and upper apron reinforcement—Mazda3 with the 2.0L (LF) and 2.3L (L3) engines

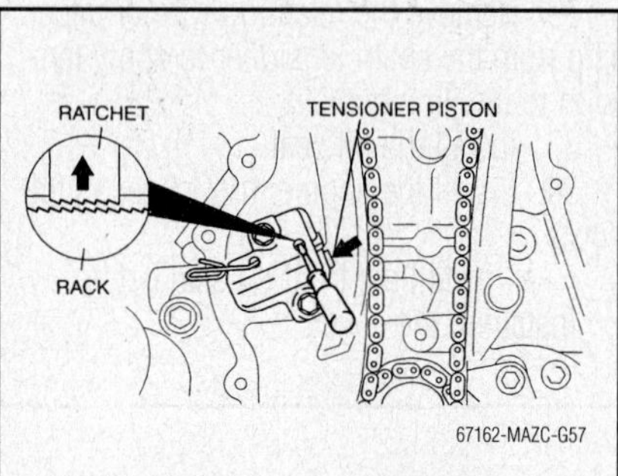

Fig. 349 Compressing and retaining the chain tensioner piston—Mazda3 with the 2.0L (LF) and 2.3L (L3) engines

24. Remove the tensioner arm, chain guide and timing chain.

To install:

25. Install tool 49 UN30 3465 as illustrated on the camshaft.
26. Install the timing chain and remove the paper clip or wire retaining the tensioner piston to apply tension to the chain.
27. Install the timing chain guide and tighten the bolts to 71–101 inch lbs. (8–11.5 Nm).
28. Install the tensioner arm and tighten the bolts to 71–101 inch lbs. (8–11.5 Nm).
29. Install a new oil seal in the front cover as follows:
    b. Coat the new seal with clean engine oil.
    c. Push the seal in by hand to get it started.
    d. Use seal installer tool 49 H010 401 and a hammer to install the seal so that it is recessed 0–0.019 inch (0–0.5mm) as shown in the accompanying illustration.
30. Apply sealant to the engine front cover. At point A the bead should be 0.087–0.125 inch (2.2–3.2mm) thick and at point B the bead should be 0.059–0.098 inch (1.5–2.5mm) thick. Refer to the

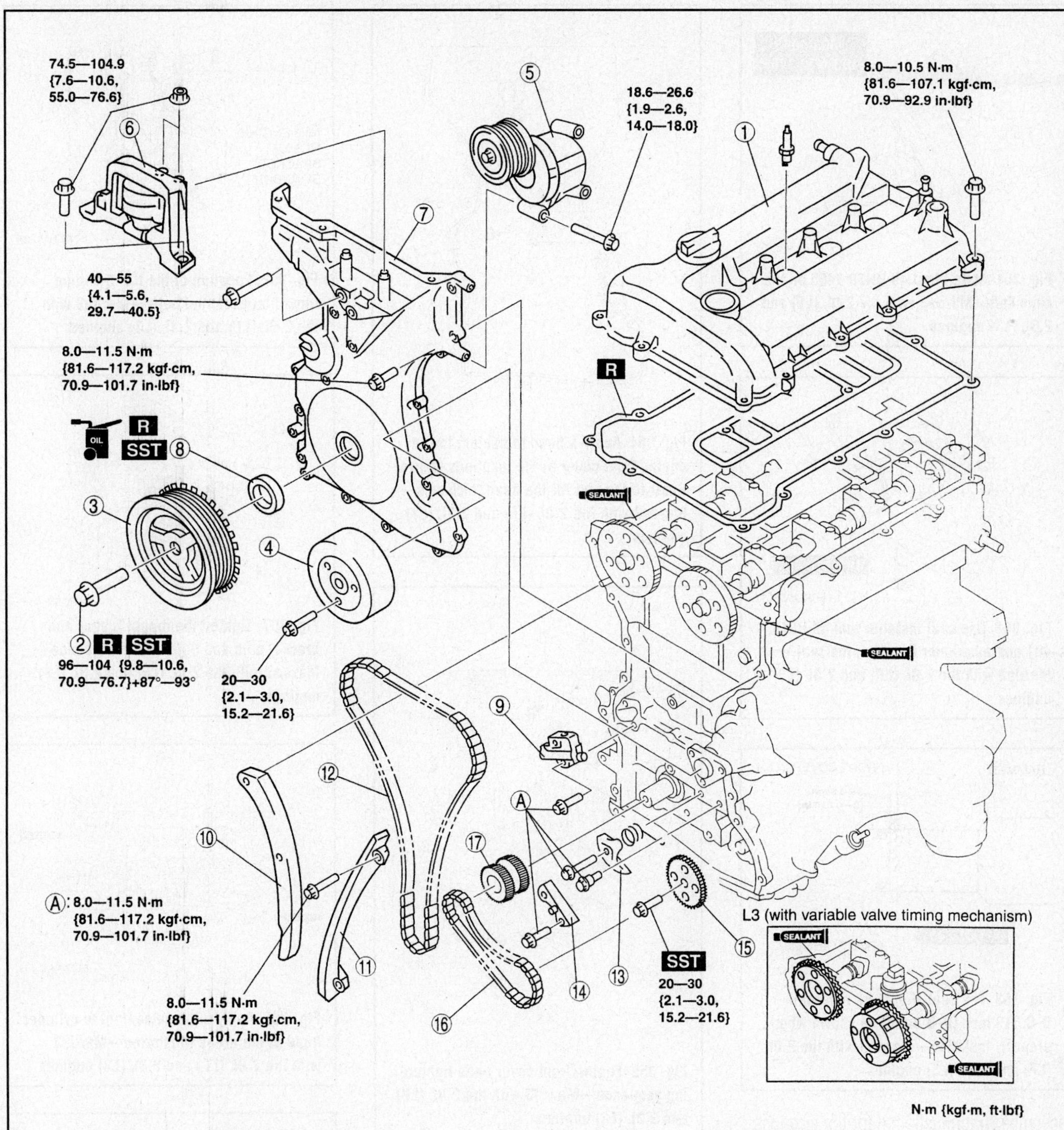

1 Cylinder head cover
2 Crankshaft pulley lock bolt
3 Crankshaft pulley
4 Water pump pulley
5 Drive belt auto tensioner
6 No.3 engine mount rubber and No.3 engine joint bracket
7 Engine front cover
8 Front oil seal
9 Chain tensioner
10 Tensioner arm
11 Chain guide
12 Timing chain
13 Oil pump chain tensioner
14 Oil pump chain guide
15 Oil pump sprocket
16 Oil pump chain
17 Crankshaft sprocket

67162-MAZC-G55

Fig. 350 Exploded view of the timing chain assembly and related components—Mazda3 with the 2.0L (LF) and 2.3L (L3) engines

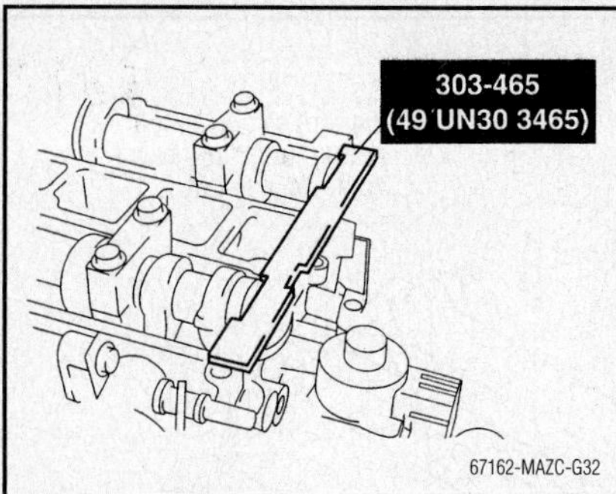

Fig. 351 Install tool 49 UN30 3465 on the camshaft—Mazda3 with the 2.0L (LF) and 2.3L (L3) engines

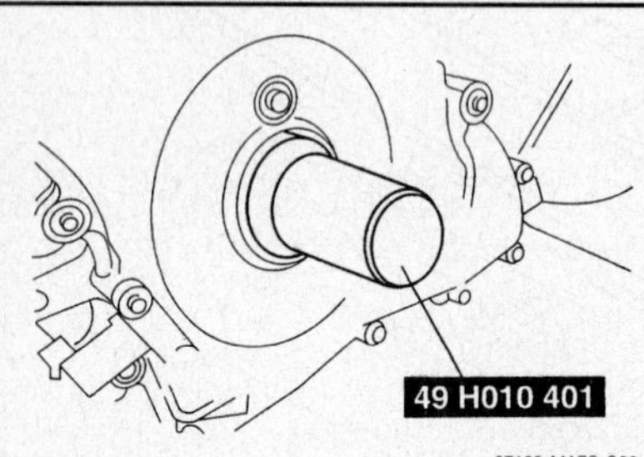

Fig. 352 Use seal installer tool 49 H010 401 and a hammer to install the seal—Mazda3 with the 2.0L (LF) and 2.3L (L3) engines

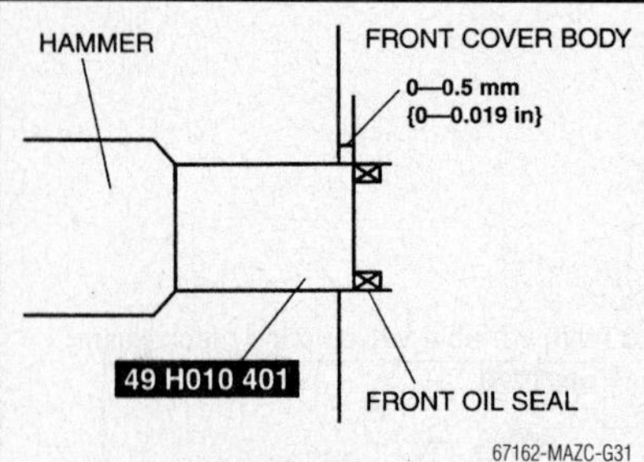

Fig. 353 The seal should be recessed 0–0.019 inch (0–0.5mm) as shown when properly installed—Mazda3 with the 2.0L (LF) and 2.3L (L3) engines

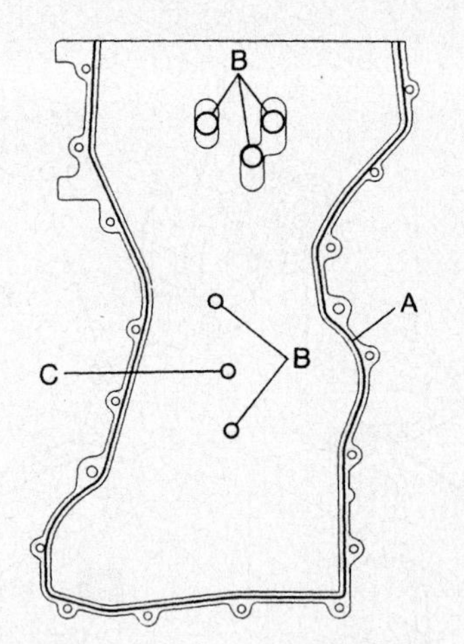

Fig. 354 Apply a bead of sealant to the engine front cover at the locations shown. Refer to the text for the bead thickness—Mazda3 with the 2.0L (LF) and 2.3L (L3) engines

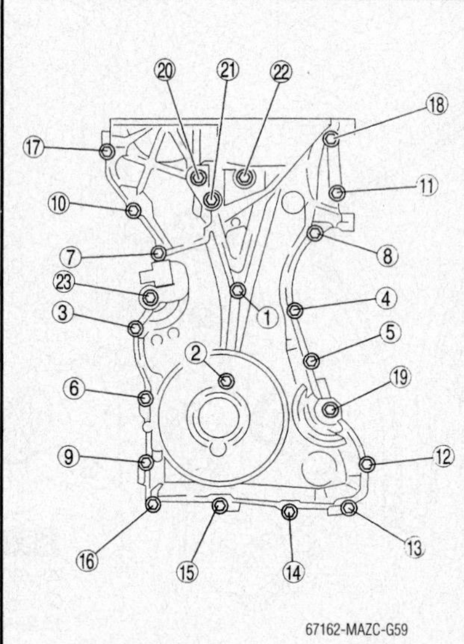

Fig. 355 Engine front cover bolts tightening sequence—Mazda3 with the 2.0L (LF) and 2.3L (L3) engines

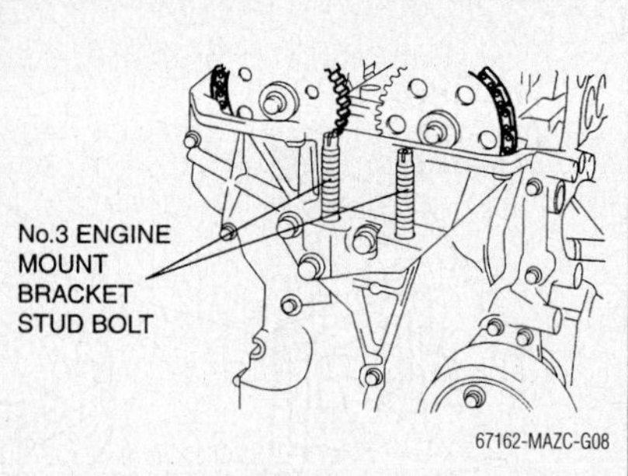

Fig. 356 Location of the No. 3 engine mount bracket stud bolts—Mazda3 with the 2.0L (LF) and 2.3L (L3) engines

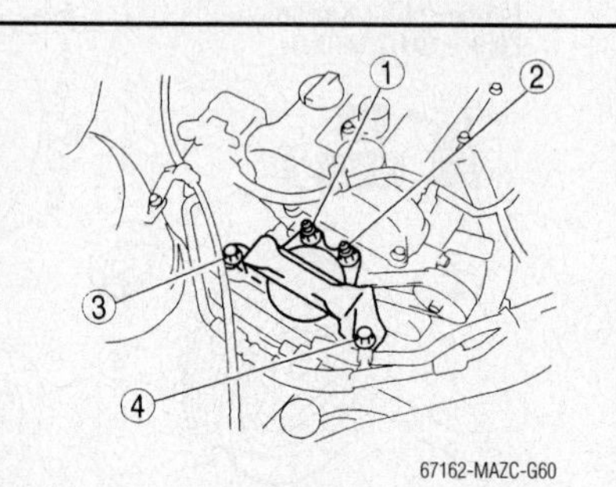

Fig. 357 Tighten the mount rubber and bracket nuts and bolts in the sequence—Mazda3 with the 2.0L (LF) and 2.3L (L3) engines

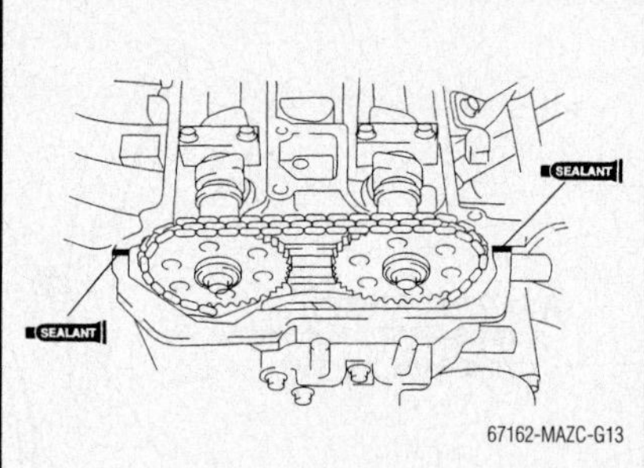

Fig. 358 Apply silicone sealant to cylinder head at the areas illustrated—Mazda3 with the 2.0L (LF) and 2.3L (L3) engines

accompanying illustration for the locations of points A and B. No sealant is needed at the points marked C on 2.3L (L3) engines with variable valve timing.

31. Install the cover within 10 minutes of apply the sealant. Tighten the cover bolts as follows:

a. Bolts 1 through 18: In sequence to 71–101 inch lbs. (8–11.5 Nm).

b. Bolts 19 through 22 : In sequence to 29.7–40.5 ft. lbs. (40–55 Nm).

c. Bolt 23 : 14.8–22 ft. lbs. (20–30 Nm).

32. Install the No. 3 engine mount and joint bracket as follows:

a. Tighten the No. 3 mount bracket stud bolts to 62–115 inch lbs. (7–13 Nm). Tighten the stud bolt with the mount nut loosened.

b. Hand tighten the mount rubber and bracket nuts and bolts, then tighten in the sequence illustrated to 55–77 ft. lbs. (74–105 Nm).

33. Install tool 49 UN30 3465 as illustrated on the camshaft.

34. Install and hand tighten the M6 x 1.0 bolt as illustrated.

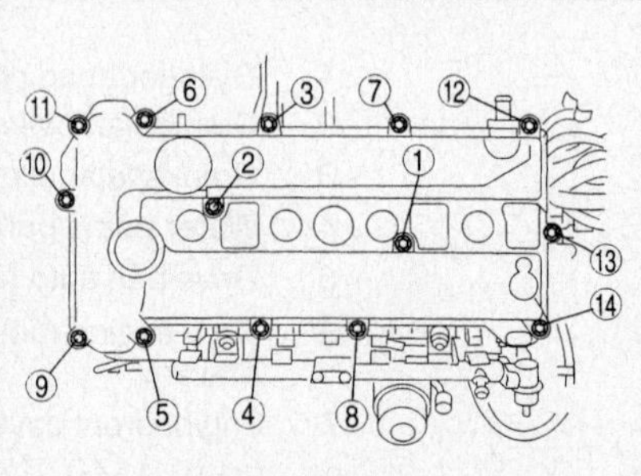

Fig. 359 Tighten the cylinder head cover bolts in the sequence shown—Mazda3 with the 2.0L (LF) and 2.3L (L3) engines

35. Turn the crankshaft clockwise until the crankshaft is in number 1 TDC (the balance weight should be attached to the tool).
36. Hold the crankshaft pulley and tighten the lock bolt in two steps. First tighten to 71–77 ft. lbs. (96–104 Nm). Then final tighten an additional 87–93 degrees.
37. Remove the M6 x 1.0 bolt.
38. Remove the tools from the camshaft and the cylinder block lower blind plug.
39. Rotate the crankshaft clockwise 2 turns until you reach TDC. If not aligned properly, loosen the crankshaft pulley lock bolt and reinstall the lock bolt again using the above procedure and special tools.
40. Install the cylinder block blind plug and tighten to 13–16 ft. lbs. (18–22 Nm).
41. Apply a 0.16–0.24 inch (4–7mm) bead of silicone sealant to cylinder head at the areas illustrated. Make sure to install the cover within 10 minutes of applying the sealant.
42. Install the cylinder head cover with a new gasket. Torque the bolts in the sequence illustrated to 71–93 inch lbs. (8–11 Nm).
43. Install the remaining components in the reverse order of removal.
44. When installing the CKP sensor perform the following:
   a. Remove the cylinder block lower blind plug.
   b. Install tool 49 JE01 061.
   c. Turn the crankshaft clockwise until the crankshaft is in the number 1 cylinder Top Dead Center position (the balance weight should be attached to tool 49 JE01 061).
   d. Using a ruler, mark the center line on the pulse wheel teeth on the crank pulley which is located at the 9th tooth counting counterclockwise from the empty space. Refer to the illustration for more detail.

CAUTION

If you do not mark the center line correctly this will cause improper engine control for the ignition and fuel system, so be sure to mark the line carefully.

   e. Install the CKP sensor making sure the mark on the sensor is aligned with the mark on the pulse wheel. Tighten the sensor bolt to 49–66 inch lbs. (5.5–7.5 Nm)and attach the electrical connector.
   f. Remove the tool from the cylinder block lower blind plug.
   g. Rotate the crankshaft clockwise 2 turns until you reach TDC. If not aligned properly, loosen the crankshaft pulley lock bolt and reinstall the lock bolt.
   h. Install the cylinder block blind plug and tighten to 13–16 ft. lbs. (18–22 Nm).
45. Change the engine oil.
46. Start the engine.
47. Inspect for the following:
   - Pulley and belt for run–out and contact
   - Any leaking fluids.
   - Ignition timing, idle speed and exhaust emissions
   - All remaining components for proper operation.

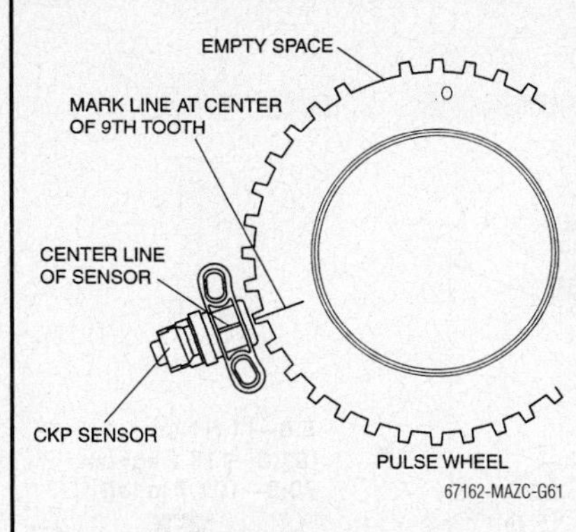

Fig. 360 Using a ruler, mark the center line on the pulse wheel teeth on the crank pulley which is located at the 9th tooth counting counterclockwise from the empty space—Mazda3 with the 2.0L (LF) and 2.3L (L3) engines

Mazda6

2.3L (L3) Engines

See Figures 361 through 371.

1. Before servicing the vehicle, refer to the precautions section.
2. Disconnect the negative battery cable.
3. Remove the cylinder head cover.
4. Remove the cylinder block lower blind plug.
5. Install tool 49 JE01 061.
6. Turn the crankshaft clockwise until the crankshaft is in the number 1 cylinder Top Dead Center position (the balance weight should be attached to tool 49 JE01 061).
7. Hold the crankshaft pulley using the tools illustrated and remove the bolt.
8. Remove the crankshaft pulley.
9. Remove the water pump pulley.
10. Remove the drive belt idler pulley.
11. Install an engine support device such as 49 E017 5A0.
12. Remove the No. 3 engine mount and joint.

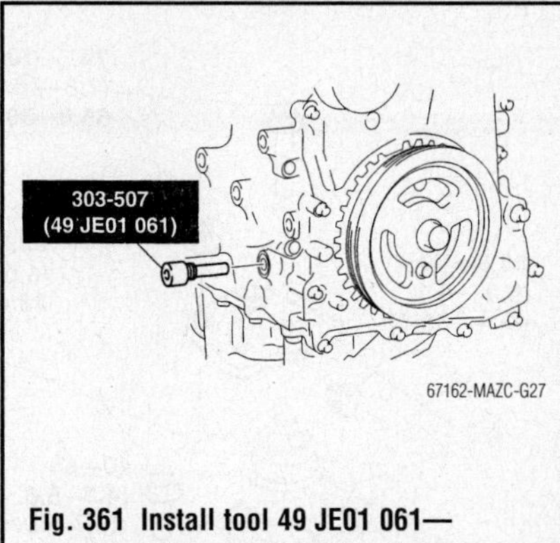

Fig. 361 Install tool 49 JE01 061—Mazda6 with the 2.3L (L3) engine

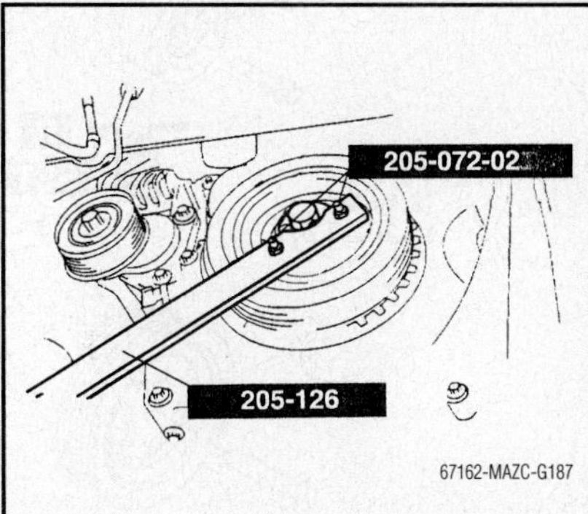

Fig. 362 Install tools 205-072-02 and 205-126 to hold the crankshaft pulley using the tools illustrated and remove the bolt—Mazda6 with the 2.3L (L3) engine

13. Remove the engine front cover and oil seal.
14. Unlock the chain tensioner using a suitable tool to slowly compress the tensioner piston. Insert a 0.059 inch (1.5mm) wire or a paper clip to hold the piston in its compressed position.
15. Remove the tensioner arm, chain guide and timing chain.

To install:

16. Install tool 49 UN30 3465 as illustrated on the camshaft.

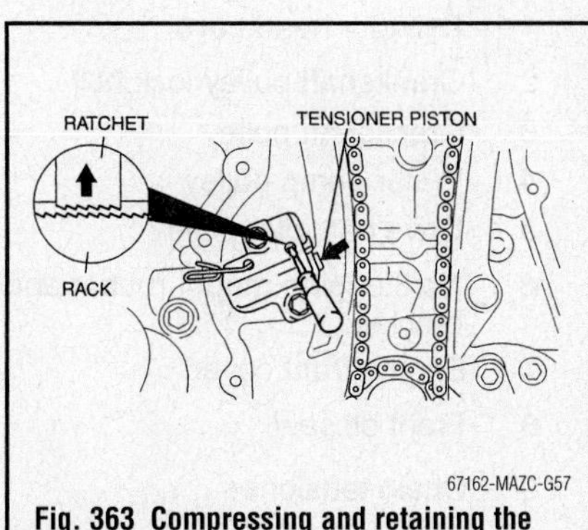

Fig. 363 Compressing and retaining the chain tensioner piston—Mazda6 with the 2.3L (L3) engine

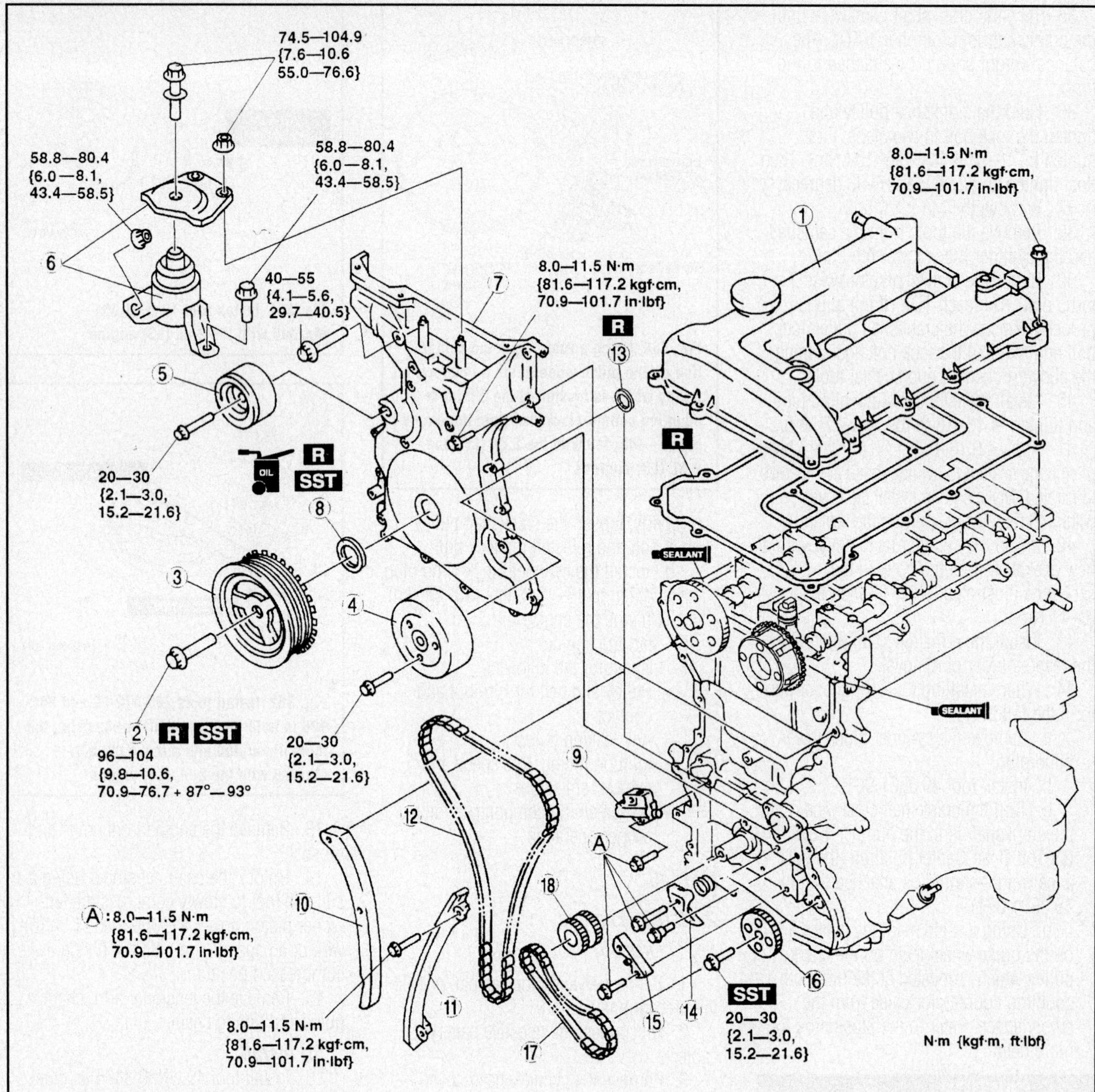

1	Cylinder head cover	10	Tensioner arm
2	Crankshaft pulley lock bolt	11	Chain guide
3	Crankshaft pulley	12	Timing chain
4	Water pump pulley	13	Seal
5	Drive belt idler pulley	14	Oil pump chain tensioner
6	No.3 engine mount rubber and No.3 engine joint bracket	15	Oil pump chain guide
7	Engine front cover	16	Oil pump sprocket
8	Front oil seal	17	Oil pump chain
9	Chain tensioner	18	Crankshaft sprocket

67162-MAZC-G186

Fig. 364 Exploded view of the timing chain assembly and related components—Mazda6 with the 2.3L (L3) engine

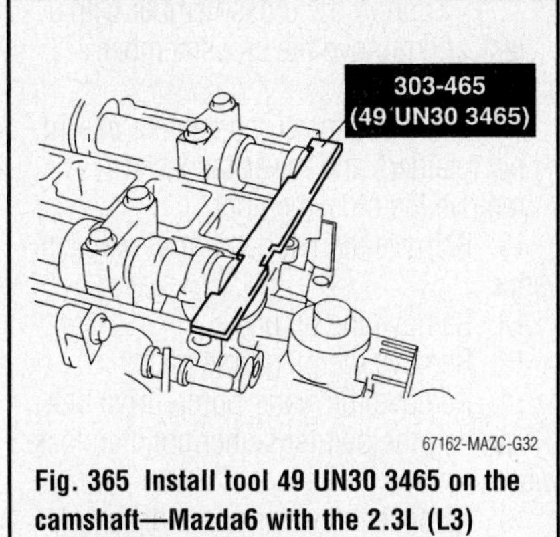

Fig. 365 Install tool 49 UN30 3465 on the camshaft—Mazda6 with the 2.3L (L3) engine

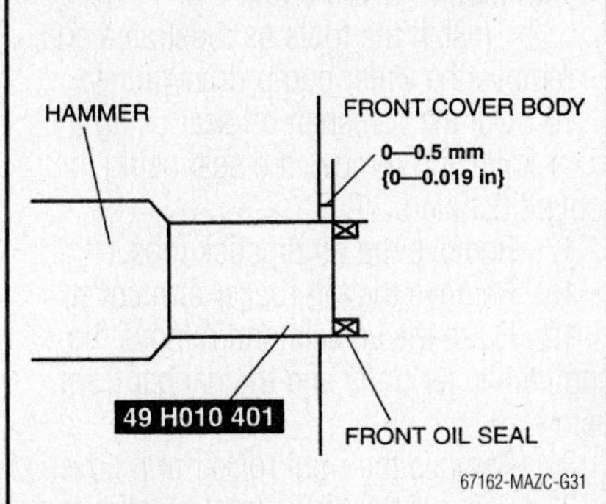

Fig. 366 The seal should be recessed 0–0.019 inch (0–0.5mm) as shown when properly installed—Mazda6 with the 2.3L (L3) engine

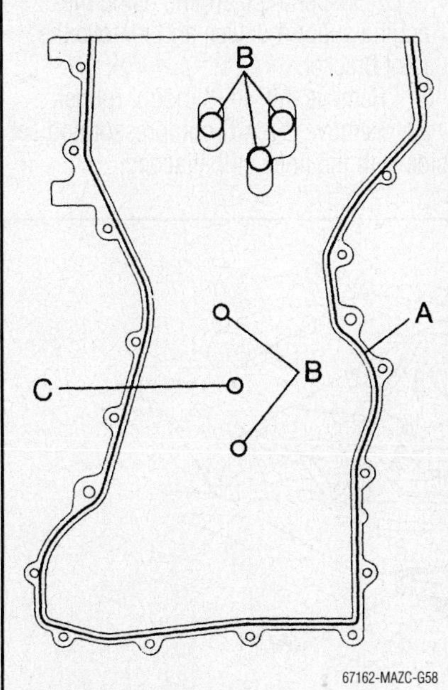

Fig. 367 Apply a bead of sealant to the engine front cover at the locations shown. Refer to the text for the bead thickness—Mazda6 with the 2.3L (L3) engine

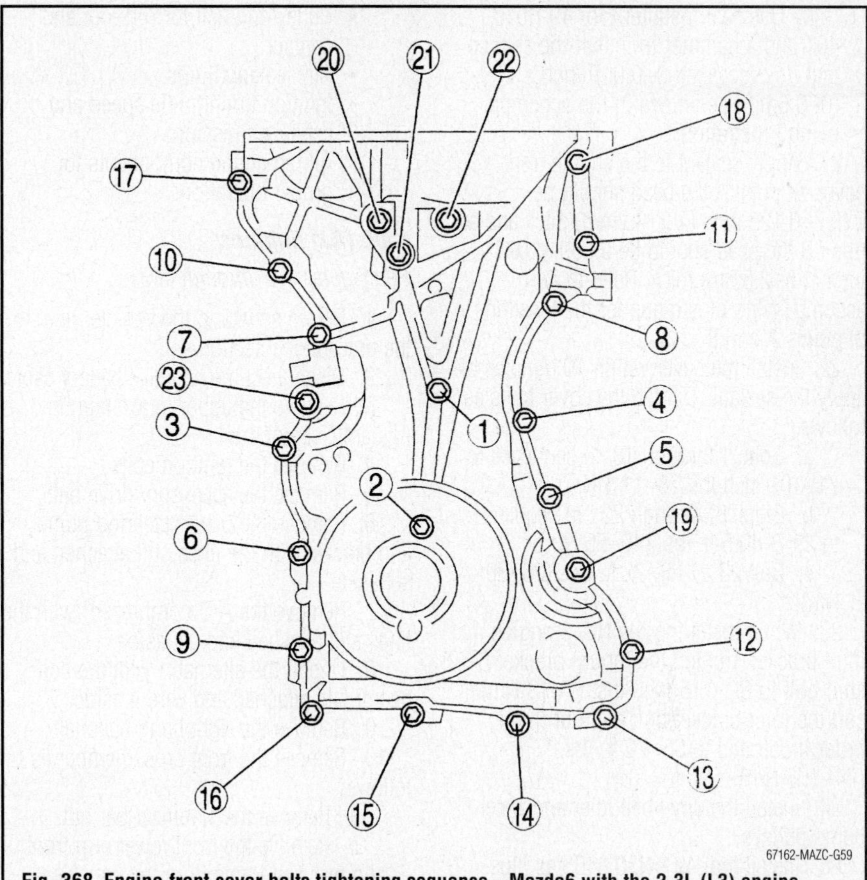

Fig. 368 Engine front cover bolts tightening sequence—Mazda6 with the 2.3L (L3) engine

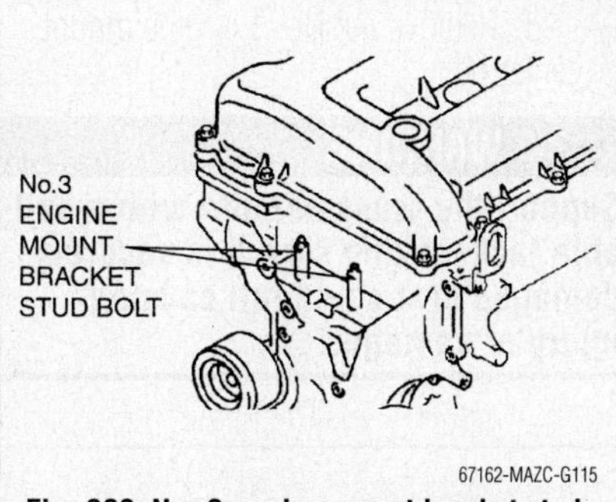

Fig. 369 No. 3 engine mount bracket stud bolt locations—Mazda6 models with the 2.3L (L3) engine

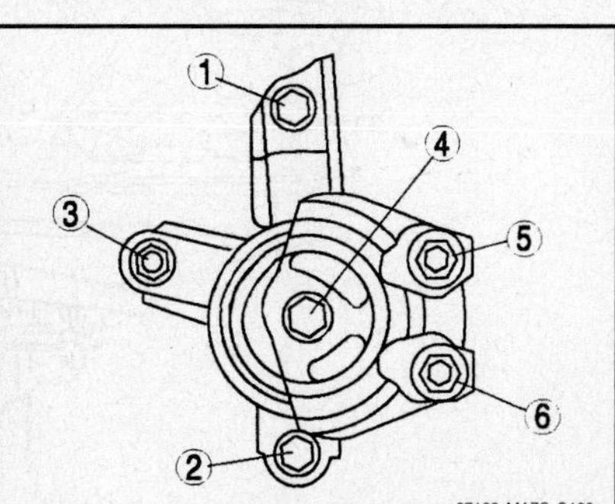

Fig. 370 Tighten the No. 3 joint bracket bolts and nuts in the order shown—Mazda6 models with the 2.3L (L3) engine

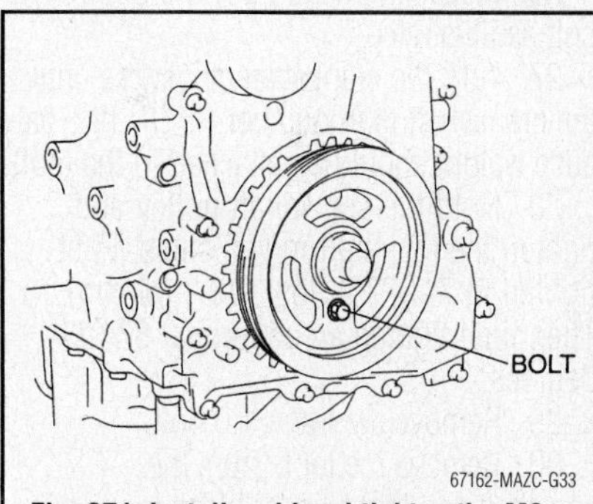

Fig. 371 Install and hand tighten the M6 x 1.0 bolt—Mazda6 with the 2.3L (L3) engine

17. Install the timing chain and remove the paper clip or wire retaining the tensioner piston to apply tension to the chain.
18. Install the timing chain guide and tighten the bolts to 71–101 inch lbs. (8–11.5 Nm).
19. Install the tensioner arm and tighten the bolts to 71–101 inch lbs. (8–11.5 Nm).
20. Install a new oil seal in the front cover as follows:
    a. Coat the new seal with clean engine oil.
    b. Push the seal in by hand to get it started.

c. Use seal installer tool 49 H010 401 and a hammer to install the seal so that it is recessed 0–0.019 inch (0–0.5mm) as shown in the accompanying illustration.

21. Apply sealant to the engine front cover. At point A the bead should be 0.087–0.125 inch (2.2–3.2mm) thick and at point B the bead should be 0.059–0.098 inch (1.5–2.5mm) thick. Refer to the accompanying illustration for the locations of points A and B.

22. Install the cover within 10 minutes of apply the sealant. Tighten the cover bolts as follows:

a. Bolts 1 through 18: In sequence to 71–101 inch lbs. (8–11.5 Nm).

b. Bolts 19 through 22 : In sequence to 29.7–40.5 ft. lbs. (40–55 Nm).

c. Bolt 23 : 14.8–22 ft. lbs. (20–30 Nm).

23. When installing the No. 3 engine joint bracket, tighten the mount bracket stud bolt to 62–115 inch lbs. (7–13 Nm) and the joint bracket bolt and nut in the order illustrated to 55–76 ft. lbs. (74–105 Nm).

24. Install the drive belt idler and water pump pulleys.

25. Install tool 49 UN30 3465 as illustrated on the camshaft.

26. Install and hand tighten the M6 x 1.0 bolt as illustrated.

27. Turn the crankshaft clockwise until the crankshaft is in number 1 TDC (the balance weight should be attached to the tool).

28. Hold the crankshaft pulley and tighten the lock bolt in two steps. First tighten to 71–77 ft. lbs. (96–104 Nm). Then final tighten an additional 87–93 degrees.

29. Remove the M6 x 1.0 bolt.

30. Remove the tools from the camshaft and the cylinder block lower blind plug.

31. Rotate the crankshaft clockwise 2 turns until you reach TDC. If not aligned properly, loosen the crankshaft pulley lock bolt and reinstall the lock bolt again using the above procedure and special tools.

32. Install the cylinder block blind plug and tighten to 13–16 ft. lbs. (18–22 Nm).

33. Install the cylinder head cover with a new gasket. Torque the bolts in the sequence illustrated to 71–93 inch lbs. (8–11 Nm).

34. Install the remaining components in the reverse order of removal.

35. Change the engine oil.

36. Start the engine.

37. Inspect for the following:

- Pulley and belt for run–out and contact
- Any leaking fluids.
- Ignition timing, idle speed and exhaust emissions
- All remaining components for proper operation.

3.0L (AJ) Engines

See Figures 372 through 403.

1. Before servicing the vehicle, refer to the precautions section.
2. Disconnect the negative battery cable.
3. Remove the upper intake manifold (dynamic chamber).
4. Remove the ignition coils.
5. Remove the accessory drive belt.
6. Remove the power steering pump and bracket with the lines still attached and set aside.
7. Remove the A/C compressor with the lines still attached and set aside.
8. Unbolt the alternator with the connector still attached and wire it aside.
9. Remove the right hand halfshaft.
10. Remove the front crossmember as follows:

a. Remove the stabilizer bar nut.

b. Remove the front lower arm front and rear ball joints from the knuckle.

c. Remove the front shock absorber lower bolt.

d. Remove the No. 1 engine mount center bolt.

⁂ WARNING

Support the crossmember with a suitable jack making sure it is securely damaged or it could fall causing injury or damage.

e. Support the crossmember with a jack and remove the crossmember bracket.

f. Remove the crossmember assembly retainers and lower the jack to remove the crossmember.

11. Remove the left hand three way converter.
12. Remove the oil pan.
13. Remove the plug hole plate.
14. Remove the water pump drive belt by rotating the belt tensioner counterclockwise.
15. Remove the water pump drive belt pulley as follows:

a. Replace part of tool 49 UN30 3009 with tool 49 UN30 3457.

b. Install the tools as illustrated and remove the water pump drive pulley.

16. Cut the camshaft oil seal using a razor knife and remove the seal using the tools illustrated.
17. Remove the oil dipstick tube.
18. Remove the left rocker arm cover.
19. Raise the vehicle and remove the engine hanger bolts and the hanger from below.
20. Remove the right rocker arm cover.
21. Remove the No. 3 joint bracket as follows:

a. Install the right hand engine hanger.

b. Use a M10 x 1.25 0.984 (25mm) length bolt to attach tool 49 UN30 3050 as illustrated. Tighten the bolt to 73–92 ft. lbs. (100–125 Nm).

c. Suspend the engine using the engine support device and remove the joint bracket.

22. Remove the No. 3 mount rubber.
23. Remove the A/C compressor and set aside with the lines still attached.

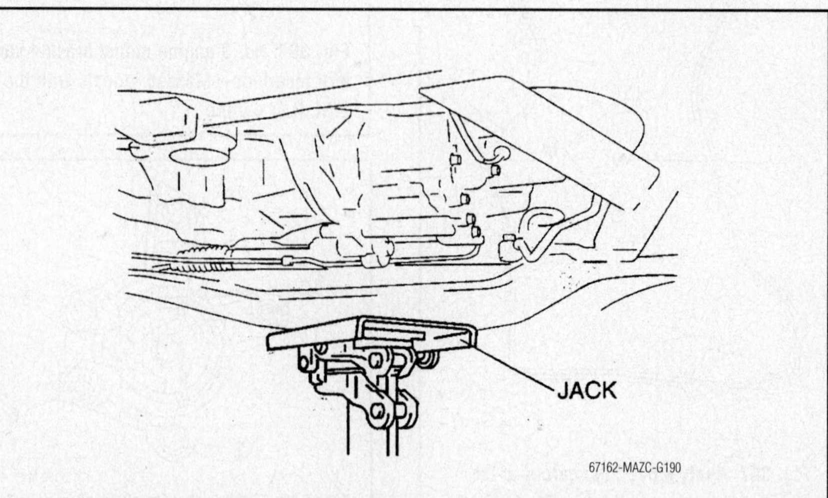

Fig. 372 Support the crossmember with a jack and remove the crossmember bracket—Mazda6 with the 3.0L (AJ) engine

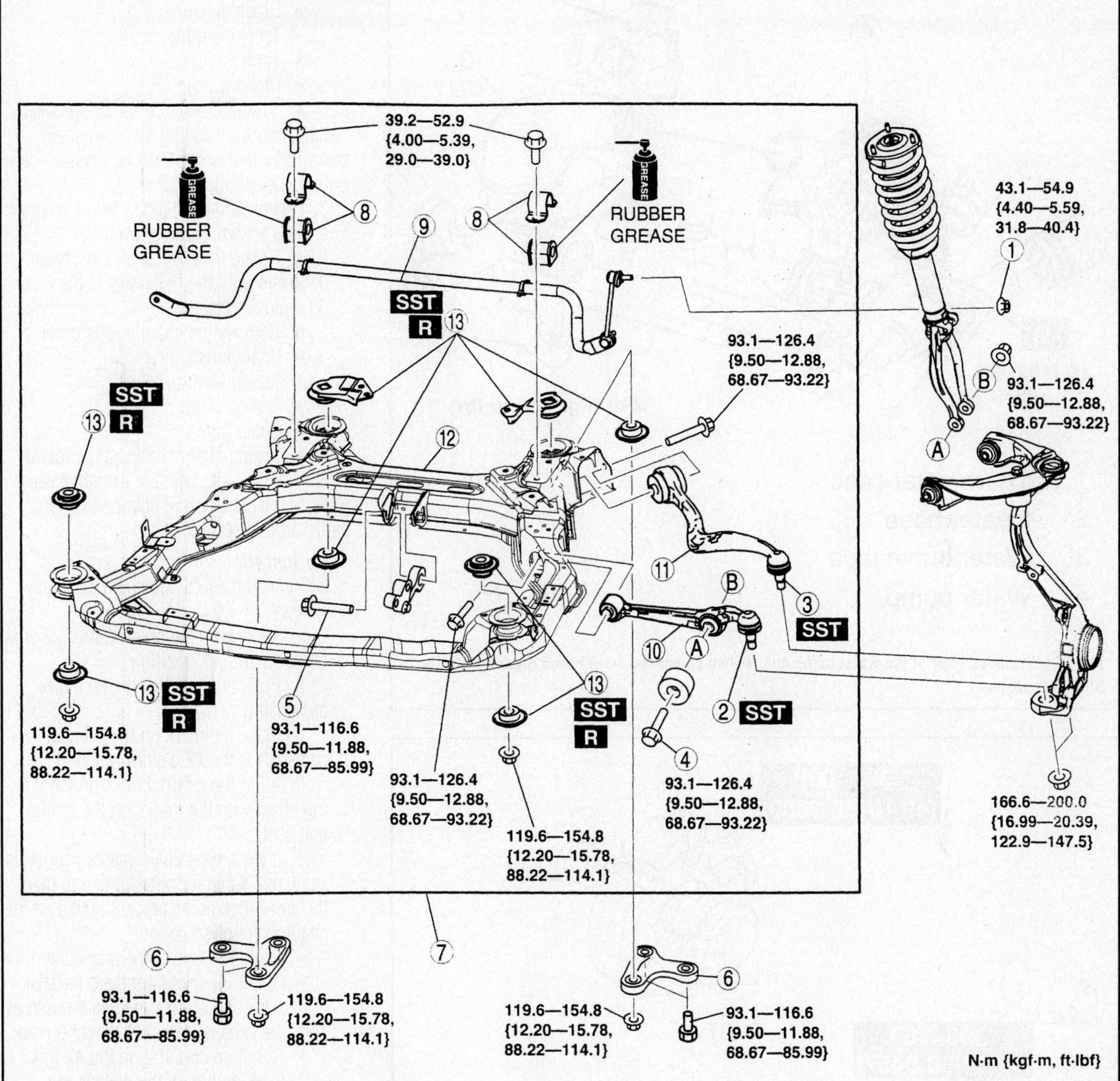

Fig. 373 Exploded view of the crossmember assembly and related components—Mazda6 with the 3.0L (AJ) engine

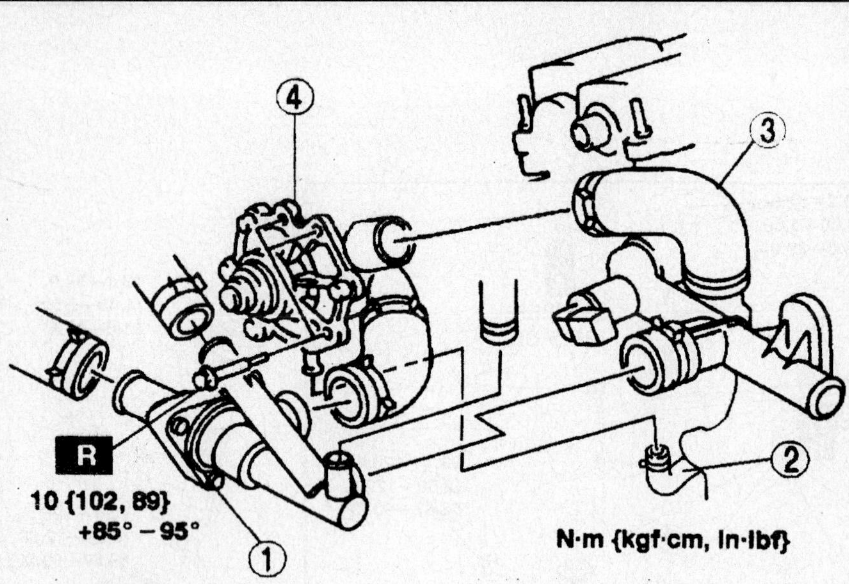

Fig. 374 Exploded view of the water pump and related components—Mazda6 models with the 3.0L (AJ) engine

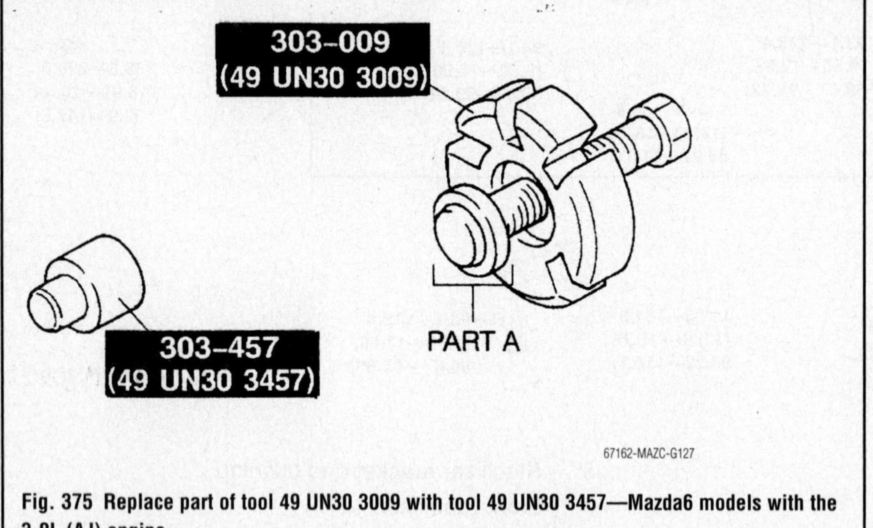

Fig. 375 Replace part of tool 49 UN30 3009 with tool 49 UN30 3457—Mazda6 models with the 3.0L (AJ) engine

24. Remove the crankshaft pulley using a suitable puller such as the one illustrated.
25. Remove the front oil seal using a pry tool as illustrated.
26. Remove the drive belt tensioner.
27. Remove the engine front cover bolts in the sequence illustrated.
28. Remove the CKP sensor pulse wheel.
29. Remove the right side chain tensioner and chain assembly as follows:
a. Rotate the crankshaft clockwise so the crankshaft keyway is at the 3 o'clock position. The camshafts should be in the neutral position.

➡Do not rotate the crankshaft counterclockwise or you may bind the chains causing engine damage.

30. Remove the right side chain in this order:
- Chain tensioner
- Tensioner arm
- Timing chain
- Chain guide
- Timing chain crankshaft sprocket

31. Remove the right side camshaft caps. Refer to the camshaft removal procedure earlier in this section
32. Remove the left side chain tensioner and chain assembly as follows:
a. Rotate the crankshaft clockwise 1⅔turns until the keyway is at the 11 o'clock position.
b. Remove the chain in this order:
- Chain tensioner
- Tensioner arm
- Timing chain
- Chain guide
- Timing chain crankshaft sprocket

33. Remove the left side camshaft caps. Refer to the camshaft removal procedure earlier in this section

To install:

34. Install the left side chain assembly as follows:
a. Make sure the crankshaft keyway is at the 11 o'clock position.
b. Position the mark on the intake camshaft to 9 o'clock.
c. Place the mark on the exhaust camshaft at the 12 o'clock position.
d. Align the colored links on the timing chain with the marks on the timing sprockets.
e. If the timing chain marks cannot be seen, use a marker or paint pen to mark the crankshaft and camshaft marks on the chain as follows:
- Mark any link as the crankshaft link
- Count 29 links from the crankshaft mark and place a mark to be used as the exhaust camshaft sprocket mark
- Continue counting to the 42 link mark and mark this link as the intake camshaft sprocket mark

35. Install the left hand chain in this order:
- Timing chain crankshaft sprocket
- Chain guide
- Timing chain
- Tensioner arm
- Chain tensioner

➡The chain guide should be installed to the actuator and allowed to hang freely when the bolts are installed. Do not hold the guide up when installing the bolts. The actuator causes wear to an O-ring and holding the guide will increase wear.

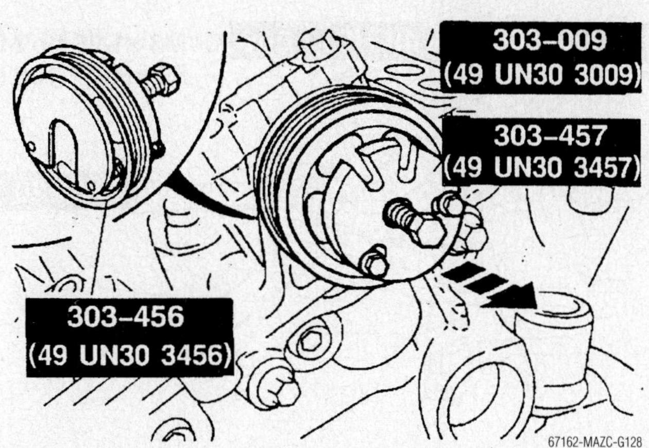

Fig. 376 Install the tools as illustrated and remove the water pump drive pulley—Mazda6 models with the 3.0L (AJ) engine

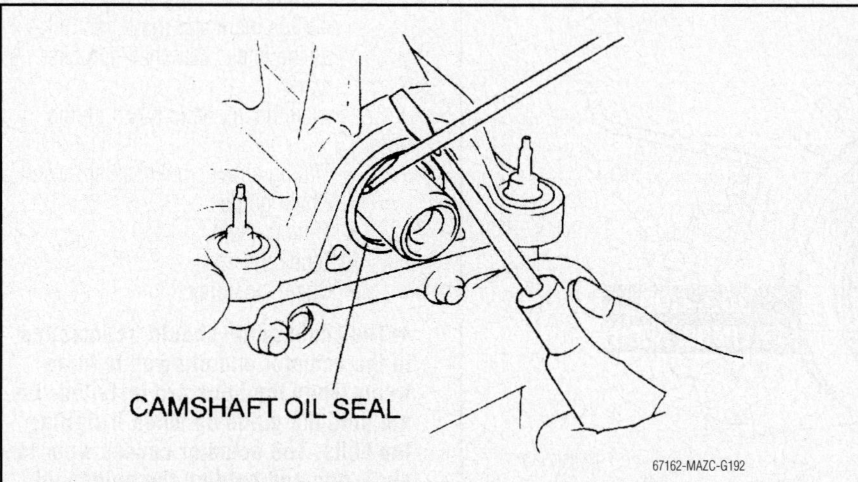

Fig. 377 Cut the camshaft oil seal using a razor knife and remove the seal as shown—Mazda6 with the 3.0L (AJ) engine

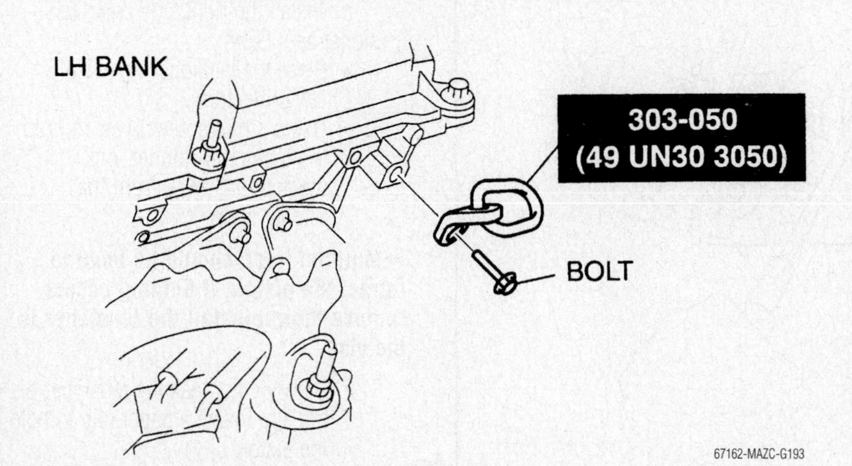

Fig. 378 Use a M10 x 1.25 0.984 (25mm) length bolt to attach tool 49 UN30 3050 as shown—Mazda6 with the 3.0L (AJ) engine

f. Install the left side camshaft caps. Refer to the camshaft removal procedure earlier in this section

g. Install the left side chain tensioner as follows:

- Place the tensioner in a vise with jaw protectors
- Use a small screwdriver to hold the tensioner ratchet lock mechanism away from the ratchet stem

➡Minimal force should be used to retract the piston, if binding occurs remove then reinstall the tensioner is the vise.

- Slowly compress the tensioner piston and install a paper clip to hold the piston
- Install the tensioner, tighten the bolts to 15–22 ft. lbs. (20–30 Nm) and remove the paper clip

36. Install the right side chain assembly as follows:

a. Make sure the crankshaft keyway is at the 3 o'clock position.

b. Place the mark on the exhaust camshaft at the 12 o'clock position.

c. Position the mark on the intake camshaft to 3 o'clock.

d. Align the colored links on the timing chain with the marks on the timing sprockets.

e. If the timing chain marks cannot be seen, use a marker or paint pen to mark the crankshaft and camshaft marks on the chain as follows:

- Mark any link as the crankshaft link
- Count 29 links from the crankshaft mark and place a mark to be used as the exhaust camshaft sprocket mark

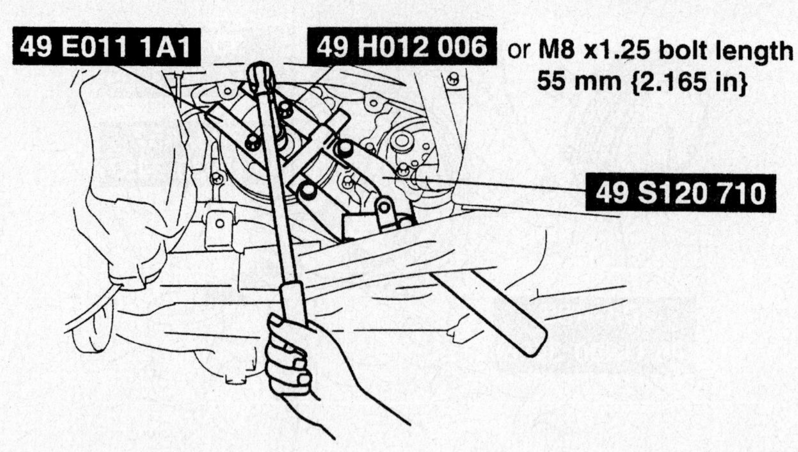

Fig. 379 Hold the crankshaft pulley using the tools illustrated and remove the pulley bolt—Mazda6 with the 3.0L (AJ) engine

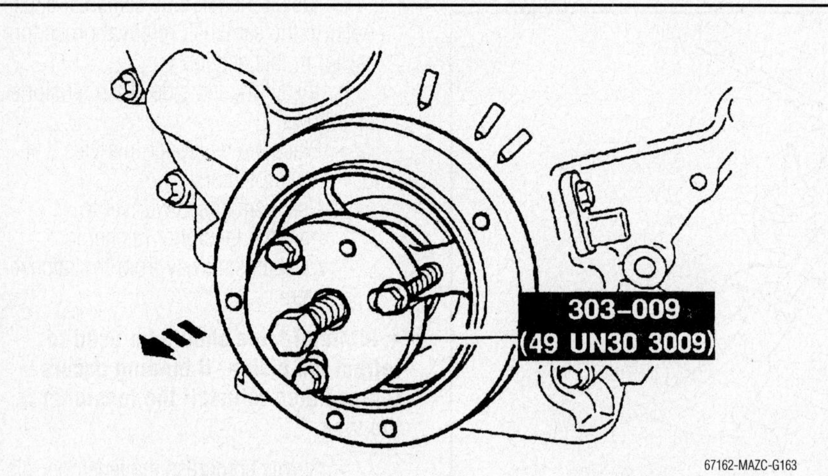

Fig. 380 Remove the crankshaft pulley using a suitable puller such as the one shown—Mazda6 with the 3.0L (AJ) engine

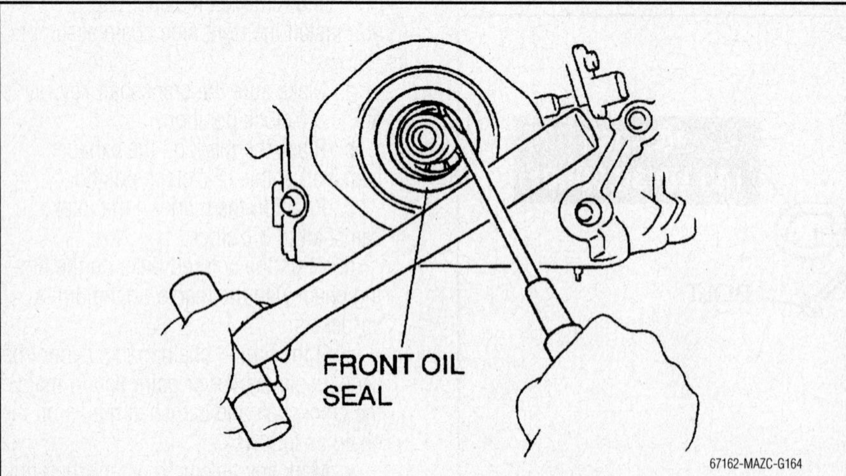

Fig. 381 Remove the front oil seal using a pry tool as shown—Mazda6 with the 3.0L (AJ) engine

- Continue counting to the 42 link mark and mark this link as the intake camshaft sprocket mark

37. Install the left hand chain in this order:
- Timing chain crankshaft sprocket
- Chain guide
- Timing chain
- Tensioner arm
- Chain tensioner

➡The chain guide should be installed to the actuator and allowed to hang freely when the bolts are installed. Do not hold the guide up when installing the bolts. The actuator causes wear to an O-ring and holding the guide will increase wear.

a. Install the right side camshaft caps. Refer to the camshaft removal procedure earlier in this section

b. Install the right side chain tensioner as follows:
- Place the tensioner in a vise with jaw protectors
- Use a small screwdriver to hold the tensioner ratchet lock mechanism away from the ratchet stem

➡Minimal force should be used to retract the piston, if binding occurs remove then reinstall the tensioner is the vise.

- Slowly compress the tensioner piston and install a paper clip to hold the piston
- Install the tensioner, tighten the bolts to 15–22 ft. lbs. (20–30 Nm) and remove the paper clip.

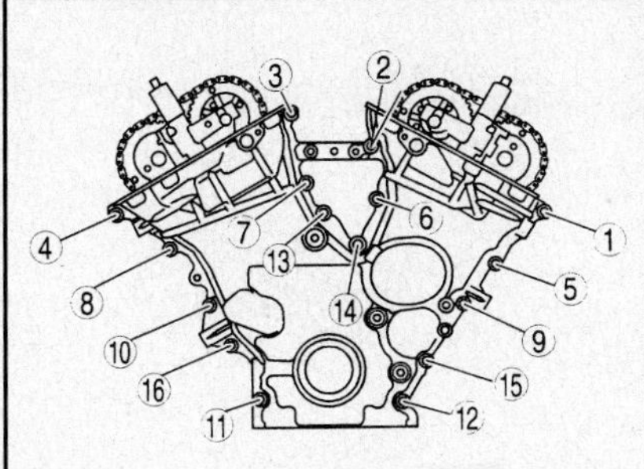

Fig. 382 Remove the engine front cover bolts in the sequence shown—Mazda6 with the 3.0L (AJ) engine

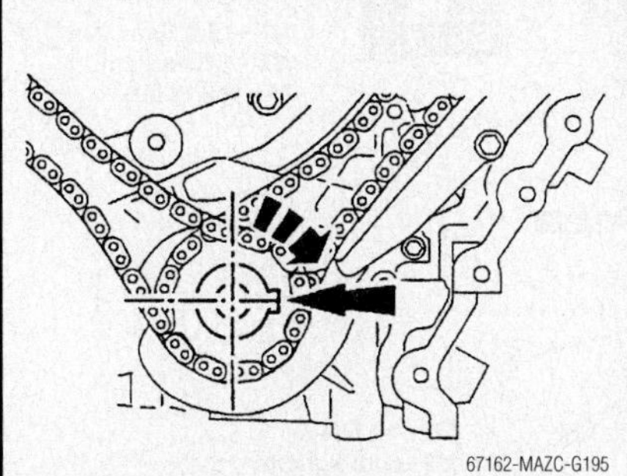

Fig. 383 Rotate the crankshaft clockwise so the crankshaft keyway is at the 3 o'clock position when removing the right side chain—Mazda6 with the 3.0L (AJ) engine

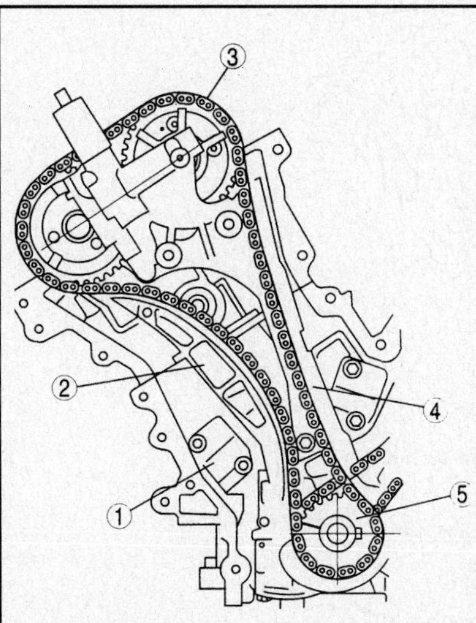

Fig. 384 Remove right side timing chain in numerical order—Mazda6 with the 3.0L (AJ) engine

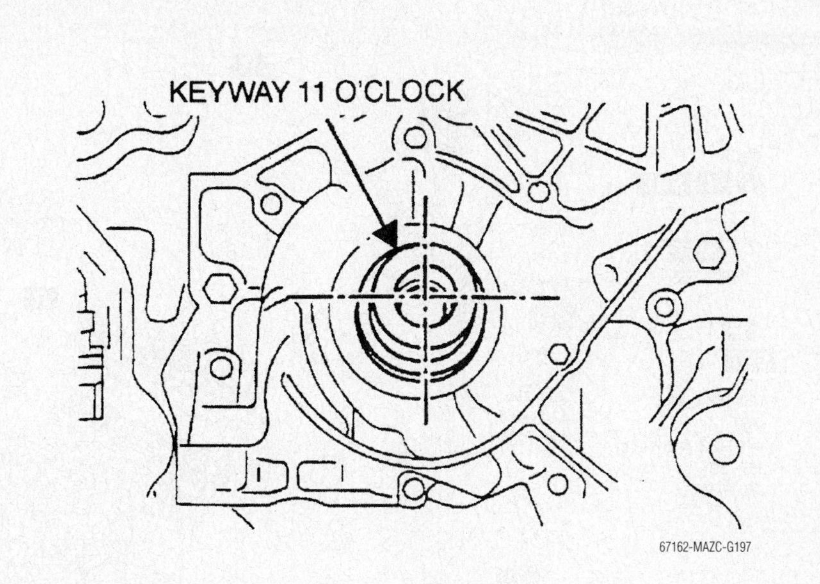

Fig. 385 Rotate the crankshaft clockwise 1⅔ turns until the keyway is at the 11 o'clock when removing the left side chain—Mazda6 with the 3.0L (AJ) engine

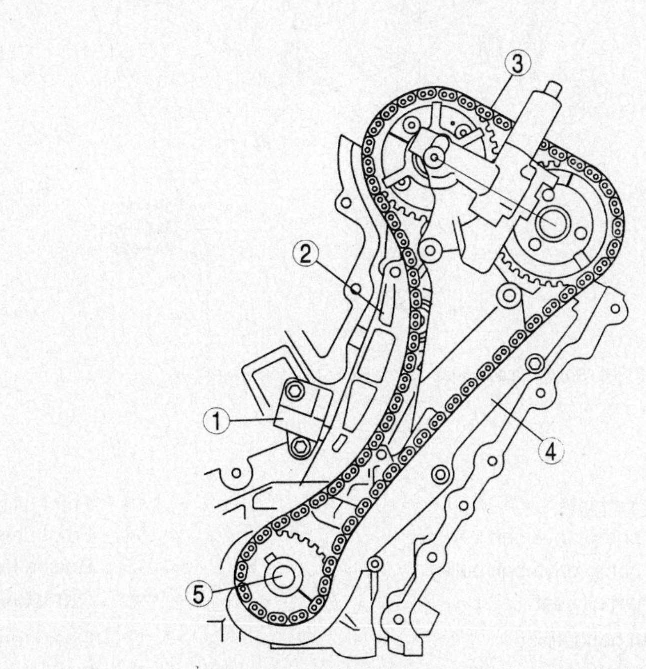

Fig. 386 Remove left side timing chain in numerical order—Mazda6 with the 3.0L (AJ) engine

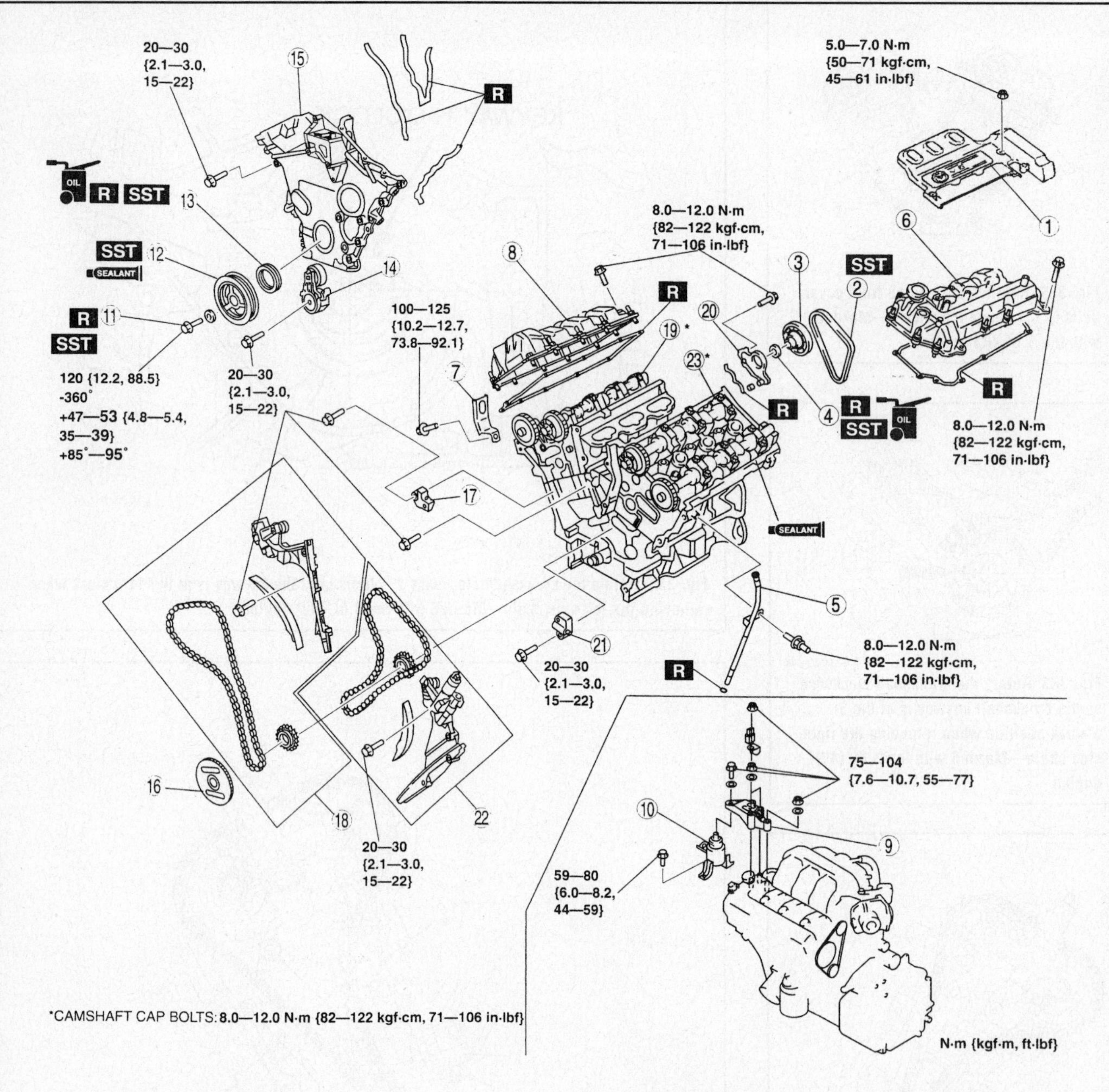

1 Plug hole plate
2 Water pump drive belt
3 Water pump drive belt pulley
4 Camshaft oil seal
5 Oil level gauge pipe
6 Cylinder head cover (LH)
7 Engine hanger (RH)
8 Cylinder head cover (RH)
9 No.3 engine joint bracket
10 No.3 engine mount rubber
11 Crankshaft pulley lock bolt
12 Crankshaft pulley
13 Front oil seal
14 Front drive belt auto tensioner
15 Engine front cover
16 CKP sensor pulse wheel
17 Chain tensioner (RH)
18 Timing chain component (RH)
19 Camshaft cap (RH)
20 Camshaft oil seal housing
21 Chain tensioner (LH)
22 Timing chain component (LH)
23 Camshaft cap (LH)

67162-MAZC-G189

Fig. 387 Exploded view of the timing chain assembly and related components—Mazda6 with the 3.0L (AJ) engine

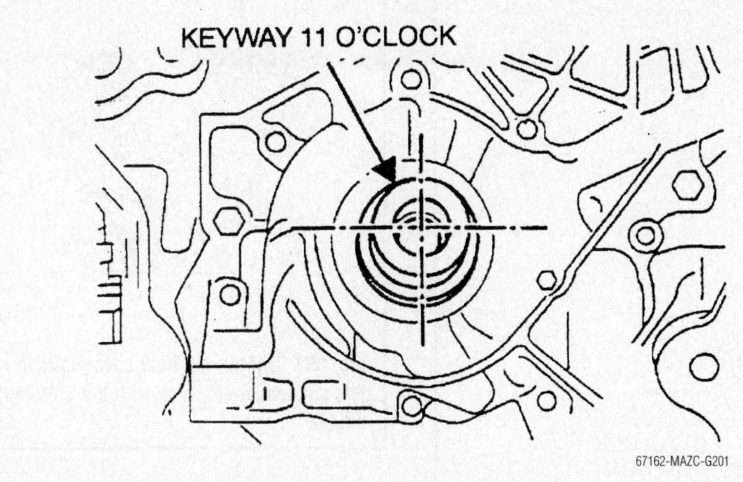

Fig. 388 When installing the left side chain, make sure the crankshaft keyway is at the 11 o'clock position—Mazda6 with the 3.0L (AJ) engine

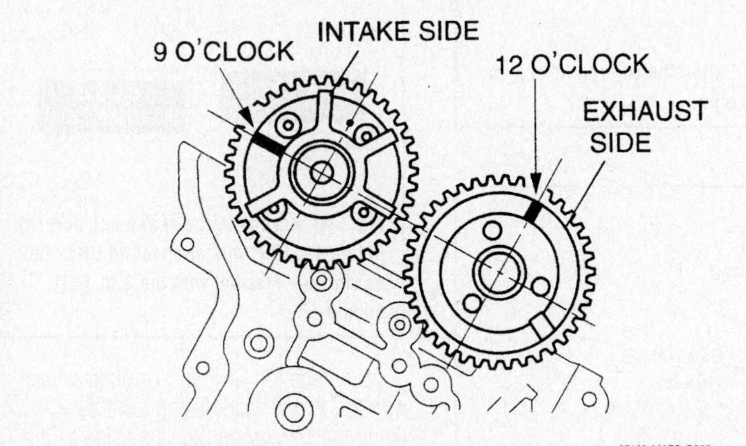

Fig. 389 When installing the left side chain, Position the mark on the intake camshaft to 9 o'clock and the mark on the exhaust camshaft at the 12 o'clock position—Mazda6 with the 3.0L (AJ) engine

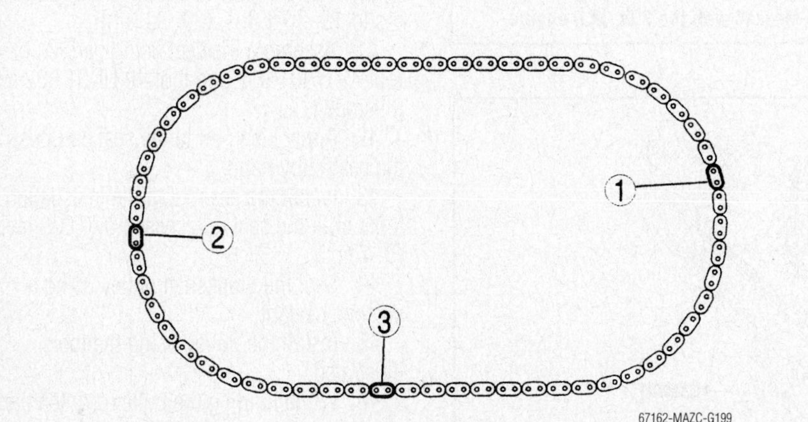

Fig. 390 If the timing chain marks cannot be seen, use a marker or paint pen to mark the crankshaft and camshaft marks on the chain. Refer to the text for marking steps.—Mazda6 with the 3.0L (AJ) engine

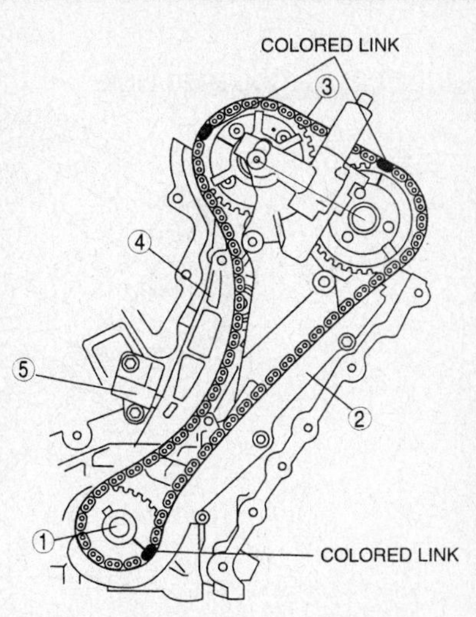

(1) Timing chain crankshaft sprocket
(2) Chain guide
(3) Timing chain
(4) Tensioner arm
(5) Chain tensioner

67162-MAZC-G200

Fig. 391 Install the left side timing chain in numerical order and align the colored links with the timing marks on the sprockets—Mazda6 with the 3.0L (AJ) engine

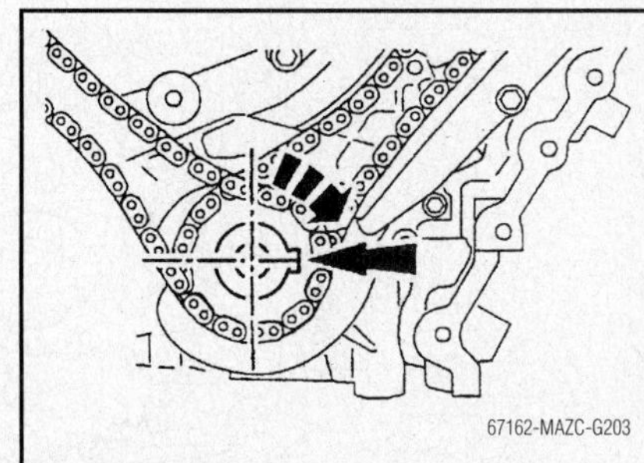

Fig. 392 When installing the right side chain, make sure the crankshaft keyway is at the 3 o'clock position—Mazda6 with the 3.0L (AJ) engine

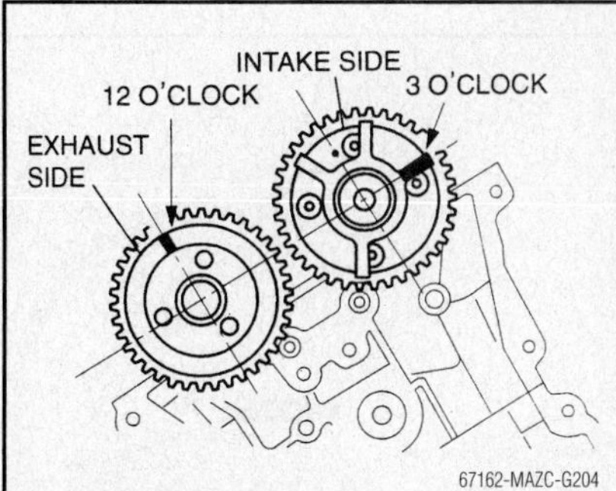

Fig. 393 When installing the right side chain, Position the mark on the exhaust camshaft to 12 o'clock and the mark on the intake camshaft at the 3 o'clock position—Mazda6 with the 3.0L (AJ) engine

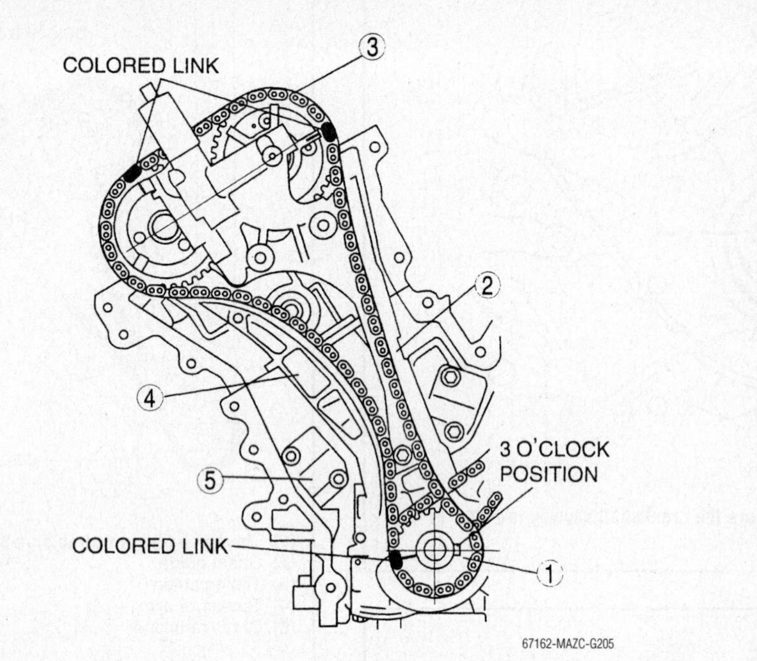

Fig. 394 Install the right side timing chain in numerical order and align the colored links with the timing marks on the sprockets—Mazda6 with the 3.0L (AJ) engine

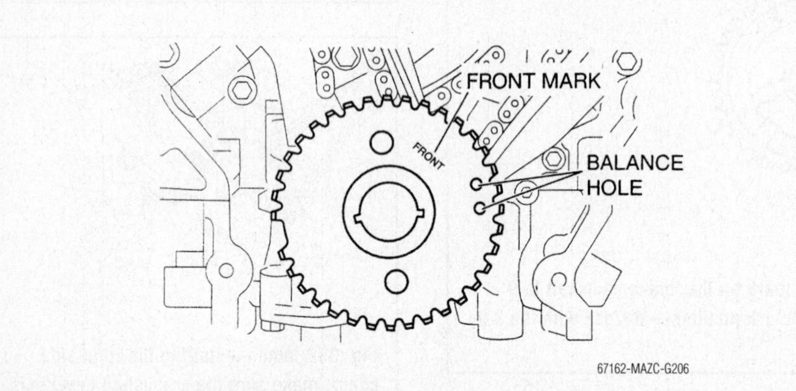

Fig. 395 Place the CKP sensor pulse wheel with the Front mark facing towards you, using the keyway on the same side as the empty space as shown—Mazda6 with the 3.0L (AJ) engine

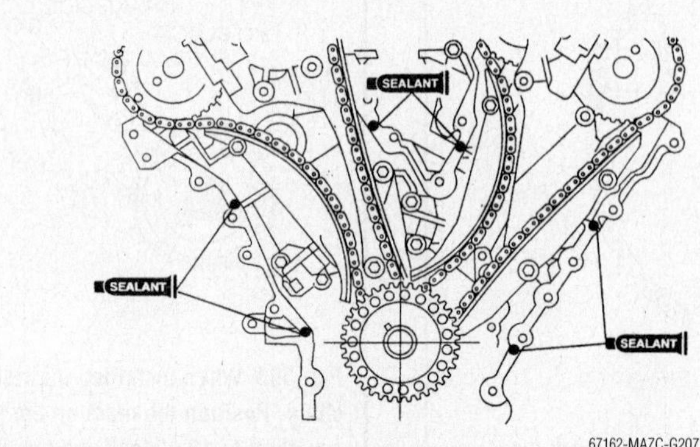

Fig. 396 Apply a 0.24 inch (6mm) bead of silicone sealant to the locations illustrated and install the front cover—Mazda6 with the 3.0L (AJ) engine

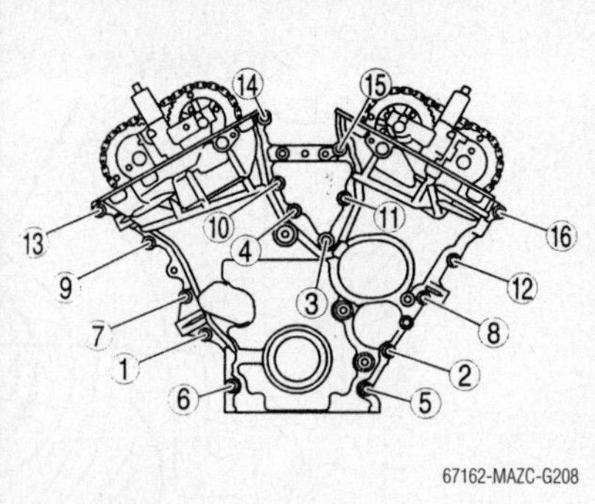

Fig. 397 Tighten the front cover bolts in this sequence—Mazda6 with the 3.0L (AJ) engine

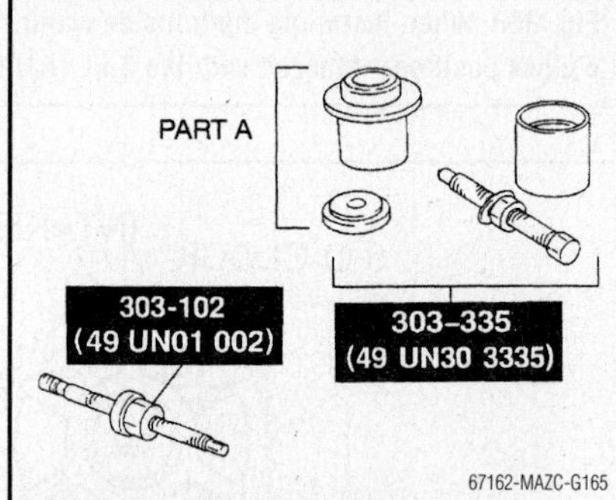

Fig. 398 Assemble the seal using part (A) of tool 49 UN01 002 and tool 49 UN01 002 as shown—Mazda6 with the 3.0L (AJ) engine

38. Place the CKP sensor pulse wheel with the **Front** mark facing towards you, using the keyway on the same side as the empty space as illustrated.
39. Apply a 0.24 inch (6mm) bead of silicone sealant to the locations illustrated and install the front cover.
40. Tighten the front cover bolts in the sequence illustrated and tighten the retainers to 15–20 ft. lbs. (20–30 Nm).
41. Assemble the seal using part (A) of tool 49 UN01 002 and tool 49 UN01 002 as illustrated.
42. Apply clean oil to the seal and push the seal in by hand.
43. Install the seal using the installation tools until the seal is recessed 0–0.039 inch (0–1mm).
44. Seal the crankshaft pulley using a silicone sealant.
45. Install the pulley using the tools illustrated.
46. While holding the pulley, tighten the crankshaft pulley bolt as follows:
    a. Tighten to 88 ft. lbs. (120 Nm).
    b. Loosen one full turn.
    c. Tighten to 35–39 ft. lbs. (47–53 Nm).
    d. Tighten 85–95 degrees.

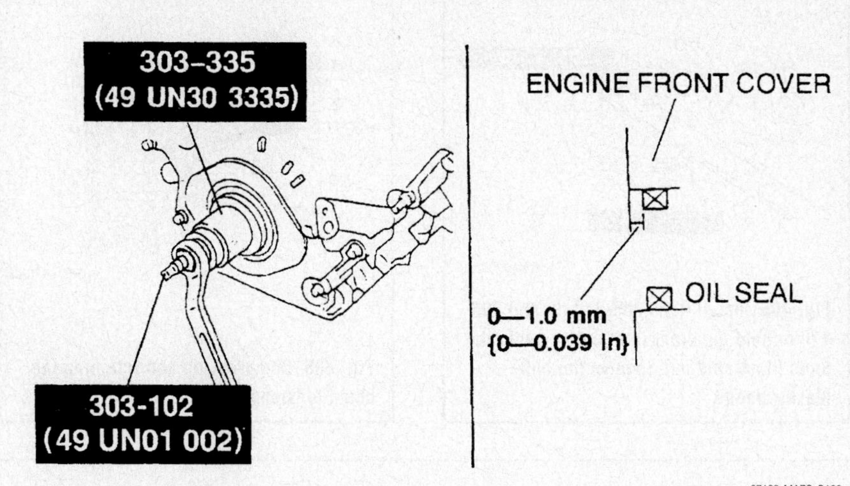

Fig. 399 Install the seal using the installation tools until the seal is recessed 0–0.039 inch (0–1mm)—Mazda6 with the 3.0L (AJ) engine

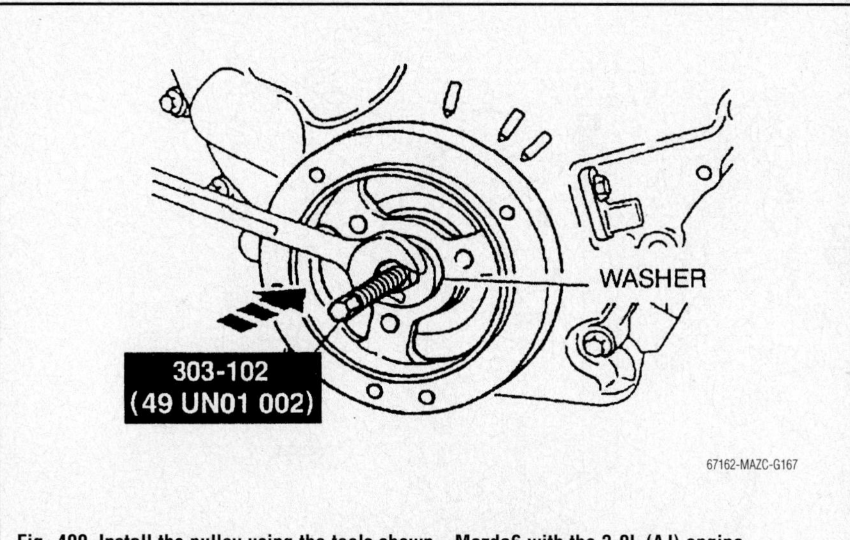

Fig. 400 Install the pulley using the tools shown—Mazda6 with the 3.0L (AJ) engine

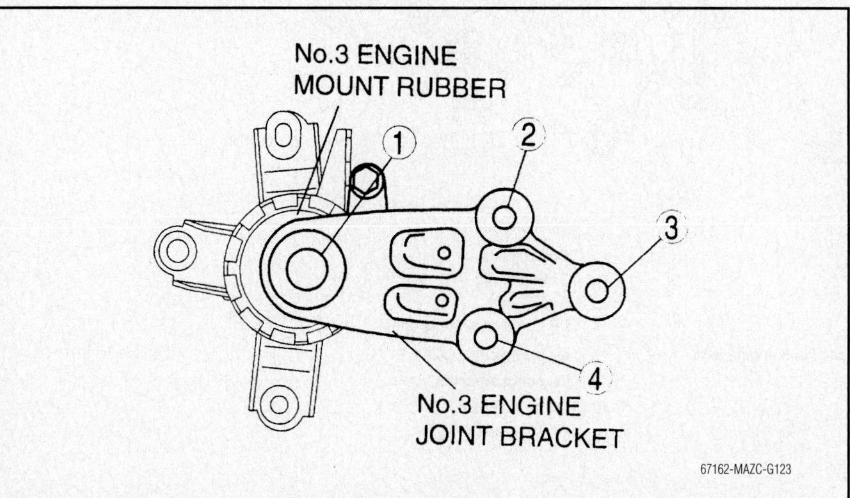

Fig. 401 Tighten the No. 3 engine joint bracket as shown—Mazda6 models with the 3.0L (AJ) engine

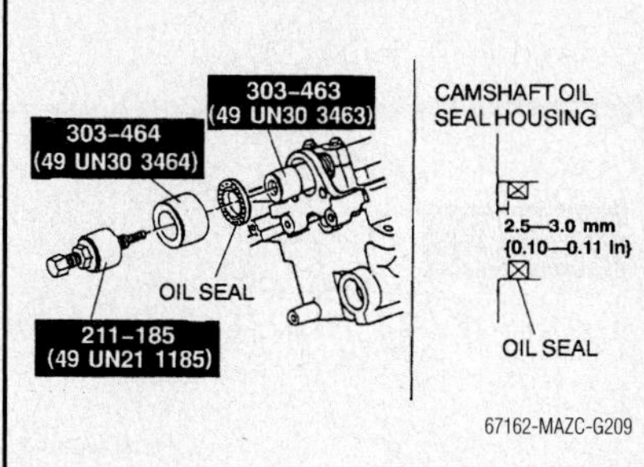

Fig. 402 Install the camshaft oil seal—Mazda6 with the 3.0L (AJ) engine

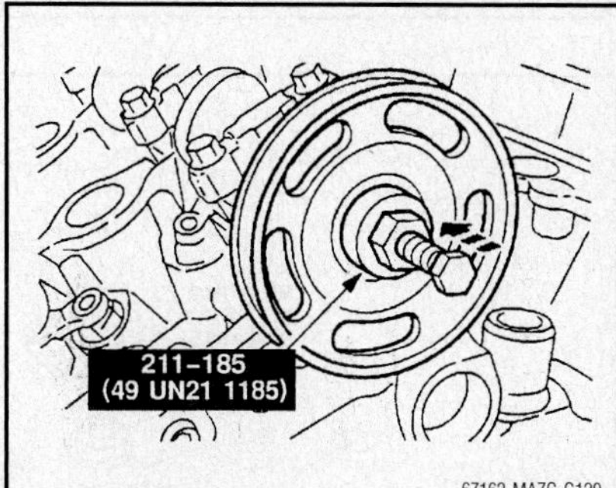

Fig. 403 Install a new water pump drive belt pulley using tool 49 UN21 1185—Mazda6 models with the 3.0L (AJ) engine

47. Install the no. 3 engine mount rubber.
48. When installing the No. 3 engine joint bracket, tighten the No.3 mount bracket bolts and nuts in the order illustrated as follows:
   a. Bolt 1: 55–77 ft. lbs. (74–105 Nm).
   b. Bolts 2, 3 and 4: 31–44 ft. lbs. (43–60 Nm).
49. Install the right hand valve cover.
50. Install the right hand engine hanger.
51. Install the left hand valve cover.
52. Install the dipstick tube.
53. Apply clean engine oil to the camshaft seal and install using the tools illustrated. The seal should be recessed 0.10–0.11 inch (2.5–3mm).
54. Install a new water pump drive belt pulley using tool 49 UN21 1185.
55. Install the water pump drive belt.
56. Install the plug hole plate.
57. Install the remaining components in the reverse order of removal.
58. Chain the engine oil and filter.
59. Check all fluid levels and replenish as necessary.
60. Road test the vehicle.

Mazdaspeed3 and Mazdaspeed6 Models

2.3L (L3 with TC) Engines

See Figures 404 through 415.

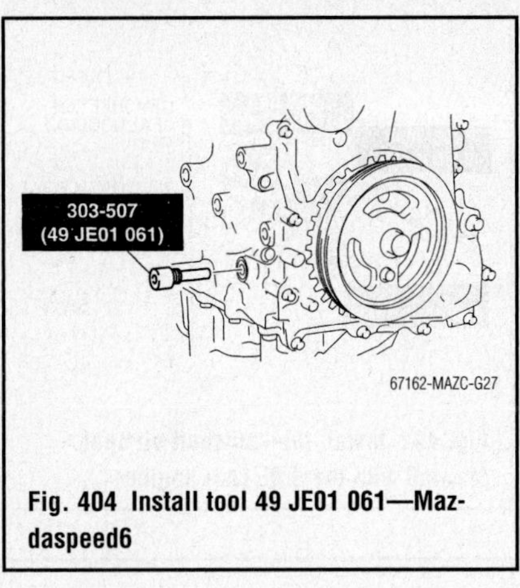

Fig. 404 Install tool 49 JE01 061—Mazdaspeed6

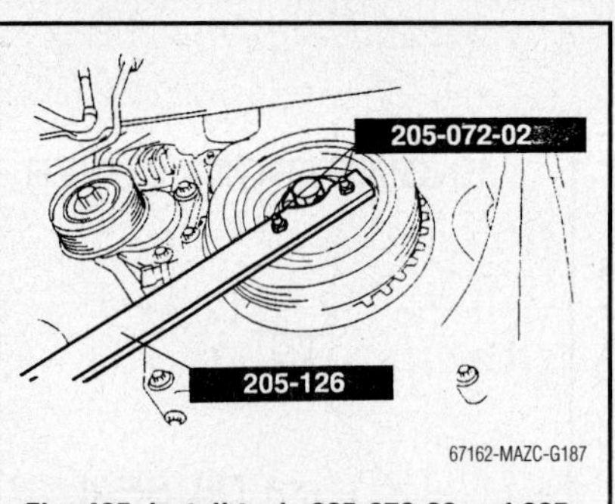

Fig. 405 Install tools 205-072-02 and 205-126 to hold the crankshaft pulley using the tools illustrated and remove the bolt—Mazdaspeed6

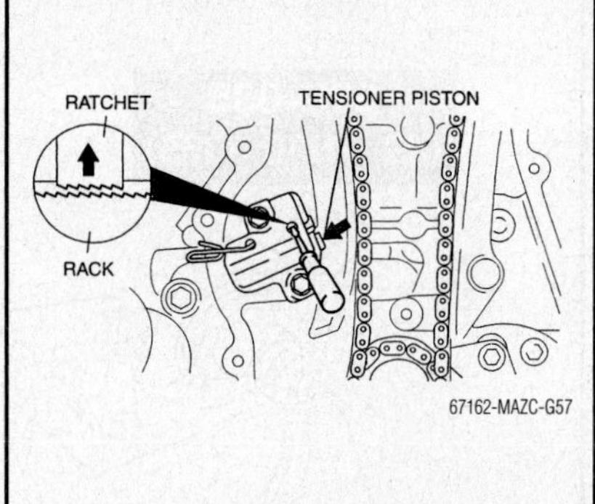

Fig. 406 Compressing and retaining the chain tensioner piston—Mazdaspeed6

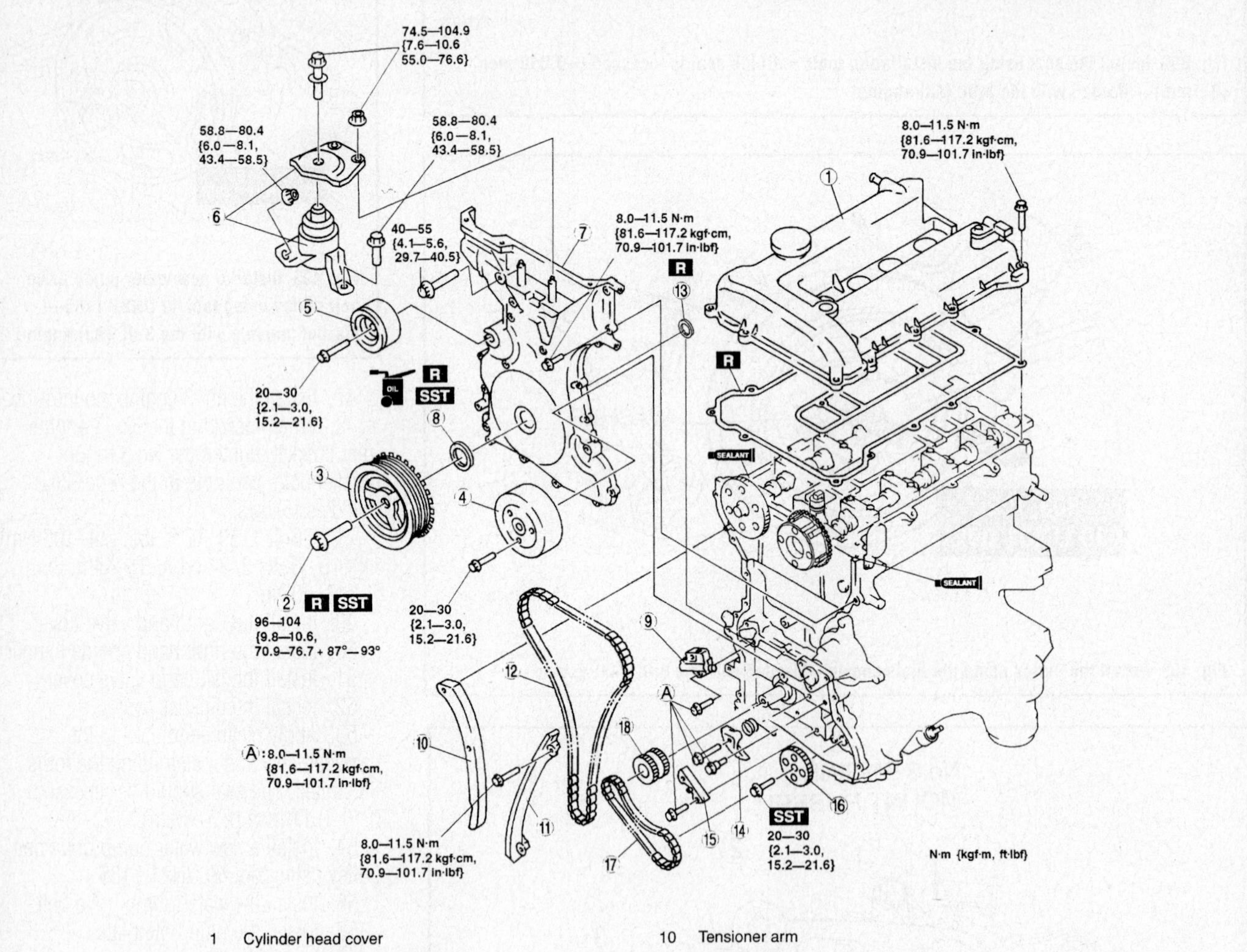

1	Cylinder head cover	10	Tensioner arm
2	Crankshaft pulley lock bolt	11	Chain guide
3	Crankshaft pulley	12	Timing chain
4	Water pump pulley	13	Seal
5	Drive belt idler pulley	14	Oil pump chain tensioner
6	No.3 engine mount rubber and No.3 engine joint bracket	15	Oil pump chain guide
7	Engine front cover	16	Oil pump sprocket
8	Front oil seal	17	Oil pump chain
9	Chain tensioner	18	Crankshaft sprocket

67162-MAZC-G186

Fig. 407 Exploded view of the timing chain assembly and related components—Mazdaspeed6

1. Before servicing the vehicle, refer to the precautions section.
2. Disconnect the negative battery cable.
3. Remove the right side front tire.
4. Remove the undercover.
5. Remove the right side splash shield.

➡If the high pressure fuel pump is removed, replace the O-ring with a new one.

6. Remove the charge air cooler cover.
7. Disconnect the spill valve control solenoid valve connector.

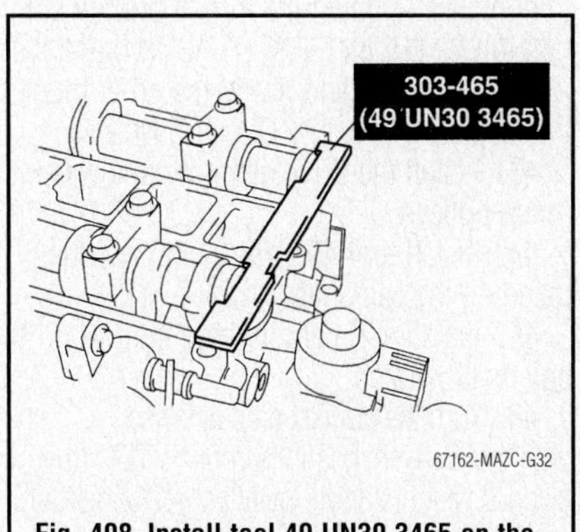

Fig. 408 Install tool 49 UN30 3465 on the camshaft—Mazdaspeed6

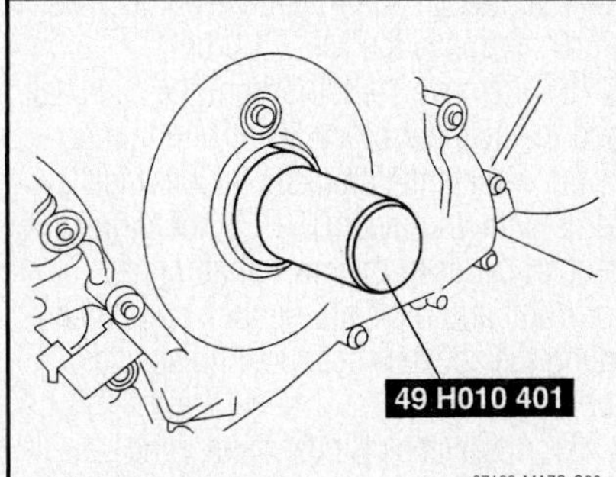

Fig. 409 Use seal installer tool 49 H010 401 and a hammer to install the seal—Mazdaspeed6

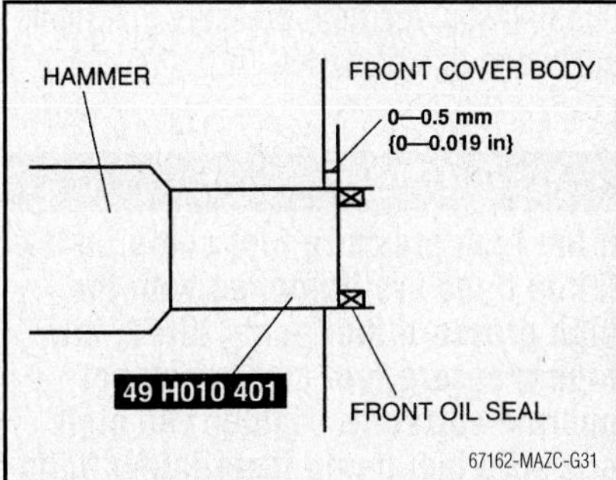

Fig. 410 The seal should be recessed 0–0.019 inch (0–0.5mm) as shown when properly installed—Mazdaspeed6

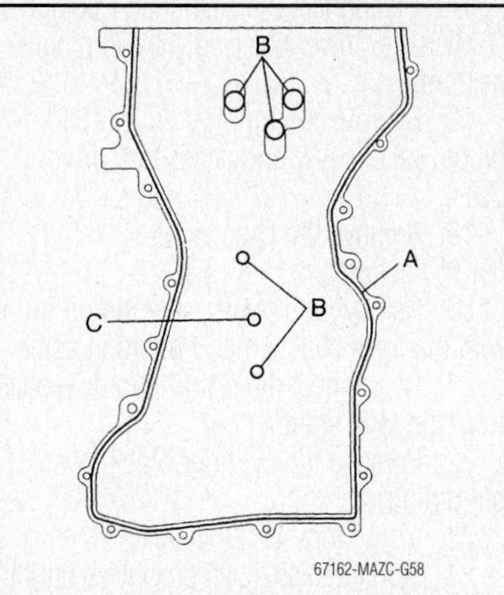

Fig. 411 Apply a bead of sealant to the engine front cover at the locations shown. Refer to the text for the bead thickness—Mazdaspeed6

8. Disconnect the quick release connector on the high pressure fuel pump.
9. Remove the battery and battery tray.

✲✲ WARNING

If the high pressure fuel pump joint nut is loosened, fuel leakage may occur resulting in death or serious injury, or damage to the equipment or the vehicle. Fuel can also irritate the skin and eyes. When removing the high pressure line pipe, always tighten the high pressure line pipe installation nut while fixing the high pressure fuel pump joint nut with a wrench. If the high pressure fuel pump joint nut has rotated, replace the high pressure fuel pump with a new one.

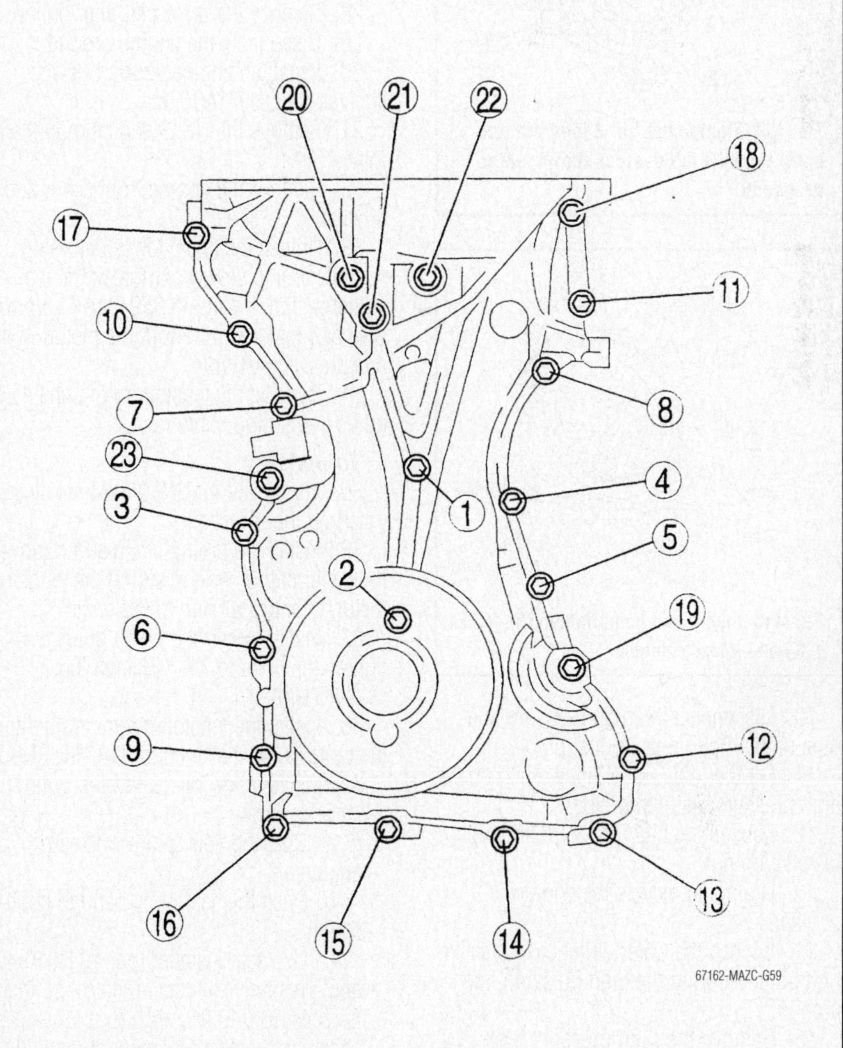

Fig. 412 Engine front cover bolts tightening sequence—Mazdaspeed6

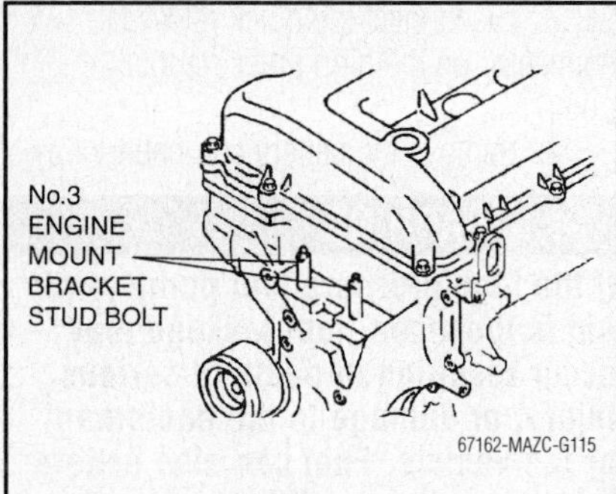

Fig. 413 No. 3 engine mount bracket stud bolt locations—Mazdaspeed6

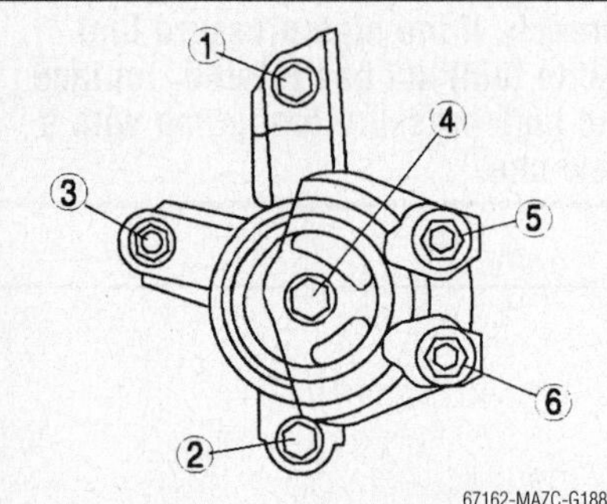

Fig. 414 Tighten the No. 3 joint bracket bolts and nuts in the order shown—Mazdaspeed6

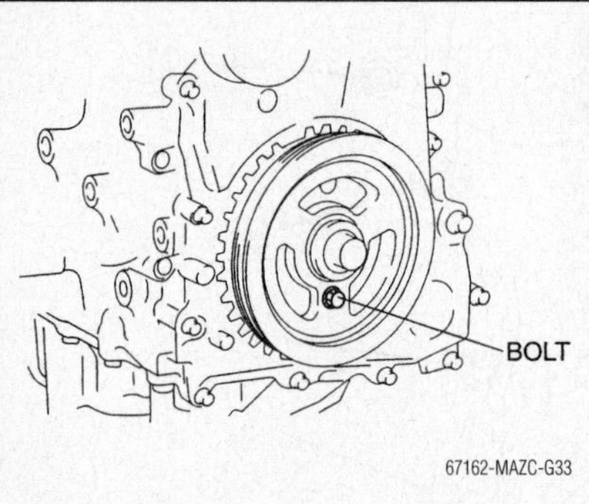

Fig. 415 Install and hand tighten the M6 x 1.0 bolt—Mazdaspeed6

10. Disconnect the high pressure line pipe of the high pressure fuel pump.
11. Fix the joint nut with a wrench on the high pressure fuel pump side.
12. Loosen the high pressure line pipe installation nut.
13. Drain and recycle the engine coolant.
14. Loosen the water outlet case installation bolts securing the high pressure line pipe.
15. Remove the high pressure fuel pump.
16. Remove the high pressure fuel pump cover.
17. Disconnect the Camshaft Position (CMP) and power steering oil pump connectors.
18. Remove the ignition coils, spark plugs, valve cover and accessory drive belt.
19. Remove the Crankshaft Position (CKP) sensor.
20. Remove the power steering oil pump with the lines attached and position aside.
21. Disconnect the right hand driveshaft from the joint shaft.
22. Remove the cylinder block lower blind plug.
23. Install tool 49 JE01 061.
24. Turn the crankshaft clockwise until the crankshaft is in the number 1 cylinder Top Dead Center position (the balance weight should be attached to tool 49 JE01 061).
25. Hold the crankshaft pulley using the tools illustrated and remove the bolt.
26. Remove the crankshaft pulley.
27. Remove the water pump pulley.
28. Remove the drive belt idler pulley.
29. Disconnect the engine ground.
30. Install an engine support device such as 49 E017 5A0.
31. Remove the No. 3 engine mount and joint.
32. Remove the engine front cover and oil seal.
33. Unlock the chain tensioner using a suitable tool to slowly compress the tensioner piston. Insert a 0.059 inch (1.5mm) wire or a paper clip to hold the piston in its compressed position.
34. Remove the tensioner arm, chain guide and timing chain.

To install:

35. Install tool 49 UN30 3465 as illustrated on the camshaft.
36. Install the timing chain and remove the paper clip or wire retaining the tensioner piston to apply tension to the chain.
37. Install the timing chain guide and tighten the bolts to 71–101 inch lbs. (8–11.5 Nm).
38. Install the tensioner arm and tighten the bolts to 71–101 inch lbs. (8–11.5 Nm).
39. Install a new oil seal in the front cover as follows:
    a. Coat the new seal with clean engine oil.
    b. Push the seal in by hand to get it started.
    c. Use seal installer tool 49 H010 401 and a hammer to install the seal so that it is recessed 0–0.019 inch (0–0.5mm) as shown in the accompanying illustration.
40. Apply sealant to the engine front cover. At point A the bead should be 0.086–0.126 inch (2.2–3.2mm) thick and at point B the bead should be 0.059–0.098 inch (1.5–2.5mm) thick. Refer to the accompanying illustration for the locations of points A and B.
41. Install the cover within 10 minutes of apply the sealant. Tighten the cover bolts as follows:
    a. Bolts 1 through 18: In sequence to 71–101 inch lbs. (8–11.5 Nm).
    b. Bolts 19 through 22 : In sequence to 29.7–40.5 ft. lbs. (40–55 Nm).
    c. Bolt 23 : 14.8–22 ft. lbs. (20–30 Nm).
42. When installing the No. 3 engine joint bracket, tighten the mount bracket stud bolt to 62–115 inch lbs. (7–13 Nm) and the joint bracket bolt and nut in the order illustrated to 55–76 ft. lbs. (74–105 Nm).
43. Install the drive belt idler and water pump pulleys.
44. Install tool 49 UN30 3465 as illustrated on the camshaft.
45. Install and hand tighten the M6 x 1.0 bolt as illustrated.
46. Turn the crankshaft clockwise until the crankshaft is in number 1 TDC (the balance weight should be attached to the tool).
47. Hold the crankshaft pulley and tighten the lock bolt in two steps. First tighten to 71–77 ft. lbs. (96–104 Nm). Then final tighten an additional 87–93 degrees.
48. Remove the M6 x 1.0 bolt.
49. Remove the tools from the camshaft and the cylinder block lower blind plug.
50. Rotate the crankshaft clockwise 2 turns until you reach TDC. If not aligned properly, loosen the crankshaft pulley lock bolt and reinstall the lock bolt again using the above procedure and special tools.
51. Install the cylinder block blind plug and tighten to 13–16 ft. lbs. (18–22 Nm).
52. Install the cylinder head cover with a new gasket. Torque the bolts in the sequence illustrated to 71–93 inch lbs. (8–11 Nm).
53. Install the high pressure fuel pump cover and tighten to 13–16 ft. lbs. (18–22 Nm).

CAUTION

If the high pressure fuel pump installation bolts are tightened with the high pressure fuel pump tilted, the high pressure fuel pump may not operate correctly. Tighten the high pressure fuel pump installation bolts in a few passes with equal torque. Tighten to 76–101 inch lbs. (8.5–11.5 Nm).

WARNING

If the high pressure fuel pump joint nut is loosened, fuel leakage may occur resulting in death or serious injury, or damage to the equipment or the vehicle. Fuel can also irritate the skin and eyes. When installing the high pressure line pipe, always tighten the high pressure line pipe installation nut while fixing the high pressure fuel pump joint nut with a wrench. If the high pressure fuel pump joint nut has rotated, replace the high pressure fuel pump with a new one.

54. Assemble the high pressure line pipe.
55. Fix the joint nut with a wrench on the high pressure fuel pump side. Tighten the high pressure line pipe installation nut to 17–26 ft. lbs. (23–35 Nm).
56. Tighten the water outlet case installation bolts to 13–16 ft. lbs. (18–22 Nm).
57. Install the quick release connector.
58. Install the remaining components in the reverse order of removal.
59. Change the engine oil.
60. Start the engine.
61. Inspect for the following:
    - Pulley and belt for run–out and contact
    - Ignition timing, idle speed and exhaust emissions
    - All remaining components for proper operation.
62. Verify that the high pressure fuel pump is assembled securely.
63. Drive the vehicle starting from a standstill and brake suddenly five to six times at a low speed.
64. Stop the vehicle and verify from outside the vehicle that there is no fuel leakage around the high pressure fuel pump.

VALVE COVERS

REMOVAL & INSTALLATION

MR-5 Miata, Mazda3 and Mazda6 Models

2.0L (LF) and 2.3L (L3) Engines

See Figures 416 and 417.

1. Before servicing the vehicle, refer to the precautions section.
2. Remove or disconnect the following:
    - Negative battery cable
    - Charge air cooler (L3 with TC models)
    - Ignition coils

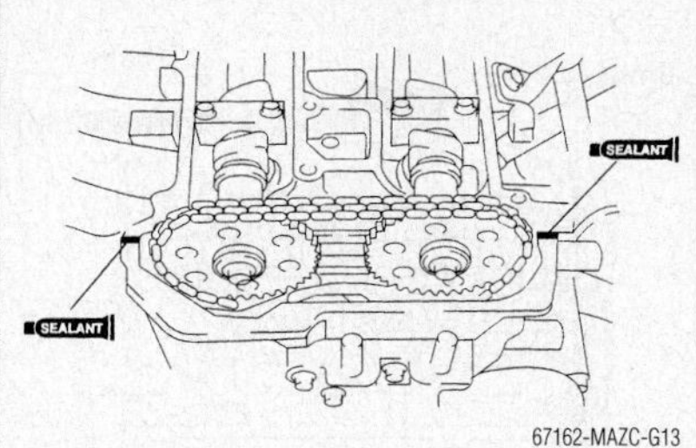

Fig. 416 Apply silicone sealant to cylinder head at the areas illustrated–All Mazda3 models, Mazda6 and Mazdaspeed6 models with the 2.3L (L3) engine models

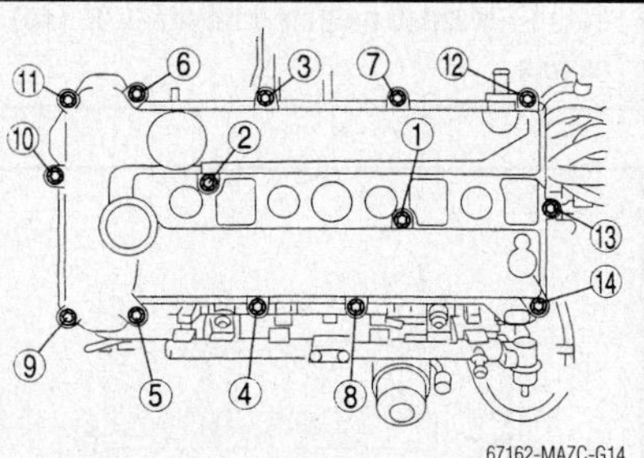

Fig. 417 Tighten the rocker arm cover bolts in the sequence shown–All Mazda3 models, Mazda6 and Mazdaspeed6 models with the 2.3L (L3) engine models

    - Vent hose
    - OCV valve connector
    - Positive Crankcase Ventilation (PCV) hose
    - Rocker arm cover and discard the gasket
3. Clean all mating surfaces of any residual gasket material.

To install:

4. Install or connect the following:
5. Apply a 0.16–0.24 inch (4–7mm) bead of silicone sealant to cylinder head at the areas illustrated. Make sure to install the cover within 10 minutes of applying the sealant.
6. Install or connect the following:
    - Rocker arm cover with a new gasket. Tighten the bolts in the sequence illustrated to 71–92 inch lbs. (8–10 Nm) on all MX-5 and Mazda3 engines, 71–101inch lbs. (8—11.5 Nm) on Mazda6 models.
    - Remaining components removed to facilitate the rocker arm cover removal
    - Negative battery cable

3.0L (AJ) Engines

See Figures 418 through 424.

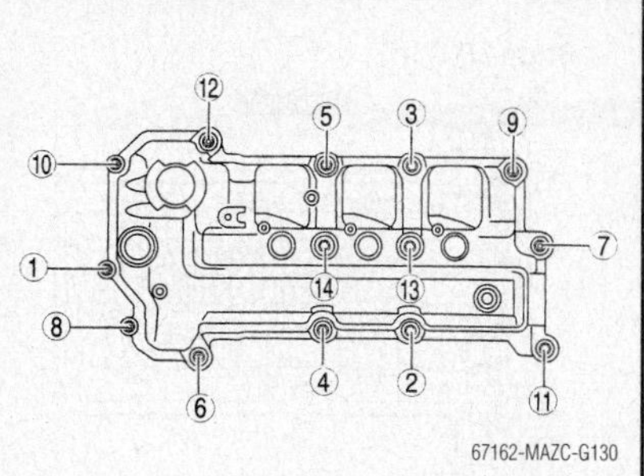

Fig. 418 Remove the left hand rocker arm cover bolts in the sequence shown—Mazda6 models with the 3.0L (AJ) engine

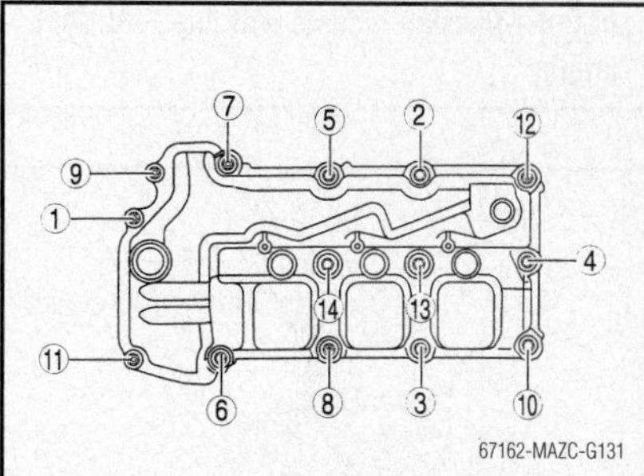

Fig. 419 Remove the right hand rocker arm cover bolts in the sequence shown—Mazda6 models with the 3.0L (AJ) engine

1. Before servicing the vehicle, refer to the precautions in the beginning of this section.
2. Remove or disconnect the following:
    - Negative battery cable
    - Any components that would interfere with cover removal
    - Rocker arm cover bolts in the sequence illustrated
    - Rocker arm cover and discard the gasket
3. Clean all mating surfaces of any residual gasket material.

To install:

4. Apply silicone sealant to cylinder head at the areas illustrated.
5. Install or connect the following:
    - Rocker arm cover with a new gasket.
    - Oil control valve with the cylinder head cover raised as shown in the accompanying illustration, being careful not to let the valve retaining bolt slip into the timing chain cover when installing and tighten the valve bolt to 71–106 INCH LBS. (8–12 Nm).

   Torque the bolts in the sequence illustrated to 71–106 inch lbs. (8–12 Nm).

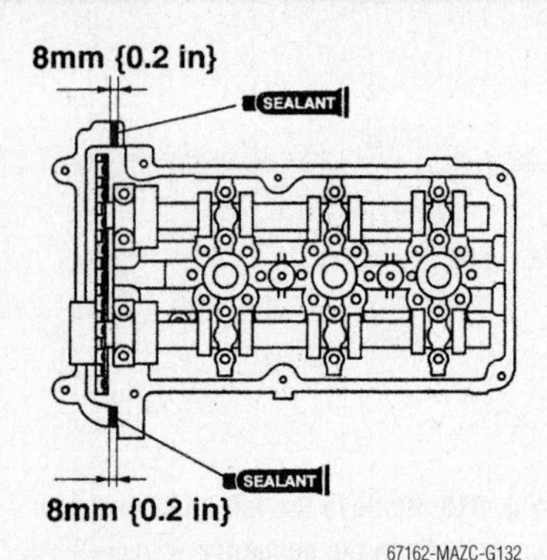

Fig. 420 Apply silicone sealant to right hand cylinder head at the areas illustrated—Mazda6 models with the 3.0L (AJ) engine

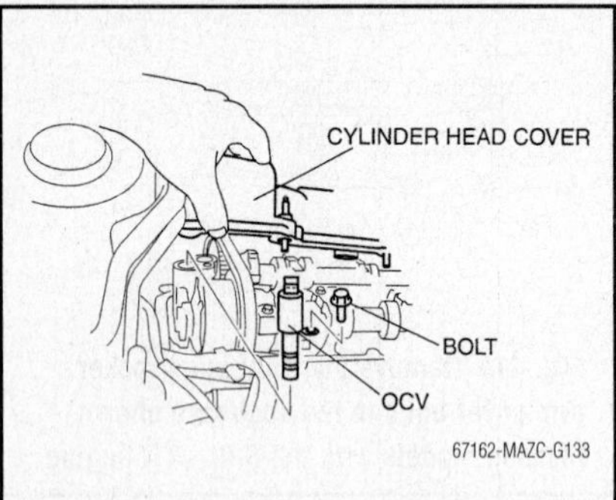

Fig. 421 Install the oil control valve on the right hand side—Mazda6 models with the 3.0L (AJ) engine

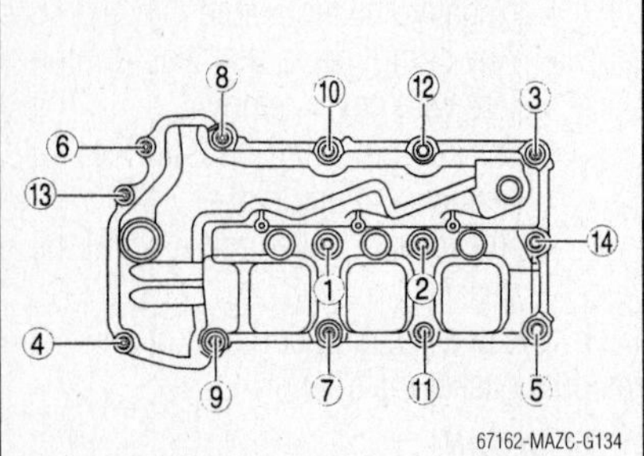

Fig. 422 Right hand cylinder head cover torque sequence—Mazda6 models with the 3.0L (AJ) engine

- Remaining components removed to facilitate the rocker arm cover removal
- Negative battery cable

VALVE LASH

ADJUSTMENT

These engines use solid cam followers with a removable adjustment shim. The valve lash clearance is measured with the original shim installed and checked against the specification. If adjustment is necessary, the original shim is removed, and a thicker or thinner shim is installed to obtain the proper clearance. Special tools are required in order to adjust the shim without removing the camshaft.

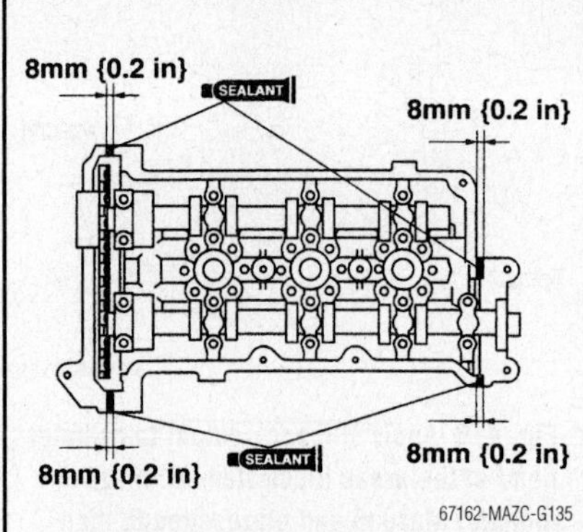

Fig. 423 Apply silicone sealant to left hand cylinder head at the areas illustrated—Mazda6 models with the 3.0L (AJ) engine

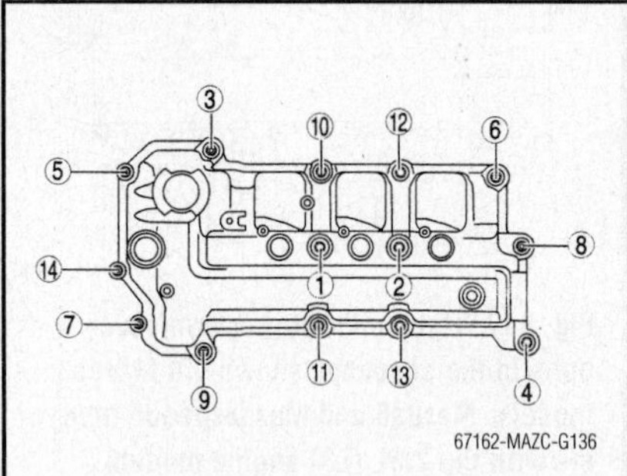

Fig. 424 Left hand cylinder head cover torque sequence—Mazda6 models with the 3.0L (AJ) engine

INSPECTION

MX-5 Miata

See Figures 425 through 432.

The valve clearance should be measured with the engine cold, the specifications are:

- Intake: 22–28mm (0.0087-0.0110 in.)
- Exhaust: 0.27–0.33 (0.0107–0.0129 in.)

1. Before servicing the vehicle, refer to the precautions section.
2. Remove the battery cover.
3. Disconnect the negative battery cable.
4. Remove the plug hole plate.
5. Disconnect the ventilation hose.
6. Remove the front suspension tower bar.
7. Remove the CMP sensor.
8. Disconnect the OCV connector.
9. Disconnect the P/S pressure switch connector.
10. Remove the ignition coils.
11. Remove the cylinder head cover.
12. Remove the drive belt.
13. Remove the engine front cover lower blind plug.
14. Remove the engine front cover upper blind plug.
15. Remove the cylinder block lower blind plug.
16. Install tool 303-507 as shown.

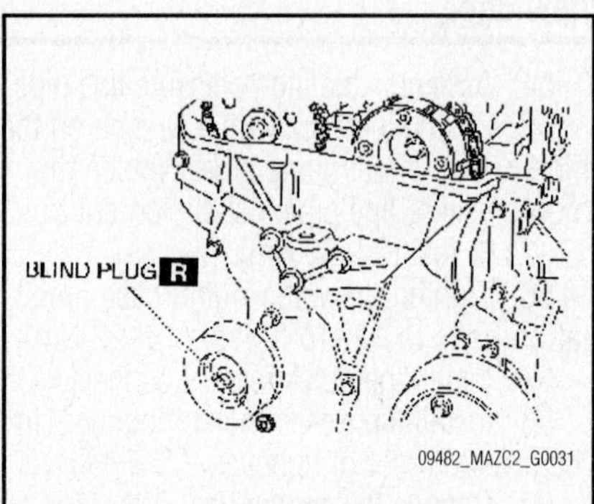

Fig. 425 Remove the engine front cover lower blind plug—MX-5 Miata

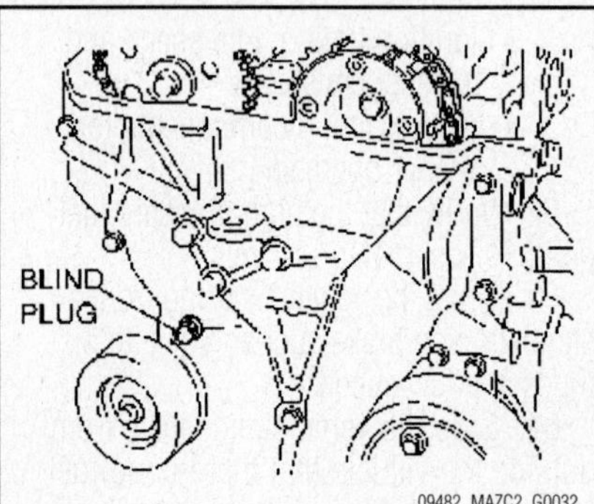

Fig. 426 Remove the engine front cover upper blind plug—MX-5 Miata

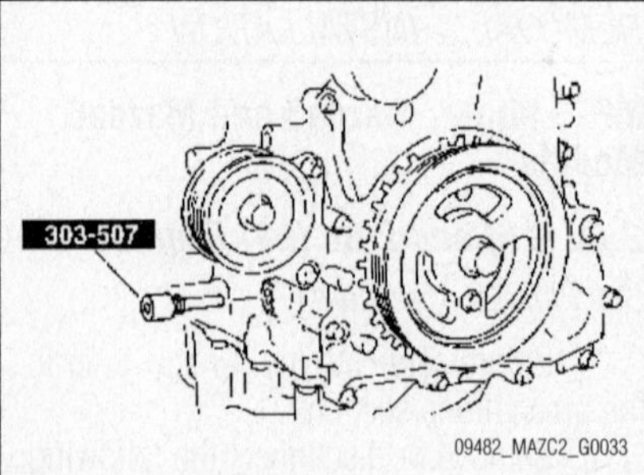

Fig. 427 Remove the cylinder block upper blind plug and install tool 303-507—MX-5 Miata

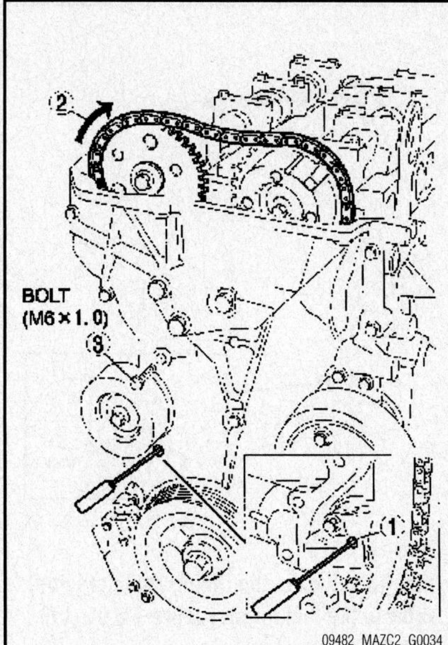

Fig. 428 Place a M6 X 1 length bolt at the engine front cover upper blind plug, secure the chain guide at the position where the tension is released—MX-5 Miata

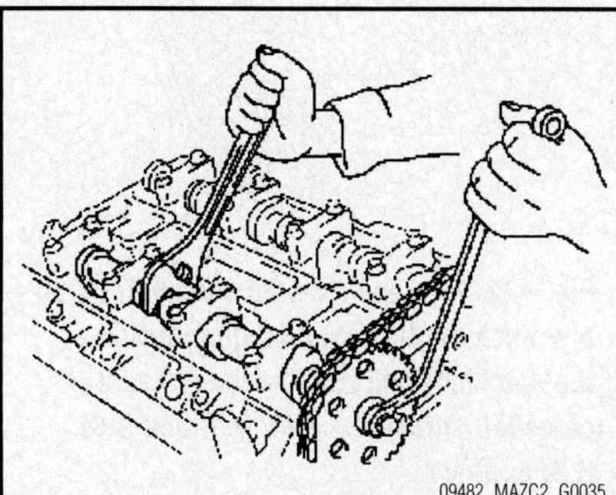

Fig. 429 Hold the exhaust camshaft using a suitable wrench on the cast hexagon—MX-5 Miata

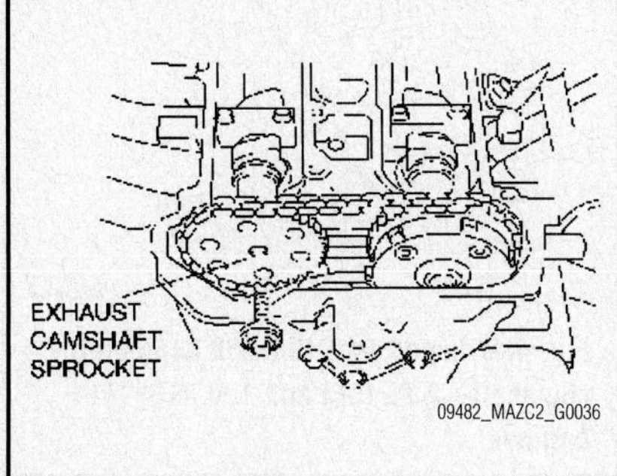

Fig. 430 Remove the exhaust camshaft sprocket—MX-5 Miata

17. Turn the crankshaft clockwise until the crankshaft is in the No.1 cylinder TDC position.

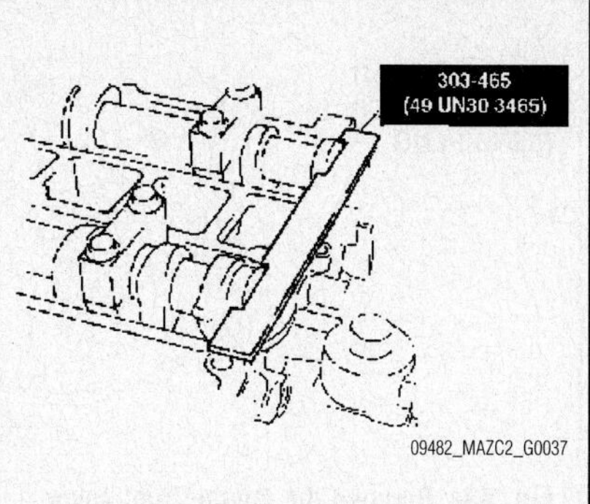

Fig. 431 Install the tool illustrated on the camshaft—MX-5 Miata

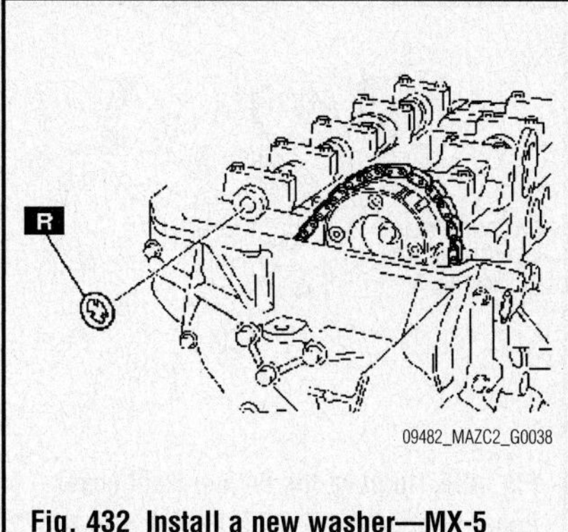

Fig. 432 Install a new washer—MX-5 Miata

18. Loosen the timing chain.
19. Using a suitable screwdriver or equivalent tool, unlock the chain tensioner ratchet.
20. Turn the exhaust camshaft clockwise using a suitable wrench on the cast hexagon and loosen the timing chain.
21. Place a M6 X 1 length bolt at the engine front cover upper blind plug, secure the chain guide at the position where the tension is released.
22. Hold the exhaust camshaft using a suitable wrench on the cast hexagon as illustrated.
23. Remove the exhaust camshaft sprocket.
24. Loosen the camshaft cap bolts in several passes in the order shown.

➡The cylinder head and the camshaft caps are numbered to make sure they are reassembled in their original position. Do not mix the caps.

25. Remove the camshafts.
26. Remove the tappet.
27. Select proper adjustment tappet using the following formula:
 a. New adjustment tappet = removed tappet thickness + Measured valve clearance - standard valve clearance. For the intake: 0.25 mm (0.0098 in.). For the exhaust: 0.30 mm (0.0118 in.).
28. Install the camshaft with No.1 cylinder camshaft lobes aligned with the TDC position.
29. Tighten the camshaft cap bolt using two steps, first to 5–9 Nm (44–80 in. lbs.), then to 14–17 Nm (10–12.5 ft. lbs.).
30. Install a new washer.
31. Install the exhaust camshaft sprocket.

➡Do not tighten the bolt for the camshaft sprocket during this step. First confirm the valve timing, then tighten the bolt.

32. Install the tool on the camshaft as shown.
33. Remove the M6 X 1.0 bolt from the engine front cover to apply tension to the timing chain.
34. Turn the crankshaft clockwise until the crankshaft is in the No.1 cylinder TDC position.
35. Hold the exhaust camshaft using a suitable wrench on the cast hexagon as shown.
36. Tighten the exhaust camshaft sprocket lock bolt to 51–55 ft. lbs. (69–75 Nm).
37. Remove the tool from the camshaft.
38. Remove the tool from the block lower blind plug.
39. Rotate the crankshaft clockwise two turns until the TDC position.
40. If not aligned, loosen the crankshaft pulley lock bolt and repeat the procedure.
41. Apply silicone sealant to the engine front cover upper blind plug.
42. Install the engine front cover upper blind plug and tighten to 71–101 in. lbs. (8–11 Nm.)
43. Install the cylinder block lower blind plug and tighten to 13–16 ft. lbs. (18–23 Nm)
44. Install the new engine front cover lower blind plug and tighten to 89–123 ft. lbs. (10–14 Nm.)
45. Install the drive belt.
46. Measure the valve clearance.
47. Turn the crankshaft clockwise so that the No.1 piston is at TDC of the compression stroke.
48. Measure the valve clearance at point A in the illustration.
49. If the valve clearance is out of the specification, adjust it.
50. Turn the crankshaft 360 degrees clockwise so that the No.4 piston is at TDC of the compression stroke.
51. Measure the valve clearance at B in the illustration.

52. If the valve clearance is out of the specification, adjust it.
53. Install the cylinder head cover.
54. Install the ignition coils.
55. Connect the OCV connector.
56. Install the CMP sensor.
57. Install the front suspension tower bar.
58. Connect the ventilation hose.
59. Install the air cleaner.
60. Install the plug hole plate.
61. Install the battery and battery tray.

2.0L (LF) Engine

See Figures 433 through 440.

1. Before servicing the vehicle, refer to the precautions section.

➡With the engine cold, standard valve clearance is 0.087–0.0110 inch (0.22–0.28mm) on the intake side and 0.0107–0.0129 inch (0.27–0.33mm) on the exhaust side.

2. Before servicing the vehicle, refer to the precautions section.
3. Remove the plug hole plate.
4. Remove the battery cover and disconnect the negative battery cable.
5. Disconnect the wiring harness.
6. Remove the right front wheel, engine undercover and splash shield.
7. Remove the ignition coils.
8. Remove the Positive Crankcase Ventilation (PCV) hose, if equipped.
9. Disconnect the oil control valve connector, if equipped.
10. Remove the cylinder head cover.
11. Measure the valve clearance by turning the crankshaft clockwise until the No. 1 piston is at Top Dead Center (TDC).
12. Measure the valve clearance at **A**. If the clearance exceeds specifications, replace the adjustment shim.
13. Turn the crankshaft clockwise 360 degrees until the No. 4 piston is at TDC.

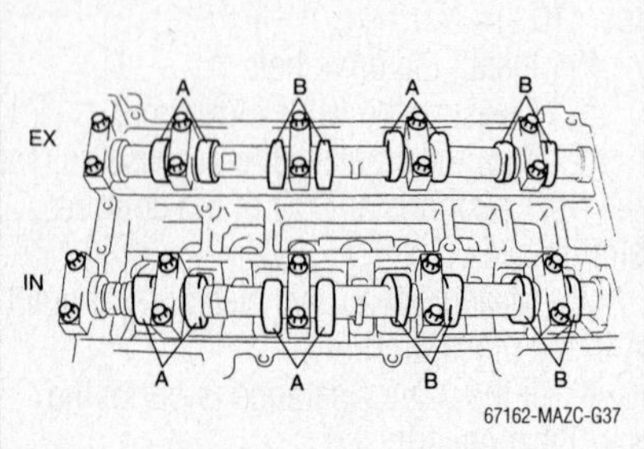

Fig. 433 Valve clearance checking positions—2.0L (LF) and 2.3L (L3) engines

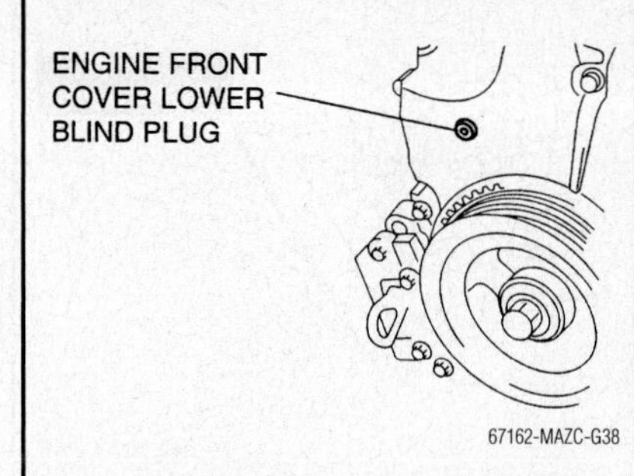

Fig. 434 Remove the engine front cover lower blind plug—2.0L (LF) and 2.3L (L3) engines

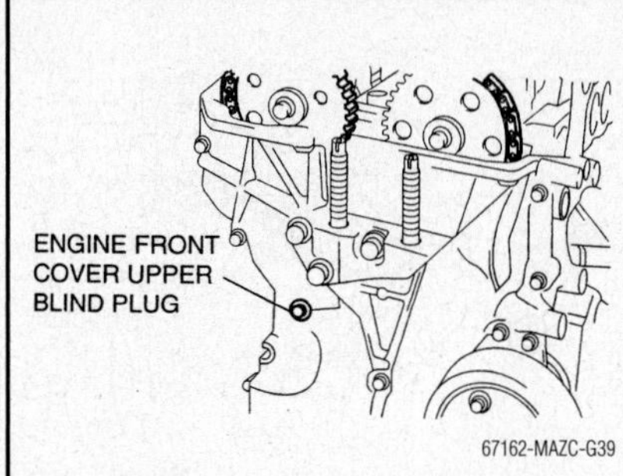

Fig. 435 Remove the engine front cover upper blind plug—2.0L (LF) and 2.3L (L3) engines

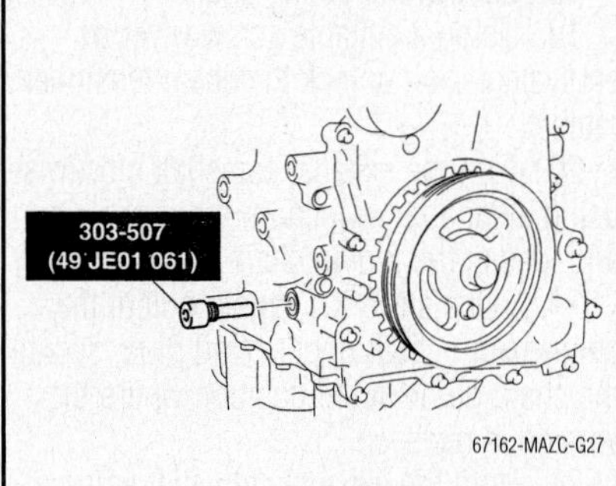

Fig. 436 Install tool 49 JE01 061—2.0L (LF) and 2.3L (L3) engines

Measure the valve clearance at **B**. If the clearance exceeds specifications, replace the adjustment shim.

14. Repeat this procedure for all the camshafts.
15. Remove the right side halfshaft from the joint shaft on Mazda6 models.
16. Remove the engine front cover lower and upper blind plugs and the cylinder block lower blind plug.
17. Install tool 49 JE01 061.
18. Turn the crankshaft clockwise until the crankshaft is in the number 1 cylinder Top Dead Center position (the balance weight should be attached to tool 49 JE01 061).
19. Loosen the timing chain as follows:
    a. Unlock the chain tensioner ratchet using a screwdriver.

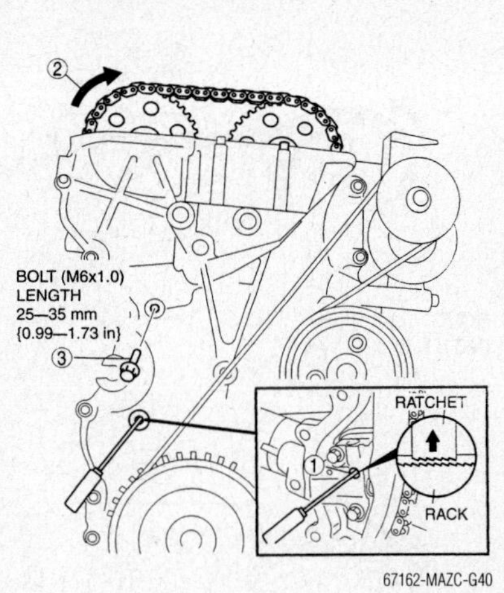

Fig. 437 Loosen the timing chain tension, refer to the text for procedure—2.0L (LF) and 2.3L (L3) engines

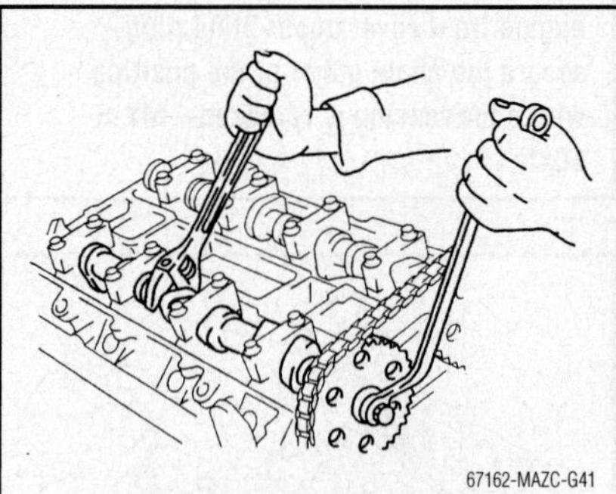

Fig. 438 Hold the exhaust camshaft using a wrench on the cast hexagon portion of the camshaft and remove the exhaust camshaft sprocket—2.0L (LF) and 2.3L (L3) engines

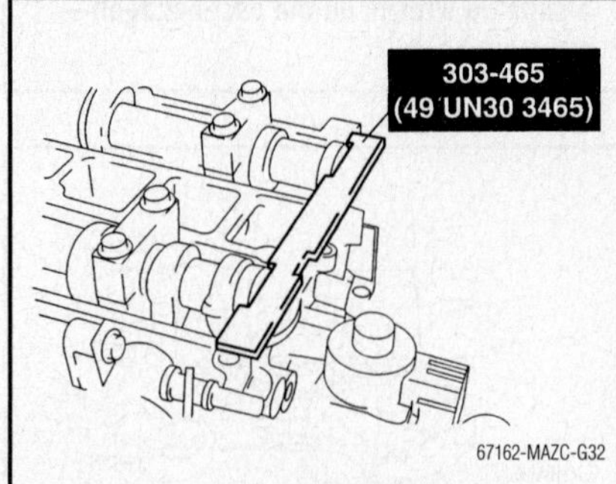

Fig. 439 Install tool 49 UN30 3465 on the camshaft—2.0L (LF) and 2.3L (L3) engines

b. Turn the exhaust camshaft clockwise using a wrench on the cast hexagon portion of the camshaft and loosen the timing chain.

c. Place a M6 x 1.0 length bolt: 0.99–1.37 inch (25–35mm) at the engine front cover upper blind plug and secure the timing chain guide at the location where the chain tension is released.

20. Hold the exhaust camshaft using a wrench on the cast hexagon portion of the camshaft and remove the exhaust camshaft sprocket.

➡The cylinder head and camshaft caps are numbered and must be reassembled in their original locations. When removing a when removed, keep the caps with the cylinder head they were removed from, do not switch the caps.

21. Loosen the camshaft cap bolts using 2–3 passes in the sequence illustrated.

22. Once all the bolts have been removed, remove the caps, keeping them in the original positions and remove the camshafts.

23. Remove the tappets and select the proper adjustment shim. The formula for selecting a shim is as follows:

a. Removed shim thickness + measured valve clearance–standard valve clearance 0.0098 inch (0.25mm) for the intake and 0.0118 inch (0.30mm) for the exhaust.

24. Install the adjustment shim.

25. Set the cam position of the number 1 cylinder at Top Dead center (TDC) and install the camshaft.

26. Temporarily install the camshaft caps in their original positions.

27. Tighten the camshaft cap bolt in two steps:

a. Step 1: 44–71 inch lbs. (5–8 Nm).

b. Step 2 : 10.4–12.5 ft. lbs. (14–17 Nm).

28. Install the exhaust camshaft sprocket and hand tighten the bolt.

29. Install tool 49 UN30 3465 as illustrated on the camshaft.

30. Remove the M6 x 1.0 bolt from the engine front cover to tension the timing chain.

31. Turn the crankshaft clockwise until the crankshaft is in the number 1 cylinder Top Dead Center position (the balance weight should be attached to tool 49 JE01 061).

32. Hold the exhaust camshaft using a wrench on the cast hexagon portion of the camshaft and tighten the exhaust camshaft sprocket bolt to 51–55 ft. lbs. (69–75 Nm).

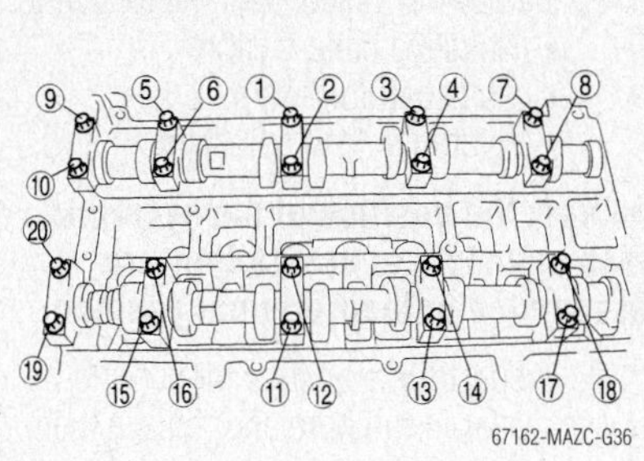

Fig. 440 Camshaft cap bolt tightening sequence—2.0L (LF) and 2.3L (L3) engines

33. Remove the tools from the camshaft and from the block lower blind plug.

34. Rotate the engine two full turns clockwise to TDC. If not aligned, remove the crankshaft pulley lockbolt, remove the tappet and repeat the adjustment shim selection and replacement procedure.

35. Apply sealant to the engine front cover upper blind plug and tighten to 71–101 inch lbs. (8–11 Nm).

36. Install the cylinder block lower blind plug and tighten to 13–16 ft. lbs. (18–22 Nm).

37. Install a new engine front cover blind plug and tighten to 7.4–10 ft. lbs. (10–14 Nm).

38. Install the right side halfshaft from the joint shaft on Mazda6 models.

39. Install the cylinder head cover.

40. Install the PCV hose, if equipped.

41. Connect the oil control valve connector, if equipped.

42. Install the ignition coils.

43. Connect the wiring harness.

44. Connect the negative battery cable and install the battery cover.

45. Install the plug hole plate.

46. Install the splash shield, undercover and wheel.

2.3L (L3) Engine

1. Before servicing the vehicle, refer to the precautions section.

➡With the engine cold, standard valve clearance is 0.087–0.0110 inch (0.22–0.28mm) on the intake side and 0.0107–0.0129 inch (0.27–0.33mm) on the exhaust side.

2. Before servicing the vehicle, refer to the precautions section.

3. Remove the plug hole plate.

4. Remove the battery cover and disconnect the negative battery cable.

5. Disconnect the wiring harness.

6. Remove the ignition coils.

7. Disconnect the Oil Control Valve (OCV) connector.

8. Remove the right front wheel, undercover and splash shield.

9. Remove the Positive Crankcase Ventilation (PCV) hose.

10. Remove the cylinder head cover.

11. Measure the valve clearance by turning the crankshaft clockwise until the No. 1 piston is at Top Dead Center (TDC).

12. Measure the valve clearance at **A**. If the clearance exceeds specifications, replace the adjustment shim.

13. Turn the crankshaft clockwise 360 degrees until the No. 4 piston is at TDC. Measure the valve clearance at **B**. If the clearance exceeds specifications, replace the adjustment shim.

14. Repeat this procedure for all the camshafts.

15. Remove the engine front cover lower and upper blind plugs and the cylinder block lower blind plug.

16. Install tool 49 JE01 061.

17. Turn the crankshaft clockwise until the crankshaft is in the number 1 cylinder Top Dead Center position (the balance weight should be attached to tool 49 JE01 061).

18. Loosen the timing chain as follows:

a. Unlock the chain tensioner ratchet using a screwdriver.

b. Turn the exhaust camshaft clockwise using a wrench on the cast hexagon portion of the camshaft and loosen the timing chain.

c. Place a M6 x 1.0 length bolt: 0.99–1.37 inch (25–35mm) at the engine front cover upper blind plug and secure the timing chain guide at the location where the chain tension is released.

19. Hold the exhaust camshaft using a wrench on the cast hexagon portion of the camshaft and remove the exhaust camshaft sprocket.

➡The cylinder head and camshaft caps are numbered and must be reassembled in their original locations. When removing a when removed, keep the caps with the cylinder head they were removed from, do not switch the caps.

20. Loosen the camshaft cap bolts using 2–3 passes in the sequence illustrated.

21. Once all the bolts have been removed, remove the caps, keeping them in the original positions and remove the camshafts.

22. Remove the tappets and select the proper adjustment shim. The formula for selecting a shim is as follows:

a. Removed shim thickness + measured valve clearance - standard valve clearance 0.0098 inch (0.25mm) for the intake and 0.0118 inch (0.30mm) for the exhaust.

23. Install the adjustment shim.
24. Set the cam position of the number 1 cylinder at Top Dead center (TDC) and install the camshaft.
25. Temporarily install the camshaft caps in their original positions.
26. Tighten the camshaft cap bolt in two steps:
    a. Step 1: 44–71 inch lbs. (5–8 Nm).
    b. Step 2 : 10.4–12.5 ft. lbs. (14–17 Nm).
27. Install the exhaust camshaft sprocket and hand tighten the bolt.
28. Install tool 49 UN30 3465 as illustrated on the camshaft.
29. Remove the M6 x 1.0 bolt from the engine front cover to tension the timing chain.
30. Turn the crankshaft clockwise until the crankshaft is in the number 1 cylinder Top Dead Center position (the balance weight should be attached to tool 49 JE01 061).
31. Hold the exhaust camshaft using a wrench on the cast hexagon portion of the camshaft and tighten the exhaust camshaft sprocket bolt to 51–55 ft. lbs. (69–75 Nm).
32. Remove the tools from the camshaft and from the block lower blind plug.
33. Rotate the engine two full turns clockwise to TDC. If not aligned, remove the crankshaft pulley lockbolt, remove the tappet and repeat the adjustment shim selection and replacement procedure.
34. Apply sealant to the engine front cover upper blind plug and tighten to 71–101 inch lbs. (8–11 Nm).
35. Install the cylinder block lower blind plug and tighten to 13–16 ft. lbs. (18–22 Nm).
36. Install a new engine front cover blind plug and tighten to 7.4–10 ft. lbs. (10–14 Nm).
37. Install the cylinder head cover.
38. Install the PCV hose.
39. Connect the OCV connector.
40. Install the ignition coils.
41. Connect the wiring harness.
42. Connect the negative battery cable and install the battery cover.
43. Install the plug hole plate.
44. Install the undercover, splash shield and wheel.

3.0L (AJ) Engine

1. Before servicing the vehicle, refer to the precautions section.
2. Remove or disconnect the following:
    - Negative battery cable
    - Camshaft followers
    - Hydraulic lash adjusters

➡Mark the position of the hydraulic lash adjusters to assure they are assembled in their original position

3. Inspect the adjusters for scoring or uneven wear in the bore and replace them as required.

To install:

4. Install or connect the following:
    - Hydraulic lash adjusters after lubricating them with clean engine oil
    - Camshaft followers
    - Negative battery cable

Mazdaspeed3 and Speed6 Models

See Figures 441 through 446.

1. Before servicing the vehicle, refer to the precautions section.

➡With the engine cold, standard valve clearance is 0.087–0.0110 inch (0.22–0.28mm) on the intake side and 0.0107–0.0129 inch (0.27–0.33mm) on the exhaust side.

2. Disconnect the negative battery cable.
3. Remove the right side front tire.
4. Remove the undercover.
5. Remove the right side splash shield.

➡If the high pressure fuel pump is removed, replace the O-ring with a new one.

6. Remove the charge air cooler cover.
7. Disconnect the spill valve control solenoid valve connector.
8. Disconnect the quick release connector on the high pressure fuel pump.
9. Remove the battery and battery tray.

⁂ WARNING

If the high pressure fuel pump joint nut is loosened, fuel leakage may occur resulting in death or serious injury, or damage to the equipment or the vehicle. Fuel can also irritate the skin and eyes. When removing the high pressure line pipe, always tighten the high pressure line pipe installation nut while fixing the high pressure fuel pump joint nut with a wrench. If the high pressure fuel pump joint nut has rotated, replace the high pressure fuel pump with a new one.

10. Disconnect the high pressure line pipe of the high pressure fuel pump.
11. Fix the joint nut with a wrench on the high pressure fuel pump side.
12. Loosen the high pressure line pipe installation nut.
13. Drain and recycle the engine coolant.
14. Loosen the water outlet case installation bolts securing the high pressure line pipe.
15. Remove the high pressure fuel pump.
16. Remove the high pressure fuel pump cover.
17. Remove the ignition coils, spark plugs, valve cover and accessory drive belt.
18. Measure the valve clearance by turning the crankshaft clockwise until the No. 1 piston is at Top Dead Center (TDC).
19. Measure the valve clearance at **A**. If the clearance exceeds specifications, replace the adjustment shim.
20. Turn the crankshaft clockwise 360 degrees until the No. 4 piston is at TDC. Measure the valve clearance at **B**. If the clearance exceeds specifications, replace the adjustment shim.
21. Repeat this procedure for all the camshafts.

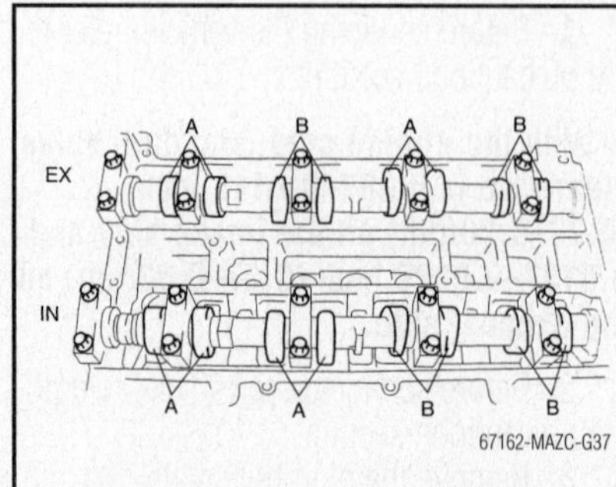

Fig. 441 Valve clearance checking positions—Mazdaspeed models

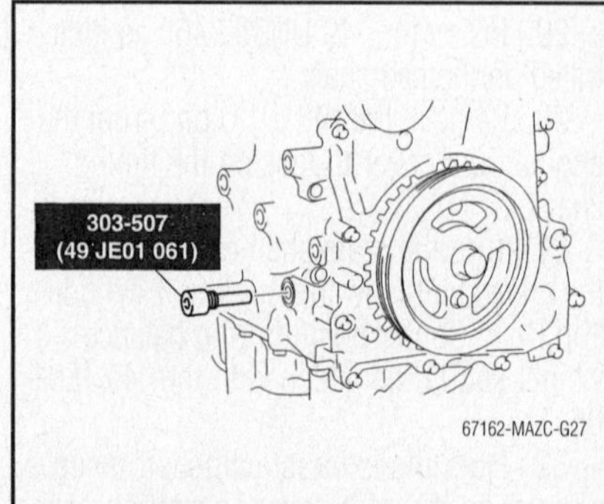

Fig. 442 Install tool 49 JE01 061—Mazdaspeed Models

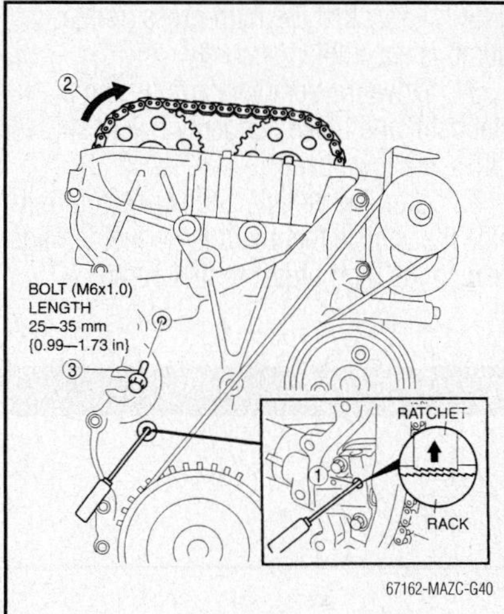

Fig. 443 Loosen the timing chain tension, refer to the text for procedure—Mazdaspeed models

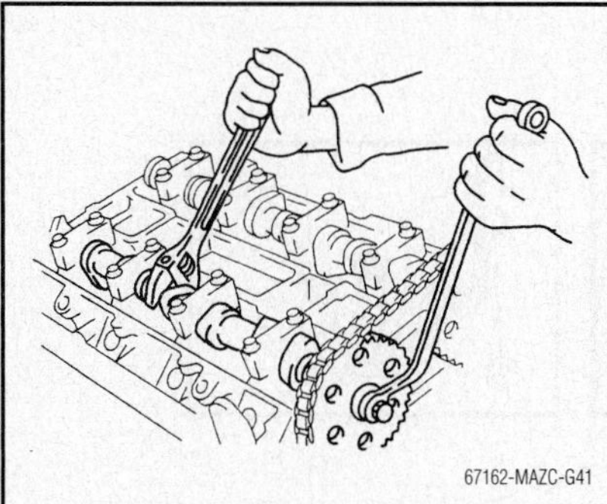

Fig. 444 Hold the exhaust camshaft using a wrench on the cast hexagon portion of the camshaft and remove the exhaust camshaft sprocket—Mazdaspeed models

22. Remove the engine front cover lower and upper blind plugs and the cylinder block lower blind plug.

23. Install tool 49 JE01 061.

24. Turn the crankshaft clockwise until the crankshaft is in the number 1 cylinder Top Dead Center position (the balance weight should be attached to tool 49 JE01 061).

25. Loosen the timing chain as follows:

a. Unlock the chain tensioner ratchet using a screwdriver.

b. Turn the exhaust camshaft clockwise using a wrench on the cast hexagon portion of the camshaft and loosen the timing chain.

c. Place a M6 x 1.0 length bolt: 0.99–1.37 inch (25–35mm) at the engine front cover upper blind plug and secure the timing chain guide at the location where the chain tension is released.

26. Hold the exhaust camshaft using a wrench on the cast hexagon portion of the camshaft and remove the exhaust camshaft sprocket.

➡The cylinder head and camshaft caps are numbered and must be reassembled in their original locations. When removing a when removed, keep the caps with the cylinder head they were removed from, do not switch the caps.

27. Loosen the camshaft cap bolts using 2–3 passes in the sequence illustrated.

28. Once all the bolts have been removed, remove the caps, keeping them in the original positions and remove the camshafts.

29. Remove the tappets and select the proper adjustment shim. The formula for selecting a shim is as follows:

a. Removed shim thickness + measured valve clearance - standard valve clearance 0.0098 inch (0.25mm) for the intake and 0.0118 inch (0.30mm) for the exhaust.

30. Install the adjustment shim.

31. Set the cam position of the number 1 cylinder at Top Dead center (TDC) and install the camshaft.

32. Temporarily install the camshaft caps in their original positions.

33. Tighten the camshaft cap bolt in two steps:

a. Step 1: 44–71 inch lbs. (5–8 Nm).

b. Step 2 : 10.4–12.5 ft. lbs. (14–17 Nm).

34. Install the exhaust camshaft sprocket and hand tighten the bolt.

35. Install tool 49 UN30 3465 as illustrated on the camshaft.

36. Remove the M6 x 1.0 bolt from the engine front cover to tension the timing chain.

37. Turn the crankshaft clockwise until the crankshaft is in the number 1 cylinder Top Dead Center position (the balance weight should be attached to tool 49 JE01 061).

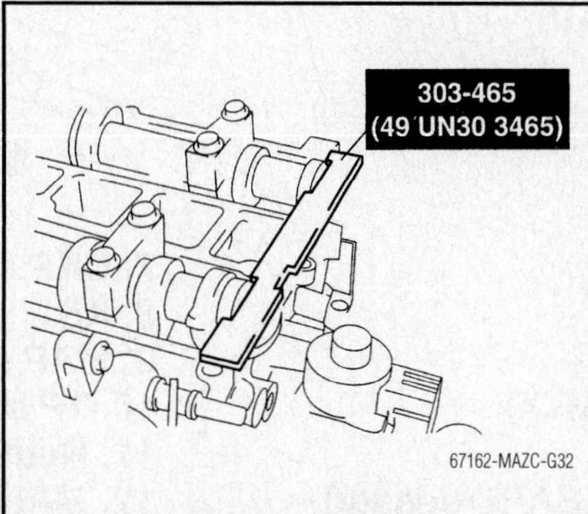

Fig. 445 Install tool 49 UN30 3465 on the camshaft—Mazdaspeed models

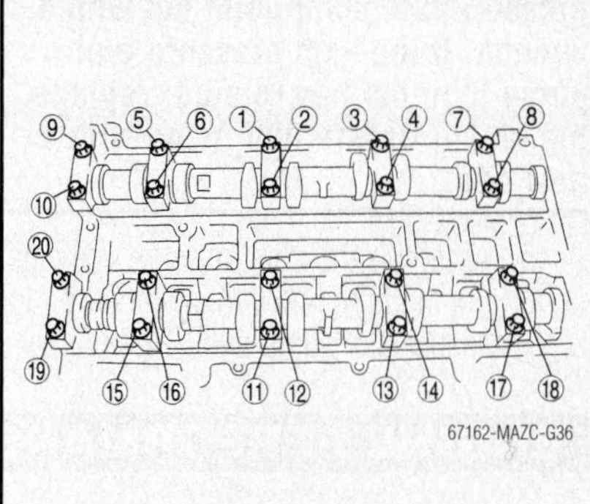

Fig. 446 Camshaft cap bolt tightening sequence—Mazdaspeed models

38. Hold the exhaust camshaft using a wrench on the cast hexagon portion of the camshaft and tighten the exhaust camshaft sprocket bolt to 51–55 ft. lbs. (69–75 Nm).

39. Remove the tools from the camshaft and from the block lower blind plug.

40. Rotate the engine two full turns clockwise to TDC. If not aligned, remove the crankshaft pulley lockbolt, remove the tappet and repeat the adjustment shim selection and replacement procedure.

41. Apply sealant to the engine front cover upper blind plug and tighten to 71–101 inch lbs. (8–11 Nm).

42. Install the cylinder block lower blind plug and tighten to 13–16 ft. lbs. (18–22 Nm).

43. Install a new engine front cover blind plug and tighten to 7.4–10 ft. lbs. (10–14 Nm).

44. Install the high pressure fuel pump cover and tighten to 13–16 ft. lbs. (18–22 Nm).

CAUTION

If the high pressure fuel pump installation bolts are tightened with the high pressure fuel pump tilted, the high pressure fuel pump may not operate correctly. Tighten the high pressure fuel pump installation bolts in a few passes with equal torque. Tighten to 76–101 inch lbs. (8.5–11.5 Nm).

WARNING

If the high pressure fuel pump joint nut is loosened, fuel leakage may occur resulting in death or serious injury, or damage to the equipment or the vehicle. Fuel can also irritate the skin and eyes. When installing the high pressure line pipe, always tighten the high pressure line pipe installation nut while fixing the high

pressure fuel pump joint nut with a wrench. If the high pressure fuel pump joint nut has rotated, replace the high pressure fuel pump with a new one.

45. Assemble the high pressure line pipe.
46. Fix the joint nut with a wrench on the high pressure fuel pump side. Tighten the high pressure line pipe installation nut to 17–26 ft. lbs. (23–35 Nm).
47. Tighten the water outlet case installation bolts to 13–16 ft. lbs. (18–22 Nm).
48. Install the quick release connector.
49. Install the remaining components in the reverse order of removal.
50. Verify that the high pressure fuel pump is assembled securely.
51. Drive the vehicle starting from a standstill and brake suddenly five to six times at a low speed.
52. Stop the vehicle and verify from outside the vehicle that there is no fuel leakage around the high pressure fuel pump.

ENGINE PERFORMANCE & EMISSION CONTROLS

2007—08 Mazda3 Models

See Figure 447.

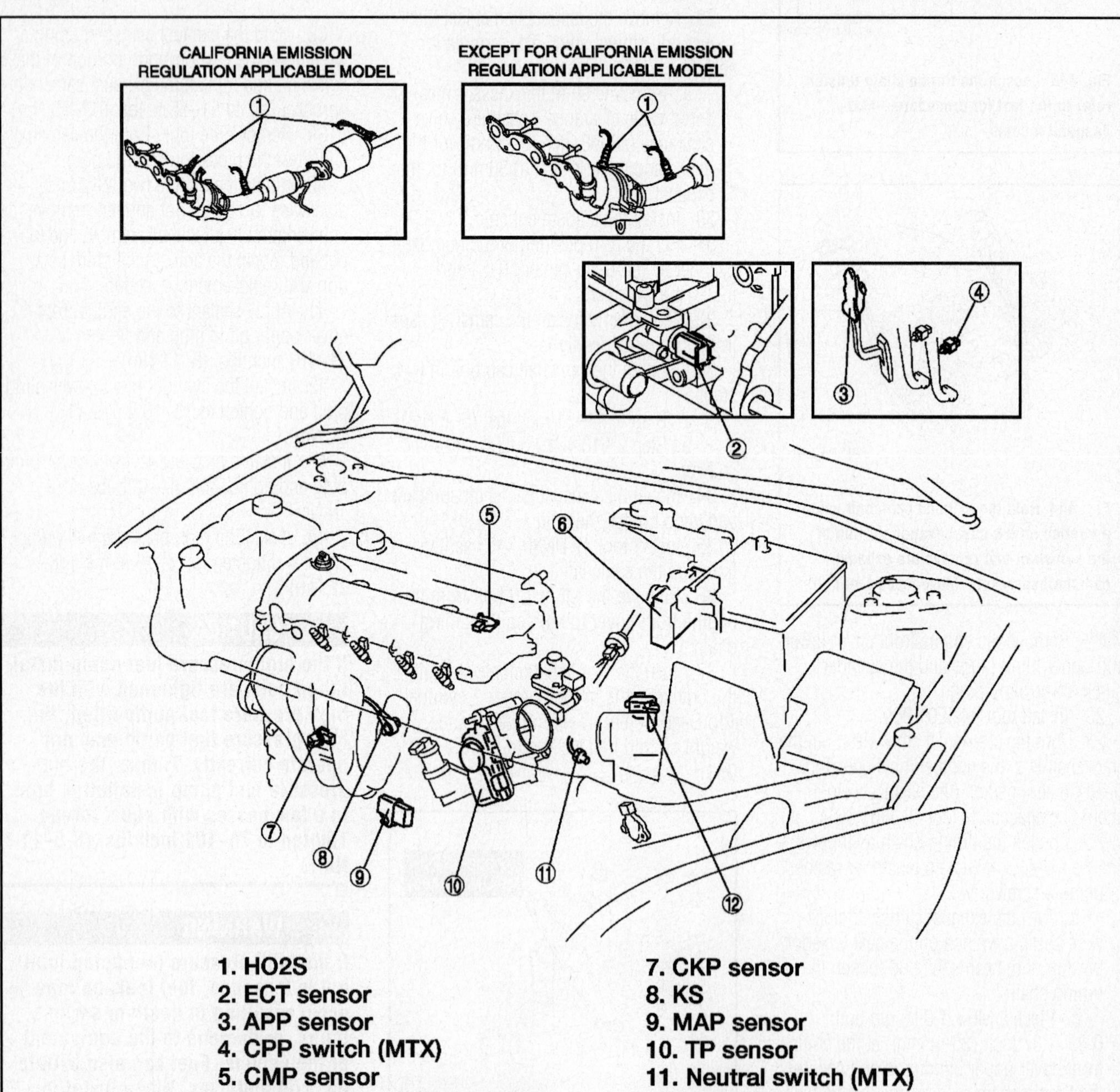

Fig. 447 Vehicle component locations 2.0L (LF) and 2.3L (L3) engines

2007—08 MX5-Miata

See Figure 448.

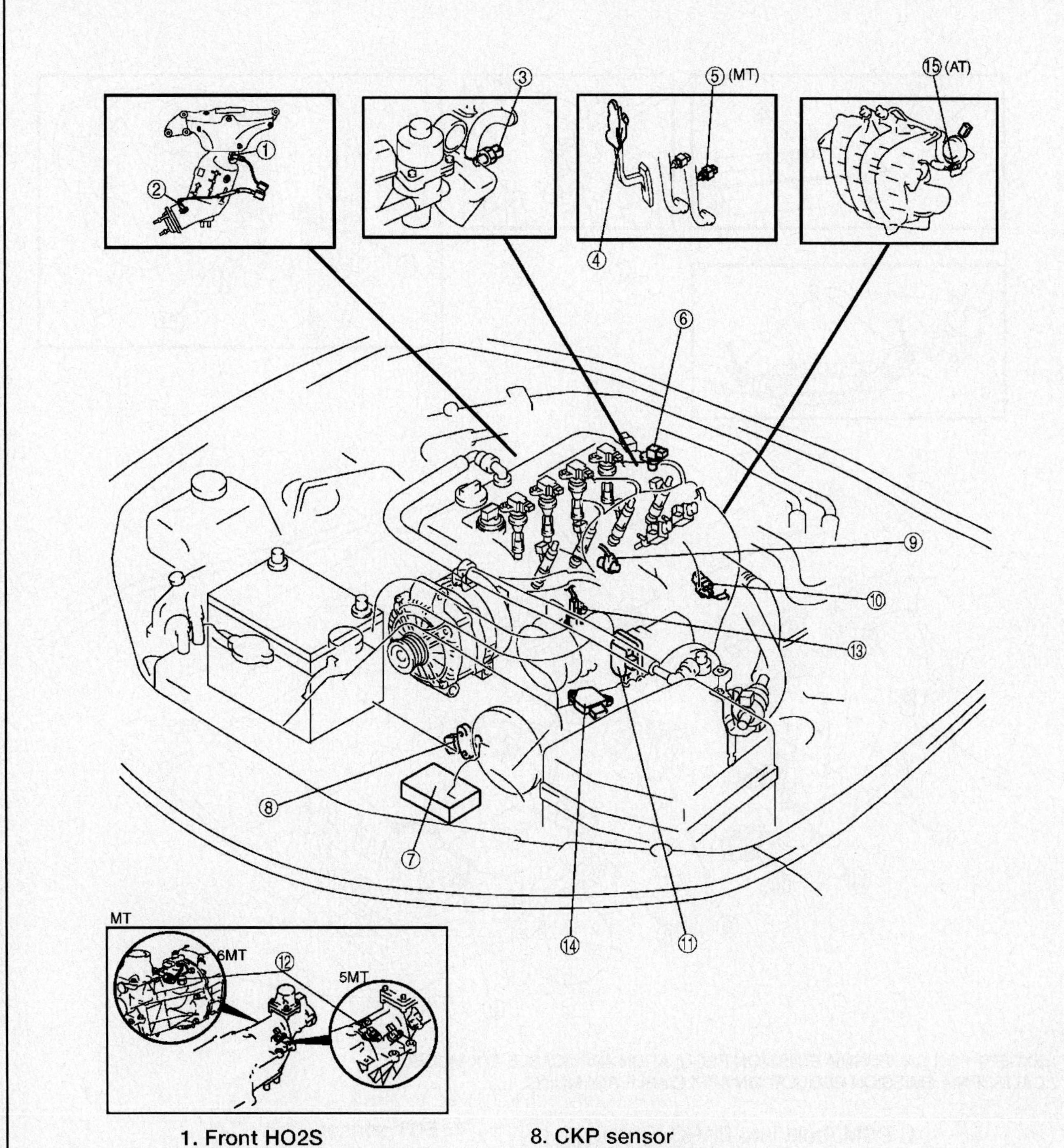

Fig. 448 Vehicle component locations 2007—08 MX5-Miata

2007–08 Mazda 6

See Figure 449 and 450.

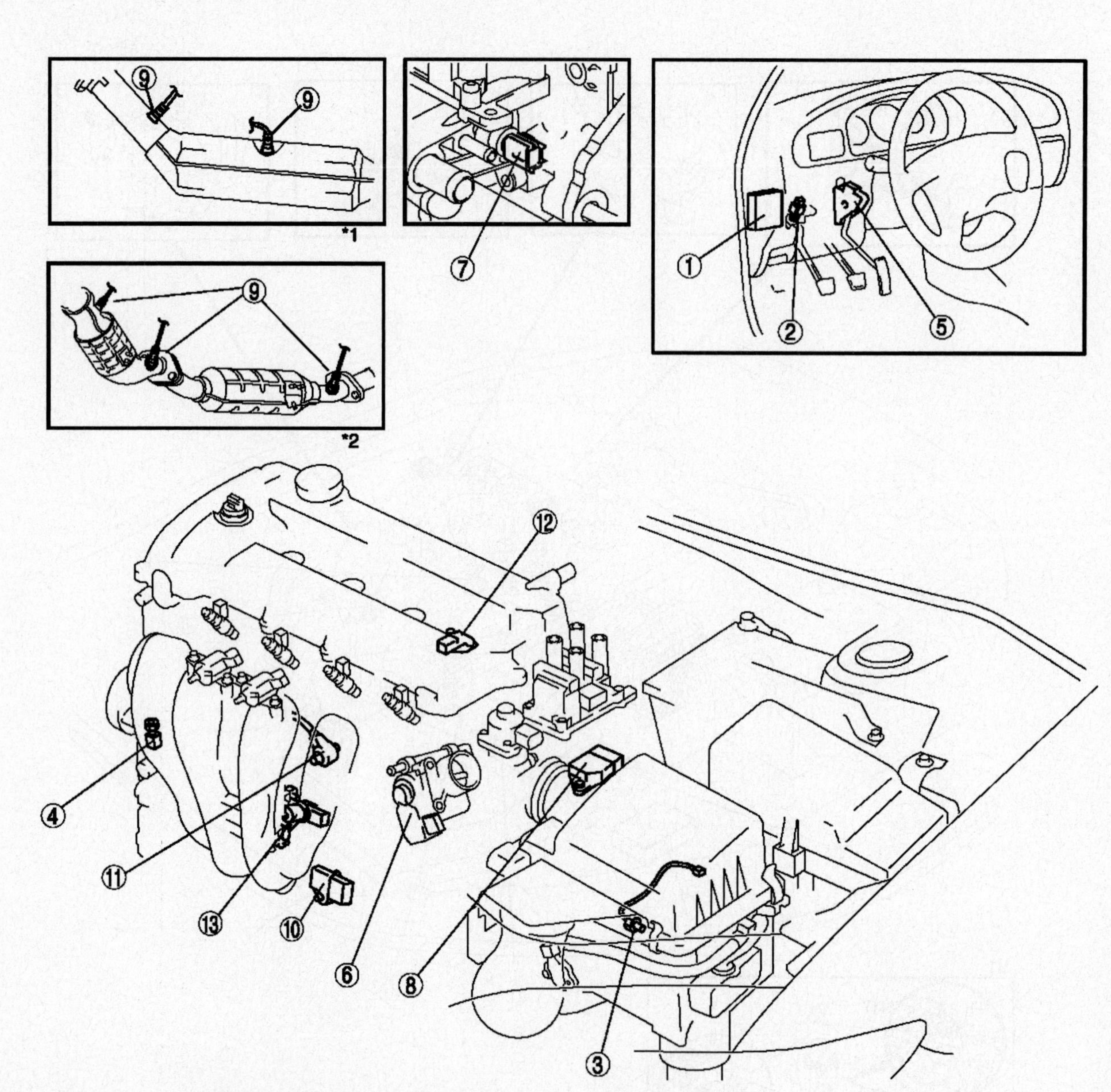

Fig. 449 Vehicle component locations 2.3L (L3) engine

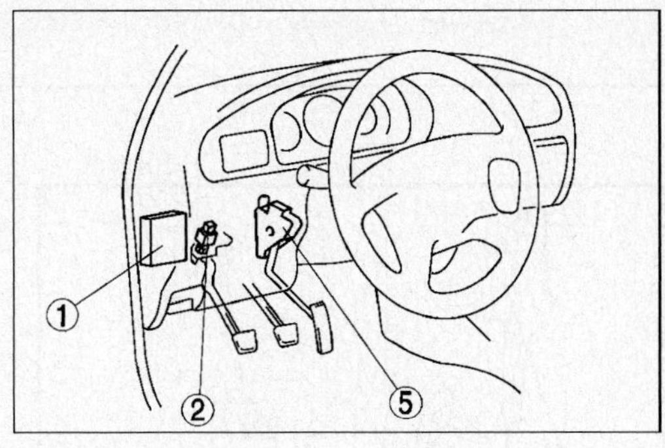

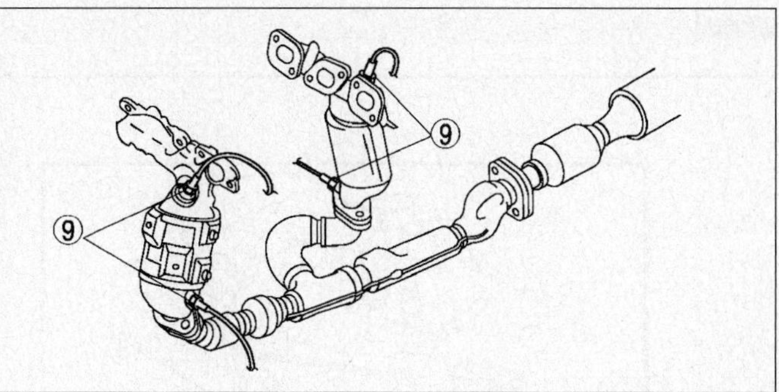

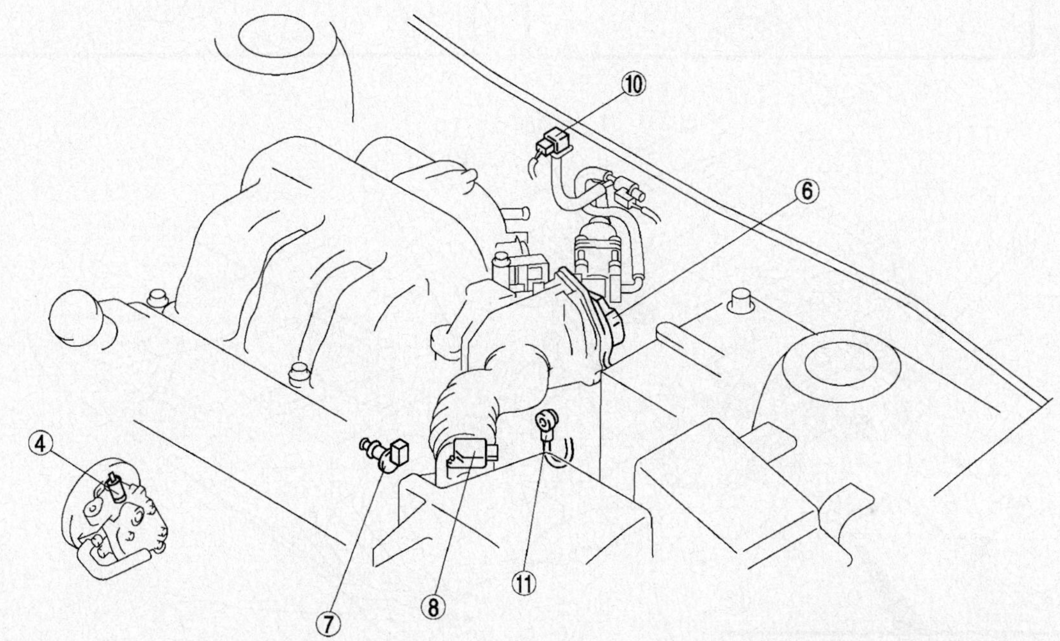

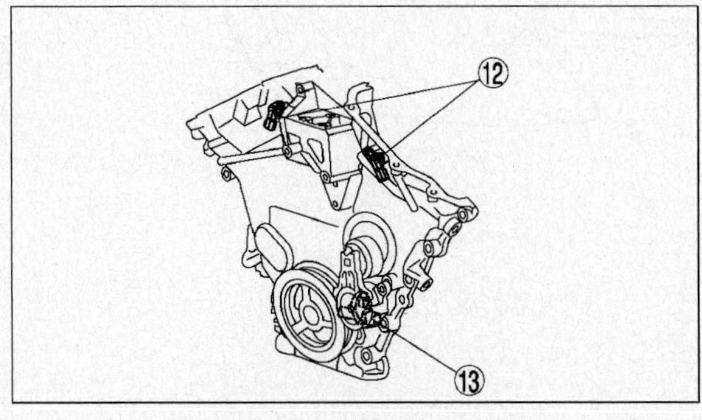

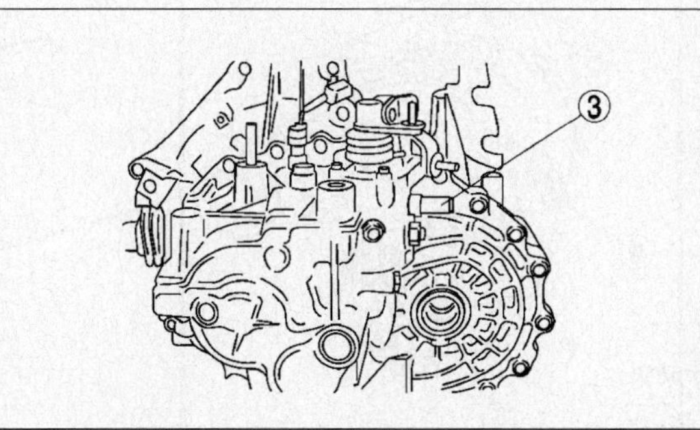

1. PCM
2. Clutch switch (MTX)
3. Neutral switch (MTX)
4. PSP switch
5. APP sensor
6. TP sensor and throttle actuator
7. ECT sensor
8. MAF/IAT sensor
9. HO2S
10. EGR boost sensor
11. KS
12. CMP sensor
13. CKP sensor

22140_MAZD_G0192

Fig. 450 Vehicle component locations 3.0L (AJ) engine

2007—08 MazdaSpeed3 and Speed6 Models

See Figure 451.

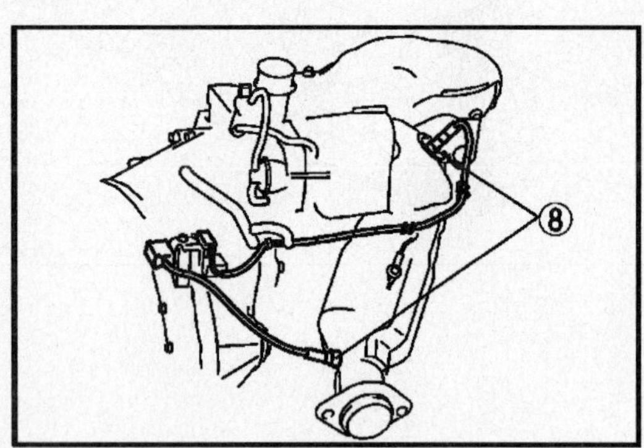

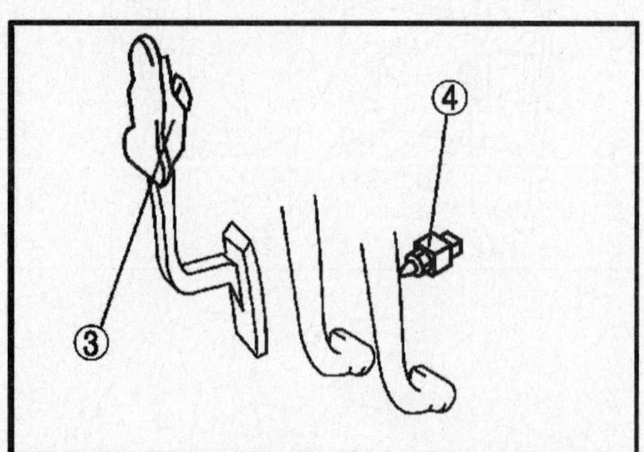

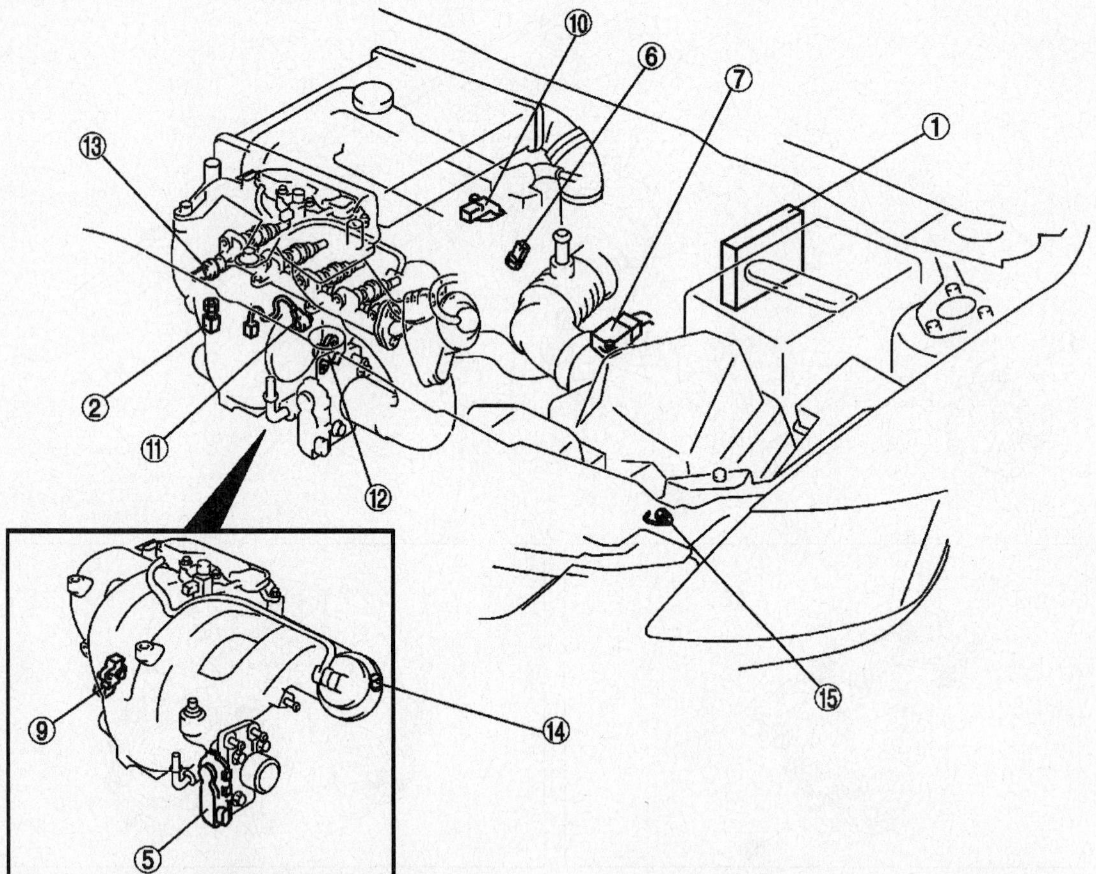

1. PCM (Built-in BARO sensor)
2. PSP switch
3. APP sensor
4. CPP sensor
5. TP sensor
6. ECT sensor
7. MAF/IAT sensor
8. HO2S
9. MAP/boost air temperature sensor
10. CMP sensor
11. KS
12. CKP sensor
13. Fuel pressure sensor
14. Variable swirl shutter valve switch
15. Neutral switch

22140_MAZD_G0193

Fig. 451 Vehicle component locations 2.3L (LF with TC) engine

ACCELERATOR PEDAL POSITION (APP) SENSOR

LOCATION

See Figure 452.

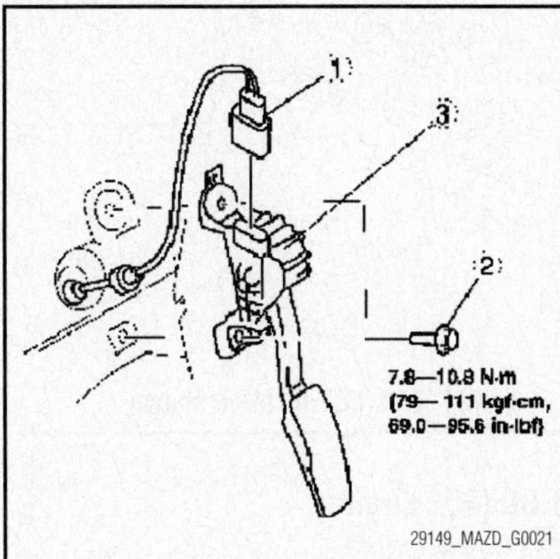

Fig. 452 Accelerator pedal position sensor location

REMOVAL & INSTALLATION

1. Remove the battery cover.
2. Disconnect the negative battery cable.
3. Remove the connector.
4. Remove the bolts.
5. Remove the
6. Remove the APP sensor.
7. Install in the reverse order of removal, and tighten the mounting bolts to 70—95 inch. lbs. (8—11 Nm).

CAMSHAFT POSITION (CMP) SENSOR

LOCATION

All L4 Engines

See Figure 453.

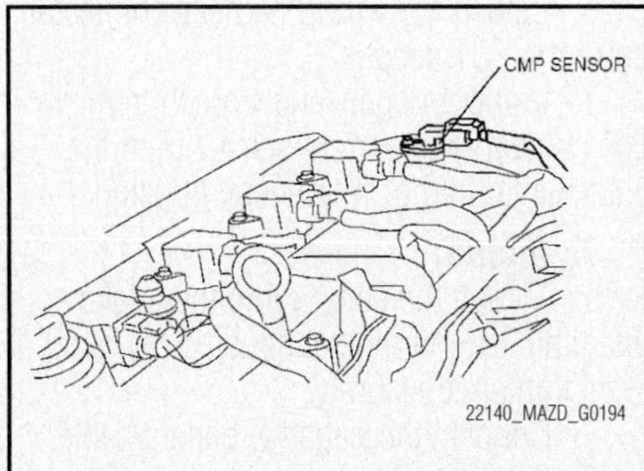

Fig. 453 2.3L and 2.0L Engines CMP sensor location

3.0L (AJ) Engine

See Figure 454.

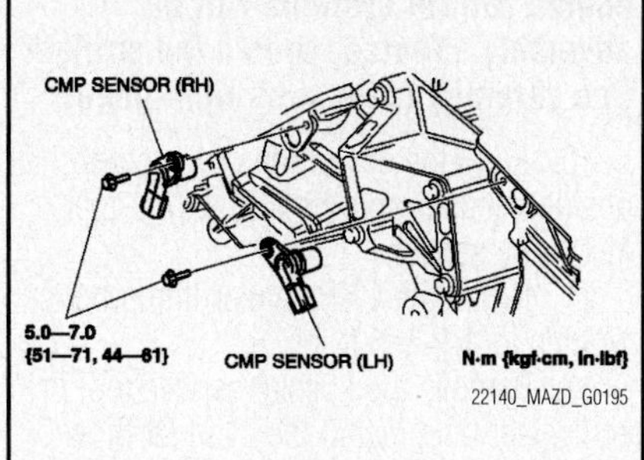

Fig. 454 3.0L (AJ) V6 Engine CMP Sensor location

REMOVAL & INSTALLATION

All L4 Engines

See Figure 455.

1. Disconnect the negative battery cable.
2. Remove the battery cover.
3. Remove the plug hole cover.
4. Remove the charge air cooler. (For turbo models)
5. Disconnect the CMP sensor connector.
6. Remove the CMP sensor installation bolt.
7. Remove the CMP sensor.
8. Make sure that the CMP sensor is free of any metallic shavings or particles.

➡If metallic shavings or particles are found on the sensor, clean them off.

To install:

9. Install the CMP sensor in the reverse order of removal.
10. Tightening torque is 49–66 inch. lbs. (5.5—7.5 Nm).

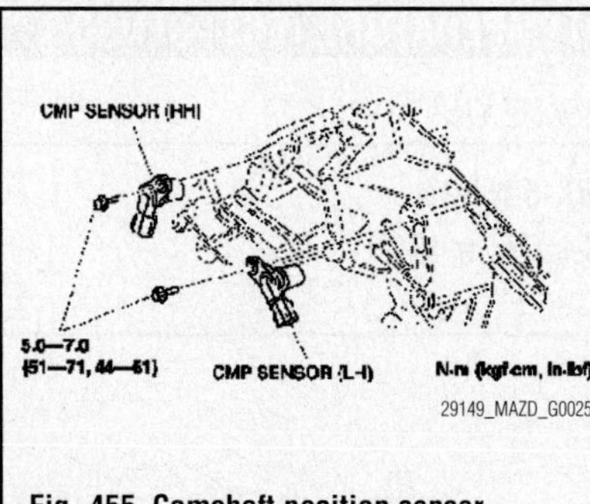

Fig. 455 Camshaft position sensor removal

3.0L (AJ) Engine

1. Disconnect the negative battery cable.
2. Disconnect the CMP sensor connector.

➡Do not forcefully pull the wiring harness of the CMP sensor. Doing so will break the harness.

3. Remove RH or LH CMP sensor retaining bolt.
4. Remove the CMP sensor.

✲✲ WARNING

When foreign material such as an iron chip is on the CMP sensor, it can cause abnormal output from the sensor because of flux turbulence and adversely affect the engine control. Be sure there is no foreign material on the CMP sensor when replacing.

To install:

5. Install the removed CMP sensor.
6. Tighten the CMP sensor mounting bolt to 44—61 inch. lbs. (5—7 Nm).
7. Install the CMP sensor connector
8. Connect the negative battery cable.

CRANKSHAFT POSITION (CKP) SENSOR

LOCATION

All L4 Engines

See Figure 456.

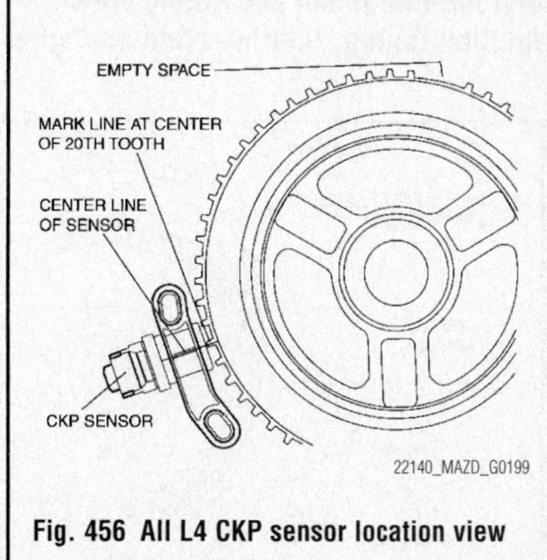

Fig. 456 All L4 CKP sensor location view

3.0L (AJ) Engine

See Figure 457.

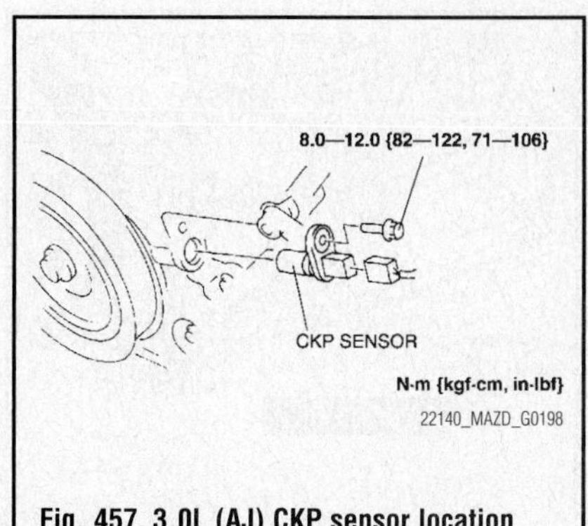

Fig. 457 3.0L (AJ) CKP sensor location view

REMOVAL & INSTALLATION

All L4 Engines

See Figures 458 and 459.

1. Remove the battery cover.
2. Disconnect the negative battery cable.
3. Remove the right front wheel. (Front wheel drive models)
4. Remove the splash shield.
5. Disconnect the CKP sensor connector.
6. Remove the installation bolts to remove the CKP sensor.

➡Perform the following procedure so that piston No.1 is at the top dead center.

7. Remove the cylinder block lower blind plug and install the special service tool or equivalent.
8. Turn the crankshaft pulley to the clockwise until it stops.
9. Using a straight edge, draw a straight line directly in the center of the ninth tooth of the crankshaft pulley pulse wheel (counting counterclockwise from the empty space).

➡If the line is not accurately drawn, ignition timing, fuel injection and other engine control systems will be adversely affected. Drawn the straight line carefully using a straight edge.

10. Align the centerline of the crankshaft position sensor and the line drawn, then install the sensor.
11. Install the CKP sensor fitting bolts. Tighten to 4.1–5.5 ft lbs
12. Remove the special service tool or equivalent then install the cylinder block lower blind plug. Tightening torque 15 ft lbs
13. Install the splash shield.
14. Install the right front wheel

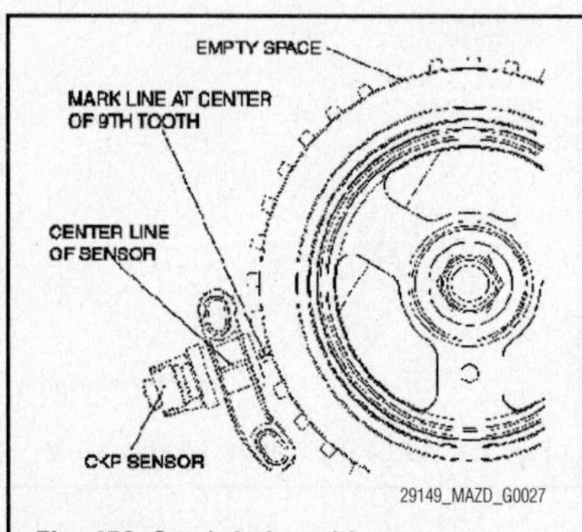

Fig. 458 Crankshaft position sensor removal and installation–L4 Engines

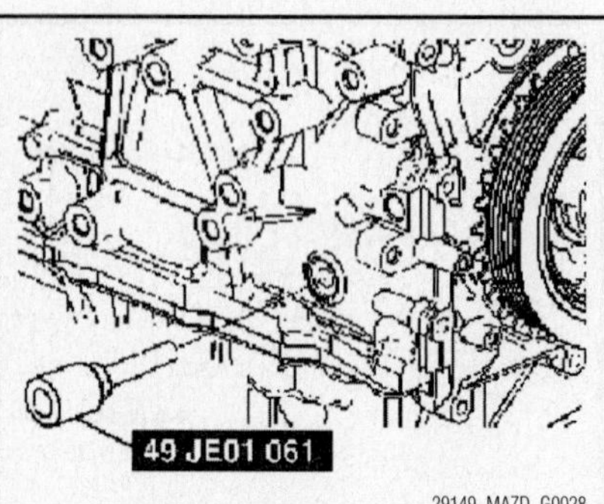

Fig. 459 Crankshaft position sensor TDC locator–L4 Engines

3.0L (AJ) Engine

1. Disconnect the negative battery cable.
2. Disconnect the CKP sensor connector.
3. Remove the CKP sensor.

To install:

⁂ WARNING

When foreign material such as an iron chip is on the CKP sensor, it can cause abnormal output from the sensor because of flux turbulence and adversely affect the engine control. Be sure there is no foreign material on the CKP sensor when replacing.

4. Install the CKP sensor.
5. Tighten the Sensor mounting bolt to 71—106 inch. lbs. (8—12 Nm).
6. Reconnect the CKP sensor connector.
7. Connect the negative battery cable.

ENGINE COOLANT TEMPERATURE (ECT) SENSOR

LOCATION

MX-5 Miata

See Figure 460.

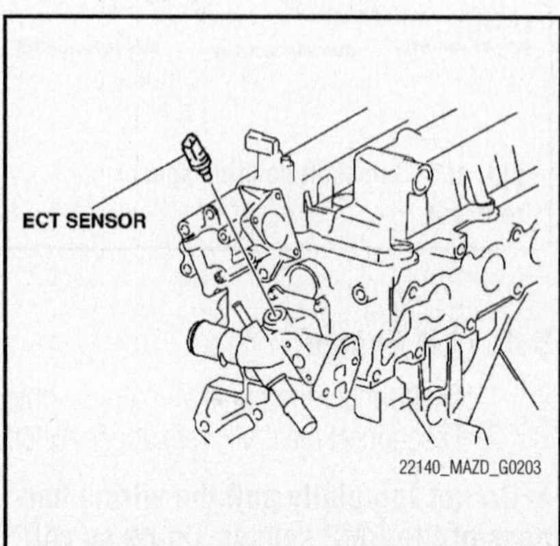

Fig. 460 MX-5 Miata ECT location shown

2.3L and 2.0L Engines

See Figure 461.

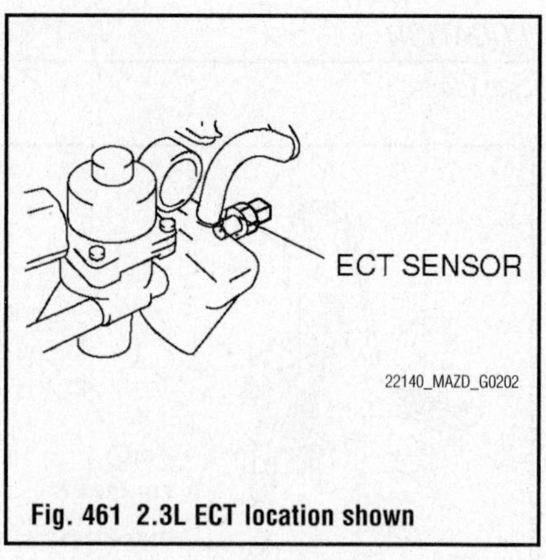

Fig. 461 2.3L ECT location shown

3.0L (AJ) Engine

See Figure 462.

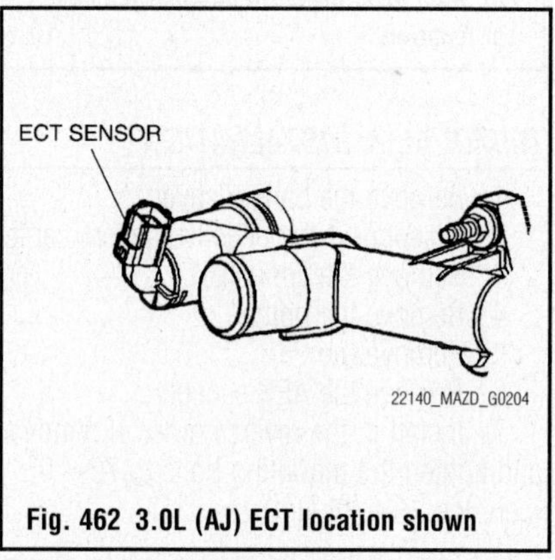

Fig. 462 3.0L (AJ) ECT location shown

MOVAL & INSTALLATION

All L4 Engines

1. Partially drain the engine cooling system until the coolant level is below the ECT sensor-mounting hole.
2. Disconnect the negative battery cable.
3. Detach the wiring harness connector from the ECT sensor.
4. Using an open-end wrench, remove the coolant temperature sensor from the intake manifold or thermostat housing.

To install:

5. Thread the sensor into the intake manifold, or thermostat housing, by hand, then tighten it securely.
6. Connect the negative battery cable.
7. Refill the engine cooling system.
8. Start the engine, check for coolant leaks and top off the cooling system.

3.0L (AJ) Engine

1. Disconnect the negative battery cable.
2. Drain the engine coolant.

3. Disconnect the ECT sensor connector.
4. Remove the ECT sensor.

To install:

5. Replace the gasket.
6. Install the ECT sensor.
7. Reconnect the ECT sensor connector.
8. Refill cooling system.
9. Connect the negative battery cable.

HEATED OXYGEN SENSOR (HO2S)

LOCATION

All L4 Engines

See Figures 463 and 464.

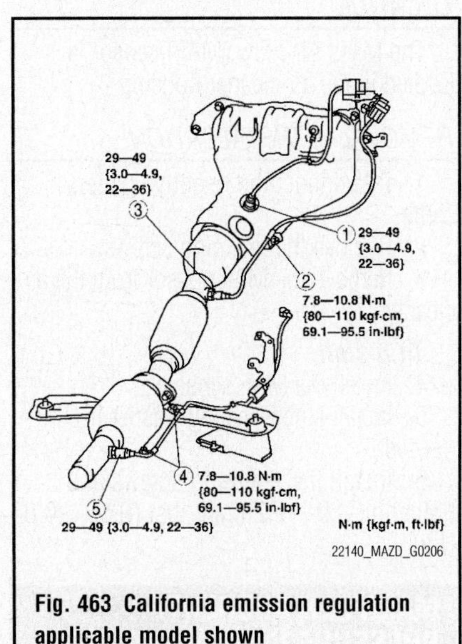

Fig. 463 California emission regulation applicable model shown

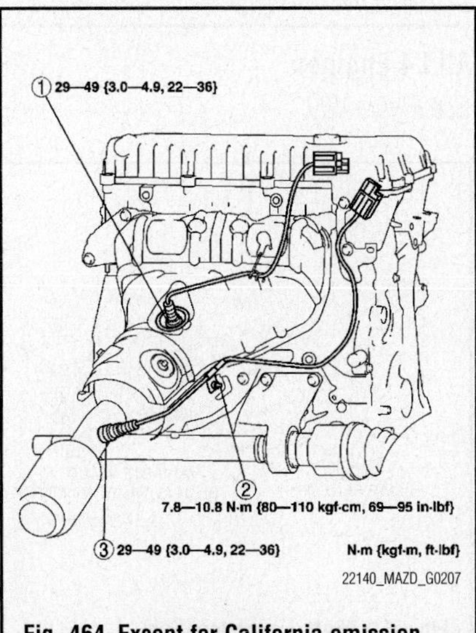

Fig. 464 Except for California emission regulation applicable model shown

3.0L (AJ) Engine

See Figure 465.

3.0L (AJ) Heated Oxygen Sensor (HO2S) location.

REMOVAL & INSTALLATION

CAUTION

The temperature of the exhaust system is extremely high after the engine has been run. To prevent personal injury, allow the exhaust system to cool completely before removing sensor from the exhaust system.

1. Disconnect the negative battery cable.
2. Raise and safely support the vehicle.
3. Disconnect the HO2S from the engine control sensor wiring.

➡If excessive force is needed to remove the sensors lubricate the sensor with penetrating oil prior to removal.

4. Remove the sensors with a sensor removal tool, such as a 49-L018-001 or equivalent.

To install:

5. Install the sensor, and then tighten it to 22–36 ft. lbs. (29—49 Nm).
6. Connect the sensor electrical wiring connector to the engine wiring harness.
7. Lower the vehicle.
8. Connect the negative battery cable.

1. Main silencer
2. Presilencer
3. TWC
4. Front pipe
5. HO2S (RR)
6. HO2S (LR)
7. HO2S (RF)
8. HO2S (LF)
9. EGR pipe
10. Exhaust manifold, WU-TWC (RH)
11. Exhaust manifold, WU-TWC (LH)

22140_MAZD_G0208

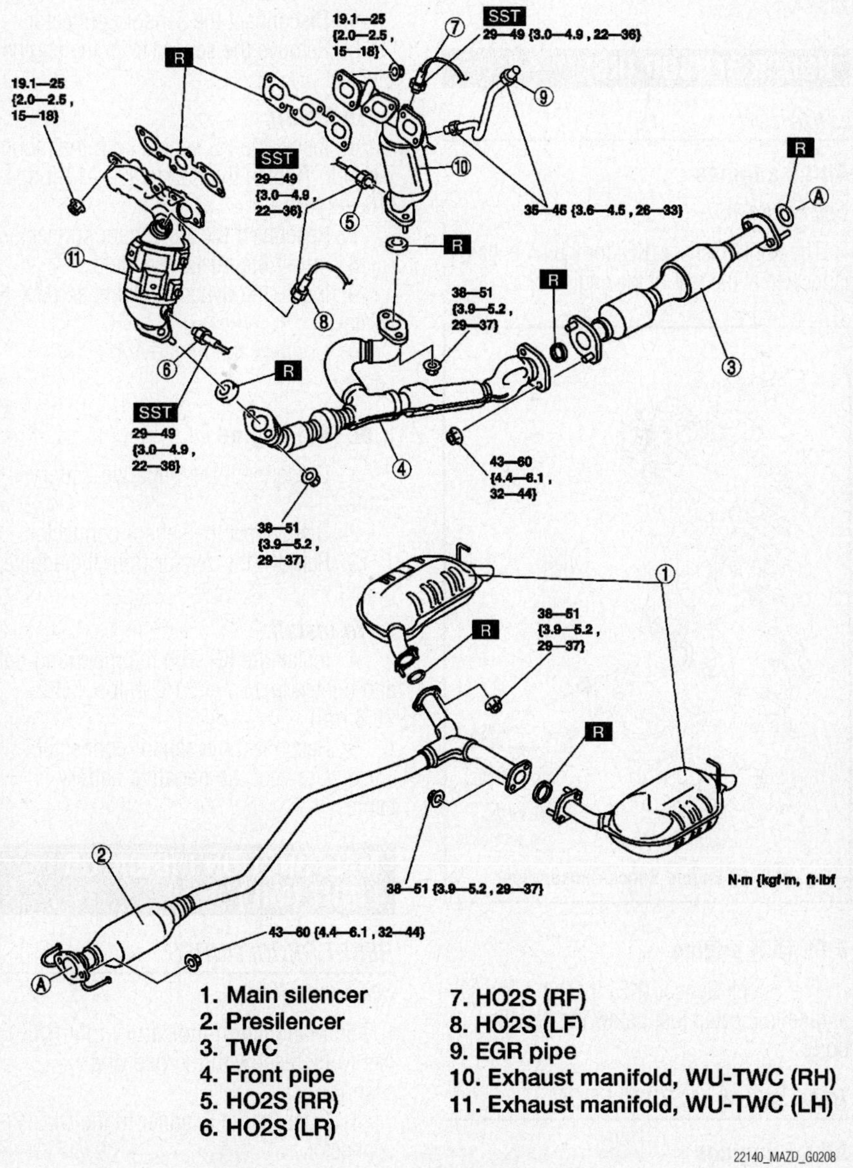

Fig. 465 3.0L (AJ) Heated Oxygen Sensor (HO2S) location view

INTAKE AIR TEMPERATURE (IAT) SENSOR

LOCATION

The Intake Air Temperature (IAT) Sensor is integral to the Mass Air Flow (MAF) sensor. It is located in the air cleaner housing.

REMOVAL & INSTALLATION

1. Disconnect the negative battery cable.
2. Remove the retaining screws.
3. Remove the IAT/MAF sensor from the air cleaner.

To install:

4. Install the IAT/MAF sensor
5. Install a new sealing washer if needed.
6. Install the retaining screws and tighten to 4.9—7.2 inch. lbs. (0.55—0.82 Nm).

KNOCK SENSOR (KS)

LOCATION

All L4 engines

See Figure 466.

The Knock Sensor (KS) for the L4 engines, is located at the rear of the engine block.

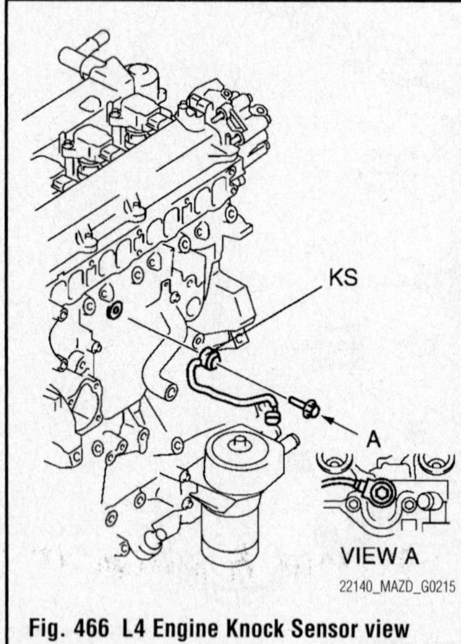

Fig. 466 L4 Engine Knock Sensor view

3.0L (AJ) engine

The Knock Sensor (KS) for the 3.0L (AJ) engine is located just below the throttle body.

REMOVAL & INSTALLATION

All L4 Engines

See Figure 467.

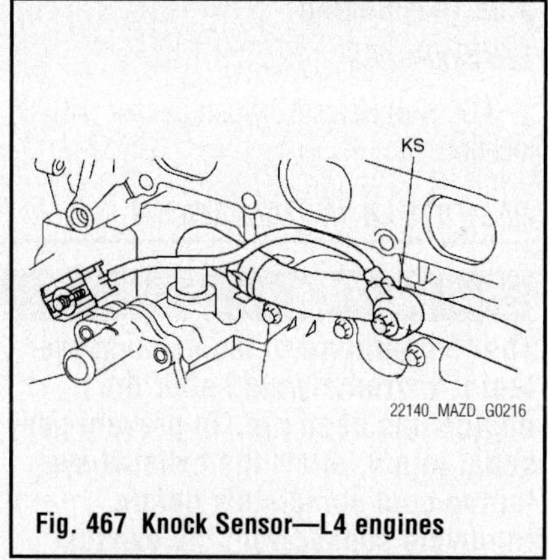

Fig. 467 Knock Sensor—L4 engines

1. Disconnect the negative battery cable.
2. Remove the dynamic chamber. (MX-5 Miata)
3. Remove the intake manifold.
4. Disconnect the sensor connector.
5. Remove the sensor from the engine block.

To install:

6. Install the KS sensor with the mounting bolt. Tighten the bolt to 12—17 ft. lbs. (16—24 Nm).
7. Reconnect the KS sensor connector.
8. Install the intake manifold.
9. Install the dynamic chamber. (MX-5 Miata)
10. Connect the negative battery cable.

3.0L (AJ) Engine

1. Disconnect the negative battery cable.
2. Disconnect the sensor connector.
3. Remove the sensor from the engine block.

To install:

4. Install the KS with the mounting bolt and tighten to 15.7—21.2 ft. lbs. (21.2—28.8 Nm).
5. Reconnect the sensor connector.
6. Connect the negative battery cable.

MALFUNCTION INDICATOR LIGHT (MIL)

RESET PROCEDURES

See Figure 468.

The Malfunction Indicator Light (MIL) has to be reset with the use of a scanner.

1. Connect the Scanner to the DLC-2 connector.
2. Clear all DTCs stored in the PCM.

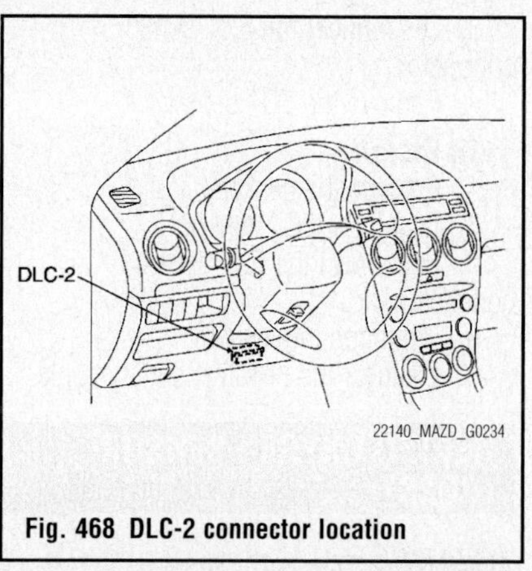

Fig. 468 DLC-2 connector location

MASS AIR FLOW (MAF) SENSOR

LOCATION

The Mass Air Flow (MAF) sensor is located in the air cleaner housing.

REMOVAL & INSTALLATION

1. Disconnect the negative battery cable.
2. Remove the retaining screws.
3. Remove the MAF sensor from the air cleaner.

To install:

4. Install the MAF sensor
5. Install a new sealing washer if needed.
6. Install the retaining screws and tighten to 4.9—7.2 inch. lbs. (0.55—0.82 Nm).

MANIFOLD ABSOLUTE PRESSURE (MAP) SENSOR

LOCATION

All L4 Engines

See Figure 469.

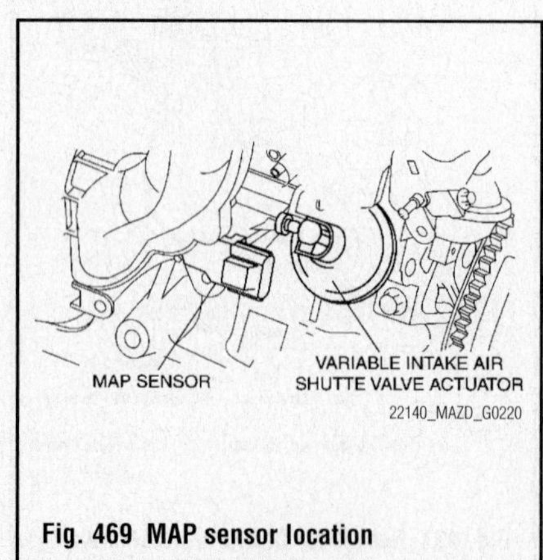

Fig. 469 MAP sensor location

REMOVAL & INSTALLATION

All L4 Engines

1. Remove the battery cover.
2. Disconnect the negative battery cable.
3. Remove the oil level gauge pipe. (Mazda3 Models)
4. Remove the plughole cover.
5. Remove the dynamic chamber. (MX-5 Miata)
6. Disconnect the vacuum hose.
7. Disconnect sensor connector.
8. Remove sensor mounting screw.
9. Remove the sensor from intake manifold.
10. Installation is the reverse of the removal procedure. Tightening to 24—32 inch. lbs. (2.7—3.7 Nm).

POWERTRAIN CONTROL MODULE (PCM)

LOCATION

The PCM for the Mazda3 is located next to the battery box.

The PCM for the MX-5 Miata is located at the front of the engine compartment to the left of the air filter housing.

The PCM location for the Mazda6 is located inside the vehicle, just to the left of the drivers compartment behind the dash.

REMOVAL & INSTALLATION

Mazda3 Models

See Figure 470.

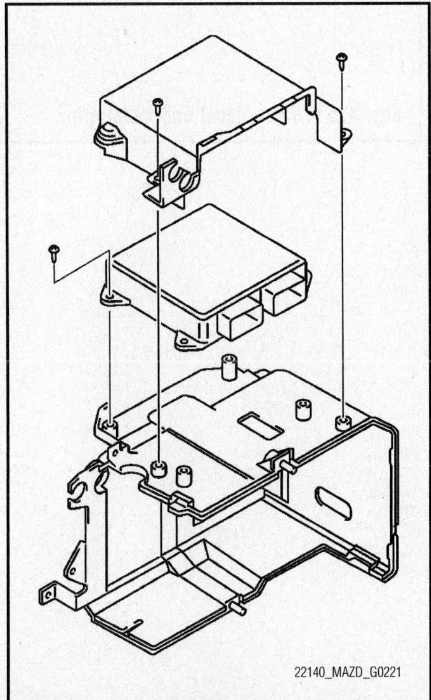

Fig. 470 Mazda3 PCM removal shown

1. Disconnect the negative battery cable.
2. Remove the battery cover.
3. Remove the battery duct.
4. Remove the battery and battery tray with the PCM.
5. Remove the PCM harness connectors.
6. Remove the PCM from the battery tray.
7. Installation is the reverse of the removal procedure.

MX-5 Miata

See Figure 471.

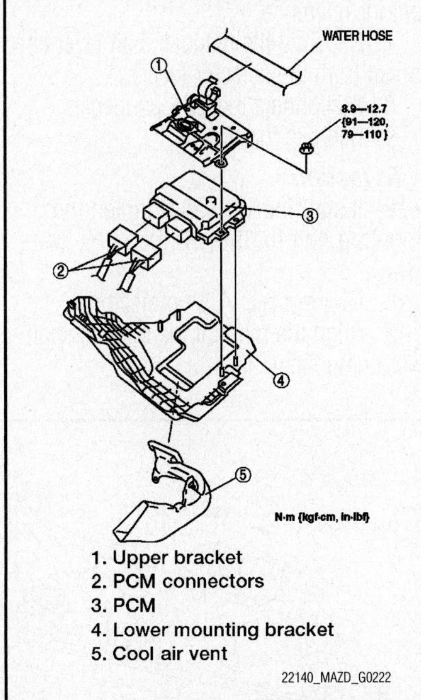

Fig. 471 MX-5 Miata PCM removal shown

1. Disconnect the negative battery cable.
2. Remove the battery cover.
3. Remove the battery duct.
4. Remove the air cleaner case.
5. Move the water hose from the PCM cover slightly out of the way.
6. Remove the PCM cover.
7. Remove the PCM connectors.
8. Remove the PCM.
9. Installation is the reverse of the removal procedure.

Mazda6 Models

See Figure 472.

1. Disconnect the negative battery cable.
2. Remove the PCM harness connectors.
3. Remove the mounting fasteners and remove the PCM from the vehicle.
4. Installation is the reverse of the removal procedure. Tighten mounting fasteners to 70–95 inch. lbs. (7.8–10.8 Nm).

THROTTLE POSITION SENSOR (TPS)

LOCATION

The TPS sensor is integral to the throttle body assembly.

REMOVAL & INSTALLATION

1. Disconnect the negative battery cable.
2. Remove the necessary air intake components to access the throttle body.
3. Detach the electrical wire harness plug from the throttle body.
4. Remove the sensor attaching bolts.
5. Remove the throttle body.
6. Installation is the reverse of the removal procedure. Tighten the mounting bolts to 71—101 inch. lbs. (8—11.5 Nm).

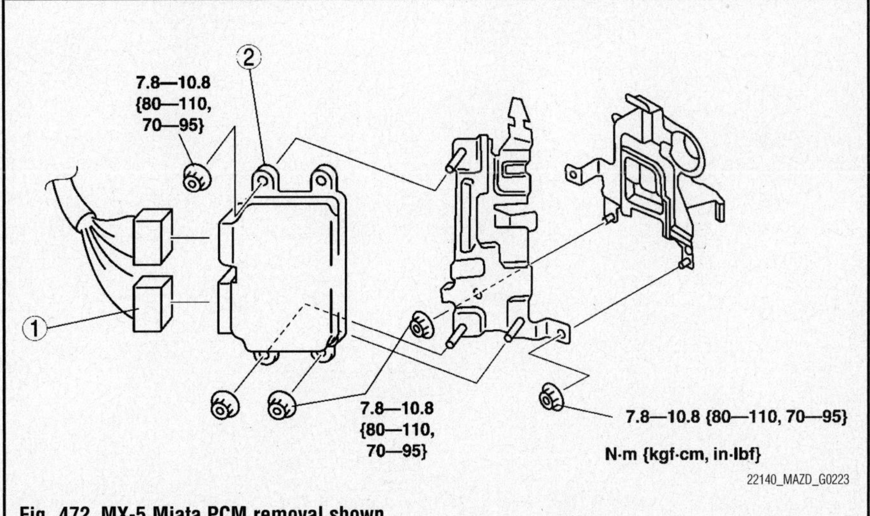

Fig. 472 MX-5 Miata PCM removal shown

VEHICLE SPEED SENSOR (VSS)

LOCATION

The Vehicle Speed Sensor (VSS) is located to the right rear of the transaxle or transmission.

REMOVAL & INSTALLATION

MX-5 Miata

P66M-D and 5M15-D Transmissions

See Figure 473.

1. Remove the battery cover.
2. Disconnect the negative battery cable.
3. Disconnect the VSS connector.
4. Remove the transmission.
5. Remove the VSS.

To install:

6. Apply the specified oil to a new O-ring and install it on a new VSS.
7. Install the VSS and tighten the mounting bolt to 79—97 inch. lbs. (8—11 Nm).
8. Install the transmission.
9. Connect the VSS connector.
10. Connect the negative battery cable.
11. Install the battery cover.

SJ6A-EL Transmission

See Figure 474.

1. Remove the battery cover.
2. Disconnect the negative battery cable.
3. Remove the tunnel member component.
4. Remove the middle exhaust pipe.
5. Remove the insulator.
6. Mark the manual shaft lever component for reference.
7. Separate the manual shaft lever component from the selector lever.
8. Disconnect the VSS connector.
9. Remove the VSS.

To install:

10. Install the VSS and tighten the mounting bolt to 35—60 inch. lbs. (4—7 Nm).
11. Connect the VSS connector.
12. Align the mark of the manual shaft lever component.
13. Install the manual shaft lever component installation nut. Tighten to 72—97 inch. lbs. (8—11 Nm).
14. Install the middle exhaust pipe.
15. Install the tunnel member component.
16. Connect the negative battery cable.
17. Install the battery cover.

All Models Except MX-5 Miata

See Figure 475.

1. Disconnect the negative battery cable.
2. Remove the undercover.
3. Disconnect the VSS connector.
4. Remove the VSS.

To install:

5. Apply ATF to a new O-ring and install it on a VSS.
6. Install the VSS and tighten the mounting bolt to 71—97 inch. lbs. (8—11 Nm).
7. Connect the VSS connector.
8. Connect the negative battery cable.

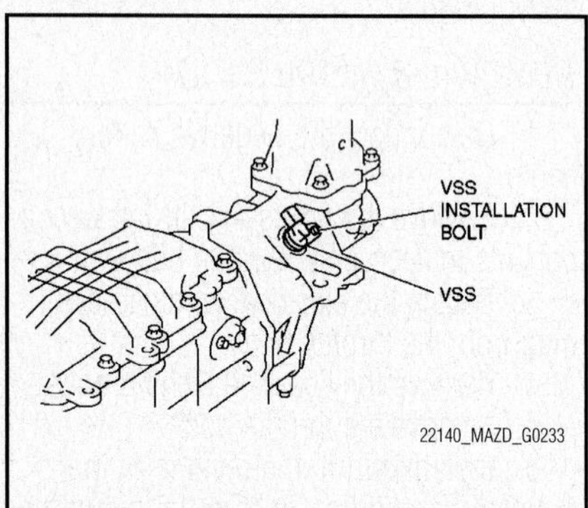

Fig. 473 VSS (P66M-D) and (5M15-D) transmission location view

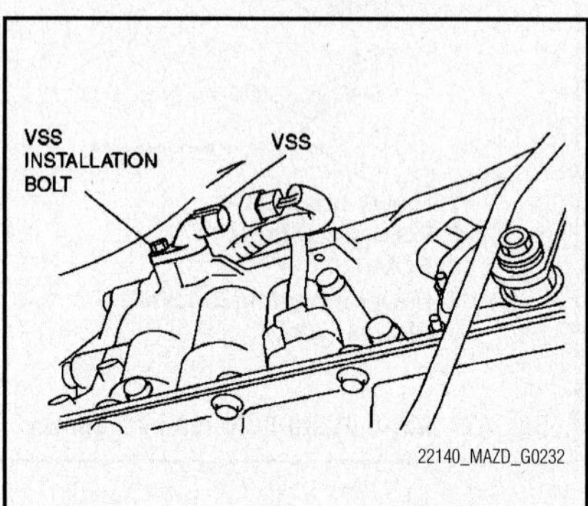

Fig. 474 VSS (SJ6A-EL) transmission location view

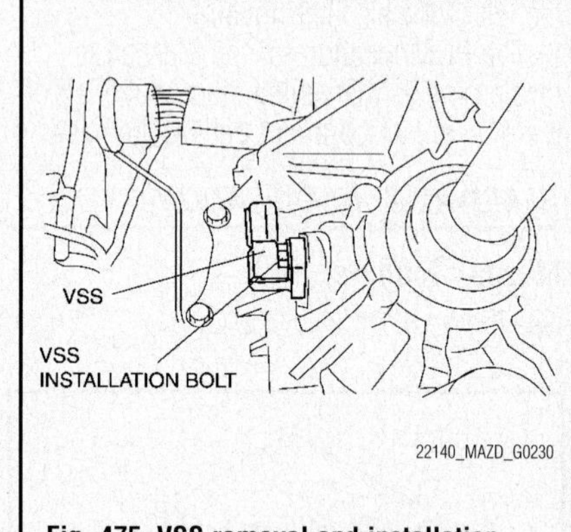

Fig. 475 VSS removal and installation

FUEL SYSTEM SERVICE PRECAUTIONS

Safety is the most important factor when performing not only fuel system maintenance but any type of maintenance. Failure to conduct maintenance and repairs in a safe manner may result in serious personal injury or death. Maintenance and testing of the vehicle's fuel system components can be accomplished safely and effectively by adhering to the following rules and guidelines.

- To avoid the possibility of fire and personal injury, always disconnect the negative battery cable unless the repair or test procedure requires that battery voltage be applied.
- Always relieve the fuel system pressure prior to disconnecting any fuel system component (injector, fuel rail, pressure regulator, etc.), fitting or fuel line connection. Exercise extreme caution whenever relieving fuel system pressure to avoid exposing skin, face and eyes to fuel spray. Please be advised that fuel under pressure may penetrate the skin or any part of the body that it contacts.
- Always place a shop towel or cloth around the fitting or connection prior to loosening to absorb any excess fuel due to spillage. Ensure that all fuel spillage (should it occur) is quickly removed from engine surfaces. Ensure that all fuel soaked cloths or towels are deposited into a suitable waste container.
- Always keep a dry chemical (Class B) fire extinguisher near the work area.
- Do not allow fuel spray or fuel vapors to come into contact with a spark or open flame.
- Always use a back-up wrench when loosening and tightening fuel line connection fittings. This will prevent unnecessary stress and torsion to fuel line piping.
- Always replace worn fuel fitting O-rings with new. Do not substitute fuel hose or equivalent where fuel pipe is installed.

Before servicing the vehicle, make sure to also refer to the precautions in the beginning of this section as well.

RELIEVING FUEL SYSTEM PRESSURE

Mazda3 Models

See Figure 476.

1. Before servicing the vehicle, refer to the precautions section.

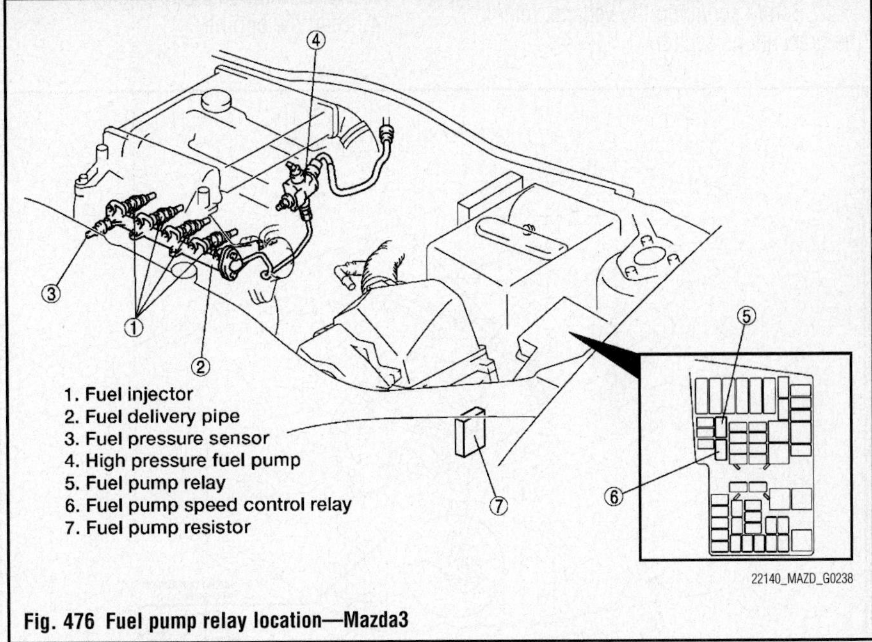

Fig. 476 Fuel pump relay location—Mazda3

2. Remove the fuel-filler cap to release the pressure inside the fuel tank.
3. Remove the fuel pump relay
4. Start the engine.
5. After the engine stalls, crank the engine several times.
6. Turn the ignition switch to the LOCK position.
7. Install the fuel pump relay.

Mazdaspeed3 Models

See Figure 477.

1. Before servicing the vehicle, refer to the precautions section.
2. Remove the fuel-filler cap to release the pressure inside the fuel tank.
3. Remove the fuel pump relay
4. Start the engine.
5. After the engine stalls, crank the engine several times.
6. Turn the ignition switch to the LOCK position.
7. Install the fuel pump relay.

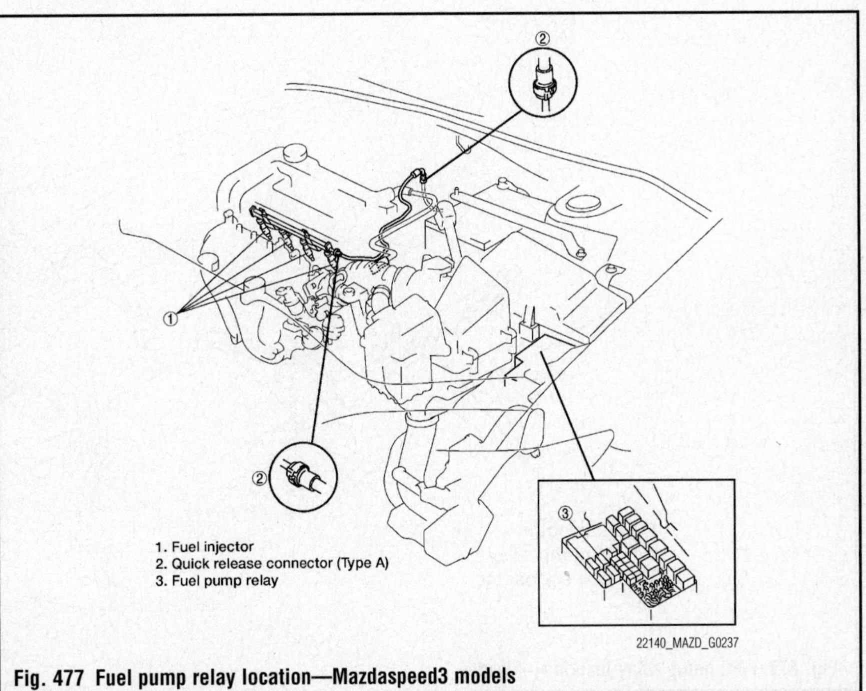

Fig. 477 Fuel pump relay location—Mazdaspeed3 models

MX-5 Miata

See Figure 478.

1. Before servicing the vehicle, refer to the precautions section.
2. Remove the fuel-filler cap to release the pressure inside the fuel tank.
3. Remove the fuel pump relay
4. Start the engine.
5. After the engine stalls, crank the engine several times.
6. Turn the ignition switch to the LOCK position.
7. Install the fuel pump relay.

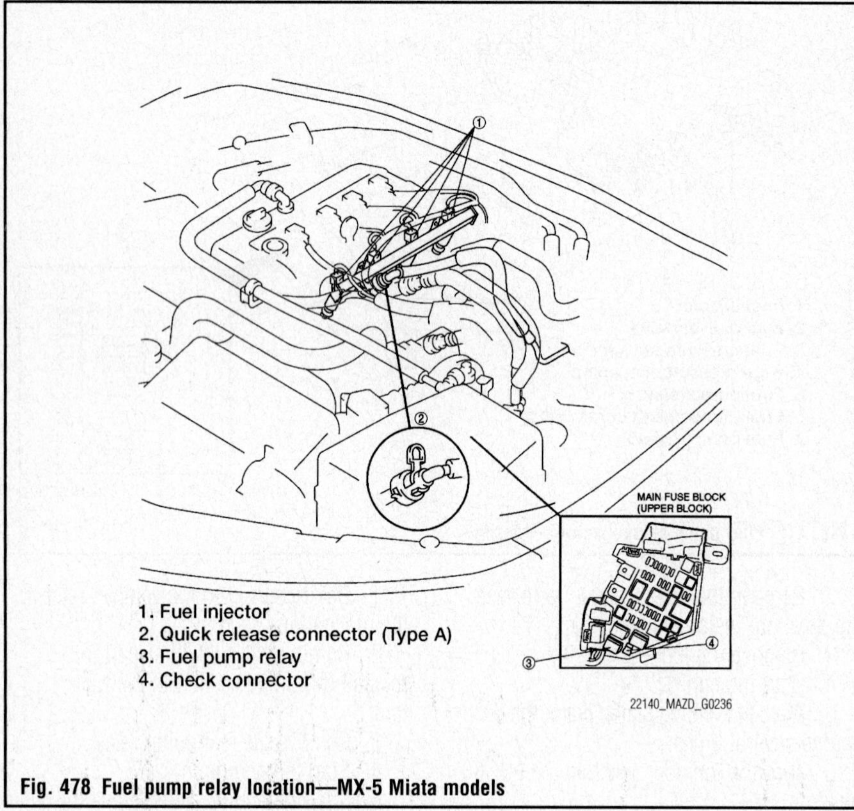

Fig. 478 Fuel pump relay location—MX-5 Miata models

Mazda6 Models

See Figure 479.

1. Before servicing the vehicle, refer to the precautions section.
2. Remove the filler cap.
3. Remove the fuel pump relay from the relay box, located in the engine compartment.
4. Start the engine.
5. After the engine stalls, turn the ignition switch **OFF**.
6. After servicing the vehicle, reinstall the relay.

Mazdaspeed6

See Figure 480.

1. Before servicing the vehicle, refer to the precautions section.
2. Remove the filler cap.
3. Remove the fuel pump relay from the relay box, located in the engine compartment.
4. Start the engine.
5. After the engine stalls, turn the ignition switch **OFF**.
6. After servicing the vehicle, reinstall the relay.

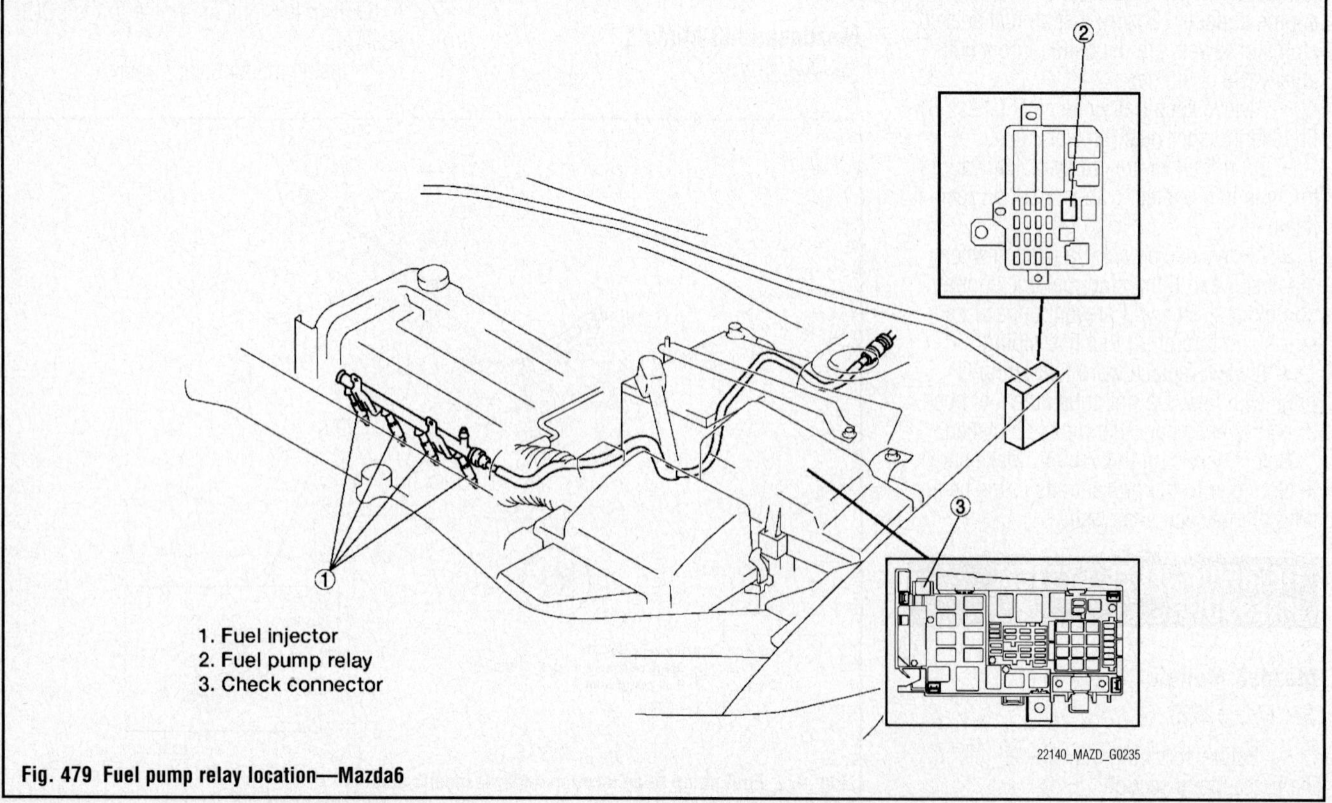

Fig. 479 Fuel pump relay location—Mazda6

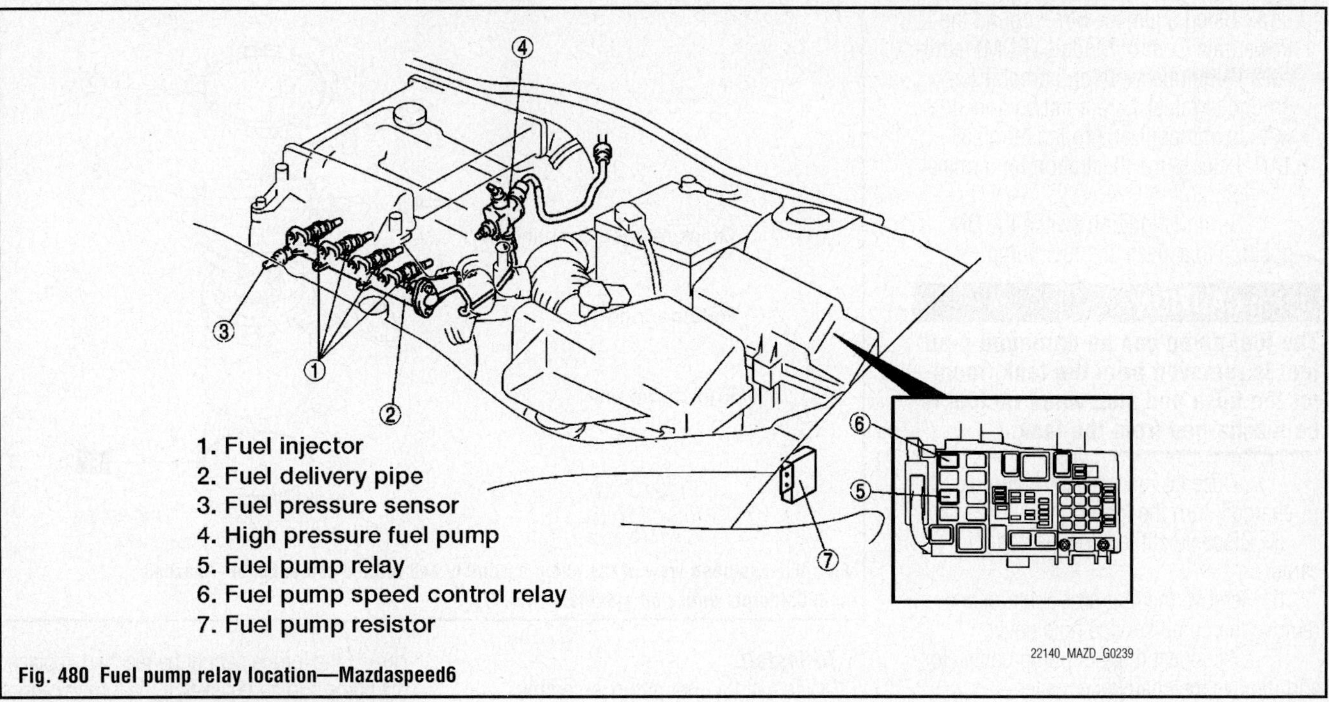

Fig. 480 Fuel pump relay location—Mazdaspeed6

FUEL FILTER

REMOVAL & INSTALLATION

The fuel filter is part of the pump assembly. Once the fuel pump is removed, remove the filter retainer and filter. refer to the fuel pump removal and installation.

FUEL LEVEL SENDING UNIT

REMOVAL & INSTALLATION

The fuel level sending unit is an integral part of the fuel pump assembly and is monitored by the PCM, which then relays this information to the fuel gauge. If the sending unit is defective, the whole pump assembly must be replaced as the pump cannot be disassembled.

FUEL PUMP

REMOVAL & INSTALLATION

MX-5 Miata

See Figure 481.

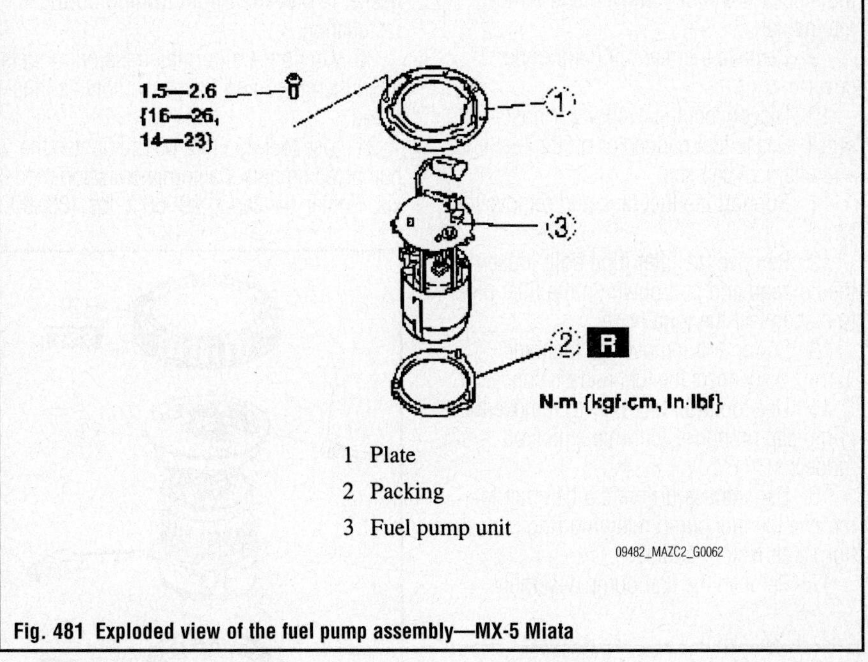

Fig. 481 Exploded view of the fuel pump assembly—MX-5 Miata

1. Before servicing the vehicle, refer to the precautions section.
2. Properly relieve the fuel system pressure.
3. Remove the battery cover.
4. Disconnect the negative battery cable.
5. Perform the following procedure to remove the service hole cover.
    a. Remove the console.
    b. Remove the quarter trim.
    c. Remove the scuff plate.
    d. Remove the tire house trim.
    e. Remove the aeroboard.
    f. Remove the front seat back bar garnish.
    g. Remove the remove the back trim.
    h. Remove the service hole cover.
6. Disconnect the quick release connector from the fuel pump unit.
7. Disconnect the fuel pump unit connector.
8. Remove plate, packing and pump assembly.

To install:

9. Installation is the reverse of removal.

Mazda3

See Figures 482 through 489.

1. Before servicing the vehicle, refer to the precautions section.
2. Relive the fuel system pressure.
3. Disconnect the quick release connector in the engine compartment.
4. Attach a long hose to the disconnect fuel pipe and drain the fuel into a suitable container as follows:

a. Using a jumper wire, ground the Powertrain Control Module (PCM) terminals. If equipped with an immobilizer, ground terminal 1AR, if not equipped with an immobilizer, ground terminal 1AQ. Refer to the illustration for terminal location.

b. Turn the ignition switch the ON position to activate the fuel pump.

✱✱ CAUTION

The fuel pump can be damaged if all fuel is removed from the tank, monitor the hose and stop when no fuel is be discharged from the tank.

c. Once no more fuel is being discharged, turn the ignition OFF.

5. Disconnect the negative battery cable.
6. Remove the rear seat cushion and remove the pump service hole cover.
7. Disconnect the fuel pump connector and remove the charcoal canister.
8. Lower the exhaust main silencer so the insulator can be removed and remove the insulator.
9. Remove the rear left hand undercover.
10. Disconnect the evaporator hose, quick release fuel connector on the fuel line and charcoal canister.
11. Support the fuel tank and remove the strap.
12. Remove the filler pipe bolt, loosen the tie band and pull down on the filler pipe to disconnect the joint hose.
13. Lower and remove the fuel tank.
14. Disconnect the fuel release connector.
15. Use tool 49 F042 001 to remove the pump cap on non–California emission models.
16. Use a brass drift and a hammer to remove the fuel pump retaining ring on California emission models.
17. Remove the fuel pump assembly.

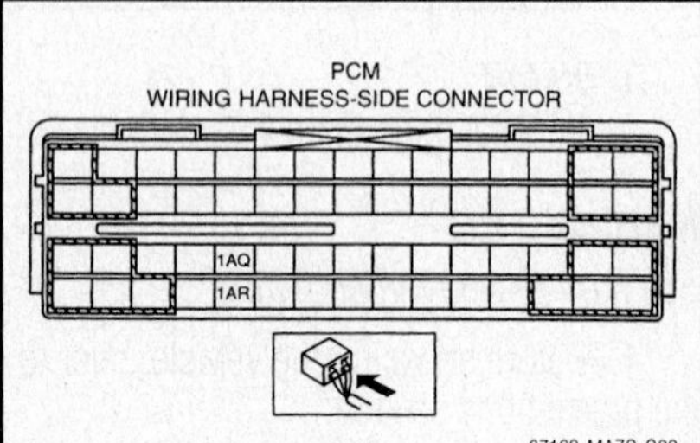

Fig. 482 If equipped with an immobilizer, ground PCM terminal 1AR, if not equipped with an immobilizer, ground PCM terminal 1AQ—Mazda3 models

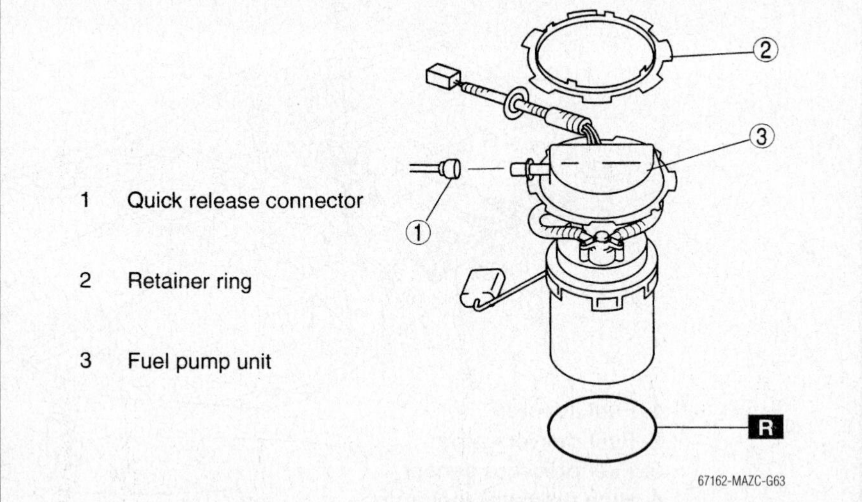

Fig. 483 Exploded view of the pump assembly and related components—Mazda3 non–California emission models

To install:

18. Install the fuel pump assembly.
19. Clean all gasoline from the pump gasket to prevent it from turning during installation.
20. On non–California emission models, align the pump and tank assembly as illustrated.
21. Use tool 49 F042 001 to tighten the pump cap on non–California emission models. Tighten the cap to 59–66 ft. lbs. (80–90 Nm). If the torque cannot be reached, replace the pump cap and gasket. If the torque cannot still be reached, replace the fuel tank.
22. On California emission models, use a brass drift and a hammer to install the fuel pump retaining ring
23. Before installing the tank apply 1.7 inch Hg (5.9 kPa) of pressure to the tank to check for leakage around the pump.

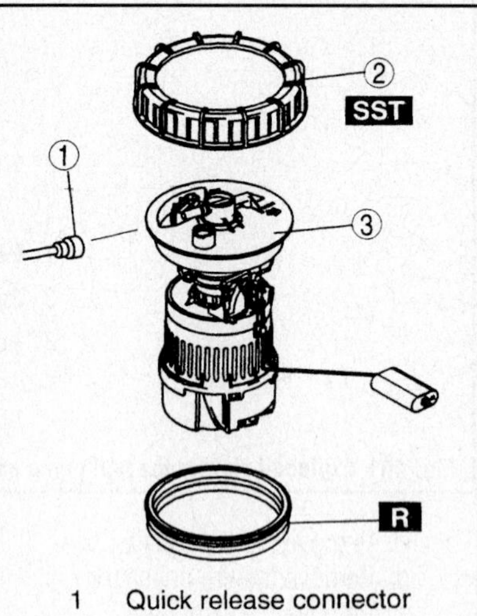

Fig. 484 Exploded view of the pump assembly and related components—Mazda3 California emission models

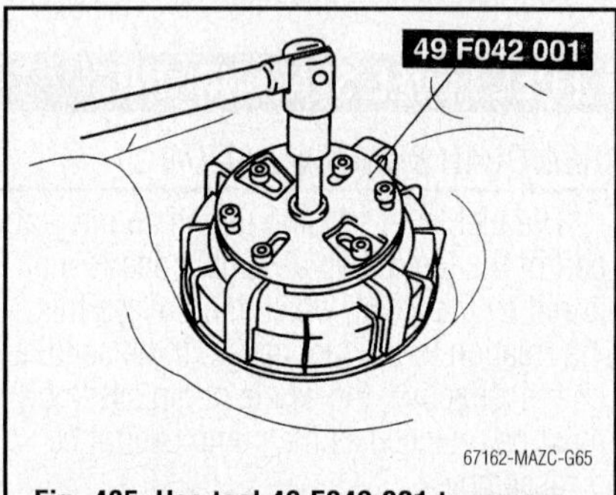

Fig. 485 Use tool 49 F042 001 to remove the pump cap—Mazda3 non–California emission models

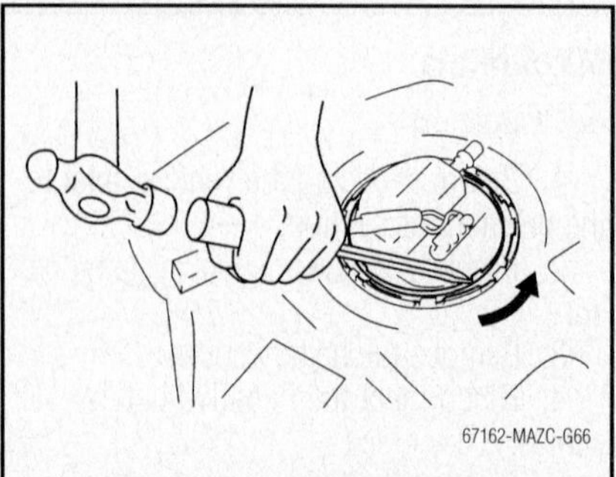

Fig. 486 Use a brass drift and a hammer to remove the fuel pump retaining ring—Mazda3 non–California emission models

24. Connect the fuel release connector.
25. Raise the fuel tank into position.
26. Install the joint hose and align the hose and clamp as illustrated.
27. Install the strap. Tighten the bolts to 16–22 ft. lbs. (22–30 Nm).
28. Connect the evaporator hose, quick release fuel connector on the fuel line and charcoal canister.
29. Install the rear left hand undercover.
30. Install insulator and the exhaust main silencer.
31. Install the charcoal canister and connect the fuel pump connector.
32. Install the pump service hole cover and rear seat.
33. Disconnect the negative battery cable.
34. Start the vehicle and check for leaks as follows:
a. Using a jumper wire, ground the Powertrain Control Module (PCM) terminals. If equipped with an immobilizer, ground terminal 1AR, if not equipped with an immobilizer, ground terminal 1AQ. Refer to the illustration for terminal location.
b. Turn the ignition switch the ON position to activate the fuel pump.
c. Check the hoses, clips and other fuel system components for leaks.
d. If there are any leaks, replace the fuel hoses and clips. If the is damage to the seal on the fuel pipe side, replace the pipe.
e. The system must be leak free for five minutes with the terminal grounded. If any component is replaced because of a system, leak, turn the ignition key OFF; remove the jumper wire from the terminal. reapply the jumper wire, turn the ignition On and check for leaks.

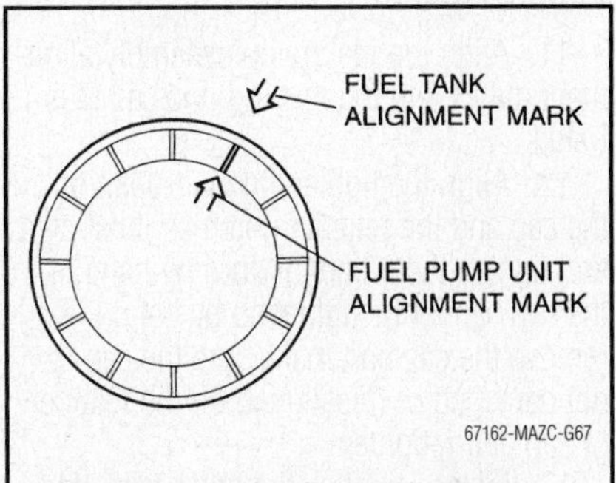

Fig. 487 Align the pump and tank assembly as shown—Mazda3 non-California emission models

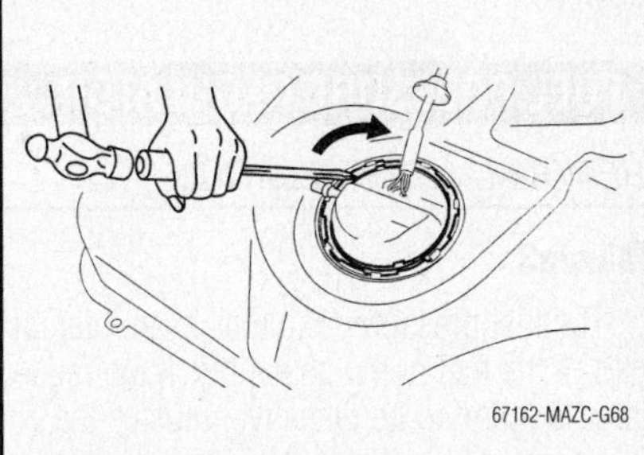

Fig. 488 Use a brass drift and a hammer install the fuel pump retaining ring—Mazda3 non-California emission models

Mazda6 Models

See Figures 490 through 493.

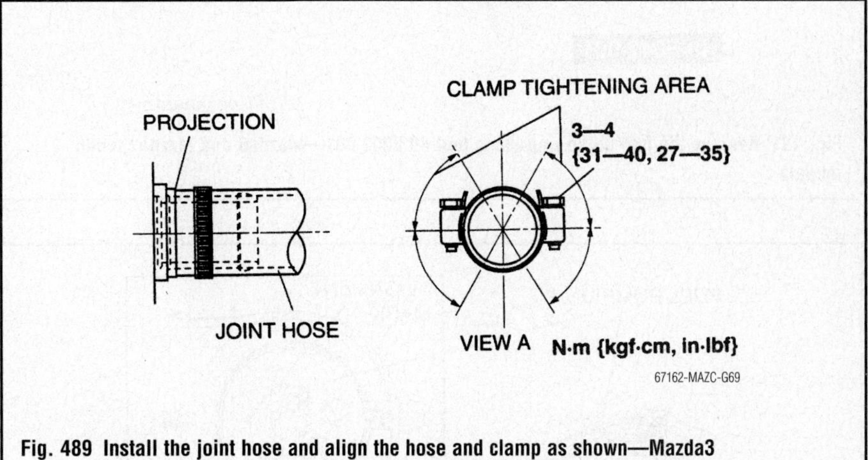

Fig. 489 Install the joint hose and align the hose and clamp as shown—Mazda3

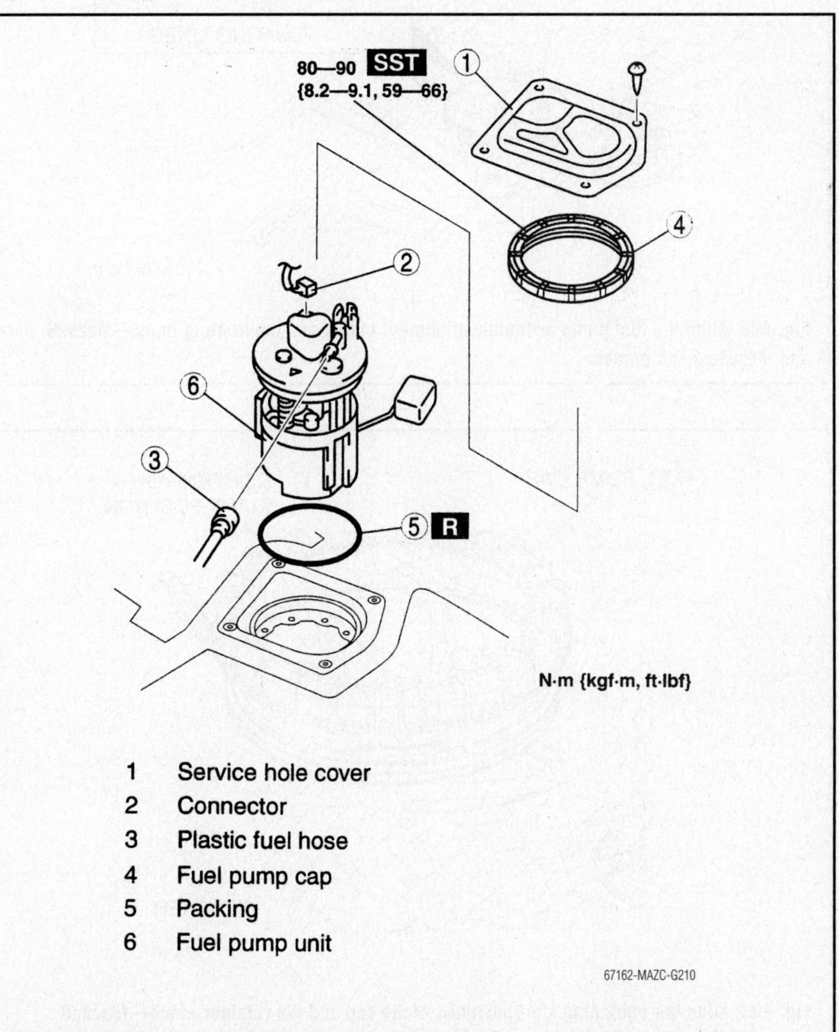

Fig. 490 Exploded view of the fuel pump assembly—Mazda6 and Mazdaspeed6 models

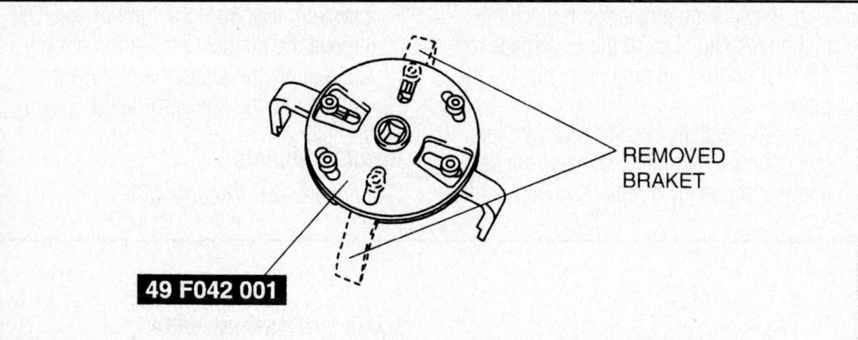

Fig. 491 Remove the fuel pump ring using tool 49 F042 001—Mazda6 and Mazdaspeed6 models

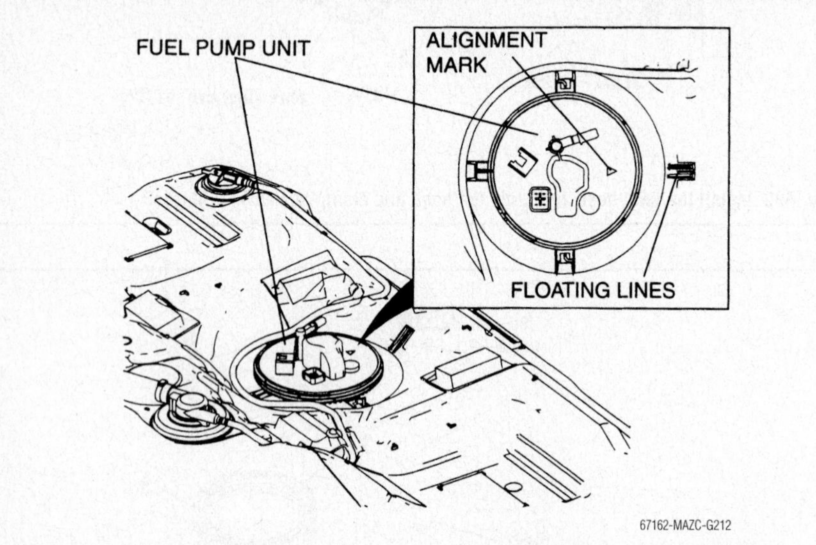

Fig. 492 Align the fuel pump assembly alignment marks and the floating lines—Mazda6 and Mazdaspeed6 models

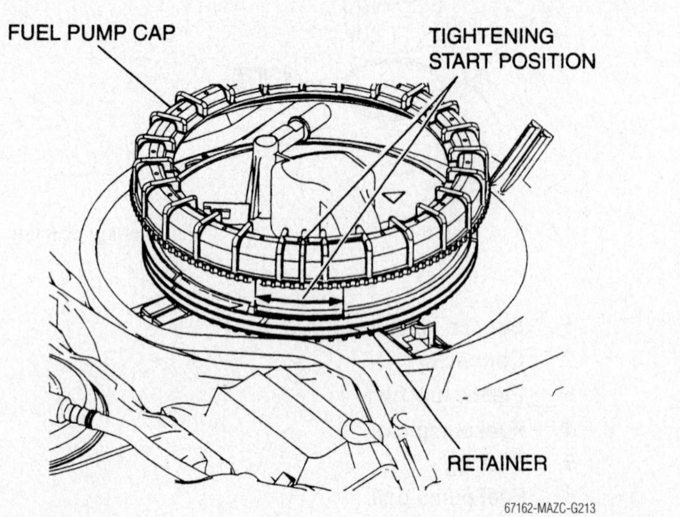

Fig. 493 Align the tightening start positions of the cap and the retainer notch—Mazda6 and Mazdaspeed6 models

1. Before servicing the vehicle, refer to the precautions section.
2. Relieve the fuel system pressure.
3. Disconnect the negative battery cable.
4. Remove the rear seat cushion.
5. Remove the service hole cover.
6. Disconnect the fuel pump electrical connector.
7. Disconnect the hoses from the fuel tank.

➡The fuel pump cap can be damaged if there is a gap between the removal/installation tool 49 F042 001. Make sure the tool is attached securely to the cap and that no gap exists between them.

8. Remove the fuel pump ring using tool 49 F042 001.
9. Remove the fuel pump and gaskets from the fuel tank.

To install:

10. Install the fuel pump using a new gasket.

➡The fuel pump cap can be damaged if there is a gap between the removal/installation tool 49 F042 001. Make sure the tool is attached securely to the cap and that no gap exists between them.

11. Align the fuel pump assembly alignment marks and the floating lines as illustrated.
12. Align the tightening start positions of the cap and the retainer notch as illustrated and tighten them one full turn by hand. If the cap cannot be tightened by hand , remove the cap and make sure the cap is not damaged or misaligned on the retainer or cap and retighten.
13. Make sure the alignment mark and the floating lines are still aligned, tighten the cap to 59–99 ft. lbs. (80–135 Nm) or a rotation angle of 50–140 degrees with the total angle of step 2 and 3 being 410–500 degrees.

FUEL PRESSURE REGULATOR

REMOVAL & INSTALLATION

Mazda3

The fuel pressure regulator is an integral part of the fuel pump assembly. If the regulator is found to be defective, replace the fuel pump assembly as the pump cannot be disassembled. If the fuel pressure exceeds 50.8–59.4 psi (350–410 kPa) the regulator is defective. If the pressure is below this specification, the pump is defective.

MX5-Miata

The fuel pressure regulator is an integral part of the fuel pump assembly. If the regulator is found to be defective, replace the fuel pump assembly as the pump cannot be disassembled. If the fuel pressure exceeds 50.8–59.4 psi (350–410 kPa) the regulator is defective. If the pressure is below this specification, the pump is defective.

Mazda6

2.3L (L3) Engines

1. Before servicing the vehicle, refer to the precautions section.

The fuel pressure regulator is an integral part of the fuel pump assembly. If the regulator is found to be defective, replace the fuel pump assembly as the pump cannot be disassembled. If the fuel pressure exceeds 50.8–59.4 psi (350–410 kPa) the regulator is defective. If the pressure is below this specification, the pump is defective.

3.0L (AJ) Engines

1. Before servicing the vehicle, refer to the precautions section.

The fuel pressure regulator is an integral part of the fuel pump assembly. If the regulator is found to be defective, replace the fuel pump assembly as the pump cannot be disassembled. If the fuel pressure exceeds 63–73 psi (430–510 kPa) the regulator is defective. If the pressure is below this specification, the pump is defective.

Mazdaspeed3 and Mazdaspeed6

1. Before servicing the vehicle, refer to the precautions section.
2. Before servicing the vehicle, refer to the precautions section.

The fuel pressure regulator is an integral part of the fuel pump assembly. If the regulator is found to be defective, replace the fuel pump assembly as the pump cannot be disassembled. If the fuel pressure exceeds 50.8–59.4 psi (350–410 kPa) the regulator is defective. If the pressure is below this specification, the pump is defective.

FUEL RAIL & INJECTORS

REMOVAL & INSTALLATION

CAUTION

Fuel injection systems remain under pressure after the engine has been turned OFF. Properly relieve fuel pressure before disconnecting any fuel lines. Failure to do so may result in fire or personal injury. Do not allow fuel spray or fuel vapors to come in contact with a spark or open flame. Keep a dry chemical fire extinguisher nearby. Never store fuel in an open container due to risk of fire or explosion.

MX5-Miata

See Figures 494 through 496.

1. Before servicing the vehicle, refer to the precautions section.
2. Properly relieve the fuel system pressure.
3. Remove the plug hole plate by lifting off and removing the plug hole plate from the installation areas (rubber and clips) as shown in the illustration.
4. Remove the battery cover.
5. Disconnect the negative battery cable.
6. Disconnect the fuel injector connector and move the harness slightly out of the way.
7. Disconnect the quick release connector.
8. Remove the fuel rail bolts and the injectors as an assembly.
9. Remove the injector clip.

CAUTION

Use of a deformed injector clip will cause the fuel injector to be connected incorrectly and could result in fuel leakage. It will also cause the injector to rotate. Therefore, always replace the clip when the injector is removed.

10. Insert a flathead screwdriver between the injector cup and clip finger.

➡When pushing the clip finger outward, deform the finger until it is removed completely from the cup notch.

11. Push the clip finger outward using a flathead screwdriver.
12. Remove the injector with the clip.
13. Hold the clip using pliers.
14. Pull the clip parallel to the injector groove and remove it from the injector. Discard the clip.

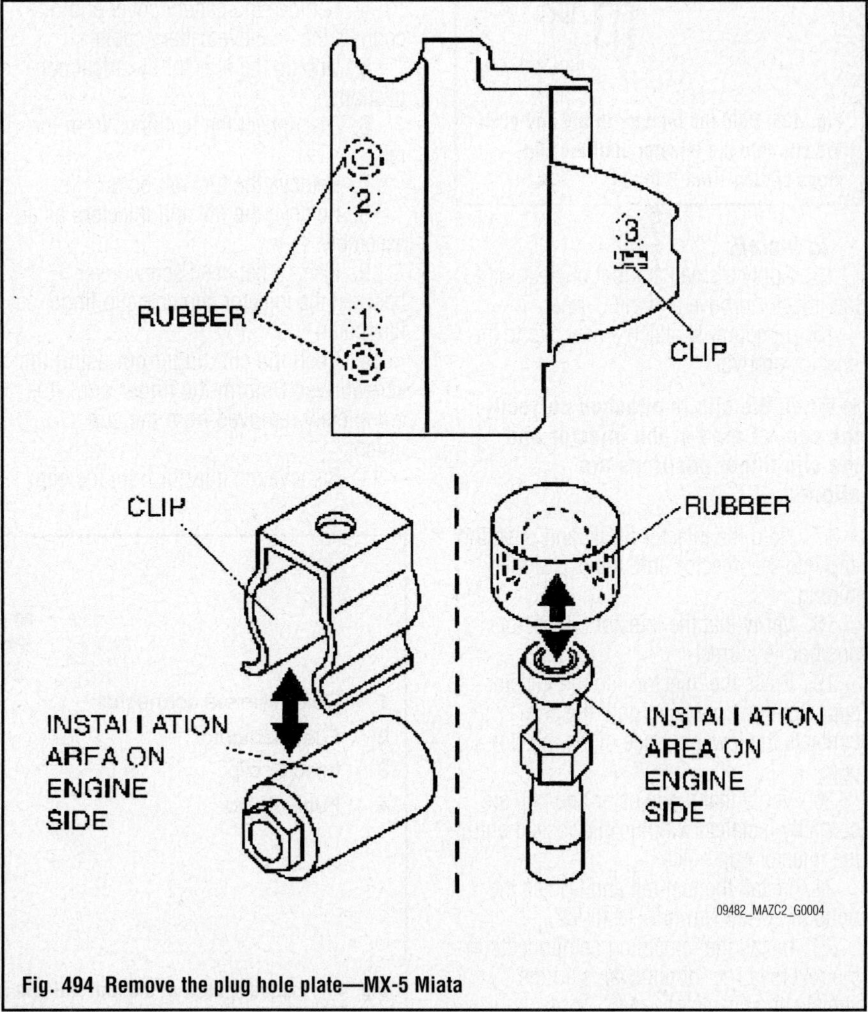

Fig. 494 Remove the plug hole plate—MX-5 Miata

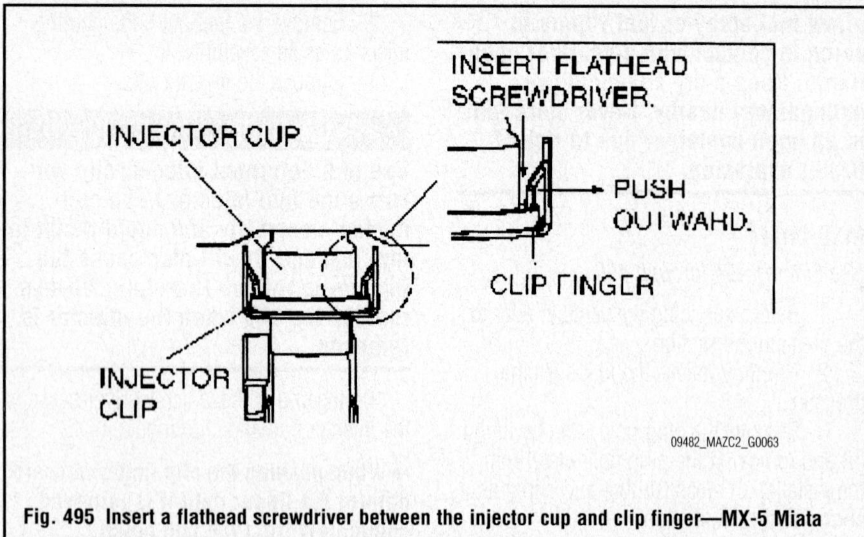

Fig. 495 Insert a flathead screwdriver between the injector cup and clip finger—MX-5 Miata

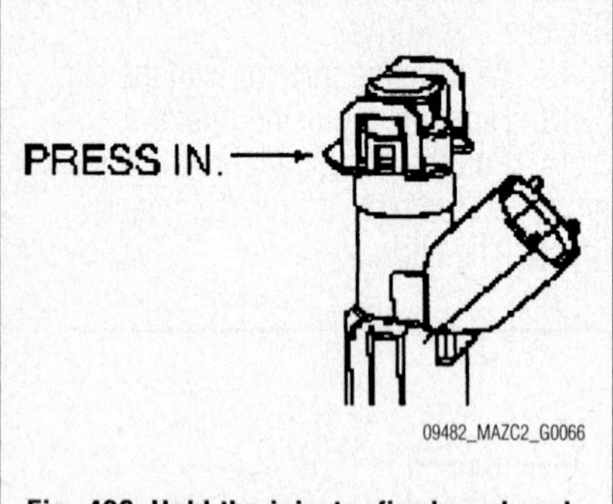

Fig. 496 Hold the injector firmly and push the clip into the injector until the clip stops sliding—MX-5 Miata

To install:

15. Apply a small amount of clean oil to the injector groove and the O-ring.
16. Temporarily attach a new clip to the injector groove.

➡When the clip is attached correctly, the central area of the injector and the clip finger positions are aligned.

17. Hold the injector firmly and push the clip into the injector until the clip stops sliding.
18. Verify that the injector connector position is correct.
19. Press the injector into the injector cup. Continue pressing until the clip contacts the lower surface of the injector cup.
20. Verify that the injector and clip are correctly installed with the clip locked onto the injector cup notch.
21. Install the fuel rail and tighten the bolts to 20–26 Nm (15–19 ft. lbs.).
22. Install the remaining components in the reverse order of removal, start the vehicle and check for leaks.

Mazda3

See Figure 497.

1. Before servicing the vehicle, refer to the precautions section.
2. Properly relieve the fuel system pressure.
3. Remove the plug hole plate.
4. Remove the battery cover and disconnect the negative battery cable.
5. Unplug the injector electrical connections.
6. Disconnect the fuel lines from the rail.
7. Remove the fuel rail bolts.
8. Remove the rail and injectors as an assembly.
9. Insert a flat head screwdriver between the injector cup and clip finger as illustrated.
10. Push the clip finger out using the screwdriver. Deform the finger until it is completely removed from the cup notch.
11. Remove the injector from the clip.
12. Use pliers to pull the clip parallel to the injector groove and remove the clip from the injector. Discard the clip.

To install:

13. Use new O–rings and coat them with clean engine oil.
14. Temporarily attach a new clip to the injector groove.

➡When the clip is correctly installed, the central area of the injector and clip finger positions are aligned.

15. While firmly holding the injector, push the clip into the injector until the clip stops. Make sure the injector connector is correctly positioned.
16. Press the injector into the cup until the cup contacts the lower surface of the cup. Make sure the injector and clip are installed properly and the clip is hooked into the injector cup notch.
17. Install the injectors and fuel rail as an assembly.
18. Install the fuel rail bolts and tighten to 15–19 ft. lbs. (20–26 Nm).
19. Attach the fuel lines to the rail.
20. Start the vehicle and check for leaks as follows:
    a. Using a jumper wire, ground the Powertrain Control Module (PCM) terminals. If equipped with an immobilizer, ground terminal 1AR, if not equipped with an immobilizer, ground terminal 1AQ. Refer to illustration 67162-mazc-g22 for terminal location.
    b. Turn the ignition switch the ON position to activate the fuel pump.
    c. Check the hoses, clips and other fuel system components for leaks.
    d. If there are any leaks, replace the fuel hoses and clips. If the is damage to the seal on the fuel pipe side, replace the pipe.
    e. The system must be leak free for five minutes with the terminal grounded.

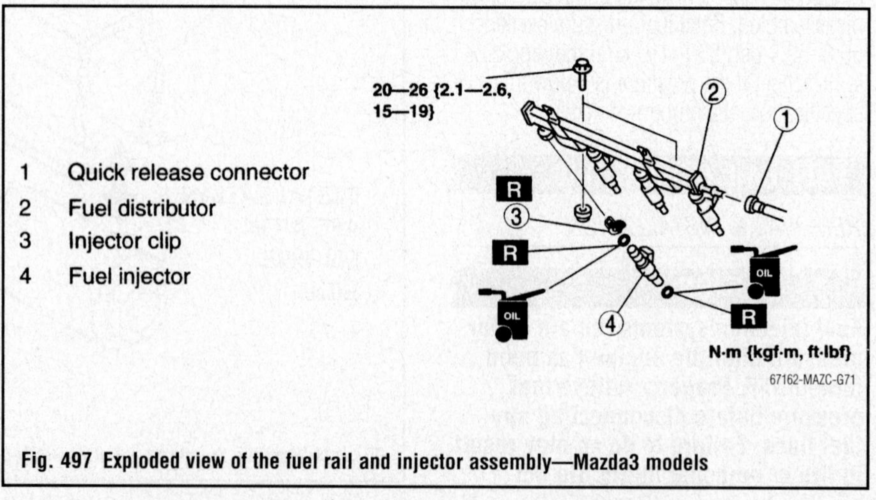

Fig. 497 Exploded view of the fuel rail and injector assembly—Mazda3 models

If any component is replaced because of a system, leak, turn the ignition key OFF; remove the jumper wire from the terminal. reapply the jumper wire, turn the ignition On and check for leaks.

21. Install the remaining components.

Mazda6

2.3L (L3) Engines

See Figures 498 through 501.

1. Before servicing the vehicle, refer to the precautions section.
2. Properly relieve the fuel system pressure.
3. Disconnect the negative battery cable.
4. Unplug the injector electrical connections.
5. Disconnect the fuel lines from the rail.
6. Remove the fuel rail bolts.
7. Remove the rail and injectors as an assembly.
8. Insert a flat head screwdriver between the injector cup and clip finger as illustrated.
9. Push the clip finger out using the screwdriver. Deform the finger until it is completely removed from the cup notch.
10. Remove the injector from the clip.
11. Use pliers to pull the clip parallel to the injector groove and remove the clip from the injector. Discard the clip.

To install:

12. Use new O–rings and coat them with clean engine oil.
13. Temporarily attach a new clip to the injector groove.

➡When the clip is correctly installed, the central area of the injector and clip finger positions are aligned.

14. While firmly holding the injector, push the clip into the injector until the clip stops. Make sure the injector connector is correctly positioned.
15. Press the injector into the cup until the cup contacts the lower surface of the cup. Make sure the injector and clip are installed properly and the clip is hooked into the injector cup notch.
16. Install the injectors and fuel rail as an assembly.
17. Install the fuel rail bolts and tighten to 15–19 ft. lbs. (20–26 Nm).
18. Attach the fuel lines to the rail.
19. Start the vehicle and check for leaks as follows:
    a. Using a jumper wire, short the check connector to terminal F/P to a body ground.
    b. Turn the ignition switch the ON position to activate the fuel pump.
    c. Check the hoses, clips and other fuel system components for leaks.
    d. If there are any leaks, replace the fuel hoses and clips. If the is damage to the seal on the fuel pipe side, replace the pipe.
    e. The system must be leak free for five minutes with the terminal grounded. If any component is replaced because of a system, leak, turn the ignition key OFF; remove the jumper wire from the terminal. reapply the jumper wire, turn the ignition On and check for leaks.
20. Install the remaining components.

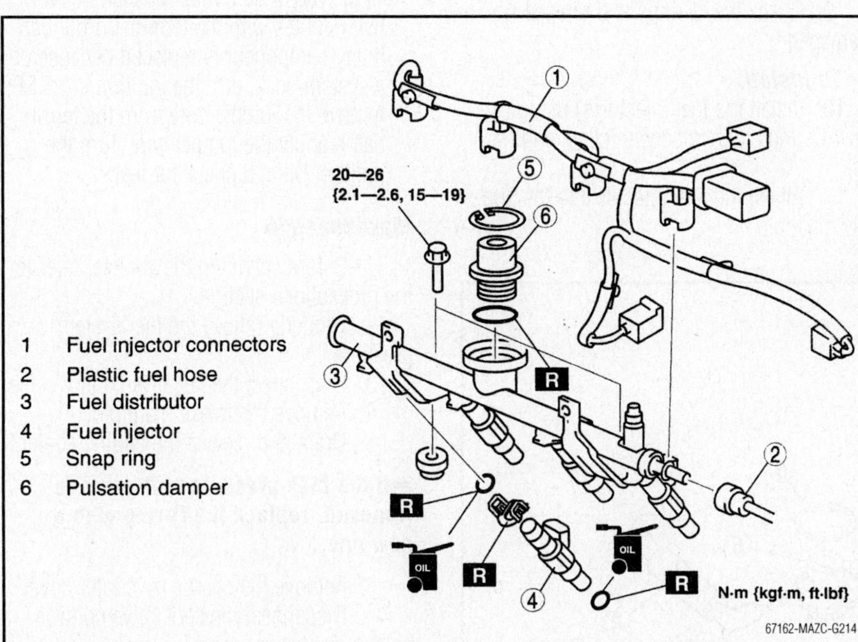

Fig. 498 Exploded view of the fuel rail and injector assembly—Mazda6 models with the 2.3L (L3) engine

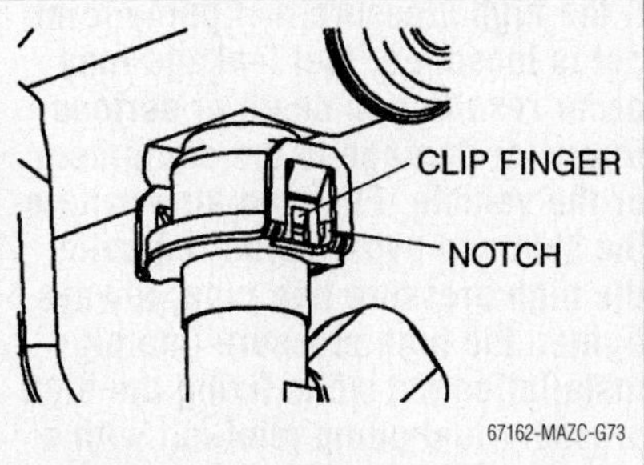

Fig. 499 Deform the finger until it is completely removed from the cup notch—Mazda3 and 6 models with the 2.3L (L3) engine

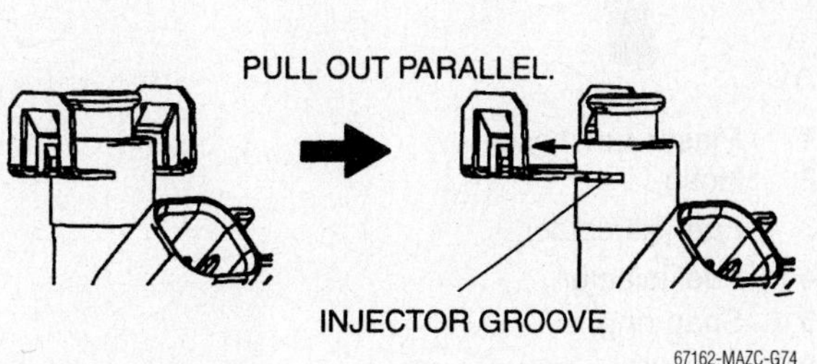

Fig. 500 Use pliers to pull the clip parallel to the injector groove and remove the clip from the injector—Mazda3 and 6 models with the 2.3L (L3) engine

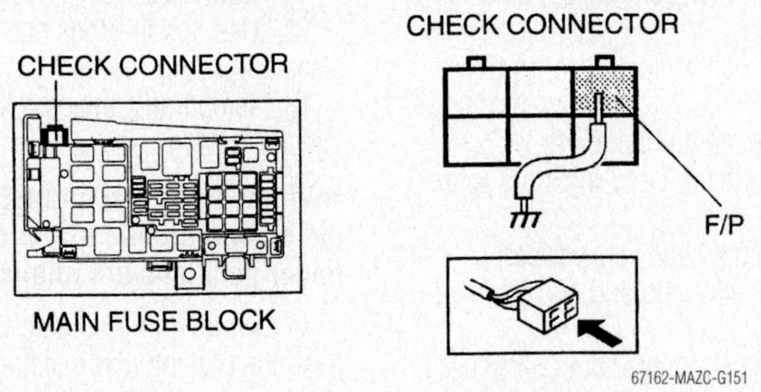

CHECK CONNECTOR
CHECK CONNECTOR
MAIN FUSE BLOCK
F/P
67162-MAZC-G151

Fig. 501 Using a jumper wire, short the check connector to terminal F/P to a body ground—Mazda6 models

3.0L (AJ) Engines

See Figure 502.

1. Before servicing the vehicle, refer to the precautions section.
2. Release the fuel system pressure.
3. Disconnect the negative battery cable.
4. Remove the upper intake manifold (dynamic chamber).
5. Disconnect the fuel injector connectors.
6. Disconnect the fuel hoses from the rail.
7. Remove the rail and injectors as an assembly.
8. Gently twist the fuel injector out of the manifold
9. Check the O-rings and replace if damaged.

To install:

10. Install the fuel injector(s) using new O-rings lubricated with clean engine oil.
11. Attach the fuel injector into the supply manifold.
12. Install the fuel rail and injectors as an assembly. Tighten the rail bolts to 72–101 inch . lbs. (8–11 Nm).
13. Connect the fuel hoses to the rail.
14. Install the injector connectors.
15. Install the upper intake manifold.
16. Connect the negative battery cable.
17. Start the vehicle and check for leaks as follows:
    a. Using a jumper wire, short the check connector to terminal F/P to a body ground.
    b. Turn the ignition switch the ON position to activate the fuel pump.
    c. Check the hoses, clips and other fuel system components for leaks.
    d. If there are any leaks, replace the fuel hoses and clips. If the is damage to the seal on the fuel pipe side, replace the pipe.
    e. The system must be leak free for five minutes with the terminal grounded. If any component is replaced because of a system, leak, turn the ignition key OFF; remove the jumper wire from the terminal. reapply the jumper wire, turn the ignition On and check for leaks.

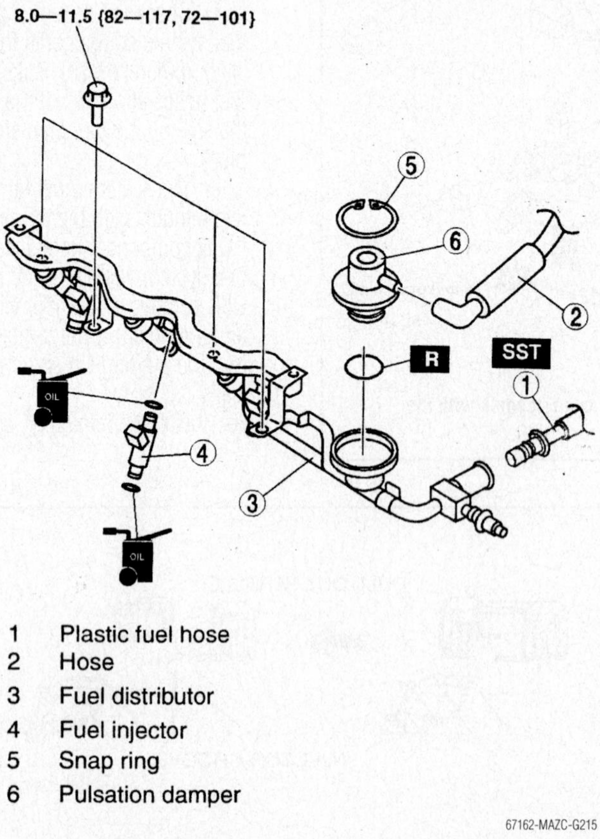

8.0—11.5 {82—117, 72—101}
OIL
R
SST
OIL

1 Plastic fuel hose
2 Hose
3 Fuel distributor
4 Fuel injector
5 Snap ring
6 Pulsation damper

67162-MAZC-G215

Fig. 502 Exploded view of the fuel rail and injector assembly—Mazda6 models with the 3.0L (AJ) engine

Mazdaspeed6

1. Before servicing the vehicle, refer to the precautions section.
2. Properly relieve the fuel system pressure.
3. Disconnect the negative battery cable.
4. Remove the intake manifold.
5. Drain and recycle the engine coolant.

➡If the high pressure fuel pump is removed, replace the O-ring with a new one.

6. Remove the charge air cooler cover.
7. Disconnect the spill valve control solenoid valve connector.
8. Disconnect the quick release connector on the high pressure fuel pump.
9. Remove the battery and battery tray.

WARNING

If the high pressure fuel pump joint nut is loosened, fuel leakage may occur resulting in death or serious injury, or damage to the equipment or the vehicle. Fuel can also irritate the skin and eyes. When removing the high pressure line pipe, always tighten the high pressure line pipe installation nut while fixing the high pressure fuel pump joint nut with a wrench. If the high pressure fuel pump joint nut has rotated, replace the high pressure fuel pump with a new one.

10. Disconnect the high pressure line pipe of the high pressure fuel pump.
11. Fix the joint nut with a wrench on the high pressure fuel pump side.
12. Loosen the high pressure line pipe installation nut.
13. Disconnect the fuel delivery pipe.
14. Unplug the injector electrical connections.
15. Remove the fuel injector bract.
16. Remove the injectors as follows:
a. Install the service tool onto the fuel injector.

✲✲ CAUTION

If the tool slips while ratcheting up the fuel injector, the fuel injector or surrounding parts could be damaged. Press fit the tool to the fuel injector firmly and operate carefully.

b. When ratcheting up the fuel injector, the fuel injector connector may contact the cylinder head and damage the fuel injector. Ratchet up the fuel injector so that the fuel injector connector does not contact the cylinder head.

➡If fuel injector No.3 contacts the oil separator, cut the tab on the oil separator. Carefully cut the tab so that the oil separator is not deformed or damaged, with no clearance on the mating surfaces of the oil separator and engine. Keep ratcheting the tool so that the fuel injector becomes free enough to ratchet up without using the tool.

✲✲ CAUTION

Do not apply excessive force to the fuel injector connector because the fuel injector could be damaged. Pull out the fuel injector by ratcheting it upright.

c. Verify that there are no gasket in the cylinder heads after removing the fuel injectors.

✲✲ WARNING

If foreign material such as metal shavings penetrates the fuel injector installation hole on the cylinder block, the engine could be damaged. Remove all foreign material and cap the fuel injector installation hole after removing the fuel injector.

d. Clean the fuel injector and around the insertion hole using a vacuum cleaner.

To install:

17. Installation is the reverse of removal, note the following:
a. Use new O–rings and coat them with clean engine oil.
b. Tighten the fuel pipe delivery bolts to 13–16 ft. lbs. (17–23 Nm).
c. Assemble the high pressure line pipe.
d. Fix the joint nut with a wrench on the high pressure fuel pump side. Tighten the high pressure line pipe installation nut to 17–26 ft. lbs. (23–35 Nm).
e. Tighten the water outlet case installation bolts to 13–16 ft. lbs. (18–22 Nm).
f. Install the quick release connector.
18. Verify that the high pressure fuel pump is assembled securely.
19. Drive the vehicle starting from a standstill and brake suddenly five to six times at a low speed.
20. Stop the vehicle and verify from outside the vehicle that there is no fuel leakage around the high pressure fuel pump.

IDLE SPEED

ADJUSTMENT

The idle speed is controlled by the PCM.

THROTTLE BODY

REMOVAL & INSTALLATION

MX-5 Miata

See Figures 503 and 504.

1. Before servicing the vehicle, refer to the precautions section.
2. Properly relieve the fuel system pressure.
3. Remove the battery cover.
4. Disconnect the negative battery cable.
5. Remove the air cleaner cover, this involves first removing the MAF/IAT sensor.
6. Remove the air cleaner element.
7. Remove the air cleaner case.
8. Disconnect the quick connect fitting from the air cleaner hose.
9. Remove the air hose by moving the purge solenoid valve slightly out of the way.
10. Remove the fresh-air duct by first removing the bumper.
11. Drain the cooling system.
12. Remove the throttle body.

To install:

13. Installation is the reverse order of removal, make sure to tighten the throttle body bolts in the sequence shown to 71–100 inch. lbs. (8–11 Nm).
14. Check and adjust all fluid levels, start the vehicle and check for proper operation.

Mazda3

See Figures 505 through 508.

1. Before servicing the vehicle, refer to the precautions section.
2. Drain the cooling system.
3. Relieve the fuel system pressure.
4. Remove the plug hole plate by lifting off and removing the plug hole plate from the areas shown in the accompanying illustration.
5. Remove the battery cover and battery duct.
6. Remove the undercover and disconnect the negative battery cable.
7. Remove the intake air cover, the air hose and the air cleaner element.
8. Remove the resonance chamber and air filter element.
9. Remove the strap and the air cleaner case.
10. Remove the fresh air duct.
11. Remove the throttle body.

To install:

12. Install the throttle body and tighten the bolts to 71–101 inch lbs. (8–11.5 Nm).
13. Install the fresh air duct.
14. Verify the rubber mounts on the battery support bracket are still in place. Insert the air cleaner case into the rubber mounts, using soapy water if necessary to ease installation.
15. Use the strap to secure the shroud panel and the air cleaner case as shown in the accompanying illustration.
16. Install the air cleaner element and the resonance chamber.
17. Install the air cleaner cover.
18. Install the air hose, make sure to align the alignment marks on the throttle body and the air hose.
19. Install the intake air cover.
20. Install the undercover.
21. Connect the negative battery cable.
22. Install the battery duct and cover.
23. Install the plug hole plate.
24. Fill the cooling system.

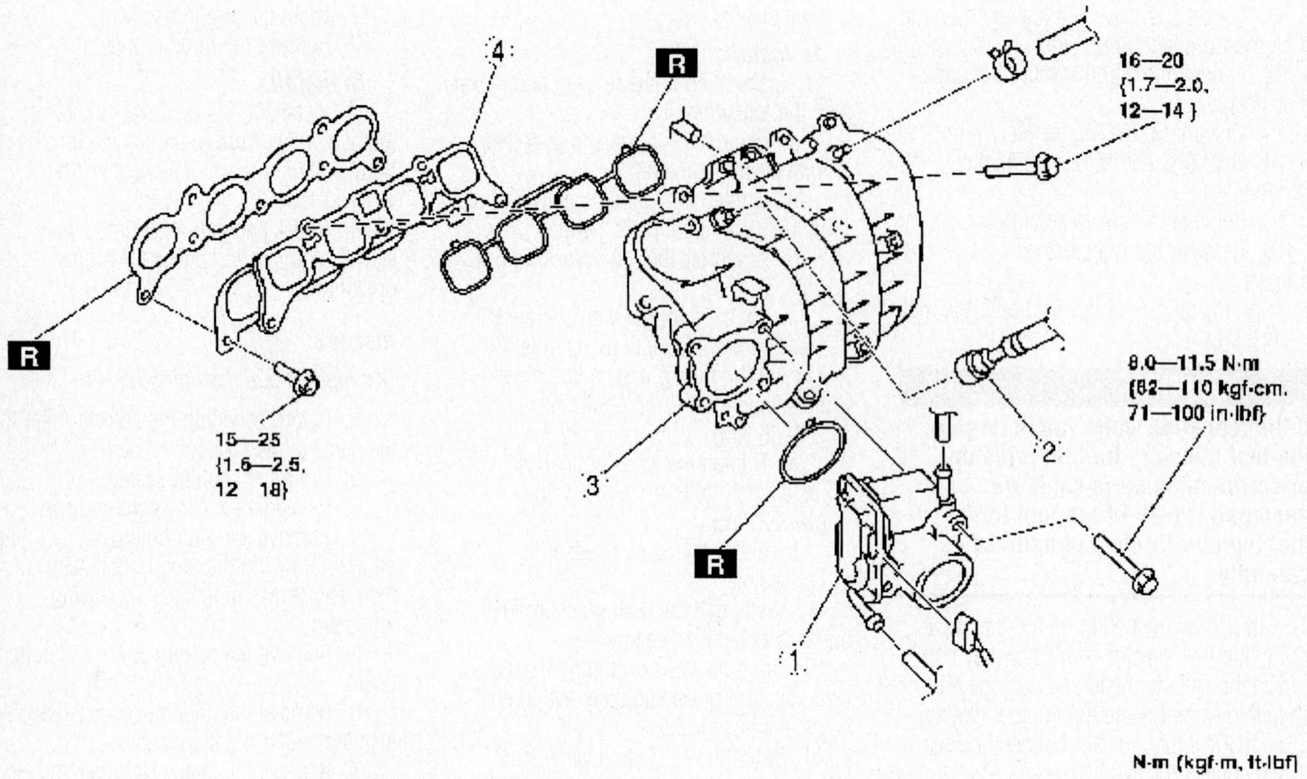

16—20 {1.7—2.0, 12—14}

8.0—11.5 N·m {82—110 kgf·cm, 71—100 in·lbf}

15—25 {1.6—2.5, 12 18}

N·m {kgf·m, ft·lbf}

1 Throttle body
2 Quick release connector (Type A)
3 Dynamic chamber
4 Intake manifold

09482_MAZC2_G0022

Fig. 503 Exploded view of the dynamic chamber and intake manifold—MX-5 Miata

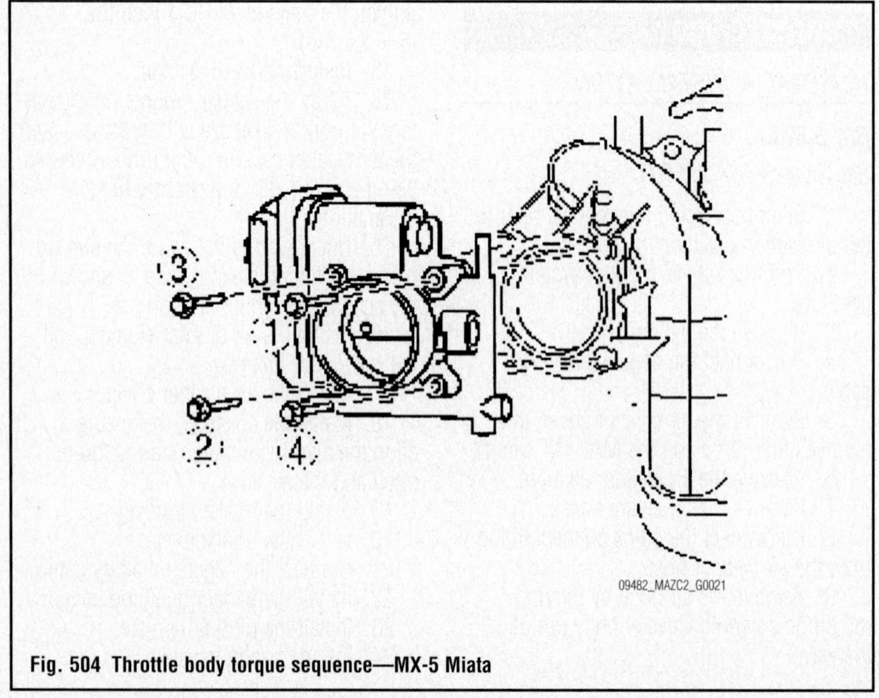

Fig. 504 Throttle body torque sequence—MX-5 Miata

25. Start the vehicle and check for leaks as follows:

a. Using a jumper wire, ground the Powertrain Control Module (PCM) terminals. If equipped with an immobilizer, ground terminal 1AR, if not equipped with an immobilizer, ground terminal 1AQ. Refer to the illustration for terminal location.

b. Turn the ignition switch the ON position to activate the fuel pump.

c. Check the hoses, clips and other fuel system components for leaks.

d. If there are any leaks, replace the fuel hoses and clips. If the is damage to the seal on the fuel pipe side, replace the pipe.

e. The system must be leak free for five minutes with the terminal grounded. If any component is replaced because of a system, leak, turn the ignition key OFF; remove the jumper wire from the terminal. reapply the jumper wire, turn the ignition On and check for leaks.

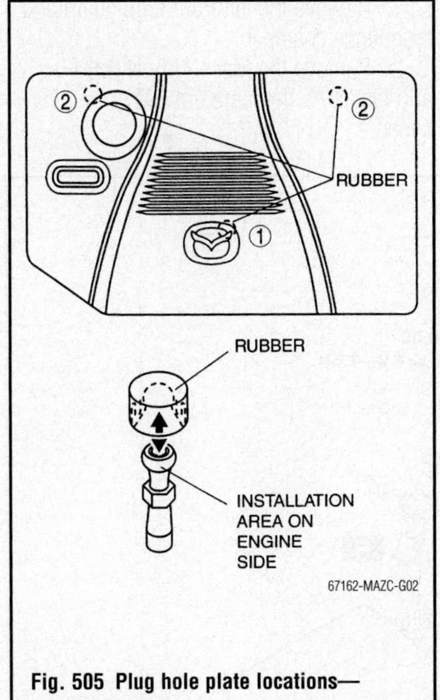

Fig. 505 Plug hole plate locations—Mazda3 models

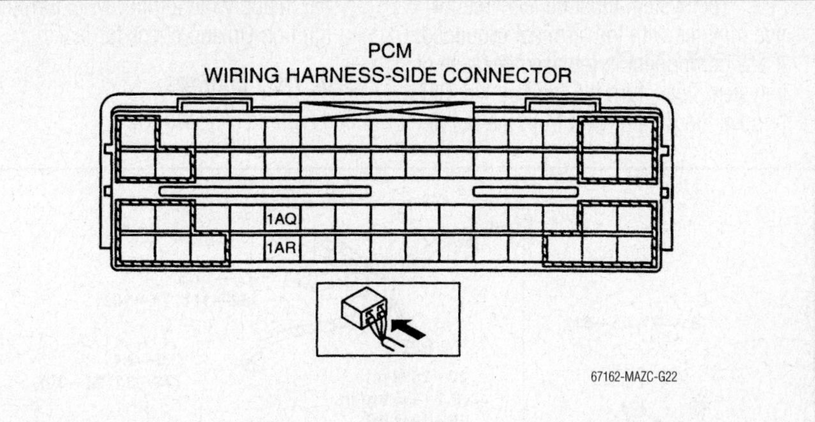

Fig. 508 If equipped with an immobilizer, ground PCM terminal 1AR, if not equipped with an immobilizer, ground PCM terminal 1AQ—Mazda3 models

Mazda6

2.3L (L3) Engines

See Figures 509 and 510.

1. Before servicing the vehicle, refer to the precautions section.
2. Drain the cooling system.
3. Relieve the fuel system pressure.
4. Disconnect the negative battery cable.
5. Remove the intake air cover and element.
6. Remove the air cleaner case.
7. Remove the VAD check valve.
8. Remove the left front mud guard and resonance chamber.
9. Remove the MAF/IAT sensor.
10. Remove the air hose.
11. Remove the water hose.
12. Remove the throttle body.

To install:

13. Install the throttle body and tighten the bolts to 71–101 inch lbs. (8–11.5 Nm).
14. Install the water hose.
15. Install the air hose.
16. Verify the rubber mounts on the air cleaner bracket are still in place. Insert the air cleaner case into the rubber mounts, using soapy water if necessary to ease installation.
17. Install the remaining components in the reverse order of removal.
18. Fill the cooling system.
19. Start the vehicle and check for leaks as follows:
    a. Using a jumper wire, short the check connector to terminal F/P to a body ground.
    b. Turn the ignition switch the ON position to activate the fuel pump.
    c. Check the hoses, clips and other fuel system components for leaks.
    d. If there are any leaks, replace the fuel hoses and clips. If the is damage to the seal on the fuel pipe side, replace the pipe.

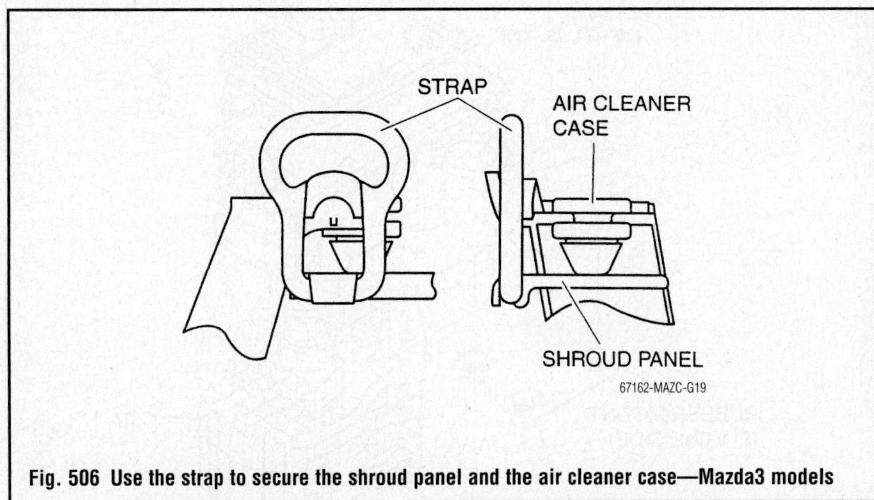

Fig. 506 Use the strap to secure the shroud panel and the air cleaner case—Mazda3 models

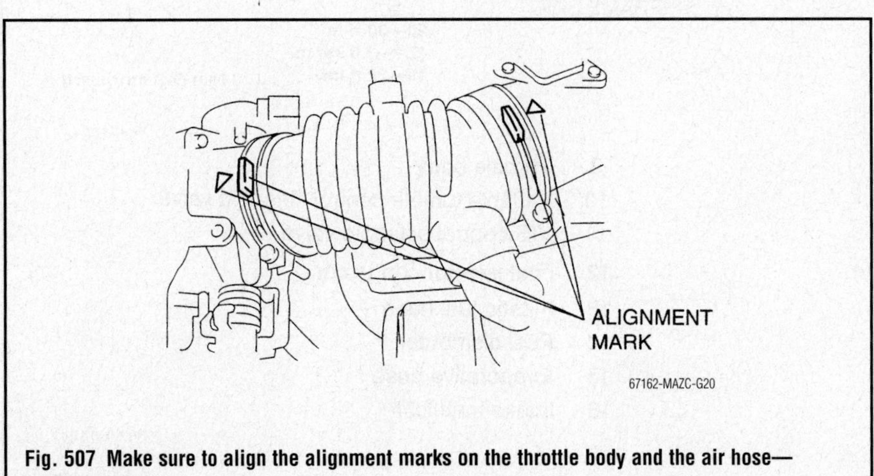

Fig. 507 Make sure to align the alignment marks on the throttle body and the air hose—Mazda3 models

e. The system must be leak free for five minutes with the terminal grounded. If any component is replaced because of a system, leak, turn the ignition key OFF; remove the jumper wire from the terminal. reapply the jumper wire, turn the ignition On and check for leaks.

3.0L (AJ) Engines

See Figures 511 and 512.

1. Remove the left front mud guard and resonance chamber.
2. Remove the Mass Airflow (MAF) and Intake Air Temperature (IAT) sensors.

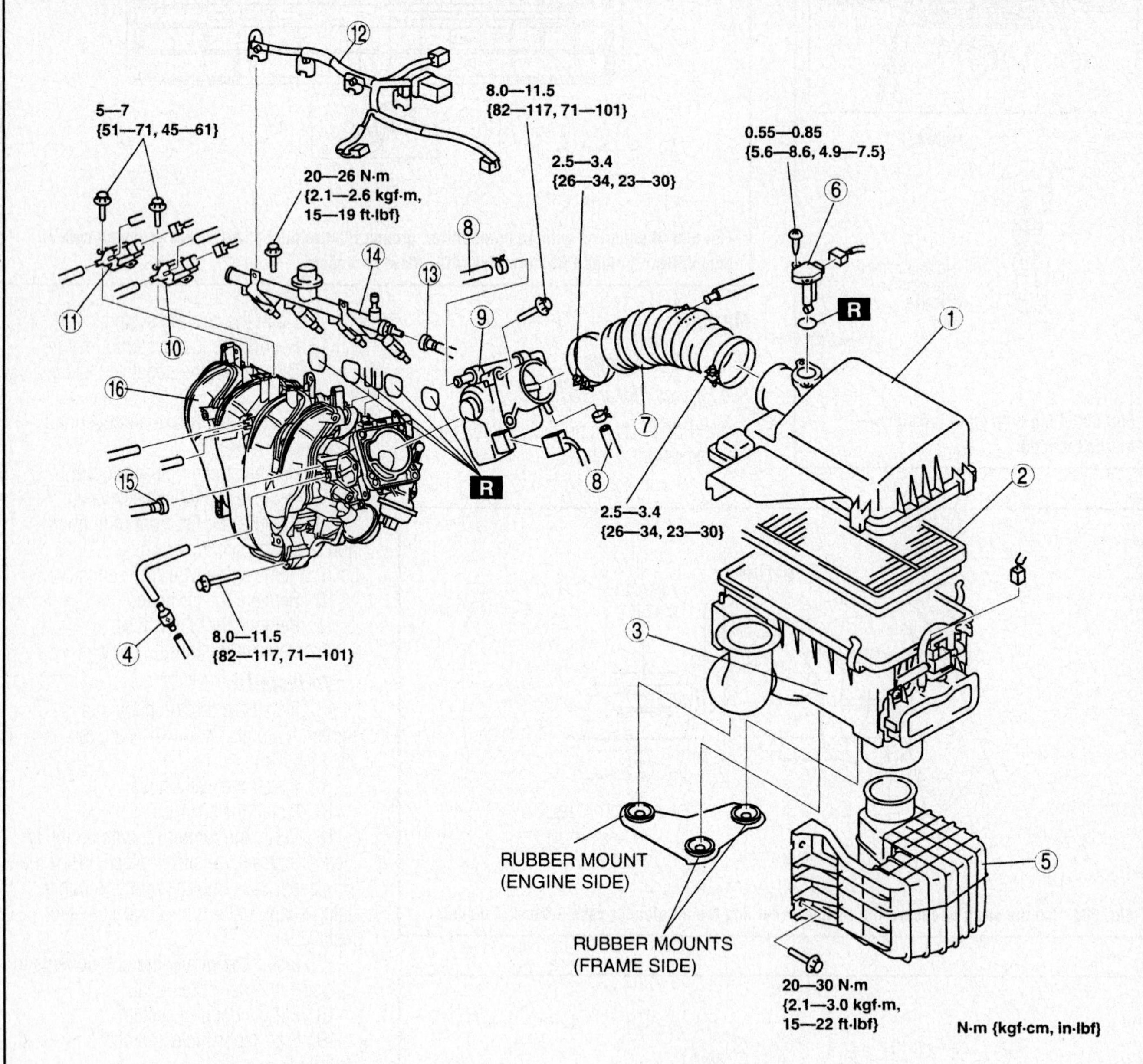

Fig. 509 Exploded view of the intake manifold assembly and related components—Mazda6 with the 2.3L (L3) engine

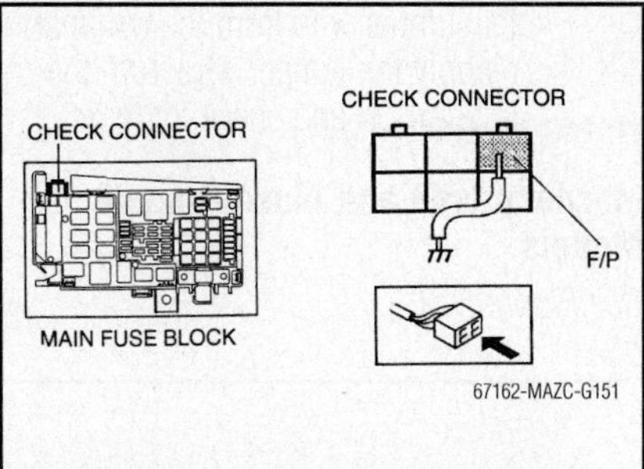

Fig. 510 Using a jumper wire, short the check connector to terminal F/P to a body ground—Mazda6 models

3. Remove the purge solenoid valve vacuum hose.
4. Remove the air hose.
5. Remove the water hose.
6. Remove the throttle body.

To install:

7. Install the throttle body and tighten the bolts to 71–101 inch lbs. (8–11.5 Nm).
8. Fill the cooling system.
9. Start the vehicle and check for leaks as follows:
   - Using a jumper wire, short the check connector to terminal F/P to a body ground.

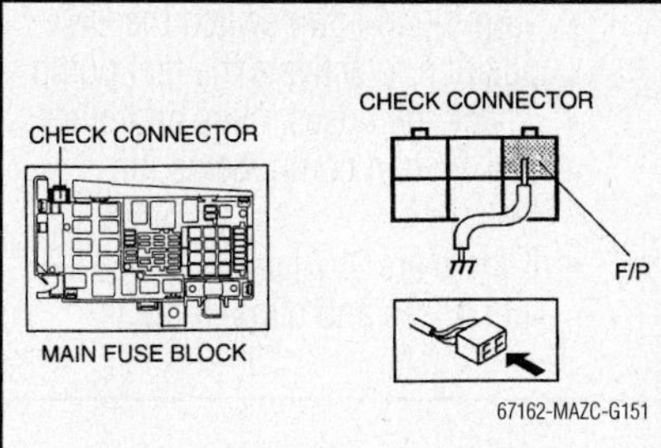

Fig. 512 Using a jumper wire, short the check connector to terminal F/P to a body ground—Mazda6 models

N·m {kgf·cm, in·lbf}

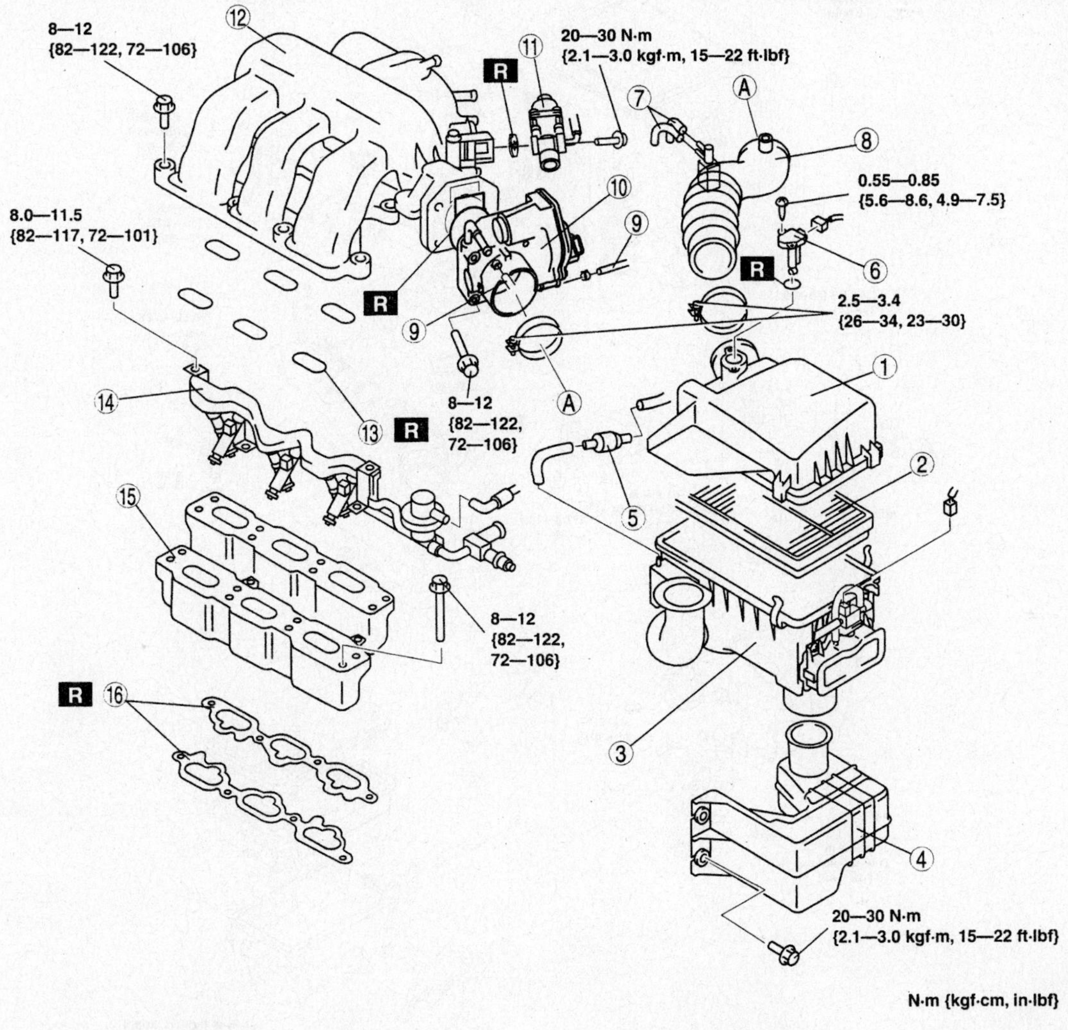

1	Air cleaner cover	9	Water hose
2	Air cleaner element	10	Throttle body
3	Air cleaner case	11	EGR valve
4	Resonance chamber	12	Dynamic chamber
5	VAD check valve (one-way)	13	Dynamic chamber gasket
6	MAF/IAT sensor	14	Fuel distributor
7	Vacuum hose (purge solenoid valve)	15	Intake manifold
8	Air hose	16	Intake manifold gasket

67162-MAZC-G152

Fig. 511 Exploded view of the intake manifold assembly and related components—Mazda6 with the 3.0L (AJ) engine

- Turn the ignition switch the ON position to activate the fuel pump.
- Check the hoses, clips and other fuel system components for leaks.
- If there are any leaks, replace the fuel hoses and clips. If the is damage to the seal on the fuel pipe side, replace the pipe.
- The system must be leak free for five minutes with the terminal grounded. If any component is replaced because of a system, leak, turn the ignition key OFF; remove the jumper wire from the terminal. reapply the jumper wire, turn the ignition On and check for leaks.

Mazdaspeed3 and Mazdaspeed6 Models

See Figure 513.

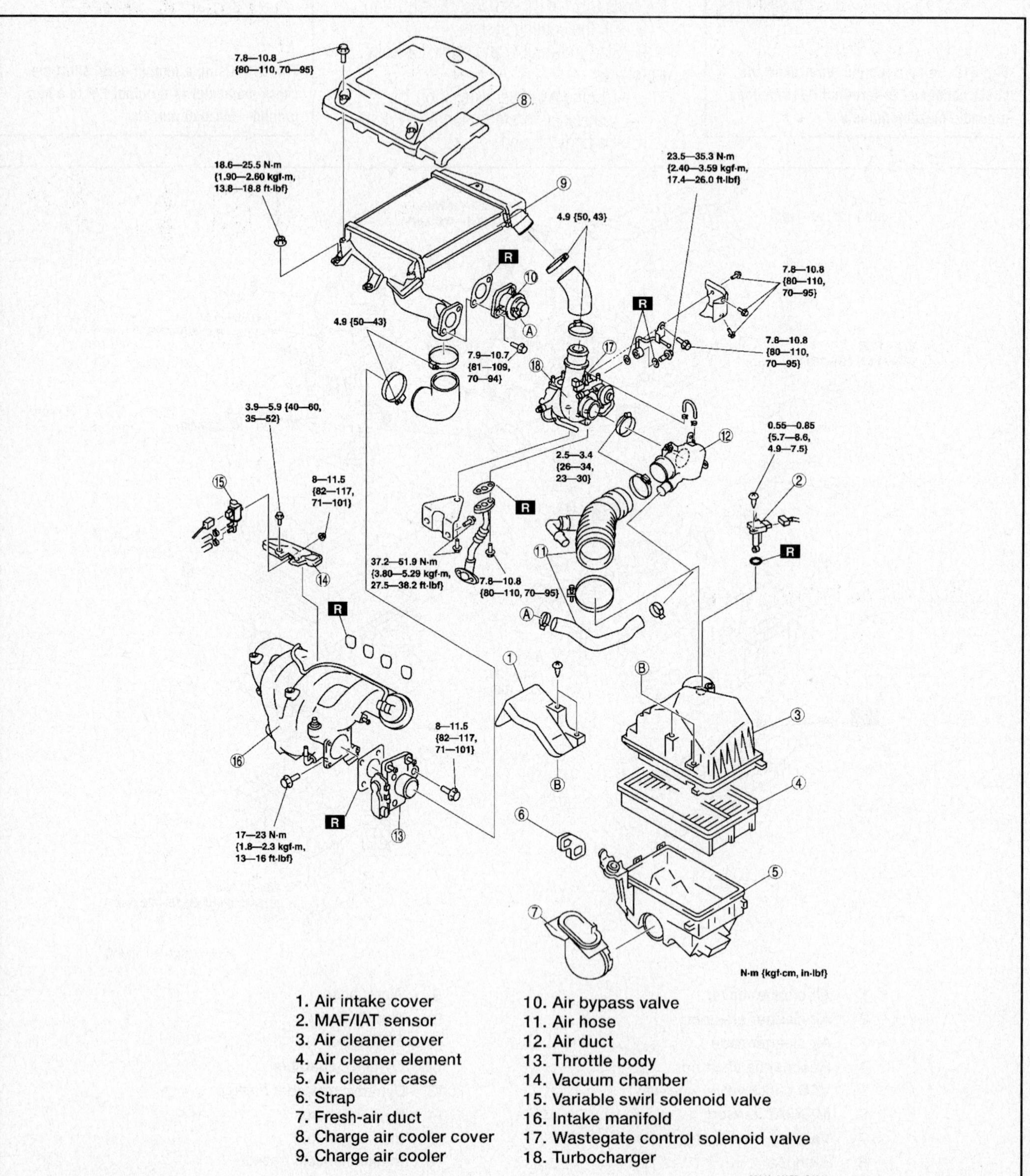

Fig. 513 Exploded view of the intake manifold assembly and related components—Mazdaspeed6 with the 2.3L (L3 with TC) engine

1. Before servicing the vehicle, refer to the precautions section.
2. Disconnect the negative battery cable.
3. Remove the charge air cooler cover.
4. Remove the battery and tray.
5. Remove the air cleaner.
6. Remove the resonance chamber, remove the left hand front mudguard before removing the resonance chamber.
7. Disconnect the MAF/IAT sensor.
8. Remove the duct.
9. Remove the charge air cooler cover.
10. Remove the charge air cooler.
11. Remove the charge air cooler bracket.
12. Remove the air bypass valve.
13. Drain and recycle the engine coolant.
14. Remove the throttle body.

To install:

15. Installation is the reverse of removal, note the following:
    a. Tighten the throttle body installation bolts to 71–101 inch lbs. (8–11.5 Nm).
    b. Perform the following when installing the air cleaner case:
    - Verify that the rubber mounts are set in the air cleaner bracket
    - Install the projections on the frame side.
    - Verify that the projections on the frame side are installed securely.
    - Install the projection on the engine side.
    - Verify that the projection on the engine side installed securely.

HEATING & AIR CONDITIONING SYSTEM

BLOWER MOTOR

REMOVAL & INSTALLATION

Mazda3 and Mazdaspeed3

See Figures 514 through 517.

1. Set the air intake mode to FRESH.
2. Disconnect the negative battery cable.
3. Remove or disconnect the following:
4. Right side front side trim.
5. Right side front scuff plate.
6. Decoration panel.
7. Glove compartment.
8. Passenger junction box and bracket.
9. Cut off the temporary steering stay on the passenger's side.
10. Detach the hood release lever from the lower panel.
11. Remove the front scuff plate.
12. Remove the front side trim.
13. Remove the lower panel.

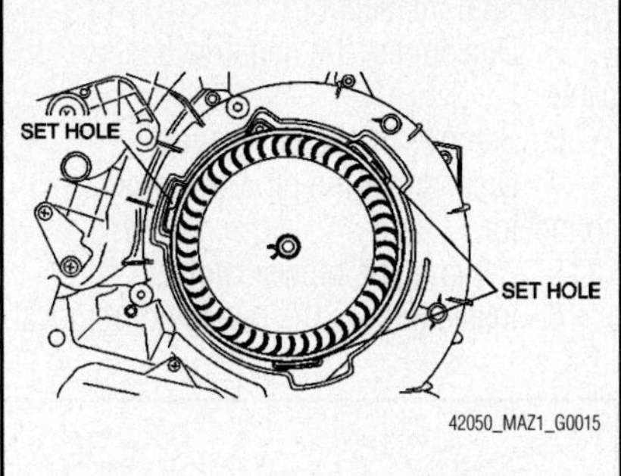

Fig. 515 Install the tool 49 B061 015 to the blower motor—Mazda3

14. Remove the screws and slide the blower case out.
15. Disconnect the air intake actuator connector.
16. Remove the blower case.
17. Remove the air guide.
18. Install the tool 49 B061 015 to the blower motor. Align the SST guide with the sirocco fan clip position and press the tool tabs into the three set holes on the blower motor until they are inserted. Rotate the tool clockwise to lock the tool and blower motor.
19. Disconnect the resistor connector if equipped with a manual air conditioner.
20. Disconnect the blower motor cooling pipe connected to the blower motor.
21. Remove the airflow mode actuator.
22. Disconnect the blower motor connector.
23. Remove the blower motor cover by pulling the lock on the top of the blower motor cover and rotating the blower motor cover.

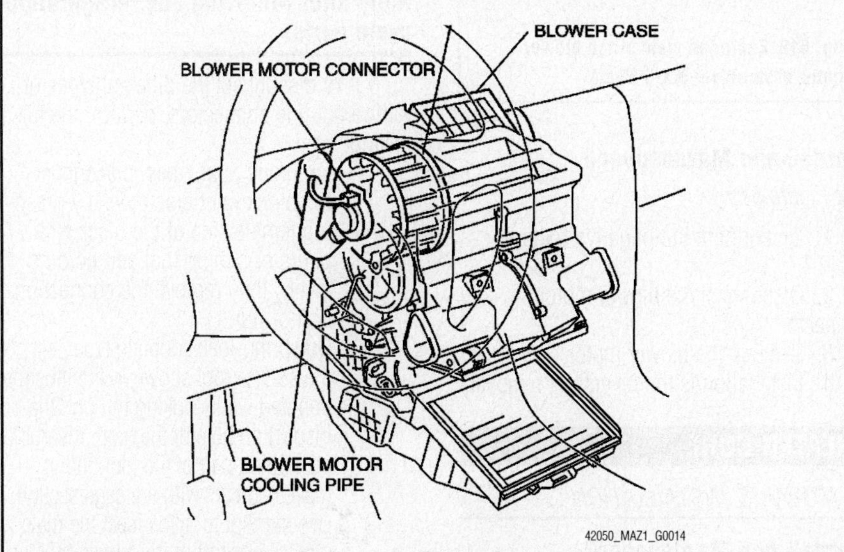

Fig. 514 Exploded view of the blower motor assembly—Mazda3

✱✱ CAUTION

When the blower motor cover is removed, the blower motor could fall in the A/C unit case causing the sirocco fan to be damaged. Therefore another person must hold the blower motor at the installation position. To prevent damage to the sirocco fan, pull the blower motor out being careful that the blower motor does not interfere with the A/C unit.

24. Remove the blower motor by pulling it out.

To install:

25. Installation is the reverse of removal, please note the following:

➡Position the blower motor projection upward and install the blower motor to the A/C unit.

✱✱ CAUTION

To prevent damage to the sirocco fan, install the blower motor being careful that the blower motor does not interfere with the A/C unit. Also, another person must hold the blower motor at the installation position.

- Install the tool to the blower motor.

- Install the blower motor with the tool installed, to the A/C unit.
- Install the blower motor cover from the driver's side. Rotate the blower motor cover until a click is heard.
- Connect the blower motor connector.
- Install the blower motor cooling pipe.
- Remove the tool from the blower motor.

26. If not replacing the blower case, replace the adhesive polyurethane on the fresh-air inlet of the blower case.

✻✻ CAUTION

To adhere new polyurethane properly, be sure to remove the adhesive agent and adhesive polyurethane completely.

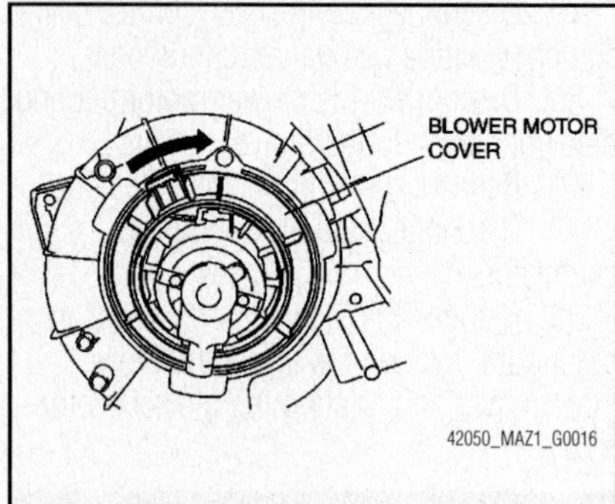

Fig. 516 Rotate the blower motor cover until a click is heard—Mazda3

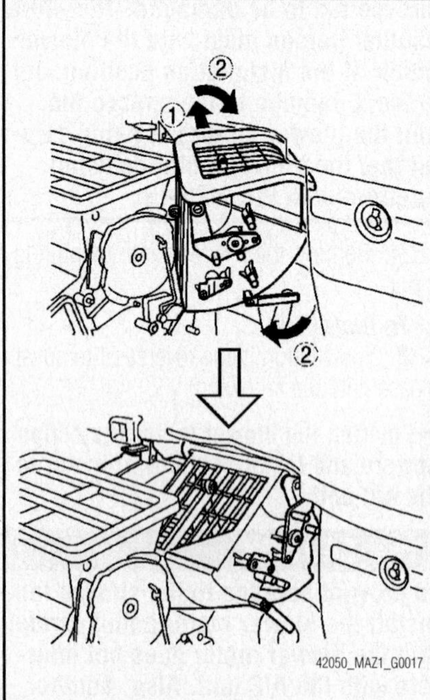

Fig. 517 Install the blower case by inserting and rotating it in the directions of the arrows—Mazda3

➡If the blower case is removed or installed, the adhesive polyurethane can be damaged. Damaged adhesive polyurethane could cause abnormal noise or other malfunctions, therefore replace it.

- Insert the screw into the blower case and install the case to the A/C unit.
- Install the blower case by inserting and rotating it in the directions of the arrows shown in the illustration

MX-5-Miata

See Figure 518.

1. Before servicing the vehicle, refer to the precautions section.
2. Disconnect the negative battery cable.
3. Remove the battery cover.
4. Disconnect the blower motor connector.
5. Remove the blower motor.
6. Installation is the reverse of removal.

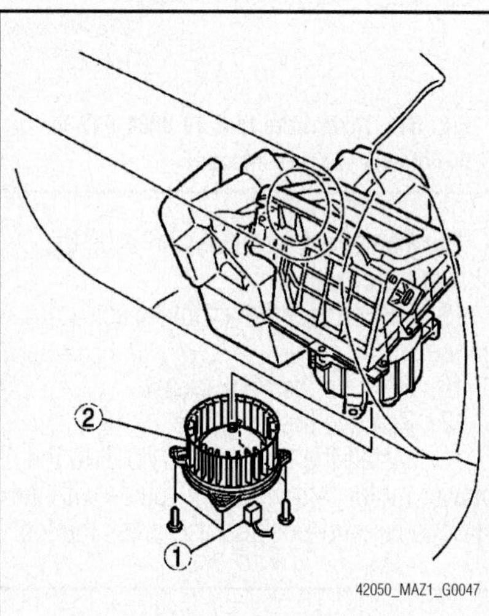

Fig. 518 Exploded view of the blower motor assembly—MX-5 Miata

Mazda6 and Mazdaspeed6

See Figure 519.

1. Disconnect the negative battery cable.
2. Disconnect the blower motor connector.
3. Remove the blower motor.
4. Installation is the reverse of removal.

HEATER CORE

REMOVAL & INSTALLATION

Mazda3 And Mazdaspeed3

See Figures 520 through 546.

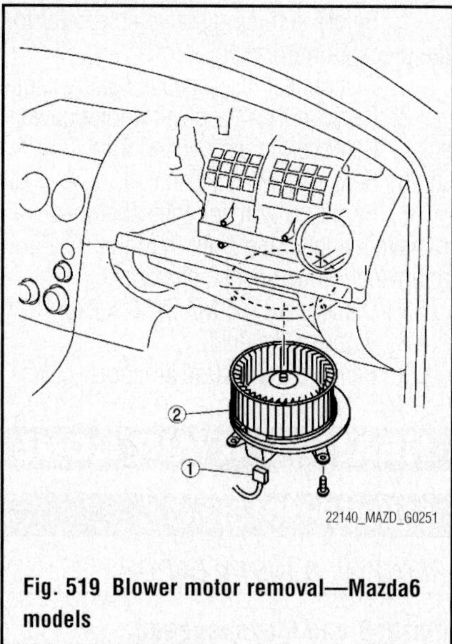

Fig. 519 Blower motor removal—Mazda6 models

1. Before servicing the vehicle, refer to the precautions section.

➡Refer the exploded view illustration for component locations and if applicable, their retainer torque specifications.

2. Remove the battery cover.
3. Disconnect the negative battery cable.
4. Discharge the refrigerant from the system.
5. Drain and recycle the engine coolant.
6. Remove the front doors.

✻✻ CAUTION

If moisture or foreign material enters the refrigeration cycle, cooling ability will be lowered and abnormal noise or other malfunction could occur. Always plug open fittings immediately after removing any refrigeration cycle parts.

7. To disconnect the different types of refrigerant line connectors, perform the following:
   - On block joint types. disconnect the block joint type pipes by grasping female side of the block with pliers or similar tool and holding firmly, then remove the connection bolt or nut.
   - On spring-lock coupling type , set the service tool shown in the illustration and while looking through the inspection hole of the tool , insert the protruding part of the tool until it makes contact with the cage section. Use the tool to disconnect the male pipe or hose from the female by pulling the male pipe or hose.

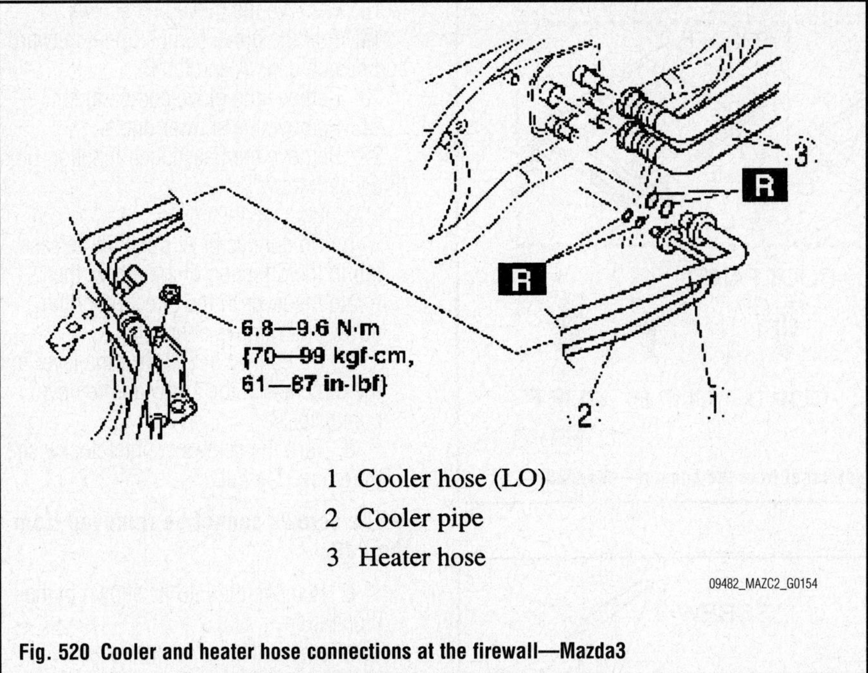

Fig. 520 Cooler and heater hose connections at the firewall—Mazda3

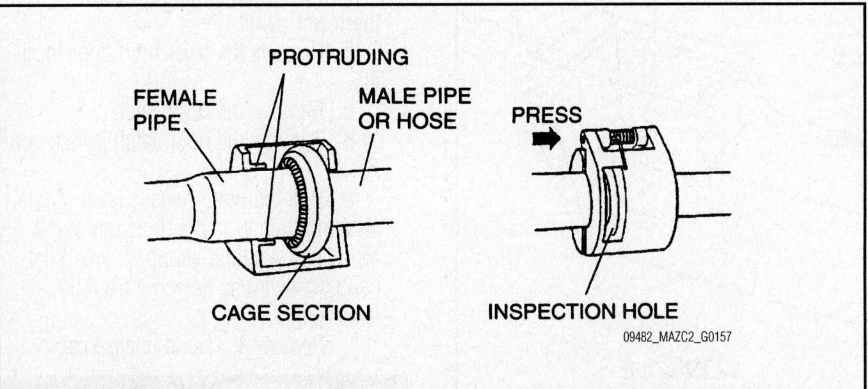

Fig. 521 Looking through the inspection hole of the tool, insert the protruding part of the tool until it makes contact with the cage on spring lock type refrigerant joints—Mazda3

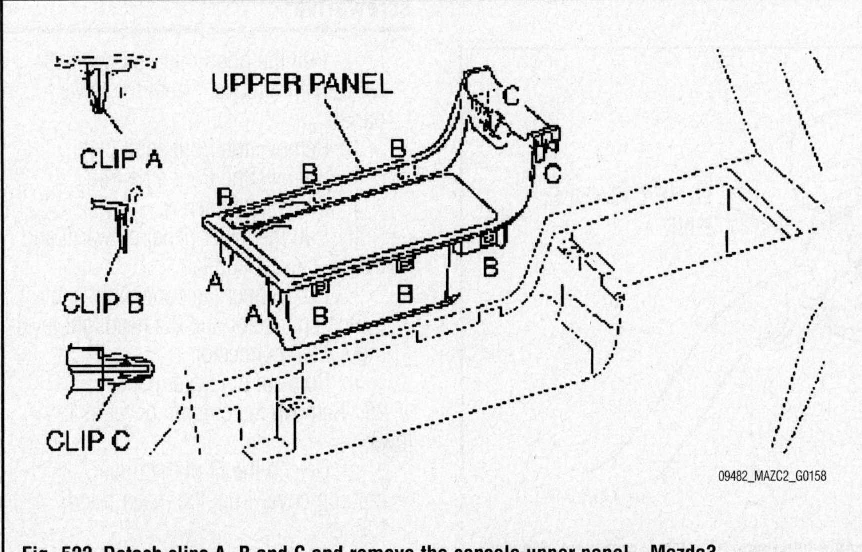

Fig. 522 Detach clips A, B and C and remove the console upper panel—Mazda3

➡**The male pipe or hose can be disconnected easily from the female pipe by pulling from the male pipe or hose while maintaining the pressure of the protruding part of the tool.**

8. Disconnect the heater hose at the firewall.
9. Remove the console as follows:
    - Detach clips A, B and C and remove the upper panel.
    - Detach clips D, E and F and remove the boot panel.
    - Remove screws.
    - Detach clips G.
    - Disconnect the cigarette lighter connector.
    - Remove the ashtray illumination, then remove the ashtray panel.
    - Remove the screws B.
    - Detach tabs H and remove the console.
10. Remove the shift lever component on models with a manual transmission as follows:
    - Remove the battery and battery tray
    - Remove the center console.
    - Remove the front heat insulator.
    - Remove the shift lever knob.
    - Remove the boot panel.
    - Remove the nuts.
    - Remove the nut.
    - Remove the bracket.
    - Remove the seal plate.
    - Remove both the shift cable end and select cable end using a fastener remover.
    - Remove the nuts.
    - Remove the shift lever component.
11. Remove the shift lever component on models with an automatic transmission as follows:
    - Remove the air cleaner component.
    - Remove the console.
    - Remove the front and center heat insulator.
    - Remove the selector lever component connector.
    - Remove the clip and selector cable.
12. Remove the selector lever component.
13. Pull the decoration panel outward and detach the clips A and tab B.
14. Remove the decoration panel.
15. Pull the front scuff plate upward, detach clips A, locator pins B and C from the body, and then remove the front scuff plate.
16. Remove the front side trim fastener.
17. Pull the front side trim in the direction of the arrow and detach clip A and locator pin B.

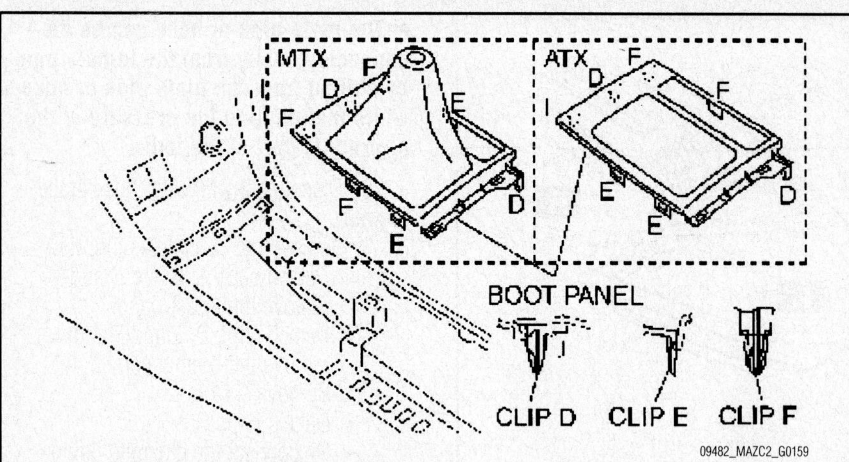

Fig. 523 Detach clips D, E and F and remove the boot panel from the console—Mazda3

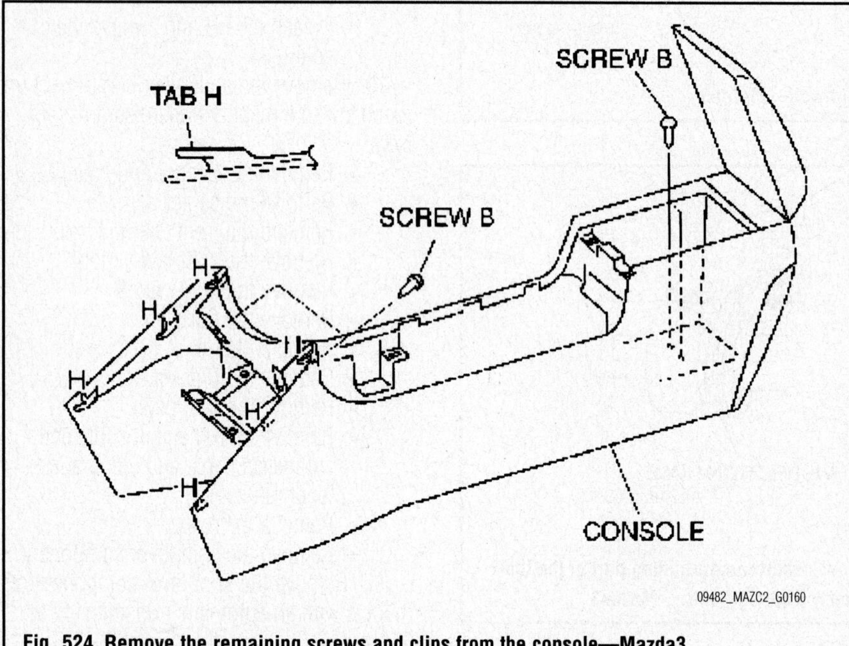

Fig. 524 Remove the remaining screws and clips from the console—Mazda3

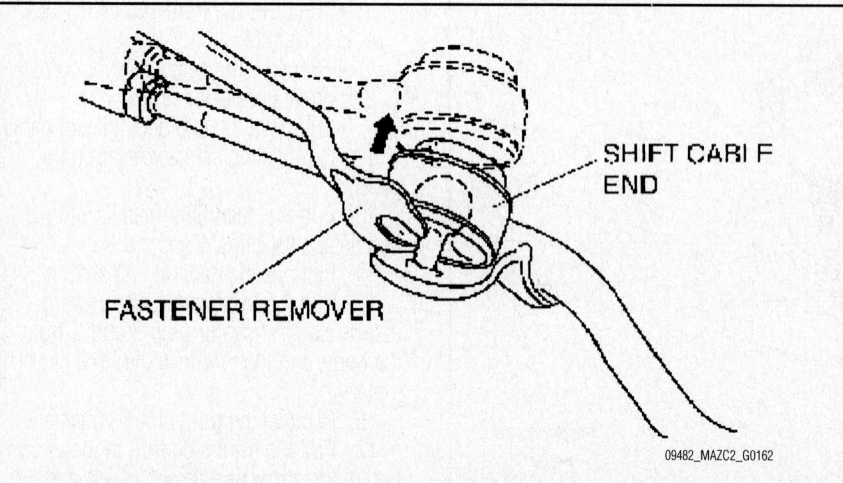

Fig. 525 Remove both the shift cable end and select cable end using a fastener remover—Mazda3

18. Remove the glove box screws.
19. Pull the glove compartment outward and detach clips A and tab B.
20. Remove the glove compartment.
21. Remove the shower ducts.
22. Remove the Passenger Junction Box (PJB) as follows:
   a. Remove the cover.
   b. On connector A, push the release tab in the direction of the arrow, then rotate the lever in the direction of the arrow and remove connector A.
   c. On connector B, rotate the lever in the direction of the arrow and remove connector B.
   d. Turn the screws counterclockwise to remove the PJB.

➡The screws cannot be removed from the PJB.

   e. Remove the PJB as shown in the illustration.
23. Remove the car navigation unit.
24. Remove the lower panel as follows:
   a. Detach the hood release lever from the lower panel.
   b. Remove the lever on 5 door models.
   c. Remove the hood latch.
   d. Remove the hood latch switch connector.
   e. Pull the hood release lever. While pushing the tab in the direction of the arrow using a tape-wrapped, small flathead screwdriver, detach it from the lower panel.
   f. Remove the hood release cable.

✲✲ CAUTION

Be careful not to damage the hood release cable when removing the hood release lever with the flathead screwdriver.

   g. Pull the hood release lever outward, then remove it from the lower panel.
   h. Remove the front scuff plate.
   i. Remove the front side trim.
   j. Remove the screw.
   k. Pull the lower panel outward, and detach clips A and tab.
   l. Disconnect the panel light control switch connector and the headlight leveling switch connector.
   m. Remove the lower panel.
25. Remove the column cover as follows:
   a. Detach the fit of the upper column cover from the meter hood rubber.
   b. Remove the upper column cover.

c. Remove the ignition key illumination.

d. Remove the screws.

e. Remove the lower column cover.

26. Remove the steering shaft as follows:

⁂ WARNING

Handling the air bag module improperly can accidentally operate (deploy) the air bag module, which may seriously injure you.

a. Remove the air bag cover.

b. Remove the bolt.

c. Using a flathead screwdriver, pry out the connector stopper plate and unplug the connector.

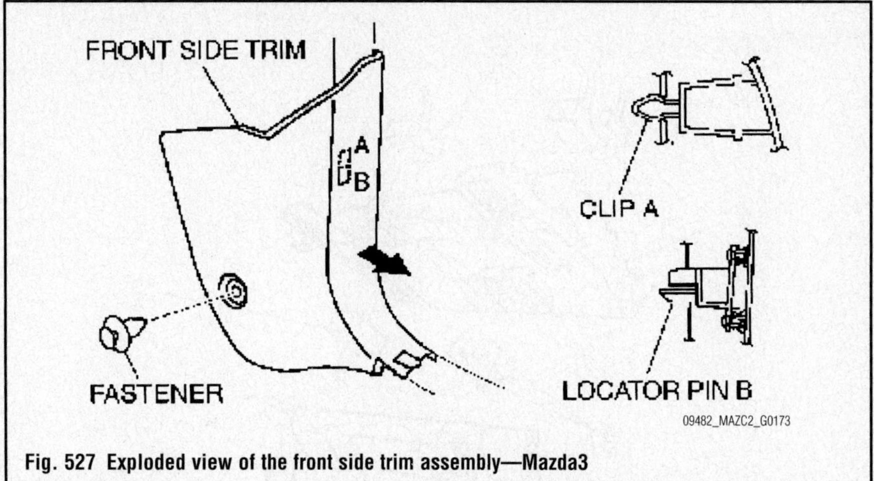

Fig. 527 Exploded view of the front side trim assembly—Mazda3

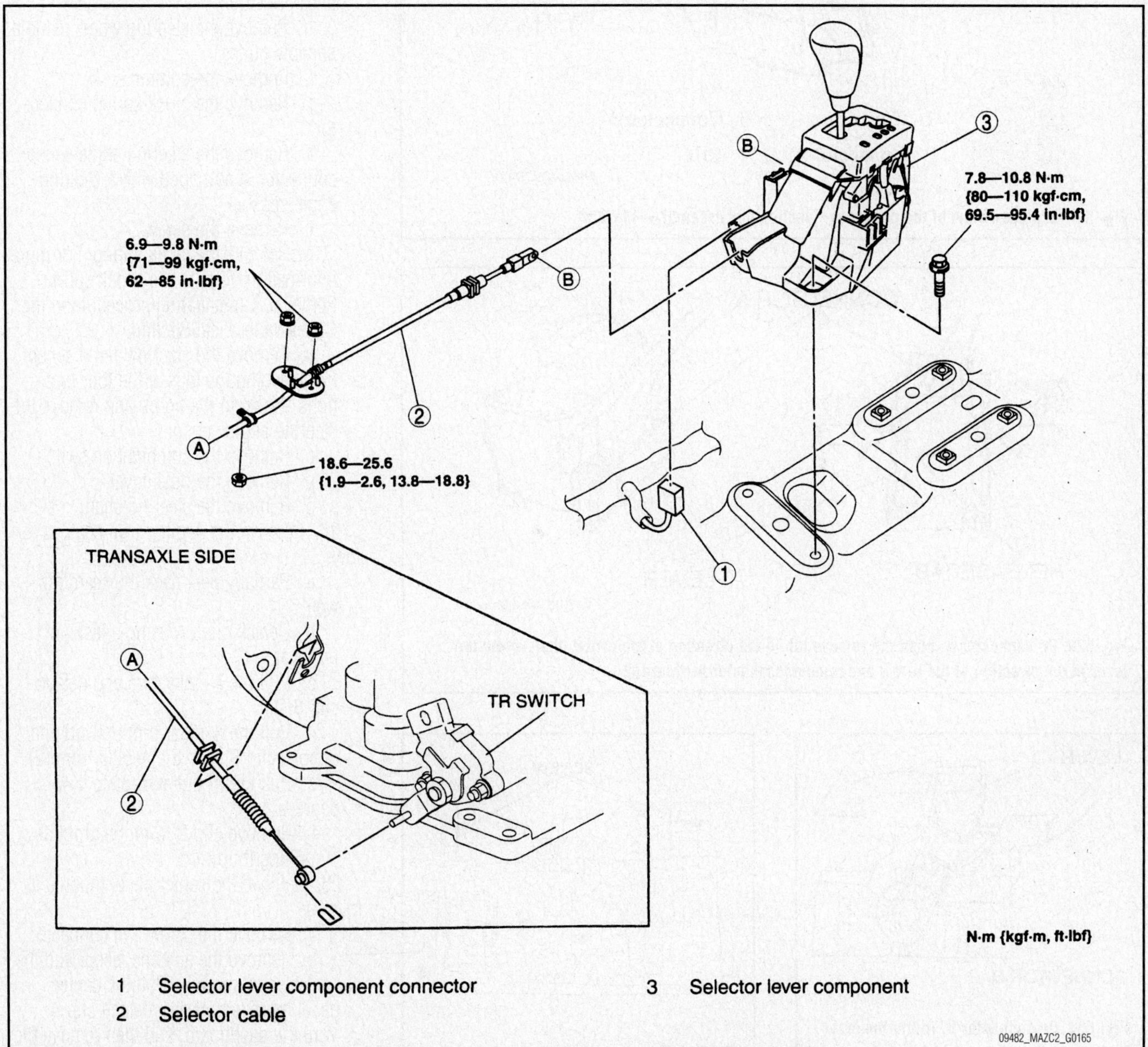

N·m {kgf·m, ft·lbf}

1 Selector lever component connector

2 Selector cable

3 Selector lever component

09482_MAZC2_G0165

Fig. 526 Exploded view of the shift lever component on models with an automatic transmission—Mazda3

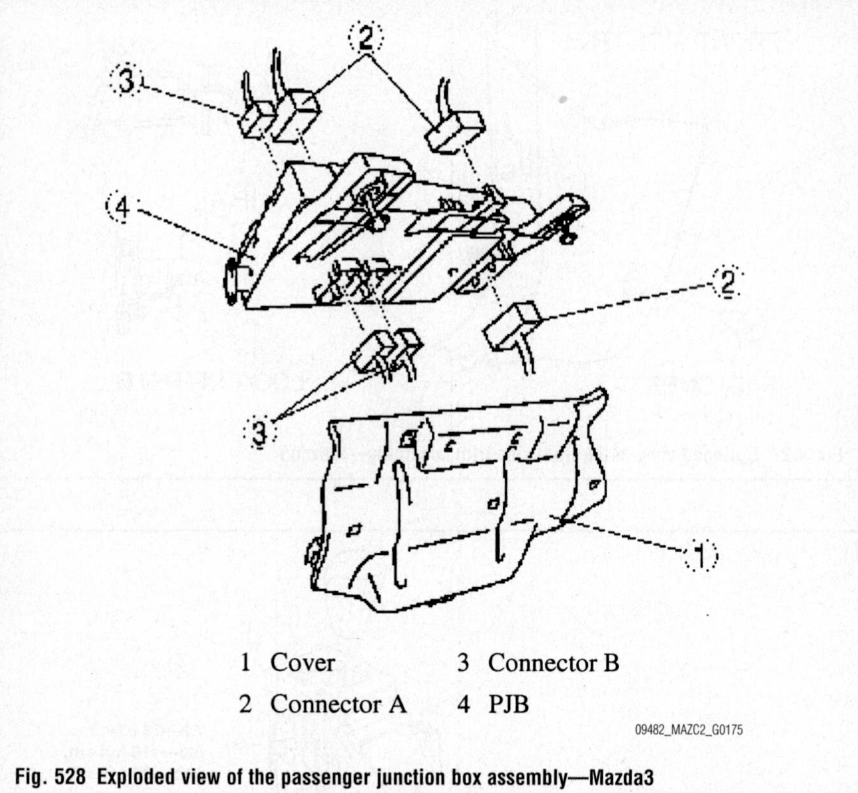

Fig. 528 Exploded view of the passenger junction box assembly—Mazda3

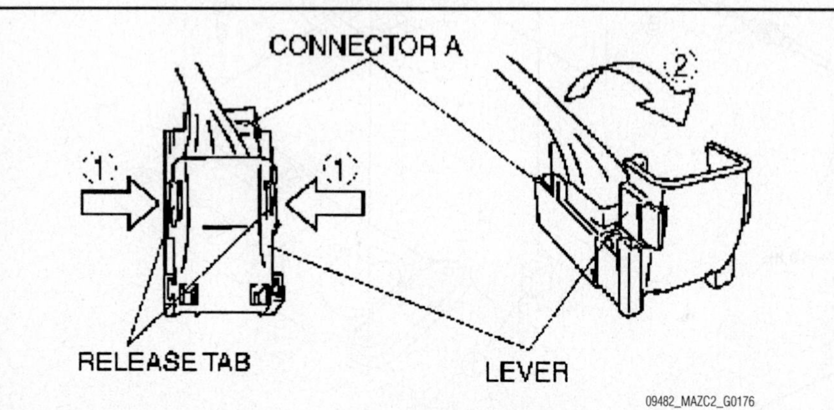

Fig. 529 On connector A, push the release tab in the direction of the arrow, then rotate the lever in the direction of the arrow and remove connector A—Mazda3

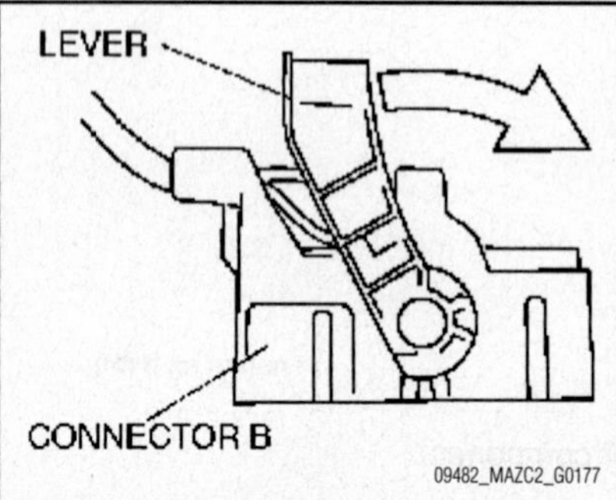

Fig. 530 On connector B, rotate the lever in the direction of the arrow and remove connector B—Mazda3

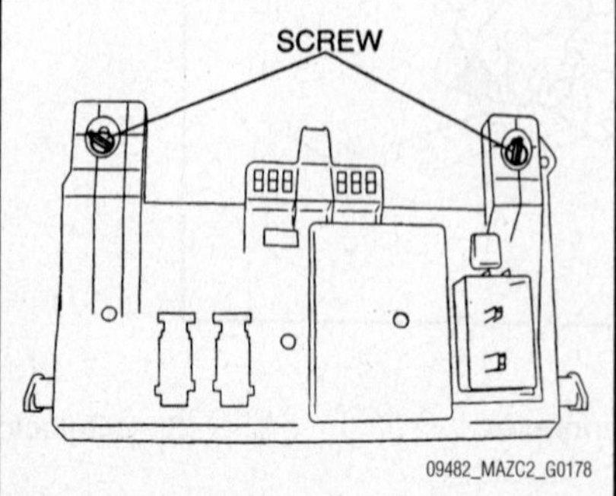

Fig. 531 Turn the screws counterclockwise to remove the PJB—Mazda3

d. Remove the driver's side air bag module.

e. Remove the lockbolt.

⁂ WARNING

Handling the air bag module improperly can accidentally deploy the air bag module, which may seriously injure you.

f. Remove the locknut.

⁂ CAUTION

Do not try to remove the steering wheel by hitting the shaft with a hammer. The column will collapse.

g. Set the vehicle in the straight-ahead position.

h. Remove the steering wheel using a suitable puller.

i. Remove the column cover.

j. Remove the clock spring connector.

k. Remove the steering angle sensor connector, if equipped with a steering angle sensor.

l. Remove the screw.

m. Remove the clock spring. For more information, refer to the Clock spring Removal & Installation procedure in the Chassis Electrical Section.

n. Remove the steering angle sensor by detaching the tabs at the four locations shown in the figure and remove the steering angle sensor.

o. Remove the combination switch.

p. Remove the dust cover.

q. Remove the steering shaft.

27. Remove the A-pillar trim shaft as follows:

a. Partially peel back the seaming welt.

b. Detach clips A using a fastener remover.

c. Pull the A-pillar trim and detach clip B (1).

d. Pull the A-pillar trim upward and remove clip B from the A-pillar trim (2).

e. Pull clip B out and rotate it 45 degrees.

f. Remove clip B from the grommet by pulling it upward.

28. Remove the center panel module as follows:

a. Remove the screw and connector.

b. Remove the antenna feeder and the center panel module. Pull the center panel module outward, detach clip A from the dashboard, and then remove the center panel module.

29. Remove the wiper arm and blade.

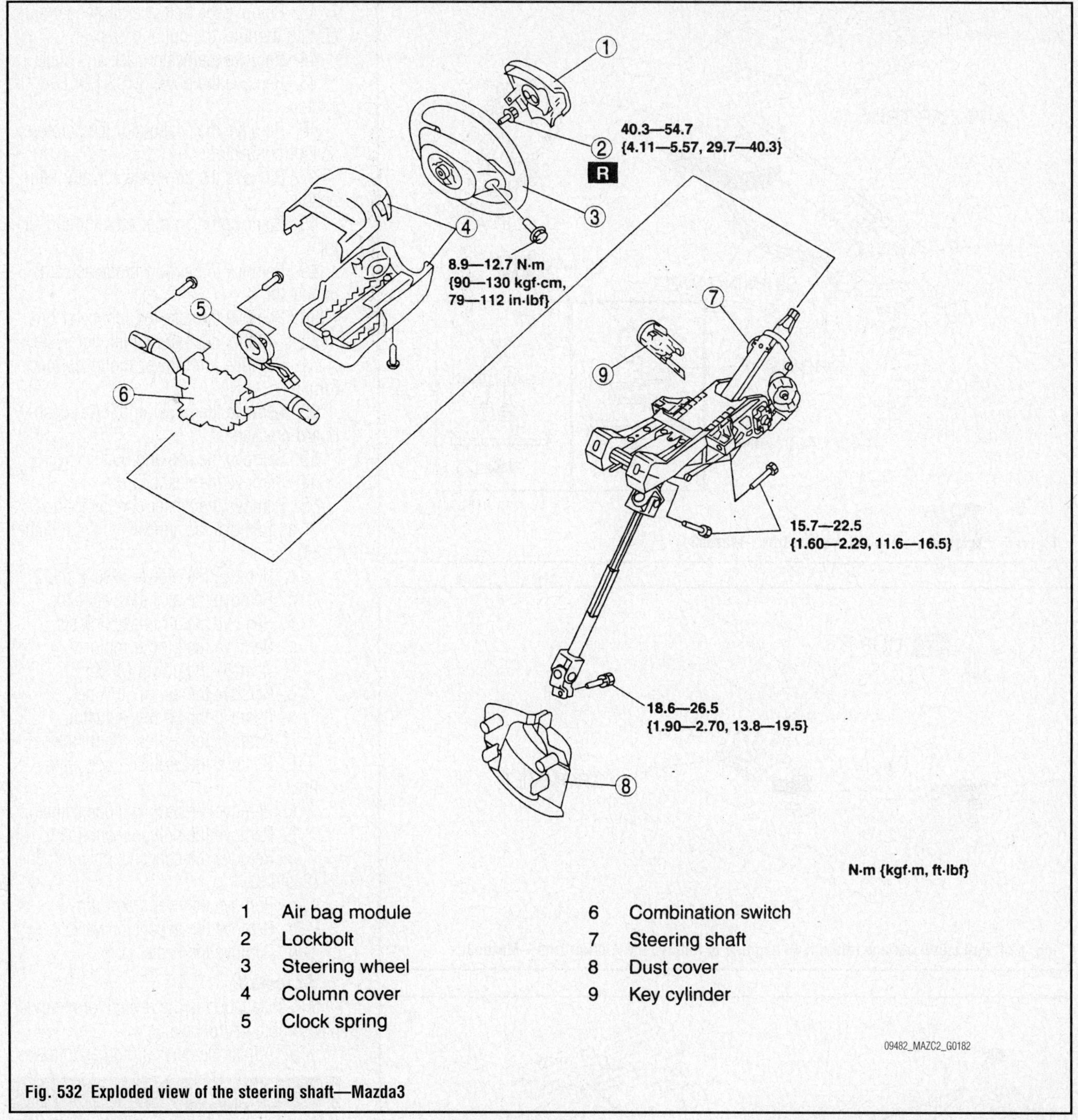

Fig. 532 Exploded view of the steering shaft—Mazda3

30. Remove the cowl grille.
31. Remove the cowl panel.
32. Remove the wiper motor bolts.
33. Move the wiper motor in the direction of the arrow, remove the securing rubber from the stud pin (for connecting the motor securing rubber), and then remove the windshield wiper motor.
34. Disconnect the windshield wiper motor connector.
35. Remove the A/C unit installation nut from the engine compartment, then remove the A/C unit.
36. Remove the rear heat duct.
37. Disconnect the drain hose connected to the A/C unit.
38. Remove the nuts and bolts for installing the dashboard to the body.
39. Remove the climate control unit.
40. Remove the LCD unit.

CAUTION

Handling the air bag module improperly can accidentally deploy the air bag module, which may seriously injure you. Due to the adoption of 2-step deployment control in the passenger-side air bag module, depending on the impact force, it is possible that inflator No.2 might not deploy. In such cases, before disposing of the air bag module, make sure to follow the inflator deployment procedures and verify complete deployment of inflators No.1 and 2.

41. Turn the ignition switch to the LOCK position.
42. Remove the dashboard garnish.

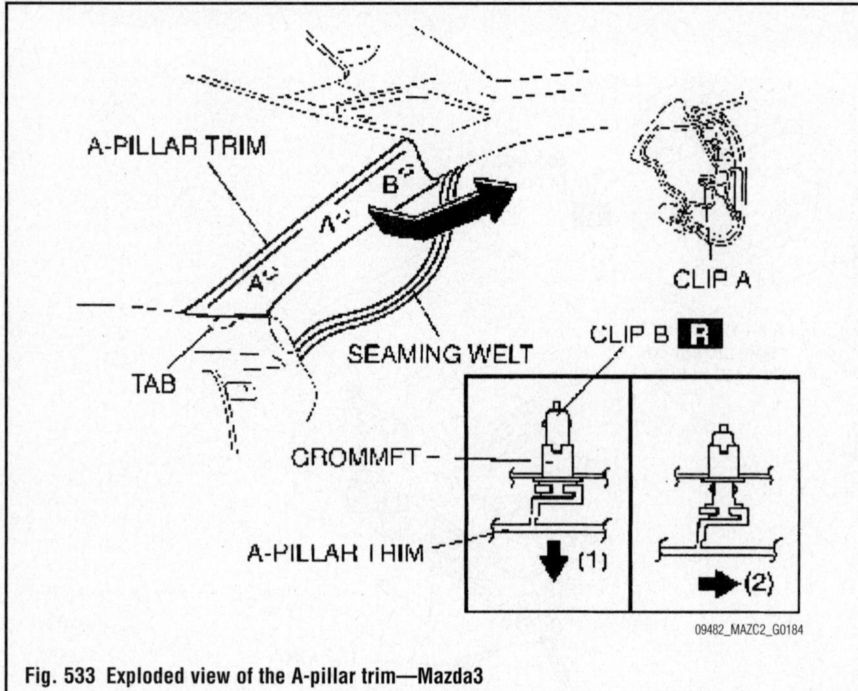

Fig. 533 Exploded view of the A-pillar trim—Mazda3

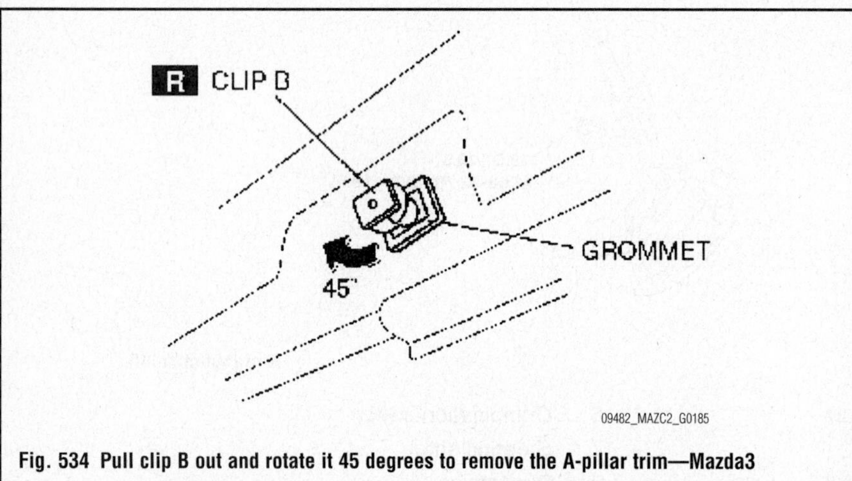

Fig. 534 Pull clip B out and rotate it 45 degrees to remove the A-pillar trim—Mazda3

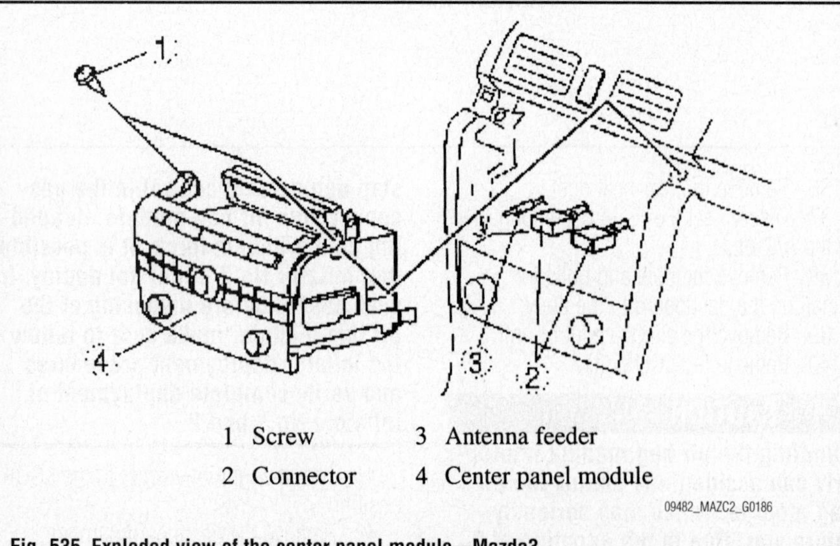

Fig. 535 Exploded view of the center panel module—Mazda3

43. Remove the bolt, the passenger side air bag module, the nut and bracket.
44. Remove the heater case and duct.
45. Remove the power MOS FET connector.
46. Remove the evaporator temperature sensor connector.
47. Remove the air intake actuator connector.
48. Remove the air mix actuator connector.
49. Remove the airflow mode actuator connector.
50. Remove the resistor connector on models with manual air conditioner system.
51. Remove the blower motor connector.
52. Remove the upper and lower dashboard brackets.
53. Remove the cap and bolt.
54. Remove the dashboard.
55. Remove the heater core as follows:
   a. Remove the adhesive polyurethane (1).
   b. Remove the blower case 1 and 2.
   c. Remove the air intake actuator.
   d. Remove the air intake link set.
   e. Remove the blower motor.
   f. Remove the power MOS FET.
   g. Remove the air mix link set.
   h. Remove the air mix actuator.
   i. Remove the airflow mode link set.
   j. Remove the airflow mode main link.
   k. Remove the airflow mode actuator.
   l. Remove the polyurethane foam.
   m. Remove the adhesive polyurethane (2 and 3).
   n. Remove the evaporator pipe.
   o. Remove the expansion valve.
   p. Remove the heater core.

To install:

56. Installation is the reverse of removal, please note the following:
   a. Adjust the clock spring as follows:

**** CAUTION**

If the clock spring is not adjusted, the spring wire in the clock spring could over-wind and break when the steering wheel is turned. Always adjust the clock spring after installing it.

- Set the front tires straight-ahead

**** CAUTION**

The clock spring will break if overwound. Do not forcibly turn the clock spring.

- Turn the clock spring clockwise until it stops

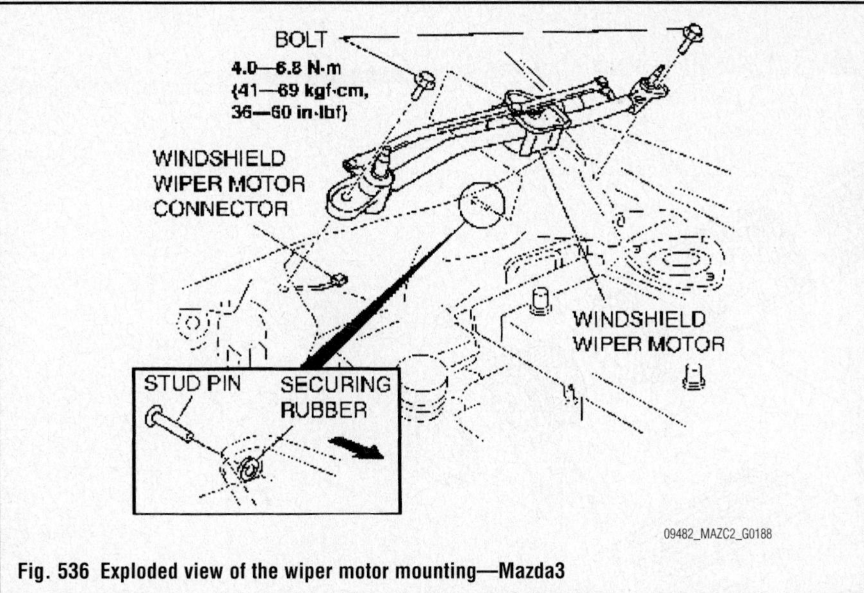

Fig. 536 Exploded view of the wiper motor mounting—Mazda3

- From the stopped position, turn the clock spring counterclockwise 2 ¾ turns
- Align the marks
- Verify that the tilt / telescope lever is in the LOCK position

b. Tighten the steering shaft bolts in alphabetical order.

c. Set the wheels in the straight-ahead position and install the steering wheel

⁂ CAUTION

When installing the center panel module, make sure that the wiring harness and antenna feeder are not caught between the unit and dashboard. If the wiring harness or the antenna feeder is caught between the unit and dashboard, it may cause malfunctions.

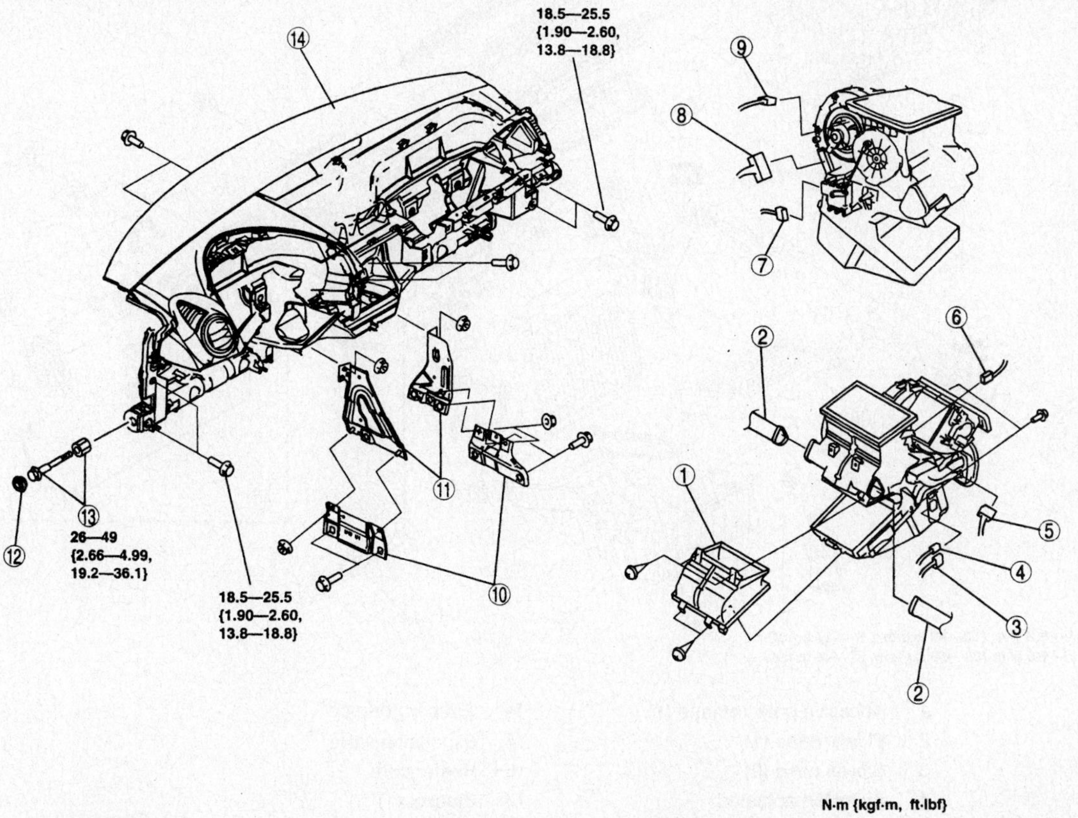

N·m {kgf·m, ft·lbf}

1	Heater case	8	Resistor connector (with manual air conditioner system)
2	Duct	9	Blower motor connector
3	Power MOS FET connector (with full-auto air conditioner system)	10	Dashboard bracket (lower)
4	Evaporator temperature sensor connector	11	Dashboard bracket (upper)
5	Air intake actuator connector	12	Cap
6	Air mix actuator connector (with full-auto air conditioner system)	13	Bolt
7	Airflow mode actuator connector (with full-auto air conditioner system)	14	Dashboard

09482_MAZC2_G0189

Fig. 537 Exploded view of the dashboard mounting—Mazda3

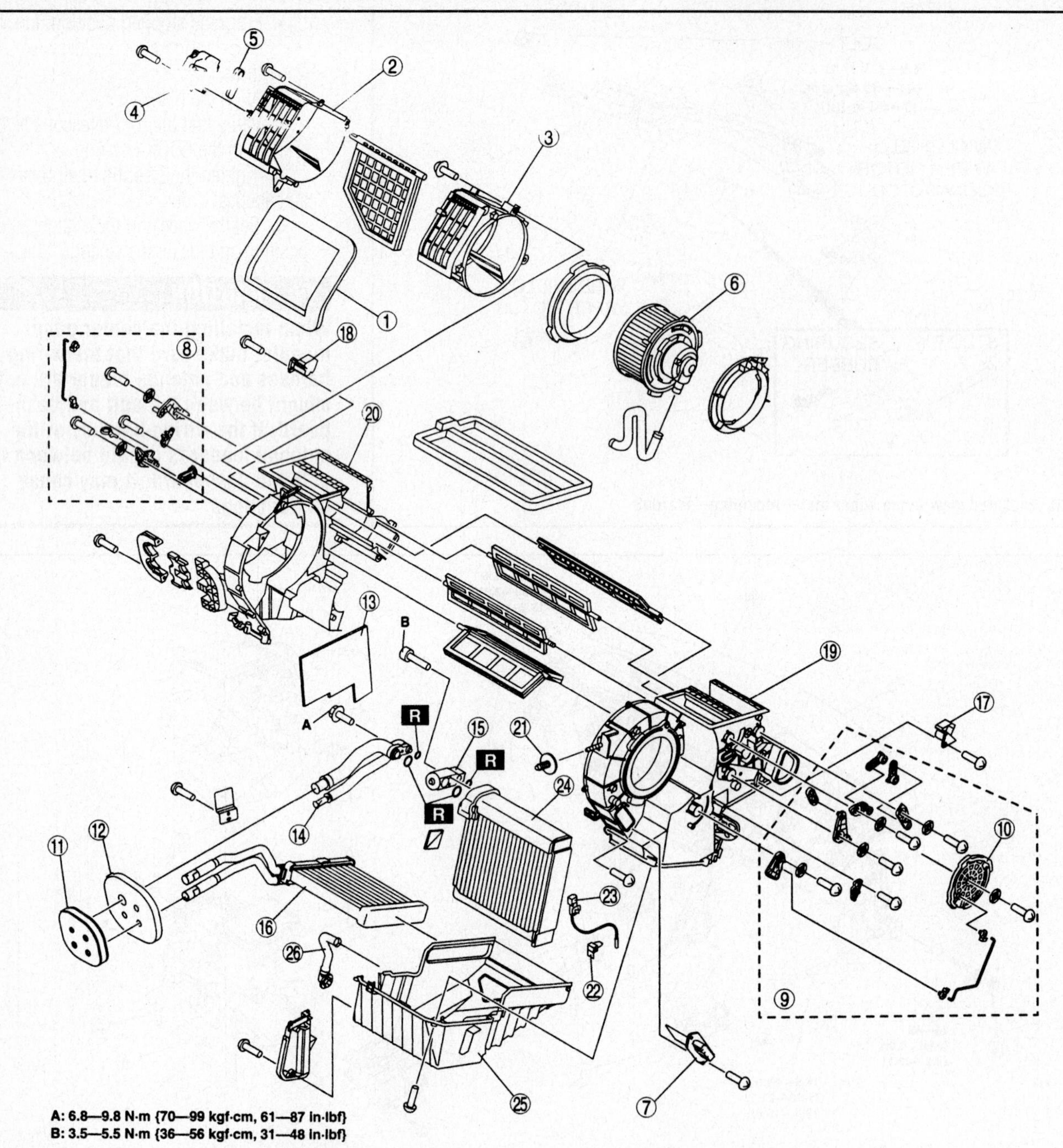

1	Adhesive polyurethane (1)	14	Evaporator pipe
2	Blower case (1)	15	Expansion valve
3	Blower case (2)	16	Heater core
4	Air intake actuator	17	Bracket (1)
5	Air intake link set	18	Bracket (2)
6	Blower motor	19	A/C case (1)
7	Resistor	20	A/C case (2)
8	Air mix link set	21	Bolt
9	Airflow mode link set	22	Sensor clamp
10	Airflow mode main link	23	Evaporator temperature sensor
11	Polyurethane foam	24	Evaporator
12	Adhesive polyurethane (2)	25	A/C case (3)
13	Adhesive polyurethane (3)	26	Drain hose

09482_MAZC2_G0190

Fig. 538 Exploded view of Type (A) A/C unit assembly—Mazda3

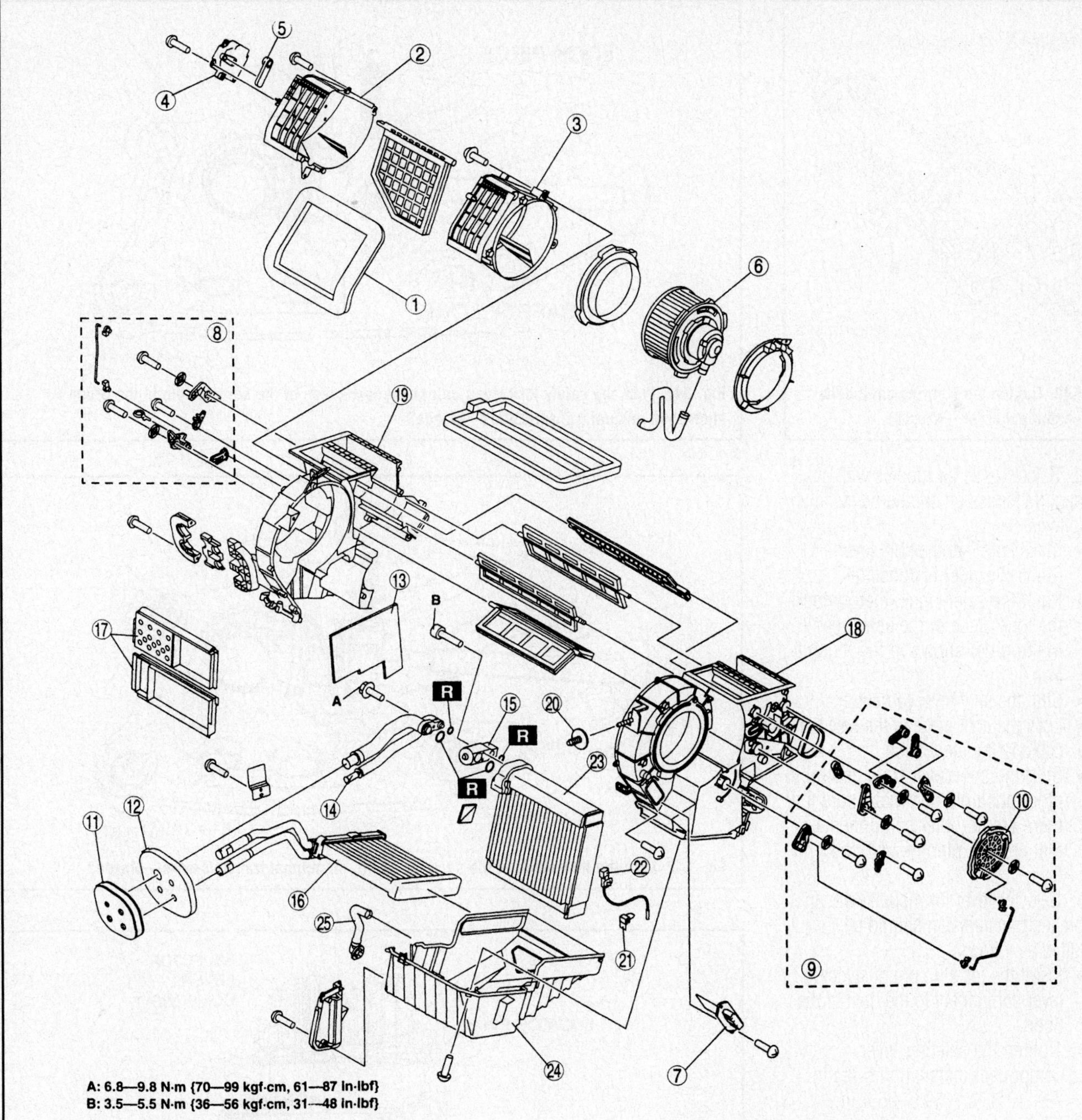

1	Adhesive polyurethane (1)	14	Evaporator pipe
2	Blower case (1)	15	Expansion valve
3	Blower case (2)	16	Heater core
4	Air intake actuator	17	Air filter cover
5	Air intake link set	18	A/C case (1)
6	Blower motor	19	A/C case (2)
7	Resistor	20	Bolt
8	Air mix link set	21	Sensor clamp
9	Airflow mode link set	22	Evaporator temperature sensor
10	Airflow mode main link	23	Evaporator
11	Polyurethane foam	24	A/C case (3)
12	Adhesive polyurethane (2)	25	Drain hose
13	Adhesive polyurethane (3)		

Fig. 539 Exploded view of Type (B) A/C unit assembly—Mazda3

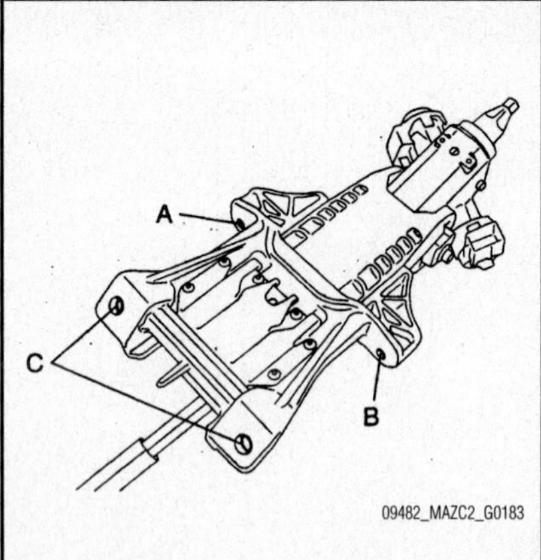

Fig. 540 Tighten the steering shaft bolts in alphabetical order—Mazda3

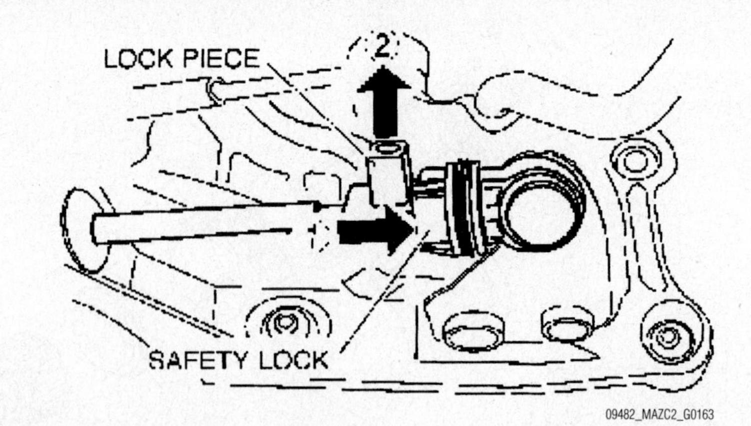

Fig. 541 Push the safety lock, then unlock the lock piece of the select cable in the order shown, on manual transmissions—Mazda3

d. Select Cable for models with an manual transmission should be installed as follows:

- Make sure that the shift lever (transaxle side) is in neutral
- Push the safety lock, then unlock the lock piece of the select cable in the order shown in the illustration.
- Shift the shift lever to neutral.
- Lock the lock piece of the selector cable in the illustration.
- Shift the shift lever from neutral to other position, and make sure that there are no other components in that area to interfere with the lever.

e. Selector lever for models with an automatic transmission should be installed as follows:

- Insert the location pin of selector lever component to the hole of the floor
- Tighten the selector lever component installation bolts to 7.8–10.8 Nm (69.5–95.4 in. lbs.).

f. Selector Cable for models with an automatic transmission should be installed as follows:

- Install the selector cable to the selector lever securely.
- Install the selector cable to the bracket securely.

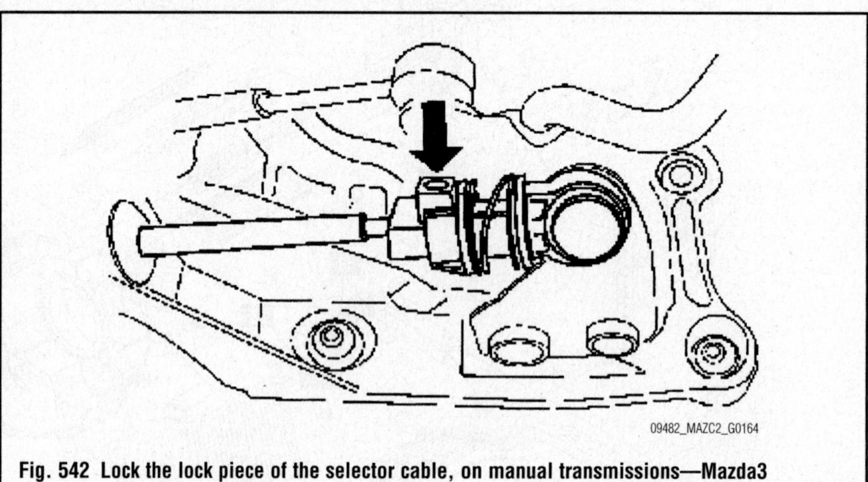

Fig. 542 Lock the lock piece of the selector cable, on manual transmissions—Mazda3

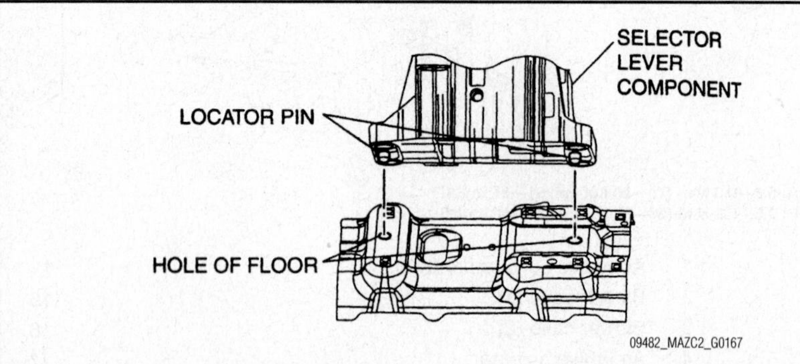

Fig. 543 Insert the location pin of selector lever component to the hole of the floor on models with an automatic transmission—Mazda3

CAUTION

Bending the selector cable in the manner shown in the illustration will damage the cable and it may become loose when shifted. When installing the selector cable, hold it straight.

- Install the clip as shown in the illustration.

➡Install the selector cable to the manual shaft cable with the clip side of the selector cable end facing the front of the vehicle.

- Install the selector cable to the manual shaft lever in such a way that the selector cable does not bear a load.
- Confirm that the end of the manual shift lever projects from the end of the selector cable.
- Install the selector cable to the selector cable bracket securely.

g. Install the connector B as shown in the illustration.

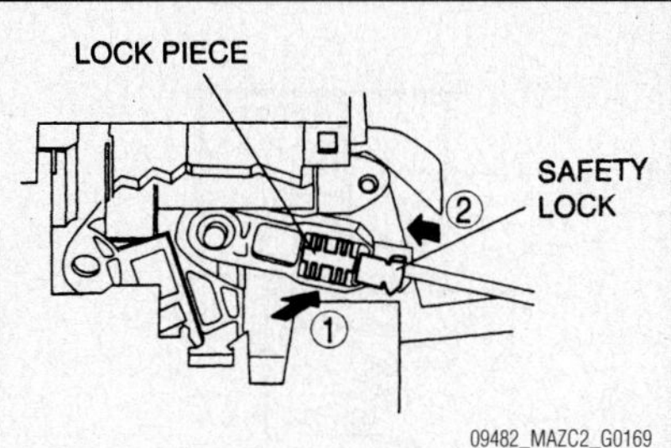

Fig. 544 Lock the lock piece and safety lock of the selector cable in this order on models with an automatic transmission—Mazda3

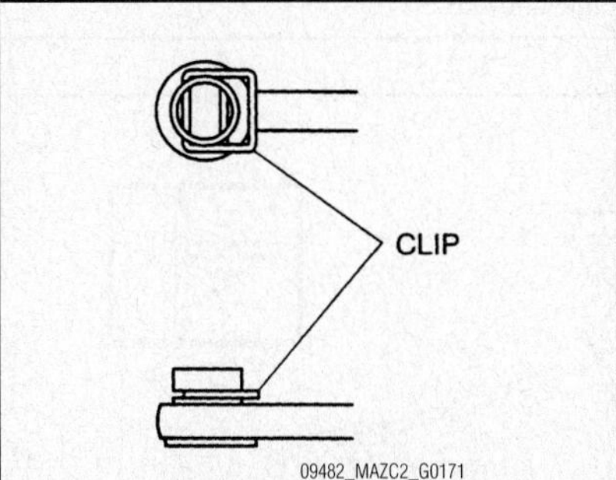

Fig. 545 Install the selector cable to the manual shaft cable with the clip side of the selector cable end facing the front of the vehicle on models with an automatic transmission—Mazda3

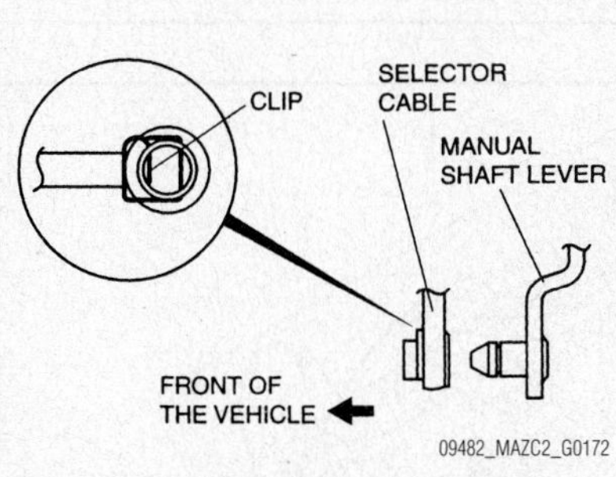

Fig. 546 Install the selector cable to the manual shaft lever in such a way that the selector cable does not bear a load on models with an automatic transmission—Mazda3

h. After connecting connector A, rotate the lever in the direction of the arrow to install connector A.

i. Turn the ignition switch to the ON position.

j. Verify that the air bag system warning light goes out.

k. If the air bag system warning light does not operate normally, inspect of the system.

MX5-Miata

See Figures 547 through 565.

1. Before servicing the vehicle, refer to the precautions section.
2. Disconnect the negative battery cable.

✻✻ CAUTION

After disconnecting the battery, wait for more than 1 minute for the SAS to deplete its stored energy.

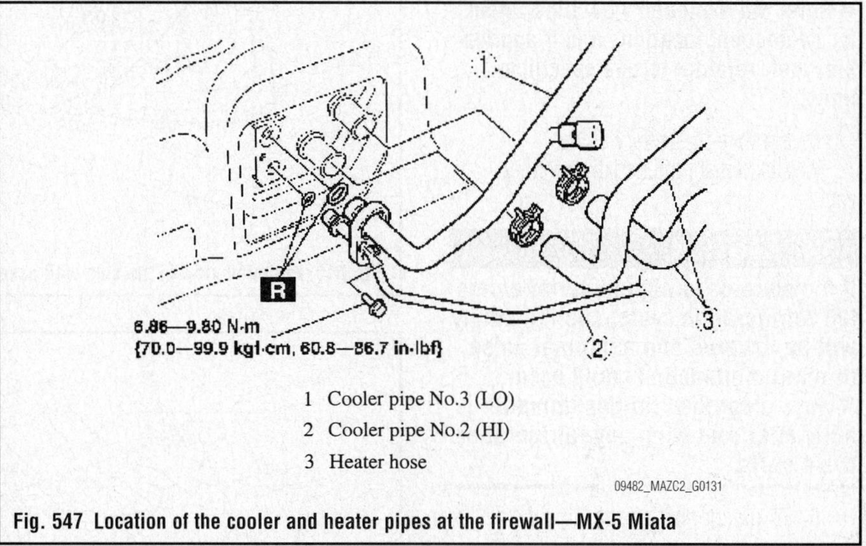

Fig. 547 Location of the cooler and heater pipes at the firewall—MX-5 Miata

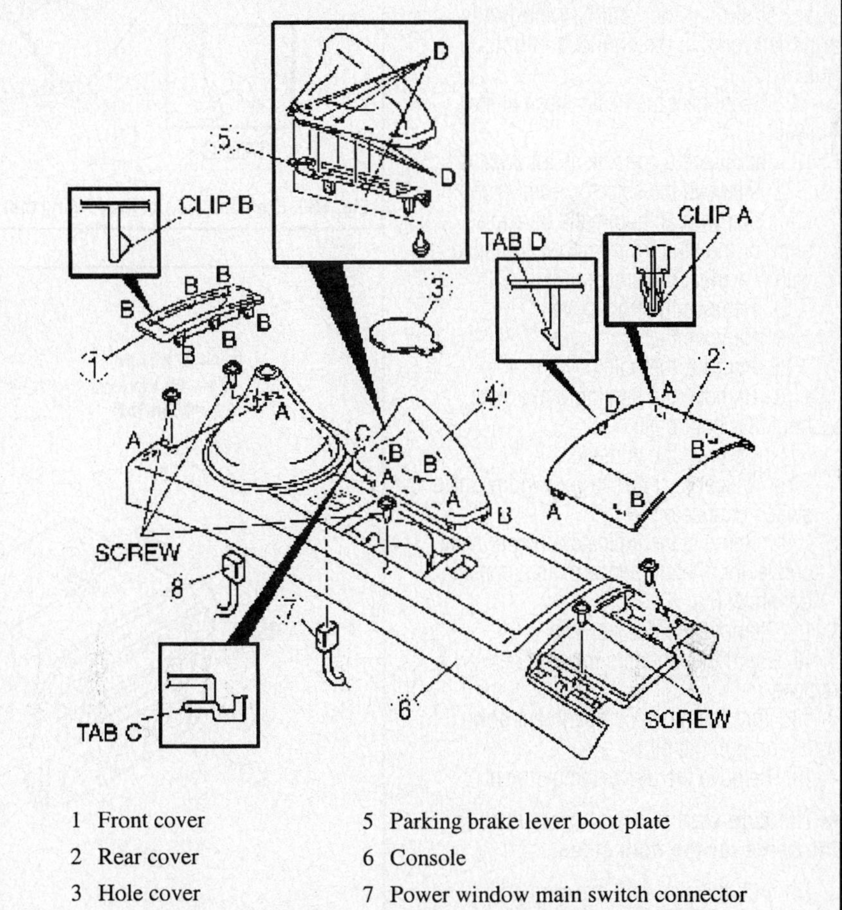

Fig. 548 Exploded view of the console assembly—MX-5 Miata

3. Drain the cooling system into a clean container for reuse.
4. Disconnect the heater hoses from the heater core.
5. Discharge and recover the air conditioning system refrigerant.

➡Refer the exploded view illustration for component locations and if applicable, their retainer torque specifications.

6. Remove the battery cover.
7. Disconnect the negative battery cable.

CAUTION

If moisture or foreign material enters the refrigeration cycle, cooling ability will be lowered and abnormal noise or other malfunction could occur. Always plug open fittings immediately after removing any refrigeration cycle parts.

8. To disconnect cooler pipe # 3, disconnect the block joint type pipes by grasping the female side of the block with pliers or similar tool and holding firmly, and then remove the connection bolt or nut.
9. Disconnect the heater hose at the firewall.
10. Remove the console as follows:
    a. Remove the selector lever knob if equipped with an automatic transmission, or the shift lever knob if equipped with a manual transmission.
    b. Remove the front cover.
    c. Remove the rear cover.
    d. Remove the hole cover.
    e. Remove the parking brake lever boot and boot plate.
    f. Remove the console.
    g. Disconnect the power window main switch connector.
    h. Remove the indicator compartment connector, if equipped with an automatic transmission.
11. Remove the dumper clip.
12. Bend the stoppers inward, then remove.
13. Turn the glove compartment downward and pull the pins.
14. Remove the glove compartment.

➡The side wall removal procedure is the same for the both sides.

15. Pull the side wall rearward and detach clips A, B pins C and tab D.
16. Remove the side wall.
17. Remove the console panel as follows:

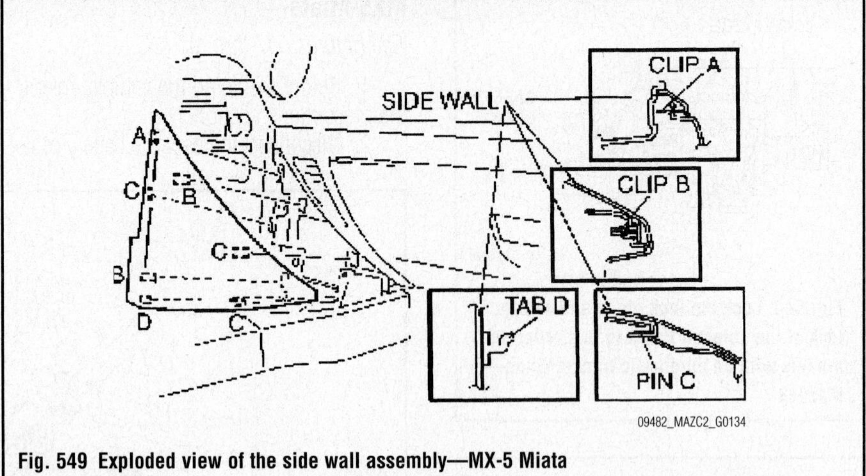

Fig. 549 Exploded view of the side wall assembly—MX-5 Miata

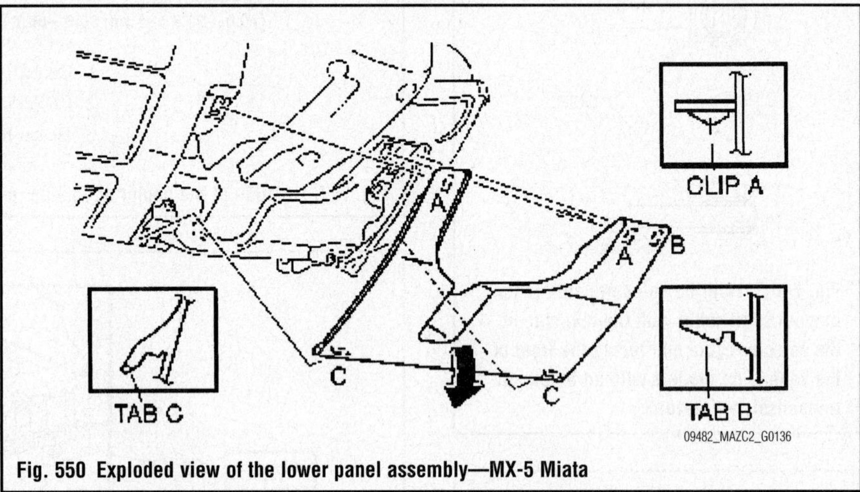

Fig. 550 Exploded view of the lower panel assembly—MX-5 Miata

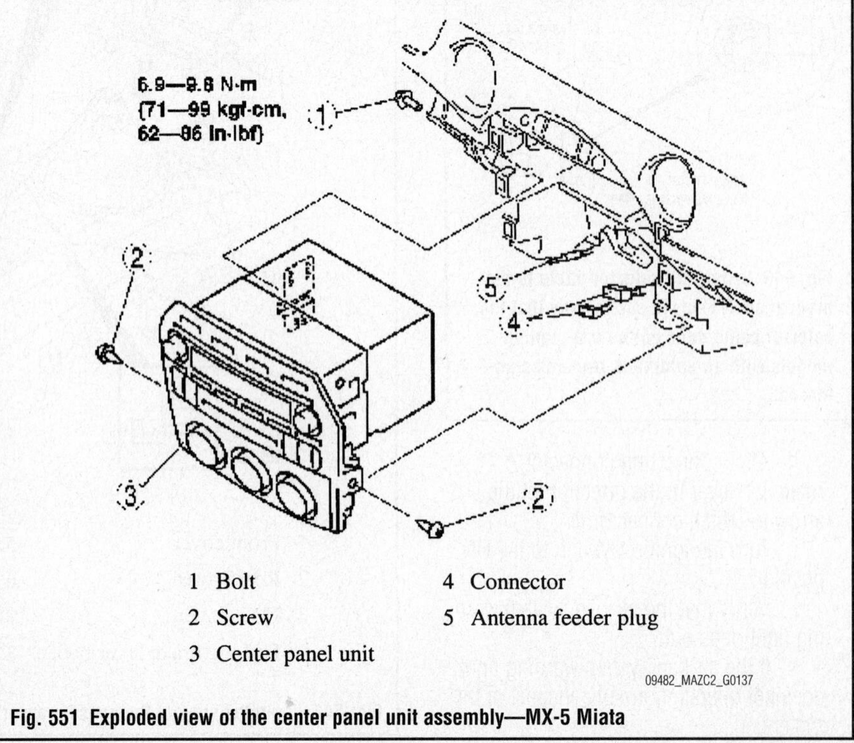

Fig. 551 Exploded view of the center panel unit assembly—MX-5 Miata

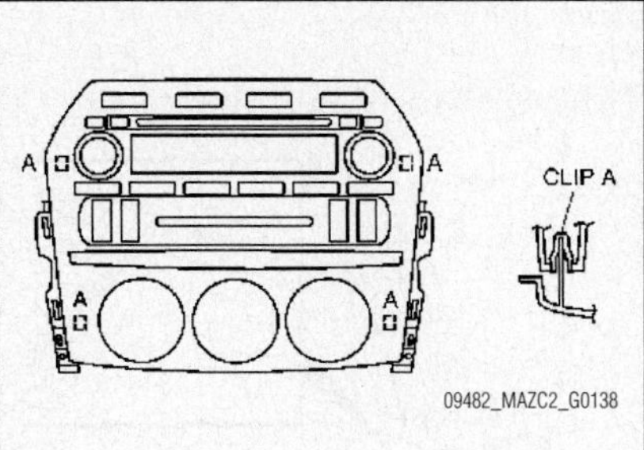

Fig. 552 Pull the center panel unit outward, detach clip A from the dashboard, and then remove the center panel unit—MX-5 Miata

a. Remove the screws.
b. Pull the console panel outward and detach tabs A and pins B.
c. Disconnect the seat warmer switch connector, if equipped.
d. Disconnect the accessory socket connector.
e. Remove the console panel.

18. Remove the center panel as follows:
a. Pull the lower panel outward and detach clips A and tab B.
b. Turn the lower panel downward and remove the tabs C.
c. Remove the lower panel.
d. Remove the knee bolster.
e. Remove the bolt and screw.
f. Pull the center panel unit outward, detach clip A from the dashboard, and then remove the center panel unit.

➡Verify that bolt has been removed when pulling the center panel unit outward. If the center panel unit is pulled with bolt installed, stress will be applied to the connectors inside the panel which may cause a malfunction. When removing the center panel unit, disconnect the audio unit connector (24-pin) first to aid in disconnecting other connectors and the antenna plug.

g. Disconnect the connector and antenna feeder plug.

19. Remove the column cover as follows:
a. Remove the tab A, then remove the upper column cover.
b. Remove the screws, then remove the lower column cover.
c. Remove the ignition key illumination.

20. Remove the driver's side air bag as follows:

✲✲ WARNING

Handling the air bag module improperly can accidentally deploy the air bag module, which may seriously injure you.

a. Turn the ignition switch to the LOCK position.
b. Remove the cover, and bolt.
c. Using a flathead screwdriver, pry out the connector stopper plate.
d. Disconnect the connector.
e. Remove the air bag.

21. Remove the steering wheel as follows:

✲✲ WARNING

Handling the air bag module improperly can accidentally deploy the air bag module, which may seriously injure you.

a. Remove the locknut.

✲✲ CAUTION

Do not try to remove the steering wheel by hitting the shaft with a hammer. The column will collapse.

b. Set the vehicle in the straight-ahead position.
c. Remove the steering wheel using a suitable puller.
d. Remove the column cover.
e. Remove the clock spring screw, connector and clock spring.

➡For vehicles with DSC, if the negative battery cable or the steering angle sensor connector or ROOM 15 A fuse is disconnected, the stored initial position of the steering angle sensor will be cleared and the DSC will not operate properly, making the vehicle unsafe to drive. Perform the steering angle sensor initialization procedure after connecting the negative battery cable.

f. Disconnect the combination switch connector.
g. Disconnect the steering angle sensor connector, on vehicles with DSC.

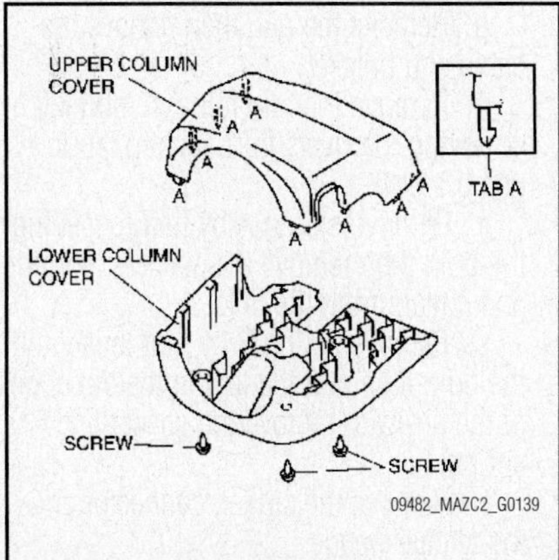

Fig. 553 Exploded view of the column cover assembly—MX-5 Miata

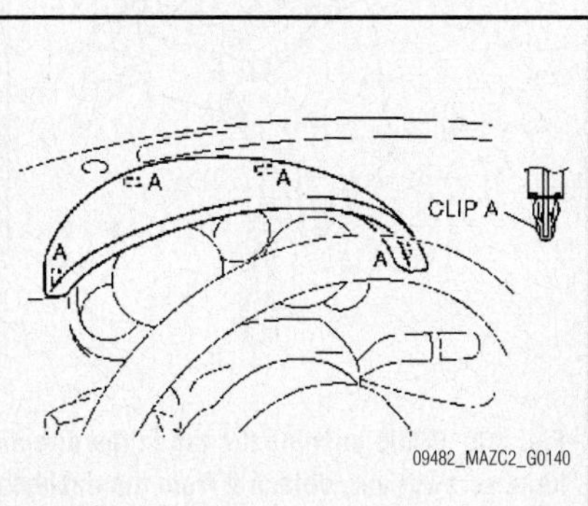

Fig. 554 Exploded view of the meter hood assembly—MX-5 Miata

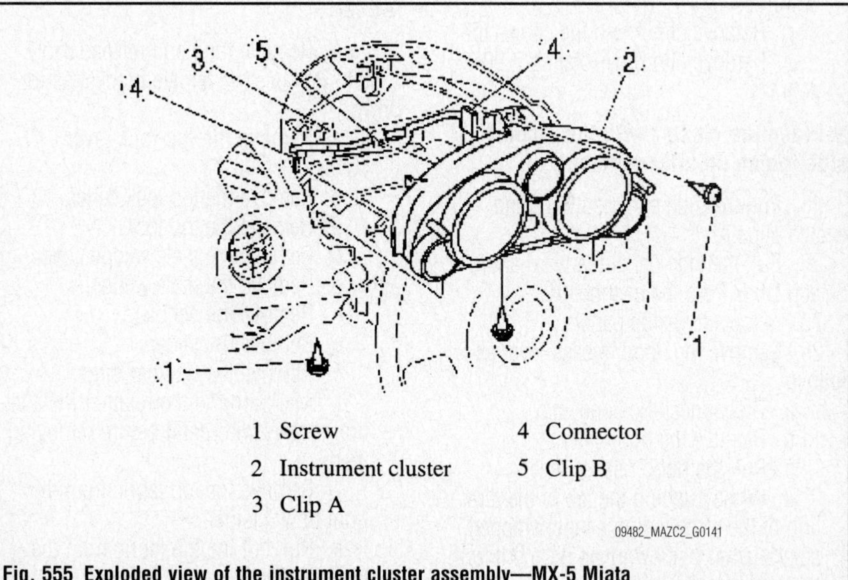

Fig. 555 Exploded view of the instrument cluster assembly—MX-5 Miata

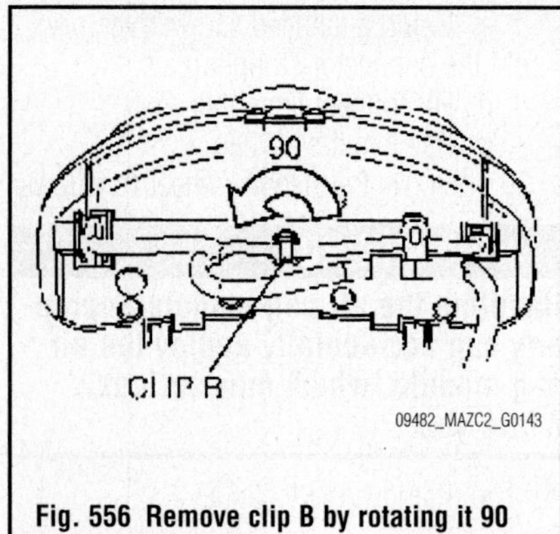

Fig. 556 Remove clip B by rotating it 90 degrees—MX-5 Miata

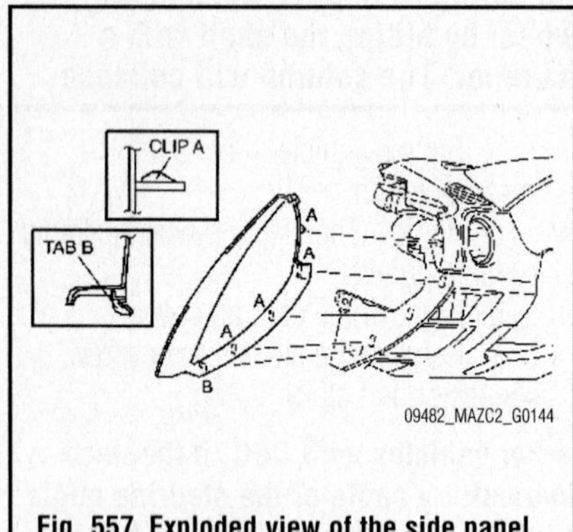

Fig. 557 Exploded view of the side panel assembly—MX-5 Miata

h. Remove the screws and then remove the combination switch.

22. Remove the steering shaft.

23. Pull the meter hood upward and detach clips A.

24. Remove the meter hood.

25. Remove the instrument cluster as Follows;

a. Remove the screw.

b. Remove the cluster by rotating it upwards.

c. Remove clip A and the connector

d. Remove clip B by rotating it 90 degrees.

➡Place the cluster with the display side facing up after removal.

26. Pull the side panel outward and detach clips A.

27. Pull the side panel rearward and detach tab B from the dashboard.

28. Remove the side panel.

29. Remove the hood release lever as follows:

a. Disconnect the connector.

b. Remove the hood latch.

c. Pull the hood release lever.

d. While pushing the tab in the direction of the arrow using a tape-wrapped, small flathead screwdriver, detach it from the dashboard.

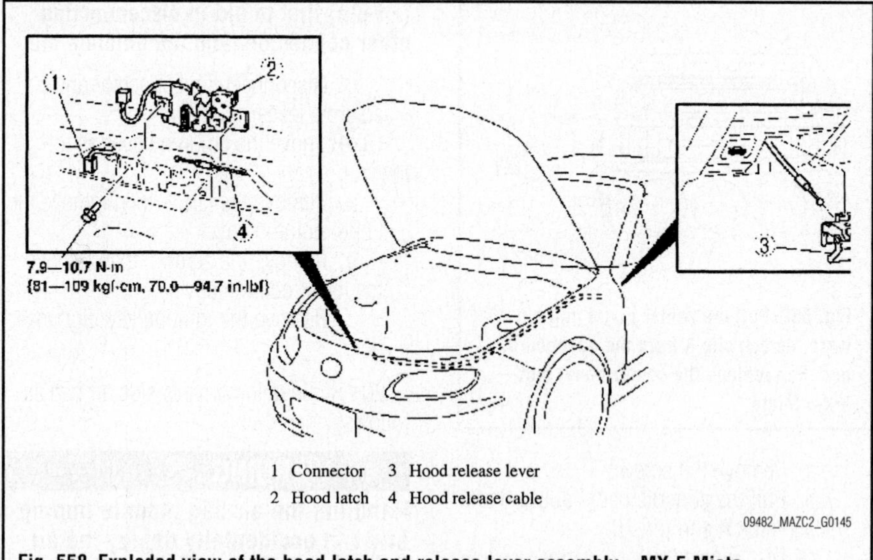

Fig. 558 Exploded view of the hood latch and release lever assembly—MX-5 Miata

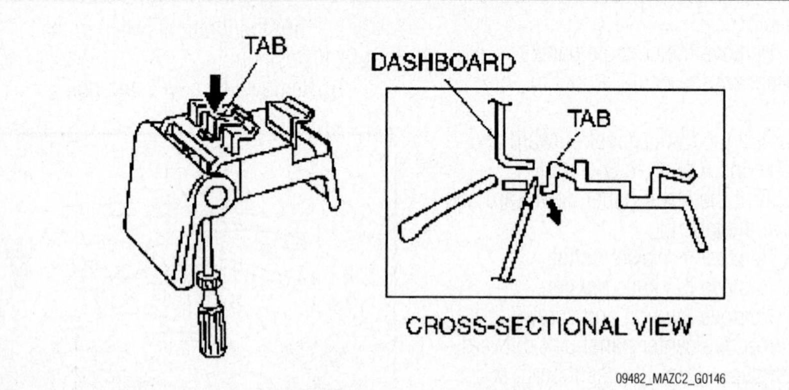

Fig. 559 While pushing the tab in the direction of the arrow using a tape-wrapped, small flat-head screwdriver, detach it from the dashboard—MX-5 Miata

➡Remove the hood release lever while taking care not to damage the hood release cable with the flathead screwdriver.

e. Remove the hood release cable.

30. Remove the female bracket as follows:

a. Remove the top lock lever cover.

b. Remove the top lock cover.

c. Remove the top lock.

d. Remove the male wedge cover.

e. Remove the male wedge.

f. Remove the set plate.

g. Remove the slider.

h. Remove the weather-strip.

i. Mark around the retainer installation screws with paint before removing them.

j. Remove the top fabric from the front bow retainer.

k. Pull out the top fabric from the front header.

l. Remove the covers.

m. Remove the cable installation rivet using a drill.

n. Remove the cables from the cable guide.

o. Remove the band installation rivet using a drill.

p. Remove the nut, then remove the cable end bracket.

q. Remove the rain rail assembly by removing the rivets from the top fabric using a drill.

r. Remove spring A by first loosening the bolt, then remove the spacers, spring and collars from the link.

s. Remove spring B by first loosening the bolt, then pull the link in the direction of the arrow to remove the spacer and spring.

t. Remove the striker, connector and the female wedge.

u. After installing the female bracket, keep in mind the following:

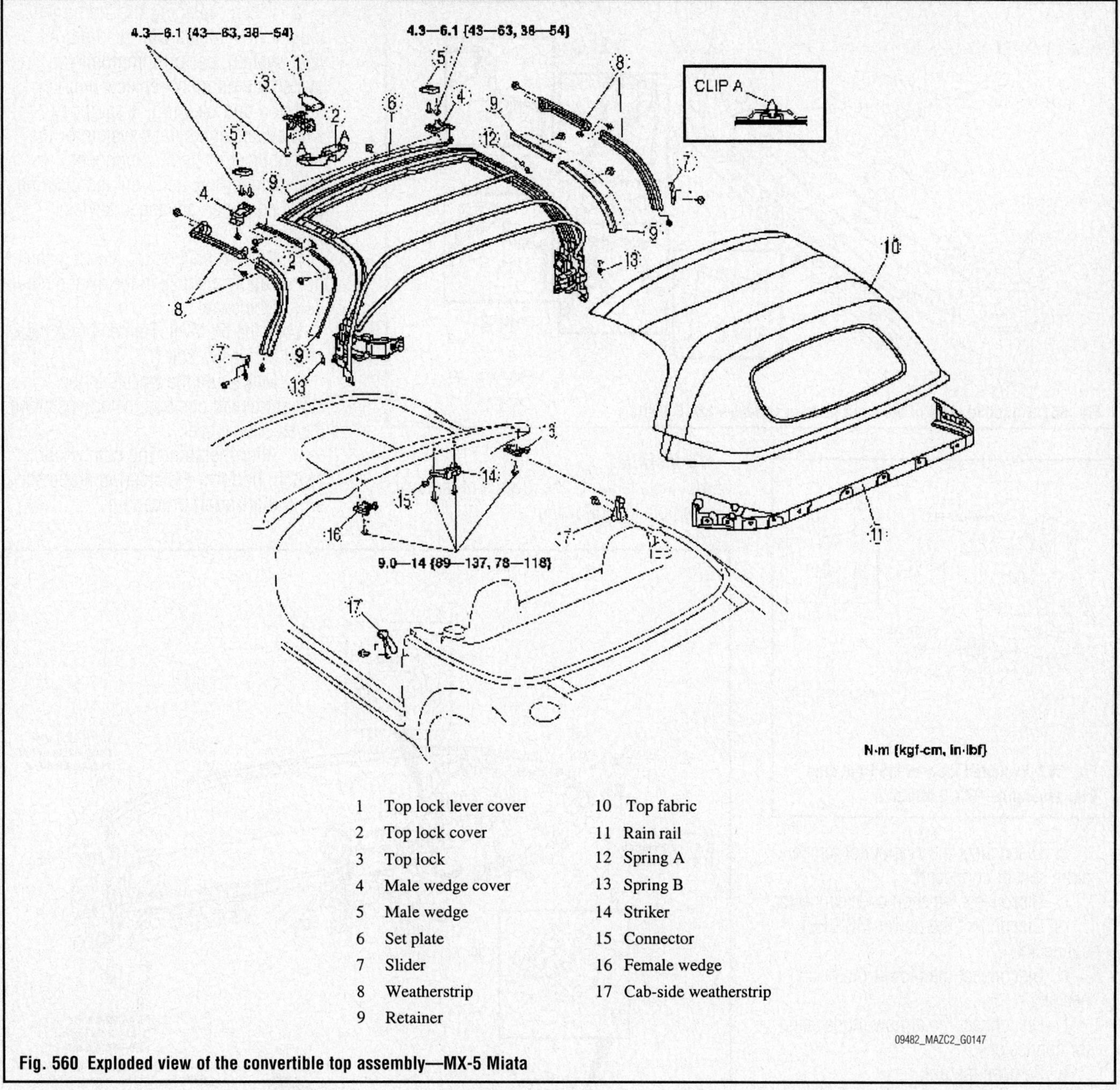

Fig. 560 Exploded view of the convertible top assembly—MX-5 Miata

v. Degrease the rain rail using white gasoline.

w. Install the insulation tape to the rivet installation hole of the rain rail.

x. Secure the top fabric and rain rail with the rivet.

y. Flatten the stem using hammer.

z. Place the link onto the top fabric.

aa. Align the link with the set plate installation hole of the top fabric, and install the top fabric to the front header.

bb. Thread the cable into the cable guide.

cc. Set the aluminum rivet to the riveter, and then secure the cable to the link with the rivet.

dd. Install the covers.

ee. Install the top fabric to the front bow.

ff. Secure the top fabric to the front bow retainer using a rubber hammer.

gg. Set the aluminum rivet to the riveter, and then secure the band to the link with the rivet.

Install the cable end bracket.

When installing the retainer assembly, install the retainers to the link, aligning the retainer marks with the retainer installation screws.

31. Pull the A-pillar trim, then disengage clips A and B.

32. Pull the A-pillar trim upward, then disengage tabs C from the body.

33. Remove the A-pillar trim.

34. Pull the scuff plate upward while detaching tabs C, and then detach clips A and D, and pins B (4), E (1) from the body, and remove the scuff plate.

35. Install in the reverse order of removal.

36. Remove the seaming welt.

37. Remove the fastener.

38. Pull the front side trim toward you, then disengage clip A and pin B from the body.

39. Remove the front side trim.

40. Fuse box No.1

41. Remove the dashboard as follows:

a. Disconnect the dashboard harness connectors.

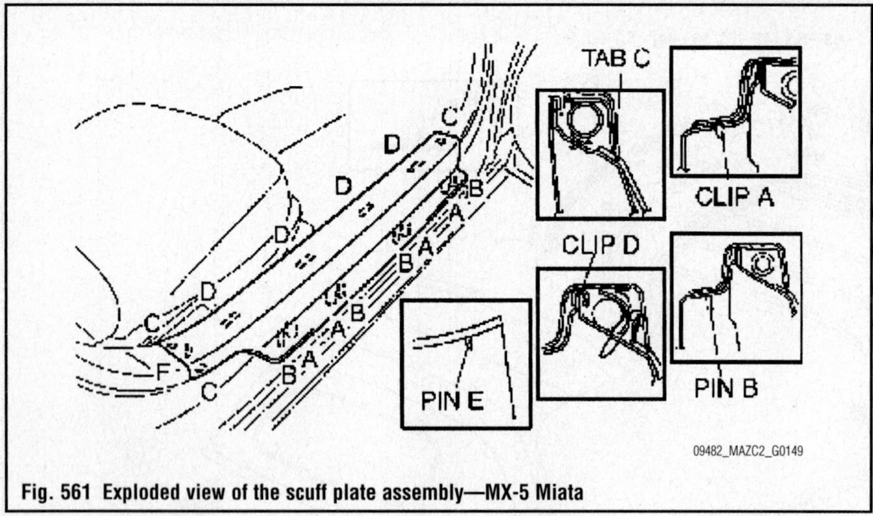

Fig. 561 Exploded view of the scuff plate assembly—MX-5 Miata

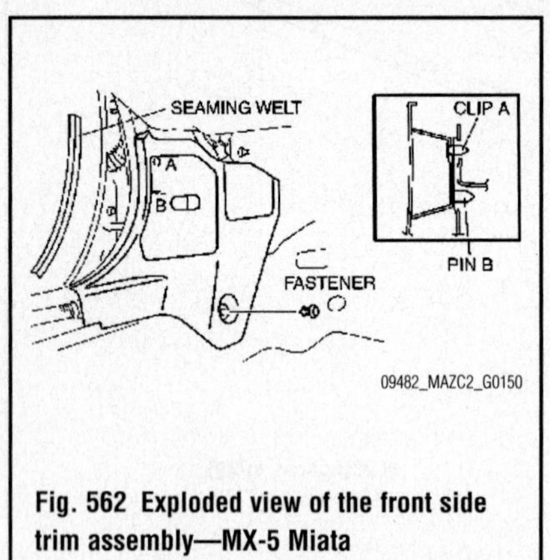

Fig. 562 Exploded view of the front side trim assembly—MX-5 Miata

b. Disconnect the evaporator temperature sensor connector.
c. Disconnect the short code connector.
d. Disconnect the power MOS FET connector.
e. Disconnect the blower motor connector.
f. Disconnect the airflow mode actuator connector.
g. Remove the nut.
h. Remove the cover.
i. Remove bolts A through D.
j. Remove the dashboard.

42. Remove the air dist unit and A/C unit.
43. Remove the heater core as follows:

a. Remove the blower motor.
b. Remove the power MOS FET.
c. Remove the polyurethane foam 1 and 2.
d. Remove the cover.
e. Remove the adhesive polyurethane.
f. Remove the air intake box.
g. Remove the air intake actuator.
h. Remove the air intake crank.
i. Remove the air intake door.
j. Remove the harness.
k. Remove the air mix actuator.
l. Remove the heater core.

To install:

44. Installation is the reverse of removal, please note the following:

a. When replacing the instrument cluster of vehicles with the immobilizer system, perform immobilizer system component replacement/key addition and clearing, if equipped with advanced keyless system or the immobilizer system component replacement/key addition and clearing, if not equipped advanced keyless system .
b. When installing the steering shaft, do not apply a shock in the axial direction of the shaft.
Lock the tilt lever. Tighten nut A then nut B and lastly bolt C.
c. Make sure the wheels in the straight-ahead position, before installing the steering wheel.
d. After installing the combination switch, perform the steering angle sensor initialization procedure

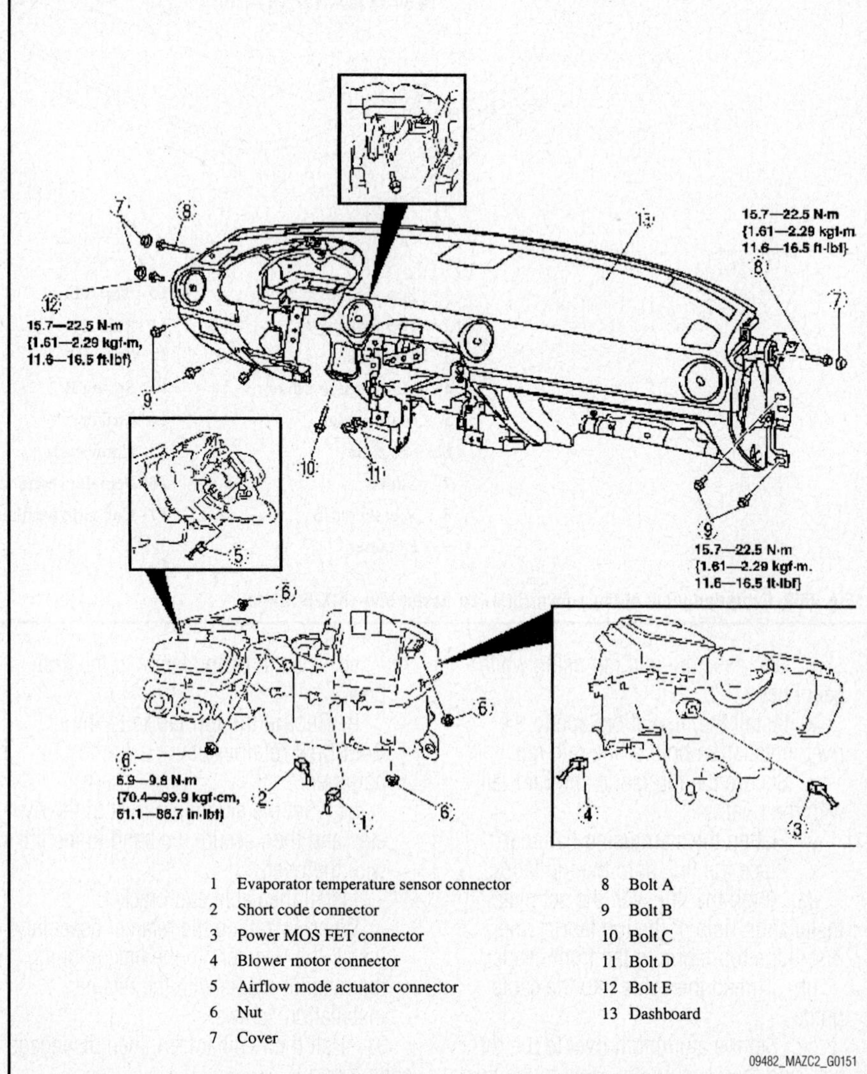

Fig. 563 Exploded view of the dashboard mounting—MX-5 Miata

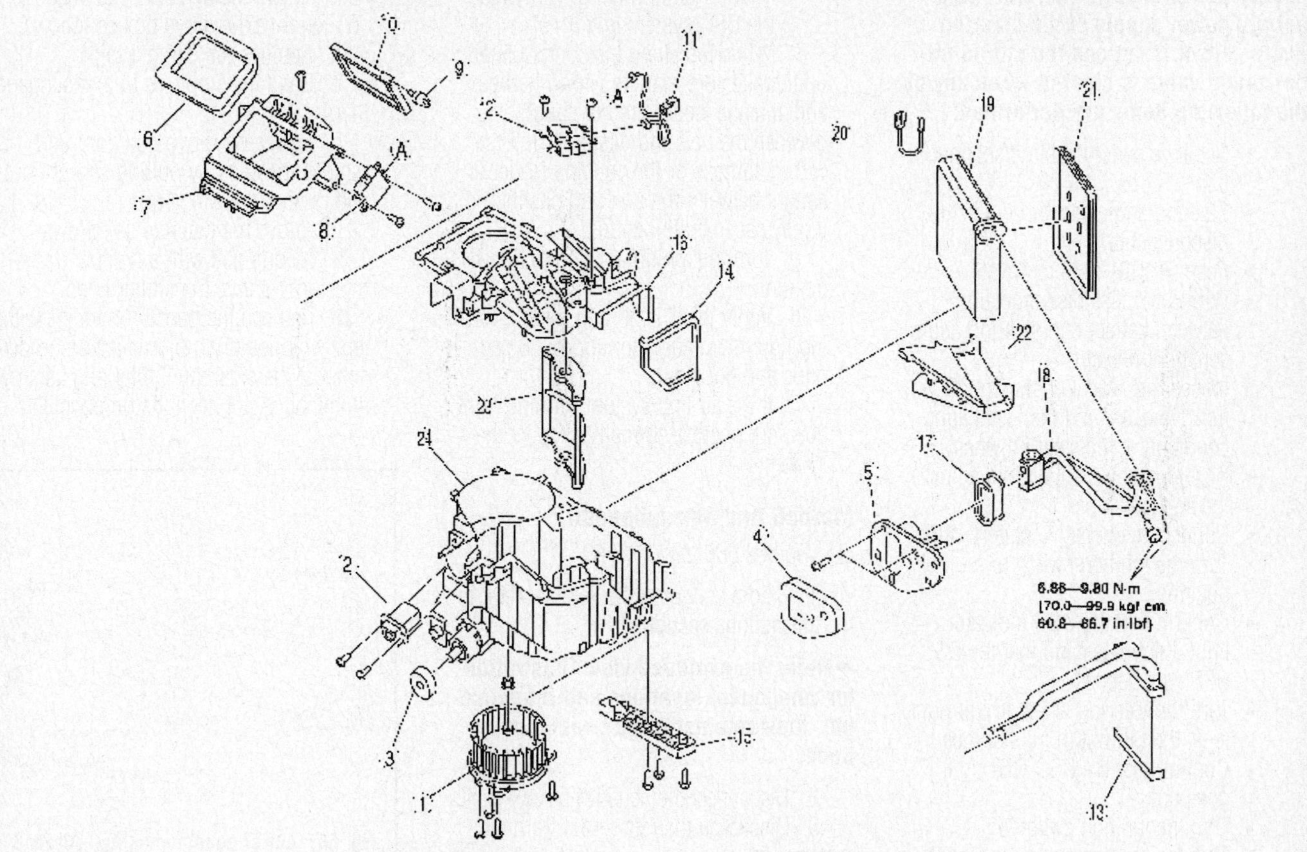

6.86—9.80 N·m {70.0—99.9 kgf cm, 60.8—86.7 in·lbf}

1 Blower motor	13 Heater core
2 Power MOS FET	14 Adhesive polyurethane (2)
3 Polyurethane foam (1)	15 Cover (2)
4 Polyurethane foam (2)	16 A/C case (1)
5 Cover (1)	17 Cover (3)
6 Adhesive polyurethane (1)	18 Expansion valve
7 Air intake box	19 Evaporator
8 Air intake actuator	20 Evaporator temperature sensor
9 Air intake crank	21 Cover (4)
10 Air intake door	22 Insulator
11 Harness	23 Air mix damper
12 Air mix actuator	24 A/C case (2)

09482_MAZC2_G0152

Fig. 564 Exploded view of the air distribution and A/C units—MX-5 Miata

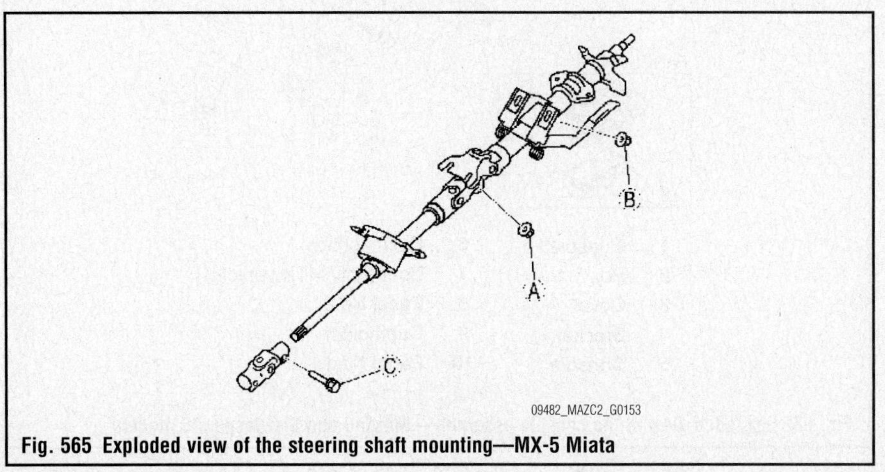

Fig. 565 Exploded view of the steering shaft mounting—MX-5 Miata

⁂ WARNING

Unless the initialization procedure of the steering angle sensor is completed, the DSC will not operate, causing an unexpected accident. Therefore, always perform the initialization procedure to ensure DSC operation if the power supply to the steering angle sensor has been cut off due to disconnection of the steering angle sensor connector or negative battery cable, or any other cause.

➡The initialization value of the steering angle sensor is stored using the

battery power supply. Therefore, the battery power supply of the steering angle sensor is cut and the stored initialization value is cleared when any of the following items are performed.

- Negative battery cable disconnection
- Steering angle sensor connector disconnection
- Fuse (ROOM 15A) removal
- Wiring harness disconnection between battery and steering angle sensor connector
- Inspect the wheel alignment, inflation pressure, and the installation condition of the steering wheel
- If there is any malfunction, adjust the applicable part
- Connect the negative battery cable
- Turn the ignition switch to the ON position
- Confirm that the DSC indicator light illuminates and that the DSC OFF light flashes
- Turn the steering wheel to full right lock, then turn it to full left lock
- Confirm that the DSC OFF light goes out
- Turn the ignition switch off
- Turn the ignition switch to the ON position again, and confirm that the DSC indicator light goes out
- If the DSC indicator light does not go out, disconnect the negative battery cable, and perform the procedure again
- Drive the vehicle for about 10 minutes and confirm that the ABS warning and DSC indicator lights do not illuminate

e. If the initialization procedure of the steering angle signal is not completed, the DSC will not operate properly and may cause an accident. Therefore, always perform initialization of the DSC HU/CM steering angle signal to ensure proper DSC operation when any of the following items are performed.

- Steering angle sensor replacement DSC HU/CM replacement
- Inspect the wheel alignment and inflation pressure.
- Park the vehicle on level ground
- Turn the ignition switch off
- Connect the WDS or equivalent to the DLC-2
- Access the active command mode, select the following commands, and then follow the indication on the monitor
- Drive the vehicle forward
- After 5 min of driving, verify that the DSC system is normal

f. When installing the center panel unit, make sure that the wiring harness and antenna feeder are not caught between the unit and dashboard. If the wiring harness or the antenna feeder is caught between the unit and dashboard, it may cause malfunctions.

g. Turn the ignition switch to the ON position.

h. Verify that the air bag system warning light illuminates for approx. 6 seconds and goes out.

i. If the air bag system warning light does not operate normally, inspect the system.

Mazda6 And Mazdaspeed6

See Figures 566 through 580.

1. Before servicing the vehicle, refer to the precautions section.

➡Refer the exploded view illustration for component locations and if applicable, their retainer torque specifications.

2. Disconnect the negative battery cable.
3. Discharge the refrigerant from the system.
4. Drain and recycle the engine coolant.
5. Remove the dynamic chamber.
6. Remove the glove box as follows:

a. Remove the damper clip.

b. Bend the stoppers in to disengage them.

c. Lower the glove box door and remove the clips by pulling straight out and remove the glove box.

7. Remove the console as follows:

a. If equipped with a manual transmission, remove the shifter knob.

b. Remove the panels (1 and 2) with a tape wrapped pry tool, make sure to disconnect the cigarette lighter and ashtray illumination connections on panel 2.

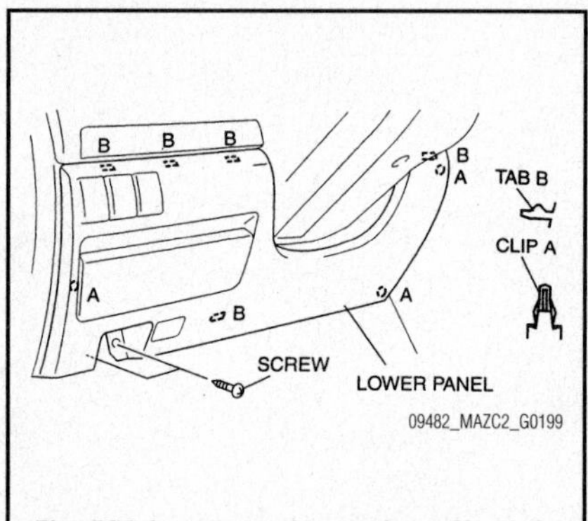

Fig. 567 Lower panel mounting—Mazda6 and Mazdaspeed6 models

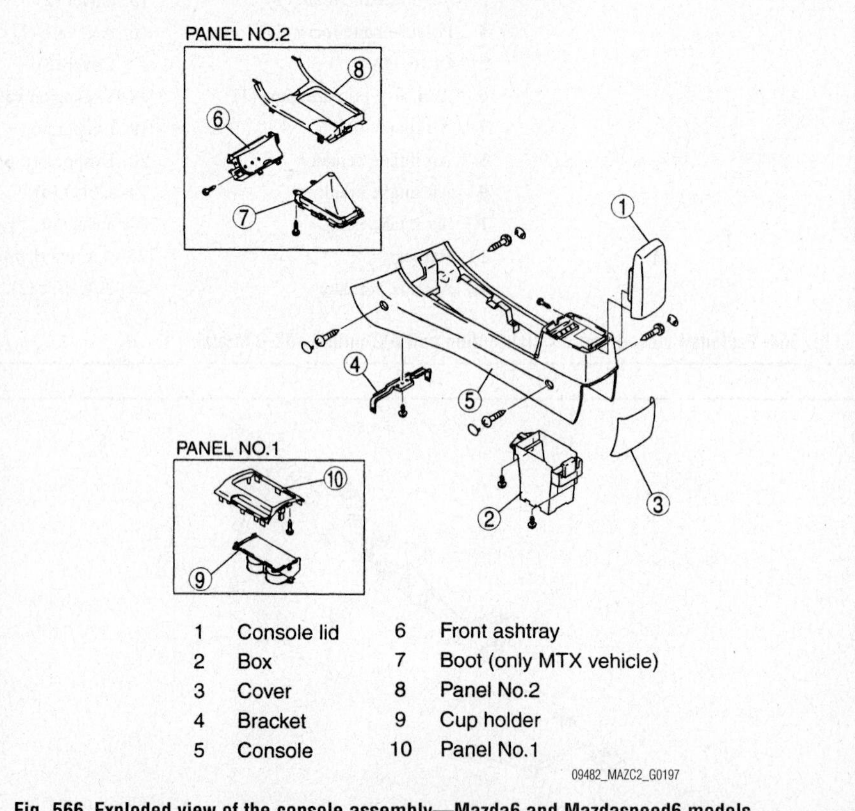

Fig. 566 Exploded view of the console assembly—Mazda6 and Mazdaspeed6 models

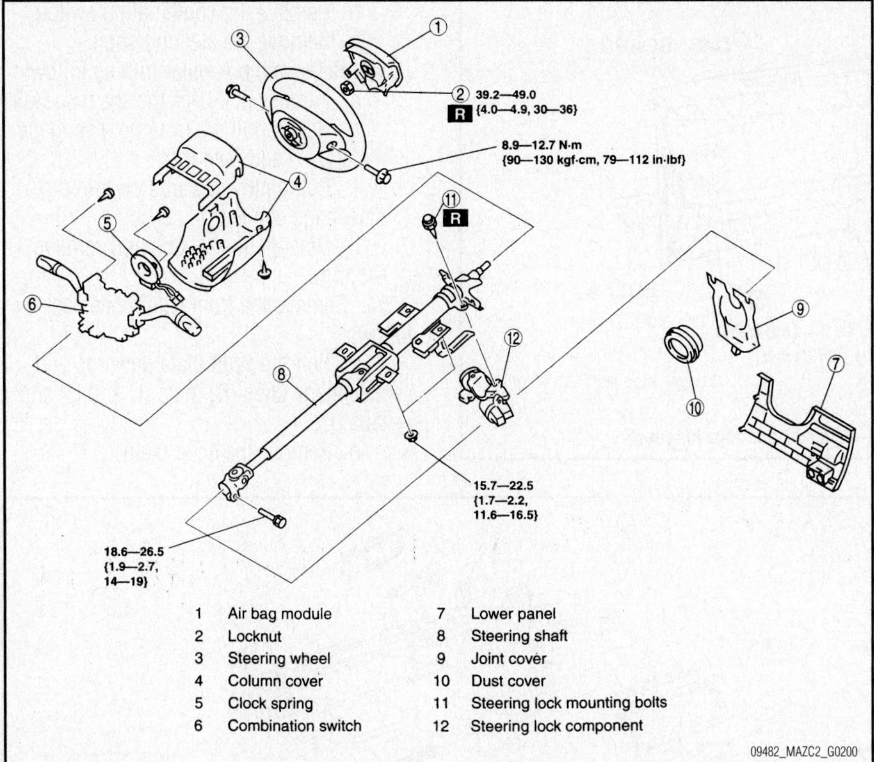

Fig. 568 Exploded view of the steering column and shaft assembly—Mazda6 and Mazdaspeed6 models

c. Remove the bolts, disconnect the socket connector and remove the console.

8. Remove the meter hood as follows:
   a. Remove the screws, pull the meter hood outwards and detach the column cover.
9. Remove the instrument cluster as follows:
   a. Lower the tilt steering wheel to its lowest point.
   b. Pull the steering wheel towards you.
   c. Remove the screws, connectors and cluster.
10. Remove the column cover.

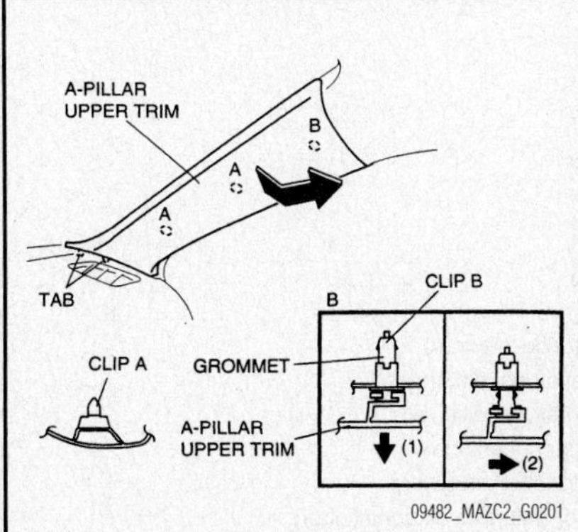

Fig. 569 Exploded view of the A-trim mounting—Mazda6 and Mazdaspeed6 models

11. Remove the lower panel as follows:
    a. Pull the hood release lever.
    b. While pushing the tab in the direction illustrated, using a tape wrapped small flat pry tool to pull the release lever outwards from the lower panel.
    c. Remove the lower panel screws.
    d. Pull the panel out to disengage the clips A from the tabs B and remove the panel.
12. Remove the steering shaft as follows:

WARNING

Handling the air bag module improperly can accidentally operate (deploy) the air bag module, which may seriously injure you.

a. Remove the air bag cover.
b. Remove the bolt.
c. Using a flathead screwdriver, pry out the connector stopper plate and unplug the connector.
d. Remove the driver's side air bag module.
e. Remove the lockbolt.

WARNING

Handling the air bag module improperly can accidentally deploy the air bag module, which may seriously injure you.

f. Remove the locknut.

CAUTION

Do not try to remove the steering wheel by hitting the shaft with a hammer. The column will collapse.

g. Set the vehicle in the straight-ahead position.
h. Remove the steering wheel using a suitable puller.
i. Remove the column cover.
j. Remove the clock spring connector.
k. Remove the screw.
l. Remove the clock spring.

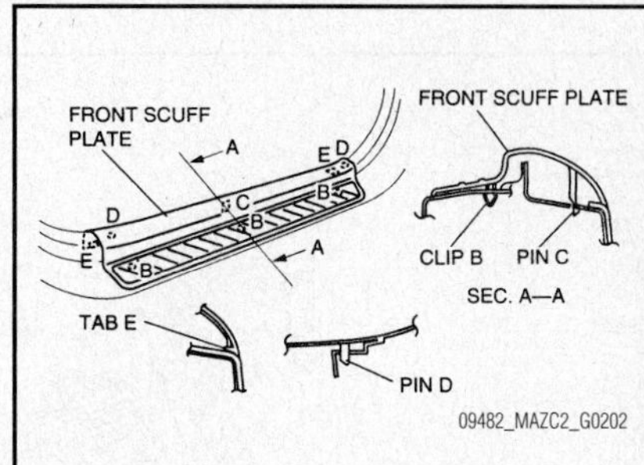

Fig. 570 Exploded view of the front scuff plate mounting—Mazda6 and Mazdaspeed6 models

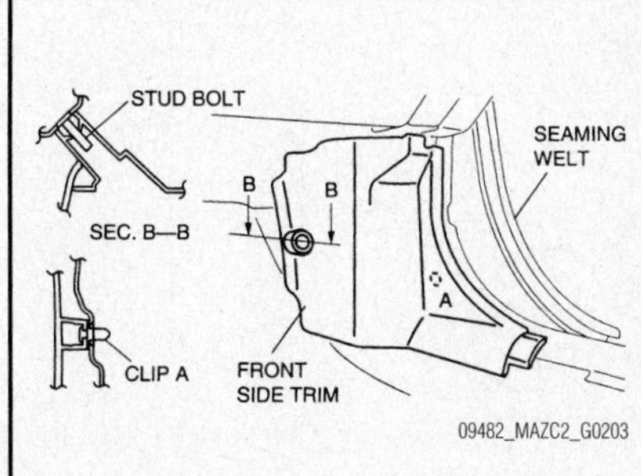

Fig. 571 Exploded view of the front side trim mounting—Mazda6 and Mazdaspeed6 models

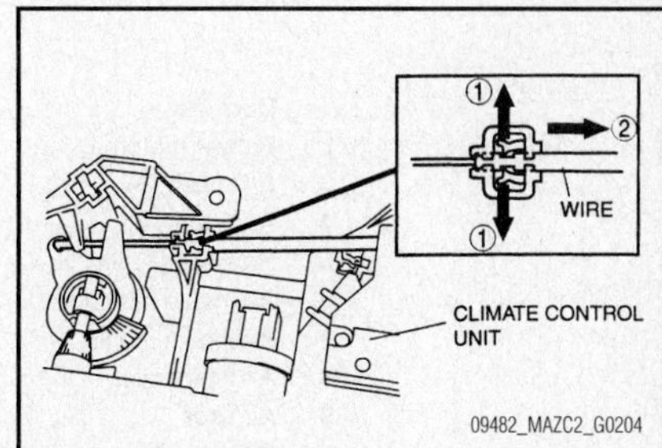

Fig. 572 If equipped with a wire type climate control unit, disconnect the wires for the front A/C unit as shown—Mazda6 and Mazdaspeed6 models

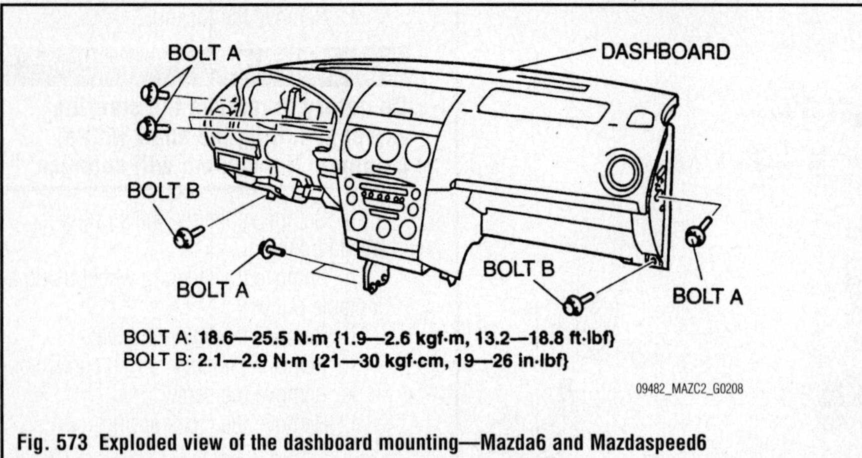

Fig. 573 Exploded view of the dashboard mounting—Mazda6 and Mazdaspeed6

m. Remove the combination switch.

n. Remove the steering shaft.

13. Remove the A-pillar trim as follows:

a. Partially pull back the seaming welt.

b. Gently pull the trim the disengage the clips A and then B (1).

c. Pull gently upwards on the trim to disengage clip B (2).

d. Disengage the tabs and remove the trim.

14. Remove the front scuff plates as follows:

a. Pull the scuff plate upwards and disengage clips (B), pins (C and D) and tabs (E).

b. Remove the scuff plate.

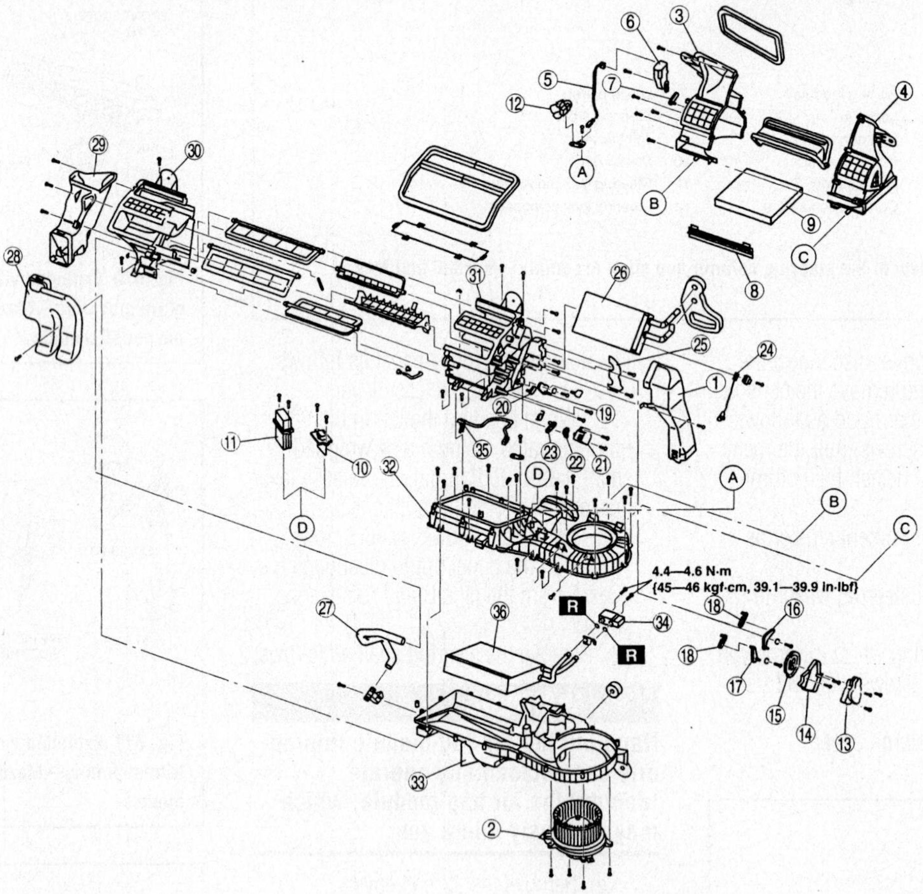

4.4—4.6 N·m {45—46 kgf·cm, 39.1—39.9 in·lbf}

1 Duct (1)
2 Blower motor
3 A/C case (3)
4 A/C case (4)
5 Harness (1)
6 Air intake actuator
7 Air intake crank
8 Air filter cover
9 Air filter
10 Resistor (manual air conditioner)
11 PWM unit (full-auto air conditioner)
12 Harness (2) (full-auto air conditioner)
13 Airflow mode actuator
14 Airflow mode bracket
15 Airflow mode main link
16 Airflow mode sub link (1)
17 Airflow mode sub link (2)
18 Airflow mode crank
19 Air mix link (1) (manual air conditioner)
20 Air mix crank (1) (manual air conditioner)
21 Air mix actuator (full-auto air conditioner)
22 Air mix link (2) (full-auto air conditioner)
23 Air mix crank (2) (full-auto air conditioner)
24 Water temperature sensor (full-auto air conditioner)
25 Heater core cover
26 Heater core

09482_MAZC2_G0206

Fig. 574 Exploded view of the A/C unit assembly—Mazda6

15. Remove the front side trims as follows:
  a. Partially pull back the seaming welt.
  b. Pull the trim outwards to disengage clip (A) and the stud bolt, then remove the trim.
16. Remove the side panels using a flat bladed pry tool to disengage the clips.
17. If equipped with a wire type climate control unit, disconnect the wires for the front A/C unit as follows:
  a. Remove the illumination bulb.
  b. Remove the unit.
  c. Remove the fan switch.
  d. Disassemble the wire as illustrated.
18. Disconnect the dashboard wiring harness connectors.
19. Remove the dashboard bolts and pull the dashboard pins out from the body.

⁂ WARNING

When removing the dashboard, make sure to support the dashboard is properly supported to avoid injury and also to avoid damage to the dashboard.

20. Remove the dashboard through the driver's side door.
21. Remove the theft control module bolt, connector and the module.

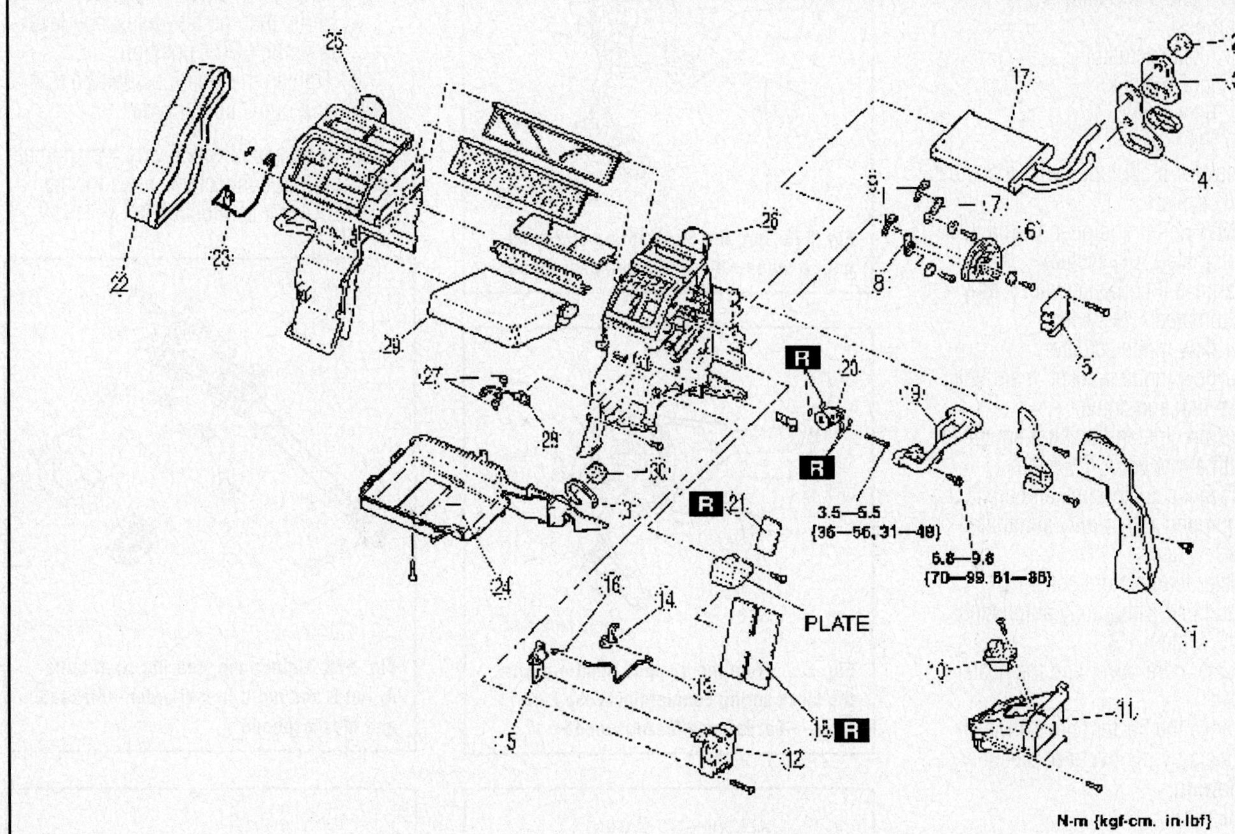

1 Duct (1)
2 Polyurethane protector (1)
3 Polyurethane protector (2)
4 Polyurethane protector (3)
5 Airflow mode actuator
6 Airflow mode main link
7 Airflow mode sub link (2)
8 Airflow mode sub link (3)
9 Airflow mode crank
10 Power MOS FET
11 Duct (2)
12 Air mix actuator
13 Air mix rod
14 Air mix crank (1)
15 Air mix crank (2)
17 Heater core
18 Adhesive polyurethane (1)
19 Outlet pipe
20 Expansion valve
21 Adhesive polyurethane (2)
22 Duct (3)
23 A/C water temperature sensor
24 A/C case (3)
25 A/C case (1)
26 A/C case (2)
27 Sensor clamp
28 Evaporator temperature sensor
29 Evaporator
30 Polyurethane protector (4)
31 Adhesive polyurethane (3)

09482_MAZC2_G0207

Fig. 575 Exploded view of the A/C unit assembly—Mazdaspeed6

22. Disconnect the heater hose at the firewall.

23. Disconnect the block joint type pipes by grasping female side of the block with pliers or similar tool and holding firmly, then remove the connection bolt or nut.

24. Remove the heater core on Mazd6 models by removing or disconnecting the following:

- Duct
- Blower motor.
- A/C case 3 and case 4
- Harness
- Air intake actuator
- Air intake crank
- Air filter cover
- Air filter
- Resistor if equipped with a manual A/C system
- PMW unit if equipped with a fully automated A/C system
- Harness if equipped with a fully automated A/C system
- Air flow mode actuator
- Air flow mode bracket, main link, sub link and crank
- Air mix link and crank if equipped with a manual A/C system
- Air mix actuator, link and crank if equipped with a fully automated A/C system
- Water temperature sensor if equipped with a fully automated A/C system
- Heater core cover and the heater core

25. Remove the heater core on Mazdspeed6 models by removing or disconnecting the following:

- Duct (1)
- Polyurethane protector 1, 2 and 3
- Airflow mode actuator, main link, sub link (2) and sub link (3) and mode crank
- Power MOS FET
- Duct (2)
- Air mix actuator, rod, crank (1 and 2) and air mix rod holder
- Heater core

To install:

26. Installation is the reverse of removal, keep in mind the following>

a. Adjust the clock spring as follows:

**** CAUTION**

If the clock spring is not adjusted, the spring wire in the clock spring could over-wind and break when the steering wheel is turned. Always adjust the clock spring after installing it.

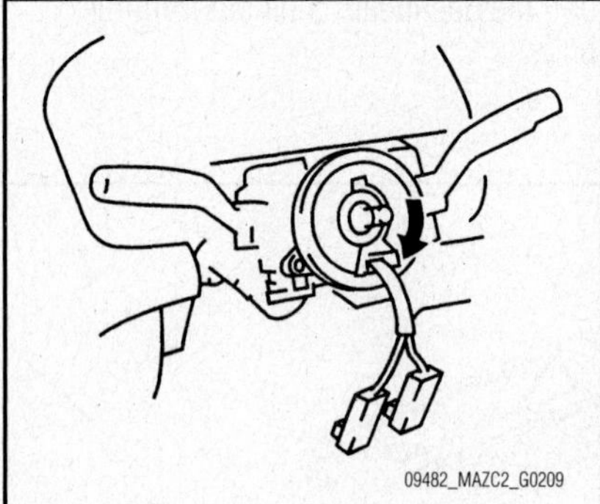

09482_MAZC2_G0209

Fig. 576 Turn the clock spring clockwise until it stops—Mazda6 and Mazdaspeed6

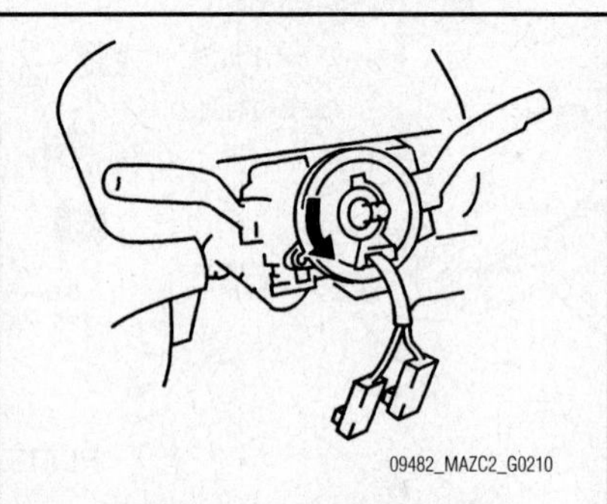

09482_MAZC2_G0210

Fig. 577 From the stopped position, turn the clock spring counterclockwise 2¾ turns—Mazda6 and Mazdaspeed6

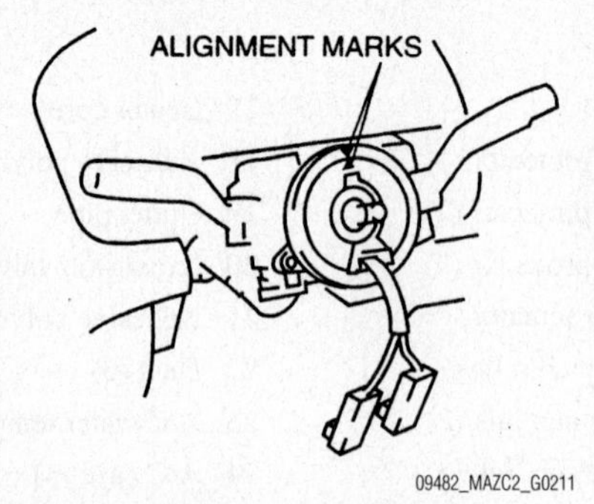

09482_MAZC2_G0211

Fig. 578 Align the marks on the clock spring—Mazda6 and Mazdaspeed6

- Set the front tires straight-ahead

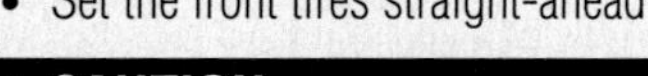

**** CAUTION**

The clock spring will break if overwound. Do not forcibly turn the clock spring.

- Turn the clock spring clockwise until it stops
- From the stopped position, turn the clock spring counterclockwise 2 ¾ turns
- Align the marks
- Verify that the tilt / telescope lever is in the LOCK position

b. Tighten the steering shaft bolts A, nut B and nut C in that order.

c. If equipped with a wire type climate control unit, connect the wires for the front A/C unit as illustrated.

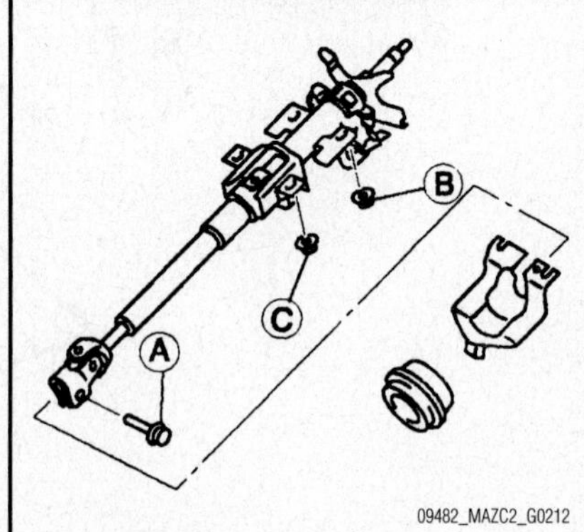

09482_MAZC2_G0212

Fig. 579 Tighten the steering shaft bolts A, nut B and nut C in that order—Mazda6 and Mazdaspeed6

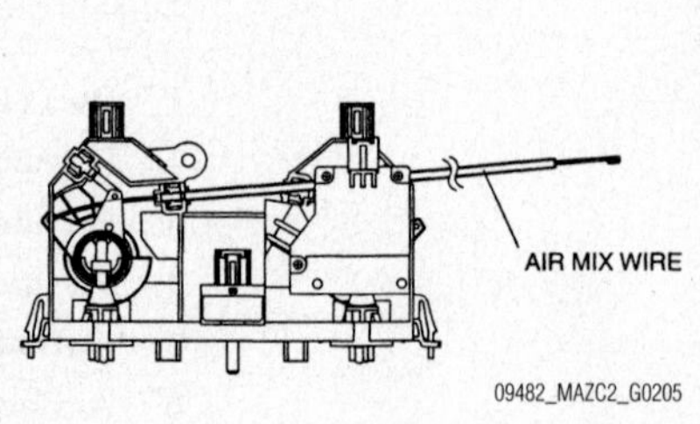

09482_MAZC2_G0205

Fig. 580 If equipped with a wire type climate control unit, connect the wires for the front A/C unit as shown—Mazda6 and Mazdaspeed6 models

STEERING

POWER RACK & PINION STEERING GEAR

REMOVAL & INSTALLATION

Mazda3

See Figures 581 and 582.

**** WARNING**

Performing the following procedures without first removing the ABS wheel-speed sensor may possibly cause an open circuit in the wiring harness if it is pulled by mistake.

Before performing the following procedures, disconnect the ABS wheel-speed sensor connector (axle side) and fix the wiring harness to an appropriate place where it will not be pulled by mistake while servicing the vehicle.

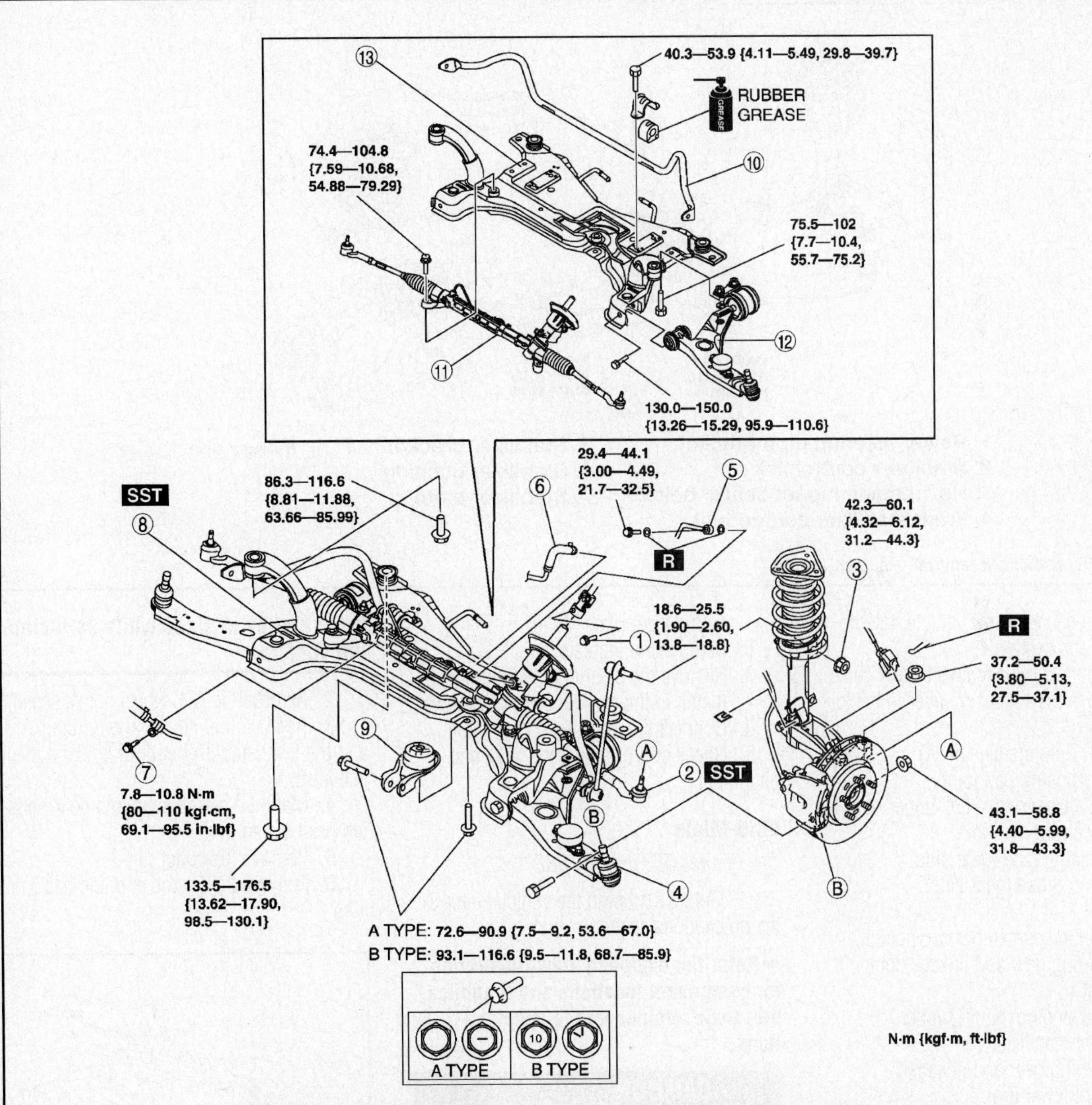

Fig. 581 Front crossmember exploded view—Mazda3

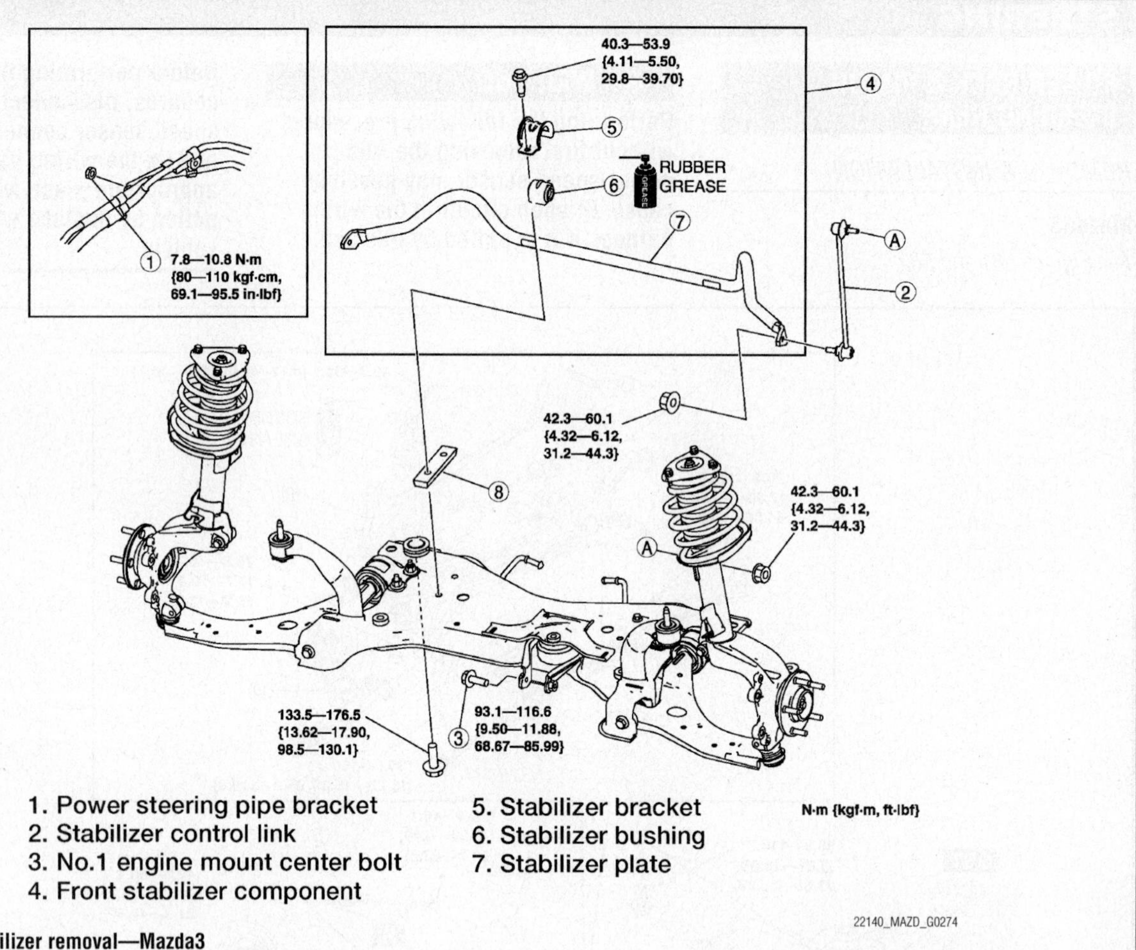

Fig. 582 Front stabilizer removal—Mazda3

1. Before servicing the vehicle, refer to the precautions section.
2. Disconnect the negative battery cable.
3. Remove front crossmember in the following order:
   - Bolt (intermediate shaft)
   - Tie-rod end ball joint
   - Stabilizer control link upper nut
   - Lower arm ball joint
   - Pressure pipe (gear side)
   - Return hose (gear side)
   - Bolt
   - Front crossmember component, steering gear and linkage component
   - No.1 engine mount rubber
   - Front stabilizer
   - Steering gear and linkage
   - Front lower arm
   - Front crossmember
4. Remove front stabilizer in the following order:
   - Power steering pipe bracket
   - Stabilizer control link
   - No.1 engine mount center bolt
   - Front stabilizer component
   - Stabilizer bracket
   - Stabilizer bushing
   - Stabilizer plate
5. Remove the insulator.
6. Remove the steering gear and linkage.
7. Install in the reverse order of removal.
8. Connect the negative battery cable.
9. Check and/or adjust the front end alignment.

MX5-Miata

See Figures 583 through 587.

1. Before servicing the vehicle, refer to the precautions section.

➡Refer the exploded view illustration for component locations and if applicable, their retainer torque specifications.

⁂ CAUTION

Performing the following procedures without first removing the ABS wheel-speed sensor may possibly cause an open circuit in the harness if it is pulled by mistake. Before performing the following procedures, remove the ABS wheel-speed sensor (axle side) and fix it to an appropriate place where the sensor will not be pulled by mistake while servicing the vehicle.

2. Remove the ABS wheel-speed sensor.
3. Remove the radiator mount bracket and the front stabilizer control link and the stabilizer.
4. Mark the pinion shaft and gear housing prior to removal.
5. Remove the cotter pin.
6. Remove the tie-rod end and ball joint nuts.

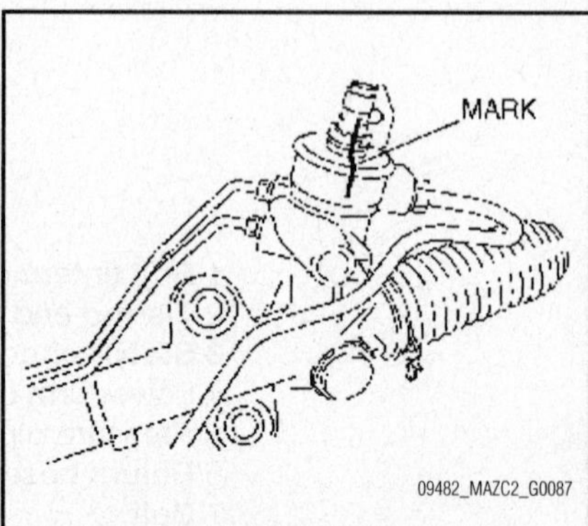

Fig. 583 Mark the pinion shaft and gear housing prior to removal—MX-5 Miata

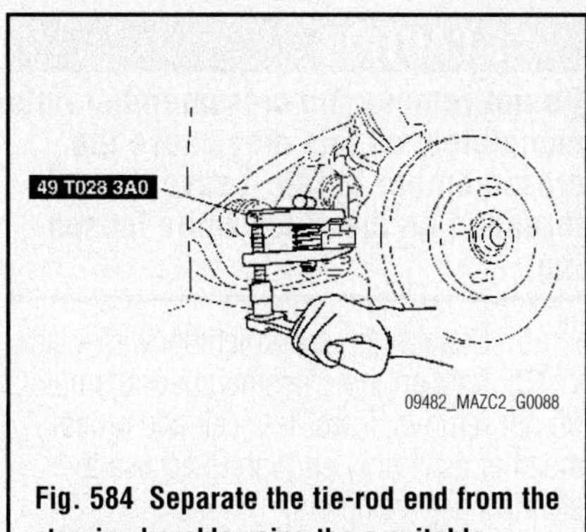

Fig. 584 Separate the tie-rod end from the steering knuckle using the a suitable puller—MX-5 Miata

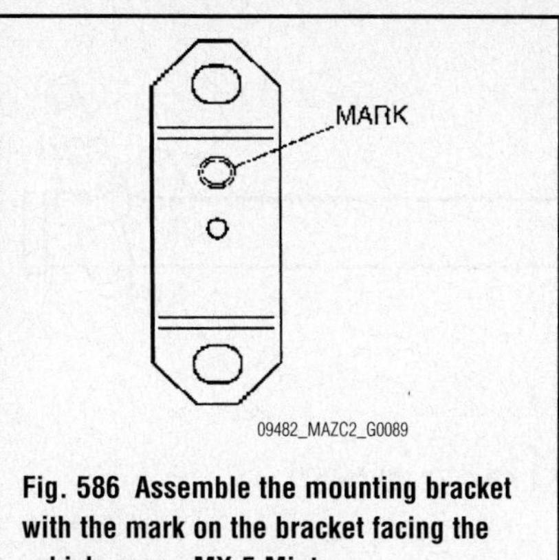

Fig. 586 Assemble the mounting bracket with the mark on the bracket facing the vehicle rear—MX-5 Miata

7. Remove the tie-rod nut.
8. Separate the tie-rod end from the steering knuckle using the a suitable puller.
9. Remove the lower mounting rubber bracket.
10. Remove the pressure pipe and return hose
11. Remove the steering gear and linkage retainers.
12. Remove the steering gear and linkage by pulling it from the right side of the vehicle.

To install:

13. Install the gear and linkage assembly and loosely tighten the bolts.
14. Assemble the mounting bracket with the mark on the bracket facing the vehicle rear.
15. Tighten the mounting bracket bolts to the 74.4–104.8 Nm (54.88–77.29 ft. lbs.) in the order shown.
16. Install the remaining components in the reverse order of removal using the accompanying illustration for torque values.
17. When installing the intermediate shaft. Align the marks and install the

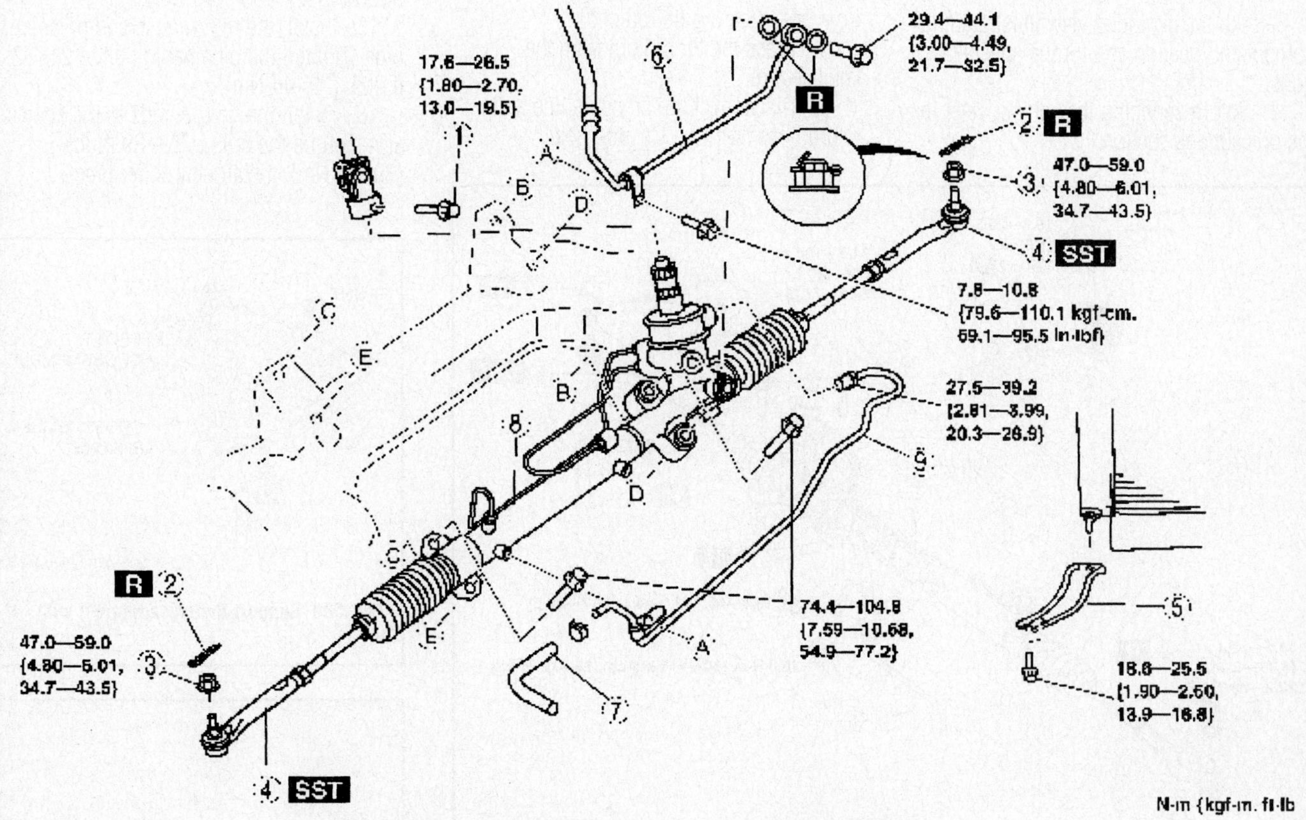

N·m {kgf·m, ft·lb}

1 Bolt (intermediate shaft)
2 Cotter pin
3 Nuts (tie-rod end ball joint)
4 Tie-rod end ball joint
5 Lower mounting rubber bracket
6 Pressure pipe
7 Return hose
8 Steering gear and linkage
9 Return pipe

09482_MAZC2_G0086

Fig. 585 Exploded view of the steering gear and linkage assembly—MX-5 Miata

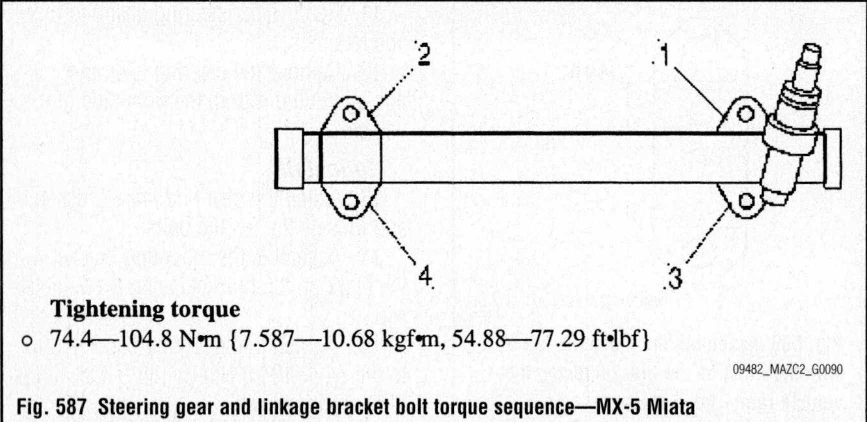

Fig. 587 Steering gear and linkage bracket bolt torque sequence—MX-5 Miata

intermediate shaft and bolt. Tighten to 17–26 Nm (13–19 ft. lbs.).

18. After installation, adjust alignment.

Mazda6 and Mazdaspeed6

2.3L (L3) Engines

See Figures 588 through 590.

Refer to the exploded view illustration for component location and torque specifications.

1. Before servicing the vehicle, refer to the precautions section.
2. Remove the ABS speed sensor.
3. Disconnect the negative battery cable.
4. Remove the front wheels.
5. Remove the intermediate shaft to steering gear pinion shaft bolt. Mark the shaft-to-gear location.
6. Remove the cotter pins and nuts from both steering tie rod ends
7. Press the tie rod out from the knuckle arm.
8. Disconnect the pressure line and return hose from the steering gear.

⁂ CAUTION

Do not remove the crossmember nuts completely as this may cause the crossmember to fall. Leave the nuts threaded on the studs while loosening.

9. Support the crossmember with a jack.
10. Loosen the crossmember nuts but do not remove, lower the jack and crossmember assembly enough to access to steering gear retainers.
11. Remove the steering gear and linkage from the left side.

To install:

12. Install the steering gear and linkage.
13. Tighten the gear mounting bolts in the sequence illustrated to 55–77 ft. lbs. (74–104 Nm).
14. Raise the crossmember assembly into position and tighten the nuts to 89–114 ft. lbs. (119–154 Nm) and the bolts to 68–86 ft. lbs. (93–116 Nm).
15. Install the return hose and pressure pipe. Tighten the pipe banjo bolt to 21–32 ft. lbs. (29–44 Nm).
16. Install the tie rod ends to the knuckle arm. Torque the nuts to 29–39 ft. lbs. (39–53 Nm). Install new cotter pins .

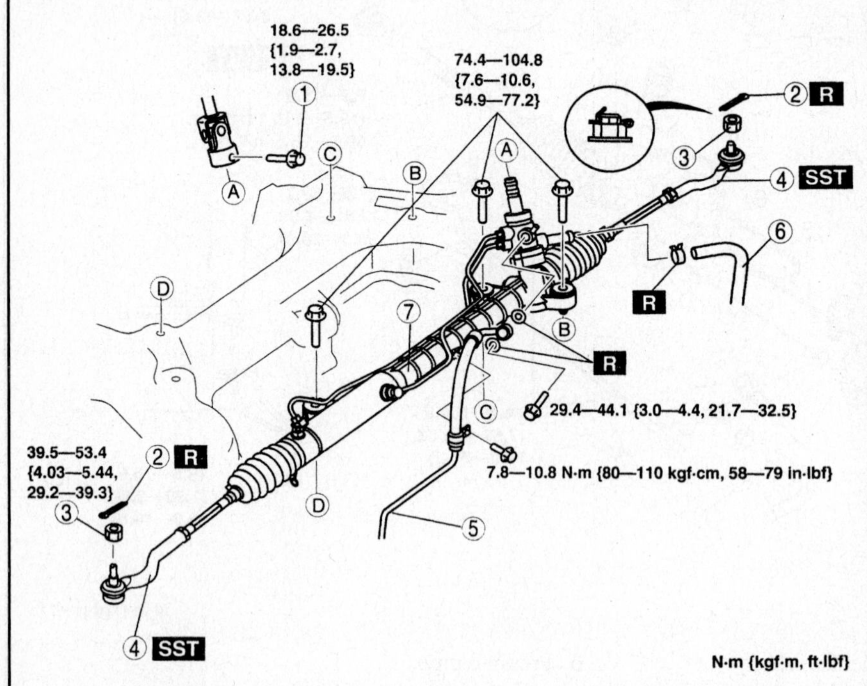

Fig. 588 Exploded view of the power steering gear assembly–Mazda6and Mazdaspeed6 with the 2.3L (L3) engine

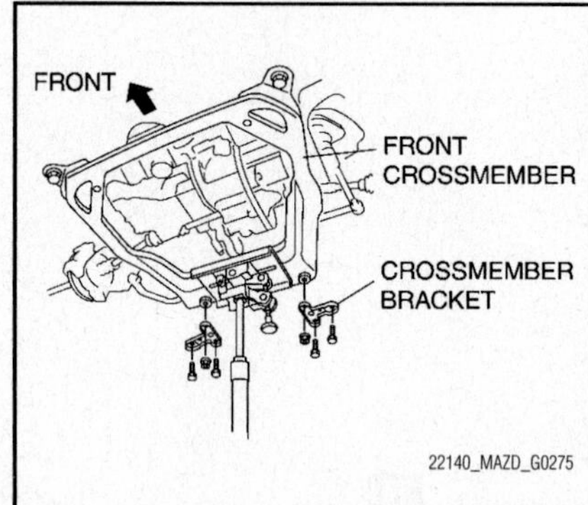

Fig. 589 Support the crossmember with a jack

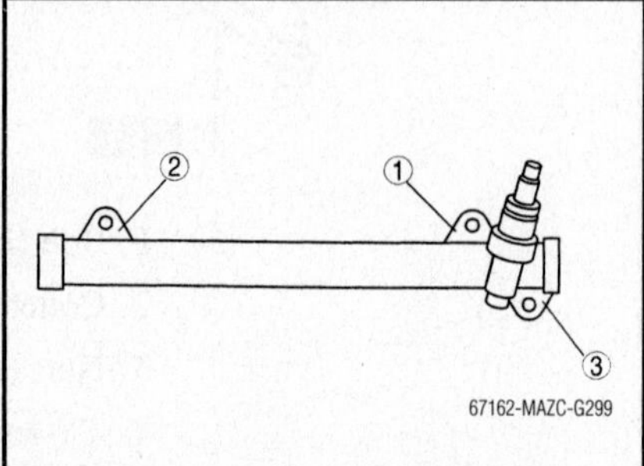

Fig. 590 Exploded view of the power steering gear assembly–Mazda6 and Mazdaspeed6 models

17. Align the intermediate shaft–to–gear marks made during removal and tighten the bolt to 13–19 ft. lbs. (18–26 Nm). Check the power steering fluid level.

18. Install the abs sensor and the wheels.

19. Connect the negative battery cable.

20. Check and/or adjust the front end alignment.

3.0L (AJ) Engines

See Figure 591.

Refer to the exploded view illustration for component location and torque specifications.

1. Before servicing the vehicle, refer to the precautions section.

2. Disconnect the negative battery cable.

3. Remove the ABS speed sensor.

4. Remove the front wheels, undercover and splash shield.

5. Separate the stabilizer bar link at the shock absorber side.

6. Separate the front lower arm front and rear ball joints from the knuckle.

7. Remove the lower side shock absorber bolt.

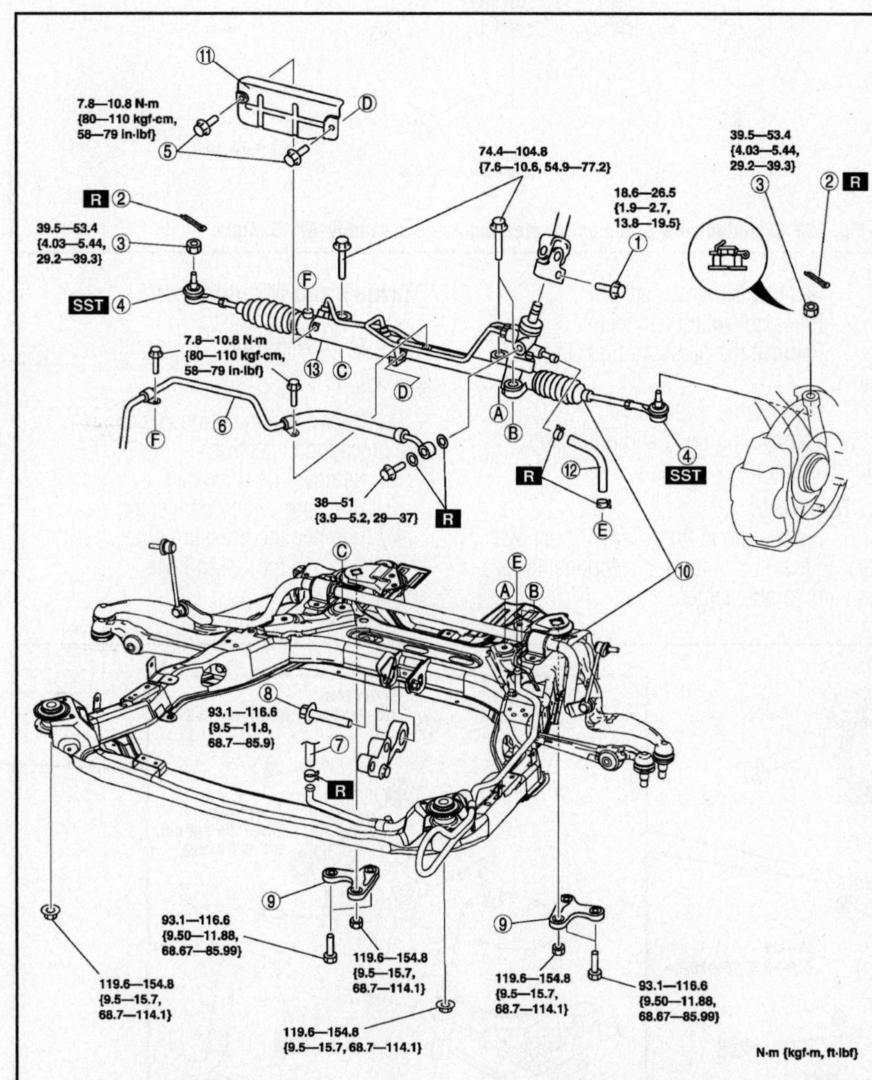

1. Bolt (intermediate shaft)
2. Cotter pin
3. Nuts (tierod end ball joint)
4. Tierod end ball joint
5. Insulator bolts
6. Pressure pipe
7. Return hose
8. No.1 engine mount center bolt
9. Crossmember brack
10. Crossmember component, steering gear and linkage
11. Insulator
12. Return hose
13. Steering gear and linkage

22140_MAZD_G0276

Fig. 591 Exploded view of the power steering gear assembly–Mazda6 with the 3.0L (AJ) engine

8. Remove the intermediate shaft to steering gear pinion shaft bolt. Mark the shaft-to-gear location.

9. Remove the cotter pins and nuts from both steering tie rod ends

10. Press the tie rod out from the knuckle arm.

11. Remove the heat shield bolts.

12. Disconnect the pressure line and return hose from the steering gear.

13. Remove the No. 1 engine mount center bolt.

⁂ CAUTION

Support the crossmember with a jack and make sure the jack is attached securely to the crossmember before removing the bracket.

14. Support the crossmember with a jack.

15. Remove the crossmember bracket.

16. Remove the crossmember bolts and lower the crossmember assembly with the gear and linkage attached.

17. Remove the heat shield and return hose.

18. Remove the steering gear and linkage.

To install:

19. Install the steering gear and linkage to the crossmember assembly.

20. Tighten the gear mounting bolts in the sequence illustrated to 55–77 ft. lbs. (74–104 Nm).

21. Install the return hose and heat shield but do not tighten the shield bolts.

22. Raise the crossmember assembly into position and tighten the retainers to the specifications shown in the accompanying illustration.

23. Install the crossmember bracket and tighten the retainers to the specifications shown in the accompanying illustration.

24. Install the No. 1 engine mount center bolt and tighten to 68–86 ft. lbs. (93–116 Nm).

25. Install the return hose and pressure pipe. Tighten the pipe banjo bolt to 21–32 ft. lbs. (29–44 Nm).

26. Install the heat shield bolts and tighten to 79 inch lbs. (10 Nm).

27. Install the tie rod ends to the knuckle arm. Torque the nuts to 29–39 ft. lbs. (39–53 Nm). Install new cotter pins .

28. Align the intermediate shaft–to–gear marks made during removal and tighten the bolt to 13–19 ft. lbs. (18–26 Nm). Check the power steering fluid level.

29. Install the lower side shock absorber bolt.

30. Attach the front lower arm front and rear ball joints to the knuckle.

31. Attach the stabilizer bar link to the shock absorber side.
32. Install the abs sensor, wheels, splash shield and undercover.
33. Connect the negative battery cable.
34. Check and/or adjust the front end alignment.

POWER STEERING PUMP

REMOVAL & INSTALLATION

Mazda3 and Mazdaspeed3

See Figure 592.

1. Remove the undercover, splash shield and mudguard.
2. Remove the front bumper.
3. Remove the washer tank.
4. Unplug the electrical connector from the pump.
5. Disconnect the pressure pipe.
6. Remove the power steering fluid reserve tank.
7. Remove the electric power steering oil pump and bracket assembly.
8. Remove the bracket and disconnect to the return hose.
9. Remove the pump.

To install:

10. Installation is the reverse of removal, refer to the illustration for components location and torque values.

MX5-Miata

See Figure 593.

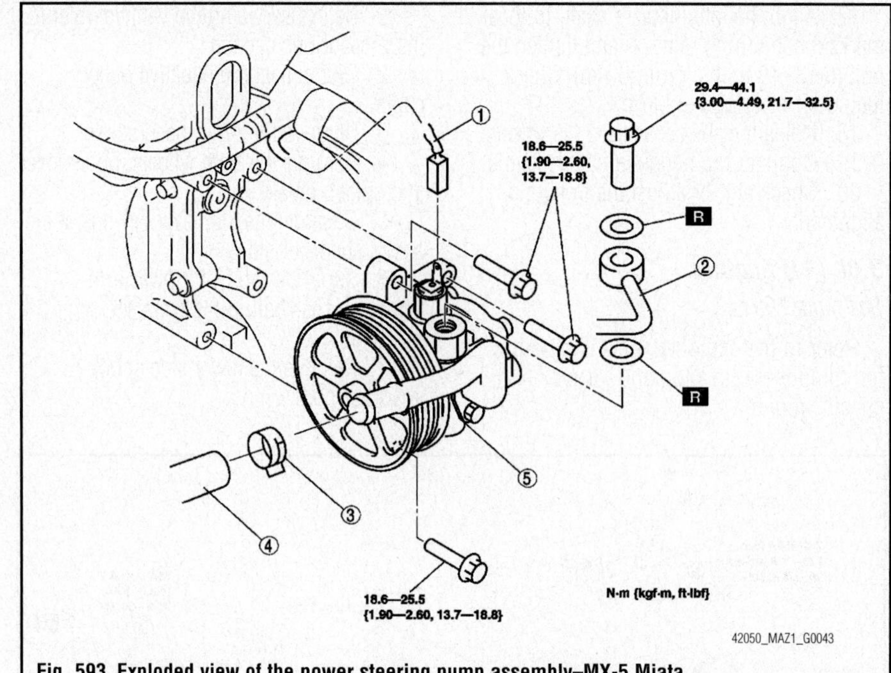

Fig. 593 Exploded view of the power steering pump assembly–MX-5 Miata

1. Remove the drive belt.
2. Pressure switch connector.
3. Remove the pressure pipe and hose band.
4. Remove the return hose.
5. Remove the bolts and the power steering pump.

To install:

6. Installation is the reverse of removal, refer to the illustration for components location and torque values.

Mazda6 and Mazdaspeed6

2.3L (L3) Engines

See Figure 594.

1. Before servicing the vehicle, refer to the precautions section.
2. Remove the drive belt.
3. Pressure switch connector.
4. Remove the pressure pipe.
5. Remove the return hose.

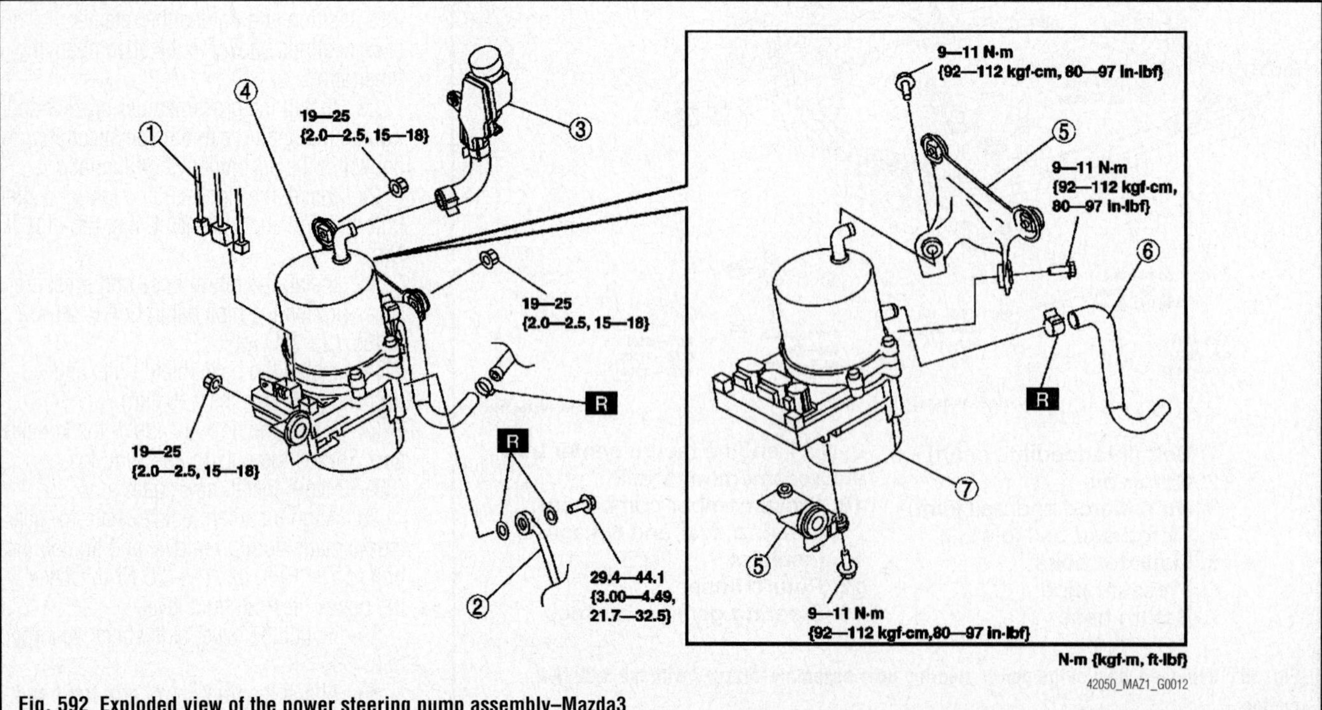

Fig. 592 Exploded view of the power steering pump assembly–Mazda3

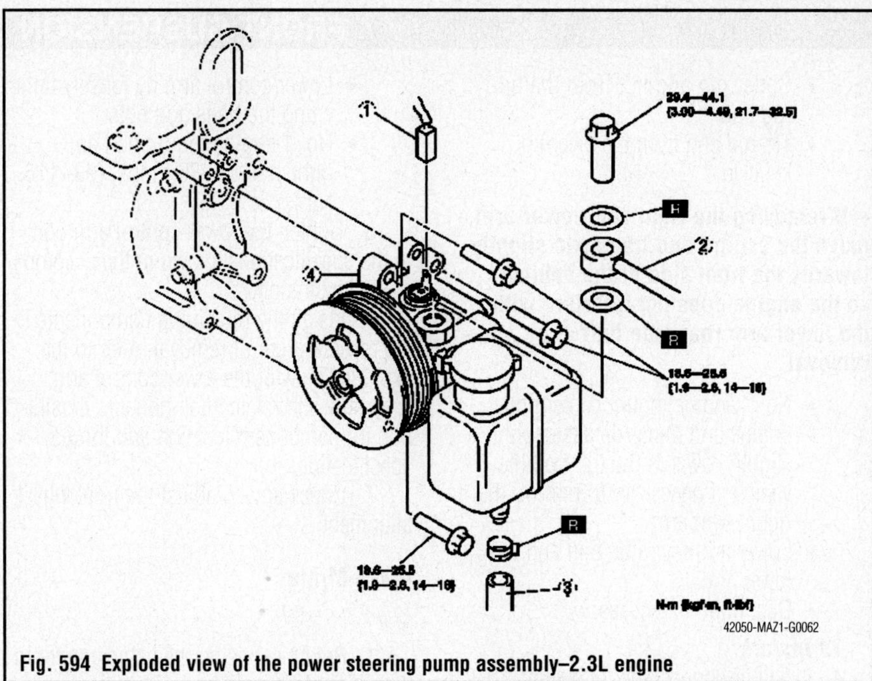

Fig. 594 Exploded view of the power steering pump assembly–2.3L engine

6. Remove the bolts and the power steering pump.

To install:

7. Installation is the reverse of removal, refer to the illustration for components location and torque values.

3.0L (AJ) Engines

See Figure 595.

1. Before servicing the vehicle, refer to the precautions section.
2. Remove the right side splash shield
3. Remove the undercover.
4. Remove the front drive belt.
5. Remove the dipstick tube.
6. Remove the A/C compressor bolts.
7. Remove the left hand exhaust manifold.
8. Pressure switch connector.
9. Remove the pressure pipe.
10. Remove the return hose.
11. Remove the bolts and the power steering pump bracket.
12. Remove the bolts and the power steering pump.

To install:

13. Installation is the reverse of removal, refer to the illustration for components location and torque values.

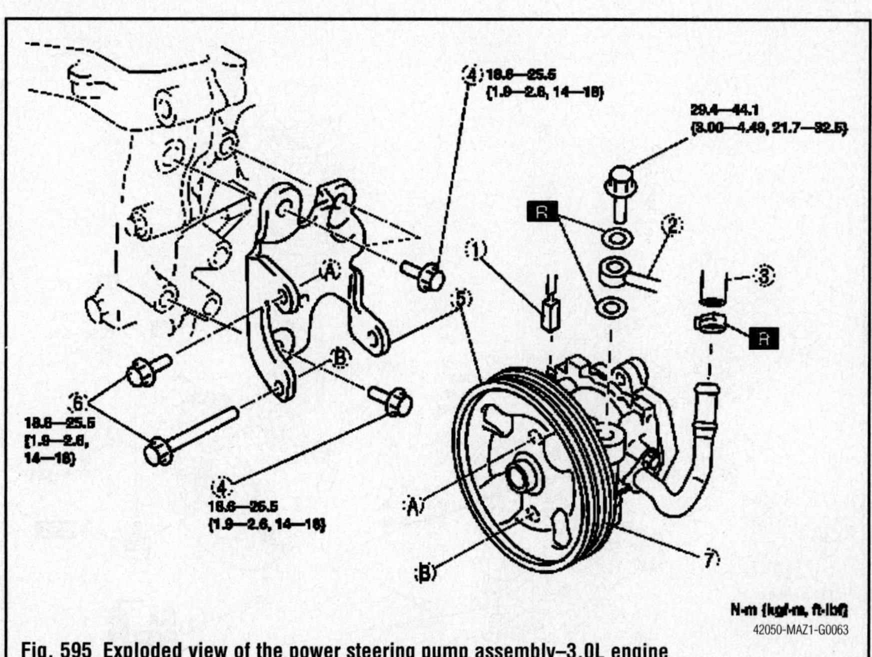

Fig. 595 Exploded view of the power steering pump assembly–3.0L engine

BLEEDING

CAUTION

Do not hold the steering wheel fully turned for 5 seconds or more. It is possible that oil temperature can rise and this will negatively affect the oil pump.

1. Check and top off the fluid level.
2. Turn the steering wheel fully to the left and right several times.
3. Inspect the fluid level and add fluid as needed.
4. Turn the steering wheel fully to the left and right several times until the fluid level stabilizes.
5. Start the engine and let it idle.
6. Turn the steering wheel fully to the left and right several times.
7. Turn the steering wheel fully to the left and right several times until the fluid is not foamy and the fluid level has not dropped.
8. Check and top off the fluid level.

LOWER BALL JOINT

REMOVAL & INSTALLATION

The lower ball joint is an integral part of the lower control and cannot be replaced separately. If the lower ball joint is defective, the entire lower control arm must be replaced. Refer to the lower control arm procedure.

LOWER CONTROL ARM

REMOVAL & INSTALLATION

Mazda3 Models

See Figure 596.

1. Before servicing the vehicle, refer to the precautions section.
2. Remove or disconnect the following:
   - Wheel
   - Cotter pin
   - Lower ball joint by loosening it
3. With the nut protecting the ball joint stud, separate the stud from the knuckle. Remove the nut.
   - Cotter pin and nut from the tie-rod end
   - Tie-rod end from the steering knuckle

➡If removing the right side lower arm, move the engine and transaxle slightly towards the front side of the vehicle so the engine does not interfere with the lower arm rear side bolt removal.

   - No. 1 engine mount center bolt.
   - Engine and transaxle assembly slightly towards the front off the vehicle if necessary to remove the right lower arm
   - Lower arm rear side bolt and the lower arm.
   - Dust boot, if necessary

To install:

4. Install or connect the following:
   - Dust boot, if removed using a press. Always fill the inside of the new boot with grease prior to installation.
   - Lower control arm by loosely tightening the bolts and nuts
   - No. 1 engine mount bolt and tighten to 68–86 ft. lbs. (93–116 Nm).
5. Tighten the lower control arm bolts to the specifications shown in the accompanying illustration.
6. Install the remaining components in the reverse order of removal, refer to the exploded view of the lower control arm assembly and related components illustration for component location and torque specifications.
7. Check and/or adjust the front wheel alignment.

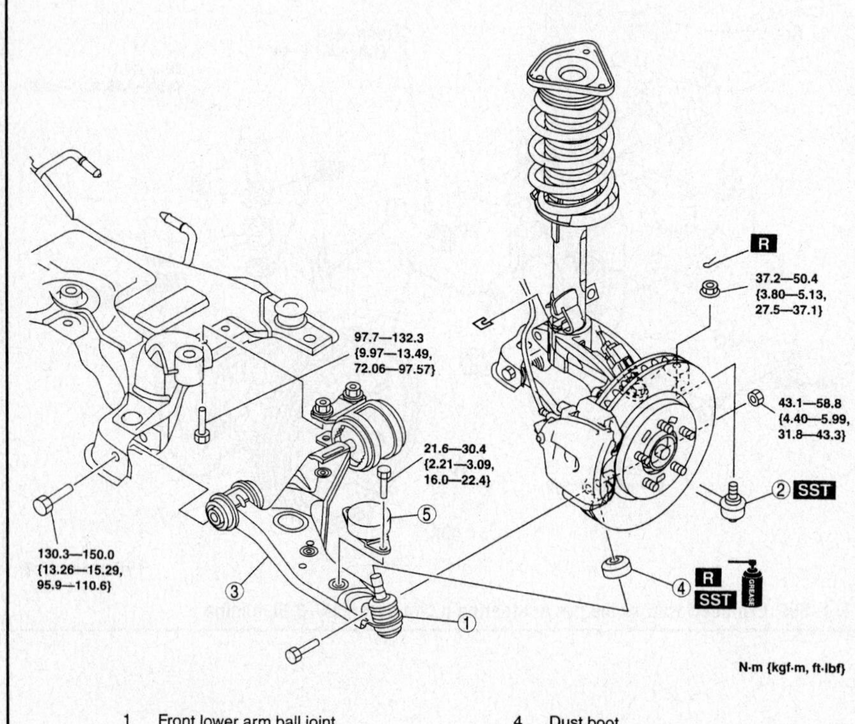

Fig. 596 Exploded view of the front lower control arm assembly and related components—Mazda3

MX-5 Miata

See Figure 597.

1. Before servicing the vehicle, refer to the precautions section.

➡Refer the exploded view illustration for component locations and if applicable, their retainer torque specifications.

✲✲ CAUTION

Performing the following procedures without first removing the ABS wheel-speed sensor may possibly cause an open circuit in the wiring harness if it is pulled by mistake. Before performing the following procedures, remove the ABS wheel-speed sensor (axle side) and fix it to an appropriate place where the sensor will not be pulled while servicing the vehicle.

2. Remove the caliper and mounting support from the steering knuckle and suspend it with a cable in a location out of the way.
3. Separate the front lower arm ball joint from the knuckle.

➡When removing the front lower arm ball joint, the steering knuckle bushing may also come off. If it comes off, replace the steering knuckle.

4. Disconnect the tie-rod end.
5. Separate the front upper arm ball joint from the knuckle
6. Remove the front hub and steering knuckle assembly.
7. Remove the stabilizer control link nut on the front lower arm side.
8. Remove the front lower arm.

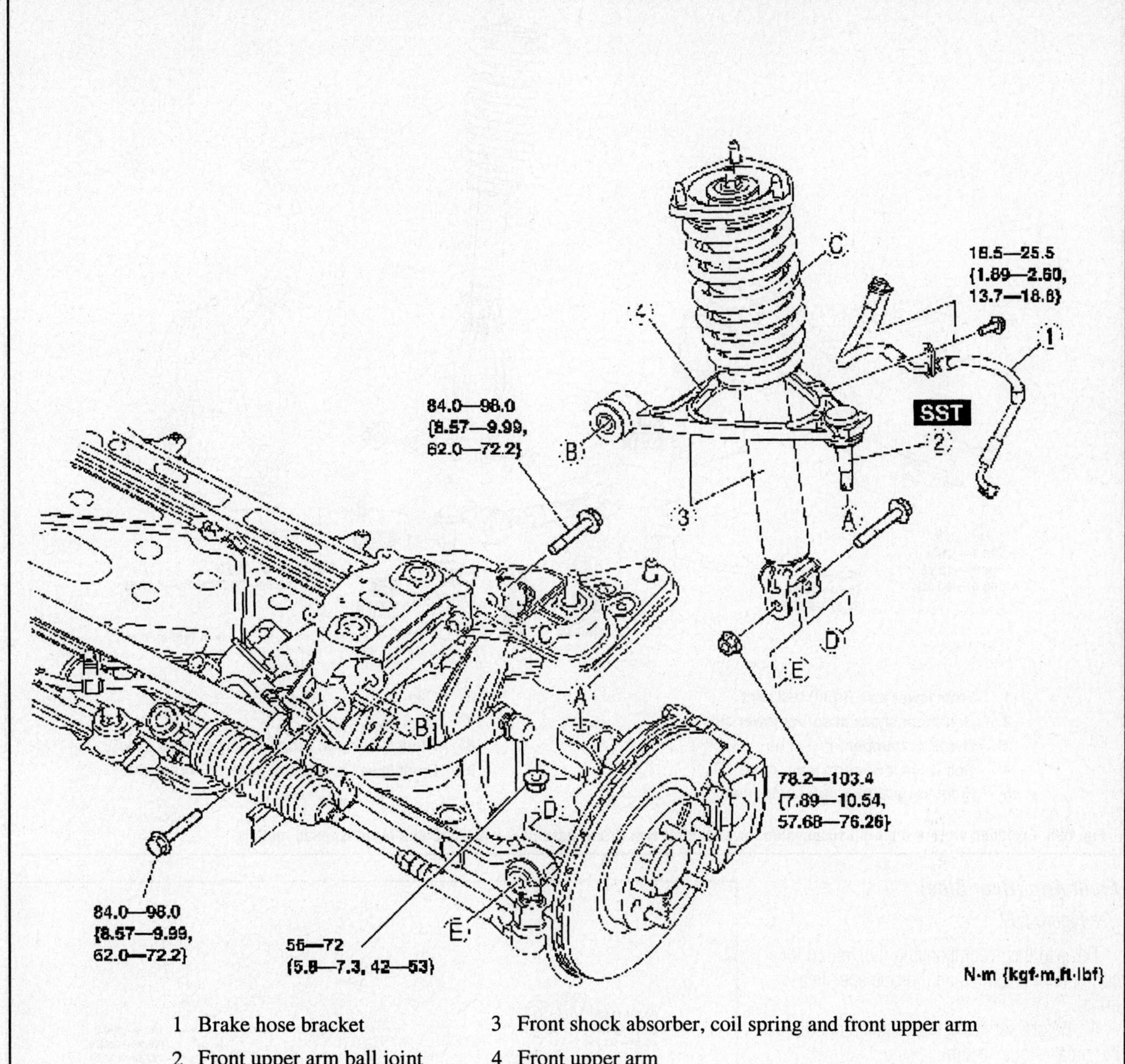

Fig. 597 Exploded view of the front upper control arm assembly—MX-5 Miata

To install:

9. Install all components in the reverse of removal.

10. Refer to illustration for tightening specifications.

Mazda6 Models

Front Arm (Front Side)

See Figure 598.

Refer to the accompanying illustration for component locations and torque specifications.

1. Before servicing the vehicle, refer to the precautions section.

2. Remove the three way converter on models with the 3.0L (AJ) engine.

3. Separate the lower control arm (front) ball joint from the knuckle.

4. Remove the front shock lower bolt and the dynamic damper.

5. Remove the lower inner side arm bolt.

6. Remove the front lower arm (front).

To install:

7. Installation is the reverse of removal. refer to the illustration for bolt torque specifications.

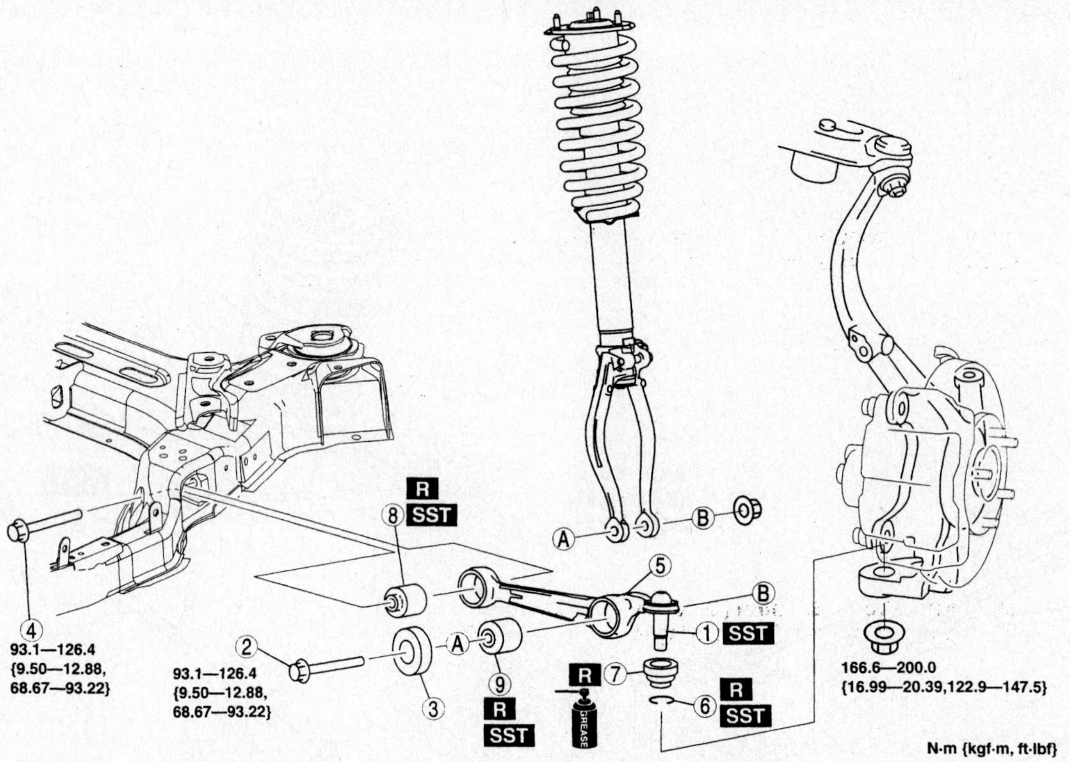

8 R SST
A
B
5
B
1 SST
4
93.1—126.4
{9.50—12.88,
68.67—93.22}
2
93.1—126.4
{9.50—12.88,
68.67—93.22}
A
9
3
R
SST
R 7
GREASE
R
6 SST
166.6—200.0
{16.99—20.39,122.9—147.5}
N·m {kgf·m, ft·lbf}

1	Front lower arm (front) ball joint	6	Clip
2	Bolt (front shock absorber lower side)	7	Dust boot
3	Dynamic damper	8	Front lower arm (front) bushing (inner side)
4	Bolt (front lower arm inner side)	9	Front lower arm (front) bushing (outer side)
5	Front lower arm (front) component		

67162-MAZC-G250

Fig. 598 Exploded view of the front suspension lower control arm (front) assembly—Mazda6 and Mazdaspeed6 models

Front Arm (Rear Side)

See Figure 599.

Refer to the accompanying illustration for component locations and torque specifications.

1. Before servicing the vehicle, refer to the precautions section.
2. Remove the wheels, engine undercover and splash shield.
3. Remove the steering gear and linkage bolts, pipe assembly from the crossmember and wire the assembly aside.
4. Remove the stabilizer bar link nut.
5. Separate the lower control arm (rear) ball joint from the knuckle.
6. Remove the No. 1 center bolt.
7. Support the crossmember with a jack and remove the crossmember bracket.
8. Remove the front lower arm (rear).

To install:

9. Installation is the reverse of removal. refer to the illustration for bolt torque specifications.

93.1—116.6
{9.50—11.88,
68.67—85.99}
3
166.6—200.0
{16.99—20.39,
122.9—147.5}
1
43.1—54.9
{4.40—5.59,
21.8—40.4}
93.1—126.4
{9.50—12.88,
68.67—93.22}
2 SST
4
119.6—154.8
{12.20—15.78,
88.22—114.1}
93.1—116.6
{9.50—11.88,
68.67—85.99}
8
SST
R
5
6
R
SST
7 R
GREASE
N·m {kgf·m, ft·lbf}

1	Nut (stabilizer control link lower side)	5	Front lower arm (rear)
2	Front lower arm (rear) ball joint	6	Clip
3	No.1 engine mount center bolt	7	Dust boot
4	Crossmember bracket	8	Front lower arm (rear) bushing

67162-MAZC-G251

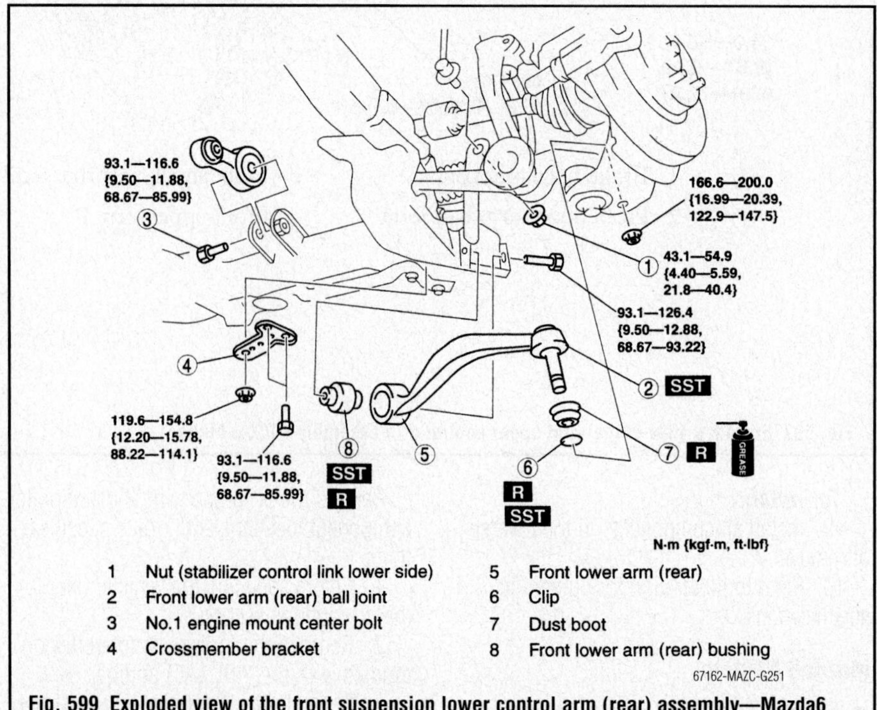

Fig. 599 Exploded view of the front suspension lower control arm (rear) assembly—Mazda6 and Mazdaspeed6 models

CONTROL ARM BUSHING REPLACEMENT

MX-5 Miata

See Figures 600 through 610.

1. Before servicing the vehicle, refer to the precautions section.
2. Remove the lower control arm.
3. Remove the clip.
4. Remove the dust boot.

➡Be careful not to damage the front lower arm. If it is damaged, replace it.

5. Mark the front upper arm as shown in the illustration.

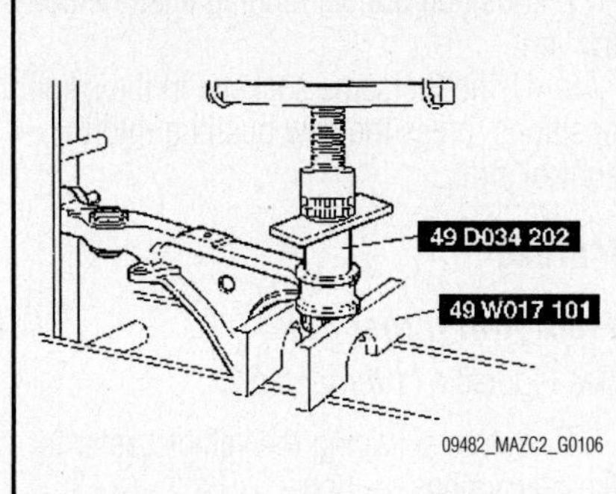

Fig. 603 Remove the front side bushing—MX-5 Miata

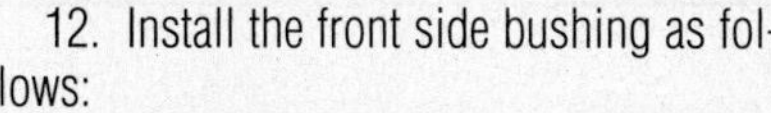

12. Install the front side bushing as follows:
   a. Press the bushing in using the tools illustrated.
   b. Insert the stopper plate rubber into the inner pipe of the bushing (front side).
13. Install the rear side bushing as follows:
   a. Align the marks placed during removal of the bushing.
   b. Press the bushing in using the tools illustrated.
14. Install the clip as follows:
   a. Wipe the grease off the ball joint stud.

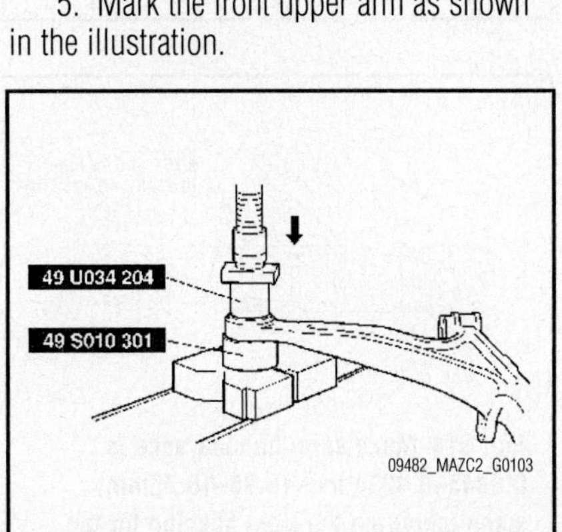

Fig. 600 Remove the rear side bushing—MX-5 Miata

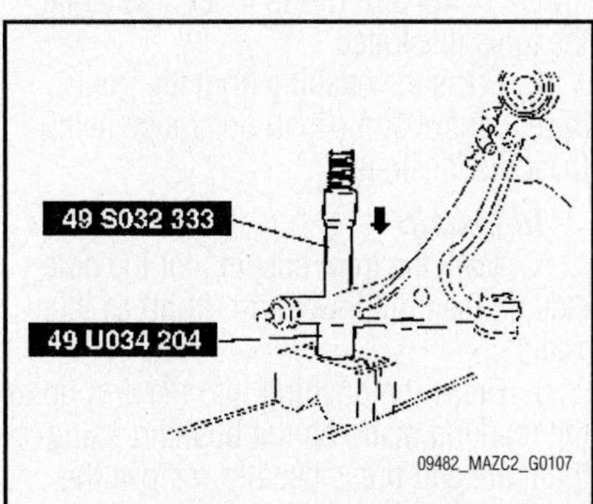

Fig. 604 Remove the shock absorber lower side bushing—MX-5 Miata

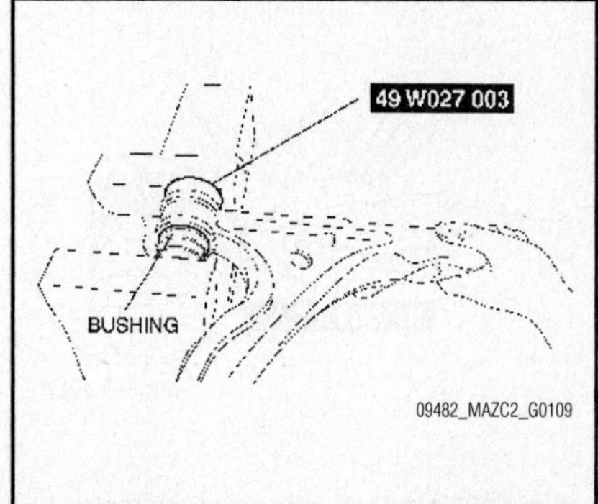

Fig. 606 Press the front side bushing in—MX-5 Miata

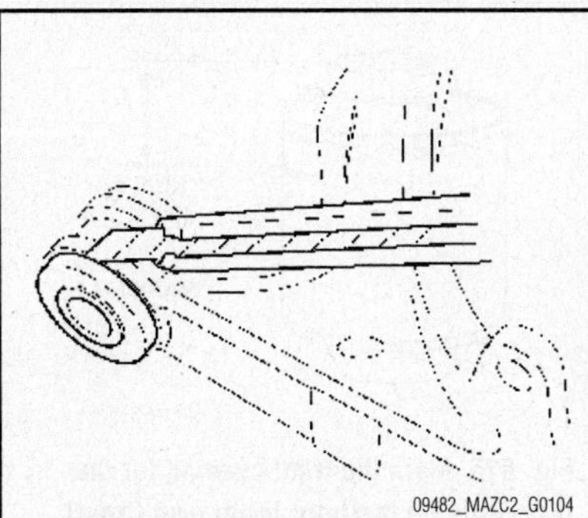

Fig. 601 Cut off the stopper plate rubber using a razor—MX-5 Miata

6. Remove the rear side bushing using the tools illustrated
7. Remove the front side bushing by cutting off the stopper plate rubber using a razor.
8. Cut off 5–6mm each side of the knob end of the front side bushing using a hacksaw.
9. Remove the bushing using the tools illustrated.
10. Remove the shock absorber lower side bushing using the tools illustrated.

To install:

11. Install the shock absorber lower side bushing by compressing the bushing using the tools illustrated.

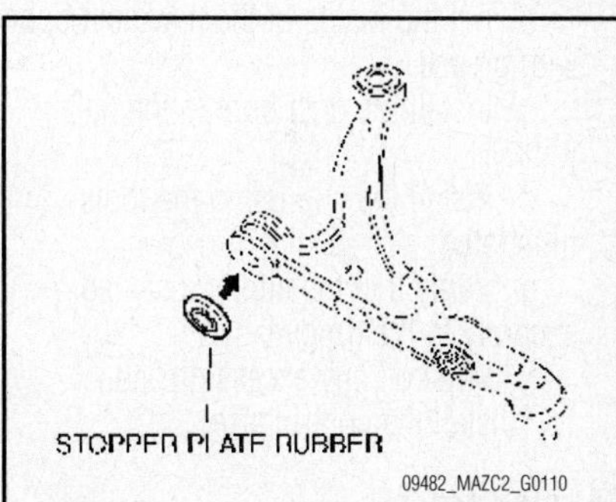

Fig. 607 Insert the stopper plate rubber into the inner pipe of the bushing—MX-5 Miata

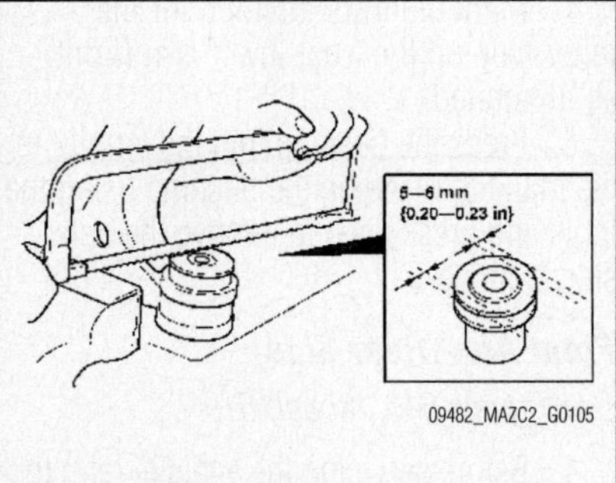

Fig. 602 Cut off 5–6mm each side of the knob end of the front side bushing—MX-5 Miata

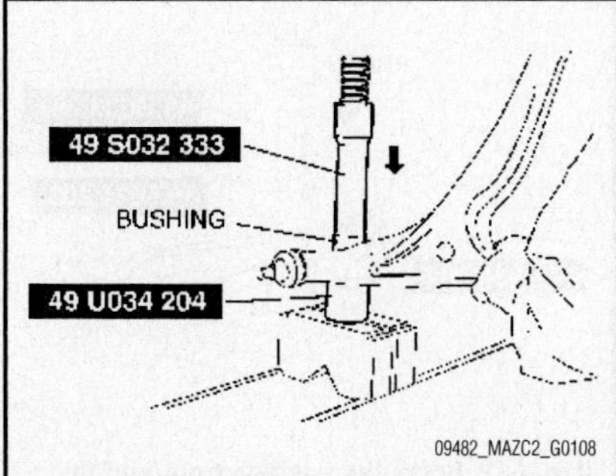

Fig. 605 Install the shock absorber lower side bushing by compressing the bushing as shown—MX-5 Miata

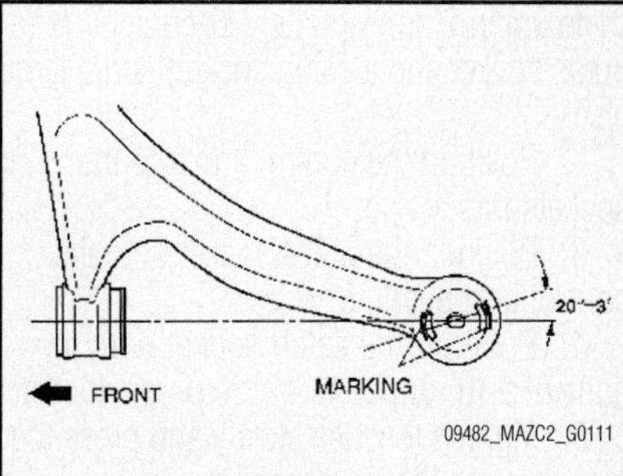

Fig. 608 Align the marks placed during removal of the rear side bushing—MX-5 Miata

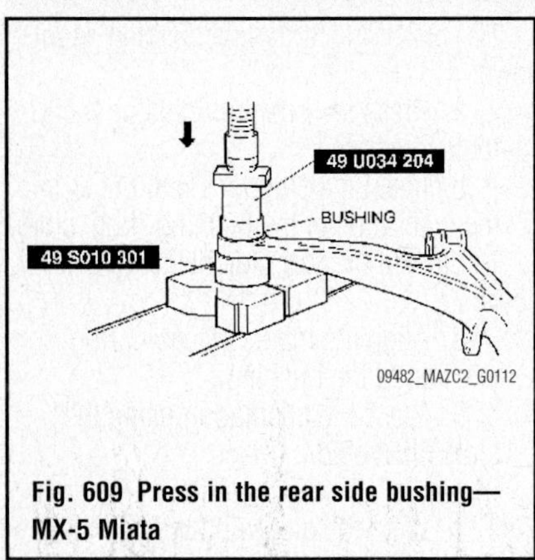

Fig. 609 Press in the rear side bushing—MX-5 Miata

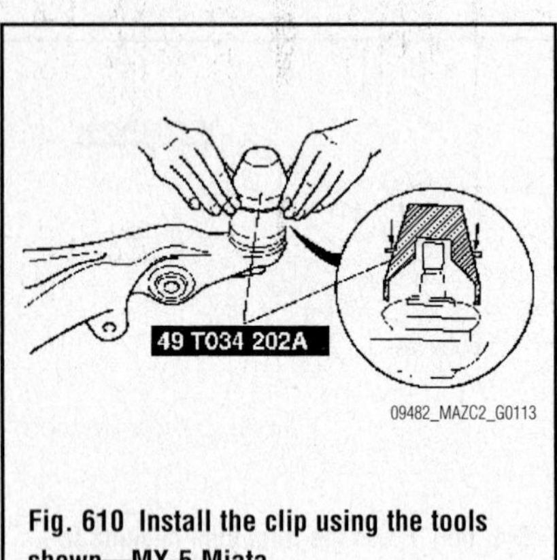

Fig. 610 Install the clip using the tools shown—MX-5 Miata

b. Fill the inside of the new dust boot with grease.

c. Install the dust boot to the ball joint.

d. Install the clip using the tools illustrated.

e. Verify that the clip is installed securely to the groove.

f. Wipe off any excess grease.

15. Install the control arm.

Mazda3 Models

1. Before servicing the vehicle, refer to the precautions section.

All Mazda's use a pressed in control arm bushing, and the pressing can usually be done using 2 appropriately sized sockets (a press socket and a catch socket) and a large vise.

2. Position the control arm and the 2 sockets into a vise.

3. Position the press socket onto the control arm bushing.

4. Position the catch socket onto the control arm, opposite of the press socket.

5. Tighten the vise slowly and press the bushing into the catch socket.

To install:

6. Apply soapy water to the new control arm bushing.

7. Position the bushing against the control arm.

8. Using the same sockets, in the same positions, press the new bushing into the control arm.

Mazda6

Front Arm (Front Side)

See Figures 611 through 615.

1. Before servicing the vehicle, refer to the precautions section.

2. Remove the front lower control arm (front).

3. Press the bushing from the front lower control arm (front) inner side using the tools illustrated.

4. Press the bushing from the front lower control arm (front) outer side using the tools illustrated.

To install:

5. Mark the front bushing for the outer side on the front lower arm (front) as illustrated.

6. Press the bushing into the arm up to the marking made on the bushing using the tools and press used to remove the bushing. Make sure the clearance is 0.3445–0.4234 inch (8.75–10.75mm) as illustrated.

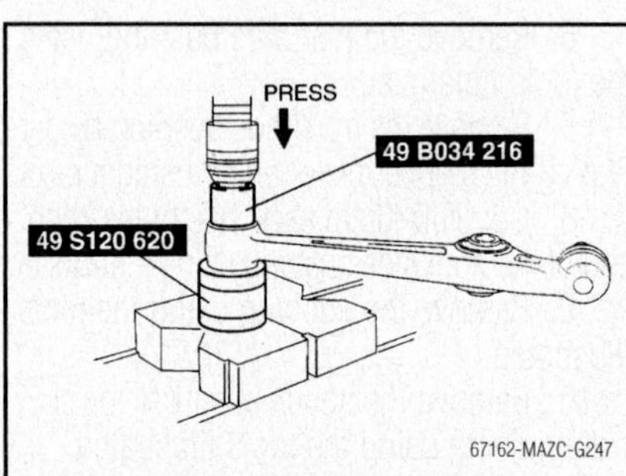

Fig. 611 Press the bushing from/into the front lower control arm (front) inner side using the removal/installer tools shown—Mazda6 and Mazdaspeed6 models

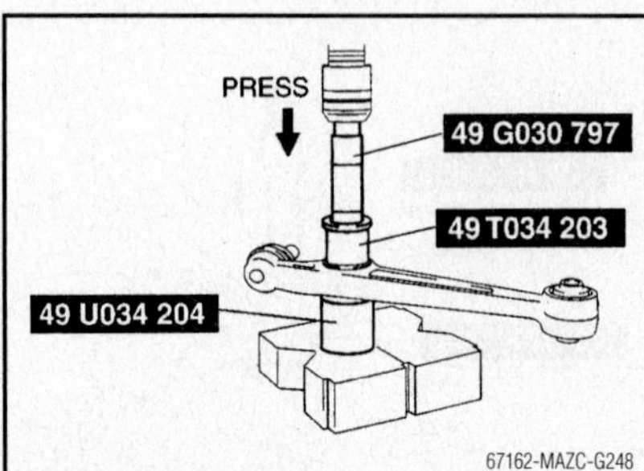

Fig. 612 Press the bushing from/into the front lower control arm (front) outer side using the removal/installer tools shown—Mazda6 and Mazdaspeed6 models

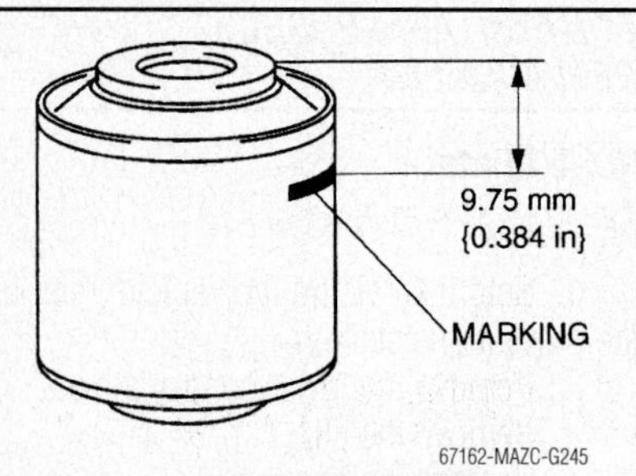

Fig. 613 Mark the front bushing for the outer side on the front lower arm (front) as shown—Mazda6 and Mazdaspeed6 models

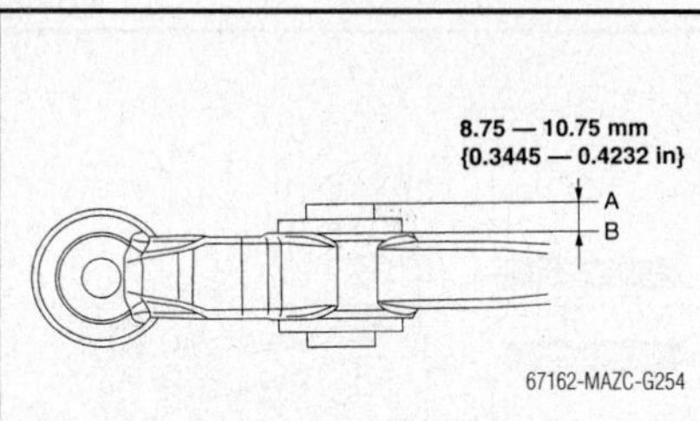

Fig. 614 Make sure the clearance is 0.3445–0.4234 inch (8.75–10.75mm) when installing the front bushing for the outer side on the front lower arm (front)—Mazda6 and Mazdaspeed6 models

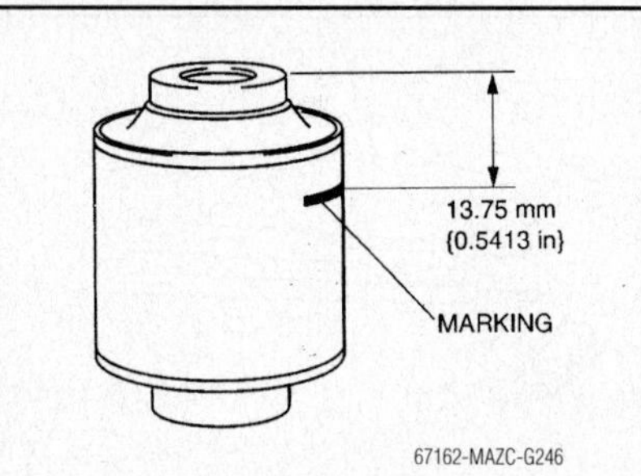

Fig. 615 Mark the front bushing for the inner side on the front lower arm (front) as shown—Mazda6 and Mazdaspeed6 models

7. Mark the front bushing for the inner side on the front lower arm (front) as illustrated.

8. Press the bushing into the arm up to the marking made on the bushing using the tools and press used to remove the outer side bushing.

Front Arm (Rear Side)

See Figures 616 through 619.

1. Before servicing the vehicle, refer to the precautions section.

2. Remove the front lower control arm (rear).

3. Press the bushing from the front lower control arm (rear) using the tools illustrated. remove the arm from the press and use a hammer to remove the bushing completely.

To install:

4. Mark the rear bushing on the front lower arm (rear) as illustrated.
5. Press the bushing into the arm up to the marking made on the bushing using the tools and press used to remove the bushing.
6. The clearance should be 0.936–1.013 inch (23–25mm) as illustrated once the bushing is seated properly.

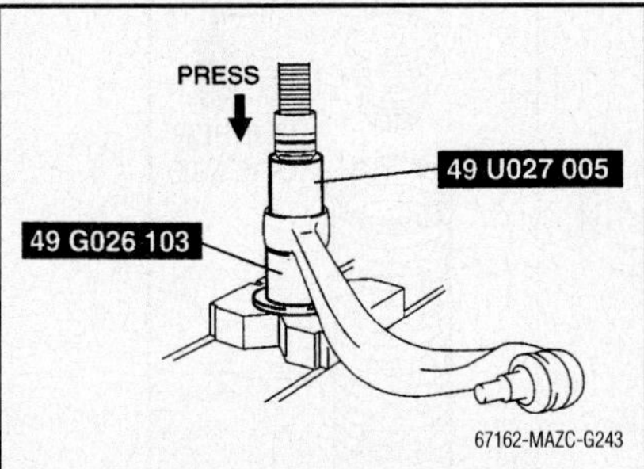

Fig. 616 Press the bushing from/into the front lower control arm (rear) using the removal/installer tools shown—Mazda6 and Mazdaspeed6 models

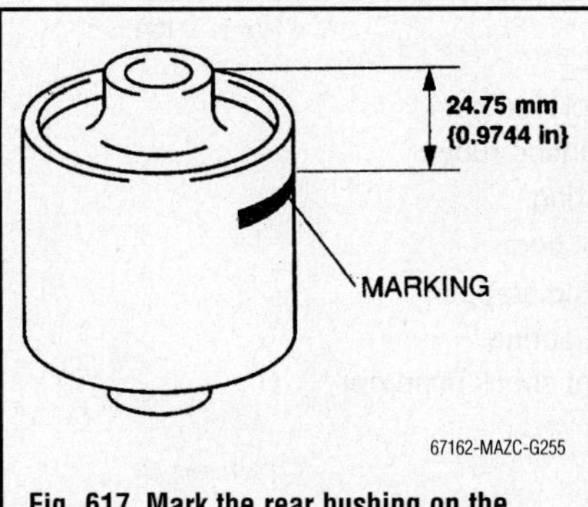

Fig. 617 Mark the rear bushing on the front lower arm (rear) as shown—Mazda6 and Mazdaspeed6 models

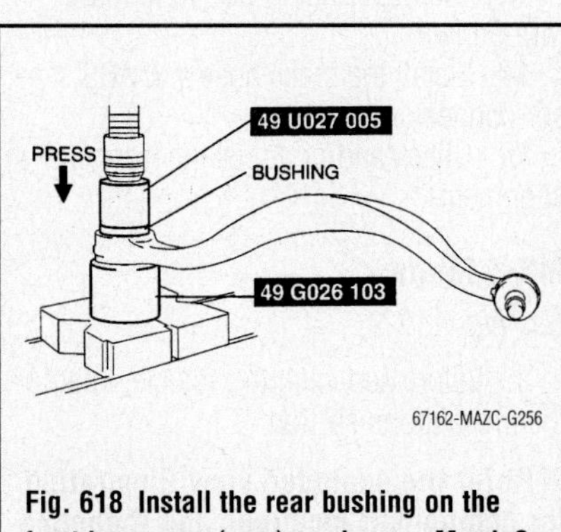

Fig. 618 Install the rear bushing on the front lower arm (rear) as shown—Mazda6 and Mazdaspeed6 models

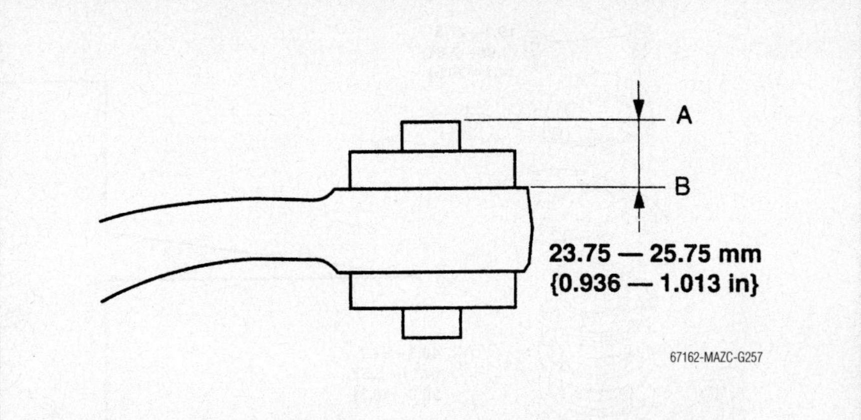

Fig. 619 The rear bushing on the front lower arm (rear) should be 0.936–1.013 inch (23–25mm) once the bushing is seated properly—Mazda6 and Mazdaspeed6 models

Rear Lower Control Arm

See Figures 620 and 621.

1. Before servicing the vehicle, refer to the precautions section.
2. Remove the lower control arm.
3. Press the bushing from the arm using the tools illustrated.

To install:

4. Press the bushing into the arm using the tools illustrated.
5. Install the lower control arm.

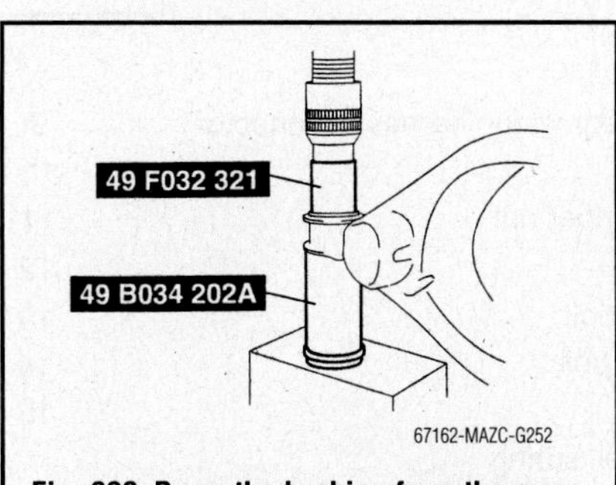

Fig. 620 Press the bushing from the rear suspension lower control arm using the tools shown—Mazda6 and Mazdaspeed6 models

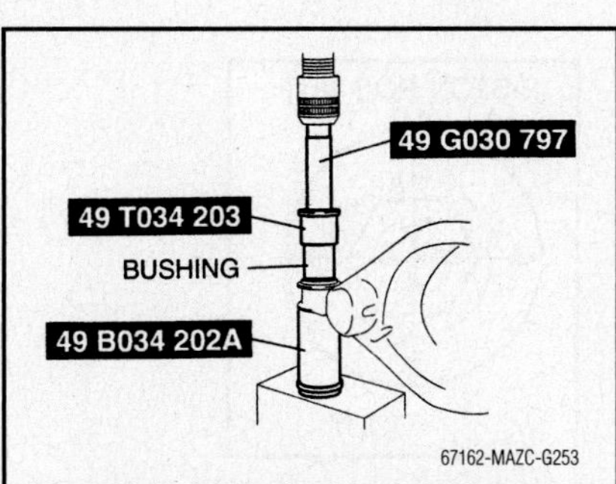

Fig. 621 Press the bushing into the rear suspension lower control arm using the tools shown—Mazda6 and Mazdaspeed6 models

MACPHERSON STRUT

REMOVAL & INSTALLATION

Mazda3 Models

See Figures 622 and 623.

1. Before servicing the vehicle, refer to the precautions section.
2. Disconnect the ABS sensor connector.
3. Disconnect the brake hose from the bracket on the strut assembly.
4. Remove the stabilizer bar link upper nut and disconnect the bar from the strut assembly.
5. Disconnect the tie rod from the knuckle.
6. Remove the shock absorber lower bolt.
7. Loosen the front lower arm inner bolt, the separate the shock from the hub assembly by tapping the knuckle with a hammer or mallet being careful not to damage any components.
8. Remove the shock absorber assembly.

To install:

9. Align the piston rod nut with the center part where the shock absorber is installed by placing the piston rod nut with lengths (A) all the same and tighten the shock absorber upper bolts. Refer to the accompanying illustration.
10. Use a jack to raise the lower control arm, attach the shock absorber and tighten the bolts
11. Tighten the upper shock nuts to 14–20 ft. lbs. (19–28 Nm) and the lower shock nut and bolt to 40–54 ft. lbs. (54–74 Nm).
12. Tighten the tie rod end nut to 27–37 ft. lbs. (37–50 Nm), install a new cotter pin. Tighten the nut, if necessary, to align the ball stud hole with the nut castellation.

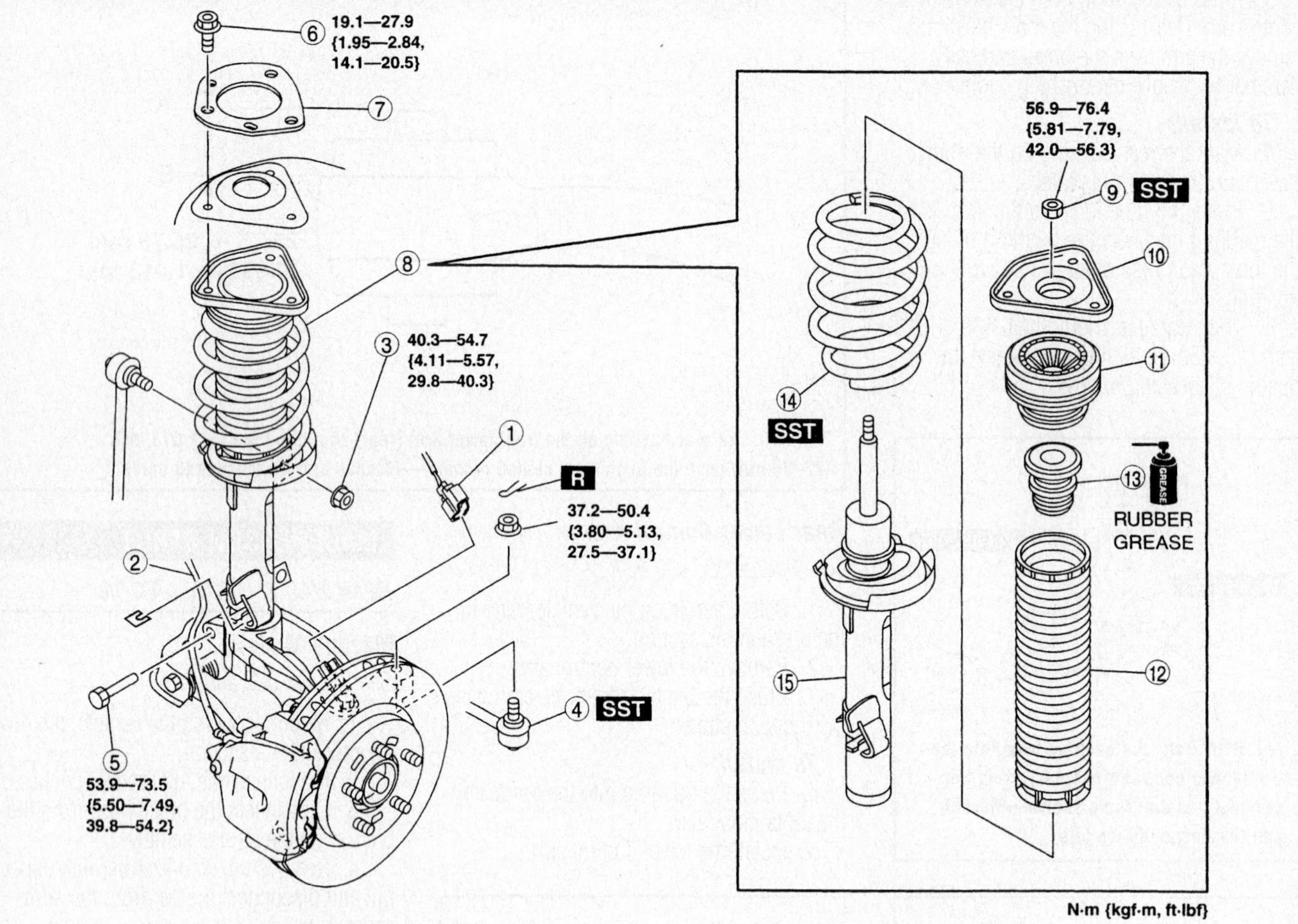

19.1—27.9
{1.95—2.84,
14.1—20.5}
56.9—76.4
{5.81—7.79,
42.0—56.3}
SST
40.3—54.7
{4.11—5.57,
29.8—40.3}
SST
R
37.2—50.4
{3.80—5.13,
27.5—37.1}
RUBBER
GREASE
GREASE
SST
53.9—73.5
{5.50—7.49,
39.8—54.2}
N·m {kgf·m, ft·lbf}

1	ABS wheel-speed sensor wiring harness connector	9	Piston rod nut
2	Brake hose	10	Mounting rubber
3	Stabilizer control link upper nut	11	Bearing
4	Tie-rod end ball joint	12	Dust boot
5	Shock absorber lower bolt	13	Bound stopper
6	Shock absorber upper bolt	14	Coil spring
7	Stiffener	15	Front shock absorber
8	Shock absorber and coil spring		

67162-MAZC-G88

Fig. 622 Exploded view of the front strut and spring assembly components—Mazda3

SHOCK
ABSORBER
UPPER BOLT
PISTON ROD NUT
A
A
A
A
BODY
OUTER SIDE

67162-MAZC-G89

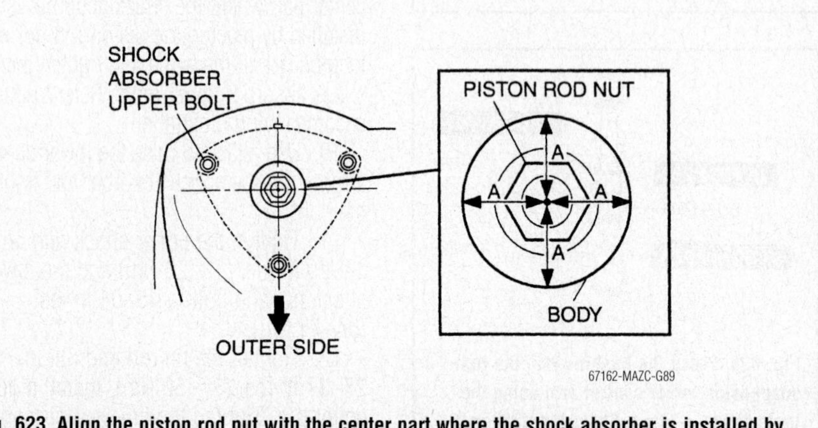

Fig. 623 Align the piston rod nut with the center part where the shock absorber is installed by placing the piston rod nut with lengths (A) all the same—Mazda3

13. Install the stabilizer bar link and tighten the upper nut to 30–40 ft. lbs. (40–54 Nm).

14. Install the brake hose and ABS sensor connector.

15. Check and/or adjust the front end alignment.

MX-5 Miata

See Figure 624.

1. Before servicing the vehicle, refer to the precautions section.

➡Refer the exploded view illustration for component locations and if applicable, their retainer torque specifications.

CAUTION

Performing the following procedures without first removing the ABS wheel-speed sensor may possibly cause an open circuit in the wiring harness if it is pulled by mistake. Before performing the following procedures, remove the ABS wheel-speed sensor (axle side) and fix it to an appropriate place where the sensor will not be pulled while servicing the vehicle.

2. Remove the front suspension tower bar.
3. Remove the brake hose bracket.
4. Remove the front upper arm ball joint from the knuckle.
5. Remove the front shock absorber, coil spring and front upper arm.
6. Remove the front shock absorber and coil spring.

To install:

7. Install the components in the reverse order of removal using the accompanying illustration for torque values.

Mazda6 Models

See Figures 625 and 626.

1. Before servicing the vehicle, refer to the precautions section.
2. Disconnect the ABS sensor.
3. Disconnect the brake hose from the bracket on the strut assembly.
4. Remove the stabilizer bar link upper nut and disconnect the bar from the strut assembly.
5. Remove the shock absorber lower bolt, the dynamic damper and the damper fork.
6. Remove the upper shock absorber nuts.
7. Remove the shock absorber assembly.

To install:

8. Install the shock absorber assembly. Position the stud bolts at a 27–33 degree angle from where the stabilizer bar is installed (center line), towards the inner side of the vehicle. Refer to the accompanying illustration.
9. Tighten the upper nuts to 21–29 ft. lbs. (29–39 Nm).
10. Install the damper fork and align the gap of the fork with the projection of the damper as illustrated and install and tighten the bolt to 31–38 ft. lbs. (43–52 Nm)
11. Install the dynamic damper and the lower shock bolt and tighten to 68–93 ft. lbs. (93–126 Nm).
12. Install the stabilizer bar link and tighten the nut to 31–40 ft. lbs. (43–54 Nm).
13. Install the brake hose, tighten the bolt to 13–18 ft. lbs. (18–25 Nm).

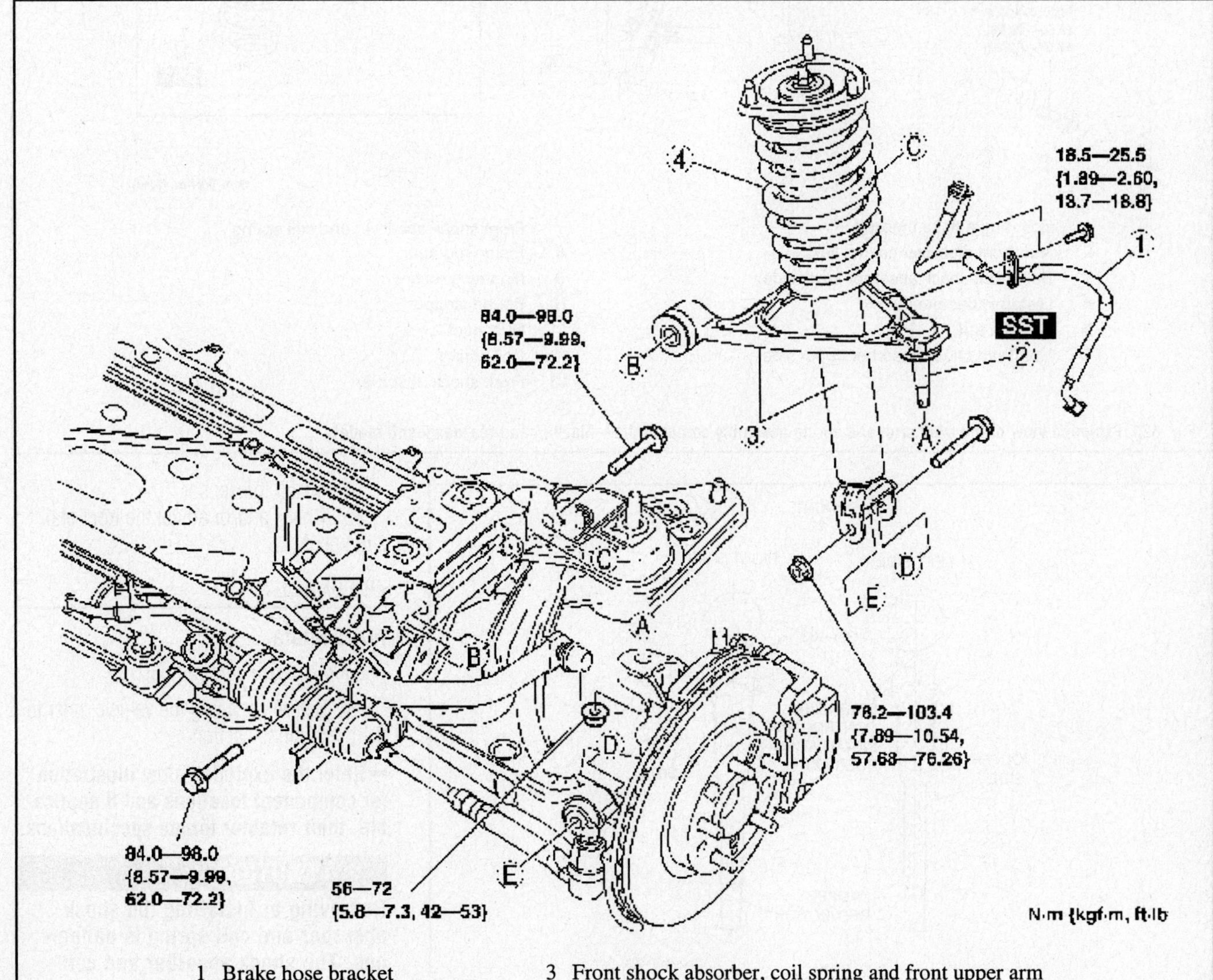

Fig. 624 Exploded view of the front strut assembly and related components—MX-5 Miata

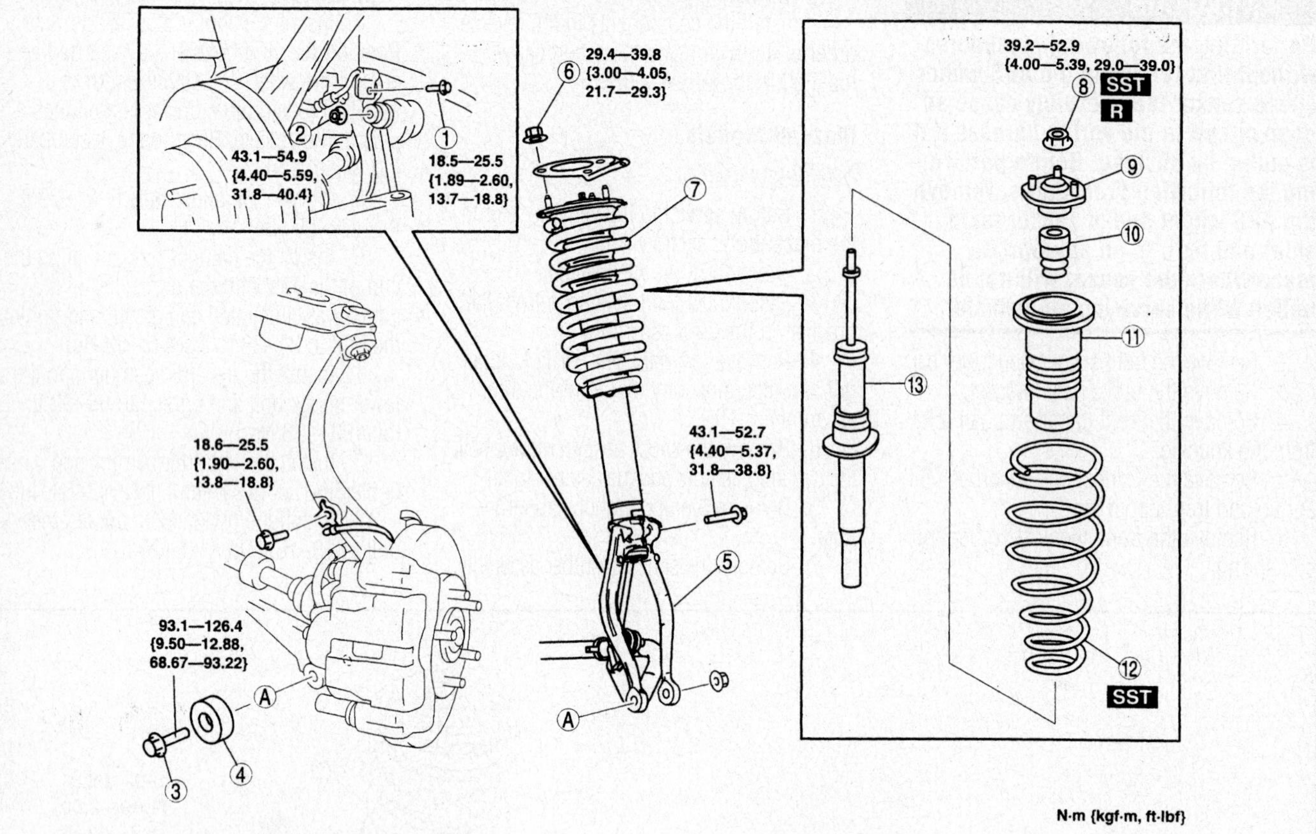

29.4—39.8
{3.00—4.05,
21.7—29.3}
43.1—54.9
{4.40—5.59,
31.8—40.4}
18.5—25.5
{1.89—2.60,
13.7—18.8}
39.2—52.9
{4.00—5.39, 29.0—39.0}
SST
R
18.6—25.5
{1.90—2.60,
13.8—18.8}
43.1—52.7
{4.40—5.37,
31.8—38.8}
93.1—126.4
{9.50—12.88,
68.67—93.22}
A
SST
N·m {kgf·m, ft·lbf}

1 Bolt (brake hose bracket)
2 Nut (front stabilizer control link)
3 Bolt (front shock absorber lower side)
4 Dynamic damper
5 Damper fork
6 Nut (front shock absorber upper side)
7 Front shock absorber and coil spring
8 Piston rod nut
9 Mounting rubber
10 Bound stopper
11 Dust boot
12 Coil spring
13 Front shock absorber

67162-MAZC-G235

Fig. 625 Exploded view of the front strut and spring assembly components—Mazda6 and Mazdaspeed6 models

FRONT
LEFT SIDE
RIGHT SIDE
27°—33°
STABILIZER
CONTROL LINK
INSTALLATION
POSITION
OUTER
SIDE
OUTER
SIDE
DAMPER
PROJECTION

67162-MAZC-G236

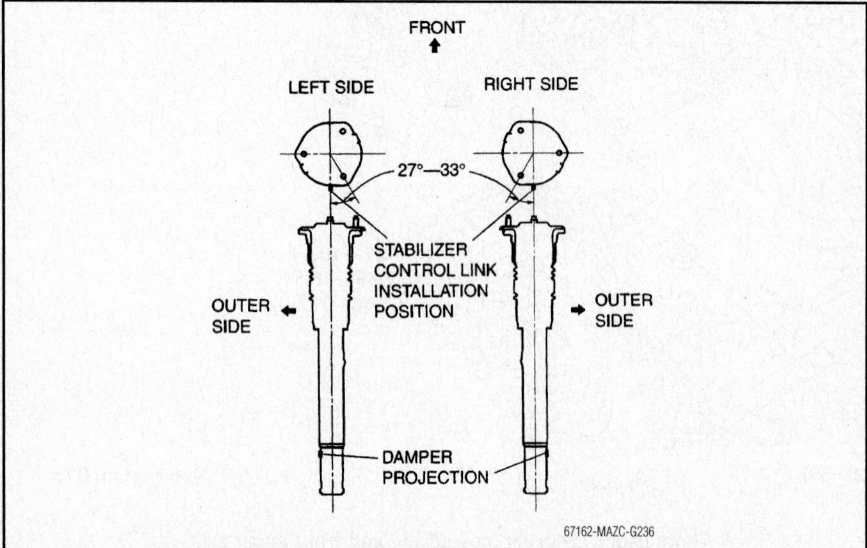

Fig. 626 Position the stud bolts at a 27–33 degree angle from where the stabilizer bar is installed (center line), towards the inner side of the vehicle when installing the front shock assembly—Mazda6 and Mazdaspeed6 models

14. Install the sensor.
15. Check and/or adjust the front end alignment.

OVERHAUL

MX-5 Miata

See Figures 627 through 630.

1. Before servicing the vehicle, refer to the precautions section.

➡Refer the exploded view illustration for component locations and if applicable, their retainer torque specifications.

✲✲ CAUTION

Removing or installing the shock absorber and coil spring is dangerous. The shock absorber and coil spring could fly off and cause serious injury or death, and damage the vehicle.

2. Remove the front shock absorber and coil spring.

⁂ WARNING

Before removing the piston rod nut, secure the shock absorber and spring in a suitable spring compressor. Otherwise, the shock absorber and spring could fly off under tremendous pressure and cause serious injury or death, or damage to vehicle parts.

3. Protect the coil spring from scratches using a piece of cloth and install the spring compressor.
4. Compress the coil spring and remove the piston rod nut.
5. Remove the following:
   - Retainer
   - Bushing
   - Upper spring seat
   - Dust boot
   - Spacer
   - Bushing
   - Stopper casing and bound stopper
   - Bound stopper
   - Stopper casing
   - Coil spring
   - Shock absorber

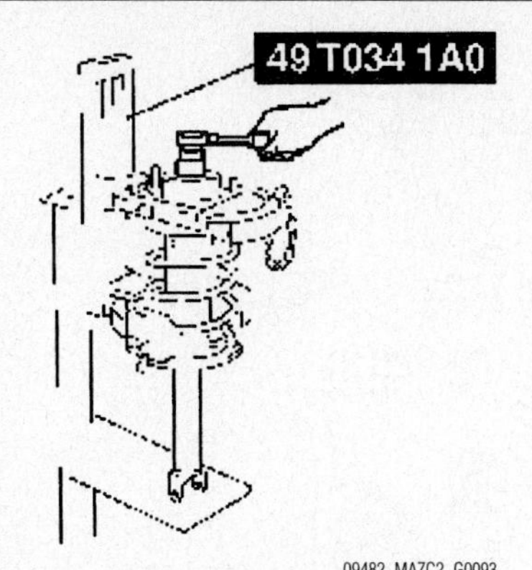

Fig. 628 Compress the front coil spring and remove the piston rod nut—MX-5 Miata

To install:

6. Install the following:
   - Shock absorber
   - Coil spring
   - Stopper casing
   - Bound stopper
   - Stopper casing and bound stopper
   - Bushing
   - Spacer
   - Dust boot

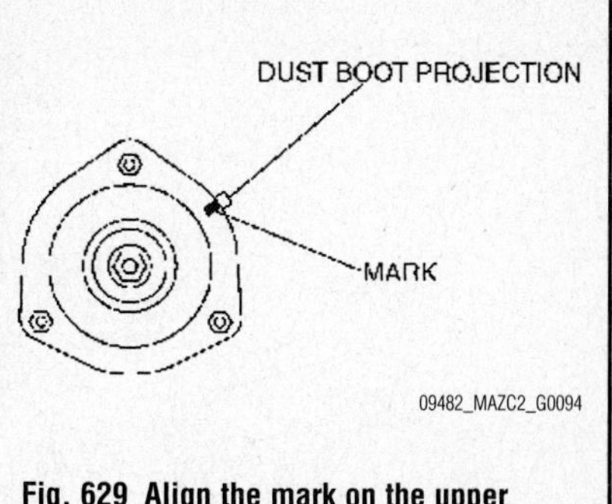

Fig. 629 Align the mark on the upper spring seat with the dust boot projection—MX-5 Miata

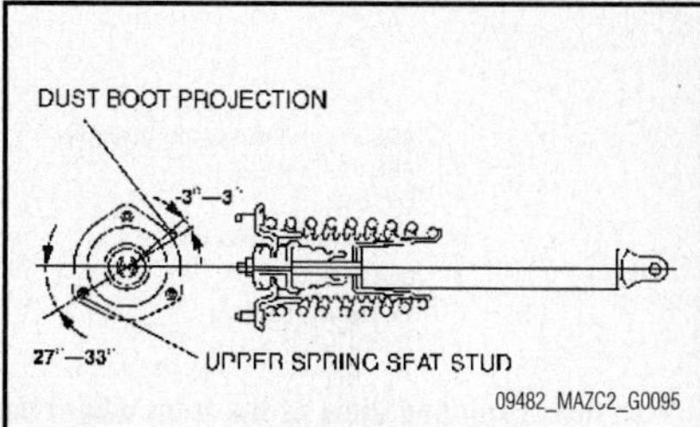

Fig. 630 Install the front strut upper spring seat so that the upper spring seat stud is at a 27–33 degree angle to the shock absorber installation shaft—MX-5 Miata

R
SST 38.1—46.1 {3.89—4.70, 28.2—34.0}
SST

N·m {kgf·m, ft·lbf}

1	Piston rod nut	7	Bushing
2	Retainer	8	Stopper casing and bound stopper
3	Bushing	9	Bound stopper
4	Upper spring seat	10	Stopper casing
5	Dust boot	11	Coil spring
6	Spacer	12	Front shock absorber

09482_MAZC2_G0092

Fig. 627 Exploded view of the front coil spring assembly—MX-5 Miata

   - Upper spring seat
   - Bushing
   - Retainer

7. Protect the coil spring from scratches using a piece of cloth and install the spring compressor.
8. Compress the coil spring using the spring compressor.
9. Install the shock absorber so that the lower end of the coil spring is seated on the step of the lower spring seat.
10. When installing the upper spring seat:
    a. Align the mark on the upper spring seat with the dust boot projection.
    b. Install the upper spring seat so that the upper spring seat stud is at a 27–33 degree angle to the shock absorber installation shaft (lower side).
11. Install the piston rod nut and tighten to 38–46 Nm (28–34 ft. lbs.).
12. Install the strut assembly.

STEERING KNUCKLE

REMOVAL & INSTALLATION

Mazda3

See Figure 631.

1. Before servicing the vehicle, refer to the precautions section.

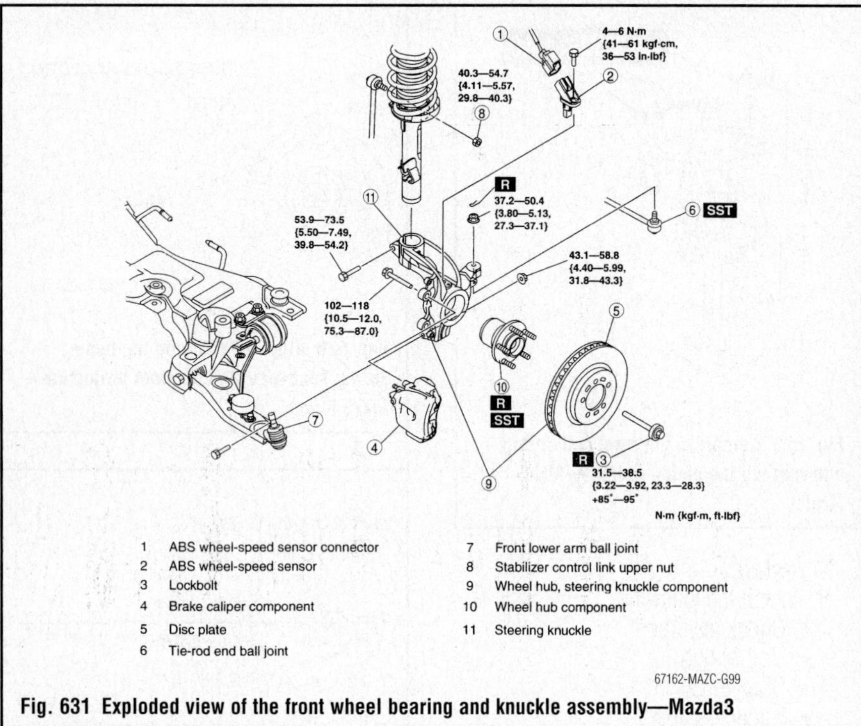

Fig. 631 Exploded view of the front wheel bearing and knuckle assembly—Mazda3

➡**Refer the exploded view illustration for component locations and if applicable, their retainer torque specifications.**

CAUTION

Performing the following procedures without first removing the ABS wheel-speed sensor may possibly cause an open circuit in the wiring harness if it is pulled by mistake. Before operations, remove the ABS wheel-speed sensor (axle side) and move the sensor away from the harnesses.

2. Disconnect the ABS wheel-speed sensor.
3. Remove the caliper with the hose still attached and support the assembly with wire so as to not place strain on the hose.
4. Remove the rotor.
5. Disconnect the tie-rod end from the knuckle.

2. Refer to the illustration for component location and torque specifications.
3. Remove or disconnect the following:
   - Wheels
   - ABS sensor connector and the sensor
   - Halfshaft lockbolt
   - Brake caliper and rotor
   - Tie rod end from the knuckle
   - Lower control arm ball joint from the knuckle
   - Stabilizer bar link nut
   - Knuckle assembly

To install:

4. Install or connect the following:
   - Knuckle assembly, tighten the bolts to the specifications shown in the accompanying illustration
   - Control arm ball joint to the knuckle tighten, the bolts to the specifications shown in the accompanying illustration
   - Tie rod end to the knuckle, tighten the bolts to the specifications shown in the accompanying illustration
   - Brake rotor and caliper
   - New halfshaft lockbolt and tighten to 23–28 ft. lbs. (31–38 Nm), plus an additional 85–95 degrees
   - ABS sensor and connector
   - Wheels

MX-5 Miata

See Figure 632.

1. Before servicing the vehicle, refer to the precautions section.

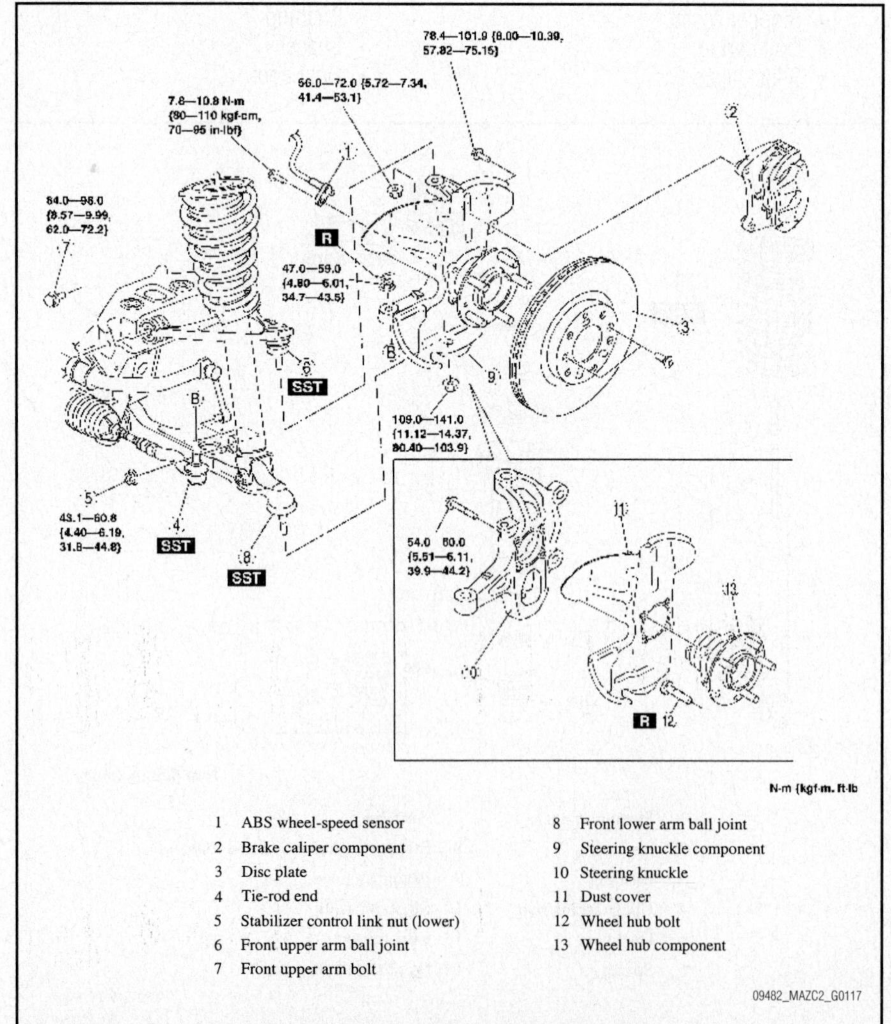

Fig. 632 Exploded view of front hub and knuckle assembly—MX-5 Miata

6. Disconnect the stabilizer control link nut (lower).
7. Disconnect the front upper arm ball joint from the knuckle.
8. Remove the front upper arm bolt.
9. Disconnect the front lower arm ball joint from the knuckle.
10. Remove the Steering knuckle assembly

To install:

11. Install the remaining components in the reverse order of removal using the accompanying illustration for torque values.
12. Inspect the wheel alignment.

Mazda6 Models

See Figures 633 through 636.

1. Before servicing the vehicle, refer to the precautions section.
2. Refer to the illustration for component location and torque specifications.
3. Remove or disconnect the following:
   - Wheels
   - ABS sensor
   - Halfshaft axle nut, unstake the nut prior to removal.
   - Brake caliper and rotor
   - Tie rod end from the knuckle
   - Damper fork–to–control arm bolt
   - Front lower control arm ball joints
   - Front upper arm ball joint
   - Wheel hub dust cover
   - Hub bolts and the hub
   - Steering knuckle

To install:

4. Install or connect the following:
   - Steering knuckle. Refer to the illustration for torque values.
   - Wheel hub dust cover
   - Wheel bearing assembly.
   - Front upper arm ball joint and tighten the nut to 29–39 ft. lbs. (39–53 Nm)
   - Front lower arm ball joints and tighten the nuts to 122–147 ft. lbs. (166–200 Nm)
   - Damper fork bolt and tighten to 68–93 ft. lbs. (93–126 Nm)
   - Tie rod end to the knuckle and tighten the nut to 29–39 ft. lbs. (39–53 Nm) and install a new cotter pin
   - Brake caliper and rotor
   - Halfshaft axle nut, tighten the nut to 173–202 ft. lbs. (235–274 Nm) and stake the nut
   - Wheels

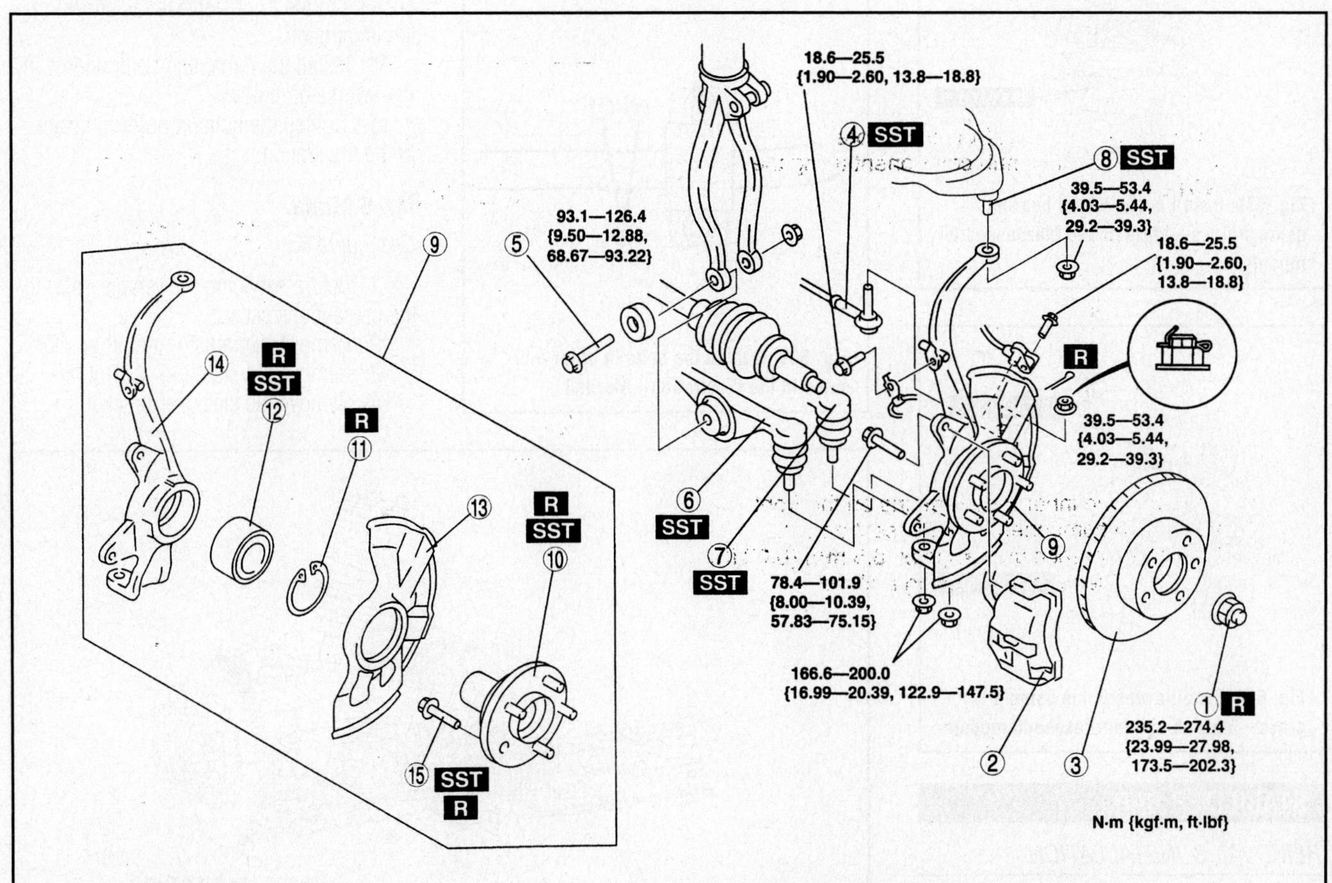

Fig. 633 Exploded view of the front wheel bearing and knuckle assembly—Mazda6 and Mazdaspeed6 models

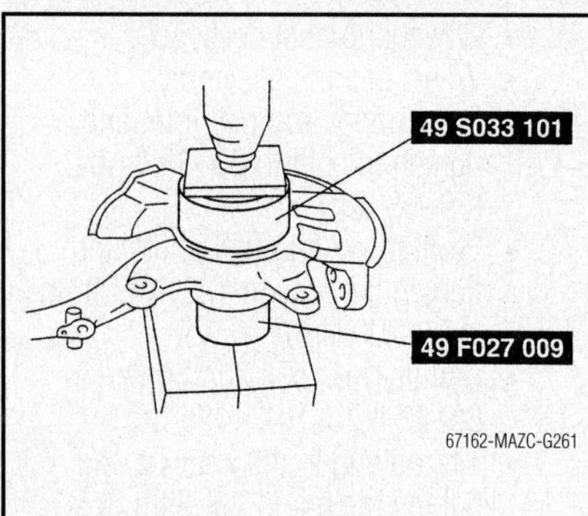

Fig. 634 Install the dust shield—Mazda6 and Mazdaspeed6 models

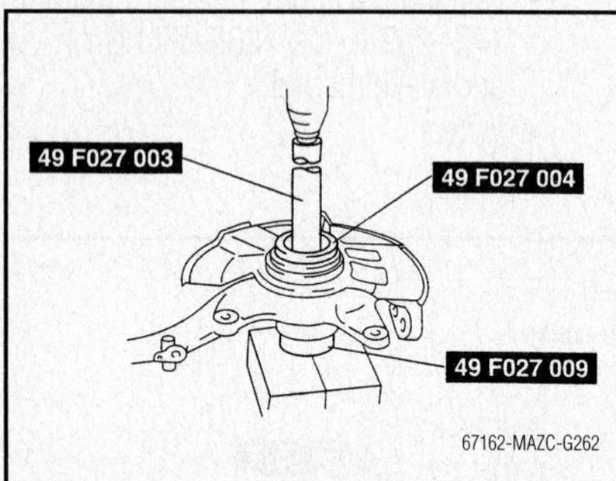

Fig. 635 Install a new wheel bearing using a press—Mazda6 and Mazdaspeed6 models

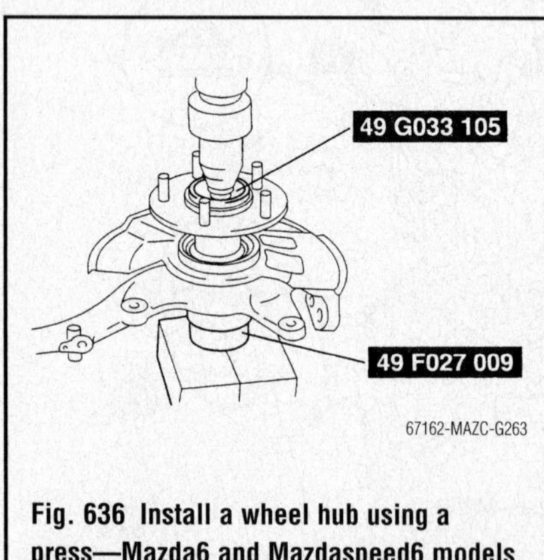

Fig. 636 Install a wheel hub using a press—Mazda6 and Mazdaspeed6 models

STABILIZER BAR

REMOVAL & INSTALLATION

Mazda3 Models

See Figures 637 and 638.

1. Before servicing the vehicle, refer to the precautions section.
2. Detach the steering shaft from the steering gear and linkage.
3. Remove the power steering pipe bracket.
4. Remove the stabilizer control link.

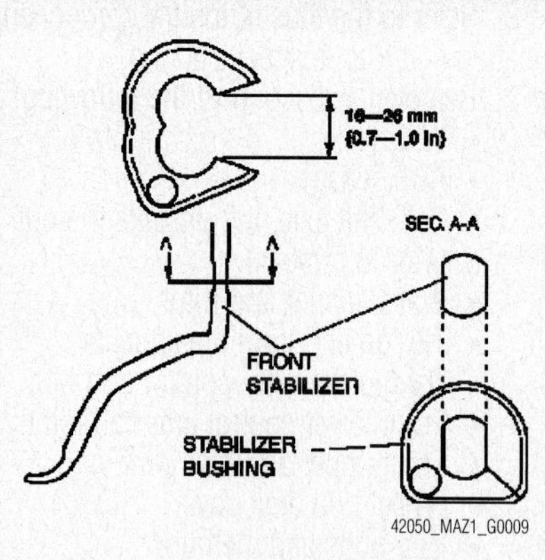

Fig. 637 Widen the stabilizer bushing opening 0.7–1 inch (16–26 mm) and install the bushing to the front stabilizer —Mazda3

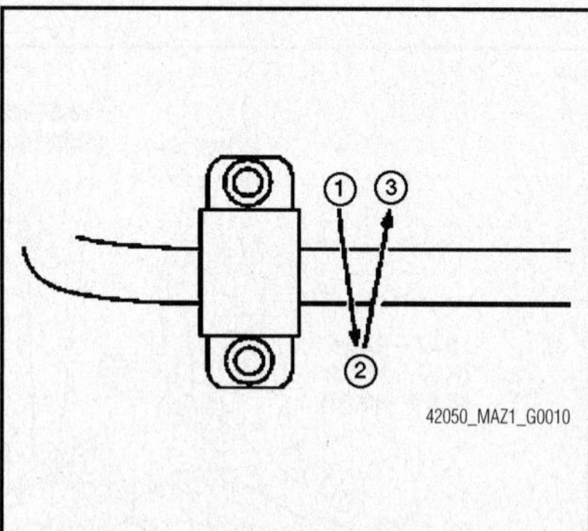

Fig. 638 Tighten the bolts in order indicated in the illustration—Mazda3

5. Support the front crossmember using a jack and remove the front crossmember bracket.
6. Remove the No.1 engine mount center bolt.
7. Detach the silencer hangers on the middle pipe from the front crossmember.
8. Lower the front crossmember slowly approx. 90 mm {3.5 in} and remove the front stabilizer .component.
9. Secure the stabilizer bracket flange using a vise.
10. Remove the front stabilizer.

To install:

11. Apply grease to the stabilizer bushing.
12. Install the stabilizer bracket using a vise
13. Widen the stabilizer bushing opening 0.7–1 inch (16–26 mm) and install the bushing to the front stabilizer as shown in the illustration.
14. Install the remaining components in the reverse of removal
15. Tighten the bolts in order indicated in the illustration.

MX-5 Miata

See Figure 639.

1. Before servicing the vehicle, refer to the precautions section.
2. Remove the radiator mount bracket.
3. Remove the stabilizer control link.
4. Remove the stabilizer bracket.

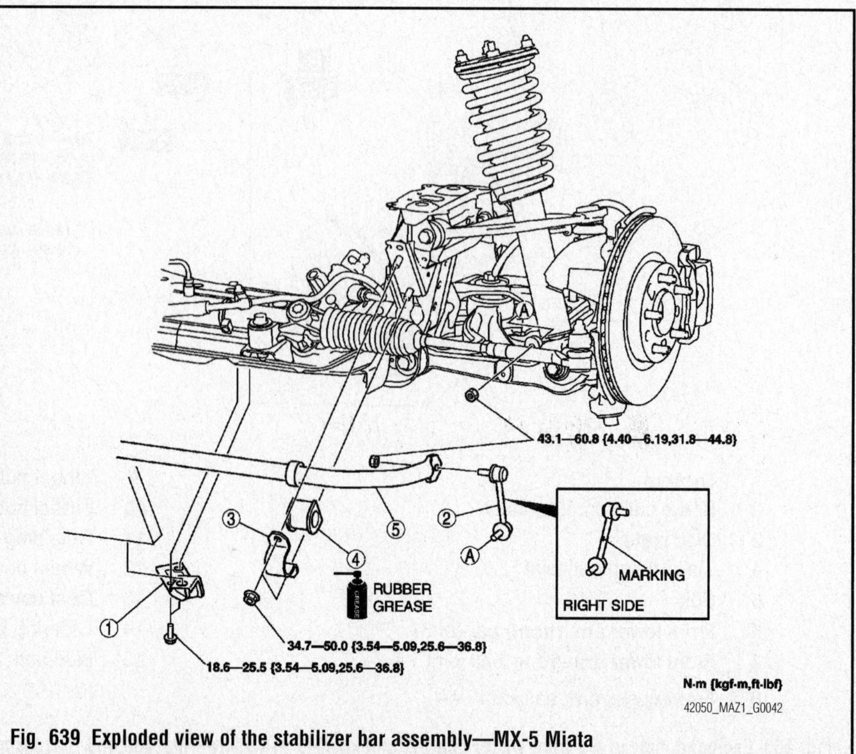

Fig. 639 Exploded view of the stabilizer bar assembly—MX-5 Miata

5. Remove the stabilizer front stabilizer bushing and stabilizer,

To install:

6. Apply grease to the stabilizer bushing.
7. Align the outer side of the stabilizer slide stopper with the stabilizer bushing.
8. Install the stabilizer bracket.
9. Install the remaining components in the reverse of removal
10. Tighten the bolts in order indicated in the illustration.

Mazda6 Models

See Figures 640 and 641.

1. Before servicing the vehicle, refer to the precautions section.
2. Remove the undercover and splash shield.
3. Remove the transverse member bolts and the transverse member.
4. Remove the front crossmember as follows:
   a. Remove the stabilizer bar nut.
   b. Remove the front lower arm front and rear ball joints from the knuckle.
   c. Remove the front shock absorber lower bolt.
   d. Remove the No. 1 engine mount center bolt.

⁂ WARNING

Support the crossmember with a suitable jack making sure it is securely damaged or it could fall causing injury or damage.

   e. Support the crossmember with a jack and remove the crossmember bracket.
   f. Remove the crossmember assembly retainers and lower the jack to remove the crossmember.
5. Remove the stabilizer control link.

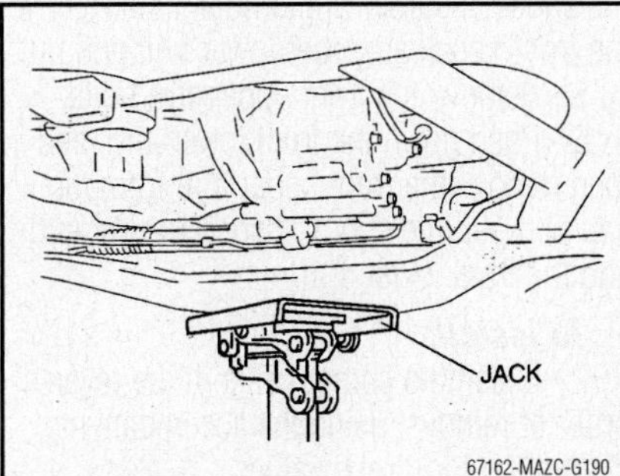

Fig. 640 Support the crossmember with a jack and remove the crossmember bracket—Mazda6 with the 3.0L (AJ) engine

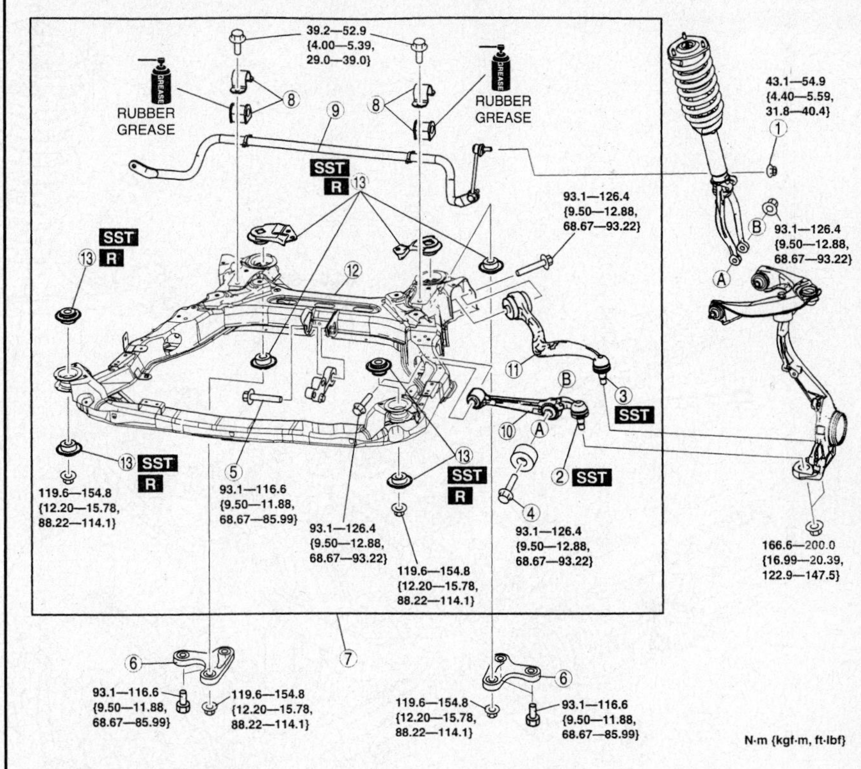

1	Nut (stabilizer control link)	8	Stabilizer bracket and bushing
2	Front lower arm (front) ball joint	9	Front Stabilizer
3	Front lower arm (rear) ball joint	10	Front lower arm (front)
4	Bolt (front shock absorber lower side)	11	Front lower arm (rear)
5	No.1 engine mount center bolt	12	Front crossmember
6	Crossmember bracket	13	Front crossmember bushing
7	Crossmember component		

67162-MAZC-G124

Fig. 641 Exploded view of the crossmember assembly and related components—Mazda6 with the 3.0L (AJ) engine

6. Remove the stabilizer bracket.
7. Remove the stabilizer front stabilizer bushing and stabilizer,

To install:

8. Apply grease to the stabilizer bushing.
9. Align the bushing with the inside of positioning plate on the stabilizer bar.
10. Install the stabilizer bracket.
11. Install the remaining components in the reverse of removal
12. Tighten the bolts in order indicated in the illustration.

UPPER BALL JOINT

REMOVAL & INSTALLATION

The upper ball joint is an integral part of the upper control and cannot be replaced separately. If the upper ball joint is defective, the entire upper control arm must be replaced. Refer to the upper control arm procedure.

UPPER CONTROL ARM

REMOVAL & INSTALLATION

MX-5 Miata

See Figures 642 and 643.

1. Before servicing the vehicle, refer to the precautions section.

➡Refer the exploded view illustration for component locations and if applicable, their retainer torque specifications.

⁂ CAUTION

Performing the following procedures without first removing the ABS wheel-speed sensor may possibly cause an open circuit in the wiring harness if it is pulled by mistake. Before performing the following procedures, remove the ABS wheel-speed sensor (axle side) and fix it to an appropriate place where the

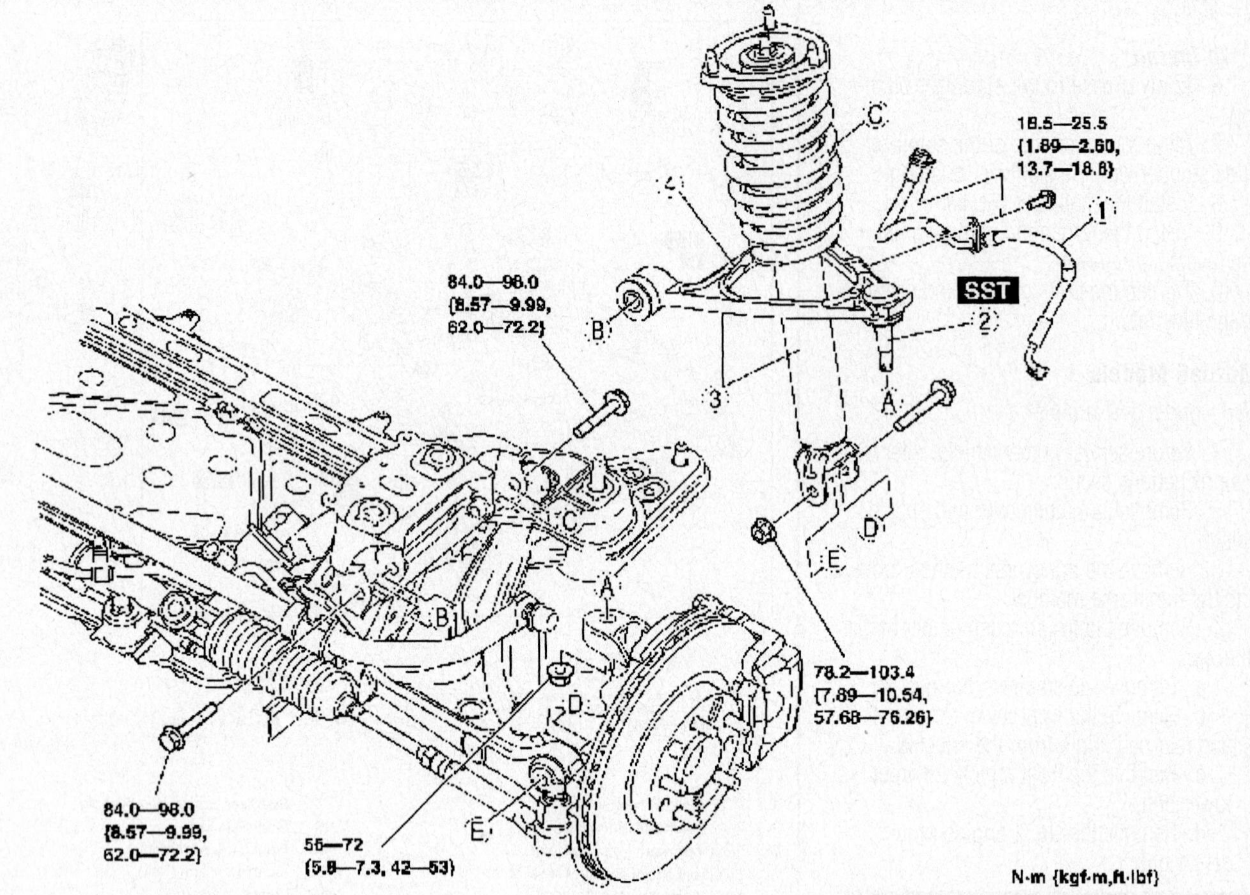

1 Brake hose bracket
2 Front upper arm ball joint
3 Front shock absorber, coil spring and front upper arm
4 Front upper arm

09482_MAZC2_G0101

Fig. 642 Exploded view of the front upper control arm assembly—MX-5 Miata

1 Rear lower arm outer bolt
2 Rear coil spring component
3 Rear upper arm

67162-MAZC-G95

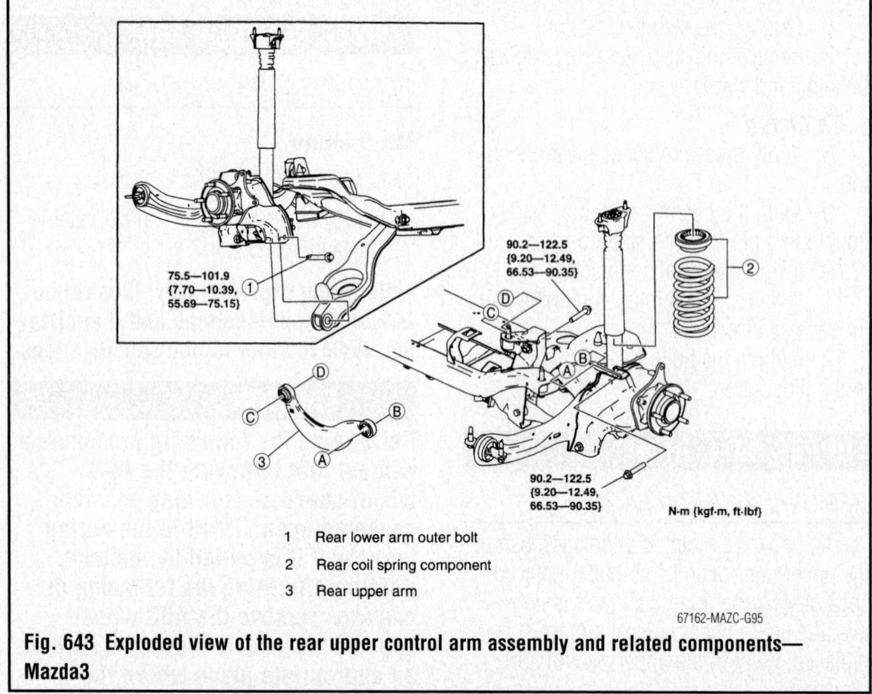

Fig. 643 Exploded view of the rear upper control arm assembly and related components—Mazda3

sensor will not be pulled while servicing the vehicle.

2. Remove the brake hose bracket.
3. Separate the front upper arm ball joint from the knuckle.
4. Remove the front shock absorber, coil spring and front upper arm, by loosening the shock absorber upper nuts, then remove the front shock absorber lower bolt and nut.
5. Remove the front upper arm bolts.
6. Push down the front lower arm, and then remove the front upper arm from the gap between the shock absorber lower end and the front lower arm.

To install:

7. Install the components in the reverse order of removal using the accompanying illustration for torque values.
8. Inspect the front wheel alignment and adjust as necessary.

Mazda6 Models

See Figures 644 and 645.

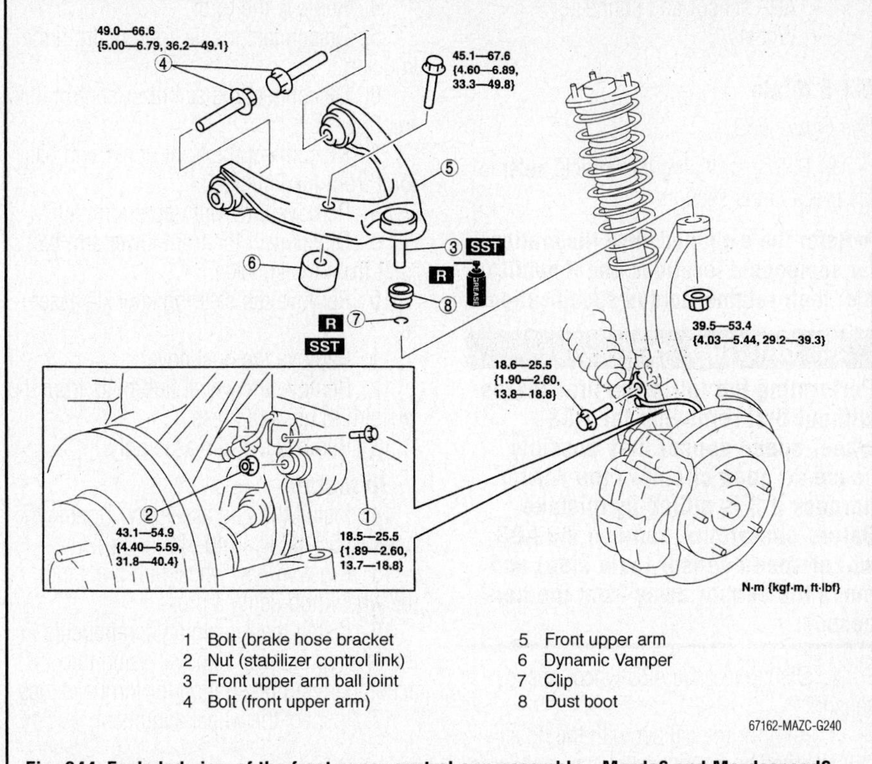

Fig. 644 Exploded view of the front upper control arm assembly—Mazda6 and Mazdaspeed6 models

1. Before servicing the vehicle, refer to the precautions section.
2. Remove the wheel.
3. Remove the ABS sensor.
4. Remove the brake hose bracket bolt.
5. Remove the stabilizer bar link nut.
6. Support the lower control arm with a jack.
7. Separate the upper arm ball joint from the knuckle using a separator tool such as 49 T028 3A0.
8. Remove the upper arm rear bolts and the arm.

To install:

- Install the upper arm and the bolts.

9. Attach the ball joint to the knuckle. Tighten the ball joint bolt to 29–39 ft. lbs.

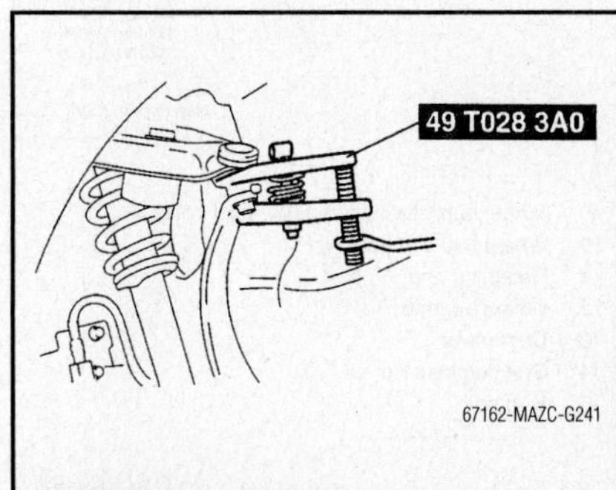

Fig. 645 Separate the upper arm ball joint from the knuckle using a separator tool—Mazda6 and Mazdaspeed6 models

(39–53 Nm) and the rear upper arm bolts to 36–49 ft. lbs. (49–66 Nm).

10. Install the stabilizer bar link nut and tighten to 31–40 ft. lbs. (43–54 Nm).
11. Install the brake hose bracket bolt and tighten to 13–18 ft. lbs. (18–25 Nm).
12. Install the ABS sensor and in the wheel.

WHEEL HUB AND BEARING

REMOVAL & INSTALLATION

Mazda3 Models

See Figures 646 through 648.

1. Before servicing the vehicle, refer to the precautions section.
2. Refer to the illustration for component location and torque specifications.
3. Remove or disconnect the following:
   - Wheels
   - ABS sensor connector and the sensor
   - Halfshaft lockbolt
   - Brake caliper and rotor
   - Tie rod end from the knuckle
   - Lower control arm ball joint from the knuckle

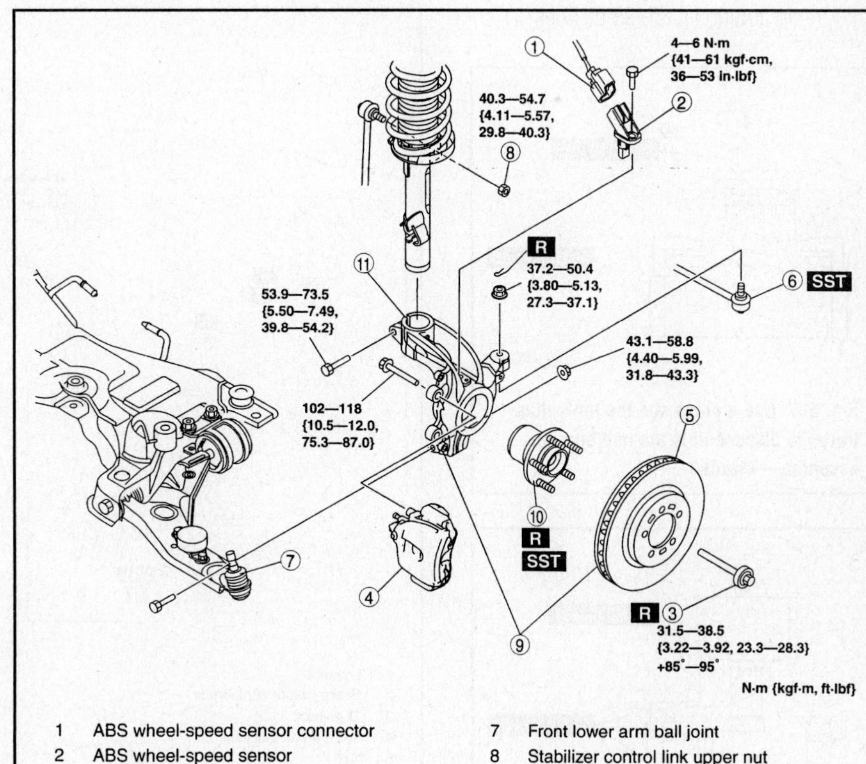

Fig. 646 Exploded view of the front wheel bearing and knuckle assembly—Mazda3

- Stabilizer bar link nut
- Knuckle assembly
- Hub bearing using a press and Mazda tools 49 G030 795 and 49 B033 1A0

4. Clean and inspect all parts but do not wash or clean the wheel bearing. The bearing must be replaced.

To install:

5. Using Mazda Press tools 49 H034 201 and 49 B033 1A0, press a new wheel bearing into the knuckle assembly. Make sure the installation tool engages to the bearing outer race properly to avoid damage.
6. Install or connect the following:
   - Press the hub onto the knuckle
   - Knuckle assembly, tighten the bolts to the specifications shown in the accompanying illustration
   - Control arm ball joint to the knuckle tighten, the bolts to the specifications shown in the accompanying illustration
   - Tie rod end to the knuckle, tighten the bolts to the specifications shown in the accompanying illustration
   - Brake rotor and caliper
   - New halfshaft lockbolt and tighten to 23–28 ft. lbs. (31–38 Nm), plus an additional 85–95 degrees
   - ABS sensor and connector
   - Wheels

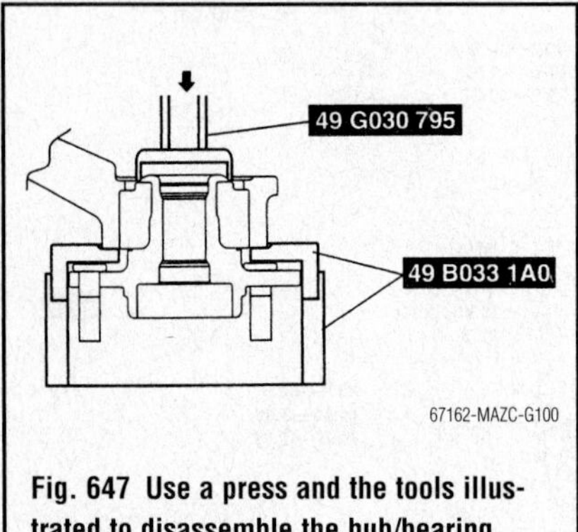

Fig. 647 Use a press and the tools illustrated to disassemble the hub/bearing assembly—Mazda3

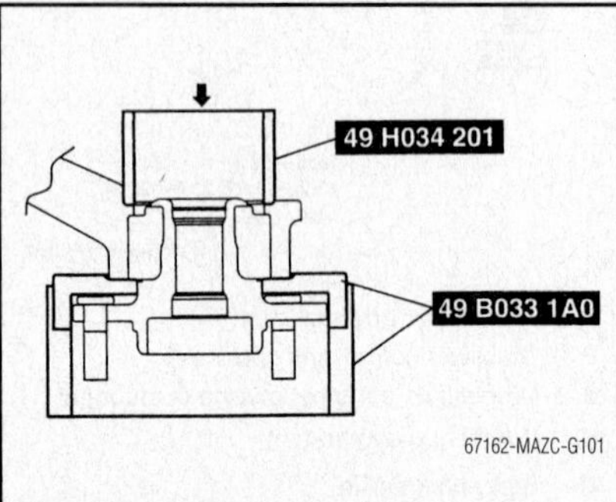

Fig. 648 Use a press and the tools illustrated to assemble the hub/bearing assembly—Mazda3

MX-5 Miata

See Figure 632.

1. Before servicing the vehicle, refer to the precautions section.

➡Refer the exploded view illustration for component locations and if applicable, their retainer torque specifications.

⁂ CAUTION

Performing the following procedures without first removing the ABS wheel-speed sensor may possibly cause an open circuit in the wiring harness if it is pulled by mistake. Before operations, remove the ABS wheel-speed sensor (axle side) and move the sensor away from the harnesses.

2. Disconnect the ABS wheel-speed sensor.
3. Remove the caliper with the hose still attached and support the assembly with wire so as to not place strain on the hose.
4. Remove the rotor.
5. Disconnect the tie-rod end from the knuckle.
6. Disconnect the stabilizer control link nut (lower).
7. Disconnect the front upper arm ball joint from the knuckle.
8. Remove the front upper arm bolt.
9. Disconnect the front lower arm ball joint from the knuckle.
10. Remove the steering knuckle assembly.
11. Remove the dust cover.
12. Remove the wheel hub bolts from the wheel hub using a press.
13. Remove the hub assembly.

To install:

14. Install the hub assembly. Tighten the bolts to 54–60 Nm (40–44 ft. lbs.).
15. Press in new wheel hub bolts into the wheel hub using a press.
16. Install the remaining components in the reverse order of removal using the accompanying illustration for torque values.
17. Inspect the wheel alignment.

Mazda6 Models

See Figures 649 through 654.

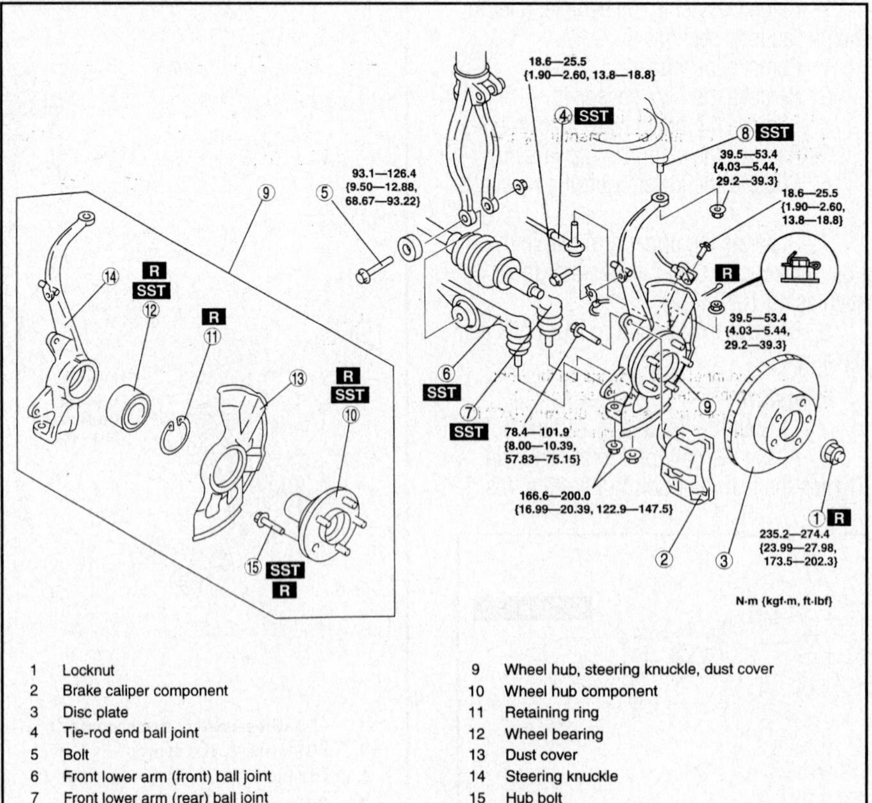

Fig. 649 Exploded view of the front wheel bearing and knuckle assembly—Mazda6 and Mazdaspeed6 models

1. Before servicing the vehicle, refer to the precautions section.
2. Refer to the illustration for component location and torque specifications.
3. Remove or disconnect the following:
   - Wheels
   - ABS sensor
   - Halfshaft axle nut, unstake the nut prior to removal.
   - Brake caliper and rotor
   - Tie rod end from the knuckle
   - Damper fork–to–control arm bolt
   - Front lower control arm ball joints
   - Front upper arm ball joint
   - Wheel hub dust cover
   - Hub bolts and the hub

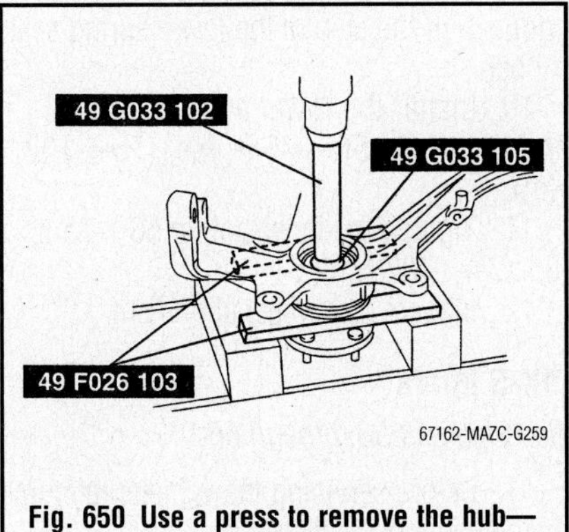

Fig. 650 Use a press to remove the hub—Mazda6 and Mazdaspeed6 models

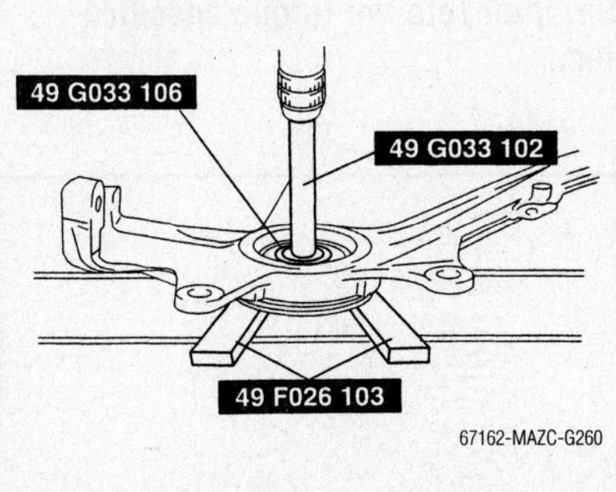

Fig. 651 Use a press to remove the wheel bearing—Mazda6 and Mazdaspeed6 models

   - Hub using a press and Mazda tools 49 F026 10, 49 G033 102 and 49 G033 105. If the bearing inner race remains in the hub, grind a section of the bearing inner race until 0.02 inch (0.5mm) remains and use a chisel to remove it.

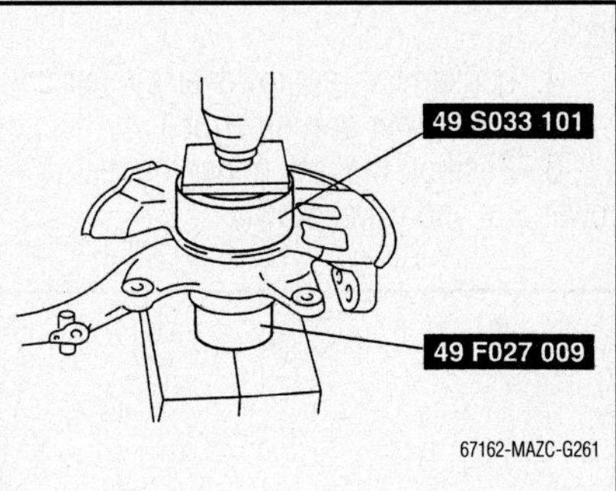

Fig. 652 Install the dust shield—Mazda6 and Mazdaspeed6 models

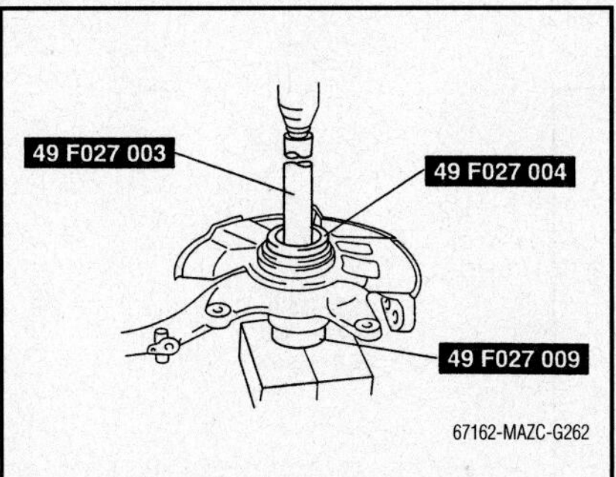

Fig. 653 Install a new wheel bearing using a press—Mazda6 and Mazdaspeed6 models

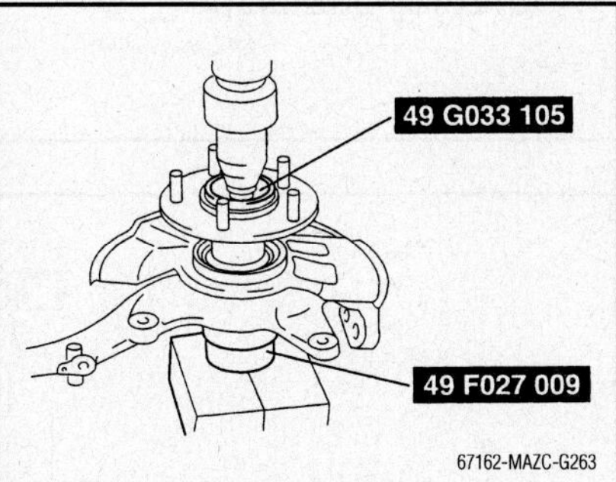

Fig. 654 Install a wheel hub using a press—Mazda6 and Mazdaspeed6 models

   - Snap ring
   - Bearing from the hub using a press and the tools illustrated
   - Brake dust shield, if it is being replaced. Mark the cover and knuckle for replacement purposes and use a chisel to remove the shield.
4. Clean and inspect all parts but do not wash or clean the wheel bearing. The bearing must be replaced.

To install:

5. Using the tools illustrated, install a new dust shield cover assembly to the knuckle, if removed.
6. Using the tools illustrated, press a new wheel bearing into the knuckle assembly and install the snap ring.
7. Install or connect the following:
   - Wheel bearing retaining ring
8. Using the tools illustrated, press in the hub assembly.
   - Wheel hub dust cover
   - Front upper arm ball joint and tighten the nut to 29–39 ft. lbs. (39–53 Nm)
   - Front lower arm ball joints and tighten the nuts to 122–147 ft. lbs. (166–200 Nm)
   - Damper fork bolt and tighten to 68–93 ft. lbs. (93–126 Nm)
   - Tie rod end to the knuckle and tighten the nut to 29–39 ft. lbs. (39–53 Nm) and install a new cotter pin
   - Brake caliper and rotor
   - Halfshaft axle nut, tighten the nut to 173–202 ft. lbs. (235–274 Nm) and stake the nut
   - Wheels

ADJUSTMENT

The front wheel bearings are not adjustable. If the bearings become loose or make noise, they must be replaced.

SUSPENSION

REAR SUSPENSION

COIL SPRING

REMOVAL & INSTALLATION

Mazda3

See Figures 655 through 657.

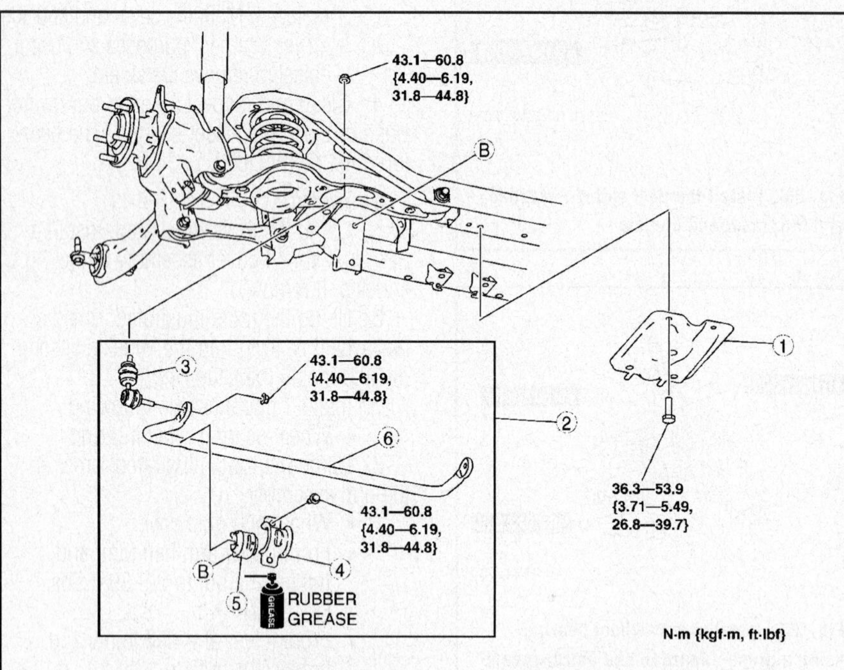

Fig. 655 Stabilizer bar and related components

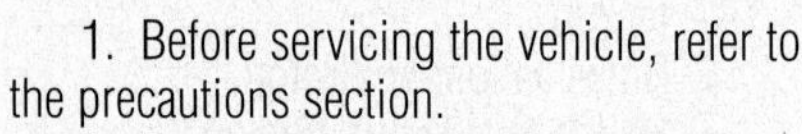

1. Before servicing the vehicle, refer to the precautions section.
2. Remove the rear stabilizer bar.
3. Support the rear lower arm using a jack.
4. Loosen the rear lower arm inner bolt.
5. Rear lower arm outer bolt.
6. Position the jack under the rear lower arm and lower slowly.
7. Remove the rear coil spring.

To install:

➡Before installation of the coil spring check the spring seat rubber. Replace as needed.

⁂ CAUTION

Installing the coil spring is dangerous. The coil spring could fly off and cause serious injury or death, and damage the vehicle.

8. Align the upper end of the rear coil spring with the step of the upper spring seat rubber.
9. Align the lower end of the rear coil spring with the step of the lower spring seat rubber.
10. Install the lower arm outer bolt and tighten to 56—75 ft. lbs. (75—102 Nm).
11. Tighten the inner bolt to 56—75 ft. lbs. (75—102 Nm).
12. Inspect rear wheel alignment.

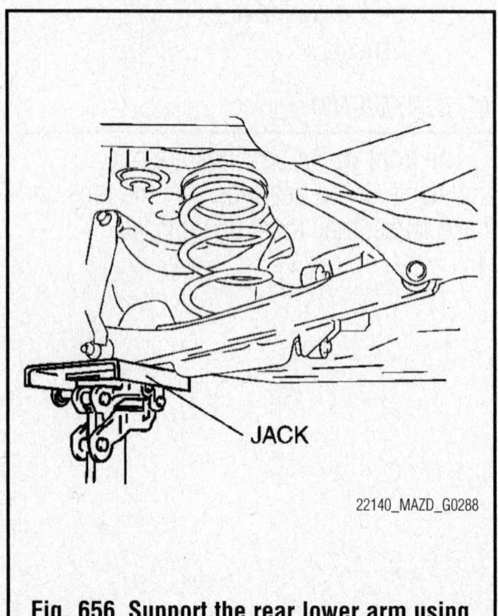

Fig. 656 Support the rear lower arm using a jack

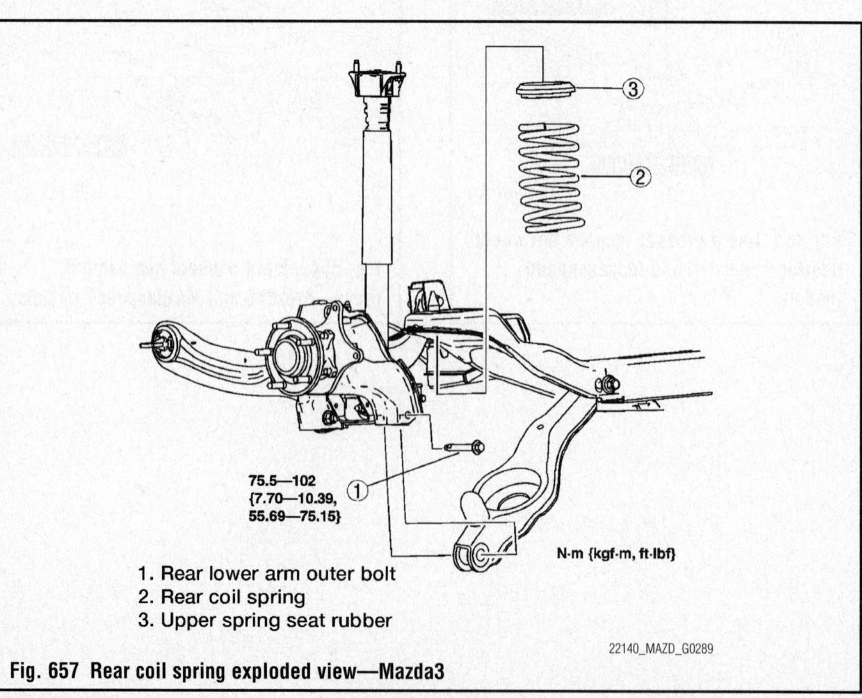

Fig. 657 Rear coil spring exploded view—Mazda3

MX-5 Miata

See Figures 658 through 662.

1. Before servicing the vehicle, refer to the precautions section.

➡Refer the exploded view illustration for component locations and if applicable, their retainer torque specifications.

⁂ CAUTION

Removing or installing the shock absorber and coil spring is dangerous. The shock absorber and coil spring could fly off and cause serious injury or death, and damage the vehicle.

2. Remove the rear shock absorber and coil spring.

⁂ WARNING

Before removing the piston rod nut, secure the shock absorber and spring in a suitable spring compressor. Otherwise, the shock absorber and spring could fly off under tremendous pressure and cause serious injury or death, or damage to vehicle parts.

3. Protect the coil spring from scratches using a piece of cloth and install the spring compressor.
4. Compress the coil spring and remove the piston rod nut.
5. Remove the following:
   - Retainer
   - Bushing
   - Upper spring seat
   - Spring seat rubber
   - Spacer
   - Bushing
   - Stopper casing and bound stopper
   - Bound stopper
   - Collar
   - Stopper casing
   - Coil spring
   - Shock absorber

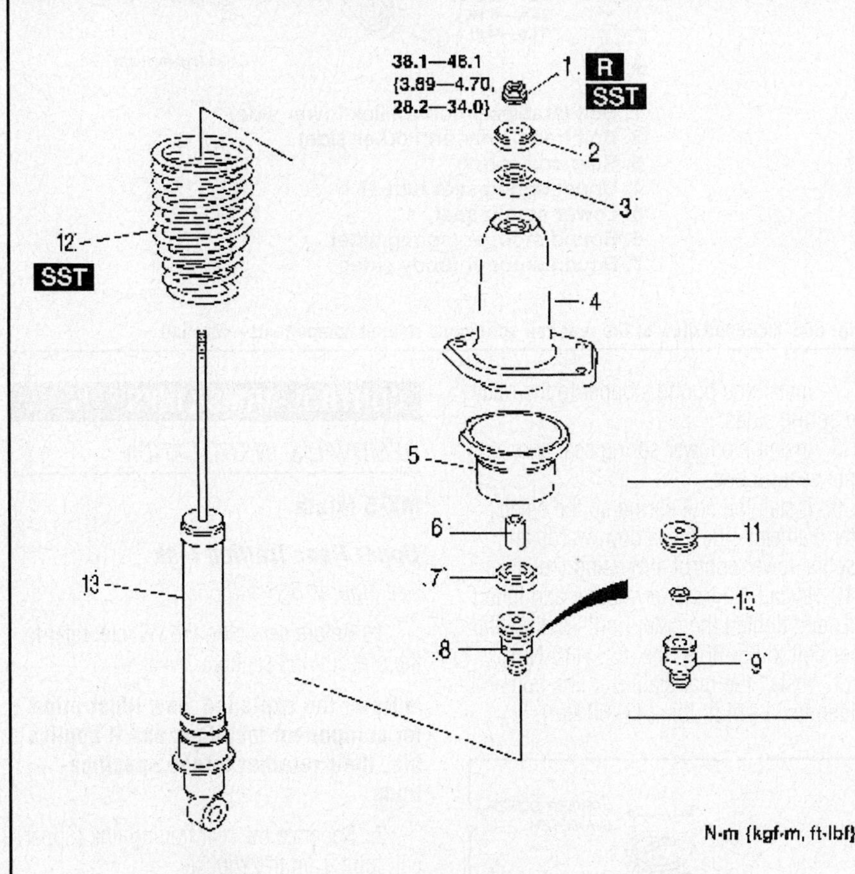

1 Piston rod nut
2 Retainer
3 Bushing
4 Upper spring seat
5 Spring seat rubber
6 Spacer
7 Bushing
8 Bound stopper and stopper casing
9 Bound stopper
10 Collar
11 Stopper casing
12 Coil spring
13 Rear shock absorber

Fig. 658 Exploded view of the rear coil spring assembly—MX-5 Miata

To install:

6. Install the following:
   - Shock absorber so that the lower end of the coil spring is seated on the step of the lower spring seat
   - Coil spring lead lower end of the lower spring seat facing the direction shown in the illustration
   - Stopper casing
   - Collar so that the tapered side is facing downward as shown in the illustration
   - Bound stopper
   - Stopper casing and bound stopper
   - Bushing

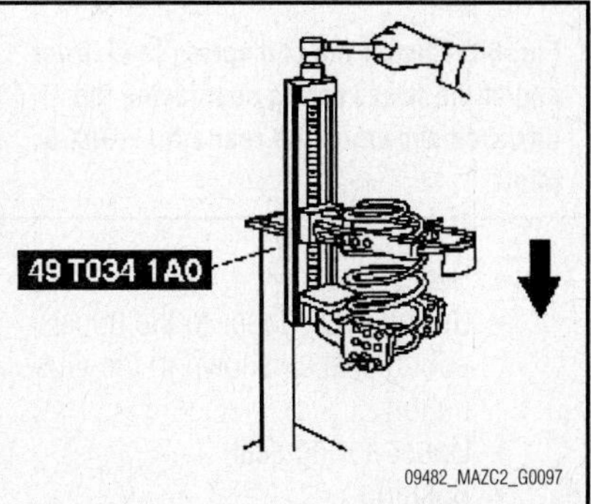

Fig. 659 Installing the coil spring—MX-5 Miata

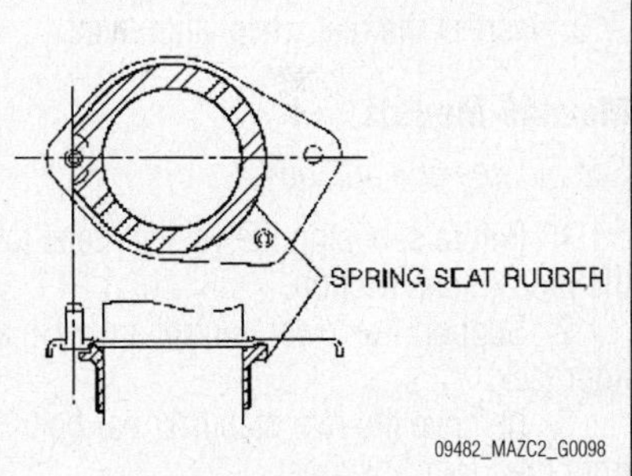

Fig. 660 Install the collar on the rear strut so that the tapered side is facing downward—MX-5 Miata

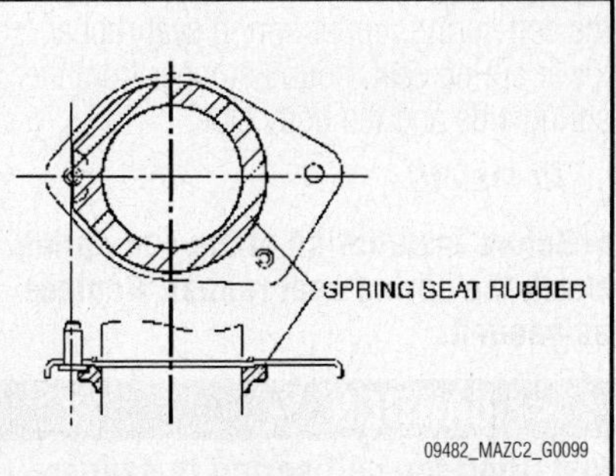

Fig. 661 Install the spring seat rubber on the rear strut to the upper spring seat as shown—MX-5 Miata

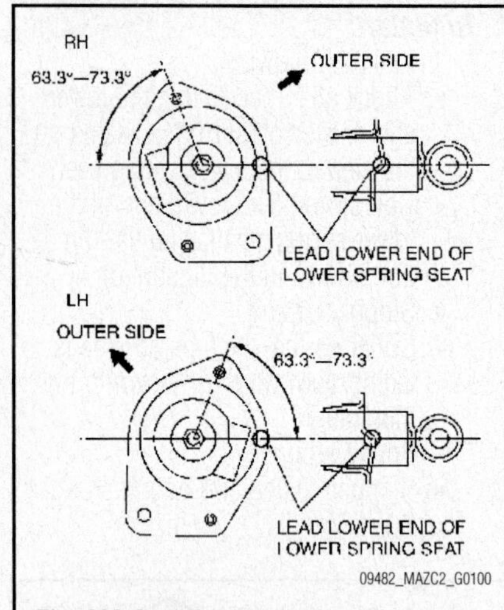

Fig. 662 Install the coil spring lead lower end of the lower spring seat facing the direction shown on the rear strut —MX-5 Miata

- Spacer
- Spring seat rubber to the upper spring seat as shown in the illustration
- Upper spring seat
- Bushing
- Retainer

7. Install the piston rod nut and tighten to 38–46 Nm (28–34 ft. lbs.).
8. Install the strut assembly.
9. Inspect the rear wheel alignment.

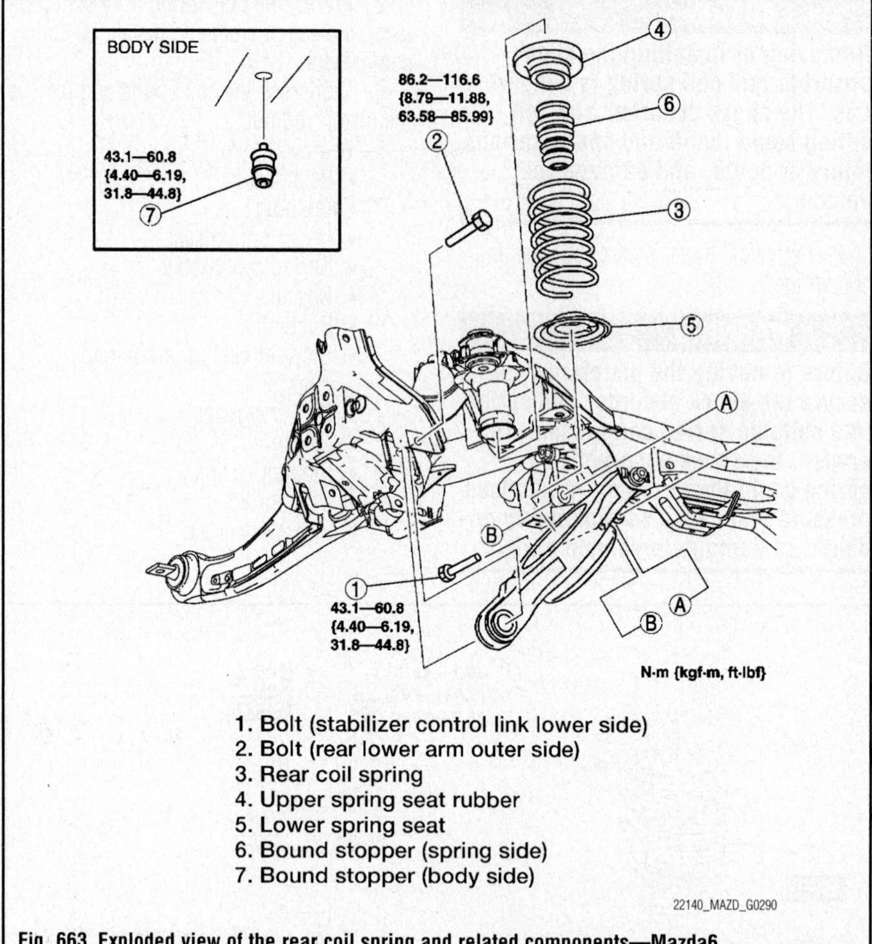

Fig. 663 Exploded view of the rear coil spring and related components—Mazda6

Mazda6 Models

See Figures 663 and 664.

1. Before servicing the vehicle, refer to the precautions section.
2. Support the lower control arm with a floor jack.
3. Remove the rear stabilizer bar bolt from the lower link side.
4. Loosen the inner bolt of the rear lower control arm.
5. Remove the outer bolts of the rear lower control arm.
6. Lower the control arm and remove the coil spring, upper spring seat rubber, lower spring seat, bound stopper from the spring side and the body side.

To install:

➡Before installation of the coil spring check the spring seat rubber. Replace as needed.

⁂ CAUTION

Installing the coil spring is dangerous. The coil spring could fly off and cause serious injury or death, and damage the vehicle.

7. Install the bound stopper to the body and spring sides.
8. Install the lower spring seat and upper seat rubber.
9. Install the coil spring so the small outer diameter side faces downwards and raise the lower control arm using the jack.
10. Install the lower arm outer and inner bolts and tighten the outer bolt first then the inner bolt to 63–86 ft. lbs. (86–116 Nm).
11. Install the rear stabilizer link bolt and tighten to 31–44 ft. lbs. (43–60 Nm).

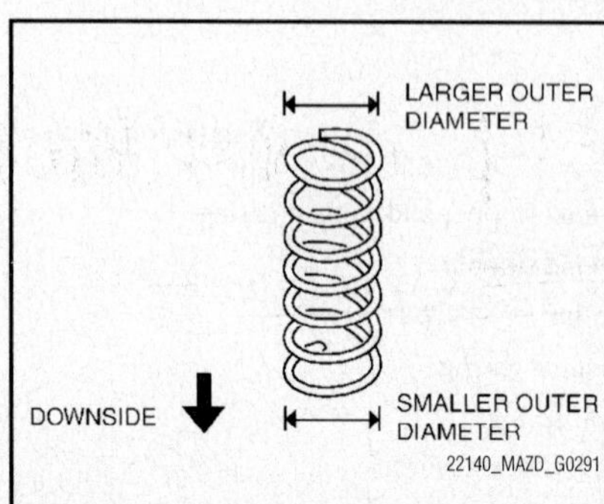

Fig. 664 Install the coil spring so the small outer diameter side faces downwards

CONTROL ARMS/LINKS

REMOVAL & INSTALLATION

MX-5 Miata

Upper Rear Trailing Link

See Figures 665 and 666.

1. Before servicing the vehicle, refer to the precautions section.

➡Refer the exploded view illustration for component locations and if applicable, their retainer torque specifications.

2. Separate the rear trailing link (upper) ball joint from the knuckle.

➡When removing the rear trailing link (upper) ball joint, the rear knuckle bushing may also come off. If it comes off, replace the rear knuckle.

3. Remove the rear upper trailing link.
4. Remove the dust boot.

To install:

5. Wipe the grease off the ball joint stud.

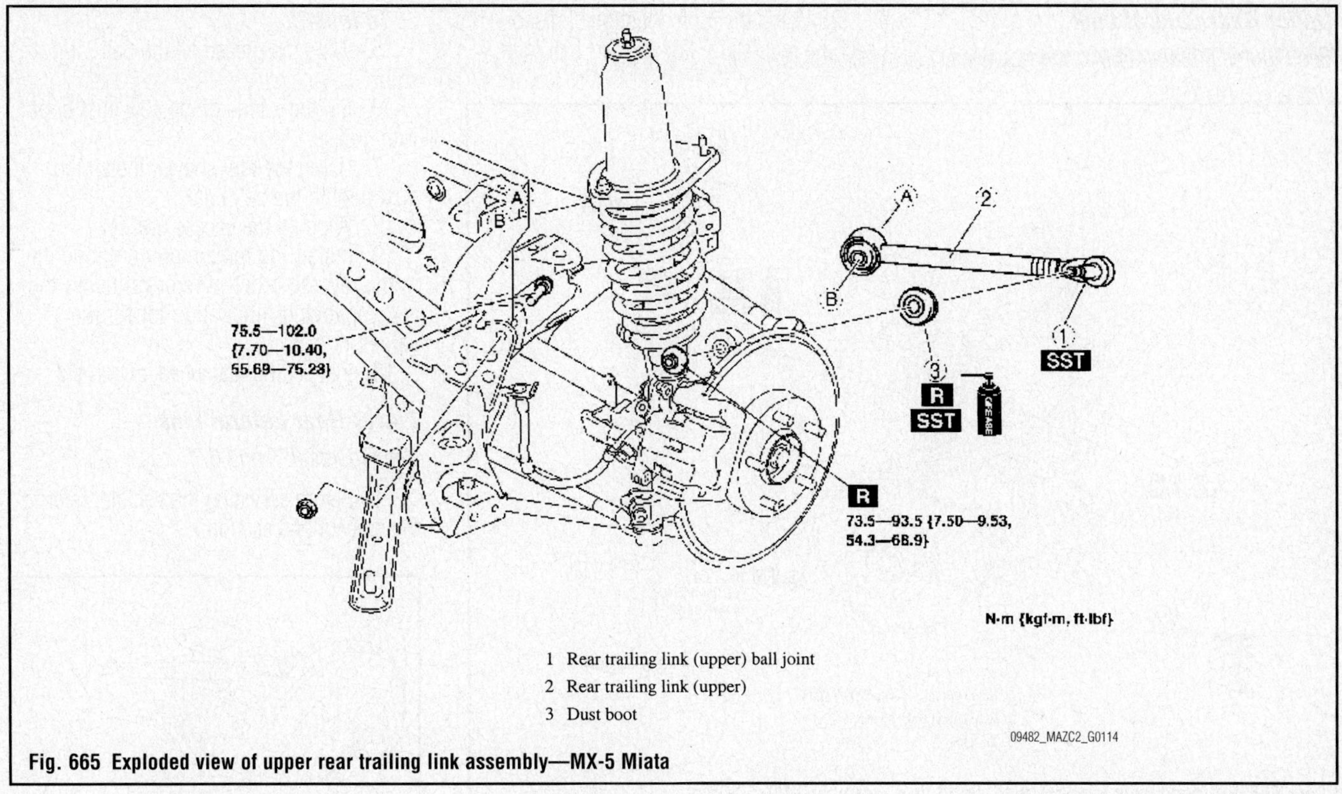

Fig. 665 Exploded view of upper rear trailing link assembly—MX-5 Miata

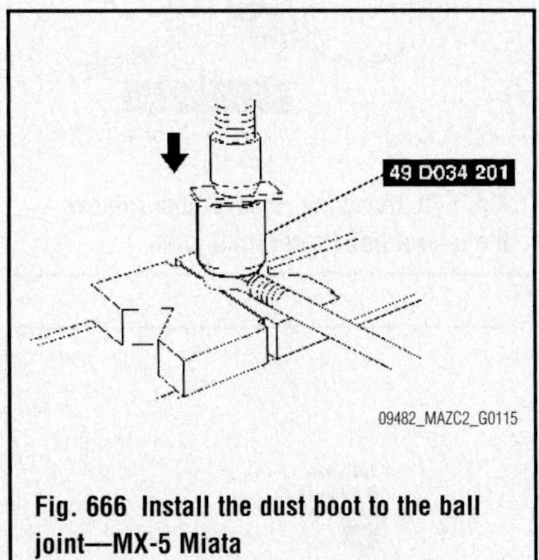

Fig. 666 Install the dust boot to the ball joint—MX-5 Miata

6. Fill the inside of the new dust boot with grease.
7. Using the tool shown, install the dust boot to the ball joint.
8. Wipe off the excess grease.
9. Install the remaining components in the reverse order of removal using the accompanying illustration for torque values.
10. Inspect the rear wheel alignment.

Lower Rear Trailing Link

See Figure 667.

1. Remove the toe control link outer bolt.
2. Remove rear trailing link (lower) outer bolt.
3. Remove rear trailing link (lower) inner bolt.
4. Remove rear trailing link (lower).

To install:

5. Install the rear trailing link (lower).
6. Install the rear trailing link (lower) inner bolt and tighten to 23—75 ft. lbs. (75—102 Nm).
7. Install the rear trailing link (lower) outer bolt and tighten to 56—75 ft. lbs. (75—102 Nm).
8. Install the toe control link outer bolt and tighten to 23—31 ft. lbs. (31—42 Nm).

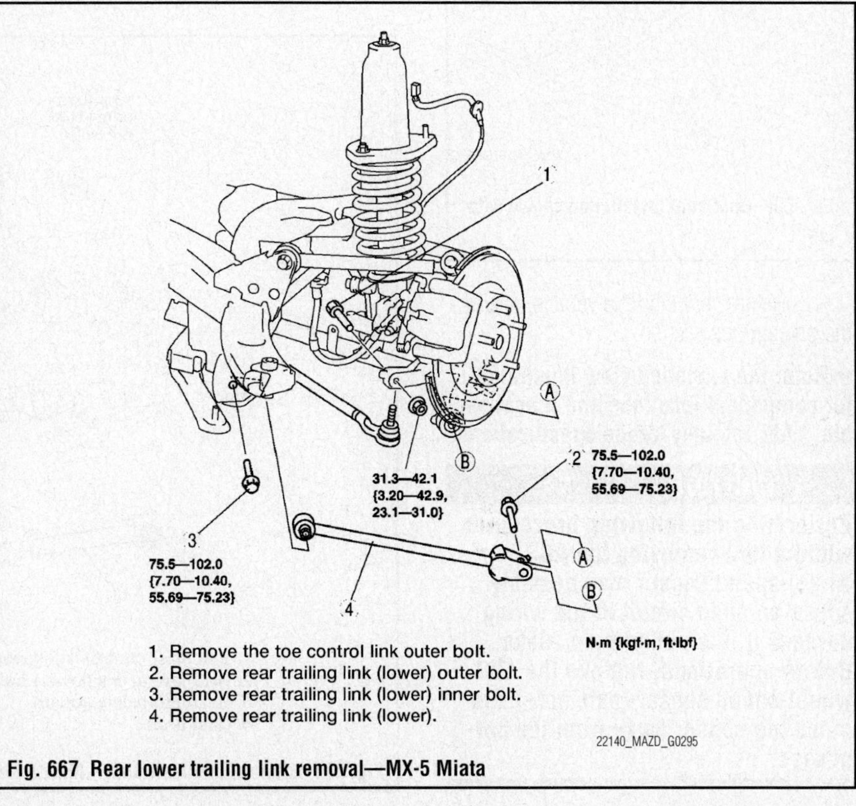

Fig. 667 Rear lower trailing link removal—MX-5 Miata

Upper Rear Lateral Link

See Figures 668 and 669.

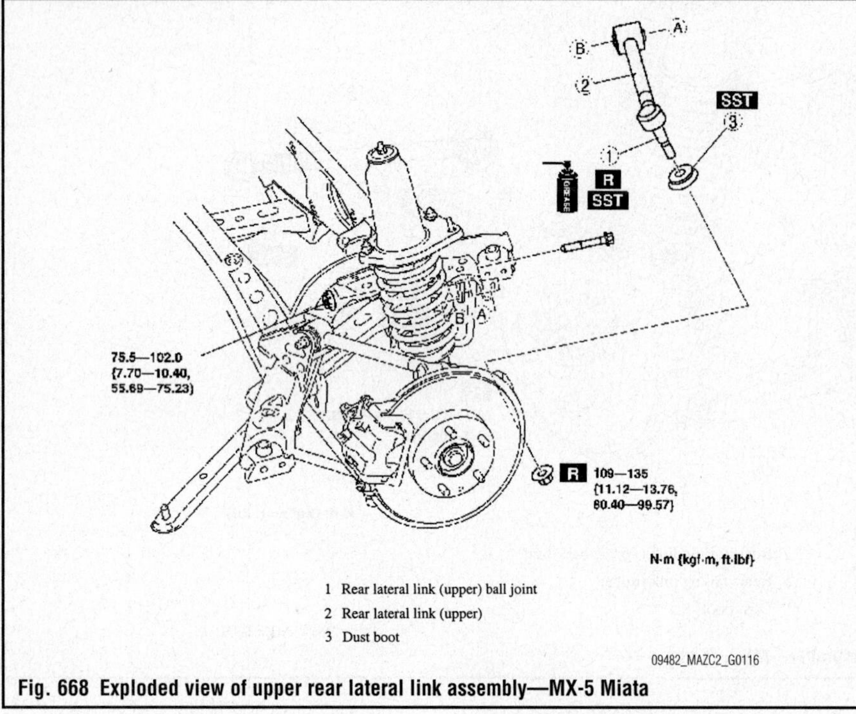

Fig. 668 Exploded view of upper rear lateral link assembly—MX-5 Miata

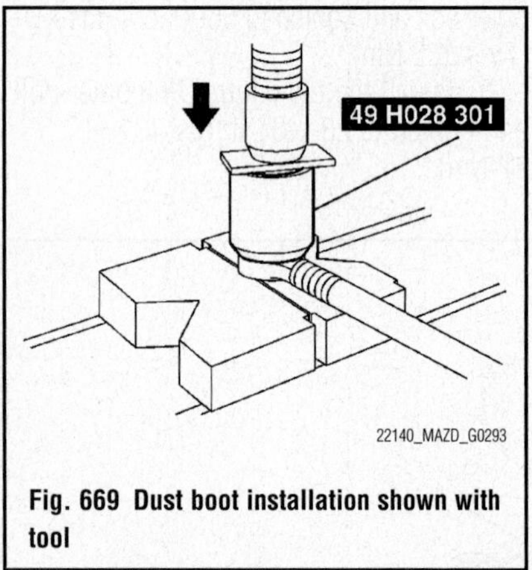

Fig. 669 Dust boot installation shown with tool

1. Before servicing the vehicle, refer to the precautions section.

➡Refer the exploded view illustration for component locations and if applicable, their retainer torque specifications.

✲✲ CAUTION

Performing the following procedures without first removing the ABS wheel-speed sensor may possibly cause an open circuit in the wiring harness if it is pulled by mistake. Before operations, remove the ABS wheel-speed sensor (axle side) and move the sensor away from the harnesses.

2. Using the tools illustrated, disconnect the rear lateral link (upper) ball joint.

➡When removing the rear lateral link (upper) ball joint, the rear knuckle bushing may also come off. If it comes off, replace the rear knuckle.

3. Remove the rear lateral link (upper).
4. Remove the dust boot.

To install:

5. Wipe the grease off the ball joint stud.
6. Fill the inside of the new dust boot with grease.
7. Using the tool shown, install the dust boot to the ball joint.
8. Wipe off the excess grease.
9. Install the remaining components in the reverse order of removal using the accompanying illustration for torque values.
10. Inspect the rear wheel alignment.

Lower Rear Lateral Link

See Figures 670 and 671.

1. Before servicing the vehicle, refer to the precautions section.

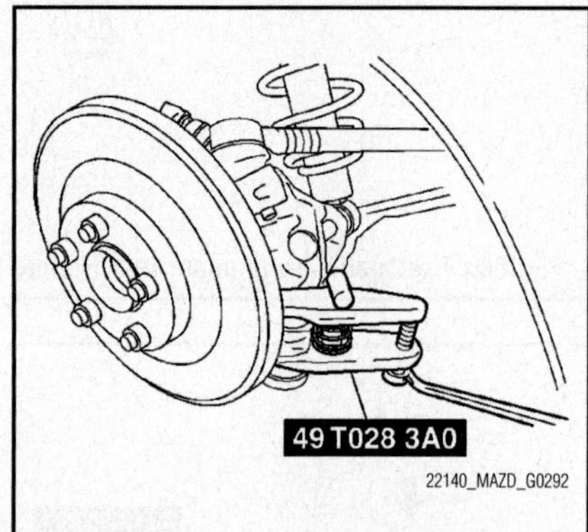

Fig. 670 Using the removal tool remove the lateral link (lower) ball joint

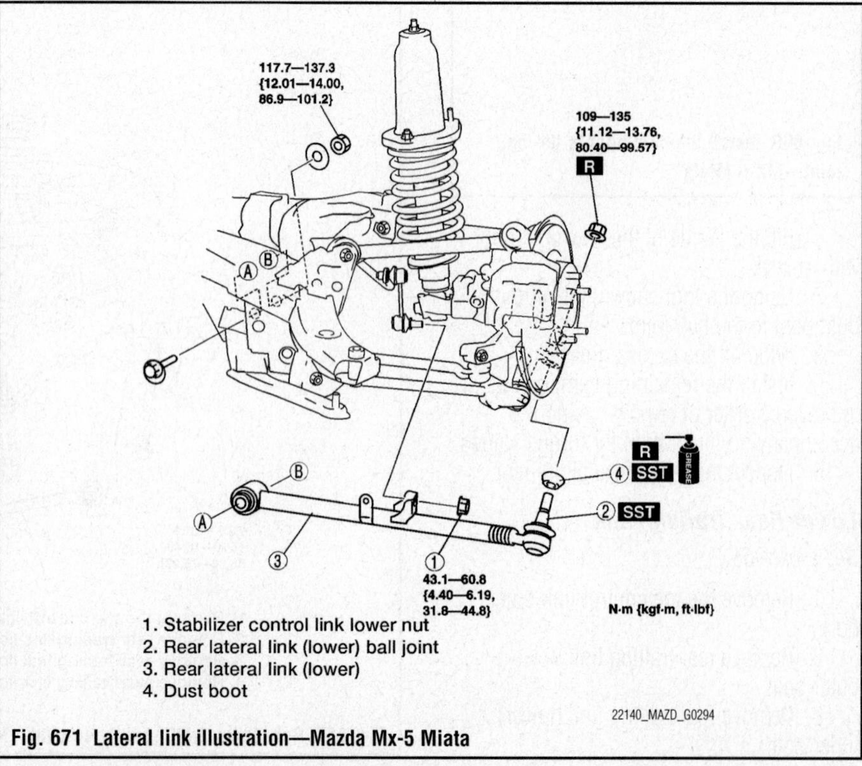

Fig. 671 Lateral link illustration—Mazda Mx-5 Miata

➡Refer the exploded view illustration for component locations and if applicable, their retainer torque specifications.

⁂ CAUTION

Performing the following procedures without first removing the ABS wheel-speed sensor may possibly cause an open circuit in the wiring harness if it is pulled by mistake. Before operations, remove the ABS wheel-speed sensor (axle side) and move the sensor away from the harnesses.

2. remove the lower rear lateral link as follow:
   - Stabilizer control link lower nut
   - Rear lateral link (lower) ball joint
   - Rear lateral link (lower)
   - Dust boot

To install:

3. Wipe the grease off the ball joint stud.
4. Fill the inside of the new dust boot with grease.
5. Install the dust boot to the ball joint.
6. Wipe off the excess grease.
7. Install the remaining components in the reverse order of removal using the accompanying illustration for torque values.
8. Inspect the rear wheel alignment.

Mazda6 Models

Upper Control Arm

See Figure 672.

1. Before servicing the vehicle, refer to the precautions section.
2. Remove the wheel.
3. Remove the ABS sensor.
4. Disconnect the parking brake cable.
5. Support the rear trailing link with a jack.
6. Remove the rear shock absorber lower bolt.
7. Remove the front trailing link bolt.
8. Remove the rear upper arm bolt and the arm.

To install:

9. Install the arm and the rear bolt.
10. Attach the bolt to the trailing link front side.
11. Tighten the rear upper bolt to 68–93 ft. lbs. (93–126 Nm) and the trailing link bolt to 63–86 ft. lbs. (86–116 Nm).
12. Install the lower shock absorber bolt and tighten to 68–93 ft. lbs. (93–126 Nm).
13. Attach the parking brake cable.
14. Install the ABS sensor.
   - Install the wheel.

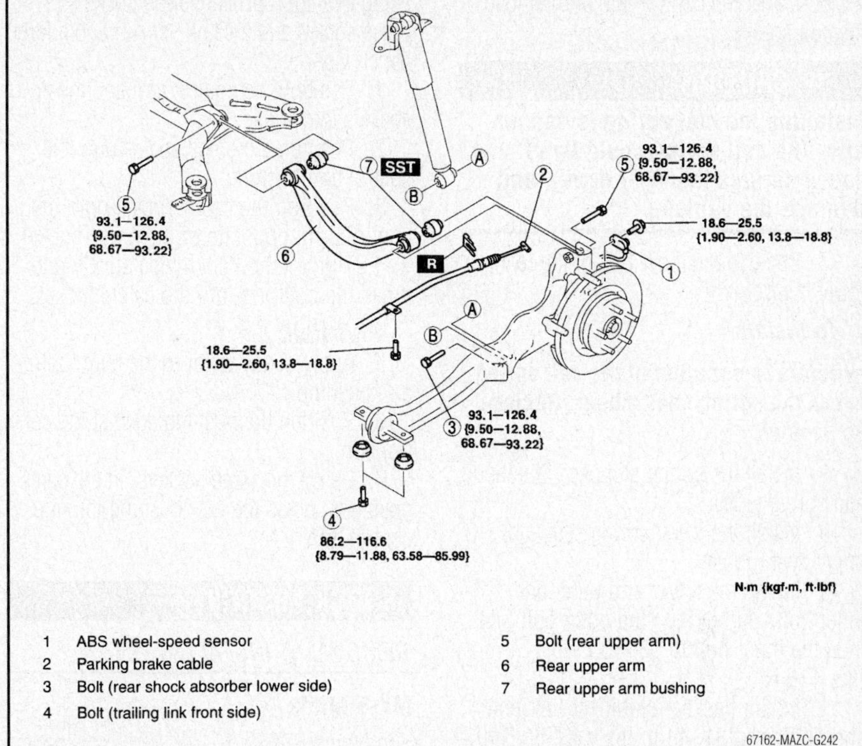

Fig. 672 Exploded view of the rear upper control arm assembly—Mazda6 and Mazdaspeed6 models

Lower Control Arm

See Figure 673.

1. Before servicing the vehicle, refer to the precautions section.
2. Support the lower control arm with a floor jack.
3. Remove the rear stabilizer bar bolt from the lower link side.
4. Loosen the inner bolt of the rear lower control arm.
5. Remove the outer bolts of the rear lower control arm.

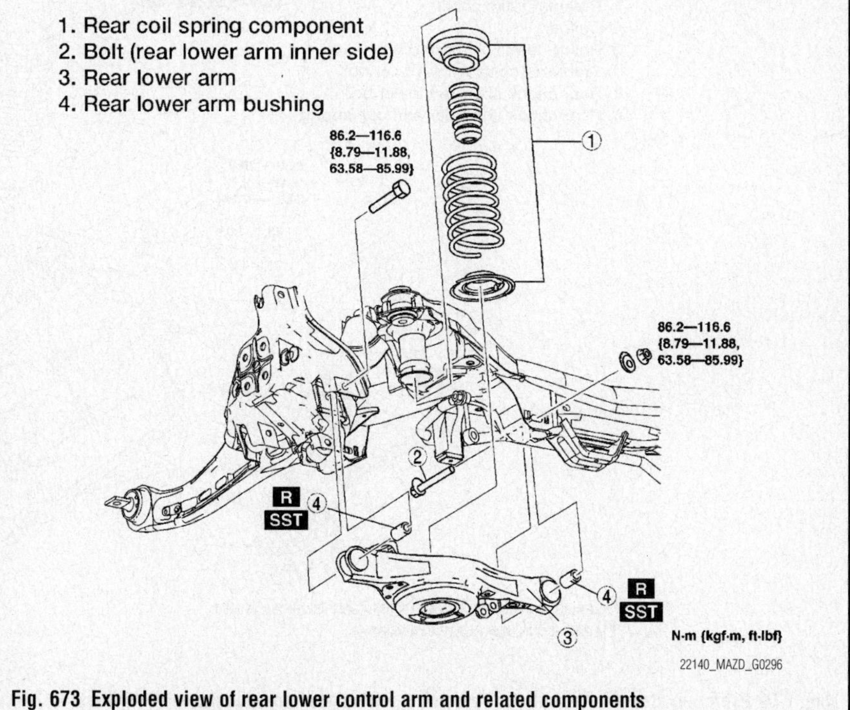

Fig. 673 Exploded view of rear lower control arm and related components

6. Lower the control arm and remove the coil spring.

⁂ CAUTION

Installing the coil spring is dangerous. The coil spring could fly off and cause serious injury or death, and damage the vehicle.

7. Remove the inner bolt of the rear lower control arm.

To install:

➡Before installation of the coil spring check the spring seat rubber. Replace as needed.

8. Install the bound stopper to the body and spring sides.
9. Install the lower spring seat and upper seat rubber.
10. Install the lower arm outer and inner bolts and tighten the outer bolt first then the inner bolt to 63–86 ft. lbs. (86–116 Nm).
11. Install the rear stabilizer link bolt and tighten to 31–44 ft. lbs. (43–60 Nm).

CONTROL ARM BUSHING REPLACEMENT

All Mazda's use a pressed in control arm bushing, and the pressing can be done using two appropriately sized sockets (a press socket and a catch socket) and a large bench vise.

1. Position the control arm and the 2 sockets into a vise.
2. Position the press socket onto the control arm bushing.
3. Position the catch socket onto the control arm, opposite of the press socket.
4. Tighten the bench vise slowly and press the bushing into the catch socket.

To install:

5. Apply soapy water to the new control arm bushing.
6. Position the bushing against the control arm.
7. Using the same sockets, in the same positions, press the new bushing into the control arm.

SHOCK ABSORBER

REMOVAL & INSTALLATION

MX-5 Miata

See Figure 674.

1. Before servicing the vehicle, refer to the precautions section.

➡Refer the exploded view illustration for component locations and if applicable, their retainer torque specifications.

⁂ CAUTION

Performing the following procedures without first removing the ABS wheel-speed sensor may possibly cause an open circuit in the wiring harness if it is pulled by mistake. Before performing the following procedures, remove the ABS wheel-speed sensor (axle side) and fix it to an appropriate place where the sensor will not be pulled while servicing the vehicle.

2. If removing the left side, remove the fuel tank protector.
3. Remove the parking brake cable.
4. Remove the caliper without disconnecting the hose, and support the caliper with wire.
5. Remove the rear lateral link upper bolt.
6. Remove the stabilizer control link upper nut.
7. Remove the rear shock absorber lower bolt.
8. Remove the rear shock absorber and coil spring.

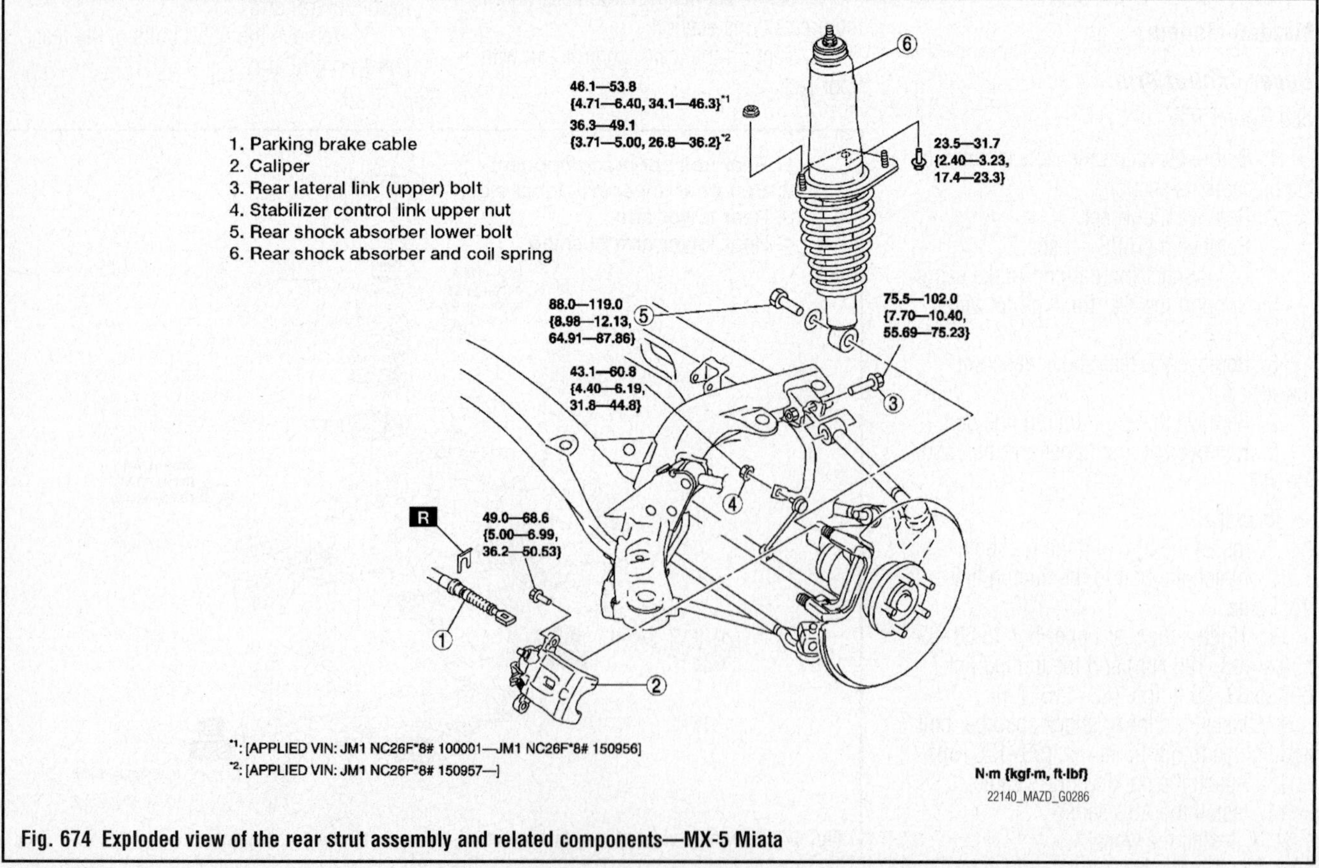

Fig. 674 Exploded view of the rear strut assembly and related components—MX-5 Miata

To install:

9. Install the components in the reverse order of removal using the accompanying illustration for torque values.

Mazda3 Models

See Figure 675.

1. Before servicing the vehicle, refer to the precautions section.
2. Support the rear axle assembly with a jack.
3. Remove or disconnect the following:
   - Rear wheel(s)
   - Top shock absorber nuts
   - Bottom nut and bolt
   - Shock absorber

To install:

4. Install or connect the following:
   - Shock absorber
   - Bottom nut and bolt and tighten to 56—75 ft. lbs. (76—102Nm).
   - Upper mounting nuts and tighten to 16–21 ft. lbs. (21–29 Nm)
   - Rear wheel(s)

Mazda6 Models

See Figures 676 and 677.

1. Before servicing the vehicle, refer to the precautions section.
2. Support the rear axle assembly with a jack.
3. Remove or disconnect the following:
   - Rear wheel(s)
   - ABS sensor
   - Top shock and bracket retainers
   - Bracket

➡The suspension will drop when the weight lifts off the wheels.

   - Bottom shock mount bolt
   - Shock absorber

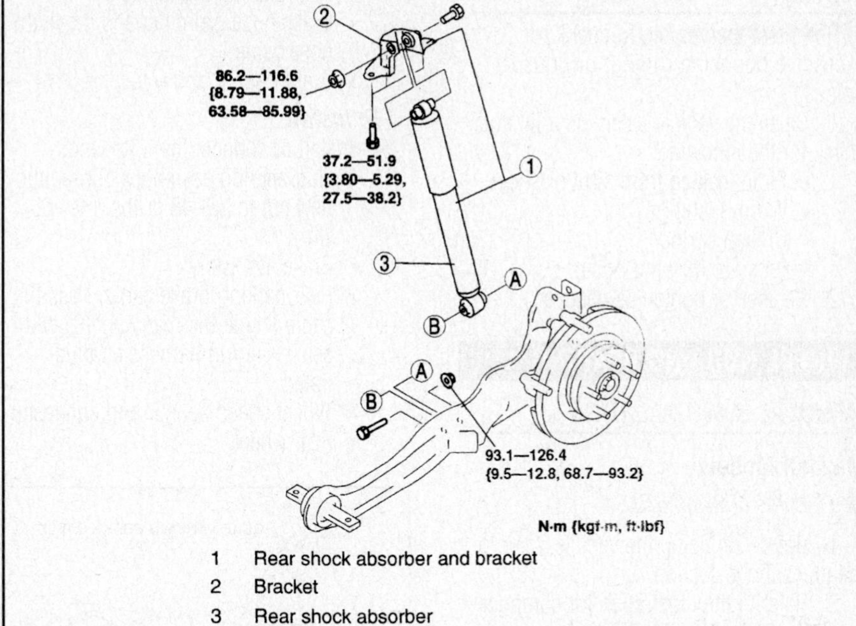

Fig. 676 Exploded view of the rear shock absorber assembly—Mazda6 and Mazdaspeed6 models

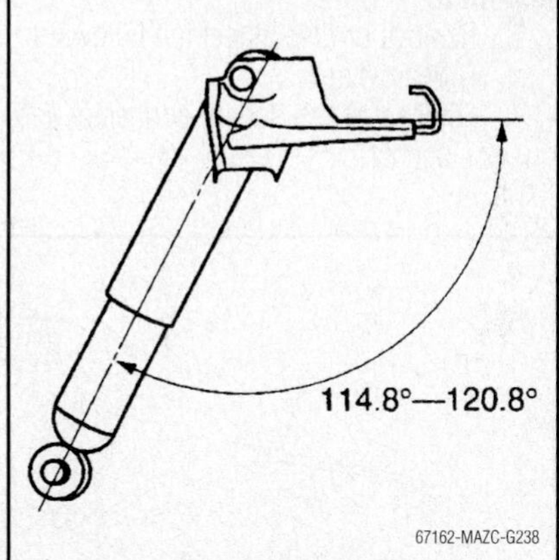

Fig. 677 Position the bracket and rear shock absorber assembly as shown when installing—Mazda6 and Mazdaspeed6 models

To install:

4. Install or connect the following:
   - Shock absorber
   - Bracket and position the strut and bracket as illustrated. Tighten the bracket bolts to 27–38 ft. lbs. (37–51 Nm).
   - Bottom shock mount bolt(s) and tighten to 68–93 ft. lbs. (93–126 Nm)
   - Upper mounting retainers and tighten to 63–85 ft. lbs. (86–116 Nm)
   - ABS sensor
   - Rear wheel(s)

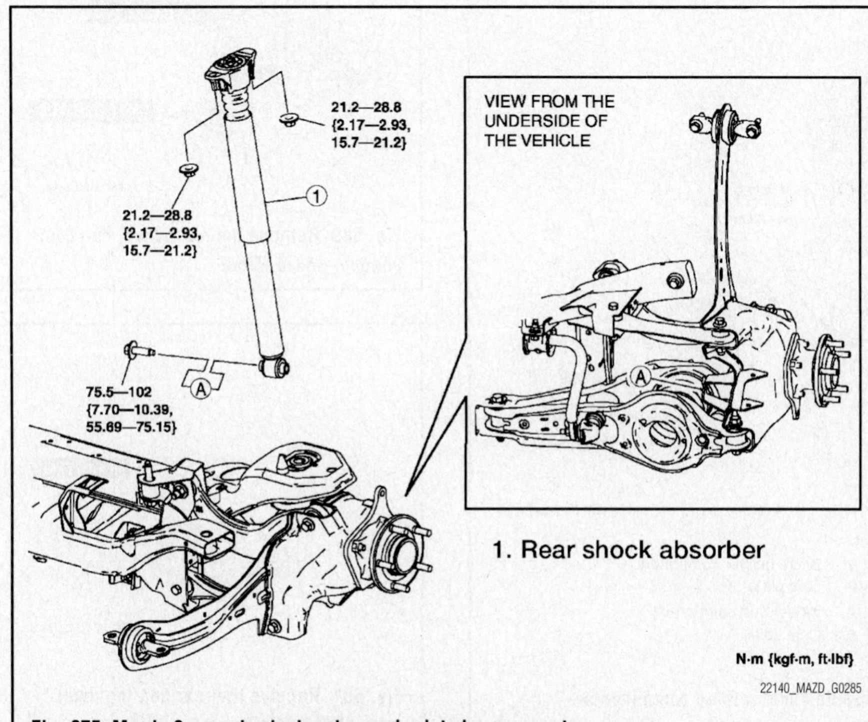

Fig. 675 Mazda 3 rear shock absorber and related components

TESTING

Road test the vehicle to check for excessive bouncing rolling, or unusual noises.

1. Raise the vehicle safely on a lift and check for the following:
   - Fluid leaking from strut or shock
   - Worn bushings
   - Broken spring
   - loose or missing bolts
2. Replace or tighten as needed.

WHEEL HUB AND BEARING

REMOVAL & INSTALLATION

Mazda3 Models

See Figures 678 and 679.

1. Before servicing the vehicle, refer to the precautions section.
2. Refer to the illustration for component location and torque specifications.

➡The wheel bearings are not serviceable. If the bearings are bad, a new hub/bearing assembly must be installed.

3. Remove or disconnect the following:
   - Rear wheels
   - Wheel speed sensor connector and sensor
   - Rear parking brake cable
   - Brake hose grommet and move the hose aside
   - Brake caliper and rotor

To install:

4. Install or connect the following:
   - Hub/bearing assembly. Torque the new nut to 36–48 ft. lbs. (49–65 Nm).
   - Brake assembly
   - Rear parking brake cable. Pass the cable inside the rear wheel speed sensor wiring harness as illustrated.
   - Wheel speed sensor and connector
   - Rear wheel

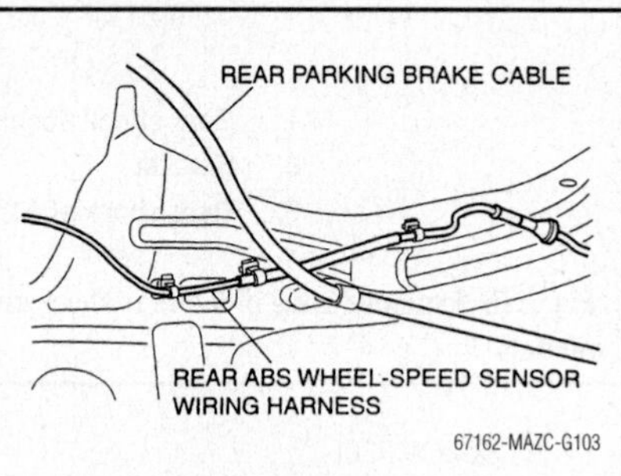

Fig. 679 Pass the cable inside the rear wheel speed sensor wiring harness when installing—Mazda3

MX-5 Miata

See Figures 680 through 684.

1. Before servicing the vehicle, refer to the precautions section.

➡Refer the exploded view illustration for component locations and if applicable, their retainer torque specifications.

CAUTION

Performing the following procedures without first removing the ABS wheel-speed sensor may possibly cause an open circuit in the wiring harness if it is pulled by mistake. Before operations, remove the ABS wheel-speed sensor (axle side) and move the sensor away from the harnesses.

2. Disconnect the ABS wheel-speed sensor.
3. Lock the disc plate by applying the brakes.
4. Knock the crimped portion of the locknut outward using a chisel and a hammer.
5. Remove the locknut.
6. Disconnect the parking brake cable.

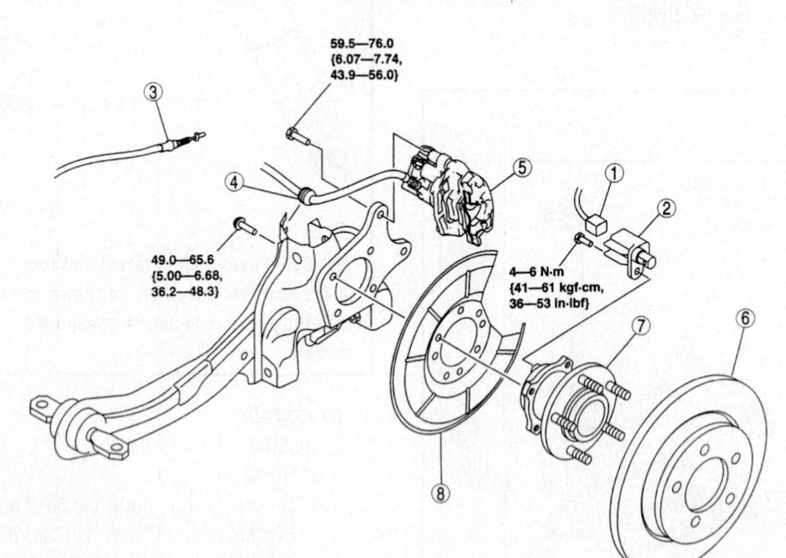

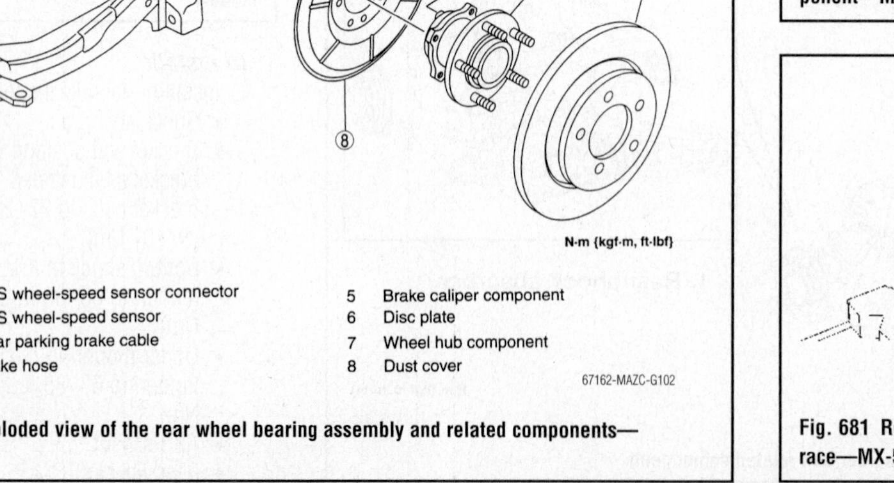

Fig. 678 Exploded view of the rear wheel bearing assembly and related components—Mazda3

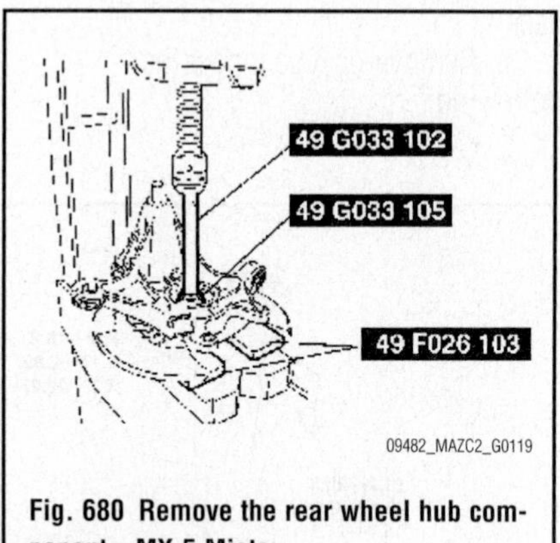

Fig. 680 Remove the rear wheel hub component—MX-5 Miata

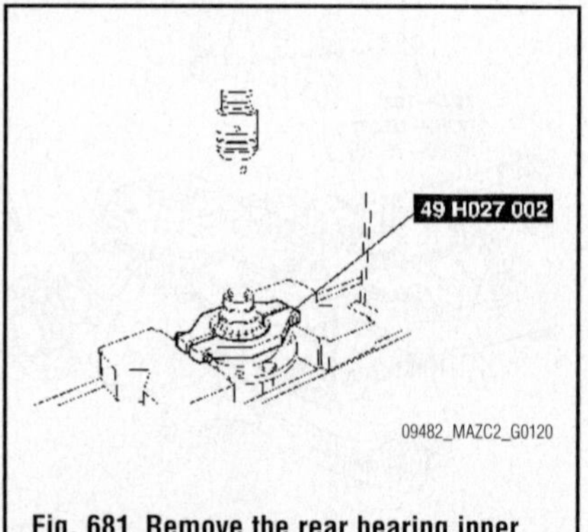

Fig. 681 Remove the rear bearing inner race—MX-5 Miata

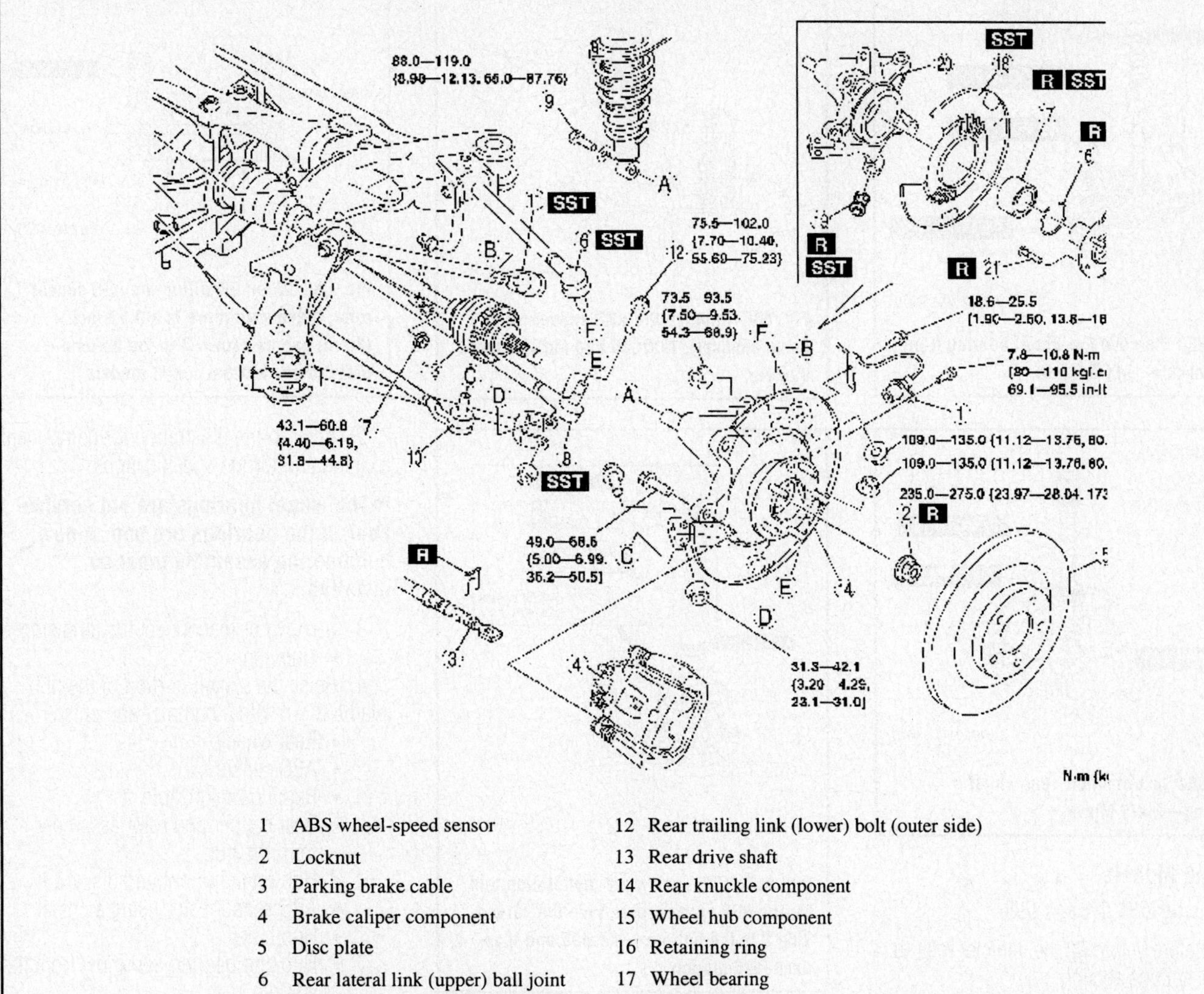

1 ABS wheel-speed sensor
2 Locknut
3 Parking brake cable
4 Brake caliper component
5 Disc plate
6 Rear lateral link (upper) ball joint
7 Stabilizer control link nut (lower)
8 Rear lateral link (lower) ball joint
9 Shock absorber bolt (lower)
10 Toe control link ball joint
11 Rear trailing link (upper) ball joint
12 Rear trailing link (lower) bolt (outer side)
13 Rear drive shaft
14 Rear knuckle component
15 Wheel hub component
16 Retaining ring
17 Wheel bearing
18 Dust cover
19 Bushing
20 Rear knuckle
21 Wheel hub bolt

09482_MAZC2_G0118

Fig. 682 Exploded view of rear hub and knuckle assembly and related components—MX-5 Miata

7. Remove the caliper with the hose still attached and support the assembly with wire so as to not place strain on the hose.
8. Remove the rotor.
9. Disconnect the rear lateral link (upper) ball joint from the knuckle.
10. Remove the stabilizer control link lower nut.
11. Disconnect the rear lateral link lower ball joint from the knuckle.
12. Remove the lower shock absorber bolt.
13. Disconnect the toe control link ball joint.
14. Disconnect the rear trailing link (upper) ball joint.
15. Remove the rear lower trailing link bolt on the outer side.
16. Temporarily install a spare nut onto the end of the rear drive shaft.
17. Tap the nut with a copper hammer to loosen the drive shaft from the wheel hub.
18. Separate the rear drive shaft from the wheel hub.
19. Remove the rear knuckle component.
20. Remove the wheel Hub Component as follows:
    a. Wind the tool illustrated and backing plate contact area with packing tape two times.
    b. Remove the wheel hub component using the tools illustrated.
    c. If the bearing inner race remains on the wheel hub component, use a chisel to secure a sufficient space for installing the service tool between wheel hub components and bearing inner race.
21. Remove the bearing inner race using the tool illustrated.

To install:

22. Install a new wheel bearing using the tool illustrated.
23. Install the wheel hub component using the tool illustrated.
24. Install the remaining components in the reverse order of removal using the accompanying illustration for torque values.
25. Inspect the wheel alignment.

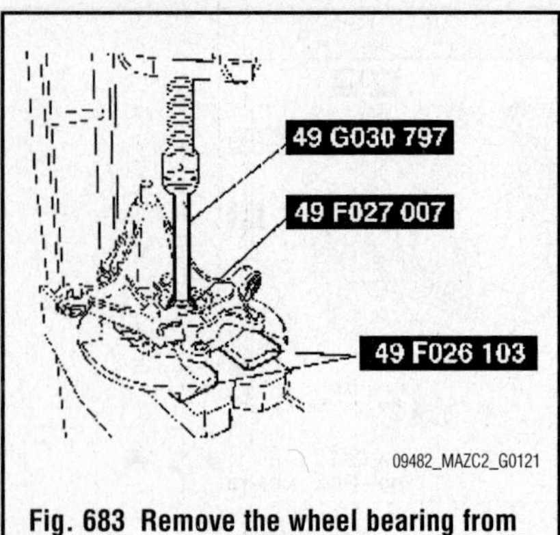

Fig. 683 Remove the wheel bearing from the knuckle—MX-5 Miata

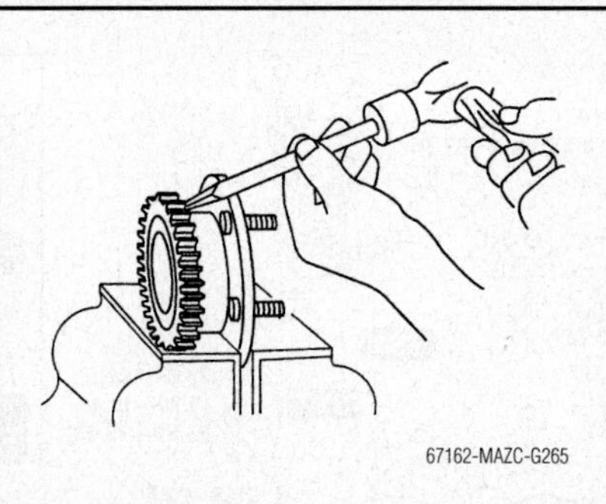

Fig. 686 Remove the ABS sensor rotor using a chisel—Mazda6 and Mazdaspeed6 models

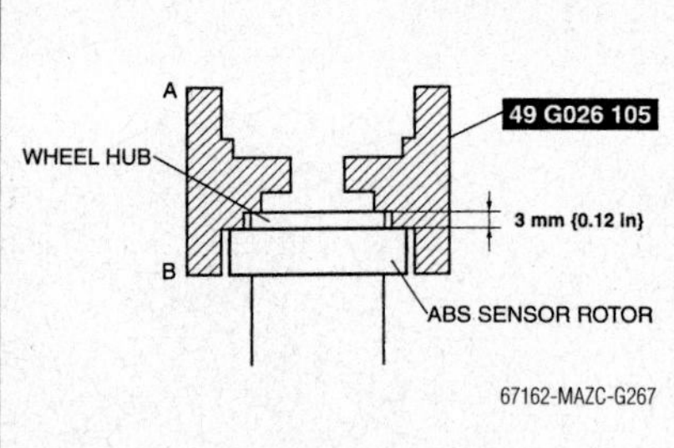

Fig. 688 When installing the ABS sensor rotor, make sure there is a 0.12 inch (3mm) gap as shown B to the bottom—Mazda6 and Mazdaspeed6 models

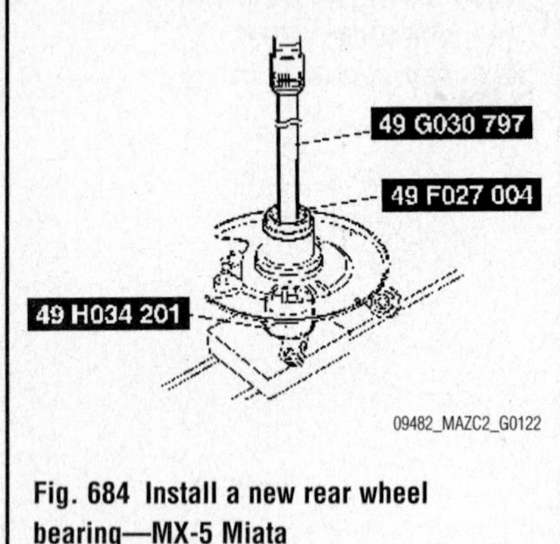

Fig. 684 Install a new rear wheel bearing—MX-5 Miata

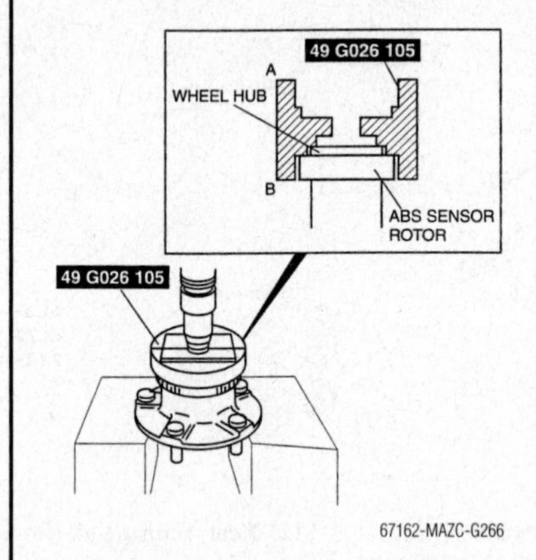

Fig. 687 When using the installation tool for the ABS sensor rotor, face the carved side B to the bottom—Mazda6 and Mazdaspeed6 models

Mazda6 Models

See Figures 685 through 688.

1. Before servicing the vehicle, refer to the precautions section.
2. Refer to the illustration for component location and torque specifications.

➡**The wheel bearings are not serviceable. If the bearings are bad, a new hub/bearing assembly must be installed.**

3. Remove or disconnect the following:
   - Hub cap
4. Raise the staked portion of the hub retaining nut with a hammer and chisel.
   - Rear wheels
   - ABS sensor
   - Parking brake cable
   - Rear caliper and rotor assembly from the hub
   - Hub retaining nut and discard it
   - ABS sensor rotor using a chisel
   - Hub bolts
   - Hub and bearing assembly from the spindle

To install:

- Hub and bearing assembly on the spindle
- Hub bolts and tighten to the specifications shown in the illustration

CAUTION

When using the installation tool for the ABS sensor rotor, face the carved sideBto the rotor.

- ABS sensor rotor using a press and tools illustrated. Make sure there is a 0.12 inch (3mm) gap as illustrated
- Remaining components in the reverse order of removal, referring to the illustration for component location and torque specifications

ADJUSTMENT

The rear wheel bearings are not adjustable. If the bearings become loose or make noise, they must be replaced.

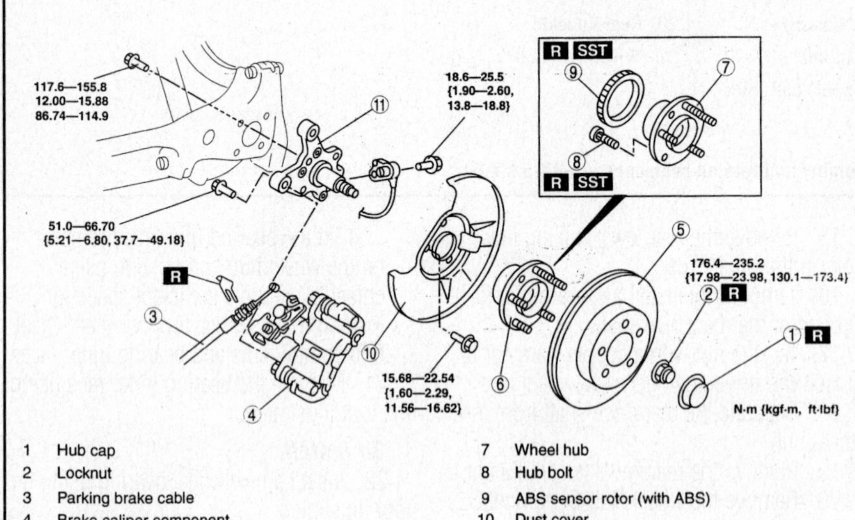

Fig. 685 Exploded view of the rear hub and spindle assembly—Mazda6 and Mazdaspeed6 models

MAZDA

B2300 • B3000 • B4000

2

SPECIFICATIONS AND MAINTENANCE CHARTS

ENGINE AND VEHICLE IDENTIFICATION

Engine							Model Year	
Code ①	Liters (cc)	Cu. In.	Cyl.	Fuel Sys.	Type	Eng. Mfg.	Code ②	Year
D	2.3 (2261)	138	4	EFI	DOHC	Ford	7	2007
U	3.0 (2999)	183	6	EFI	DOHC	Ford	8	2008
E	4.0 (3998)	244	6	EFI	DOHC	Ford		

EFI: Electronic fuel injection

DOHC: Dual Overhead Camshafts

① 8th digit of the Vehicle Identification Number (VIN)

② 10th digit of the Vehicle Identification Number (VIN)

22140_BPUP_C0001

GENERAL ENGINE SPECIFICATIONS

Year	Model	Engine Displacement Liters	Engine (VIN)	Net Horsepower @ rpm	Net Torque @ rpm (ft. lbs.)	Bore x Stroke (in.)	Com-pression Ratio	Oil Pressure @ rpm
2007	B2300	2.3	D	143@5250	154@3750	3.78X3.90	9.1:1	40-60@2000
	B3000	3.0	U	148@5000	180@3250	3.50x3.14	9.3:1	40-60@2500
	B4000	4.0	X	207@5250	238@3000	3.95x3.35	9.0:1	40-60@2000
2008	B2300	2.3	D	143@5250	154@3750	3.45X3.70	9.7:1	40-60@2000
	B4000	4.0	X	207@5250	238@3000	3.95x3.35	9.0:1	40-60@2000

22140_BPUP_C0002

GASOLINE ENGINE TUNE-UP SPECIFICATIONS

Year	Engine Displacement Liters	Engine VIN	Spark Plug Gap (in.)	Ignition Timing (deg.) ①		Fuel Pump (psi)	Idle Speed (rpm)		Valve Clearance	
				MT	AT		MT	AT	In.	Ex.
2007	2.3	D	0.041-0.045	10B ①	10B ①	56-72	①	①	HYD	HYD
	3.0	U	0.042-0.046	10B ①	10B ①	35-45	①	①	HYD	HYD
	4.0	E	0.061-0.068	10B ①	10B ①	35-45	①	①	HYD	HYD
2008	2.3	D	0.041-0.045	10B ①	10B ①	56-72	①	①	HYD	HYD
	4.0	E	0.061-0.068	10B ①	10B ①	35-45	①	①	HYD	HYD

NOTE: The Vehicle Emission Control Information label often reflects specification changes changes made during production. The label figures must be used if they differ from those in this chart.

B: Before top dead center

HYD: Hydraulic

① Electronically controlled and cannot be adjusted

22140_BPUP_C0003

CAPACITIES

Year	Model	Engine Displacement Liters	Engine VIN	Engine Oil with Filter (qts.)	Transmission (pts.) 5-Spd	Transmission (pts.) Auto.	Transfer Case (pts.)	Drive Axle Front (pts.)	Drive Axle Rear (pts.)	Fuel Tank (gal.)	Cooling System (qts.)
2007	B2300	2.3	D	4.0	5.6	19.8	NA	NA	5.0	①	②
	B3000	3.0	U	4.5	5.6	③	2.50	3.25	5.0	①	④
	B4000	4.0	E	5.0	5.6	③	2.50	3.25	5.0	①	④
2008	B2300	2.3	D	4.0	5.6	19.8	NA	NA	5.0	①	②
	B4000	4.0	E	5.0	5.6	③	2.50	3.25	5.0	①	④

NOTE: All capacities are approximate. Add fluid gradually and check to be sure a proper fluid level is obtained.

NA: not available

① Regular Cab short wheel base: 17 gal.
Regular Cab long wheel base: 20.3 gal.
Super Cab: 19.5 gal.

② w/MT: 11.2
w/AT: 10.9

③ 2WD: 20.0
4WD: 20.6

④ w/MT: 15.2
w/AT: 14.8

22140_BPUP_C0004

FLUID SPECIFICATIONS

Year	Model	Engine Displacement Liters	Engine ID/VIN	Engine Oil	Auto. Trans.	Drive Axle	Power Steering Fluid	Brake Master Cylinder
2007	B-Series	2.3	D	5W-20 ①	Mercon® V	80W-90 ② ③	Mercon®	DOT 3
		3.0	U	5W-20 ①	Mercon® V	80W-90 ② ③	Mercon®	DOT 3
		4.0	E	5W-20 ①	Mercon® V	80W-90 ② ③	Mercon®	DOT 3
2008	B-Series	2.3	D	5W-20 ①	Mercon® V	80W-90 ② ③	Mercon®	DOT 3
		4.0	E	5W-20 ①	Mercon® V	80W-90 ② ③	Mercon®	DOT 3

DOT: Department Of Transportation

® Registered trademark

① Synthetic motor oil is recommended

② 8.8 inch. ring gear high torque 75W-140 synthetic lubricant

③ Friction modifier XL-3

22140_BPUP_C0015

VALVE SPECIFICATIONS

Year	Engine Displacement Liters	Engine VIN	Seat Angle (deg.)	Face Angle (deg.)	Spring Test Pressure (lbs. @ in.)	Spring Installed Height (in.)	Stem-to-Guide Clearance (in.) Intake	Stem-to-Guide Clearance (in.) Exhaust	Stem Diameter (in.) Intake	Stem Diameter (in.) Exhaust
2007	2.3	D	45	45	①	1.129	0.0009-0.0027	0.0012-0.0029	0.2154-0.2159	0.2152-0.2157
	3.0	U	44.5	45	NA	1.736-1.650	0.0010-0.0027	0.0015-0.0032	0.2744-0.2752	0.2739-0.2747
	4.0	E	45	45	202-225@ 1.413-1.445	1.569-1.601	0.0010-0.0020	0.0010-0.0030	0.2740-0.2750	0.2740
2008	2.3	D	45	45	①	1.129	0.0009-0.0027	0.0012-0.0029	0.2154-0.2159	0.2152-0.2157
	4.0	E	45	45	202-225@ 1.413-1.445	1.569-1.608	0.0010-0.0020	0.0010-0.0030	0.2740-0.2750	0.2740

NA: not available

① Intake: 97.0@1.201
Exhaust: 93.3@1.201

22140_BPUP_C0005

CAMSHAFT SPECIFICATIONS CHART

All measurements are given in inches.

Year	Engine Displ. Liters	Engine VIN	Journal Dia.	Brg. Oil Clearance	Shaft End-play	Runout	Lobe Height Intake	Lobe Height Exhaust
2007	2.3	D	0.9827-0.9834	0.0014-0.0031	0.0035-0.0094	0.0012	1.6710 0.1569	1.6210 0.1569
	3.0	U	2.0074-2.0084	0.0010-0.0030	①	②	0.2600	0.2600
	4.0	E	1.0990-1.1010	0.0020-0.0040	0.0003-0.0070	0.0012	0.2590	0.2590
2008	2.3	D	0.9827-0.9834	0.0014-0.0031	0.0035-0.0094	0.0020	1.6710 0.1569	1.6210 0.1569
	4.0	E	1.0990-1.1010	0.0020-0.0040	0.0003-0.0070	0.0012	0.2590	0.2590

NA - Not available

① 0.007 service limit

② 0.002 limit. Runout no.2 or no.3 journal relative to no.1 and no.4 journals

22140_BPUP_C0007

CRANKSHAFT AND CONNECTING ROD SPECIFICATIONS

All measurements are given in inches.

Year	Engine Displacement Liters	Engine VIN	Crankshaft Main Brg. Journal Dia.	Crankshaft Main Brg. Oil Clearance	Crankshaft Shaft End-play	Crankshaft Thrust on No.	Connecting Rod Journal Diameter	Connecting Rod Oil Clearance	Connecting Rod Side Clearance
2007	2.3	D	2.0460-2.0470	0.0007-0.0013	0.0080-0.0160	3	1.9670-1.9680	0.0010-0.0020	0.0760-0.1200
	3.0	U	2.5190-2.5198	0.0010-0.0014	0.0040-0.0080	3	2.1253-2.1261	0.0010-0.0014	0.0060-0.0114
	4.0	E	2.2430-2.2440	0.0008-0.0015	0.0020-0.0126	3	2.1250-2.1260	0.0003-0.0024	0.0036-0.0106
2008	2.3	D	2.0460-2.0470	0.0007-0.0013	0.0080-0.0160	3	1.9670-1.9680	0.0010-0.0020	0.0760-0.1200
	4.0	E	2.2430-2.2440	0.0008-0.0015	0.0020-0.0126	3	2.1250-2.1260	0.0003-0.0024	0.0036-0.0106

22140_BPUP_C0008

PISTON AND RING SPECIFICATIONS

All measurements are given in inches.

Year	Engine Displacement Liters	Engine VIN	Piston Clearance	Ring Gap Top Compression	Ring Gap Bottom Compression	Ring Gap Oil Control	Ring Side Clearance Top Compression	Ring Side Clearance Bottom Compression	Ring Side Clearance Oil Control
2007	2.3	D	0.0009-0.0017	0.0060-0.0012	0.0120-0.0180	0.0070-0.0270	0.0008-0.0013	0.0004-0.0011	0.0025-0.0054
	3.0	U	0.0012-0.0022	0.010-0.020	0.010-0.020	0.010-0.049	0.0602-0.0612	0.0602-0.0612	0.1587-0.1596
	4.0	E	0.0008-0.0019	0.015-0.023	0.015-0.023	0.015-0.055	0.0010-0.0030	0.0010-0.0030	SNUG
2008	2.3	D	0.0009-0.0017	0.0060-0.0012	0.0120-0.0180	0.0070-0.0270	0.0008-0.0013	0.0004-0.0011	0.0025-0.0054
	4.0	E	0.0008-0.0019	0.015-0.023	0.015-0.023	0.015-0.055	0.0010-0.0030	0.0010-0.0030	SNUG

22140_BPUP_C0006

TORQUE SPECIFICATIONS

All readings in ft. lbs.

Year	Engine Displacement Liters	Engine VIN	Cylinder Head Bolts	Main Bearing Bolts	Rod Bearing Bolts	Crankshaft Damper Bolts	Flywheel Bolts	Manifold Intake *	Manifold Exhaust	Spark Plugs	Oil Pan Drain Plug
2007	2.3	D	①	L	M	②	③	13	40	9	21
	3.0	U	④	59	26	107	59	21	⑤	11	NA
	4.0	E	⑥	⑦	⑧	⑨	⑩	7	16	13	18
2008	2.3	D	①	L	M	②	③	13	40	9	21
	4.0	E	⑪	⑦	⑧	⑨	⑩	7	16	13	18

NA: Information not available

* NOTE: Applies to Lower Manifold only.

① Step 1: 44 inch lbs.
Step 2: 11 ft. lbs.
Step 3: 33 ft. lbs.
Step 4: +90 degrees
Step 5: +90 degrees

② Step 1: 74 ft. lbs.
Step 2: +90 degrees

③ Step 1: 37 ft. lbs.
Step 2: 59 ft. lbs.
Step 3: 83 ft. lbs.

④ Step 1: 59 ft. lbs.
Step 2: Back off 1 full turn
Step 3: 34-40 ft. lbs.
Step 4: 63-73 ft. lbs.

⑤ Step 1: 89 inch lbs.
Step 2: 17 ft. lbs.

⑥ 8mm bolts: 24 ft. lbs.
12mm bolts: 24 ft. lbs. +80 degrees, +80 degrees

⑦ Step 1: 26 ft. lbs.
Step 2: +57 degrees

⑧ Step 1: 15 ft. lbs.
Step 2: +90 degrees

⑨ Step 1: 37 ft. lbs.
Step 2: +90 degrees

⑩ Step 1: 10 ft.lbs.
Step 2: 52 ft. lbs.

⑪ Step 1: 12mm bolts 9 ft. lbs.
Step 2: 12mm bolts 18 ft. lbs.
Step 3: 8mm bolts 24 ft. lbs.
Step 4: 12 mm bolts +90 degrees
Step 5: 12mm bolts +90 degrees

⑪ Step 1: 26 ft. lbs.
Step 2: +57 degrees

22140_BPUP_C0009

WHEEL ALIGNMENT

Year	Model		Caster Range (+/-Deg.)	Caster Preferred Setting (Deg.)	Camber Range (+/-Deg.)	Camber Preferred Setting (Deg.)	Toe-in (in.)
2007	B Series	2WD	1.0	①	0.70	-0.50	0.06+/-0.25
		4WD	1.0	②	0.70	-0.50	0.12+/-0.25
		Rear	—	—	0.75	0	0
2008	B Series	2WD	1.0	①	0.70	-0.50	0.06+/-0.25
		4WD	1.0	②	0.70	-0.50	0.12+/-0.25
		Rear	—	—	0.75	0	0

① Left: +3.5
Right: +3.9

② Left: +2.8
Right: +3.5

22140_BPUP_C0012

TIRE, WHEEL AND BALL JOINT SPECIFICATIONS

Year	Model	OEM Tires Standard	OEM Tires Optional	Tire Pressures (psi) Front	Tire Pressures (psi) Rear	Wheel Size	Ball Joint Inspection	Lug Nut Torque (ft. lbs.)
2007	B2300	P225/70R15	none	①	①	7-JJ	②	100
	B3000	P235/75R15	none	①	①	7-JJ	②	100
	B4000	P235/75R15	P255/70R16	①	①	7-JJ	②	100
2008	B2300	P225/70R15	none	①	①	7-JJ	②	100
	B4000	P235/75R15	③ ④	①	①	7-JJ	②	100

OEM: Original Equipment Manufacturer

PSI: Pounds Per Square Inch

① See placard on door post

② Upper 0 - 0.032
Lower 0 - 0.008

22140_BPUP_C0011

BRAKE SPECIFICATIONS

All measurements in inches unless noted

Year	Model	Brake Disc Original Thickness	Brake Disc Minimum Thickness	Brake Disc Maximum Runout	Brake Drum Diameter Original Inside Diameter	Brake Drum Diameter Max. Wear Limit	Brake Drum Diameter Maximum Machine Diameter	Minimum Lining Thickness	Brake Caliper Bracket Bolts (ft. lbs.)	Brake Caliper Mounting Bolts (ft. lbs.)
2007	B-Series	NA	0.965	0.0030	9.00	NA	①	0.0030	85	27
2008	B-Series	NA	0.988	0.0030	10.00	NA	①	0.0030	85	27

NOTE: Due to changes made during production, refer to manufacturer's specifications if they differ from those in this chart

NA: Not Available

① Molded into the drum

22140_BPUP_C0010

SCHEDULED MAINTENANCE INTERVALS

TO BE SERVICED	TYPE OF SERVICE	VEHICLE MILEAGE INTERVAL (x1000)												
		5	10	15	20	25	30	35	40	45	50	55	60	65
Engine oil & filter	R	✓	✓	✓	✓	✓	✓	✓	✓	✓	✓	✓	✓	✓
Tires	Rotate	✓	✓	✓	✓	✓	✓	✓	✓	✓	✓	✓	✓	✓
Auto trans. fluid	I									✓				
Brake pads/shoes	I			✓			✓			✓			✓	
Wheel ends ①	I			✓			✓			✓			✓	
Coolant hoses	S/I			✓			✓			✓			✓	
Steering linkage	I/L			✓			✓			✓			✓	
Suspension and driveshaft	I			✓			✓			✓			✓	
Cabin air filter	R			✓			✓			✓			✓	
Ball joints (2WD)	I/L			✓			✓			✓			✓	
Exhaust system	I						✓						✓	
Engine air filter	R						✓						✓	
Fuel filter	R						✓						✓	
Manual trans fluid	R												✓	
Auto trans fluid	R						✓						✓	
Wheel bearings (2WD) ②	I/L												✓	
Premium Gold coolant	I												✓	
Spark plugs	R	every 100,000 miles												
PCV valve ③	R	every 60,000 miles												
Rear axle fluid	R	every 150,000 miles												
Accessory drive belts	I	every 100,000 miles												
Accessory drive belts	R	every 150,000 miles, if not previously replaced												

R: Replace S: Service I: Inspect L: Lubricate

① Check for play and noise

② Replace after 150,000 miles

③ At 60,000 miles (96,000 km), the dealer will replace the PCV valve at no cost, except Canada and CA

Special Operating Condition Requirements

When towing a trailer or using a camper or car-top carrier:

Change engine oil and install a new oil filter every 4,800 km (3,000 miles), 3 months or 200 hours of engine operation (whichever occurs first).

Change transfer case fluid every 96,000 km (60,000 miles).

Change manual transmission fluid as required.

Change automatic transmission fluid, lubricate 4x2 wheel bearings, install new grease seals and adjust bearings every

During extensive idling and/or low speed driving for long distances, as in heavy commercial use such as delivery, taxi or livery:

Change engine oil and install a new oil filter every 4,800 km (3,000 miles), 3 months or 200 hours of engine operation (whichever occurs first).

Lube front lower control arm and steering linkage ball joints with zerk fittings (if equipped) every 4,800 km (3,000 miles) or 3 months.

Inspect brake system and check battery electrolyte level (Patrol cars) every 8,000 km (5,000 miles).

Install a new fuel filter every 24,000 km (15,000 miles).

Change automatic transmission fluid, lubricate 4x2 wheel bearings, install new grease seals and adjust bearings every 48,000 km (30,000 miles). If equipped, change the in-line service installed transmission fluid filter.

Install new spark plugs and change transfer case fluid every 96,000 km (60,000 miles).

Change transfer case fluid every 96,000 km (60,000 miles).

Install a new cabin air filter as required.

When operating in dusty conditions such as unpaved or dusty roads:

Change engine oil and install a new oil filter every 3,000 miles (4,800 km) or 3 months.

Install a new fuel filter every 24,000 km (15,000 miles).

Change automatic transmission fluid every 48,000 km (30,000 miles). If equipped, change the in-line service installed transmission fluid filter.

Install a new engine air filter as required.

Install a new cabin air filter as required.

When operating in off-road conditions:

Change automatic transmission fluid every 48,000 km (30,000 miles). If equipped, change the in-line service installed transmission fluid filter.

Change transfer case fluid every 96,000 km (60,000 miles).

Install a new cabin air filter as required.

Inspect and lubricate U-joints.

22140_BPUP_C0013

PRECAUTIONS

Before servicing any vehicle, please be sure to read all of the following precautions, which deal with personal safety, prevention of component damage, and important points to take into consideration when servicing a motor vehicle:

- Never open, service or drain the radiator or cooling system when the engine is hot; serious burns can occur from the steam and hot coolant.
- Observe all applicable safety precautions when working around fuel. Whenever servicing the fuel system, always work in a well-ventilated area. Do not allow fuel spray or vapors to come in contact with a spark, open flame, or excessive heat (a hot drop light, for example). Keep a dry chemical fire extinguisher near the work area. Always keep fuel in a container specifically designed for fuel storage; also, always properly seal fuel containers to avoid the possibility of fire or explosion. Refer to the additional fuel system precautions later in this section.
- Fuel injection systems often remain pressurized, even after the engine has been turned **OFF**. The fuel system pressure must be relieved before disconnecting any fuel lines. Failure to do so may result in fire and/or personal injury.
- Brake fluid often contains polyglycol ethers and polyglycols. Avoid contact with the eyes and wash your hands thoroughly after handling brake fluid. If you do get brake fluid in your eyes, flush your eyes with clean, running water for 15 minutes. If eye irritation persists, or if you have taken brake fluid internally, IMMEDIATELY seek medical assistance.
- The EPA warns that prolonged contact with used engine oil may cause a number of skin disorders, including cancer. You should make every effort to minimize your exposure to used engine oil. Protective gloves should be worn when changing oil. Wash your hands and any other exposed skin areas as soon as possible after exposure to used engine oil. Soap and water, or waterless hand cleaner should be used.
- All new vehicles are now equipped with an air bag system, often referred to as a Supplemental Restraint System (SRS) or Supplemental Inflatable Restraint (SIR) system. The system must be disabled before performing service on or around system components, steering column, instrument panel components, wiring and sensors. Failure to follow safety and disabling procedures could result in accidental air bag deployment, possible personal injury and unnecessary system repairs.
- Always wear safety goggles when working with, or around, the air bag system. When carrying a non-deployed air bag, be sure the bag and trim cover are pointed away from your body. When placing a non-deployed air bag on a work surface, always face the bag and trim cover upward, away from the surface. This will reduce the motion of the module if it is accidentally deployed. Refer to the additional air bag system precautions later in this section.
- Clean, high quality brake fluid from a sealed container is essential to the safe and proper operation of the brake system. You should always buy the correct type of brake fluid for your vehicle. If the brake fluid becomes contaminated, completely flush the system with new fluid. Never reuse any brake fluid. Any brake fluid that is removed from the system should be discarded. Also, do not allow any brake fluid to come in contact with a painted surface; it will damage the paint.
- Never operate the engine without the proper amount and type of engine oil; doing so WILL result in severe engine damage.
- Timing belt maintenance is extremely important. Many models utilize an interference-type, non-freewheeling engine. If the timing belt breaks, the valves in the cylinder head may strike the pistons, causing potentially serious (also time-consuming and expensive) engine damage. Refer to the maintenance interval charts for the recommended replacement interval for the timing belt, and to the timing belt section for belt replacement and inspection.
- Disconnecting the negative battery cable on some vehicles may interfere with the functions of the on-board computer system(s) and may require the computer to undergo a relearning process once the negative battery cable is reconnected.
- When servicing drum brakes, only disassemble and assemble one side at a time, leaving the remaining side intact for reference.
- Only an MVAC-trained, EPA-certified automotive technician should service the air conditioning system or its components.

BRAKES — ANTI-LOCK BRAKE SYSTEM (ABS)

GENERAL INFORMATION

The anti-lock control system consists of the following components:

- Hydraulic control unit (HCU)
- Anti-lock brake control module
- Rear anti-lock brake sensor
- Rear anti-lock brake sensor indicator
- Front anti-lock brake sensor
- Front anti-lock brake sensor indicator
- Yellow ABS warning indicator
- Red brake warning indicator

The anti-lock brake control module manages anti-lock braking. When the ignition switch is placed in the RUN position, the module carries out a preliminary electrical check. At approximately 20 km/h (12 MPH), the hydraulic pump motor is turned on for approximately 0.5 second. Any malfunction of the ABS disables the ABS and the anti-lock brake warning indicator illuminates. However, the power-assist braking system still functions normally.

The ABS module needs to be reconfigured whenever a new ABS module is installed. Follow the WDS or equivalent Tester directions for the configuration procedure.

The electronic brake distribution force controls the rear brake pressure and acts as an electronic proportioning valve. It is controlled by the ABS module. When the electronic brake force distribution is disabled, the amber ABS warning indicator illuminates.

Using a specially developed pressure/pedal movement mechanism, the panic assist braking functions by fully applying the brakes during a panic stop. Depending on the vehicle speed, if a pedal stroke is more rapid than normal, the brake booster automatically applies full brake boost. Panic assist is disabled when the driver releases the brake pedal.

WHEEL SPEED SENSORS

REMOVAL & INSTALLATION

Front

See Figure 1.

1. Before servicing the vehicle, refer to the Precautions Section.
2. Disconnect the negative battery cable.
3. Remove the front wheel and tire assembly.

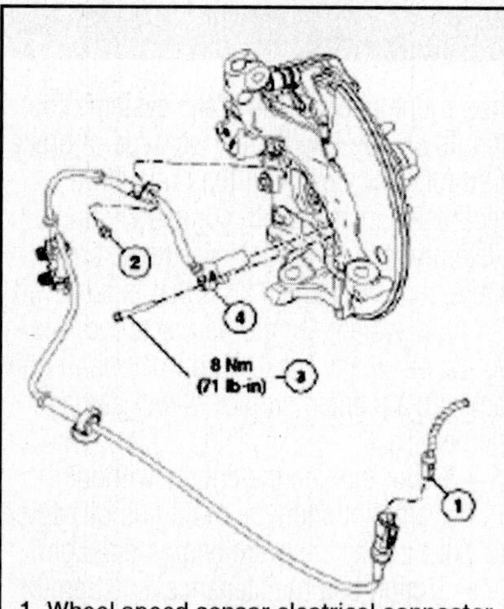

1. Wheel speed sensor electrical connector
2. Wheel speed sensor harness bolt
3. Wheel speed sensor bolt
4. Wheel speed sensor

22140_BPUP_G0026

Fig. 1 Front wheel sensor and wiring exploded view

➡The harness connector is located in the engine compartment.

4. Disconnect the harness electrical connector.
5. Remove the brake disc (4WD models)
6. Remove the 3 brake disc shield bolts and the brake disc shield. (4WD models)
7. Release the grommet from the body and pull the connector through.
8. Release the sensor harness retainer.

➡Clean the area surrounding the wheel speed sensor and bolt before removal.

9. Remove the wheel speed sensor bolt.

➡Thoroughly clean the wheel speed sensor and wheel speed sensor mounting surface. Apply wheel bearing grease to the mounting surface.

To install:

10. Install the wheel speed sensor bolt.
11. Tighten wheel speed sensor bolt to 9 ft. lbs. (12 Nm).
12. Install the sensor harness retainer.
13. Tighten wheel speed sensor bolt to 11 ft. lbs. (15 Nm).
14. Connect the harness electrical connector.
15. Install the3 brake disc shield bolts and the brake disc shield. (4WD models).
16. Tighten the3 brake disc shield bolts to 9 ft. lbs. (12 Nm).
17. Install the brake disc (4WD models)
18. Connect the negative battery cable.

Rear

➡On 4WD vehicles, the rear sensor ring is integral to the rear axle halfshaft and cannot be repaired separately.

1. Before servicing the vehicle, refer to the Precautions Section.
2. Disconnect the negative battery cable.
3. Raise the vehicle on a hoist.
4. Release the grommet and pull the connector through the body.
5. Disconnect the wheel speed sensor harness electrical connector.
6. Remove the wheel speed sensor harness bolts.

➡Clean the area surrounding the wheel speed sensor and bolt before removal.

7. Remove the wheel speed sensor bolt and wheel speed sensor.
8. On FWD vehicles, remove the rear ABS sensor nut and ring.

➡Thoroughly clean the wheel speed sensor and mounting surface before installation.

9. To install, reverse the removal procedure. Tighten ABS sensor and ring to 214 ft. lbs. (290 Nm). Tighten all remaining fasteners to 80 inch lbs. (9 Nm).

WHEEL SPEED SENSOR RINGS

REMOVAL & INSTALLATION

The front tone ring for 2WD models is part of the front brake rotor. Refer to the Brake Rotor removal and installation. On 4WD models, the tone ring is incorporated into the hub and bearing assembly. Refer to the Front Hub and Bearing replacement. The rear (exciter) tone ring is inside the rear axle housing affixed to the ring gear.

BRAKES

BLEEDING THE BRAKE SYSTEM

BLEEDING PROCEDURE

**** WARNING**

Use of any other than the approved DOT 3 brake fluid will cause permanent damage to brake components and will render the brakes inoperative. Failure to follow these instructions may result in personal injury.

**** WARNING**

Brake fluid contains polyglycol ethers and polyglycols. Avoid contact with eyes. Wash hands thoroughly after handling. If brake fluid contacts eyes, flush eyes with running water for 15 minutes. Get medical attention if irritation persists. If taken internally, drink water and induce vomiting. Get medical attention immediately. Failure to follow these instructions may result in personal injury.

**** CAUTION**

Do not allow the brake master cylinder reservoir to run dry during the bleeding operation. Keep the master cylinder reservoir filled with the specified brake fluid. Never reuse the brake fluid that has been drained from the hydraulic system.

**** CAUTION**

Brake fluid is harmful to painted and plastic surfaces. If brake fluid is spilled onto a painted or plastic surface, immediately wash it with water.

➡When any part of the hydraulic system has been disconnected or a new component is installed, air may enter the system, causing spongy brake pedal action. This requires the bleeding of the hydraulic system after it has been correctly connected.

**** CAUTION**

Be sure to check the brake fluid level in the brake master cylinder reservoir often. Do not let it run dry.

1. Fill the brake master cylinder reservoir with the specified brake fluid.
2. Begin bleeding the system, going in order from the RH rear wheel, to the LH rear wheel, then to the RH front wheel and ending with the LH front wheel.
3. Attach a rubber drain hose to the rear bleeder screw and submerge the free end in a container partially filled with clean brake fluid.
4. Have and assistant pump the brake pedal 10 times and then hold firm pressure on the brake pedal.
5. Loosen the bleeder screw until the fluid flow stops. Maintain pressure on the brake pedal and tighten the bleeder screw.
6. Repeat steps 4 and 5 until clear, bubble-free brake fluid flows.

7. Tighten the bleeder screw and refill the brake master cylinder reservoir as necessary.
8. Continue bleeding the brake hydraulic system at each wheel.
9. Fill the brake master cylinder reservoir with the specified brake fluid.

BLEEDING THE ABS SYSTEM

PRECAUTIONS

CAUTION

Use of any other than approved DOT 3 brake fluid will cause permanent damage to brake components and will render the brakes inoperative. Failure to follow these instructions may result in personal injury.

WARNING

Brake fluid contains polyglycol ethers and polyglycols. Avoid contact with eyes. Wash hands thoroughly after handling. If brake fluid contacts eyes, flush eyes with running water for 15 minutes. Get medical attention if irritation persists. If taken internally, drink water and induce vomiting. Get medical attention immediately. Failure to follow these instructions may result in personal injury.

CAUTION

Brake fluid is harmful to painted and plastic surfaces. If brake fluid is spilled onto a painted or plastic surface, immediately wash it with water.

CAUTION

Do not allow the brake master cylinder reservoir to run dry during the bleeding operation. Keep the brake master cylinder reservoir filled with the specified brake fluid. Never reuse the brake fluid that has been drained from the hydraulic system.

➡When any part of the hydraulic system has been disconnected for repair or replacement, air can get into the system and cause spongy brake pedal action. This requires bleeding of the hydraulic system after it has been properly connected. The hydraulic system can be bled manually or with pressure bleeding equipment.

➡Bleeding the Hydraulic Control Unit (HCU) is required only when removing or installing the HCU or master cylinder, or opening the lines to the HCU.

➡Carrying out the System Bleed function drives trapped air from the HCU. Subsequent bleeding removes the air from the brake hydraulic system through the bleeder screws.

➡Adequate voltage to the HCU module is required during the anti-lock control portion of the system bleed.

1. Connect the WDS or equivalent Tester.
2. Access the SYSTEM BLEED FUNCTION. Go to the Tool Tab-Chassis-Braking-ABS Service Bleed and follow the directions on the diagnostic tool.
3. Manually bleed the brake hydraulic system.
4. Repeat the procedure carrying out a total of two diagnostic tool cycles and two manual bleed cycles.

MANUAL BLEEDING PROCEDURE

CAUTION

Use of any other than approved DOT 3 brake fluid will cause permanent damage to brake components and will render the brakes inoperative. Failure to follow these instructions may result in personal injury.

WARNING

Brake fluid contains polyglycol ethers and polyglycols. Avoid contact with eyes. Wash hands thoroughly after handling. If brake fluid contacts eyes, flush eyes with running water for 15 minutes. Get medical attention if irritation persists. If taken internally, drink water and induce vomiting. Get medical attention immediately. Failure to follow these instructions may result in personal injury.

CAUTION

Brake fluid is harmful to painted and plastic surfaces. If brake fluid is spilled onto a painted or plastic surface, immediately wash it with water.

CAUTION

Do not allow the brake master cylinder reservoir to run dry during the bleeding operation. Keep the brake master cylinder reservoir filled with the specified brake fluid. Never reuse the brake fluid that has been drained from the hydraulic system.

➡When any part of the hydraulic system has been disconnected for repair or replacement, air can get into the system and cause spongy brake pedal action. This requires bleeding of the hydraulic system after it has been properly connected. The hydraulic system can be bled manually or with pressure bleeding equipment.

➡Bleeding the Hydraulic Control Unit (HCU) is required only when removing or installing the HCU or master cylinder, or opening the lines to the HCU.

➡Carrying out the System Bleed function drives trapped air from the HCU. Subsequent bleeding removes the air from the brake hydraulic system through the bleeder screws.

1. Clean all dirt from and remove the brake master cylinder filler cap and fill the brake master cylinder reservoir with the specified brake fluid.
2. Place a box end wrench on the RH rear bleeder screw. Attach a rubber drain tube to the RH rear bleeder screw and submerge the free end of the tube in a container partially filled with clean brake fluid.
3. Have an assistant pump, and then hold firm pressure on, the brake pedal.
4. Loosen the RH rear bleeder screw until a stream of brake fluid comes out. While an assistant maintains pressure on the brake pedal, tighten the RH rear bleeder screw.
5. Repeat until clear, bubble-free fluid comes out.
6. Refill the brake master cylinder reservoir as necessary.
7. Tighten the RH rear bleeder screw.
8. Tighten to 71 inch lbs. (8 Nm).
9. Repeat Steps 2-5 for the LH rear bleeder screw.
10. Place a box end wrench on the RH front disc brake caliper bleeder screw. Attach a rubber drain tube to the RH front disc brake caliper bleeder screw, and submerge the free end of the tube in a container partially filled with clean brake fluid.
11. Have an assistant pump, and then hold firm pressure on, the brake pedal.
12. Loosen the RH front disc brake caliper bleeder screw until a stream of brake fluid comes out. While an assistant maintains pressure on the brake pedal, tighten the RH front disc brake caliper bleeder screw.

13. Repeat until clear, bubble-free fluid comes out.

14. Refill the brake master cylinder reservoir as necessary.

15. Tighten the RH front disc brake caliper bleeder screw.

16. Tighten to 11 ft. lbs. (15 Nm).

17. Repeat Steps 7-10 for the LH front disc brake caliper bleeder screw.

PRESSURIZED BLEEDING PROCEEDURE

➡**Master cylinder pressure bleeder adapter tools are available from various manufacturers of pressure bleeding equipment. Follow the instructions of the manufacturer when installing the adapter. Install the bleeder adapter to the brake master cylinder reservoir, and attach the bleeder tank hose to the fitting on the adapter.**

1. Bleed the longest line first. Make sure the bleeder tank contains enough specified brake fluid to complete the bleeding operation. Place a box-end wrench on the RH rear bleeder screw. Attach a rubber drain tube to the RH rear bleeder screw, and submerge the free end of the tube in a container partially filled with clean brake fluid.

2. Open the valve on the bleeder tank.

3. Loosen the RH rear bleeder screw. Leave open until clear, bubble-free brake fluid flows, then tighten the RH rear bleeder screw and remove the rubber hose.

4. Tighten to 71 inch lbs. (6 Nm).

5. Continue bleeding the rest of the system, going in order from the LH rear bleeder screw to the RH front disc brake caliper bleeder screw, ending with the LH front disc brake caliper bleeder screw.

6. Tighten the brake caliper bleeder screws to 11 ft. lbs. (15 Nm).

7. Close the bleeder tank valve. Remove the tank hose from the adapter, and remove the adapter.

⁂ CAUTION

Dust and dirt accumulating on brake parts during normal use may contain asbestos fibers from production or aftermarket brake linings. Breathing excessive concentrations of asbestos fibers can cause serious bodily harm. Exercise care when servicing brake parts. Do not sand or grind brake lining unless equipment used is designed to contain the dust residue. Do not clean brake parts with compressed air or by dry brushing. Cleaning should be done by dampening the brake components with a fine mist of water, then wiping the brake components clean with a dampened cloth. Dispose of cloth and all residue containing asbestos fibers in an impermeable container with the appropriate label. Follow practices prescribed by the Occupational Safety and Health Administration (OSHA) and the Environmental Protection Agency (EPA) for the handling, processing, and disposing of dust or debris that may contain asbestos fibers.

BRAKE CALIPER

REMOVAL & INSTALLATION

See Figure 2.

1. Before servicing the vehicle, refer to the Precautions Section.

2. Loosen the wheel lug nuts.

3. Raise and safely support the front of the vehicle. Remove the wheel.

4. Remove the brake caliper flow bolt and position the brake hose aside.

5. Discard the 2 copper washers.

➡**Plug the brake hose.**

6. Remove the two caliper slide pin bolts and lift the caliper from the anchor plate.

➡**Use care to retain as much of the original caliper slide pin grease as possible.**

7. Position the caliper on a frame member or suspend it with some wire. Do not allow the caliper to hang by the brake hose.

8. Disconnect and plug the brake hose at the caliper. Remove the caliper from the rotor.

To install:

9. Position the caliper over the brake pads and align the slide pin mounting holes.

10. Install the slide pin bolts.

11. Tighten the bolts to 27 ft. lbs. (36 Nm)

12. Install the caliper brake hose using new washers. Tighten the bolt to 25 ft. lbs. (34 Nm).

13. Bleed the brake caliper.

14. Install the wheel and snug the lug nuts.

15. Lower the vehicle and tighten the lug nuts to 100 ft. lbs. (135 Nm).

➡**The first couple of times you apply the brakes, the pedal may go to the floor. Continue to pump the brake pedal until it feels firm.**

16. Start the engine and apply the brakes several times to readjust the caliper pistons. Ensure that the pedal feels firm before operating the vehicle.

17. Check and adjust the brake fluid level, as required.

DISC BRAKE PADS

REMOVAL & INSTALLATION

See Figure 3.

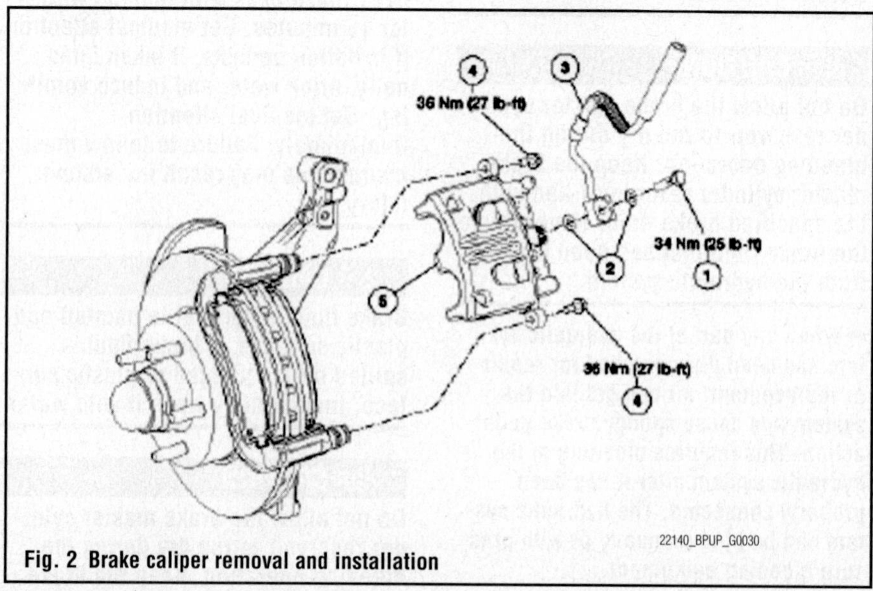

Fig. 2 Brake caliper removal and installation

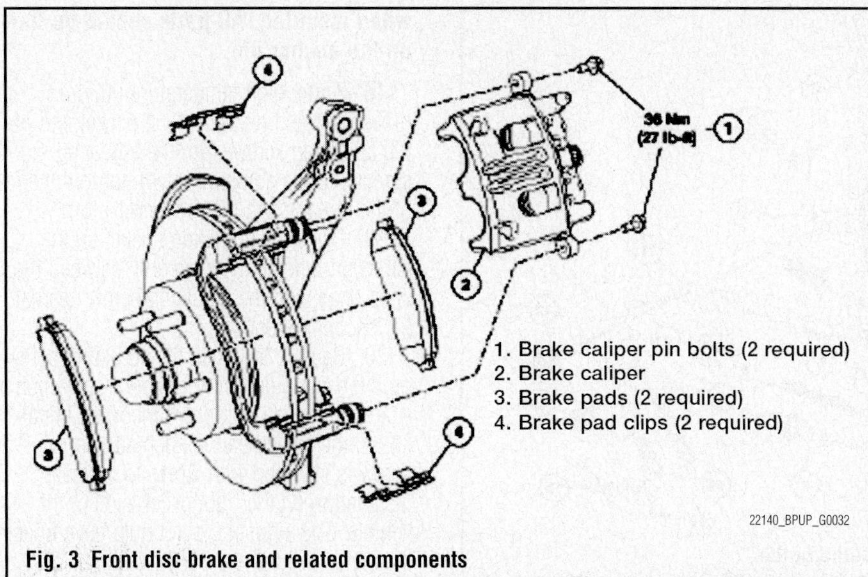

Fig. 3 Front disc brake and related components

1. Before servicing the vehicle, refer to the Precautions Section.
2. Raise and safely support the front of the vehicle. Remove the wheels.

✻✻ CAUTION

Do not pry in the caliper sight hole to retract the pistons, as this can damage the pistons and boots

➡Do not remove the anchor pins unless installing new pins. Inspect the anchor plate guide pins. They should slide in and out of the anchor plates with no binding. Check the boots for cracks or tears. If the pins do not slide easily or the boots are cracked or torn, install new boots and pins. Lubricate the pins using silicone brake caliper grease.

3. Remove the two caliper slide pin bolts and lift the caliper from the anchor plate.

➡Use care to retain as much of the original caliper slide pin grease as possible.

4. Position the caliper on a frame member or suspend it with some wire. Do not allow the caliper to hang by the brake hose.
5. Remove the brake pads and, if necessary, the anti-rattle clips from the anchor plate.
6. Remove the shims, if any, from the brake pads for re-use.

To install:

✻✻ CAUTION

Do not allow grease, oil, brake fluid or other contaminants to contact the pad lining material. Do not install contaminated pads.

7. If removed, install new anti-rattle clips.
8. Install the brake pads to the anchor plate.
9. Position the caliper over the brake pads and align the slide pin mounting holes.
10. Tighten the bottom caliper bolt before tightening the top caliper bolt.
11. Install the slide pin bolts and tighten them to 27 ft. lbs. (36 Nm).
12. Install the wheel and snug the lug nuts.
13. Lower the vehicle and tighten the lug nuts to 100 ft. lbs. (135 Nm).

➡The first couple of times you apply the brakes, the pedal may go to the floor. Continue to pump the brake pedal until it feels firm.

14. Start the engine and apply the brakes several times to readjust the caliper pistons. Ensure that the pedal feels firm before operating the vehicle.
15. Check and adjust the brake fluid level, as required.

BRAKES — REAR DRUM BRAKES

✻✻ CAUTION

Dust and dirt accumulating on brake parts during normal use may contain asbestos fibers from production or aftermarket brake linings. Breathing excessive concentrations of asbestos fibers can cause serious bodily harm. Exercise care when servicing brake parts. Do not sand or grind brake lining unless equipment used is designed to contain the dust residue. Do not clean brake parts with compressed air or by dry brushing. Cleaning should be done by dampening the brake components with a fine mist of water, then wiping the brake components clean with a dampened cloth. Dispose of cloth and all residue containing asbestos fibers in an impermeable container with the appropriate label. Follow practices prescribed by the Occupational Safety and Health Administration (OSHA) and the Environmental Protection Agency (EPA) for the handling, processing, and disposing of dust or debris that may contain asbestos fibers.

BRAKE DRUM

REMOVAL & INSTALLATION

1. Before servicing the vehicle, refer to the Precautions Section.
2. Raise and safely support the vehicle. Remove the wheel and tire assembly.
3. Remove the retaining nuts, if equipped, and remove the brake drum.
4. Inspect the brake drum surface for wear, scoring and runout. Machine or replace, as necessary.

To install:

5. Install the brake drum and secure in place with the retainer nuts, if equipped.
6. Adjust the rear brakes.
7. Install the wheel. Lower the vehicle.

BRAKE SHOES

REMOVAL & INSTALLATION

See Figure 4.

1. Before servicing the vehicle, refer to the Precautions Section.
2. Raise and safely support the vehicle. Remove the wheel and tire assembly and the brake drum.
3. Pull backward on the adjusting lever cable to disengage the adjusting lever from the adjusting screw. Move the outboard side of the adjusting screw upward and back off the pivot nut as far as it will go.

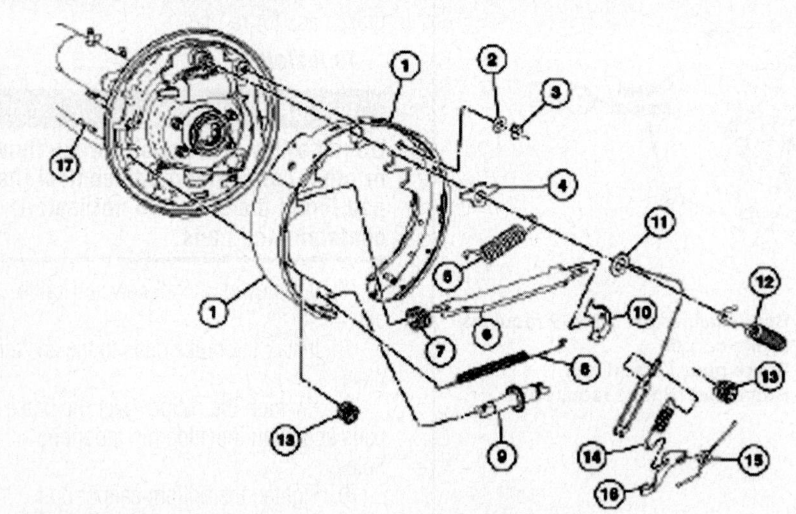

1. Brake shoes and linings (2 required)
2. Washer
3. Parking brake lever pin retainer
4. Brake shoe guide plate
5. Primary brake shoe retracting spring
6. Parking brake strut
7. Parking brake strut spring
8. Brake shoe adjusting screw spring
9. Brake shoe adjusting screw
10. Cable guide
11. Brake shoe adjusting lever cable
12. Secondary brake shoe retracting spring
13. Brake shoe hold-down spring
14. Brake shoe adjusting lever cable spring
15. Adjusting lever return spring
16. Brake shoe adjusting lever
17. Brake shoe hold-down spring pin

22140_BPUP_G0031

Fig. 4 Rear brake assembly and related components

4. Pull the adjusting lever, cable and automatic adjuster spring down and toward the rear to unhook the pivot hook from the large hole in the secondary shoe web. Do not pry the pivot hook from the hole.
5. Remove the automatic adjuster spring and adjusting lever.
6. Remove the secondary shoe-to-anchor spring using a suitable brake spring removal/installation tool. Using the tool, remove the primary shoe-to-anchor spring and unhook the cable anchor. Remove the anchor pin plate, if equipped.
7. Remove the cable guide from the secondary shoe.
8. Remove the shoe hold-down springs, shoes, adjusting screw, pivot nut and socket. Note the color and position of each hold-down spring so they can be reassembled in the same position.
9. Remove the parking brake link and spring. Disconnect the parking brake cable from the parking brake lever.
10. Remove the secondary brake shoe. On 9 inch rear brakes, remove the parking brake lever from the shoe. On 10 inch rear brakes, remove the retainer clip and spring washer and remove the parking brake lever.

To install:

11. Clean the backing plate ledge pads and sand lightly. Apply a light coating of high temperature lithium grease to the points where the brake shoes touch the backing plate. Lubricate the adjusting cable eye and the anchor pin area.
12. Install the parking brake lever on the secondary shoe. On 10 inch brakes, secure with the spring washer and retaining clip.
13. Position the brake shoes on the backing plate and install the hold-down spring pins, springs and cups. Install the parking brake link, spring and washer. Connect the parking brake cable to the parking brake lever.
14. Install the anchor pin plate, if equipped, and place the cable anchor over the anchor pin with the crimped side toward the backing plate.
15. Install the primary shoe-to-anchor spring using the brake spring removal/installation tool.
16. Install the cable guide on the secondary shoe with the flanged hole fitted into the hole in the secondary shoe. Thread the cable around the cable guide groove.

➡Make sure the cable is positioned in the groove and not between the guide and shoe web.

17. Install the secondary shoe-to-anchor (long) spring.

➡Make sure the cable end is not cocked or binding on the anchor pin when installed. All parts should be flat on the anchor pin.

18. Apply high temperature lithium grease to the threads and the socket end of the adjusting screw. Turn the adjusting screw into the adjusting pivot nut to the end of the threads and then loosen,½ turn.
19. Place the adjusting socket on the screw and install the assembly between the shoe ends with the adjusting screw nearest the secondary shoe.
20. Be sure to install the adjusting screw on the same side of the vehicle from which it was removed. To prevent incorrect installation, the socket end of each adjusting screw is stamped with **R**or**L**, to indicate installation on the right or left side of the vehicle. The adjusting pivot nuts have lines machined around the body of the nut, 2 lines indicating the right side nut and 1 line indicating the left side nut.
21. Hook the cable hook into the hole in the adjusting lever from the outboard plate side. The adjusting levers are also stamped with an **R**or**L** to indicate right or left side installation.
22. Place the hooked end of the adjuster spring in the large hole in the primary shoe web and connect the loop end of the spring to the adjuster lever hole.
23. Pull the adjuster lever, cable and automatic adjuster spring down toward the rear to engage the pivot hook in the large hole in the secondary shoe web.
24. After installation, check the action of the adjuster by pulling the section of the cable between the cable guide and the adjusting lever toward the secondary shoe web far enough to lift the lever past a tooth on the adjusting screw wheel. The lever should snap into position behind the next tooth and releasing the cable should cause the adjuster spring to return the lever to its original position. This return action will turn the adjusting screw 1 tooth.
25. If pulling the cable does not produce the action described previously, or if lever action is sluggish instead of positive and sharp, check the position of the lever on the adjusting screw toothed wheel. With the brake in a vertical position, anchor at the top, the lever should contact the adjusting wheel 1 tooth above the centerline of the adjusting screw. If the contact point is below the centerline, the lever will not lock on the adjusting screw wheel teeth and the screw will not turn, since the lever is actuated by the cable.
26. Adjust the brake shoes using either a brake adjustment gauge or manually with the drums installed.

27. Install the wheels, and lower the vehicle.

28. Check and adjust the brake fluid level, as required.

WHEEL CYLINDER

REMOVAL & INSTALLATION

See Figure 5.

1. Before servicing the vehicle, refer to the Precautions Section.
2. Raise and support the vehicle safely.
3. Remove the wheel and tire assembly.
4. Remove the brakes shoes.
5. Disconnect the brake line at the wheel cylinder.
6. Remove the wheel cylinder bolts.

To install:

7. Installation is the reverse of the removal procedure.
8. Torque the brake tube-to-wheel cylinder fitting to 13 ft. lbs. (18 Nm).

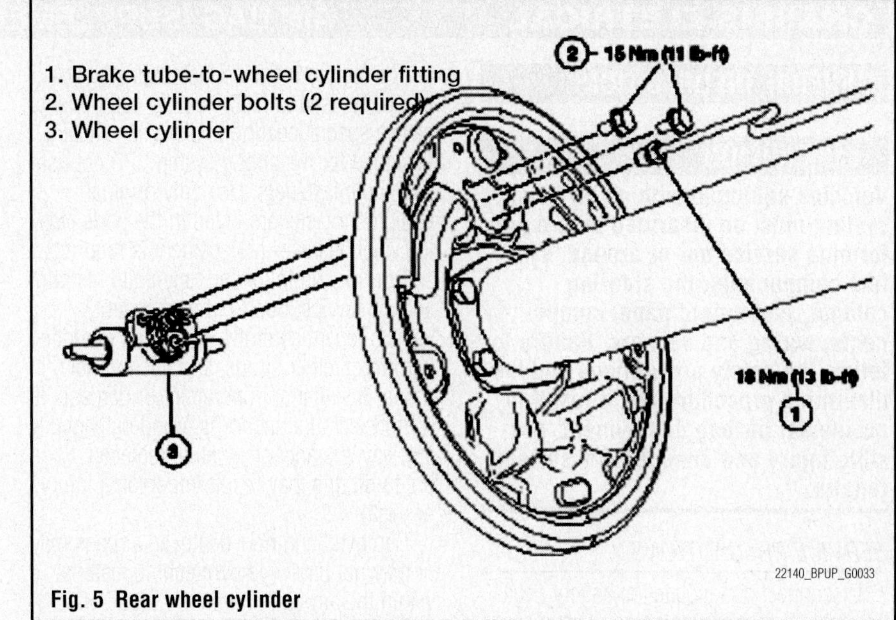

Fig. 5 Rear wheel cylinder

9. Torque the retaining bolts to 11 ft. lbs. (15 Nm).
10. Bleed the brake system.
11. Correct and adjust the brake fluid level.

PARKING BRAKE CABLES

ADJUSTMENT

See Figure 6.

1. After installation of the rear brake shoes, check the action of the adjuster by pulling the section of the cable between the cable guide and the adjusting lever toward the secondary shoe web far enough to lift the lever past a tooth on the adjusting screw wheel. The lever should snap into position behind the next tooth and releasing the cable should cause the adjuster spring to return the lever to its original position. This return action will turn the adjusting screw 1 tooth.
2. If pulling the cable does not produce the action described previously, or if lever action is sluggish instead of positive and sharp, check the position of the lever on the adjusting screw toothed wheel. With the brake in a vertical position, anchor at the top, the lever should contact the adjusting wheel 1 tooth above the centerline of the adjusting screw. If the contact point is below the centerline, the lever will not lock on the adjusting screw wheel teeth and the screw will not turn, since the lever is actuated by the cable.

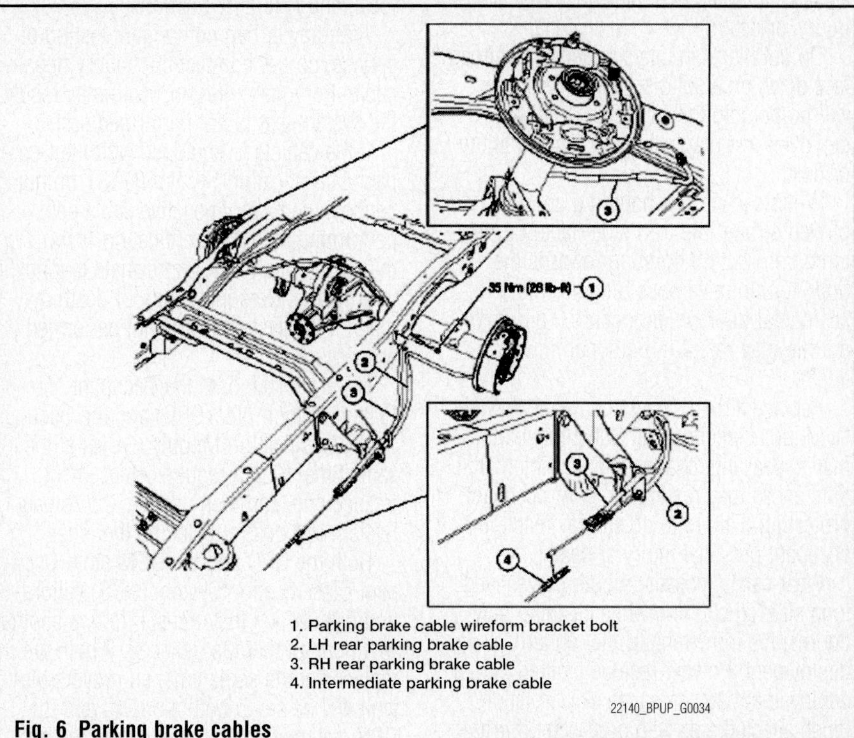

Fig. 6 Parking brake cables

3. Adjust the brake shoes using either a brake adjustment gauge or manually with the drums installed.
4. Install the wheels, and lower the vehicle.

PARKING BRAKE SHOES

REMOVAL & INSTALLATION

The rear drum brake shoes serve as the parking brakes. Refer to the procedures under Rear Drum Brakes.

GENERAL INFORMATION

CAUTION

Vehicles equipped with an air bag system must be disarmed before performing service on, or around, system components, the steering column, instrument panel components, wiring and sensors. Failure to follow the safety precautions and the disarming procedure could result in accidental air bag deployment, possible injury and unnecessary system repairs.

SERVICE PRECAUTIONS

Disconnect and isolate the battery negative cable before beginning any airbag system component diagnosis, testing, removal, or installation procedures. Allow system capacitor to discharge for two minutes before beginning any component service. This will disable the airbag system. Failure to disable the airbag system may result in accidental airbag deployment, personal injury, or death.

Do not place an intact undeployed airbag face down on a solid surface. The airbag will propel into the air if accidentally deployed and may result in personal injury or death.

When carrying or handling an undeployed airbag, the trim side (face) of the airbag should be pointing towards the body to minimize possibility of injury if accidental deployment occurs. Failure to do this may result in personal injury or death.

Replace airbag system components with OEM replacement parts. Substitute parts may appear interchangeable, but internal differences may result in inferior occupant protection. Failure to do so may result in occupant personal injury or death.

Wear safety glasses, rubber gloves, and long sleeved clothing when cleaning powder residue from vehicle after an airbag deployment. Powder residue emitted from a deployed airbag can cause skin irritation. Flush affected area with cool water if irritation is experienced. If nasal or throat irritation is experienced, exit the vehicle for fresh air until the irritation ceases. If irritation continues, see a physician.

Do not use a replacement airbag that is not in the original packaging. This may result in improper deployment, personal injury, or death.

The factory installed fasteners, screws and bolts used to fasten airbag components have a special coating and are specifically designed for the airbag system. Do not use substitute fasteners. Use only original equipment fasteners listed in the parts catalog when fastener replacement is required.

During, and following, any child restraint anchor service, due to impact event or vehicle repair, carefully inspect all mounting hardware, tether straps, and anchors for proper installation, operation, or damage. If a child restraint anchor is found damaged in any way, the anchor must be replaced. Failure to do this may result in personal injury or death.

Deployed and non-deployed airbags may or may not have live pyrotechnic material within the airbag inflator.

Do not dispose of driver/passenger/curtain airbags or seat belt tensioners unless you are sure of complete deployment. Refer to the Hazardous Substance Control System for proper disposal.

Dispose of deployed airbags and tensioners consistent with state, provincial, local, and federal regulations.

After any airbag component testing or service, do not connect the battery negative cable. Personal injury or death may result if the system test is not performed first.

If the vehicle is equipped with the Occupant Classification System (OCS), do not connect the battery negative cable before performing the OCS Verification Test using the scan tool and the appropriate diagnostic information. Personal injury or death may result if the system test is not performed properly.

Never replace both the Occupant Restraint Controller (ORC) and the Occupant Classification Module (OCM) at the same time. If both require replacement, replace one, and then perform the Airbag System test before replacing the other.

Both the ORC and the OCM store Occupant Classification System (OCS) calibration data, which they transfer to one another when one of them is replaced. If both are replaced at the same time, an irreversible fault will be set in both modules and the OCS may malfunction and cause personal injury or death.

If equipped with OCS, the Seat Weight Sensor is a sensitive, calibrated unit and must be handled carefully. Do not drop or handle roughly. If dropped or damaged, replace with another sensor. Failure to do so may result in occupant injury or death.

If equipped with OCS, the front passenger seat must be handled carefully as well. When removing the seat, be careful when setting on floor not to drop. If dropped, the sensor may be inoperative, could result in occupant injury, or possibly death.

If equipped with OCS, when the passenger front seat is on the floor, no one should sit in the front passenger seat. This uneven force may damage the sensing ability of the seat weight sensors. If sat on and damaged, the sensor may be inoperative, could result in occupant injury, or possibly death.

CAUTION

All vehicles are equipped with an air bag system. The system MUST BE disabled before performing service on or around system components, steering column, instrument panel components, wiring and sensors. Failure to follow safety and disabling procedures could result in accidental air bag deployment, possible personal injury and unnecessary system repairs.

DISARMING THE SYSTEM

➡This procedure may also be referred to as Depowering the SRS.

CAUTION

The Supplemental Inflatable Restraint (SIR) system must be disarmed before performing service around SIR system components or SIR system wiring. Failure to do so may cause accidental deployment of the air bag, resulting in unnecessary SIR system repairs and/or personal injury.

The positive battery cable must be disconnected for a minimum of 1 minute before beginning any air bag work to deenergize the back-up power supply. It is a good idea to disengage both the positive and negative battery cables to ensure that the Air Bag system is definitely discharged.

1. Before servicing the vehicle, refer to the Precautions Section.
2. Turn all vehicle accessories.**OFF**
3. Turn the ignition switch to **OFF**.
4. At the central junction box (SJB), located below the left side of the instrument panel, remove the cover and the restraints control module (RCM) fuse(s) from the SJB. See the Owner's Manual.

5. Turn the ignition **ON** and visually monitor the air bag indicator for at least 30 seconds. The air bag indicator will remain lit continuously (no flashing) if the correct RCM fuse has been removed. If the air bag indicator does not remain lit continuously, remove the correct RCM fuse before proceeding.

6. Turn the ignition **OFF**.

CAUTION

To avoid accidental deployment and possible personal injury, the backup power supply must be depleted before repairing or replacing any front or side air bag supplemental restraint system (SRS) components and before servicing, replacing, adjusting or striking components near the front or side air bag sensors, such as doors, instrument panel, console, door latches, strikers, seats and hood latches.

The front impact severity sensor is located on the radiator support bracket.

The first row side impact sensors (if equipped) are located at or near the base of the B-pillars.

The second row side impact sensors (if equipped) are located on the C-pillar.

➡**To deplete the backup power supply energy, disconnect the battery ground cable and wait at least one minute. Be sure to disconnect auxiliary batteries and power supplies (if equipped).**

7. Disconnect the battery ground cable and wait at least one minute.

ARMING THE SYSTEM

➡**This procedure may also be referred to as Repowering the SRS.**

1. Before servicing the vehicle, refer to the Precautions Section.

CAUTION

The restraint system diagnostic tool is for restraint system service only. Remove from vehicle prior to road use. Failure to remove could result in injury and possible violation of vehicle safety standards.

2. Make sure all restraint system diagnostic tool(s) that may have been installed during the repair have been removed from the vehicle and all SRS components are connected.

3. Turn the ignition switch from **OFF** to **ON**.

4. Install RCM fuse(s) to the SJB and close the cover.

CAUTION

Be sure that nobody is in the vehicle and that there is nothing blocking or set in front of any air bag module when the battery ground cable is connected.

5. Connect the battery ground cable.

6. Prove out the supplemental restraint system (SRS) as follows:

7. Turn the ignition key from **ON** to **OFF**. Wait 10 seconds, and then turn the key back to ON and visually monitor the air bag indicator with the air bag modules installed. The air bag indicator will light continuously for approximately six seconds and then turn off. If an air bag supplemental restraint system (SRS) fault is present, the air bag indicator will either:

- Fail to light
- Remain lit continuously
- Flash

8. The flashing might not occur until approximately 30 seconds after the ignition switch has been turned from the **OFF** to the **ON** position. This is the time required for the restraints control module (RCM) to complete the testing of the SRS. If the air bag indicator is inoperative and a SRS fault exists, a chime will sound in a pattern of five sets of five beeps. If this occurs, the air bag indicator and any SRS fault discovered must be diagnosed and repaired.

9. Clear all continuous DTC's from the restraints control module using a scan tool.

CLOCKSPRING CENTERING

See Figures 7 and 8.

1. Before servicing the vehicle, refer to the Precautions Section.

CAUTION

Incorrect centering of the clock spring may result in premature component failure. If in doubt when centering the clockspring, repeat the centering procedure. Failure to follow this instruction may result in personal injury.

➡**Make sure the road wheels are in the straight-ahead position.**

2. If the vehicle's clockspring has rotated out of center, follow these steps to center the clockspring.

3. Hold the clockspring outer housing stationary.

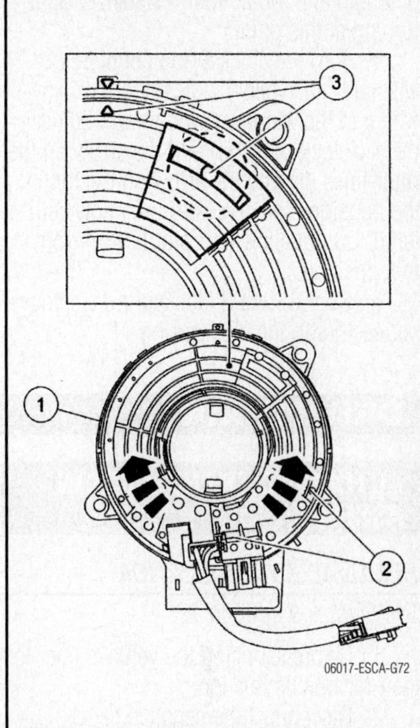

Fig. 7 Outer housing (1), rotor (2), clockspring aligning

WARNING

Overturning will destroy the clockspring. The internal ribbon wire acts as the stop and can be broken from its internal connection.

4. While turning the rotor clockwise, carefully feel for the ribbon wire to run out

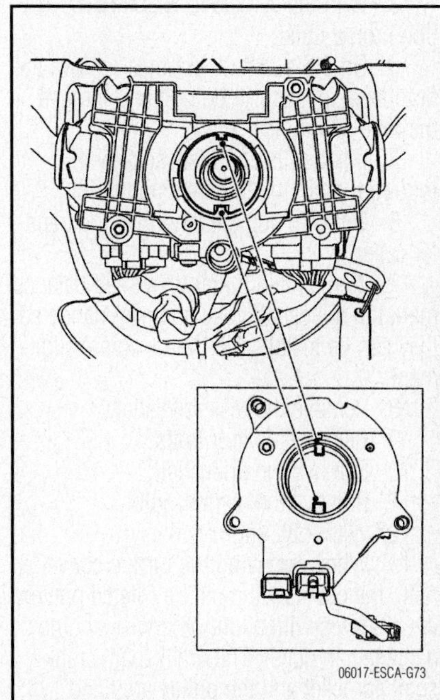

Fig. 8 Clockspring installation

of length and for a slight resistance. Stop turning at this point.

5. Turn the clockspring counterclockwise until the yellow indicator shows anywhere in the window (window will be near the 1 o'clock position) and the arrow on the rotor lines up with the arrow on the top of the housing. The clockspring is now centered. Do not allow the rotor to turn from this position.

6. Connect the 2 clockspring electrical connectors to the clockspring.

➡**Slight turning of the clockspring rotor is allowable for alignment purposes to the steering column.**

7. Align the clockspring for installation.

8. Align the large slot to the large tab in the clockspring.

9. Align the small slot to the small tab in the clockspring.

10. Install the 3 clockspring screws.

11. For vehicles reusing a clockspring that was removed, remove the tape. For vehicles installing a new clockspring, remove the retaining pin.

12. Install the lower steering column shroud and the 3 screws.

13. Position the steering column completely downward and lock in place.

14. Install the upper steering column shroud and engage the retaining tabs.

15. Install the steering wheel. Torque the steering wheel pinion bolt to 9 ft. lbs. (12 Nm.).

16. Repower the system.

DRIVETRAIN

AUTOMATIC TRANSMISSION ASSEMBLY

REMOVAL & INSTALLATION

See Figures 9 through 17.

1. Before servicing the vehicle, refer to the Precautions Section.

2. Disconnect the negative battery cable.

➡**On some vehicles, when the battery cable is disconnected and reconnected, some abnormal drive symptoms may occur while the vehicle relearns its adaptive strategy. The vehicle may need to be driven to relearn its strategy.**

➡**If the transmission is to be removed for a period, support the engine with a safety stand and a wood block.**

3. With the vehicle in **NEUTRAL**, position it on a hoist.

4. On 3.0 and 4.0L engines remove the fluid level indicator tube bolt and remove the tube and indicator.

5. If transmission disassembly is required, drain the transmission fluid.

6. On 4WD vehicles, remove the transfer case.

7. To maintain initial driveshaft balance, mark the driveshaft yoke and axle flange so they can be installed in their original alignment.

8. Remove the rear driveshaft.

9. Remove the four bolts.

10. Remove the driveshaft.

11. Remove the starter motor.

12. With 2.3L engine:

13. When removing the torque converter nuts, the crankshaft must be rotated only in the clockwise direction, otherwise engine damage can occur. The crankshaft, crankshaft sprocket and the pulley are fitted together by friction between the flange faces on each part. For that reason, the crankshaft sprocket can also be moved when the crankshaft pulley is turned in the counterclockwise direction.

14. It may be necessary to gain access to the flexplate nuts through the wheel well.

15. Mark the torque converter and the flexplate for correct alignment at reinstallation.

16. Remove and discard the four torque converter nuts. Rotate the flexplate to access to all the nuts.

17. With 3.0L and 4.0L engine:

18. Mark the torque converter and the flexplate for correct alignment at reinstallation. Remove the four nuts. Rotate the flexplate to access to all the nuts.

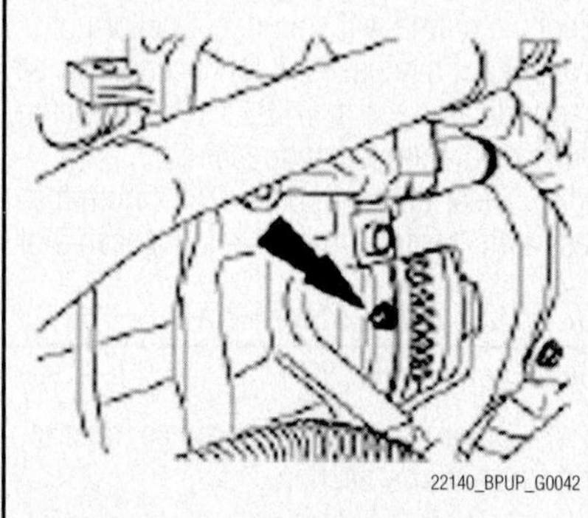

Fig. 9 Four torque converter to flexplate nuts

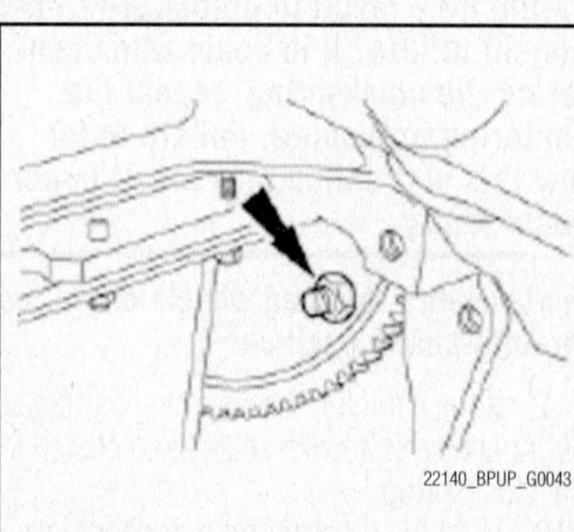

Fig. 10 Four torque converter to flexplate nuts

19. Remove or disconnect the following:
   - Shift cable.
   - Wiring harness from the case
   - Wiring harness
   - Transmission selector lever bracket
   - Transmission selector lever cable retainer from the case
   - Transmission electrical connectors
   - Intermediate shaft speed sensor connector
   - Digital Transmission Range (TR) sensor connector
   - Output Shaft Speed (OSS) sensor connector

20. Remove the three way catalytic converter.

21. Remove the rear engine cover plate.

➡**Care should be taken not to bend or damage the cooler lines.**

22. Hold the case fitting and remove the transmission cooler lines.

23. Position a transmission jack under the transmission. Raise and support the transmission.

24. Remove the crossmember.

25. Remove the 6 bolts (3 each side).

26. Remove the transmission mount.

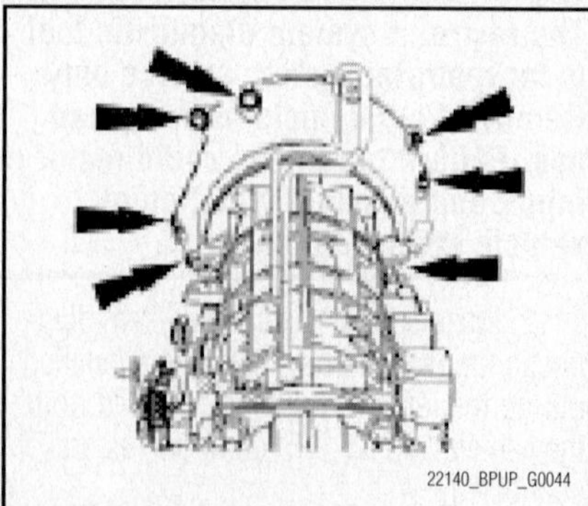

Fig. 11 Transmission upper retaining bolts

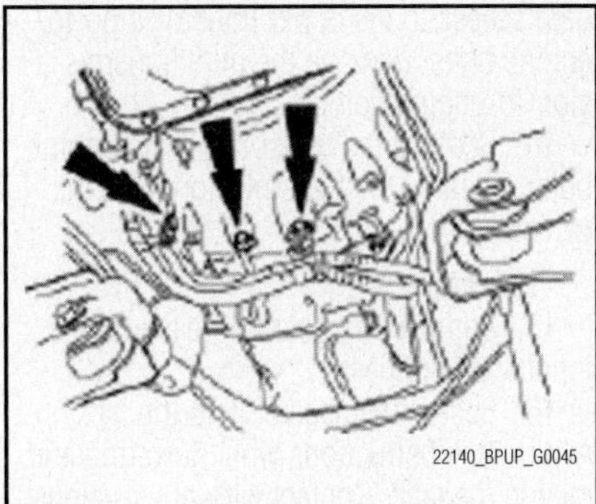

Fig. 12 Transmission lower retaining bolts

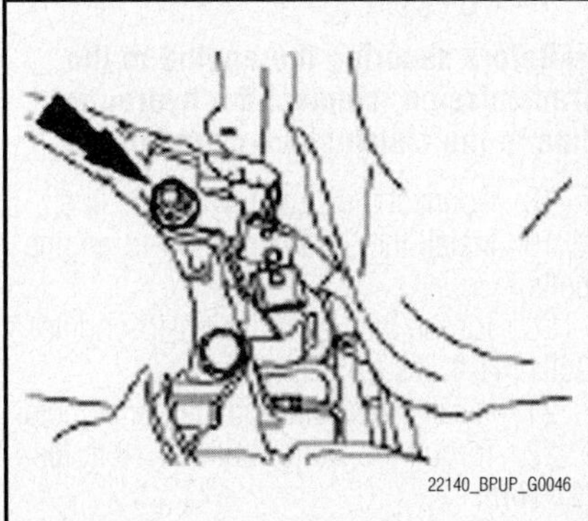

Fig. 13 O2 Sensor bracket location

27. If equipped, remove the rear vibration damper.

28. (2.3L engine) remove the transmission upper fill tube.

29. Lower the jack to gain access to screws. Remove the transmission-to-engine bolts.

30. Remove the lower screws.

31. Remove the HO2S connector bracket from the transmission.

32. On 4WD vehicles, remove the vent tube assembly.

⁂ CAUTION

The torque converter is heavy and may result in injury if it falls out of the transmission. Secure the torque converter in the transmission. Failure to follow these instructions may result in personal injury. Install the SST T979-7902-A or equivalent before lowering the transmission from the vehicle.

⁂ CAUTION

Secure the transmission to the transmission jack with a safety chain. Failure to follow these instructions may result in personal injury.

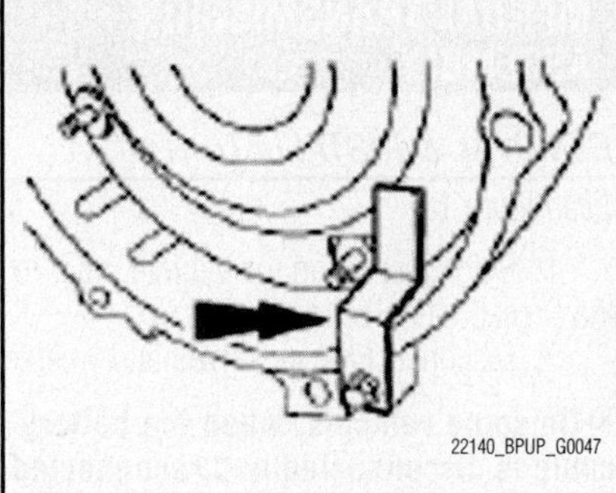

Fig. 14 Torque converter retaining bracket

33. Lower the transmission.

34. If the transmission is being overhauled or if installing a new or remanufactured transmission, carry out transmission fluid cooler backflushing and cleaning.

To install:

35. On 4WD vehicles, install the vent tube assembly.

⁂ CAUTION

Secure the transmission to the transmission jack with a safety chain. Failure to follow these instructions may result in personal injury.

36. Raise and position the transmission.

37. Remove the holding tool.

38. Install the transmission.

39. Align the flexplate to the converter marks made at removal.

➡On 2.3L engine when aligning or installing the torque converter nuts, the crankshaft must be rotated only in the clockwise direction, otherwise engine damage can occur. The crankshaft, the crankshaft sprocket and the pulley are fitted together by friction between the flange faces on each part. For that reason, the crankshaft sprocket can also be moved when the crankshaft pulley is turned in the counterclockwise direction.

40. Install the transmission-to-engine upper bolts. See illustration for proper installation.

41. With the 2.3L engine and 3.0L engine, install the upper transmission bolts See illustration for proper installation.

42. Torque to 41 ft. lbs. (55 Nm).

43. Install lower screws

44. Install four new torque converter to flywheel nuts and tighten to 26 ft. lbs. (35 Nm).

45. Install the upper fluid filler tube and bracket screw and tighten to 41 ft. lbs. (55 Nm).

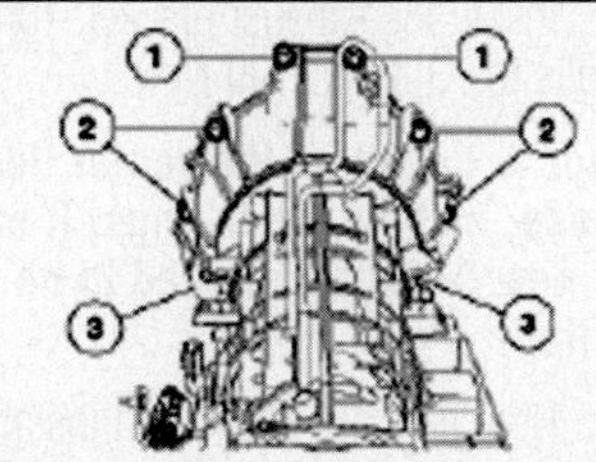

1. Install 1 M10 x 40 mm (1.57 in) bolt.
2. Install 4 M10 x 50 mm (1.96 in) bolts.
3. Install 2 M10 x 75 mm (2.95 in) bolts.

22140_BPUP_G0048

Fig. 15 Bolt size installation guide—4.0L engine

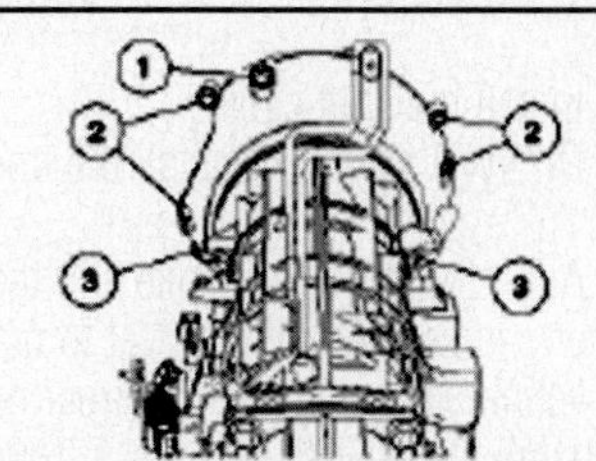

1. Install 1 M10 x 40 mm (1.57 in) bolt.
2. Install 4 M10 x 50 mm (1.96 in) bolts.
3. Install 2 M10 x 75 mm (2.95 in) bolts.

22140_BPUP_G0049

Fig. 16 Bolt size installation guide2.3L engine—3.0L engine

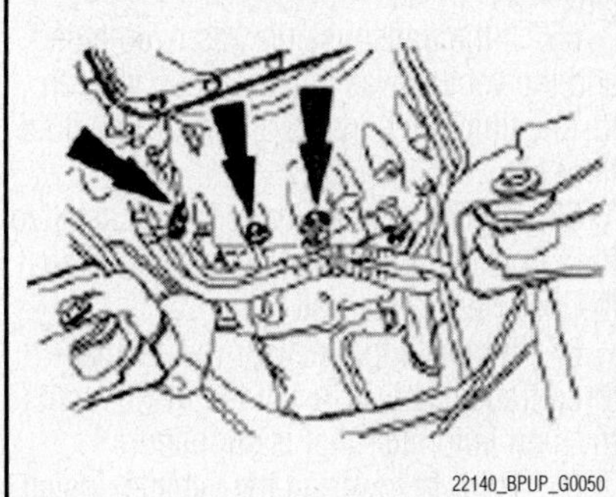

Fig. 17 Lower screw installation guide—all engines

46. Install the O2 sensor connector bracket.

47. Install the rear engine cover plate.

48. Install the crossmember.

49. Tighten the bolts to 74 ft. lbs. (101 Nm).

50. Install the transmission mount into the crossmember and torque the nuts to 66 ft. lbs. (90 Nm) .

51. On 4WD vehicles, install the 6 crossmember bolts (3 each side).

52. Tighten the bolts to 73 ft. lbs. (99 Nm).

53. If equipped, install the rear vibration damper and tight to 22 ft. lbs. (30 Nm).

54. Install the exhaust bracket. Torque the bolts to 73 ft. lbs. (99 Nm).

➡Prior to installing the cooler lines to the case, inspect the O-rings. If damaged new O-rings will need to be installed.

55. Hold the case fitting and install the transmission cooler lines. Torque to 21 ft. lbs. (28 Nm).
56. Install the starter motor.
57. Install the catalytic converter assembly.
58. Position the transmission wiring harness in place.
59. Connect the transmission wiring harness.
60. Install the shift cable.
61. On 4WD vehicles, install the transfer case.
62. Align the driveshaft yoke and the axle shaft marks made at removal to maintain driveline balance. Install the rear driveshaft. Install the driveshaft bolts. Torque to 83 ft. lbs. (112 Nm).
63. Use the following guidelines for installing the in-line transmission fluid filter:
64. If the transmission was overhauled and the vehicle was equipped with an in-line fluid filter, install a new in-line fluid filter.
65. If the transmission was overhauled and the vehicle was not equipped with an in-line fluid filter, install a new in-line fluid filter kit.
66. If the transmission is being installed for a non-internal repair, do not install an in-line filter or filter kit.
67. If installing a new or re-manufactured transmission, install the in-line transmission fluid filter that is supplied.
68. Prior to lowering the vehicle, install a new in-line transmission filter or a filter kit.
69. Install the transmission fill tube and indicator as an assembly.

➡When the battery has been disconnected and reconnected, some abnormal drive symptoms can occur while the vehicle relearns its adaptive strategy.

70. Connect the battery ground cable.
71. Fill the transmission with clean automatic transmission fluid to the specified level.
72. Check the transmission for correct operation.
73. Verify that the shift cable is correctly adjusted.

MANUAL TRANSMISSION ASSEMBLY

REMOVAL & INSTALLATION

See Figure 18.

1. Before servicing the vehicle, refer to the Precautions Section.
2. Disconnect the negative battery cable.

➡On some vehicles, when the battery cable is disconnected and reconnected, some abnormal drive symptoms may occur while the vehicle relearns its adaptive strategy. The vehicle may need to be driven to relearn its strategy.

3. Remove the upper gearshift lever, the outer gearshift lever boot and the console as an assembly.
4. With the vehicle in neutral raise and support the vehicle safely.
5. If transmission disassembly is required, remove the drain plug and drain the transmission fluid. Install the drain plug after draining all the fluid.
6. To maintain initial driveshaft balance, index-mark the driveshaft yoke to the axle flange, so they can be installed in their original positions. Remove the rear driveshaft.
7. If equipped with 4WD, remove the transfer case.
8. Using special tool 308-182, disconnect the clutch hydraulic line.
9. Remove the starter motor.
10. Place a suitable jack under the transmission. Secure the transmission to the jack with a safety strap.
11. Remove the transmission mount bolts and the transmission mount.
12. Remove the transmission bolt and reposition the exhaust bracket. Remove the exhaust inlet crossover pipe on 3.0L and 4.0L engines.
13. Lower the transmission enough to gain access to the upper transmission-to-engine bolts. Remove the nine transmission-to-engine bolts.
14. Remove the transmission to engine bolts. Remove the transmission from the vehicle.

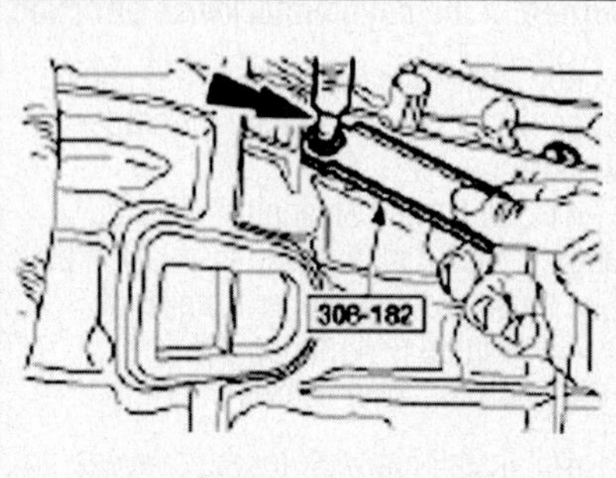

Fig. 18 Special disconnect tool for the clutch hydraulic line 308-182.

To install:

15. Apply a thin coat of grease on the input shaft splines.
16. Secure the assembly to the jack. Avoid any obstructions while lowering and raising the jack. Contact with obstructions may cause the assembly to fall off the jack, which may result in serious personal injury.
17. Raise and position the transmission to the engine.

➡Before securing the engine to the transmission, connect the hydraulic line to the clutch slave cylinder.

18. Connect the clutch hydraulic line.
19. Install the 9 transmission-to-engine bolts.
20. Torque the transmission-to-engine bolts: 41 ft. lbs. (55 Nm).
21. Install the 3 bolts and the starter motor.
22. Torque the starter bolts to 18 ft. lbs. (25 Nm).
23. Connect the vehicle speed sensor (VSS) electrical connector and the reverse lamp switch electrical connector. Then clip the wiring harness to the transmission.
24. For 4.0L engine, install the catalytic converter Y-pipe.
25. Install the transmission mount and the 2 transmission mount bolts.
26. Tighten to 72 ft. lbs. (98 Nm).
27. Install the heated oxygen sensor (HO2S) bracket and nut to the extension housing.
28. Tighten to 29 ft. lbs. (39 Nm).
29. Install the crossmember and the 2 transmission mount nuts. Do not tighten the nuts at this time.
30. Install the 6 crossmember bolts.
31. Tighten the transmission mount nuts to 72 ft. lbs. (98 Nm).
32. Tighten the crossmember bolts to 74 ft. lbs. (101 Nm).
33. Remove the transmission jack.
34. Connect the wire harness to the crossmember.
35. On 4WD vehicles, install the transfer case.
36. Align the index marks when installing the front and rear driveshafts.
37. Install the upper gearshift lever and gearshift lever boot as an assembly.
38. Check and, if necessary, fill the transmission with the specified type and quantity of fluid.

39. Warm up the engine and transmission, inspect for oil leakage and inspect transmission operation.

CLUTCH DRIVEN DISC & PRESSURE PLATE

REMOVAL & INSTALLATION

See Figures 19 and 20.

1. Before servicing the vehicle, refer to the precautions in the beginning of this section.
2. Remove or disconnect the following:
   - Negative battery cable
   - Transmission see manual transmission removal in this section.

➡If the clutch disc and pressure plate are to be reinstalled, bolts must be removed evenly or permanent damage to the diaphragm spring will occur resulting in complete clutch release.

   - Bolts, clutch pressure plate and the clutch disc.

➡If the parts are to be reused, index-mark the clutch pressure plate to the flywheel.

To installation:

3. Lubricate the transmission input shaft pilot bearing with front axle grease.
4. Using a suitable press, press downward on the pressure plate fingers until the adjusting ring moves freely.
5. Rotate the adjusting ring counterclockwise to compress the tension springs. Hold the adjusting ring in this position.
6. Release the pressure on the fingers, the adjusting ring will stay in the reset position.

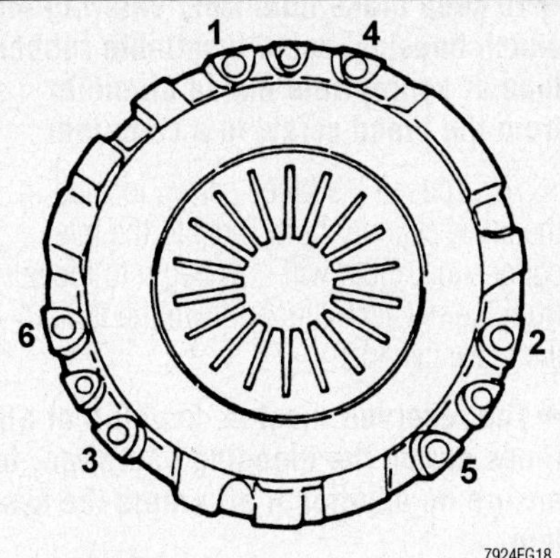

Fig. 20 Tighten the bolts gradually in the correct sequence to avoid warping the pressure plate

7. Position the clutch disc on the flywheel.

➡If reusing the clutch pressure plate and flywheel, align the marks made during removal.

8. Align the clutch disc and the clutch pressure plate. Install the bolts and tighten in a star pattern sequence to 24 ft. lbs. (35 Nm).
   - Install the transmission.

ADJUSTMENTS

Because the clutch is hydraulically driven, there is no adjustment required.

In the event the clutch pedal develops a squeak or uneven feel when depressing, spray the pedal bushing assembly with penetrating oil and work the pedal back-and-forth.

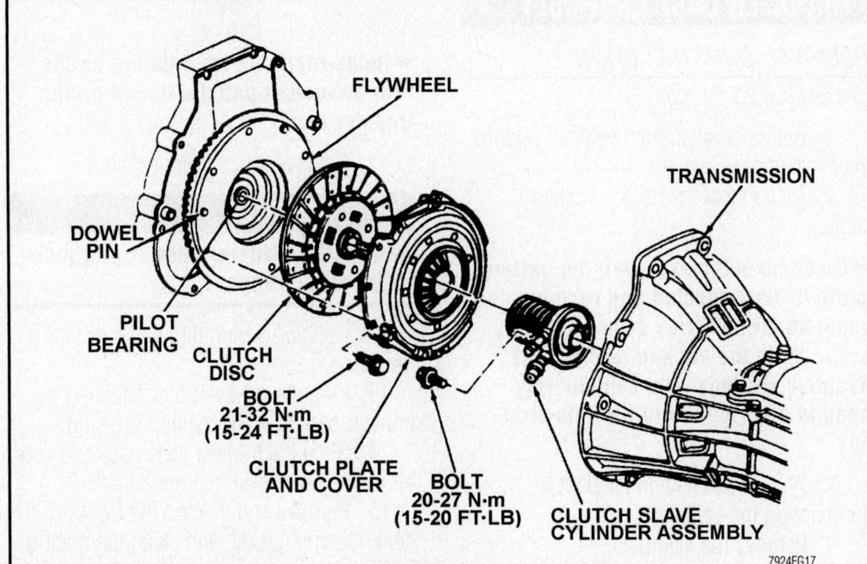

Fig. 19 Clutch disc, pressure plate and bearing assembly

CLUTCH MASTER CYLINDER

REMOVAL & INSTALLATION

1. Before servicing the vehicle, refer to the Precautions Section.
2. Disconnect the negative battery cable.
3. Disconnect the clutch master cylinder rod from the clutch pedal.
4. Remove the clutch pedal bracket nut.

✻✻ CAUTION

Brake fluid contains polyglycol ethers and polyglycols. Avoid contact with eyes. Wash hands thoroughly after handling. If brake fluid contacts eyes, flush eyes with running water for 15 minutes. Get medical attention if irritation persists. If taken internally, drink water and induce vomiting. Get medical attention immediately. Failure to follow these instructions may result in personal injury.

✻✻ WARNING

Brake fluid is harmful to painted and plastic surfaces. If brake fluid is spilled onto a painted or plastic surface, wash it immediately with water.

5. Disconnect the clutch master cylinder hose from the clutch master cylinder and plug it to prevent excess fluid loss.
6. Disconnect the clutch slave cylinder tube from the clutch master cylinder.
7. Remove the clutch master cylinder nut.
8. Remove the clutch master cylinder.
9. To install, reverse the removal procedure. Torque all fasteners to specification.
10. Bleed the air from the system.

BENCH BLEEDING PROCEDURE

1. Position the clutch master cylinder reservoir above the clutch master cylinder and the quick connect fitting below the clutch master cylinder.
2. Press the internal mechanism of the quick connect fitting to open the valve.
3. Compress and hold the clutch master cylinder push rod.
4. Release the quick connect fitting valve.
5. Release the clutch master cylinder push rod.
6. Fill the clutch master cylinder reservoir.
7. Repeat steps 2 and 3 four times.

CLUTCH SLAVE CYLINDER

REMOVAL & INSTALLATION

✱✱ CAUTION

Brake fluid contains polyglycol ethers and polyglycols. Avoid contact with eyes. Wash hands thoroughly after handling. If brake fluid contacts eyes, flush eyes with running water for 15 minutes. Get medical attention if irritation persists. If taken internally, drink water and induce vomiting. Get medical attention immediately. Failure to follow these instructions may result in personal injury.

✱✱ WARNING

Brake fluid is harmful to painted and plastic surfaces. If brake fluid is spilled onto a painted or plastic surface, wash it immediately with water. Remove the clutch slave cylinder to clutch line adapter.

1. Before servicing the vehicle, refer to the Precautions Section.
2. Disconnect the negative battery cable.
3. Remove the transmission.
4. Remove the clutch slave cylinder bolts.
5. Remove the clutch slave cylinder.
6. To install, reverse the removal procedure. Torque slave cylinder bolts to 15 ft. lbs. (20 Nm).
7. Bleed the air from the system.

CLUTCH HYDRAULIC SYSTEM BLEEDING

BLEEDING PROCEDURE

The following procedure is recommended for bleeding the clutch hydraulic system installed on the vehicle. It is recommended that the original clutch tube, with quick-connect fitting be replaced when servicing the hydraulic system, because air can be trapped in the quick-connect fitting and prevent complete bleeding of the system. The replacement tube does not include a quick-connect fitting.

1. Before servicing the vehicle, refer to the precautions in the beginning of this section.
2. Clean the dirt and grease from the dust cap.
3. Remove the cap and diaphragm and fill the reservoir to the top with approved brake fluid C6AZ-19542-AA or BA, (ESA-M6C25-A).

➡To keep brake fluid from entering the clutch housing, route a suitable rubber tube of appropriate inside diameter from the bleed screw to a container.

4. Loosen the bleed screw, located in the slave cylinder body, next to the inlet connection. Fluid will now begin to move from the master cylinder down the tube to the slave cylinder.

➡The reservoir must be kept full at all times during the bleeding operation, to ensure no additional air enters the system.

5. Observe the bleed screw outlet. When the slave cylinder is full, a steady stream of fluid will flow from the outlet port. Tighten the bleed screw.
6. Depress the clutch pedal to the floor and hold for 1–2 seconds. Release the pedal as rapidly as possible. The pedal must be released completely. Pause for 1–2 seconds. Repeat 10 times.
7. Check the fluid level in the reservoir. The fluid should be level with the step when the diaphragm is removed.
8. Hold the pedal to the floor, slightly open the bleed screw to allow any additional air to escape. Close the bleed screw, then release the pedal.
9. Check the fluid in the reservoir. The hydraulic system should now be fully bled, and should actuate the clutch.
10. Check the vehicle by starting, pushing the clutch pedal to the floor and selecting reverse gear. There should be no grating of gears. If there is, and the hydraulic system still contains air; repeat the bleeding procedure.

TRANSFER CASE ASSEMBLY

REMOVAL & INSTALLATION

See Figures 21 through 23.

1. Before servicing the vehicle, refer to the Precautions Section.
2. Disconnect the negative battery cable.

➡On some vehicles, when the battery cable is disconnected and reconnected, some abnormal drive symptoms may occur while the vehicle relearns its adaptive strategy. The vehicle may need to be driven to relearn its strategy.

3. With the vehicle in **NEUTRAL**, raise and support the vehicle.
4. Remove the skid plate.
5. Remove the damper.

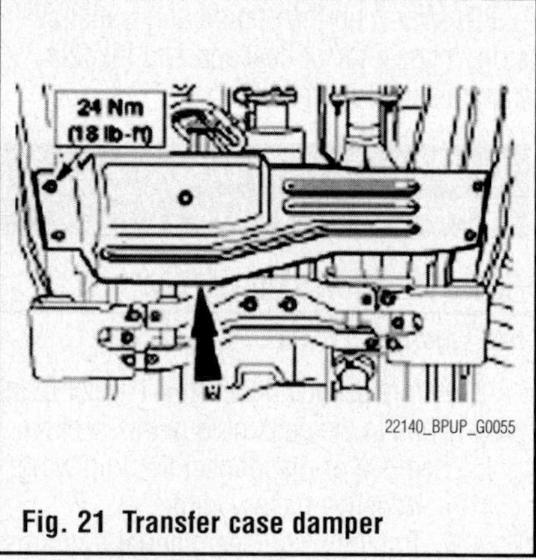

Fig. 21 Transfer case damper

6. Disconnect the transfer case harness connector and position it aside.
7. If transfer case disassembly is necessary, remove the drain plug and drain the fluid. Install the drain plug when all of the fluid has drained.

➡Index-mark the front output shaft assembly and the front driveshaft constant velocity (CV) joint.

✱✱ WARNING

Always disconnect the front driveshaft from the transfer case first. Otherwise, the weight of the driveshaft can pinch the boot between the shaft and the boot can and cause the boot to tear.

8. Index-mark the front output shaft assembly and the front driveshaft constant velocity (CV) joint.
9. Remove and discard the bolts and washers.
10. Disconnect the front driveshaft from the transfer case and position the driveshaft aside.

➡Index-mark the front flange on the rear driveshaft and the flange on the transfer case.

11. Remove the rear driveshaft.

✱✱ CAUTION

Secure the transfer case to the jack with safety straps.

12. Position a high lift jack under the transfer case.
13. Remove the five bolts retaining the transfer case to the extension housing.
14. Slide the transfer case rearward and off of the transmission output shaft.
15. Remove and discard the front extension housing gasket, and clean the mating surfaces.

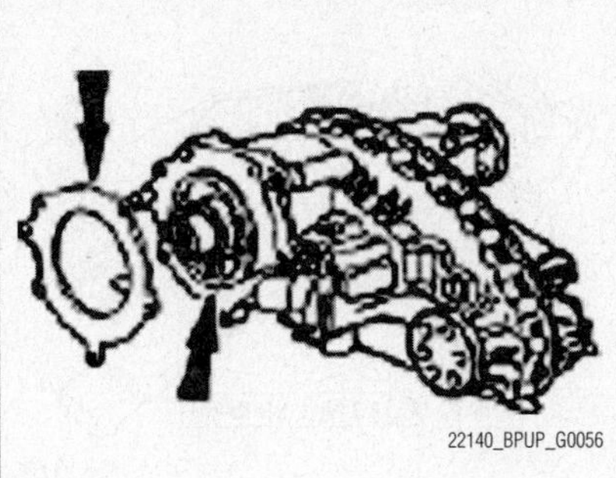

Fig. 22 New gasket on the front of the transfer case.

To install:

16. Install the transfer case with a new gasket.

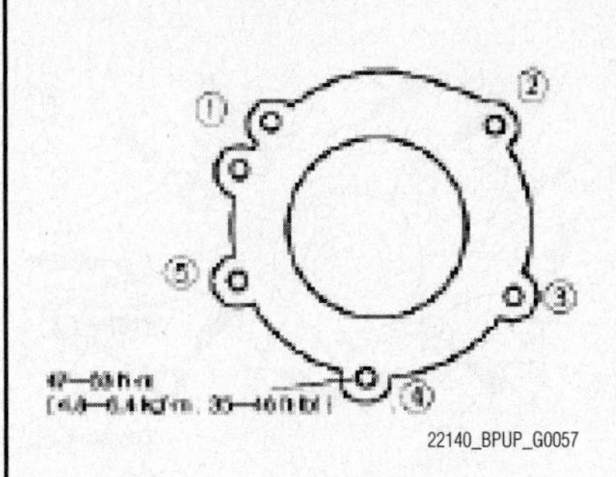

Fig. 23 Install procedure of the axle to the transfer case

17. Tighten the bolts that retain the transfer case to the extension housing in a clockwise direction beginning with the upper LH bolt. Torque to 40 ft. lbs. (54 Nm).

18. Install the front driveshaft with new bolts and washers and the rear driveshaft with new bolts. If new bolts are not available, coat the threads of the original bolts with Threadlock and Sealer E0AZ-19554-AA, or equivalent.

➡When installing the front driveshaft, always connect it to the axle first and then connect it to the transfer case.

➡Align the index marks when installing the front and rear driveshafts.

19. The remainder of installation is the reverse of the removal procedure.

20. Check and, if necessary, fill the transfer case with the specified type and quantity of fluid.

ENGINE COOLING

✲✲ CAUTION

Never remove the pressure relief cap or coolant recovery reservoir cap while the engine is operating or when the cooling system is hot. Failure to follow these instructions can result in damage to the cooling system or engine or personal injury. To avoid having scalding hot coolant or steam blow out of the coolant reservoir when removing the pressure relief cap, wait until the engine has cooled, then wrap a thick cloth around the pressure relief cap and turn it slowly. Step back while the pressure is released from the cooling system. When you are sure all the pressure has been released, (still with a cloth) turn and remove the pressure relief cap. Failure to follow these instructions may result in personal injury.

✲✲ WARNING

Engine coolant provides freeze protection, boil protection, cooling efficiency and corrosion protection to the engine and cooling components. In order to obtain these protections, the engine coolant must be maintained at the correct concentration and fluid level. When adding engine coolant, use a 50/50 mixture of engine coolant and clean, drinkable water.

✲✲ WARNING

Do not add Mazda Specialty Orange Engine Coolant, VC-2. Mixing coolants may degrade the coolant's corrosion protection.

COOLANT TEMPERATURE SENSOR

REMOVAL & INSTALLATION

1. See all precautions in this section.
2. Disconnect the negative battery cable.
3. Drain the cooling system

➡ The coolant must be recovered in a suitable, clean container for reuse. If the coolant is contaminated it must be recycled or disposed of properly.

4. Disconnect the electrical connector.
5. Remove the engine coolant temperature sensor.
6. To install, reverse the removal procedure.
7. Coat the threads with pipe thread sealant.
8. Fill with Mazda Premium Gold Engine Coolant VC-7-A (in Oregon VC-7-B) (yellow color). Do not mix coolant types.
9. See Filling and Bleeding Procedure in this section.

WATER PUMP

REMOVAL & INSTALLATION

2.3L Engine

See Figure 24.

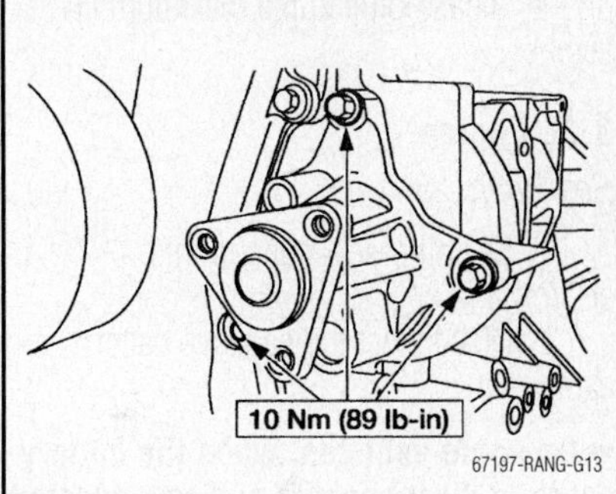

Fig. 24 Water pump mounting bolts—2.3L engine

1. See all precautions in this section.
2. Disconnect the negative battery cable.

➡On some vehicles, when the battery cable is disconnected and reconnected, some abnormal drive symptoms may occur while the vehicle relearns its adaptive strategy. The vehicle may need to be driven to relearn its strategy.

3. Remove the air cleaner outlet pipe.
4. Drain the cooling system.
5. Remove the drive belt.
6. Remove the fan if equipped with a clutch driven fan, see cooling fan removal in this section.
7. Remove the water pump pulley.
8. Remove the water pump.

To install:

9. Clean the mating surfaces where the water pump attaches to the engine.

➡Lubricate the water pump O-ring.

10. To install, reverse the removal procedure. Torque the water pump mount bolts to 89 inch lbs. (10 Nm). Torque the pulley bolts to 18 ft. lbs. (25 Nm).
11. See all Filling and Bleeding Procedure in this section.
12. Maintain the engine speed at 2,500 RPM for an additional three minutes.
13. Increase engine speed to 4,000 RPM and hold for five seconds.
14. Return the engine speed to 2,500 RPM and hold for an additional three minutes.
15. Repeat the previous two steps.
16. Stop the engine and check for leaks.
17. Verify the correct fluid level after the engine cools for 20 minutes. Top off the degas bottle to the maximum fill level.

3.0L Engine

See Figure 25.

1. See all precautions in this section.
2. Disconnect the negative battery cable.

➡On some vehicles, when the battery cable is disconnected and reconnected, some abnormal drive symptoms may occur while the vehicle relearns its adaptive strategy. The vehicle may need to be driven to relearn its strategy.

3. Drain the cooling system.
4. Remove or disconnect the following:
   - Air cleaner outlet tube
   - Fan and radiator shroud, see cooling fan removal in this section.
   - Water bypass tube
   - Drive belt
   - Heater hose
   - Water pump pulley
   - Lower radiator hose
   - Air conditioning compressor and bracket assembly and move them aside
   - Water pump

To install:

5. Clean the mating surfaces where the water pump attaches to the engine.
6. Install or connect the following:
   - Water pump. Torque the bolts to 89 inch lbs. (10 Nm).

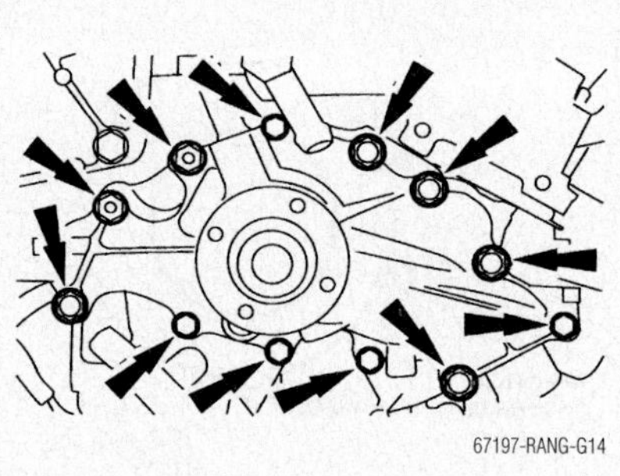

Fig. 25 Water pump mounting bolts—3.0L engine

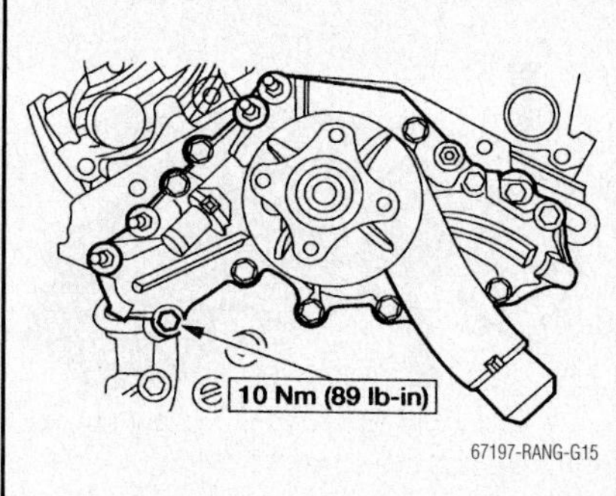

Fig. 26 Water pump mounting bolts—4.0L engine

   - Air conditioning compressor mounting bracket. Torque the bolts to 44 ft. lbs. (61 Nm).
   - Water pump pulley. Torque the bolts to 18 ft. lbs. (25 Nm).
   - Drive belt
   - Heater hose
   - Lower radiator hose
   - Fan and shroud
   - Air cleaner outlet tube
   - Negative battery cable
7. Fill the cooling system.
8. See all Filling and Bleeding Procedure in this section.
9. Maintain the engine speed at 2,500 RPM for an additional three minutes.
10. Increase engine speed to 4,000 RPM and hold for five seconds.
11. Return the engine speed to 2,500 RPM and hold for an additional three minutes.
12. Repeat the previous two steps.
13. Stop the engine and check for leaks.
14. Verify the correct fluid level after the engine cools for 20 minutes. Top off the degas bottle to the maximum fill level.

4.0L Engine

See Figure 26.

1. See all precautions in this section.
2. Disconnect the negative battery cable.

➡On some vehicles, when the battery cable is disconnected and reconnected, some abnormal drive symptoms may occur while the vehicle relearns its adaptive strategy. The vehicle may need to be driven to relearn its strategy.

3. Drain the cooling system.
4. Remove or disconnect the following:
   - Fan shroud, see cooling fan removal in this section
   - Accessory drive belt
   - Idler pulley
   - Water bypass hose
   - Heater hose
   - Lower radiator hose
   - Water pump pulley
   - Water pump

To install:

5. Clean the mating surfaces where the water pump attaches to the engine.

WARNING

Use care when scraping the water pump-to-engine block mating surfaces. Gouges in the aluminum could form leak paths.

6. Clean all the sealing surfaces.
7. To install, reverse the removal procedure. Torque the water pump bolts to 89 inch lbs. (10 Nm). Torque the pulley bolts to 18 ft. lbs. (25 Nm).
8. Fill the cooling system.
9. See all Filling and Bleeding Procedures in this section.
10. Maintain the engine speed at 2,500 RPM for an additional three minutes.
11. Increase engine speed to 4,000 RPM and hold for five seconds.
12. Return the engine speed to 2,500 RPM and hold for an additional three minutes.
13. Repeat the previous two steps.
14. Stop the engine and check for leaks.
15. Verify the correct fluid level after the engine cools for 20 minutes. Top off the degas bottle to the maximum fill level.

THERMOSTAT

REMOVAL & INSTALLATION

1. See all precautions in this section.
2. Disconnect the negative battery cable.

3. Drain and recycle the engine coolant.
4. Remove the upper and lower coolant hoses on 2.3L engine.
5. Remove the upper coolant hose on 4.0L engine.
6. Remove the thermostat housing.
7. Remove the bolts and separate the water outlet adapter from the thermostat housing.

➡The thermostat is indexed and must be installed correctly.

8. Remove the thermostat and the O-ring seal.

➡Clean and inspect the sealing surfaces.

9. Fill the cooling system.
10. See Filling and Bleeding Procedure in this section.

ENGINE ELECTRICAL

CHARGING SYSTEM

✲✲ WARNING

Do not allow any metal object to come in contact with the housing and the internal diode cooling fins with key on or off. A short circuit will result and burn out the diodes.

ALTERNATOR

REMOVAL & INSTALLATION

2.3L Engine

See Figures 27 and 28.

1. See all precautions in this section.
2. Disconnect the negative battery cable.
3. Remove the air cleaner outlet tube.
4. Remove the accessory drive belt.
5. Remove the bolts and position the alternator aside.
6. Remove the alternator.
7. Remove the nut and disconnect the electrical connectors.
8. To install, reverse the removal procedure.
9. Tighten alternator to 18 ft. lbs. (25 Nm).

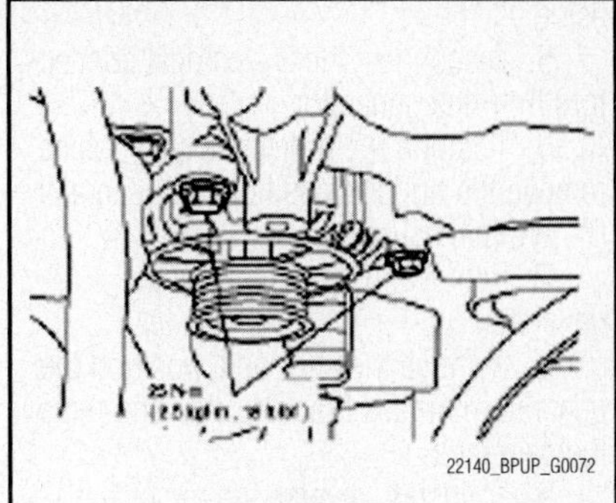

Fig. 28 Alternator and system components—2.3L engine

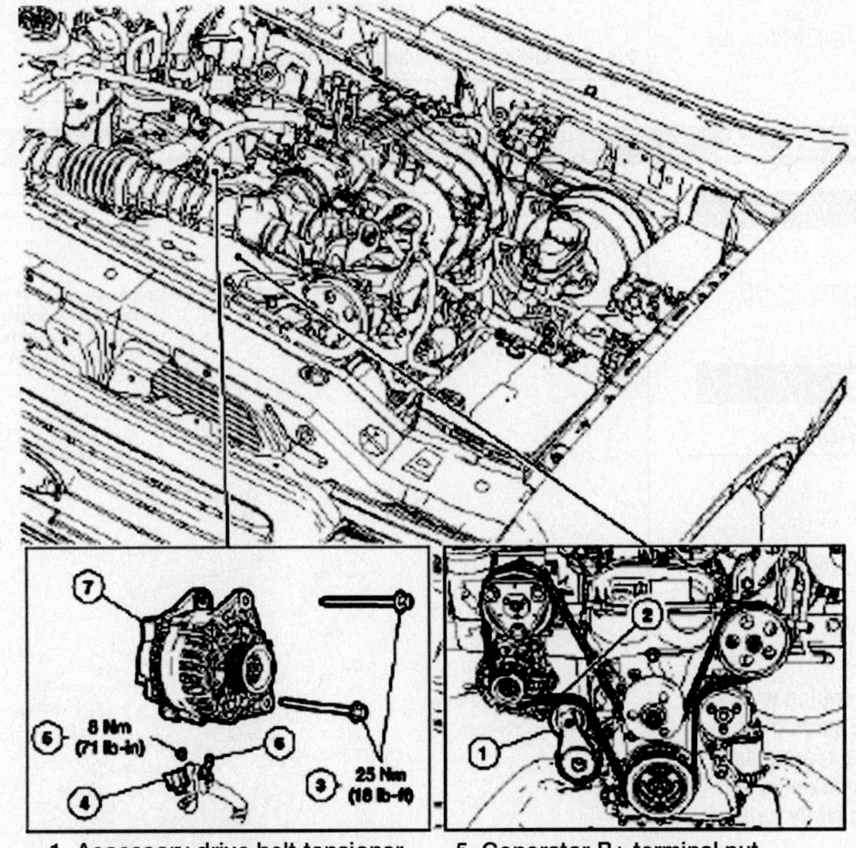

Fig. 27 Alternator bolts and positioning—2.3L engine

3.0L and 4.0L Engines

See Figure 29.

1. See all precautions in this section.
2. Disconnect the negative battery cable.
3. Remove the air cleaner outlet tube.
4. Rotate the front end accessory drive belt tensioner counterclockwise and position the accessory drive belt aside.
5. Disconnect the 2 electrical connectors from the generator.
6. Position the protective cover aside, remove the nut and position the generator B+ terminal aside.
7. Remove the 2 bolts from the generator.
8. Remove the stud bolt, position the generator harness bracket aside and remove the generator.
9. To install, reverse the removal procedure.
10. Tightening the B+ terminal to 71 inch lbs. (8 Nm).
11. Tightening the alternator bolts to 35 ft. lbs. (47 Nm).
12. Tightening the stud bolt to 35 ft. lbs. (47 Nm).

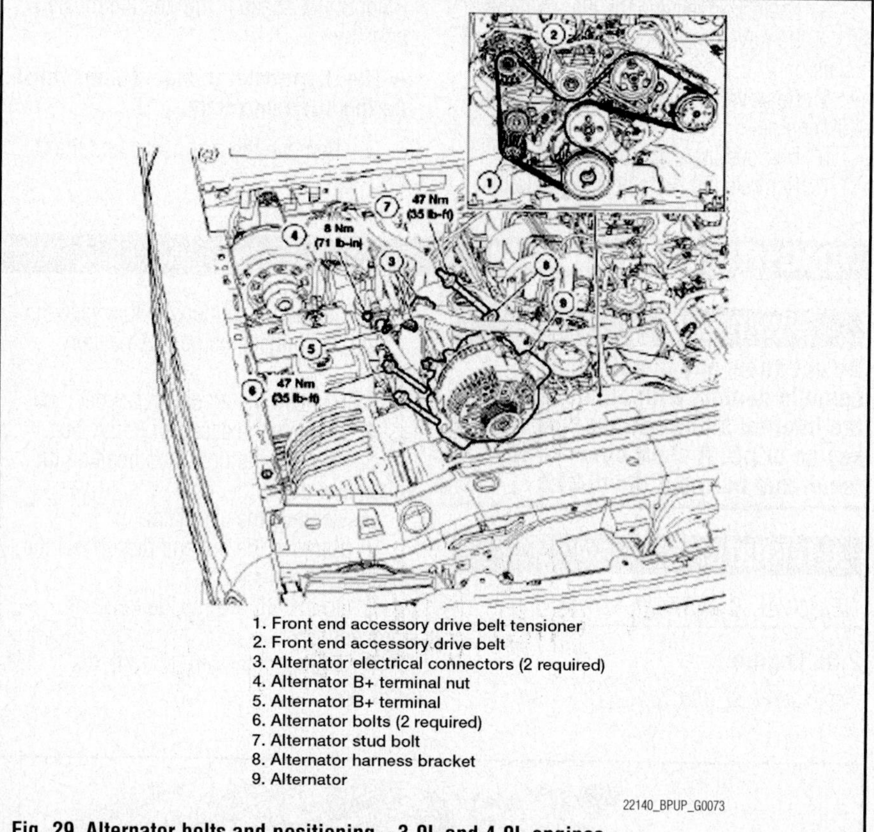

Fig. 29 Alternator bolts and positioning—3.0L and 4.0L engines

ENGINE ELECTRICAL

DISTRIBUTORLESS IGNITION SYSTEM

ADJUSTMENT

The ignition timing is preset to 10 degrees Before Top Dead Center (BTDC) and is not adjustable.

IGNITION COIL PACK

REMOVAL & INSTALLATION

2.3L Engine

See Figure 30.

1. Remove or disconnect the following:
   - Negative battery cable.
   - Ignition wires

➡Do not pull on the spark plug wire as it may separate from the spark plug wire connector inside the spark plug wire boot. Remove the spark plug wires by slightly twisting while pulling upwards.

- Ignition coil electrical connectors
- Remove the bolts and remove the ignition coil.

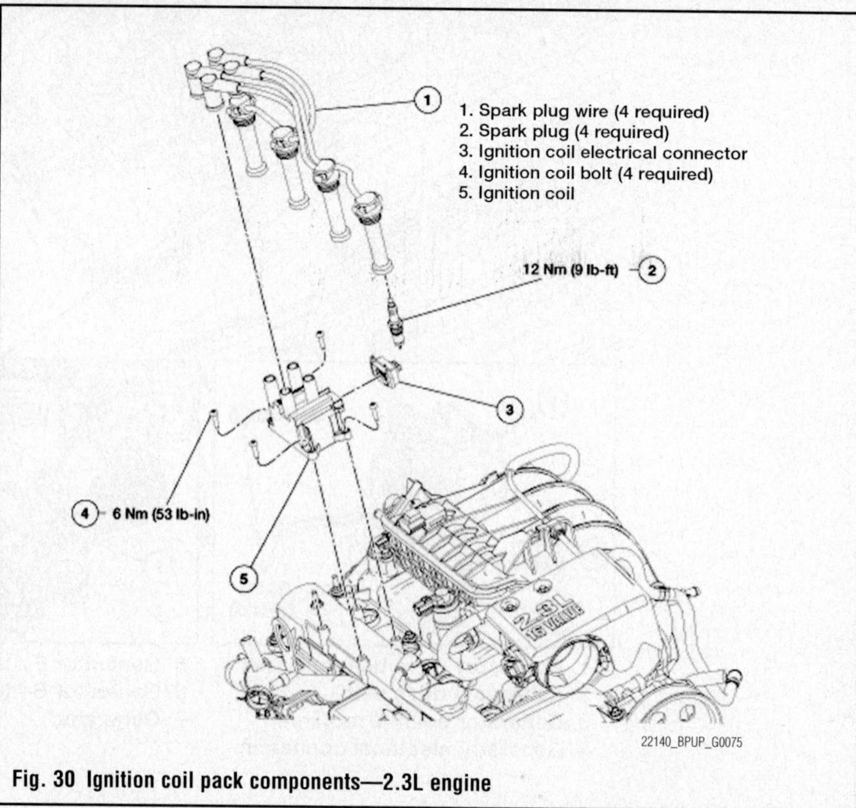

Fig. 30 Ignition coil pack components—2.3L engine

✱✱ CAUTION

Correct installation of the ignition wires is critical to engine operation. If one spark plug wire is not correctly installed at either the spark plugs or the ignition coil, both spark plugs connected to the ignition coil may not fire under load.

2. To install, reverse the removal procedure.
3. Wipe the coil towers with a clean cloth dampened with soap and water. Remove any soap film and dry with compressed air. Inspect for cracks, carbon tracking and dirt. To install, reverse the removal procedure.
4. Apply silicone dielectric compound to the inside of the spark plug wire boots.
5. Torque coil bolts to 53 inch lbs. (6 Nm).

3.0L and 4.0L Engines

See Figure 31.

1. Remove or disconnect the following:
   - Negative battery cable.
   - Ignition wires

➡Do not pull on the spark plug wire as it may separate from the spark plug wire connector inside the spark plug wire boot. Remove the spark plug wires by slightly twisting while pulling upwards.

   - Ignition coil electrical connectors
   - Remove the bolts and remove the ignition coil.

➡Correct installation of the ignition wires is critical to engine operation. If one spark plug wire is not correctly installed at either the spark plugs or the ignition coil, both spark plugs connected to the ignition coil may not fire under load.

2. To install, reverse the removal procedure.
3. Wipe the coil towers with a clean cloth dampened with soap and water. Remove any soap film and dry with compressed air. Inspect for cracks, carbon tracking and dirt. To install, reverse the removal procedure.
4. Apply silicone dielectric compound to the inside of the spark plug wire boots.
5. Torque coil bolts to 53 inch lbs. (6 Nm).

FIRING ORDER

The 2.3L engine firing order is: 1–3–4–2. The 3.0L and 4.0L engine firing order is :1—4—2—5—6.

IGNITION TIMING

The ignition timing is controlled by the Powertrain Control Module and is not adjustable.

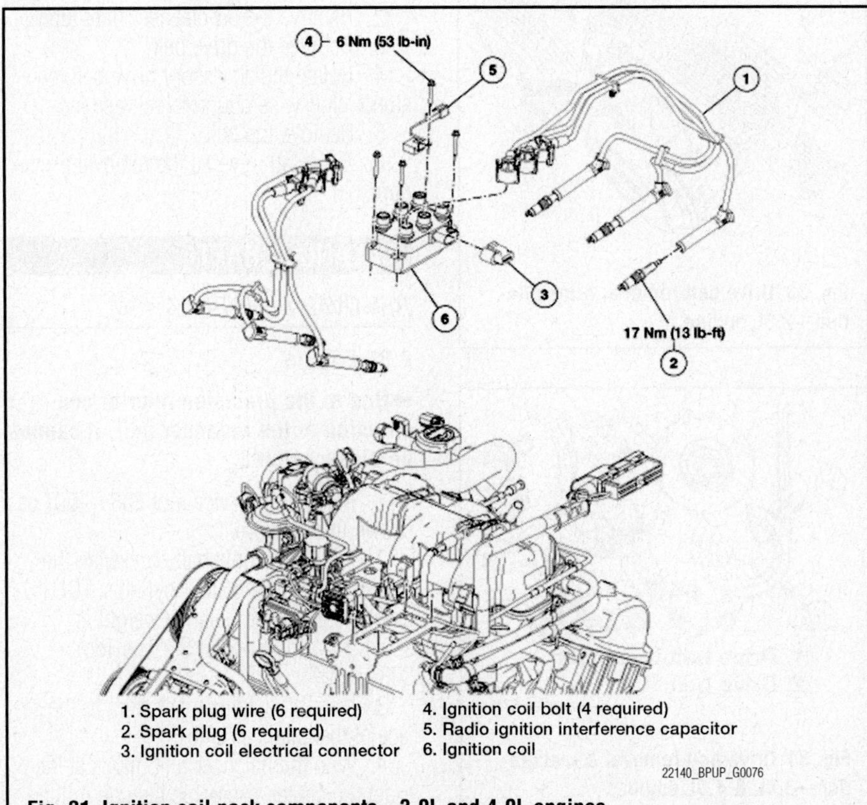

Fig. 31 Ignition coil pack components—3.0L and 4.0L engines

SPARK PLUGS

REMOVAL & INSTALLATION

2.3L Engine

1. Disconnect the negative battery cable.
2. Remove or disconnect the following:
   - 4 ignition coil-on-plug electrical connectors.
   - 4 ignition coil-to-valve cover bolts.
   - 4 ignition coils.

➡Use compressed air to remove any foreign material in the spark plug well before removing the spark plugs.

   - Remove the 4 spark plugs.
3. Inspect the spark plugs. Replace as necessary.
4. Apply a small amount of dielectric grease to the inside of the ignition coil boots before attaching to the spark plugs.
5. To install, reverse removal procedure.
6. Torque spark plugs to 144 inch lbs. (15 Nm).

3.0L and 4.0L Engines

➡All B-Series trucks with 4.0L engine have factory-installed, color-coded spark plugs. These spark plugs are not interchangeable between the right and left bank.

1. Disconnect the negative battery cable.
2. Remove the right front tire and wheel.
3. Remove the right front fender splash shield.
4. Remove or disconnect the following:
   - 6 ignition coil-on-plug electrical connectors
   - 6 ignition coil-to-valve cover bolts
   - 6 ignition coils

➡Use compressed air to remove any foreign material in the spark plug well before removing the spark plugs.

   - Remove the 6 spark plugs.
5. Inspect the spark plugs. Replace as necessary.

➡Platinum-enhanced spark plugs should not require adjustment throughout their service life.

6. Apply a small amount of dielectric grease to the inside of the ignition coil boots before attaching to the spark plugs.
7. To install, reverse removal procedure.
8. Torque spark plugs to 144 inch lbs. (15 Nm).

ENGINE ELECTRICAL

CAUTION

When carrying out maintenance on the starting system be aware that heavy gauge leads are connected directly to the battery. Make sure protective caps are in place when maintenance is completed. Failure to follow these instructions may result in personal injury.

STARTER

REMOVAL & INSTALLATION

2.3L Engine

See Figure 32.

1. Before servicing the vehicle, refer to the Precautions Section.
2. Disconnect the negative battery cable.

➡On some vehicles, when the battery cable is disconnected and reconnected, some abnormal drive symptoms may occur while the vehicle relearns its adaptive strategy. The vehicle may need to be driven to relearn its strategy.

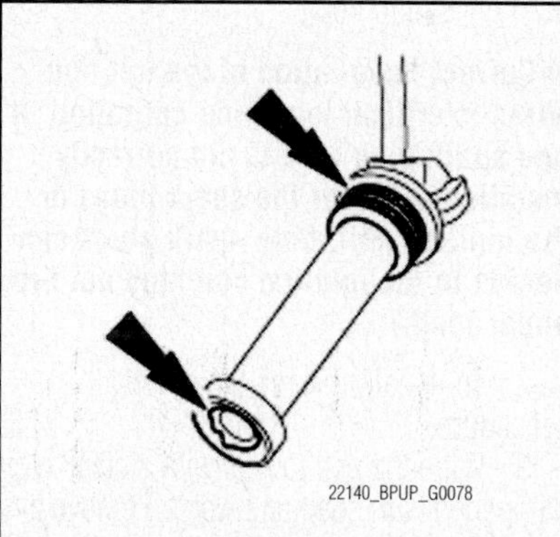

Fig. 32 Exploded view of the starter

3. Raise and support the vehicle safely.
4. Disconnect the electrical connections from the starter.
5. Disconnect the positive battery cable.
6. Remove the starter retaining bolts. Remove the starter from the engine.

To install:

7. Installation is the reverse of the removal procedure.
8. Torque the starter retaining bolts to 9 ft. lbs. (12 Nm).

3.0L and 4.0L Engines

1. Before servicing the vehicle, refer to the precautions in the beginning of this section.
2. Remove or disconnect the following:
   - Negative battery cable
   - Starter harness connectors
   - Starter motor

To install:

3. Install or connect the following:
   - Starter motor. Tighten the bolts to 24–33 ft. lbs. (32–46 Nm).
   - Starter harness connectors. Tighten the battery cable nut to 87–104 inch lbs. (10–12 Nm).
   - Negative battery cable

ENGINE MECHANICAL

ACCESSORY DRIVE BELTS

INSPECTION

Inspect the drive belt for signs of glazing or cracking. A glazed belt will be perfectly smooth from slippage, while a good belt will have a slight texture of fabric visible. Cracks will usually start at the inner edge of the belt and run outward. All worn or damaged drive belts should be replaced immediately.

ADJUSTMENT

Belt tensioners are preloaded and no adjustments can be made.

REMOVAL & INSTALLATION

2.3L Engine

See Figure 33.

1. Disconnect the negative battery cable.
2. Rotate the accessory drive belt tensioner clockwise and remove the accessory drive belt.
3. To install, reverse the removal procedure.

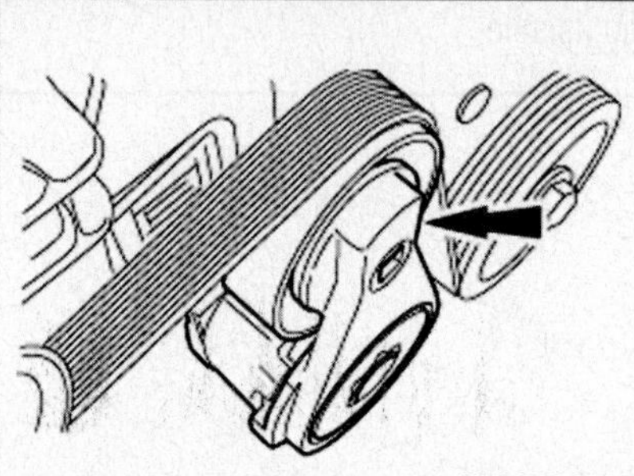

Fig. 33 Drive belt removal & installation—2.3L engine

3.0L and 4.0L Engines

See Figure 34.

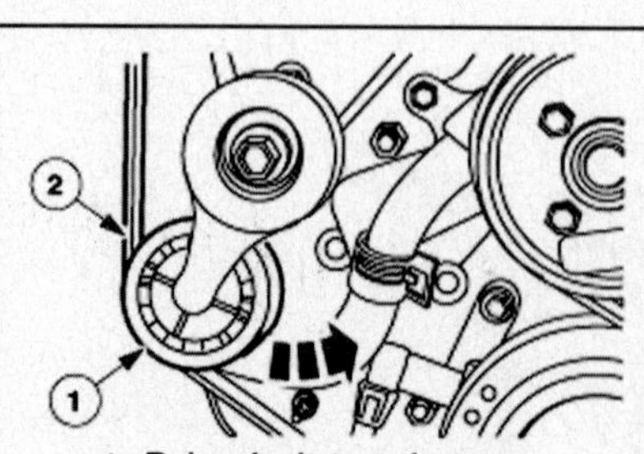

Fig. 34 Drive belt removal & installation—3.0L & 4.0L engines

1. Disconnect the negative battery cable.
2. Remove the air cleaner outlet tube.
3. Remove the drive belt.
4. Rotate the accessory drive belt tensioner clockwise counterclockwise for
5. Remove the belt.
6. To install, reverse the removal procedure.

BALANCE SHAFT

REMOVAL & INSTALLATION

2.3L Engine

➡Due to the precision interior construction of the balancer unit, it cannot be disassembled.

1. Install the service tool 303 —507 as shown in the figure.
2. Turn the crankshaft clockwise the crankshaft is in the No.1 cylinder TDC position (until the balance weight is attached to the Special Service Tool 303 —507.
3. Install the adjustment shim to the seat face of the balancer unit.
4. With the balancer unit marks at the exact top center, assemble the unit to the

cylinder block. Set the SST as shown, then measure the gear backlash using a dial gauge.

➡For an accurate measurement of gear backlash, insert a screwdriver into the crankshaft No. 1 balance weight area and set both the rotation and the thrust direction with the screwdriver, using a prying action,as shown in the figure.

➡If the backlash exceeds the specified range, re-measure the backlash and, using the adjustment shim selection table, select the proper shim, according to the following procedure.

CAUTION

When measuring the backlash, rotate the crankshaft one full rotation and verify that it is within the specified range at all of the following six positions: 10°, 30°, 100°, 190°, 210°, 280° ATDC.

5. Value range is 0.005–0.101 mm {0.00019–0.0039 in}
6. Using master adjustment shim (No.50), assemble the balancer unit to the cylinder block, and then measure the backlash.
7. Select the proper adjustment shim according to the measured value.
8. Install the selected adjustment shim to the balancer unit, then assemble the balancer unit to the cylinder

CAMSHAFT AND VALVE LIFTERS

REMOVAL & INSTALLATION

➡Although Mazda suggests that this component is removable while the engine is installed in the vehicle, depending on the particular options with which your truck is equipped, working clearance may be extremely tight and this procedure may be much easier to perform with the engine removed. Before commencing, read through this procedure and make certain enough clearance, or working room, exists with the engine in the vehicle; if there is not enough space, the engine should be removed.

2.3L Engine

See Figures 35 through 44.

➡On some engines the crankshaft, crankshaft sprocket and the pulley are fitted together by friction, with diamond washers between the flange on each part. For that reason, the crankshaft sprocket is also unfastened if you loosen the pulley. Therefore, the engine must be retimed each time the damper is removed. Otherwise severe damage can occur.

1. Before servicing the vehicle, refer to the Precautions Section.
2. Relieve the fuel system pressure.
3. Disconnect the negative battery cable.

➡On some vehicles, when the battery cable is disconnected and reconnected, some abnormal drive symptoms may occur while the vehicle relearns its adaptive strategy. The vehicle may need to be driven to relearn its strategy.

4. Drain the cooling system.
5. Properly discharge the air conditioning system.
6. Remove or disconnect the following:
   - Drive belt
   - Accessory drive belt tensioner
   - Cooling fan drive pulley
   - Coolant pump pulley
   - Power steering pump
   - Engine oil level indicator assembly
   - Engine oil level indicator.
   - Engine oil level indicator tube
   - Water outlet tube
   - Water outlet tube.
   - Air conditioning compressor.

➡The alternator will be removed with the accessory bracket.

   - Accessory bracket
   - Right motor mount
   - Coolant hose from the thermostat
   - Coolant hose from the EGR valve
   - Coolant tube assembly
   - Exhaust manifold and gasket
   - Block heater (if so equipped)
   - Water outlet
   - EGR valve
   - Power steering pump and reservoir as an assembly
   - Idle Air Control (IAC) valve
   - Throttle Position Sensor (TPS)
   - Manifold Absolute Pressure (MAP) sensor
   - Swirl control valve monitor electrical connector
   - Crankshaft Position (CKP) sensor and the wiring harness pin-type retainers
   - Knock Sensor (KS)
   - Electric thermostat
   - Swirl control valve
   - Camshaft Position (CMP) sensor electrical connector and disconnect the PCV hose from the intake manifold
   - Engine wiring harness pin-type retainers from the intake manifold
   - Engine wiring harness connector bracket. Position the engine wiring harness aside
   - EGR tube
   - Fuel supply line clip from the front of the intake manifold. Disconnect the vacuum hose from the intake manifold
   - Intake manifold assembly
   - Fuel injector electrical connectors. Detach the wiring harness pin-type retainers
   - Ignition coil and the Cylinder Head Temperature (CHT) sensor electrical connectors
   - Engine wiring harness anchors from the valve cover studs. Remove the engine wiring harness
   - Ignition coil
   - Bypass hose
   - Thermostat housing
   - Knock sensor and the engine vent cover
   - Left motor mount
   - Fuel injector supply manifold with the injectors and the ground strap
   - Water pump pulley
   - Water pump
   - CMP sensor
   - CHT sensor
   - Spark plugs
   - Valve cover
   - CKP sensor
   - Crankshaft vibration damper

➡There is one front cover bolt behind the cooling fan drive pulley. To remove this bolt, align one of the cooling fan drive pulley access holes with the bolt head to access the bolt.

   - Front cover
   - Timing chain tensioner
   - Timing chain guides
   - Timing chain assembly

➡Use a wrench on the flats between cylinders No. 1 and No. 2 to hold the camshaft in place.

   - Camshaft drive sprockets
   - Oil pump chain tensioner and guide

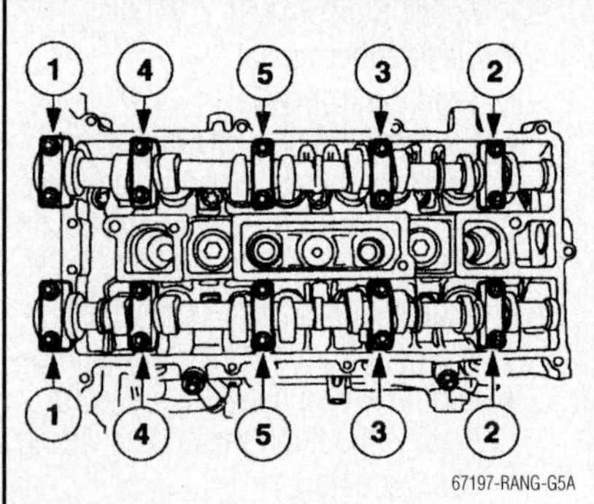

Fig. 35 Camshaft cap loosening sequence—2.3L engine

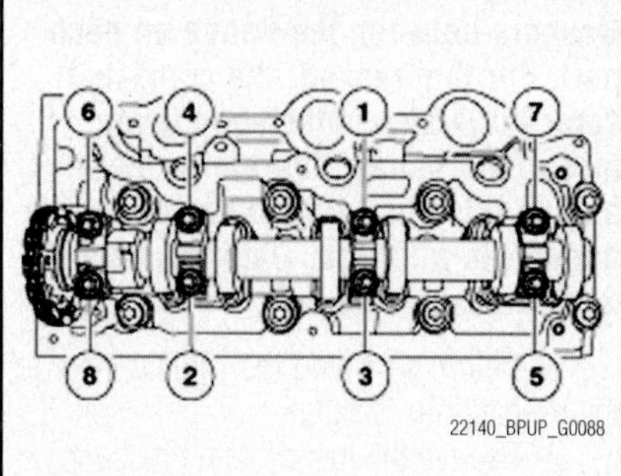

Fig. 37 Position of the camshaft bearing caps tightening sequence

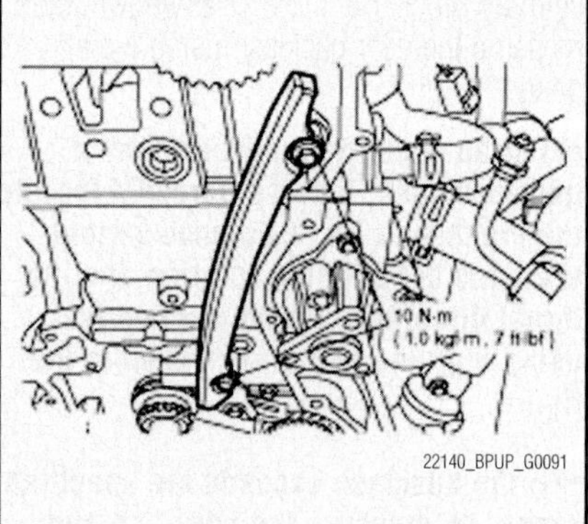

Fig. 39 LH timing chain guide and bolts

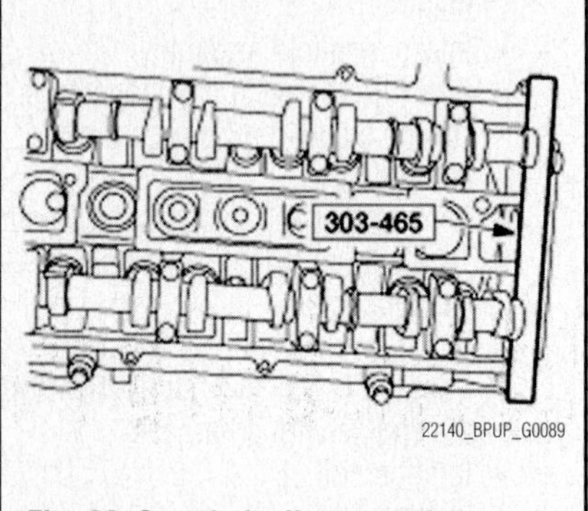

Fig. 36 Camshaft alignment plate 303-465

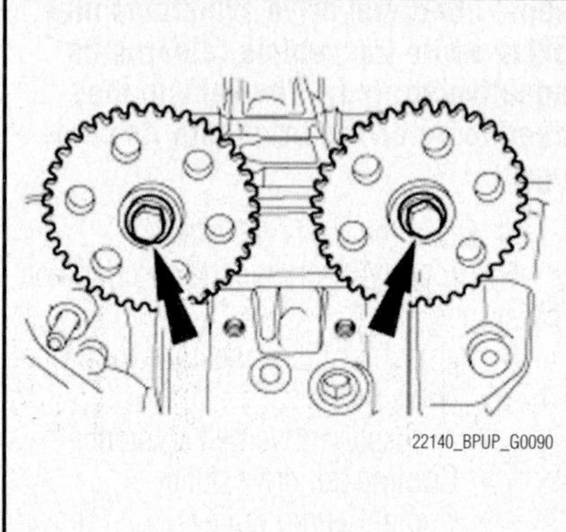

Fig. 38 Camshaft drive gears

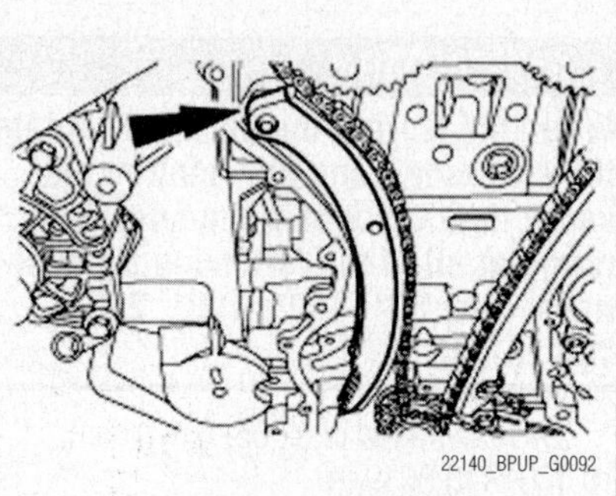

Fig. 40 RH timing chain guide and bolts

➡The oil pump chain sprocket must be held in place

- Oil pump chain and sprockets

➡Note the position of the lobes on the No. 1 cylinder before removing the camshafts for assembly reference.

✲✲ WARNING

Failure to follow the camshaft loosening procedure can result in damage to the camshafts.

7. Loosen the camshaft bearing cap bolts in sequence, one turn at a time. Repeat the first step until all tension is released from the camshaft bearing caps. Remove the camshaft bearing caps.
8. Remove the camshaft alignment plate 303-465
9. Remove the camshafts.

To install:

➡Install the camshafts with the alignment notches in the camshaft lined up so the camshaft alignment plate can be installed without rotating the camshafts. Make sure the lobes on the No. 1 cylinder are in the same position as noted in the disassembly procedure. Rotating the camshafts, or installing the camshafts 180 degrees out of position can cause severe damage to the valves and pistons.

➡Lubricate the camshaft journals and bearing caps with clean engine oil. Install the camshafts and bearing caps.

✲✲ WARNING

Do not rely on the camshaft alignment plate to prevent camshaft rotation. Damage to the tool or the camshaft can occur. If necessary, remove the bolts and the camshaft sprockets. Use the flats on the camshaft to prevent camshaft rotation.

10. Remove the camshaft alignment plate 303-465

➡Tighten the bolts in the sequence shown in three stages.

- Step 1: Tighten the camshaft bearing caps one turn at a time until tight.
- Step 2: Tighten the bolts to 7 Nm (62 inch lbs.)
- Step 3: Tighten the bolts to 16 Nm (12 ft. lbs.)

11. Install the camshaft drive gears and hand-tighten the bolts.

➡Do not tighten the bolts at this time.

12. Install the LH timing chain guide and bolts.
13. Tighten the LH timing chain guide bolts to 89 inch lbs. (10 Nm)
14. Install the timing chain.
15. Install the RH timing chain guide.
16. Tighten the RH timing chain guide bolts to 89 inch lbs. (10 Nm)
17. Install the timing chain tensioner and the bolts.
18. Tighten the timing chain tensioner bolts to 89 inch lbs. (10 Nm).
19. Remove the paper clip to release the piston.
20. Install the camshaft alignment plate.

✲✲ WARNING

Do not rely on the camshaft alignment plate to prevent camshaft rotation. Damage to the tool or the camshafts can result. Using the flats on the camshafts to prevent camshaft rotation.

21. Tighten the camshaft sprockets bolts to 53 ft. lbs. (72 Nm)
22. Install the front cover
23. Tighten the bolts in the sequence shown, to the following specifications:

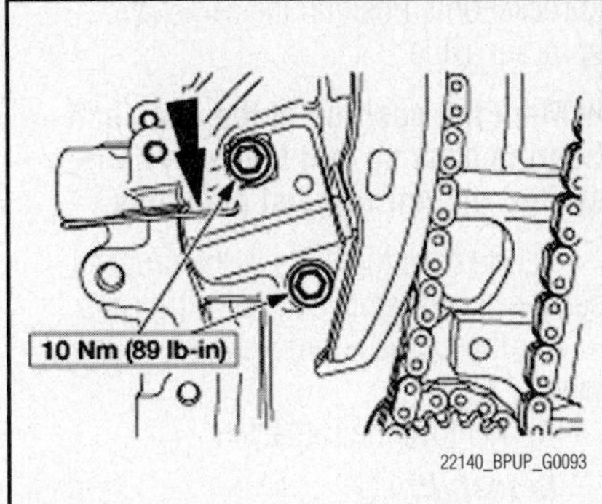

Fig. 41 Timing chain tensioner and the bolts—2.3L engine

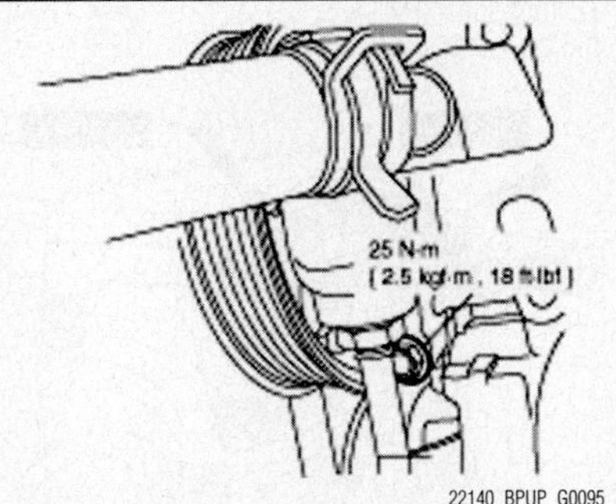

Fig. 43 Power steering pump and install the lower retaining bolt—2.3L engine

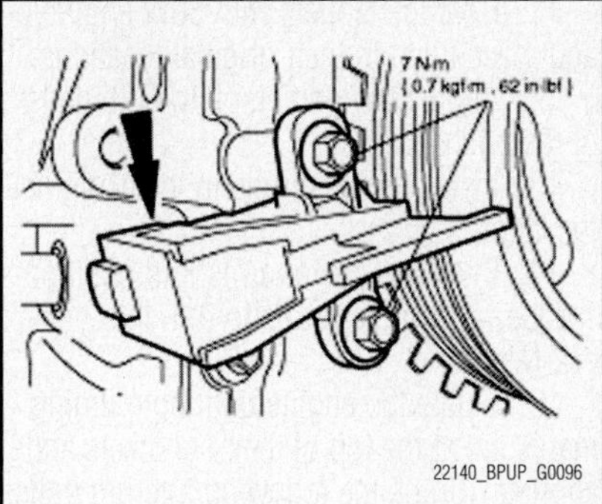

Fig. 44 CKP sensor with the alignment tool—2.3L engine

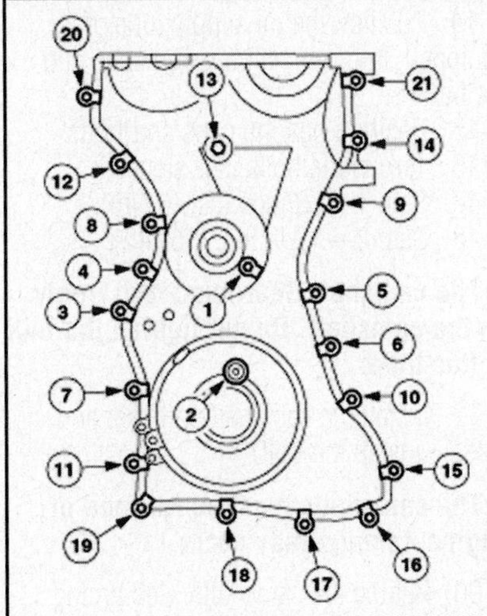

Fig. 42 Front cover tightening sequence—2.3L engine

- 8 mm bolts —7 ft. lbs. (10 Nm).
- 10 mm bolts—18 ft. lbs. (25 Nm).
- 13 mm bolts—35 ft. lbs. (48 Nm).

✱✱ WARNING

There is one bolt behind the cooling fan drive pulley. This bolt can be accessed by lining up one of the holes in the pulley with the bolt.

24. Position the power steering pump and install the lower retaining bolt.
25. Tighten the power steering pump bolts to 18 ft. lbs. (25 Nm).
26. Connect the PSP switch electrical connector.
27. Install the water pump pulley.
28. Tighten the water pump pulley bolts to 15 ft. lbs. (20 Nm).
29. Install the belt tensioner.
30. Tighten the belt tensioner bolts to 36 ft. lbs. (50 Nm).

✱✱ CAUTION

The crankshaft, the crankshaft sprocket and the pulley are fitted together by friction, with diamond washers between the flange faces on each part. For that reason, the crankshaft sprocket is also unfastened if you loosen the pulley. Therefore, the engine must be retimed each time the damper is removed. Otherwise severe engine damage can occur.

31. Install the crankshaft vibration damper:

➡Do not reuse the crankshaft pulley bolt. Tighten the bolt in two stages.

32. Tighten the water pump pulley bolts to 15 ft. lbs. (20 Nm).

✱✱ CAUTION

A new CKP sensor must be installed whenever the old sensor is removed. Install a new CKP sensor, do not tighten the bolts at this time

33. Tighten the CKP sensor bolts to 82 inch. lbs. (7 Nm).
34. Connect the CKP sensor electrical connector and the wiring harness pin-type retainers.
35. Connect the battery ground cable.

3.0L Engine

1. Before servicing the vehicle, refer to the Precautions Section.
2. Properly relieve the fuel system pressure.
3. Disconnect the negative battery cable.

➡On some vehicles, when the battery cable is disconnected and reconnected, some abnormal drive symptoms may occur while the vehicle relearns its adaptive strategy. The vehicle may need to be driven to relearn its strategy.

4. Drain the engine oil.
5. Remove or disconnect the following:
   - Spark plug wires from the plugs
   - Rocker arm covers
   - Intake manifold
   - Remove the lower intake manifold. See removal and installation in the intake manifold section
   - Remove the rocker arms and the push rods.
   - Remove the tappet guide plate retainer.
6. Rotate the crankshaft so that No. 1 piston is at Top Dead Center (TDC) on the compression stroke.
   - Loosen the rocker arm bolts enough to pivot the rocker arms out of the way and remove the pushrods. Identify them for installation
   - Camshaft gear attaching bolt and washer, then slide the gear off the camshaft
7. Carefully slide the camshaft out of the engine block, using caution to avoid any damage to the camshaft bearings.

To install:

8. Oil the camshaft journals and cam lobes with heavy SJ engine oil (50W). Install the spacer ring with the chamfered side toward the camshaft, then insert the camshaft key.
9. Install or connect the following:
   - Camshaft using caution to avoid any damage to the camshaft bearings
   - Thrust plate. Torque the screws to 84 inch lbs. (10 Nm).
10. Rotate the camshaft and crankshaft as necessary to align the timing marks. Install the camshaft gear and chain. Torque the bolt to 46 ft. lbs. (62 Nm).

11. Coat the tappets with 50W engine oil and place them in their original locations.
12. Apply 50W engine oil to both ends of the pushrods.
13. Install the pushrods in their original locations.
14. Pivot the rocker arms into position. Torque the fulcrum bolts to 9 ft. lbs. (12 Nm).
15. Rotate the engine until both timing marks are at the top of their sprockets and aligned. Torque the following fulcrum bolts to 18 ft. lbs. (24 Nm):
16. No.1 intake
17. No.2 exhaust
18. No.4 intake
19. No.5 exhaust
20. Rotate the engine until the camshaft timing mark is at the bottom of the sprocket and the crankshaft timing mark is at the top of the sprocket, and both are aligned. Torque the following fulcrum bolts to 18 ft. lbs. (24 Nm):
21. No.1 exhaust
22. No.2 intake
23. No.3 intake and exhaust
24. No.4 exhaust
25. No.5 intake
26. No.6 intake and exhaust
27. Torque all the bolts to 24 ft. lbs. (33 Nm).
28. Turn the engine by hand to 0 degrees Before Top Dead center (BTDC) of the power stroke on No. 1 cylinder.
29. Install or connect the following:
- Intake manifold
- Rocker arm covers
- Spark plug wires
- Fuel lines to the fuel supply manifold
- Negative battery cable

30. Replace the oil filter and refill the engine with the specified amount of engine oil.
31. Start the engine and check the ignition timing and idle speed. Adjust if necessary. Run the engine at fast idle and check for coolant, fuel, vacuum or oil leaks.

4.0L Engine—Right Side

See Figures 45 through 48.

1. Before servicing the vehicle, refer to the Precautions Section.
2. Properly relieve the fuel system pressure.
3. Disconnect the negative battery cable.

➡On some vehicles, when the battery cable is disconnected and reconnected, some abnormal drive symptoms may occur while the vehicle relearns its adaptive strategy. The vehicle may need to be driven to relearn its strategy.

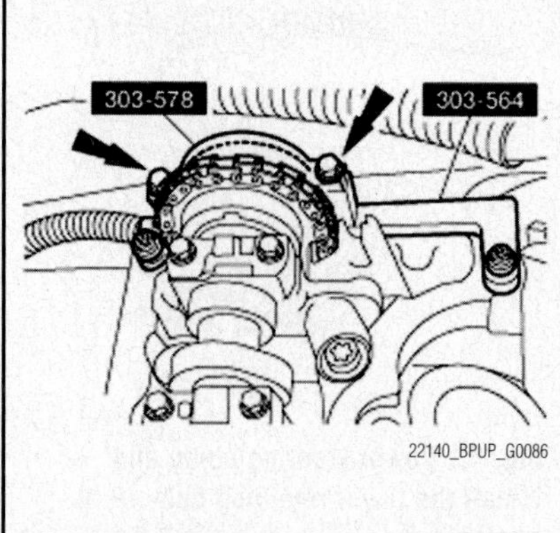

Fig. 45 RH hydraulic chain tensioner.

4. Remove the valve covers. Remove the fuel supply manifold.

➡Mark each camshaft roller follower to ensure its original position during reassembly.

5. Using tool 303-581, remove the roller followers.
6. Remove the RH hydraulic chain tensioner.
7. Install the camshaft sprocket holding tool and the adaptor 303-575 on the rear of the RH cylinder head.

➡The camshaft sprocket is a left hand threaded bolt pattern.

8. Loosen and remove the camshaft sprocket bolt. Position the camshaft sprocket aside.

➡Mark the position of the camshaft bearing caps so that they can be reinstalled in their original positions.

9. Remove the bolts in the proper sequence and remove the bearing caps.
10. If equipped, remove the oil supply tube.
11. Remove the camshaft.

To install:

12. Lubricate all of the moving parts with SAE 50W engine oil.
13. Install camshaft onto the cylinder head.
14. Position the oil supply tube, if equipped, the camshaft bearing caps and the bolts.
15. Position bearing caps and bolts.
16. Torque the bolts in 2 steps:
17. Step 1—53 inch lbs. (6 Nm).
18. Step 2—12 ft. lbs. (16 Nm).

➡The camshaft Gear must turn freely on the camshaft. Do not tighten the bolt at this time.

19. Install the camshaft sprocket and loosely install the bolt.

➡The camshafts must be retimed or engine damage may occur.

20. Retime the camshafts. See timing chain installation section.
21. Continue the installation in the reverse order of the removal procedure.

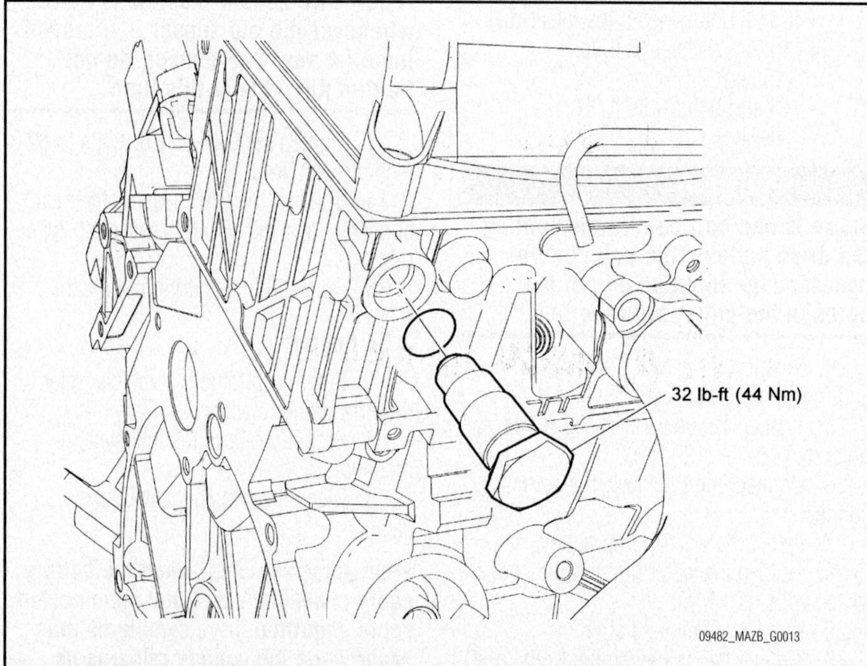

Fig. 46 Right side hydraulic chain tensioner location—4.0L engine

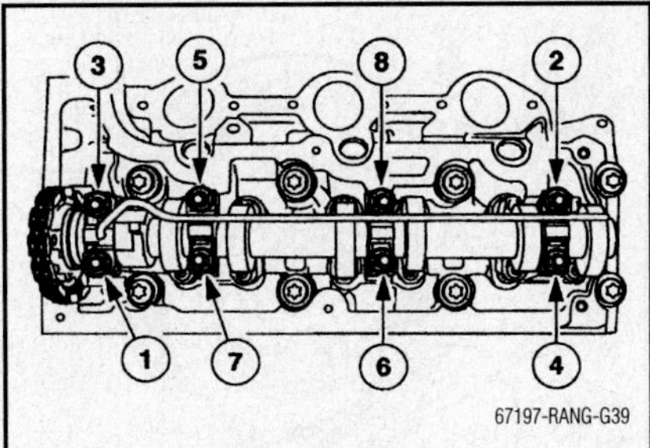

Fig. 47 Camshaft bolt removal sequence—4.0L engine

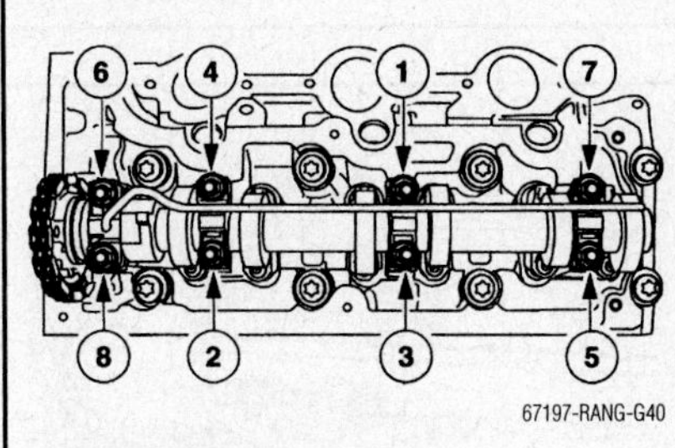

Fig. 48 Camshaft bolt torque sequence—4.0L engine

4.0L Engine–Left Side

See Figures 49 through 51.

1. Before servicing the vehicle, refer to the Precautions Section.
2. Properly relieve the fuel system pressure.
3. Disconnect the negative battery cable.

➡On some vehicles, when the battery cable is disconnected and reconnected, some abnormal drive symptoms may occur while the vehicle relearns its adaptive strategy. The vehicle may need to be driven to relearn its strategy.

4. Remove the valve covers. Remove the fuel supply manifold.

➡Mark each camshaft roller follower to ensure its original position during reassembly.

5. Using tool 303-581, remove the roller followers.
6. Remove the intake manifold. Remove the thermostat housing.
7. Remove the hydraulic chain tensioner.
8. Install camshaft sprocket holding tool 303-564 and 303-578 on the front of the camshaft.
9. Loosen and remove the camshaft sprocket bolt.
10. Position the camshaft sprocket aside.

➡Mark the position of the camshaft bearing caps so that they can be reinstalled in their original positions.

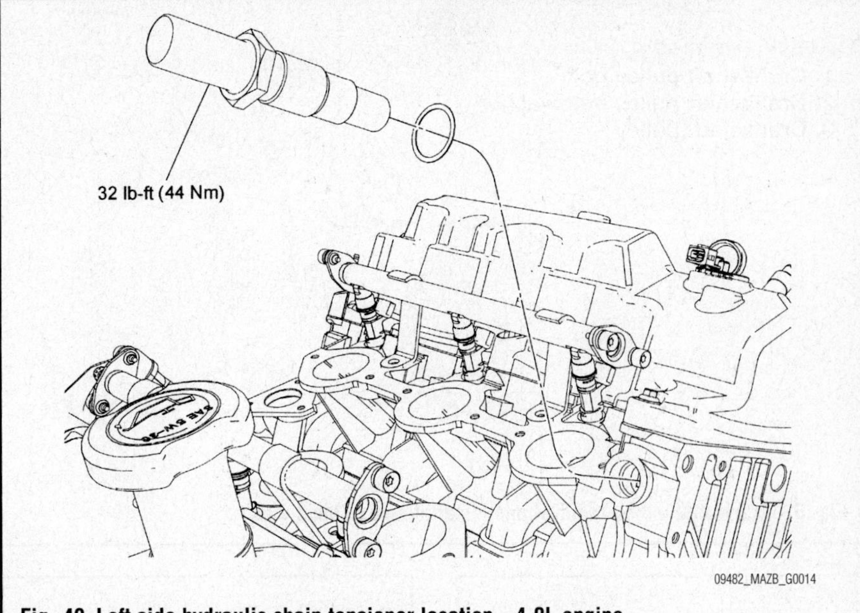

Fig. 49 Left side hydraulic chain tensioner location—4.0L engine

22140_BPUP_G0087

Fig. 50 Position of the camshaft bearing caps

11. Remove the bolts in the proper sequence and remove the bearing caps.
12. If equipped, remove the oil supply tube.
13. Remove the camshaft.

To install:

14. Lubricate all of the moving parts with SAE 50W engine oil.
15. Install camshaft onto the cylinder head.
16. Position the oil rail, if equipped and install the bearing caps and bolts. Torque the bolts in 2 steps:
17. Step 1—53.5 inch lbs. (6 Nm).
18. Step 2—12 ft. lbs. (16 Nm).

➡The camshaft Gear must turn freely on the camshaft. Do not tighten the bolt at this time.

19. Install the camshaft sprocket and loosely install the bolt.

➡The camshafts must be retimed or engine damage may occur.

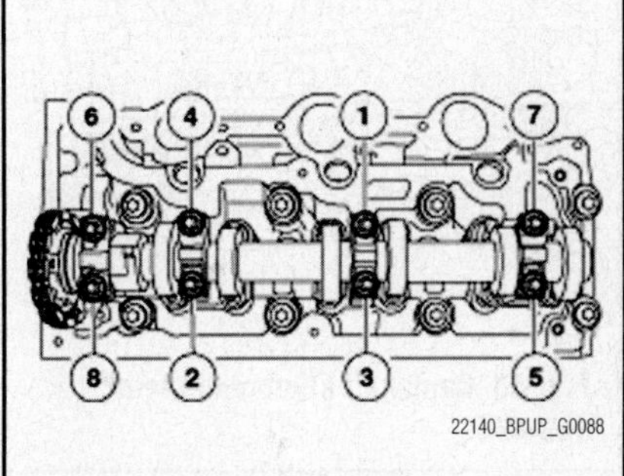

Fig. 51 Position of the camshaft bearing caps tightening sequence

20. Retime the camshafts. See timing chain installation section.
21. Continue the installation in the reverse order of the removal procedure.

CRANKSHAFT DAMPER

REMOVAL & INSTALLATION

2.3L Engine

See Figures 52 through 58.

CAUTION

The crankshaft, the crankshaft sprocket and the pulley are fitted together by friction, with diamond washers between the flange faces on each part. For that reason, the crankshaft sprocket is also unfastened if you loosen the pulley. Therefore, the engine must be retimed each time the damper is removed. Otherwise severe damage can occur.

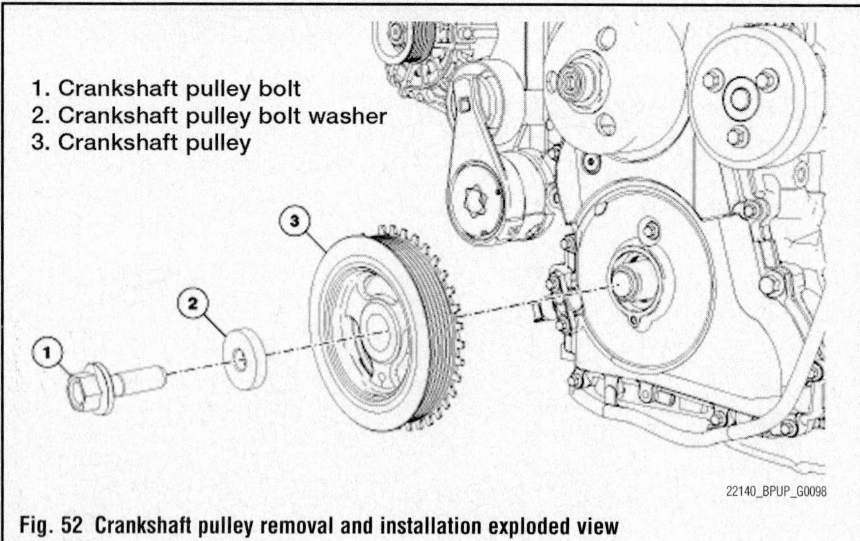

Fig. 52 Crankshaft pulley removal and installation exploded view

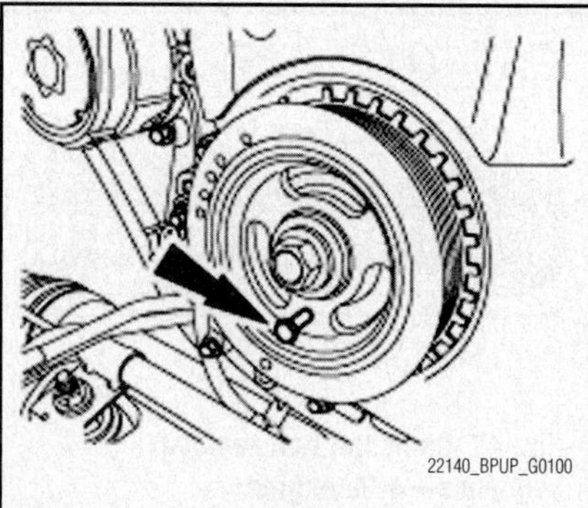

Fig. 56 M6 X 1.0 X 2.25 bolt (Mazda part No. 1F20-12-428)

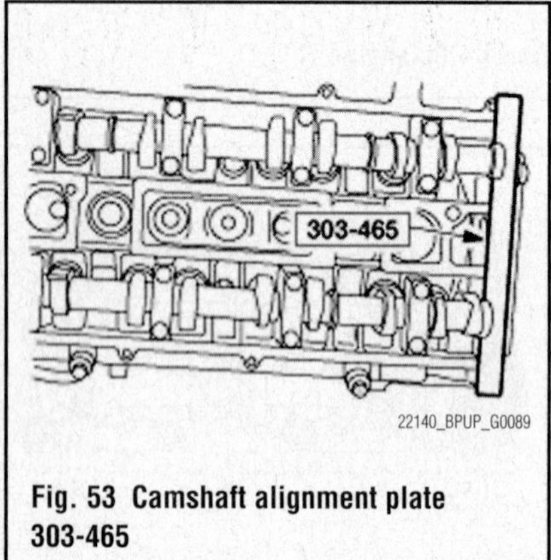

Fig. 53 Camshaft alignment plate 303-465

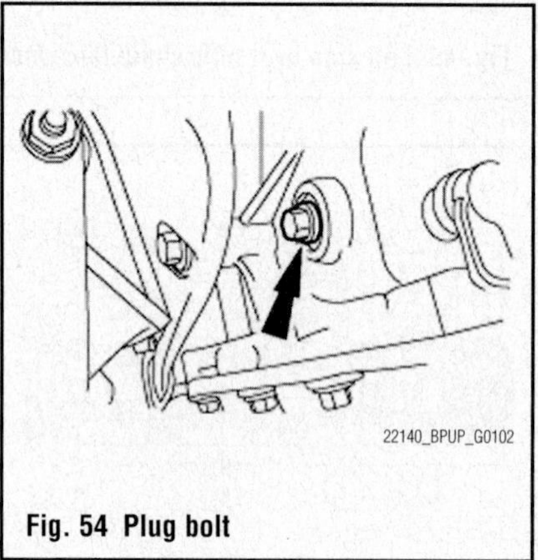

Fig. 54 Plug bolt

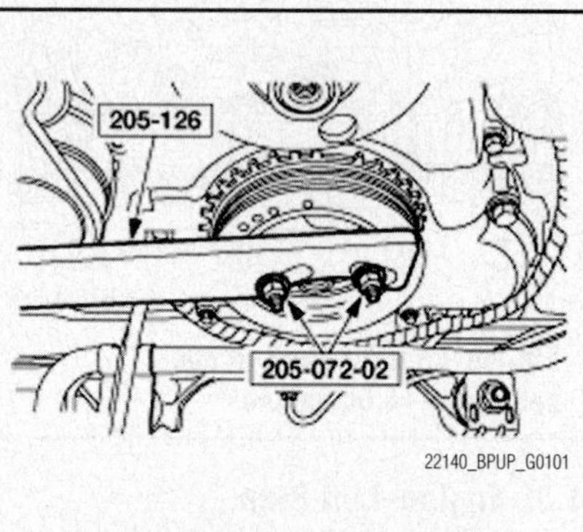

Fig. 57 Crank pulley holding fixture 205—072—02—and 205—126

1. Before servicing the vehicle, refer to the Precautions Section.
2. Disconnect the negative battery cable.

➡On some vehicles, when the battery cable is disconnected and reconnected, some abnormal drive symptoms may occur while the vehicle relearns its adaptive strategy. The vehicle may need to be driven to relearn its strategy.

3. Install the camshaft alignment plate 303-465 in the slots on the rear of both camshafts.
4. Remove the plug bolt.

➡Only turn the engine in the normal direction of rotation. Installing the special tool in this step will prevent the engine from being rotated in the clockwise direction. However, the engine can still be rotated in the counterclockwise direction.

5. Install the crankshaft TDC timing peg 303—507—2.3L engine.

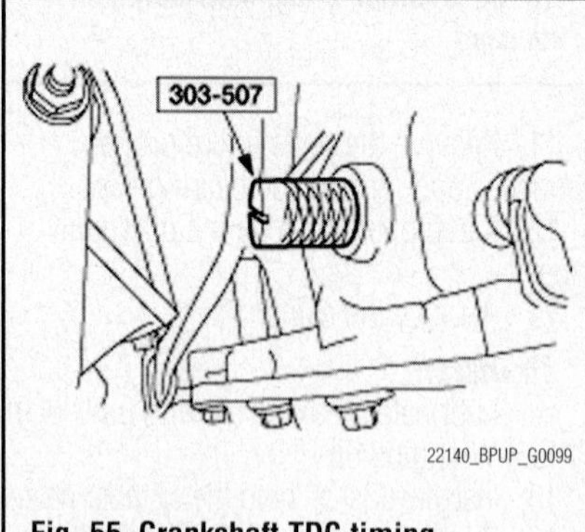

Fig. 55 Crankshaft TDC timing peg303—507

6. Install a M6 X 1.0 X 2.25 bolt (Mazda part No. 1F20-12-428) to verify ignition timing.
7. Install the holding fixture 205—072—02 and 205—126.

WARNING

Failure to hold the crankshaft pulley in place during bolt loosening can cause damage to the engine.

Remove the bolt and the crankshaft pulley.

8. Remove the crankshaft bolt.
9. Remove the holding fixture 205—072—02 and 205—126.
10. Remove the M6 bolt.
11. Remove the crankshaft pulley.

To install:

➡Do not reuse the crankshaft damper bolt or washer.

➡Apply clean engine oil on seal area before installing.

12. Install the new crankshaft pulley and hand-tighten the bolt.

WARNING

Only hand-tighten the bolt or damage to the front cover can occur.

➡This step will correctly align the crankshaft pulley to the crankshaft.

13. Install a standard 0.23 inch (6mm) x 0.7 in (18mm) bolt through the crankshaft pulley and thread it into the front cover.
14. Rotate the crankshaft pulley as necessary to align the bolt holes.

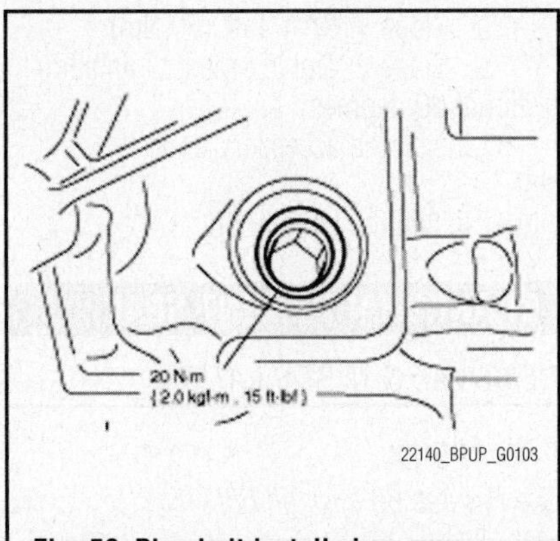

Fig. 58 Plug bolt install view

✲✲ WARNING

Failure to hold the crankshaft pulley in place during bolt tightening can cause damage to the engine front cover. Using the holding fixture 205—072—02 and 205—126 hold the crankshaft pulley in place and tighten the crankshaft pulley bolt in two stages.

15. Step 1: Tighten the crankshaft pulley bolt to 74 ft. lbs. (100 Nm).
16. Step 2: Tighten the crankshaft pulley bolt an additional 90 °.
17. Remove the holding fixture205—072—02 and 205—126
18. Remove the 0.23 inch (6mm) x 0.7 in (18mm) bolt.
19. Remove the crankshaft TDC timing peg.

✲✲ WARNING

Only hand-tighten the bolt or damage to the front cover can occur. Using 0.23 inch (6mm) x 0.7 in (18mm) bolt, check the position of the crankshaft pulley.

20. If it is not possible to install the 0.23 inch (6mm) x 0.7 in (18mm) bolt, correct the engine timing.
21. Using the camshaft alignment plate 303—465, check the position of the camshafts.
22. If it is not possible to install the camshaft alignment plate 303—465, correct the engine timing.
23. Remove the camshaft alignment plate.
24. Remove the 6 mm 0.23 inch (6mm) x 0.7 in (18mm) bolt.
25. Remove the crankshaft TDC timing peg.
26. Install the plug bolt.
27. Tighten the Plug Bolt to 15 ft. lbs. (20 Nm).
28. Install the valve cover.
29. Install the accessory drive belt.
30. Install the fan and shroud.

3.0L Engine

See Figures 59 through 61.

1. Before servicing the vehicle, refer to the Precautions Section.
2. Disconnect the negative battery cable.

➡On some vehicles, when the battery cable is disconnected and reconnected, some abnormal drive symptoms may occur while the vehicle relearns its adaptive strategy. The vehicle may need to be driven to relearn its strategy.

3. Remove the fan shroud. See removal and installation of the cooling fan.
4. Remove the accessory drive belt. See removal and installation of drive belts.
5. Remove the air cleaner outlet tube.
6. Remove the crankshaft pulley.
7. Remove the vibration damper bolt and the washer.
8. Using Crankshaft Damper Remover 303—009, remove the crankshaft damper.
9. Remove the crankshaft key, if required.

To install:

10. Install the crankshaft front oil seal.
11. Lubricate the seal lip with clean engine oil.
12. Using the vibration damper and seal replacer 303—175, install the front oil seal.
13. Using the vibration damper and seal replacer 303—175or equivalent, install the crankshaft damper.
14. Tighten to 107 ft. lbs. (145 Nm).
15. Coat the sealing surfaces of the crankshaft damper with clean engine oil.
16. Coat the crankshaft damper keyway with silicone.

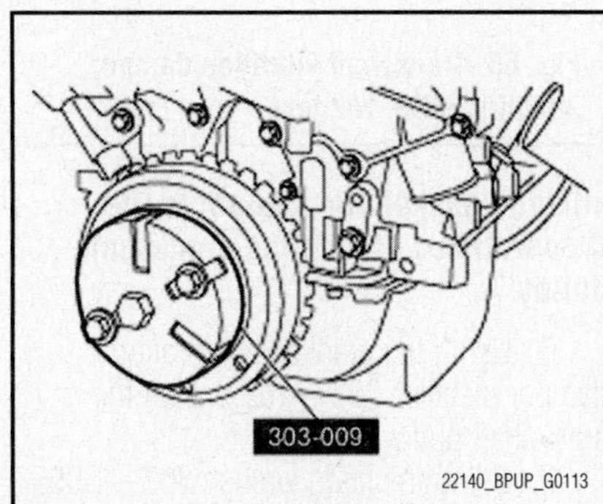
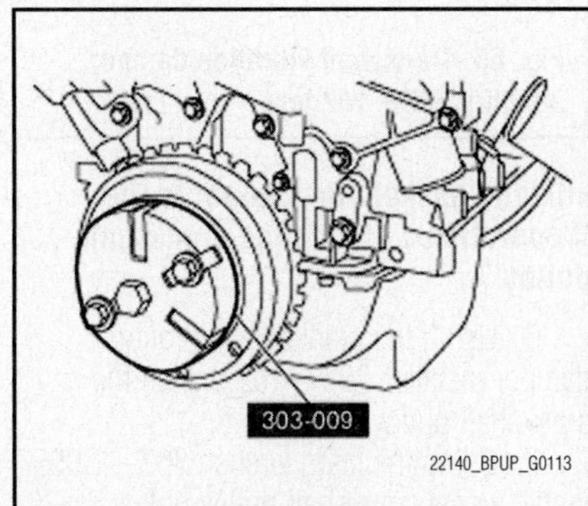

Fig. 59 Crankshaft Damper Remover 303—009—3.0L engine

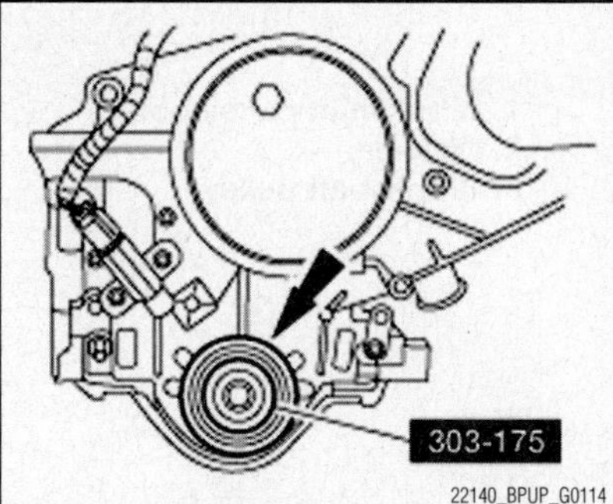

Fig. 60 Crankshaft vibration damper and seal replacer 303—175—3.0L engine

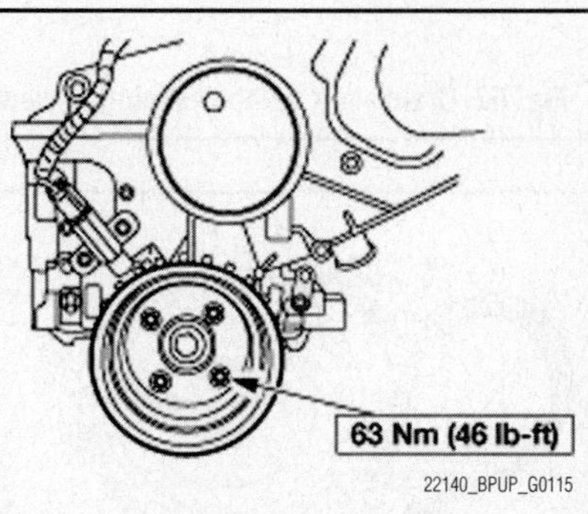

Fig. 61 Installation of the crankshaft pulley—3.0L engine

17. Install the crankshaft damper.
18. Install the crankshaft pulley.
19. Tighten the crankshaft pulley bolts to 46 ft. lbs. (63 Nm).
20. Install the drive belt.
21. Install the cooling fan blade and fan.
22. Install the air cleaner outlet tube.
23. Connect the negative battery cable.

4.0L Engine

See Figures 62 through 65.

1. Before servicing the vehicle, refer to the Precautions Section.
2. Disconnect the negative battery cable.

➡On some vehicles, when the battery cable is disconnected and reconnected, some abnormal drive symptoms may occur while the vehicle relearns its adaptive strategy. The vehicle may need to be driven to relearn its strategy.

3. Remove the fan shroud. See removal and installation of the cooling fan.
4. Remove the accessory drive belt. See removal and installation of drive belts.

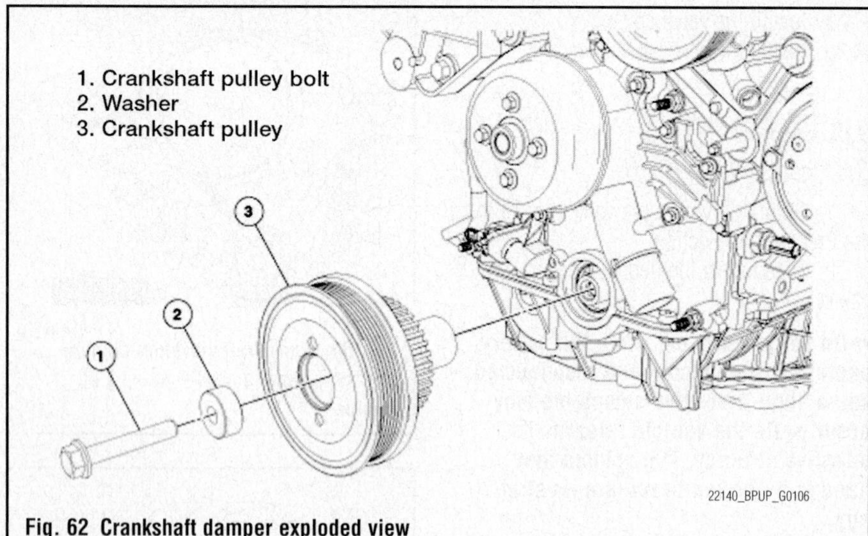

Fig. 62 Crankshaft damper exploded view

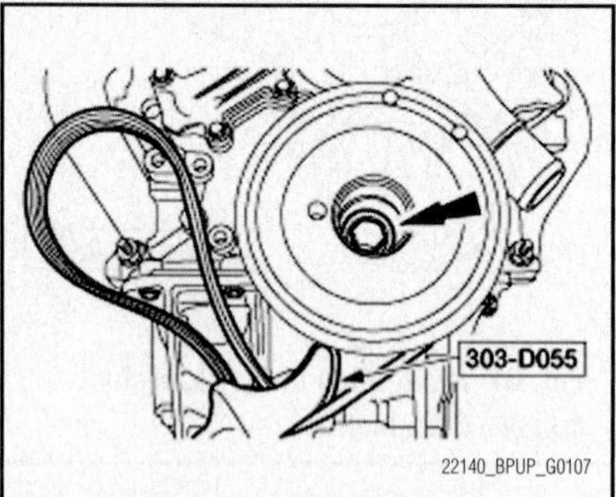

Fig. 63 Strap wrench 303—D055, and crankshaft pulley bolt to be removed and installed.

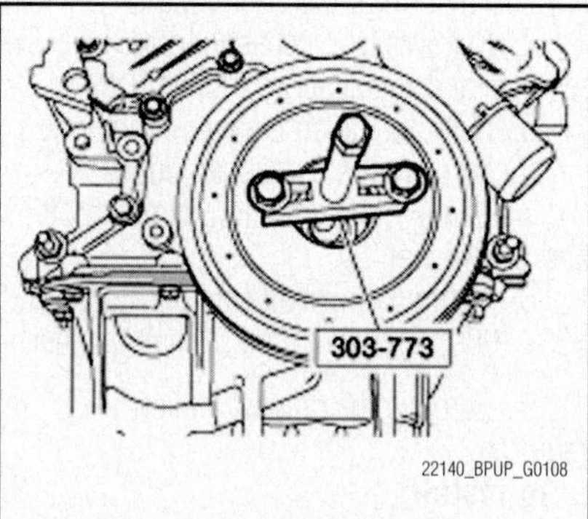

Fig. 64 Crankshaft vibration damper remover 303—773 tool

⁂ CAUTION

This bolt is a torque-to-yield design and cannot be reused.

5. Using the strap wrench 303—D055, remove the crankshaft pulley bolt.

➡If the crankshaft pulley bolt is not installed, the crankshaft vibration damper remover 303—773 will bottom out before the pulley is off. Install the crankshaft pulley bolt 2-3 turns. Using the crankshaft vibration damper remover 303—773 and the 8 mm x 1.25 x 100 mm bolts, remove the pulley.

➡If not secured within four minutes, the sealant must be removed and the sealing area cleaned. To clean the sealing area, use silicone gasket remover and metal surface prep. Follow the directions on the packaging. Failure to follow this procedure can cause future oil leakage. Apply silicone gasket and sealant to the Woodruff key slot on the crankshaft pulley.

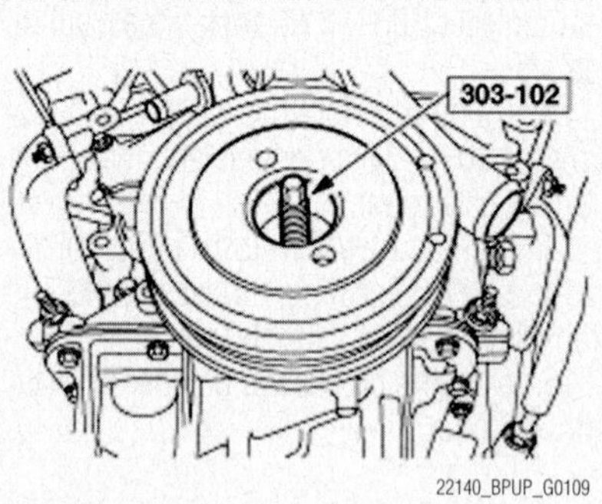

Fig. 65 Crankshaft vibration damper installer 303—102 tool

6. Using the crankshaft vibration damper installer 303—102, install the crankshaft pulley.
7. Using the strap wrench 303—D055, install a new crankshaft pulley bolt.
8. Tighten the bolt in two stages:
   a. Stage 1: 37 ft. lbs. (50 Nm).
   b. Stage 2: Tighten the bolt an additional 90 degrees.
9. Install the accessory drive belt.
10. Install the fan shroud.

CRANKSHAFT FRONT SEAL

REMOVAL & INSTALLATION

2.3L Engine

See Figures 66 through 68.

⁂ WARNING

The crankshaft, the crankshaft sprocket and the pulley are fitted together by friction, with diamond washers between the flange faces on each part. For that reason, the crankshaft sprocket is also unfastened if you loosen the pulley. Therefore, the engine must be retimed each time the damper is removed. Otherwise severe damage can occur. Remove the crankshaft pulley.

⁂ CAUTION

Use care not to damage the engine front cover or the crankshaft when removing the seal.

1. Remove the crankshaft damper. See removal and installation of crankshaft damper in this section.
2. Remove the crankshaft front oil seal using 303—409 seal remover or similar seal removal tool.
3. Lubricate the oil seal with clean engine oil.
4. Using the crankshaft front oil seal installer 303—096, install the crankshaft front oil seal.
5. Install the crank shaft pulley.

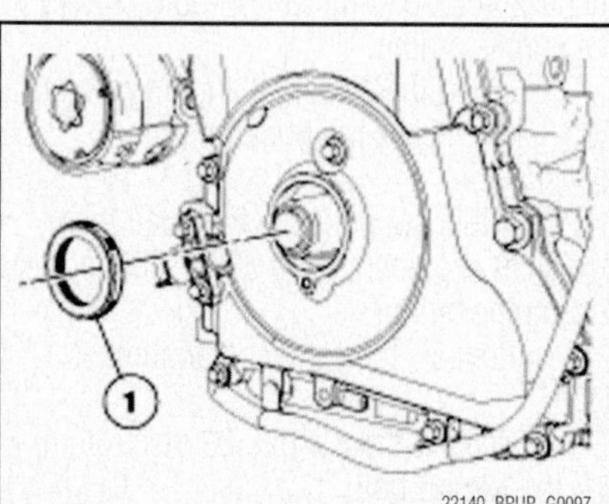

Fig. 66 Crankshaft front seal removal / installation exploded view—2.3L engine

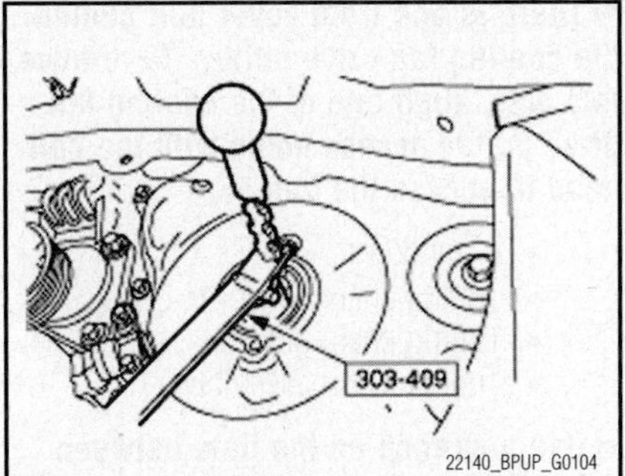

Fig. 67 Removal of the crankshaft front oil seal with tool 303—409 for 2.3L engine

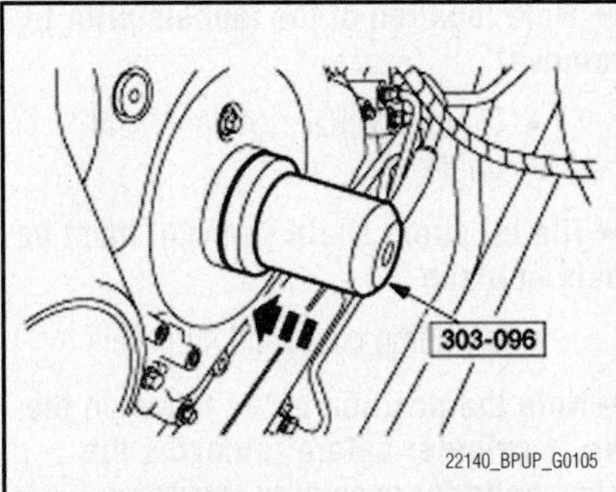

Fig. 68 Installation of the crankshaft front oil seal with tool 303—096 for 2.3L engine

3.0L Engine

See Figures 69 through 71.

1. Before servicing the vehicle, refer to the Precautions Section.
2. Disconnect the negative battery cable.

➡On some vehicles, when the battery cable is disconnected and reconnected, some abnormal drive symptoms may occur while the vehicle relearns its adaptive strategy. The vehicle may need to be driven to relearn its strategy.

3. Remove the fan shroud. See removal and installation of the cooling fan.
4. Remove the accessory drive belt. See removal and installation of drive belts.
5. Remove the air cleaner outlet tube.
6. Remove the crankshaft pulley.
7. Remove the vibration damper bolt and the washer.
8. Using crankshaft damper remover 303—009, remove the crankshaft damper.
9. Remove the crankshaft key, if required.

To install:

10. Install the crankshaft front oil seal.

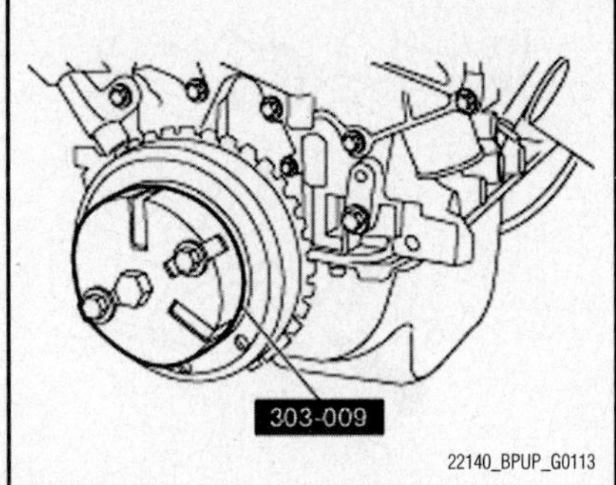

Fig. 69 Crankshaft damper remover 303—009

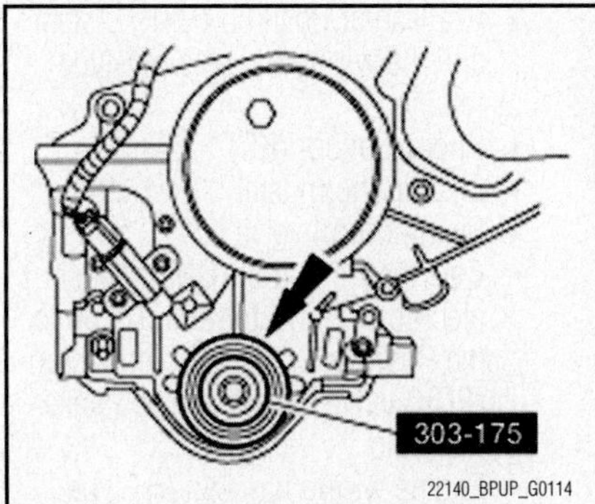

Fig. 70 Crankshaft vibration damper and seal replacer 303—175

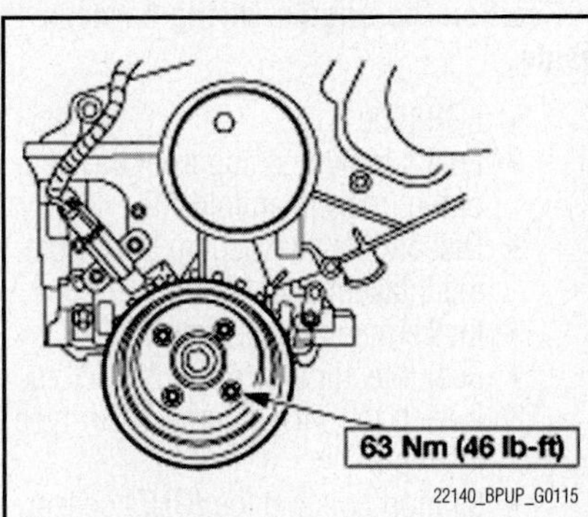

Fig. 71 Installation of the crankshaft pulley

11. Lubricate the seal lip with clean engine oil.
12. Using the vibration damper and seal replacer 303—175, install the front oil seal.
13. Using the vibration damper and seal replacer 303—175 or equivalent, install the crankshaft damper.
14. Tighten to 107 ft. lbs. (145 Nm).
15. Coat the sealing surfaces of the crankshaft damper with clean engine oil.
16. Coat the crankshaft damper keyway with silicone.
17. Install the crankshaft damper.
18. Install the crankshaft pulley.
19. Tighten the crankshaft pulley bolts to 46 ft. lbs. (63 Nm).
20. Install the drive belt.
21. Install the cooling fan blade and fan.
22. Install the air cleaner outlet tube.
23. Connect the negative battery cable.

4.0L Engine

See Figures 72 through 74.

1. Remove the crankshaft pulley. See crankshaft pulley removal and installation in this section.
2. Using the front cover seal remover 303—107, remove the crankshaft front oil seal.

To install:

➡Lubricate the seal lip with clean engine oil. Using the front cover seal remover 303—107and front cover aligner replacer 303—093 and 303—102, install the crankshaft front oil seal.

3. Install the crankshaft pulley.

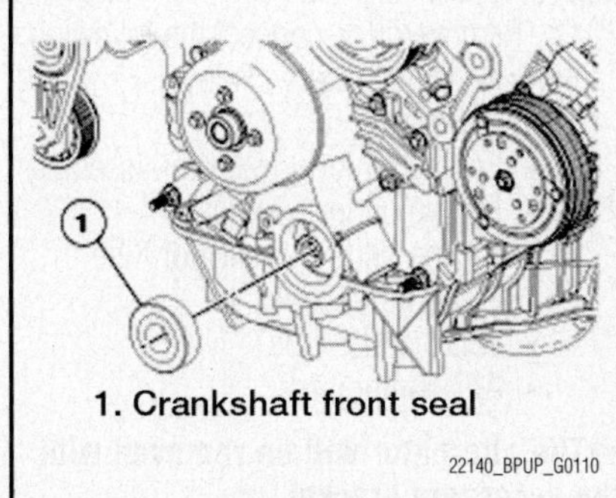

Fig. 72 Crankshaft front oil seal removal and installation exploded view

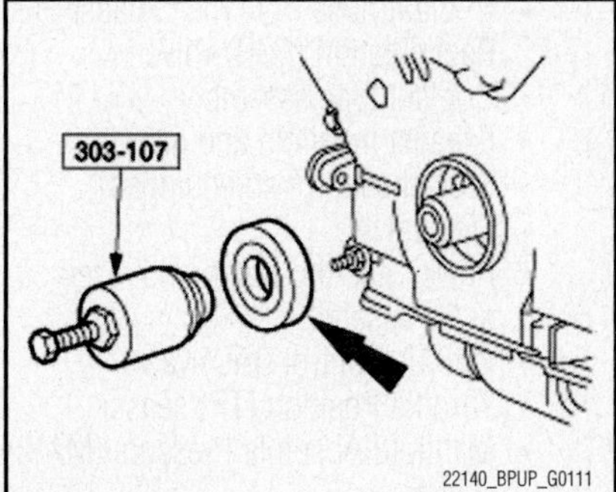

Fig. 73 Front cover seal removal tool 303—107

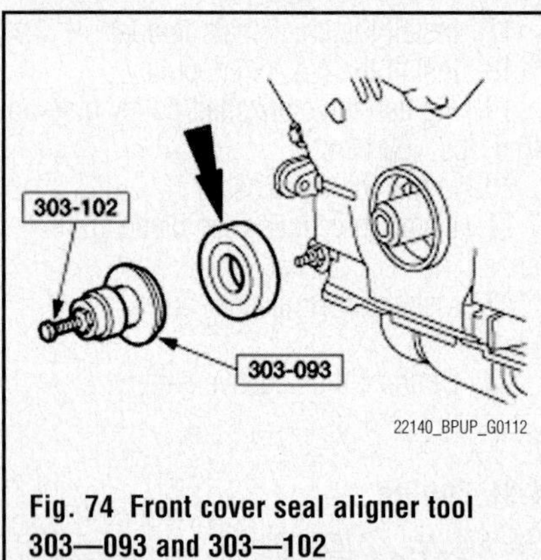

Fig. 74 Front cover seal aligner tool 303—093 and 303—102

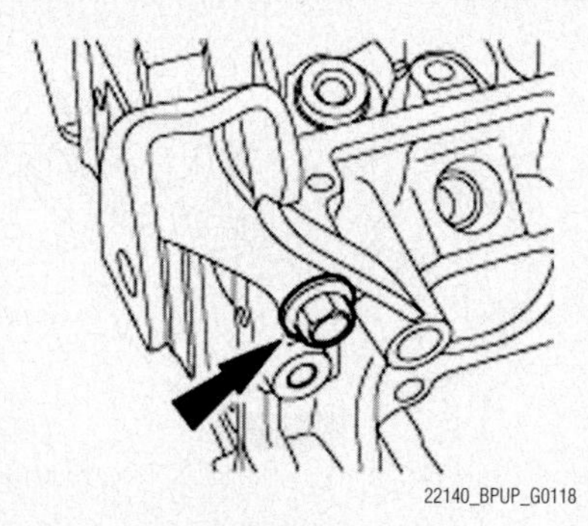

Fig. 75 Rear lifting eye and the front bracket mounting location

CYLINDER HEAD

REMOVAL & INSTALLATION

2.3L Engine

See Figures 75 through 77.

1. Before servicing the vehicle, refer to the precautions in the beginning of this section.
2. Relieve the fuel system pressure.
3. Drain the cooling system.
4. Properly discharge the A/C system.
5. Remove the rear lifting eye and the front bracket.
6. Remove the Cylinder Head Temperature (CHT) sensor.
7. Remove or disconnect the following:
   - Negative battery cable
   - Drive belt
   - Engine oil level indicator assembly
   - Engine oil level indicator
   - Engine oil level indicator tube
   - Water outlet tube
   - Water outlet tube
   - A/C compressor

➡The alternator will be removed with the accessory bracket.

   - Accessory bracket
   - Right motor mount
   - Coolant hose from the thermostat
   - Coolant hose from the Exhaust Gas Recirculation (EGR) valve
   - Coolant tube assembly
   - Exhaust manifold and gasket
   - Block heater (if so equipped)
   - EGR valve
   - Power steering pump and reservoir as an assembly
   - Idle Air Control (IAC) valve
   - Throttle Position (TP) sensor
   - Manifold Absolute Pressure (MAP) sensor
   - Swirl control valve monitor electrical connector
   - Crankshaft Position (CKP) sensor and the wiring harness pin-type retainers
   - Knock Sensor (KS)
   - Electric thermostat
   - Swirl control valve
   - Camshaft Position (CMP) sensor electrical connector and disconnect the Positive Crankcase Ventilation (PCV) valve hose from the intake manifold
   - Engine wiring harness pin-type retainers from the intake manifold
   - Engine wiring harness connector bracket.

➡Position the engine wiring harness aside.

   - EGR tube
   - Fuel supply line clip from the front of the intake manifold
   - Disconnect the vacuum hose from the intake manifold
   - Intake manifold assembly
   - Fuel injector electrical connectors. Detach the wiring harness pin-type retainers
   - Ignition coil and the (CHT) sensor electrical connectors
   - Engine wiring harness anchors from the valve cover studs
   - Remove the engine wiring harness
   - Ignition coil
   - Bypass hose
   - Thermostat housing
   - Knock Sensor (KS) and the engine vent cover
   - Left motor mount
   - Fuel injector supply manifold with the injectors and the ground strap
   - Water pump pulley
   - Water pump
   - Spark plugs
   - Valve cover
   - Crankshaft vibration damper

➡There is one front cover bolt behind the cooling fan drive pulley. To remove this bolt, align one of the cooling fan drive pulley access holes with the bolt head to access the bolt.

   - Front cover
   - Timing chain tensioner
   - Timing chain guides
   - Timing chain assembly

➡Use a wrench on the flats between cylinders No. 1 and No. 2 to hold the camshaft in place

   - Camshaft drive sprockets
   - Remove the 16 tappets

➡ Note location of the tappets prior to removal.

   - Oil pump chain tensioner and guide

➡The oil pump chain sprocket must be held in place

   - Oil pump chain and sprockets

➡Note the position of the lobes on the No. 1 cylinder before removing the camshafts for assembly reference.

8. Loosen the camshaft bearing cap bolts in sequence, one turn at a time. Repeat the first step until all tension is released from the camshaft bearing caps. Remove the camshaft bearing caps.
9. Remove or disconnect the following:
   - Camshafts
   - Cylinder head bolts and the cylinder head
   - Cylinder head gasket
10. Installation is the reverse of removal. Apply RTV sealer to the places shown. The head must be installed within 4 minutes of application. Observe the following torques for the cylinder heads:
   - Step 1: Tighten the bolts to 5 Nm (44 inch lbs.)
   - Step 2: Tighten the bolts to 15 Nm (11 ft. lbs.)
   - Step 3: tighten the bolts to 45 Nm (33 ft. lbs.)
   - Step 4: Tighten the bolts an additional 90 degrees (1/4 turn)
   - Step 5: Tighten the bolts an additional 90 degrees (1/4 turn)

➡Install the camshafts with the alignment notches in the camshaft lined up so the camshaft alignment plate can be installed without rotating the camshafts. Make sure the lobes on the No. 1 cylinder are in the same position as noted in the disassembly procedure. Rotating the camshafts, or installing

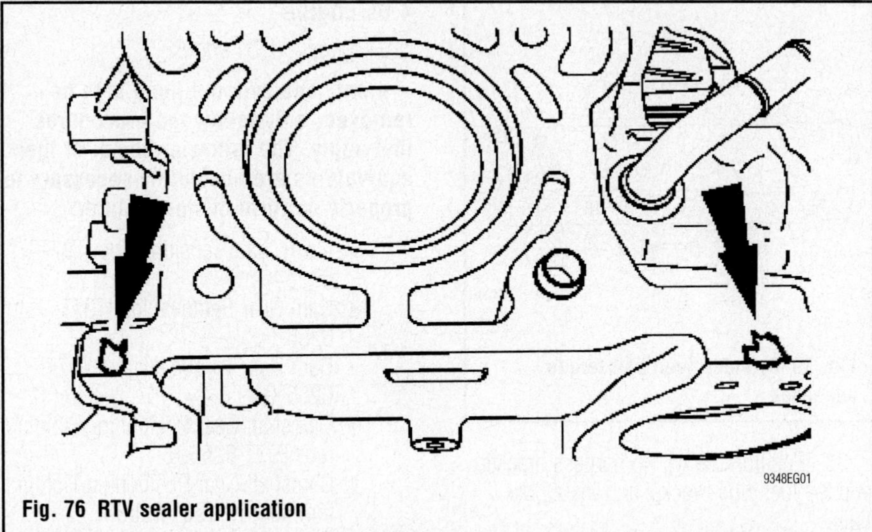

Fig. 76 RTV sealer application

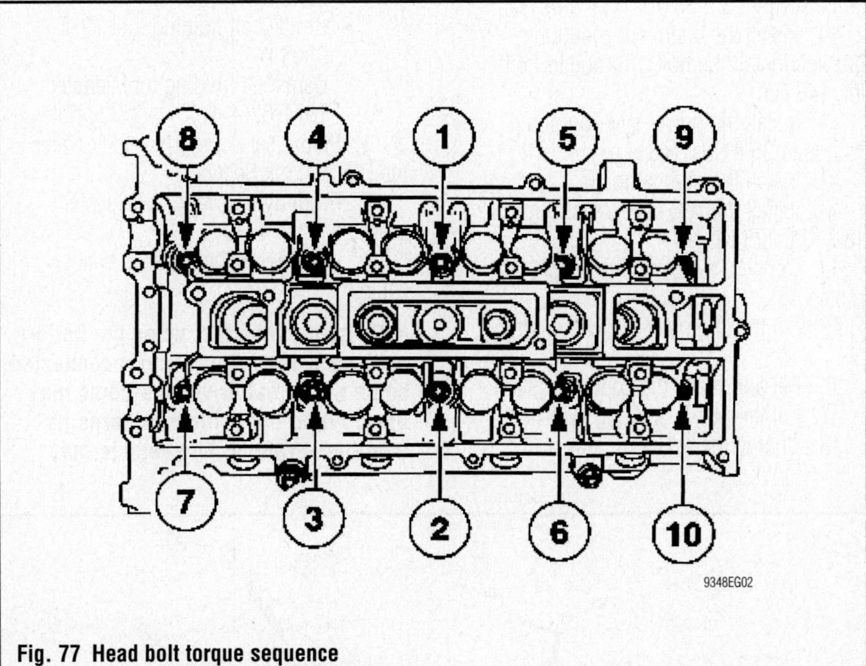

Fig. 77 Head bolt torque sequence

the camshafts 180 degrees out of position can cause severe damage to the valves and pistons. Lubricate the camshaft journals and bearing caps with clean engine oil. Install the camshafts and bearing caps. Tighten the bolts in the sequence shown in three stages.

11. Observe the following torques for the camshafts.

- Step 1: Tighten the camshaft bearing caps one turn at a time until tight
- Step 2: Tighten the bolts to 7 Nm (62 inch lbs.)
- Step 3: Tighten the bolts to 16 Nm (12 ft. lbs.)

12. Observe the following torques for the crankshaft vibration damper.

➡Do not reuse the crankshaft pulley bolt. Tighten the bolt in two stages.

- Step 1: Tighten the bolt to 40 Nm (30 ft. lbs.)
- Step 2: Tighten the bolt and additional 90 degrees (1/4 turn).

3.0L Engine

See Figures 78 and 79.

1. Before servicing the vehicle, refer to the Precautions Section.
2. With the vehicle in **NEUTRAL**, position it on a hoist.
3. Disconnect the negative battery cable.

➡On some vehicles, when the battery cable is disconnected and reconnected, some abnormal drive symptoms may occur while the vehicle relearns its adaptive strategy. The vehicle may need to be driven to relearn its strategy.

4. Remove the lower intake manifold. See Intake manifold removal and installation.
5. Remove the drive belt from the A/C compressor pulley.
6. Remove the drive belt from the A/C compressor pulley.
7. Remove the A/C compressor. See A/C compressor removal and installation.

⁂ WARNING

Refrigerant compressor oil (mineral oil) F73Z-19577-AA (Motorcraft YN-9-A) should be used to lubricate R-134a refrigerant system O-ring seals only and should not be added to the R-134a refrigerant system as an A/C compressor lubricant. PAG refrigerant compressor oil F7AZ-19D589-DA (Motorcraft YN-12-C) or equivalent meeting Ford specification WSH-M1C231-B only should be used as an A/C compressor lubricant.

➡Installation of a new suction accumulator is not required when repairing the air conditioning system except when there is physical evidence of system contamination from a failed A/C compressor or damage to the suction accumulator.

8. If flushing of the refrigerant system has not been carried out, recover the refrigerant.
9. Position the air cleaner outlet tube aside.
10. Disconnect the electrical connector.
11. Remove the clamp nut on the compressor manifold and tube assembly.
12. Loosen the bolt and disconnect the compressor manifold and tube assembly from the A/C compressor.
13. Remove the bolts and the A/C compressor. Discard the O-ring seals.
14. Remove the alternator.
15. Remove the power steering pump pulley.
16. Disconnect the power steering pressure hose. Remove and discard the seal.
17. Disconnect the power steering return hose.
18. Disconnect the power steering return line hose.

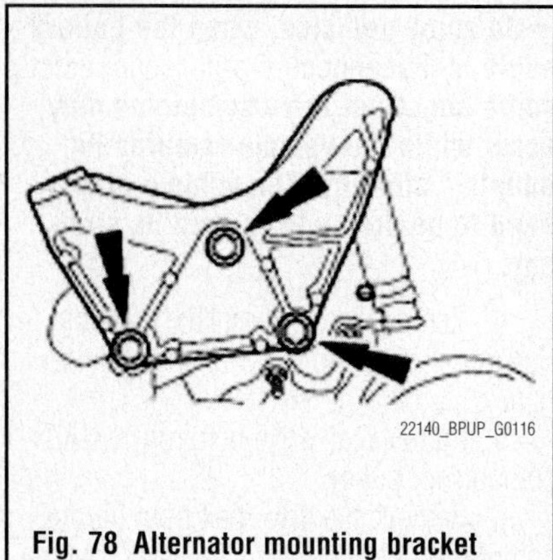
Fig. 78 Alternator mounting bracket

19. Remove the bolts and the power steering pump.
20. Remove the alternator mounting bracket.
21. Remove the rocker arms and the push rods.
22. Remove the nuts and detach the fuel tube bracket and the wiring harness bracket.
23. Remove the two studs and bolts from the rear of the A/C compressor mounting bracket.
24. Remove the bolts and remove the front of the A/C compressor mounting bracket.
25. Remove the exhaust manifolds.
26. Remove and discard the eight cylinder head bolts.
27. Remove the cylinder heads.
28. Remove and discard the cylinder head gaskets.

To install:

29. Clean all the sealing surfaces.
30. Check the cylinder head for flatness.

➡The "V" notch in the head gasket faces the front of the engine.

31. Install new cylinder head gaskets.
32. Install the cylinder heads.
33. Install the bolts. Tighten the bolts in five steps in the sequence shown.
    - Step 1: Tighten to 37 ft. lbs. (50 Nm)
    - Step 2: Loosen the bolts one full turn
    - Step 3: Tighten to 22 ft. lbs. (30 Nm)
    - Step 4: Tighten the bolts 90 degrees
    - Step 5: Tighten the bolts an additional 90 degrees
34. Install the lower intake manifold.
35. Install the exhaust manifolds.
36. Install the A/C compressor mounting bracket. Torque to 34 ft. lbs. (46 Nm).
37. Install the studs and bolts to the A/C compressor mounting bracket. Torque to 35 ft. lbs. (48 Nm).

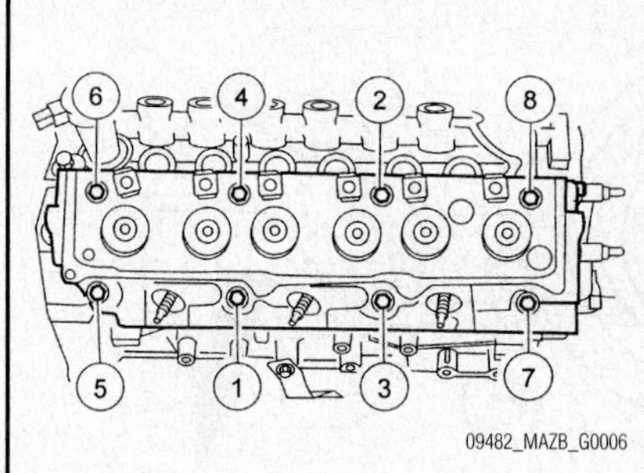

Fig. 79 Cylinder head bolt torque sequence

38. Position the wiring harness bracket and the fuel tube bracket and install the nuts.
39. Install the push rods and the rocker arms. Torque to 24 ft. lbs. (32 Nm).
40. Install the alternator mounting bracket. Install the bolts. Torque to 34 ft. lbs. (46 Nm).
41. Install the power steering pump. Torque to 35 ft. lbs. (48 Nm).
42. Install the alternator.
43. Install the A/C compressor. Torque to 18 ft. lbs. (25 Nm).
44. Connect the battery ground cable.
45. Fill the engine with clean engine oil.
46. Fill and bleed the cooling system.
47. Fill the power steering system.
48. Charge the A/C system.

4.0L Engine

See Figures 80, 81 and 82

➡If only one cylinder head is to be removed, only follow the procedures that apply. The following tools, or their equivalents are absolutely necessary to properly perform this procedure:

- Cam Chain Tensioner tool T97T-6K254-A
- Cam Gear Removal tool T97T-6256-F
- Cam Gear Torque adapter T97T-6256-G
- Camshaft Gear Positioning/Holding tool T97T-6256-B
- Camshaft Gear Positioning/Holding tool adapter T97T-6256-A
- Camshaft holding tool T97T-6256-C
- Crankshaft holding tool T97T-6303-A
- Camshaft holding tool adapter T97T-6256-D

1. Before servicing the vehicle, refer to the Precautions Section.
2. Properly relieve the fuel system pressure.
3. Disconnect the negative battery cable.

➡On some vehicles, when the battery cable is disconnected and reconnected, some abnormal drive symptoms may occur while the vehicle relearns its adaptive strategy. The vehicle may

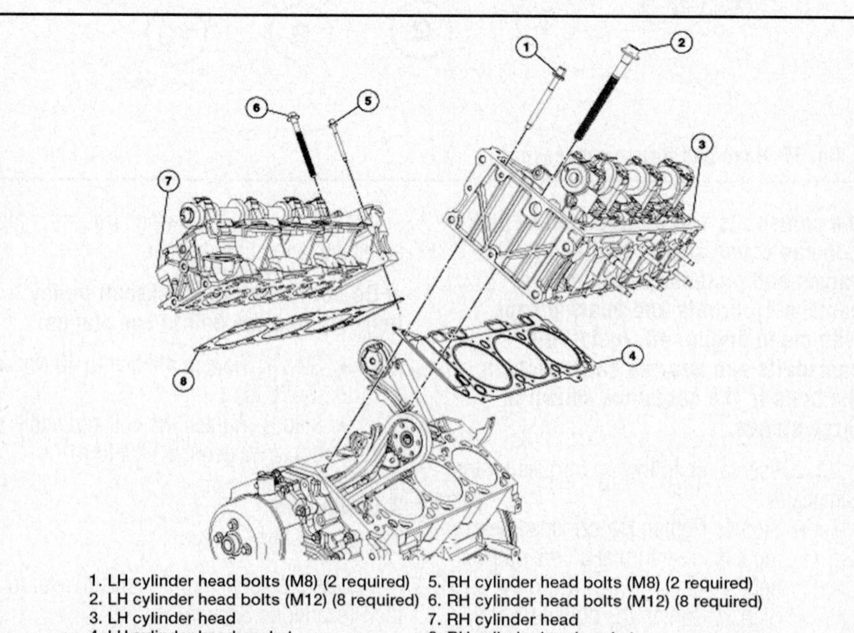

Fig. 80 Cylinder head removal exploded view

need to be driven to relearn its strategy.

4. Drain the cooling system.
5. Remove or disconnect the following:
   - Lower intake manifold
   - Fan blade and shroud
   - Valve cover
   - Roller followers, if equipped
   - Drive belt
   - Upper radiator hose and tube
   - Alternator electrical connectors
   - Alternator mounting bracket
   - Engine accessory bracket and move it aside
   - Camshaft Position (CMP) electrical connector
   - Crankshaft Position (CKP) sensor electrical connector
   - Engine Coolant Temperature (ECT) sensor electrical connector
   - Coil pack electrical connector
   - Exhaust Gas Recirculation (EGR) valve electrical connector
   - EGR valve bracket and move it aside
   - Heater hoses
   - Fuel injector electrical connectors
   - Water bypass hose
   - Thermostat housing
   - Spark plug wires
   - Fuel injection supply manifold
   - Fuel injectors
   - Crankcase vent separator spring
   - Oil dipstick housing
   - Exhaust manifold
   - Hydraulic chain tensioner
   - Cassette retaining bolt
   - Camshaft sprocket

⁂ WARNING

Remove the camshaft sprocket to prevent breaking the cassette and to gain clearance to remove the cylinder head.

➡Hold the chain and cassette with a rubber band to aid in removal and prevent the chain from falling into the cylinder block.

- Cylinder head and discard the gasket

➡Discard the head bolts.

To install:

6. Thoroughly clean all gasket mating surfaces. Remove all traces of old gasket material, oil, grease or dirt.
7. Insure that the rubber band is holding the right-hand chain to the cassette.
8. Install a new head gasket and the cylinder head. Be sure to use new cylinder head bolts.

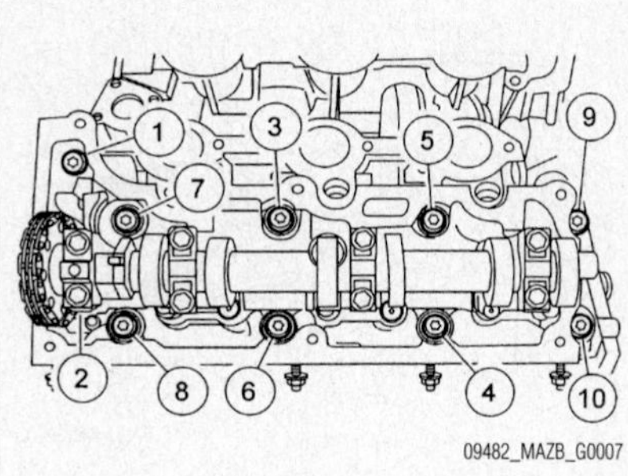

Fig. 81 Cylinder head bolt loosening sequence

Fig. 82 Cylinder head bolt torque sequence

⁂ WARNING

Cylinder head bolts are a torque-to-yield design. New bolts must be installed.

9. Torque the new cylinder head bolts in sequence as follows:
10. Install bolts 8 (12mm) and torque, in sequence to:
    - Step 1: 9 ft. lbs. (12 Nm)
    - Step 2: 18 ft. lbs. (25 Nm)
    - Step 3: Install bolts 5 and 6 (8mm), and torque to 24 ft. lbs. (32 Nm)
    - Step 4: 8 (12mm) bolts plus 90 degrees
    - Step 5: 8 (12mm) bolts plus an additional 90 degrees
11. Install or connect the following:
    - Camshaft sprocket in the cassette and make certain that the camshaft sprocket turns freely on the camshaft
    - Cassette retaining bolt. Torque the bolt to 89 inch lbs. (10 Nm)
    - Exhaust manifold
    - Oil level indicator tube. Torque the bolt to 18 ft. lbs. (25 Nm)
    - Crankcase vent separator and spring
    - Thermostat housing. Torque the bolts to 8 ft. lbs. (11 Nm)
    - Water bypass hose
    - Heater hoses
    - EGR bracket. Torque the bolt to 89 inch lbs. (10 Nm)
    - EGR tube. Torque the nut to 30 ft. lbs. (40 Nm)
    - ECT sensor electrical connector
    - Electrical harness retainer. Torque the bolt to 89 inch lbs. (10 Nm)
    - CKP and CMP electrical connectors
    - Accessory bracket. Torque the bolts to 31 ft. lbs. (42 Nm)
    - Alternator mounting bracket. Torque the bolts to 31 ft. lbs. (42 Nm)
    - Alternator and electrical connectors
    - Drive belt
    - Fan shroud
    - Roller followers
    - Valve cover
    - Lower intake manifold
    - Negative battery cable
12. Change the engine oil and filter.
13. Refill the cooling system.
14. Start the engine and check for leaks, repair if necessary.

ENGINE ASSEMBLY

REMOVAL & INSTALLATION

2.3L Engine

See Figure 83.

1. Before servicing the vehicle, refer to the Precautions Section.
2. Relieve the fuel system pressure.
3. Disconnect the negative battery cable.

➡On some vehicles, when the battery cable is disconnected and reconnected, some abnormal drive symptoms may occur while the vehicle relearns its adaptive strategy. The vehicle may need to be driven to relearn its strategy.

4. Drain the cooling system.
5. Drain the engine oil.
6. Properly discharge the air conditioning system.
7. Remove or disconnect the following:
   - Hood
   - Accelerator control snow shield
   - Air cleaner tube
   - Upper radiator hose
   - Lower radiator hose
   - Fan and shroud
   - PCM electrical connector. Remove the retaining nut on the harness clamp. Position the harness on the engine.

- Ground stud for the PCM
- Heater hoses
- All vacuum hoses
- Coolant reservoir hoses
- Air conditioning compressor clutch
- MAF electrical connector
- Air conditioning compressor manifold, plug the lines and the compressor ports
- Accelerator and speed control cables
- Power steering return hose
- Power steering switch electrical connector
- High pressure power steering hose
- Fuel supply hose
- 42-pin electrical connector
- Vacuum regulator solenoid supply hose
- Evaporative purge hose
- Brake booster vacuum hose and the engine ground strap
- Solenoid control wire at the starter
- Starter wiring harness clamp bolt and position it out of the way.
- RH splash shield
- Alternator electrical connections
- Block heater electrical connector
- Front heated oxygen sensor electrical connector at the bell housing
- Oil pressure sensor electrical connector
- Engine wiring pushpins and position the engine wiring harnesses out of the way.
- Oil filter
- With automatic transmission, the bolt retaining the transmission cooling tubes to the engine. Remove the bracket.
- Transmission dust shield
- Starter motor
- Heated oxygen sensor electrical connector at the rear of the transmission
- Transmission wiring harness
- Vehicle speed sensor, transmission range sensor, backup light switch and the transmission electrical connectors. Disconnect the pushpins and position the harness forward to the engine.
- Oil filter adapter

➡Leave two side bolts in until the engine is ready to be removed.

- Nine of the transmission-to-engine bolts
- With automatic transmission, the transmission fluid indicator and tube assembly
- Starter dust shield

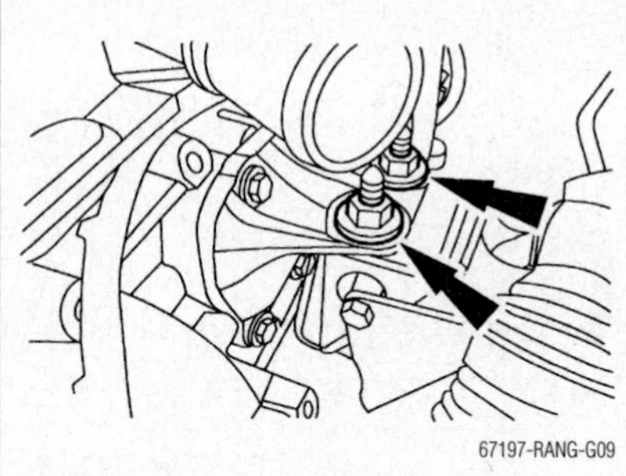

Fig. 83 Engine support insulator nuts, left side shown, right side similar—2.3L engine

➡Mark one stud and the flexplate for assembly reference.

- With automatic transmission, the four torque converter nuts

8. Support the transmission with a floor jack.
9. Support the engine with a floor crane using a spreader bar.
10. Remove the two side transmission-to-engine bolts.
11. Remove the four engine support insulator.
12. Remove the engine from the vehicle.

To install:

13. Install or connect the following:
    - Automatic transmission, the four torque converter nuts
    - Flexplate
    - Starter dust shield
    - Transmission fluid indicator and tube assembly
    - Nine of the transmission-to-engine bolts
    - Vehicle speed sensor, transmission range sensor, backup light switch and the transmission electrical connectors
    - Connect the pushpins and position the harness forward to the engine.
    - Oil filter adapter
    - Transmission dust shield
    - Starter motor
    - Heated oxygen sensor electrical connector at the rear of the transmission
    - Transmission wiring harness
    - With automatic transmission, Install the bolt and bracket retaining the transmission cooling tubes to the engine.
    - Front heated oxygen sensor electrical connector at the bell housing
    - Oil pressure sensor electrical connector
    - Engine wiring pushpins and position the engine wiring harnesses out of the way.
    - Oil filter
    - Alternator electrical connections
    - Block heater electrical connector
    - RH splash shield
    - Starter wiring harness
    - Solenoid control wire
    - Evaporative purge hose
    - Brake booster vacuum hose and the engine ground strap
    - Vacuum regulator solenoid supply hose
    - Accelerator and speed control cables
    - Power steering return hose
    - Power steering switch electrical connector
    - High pressure power steering hose
    - Fuel supply hose
    - 42-pin electrical connector
    - Ground stud for the PCM
    - Heater hoses
    - All vacuum hoses
    - Coolant reservoir hoses
    - Air conditioning compressor clutch
    - MAF electrical connector
    - Air conditioning compressor manifold
    - PCM electrical connector.
    - Accelerator control snow shield
    - Air cleaner tube
    - Upper radiator hose
    - Lower radiator hose
    - Fan and shroud
    - Hood
14. Observe the following torques:
    - Torque converter bolts: 26 ft. lbs. (35 Nm)
    - Nine transmission-to-engine bolts 35 ft. lbs. (48 Nm)
    - Oil filter adapter: 18 ft. lbs. (25 Nm)
    - Starter: 30 ft. lbs. (40 Nm)
    - Engine support nuts: 75 ft. lbs. (102 Nm)

3.0L Engine

See Figure 84.

1. Before servicing the vehicle, refer to the Precautions Section.
2. Relieve the fuel system pressure.
3. Disconnect the negative battery cable.

➡On some vehicles, when the battery cable is disconnected and reconnected, some abnormal drive symptoms may occur while the vehicle relearns its

adaptive strategy. The vehicle may need to be driven to relearn its strategy.

4. Drain the cooling system.
5. Drain the engine oil.
6. Properly discharge the air conditioning system.
7. Remove or disconnect the following:
   - Hood
   - Air cleaner outlet tube
   - Upper and the lower radiator hoses

WARNING

The fan clutch has left-hand threads.

   - The fan clutch and blade as an assembly
   - Drive belt
   - Fan shroud
   - Radiator
   - Air conditioning manifold and tube. Remove the nut and position the line aside.
   - Air conditioning compressor wiring
   - Air conditioning compressor and the air conditioning compressor mounting bracket
   - Heater hoses
   - Ground cable
   - Fuel lines
   - Snow shield
   - Accelerator cable and the speed control actuator cable
   - All vacuum lines
   - 42-pin connector
   - Powertrain Control Module (PCM) connector
   - Nut from the PCM harness
   - Stud bolt and the PCM ground strap
   - Alternator wiring and position aside
   - Both Heated Oxygen Sensors (HO2S)
   - Transmission harness connectors
   - MAF sensor
   - LH HO2S
   - Dual converter Y pipe
   - Starter motor and the starter grounding stud bolt
   - Torque converter nuts
   - 8 transmission-to-engine bolts
8. Install the lifting eyes.
9. Remove the four nuts.
10. Support the transmission.
11. Remove the engine from the vehicle.

To install:

12. Install or connect the following:
   - Torque converter nuts
   - 8 transmission-to-engine bolts
   - Dual converter Y pipe

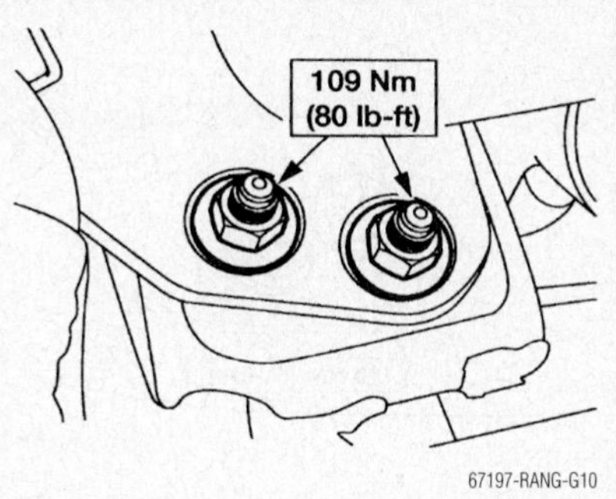

Fig. 84 Engine support nuts, one side shown—3.0L engine

   - Starter motor and the starter grounding stud bolt
   - Stud bolt and the PCM ground strap
   - Alternator wiring and position aside
   - Both Heated Oxygen Sensors (HO2S)
   - Transmission harness connectors
   - MAF sensor
   - LH HO2S
   - Heater hoses
   - Ground cable
   - Fuel lines
   - Snow shield
   - Accelerator cable and the speed control actuator cable
   - All vacuum lines
   - 42-pin connector
   - Powertrain Control Module (PCM) connector
   - Air conditioning compressor
   - Air conditioning manifold and tube
   - Air conditioning compressor wiring
   - Radiator
   - The fan clutch and blade as an assembly

WARNING

The fan clutch has left-hand threads.

   - Fan shroud
   - Drive belt
   - Upper and the lower radiator hoses
   - Air cleaner outlet tube
   - Hood
13. Fill the cooling system.
14. Fill the engine oil.
15. Observe the following torques:
   - Engine mount nuts: 80 ft. lbs. (109 Nm)
   - Transmission-to-engine bolts: 33 ft. lbs. (45 Nm)
   - Torque converter nuts 35 ft. lbs. (47 Nm).

4.0L Engine

See Figures 85 and 86.

1. Before servicing the vehicle, refer to the Precautions Section.
2. Relieve the fuel system pressure.
3. Disconnect the negative battery cable.

➡On some vehicles, when the battery cable is disconnected and reconnected, some abnormal drive symptoms may occur while the vehicle relearns its adaptive strategy. The vehicle may need to be driven to relearn its strategy.

CAUTION

If the fuel supply manifold is used as a leverage device, damage may occur to the supply manifold. Care must be taken when working around the fuel supply manifold.

4. Remove or disconnect the following:
   - Accelerator cable from engine
   - Speed control cable from engine
   - Radiator, the fan blade, and the fan shroud
   - Accessory bracket bolts and position bracket aside
   - Alternator wiring
   - Wiring harness retainer and position alternator wiring away from engine
   - Engine electrical connector
   - PCM connector
   - PCM ground wire
   - Engine ground wire
   - Brake booster vacuum hose
   - Air conditioning high pressure switch electrical connector
   - Bolt and position the air conditioning lines aside

➡Heater hose will be removed with engine.

   - Heater hoses
   - Fuel line
   - Starter motor
   - Engine oil
   - Oil drain plug
   - Transmission portion of wiring harness
   - RH and LH HO2S connectors
   - Transmission control connector
   - Output shaft speed sensor connector
   - Digital transmission range sensor connector
   - Catalyst monitor sensor electrical connector

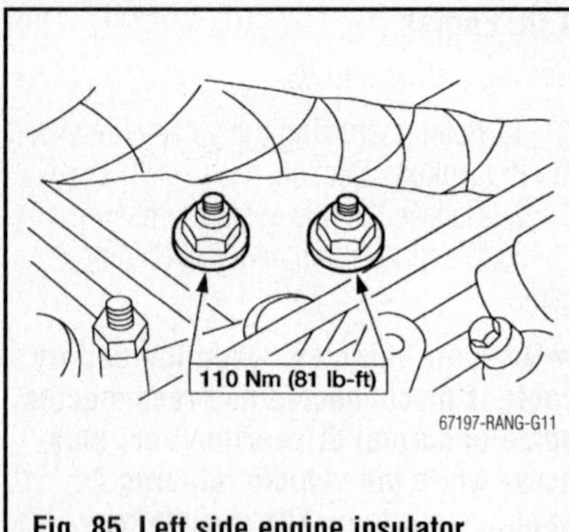

Fig. 85 Left side engine insulator nuts—4.0L engine

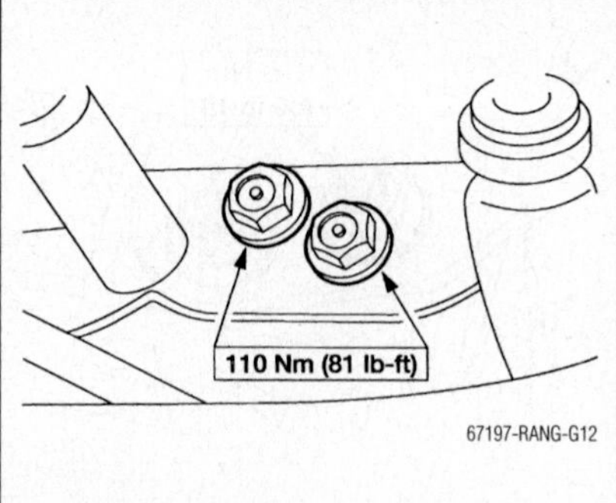

Fig. 86 Right side engine insulator nuts—4.0L engine

- Transmission/transfer case portion of the wiring harness from any routing clips or pushpins. Route transmission/transfer case portion of the wiring harness to top of engine.
- Bolt, and position the transmission cooling line bracket aside
- Air conditioning line bracket nut and position it aside
- Power steering return hose
- Power steering pressure hose
- Vapor management valve hose connector
- Eight bolts and the LH and the RH engine support insulator nuts

➡The lifting eyes should be installed on the exhaust manifold studs for number three and number four cylinders

5. Install the lifting eyes.
6. Install the spreader bar to the lifting eyes.
7. Attach a floor crane to the spreader bar and remove the engine.

To install:

➡Remove the lifting eyes on the exhaust manifold studs.

8. Install the eight bolts and the LH and the RH engine support insulator nuts.
9. Install the vapor management valve hose connector.
10. Install the power steering pressure hose.
11. Install the air conditioning line bracket nut and position it aside.
12. Install the bolt, and position the transmission cooling line bracket into place.
13. Install the transmission/transfer case portion of the wiring harness from any routing clips or pushpins. Route transmission/transfer case portion of the wiring harness to top of engine.
14. Install the catalyst monitor sensor electrical connector.
15. Install the digital transmission range sensor connector.
16. Install the output shaft speed sensor connector.
17. Install the transmission control connector.
18. Install the RH and LH HO2S connectors.
19. Install the heater hoses.
20. Install the starter motor.
21. Install the fuel line.
22. Install the engine oil.
23. Install the oil drain plug.
24. Install the air conditioning high pressure switch electrical connector.
25. Install the bolt and position the air conditioning lines aside.
26. Install the brake booster vacuum hose.
27. Install the engine ground wire.
28. Install the PCM ground wire.
29. Install the PCM connector.
30. Install the alternator wiring.
31. Install the radiator, the fan blade, and the fan shroud.
32. Install the speed control cable.
33. Install the accelerator cable.
34. Observe the following torques:
    - Left and right engine insulator nuts: 81 ft. lbs. (110 Nm)
    - Engine mount nuts: 59 ft. lbs. (80 Nm)
    - Transmission-to-engine bolts: 35 ft. lbs. (47 Nm)
    - Torque converter nuts: 35 ft. lbs. (47 Nm).

EXHAUST MANIFOLD

REMOVAL & INSTALLATION

2.3L Engine

See Figure 87.

1. Before servicing the vehicle, refer to the Precautions Section.
2. Disconnect the negative battery cable.

➡On some vehicles, when the battery cable is disconnected and reconnected, some abnormal drive symptoms may occur while the vehicle relearns its adaptive strategy. The vehicle may need to be driven to relearn its strategy.

3. Remove or disconnect the following:
    - Exhaust flange nuts
    - Drive belt
    - Coolant
    - Upper radiator hose and the engine reservoir hose
    - Air conditioning compressor
    - Heater hose
    - Oil indicator and the upper bolt for the tube assembly
    - Lower bolt and remove the oil indicator tube assembly
    - Front radiator tube
4. Remove the pushpins and position the right inner fender splash shield out of the way.
5. Remove or disconnect the following:
    - Alternator electrical connectors
    - Lower Front End Accessory Drive (FEAD) mounting bolts
    - Upper mounting bolt and the FEAD assembly
    - Two nuts and position the coolant tube out of the way
    - Exhaust manifold
    - Exhaust manifold gasket

To install:

6. Install or connect the following:
    - Exhaust manifold gasket
    - Exhaust manifold and the nuts
    - Coolant tube and the nuts
    - FEAD assembly and the upper mounting bolts, finger tight.
    - Lower FEAD mounting bolts, finger tight. Then, torque all FEAD bolts, in the sequence shown, to 35 ft. lbs. (47 Nm).
    - Alternator electrical connectors
    - Right inner splash shield and pushpins
    - Upper radiator tube and install the bolts
    - Oil indicator tube assembly and the lower bolt
    - Oil indicator tube upper bolt and the oil indicator
    - Heater water hose
    - Air conditioning compressor
    - Upper radiator hose and the engine reservoir hose
7. Fill the cooling system.
8. Install the serpentine drive belt.
9. Install the exhaust flange nuts.
10. Connect the battery ground cable.

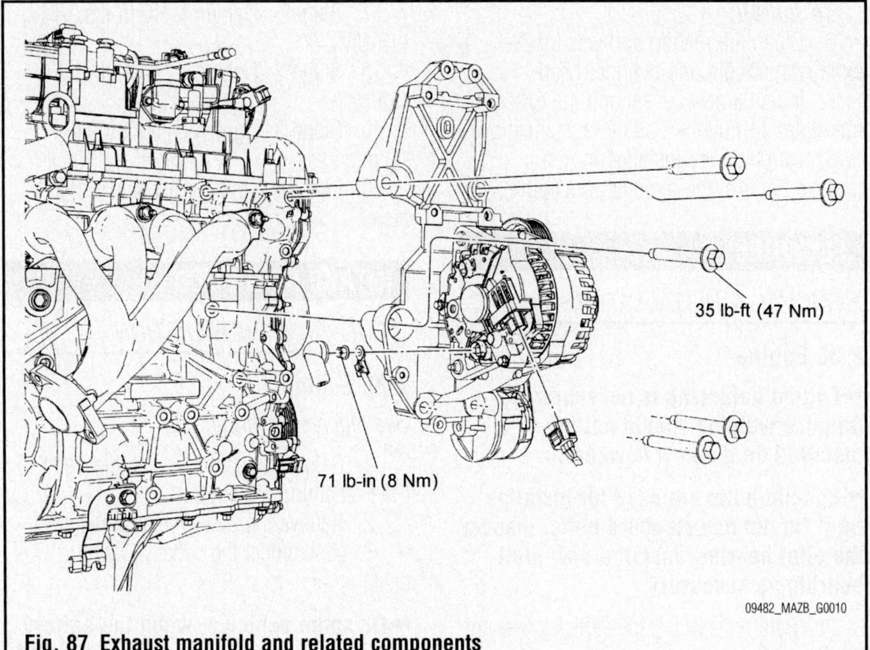

Fig. 87 Exhaust manifold and related components

3.0L Engine–Left Side

See Figure 88.

1. Before servicing the vehicle, refer to the Precautions Section.
2. Disconnect the negative battery cable.

➡On some vehicles, when the battery cable is disconnected and reconnected, some abnormal drive symptoms may occur while the vehicle relearns its adaptive strategy. The vehicle may need to be driven to relearn its strategy.

3. Install or connect the following:
   - Exhaust flange nuts
   - Exhaust Gas Recirculation (EGR) valve from the exhaust manifold tube
   - Oil lever indicator and bracket
   - Exhaust manifold and discard the gasket

To install:

4. Clean the mating surfaces for the exhaust manifold and cylinder head.
5. Install a new gasket and the exhaust manifold. Torque the bolts in sequence to:
   a. 89 inch lbs. (10 Nm).
   b. 15 ft. lbs. (20 Nm).
6. Install or connect the following:
   - Oil lever indicator tube and bracket. Torque the bolt to 12 ft. lbs. (16 Nm).
   - EGR valve to the exhaust manifold tube. Torque the fastener to 26 ft. lbs. (35 Nm).
   - Exhaust flange. Torque the nuts to 25 ft. lbs. (34 Nm).
   - Negative battery cable

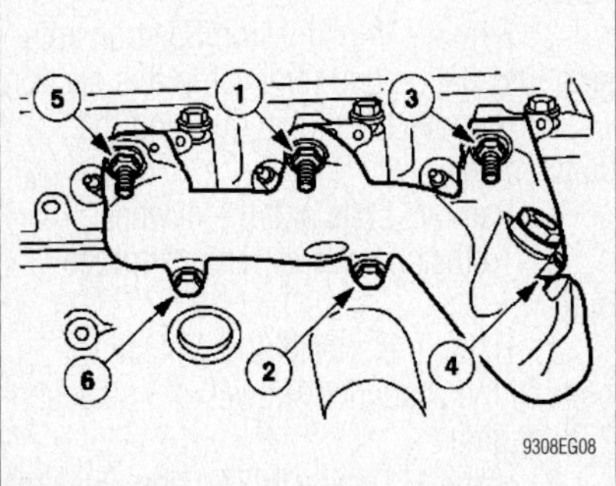

Fig. 88 Left side exhaust manifold bolt torque sequence–3.0L engine

7. Start the vehicle and check for leaks, repair if necessary.

3.0L Engine—Right Side

See Figure 89.

1. Before servicing the vehicle, refer to the Precautions Section.
2. Disconnect the negative battery cable.

➡On some vehicles, when the battery cable is disconnected and reconnected, some abnormal drive symptoms may occur while the vehicle relearns its adaptive strategy. The vehicle may need to be driven to relearn its strategy.

3. Remove or disconnect the following:
   - Exhaust manifold flange
   - Ignition coil support bracket
   - Exhaust manifold and discard the gasket

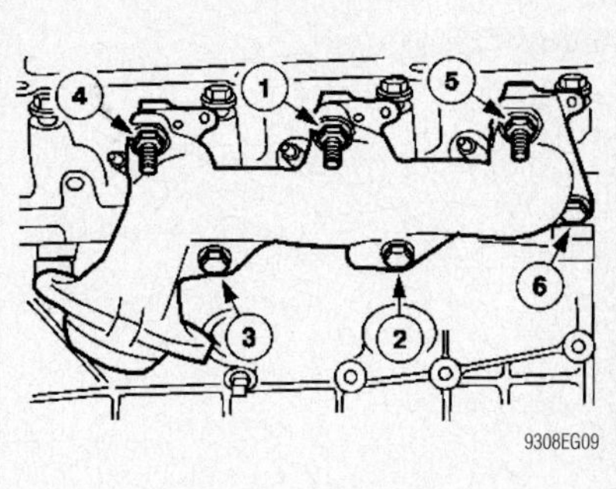

Fig. 89 Right side exhaust manifold bolt torque sequence–3.0L engine

To install:

4. Clean the mating surfaces for the exhaust manifold and cylinder head.
5. Install a new gasket and the exhaust manifold.
6. Torque the bolts in sequence to 18 ft. lbs. (25 Nm).
7. Install or connect the following:
   - Ignition coil support bracket. Torque the bolts to 15 ft. lbs. (20 Nm).
   - Exhaust flange nuts. Torque the nuts to 33 ft. lbs. (46 Nm).
   - Negative battery cable
8. Start the vehicle and check for leaks, repair if necessary.

4.0L Engine—Left Side

See Figure 90.

1. Before servicing the vehicle, refer to the Precautions Section.
2. Disconnect the negative battery cable.

➡On some vehicles, when the battery cable is disconnected and reconnected, some abnormal drive symptoms may occur while the vehicle relearns its adaptive strategy. The vehicle may need to be driven to relearn its strategy.

3. Raise and support the vehicle safely. Remove the exhaust pipe flange retaining bolts.
4. Lower the vehicle.
5. Disconnect the Differential Pressure Feedback Exhaust Gas Recirculation (DPFE) transducer hoses.
6. Disconnect the Exhaust Gas Recirculation (EGR) valve to exhaust manifold tube.
7. Disconnect the EGR valve to exhaust manifold tube from the EGR valve and remove the EGR valve to exhaust manifold tube.
8. Remove the exhaust manifold retaining nuts. Remove the exhaust manifold from the engine.

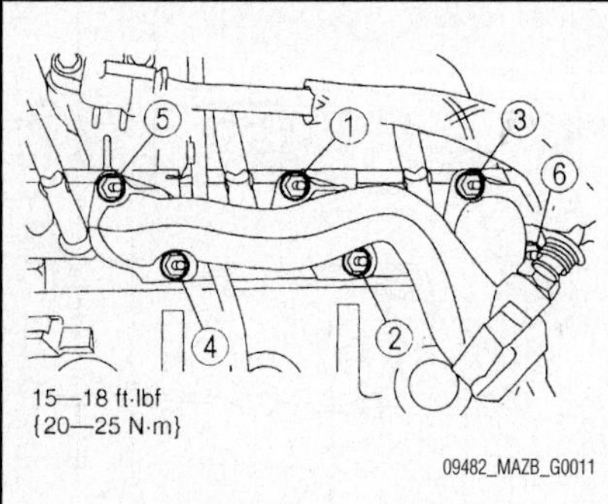

Fig. 90 Left side exhaust manifold bolt torque sequence—4.0L engine

To install:

9. Clean the mating surfaces for the exhaust manifold and cylinder head
10. Install a new gasket and the exhaust manifold. Torque the exhaust flange nuts to 33 ft. lbs. (46 Nm).
11. Continue the installation in the reverse order of the removal procedure.

4.0L Engine—Right Side

See Figure 91.

1. Before servicing the vehicle, refer to the Precautions Section.
2. Disconnect the negative battery cable.

➡On some vehicles, when the battery cable is disconnected and reconnected, some abnormal drive symptoms may occur while the vehicle relearns its adaptive strategy. The vehicle may need to be driven to relearn its strategy.

3. Raise and support the vehicle safely. Remove the exhaust pipe flange retaining bolts.
4. Lower the vehicle.
5. Remove the exhaust manifold retaining nuts. Remove the exhaust manifold from the engine.

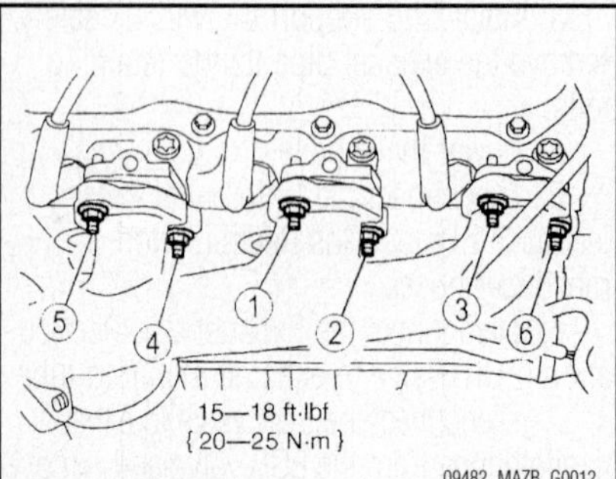

Fig. 91 Right side exhaust manifold bolt torque sequence—4.0L engine

To install:

6. Clean the mating surfaces for the exhaust manifold and cylinder head
7. Install a new gasket and the exhaust manifold. Torque the bolts to specification.
8. Continue the installation in the reverse order of the removal procedure.

FLYWHEEL

REMOVAL & INSTALLATION

2.3L Engine

➡Engine balancing is not required. Balance weights should not be installed on the new flywheel.

➡Special bolts are used for installation. Do not use standard bolts. Inspect the pilot bearing. Install a new pilot bearing as necessary

1. Before servicing the vehicle, refer to the Precautions Section.
2. Disconnect the negative battery cable.
3. Remove transmission See transmission installation and removal in this section.
4. Remove the clutch disc and the clutch cover.
5. Remove bolts and the flywheel.
6. To install, reverse removal procedure.
7. Using the special tool, install the bolts in the sequence shown, in 3 stages.—2.3L engine
8. Stage 1: Tighten to 37 ft. lbs. (50 Nm).
9. Stage 2: Tighten to 59 ft. lbs. (80 Nm).
10. Stage 3: Tighten to 83 ft. lbs. (112 Nm).

3.0L and 4.0L Engines

➡Engine balancing is not required. Balance weights should not be installed on the new flywheel.

➡Special bolts are used for installation. Do not use standard bolts. Inspect the pilot bearing. Install a new pilot bearing as necessary

1. Disconnect the negative battery cable.
2. Remove transmission See transmission installation and removal in this section.
3. Remove the clutch disc and the clutch cover.
4. Remove bolts and the flywheel.
5. To install, reverse removal procedure.
6. Using the special tool, install the bolts in the sequence shown, in 3 stages.
7. Stage 1: Tighten to 10 ft. lbs. (13 Nm).
8. Stage 2: Tighten to 37 ft. lbs. (50 Nm).
9. Stage 3: Tighten an additional 90 degrees.
10. Install the clutch disc and clutch cover.

INTAKE MANIFOLD

REMOVAL & INSTALLATION

2.3L Engine

See Figures 92 through 95.

1. Before servicing the vehicle, refer to the Precautions Section.
2. Relieve the fuel system pressure.
3. Disconnect the negative battery cable.

➡On some vehicles, when the battery cable is disconnected and reconnected, some abnormal drive symptoms may occur while the vehicle relearns its adaptive strategy. The vehicle may need to be driven to relearn its strategy.

4. Remove the accelerator control splash shield. Remove the air cleaner outlet tube.
5. Disconnect the throttle cables from the intake manifold.
6. Disconnect the Throttle Position (TP) sensor.
7. Disconnect breather hose.
8. Disconnec the Manifold Absolute Pressure (MAP) sensor electrical connector.
9. Disconnect the vacuum hose and the Idle Air Control (IAC) valve electrical connector.
10. Disconnect the engine vacuum harness and brake booster hose and the Positive Crankcase Ventilation (PCV) valve.
11. Disconnect and plug the heater hoses.
12. Disconnect the two engine wiring harness pin type retainers from the intake manifold. Disconnect the pin type retainer from the rear of the intake manifold.
13. Remove the Exhaust Gas Recirculation (EGR) tube bracket bolt. Disconnect the fuel hose from the clip. Remove the bolts from the EGR tube flange.
14. Detach the 42 pin electrical connector from the bracket and disconnect the vapor purge hose.
15. Disconnect the Knock Sensor (KS) sensor electrical connector and disconnect the connector from the intake manifold.

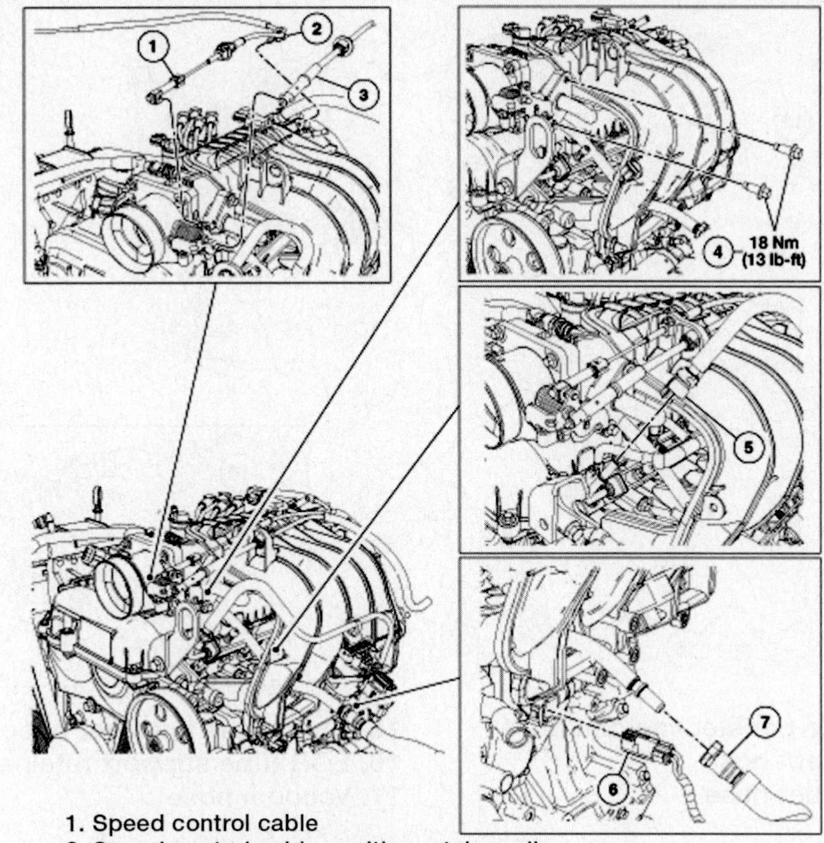

Fig. 92 Intake manifold removal and installation exploded view and component identification—2.3L engine

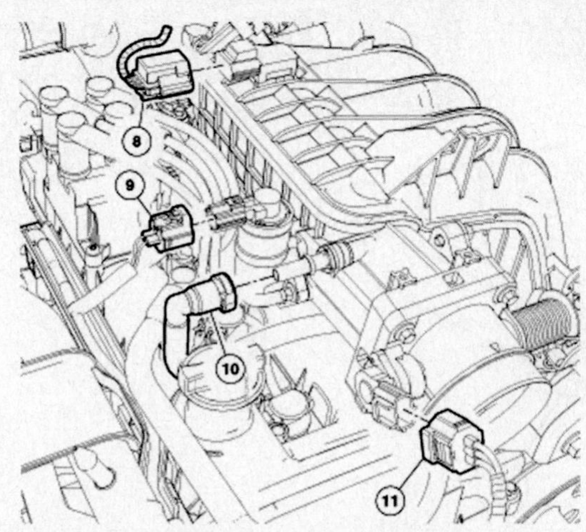

Fig. 93 Intake manifold component identification part 1—2.3L engine

16. Remove the wiring harness push pin from the bottom of the intake manifold. Remove the bracket. Disconnect the fuel supply line from the manifold.
17. Remove the intake manifold retaining bolts. Remove the intake manifold from the engine.

To install:

18. Be sure to clean all old gasket material from the sealing surfaces, using the proper tools.
19. Installation is the reverse of the removal procedure.
20. Be sure to use new gaskets.
21. Torque the bolts to 13 ft. lbs. (18 Nm).
22. Be sure to fill the cooling system with the proper grade and type engine coolant.
23. Start the engine and check for leaks, correct as required.

12
13
14
15
17
16
18 Nm
(13 lb-ft)

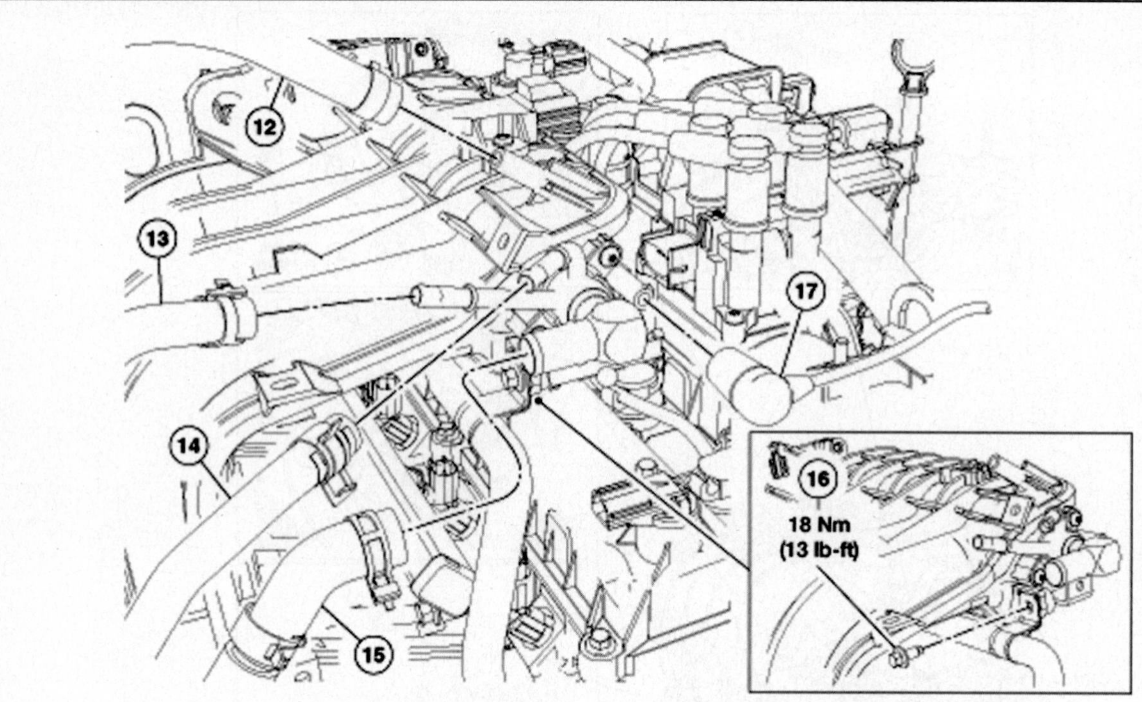

12. Brake booster vacuum hose
13. Coolant hose
14. Coolant hose
15. Crankcase breather hose
16. EGR tube support retainer bolt
17. Vacuum hose

22140_BPUP_G0121

Fig. 94 Intake manifold component identification part 2—2.3L engine

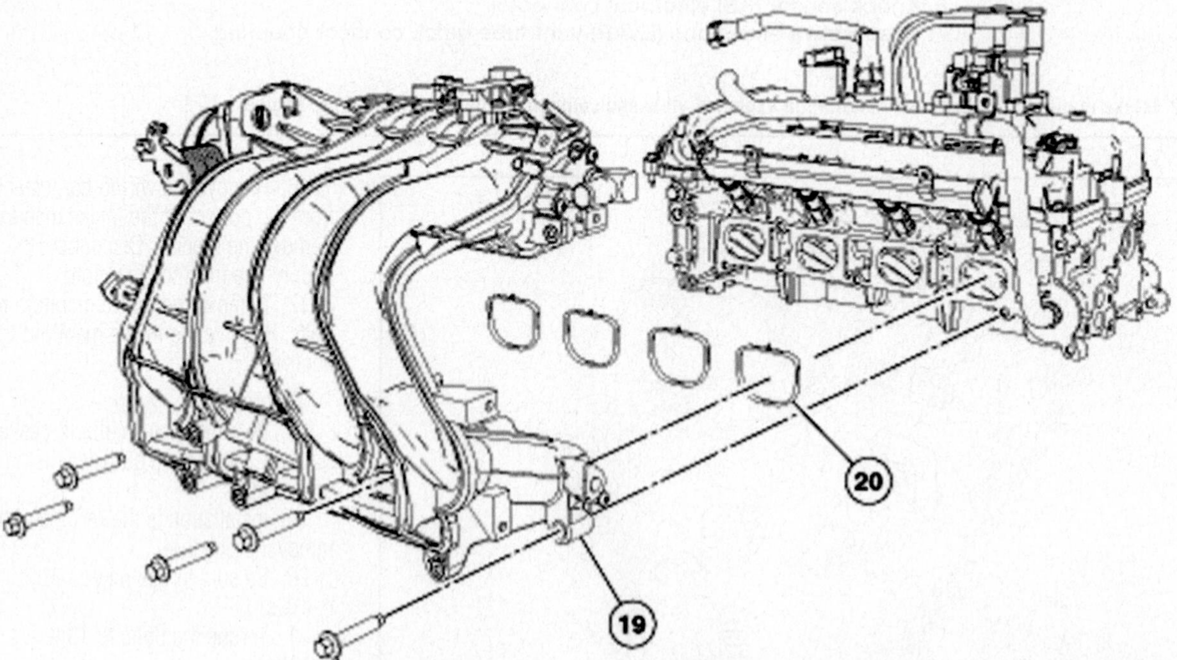

20
19
18 Nm (13 lb-ft) 18

18. Intake manifold bolts (5 required)
19. Intake manifold
20. Intake manifold gaskets (4 required)

22140_BPUP_G0122

Fig. 95 Intake manifold component identification part 3—2.3L engine

3.0L Engine —Upper Manifold

See Figures 96 through 100.

1. Before servicing the vehicle, refer to the Precautions Section.
2. Disconnect the negative battery cable.

➡On some vehicles, when the battery cable is disconnected and reconnected, some abnormal drive symptoms may occur while the vehicle relearns its adaptive strategy. The vehicle may need to be driven to relearn its strategy.

3. Remove the air cleaner outlet pipe.
4. Remove the bolt and the splash shield.
5. Disconnect the accelerator cable, speed control cable and the return spring.
6. Disconnect the accelerator cable and speed control cable from the accelerator bracket.
7. Disconnect the Throttle Position (TP) sensor and Idle Air Control (IAC) valve electrical connectors.
8. Disconnect engine vacuum harness-to-vacuum reservoir hose connection.
9. Disconnect the exhaust manifold-to-Exhaust Gas Recirculation (EGR) valve tube from the EGR valve.
10. Disconnect the fuel vapor hose from the intake manifold.
11. Disconnect the Exhaust Gas Recirculation (EGR) vacuum regulator solenoid electrical connector.
12. Disconnect the brake booster vacuum hose and the Positive Crankcase Ventilation (PCV) valve tube.
13. Remove the oil level indicator tube bracket nut and position the tube and bracket out of the way.
14. Remove the nut and the upper intake manifold brace.
15. Remove the bolts, the stud bolt and the upper intake manifold. Remove and discard the gaskets.

To install:

16. Be sure to clean all old gasket material from the sealing surfaces, using the proper tools.
17. Be sure to use new gaskets.
18. Install the nut and the upper intake manifold brace.
19. Install the oil level indicator tube bracket nut and position the tube and bracket out of the way.
20. Install the brake booster vacuum hose and the Positive Crankcase Ventilation (PCV) valve tube.
21. Install the EGR vacuum regulator solenoid electrical connector.

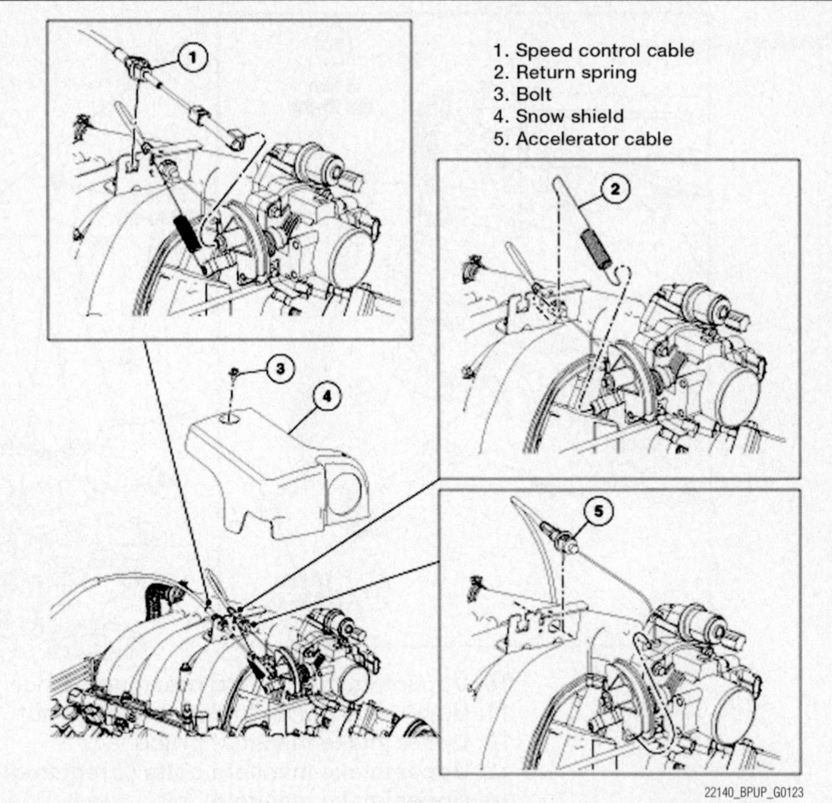

Fig. 96 Intake manifold removal and installation exploded view and component identification—3.0L engine

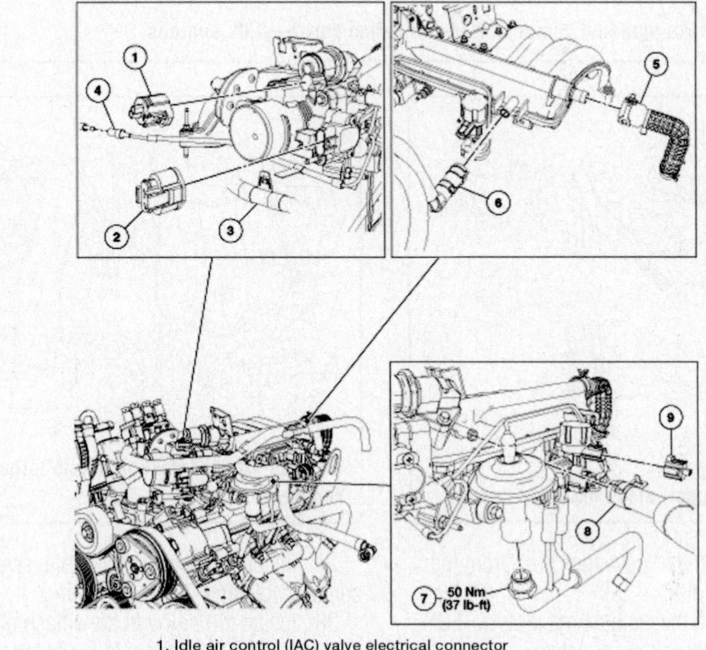

Fig. 97 Intake manifold component identification part 1—3.0L engine

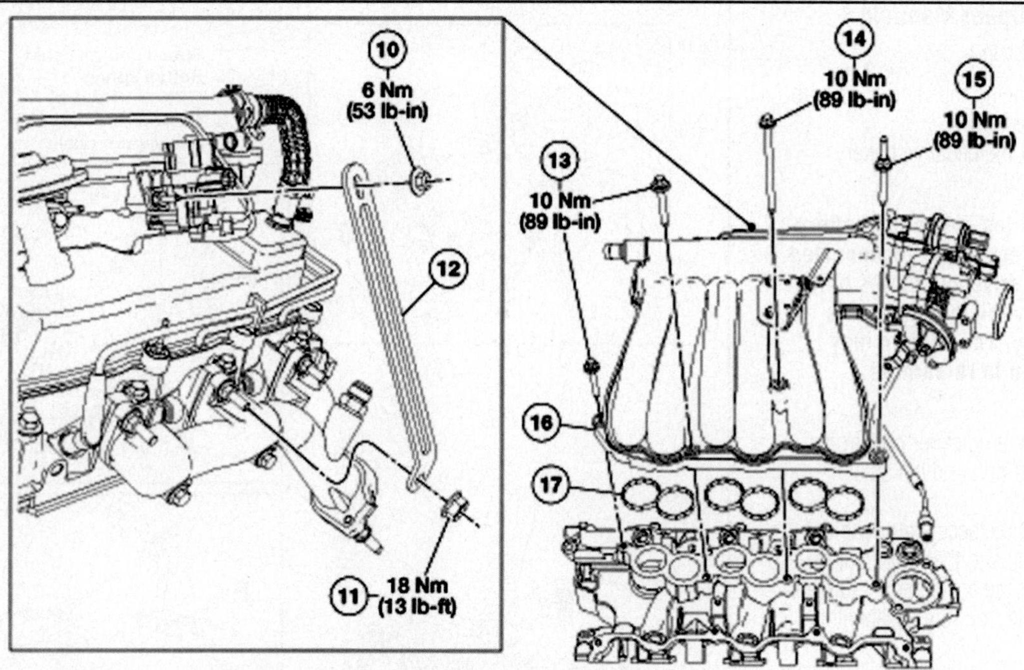

10. Upper intake manifold brace upper nut
11. Upper intake manifold brace lower nut
12. Upper intake manifold brace
13. Upper intake manifold bolts (2 required)
14. Upper intake manifold bolt
15. Upper intake manifold stud bolt
16. Upper intake manifold
17. Upper intake manifold gaskets (3 required)

22140_BPUP_G0125

Fig. 98 Intake manifold component identification part 2—3.0L engines

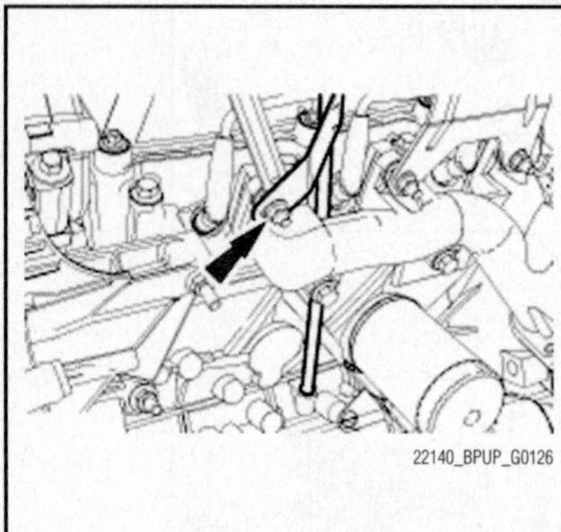

Fig. 99 Upper intake manifold brace.

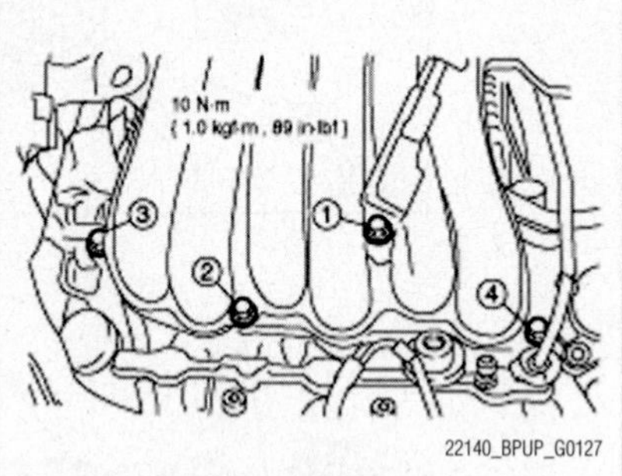

Fig. 100 Upper intake manifold torque sequence

22. Install the fuel vapor hose from the intake manifold..
23. Install the exhaust manifold-to-EGR valve tube from the EGR valve.
24. Install engine vacuum harness-to-vacuum reservoir hose connection.
25. Install the (TP) sensor and (IAC) valve electrical connectors.
26. Install the accelerator cable and speed control cable from the accelerator bracket.
27. Install the accelerator cable, speed control cable and the return spring.
28. Install the bolt and the splash shield.
29. Install the air cleaner outlet pipe.
30. Torque the bolts to 89 inch lbs. (10 Nm).
31. Connect the negative battery cable.

3.0L Engine–Lower Manifold

See Figures 101 through 104.

1. Before servicing the vehicle, refer to the Precautions Section.
2. Relieve the fuel system pressure.
3. Disconnect the negative battery cable.

➡On some vehicles, when the battery cable is disconnected and reconnected, some abnormal drive symptoms may occur while the vehicle relearns its adaptive strategy. The vehicle may need to be driven to relearn its strategy.

4. Drain the engine cooling system.
5. Remove both the valve covers.
6. Disconnect the upper radiator hose, heater hose and coolant bypass hose.
7. Disconnect the Engine Coolant Temperature (ECT) sensor and coolant temperature sender electrical connectors.
8. Disconnect the Crankshaft Position (CKP) sensor and detach the wiring from the stud bolt.
9. Disconnect the fuel injector electrical connectors and position the engine control wiring aside.
10. Disconnect the fuel tube.
11. Loosen the rocker arm bolts.

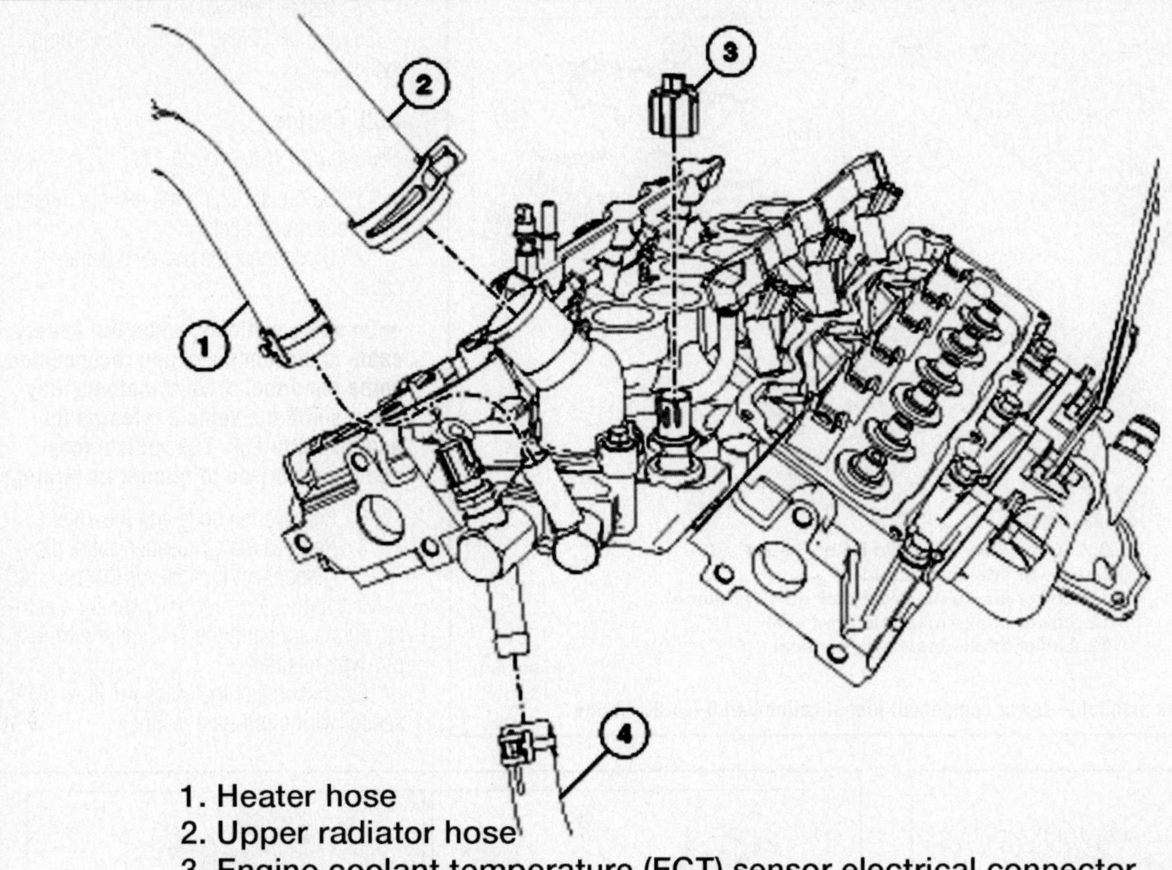

Fig. 101 Intake manifold—lower component identification part 1—3.0L engine

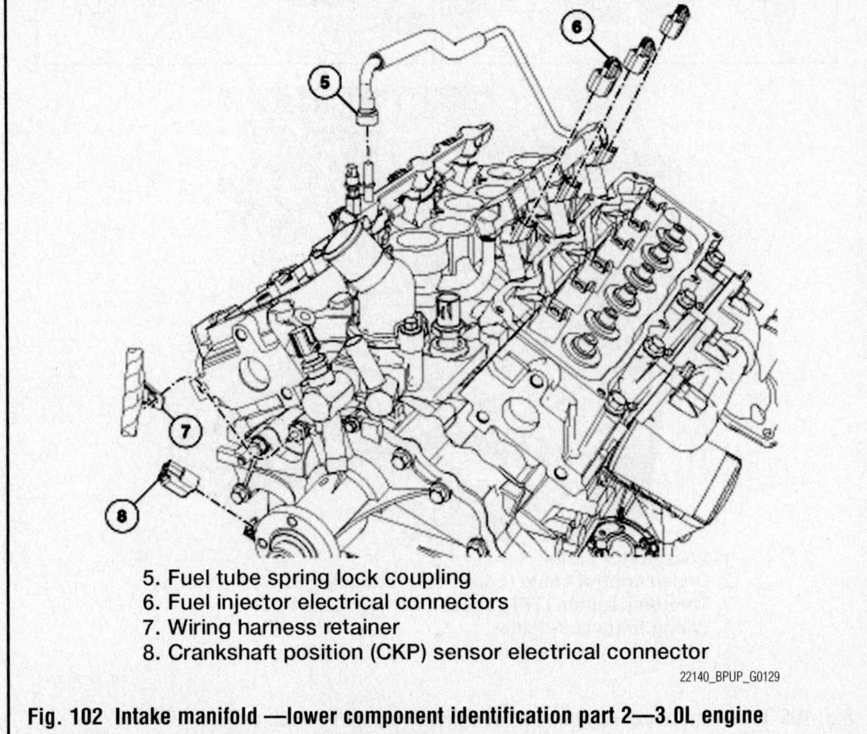

Fig. 102 Intake manifold —lower component identification part 2—3.0L engine

➡Identify the location of each pushrod. Each pushrod is to be installed in the original location to prevent premature wear.

12. Remove all the pushrods.
13. Remove the bolts from the lower intake manifold.

➡Gently loosen the intake manifold to separate the silicone sealant from the cylinder block.

14. Remove the lower intake manifold.
15. Remove and discard the intake manifold gaskets and end seals.

To install:

➡Clean and inspect all mating surfaces.

➡If the lower intake manifold is not installed within four minutes, remove the sealant and reapply.

16. Apply a drop of silicone gasket and sealant at the four cylinder block-to-cylinder head seams.

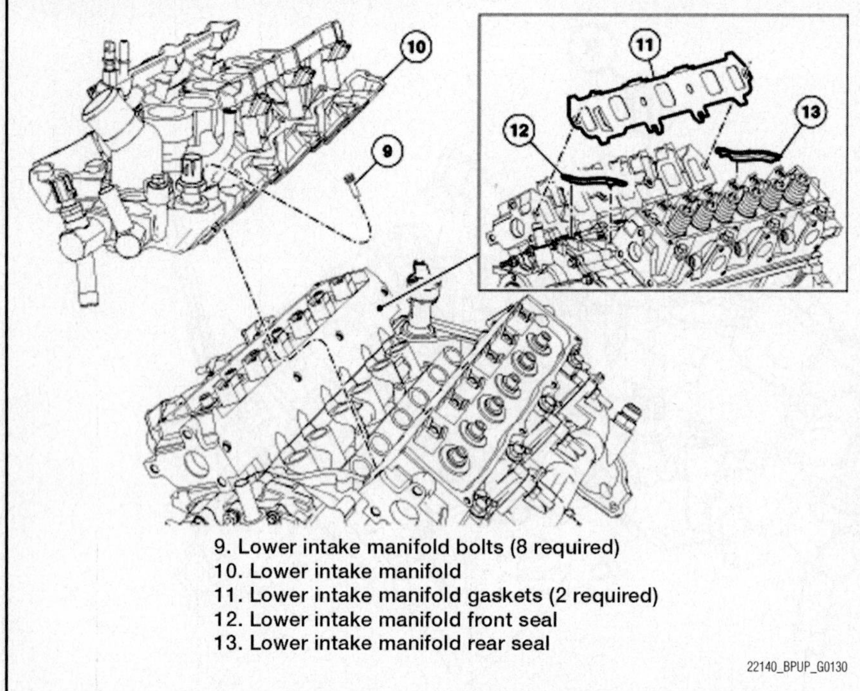

Fig. 103 Intake manifold—lower component identification part 3—3.0L engine

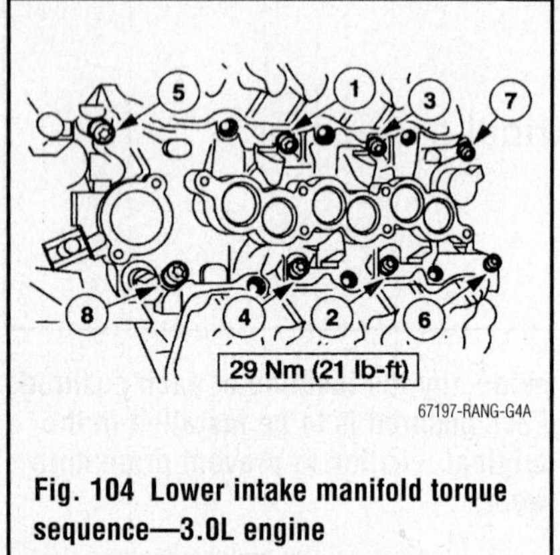

Fig. 104 Lower intake manifold torque sequence—3.0L engine

17. Install the intake manifold gaskets and end seals.
18. Position the lower intake manifold.
19. Install the bolts and tighten in the sequence shown to 21 ft. lbs. (29 Nm).
20. Install the pushrods.
21. Tighten the rocker arms to 24 ft. lbs. (32 Nm).
22. Connect the fuel tube.
23. Connect the fuel injector electrical connectors.
24. Connect the Crankshaft Position (CKP) sensor and attach the wiring from the stud bolt.
25. Connect the Engine Coolant Temperature (ECT) sensor and coolant temperature sender electrical connectors.
26. Connect the upper radiator hose, heater hose and coolant bypass hose.
27. Install both valve covers.
28. Fill and bleed the engine cooling system.

4.0L Engine

See Figures 105 through 111.

1. Before servicing the vehicle, refer to the Precautions Section.
2. Disconnect the negative battery cable.

➡On some vehicles, when the battery cable is disconnected and reconnected, some abnormal drive symptoms may occur while the vehicle relearns its adaptive strategy. The vehicle may need to be driven to relearn its strategy.

3. Remove the bolts and the shield.
4. Remove the air cleaner outlet pipe.
5. Disconnect the Idle Air Control (IAC) valve, Throttle Position (TP) sensor electrical connectors and the TP sensor wiring pin-type retainer.
6. Disconnect the Mass Air Flow (MAF) sensor wiring pin-type retainer.

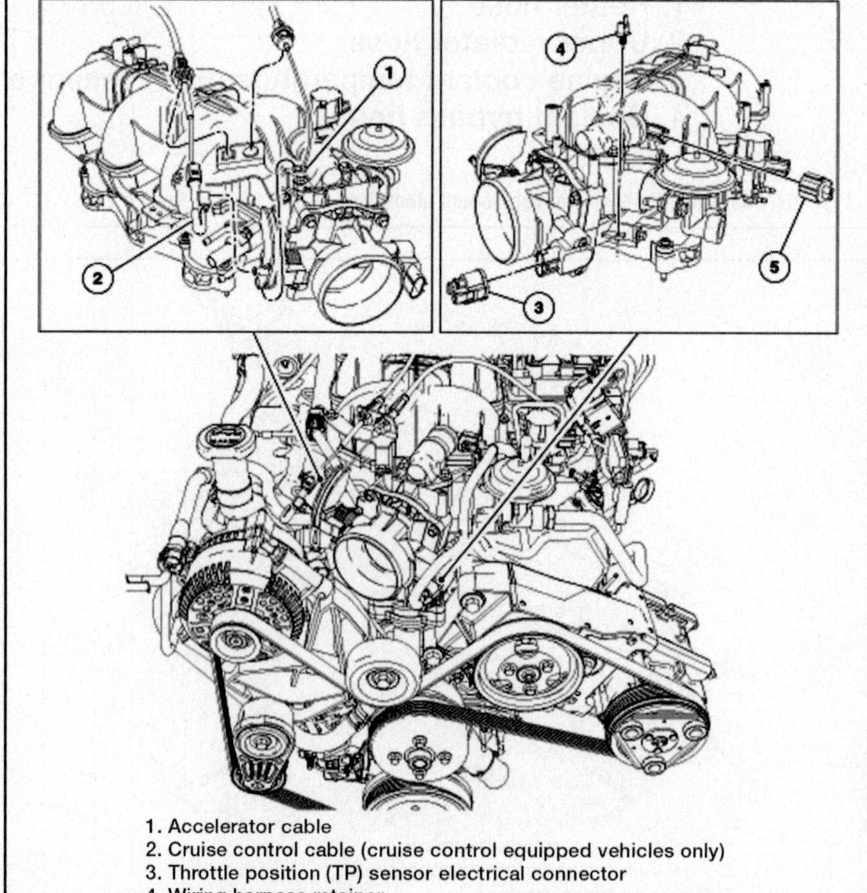

Fig. 105 Intake manifold component identification part 1—4.0L engine

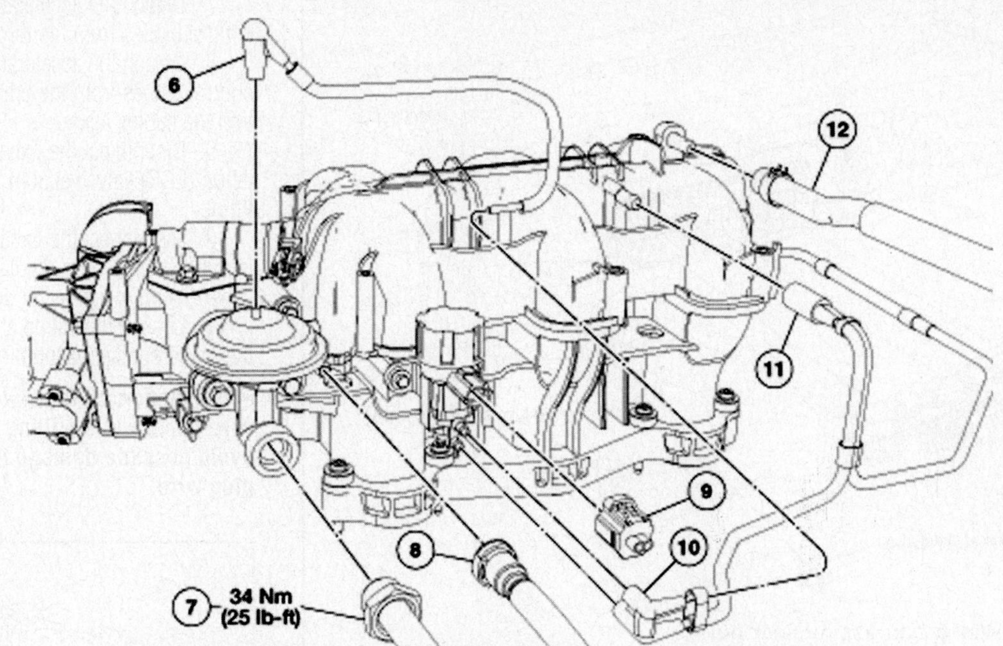

Fig. 106 Intake manifold component identification part 2—4.0L engine

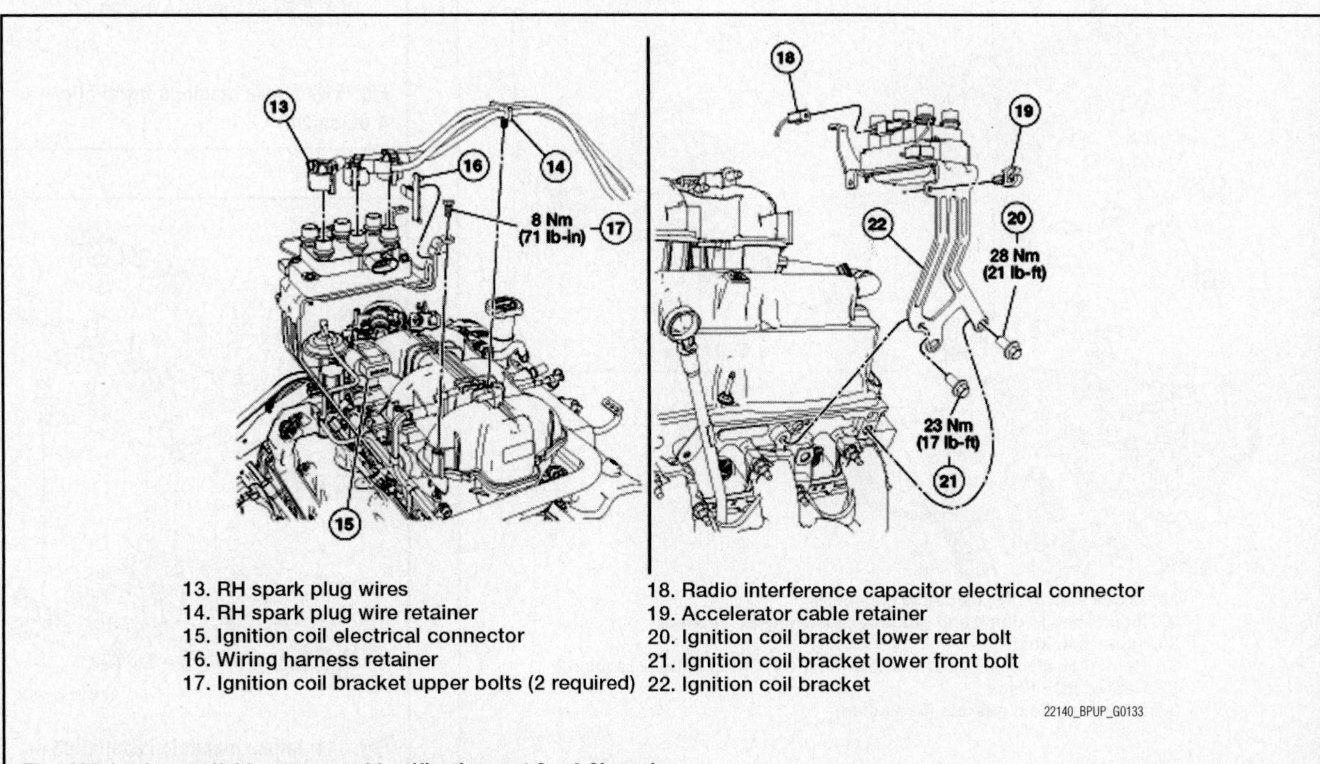

Fig. 107 Intake manifold component identification part 3—4.0L engine

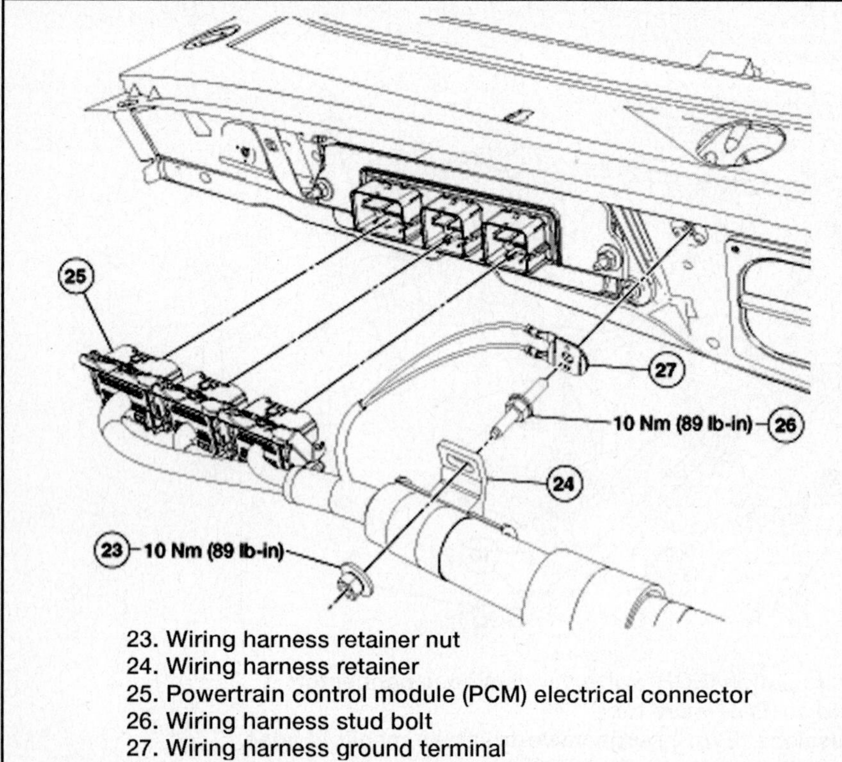

23. Wiring harness retainer nut
24. Wiring harness retainer
25. Powertrain control module (PCM) electrical connector
26. Wiring harness stud bolt
27. Wiring harness ground terminal

22140_BPUP_G0134

Fig. 108 Powertrain Control Module (PCM) electrical connections part 4—4.0L engine

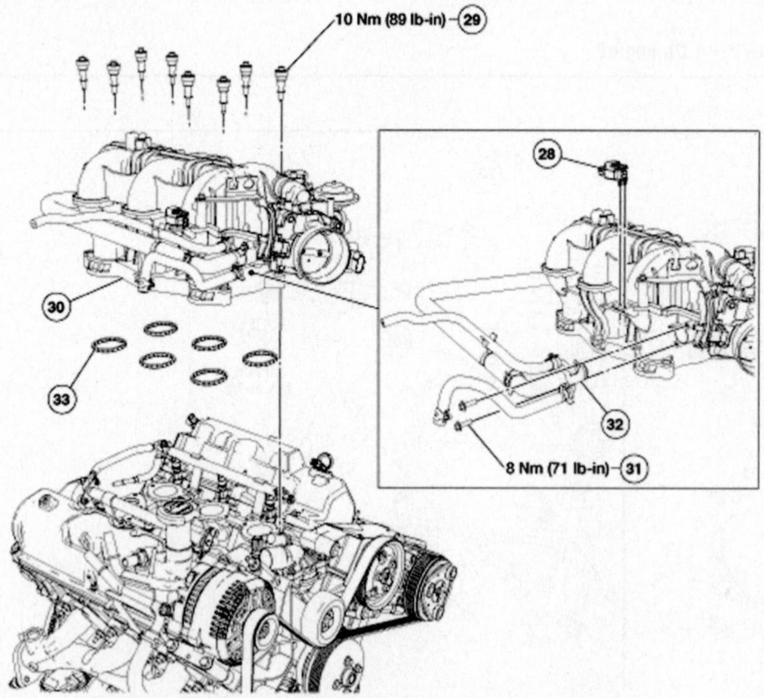

10 Nm (89 lb-in) 29

28. Knock sensor (KS) electrical connector
29. Intake manifold bolt and grommet assembly (8 required)
30. Intake manifold
31. Heated positive crankcase ventilation (PCV) fitting bolts (2 required)
32. Heated PCV fitting
33. Intake manifold gaskets (6 required)

22140_BPUP_G0135

Fig. 109 Intake manifold component identification part 5—4.0L engine

7. Detach the accelerator and speed control cables from the throttle body.
8. Detach the accelerator and speed control cables from the bracket, and position the cables aside.
9. Disconnect the Exhaust Gas Recirculation (EGR) valve vacuum hose and tube fitting.
10. Disconnect the Exhaust Gas Recirculation (EGR) vacuum regulator solenoid valve electrical connector and vacuum hose.
11. Loosen the clamp and disconnect the brake booster vacuum hose.

➡It is important to twist the spark plug wire boots while pulling upward to avoid possible damage to the spark plug wire.

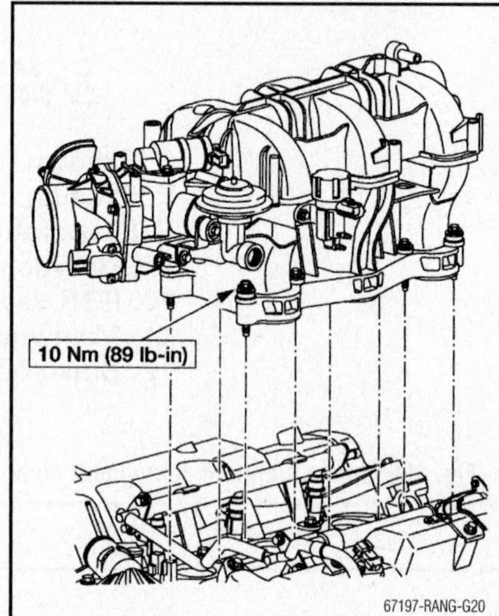

67197-RANG-G20

Fig. 110 Intake manifold installation—4.0L engine

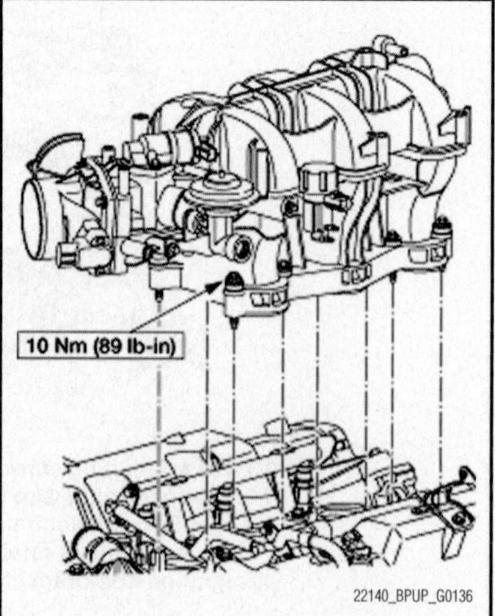

22140_BPUP_G0136

Fig. 111 Intake manifold installation—4.0L engine

➡Mark the spark plug wire locations before removing them.

12. Disconnect the RH spark plug wires from the coil. Remove the spark plug wire routing clip pin-type retainer and position the wires aside.
13. Remove the wiring harness bracket retainer, then position the wiring harness aside.
14. Remove the accelerator cable routing clip pin-type retainer and the wiring harness pin-type retainer.
15. Remove the bolts.
16. Remove the bolts and position the coil and bracket aside.
17. Disconnect the vacuum hoses.
18. Remove the nut.
19. Disconnect the Powertrain Control Module (PCM) electrical connector.
20. Remove the retainer and position the ground wires aside.
21. Detach the electrical connector retainer.
22. Remove the intake manifold bolts and lift up the intake manifold.
23. Remove the heated Positive Crankcase Ventilation (PCV) hose retainers and remove the heated PCV fitting.
24. Remove the intake manifold retaining bolts. Remove the intake manifold from the engine.

To install:

25. Be sure to clean all old gasket material from the sealing surfaces, using the proper tools.
26. Installation is the reverse of the removal procedure.
27. Be sure to use new gaskets.
28. Remove the bolts.
29. Remove the accelerator cable routing clip pin-type retainer and the wiring harness pin-type retainer.
30. Remove the wiring harness bracket retainer, then position the wiring harness aside.
31. Disconnect the RH spark plug wires from the coil. Remove the spark plug wire routing clip pin-type retainer and position the wires aside.
32. Connect the brake booster vacuum hose.
33. Disconnect the EGR vacuum regulator solenoid valve electrical connector and vacuum hose.
34. Disconnect the EGR valve vacuum hose and tube fitting.
35. Detach the accelerator and speed control cables from the bracket, and position the cables aside.
36. Detach the accelerator and speed control cables from the throttle body.
37. Disconnect the MAF sensor wiring pin-type retainer.
38. Disconnect the IAC valve, TP sensor electrical connectors and the TP sensor wiring pin.
39. Remove the air cleaner outlet pipe
40. Torque the intake manifold bolts to 89 inch lbs. (10 Nm).

OIL PRESSURE SENSOR

REMOVAL & INSTALLATION

1. Disconnect the negative battery cable.
2. Disconnect the wiring at the sender/switch.
3. Remove the oil pressure sender/switch from the engine.
4. To install, reverse removal procedure.
5. Apply sealant to oil pressure sender threads.
6. Torque the oil pressure sender to 11 ft. lbs. (15 Nm).

OIL PAN

REMOVAL & INSTALLATION

2.3L Engine

See Figures 112 through 117.

1. Before servicing the vehicle, refer to the precautions in the beginning of this section.
2. Drain the engine oil.
3. Only the LH side of the engine will be raised. Remove the LH engine support insulator-to-engine bracket nuts.
4. Install the lifting eyes 301—D030 or similar hoist ring.
5. Install the 3–Bar Engine Support Kit 301—F072 and raise the LH side of the engine approximately 25 mm (1 inch).
6. Detach the wiring harness retainers from the engine front cover.
7. Remove the 4 engine front cover-to-oil pan bolts.
8. Remove the transmission bracket-to-transmission bolts.
9. Remove the 4 transmission-to-oil pan bolts.

⁂ WARNING

To prevent damage to the transmission, do not loosen the transmission-to-engine bolts more than 5 mm (0.19 inch).

10. Loosen the 7 transmission-to-engine bolts 5 mm (0.19 inch).

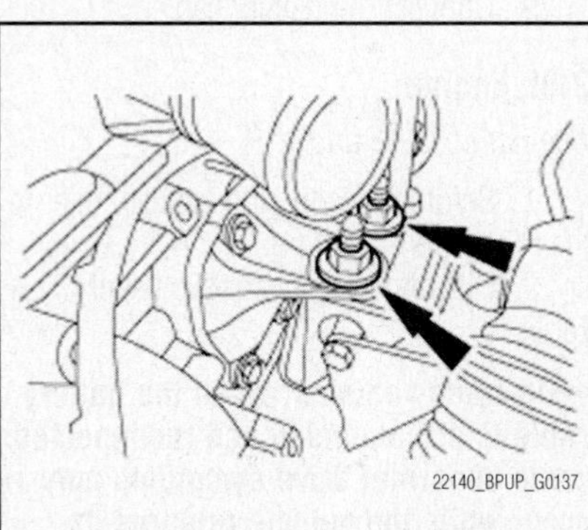

Fig. 112 LH engine support insulator-to-engine bracket nuts—2.3L engine

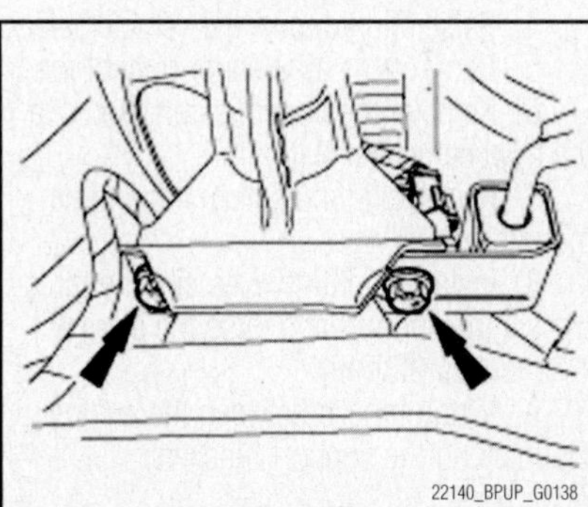

Fig. 113 Transmission bracket-to-transmission bolts—2.3L engine

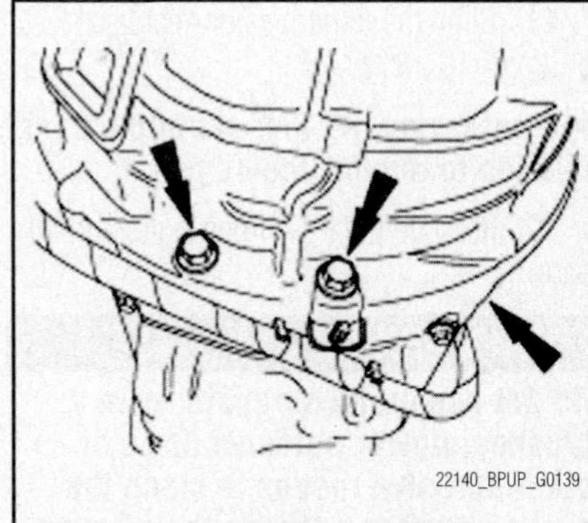

Fig. 114 4 transmission-to-oil pan bolts—2.3L engine

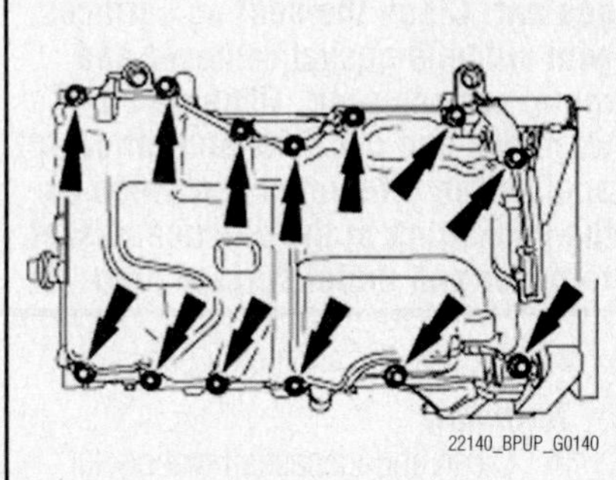

Fig. 115 13 oil pan bolts and oil pan—2.3L engine

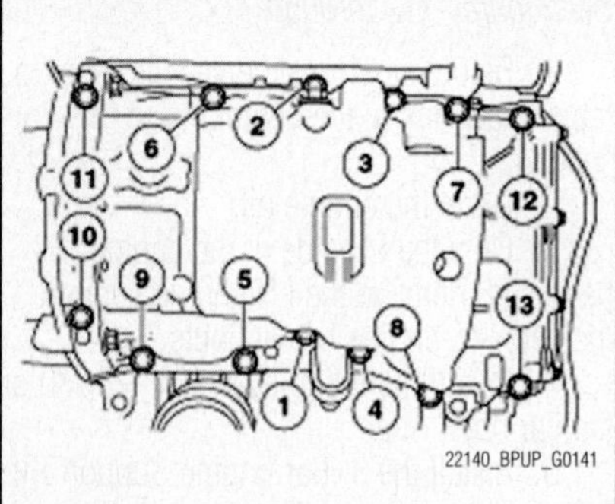

Fig. 116 Oil pan bolt tightening sequence—2.3L engine

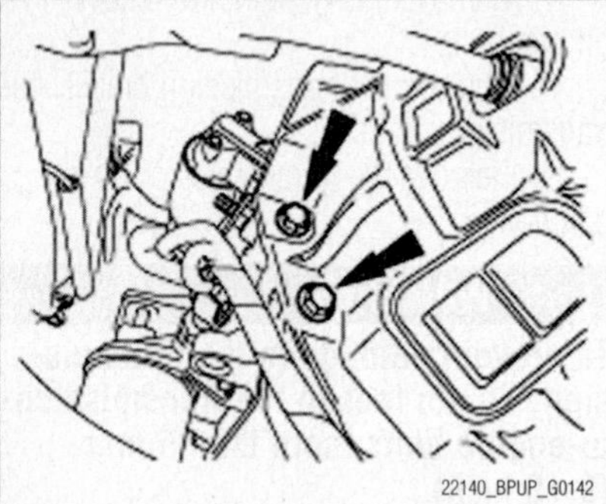

Fig. 117 Transmission-to-engine bolts location—2.3L engine

11. Slide the transmission rearward 5 mm (0.19 inch).

➡If necessary, lift the rear of the transmission to remove the oil pan.

12. Remove the 13 oil pan bolts and oil pan.

⁂ WARNING

Do not use metal scrapers, wire brushes, power abrasive discs or other abrasive means to clean the sealing surfaces. These tools cause scratches and gouges, which make leak paths. Use a plastic scraping tool to remove all traces of old sealant. Clean the sealing surfaces with silicone gasket remover and metal surface prep. Observe all warnings and cautions and follow all application directions contained on the packaging of the silicone gasket remover and metal surface prep.

13. Remove the engine oil pan.

To install:

14. Clean and inspect all mating surfaces.

➡The oil pan must be installed and the bolts tightened with four minutes of applying the silicone gasket and sealant.

15. Apply a 2.5 mm bead of silicone gasket and sealant to the oil pan. Install the oil pan. Tighten the oil pan in the sequence shown.
16. Lubricate the O-ring with clean engine oil and install the engine oil level indicator assembly.
17. Install the engine front cover-to-oil pan bolts.
18. Torque to 89 inch lbs. (10 Nm).
19. Tighten the oil pan bolts in the sequence shown.
20. Torque to 18 ft. lbs. (25 Nm).
21. Attach the wiring harness retainers to the engine front cover.
22. Slide the transmission forward and tighten the 7 transmission-to-engine bolts.
23. Torque to 35 ft. lbs. (48 Nm).
24. Install the 4 transmission-to-oil pan bolts. Torque to 35 ft. lbs. (48 Nm).
25. Install the transmission bracket-to-transmission bolts. Torque to 66 ft. lbs. (90 Nm).
26. Lower the engine and remove the 3 bar engine support kit and the lifting eyes.
27. Install the engine support insulator nuts and torque to 75 ft. lbs. (102 Nm).
28. Install the oil level tube assembly.
29. Connect the battery cable.

3.0L Engine

See Figures 118 and 119.

1. Before servicing the vehicle, refer to the Precautions Section.
2. Disconnect the negative battery cable.

➡On some vehicles, when the battery cable is disconnected and reconnected, some abnormal drive symptoms may occur while the vehicle relearns its adaptive strategy. The vehicle may need to be driven to relearn its strategy.

3. Drain the engine oil.
4. Raise and support the vehicle safely.
5. Remove the air cleaner outlet tube.
6. Remove the transmission. Remove the flywheel or flexplate.
7. Remove the nuts from the engine mounts.
8. Install the lifting eyes and the three-bar engine support and raise the engine 1.72 inches (43mm).
9. Install blocks between the engine mounts and the support brackets.

➡If equipped with coil springs, remove the air cleaner outlet tube. Remove the nuts from the engine mounts. Using an engine lifting device, raise the engine about 1.72 inches. Install blocks between the engine mounts and the support brackets.

10. Remove the oil pan retaining bolts. Remove the oil pan. Discard the oil pan gasket.

To install:

11. Apply a ⅛ (4mm) bead of RTV sealer to the junctions of the rear main bearing cap and block, and the front cover and block. The oil pan must be installed within 15 minutes, after applying the sealant.
12. Position a new pan gasket on the pan. Be sure to align the fastener holes and secure the gasket with gasket clips. Install the oil pan.

➡Make sure the fasteners are clean and dry prior to installation.

13. Tighten the oil pan bolts in the sequence shown.
14. Tighten the oil pan bolts to 9 ft. lbs. (12 Nm).
15. Remove the blocks from between the engine mounts and the support brackets.
16. Lower the engine and remove the three-bar engine support and the lifting eyes.

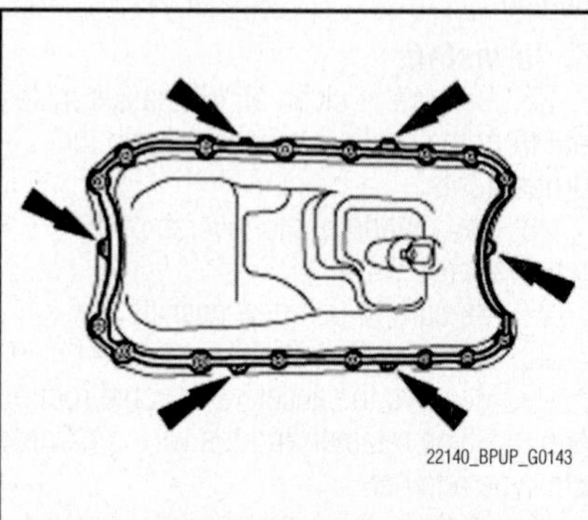

Fig. 118 Securing the gasket with the gasket clips—3.0L engines

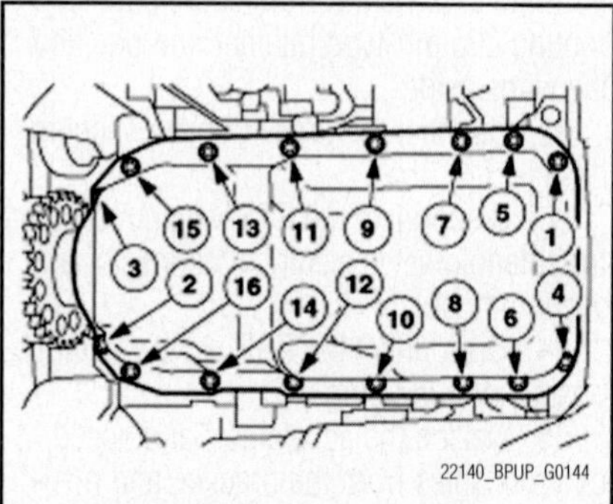

Fig. 119 Oil pan tightening sequence—3.0L engines

17. Install the nuts.
18. Install the air cleaner outlet tube.
19. Install the flywheel or flexplate.
20. Fill the engine with clean engine oil.
21. Connect the battery cable.
22. Start the engine and check for leaks.

4.0L Engine

See Figure 120.

1. Before servicing the vehicle, refer to the Precautions Section.
2. Disconnect the negative battery cable.

➡On some vehicles, when the battery cable is disconnected and reconnected, some abnormal drive symptoms may occur while the vehicle relearns its adaptive strategy. The vehicle may need to be driven to relearn its strategy.

3. With the vehicle in neutral, position. raise and support the vehicle safely.
4. Drain the engine oil.
5. Remove the oil pan retaining bolts.
6. Remove the oil pan and discard the gasket

To install:

7. Torque the bolts to 19 ft. lbs. (26 Nm).
8. Install the negative battery cable.
9. Fill the engine with clean oil.
10. Start the vehicle and check for leaks, repair if necessary.

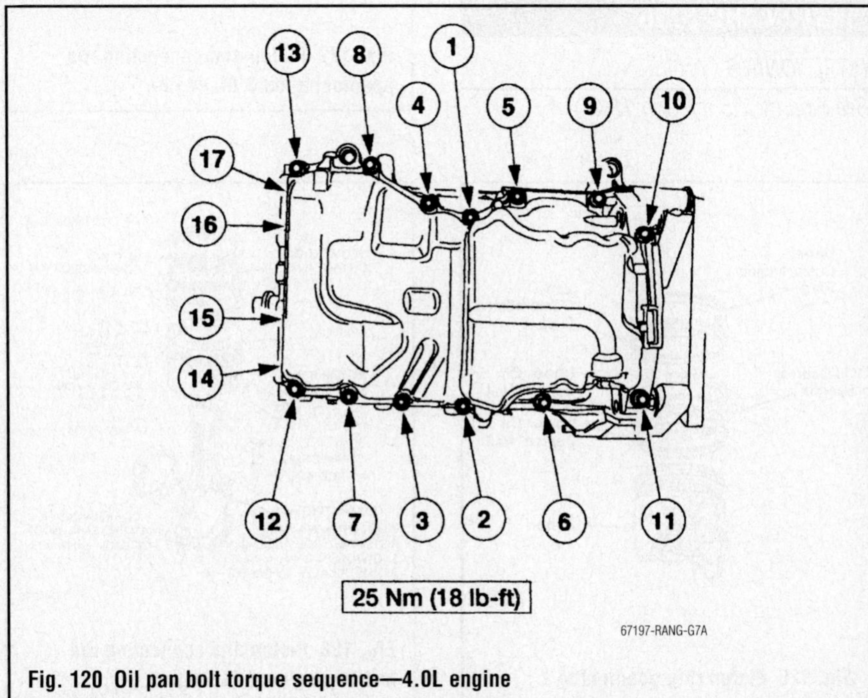

Fig. 120 Oil pan bolt torque sequence—4.0L engine

OIL PUMP

REMOVAL & INSTALLATION

2.3L Engine

See Figures 121 and 122.

➡The oil pump is located on the front of the engine and is turned by the timing belt.

1. Before servicing the vehicle, refer to the precautions in the beginning of this section.
2. Remove or disconnect the following:
   - Negative battery cable
   - Timing chain
   - Oil pump chain and sprockets
   - Oil pan
   - Oil pump pickup tube and gasket
   - Oil pump assembly and gasket

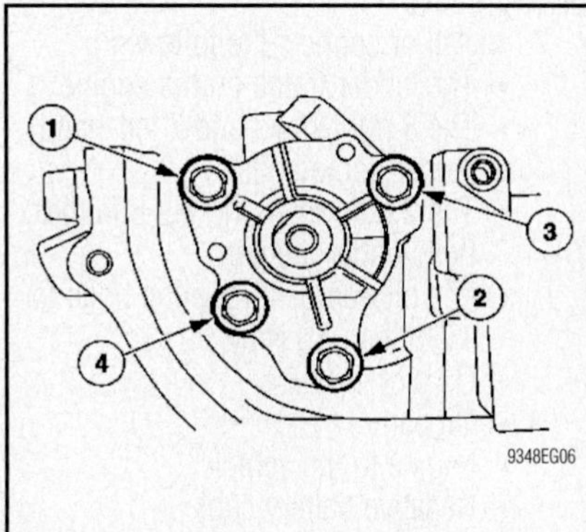

Fig. 121 Oil pump torque sequence—2.3L engine

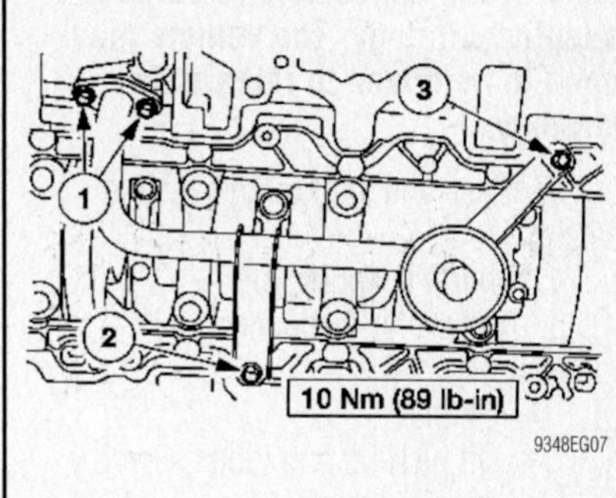

Fig. 122 Oil pump pickup tube torque sequence—2.3L engine

To install:

3. Turn the crankshaft clockwise to position the No. 1 piston.
4. Remove the plug bolt.
5. Install the engine timing peg 303-507.

➡Clean the gasket surface with metal surface cleaner.

6. Install a new gasket and the oil pump assembly. Tighten the bolts in the sequence shown in two stages.
   - Step 1: Tighten the bolts to 80 inch lbs. (10 Nm)
   - Step 2: Tight the bolts to 17 ft. lbs. (23 Nm)
7. Install a new oil pump pickup tube gasket and the pickup tube. Tighten the bolts in the sequence shown

3.0L Engine

See Figure 123.

1. Before servicing the vehicle, refer to the Precautions Section.
2. Disconnect the negative battery cable.

➡On some vehicles, when the battery cable is disconnected and reconnected, some abnormal drive symptoms may

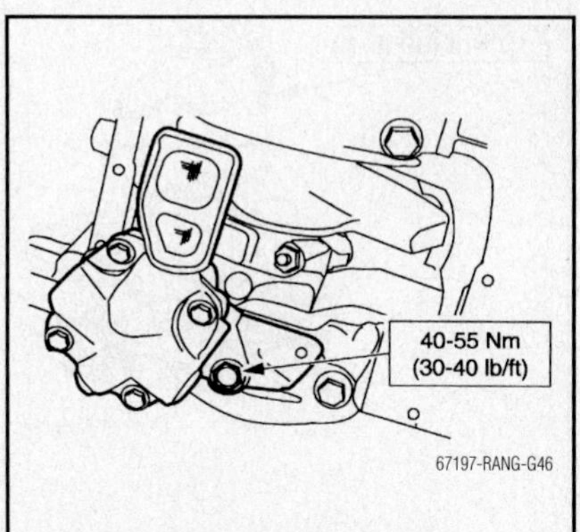

Fig. 123 Oil pump installed—3.0L engines

occur while the vehicle relearns its adaptive strategy. The vehicle may need to be driven to relearn its strategy.

3. Raise and support the vehicle safely.
4. Drain the engine oil.
5. Remove or disconnect the following:
    - Oil pan
    - Oil pick-up and tube assembly from the pump
    - Oil pump retainer bolts and the oil pump

To install:

6. Prime the oil pump with clean engine oil by filling either the inlet or outlet port. Rotate the pump shaft to distribute the oil within the pump body.
7. Install the oil pump and tighten the mounting bolts to 35 ft. lbs. (46 Nm).
8. Install or connect the following:
    - Oil pick-up and tube assembly
    - Oil pan
9. Fill the engine with clean oil.
10. Start the vehicle and check for leaks, repair if necessary.

4.0L Engine

See Figure 124.

➡The oil pump cannot be removed with the engine in the vehicle.

1. Before servicing the vehicle, refer to the Precautions Section.
2. Disconnect the negative battery cable.

➡On some vehicles, when the battery cable is disconnected and reconnected, some abnormal drive symptoms may occur while the vehicle relearns its adaptive strategy. The vehicle may need to be driven to relearn its strategy.

3. Raise and support the vehicle safely.
4. Drain the engine oil.
5. Remove or disconnect the following:
    - Engine from the vehicle
    - Oil pan
    - Unbolt the oil pick-up tube
    - The 8 ladder frame bolts that were under the oil pan
    - The 2 rear outer ladder frame bolts
    - The 7 left-hand and the 8 right-hand ladder frame bolts
    - The ladder frame from the engine
    - The 2 oil pump attaching bolts and the pump

To install:

6. Submerge the pump in clean engine oil to prime it.
7. Install or connect the following:
    - The ladder frame on the engine
    - The 8 right-hand and 7 left-hand ladder frame bolts
    - The 2 rear outer and the 8 frame bolts under the pan
    - The oil pump. Torque the bolts to 14 ft. lbs. (19 Nm).
    - Oil pick-up tube
    - Oil pan
    - Engine to the vehicle
    - Negative battery cable
8. Fill the engine with clean oil.
9. Start the vehicle and check for leaks, repair if necessary.

PISTON AND RING

POSITIONING

See Figures 125 through 128.

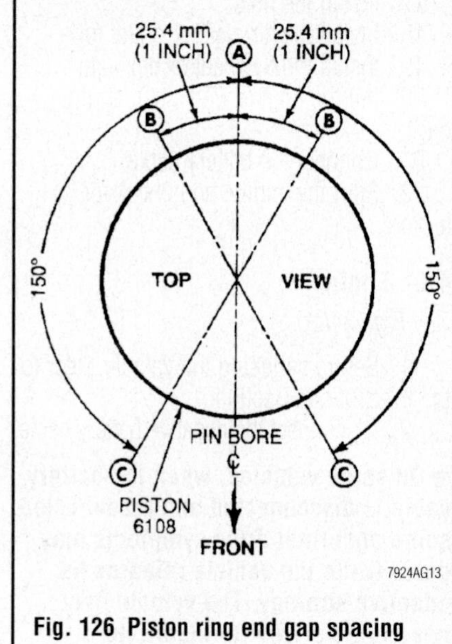

Fig. 126 Piston ring end gap spacing

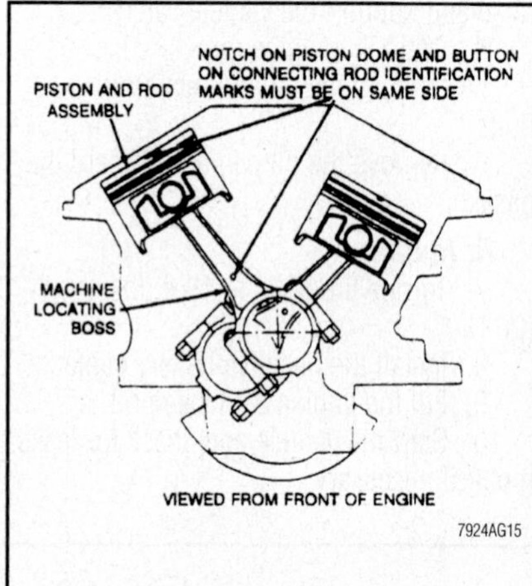

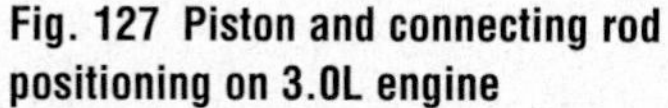

Fig. 127 Piston and connecting rod positioning on 3.0L engine

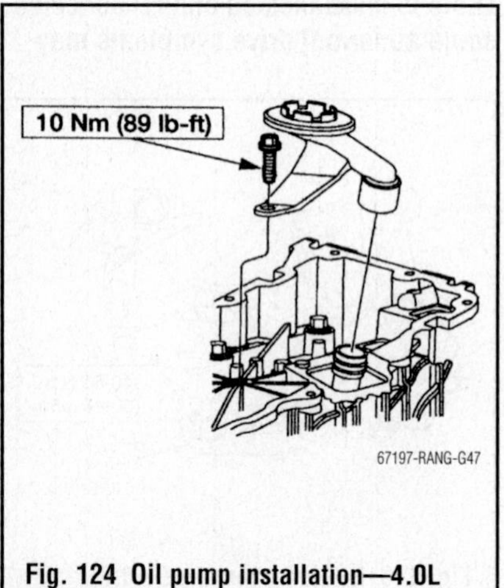

Fig. 124 Oil pump installation—4.0L engine

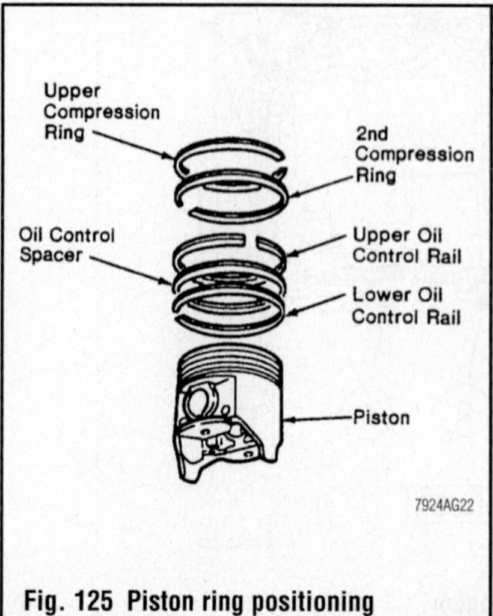

Fig. 125 Piston ring positioning

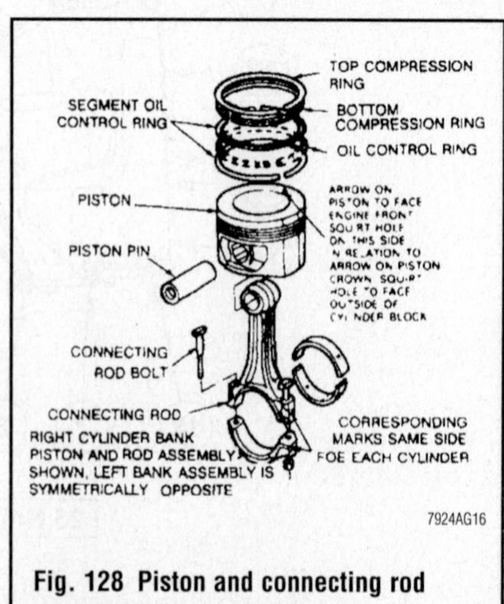

Fig. 128 Piston and connecting rod positioning on 4.0L engine

REAR MAIN SEAL

REMOVAL & INSTALLATION

2.3L Engine

See Figure 129.

1. Before servicing the vehicle, refer to the Precautions Section.
2. Disconnect the negative battery cable.

➡On some vehicles, when the battery cable is disconnected and reconnected, some abnormal drive symptoms may occur while the vehicle relearns its adaptive strategy. The vehicle may need to be driven to relearn its strategy.

3. Raise and support the vehicle safely.
4. Remove the flywheel or flexplate.
5. Remove the oil pan.
6. Remove the bolts and the crankshaft rear oil seal.

To install:

7. Install the crankshaft rear main oil seal.
8. Tighten the bolts in the sequence shown to 7 ft. lbs. (10 Nm).
9. Install the oil pan.
10. Install the flywheel or flexplate.
11. Connect the negative battery cable.

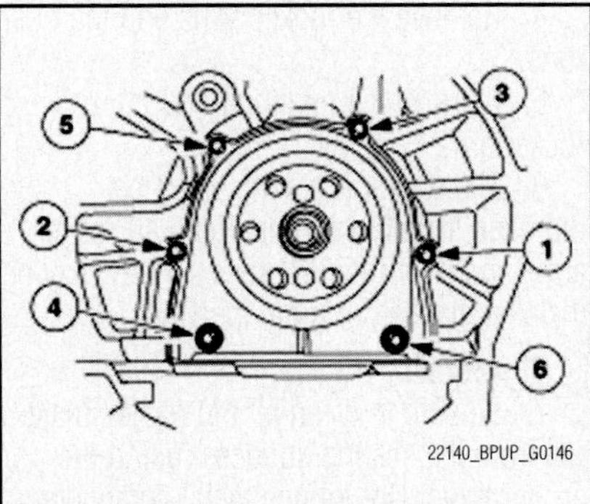

Fig. 129 Rear main seal tightening sequence—2.3L engine

3.0L Engine

See Figures 130 and 131.

1. Before servicing the vehicle, refer to the Precautions Section.
2. Disconnect the negative battery cable.
3. Remove the flexplate or flywheel.

⁂ WARNING

Use care to avoid scratching or damaging the oil seal surface or leakage may occur.

4. Using a sharp awl, punch one hole into the crankshaft rear oil seal metal surface between the seal lip and the cylinder block.

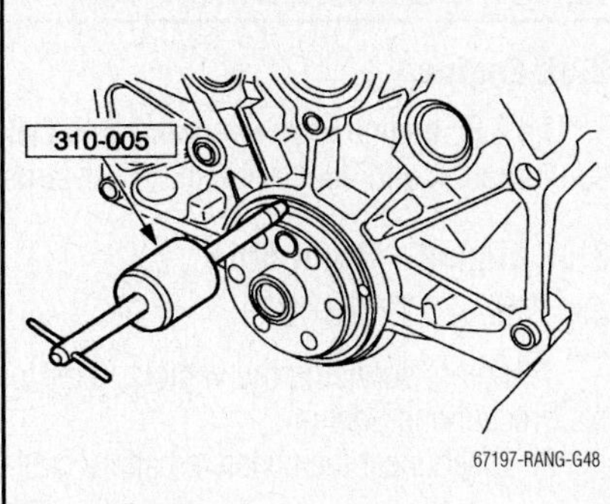

Fig. 130 Rear main seal removal—3.0L engine

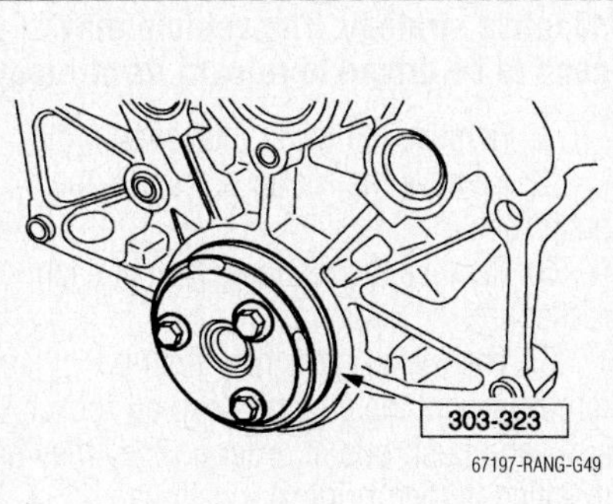

Fig. 131 Rear main seal installation—3.0L engine

5. Screw the threaded end of the special tool into the oil seal. Use the special tool to remove the crankshaft rear oil seal.

To install:

6. Lubricate the outer lips and the inner seal on the crankshaft rear oil seal with clean engine oil.
7. Using the special tool, install the crankshaft rear oil seal.
8. Alternate bolt tightening to correctly seat the crankshaft rear oil seal.
9. Install the flexplate or flywheel.
10. Connect the negative battery cable.

4.0L Engine

See Figures 132 through 135.

1. Before servicing the vehicle, refer to the Precautions Section.
2. Remove the flexplate or flywheel.
3. Disconnect the negative battery cable.

⁂ WARNING

Avoid scratching or damaging the oil crankshaft seal running surface during removal of the crankshaft rear oil seal.

4. Using the special tool303—409 remove the crankshaft rear oil seal.

To install:

➡Be sure the crankshaft rear sealing surface is clean and free of any rust or corrosion. To clean the crankshaft rear

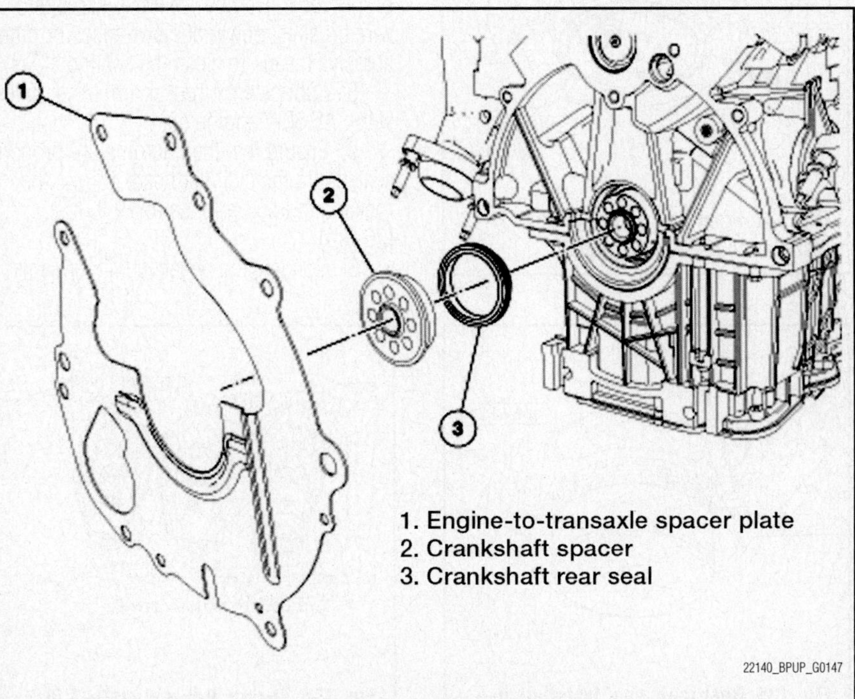

Fig. 132 Rear main seal exploded view and component locations—4.0L engine

sealing surface, use extra-fine emery cloth or extra-fine 0000 steel wool with metal surface cleaner.

5. Lubricate the crankshaft rear oil seal with clean engine oil and install on the rear main seal protector, which is part of the crankshaft rear oil seal service set 49 UN—01—1360.
6. Using the special tool 49 UN—01—1360, install the crankshaft rear oil seal.
7. Remove all the special tools.
8. Install the crankshaft spacer.
9. Install the crankshaft spacer plate.

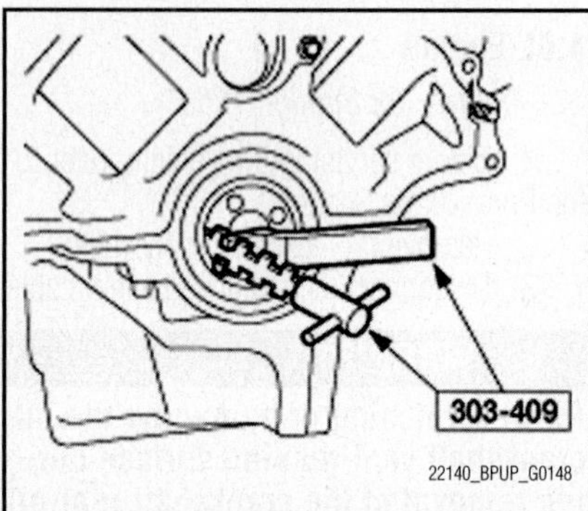

Fig. 133 Rear main seal exploded view and component locations—4.0L engine

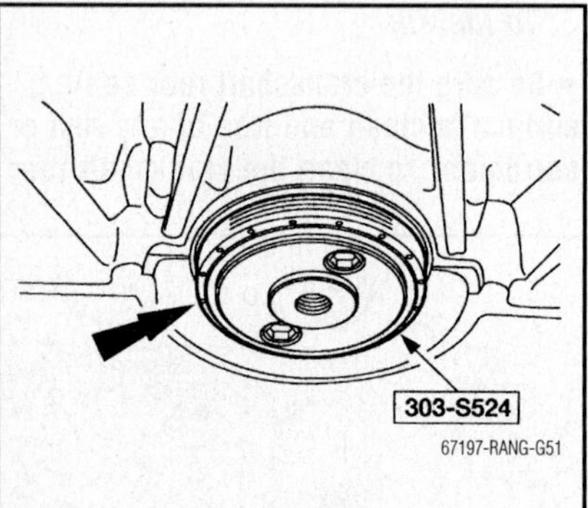

Fig. 134 Front part of rear main seal installation tool—4.0L engine

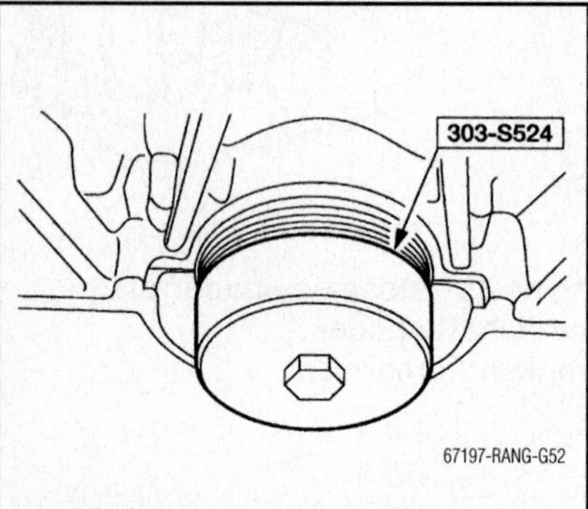

Fig. 135 Rear main seal installation—4.0L engine

10. Install the flexplate or flywheel.
11. Connect the negative battery cable.

ROCKER ARMS/SHAFTS

REMOVAL & INSTALLATION

2.3L Engine

The 2.3L engine is an overhead camshaft configuration and does not use rocker arms.

3.0L Engine—Right Side

See Figures 136 and 137.

1. Before servicing the vehicle, refer to the Precautions Section.
2. Disconnect the negative battery cable.

➡On some vehicles, when the battery cable is disconnected and reconnected, some abnormal drive symptoms may occur while the vehicle relearns its adaptive strategy. The vehicle may need to be driven to relearn its strategy.

3. Remove the lower intake manifold.
4. Remove the rocker arms and the push rods.
5. Remove the retaining bolt at each rocker arm.
6. The rocker arms may then be removed from the engine. Keep all rocker arms and pushrods in order so they may be installed in their original locations.

To install:

7. Be sure to clean all old gasket material from the sealing surfaces, using the proper tools. Do not use metal scrappers, wire brushes, power abrasive discs or other abrasive means to clean the sealing areas.
8. Lubricate the rocker arm assemblies with SAE 50W engine oil.
9. Ensure that the fulcrums are properly seated into the cylinder head. Torque the rocker arm fulcrum bolts to 19 ft. lbs. (26 Nm).
10. Connect the negative battery cable.

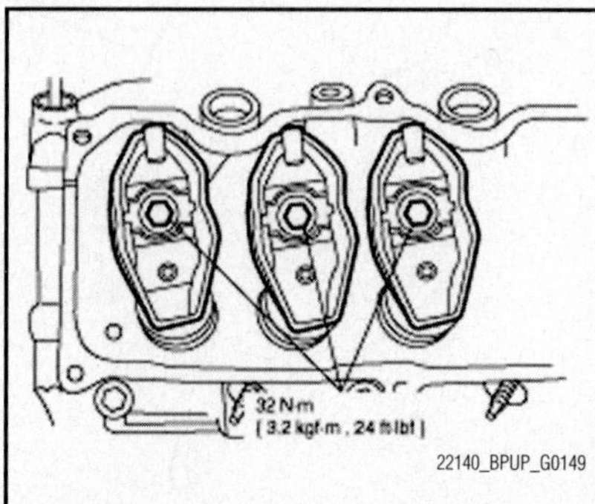

Fig. 136 Rocker arm removal—3.0L engine

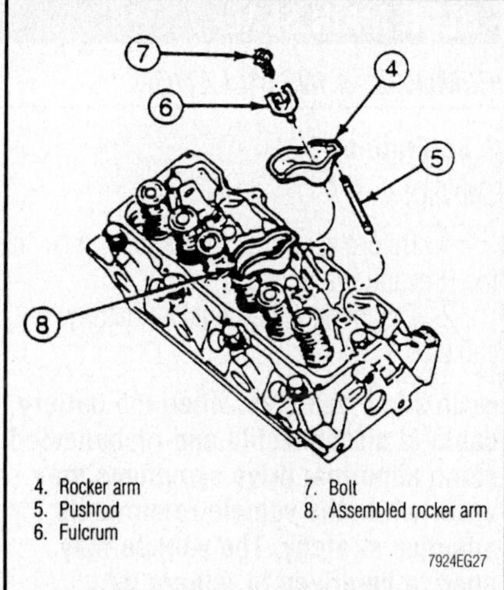

Fig. 137 Rocker arm and related components—3.0L engine

4.0L Engine

1. Before servicing the vehicle, refer to the Precautions Section.
2. Disconnect the negative battery cable.

➡On some vehicles, when the battery cable is disconnected and reconnected, some abnormal drive symptoms may occur while the vehicle relearns its adaptive strategy. The vehicle may need to be driven to relearn its strategy.

3. Remove the lower intake manifold.
4. Remove the rocker arms and the push rods.
5. Remove the retaining bolt at each rocker arm.
6. The rocker arms may then be removed from the engine. Keep all rocker arms and pushrods in order so they may be installed in their original locations.

To install:

7. Be sure to clean all old gasket material from the sealing surfaces, using the proper tools. Do not use metal scrappers, wire brushes, power abrasive discs or other abrasive means to clean the sealing areas.
8. Lubricate the rocker arm assemblies with SAE 50W engine oil.
9. Ensure that the fulcrums are properly seated into the cylinder head. Torque the rocker arm fulcrum bolts to 19 ft. lbs. (26 Nm).

TIMING CHAIN COVER AND SEAL

REMOVAL & INSTALLATION

2.3L Engine

See Figures 138 through 140.

➡On some engines the crankshaft, crankshaft sprocket and the pulley are fitted together by friction, with diamond washers between the flange on each part. For that reason, the crankshaft sprocket is also unfastened if you loosen the pulley. Therefore, the engine must be retimed each time the damper is removed. Otherwise severe damage can occur.

1. Before servicing the vehicle, refer to the Precautions Section.
2. Relieve the fuel system pressure.
3. Disconnect the negative battery cable.

➡On some vehicles, when the battery cable is disconnected and reconnected, some abnormal drive symptoms may occur while the vehicle relearns its adaptive strategy. The vehicle may need to be driven to relearn its strategy.

4. Drain the cooling system.
5. Properly discharge the air conditioning system.
6. Remove or disconnect the following:
   - Drive belt
   - Accessory drive belt tensioner
   - Cooling fan drive pulley
   - Coolant pump pulley
   - Power steering pump
   - Engine oil level indicator assembly
   - Engine oil level indicator
   - Engine oil level indicator tube
   - Water outlet tube
   - Water outlet tube
   - Air conditioning compressor

➡The alternator will be removed with the accessory bracket.

   - Accessory bracket
   - Right motor mount.
   - Coolant hose from the thermostat
   - Coolant hose from the Exhaust Gas Recirculation (EGR) valve
   - Coolant tube assembly
   - Exhaust manifold and gasket
   - Block heater (if so equipped)
   - Water outlet
   - EGR valve
   - Power steering pump and reservoir as an assembly
   - Idle Air Control (IAC) valve
   - Throttle Position (TP) sensor
   - Manifold Absolute Pressure (MAP) sensor
   - Swirl control valve monitor electrical connector
   - Crankshaft Position (CKP) sensor and the wiring harness pin-type retainers
   - Knock Sensor (KS)
   - Electric thermostat
   - Swirl control valve
   - Camshaft Position (CMP) sensor electrical connector and disconnect the Positive Crankcase Ventilation (PCV) hose from the intake manifold
   - Engine wiring harness pin-type retainers from the intake manifold
   - Engine wiring harness connector bracket. Position the engine wiring harness aside
   - Exhaust Gas Recirculation (EGR) tube
   - Fuel supply line clip from the front of the intake manifold. Disconnect the vacuum hose from the intake manifold
   - Intake manifold assembly
   - Fuel injector electrical connectors. Detach the wiring harness pin-type retainers
   - Ignition coil and the Cylinder Head Temperature (CHT) sensor electrical connectors
   - Engine wiring harness anchors from the valve cover studs. Remove the engine wiring harness
   - Ignition coil
   - Bypass hose
   - Thermostat housing
   - Knock sensor and the engine vent cover
   - Left motor mount
   - Fuel injector supply manifold with the injectors and the ground strap
   - Water pump pulley
   - Water pump
   - CMP sensor
   - CHT sensor
   - Spark plugs
   - Valve cover
   - CKP sensor
   - Crankshaft vibration damper

Fig. 138 Front cover tightening sequence—2.3L engine

➡There is one front cover bolt behind the cooling fan drive pulley. To remove this bolt, align one of the cooling fan drive pulley access holes with the bolt head to access the bolt.

7. Remove the front cover.

To install:

8. Install the front cover
9. Tighten the bolts in the sequence shown, to the following specifications:
   - 8 mm bolts—7 ft. lbs. (10 Nm).
   - 10 mm bolts—18 ft. lbs. (25 Nm).
   - 13 mm bolts—35 ft. lbs. (48 Nm).

⁂ WARNING

There is one bolt behind the cooling fan drive pulley. This bolt can be accessed by lining up one of the holes in the pulley with the bolt.

10. Position the power steering pump and install the lower retaining bolt.
11. Tighten the power steering pump bolts to 18 ft. lbs. (25 Nm).
12. Connect the power steering pump switch electrical connector.
13. Install the water pump pulley.
14. Tighten the water pump pulley bolts to 15 ft. lbs. (20 Nm).
15. Install the belt tensioner.
16. Tighten the belt tensioner bolts to 36 ft. lbs. (50 Nm).

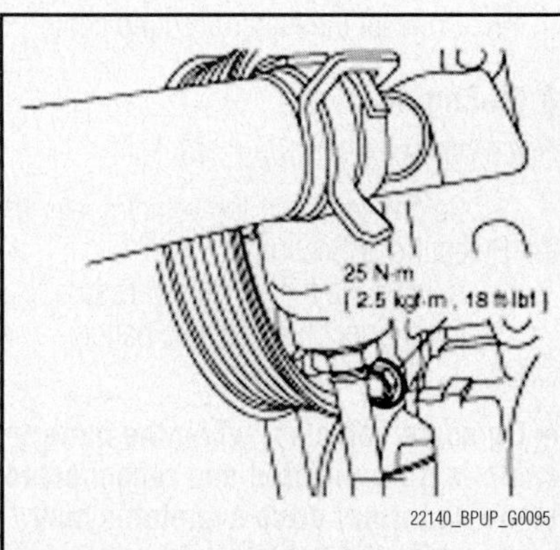

Fig. 139 Power steering pump and install the lower retaining bolt—2.3L engine

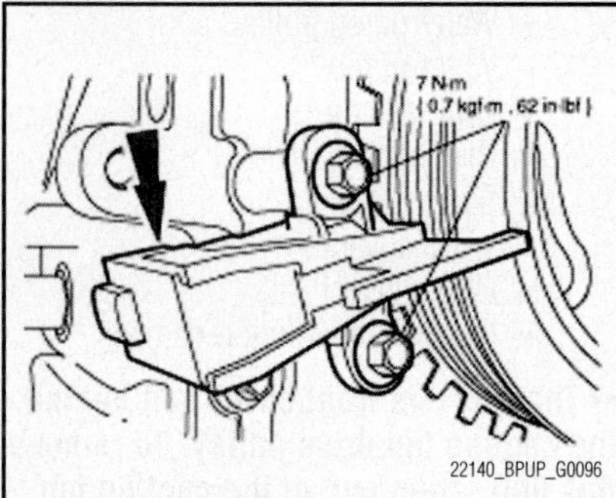

Fig. 140 CKP sensor with the alignment tool—2.3L engine

WARNING

The crankshaft, the crankshaft sprocket and the pulley are fitted together by friction, with diamond washers between the flange faces on each part. For that reason, the crankshaft sprocket is also unfastened if you loosen the pulley. Therefore, the engine must be retimed each time the damper is removed. Otherwise severe engine damage can occur.

17. Install the crankshaft vibration damper:

➡Do not reuse the crankshaft pulley bolt. Tighten the bolt in two stages.

18. Tighten the water pump pulley bolts to 15 ft. lbs. (20 Nm).

➡A new CKP sensor must be installed whenever the old sensor is removed. Install a new CKP sensor, do not tighten the bolts at this time.

19. Tighten the CKP sensor bolts to 82 inch. lbs. (7 Nm).
20. Connect the CKP sensor electrical connector and the wiring harness pin-type retainers.
21. Connect the battery ground cable.

3.0L Engine

See Figures 141 through 148.

1. Before servicing the vehicle, refer to the Precautions Section.
2. Relieve the fuel system pressure.
3. Disconnect the negative battery cable.

➡On some vehicles, when the battery cable is disconnected and reconnected, some abnormal drive symptoms may occur while the vehicle relearns its adaptive strategy. The vehicle may need to be driven to relearn its strategy.

4. Drain the cooling system.
5. Remove the fan and shroud.
6. Remove the coolant pump pulley.
7. Remove the nut and detach the wiring harness bracket.
8. Remove the nut and position the bracket aside.
9. Remove the stud bolts from the A/C compressor mounting bracket.
10. Remove the bolt and the accessory drive belt tensioner.
11. Remove the bolts and position the power steering bracket and compressor aside.
12. Disconnect the heater hose at the water pump.
13. Disconnect the radiator lower hose at the coolant pump outlet.
14. Remove the crankshaft front oil seal.
15. Remove the oil pan.
16. Remove the Crankshaft Position (CKP) sensor.
17. Disconnect the CKP sensor electrical connector.
18. Remove the 2 bolts.
19. Remove the CKP sensor.
20. Remove the vibration damper bolt and the washer.
21. Using Crankshaft Damper Remover 303—009, remove the crankshaft damper.
22. Remove the crankshaft key, if required
23. Remove the engine front cover and the coolant pump as an assembly.
24. Remove the bolts.
25. Remove and discard the front cover gasket.

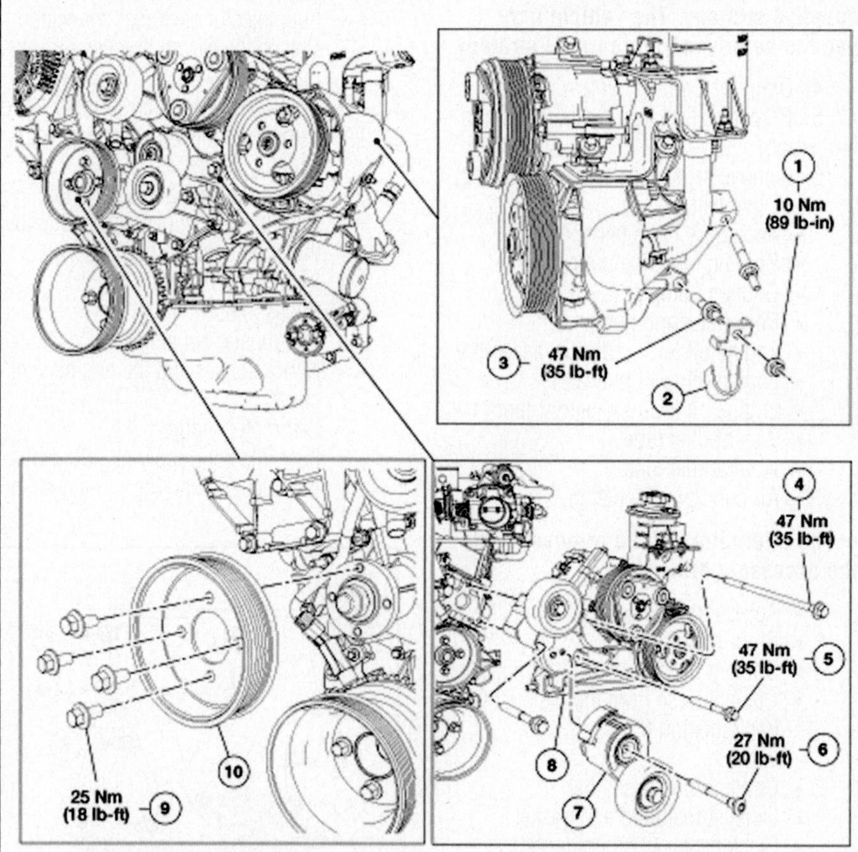

Fig. 141 Engine front cover removal and installation exploded view of accessories part 1—3.0L engine

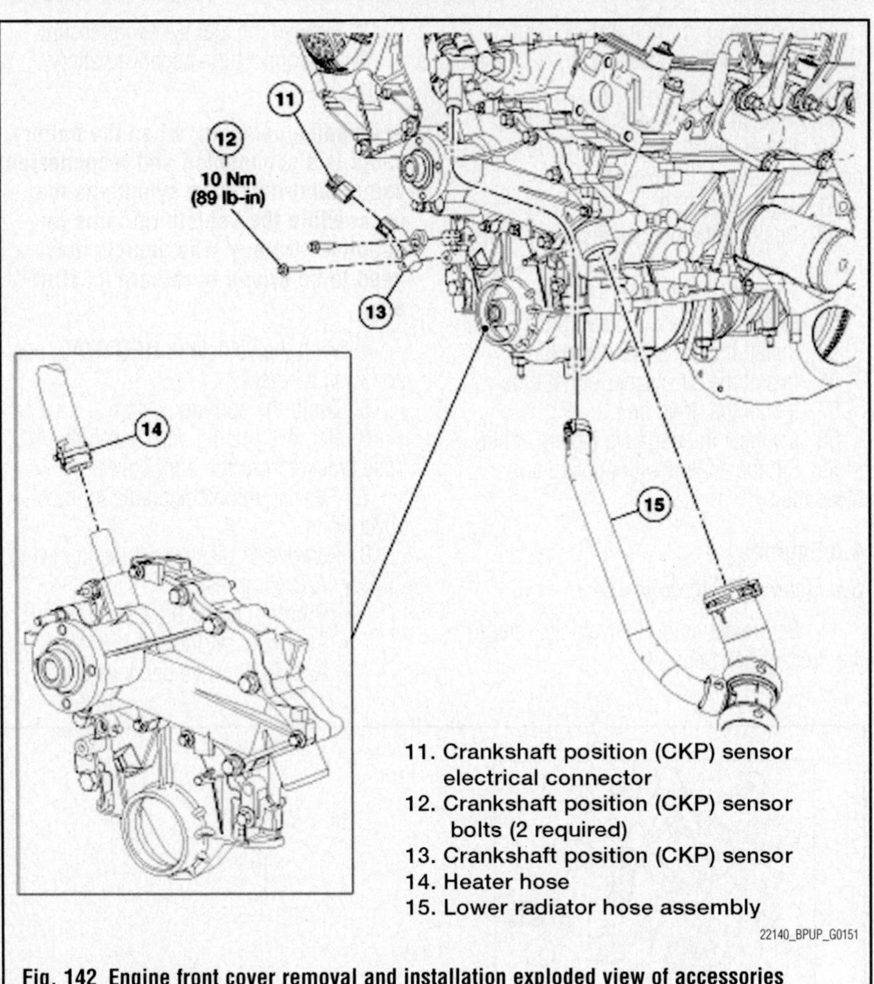

Fig. 142 Engine front cover removal and installation exploded view of accessories part 2—3.0L engine

Fig. 144 Crankshaft damper remover 303—009

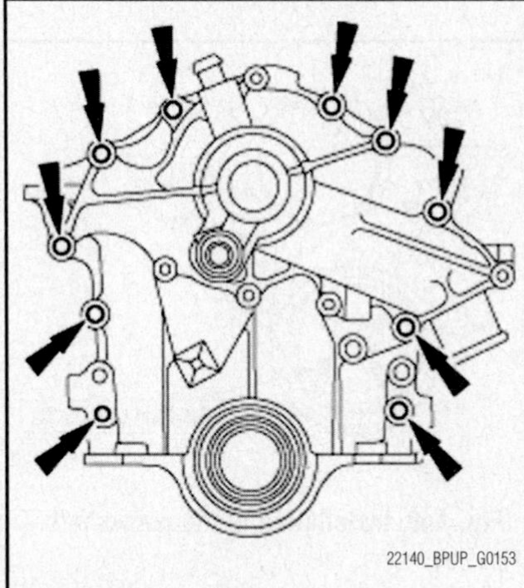

Fig. 145 Engine front cover and the coolant pump as an assembly—3.0L engine

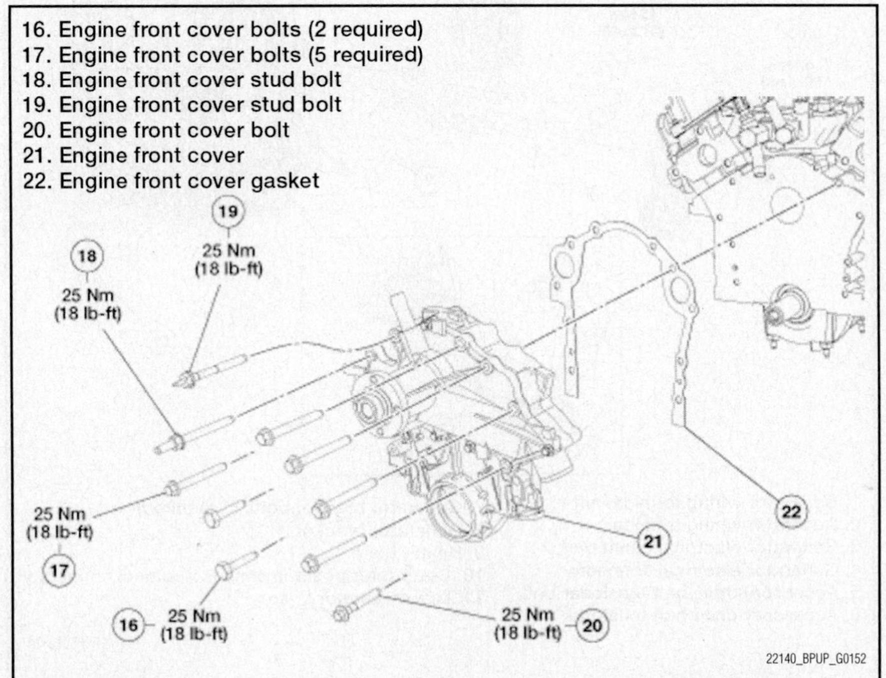

Fig. 143 Engine front cover removal and installation exploded view of accessories part 3—3.0L engine

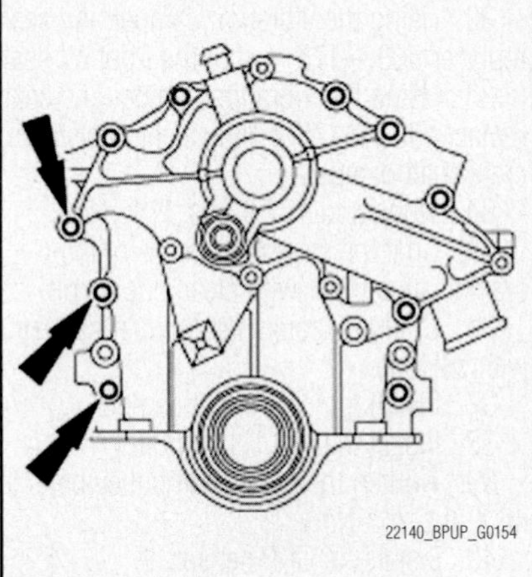

Fig. 146 Engine front cover and the bolts to be installed with apply pipe sealant with Teflon—3.0L engine

To install:

26. Clean all sealing surfaces and position a new front cover gasket.
27. Position the engine front cover.
28. Apply pipe sealant with Teflon® to the bolts to be installed in the locations shown.

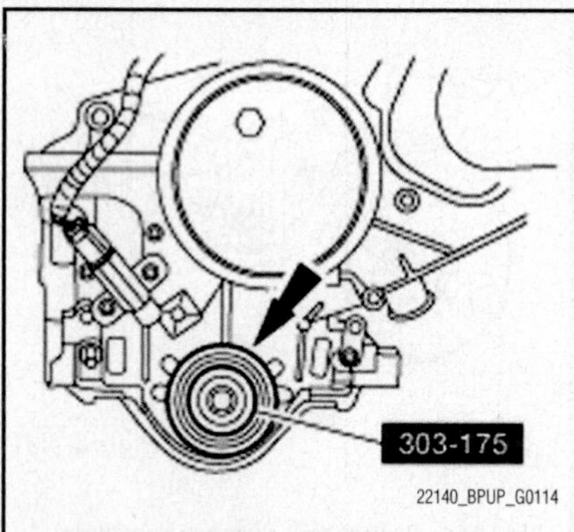

Fig. 147 Crankshaft vibration damper and seal replacer 303—175

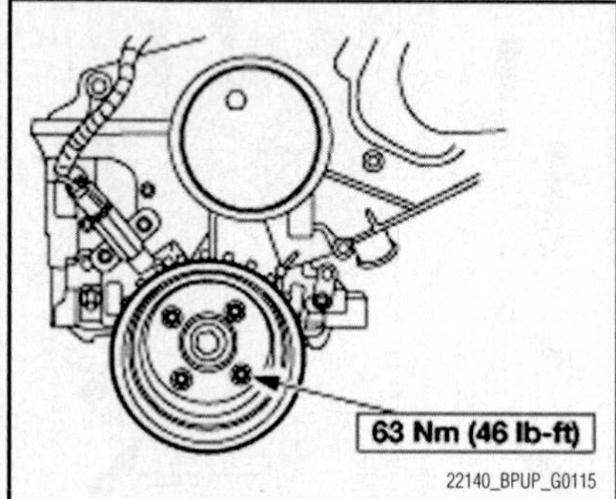

Fig. 148 Installation of the crankshaft pulley

29. Install the front cover bolts.
30. Install the crankshaft front oil seal.
31. Lubricate the seal lip with clean engine oil.
32. Using the vibration damper and seal replacer 303—175, install the front oil seal.
33. Using the vibration damper and seal replacer 303—175or equivalent, install the crankshaft damper.
34. Tighten to 107 ft. lbs. (145 Nm).
35. Coat the sealing surfaces of the crankshaft damper with clean engine oil.
36. Coat the crankshaft damper keyway with silicone.
37. Install the crankshaft damper.
38. Install the crankshaft pulley.
39. Tighten the crankshaft pulley bolts to 46 ft. lbs. (63 Nm).
40. Install the CKP sensor.
41. Tighten to 89 inch lbs. (10 Nm).
42. Connect the CKP sensor electrical connector.
43. Install the oil pan.
44. Connect the lower radiator hose to the coolant pump outlet.
45. Connect the heater hose at the coolant pump.
46. Position back the A/C compressor and power steering bracket and install the bolts.
47. Tighten to 35 inch lbs. (47 Nm).
48. Position the accessory drive belt tensioner and install the bolt.
49. Tighten to 20 ft. lbs. (27 Nm).
50. Install the stud bolts to the A/C compressor mounting bracket.
51. Tighten to 35 ft. lbs. (47 Nm).
52. Position the bracket and install the nut.
53. Tighten to 8 ft. lbs. (11 Nm).
54. Install the coolant pump pulley.
55. Install the fan shroud and fan.
56. Install the air cleaner outlet tube.
57. Install the drive belt.
58. Connect the negative battery cable.
59. Fill the engine with coolant and bleed the system.

4.0 Engines

See Figures 149 through 154.

1. Before servicing the vehicle, refer to the Precautions Section.
2. Relieve the fuel system pressure.
3. Disconnect the negative battery cable.

➡On some vehicles, when the battery cable is disconnected and reconnected, some abnormal drive symptoms may occur while the vehicle relearns its adaptive strategy. The vehicle may need to be driven to relearn its strategy.

4. With the vehicle in **NEUTRAL**, position it on a hoist.
5. Drain the cooling system.
6. Remove the nut and detach the A/C hose bracket from the front cover.
7. Remove the block cradle to front cover bolts.
8. Remove or disconnect the following:
   - A/C compressor
   - Power steering bracket bolts
   - Alternator wiring
   - Accessory drive belt tensioner

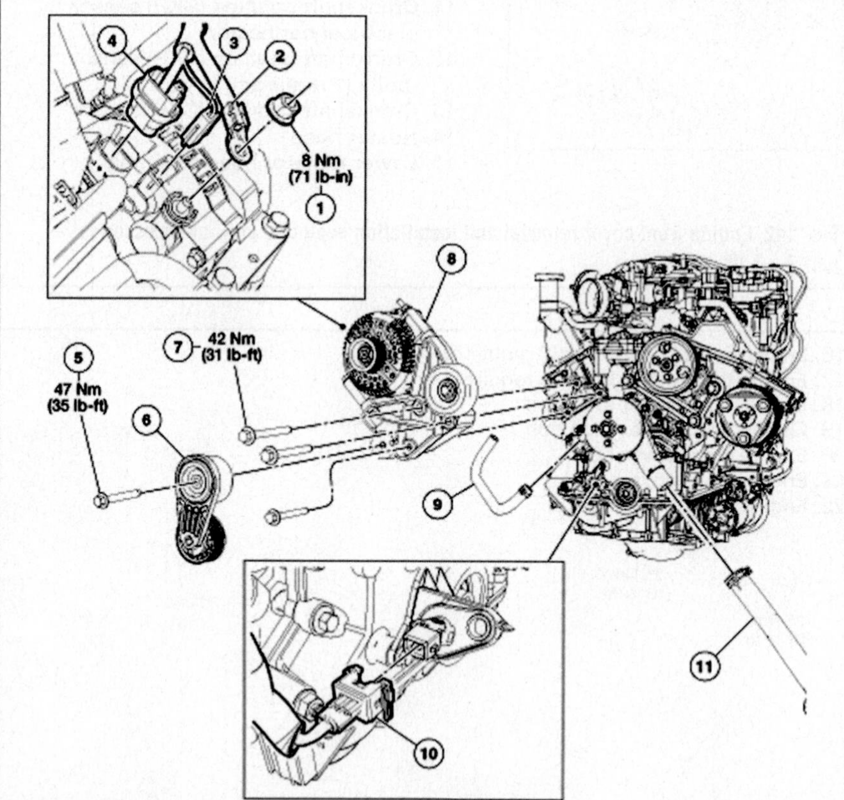

Fig. 149 Engine front cover removal and installation exploded view of accessories part 1—4.0L engine

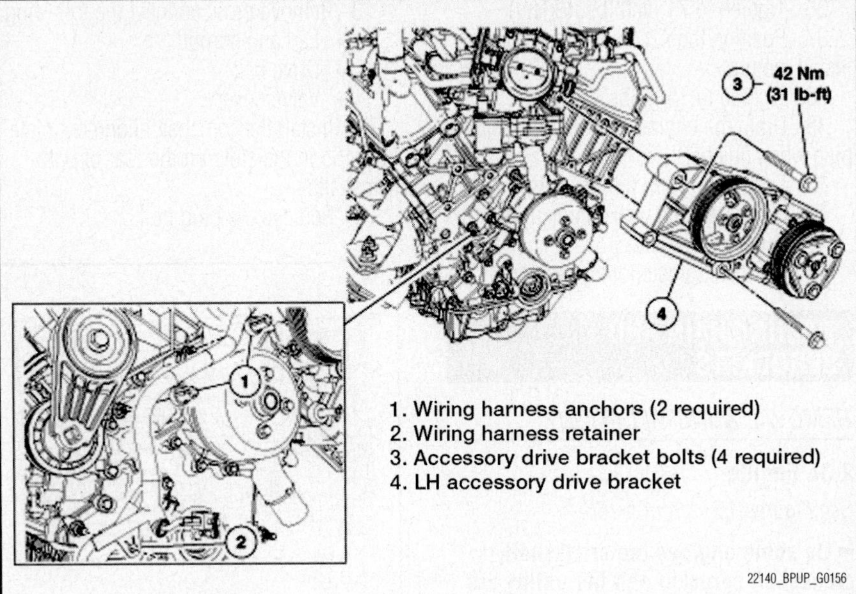

Fig. 150 Engine front cover removal and installation exploded view of accessories part 2—4.0L engine

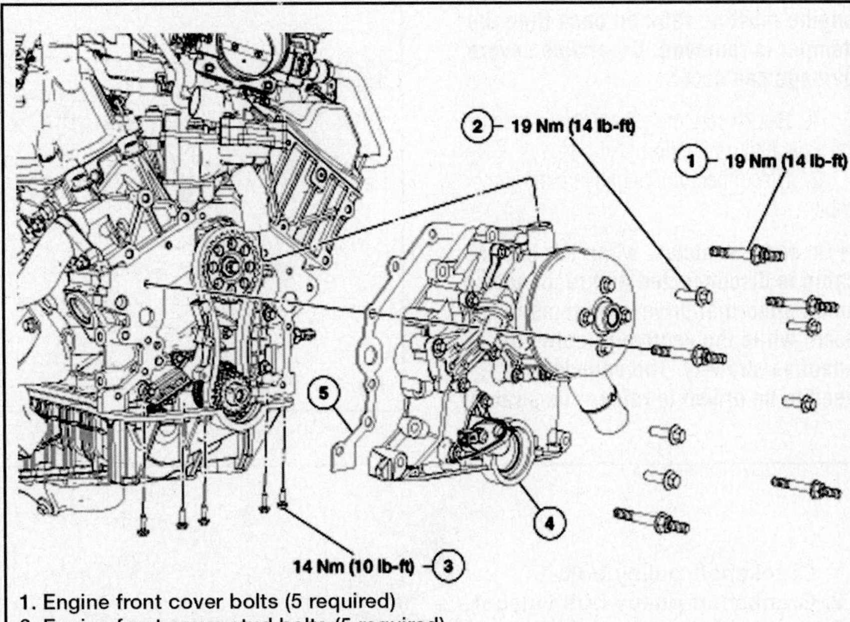

Fig. 151 Engine front cover removal and installation exploded view of accessories part 3—4.0L engine

- Generator mounting bracket
- Crankshaft Position (CKP) sensor electrical connector
- CKP wiring harness retainer.
- Wiring harness anchors
- Bypass hose clamp
- Heater hose from the coolant pump
- Lower radiator hose from the coolant pump

➡This bolt is a torque-to-yield design and cannot be reused.

9. Using the strap wrench 303—D055, remove the crankshaft pulley bolt.

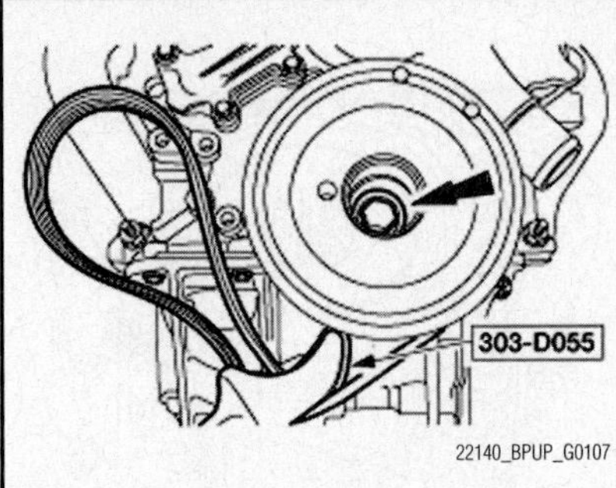

Fig. 152 Strap wrench 303—D055, and crankshaft pulley bolt to be removed and installed

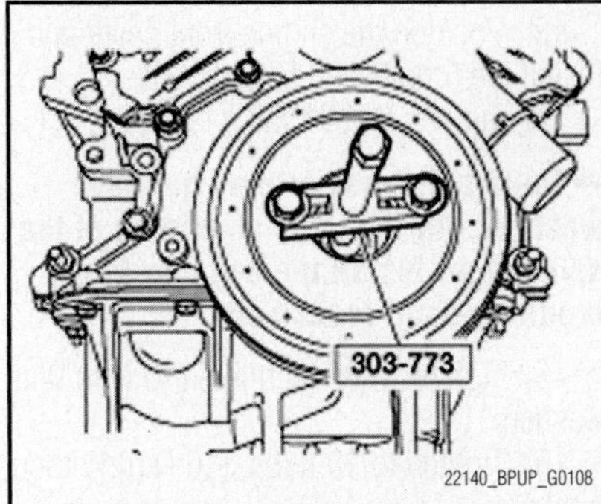

Fig. 153 Crankshaft vibration damper remover 303—773 tool

➡If the crankshaft pulley bolt is not installed, the crankshaft vibration damper remover 303—773 will bottom out before the pulley is off. Install the crankshaft pulley bolt 2-3 turns. Using the crankshaft vibration damper remover 303—773 and the 8 mm x 1.25 x 100 mm bolts, remove the pulley.

10. Remove the crankshaft front seal.

➡Note the positions of the stud bolts for installation reference.

11. Remove the 5 bolts, the 5 stud bolts, the engine front cover and the gasket.

To install:

➡If not secured within four minutes, the sealant must be removed and the sealing area cleaned. To clean the sealing area, use silicone gasket remover and metal surface prep. Follow the directions on the packaging. Failure to follow this procedure can cause future oil leakage. Apply silicone gasket and sealant to the Woodruff key slot on the crankshaft pulley.

12. Make sure the stud bolts are installed in their original positions.

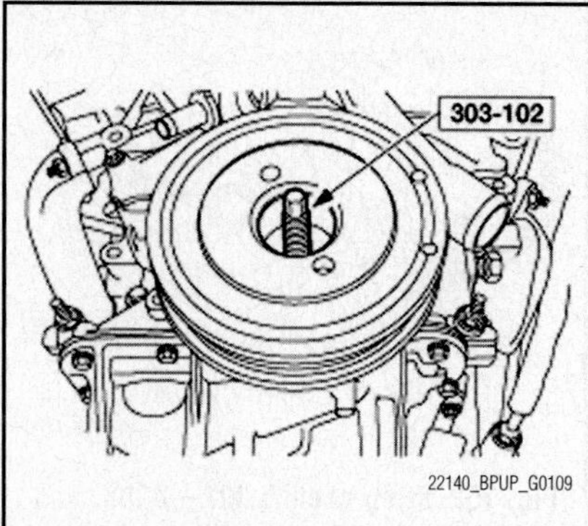

Fig. 154 Crankshaft vibration damper installer 303—102 tool

13. Position the engine front cover and install the 5 bolts and 5 stud bolts.
14. Tighten to 14 ft lbs. (19 Nm).

➡The 5 bolts may or may not have washers. All the bolts should be of the same type. Install the cylinder block cradle-to-front cover bolts.

15. Tighten to 89 inch lbs. (10 Nm).With washers.
16. Tighten to 10 ft lbs. (10 Nm).Without washers.
17. Using the crankshaft vibration damper installer 303—102, install the crankshaft pulley.
18. Using the strap wrench 303—D055, install a new crankshaft pulley bolt.
19. Tighten the bolt in two stages.
20. Torque Stage 1: 37 ft. lbs. (50 Nm).
21. Torque Stage 2: Tighten the bolt an additional 90 degrees
22. Position the bypass hose clamp.
23. Connect the lower radiator hose to the coolant pump.
24. Connect the heater hose to the coolant pump.
25. Position the wiring harness and install the bolt.
26. Tighten the wiring harness to 89 inch lbs. (10 Nm).
27. Install new wiring harness anchors and position the CKP sensor wiring.
28. Attach the wiring harness retainer.
29. Connect the CKP sensor electrical connector.
30. Position the A/C compressor and power steering bracket and install the 4 bolts.
31. Tighten to 31 ft. lbs. (42 Nm).
32. Position the generator bracket and install the 3 bolts.
33. Tighten to 35 ft. lbs. (47 Nm).
34. Connect the generator electrical connectors and the wiring harness retainer.
35. Attach the generator terminal and install the nut.
36. Tighten to 71 inch lbs. (8 Nm).
37. Position the A/C tube bracket and install the nut.
38. Tighten to 15 ft. lbs. (20 Nm).
39. Drain the engine oil. Install the drain plug when finished.
40. Tighten to 19 ft. lbs. (26 Nm).
41. Fill the engine with clean engine oil.
42. Connect the battery.
43. Fill the engine cooling system.

TIMING CHAIN AND SPROCKETS

REMOVAL & INSTALLATION

2.3L Engine

See Figures 155 through 166.

➡On some engines the crankshaft, crankshaft sprocket and the pulley are fitted together by friction, with diamond washers between the flange on each part. For that reason, the crankshaft sprocket is also unfastened if you loosen the pulley. Therefore, the engine must be retimed each time the damper is removed. Otherwise severe damage can occur.

1. Before servicing the vehicle, refer to the Precautions Section.
2. Disconnect the negative battery cable.

➡On some vehicles, when the battery cable is disconnected and reconnected, some abnormal drive symptoms may occur while the vehicle relearns its adaptive strategy. The vehicle may need to be driven to relearn its strategy.

3. Remove or disconnect the following:
   - Fan and shroud
   - Drive belt
   - Valve cover
4. Install the camshaft alignment plate 303-465 in the slots on the rear of both camshafts.
5. Remove the plug bolt.

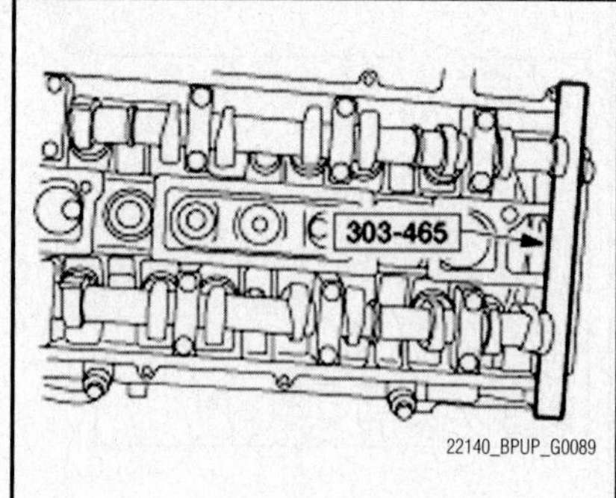

Fig. 156 Camshaft alignment plate 303-465

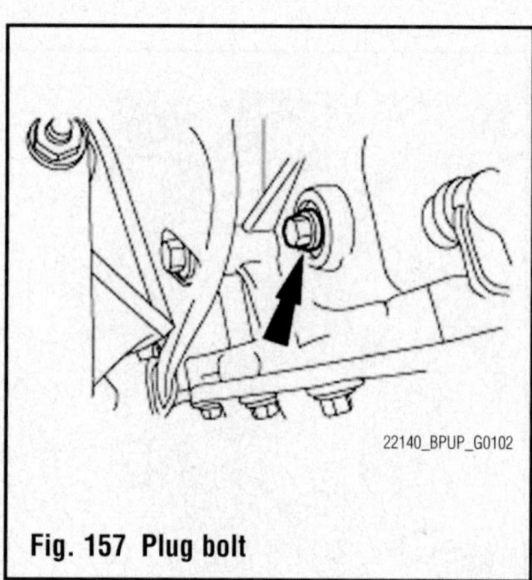
Fig. 157 Plug bolt

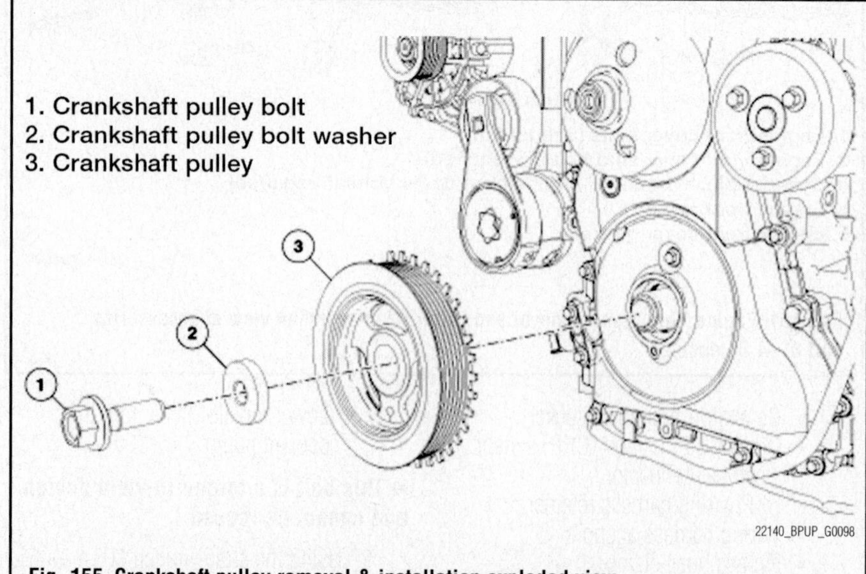

Fig. 155 Crankshaft pulley removal & installation exploded view

➡ Only turn the engine in the normal direction of rotation. Installing the special tool in this step will prevent the engine from being rotated in the clockwise direction. However, the engine can still be rotated in the counterclockwise direction.

6. Install the crankshaft TDC timing peg (2.3L engine) 303—507.
7. Install a M6 X 1.0 X 2.25 bolt (Mazda part No. 1F20-12-428) to verify ignition timing.
8. Install the holding fixture 205—072—02 and 205—126.

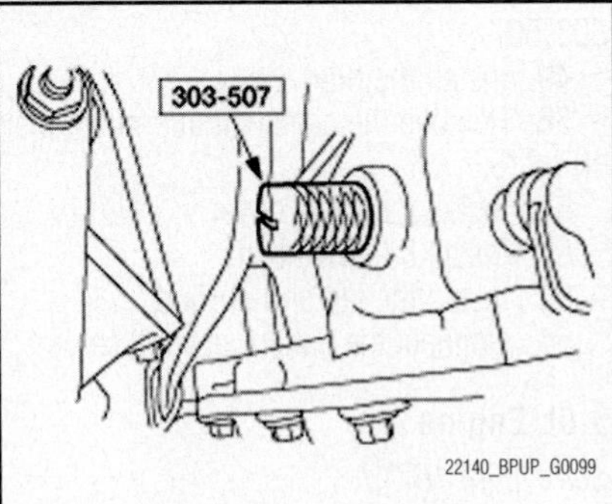

Fig. 158 Crankshaft TDC timing peg303—507

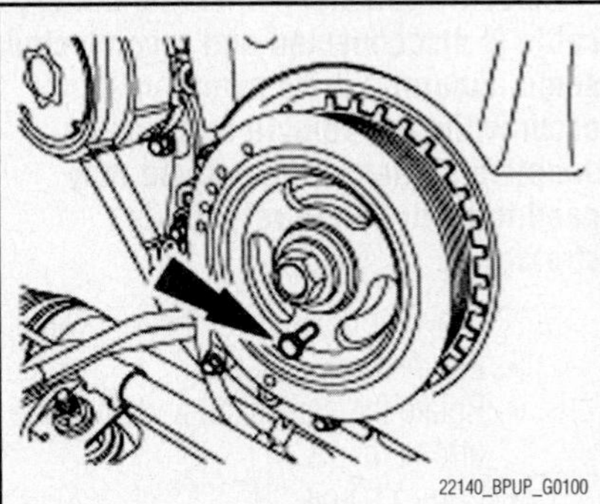

Fig. 159 M6 X 1.0 X 2.25 bolt (Mazda part No. 1F20-12-428)

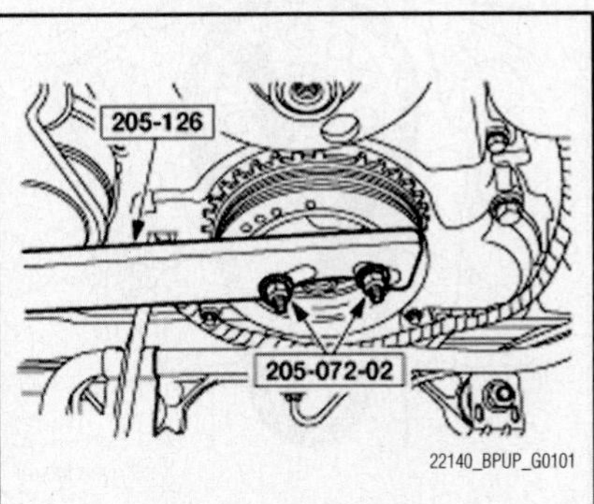

Fig. 160 Crank pulley holding fixture 205-072-02-and 205-126

WARNING

Failure to hold the crankshaft pulley in place during bolt loosening can cause damage to the engine. Remove the bolt and the crankshaft pulley.

9. Remove the crankshaft bolt.
10. Remove the holding fixture 205—072—02 and 205—126.
11. Remove the M6 bolt.
12. Remove the crankshaft pulley.
13. Remove or disconnect the following:
   - Camshaft pulley
   - Crankshaft position sensor
   - Belt tensioner
   - Water pump pulley
   - Power steering high pressure hose. Remove the nylon O-ring.
   - Power steering return hose
   - Power steering pump

➡This step is needed only if a new front cover is being installed.

14. Using a three-jaw puller, remove the fan drive pulley.

➡There is one bolt behind the cooling fan drive pulley. This bolt can be accessed by lining up one of the holes in the pulley with the bolt.

15. Remove the bolts and the engine front cover.
16. Compress the timing chain tensioner and remove the tensioner.
17. Remove the right-hand timing chain guide.
18. Remove the timing chain.
19. Remove the bolts and the left-hand timing chain guide.

WARNING

Do not rely on the Camshaft Alignment Plate to prevent camshaft rotation. Damage to the tool or the camshaft can occur.

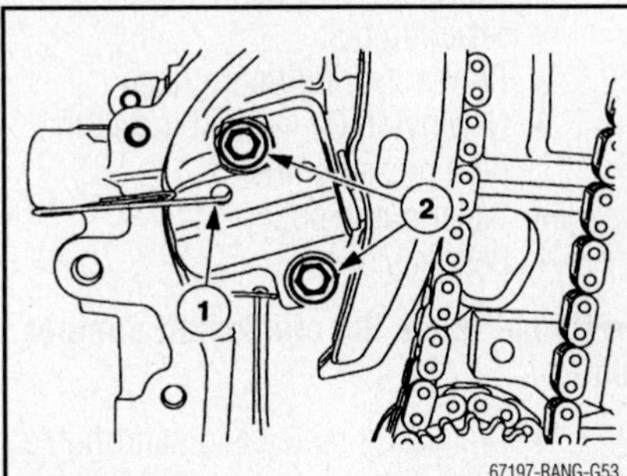

Fig. 161 Timing chain tensioner removal; 1-paper clip, 2 bolts—2.3L engine

Fig. 162 Right side timing chain guide—2.3L engine

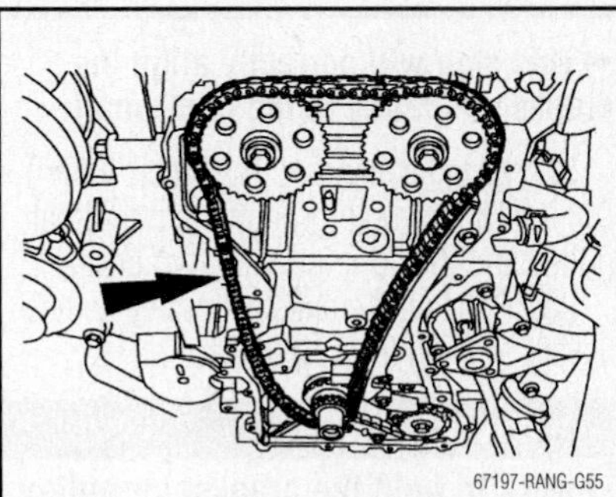

Fig. 163 Timing chain removal—2.3L engine

Fig. 164 Left timing chain guide—2.3L engine

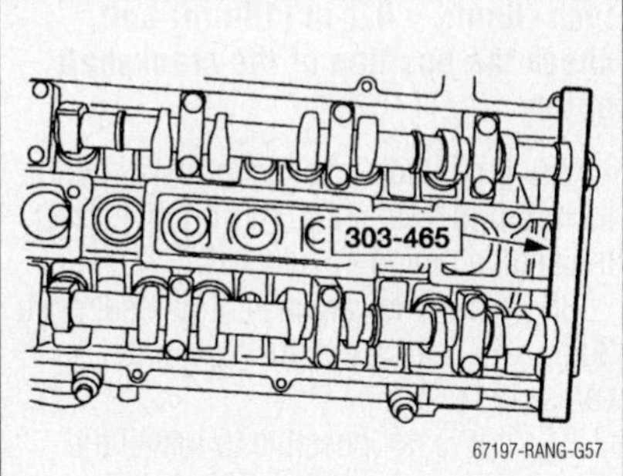

Fig. 165 Camshaft alignment plate installed—2.3L engine

20. If necessary, remove the bolts and the camshaft sprockets. Use the flats on the camshaft to prevent camshaft rotation.

To install:

➡Do not reuse the crankshaft damper bolt or washer.

➡Apply clean engine oil on seal area before installing.

21. Install the new crankshaft pulley and hand-tighten the bolt.

⁂ WARNING

Only hand-tighten the bolt or damage to the front cover can occur.

➡This step will correctly align the crankshaft pulley to the crankshaft.

22. Install a standard 0.23 inch (6mm) x 0.7 in (18mm) bolt through the crankshaft pulley and thread it into the front cover.
23. Rotate the crankshaft pulley as necessary to align the bolt holes.

⁂ WARNING

Failure to hold the crankshaft pulley in place during bolt tightening can cause damage to the engine front cover. Using the holding fixture 205—072—02 and 205—126 hold the crankshaft pulley in place and tighten the crankshaft pulley bolt in two stages.

24. Tighten the crankshaft pulley bolt to 74 ft. lbs. (100 Nm).
25. Tighten the crankshaft pulley bolt an additional 90°.
26. Remove the holding fixture 205—072—02 and 205—126
27. Remove the 0.23 inch (6mm) x 0.7 in (18mm) bolt.
28. Remove the crankshaft TDC timing peg

➡Only hand-tighten the bolt or damage to the front cover can occur. Using 0.23 inch (6mm) x 0.7 in (18mm) bolt, check the position of the crankshaft pulley.

29. If it is not possible to install the 0.23 inch (6mm) x 0.7 in (18mm) bolt, correct the engine timing.
30. Using the camshaft alignment plate 303—507, check the position of the camshafts.
31. If it is not possible to install the camshaft alignment plate 303—507, correct the engine timing.
32. Remove the camshaft alignment plate.

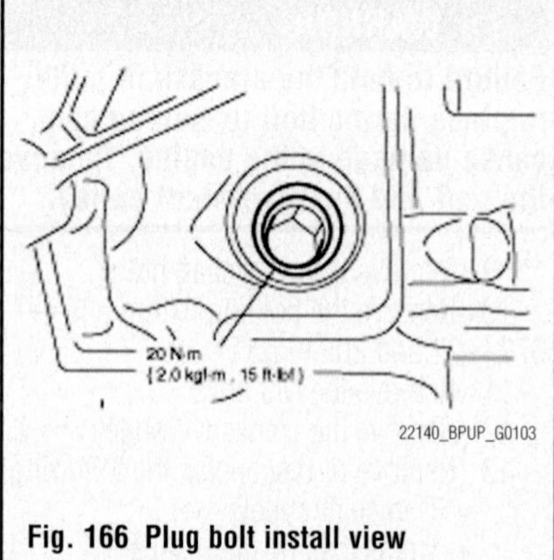

Fig. 166 Plug bolt install view

33. Remove the 6mm 0.23 inch (6mm) x 0.7 in (18mm) bolt.
34. Remove the crankshaft TDC timing peg.
35. Install the plug bolt.
36. Tighten the Plug Bolt to 15 ft. lbs. (20 Nm).
37. Install the fan drive pulley using a nut and bolt with flat washers.
38. Clean and inspect the mounting surfaces of the engine and the front cover.

➡The engine front cover must be installed and the bolts tightened within four minutes of applying the silicone gasket and sealant.

39. Apply a 2.5 mm bead of silicone gasket and sealant to the cylinder head and oil pan joint areas.
40. Apply a 2.5 mm bead of silicone gasket and sealant to the front cover.
41. Install the front cover. Tighten the bolts in the sequence shown, to the following specifications:
   - Step 1: 8 mm bolts to 10 Nm (89 inch lbs.)
   - Step 2: 10 mm bolts to 25 Nm (18 ft. lbs.)
   - Step 3: 13 mm bolts to 48 Nm (35 ft. lbs.)
42. Install or connect the following:
   - Power steering pump and lower retaining bolt
   - Power steering return hose
   - New nylon O-ring and install the high pressure line.
   - Water pump pulley
   - Belt tensioner

➡Do not reuse the crankshaft damper bolt.

   - Crankshaft pulley and hand-tighten the bolt
43. Install an M6 bolt in the crankshaft pulley. Tighten the crankshaft retaining bolt in two stages.
   - Step 1: 40 Nm (30 ft. lbs.)
   - Step 2: Rotate the bolt an additional 90 degrees.
44. Install the crankshaft position sensor, do not tighten the bolts at this time.

➡A new sensor must be installed whenever the old one is removed.

45. Adjust the crankshaft position sensor with the Alignment Tool, and tighten the mounting bolts.
46. Connect the crankshaft position sensor electrical connector.
47. Remove the M6 bolt from the crankshaft pulley.
48. Remove the Crankshaft Timing Peg 303-507.
49. Install the plug.
50. Remove the camshaft alignment plate 303-376.
51. Install the valve cover.
52. Install the drive belt.
53. Install the fan and shroud.
54. Connect the battery ground cable.

3.0L Engine

See Figure 167.

1. Before servicing the vehicle, refer to the Precautions Section.
2. Disconnect the negative battery cable.

➡On some vehicles, when the battery cable is disconnected and reconnected, some abnormal drive symptoms may occur while the vehicle relearns its adaptive strategy. The vehicle may need to be driven to relearn its strategy.

3. Remove or disconnect the following:
   - Engine front cover
   - Rotate the crankshaft and align the timing marks
   - Sprocket bolt

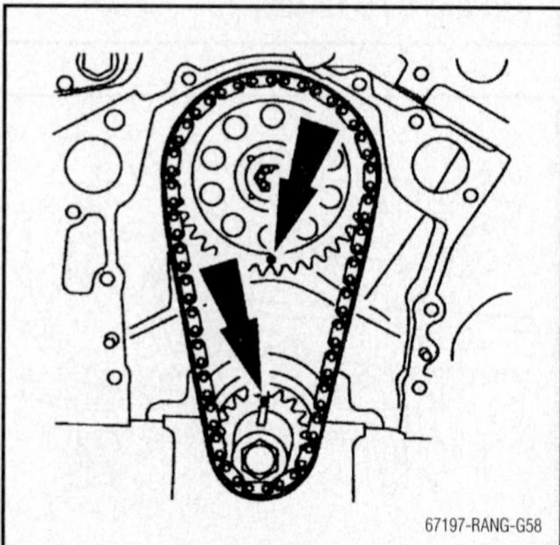

Fig. 167 Timing mark alignment—3.0L engine

- Timing chain, camshaft sprocket and crankshaft sprocket as an assembly

To install:

4. Install or connect the following:
   - Timing chain, camshaft and crankshaft sprockets as an assembly
5. Align the timing marks.
6. Install or connect the following:
   - Sprocket bolt. Torque the bolt to 46 ft. lbs. (63 Nm).
   - Engine front cover
   - Negative battery cable

4.0L Engine

See Figures 168 through 180.

1. Before servicing the vehicle, refer to the Precautions Section.
2. Disconnect the negative battery cable.

➡On some vehicles, when the battery cable is disconnected and reconnected, some abnormal drive symptoms may occur while the vehicle relearns its adaptive strategy. The vehicle may need to be driven to relearn its strategy.

3. With the vehicle in neutral, position it on a hoist.
4. Remove the intake manifold.
5. Remove the fuel supply manifold.
6. Remove the accessory drive belt.
7. Remove the thermostat housing.
8. Remove the roller followers.
9. The crankshaft timing tool 303—573 must be installed on the damper and should contact the engine block, this positions the engine at TDC.

WARNING

Do not rotate the engine counterclockwise. Rotating the engine counterclockwise will result in incorrect timing of the engine.

➡You must retime the LH and RH camshafts when either camshaft is disturbed. Turn the crankshaft clockwise to position the number 1 cylinder at top dead center (TDC).

10. Remove the LH hydraulic chain tensioner.
11. Remove the LH camshaft sprocket bolt.
12. Install the camshaft sprocket holding tool 303—564 and the adapter
13. Remove the LH camshaft sprocket bolt.
14. Using the crankshaft holding tool 303—674 to prevent the crankshaft from turning, remove the jackshaft sprocket bolt.
15. Remove the bolts and the primary chain tensioner.
16. Remove the primary chain and sprockets as an assembly.
17. Remove the LH cassette upper bolt
18. Remove the LH cassette lower bolt and the LH cassette.

To install:

19. The camshaft chain sprockets must be oriented correctly. Position the LH cassette.
20. Install the LH cassette lower bolt.
21. Torque the LH cassette lower bolt to 14 ft. lbs. (19 Nm).
22. Install the LH cassette upper bolt.
23. Torque the LH cassette upper bolt to 8 ft. lbs. (12 Nm).
24. Install the primary chain and sprockets as an assembly.
25. Install the primary chain tensioner and the bolts.
26. Torque the primary chain tensioner bolts to 80 inch lbs. (9 Nm).
27. Using the crankshaft holding tool303—674 to prevent the crankshaft from turning, tighten the jackshaft sprocket bolt in 2 stages as follows:
    - Torque Stage 1: 33 ft. lbs. (45 Nm)
    - Torque Stage 2: Tighten an additional 90 degrees
28. Loosely install the camshaft sprocket bolt.
29. Install the LH hydraulic chain tensioner during camshaft timing.
30. Retime the LH and RH camshafts as follows.
31. The crankshaft timing tool 303—578 and 303—564 must be installed on the damper and should contact the engine block, this positions the engine at TDC.
32. Install the crankshaft timing tool.
33. Install the camshaft sprocket holding tool 303—575 and the adaptor to the RH cylinder head and tighten the 2 top clamp bolts.

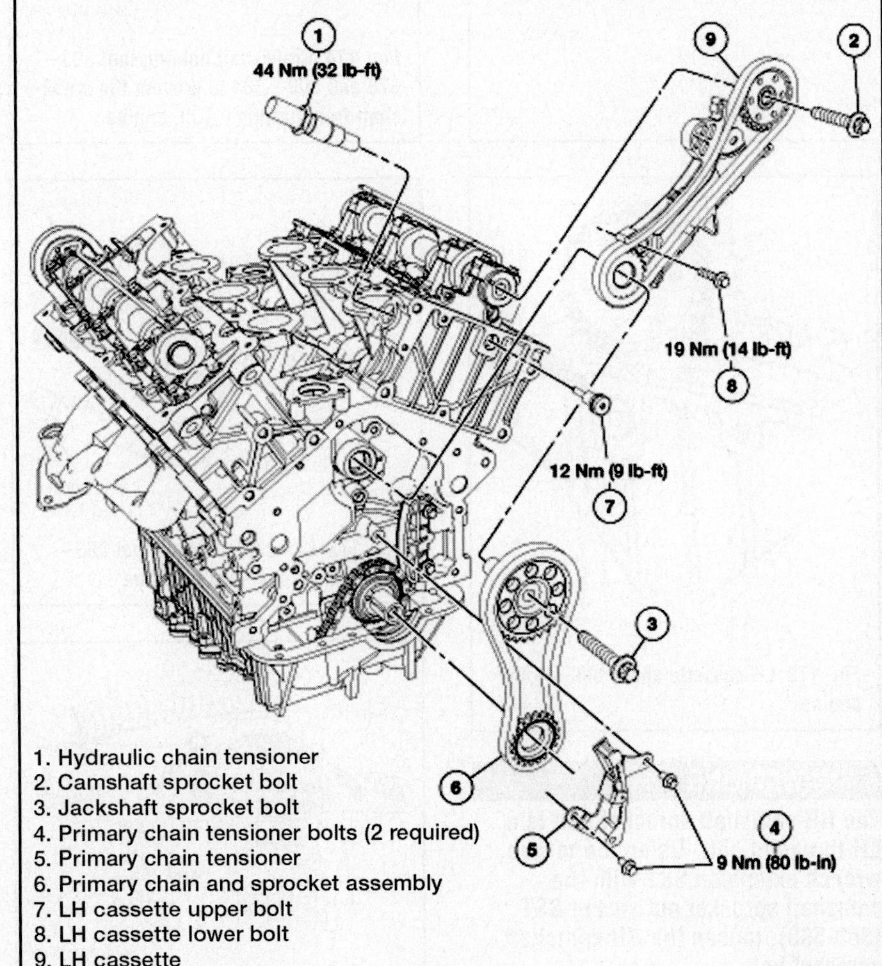

Fig. 168 Timing chain components exploded view—4.0L engine

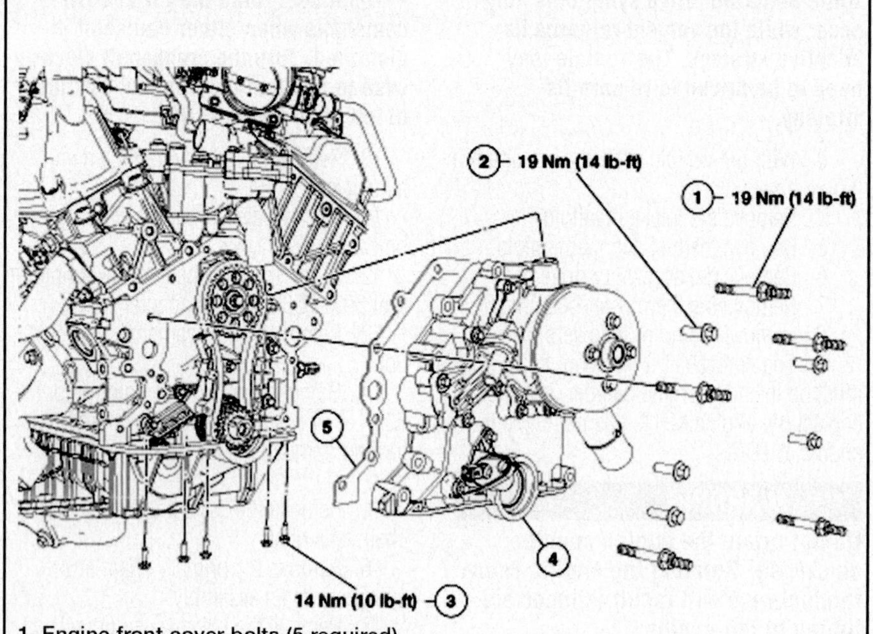

1. Engine front cover bolts (5 required)
2. Engine front cover stud bolts (5 required)
3. Cylinder block cradle-to-engine front cover bolts (5 required)
4. Engine front cover
5. Engine front cover gasket

Fig. 169 Engine front cover—4.0L engine

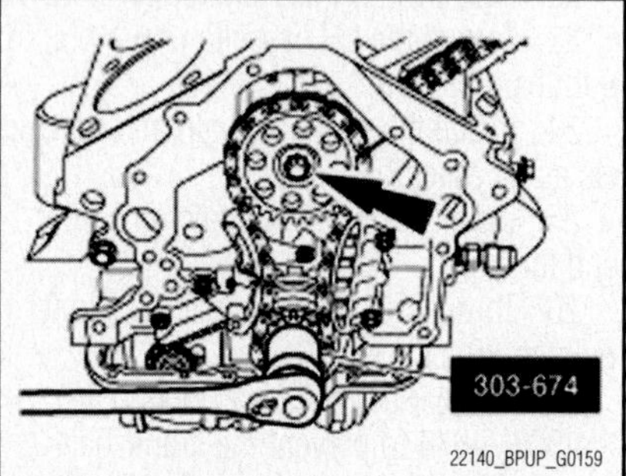

Fig. 170 Crankshaft holding tool 303—674 to prevent the crankshaft from turning—4.0L engine

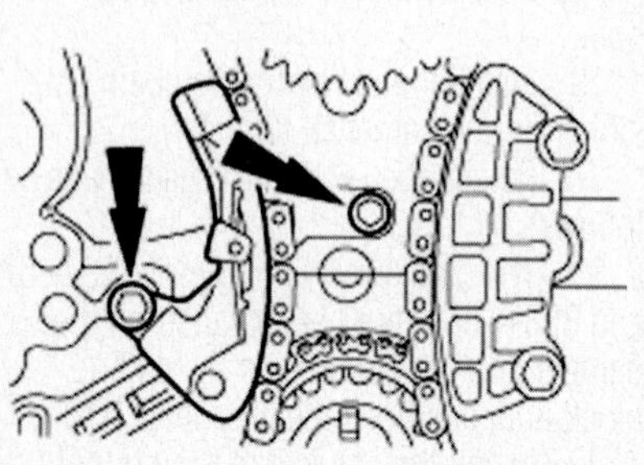

Fig. 171 Primary chain tensioner location and bolt locations—4.0L engine

Fig. 172 LH cassette upper bolt—4.0L engine

WARNING

The RH camshaft sprocket bolt is a LH-threaded bolt. Using the torque wrench extension SST with the camshaft sprocket nut socket SST (303-565), loosen the RH camshaft sprocket bolt.

34. Loosen the top 2 special tool clamp bolts.

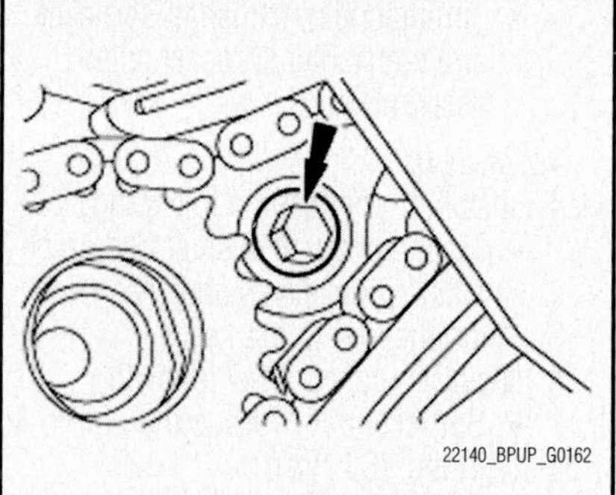

Fig. 173 LH cassette lower bolt and the LH cassette.— 4.0L engine

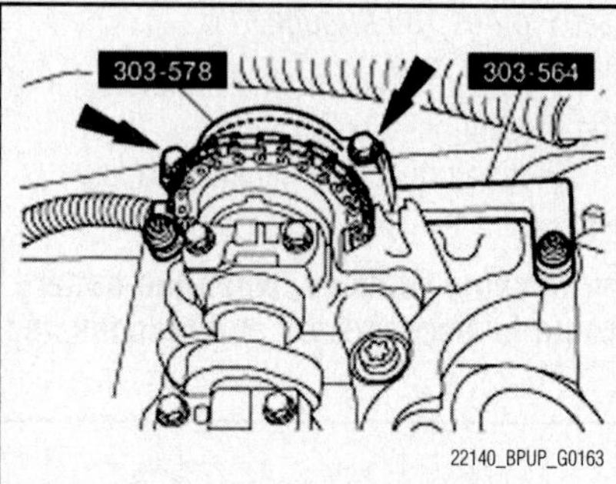

Fig. 174 Crankshaft holding tool 303—578 and 303—564 to prevent the crankshaft from turning—4.0L engine

Fig. 175 Camshaft holding tool 303—575 and adaptor—4.0L engine

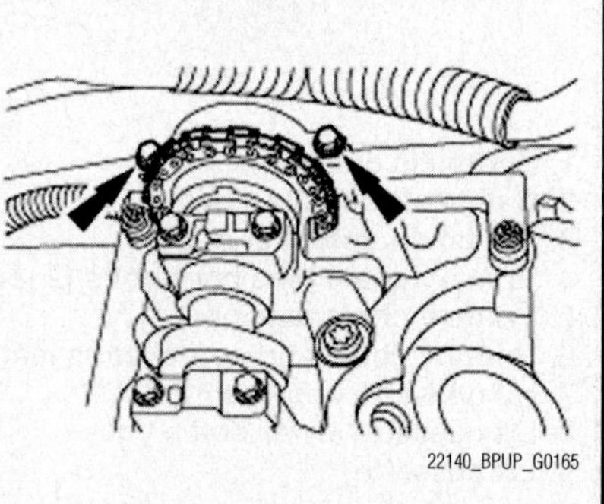

Fig. 176 RH camshaft sprocket bolt.—4.0L engine

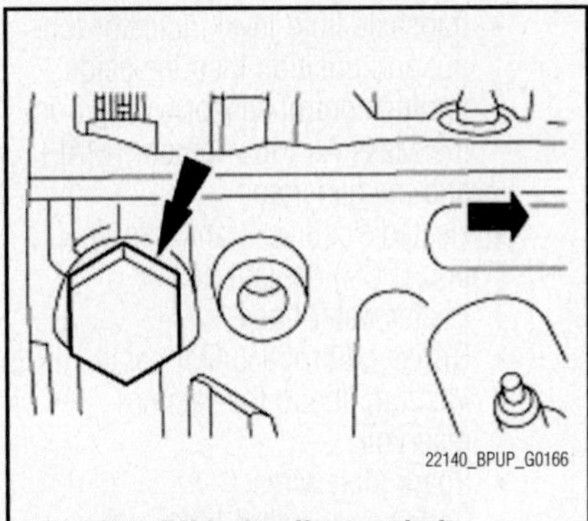
Fig. 177 RH hydraulic camshaft tensioner—4.0L engine

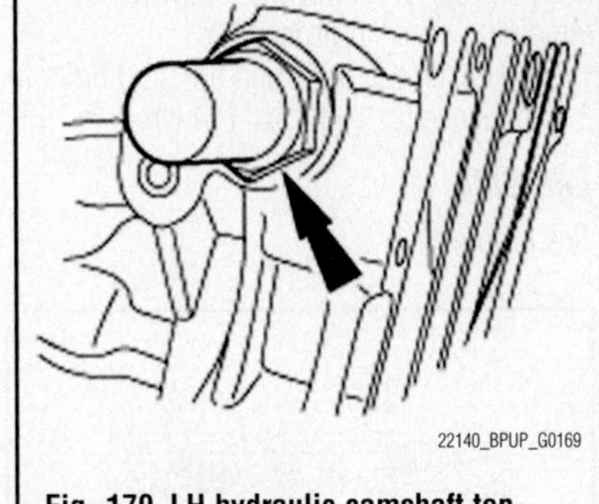
Fig. 179 LH hydraulic camshaft tensioner—4.0L engine

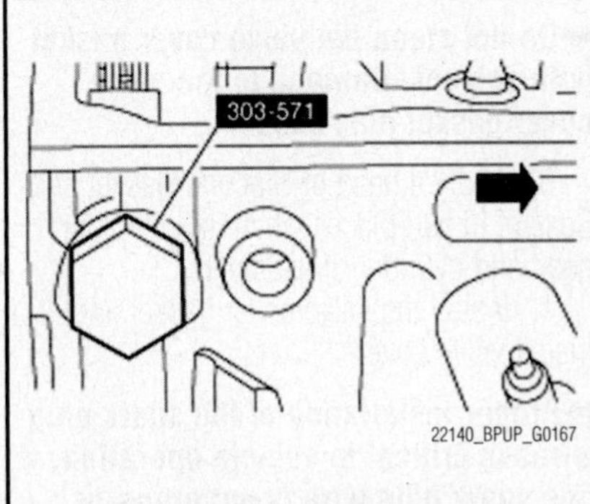

Fig. 178 Timing chain tensioner service tool 303-571—4.0L engine

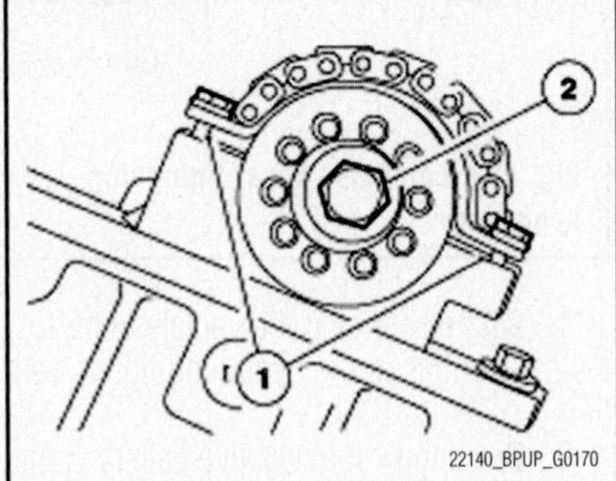

Fig. 180 Top 2 clamp bolts and the LH camshaft bolt—4.0L engine

35. The camshaft timing slots are off-center. Position the camshaft timing slots below the centerline of the camshaft to correctly fit the special tools and install the camshaft holding tool 303—576 and adaptor 303—577 on the front of the RH cylinder head.
36. Remove the RH lower splash shield.
37. Remove the RH hydraulic camshaft tensioner.
38. Install the timing chain tensioner 303—571.

WARNING

The RH camshaft sprocket bolt is a LH-threaded bolt. Tighten the bolts.

39. Tighten the special tool top 2 clamp bolts to 89 inch lbs. (10 Nm).
40. Using the torque wrench extension service tool with the camshaft sprocket nut socket 303—565, tighten the camshaft bolt.
41. Tighten camshaft bolts to 45 ft. lbs. (61 Nm).
42. Remove the timing chain tensioner 303—571
43. Install a new O-ring seal on the tensioner and lubricate with clean engine oil. Install the RH hydraulic camshaft tensioner.
44. Tighten camshaft bolts to 32 ft. lbs. (44 Nm).
45. Install the RH lower splash shield.
46. Remove the special tools from the RH cylinder head.
47. Install the camshaft sprocket holding tool 303—564 and—578 the adaptor SST on the front of the LH cylinder head and tighten the top 2 clamp bolts.
48. Tighten camshaft sprocket bolts to 89 inch lbs. (10 Nm).
49. Loosen the LH camshaft sprocket bolt.
50. Loosen the top 2 clamp bolts on the camshaft sprocket holding tool to allow the camshaft sprocket to rotate freely.
51. The camshaft timing slots are off-center. Position the camshaft timing slots below the centerline of the camshaft to correctly fit the special tools and install the camshaft holding tool 303—576 and adaptor 303—577 on the rear of the LH cylinder head.
52. Remove the LH hydraulic camshaft tensioner.
53. Install the timing chain tensioner 303—571.
54. Tighten camshaft top 2 clamp bolts to 89 inch lbs. (10 Nm).
55. Tighten LH camshaft bolt to 63 ft. lbs. (85 Nm).
56. Remove the timing chain tensioner 303—571.
57. Install a new O-ring seal on the tensioner and lubricate with clean engine oil. Install the LH hydraulic camshaft tensioner.
58. Tighten camshaft sprocket bolts to 32 ft. lbs. (44 Nm).
59. Remove the crankshaft timing tool 303—573.
60. Position the A/C hose and install the nut.
61. Install the camshaft roller followers.
62. Install the thermostat housing.
63. Install the engine front cover.
64. Install the fuel rail.
65. Install the valve covers.
66. Connect the battery cable.

VALVE COVERS

REMOVAL & INSTALLATION

2.3L Engine

See Figure 181.

1. Before servicing the vehicle, refer to the precautions in the beginning of this section.
2. Disconnect the negative battery cable.
3. Remove or disconnect the following:
   - Intake manifold assembly
   - Cylinder Head Temperature (CHT) sensor boot and electrical connector
   - Engine wiring harness from the valve cover studs
   - Spark plug wires and ignition coil
   - Camshaft position sensor
   - Crankcase ventilation tube
   - Bolts and the valve cover
   - If necessary, valve cover gaskets.

To install:

4. Install new valve cover gaskets if removed.

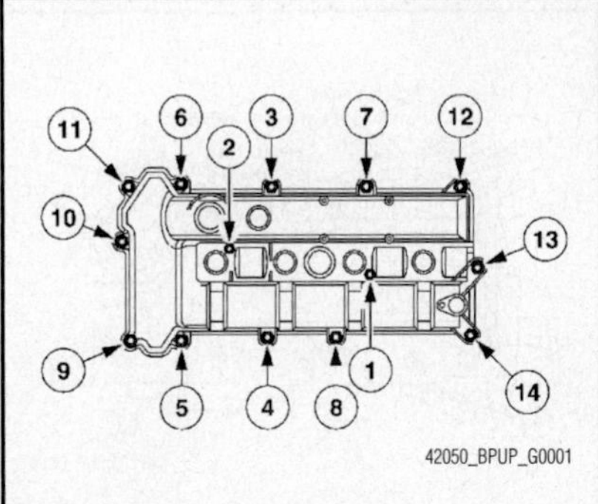

Fig. 181 Valve cover tightening sequence

5. Tighten valve cover bolts in sequence to: 7 ft. lbs. (10 Nm).

6. To complete installation, reverse remaining removal procedure.

7. Tighten the Camshaft Position (CMP) sensor and the bolt to: 62 inch lbs. (7 Nm).

8. Tighten valve cover bolts in sequence to: 53 inch lbs. (6 Nm).

3.0L Engine

Right Side

See Figure 182.

1. Before servicing the vehicle, refer to the precautions in the beginning of this section.
2. Disconnect the negative battery cable.
3. Remove or disconnect the following:
   - Engine control sensor wiring from the ignition coil
   - Plug wires from the RH spark plugs and position the wires aside
   - Engine control sensor wiring from the Intake Air Temperature (IAT) sensor
   - Engine control sensor wiring from the Mass Air Flow (MAF) sensor
   - Vacuum line
   - Ignition coil bracket
   - Crankcase ventilation tube
   - Two bolts and the six stud bolts and remove the valve cover
   - Valve cover gasket

To install:

4. Clean all the mating surfaces.

➡Do not clean the valve cover gasket with solvent. Damage to the valve cover gasket may occur.

5. Apply a bead of silicone gasket and sealant in two places where the cylinder head and cylinder block meet.
6. Install the valve cover gasket and install valve cover.

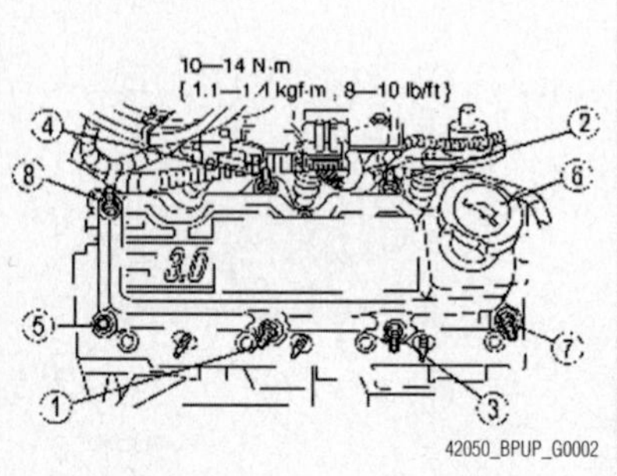

Fig. 182 Right valve cover tightening sequence

7. To complete installation, reverse remaining removal procedure.
8. Tighten the bolts and the stud bolts in the sequence to: 10 ft. lbs. (14 Nm).

Left Side

See Figure 183.

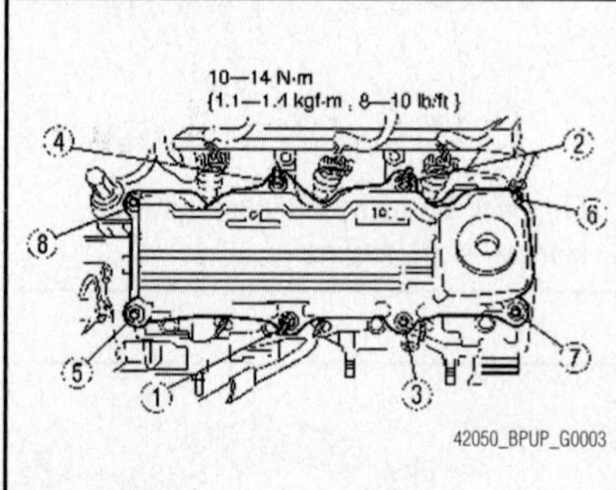

Fig. 183 Left valve cover tightening sequence

1. Before servicing the vehicle, refer to the precautions in the beginning of this section.
2. Disconnect the negative battery cable.
3. Remove or disconnect the following:
   - Upper intake manifold
   - LH spark plug wires at the spark plugs
   - Five bolts and the three stud bolts
   - Valve cover and valve cover gasket

To install:

4. Clean all the mating surfaces.

➡Do not clean the valve cover gasket with solvent. Damage to the valve cover gasket may occur.

5. Apply a bead of Silicone Gasket and Sealant in two places where the cylinder head and cylinder block meet.
6. Install the valve cover gasket and install valve cover.
7. To complete installation, reverse remaining removal procedure.
8. Tighten the bolts and the stud bolts in the sequence to: 10 ft. lbs. (14 Nm)

4.0L Engine

Right Side

See Figure 184.

1. Before servicing the vehicle, refer to the precautions in the beginning of this section.
2. Disconnect the negative battery cable.
3. Drain and recycle the engine coolant.
4. Remove or disconnect the following:
   - Air cleaner outlet tube
   - Upper radiator hose
   - transaxle fluid level indicator tube nut and position the tube aside.
   - Engine control sensor wiring from the Mass Air Flow Sensor (MAF) and the generator
   - Heated Positive Crankcase Ventilation (PCV) coolant hose
   - Heater water hose
   - Cruise control actuator cable
   - Vacuum line to the vacuum reservoir
   - Spark plug wires
   - Bolts and the stud-bolt
   - Valve cover and valve cover gasket

To install:

5. Clean all the mating surfaces.

➡Do not clean the valve cover gasket with solvent. Damage to the valve cover gasket may occur.

6. Apply a bead of Silicone Gasket and Sealant in two places where the cylinder head and cylinder block meet.
7. Install the valve cover gasket and install valve cover.

➡Proper installation of the spark plug wires is critical to vehicle operation. If one spark plug wire is not properly installed on the spark plug or ignition coil, both spark plugs connected to that segment of the ignition coil may not fire under load.

➡Wipe the spark plug wires with a clean, damp cloth prior to inspection. When a spark plug wire is removed for any reason from a spark plug or ignition coil, or a new spark plug wire is installed, dielectric compound or equivalent must be applied to the spark plug wire boot prior to installation. Use a small, clean tool to coat the entire interior surface of the boot with dielectric compound or equivalent.

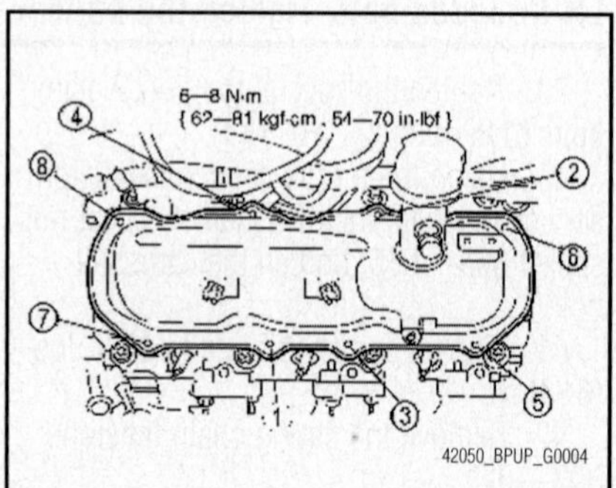

Fig. 184 Right valve cover tightening sequence

8. To complete installation, reverse remaining removal procedure.

9. Tighten the bolts and the stud bolts in the sequence to: 70 inch lbs. (8 Nm).

Left Side

See Figure 185.

1. Before servicing the vehicle, refer to the precautions in the beginning of this section.
2. Disconnect the negative battery cable.
3. Remove or disconnect the following:
   - Upper intake manifold
   - Vacuum lines from the Exhaust Gas Recirculation (EGR) valve
   - Valve cover nuts and stud bolts
   - Valve cover and gasket

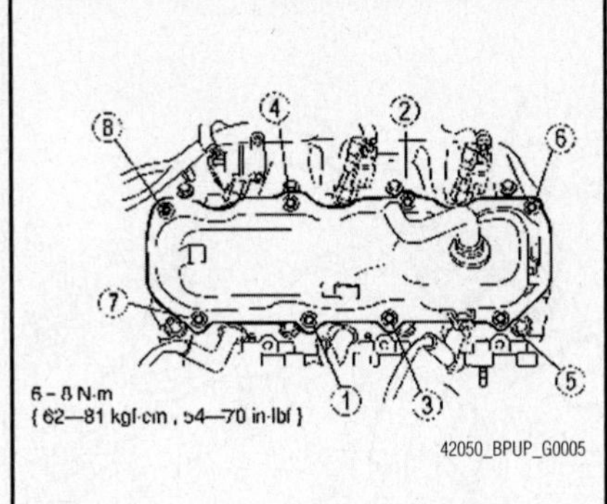

Fig. 185 Left valve cover tightening sequence

To install:

4. Clean all the mating surfaces.

➡**Do not clean the valve cover gasket with solvent. Damage to the valve cover gasket may occur.**

5. Apply a bead of Silicone Gasket and Sealant in two places where the cylinder head and cylinder block meet.
6. Install the valve cover gasket and install valve cover.
7. To complete installation, reverse remaining removal procedure.
8. Tighten the bolts and the stud bolts in the sequence to: 70 inch lbs. (8 Nm).

ENGINE PERFORMANCE & EMISSION CONTROL

ACCELERATOR PEDAL POSITION (APP) SENSOR

LOCATION

The APP sensor is located above the accelerator pedal.

REMOVAL & INSTALLATION

See Figure 186.

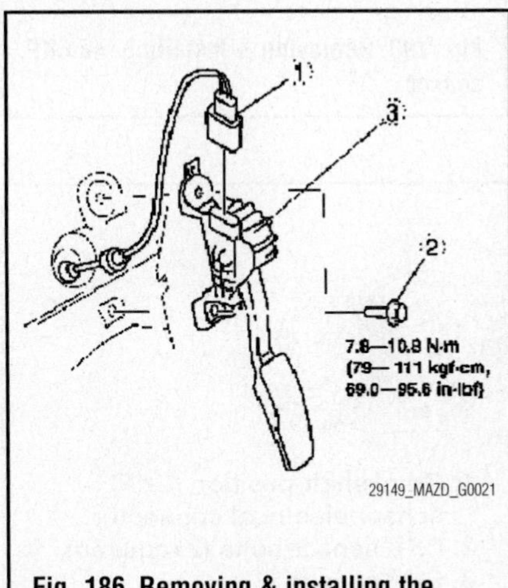

Fig. 186 Removing & installing the APP sensor

Refer to the accompanying illustration for Accelerator Pedal Position (APP) Sensor removal and installation.

CAMSHAFT POSITION (CMP) SENSOR

LOCATION

See Figures 187 and 188.

Refer to the accompanying illustrations for CMP sensor location.

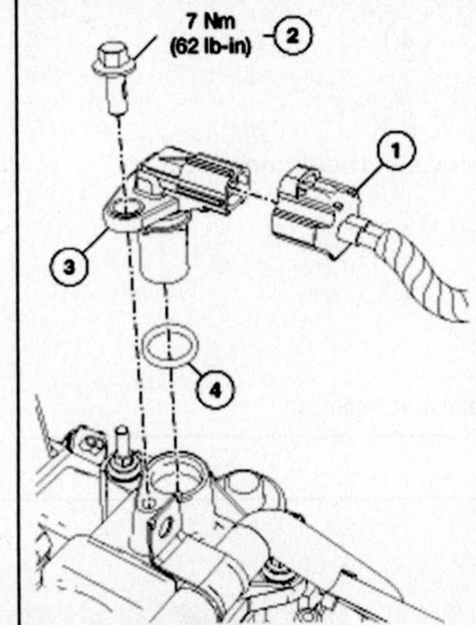

Fig. 187 Camshaft Position Sensor (CMP) 2.3L engine

REMOVAL & INSTALLATION

See Figure 189.

1. Disconnect the negative battery cable.
2. Disconnect the CMP sensor connector.
3. Remove the CMP sensor installation bolt.
4. Remove the CMP sensor.
5. Install the CMP sensor in the reverse order of removal. Tighten to 71 inch lbs. (8 Nm).

CRANKSHAFT POSITION (CKP) SENSOR

LOCATION

See Figures 190 and 191.

Refer to the accompanying illustrations for CKP sensor location.

REMOVAL & INSTALLATION

See Figures 192 and 193.

1. Remove the right front wheel.
2. Remove the splash shield.
3. Disconnect the CKP sensor connector.
4. Remove the installation bolts to remove the CKP sensor.

To install:

➡**Perform the following procedure so that piston No.1 is at the top dead center.**

5. Remove the right side driveshaft.
6. Remove the cylinder block lower blind plug and install the special service tool or equivalent.
7. Turn the crankshaft pulley to the clockwise until it stops.
8. Using a straight edge, draw a straight line directly in the center of the ninth tooth of the crankshaft pulley pulse wheel (counting counterclockwise from the empty space).

➡**If the line is not accurately drawn, ignition timing, fuel injection and other engine control systems will be adversely affected. Drawn the straight line carefully using a straight edge.**

9. Align the centerline of the crankshaft position sensor and the line drawn, then install the sensor.

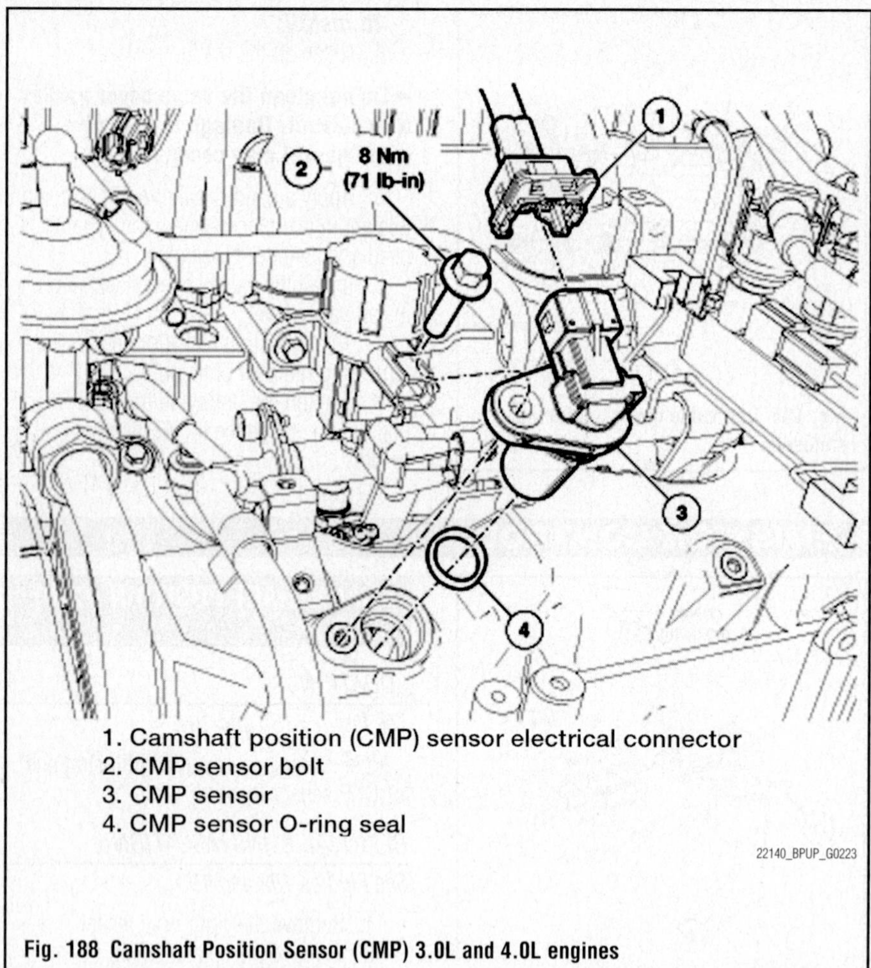

Fig. 188 Camshaft Position Sensor (CMP) 3.0L and 4.0L engines

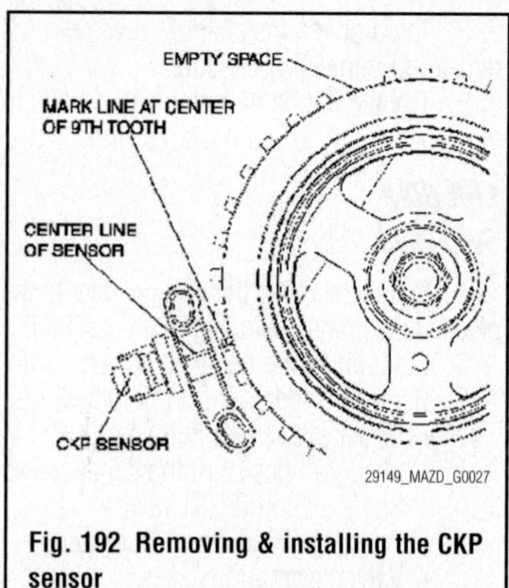

Fig. 192 Removing & installing the CKP sensor

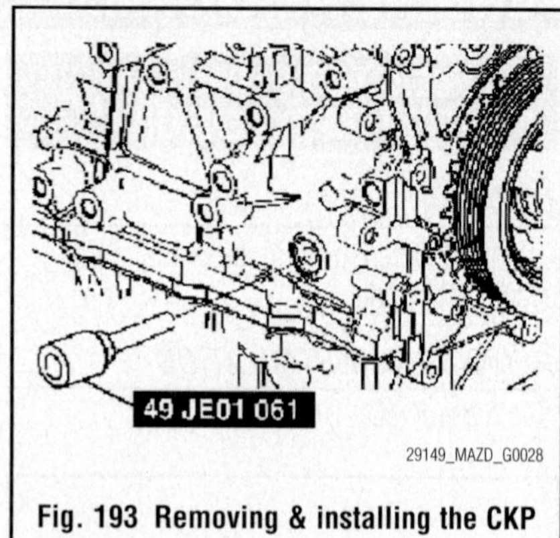

Fig. 193 Removing & installing the CKP sensor

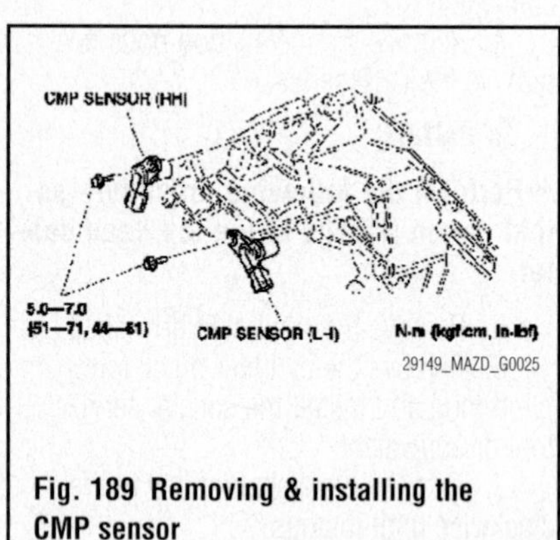

Fig. 189 Removing & installing the CMP sensor

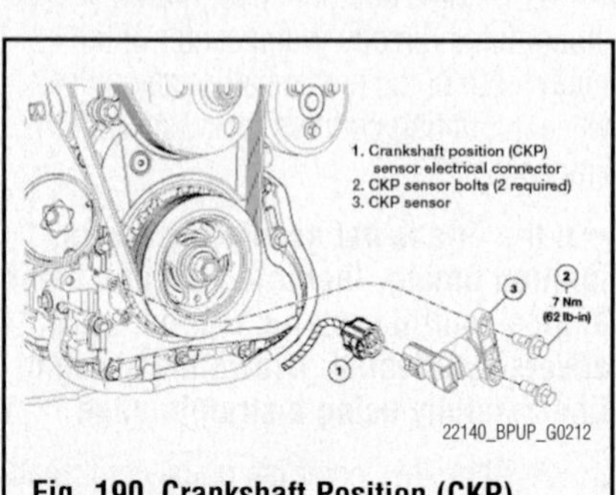

Fig. 190 Crankshaft Position (CKP) sensor—2.3L engine

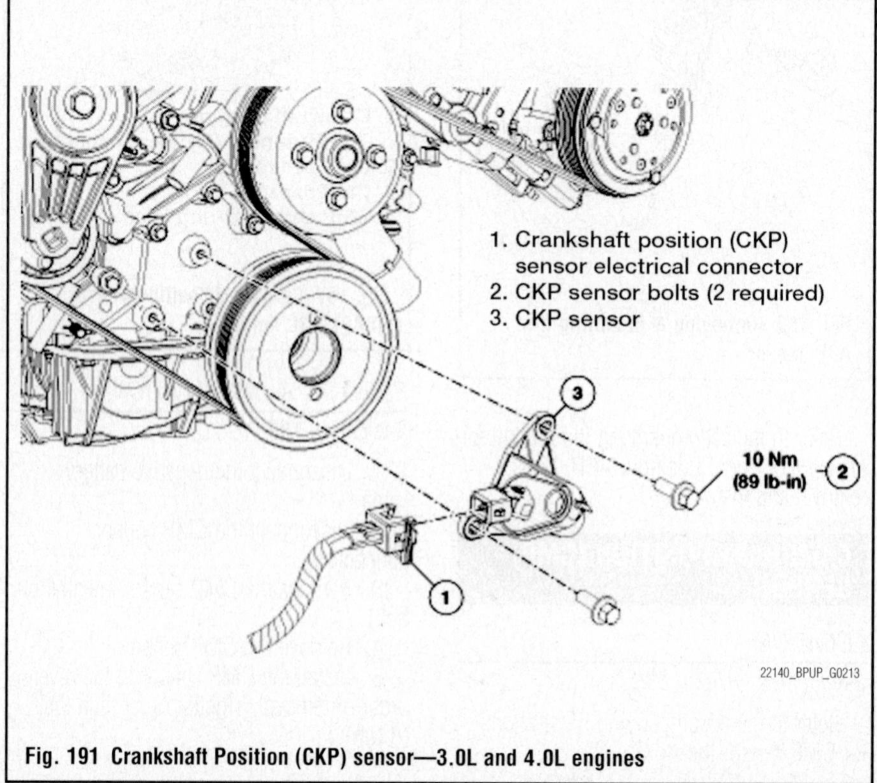

Fig. 191 Crankshaft Position (CKP) sensor—3.0L and 4.0L engines

10. Install the CKP sensor fitting bolts. Tightening torque 4.1–5.5 ft lbs.

11. Remove the special service tool or equivalent then install the cylinder block lower blind plug.

12. Tightening torque 15 ft lbs. (20 Nm).

13. Install the right side driveshaft.

14. Install the splash shield.

15. Install the right front wheel

ENGINE COOLANT TEMPERATURE (ECT) SENSOR

LOCATION

See Figures 194 and 195.

The Engine Coolant Temperature (ECT) sensor is located in the cylinder head.

REMOVAL & INSTALLATION

1. Partially drain the engine cooling system until the coolant level is below the ECT sensor-mounting hole.
2. Disconnect the negative battery cable.
3. Detach the wiring harness connector from the ECT sensor.
4. Using an open-end wrench, remove the coolant temperature sensor from the intake manifold or thermostat housing.

To install:

5. Thread the sensor into the intake manifold, or thermostat housing, by hand, then tighten it securely.
6. Connect the negative battery cable.
7. Refill the engine cooling system.
8. Start the engine, check for coolant leaks and top off the cooling system.

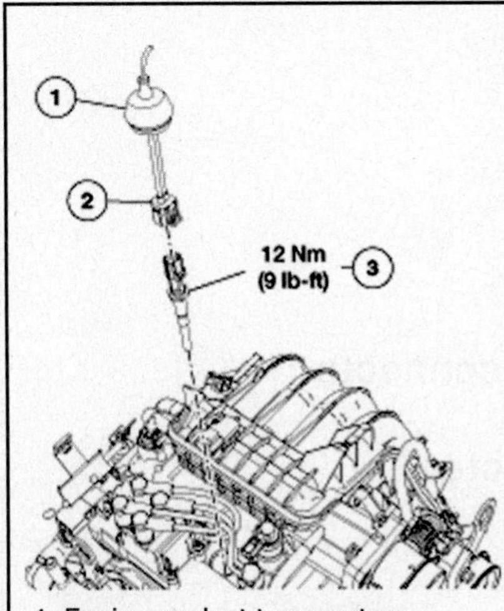

Fig. 194 Engine Coolant Temperature (ECT) sensor—2.3L engine

HEATED OXYGEN (HO2S) SENSOR

LOCATION

See Figures 196 and 197.

The oxygen sensors are located in the exhaust manifold and downstream from the catalytic converters.

REMOVAL & INSTALLATION

⁂ CAUTION

The temperature of the exhaust system is extremely high after the engine has been run. To prevent personal injury, allow the exhaust system to cool completely before removing sensor from the exhaust system.

1. Disconnect the negative battery cable.
2. Raise and safely support the vehicle on jack stands.
3. Disconnect the HO2S from the engine control sensor wiring.

➡If excessive force is needed to remove the sensors lubricate the sensor with penetrating oil prior to removal.

4. Remove the sensors with a sensor removal tool, such as Ford Tool T94P-9472-A or equivalent.

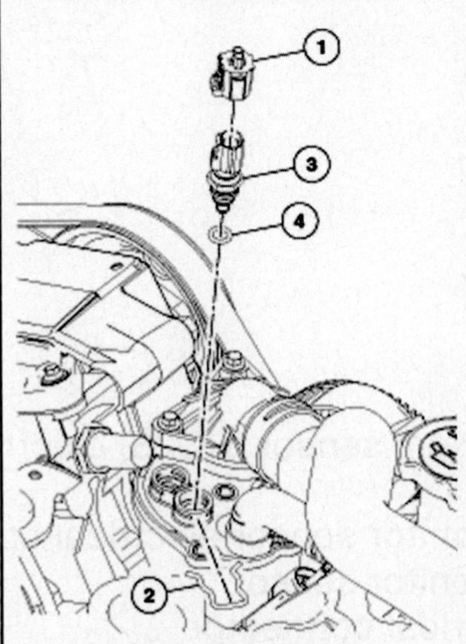

Fig. 195 Engine Coolant Temperature (ECT) sensor—3.0L and 4.0L engines

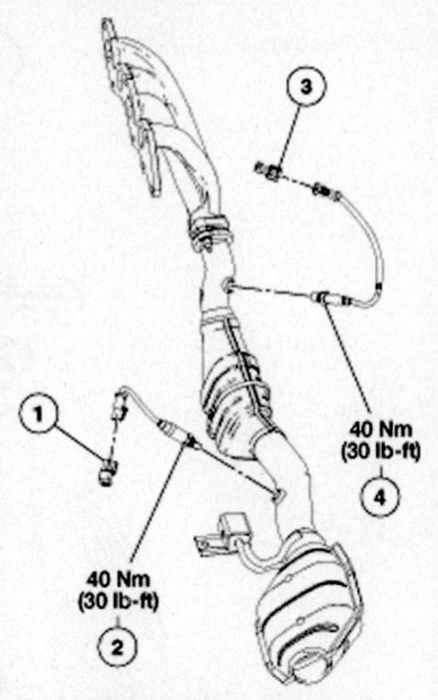

Fig. 196 Heated Oxygen Sensor (HO2S)—2.3L engine

To install:

5. Install the sensor in the mounting boss, and then tighten it to 27–33 ft. lbs.
6. Connect the sensor electrical wiring connector to the engine wiring harness.
7. Lower the vehicle.
8. Connect the negative battery cable.

IDLE AIR CONTROL (IAC) VALVE

LOCATION

See Figures 198 and 199.

Refer to the accompanying illustrations for IAC valve location.

REMOVAL & INSTALLATION

1. Disconnect the negative battery cable.
2. Detach the electrical connector at the sensor.
3. Remove the sensor from the throttle body.

To install:

4. Install the sensor in the opening in the throttle body and tighten the sensor.
5. Attach the electrical connector to the sensor.
6. Connect the negative battery cable.

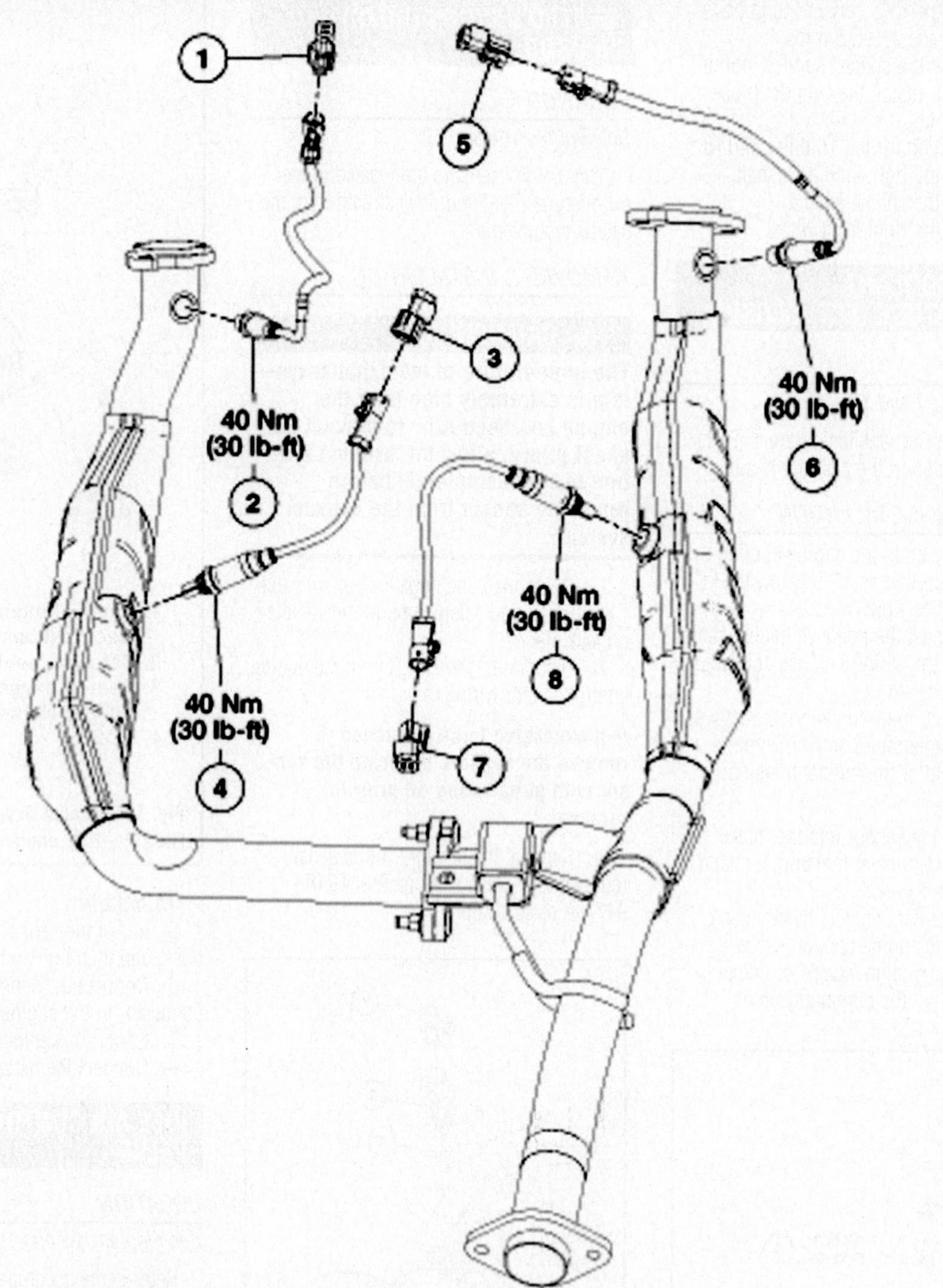

1. Heated oxygen sensor (HO2S) electrical connector
2. HO2S
3. Catalyst monitor sensor electrical connector
4. Catalyst monitor sensor
5. HO2S electrical connector
6. HO2S
7. Catalyst monitor sensor electrical connector
8. Catalyst monitor sensor

22140_BPUP_G0211

Fig. 197 Heated Oxygen Sensor (HO2S)—3.0L and 4.0L engines

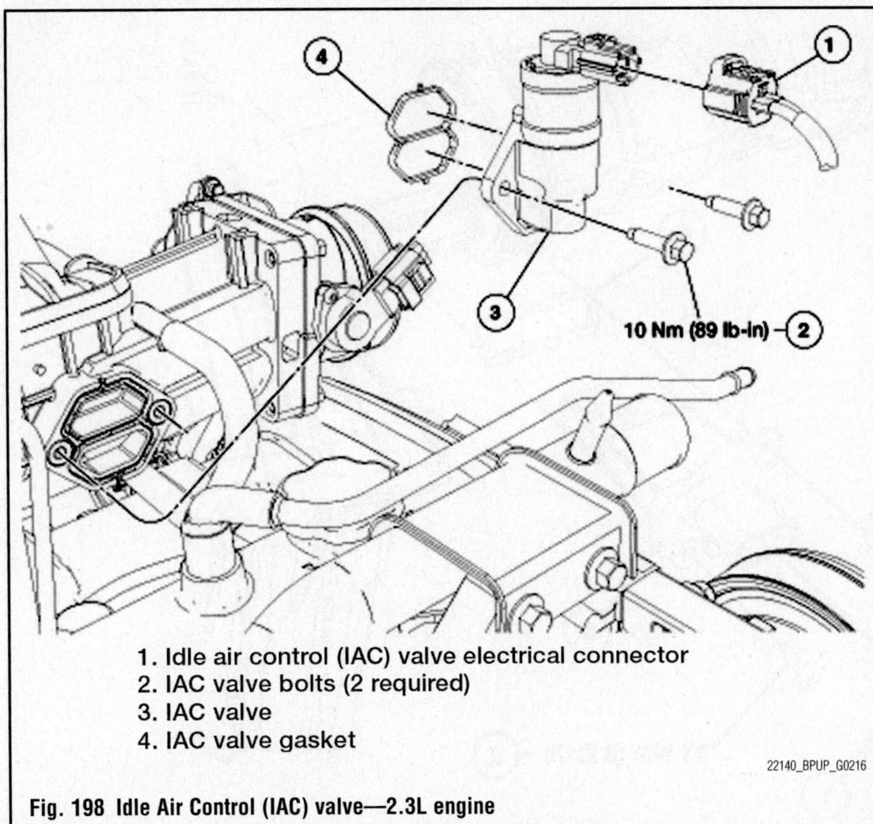

Fig. 198 Idle Air Control (IAC) valve—2.3L engine

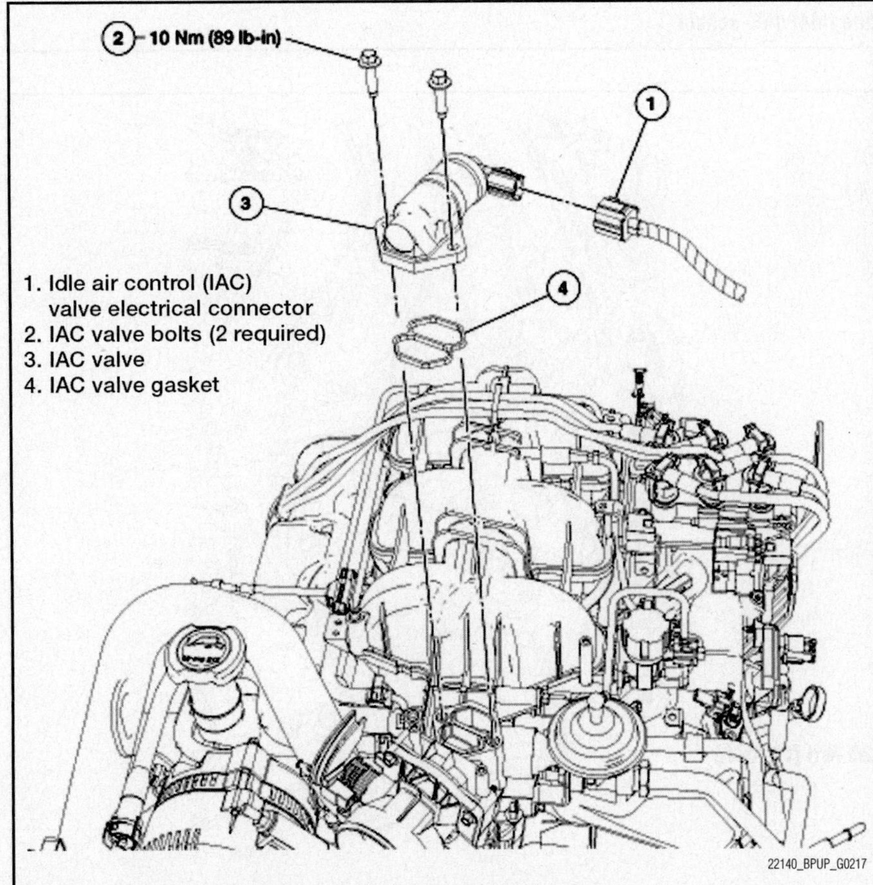

Fig. 199 Idle Air Control (IAC) valve—3.0L and 4.0L engines

INTAKE AIR TEMPERATURE (IAT) SENSOR

LOCATION

See Figure 200.

The Intake Air Temperature (IAT) sensor is integrated with the Mass Air Flow (MAF) sensor.

REMOVAL & INSTALLATION

1. Disconnect the negative battery cable.
2. Remove the IAT sensor from the air cleaner housing or intake air pipe.

To install:

3. Install a new sealing washer to the thermo sensor.
4. Install the IAT sensor from the air cleaner housing.
5. Connect the negative battery cable.

KNOCK SENSOR (KS)

LOCATION

See Figures 201 and 202.

Refer to the accompanying illustrations for KS location.

REMOVAL & INSTALLATION

1. Disconnect the negative battery cable.
2. Detach the electrical connector at the sensor.
3. Remove the sensor from the throttle body.

To install:

4. Install the sensor in the opening in the throttle body and tighten the sensor.
5. Attach the electrical connector to the sensor.
6. Connect the negative battery cable.

MALFUNCTION INDICATOR LIGHT (MIL)

RESET PROCEDURES

To turn off the Malfunction Indictor Light(MIL) after a repair, a reset command from the IDS or equivalent tester must be sent, or three consecutive drive cycles must be completed without a fault.

MASS AIR FLOW (MAF) SENSOR

LOCATION

See Figure 203.

The Mass Air Flow (MAF) sensor is located in the intake air duct leading to the throttle housing.

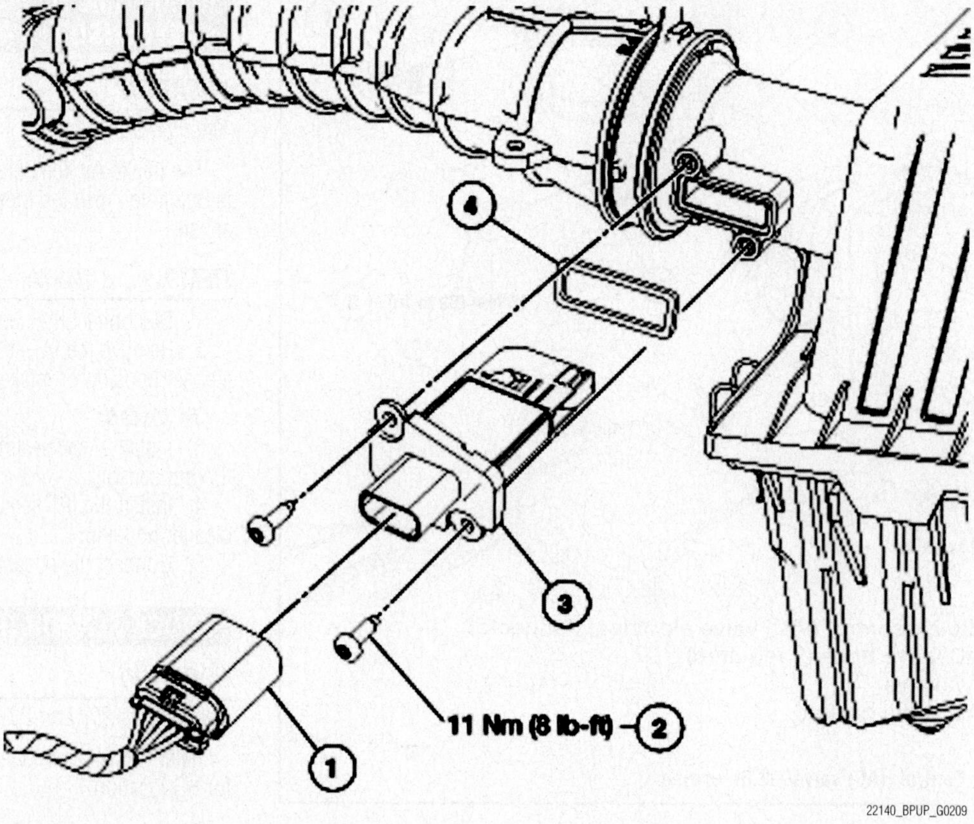

Fig. 200 Mass Air Flow and Intake Air Temperature (MAF/IAT) sensor

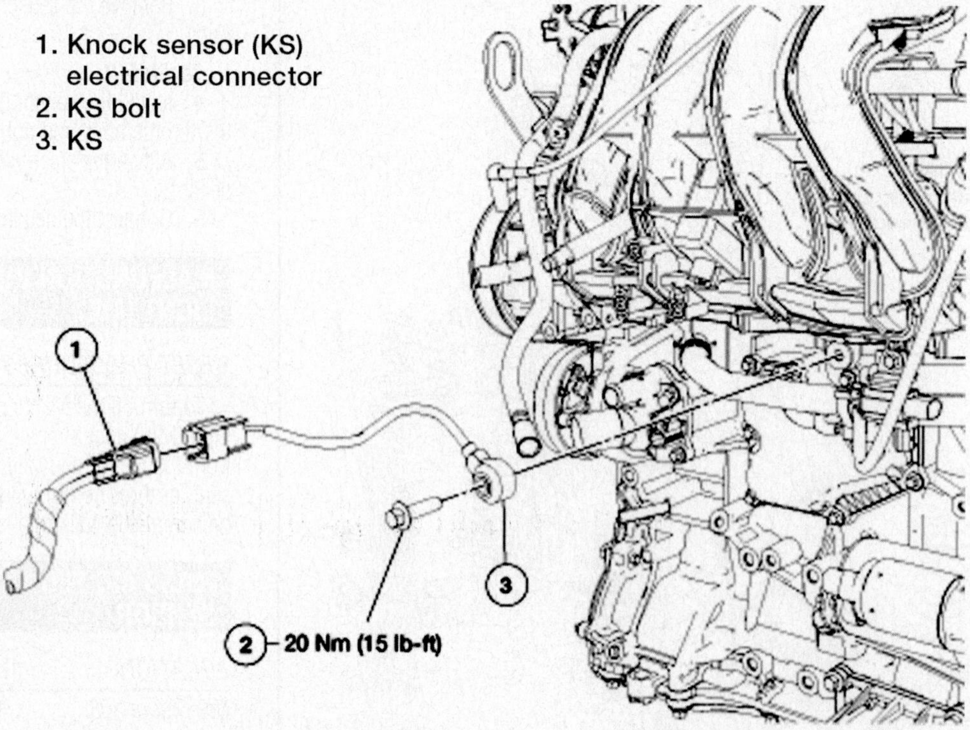

Fig. 201 Knock Sensor (KS)—2.3L engine

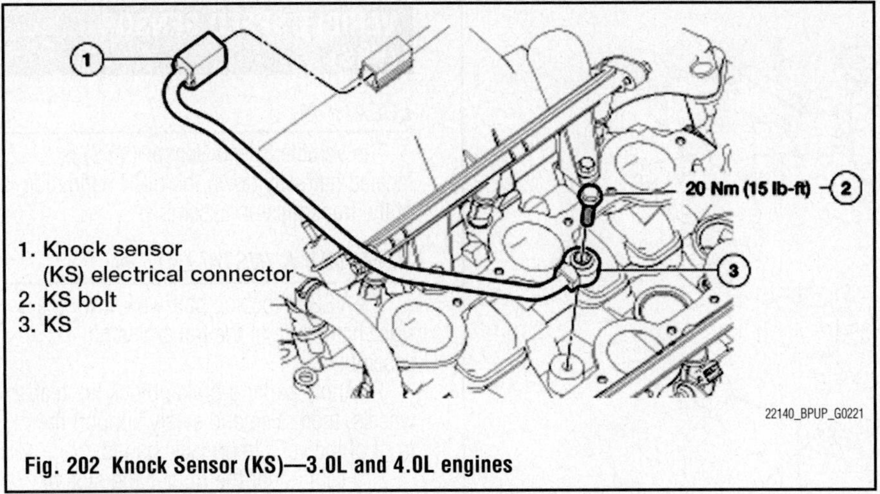

Fig. 202 Knock Sensor (KS)—3.0L and 4.0L engines

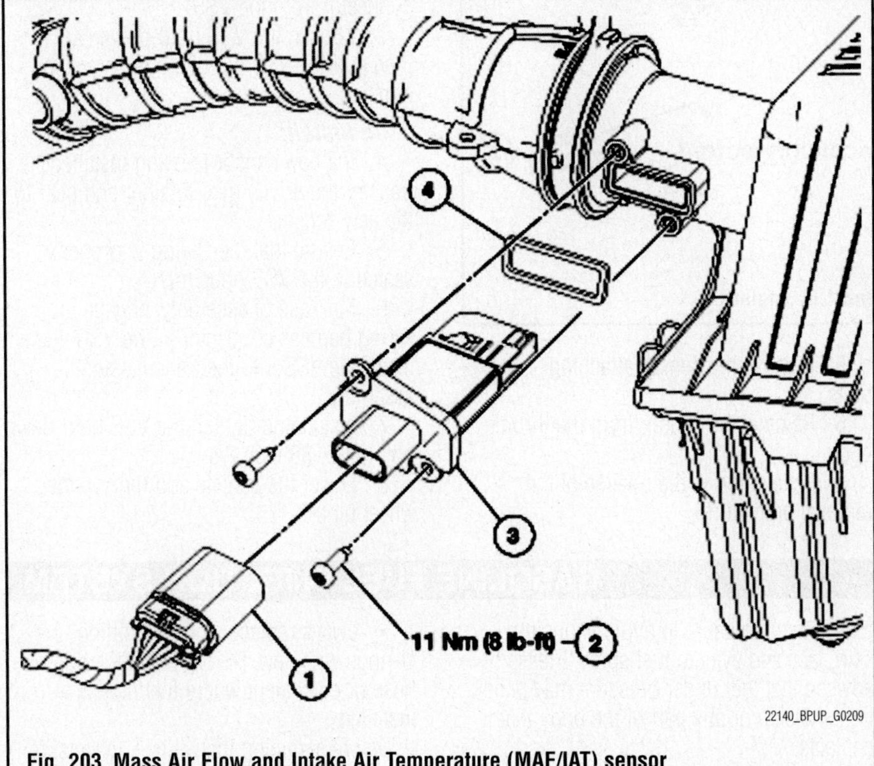

Fig. 203 Mass Air Flow and Intake Air Temperature (MAF/IAT) sensor

REMOVAL & INSTALLATION

See Figure 204.

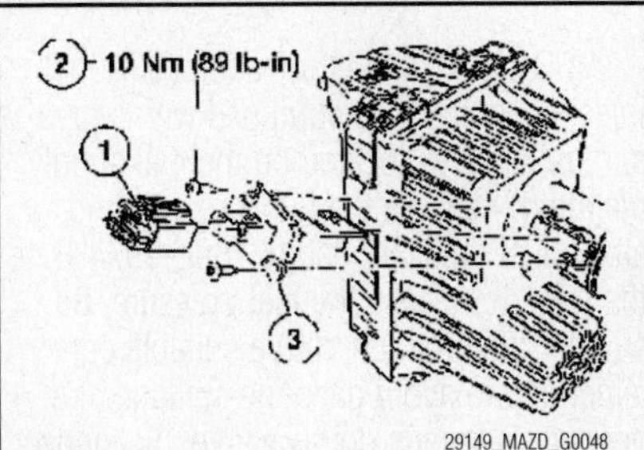

Fig. 204 Mass Air Flow Sensor removal & installation

1. Remove the air cleaner outlet tube.
2. Disconnect the Mass Air Flow (MAF) sensor electrical connector.
3. Remove the attaching four nuts and the MAF sensor.
4. Installation is the reverse of the removal procedure.

POWERTRAIN CONTROL MODULE (PCM)

REMOVAL & INSTALLATION

1. Disconnect the negative battery cable.
2. Disengage the wiring harness connector from the PCM by loosening the connector retaining bolt, then pulling the connector from the module.
3. Remove the two nuts and the PCM cover.
4. Remove the PCM from the bracket by pulling the unit outward.

To install:

5. Install the PCM in the mounting bracket.
6. Install the PCM cover and tighten the two nuts.
7. Attach the wiring harness connector to the module, then tighten the connector-retaining bolt.
8. Connect the negative battery cable.

THROTTLE POSITION SENSOR (TPS)

LOCATION

See Figures 205 and 206.

The Throttle Position Sensor (TPS) is located on the throttle hosing.

REMOVAL & INSTALLATION

1. Disconnect the negative battery cable.

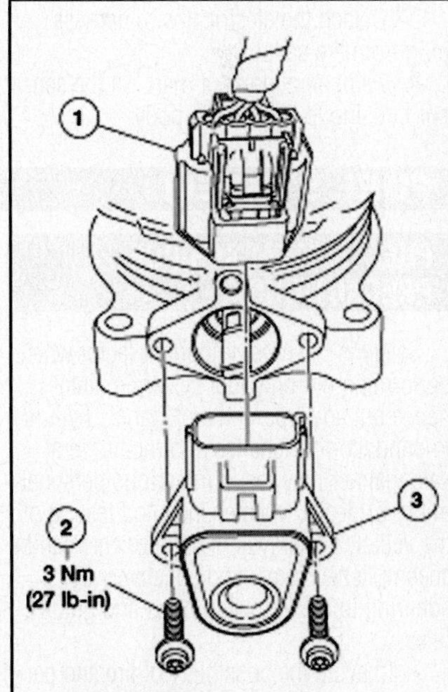

Fig. 205 Throttle Position Sensor (TPS)—2.3L engine

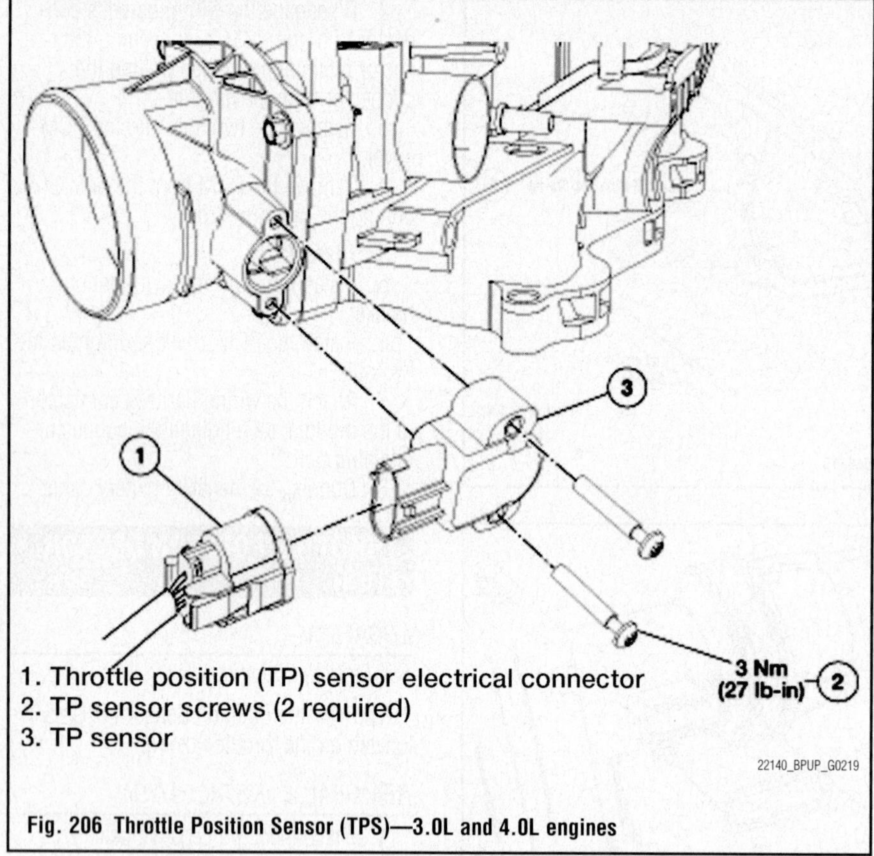

Fig. 206 Throttle Position Sensor (TPS)—3.0L and 4.0L engines

2. Remove the necessary air intake components to access the TPS.
3. Detach the electrical wire harness plug from the sensor.
4. Paint an alignment mark on the sensor housing to the throttle body.
5. Remove the sensor attaching bolts.
6. Remove the sensor from the throttle body.
7. Installation is the reverse of the removal procedure.

VEHICLE SPEED SENSOR (VSS)

LOCATION

The Vehicle Speed Sensor (VSS) is located half-way down the right-hand side of the transmission assembly.

REMOVAL & INSTALLATION

The VSS is located half-way down the right-hand side of the transmission assembly.

1. Apply parking brake, block the rear wheels, then raise and safely support the front of the vehicle on jack stands.
2. From under the right-hand side of the vehicle, disengage the wiring harness connector from the VSS.
3. Loosen the VSS hold-down bolt, then pull the VSS out of the transmission housing.

To install:

4. If a new sensor is being installed, transfer the driven gear retainer and gear to the new sensor.
5. Ensure that the O-ring is properly seated in the VSS housing.
6. For ease of assembly, engage the wiring harness connector to the VSS, then insert the VSS into the transmission assembly.
7. Install and tighten the VSS hold-down bolt to 62–88 inch lbs.
8. Lower the vehicle and remove the wheel blocks.

FUEL SYSTEM

GASOLINE FUEL INJECTION SYSTEM

FUEL SYSTEM SERVICE PRECAUTIONS

Safety is the most important factor when performing not only fuel system maintenance but any type of maintenance. Failure to conduct maintenance and repairs in a safe manner may result in serious personal injury or death. Maintenance and testing of the vehicle's fuel system components can be accomplished safely and effectively by adhering to the following rules and guidelines.

- To avoid the possibility of fire and personal injury, always disconnect the negative battery cable unless the repair or test procedure requires that battery voltage be applied.
- Always relieve the fuel system pressure prior to disconnecting any fuel system component (injector, fuel rail, pressure regulator, etc.), fitting or fuel line connection. Exercise extreme caution whenever relieving fuel system pressure to avoid exposing skin, face and eyes to fuel spray. Please be advised that fuel under pressure may penetrate the skin or any part of the body that it contacts.
- Always place a shop towel or cloth around the fitting or connection prior to loosening to absorb any excess fuel due to spillage. Ensure that all fuel spillage (should it occur) is quickly removed from engine surfaces. Ensure that all fuel soaked cloths or towels are deposited into a suitable waste container.
- Always keep a dry chemical (Class B) fire extinguisher near the work area.
- Do not allow fuel spray or fuel vapors to come into contact with a spark or open flame.
- Always use a back-up wrench when loosening and tightening fuel line connection fittings. This will prevent unnecessary stress and torsion to fuel line piping.
- Always replace worn fuel fitting O-rings with new. Do not substitute fuel hose or equivalent where fuel pipe is installed.

Before servicing the vehicle, make sure to also refer to the precautions in the beginning of this section as well.

RELIEVING FUEL SYSTEM PRESSURE

All Sequential Fuel Injection (SFI) fuel injected engines are equipped with a pressure relief valve located on the fuel supply manifold. Remove the fuel tank cap and attach fuel pressure gauge T80L-9974-B, to the valve to release the fuel pressure. Be sure to drain the fuel into a suitable container and to avoid gasoline spillage. If a pressure gauge is not available, disconnect the vacuum hose from the fuel pressure regulator and attach a hand-held vacuum pump. Apply about 25 in. Hg (84 kPa) of vacuum to the regulator to vent the fuel system pressure into the fuel tank through the fuel return hose. Note that this procedure will remove the fuel pressure from the lines, but not the fuel. Take precautions to avoid the risk of fire and use clean rags to soak up any spilled fuel when the lines are disconnected.

An alternate method of relieving the fuel system pressure involves disconnecting the inertia switch.

FUEL FILTER

REMOVAL & INSTALLATION

See Figure 207.

** CAUTION

Do not smoke or carry lighted tobacco or open flame of any type when working on or near any fuel-related component. Highly flammable mixtures are always present and can be ignited, resulting in possible personal injury.

** CAUTION

Fuel in the fuel system remains under high pressure even when the engine is not running. Before repairing or disconnecting any of the fuel lines or fuel system components, the fuel system pressure must be relieved to prevent accidental spraying of fuel, causing personal injury or a fire hazard.

1. Before servicing the vehicle, refer to the precautions in the beginning of this section.
2. Properly relieve the fuel system pressure.
3. Remove or disconnect the following:
   - Negative battery cable
   - Push connect and R-clip fittings from the fuel filter
   - Fuel filter

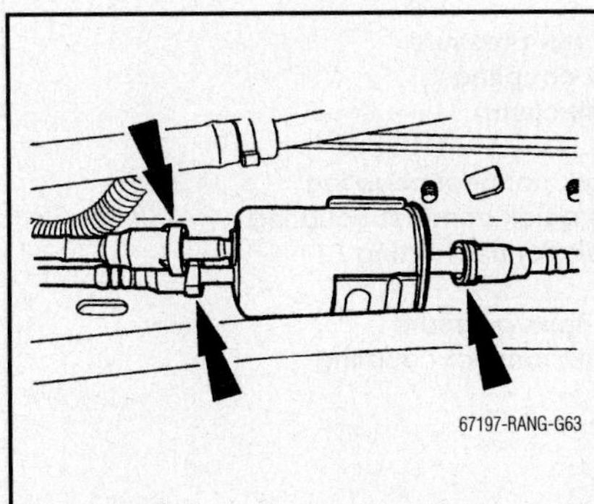

Fig. 207 Fuel filter location and connections

To install:

4. Install or connect the following:
   - Fuel filter. Torque the nut to 17 ft. lbs. (23 Nm).
   - R-clip and push connect fittings
   - Negative battery cable
5. Start the vehicle, check for leaks and repair if necessary.

FUEL LEVEL SENDING UNIT

REMOVAL & INSTALLATION

See Figures 208 and 209.

1. Disconnect the negative battery cable.

➡On some vehicles, when the battery cable is disconnected and reconnected, some abnormal drive symptoms may occur while the vehicle relearns its adaptive strategy. The vehicle may need to be driven to relearn its strategy.

The fuel pump must be handled carefully to avoid damage to the float arm and filter. Remove the fuel pump assembly.

2. Properly relieve the fuel system pressure
3. Clean the area around the fuel pump mounting flange.
4. Using the fuel tank sender unit wrench 310—123, remove the fuel tank sending unit assembly locking retainer ring.

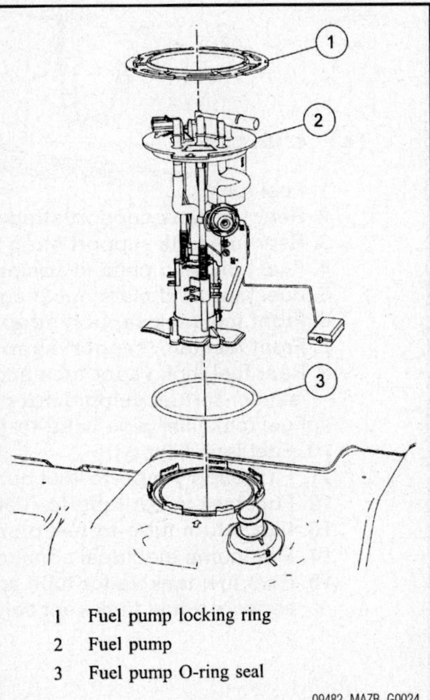

Fig. 208 Fuel pump and related components

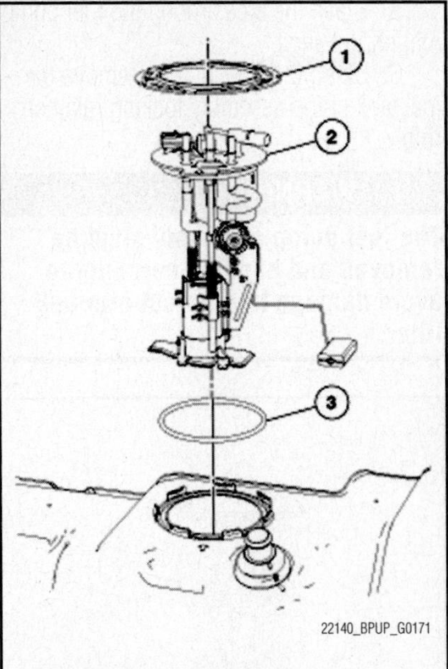

Fig. 209 Fuel sending and pump unit removal tool

5. Remove and discard the fuel pump mounting gasket.

To install:

6. Clean the fuel pump mounting flange and the fuel tank mounting surface.
7. Install a new fuel pump mounting gasket
8. Install the fuel tank sending unit assembly with the float toward the rear of the fuel tank. Align the arrows molded into the fuel tank flange.
9. Install the locking retainer ring while compressing the fuel tank sending unit assembly into the tank.
10. Using the fuel tank sender unit wrench, tighten the fuel tank sending unit assembly locking retainer ring until it locks in place.
11. Install the fuel tank.

FUEL PUMP

REMOVAL & INSTALLATION

See Figure 210.

1. Before servicing the vehicle, refer to the Precautions Section.
2. Properly relieve the fuel system pressure.
3. Disconnect the negative battery cable.

➡On some vehicles, when the battery cable is disconnected and reconnected, some abnormal drive symptoms may occur while the vehicle relearns its adaptive strategy. The vehicle may need to be driven to relearn its strategy.

4. Remove the fuel tank.

5. Clean the area around the fuel pump mounting flange.

6. Using the special tool, remove the fuel tank pump assembly locking retainer ring.

✲✲ WARNING

The fuel pump assembly must be removed and handled carefully to avoid damage to the float arm and filter.

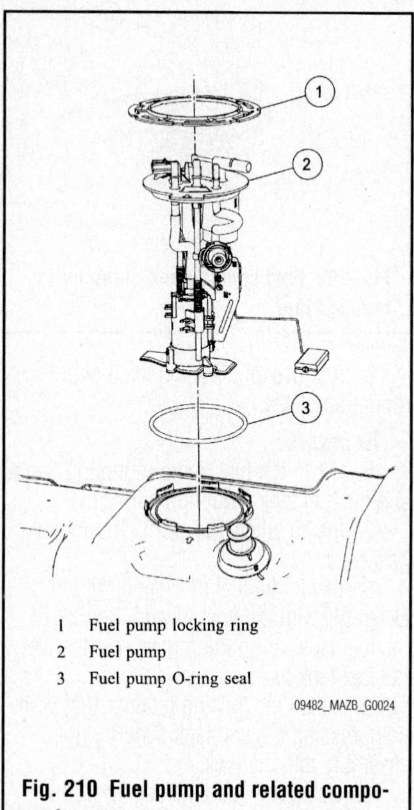

Fig. 210 Fuel pump and related components

7. Remove the fuel pump assembly.

8. Remove and discard the fuel pump mounting gasket.

To install:

9. Clean the fuel pump mounting flange and the fuel tank mounting surface.

10. Install a new fuel pump mounting gasket.

11. Install the fuel pump and sender assembly with the float toward the rear of the tank. Align the arrows molded into the tank and flange.

12. Install the locking ring while compressing the pump assembly into the tank.

13. Using the special tool, tighten the fuel pump assembly locking ring retainer ring until it locks in place.

14. Install the fuel tank.

FUEL TANK

REMOVAL & INSTALLATION

See Figure 212.

1. Relieve the fuel pressure.

2. Disconnect the fill pipe main hose from the main fill pipe.

3. Drain the fuel tank.

4. If equipped, remove the 4 nuts (two each side) and remove the skid plate.

5. Support the fuel tank.

6. Remove the bolts and pivot the fuel tank support straps away from the fuel tank.

7. Disconnect the fuel pump connector.

8. Lower the fuel tank slightly.

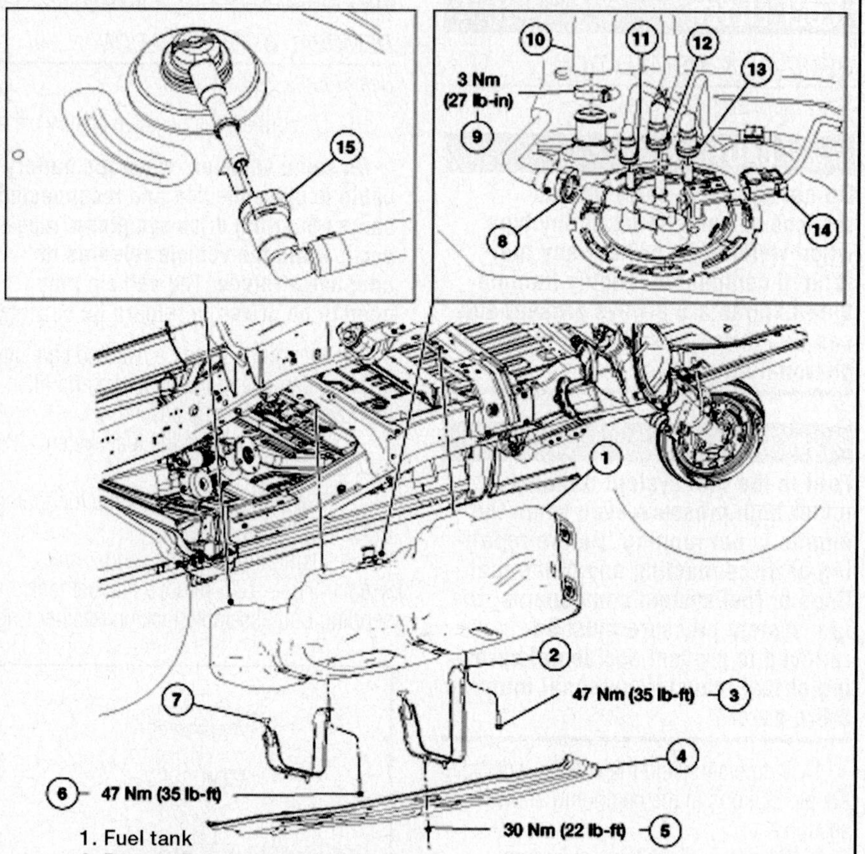

3 Nm (27 lb-in) (9)

47 Nm (35 lb-ft) (3)

(6) 47 Nm (35 lb-ft)

30 Nm (22 lb-ft) (5)

1. Fuel tank
2. Rear fuel tank support strap
3. Rear fuel tank support strap bolt
4. Fuel tank skid plate (if equipped)
5. Fuel tank skid plate nut (if equipped)
6. Front fuel tank support strap bolt
7. Front fuel tank support strap
8. Rear fuel tank vapor tube and fuel tank pressure sensor-to-fuel pump quick connect coupling
9. Fuel tank filler pipe hose-to-fuel tank clamp
10. Fuel tank filler pipe
11. Fuel supply tube-to-fuel pump quick connect coupling
12. Fuel tank vapor tube-to-fuel pump quick connect coupling
13. Fuel return tube-to-fuel pump quick connect fitting
14. Fuel pump electrical connector
15. Rear fuel tank vapor tube and fuel tank pressure sensor-to-fuel tank vent valve quick connect coupling

22140_BPUP_G0172

Fig. 212 Exploded view of the fuel tank removal and installation

9. Disconnect the fuel pump electrical connector.
10. Disconnect the fuel supply tube, vapor tube and the fill tube vent hose from the fuel pump.
11. Disconnect the vapor tube fitting at the rear of the fuel tank.
12. Lower the fuel tank.
13. To install, reverse the removal procedure.
14. Tighten the skid plate to 22 ft. lbs. (30 Nm).
15. Tighten the fuel tank support straps to 30 ft. lbs. (40 Nm).

FUEL RAIL & INJECTORS

REMOVAL & INSTALLATION

2.3L Engine

See Figures 213 through 215.

1. Before servicing the vehicle, refer to the precautions in the beginning of this section.
2. Properly relieve the fuel system pressure.
3. Remove or disconnect the following:
4. Remove the intake manifold.
5. Disconnect the fuel tube spring lock coupling.
6. Disconnect the fuel injector electrical connectors.
7. Remove or disconnect the following:
    - Negative battery cable
    - Upper intake manifold
    - Fuel injector connectors
    - Fuel injector harness from the fuel injector supply manifold
    - Fuel line spring lock
    - Fuel line
    - Fuel injection supply manifold
    - Fuel injector retaining clip
    - Fuel injector

WARNING

Use O-ring seals that are made of special fuel-resistant material. Use of ordinary O-ring seals can cause the fuel system to leak. Do not reuse the O-ring seals.

8. Installation is the reverse of removal. Install new O-rings. Lubricate the O-rings with clean engine oil.

3.0L and 4.0L Engines

See Figures 216 through 221.

1. Before servicing the vehicle, refer to the Precautions Section.
2. Disconnect the negative battery cable.

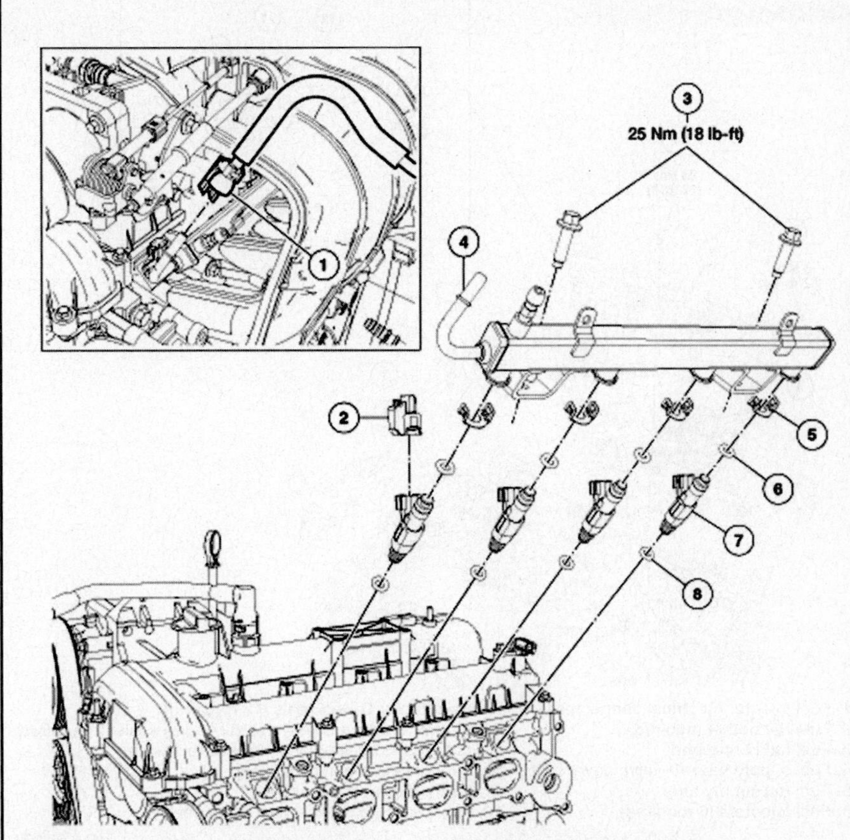

Fig. 213 Fuel injection supply manifold and fuel injector removal and installation exploded component view—2.3L engine

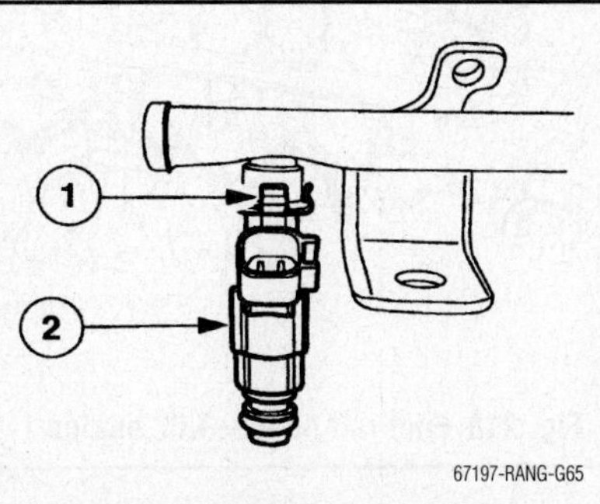

Fig. 214 Fuel injector-to-fuel rail installation. 1 retaining clip; 2 injector—2.3L engine

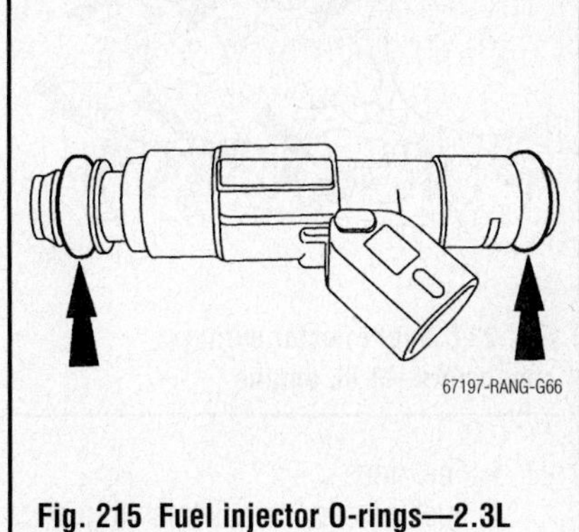

Fig. 215 Fuel injector O-rings—2.3L engine

➡On some vehicles, when the battery cable is disconnected and reconnected, some abnormal drive symptoms may occur while the vehicle relearns its adaptive strategy. The vehicle may need to be driven to relearn its strategy.

3. Properly relieve the fuel system pressure.
4. Remove or disconnect the following:
    - Upper intake manifold
    - Engine control sensor wiring from the fuel injectors

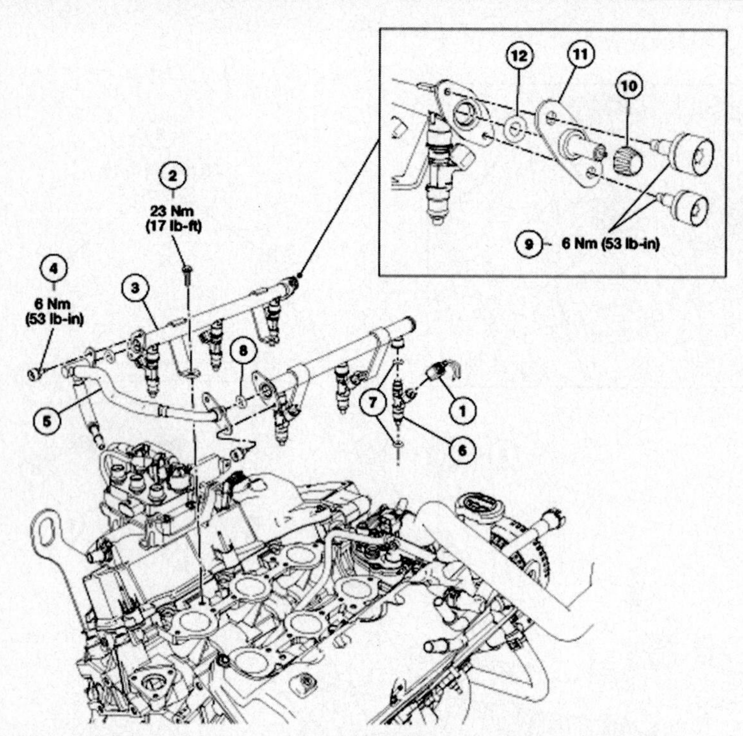

Fig. 216 Fuel injection supply manifold and fuel injector removal and installation exploded component view—3.0L and 4.0L engines

Fig. 217 Fuel injector wiring connectors—3.0L engine

Fig. 218 Fuel rail bolts—3.0L engine

- Fuel lines
- Fuel injection supply manifold and injectors as an assembly
- Vacuum line
- Fuel injectors from the supply manifold
- Inspect the O-rings and replace them as needed

To install:

5. Install or connect the following:
   - Fuel injectors
   - Vacuum line
   - Fuel injection supply manifold. Torque the bolts to 89 inch lbs. (10 Nm).
   - Fuel line
   - Engine control sensor wiring to the fuel injectors
   - Upper intake manifold
   - Negative battery cable
6. Start the vehicle, check for leaks and repair if necessary.

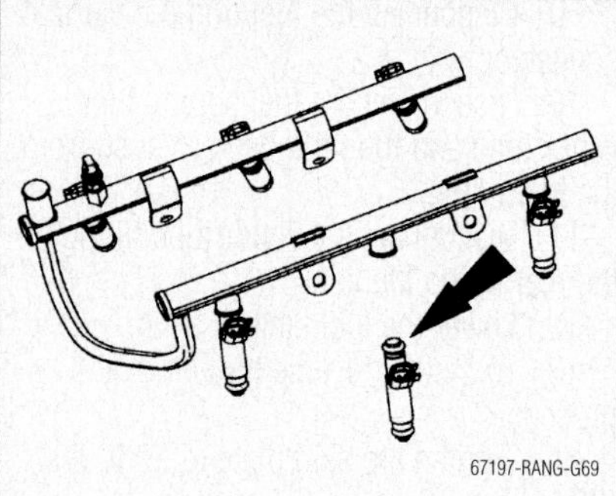

Fig. 219 Fuel rail and injectors—3.0L engine

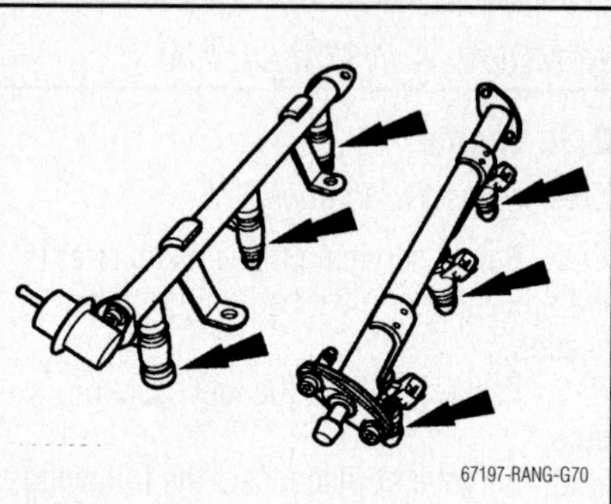

Fig. 220 Fuel rail and injectors—4.0L engine

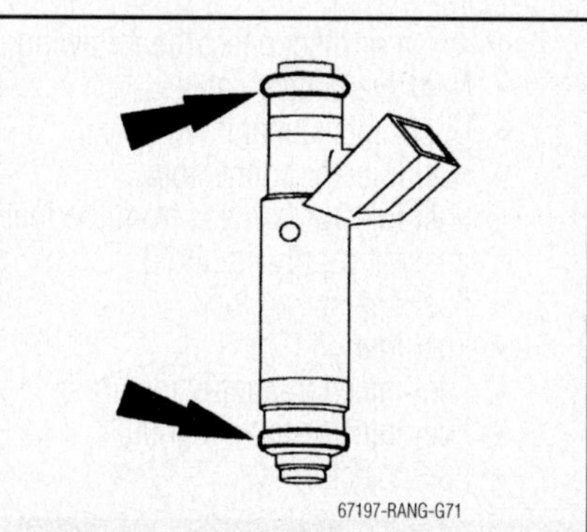

Fig. 221 Fuel injector O-rings—4.0L engine

IDLE SPEED

ADJUSTMENT

No adjustments can be made The Electronic Control Module (ECM) controls idle speed dependent on input.

THROTTLE BODY

REMOVAL & INSTALLATION

2.3L Engine

See Figure 222.

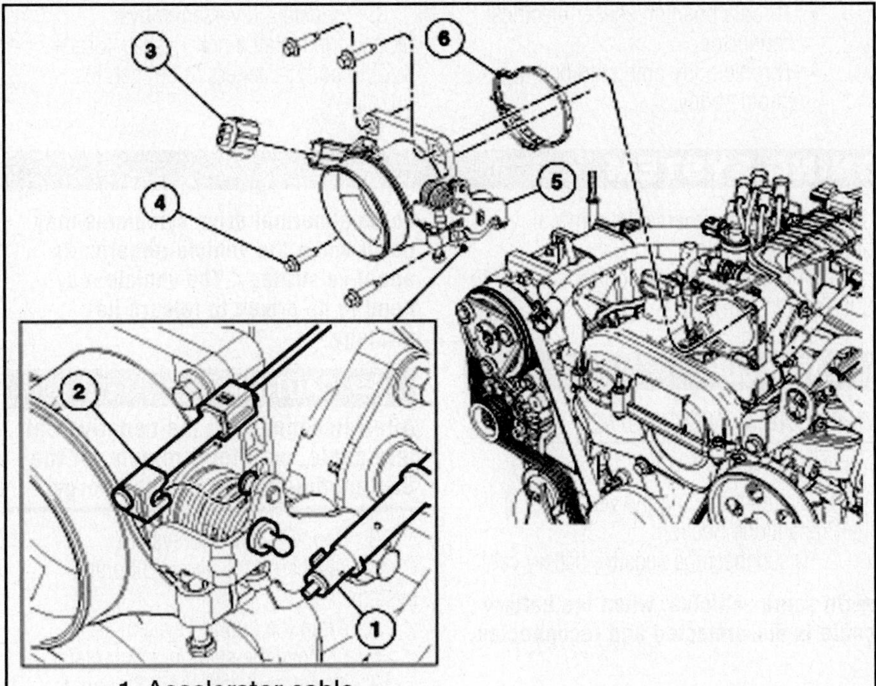

Fig. 222 Exploded view of the throttle body—2.3L engine

✱✱ CAUTION

Do not smoke or carry lighted tobacco or open flame of any type when working on or near any fuel-related components. Highly flammable mixtures are always present and may be ignited. Failure to follow these instructions may result in personal injury.

✱✱ WARNING

The throttle body bore and plate area have a special coating and should not be cleaned.

1. Before servicing the vehicle, refer to the Precautions Section.
2. Remove or disconnect the following:
    - Negative battery cable.
    - Air cleaner outlet pipe
    - If equipped, transmission vent tube and bracket bolt and set aside
    - Accelerator cable and, if equipped, the cruise control actuator cable
    - Throttle position sensor electrical connector
    - Screws and throttle body

➡The throttle body sealing surfaces are soft materials. Use care if cleaning is necessary.

3. To install, reverse the removal procedure.
4. Torque throttle body screws to 89 inch lbs. (10 Nm).

3.0L and 4.0L Engines

See Figure 223.

✱✱ WARNING

Do not smoke or carry lighted tobacco or open flame of any type when working on or near any fuel-related components. Highly flammable mixtures are always present and may be ignited. Failure to follow these instructions may result in personal injury.

✱✱ CAUTION

Throttle body bore and plate area have a special coating and should not be cleaned.

1. Before servicing the vehicle, refer to the Precautions Section.
2. Remove or disconnect the following:
    - Negative battery cable.

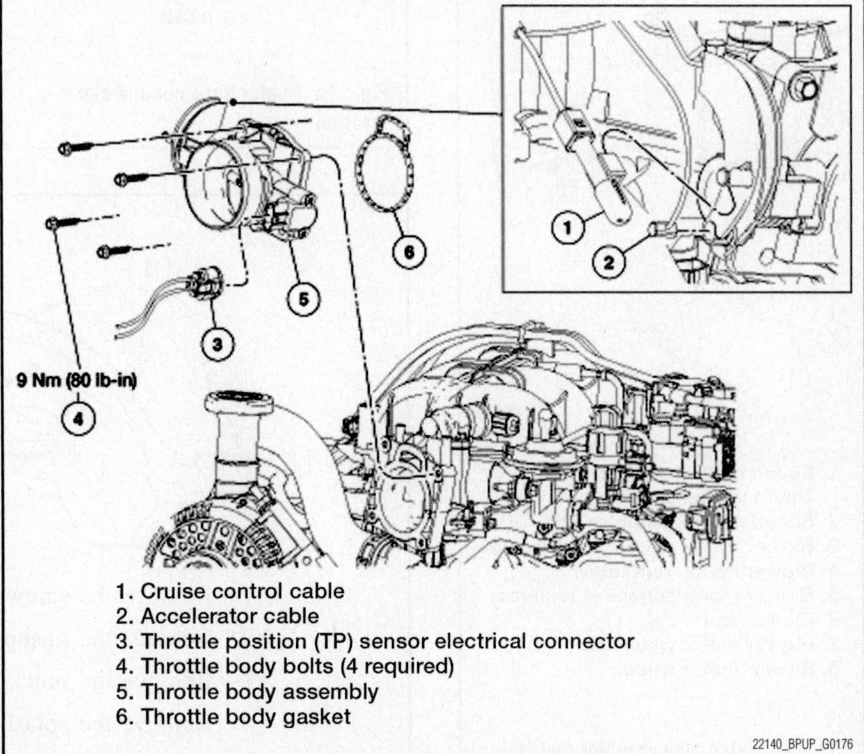

Fig. 223 Throttle body removal and installation exploded component view—3.0L and 4.0L engines

- Air cleaner outlet pipe
- Accelerator cable and, if equipped, the cruise control actuator cable
- Throttle position sensor electrical connector
- Throttle body bolt, stud bolt and throttle body

3. To install, reverse removal procedure. Install a new gasket. Torque throttle body fasteners to 89 inch lbs. (10 Nm).

HEATING & AIR CONDITIONING SYSTEM

BLOWER MOTOR

REMOVAL & INSTALLATION

See Figure 224.

1. Before servicing the vehicle, refer to the Precautions Section.
2. Disconnect the negative battery cable.
3. Disconnect the speed control actuator electrical connector.
4. Remove the speed control actuator bolt and position the speed control actuator aside.
5. Disconnect the blower motor electrical connector.
6. Remove the blower motor vent tube.
7. Remove the 4 blower motor screws and the blower motor.
8. Remove the blower motor wheel clip.
9. Remove the blower motor wheel.
10. To install, reverse the removal procedure.
11. Tighten the speed control actuator to 8 ft. lbs. (11 Nm).

2 - 11 Nm (8 lb-ft)

1. Speed control actuator electrical connector
2. Speed control actuator bolt
3. Blower motor electrical connector
4. Blower motor vent tube
5. Blower motor screws (4 required)
6. Blower motor
7. Blower motor wheel clip
8. Blower motor wheel

22140_BPUP_G0177

Fig. 224 Cabin blower motor removal and installation exploded component view

HEATER CORE

REMOVAL & INSTALLATION

See Figures 225 through 233.

1. Before servicing the vehicle, refer to the Precautions Section.
2. Disconnect the negative battery cable.

➡On some vehicles, when the battery cable is disconnected and reconnected, some abnormal drive symptoms may occur while the vehicle relearns its adaptive strategy. The vehicle may need to be driven to relearn its strategy.

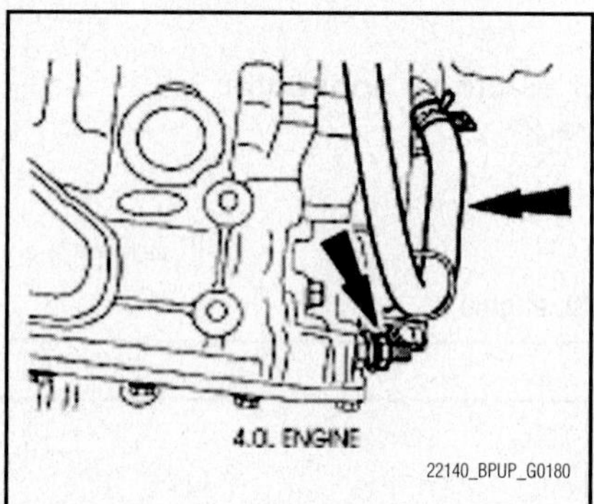

Fig. 225 Heater hose nut and clip location

⁂ CAUTION

After disconnecting the negative battery cable, wait for 1 minute for the SRS module to deplete its energy.

3. Depower the SRS system.
4. Discharge the air conditioning system.
5. Drain the cooling system.
6. Remove the suction accumulator.
7. If equipped with the 2.3L engine, remove the A/C compressor and the engine oil indicator and tube.
8. If equipped with the 3.0L or 4.0L engine, position the coolant reservoir and windshield washer reservoir aside.
9. Detach the speed control servo, on vehicles equipped with cruise control.
10. Disconnect the blower motor and blower motor resistor electrical connectors.
11. On 2.3L engine, remove the heater hoses from the heater core.
12. Disconnect the vacuum line from the water control valve.
13. Remove the heater hose from the rear coolant tube.

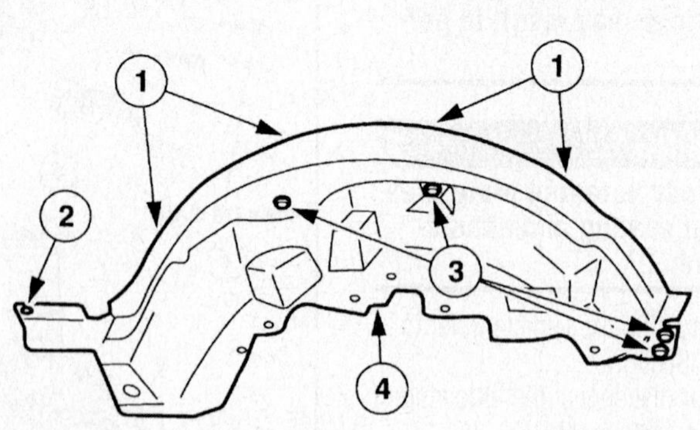

1 Remove the screws.
2 Remove the pushpin.
3 Remove the bolts.
4 Remove the splash shield.

09482_MAZB_G0003

Fig. 226 Splash shield removal points

14. For all engines, remove the heater hoses from between the heater core and the water control valve.

15. On 2.3L engine, disconnect the heater hose from the water outlet tube and remove the heater hose assembly.

16. For all engines, remove the heater hose from between the water control valve and the water pump.

17. On 4.0L engine, remove the nut from the clip.

18. Remove the heater hose from between the water control valve and the water outlet.

19. Detach the pin-type retainer and position aside the windshield washer hose.

20. Disconnect and detach the heater control valve vacuum hose.

21. Disconnect the vacuum supply hose near the evaporator core housing.

22. Disconnect the condenser to evaporator line spring lock coupling from the evaporator core inlet. Discard the O-ring seals.

➡This step is performed at the lower passenger side dash panel, inside the passenger compartment.

23. Disconnect the vacuum hose connector and remove the nut.

24. If equipped with the 4.0L engine, remove the vehicle splash guard. Remove the splash shield. On the right splash shield, disconnect the vacuum lines from the vacuum storage reservoir.

➡Remove the electrical connector locators from the splash shield prior to removal.

25. Remove the nuts and the evaporator core housing.

26. Lock the steering column.

27. Remove the front seats.

28. Remove the screws and position the parking brake release handle aside.

29. Remove the left and right lower cowl kick panels.

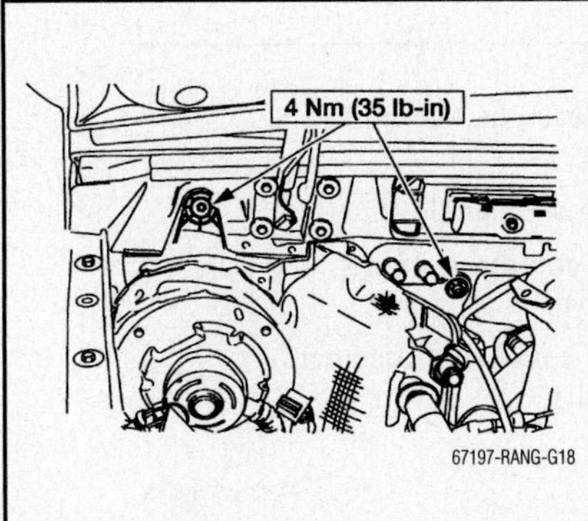

Fig. 227 Evaporator core housing bolts

30. Position the parking brake assembly aside.

31. Remove the screws and position the hood release handle aside.

32. Remove the instrument panel steering column cover.

33. Remove the instrument panel steering column opening cover reinforcement.

34. Disconnect the Brake Pedal Position (BPP) switch electrical connector from the steering column shaft.

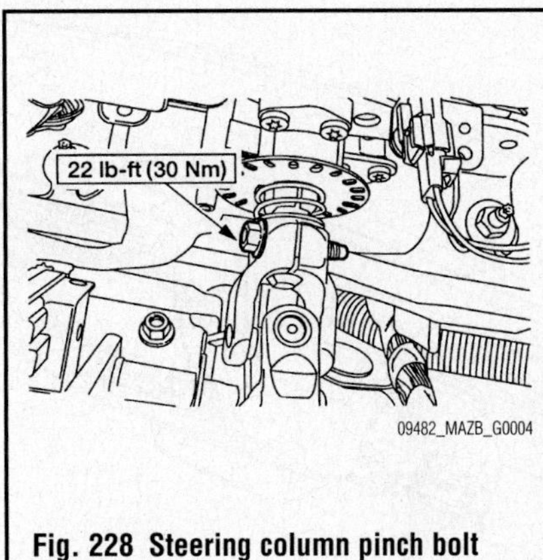

Fig. 228 Steering column pinch bolt

35. If equipped, disconnect the Clutch Pedal Position (CPP) switch electrical connector.

36. If equipped, disconnect the shift cable from the steering column.

37. Remove the steering column pinch bolt and disconnect the steering column intermediate shaft.

➡To avoid damage to the clockspring, do not allow the steering shaft to rotate while the intermediate shaft is disconnected.

38. Remove the left and right side garnish moldings.

39. Remove the bolt covers and remove the bolts.

40. Remove the assist handle.

41. Remove the windshield side garnish molding.

42. Remove the door moldings.

43. On regular cab vehicles

44. Remove the screws.

45. Remove the scuff plate.

46. Disconnect the electrical connectors and the ground wire from the RH side lower kick panel.

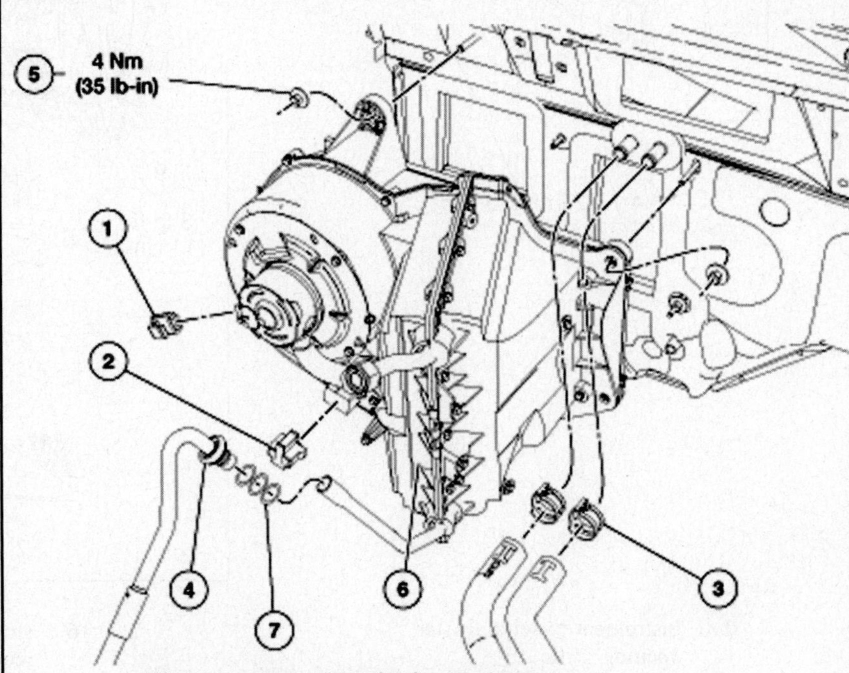

Fig. 229 Evaporator core housing

➡To avoid damaging the bulkhead electrical connectors, be sure the release tab is fully depressed before pulling release lever into the disconnect position.

47. Disconnect the LH side bulkhead electrical connector. Press the release tab and pull the release lever.

48. Remove the audio unit. Insert the removal tool. Remove and support the audio unit.

49. Disconnect the audio unit electrical connector and antenna cable.

50. Lower the glove compartment. Press the release tabs inward while lowering the compartment.

51. Through the glove compartment

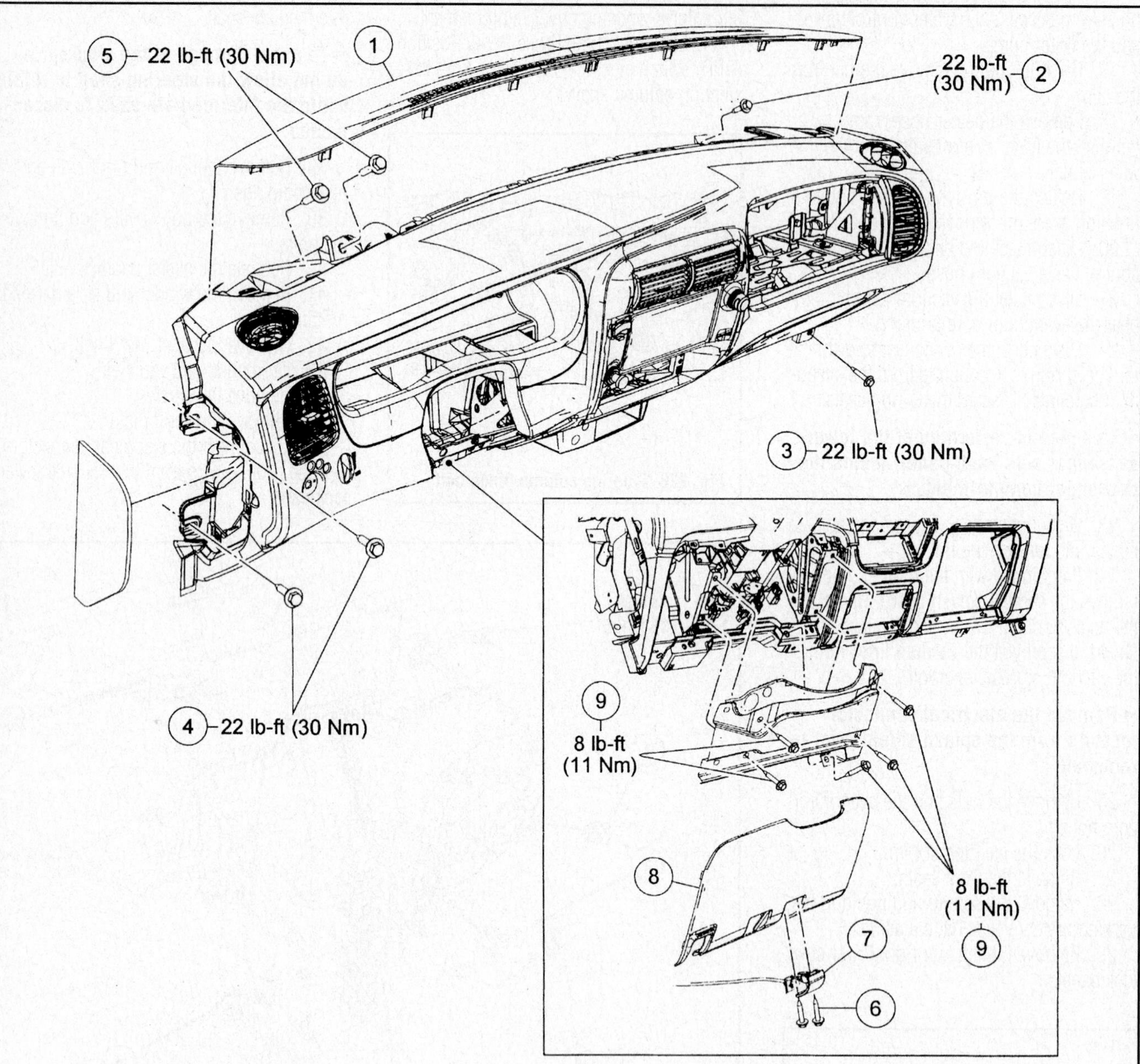

Fig. 230 Instrument panel and related components

opening, disconnect the blend door actuator electrical connector.

52. Through the glove compartment opening, disconnect the climate control vacuum harness connector.

53. Raise and secure the glove compartment. Press the release tabs inward while raising the glove compartment.

54. Remove the instrument panel defroster opening grille.

55. Remove the instrument panel cowl top bolts.

56. If equipped, remove the floor console.

57. If not equipped with the high-series floor console, remove the cup holders.

58. Release the clips and remove the Restraints Control Module (RCM) cover.

59. If not equipped with high-series floor console, remove the consolette mat.

60. Remove the screws and remove the RCM cover.

61. Remove the consolette base.

62. Remove the gearshift lever.

63. Remove the screws and remove the manual transmission consolette.

64. Disconnect the RCM electrical connector.

65. Pull the floor carpeting back.

66. On 2WD vehicles, disconnect the instrument panel main harness.

67. On 4WD vehicles disconnect the instrument panel main harness. From underneath the vehicle, release the instrument panel main harness at the transfer case.

68. Remove the instrument panel side finish panel.

69. Disconnect the door harness electrical connector.

70. Remove the RH side instrument panel bolt. If necessary, transfer the components to the new instrument panel.

71. Remove the LH instrument panel cowl side bolts.

72. Remove the instrument panel.

73. Remove the evaporator core housing.

74. Remove the Powertrain Control Module (PCM).

75. Remove the PCM heat sink.

76. Remove the four nuts from the engine side of the dash panel. Position the plenum chamber on the vehicle floor.

77. Remove the heater core cover.

78. Remove the heater core.

To install:

79. During installation, be sure to install a new oval foam seal around the heater core inlet and outlet tubes.

80. Torque the steering column pinch bolt to 21 ft. lbs.

➡To avoid damage to the clockspring, do not allow the steering shaft to rotate while the intermediate shaft is disconnected.

81. Lubricate the refrigerant system with the correct amount of clean PAG oil or equivalent. Install new O-ring seals lubricated in clean mineral oil. Lubricate the coolant hoses with plain water only, if needed.

82. Evacuate, leak test, and charge the refrigerant system.

83. Install the heater core.

84. Install the heater core cover.

85. Install the four nuts from the engine side of the dash panel. Position the plenum chamber on the vehicle.

86. Install the Powertrain Control Module (PCM) heat sink.

87. Install the PCM.

88. Install the evaporator core housing.

89. Install the instrument panel.

90. Install gearshift lever.

91. Install consolette base.

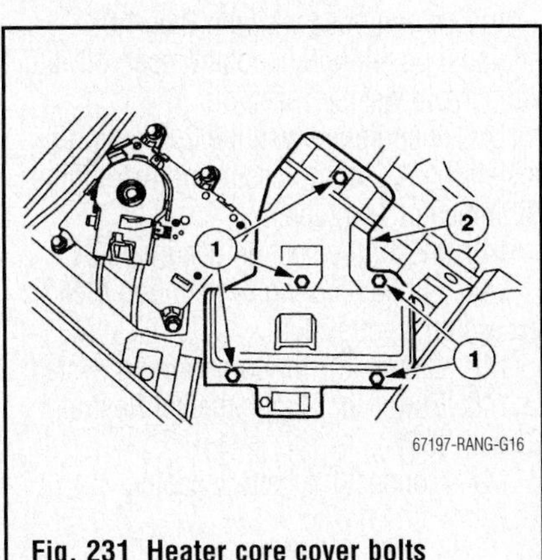

Fig. 231 Heater core cover bolts

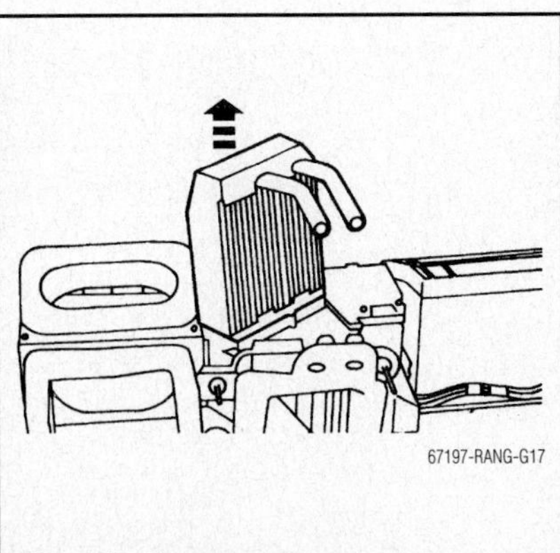

Fig. 232 Heater core removal

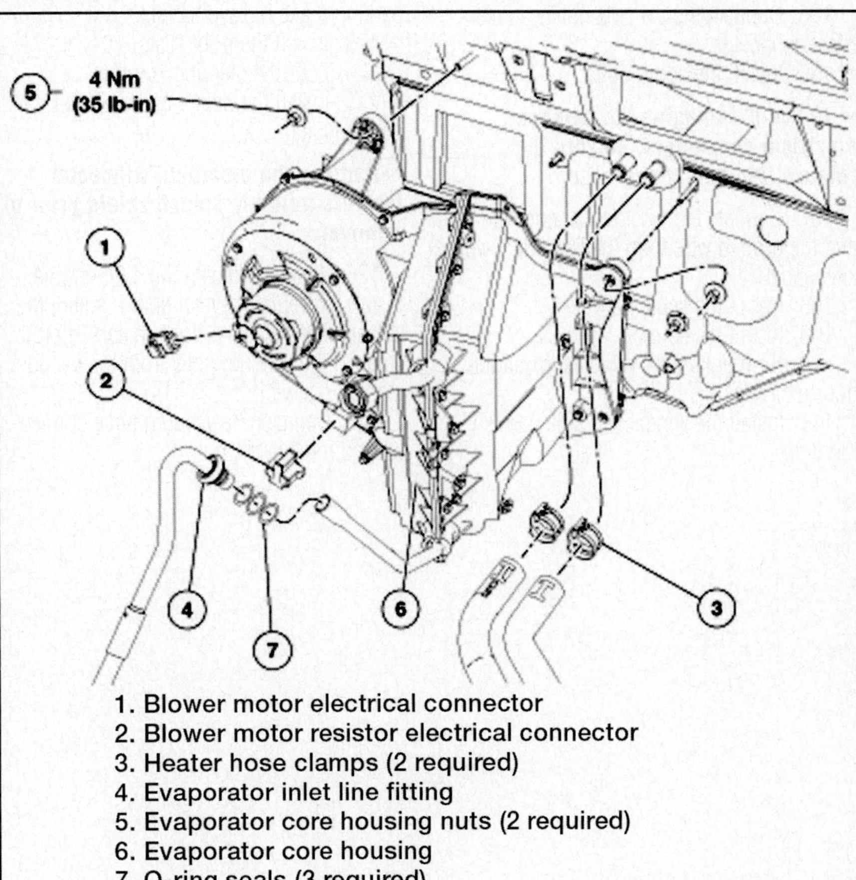

Fig. 233 Evaporator core housing

92. Install screws and install the Restraints Control Module (RCM) cover.

93. Install, if equipped with high-series floor console, the consolette mat.

94. Install the clips and the RCM cover.

95. Install, if not equipped with the high-series floor console, the cup holders.

96. Install the floor console.

97. Install the instrument panel cowl top bolts.

98. Install the instrument panel defroster opening grille.

99. Install the glove compartment. Press the release tabs inward while raising the glove compartment.

100. Through the glove compartment opening, connect the climate control vacuum harness connector.

101. Through the glove compartment opening, connect the blend door actuator electrical connector.

102. Install the glove compartment. Press the release tabs inward while lowering the compartment.

103. Connect the audio unit electrical connector and antenna cable.

104. Install the audio unit. Insert the tool. and support the audio unit.

105. Connect the LH side bulkhead electrical connector.

106. Install the audio unit.

➡To avoid damaging the bulkhead electrical connectors, be sure the release tab is fully depressed.

107. Connect the electrical connectors and the ground wire from the RH side lower kick panel.

108. Install the scuff plate.

109. Install the screws.

110. On regular cab, vehicles install the door moldings.

111. Install the windshield side garnish molding.

112. Install the assist handle.

113. Install the bolt covers and the bolts.

114. Install the left and right side garnish moldings.

➡To avoid damage to the clockspring, do not allow the steering shaft to rotate while the intermediate shaft is disconnected.

115. Install the steering column pinch bolt and disconnect the steering column intermediate shaft.

116. If equipped, connect the shift cable to the steering column.

117. If equipped, connect the Clutch Pedal Position (CPP) switch electrical connector.

118. Connect the Brake Pedal Position (BPP) switch electrical connector from the steering column shaft.

119. Install the instrument panel steering column opening cover reinforcement.

120. Install the screws and position the hood release handle.

121. Install the left and right lower cowl kick panels.

122. Install the screws and position the parking brake release handle.

123. Install the front seats.

124. Lock the steering column.

125. Install the nuts and the evaporator core housing.

➡Remove the electrical connector locators from the splash shield prior to removal.

126. If equipped with the 4.0L engine, install the vehicle splash guard. Install the splash shield. On the right splash shield, connect the vacuum lines from the vacuum storage reservoir.

127. Connect the vacuum hose connector and remove the nut.

➡This step is performed at the lower passenger side dash panel, inside the passenger compartment.

128. Connect the vacuum hose connector and remove the nut.

129. Connect the vacuum supply hose near the evaporator core housing.

130. Connect and detach the heater control valve vacuum hose.

131. Install the pin-type retainer and position the windshield washer hose into place.

132. Install the heater hose from between the water control valve and the water outlet.

133. On 4.0L engine, install the nut from the clip.

134. Install the heater hose from between the water control valve and the water pump.

135. On 2.3L engine, connect the heater hose from the water outlet tube and remove the heater hose assembly.

136. Install the heater hoses from between the heater core and the water control valve.

137. Install the heater hose from the rear coolant tube.

138. On 2.3L engine, install the heater hoses from the heater core.

139. Connect the blower motor and blower motor resistor electrical connectors.

140. Install the speed control servo, on vehicles equipped with cruise control.

141. If equipped with the 3 or 4.0L engines, position the coolant reservoir and windshield washer reservoir.

142. If equipped with the 2.3L engine, install the A/C compressor and the engine oil indicator and tube.

143. Install the suction accumulator.

144. Fill the cooling system and bleed excess air from system.

145. Charge the air conditioning system

146. Power the Supplemental Restraint System (SRS).

147. Connect the battery cable.

STEERING

POWER RACK & PINION STEERING GEAR

REMOVAL & INSTALLATION

2WD Models

See Figure 234.

1. Before servicing the vehicle, refer to the Precautions Section.
2. Disconnect the negative battery cable.
3. Place the steering wheel in the straight-ahead position and remove the ignition key.
4. Rotate the steering wheel until the steering column locks into position.

WARNING

Do not rotate the steering wheel when the shaft is disconnected from the steering gear as damage to the clock spring could occur.

5. Drain the power steering fluid reservoir.
6. Remove or disconnect the following:
   - Negative battery cable
   - Bolt retaining the lower steering column shaft to the steering gear input shaft
   - Stabilizer bar
   - Quick-connect fittings for the power steering pressure and return hoses at the steering gear housing
   - Nuts securing the power steering cooler and remove the cooler
   - Outer tie rod ends
   - Nuts, bolts and washer assemblies retaining the steering gear housing to the front crossmember
   - Steering gear from the vehicle

To install:

7. Install or connect the following:
   - Position the steering gear to the front crossmember and install the nuts, bolts and washer assemblies. Torque to 94–127 ft. lbs. (128–172 Nm).
   - Power steering cooler retaining bolts
   - Power steering lines to the steering gear housing and torque the fittings to 20–26 ft. lbs. (27–35 Nm).
   - Outer tie rod ends and ensure that the steering shaft or gear input shaft has not been rotated
   - Intermediate shaft-to-steering input shaft retaining (pinch) bolt and torque the bolt to 41 ft. lbs. (55 Nm).
   - Negative battery cable
8. Fill the power steering pump reservoir.
9. Bleed the air from the power steering system.
10. Ensure that there are no leaks and the fluid is maintained at the proper level.
11. Check the alignment.

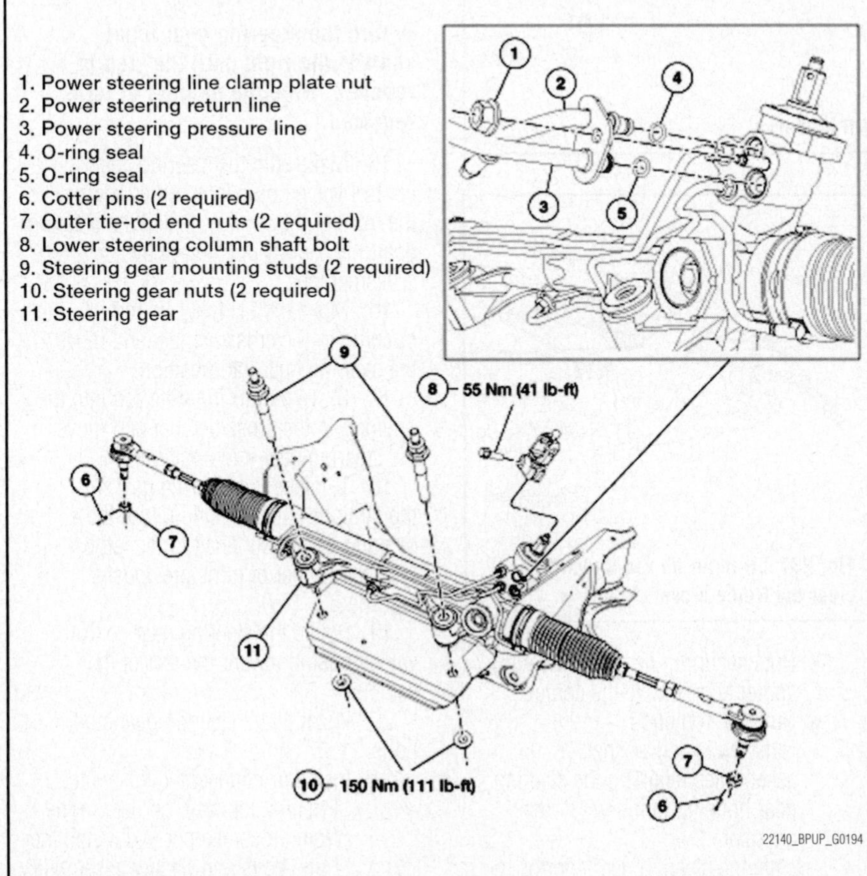

Fig. 234 Exploded view of the steering gear—2WD models

4WD Models

See Figures 235 through 238.

1. Before servicing the vehicle, refer to the Precautions Section.
2. Disconnect the negative battery cable.

WARNING

Do not rotate the steering wheel when the shaft is disconnected from the steering gear as damage to the clock spring could occur.

3. Place the steering wheel in the straight-ahead position and remove the ignition key.
4. Rotate the steering wheel until the steering column locks into position.
5. Drain the power steering fluid reservoir.
6. On vehicles with 4.0L engine, remove the power steering fluid cooler.
7. Remove or disconnect the following:
   - Negative battery cable
   - Bolt retaining the lower steering column shaft to the steering gear input shaft
   - Stabilizer bar

➡Do not allow the steering column shaft to rotate while it is disconnected from the steering gear or damage to the clockspring can result. If there is evidence that the steering column shaft has rotated, the clockspring must be removed and reentered.

8. Remove the lower steering column shaft to steering gear bolt and disconnect the shaft from the gear. Discard the bolt.

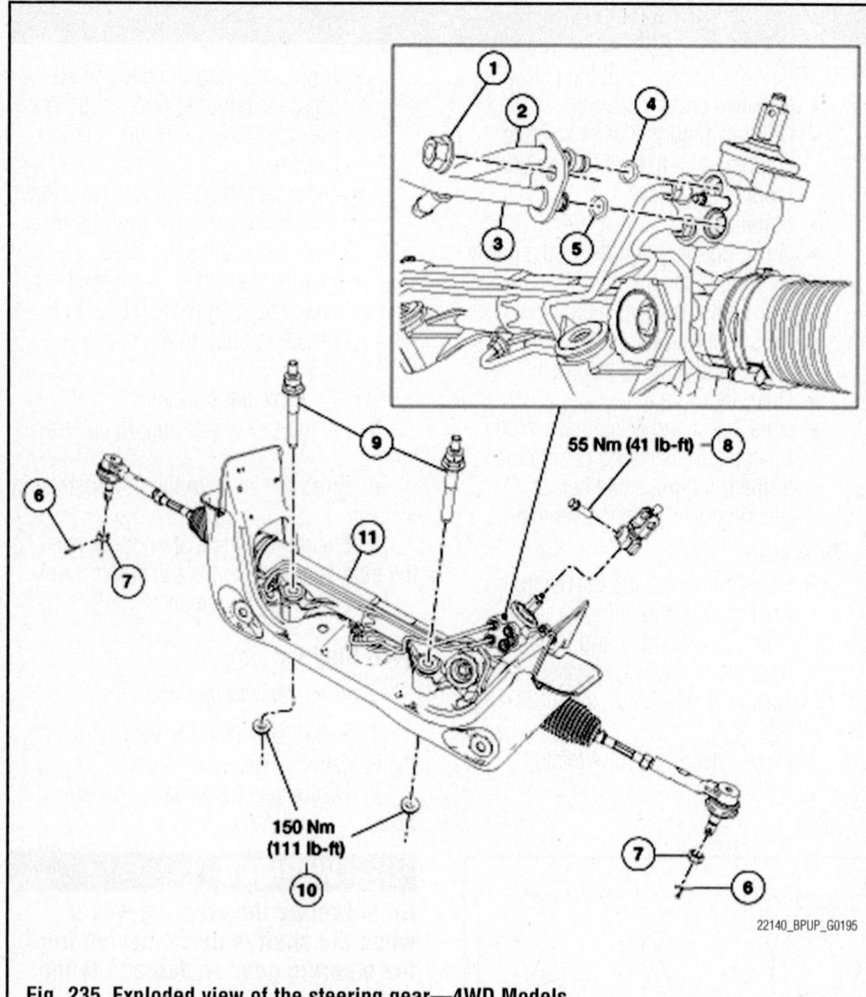

Fig. 235 Exploded view of the steering gear—4WD Models

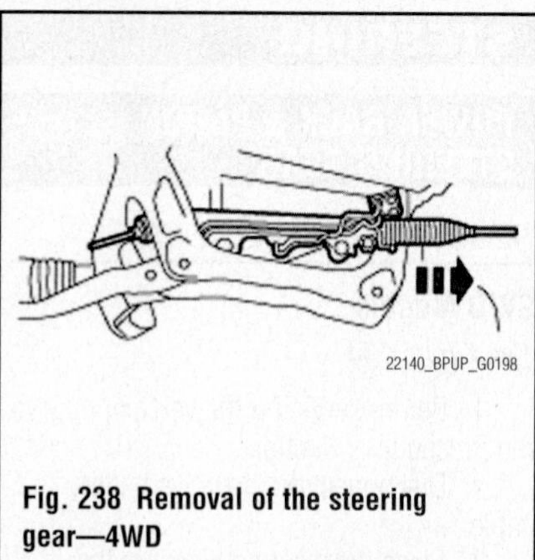

Fig. 238 Removal of the steering gear—4WD

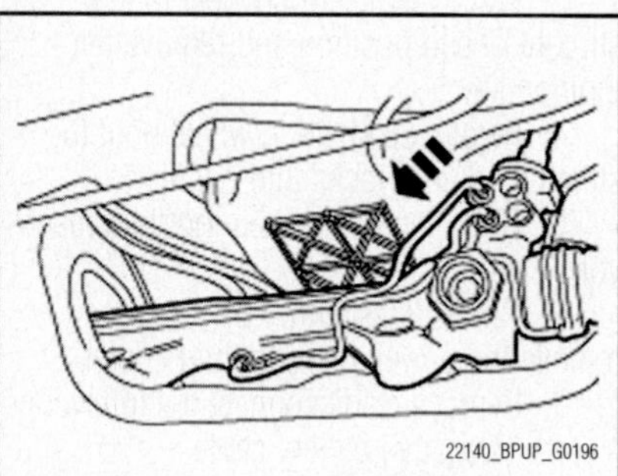

Fig. 236 Rotating the steering gear control valve housing toward the front of the vehicle.

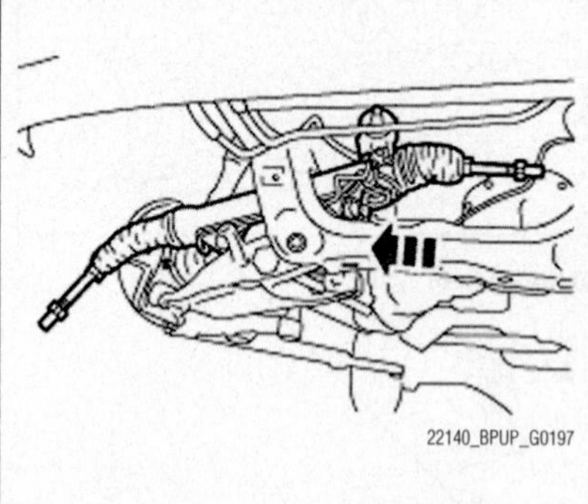

Fig. 237 LH inner tie rod forward to clear the frame crossmember

9. Remove the steering line clamp plate nut and disconnect the pressure and return lines. Discard the O-ring seals.
10. Remove or disconnect the following:
   - Quick-connect fittings for the power steering pressure and return hoses at the steering gear housing
   - Nuts securing the power steering cooler and remove the cooler
   - Outer tie rod ends
   - Nuts, bolts and washer assemblies retaining the steering gear housing to the front cross-member
11. Rotate the steering gear control valve housing toward the front of the vehicle.
12. Turn the steering gear input shaft to the right until the stop is reached.
13. Move the steering gear as far to the RH side of the vehicle as possible.
14. Remove the steering gear.

To install:

➡Handle the steering gear with caution to avoid damage to fluid transfer tubes and to avoid dimples in the steering gear bellows boot.

➡Make sure the steering gear input shaft is turned to the left until the stop is reached.

➡Turn the steering gear input shaft to the right until the stop is reached. Note the number of turns required.

15. Make sure the steering gear control valve housing is turned toward the front of the vehicle. Install the steering gear into the RH opening of the crossmember.
16. Move the LH inner tie rod into the opening in the crossmember and move the steering gear into position.
17. Move the LH inner tie rod into the opening in the crossmember and move the steering gear into position.
18. To place the steering gear in the straight-ahead position, turn the steering gear input shaft to the left by 1/2 the number of turns previously recorded.
19. Rotate the steering gear control valve housing toward the rear of the vehicle.
20. Install the 2 steering gear stud bolts.
21. Install or connect the following:
   - Position the steering gear to the front crossmember and install the nuts, bolts and washer assemblies. Torque to 94–110 ft. lbs. (150–172 Nm).

- Power steering cooler retaining bolts
- Power steering lines to the steering gear housing and torque the fittings to 20–26 ft. lbs. (27–35 Nm)
- Outer tie rod ends and ensure that the steering shaft or gear input shaft has not been rotated
- Intermediate shaft-to-steering input shaft retaining (pinch) bolt and torque the bolt to 41 ft. lbs. (55 Nm)

22. Position the outer tie rods and install the nuts and new cotter pins.
23. Check that the brake disc shields are not bent and are not in contact with the outer tie-rod end boots.
24. Tighten the tie rod nuts to 59 ft. lbs. (80 Nm).
25. On vehicles with 4.0L engine, install the power steering fluid cooler.
26. Connect the negative battery cable.
27. Fill the power steering pump reservoir.
28. Bleed the air from the power steering system.
29. Ensure that there are no leaks and the fluid is maintained at the proper level.
30. Check the alignment.

POWER STEERING PUMP

REMOVAL & INSTALLATION

2.3L Engine

See Figure 239.

1. Before servicing the vehicle, refer to the Precautions Section.
2. Disconnect the negative battery cable.

✲✲ WARNING

Do not allow power steering fluid to contact the accessory drive belt or the belt may be damaged. Using a suitable suction device, remove the power steering fluid from the reservoir.

3. Remove the power steering pump pulley.
4. Compress the clamp and disconnect the supply hose.
5. Disconnect the pressure line-to-pump fitting.
6. Remove and discard the Teflon seal.
7. Remove the 4 power steering pump bolts and the power steering pump.

✲✲ WARNING

A new Teflon seal must be installed any time the pressure line fitting is disconnected from the power steering pump or a fluid leak may occur.

8. To install, reverse the removal procedure.
9. Tighten the pressure line-to-pump fitting to 48 ft. lbs. (65 Nm).
10. Tighten the 4 power steering pump bolts and the power steering pump to 17 ft. lbs. (23 Nm).

✲✲ WARNING

A new Teflon seal must be installed any time the pressure line fitting is disconnected from the power steering pump or a fluid leak may occur.

11. Fill the power steering system.

3.0L Engine

See Figure 240.

✲✲ WARNING

Do not allow power steering fluid to contact the accessory drive belt or the belt may be damaged.

1. Using a suitable suction device, remove the power steering fluid from the reservoir.
2. Remove the pump pulley.
3. Remove and discard the O-ring seal.
4. Compress the clamp and disconnect the power steering supply hose.
5. Remove the 3 bolts and the power steering pump.

✲✲ WARNING

A new Teflon O-ring seal must be installed any time the power steering pressure line is disconnected from the power steering pump.

6. To install, reverse the removal procedure.

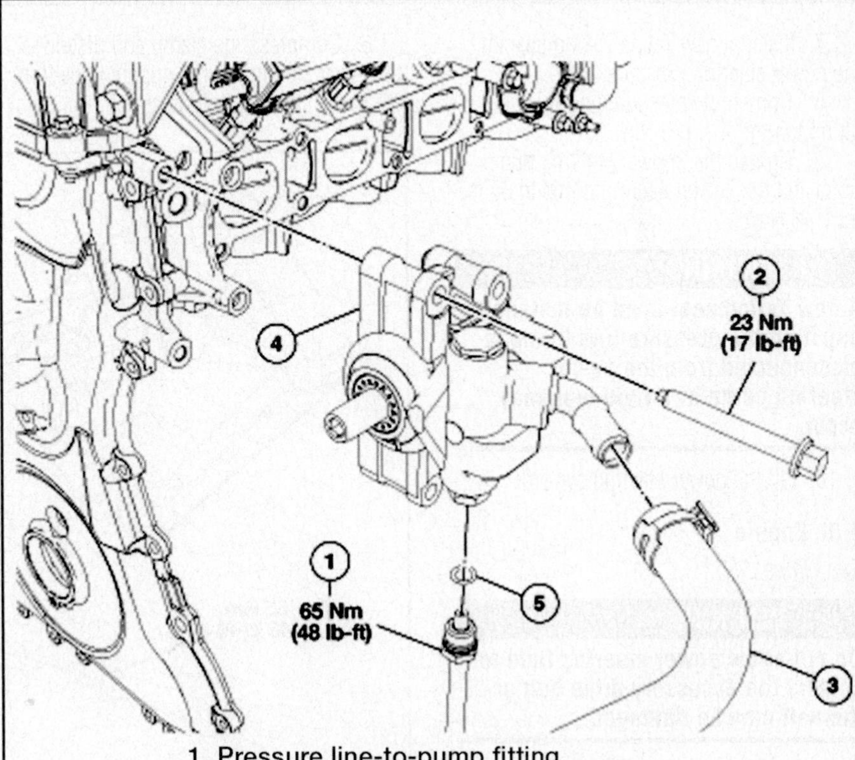

Fig. 239 Power steering pump and components exploded view—2.3L engine

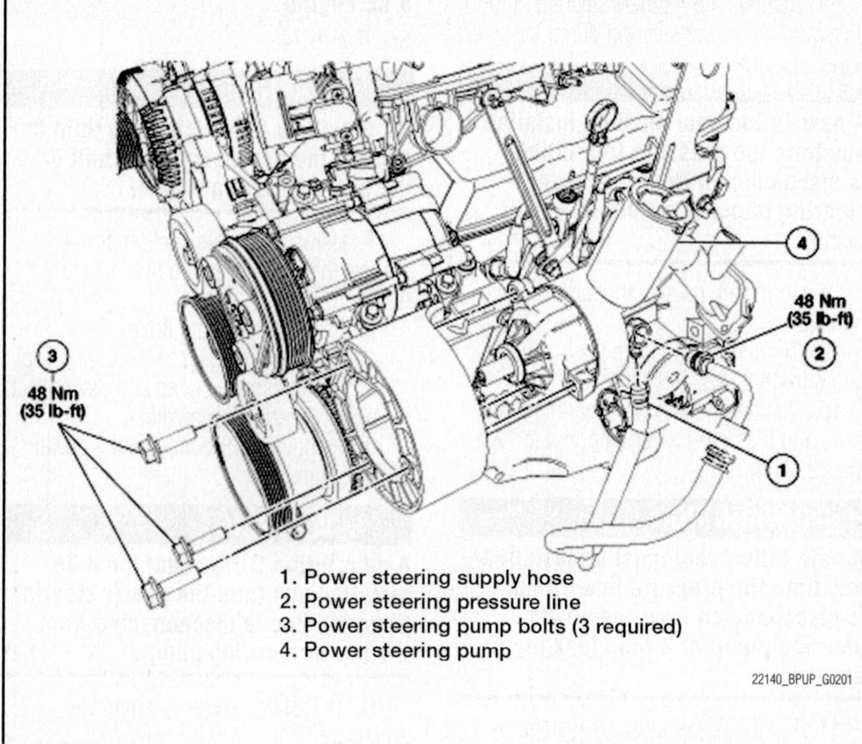

Fig. 240 Power steering pump and components exploded view—3.0L engines

7. Install a new Teflon O-ring seal on the power steering pressure hose
8. Tighten the pressure line-to-pump fitting to48 ft. lbs. (65 Nm).
9. Tighten the 3power steering pump bolts and the power steering pump to 35 ft. lbs. (48 Nm).

※※ WARNING

A new Teflon seal must be installed any time the pressure line fitting is disconnected from the power steering pump or a fluid leak may occur.

10. Fill the power steering system.

4.0L Engine

See Figure 241.

※※ WARNING

Do not allow power steering fluid to contact the accessory drive belt or the belt may be damaged.

1. Using a suitable suction device, remove the power steering fluid from the reservoir.
2. Rotate the accessory drive belt tensioner counterclockwise and remove the accessory drive belt.
3. Remove the engine cooling fan.
4. Remove the power steering pump pulley.
5. Compress the clamp and disconnect the power steering pump supply hose from the power steering fluid reservoir.
6. Disconnect the power steering pressure line to pump fitting.
7. Compress the clamp and disconnect the power steering pump supply hose from the power steering pump.
8. Remove the 3 bolts and the power steering pump.

※※ WARNING

A new Teflon O-ring seal must be installed any time the power steering pressure line is disconnected from the power steering pump or a fluid leak may occur.

9. To install, reverse removal procedure
10. Install a new Teflon O-ring seal on the power steering pressure line
11. Tighten the pressure line-to-pump fitting to 48 ft. lbs. (65 Nm).
12. Tighten the 3 power steering pump bolts and the power steering pump to 18 ft. lbs. (25 Nm).

※※ WARNING

A new Teflon seal must be installed any time the pressure line fitting is disconnected from the power steering pump or a fluid leak may occur.

13. Fill the power steering system.

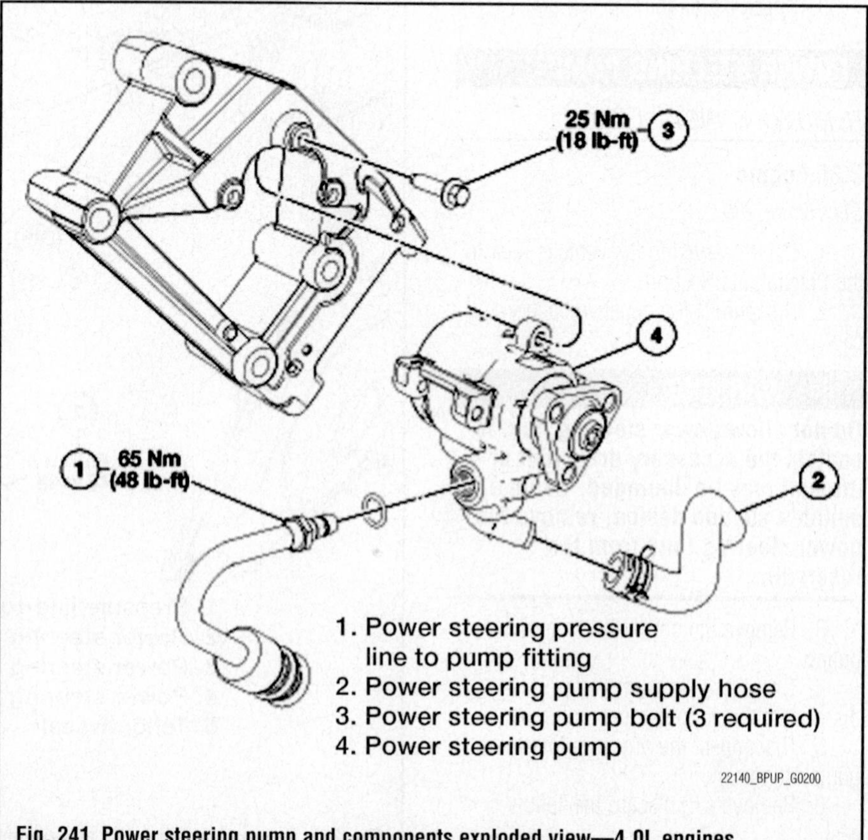

Fig. 241 Power steering pump and components exploded view—4.0L engines

COIL SPRING

REMOVAL & INSTALLATION

See Figures 242 and 243.

⁂ WARNING

Do not apply heat or flame to the shock absorber or strut tube. The shock absorber and strut tube are gas pressurized and could explode if heated. Failure to follow this instruction may result in serious personal injury.

⁂ WARNING

Keep all body parts clear of shock absorbers or strut rods. Shock absorbers or struts can extend unassisted. Failure to follow this instruction may result in serious personal injury.

⁂ WARNING

Suspension fasteners are critical parts because they affect performance of vital components and systems and their failure may result in major service expense. New parts must be installed with the same part number or an equivalent part if replacement is necessary. Do not use a replacement part of lesser quality or substitute design. Torque values must be used as specified during reassembly to make sure of correct retention of these parts.

1. Mark the lower arm to the frame for reassembly.
2. Before servicing the vehicle, refer to the precautions in the beginning of this section.

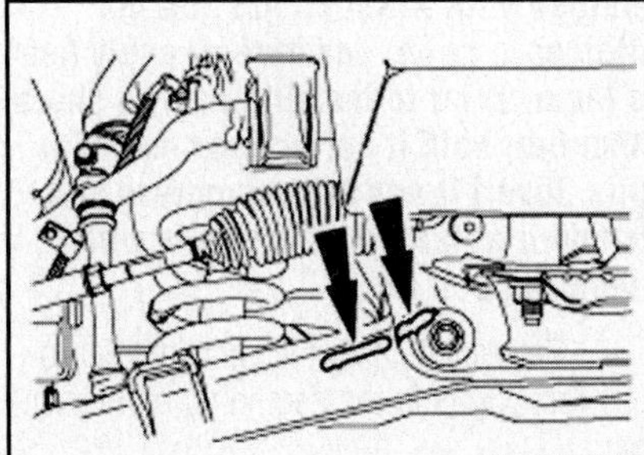

Fig. 242 Index marking on the lower arm to the frame with the vehicle in a static ground position (curb height).

3. Remove or disconnect the following:
   - Wheel and tire assembly
   - Shock absorber
   - Front stabilizer bar link nut
4. Use a coil spring compressor to compress the coil spring.
5. Using a suitable jack, support the lower control arm.
6. Remove the cotter pin and castellated nut.
7. Separate the lower ball joint from the front wheel spindle.
8. Position the front wheel spindle out of the way and remove the coil spring.

To install:

➡The end of the coil spring must cover the first hole and should not be visible in the second hole.

9. Install the coil spring in the lower arm.

⁂ WARNING

Always install the cotter pin into the lower ball joint castellated nut from outboard to inboard. Failure to do so will result in damage to the wheel and tire assembly.

10. Install the lower ball joint.
11. Tighten the lower ball joint to 98 ft. lbs. (133 Nm).
12. Install a new cotter pin.
13. Install the front stabilizer bar link nut.
14. Install a new stabilizer bar link nut and grommet.
15. Tighten the new stabilizer bar link nut to 26 ft. lbs. (35 Nm).

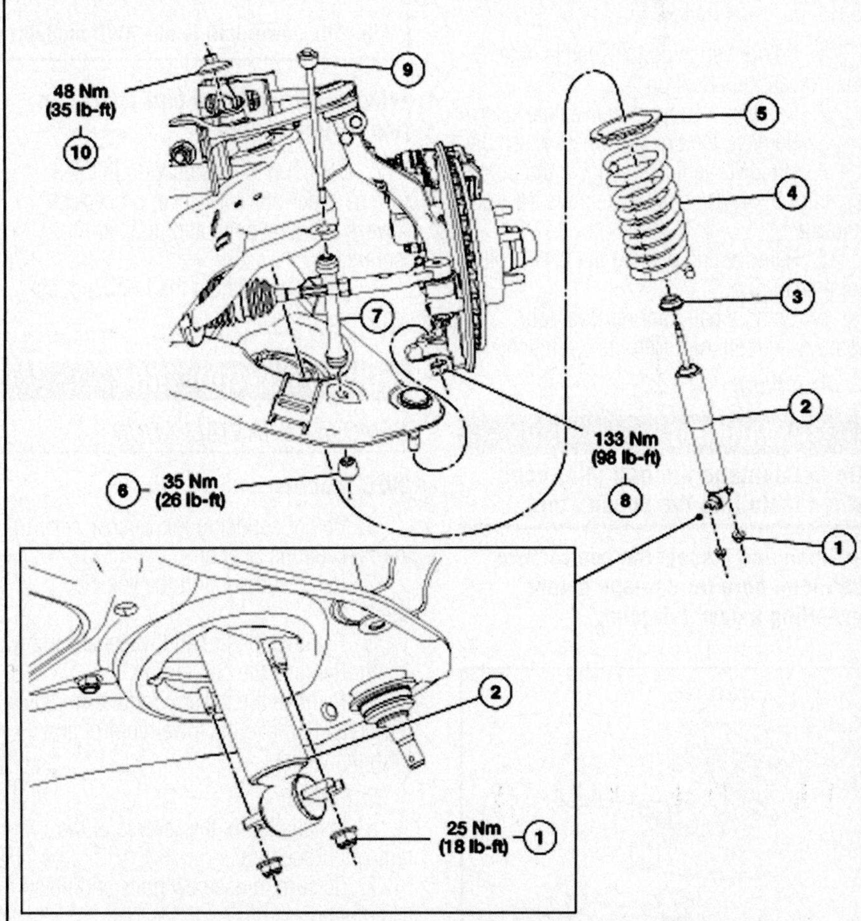

1. Lower shock nut (2 required)
2. Shock absorber
3. Shock absorber insulator
4. Coil spring
5. Coil spring insulator
6. Stabilizer bar link grommet and nut
7. Stabilizer bar link and grommet
8. Lower ball joint nut
9. Stabilizer bar link stud and grommet
10. Shock absorber rod nut

Fig. 243 Front coil spring and components exploded view

16. Remove the coil spring compressor.
17. Install the front shock absorber and the two lower nuts.
18. Tighten the front shock absorber lower nuts to 18 ft. lbs. (25 Nm).
19. Install the upper shock absorber bushing and nut/washer assembly.
20. Tighten the front shock absorber upper nuts to 35 ft. lbs. (35 Nm).
21. Position a suitable jack under the lower arm and raise the suspension until the index marks on the lower arm and the frame are aligned (curb height).
22. Tighten the lower arm to 148 ft. lbs. (200 Nm).
23. Remove the jack from under the lower control arm.
24. Install the wheel and tire assembly.

LOWER BALL JOINT

REMOVAL & INSTALLATION

See Figures 244 and 245.

1. Before servicing the vehicle, refer to the Precautions Section.
2. Raise and support the vehicle safely.
3. Remove the tire and wheel assembly.
4. On 2WD vehicles, remove the front spindle. On 4WD vehicles, remove the front knuckle.
5. Remove and discard the lower ball joint snap ring.
6. Using a ball joint removal too, remove the ball joint from its mounting.

To install:

WARNING

Do not damage the ball joint boot when installing the special tool.

➡Clean and inspect the control arm ball joint bore for damage before installing a new ball joint.

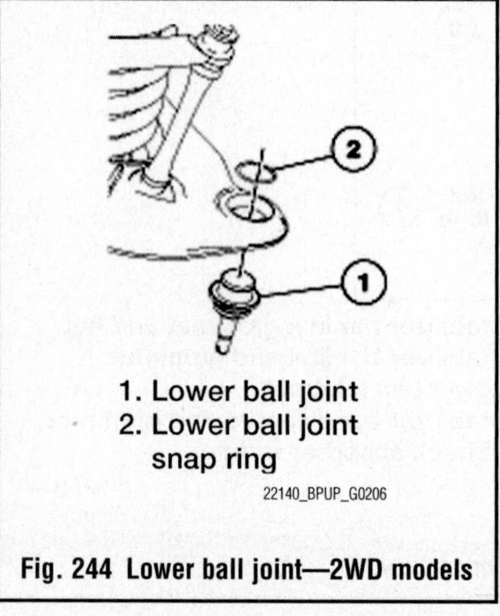

Fig. 244 Lower ball joint—2WD models

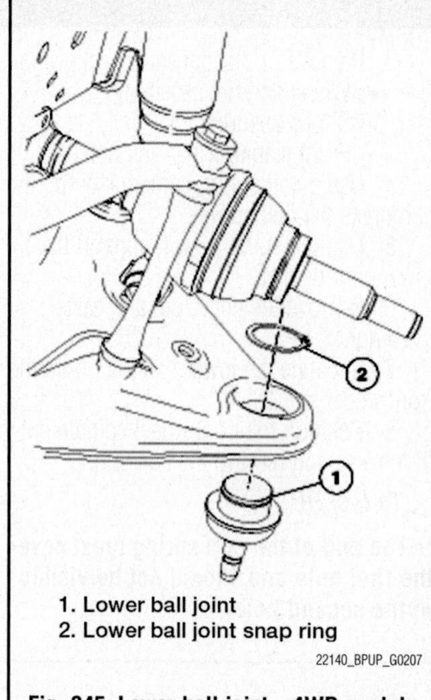

Fig. 245 Lower ball joint—4WD models

➡Make sure the new ball joint snap ring is fully seated.

7. Continue the installation in the reverse order of the removal procedure. Always use new bolts and nuts, as required.
8. Check and adjust the front end, as required.

LOWER CONTROL ARM

REMOVAL & INSTALLATION

2WD Models

1. Before servicing the vehicle, refer to the Precautions Section.
2. Raise and support the vehicle safely.
3. Remove the tire and wheel assembly.
4. Remove the coil spring.
5. Remove the lower control arm retaining bolts. Remove the lower control arm from the vehicle.

To install:

6. Installation is the reverse of the removal procedure.
7. Be sure to use new bolts and nuts, as required.
8. Torque the left side lower control arm rearward bolt to 148 ft. lbs. (200 Nm).
9. Torque the right side lower control arm rearward bolt to 129 ft. lbs. (174 Nm).
10. Torque the lower control arm forward bolts to 148 ft. lbs. (200 Nm).
11. Check and adjust the front end alignment, as required.

4WD Models

1. Before servicing the vehicle, refer to the Precautions Section.
2. Raise and support the vehicle safely.
3. Remove the tire and wheel assembly.
4. Remove the torsion bar.
5. Remove and discard the stabilizer bar link nut and grommet.
6. Remove the stabilizer link stud and grommet and the stabilizer link assembly.
7. Remove and discard the two lower shock absorber retaining nuts.
8. Remove and discard the lower ball joint cotter pin and nut.

➡Do not use a hammer to separate the ball joint from the knuckle or damage to the knuckle will result.

9. Using the proper tool, separate the ball joint from the knuckle.
10. Remove the lower control arm retaining bolts. Remove the lower control arm from the vehicle.

To install:

11. Installation is the reverse of the removal procedure.
12. Be sure to use new bolts and nuts, as required.

WARNING

Always install the cotter pin into the lower ball joint castellated nut from outboard to inboard. Failure to do so will result in damage to the wheel and tire assembly.

13. Torque the lower control arm bolts to 148 ft. lbs. (200 Nm).
14. Check and adjust the front end alignment, as required.

SHOCK ABSORBERS

REMOVAL & INSTALLATION

➡Low pressure gas shocks are charged with Nitrogen gas. Do not attempt to open, puncture or apply heat to them. Prior to installing a new shock absorber, hold it upright and extend it fully. Invert it and fully compress and extend it at least 3 times. This will bleed trapped air.

1. Before servicing the vehicle, refer to the precautions in the beginning of this section.
2. Remove or disconnect the following:
   - Negative battery cable
   - Upper shock-to-frame attaching nut, washer and insulator assembly

- Lower shock-to-control arm attaching nuts
- Slightly compress the shock absorber by hand and remove it from the vehicle

To install:

3. Install or connect the following:
   - Position the lower washer and insulator on the shock absorber rod and position the shock absorber to the upper frame bracket mount
   - Position the upper insulator and washer on the shock absorber rod and install the attaching nut loosely.
   - Position the lower shock absorber mounting studs into the control arm and install the attaching nuts loosely.
   - Torque the lower shock attaching nuts to 15–21 ft. lbs. (21–29 Nm), and the upper shock attaching bolts to 30–40 ft. lbs. (40–55 Nm).
   - Negative battery cable

STEERING KNUCKLE

REMOVAL & INSTALLATION

2WD Models

1. Before servicing the vehicle, refer to the Precautions Section.
2. Raise and support the vehicle safely.
3. Remove the tire and wheel assembly.
4. Remove the front disc brake caliper and wire it to the side out of the way.

➡Do not allow the caliper to hang by the brake line. Damage may result.

5. Remove the rotor.
6. Remove the hub nut and washer.

➡Do not use a hammer to separate the outer CV joint from the hub. Damage to the outboard CV threads and to internal components may result.

7. Use special tool 205-D070, or equivalent and separate the outer CV joint from the hub.
8. Remove the brake dust shield. Remove the bolt and detach the wheel speed sensor from the wheel hub.

➡Do not overextend the CV joint and boots when removing the hub and bearing assembly.

9. Remove the three retaining bolts. Remove the wheel and hub.
10. Remove the torsion bar.

➡Secure the front axle shaft to prevent it from overextending. Failure to do so can cause damage to the front axle shaft. Suspend the front axle shaft with wire.

11. Remove the tie rod end cotter pin and nut.
12. Using the proper tool separate the tie rod end from the front wheel knuckle.
13. Remove the lower ball joint cotter pin and nut.
14. Using the proper tool separate the front wheel knuckle from the front lower suspension arm.
15. Remove the pinch bolt, nut and the front wheel knuckle.

To install:

➡The knuckle main seal should only be removed from the knuckle if it is damaged. The main seal cannot be reused once it has been removed from the knuckle. Failure to lubricate the new seal prior to installation can cause premature wear of the seal, resulting in damage to the axle shaft and or wheel knuckle.

16. Installation is the reverse of the removal procedure. Be sure to use new bolts and nuts, as required.
17. Torque the hub nut to 162 ft. lbs..
18. Check and adjust the front end alignment, as required.

SPINDLE

REMOVAL & INSTALLATION

2WD Models

1. Before servicing the vehicle, refer to the Precautions Section.
2. Raise and support the vehicle safely.
3. Remove the tire and wheel assembly.
4. Remove the brake disc shield. Remove and discard the wheel speed sensor harness bracket bolt.
5. Remove and discard the outer tie rod end nut.

➡Do not use a hammer to separate the tie rod end from the wheel spindle or damage to the wheel spindle will result.

6. Using the proper tool separate the outer tie rod end from the wheel spindle.
7. Using a floor jack support the front suspension under the lower control arm.
8. Remove and discard the upper ball joint nut and pinch bolt. Using the proper tool disconnect the upper ball joint from the wheel spindle.
9. Remove and discard the lower ball joint cotter pin and nut. Using the proper tool disconnect the lower ball joint from the wheel spindle.

➡Do not use a hammer to separate the ball joint from the wheel spindle or damage to the wheel spindle will result.

10. Remove the front spindle from the vehicle.

To install:

11. Installation is the reverse of the removal procedure.
12. Be sure to use new bolts and nuts, as required. Torque the upper ball joint to wheel spindle retaining nut to 46 ft. lbs. Torque the lower ball joint to wheel spindle retaining nut to 98 ft. lbs..
13. Torque the outer tie rod nut to 59 ft. lbs..

✻✻ WARNING

Always install the cotter pin into the lower ball joint castellated nut from outboard to inboard. Failure to do so will result in damage to the wheel and tire assembly.

14. Check and adjust the front end alignment.

STABILIZER BAR

REMOVAL & INSTALLATION

1. Before servicing the vehicle, refer to the Precautions Section.
2. Raise and support the vehicle safely.
3. Remove the wheel and tire assembly.
4. Remove the front stabilizer bar link nuts from the front suspension lower arms. Discard the nuts.
5. Remove the front stabilizer bar link studs and the front stabilizer bar links.
6. Remove the four bolts and two brackets. Discard the bolts.
7. Remove the front stabilizer bar.
8. Remove the stabilizer bar insulator.

To install:

9. Installation is the reverse of removal. Be sure to use new bolts and nuts. Observe the following torques:
   - End link nuts: 18 ft. lbs. (25 Nm)
   - Bracket bolts: 30 ft. lbs. (40 Nm)

➡In the event the self-tapping bolts cannot be installed in the frame, there is a kit available with flag nuts.

TORSION BAR

REMOVAL & INSTALLATION

WARNING

The electrical power to the air suspension system must be shut off prior to hoisting, jacking or towing an air suspension vehicle. This can be accomplished by turning off the air suspension switch located in the rear jack storage area. Failure to do so can result in unexpected inflation or deflation of the air springs or shocks, which can result in shifting of the vehicle during these operations.

1. Before servicing the vehicle, refer to the precautions in the beginning of this section.
2. Remove or disconnect the following:
3. Remove the torsion bar cover plate

➡Before relieving the torsion bar tension, measure and record the measurement of the torsion bar adjustment bolt. This measurement will be used as the preset depth for the new torsion bar adjustment bolt during installation.

4. Relieve the torsion bar tension.
5. position the torsion bar tool and adapters.
6. Tighten the torsion bar tool until the torsion bar adjuster lifts off the adjustment bolt.

WARNING

The torsion bar adjustment bolt is coated with dry adhesive; and must be replaced if it is backed off or removed. Failure to do so can cause the adjustment bolt to loosen during operation and cause a loss of vehicle alignment.

7. Remove the torsion bar adjustment bolt and nut.
8. Loosen the torsion bar tool until the tension is removed from the torsion bar.
9. Mark the torsion bar and the adjuster for proper installation.
10. Remove the torsion bar insulator.
11. Grasp the torsion bar, and pull it free from the front suspension lower arm.

To install:

12. Position the torsion bar and the torsion bar adjuster.
13. Align the marks on the torsion bar and the torsion bar adjuster, then install the torsion bar adjuster.
14. Position the torsion bar insulator.
15. Install the torsion bar tool and the adapters.
16. Tighten the torsion bar tool until the new adjustment bolt and nut can be installed.
17. Turn the adjustment bolt until the preliminary adjustment measurement (recorded length of the old adjustment bolt) is reached.
18. Install the torsion bar cover plate. Torque the bolts to 46 ft. lbs. (63 Nm).
19. If equipped with air suspension, reactivate the system by turning on the air suspension switch.
20. Lower the vehicle.
21. Adjust the ride height.
22. Check the alignment.

UPPER BALL JOINT

REMOVAL & INSTALLATION

The ball joints are integral with the control arm. If the ball joint is defective, the entire control arm must be replaced.

UPPER CONTROL ARM

REMOVAL & INSTALLATION

1. Before servicing the vehicle, refer to the Precautions Section.
2. Raise and support the vehicle safely.
3. Remove the tire and wheel assembly.
4. Using a suitable jack, support the lower control arm assembly.

➡To avoid possible damage to the wheel spindle (2WD) or knuckle (4WD) secure the assembly to keep it from tilting before removing the pinch bolt and nut.

5. Remove and discard the upper ball joint nut and pinch bolt. Separate the ball joint from the wheel spindle (2WD) or the knuckle (4WD).
6. Remove and discard the two upper control arm nuts.
7. Remove the upper control arm from the vehicle.

To install:

8. Installation is the reverse of the removal procedure. Be sure to use new nuts and bolts, as required.
9. The upper control arm retaining nuts must be fully tightened at curb height. Torque them to 98 ft. lbs.
10. Make sure that the shims are installed in their original locations.
11. Torque the upper ball joint nut and pinch bolt to 46 ft. lbs.
12. Check and adjust the front end alignment, as required.

WHEEL BEARINGS

REMOVAL & INSTALLATION

2WD Models

1. Before servicing the vehicle, refer to the precautions in the beginning of this section.
2. Remove or disconnect the following:
   - Disc brake caliper anchor plate
   - Hub grease cap
   - Cotter pin
   - Nut retainer
   - Spindle nut
   - Wheel outer bearing retainer washer
   - Outer front wheel bearing
   - Brake disc and hub
   - Hub grease seal
   - Inner wheel bearing

To install:

3. Thoroughly clean and inspect the front wheel bearings and the brake disc and hub.
4. Lubricate the front wheel bearings.
5. Install the inner front wheel bearing.
6. Install a new wheel hub grease seal.
7. Position the brake disc and hub.
8. Assemble all parts and adjust the bearings.
9. Tighten lug nuts to 100 ft. lbs. (135 Nm).

CAUTION

Before driving the vehicle, pump the brake pedal several times to restore normal brake pedal travel.

4WD Models

1. Before servicing the vehicle, refer to the precautions in the beginning of this section.
2. Remove or disconnect the following:
   - Negative battery cable
   - Wheel assembly
   - Disc brake caliper anchor plate
   - Hub grease cap
   - Cotter pin
   - Nut retainer
   - Spindle nut
   - Wheel bearing
   - Pull the hub and rotor assembly from spindle.
   - Grease seal—discard seal

To install:

3. Use a bearing packer tool and properly repack the wheel bearings with the proper grade and type of grease. If a bearing packer is not available, work as much of the grease as possible between the rollers and cages. Also, grease the cone surfaces.

4. Install or connect the following:
   - Inner bearing cone and roller assembly in the inner cup. A light film of grease should be included between the lips of the new grease seal.
   - Grease seal by driving in place with Hub Seal Replacer tool T83T-1175-B and Driver Handle T80T-4000-W
   - Hub and rotor assembly onto the spindle. Keep the hub centered on the spindle to prevent damage to the spindle and the retainer
   - Rotor onto the spindle
   - Install retainer, new cotter pin and grease cap.
   - Axle shaft spacer
   - Install caliper.
   - Wheel assembly
   - Negative battery cable
5. Tighten lug nuts to 100 ft. lbs. (135 Nm).

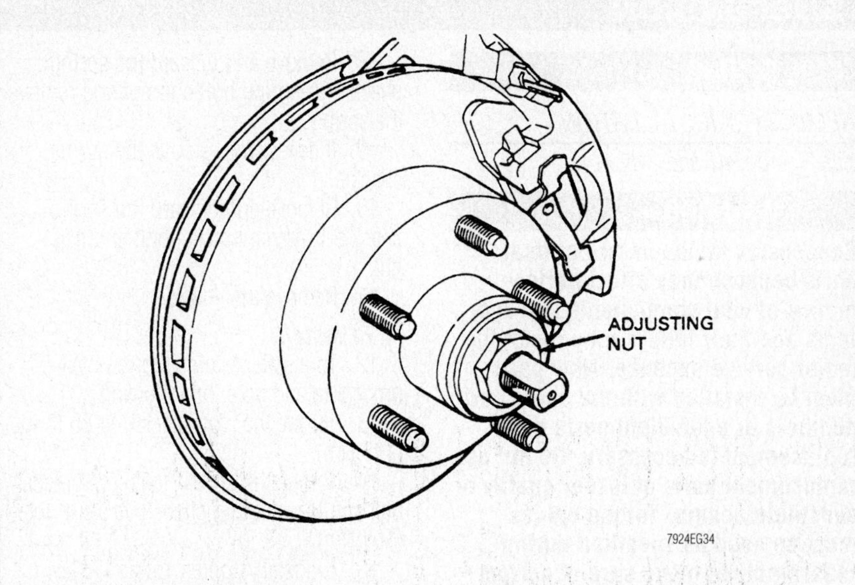

Fig. 246 Loosen the adjusting nut 3 turns, then rock the entire wheel assembly in-and-out to spread the brake pads before attempting to adjust the bearing—2WD vehicles

**** CAUTION**

Before driving the vehicle, pump the brake pedal several times to restore normal brake pedal travel.

ADJUSTMENT

2WD Models

See Figure 246.

1. Before servicing the vehicle, refer to the precautions in the beginning of this section.
2. Remove the grease cap from the hub and wipe the excess grease from the end of the spindle. Remove the cotter pin and retainer. Discard the cotter pin.
3. Loosen the adjusting nut 3 turns.

**** WARNING**

Obtain running clearance between the disc brake rotor surface and shoe linings by rocking the entire wheel assembly in and out several times in order to push the caliper and brake pads away from the rotor. An alternate method to obtain proper running clearance is to tap lightly on the caliper housing. Be sure not to tap on any other area that may damage the disc brake rotor or the brake lining surfaces. Do not pry on the phenolic caliper piston. The running clearance must be maintained throughout the adjustment procedure. If proper clearance cannot be maintained, the caliper must be removed from its mounting.

4. While rotating the wheel assembly, tighten the adjusting nut to 17–25 ft. lbs. (23–34 Nm) in order to seat the bearings. Loosen the adjusting nut a half turn. Retighten the adjusting nut 18–20 inch lbs. (2.0–2.2 Nm).
5. Place the retainer on the adjusting nut. The castellations on the retainer must be in alignment with the cotter pin holes in the spindle. Once this is accomplished, install a new cotter pin and bend the ends to insure its being locked in place.
6. Check for proper wheel rotation. If correct, install the grease cap.
7. Lower the vehicle and tighten the lug nuts to 100 ft. lbs.., (136 Nm) if the wheel was removed. Before driving the vehicle, pump the brake pedal several times to restore normal brake pedal travel.

**** CAUTION**

If the wheel was removed, retighten the wheel lug nuts to specification after about 500 miles (804km) of driving. Failure to do this could result in the wheel coming off while the vehicle is in motion causing loss of vehicle control or collision.

4WD Models

1. If a packer is not available, work as much grease as possible between the rollers and cages. Grease the cone surfaces. Pack the bearing cone and roller assemblies with appropriate grease.
2. Place the bearing cone and roller assembly in the inner cup. A light film of grease should be included between the lips of the new grease seal. Additional grease may be added to the new grease seal if required.
3. When installing seal, be sure it is properly seated.
4. Install the seal.

COIL SPRING

REMOVAL & INSTALLATION

See Figures 247 and 248.

✲✲ WARNING

Suspension fasteners are critical parts because they affect performance of vital components and systems and their failure may result in major service expense. New parts must be installed with the same part numbers or equivalent parts if replacement is necessary. Do not use replacement parts of lesser quality or substitute design. Torque values must be used as specified during reassembly to make sure of correct retention of these parts.

1. With the vehicle in neutral, position it on a hoist.
2. Using a suitable jack, support the rear axle.
3. Remove and discard the lower shock absorber bolt and nut.

➡When installing new U-bolts, the distance between the U-bolts should be 93 mm (3.66 in) ± 3 mm (0.12 in).This is important for providing the correct U-bolt clamp load and retention of parts. Remove and discard the 4 rear U-bolt nuts.

4. Remove the U-bolts and the spring plate.
5. Carefully lower the rear axle and remove the rear spring spacer, if equipped.
6. Remove and discard the spring-to-frame bracket bolt and flag nut.
7. To install, tighten to 148 ft. lbs. (200 Nm).
8. Remove and discard the spring shackle-to-spring bolt and nut and remove the spring.
9. If necessary, remove the spring shackle.
10. Remove and discard the spring shackle-to-frame bracket bolt and flag nut.
11. Remove the shackle

To install:

12. To install, reverse the removal procedure and note the following.
13. Torque the rear shackle to 85 ft. lbs. (115 Nm).
14. Torque the spring-to-frame bracket bolt and flag nut rear shackle to 76 ft. lbs. (103 Nm).
15. To install, tighten Torsion axle U-bolt nuts in 4 stages.
    a. Stage 1: Tighten in a cross pattern to 18 ft. lbs. (25 Nm).
    b. Stage 2: Tighten in a cross pattern to 37 ft. lbs. (50 Nm).
    c. Stage 3: Tighten in a cross pattern to 55 ft. lbs. (75 Nm).
    d. Stage 4: Tighten in a cross pattern to 76 ft. lbs. (103 Nm).
16. To install, tighten standard axle U-bolt nuts in 4 stages:
    a. Stage 1: Tighten in a cross pattern to 22 ft. lbs. (30 Nm).
    b. Stage 2: Tighten in a cross pattern to 44 ft. lbs. (60 Nm).
    c. Stage 3: Tighten in a cross pattern to 66 ft. lbs. (90 Nm).
    d. Stage 4: Tighten in a cross pattern to 85 ft. lbs. (115 Nm).

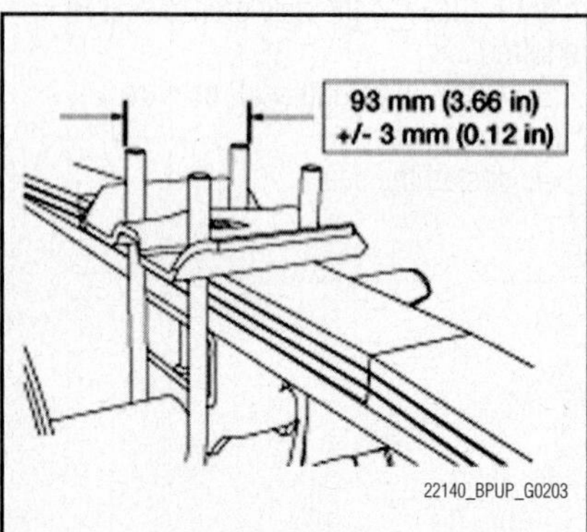

Fig. 247 Standard axle U-bolt nuts

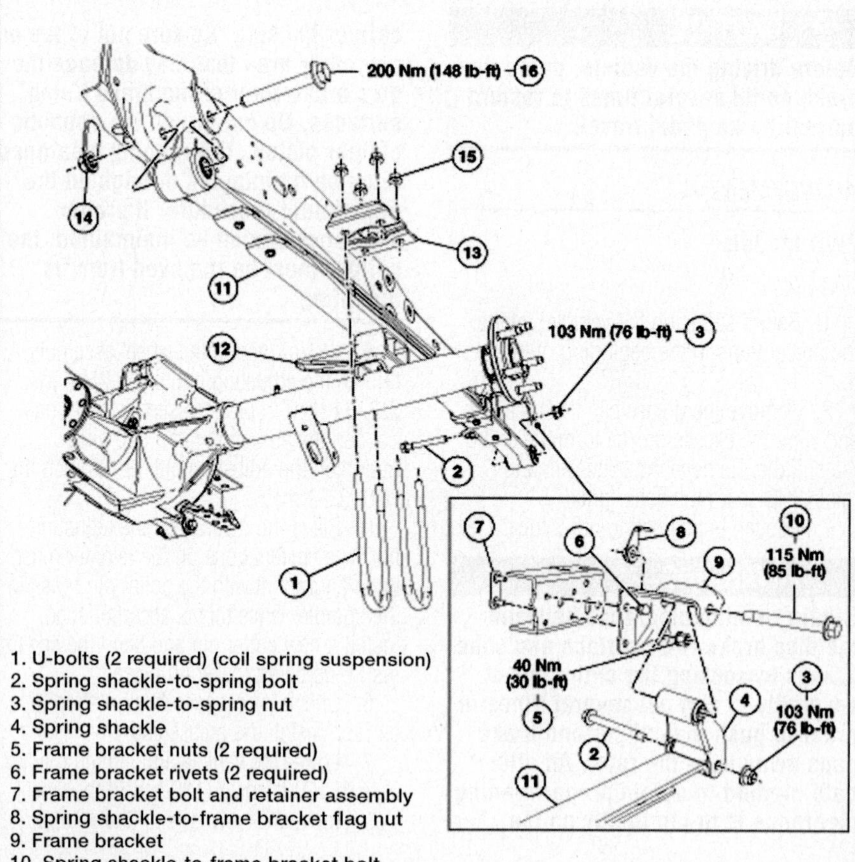

1. U-bolts (2 required) (coil spring suspension)
2. Spring shackle-to-spring bolt
3. Spring shackle-to-spring nut
4. Spring shackle
5. Frame bracket nuts (2 required)
6. Frame bracket rivets (2 required)
7. Frame bracket bolt and retainer assembly
8. Spring shackle-to-frame bracket flag nut
9. Frame bracket
10. Spring shackle-to-frame bracket bolt
11. Leaf spring
12. Rear spring spacer (torsion bar suspension only)
13. Spring plate
14. Spring-to-frame bracket flag nut
15. U-bolt nuts (4 required)
16. Spring-to-frame bracket bolt

Fig. 248 Rear suspension springs exploded view

LEAF SPRING

REMOVAL & INSTALLATION

See Figure 248.

1. Before servicing the vehicle, refer to the precautions in the beginning of this section.
2. Remove or disconnect the following:
   - Negative battery cable
   - Rear wheels
   - U-bolts from the rear spring plate
   - Hardware from the spring to bracket at the front of the rear spring
   - Upper and lower shackle bolts at the rear of the spring
   - Spring and shackle from the bracket

To install:

3. Install or connect the following:
   - Spring and shackle to the bracket
   - Upper and lower shackle bolts at the rear of the spring. Torque the nuts to 87 ft. lbs. (118 Nm).
   - U-bolts to the spring plate. Torque the nuts 87 ft. lbs. (113 Nm).
   - Rear wheels
   - Negative battery cable

SHOCK ABSORBER

REMOVAL & INSTALLATION

➡**Low pressure gas shocks are charged with nitrogen gas. Do not attempt to open, puncture or apply heat to them. Prior to installing a new shock absorber, hold it upright and extend it fully. Invert it and fully compress and extend it at least 3 times. This will bleed trapped air.**

1. Before servicing the vehicle, refer to the Precautions Section.
2. Raise and support the vehicle safely. Remove the tire and wheel assembly.
3. Remove or disconnect the following:
   - Upper shock-to-frame attaching nut
   - Lower shock nut
   - Slightly compress the shock absorber by hand and remove it from the vehicle

To install:

4. Install or connect the following:
   - Shock absorber upper end and nut
   - Shock absorber lower end and nut
   - Torque the upper shock attaching nuts to 46 ft. lbs. (63 Nm)
   - Torque the lower shock attaching nuts to 59 ft. lbs. (80 Nm)

STABILIZER BAR

REMOVAL & INSTALLATION

See Figure 249.

Suspension fasteners are critical parts because they affect performance of vital components and systems and their failure can result in major service expense. A new part with the same part number or an equivalent part must be installed, if installation is necessary. Do not use a part of lesser quality or substitute design. Torque values must be used as specified during reassembly to make sure of correct retention of these parts.

1. Before servicing the vehicle, refer to the Precautions Section.
2. With the vehicle in **NEUTRAL**, position it on a hoist.
3. Remove the wheel and tire assembly.
4. Remove the rear stabilizer bar link nuts and bolt. Discard the bolts and nuts.
5. Remove the rear stabilizer bar link.
6. Remove the rear stabilizer bar mounting bracket bolts and remove the brackets. Discard the bolts and nuts.
7. Remove the 4 stabilizer bar bracket bolts and then remove the stabilizer bar, bushings and brackets.

To install:

8. Installation is the reverse of the removal procedure.
9. Observe the following torques:
   - Link-to-stabilizer bar bolts and nuts: 52 ft. lbs. (70 Nm)
   - Bar link-to-frame bolts and nuts: 52 ft. lbs. (70 Nm)
   - 4 stabilizer bar bracket bolts: 30 ft. lbs. (40 Nm)

WHEEL BEARINGS

REMOVAL & INSTALLATION

1. Before servicing the vehicle, refer to the Precautions Section.
2. Remove the axle shaft.

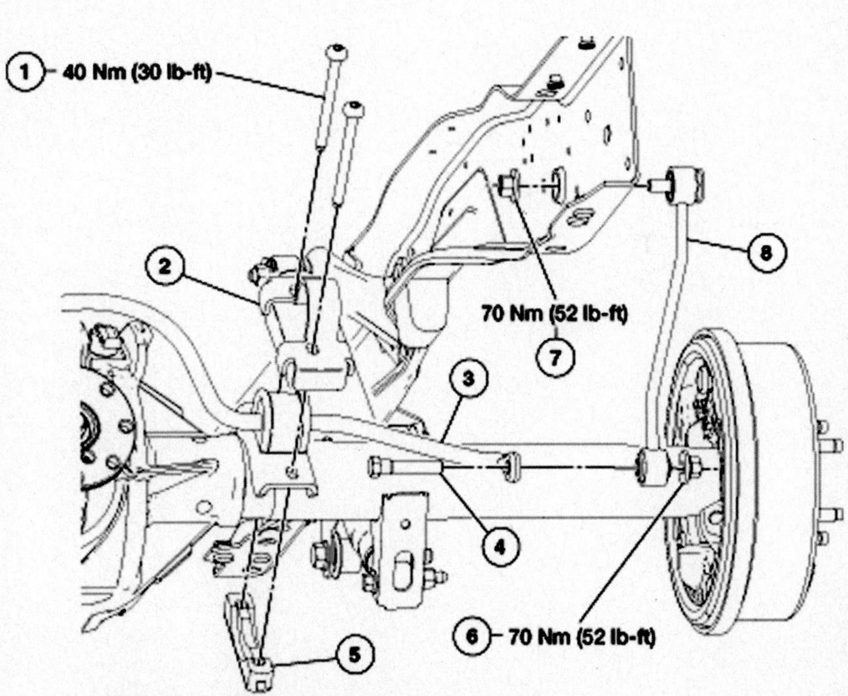

Fig. 249 Stabilizer bar and link

➡If only a new seal needs to be installed, use care to avoid damaging the seal bore.

3. Using a suitable seal remover, remove the axle shaft oil seal. Discard the oil seal.
4. Inspect the rear wheel bearing and axle shaft for wear or damage.
5. If necessary, using the impact slide hammer remove the rear wheel bearing.

To install:

6. Lubricate the new rear wheel bearing with lubricant.
7. Use driver handle and axle tube bearing replacer to install the rear wheel bearing.
8. Lubricate the lip of the new wheel bearing oil seal with appropriate grease
9. Use driver handle axle tube bearing replacer to install the inner wheel bearing oil seal.
10. Install the axle shaft.

ADJUSTMENT

There is no adjustment on in the rear bearings.

REPACKING

The rear wheel bearings are not serviceable and are lubricated by differential fluid.

MAZDA

CX-7

3

SPECIFICATIONS AND MAINTENANCE CHARTS

ENGINE AND VEHICLE IDENTIFICATION

Engine							Model Year	
Code ①	Liters (cc)	Cu. In.	Cyl.	Fuel Sys.	Engine Type	Eng. Mfg.	Code ②	Year
L3	2.3 (2260)	137.9	4	MPFI	DOHC	Mazda	7	2007
							8	2008

MPFI: Multi-Point Fuel Injection

DOHC: Double Over Head Cam

① Located above the starter

② 10th digit of the Vehicle Identification Number (VIN)

22140_MCX7_C0001

GENERAL ENGINE SPECIFICATIONS

Year	Model	Engine Displacement Liters (VIN)	Net Horsepower @ rpm	Net Torque @ rpm (ft. lbs.)	Bore x Stroke (in.)	Com-pression Ratio	Oil Pressure @ rpm
2007	CX-7	2.3 (L3)	244@5000	258@2500	344x3.70	9.5:1	43-79@3000
2008	CX-7	2.3 (L3)	244@5000	258@2500	344x3.70	9.5:1	43-79@3000

NA: Not Available

EFI: Electronic Fuel Injection

22140_MCX7_C0003

ENGINE TUNE-UP SPECIFICATIONS

Year	Engine Displacement Liters (VIN)	Spark Plug Gap (in.)	Ignition Timing (deg.)		Fuel Pump (psi)	Idle Speed (rpm)		Valve Clearance	
			MT	AT		MT	AT	In.	Ex.
2007	2.3 (L3)	28-31	—	10B	60-71	—	650-750	0.008-0.011	0.010-0.012
2008	2.3 (L3)	28-31	—	10B	60-71	—	650-750	0.008-0.011	0.010-0.012

NOTE: The Vehicle Emission Control Information label often reflects specification changes made during production.

The label figures must be used if they differ from those in this chart.

B: Before top dead center

22140_MCX7_C0002

CAPACITIES

Year	Model	Engine Displacement Liters (VIN)	Engine Oil with Filter (qts.)	Transmission (pts.) Man	Transmission (pts.) Auto.	Drive Axle Front (pts.)	Drive Axle Rear (pts.)	Fuel Tank (gal.)	Cooling System (qts.)
2007	CX-7	2.3 (L3)	6.0	—	14.8	①	2.12	18.2	9.5
2008	CX-7	2.3 (L3)	6.0	—	14.8	①	2.12	18.2	9.5

NOTE: All capacities are approximate. Add fluid gradually and check to be sure a proper fluid level is obtained.

① Included in transaxle

22140_MCX7_C0004

VALVE SPECIFICATIONS

Year	Engine Displacement Liters (VIN)	Seat Angle (deg.)	Face Angle (deg.)	Maximum out of Square (in.)	Spring Free Length (in.)	Stem-to-Guide Clearance (in.) Intake	Stem-to-Guide Clearance (in.) Exhaust	Stem Diameter (in.) Intake	Stem Diameter (in.) Exhaust
2007	2.3 (L3)	NA	NA	NA	NA	NA	NA	NA	NA
2008	2.3 (L3)	NA	NA	NA	NA	NA	NA	NA	NA

NA: Not Available

22140_MCX7_C0005

CRANKSHAFT AND CONNECTING ROD SPECIFICATIONS

All measurements are given in inches.

Year	Engine Displacement Liters (VIN)	Crankshaft Main Brg. Journal Dia.	Crankshaft Main Brg. Oil Clearance	Crankshaft Shaft End-play	Crankshaft Thrust on No.	Connecting Rod Journal Diameter	Connecting Rod Oil Clearance	Connecting Rod Side Clearance
2007	2.3 (L3)	NA	NA	NA	NA	NA	NA	NA
2008	2.3 (L3)	NA	NA	NA	NA	NA	NA	NA

NA: Not Avilable

22140_MCX7_C0006

PISTON AND RING SPECIFICATIONS

All measurements are given in inches.

Year	Engine Displacement Liters (VIN)	Piston Clearance	Ring Gap Top Compression	Ring Gap Bottom Compression	Ring Gap Oil Control	Ring Side Clearance Top Compression	Ring Side Clearance Bottom Compression	Ring Side Clearance Oil Control
2007	2.3 (L3)	NA	NA	NA	NA	NA	NA	NA
2008	2.3 (L3)	NA	NA	NA	NA	NA	NA	NA

NA: Not Available

22140_MCX7_C0007

TORQUE SPECIFICATIONS

All readings in ft. lbs.

Year	Engine Displacement Liters (VIN)	Cylinder Head Bolts	Main Bearing Bolts	Rod Bearing Bolts	Crankshaft Damper Bolts	Flywheel Bolts	Manifold Intake	Manifold Exhaust	Spark Plugs	Oil Pan Drain Plug
2007	2.3 (L3)	①	NA	NA	②	71-76	13-16	32-47	8-10	22-30
2008	2.3 (L3)	①	NA	NA	②	71-76	13-16	32-47	8-10	22-30

NA: Not Available

① Step 1: 97 inch lbs.
Step 2: Tighten 12.5 ft. lbs
Step 3: Tighten 33 ft. lbs.
Step 4: Tighten 90 degrees
Step 5: Tighten 90 degrees

② 71-77 ft. lbs. plus 90 degrees

22140_MCX7_C0008

WHEEL ALIGNMENT

Year	Model		Caster Range (+/-Deg.)	Caster Preferred Setting (Deg.)	Camber Range (+/-Deg.)	Camber Preferred Setting (Deg.)	Toe-in (in.)
2007	CX-7	F	1.00	+2.58	1.00	-0 17' +/- 1.00	0 +/- 0 22'
		R	—	—	1.00	-0 09' +/- 1.00	0 +/- 0 22'
2008	CX-7	F	1.00	+2.58	1.00	-0 17' +/- 1.00	0 +/- 0 22'
		R	—	—	1.00	-0 09' +/- 1.00	0 +/- 0 22'

22140_MCX7_C0010

TIRE, WHEEL AND BALL JOINT SPECIFICATIONS

		OEM Tires		Tire Pressures (psi)		Wheel	Ball Joint	Lug Nut
Year	Model	Standard	Optional	Front	Rear	Size	Inspection	(ft. lbs.)
2007	CX-7	P235/60R18 102H	235/60R18 103H	①	①	18 × 7 1/2J	②	65-87
2008	CX-7	P235/60R18 102H	235/60R18 103H	①	①	18 × 7 1/2J	②	65-87

OEM: Original Equipment Manufacturer

PSI: Pounds Per Square Inch

① P235/60R18 102H: 32 psi.
235/60R18 103H: 34 psi.

② Lower arm ball rotation torque: 13-17 inch lbs.

22140_MCX7_C0011

BRAKE SPECIFICATIONS

All measurements in inches unless noted

						Brake Drum				Brake Caliper	
			Brake Disc			Original	Max.	Maximum	Minimum	Bracket	Mounting
			Original	Minimum	Maximum	Inside	Wear	Machine	Lining	Bolts	Bolts
Year	Model		Thickness	Thickness	Runout	Diameter	Limit	Diameter	Thickness	(ft. lbs.)	(ft. lbs.)
2007	CX-7	F	NA	1.030	0.002	—	—	—	0.080	93-96	62-69
		R	NA	0.630	0.002	—	—	—	0.080	58-75	16-23
2008	CX-7	F	NA	1.030	0.002	—	—	—	0.080	93-96	62-69
		R	NA	0.630	0.002	—	—	—	0.080	58-75	16-23

NA: Not Avilable

F: Front

R: Rear

22140_MCX7_C0012

SCHEDULED MAINTENANCE INTERVALS

Mazda—CX-7

		VEHICLE MILEAGE INTERVAL (x1000)												
TO BE SERVICED	TYPE OF SERVICE	7.5	15	22.5	30	37.5	45	52.5	60	67.5	75	82.5	90	97.5
Engine oil & filter	R	✓	✓	✓	✓	✓	✓	✓	✓	✓	✓	✓	✓	✓
Air cleaner element	R					✓					✓			
Engine coolant	R													
Spark plugs	R										✓			
Bolts & nuts on chassis & body	S/I				✓				✓				✓	
Brake lines, hoses & connections	S/I				✓				✓				✓	
Cooling system	S/I				✓				✓				✓	
Disc brakes	S/I				✓				✓				✓	
Drive belts	S/I					✓					✓			
Drive shaft dust boots	S/I				✓				✓				✓	
Exhaust system heat shield	S/I				✓				✓				✓	
Front & rear suspension ball joints	S/I				✓				✓				✓	
Fuel lines & hoses	S/I				✓				✓				✓	
Steering operation & linkages	S/I				✓				✓				✓	
Fuel filter	R								✓					
Valve clearance	S/I										✓			
Hose & tube for emission	S/I								✓					

R: Replace S/I: Service or Inspect

FREQUENT OPERATION MAINTENANCE (SEVERE SERVICE)

If a vehicle is operated under any of the following conditions it is considered severe service

- Extremely dusty areas.

- 50% or more of the vehicle operation is in 32°C (90°F) or higher temperatures, or constant operation in temperatures below 0°C (32°F).

- Prolonged idling (vehicle operation in stop and go traffic).

- Frequent short running periods (engine does not warm to normal operating temperatures).

- Police, taxi, delivery usage or trailer towing usage.

Oil & oil filter: change every 5000 miles.

Air cleaner element: service or inspect every 15,000 miles

Automatic transaxle fluid: service or inspect every 15,000 miles.

Bolts & nuts on chassis & body: tighten every 15,000 miles.

Disc brakes: service or inspect every 15,000 miles.

22140_MCX7_C0009

PRECAUTIONS

Before servicing any vehicle, please be sure to read all of the following precautions, which deal with personal safety, prevention of component damage, and important points to take into consideration when servicing a motor vehicle:

- Never open, service or drain the radiator or cooling system when the engine is hot; serious burns can occur from the steam and hot coolant.
- Observe all applicable safety precautions when working around fuel. Whenever servicing the fuel system, always work in a well-ventilated area. Do not allow fuel spray or vapors to come in contact with a spark, open flame, or excessive heat (a hot drop light, for example). Keep a dry chemical fire extinguisher near the work area. Always keep fuel in a container specifically designed for fuel storage; also, always properly seal fuel containers to avoid the possibility of fire or explosion. Refer to the additional fuel system precautions later in this section.
- Fuel injection systems often remain pressurized, even after the engine has been turned **OFF**. The fuel system pressure must be relieved before disconnecting any fuel lines. Failure to do so may result in fire and/or personal injury.
- Brake fluid often contains polyglycol ethers and polyglycols. Avoid contact with the eyes and wash your hands thoroughly after handling brake fluid. If you do get brake fluid in your eyes, flush your eyes with clean, running water for 15 minutes. If eye irritation persists, or if you have taken brake fluid internally, IMMEDIATELY seek medical assistance.
- The EPA warns that prolonged contact with used engine oil may cause a number of skin disorders, including cancer. You should make every effort to minimize your exposure to used engine oil. Protective gloves should be worn when changing oil. Wash your hands and any other exposed skin areas as soon as possible after exposure to used engine oil. Soap and water, or waterless hand cleaner should be used.
- All new vehicles are now equipped with an air bag system, often referred to as a Supplemental Restraint System (SRS) or Supplemental Inflatable Restraint (SIR) system. The system must be disabled before performing service on or around system components, steering column, instrument panel components, wiring and sensors. Failure to follow safety and disabling procedures could result in accidental air bag deployment, possible personal injury and unnecessary system repairs.
- Always wear safety goggles when working with, or around, the air bag system. When carrying a non-deployed air bag, be sure the bag and trim cover are pointed away from your body. When placing a non-deployed air bag on a work surface, always face the bag and trim cover upward, away from the surface. This will reduce the motion of the module if it is accidentally deployed. Refer to the additional air bag system precautions later in this section.
- Clean, high quality brake fluid from a sealed container is essential to the safe and proper operation of the brake system. You should always buy the correct type of brake fluid for your vehicle. If the brake fluid becomes contaminated, completely flush the system with new fluid. Never reuse any brake fluid. Any brake fluid that is removed from the system should be discarded. Also, do not allow any brake fluid to come in contact with a painted surface; it will damage the paint.
- Never operate the engine without the proper amount and type of engine oil; doing so WILL result in severe engine damage.
- Timing belt maintenance is extremely important. Many models utilize an interference-type, non-freewheeling engine. If the timing belt breaks, the valves in the cylinder head may strike the pistons, causing potentially serious (also time-consuming and expensive) engine damage. Refer to the maintenance interval charts for the recommended replacement interval for the timing belt, and to the timing belt section for belt replacement and inspection.
- Disconnecting the negative battery cable on some vehicles may interfere with the functions of the on-board computer system(s) and may require the computer to undergo a relearning process once the negative battery cable is reconnected.
- When servicing drum brakes, only disassemble and assemble one side at a time, leaving the remaining side intact for reference.
- Only an MVAC-trained, EPA-certified automotive technician should service the air conditioning system or its components.

BRAKES — ANTI-LOCK BRAKE SYSTEM (ABS)

GENERAL INFORMATION

PRECAUTIONS

- Certain components within the ABS system are not intended to be serviced or repaired individually.
- Do not use rubber hoses or other parts not specifically specified for and ABS system. When using repair kits, replace all parts included in the kit. Partial or incorrect repair may lead to functional problems and require the replacement of components.
- Lubricate rubber parts with clean, fresh brake fluid to ease assembly. Do not use shop air to clean parts; damage to rubber components may result.
- Use only DOT 3 brake fluid from an unopened container.
- If any hydraulic component or line is removed or replaced, it may be necessary to bleed the entire system.
- A clean repair area is essential. Always clean the reservoir and cap thoroughly before removing the cap. The slightest amount of dirt in the fluid may plug an orifice and impair the system function. Perform repairs after components have been thoroughly cleaned; use only denatured alcohol to clean components. Do not allow ABS components to come into contact with any substance containing mineral oil; this includes used shop rags.
- The Anti-Lock control unit is a microprocessor similar to other computer units in the vehicle. Ensure that the ignition switch is **OFF** before removing or installing controller harnesses. Avoid static electricity discharge at or near the controller.
- If any arc welding is to be done on the vehicle, the control unit should be unplugged before welding operations begin.

WHEEL SPEED SENSORS

REMOVAL & INSTALLATION

Front

See Figure 1.

1. Before servicing the vehicle, refer to the Precautions Section.

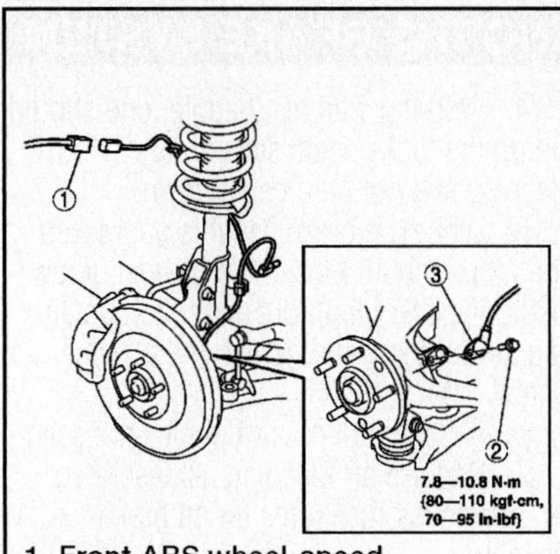

Fig. 1 Front wheel speed sensor

2. If removing the left front sensor, remove the battery, battery tray, air cleaner and hose.
3. Disconnect the wiring harness.
4. Unbolt and remove the wheel speed sensor.
5. Installation is the reverse of the removal procedure.

Rear

See Figures 2 and 3.

1. Before servicing the vehicle, refer to the Precautions Section.
2. Remove the rear seat.
3. Remove the rear scuff plate inner.
4. Remove the sub trunk box.
5. Remove the trunk end trim.
6. Remove the trunk side trim.
7. Remove the wiring harness connector.
8. Unbolt and remove the rear wheel-speed sensor
9. Install in the reverse order of removal.

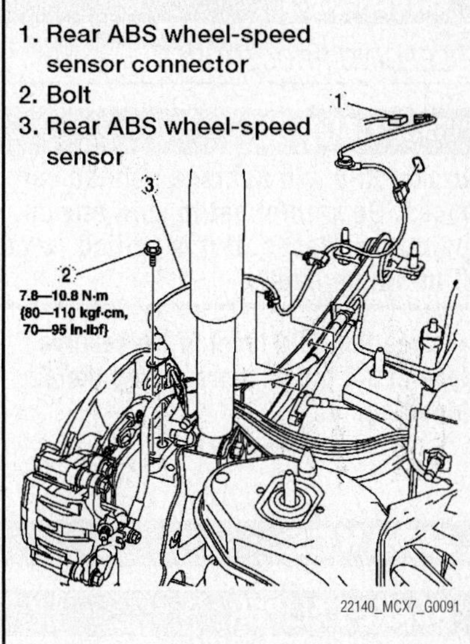

Fig. 3 Rear wheel speed sensor—AWD

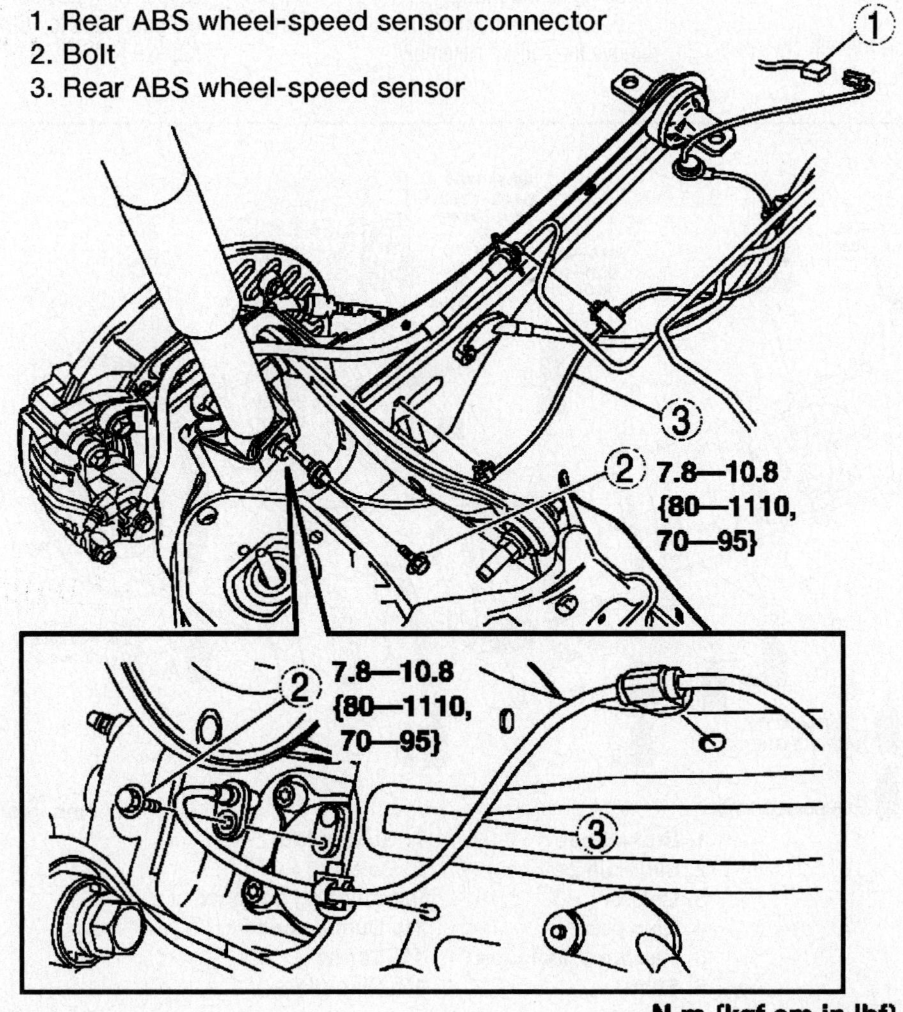

Fig. 2 Rear wheel speed sensor—2WD

BLEEDING PROCEDURE

CAUTION

Brake fluid will damage painted surfaces. Be careful not to spill any on painted surfaces. If it is spilled, wipe it off immediately.

➡Keep the fluid level in the reserve tank at 3/4 full or more during the air bleeding.

➡Begin air bleeding with the brake caliper that is furthest from the master cylinder.

➡Use DOT-3 brake fluid.

1. Before servicing the vehicle, refer to the Precautions Section.
2. Remove the bleeder cap on the brake caliper, and attach a vinyl tube to the bleeder screw.
3. Place the other end of the vinyl tube in a clear container and fill the container with fluid during air bleeding.
4. Working with two people, one should pump the brake pedal several times and depress and hold the pedal down.
5. While the brake pedal is depressed, the other should loosen the bleeder screw using the SST, drain out any fluid containing air bubbles, and tighten the bleeder screw.
6. Repeat until no air bubbles are seen.
7. Perform air bleeding as described in the above procedures for all brake calipers.

BRAKES FRONT DISC BRAKES

BRAKE CALIPER

REMOVAL & INSTALLATION

See Figure 4.

1. Before servicing the vehicle, refer to the Precautions Section.
2. Raise and support the vehicle safely.
3. Remove the wheels.
4. Disconnect the brake hose from the caliper by removing the union bolt and 2 gaskets. Plug the end of the hose to prevent loss of fluid.
5. Remove the bolts that attach the caliper to its mounting.
6. Lift the bottom of the caliper up and remove the caliper assembly.

To install:

7. Grease the caliper slides and bolts with lithium grease or equivalent. Install the caliper and secure with the bolts.
8. Connect the brake hose to the caliper, using 2 new washers. Torque the union bolt to 16–21 ft. lbs. (22–29 Nm).

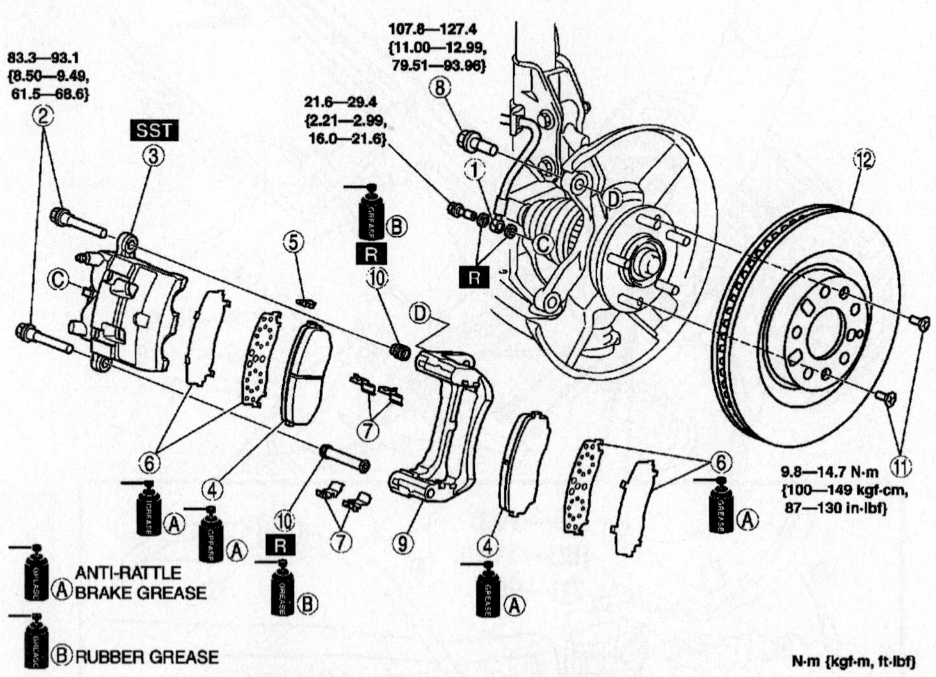

Fig. 4 Front brake exploded view

9. Fill the brake system to the proper level and bleed the brake system.

10. Install the tire and wheel assembly.

11. Top off the brake fluid level in the master cylinder. Check for leaks and proper brake operation.

DISC BRAKE PADS

REMOVAL & INSTALLATION

See Figure 5.

1. Before servicing the vehicle, refer to the Precautions Section.
2. Raise the vehicle and support it safely.
3. Remove the wheel and tire assembly.
4. When servicing the front pads, loosen the brake caliper upper side mounting bolt. Loosen and remove the lower side mounting bolt. Lift the caliper and suspend it so the hose is not stretched.
5. If equipped, remove the anti-squeal spring.
6. Remove the brake pads.

To install:

7. Siphon a small amount of brake fluid from the reservoir. Press in the brake caliper piston with the proper tool.
8. Before installing the new pads, check the disc thickness and disc runout.
9. Install the pad support plates.
10. Install the anti-squeal shims to each pad.

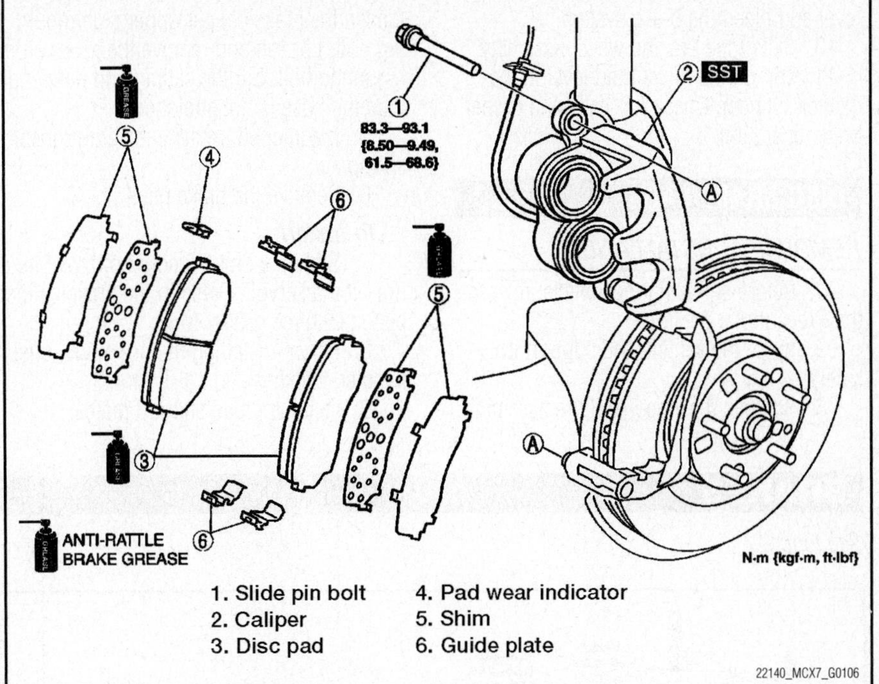

Fig. 5 Brake pad replacement exploded view

➡Apply disc brake grease to both sides of the inner anti-squeal shims.

11. Install the disc pads so the wear indicator plate is facing downward.
12. If removed, install the anti-squeal springs.
13. Carefully install the brake caliper so the boot is not wedged.
14. Install the wheel and tire assembly.
15. Check and adjust the fluid level. Apply the brake pedal several times.
16. Road test the vehicle for proper operation.

BRAKES — REAR DISC BRAKES

BRAKE CALIPER

REMOVAL & INSTALLATION

See Figure 6.

1. Before servicing the vehicle, refer to the Precautions Section.
2. Raise and support the vehicle safely.
3. Remove the wheels.
4. Disconnect the brake hose from the caliper by removing the union bolt and 2 gaskets. Plug the end of the hose to prevent loss of fluid.
5. Remove the bolts that attach the caliper to its mounting.
6. Lift the bottom of the caliper up and remove the caliper assembly.

To install:

7. Grease the caliper slides and bolts with lithium grease or equivalent. Install the caliper and secure with the bolts.
8. Connect the brake hose to the caliper, using 2 new washers. Torque the union bolt to 16–21 ft. lbs. (22–29 Nm).

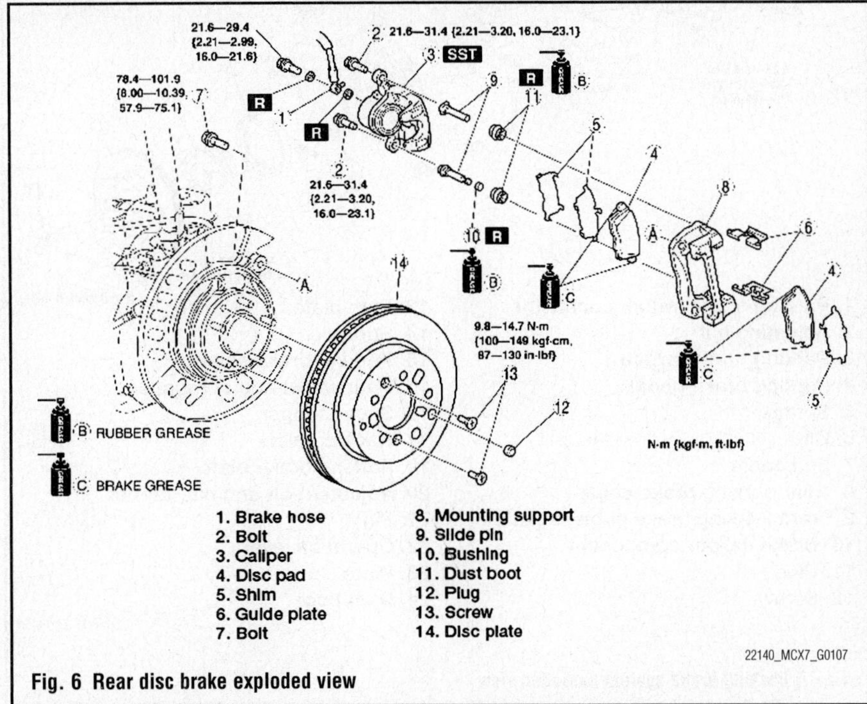

Fig. 6 Rear disc brake exploded view

9. Fill the brake system to the proper level and bleed the brake system.

10. Install the tire and wheel assembly.

11. Top off the brake fluid level in the master cylinder. Check for leaks and proper brake operation.

DISC BRAKE PADS

REMOVAL & INSTALLATION

1. Before servicing the vehicle, refer to the Precautions Section.
2. Raise the vehicle and support it safely.
3. Remove the wheel and tire assembly.
4. When servicing the front pads, loosen the brake caliper upper side mounting bolt. Loosen and remove the lower side mounting bolt. Lift the caliper and suspend it so the hose is not stretched.
5. If equipped, remove the anti-squeal spring.
6. Remove the brake pads.

To install:

7. Siphon a small amount of brake fluid from the reservoir. Press in the brake caliper piston with the proper tool.
8. Before installing the new pads, check the disc thickness and disc runout.
9. Install the pad support plates.
10. Install the anti-squeal shims to each pad.

➡Apply disc brake grease to both sides of the inner anti-squeal shims.

11. Install the disc pads so the wear indicator plate is facing downward.
12. If removed, install the anti-squeal springs.
13. Carefully install the brake caliper so the boot is not wedged.
14. Install the wheel and tire assembly.
15. Check and adjust the fluid level. Apply the brake pedal several times.
16. Road test the vehicle for proper operation.

BRAKES

See Figure 7.

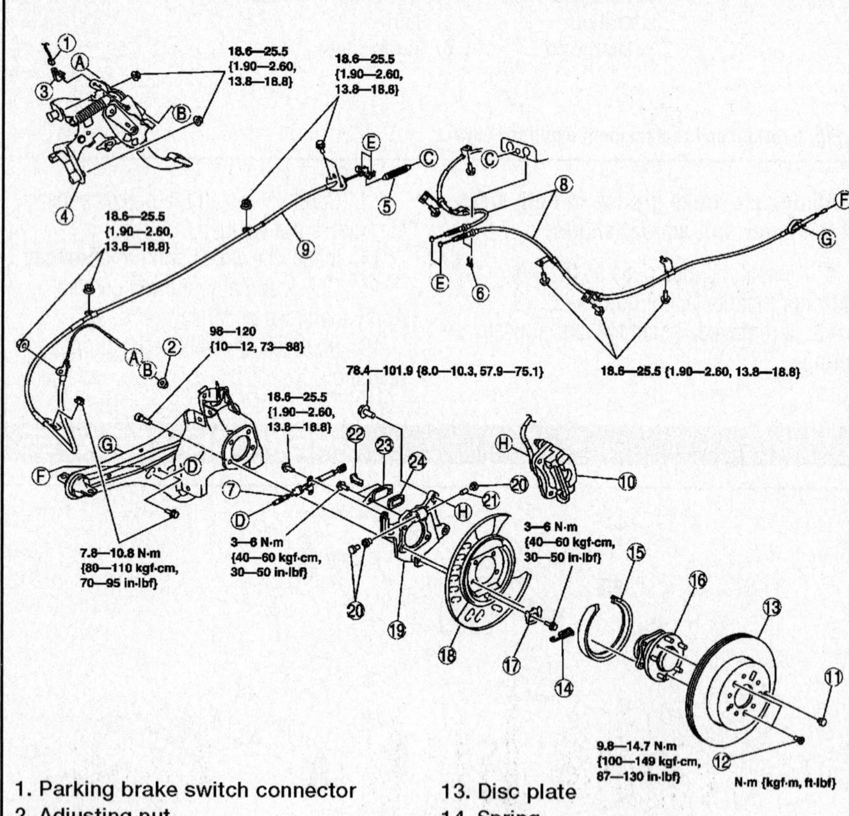

1. Parking brake switch connector
2. Adjusting nut
3. Parking brake switch
4. Parking brake pedal
5. Spring
6. Clip
7. End cable
8. Rear parking brake cable
9. Front Parking brake cable, equalizer
10. Brake caliper component
11. Plug
12. Screw
13. Disc plate
14. Spring
15. Parking brake shoe
16. Rear wheel hub component
17. Shoe stopper
18. Backing plate
19. Parking brake plate
20. Adjuster bolt and nut, tappet
21. Pin
22. Operation lever
23. Plate
24. Dust boot

22140_MCX7_G0108

Fig. 7 Parking brake system exploded view

PARKING BRAKE

PARKING BRAKE CABLES

ADJUSTMENT

See Figure 8.

1. Before servicing the vehicle, refer to the Precautions Section.
2. Start the engine and depress the parking brake pedal several times.
3. Stop the engine.
4. Turn the adjusting nut at the front of the parking cable.
5. After adjustment, inspect the following points:
6. Turn the ignition switch on, depress the parking brake pedal one notch, and verify that the brake system warning light illuminates.
7. Verify that the rear brakes do not drag.

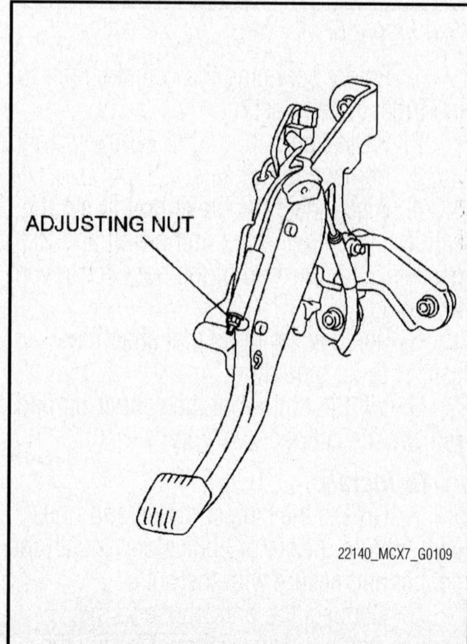

Fig. 8 Parking brake adjustment nut

CHASSIS ELECTRICAL

AIR BAG (SUPPLEMENTAL RESTRAINT SYSTEM)

GENERAL INFORMATION

SERVICE PRECAUTIONS

Disassembling the air bag system components could cause it to not operate (deploy) normally. Never disassemble any air bag system components.

Oil, grease, or water on the air bag modules may cause the air bags and pre-tensioner seat belts to fail to operate (deploy) in an accident. Never allow oil, grease, or water to get on the air bag modules or pre-tensioner seat belts.

Inserting a screwdriver or similar object into the connector of an air bag module or a pre-tensioner seat belt may damage the connector and cause the air bag module or the pre-tensioner seat belt to operate (deploy) improperly, which may cause serious injury. Never insert any foreign objects into the air bag module or seat belt connectors.

The seat weight sensor has a built-in strain gauge which may operate improperly if the sensor is dropped by itself or when installed to the seat. If it is dropped, replace the seat weight sensor with a new one.

Oil, grease, or water on the seat weight sensor may cause the system to operate (deploy) improperly. Never allow oil, grease, or water to get on the seat weight sensor.

Foreign material in the seat weight sensor components may cause the system to operate (deploy) improperly. Always make sure that no foreign material can get into the seat weight sensor.

Disassembling the seat weight sensor, or tightening any of the nuts and bolts installed to the sensor body may cause it to operate (deploy) improperly. Never disassemble the seat weight sensor or tighten any of the nuts or bolts installed to the body of the sensor.

Even if an air bag module or a pre-tensioner seat belt does not operate (deploy) in a collision and does not have any external signs of damage, it may have been damaged internally, which may cause improper operation. Before reusing a live (undeployed) air bag module and the pre-tensioner seat belts, always use the on-board diagnostic to diagnose the air bag module and the pre-tensioner seat belts to verify that they have no malfunction.

Incorrectly repairing an air bag wiring harness can accidentally operate (deploy) the air bag module and pre-tensioner seat belts. If a problem is found in the air bag wiring harness, always replace the wiring harness with a new one.

DISARMING THE SYSTEM

1. Disconnect the negative battery cable and wait for 1 minute or more.

ARMING THE SYSTEM

1. Connect the negative battery cable. Check for proper operation of the air bag warning light.
2. Perform the steering angle sensor reference point setting.

DRIVETRAIN

AUTOMATIC TRANSAXLE ASSEMBLY

REMOVAL & INSTALLATION

See Figures 9 through 13.

1. Before servicing the vehicle, refer to the Precautions Section.
2. Disconnect the negative battery cable.
3. Remove or disconnect the following:
   - Remove the following parts.
   - Battery and battery tray
   - Air cleaner component
   - Engine cover
   - Charge air cooler
   - Windshield wiper arm and blade
   - Cowl grille
   - Front wheel
   - Splash shield
   - Undercover
   - Side cover
   - Propeller shafts
   - Transfer oil cooler
4. Drain the ATF.
5. Support the engine from above.

**** WARNING**

Improperly jacking a transaxle is dangerous. It can slip off the jack and may cause serious injury.

**** CAUTION**

To prevent the torque converter and transaxle from separating, remove the transaxle without tilting it toward the torque converter.

6. Remove in the order shown in the figure.
7. Install in the reverse order of removal.
8. Add ATF to the specified level.

TRANSFER CASE ASSEMBLY

REMOVAL & INSTALLATION

See Figure 14.

1. Before servicing the vehicle, refer to the precautions in the beginning of this section.
2. Disconnect the negative battery cable.
3. Remove or disconnect the following:
   - Battery, battery tray
   - Engine cover
   - Intercooler
   - Front exhaust pipe
4. Disconnect the propeller shaft from the transfer side.
5. Remove the transfer oil cooler with the hose still connected.
6. Remove in the order indicated in the table.
7. Install in the reverse order of removal.
8. Warm up the engine and transaxle, inspect for oil leakage, and inspect the transfer operation.

FRONT HALFSHAFTS

REMOVAL & INSTALLATION

See Figure 15.

1. Before servicing the vehicle, refer to the Precautions Section.
2. Remove the side cover and under-cover.
3. Remove the front ABS wheel-speed sensor.
4. Remove the hub locknut.
5. Remove the tie rod end.
6. Remove the lower ball joint.
7. Separate the outer stub shaft from the wheel hub using a plastic-faced hammer.
8. Separate the inner CV joint from the transaxle using a suitable prytool.
9. Install in the reverse order of removal.

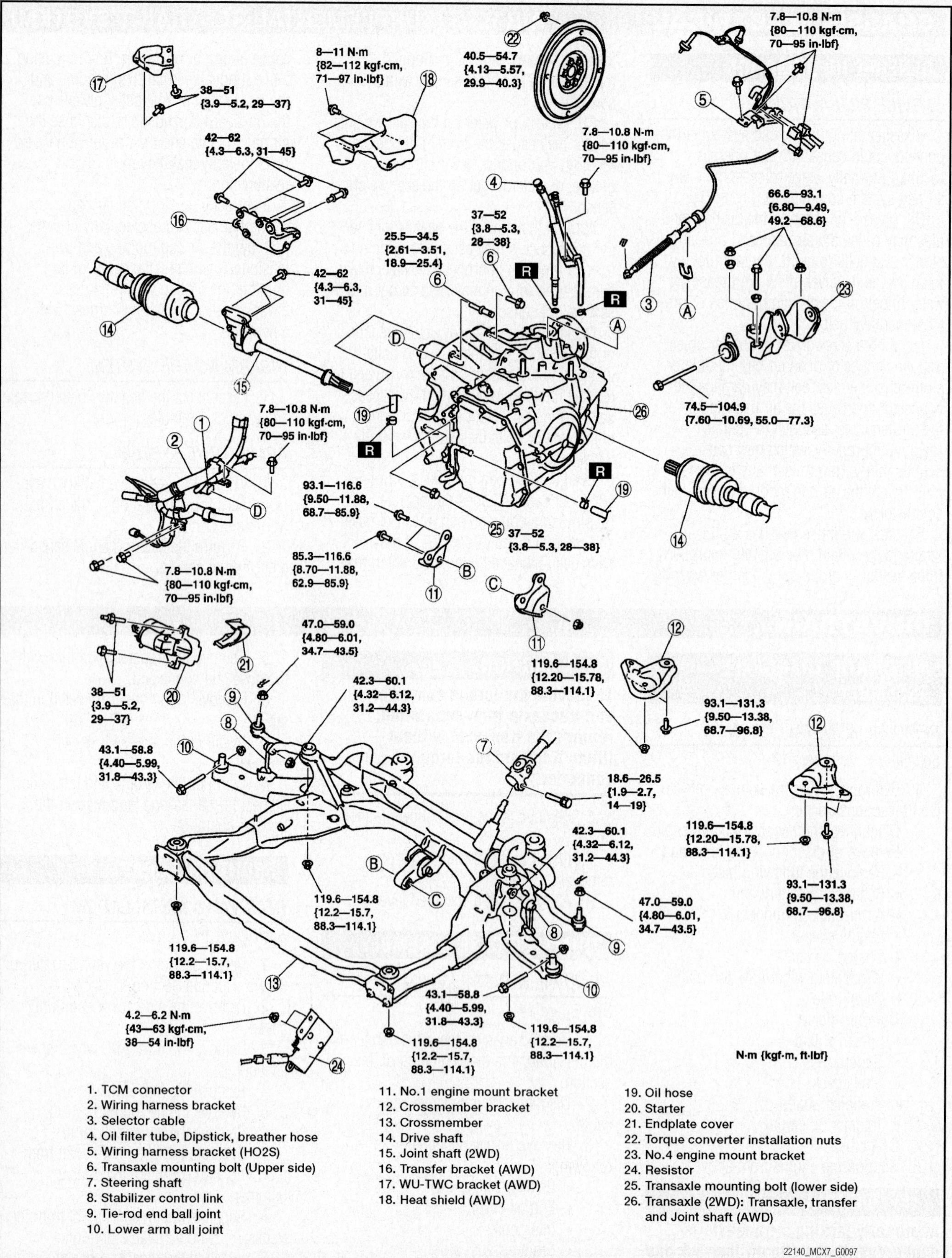

1. TCM connector
2. Wiring harness bracket
3. Selector cable
4. Oil filter tube, Dipstick, breather hose
5. Wiring harness bracket (HO2S)
6. Transaxle mounting bolt (Upper side)
7. Steering shaft
8. Stabilizer control link
9. Tie-rod end ball joint
10. Lower arm ball joint
11. No.1 engine mount bracket
12. Crossmember bracket
13. Crossmember
14. Drive shaft
15. Joint shaft (2WD)
16. Transfer bracket (AWD)
17. WU-TWC bracket (AWD)
18. Heat shield (AWD)
19. Oil hose
20. Starter
21. Endplate cover
22. Torque converter installation nuts
23. No.4 engine mount bracket
24. Resistor
25. Transaxle mounting bolt (lower side)
26. Transaxle (2WD): Transaxle, transfer and Joint shaft (AWD)

22140_MCX7_G0097

Fig. 9 Transaxle removal and installation—1 of 2

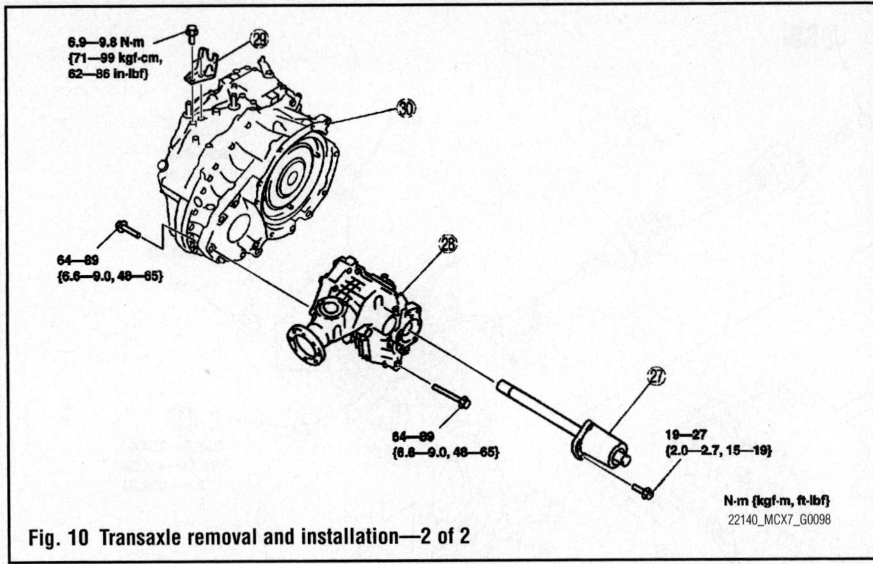

Fig. 10 Transaxle removal and installation—2 of 2

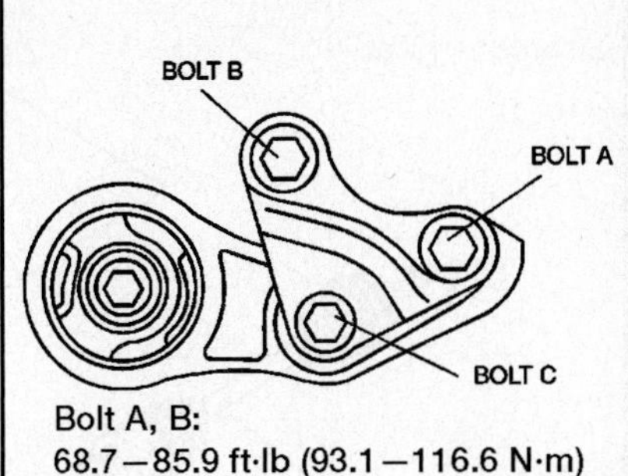

Fig. 13 No. 4 Engine mount bracket mounting fastener locations and torque specifications

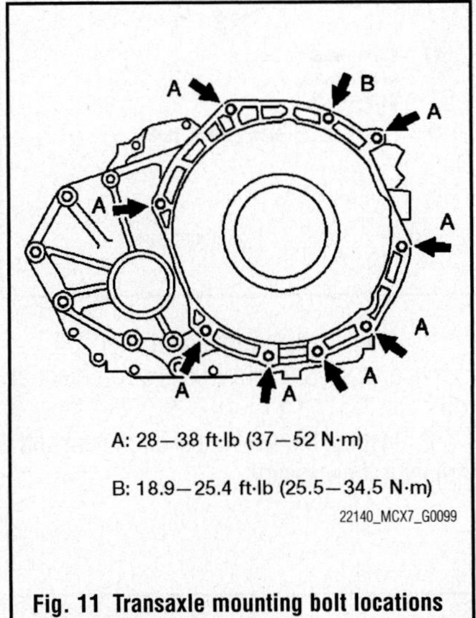

Fig. 11 Transaxle mounting bolt locations and torque specifications

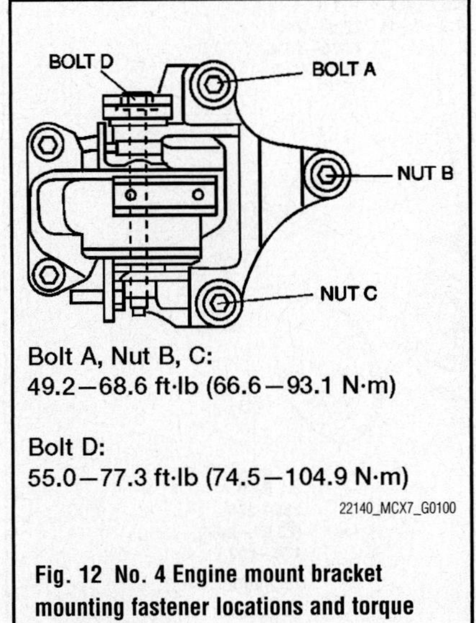

Fig. 12 No. 4 Engine mount bracket mounting fastener locations and torque specifications

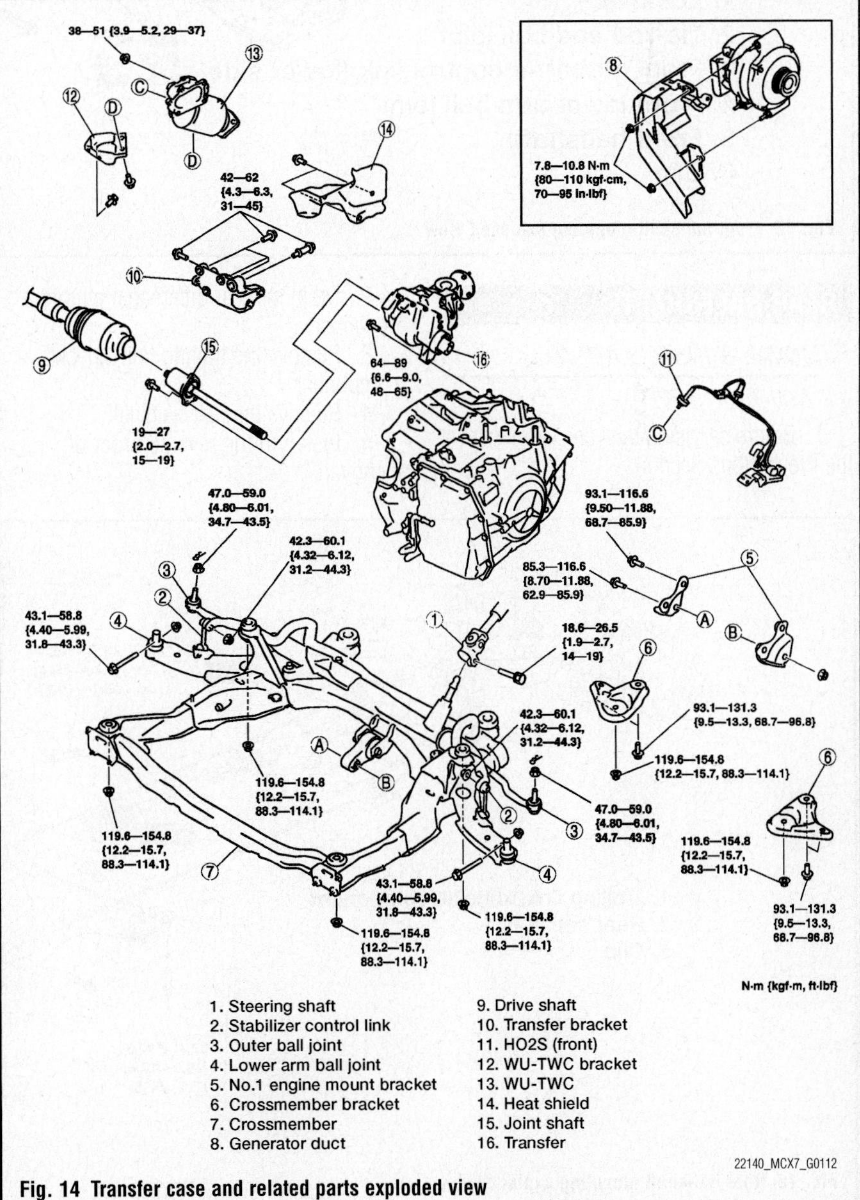

Fig. 14 Transfer case and related parts exploded view

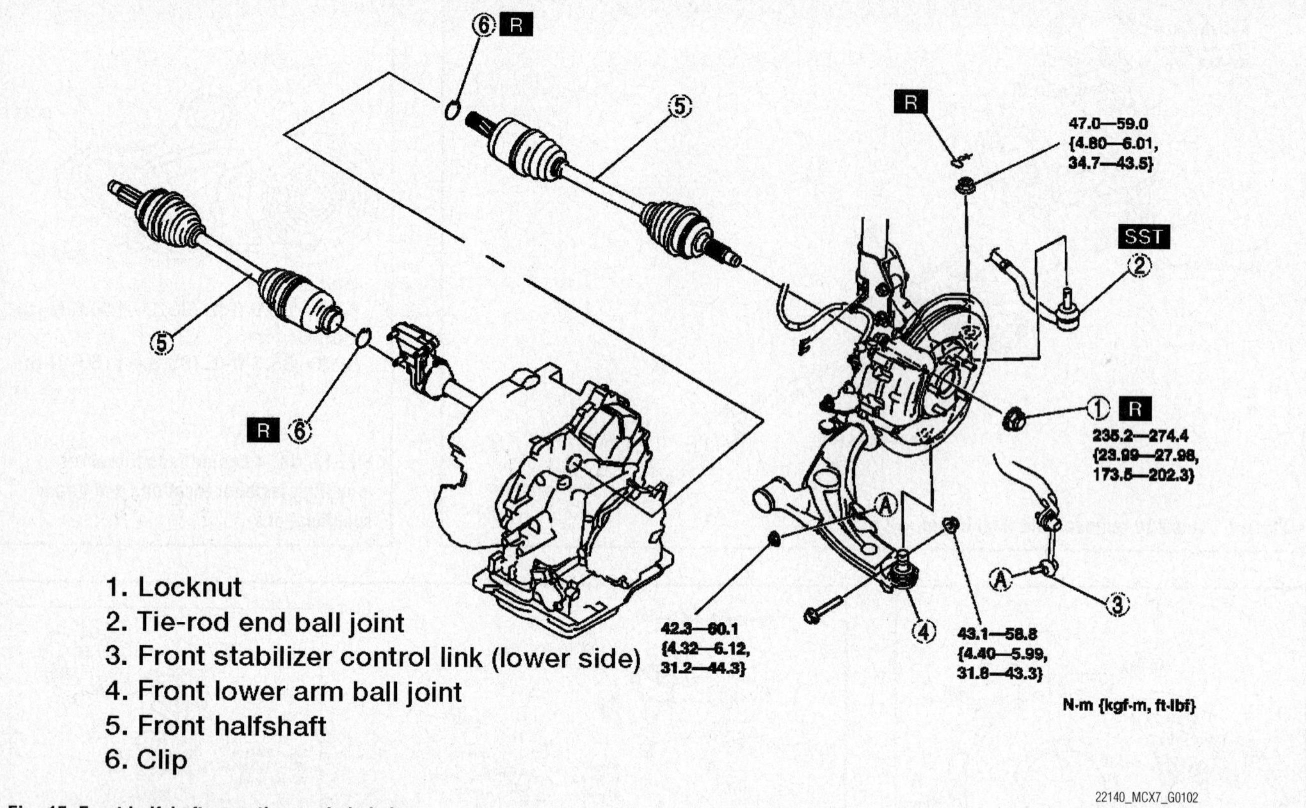

47.0—59.0 {4.80—6.01, 34.7—43.5}
SST
235.2—274.4 {23.99—27.98, 173.5—202.3}
42.3—60.1 {4.32—6.12, 31.2—44.3}
43.1—58.8 {4.40—5.99, 31.8—43.3}
N·m {kgf·m, ft·lbf}
1. Locknut
2. Tie-rod end ball joint
3. Front stabilizer control link (lower side)
4. Front lower arm ball joint
5. Front halfshaft
6. Clip
22140_MCX7_G0102

Fig. 15 Front halfshaft mounting exploded view

REAR HALFSHAFTS

REMOVAL & INSTALLATION

See Figure 16.

1. Before servicing the vehicle, refer to the Precautions Section.
2. Drain the rear differential oil into a container.
3. Remove the trailing link, knuckle component.
4. Remove the rear halfshaft.
5. Install in the reverse order of removal.
6. After installation, add 1 liter API service GL-5 SAE 80W-90 rear differential oil.
7. Inspect the rear wheel alignment and adjust it if necessary.

78.4—101.9 {8.0—10.3, 57.9—75.1}
75.5—102 {7.70—10.4, 55.7—75.2}
75.5—102 {7.70—10.4, 55.7—75.2}
SST
97.7—132.3 {10.0—13.4, 72.1—97.5}
75.5—102 {7.70—10.4, 55.7—75.2}
235—275 {23.97—28.0, 174—202.}
N·m {kgf·m, ft·lbf}
1. Trailing link, knuckle component
2. Rear halfshaft
3. Clip
22140_MCX7_G0103

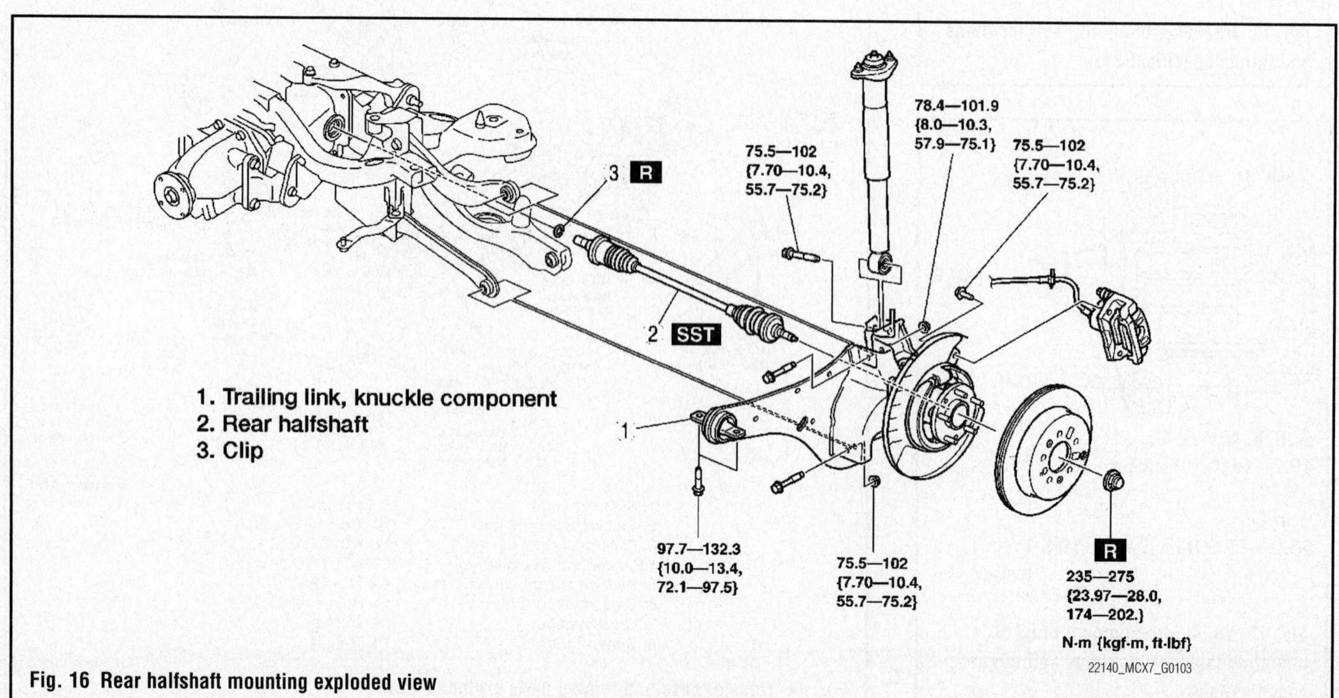

Fig. 16 Rear halfshaft mounting exploded view

ENGINE COOLING

⁂ WARNING

Never remove the cooling system cap or loosen the radiator drain plug while the engine is running, or when the engine and radiator are hot. Scalding engine coolant and steam may shoot out and cause serious injury. It may also damage the engine and cooling system. Turn off the engine and wait until it is cool. Even then, be very careful when removing the cap. Wrap a thick cloth around it and slowly turn it counterclockwise to the first stop. Step back while the pressure escapes. When you are sure all the pressure is gone, press down on the cap using the cloth, turn it, and remove it.

THERMOSTAT

REMOVAL & INSTALLATION

See Figure 17.

1. Before servicing the vehicle, refer to the precautions in the beginning of this section.
2. Disconnect the negative battery cable.
3. Drain the engine coolant.
4. Remove the front splash shield (RH).
5. Remove the charge air cooler duct.
6. Remove the drive belt.
7. Remove the P/S oil pump with hose and pipe still connected. Position the P/S oil pump out of the way.
8. Remove or disconnect the following:
   - Water hose
   - Lower radiator hose
   - Thermostat component
   - Gasket
9. Install in the reverse order of removal.
10. Refill the engine coolant.
11. Inspect for engine coolant leakage.

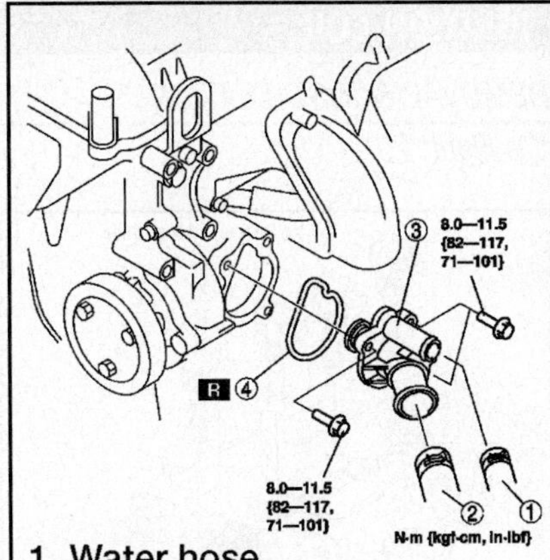

Fig. 17 Thermostat exploded view

WATER PUMP

REMOVAL & INSTALLATION

See Figure 18.

1. Before servicing the vehicle, refer to the precautions in the beginning of this section.

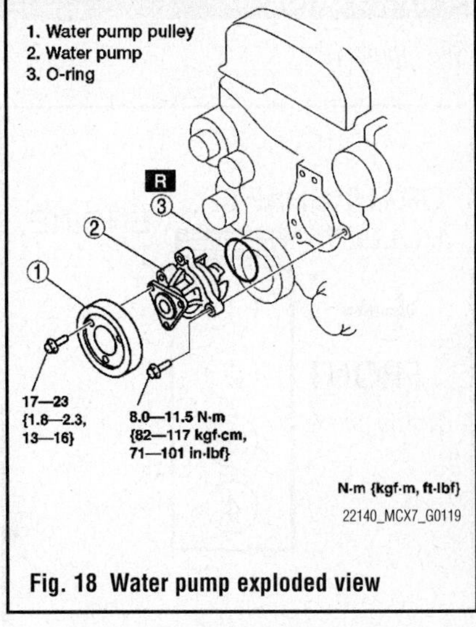

Fig. 18 Water pump exploded view

2. Disconnect the negative battery cable.
3. Drain the engine coolant.
4. Remove the front splash shield (RH).
5. Remove the charge air cooler duct.
6. Loosen the water pump pulley bolts before removing the drive belt.
7. Remove the drive belt.
8. Remove the water pump pulley.
9. Remove the water pump and O-ring seal.

➡Use a new O-ring seal for assembly. Lubricate the seal with engine coolant.

10. Install in the reverse order of removal.
11. Refill the engine coolant.
12. Inspect for engine coolant leakage.

ENGINE ELECTRICAL

ALTERNATOR

REMOVAL & INSTALLATION

⁂ WARNING

Remove and install all parts when the engine is cold, otherwise they can cause severe burns or serious injury.

⁂ WARNING

When the battery cables are connected, touching the vehicle body with alternator terminal B will generate sparks. This can cause personal injury, fire, and damage to the electrical components. Always disconnect the negative battery cable before performing the following operation.

⁂ CAUTION

The alternator can be damaged by the heat from the exhaust manifold. Make sure the alternator duct is installed securely.

CHARGING SYSTEM

1. Before servicing the vehicle, refer to the precautions in the beginning of this section.
2. Disconnect the negative battery cable.
3. Remove the charge air cooler duct and charge air cooler cover.
4. Remove the front splash shield (RH).
5. Remove the drive belt.
6. Remove the alternator duct
7. Remove the wiring harness bracket
8. Disconnect the terminal B cable
9. Disconnect the alternator connector
10. Alternator
11. Install in the reverse order of removal.

ENGINE ELECTRICAL — DISTRIBUTORLESS IGNITION SYSTEM

FIRING ORDERS

See Figure 19.

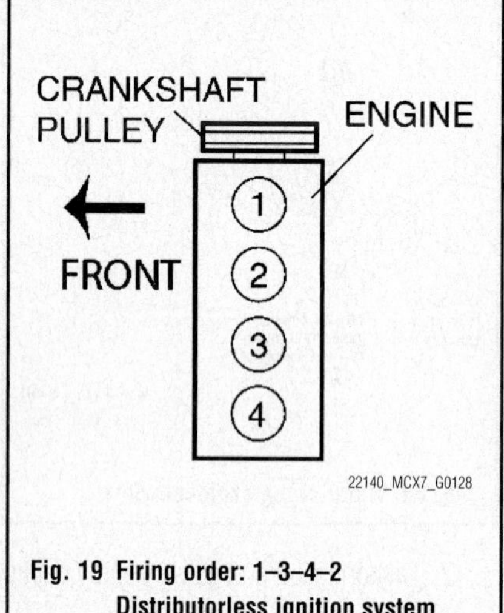

Fig. 19 Firing order: 1–3–4–2 Distributorless ignition system

IGNITION COIL

REMOVAL & INSTALLATION

See Figure 20.

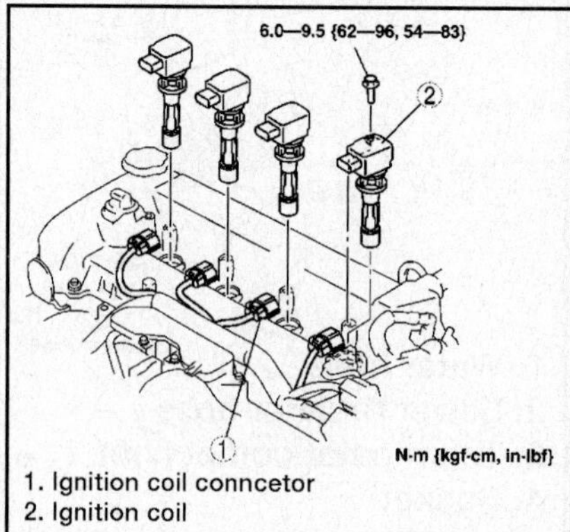

Fig. 20 Ignition coil mounting exploded view

1. Before servicing the vehicle, refer to the precautions in the beginning of this section.
2. Disconnect the negative battery cable.
3. Remove the charge air cooler.
4. Disconnect the ignition coil connectors.
5. Remove the ignition coils.
6. Install in the reverse order of removal.

SPARK PLUGS

REMOVAL & INSTALLATION

1. Before servicing the vehicle, refer to the precautions in the beginning of this section.
2. Disconnect the negative battery cable.
3. Remove the charge air cooler.
4. Remove the ignition coils.
5. Remove the spark plugs using a plug-wrench.
6. Install in the reverse order of removal. Tighten the spark plugs to 96–120 inch lbs. (10–14 Nm).

ENGINE ELECTRICAL — STARTING SYSTEM

STARTER

REMOVAL & INSTALLATION

See Figure 21.

WARNING

Remove and install all parts when the engine is cold, otherwise they can cause severe burns or serious injury.

WARNING

When the battery cables are connected, touching the vehicle body with starter terminal B will generate sparks. This can cause personal injury, fire, and damage to the electrical components. Always disconnect the negative battery cable before performing the following operation.

1. Before servicing the vehicle, refer to the precautions in the beginning of this section.
2. Remove the battery and battery tray.
3. Remove the charge air cooler duct, air cleaner and fresh air duct component, and air hose.
4. Remove the undercover.
5. Remove the wiring harness and bracket.

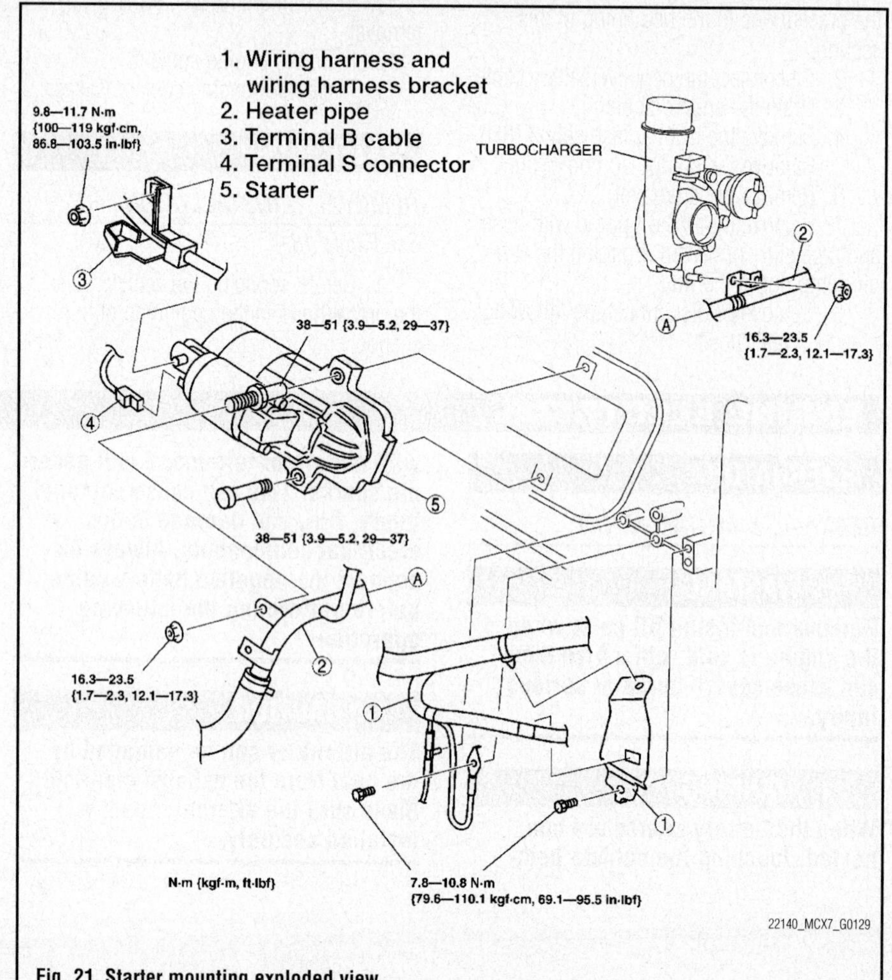

Fig. 21 Starter mounting exploded view

6. Remove the heater pipe with the water hoses still connected. Position the heater pipe so that it is out of the way.

7. Remove the terminal B cable and the terminal S connector.

8. Remove the starter.

9. Install in the reverse order of removal.

SOLENOID OR RELAY REPLACEMENT

See Figure 22.

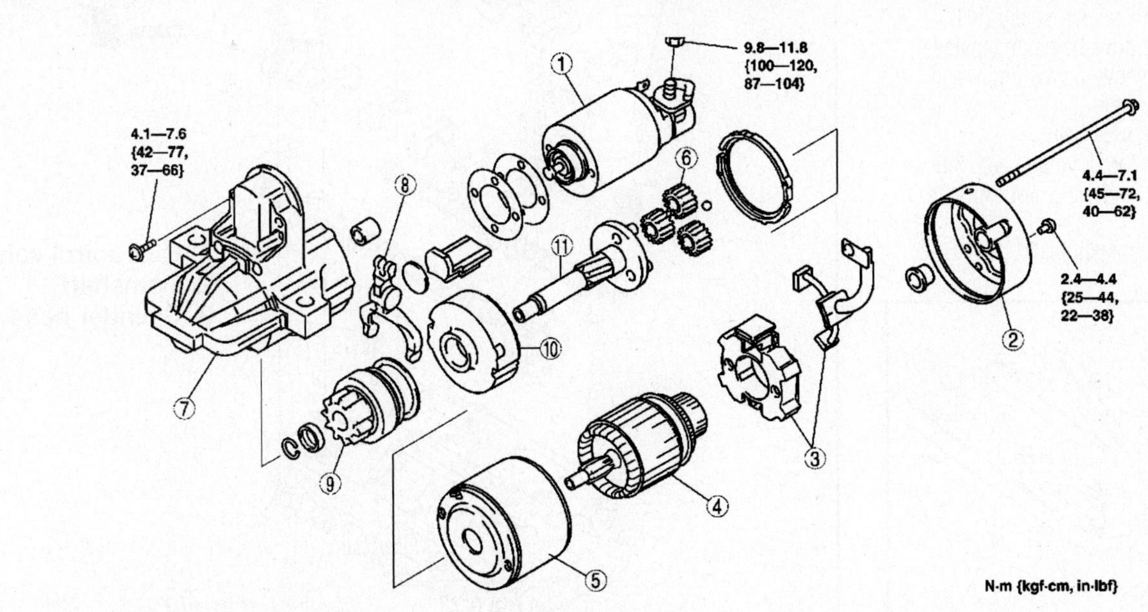

Fig. 22 Starter exploded view

ENGINE MECHANICAL

ACCESSORY DRIVE BELTS

ACCESSORY BELT ROUTING

See Figure 23.

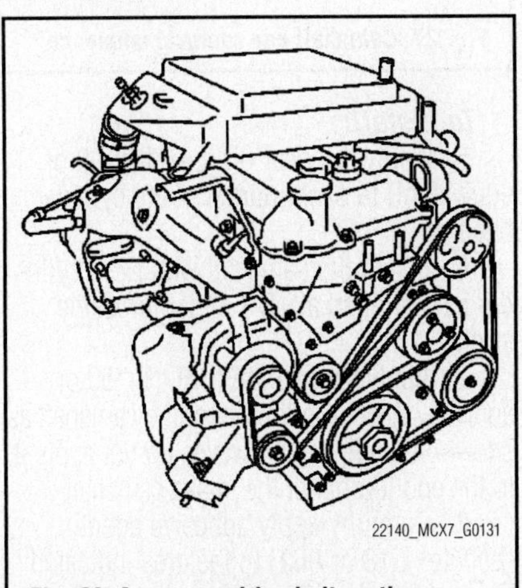

Fig. 23 Accessory drive belt routing

INSPECTION

See Figure 24.

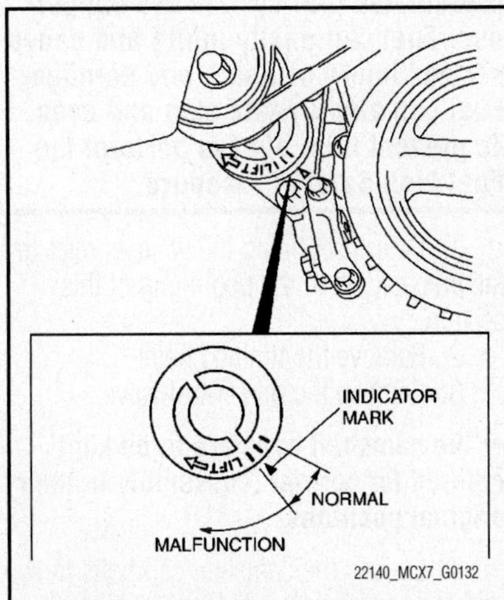

Fig. 24 Auto belt tensioner wear indicator

Inspect the drive belt for signs of glazing or cracking. A glazed belt will be perfectly smooth from slippage, while a good belt will have a slight texture of fabric visible. Cracks will usually start at the inner edge of the belt and run outward. All worn or damaged drive belts should be replaced immediately.

➡Drive belt deflection/tension inspection is not necessary because of the use of the drive belt auto tensioner.

1. Verify that the drive belt auto tensioner indicator mark does not exceed the limit.

2. If it exceeds the limit, replace the drive belt.

ADJUSTMENT

Drive belt adjustment is not necessary because of the use of the drive belt auto tensioner.

REMOVAL & INSTALLATION

See Figure 25.

1. Before servicing the vehicle, refer to the precautions in the beginning of this section.
2. Remove the splash shield (RH).
3. Rotate the drive belt auto tensioner in the direction shown in the figure and remove the drive belt.
4. Install the drive belt.
5. Verify that the drive belt auto tensioner indicator mark does not exceed the limit.
6. Install the splash shield (RH).

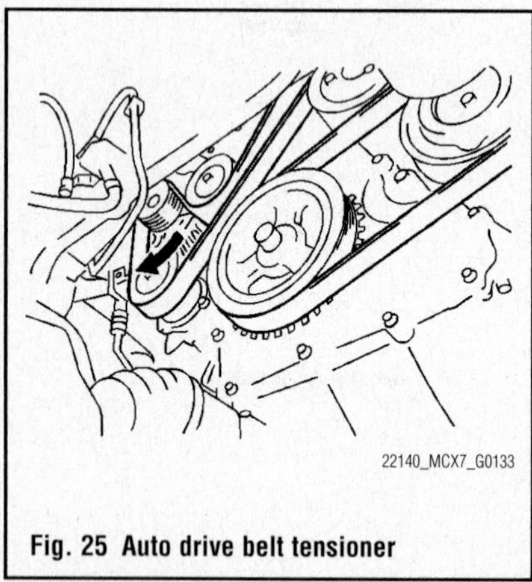

Fig. 25 Auto drive belt tensioner

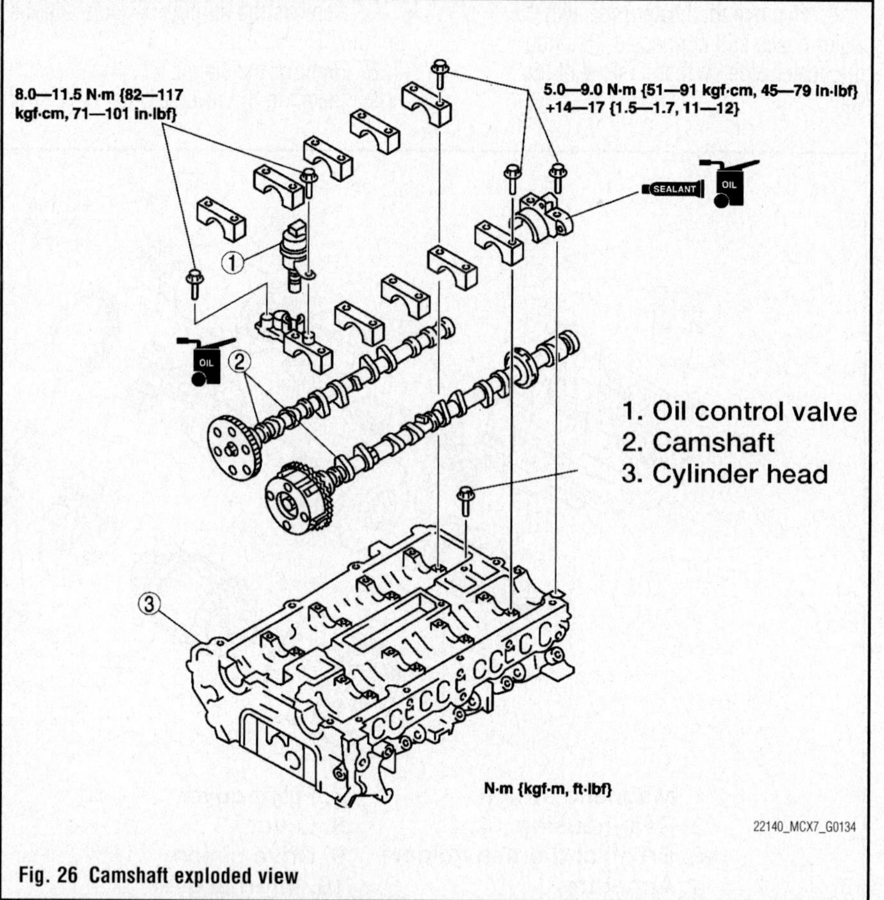

Fig. 26 Camshaft exploded view

CAMSHAFT

INSPECTION

1. Before servicing the vehicle, refer to the Precautions Section.
2. Remove the camshaft from the engine.
3. Check the camshaft bearing journals for damage and binding.
4. If the journals are binding, check the cylinder head for damage.
5. Check the cylinder head for clogged oil holes.
6. Check the camshaft surface for abnormal wear and damage. Replace the camshaft, as required.
7. Measure the camshaft lobe surface and replace the camshaft if not within specification.
8. Measure the camshaft journal diameter and replace the camshaft if not within specification.
9. Measure the camshaft run out and replace the camshaft if not within specification.

REMOVAL & INSTALLATION

See Figures 26 through 30.

**** WARNING**

Fuel vapor is hazardous. It can very easily ignite, causing death, serious injury, or damage. Always keep sparks and flames away from fuel.

**** WARNING**

Fuel line spills and leakage from the pressurized fuel system are dangerous. Fuel can easily ignite and cause serious injury or death and damage. Fuel can also irritate skin and eyes. To prevent this, always perform the Fuel Line Safety Procedure.

1. Before servicing the vehicle, refer to the precautions in the beginning of this section.
2. Remove the timing chain.
3. Remove the oil control valve.

➡The camshaft caps are to be kept ordered for correct reassembly in their original positions.

4. Loosen the camshaft cap bolts in two or three passes in the order shown, and remove them.

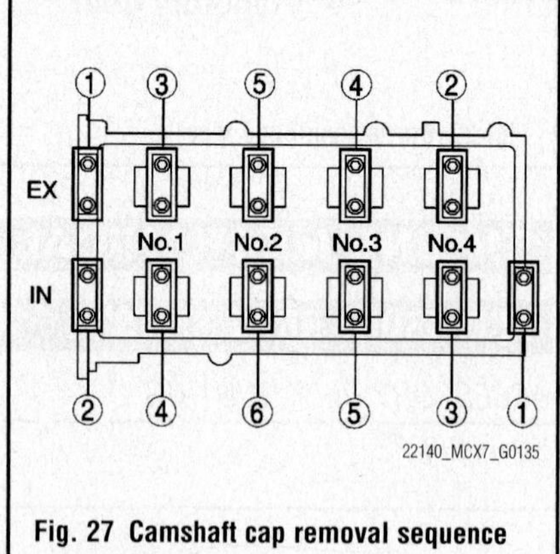

Fig. 27 Camshaft cap removal sequence

To install:

5. Apply the gear oil (SAE No. 90 or equivalent) to each journal of the cylinder head.
6. Install the camshaft with No.3 cylinder cam aligned at TDC of compression stroke.
7. Apply the gear oil (SAE No. 90 or equivalent) to each journal of the camshaft as shown in the figure. However, do not apply it to the end journal of the intake camshaft.
8. Carefully apply adhesive agent (Loctite® 518 or 962) to the area indicated

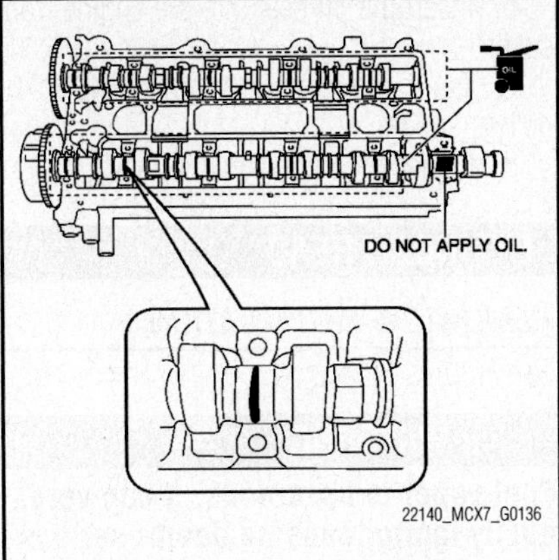

Fig. 28 Apply oil to the camshaft journals

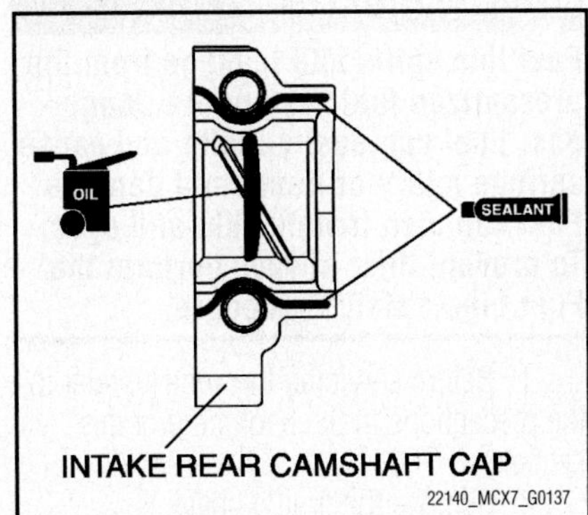

Fig. 29 Apply sealant and oil to the rear intake camshaft cap

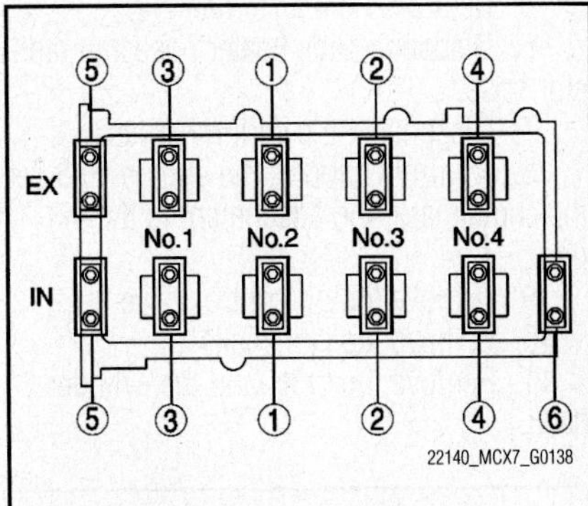

Fig. 30 Camshaft bearing cap torque sequence

in the figure so that it does not leak into the sliding part then, apply the gear oil (SAE No. 90 or equivalent) to the journal.

9. Install the camshaft caps and temporarily tighten the camshaft cap bolts evenly in two or three passes until snug, then tighten as follows:
   a. Step 1: 45–79 inch lbs. (5–9 Nm)
   b. Step 2: 11–12 ft. lbs. (14–17 Nm)
10. The remainder of the installation is the reverse of the removal procedure.

CRANKSHAFT DAMPER

REMOVAL & INSTALLATION

See Figures 31 through 36.

1. Before servicing the vehicle, refer to the precautions in the beginning of this section.
2. Disconnect the negative battery cable.
3. Remove the undercover.
4. Remove the splash shield (RH).
5. Remove the charge air cooler.
6. Remove the high pressure fuel pump.
7. Remove the drive belt.
8. Remove the ignition coils.
9. Disconnect the wiring harness.
10. Remove the ventilation hose.
11. Remove cylinder head cover.
12. Remove the crankshaft position sensor.
13. Remove the bracket shown in the figure, and place the power steering pressure hose outside of the vehicle.
14. Rotate the crankshaft in the direction of the engine rotation and remove the

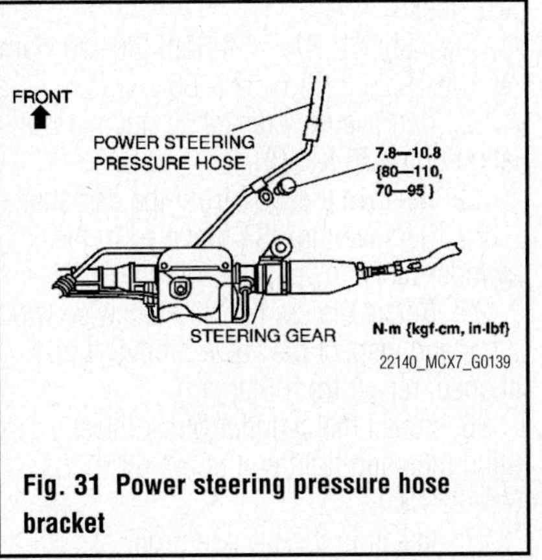

Fig. 31 Power steering pressure hose bracket

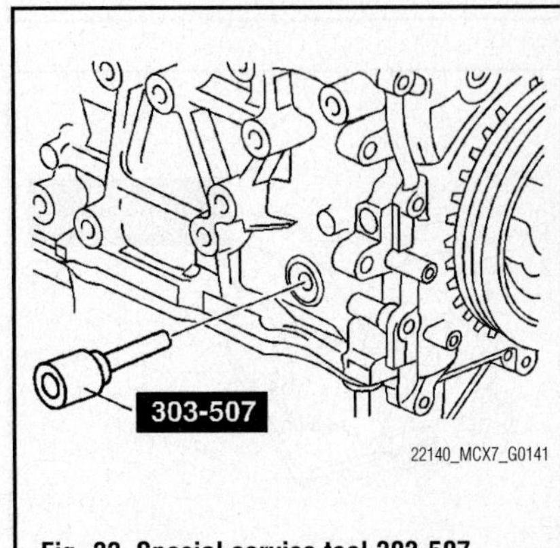

Fig. 33 Special service tool 303-507

1. Crankshaft pulley lock bolt
2. Crankshaft pulley
3. Front oil seal

R SST
R SST
OIL
N·m { kgf·m, ft·lbf}
96—104 {9.8—10.6 , 70.9—76.7} +87°—93°
22140_MCX7_G0140

Fig. 32 Crankshaft pulley exploded view

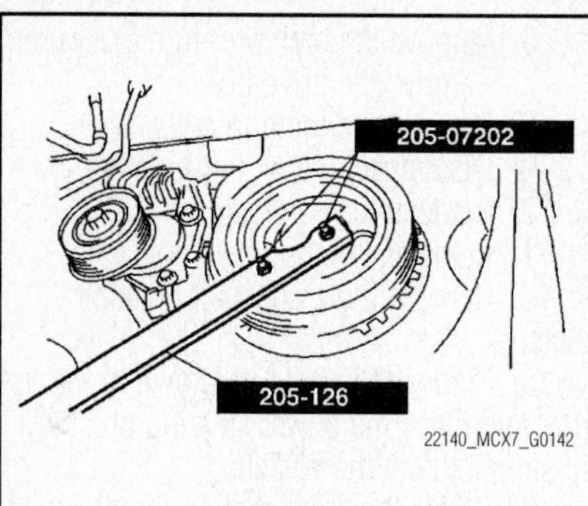

Fig. 34 Special service tools 205-07202 and 205-126

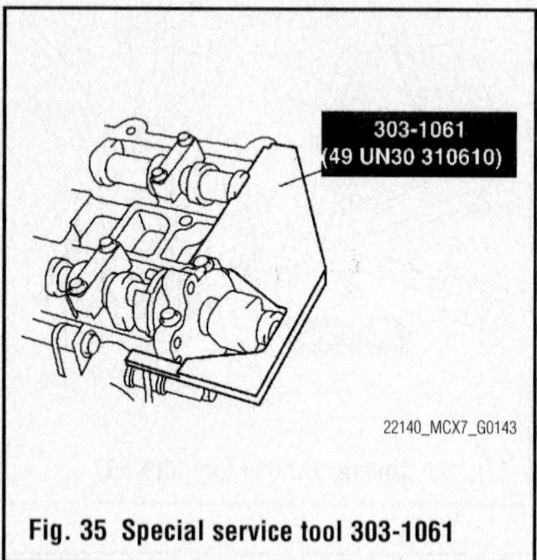

Fig. 35 Special service tool 303-1061

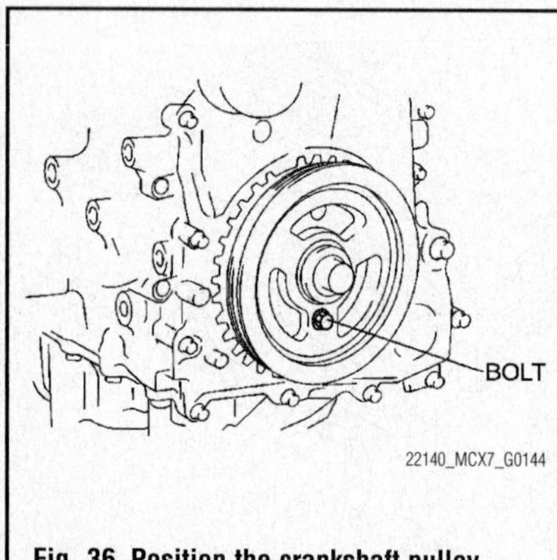

Fig. 36 Position the crankshaft pulley

cylinder block lower blind plug when the No. 1 cylinder is at the point prior to top dead center (TDC) of compression, then install SST 303-507.

15. Rotate the crankshaft in the direction of the engine rotation so that the No.1 piston is at TDC of the compression stroke. (Until the crank weight contacts SST and stops.)

16. Install SSTs 205-07202 and 205-126 to the crankshaft pulley and lock the crankshaft against rotation.

17. Remove the crankshaft pulley lock bolt and the crankshaft pulley.

To install:

18. Install SST 303-1061 to the camshaft as shown in the figure.

19. Verify that cylinder No.1 is at TDC of the compression stroke. (Position crank weight contacts SST 303-507.)

20. To position the crankshaft pulley, temporarily tighten it and, using a suitable bolt (M6 x 1.0), fix the crankshaft pulley to the engine front cover.

21. Install SSTs 205-07202 and 205-126 to the crankshaft pulley and lock the crankshaft against rotation, and tighten the crankshaft pulley lock bolt using the following two steps:

a. Step 1: 71–77 ft. lbs. (96–104 Nm)

b. Step 2: Plus 87°–93°

22. Remove the crankshaft pulley installation bolt (M6 X 1.0).

23. Remove the SST from the camshaft.

24. Remove the SST installed in the cylinder block lower blind plug hole.

25. Rotate the crankshaft clockwise two turns and inspect the valve timing. If not aligned, repeat from Step 1.

26. Install the cylinder block lower blind plug and tighten it to 14–16 ft. lbs. (18–22 Nm).

27. Install in the reverse order of removal.

CRANKSHAFT FRONT SEAL

REMOVAL & INSTALLATION

See Figure 37.

1. Before servicing the vehicle, refer to the precautions in the beginning of this section.

2. Remove the crankshaft pulley.

3. Remove the oil seal lip using a razor.

4. Remove the oil seal using a suitable prytool.

To install:

5. Apply clean engine oil to a new oil seal.

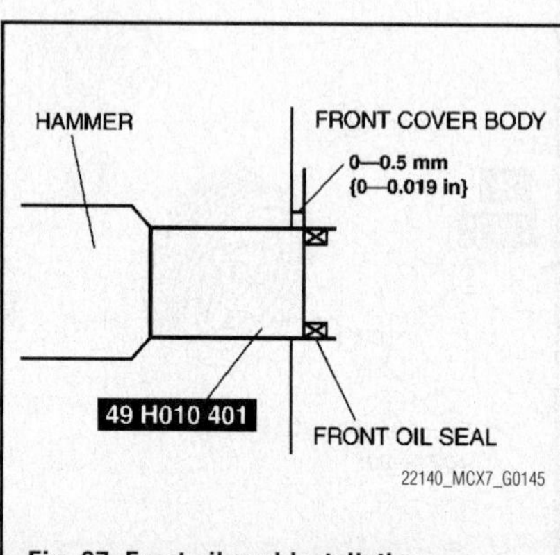

Fig. 37 Front oil seal installation

6. Insert the oil seal into the engine front cover.

7. Tap in the oil seal using SST 49 H010 401 as shown.

8. Replace the crankshaft pulley.

CYLINDER HEAD

REMOVAL & INSTALLATION

See Figures 38 and 39.

**** WARNING**

Fuel vapor is hazardous. It can very easily ignite, causing death, serious injury, or damage. Always keep sparks and flames away from fuel.

**** WARNING**

Fuel line spills and leakage from the pressurized fuel system are dangerous. Fuel can easily ignite and cause serious injury or death and damage. Fuel can also irritate skin and eyes. To prevent this, always perform the Fuel Line Safety Procedure.

1. Before servicing the vehicle, refer to the precautions in the beginning of this section.

2. Remove the timing chain.

3. Remove the generator. Place the generator out of the way with the wiring harnesses connected.

4. Remove the exhaust manifold.

5. Remove the intake manifold.

6. Disconnect the heater hose and radiator hose.

7. Disconnect the wiring harness.

8. To firmly support the engine, first set the engine jack and attachment to the oil pan.

9. Remove the oil control valve

10. Remove the camshafts.

11. Remove the bolts and the cylinder head.

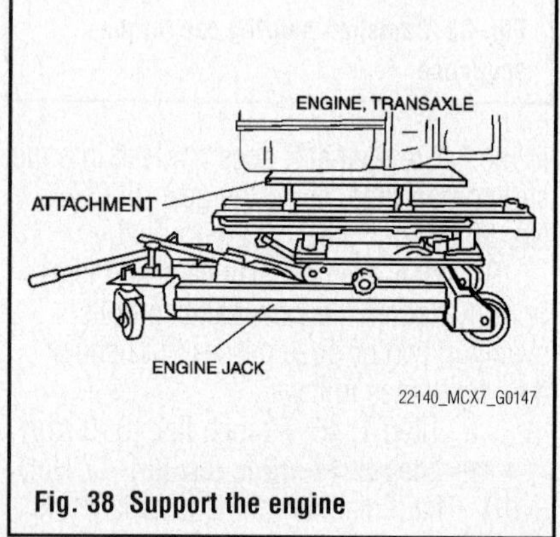

Fig. 38 Support the engine

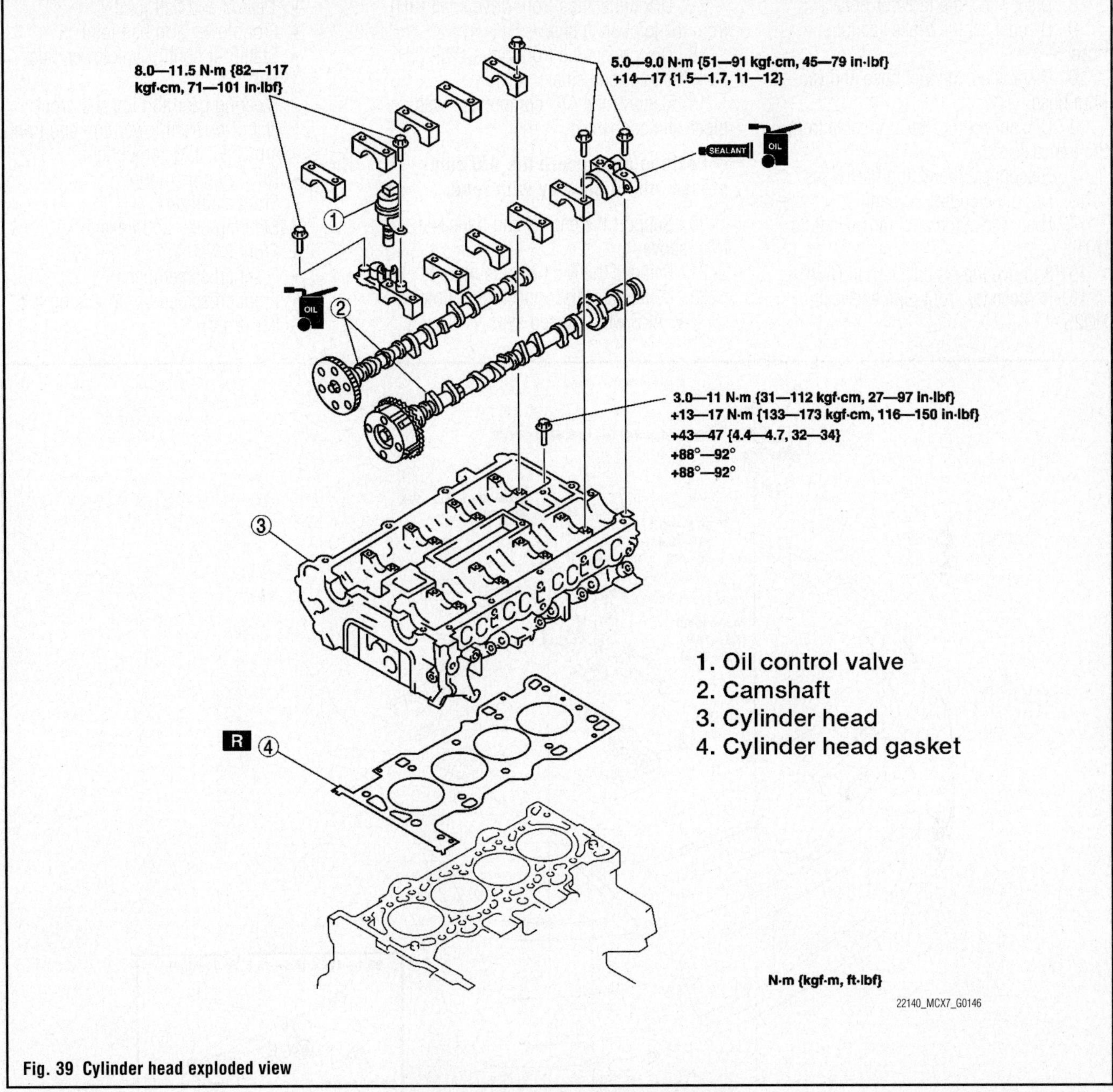

Fig. 39 Cylinder head exploded view

To install:

12. Measure the cylinder head bolts. Replace any bolts that exceed 5.740 inches (146 mm).
13. Install the cylinder head with a new gasket. Tighten the bolts in sequence as follows:
    a. Step 1: 27–97 inch lbs. (3–11 Nm)
    b. Step 2: 116–150 inch lbs. (13–17 Nm)
    c. Step 3: 32–34 ft. lbs. (43–47 Nm)
    d. Step 4: Plus 88°–92°
    e. Step 5: Plus 88°–92°
14. Install in the reverse order of removal.
15. Bleed the air from the cooling system.
16. Inspect the compression pressure.

ENGINE ASSEMBLY

REMOVAL & INSTALLATION

See Figures 40 and 41.

**** WARNING**

Fuel vapor is hazardous. It can very easily ignite, causing death, serious injury, or damage. Always keep sparks and flames away from fuel.

**** WARNING**

Fuel line spills and leakage from the pressurized fuel system are dangerous. Fuel can easily ignite and cause serious injury or death and damage. Fuel can also irritate skin and eyes. To prevent this, always perform the Fuel Line Safety Procedure.

1. Before servicing the vehicle, refer to the precautions in the beginning of this section.

Remove the battery and battery tray.

2. Remove the charge air cooler duct.
3. Remove the air cleaner.
4. Remove the front tires.
5. Remove the undercover, splash shield and mudguard.
6. Remove the drive belt.
7. Disconnect the heater hose.

8. Disconnect the radiator hose.
9. Disconnect the brake vacuum hose.
10. Disconnect the fuel hose and vacuum hose.
11. Disconnect the selector cable from the transaxle side.
12. Disconnect the wiring harnesses.
13. Drain the engine coolant
14. Drain the automatic transaxle fluid (ATF).
15. Remove the propeller shaft (AWD).
16. Remove the front pipe and rear HO2S.
17. Disconnect the front drive shaft (RH) from the joint shaft side.
18. Disconnect the front drive shaft (LH) from the transaxle side.
19. Remove the A/C compressor with the pipes still connected.

➡Position and secure the A/C compressor out of the way with rope.

20. Support the engine and transaxle from above.
21. Remove the No.1 engine mount.
22. Remove or disconnect the following:
  - ABS wheel speed sensor
  - Tie-rod end ball joint
  - Front lower arm ball joint
  - Stabilizer control link lower side nut
  - Steering gear and linkage, front stabilizer, front lower arm and front crossmember component
  - No. 1 engine mount
  - Front stabilizer
  - Steering gear and linkage
  - Front lower arm
  - Front crossmember
  - Front crossmember mounting rubber (front)

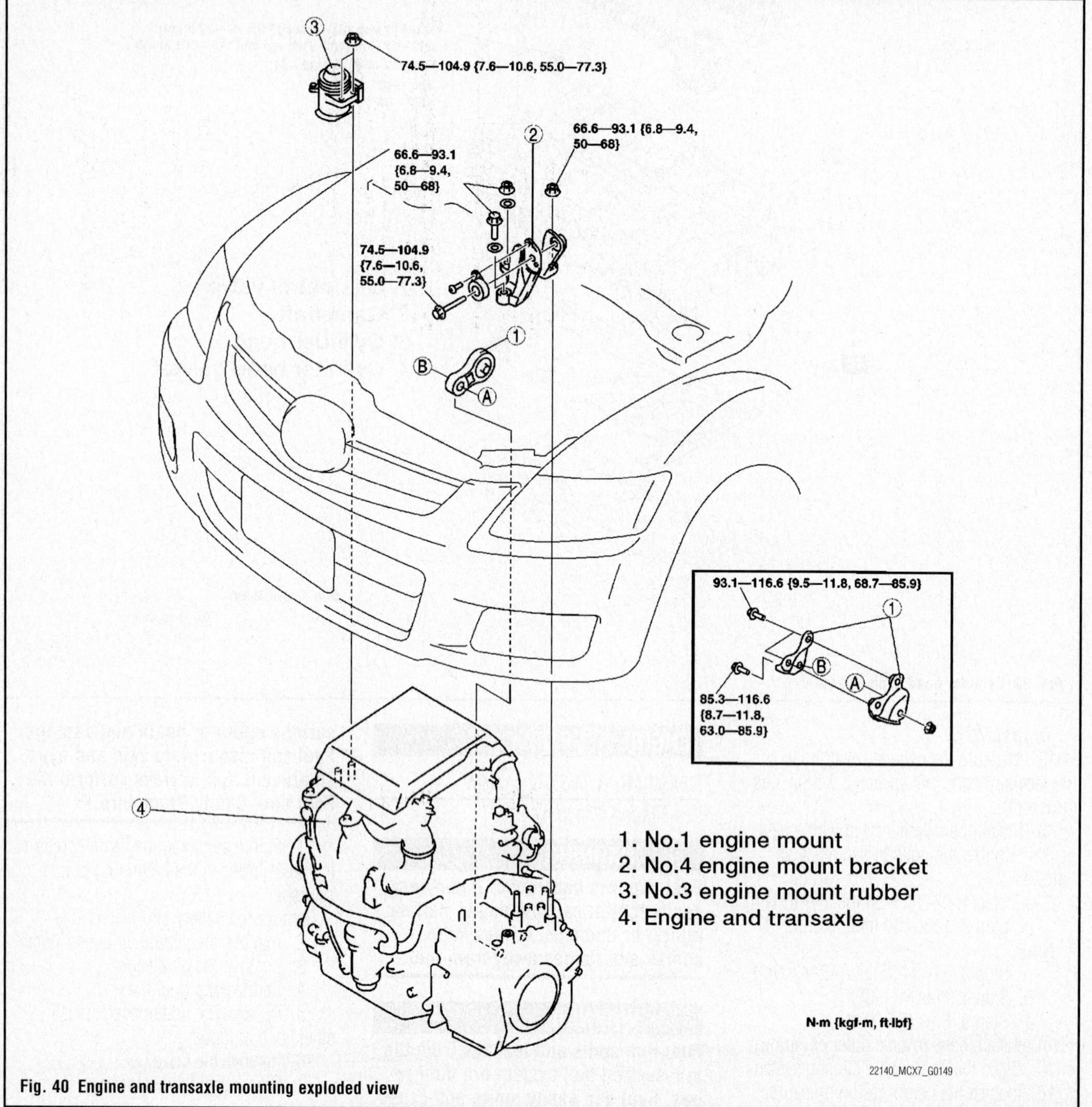

Fig. 40 Engine and transaxle mounting exploded view

- Front crossmember mounting rubber (rear)
- Engine and transaxle

To install:

23. Install in the reverse order of removal.
24. Start the engine, and inspect and adjust the following:
    - Air bleeding
    - Front wheel alignment
    - Bleed the air from the cooling system.
    - Runout and contact on pulley and belt
    - Leakage of engine oil, engine coolant, automatic transmission fluid, or fuel.
    - Ignition timing, idle speed and idle mixture (CO and HC)
    - Engine accessories operation
25. Perform a road test and verify that there is no abnormal vibration or noise.

EXHAUST MANIFOLD

REMOVAL & INSTALLATION

See Figure 42.

✱✱ WARNING

A hot engine and exhaust system can cause severe burns. Turn off the engine and wait until they are cool before removing the exhaust system.

1. Before servicing the vehicle, refer to the precautions in the beginning of this section.
2. Disconnect the negative battery cable.
3. Remove the charge air cooler duct.
4. Remove the charge air cooler cover.

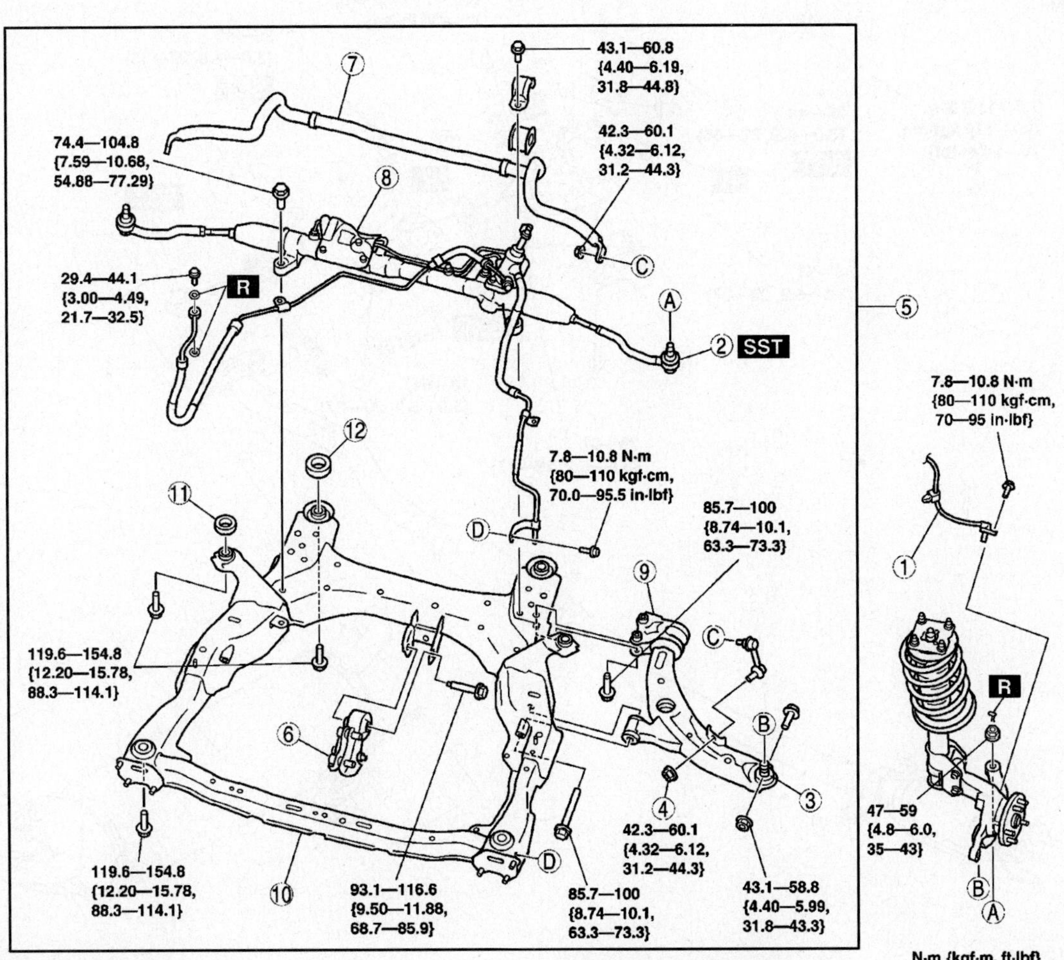

Fig. 41 Front crossmember exploded view

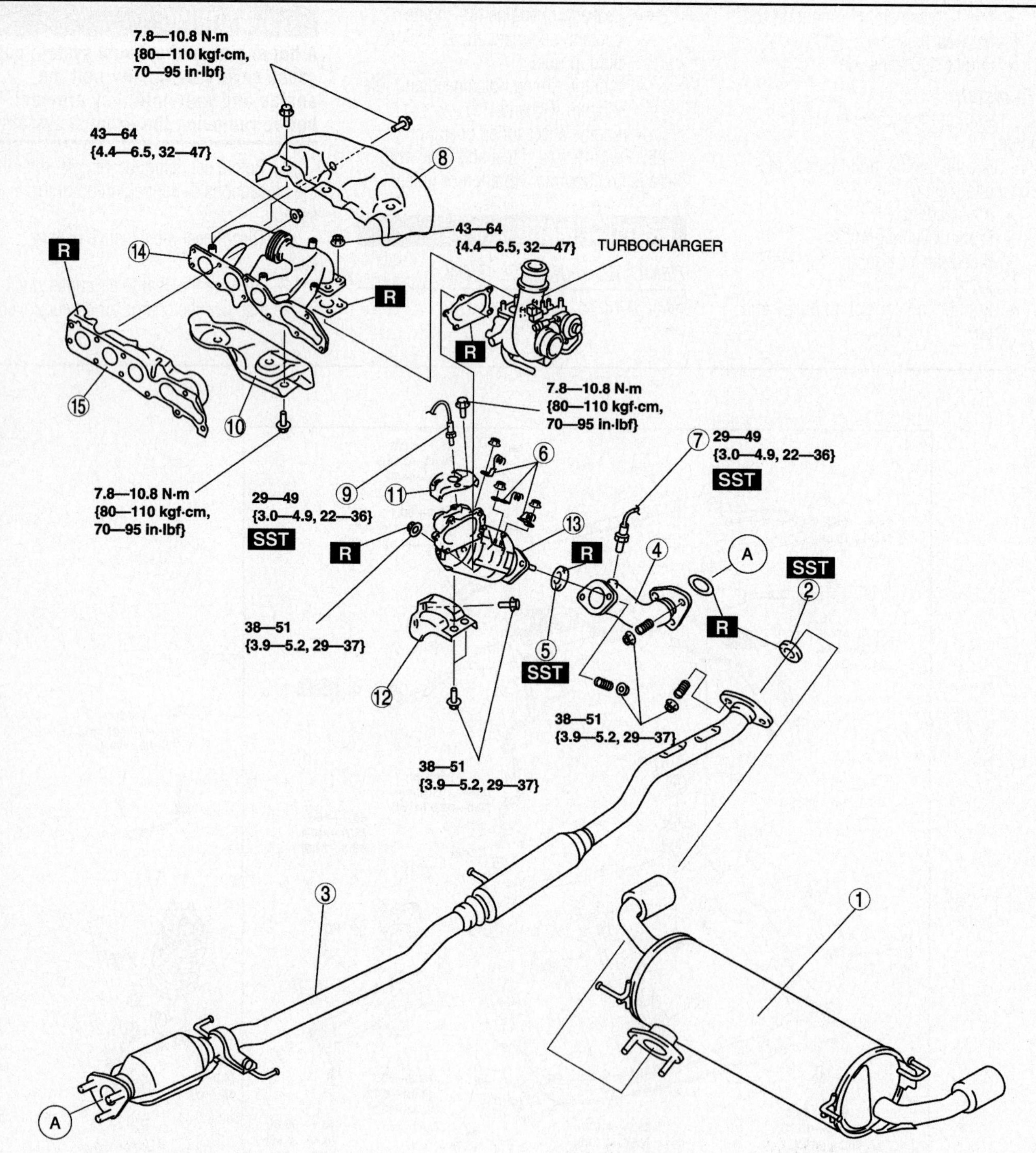

1. Main silencer
2. Seal ring (middle pipe side)
3. Middle pipe
4. Front pipe
5. Seal ring (WU-TWC side)
6. Clip
7. Rear HO2S
8. Exhaust manifold insulator (Upper)
9. Front HO2S
10. Exhaust manifold insulator (Lower)
11. WU-TWC insulator
12. WU-TWC bracket
13. WU-TWC
14. Exhaust manifold
15. Exhaust manifold gasket

N·m {kgf·m, ft·lbf}

22140_MCX7_G0151

Fig. 42 Exhaust system exploded view

5. Remove the charge air cooler.
6. Remove the main silencer
7. Remove the seal ring (middle pipe side)
8. Remove the middle pipe
9. Remove the front pipe
10. Remove the seal ring (WU-TWC side)
11. Remove the clip
12. Remove the rear HO2S
13. Remove the exhaust manifold insulator (Upper)
14. Remove the front HO2S
15. Remove the exhaust manifold insulator (Lower)
16. Remove the WU-TWC insulator
17. Remove the WU-TWC bracket
18. Remove the WU-TWC
19. Remove the exhaust manifold
20. Remove the exhaust manifold gasket
21. Install in the reverse order of removal.

FLEXPLATE

REMOVAL & INSTALLATION

See Figures 43 through 45.

1. Before servicing the vehicle, refer to the precautions in the beginning of this section.
2. Remove the transaxle.
3. Install SST 49 E011 1A0 or equivalent as shown to hold the flexplate from turning.
4. Remove the bolts and remove the flexplate.

To install:

5. Remove the sealant from the bolt holes in the crankshaft and from the drive plate mounting bolts.
6. Install the drive plate.
7. Install the backing plate.
8. Set the SST 49 E011 1A0 or equivalent against the flexplate as shown.
9. Tighten the flexplate bolts in sequence and in two or three passes to 80–82 ft. lbs. (108–116 Nm).
10. Install the transaxle.

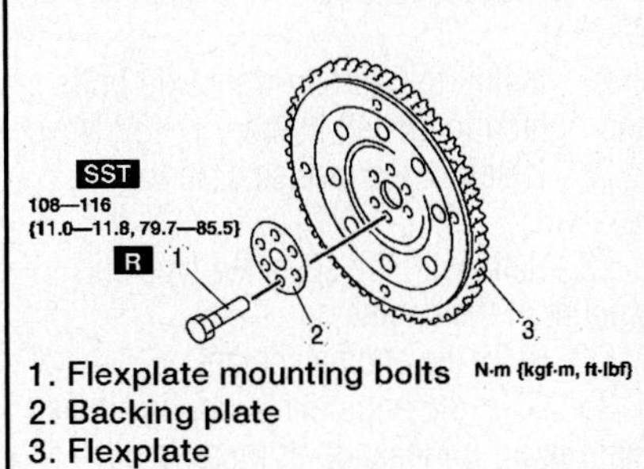

Fig. 43 Flexplate exploded view

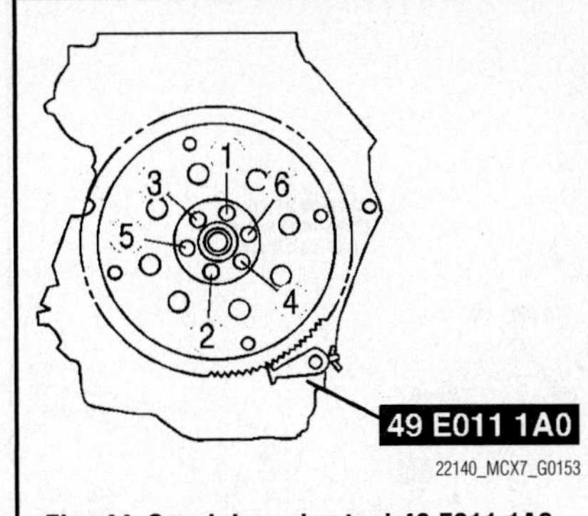

Fig. 44 Special service tool 49 E011 1A0 in place for bolt removal

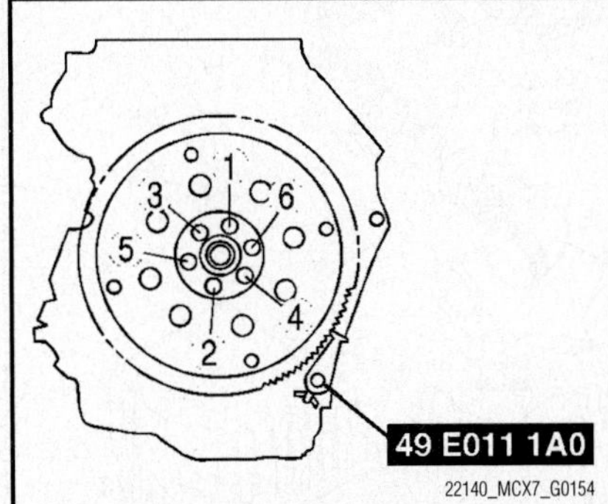

Fig. 45 Special service tool 49 E011 1A0 in place for bolt installation

INTAKE MANIFOLD

REMOVAL & INSTALLATION

See Figures 46 and 47.

**** WARNING**

A hot engine and intake air system can cause severe burns. Turn off the engine and wait until they are cool before removing the intake air system.

**** WARNING**

Fuel vapor is hazardous. It can easily ignite, causing serious injury and damage. Always keep sparks and flames away from fuel.

**** WARNING**

Fuel line spills and leakage are dangerous. Fuel can ignite and cause serious injuries or death and damage. Fuel can also irritate skin and eyes.

1. Before servicing the vehicle, refer to the precautions in the beginning of this section.
2. Disconnect the negative battery cable.
3. Drain the cooling system.
4. Remove the charge air cooler duct.
5. Remove the battery and battery tray.
6. Remove the undercover.
7. Remove the Mass Airflow sensor (MAF).
8. Remove the fresh air duct.
9. Remove the air cleaner cover and case.
10. Remove the resonance chamber.
11. Remove the charge air cooler.
12. Remove the air bypass valve.
13. Remove the air hose and duct.
14. Remove the throttle body.
15. Remove the vacuum chamber.
16. Remove the variable swirl solenoid valve.
17. Remove the fuel delivery pipe cover.
18. Disconnect the quick connector connected to the intake manifold.
19. Remove the EGR pipe.
20. Disconnect the OCV connector.
21. Disconnect the PSP switch connector.
22. Remove the oil level gauge pipe.
23. Remove the splash shield (RF).
24. Remove the drive belt.
25. Set the power steering oil pump out of the way.
26. Disconnect the fuel pressure sensor connector.
27. Disconnect the vacuum hose connected between the intake manifold and the master cylinder from the intake manifold.
28. Disconnect the MAP sensor connector.
29. Remove the intake manifold installation bolts.
30. Disconnect the evaporative hose connected between the intake manifold and the PCV valve from the intake manifold.
31. Remove the intake manifold.
32. Install in the reverse order of removal. Tighten the intake manifold bolts to 13–16 ft. lbs. (17–23 Nm).

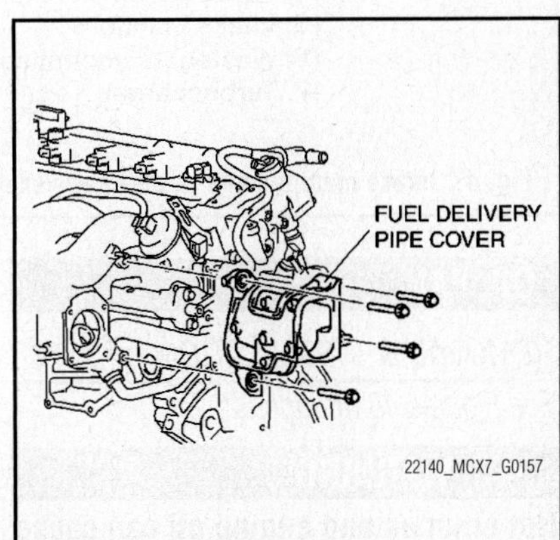

Fig. 46 Fuel delivery pipe cover

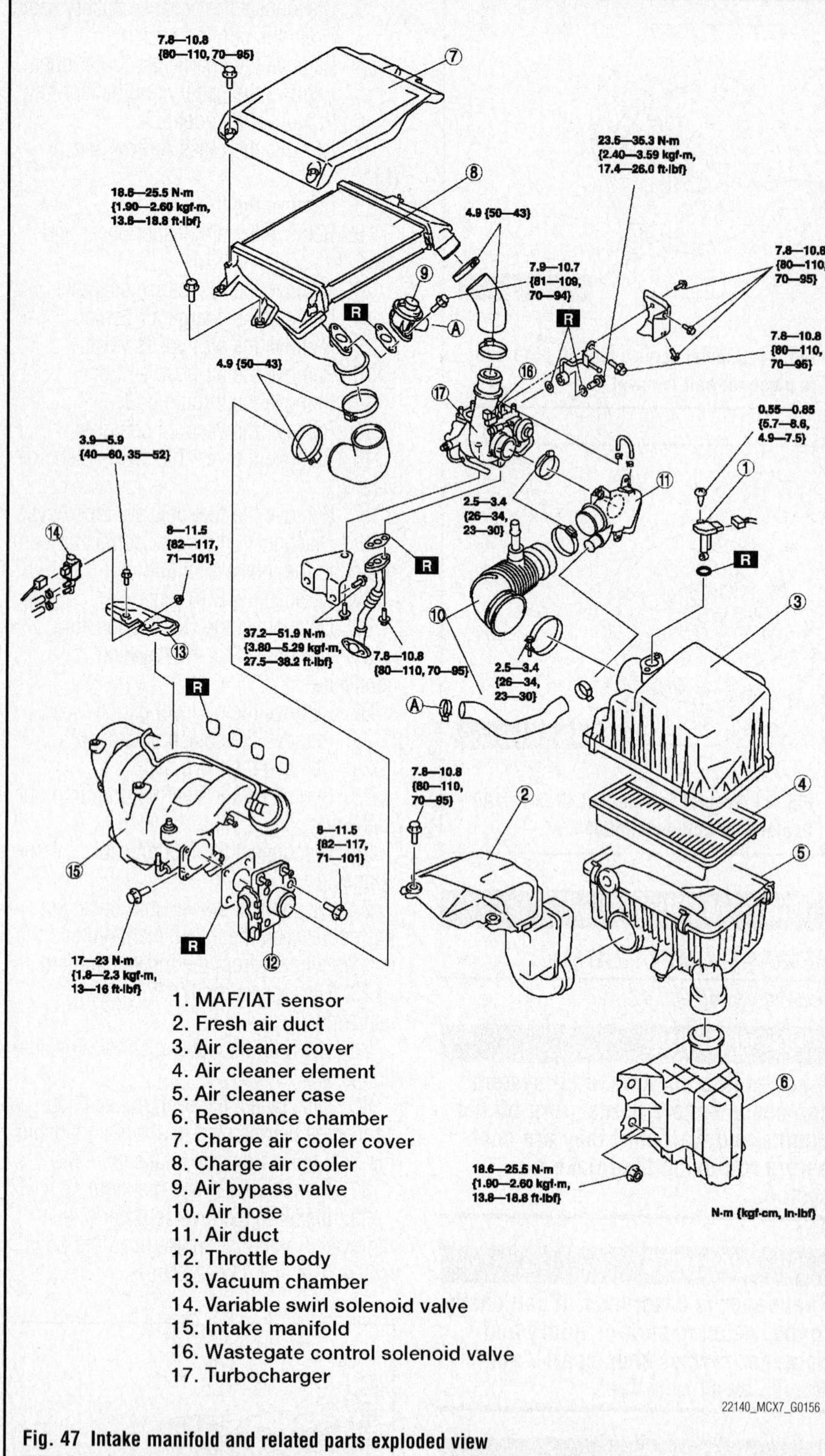

Fig. 47 Intake manifold and related parts exploded view

OIL PAN

REMOVAL & INSTALLATION

See Figures 48 through 52.

✲✲ WARNING

Hot engines and engine oil can cause severe burns. Turn off the engine and wait until it and the engine oil have cooled.

✲✲ WARNING

A vehicle that is lifted but not securely supported on safety stands is dangerous. It can slip or fall, causing death or serious injury. Never work around or under a lifted vehicle if it is not securely supported on safety stands.

✲✲ WARNING

Continuous exposure to USED engine oil has caused skin cancer in laboratory mice. Protect your skin by washing with soap and water immediately after working with engine oil.

1. Before servicing the vehicle, refer to the precautions in the beginning of this section.
2. Relieve the fuel system pressure.
3. Remove the battery and battery tray.
4. Remove the undercover.
5. Remove the splash shield (RH).
6. Drain the engine oil.
7. Drain the engine coolant.
8. Remove the charge air cooler, air cleaner and fresh air duct component, and air hose.
9. Disconnect the quick release connector on the high pressure fuel pump.
10. Remove the high pressure fuel pump.
11. Remove the ignition coils.
12. Loosen the water pump pulley bolts before removing the drive belt.
13. Remove the drive belt.
14. Remove the P/S oil pump with hose and pipe still connected. Position the P/S oil pump out of the way.
15. Remove the crankshaft position (CKP) sensor.
16. Remove the engine front cover.
17. Remove the oil dipstick tube.
18. Remove the oil pan.

To install:

19. Apply silicone sealant to the oil pan along the inside of the bolt holes as shown in the figure.
20. Install the oil pan. Tighten the bolts in sequence to 13–16 ft. lbs. (17–23 Nm).
21. Install the oil pan-transaxle bolts and tighten to 28–38 ft. lbs. (37–52 Nm).
22. Install in the reverse order of removal.
23. Refill with the specified type and amount of the engine oil.
24. Refill the engine coolant.
25. Start the engine and confirm that there is no oil leakage. If there is oil leakage, repair or replace the applicable part.
26. Inspect the oil level.
27. Inspect for engine coolant leakage.

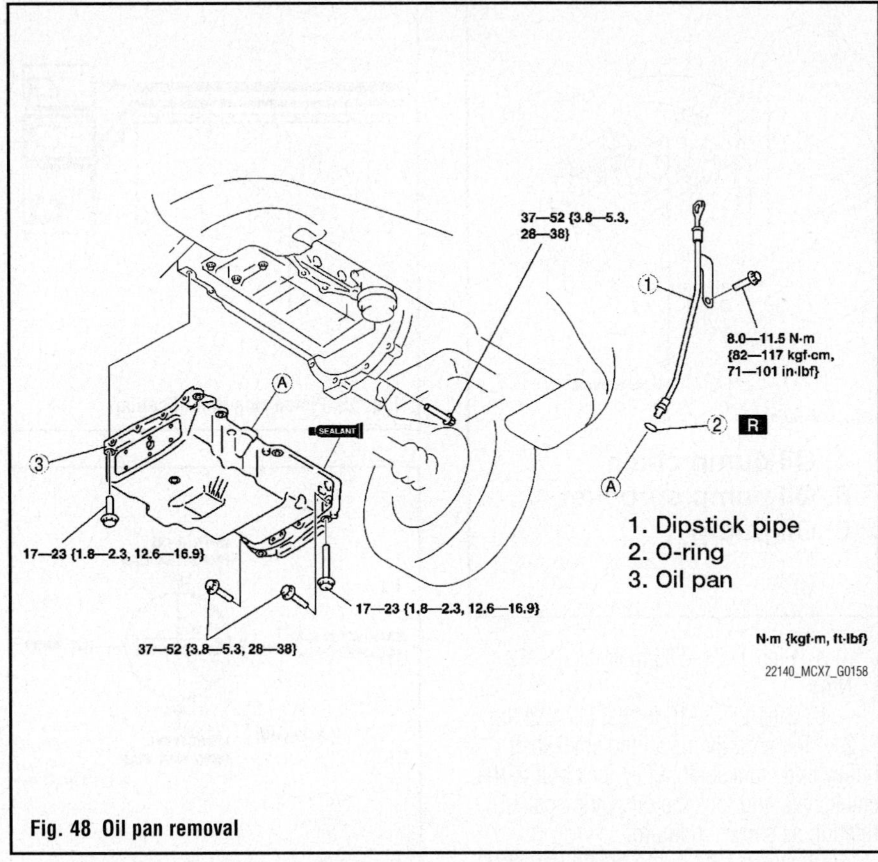

Fig. 48 Oil pan removal

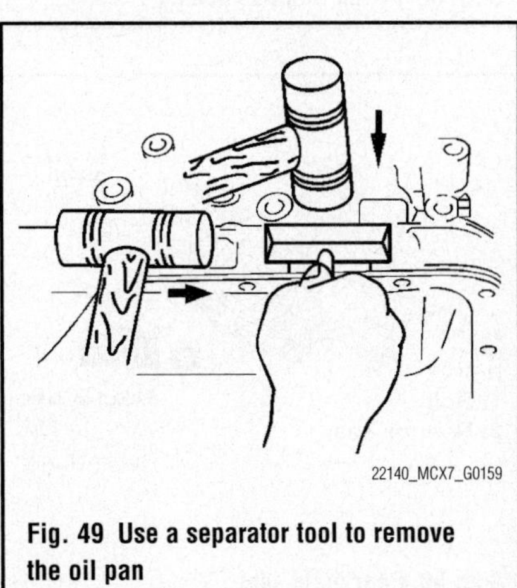

Fig. 49 Use a separator tool to remove the oil pan

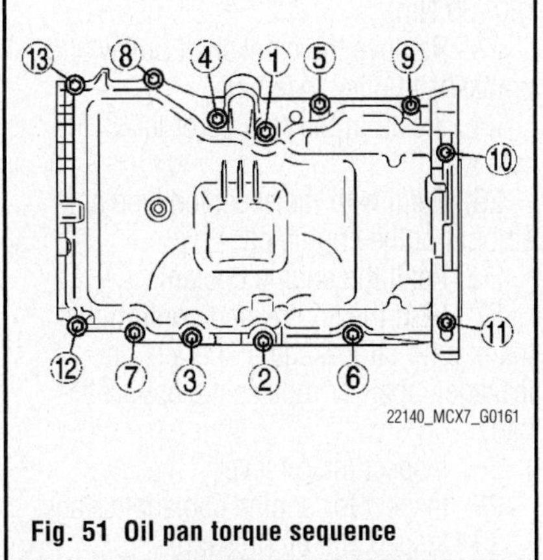

Fig. 51 Oil pan torque sequence

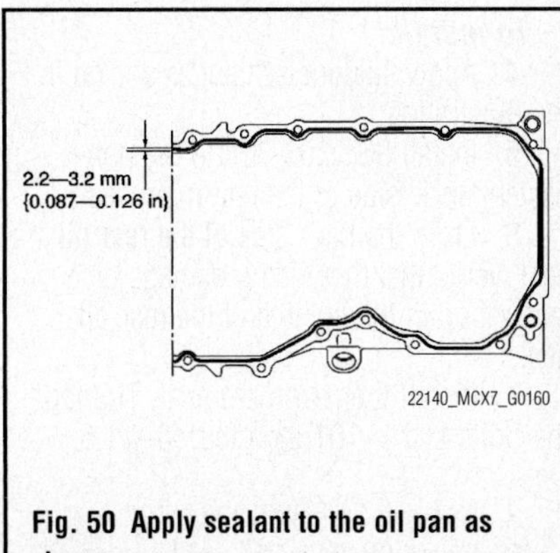

Fig. 50 Apply sealant to the oil pan as shown

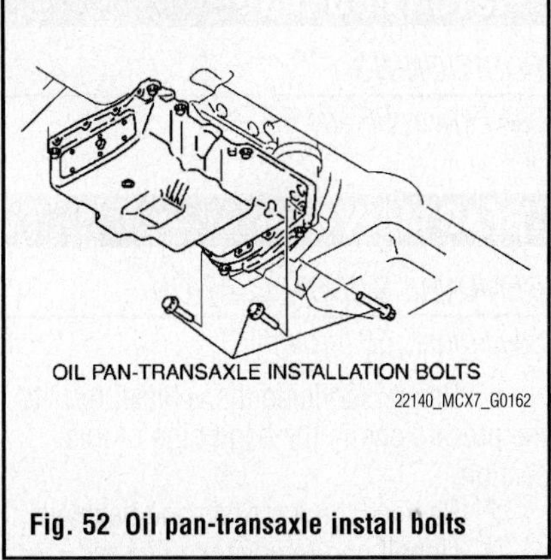

Fig. 52 Oil pan-transaxle install bolts

OIL PUMP

REMOVAL & INSTALLATION

See Figures 53 through 55.

WARNING

Hot engines and engine oil can cause severe burns. Turn off the engine and wait until it and the engine oil have cooled.

WARNING

A vehicle that is lifted but not securely supported on safety stands is dangerous. It can slip or fall, causing death or serious injury. Never work around or under a lifted vehicle if it is not securely supported on safety stands.

WARNING

Continuous exposure to USED engine oil has caused skin cancer in laboratory mice. Protect your skin by washing with soap and water immediately after working with engine oil.

1. Before servicing the vehicle, refer to the precautions in the beginning of this section.
2. Remove the battery and battery tray.
3. Remove the undercover.
4. Remove the splash shield (RH).
5. Drain the engine oil.
6. Drain the engine coolant.
7. Remove the charge air cooler, air cleaner and fresh air duct component, and air hose.
8. Disconnect the quick release connector on the high pressure fuel pump.
9. Remove the high pressure fuel pump.
10. Remove the ignition coils.
11. Loosen the water pump pulley bolts before removing the drive belt.
12. Remove the drive belt.
13. Remove the P/S oil pump with hose and pipe still connected. Position the P/S oil pump out of the way.
14. Remove the crankshaft position (CKP) sensor.
15. Remove the engine front cover.
16. Remove the oil pan.
17. Temporarily install the crankshaft pulley and crankshaft pulley lock bolt to the crankshaft, and lock the oil pump against rotation as shown in figure.
18. Temporarily install the crankshaft pulley and crankshaft pulley lock bolt to the

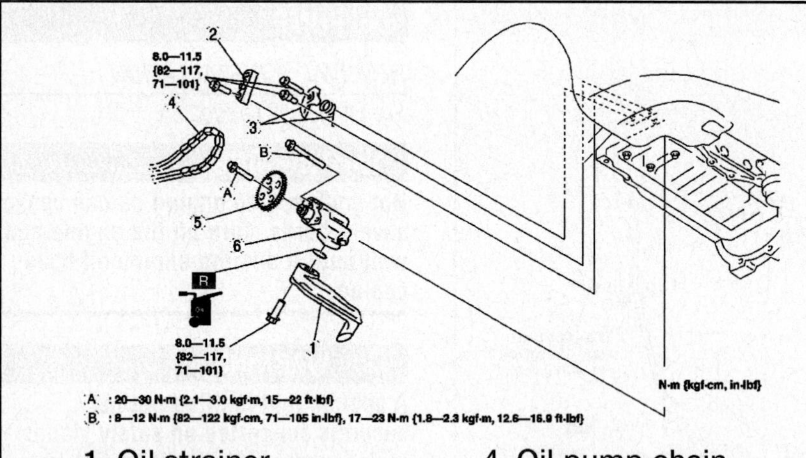

1. Oil strainer
2. Oil pump chain guide
3. Oil pump chain tensioner
4. Oil pump chain
5. Oil pump sprocket
6. Oil pump

22140_MCX7_G0163

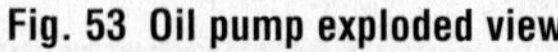

Fig. 53 Oil pump exploded view

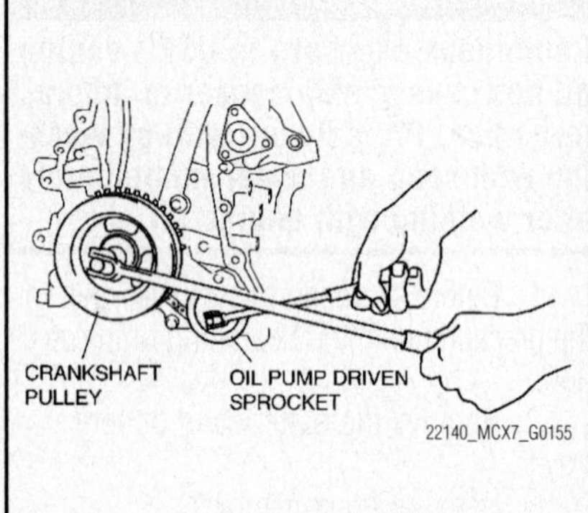

22140_MCX7_G0155

Fig. 54 Oil pump driven sprocket removal and installation

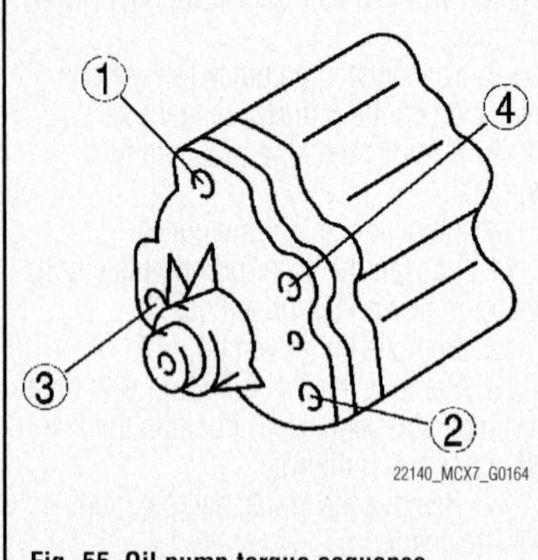

22140_MCX7_G0164

Fig. 55 Oil pump torque sequence

crankshaft, and lock the oil pump against rotation as shown in figure.

19. Remove the oil pump.

To install:

20. Install the oil pump. Tighten the oil pump bolts as follows:

a. Step 1: 71–105 inch lbs. (8–12 Nm)

b. Step 2: 13–16 ft. lbs. (17–23 Nm)

21. Temporarily install the crankshaft pulley and crankshaft pulley lock bolt to the crankshaft, and lock the oil pump against rotation as shown in figure.

22. Install the oil pump driven sprocket and tighten the bolt to 15–22 ft. lbs. (20–30 Nm).

23. Remove the crankshaft pulley and crankshaft pulley lock bolt.

24. Install in the reverse order of removal.

25. Refill with the specified type and amount of the engine oil.

26. Refill the engine coolant.

27. Start the engine and confirm that there is no oil leakage. If there is oil leakage, repair or replace the applicable part.

28. Inspect the oil level.

29. Inspect for engine coolant leakage.

30. Inspect the oil pressure.

PISTON AND RING

POSITIONING

See Figures 56 and 57.

REAR MAIN SEAL

REMOVAL & INSTALLATION

See Figures 58 through 61.

1. Before servicing the vehicle, refer to the precautions in the beginning of this section.

2. Remove the transaxle and flexplate.

3. Unbolt and remove the oil seal.

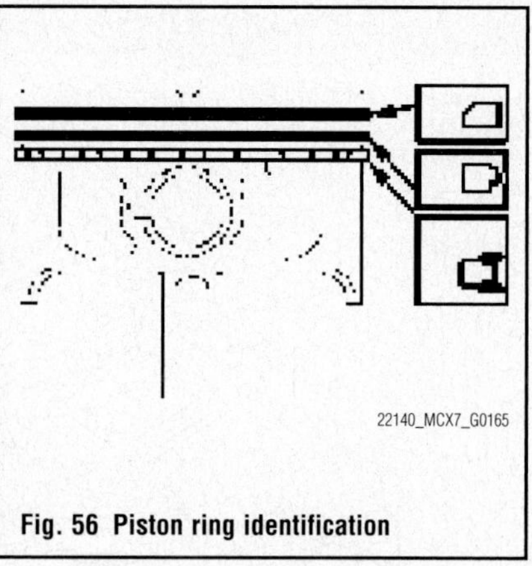

22140_MCX7_G0165

Fig. 56 Piston ring identification

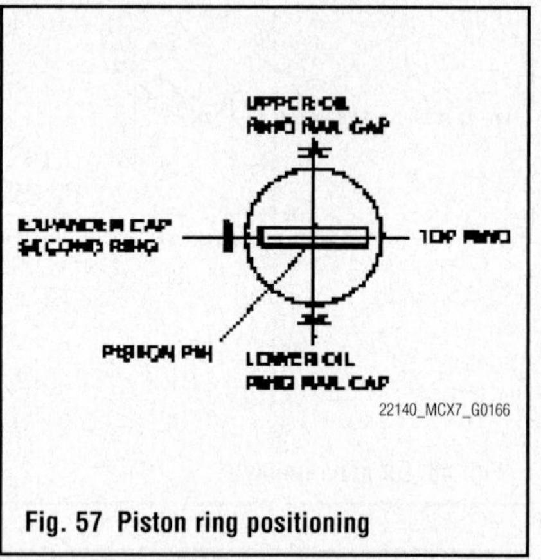

22140_MCX7_G0166

Fig. 57 Piston ring positioning

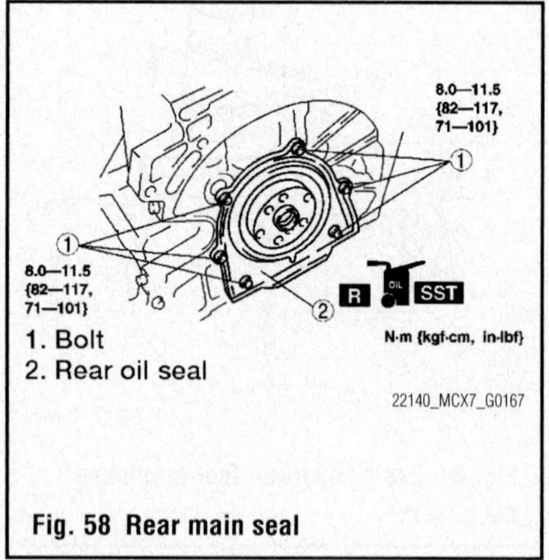

22140_MCX7_G0167

Fig. 58 Rear main seal

To install:

4. Apply silicone sealant as shown in the illustration.

5. Install SST 303-328 to the non-woven fabric side of the rear main seal.

6. From the back side of the rear oil seal, verify that there is no damage or separation in the lip area of the rear oil seal.

7. Install the rear main seal. Tighten the bolts to 71–101 inch lbs. (8–11.5 Nm).

8. Remove SST 303-328.

9. Install the flexplate and transaxle.

SEALANT
22140_MCX7_G0168

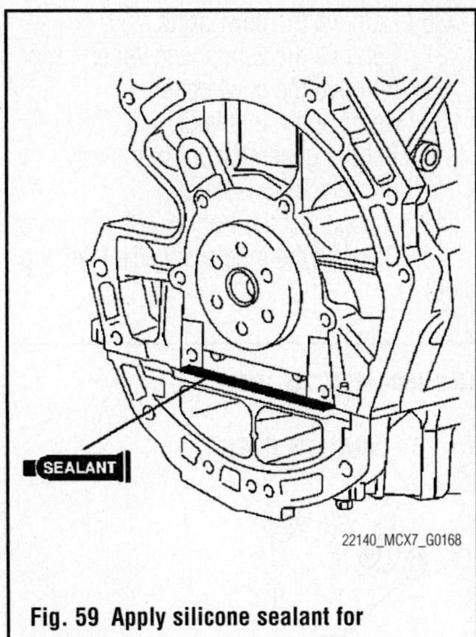

Fig. 59 Apply silicone sealant for installation

NON-WOVEN FABRIC
303-328
(49 UN30 3328)
22140_MCX7_G0169

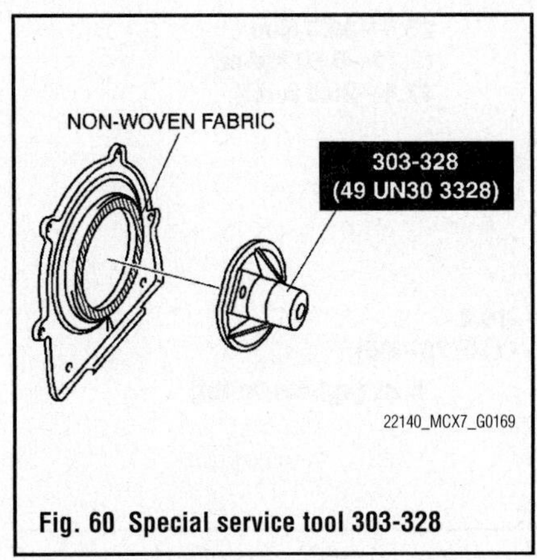

Fig. 60 Special service tool 303-328

3
5
1
6
2
4
22140_MCX7_G0170

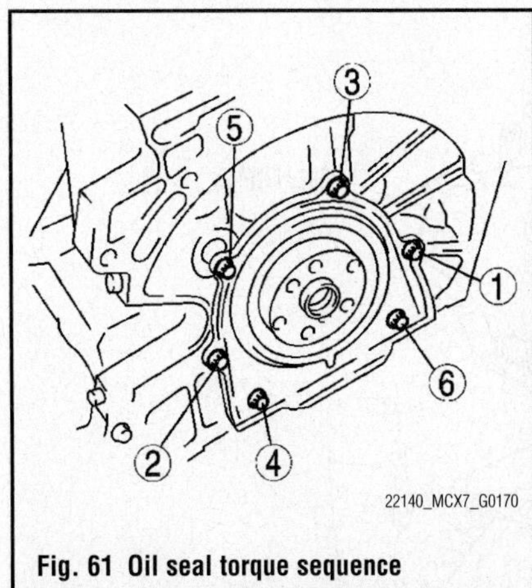

Fig. 61 Oil seal torque sequence

TURBOCHARGER

REMOVAL & INSTALLATION

See Figures 62 through 66.

⁂ WARNING

A hot engine and intake air system can cause severe burns. Turn off the engine and wait until they are cool before removing the intake air system.

⁂ WARNING

Fuel vapor is hazardous. It can easily ignite, causing serious injury and damage. Always keep sparks and flames away from fuel.

⁂ WARNING

Fuel line spills and leakage are dangerous. Fuel can ignite and cause serious injuries or death and damage. Fuel can also irritate skin and eyes.

1. Before servicing the vehicle, refer to the precautions in the beginning of this section.

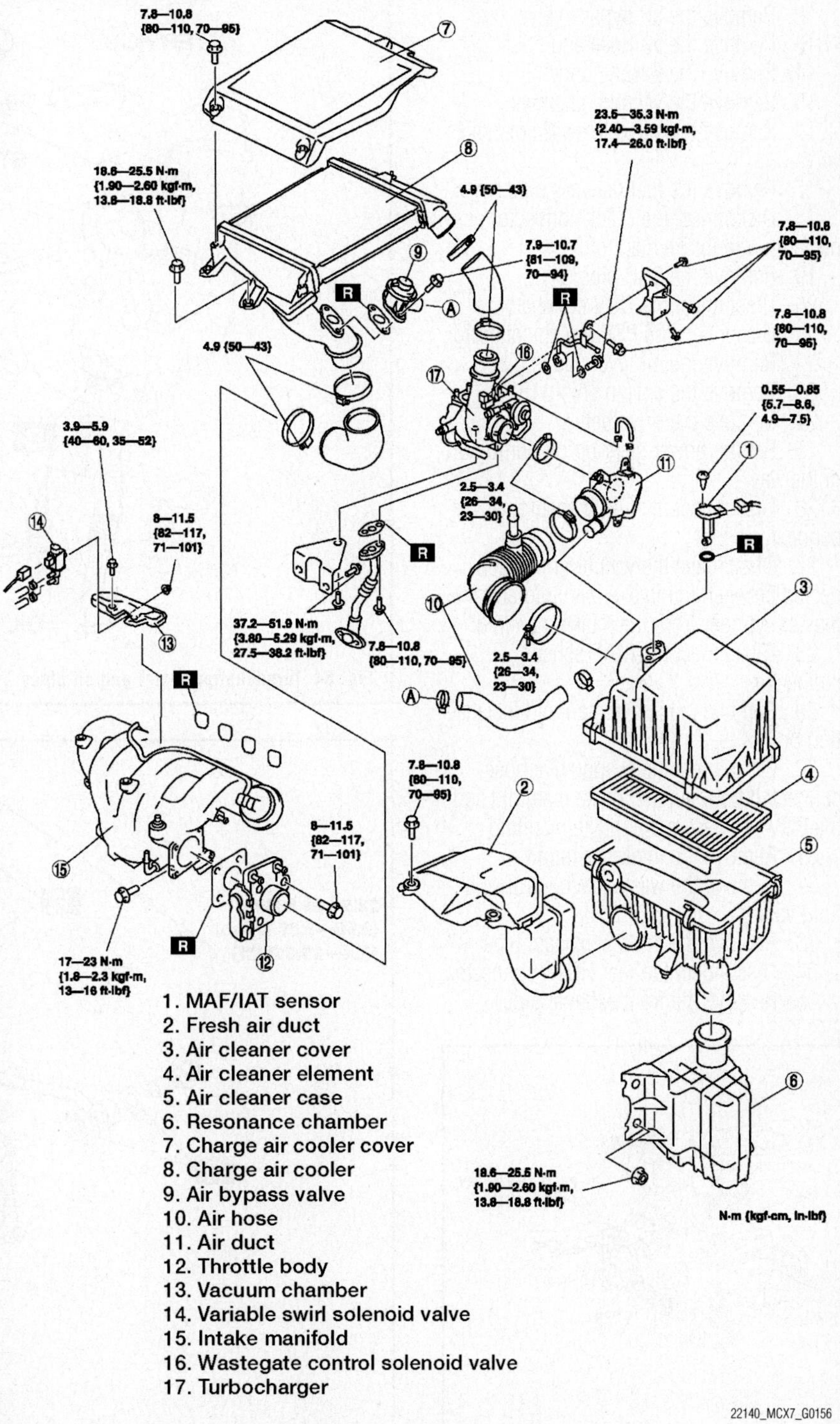

7.8—10.8
{80—110, 70—95}
18.6—25.5 N·m
{1.90—2.60 kgf·m,
13.8—18.8 ft·lbf}
4.9 {50—43}
7.9—10.7
{81—109,
70—94}
23.5—35.3 N·m
{2.40—3.59 kgf·m,
17.4—26.0 ft·lbf}
7.8—10.8
{80—110,
70—95}
0.55—0.85
{5.7—8.6,
4.9—7.5}
3.9—5.9
{40—60, 35—52}
8—11.5
{82—117,
71—101}
2.5—3.4
{26—34,
23—30}
37.2—51.9 N·m
{3.80—5.29 kgf·m,
27.5—38.2 ft·lbf}
17—23 N·m
{1.8—2.3 kgf·m,
13—16 ft·lbf}
R
A
1. MAF/IAT sensor
2. Fresh air duct
3. Air cleaner cover
4. Air cleaner element
5. Air cleaner case
6. Resonance chamber
7. Charge air cooler cover
8. Charge air cooler
9. Air bypass valve
10. Air hose
11. Air duct
12. Throttle body
13. Vacuum chamber
14. Variable swirl solenoid valve
15. Intake manifold
16. Wastegate control solenoid valve
17. Turbocharger
N·m {kgf·cm, in·lbf}
22140_MCX7_G0156

Fig. 62 Turbocharger and related parts exploded view

2. Disconnect the negative battery cable.
3. Drain the cooling system.
4. Remove the charge air cooler duct.
5. Remove the battery and battery tray.
6. Remove the undercover.
7. Remove the Mass Airflow sensor (MAF).
8. Remove the fresh air duct.
9. Remove the air cleaner cover and case.
10. Remove the resonance chamber.
11. Remove the charge air cooler.
12. Remove the air bypass valve.
13. Remove the air hose and duct.
14. Remove the throttle body.
15. Remove the vacuum chamber.
16. Remove the variable swirl solenoid valve.
17. Remove the fuel delivery pipe cover.
18. Disconnect the quick connector connected to the intake manifold.
19. Remove the EGR pipe.
20. Disconnect the OCV connector.
21. Disconnect the PSP switch connector.
22. Remove the oil level gauge pipe.
23. Remove the splash shield (RF).
24. Remove the drive belt.
25. Set the power steering oil pump out of the way.
26. Disconnect the fuel pressure sensor connector.
27. Disconnect the vacuum hose connected between the intake manifold and the master cylinder from the intake manifold.
28. Disconnect the MAP sensor connector.
29. Remove the intake manifold installation bolts.
30. Disconnect the evaporative hose connected between the intake manifold and the PCV valve from the intake manifold.
31. Remove the intake manifold.
32. Remove the wastegate control solenoid valve.
33. Remove the middle exhaust pipe
34. Disconnect the rear HO2S connector.
35. Remove the front exhaust pipe.

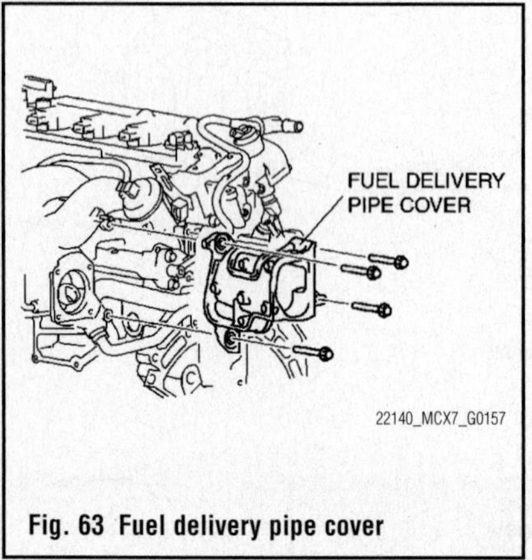

Fig. 63 Fuel delivery pipe cover

✻✻ CAUTION

When removing the cowl grille, a part or tool may hit the edge of the windshield and could damage it. Protect the windshield by covering it with a clean rag to prevent damage to the windshield.

36. Remove the cowl grille.
37. Remove the windshield wiper motor.
38. Remove the cowl panel.
39. Remove the insulator
40. Set the generator out of the way.
41. Remove the insulator
42. Disconnect the front HO2S connector.
43. Remove the insulator (WU-TWC top side).

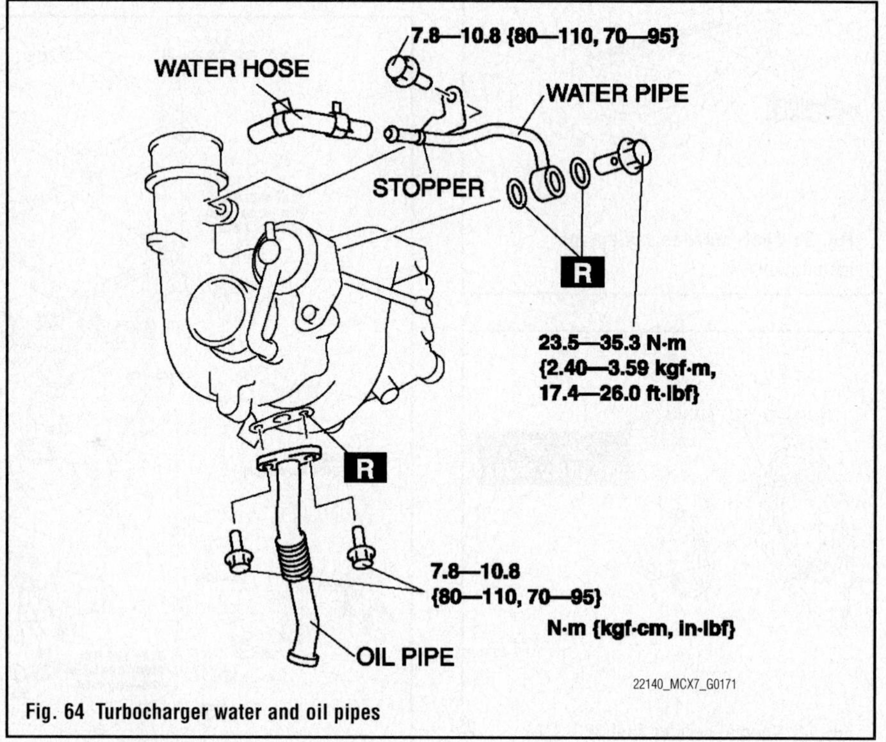

Fig. 64 Turbocharger water and oil pipes

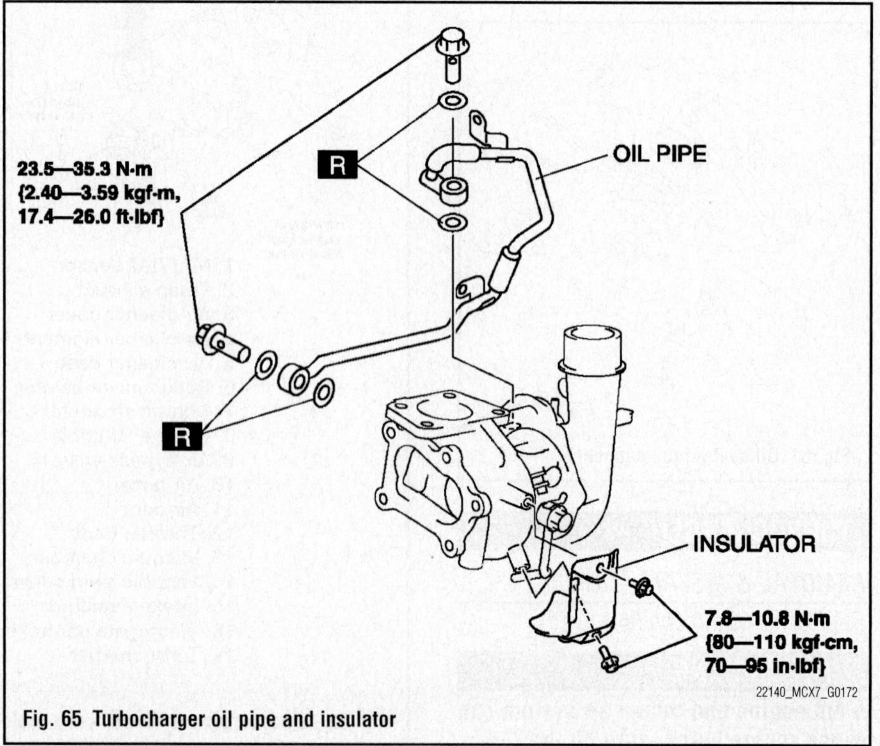

Fig. 65 Turbocharger oil pipe and insulator

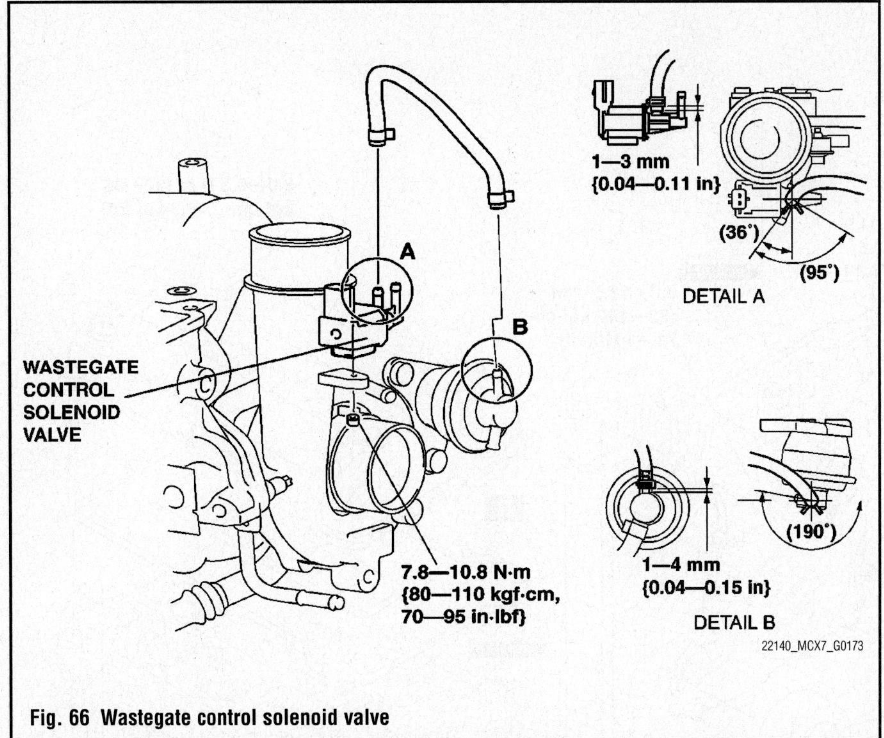

Fig. 66 Wastegate control solenoid valve

44. Remove the vacuum hose (Brake master back side).
45. Remove the front HO2S.
46. Remove the WU-TWC.
47. Remove the oil pipes, water pipe, and insulator.
48. Remove the turbocharger.
49. Install in the reverse order of removal. Tighten the intake manifold bolts to 13–16 ft. lbs. (17–23 Nm).

TIMING CHAIN AND SPROCKETS

REMOVAL & INSTALLATION

See Figures 67 through 80.

1. Before servicing the vehicle, refer to the precautions in the beginning of this section.

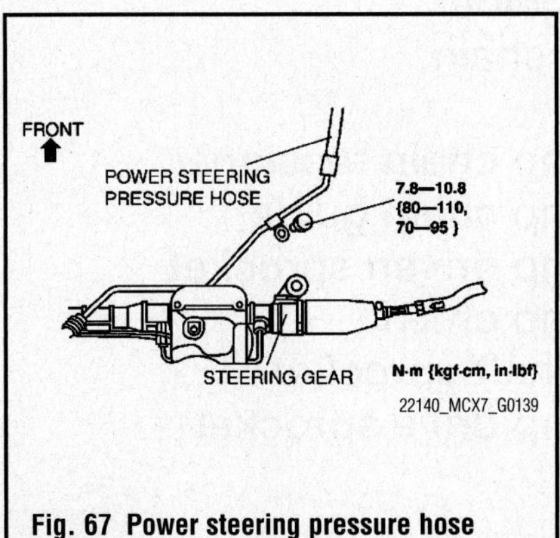

Fig. 67 Power steering pressure hose bracket

2. Disconnect the negative battery cable.
3. Remove the undercover.
4. Remove the splash shield.
5. Remove the high pressure fuel pump.

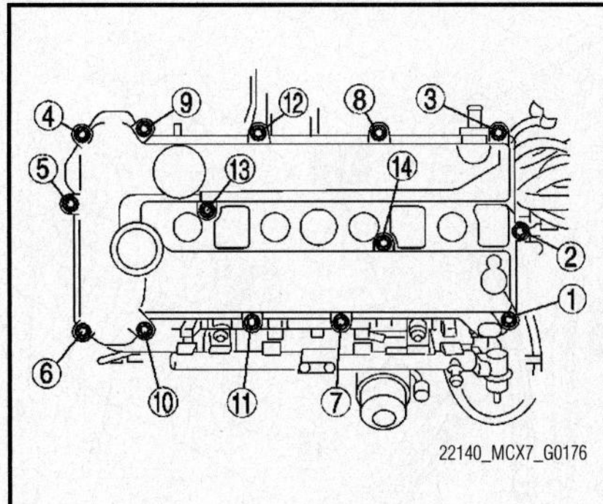

Fig. 68 Valve cover bolt removal sequence

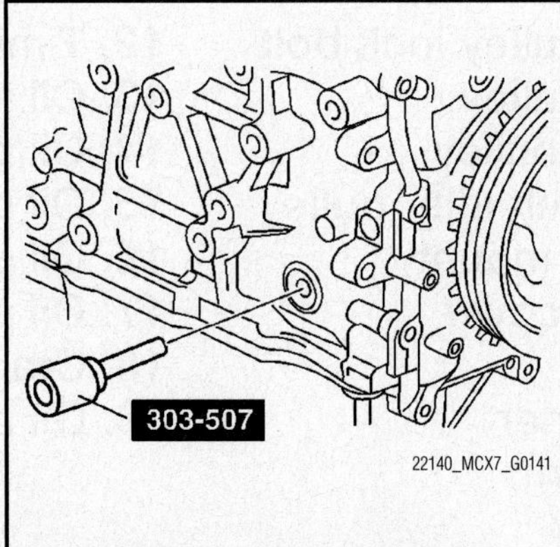

Fig. 69 Special service tool 303-507

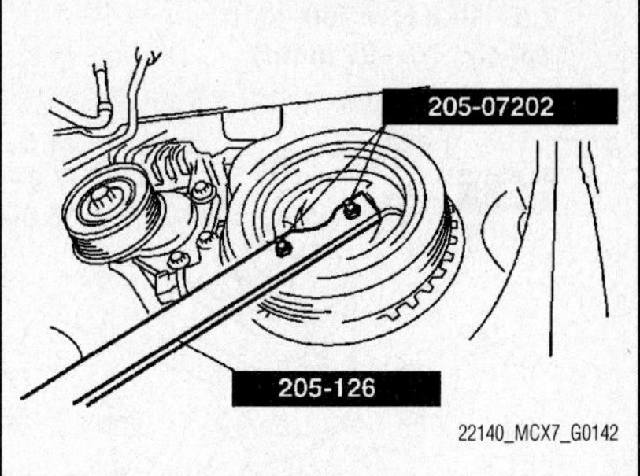

Fig. 70 Special service tools 205-07202 and 205-126

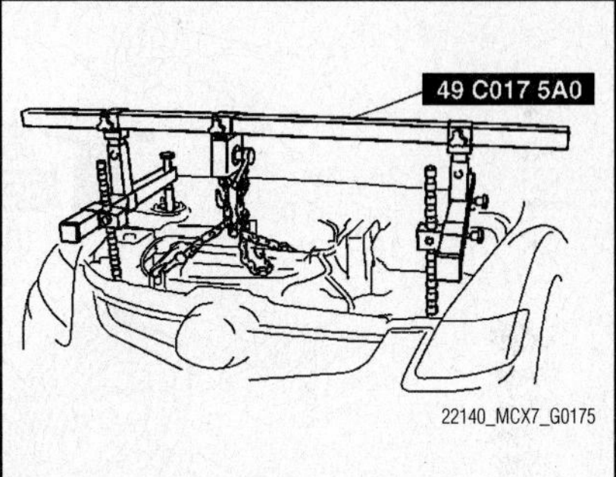

Fig. 71 Engine support fixture 49 C017 5A0

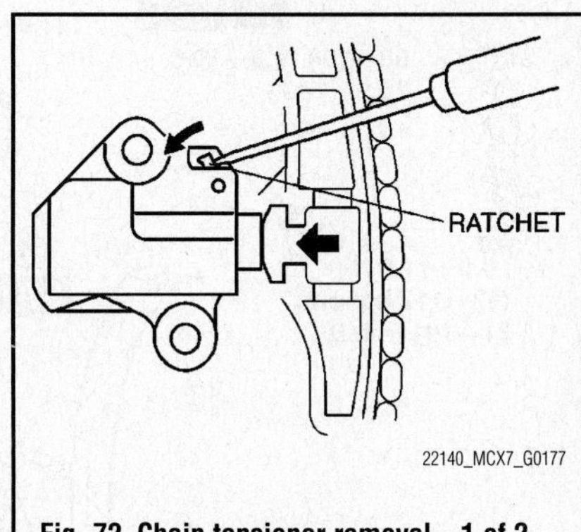

Fig. 72 Chain tensioner removal—1 of 2

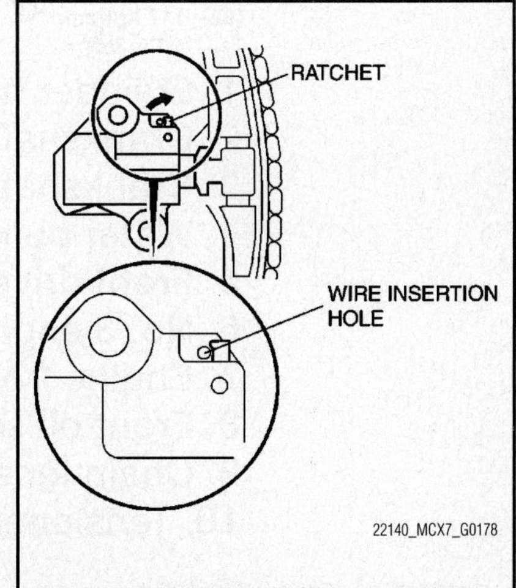

Fig. 73 Chain tensioner removal—2 of 2

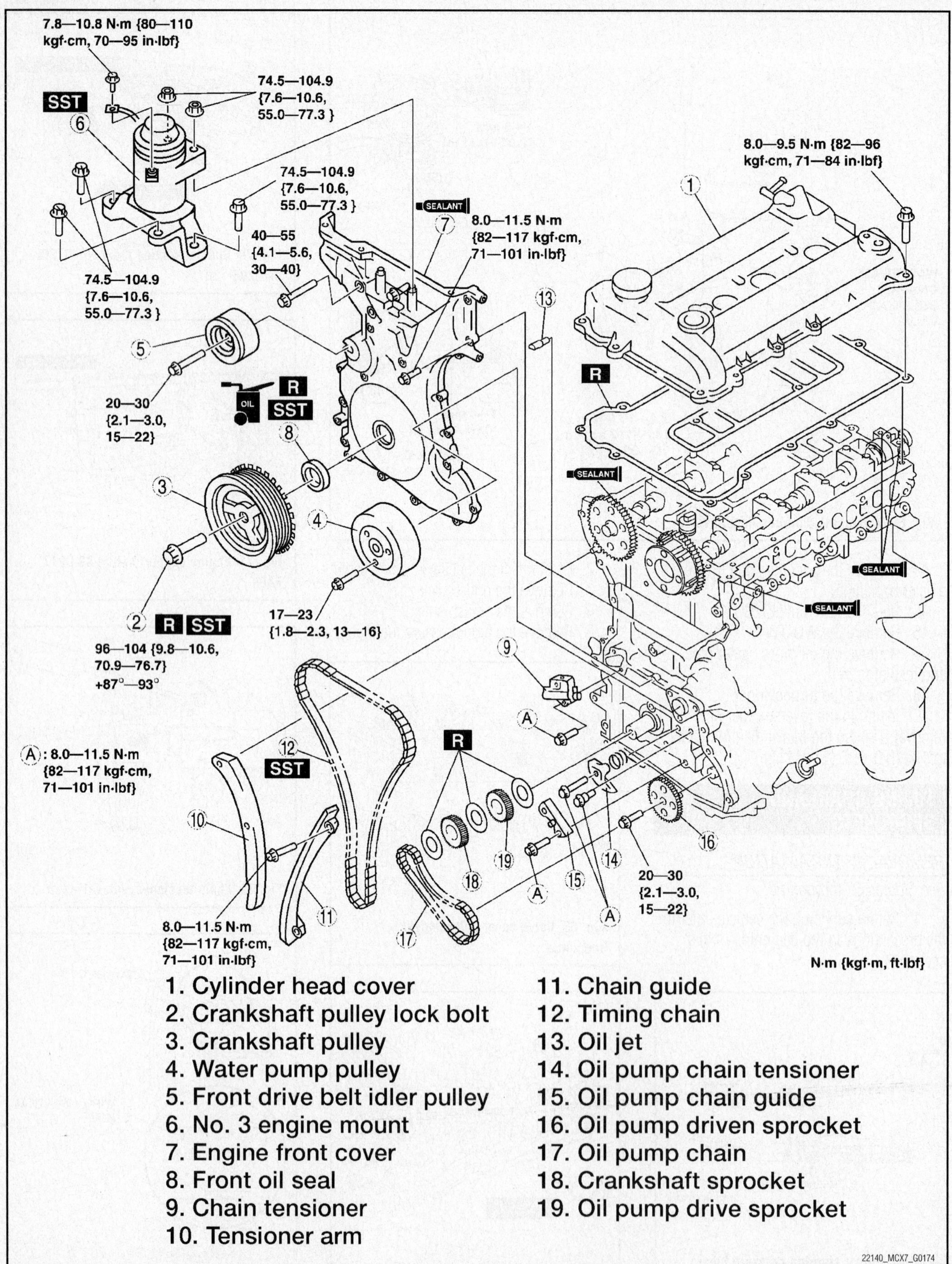

Fig. 74 Timing chain exploded view

6. Remove the ignition coils.
7. Disconnect the wiring harness.
8. Remove the ventilation hose.
9. Remove the drive belt.
10. Remove the crankshaft position (CKP) sensor.
11. Remove the P/S oil pump with the hoses and pipes still connected.

➡Position and secure the P/S oil pump out of the way with a rope or wire.

12. Remove the bracket shown in the figure, and place the power steering pressure hose outside of the vehicle.
13. Remove the valve cover.
14. Rotate the crankshaft in the direction of the engine rotation and remove the cylinder block lower blind plug when the No. 1 cylinder is at the point prior to top dead center (TDC) of compression, then install SST 303-507.
15. Rotate the crankshaft in the direction of the engine rotation so that the No.1 piston is at TDC of the compression stroke. (Until the crank weight contacts SST and stops.)
16. Install SSTs 205-07202 and 205-126 to the crankshaft pulley and lock the crankshaft against rotation.
17. Remove the crankshaft pulley lock bolt and the crankshaft pulley.
18. Remove the water pump pulley.
19. Remove the front drive belt idler pulley.
20. Install engine support fixture 49 C017 5A0 or equivalent.
21. Remove the No. 3 engine mount
22. Remove the engine front cover
23. Remove the front oil seal
24. Re-set the chain tensioner as follows:
    a. Press the timing chain tensioner ratchet to the left using a thin flathead screwdriver (precision screwdriver) to unlock the plunger.
    b. Slowly press the plunger back in the direction shown in the figure while pressing the ratchet.
    c. Release the ratchet with the plunger still pressed down.
    d. Press-in the plunger until the ratchet position is as indicated in the figure, and then insert a wire to lock the plunger.
25. Remove the tensioner arm.
26. Remove the chain guide.
27. Remove the timing chain.
28. Remove the oil jet.
29. Remove the oil pump chain tensioner.
30. Remove the oil pump chain guide.
31. Remove the oil pump driven sprocket.
32. Remove the oil pump chain.
33. Remove the crankshaft sprocket.
34. Remove the oil pump drive sprocket and the oil pump chain.

To install:

35. Replace the oil pump drive sprocket and the oil pump chain.
36. Replace the crankshaft sprocket.
37. Replace the oil pump chain.
38. Replace the oil pump driven sprocket.
39. Replace the oil pump chain guide.

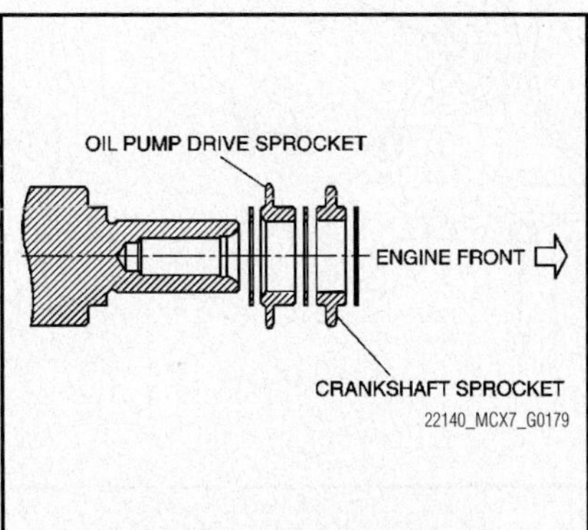

Fig. 75 Oil pump drive sprocket and crankshaft sprocket

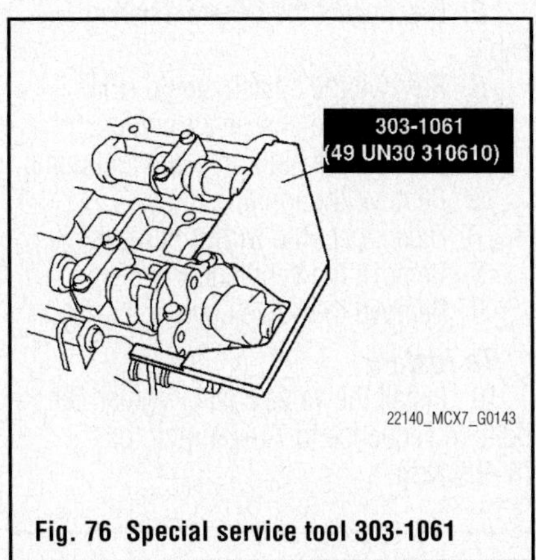

Fig. 76 Special service tool 303-1061

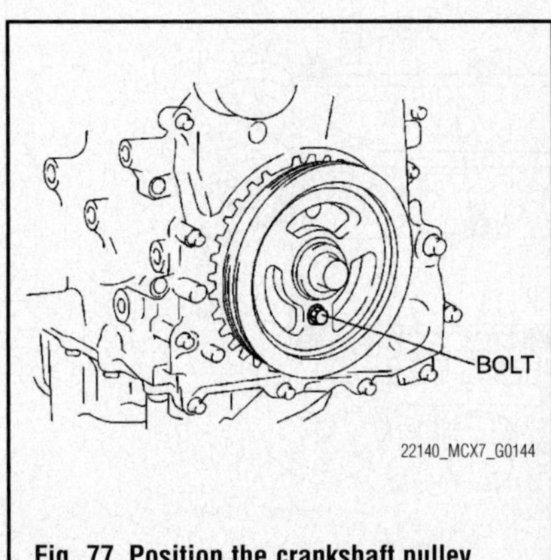

Fig. 77 Position the crankshaft pulley

40. Replace the oil pump chain tensioner.
41. Replace the oil jet.
42. Install the chain guide and the tensioner arm.
43. Install SST 303-1061 to the camshaft as shown in the figure.
44. Verify that cylinder No.1 is at TDC of the compression stroke. (Position crank weight contacts SST 303-507.)
45. Install the timing chain.
46. Remove the wire or paper clip from the chain tensioner piston and apply tension to the timing chain.

➡Install the engine front cover within 10 min of applying the silicone sealant.

47. Apply silicone sealant to the engine front cover as shown. Sealant is not needed in area C.
48. Install the front cover. Tighten the bolts in sequence as follows:
    - Bolts 1–18: 71–101 inch lbs. (8–11.5 Nm)
    - Bolts 19–22: 30–40 ft. lbs. (40–55 Nm)
49. Replace the front oil seal
50. Replace the engine front cover

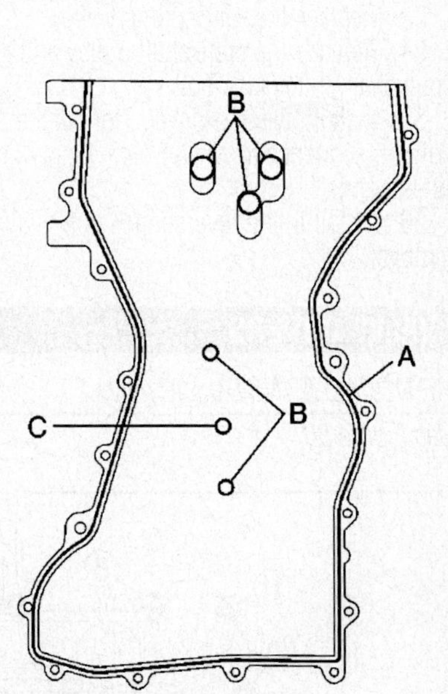

Fig. 78 Applying sealant to the front cover

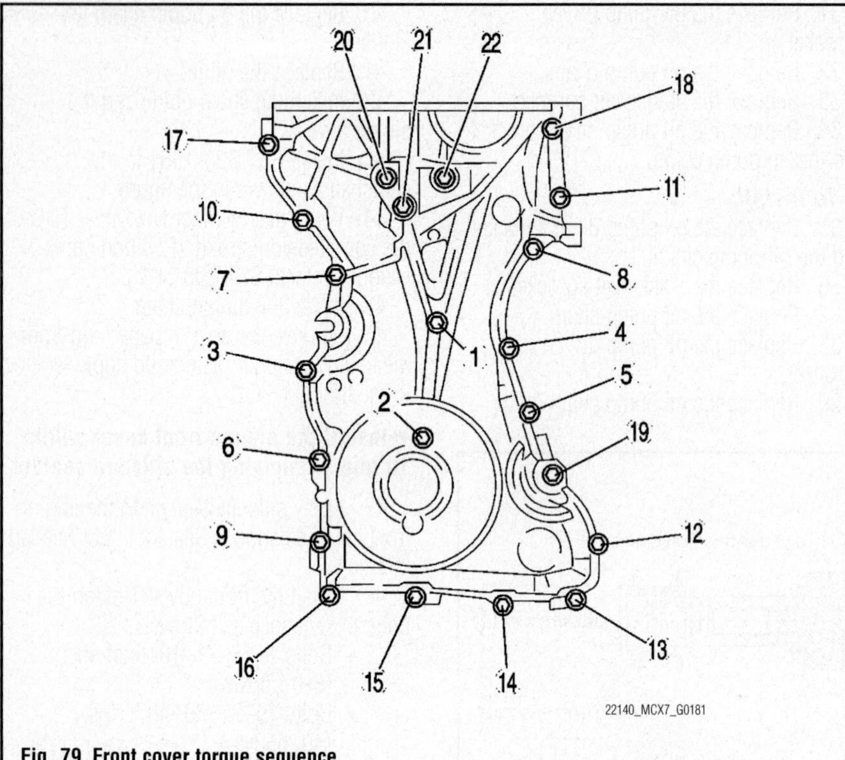

Fig. 79 Front cover torque sequence

51. Replace the No. 3 engine mount
52. Replace the front drive belt idler pulley.
53. Replace the water pump pulley.
54. Replace the crankshaft pulley and the crankshaft pulley lock bolt.
55. Install the valve cover. Tighten the bolts in sequence to 71–84 inch lbs. (8–9.5 Nm).
56. Install in the reverse order of removal.

VALVE COVERS

REMOVAL & INSTALLATION

See Figure 80.

1. Before servicing the vehicle, refer to the precautions in the beginning of this section.
2. Disconnect the negative battery cable.
3. Remove the splash shield (RH).
4. Remove the charge air cooler.
5. Remove the high pressure fuel pump.
6. Remove the ignition coils.
7. Disconnect the wiring harness.
8. Remove the ventilation hose.
9. Remove the valve cover.

To install:

10. Install the valve cover. Tighten the bolts in sequence to 71–84 inch lbs. (8–9.5 Nm).
11. Replace the ventilation hose.
12. Connect the wiring harness.
13. Replace the ignition coils.
14. Replace the high pressure fuel pump.
15. Replace the charge air cooler.
16. Replace the splash shield (RH).
17. Connect the negative battery cable.

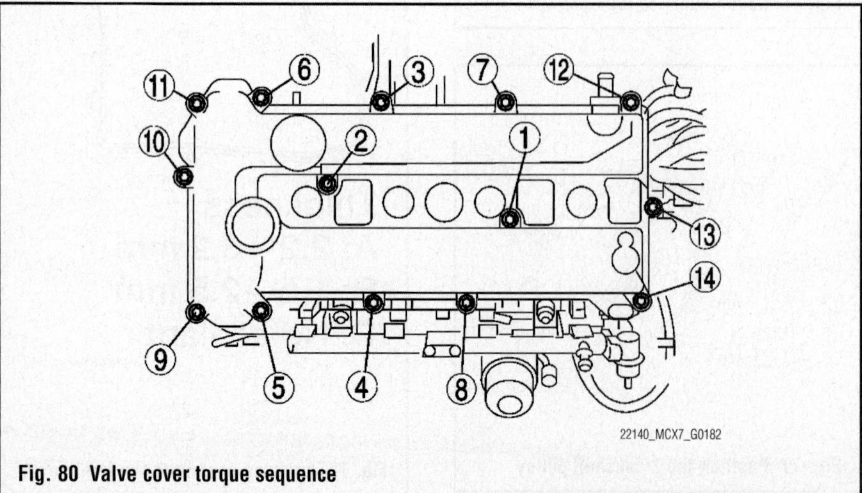

Fig. 80 Valve cover torque sequence

VALVE LASH

ADJUSTMENT

See Figure 81.

➡Check valve clearance with the engine cold.

1. Before servicing the vehicle, refer to the precautions in the beginning of this section.
2. Remove the valve cover.
3. Set the No. 1 cylinder to Top Dead Center (TDC) of the compression stroke.
4. Measure and note the valve clearances of location **A** shown in the illustration.
5. Rotate the crankshaft 1 full turn so that cylinder No. 4 is at TDC of the compression stroke.
6. Measure and note the valve clearances of location **B** shown in the illustration.
7. Specified valve clearance is:
    - Intake: 0.0087–0.0110 inch (0.22–0.28 mm)
    - Exhaust: 0.0110–0.0120 inch (0.27–0.33 mm).
8. Replace the tappets of any valve found outside the specified values and remeasure the valve clearance.

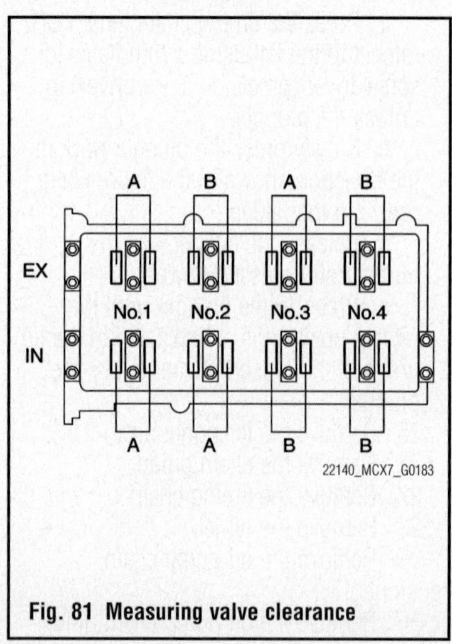

Fig. 81 Measuring valve clearance

ENGINE PERFORMANCE & EMISSION CONTROL

See Figures 82 and 83.

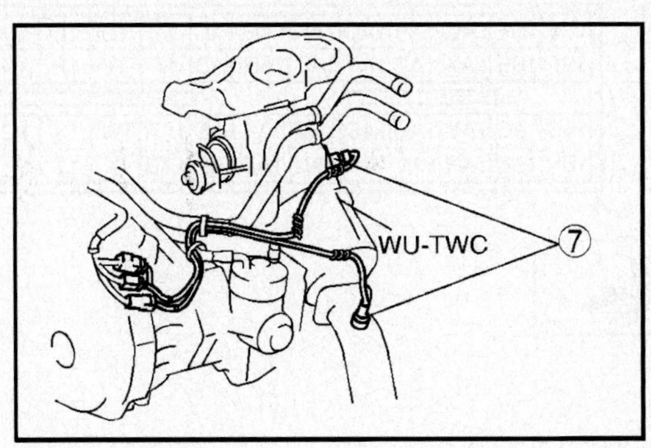

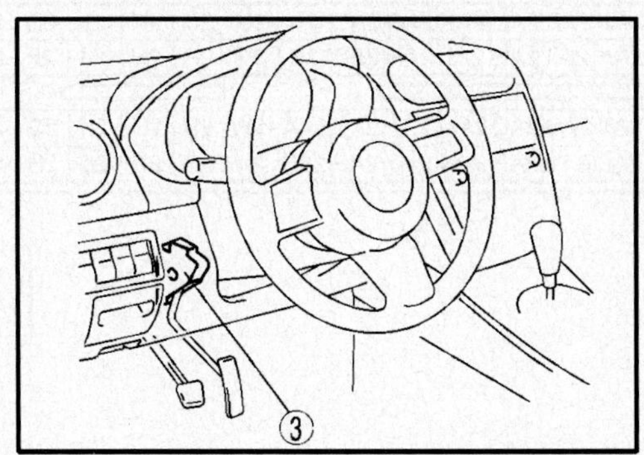

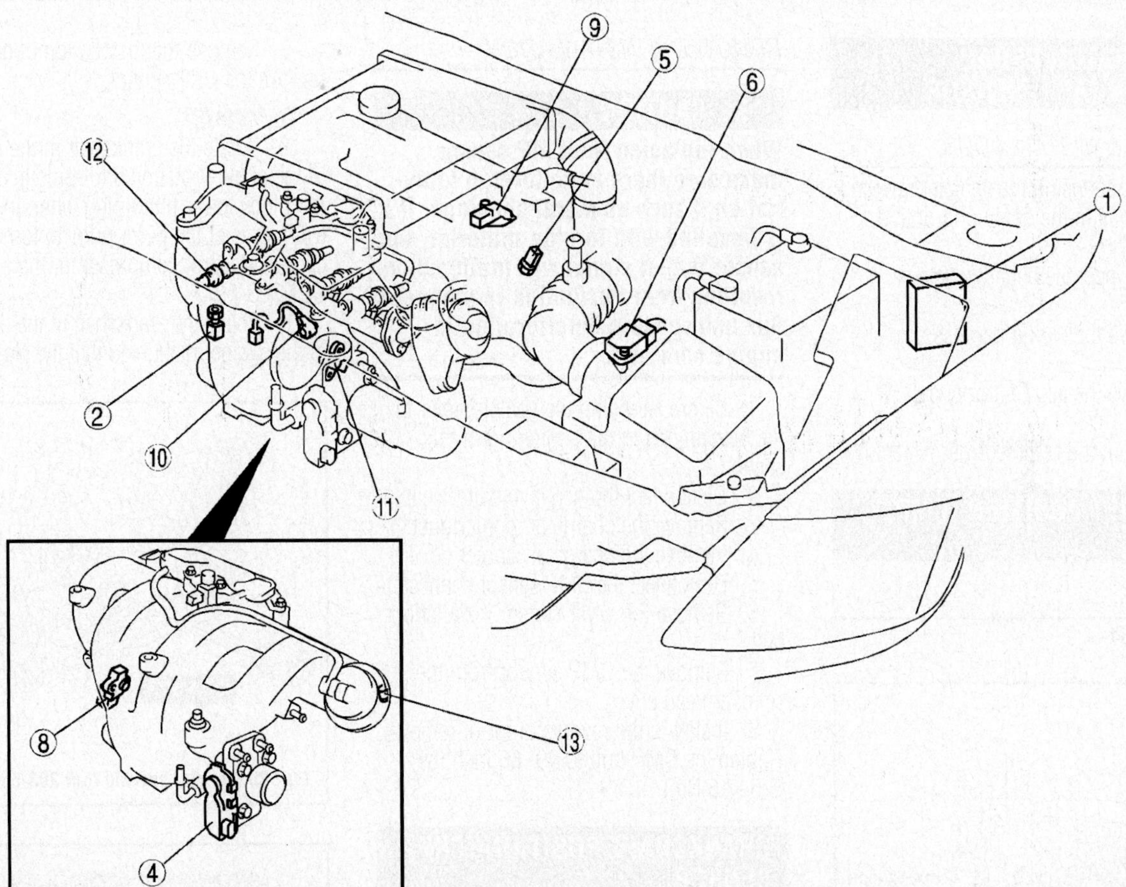

1. Powertrain control module
2. Power Steering pressure switch
3. Acclerator pedal position sensor
4. Throttle position sensor
5. Engine coolant temperature sensor
6. Mass air flow/Intake air temperature sensor
7. Heated oxygen sensor
8. Manifold absolute pressure/boost air temperature sensor
9. Camshaft position sensor
10. Knock Sensor
11. Crankshaft position sensor
12. Fuel pressure sensor
13. Variable swirl shutter valve switch

22140_MCX7_G0184

Fig. 82 Control system component locations

PCM
WIRING HARNESS-SIDE CONNECTOR

2BE	2BA	2AW	2AS	2AO	2AK	2AG	2AC	2Y	2U	2Q	2M	2I	2E	2A
2BF	2BB	2AX	2AT	2AP	2AL	2AH	2AD	2Z	2V	2R	2N	2J	2F	2B
2BG	2BC	2AY	2AU	2AQ	2AM	2AI	2AE	2AA	2W	2S	2O	2K	2G	2C
2BH	2BD	2AZ	2AV	2AR	2AN	2AJ	2AF	2AB	2X	2T	2P	2L	2H	2D

1BE	1BA	1AW	1AS	1AO	1AK	1AG	1AC	1Y	1U	1Q	1M	1I	1E	1A
1BF	1BB	1AX	1AT	1AP	1AL	1AH	1AD	1Z	1V	1R	1N	1J	1F	1B
1BG	1BC	1AY	1AU	1AQ	1AM	1AI	1AE	1AA	1W	1S	1O	1K	1G	1C
1BH	1BD	1AZ	1AV	1AR	1AN	1AJ	1AF	1AB	1X	1T	1P	1L	1H	1D

22140_MCX7_G0185

Fig. 83 PCM pin identification

ACCELERATOR PEDAL POSITION (APP) SENSOR

REMOVAL & INSTALLATION

1. Before servicing the vehicle, refer to the precautions in the beginning of this section.
2. Disconnect the negative battery cable.
3. Remove the wiring harness connector.
4. Remove the accelerator pedal assembly.
5. Install in the reverse order of removal.

CAMSHAFT POSITION (CMP) SENSOR

LOCATION

See Figure 84.

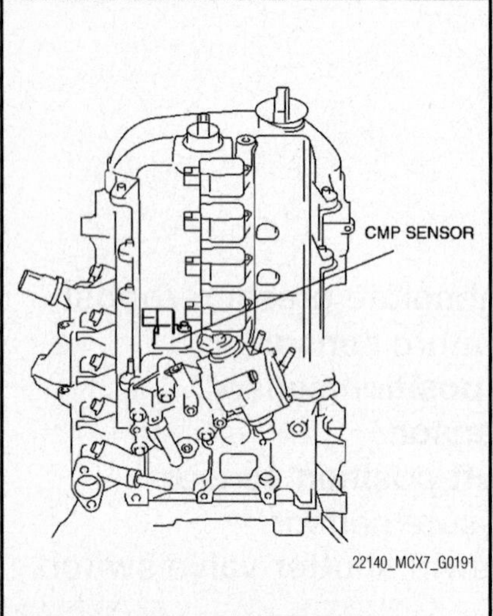

22140_MCX7_G0191

Fig. 84 Camshaft position sensor

REMOVAL & INSTALLATION

✱✱ CAUTION

When replacing the CMP sensor, make sure there is no foreign material on it such as metal shavings. If it is installed with foreign material, the sensor output signal will malfunction resulting from fluctuation in magnetic flux and cause a deterioration in engine control.

1. Before servicing the vehicle, refer to the precautions in the beginning of this section.
2. Disconnect the negative battery cable.
3. Remove the charge air cooler duct.
4. Remove the charge air cooler cover.
5. Disconnect the CMP sensor connector.
6. Remove the CMP sensor installation bolt.
7. Remove the CMP sensor from the cylinder head cover.
8. Install in the reverse order of removal. Tighten the CMP bolt to 49–66 inch lbs. (5.5–7.5 Nm).

CRANKSHAFT POSITION (CKP) SENSOR

REMOVAL & INSTALLATION

See Figures 85 and 86.

1. Before servicing the vehicle, refer to the precautions in the beginning of this section.
2. Disconnect the negative battery cable.
3. Remove the charge air cooler duct.
4. Remove the undercover.
5. Remove the splash shield (RH).
6. Disconnect the CKP sensor connector.
7. Remove the installation bolts to remove the CKP sensor.

To install:

8. Rotate the crankshaft in the direction of the engine rotation and remove the cylinder block lower blind plug when the No. 1 cylinder is at the point prior to top dead center (TDC) of compression, then install SST 303-507.
9. Rotate the crankshaft in the direction of the engine rotation so that the No.1 piston

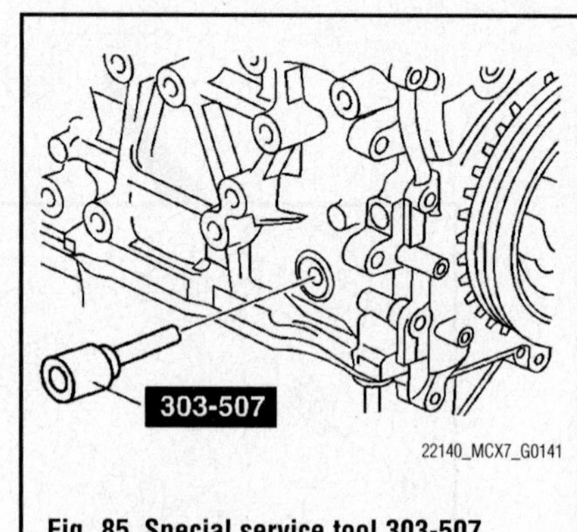

22140_MCX7_G0141

Fig. 85 Special service tool 303-507

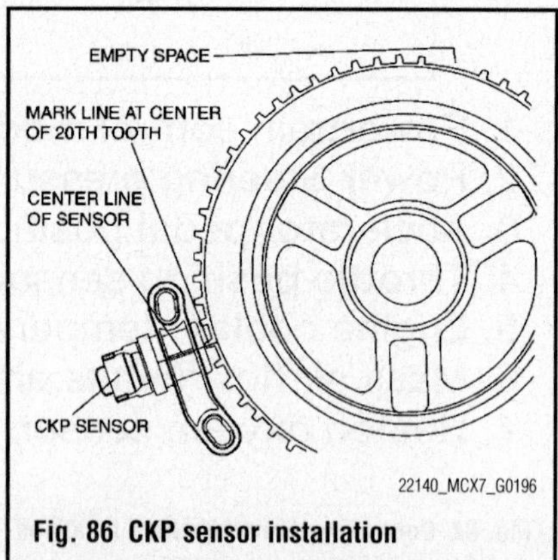

22140_MCX7_G0196

Fig. 86 CKP sensor installation

is at TDC of the compression stroke. (Until the crank weight contacts SST and stops.)

10. Using a straight edge, draw a straight line directly in the center of the twentieth tooth of the crankshaft pulley pulse wheel (counting counterclockwise from the empty space).

✻✻ CAUTION

If the line is not accurately drawn, ignition timing, fuel injection and other engine control systems will be adversely affected. Draw the straight line carefully using a straight edge.

11. Align the center line of the crankshaft position sensor and the line drawn on the tooth, then install the sensor. Tighten the bolts to 49–66 inch lbs. (5.5–7.5 Nm).

12. Remove the SST then install the cylinder block lower blind plug. Tighten to 14–16 ft. lbs. (18–22 Nm).

ENGINE COOLANT TEMPERATURE (ECT) SENSOR

LOCATION

See Figure 87.

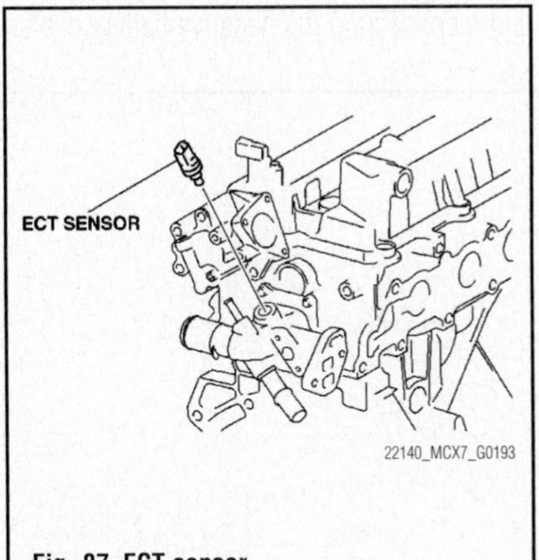

Fig. 87 ECT sensor

REMOVAL & INSTALLATION

✻✻ CAUTION

A hot engine can cause severe burns. Turn off the engine and wait until it is cool before removing the ECT sensor.

1. Before servicing the vehicle, refer to the precautions in the beginning of this section.
2. Disconnect the negative battery cable.
3. Remove the charge air cooler duct.
4. Remove the battery and battery tray.
5. Remove the air cleaner and air hose.
6. Drain the engine coolant.
7. Disconnect the wiring harness.
8. Remove the ECT sensor.
9. Install in the reverse order of removal. Tighten to 89–123 inch lbs. (10–14 Nm).
10. Refill the engine coolant.

FUEL LEVEL SENDER

LOCATION

In the fuel tank.

REMOVAL & INSTALLATION

See Figures 88 through 90.

✻✻ CAUTION

Because the fuel tank is constructed such that the fuel level is higher than the installation surface of the fuel pump, fuel leakage could occur. If the fuel gauge indicates a fuel level of half or more, perform the following Steps 1-6 to drain approx. 10 L (11 US gal, 8.8 imp gal) of fuel.

➡Disconnecting/connecting the quick release connector without cleaning it may possibly cause damage to the fuel pipe and quick release connector. Always clean the quick release connector joint area before disconnecting/connecting using a cloth or soft brush, and make sure that it is free of foreign material.

1. Before servicing the vehicle, refer to the precautions in the beginning of this section.
2. Remove the rear seat cushion.
3. Remove the service hole cover.
4. Disconnect the quick release connector connected to the fuel pump unit.

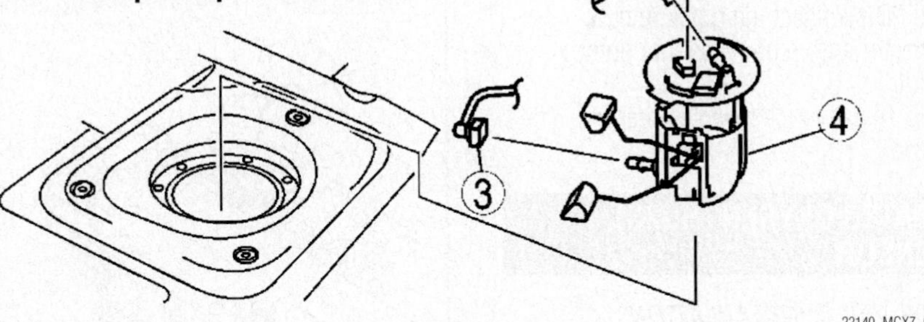

Fig. 88 Fuel pump module exploded view

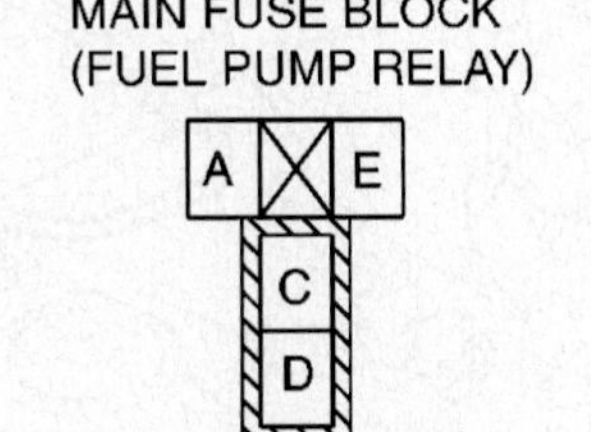

Fig. 89 Fuel pump relay terminal identification

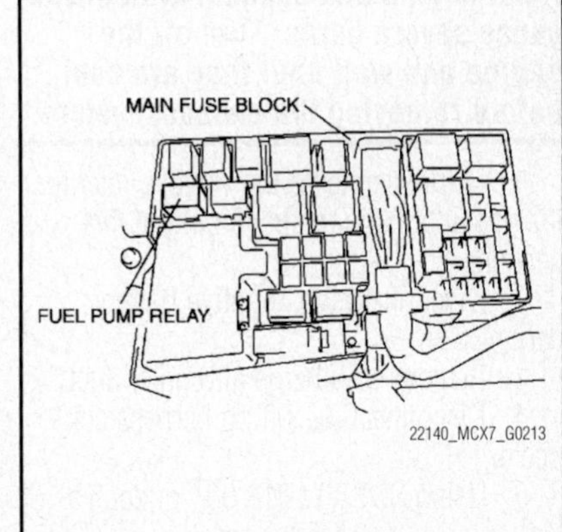

Fig. 90 Fuel pump relay location

5. Connect a long hose to the disconnected quick release connector and drain the fuel into a container used for collecting gasoline.

6. Remove the fuel pump relay.

7. Using a jumper wire, short fuel pump relay terminals C and D in the main fuse block.

8. Turn the ignition key **ON** and operate the fuel pump.

**** CAUTION**

The fuel pump could be damaged if it is operated (fuel pump idling) while there is no fuel in the fuel tank. Verify the amount of fuel being discharged from the hose and stop operation of the fuel pump when essentially no fuel is being discharged.

9. When essentially no fuel is being discharged, stop the fuel pump.

10. Turn the ignition switch to LOCK position to stop the fuel pump.

11. Disconnect the jumper wire from the check connector.

12. Disconnect the negative battery cable.

13. Remove the fuel pump bracket installation screws.

14. Remove the connector.

15. Disconnect the quick release connector and remove the fuel pump module.

16. Install in the reverse order of removal.

HEATED OXYGEN (HO2S) SENSOR

REMOVAL & INSTALLATION

See Figures 91 and 92.

**** WARNING**

A hot engine and exhaust system can cause severe burns. Turn off the engine and wait until they are cool before removing the exhaust system.

1. Before servicing the vehicle, refer to the precautions in the beginning of this section.

2. Disconnect the negative battery cable.

3. Remove the charge air cooler duct.

4. Disconnect the wiring harness connector.

5. Using SST 49 L018 001 or equivalent, remove the HO2S.

6. Install in the reverse order of removal.

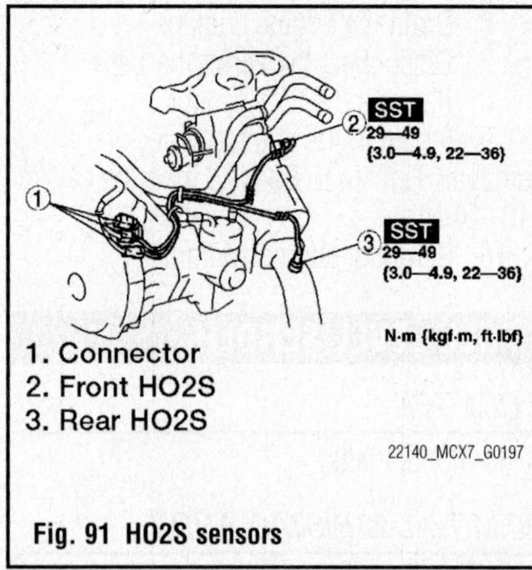

Fig. 91 HO2S sensors

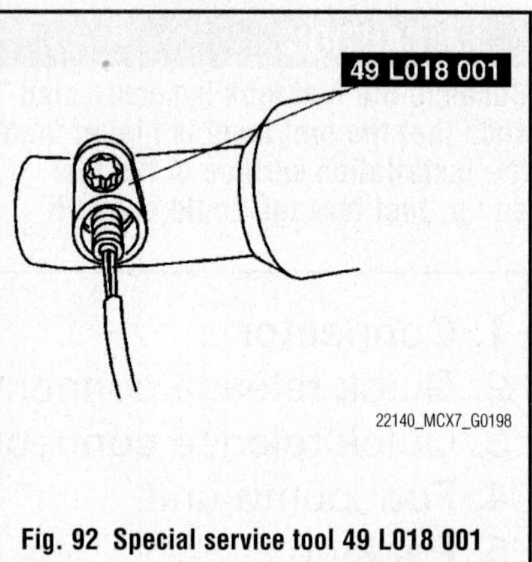

Fig. 92 Special service tool 49 L018 001

INTAKE AIR TEMPERATURE (IAT) SENSOR

LOCATION

Part of the Mass Air Flow sensor.

KNOCK SENSOR (KS)

REMOVAL & INSTALLATION

See Figure 93.

**** CAUTION**

Be careful not to impact the sensor.

1. Disconnect the negative battery cable.

2. Remove the intake manifold.

3. Remove the KS attachment bolt to remove the KS.

4. Install in the reverse order of removal. Tighten the bolt to 12–17 ft. lbs. (16–24 Nm).

MALFUNCTION INDICATOR LIGHT (MIL)

RESET PROCEDURES

Reset the Malfunction Indicator Lamp (MIL) with a scan tool or by disconnecting and reconnecting the negative battery cable.

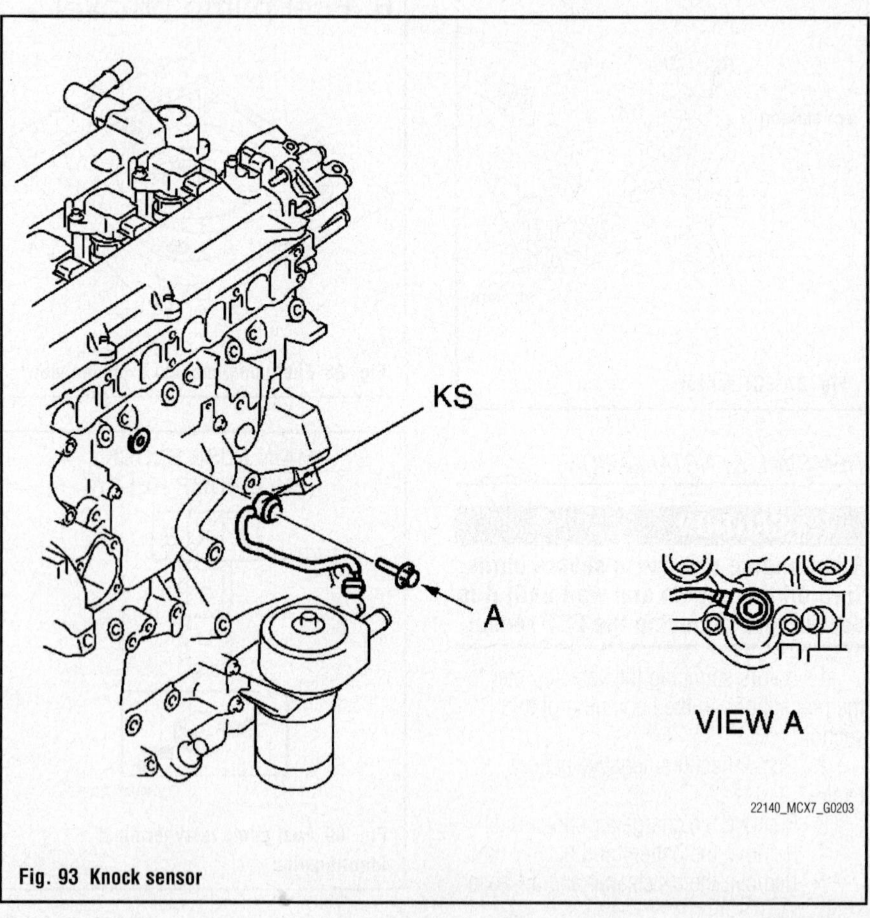

Fig. 93 Knock sensor

MANIFOLD ABSOLUTE PRESSURE (MAP) SENSOR

REMOVAL & INSTALLATION

See Figure 94.

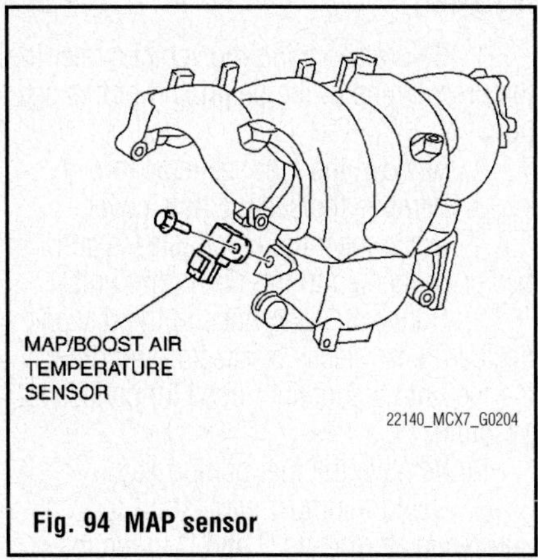

Fig. 94 MAP sensor

1. Disconnect the negative battery cable.
2. Remove the charge air cooler duct.
3. Remove the oil level gauge pipe.
4. Disconnect the MAP/boost air temperature sensor connector.
5. Remove MAP/boost air temperature sensor from the intake manifold.
6. Install in the reverse order of removal. Tighten the MAP bolt to 45–61 inch lbs. (5–7 Nm).

OIL PRESSURE SWITCH

LOCATION

On the oil filter adapter.

REMOVAL & INSTALLATION

1. Before servicing the vehicle, refer to the precautions in the beginning of this section.
2. Disconnect the negative battery cable.
3. Remove the undercover.
4. Remove the oil pressure switch.

To install:

5. Apply silicone sealant to the oil pressure switch threads.
6. Tighten the switch to 9–13 ft. lbs. (12–18 Nm).
7. Install in reverse of removal.

POWERTRAIN CONTROL MODULE (PCM)

REMOVAL & INSTALLATION

See Figure 95.

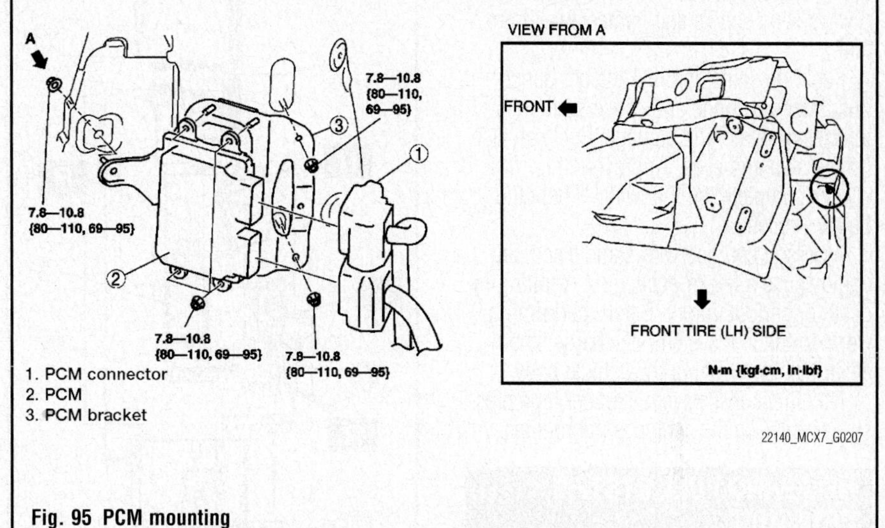

Fig. 95 PCM mounting

➡If the PCM is replaced, the replacement must be programmed with the vehicle parameters..

1. Disconnect the negative battery cable.
2. Remove the charge air cooler duct.
3. Remove the Battery and battery tray.
4. Remove the mudguard (LH).
5. Disconnect the PCM connector.
6. Remove the PCM
7. Install in the reverse order of removal.

VARIABLE CAMSHAFT TIMING OIL CONTROL SOLENOID

LOCATION

Between the camshafts.

REMOVAL & INSTALLATION

See Figure 96.

1. Disconnect the negative battery cable.
2. Remove the charge air cooler.
3. Remove the ignition coils.
4. Disconnect the wiring harness.
5. Remove the ventilation hose.
6. Remove the cylinder head cover.
7. Remove in the order indicated in the table.
8. Install in the reverse order of removal. Tighten the OCV bolt to 71–101 inch lbs. (8–11.5 Nm).

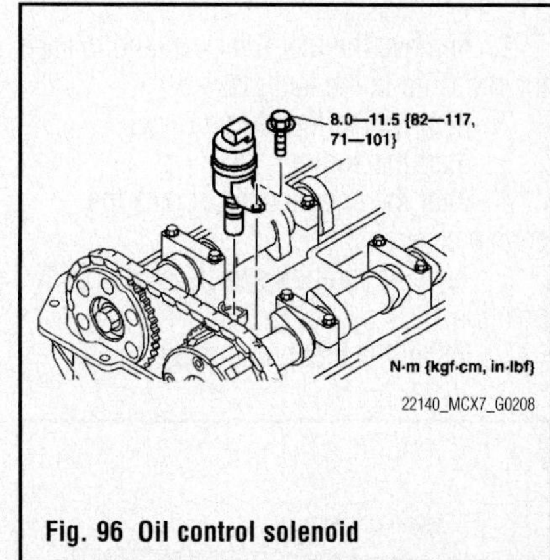

Fig. 96 Oil control solenoid

VEHICLE SPEED SENSOR (VSS)

LOCATION

Inside the transaxle.

REMOVAL & INSTALLATION

1. Remove the transaxle.
2. Remove the oil pan and valve body.
3. Replace the VSS.
4. Reassemble the transaxle and install it.

FUEL SYSTEM SERVICE PRECAUTIONS

Fuel vapor is hazardous. It can easily ignite, causing serious injury and damage. Always keep sparks and flames away from fuel.

Fuel line spills and leakage are dangerous. Fuel can ignite and cause serious injuries or death and damage. Fuel can also irritate skin and eyes. To prevent this, always complete the following "Fuel Line Safety Procedure".

A person charged with static electricity could cause a fire or explosion, resulting in death or serious injury. Before performing work on the fuel system, discharge static electricity by touching the vehicle body.

Fuel in the fuel system is under high pressure even when the engine is not running.

RELIEVING FUEL SYSTEM PRESSURE

See Figure 97.

1. Remove the fuel-filler cap and release the pressure in the fuel tank.
2. Remove the fuel pump relay.
3. Start the engine.
4. After the engine stalls, crank the engine several times.
5. Turn the ignition switch to the LOCK position.
6. Install the fuel pump relay.

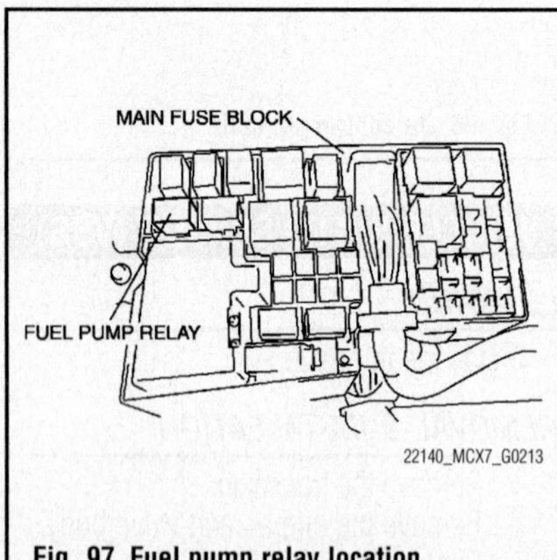

Fig. 97 Fuel pump relay location

FUEL FILTER

REMOVAL & INSTALLATION

See Figure 98.

1. Before servicing the vehicle, refer to the precautions in the beginning of this section.

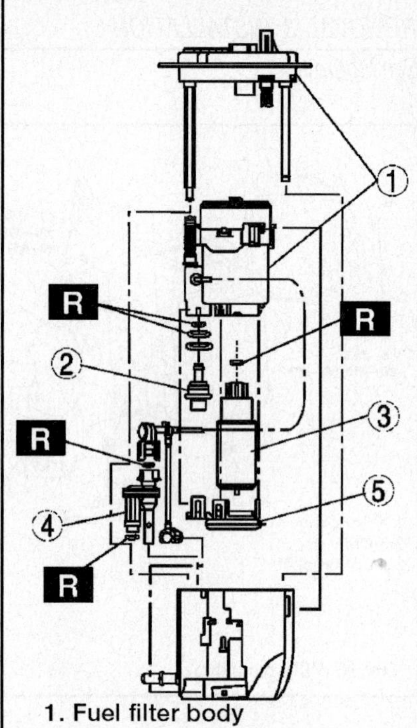

Fig. 98 Fuel pump module exploded view

2. Relieve the fuel system pressure.
3. Remove the fuel pump module.
4. Disassemble as shown and replace the fuel filter.
5. Assemble in reverse order.

FUEL PUMP

REMOVAL & INSTALLATION

In-Tank Fuel Pump

See Figures 99 through 101.

CAUTION

Because the fuel tank is constructed such that the fuel level is higher than the installation surface of the fuel pump, fuel leakage could occur. If the fuel gauge indicates a fuel level of half or more, perform the following Steps 1-6 to drain approx. 10 L (11 US gal, 8.8 imp gal) of fuel.

➡Disconnecting/connecting the quick release connector without cleaning it may possibly cause damage to the fuel pipe and quick release connector. Always clean the quick release connector joint area before disconnecting/connecting using a cloth or soft brush, and make sure that it is free of foreign material.

1. Before servicing the vehicle, refer to the precautions in the beginning of this section.
2. Remove the rear seat cushion.
3. Remove the service hole cover.
4. Disconnect the quick release connector connected to the fuel pump unit.
5. Connect a long hose to the disconnected quick release connector and drain the fuel into a container used for collecting gasoline.
6. Remove the fuel pump relay.
7. Using a jumper wire, short fuel pump relay terminals C and D in the main fuse block.
8. Turn the ignition key **ON** and operate the fuel pump.

CAUTION

The fuel pump could be damaged if it is operated (fuel pump idling) while there is no fuel in the fuel tank. Verify the amount of fuel being discharged from the hose and stop operation of the fuel pump when essentially no fuel is being discharged.

9. When essentially no fuel is being discharged, stop the fuel pump.
10. Turn the ignition switch to LOCK position to stop the fuel pump.
11. Disconnect the jumper wire from the check connector.
12. Disconnect the negative battery cable.
13. Remove the fuel pump bracket installation screws.
14. Remove the connector.
15. Disconnect the quick release connector and remove the fuel pump module.
16. Install in the reverse order of removal.

High Pressure Fuel Pump

See Figure 102.

CAUTION

Do not disassemble the high pressure fuel pump.

CAUTION

Do not scratch or damage the fuel sealing surface of the high and low fuel ports.

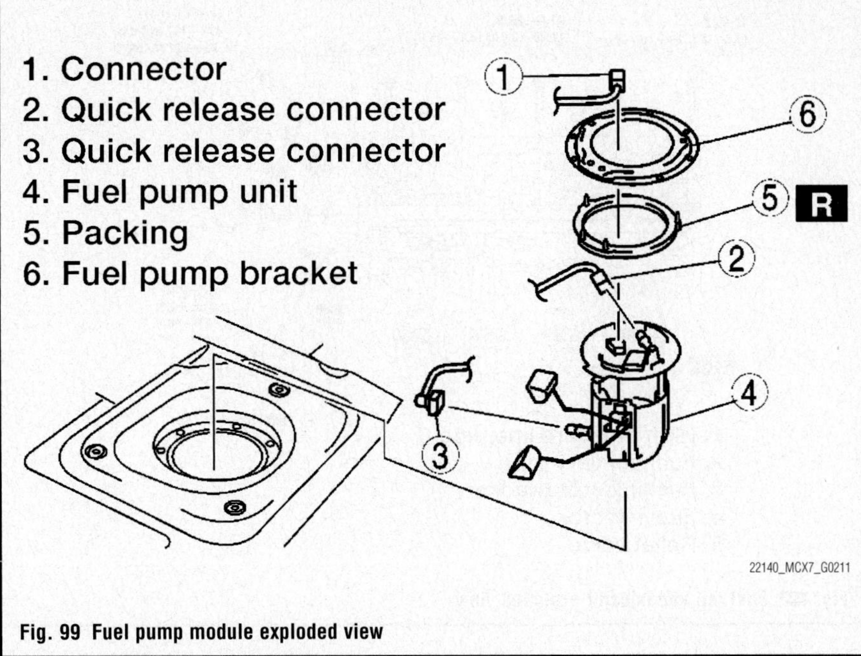

Fig. 99 Fuel pump module exploded view

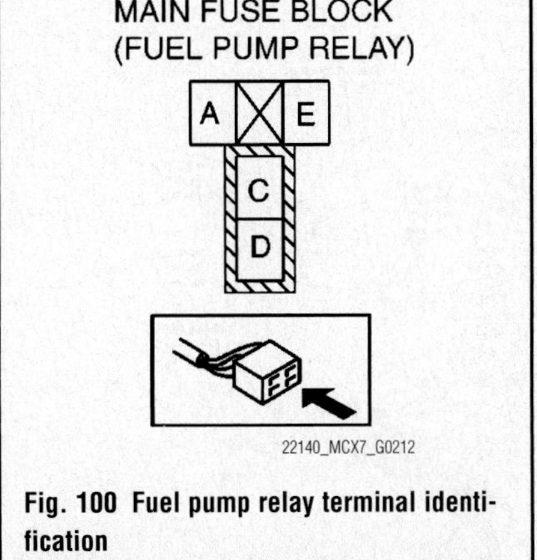

Fig. 100 Fuel pump relay terminal identification

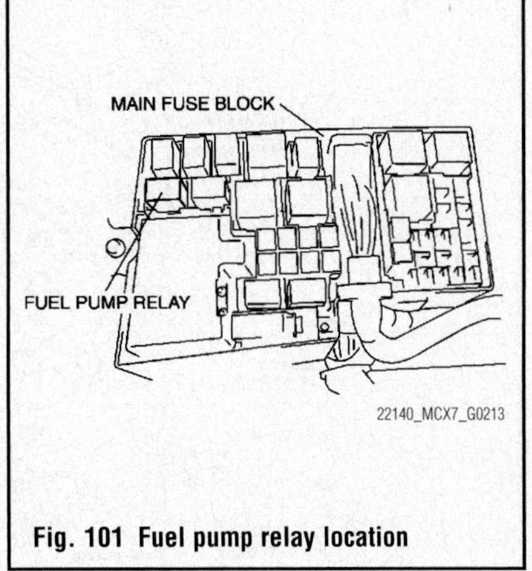

Fig. 101 Fuel pump relay location

⁂ CAUTION

When removing the high pressure fuel pipe, secure the joint (pump side) so that it does not rotate, and loosen the screw (pipe side).

➡If the high pressure fuel pump is removed, replace the O-ring with a new one.

1. Before servicing the vehicle, refer to the precautions in the beginning of this section.
2. Disconnect the negative battery cable.
3. Remove the charge air cooler duct.
4. Disconnect the spill valve control solenoid valve connector.
5. Disconnect the quick release connector on the high pressure fuel pump.
6. Remove the battery and battery tray.
7. Remove the air duct.

⁂ WARNING

If the high pressure fuel pump joint nut is loosened, fuel leakage may occur resulting in death or serious injury, or damage to the equipment or the vehicle. Fuel can also irritate the skin and eyes.

⁂ WARNING

When removing the high pressure line pipe, always tighten the high pressure line pipe installation nut while fixing the high pressure fuel pump joint nut with a wrench. If the high pressure fuel pump joint nut has rotated, replace the high pressure fuel pump with a new one.

8. Fix the joint nut with a wrench on the high pressure fuel pump side as shown in the figure.
9. Loosen the high pressure line pipe installation nut.
10. Drain engine coolant.
11. Loosen the water outlet case installation bolts securing the high pressure line pipe.
12. Remove the high pressure fuel pump.
13. Remove the high pressure fuel pump cover.

To install:

14. Install the pump cover and tighten the bolts to 13–16 ft. lbs. (17–23 Nm).

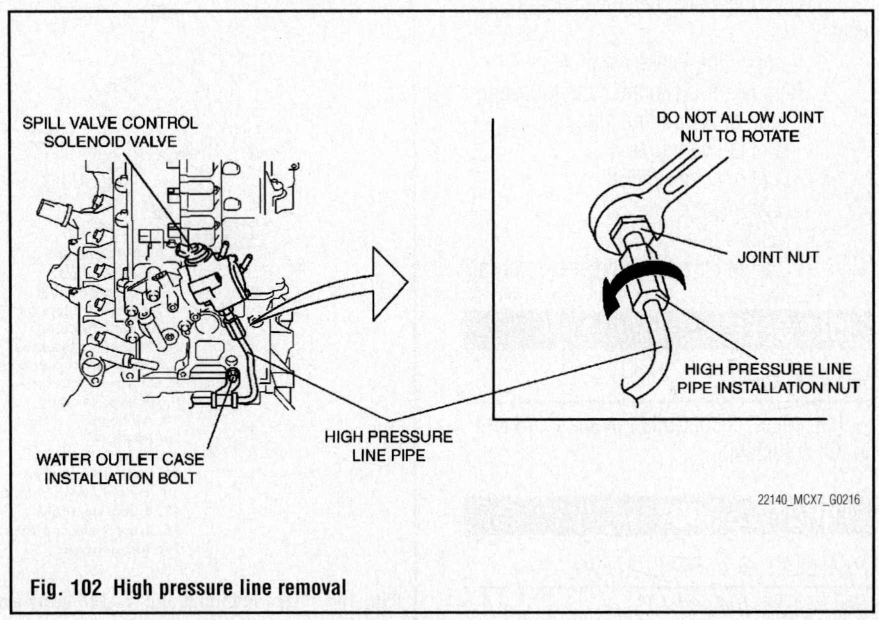

Fig. 102 High pressure line removal

⁂ CAUTION

If the high pressure fuel pump installation bolts are tightened with the high pressure fuel pump tilted, the high pressure fuel pump may not operate correctly.

15. Install the high pressure pump. Tighten the high pressure fuel pump installation bolts in a few passes to 71–101 inch lbs. (8.5–11.5 Nm).
16. Assemble the high pressure line pipe.
17. Fix the joint nut with a wrench on the high pressure fuel pump side as shown in the figure.
18. Tighten the high pressure line pipe installation nut to 18–26 ft. lbs. (23.5–35.5 Nm).
19. Tighten the water outlet case installation bolts to 71–101 inch lbs. (8.5–11.5 Nm).
20. Install the quick release connector.
21. Verify that the high pressure fuel pump is assembled securely.
22. Drive the vehicle starting from a standstill and brake suddenly five to six times at a low speed.
23. Stop the vehicle and verify from outside the vehicle that there is no fuel leakage around the high pressure fuel pump.

FUEL RAIL & INJECTORS

REMOVAL & INSTALLATION

See Figure 103.

1. Before servicing the vehicle, refer to the precautions in the beginning of this section.
2. Disconnect the negative battery cable.
3. Remove the intake manifold.
4. Remove or disconnect the following:
   - High pressure line pipe
   - Fuel delivery pipe
   - Fuel injector bracket
   - Fuel injector
   - Relief valve
5. Install in the reverse order of removal.

IDLE SPEED

ADJUSTMENT

Idle speed is controlled by the PCM and is not adjustable.

THROTTLE BODY

REMOVAL & INSTALLATION

See Figures 104 and 105.

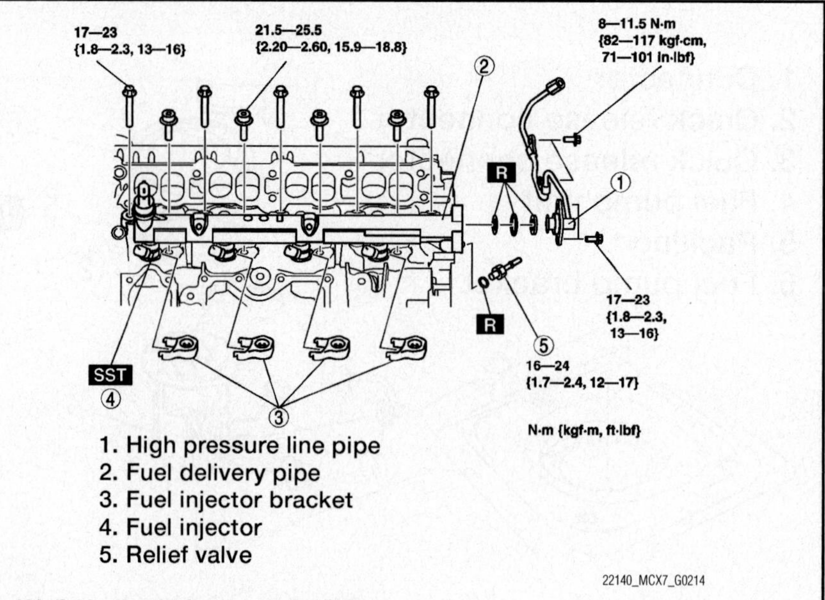

Fig. 103 Fuel rail and injector exploded view

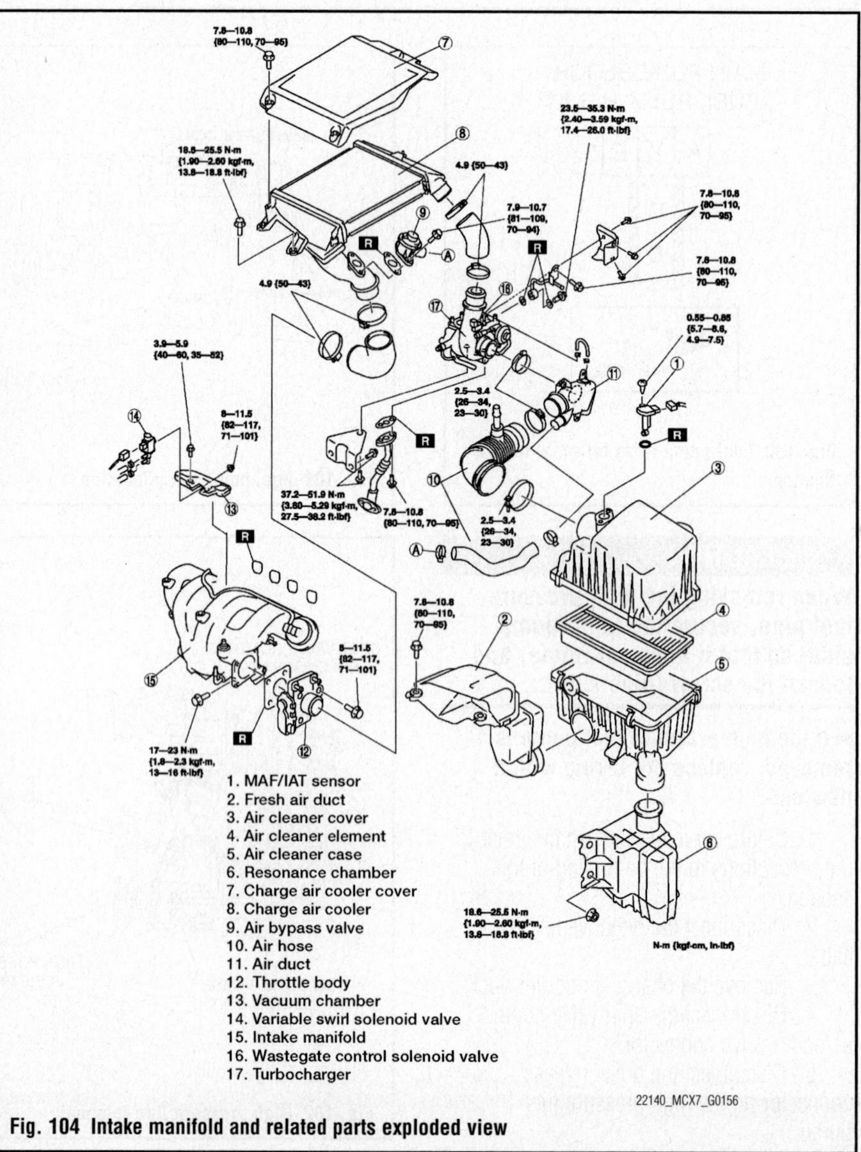

Fig. 104 Intake manifold and related parts exploded view

✻✻ WARNING

A hot engine and intake air system can cause severe burns. Turn off the engine and wait until they are cool before removing the intake air system.

✻✻ WARNING

Fuel vapor is hazardous. It can easily ignite, causing serious injury and damage. Always keep sparks and flames away from fuel.

✻✻ WARNING

Fuel line spills and leakage are dangerous. Fuel can ignite and cause serious injuries or death and damage. Fuel can also irritate skin and eyes.

1. Before servicing the vehicle, refer to the precautions in the beginning of this section.
2. Disconnect the negative battery cable.
3. Drain the cooling system.
4. Remove the charge air cooler duct.
5. Remove the battery and battery tray.
6. Remove the undercover.
7. Remove the Mass Airflow sensor (MAF).
8. Remove the fresh air duct.
9. Remove the air cleaner cover and case.
10. Remove the resonance chamber.
11. Remove the charge air cooler.
12. Remove the air bypass valve.
13. Remove the air hose and duct.

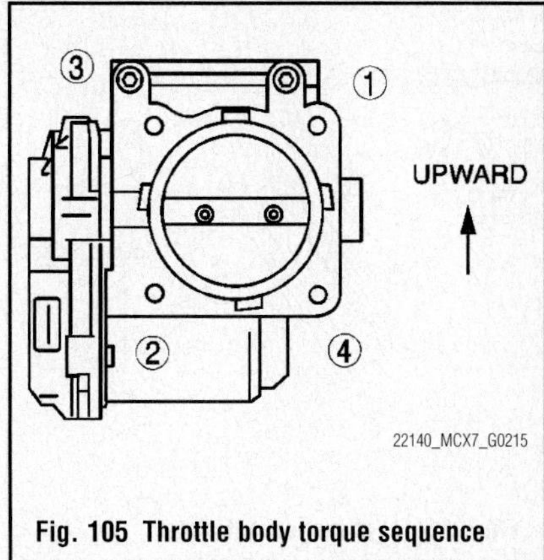

Fig. 105 Throttle body torque sequence

14. Remove the throttle body.
15. Install in the reverse order of removal. Tighten the throttle body bolts in sequence to 71–101 inch lbs. (8–11 Nm).

HEATING & AIR CONDITIONING SYSTEM

BLOWER MOTOR

REMOVAL & INSTALLATION

See Figure 106.

1. Before servicing the vehicle, refer to the precautions in the beginning of this section.
2. Disconnect the negative battery cable.
3. Remove the dashboard undercover (passenger's side).
4. Remove the blower motor connector and the blower motor
5. Install in the reverse order of removal.

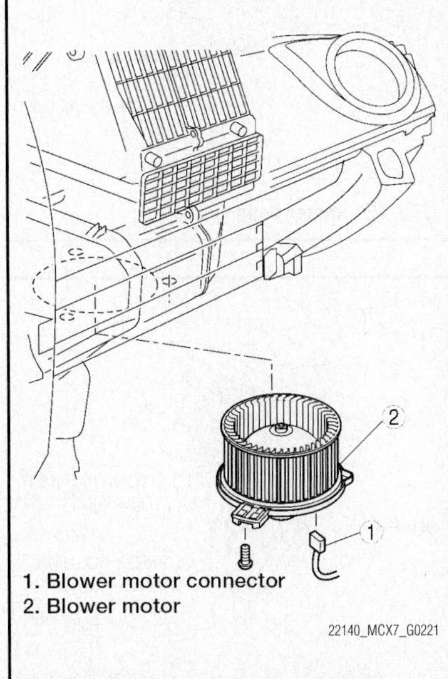

Fig. 106 Blower motor exploded view

HEATER CORE

REMOVAL & INSTALLATION

See Figures 107 through 134.

1. Before servicing the vehicle, refer to the Precautions Section.
2. Disconnect the negative battery cable.
3. Remove or disconnect the following:
   - Console side panel
   - Console
   - Front scuff plate inner
   - Front side trim
   - Dashboard undercover

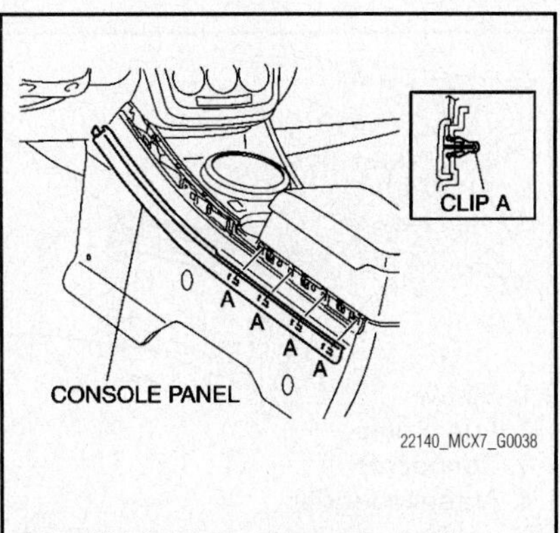

Fig. 107 Console side panel—1 of 2

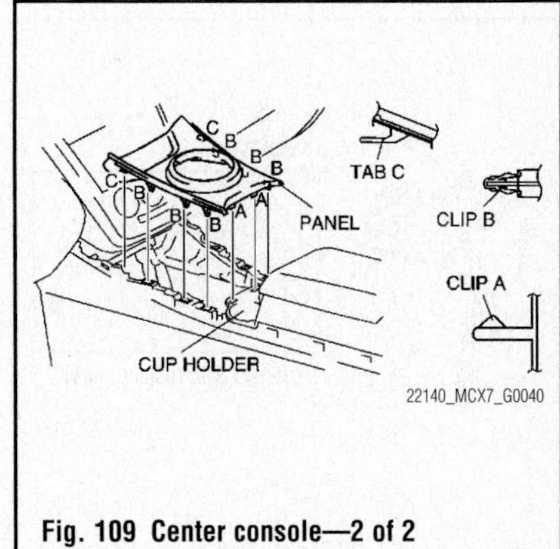

Fig. 109 Center console—2 of 2

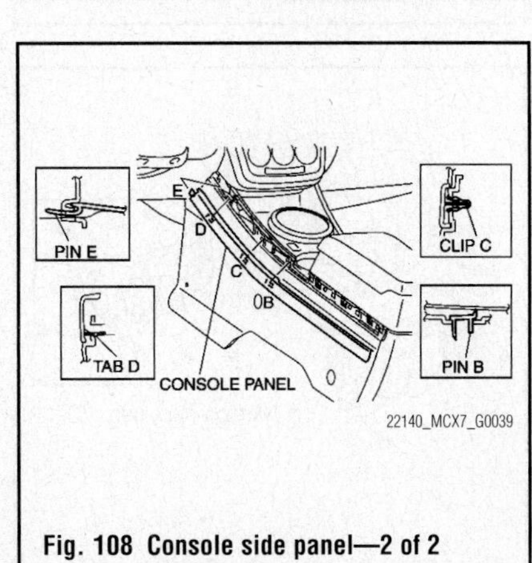

Fig. 108 Console side panel—2 of 2

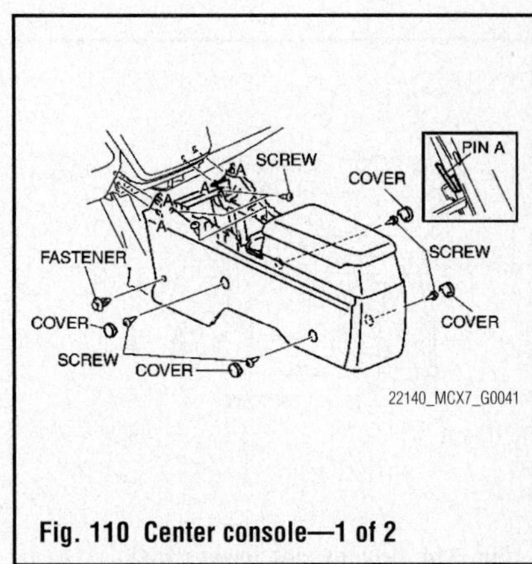

Fig. 110 Center console—1 of 2

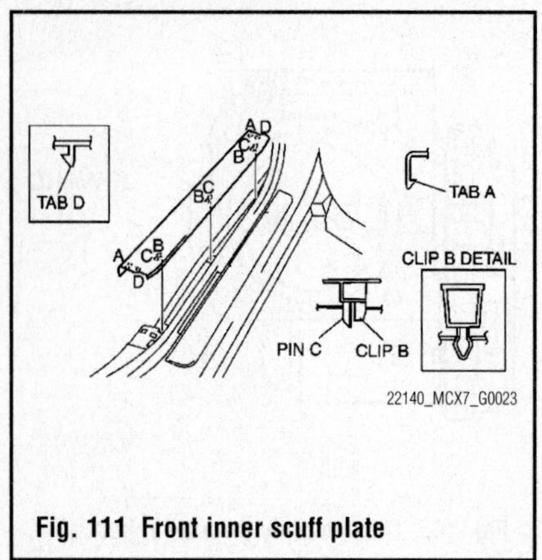

Fig. 111 Front inner scuff plate

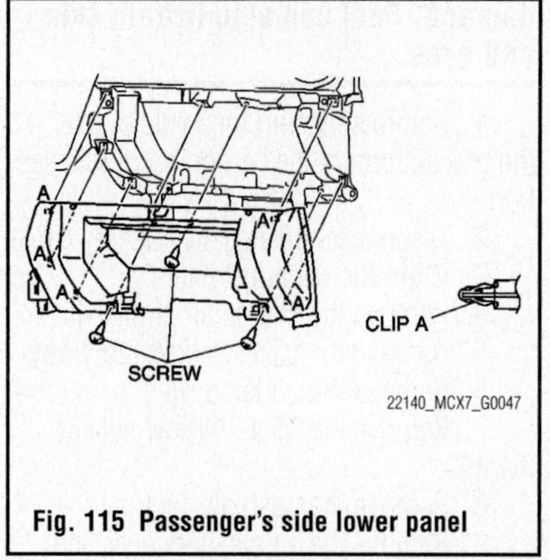

Fig. 115 Passenger's side lower panel

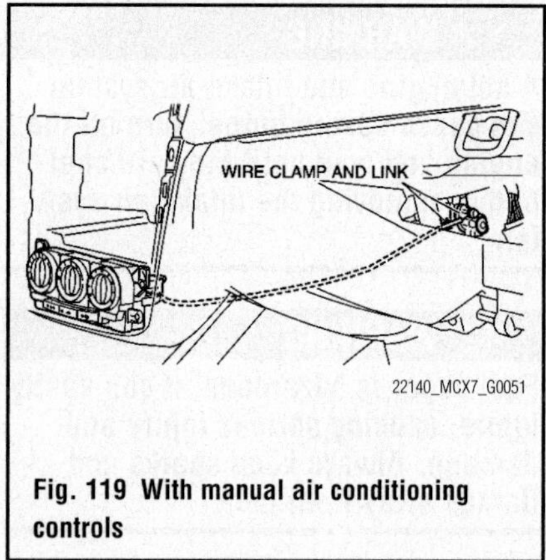

Fig. 119 With manual air conditioning controls

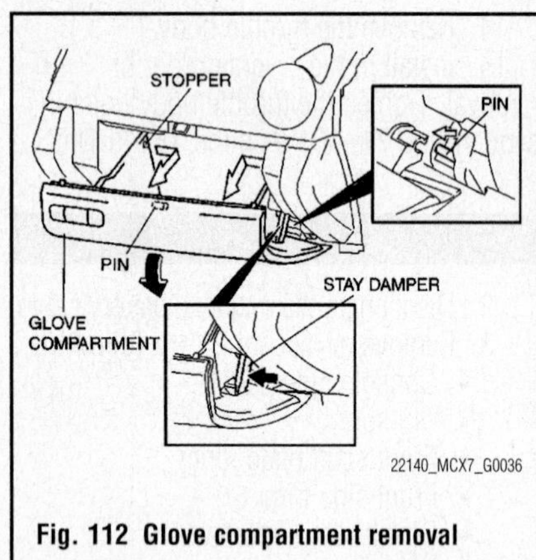

Fig. 112 Glove compartment removal

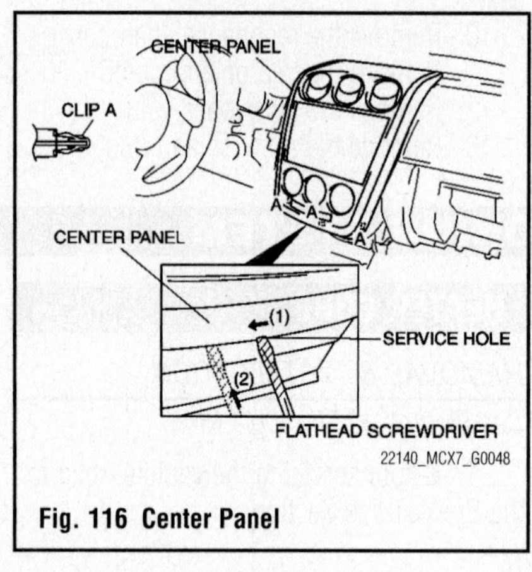

Fig. 116 Center Panel

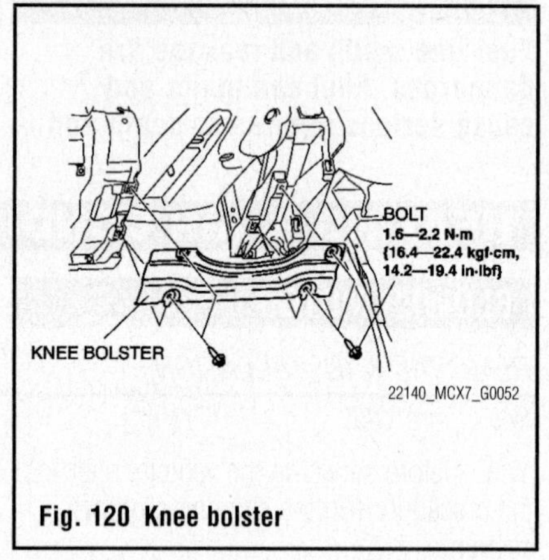

Fig. 120 Knee bolster

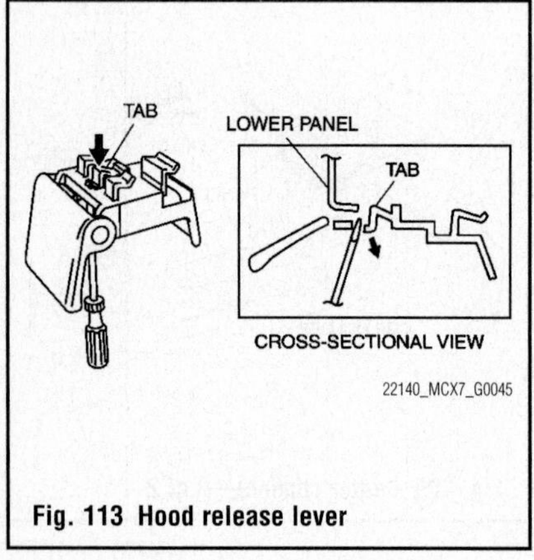

Fig. 113 Hood release lever

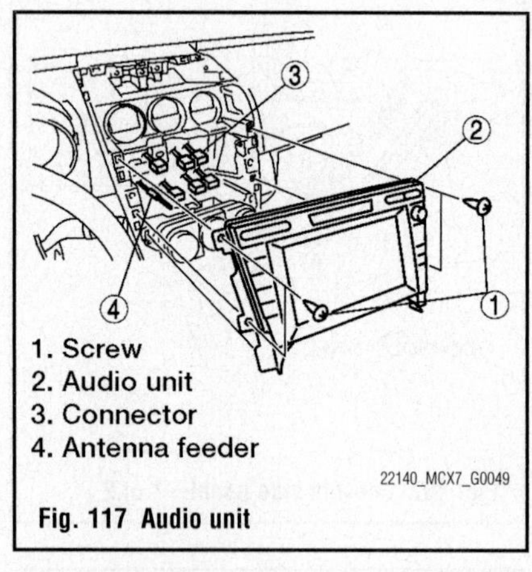

Fig. 117 Audio unit

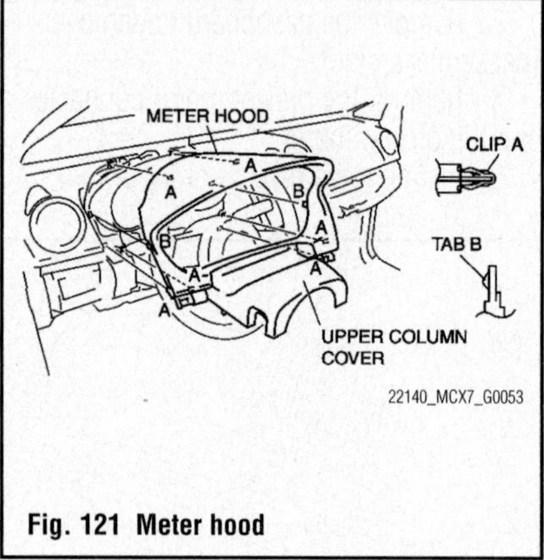

Fig. 121 Meter hood

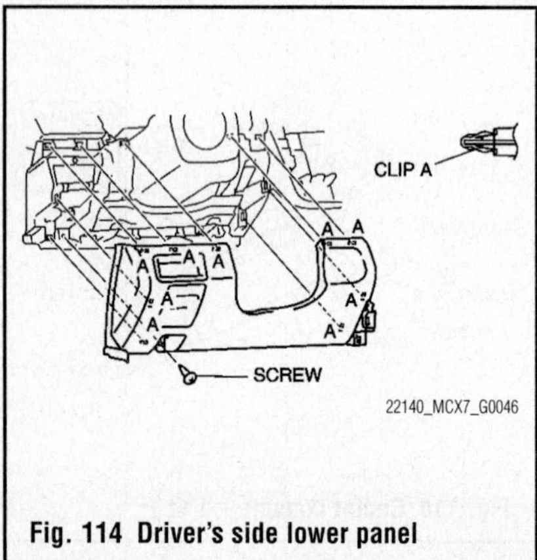

Fig. 114 Driver's side lower panel

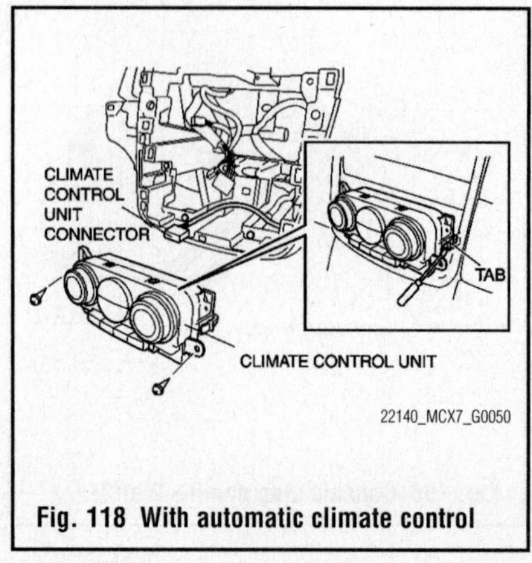

Fig. 118 With automatic climate control

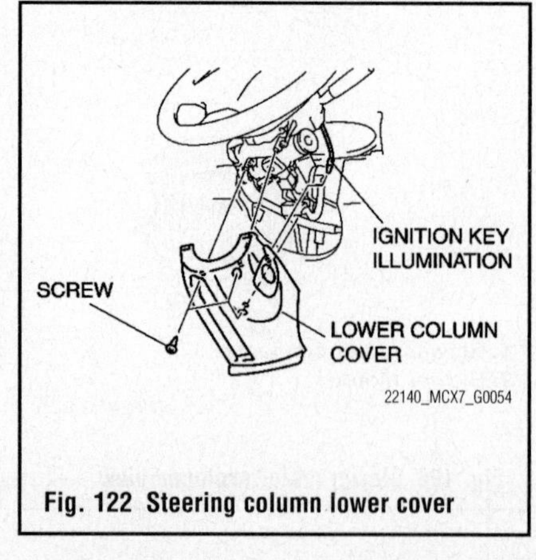

Fig. 122 Steering column lower cover

1. Screw
2. Connector
3. Instrument cluster
22140_MCX7_G0055

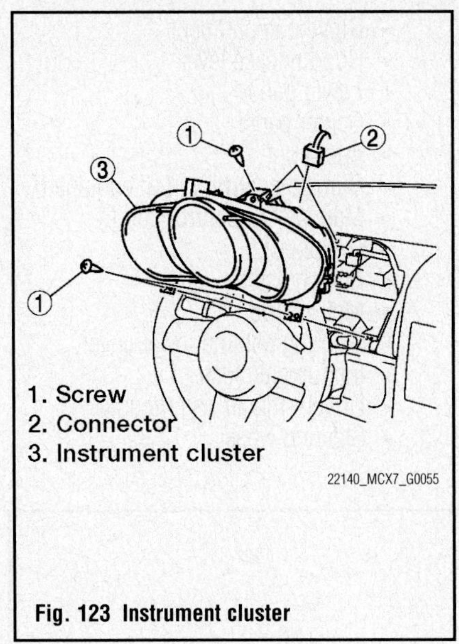

Fig. 123 Instrument cluster

8.8—12.8 {89.8—130.5, 77.8—113.2}
8.8—12.8 {89.8—130.5, 77.8—113.2}
N·m {kgf·cm, in·lbf}
1. Cover
2. Bolt
3. Connector A
4. Connector B
5. Driver-side air bag module
22140_MCX7_G0056

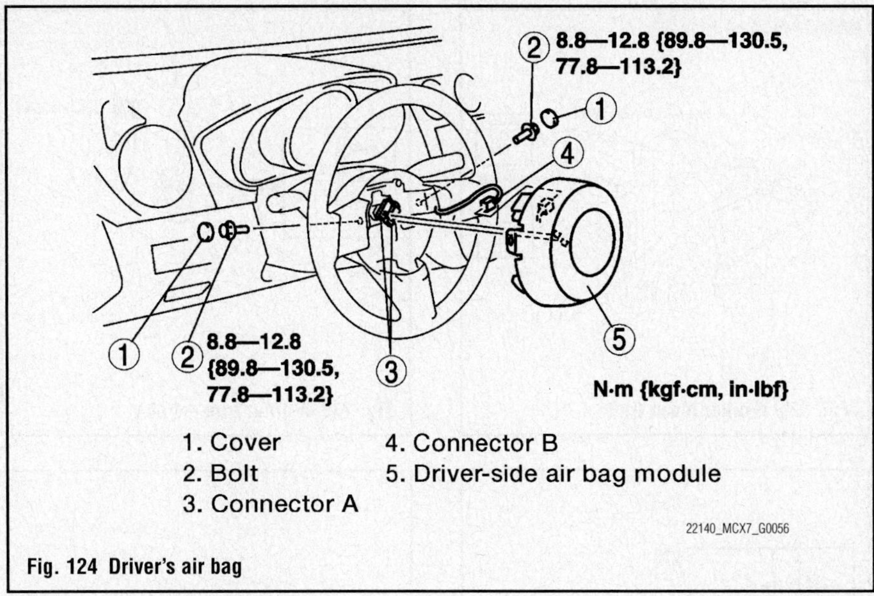

Fig. 124 Driver's air bag

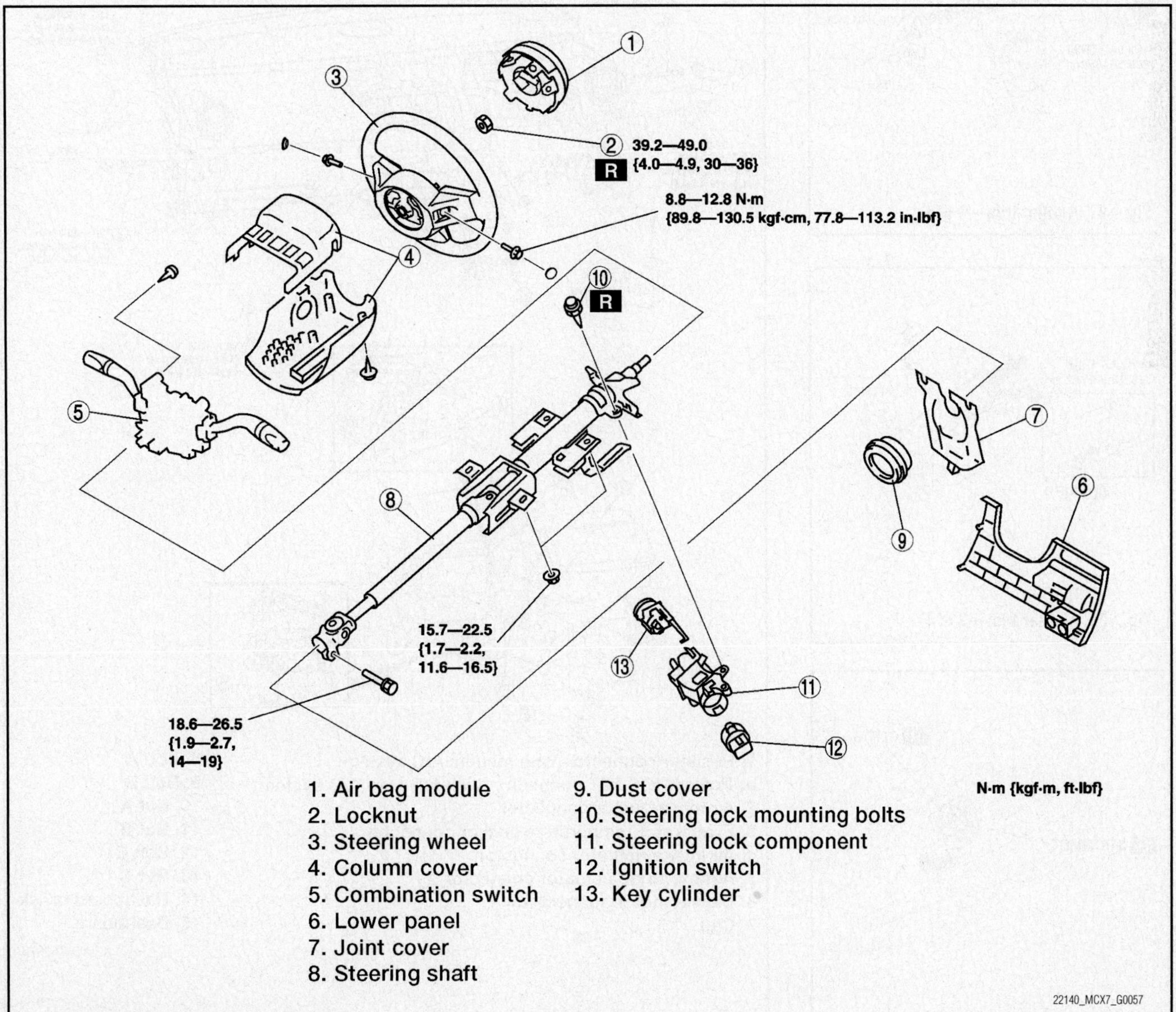

39.2—49.0 {4.0—4.9, 30—36}
8.8—12.8 N·m {89.8—130.5 kgf·cm, 77.8—113.2 in·lbf}
15.7—22.5 {1.7—2.2, 11.6—16.5}
18.6—26.5 {1.9—2.7, 14—19}
N·m {kgf·m, ft·lbf}
1. Air bag module
2. Locknut
3. Steering wheel
4. Column cover
5. Combination switch
6. Lower panel
7. Joint cover
8. Steering shaft
9. Dust cover
10. Steering lock mounting bolts
11. Steering lock component
12. Ignition switch
13. Key cylinder
22140_MCX7_G0057

Fig. 125 Steering column exploded view

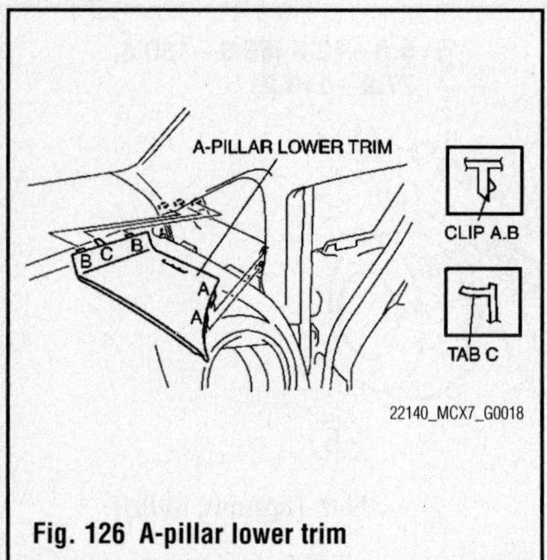

Fig. 126 A-pillar lower trim

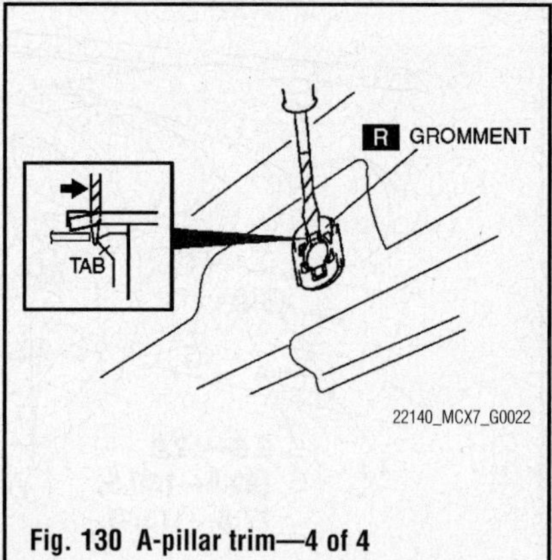

Fig. 130 A-pillar trim—4 of 4

- Glove compartment
- Hood release lever
- Lower panel
- Center panel
- Audio unit
- Climate control unit (if equipped)
- Manual A/C control unit (if equipped)
- Knee bolster
- Meter hood
- Steering column lower cover
- Instrument cluster
- Driver-side air bag module
- Steering wheel

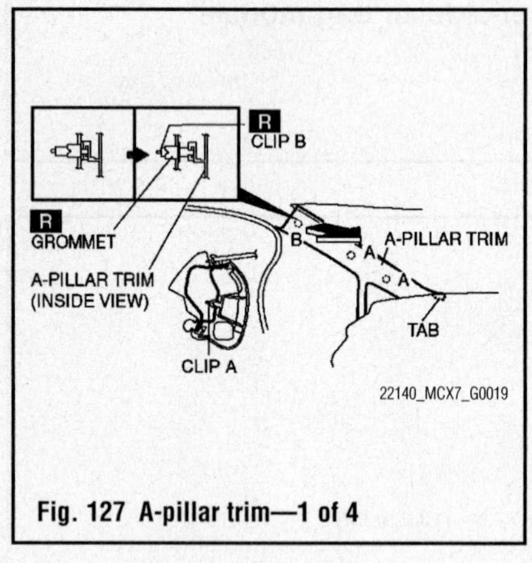

Fig. 127 A-pillar trim—1 of 4

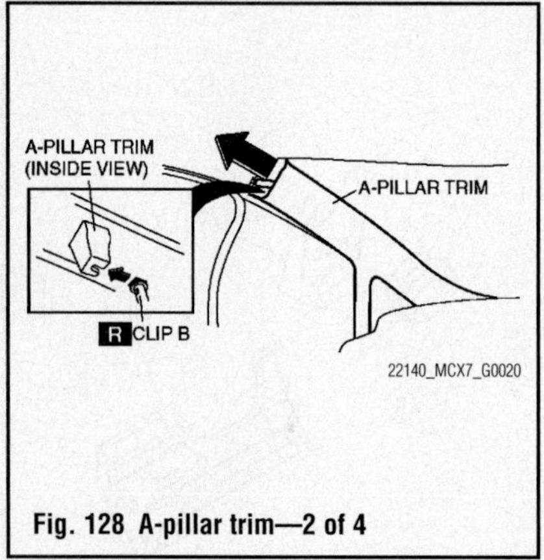

Fig. 128 A-pillar trim—2 of 4

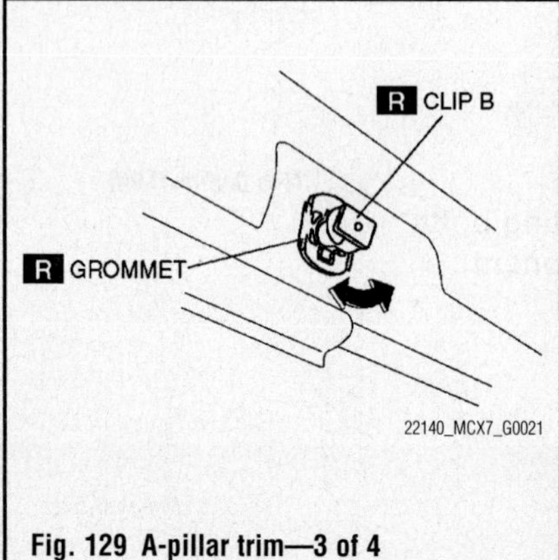

Fig. 129 A-pillar trim—3 of 4

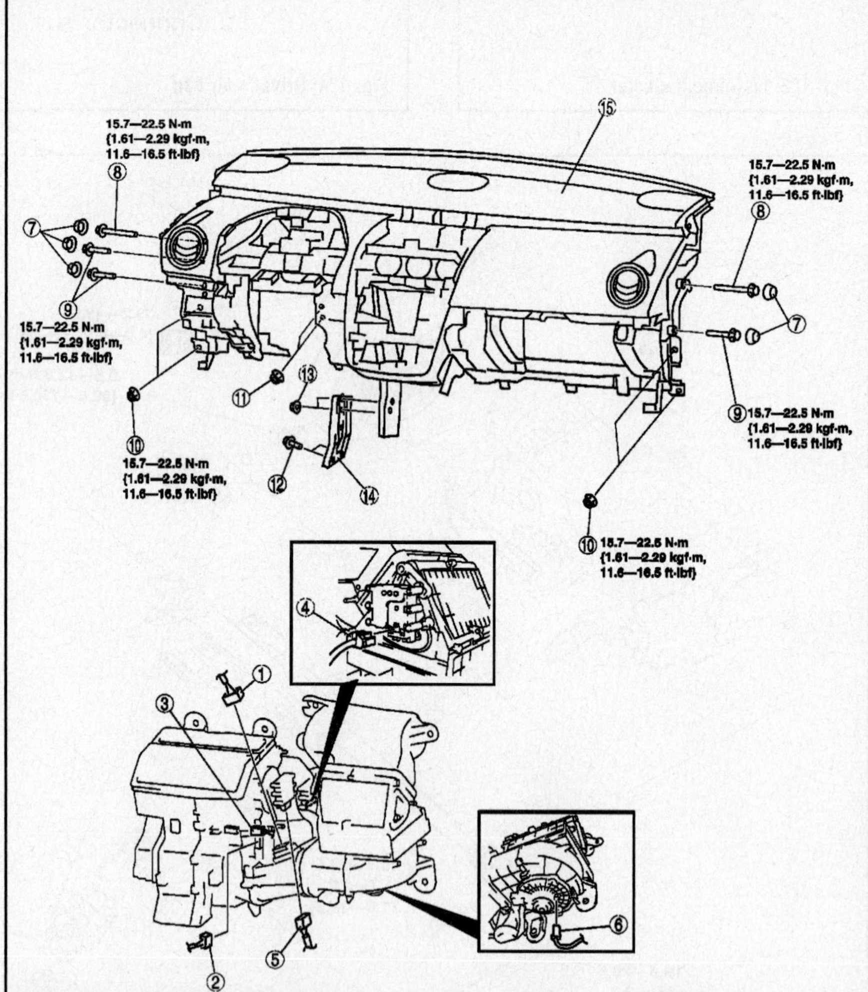

Fig. 131 Instrument panel mounting

1. Insulator
2. Heater hose
3. Cooler hose (LO)
4. Cooler pipe
5. A/C unit
22140_MCX7_G0229

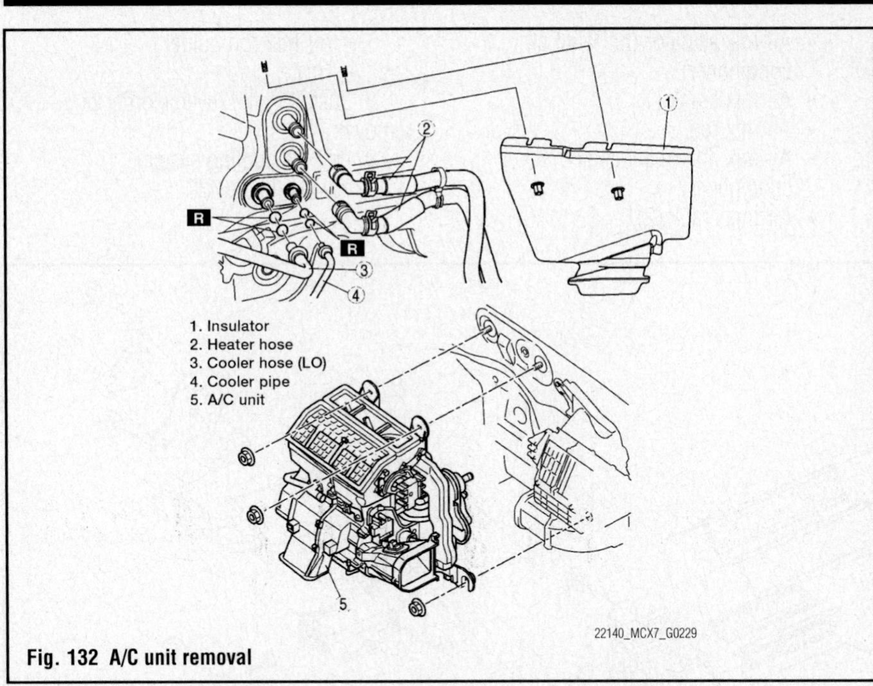

Fig. 132 A/C unit removal

- Combination switch
- Steering shaft
- A-pillar lower trim
- A-pillar trim

4. Disconnect the dashboard harness connectors.

5. Unbolt and remove the instrument panel.

6. Remove or disconnect the following:

- Insulator
- Heater hose
- Cooler hose (LO)
- Cooler pipe
- A/C unit

7. To disassemble the A/C unit, remove or disconnect the following:

- Duct (1)
- Polyurethane protector (1)
- Polyurethane protector (2)
- Polyurethane protector (3)
- Airflow mode actuator
- Airflow mode main link

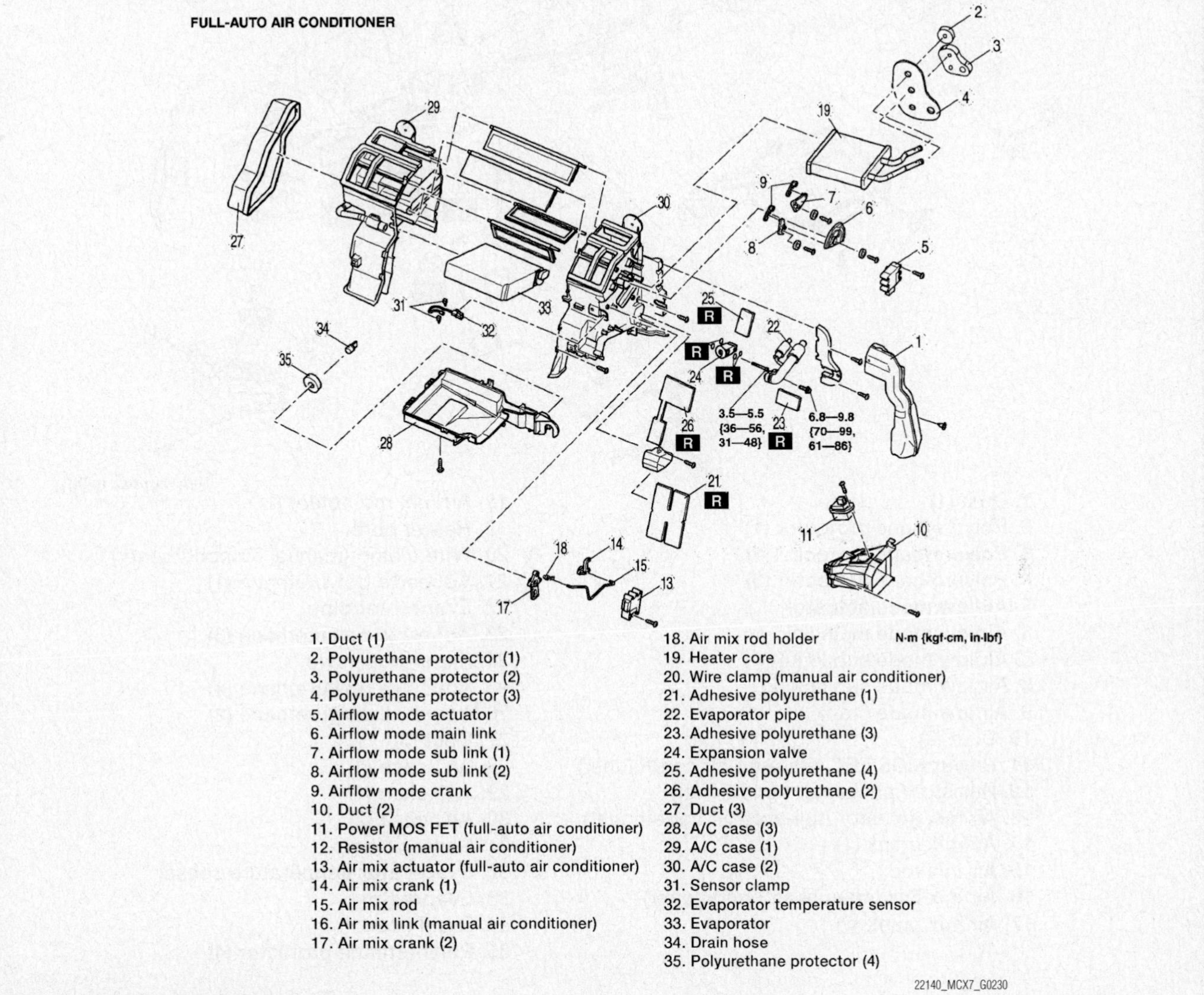

FULL-AUTO AIR CONDITIONER
3.5—5.5 {36—56, 31—48}
6.8—9.8 {70—99, 61—86}
N·m {kgf-cm, in-lbf}
1. Duct (1)
2. Polyurethane protector (1)
3. Polyurethane protector (2)
4. Polyurethane protector (3)
5. Airflow mode actuator
6. Airflow mode main link
7. Airflow mode sub link (1)
8. Airflow mode sub link (2)
9. Airflow mode crank
10. Duct (2)
11. Power MOS FET (full-auto air conditioner)
12. Resistor (manual air conditioner)
13. Air mix actuator (full-auto air conditioner)
14. Air mix crank (1)
15. Air mix rod
16. Air mix link (manual air conditioner)
17. Air mix crank (2)
18. Air mix rod holder
19. Heater core
20. Wire clamp (manual air conditioner)
21. Adhesive polyurethane (1)
22. Evaporator pipe
23. Adhesive polyurethane (3)
24. Expansion valve
25. Adhesive polyurethane (4)
26. Adhesive polyurethane (2)
27. Duct (3)
28. A/C case (3)
29. A/C case (1)
30. A/C case (2)
31. Sensor clamp
32. Evaporator temperature sensor
33. Evaporator
34. Drain hose
35. Polyurethane protector (4)
22140_MCX7_G0230

Fig. 133 Full-auto A/C unit exploded view

- Airflow mode sub link (1)
- Airflow mode sub link (2)
- Airflow mode crank
- Duct (2)
- Power MOS FET (full-auto air conditioner)
- Resistor (manual air conditioner)
- Air mix actuator (full-auto air conditioner)
- Air mix crank (1)
- Air mix rod
- Air mix link (manual air conditioner)
- Air mix crank (2)
- Air mix rod holder
- Heater core

8. Install in the reverse order of removal.
9. Fill the cooling system.
10. Recharge the A/C.

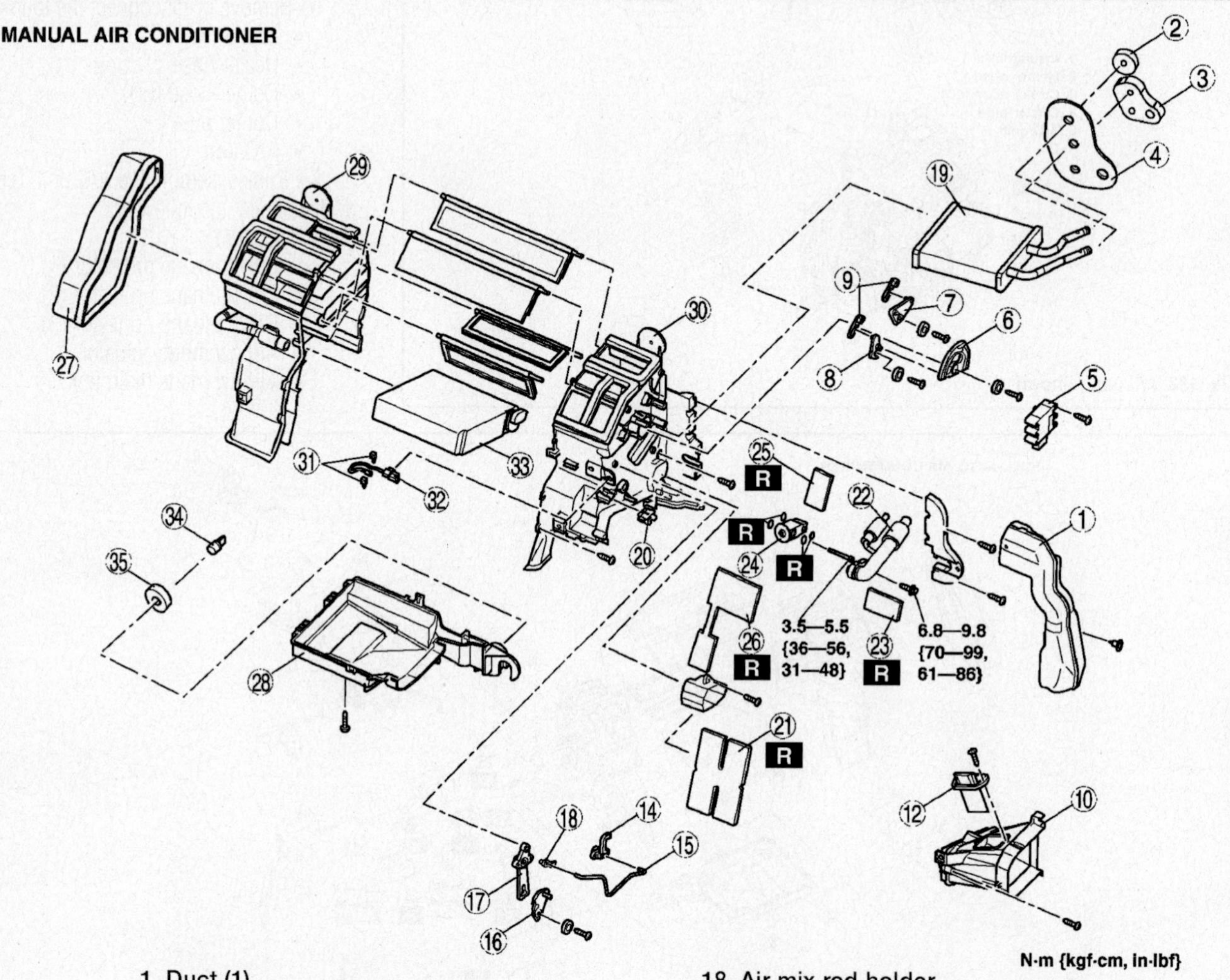

1. Duct (1)
2. Polyurethane protector (1)
3. Polyurethane protector (2)
4. Polyurethane protector (3)
5. Airflow mode actuator
6. Airflow mode main link
7. Airflow mode sub link (1)
8. Airflow mode sub link (2)
9. Airflow mode crank
10. Duct (2)
11. Power MOS FET (full-auto air conditioner)
12. Resistor (manual air conditioner)
13. Air mix actuator (full-auto air conditioner)
14. Air mix crank (1)
15. Air mix rod
16. Air mix link (manual air conditioner)
17. Air mix crank (2)
18. Air mix rod holder
19. Heater core
20. Wire clamp (manual air conditioner)
21. Adhesive polyurethane (1)
22. Evaporator pipe
23. Adhesive polyurethane (3)
24. Expansion valve
25. Adhesive polyurethane (4)
26. Adhesive polyurethane (2)
27. Duct (3)
28. A/C case (3)
29. A/C case (1)
30. A/C case (2)
31. Sensor clamp
32. Evaporator temperature sensor
33. Evaporator
34. Drain hose
35. Polyurethane protector (4)

22140_MCX7_G0231

Fig. 134 Manual A/C unit exploded view

STEERING

POWER RACK & PINION STEERING GEAR

REMOVAL & INSTALLATION

See Figures 135 and 136.

✻✻ CAUTION

Performing the following procedures without first removing the ABS wheel-speed sensor may possibly cause an open circuit in the wiring harness if it is pulled by mistake. Before performing the following procedures, disconnect the ABS wheel-speed sensor connector (axle side) and fix the wiring harness to an appropriate place where it will not be pulled by mistake while servicing the vehicle.

1. Before servicing the vehicle, refer to the Precautions Section.
2. Remove the transverse member.
3. Remove the front crossmember, lower arm, front stabilizer, and steering gear and linkage as a single unit.
4. Remove the front stabilizer from the crossmember component.
5. Remove insulator and the power steering gear.
6. Install in the reverse order of removal.
7. After installation, inspect the front wheel alignment and adjust it if necessary.

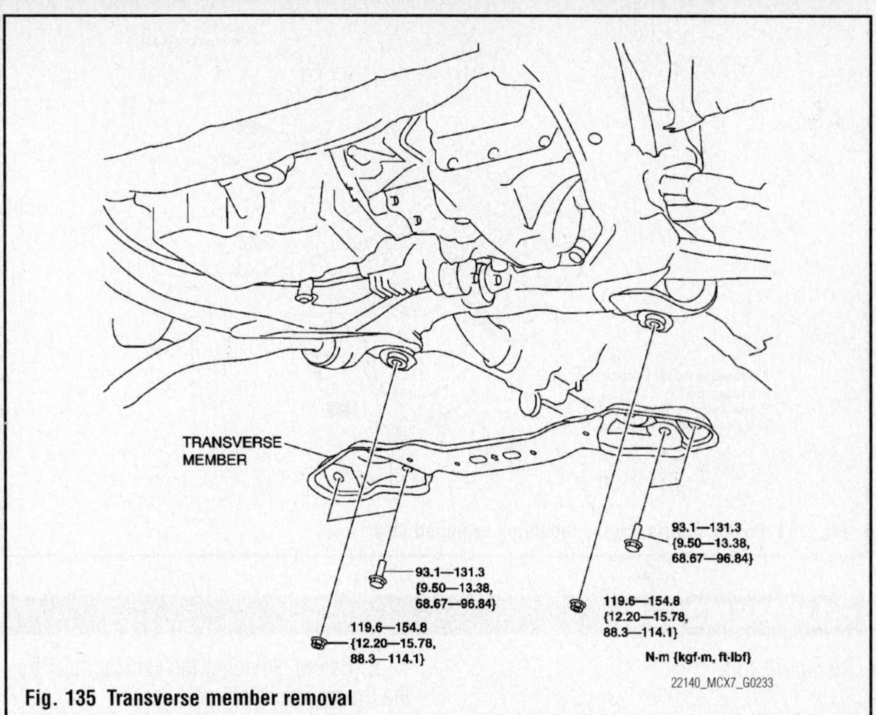

Fig. 135 Transverse member removal

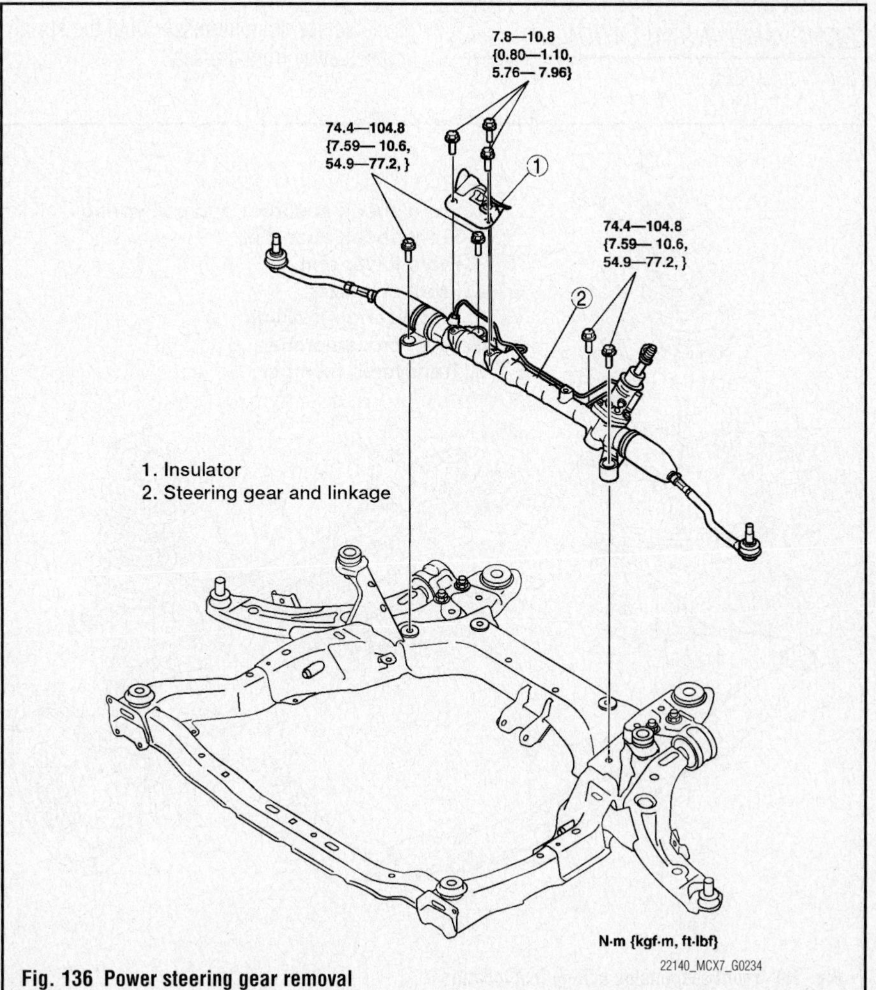

Fig. 136 Power steering gear removal

POWER STEERING PUMP

REMOVAL & INSTALLATION

See Figure 137.

1. Before servicing the vehicle, refer to the Precautions Section.
2. Remove the accessory drive belt.
3. Remove the pressure switch connector.
4. Remove the high pressure line.
5. Disconnect the return hose.
6. Unbolt and remove the power steering pump.
7. Install in reverse of removal. Tighten the power steering pump bolts to 14–18 ft. lbs. (29.4–44.1 Nm).

BLEEDING

✻✻ CAUTION

Do not turn the steering wheel during the fluid level inspection, otherwise the fluid level changes and cannot be inspected correctly.

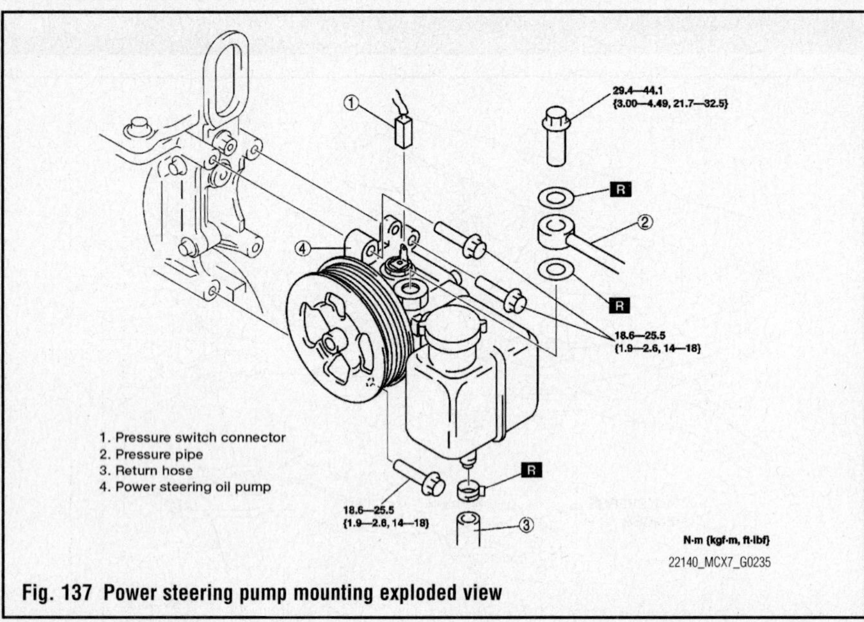

Fig. 137 Power steering pump mounting exploded view

1. Before servicing the vehicle, refer to the Precautions Section.
2. Inspect the fluid level.
3. Jack up the front of the vehicle and support it on safety stands.
4. Turn the steering wheel fully to the left and right several times with the engine not running.
5. Re-inspect the fluid level.
6. If it has dropped, add fluid.
7. Repeat until the fluid level stabilizes.
8. Lower the vehicle.
9. Start the engine and let it idle.
10. Turn the steering wheel fully to the left and right several times.
11. Verify that the fluid is not foamy and that the fluid level has not dropped.
12. If the fluid level has dropped, add fluid as necessary and repeat the last two steps.

SUSPENSION

FRONT SUSPENSION

See Figure 138.

COIL SPRING

REMOVAL & INSTALLATION

See Figure 139.

1. Before servicing the vehicle, refer to the Precautions Section.
2. Remove the strut from the vehicle.
3. Compress the coil spring using a suitable spring compressor until the spring comes away from the seat.

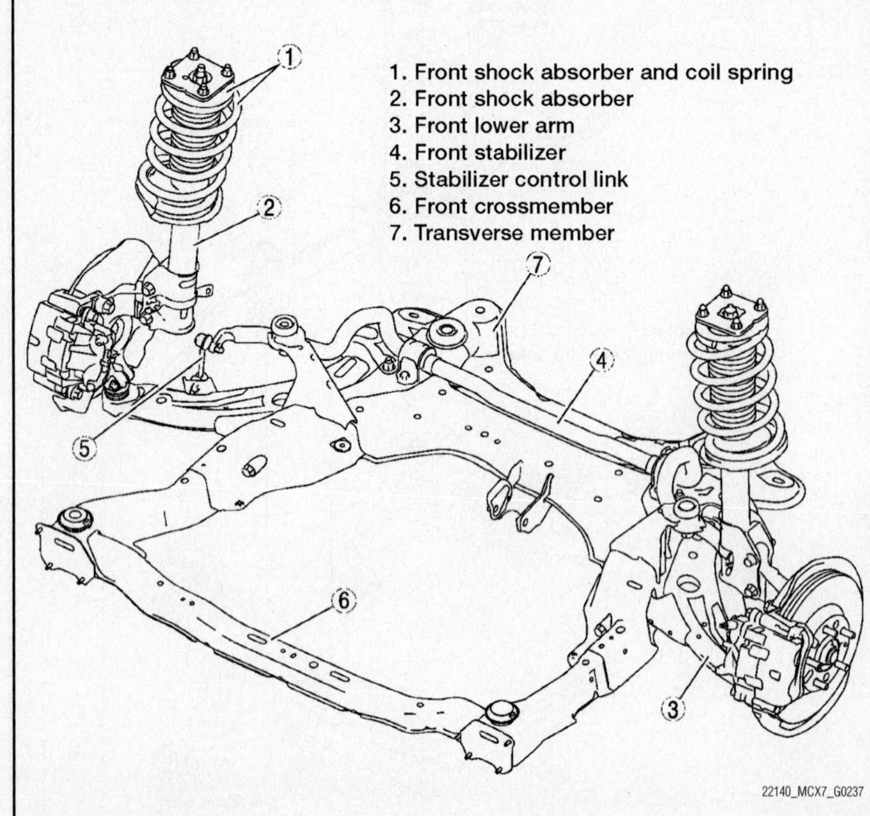

Fig. 138 Front suspension component locations

4. Remove the large center nut and slowly release the spring compressor.

To install:

5. Compress the spring and install it on the strut.
6. Install the lower washer and mounting bracket.
7. Install the upper washer and a new nut.
8. Install the strut assembly in the vehicle.

LOWER BALL JOINT

REMOVAL & INSTALLATION

The lower ball joint is serviced with the lower control arm as an assembly.

LOWER CONTROL ARM

REMOVAL & INSTALLATION

See Figure 140.

1. Before servicing the vehicle, refer to the Precautions Section.
2. Remove the undercover.
3. Remove or disconnect the following:
   - ABS wheel speed sensor
   - Tie-rod end ball joint
   - Stabilizer control link lower nut
   - Front lower arm ball joint
   - Front lower arm
4. Install in reverse of removal.

MACPHERSON STRUT

REMOVAL & INSTALLATION

See Figures 141 and 142.

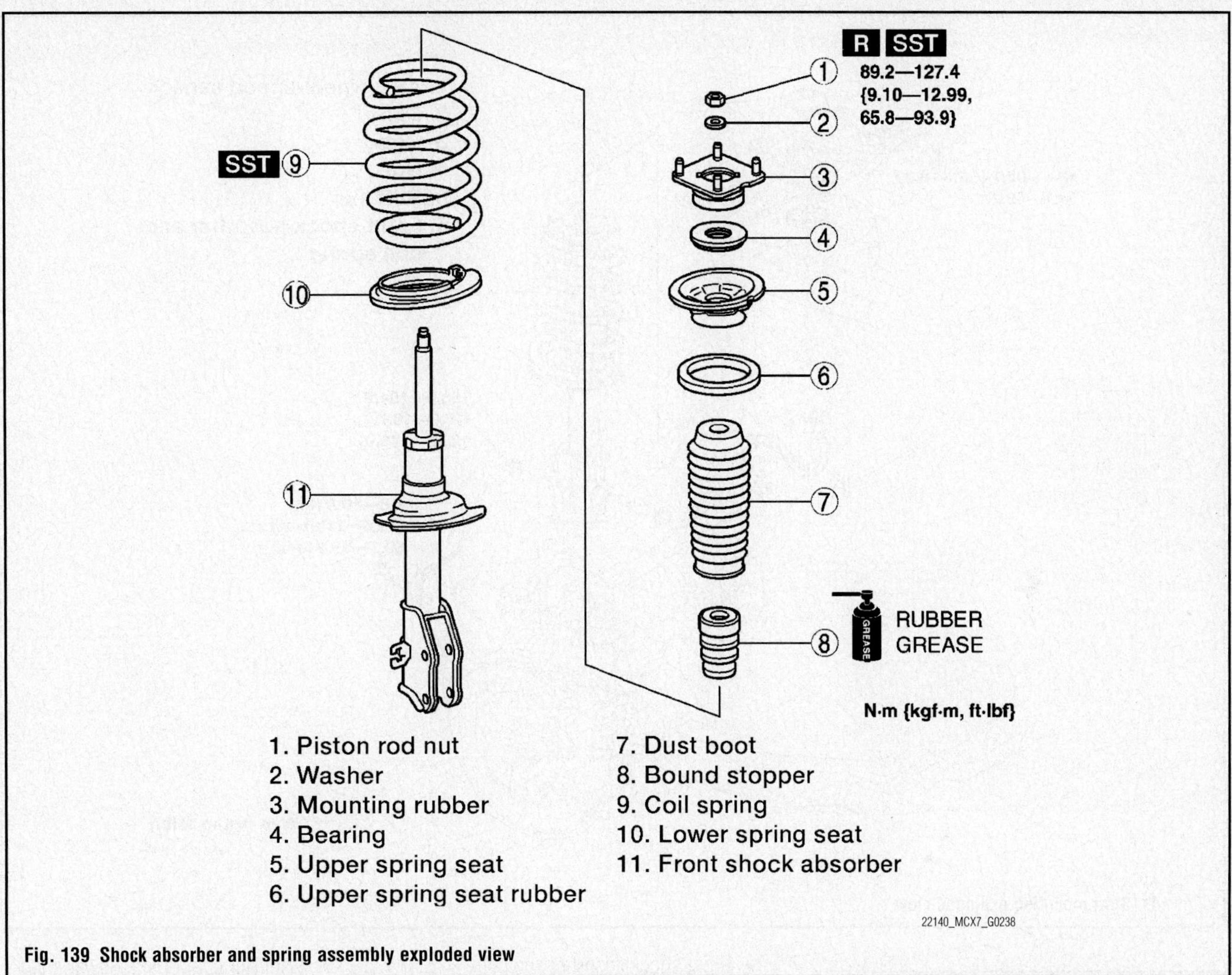

Fig. 139 Shock absorber and spring assembly exploded view

1. ABS wheel speed sensor
2. Tie-rod end ball joint
3. Stabilizer control link lower nut
4. Front lower arm ball joint
5. Front lower arm

47.0—59.0
{4.80—6.01,
34.7—43.5}

7.8—10.8 N·m
{80—110 kgf·cm,
69.1—95.5 in·lbf}

R

85.7—100.0
{8.74—10.19,
63.21—73.75}

85.7—100.0
{8.74—10.19,
63.21—73.75}

43.1—58.8
{4.40—5.99,
31.8—43.3}

43.1—60.8
{4.32—6.12,
31.2—44.3}

SST

N·m {kgf·m, ft·lbf}

22140_MCX7_G0239

Fig. 140 Lower control arm mounting exploded view

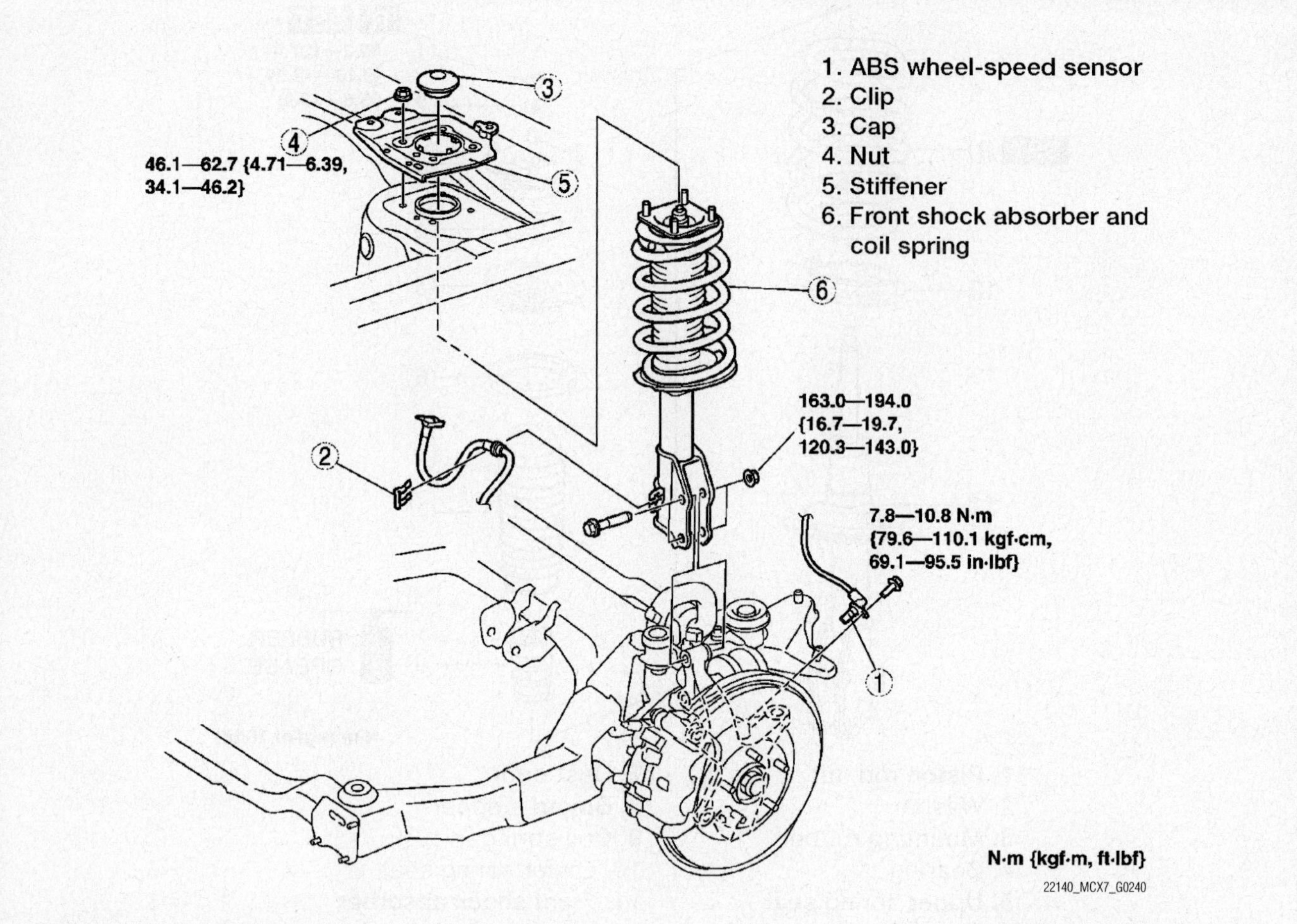

1. ABS wheel-speed sensor
2. Clip
3. Cap
4. Nut
5. Stiffener
6. Front shock absorber and coil spring
46.1—62.7 {4.71—6.39, 34.1—46.2}
163.0—194.0 {16.7—19.7, 120.3—143.0}
7.8—10.8 N·m {79.6—110.1 kgf·cm, 69.1—95.5 in·lbf}
N·m {kgf·m, ft·lbf}
22140_MCX7_G0240

Fig. 141 Strut mounting exploded view

FRONT
LH
RH
IDENTIFICATION MARK
22140_MCX7_G0241

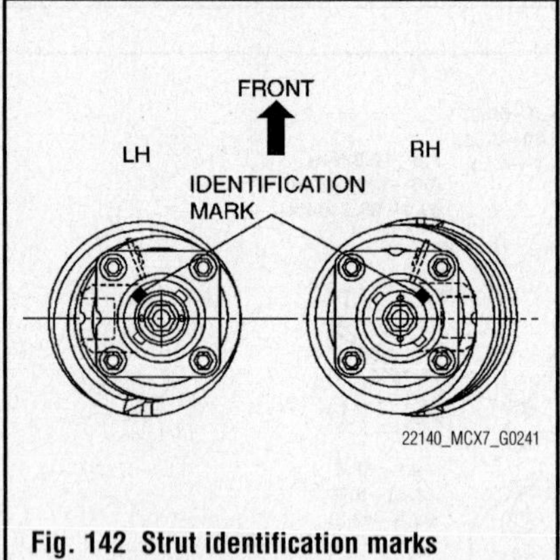

Fig. 142 Strut identification marks

1. Before servicing the vehicle, refer to the Precautions Section.
2. Remove the windshield wiper arm.
3. Remove the cowl grille.
4. Remove or disconnect the following:
   - ABS wheel-speed sensor
   - Clip
   - Cap
   - Nut
   - Stiffener
   - Front shock absorber and coil spring
5. Install in the reverse order of removal.

➡Install the front shock absorber and coil spring so that the identification mark on the mounting rubber is facing to the position indicated in the figure.

6. Inspect for front wheel alignment, and adjust it as necessary.

STEERING KNUCKLE

REMOVAL & INSTALLATION

See Figure 143.

1. Before servicing the vehicle, refer to the Precautions Section.
2. Remove the front ABS wheel-speed sensor.
3. Remove or disconnect the following:
   - Brake hose clip
   - Locknut
   - Brake caliper component
   - Disc plate
   - Stabilizer control link (lower side)
   - Tie-rod end ball joint
   - Nut (front shock absorber lower side)
   - Front lower arm ball joint
   - Wheel hub, steering knuckle component
   - Wheel hub
   - Retaining ring
   - Wheel bearing
   - Dust cover
   - Steering knuckle
4. Install in the reverse order of removal.
5. After installation, inspect the front wheel alignment and adjust it if necessary.

STABILIZER BAR

REMOVAL & INSTALLATION

See Figures 144 through 146.

1. Before servicing the vehicle, refer to the Precautions Section.
2. Remove the side cover and undercover.
3. Drain the power steering fluid.

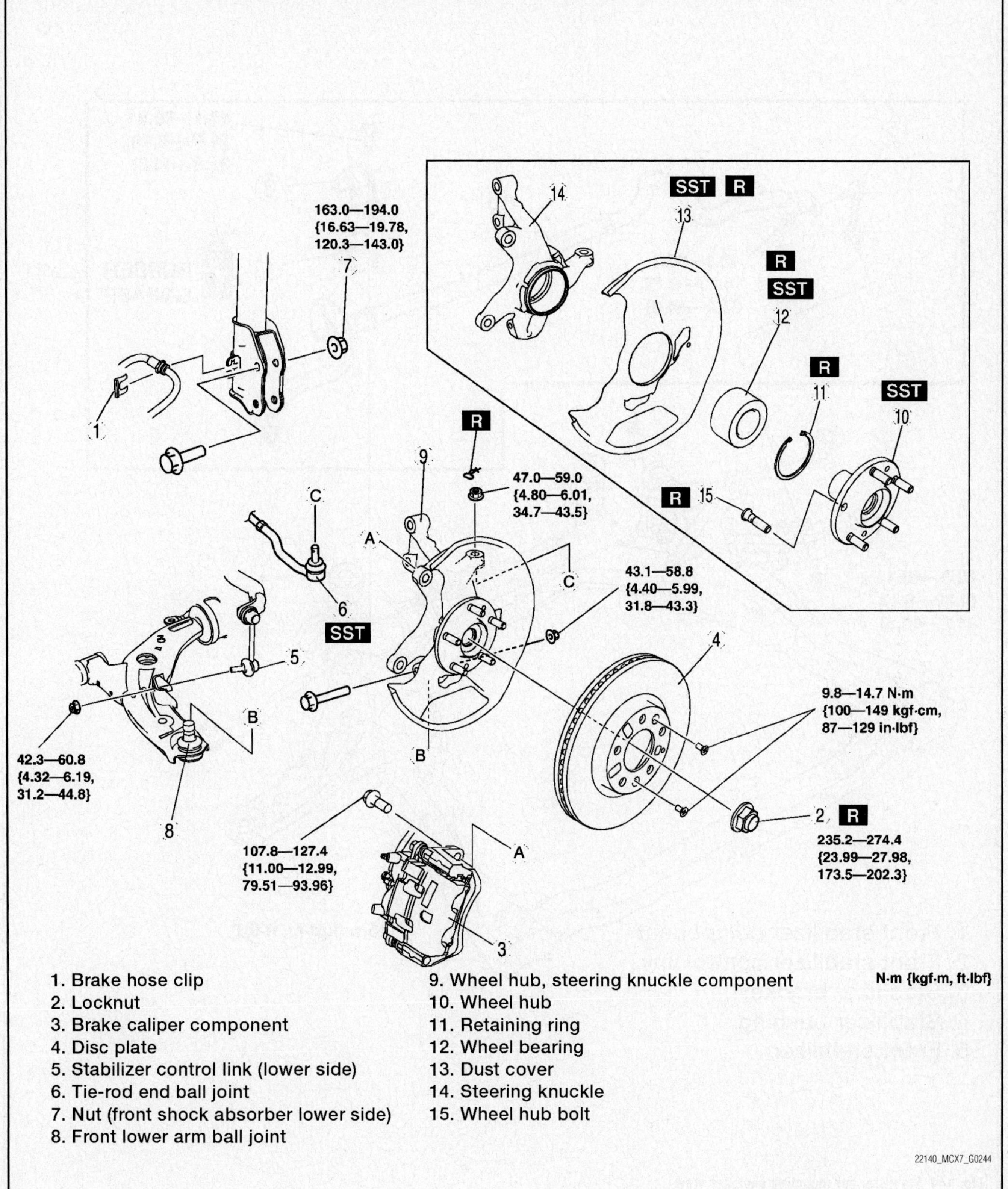

Fig. 143 Steering knuckle, hub, and wheel bearing exploded view

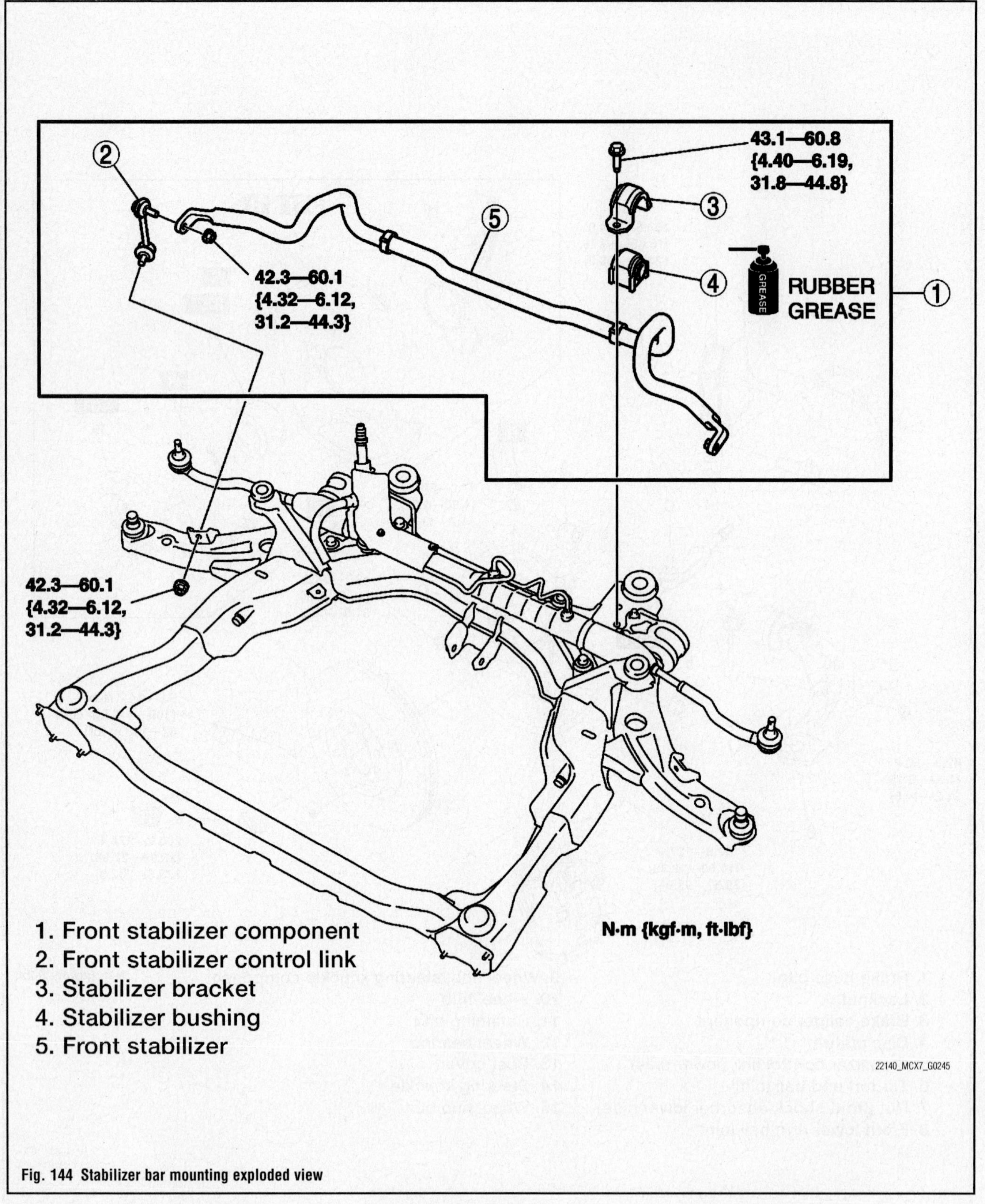

Fig. 144 Stabilizer bar mounting exploded view

4. Remove the transverse member.
5. Remove the front crossmember component.
6. Remove or disconnect the following:

- Front stabilizer component
- Front stabilizer control link
- Stabilizer bracket
- Stabilizer bushing
- Front stabilizer

7. Install in the reverse order of removal.

➡Install the stabilizer bar so that the identification mark is on the left side of the vehicle.

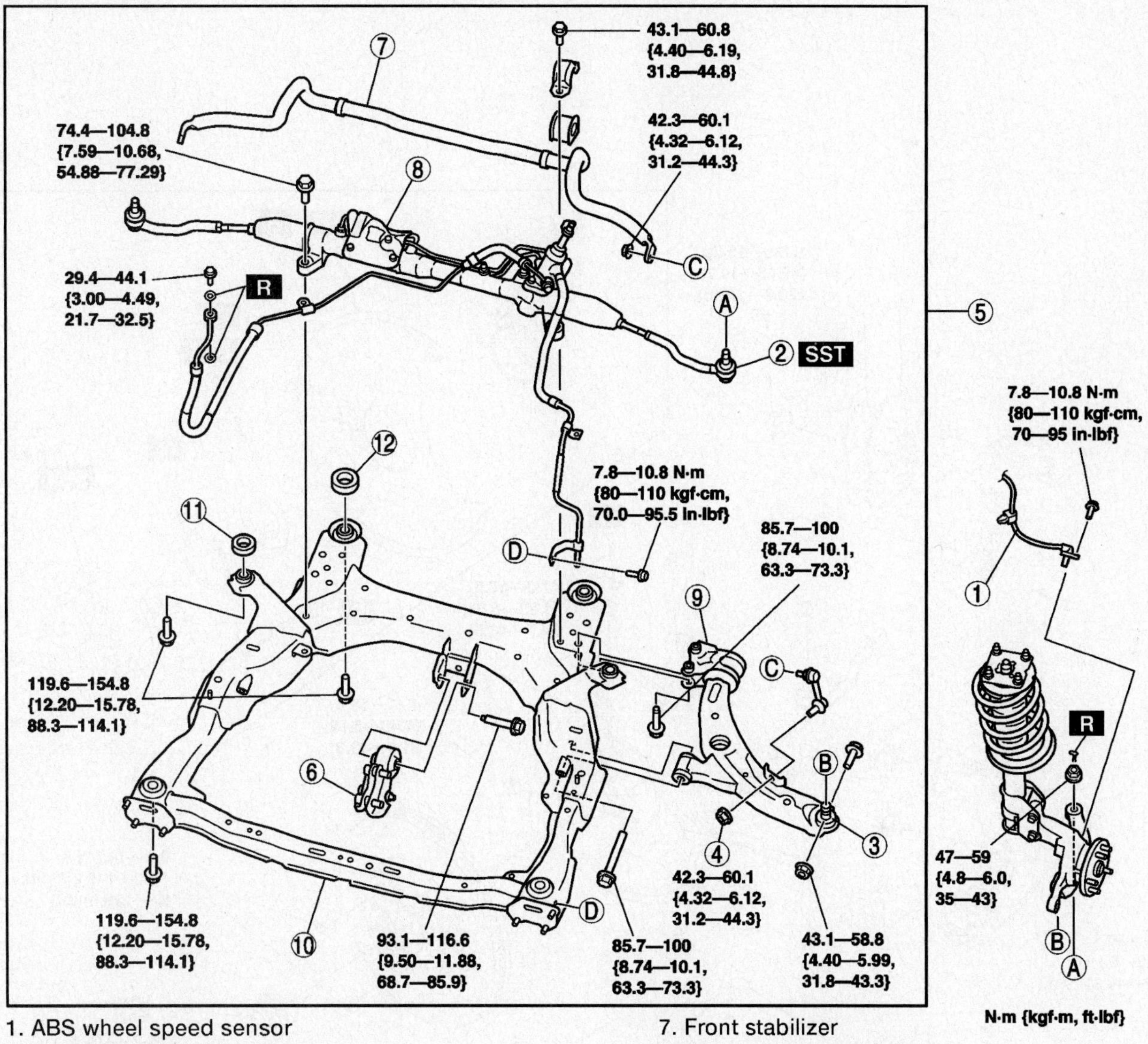

1. ABS wheel speed sensor
2. Tie-rod end ball joint
3. Front lower arm ball joint
4. Stabilizer control link lower side nut
5. Steering gear and linkage, front stabilizer, front lower arm and front crossmember component
6. No. 1 engine mount
7. Front stabilizer
8. Steering gear and linkage
9. Front lower arm
10. Front crossmember
11. Front crossmember mounting rubber (front)
12. Front crossmember mounting rubber (rear)

N·m {kgf·m, ft·lbf}

22140_MCX7_G0246

Fig. 145 Front crossmember exploded view

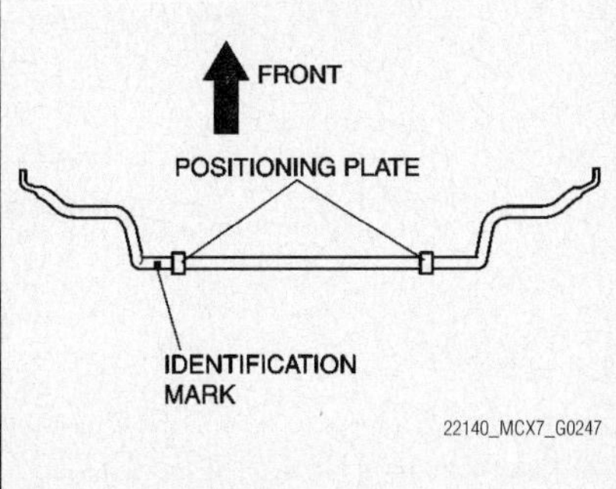

Fig. 146 Stabilizer bar identification mark

WHEEL BEARINGS

REMOVAL & INSTALLATION

See Figure 147.

1. Before servicing the vehicle, refer to the Precautions Section.
2. Remove the front ABS wheel-speed sensor.
3. Remove or disconnect the following:
   - Brake hose clip
   - Locknut
   - Brake caliper component
   - Disc plate
   - Stabilizer control link (lower side)
   - Tie-rod end ball joint
   - Nut (front shock absorber lower side)
   - Front lower arm ball joint
   - Wheel hub, steering knuckle component
   - Wheel hub
   - Retaining ring
   - Wheel bearing
   - Dust cover
   - Steering knuckle
4. Using a hydraulic press, remove the wheel hub.

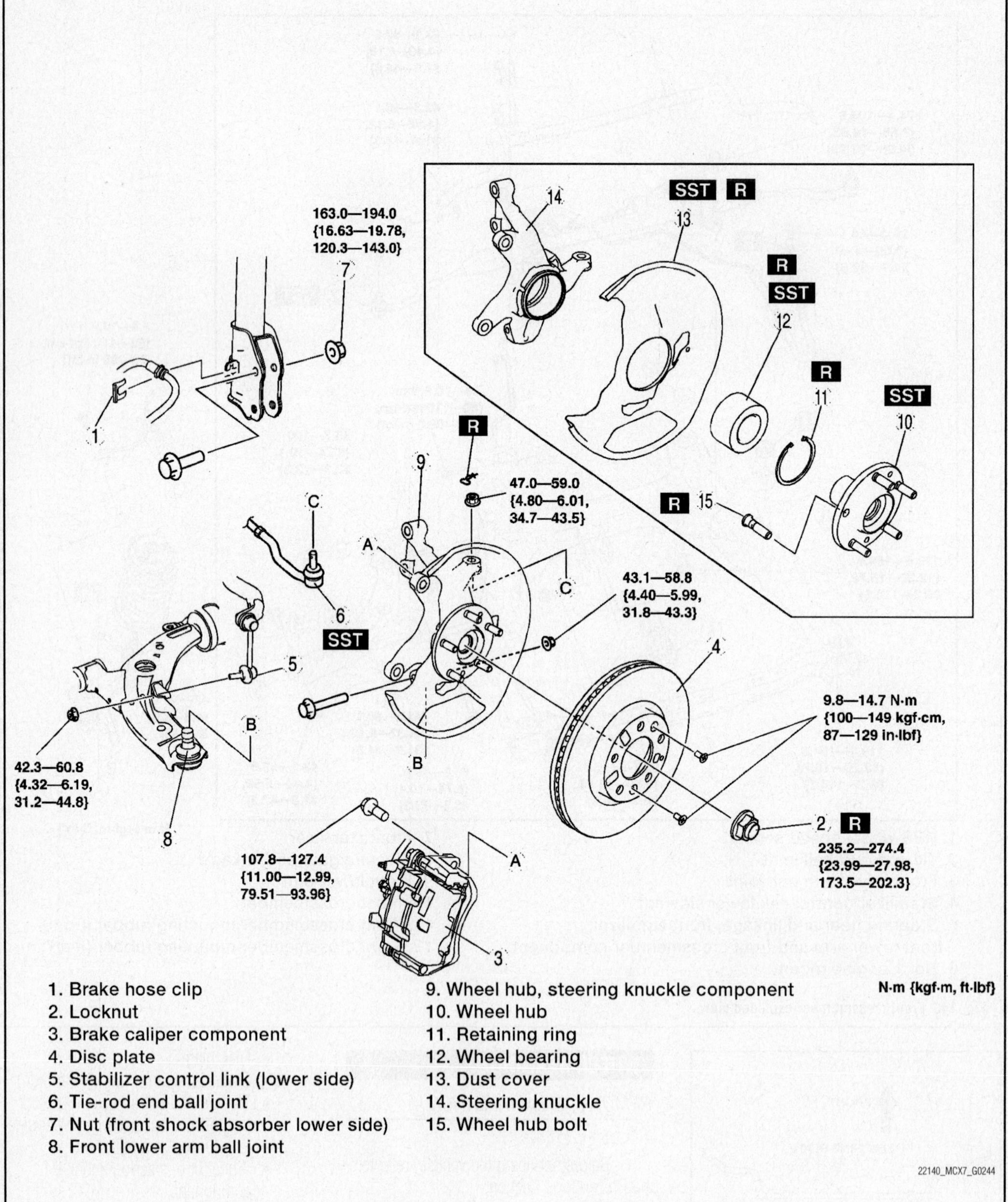

Fig. 147 Steering knuckle, hub, and wheel bearing exploded view

5. Remove the retainer and press the wheel bearing from the knuckle.
6. Install in the reverse order of removal.
7. After installation, inspect the front wheel alignment and adjust it if necessary.

REAR SUSPENSION

See Figures 148 and 149.

COIL SPRING

REMOVAL & INSTALLATION

See Figure 150.

1. Before servicing the vehicle, refer to the Precautions Section.
2. Remove the rear wheel.
3. Remove the middle exhaust pipe.
4. Remove the rear stabilizer.
5. Raise the rear trailing link to the unloaded condition with a jack.
6. Loosen the inner and outer bolts of the lateral link.
7. Loosen the outer bolt of the upper arm.
8. Support the lower arm with a jack.
9. Loosen the inner bolt of the lower arm.

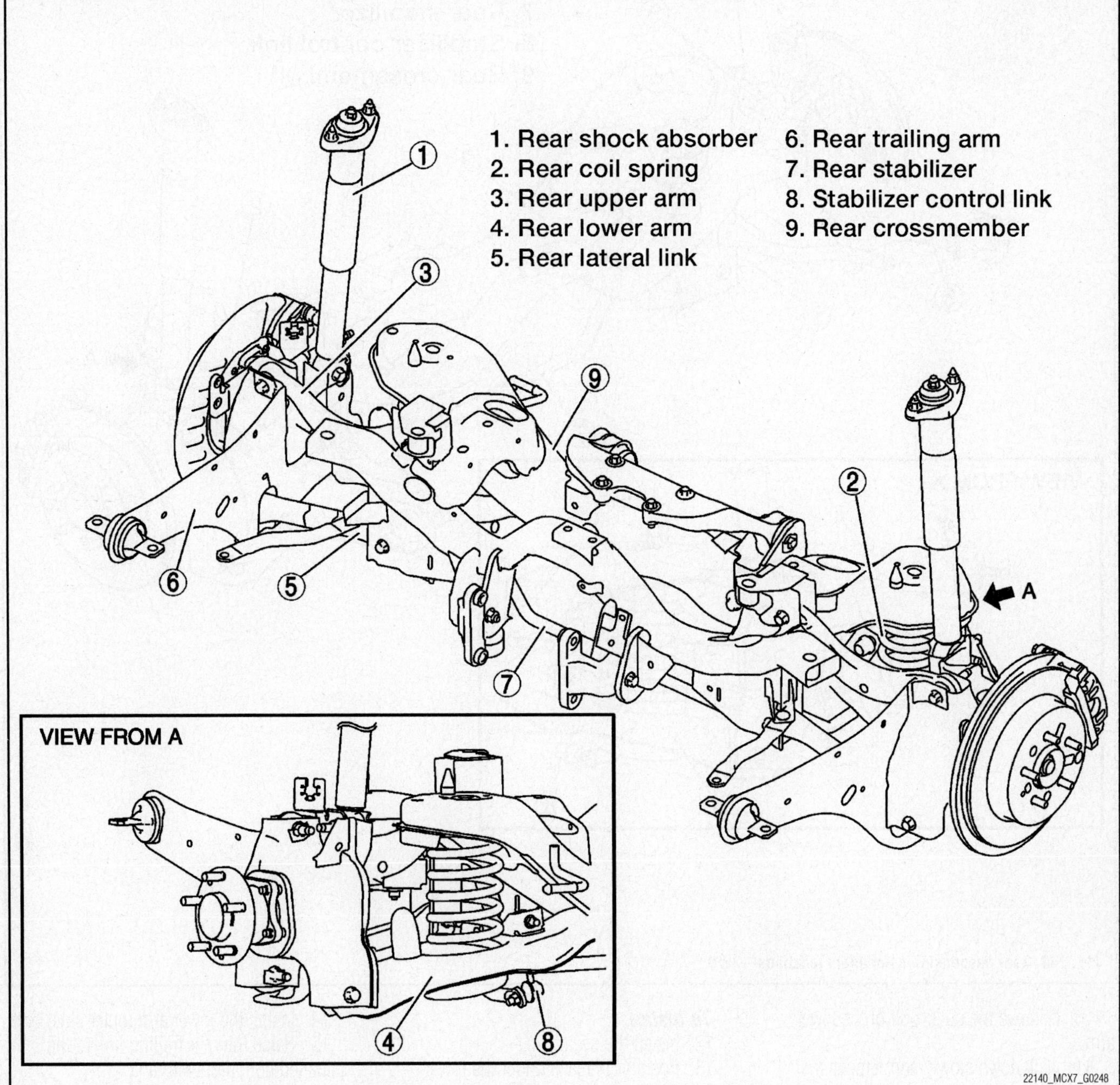

Fig. 148 Rear suspension component locations—AWD

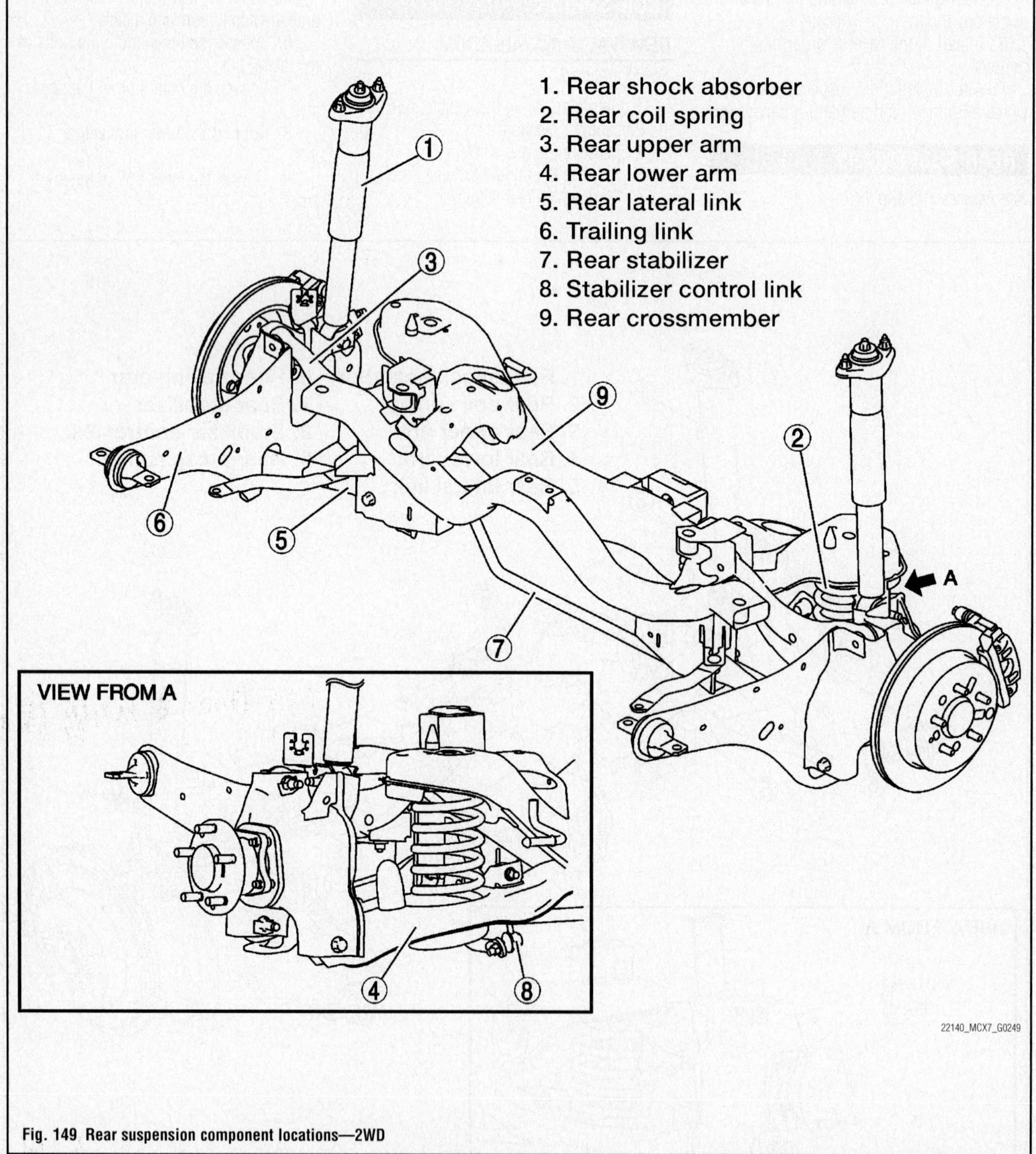

Fig. 149 Rear suspension component locations—2WD

10. Remove the outer bolt of the lower arm.
11. Jack down slowly and remove the coil spring.

⁂ WARNING

Removing the coil spring is dangerous. The coil spring could fly off, the cause serious injury or death.

To install:

12. Install the spring.
13. Position the jack under the lower arm and jack up slowly.

⁂ WARNING

Installing the coil spring is dangerous. The coil spring could fly off, and cause serious injury or death.

14. Install the lower arm (outer side) bolt.
15. Raise the rear trailing link to the unloaded condition with a jack.
16. Tighten the outer bolt of the upper arm to 72.3–75.2 ft. lbs. (75.5–102 Nm).
17. Tighten the inner and outer bolts of the lateral link to 72.3–75.2 ft. lbs. (75.5–102 Nm).
18. Install the rear stabilizer.

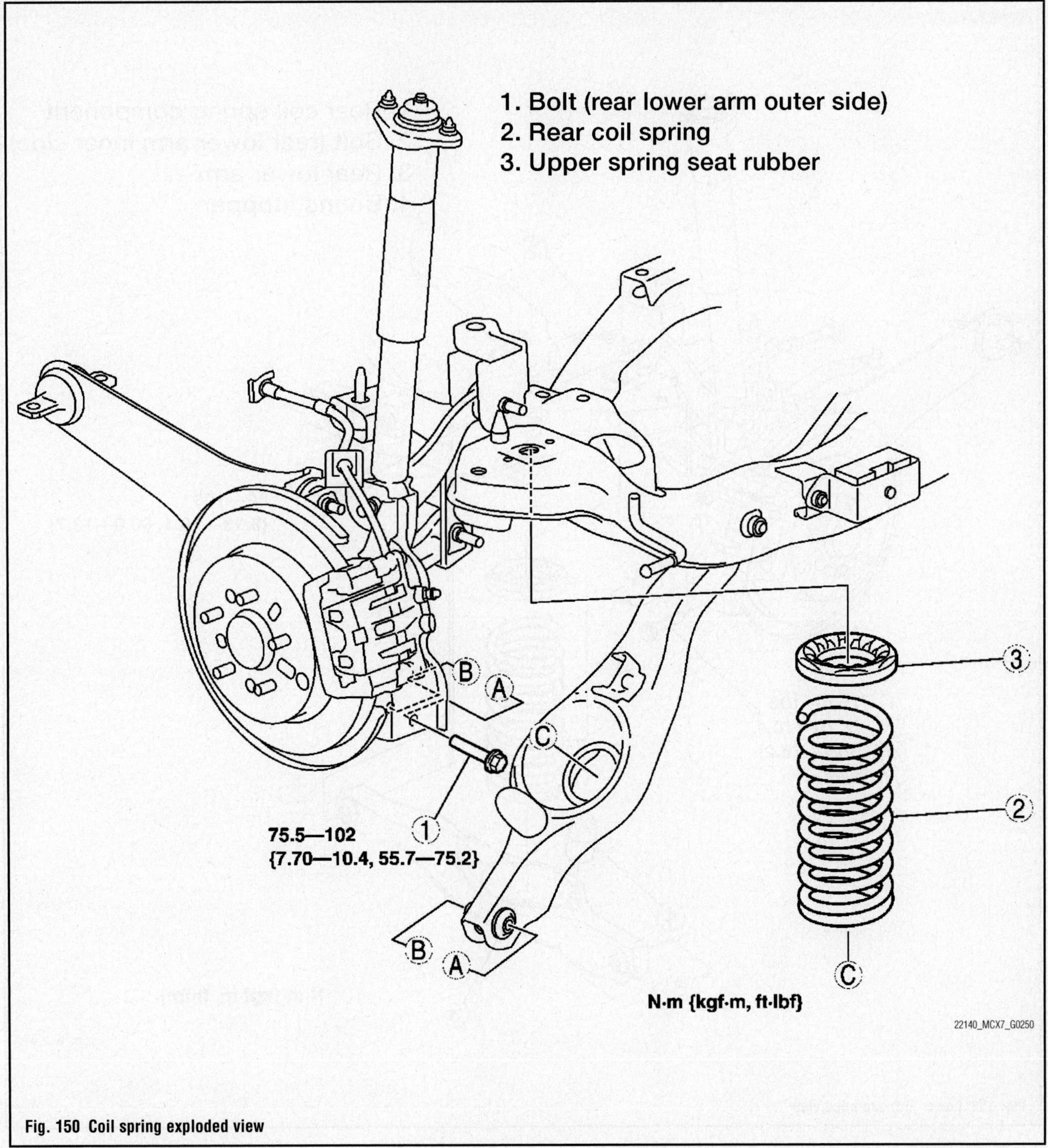

Fig. 150 Coil spring exploded view

19. Install the middle exhaust pipe.
20. Install the rear wheel.
21. Check the rear wheel alignment and adjust if necessary.

CONTROL ARMS/LINKS

REMOVAL & INSTALLATION

Lower Arm

See Figure 151.

1. Before servicing the vehicle, refer to the Precautions Section.
2. Remove the rear wheel.
3. Remove the middle exhaust pipe.
4. Remove the rear stabilizer.
5. Raise the rear trailing link to the unloaded condition with a jack.
6. Loosen the inner and outer bolts of the lateral link.
7. Loosen the outer bolt of the upper arm.
8. Support the lower arm with a jack.
9. Loosen the inner bolt of the lower arm.
10. Remove the outer bolt of the lower arm.
11. Jack down slowly and remove the coil spring.

WARNING

Removing the coil spring is dangerous. The coil spring could fly off, the cause serious injury or death.

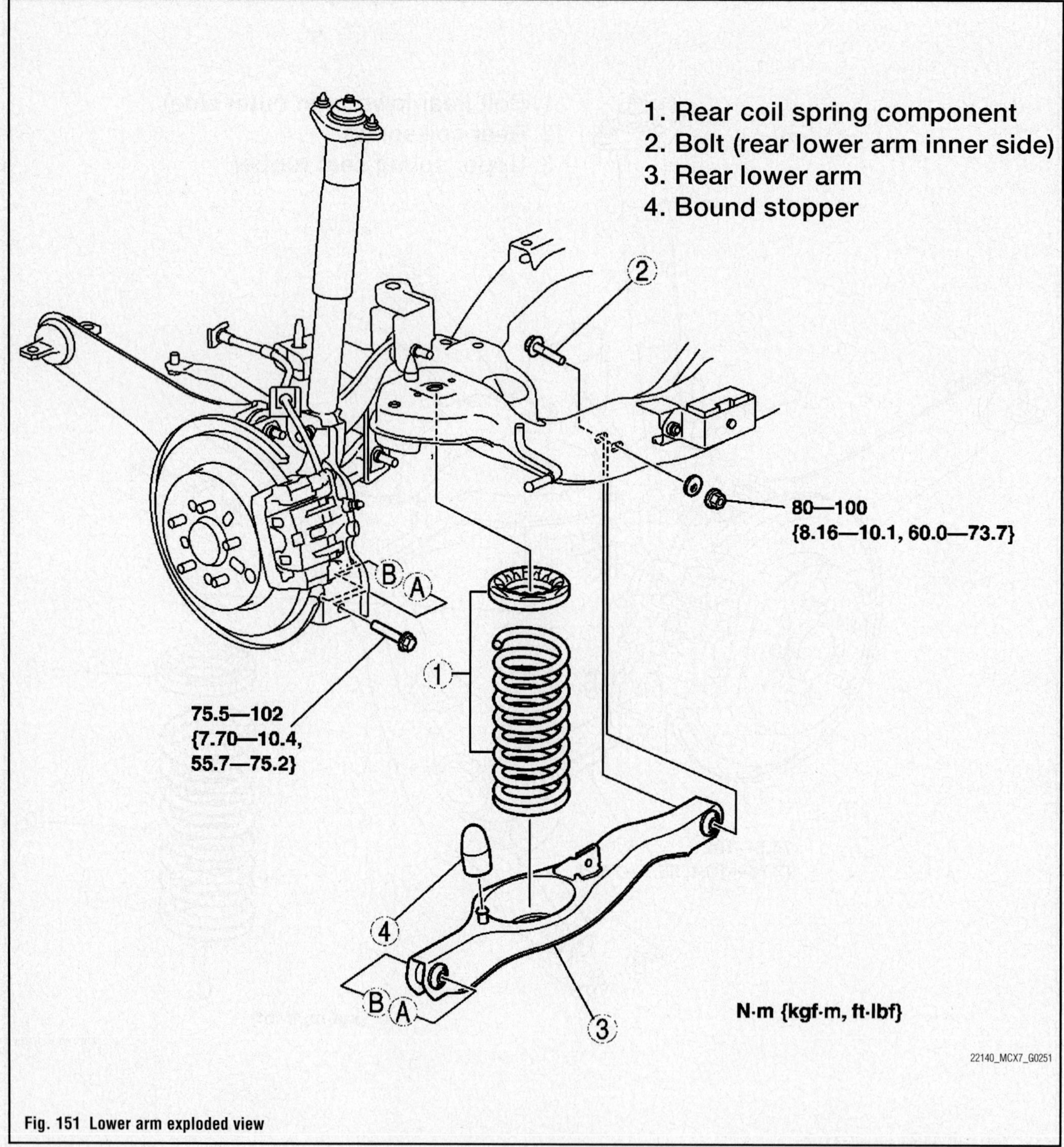

Fig. 151 Lower arm exploded view

12. Remove the bolt and remove the lower arm.

To install:

13. Install the lower arm
14. Install the spring.
15. Position the jack under the lower arm and jack up slowly.

WARNING

Installing the coil spring is dangerous. The coil spring could fly off, and cause serious injury or death.

16. Install the lower arm (outer side) bolt.
17. Raise the rear trailing link to the unloaded condition with a jack.
18. Tighten the outer bolt of the upper arm to 72.3–75.2 ft. lbs. (75.5–102 Nm).
19. Tighten the inner and outer bolts of the lateral link to 72.3–75.2 ft. lbs. (75.5–102 Nm).
20. Install the rear stabilizer.
21. Install the middle exhaust pipe.
22. Install the rear wheel.
23. Check the rear wheel alignment and adjust if necessary.

Upper Arm

See Figure 152.

1. Before servicing the vehicle, refer to the Precautions Section.
2. Remove the rear wheel.

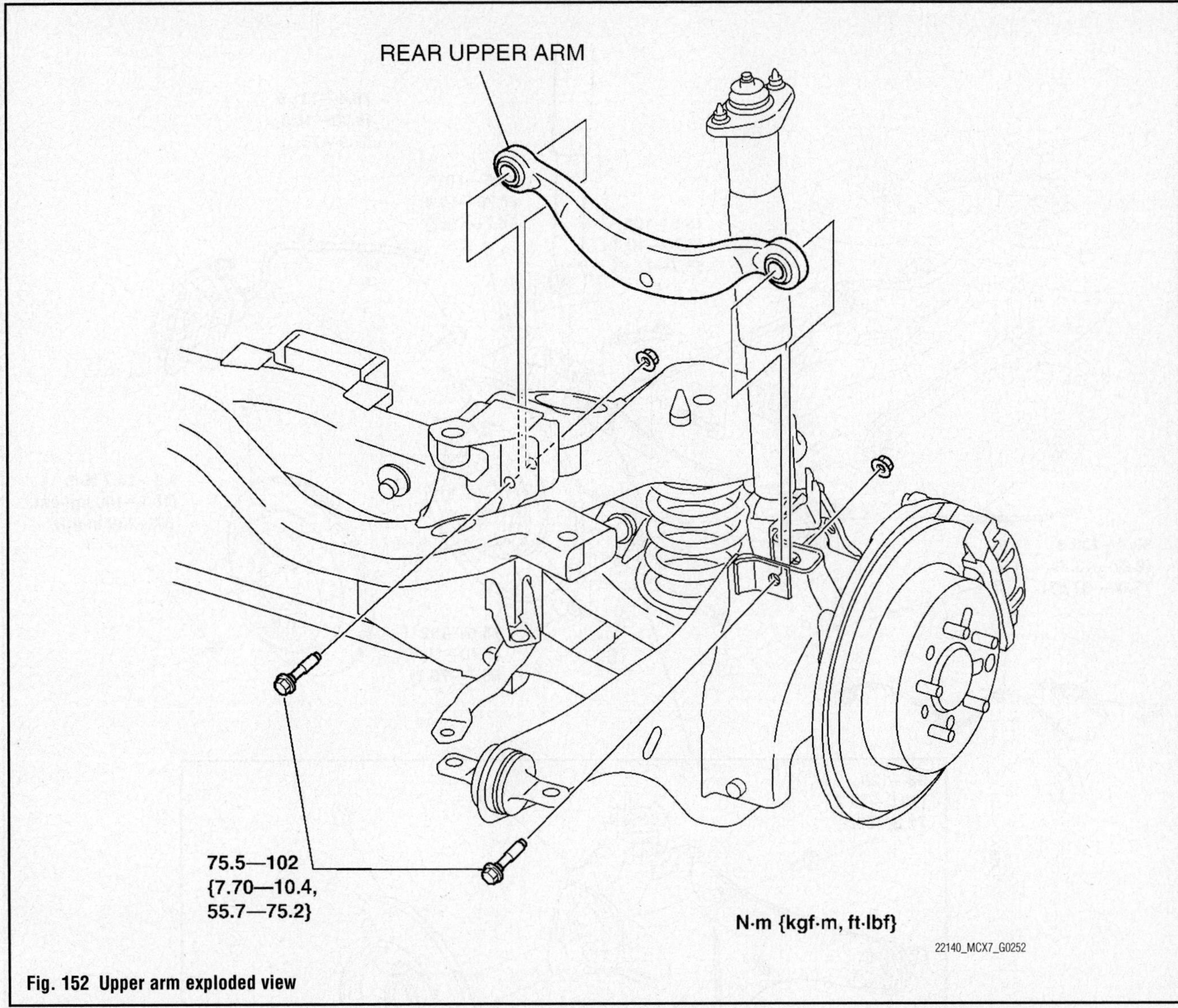

Fig. 152 Upper arm exploded view

3. Raise the trailing link to the unloaded condition with the jack.

➡Jacking up the rear suspension to the no-occupant position will lighten the force on the bushing and make it easier to perform the procedure.

4. Remove the rear upper arm.

To install:

5. Install the upper arm. Tighten the bolts to 55.7–75.2 ft. lbs. (75.5–102 Nm).
6. Install the rear wheel.
7. Check the rear wheel alignment and adjust if necessary.

Lateral Link

See Figure 153.

1. Before servicing the vehicle, refer to the Precautions Section.
2. Remove the rear wheel.
3. Raise the trailing link to the unloaded condition with the jack.

➡Jacking up the rear suspension to the no-occupant position will lighten the force on the bushing and make it easier to perform the procedure.

4. Remove the rear lateral link.

To install:

5. Install the lateral link. Tighten the bolts to 55.7–75.2 ft. lbs. (75.5–102 Nm).
6. Install the rear wheel.
7. Check the rear wheel alignment and adjust if necessary.

Trailing Link

2WD Models

See Figure 154.

1. Before servicing the vehicle, refer to the Precautions Section.
2. Remove the middle pipe.
3. Remove the rear stabilizer.
4. Remove the rear coil spring.
5. Remove or disconnect the following:
   - ABS wheel-speed sensor
   - Parking brake cable
   - Caliper component
   - Disc plate
   - Bolt (rear shock absorber lower side)
   - Bolt (rear lateral link outer side)
   - Bolt (rear upper arm outer side)
   - Trailing link, rear wheel hub component and parking brake component
   - Rear wheel hub component
   - Parking brake component
   - Trailing link
6. Install in the reverse order of removal.
7. Check the rear wheel alignment and adjust if necessary.

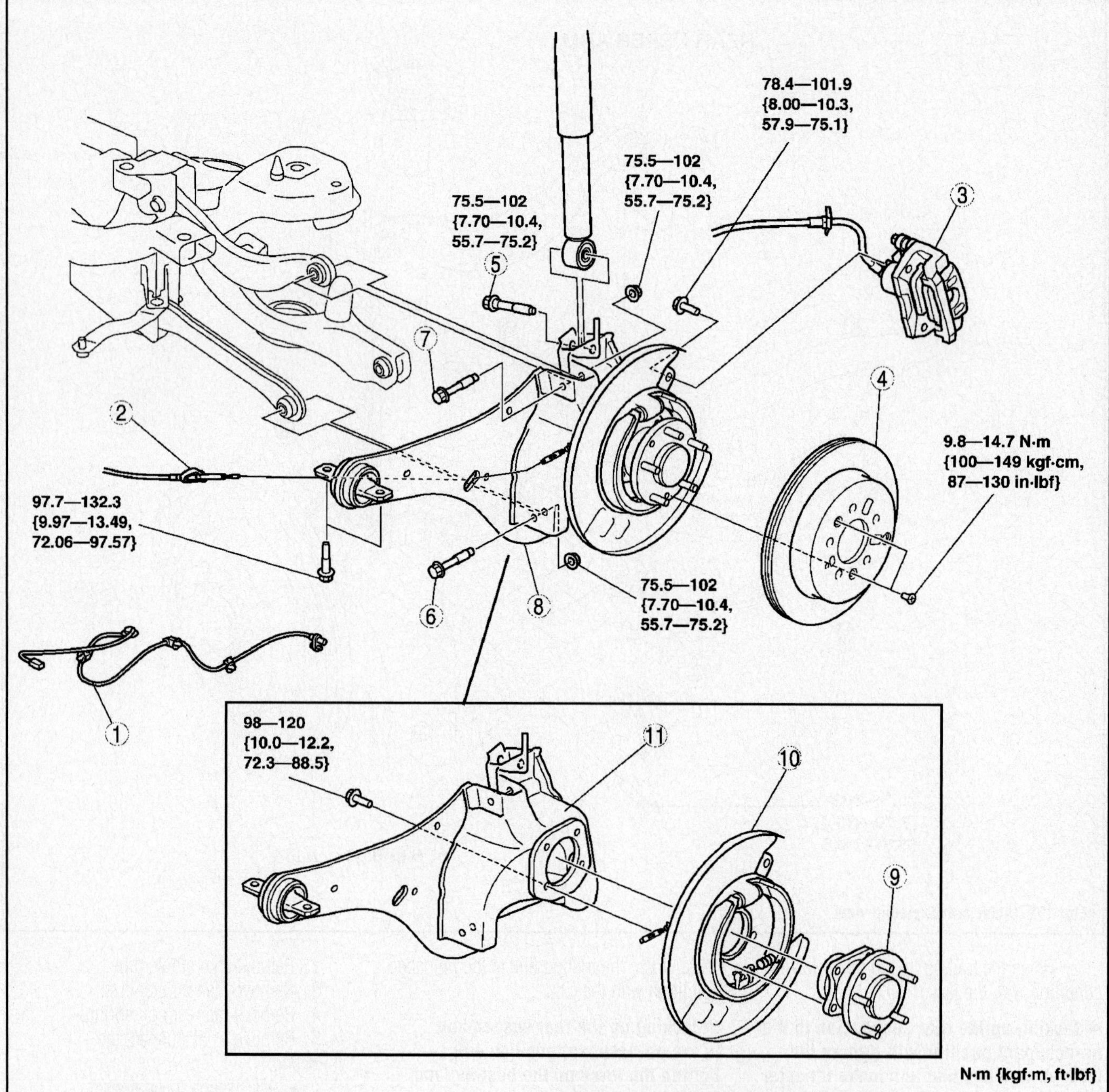

Fig. 153 Lateral link removal

AWD Models

See Figure 155.

1. Before servicing the vehicle, refer to the Precautions Section.
2. Remove the middle pipe.
3. Remove the rear stabilizer.
4. Remove the rear coil spring.
5. Remove or disconnect the following:
   - Rear ABS wheel-speed sensor
   - Parking brake cable
   - Locknut
   - Brake caliper component

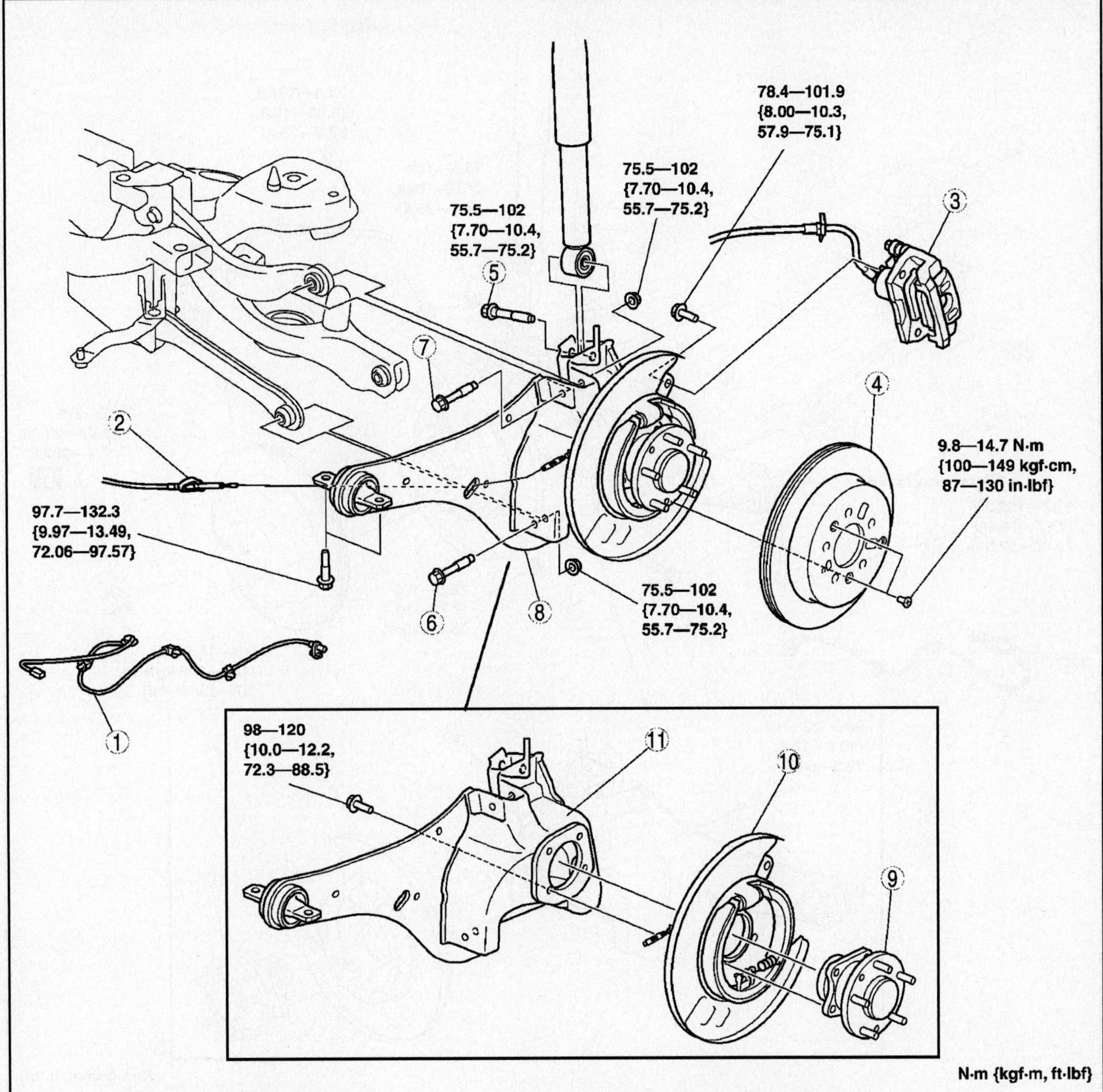

Fig. 154 Trailing link exploded view—2WD models

- Disc plate
- Bolt (rear shock absorber lower side)
- Bolt (rear lateral link outer side)
- Bolt (rear upper arm outer side)
- Trailing link, rear wheel hub component and parking brake component
- Rear wheel hub component
- Parking brake component
- Trailing link

6. Install in the reverse order of removal.

7. Check the rear wheel alignment and adjust if necessary.

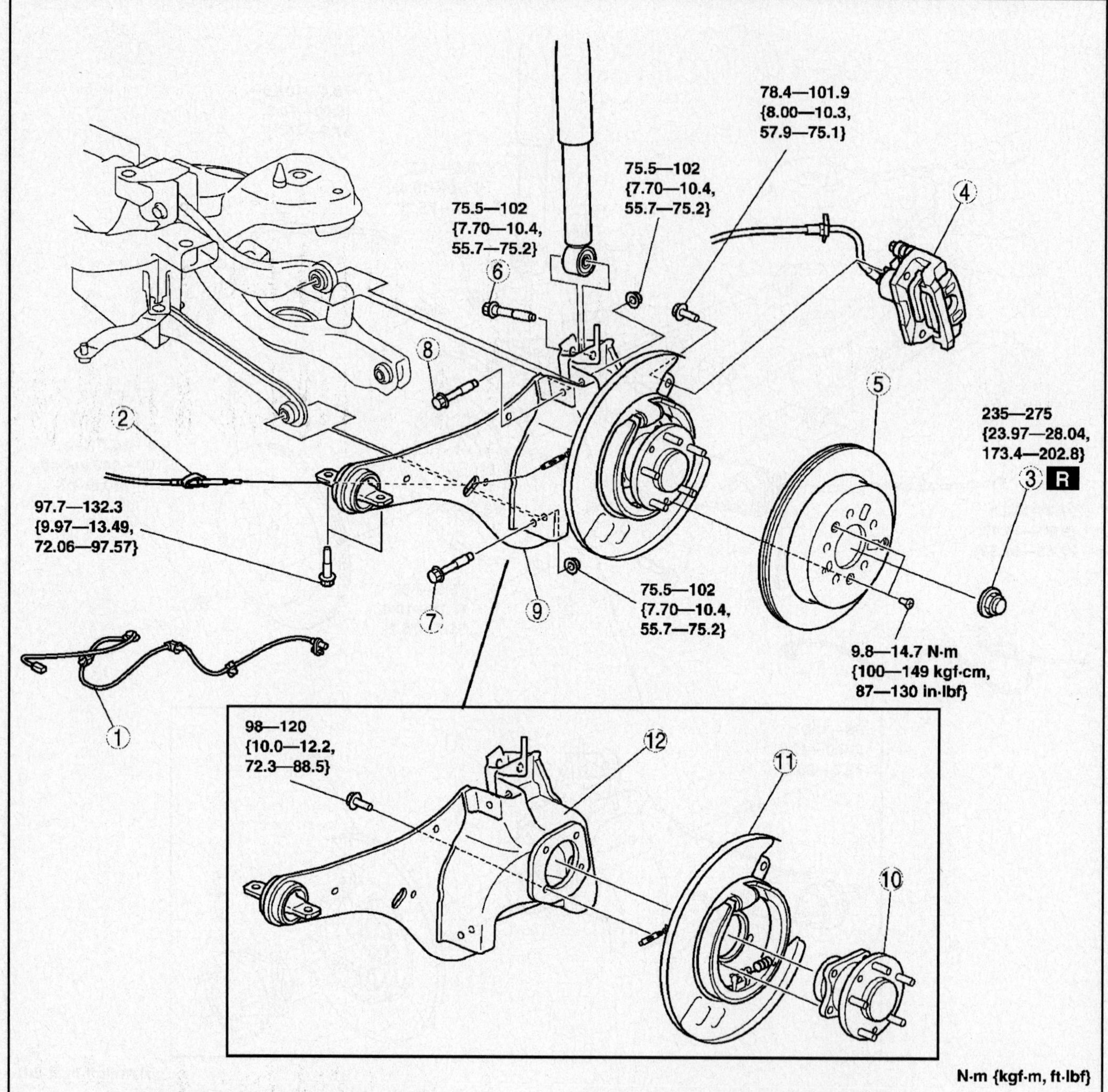

1. Rear ABS wheel-speed sensor
2. Parking brake cable
3. Locknut
4. Brake caliper component
5. Disc plate
6. Bolt (rear shock absorber lower side)
7. Bolt (rear lateral link outer side)
8. Bolt (rear upper arm outer side)
9. Trailing link, rear wheel hub component and parking brake component
10. Rear wheel hub component
11. Parking brake component
12. Trailing link

22140_MCX7_G0254

Fig. 155 Trailing link exploded view—AWD models

SHOCK ABSORBER

REMOVAL & INSTALLATION

See Figures 156 and 157.

1. Before servicing the vehicle, refer to the Precautions Section.
2. Remove the rear wheel.
3. Support the rear axle with the jack.
4. Remove the service hole cover on the trunk side trim.
5. Remove the nuts.
6. Remove or disconnect the following:
   - Nut

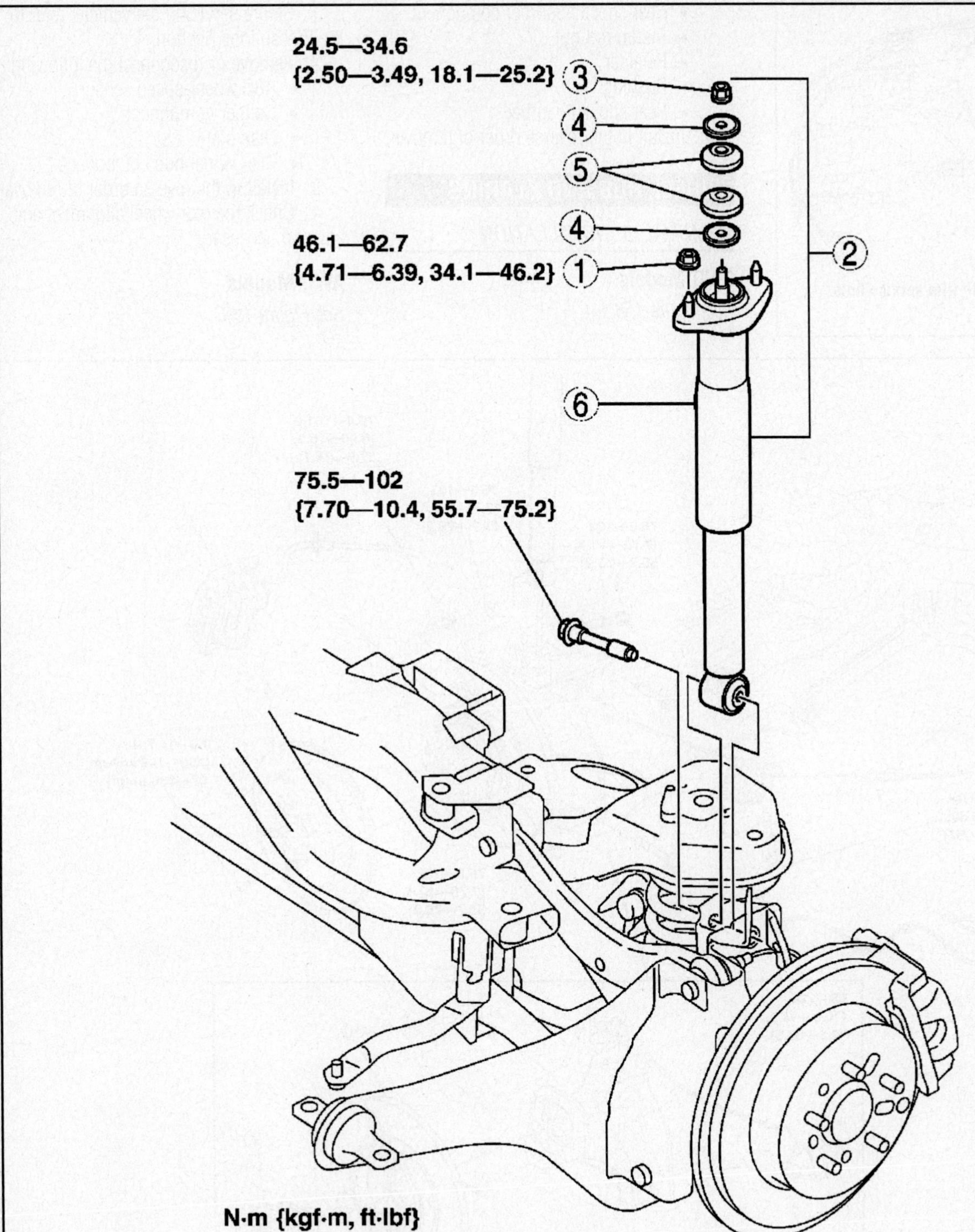

Fig. 156 Shock absorber exploded view

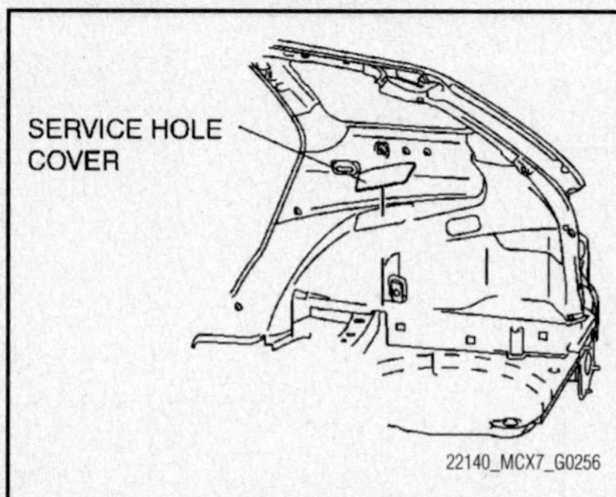

Fig. 157 Trunk side trim service hole cover

- Rear shock absorber component
- Piston rod nut
- Retainer
- Bushing
- Rear shock absorber

7. Install in the reverse order of removal.

WHEEL HUB AND BEARING

REMOVAL & INSTALLATION

2WD Models

See Figure 158.

1. Before servicing the vehicle, refer to the Precautions Section.
2. Remove or disconnect the following:
   - ABS wheel-speed sensor
   - Caliper component
   - Disc plate
   - Rear wheel hub component
3. Install in the reverse order of removal.
4. Check the rear wheel alignment and adjust if necessary.

AWD Models

See Figure 159.

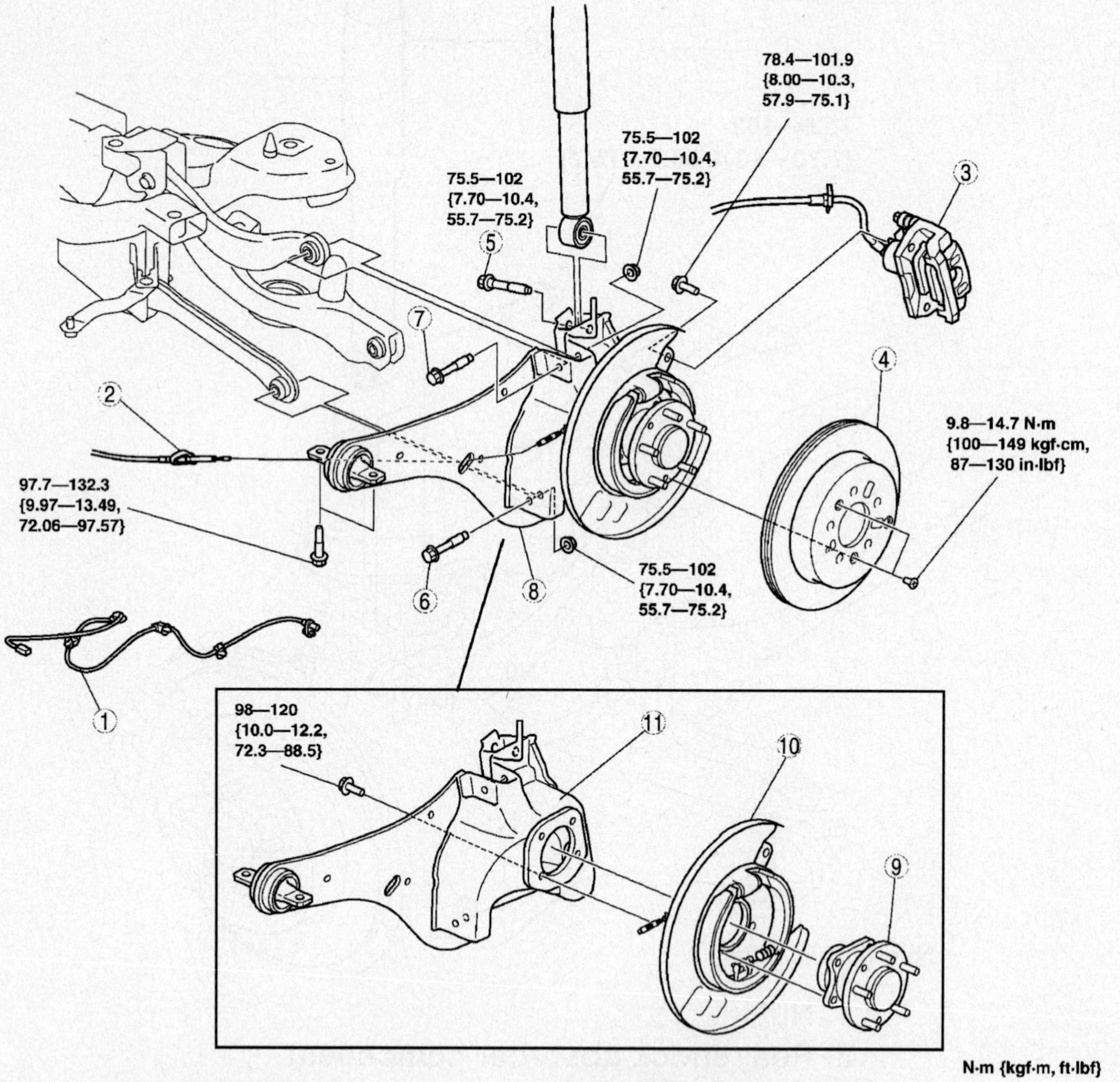

N·m {kgf·m, ft·lbf}

1. ABS wheel-speed sensor
2. Parking brake cable
3. Caliper component
4. Disc plate
5. Bolt (rear shock absorber lower side)
6. Bolt (rear lateral link outer side)
7. Bolt (rear upper arm outer side)
8. Trailing link, rear wheel hub component and parking brake component
9. Rear wheel hub component
10. Parking brake component
11. Trailing link

22140_MCX7_G0253

Fig. 158 Rear hub removal exploded view—2WD models

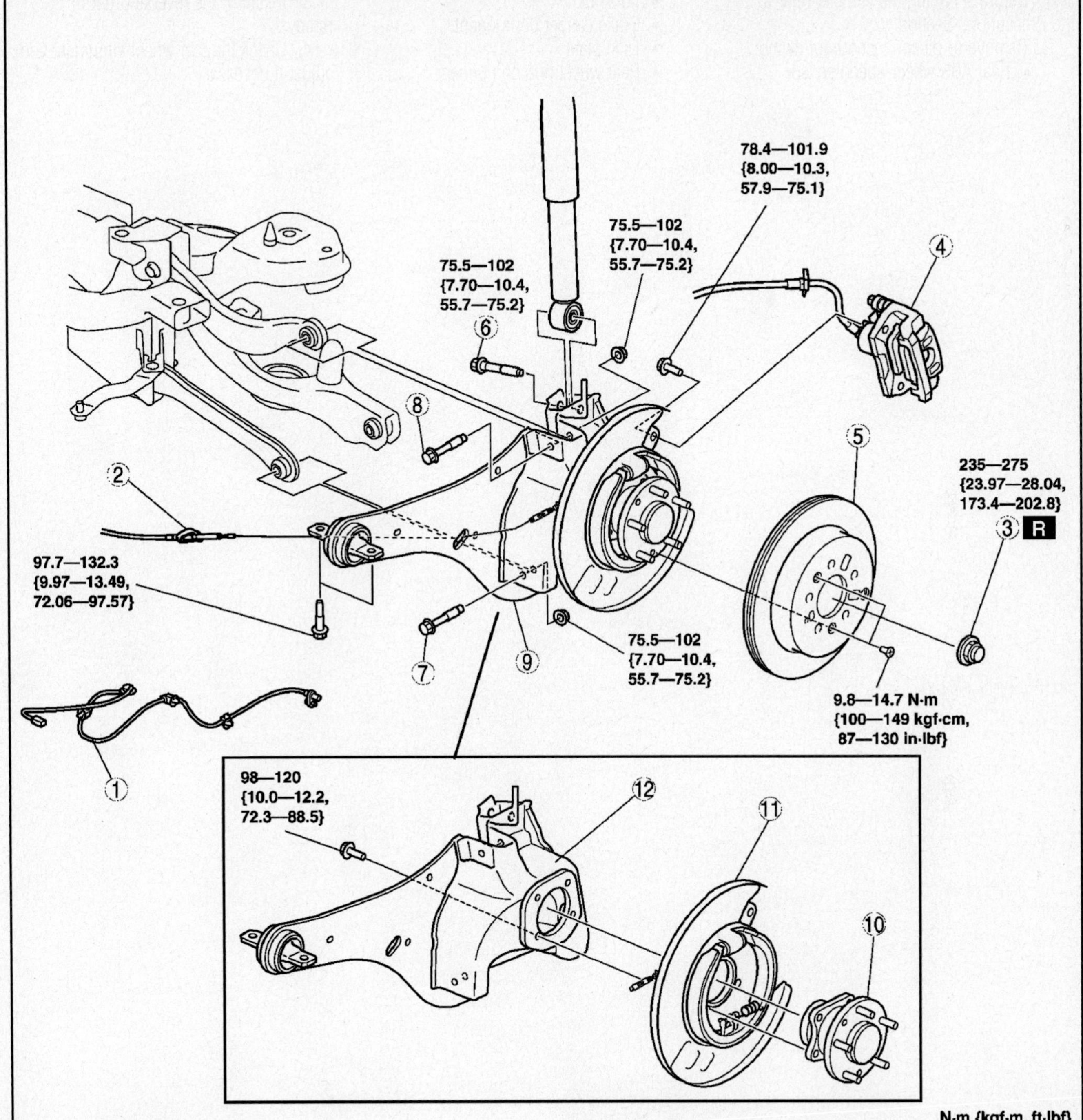

Fig. 159 Rear hub removal exploded view—AWD models

1. Before servicing the vehicle, refer to the Precautions Section.
2. Remove or disconnect the following:
   - Rear ABS wheel-speed sensor
   - Locknut
   - Brake caliper component
   - Disc plate
   - Rear wheel hub component
3. Install in the reverse order of removal.
4. Check the rear wheel alignment and adjust if necessary.

MAZDA

CX-9

4

SPECIFICATIONS AND MAINTENANCE CHARTS

ENGINE AND VEHICLE IDENTIFICATION CHART

Engine Code							Model Year	
Code	Liters (cc)	Cu. In.	Cyl.	Fuel Sys.	Engine Type	Eng. Mfg.	Code ①	Year
A ②	3.7 (3721)	227	6	SMFI	DOHC	Ford Motor Co.	7	2007
C ③	3.5 (3500)	214	6	SMFI	DOHC	Ford Motor Co.	8	2008
V ③	3.7 (3721)	227	6	SMFI	DOHC	Ford Motor Co.		
Y ②	3.5 (3500)	214	6	SMFI	DOHC	Ford Motor Co.		

DOHC: Double Overhead Cam

SMFI: Sequential Multi-port Fuel Injection

① 10th position of VIN

② Federal/Canada

③ California

22140_MCX9_C0001

GENERAL ENGINE SPECIFICATIONS

Year	Engine Displacement Liters (VIN)	Net Horsepower @ rpm	Net Torque @ rpm (ft. lbs.)	Bore x Stroke (in.)	Com-pression Ratio	Oil Pressure @ rpm
2007	3.5 (C)	265@6250	250@4500	3.64x3.41	10.3:1	NA
	3.5 (Y)	265@6250	250@4500	3.64x3.41	10.3:1	NA
2008	3.7 (A)	273@6250	270@4250	3.76x3.41	10.0:1	NA
	3.7 (V)	273@6250	270@4250	3.76x3.41	10.0:1	NA

NA - Not Available

22140_MCX9_C0002

ENGINE TUNE-UP SPECIFICATIONS

Year	Engine Displacement Liters (VIN)	Spark Plug Gap (in.)	Ignition Timing (deg.)		Fuel Pump (psi)	Idle Speed (rpm)		Valve Clearance (in.)	
			MT	AT		MT	AT	Intake	Exhaust
2007	3.5 (C)	0.051-0.057	—	13 BTDC	48-70	—	570-670	0.006-0.010	0.012-0.016
	3.5 (Y)	0.051-0.057	—	13 BTDC	48-70	—	570-670	0.006-0.010	0.012-0.016
2008	3.7 (A)	0.051-0.057	—	13 BTDC	48-70	—	570-670	0.006-0.010	0.012-0.016
	3.7 (V)	0.051-0.057	—	13 BTDC	48-70	—	570-670	0.006-0.010	0.012-0.016

BTDC - Before Top Dead Center

22140_MCX9_C0003

CAPACITIES

Year	Model	Engine Displacement Liters (VIN)	Engine Oil with Filter (qts.)	Transmission (pts.) 5-Spd	Transmission (pts.) Auto.	Transfer Case (pts.)	Drive Axle Front (pts.)	Drive Axle Rear (pts.)	Fuel Tank (gal.)	Cooling System (qts.)
2007	CX9	3.5 (C)	5.5	—	14.8	1.12	—	2.2	20.1	①
	CX9	3.5 (Y)	5.5	—	14.8	1.12	—	2.2	20.1	①
2008	CX9	3.7 (A)	5.5	—	14.8	1.12	—	2.2	20.1	②
	CX9	3.7 (V)	5.5	—	14.8	1.12	—	2.2	20.1	②

NOTE: All capacities are approximate. Add fluid gradually and check to be sure a proper fluid level is obtained.

① Single fan: 11.7 qt., Dual fan: 12.4 qt.

② Single fan: 12.3 qt., Dual fan: 12.9 qt.

22140_MCX9_C0004

FLUID SPECIFICATIONS

Year	Engine Displacement Liters (VIN)	Engine Oil	Auto. Trans.	Transfer Case	Drive Axle	Power Steering Fluid	Brake Master Cylinder
2007	3.5 (C)	①	JWS3309	75W-140	80W-90	Mercon® ATF Fluid	DOT 3
	3.5 (Y)	①	JWS3309	75W-140	80W-90	Mercon® ATF Fluid	DOT 3
2008	3.5 (C)	①	JWS3309	75W-140	80W-90	Mercon® ATF Fluid	DOT 3
	3.5 (Y)	①	JWS3309	75W-140	80W-90	Mercon® ATF Fluid	DOT 3

DOT: Department Of Transpotation

① 5W-20 Premium Synthetic Blend Motor Oil (US) or 5W-20 Super Premium Motor Oil (Canada)

22140_MCX9_C0005

VALVE SPECIFICATIONS

Year	Engine Displacement Liters (VIN)	Seat Angle (deg.)	Face Angle (deg.)	Spring Test Pressure (lbs. @ in.)	Spring Installed Height (in.)	Stem-to-Guide Clearance (in.) Intake	Stem-to-Guide Clearance (in.) Exhaust	Stem Diameter (in.) Intake	Stem Diameter (in.) Exhaust
2007	3.5 (C)	89.0-91.0	90.50-91.50	115 @ 1.08	1.4500	0.0008-0.0027	0.0013-0.0320	0.2157-0.2164	0.2151-0.2159
	3.5 (Y)	89.0-91.0	90.50-91.50	115 @ 1.08	1.4500	0.0008-0.0027	0.0013-0.0320	0.2157-0.2164	0.2151-0.2159
2008	3.7 (A)	89.0-91.0	90.50-91.50	115 @ 1.08	1.4500	0.0008-0.0027	0.0013-0.0320	0.2157-0.2164	0.2151-0.2159
	3.7 (V)	89.0-91.0	90.50-91.50	115 @ 1.08	1.4500	0.0008-0.0027	0.0013-0.0320	0.2157-0.2164	0.2151-0.2159

22140_MCX9_C0006

CAMSHAFT SPECIFICATIONS

All measurements are given in inches

Year	Engine Displacement Liters (VIN)	Journal Diameter	Bearing Oil Clearance	Shaft End-play	Runout	Lobe Height Intake	Lobe Height Exhaust
2007	3.5 (C)	①	②	0.0012-0.0066	0.0015	0.3800	0.3800
	3.5 (Y)	①	②	0.0012-0.0066	0.0015	0.3800	0.3800
2008	3.7 (A)	①	②	0.0012-0.0066	0.0015	0.3800	0.3800
	3.7 (V)	①	②	0.0012-0.0066	0.0015	0.3800	0.3800

① 1st journal: 1.2202-1.2209 in.
Intermediate journals: 1.021-1.022 in.

② 1st journal: 0.0027 MAX
Intermediate journals: 0.0029 MAX

22140_MCX9_C0007

CRANKSHAFT AND CONNECTING ROD SPECIFICATIONS

All measurements are given in inches

Year	Engine Displacement Liters (VIN)	Crankshaft Main Brg. Journal Dia.	Crankshaft Main Brg. Oil Clearance	Crankshaft Shaft End-play	Crankshaft Thrust on No.	Connecting Rod Journal Diameter	Connecting Rod Oil Clearance	Connecting Rod Side Clearance
2007	3.5 (C)	2.657	NA	0.0039-0.0114	NA	2.204-2.205	NA	0.0068-0.0167
	3.5 (Y)	2.657	NA	0.0039-0.0114	NA	2.204-2.205	NA	0.0068-0.0167
2008	3.7 (A)	2.657	NA	0.0039-0.0114	NA	2.204-2.205	NA	0.0068-0.0167
	3.7 (V)	2.657	NA	0.0039-0.0114	NA	2.204-2.205	NA	0.0068-0.0167

NA - Not Available

22140_MCX9_C0008

PISTON AND RING SPECIFICATIONS

All measurements are given in inches

Year	Engine Displacement Liters (VIN)	Piston Clearance	Ring Gap Top Compression	Ring Gap Bottom Compression	Ring Gap Oil Control	Ring Side Clearance Top Compression	Ring Side Clearance Bottom Compression	Ring Side Clearance Oil Control
2007	3.5 (C)	0.0003-0.0017	0.0059-0.0098	0.0118-0.0216	0.0059-0.0177	NA	NA	NA
	3.5 (C)	0.0003-0.0017	0.0059-0.0098	0.0118-0.0216	0.0059-0.0177	NA	NA	NA
2008	3.7 (A)	0.0003-0.0017	0.0059-0.0098	0.0118-0.0216	0.0059-0.0177	NA	NA	NA
	3.7 (V)	0.0003-0.0017	0.0059-0.0098	0.0118-0.0216	0.0059-0.0177	NA	NA	NA

NA: Not Available

22140_MCX9_C0009

TORQUE SPECIFICATIONS

All readings in ft. lbs.

Year	Engine Displacement Liters (VIN)	Cylinder Head Bolts	Main Bearing Bolts	Rod Bearing Bolts	Crankshaft Damper Bolts	Flywheel Bolts	Manifold Intake	Manifold Exhaust	Spark Plugs	Oil Pan Drain Plug
2007	3.5 (C)	①	NA	NA	②	59	③	19	11	20
	3.5 (Y)	①	NA	NA	②	59	③	19	11	20
2008	3.7 (A)	①	NA	NA	②	59	③	19	11	20
	3.7 (V)	①	NA	NA	②	59	③	19	11	20

① Step 1: Tigthen bolts 1-16 to 15 ft. lbs.
Step 2: Tigthen bolts 1-16 to 26 ft. lbs.
Step 3: Tigthen bolts 1-16 to +90 degrees
Step 4: Tigthen bolts 1-16 to +90 degrees
Step 4: Tigthen bolts 1-16 to +90 degrees
Step 5: Tigthen bolts 17-18 to 76-101 inch lbs.

② Step 1: 89 ft. lbs.
Step 2: Loosen one full turn
Step 3: 37 ft. lbs.
Step 4: +90 degrees

③ Upper intake manifold: 89 inch lbs.
Lower intake manfold: 89 inch lbs.

22140_MCX9_C0010

WHEEL ALIGNMENT

Year	Model		Caster Range (+/-Deg.)	Caster Preferred Setting (Deg.)	Camber Range (+/-Deg.)	Camber Preferred Setting (Deg.)	Toe-in (in.)
2007	CX9	F	±1°	3°04′	±1°	−0°21′	0.08±0.16
		R	±1°	—	±1°	-0°33′	0.08±0.16
2008	CX9	F	±1°	3°04′	±1°	−0°21′	0.08±0.16
		R	±1°	—	±1°	-0°33′	0.08±0.16

NOTE: Perform wheel alignment with the fuel tank full.

22140_MCX9_C0014

TIRE, WHEEL AND BALL JOINT SPECIFICATIONS

Year	Model	OEM Tires		Tire Pressures (psi)		Wheel	Ball Joint	Lug Nut
		Standard	Optional	Front	Rear	Size	Inspection	(ft. lbs.)
2007	CX9	P245/60R18	P245/50R20	①	①	②	NS	94
2008	CX9	P245/60R18	P245/50R20	①	①	②	NS	94

OEM: Original Equipment Manufacturer

PSI: Pounds Per Square Inch

NS: Not specified by manufacturer

① See the safety certification label on the driver side door jamb for tire pressures.

② P245/60R18: 18x7.5

P245/50R20: 20x7.5

22140_MCX9_C0011

BRAKE SPECIFICATIONS

All measurements in inches unless noted

Year	Model		Brake Disc			Brake Drum Diameter			Minimum Lining Thickness		Brake Caliper	
			Original Thickness	Minimum Thickness	Maximum Runout	Original Inside Diameter	Max. Wear Limit	Maximum Machine Diameter	Front	Rear	Bracket Bolts (ft. lbs.)	Mounting Bolts (ft. lbs.)
2007	CX9	F	1.102	1.030	0.002	—	—	—	0.080	—	98	65
		R	0.708	0.630	0.002	—	—	—	—	0.08	66	19
2008	CX9	F	1.102	1.030	0.002	—	—	—	0.080	—	98	65
		R	0.708	0.630	0.002	—	—	—	—	0.08	66	19

F: Front

R: Rear

22140_MCX9_C0012

SCHEDULED MAINTENANCE INTERVALS

MAZDA CX9

TO BE SERVICED	TYPE OF SERVICE	VEHICLE MILEAGE INTERVAL (x1000)											
		10	20	30	40	50	60	70	80	90	100	110	120
Accessory drive belts	I & A			✓			✓			✓			✓
Air cleaner element	R			✓			✓			✓			✓
Air conditioning filter	R			✓			✓			✓			✓
Brake fluid	R											✓	
Brake hoses & lines (including ABS)	I		✓		✓		✓		✓		✓		
Cooling system hoses & connections	I		✓		✓		✓		✓		✓		
Engine coolant	R												✓
Engine oil	R	✓	✓	✓	✓	✓	✓	✓	✓	✓	✓	✓	✓
Engine oil and coolant levels	I	Inspect at each fuel stop											
Engine oil filter	R		✓		✓		✓		✓		✓		
Exhaust system	I		✓		✓		✓		✓		✓		
Fluid levels and condition	I		✓		✓		✓		✓		✓		
Front and rear brakes	I		✓		✓		✓		✓		✓		
Fuel lines & connection	I		✓		✓		✓		✓		✓		
Halfshaft boots	I		✓		✓		✓		✓		✓		
Idle speed	I & A											✓	
Parking brake system	I & A		✓		✓		✓		✓		✓		
Rear differential fluid	R										✓		
Rotate and inspect tires	I	✓	✓	✓	✓	✓	✓	✓	✓	✓	✓	✓	✓
Spark plugs	R											✓	
Suspension components	I		✓		✓		✓		✓		✓		
Tie rod ends, steering gear box & boots	I		✓		✓		✓		✓		✓		
Transmission fluid	R												✓
Valve clearance	I											✓	

R: Replace I: Inspect A: Adjust

FREQUENT OPERATION MAINTENANCE (SEVERE SERVICE)

If a vehicle is operated under any of the following conditions it is considered severe service:

- Towing a trailer or using a camper or car-top carrier.

- Repeated short trips of less than 5 miles in temperatures below freezing, or trips of less than 10 miles in any temperature.

- Extensive idling or low-speed driving for long distances as in heavy commercial use, such as delivery, taxi or police cars.

- Operating on rough, muddy or salt-covered roads.

- Operating on unpaved or dusty roads.

- Driving in extremely hot (over 90°) conditions.

Air cleaner element: replace every 15,000 miles

Engine oil and filter: replace every 3750 miles or 6 months, whichever occurs first.

Timing belt: replace every 60,000 miles if the vehicle is regularly driven in temperatures above 110°F or below -20°F.

Transmission fluid: replace every 30,000 miles.

Rear differential fluid: replace every 60,000 miles.

Front and rear brakes: inspect every 7500 miles or 6 months, whichever occurs first.

Locks and hinges: lubricate every 15,000 miles.

Tie rods, steering gear box, boots: inspect every 7500 miles or 6 months, whichever occurs first.

Suspension components: inspect every 7500 miles or 6 months, whichever occurs first.

Halfshaft boots: inspect every 7500 miles or 6 months, whichever occurs first.

22140_MCX9_C0013

PRECAUTIONS

Before servicing any vehicle, please be sure to read all of the following precautions, which deal with personal safety, prevention of component damage, and important points to take into consideration when servicing a motor vehicle:

- Never open, service or drain the radiator or cooling system when the engine is hot; serious burns can occur from the steam and hot coolant.
- Observe all applicable safety precautions when working around fuel. Whenever servicing the fuel system, always work in a well-ventilated area. Do not allow fuel spray or vapors to come in contact with a spark, open flame, or excessive heat (a hot drop light, for example). Keep a dry chemical fire extinguisher near the work area. Always keep fuel in a container specifically designed for fuel storage; also, always properly seal fuel containers to avoid the possibility of fire or explosion. Refer to the additional fuel system precautions later in this section.
- Fuel injection systems often remain pressurized, even after the engine has been turned **OFF**. The fuel system pressure must be relieved before disconnecting any fuel lines. Failure to do so may result in fire and/or personal injury.
- Brake fluid often contains polyglycol ethers and polyglycols. Avoid contact with the eyes and wash your hands thoroughly after handling brake fluid. If you do get brake fluid in your eyes, flush your eyes with clean, running water for 15 minutes. If eye irritation persists, or if you have taken brake fluid internally, IMMEDIATELY seek medical assistance.
- The EPA warns that prolonged contact with used engine oil may cause a number of skin disorders, including cancer. You should make every effort to minimize your exposure to used engine oil. Protective gloves should be worn when changing oil. Wash your hands and any other exposed skin areas as soon as possible after exposure to used engine oil. Soap and water, or waterless hand cleaner should be used.
- All new vehicles are now equipped with an air bag system, often referred to as a Supplemental Restraint System (SRS) or Supplemental Inflatable Restraint (SIR) system. The system must be disabled before performing service on or around system components, steering column, instrument panel components, wiring and sensors. Failure to follow safety and disabling procedures could result in accidental air bag deployment, possible personal injury and unnecessary system repairs.
- Always wear safety goggles when working with, or around, the air bag system. When carrying a non-deployed air bag, be sure the bag and trim cover are pointed away from your body. When placing a non-deployed air bag on a work surface, always face the bag and trim cover upward, away from the surface. This will reduce the motion of the module if it is accidentally deployed. Refer to the additional air bag system precautions later in this section.
- Clean, high quality brake fluid from a sealed container is essential to the safe and proper operation of the brake system. You should always buy the correct type of brake fluid for your vehicle. If the brake fluid becomes contaminated, completely flush the system with new fluid. Never reuse any brake fluid. Any brake fluid that is removed from the system should be discarded. Also, do not allow any brake fluid to come in contact with a painted surface; it will damage the paint.
- Never operate the engine without the proper amount and type of engine oil; doing so WILL result in severe engine damage.
- Timing belt maintenance is extremely important. Many models utilize an interference-type, non-freewheeling engine. If the timing belt breaks, the valves in the cylinder head may strike the pistons, causing potentially serious (also time-consuming and expensive) engine damage. Refer to the maintenance interval charts for the recommended replacement interval for the timing belt, and to the timing belt section for belt replacement and inspection.
- Disconnecting the negative battery cable on some vehicles may interfere with the functions of the on-board computer system(s) and may require the computer to undergo a relearning process once the negative battery cable is reconnected.
- When servicing drum brakes, only disassemble and assemble one side at a time, leaving the remaining side intact for reference.
- Only an MVAC-trained, EPA-certified automotive technician should service the air conditioning system or its components.

BRAKES

ANTI-LOCK BRAKE SYSTEM (ABS)

GENERAL INFORMATION

As part of the Anti-Lock Brake System (ABS), this vehicle is equipped with Electronic Stability Control (ESC). ESC is a computerized technology that improves the safety of a vehicle's handling by detecting and preventing skids. When ESC detects loss of steering control, ESC automatically applies individual brakes to help "steer" the vehicle where the driver wants to go. Braking is automatically applied to individual wheels, such as the outer front wheel to counter oversteer, or the inner rear wheel to counter under steer. Some ESC systems also reduce engine power until control is regained.

ESC compares the driver's intended direction (by measuring steering angle) to the vehicle's actual direction (by measuring lateral acceleration, vehicle rotation (yaw) and individual road wheel speeds). If the vehicle is not going where the driver is steering, ESC then brakes individual front or rear wheels and/or reduces excess engine power as needed to help correct under steer (plowing) or oversteer (fishtailing).

ESC incorporates yaw rate control into the ABS. Yaw is rotation around the vertical axis; i.e. spinning left or right. Anti-lock brakes enable ESC to brake individual wheels. The ESC used on this vehicle also incorporates a Traction Control System (TCS), which senses drive-wheel slip under acceleration and individually brakes the slipping wheel or wheels and/or reduces excess engine power until control is regained.

WHEEL SPEED SENSORS

REMOVAL & INSTALLATION

Front

See Figure 1.

1. Before servicing the vehicle, refer to the Precautions Section.
2. When removing the right front ABS wheel-speed sensor, perform the following:.
    a. Remove the coolant reserve tank installation bolts and move the coolant reserve tank.
    b. Remove the power steering reserve tank installation bolt and nut, and move the power steering reserve tank.

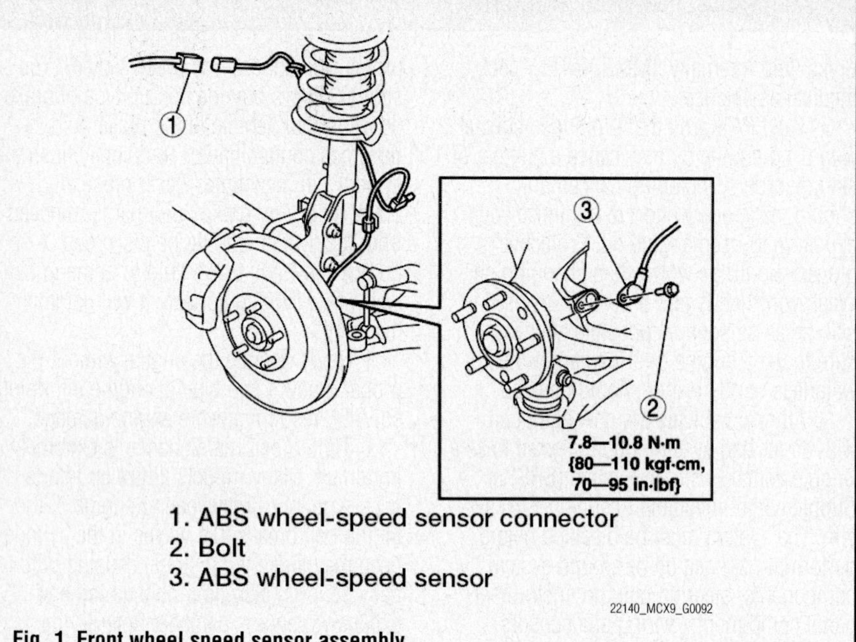

Fig. 1 Front wheel speed sensor assembly

3. When removing the left front ABS wheel-speed sensor, perform the following procedure:
   a. Remove the air cleaner case.
4. Disconnect the front ABS wheel-speed sensor connector.
5. Remove the attaching bolt.
6. Remove the front ABS wheel-speed sensor.

To install:

7. Install the front ABS wheel-speed sensor.
8. Install the attaching bolt and tighten to 70–95 inch lbs. (8–11 Nm).
9. Disconnect the front ABS wheel-speed sensor connector.
   a. Install the air cleaner case.
10. When removing the left front ABS wheel-speed sensor, perform the following procedure:
    a. Install the power steering reserve tank and tighten bolts to 14–19 ft. lbs. (19–26 Nm).
    b. Install the coolant reserve tank and tighten bolts to 62–104 inch lbs. (7–12 Nm).
11. When removing the right front ABS wheel-speed sensor, perform the following:.

Rear

See Figures 2 through 8.

1. Before servicing the vehicle, refer to the Precautions Section.
2. Remove the trunk box.
3. Remove the seat side box.
4. Remove the third-row seat.
5. Remove the rear scuff plate inner.
6. Remove the third-row seat belt lower anchor installation bolt.
7. Remove the trunk side trim.

Fig. 2 Removing the trunk box

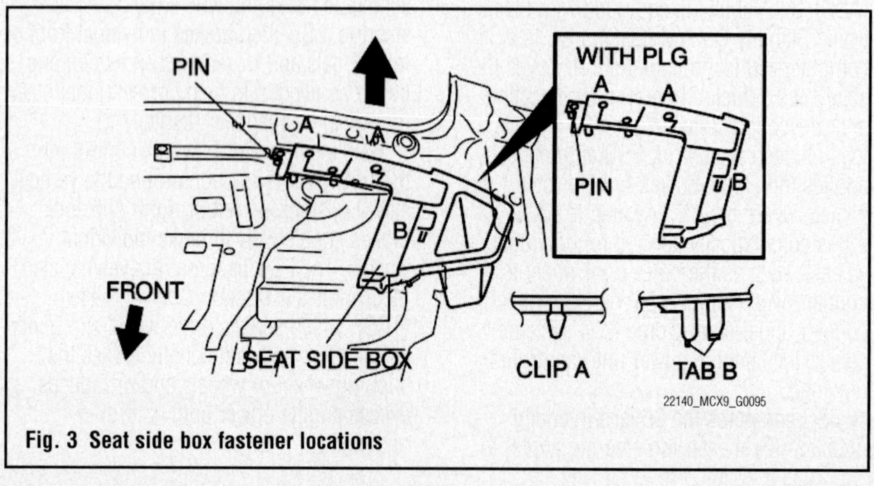

Fig. 3 Seat side box fastener locations

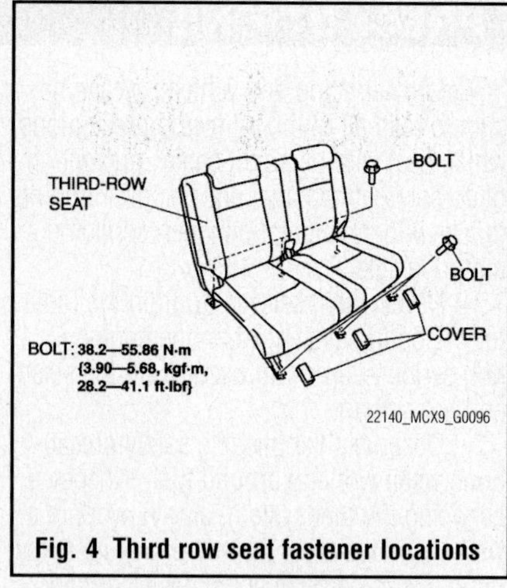

Fig. 4 Third row seat fastener locations

8. Disconnect the rear ABS wheel-speed sensor connector.
9. Remove the attaching bolt.
10. Remove the rear ABS wheel-speed sensor.

To install:

11. Install the rear ABS wheel-speed sensor.
12. Install the attaching bolt and tighten to 70–95 inch lbs. (8–11 Nm).
13. Connect the rear ABS wheel-speed sensor connector.
14. Install the trunk side trim.
15. Install the third-row seat belt lower anchor bolt. Tighten to 28–58 ft. lbs. (38–78 Nm).
16. Install the rear scuff plate inner.
17. Install the third-row seat.
18. Install the seat side box.
19. Install the trunk box.

WHEEL SPEED SENSOR RINGS

REMOVAL & INSTALLATION

The speed sensor tone rings are integral either the halfshafts or wheel hubs and cannot be serviced separately.

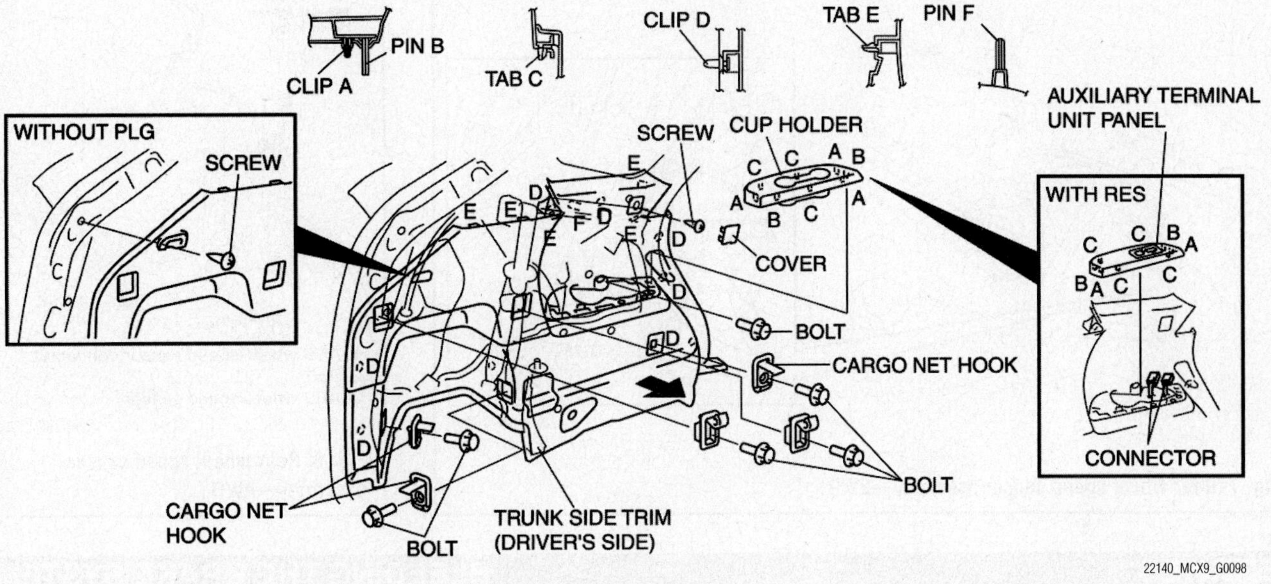

22140_MCX9_G0098

Fig. 5 Trunk side trim fastener locations—driver's side

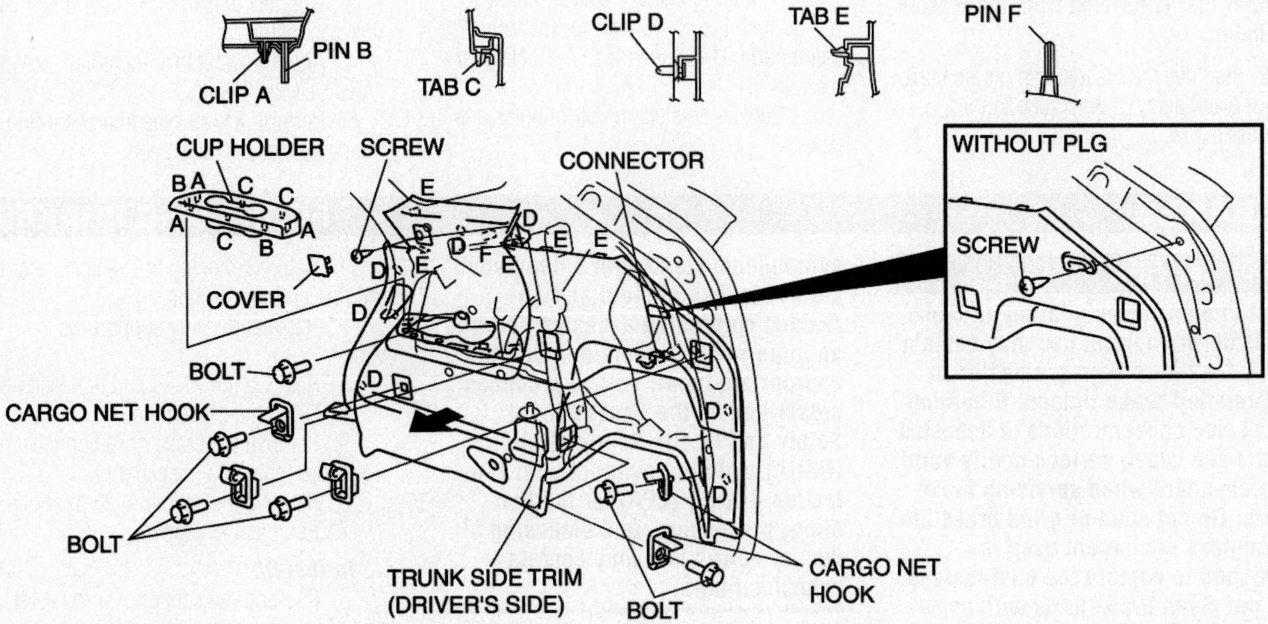

22140_MCX9_G0099

Fig. 6 Trunk side trim fastener locations—passenger's side

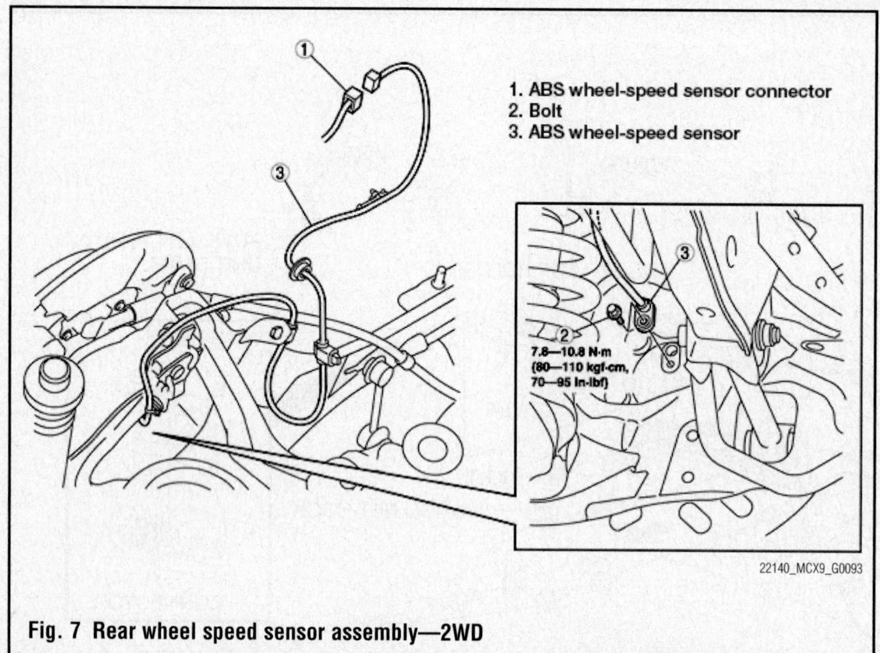

Fig. 7 Rear wheel speed sensor assembly—2WD

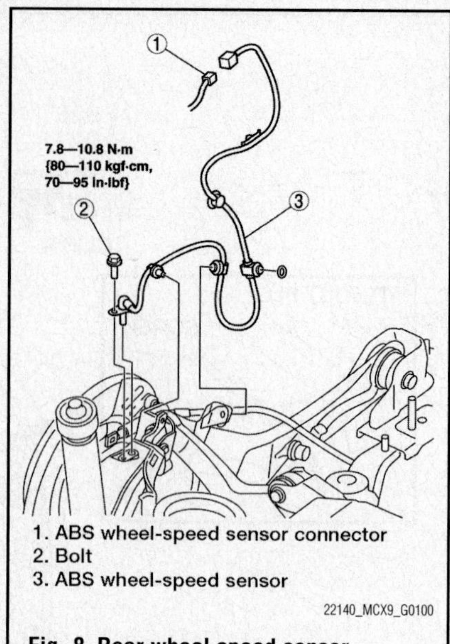

Fig. 8 Rear wheel speed sensor assembly—AWD

BRAKES — BLEEDING THE BRAKE SYSTEM

BLEEDING PROCEDURE

1. Before servicing the vehicle, refer to the Precautions Section.

✻✻ WARNING

Keep the fluid level in the reserve tank at 3/4 full or more during the air bleeding.

➡Begin air bleeding with the brake caliper that is furthest from the master cylinder.

2. Remove the bleeder cap on the brake caliper, and attach a vinyl tube to the bleeder screw.
3. Place the other end of the vinyl tube in a clear container and fill the container with fluid during air bleeding.
4. Working with two people, one should pump the brake pedal several times and depress and hold the pedal down.
5. While the brake pedal is depressed, the other should loosen the bleeder screw using a commercially available flare nut wrench, drain out any fluid containing air bubbles, and tighten the bleeder screw to 53–86 inch lbs. 6–10 Nm) for the front caliper and 61–86 inch lbs. 7–10 Nm) for the rear.
6. Repeat the steps no air bubbles are seen.
7. Perform air bleeding as described for all brake calipers.
8. After air bleeding, check for proper brake operation, any fluid leakage and correct fluid level.

BLEEDING THE ABS SYSTEM

1. Before servicing the vehicle, refer to the Precautions Section.
2. Turn the ignition switch **OFF**
3. Connect a scan tool to the diagnostic link connector.
4. After the vehicle is identified, select “ABS Service Bleed”.
5. Perform the air bleeding according to the directions on the screen.

BRAKES — FRONT DISC BRAKES

✻✻ CAUTION

Dust and dirt accumulating on brake parts during normal use may contain asbestos fibers from production or aftermarket brake linings. Breathing excessive concentrations of asbestos fibers can cause serious bodily harm. Exercise care when servicing brake parts. Do not sand or grind brake lining unless equipment used is designed to contain the dust residue. Do not clean brake parts with compressed air or by dry brushing. Cleaning should be done by dampening the brake components with a fine mist of water, then wiping the brake components clean with a dampened cloth. Dispose of cloth and all residue containing asbestos fibers in an impermeable container with the appropriate label. Follow practices prescribed by the Occupational Safety and Health Administration (OSHA) and the Environmental Protection Agency (EPA) for the handling, processing, and disposing of dust or debris that may contain asbestos fibers.

BRAKE CALIPER

REMOVAL & INSTALLATION

See Figure 9.

1. Before servicing the vehicle, refer to the Precautions Section.
2. Raise and safely support the vehicle.
3. Remove the wheel and tire assembly.
4. Drain the brake fluid.
5. Remove the brake hose connection, if you are replacing the caliper.
6. Remove the caliper slide pin bolts.
7. Remove the caliper.

To install:

8. Position the caliper over the rotor so the caliper engages the adapter correctly.
9. Install the caliper. Tighten the front caliper slide pin bolts to 62–69 ft. lbs. (83–93 Nm).

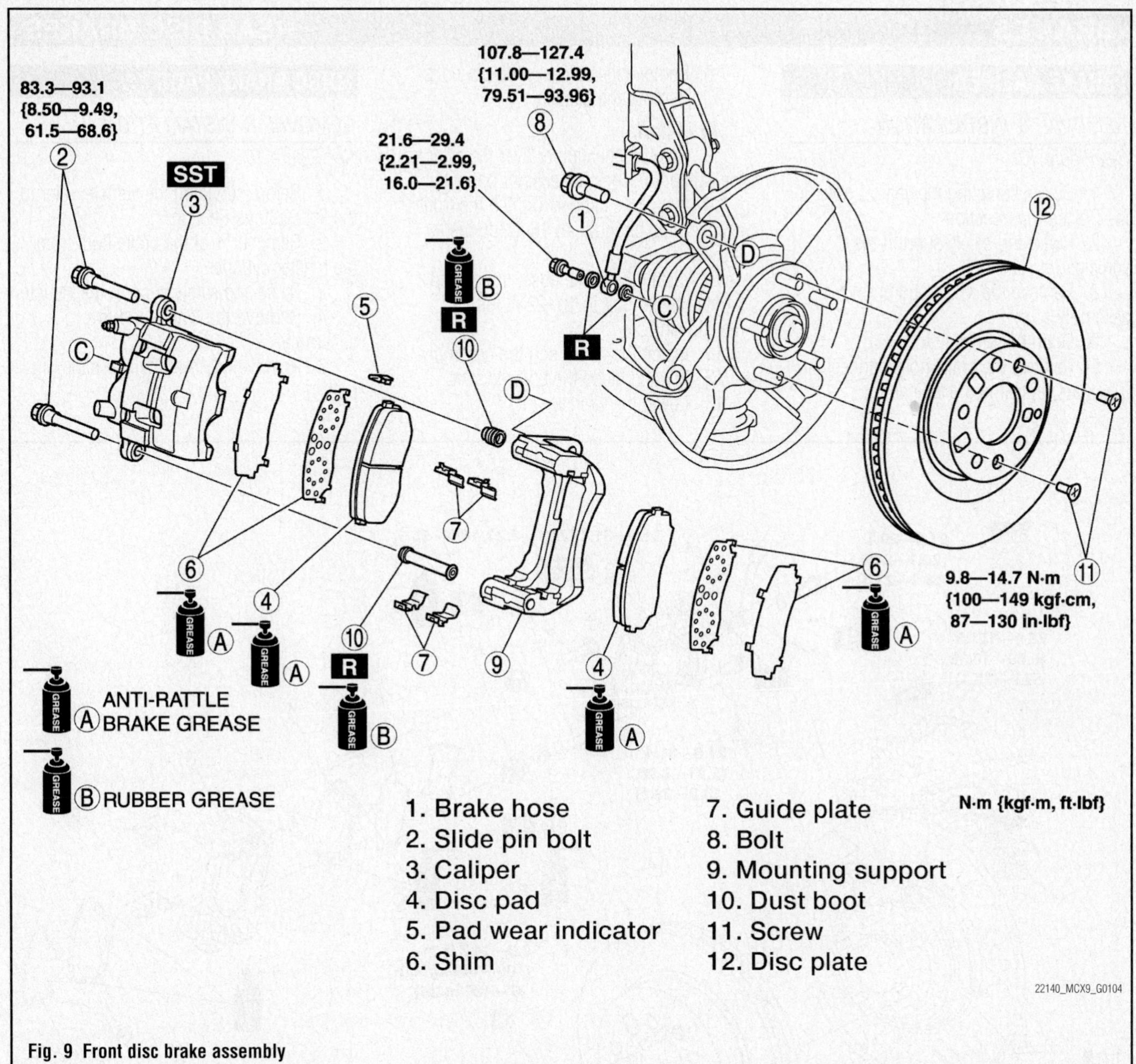

Fig. 9 Front disc brake assembly

10. Install the brake hose, if removed. Tighten banjo bolt to 16–22 ft. lbs. (22–30 Nm).
11. Install the wheel and tire assembly.
12. Fill and bleed the brake system.
13. Lower the vehicle.

DISC BRAKE PADS

REMOVAL & INSTALLATION

See Figure 9.

1. Before servicing the vehicle, refer to the Precautions Section.
2. Remove ½ of the brake fluid from the master cylinder.
3. Raise and safely support the vehicle.
4. Remove the wheel and tire assembly.
5. Remove the lower caliper slide pin bolt.
6. Remove the caliper from the caliper support but do NOT disconnect the fluid line. Suspend the caliper from the suspension with a piece of wire.
7. Remove the disc pad, pad wear indicator, shim and guide plate.

To install:

8. Clean the exposed portion of the caliper piston, then press the piston back into the caliper bore using the old inner brake pad and a C-clamp.
9. Remove the disc pad, pad wear indicator, shim and guide plate making sure the shims and clips are properly positioned.
10. Install the caliper over the rotor so the caliper engages the adapter correctly.
11. Install the caliper slide pin bolt and tighten to 62–69 ft. lbs. (83–93 Nm).
12. Install the wheel and tire assembly and lower the vehicle.
13. Apply the brake pedal several times until a firm pedal is obtained.
14. Check the fluid level in the master cylinder and add fluid, as necessary.

BRAKE CALIPER

REMOVAL & INSTALLATION

See Figure 10.

1. Before servicing the vehicle, refer to the Precautions Section.
2. Raise and safely support the vehicle.
3. Remove the wheel and tire assembly.
4. Drain the brake fluid.
5. Remove the brake hose connection, if you are replacing the caliper.
6. Remove the caliper slide pin bolts.
7. Remove the caliper.

To install:

8. Position the caliper over the rotor so the caliper engages the adapter correctly.
9. Install the caliper. Tighten the front caliper slide pin bolts to 16–23 ft. lbs. (22–31 Nm).
10. Install the brake hose, if removed. Tighten banjo bolt to 16–22 ft. lbs. (22–30 Nm).
11. Install the wheel and tire assembly.
12. Fill and bleed the brake system.
13. Lower the vehicle.

DISC BRAKE PADS

REMOVAL & INSTALLATION

See Figure 10.

1. Before servicing the vehicle, refer to the Precautions Section.
2. Remove ½ of the brake fluid from the master cylinder.
3. Raise and safely support the vehicle.
4. Remove the wheel and tire assembly.
5. Remove the lower caliper slide pin bolt.

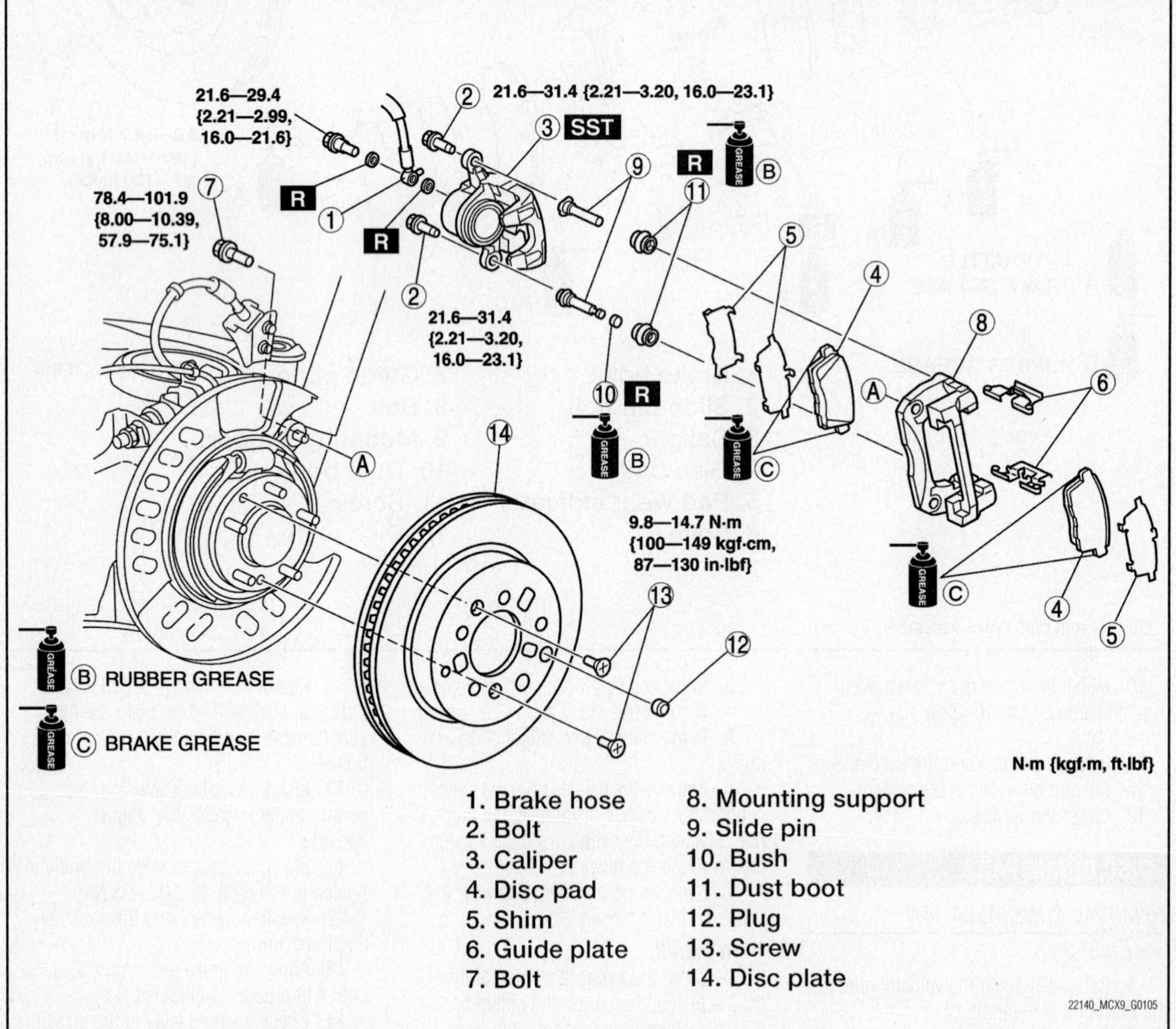

Fig. 10 Rear disc brake assembly

6. Remove the caliper from the caliper support but do NOT disconnect the fluid line. Suspend the caliper from the suspension with a piece of wire.

7. Remove the disc pad, shim and guide plate.

To install:

8. Clean the exposed portion of the caliper piston, then press the piston back into the caliper bore using the old inner brake pad and a C-clamp.

9. Remove the disc pad, shim and guide plate making sure the shims and clips are properly positioned.

10. Install the caliper over the rotor so the caliper engages the adapter correctly.

11. Install the caliper slide pin bolt and tighten to 16–23 ft. lbs. (22–31 Nm).

12. Install the wheel and tire assembly and lower the vehicle.

13. Apply the brake pedal several times until a firm pedal is obtained.

14. Check the fluid level in the master cylinder and add fluid, as necessary.

PARKING BRAKE CABLES

ADJUSTMENT

See Figure 11.

1. Before servicing the vehicle, refer to the Precautions Section.

2. Turn the adjusting nut at the front of the parking cable.

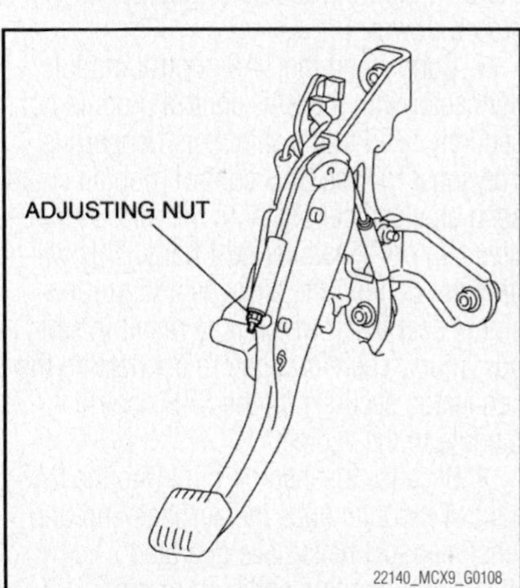

Fig. 11 The adjusting nut is located at the front of the parking brake cable

3. After adjustment, inspect the following points:

a. Turn the ignition switch on, depress the parking brake pedal one notch, and verify that the brake system warning light illuminates.

b. Verify that the rear brakes do not drag.

PARKING BRAKE SHOES

REMOVAL & INSTALLATION

See Figures 12 through 14.

1. Before servicing the vehicle, refer to the Precautions Section.

2. Raise and safely support the vehicle securely on jackstands.

3. Remove the wheels.

4. Insert a flathead screwdriver into the service hole and turn the adjuster in the direction of the arrow to compress the parking brake shoe.

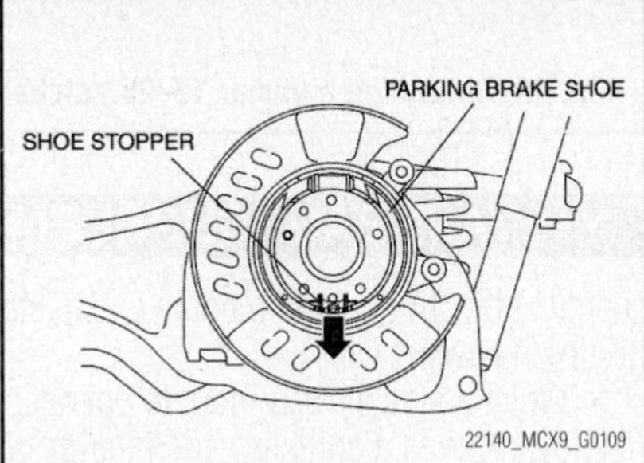

Fig. 12 Pull the parking brake shoe downward and disengage it from the shoe stopper

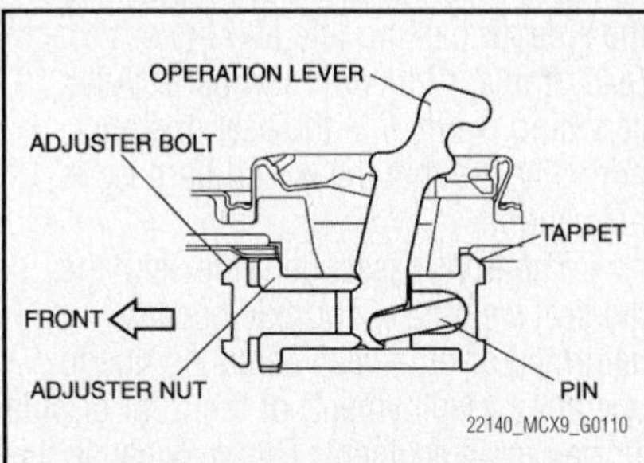

Fig. 13 Install the operation lever, pin, adjuster bolt and nut, and tappet so that the adjuster nut is facing toward the vehicle front

5. Remove the brake rotor.

6. Pull the parking brake shoe downward and disengage it from the shoe stopper.

7. Press the adjuster bolt and tappet by hand, and slowly remove the parking brake shoe.

➡When removing the parking brake shoe, firmly press the adjuster bolt and tappet by hand and slowly remove the parking brake shoe to prevent the adjuster bolt, tappet, operation lever and other parts from flying off.

To install:

8. Install the operation lever, pin, adjuster bolt and nut, and tappet so that the adjuster nut is facing toward the vehicle front.

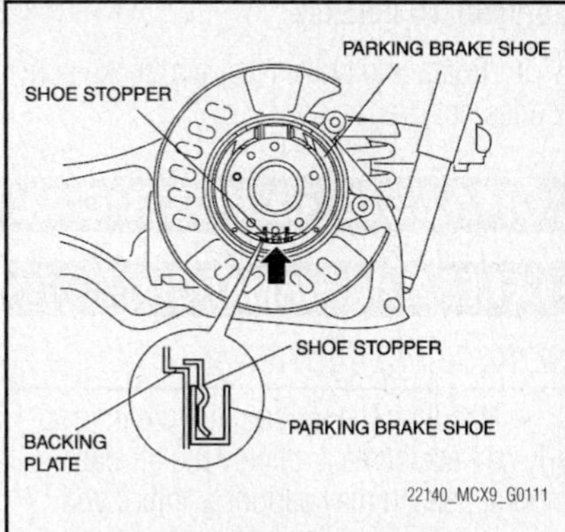

Fig. 14 After installing the opening of the parking brake shoe to the adjuster bolt and tappet, push the brake shoe upward and attach it to the shoe stopper

9. Completely tighten the adjuster bolt and nut.

10. Move the operation lever by hand and verify that it operates properly.

➡If proper operation cannot be verified, reinstall.

11. Install the brake shoes.

a. Measure the parking brake lining thickness with a Vernier caliper. If it is less than 0.04 in. (1.0 mm), install a new parking brake shoe.

b. Apply grease to the contact surface of the parking brake shoe and shoe stopper.

c. After installing the opening of the parking brake shoe to the adjuster bolt and tappet, push the brake shoe upward and attach it to the shoe stopper.

12. Install the brake rotor.

a. Measure the inner diameter of the brake rotor with a Vernier caliper.

b. If it exceeds 7.52 in (191.0 mm) in diameter, install a new brake rotor.

c. Install the brake rotor and screw. Tighten the screws to 87–130 inch lbs. (10–15)

13. Adjust the parking brakes.

14. Install the wheels and lower the vehicle.

ADJUSTMENT

See Figure 15.

1. Insert a flathead screwdriver into the service hole and turn the adjuster in the direction of the arrow to expand the parking brake shoe until the brake rotor cannot rotate.
2. Return the adjuster 13–25 notches in the direction of the arrow.

➡Shoe clearance can be adjusted to 0.15 mm (0.006 in) by returning the adjuster 15 notches.

3. Rotate the brake rotor and make sure it does not drag.

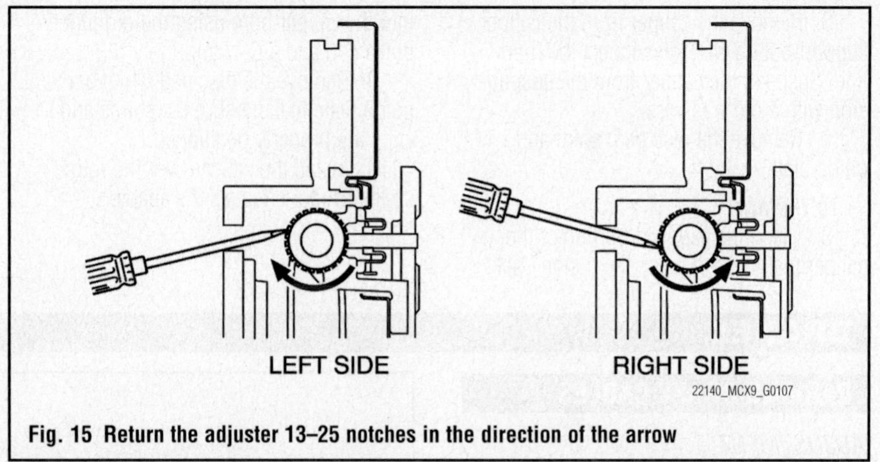

Fig. 15 Return the adjuster 13–25 notches in the direction of the arrow

CHASSIS ELECTRICAL

AIR BAG (SUPPLEMENTAL RESTRAINT SYSTEM)

GENERAL INFORMATION

SERVICE PRECAUTIONS

• Handling the air bag module improperly can accidentally deploy the air bag module, which may seriously injure you.

• Inspecting an air bag module using a tester can operate (deploy) the air bag module, which may cause serious injury. Do not use a tester to inspect an air bag module. Always use the on-board diagnostic function to diagnose the air bag module for malfunctions.

• Before removing the air bag module or disconnecting the air bag module connector, always turn the ignition switch to the LOCK position, disconnect the negative battery cable, and then wait for 1 min or more to allow the backup power supply of the SAS control module to deplete its stored power.

Handling a live (undeployed) air bag module that is pointed toward your body could result in serious injury if the air bag module were to accidentally operate (deploy). When carrying a live (undeployed) air bag module, point the deployment surface away from your body to lessen the chance of injury in case it operates (deploys).

• A live (undeployed) air bag module placed with its deployment surface to ground is dangerous. If the air bag module were to accidentally operate (deploy), it could cause serious injury. Always place a live (undeployed) air bag module with its deployment surface up.

• Before removing the side air bag module or disconnecting the side air bag module connector, always turn the ignition switch to the LOCK position, disconnect the negative battery cable, and then wait for 1 min or more to allow the backup power supply of the SAS control module to deplete its stored power.

• When a side air bag module operates (deploys) due to a collision, the interior of the seat back (pad, frame, trim) may become damaged. If a side air bag does not operate (deploy) normally from a seat back that has been reused, a serious accident may result. After a side air bag has operated (deployed), always replace both the side air bag module and the seat back (pad, frame, trim) with new parts. After servicing, verify that the seat operates normally and that the wiring harness is not caught.

• The front passenger-side seat and the seat weight sensor may become deformed or otherwise damaged due to operation (deployment) of the front or side air bag in an accident. This may cause the passenger sensing function to operate improperly and result in a serious accident. Always replace the front passenger seat and seat weight sensor with new ones after the front or side air bags have operated (deployed). After servicing, verify that the seat operates normally and that the wiring harness is not caught. If the collision is not hard enough to cause the front or side air bags to operate (deploy), inspect the seat weight sensor and replace it if there is any malfunction.

• Removing the SAS control module or disconnecting the SAS control module connector with the ignition switch at the ON position can activate the sensor in the SAS control module and operate (deploy) the air bags and pre-tensioner seat belts, which may cause serious injury. Before removing the SAS control module or disconnecting the SAS control module connector, always turn the ignition switch to the LOCK position, disconnect the negative battery cable, and then wait for 1 min or more to allow the backup power supply of the SAS control module to deplete its stored power.

• Connecting the SAS control module connector with the SAS control module not securely fixed to the vehicle is dangerous. The sensor in the SAS control module could send an electrical signal to the air bag modules and pre-tensioner seat belts. This will operate (deploy) the air bags and pre-tensioner seat belts, which may result in serious injury. Therefore, before connecting the connector, securely fix the SAS control module to the vehicle.

• Because a sensor is built into the SAS control module, once the air bags and pre-tensioner seat belts have operated (deployed) due to a collision or other causes, the SAS control module must be replaced with a new one even if the used one does not have any visible external damage or deformation. The used SAS control module may have been damaged internally, which may cause improper operation. If the SAS control module is reused, the air bags and pre-tensioner seat belts may not operate (deploy) normally, which could result in a serious accident. Always replace the SAS control module with a new one. The SAS control module cannot be bench-checked or self-checked.

• Removing the crash zone sensor or disconnecting the crash zone sensor connector with the ignition switch at the ON position can activate the crash zone sensor and operate (deploy) the air bags and pre-tensioner seat belts, which may cause serious injury. Before removing the crash zone sensor or disconnecting the crash zone sensor connector, always turn the ignition switch to the LOCK position, disconnect the negative battery cable, and then wait for 1 min or more to allow the backup power supply of the SAS control module to deplete its stored power.

- If the crash zone sensor is subjected to shock or the sensor is disassembled, the air bags and pre-tensioner seat belts may accidentally operate (deploy) and cause injury, or the system may fail to operate normally and cause a serious accident. Do not subject the crash zone sensor to shock or disassemble the sensor.
- Because a sensor is built into the crash zone sensor, once the air bags and pre-tensioner seat belts have operated (deployed) due to a collision or other causes, the crash zone sensor must be replaced with a new one even if the used one does not have any visible external damage or deformation. If the crash zone sensor is reused, the air bags and pre-tensioner seat belts may not operate (deploy) normally, which could result in a serious accident. Always replace the crash zone sensor with a new one. The crash zone sensor cannot be bench-checked or self-checked.
- Removing the side air bag sensor or disconnecting the side air bag sensor connector with the ignition switch at the ON position can activate the side air bag sensor and operate (deploy) the side air bag, which may cause serious injury. Before removing the side air bag sensor or disconnecting the side air bag sensor connector, always turn the ignition switch to the LOCK position, disconnect the negative battery cable, and then wait for 1 min or more to allow the backup power supply of the SAS control module to deplete its stored power.
- If the side air bag sensor is subjected to shock or the sensor is disassembled, the side air bag may accidentally operate (deploy) and cause injury, or the system may fail to operate normally and cause a serious accident. Do not subject the side air bag sensor to shock or disassemble the sensor.
- Because a sensor is built into the side air bag sensor, once the air bag has operated (deployed) due to a collision or other causes, the side air bag sensor must be replaced with a new one even if the used one does not have any visible external damage or deformation. If the side air bag sensor is reused, the side air bag may not operate (deploy) normally, which could result in a serious accident. Always replace the side air bag sensor with a new one. The side air bag sensor cannot be bench-checked or self-checked.
- Inspecting a pre-tensioner seat belt using a tester can operate (deploy) the pre-tensioner seat belt, which may cause serious injury. Do not use a tester to inspect a pre-tensioner seat belt. Always use the on-board diagnostic function to diagnose the pre-tensioner seat belt for malfunctions.
- Disassembling the air bag system components could cause it to not operate (deploy) normally. Never disassemble any air bag system components.
- Oil, grease, or water on the air bag modules may cause the air bags and pre-tensioner seat belts to fail to operate (deploy) in an accident. Never allow oil, grease, or water to get on the air bag modules or pre-tensioner seat belts.
- Inserting a screwdriver or similar object into the connector of an air bag module or a pre-tensioner seat belt may damage the connector and cause the air bag module or the pre-tensioner seat belt to operate (deploy) improperly, which may cause serious injury. Never insert any foreign objects into the air bag module or seat belt connectors.
- The seat weight sensor has a built-in strain gauge which may operate improperly if the sensor is dropped by itself or when installed to the seat. If it is dropped, replace the seat weight sensor with a new one.
- Oil, grease, or water on the seat weight sensor may cause the system to operate (deploy) improperly. Never allow oil, grease, or water to get on the seat weight sensor.
- Foreign material in the seat weight sensor components may cause the system to operate (deploy) improperly. Always make sure that no foreign material can get into the seat weight sensor.
- Disassembling the seat weight sensor, or tightening any of the nuts and bolts installed to the sensor body may cause it to operate (deploy) improperly. Never disassemble the seat weight sensor or tighten any of the nuts or bolts installed to the body of the sensor.
- Even if an air bag module or a pre-tensioner seat belt does not operate (deploy) in a collision and does not have any external signs of damage, it may have been damaged internally, which may cause improper operation. Before reusing a live (undeployed) air bag module and the pre-tensioner seat belts, always use the on-board diagnostic to diagnose the air bag module and the pre-tensioner seat belts to verify that they have no malfunction.
- Incorrectly repairing an air bag wiring harness can accidentally operate (deploy) the air bag module and pre-tensioner seat belts. If a problem is found in the air bag wiring harness, always replace the wiring harness with a new one.

DISARMING THE SYSTEM

To avoid personal injury when working on vehicles equipped with an air bag, the negative battery cable must be disconnected and at least 1 minute must elapse before working on the system. Failure to do so may result in deployment of the air bag. You should also wrap or isolate the negative battery cable with electrical or other non-conductive tape.

ARMING THE SYSTEM

To arm the system after service is completed, connect the negative battery cable.

DRIVETRAIN

AUTOMATIC TRANSAXLE ASSEMBLY

REMOVAL & INSTALLATION

See Figures 16 through 25.

1. Before servicing the vehicle, refer to the Precautions Section.
2. Disconnect the negative battery cable.
3. Remove the battery and battery tray.
4. Remove the air cleaner.
5. Remove the engine cover.
6. Remove the windshield wiper arm and blade.
7. Remove the cowl grille.
8. Remove the cowl panel.
9. Raise and safely support the vehicle securely on jackstands.
10. Remove the front wheel.
11. Remove the splash shield.
12. Remove the side cover.
13. Remove the propeller shaft (transfer side).
14. Drain the transmission fluid.

✻✻ CAUTION

Improperly jacking a transaxle is dangerous. It can slip off the jack and may cause serious injury.

✻✻ WARNING

To prevent the torque converter and transaxle from separating, remove the transaxle without tilting it toward the torque converter.

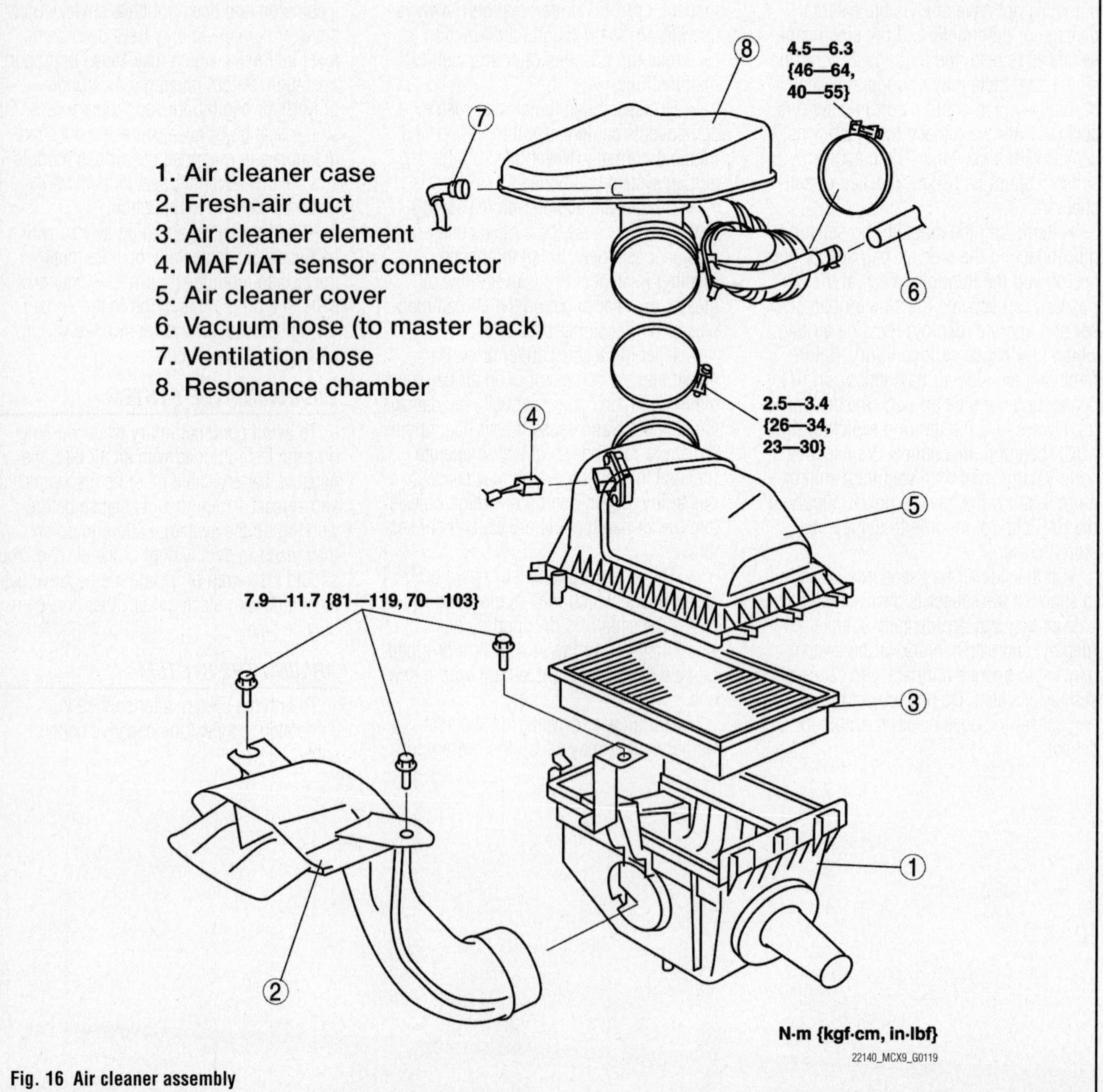

Fig. 16 Air cleaner assembly

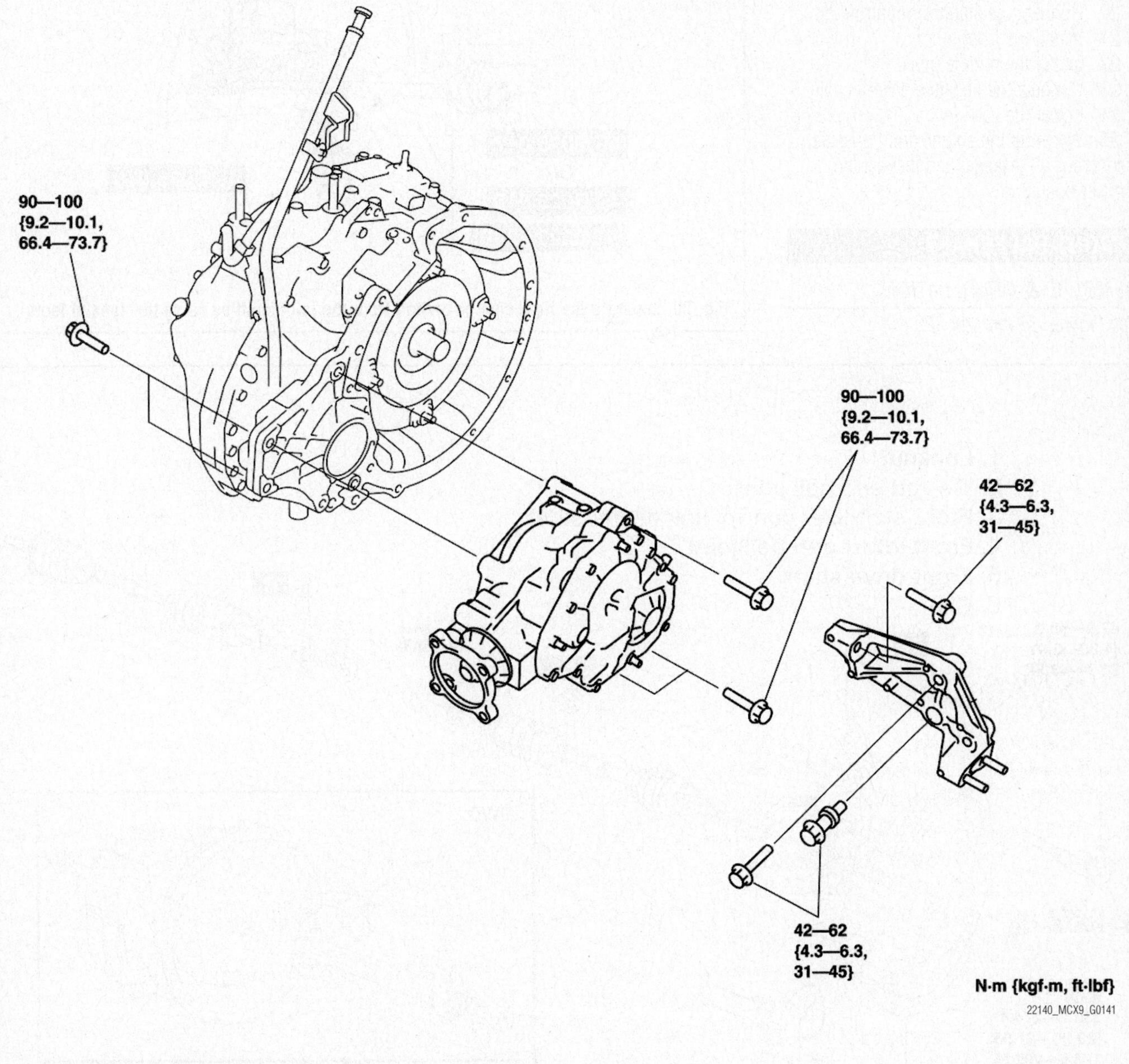

Fig. 28 Transfer case and related components

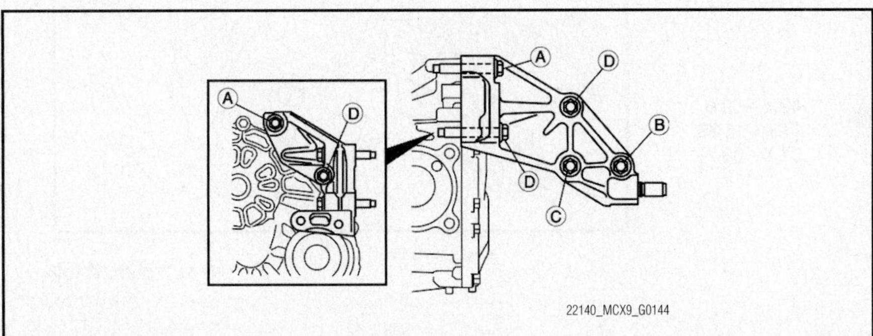

Fig. 29 Tighten bolt A, B and C in the order of C–A–B to 31–45 ft. lbs. (42–62 Nm). Then tighten bolt D to 31–45 ft. lbs. (42–62 Nm)

14. Install the transfer case bracket.
    a. Install the transfer bracket to the transfer case, then temporarily tighten bolt C.
    b. Temporarily tighten bolt A and B.
    c. Tighten bolt A, B and C in the order of C–A–B to 31–45 ft. lbs. (42–62 Nm).
    d. Tighten bolt D to 31–45 ft. lbs. (42–62 Nm).
15. Connect the propeller shaft from the transfer side.
16. Install the joint shaft.
17. Install the right side front drive shaft.

18. Connect the front lower arm ball joint.
19. Connect the tie-rod end ball joint.
20. Connect the stabilizer control link.
21. Install the front pipe.
22. Install the middle pipe.
23. Connect the negative battery cable.
24. Lower the vehicle.
25. Warm up the engine and transaxle, inspect for oil leakage, and inspect the transfer operation.

FRONT HALFSHAFT

REMOVAL & INSTALLATION

See Figures 30 through 32.

49 B025 017
OR
49 B025 016
49 0223 630B
49 B025 010
22140_MCX9_G0130

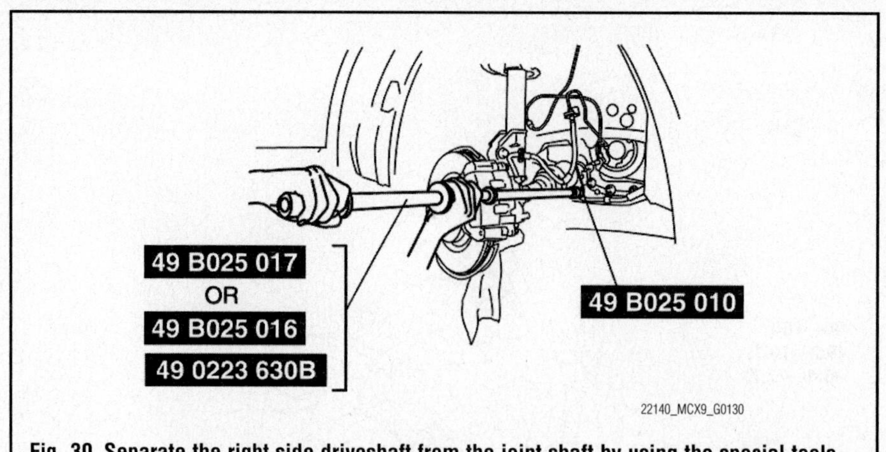

Fig. 30 Separate the right side driveshaft from the joint shaft by using the special tools

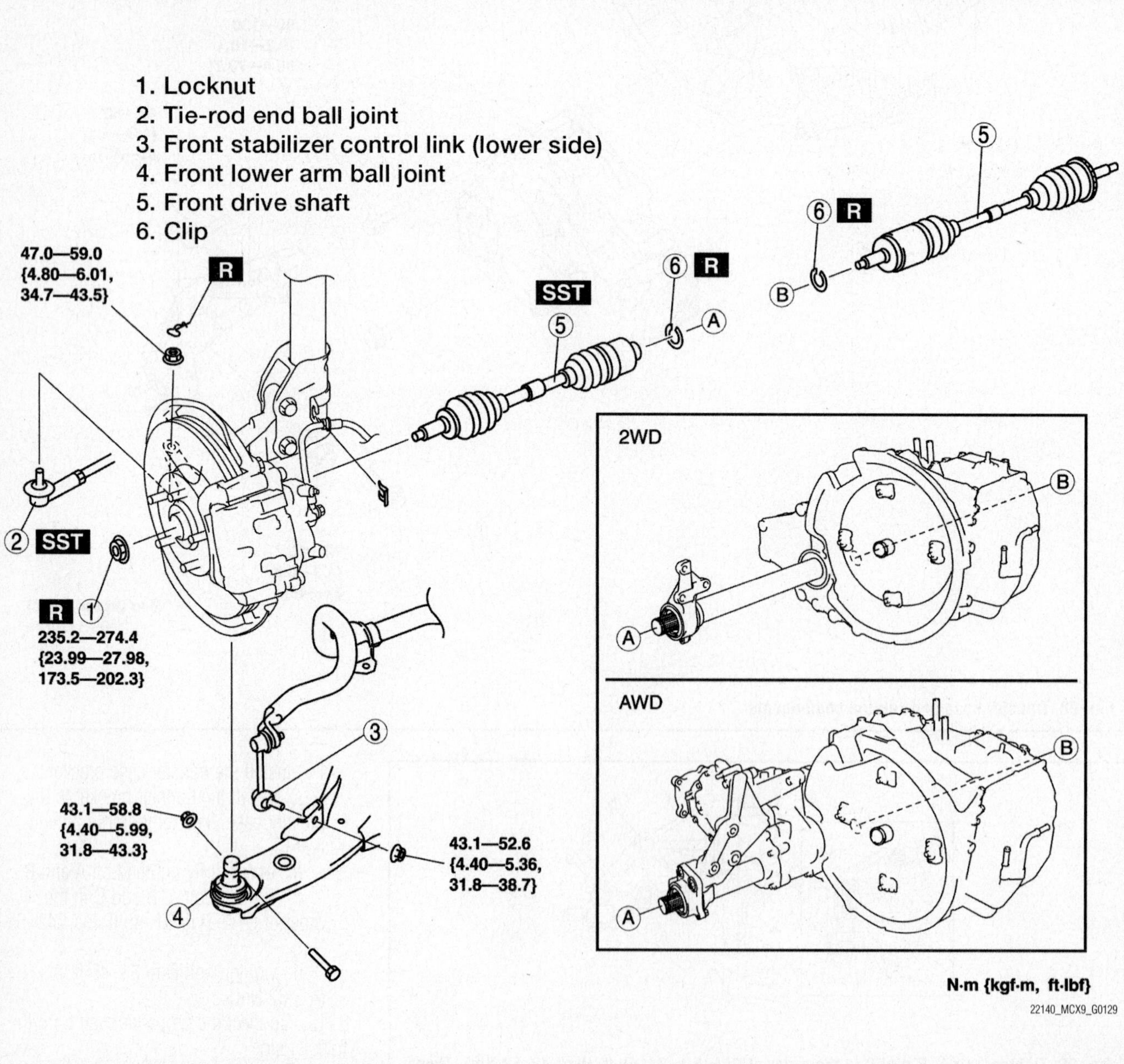

1. Locknut
2. Tie-rod end ball joint
3. Front stabilizer control link (lower side)
4. Front lower arm ball joint
5. Front drive shaft
6. Clip
47.0—59.0 {4.80—6.01, 34.7—43.5}
R
SST
235.2—274.4 {23.99—27.98, 173.5—202.3}
43.1—58.8 {4.40—5.99, 31.8—43.3}
43.1—52.6 {4.40—5.36, 31.8—38.7}
2WD
AWD
A
B
N·m {kgf·m, ft·lbf}
22140_MCX9_G0129

Fig. 31 Front halfshaft and related components

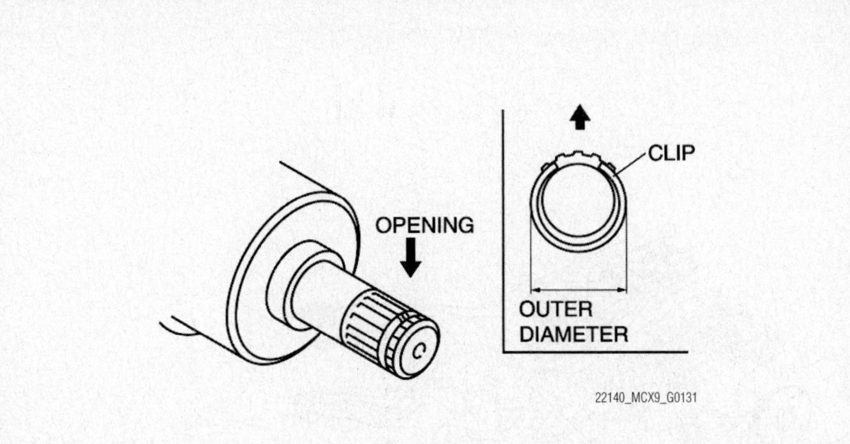

Fig. 32 Install a new clip onto the driveshaft with the opening facing upward. Ensure that the diameter of the clip does not exceed 1.34 in. (34 mm)

⁂ WARNING

Performing the following procedures without first removing the ABS wheel speed sensor may possibly cause an open circuit in the harness if it is pulled by mistake. Before performing the following procedures, remove the ABS wheel speed sensor (axle side) and fix it to an appropriate place where the sensor will not be pulled by mistake while the vehicle is being serviced.

1. Before servicing the vehicle, refer to the Precautions Section.
2. Raise and safely support the vehicle securely on jackstands.
3. Remove the side cover.
4. Remove the front ABS wheel-speed sensor.
5. Remove the locknut.
   a. Lock the hub by applying the brakes.
   b. Knock the crimped portion of the locknut outward using a small chisel and a hammer.
   c. Remove the locknut.
6. Disconnect the tie-rod end ball joint.
7. Disconnect the front stabilizer control link (lower side).
8. Disconnect the front lower arm ball joint.
9. Remove the front halfshaft.
   a. Install a spare nut onto the driveshaft so that the nut is flush with the end of the driveshaft.
   b. Tap the nut with a copper hammer to loosen the driveshaft from the front wheel hub.
   c. Separate the driveshaft from the wheel hub.

⁂ WARNING

The sharp edges of the driveshaft can slice or puncture the oil seal. Be careful when removing the driveshaft from the transaxle.

   d. Separate the left side driveshaft from the transaxle by prying with a bar inserted between the outer ring and the transaxle, as shown.
   e. Separate the right side driveshaft from the joint shaft by using the special tools.
10. Remove the clip.

To install:

11. Install the clip.
   a. Install a new clip onto the driveshaft with the opening facing upward. Ensure that the diameter of the clip does not exceed the specification on installation.
   b. After installation, measure the outer diameter. If it exceeds 1.34 in. (34 mm), repeat steps using a new clip.
12. Install the left side halfshaft as follows:
   a. Insert the driveshaft into the wheel hub.
   b. Apply transaxle oil to the oil seal lip.
   c. Push the driveshaft into the transaxle.
   d. After installation, pull the transaxle side outer ring forward to confirm that the driveshaft is securely held by the clip.
13. Install the right side halfshaft as follows:
   a. Insert the driveshaft into the wheel hub.
   b. Insert the driveshaft into the joint shaft.
   c. After installation, pull the transaxle side outer ring forward to confirm that the driveshaft is securely held by the clip.
14. Connect the front lower arm ball joint.
15. Connect the front stabilizer control link (lower side).
16. Connect the tie-rod end ball joint.
17. Install a new locknut and stake it into place.
18. Install the front ABS wheel-speed sensor.
19. Install the side cover.
20. Lower the vehicle.

REAR HALFSHAFT

REMOVAL & INSTALLATION

See Figure 33.

⁂ WARNING

Performing the following procedures without first removing the ABS wheel speed sensor may possibly cause an open circuit in the harness if it is pulled by mistake. Before performing the following procedures, remove the ABS wheel speed sensor (wheel side) and fix it to an appropriate place where the sensor will not be pulled by mistake while the vehicle is being serviced.

1. Before servicing the vehicle, refer to the Precautions Section.
2. Raise and safely support the vehicle securely on jackstands.
3. Remove the rear ABS wheel-speed sensor.
4. Drain the rear differential oil into a container.
5. Remove the rear lateral link.
6. Remove the rear coil spring.
7. Remove the rear lower arm.
8. Remove the rear trailing link and rear wheel hub.
9. Remove the rear halfshaft.
   a. Disengage the rear drive shaft using the special tools.

⁂ WARNING

Be careful not to damage the rear differential oil seal.

10. Remove the oil seal.
11. Remove the clip.

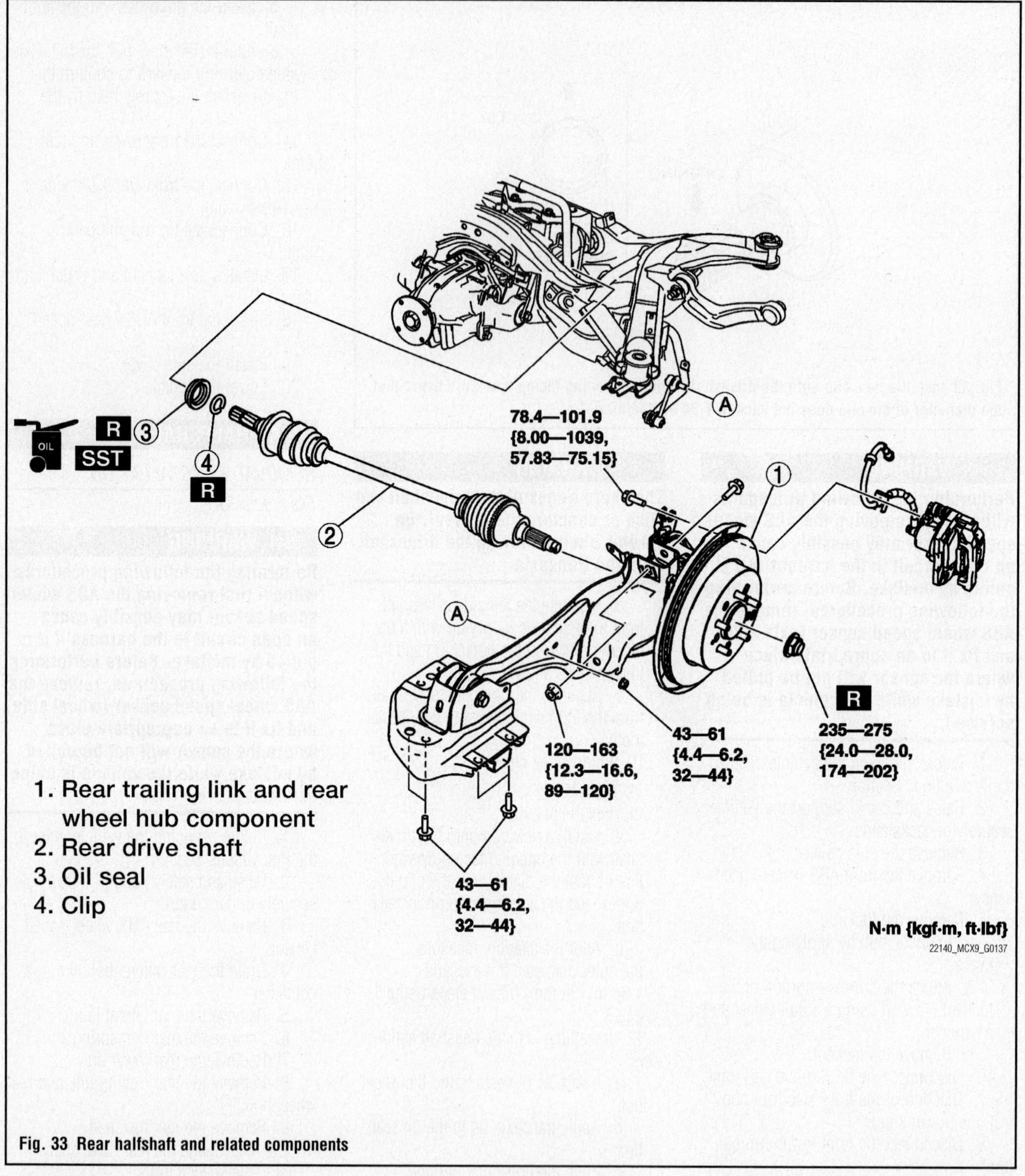

Fig. 33 Rear halfshaft and related components

To install:

12. Install the clip.

a. Install a new clip onto the joint shaft with the opening facing upward. Ensure that the diameter of the clip does not exceed 1.16 in. (29.5 mm) on installation.

b. After installation, measure the outer diameter. If it exceeds the specification, repeat steps using a new clip.

13. Install the oil seal.
14. Install the rear halfshaft.
15. Install the rear trailing link and rear wheel hub.
16. Install the rear lower arm.
17. Install the rear coil spring.
18. Install the rear lateral link.
19. Drain the rear differential oil into a container.
20. Install the rear ABS wheel-speed sensor.
21. After installation, add the specified rear differential oil.
22. Inspect the rear wheel alignment and adjust it if necessary.
23. Lower the vehicle.

ENGINE COOLING

⁂ WARNING

Never remove the cooling system cap or loosen the radiator drain plug while the engine is running, or when the engine and radiator are hot. Scalding engine coolant and steam may shoot out and cause serious injury. It may also damage the engine and cooling system. Turn off the engine and wait until it is cool. Even then, be very careful when removing the cap. Wrap a thick cloth around it and slowly turn it counter-clockwise to the first stop. Step back while the pressure escapes. When you are sure all the pressure is gone, press down on the cap using the cloth, turn it, and remove it.

THERMOSTAT

REMOVAL & INSTALLATION

See Figures 34 and 35.

1. Before servicing the vehicle, refer to the Precautions Section.
2. Disconnect the negative battery cable.
3. Drain the engine coolant.
4. Remove the air cleaner and fresh air duct.
5. Remove the thermostat cover.
6. Remove the O-ring.
7. Remove the thermostat.

To install:

8. Install the thermostat into the thermostat case with the jiggle pin at the top.
9. Install the O-ring.
   a. Clean and inspect the O-ring. Install a new O-ring if necessary.
   b. Apply clean engine coolant to the O-ring.
   c. Install the O-ring.

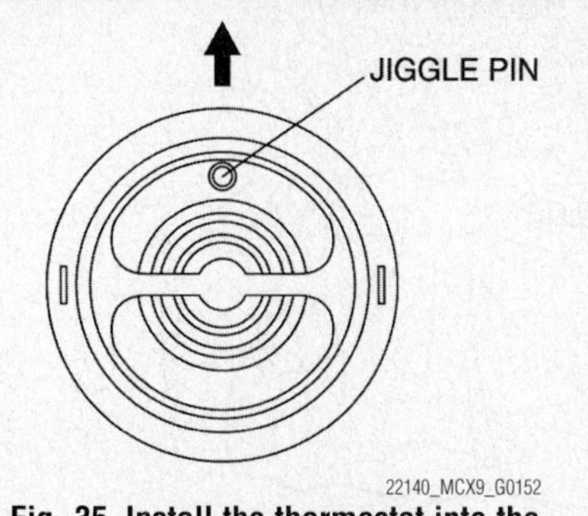

Fig. 35 Install the thermostat into the thermostat case with the jiggle pin at the top

10. Install the thermostat cover.
11. Install the air cleaner and fresh air duct.
12. Connect the negative battery cable.
13. Refill the engine coolant.
    a. After the engine warms up, perform the following steps.
    b. Run the engine at approx. 4,000 RPM for 1 min.
    c. Run the engine at idle for 1 min.
    d. Repeat the following steps 2 times.
    e. Operate the heater at the maximum temperature and airflow, and verify that hot air blows from vent.
    f. Stop the engine, and inspect the engine coolant level after the engine coolant temperature decreases. If it is low, repeat steps.
14. Inspect for engine coolant leakage.

WATER PUMP

REMOVAL & INSTALLATION

See Figures 36 through 39.

Mazda's official procedure for water pump removal and installation requires the engine and transaxle to be removed from the vehicle. It may be possible to perform this procedure with the engine and transaxle in the vehicle.

1. Before servicing the vehicle, refer to the Precautions Section.
2. Drain the engine oil.
3. Remove the engine and transaxle.
4. Secure the engine and transaxle using a hoist and an engine stand.
5. Remove the dynamic chamber and throttle body as a single unit.
6. Remove the ignition coils.
7. Remove the dipstick.
8. Remove the power steering oil pump drive belt.
9. Remove the power steering oil pump.
10. Remove the timing chain.
11. Remove the upper timing chain guides.
    a. Loosen the upper timing chain guide (RH) bolt A.
    b. Remove the upper timing chain guide (RH) bolt B.
    c. Move the upper timing chain guide (RH) up and to the right.
    d. Retighten the bolt A.
12. Remove the water pump and gasket.

⁂ WARNING

When the water pump is removed, the engine coolant may flow through the cylinder block and penetrate the oil pan. To prevent the engine coolant from accumulating in the oil pan, remove the oil pan drain plug before removing the water pump.

   a. Remove the oil pan drain plug.
   b. To prevent engine coolant from penetrating the oil pan, line the cylinder block with plastic sheeting as shown. before removing the water pump.

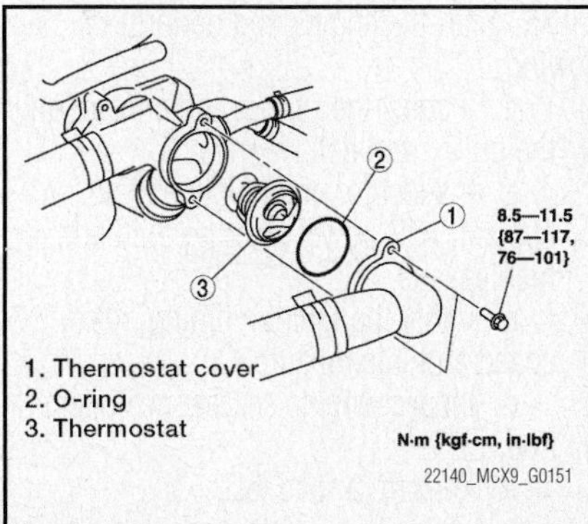

Fig. 34 Thermostat and related components

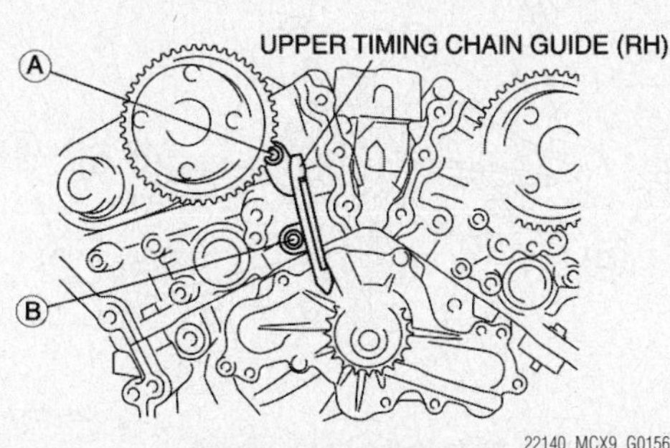

Fig. 36 Loosen the upper timing chain guide (RH) bolt A. Remove the upper timing chain guide (RH) bolt B. Move the upper timing chain guide (RH) up and to the right, then retighten the bolt A

Fig. 37 To prevent engine coolant from penetrating the oil pan, line the cylinder block with plastic sheeting as shown. before removing the water pump

To install:

13. Install the water pump and gasket.
14. Tighten the water pump bolts in the order shown. 76–101 ft. lbs. (8.5–11.5 Nm).
15. Install the upper timing chain guides.
16. Install the timing chain.
17. Install the power steering oil pump.
18. Install the power steering oil pump drive belt.
19. Install the dipstick.
20. Install the ignition coils.
21. Install the dynamic chamber and throttle body as a single unit.
22. Secure the engine and transaxle using a hoist and the special tool.

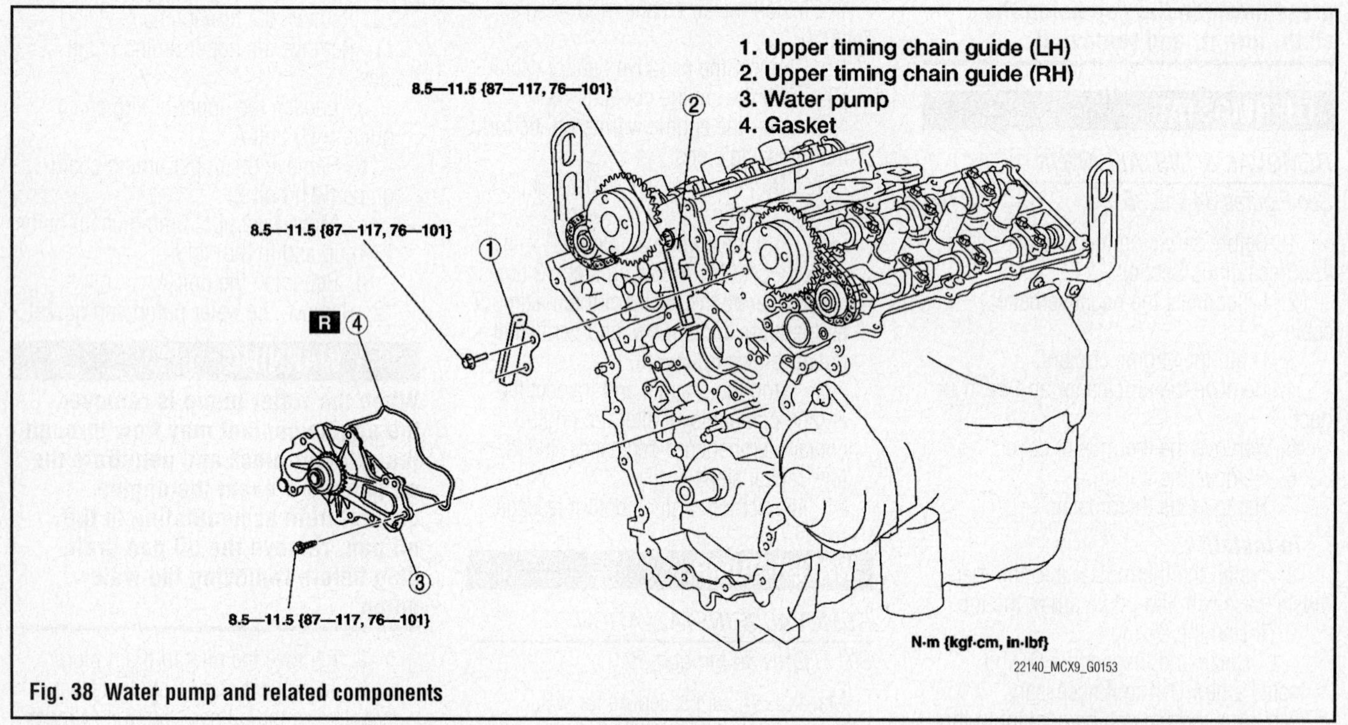

Fig. 38 Water pump and related components

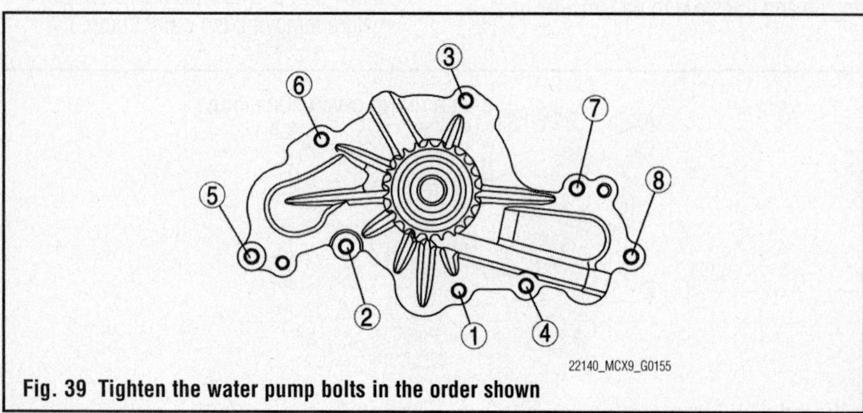

Fig. 39 Tighten the water pump bolts in the order shown

23. Install the engine and transaxle.
24. Fill the engine with oil.
25. Start the engine and perform the following:
    a. Inspect the runout and contact on the pulley and belt.
    b. Inspect for engine oil, engine coolant, ATF, power steering fluid and fuel leakage.
    c. Verify the ignition timing, idle speed and idle mixture.
    d. Inspect engine accessories operation.
    e. Perform a road test.

ENGINE ELECTRICAL — CHARGING SYSTEM

ALTERNATOR

REMOVAL & INSTALLATION

See Figure 40.

**** WARNING**

Remove and install all parts when the engine is cold, otherwise they can cause severe burns or serious injury.

**** WARNING**

When the battery cables are connected, touching the vehicle body with alternator terminal B will generate sparks. This can cause personal injury, fire, and damage to the electrical components. Always disconnect the negative battery cable before performing the following operation.

1. Before servicing the vehicle, refer to the Precautions Section.
2. Disconnect the negative battery cable.
3. Drain the engine coolant.
4. Remove the air cleaner and fresh air duct.
5. Remove the cooling fan.
6. Remove the front splash shield (RH).
7. Remove the alternator and A/C drive belt.
8. Remove the terminal B cable.
9. Remove the alternator connector.
10. Remove the alternator lower bolt.
    a. The alternator lower bolt cannot be fully removed from the engine because it contacts the body frame. However, the alternator can be removed/installed without fully removing the lower bolt because there is a notch at the lower bolt installation part of the alternator.
    b. Fully loosen the alternator lower bolt.
11. Remove the alternator from the top.

To install:

12. Install the alternator from the top.
13. Install the alternator upper bolt.
    a. Tighten the upper and lower bolts temporarily, then tighten to 35 ft. lbs. (48 Nm).
14. Install the alternator connector.
15. Install the terminal B cable.
16. Install the alternator and A/C drive belt.
17. Install the front splash shield (RH).
18. Install the cooling fan.
19. Install the air cleaner and fresh air duct.
20. Connect the negative battery cable.
21. Refill the engine coolant.
    a. After the engine warms up, perform the following steps.
    b. Run the engine at approx. 4,000 RPM for 1 min.
    c. Run the engine at idle for 1 min.
    d. Repeat the following steps 2 times.
    e. Operate the heater at the maximum temperature and airflow, and verify that hot air blows from vent.
    f. Stop the engine, and inspect the engine coolant level after the engine coolant temperature decreases. If it is low, repeat steps.
22. Inspect for engine coolant leakage.

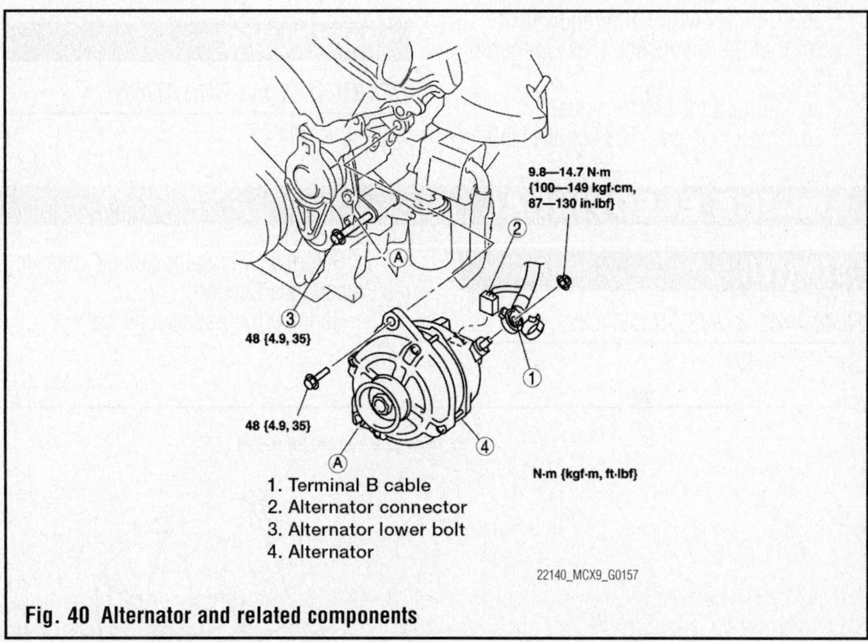

Fig. 40 Alternator and related components

ENGINE ELECTRICAL — DISTRIBUTORLESS IGNITION SYSTEM

IGNITION COIL

REMOVAL & INSTALLATION

See Figures 41 and 42.

1. Before servicing the vehicle, refer to the Precautions Section.
2. Disconnect the negative battery cable.
3. Remove the engine cover.
4. Remove the intake manifold dynamic chamber.
5. Disconnect the coil connector.
6. Remove the ignition coil.

To install:

7. Apply small amount of Motorcraft XG-3A dielectric grease to the inside of the

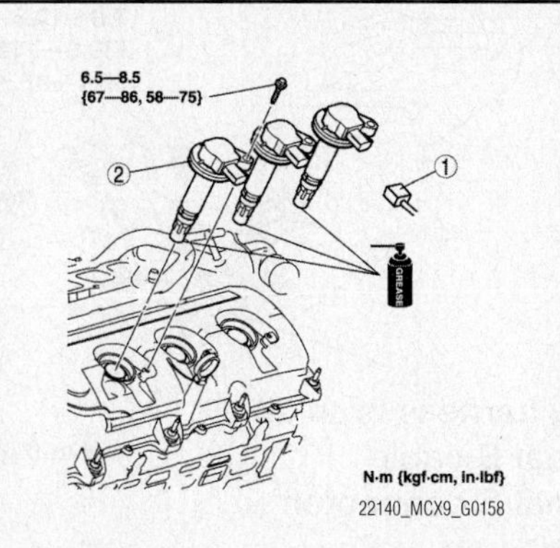

Fig. 41 Ignition coils and related components

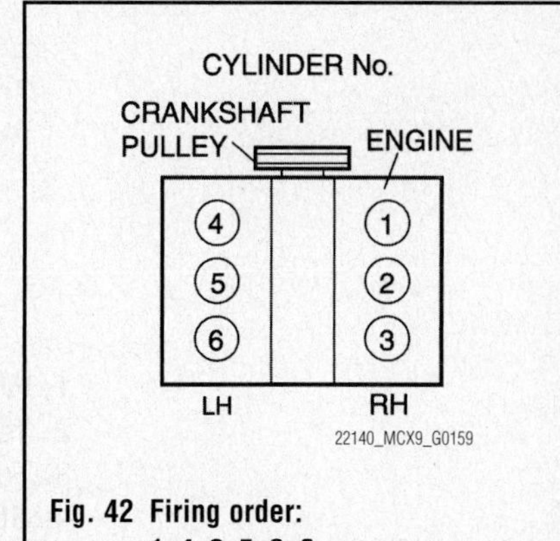

Fig. 42 Firing order: 1-4-2-5-3-6 Direct ignition system

ignition coils, then install the ignition coils to the spark plugs.

8. Tighten coils to 58–75 inch lbs. (7–9 Nm).

9. Install the intake manifold dynamic chamber.

10. Install the engine cover.

11. Connect the negative battery cable.

FIRING ORDER

See Figure 42.

IGNITION TIMING

INSPECTION

1. Perform the engine tune-up preparation as follows:
   a. Verify that the selector lever is in the P position.
   b. Connect M-MDS or equivalent scan tool the Diagnostic Link Connector (DLC).
   c. Turn off the electrical loads.
   d. Verify that no DTCs are available.
2. Warm up the engine as follows.
   a. Start the engine.
   b. Perform no-load racing at 2,5003,000 RPM for 3 min. (ECT is approximately 176°F (80°C)} or more)
   c. Release the accelerator pedal.
   d. Wait until the cooling fans stop.
3. Inspect the ignition timing as follows:
   a. Verify that the ignition timing is within the specification using M-MDS or equivalent scan tool. Ignition timing should be 13° BTDC
   b. Verify that ignition timing advances when the engine speed increases gradually.

ADJUSTMENT

Ignition timing is control by the Engine Control Module (ECM) and is not adjustable.

SPARK PLUGS

REMOVAL & INSTALLATION

See Figure 41.

1. Before servicing the vehicle, refer to the Precautions Section.
2. Disconnect the negative battery cable.
3. Remove the engine cover.
4. Remove the intake manifold dynamic chamber.
5. Remove the ignition coils.
6. Remove the spark plugs using a plug-wrench.

To install:

7. Install the spark plugs and tighten to 79–177 inch lbs. (9–20 Nm).
8. Apply small amount of Motorcraft XG-3A dielectric grease to the inside of the ignition coils, then install the ignition coils to the spark plugs.
9. Tighten coils to 58–75 inch lbs. (7–9 Nm).
10. Install the intake manifold dynamic chamber.
11. Install the engine cover.
12. Connect the negative battery cable.

ENGINE ELECTRICAL

STARTING SYSTEM

STARTER

REMOVAL & INSTALLATION

See Figure 43.

1. Before servicing the vehicle, refer to the Precautions Section.
2. Remove the battery and battery tray.
3. Position the selector cable out of the way.
4. Remove the wiring harness bracket.
5. Disconnect the terminal B cable.

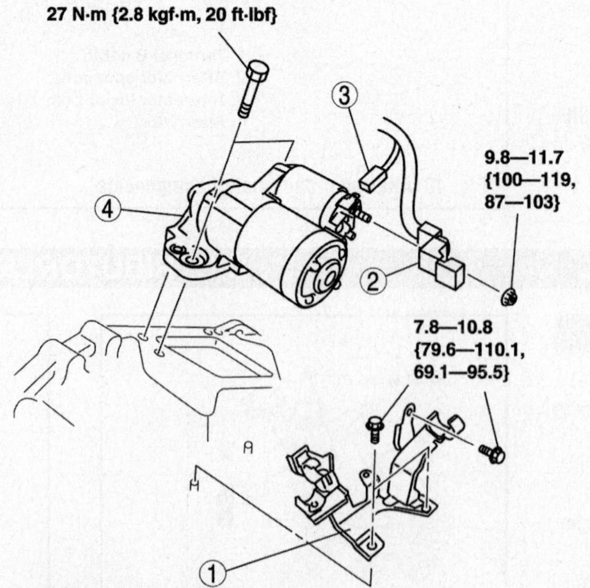

Fig. 43 Starter and related components

6. Disconnect the terminal S connector.
7. Remove the starter.

To install:

8. Install the starter. Tighten bolts to 20 ft. lbs. (27 Nm).
9. Connect the terminal S connector.
10. Connect the terminal B cable. Tighten to 87–103 inch lbs. (10–12 Nm).
11. Install the wiring harness bracket. Tighten to 69–96 inch lbs. (8–11 Nm).
12. Position the selector cable out of the way.
13. Install the battery and battery tray.

SOLENOID OR RELAY REPLACEMENT

See Figure 44.

1. Before servicing the vehicle, refer to the Precautions Section.
2. Remove the starter from the vehicle.

➡**Do not clamp the yoke assembly with a vise.**

3. Disconnect the lead from the M terminal of the magnetic switch.
4. Remove the screw and bracket from the solenoid.
5. Remove the solenoid from the starter.

To install:

6. Installation is the reverse of removal.
7. Tighten the solenoid screws to 37–66 inch lbs. (4–8 Nm).

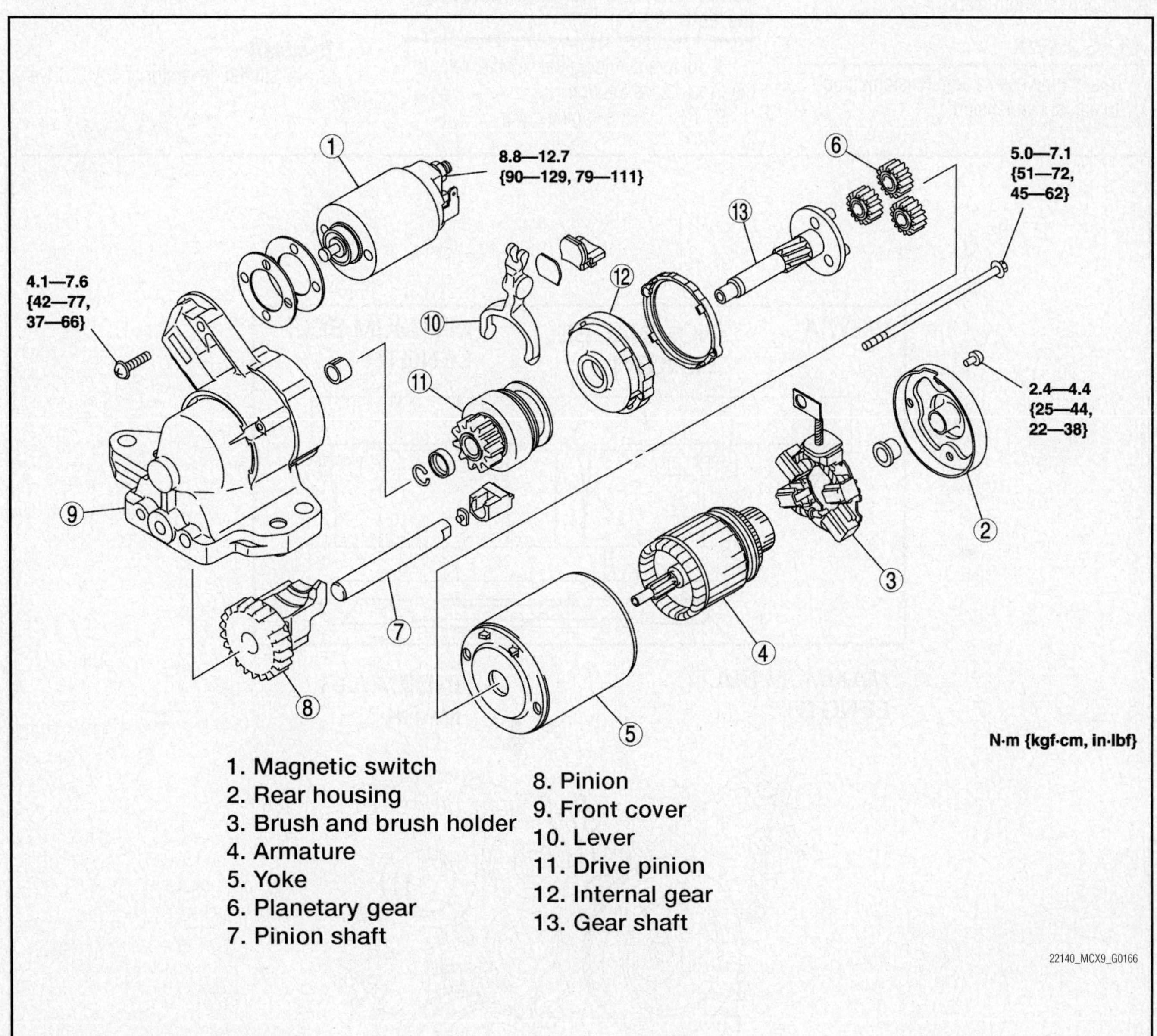

Fig. 44 Starter solenoid replacement

ENGINE MECHANICAL

ACCESSORY DRIVE BELTS

INSPECTION

Verify that the drive belt auto tensioner indicator mark is within the maximum and minimum belt length.

When the generator and A/C drive belt is replaced with a new one, the drive belt auto tensioner indicator mark is aligned with the nominal belt length mark.

ADJUSTMENT

This engine uses a auto tensioning pulley to adjust belt tension.

REMOVAL & INSTALLATION

Generator and A/C Drive belt

See Figure 45.

⁂ CAUTION

When removing the generator and A/C drive belt, perform the procedure with two people, one releasing the belt tension and the other removing the belt.

1. Before servicing the vehicle, refer to the Precautions Section.
2. Remove the engine cover.
3. Remove the front wheel and tire (RH).
4. Remove the splash shield (RH).
5. Set a breaker bar on the center of the tensioner pulley.
6. Using the breaker bar, turn the center of the tensioner pulley clockwise to release the tension on the generator and A/C drive belt.
7. Remove the generator and A/C drive belt.

To install:

8. Install the generator and A/C drive belt.

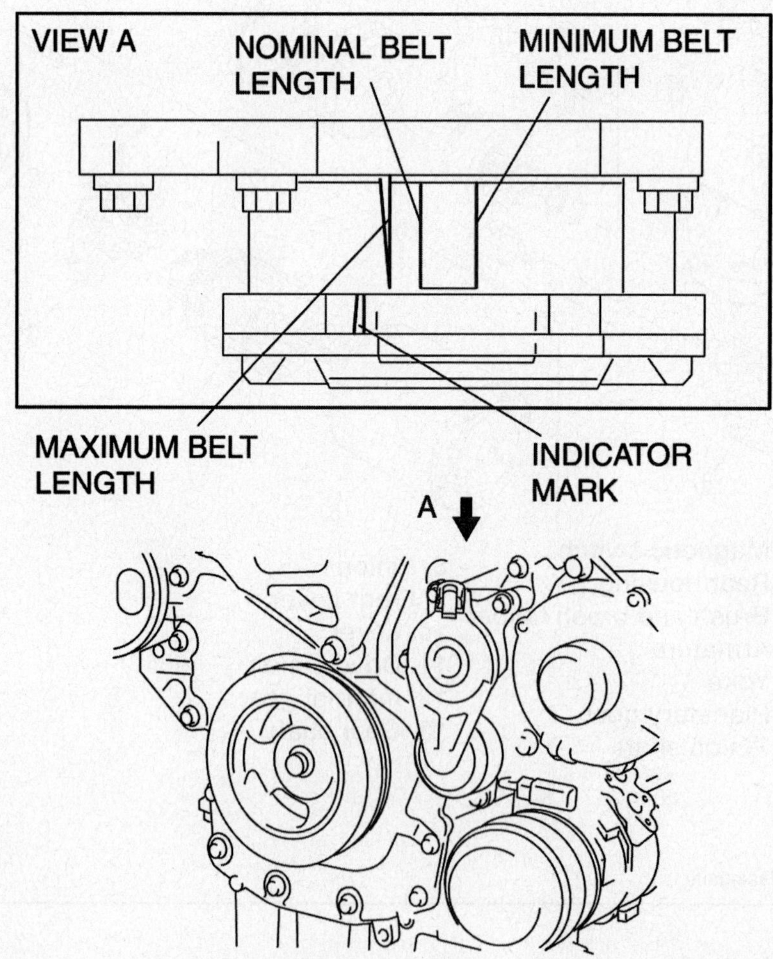

Fig. 45 Verify that the drive belt auto tensioner indicator mark is within the maximum and minimum belt length

➡When the generator and A/C drive belt is replaced with a new one, the drive belt auto tensioner indicator mark is aligned with the nominal belt length mark.

9. Verify that the drive belt auto tensioner indicator mark is within the maximum and minimum belt length.
10. If the drive belt auto tensioner indicator mark is not within the limit, replace the generator and A/C drive belt.
11. Install the splash shield (RH).
12. Install the front wheel and tire (RH).
13. Install the engine cover.

Power Steering Oil Pump Drive Belt

See Figures 46 through 48.

1. Before servicing the vehicle, refer to the Precautions Section.
2. Remove the generator and A/C drive belt.
3. Remove the power steering oil pump drive belt.
    a. Install the special tool between the power steering oil pump drive belt and the power steering oil pump pulley.
    b. Hold the special tool by hand until it is lodges between the power steering oil pump pulley and power steering oil pump drive belt.
    c. Turn the crankshaft clockwise to remove the power steering oil pump drive belt.
    d. If there is power steering oil pump drive belt damage and cracks, replace the power steering oil pump drive belt.
4. Install the power steering oil pump drive belt.
    a. Install the special tool according to the following steps.
    b. Loosen the clamp adapter nut of the special tool.
    c. There is no need to remove the clamp adapter nut completely when positioning the special tool on the power steering oil pump pulley.
    d. Position the special tool on the power steering oil pump pulley with the clamp adapter in one of the holes of the pulley.
    e. Tighten the clamp adapter nut by hand.
    f. Hand tightening of the clamp adapter nut is sufficient.
    g. Position the power steering oil pump drive belt around the special tool and the power steering oil pump pulley.

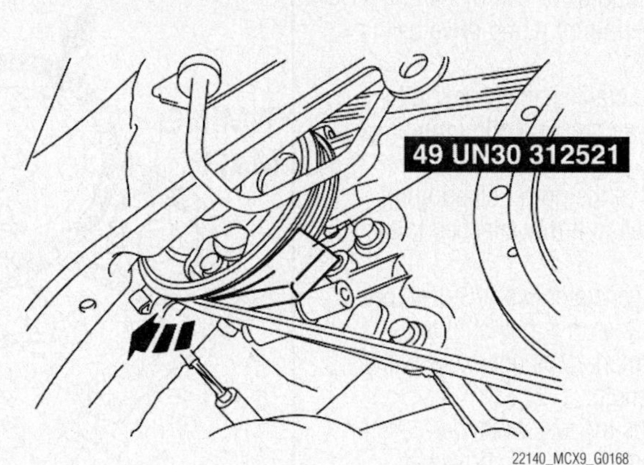

Fig. 46 Install the special tool between the power steering oil pump drive belt and the power steering oil pump pulley

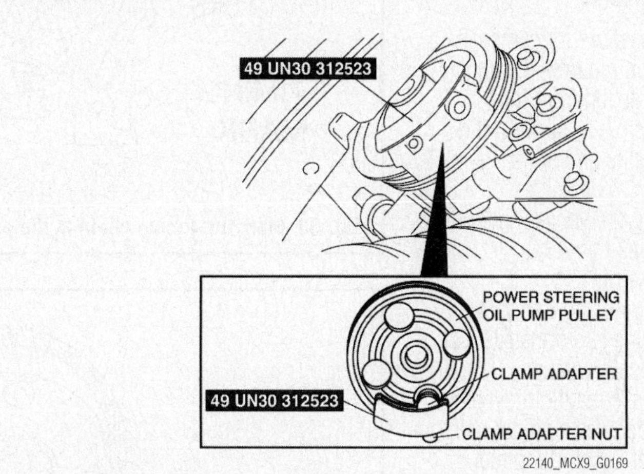

Fig. 47 Position the special tool on the power steering oil pump pulley with the clamp adapter in one of the holes of the pulley

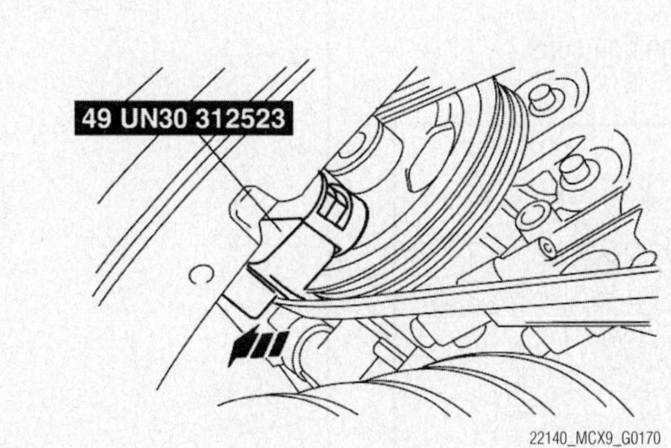

Fig. 48 Position the power steering oil pump drive belt around the special tool and the power steering oil pump pulley

h. Hold the special tool and power steering oil pump drive belt by hand until the power steering oil pump drive belt is properly seated.

i. Turn the crankshaft clockwise to install the power steering oil pump drive belt.

j. Verify that the power steering oil pump drive belt is firmly attached to the pulleys.

5. Install the generator and A/C drive belt.
6. Verify that the drive belt auto tensioner indicator mark does not exceeds the maximum belt length.
7. If it exceeds the maximum belt length, replace the generator and A/C drive belt.

CAMSHAFT AND VALVE LIFTERS

REMOVAL & INSTALLATION

See Figures 49 through 66.

Mazda's official procedure for camshaft removal and installation requires the engine and transaxle to be removed from the vehicle. It may be possible to perform this procedure with the engine and transaxle in the vehicle.

1. Before servicing the vehicle, refer to the Precautions Section.
2. Drain the engine oil.
3. Remove the engine and transaxle.
4. Secure the engine and transaxle using a hoist and an engine stand.
5. Remove the intake manifold dynamic chamber and throttle body as a single unit.
6. Remove the fuel injector and fuel distributor together as a single unit.
7. Remove the thermostat and thermostat housing together as a single unit.
8. Remove the Intake manifold.
9. Remove both catalytic converters together with both exhaust manifolds as a single unit.
10. Remove the ignition coils.

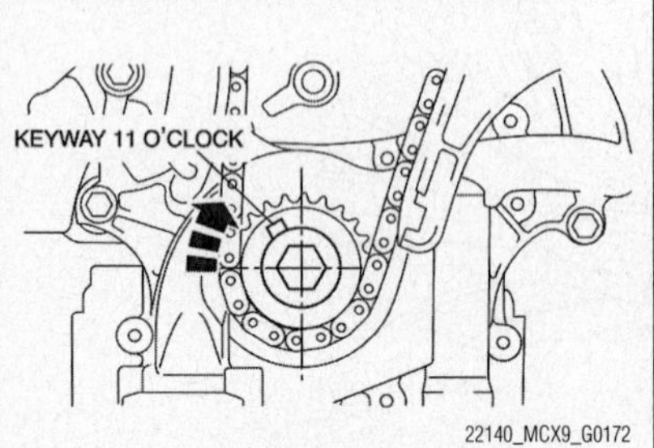

Fig. 49 Turn the crankshaft clockwise so that the crankshaft keyway is in the 11 o'clock position. This will position the No.1 cylinder at TDC

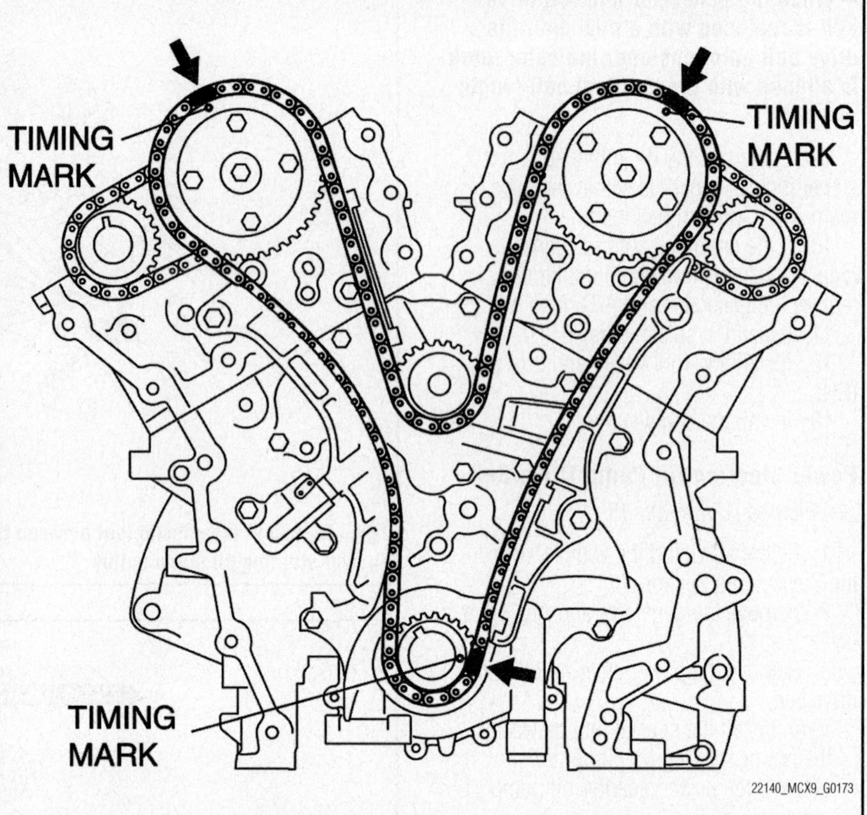

Fig. 50 Mark the timing chain at the position of each timing sprocket timing mark

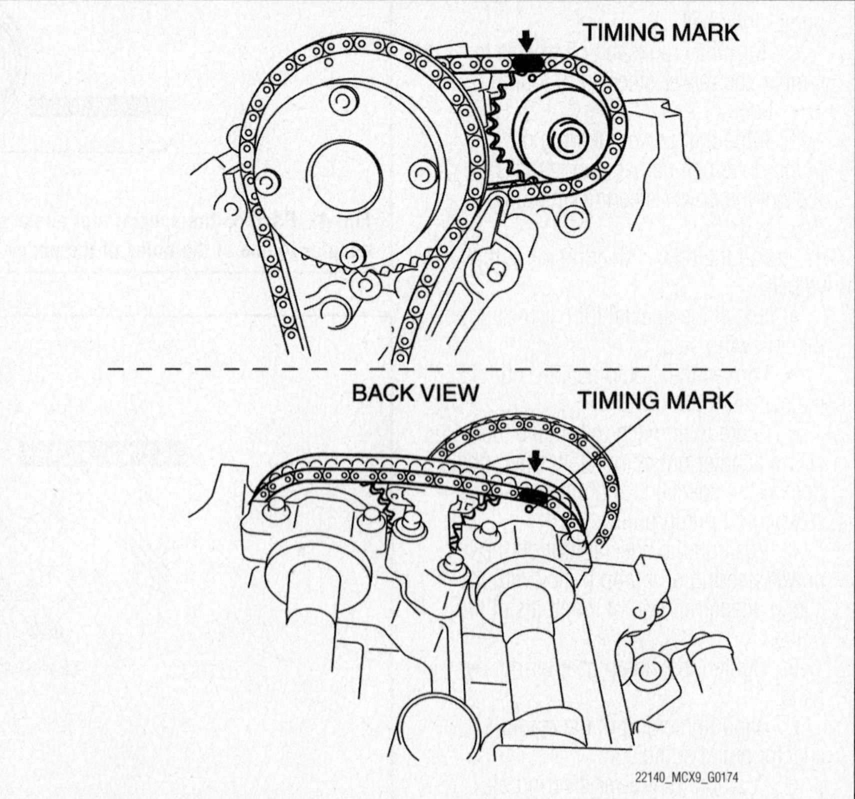

Fig. 51 Mark the camshaft timing chain at the positions where it is aligned with each of the camshaft sprocket on both banks

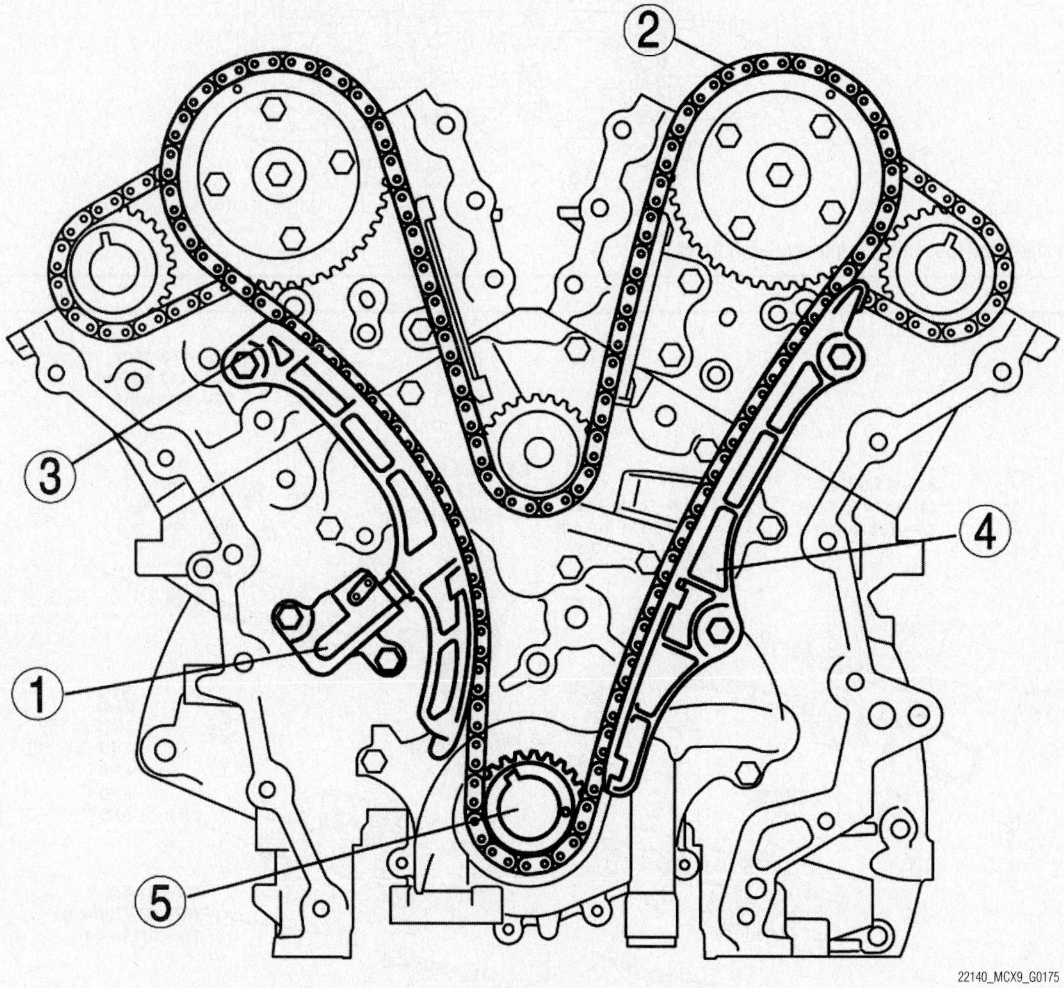

Fig. 52 Remove the timing chain in the following order: (1) Chain tensioner, (2) Timing chain, (3) Tensioner arm, (4) Chain guide and (5) Crankshaft sprocket

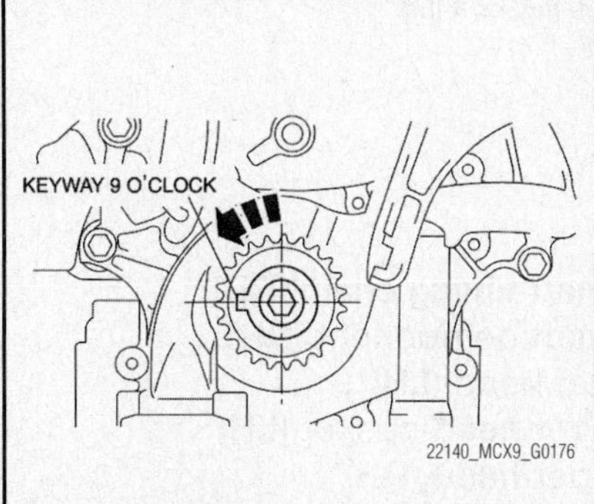

Fig. 53 Rotate the crankshaft counter-clockwise until the keyway is in the 9 o'clock position

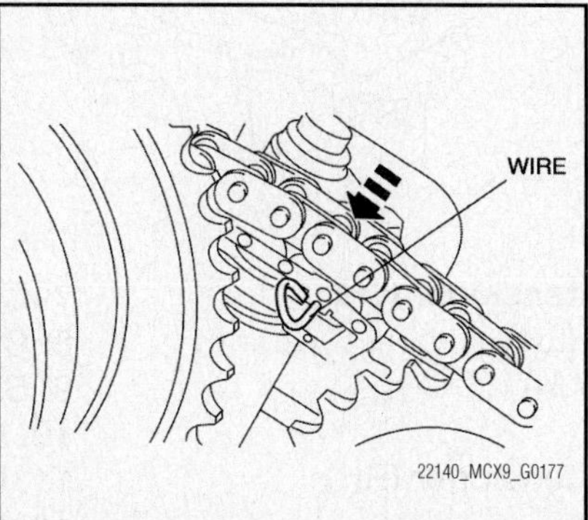

Fig. 54 Insert a paper clip into the camshaft timing chain tensioner to hold the tensioner piston

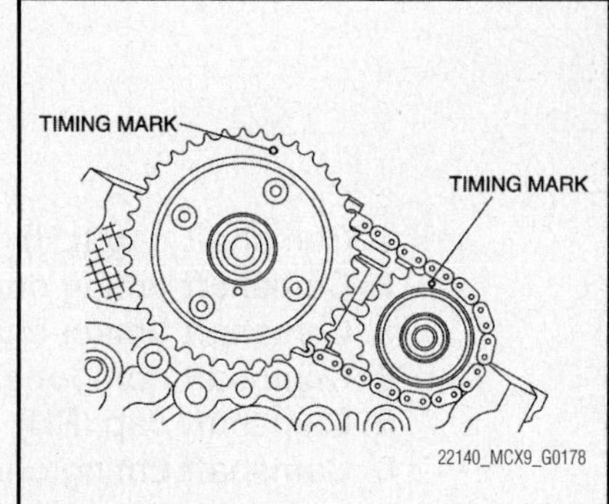

Fig. 55 Camshafts in the neutral position—left hand

ENGINE FRONT SIDE
14 13 12 11 10 9
15
16
8
7
6 5 4 3 2 1
22140_MCX9_G0179

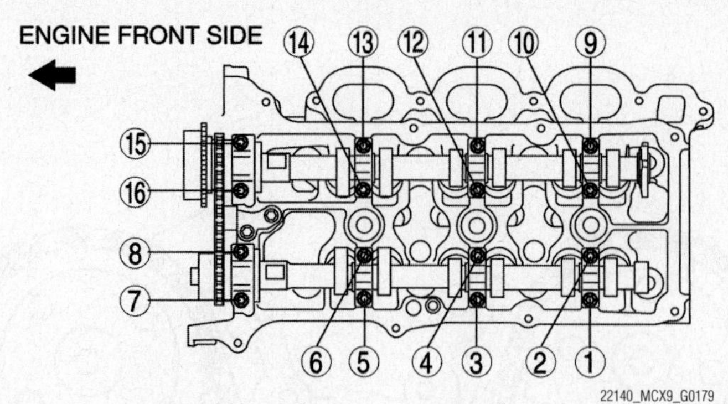

Fig. 56 Camshaft cap loosening sequence—left hand

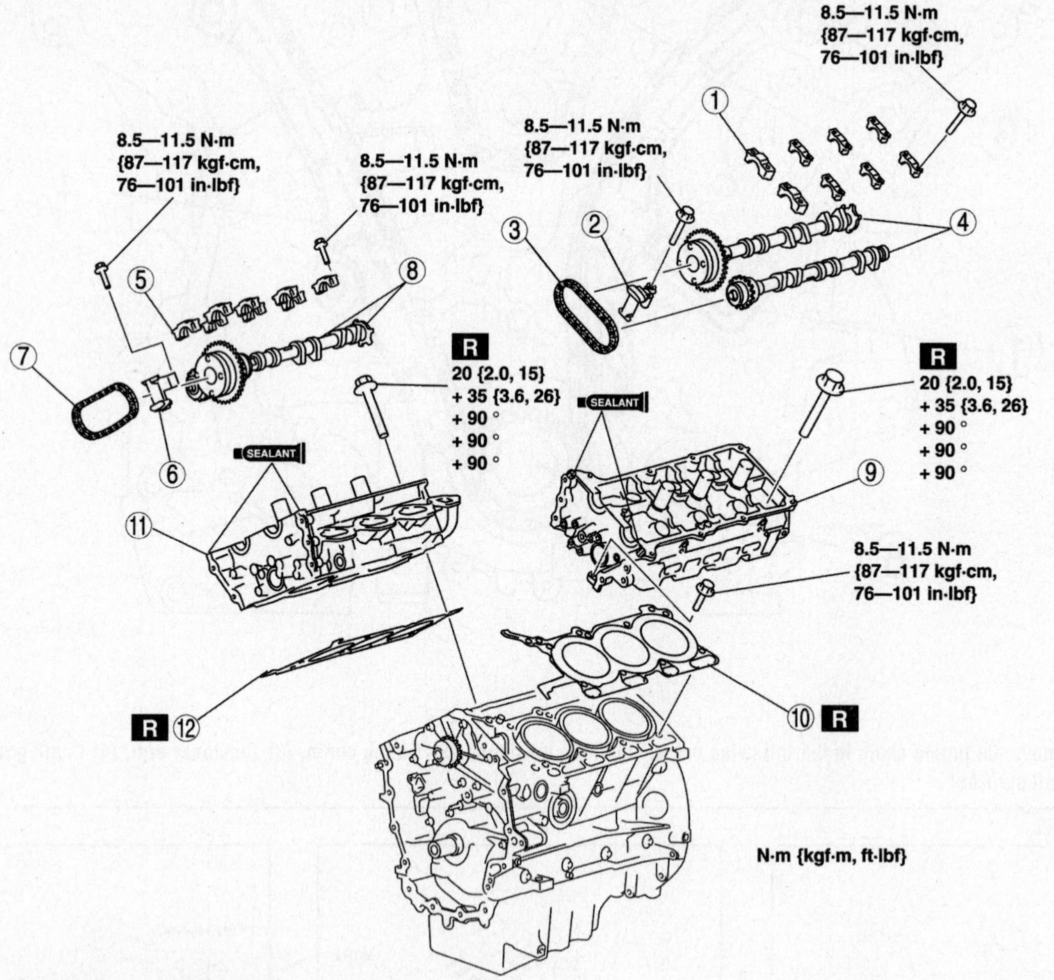

8.5—11.5 N·m {87—117 kgf·cm, 76—101 in·lbf}
8.5—11.5 N·m {87—117 kgf·cm, 76—101 in·lbf}
8.5—11.5 N·m {87—117 kgf·cm, 76—101 in·lbf}
8.5—11.5 N·m {87—117 kgf·cm, 76—101 in·lbf}
R
20 {2.0, 15} + 35 {3.6, 26} + 90 ° + 90 ° + 90 °
R
20 {2.0, 15} + 35 {3.6, 26} + 90 ° + 90 ° + 90 °
SEALANT
SEALANT
8.5—11.5 N·m {87—117 kgf·cm, 76—101 in·lbf}
R 12
10 R
N·m {kgf·m, ft·lbf}

1. Camshaft cap (LH)
2. Camshaft timing chain tensioner (LH)
3. Camshaft timing chain (LH)
4. Camshaft component (LH)
5. Camshaft cap (RH)
6. Camshaft timing chain tensioner (RH)
7. Camshaft timing chain (RH)
8. Camshaft component (RH)
9. Cylinder head (LH)
10. Cylinder head gasket (LH)
11. Cylinder head (RH)
12. Cylinder head gasket (RH)

22140_MCX9_G0171

Fig. 57 Exploded view of the cylinder head assembly

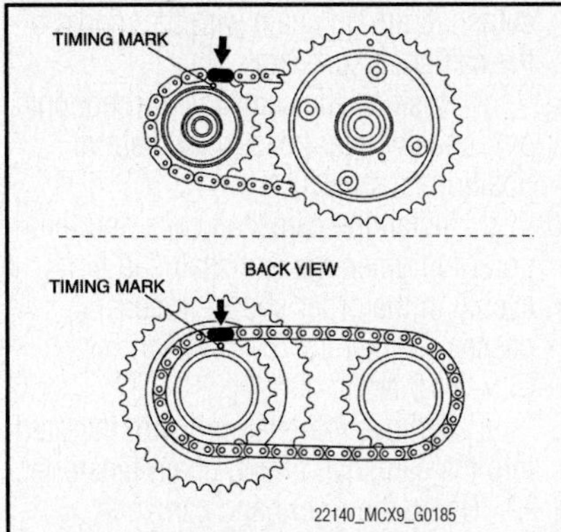

Fig. 58 Install the camshaft timing chain by aligning the colored links on the camshaft timing chain with the marks on the camshaft sprockets—right hand

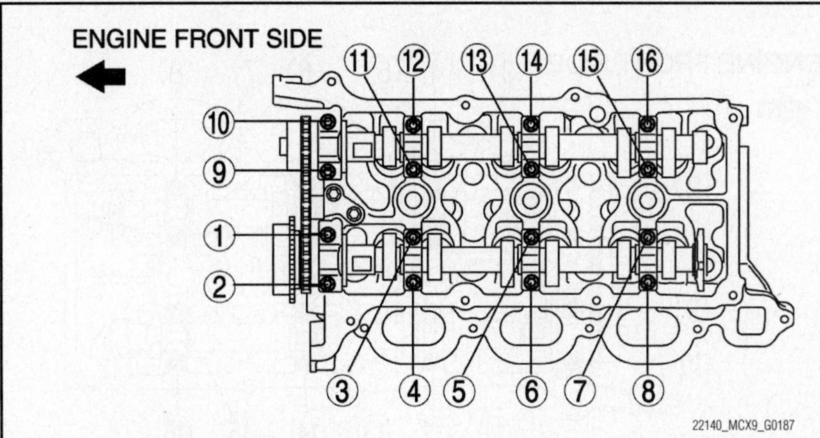

Fig. 62 Tighten the camshaft cap bolts to the specified torque, in several passes, in the order shown—right hand

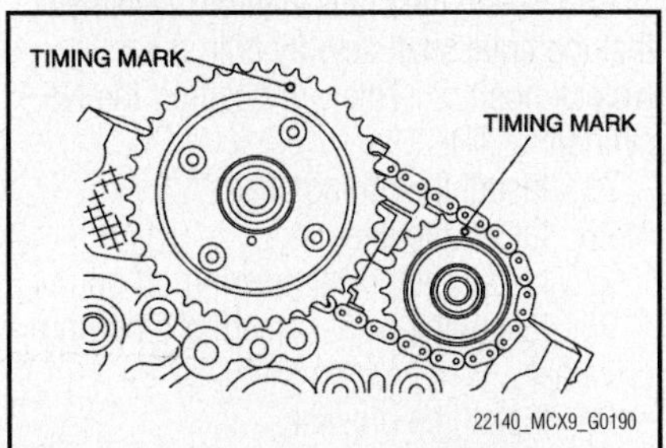

Fig. 59 Position the camshaft component onto the cylinder head in the neutral position as shown—left hand

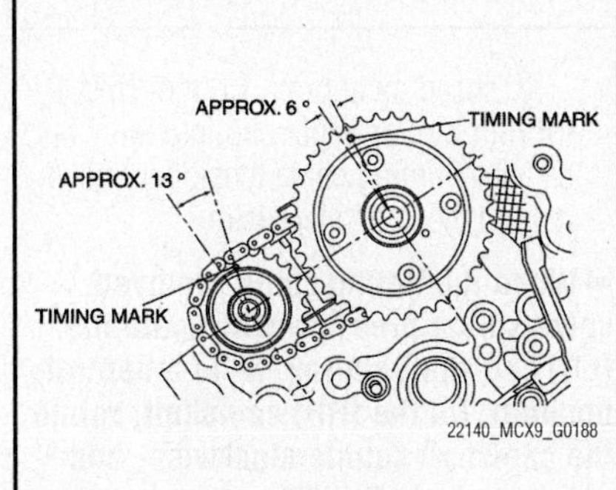

Fig. 63 Using the right hand camshaft sprocket installation bolt, rotate the right hand camshaft clockwise so the timing marks align as shown

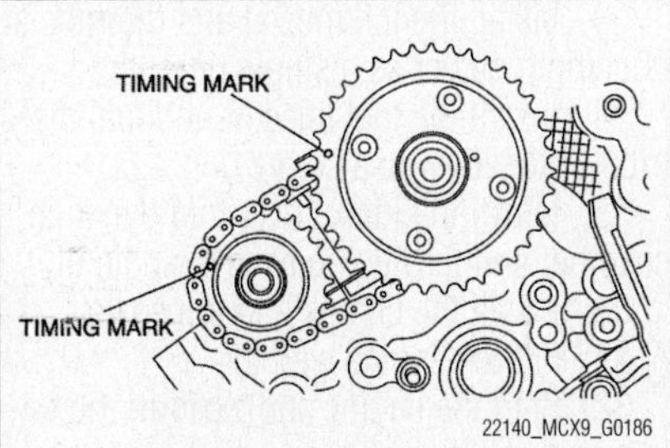

Fig. 60 Position the camshaft component onto the cylinder head in the neutral position as shown—right hand

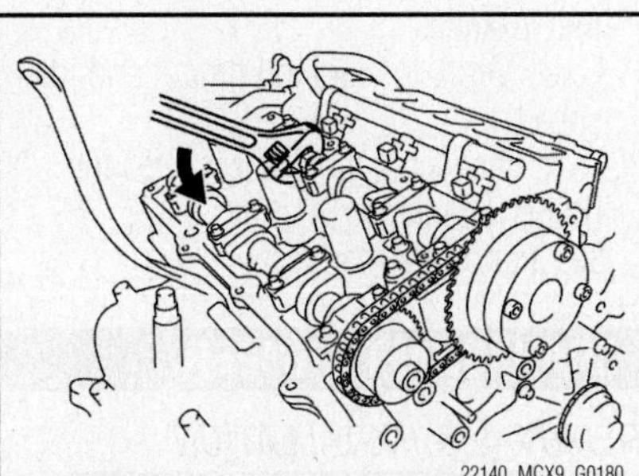

Fig. 61 Put a wrench as shown on the camshaft, rotate the camshaft counterclockwise, and set the camshaft to the neutral position—right hand

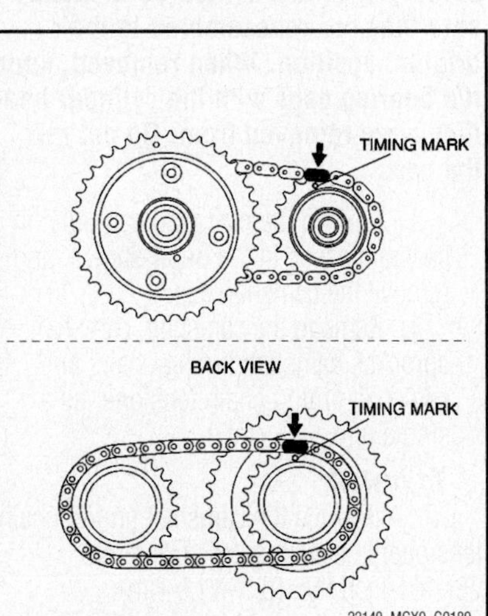

Fig. 64 Install the camshaft timing chain by aligning the colored links on the camshaft timing chain with the marks on the camshaft sprockets—left hand

11. Remove the dipstick.
12. Remove the power steering oil pump drive belt.
13. Remove the power steering oil pump.
14. Remove the alternator.

➡When removing the timing chain and marking the timing marks on the chain, mark the camshaft timing chain as well.

15. Remove the timing chain.
16. Remove the camshaft caps.
17. Remove the camshaft timing chain tensioners.
18. Remove the camshaft timing chains.
19. Remove the camshafts.

✻✻ WARNING

Do not rotate the crankshaft counterclockwise. The timing chains may bind, causing engine damage.

a. Turn the crankshaft clockwise so that the crankshaft keyway is in the 11 o'clock position. This will position the No.1 cylinder at TDC.

➡Verify that there are timing marks in three locations (Yellow 1, Black 2) on the timing chain. If any timing marks are missing, mark the timing chain. When marking the crankshaft sprocket side timing chain, change the mark color. When the timing chain is replaced with a new one, mark the new timing chain at the same positions as the removed timing chain.

b. Mark the timing chain at the position of each timing sprocket timing mark.

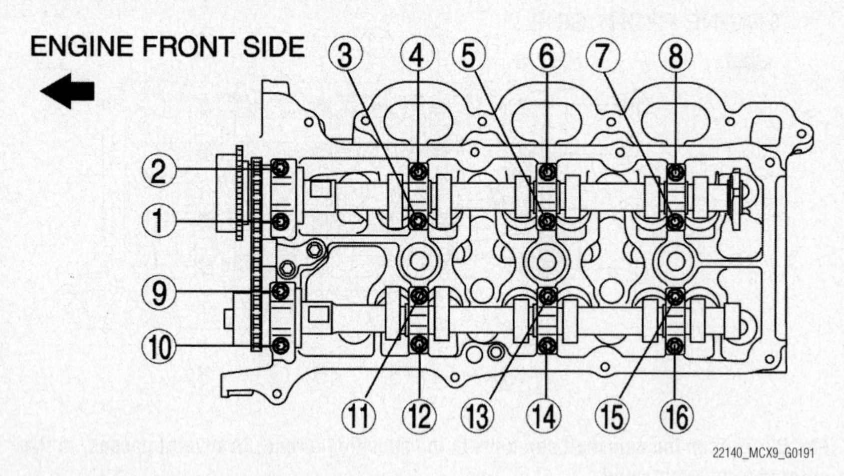

Fig. 65 Tighten the camshaft cap bolts to the specified torque, in several passes, in the order shown—left hand

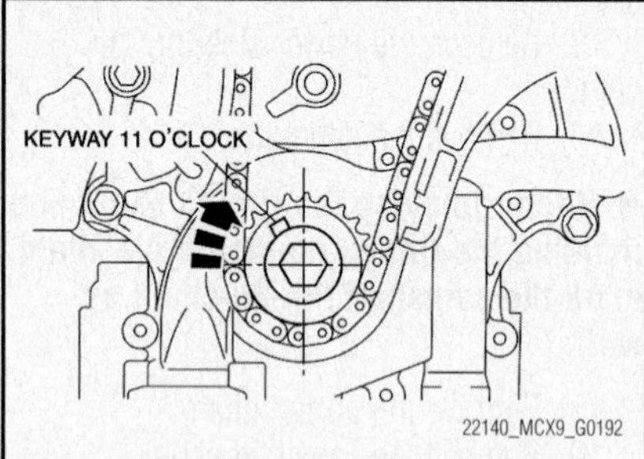

Fig. 66 Turn the crankshaft clockwise so that the crankshaft keyway is in the 11 o'clock position. This will position the No.1 cylinder at TDC.

➡Verify that there are timing marks in two locations on the camshaft timing chain. If any timing marks are missing, mark the camshaft timing chain. If replacing with a new camshaft timing chain, place alignment marks in the same positions as those prior to the replacement.

c. Mark the camshaft timing chain at the positions where it is aligned with each of the camshaft sprocket on both banks.

d. Remove the timing chain in the following order.

- Chain tensioner
- Timing chain
- Tensioner arm
- Chain guide
- Crankshaft sprocket

e. Rotate the crankshaft counterclockwise until the keyway is in the 9 o'clock position.

f. Slowly compress the camshaft timing chain tensioner (LH) piston by hand.

g. Insert an approx. 1.0 mm {0.039 in} thin wire or paper clip into the camshaft timing chain tensioner shown. to hold the tensioner piston.

➡When the timing chain removed, valve spring pressure will rotate the (LH) camshaft approx. 3° to a neutral position. On the (RH) camshaft, rotate the camshaft counterclockwise, and set the camshaft to the neutral position.

h. Verify that the camshafts are in the neutral position.

➡The cylinder head and the camshaft bearing caps are numbered to make sure they are reassembled in their original position. When removed, keep the bearing caps with the cylinder head they were removed from. Do not mix the caps.

i. Loosen the camshaft cap bolts in several passes in the order shown. and remove the camshaft cap.

j. Remove the camshaft, camshaft sprocket, camshaft timing chain, and camshaft timing chain tensioner as a single unit.

To install:

20. Assemble the camshaft timing chain tensioners on both sides. Tighten to 76–101 inch lbs. (8.5–11.5 Nm).

21. Install the right hand camshaft.

➡Install the camshaft, camshaft sprocket and camshaft timing chain of the both bank as a single unit.

a. Install the camshafts.

b. Install the camshaft timing chain by aligning the colored links on the camshaft timing chain with the marks on the camshaft sprockets.

c. Position the camshaft component onto the cylinder head in the neutral position as shown.

d. Install the camshaft caps and temporarily tighten the camshaft cap bolts evenly in the order shown in several passes. Tighten to 76–101 inch lbs. (8.5–11.5 Nm).

e. Remove the retaining wire inserted into the camshaft timing chain tensioner.

22. Using the right hand camshaft sprocket installation bolt, rotate the right hand camshaft clockwise so the timing marks align as shown.

23. Install the left hand camshaft in the same manner as the right hand camshaft.

24. Rotate the crankshaft clockwise so that the crankshaft keyway is in the 11 o'clock position. This will position the No.1 cylinder at TDC.

25. Install the timing chain.

26. Install the alternator.

27. Install the power steering oil pump.

28. Install the power steering oil pump drive belt.

29. Install the dipstick.

30. Install the ignition coils.

31. Install both catalytic converters together with both exhaust manifolds as a single unit.

32. Install the Intake manifold.

33. Install the thermostat and thermostat housing together as a single unit.

34. Install the fuel injector and fuel distributor together as a single unit.

35. Install the intake manifold dynamic chamber and throttle body as a single unit.

36. Install the engine and transaxle.

37. Fill the engine with oil.

38. Start the engine and perform the following inspections:

a. Inspect the runout and contact on the pulley and belt.

b. Inspect for engine oil, engine coolant, ATF, power steering fluid and fuel leakage.

c. Verify the ignition timing and idle speed.

d. Inspect engine accessories for proper operation.

39. Perform a road test.

CRANKSHAFT DAMPER

REMOVAL & INSTALLATION

See Figures 67 through 71.

Mazda's official procedure for the crankshaft damper removal and installation requires the engine and transaxle to be removed from the vehicle. It may be possible to perform this procedure with the engine and transaxle in the vehicle.

1. Before servicing the vehicle, refer to the Precautions Section.
2. Drain the engine oil.
3. Remove the engine and transaxle.
4. Secure the engine and transaxle using a hoist and an engine stand.
5. Remove the intake manifold dynamic chamber and throttle body as a single unit.
6. Remove the ignition coils.

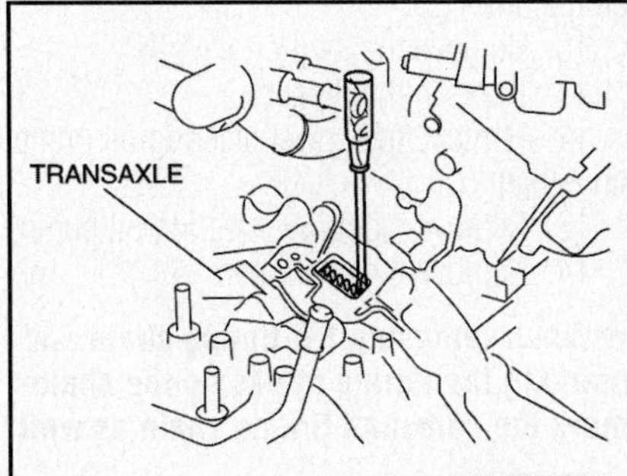

Fig. 67 Set a flathead screwdriver to the drive plate as shown to lock the crankshaft rotation

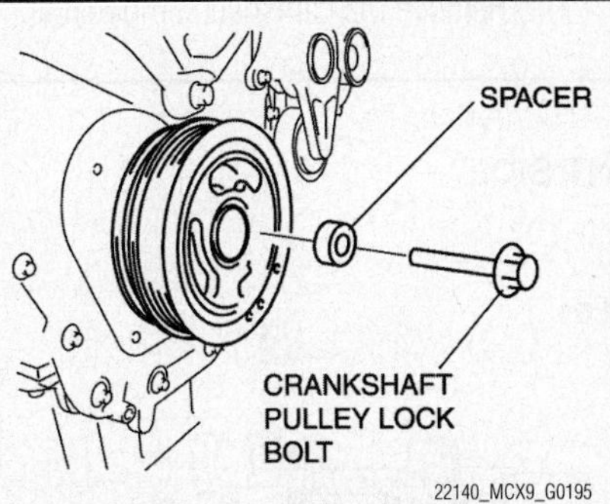

Fig. 68 Install a suitable spacer 0.55 in. (14 mm) in thickness and 1.18 in. (30 mm) in diameter to the crankshaft pulley lock bolt, and install the crankshaft pulley lock bolt to the crankshaft

7. Remove the dipstick.
8. Remove the power steering oil pump drive belt.
9. Remove the power steering oil pump.
10. Remove the drive belt auto tensioner.
11. Remove the crankshaft pulley lock bolt.
    a. Remove the starter.
    b. Set a flathead screwdriver to the drive plate as shown to lock the crankshaft rotation.
    c. Remove the crankshaft pulley lock bolt and washer.
12. Remove the crankshaft pulley.
    a. Install a suitable spacer 0.55 in. (14 mm) in thickness and 1.18 in. (30 mm) in diameter to the crankshaft pulley lock bolt, and install the crankshaft pulley lock bolt to the crankshaft.
    b. Remove the crankshaft pulley using a gear puller.
    c. Remove the crankshaft pulley lock bolt and spacer.

To install:

13. Install the crankshaft pulley
14. Install the crankshaft pulley lock bolt.
    a. Set a flathead screwdriver to the drive plate in the position indicated. to lock the crankshaft rotation.
    b. Tighten the new crankshaft pulley lock bolt in four steps.
       - Step 1: Tighten to 103 ft. lbs. (140 Nm)
       - Step 2: Loosen 360° (one full turn)
       - Step 3: Tighten to 35–39 ft. lbs (47–53 Nm)
       - Step 4: Tighten an additional 85–95° of rotation
15. Install the power steering oil pump.
16. Install the power steering oil pump drive belt.
17. Install the dipstick.
18. Install the ignition coils.
19. Install the intake manifold dynamic chamber and throttle body as a single unit.
20. Install the engine and transaxle.
21. Start the engine and perform the following inspections:
    a. Inspect the runout and contact on the pulley and belt.
    b. Inspect engine accessories for proper operation.

CRANKSHAFT FRONT SEAL

REMOVAL & INSTALLATION

See Figures 67 through 73.

Mazda's official procedure for the crankshaft damper removal and installation requires the engine and transaxle to be

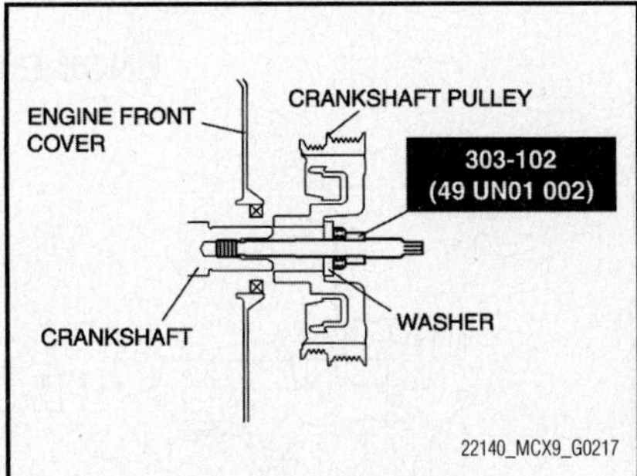

Fig. 70 Install the crankshaft pulley, washer, and special tool to the crankshaft

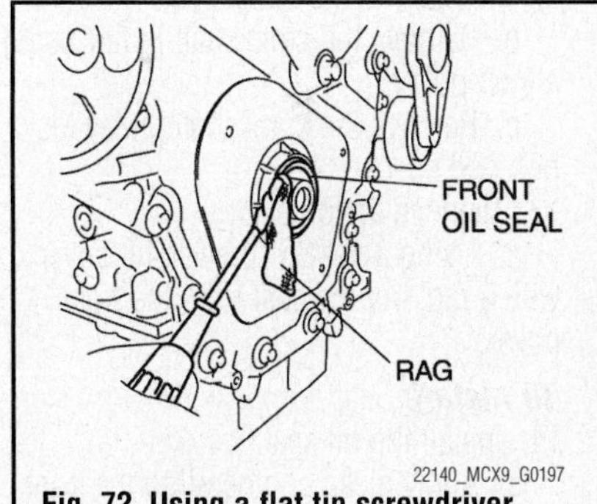

Fig. 72 Using a flat tip screwdriver cover with a rag, pry the seal from the front cover

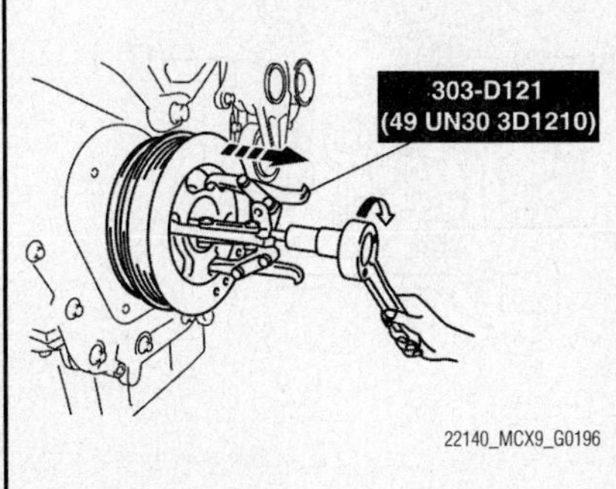

Fig. 69 Remove the crankshaft pulley using a gear puller

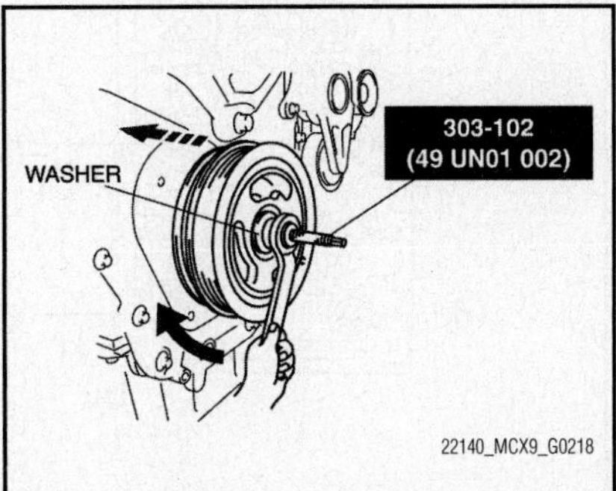

Fig. 71 Tighten the special tool nut and install the crankshaft pulley

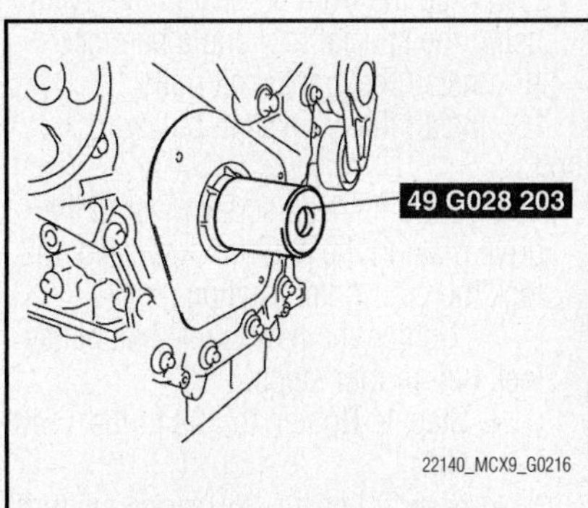

Fig. 73 Tap the front oil seal in evenly using the special tool and a hammer

removed from the vehicle. It may be possible to perform this procedure with the engine and transaxle in the vehicle.

1. Before servicing the vehicle, refer to the Precautions Section.
2. Drain the engine oil.
3. Remove the engine and transaxle.
4. Secure the engine and transaxle using a hoist and an engine stand.
5. Remove the intake manifold dynamic chamber and throttle body as a single unit.
6. Remove the ignition coils.
7. Remove the dipstick.
8. Remove the power steering oil pump drive belt.
9. Remove the power steering oil pump.
10. Remove the drive belt auto tensioner.
11. Remove the crankshaft pulley lock bolt.
   a. Remove the starter.
   b. Set a flathead screwdriver to the drive plate as shown to lock the crankshaft rotation.
   c. Remove the crankshaft pulley lock bolt and washer.
12. Remove the crankshaft pulley.
   a. Install a suitable spacer 0.55 in. (14 mm) in thickness and 1.18 in. (30 mm) in diameter to the crankshaft pulley lock bolt, and install the crankshaft pulley lock bolt to the crankshaft.
   b. Remove the crankshaft pulley using a gear puller.
   c. Remove the crankshaft pulley lock bolt and spacer.
13. Remove the oil seal.
   a. Using a flat tip screwdriver cover with a rag, pry the seal from the front cover

To install:

14. Install the oil seal.
   a. Apply clean engine oil to the front oil seal bore in the engine front cover.
   b. Push the front oil seal slightly in by hand.
   c. Tap the front oil seal in evenly using the special tool and a hammer.
15. Install the crankshaft pulley.
16. Install the crankshaft pulley lock bolt.
   a. Set a flathead screwdriver to the drive plate in the position indicated. to lock the crankshaft rotation.
   b. Tighten the new crankshaft pulley lock bolt in four steps.
      - Step 1: Tighten to 103 ft. lbs. (140 Nm)
      - Step 2: Loosen 360° (one full turn)
      - Step 3: Tighten to 35–39 ft. lbs (47–53 Nm)
      - Step 4: Tighten an additional 85–95° of rotation
17. Install the power steering oil pump.
18. Install the power steering oil pump drive belt.
19. Install the dipstick.
20. Install the ignition coils.
21. Install the intake manifold dynamic chamber and throttle body as a single unit.
22. Install the engine and transaxle.
23. Start the engine and perform the following inspections:
   a. Inspect the runout and contact on the pulley and belt.
   b. Inspect engine accessories for proper operation.

CYLINDER HEAD

REMOVAL & INSTALLATION

See Figures 49 through 56, 61 and 74 through 84.

Mazda's official procedure for cylinder head removal and installation requires the engine and transaxle to be removed from the vehicle. It may be possible to perform this procedure with the engine and transaxle in the vehicle.

1. Before servicing the vehicle, refer to the Precautions Section.
2. Drain the engine oil.
3. Remove the engine and transaxle.
4. Secure the engine and transaxle using a hoist and an engine stand.
5. Remove the intake manifold dynamic chamber and throttle body as a single unit.
6. Remove the fuel injector and fuel distributor together as a single unit.
7. Remove the thermostat and thermostat housing together as a single unit.
8. Remove the Intake manifold.
9. Remove both catalytic converters together with both exhaust manifolds as a single unit.
10. Remove the ignition coils.
11. Remove the dipstick.
12. Remove the power steering oil pump drive belt.
13. Remove the power steering oil pump.
14. Remove the alternator.

➡When removing the timing chain and marking the timing marks on the chain, mark the camshaft timing chain as well.

15. Remove the timing chain.
16. Remove the camshaft caps.
17. Remove the camshaft timing chain tensioners.
18. Remove the camshaft timing chains.
19. Remove the camshaft components.

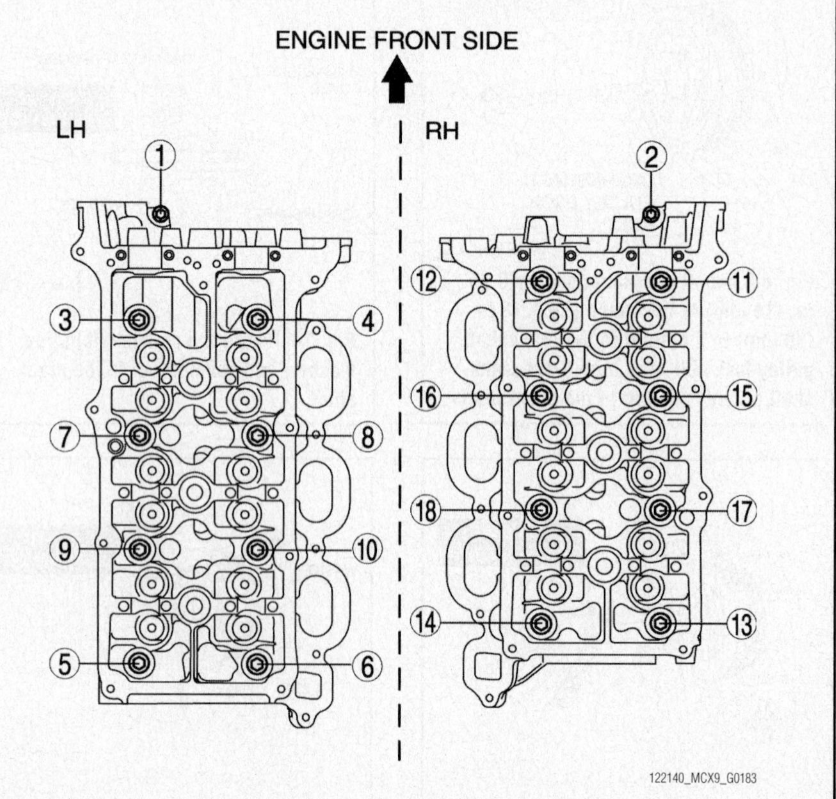

Fig. 74 Loosen the cylinder head bolts in several passes in the order shown

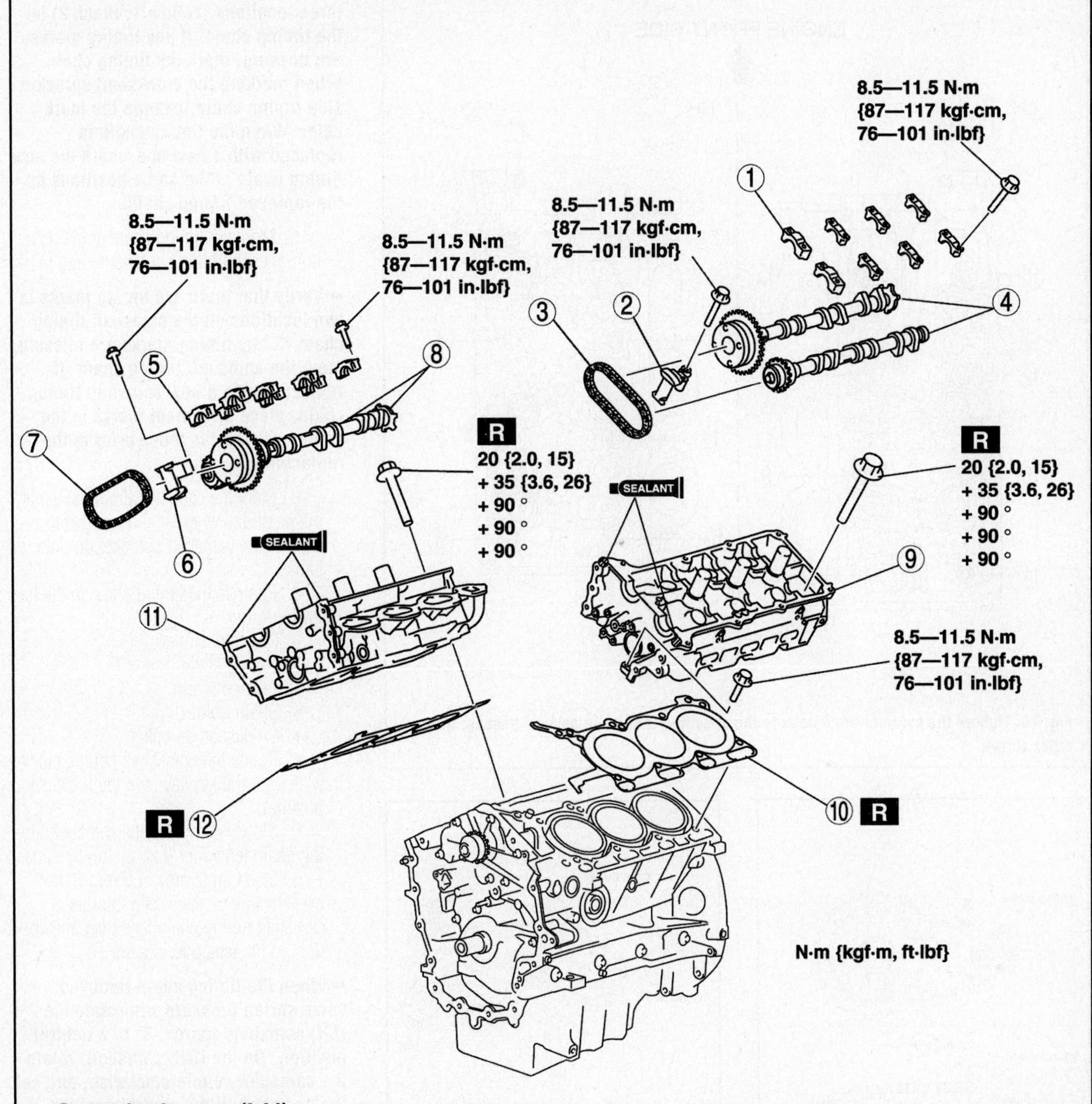

1. Camshaft cap (LH)
2. Camshaft timing chain tensioner (LH)
3. Camshaft timing chain (LH)
4. Camshaft component (LH)
5. Camshaft cap (RH)
6. Camshaft timing chain tensioner (RH)
7. Camshaft timing chain (RH)
8. Camshaft component (RH)
9. Cylinder head (LH)
10. Cylinder head gasket (LH)
11. Cylinder head (RH)
12. Cylinder head gasket (RH)

22140_MCX9_G0171

Fig. 75 Exploded view of the cylinder head assembly

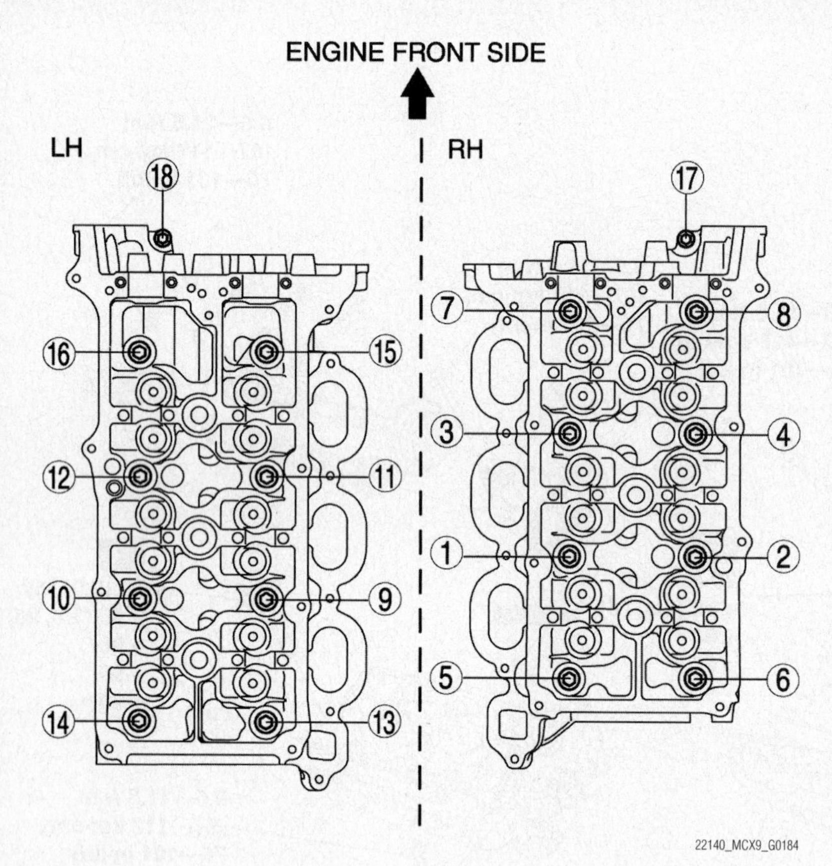

Fig. 76 Tighten the cylinder head bolts to the specified torque, in several passes, in the order shown

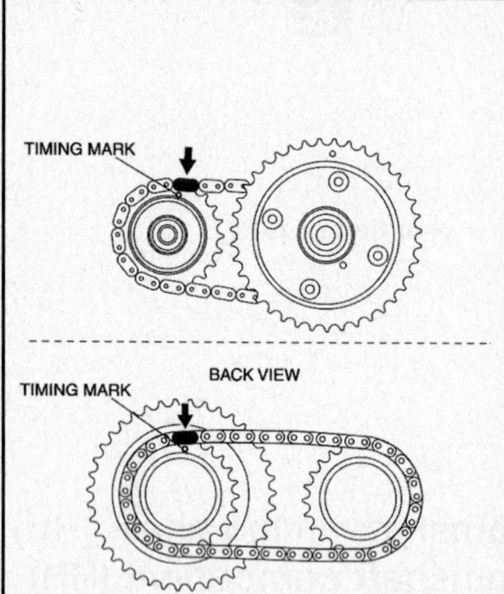

Fig. 77 Install the camshaft timing chain by aligning the colored links on the camshaft timing chain with the marks on the camshaft sprockets—right hand

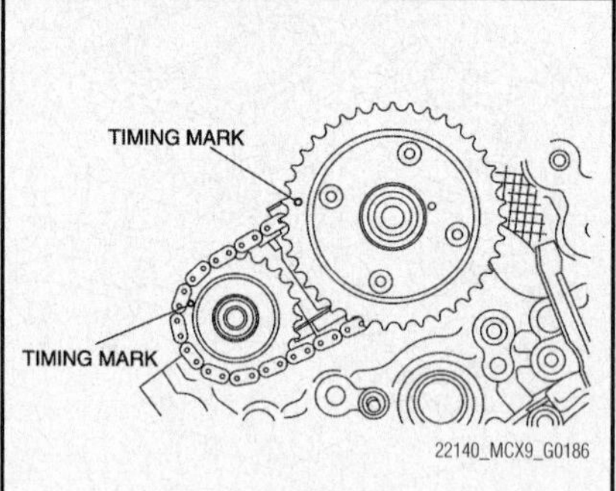

Fig. 78 Position the camshaft component onto the cylinder head in the neutral position as shown—right hand

✲✲ WARNING

Do not rotate the crankshaft counterclockwise. The timing chains may bind, causing engine damage.

a. Turn the crankshaft clockwise so that the crankshaft keyway is in the 11 o'clock position. This will position the No.1 cylinder at TDC.

➡Verify that there are timing marks in three locations (Yellow 1, Black 2) on the timing chain. If any timing marks are missing, mark the timing chain. When marking the crankshaft sprocket side timing chain, change the mark color. When the timing chain is replaced with a new one, mark the new timing chain at the same positions as the removed timing chain.

b. Mark the timing chain at the position of each timing sprocket timing mark.

➡Verify that there are timing marks in two locations on the camshaft timing chain. If any timing marks are missing, mark the camshaft timing chain. If replacing with a new camshaft timing chain, place alignment marks in the same positions as those prior to the replacement.

c. Mark the camshaft timing chain at the positions where it is aligned with each of the camshaft sprocket on both banks.

d. Remove the timing chain in the following order.

- Chain tensioner
- Timing chain
- Tensioner arm
- Chain guide
- Crankshaft sprocket

e. Rotate the crankshaft counterclockwise until the keyway is in the 9 o'clock position.

f. Slowly compress the camshaft timing chain tensioner (LH) piston by hand.

g. Insert an approx. 1.0 mm {0.039 in} thin wire or paper clip into the camshaft timing chain tensioner shown. to hold the tensioner piston.

➡When the timing chain removed, valve spring pressure will rotate the (LH) camshaft approx. 3° to a neutral position. On the (RH) camshaft, rotate the camshaft counterclockwise, and set the camshaft to the neutral position.

h. Verify that the camshafts are in the neutral position.

➡The cylinder head and the camshaft bearing caps are numbered to make sure they are reassembled in their original position. When removed, keep the bearing caps with the cylinder head they were removed from. Do not mix the caps.

i. Loosen the camshaft cap bolts in several passes in the order shown. and remove the camshaft cap.

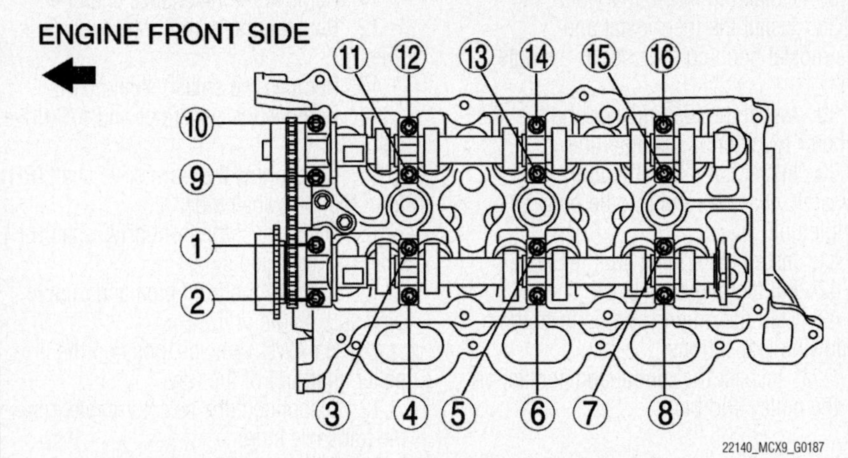

Fig. 79 Tighten the camshaft cap bolts to the specified torque, in several passes, in the order shown—right hand

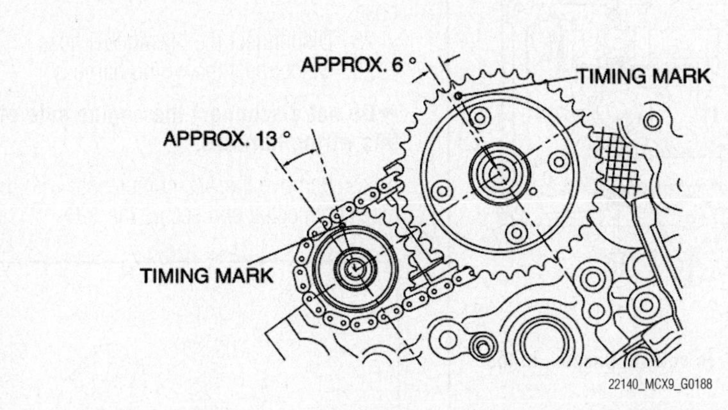

Fig. 80 Using the right hand camshaft sprocket installation bolt, rotate the right hand camshaft clockwise so the timing marks align as shown

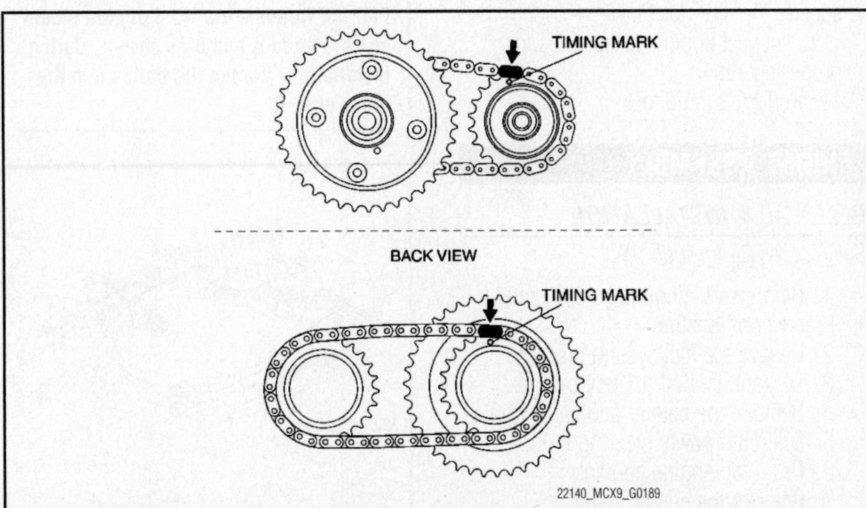

Fig. 81 Install the camshaft timing chain by aligning the colored links on the camshaft timing chain with the marks on the camshaft sprockets—left hand

j. Remove the camshaft, camshaft sprocket, camshaft timing chain, and camshaft timing chain tensioner as a single unit.

20. Remove the cylinder head.
21. Remove the cylinder head gasket.

To install:

22. Install the cylinder head gasket.
23. Install the cylinder head.
24. Install the cylinder head bolts and tighten to the following torque in the order shown:
    - Step 1: Tighten bolts 1–16 to 15 ft. lbs. (20 Nm).
    - Step 2: Tighten bolts 1–16 to 26 ft. lbs. (35 Nm).
    - Step 3: Tighten bolts 1–16 an additional 90° rotation.
    - Step 4: Tighten bolts 1–16 an additional 90° rotation.
    - Step 5: Tighten bolts 1–16 an additional 90° rotation.
    - Step 6: Tighten bolts 17–18 to 76–101 inch lbs. (9–12 Nm).
25. Assemble the camshaft timing chain tensioners on both sides. Tighten to 76–101 inch lbs. (8.5–11.5 Nm).
26. Install the right hand camshaft.

➡Install the camshaft, camshaft sprocket and camshaft timing chain of the both bank as a single unit.

a. Install the camshafts.

b. Install the camshaft timing chain by aligning the colored links on the camshaft timing chain with the marks on the camshaft sprockets.

c. Position the camshaft component onto the cylinder head in the neutral position as shown.

d. Install the camshaft caps and temporarily tighten the camshaft cap bolts evenly in the order shown in several passes. Tighten to 76–101 inch lbs. (8.5–11.5 Nm).

e. Remove the retaining wire inserted into the camshaft timing chain tensioner.

27. Using the right hand camshaft sprocket installation bolt, rotate the right hand camshaft clockwise so the timing marks align as shown.
28. Install the left hand camshaft in the same manner as the right hand camshaft.
29. Rotate the crankshaft clockwise so that the crankshaft keyway is in the 11 o'clock position. This will position the No.1 cylinder at TDC.
30. Install the timing chain.
31. Install the alternator.
32. Install the power steering oil pump.

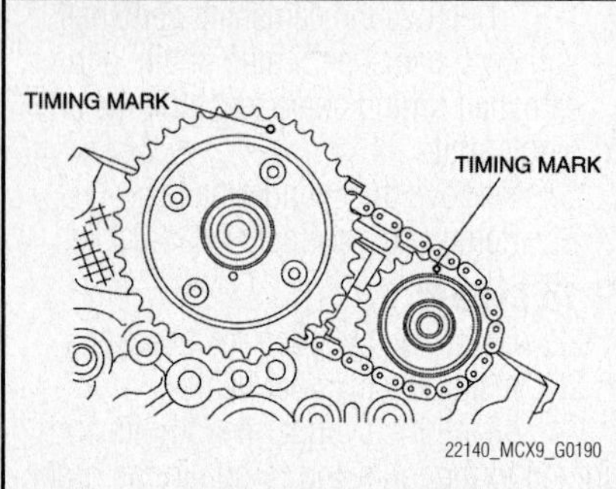

Fig. 82 Position the camshaft component onto the cylinder head in the neutral position as shown—left hand

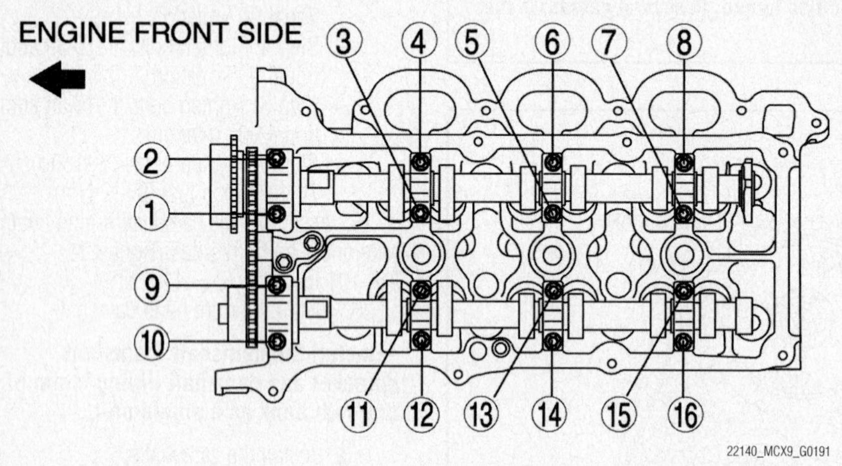

Fig. 83 Tighten the camshaft cap bolts to the specified torque, in several passes, in the order shown—left hand

33. Install the power steering oil pump drive belt.
34. Install the dipstick.
35. Install the ignition coils.
36. Install both catalytic converters together with both exhaust manifolds as a single unit.

KEYWAY 11 O'CLOCK

22140_MCX9_G0192

Fig. 84 Turn the crankshaft clockwise so that the crankshaft keyway is in the 11 o'clock position. This will position the No.1 cylinder at TDC.

37. Install the intake manifold.
38. Install the thermostat and thermostat housing together as a single unit.
39. Install the fuel injector and fuel distributor together as a single unit.
40. Install the intake manifold dynamic chamber and throttle body as a single unit.
41. Install the engine and transaxle.
42. Fill the engine with oil.
43. Start the engine and perform the following inspections:
    a. Inspect the runout and contact on the pulley and belt.
    b. Inspect for engine oil, engine coolant, ATF, power steering fluid and fuel leakage.
    c. Verify the ignition timing and idle speed.
    d. Inspect engine accessories for proper operation.
44. Perform a road test.

ENGINE ASSEMBLY

REMOVAL & INSTALLATION

See Figures 85 through 95.

1. Before servicing the vehicle, refer to the Precautions Section.
2. Relieve the fuel system pressure.
3. Disconnect both battery cables.
4. Remove the battery and battery tray.
5. Drain the power steering fluid.
6. Drain the engine coolant.
7. Remove the engine cover.
8. Remove the ventilation hose.
9. Disconnect the vacuum hose.
10. Remove the air cleaner.
11. Remove the resonance chamber.
12. Remove the both front wheels and tires.
13. Remove the splash shield (RH).
14. Remove the generator and A/C drive belt.
15. Disconnect the front drive shaft (RH) from the joint shaft side.
16. Disconnect the front drive shaft (LH) from the transaxle side.
17. Remove the front pipe and middle pipe as a single unit.
18. On AWD vehicles, position the propeller shaft out of the way.
19. Disconnect the selector cable from the transaxle side.
20. Disconnect the heater hose.
21. Disconnect the power steering return hose.
22. Disconnect the radiator hose.
23. Disconnect the water hose.
24. Disconnect the ATF oil refrigerant hose.
25. Disconnect the plastic fuel hose.
26. Disconnect the wiring harness.

➡Do not disconnect the engine side of this wiring harness.

27. Remove the A/C compressor with the pipes connected and secure the A/C

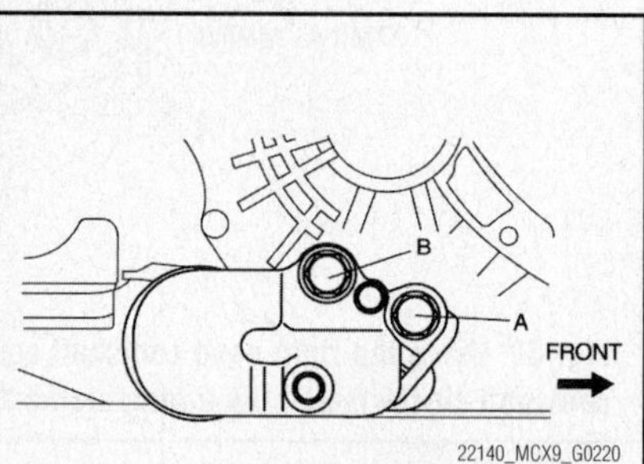

Fig. 85 Remove the No.1 engine mount bracket bolts A and B as shown. During installation, tighten the bolts on in the order of A, B

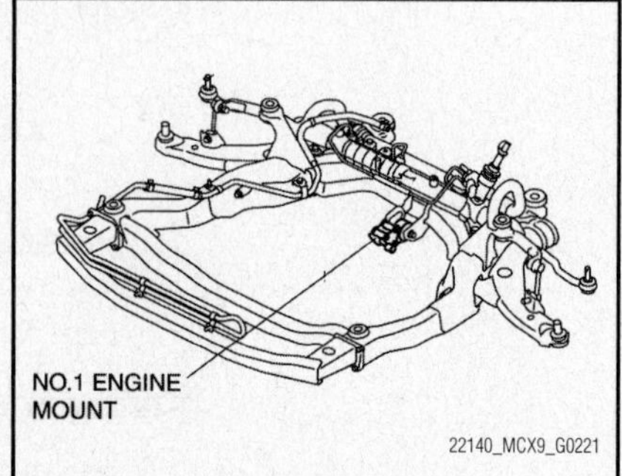

Fig. 86 Remove the No.1 engine mount, No.1 engine mount bracket and the front crossmember as a single unit

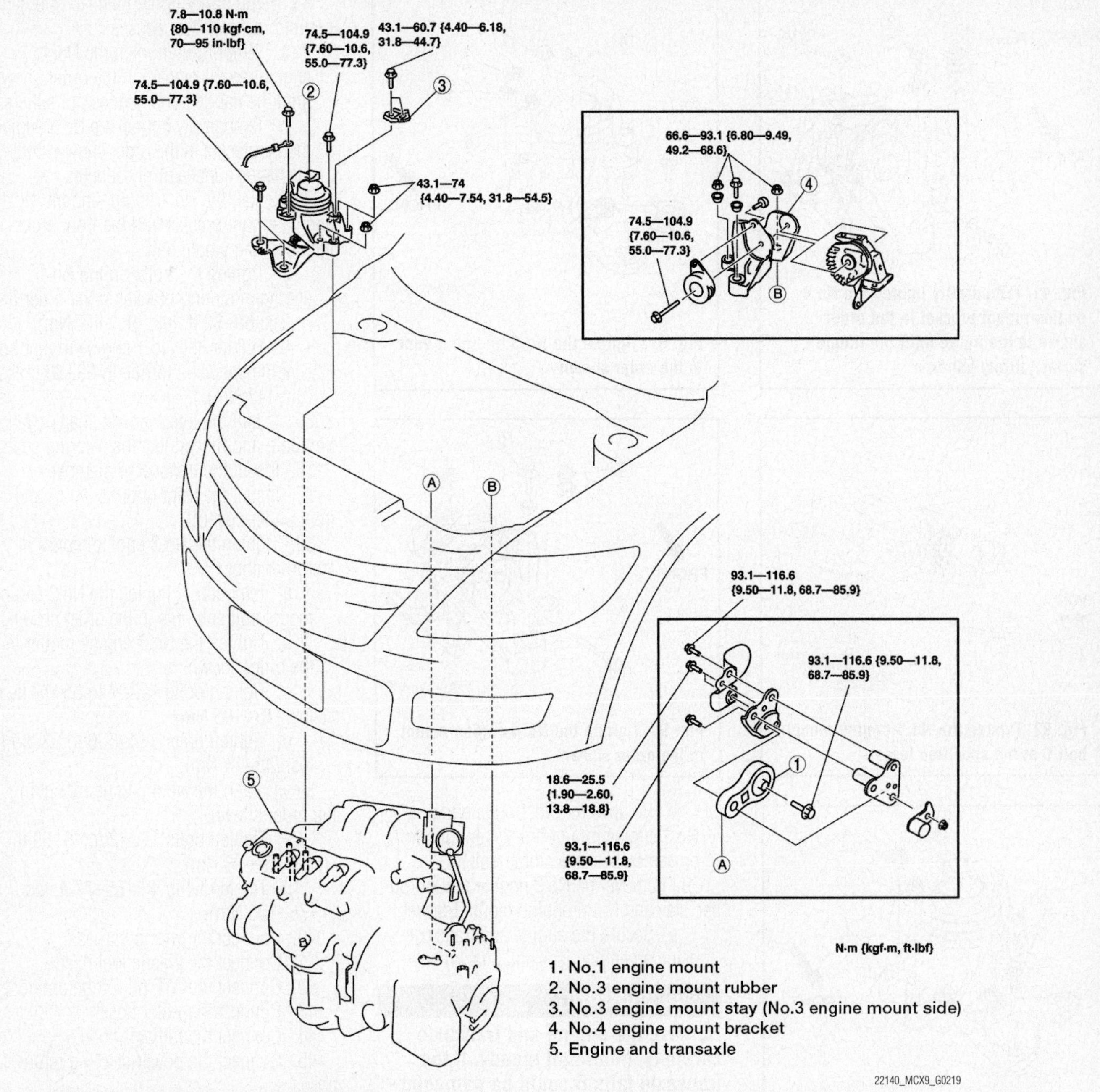

7.8—10.8 N·m {80—110 kgf·cm, 70—95 in·lbf}
74.5—104.9 {7.60—10.6, 55.0—77.3}
43.1—60.7 {4.40—6.18, 31.8—44.7}
74.5—104.9 {7.60—10.6, 55.0—77.3}
43.1—74 {4.40—7.54, 31.8—54.5}
66.6—93.1 {6.80—9.49, 49.2—68.6}
74.5—104.9 {7.60—10.6, 55.0—77.3}
93.1—116.6 {9.50—11.8, 68.7—85.9}
93.1—116.6 {9.50—11.8, 68.7—85.9}
18.6—25.5 {1.90—2.60, 13.8—18.8}
93.1—116.6 {9.50—11.8, 68.7—85.9}
N·m {kgf·m, ft·lbf}
1. No.1 engine mount
2. No.3 engine mount rubber
3. No.3 engine mount stay (No.3 engine mount side)
4. No.4 engine mount bracket
5. Engine and transaxle
22140_MCX9_G0219

Fig. 87 Engine mounts and related components

NO.3 ENGINE MOUNT BRACKET STUD BOLT
22140_MCX9_G0222

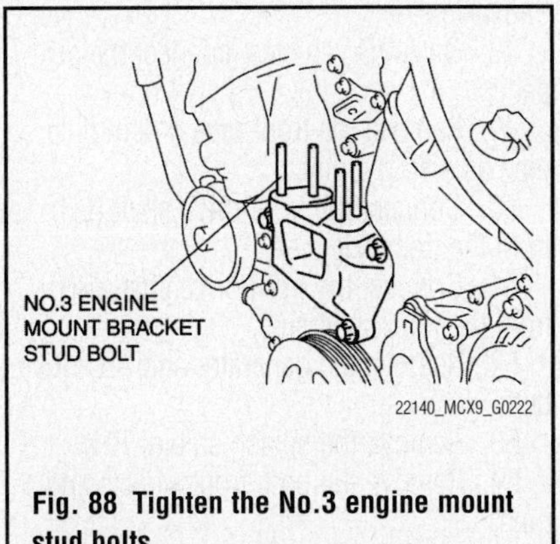

Fig. 88 Tighten the No.3 engine mount stud bolts

FRONT
22140_MCX9_G0223

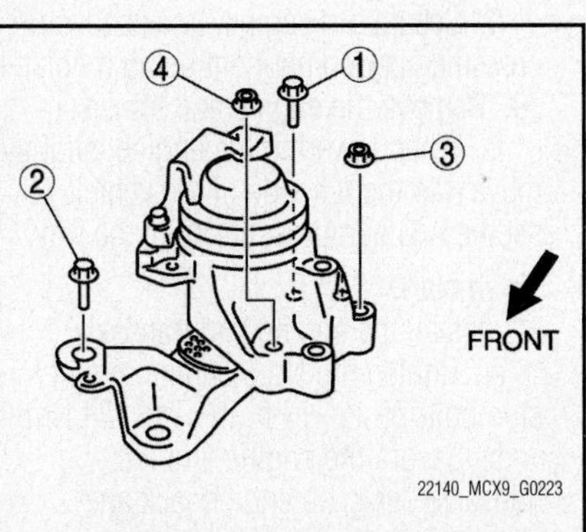

Fig. 89 Tighten the No.3 engine mount bolts and nuts in the order shown

FRONT
22140_MCX9_G0224

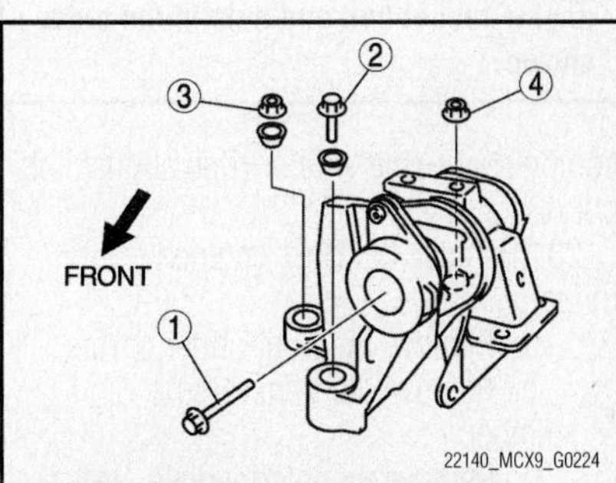

Fig. 90 Temporarily tighten the No.4 engine mount bracket in the order shown in the figure until the flange surface does not adhere

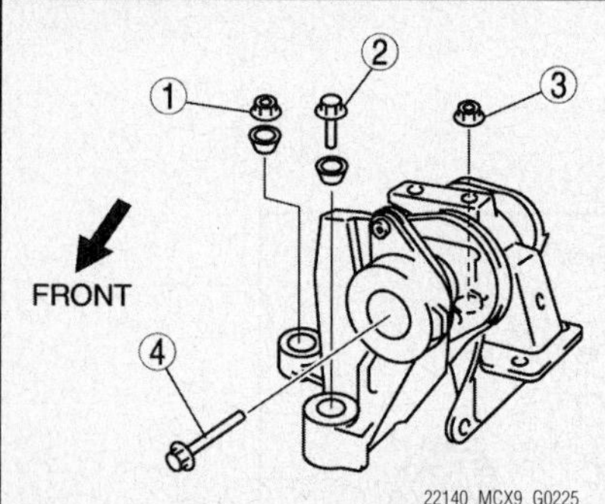

Fig. 91 Temporarily tighten the No.4 engine mount bracket in the order shown in the figure until the flange surface firmly adheres

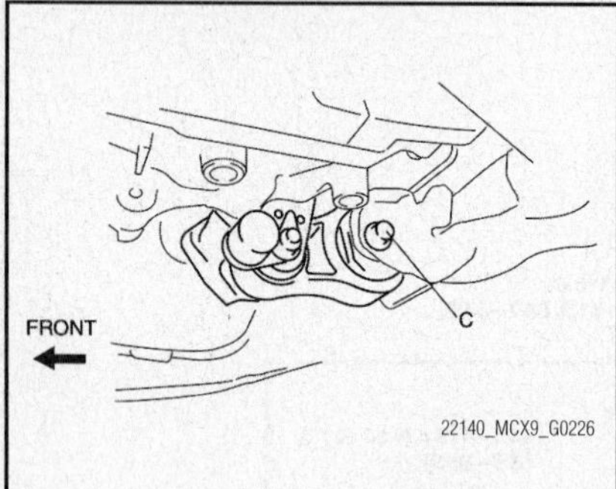

Fig. 92 Tighten the No.1 engine mount bolt C at the specified torque

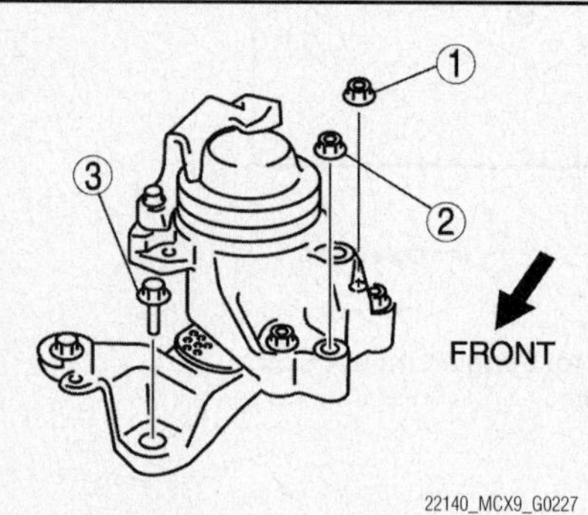

Fig. 93 Temporarily tighten the No.3 engine mount bolt and nuts in the order shown

compressor using wire or rope so that it is out of the way.

28. Remove the No.1 engine mount.
    a. Remove the front under covers.
    b. Remove the transverse member.
    c. Remove the intermediate shaft installation bolt, and disconnect the steering shaft.
    d. Remove the No.1 engine mount bracket bolts A and B as shown.

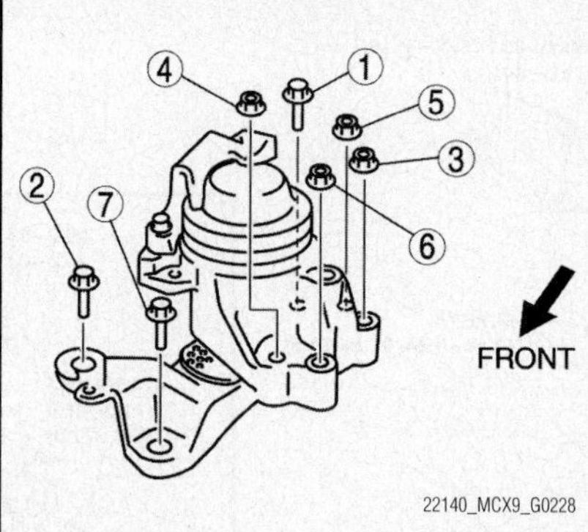

Fig. 94 Tighten the No.3 engine mount in the order shown

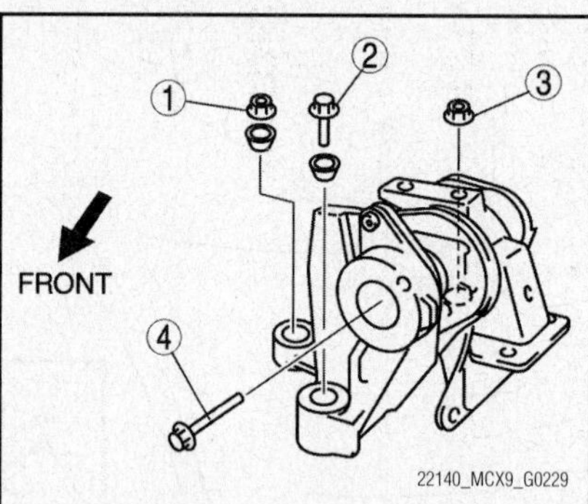

Fig. 95 Tighten the No.4 engine mount in the order shown

    e. Remove the No.1 engine mount, No.1 engine mount bracket and the front crossmember as a single unit.
29. Remove the No.3 engine mount rubber, stay and No. 4 engine mount bracket.
    a. Secure the engine and transaxle using an engine jack and attachment.

**** WARNING**

Remove the engine and transaxle carefully, holding it steady. If the transaxle falls it could be damaged or cause injury.

    b. Secure the engine, transaxle, and crossmember component using a hoist.
30. Remove the engine and transaxle.
    a. When lowering the engine, slightly move it to the left side of the vehicle to set the A/C compressor out of the way.

To install:

31. Install the engine and transaxle.
    a. Tighten the No.3 engine mount stud bolts to 12–17 ft. lbs. (16–24 Nm).
    b. Secure the engine and the transaxle using an engine jack and attachment.
32. Temporarily tighten the No.3 engine mount bolts and nuts in the order shown.
33. Temporarily tighten the No. 4 engine mount bracket in two passes.
    a. Temporarily tighten the No.4 engine mount bracket in the order shown until the flange surface does not adhere.
    b. Temporarily tighten the No.4 engine mount bracket in the order shown. until the flange surface firmly adheres.
34. Install the No.1 engine mount, No.1 engine mount bracket and the front crossmember as a single unit.
    a. Tighten the bolts on the No.1 engine mounting bracket in the order of A, B to 69–86 ft. lbs. (93–117 Nm).
    b. Tighten the No.1 engine mount bolt C at the specified torque to 69–86 ft. lbs. (93–117 Nm).
35. Install the intermediate shaft installation bolt, and disconnect the steering shaft.
36. Install the transverse member.
37. Install the front under cover A and front under cover B.
38. Tighten the No.3 engine mount in the order shown.
    a. Temporarily tighten the No.3 engine mount bolt and nuts in the order shown.
    b. Tighten the No.3 engine mount in the order shown.
    c. Tighten bolts 1–2–7 to 55–77 ft. lbs. (75–105 Nm).
    d. Tighten bolts 3–4–5–6 to 32–55 ft. lbs. (43–74 Nm).
39. Tighten the No.4 engine mount in the order shown.
    a. Tighten bolts 1–2–3 to 49–69 ft. lbs. (67–93 Nm).
    b. Tighten bolts 4 to 55–77 ft. lbs. (75–105 Nm).
40. Connect the wiring harness.
41. Connect the plastic fuel hose.
42. Connect the ATF oil refrigerant hose.
43. Connect the water hose.
44. Connect the radiator hose.
45. Connect the power steering return hose.
46. Connect the heater hose.
47. Connect the selector cable from the transaxle side.
48. On AWD vehicles, connect the propeller.
49. Remove the front pipe and middle pipe as a single unit.
50. Connect the front drive shaft (LH) from the transaxle side.
51. Connect the front drive shaft (RH) from the joint shaft side.
52. Remove the generator and A/C drive belt.
53. Remove the splash shield (RH).
54. Remove the both front wheels and tires.
55. Remove the resonance chamber.

56. Remove the air cleaner.
57. Connect the vacuum hose.
58. Remove the ventilation hose.
59. Remove the engine cover.
60. Drain the engine coolant.
61. Drain the power steering fluid.
62. Install the battery and battery tray.
63. Connect both battery cables.
64. Start the engine and perform the following inspections:
    a. Inspect the runout and contact on the pulley and belt.
    b. Inspect for engine oil, engine coolant, ATF, power steering fluid and fuel leakage.
    c. Verify the ignition timing, idle speed.
    d. Inspect engine accessories for proper operation.
65. Perform a road test.

EXHAUST MANIFOLD

REMOVAL & INSTALLATION

Right Side

See Figure 96.

1. Before servicing the vehicle, refer to the Precautions Section.

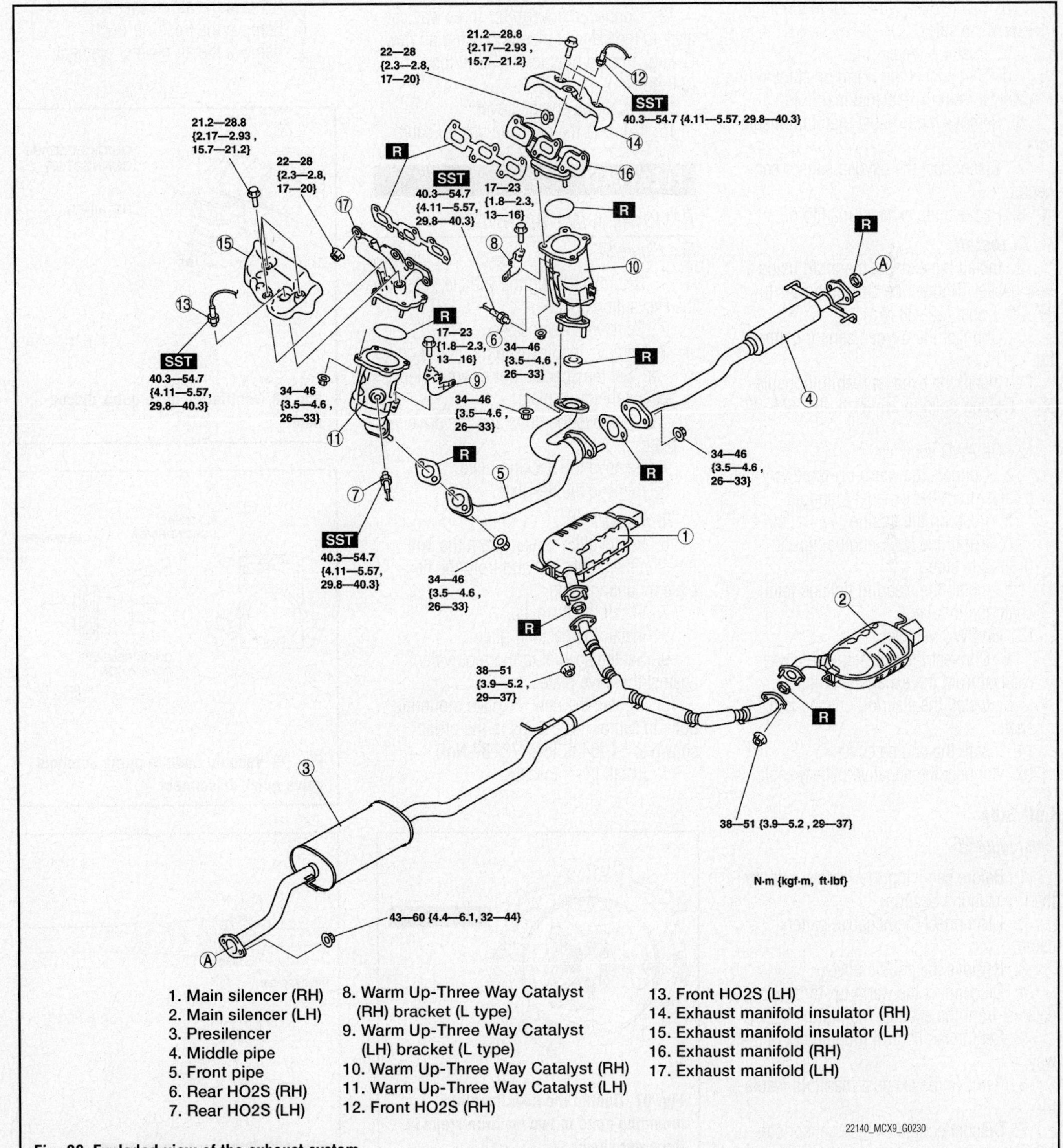

Fig. 96 Exploded view of the exhaust system

2. Disconnect the negative battery cable.
3. Remove the engine cover.
4. On 2WD vehicles:
a. Remove the steering linkage joint shaft.
b. Disconnect the warm up-three way catalyst from the exhaust manifold.
5. On AWD vehicles:
a. Remove the steering linkage joint shaft and insulator.
b. Remove the No.1 engine mount (transaxle side).
c. Push up the engine.
d. Disconnect the warm up-three way catalyst from the exhaust manifold.
6. Remove the exhaust manifold insulators.
7. Disconnect the oxygen sensor connector.
8. Remove the exhaust manifold.

To install:

9. Install the exhaust manifold using a new gasket. Tighten the attaching nuts to 17–20 ft. lbs. (22–28 Nm).
10. Connect the oxygen sensor connector.
11. Install the exhaust manifold insulators. Tighten bolts to 16–21 ft. lbs. (21–29 Nm).
12. On AWD vehicles:
a. Connect the warm up-three way catalyst from the exhaust manifold.
b. Push up the engine.
c. Install the No.1 engine mount (transaxle side).
d. Install the steering linkage joint shaft and insulator.
13. On 2WD vehicles:
a. Connect the warm up-three way catalyst from the exhaust manifold.
b. Install the steering linkage joint shaft.
14. Install the engine cover.
15. Connect the negative battery cable.

Left Side

See Figure 96.

1. Before servicing the vehicle, refer to the Precautions Section.
2. Disconnect the negative battery cable.
3. Remove the engine cover.
4. Disconnect the warm up-three way catalyst from the exhaust manifold.
5. Set the fan control module out of the way.
6. Remove the exhaust manifold insulator.
7. Disconnect the oxygen sensor connector.
8. Remove the exhaust manifold and gasket.

To install:

9. Install the exhaust manifold using a new gasket. Tighten the attaching nuts to 17–20 ft. lbs. (22–28 Nm).
10. Connect the oxygen sensor connector.
11. Install the exhaust manifold insulator. Tighten bolts to 16–21 ft. lbs. (21–29 Nm).
12. Replace the fan control module.
13. Connect the warm up-three way catalyst to the exhaust manifold using an new O-ring. Tighten nuts to 26–33 ft. lbs. (34–46 Nm).
14. Install the engine cover.
15. Connect the negative battery cable.

FLEXPLATE

REMOVAL & INSTALLATION

See Figure 97.

1. Before servicing the vehicle, refer to the Precautions Section.
2. Remove the transaxle.
3. Remove the flexplate mounting bolts.
a. Set the special tool or equivalent against the drive plate.
b. Remove the bolts and the drive plate.
4. Remove the backing plate.
5. Remove the flexplate.

To install:

6. Remove the sealant from the bolt holes in the crankshaft and from the flexplate mounting bolts.
7. Install the flexplate.
8. Install the backing plate.
9. Set the special tool or equivalent against the drive plate.
10. Tighten the new flexplate mounting bolts in two or three steps in the order shown to 54–64 ft. lbs. (73–87 Nm).
11. Install the transaxle.

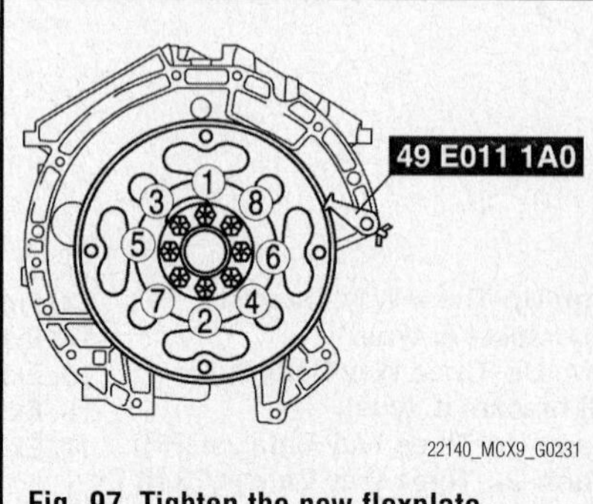

Fig. 97 Tighten the new flexplate mounting bolts in two or three steps in the order shown

INTAKE MANIFOLD

REMOVAL & INSTALLATION

See Figures 98 through 107.

1. Before servicing the vehicle, refer to the Precautions Section.
2. Properly relieve the fuel system pressure.
3. Remove the battery.
4. Remove the engine cover.
5. Remove the air cleaner case.
6. Remove the fresh-air duct.
7. Remove the air cleaner element.

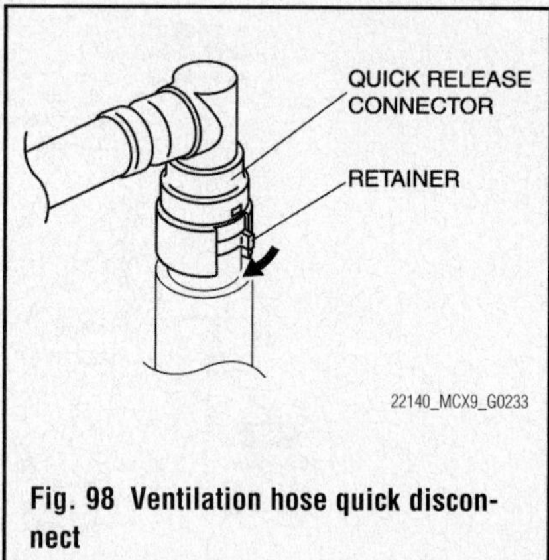

Fig. 98 Ventilation hose quick disconnect

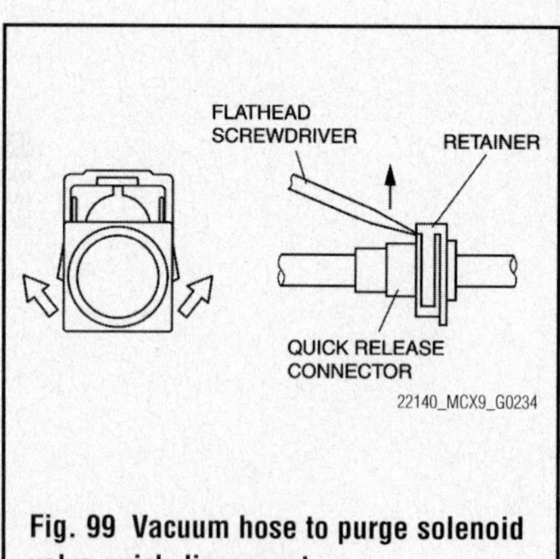

Fig. 99 Vacuum hose to purge solenoid valve quick disconnect

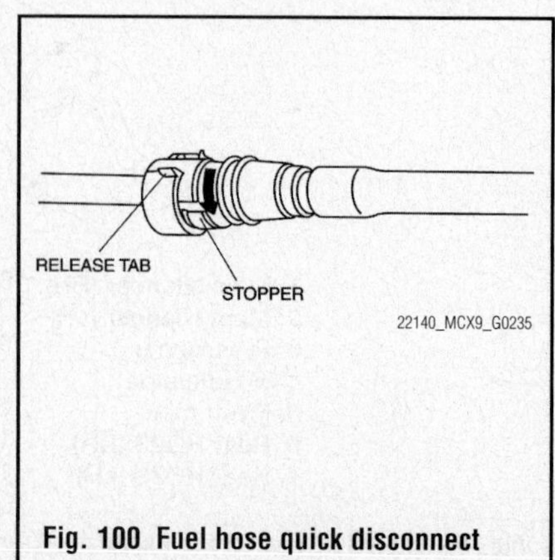

Fig. 100 Fuel hose quick disconnect

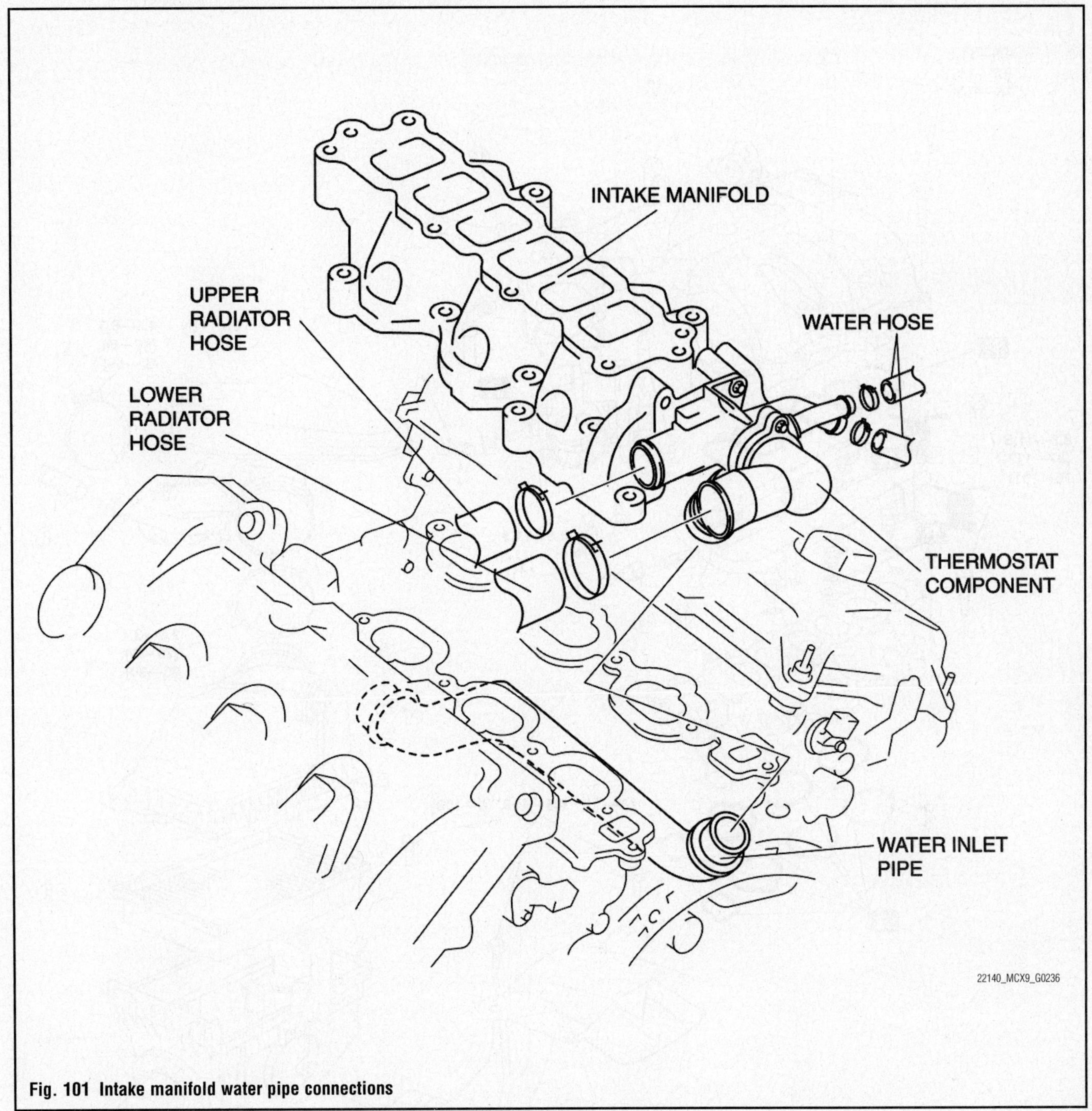

Fig. 101 Intake manifold water pipe connections

8. Disconnect the MAF/IAT sensor connector.
9. Remove the air cleaner cover.
10. Disconnect the vacuum hose to the master back.
11. Disconnect the ventilation hose.
   a. Release the retainer shown in the figure.
   b. Disconnect the quick release connector.
   c. Cover the disconnected quick release connector and fuel pipe with vinyl sheeting or a similar material to prevent it from scratches or dirt.
12. Remove the resonance chamber.
13. Disconnect the TP sensor connector.
14. Remove the throttle body.
15. Disconnect the vacuum hose to the purge solenoid valve.
   a. Move the retainer upward using a small flathead screwdriver or a similar tool.
   b. Pull out the fuel hose straight from the fuel pipe and disconnect it.
   c. Cover the disconnected quick release connector and fuel pipe with vinyl sheeting or a similar material to prevent it from scratches or dirt.
16. Disconnect the vacuum hose to the master back.
17. Remove the dynamic chamber.
18. Disconnect the fuel hose.
   a. Rotate the release tab on the quick release connector to the stopper position.
   b. Pull out the fuel hose straight from the fuel pipe and disconnect it.
   c. Cover the disconnected quick release connector and fuel pipe with vinyl sheeting or a similar material to prevent it from scratches or dirt.
19. Remove the fuel distributor
20. Remove the intake manifold.

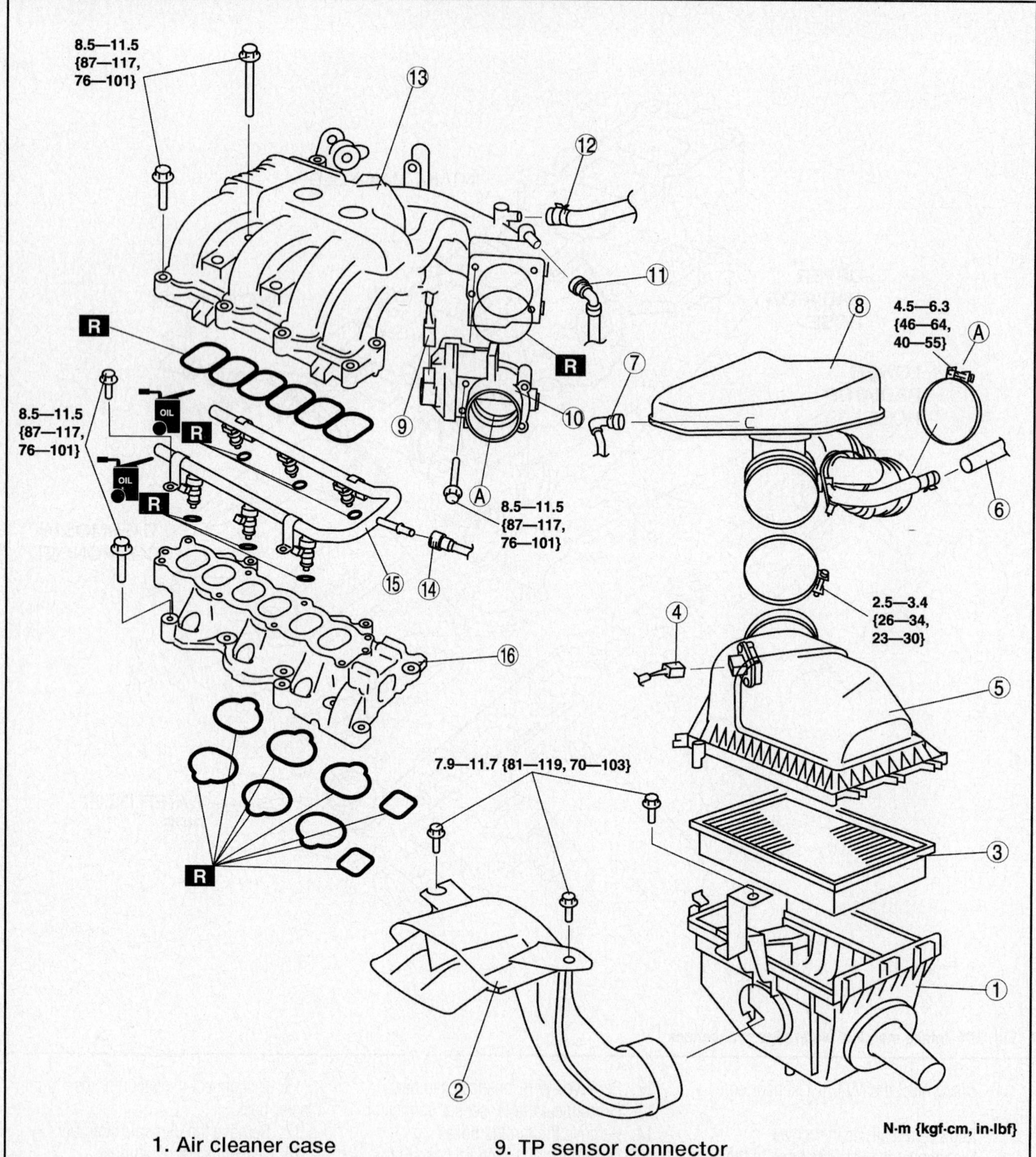

Fig. 102 Exploded view of the air intake assembly

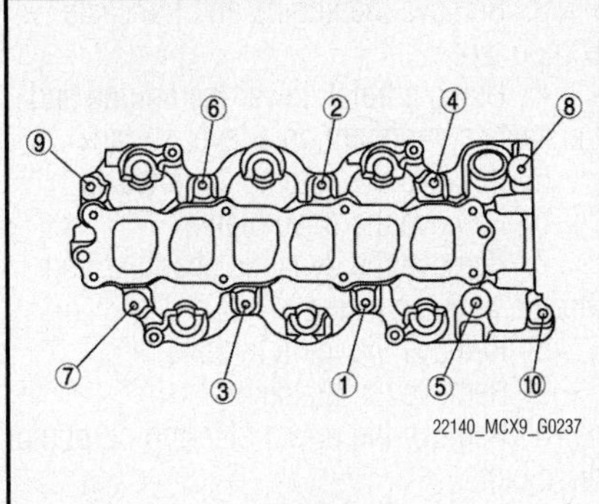

Fig. 103 Tighten the intake manifold installation bolts in the order shown

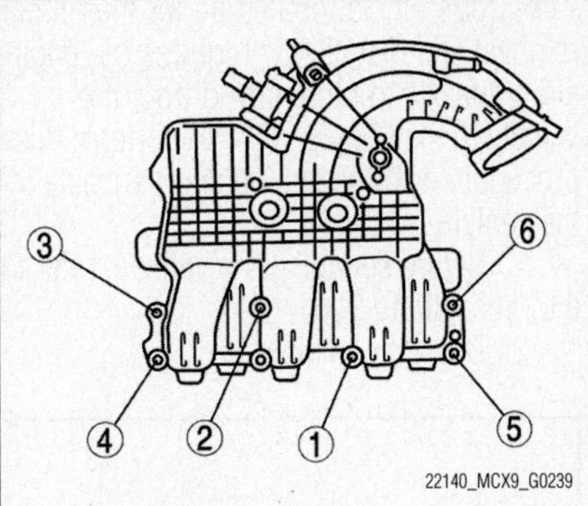

Fig. 105 Tighten the dynamic chamber installation bolts in the order shown

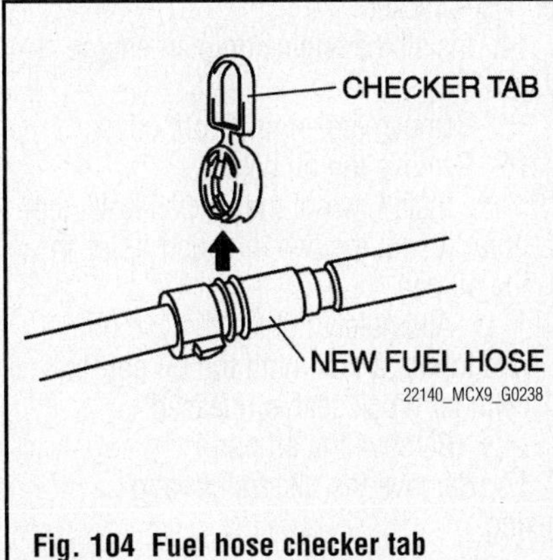

Fig. 104 Fuel hose checker tab

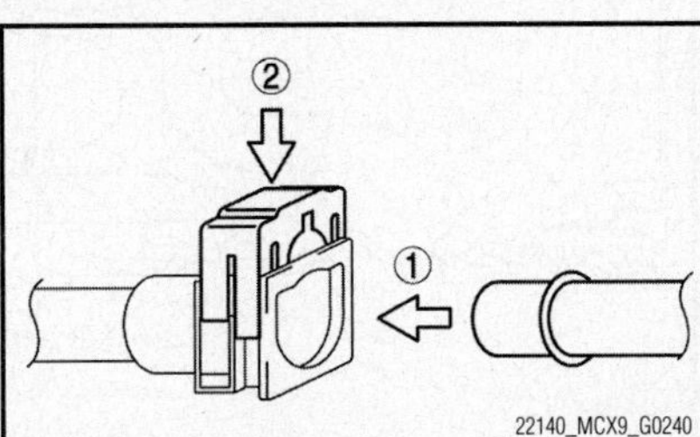

Fig. 106 To connect the vacuum hose to the purge solenoid valve, insert the fuel pipe straight to the end of the quick release connector and push down the retainer

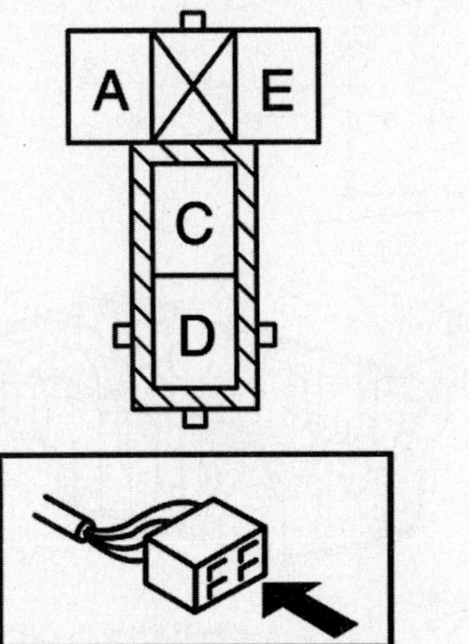

Fig. 107 Using a jumper wire, short fuel pump relay terminals C and D in the main fuse block to check for fuel leaks

a. Drain the engine coolant.
b. Disconnect the upper radiator hose.
c. Disconnect the lower radiator hose.
d. Disconnect the two water hose from the thermostat component.
e. Disconnect the thermostat component side of the water inlet pipe which is under the intake manifold.
f. Remove the intake manifold and thermostat component as a single unit.
g. Remove the thermostat component from the intake manifold.

To install:

21. Install the intake manifold.
    a. Tighten the intake manifold installation bolts in the order shown.
    b. Tighten the bolts to 76–101 inch lbs. (9–12 Nm).
22. Install the fuel distributor
23. Connect the fuel hose.
    a. If the quick release connector O-ring is damaged or has slipped, replace the fuel hose.
    b. A checker tab is integrated with the quick release connector for new fuel hoses and evaporative hoses. Remove the checker tab from the quick release connector after the connector is completely engaged with the fuel pipe.
    c. Remove the retainer remaining on the charcoal canister pipe.
    d. Install a new retainer to the quick release connector.
    e. Reconnect the hose straight to the pipe until a click is heard.
    f. Lightly pull and push the quick release connector a few times by hand, and then verify that it is connected securely.
24. Install the dynamic chamber.
    a. Tighten the dynamic chamber installation bolts in the order shown.
    b. Tighten the bolts to 76–101 inch lbs. (9–12 Nm).
25. Connect the vacuum hose to the master back.
26. Connect the vacuum hose to the purge solenoid valve.
    a. If the quick release connector O-ring is damaged or has slipped, replace the fuel hose.
    b. Inspect the fuel hose and fuel pipe sealing surface for damage and deformation. If there is any malfunction, replace it with a new one.
    c. Install the quick release connector.
    d. Insert the fuel pipe straight to the end of the quick release connector.

e. Push down the retainer using a finger.

f. If the retainer cannot be pushed down, push the fuel pipe further to the quick release connector.

g. Lightly pull and push the quick release connector a few times by hand, and then verify that it is connected securely.

27. Install the throttle body.
28. Connect the TP sensor connector.
29. Install the resonance chamber.
30. Connect the ventilation hose.

a. Inspect the fuel hose and fuel pipe sealing surface for damage and deformation. If there is any malfunction, replace it with a new one.

b. Connect the quick release connector.

31. Connect the vacuum hose to the master back.
32. Install the air cleaner cover.
33. Connect the MAF/IAT sensor connector.
34. Install the air cleaner element.
35. Install the fresh-air duct.
36. Install the air cleaner case.
37. Install the engine cover.
38. Install the battery but leave the negative battery cable disconnected.
39. Perform a fuel leakage test as follows:

a. Remove the fuel pump relay.

**** WARNING**

Short the specified terminals because shorting the wrong terminal of the main fuse block may cause malfunctions.

b. Using a jumper wire, short fuel pump relay terminals C and D in the main fuse block.

c. Connect the negative battery cable and operate the fuel pump.

d. Verify that there is no fuel leakage from the pressurized parts.

e. If there is leakage, replace the fuel hoses and clips.

f. If there is damage on the seal on the fuel pipe side, replace the fuel pipe.

g. There should be no leakage after 5 minutes.

h. After making any repairs, assemble the system and repeat the leakage test.

OIL PAN

REMOVAL & INSTALLATION

See Figures 108 through 113.

Mazda's official procedure for the oil pan removal and installation requires the engine and transaxle to be removed from the vehicle. It may be possible to perform this procedure with the engine and transaxle in the vehicle.

1. Before servicing the vehicle, refer to the Precautions Section.
2. Drain the engine oil.

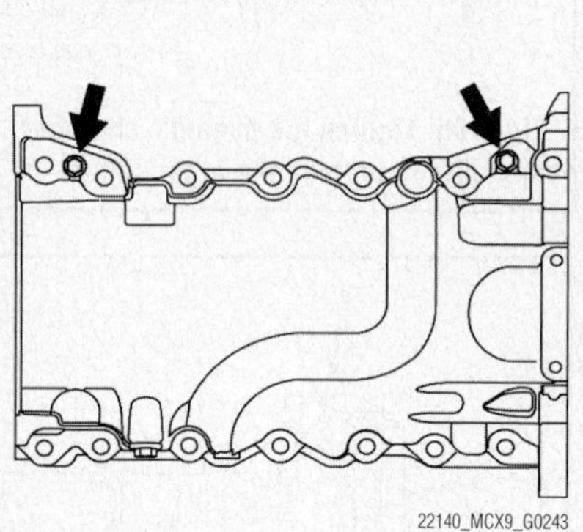

Fig. 108 Install two of the oil pan bolts temporarily into the two threaded holes in the oil pan. Alternately tighten the two bolts one turn at a time until the oil pan-to-cylinder block seal is released

3. Remove the engine and transaxle component.
4. Using a hoist, lower the engine and transaxle component on a level surface.
5. Remove the automatic transaxle.
6. Remove the drive plate.
7. Remove the dynamic chamber and throttle body as a single unit.
8. Remove the ignition coils.
9. Remove the dipstick.
10. Remove the power steering oil pump drive belt.
11. Remove the power steering oil pump.
12. Remove the exhaust manifold (RH).
13. On AWD vehicles, remove the transfer case bracket.
14. Install the engine onto an engine stand.
15. Remove the engine front cover.
16. Remove the oil pan.

a. Install two of the oil pan bolts temporarily into the two threaded holes in the oil pan.

b. Alternately tighten the two bolts one turn at a time until the oil pan-to-cylinder block seal is released.

c. Remove the oil pan.

17. Remove the oil strainer and O-ring.

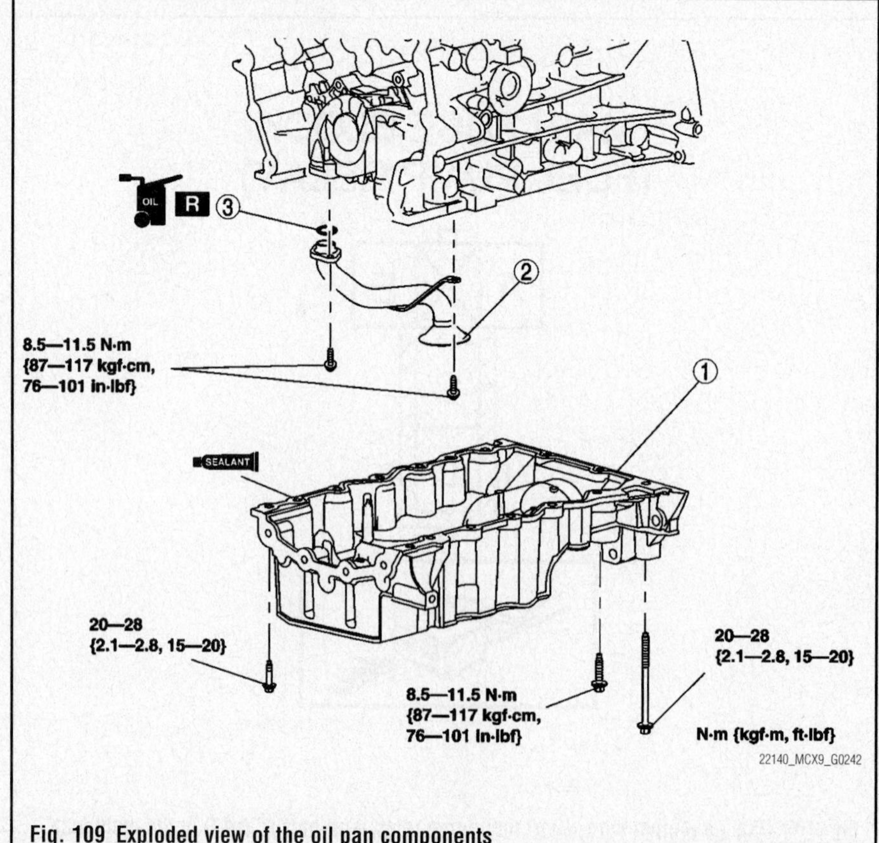

Fig. 109 Exploded view of the oil pan components

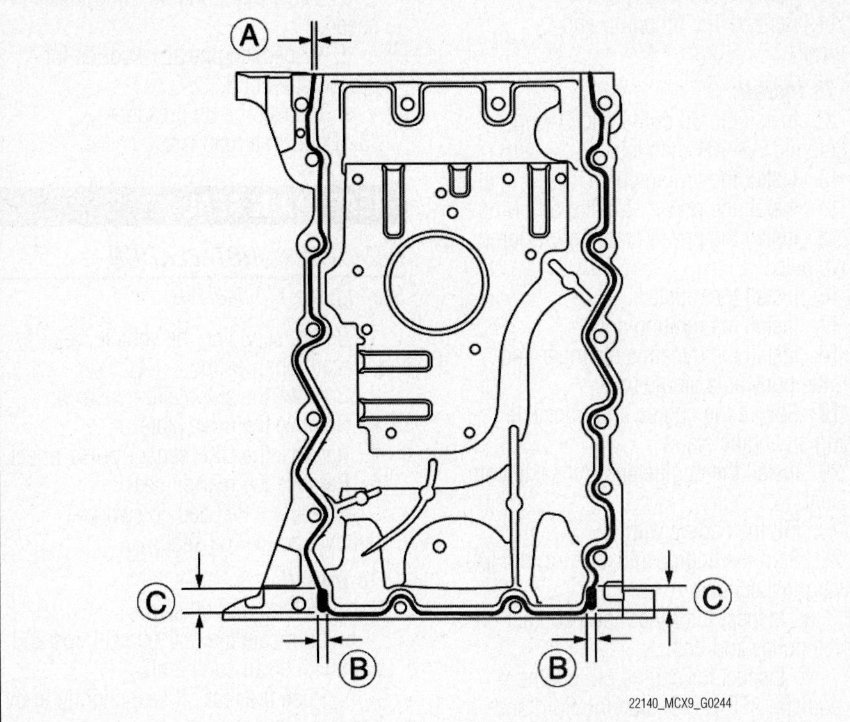

Fig. 110 Apply Loctite® 5900 silicone sealant to the oil pan along the inside of the bolt holes as shown

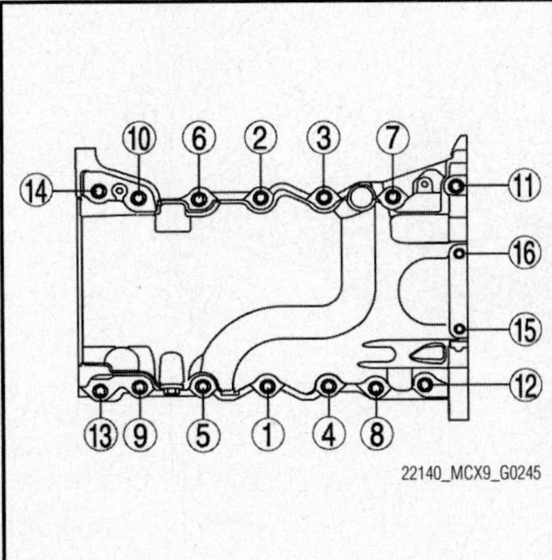

Fig. 111 Tighten the oil pan bolts in the order shown

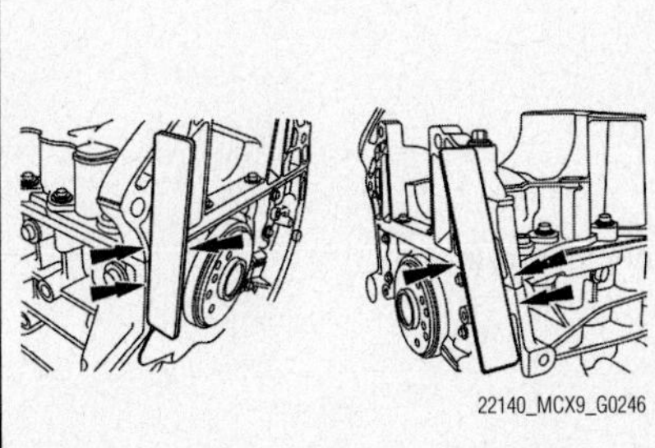

Fig. 112 Using a straightedge, align the oil pan flush with the rear of the cylinder block at the two areas as shown

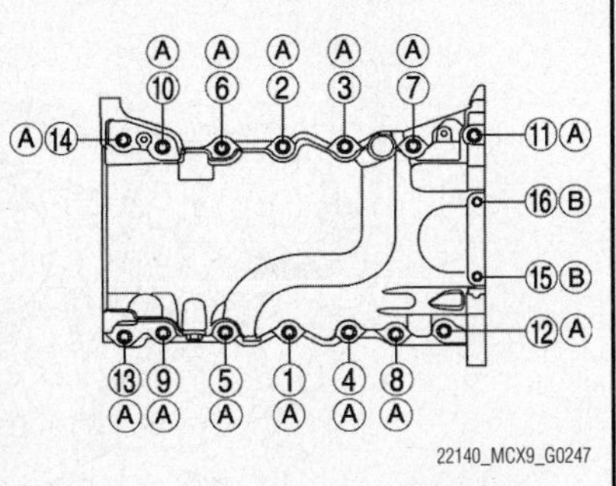

Fig. 113 Final tighten the oil pan bolts in the order shown and to the proper torque

To install:

18. Install the oil strainer using a new O-ring. Tighten to 76–101 inch lbs. (9–12 Nm).
19. Install the oil pan.

***** WARNING**

Apply silicon sealant in a single, unbroken line around the whole perimeter. The oil pan and bolts must be installed and the oil pan aligned to the cylinder block within 5 min of sealant application. Final tightening of the oil pan bolts must be carried out within 60 min of sealant application. Using bolts with the old seal adhering could cause cracks in the cylinder block.

a. Completely clean and remove any oil, dirt, sealant or other foreign material that may be adhering to the cylinder block and oil pan.

b. When reusing the oil pan installation bolts, clean any old sealant from the bolts.

c. Apply Loctite® 5900 silicone sealant to the oil pan along the inside of the bolt holes as shown.

d. Silicone sealant thickness should be as follows:

- Area A: 0.12 in. (3 mm)
- Area B: 0.20–0.23 in. 0.12 in (5.0–6.0 mm)
- Area C: 0.39 in (10 mm)

e. Install the oil pan and the bolts to the cylinder block.

f. Tighten the bolts in the order shown as follows:

- Step 1: Tighten bolts to 27 inch lbs. (3 Nm).
- Step 2: Loosen the bolts 180°.
- Step 3: Using a straightedge, align the oil pan flush with the rear of the cylinder block at the two areas as shown.
- Step 4: Tighten Bolt A to 15–20 ft. lbs. (20–28 Nm)
- Step 4: Tighten Bolt B to 76–101 inch lbs. (9–12 Nm)

20. Remove the engine stand.
21. On AWD vehicles, install the transfer case bracket.
22. Install the exhaust manifold (RH).
23. Install the power steering oil pump.
24. Install the power steering oil pump drive belt.
25. Install the dipstick.
26. Install the ignition coils.
27. Install the dynamic chamber and throttle body as a single unit.
28. Install the drive plate.
29. Install the automatic transaxle.
30. Using a hoist, lower the engine and transaxle component on a level surface.
31. Install the engine and transaxle component.
32. Fill the engine with oil.
33. Install the engine front cover.
34. Start the engine and perform the following inspections:

a. Inspect the runout and contact on the pulley and belt.

b. Inspect for engine oil, engine coolant, ATF, power steering fluid and fuel leakage.

c. Verify the ignition timing, idle speed.

d. Inspect engine accessories for proper operation.

35. Perform a road test.

OIL PUMP

REMOVAL & INSTALLATION

See Figure 114.

Mazda's official procedure for the oil pump removal and installation requires the engine and transaxle to be removed from the vehicle. It may be possible to perform this procedure with the engine and transaxle in the vehicle.

1. Before servicing the vehicle, refer to the Precautions Section.
2. Drain the engine oil.
3. Remove the engine and transaxle component.
4. Secure the engine and transaxle using an engine stand.
5. Remove the dynamic chamber and throttle body as a single unit.
6. Remove the ignition coils.
7. Remove the dipstick.
8. Remove the power steering oil pump drive belt.
9. Remove the power steering oil pump.
10. Remove the timing chain.
11. Remove the oil pump and O-ring.

To install:

12. Install the oil pump and O-ring. Tighten to 76–101 inch lbs. (9–12 Nm)
13. Install the timing chain.
14. Install the power steering oil pump.
15. Install the power steering oil pump drive belt.
16. Install the dipstick.
17. Install the ignition coils.
18. Install the dynamic chamber and throttle body as a single unit.
19. Secure the engine and transaxle using an engine stand.
20. Install the engine and transaxle component.
21. Fill the engine with oil.
22. Start the engine and perform the following inspections:
    a. Inspect the runout and contact on the pulley and belt.
    b. Inspect for engine oil, engine coolant, ATF, power steering fluid and fuel leakage.
    c. Verify the ignition timing, idle speed.
    d. Inspect engine accessories for proper operation.
    e. Inspect the oil pressure.
23. Perform a road test.

REAR MAIN SEAL

REMOVAL & INSTALLATION

See Figures 115 and 116.

1. Before servicing the vehicle, refer to the Precautions Section.
2. Remove the automatic transaxle.
3. Remove the drive plate.
4. Remove the CKP sensor pulse wheel.
5. Remove the rear oil seal.
    a. Using a flathead screwdriver, remove the rear oil seal.

To install:

6. Install the rear oil seal.
    a. Lubricate the rear oil seal lips and bore with clean engine oil.
    b. Push the rear oil seal slightly in by hand.

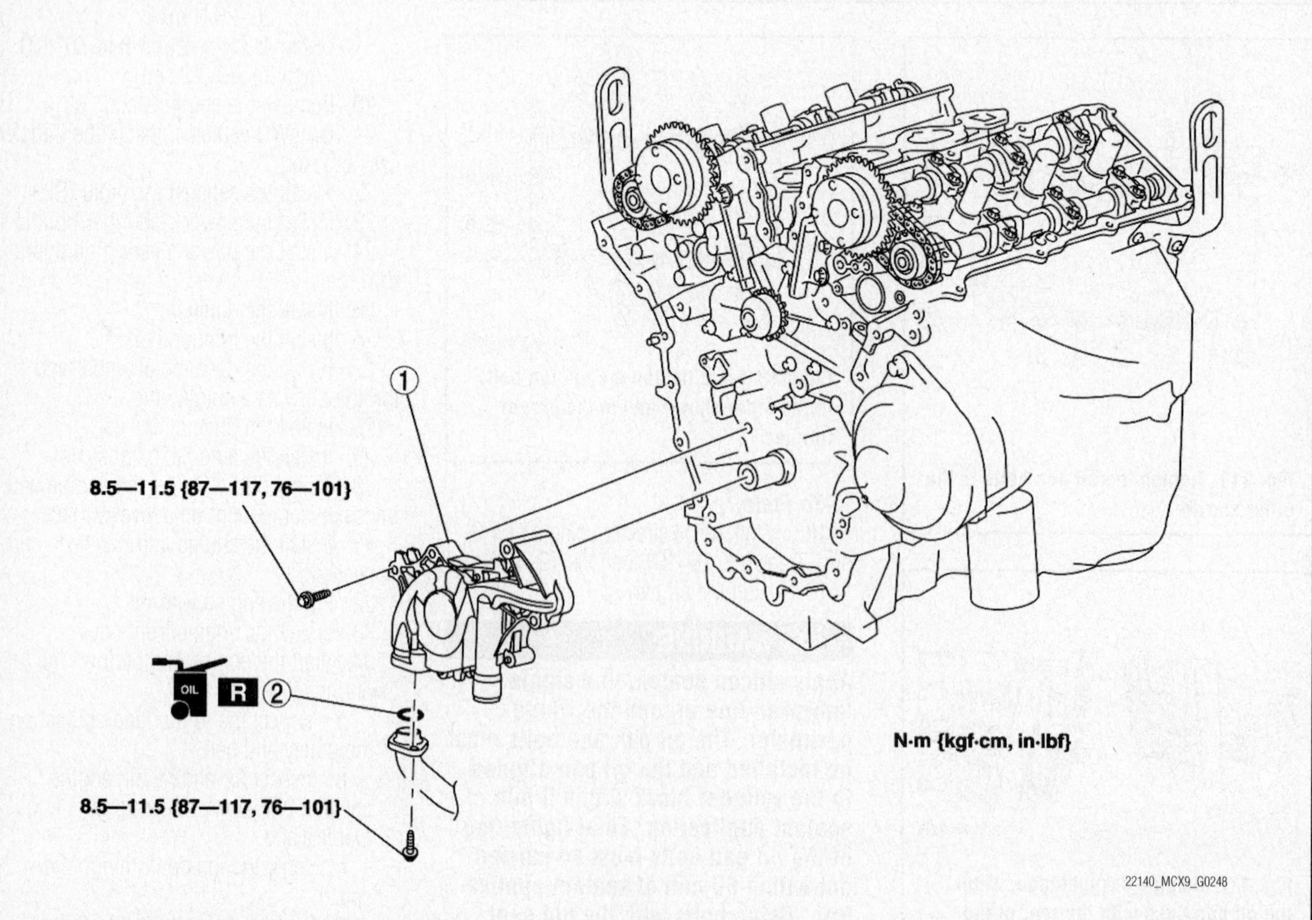

Fig. 114 Exploded view of the oil pump components

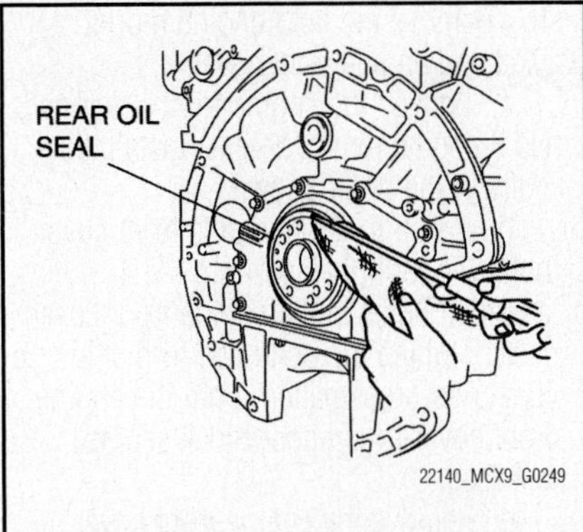

Fig. 115 Using a flathead screwdriver, remove the rear oil seal

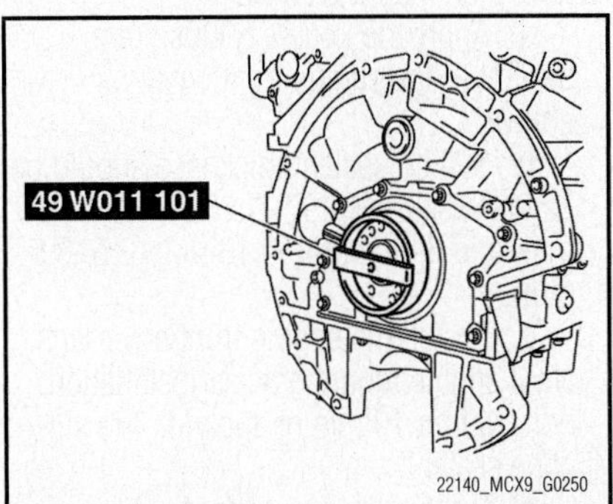

Fig. 116 Tap the rear oil seal in evenly using the special tool and a hammer

c. Tap the rear oil seal in evenly using the special tool and a hammer.
7. Install the CKP sensor pulse wheel.
8. Install the drive plate.
9. Install the automatic transaxle.

TIMING CHAIN COVER AND SEAL

REMOVAL & INSTALLATION

See Figures 117 through 127.

Mazda's official procedure for the timing chain cover removal and installation requires the engine and transaxle to be removed from the vehicle. It may be possible to perform this procedure with the engine and transaxle in the vehicle.

1. Before servicing the vehicle, refer to the Precautions Section.
2. Drain the engine oil.
3. Remove the engine and transaxle.
4. Secure the engine and transaxle using a hoist and an engine stand.
5. Remove the intake manifold dynamic chamber and throttle body as a single unit.
6. Remove the ignition coils.
7. Remove the dipstick.
8. Remove the power steering oil pump drive belt.

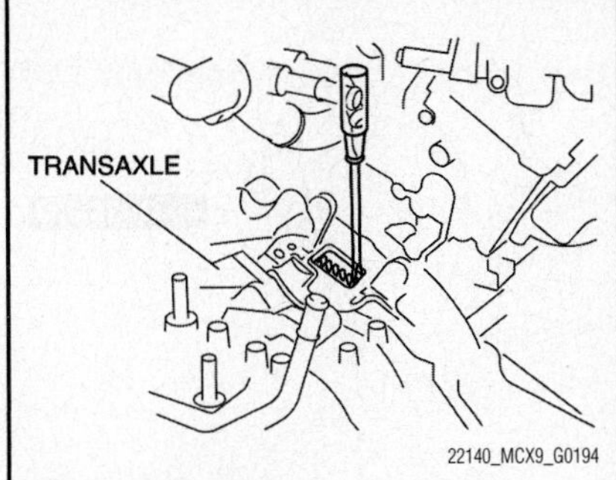

Fig. 117 Set a flathead screwdriver to the drive plate as shown to lock the crankshaft rotation

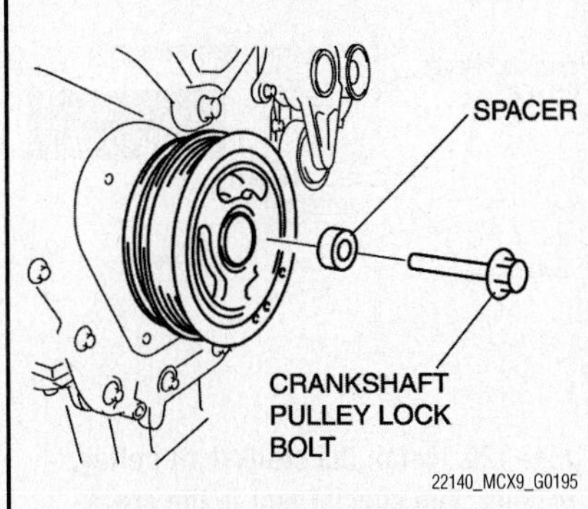

Fig. 118 Install a suitable spacer 0.55 in. (14 mm) in thickness and 1.18 in. (30 mm) in diameter to the crankshaft pulley lock bolt, and install the crankshaft pulley lock bolt to the crankshaft

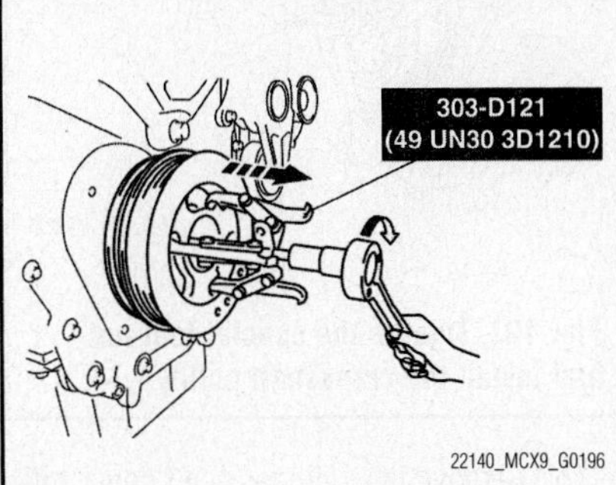

Fig. 119 Remove the crankshaft pulley using a gear puller

9. Remove the power steering oil pump.
10. Remove the drive belt auto tensioner.
11. Remove the crankshaft pulley lock bolt.
    a. Remove the starter.
    b. Set a flathead screwdriver to the drive plate as shown to lock the crankshaft rotation.

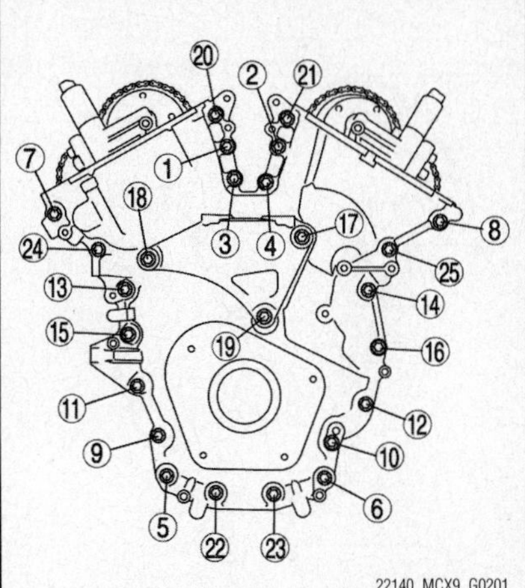

Fig. 120 Loosen the engine front cover and No.3 engine mount bracket installation bolts in the order shown

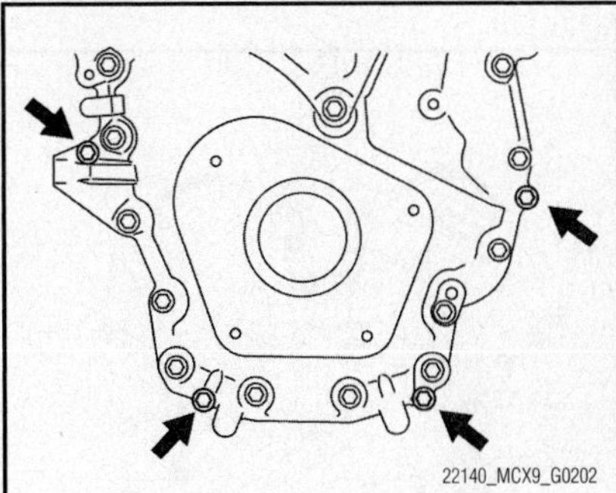

Fig. 121 Install 4 of the engine front cover bolts (finger tightened) into the 4 threaded holes in the engine front cover. Tighten the bolts one turn at a time in a criss-cross pattern until the engine front cover-to-cylinder block seal is released

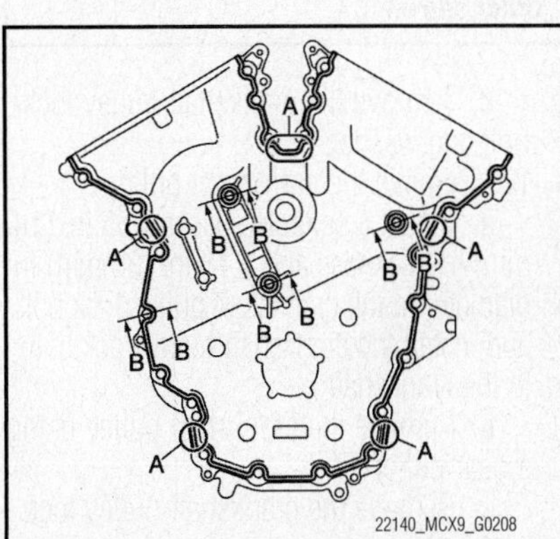

Fig. 122 Apply the Loctite® 5900 silicon sealant to the engine front cover as shown. Silicon sealant thickness should be as follows: A—0.197–0.236 in. (5.0–6.0 mm) and B— 0.099–0.137 in. (2.5–3.5 mm)

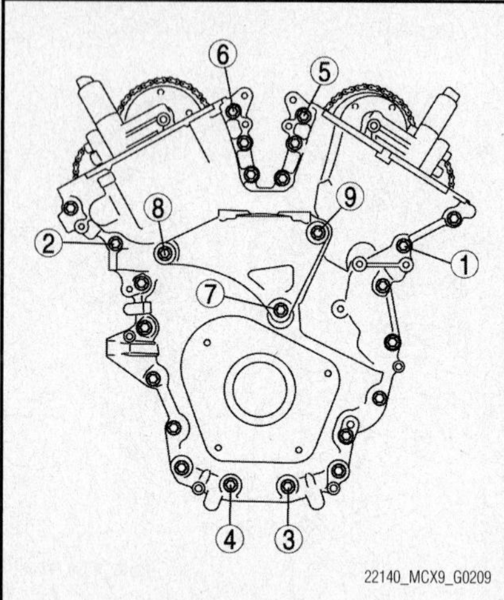

Fig. 123 Tighten the engine front cover and No.3 engine mount bracket installation bolts in the order shown

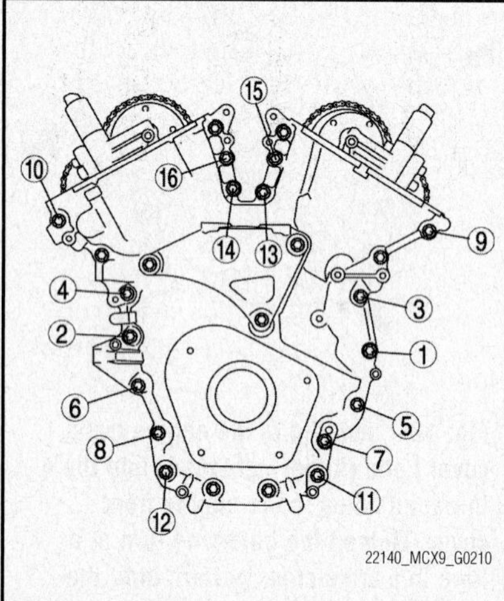

Fig. 124 Tighten the engine front cover installation bolts in the order in the order shown

c. Remove the crankshaft pulley lock bolt and washer.

12. Remove the crankshaft pulley.

a. Install a suitable spacer 0.55 in. (14 mm) in thickness and 1.18 in. (30 mm) in diameter to the crankshaft pulley lock bolt, and install the crankshaft pulley lock bolt to the crankshaft.

b. Remove the crankshaft pulley using a gear puller.

c. Remove the crankshaft pulley lock bolt and spacer.

13. Remove the front oil seal.

a. Remove the front oil seal using a flathead screwdriver.

14. Remove the cylinder head covers, as outlined later in this section. The bolts must be removed in the proper sequence.

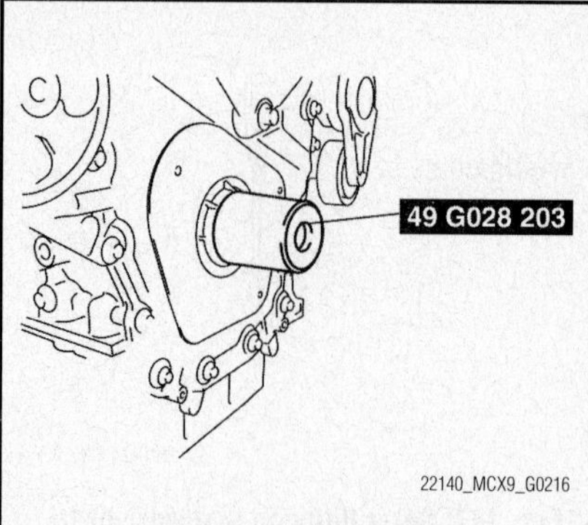

Fig. 125 Tap the front oil seal in evenly using the special tool and a hammer

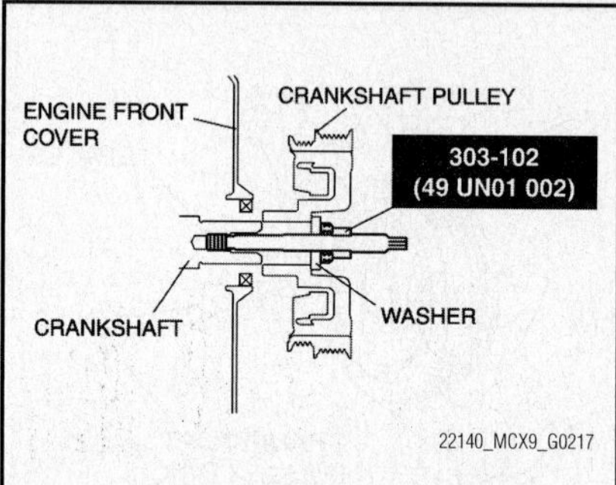

Fig. 126 Install the crankshaft pulley, washer, and special tool to the crankshaft

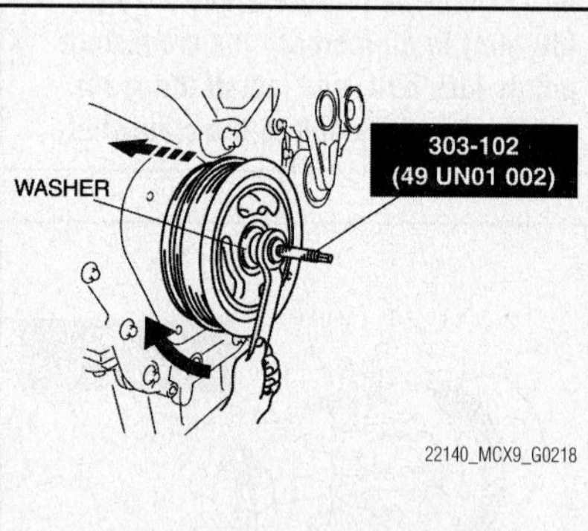

Fig. 127 Tighten the special tool nut and install the crankshaft pulley

15. Remove the cylinder head cover oil seals.

a. Inspect the OCV attachment hole seals and spark plug tube attachment hole seals. Remove the any damaged seals.

b. The OCV attachment hole seals removal is shown, the spark plug tube attachment hole seals removal procedure is the same.

c. To remove, tap the OCV attachment hole seals and spark plug tube attachment hole seals using the special tool and hammer.

16. Remove the no.3 engine mount bracket and engine front cover.

a. Loosen the engine front cover and No.3 engine mount bracket installation bolts in the order shown.

b. Install 4 of the engine front cover bolts (finger tightened) into the 4 threaded holes in the engine front cover.

c. Tighten the bolts one turn at a time in a criss-cross pattern until the engine front cover-to-cylinder block seal is released.

d. Remove the engine front cover.

To install:

17. Install the engine front cover and No.3 engine mount bracket.

a. Apply the Loctite 5900silicon sealant to the engine front cover as shown.

b. Silicon sealant thickness should be as follows: A—0.197–0.236 in. (5.0–6.0 mm) and B— 0.099–0.137 in. (2.5–3.5 mm).

c. Install the engine front cover and No.3 engine mount bracket installation bolts within 10 min of applying the silicone sealant.

d. Tighten the engine front cover and No.3 engine mount bracket installation bolts in 4 steps in the order shown.

- Step 1: Tighten bolts 1–6 to 89 inch lbs. (10 Nm).
- Step 2: Tighten bolts 7–9 to 11 ft. lbs. (15 Nm).
- Step 3: Tighten bolts 1–6 to 15–20 ft. lbs. (20–28 Nm).
- Step 4: Tighten bolts 7–9 to 52–59 ft. lbs. (70–80 Nm).

e. Install the engine front cover installation bolts within 60 min of applying the silicone sealant.

f. Tighten the engine front cover installation bolts in the order in 2 steps in the order shown.

- Step 1: Tighten to 89 inch lbs. (10 Nm).
- Step 2: Tighten to 15–20 ft. lbs. (20–28 Nm).

18. Install the cylinder head cover oil seal.

a. Installation of new seals is only required if damaged seals were removed during disassembly of the engine.

b. Push the OCV attachment hole seals and spark plug tube attachment hole seals slightly in by hand.

c. Tap the OCV attachment hole seals and spark plug tube attachment hole seals using the special tool and hammer.

19. Install the cylinder head cover, as outlined later in this section. Note the following:
   a. Make sure to apply Loctite® 5900 silicone sealant to the mating faces, as shown in the procedure.
   b. Install the cylinder head cover with a new gasket.
   c. Install the cylinder head cover installation bolt and studs within 5 min of applying the silicone sealant.
   d. Tighten the bolts in the order shown to 76–101 inch lbs. (9–12 Nm).
20. Install the front oil seal
   a. Apply clean engine oil to the front oil seal bore in the engine front cover.
   b. Push the front oil seal slightly in by hand.
   c. Tap the front oil seal in evenly using the special tool and a hammer.
21. Install the crankshaft pulley
22. Install the crankshaft pulley lock bolt.
   a. Set a flathead screwdriver to the drive plate in the position indicated. to lock the crankshaft rotation.
   b. Tighten the new crankshaft pulley lock bolt in four steps.
      - Step 1: Tighten to 103 ft. lbs. (140 Nm)
      - Step 2: Loosen 360° (one full turn)
      - Step 3: Tighten to 35–39 ft. lbs (47–53 Nm)
      - Step 4: Tighten an additional 85–95° of rotation
23. Install the power steering oil pump.
24. Install the power steering oil pump drive belt.
25. Install the dipstick.
26. Install the ignition coils.
27. Install the intake manifold dynamic chamber and throttle body as a single unit.
28. Install the engine and transaxle.
29. Fill the engine with oil.
30. Start the engine and perform the following inspections:
   a. Inspect the runout and contact on the pulley and belt.
   b. Inspect for engine oil, engine coolant, ATF, power steering fluid and fuel leakage.
   c. Verify the ignition timing and idle speed.
   d. Inspect engine accessories for proper operation.
31. Perform a road test.

TIMING CHAIN AND SPROCKETS

REMOVAL & INSTALLATION

See Figures 128 through 135.

Mazda's official procedure for timing chain removal and installation requires the engine and transaxle to be removed from the vehicle. It may be possible to perform this procedure with the engine and transaxle in the vehicle.

Fig. 128 Loosen the OCV component installation bolts in the order shown—right hand

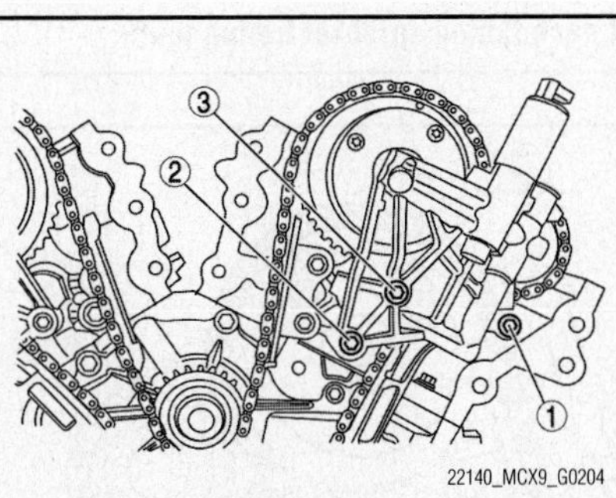

Fig. 129 Loosen the OCV component installation bolts in the order shown—left hand

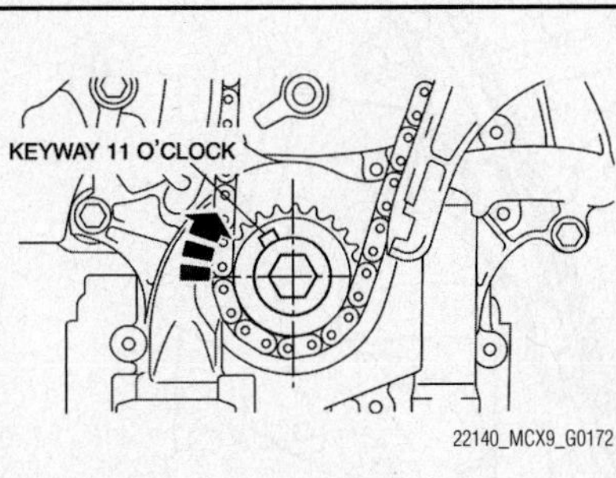

Fig. 130 Turn the crankshaft clockwise so that the crankshaft keyway is in the 11 o'clock position. This will position the No.1 cylinder at TDC

1. Before servicing the vehicle, refer to the Precautions Section.
2. Drain the engine oil.
3. Remove the engine and transaxle.
4. Secure the engine and transaxle using a hoist and an engine stand.
5. Remove the Timing Chain Cover and Seal, as outlined earlier in this section.
6. Remove the OCV component.
   a. Loosen the OCV component installation bolts in the order shown
7. Remove the chain tensioner.
8. Remove the timing chain.

➡When removing the timing chain and marking the timing marks on the chain, mark the camshaft timing chain as well.

⁂ WARNING

Do not rotate the crankshaft counterclockwise. The timing chains may bind, causing engine damage.

   a. Turn the crankshaft clockwise so that the crankshaft keyway is in the 11 o'clock position. This will position the No.1 cylinder at TDC.

➡Verify that there are timing marks in three locations (Yellow 1, Black 2) on the timing chain. If any timing marks are missing, mark the timing chain. When marking the crankshaft sprocket side timing chain, change the mark color. When the timing chain is replaced with a new one, mark the new timing chain at the same positions as the removed timing chain.

   b. Mark the timing chain at the position of each timing sprocket timing mark.

➡Verify that there are timing marks in two locations on the camshaft timing chain. If any timing marks are missing, mark the camshaft timing chain. If replacing with a new camshaft timing chain, place alignment marks in the same positions as those prior to the replacement.

   c. Mark the camshaft timing chain at the positions where it is aligned with each of the camshaft sprocket on both banks.
   d. Remove the timing chain in the following order:
      - Chain tensioner
      - Timing chain
      - Tensioner arm
      - Chain guide
      - Crankshaft sprocket

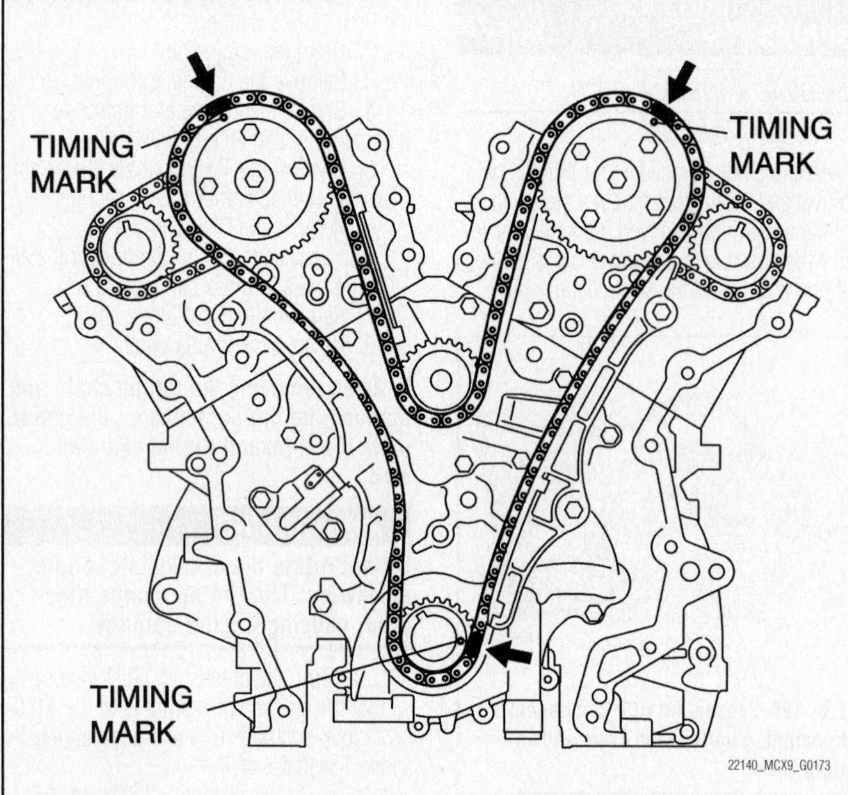

Fig. 131 Mark the timing chain at the position of each timing sprocket timing mark

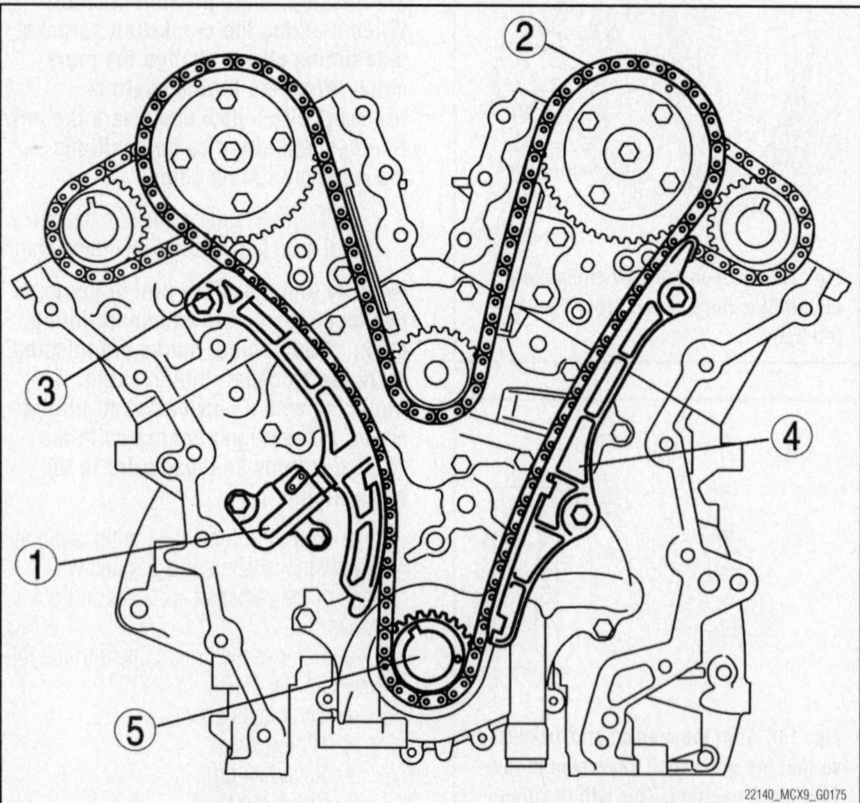

Fig. 132 Remove the timing chain in the following order: (1) Chain tensioner, (2) Timing chain, (3) Tensioner arm, (4) Chain guide and (5) Crankshaft sprocket

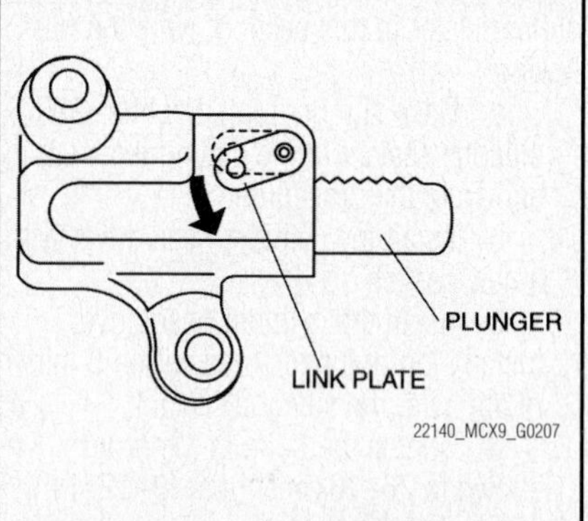

Fig. 133 Insert a paper clip where the link plate hole and the tensioner body hole overlap to fix the link plate and lock the plunger

To install:

9. Install the timing chain
   a. Verify that the crankshaft keyway is at 11 o'clock position.

➡Of the three marked locations on the timing chain, align the mark that has a different color to the crankshaft sprocket side timing mark.

   b. Place alignment marks on the timing chain corresponding to each of the alignment marks on the timing sprocket.
   c. Push down the link plate of the timing chain tensioner and release the plunger lock.

➡The plunger should retract with minimal force. If binding occurs, remove the chain tensioner from the vise and reset it in the vise.

   d. Secure the chain tensioner using a vice attached with a soft base, and slowly press the plunger back shown. while pressing down the link plate.
   e. Release the pressure slightly from the plunger, and move the plunger back and forth 0.08–0.11 in (2–3 mm).
   f. Insert a paper clip where the link plate hole and the tensioner body hole overlap to fix the link plate and lock the plunger.
   g. Install the timing chain in the following order:
   - Crankshaft sprocket
   - Chain guide
   - Tensioner arm
   - Timing chain
   - Chain tensioner

   h. Tighten components to 76–101 inch lbs. (9–12 Nm).
   i. Remove the retaining wire.

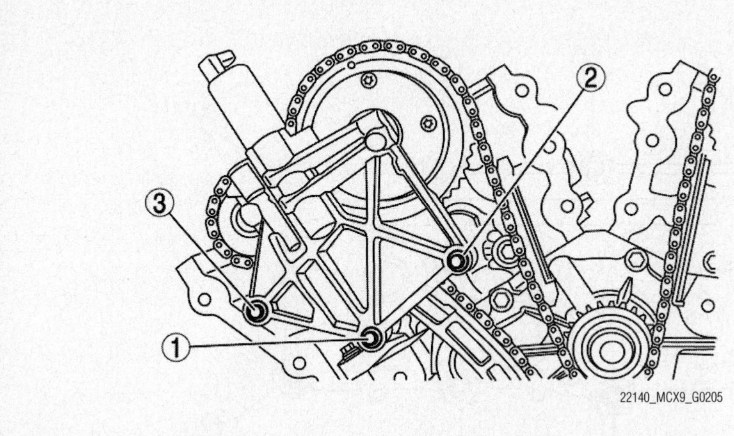

Fig. 134 Tighten the OCV component installation bolts in the order shown—right hand

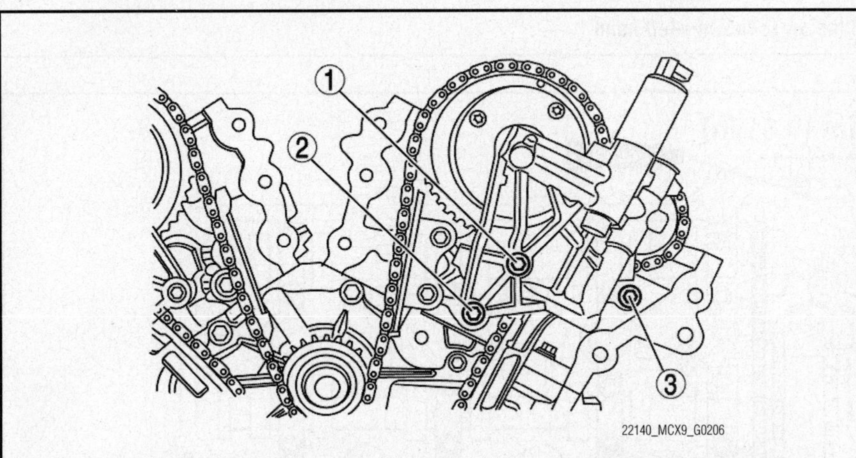

Fig. 135 Tighten the OCV component installation bolts in the order shown—left hand

10. Install the chain tensioner
11. Install the OCV components.
    a. Tighten the OCV component installation bolts in the order shown
12. Install the Timing Chain Cover and Seal, as outlined earlier in this section.
13. Install the engine and transaxle.
14. Fill the engine with oil.
15. Start the engine and perform the following inspections:
    a. Inspect the runout and contact on the pulley and belt.
    b. Inspect for engine oil, engine coolant, ATF, power steering fluid and fuel leakage.
    c. Verify the ignition timing and idle speed.
    d. Inspect engine accessories for proper operation.
16. Perform a road test.

VALVE COVERS

REMOVAL & INSTALLATION

See Figures 136 through 141.

Mazda's official procedure for timing chain removal and installation requires the engine and transaxle to be removed from the vehicle. It may be possible to perform this procedure with the engine and transaxle in the vehicle.

1. Before servicing the vehicle, refer to the Precautions Section.
2. Remove the engine and transaxle.
3. Secure the engine and transaxle using a hoist and an engine stand.
4. Disconnect the ignition coil connectors.
5. Remove the ignition coils.
6. Disconnect the control wiring harness.
7. Loosen the valve cover retaining bolts in the order shown.
8. Remove the valve cover and gasket.

To install:

9. Apply Loctite® 5900 silicone sealant to the mating faces as shown.
10. Install the valve cover using a new gasket.
11. Install the valve cover retaining bolts and tighten in the order shown to 47–63 inch lbs (5–7 Nm).
12. Connect the control wiring harness.
13. Install the ignition coils.
14. Connect the ignition coil connectors.
15. Install the engine and transaxle.
16. Run the engine and check for leaks.

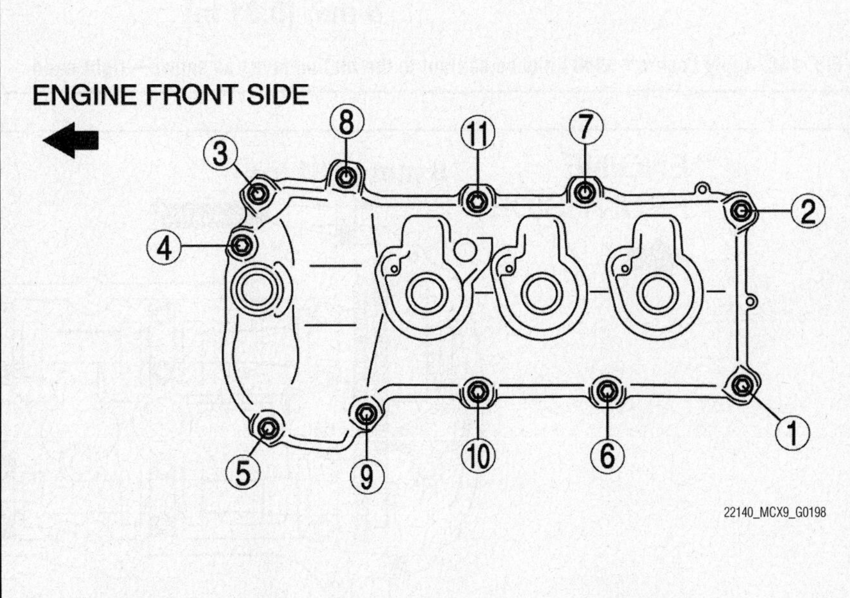

Fig. 136 Remove the cylinder head cover bolts in the order shown—right hand

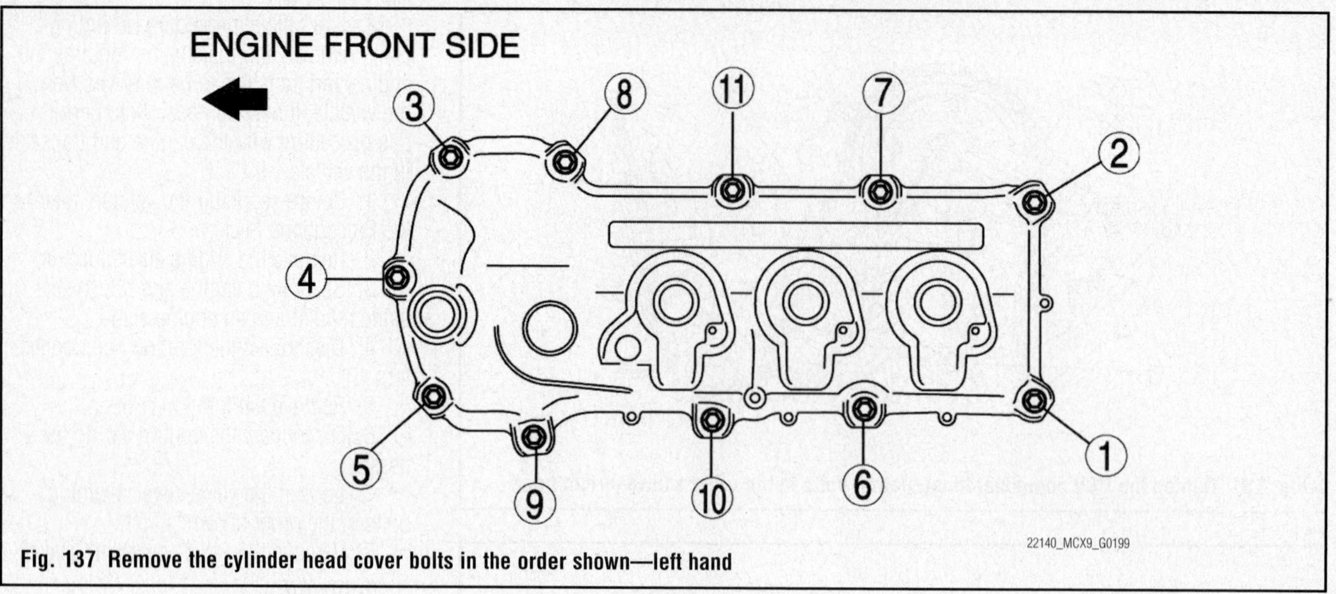

Fig. 137 Remove the cylinder head cover bolts in the order shown—left hand

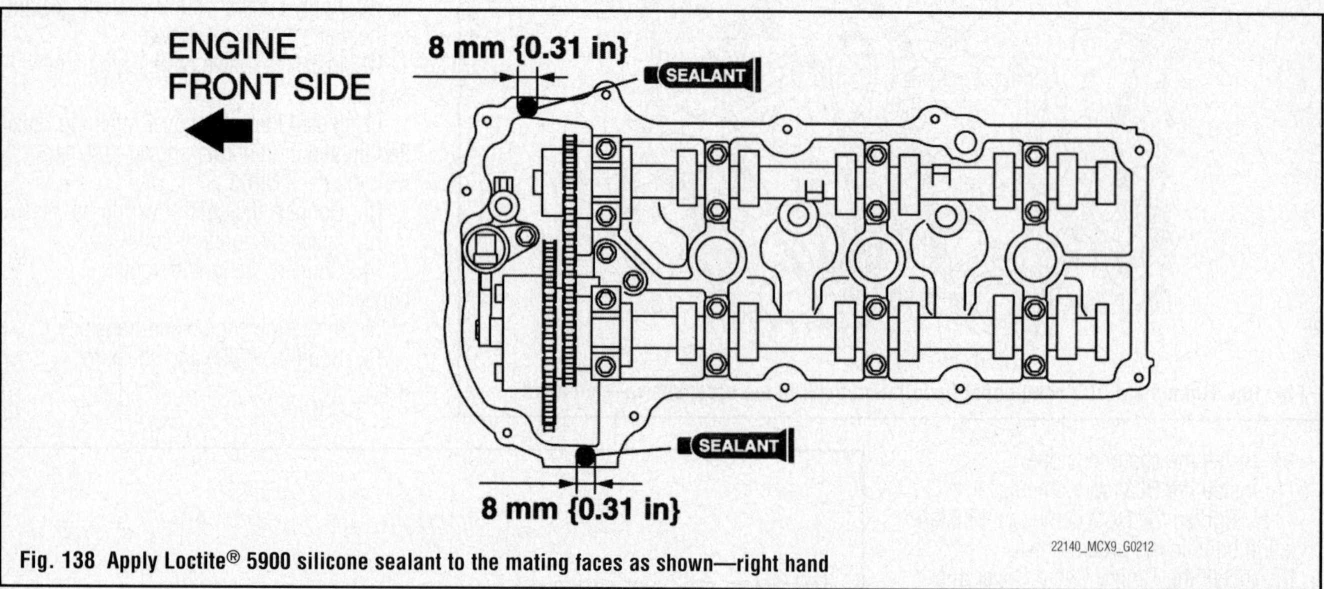

Fig. 138 Apply Loctite® 5900 silicone sealant to the mating faces as shown—right hand

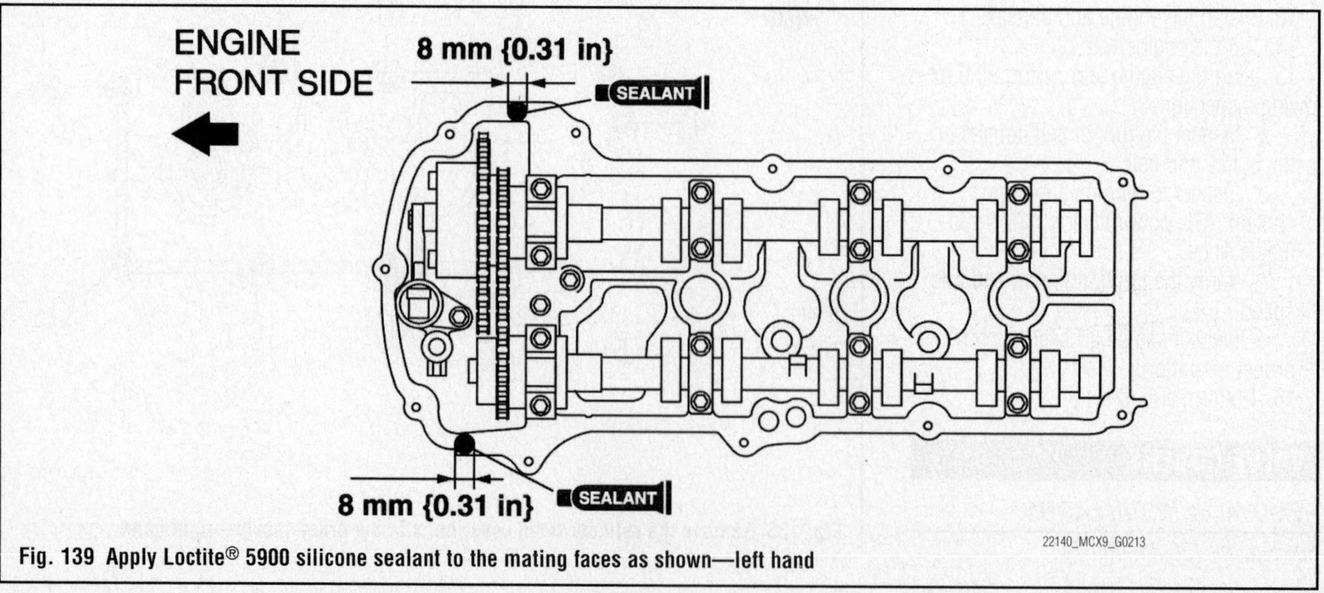

Fig. 139 Apply Loctite® 5900 silicone sealant to the mating faces as shown—left hand

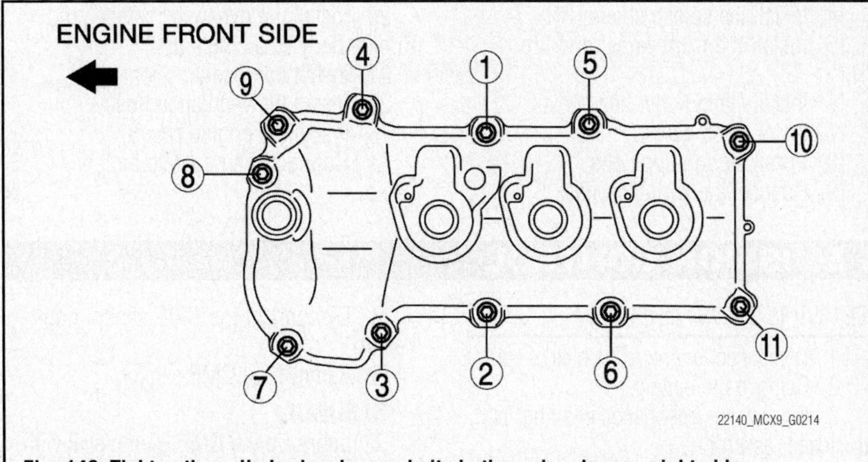

Fig. 140 Tighten the cylinder head cover bolts in the order shown—right side

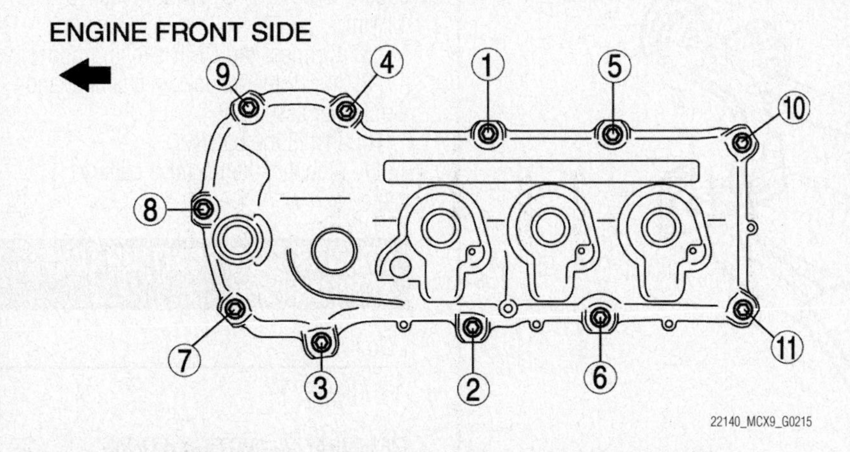

Fig. 141 Tighten the cylinder head cover bolts in the order shown—left side

VALVE LASH

ADJUSTMENT

See Figures 142 and 143.

1. Disconnect the negative battery cable.
2. Remove the engine cover.
3. Remove the ventilation hose.

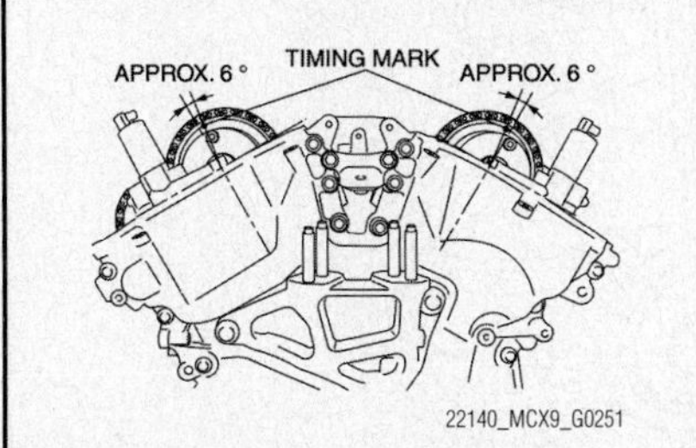

Fig. 142 Rotate the crankshaft clockwise and verify that the timing mark of the intake side camshaft sprocket is in the position shown in the figure. In this position, the No.1 cylinder is at the TDC of the compression stroke

4. Remove the resonance chamber.
5. Remove the dynamic chamber and throttle body as a single unit.
6. Disconnect the wiring harness.
7. Remove the ignition coils.
8. Remove the dipstick.
9. Remove the cylinder head cover.
10. Remove the front wheel and tire (RH).
11. Remove the splash shield (RH).
12. Measure the valve clearance.
    a. Rotate the crankshaft clockwise and verify that the timing mark of the intake side camshaft sprocket is in the position shown in the figure. In this position, the No.1 cylinder is at the TDC of the compression stroke.
    b. Measure the valve clearance of location A shown in the figure.

➡Make sure to note down the measured values for choosing the suitable replacement tappets.

    c. Mark the crank pulley and engine front cover as shown in the figure.
    d. Rotate the crankshaft clockwise 360° so that the No.5 cylinder is at TDC of the compression stroke.
    e. Measure the valve clearance of location B shown in the figure.
13. If it is not within the specification, replace the tappet and adjust the valve clearance to the median value of the standard.
    a. Remove the camshafts and camshaft sprocket as a single unit.

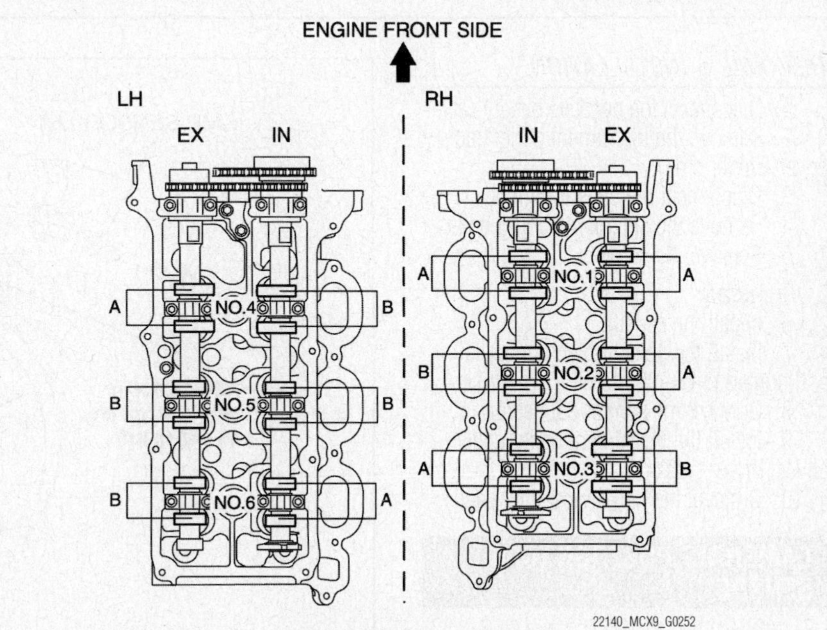

Fig. 143 Measure the valve clearance of location A, rotate the crankshaft clockwise 360° so that the No.5 cylinder is at TDC of the compression stroke and then measure the valve clearance of location B as shown

b. Remove the tappet.

c. Install an appropriate tappet based on the results of the valve clearance inspection.

Selected tappet = Removed tappet thickness + Measured valve clearance - Standard valve clearance

14. Install the splash shield (RH).
15. Install the front wheel and tire (RH).
16. Install the cylinder head cover.
17. Install the dipstick.
18. Install the ignition coils.
19. Connect the wiring harness.
20. Install the dynamic chamber and throttle body as a single unit.
21. Install the resonance chamber.
22. Install the ventilation hose.
23. Install the engine cover.
24. Connect the negative battery cable.

ENGINE PERFORMANCE & EMISSION CONTROL

ACCELERATOR PEDAL POSITION (APP) SENSOR

LOCATION

See Figure 144.

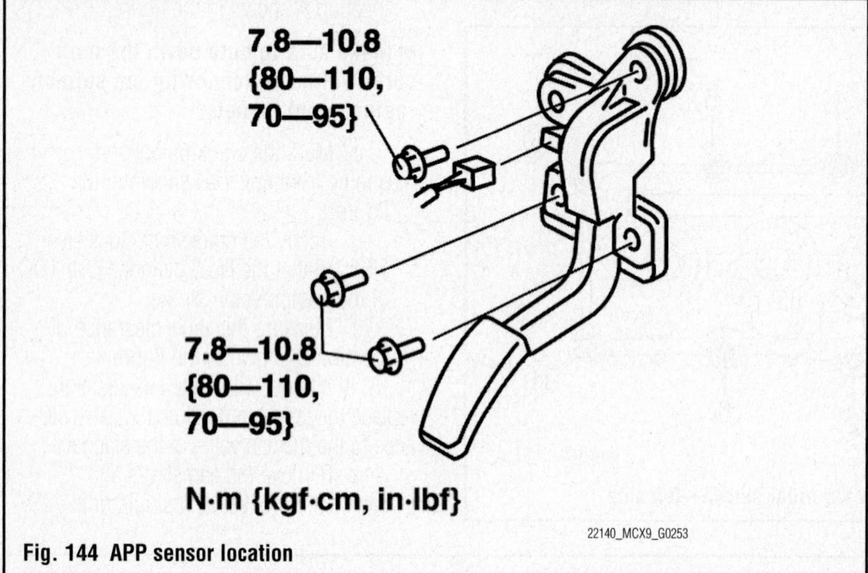

Fig. 144 APP sensor location

REMOVAL & INSTALLATION

1. Disconnect the negative battery cable.
2. Remove the instrument panel under cover (Driver side).
3. Disconnect the sensor connector
4. Remove the sensor attaching bolts.
5. Remove the sensor.

To install:

6. Install the sensor.
7. Install the sensor attaching bolts and tighten to 70–95 inch lbs. (8–11 Nm).
8. Connect the sensor connector
9. Install the instrument panel under cover (Driver side).
10. Connect the negative battery cable.

CAMSHAFT POSITION (CMP) SENSOR

LOCATION

See Figure 145.

REMOVAL & INSTALLATION

1. Disconnect the negative battery cable.
2. Remove the battery.
3. Remove the resonance chamber and air cleaner assembly.
4. Disconnect the CMP sensor connector
5. Remove the CMP sensor.

To install:

6. Lubricate the CMP O-ring seal with clean engine oil.
7. Install the CMP sensor. Tighten attaching bolt to 70–85 inch lbs. (8–11 Nm).
8. Connect the CMP sensor connector
9. Install the resonance chamber and air cleaner assembly.
10. Install the battery.
11. Connect the negative battery cable.

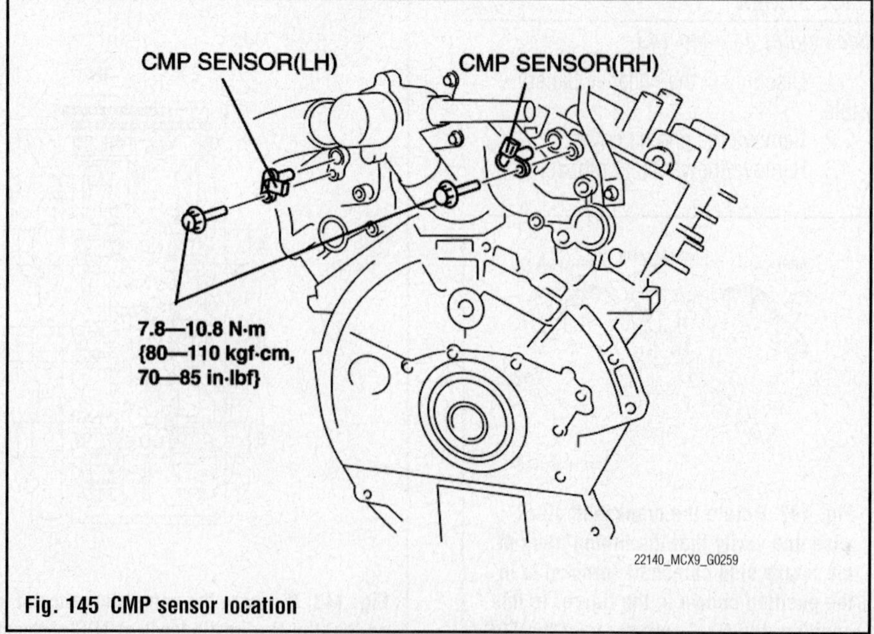

Fig. 145 CMP sensor location

CRANKSHAFT POSITION (CKP) SENSOR

LOCATION

See Figure 146.

REMOVAL & INSTALLATION

1. Disconnect the negative battery cable.

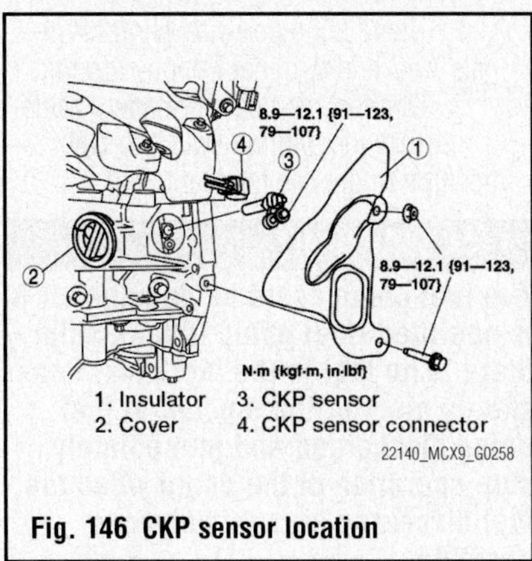

Fig. 146 CKP sensor location

2. Remove the left side catalytic converter.
3. Remove the insulator.
4. Remove the cover.
5. Remove the CKP sensor.
6. Disconnect the CKP sensor connector.

To install:

7. Connect the CKP sensor connector.
8. Install the CKP sensor. Tighten to 79–107 inch lbs. (9–12 Nm).
9. Install the cover. Tighten fasteners to 79–107 inch lbs. (9–12 Nm).
10. Install the insulator.
11. Install the left side catalytic converter.
12. Connect the negative battery cable.

CYLINDER HEAD TEMPERATURE (CHT) SENSOR

LOCATION

See Figure 147.

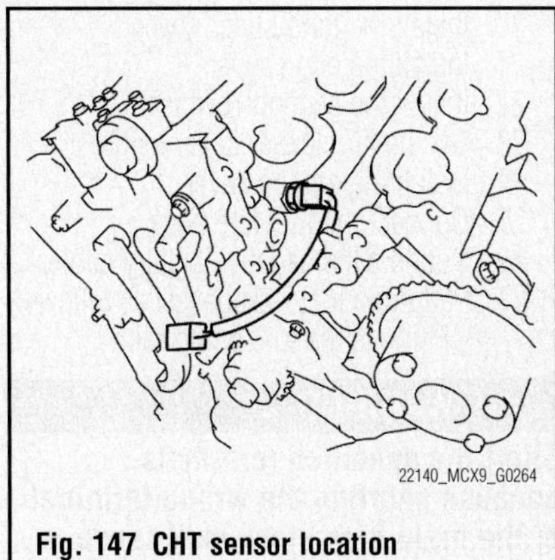

Fig. 147 CHT sensor location

REMOVAL & INSTALLATION

⁂ WARNING

Do not reuse the CHT sensor. If the sensor is removed, install a new sensor.

1. Disconnect the negative battery cable.
2. Remove the intake manifold.
3. Disconnect the CHT sensor connector.
4. Remove the CHT sensor.

To install:

5. Install the CHT sensor. Tighten to 79–93 inch lbs. (9–11 Nm).
6. Connect the CHT sensor connector.
7. Install the intake manifold.
8. Connect the negative battery cable.

ELECTRIC FAN SWITCH

LOCATION

See Figure 148.

REMOVAL & INSTALLATION

1. Before servicing the vehicle, refer to the Precautions Section.
2. Disconnect the negative battery cable.
3. Drain the engine coolant.
4. Remove the air cleaner and fresh air duct.
5. Remove the engine cover.
6. Remove the dipstick.
7. Disconnect the fan electrical connector.
8. Disconnect the fan wiring harness.
9. Remove the relay box.
10. Remove the upper radiator hose.
11. Remove the oil cooler water hose.
12. Remove the transmission oil refrigerant pipe

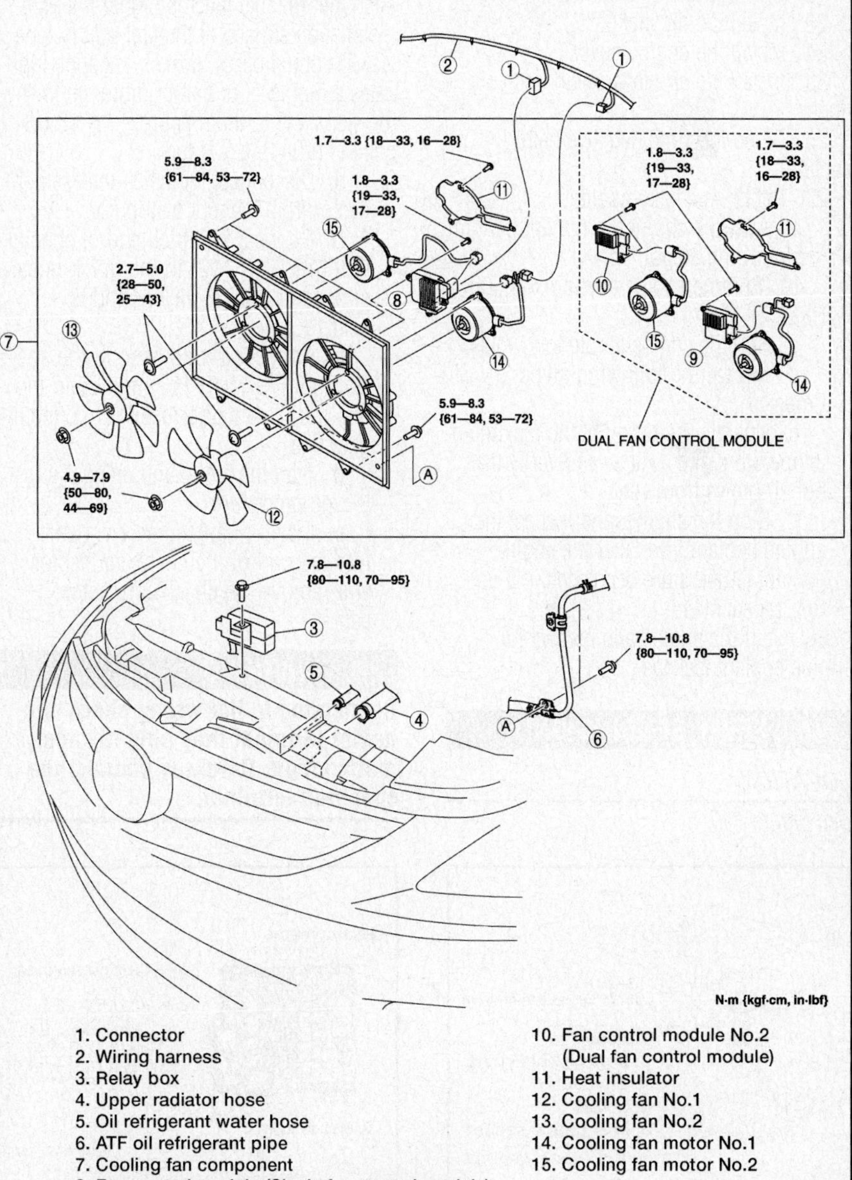

Fig. 148 Engine fan and related components

a. Remove the transmission oil refrigerant pipe with the hoses still connected. Position the transmission oil refrigerant pipe so that it is out of the way.

13. Remove the cooling fan from the top.
14. Remove the fan control module(s).

To install:

15. Install the fan control module(s).
16. Install the cooling fan from the top. Tighten to 53–72 inch lbs. (6–8 Nm).
17. Install the transmission oil refrigerant pipe.
18. Install the oil cooler water hose.
19. Install the upper radiator hose.
20. Install the relay box.
21. Connect the fan wiring harness.
22. Connect the fan electrical connector.
23. Install the dipstick.
24. Install the engine cover.
25. Install the air cleaner and fresh air duct.
26. Disconnect the negative battery cable.
27. Refill the engine coolant.
    a. After the engine warms up, perform the following steps.
    b. Run the engine at approx. 4,000 RPM for 1 min.
    c. Run the engine at idle for 1 min.
    d. Repeat the following steps 2 times.
    e. Operate the heater at the maximum temperature and airflow, and verify that hot air blows from vent.
    f. Stop the engine, and inspect the engine coolant level after the engine coolant temperature decreases. If it is low, repeat steps.
28. Start the engine and inspect for engine coolant leakage.

FUEL LEVEL SENDING UNIT

LOCATION

See Figure 149.

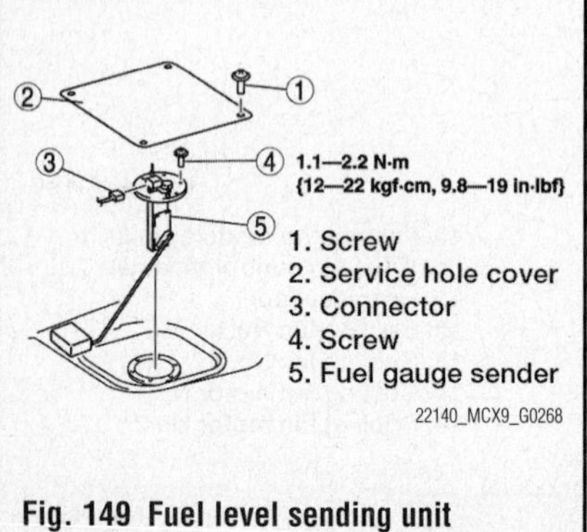

Fig. 149 Fuel level sending unit assembly

REMOVAL & INSTALLATION

See Figure 150.

1. Level the vehicle.
2. Fuel in the fuel system is under high pressure also when the engine is not running.
    a. Remove the fuel filler cap and release the pressure in the fuel tank.
    b. When disconnecting a fuel line hose, wrap a rag around it to protect against fuel leakage.
    c. Plug the hose after removal.
3. Disconnect the negative battery cable.
4. Remove the engine cover.
5. Set the air cleaner cover aside.
6. Because the fuel tank is constructed such that the fuel level is higher than the installation surface of the fuel pump, fuel leakage could occur. If the fuel gauge indicates a fuel level of half or more, perform the steps 1-6 to drain approx. 11–15 US gallons (10–15 L) of fuel.
    a. Disconnect the quick release connector from the fuel distributor.
    b. Connect the thick end of special tool (49 T013 102) to the quick release connector until a click sound is heard.
    c. Connect a long hose to the special tool (49 T013 102) and drain the fuel into a container used for collecting gasoline.
    d. Start the fuel pump using the following procedure.
    e. Insert a flathead screwdriver into tab part of the check connector and remove the check connector cap.

WARNING

Connecting to the wrong check connector terminal may only to cause malfunction. Carefully connect the specified terminal.

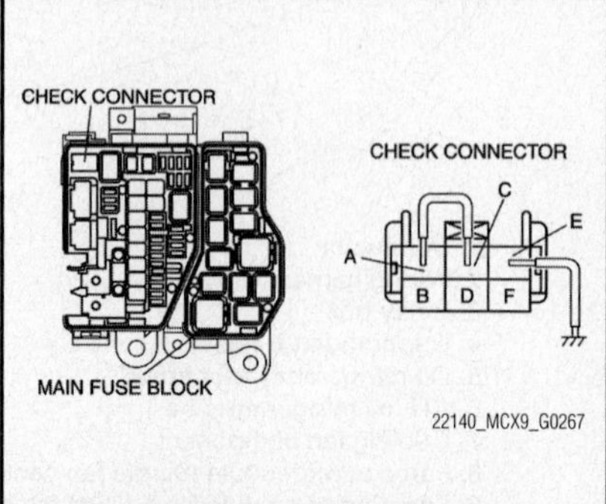

Fig. 150 Using a jumper wire, short the check connector terminals

    f. Using a jumper wire, short terminals A–C and terminal E–body ground.
    g. Connect the negative battery cable.
    h. Turn the ignition switch to **ON** position to operate the fuel pump.

WARNING

The fuel pump could be damaged if it is operated (fuel pump idling) while there is no fuel in the fuel tank. Constantly monitor the amount of fuel being discharged and immediately stop operation of the pump when the fuel discharge amount becomes unstable.

    i. When essentially no fuel is being discharged, stop operation of the fuel pump.
    j. Disconnect the negative battery cable.
7. Remove the second-row seat.
8. Remove the edge cover.
9. Remove the long slider cover.
10. Remove the rear heat duct No.4.
11. Remove the screw.
12. Remove the service hole cover.
13. Disconnect the fuel level sender connector
14. Remove the screw and the fuel gauge sender sub-unit.

To install:

15. Install the fuel gauge sender sub-unit. Tighten the attaching screw to 10–19 inch lbs. (1–2 Nm).
16. Connect the fuel level sender connector.
17. Install the service hole cover.
18. Install the screw.
19. Install the rear heat duct No.4.
20. Install the long slider cover.
21. Install the edge cover.
22. Install the second-row seat.
23. Set the air cleaner cover aside.
24. Install the engine cover.
25. Connect the fuel line hose.
26. Connect the negative battery cable.
27. Perform a fuel leakage test as follows:
    a. Remove the fuel pump relay.

WARNING

Short the specified terminals because shorting the wrong terminal of the main fuse block may cause malfunctions.

    b. Using a jumper wire, short fuel pump relay terminals C and D in the main fuse block.
    c. Connect the negative battery cable and operate the fuel pump.
    d. Verify that there is no fuel leakage from the pressurized parts.

e. If there is leakage, replace the fuel hoses and clips.

f. If there is damage on the seal on the fuel pipe side, replace the fuel pipe.

g. There should be no leakage after 5 minutes.

h. After making any repairs, assemble the system and repeat the leakage test.

HEATED OXYGEN (HO2S) SENSOR

LOCATION

See Figures 151 and 152.

REMOVAL & INSTALLATION

1. Disconnect the negative battery cable.
2. On the left side, remove the engine cover.
3. On the right side, remove the cowl panel.
4. Disconnect the HO2S connector.

➡If necessary, lubricate the HO2S with Penetrating and Lock Lubricant loosen to aid in removal.

5. Remove the HO2S using the special tool.

To install:

6. Apply a light coat of anti-seize lubricant to the threads of the HO2S.
7. Install the HO2S using the special tool. Tighten to 30–40 ft. lbs. (40–55 Nm).
8. Connect the HO2S connector.
9. On the right side, Install the cowl panel.
10. On the left side, Install the engine cover.
11. Connect the negative battery cable.

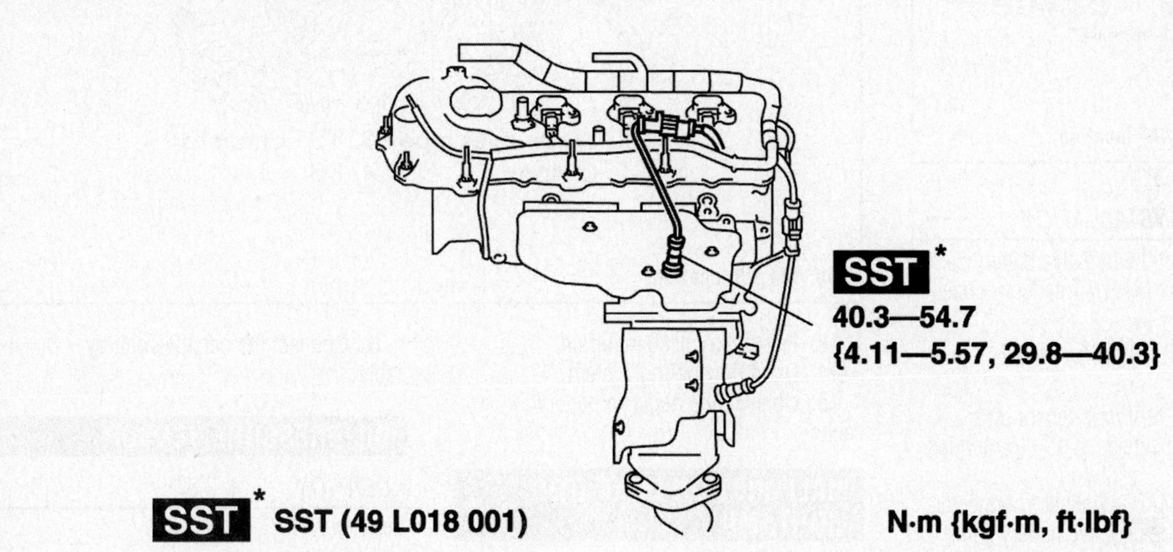

Fig. 151 Front HO2S location

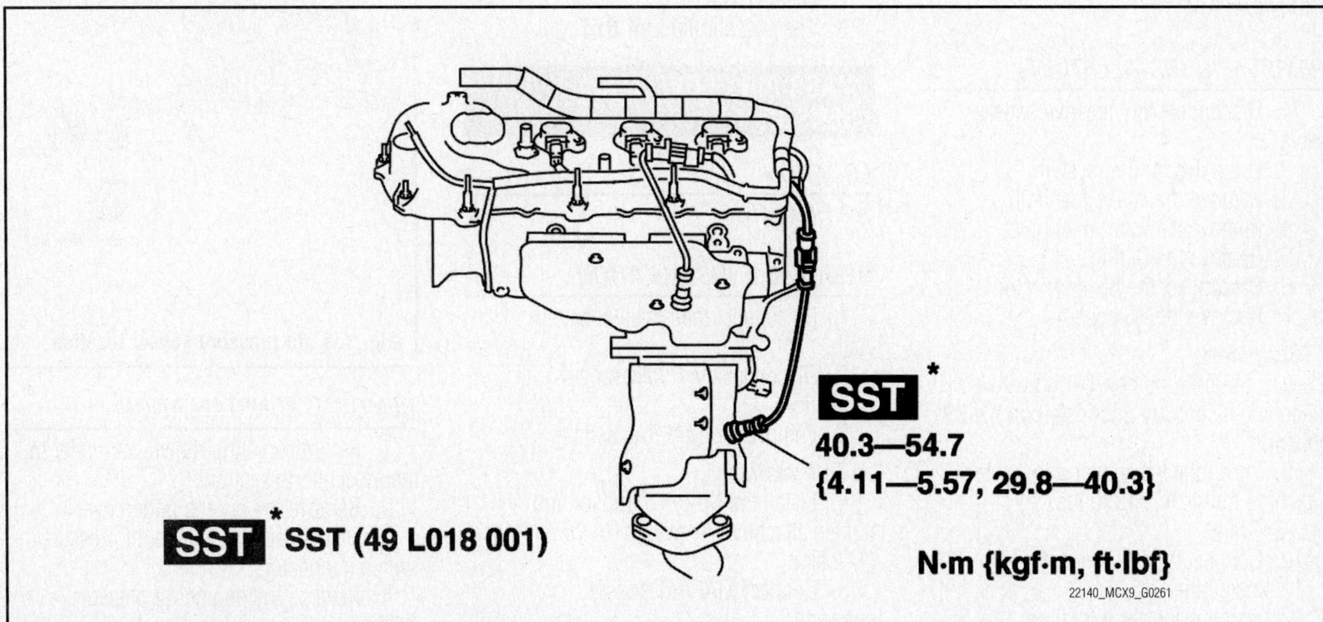

Fig. 152 Rear HO2S location

INTAKE AIR TEMPERATURE (IAT) SENSOR

LOCATION

See Figure 153.

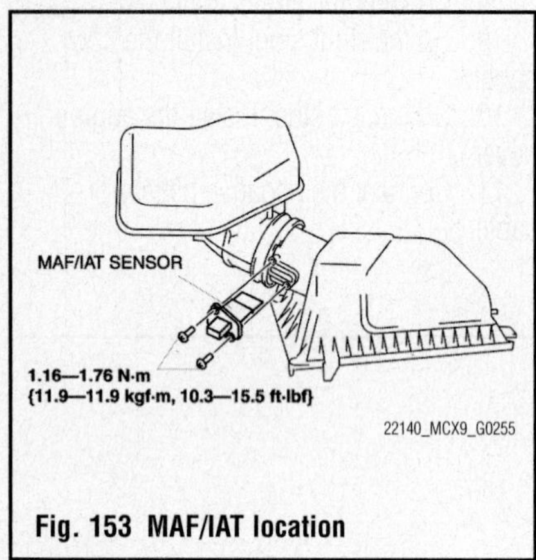

Fig. 153 MAF/IAT location

REMOVAL & INSTALLATION

1. Disconnect the negative battery cable.
2. Disconnect MAF/IAT sensor connector.
3. Remove the MAF/IAT sensor.

To install:

4. Install the MAF/IAT sensor and tighten attaching screws to 10–16 inch lbs. (1–2 Nm).
5. Connect MAF/IAT sensor connector.
6. Connect the negative battery cable.

KNOCK SENSOR (KS)

LOCATION

See Figure 154.

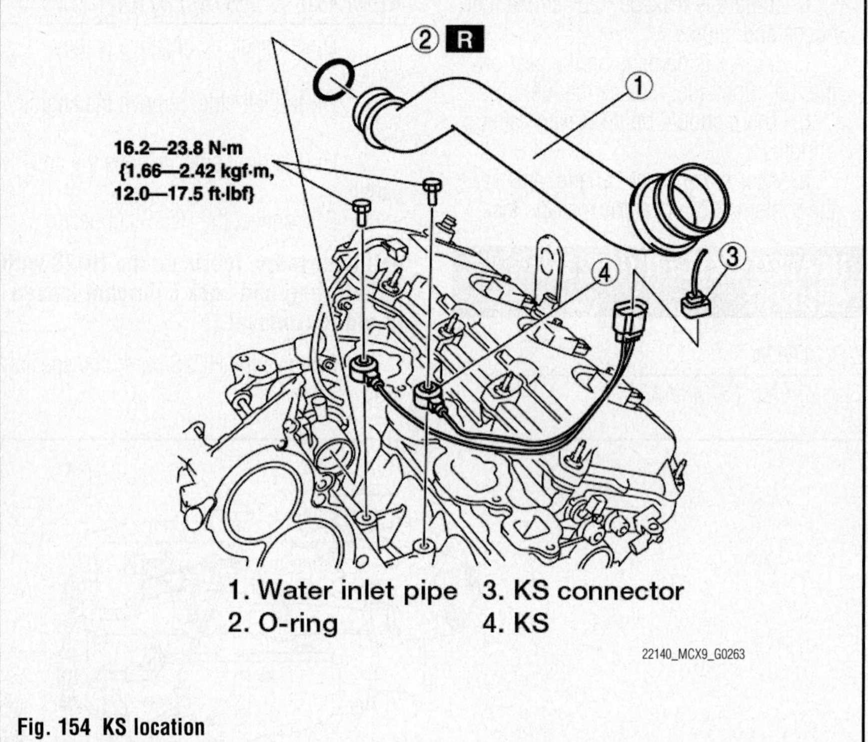

Fig. 154 KS location

REMOVAL & INSTALLATION

1. Disconnect the negative battery cable.
2. Drain the cooling system.
3. Remove the intake manifold.
4. Remove the water inlet pipe.
5. Remove the O-ring.
6. Disconnect the KS connector.
7. Remove the KS sensor

To install:

8. Lubricate the new O-ring with clean engine coolant and install on the KS sensor.
9. Install the KS sensor and tighten attaching bolt to 12–18 ft. lbs. (16–24 Nm).
10. Connect the KS connector.
11. Install the O-ring.
12. Install the water inlet pipe.
13. Install the intake manifold.
14. Drain the cooling system.
15. Connect the negative battery cable.

MALFUNCTION INDICATOR LIGHT (MIL)

RESET PROCEDURES

1. Connect a scan tool to the diagnostic connector.
2. Clear DTCs.
3. The MIL should turn **OFF**.

MASS AIR FLOW (MAF) SENSOR

LOCATION

See Figure 153.

REMOVAL & INSTALLATION

1. Disconnect the negative battery cable.
2. Disconnect MAF/IAT sensor connector.
3. Remove the MAF/IAT sensor.

To install:

4. Install the MAF/IAT sensor and tighten attaching screws to 10–16 inch lbs. (1–2 Nm).
5. Connect MAF/IAT sensor connector.
6. Connect the negative battery cable.

OIL PRESSURE SENSOR

LOCATION

See Figure 155.

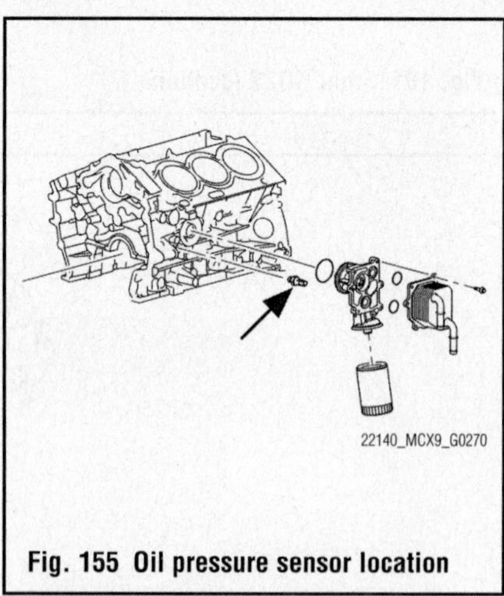

Fig. 155 Oil pressure sensor location

REMOVAL & INSTALLATION

1. Before servicing the vehicle, refer to the precautions section.
2. Remove the engine undercover.
3. Disconnect the engine oil pressure switch connector.
4. Remove the engine oil pressure switch.

To install:

5. Apply Three bond 1215 or equivalent sealant to 2 or 3 threads of the sensor.

6. Install the sensor and tighten to 89 inch lbs. (10 Nm).

WARNING

Do not start the engine for 2 hour or more after the engine oil pressure switch installation.

7. Connect the engine oil pressure switch connector.

8. Install the engine undercover.

POWERTRAIN CONTROL MODULE (PCM)

LOCATION

See Figure 156.

REMOVAL & INSTALLATION

1. Before servicing the vehicle, refer to the precautions section.

2. Using a scan tool, perform the PCM configuration.

3. Disconnect the negative battery cable.

4. Remove the battery and battery tray.

5. Disconnect the PCM connectors.

6. Remove PCM and bracket as and assembly.

7. Remove PCM from bracket.

To install:

➡When replacing the PCM on the vehicles, perform the PCM parameter reset.

8. Install PCM from bracket. Tighten to 70–95 inch lbs. (8–11 Nm).

9. Install PCM and bracket as and assembly. Tighten to 70–95 inch lbs. (8–11 Nm).

10. Connect the PCM connectors.

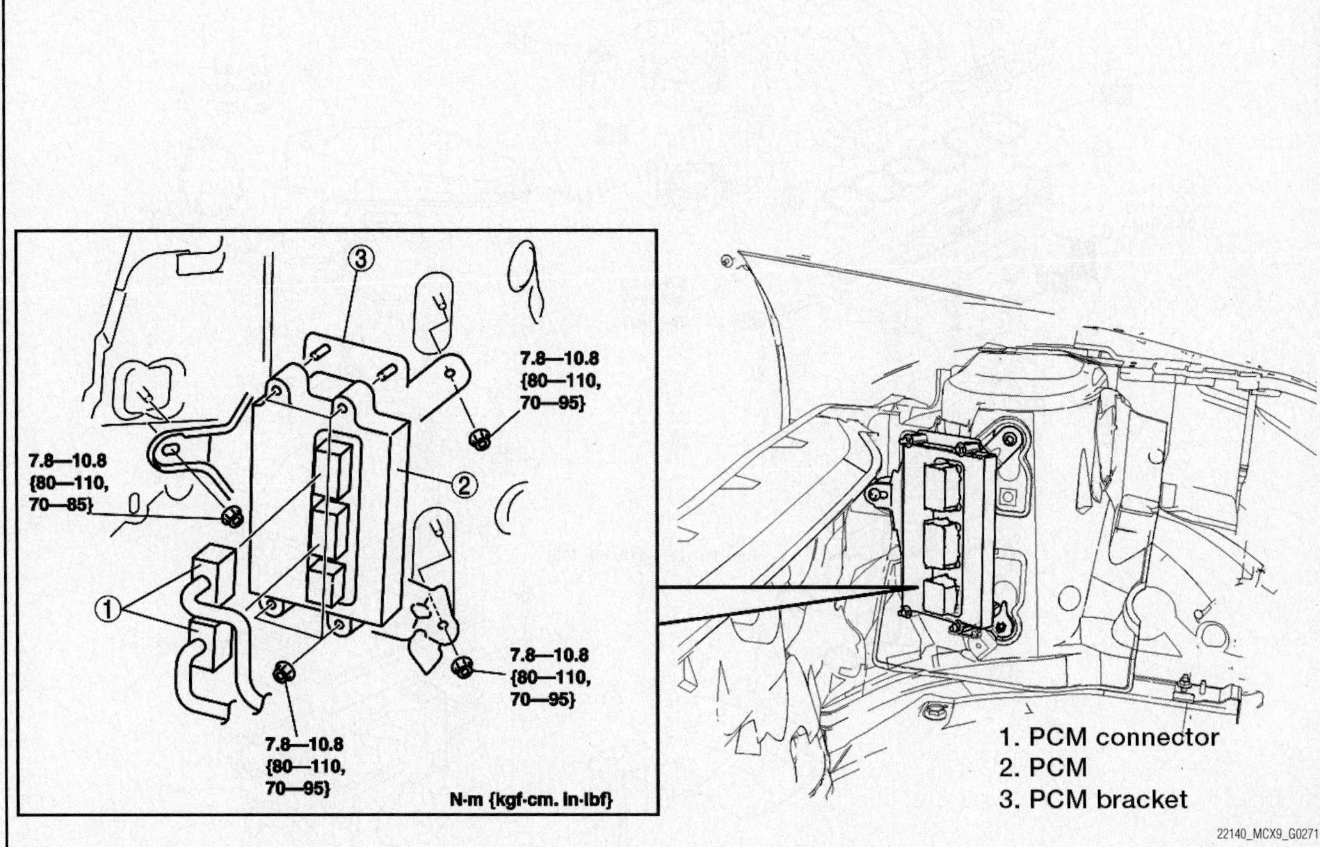

Fig. 156 PCM location

11. Install the battery and battery tray.
12. Connect the negative battery cable.
13. Using a scan tool, perform the PCM configuration.

THROTTLE POSITION SENSOR (TPS)

LOCATION

See Figure 157.

REMOVAL & INSTALLATION

1. Before servicing the vehicle, refer to the precautions section.
2. Disconnect the negative battery cable.
3. Remove the air cleaner case.
4. Remove the fresh-air duct.
5. Remove the air cleaner element.
6. Disconnect the MAF/IAT sensor connector.
7. Remove the air cleaner cover.
8. Disconnect the vacuum hose to master back.
9. Disconnect the ventilation hose.
10. Remove the resonance chamber.
11. Disconnect the TPS sensor connector.
12. Remove the throttle body.

To install:

13. Install the throttle body. Tighten attaching bolts to 76–101 inch lbs. (9–12 Nm).

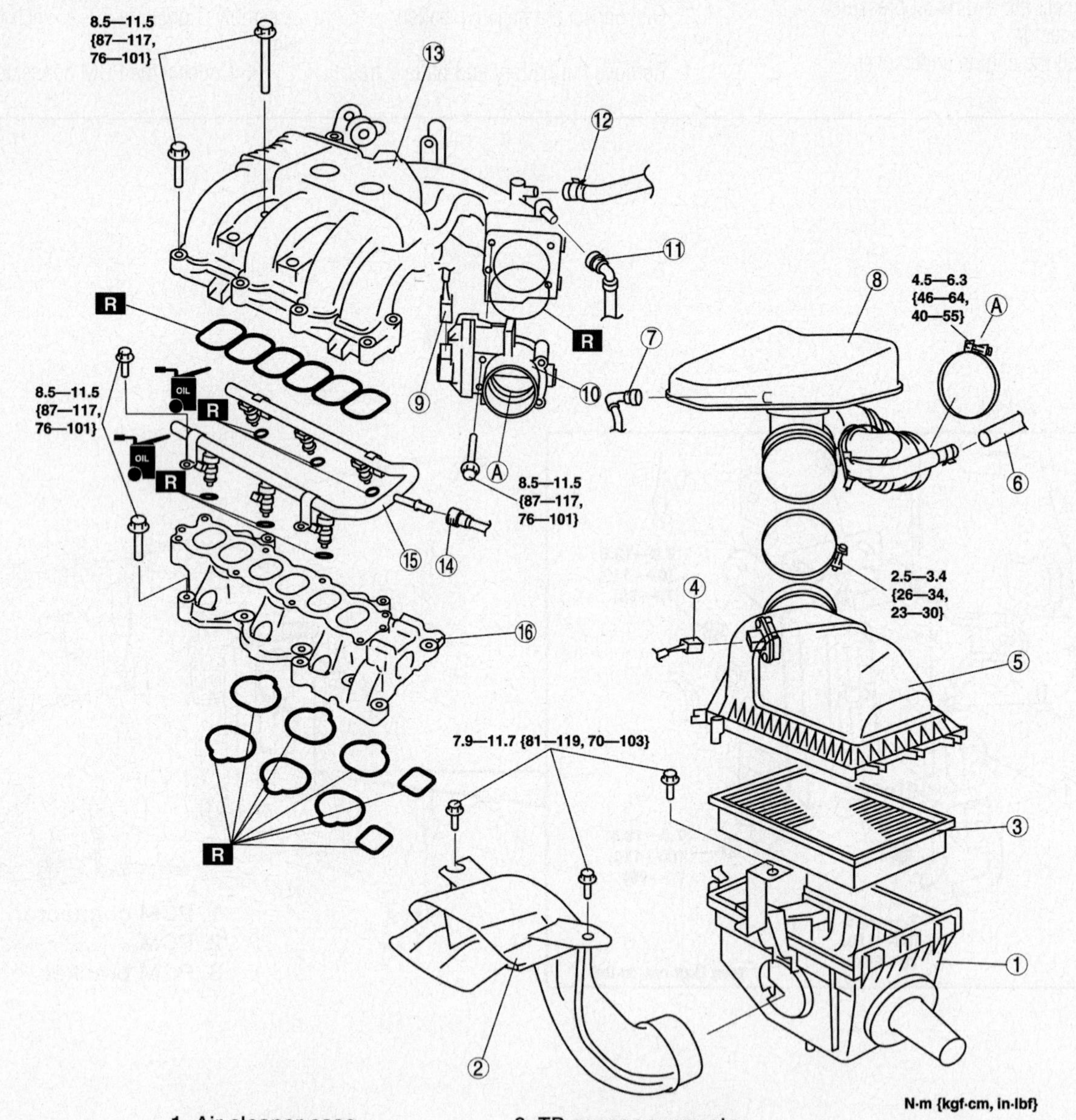

Fig. 157 Throttle Position (TP) sensor and related components

WHEEL ALIGNMENT

Year	Model		Caster Range (+/-Deg.)	Caster Preferred Setting (Deg.)	Camber Range (+/-Deg.)	Camber Preferred Setting (Deg.)	Toe-in (Deg.)
2007	RX-8	Front	1.0	① ②	1.0	③ ④	0° 11′±21′
		Rear	—	—	1.0	⑤ ⑥	0° 16′±20′
2008	RX-8	Front	1.0	① ②	1.0	③ ④	0° 11′±21′
		Rear	—	—	1.0	⑤ ⑥	0° 16′±20′

Note Unloaded vehicle: Fuel tank is full. Engine coolant and engine oil are at specified level. Jack and tools are in designated position.

The following specifications are for the front Caster settings for Standard Suspension

① 1: 6°31′±1° - Vehicle height* - 14.2—14.5 inches

2: 6°18′±1 - Vehicle height* - 14.6—14.9 inches

3: 6°06′±1° - Vehicle height* - 15.0—15.3 inches

4: 5°53′±1° - Vehicle height* - 15.4—15.7 inches

5: 5°40′±1° - Vehicle height* - 15.8—16.1 inches

The following specifications are for the front Caster settings for Sport Suspension

② 1: 6°41′±1° - Vehicle height* - 13.9—14.2 inches

2: 6°28′±1 - Vehicle height* - 14.3—14.6inches

3: 6°16′±1° - Vehicle height* - 14.7—15.0 inches

4: 6°03′±1° - Vehicle height* - 15.1—15.4 inches

5: 5°50′±1° - Vehicle height* - 15.5—15.8 inches

The following specifications are for the front Camber settings on Standard Suspension

③ 1: -0°33′±1° - Vehicle height* - 14.4—14.8 inches

2: -0°13′±1 - Vehicle height* - 14.9—15.1 inches

3: 0°04′±1° - Vehicle height* - 15.2—15.5 inches

4: -0°20′±1° - Vehicle height* - 15.6—15.9 inches

5: -0°33′±1° - Vehicle height* -16.0—16.3 inches

The following specifications are for the front Camber settings on Sport Suspension

④ 1: 0°45′±1° - Vehicle height* - 1 4.2—14.5 inches

2: 0°25′±1 - Vehicle height* - 14.6—14.9 inches

3: 0°06′±1° - Vehicle height* - 15.0—15.3 inches

4: -0°11′±1° - Vehicle height* - 15.4—15.7 inches

5: -0°26′±1° - Vehicle height* - 15.8—16.1 inches

The following specifications are for the rear Camber settings for Standard Suspension

⑤ 1: -1°30′±1° - Vehicle height* - 14.2—14.5 inches

2: -1°12′±1 - Vehicle height* - 14.6—14.9 inches

3: 0°56′±1° - Vehicle height* - 15.0—15.3 inches

4: -0°43′±1° - Vehicle height* - 15.4—15.7 inches

5: -0°33′±1° - Vehicle height* - 15.8—16.1 inches

The following specifications are for the rear Camber settings on the Sport Suspension

⑥ 1: -1°44′±1° - Vehicle height* - 14.2—14.5 inches

2: -1°24′±1 - Vehicle height* - 14.6—14.9 inches

3: 0°07′±1° - Vehicle height* - 15.0—15.3 inches

4: -0°52′±1° - Vehicle height* - 15.4—15.7 inches

5: -0°40′±1° - Vehicle height* - 15.8—16.1 inches

* : From the end of the front fender to the center of the wheel inches

22140_MRX8_C0011

TIRE AND WHEEL SPECIFICATIONS

Year	Model	OEM Tires		Tire Pressures (psi)		Wheel	Lug Nut
		Standard	Optional	Front	Rear	Size	Torque (ft. lbs.)
2007	RX-8	P225/55R16	P225/45R18	32	32	① ②	65-87
2008	RX-8	P225/55R16	P225/45R18	32	32	① ②	65-87

OEM: Original Equipment Manufacturer

PSI: Pounds Per Square Inch

① 16 X 7 inches standard

② 18X8 optional

22140_MRX8_C0013

BRAKE SPECIFICATIONS

All measurements in inches unless noted

Year	Model		Brake Disc			Minimum Lining Thickness		Brake Caliper	
			Original Thickness	Minimum Thickness	Maximum Runout	Front	Rear	Bracket Bolts (ft. lbs.)	Mounting Bolts (ft. lbs.)
2007	RX-8	F	①	0.870	0.002	0.790	—	58-75	16-23
		R	①	0.630	0.002	—	0.079	36-51	16-23
2008	RX-8	F	①	0.870	0.002	0.790	—	58-75	16-23
		R	①	0.630	0.002	—	0.079	36-51	16-23

① Not available

22140_MRX8_C0012

SCHEDULED MAINTENANCE INTERVALS

MAZDA RX-8 2007- 2008

TO BE SERVICED	TYPE OF SERVICE	VEHICLE MILEAGE INTERVAL (x1000)											
		7.5	15	22.5	30	37.5	45	52.5	60	67.5	75	82.5	90
Engine oil & filter	R	✓	✓	✓	✓	✓	✓	✓	✓	✓	✓	✓	✓
Cabin air filter	R			✓			✓				✓		
Engine coolant strength hoses & clamps	S/I				✓				✓				✓
Air cleaner filter	R					✓					✓		
Brake fluid	R				✓				✓				✓
Engine coolant ① ②	R								✓				
Spark plugs	R					✓					✓		
Drive belts	S/I				✓				✓				✓
Exhaust system & heat shields	S/I				✓				✓				✓
Manual transmission oil	R								✓				
Rear differential oil	R								✓				
Front & rear brakes	S/I			✓			✓			✓			✓
Fuel filter	R												

R: Replace S/I: Service or Inspect

① Engine coolant: Use FL22 type coolant in vehicles with the inscription "FL22" on the radiator cap itself or the surrounding area. Use FL22 when replacing the coolant. For F22 change initially at 120,000 miles or 10 ten years and every 60,000 miles or years thereafter.

② All other coolants must initally be changes at 60,000 miles or 4 years or 10 ten years and every 2 years years thereafter.

22140_MRX8_C0014

SCHEDULED MAINTENANCE INTERVALS

MAZDA RX-8 2007- 2008

TO BE SERVICED	TYPE OF SERVICE	VEHICLE MILEAGE INTERVAL (x1000)											
		7.5	15	22.5	30	37.5	45	52.5	60	67.5	75	82.5	90
Brake lines, hoses and connections	I				✓				✓				✓
Disc brakes	S/I		✓		✓		✓		✓		✓		✓
Tire (Rotation)	S/I	✓	✓	✓	✓	✓	✓	✓	✓	✓	✓	✓	✓
Steering operation and linkages	S/I				✓				✓				✓
Front and rear suspension, ball joints and wheel bearing axial play	S/I				✓				✓				✓
Manual transmission oil	R								✓				
Rear differential oil	R								✓				
Driveshaft dust boots	I				✓				✓				✓
Exhaust system and heat shields	I						✓		✓				✓
All locks and hinges	S/I	✓	✓	✓	✓	✓	✓	✓	✓	✓	✓	✓	✓
Chassis/body bolts and nuts	S/I			✓			✓			✓			✓

R: Replace S/I: Service or Inspect

FREQUENT OPERATION MAINTENANCE (SEVERE SERVICE)

If a vehicle is operated under any of the following conditions it is considered severe service:

- Extremely dusty areas.

- 50% or more of the vehicle operation is in 32°C (90°F) or higher temperatures, or constant operation in temperatures below 0°C (32°F).

- Prolonged idling (vehicle operation in stop and go traffic.

- Frequent short running periods (engine does not warm to normal operating temperatures).

- Police, taxi, delivery usage or trailer towing usage.

Engine oil & filter: replace every 5000 miles.

Air cleaner filter: change every 35,000 miles.

Engine coolant: replace every 60,000 miles.

Exhaust system: check every 30,000 miles.

22140_MRX8_C0015

PRECAUTIONS

Before servicing any vehicle, please be sure to read all of the following precautions, which deal with personal safety, prevention of component damage, and important points to take into consideration when servicing a motor vehicle:

- Never open, service or drain the radiator or cooling system when the engine is hot; serious burns can occur from the steam and hot coolant.
- Observe all applicable safety precautions when working around fuel. Whenever servicing the fuel system, always work in a well-ventilated area. Do not allow fuel spray or vapors to come in contact with a spark, open flame, or excessive heat (a hot drop light, for example). Keep a dry chemical fire extinguisher near the work area. Always keep fuel in a container specifically designed for fuel storage; also, always properly seal fuel containers to avoid the possibility of fire or explosion. Refer to the additional fuel system precautions later in this section.
- Fuel injection systems often remain pressurized, even after the engine has been turned **OFF**. The fuel system pressure must be relieved before disconnecting any fuel lines. Failure to do so may result in fire and/or personal injury.
- Brake fluid often contains polyglycol ethers and polyglycols. Avoid contact with the eyes and wash your hands thoroughly after handling brake fluid. If you do get brake fluid in your eyes, flush your eyes with clean, running water for 15 minutes. If eye irritation persists, or if you have taken brake fluid internally, IMMEDIATELY seek medical assistance.
- The EPA warns that prolonged contact with used engine oil may cause a number of skin disorders, including cancer. You should make every effort to minimize your exposure to used engine oil. Protective gloves should be worn when changing oil. Wash your hands and any other exposed skin areas as soon as possible after exposure to used engine oil. Soap and water, or waterless hand cleaner should be used.
- All new vehicles are now equipped with an air bag system, often referred to as a Supplemental Restraint System (SRS) or Supplemental Inflatable Restraint (SIR) system. The system must be disabled before performing service on or around system components, steering column, instrument panel components, wiring and sensors. Failure to follow safety and disabling procedures could result in accidental air bag deployment, possible personal injury and unnecessary system repairs.
- Always wear safety goggles when working with, or around, the air bag system. When carrying a non-deployed air bag, be sure the bag and trim cover are pointed away from your body. When placing a non-deployed air bag on a work surface, always face the bag and trim cover upward, away from the surface. This will reduce the motion of the module if it is accidentally deployed. Refer to the additional air bag system precautions later in this section.
- Clean, high quality brake fluid from a sealed container is essential to the safe and proper operation of the brake system. You should always buy the correct type of brake fluid for your vehicle. If the brake fluid becomes contaminated, completely flush the system with new fluid. Never reuse any brake fluid. Any brake fluid that is removed from the system should be discarded. Also, do not allow any brake fluid to come in contact with a painted surface; it will damage the paint.
- Never operate the engine without the proper amount and type of engine oil; doing so WILL result in severe engine damage.
- Timing belt maintenance is extremely important. Many models utilize an interference-type, non-freewheeling engine. If the timing belt breaks, the valves in the cylinder head may strike the pistons, causing potentially serious (also time-consuming and expensive) engine damage. Refer to the maintenance interval charts for the recommended replacement interval for the timing belt, and to the timing belt section for belt replacement and inspection.
- Disconnecting the negative battery cable on some vehicles may interfere with the functions of the on-board computer system(s) and may require the computer to undergo a relearning process once the negative battery cable is reconnected.
- When servicing drum brakes, only disassemble and assemble one side at a time, leaving the remaining side intact for reference.
- Only an MVAC-trained, EPA-certified automotive technician should service the air conditioning system or its components.

BRAKES

ANTI-LOCK BRAKE SYSTEM (ABS)

GENERAL INFORMATION

The Anti-lock Brake System (ABS) module simultaneously manages the anti-lock braking, traction control and engine control systems to maintain vehicle control during deceleration and acceleration.

When the ignition switch is in the **RUN** position, the module carries out a preliminary electrical check and, at approximately 12 MPH (20 km/h), the hydraulic pump motor is turned on for approximately one half-second. Any malfunction of the anti-lock brake system disables the traction control and stability assist (if equipped) and the anti-lock brake warning indicator illuminates. However, the power-assist braking system functions normally.

WHEEL SPEED SENSORS

REMOVAL & INSTALLATION

Front

See Figure 1.

➡If there is any malfunction in the front ABS wheel-speed sensor unit, replace the wheel hub, as outlined in the Suspension Section.

1. Raise and safely support the vehicle.
2. Remove the wheel and tire assembly.
3. Remove the mudguard.
4. Detach the ABS speed sensor connector.
5. Remove the bolt and the wire harness.
6. Installation is the reverse of the removal procedure. Refer to the tightening specifications on the accompanying illustration.

Rear

See Figure 2.

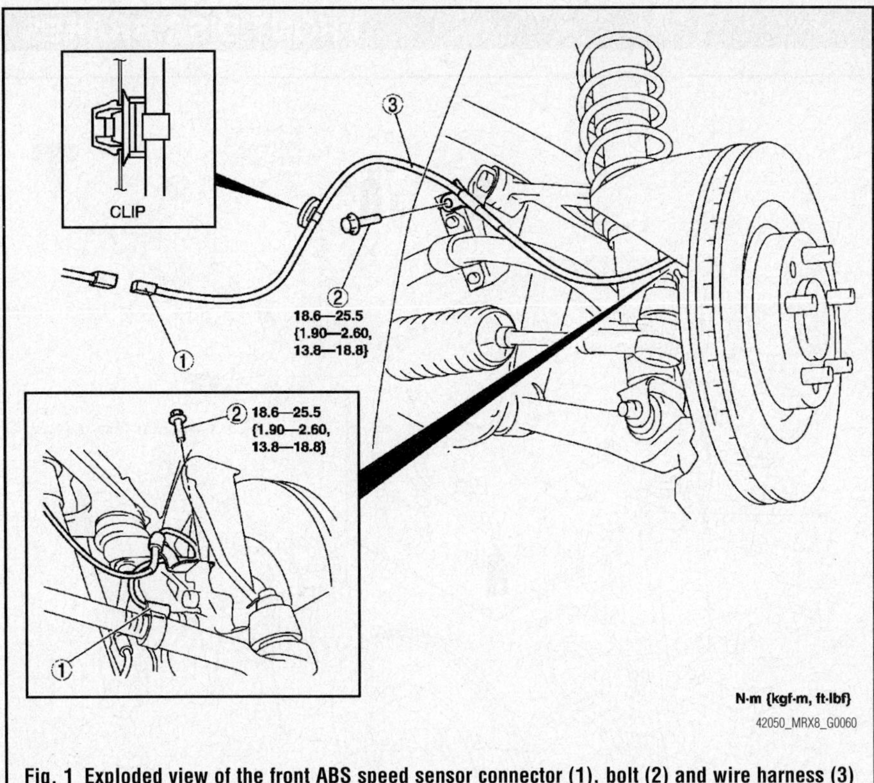

Fig. 1 Exploded view of the front ABS speed sensor connector (1), bolt (2) and wire harness (3)

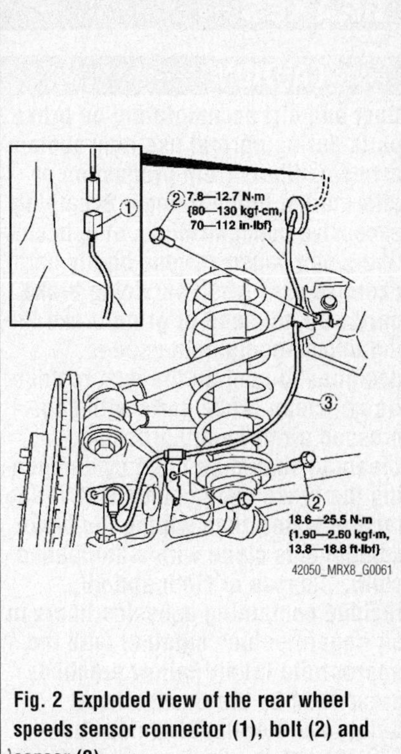

Fig. 2 Exploded view of the rear wheel speeds sensor connector (1), bolt (2) and sensor (3)

1. Raise and safely support the vehicle.
2. Remove the wheel and tire assembly.
3. Remove the trunk side trim.
4. Detach the sensor electrical connector.
5. Unfasten the retaining bolt.
6. Remove the rear wheel speed sensor.
7. Installation is the reverse of the removal procedure. Refer to the tightening specifications on the accompanying illustration.

BRAKES — BLEEDING THE BRAKE SYSTEM

BLEEDING PROCEDURE

See Figure 3.

✲✲ WARNING

Brake fluid will damage painted surfaces. Be careful not to spill any on painted surfaces. If it is spilled, wipe it off immediately.

✲✲ WARNING

Keep the fluid level in the reserve tank at 3/4 full or more during the air bleeding.

1. Begin air bleeding with the master cylinder and then continue with the brake caliper that is furthest away from the master cylinder. Finish by bleeding air from the master cylinder again.

➡Only use SAE J1703, FMVSS 116 DOT-3 Brake Fluid.

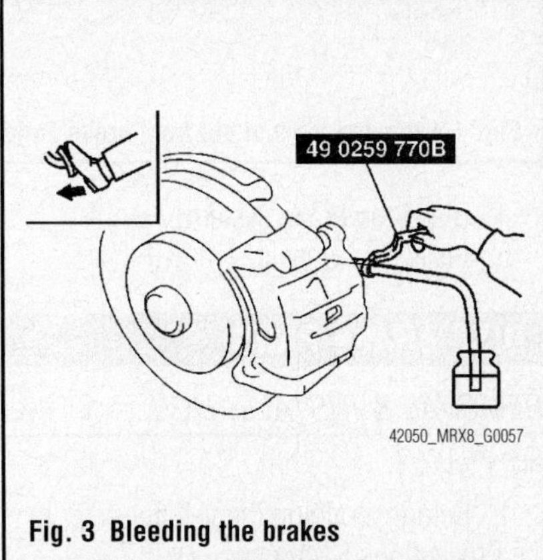

Fig. 3 Bleeding the brakes

2. Remove the bleeder cap from the brake caliper, and connect a vinyl tube to the bleeder screw.
3. Place the other end of the vinyl tube in a clear container, and fill the container with fluid during air bleeding.
4. Working with two people, one should depress the brake pedal a few times and then depress and hold the pedal down.
5. While the brake pedal is being held down, the other person should loosen the bleeder screw using the SST, and bleed any fluid containing air bubbles. Once completed, tighten the bleeder screw to 62–86 inch lbs. (6.9–9.8 Nm).
6. Repeat Steps 3 and 4 until no air bubbles are seen.
7. Perform the bleeding procedure outlined above for each of the brake calipers.
8. After air bleeding, inspect the following:
   a. Brake operation
   b. Fluid leakage
   c. Fluid level

BLEEDING THE ABS SYSTEM

There is no special bleeding procedure for the ABS system. The system is bled using the bleeding procedure located at the beginning of this section.

✻✻ CAUTION

Dust and dirt accumulating on brake parts during normal use may contain asbestos fibers from production or aftermarket brake linings. Breathing excessive concentrations of asbestos fibers can cause serious bodily harm. Exercise care when servicing brake parts. Do not sand or grind brake lining unless equipment used is designed to contain the dust residue. Do not clean brake parts with compressed air or by dry brushing. Cleaning should be done by dampening the brake components with a fine mist of water, then wiping the brake components clean with a dampened cloth. Dispose of cloth and all residue containing asbestos fibers in an impermeable container with the appropriate label. Follow practices prescribed by the Occupational Safety and Health Administration (OSHA) and the Environmental Protection Agency (EPA) for the handling, processing, and disposing of dust or debris that may contain asbestos fibers.

BRAKE CALIPER

REMOVAL & INSTALLATION

See Figure 4.

1. Before servicing the vehicle, refer to the Precautions Section.
2. Remove or disconnect the following:
   - Wheels
   - Flexible brake hose from the caliper
   - Caliper bolt
   - Caliper

To install:

3. Install or connect the following:
   - Caliper on the brake disc
   - Caliper mounting bolts and tighten the bolts 16–23 ft. lbs. (22–31 Nm)
   - Brake hose to the caliper and tighten the hose nut to 16–21 ft. lbs. (22–29 Nm)

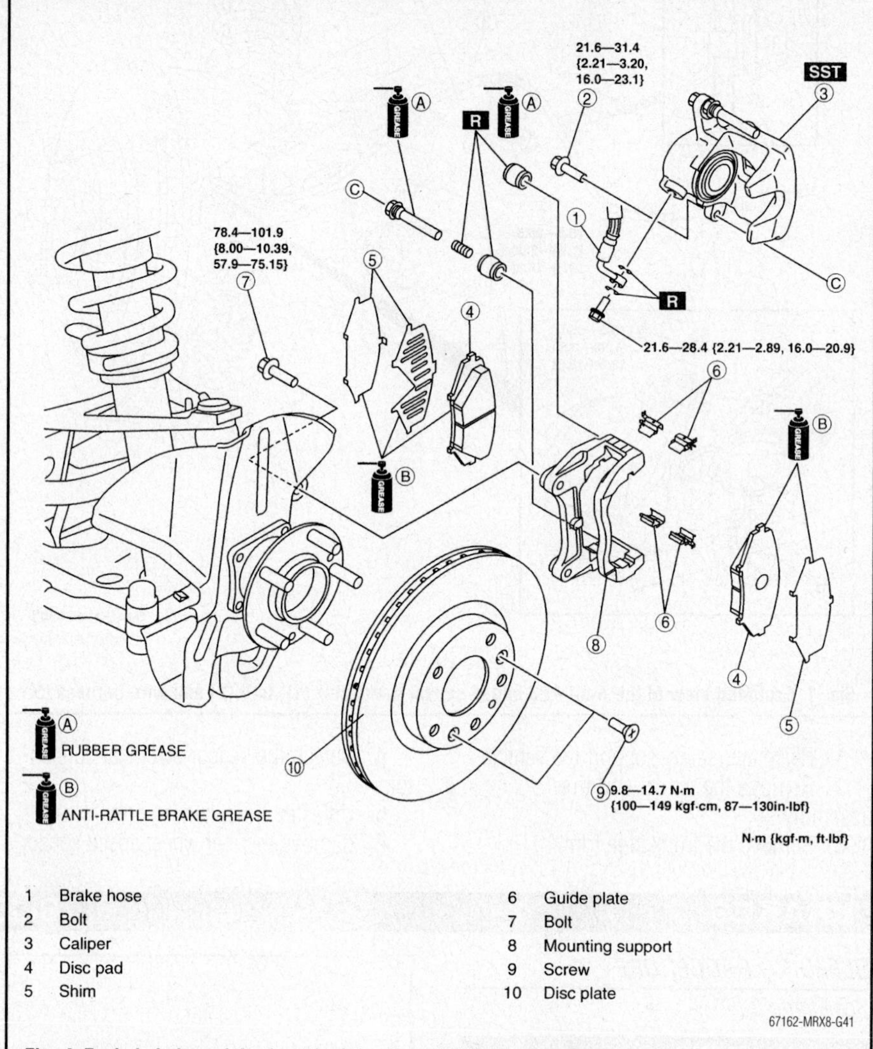

Fig. 4 Exploded view of the front brake caliper–RX-8

4. Bleed the brake system.
5. Install the wheels.

DISC BRAKE PADS

REMOVAL & INSTALLATION

See Figure 4.

1. Before servicing the vehicle, refer to the Precautions Section.
2. Remove or disconnect the following:
   - Wheels
   - Caliper
   - Brake pads
   - Shim
   - Guide plate

To install:

3. Install or connect the following:
   - Guide plate
   - Shim
   - Brake pads
   - Caliper and tighten the bolts to 16–23 ft. lbs. (22–32 Nm).
4. Install the wheels.

⁂ CAUTION

Dust and dirt accumulating on brake parts during normal use may contain asbestos fibers from production or aftermarket brake linings. Breathing excessive concentrations of asbestos fibers can cause serious bodily harm. Exercise care when servicing brake parts. Do not sand or grind brake lining unless equipment used is designed to contain the dust residue. Do not clean brake parts with compressed air or by dry brushing. Cleaning should be done by dampening the brake components with a fine mist of water, then wiping the brake components clean with a dampened cloth. Dispose of cloth and all residue containing asbestos fibers in an impermeable container with the appropriate label. Follow practices prescribed by the Occupational Safety and Health Administration (OSHA) and the Environmental Protection Agency (EPA) for the handling, processing, and disposing of dust or debris that may contain asbestos fibers.

BRAKE CALIPER

REMOVAL & INSTALLATION

See Figure 5.

1. Before servicing the vehicle, refer to the Precautions Section.
2. Remove or disconnect the following:
   - Wheels
   - Flexible brake line from the caliper assembly
   - Caliper mounting bolts
   - Caliper

To install:

3. Install or connect the following:
   - Caliper. Torque the caliper mount bolts to 16–23 ft. lbs. (22–32 Nm).
   - Brake hose. Torque the line bolt to 16–22 ft. lbs. (22–30 Nm).
   - Parking brake cable

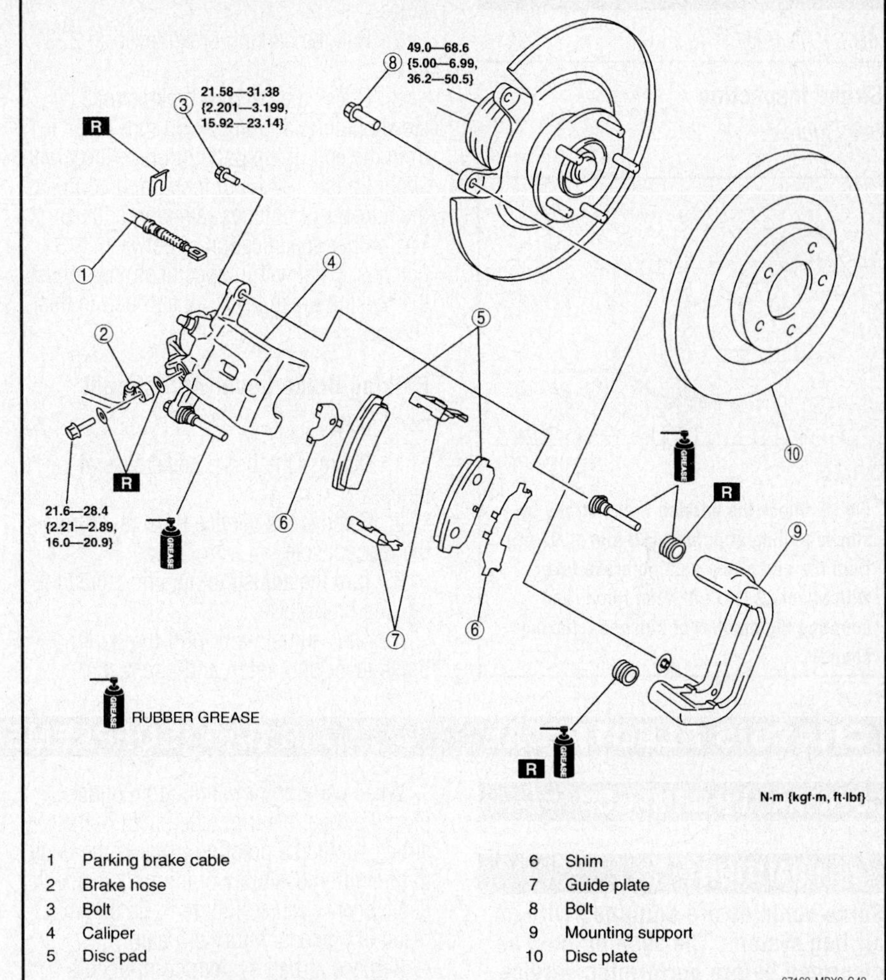

Fig. 5 Exploded view of rear brake system–RX-8

4. Install the wheels and bleed the brake system.

DISC BRAKE PADS

REMOVAL & INSTALLATION

See Figure 5.

1. Before servicing the vehicle, refer to the Precautions Section.
2. Remove or disconnect the following:
   - Wheels
   - Caliper
   - Brake pads
   - Shim
   - Guide plate

To install:

3. Install or connect the following:
   - Guide plate
   - Shim
   - Brake pads
   - Caliper and tighten the bolt to 16–23 ft. lbs. (22–32 Nm).
4. Install the wheels.

BRAKES

PARKING BRAKE

PARKING BRAKE CABLES

ADJUSTMENT

Stroke Inspection

See Figure 6.

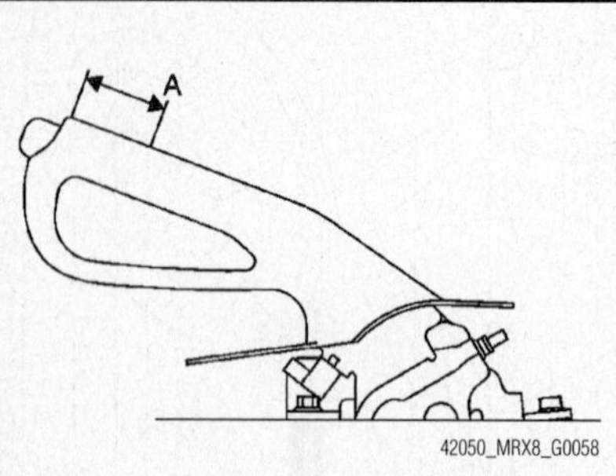

Fig. 6 Check the parking brake stroke by slowly pulling at point A (50 mm {1.97 in}) from the end of the parking brake lever with about 22 lbs. (98 N) of force, and counting the number of notches (clicking sound)

1. Depress the brake pedal several times.
2. Pull the parking brake lever up 2–3 times.
3. Check the parking brake stroke by slowly pulling at point A (50 mm {1.97 in}) from the end of the parking brake lever with about 22 lbs. (98 N) of force, and counting the number of notches (clicking sound). The proper specification is between 1–3 notches. If not within specifications, adjust the parking brake lever, as outlined in this section.

Parking Brake Lever Adjustment

See Figure 7.

1. Depress the brake pedal several times.
2. Remove the service hole cover from the rear console.
3. Turn the adjusting nut and adjust the parking brake lever.
4. After adjustment, pull the parking brake lever one notch and check that the parking brake warning light illuminates.
5. Make sure that the rear brakes do not drag.

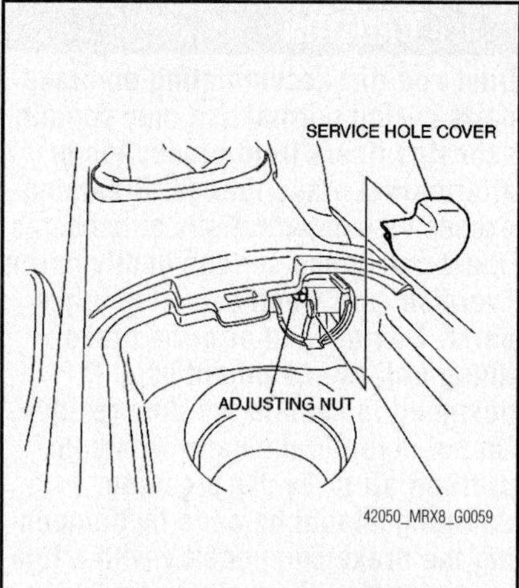

Fig. 7 Remove the service hole cover from the rear console to adjust the parking brake lever

CHASSIS ELECTRICAL

AIR BAG (SUPPLEMENTAL RESTRAINT SYSTEM)

GENERAL INFORMATION

⁂ CAUTION

Some vehicles are equipped with an air bag system. The system must be disarmed before performing service on, or around, system components, the steering column, instrument panel components, wiring and sensors. Failure to follow the safety precautions and the disarming procedure could result in accidental air bag deployment, possible injury and unnecessary system repairs.

SERVICE PRECAUTIONS

Disconnect and isolate the battery negative cable before beginning any airbag system component diagnosis, testing, removal, or installation procedures. Allow system capacitor to discharge for two minutes before beginning any component service. This will disable the airbag system. Failure to disable the airbag system may result in accidental airbag deployment, personal injury, or death.

Do not place an intact undeployed airbag face down on a solid surface. The airbag will propel into the air if accidentally deployed and may result in personal injury or death.

When carrying or handling an undeployed airbag, the trim side (face) of the airbag should be pointing towards the body to minimize possibility of injury if accidental deployment occurs. Failure to do this may result in personal injury or death.

Replace airbag system components with OEM replacement parts. Substitute parts may appear interchangeable, but internal differences may result in inferior occupant protection. Failure to do so may result in occupant personal injury or death.

Wear safety glasses, rubber gloves, and long sleeved clothing when cleaning powder residue from vehicle after an airbag deployment. Powder residue emitted from a deployed airbag can cause skin irritation. Flush affected area with cool water if irritation is experienced. If nasal or throat irritation is experienced, exit the vehicle for fresh air until the irritation ceases. If irritation continues, see a physician.

Do not use a replacement airbag that is not in the original packaging. This may result in improper deployment, personal injury, or death.

The factory installed fasteners, screws and bolts used to fasten airbag components have a special coating and are specifically designed for the airbag system. Do not use substitute fasteners. Use only original equipment fasteners listed in the parts catalog when fastener replacement is required.

During, and following, any child restraint anchor service, due to impact event or vehicle repair, carefully inspect all mounting hardware, tether straps, and anchors for proper installation, operation, or damage. If a child restraint anchor is found damaged in any way, the anchor must be replaced. Failure to do this may result in personal injury or death.

Deployed and non-deployed airbags may or may not have live pyrotechnic material within the airbag inflator.

Do not dispose of driver/passenger/curtain airbags or seat belt tensioners unless you are sure of complete deployment. Refer to the Hazardous Substance Control System for proper disposal.

Dispose of deployed airbags and tensioners consistent with state, provincial, local, and federal regulations.

After any airbag component testing or service, do not connect the battery negative cable. Personal injury or death may result if the system test is not performed first.

If the vehicle is equipped with the Occupant Classification System (OCS), do not connect the battery negative cable before performing the OCS Verification Test using the scan tool and the appropriate diagnostic information. Personal injury or death may result if the system test is not performed properly.

Never replace both the Occupant Restraint Controller (ORC) and the Occupant Classification Module (OCM) at the same time. If both require replacement, replace one, then perform the Airbag System test before replacing the other.

Both the ORC and the OCM store Occupant Classification System (OCS) calibration data, which they transfer to one another when one of them is replaced. If both are replaced at the same time, an irreversible fault will be set in both modules and the OCS may malfunction and cause personal injury or death.

If equipped with OCS, the Seat Weight Sensor is a sensitive, calibrated unit and must be handled carefully. Do not drop or handle roughly. If dropped or damaged, replace with another sensor. Failure to do so may result in occupant injury or death.

If equipped with OCS, the front passenger seat must be handled carefully as well. When removing the seat, be careful when setting on floor not to drop. If dropped, the sensor may be inoperative, could result in occupant injury, or possibly death.

If equipped with OCS, when the passenger front seat is on the floor, no one should sit in the front passenger seat. This uneven force may damage the sensing ability of the seat weight sensors. If sat on and damaged, the sensor may be inoperative, could result in occupant injury, or possibly death.

DISARMING THE SYSTEM

1. Before servicing the vehicle, refer to the Precautions Section.
2. If equipped, deactivate the audio anti-theft system.
3. Turn the ignition switch to **LOCK**.
4. Disconnect and isolate the negative battery cable and wait for more than 1 minute to allow the backup power supply to deplete its stored power.

ARMING

1. Before servicing the vehicle, refer to the Precautions Section.
2. Connect the negative battery cable, turn the ignition switch **ON** and verify the air bag warning light cones on for 6 seconds. If the light does not illuminate there are problems with the system.
3. If equipped, activate the audio anti-theft system.

CLOCKSPRING CENTERING

See Figures 8 through 10.

1. Before servicing the vehicle, refer to the Precautions Section.
2. Disconnect the negative battery.

➡The adjustment procedure is also specified on the caution label of the clock spring.

3. Set the front tires straight-ahead.

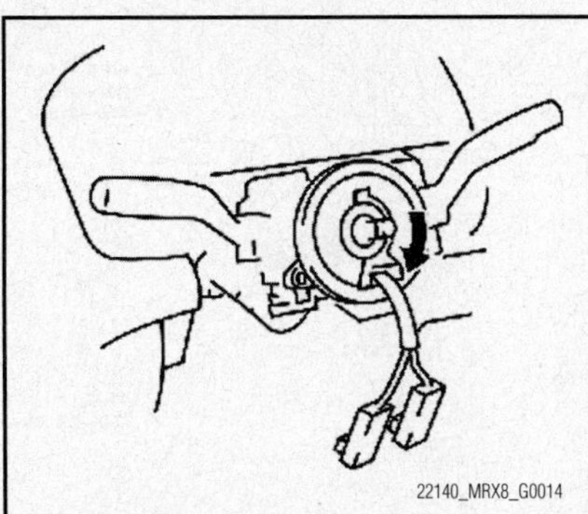

Fig. 8 Clock spring clockwise until it stops— not forcibly

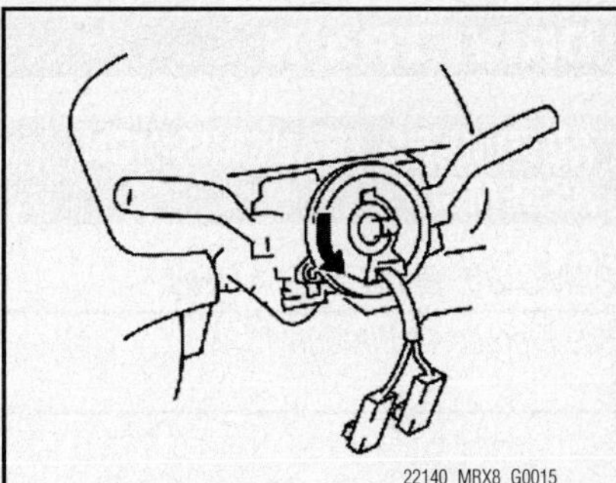

Fig. 9 Clock spring in the stopped position, turn the clock spring counterclockwise 2¾ turns.

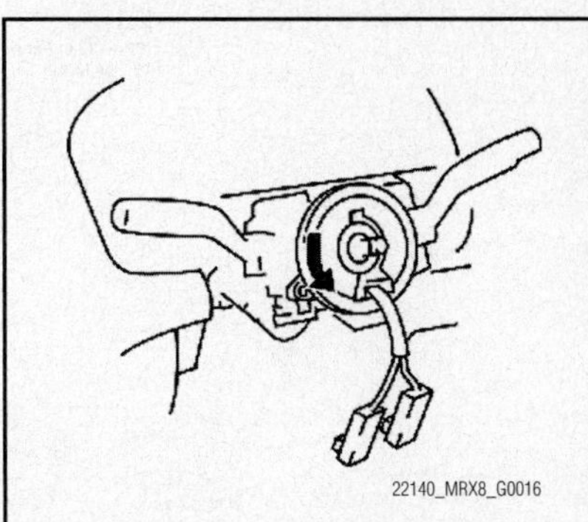

Fig. 10 Clock spring and housing marks aligned

4. The clock spring will break if overwound. Do not forcibly turn the clock spring.
5. Turn the clock spring clockwise until it stops— not forcibly.
6. From the stopped position, turn the clock spring counterclockwise 2¾ turns.
7. Align the marks.
8. Connect the negative battery cable.
9. Check for SIR. Malfunction light.

DRIVETRAIN

AUTOMATIC TRANSMISSION ASSEMBLY

REMOVAL & INSTALLATION

See Figures 11 through 13.

1. Before servicing the vehicle, refer to the Precautions Section.
2. Refer to the illustration for component location and torque specifications.
3. Drain the transaxle oil.
4. Remove or disconnect the following:
   - Engine cover
   - Battery cover.
   - Negative battery cable.
   - Front and rear tunnel supports

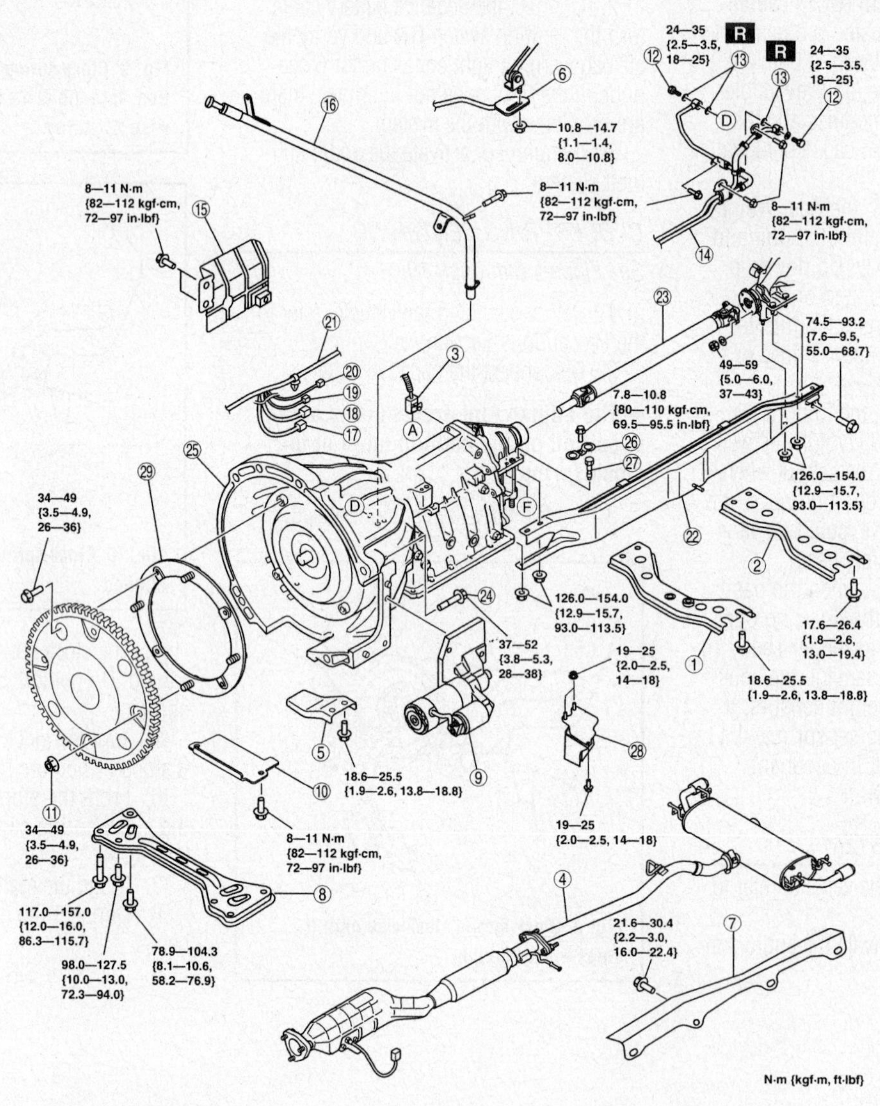

1 Front tunnel member
2 Rear tunnel member
3 Heated oxygen sensor connector
4 Catalytic converter, middle pipe, main silencer
5 Exhaust manifold stay
6 Manual shaft lever component
7 Heat insulator
8 Transverse member
9 Starter
10 Under cover
11 Torque converter installation nuts
12 Connector bolt
13 Washer
14 Oil pipe, oil hose
15 Insulator
16 Oil filter tube, Dipstick
17 TR switch connector
18 Solenoid valve connector
19 VSS connector
20 Turbine sensor connector
21 Wire
22 Power plant frame
23 Propeller shaft
24 Transmission installation bolt
25 Transmission
26 Stopper
27 Bolt
28 Dynamic dumper
29 Driven plate

67162-MRX8-G26

Fig. 11 Exploded view of the automatic transaxle assembly mounting—RX-8

- Oxygen sensor connector
- Catalytic converter
- Middle exhaust pipe
- Rear silencer
- Exhaust manifold stay
- Manual shaft lever after match marking its position
- Heat insulator
- Transverse member
- Starter
- Under cover
- Torque converter nuts
- Drive plate bolts
- Connector bolt
- Washer
- Oil pipe and hose
- Insulator
- Oil filler tube
- Transmission range switch connector
- Solenoid valve connector
- VSS connector
- Turbine sensor connector
- Connector wire loom
- Power plant frame

5. Support the transmission with a suitable jack
6. Support the rear differential with a block of wood.
7. Place match marks on the driveshaft, then remove the driveshaft.
8. Remove the transmission bolts and the transmission.

To install:

9. Install the transmission and tighten the mounting bolts to 28–38 ft. lbs. (37–52 Nm).
10. Support the transmission with a jack and install the power plant frame. Temporarily tighten the bolts in the sequence shown.
11. Raise the front end of the frame and adjust dimension A to 1.91–2.22 inches (48.4–56.4mm).
12. Tighten the power plant frame bolts in the sequence shown to 93–113 ft. lbs. (126–154 Nm) for bolts 1 and 2 and 55–69 ft. lbs. (75–93 Nm) for bolt 3.
13. Install the torque converter nuts and drive plate bolts and tighten them equally and evenly to 26–36 ft. lbs. (34–49 Nm).
14. Install or connect the following:
    - Driveshaft
    - Connector wire loom
    - Turbine sensor connector
    - VSS connector
    - Solenoid valve connector
    - Transmission range switch connector
    - Oil filler tube
    - Insulator
    - Oil pipe and hose

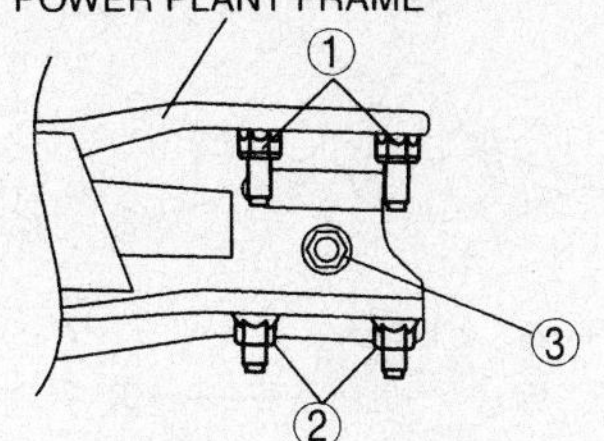

Fig. 12 Tighten the power plant frame bolts using this sequence—RX-8

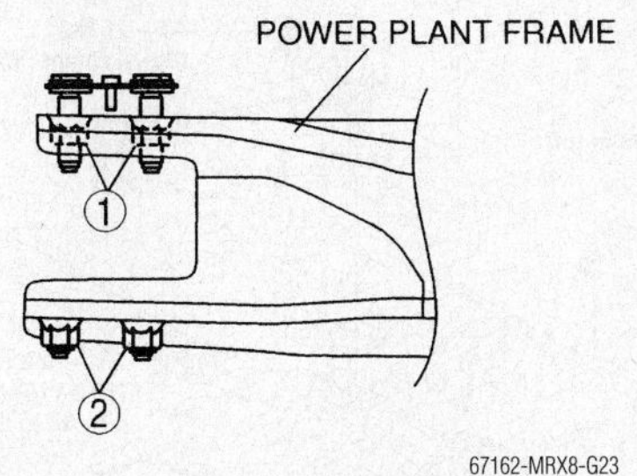

Fig. 13 Measure the distance A which should be 1.91–2.22 inch (48.4–56.4mm)—RX-8

- Washer
- Connector bolt
- Under cover
- Starter
- Transverse member
- Heat insulator
- Manual shaft lever
- Exhaust manifold stay
- Rear silencer
- Middle exhaust pipe
- Catalytic converter
- Oxygen sensor connector
- Front and rear tunnel supports
- Engine cover

15. Fill the transmission with fluid. Road test the vehicle and check for leaks. Top off all fluids as needed.
16. On models with dynamic suspension, perform the steering angle sensor initialization procedure.

MANUAL TRANSMISSION ASSEMBLY

REMOVAL & INSTALLATION

See Figures 14 through 17.

1. Before servicing the vehicle, refer to the Precautions Section.
2. Refer to the illustration for component location and torque specifications.
3. Drain the transmission oil.
4. Remove or disconnect the following:
    - Engine cover
    - Negative battery cable
    - Shifter knob
    - Shifter panel
    - Upper and lower shift insulators
    - Shift lever
    - Front and rear tunnel supports
    - Oxygen sensor connector and bracket
    - Catalytic converter
    - Middle exhaust pipe
    - Rear silencer
    - Exhaust manifold stay
    - Heat insulator
    - Starter
    - Clutch slave cylinder
5. Support the transmission with a suitable jack
6. Support the rear differential with a block of wood.
7. Place match marks on the driveshaft, then remove the driveshaft.
8. Remove or disconnect the following:
    - Power plant frame
    - Back up light switch connector
    - Neutral safety switch connector
    - Transmission bolts and the transmission

To install:

9. Install the transmission and tighten the mounting bolts to 28–38 ft. lbs. (37–52 Nm).
10. Support the transmission with a jack and install the power plant frame. Temporarily tighten the bolts in the sequence shown.
11. Raise the front end of the frame and adjust dimension A to 1.91–2.22 inches (48.4–56.4mm).
12. Tighten the power plant frame bolts in the sequence shown to 93–113 ft. lbs. (126–154 Nm) for bolts 1 and 2 and 55–69 ft. lbs. (75–93 Nm) for bolt 3.
13. Install or connect the following:
    - Neutral safety switch connector
    - Back up light switch connector
    - Clutch release cylinder
    - Drive shaft
    - Starter
    - Heat insulator

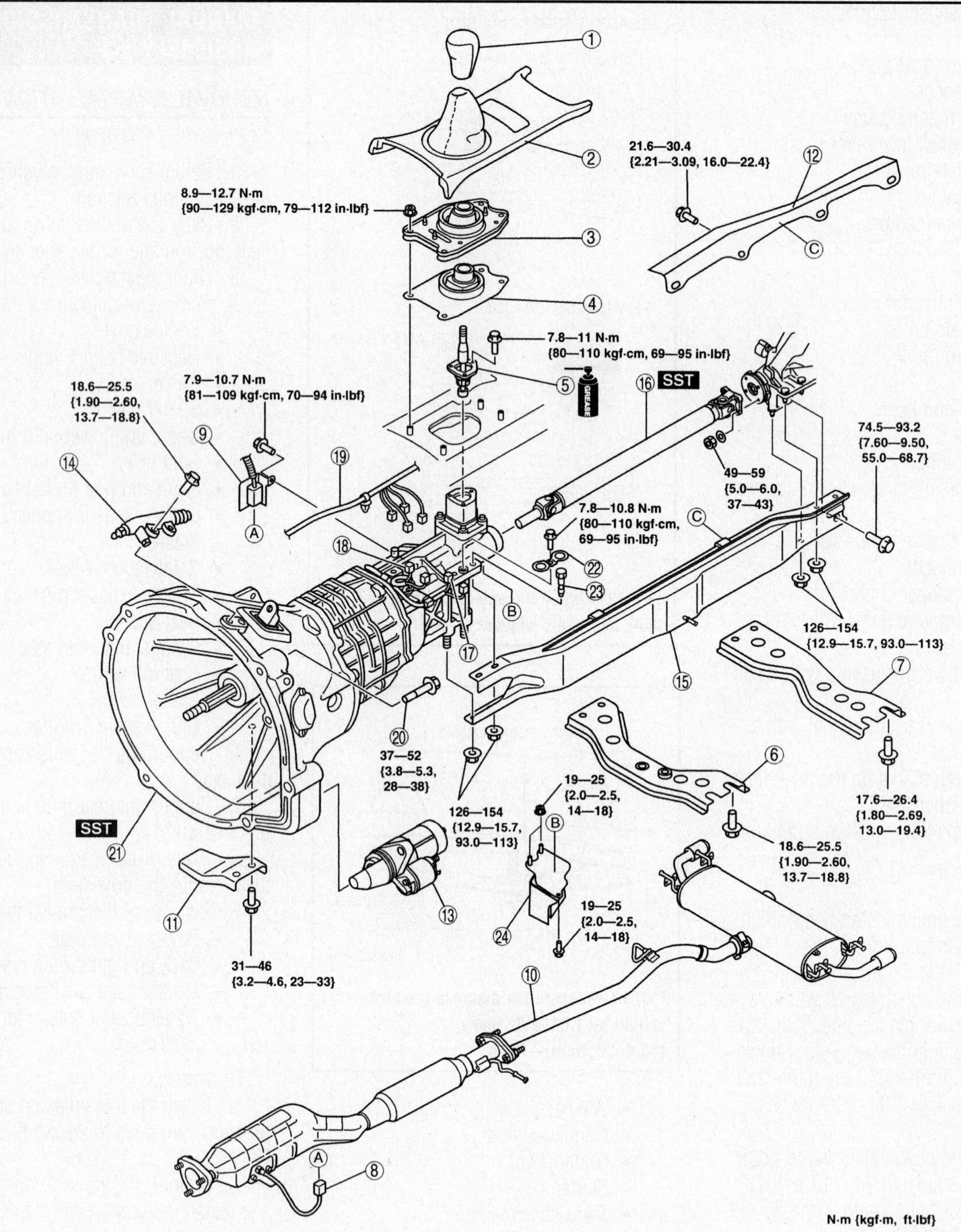

1 Shift lever knob
2 Upper panel
3 Shift insulator component (outer)
4 Shift insulator component (inner)
5 Shift lever component
6 Front tunnel member
7 Rear tunnel member
8 Heated oxygen sensor connector
9 Heated oxygen sensor connector bracket
10 Catalytic converter, middle pipe, main silencer
11 Exhaust manifold stay
12 Heat insulator
13 Starter
14 Clutch release cylinder
15 Power plant frame
16 Propeller shaft
17 Back-up light switch connector
18 Neutral switch connector
19 Wire
20 Transmission installation bolt
21 Transmission
22 Stopper
23 Bolt
24 Dynamic damper

67162-MRX8-G22

Fig. 14 Exploded view of the manual transmission mounting

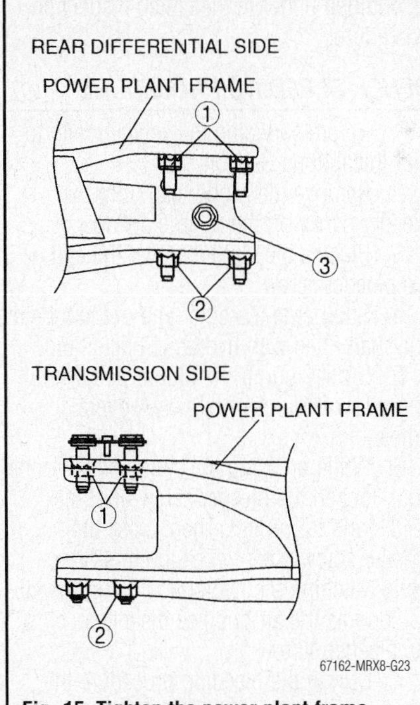

Fig. 15 Tighten the power plant frame bolts using this sequence—RX-8

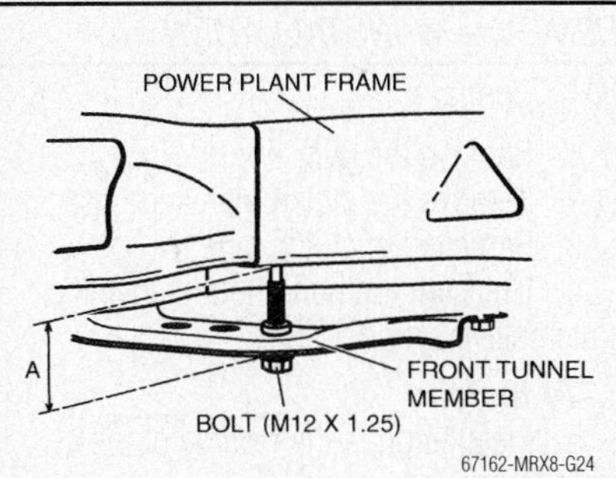

Fig. 16 Measure the distance A which should be 1.91–2.22 inch (48.4–56.4mm)—RX-8

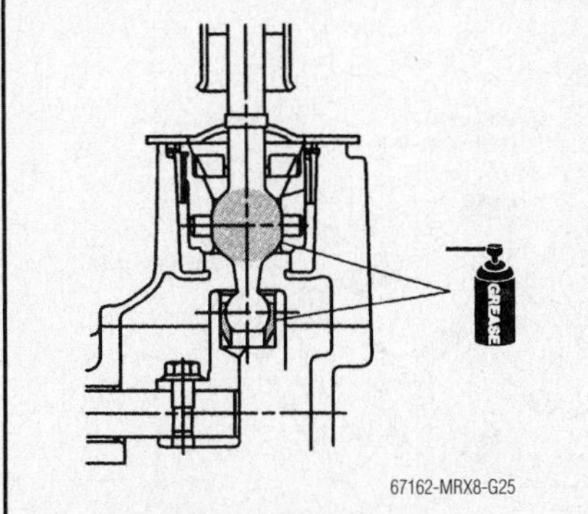

Fig. 17 Apply grease to the shift lever components—RX-8

- Exhaust manifold stay
- Rear silencer
- Middle exhaust pipe
- Catalytic converter
- Oxygen sensor connector and bracket
- Front and rear tunnel supports
- Shift lever

14. Apply grease to the shift lever components as illustrated.
15. Install or connect the following:
    - Upper and lower shift insulators
    - Shifter panel
    - Shifter knob
    - Negative battery cable
    - Engine cover
16. Fill the transmission with fluid. Road test the vehicle and check for leaks. Top off all fluids as needed.
17. On models with dynamic suspension, perform the steering angle sensor initialization procedure.

CLUTCH DRIVEN DISC & PRESSURE PLATE

REMOVAL & INSTALLATION

1. Before servicing the vehicle, refer to the Precautions Section.
2. Remove or disconnect the following:
    - Negative battery cable
    - Clutch release cylinder
    - Transmission
    - Rubber boot
    - Clutch release collar
    - Clutch release fork
    - Clutch cover
    - Pressure plate loosening the bolts one turn each in a overlapping–cross pattern
    - Clutch disc
3. Inspect the pilot bearing. If it is worn or damaged and does not turn easily by hand, remove it using a puller/slide hammer.
4. Check the flywheel surface for scoring, cracks or burning and machine or replace, as necessary.
5. Install Holder tool 49 F011 101 to keep the flywheel from turning, remove the flywheel lock bolt and remove the flywheel.
6. Inspect the clutch release bearing for wear. Replace it if it sticks or does not turn easily.
7. Inspect the release fork for wear or damage and replace as necessary.

To install:

8. Lubricate the release fork fingers and pivot with molybdenum grease and install in the release fork boot.
9. Install or connect the following:
    - Clutch release bearing on the release fork
    - New pilot bearing in the flywheel, if removed, using a installation tool
10. Be sure the flywheel mounting surface and the crankshaft or eccentric shaft mounting surfaces are clean. Remove any old sealant from the flywheel lock bolt and threads.
    - Flywheel
    - Sealant to the flywheel lock bolt threads and install it tight
    - Flywheel holding tool. Tighten the lock bolt to 289–361 ft. lbs. (392–490 Nm)..
    - Small amount of molybdenum grease to the clutch disc splines
    - Clutch disc on the flywheel with the spring side toward the transaxle
    - An alignment tool in the pilot bearing to position the clutch disc
    - Clutch pressure plate by aligning the dowel holes with the flywheel dowels
    - Pressure plate. Gradually, torque the bolts, in a crisscross pattern, to 13–19 ft. lbs. (18–27 Nm).
11. Remove the alignment tool.
    - Clutch release fork
    - Clutch release collar
    - Rubber boot
    - Transaxle
    - Clutch release cylinder
    - Negative battery cable
12. On models with dynamic suspension, perform the steering angle sensor initialization procedure.

ADJUSTMENTS

Clutch Pedal Stroke Inspection & Adjustment

See Figure 18.

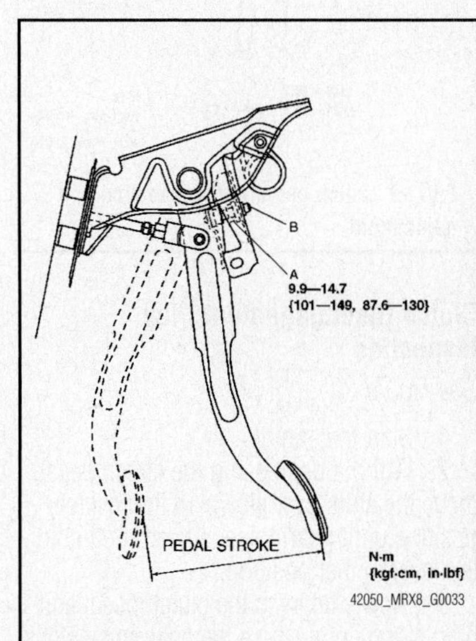

Fig. 18 Clutch pedal stroke inspection and adjustment

1. Measure the clutch pedal stroke. The stroke should be 5.12 in. (130mm).
   a. If there is any malfunction, loosen locknut A and adjust the pedal stroke with adjusting bolt B.
   b. Tighten locknut A after adjustment.

Clutch Pedal Play Inspection & Adjustment

See Figure 19.

1. Lightly depress the clutch pedal by hand until clutch resistance is felt and then measure the pedal play. It should be as follows:
   a. Clutch pedal play: 0.20–0.59 in. (5–15mm).
   b. Clutch pedal push rod play (at push rod setting line): 0.004–0.020 in. (0.2–0.5mm), (at pedal pad - reference value): 0.020–0.110 in. (0.5–2.9mm).
2. If the measurements are not within specifications, loosen locknut C and turn push rod D to adjust the pedal play.
3. Re-measure the pedal play and, if it is within the specification, tighten locknut C.

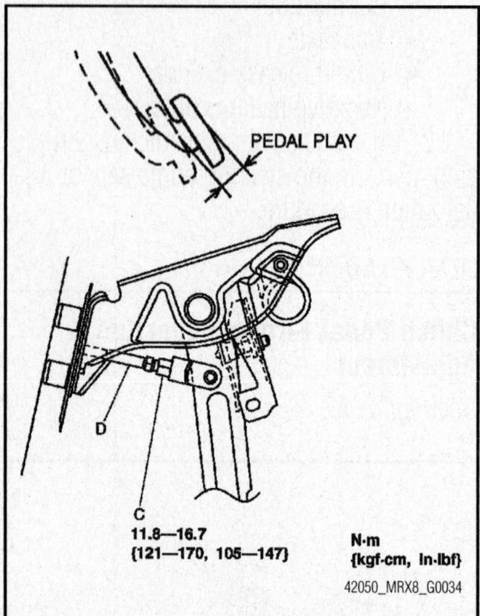

Fig. 19 Clutch pedal play inspection and adjustment

Clutch Disengagement Play Inspection

See Figure 20.

1. Start the engine.
2. Without depressing the clutch pedal, move the shift lever slowly to the reverse position until gear noise is heard and hold the lever in that position.
3. Slowly depress the clutch pedal and hold at the point where the gear noise stops (clutch disengagement point).
4. Measure distance A (as shown in illustration, from pedal not depressed to clutch disengagement point) and verify that it is within specifications.

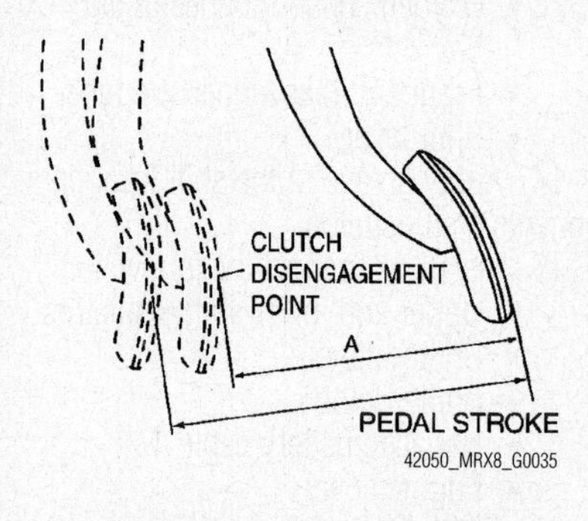

Fig. 20 Clutch disengagement stroke (reference value)—A: 4.402 in. (111.8mm)

CLUTCH MASTER CYLINDER

REMOVAL & INSTALLATION

See Figure 21.

**** WARNING**

Brake fluid will damage painted surfaces. Do NOTspill any on painted surfaces. If any is spilled, wipe it off immediately.

1. Remove the components as outlined in the accompanying illustration.

To install:

2. Installation in the reverse order of removal.
3. Bleed the air from the system, as outlined later in this section.
4. Inspect and adjust the clutch pedal, as outlined under in the clutch inspection procedure.

BENCH BLEEDING PROCEDURE

1. Before servicing the vehicle, refer to the Precautions Section.
2. Remove the rubber cap from the bleeder screw on the release cylinder.
3. Place a bleeder tube over the end of the bleeder screw.
4. Submerge the other end of the tube in a jar half filled with hydraulic brake fluid.
5. Slowly pump the clutch pedal fully and allow it to return slowly, several times.
6. While pressing the clutch pedal to the floor, loosen the bleeder screw until the fluid starts to run out. Then, close the bleeder screw. Keep repeating this Step, while watching the hydraulic fluid in the jar. As soon as the air bubbles disappear, close the bleeder screw.
7. During the bleeding procedure the reservoir must be kept at least ¾ full.

CLUTCH SLAVE CYLINDER

REMOVAL & INSTALLATION

See Figure 22.

1. Remove the clip.
2. Remove the clutch pipe and hose.
3. Remove the clutch pipe
4. Unfasten the bolt, then remove the clutch slave cylinder.

To install:

5. Installation is the reverse of the removal procedure.

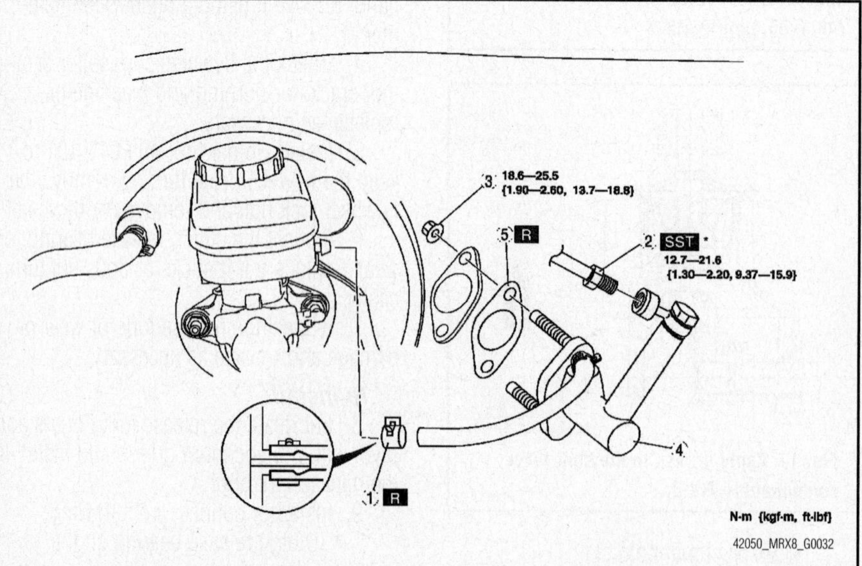

Fig. 21 Remove the hose clip (1), clutch pipe (2), nut (3), cluster master cylinder (4) and packing (5)

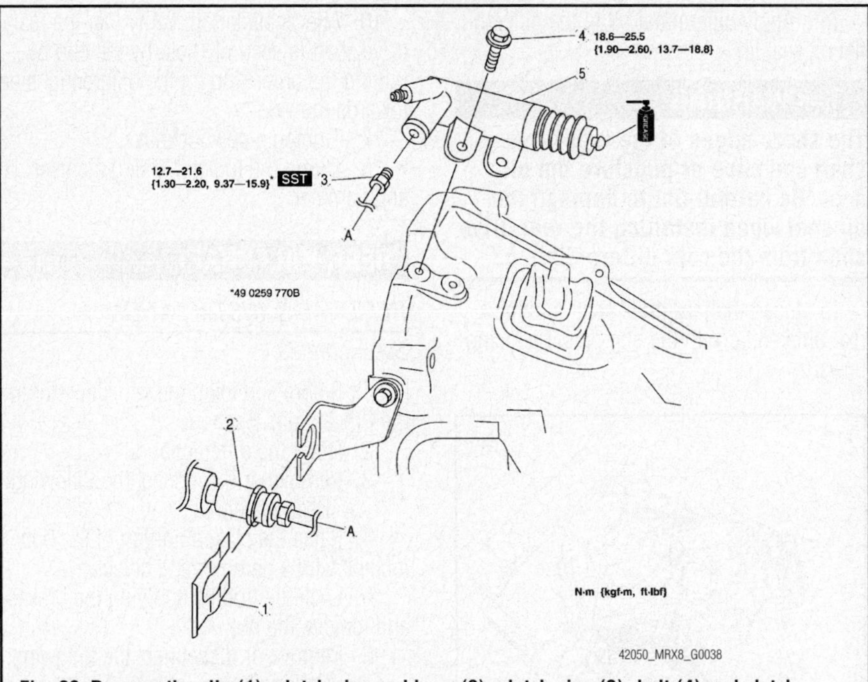

Fig. 22 Remove the clip (1), clutch pipe and hose (2), clutch pipe (3), bolt (4) and clutch slave cylinder (5)

6. Bleed the air from the system, as outlined later in this section.

CLUTCH HYDRAULIC SYSTEM BLEEDING

1. Before servicing the vehicle, refer to the Precautions Section.
2. Remove the rubber cap from the bleeder screw on the release cylinder.
3. Place a bleeder tube over the end of the bleeder screw.
4. Submerge the other end of the tube in a jar half filled with hydraulic brake fluid.
5. Slowly pump the clutch pedal fully and allow it to return slowly, several times.
6. While pressing the clutch pedal to the floor, loosen the bleeder screw until the fluid starts to run out. Then, close the bleeder screw. Keep repeating this Step, while watching the hydraulic fluid in the jar. As soon as the air bubbles disappear, close the bleeder screw.
7. During the bleeding procedure the reservoir must be kept at least ¾ full.

FRONT HALFSHAFT

REMOVAL & INSTALLATION

See Figures 23 through 28.

➡Mazda refers to the Halfshafts as Drive Shafts.

✲✲ WARNING

Performing the following procedures without first removing the ABS wheel-speed sensor may possibly cause an open circuit in the wiring harness if it is pulled by mistake. Before performing the following procedures, remove the ABS wheel-speed sensor (axle side) and fix it to an appropriate place where the sensor will not be pulled by mistake while servicing the vehicle.

1. Drain the rear differential oil.
2. Lock the disc plate by applying the brakes.
3. Knock the crimped portion of the locknut outward using a chisel and a hammer.
4. Suspend the brake caliper component using a cable or equivalent.
5. Temporarily tighten the wheel nut to prevent the disc plate from falling off.
6. Temporarily install a spare nut to the end of the rear drive shaft.

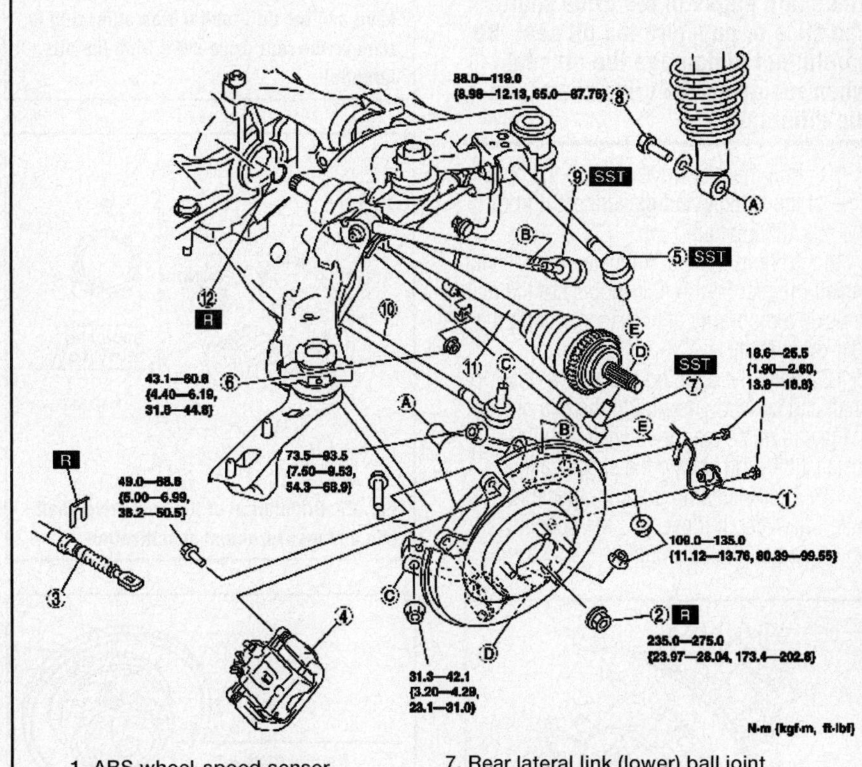

Fig. 23 Rear halfshaft/drive shaft and components —exploded view

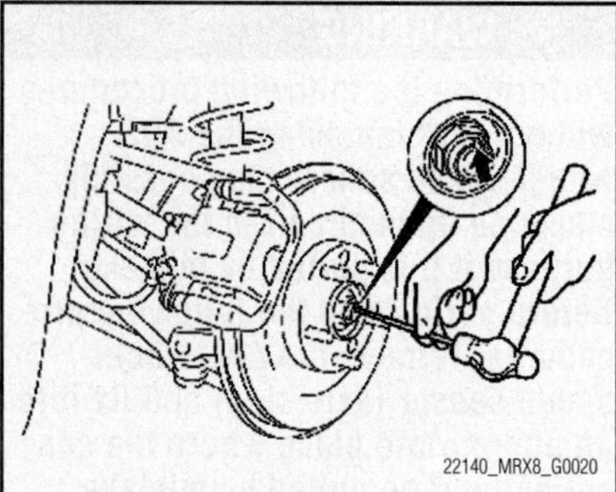

Fig. 24 Removal of the driveshaft lock nut at the rear brake rotor

7. Knock the nut with copper hammer lightly and remove the rear drive shaft from the wheel hub.
8. Separate the rear drive shaft from the wheel hub.
9. Insert a tire lever or equivalent between the rear differential and differential side outer ring, and then remove the rear drive shaft.

**** WARNING**

The sharp edges of the drive shaft can slice or puncture the oil seal. Be careful not to damage the oil seal when removing the drive shaft from the differential.

10. Pull the rear drive shaft to the outer side of the vehicle and disconnect it from the rear differential.
11. To hold the rear knuckle component, install the rear lateral link (upper) to the rear knuckle temporarily after disconnecting the rear drive shaft.
12. Point the opening of the new drive shaft clip upward, install it to the clip groove at the end of the rear drive shaft with the installation width within the specification.
13. After installing the clip, measure the outer diameter. If it exceeds the specification, reinstall the new clip.

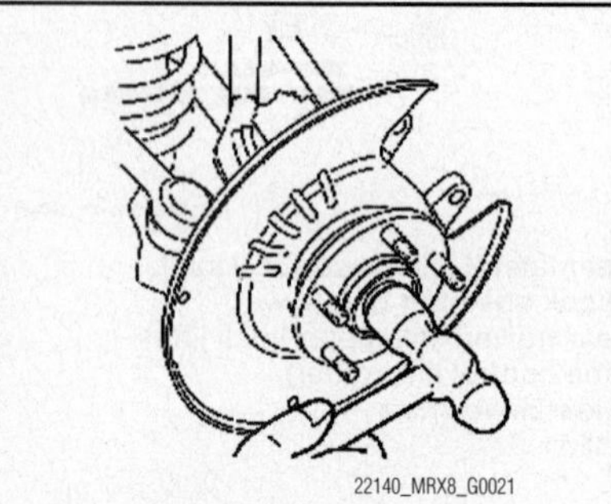

Fig. 25 Tapping the nut with a copper hammer lightly to remove the rear drive shaft from the wheel hub/rotor

14. Apply differential oil to the differential oil seal lip.

**** WARNING**

The sharp edges of the rear drive shaft can slice or puncture the oil seal. Be careful not to damage the oil seal when installing the rear drive shaft from the rear differential.

15. Insert the rear drive shaft into the rear differential with the clip opening facing upward.

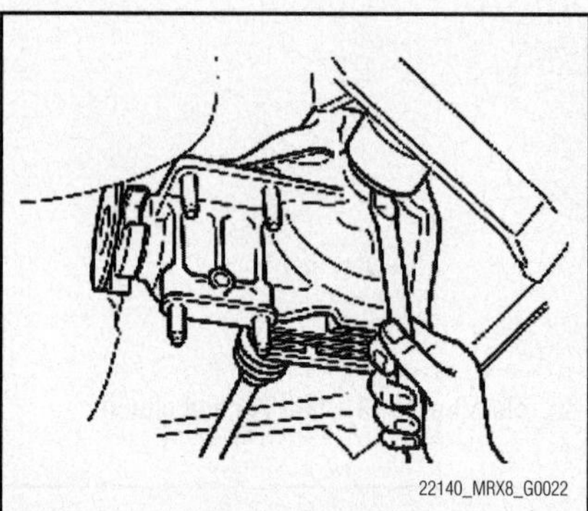

Fig. 26 Prying between the inner driveshaft and the differential side outer ring to remove the rear drive shaft from the differential

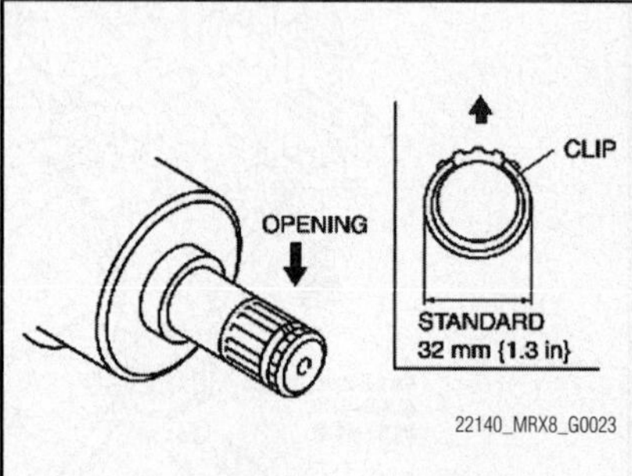

Fig. 27 Orientation of the new drive shaft clip and measurement specification

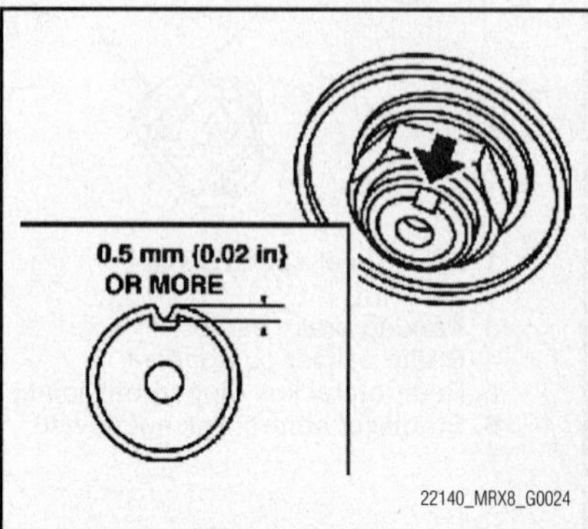
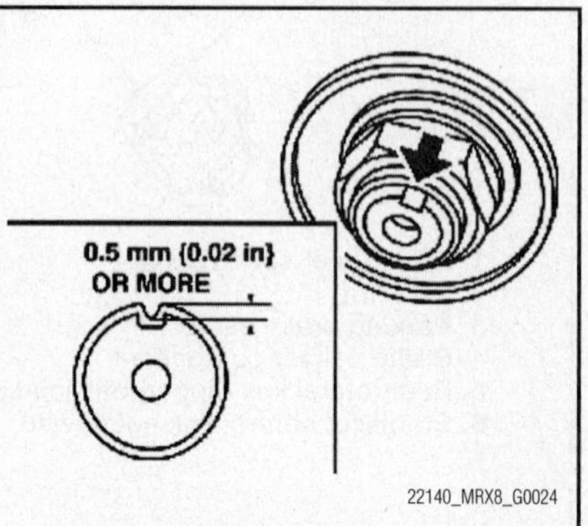

Fig. 28 Axle locknut installation and specification

16. After installation, verify that the rear drive shaft is securely held by the clip by pulling the outer ring on the differential side towards the axle
17. Tighten a new locknut.
18. Crimp the locknut, using a chisel and hammer.

REAR HALFSHAFT

REMOVAL & INSTALLATION

See Figure 29.

1. Before servicing the vehicle, refer to the Precautions Section.
2. Drain the differential oil.
3. Remove or disconnect the following:
   - Rear wheels
4. Raise the staked portion of the hub locknut with a hammer and chisel.
5. Lock the hub by applying the brakes and remove the nut.
6. Remove or disconnect the following:
   - Parking brake cable
   - Brake caliper and wire aside
   - Later link upper ball joint
   - Stabilizer link
   - Lower ball joint
   - Lower shock absorber bolt
   - Trailing link upper ball joint
   - Outer toe control link
7. Position a pry bar between the inner CV-joint and differential case. Carefully pry the halfshaft from the transaxle being careful not to damage the oil seal.
8. Pull outward on the hub/knuckle assembly, push the outer CV-joint stub shaft through the hub and remove the halfshaft. If the halfshaft is stuck in the hub, install the old hub nut to protect the stub shaft threads. Tap on the nut, using only a soft mallet, to remove the halfshaft.

➡Reinstall the rear lateral upper link to the knuckle to temporarily hold it in position.

To install:

9. Install or connect the following:
   - New circlip on the end of the halfshaft, if removed, with the end gap facing upward.
   - Differential oil to the oil seal lip
   - Halfshaft into the differential, being careful not to damage the oil seal
   - Other end of the halfshaft through the hub. Loosely install a new locknut
   - Outer toe control link
   - Trailing link upper ball joint
   - Lower shock absorber bolt
   - Lower ball joint
   - Stabilizer link

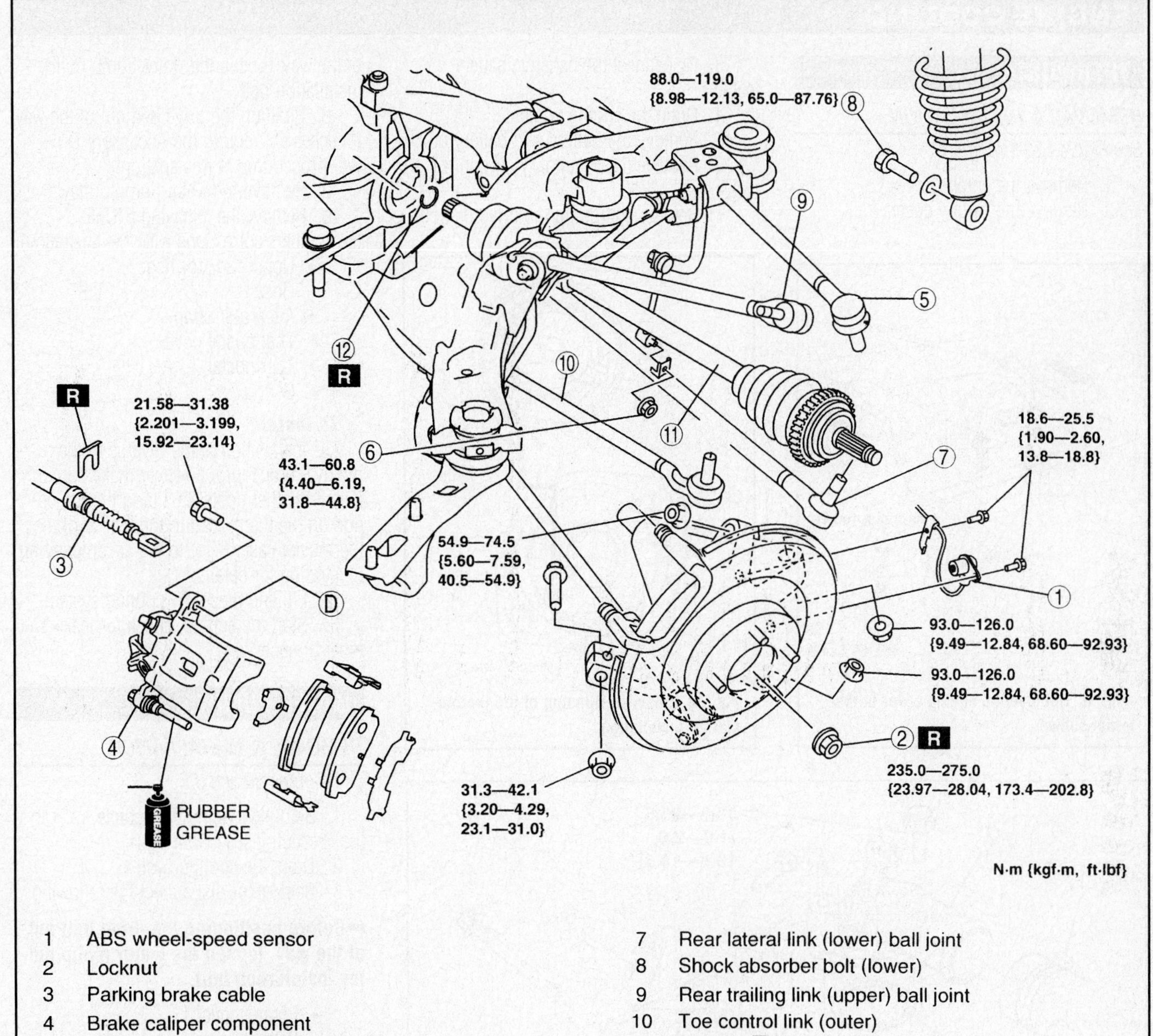

Fig. 29 Exploded view of rear halfshaft mounting—RX-8

- Later link upper ball joint
- Brake caliper and wire aside
- Parking brake cable
- Wheels
- New hub nut. Torque it to 174–203 ft. lbs. (235–275 Nm). After tightening, stake the locknut using a hammer and dull bladed chisel.

10. Fill the differential.

ENGINE COOLING

THERMOSTAT

REMOVAL & INSTALLATION

See Figures 30 through 32.

1. Remove the engine cover.
2. Remove the battery cover.
3. Disconnect the negative battery cable.
4. Drain the engine coolant.
5. Remove the battery and battery box.
6. Remove the secondary air control valve.
7. Before positioning the drive belt out of the way, loosen the water pump pulley installation bolt.
8. Position the drive belt out of the way. If necessary, refer to the Accessory Drive Belt procedures in this section.
9. Remove the water pump pulley.
10. Remove the following (NOTE: the numbers correspond with the illustration):
   - Upper radiator hose
   - Hose
   - Alternator strap
   - Thermostat cover
   - Thermostat
   - O-ring

To install:

11. Installation is the reverse of the removal procedure, however make sure to install the thermostat by fitting the projection on the thermostat to the recess of the thermostat case. Refer to the accompanying illustration for details.
12. Fill and bleed the cooling system.
13. Start the engine, check for leaks and repair if necessary.

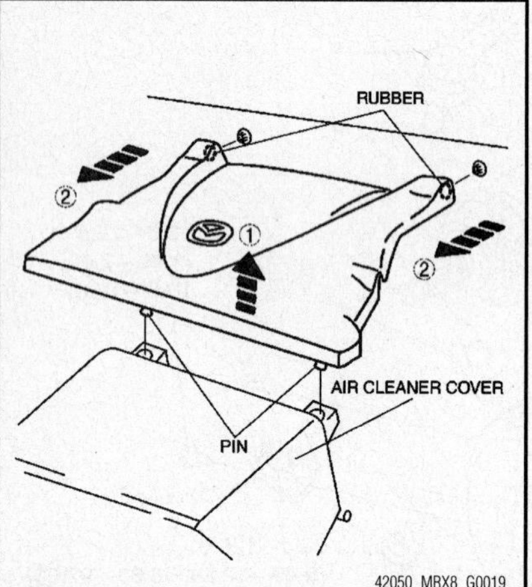

Fig. 30 Remove the engine cover in the order shown

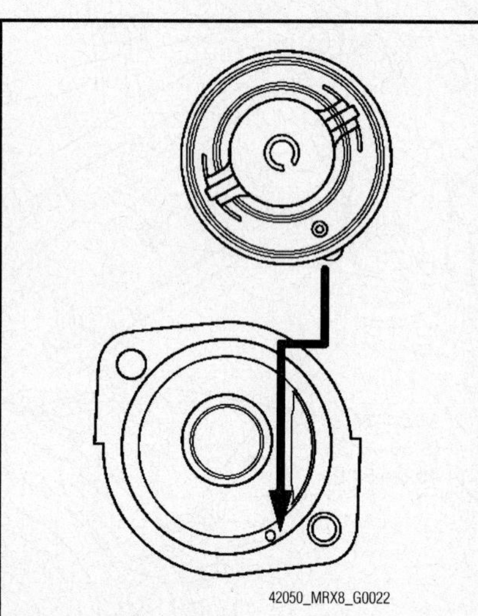

Fig. 32 Proper positioning of the thermostat for installation

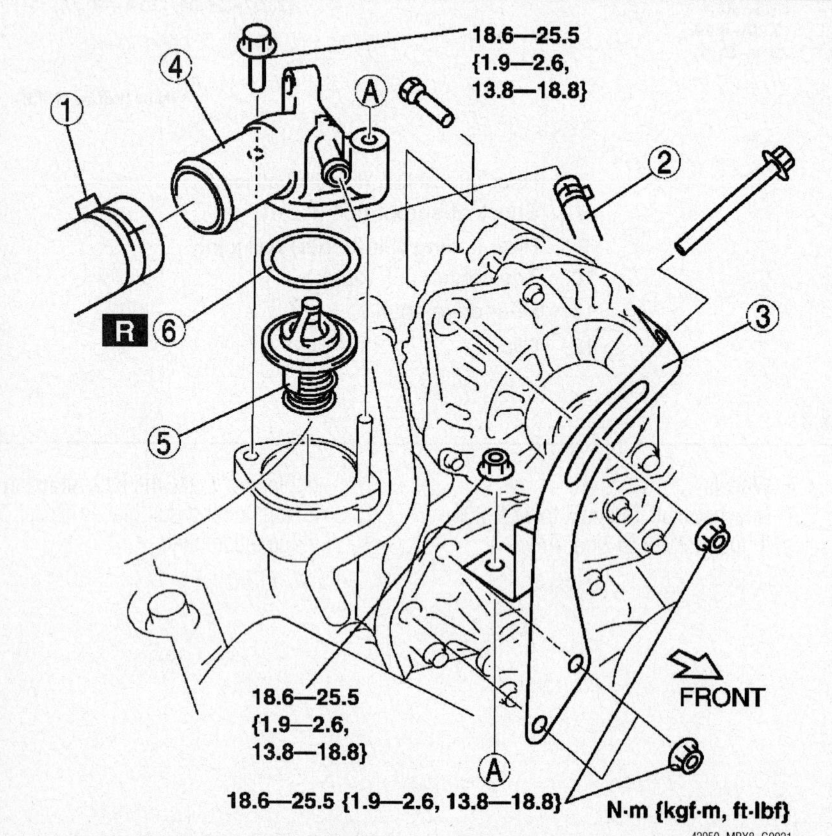

Fig. 31 Thermostat removal and installation

WATER PUMP

REMOVAL & INSTALLATION

See Figures 30 and 33.

1. Before servicing the vehicle, refer to the Precautions Section.
2. Drain the cooling system.
3. Remove or disconnect the following:

➡Before positioning the drive belt out of the way, loosen the water pump pulley installation bolt.

   - Engine cover
   - Battery cover
   - Battery cables, battery box and tray
   - Loosen the water pump pulley bolt
   - Drive belt
   - Water pump pulley
   - Front engine hangar
   - Alternator strap
   - Water pump and gasket

To install:

4. Installation is the reverse of removal. Tighten the fasteners to the specifications shown in the accompanying illustration.
5. Fill and bleed the cooling system.
6. On models with dynamic suspension, perform the steering angle sensor initialization procedure.
7. Start the engine, check for leaks and repair if necessary.

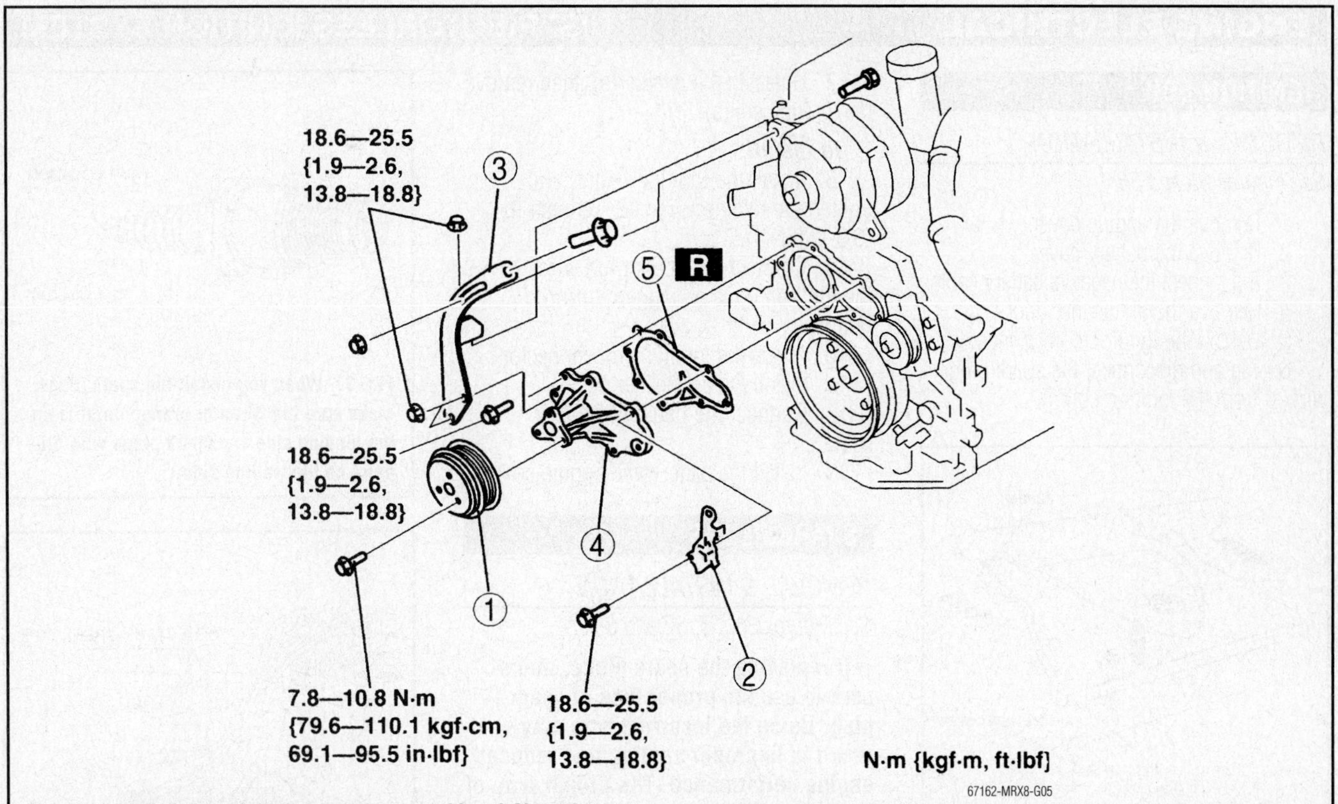

Fig. 33 Exploded view of the water pump assembly—1.3L engine

ENGINE ELECTRICAL — CHARGING SYSTEM

ALTERNATOR

REMOVAL & INSTALLATION

See Figure 34.

1. Before servicing the vehicle, refer to the Precautions Section.
2. Remove or disconnect the following:
   - Negative battery cable
   - Engine cover
   - Rear engine cross brace
   - Intake air duct
   - Accessory drive belt
   - Electrical connectors from the alternator
   - Alternator bolts
   - Alternator

To install:

3. Installation is the reverse of the removal procedure, noting the following:
   a. Tighten the left side engine cross brace nut to 40 ft. lbs. (45 Nm) and the right side nut to 16 ft. lbs. (22 Nm)

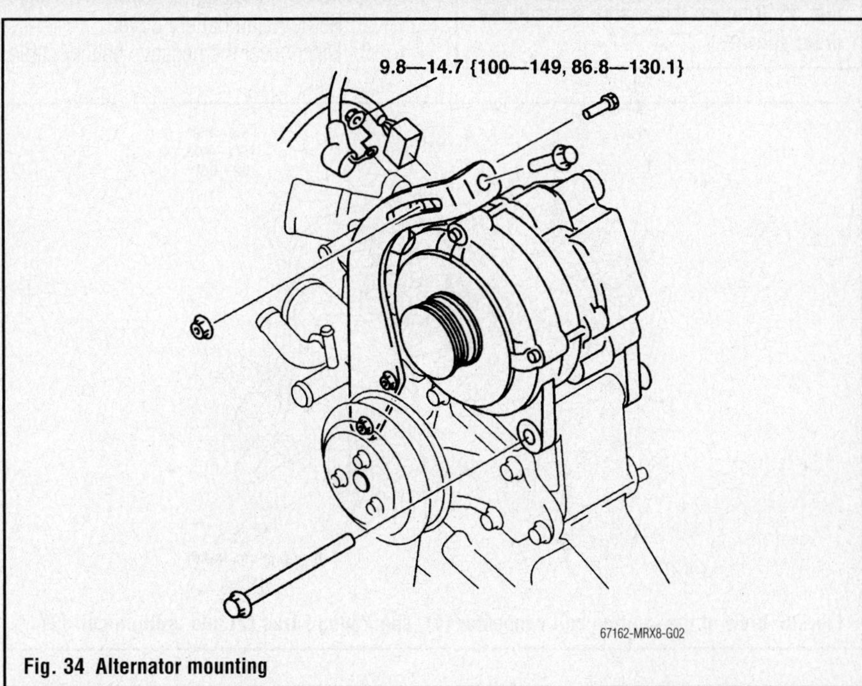

Fig. 34 Alternator mounting

IGNITION COIL

REMOVAL & INSTALLATION

See Figures 35 and 36.

1. Remove the engine cover.
2. Remove the battery cover.
3. Disconnect the negative battery cable.
4. Remove the air cleaner duct.
5. Detach the ignition coil connector.
6. Tag and disconnect the spark plug wire(s) from the ignition coil(s).

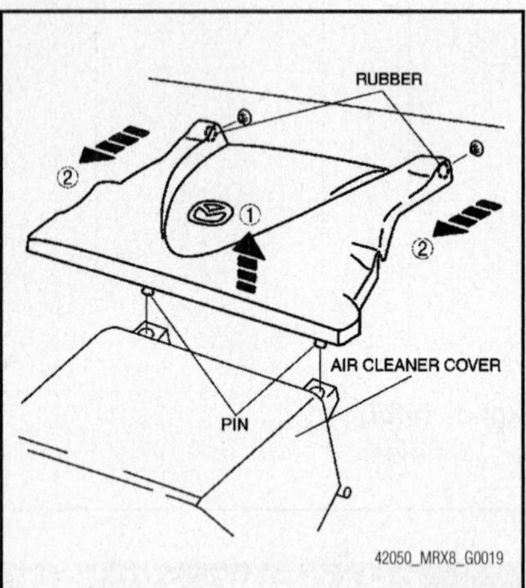

Fig. 35 Remove the engine cover in the order shown

7. Unfasten the retainer(s), then remove the ignition coil(s).

To install:

8. Install the ignition coil(s) and tighten the retainer(s) to 62–86 inch lbs. (6.9–9.8 Nm).
9. Connect the spark plug wires to the ignition coils as tagged during removal.
10. Attach the ignition coil connector.
11. Install the air cleaner duct.
12. Connect the negative battery cable.
13. Install the battery and engine covers.

SPARK PLUGS

REMOVAL & INSTALLATION

See Figures 35, 36, 37 and 38

➡If replacing the spark plugs, make sure to use the proper type of spark plug. Using the incorrect type may result in improper sealing and reduced engine performance. The proper type of spark plug for the Mazda RX-8 is NGK RE7A-L for the leading side and NGK RE9B-T for the trailing side.

1. Remove the engine cover.
2. Remove the battery cover.
3. Disconnect the negative battery cable.

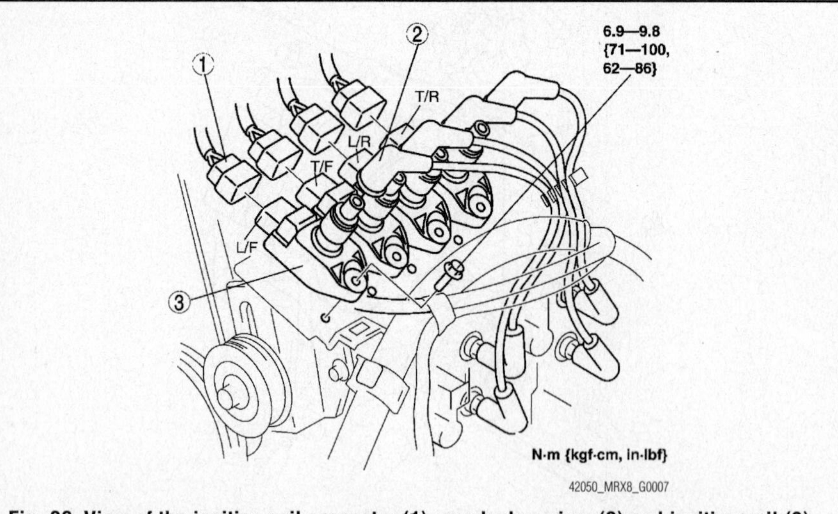

Fig. 36 View of the ignition coil connector (1), spark plug wires (2) and ignition coil (3)

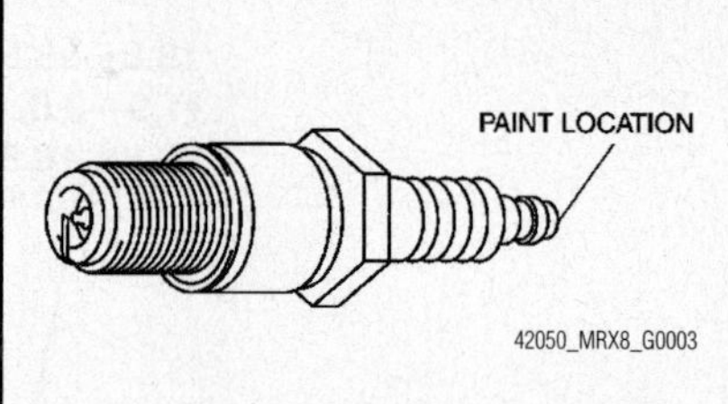

Fig. 37 When you install the spark plugs, make sure the white or orange paint is on the leading side and spark plugs with blue paint on the trailing side

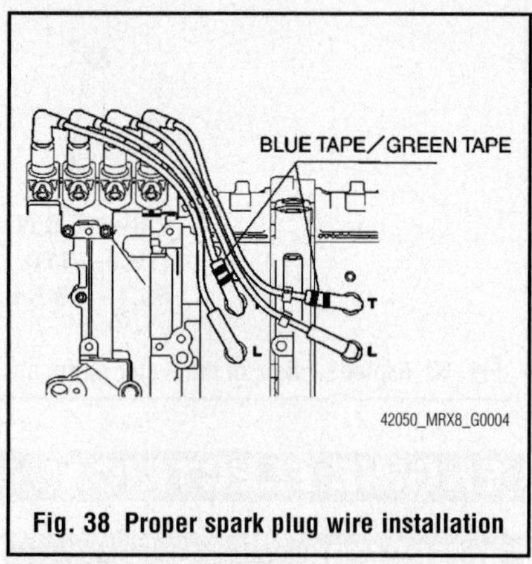

Fig. 38 Proper spark plug wire installation

4. Tag and disconnect the spark plug wires from the spark plugs.
5. Remove the spark plugs using a spark plug plug-wrench. It may be easier to remove some of the spark plugs from underneath the vehicle.

To install:

➡Install the spark plugs marked with white or orange paint on the leading side and spark plugs with blue paint on the trailing side.

6. Install the spark plug and tighten to 114–156 inch lbs. (12.8–17.7 Nm).
7. Connect the spark plug wires, as tagged during removal.
8. Connect the negative battery cable.
9. Install the battery cover.
10. Install the engine cover.

STARTER

REMOVAL & INSTALLATION

See Figure 39.

1. Remove or disconnect the following:
   - Engine cover
   - Negative battery cable
   - Air cleaner
   - Starter electrical connectors
   - Starter

To install:

2. Install or connect the following:
   - Starter and loosely tighten the lower starter mounting bolt
   - Starter electrical connectors
   - Starter bolts. Torque the bolts 14–18 ft. lbs. (19–25 Nm) on automatic transmission model, or 29–37 ft. lbs. (38–51 Nm) on manual transmission model.
   - Air cleaner
   - Negative battery cable
   - Engine cover

3. On models with dynamic suspension, perform the steering angle sensor initialization procedure.

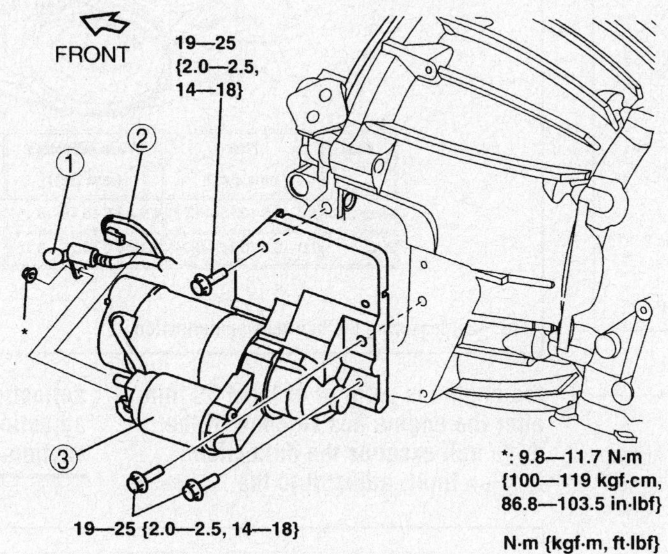

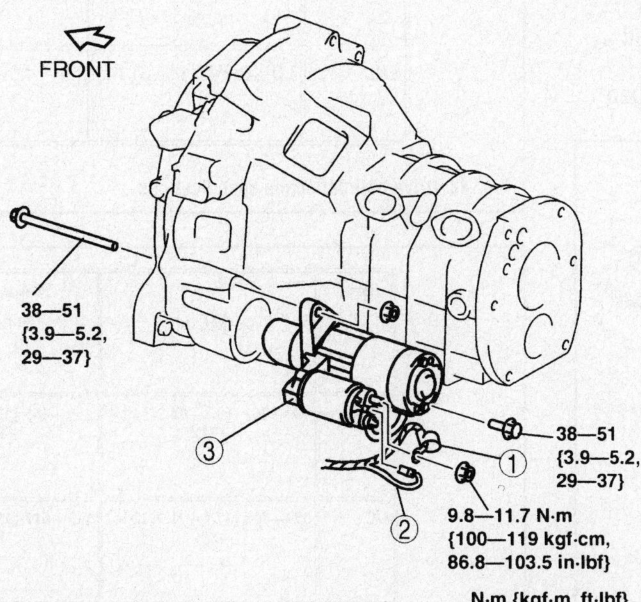

Fig. 39 Starter mounting—A/T top; M/T bottom

ENGINE MECHANICAL

ACCESSORY DRIVE BELTS

ACCESSORY BELT ROUTING

See Figure 40.

Refer to the accompanying illustration for accessory drive belt routing.

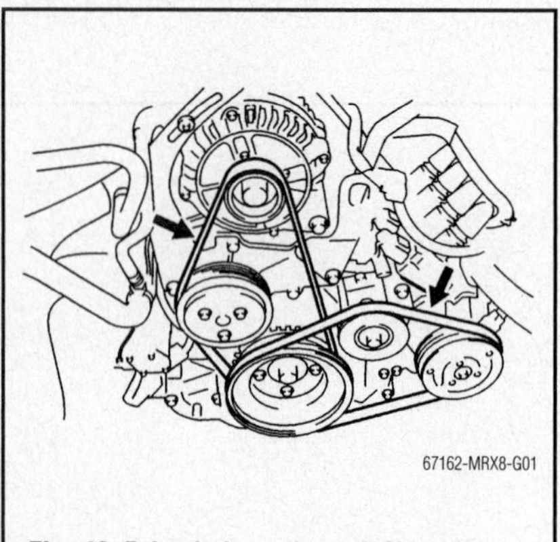

Fig. 40 Drive belt routing—1.3l engines

INSPECTION

See Figures 41 through 44.

Inspect the drive belt for signs of glazing or cracking. A glazed belt will be perfectly smooth from slippage, while a good belt will have a slight texture of fabric visible. Cracks will usually start at the inner edge of the belt and run outward. All worn or damaged drive belts should be replaced immediately.

⁂ WARNING

The drive belt deflection can be inspected only between specified pulleys. Perform the drive belt deflection/tension inspection when the engine is cold, or at least 30 min after the engine has stopped. If the drive belt exceeds the deflection/tension limit, adjust it to the deflection/tension used when adjusting, as outlined in this section..

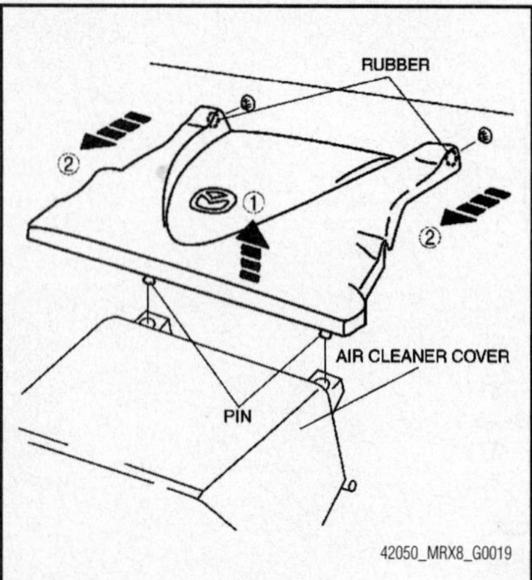

Fig. 41 Remove the engine cover in the order shown

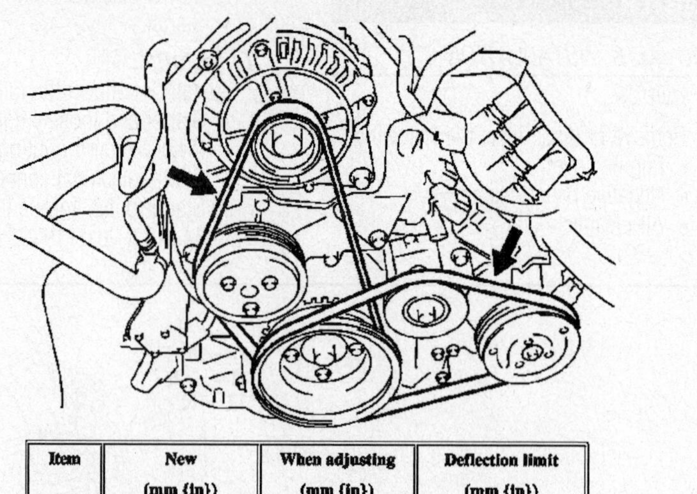

Item	New (mm {in})	When adjusting (mm {in})	Deflection limit (mm {in})
Generator	4.0—4.5 {0.16—0.17}	4.5—5.0 {0.18—0.19}	6.0 {0.24}or more
A/C	3.0—3.8 {0.11—0.14}	3.3—4.0 {0.13—0.15}	5.5 {0.21}or more

42050_MRX8_G0014

Fig. 42 Measuring the belt tension/deflection

Item	New (mm {in})	When adjusting (mm {in})	Deflection limit (mm {in})
Generator	4.0—4.5 {0.16—0.17}	4.5—5.0 {0.18—0.19}	6.0 {0.24} or more
A/C	3.0—3.8 {0.11—0.14}	3.3—4.0 {0.13—0.15}	5.5 {0.21} or more

42050_MRX8_G0015

Fig. 43 Drive belt deflection specifications

Item	New (N {kgf, lbf})	When adjusting (N {kgf, lbf})	Tension limit (N {kgf, lbf})
Generator	620—767 {63.3—78.2, 140—172}	519—666 {53.0—67.9, 117—149}	344 {35.1, 77.3} or less
A/C	559—706 {57.1—71.9, 126—158}	519—617 {53.0—62.9, 117—138}	265 {27.1, 59.6} or less

42050_MRX8_G0016

Fig. 44 Drive belt tension specifications

After replacing with a new drive belt, assemble with the deflection/tension specified for a new drive belt. Operate the alternator drive belt for 1 min or more and the A/C drive belt for 5 min or more while idling the engine. Then adjust it to the used belt deflection/tension specifications.

1. Remove the engine cover.
2. Apply a pressure of about 22 lbs. (98 Nm) to the back of the drive belt, in the middle of the pulleys shown in the figure and inspect the deflection. Otherwise, inspect the tension using the Special Service Tool (SST), as shown in the accompanying illustration.

➡The drive belt tension can be inspected anywhere between the pulleys. The drive belt deflection can be inspected only between specified pulleys.

3. If the drive belt deflection is at the deflection limit or more, or the drive belt tension is at the tension limit or less, the belt tension must be adjusted, as outlined in this section.
4. When finished, install the engine cover.

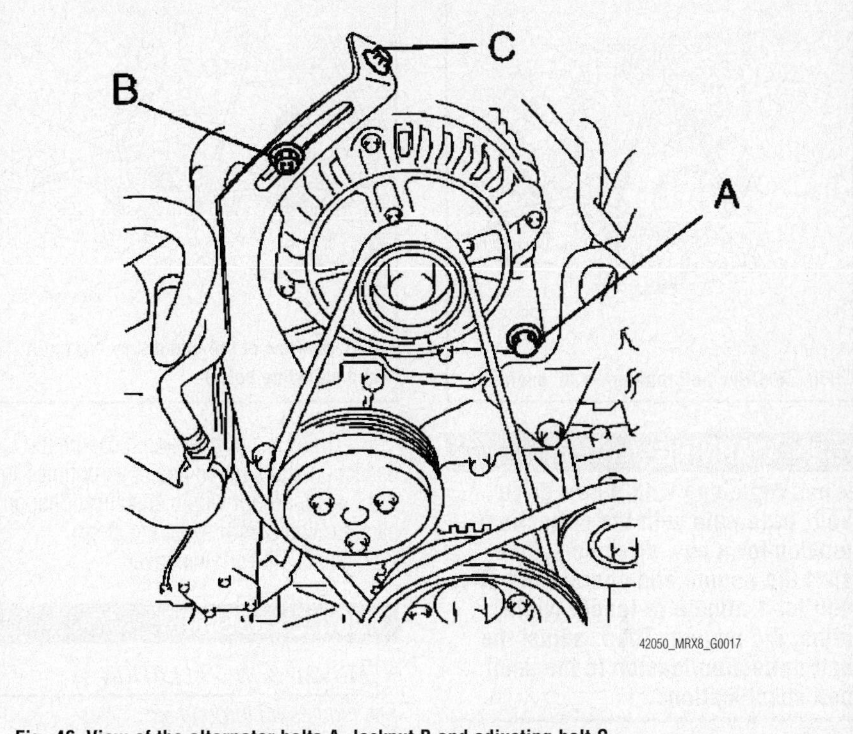

Fig. 46 View of the alternator bolts A, locknut B and adjusting bolt C

ADJUSTMENT

Alternator Belt

See Figures 45 and 46.

1. Remove the engine cover.
2. Loosen generator installation bolt A and locknut B.
3. Adjust the drive belt deflection and tension by turning adjusting bolt C to the specification:
   a. Drive belt deflection (with pressure of 22 lbs. [98 Nm]): 0.18-0.19 inch lbs. (4.5–5.0mm).
   b. Drive belt tension (when using the SST): 117–149 lbs. (519–666 Nm).
4. Tighten generator installation bolt A and locknut B, to the following:
   a. Bolt A: 29–37 ft. lbs. (38–51 Nm).
   b. Locknut B: 15–22 ft. lbs. (20–30 Nm).
5. Crank the engine and measure the deflection and tension again. If not within the specification, repeat Steps 2 & 3.
6. Install the engine cover

A/C Compressor Belt

See Figure 30, 45 and 47.

1. Remove the engine cover.
2. Loosen idle pulley locknut A.
3. Adjust the drive belt deflection and tension by turning adjusting bolt B to the specification.
   a. Drive belt deflection (with pressure of 22 lbs. [98 Nm]): 0.13-0.15 inch lbs. (3.3–4.0mm).
   b. Drive belt tension (when using the SST): 117–138 lbs. (519–617 Nm).
4. Tighten idle pulley locknut A to 27–38 ft. lbs. (37–52 Nm).
5. Crank the engine, then measure the deflection and tension again. If not within specifications, repeat Steps 2–5.
6. Install the engine cover

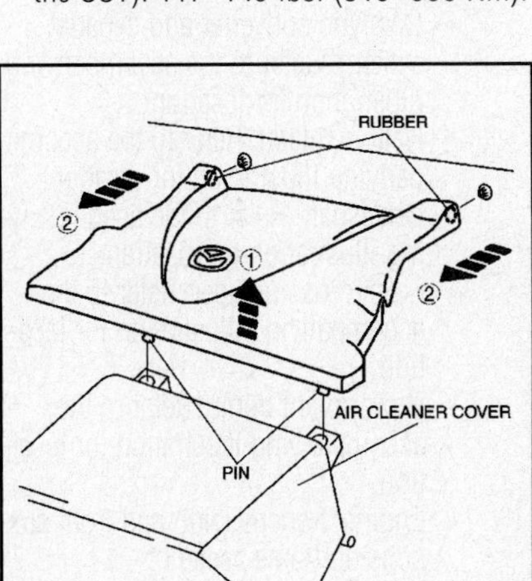

Fig. 45 Remove the engine cover in the order shown

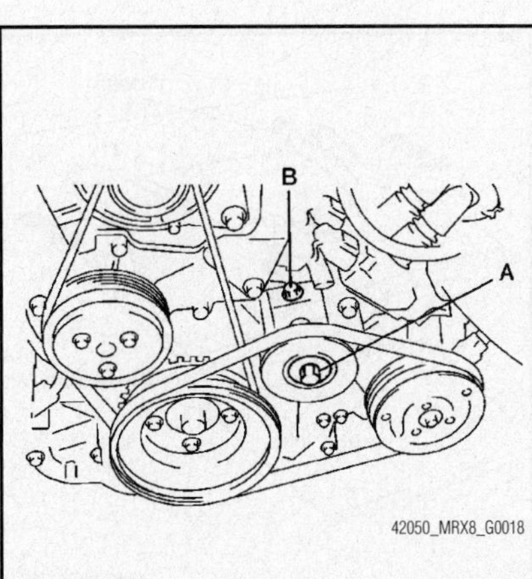

Fig. 47 View of the idle pulley locknut A and adjusting bolt B

REMOVAL & INSTALLATION

Alternator Belt

See Figures 45, 46 and 48.

1. Remove the engine cover.
2. Remove the A/C compressor belt, as outlined later in this section.
3. Loosen generator installation bolt A and locknut B.
4. Loosen adjusting bolt C, then remove the drive belt.

To install:

5. Install the drive belt and adjust the drive belt deflection by tightening adjusting bolt C to the specifications given in the Adjustment procedure in this section.

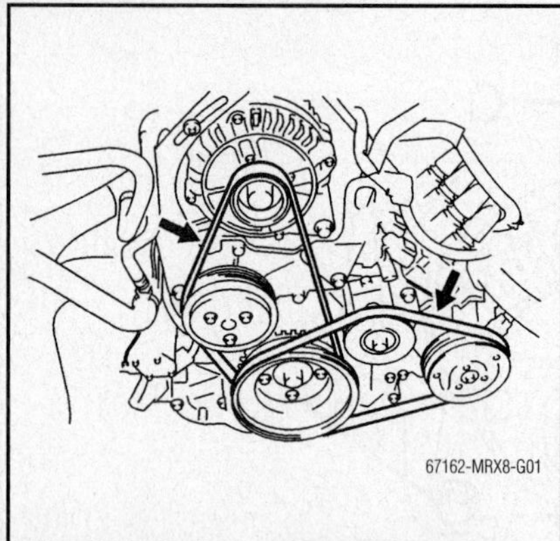

Fig. 48 Drive belt routing—1.3L engines

✻✻ WARNING

When replacing with a new drive belt, assemble with the deflection/tension for a new drive belt, then start the engine and operate the drive belt for 1 minute or longer while idling the engine. Then, adjust the belt deflection/tension to the used belt specifications.

6. Tighten generator installation bolt A and locknut B, as follows:
   a. A: 29–37 ft. lbs. (38–51 Nm)
   b. B: 15–22 ft. lbs. (20–30 Nm)
7. Crank the engine and measure the deflection and tension again, as outlined in this section. If not within specifications, go back to Step 3 and repeat the steps.
8. Install the A/C compressor belt.
9. Install the engine cover

A/C Compressor Belt

See Figure 49.

1. Remove the engine cover.
2. Loosen idle pulley locknut A.
3. Loosen adjusting bolt B, then remove the A/C compressor drive belt.

To install:

4. Install the drive belt and adjust the drive belt deflection by tightening adjusting bolt B to the specifications given in the Adjustment procedure in this section.

✻✻ WARNING

When installing a new drive belt, assemble it with the deflection/tension specifications for new drive belt, then start the engine and operate the drive belt for 5 min or longer while idling the engine. Then adjust the belt to the used deflection/tension specifications.

5. Tighten idle pulley locknut A to 27–38 ft. lbs. (37–52 Nm).

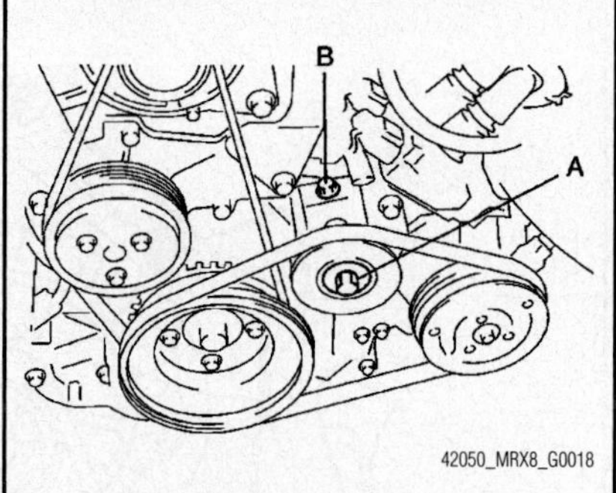

Fig. 49 View of the idle pulley locknut A and adjusting bolt B

6. Crank the engine and measure the deflection and tension again, as outlined in this section. If not within specifications, go back to Step 2 and repeat the steps.
7. Install the engine cover

ENGINE ASSEMBLY

REMOVAL & INSTALLATION

See Figures 50 through 52.

➡The procedure for pulling the engine requires removing the transaxle and front crossmember along with it. A suitable support fixture that will support the entire assembly must be used for removal.

1. Before servicing the vehicle, refer to the Precautions Section.
2. Properly relieve the fuel system pressure.
3. Drain the engine oil.
4. Drain the transaxle fluid.
5. Drain the cooling system.
6. Place the front wheels in the straight ahead position.
7. Raise and support the vehicle.

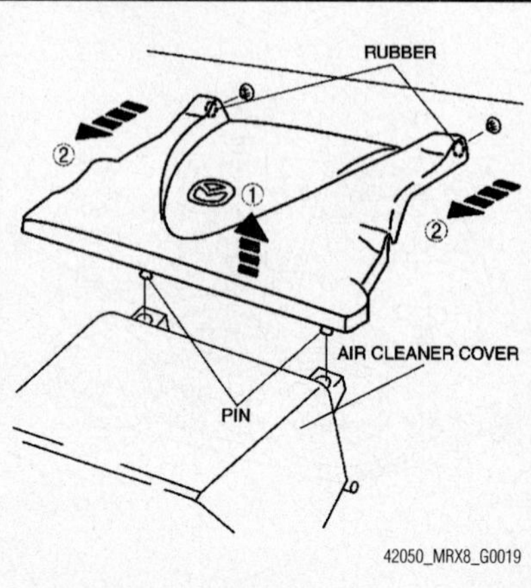

Fig. 50 Remove the engine cover in the order shown

8. Remove or disconnect the following:
   - Front wheels
   - Engine cover
   - Rear engine cross brace
   - Battery cover, battery, box and tray
   - Air cleaner, intake duct and insulator
   - Powertrain Control Module (PCM)
   - Secondary Air Injection (AIR) pump
   - Brake vacuum hose
   - Charcoal canister connector
   - Fuel lines
   - Ignition coil
   - Accessory drive belts
   - A/C compressor and wire aside
   - Engine wiring harness from main fuse block
   - Engine splash shield
   - ABS speed sensor connector
   - Radiator, heater and coolant tank hoses
   - Selector link on automatic transmission
   - Manual transmission shift lever
   - Clutch slave cylinder and wire aside
   - Steering shaft pinch bolt. Refer to the accompanying illustration for location.
   - Oil lines. Refer to the accompanying illustration for location.
   - Transmission cooler lines. Refer to the accompanying illustration for location.
   - Brake caliper. Refer to the accompanying illustration for location.
   - Lower strut bolt. Refer to the accompanying illustration for location.
   - Rear crossmembers. Refer to the accompanying illustration for location.
   - Catalytic converter and exhaust system. Refer to the accompanying illustration for location.
   - Heat insulator. Refer to the accompanying illustration for location.
   - Driveshaft. Refer to the accompanying illustration for location.
   - Front crossmember. Refer to the accompanying illustration for location.
   - Power plant frame. Refer to the accompanying illustration for location.
   - Engine, transmission and front suspension frame assembly
9. Use an engine hoist and separate the engine/transmission assembly from the front suspension crossmember using the steps shown in the accompanying illustration.

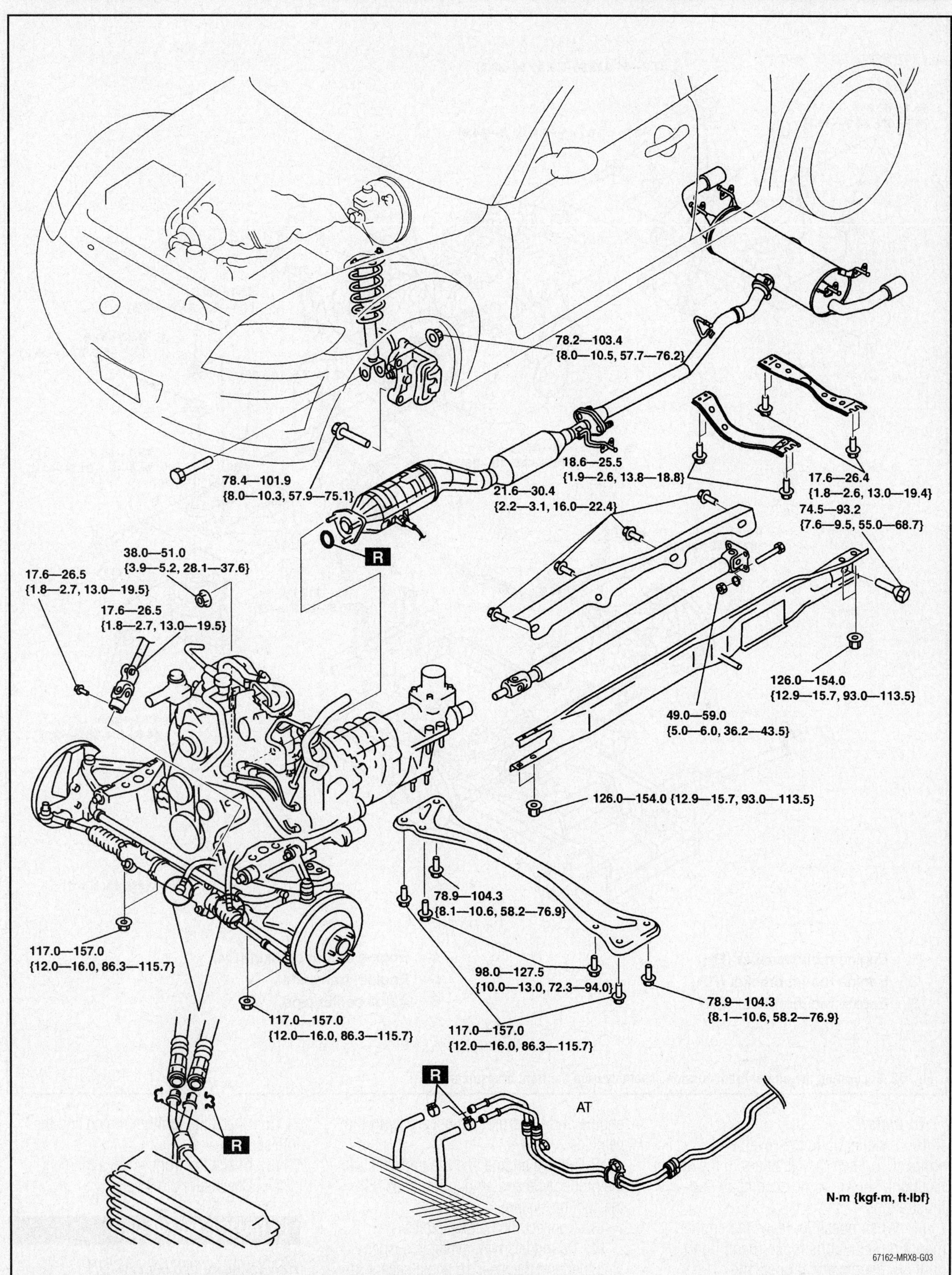

Fig. 51 Location of the engine mounting components and their torque specifications

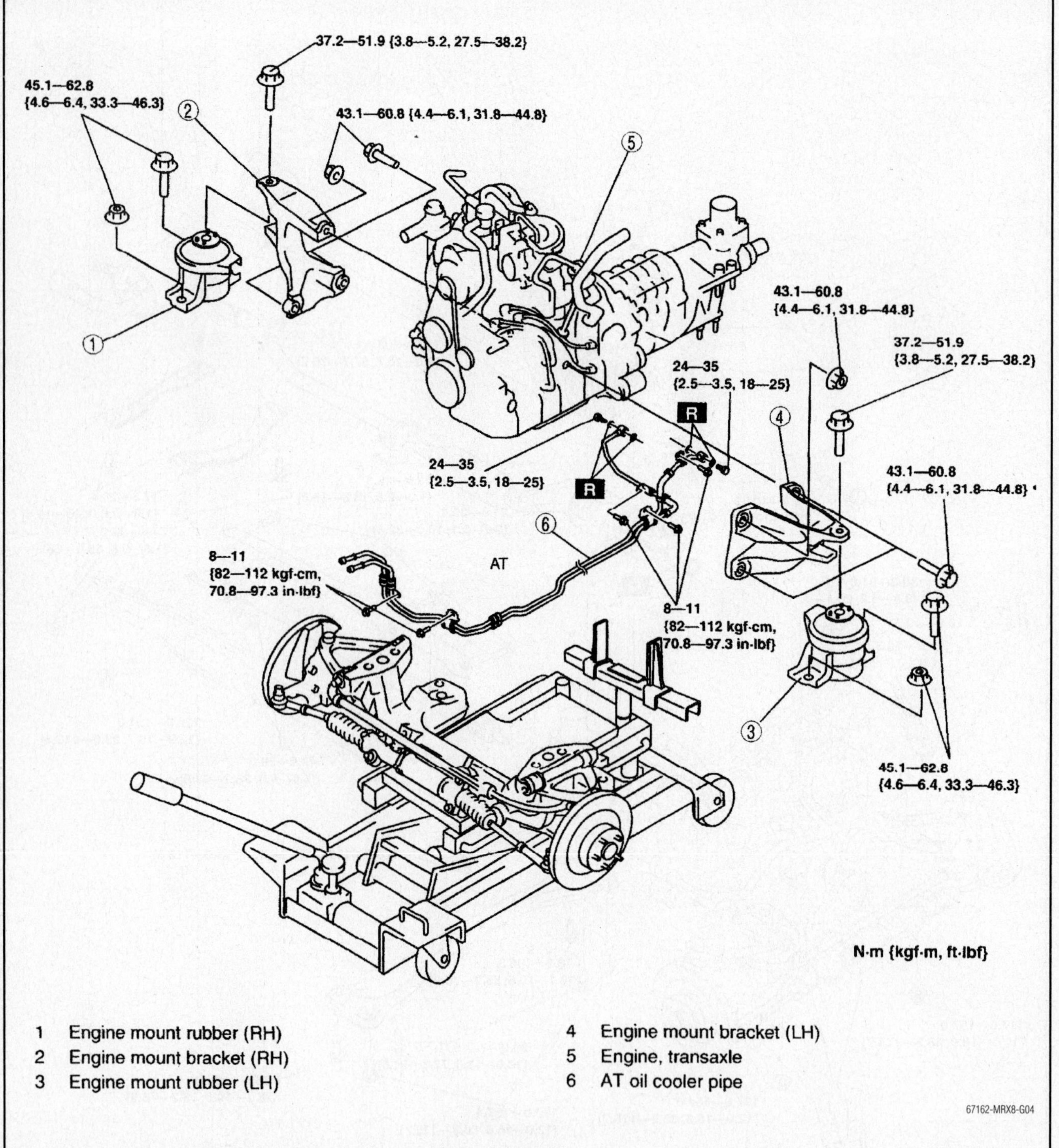

Fig. 52 Separating the engine/transmission assembly from the front crossmember

To install:

10. Installation is the reverse of removal. Tighten the fasteners to the specifications shown in the accompanying illustrations.

11. When possible, leave the engine mounting nuts/bolts loose (hand tight) until all mounts are aligned and bolted. This may help in aligning the engine and transmission assembly in the vehicle.

12. Fill the engine and the transaxle with the proper type and amount of fluids. Fill the cooling system.

13. Connect the battery cables.

14. On models with dynamic suspension, perform the steering angle sensor initialization procedure.

15. Check the ignition timing and the idle speed.

16. Check the front wheel alignment.

17. Check all fluid levels.

EXHAUST MANIFOLD

REMOVAL & INSTALLATION

See Figures 53 through 55.

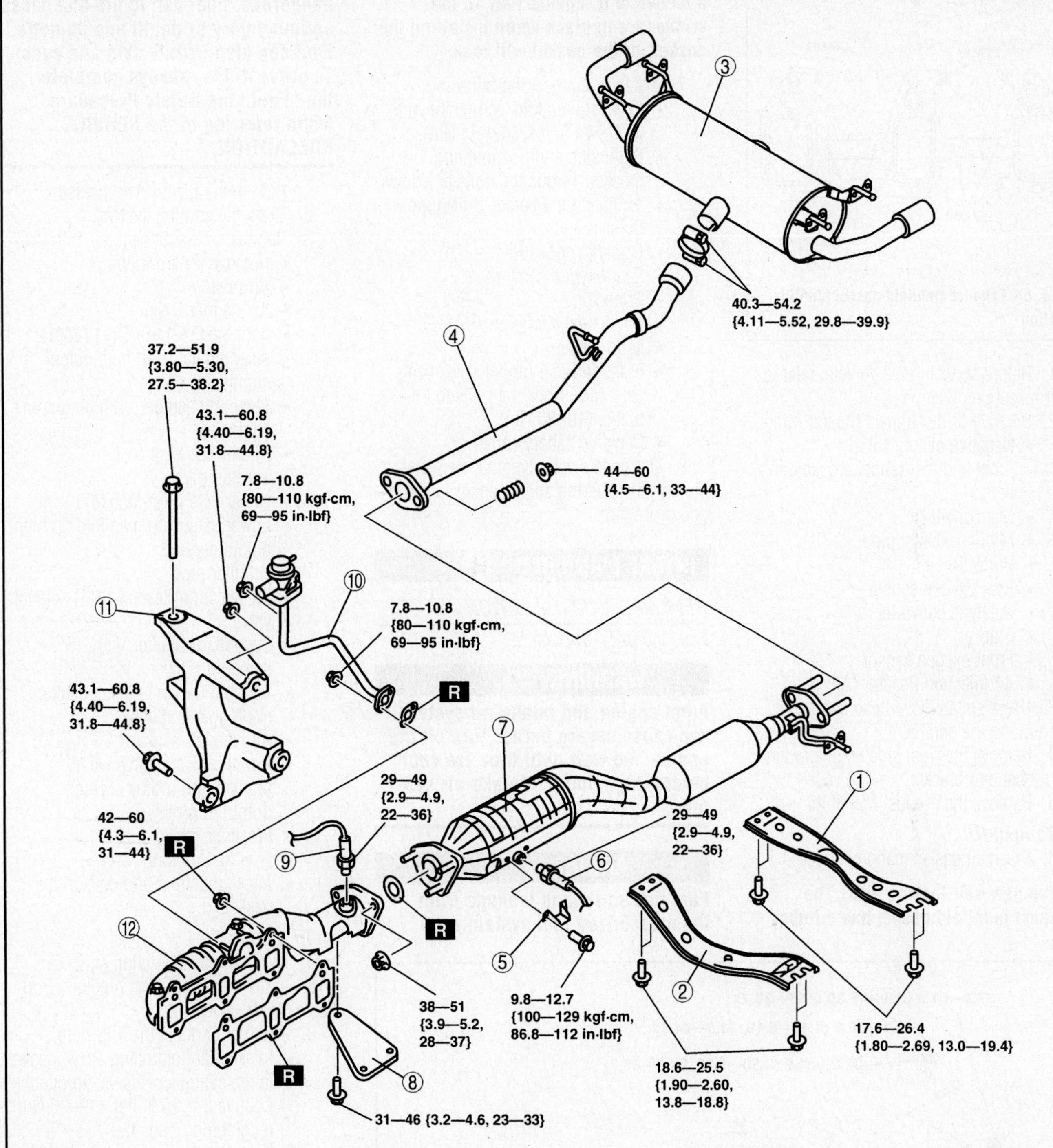

1	Rear tunnel member	7	Catalytic converter
2	Front tunnel member	8	Bracket
3	Main silencer	9	Front heated oxygen sensor
4	Middle pipe	10	AIR pipe
5	Protector	11	Engine mount bracket (RH)
6	Rear heated oxygen sensor	12	Exhaust manifold

67162-MRX8-G12

Fig. 53 Exploded view of the exhaust system

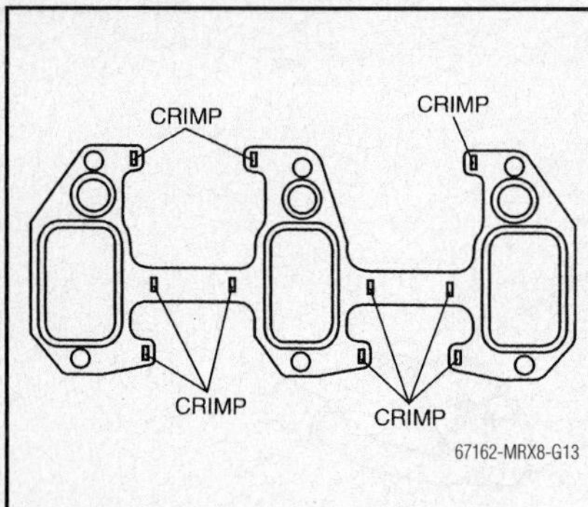

Fig. 54 Exhaust manifold gasket identification

1. Before servicing the vehicle, refer to the Precautions Section.
2. Remove or disconnect the following:
   - Negative battery cable
   - Front and rear tunnel crossmembers
   - Main silencer
   - Middle exhaust pipe
   - Protector
   - Rear oxygen sensor
   - Catalytic converter
   - Bracket
   - Front oxygen sensor
   - Air Injection Reactor (AIR pipe)
3. Use a suitable overhead engine lift and support the engine.
4. Remove the right side engine mounting rubber and bracket.
5. Remove the exhaust manifold.

To install:

6. Clean all gasket mating surfaces.

➡Use new self-locking nuts. The exhaust manifold gasket has crimps attached to it. Ensure that all the crimps are in place when installing the gasket, or the gasket will leak.

7. Install or connect the following:
   - Exhaust manifold. torque the nuts to 31–44 ft. lbs. (43–61 Nm).
   - Right side engine mounting bracket. Torque the bolts as shown.
   - Air Injection Reactor (AIR) pipe
   - Bracket
   - Catalytic converter
   - Rear oxygen sensor
   - Protector
   - Middle exhaust pipe
   - Main silencer
   - Front and rear tunnel crossmembers and torque the bolts to 14–19 ft. lbs. (19–26 Nm)
   - Negative battery cable
8. On models with dynamic suspension, perform the steering angle sensor initialization procedure.

INTAKE MANIFOLD

REMOVAL & INSTALLATION

See Figures 56 through 59.

✻✻ WARNING

A hot engine and intake-air system can cause severe burns. Turn off the engine and wait until they are cool before removing the intake-air system.

✻✻ WARNING

Fuel line spills and leakage from the pressurized fuel system are dangerous. Fuel can ignite and cause serious injury or death and damage. Fuel can also irritate skin and eyes. To prevent this, always complete the "Fuel Line Safety Procedure," while referring to the SERVICE PRECAUTION.

1. Relieve the fuel system pressure.
2. Drain the cooling system.
3. Remove or disconnect the following:
   - Remove the front bumper.
   - Air hose
   - Air cleaner cover
   - Variable Fresh Air Duct (VFAD) solenoid valve on high output engines
   - Vacuum chamber on high output engines
   - Air cleaner case
   - Throttle body
   - Upper extension manifold
   - Lower extension manifold on high output engines
   - Oil filler pipe
   - Air Injection Reactor (AIR) solenoid valve
   - Secondary Shutter Valve (SSV) solenoid
   - Variable Dynamic Effect Intake (VDI) solenoid valve.
   - Air cleaner insulator
   - Auxiliary Port Valve (APV) bracket and motor on high output engines
   - Fresh air intake duct
   - Fuel distributors
   - Intake manifold and discard the gasket

To install:

4. Clean all gasket mating surfaces.
5. Apply clean oil to the Auxiliary Port Valve (APV) valves.
6. Install or connect the following:
   - Intake manifold using a new gasket in the sequence shown. Torque the bolts to 14–19 ft. lbs. (19–26 Nm) retighten bolt no. 1.
   - Fuel distributors and torque the bolts to 14–19 ft. lbs. (19–26 Nm)
   - Fresh air intake duct and torque the bolts to 69–96 inch lbs. (8–11 Nm)
   - APV bracket and motor on high output engines
   - Air cleaner insulator and torque the bolts to 69–96 inch lbs. (8–11 Nm)
   - Variable Dynamic Effect Intake (VDI) solenoid valve.
   - Secondary Shutter Valve (SSV) solenoid
   - Air Injection Reactor (AIR) solenoid valve

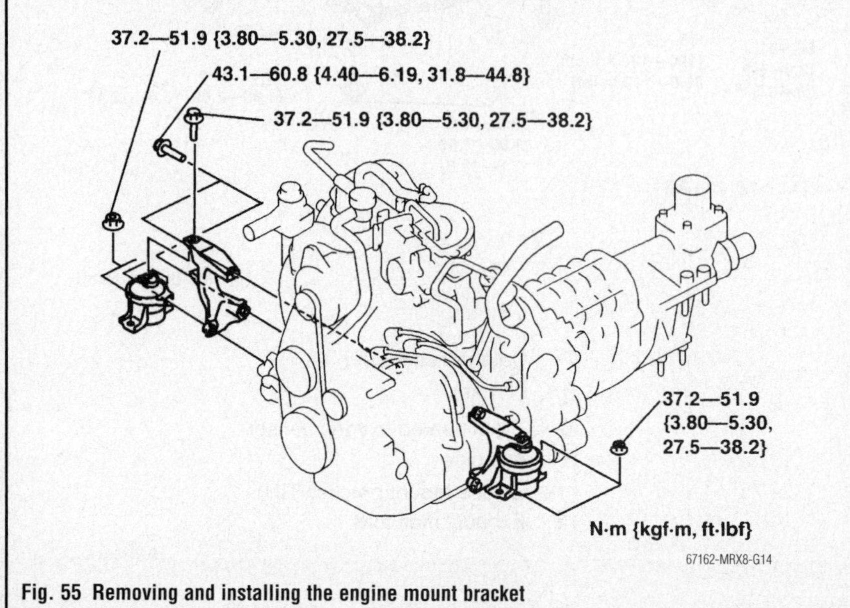

Fig. 55 Removing and installing the engine mount bracket

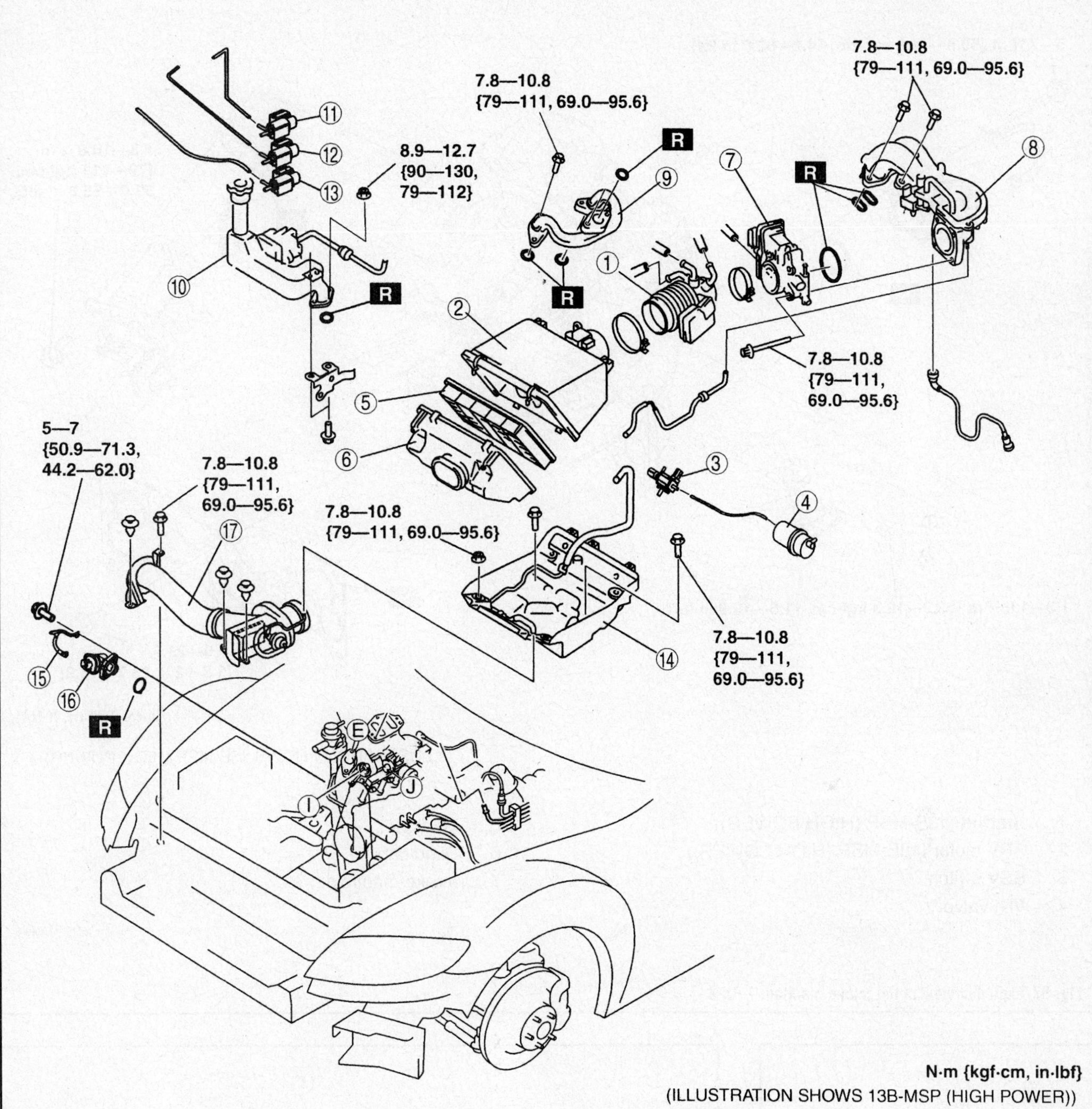

1 Air hose
2 Air cleaner cover
3 VFAD solenoid valve (13B-MSP (High power))
4 Vacuum chamber (13B-MSP (High power))
5 Air cleaner element
6 Air cleaner case
7 Throttle body
8 Extension manifold (upper)
9 Extension manifold (lower) (13B-MSP (High power))
10 Oil filler pipe
11 AIR solenoid valve
12 SSV solenoid valve
13 VDI solenoid valve
14 Air cleaner insulator
15 Bracket (13B-MSP (High power))
16 APV motor (13B-MSP (High power))
17 Fresh-air duct

67162-MRX8-G08

Fig. 56 Exploded view of the air intake system—RX-8

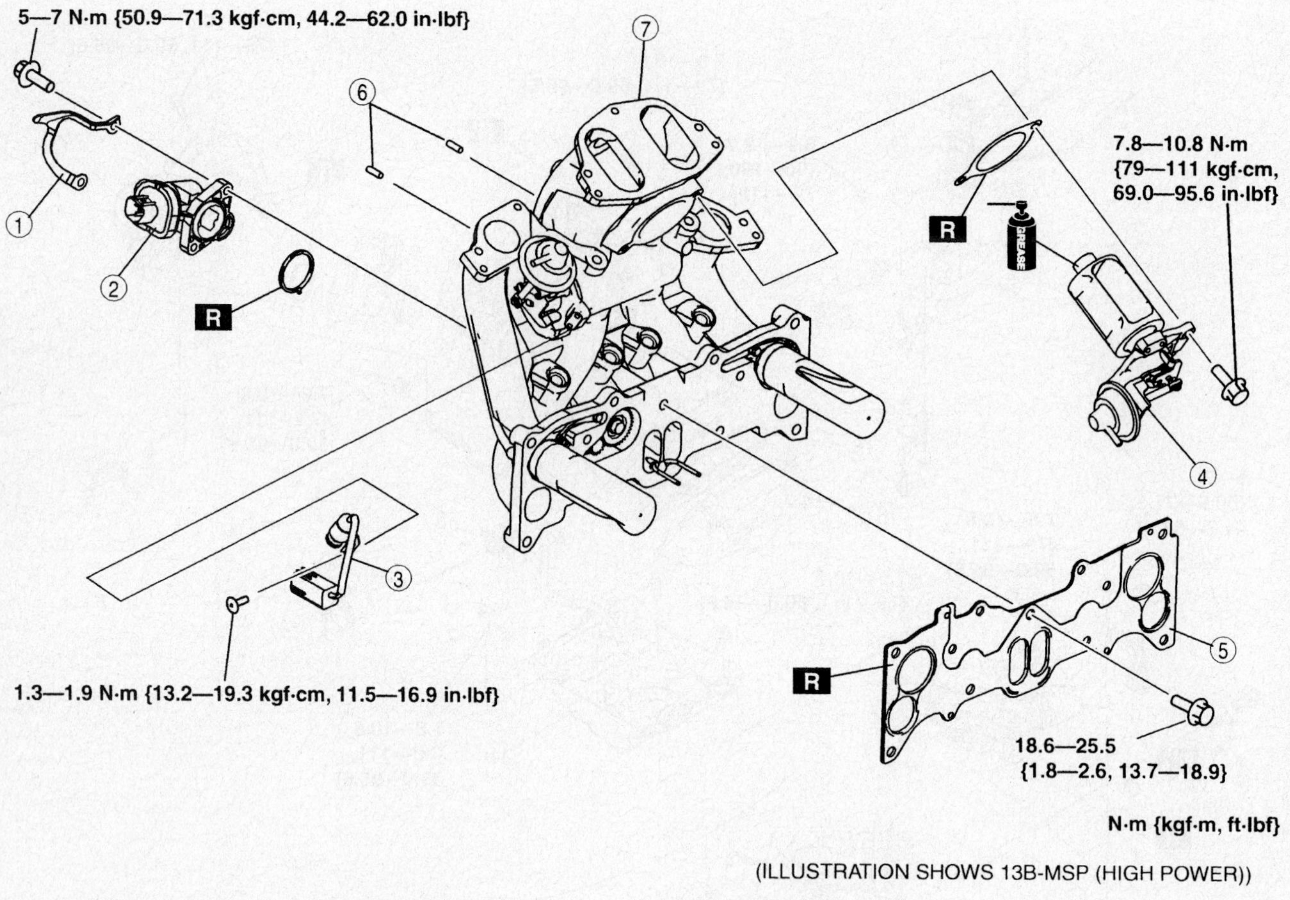

5—7 N·m {50.9—71.3 kgf·cm, 44.2—62.0 in·lbf}
7.8—10.8 N·m {79—111 kgf·cm, 69.0—95.6 in·lbf}
GREASE
R
1.3—1.9 N·m {13.2—19.3 kgf·cm, 11.5—16.9 in·lbf}
18.6—25.5 {1.8—2.6, 13.7—18.9}
N·m {kgf·m, ft·lbf}
(ILLUSTRATION SHOWS 13B-MSP (HIGH POWER))

1 Bracket (13B-MSP (HIGH POWER))
2 APV motor (13B-MSP (HIGH POWER))
3 SSV switch
4 VDI valve
5 Gasket
6 Blind cap
7 Intake manifold

67162-MRX8-G09

Fig. 57 Exploded view of the intake manifold—RX-8

APV
APV
OIL APPLICATION POINT
67162-MRX8-G10

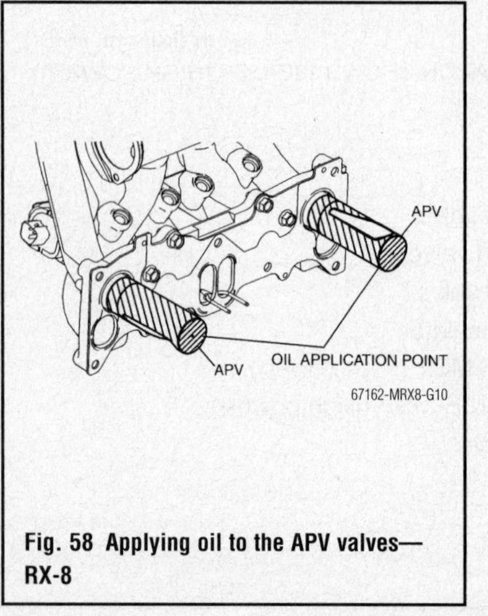

Fig. 58 Applying oil to the APV valves—RX-8

(ILLUSTRATION SHOWS 13B-MSP (HIGH POWER))
67162-MRX8-G11

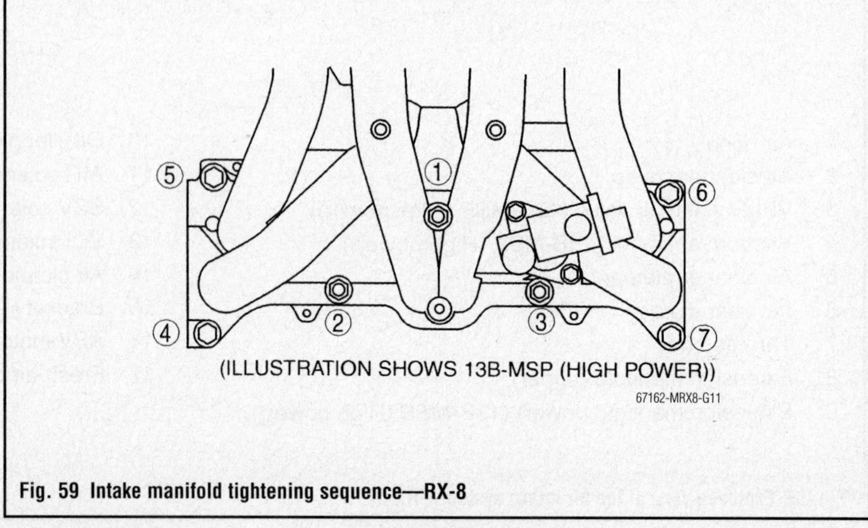

Fig. 59 Intake manifold tightening sequence—RX-8

- Oil filler pipe and torque the bolts to 79–112 inch lbs. (9–13 Nm)
- Lower extension manifold on high output engines and torque the bolts to 79–112 inch lbs. (9–13 Nm)
- Upper extension manifold and torque the bolts to 69–96 inch lbs. (8–11 Nm)
- Throttle body and torque the bolts to 69–96 inch lbs. (8–11 Nm)
- Air cleaner case
- Vacuum chamber on high output engines
- Variable Fresh Air Duct (VFAD) solenoid valve on high output engines
- Air cleaner cover
- Air hose
- Engine and transmission assembly
- Negative battery cable

7. Fill the cooling system.
8. On models with dynamic suspension, perform the steering angle sensor initialization procedure.
9. Run the engine and check for leaks.

OIL PRESSURE SENSOR

TESTING

See Figures 60 and 61.

⁂ WARNING

Remove and install all parts when the engine is cold, otherwise they can cause severe burns or serious injury. Continuous exposure to USED engine oil has caused skin cancer in laboratory mice. Protect your skin by washing with soap and water immediately after working with engine oil.

1. Disconnect the connector, and remove the oil pressure switch.

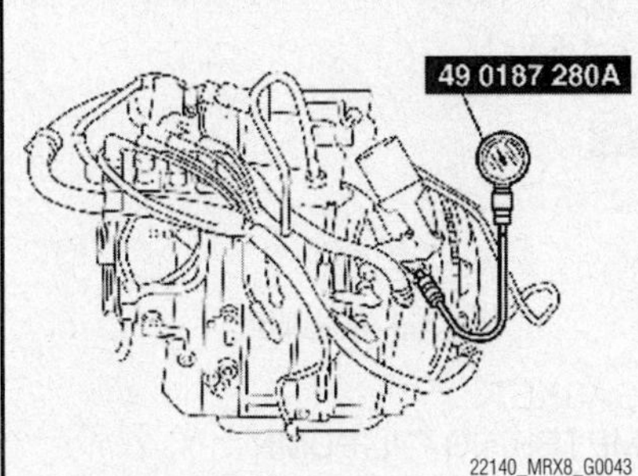

Fig. 60 View of the oil pressure switch location and the Special Service Tool 49 0187 280A

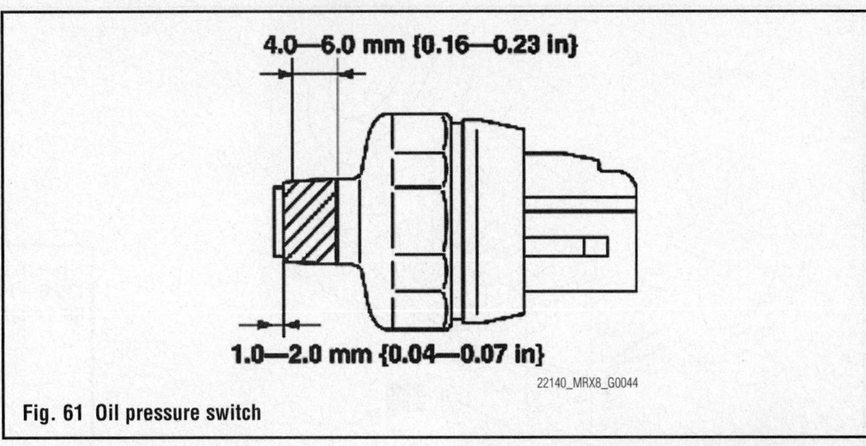

Fig. 61 Oil pressure switch

2. Connect the Special Service Tool 49 0187 280A or suitable equivalent to the oil pressure switch installation hole from underneath the vehicle.
3. Warm up the engine to normal operating temperature.
4. Run the engine at the specified speed, and note the gauge readings.

➡If not as specified, inspect for the cause and repair or replace if necessary.

➡The oil pressure can vary with oil viscosity and temperature.

5. Oil pressure reference value oil temperature: 212°F (100°C)**50.8 psi** 3,000 rpm.

⁂ CAUTION

Stop the engine and wait until it is cool.

6. Remove the pressure gauge.

⁂ WARNING

Make sure that there is no sealant between 0.04–0.07 inches (1.0–2.0 mm) from the end of the oil pressure switch to prevent a possible operation malfunction.

7. Apply silicone sealant to the oil pressure switch threads as shown in the figure.
8. Install the oil pressure switch.
9. Connect the connector.
10. Start the engine and confirm that there is no oil leakage.

⁂ WARNING

If there is any oil leakage, repair or replace the applicable part.

REMOVAL & INSTALLATION

See Figure 61.

⁂ WARNING

Remove and install all parts when the engine is cold, otherwise they can cause severe burns or serious injury. Continuous exposure to USED engine oil has caused skin cancer in laboratory mice. Protect your skin by washing with soap and water immediately after working with engine oil.

1. Disconnect the connector, and remove the oil pressure switch.
2. Apply silicone sealant to the oil pressure switch threads.

⁂ WARNING

Make sure that there is no sealant between 0.04–0.07 inches (10–2.0 mm) from the end of the oil pressure switch to prevent a possible operation malfunction.

3. Install the oil pressure switch.
4. Connect the connector.
5. Start the engine and confirm that there is no oil leakage.

⁂ WARNING

If there is any oil leakage, repair or replace the applicable part.

OIL PAN

REMOVAL & INSTALLATION

See Figure 62.

1. Before servicing the vehicle, refer to the Precautions Section.
2. Drain the engine oil.
3. Remove or disconnect the following:
   - Engine cover
   - Battery cover
   - Negative battery cable
   - Electrical connector

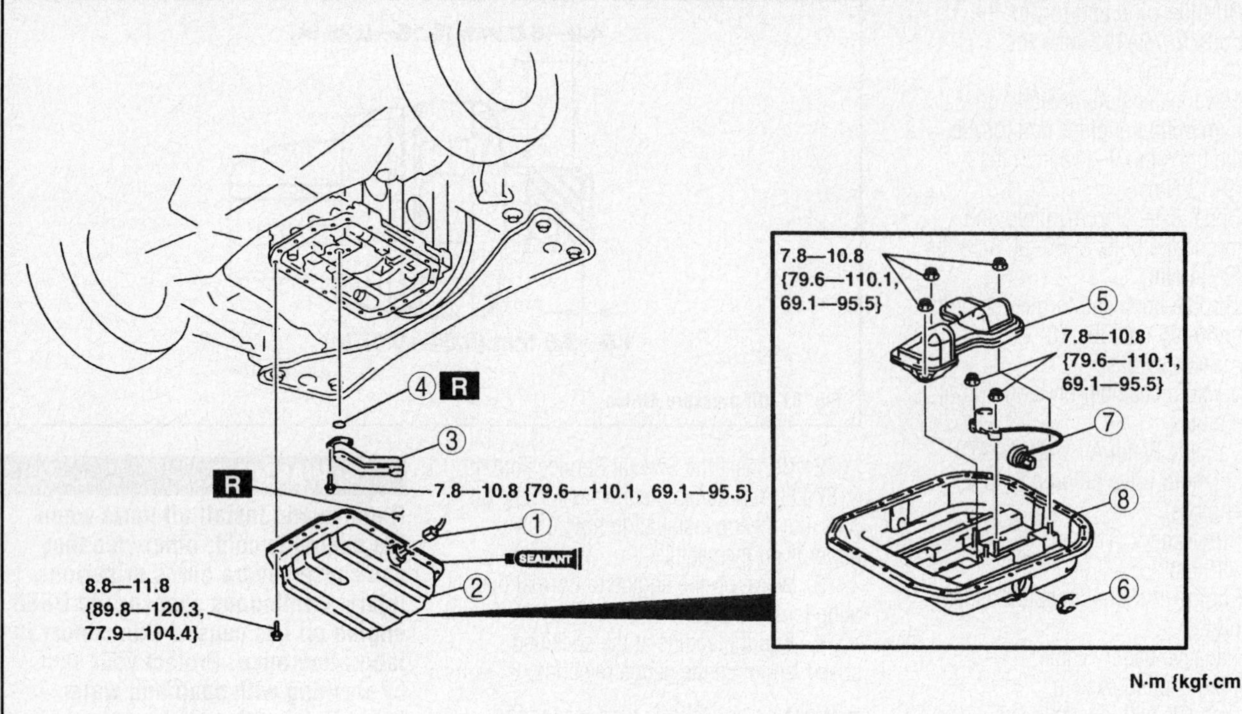

Fig. 62 Exploded view of the oil pan and related components

- Oil pan bolts and the oil pan using a seal cutter, then insert a flat pry tool into the locations illustrated.

To install:

4. Clean the oil pan. Clean all dirt, oil and old sealant from the oil pan and cylinder block contact surfaces.
5. Apply a continuous bead of silicone sealant around the perimeter of the oil pan.
6. Install the oil pan and tighten the bolts to 78–104 inch lbs. (9–12 Nm).
   - Electrical connector
   - Negative battery cable
   - Battery cover
   - Engine cover
7. Fill the engine with clean oil.
8. Start the vehicle, check for leaks and repair if necessary.
9. On models with dynamic suspension, perform the steering angle sensor initialization procedure.

OIL PUMP

REMOVAL & INSTALLATION

See Figure 63.

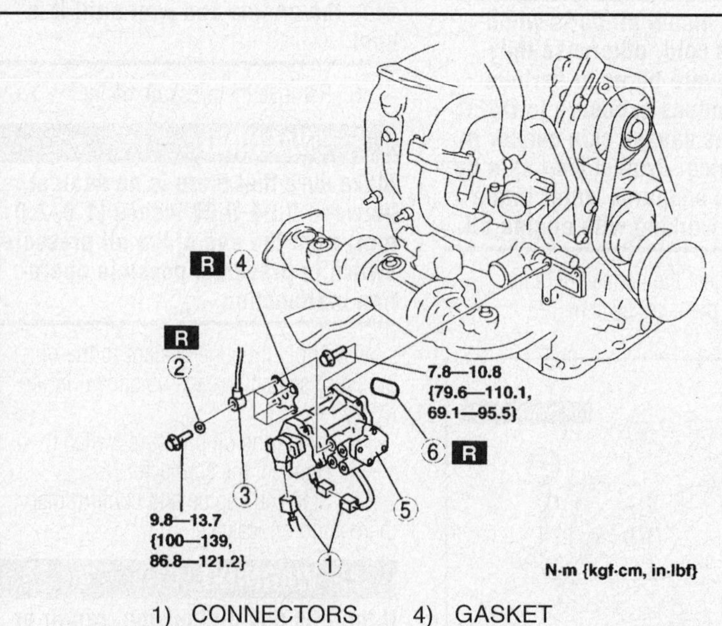

Fig. 63 Exploded view of the oil pump mounting—RX-8

1. Before servicing the vehicle, refer to the Precautions Section.
2. Drain the engine oil.
3. Remove or disconnect the following:
   - Engine cover
   - Battery cables
   - Battery cover, battery box and tray
   - Upper and lower extension manifolds
   - Electrical connectors
   - Oil pipe
   - Gasket
   - Oil pump and o-ring

To install:

4. Clean the oil, dirt and old sealant from all contact surfaces.
5. Install or connect the following:
   - New O-rings on the oil pump
   - Oil pump. Torque the bolts to 87–122 inch lbs. (10–14 Nm).
   - Gasket
   - Oil pipe
   - Electrical connectors
   - Upper and lower extension manifolds
   - Battery cover, battery box and tray
   - Battery cables
   - Engine cover
6. Fill the engine with clean oil.
7. Start the vehicle, check for leaks and repair if necessary.
8. On models with dynamic suspension, perform the steering angle sensor initialization procedure.

REAR MAIN SEAL

REMOVAL & INSTALLATION

See Figure 64.

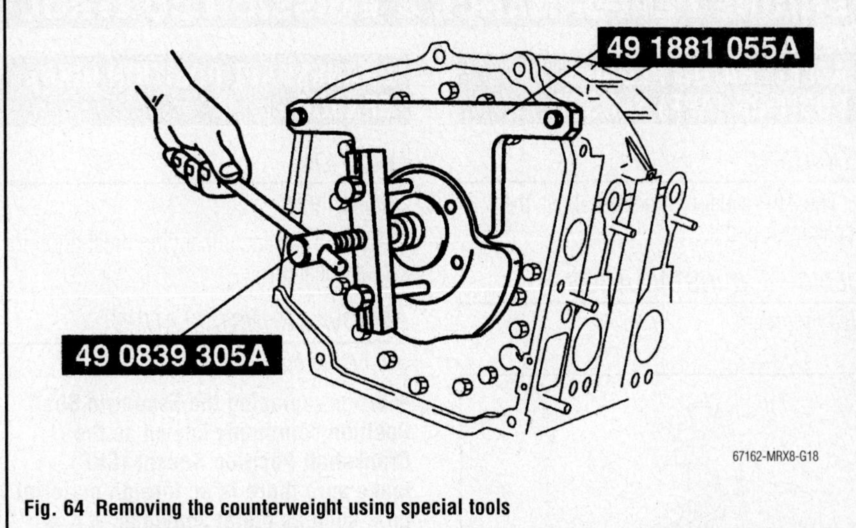

Fig. 64 Removing the counterweight using special tools

1. Before servicing the vehicle, refer to the Precautions Section.
2. Remove or disconnect the following:
   - Negative battery cable
   - Transmission assembly
   - Clutch/flywheel assembly, if equipped with a manual transaxle/transmission
   - Flexplate/shim plates, if equipped with an automatic transaxle/transmission
   - Using special tools 49 1881-055A and 49-0820-035, remove the counterweight locknut.
   - Using special tools 49 1881-055A and 49-0839-305A, remove the counterweight
   - Place a rag over the eccentric shaft and using a pry tool, carefully pry the oil seal from the oil seal housing.
3. Clean the gasket mounting surfaces.

To install:

4. Clean the oil seal housing. Coat the lip of the oil seal and the housing with clean engine oil.
5. Install or connect the following:
   - New oil seal into the housing by tapping it evenly into place with a hammer and a seal installer until it is flush with the edge of the rear cover
   - Install the key into the eccentric shaft and install the counterweight
   - Apply sealant to the seating face then install the locknut and loosely tighten.
   - Lock the counterweight using the special tools, then tighten the locknut to 290–361 ft. lbs. (392–490 Nm).
   - Clutch/flywheel assembly or the flexplate, as applicable
   - Transaxle/transmission
   - Negative battery cable
6. On models with dynamic suspension, perform the steering angle sensor initialization procedure.

ENGINE PERFORMANCE & EMISSION CONTROL

ACCELERATOR PEDAL POSITION (APP) SENSOR

LOCATION

The APP sensor is located above the accelerator pedal.

REMOVAL & INSTALLATION

See Figure 65.

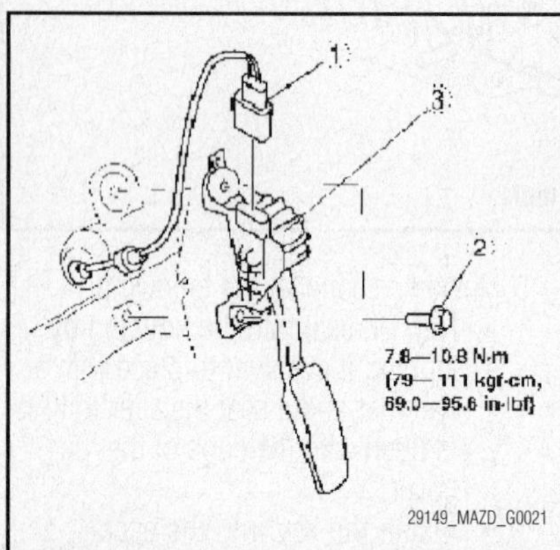

Fig. 65 Accelerator pedal position sensor location

CRANKSHAFT POSITION (CKP) SENSOR

LOCATION

See Figure 66.

Refer to the accompanying illustration for sensor location.

REMOVAL & INSTALLATION

See Figure 66.

➡When replacing the Eccentric Shaft Position commonly known as the Crankshaft Position Sensor (CKP), make sure there is no foreign material on it such as metal shavings. If it is installed with foreign material, the sensor output signal will malfunction resulting from fluctuation in magnetic flux and cause a deterioration in engine control.

1. Disconnect the eccentric shaft position sensor connector.
2. Remove the eccentric shaft position sensor installation bolt and remove the eccentric shaft position sensor.
3. Install in the reverse order of removal.
4. Tighten the eccentric shaft position sensor bolt to 72 inch lbs. (8.3 Nm).

ENGINE COOLANT TEMPERATURE (ECT) SENSOR

REMOVAL & INSTALLATION

See Figure 67.

1. Partially drain the engine cooling system until the coolant level is below the ECT sensor-mounting hole.
2. Disconnect the negative battery cable.
3. Detach the wiring harness connector from the ECT sensor.
4. Using an open-end wrench, remove the coolant temperature sensor from the intake manifold or thermostat housing.

To install:

5. Thread the sensor into the intake manifold, or thermostat housing, by hand, then tighten it securely.
6. ECT sensor tightening torque 14 ft. lbs. (20 Nm).
7. Connect the negative battery cable.
8. Refill the engine cooling system.

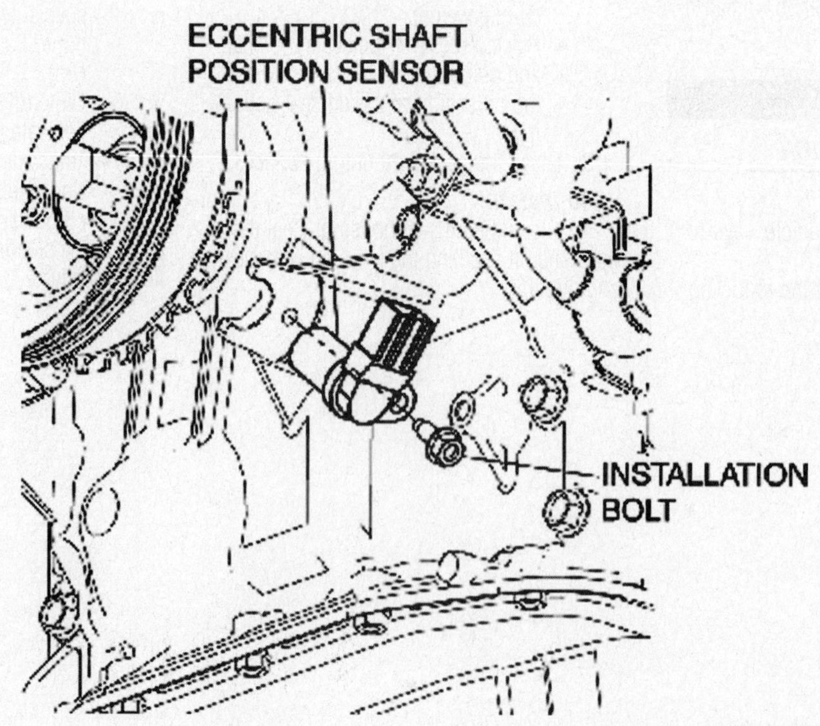

Fig. 66 Eccentric shaft position sensor

BARO SENSOR
WIRING HARNESS-SIDE CONNECTOR

PCM
WIRING HARNESS-SIDE CONNECTOR

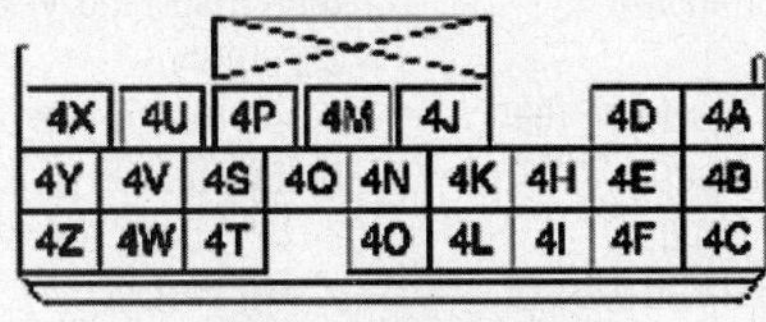

22140_MRX8_G0041

Fig. 67 Eccentric shaft position sensor

9. Start the engine, check for coolant leaks and top off the cooling system.

HEATED OXYGEN (HO2S) SENSOR

REMOVAL & INSTALLATION

CAUTION

The temperature of the exhaust system is extremely high after the engine has been run. To prevent personal injury, allow the exhaust system to cool completely before removing sensor from the exhaust system.

1. Disconnect the negative battery cable.
2. Raise and safely support the vehicle on jack stands.
3. Disconnect the HO2S from the engine control sensor wiring.

➡If excessive force is needed to remove the sensors lubricate the sensor with penetrating oil prior to removal.

4. Remove the sensors with a sensor removal tool, such as Ford Tool T94P-9472-A or equivalent.

To install:

5. Install the sensor in the mounting boss, and then tighten it to 27–33 ft. lbs.
6. Connect the sensor electrical wiring connector to the engine wiring harness.
7. Lower the vehicle.
8. Connect the negative battery cable.

IDLE AIR CONTROL (IAC) VALVE

REMOVAL & INSTALLATION

1. Disconnect the negative battery cable.
2. Detach the electrical connector at the sensor.
3. Remove the sensor from the throttle body.

To install:

4. Install the sensor in the opening in the throttle body and tighten the sensor.
5. Attach the electrical connector to the sensor.
6. Connect the negative battery cable.

INTAKE AIR TEMPERATURE (IAT) SENSOR

REMOVAL & INSTALLATION

1. Disconnect the negative battery cable.
2. Remove the IAT sensor from the air cleaner housing or intake air pipe.

To install:

3. Install a new sealing washer to the thermo sensor.

KNOCK SENSOR (KS)

OPERATION

The Knock Sensor (KS) converts the vibration of the cylinder block into a voltage and outputs it. If there is a malfunction of the knock sensor, the voltage output will not change. The Powertrain Control Module (PCM). checks whether the voltage output changes.

REMOVAL & INSTALLATION

See Figure 68.

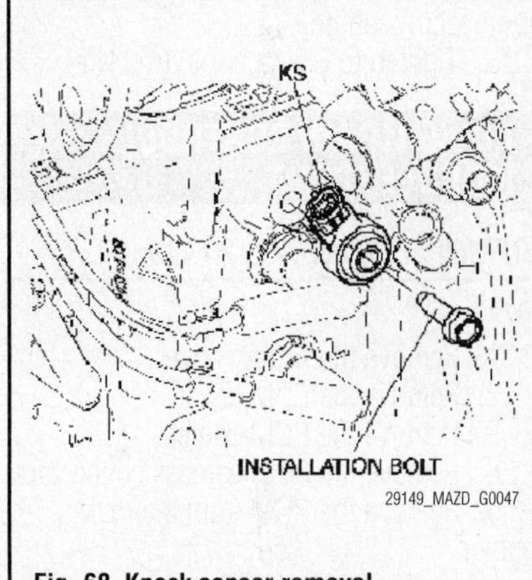

29149_MAZD_G0047

Fig. 68 Knock sensor removal

1. Disconnect the negative battery cable.
2. Disconnect the sensor connector.
3. Remove the installation bolt and sensor from the engine block.
4. Installation is the reverse of the removal procedure.
5. Tightening torque 12–18 ft. lbs.(15–25 Nm).

MALFUNCTION INDICATOR LIGHT (MIL)

RESET PROCEDURES

Use the WDS to clear the codes from the Electronic Control Module (ECM)/Powertrain Control Module (PCM).

MASS AIR FLOW (MAF) SENSOR

REMOVAL & INSTALLATION

See Figure 69.

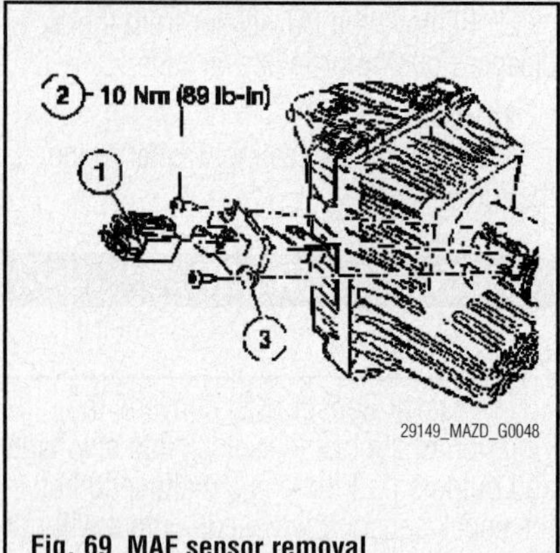

Fig. 69 MAF sensor removal

1. Remove the air cleaner outlet tube.
2. Disconnect the MAF sensor electrical connector.
3. Remove the attaching four nuts and the MAF sensor.
4. Installation is the reverse of the removal procedure.
5. Tighten to 6.5 inch lbs. (.70 Nm).

POWERTRAIN CONTROL MODULE (PCM)

REMOVAL & INSTALLATION

See Figure 70.

1. Remove the battery cover.
2. Remove the PCM cover.
3. Remove the PCM cooler.
4. Remove the PCM harness connectors.
5. Remove the PCM from the PCM cooler.
6. Installation is the reverse of the removal procedure.

THROTTLE POSITION SENSOR (TPS)

REMOVAL & INSTALLATION

1. Disconnect the negative battery cable.
2. Remove the necessary air intake components to access the TPS.
3. Detach the electrical wire harness plug from the sensor.
4. Paint an alignment mark on the sensor housing to the throttle body.
5. Remove the sensor attaching bolts.
6. Remove the sensor from the throttle body.
7. Installation is the reverse of the removal procedure.

VEHICLE SPEED SENSOR (VSS)

REMOVAL & INSTALLATION

The VSS is located half-way down the right-hand side of the transmission assembly.

1. Apply parking brake, block the rear wheels, then raise and safely support the front of the vehicle on jack stands.
2. From under the right-hand side of the vehicle, disengage the wiring harness connector from the VSS.
3. Loosen the VSS hold-down bolt, then pull the VSS out of the transmission housing.

To install:

4. If a new sensor is being installed, transfer the driven gear retainer and gear to the new sensor.
5. Ensure that the O-ring is properly seated in the VSS housing.
6. For ease of assembly, engage the wiring harness connector to the VSS, then insert the VSS into the transmission assembly.
7. Install and tighten the VSS hold-down bolt to 62–88 in lbs.
8. Lower the vehicle and remove the wheel blocks.

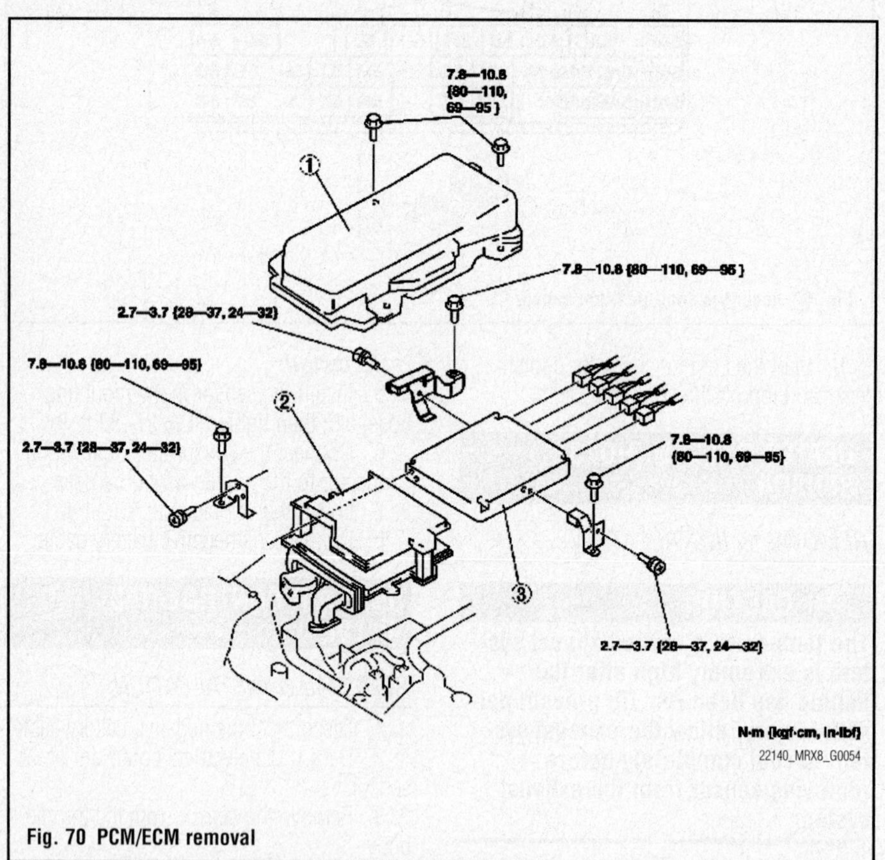

Fig. 70 PCM/ECM removal

FUEL SYSTEM SERVICE PRECAUTIONS

Safety is the most important factor when performing not only fuel system maintenance but any type of maintenance. Failure to conduct maintenance and repairs in a safe manner may result in serious personal injury or death. Maintenance and testing of the vehicle's fuel system components can be accomplished safely and effectively by adhering to the following rules and guidelines.

- To avoid the possibility of fire and personal injury, always disconnect the negative battery cable unless the repair or test procedure requires that battery voltage be applied.
- Always relieve the fuel system pressure prior to disconnecting any fuel system component (injector, fuel rail, pressure regulator, etc.), fitting or fuel line connection. Exercise extreme caution whenever relieving fuel system pressure to avoid exposing skin, face and eyes to fuel spray. Please be advised that fuel under pressure may penetrate the skin or any part of the body that it contacts.
- Always place a shop towel or cloth around the fitting or connection prior to loosening to absorb any excess fuel due to spillage. Ensure that all fuel spillage (should it occur) is quickly removed from engine surfaces. Ensure that all fuel soaked cloths or towels are deposited into a suitable waste container.
- Always keep a dry chemical (Class B) fire extinguisher near the work area.
- Do not allow fuel spray or fuel vapors to come into contact with a spark or open flame.
- Always use a back-up wrench when loosening and tightening fuel line connection fittings. This will prevent unnecessary stress and torsion to fuel line piping.
- Always replace worn fuel fitting O-rings with new. Do not substitute fuel hose or equivalent where fuel pipe is installed.

➡**Before servicing the vehicle, make sure to also refer to the precautions in the beginning of this section as well.**

RELIEVING FUEL SYSTEM PRESSURE

See Figure 71.

✱✱ WARNING

Fuel is extremely flammable. Always keep sparks and flame away from fuel. Ignition may cause death or serious injury, or damage to equipment.

✱✱ WARNING

Fuel line spills and leakage from the pressurized fuel system are dangerous. Fuel can ignite and cause serious injury or death and damage. Fuel can also irritate skin and eyes. To prevent this, always complete the "Fuel Line Safety Procedure".

✱✱ WARNING

If there is foreign material on the connecting area of the quick release connector, it might damage the connector or fuel pipe. To prevent this, when the quick release connector has been disconnected, clean the connecting area before reconnecting it.

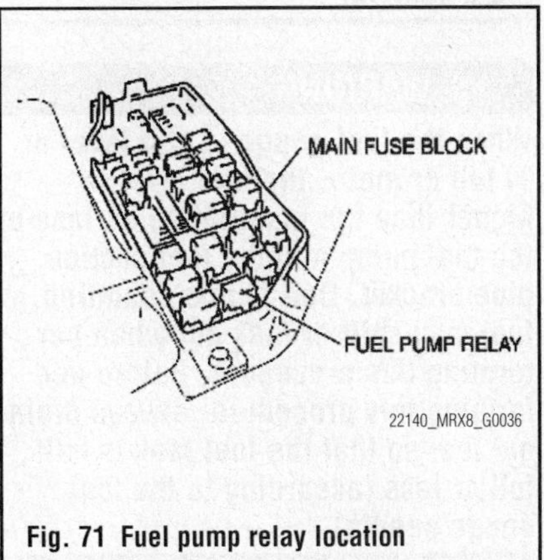

Fig. 71 Fuel pump relay location

1. Before servicing the vehicle, refer to the Precautions Section.
2. Remove the filler cap.
3. Remove the fuel pump relay from the relay box, located in the main fuse block.
4. Start the engine.
5. After the engine stalls, turn the ignition switch **OFF**.
6. After servicing the vehicle, reinstall the relay.

FUEL FILTER

REMOVAL & INSTALLATION

The fuel filter is part of the pump assembly, see the removal and installation procedure in this section.

FUEL PUMP

REMOVAL & INSTALLATION

See Figure 72.

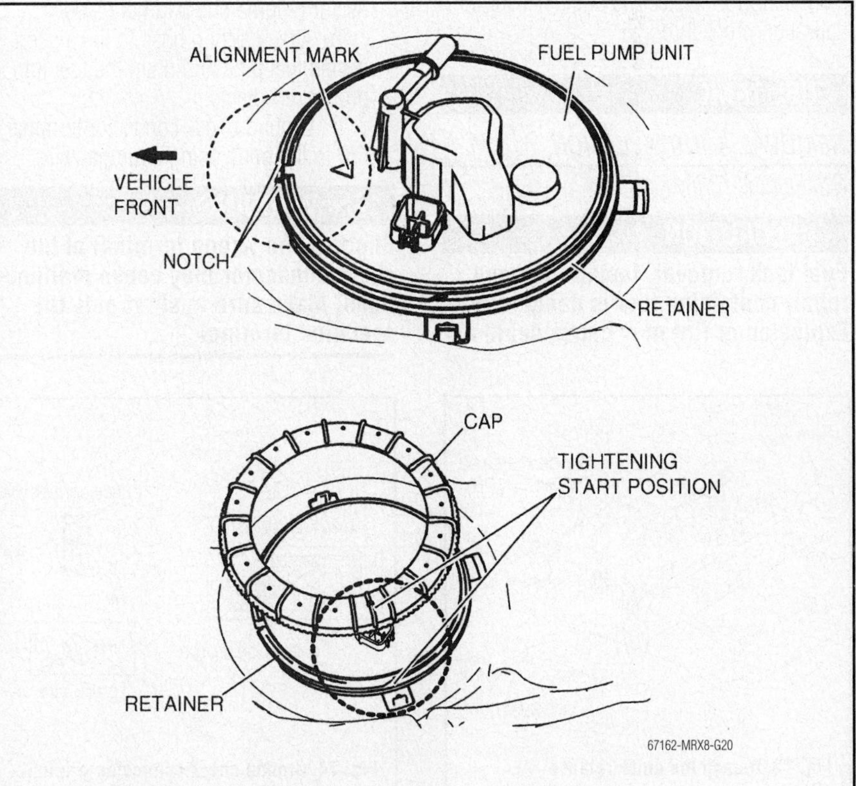

Fig. 72 Installing the fuel the fuel pump ring–RX-8

1. Before servicing the vehicle, refer to the Precautions Section.
2. Relieve the fuel system pressure.
3. Remove or disconnect the following:
   - Negative battery cable
   - Rear seat cushion
4. Drain the fuel from the tank.
   - Service hole cover
   - Fuel pump electrical connector
   - All fuel hoses from the fuel pump unit
   - Fuel pump ring using tool 49 T042 001
   - Fuel pump and gaskets from the fuel tank

To install:

5. Align the fuel pump alignment mark with the notch in the retainer and install the fuel pump using a new gasket.
6. Align the cap with the retainer and tighten one full turn by hand.
7. Tighten the cap with the special tool to 75 ft. lbs. (102 Nm).
8. Install or connect the following:
   - Fuel hoses to the fuel pump
   - Fuel pump electrical connector
   - Service hole cover
   - Rear seat cushion
   - Negative battery cable
9. Add a minimum of 10 gallons of fuel to the tank and check for leaks.
10. On models with dynamic suspension, perform the steering angle sensor initialization procedure.

FUEL TANK

REMOVAL & INSTALLATION

See Figures 73 through 76.

✱✱ CAUTION

Fuel tank removal, installation and repair containing fuel is dangerous. Explosion or fire may cause death or serious injury. Always properly steam clean a fuel tank before repairing it.

1. Park the vehicle on a level surface.
2. Relieve the fuel system pressure, as outlined in this section.
3. Remove the rear seat.
4. Drain fuel from the fuel tank, as follows:

✱✱ CAUTION

A person charged with static electricity could cause a fire or explosion, resulting in death or serious injury. Before draining fuel, make sure to discharge static electricity by touching a vehicle.

✱✱ WARNING

When the fuel gauge is at a level of ¾ full or more, the fuel level is higher than the installation surface of the fuel pump and the fuel suction pipe bracket. Due to this condition, fuel may spill or leak out when performing this procedure. Before performing this procedure, always drain out fuel so that the fuel tank is half full or less (according to the fuel gauge needle).

a. Disconnect the quick release connector (engine compartment side
b. Attach a long hose to the disconnected fuel pipe and drain the fuel into a proper receptacle.
c. Ground check connector terminal F/P to the body using a jumper wire

✱✱ WARNING

Shorting the wrong terminal of the check connector may cause malfunctions. Make sure to short only the specified terminal.

d. Turn the ignition switch to the **ON** position and operate the fuel pump for about. 20 min.

✱✱ WARNING

The fuel pump may malfunction if it is operated without any fuel in the fuel tank (fuel pump idling). Constantly monitor the amount of fuel being discharged and immediately stop operation of the pump when essentially no fuel is being discharged.

e. When essentially no fuel is being discharged from the hose, turn the ignition switch to the **LOCK** position.

➡When operating the fuel pump with a full fuel tank, fuel discharge will become erratic after approx. 10 min but will continue for approx. 10 min more and then essentially no fuel will be discharged. At this time, the fuel gauge needle will be at the halfway position.

f. Disconnect the jumper wire.
g. Disconnect the negative battery cable.

5. Remove the following components:
   a. Fuel pump unit.
   b. Main silencer and middle pipe.
   c. Power plant frame.
   d. Propeller shaft.
   e. Position the parking brake cable out of the way.
6. Remove the fuel tank and components as shown in the accompanying illustration.
7. Installation is the reverse of the removal procedure.
8. After installation is complete, inspect all parts by performing the following Fuel Line Inspection procedure:

✱✱ CAUTION

Fuel vapor is hazardous. It can very easily ignite, causing serious injury and damage. Fuel can also irritate skin and eyes. To prevent this, always complete the "Fuel Line Inspection".

✱✱ WARNING

Shorting the wrong terminal of the check connector may cause malfunctions. Make sure to short only the specified terminal.

a. Ground the check connector terminal F/P to the body using a jumper wire.

Fig. 73 Detach the quick release connector

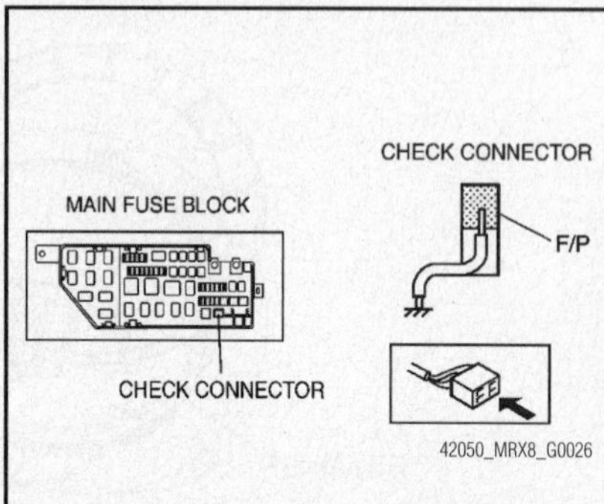

Fig. 74 Ground check connector terminal F/P to the body using a jumper wire

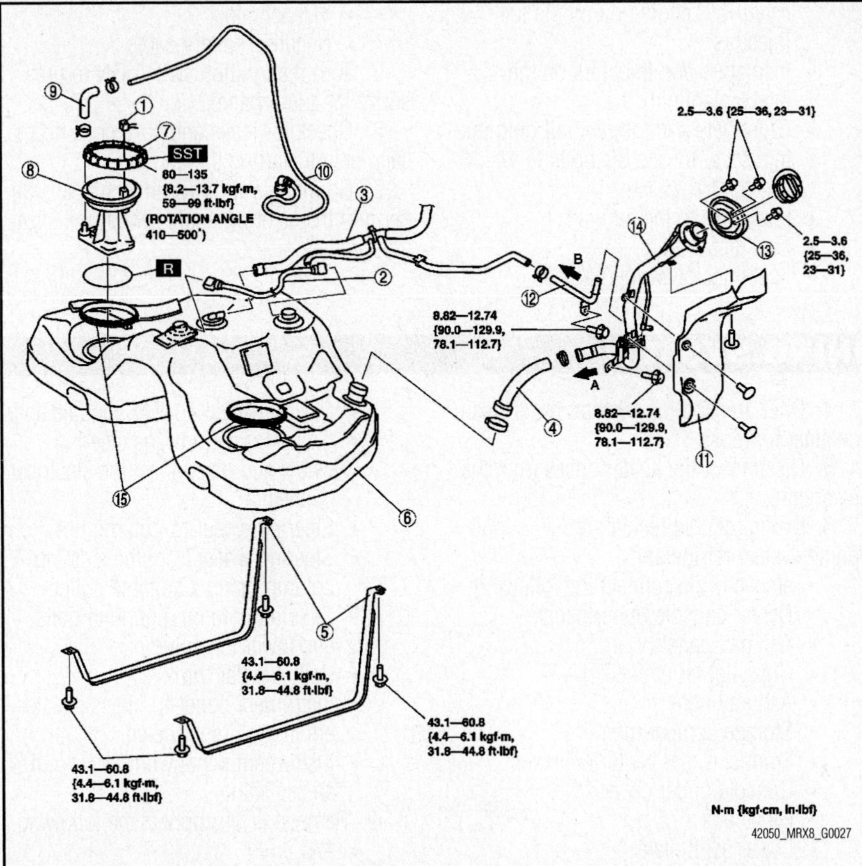

Fig. 75 Exploded view of the fuel tank and related components—Note that the removal order corresponds with the key list number

b. Turn the ignition switch to the **ON** position and operate the fuel pump.

c. Check that that there is no fuel leakage from the pressurized parts.

- If there is leakage, replace the fuel hoses.
- If there is damage to the seal on the fuel pipe side, replace the fuel pipe.
- There should not be any fuel leakage after 5 minutes.

9. After reinstallation, repeat Step a.–c. of the fuel leakage inspection.

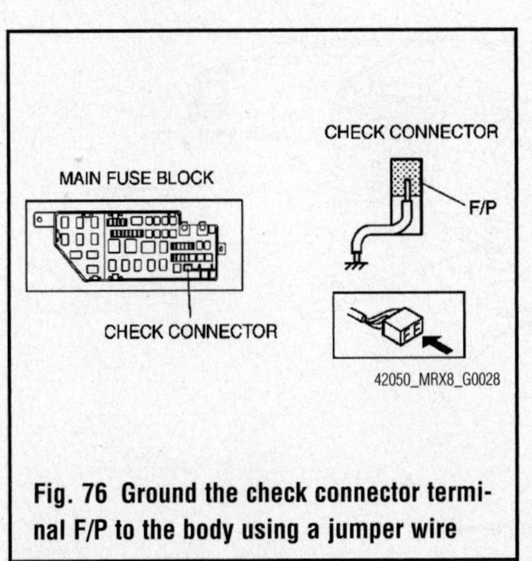

Fig. 76 Ground the check connector terminal F/P to the body using a jumper wire

FUEL RAIL & INJECTORS

REMOVAL & INSTALLATION

See Figure 77.

✲✲ CAUTION

Fuel injection systems remain under pressure after the engine has been turned OFF. Properly relieve fuel pressure before disconnecting any fuel lines. Failure to do so may result in fire or personal injury. Do not allow fuel spray or fuel vapors to come in contact with a spark or open flame. Keep a dry chemical fire extinguisher nearby. Never store fuel in an open container due to risk of fire or explosion.

1. Before servicing the vehicle, refer to the Precautions Section.
2. Relieve the fuel system pressure.
3. Remove or disconnect the following:
   - Negative battery cable
   - Upper and lower extension manifolds
   - Variable Dynamic Effect (VDI) actuator and position out of the way
   - Fuel injector wiring harness
   - Fuel lines at the fuel rail

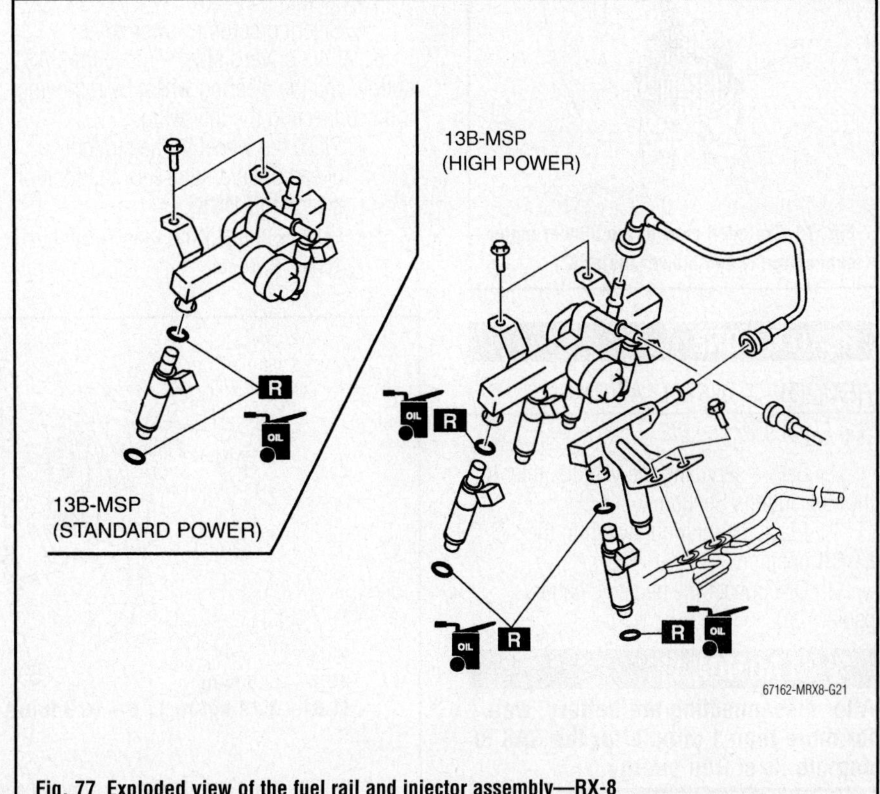

Fig. 77 Exploded view of the fuel rail and injector assembly—RX-8

- Fuel distributor from intake manifold and housing sides
- Fuel rail with the injectors attached
- Fuel injectors, grommets and O-rings from the fuel rail
- O-rings from the fuel injectors

To install:

4. Install or connect the following:
   - New O-rings and grommets lubricated with engine oil on the fuel injectors.
   - Insulators and injectors on the intake manifold
   - Grommets and the fuel rail onto the injectors. Torque the bolts to 14–18 ft. lbs. (19–25 Nm).
   - Fuel lines to the fuel rail
   - Fuel distributors
   - Fuel injector wiring harness
   - VDI actuator
   - Negative battery cable
5. Turn the ignition switch **ON** to pressurize the fuel system.
6. Check for leaks and correct as necessary, before starting the engine.
7. On models with dynamic suspension, perform the steering angle sensor initialization procedure.

HEATING & AIR CONDITIONING SYSTEM

BLOWER MOTOR

REMOVAL & INSTALLATION

See Figure 78.

1. Disconnect the negative battery cable.
2. Detach the electrical connector from the blower motor.
3. Remove the fasteners, then remove the blower motor.

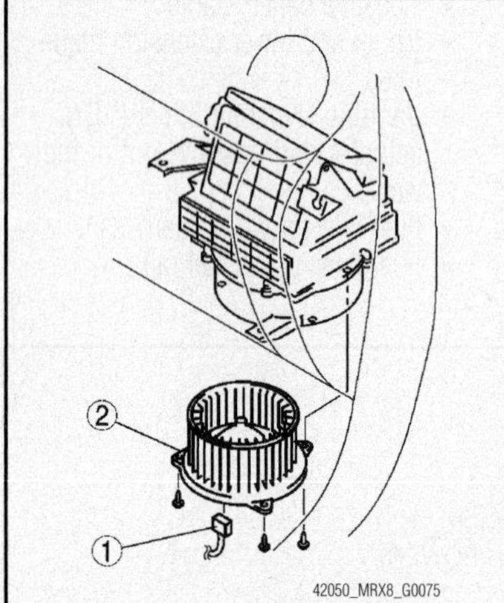

Fig. 78 Exploded view of the blower motor connector (1) and blower motor (2)

HEATER CORE

REMOVAL & INSTALLATION

See Figures 79 and 80.

1. Before servicing the vehicle, refer to the Precautions Section.
2. Place the ignition switch in the **LOCK** position.
3. Disconnect the negative battery cable.

⁂ CAUTION

After disconnecting the battery, wait for more than 1 minute for the SAS to deplete its stored energy.

4. Drain the cooling system into a clean container for reuse.
5. Disconnect the heater hoses from the heater core.
6. Discharge and recover the air conditioning system refrigerant.
7. Remove or disconnect the following:
   - Center console upper panel
   - Ash tray panel
   - Cigar lighter connector
   - Ash tray light
   - Storage compartment
   - Front and rear center consoles
   - Console under cover
   - Glove box
   - Lower scuff plate
   - Lower door side trim plate
   - Lower dashboard side panel
   - Lower dashboard front panel
   - Steering column upper cover
   - Ignition key light
   - Steering column lower cover
8. At the driver's side, remove the SAS module and the steering wheel by removing or disconnecting the following:
   - Place the wheel in the straight-ahead position and turn the ignition switch to **LOCK**.
   - Cover clips at both sides of the steering wheel
   - Steering wheel-to-SAS module bolts
   - SAS module from the steering wheel and disconnect the electrical connector
   - Steering wheel-to-column nut
   - Steering wheel from the steering column using a suitable puller
   - Steering column mounting bolts and lower the column
   - Both A pillar trims
   - Instrument panel-to-chassis fasteners in the order shown
   - Instrument panel with the help of an assistant
9. Remove or disconnect the following:
   - A/C unit
10. Separate the heater core from the A/C unit.

To install:

11. Install the heater core to the A/C unit.
12. Install or connect the following:
    - A/C unit
    - Instrument panel and tighten the fasteners as shown
    - Both A pillar trims
    - Raise the steering column and tighten the bolts to 14 ft. lbs. (19 Nm)

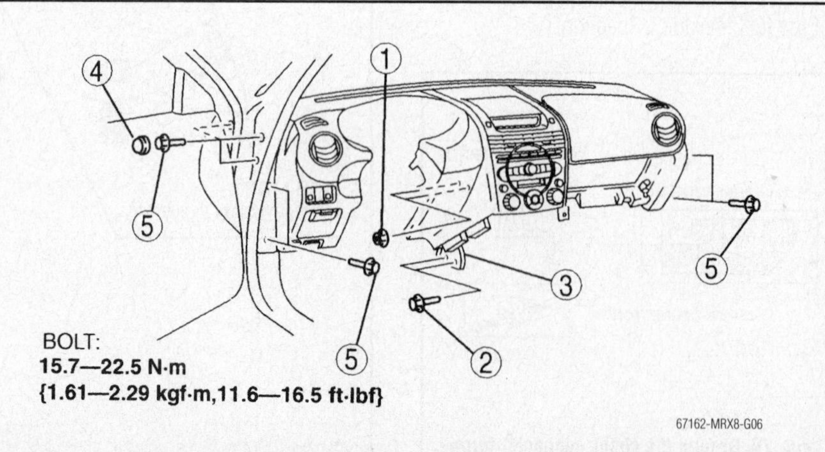

Fig. 79 Instrument panel fastener removal sequence—RX-8

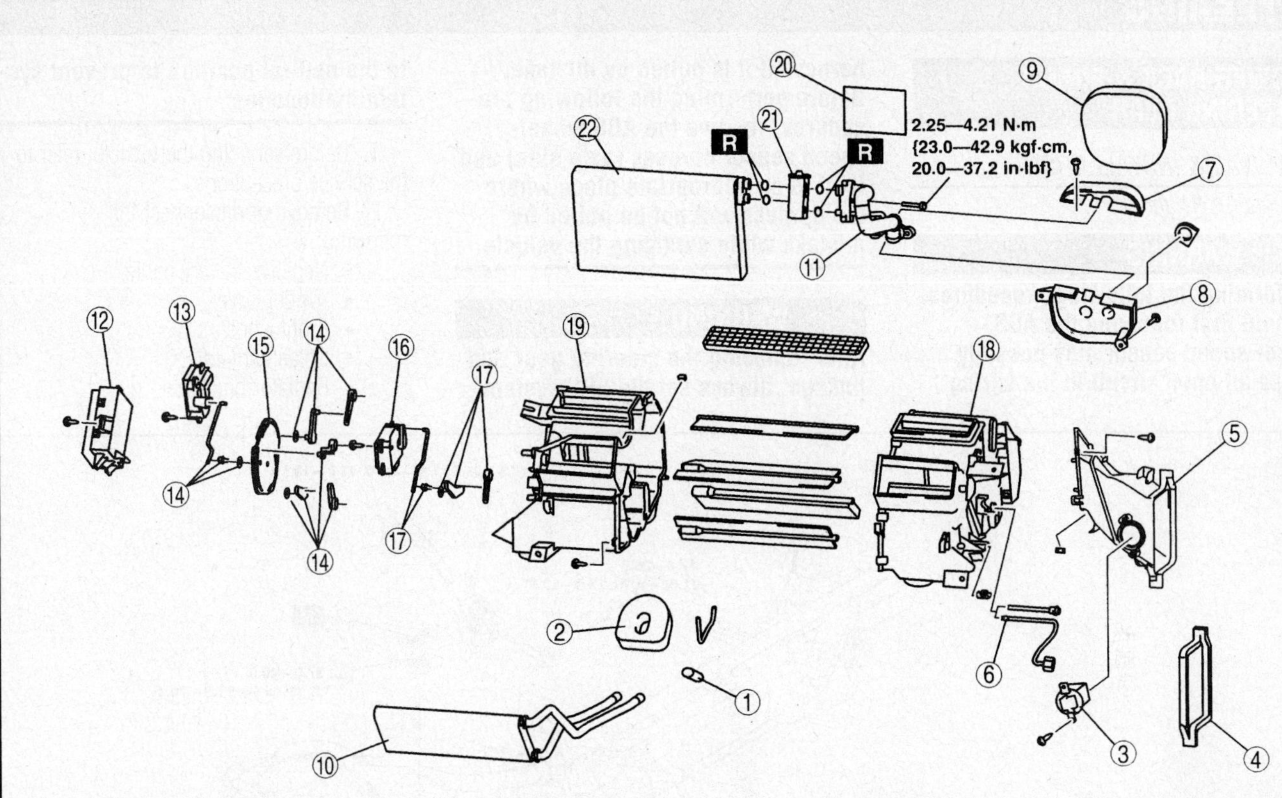

1	Drain hose	12	A/C amplifier
2	Polyurethane foam (1)	13	Airflow mode actuator
3	Resistor	14	Airflow mode link set
4	Adhesive polyurethane (1)	15	Airflow mode main link
5	Air duct	16	Air mix actuator
6	Evaporator temperature sensor	17	Air mix link set
7	Polyurethane foam (2)	18	A/C case (1)
8	Bracket (1)	19	A/C case (2)
9	Bracket (2)	20	Adhesive polyurethane (2)
10	Heater core	21	Expansion valve
11	Evaporator pipe	22	Evaporator

67162-MRX8-G07

Fig. 80 Exploded view of the A/C unit with heater core—RX8

- Steering wheel to steering column
- Steering wheel-to-column nut and tighten to 33 ft. lbs. (45 Nm)
- SAS module to the steering wheel and connect the electrical connector
- Steering wheel-to-SAS module bolts and tighten to 70–103 inch lbs. (8–12 Nm)
- Cover clips at both sides of the steering wheel
- Steering column lower cover
- Ignition key light
- Steering column upper cover
- Lower dashboard front panel
- Lower dashboard side panel
- Lower door side trim plate
- Lower scuff plate
- Glove box
- Console under cover
- Front and rear center consoles
- Storage compartment
- Ash tray light
- Cigar lighter connector
- Ash tray panel
- Center console upper panel

13. Connect the heater hoses to the heater core.

14. Refill the cooling system.

15. Connect the negative battery cable.

16. Evacuate, charge and leak test the air conditioning system refrigerant.

17. On models with dynamic suspension, perform the steering angle sensor initialization procedure.

18. Operate the engine to normal operating temperatures; then, check the climate control operation and check for leaks.

STEERING

POWER RACK & PINION STEERING GEAR

REMOVAL & INSTALLATION

See Figures 81 and 82.

WARNING

Performing the following procedures without first removing the ABS wheel-speed sensor may possibly cause an open circuit in the wiring harness if it is pulled by mistake. Before performing the following procedures, remove the ABS wheel-speed sensor harness (axle side) and fix it to an appropriate place where the harness will not be pulled by mistake while servicing the vehicle.

WARNING

After replacing the steering gear and linkage, always set the EPS system to the neutral position to prevent system malfunction

1. Before servicing the vehicle, refer to the service precautions.
2. Remove or disconnect the following:
    - Negative battery cable
    - Under cover.
    - Front wheels
    - Splash shield
    - Radiator bracket

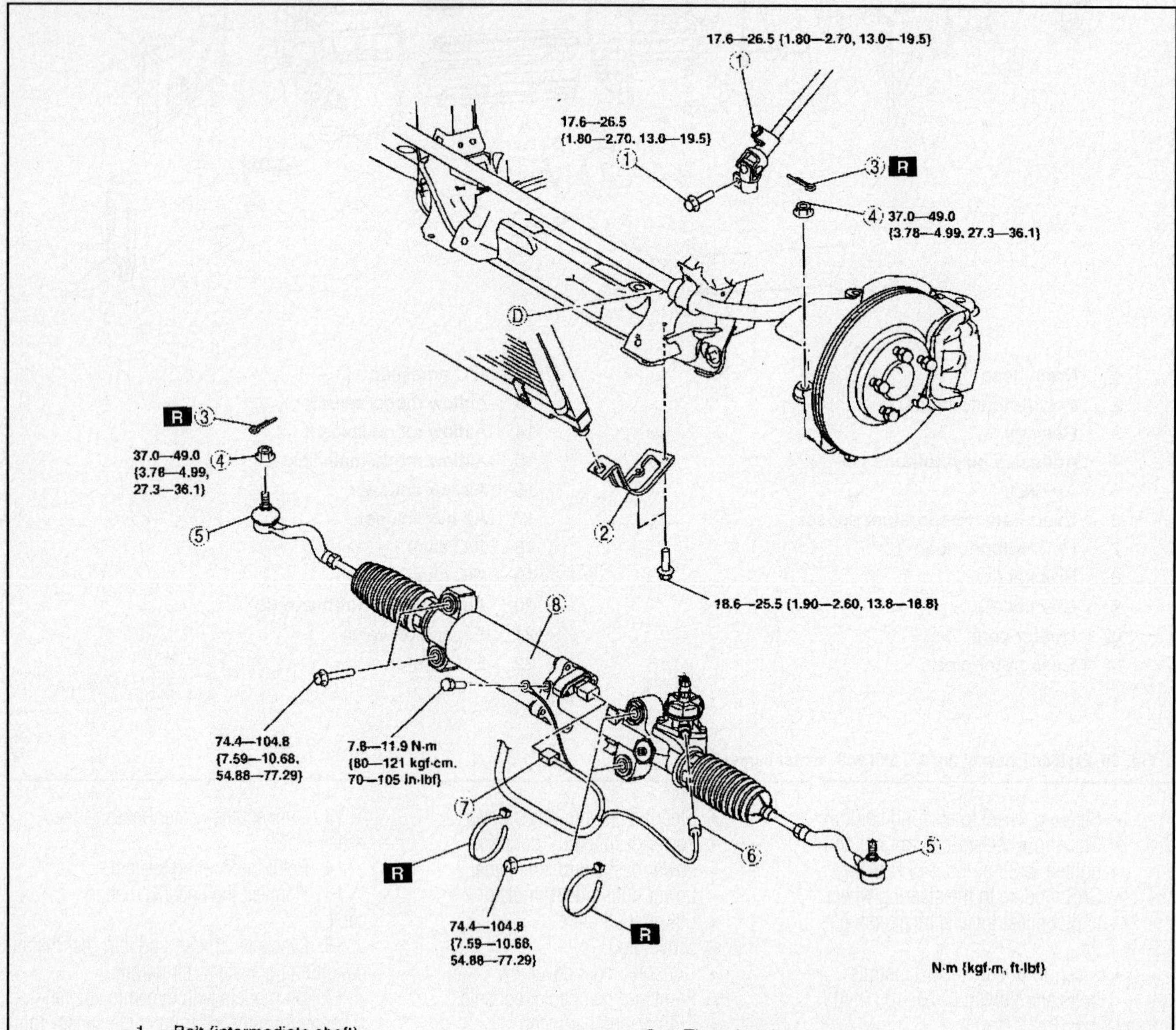

Fig. 81 Exploded view of steering gear mounting

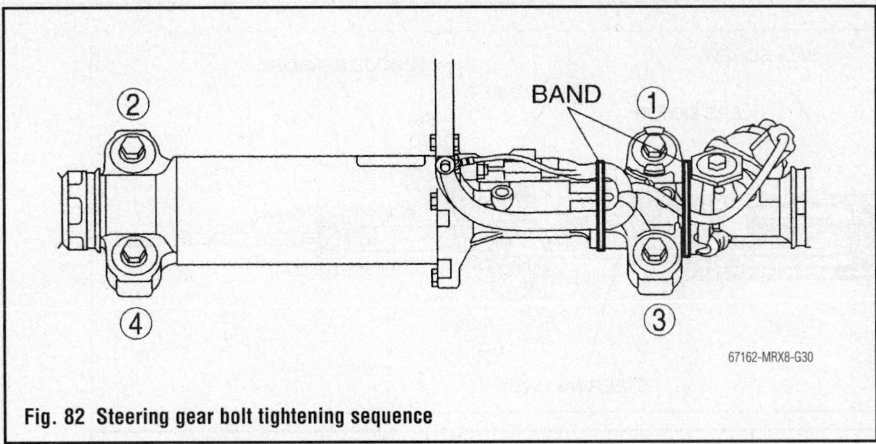

Fig. 82 Steering gear bolt tightening sequence

- Cotter pins and nuts from both steering tie rod ends

3. Press the tie rod out from the knuckle arm.

- Intermediate shaft to steering gear pinion shaft bolt. Mark the shaft-to-gear location.
- Shaft from the steering gear
- Torque sensor connector
- Electronic Power Steering (ESP) motor connector
- Steering gear mounting nuts
- Steering gear and linkage from the vehicle

To install:

4. Install or connect the following:

- Steering gear and linkage to the vehicle. Torque the bolts in sequence to 55–77 ft. lbs. (75–104 Nm).
- Steering shaft to the steering gear pinion shaft, align the marks made during removal and tighten the bolt to 19 ft. lbs. (26 Nm)
- Tie rod ends to the knuckle arm. Torque the nuts to 27–36 ft. lbs. (37–49 Nm).
- Electronic Power Steering (ESP) motor connector
- Torque sensor connector
- New cotter pins
- Radiator bracket
- Splash shield
- Wheels
- Negative battery cable

5. Check and/or adjust the front end alignment.

6. On models with dynamic suspension, perform the steering angle sensor initialization procedure.

EPS System Neutral Position Setting

See Figure 83.

1. Set the front wheels in the straight-ahead position. (Steering wheel is within 45° to the left or right of center position.)

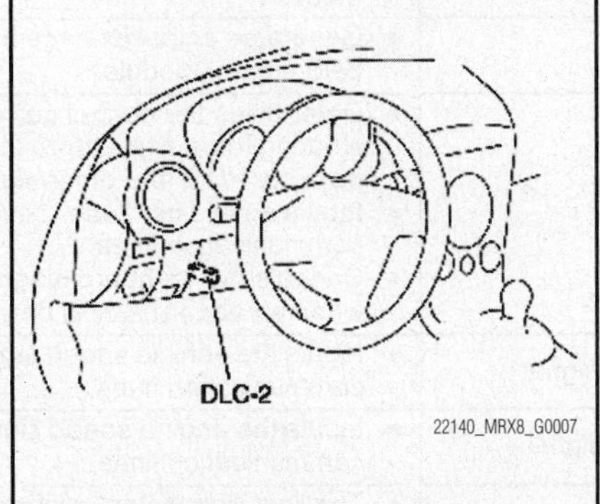

Fig. 83 Data Link Connector (DLC) location

2. Jack up both front tires so that there is no weight on them.

3. Lower the jack until the front tires touch the ground. At this time, be careful not to touch the tires or the steering wheel.

4. Connect the M-MDS to the Data Link Connector DLC-2.

5. After the vehicle is identified, select the following items from the initialization screen of the M-MDS(Scan Tool).

6. When using the IDS (laptop PC),Select "Chassis", Select "EPS, Select "Neutral Position Setting".

7. When using the PDS (Pocket PC),Select "Module Tests", Select "EPS", Select "Data Logger", Select "TRQ_S_CAL".

8. Perform the neutral position setting procedure according to the directions on the screen.

9. After setting to neutral, start the engine, rotate the steering wheel slowly in both directions within a range of 90°, and verify that the steering force does not differ in either directions.

➡If the steering force is different in either direction, inspect the power steering system following a separate troubleshooting procedure.

POWER STEERING PUMP

See Figures 84 and 85.

The conventional power steering pump is not used in this vehicle, an Electronic Power Steering (EPS)system is used.

For replacement of the EPS motor see Power Rack & Pinion Steering Gear removal and installation

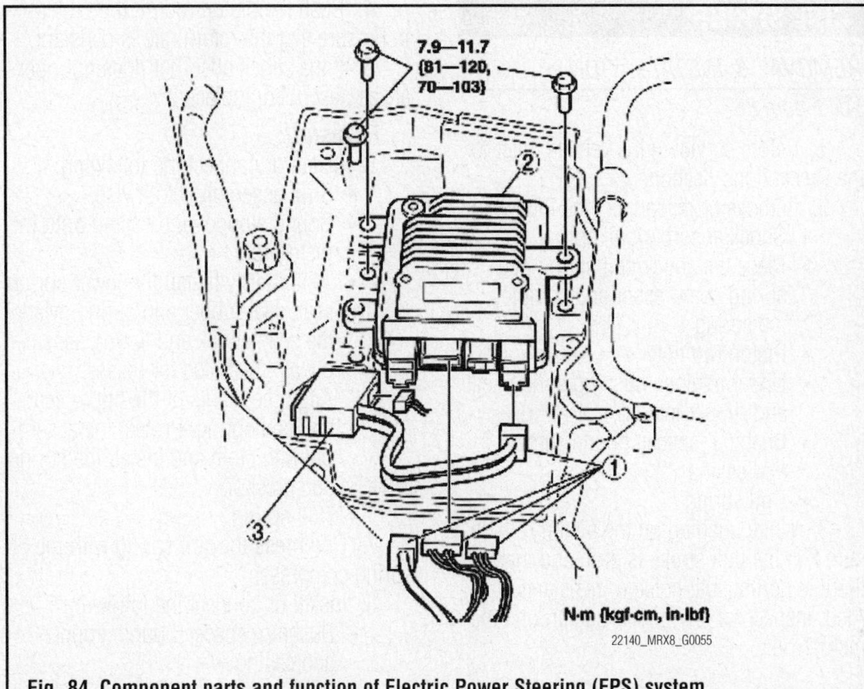

Fig. 84 Component parts and function of Electric Power Steering (EPS) system

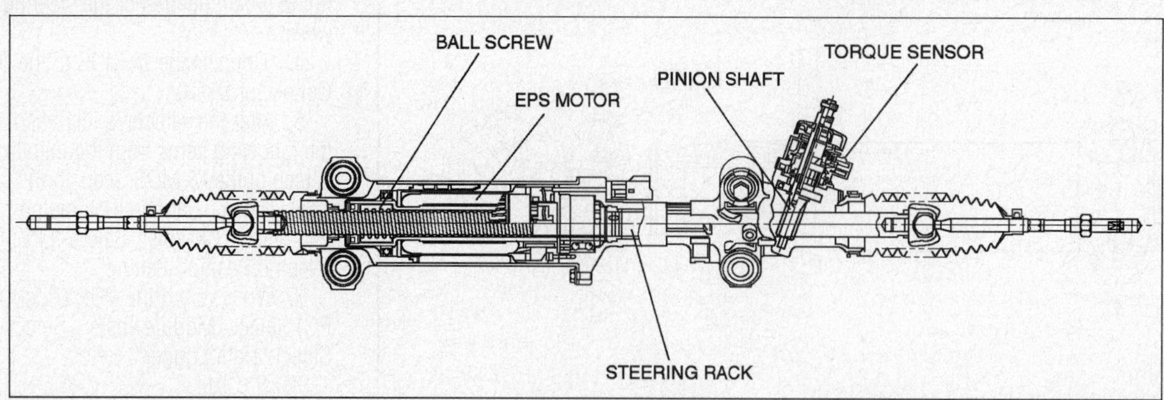

Part name		Function
Steering gear and linkage	Torque sensor	• Detects the steering force signal and inputs it to the EPS control module.
	EPS motor	• Generates an assist force based on the control current from the EPS control module.
EPS control module		• Determines the control current for the EPS motor based on the steering force signal from the torque sensor, vehicle speed signal from the PCM and other signals. • Inputs an idle increase request signal to the PCM via CAN communication lines. • Controls the on-board diagnostic system and fail-safe function when an abnormality is detected in the EPS system.
PCM	Vehicle speed signal	• Inputs the vehicle speed signal to the EPS control module via CAN communication lines.
	Engine speed signal	• Inputs the engine speed signal to the EPS control module via CAN communication lines.
Instrument cluster	EPS warning light	• The light illuminates to inform the driver when a system malfunction is detected.

22140_MRX8_G0056

Fig. 85 Electronic Power Steering (EPS) control module

SUSPENSION

FRONT SUSPENSION

COIL SPRING

REMOVAL & INSTALLATION

See Figure 86.

1. Before servicing the vehicle, refer to the Precautions Section.
2. Remove or disconnect the following:
   - Shock absorber/coil spring
   - Place the shock/coil spring in a spring compressor and compress the spring
   - Piston rod nut
   - Upper retainer, bushing spring seat and rubber insulator
   - Bushing, spacer, bound stopper and casing
   - Coil spring
3. While pushing on the piston rod, be sure that the pull stroke is even and that there is no unusual noise or resistance. Also, inspect for any oil leakage around the piston rod.
4. Push the piston rod in, then release it. Be sure that the return rate is constant.
5. If the shock absorber does not operate as described, replace it.

To install:

6. Install or connect the following:
   - Strut assembly into a vise
   - Bound stopper and casing onto the piston rod
   - Temporarily install the lower spring seat, seat rubber and spring. Mark the seat, shock and spring assembly as illustrated for reassembly. Align the marks of the upper seat and coil spring. Protect the assembly with cloth and install the spring compressor.
   - Coil spring
7. Compress the coil spring with the spring compressor
8. Install or connect the following:
   - Bushing, spacer, bound stopper and casing

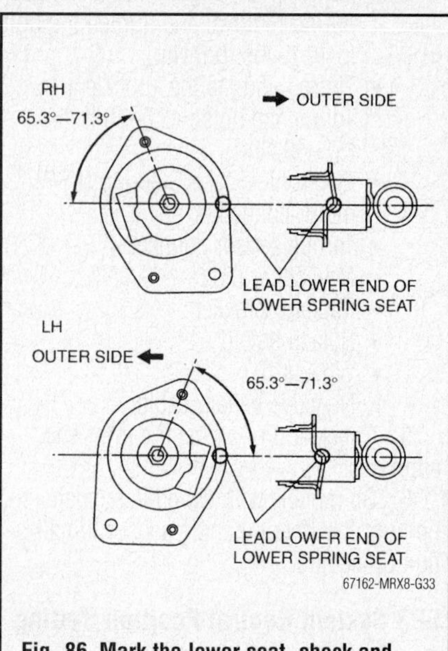

Fig. 86 Mark the lower seat, shock and spring assembly as illustrated for reassembly

- Upper retainer, bushing spring seat and rubber insulator
- Piston rod upper nut

9. Be sure that the spring upper seat notched portion is facing inward and tighten the piston rod upper nut to 23–34 ft. lbs. (32–46 Nm).

10. Be sure that the spring is well seated in the upper seats.

11. Install the shock to the vehicle.

LOWER BALL JOINT

REMOVAL & INSTALLATION

1. Before servicing the vehicle, refer to the Precautions Section.
2. Remove or disconnect the following:
   - Wheel
   - Ball joint clip
   - Ball joint using a ball joint remover

To install:

3. Install or connect the following:
   - Ball joint to lower control arm using a ball joint installer
   - Ball joint clip
   - Wheel
4. Check and/or adjust the front wheel alignment.

LOWER CONTROL ARM

REMOVAL & INSTALLATION

Performing the following procedures without first removing the ABS wheel-speed sensor may possibly cause an open circuit in the wiring harness if it is pulled by mistake. Before performing the following procedures, disconnect the ABS wheel-speed sensor harness connector (axle side) and fix it to an appropriate place where the sensor will not be pulled by mistake while servicing the vehicle.

1. Before servicing the vehicle, refer to the Precautions Section.
2. Remove the strut cross brace in the engine compartment.
3. Remove or disconnect the following:
   - Front wheel
   - Brake caliper and wire aside
   - Brake rotor
   - Tie rod end
   - Lower ball joint
   - Upper ball joint
   - Stabilizer link from the lower control arm
   - Lower control arm bolts and nuts
   - Lower control arm

To install:

4. Install or connect the following:
   - Lower control arm
   - Lower control arm bolt and nut and tighten to 62–72 ft. lbs. (84–98 Nm)
   - Stabilizer link to the lower control arm and tighten to 32–45 ft. lbs. (43–61 Nm)
   - Upper ball joint
   - Lower ball joint
   - Tie rod end
   - Brake rotor
   - Brake caliper and wire aside
   - Front wheel
5. Check and/or adjust the front wheel alignment.

SHOCK ABSORBERS

REMOVAL & INSTALLATION

See Figure 87.

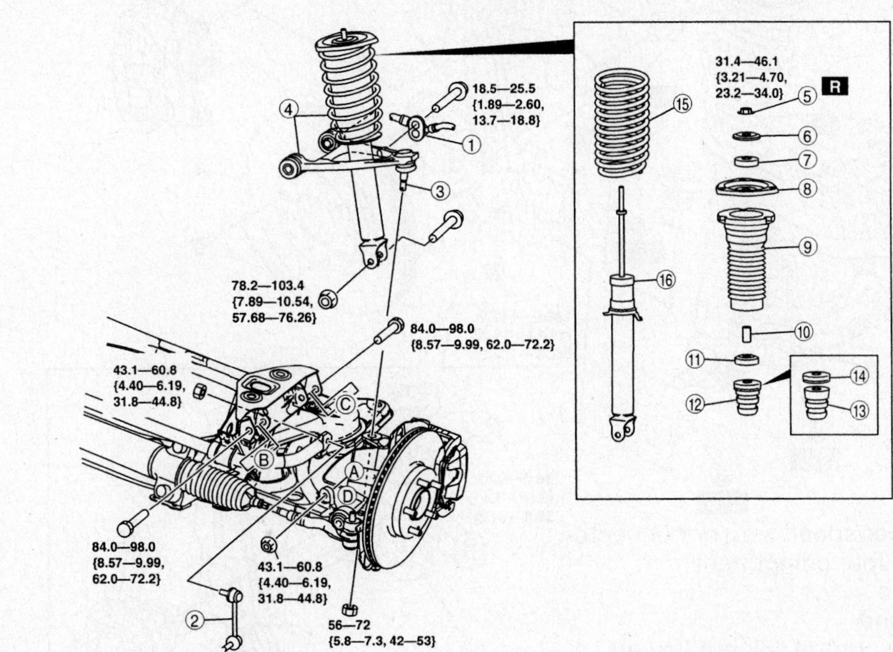

N·m {kgf-m, ft·lbf}

1	Brake hose bracket	9	Dust boot
2	Front stabilizer control link	10	Spacer
3	Front upper arm ball joint	11	Bushing
4	Front shock absorber and coil spring	12	Stopper casing and bound stopper
5	Piston rod nut	13	Bound stopper
6	Retainer	14	Stopper casing
7	Bushing	15	Coil spring
8	Upper spring seat	16	Front shock absorber

67162-MRX8-G31

Fig. 87 Exploded view of front shock absorber mounting—RX-8

1. Before servicing the vehicle, refer to the Precautions Section.
2. Remove the strut cross brace in the engine compartment and the upper shock mounting nuts.
3. Remove or disconnect the following:
   - Front wheel
   - Brake hose bracket
   - Stabilizer bar nut
   - Upper arm ball joint
4. Remove the upper and lower shock absorber mounting nuts and bolts and remove the shock absorber/coil spring assembly.

To install:

5. Installation is the reverse of removal. Tighten the upper shock nuts to 34–46 ft. lbs. (46–62 Nm) and the lower shock nut and bolt to 58–76 ft. lbs. (78–103 Nm)
6. Check and/or adjust the front end alignment.

STEERING KNUCKLE

REMOVAL & INSTALLATION

See Figures 88 and 89.

**** WARNING**

Performing the following procedures without first removing the ABS wheel-speed sensor may possibly cause an open circuit in the wiring harness if it is pulled by mistake. Before performing the following procedures, disconnect the ABS wheel-speed sensor harness connector (axle side) and fix it to an appropriate place where the sensor will not be pulled by mistake while servicing the vehicle.

1. Before servicing the vehicle, refer to the Precautions Section.
2. Remove or disconnect the following:
   - Front wheel
   - ABS wheel-speed sensor connector
   - Brake caliper component
   - Disc plate
   - Tie rod end
   - Stabilizer control link nut (lower)
   - Front upper arm ball joint
   - Front upper arm bolt
   - Front lower arm ball joint
   - Wheel hub, steering knuckle component
   - Steering knuckle
   - Dust cover
   - Wheel hub bolt
   - Wheel hub component

To install:

3. Install or connect the following:
   - Wheel hub component, steering knuckle component

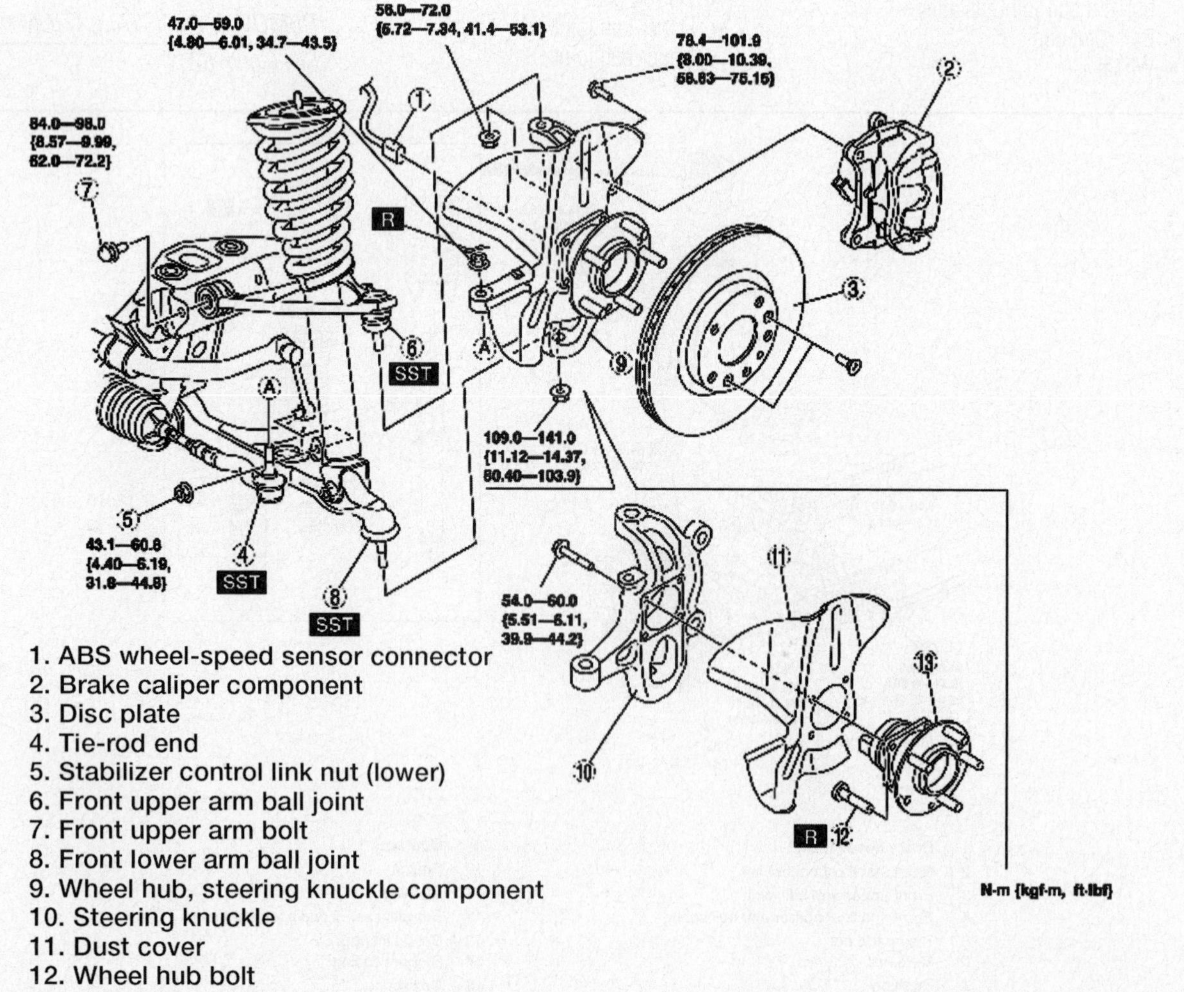

Fig. 88 Wheel hub, steering ,knuckle and associated components exploded view

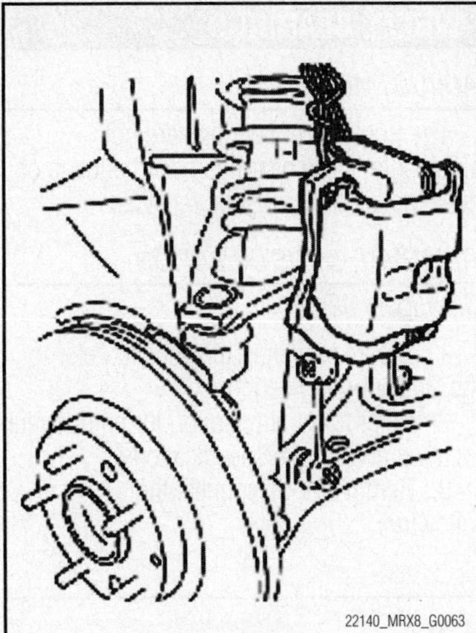

Fig. 89 Suspend the brake caliper component using a cable or equivalent

- Tighten the **UPPER** steering knuckle arm ball joint nut to 41.4–53.1 ft lbs. (56–72 Nm).
- Tighten the **LOWER** steering knuckle arm ball joint nut to 80.4–103.9 ft lbs. (109–141 Nm).
- Wheel hub bolts
- Dust cover
- Tighten to 39.9–44.2 ft lbs. (54–60 Nm).
- Stabilizer control link nut (lower)
- Tighten stabilizer control link nut (lower)to 31.8–44.8 ft lbs. (43.1–60.8 Nm).
- Tie rod end
- Tighten the tie rod end to 34.7–43.5 ft lbs. (47–59 Nm).
- Disc plate
- Brake caliper component
- Tighten Brake caliper to 58.8 –75.1 ft lbs. (78.4–101.9 Nm).
- ABS wheel-speed sensor connector
- Tighten Brake caliper to 58.8 –75.1 ft lbs. (78.4–101.9 Nm).
- Tighten disc plate with wheel to 87 ft. lbs

4. After installation, inspect the front wheel alignment.

STABILIZER BAR

REMOVAL & INSTALLATION

See Figures 90 and 91.

1. Raise and safely support the vehicle.

WARNING

Performing the following procedures without first removing the ABS wheel-speed sensor may possibly cause an open circuit in the wiring harness if it is pulled by mistake. Before performing the following procedures, remove the ABS wheel-speed sensor (axle side) and temporarily affix it out of the way, where the sensor will not be pulled while servicing the vehicle.

2. Remove the ABS wheel-speed sensor (axle side) and temporarily affix it out of the way, where the sensor will not be pulled while servicing the vehicle. For details, refer to the ABS section.
3. Remove the radiator mount bracket.
4. Remove the stabilizer control link.
5. Separate the tie rod end.
6. Remove the stabilizer bracket.
7. Remove the stabilizer bushing.
8. Remove the front stabilizer bar from the vehicle.

To install:

9. Installation is the reverse of the removal procedure. Note the tightening specifications on the illustration. When installing the stabilizer bracket, note the following:
   a. Apply rubber grease to the inner side of the stabilizer bushing.
   b. Align the outer side of the stabilizer slide stopper with the stabilizer bushing.
   c. Install the stabilizer bracket.

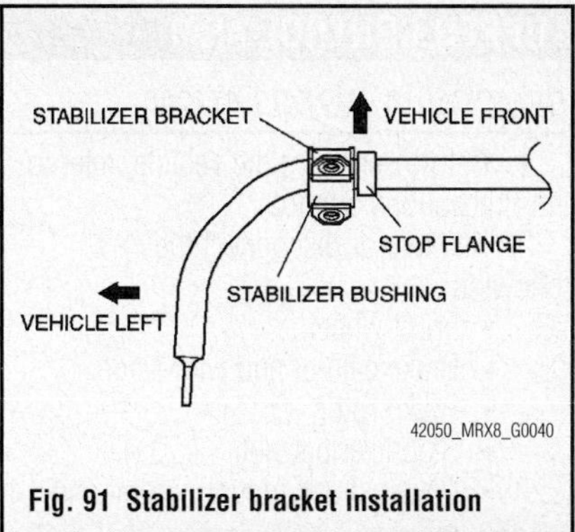

Fig. 91 Stabilizer bracket installation

UPPER BALL JOINT

REMOVAL & INSTALLATION

1. Before servicing the vehicle, refer to the Precautions Section.
2. Remove or disconnect the following:
   - Wheel
   - Lower nut on the vehicle side
   - Ball joint using a ball joint remover

To install:

3. Install or connect the following:
   - Ball joint to lower control arm using a ball joint installer
   - Tighten the nut to 42–53 ft. lbs. (56–72 Nm).
   - Wheel
4. Check and/or adjust the front wheel alignment.

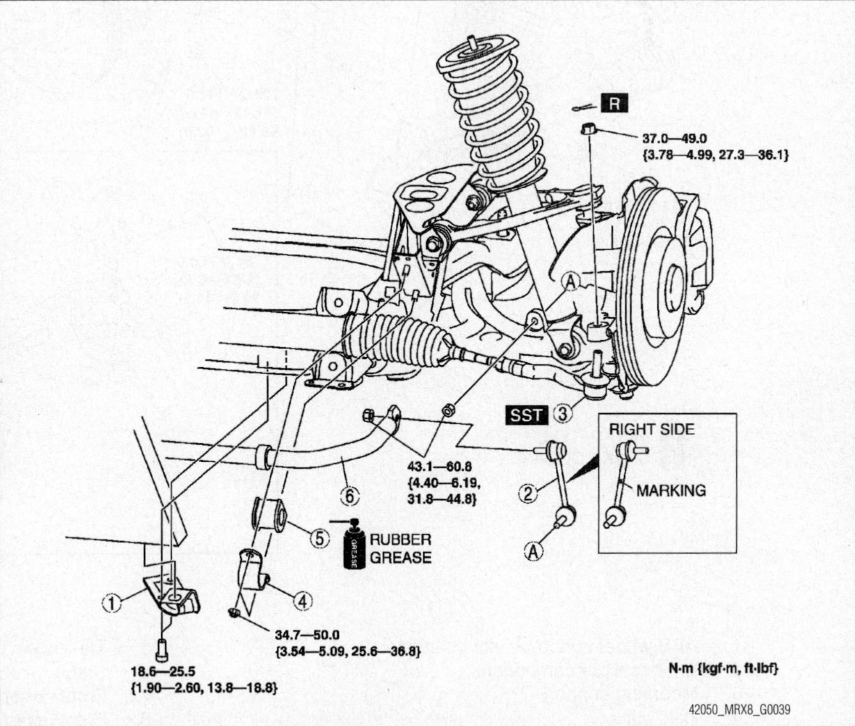

Fig. 90 Exploded view of the radiator mount bracket (1), stabilizer control link (2), tie-rod end (3), stabilizer bracket (4), stabilizer bushing (5) and front stabilizer bar (6)

UPPER CONTROL ARM

REMOVAL & INSTALLATION

1. Before servicing the vehicle, refer to the Precautions Section.
2. Remove or disconnect the following:
   - Front wheel
   - Brake caliper and wire aside
   - Brake rotor
   - Stabilizer bar nut
   - Lower shock mounting bolt and nut
   - Upper arm ball joint
3. Remove the front upper arm bolts, push down on the lower arm and remove the upper arm through the lower shock and lower arm gap.

To install:

4. Install the upper control arm.
5. Loosely tighten the bolt and nut.
6. Loosely install the lower strut mounting bolt.
   - Upper arm ball joint. Torque the nut 42–53 ft. lbs. (56–72 Nm).
   - Wheel
7. Torque upper control arm bolt to 62–72 ft. lbs. (84–98 Nm) and the lower strut mounting bolt to 58–76 ft. lbs. (78–103 Nm).
8. Check and/or adjust the front wheel alignment.

WHEEL BEARINGS

ADJUSTMENT

The front wheel bearings are not adjustable. If the bearings become loose or make noise, they must be replaced.

REMOVAL & INSTALLATION

See Figure 92.

1. Before servicing the vehicle, refer to the Precautions Section.
2. Refer to the illustration for component location and torque specifications.
3. Remove or disconnect the following:

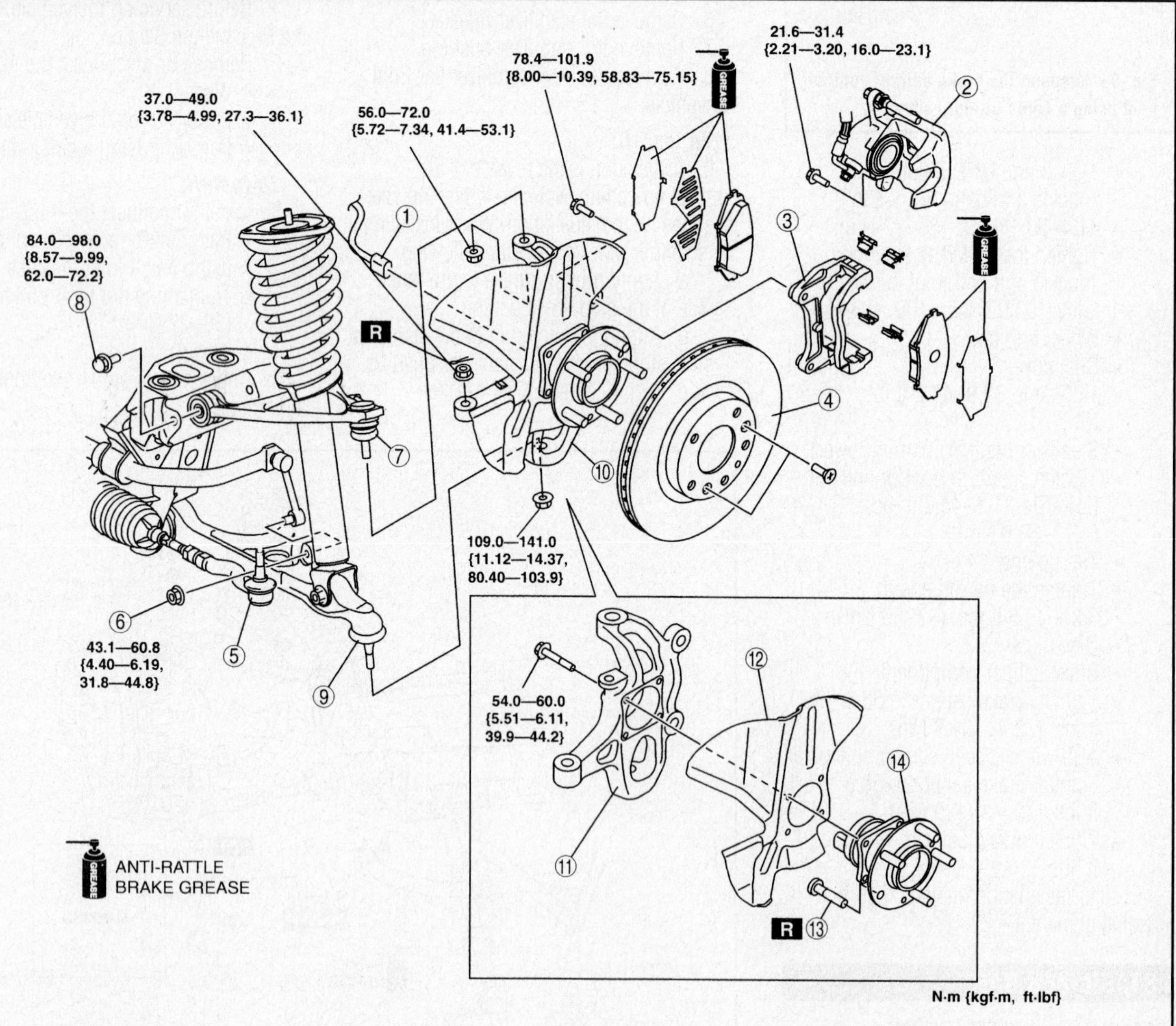

Fig. 92 Exploded view of the front wheel bearing and knuckle assembly–RX-8

- Wheels
- ABS wheel speed sensor
- Brake caliper and rotor
- Tie rod end from the knuckle
- Stabilizer bar link
- Upper control arm ball joint from the knuckle
- Upper arm bolt
- Lower ball joint
- Wheel hub and knuckle assembly
- Separate the knuckle from the wheel hub and remove the backing plate.

4. Clean and inspect all parts but do not wash or clean the wheel bearing. The wheel hub/bearing must be replaced.

To install:

- Install the backing plate to the hub, then install the knuckle to the hub and tighten the bolts to 40–44 ft. lbs. (54–60 Nm).
- Lower ball joint
- Upper arm bolt and tighten to 62–72 ft. lbs. (84–96 Nm).
- Upper control arm ball joint
- Stabilizer bar link and tighten to 32–45 ft. lbs. (43–61 Nm).
- Tie rod end
- ABS wheel speed sensor
- Brake caliper and rotor
- Wheels

SUSPENSION — REAR SUSPENSION

COIL SPRING

REMOVAL & INSTALLATION

See Figure 93.

1. Before servicing the vehicle, refer to the Precautions Section.
2. Remove or disconnect the following:
   - Shock absorber/coil spring
   - Place the shock/coil spring in a spring compressor and compress the spring
   - Piston rod nut
   - Upper retainer, bushing spring seat and rubber insulator
   - Bushing, spacer, bound stopper and casing
   - Coil spring
3. While pushing on the piston rod, be sure that the pull stroke is even and that there is no unusual noise or resistance. Also, inspect for any oil leakage around the piston rod.
4. Push the piston rod in, then release it. Be sure that the return rate is constant.
5. If the shock absorber does not operate as described, replace it.

To install:

6. Install or connect the following:
   - Strut assembly into a vise
   - Bound stopper and casing onto the piston rod
   - Temporarily install the lower spring seat, seat rubber and spring. Mark the seat, shock and spring assembly as illustrated for reassembly. Align the marks of the upper seat and coil spring. Protect the assembly with cloth and install the spring compressor.
   - Coil spring
7. Compress the coil spring with the spring compressor
8. Install or connect the following:
   - Bushing, spacer, bound stopper and casing
   - Upper retainer, bushing spring seat and rubber insulator
   - Piston rod upper nut
9. Be sure that the spring upper seat notched portion is facing inward and tighten the piston rod upper nut to 23–34 ft. lbs. (32–46 Nm).
10. Be sure that the spring is well seated in the upper seats.
11. Install the shock to the vehicle.

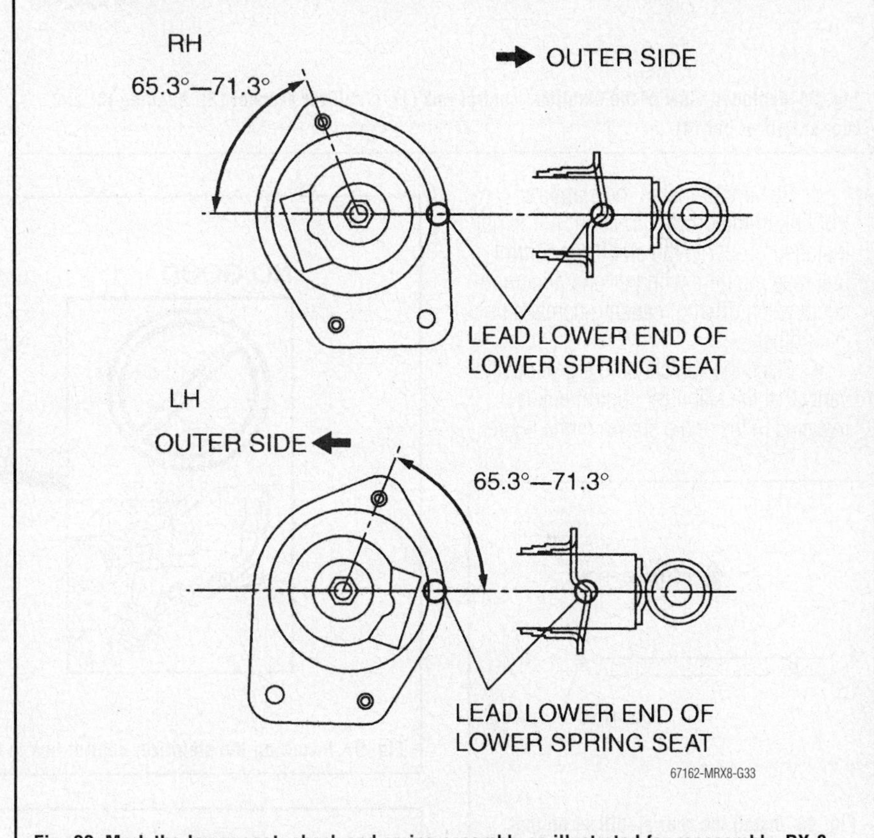

Fig. 93 Mark the lower seat, shock and spring assembly as illustrated for reassembly–RX-8

STABILIZER BAR

REMOVAL & INSTALLATION

See Figures 94 through 98.

1. Raise and safely support the vehicle.
2. Remove the stabilizer control link.
3. Remove the stabilizer bracket.
4. Remove the bushing and the rear stabilizer bar.

To install:

5. Installation is the reverse of the removal procedure, using the tightening specifications in the illustration and noting the following important steps:
   a. Install the rear stabilizer so that the identification mark is on the right side of the vehicle.
   b. Install the bushings aligned with the stop flanges, as shown in the accompanying illustration.
   c. Install the stabilizer control link in the proper angle, as shown in the accompanying illustration.

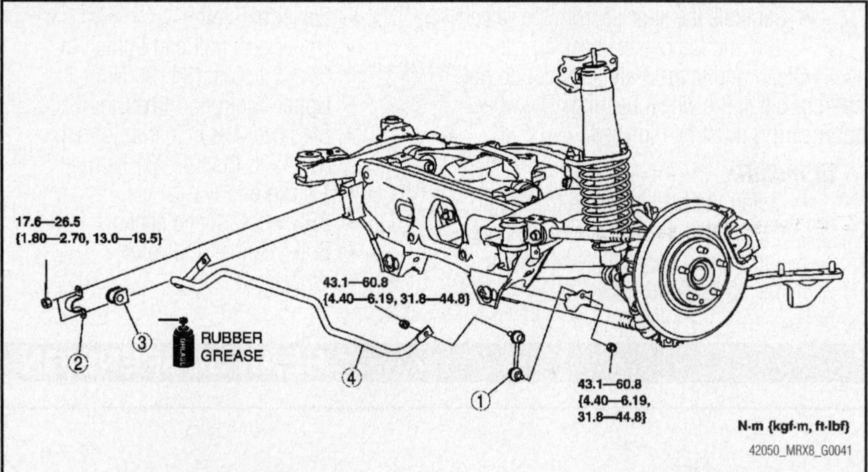

Fig. 94 Exploded view of the stabilizer control link (1), stabilizer bracket (2), bushing (3) and rear stabilizer bar (4)

d. Be sure to install the stabilizer control link in the proper position. If it is not installed properly, the stabilizer control link may interfere with peripheral components when driving, causing damage to each other.

e. Place the vehicle on the ground and verify that the stabilizer control link is installed in the angle shown in the figure.

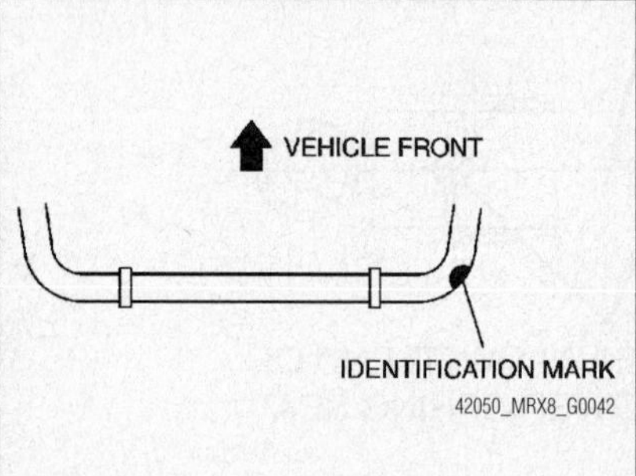

Fig. 95 Install the rear stabilizer so that the identification mark is on the right side of the vehicle

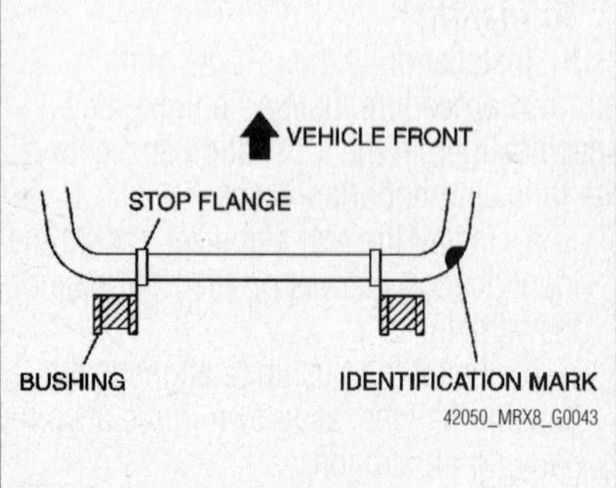

Fig. 96 Install the bushings aligned with the stop flanges

CONTROL ARMS/LINKS

REMOVAL & INSTALLATION

Rear Lateral Link (Upper)

See Figures 99 through 101.

WARNING

Performing the following procedures without first removing the ABS wheel-speed sensor may possibly cause an open circuit in the wiring harness if it is pulled by mistake. Before operations, remove the ABS wheel-speed sensor (axle side) and move the sensor away from the harnesses.

1. Raise and safely support the vehicle.

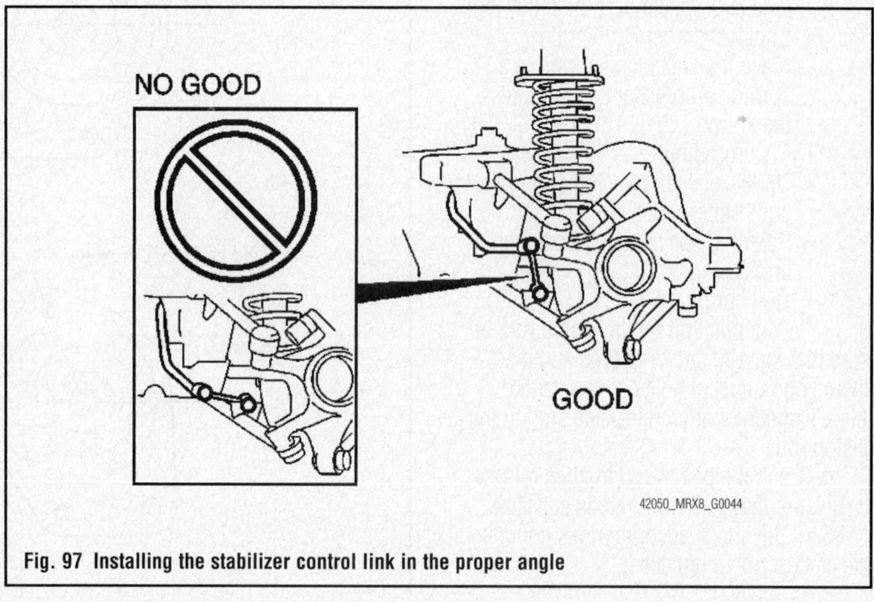

Fig. 97 Installing the stabilizer control link in the proper angle

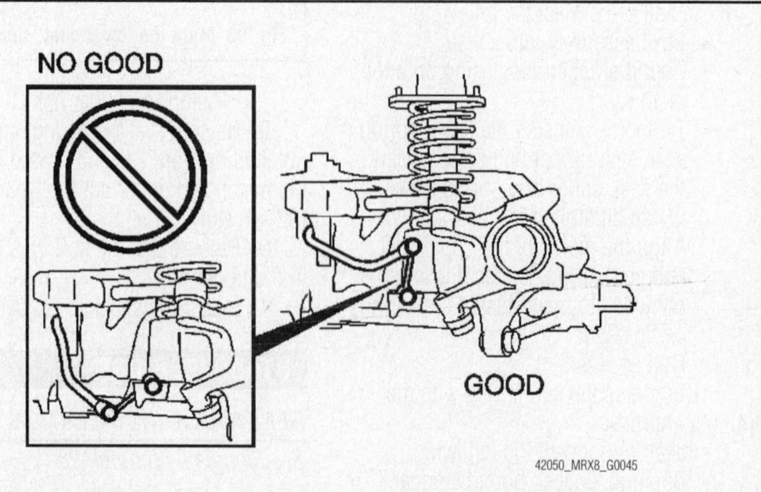

Fig. 98 Place the vehicle on the ground and verify that the stabilizer control link is installed in the angle shown

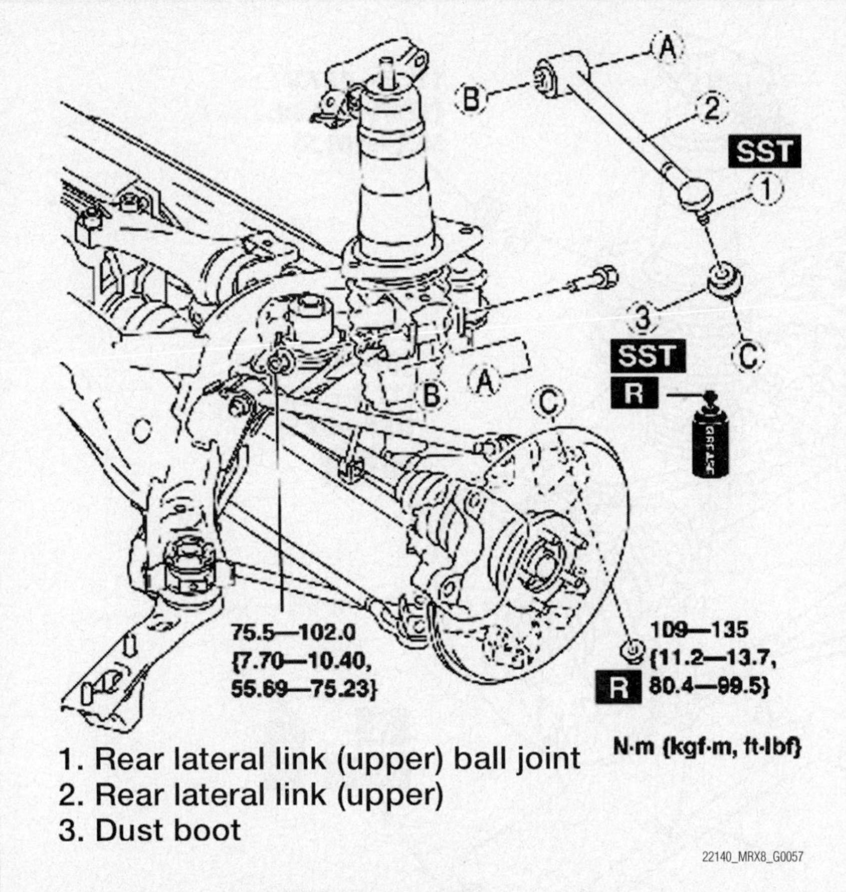

Fig. 99 Rear suspension exploded view emphasizing the upper lateral link

2. Remove the rear wheels.
3. Remove the rear lateral link (lower)bolt.
4. Remove the rear dust boot.
5. Using the Special Service Tool (SST)49 T028 3A0, disconnect the rear lateral link (upper) ball joint.

➡When removing the rear lateral link (upper) ball joint, the rear knuckle bushing may also come off. If it comes off, replace the rear knuckle.

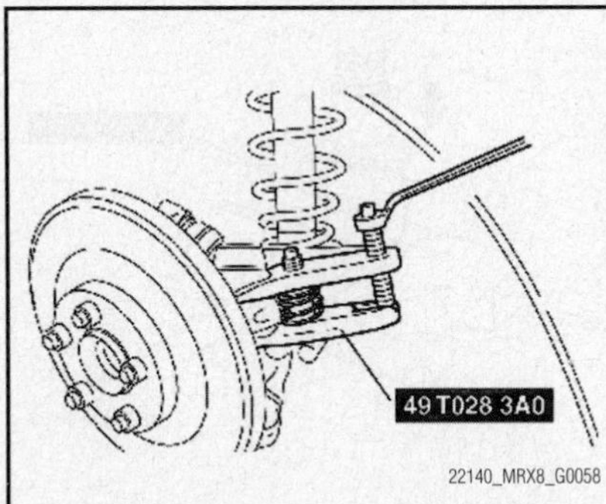

Fig. 100 Rear suspension lateral link (upper) ball joint removal with the Special Service Tool (SST)49 T028 3A0

To install:

6. Wipe the grease off the ball joint stud.
7. Fill the inside of the new dust boot with grease.
8. Using the Special Service Tool (SST) 49 H028 301, install the dust boot to the ball joint.
9. Wipe off the excess grease.
10. Install in the reverse order of removal.
11. Tighten the inner bolt and nut to 86.9–101.2 ft.lbs. (117.7–137.3 Nm).
12. Tighten the ball joint stud to 80.4–99.5 ft.lbs. (109–135 Nm).

Rear Lateral Link (Lower)

See Figures 102 through 104.

1. Raise and safely support the vehicle.
2. Remove the rear wheels.
3. Remove the stabilizer control link lower nut.
4. Using the Special Service Tool (SST)49 T028 3A0, disconnect the rear lateral link (lower) ball joint.

➡When removing the rear lateral link (lower) ball joint, the rear knuckle bushing may also come off. If it comes off, replace the rear knuckle.

5. Remove the rear lateral link (lower)bolt.
6. Remove the rear dust boot.

To install:

7. Wipe the grease off the ball joint stud.
8. Fill the inside of the new dust boot with grease.
9. Using the Special Service Tool (SST) 49 H028 301, install the dust boot to the ball joint.
10. Install in the reverse order of removal.
11. Tighten the inner bolt and nut to 86.9–101.2 ft.lbs. (117.7–137.3 Nm).
12. Tighten the ball joint stud to 80.4–99.5 ft.lbs. (109–135 Nm).
13. Tighten the stabilizer control link lower nut to 31.8–44.5 ft.lbs. (43.1–60.8 Nm).

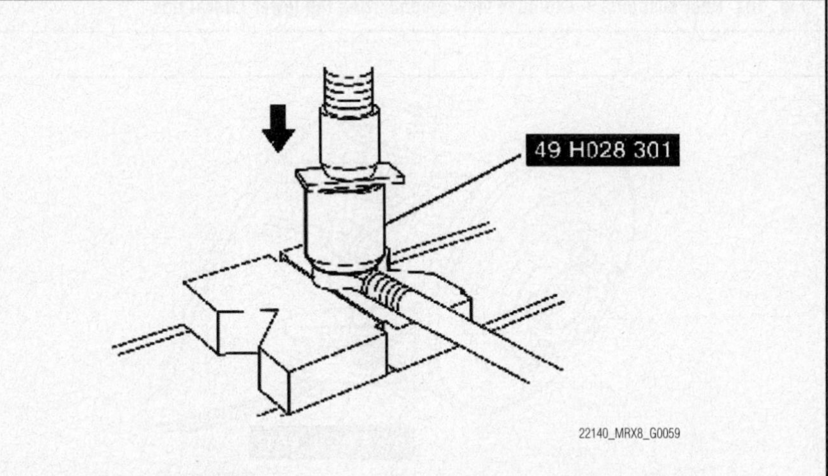

Fig. 101 Rear suspension lateral link (upper) ball joint dust boot installation with Special Service Tool (SST)49 H028 301

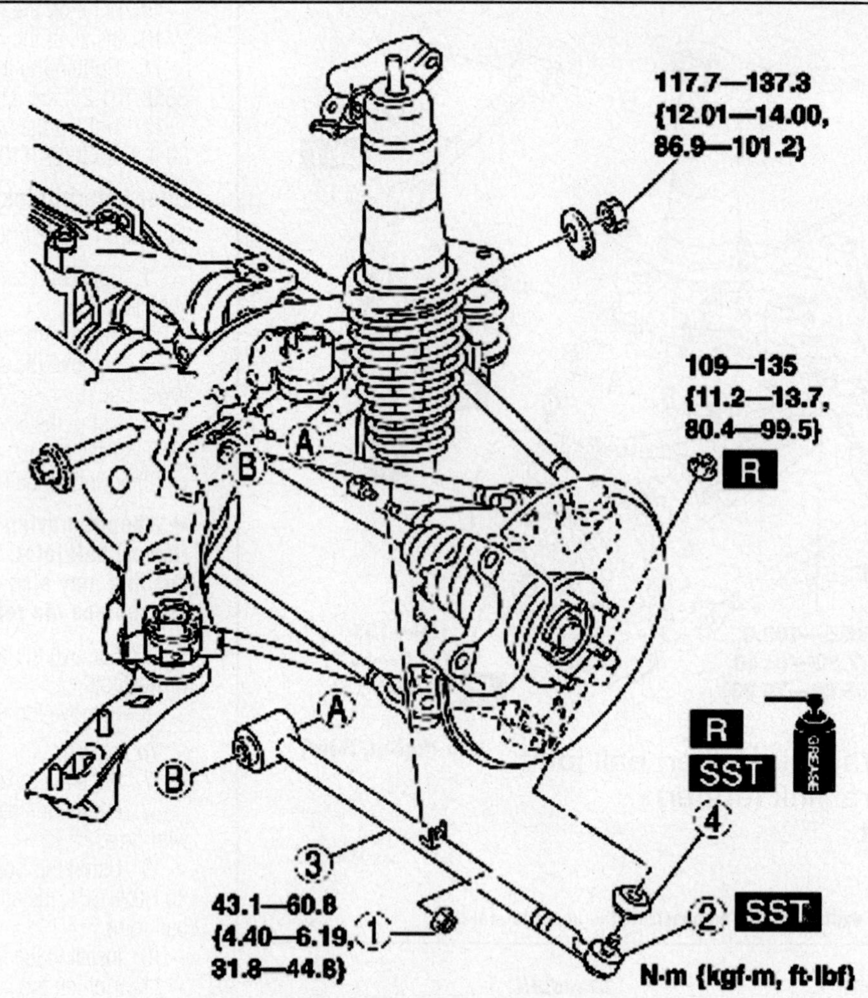

Fig. 102 Rear suspension exploded view emphasizing the lower lateral link

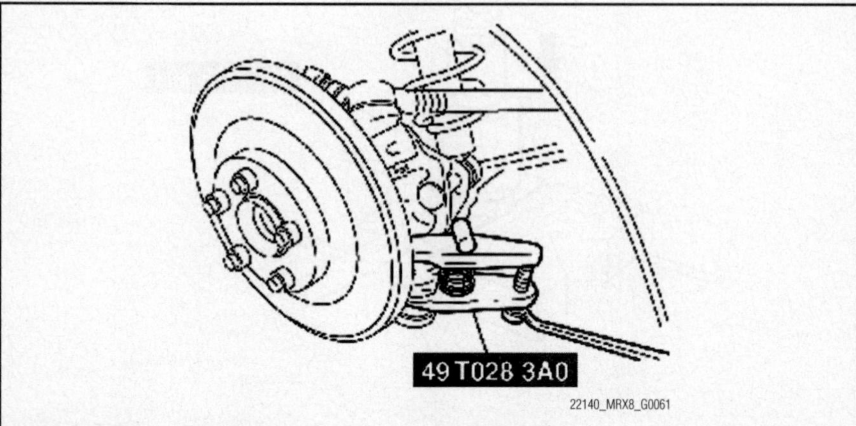

Fig. 103 Rear suspension lateral link (lower) ball joint removal with the Special Service Tool (SST)49 T028 3A0

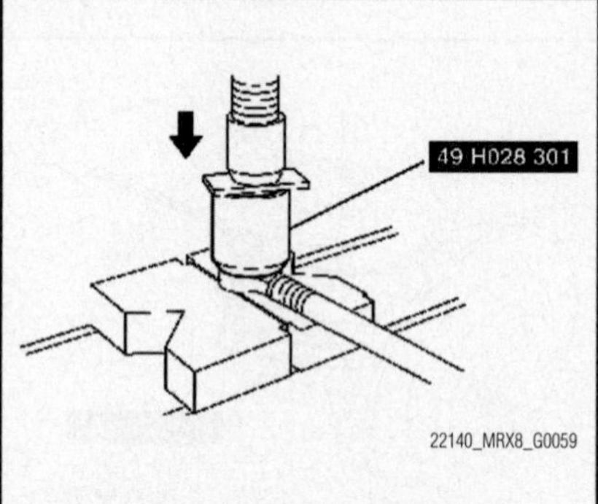

Fig. 104 Rear suspension lateral link (lower) ball joint dust boot installation with Special Service Tool (SST)49 H028 301

SHOCK ABSORBER

REMOVAL & INSTALLATION

See Figure 105.

1. Before servicing the vehicle, refer to the Precautions Section.
2. Remove or disconnect the following:
   - Rear wheel
   - Brake caliper and wire aside
   - Rear lateral link upper inner bolt
   - Stabilizer bar upper nut
   - Shock absorber lower bolt
   - Inside trunk end and side trim panels
   - Shock absorber upper bracket and mounting nuts
   - Shock absorber/coil spring assembly

To install:

3. Installation is the reverse of removal. Tighten the upper shock nuts to 34–46 ft. lbs. (46–62 Nm) and the lower shock nut and bolt to 65–88 ft. lbs. (88–119 Nm). Tighten the shock bracket bolts and nuts to 28–38 ft. lbs. (37–52 Nm).
4. Check and/or adjust the front end alignment.

WHEEL BEARINGS

ADJUSTMENT

The rear wheel bearings are not adjustable. If the bearings become loose or make noise, they must be replaced.

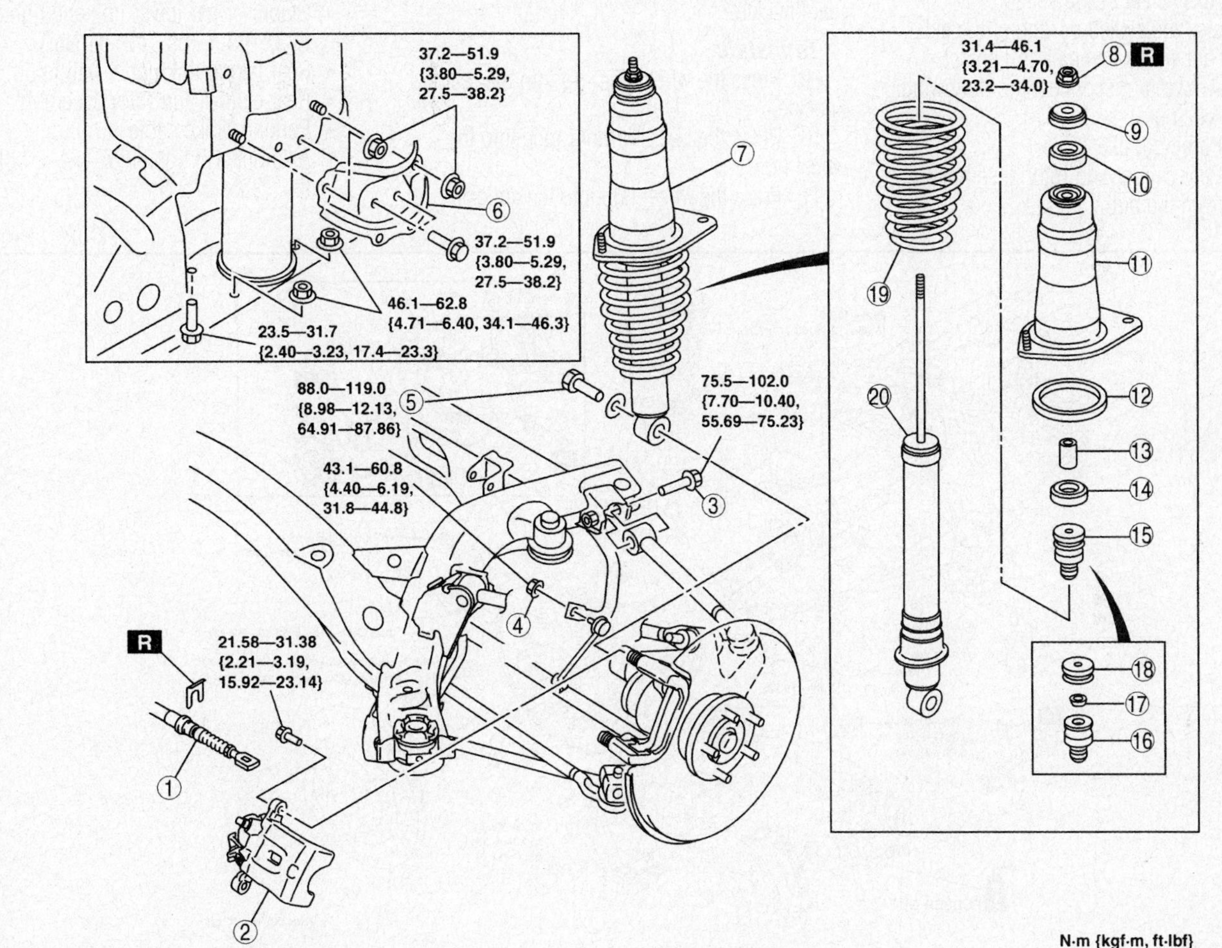

1	Parking brake cable	11	Upper spring seat
2	Caliper	12	Spring seat rubber
3	Rear lateral link (upper) inner bolt	13	Bushing
4	Stabilizer control link upper nut	14	Spacer
5	Rear shock absorber lower bolt	15	Bound stopper and stopper casing
6	Rear shock absorber bracket	16	Bound stopper
7	Rear shock absorber and coil spring	17	Collar
8	Piston rod nut	18	Stopper casing
9	Retainer	19	Coil spring
10	Bushing	20	Rear shock absorber

67162-MRX8-G32

Fig. 105 Exploded view of rear shock absorber mounting

REMOVAL & INSTALLATION

See Figure 106.

1. Before servicing the vehicle, refer to the Precautions Section.
2. Refer to the illustration for component location and torque specifications.

➡The wheel bearings are not serviceable. If the bearings are bad, a new hub/bearing assembly must be installed.

3. Remove or disconnect the following:
    - Rear wheels
    - ABS wheel speed sensor
4. Raise the staked portion of the axle retaining nut with a hammer and chisel.
5. Remove or disconnect the following:
    - Axle nut
    - Parking brake cable
    - Rear caliper and rotor assembly from the hub
    - Rear lateral link upper ball joint
    - Stabilizer bar lower link
    - Rear lateral link lower ball joint
    - Lower shock absorber bolt
    - Rear trailing link lower outside bolt
    - Rear trailing link upper ball joint
    - Toe control link outside bolt
    - Axle shaft
    - Wheel hub/knuckle assembly
6. Press the wheel hub from the knuckle.
7. Press the bearing inner race from the wheel hub.
8. Press the wheel bearing from the rear knuckle.

To install:

9. Press the wheel bearing into the rear knuckle.
10. Press the bearing inner race into the wheel hub.
11. Press the wheel hub into the knuckle.
12. Install or connect the following:
    - Wheel hub/knuckle assembly. Tighten the bolts to 23–31 ft. lbs. (31–42 Nm).
    - Axle shaft
    - Toe control link outside bolt
    - Rear trailing link upper ball joint
    - Rear trailing link lower outside bolt and tighten to 56–75 ft. lbs. (76–102 Nm).
    - Lower shock absorber bolt and tighten to 65–88 ft. lbs. (86–119 Nm).
    - Rear lateral link lower ball joint
    - Stabilizer bar lower link and tighten to 32–45 ft. lbs. (43–61 Nm).
    - Rear lateral link upper ball joint
    - Rear caliper and rotor assembly
    - Parking brake cable
    - Axle nut and tighten to 174–203 ft. lbs. (235–275 Nm).

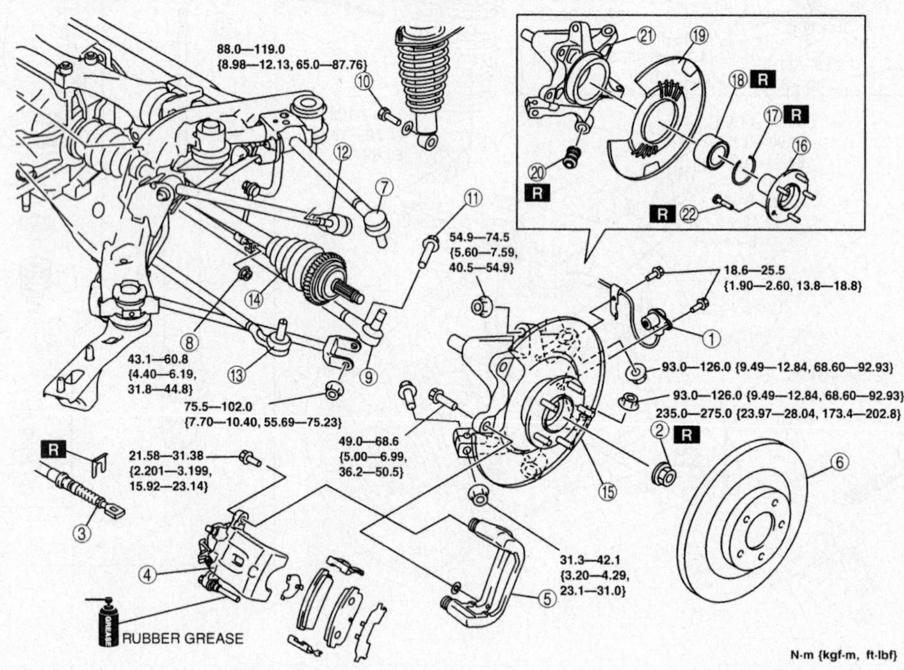

1	ABS wheel-speed sensor	12	Rear trailing link (upper) ball joint
2	Locknut	13	Toe control link outside bolt
3	Parking brake cable	14	Rear drive shaft
4	Brake caliper component	15	Rear knuckle component
5	Mounting support	16	Wheel hub component
6	Disc plate	17	Retaining ring
7	Rear lateral link (upper) ball joint	18	Wheel bearing
8	Stabilizer control link (lower)	19	Dust cover
9	Rear lateral link (lower) ball joint	20	Bushing
10	Shock absorber bolt (lower)	21	Rear knuckle
11	Rear trailing link (lower) outside bolt	22	Wheel hub bolt

67162-MRX8-G40

Fig. 106 Exploded view of the rear wheel hub and bearing–RX-8

MAZDA

Tribute

6

BRAKES 6-10

CHASSIS ELECTRICAL 6-18

DRIVETRAIN 6-20

ENGINE COOLING 6-31

ENGINE ELECTRICAL 6-34

ENGINE MECHANICAL 6-40

SPECIFICATIONS AND MAINTENANCE CHARTS

ENGINE AND VEHICLE IDENTIFICATION

Engine							Model Year	
Code ①	Liters	Cu. In.	Cyl.	Fuel Sys.	Engine Type	Eng. Mfg.	Code ②	Year
Z	2.3	137	4	MPI	DOHC	Ford	7	2007
1	3.0	182	6	MFI	DOHC	Ford	8	2008

MPI: Multi-port Fuel Injection

DOHC: Double Overhead Camshafts

① 8th digit of VIN

② 10th digit of VIN

22140_TRIB_C0001

GENERAL ENGINE SPECIFICATIONS

Year	Model	Engine Displacement Liters	Engine VIN	Net Horsepower @ rpm	Net Torque @ rpm (ft. lbs.)	Bore x Stroke (in.)	Com-pression Ratio	Oil Pressure @ rpm
2007	Tribute	2.3	Z	153@5800	152@4250	3.44x3.70	9.7:1	29-39@2000
	Tribute	3.0	1	200@6000	193@4850	3.50x3.13	10.0:1	11@1500
2008	Tribute	2.3	Z	153@5800	152@4250	3.44x3.70	9.7:1	29-39@2000
	Tribute	3.0	1	200@6000	193@4850	3.50x3.13	10.0:1	11@1500

22140_TRIB_C0002

ENGINE TUNE-UP SPECIFICATIONS

Year	Engine Displacement Liters	Engine VIN	Spark Plug Gap (in.)	Ignition Timing (deg.)		Fuel Pump (psi)	Idle Speed (rpm)		Valve Clearance	
				MT	AT		MT	AT	Intake	Exhaust
2007	2.3	Z	0.049-0.053	10 BTDC	10 BTDC	39	①	①	0.008-0.011	0.010-0.013
	3.0	1	0.052-0.056	10 BTDC	10 BTDC	39	①	①	HYD.	HYD.
2008	2.3	Z	0.049-0.053	10 BTDC	10 BTDC	39	①	①	0.008-0.011	0.010-0.013
	3.0	1	0.052-0.056	10 BTDC	10 BTDC	39	①	①	HYD.	HYD.

BTDC: Before Top Dead Center

HYD: Hydraulic lash adjusters

① Refer to Vehicle Emission Control Information Label

22140_TRIB_C0003

CAPACITIES

Year	Model	Engine Displacement Liters	Engine VIN	Engine Oil with Filter (qts.)	Transmission (pts.) Manual	Transmission (pts.) Auto.	Transfer Case (pts.)	Drive Axle Front (pts.)	Drive Axle Rear (pts.)	Fuel Tank (gal.)	Cooling System (qts.)
2007	Tribute	2.3	Z	4.25	5.0	20.0	2.95	—	3.0	16.5	①
	Tribute	3.0	1	6	5.0	20.5	2.95	—	3.0	16.5	10.5
2008	Tribute	2.3	Z	4.25	5.0	20.0	2.95	—	3.0	16.5	①
	Tribute	3.0	1	6	5.0	20.5	2.95	—	3.0	16.5	10.5

NOTE: All capacities are approximate. Add fluid gradually and check to be sure a proper fluid level is obtained.

① With manual transaxle: 6.9 qts; with automatic transaxle: 8.0 qts.

22140_TRIB_C0004

FLUID SPECIFICATIONS

Year	Model	Engine Displacement Liters	Engine ID/VIN	Engine Oil	Auto. Trans. ①	Drive Axle	Power Steering Fluid	Brake Master Cylinder
2007	Tribute	2.3	Z	5W-20 A	Mercon®V	80W-90	Mercon®V	DOT 3
		3.0	1	5W-20	Mercon®V	80W-90	Mercon®V	DOT 3
2008	Tribute	2.3	Z	5W-20	Mercon®V	80W-90	NA	DOT 3
		3.0	1	5W-20	Mercon®V	80W-90	NA	DOT 3

NA - Not Available

DOT: Department Of Transpotation

® Registerd Trademark

① SAE 5W-20 Premium Synthetic Blend Motor Oil (US only) SAE-5W-20 Super Premium Motor Oil (Canada)

22140_TRIB_C0005

VALVE SPECIFICATIONS

Year	Engine Displacement Liters	Engine VIN	Seat Angle (deg.)	Face Angle (deg.)	Spring Test Pressure (lbs. @ in.)	Spring Installed Height (in.)	Stem-to-Guide Clearance (in.) Intake	Stem-to-Guide Clearance (in.) Exhaust	Stem Diameter (in.) Intake	Stem Diameter (in.) Exhaust
2007	2.3	Z	45	45	38.6@1.49	1.496	0.0001	0.00011	0.2153-0.2159	0.2151-0.2157
	3.0	1	44.75	45.5	156@ 1.18	1.570	0.0017-0.0037	0.0017-0.0037	0.2350-0.2358	0.2343-0.2350
2008	2.3	Z	45	45	38.6@1.49	1.496	0.0001	0.00011	0.2153-0.2159	0.2151-0.2157
	3.0	1	44.75	45.5	156@ 1.18	1.570	0.0017-0.0037	0.0017-0.0037	0.2350-0.2358	0.2343-0.2350

22140_TRIB_C0006

CAMSHAFT SPECIFICATIONS CHART

All measurements are given in inches.

Year	Engine Displacement Liters	Engine VIN	Journal Dia.	Brg. Oil Clearance	Shaft End-play	Runout	Journal Bore	Lobe Height Intake	Lobe Height Exhaust
2007	2.3	Z	0.9820-0.9830	0.001-0.003	NA	0.001 ①	0.001-0.003	0.3240	0.3070
	3.0	1	1.0600-1.0610	0.001-0.00029	0.00748	NA	1.063-1.062	0.1890	0.1890
2008	2.3	Z	0.9820-0.9830	0.001-0.003	NA	0.001 ①	0.001-0.003	0.3240	0.3070
	3.0	1	1.0600-1.0610	0.001-0.00029	0.00748	NA	1.063-1.062	0.1890	0.1890

NA: Information not available

① Supported by No. 1 and No. 5 journals.

22140_TRIB_C0007

CRANKSHAFT AND CONNECTING ROD SPECIFICATIONS

All measurements are given in inches.

Year	Engine Displacement Liters	Engine VIN	Crankshaft Main Brg. Journal Dia.	Crankshaft Main Brg. Oil Clearance	Crankshaft Shaft End-play	Crankshaft Thrust on No.	Connecting Rod Journal Diameter	Connecting Rod Oil Clearance	Connecting Rod Side Clearance
2007	2.3	Z	2.0460-2.0470	0.0006-0.0015	0.0070-0.0180	NA	NA	0.001-0.002	0.0760-0.1200
	3.0	1	2.4670-2.4790	0.0009-0.0019	0.0050-0.0100	3	NA	0.001-0.0025	0.0039-0.0118
2008	2.3	Z	2.0460-2.0470	0.0006-0.0015	0.0070-0.0180	NA	NA	0.001-0.002	0.0760-0.1200
	3.0	1	2.4670-2.4790	0.0009-0.0019	0.0050-0.0100	3	NA	0.001-0.0025	0.0039-0.0118

NA: Information not available

22140_TRIB_C0008

PISTON AND RING SPECIFICATIONS

All measurements are given in inches.

Year	Engine Displacement Liters	Engine VIN	Piston Clearance	Ring Gap Top Compression	Ring Gap Bottom Compression	Ring Gap Oil Control	Ring Side Clearance Top Compression	Ring Side Clearance Bottom Compression	Ring Side Clearance Oil Control
2007	2.3	Z	0.0009-0.0017	N/A	N/A	N/A	NA	NA	NA
	3.0	1	0.0005-0.0009	0.0039-0.0098	0.0106-0.0165	0.0059-0.0255	NA	NA	snug
2008	2.3	Z	0.0009-0.0017	N/A	N/A	N/A	NA	NA	NA
	3.0	1	0.0005-0.0009	0.0039-0.0098	0.0106-0.0165	0.0059-0.0255	NA	NA	snug

NA: Information not available

22140_TRIB_C0009

TORQUE SPECIFICATIONS

All readings in ft. lbs.

Year	Engine Displacement Liters	Engine VIN	Cylinder Head Bolts	Main Bearing Bolts	Rod Bearing Bolts	Crankshaft Damper Bolts	Flywheel Bolts	Manifold Intake	Manifold Exhaust	Spark Plugs	Oil Pan Drain Plug
2007	2.3	Z	①	NA	NA	②	③	13	35	11	21
	3.0	1	④	⑤	⑥	⑦	59	⑧	15	11	19
2008	2.3	Z	①	NA	NA	②	③	13	35	11	21
	3.0	1	④	⑤	⑥	⑦	59	⑧	15	11	19

NA: Information not available

① Step 1: 44 inch lbs.
Step 2: 11 ft. lbs.
Step 3: 33 ft. lbs.
Step 4: +90 degrees
Step 5: + 90 degrees

② Step 1: 74 ft. lbs.
Step 2: plus 90 degrees

③ Step 1: 37 ft. lbs.
Step 2: 59 ft. lbs
Step 3: 83 ft. lbs.

④ Step 1: 30 ft. lbs. (40 Nm).
Step 2: Tighten the bolts 90 degrees.
Step 3: Loosen the bolts one full turn.
Step 4: 30 ft. lbs. (40 Nm).
Step 5: Tighten the bolts 90 degrees.
Step 6: Tighten the bolts 90 degrees.

⑤ Step 1: Fasteners 1-8: 18 ft. lbs.
Step 2: Fasteners 9-19: 30 ft. lbs.
Step 3: Fasteners 1-16: +90 degrees
Step 4: fasteners 17-22: 18 ft. lbs.

⑥ Step 1: 17 ft. lbs.
Step 2: 32 ft. lbs.

⑦ Step 1: 89 ft. lbs.
Step 2: Loosen 1 full turn
Step 3: 37 ft. lbs.
Step 4: 66 ft. lbs.

⑧ 89 inch lbs.

22140_TRIB_C0010

WHEEL ALIGNMENT

Year	Model		Caster Range (+/-Deg.)	Caster Preferred Setting (Deg.)	Camber Range (+/-Deg.)	Camber Preferred Setting (Deg.)	Toe-in (in.)
2007	Tribute	F	1.00	+1.60	1.00	-1.00	-0.23+/-0.32
		R	NA	NA	0.75	0.00	0.18+/-0.20
2008	Tribute	F	1.00	+1.60	1.00	-1.00	0.23+/-0.32
		R	NA	NA	0.75	0.00	0.18+/-0.20

NA: Information not available

22140_TRIB_C0011

TIRE, WHEEL AND BALL JOINT SPECIFICATIONS

Year	Model	OEM Tires Standard	OEM Tires Optional	Tire Pressures (psi) Front	Tire Pressures (psi) Rear	Wheel Size	Ball Joint Inspection	Lug Nuts (ft. lbs.)
2007	Tribute	P235/70SR16	NA	①	①	7	0.080 in.	98
2008	Tribute	P235/70R16	NA	①	①	7	0.030 in.	98

OEM: Original Equipment Manufacturer

PSI: Pounds Per Square Inch

NA: Not Available

① See certification label located on the inside driver's side front door jamb.

22140_TRIB_C0012

BRAKE SPECIFICATIONS

All measurements in inches unless noted

Year	Model		Brake Disc Original Thickness	Brake Disc Minimum Thickness	Brake Disc Maximum Run-out	Brake Drum Original Inside Diameter	Brake Drum Maximum Machine Diameter	Minimum Lining Thickness	Brake Caliper Bracket Bolts (ft. lbs.)	Brake Caliper Mounting Bolts (ft. lbs.)
2007	Tribute	F	NA	①	0.004	NA	9.05	0.118	111	②
		R	NA	0.430	0.004	—	—	③	NA	26
2008	Tribute	F	NA	①	0.004	NA	9.05	0.118	111	②
		R	NA	0.430	0.004	—	—	③	NA	26

NA: Information not available

① With disc/drum: 0.860 in.

With 4-wheel disc: 0.950 in.

② With disc/drum: 26 ft. lbs.

With 4-wheel disc: 33 ft. lbs.

③ Drum brake shoe: 0.030

Disk brake pad: 0.118

22140_TRIB_C0013

SCHEDULED MAINTENANCE INTERVALS

2007-08 Mazda Tribute

TO BE SERVICED	TYPE OF SERVICE	VEHICLE MILEAGE INTERVAL (x1000)													
		3	7.5	12	24	36	37.5	48	52.5	60	67.5	72	82.5	96	108
Air cleaner filter. Install a new engine air filter at 30,000 miles	I/R			✓		✓		✓		✓		✓		✓	
Accessory drive belt	I ⑤	✓	✓	✓	✓	✓	✓	✓	✓	✓	✓	✓	✓	✓	
Auto. Trans. fluid level ①	R			✓		✓		✓		✓		✓		✓	
Auto. Trans. Fluid	⑱					✓						✓		✓	
Ball joints	L			✓		✓		✓		✓		✓		✓	
Brake system ②	S/I			✓		✓		✓		✓		✓		✓	
Cabin air filter	R					✓				✓				✓	
Cooling system hoses and clamps	S/I			✓		✓			✓			✓		✓	
Inspect and lubricate 4WD front axle shaft U-joints	I			✓		✓		✓		✓		✓		✓	
Driveshafts & halfshafts	S/I			✓		✓		✓		✓		✓		✓	
Engine cooling system and hoses	I			✓		✓		✓		✓		✓		✓	
Engine coolant (Premium Gold)	R	Five years or 100,000 miles, then every 3 years or 50,000 miles													
Engine coolant (exc. Premium Gold)	R	At 6 years or 100,000 miles; then every 3 years or 50,000 miles													
Engine oil & filter	R	✓	✓	✓	✓	✓	✓	✓	✓	✓	✓	✓	✓	✓	✓
Front wheel bearings and seals (2WD)	R	Every 150,000 miles, if not previously done													
Fuel filter	R					✓				✓				✓	
Man. Trans. Fluid	R	Every 120,000 miles													
PCV valve	S/I	Every 100,000 miles													
Exhaust system & heat shields	S/I			✓		✓				✓		✓		✓	
Rear axle lubricant (4WD)	R	Every 150,000 miles													
Rotate tires	S/I	✓	✓	✓	✓	✓	✓	✓	✓	✓	✓	✓	✓	✓	✓
Steering linkage	S/I			✓		✓		✓		✓		✓		✓	
Spark plugs	R	Change at 100,000 miles													
Suspension components	S/I			✓		✓	✓			✓				✓	
Transfer case	R	Every 150,000 miles													
Tires inspect and rotate	I	✓		✓		✓		✓		✓		✓		✓	
Multi-Point inspection	I	✓	✓	✓	✓	✓	✓	✓	✓	✓	✓	✓	✓	✓	✓

R: Replace S/I: Inspect and service, if necessary L: Lubricate A: Adjust C: Clean

① Change automatic transaxle fluid on all vehicles equipped with the TorqShift ™ transmission and externally mounted remote filter.

② Inspect the reservoir fluid level, rotor and or drum, brake lines, hoses, calipers and or wheel cylinders

22140_TRIB_C0014

SCHEDULED MAINTENANCE INTERVALS

2007-08 Mazda Tribute

Footnotes Continued

Special Operating Condition Requirements

When towing a trailer or using a camper or car-top carrier:

Change engine oil and install a new oil filter every 4,800 km (3,000 miles) or 3 months.

Change transfer case fluid every 96,000 km (60,000 miles).

Change manual transmission fluid as required.

Inspect and lubricate U-joints as required.

During extensive idling and/or low speed driving for long distances, as in heavy commercial use such as delivery, taxi, patrol car or livery:

Change engine oil and install a new oil filter, lube front lower control arm and steering linkage ball joints with zerk fittings (if equipped) every 4,800 km (3,000 miles) or 3 months.

Inspect brake system and check battery electrolyte level (Patrol cars) every 8,000 km (5,000 miles).

Install a new fuel filter every 24,000 km (15,000 miles).

Change automatic transmission fluid, lubricate 4x2 wheel bearings, install new grease seals and adjust bearings every 48,000 km (30,000 miles).

Install new spark plugs and change transfer case fluid every 96,000 km (60,000 miles).

Install a new cabin air filter as required.

When operating in dusty conditions such as unpaved or dusty roads:

Change engine oil and install a new oil filter every 4,800 km (3,000 miles) or 3 months.

Install a new fuel filter every 24,000 km (15,000 miles).

Change automatic transmission fluid every 48,000 km (30,000 miles).

Change transfer case fluid every 96,000 km (60,000 miles).

Install a new engine air filter as required.

Install a new cabin air filter as required.

When operating in off-road conditions:

Change automatic transmission fluid every 48,000 km (30,000 miles).

Change transfer case fluid every 96,000 km (60,000 miles).

Install a new cabin air filter as required.

Inspect and lubricate U-joints.

Inspect and lubricate steering linkage ball joints with zerk fittings.

22140_TRIB_C0015

PRECAUTIONS

Before servicing any vehicle, please be sure to read all of the following precautions, which deal with personal safety, prevention of component damage, and important points to take into consideration when servicing a motor vehicle:

- Never open, service or drain the radiator or cooling system when the engine is hot; serious burns can occur from the steam and hot coolant.
- Observe all applicable safety precautions when working around fuel. Whenever servicing the fuel system, always work in a well-ventilated area. Do not allow fuel spray or vapors to come in contact with a spark, open flame, or excessive heat (a hot drop light, for example). Keep a dry chemical fire extinguisher near the work area. Always keep fuel in a container specifically designed for fuel storage; also, always properly seal fuel containers to avoid the possibility of fire or explosion. Refer to the additional fuel system precautions later in this section.
- Fuel injection systems often remain pressurized, even after the engine has been turned **OFF**. The fuel system pressure must be relieved before disconnecting any fuel lines. Failure to do so may result in fire and/or personal injury.
- Brake fluid often contains polyglycol ethers and polyglycols. Avoid contact with the eyes and wash your hands thoroughly after handling brake fluid. If you do get brake fluid in your eyes, flush your eyes with clean, running water for 15 minutes. If eye irritation persists, or if you have taken brake fluid internally, IMMEDIATELY seek medical assistance.
- The EPA warns that prolonged contact with used engine oil may cause a number of skin disorders, including cancer. You should make every effort to minimize your exposure to used engine oil. Protective gloves should be worn when changing oil. Wash your hands and any other exposed skin areas as soon as possible after exposure to used engine oil. Soap and water, or waterless hand cleaner should be used.
- All new vehicles are now equipped with an air bag system, often referred to as a Supplemental Restraint System (SRS) or Supplemental Inflatable Restraint (SIR) system. The system must be disabled before performing service on or around system components, steering column, instrument panel components, wiring and sensors. Failure to follow safety and disabling procedures could result in accidental air bag deployment, possible personal injury and unnecessary system repairs.
- Always wear safety goggles when working with, or around, the air bag system. When carrying a non-deployed air bag, be sure the bag and trim cover are pointed away from your body. When placing a non-deployed air bag on a work surface, always face the bag and trim cover upward, away from the surface. This will reduce the motion of the module if it is accidentally deployed. Refer to the additional air bag system precautions later in this section.
- Clean, high quality brake fluid from a sealed container is essential to the safe and proper operation of the brake system. You should always buy the correct type of brake fluid for your vehicle. If the brake fluid becomes contaminated, completely flush the system with new fluid. Never reuse any brake fluid. Any brake fluid that is removed from the system should be discarded. Also, do not allow any brake fluid to come in contact with a painted surface; it will damage the paint.
- Never operate the engine without the proper amount and type of engine oil; doing so WILL result in severe engine damage.
- Timing belt maintenance is extremely important. Many models utilize an interference-type, non-freewheeling engine. If the timing belt breaks, the valves in the cylinder head may strike the pistons, causing potentially serious (also time-consuming and expensive) engine damage. Refer to the maintenance interval charts for the recommended replacement interval for the timing belt, and to the timing belt section for belt replacement and inspection.
- Disconnecting the negative battery cable on some vehicles may interfere with the functions of the on-board computer system(s) and may require the computer to undergo a relearning process once the negative battery cable is reconnected.
- When servicing drum brakes, only disassemble and assemble one side at a time, leaving the remaining side intact for reference.
- Only an MVAC-trained, EPA-certified automotive technician should service the air conditioning system or its components.

BRAKES

ANTI-LOCK BRAKE SYSTEM (ABS)

GENERAL INFORMATION

The Anti-lock Brake System (ABS) module simultaneously manages the anti-lock braking, traction control and engine control systems to maintain vehicle control during deceleration and acceleration.

When the ignition switch is in the **RUN** position, the module carries out a preliminary electrical check and, at approximately 12 MPH (20 km/h), the hydraulic pump motor is turned on for approximately one half-second. Any malfunction of the anti-lock brake system disables the traction control and stability assist (if equipped) and the anti-lock brake warning indicator illuminates. However, the power-assist braking system functions normally.

The Anti-lock Brake System (ABS) consists of the following components:

- Hydraulic Control Unit (HCU)
- ABS control module
- Rear anti-lock brake sensor
- Rear anti-lock brake sensor indicator
- Front anti-lock brake sensor
- Front anti-lock brake sensor indicator
- Yellow ABS warning indicator
- Red brake warning indicator

WHEEL SPEED SENSORS

REMOVAL & INSTALLATION

Front

See Figure 1.

1. Raise and safely support the vehicle.

➡The harness connector is located in the engine compartment.

2. Disconnect the electrical connector.

⁂ WARNING

Care must be taken during the removal of the plug to prevent damage. If the plug is damaged, a new sensor may need to be installed, even though the sensor is functional in all other aspects.

3. Remove the grommet from the body.
4. When removing the body plug, rotate the plug into a position which allows the use of a small screwdriver to release the tabs on the underside of the body plug. These 2 tabs are located at right angles to the sensor wire.

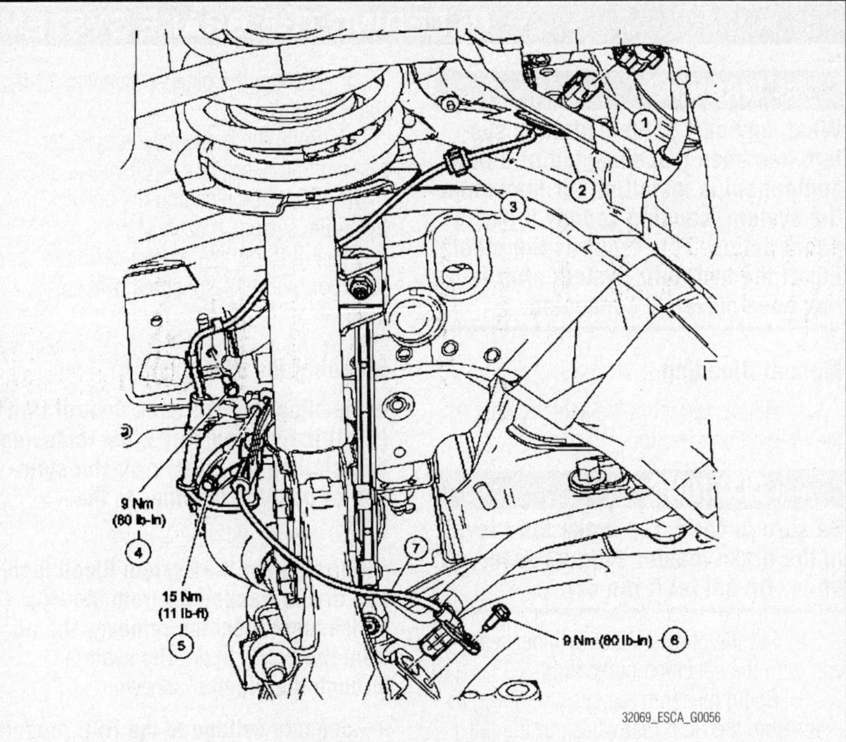

Fig. 1 View of the front wheel speed sensor wire connector (1), grommet (2), front wheel speed sensor wire retainer (3), front wheel speed sensor wire-to-body bolts (4), front wheel speed sensor bolts (5, 6) and front wheel speed sensor (7)

5. Remove the front wheel speed sensor wire from the retainer.
6. Remove the front wheel speed sensor wire-to-body bolt.
7. Remove the front wheel speed sensor wire bolt.
8. Remove the front wheel speed sensor bolt from the wheel knuckle.

➡Clean off any foreign material that may have collected around the sensor before removal.

9. Remove the front wheel speed sensor.

➡Thoroughly clean the mounting surface.

10. Installation is the reverse of the removal procedure, noting the following tightening specifications:
    a. Front wheel speed sensor-to-knuckle bolt: 80 inch lbs. (9 Nm)
    b. Front wheel speed sensor wire bolt: 11 ft. lbs. (15 Nm)
    c. Front wheel speed sensor wire-to-body bolt: 80 inch lbs. (9 Nm)

Rear

See Figure 2.

1. Remove the wheel and tire.

✻✻ WARNING

Care must be taken during the removal of the plug to prevent damage. If the plug is damaged, a new sensor may need to be installed even though the sensor is functional in all other aspects.

2. Remove the grommet from the body.
3. When removing the body plug, rotate the plug into a position which allows the use of a small screwdriver to release the tabs on the underside of the body plug. These 2 tabs are located at right angles to the sensor wire.
4. Disconnect the sensor wiring.
5. Detach the sensor wiring bolts.

➡Clean off any dirt that may have collected around the sensor before removal.

6. Remove the bolt and the sensor.

➡Thoroughly clean the mounting surface.

7. Installation is the reverse of the removal procedure. Tighten all retainers to 80 inch lbs. (9 Nm).

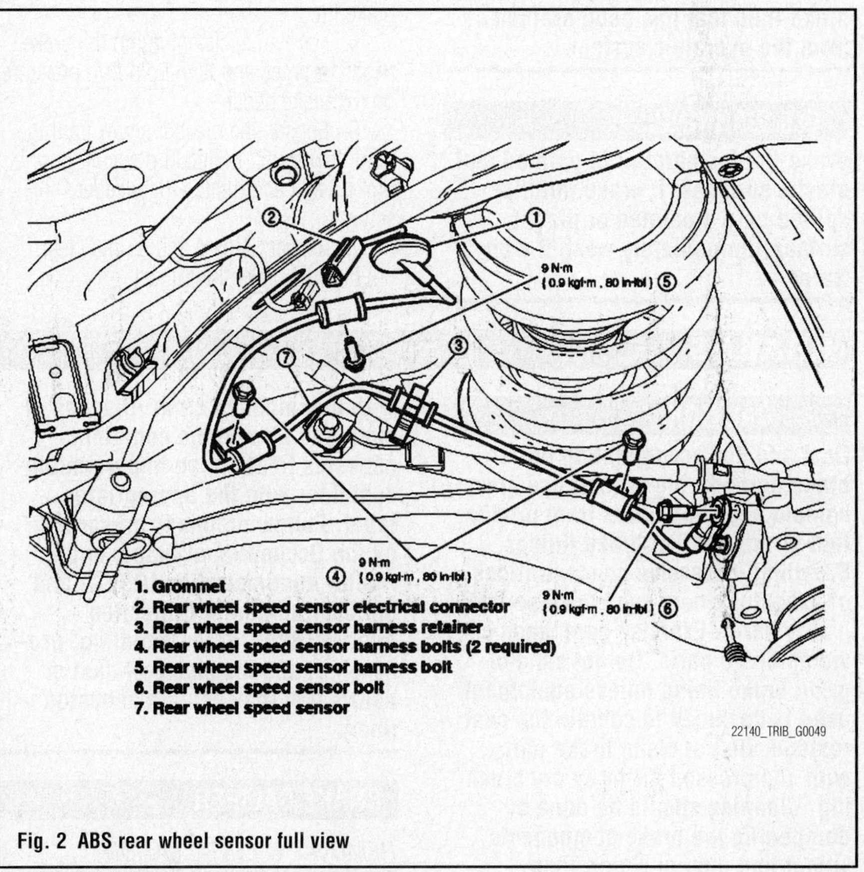

Fig. 2 ABS rear wheel sensor full view

BRAKES — BLEEDING THE BRAKE SYSTEM

BLEEDING PROCEDURE

WARNING

Use of any other than the approved DOT 3 brake fluid will cause permanent damage to brake components and will render the brakes inoperative. Failure to follow these instructions may result in personal injury.

- Brake fluid contains polyglycol ethers and polyglycols. Avoid contact with eyes. Wash hands thoroughly after handling. If brake fluid contacts eyes, flush eyes with running water for 15 minutes. Get medical attention if irritation persists. If taken internally, drink water and induce vomiting. Get medical attention immediately. Failure to follow these instructions may result in personal injury.

WARNING

Do not allow the brake master cylinder reservoir to run dry during the bleeding operation. Keep the master cylinder reservoir filled with the specified brake fluid. Never reuse the brake fluid that has been drained from the hydraulic system.

WARNING

Brake fluid is harmful to painted and plastic surfaces. If brake fluid is spilled onto a painted or plastic surface, immediately wash it with water.

WARNING

When any part of the hydraulic system has been disconnected or a new component is installed, air may enter the system, causing spongy brake pedal action. This requires the bleeding of the hydraulic system after it has been correctly connected.

Manual Bleeding

1. Before servicing the vehicle, refer to the Precautions Section.

WARNING

Be sure to check the brake fluid level in the brake master cylinder reservoir often. Do not let it run dry.

2. Fill the brake master cylinder reservoir with the specified brake fluid.
3. Begin bleeding the system, going in order from the right rear wheel, to the left rear wheel, to the right front wheel, and ending with the left front wheel.
4. Attach a rubber drain hose to the rear bleeder screw and submerge the free end in a container partially filled with clean brake fluid.
5. Have an assistant pump the brake pedal 10 times and then hold firm pressure on the brake pedal.
6. Loosen the bleeder screw until the fluid flow stops. Maintain pressure on the brake pedal and tighten the bleeder screw.
7. Repeat Steps 4 and 5 until clear, bubble-free brake fluid flows.
8. Tighten the bleeder screw to 12 ft. lbs. (16 Nm).
9. Refill the brake master cylinder reservoir as necessary.
10. Continue bleeding the brake hydraulic system at each wheel.
11. Fill the brake master cylinder reservoir with the specified brake fluid.

Bleeding the ABS System

➡Bleeding the Hydraulic Control Unit (HCU) is required only when removing or installing the HCU or master cylinder, or opening the lines to the HCU.

➡Carrying out the System Bleed function drives trapped air from the HCU. Subsequent bleeding removes the air from the brake hydraulic system through the bleeder screws.

➡Adequate voltage to the HCU module is required during the anti-lock control portion of the system bleed.

1. Connect a suitable scan/diagnostic tool.
2. Access the **SYSTEM BLEED FUNCTION**. Go to the Tool Tab-Chassis-Braking-ABS Service Bleed and follow the directions on the diagnostic tool.
3. Manually bleed the brake hydraulic system. For additional information, refer to Manual Bleed in this section.
4. Repeat the procedure carrying out a total of two diagnostic tool cycles and two manual bleed cycles.

BRAKES — FRONT DISC BRAKES

CAUTION

Dust and dirt accumulating on brake parts during normal use may contain asbestos fibers from production or aftermarket brake linings. Breathing excessive concentrations of asbestos fibers can cause serious bodily harm. Exercise care when servicing brake parts. Do not sand or grind brake lining unless equipment used is designed to contain the dust residue. Do not clean brake parts with compressed air or by dry brushing. Cleaning should be done by dampening the brake components with a fine mist of water, then wiping the brake components clean with a dampened cloth. Dispose of cloth and all residue containing asbestos fibers in an impermeable container with the appropriate label. Follow practices prescribed by the Occupational Safety and Health Administration (OSHA) and the Environmental Protection Agency (EPA) for the handling, processing, and disposing of dust or debris that may contain asbestos fibers.

BRAKE CALIPER

REMOVAL & INSTALLATION

See Figure 3.

1. Before servicing the vehicle, refer to the Precautions Section.
2. Remove the wheel and tire assembly.
3. Remove the brake caliper clip.
4. Remove the brake caliper dust boot caps.
5. Remove the brake caliper guide bolts.
6. Remove the brake caliper.

CAUTION

Do not allow the brake caliper to hang by the flexible brake hose.

7. Remove the disc brake pads.
8. Remove the brake caliper jounce hose. Loosen the jounce hose fitting prior to removing the brake caliper.

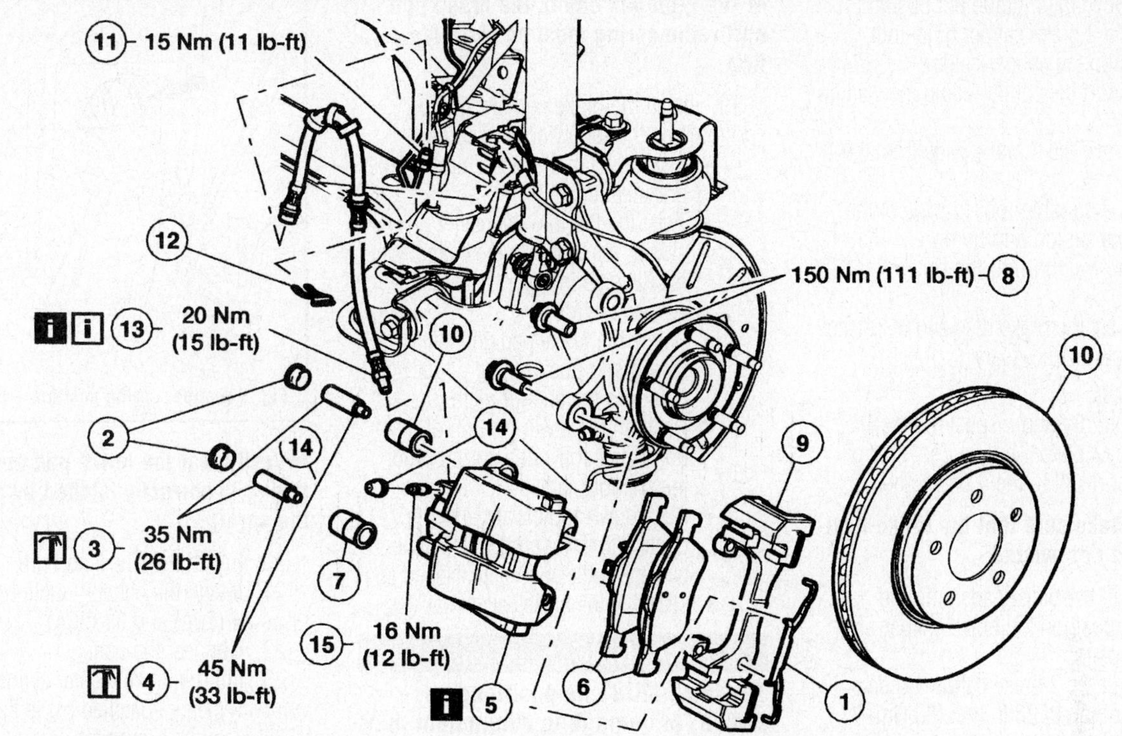

1 Brake caliper clip
2 Brake caliper dust boot caps
3 Brake caliper guide bolts (disc-drum system)
4 Brake caliper guide bolts (4-wheel disc brake system)
5 Brake caliper (RH/LH)
6 Disc brake pads (kit)
7 Brake caliper dust boots
8 Brake caliper anchor plate bolts
9 Brake caliper anchor plate
10 Brake disc
11 Brake line fitting nut
12 Brake caliper jounce hose retaining clip
13 Brake caliper jounce hose
14 Bleeder screw cap
15 Bleeder screw

67197-ESCA-G49

Fig. 3 Front caliper installation

To install:

9. To install, reverse the removal procedure.

10. Torque the caliper pin bolts to 26 ft. lbs. (35 Nm) on disc/drum systems; 33 ft. lbs. (45 Nm) on 4-wheel disc systems.

11. If the hydraulic system was opened, bleed the brake system.

➡Thread the brake caliper jounce hose onto the brake caliper before installing the brake caliper.

➡Make sure that the brake caliper jounce hose is not twisted.

12. Position the brake caliper to the anchor plate and tighten the brake caliper jounce hose.

DISC BRAKE PADS

REMOVAL & INSTALLATION

See Figures 4 and 5.

1. Before servicing the vehicle, refer to the Precautions Section.

2. With the vehicle in NEUTRAL, position it on a hoist.

➡The following steps must be followed to prevent the accumulator from charging and pressurizing the brake system.

3. Disconnect the battery.

4. Remove the battery junction box (BJB) fuses 24 (50A) and 31 (50A).

5. For the LH brake caliper, release the lower portion of the brake pad anti-rattle spring.

6. Apply force to the center of the spring and pull outward at the bottom of the spring to remove it from the lower brake caliper cavity.

7. Rotate the spring upward and remove it from the brake caliper.

8. For the RH brake caliper, release the upper portion of the brake pad anti-rattle spring.

9. Apply force to the center of the spring and pull outward at the top of the spring to remove it from the upper brake caliper cavity.

10. Rotate the spring downward and remove it from the brake caliper.

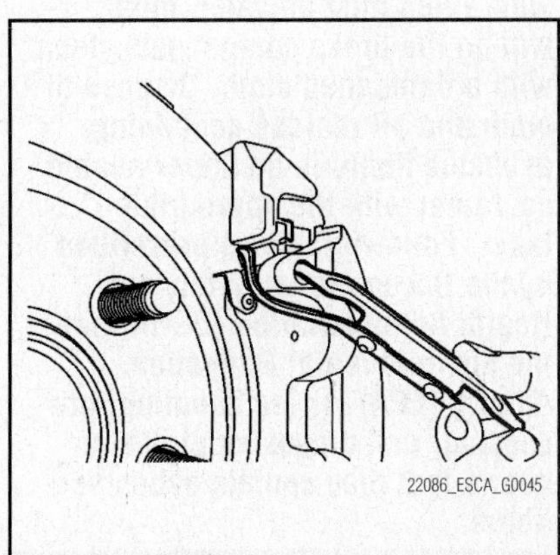

Fig. 4 LH brake caliper spring removal

11. Remove the 2 guide pin bushing caps and the 2 brake caliper guide pin bolts, position the caliper aside.
12. Support the caliper using mechanic's wire.
13. Remove the 2 brake pads from the caliper.
14. Use a suitable tool to protect the brake caliper piston and compress the brake caliper piston into the brake caliper.
15. Inspect the brake disc and resurface or install new as necessary

To install:

16. Clean, dry and inspect the brake caliper anchor plate. Apply a light coat of specified lubricant to the 4 brake pad

➡NOTE: Make sure that the brake flexible hose is not twisted.

17. Install the brake pads onto the caliper and position the brake caliper onto the anchor plate.
18. Install the 2 brake caliper guide pin bolts and tighten to 26 ft. lbs. (35 Nm) install the 2 bushing caps.

➡The 2-tabbed end of the brake pad anti-rattle spring must be installed first.

19. Install the brake pad anti-rattle spring using the following procedure:
    - Insert the tab of the spring into the brake caliper cavity.
    - Twist the tab into the cavity (LH side in the upper brake caliper cavity, RH side in the lower brake caliper cavity).
    - Rotate the brake pad anti-rattle spring and position the upper portion onto the anchor plate.
    - Position the lower portion of the brake pad anti-rattle spring onto the anchor plate.
    - Push down and inward until the upper and lower ends of the brake pad anti-rattle spring are latched and seated in the brake caliper cavities.

✱✱ WARNING

The latch MUST be positioned as shown, or damage to component may occur.

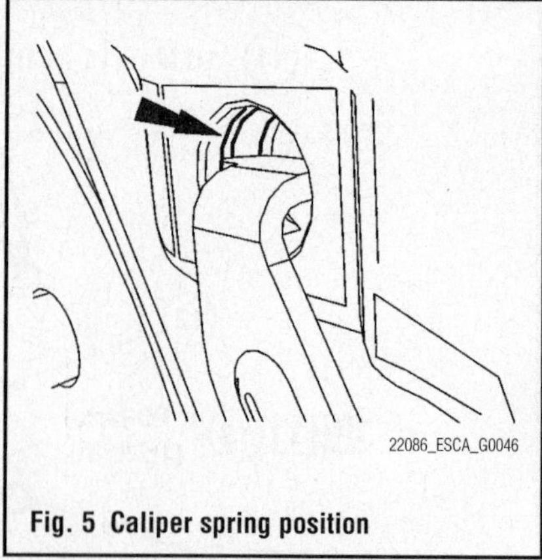

Fig. 5 Caliper spring position

➡Verify that the brake pad anti-rattle spring is correctly latched by pulling on the spring.

20. Install the wheel and tire.
21. Install the battery junction box (BJB) fuses 24 (50A) and 31 (50A).
22. Connect the battery.
23. Fill the brake master cylinder reservoir with clean, specified brake fluid.
24. Test the brakes for normal operation.

BRAKES

REAR DISC BRAKES

✱✱ CAUTION

Dust and dirt accumulating on brake parts during normal use may contain asbestos fibers from production or aftermarket brake linings. Breathing excessive concentrations of asbestos fibers can cause serious bodily harm. Exercise care when servicing brake parts. Do not sand or grind brake lining unless equipment used is designed to contain the dust residue. Do not clean brake parts with compressed air or by dry brushing. Cleaning should be done by dampening the brake components with a fine mist of water, then wiping the brake components clean with a dampened cloth. Dispose of cloth and all residue containing asbestos fibers in an impermeable container with the appropriate label. Follow practices prescribed by the Occupational Safety and Health Administration (OSHA) and the Environmental Protection Agency (EPA) for the handling, processing, and disposing of dust or debris that may contain asbestos fibers.

BRAKE CALIPER

REMOVAL & INSTALLATION

See Figure 6.

1. Before servicing the vehicle, refer to the Precautions Section.
2. Remove the wheel and tire assembly.

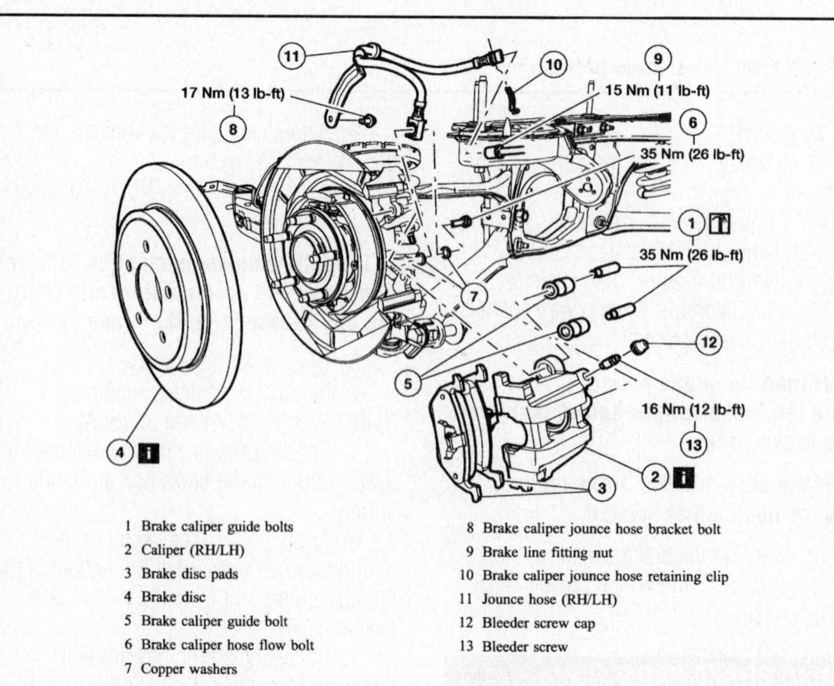

Fig. 6 Rear caliper installation

3. Remove the brake caliper guide bolts.
4. Remove the caliper.

✱✱ CAUTION

Do not allow the brake caliper to hang by the flexible brake hose.

5. Remove the brake disc pads.
6. Remove the brake caliper hose flow bolt.
7. Remove and discard the copper washers.
8. To install, reverse the removal procedure. Use new copper washers. Torque the caliper pin bolts to 26 ft. lbs. (35 Nm); torque the flow bolt to 26 ft. lbs. (35 Nm).
9. If the hydraulic system was opened, bleed the brake system.

DISC BRAKE PADS

REMOVAL & INSTALLATION

1. Before servicing the vehicle, refer to the Precautions Section.
2. Remove the wheel and tire assembly.
3. Remove the brake caliper guide bolts.
4. Remove the caliper.

✱✱ CAUTION

Do not allow the brake caliper to hang by the flexible brake hose.

5. Remove the brake disc pads.
6. To install, reverse the removal procedure.
7. If the hydraulic system was opened, bleed the brake system.

BRAKES — REAR DRUM BRAKES

✱✱ CAUTION

Dust and dirt accumulating on brake parts during normal use may contain asbestos fibers from production or aftermarket brake linings. Breathing excessive concentrations of asbestos fibers can cause serious bodily harm. Exercise care when servicing brake parts. Do not sand or grind brake lining unless equipment used is designed to contain the dust residue. Do not clean brake parts with compressed air or by dry brushing. Cleaning should be done by dampening the brake components with a fine mist of water, then wiping the brake components clean with a dampened cloth. Dispose of cloth and all residue containing asbestos fibers in an impermeable container with the appropriate label. Follow practices prescribed by the Occupational Safety and Health Administration (OSHA) and the Environmental Protection Agency (EPA) for the handling, processing, and disposing of dust or debris that may contain asbestos fibers.

BRAKE DRUM

REMOVAL & INSTALLATION

See Figure 7.

1. Before servicing the vehicle, refer to the Precautions Section.
2. Remove the tire and wheel assembly.

✱✱ CAUTION

Use of a brake drum puller or a torch is not recommended. Brake drum distortion can result.

➡If the brake drum is rusted to the axle shaft pilot diameter, tap the center of the brake drum between the wheel studs.

3. Remove the brake drum.
4. If equipped, remove the brake drum retaining clips.
5. If the brake drums will not come off, follow these steps.
6. Move the brake shoe adjusting lever off the brake adjuster screw.
7. Loosen the brake shoe adjuster screw nut by adjusting the nut upward.
8. Using the special tool, 134-R0191, measure the brake drum inside diameter.

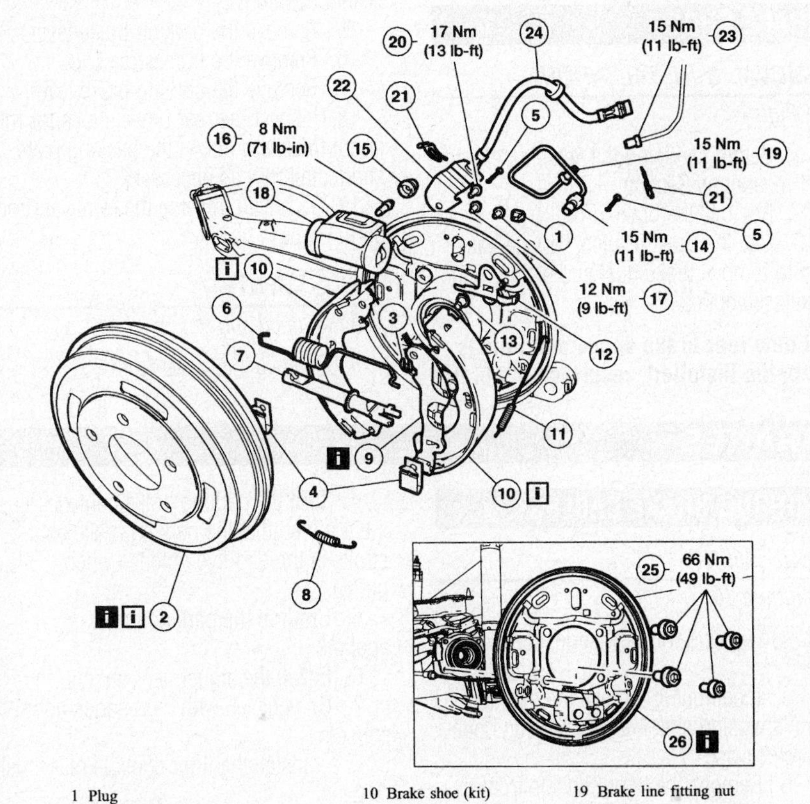

1 Plug
2 Brake drum
3 Parking brake lever clip
4 Brake shoe retaining clips
5 Brake shoe retaining pins
6 Upper return spring
7 Adjuster assembly (LH/RH)
8 Lower return spring
9 Parking brake actuator lever (LH/RH)
10 Brake shoe (kit)
11 Adjuster lever spring
12 Adjuster lever (LH/RH)
13 Pivot pin (part of 2200)
14 Brake line fitting nut
15 Bleeder screw cap
16 Bleeder screw
17 Wheel cylinder bolts
18 Wheel cylinder
19 Brake line fitting nut
20 Jounce hose bracket bolt
21 Jounce hose retaining clips
22 Jounce hose bracket
23 Brake line fitting nut
24 Jounce hose (LH/RH)
25 Backing plate bolts
26 Backing plate

67197-ESCA-G51

Fig. 7 Drum brake exploded view

9. Install a new brake drum if the maximum inside diameter exceeds the specification shown on the outside of the brake drum.

To install:

WARNING

Whenever a wheel is installed, always remove any corrosion, dirt or foreign material present on the mounting surfaces of the wheel or the surface of the wheel hub, brake drum or brake disc that contacts the wheel. Installing wheels without correct metal-to-metal contact at the wheel mounting surfaces can cause the wheel nuts to loosen and the wheel to come off while the vehicle is in motion, causing loss of control. Failure to follow these instructions may result in personal injury.

10. Clean the wheel hub mounting surface and wheel pilot.
11. Install the tire and wheel assembly.

BRAKE SHOES

REMOVAL & INSTALLATION

See Figure 7.

1. Before servicing the vehicle, refer to the Precautions Section.
2. Remove the brake drum.
3. Use the Brake/Clutch/Service Vacuum to remove brake dust and dirt from the brake assemblies.

➡If new rear brake shoes and linings are being installed, resurface the brake drums to remove glazing and to ensure an equal friction surface from side-to-side. Resurfacing will also correct out-of-round and bell conditions.

4. Using the special tool, measure the braking surface diameter. If the inside diameter measures more than the maximum specification shown on the outside of the brake drum, install a new brake drum.
5. Remove the parking brake cable from the parking brake cable lever.
6. Remove the hold-down clips and pins.
7. Remove the lower spring.
8. Remove the rear brake shoes.
9. Pull the bottom of the brake shoe forward.
10. Release the upper return spring.
11. Remove both brake shoes together.
12. Remove the self adjuster lever.
13. Remove the self adjuster and spring assembly.
14. Return the self adjuster to the fully seated position.
15. Remove the parking brake lever.
16. Remove the horseshoe clip.
17. Remove the parking brake lever.
18. Inspect the rear brake shoes for minimum thickness above the backing plate, and install new as necessary.
19. To install, reverse the removal procedure.

ADJUSTMENT

See Figures 8 and 9.

1. Remove the brake drum.
2. Using the special tool, 134-R0191, measure the brake drum inside diameter.
3. Position the special tool on the brake shoes and linings and adjust accordingly.
4. Install the brake drum

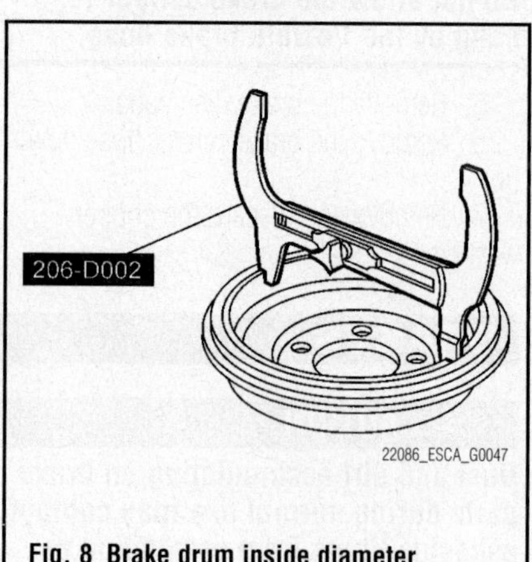

Fig. 8 Brake drum inside diameter

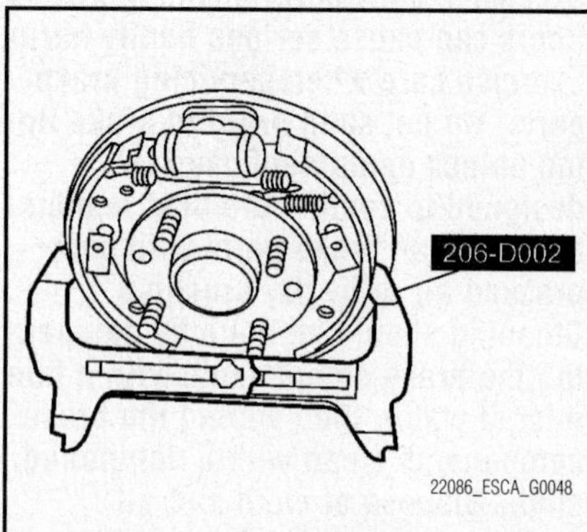

Fig. 9 Adjustment of brake shoes-to-drum inside diameter

PARKING BRAKE CABLES

ADJUSTMENT

See Figure 10.

1. On vehicle with low series floor console:
   a. If equipped, remove the manual transmission shift knob by turning counterclockwise.
   b. Remove the floor console front finish panel.
   c. Apply the parking brake, then remove the floor console rear finish panel.
2. On vehicle with high series floor console:
   a. Apply the parking brake.
   b. Remove the floor console finish panel.
3. Remove the adjustment nut clip.
4. Turn the parking brake control adjustment nut so that the parking brake control stroke is three to five notches when pulled.
5. Confirm the parking brake is applied.
6. Install the adjustment nut clip.
7. On vehicles with high series floor console:
   a. Install the floor console rear finish panel.
8. On vehicles with low series floor console:
   a. Install the floor console rear finish panel.
   b. Install the floor console front finish panel. If equipped, install the manual transmission shifter knob by turning clockwise.

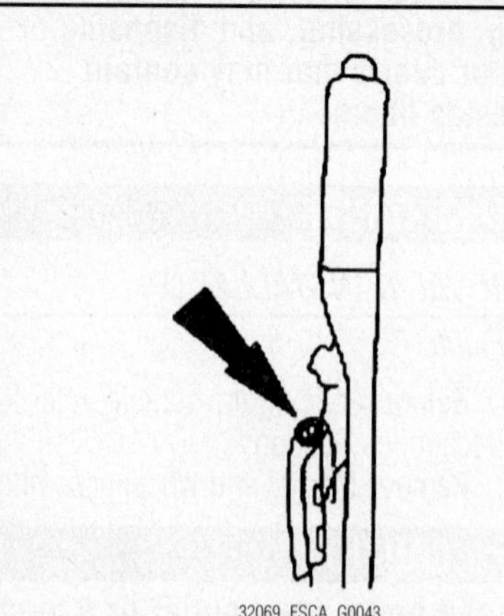

Fig. 10 Turn the parking brake control adjustment nut so that the parking brake control stroke is three to five notches when pulled

PARKING BRAKE SHOES

REMOVAL & INSTALLATION

With Rear Drum Brakes

The rear drum brake shoes serve as the parking brakes. Refer to the procedures under Rear Drum Brakes.

With Rear Disc Brakes

See Figure 11.

1. Before servicing the vehicle, refer to the Precautions Section.
2. Remove the rear brake disc.
3. Remove the parking brake shoe upper return spring.
4. Remove the 2 parking brake shoe retaining pins.
5. Remove the 2 parking brake shoe retaining springs.
6. Remove the parking brake shoe lower return spring.
7. Remove the parking brake shoe adjuster.
8. Remove the parking brake shoes.

To install:

9. To install, reverse the removal procedure.
   - Using anti-seize lubricant, lubricate the parking brake shoe contact points before installation of the rear parking brake shoes.

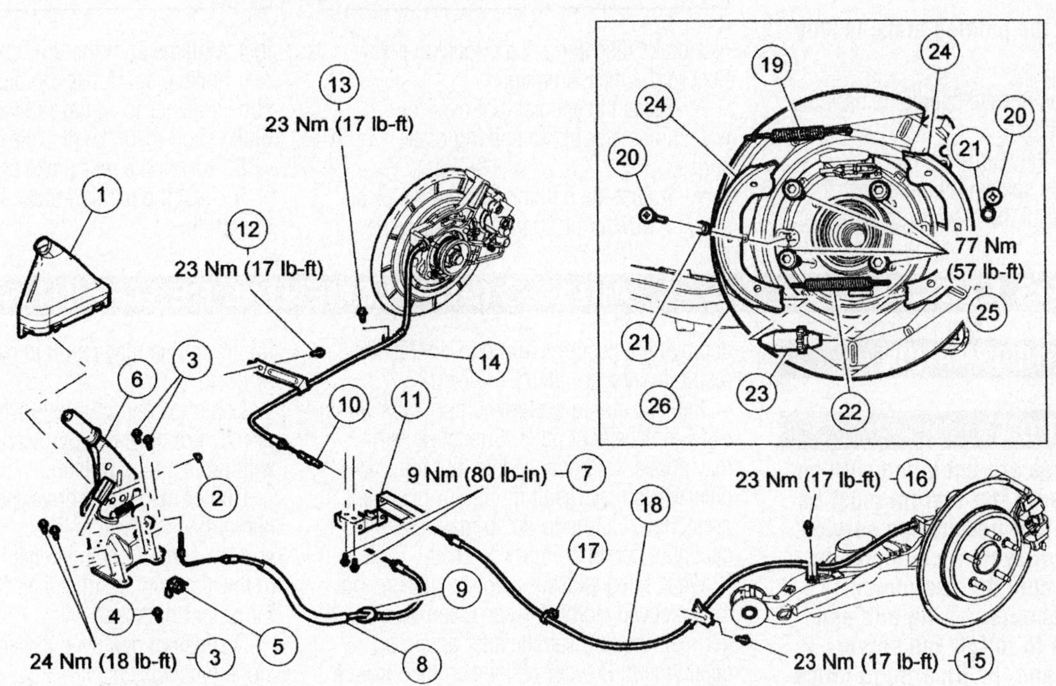

1 Parking brake control boot
2 Front cable adjuster nut
3 Parking brake control bolts (4 required)
4 Warning indicator switch screw
5 Warning indicator switch
6 Parking brake control
7 Parking brake equalizer bracket bolts (2 required)
8 Front parking brake cable
9 Grommet
10 Cable connector
11 Parking brake equalizer and bracket
12 Cable bracket-to-body bolt
13 Cable bracket-to-trailing arm bolt
14 Rear parking brake cable (RH)
15 Cable bracket-to-body bolt
16 Cable bracket-to-trailing arm bolt
17 Cable-to-fuel tank strap clip
18 Rear parking brake cable (LH)
19 Parking brake shoe upper return spring
20 Parking brake shoe retaining pins (2 required)
21 Parking brake shoe retaining springs (2 required)
22 Parking brake shoe lower return spring
23 Parking brake shoe adjuster
24 Parking brake shoe (LH/RH)
25 Support plate bolts (4 required)
26 Support plate (LH/RH)

06017-ESCA-G99

Fig. 11 Parking brake system—With rear disc brakes

- Lubricate the adjust screw threads with anti-seize lubricant.
- Adjust the parking brake shoe and lining.
- Check the parking brake for normal operation.

ADJUSTMENT

See Figures 12 and 13.

1. Before servicing the vehicle, refer to the Precautions Section.
2. With the vehicle in NEUTRAL, position it on a hoist.

➡Make sure the parking brake is fully released.

3. Using the release handle, release the parking brake control.
4. Remove the rear brake disc.
5. Using the special tool, measure the inside diameter of the drum portion of the rear brake disc and set the locking screw. Record the measurement.
6. Place the special tool over the widest diameter of the parking brake shoes.
7. Adjust the parking brake shoe clearance to 1.07 mm (0.04 in) less than the inside diameter of the drum portion of the rear brake disc. Rotate the parking brake shoe adjuster to achieve the correct parking brake shoe-to-brake disc clearance.
8. Install the rear brake disc.
9. Test the parking brake for normal operation.

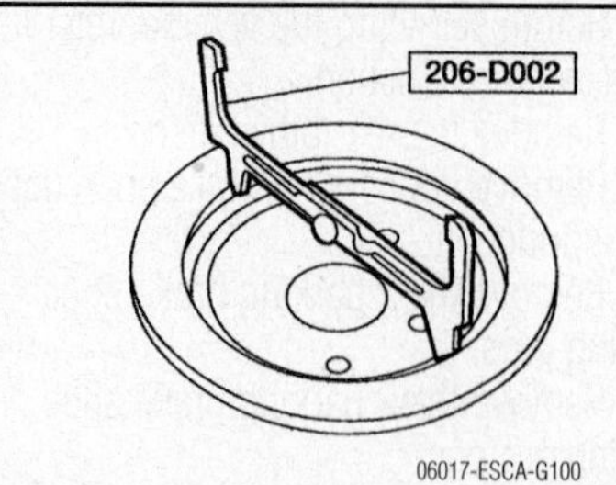

Fig. 12 Using the special tool, measure the inside diameter of the drum portion of the rear brake disc and set the locking screw

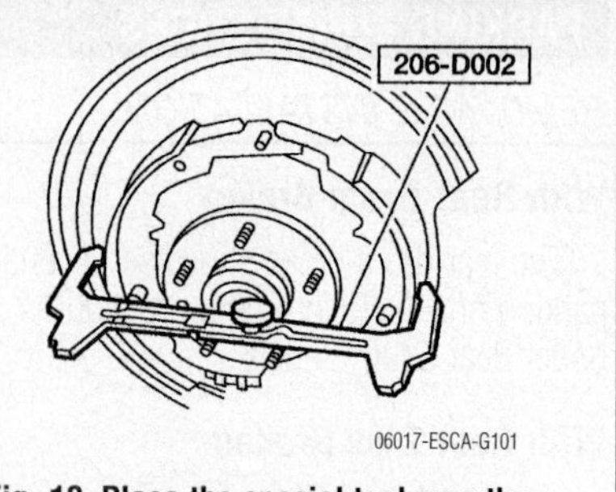

Fig. 13 Place the special tool over the widest diameter of the parking brake shoes

CHASSIS ELECTRICAL

AIR BAG (SUPPLEMENTAL RESTRAINT SYSTEM)

GENERAL INFORMATION

⁂ CAUTION

Some vehicles are equipped with an air bag system. The system must be disarmed before performing service on, or around, system components, the steering column, instrument panel components, wiring and sensors. Failure to follow the safety precautions and the disarming procedure could result in accidental air bag deployment, possible injury and unnecessary system repairs.

SERVICE PRECAUTIONS

Disconnect and isolate the battery negative cable before beginning any airbag system component diagnosis, testing, removal, or installation procedures. Allow system capacitor to discharge for two minutes before beginning any component service. This will disable the airbag system. Failure to disable the airbag system may result in accidental airbag deployment, personal injury, or death.

Do not place an intact undeployed airbag face down on a solid surface. The airbag will propel into the air if accidentally deployed and may result in personal injury or death.

When carrying or handling an undeployed airbag, the trim side (face) of the airbag should be pointing towards the body to minimize possibility of injury if accidental deployment occurs. Failure to do this may result in personal injury or death.

Replace airbag system components with OEM replacement parts. Substitute parts may appear interchangeable, but internal differences may result in inferior occupant protection. Failure to do so may result in occupant personal injury or death.

Wear safety glasses, rubber gloves, and long sleeved clothing when cleaning powder residue from vehicle after an airbag deployment. Powder residue emitted from a deployed airbag can cause skin irritation. Flush affected area with cool water if irritation is experienced. If nasal or throat irritation is experienced, exit the vehicle for fresh air until the irritation ceases. If irritation continues, see a physician.

Do not use a replacement airbag that is not in the original packaging. This may result in improper deployment, personal injury, or death.

The factory installed fasteners, screws and bolts used to fasten airbag components have a special coating and are specifically designed for the airbag system. Do not use substitute fasteners. Use only original equipment fasteners listed in the parts catalog when fastener replacement is required.

During, and following, any child restraint anchor service, due to impact event or vehicle repair, carefully inspect all mounting hardware, tether straps, and anchors for proper installation, operation, or damage. If a child restraint anchor is found damaged in any way, the anchor must be replaced. Failure to do this may result in personal injury or death.

Deployed and non-deployed airbags may or may not have live pyrotechnic material within the airbag inflator.

Do not dispose of driver/passenger/curtain airbags or seat belt tensioners unless you are sure of complete deployment. Refer to the Hazardous Substance Control System for proper disposal.

Dispose of deployed airbags and tensioners consistent with state, provincial, local, and federal regulations.

After any airbag component testing or service, do not connect the battery negative cable. Personal injury or death may result if the system test is not performed first.

If the vehicle is equipped with the Occupant Classification System (OCS), do not connect the battery negative cable before performing the OCS Verification Test using the scan tool and the appropriate diagnostic information. Personal injury or death may result if the system test is not performed properly.

Never replace both the Occupant Restraint Controller (ORC) and the Occupant Classification Module (OCM) at the same time. If both require replacement, replace one, then perform the Airbag System test before replacing the other.

Both the ORC and the OCM store Occupant Classification System (OCS) calibration data, which they transfer to one another when one of them is replaced. If both are replaced at the same time, an irreversible fault will be set in both modules and the

OCS may malfunction and cause personal injury or death.

If equipped with OCS, the Seat Weight Sensor is a sensitive, calibrated unit and must be handled carefully. Do not drop or handle roughly. If dropped or damaged, replace with another sensor. Failure to do so may result in occupant injury or death.

If equipped with OCS, the front passenger seat must be handled carefully as well. When removing the seat, be careful when setting on floor not to drop. If dropped, the sensor may be inoperative, could result in occupant injury, or possibly death.

If equipped with OCS, when the passenger front seat is on the floor, no one should sit in the front passenger seat. This uneven force may damage the sensing ability of the seat weight sensors. If sat on and damaged, the sensor may be inoperative, could result in occupant injury, or possibly death.

DISARMING THE SYSTEM

1. Before servicing the vehicle, refer to the Precautions Section.
2. Turn all vehicle accessories **OFF**.
3. Turn the ignition switch to **OFF**.
4. At the Central Junction Box (CJB), located below the left side of the instrument panel, remove the cover and the Restraints Control Module (RCM) fuse(s) from the CJB. See the Owner's Manual.
5. Turn the ignition ON and visually monitor the air bag indicator for at least 30 seconds. The air bag indicator will remain lit continuously (no flashing) if the correct RCM fuse has been removed. If the air bag indicator does not remain lit continuously, remove the correct RCM fuse before proceeding.
6. Turn the ignition **OFF**.

CAUTION

To avoid accidental deployment and possible personal injury, the backup power supply must be depleted before repairing or replacing any front or side air bag Supplemental Restraint System (SRS) components and before servicing, replacing, adjusting or striking components near the front or side air bag sensors, such as doors, instrument panel, console, door latches, strikers, seats and hood latches.

The front impact severity sensor is located on the radiator support bracket.

The first row side impact sensors (if equipped) are located at or near the base of the B-pillars.

The second row side impact sensors (if equipped) are located on the C-pillar.

➡To deplete the backup power supply energy, disconnect the battery ground cable and wait at least one minute. Be sure to disconnect auxiliary batteries and power supplies (if equipped).

7. Disconnect the battery ground cable and wait at least one minute

ARMING THE SYSTEM

1. Before servicing the vehicle, refer to the Precautions Section.

CAUTION

The restraint system diagnostic tool is for restraint system service only. Remove from vehicle prior to road use. Failure to remove could result in injury and possible violation of vehicle safety standards.

2. Make sure all restraint system diagnostic tool(s) that may have been installed during the repair have been removed from the vehicle and all SRS components are connected.
3. Turn the ignition switch from **OFF** to **ON**.
4. Install RCM fuse(s) to the CJB and close the cover.

CAUTION

Be sure that nobody is in the vehicle and that there is nothing blocking or set in front of any air bag module when the battery ground cable is connected.

5. Connect the battery ground cable.
6. Prove out the Supplemental Restraint System (SRS) as follows:
   a. Turn the ignition key from **ON** to **OFF**.Wait 10 seconds, then turn the key back to ON and visually monitor the air bag indicator with the air bag modules installed. The air bag indicator will light continuously for approximately six seconds and then turn off. If an air bag Supplemental Restraint System (SRS) fault is present, the air bag indicator will either:
      - fail to light.

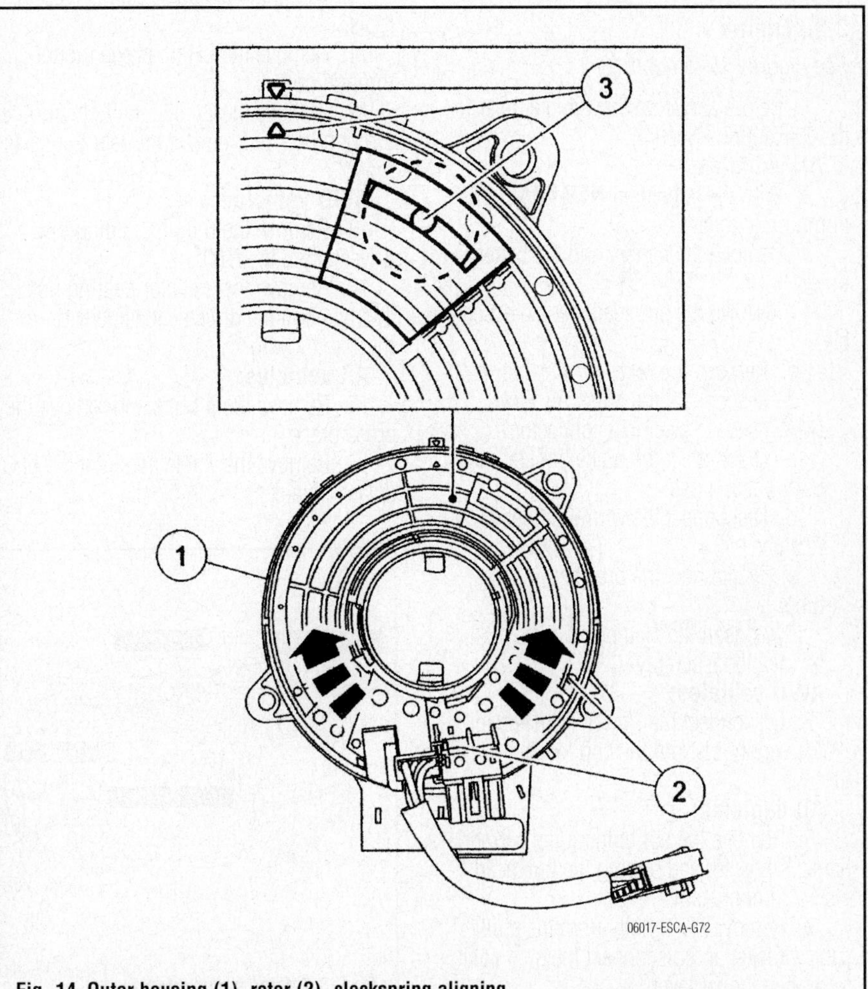

Fig. 14 Outer housing (1), rotor (2), clockspring aligning

- remain lit continuously.
- flash.

b. The flashing might not occur until approximately 30 seconds after the ignition switch has been turned from the **OFF** to the **ON** position. This is the time required for the restraints control module (RCM) to complete the testing of the SRS. If the air bag indicator is inoperative and a SRS fault exists, a chime will sound in a pattern of five sets of five beeps. If this occurs, the air bag indicator and any SRS fault discovered must be diagnosed and repaired.

7. Clear all continuous DTCs from the restraints control module using a scan tool.

CLOCKSPRING CENTERING

See Figure 14.

1. Before servicing the vehicle, refer to the Precautions Section.

⁂ CAUTION

Incorrect centralization may result in premature component failure. If in doubt when centralizing the clockspring, repeat the centralizing procedure. Failure to follow this instruction may result in personal injury.

➡Make sure the road wheels are in the straight-ahead position.

2. If the vehicle's clockspring has rotated out of center, follow these steps to center the clockspring.

a. Hold the clockspring outer housing stationary.

⁂ WARNING

Overturning will destroy the clockspring. The internal ribbon wire acts as the stop and can be broken from its internal connection.

b. While turning the rotor clockwise, carefully feel for the ribbon wire to run out of length and for a slight resistance. Stop turning at this point.

c. Turn the clockspring counterclockwise until the yellow indicator shows anywhere in the window (window will be near the 1 o'clock position) and the arrow on the rotor lines up with the arrow on the top of the housing. The clockspring is now centered. Do not allow the rotor to turn from this position.

DRIVETRAIN

AUTOMATIC TRANSAXLE ASSEMBLY

REMOVAL & INSTALLATION

2.3L Engine

See Figures 15 through 18.

1. Before servicing the vehicle, refer to the Precautions Section.

All vehicles:

2. With the vehicle in **NEUTRAL**, position it on a hoist.
3. Remove the battery and the battery tray.
4. Remove the air cleaner as an assembly.

a. Remove the bolt.

b. Disconnect the Mass Air Flow (MAF) sensor electrical connector.

c. Disconnect the brake booster vacuum hose.

d. Disconnect the wiring harness retainer

e. Disconnect the breather tube.

f. Loosen the clamp and remove the air cleaner assembly.

AWD vehicles:

5. Disconnect the Power Transfer Unit (PTU) vent hose from the clip located on the fill tube

All vehicles:

6. Remove the nut holding the wiring harness bracket and unplug the bulkhead electrical connector.
7. Remove the 2 bolts from the shift cable bracket and disconnect the shift cable from the manual lever.
8. Disconnect the Transmission Range (TR) sensor electrical connector.
9. Remove the 3 upper transaxle retaining bolts.
10. Install the suitable engine support tools.
11. Remove the LH transaxle mount through bolt.
12. Remove the 2 nuts, the bolt and the through bolt and remove the rear transaxle mount.

AWD vehicles:

13. Remove the 6 bolts holding the driveshaft to the PTO.
14. Remove the 2 center bearing nuts and position the driveshaft aside with mechanic's wire

All vehicles:

15. Remove the 4 bolts and remove the cross brace.
16. Remove the 7 retainers and the LH splash shield.

Fig. 15 Engine support tools

17. Remove the 5 retainers and the RH splash shield.
18. Remove the bolt for the mount and the 2 bolts from the cross brace.
19. Remove and discard the nut and remove the cross brace.

➡If transmission disassembly or installation of a new transmission is necessary, the transmission fluid will need to be drained.

20. Remove the drain plug and drain the fluid.
21. Remove and discard the LH front axle wheel hub nut.
22. Remove the frame bolt from the LH and RH control arms.
23. Using a suitable tool, separate the LH halfshaft from the front wheel knuckle.
24. Using a suitable tool, remove the LH halfshaft and disconnect the RH halfshaft from the intermediate shaft.
25. Remove the 2 intermediate shaft retaining nuts.
26. Remove the intermediate shaft.
27. Remove the 2 bolts which hold the exhaust bracket to the intermediate shaft bracket.
28. Remove the 2 nuts on the other end of the exhaust-to-intermediate shaft bracket and remove the bracket.
29. Remove the 3 nuts and separate the flexpipe from the exhaust manifold.

AWD vehicles:

30. Remove the 6 bolts holding the engine bracket to the PTU and remove the bracket.
31. Remove the 3 bolts and the PTU assembly.

Front wheel drive (FWD) vehicles:

32. Remove the 3 bolts and the dampener.

All vehicles:

33. Remove the 3 bolts holding the transaxle front mount plate.
34. Remove the fluid cooler line.
35. Remove the fluid cooler tube.
    a. Disconnect the Output Shaft Sensor (OSS) sensor electrical connector (black).
    b. Disconnect the Transmission Shaft Sensor (TSS) sensor electrical connector (white).
    c. Disconnect the wiring harness retainer from the transmission case and position the harness aside
    d. Remove the transmission fluid cooler line retaining bracket bolt.
    e. Remove the fluid cooler tube and position it aside.
36. Remove the OSS sensor.
37. Disconnect the starter terminals.
38. Remove the wire harness clip retainer and the ground wire from the starter bolts.
39. Remove the 3 bolts and remove the starter.
40. Remove the starter motor isolator.
41. Remove and discard the 4 torque converter nuts.
42. Using a suitable tool, lower the transmission.
43. Push the converter back from the flexplate. Use a suitable transmission jack to support the transaxle and remove the 3 rear bell housing bolts.
44. Remove the 4 remaining transaxle-to-engine bolts.

➡**The torque converter is heavy. Secure torque converter before lowering the transaxle.**

45. Lower the transaxle from the engine compartment.

To install:

⁂ WARNING

Carry out the transmission fluid cooler back flushing and cleaning if the transaxle is being overhauled or installing a new or remanufactured transaxle. Carry out the transmission fluid cooler flow test if the transaxle is being overhauled or installing a new or remanufactured transaxle.

All vehicles:

46. Lubricate the torque converter pilot hub with grease.
47. Rotate the torque converter to place the paint dot in the 6 o'clock position.
48. Position the transaxle in place.
49. Move the transaxle assembly toward the engine assembly and install the 4 bolts.
50. Tighten the transaxle mounting bolts to 30 ft. lbs. (40 Nm).
51. Install the transaxle retaining bolts and tighten to 30 ft. lbs. (40 Nm).
52. Install 4 new torque converter nuts and tighten to 30 ft. lbs. (40 Nm).
53. Install the transmission fluid cooler tube and tighten to 17 ft. lbs. (23 Nm).
54. Install the OSS sensor and tighten to 9 ft. lbs. (12 Nm).
55. Install the fluid cooler tube.
    a. Connect the OSS sensor.
    b. Connect the TSS sensor (white connector).
    c. Connect the wiring harness retainer to the transmission case.
    d. Install the fluid cooler tube and tighten to 17 ft. lbs. (23 Nm).
    e. Install fluid cooler bolt and tighten to 10 ft. lbs. (13 Nm).
56. Install the starter motor isolator.
57. Install the starter motor and mounting bolts, tighten to 26 ft. lbs. (35 Nm).
58. Connect starter terminals and tighten nuts to 89 inch lbs. (10 Nm) for battery cable. And 62 inch lbs. (7 Nm). for solenoid wire nut.
59. Install the wire harness clip retainer and the ground wire to starter bolts.
60. Install the lower front mount bracket and tighten to 41 ft. lbs. (55 Nm).
61. Using a suitable tool, raise the transaxle.
62. Install the LH upper transaxle mount through bolt and tighten to 76 ft. lbs. (103 Nm).
63. Remove engine support tools.
64. Install the upper transaxle bolts and tighten to 30 ft. lbs. (40 Nm).
65. Connect the TR sensor electrical connector.
66. Connect the shift cable to the manual lever and install the 2 bolts then, tighten to 17 ft. lbs. (23 Nm).
67. Install the wire harness bracket nut and tighten to 89 inch lbs. (10 Nm). Plug in the bulkhead electrical connector.
68. Install the rear transmission mount and tighten to 59 ft. lbs. (80 Nm).
69. Install the rear transaxle mount through bolt and tighten to 89 ft. lbs. (120 Nm).

FWD vehicles:

70. Install the dampener and tighten bolts to 30 ft. lbs. (40 Nm).

AWD vehicles:

71. Install the PTU assembly and tighten to 33 ft. lbs. (45 Nm).
72. Position the driveshaft in place and install the nuts. Tighten nuts to 35 ft. lbs. (48 Nm).
73. Install the driveshaft and tighten mounting bolts to 15 ft. lbs. (20 Nm).

All vehicles

74. Install the bracket and tighten to 41 ft. lbs. (55 Nm).
75. Install the front nuts on the exhaust-to-intermediate shaft bracket and tighten to 22 ft. lbs. (30 Nm).

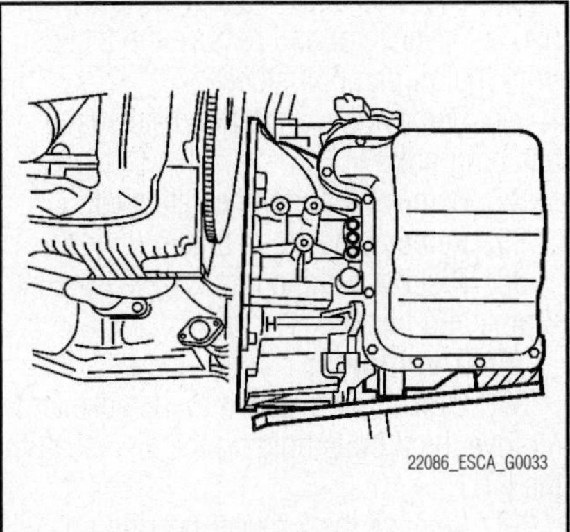

Fig. 16 Transaxle assembly removal

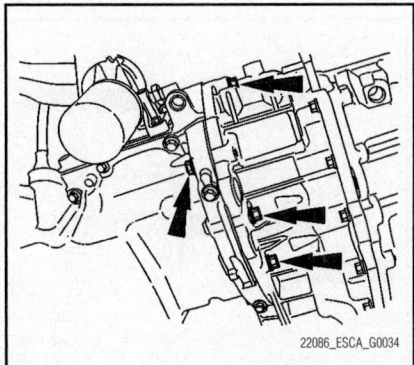

Fig. 17 Transaxle and mounting bolts shown

Fig. 18 LH upper transaxle through bolt shown

76. Install the bolts on the rear of the exhaust-to-intermediate shaft bracket and tighten to 22 ft. lbs. (30 Nm).
77. Install the flexpipe on the exhaust manifold, and tighten the mounting nuts to 18 ft. (25 Nm).
78. Install the intermediate shaft.
79. Install the intermediate shaft retaining nuts and tighten to 20 ft. lbs. (27 Nm).
80. Install the LH and RH halfshafts and frame bolts and tighten to 85 ft. lbs. (115 Nm).
81. Install the LH halfshaft into the wheel knuckle.
82. Install the LH hub nut and tighten to 214 ft. lbs. (290 Nm).
83. Install the cross brace and tighten rear nut to 129 ft. lbs. (175 Nm).
84. Tighten front cross brace bolts to 66 ft. lbs. (90 Nm).
85. Install the bolt for the mount and tighten to 85 ft. lbs. (115 Nm).
86. Install the cross brace and the 4 bolts, tighten bolts to 85 ft. lbs. (115 Nm).
87. Install the LH splash shield and the 7 retainers.
88. Install the RH splash shield and the 5 retainers.

AWD vehicles:

89. Install the vent tube to the fluid level indicator.

All vehicles:

90. Install the air cleaner assembly.
    a. Install the bolt and tighten to 89 inch lbs. (10 Nm).
    b. Reconnect the MAF sensor electrical connector.
    c. Reconnect the brake booster vacuum hose.
    d. Reconnect the wiring harness retainer.
    e. Reconnect the breather tube.
    f. Tighten the clamp and install the air cleaner assembly.
91. Install the battery tray.
92. Fill the transaxle with clean automatic transmission fluid.
93. Check the fluid level and correct as necessary.

➡Verify that the shift cable is correctly adjusted. The vehicle should start inPARKandREVERSEand the reverse lamps should illuminate inREVERSE.

3.0L Engine

See Figures 19 through 23.

1. Before servicing the vehicle, refer to the Precautions Section.

All vehicles:

2. Remove the battery and the battery tray.
3. Remove the air cleaner as an assembly.
    a. Disconnect the breather tube.
    b. Disconnect the Mass Air Flow (MAF) sensor electrical connector and the wiring harness fastener.
    c. Remove the air intake tube.
    d. Remove the air cleaner assembly retaining bolt.
    e. Remove the air cleaner assembly
4. Disconnect the Transmission Range (TR) sensor.
5. Disconnect the transaxle harness connector, remove the wire harness bracket nut and position the harness bracket aside
6. Remove the main control cover vent tube.

AWD vehicles:

7. Disconnect the Power Transfer Unit (PTU) vent hose from the transmission fluid filler tube.

All vehicles:

8. Disconnect the Power Transfer Unit (PTU) vent hose from the transmission fluid filler tube.

All vehicles:

9. Disconnect the wire harness from the battery tray hold-down bracket.
10. Disconnect the shift cable from the manual lever
11. Disconnect the wire harness retainer from the shift cable bracket, remove the 2 retaining bolts, and position the cable and bracket aside.
12. Remove the selector lever cable retainer from the transmission fluid filler tube
13. Disconnect the starter motor harness connector.
14. Disconnect the ground wire
15. Remove the 2 starter bolts and remove the starter motor.
16. Remove and disconnect both electrical connectors from the upper intake to gain access to the engine for installing the lifting bracket

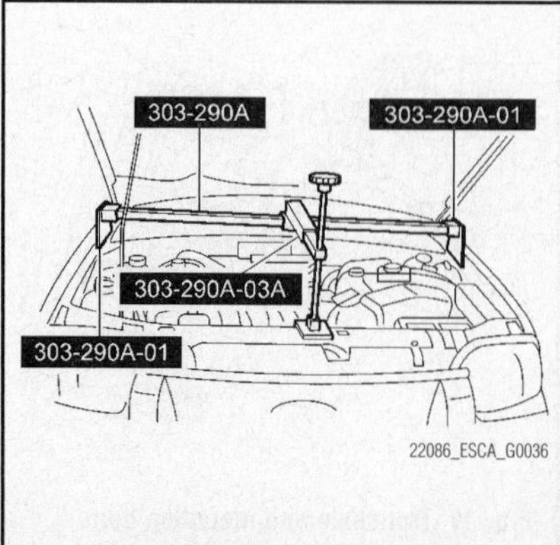

Fig. 19 Engine support system shown

17. Install suitable engine support system and secure engine with proper adapters.
18. Remove the 4 upper transaxle retaining bolts
19. Loosen, but do not remove, the 4 retaining nuts holding the bracket to the transaxle case. Remove the LH upper transaxle mount bolt.
20. Remove the RH upper engine mount bolt.
21. Remove the front wheels and tires.
22. Remove the 7 retainers and the LH splash shield.
23. Remove the 6 retainers and the RH splash shield.
24. If transaxle disassembly is necessary, remove the drain plug and drain the transmission fluid. After the fluid has drained, install the drain plug.
25. Disconnect the LH and RH suspension.
    a. Disconnect the sway bar link.
    b. Remove the tie-rod end retaining nut
    c. Remove the lower control arm knuckle bolt.
26. Using a suitable tool, disconnect the LH and RH tie-rod end from the steering knuckle.
27. Carefully pry down on the LH and RH lower control arms and disconnect the steering knuckle from the lower ball joint and position the steering knuckle aside.
28. Remove the brake hose retainer and the ABS sensor retaining bolt from the RH strut.
29. Using a suitable tool, remove the RH halfshaft from the intermediate shaft and secure the halfshaft aside.
30. Remove the brake hose retainer and the ABS sensor retaining bolt from the LH strut.
31. Using a pry bar between the transaxle case and the LH halfshaft, carefully disconnect the halfshaft from the transaxle case and secure the halfshaft aside.
32. Disconnect the heated oxygen sensor (HO2S) connector and remove the 2 clips from the oil pan bolt studs.
33. Remove the 2 intermediate shaft retaining nuts.
34. Remove the intermediate shaft.
35. Remove the cross brace.
36. Disconnect and remove the exhaust Y-pipe and hanger.

AWD vehicles:

37. Index the driveshaft to the yoke and remove the 6 bolts holding the driveshaft to the PTU
38. Remove the 2 center bearing nuts and position the driveshaft aside with mechanic's wire.

39. Remove the RH catalytic converter to gain access to the PTU bracket.

40. Remove the bolts and the PTU support bracket.

All vehicles:

41. Remove the bolt for the mount and the 2 bolts for the cross brace.

42. Remove and discard the nut and remove the cross brace.

43. Remove the electrical connectors and harness fastener from the lower mount bracket.

AWD vehicles:

44. Remove the bolts from the PTU and remove the PTU.

All vehicles:

45. Remove the through bolt from the rear transaxle mount.

46. Disconnect the electrical connector and remove the bolt and output shaft speed (OSS) sensor.

➡It is necessary to raise the engine a couple of inches in order to remove the transaxle.

47. Using the engine support system, raise the front of the engine.

➡It is necessary to lower the transaxle in order to clear the subframe to remove the transaxle.

48. Lower the transaxle enough to clear the frame.

49. Remove the access cover.

50. Remove and discard the 4 torque converter nuts.

51. Remove the fluid cooler tube and position it aside.

52. Disconnect the turbine shaft speed (TSS) sensor and the harness retainers.

53. Support the transaxle with a high-lift jack.

54. Remove the remaining transaxle mounting bolts.

⁂ WARNING

The torque converter is heavy. Secure the torque converter before lowering the transaxle.

55. Remove transaxle from the engine compartment of vehicle.

To install:

⁂ WARNING

Carry out the transmission fluid cooler back flushing and cleaning if the transaxle is being overhauled or installing a new or remanufactured transaxle. Carry out the transmission fluid cooler flow test if the transaxle is being overhauled or installing a new or remanufactured transaxle.

All vehicles:

56. Lubricate the torque converter pilot hub with grease.

57. Position the transaxle in place.

58. Move the transaxle assembly toward the engine assembly and install the bolt.

59. Install the 2 nuts and the stud tighten all the bolts to 35 ft. lbs. (48 Nm).

60. Install lower transaxle bolts and tighten to 35 ft. lbs. (48 Nm).

61. Install the mount bracket and tighten to 41 ft. lbs. (55 Nm).

62. Loosely install the rear transaxle mount bolt.

63. Install the LH transaxle mount through bolt

64. Tighten the rear transaxle mount bolt to 89 ft. lbs. (120 Nm).

65. Install the transmission fluid cooler tubes and tighten to 18 ft. lbs. (25 Nm).

66. Install the output shaft speed (OSS) sensor and tighten the bolt to 9 ft. lbs. (12 Nm).

67. Connect the OSS speed sensor and the turbine shaft speed (TSS) sensor electrical connectors and connect the wiring harness fasteners.

68. Install 4 new torque converter nuts and tighten to 30 ft. lbs. (40 Nm).

69. Install the access cover.

AWD vehicles:

70. Position the Power Transfer Unit (PTU) in place and install the bolt and tighten to 35 ft. lbs. (37 Nm).

71. Install the PTU-to-transaxle bolts and tighten to 52 ft. lbs. (70 Nm).

All vehicles:

72. Connect the electrical connector fasteners to the lower mount bracket.

73. Install the cross brace and the rear nut. Tighten the rear nut to 129 ft. lbs. (175 Nm). and the front bolts to 66 ft. lbs. (90 Nm).

74. Install the bolt for the mount and tighten to 85 ft. lbs. (115 Nm).

AWD vehicles:

75. Install the PTU support bracket and tighten mounting bolts to 41 ft. lbs. (55 Nm).

76. Install the RH catalytic converter.

77. Position the driveshaft in place and install the nuts. Tighten to 35 ft. lbs. (48 Nm).

78. Connect the rear driveshaft to the PTU. Install the PTU bolts and tighten to 27 ft. lbs. (37 Nm).

All vehicles:

79. Install the exhaust Y-pipe with new gaskets and tighten mounting nuts to 21 ft. lbs. (29 Nm).

80. Install the exhaust rubber hanger.

81. Install the cross brace and tighten to 85 ft. lbs. (115 Nm).

82. Install the intermediate shaft and retaining nuts. Tighten nuts to 20ft. lbs. (27 Nm).

83. Connect the heated oxygen sensor (HO2S) wire to the oil pan bolt studs and connect the connector.

84. Install the LH halfshaft into the transaxle and the RH halfshafts in the intermediate shaft and install the ball joints in the knuckles

85. Install the LH and RH brake hose retainer and the ABS sensor bolt. Tighten bolt to 11 ft. lbs. (15 Nm).

86. Reconnect the LH and RH suspension.

a. Reconnect the sway bar link and tighten nuts to 46 ft. lbs. (63 Nm).

b. Install the tie-rod end retaining nut and tighten to 41 ft. lbs. (55 Nm).

c. Install the lower control arm knuckle bolt and tighten to 46 ft. lbs. (63 Nm).

87. Install the LH splash shield, the retainer and the 5 bolts.

88. Install the RH splash shield, the retainer and the 5 bolts.

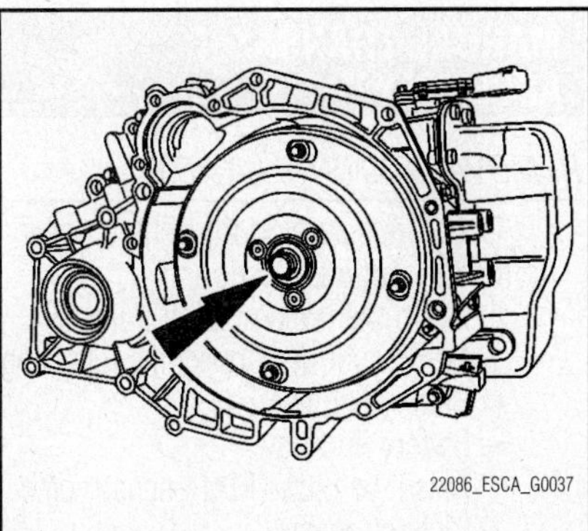

Fig. 20 Transaxle and torque converter pilot hub shown

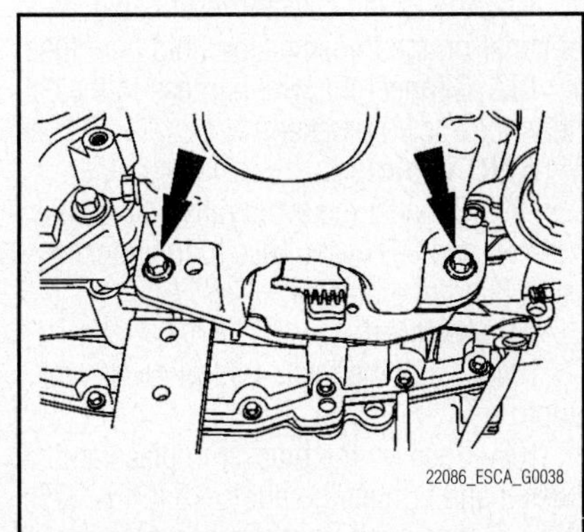

Fig. 21 Lower transaxle bolts shown

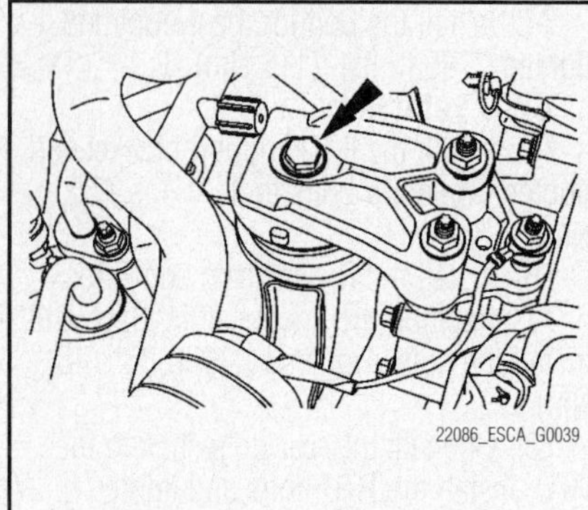

Fig. 22 RH engine mounting bolt view—3.0L engine

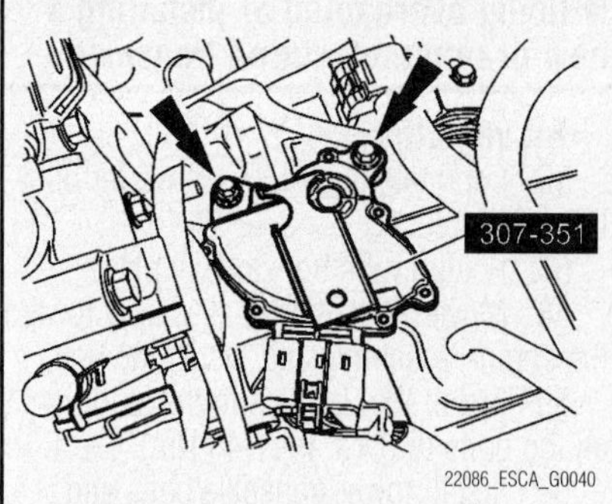

Fig. 23 Transmission range sensor tool 307-351 shown for alignment

89. Install the front wheels and tires.

90. Using the engine support system, lower the engine onto the RH engine mount.

91. Install the bolt for the RH engine mount and tighten to 89 ft. lbs. (120 Nm).

92. Tighten the LH upper transaxle mount assembly.

a. Tighten the 4 nuts for the bracket to 30 ft. lbs. (40 Nm).

b. Tighten the through bolt to 76 ft. lbs. (104 Nm).

93. Install the upper transaxle retaining bolts and tighten to 35 ft. lbs. (48 Nm).

94. Remove the engine support system tool.

95. Connect the electrical connectors together and then connect them to the upper intake manifold.

96. Install the starter motor and bolts. Tighten the bolts to 18 ft. lbs. (25 Nm).

97. Install and tighten the ground cable to 20 ft. lbs. (27 Nm).

98. Reconnect the starter cables. Tighten battery cable to 20 ft. lbs. (27 Nm), and the solenoid nut to 44 inch lbs. (5 Nm).

99. Position the shift cable and bracket in place, install the bolts and install the wiring harness retainer. Tighten bolts to 17 ft. lbs. (23 Nm).

100. Connect the shift cable to the manual lever.

101. Install the selector lever cable retainer on the transmission fluid filler tube.

102. Connect the wire harness to the battery hold-down bracket.

AWD vehicles:

103. Connect the evaporative emissions (EVAP) and PTU vent hose to the transmission filler tube bracket.

All vehicles:

104. Install the main control cover vent tube.

105. Position the bracket in place and install and tighten the nut to 18 ft. lbs. (25 Nm). Connect the transaxle harness connector.

106. Connect the Transmission Range (TR) sensor.

➡ If installing an exchange transaxle, the digital TR sensor must be aligned.

107. Using the special tool, align the digital TR sensor.

108. Install the air cleaner as an assembly.

a. Install the air cleaner assembly.

b. Connect the air intake tube.

c. Install the retaining bolt and tighten to 89 inch lbs. (10 Nm).

d. Install the breather tube

e. Connect the MAF sensor electrical connector.

➡Before installing the battery tray, check the vent tube hose for any obstructions, kinks or incorrect routing position

109. Install the battery tray.

110. Fill the transaxle with clean automatic transmission fluid.

111. Verify that the shift cable is correctly adjusted. The vehicle should start in **PARK** and **NEUTRAL** and the reverse lamps should illuminate in **REVERSE**.

112. Check the fluid level and correct as necessary.

MANUAL TRANSAXLE ASSEMBLY

REMOVAL & INSTALLATION

1. Before servicing the vehicle, refer to the Precautions Section.
2. Drain the transmission fluid.
3. Remove or disconnect the following:
   - Battery cables
   - Battery and tray
   - Mass Air Flow (MAF) sensor electrical connector
   - Accelerator cable from the air cleaner outlet tube
   - Emission management tube and hose
   - Crankcase ventilation hose
   - Air cleaner outlet tube
   - Air cleaner housing
   - Back-up lamp switch electrical connector
   - Front wire harness bracket and move it aside
   - Front wire harness bracket spacer
   - Wire harness from the rear harness bracket
   - Park Neutral Position (PNP) electrical connector
   - Rear wire harness bracket and move it aside
   - Vehicle Speed Sensor (VSS) electrical connector
   - Clutch slave cylinder line from the bracket and move it aside while properly supporting the engine
   - Left side transmission support insulator and bracket
   - Rear transmission support insulator
   - Front transmission support insulator and bracket
   - Starter and move it aside
   - Top transmission flywheel housing bolts
   - Front transmission flywheel housing bolts
   - Transfer case, if equipped
   - Left side halfshaft
   - Rear transmission support insulator bracket
   - Shifter linkage and stabilizer bar
   - Transverse crossmember
   - Front to aft crossmember
   - Left side splash shield and properly support the transmission
   - Remaining transmission flywheel housing bolts
   - Transmission and separate the right side halfshaft from the transmission

To install:

4. Align the right side half shaft to the transmission and position the transmission to the engine.
5. Install or connect the following:
   - Transmission flywheel housing bolts. Torque the bolts to 33 ft. lbs. (45 Nm) and remove the transmission support
   - Left side splash shield
   - Front-to-aft crossmember. Torque the bolts to 66 ft. lbs. (90 Nm).
   - Transverse crossmember. Torque the bolts to 85 ft. lbs. (115 Nm).
   - Shifter linkage. Torque the bolt to 15 ft. lbs. (20 Nm).

- Stabilizer bar. Torque the bolt to 30 ft. lbs. (40 Nm).
- Rear transmission support bracket. Torque the bolts to 66 ft. lbs. (90 Nm).
- Left side halfshaft
- Transfer case, if equipped
- Front transmission flywheel housing bolts. Torque the bolts to 33 ft. lbs. (45 Nm).
- Top transmission flywheel housing bolts. Torque the bolts to 33 ft. lbs. (45 Nm).
- Starter. Torque the bolts to 33 ft. lbs. (45 Nm).
- Front transmission support insulator and bracket. Torque the lower bolt to 66 ft. lbs. (90 Nm) and the 3 upper bolts to 41 ft. lbs. (55 Nm).
- Rear transmission support insulator bolt. Torque the bolt to 66 ft. lbs. (90 Nm).
- Left side transmission support insulator bracket. Torque the bolts to 66 ft. lbs. (90 Nm).
- Left side transmission support insulator. Torque the large bolt to 66 ft. lbs. (90 Nm) and the 3 remaining bolts to 41 ft. lbs. (55 Nm).
- Clutch slave cylinder. Torque the bolt to 15 ft. lbs. (20 Nm).
- Clutch slave cylinder line to the bracket and install the retaining clip
- VSS electrical connector
- Rear wire harness bracket. Torque the bolts to 80 inch lbs. (9 Nm).
- PNP switch electrical connector
- Wire harness to the rear bracket
- Front wire harness bracket spacer and bracket. Torque the bolt to 9 ft. lbs. (12 Nm).
- Back-up lamp switch electrical connector
- Air cleaner housing
- MAF sensor electrical connector
- Air cleaner outlet tube
- Crankcase ventilation hose
- Emission management tube and hose
- Accelerator cable to the air cleaner outlet tube
- Battery and tray
- Both battery cables

6. Fill the transmission to the proper level.

7. Start the vehicle and check for leaks, repair if necessary.

TRANSFER CASE ASSEMBLY

REMOVAL & INSTALLATION

See Figure 24.

1. Before servicing the vehicle, refer to the Precautions Section.
2. Disconnect the battery.
3. Drain the transfer case.
4. Remove the front right intermediate shaft.
5. Remove the driveshaft.
6. Remove the 4 bolts and the crossmember brace.
7. Remove the alternator.
8. Remove the exhaust as required.
9. Remove the heat shield.
10. Remove the transfer case-to-transaxle bolts.
11. Remove the transfer case.

✲✲ WARNING

A new transfer case driven gear seal must be installed whenever the intermediate shaft or transfer case is removed from the vehicle.

➡If necessary, replace the right differential fluid seal.

12. To install, reverse the removal procedure. Observe the following torques:
    - Crossmember bolts: 30 ft. lbs. (40 Nm).
    - Transfer case-to-transaxle bolts: 33 ft. lbs. (45 Nm).

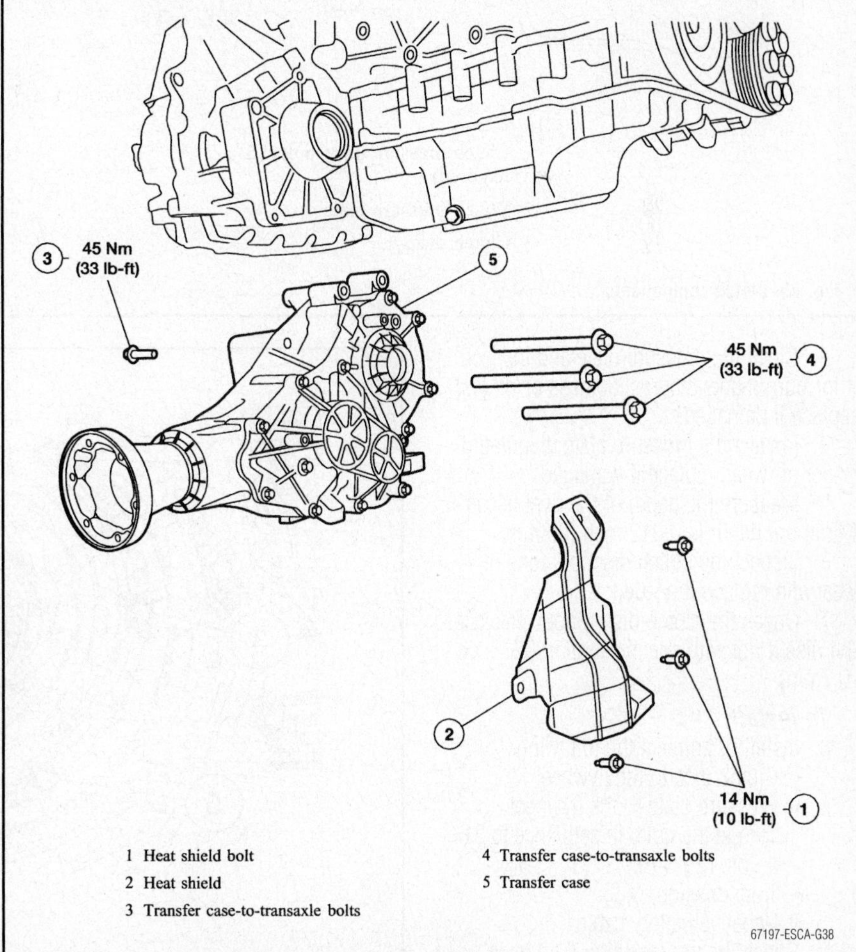

Fig. 24 Transfer case mounting with automatic transaxle

CLUTCH DRIVEN DISC & PRESSURE PLATE

REMOVAL & INSTALLATION

See Figures 25 and 26.

1. Before servicing the vehicle, refer to the Precautions Section.
2. Remove or disconnect the following:
    - Negative battery cable
    - Transmission and lock the flywheel to the engine with special tool 303–103
    - Pressure plate bolts by loosening them evenly
    - Clutch pressure plate and disc

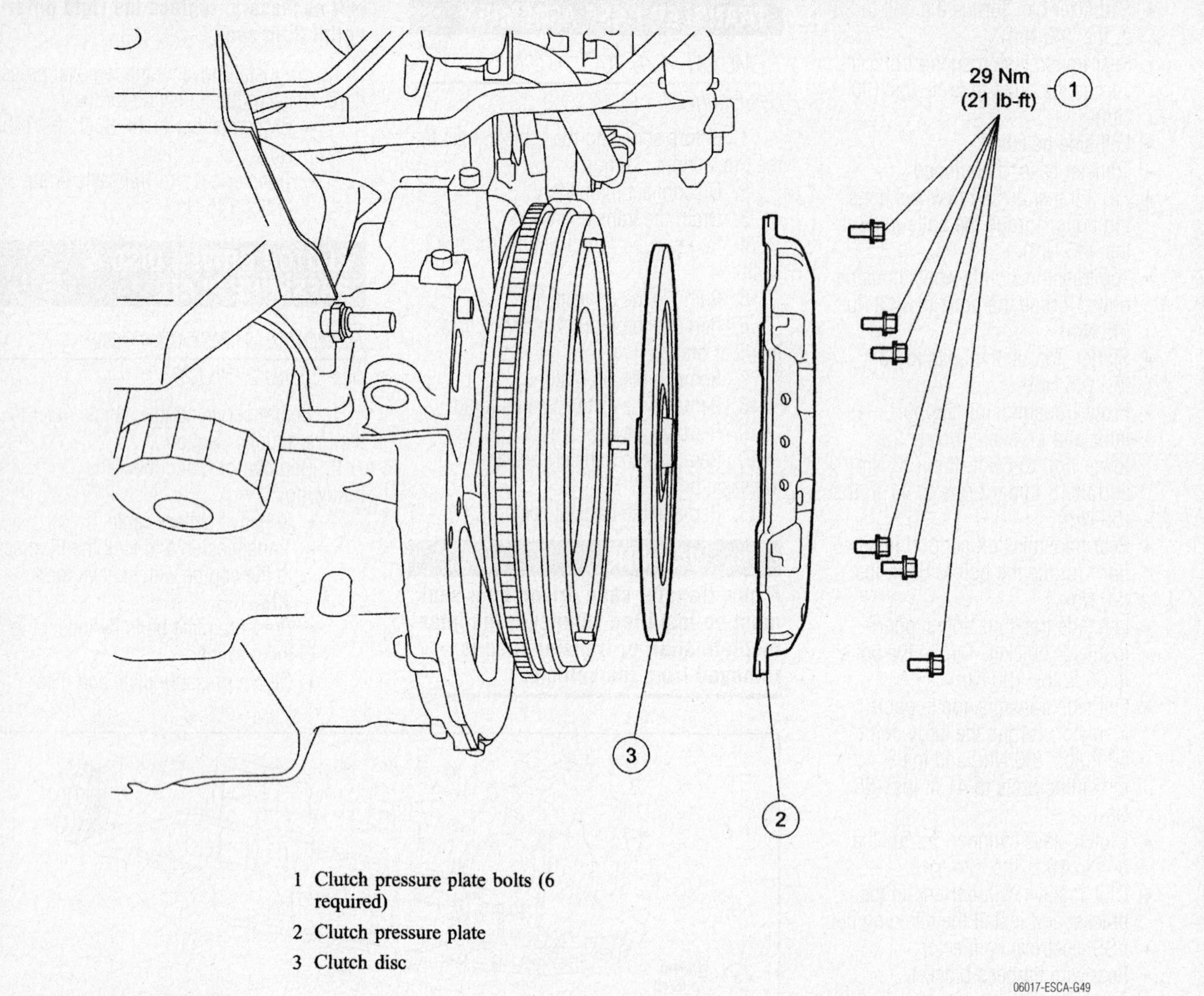

29 Nm
(21 lb-ft)
1
2
3
1 Clutch pressure plate bolts (6 required)
2 Clutch pressure plate
3 Clutch disc
06017-ESCA-G49

Fig. 25 Clutch components

3. Clean the pressure plate and inspect it for burn marks, scores, flatness or ridges, replace if damaged.
4. Inspect the pressure plate diaphragm finger for wear, replace if damaged.
5. Measure the depth of the rivet heads. Minimum depth is 0.012 inch (0.3mm).
6. Inspect the clutch disc for signs of wear and replace if needed.
7. Check the clutch disc runout. Replace the disc if not with specification: 0.027 inch (0.7mm).

To install:

8. Install or connect the following:
   - Clutch disc to the flywheel
   - Pressure plate to the flywheel. Torque the bolts in sequence to 21 ft. lbs. (29 Nm).
   - Transmission
   - Negative battery cable
9. Check the transmission fluid level and top off if necessary.

1
6
4
3
2
5
06017-ESCA-G50

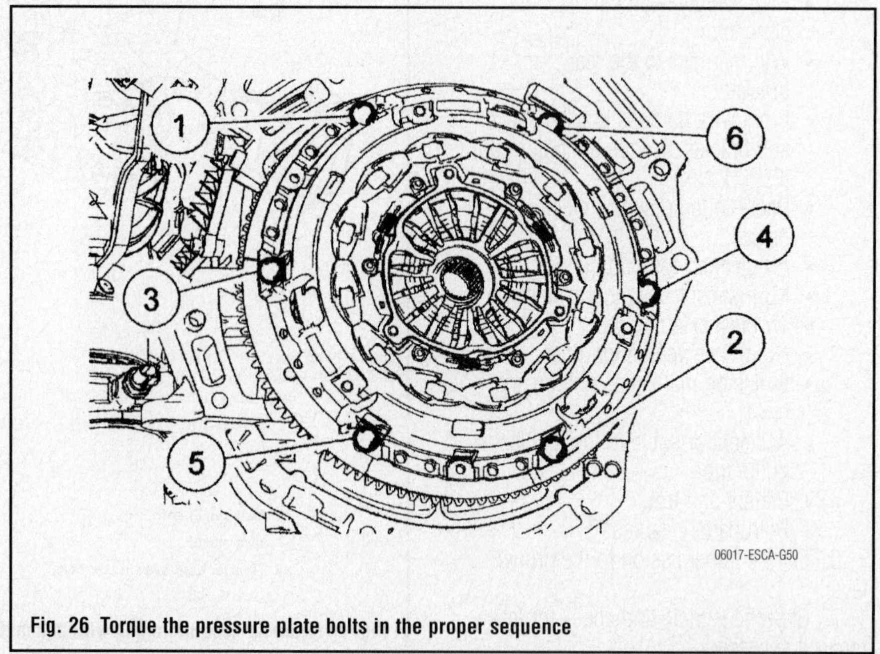

Fig. 26 Torque the pressure plate bolts in the proper sequence

ADJUSTMENTS

The clutch is hydraulically driven and therefore no adjustment is required.

CLUTCH MASTER CYLINDER

REMOVAL & INSTALLATION

See Figure 27.

1. Before servicing the vehicle, refer to the Precautions Section.

✻✻ WARNING

Brake fluid is harmful to painted and plastic surfaces. If brake fluid is spilled onto a painted or plastic surface, wash it immediately with water.

➡**If removing the clutch line, remove the engine air cleaner.**

➡**If removing the clutch slave cylinder or the clutch slave cylinder-to-clutch line adapter, remove the transaxle.**

2. Remove the clutch master cylinder hose
3. Remove the clutch master cylinder push rod
4. Remove the fitting at the master cylinder.
5. Remove the clutch master cylinder

✻✻ WARNING

Make sure the O-rings are properly positioned on the hydraulic line fittings or leaks may occur.

6. To install, reverse the removal procedure. Torque the mounting nuts to 17 ft. lbs. (23 Nm).
7. Fill and bleed the system.

BENCH BLEEDING PROCEDURE

See Figures 28 through 30.

✻✻ CAUTION

Carefully read cautionary information on product label. For additional information, see product Material Safety Data Sheet (MSDS). Failure to follow these instructions may result in personal injury.

✻✻ CAUTION

Do not spill brake fluid onto painted or plastic surfaces. If spilled, wipe up immediately. Brake fluid is harmful to painted or plastic surfaces.

1. With the vehicle in **NEUTRAL**, position it on a hoist.

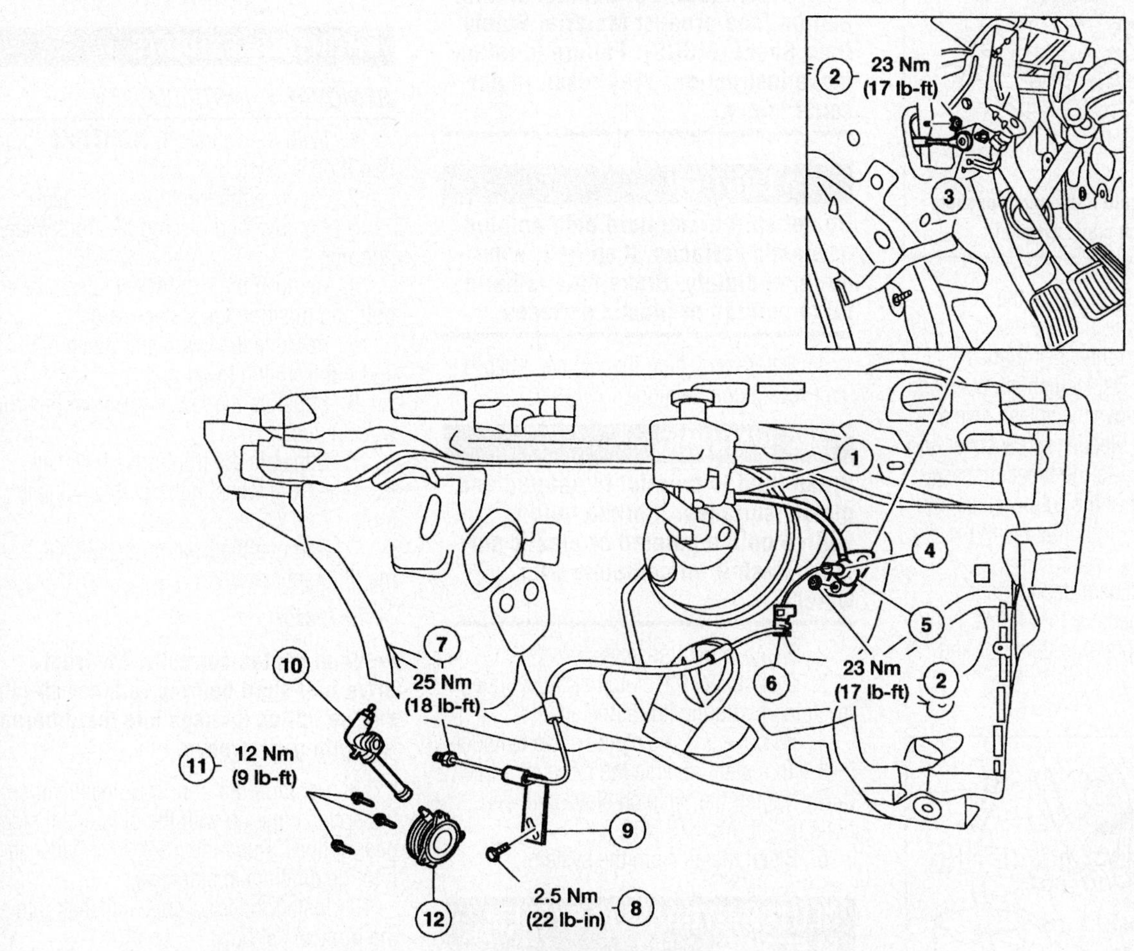

Fig. 27 Hydraulic clutch components

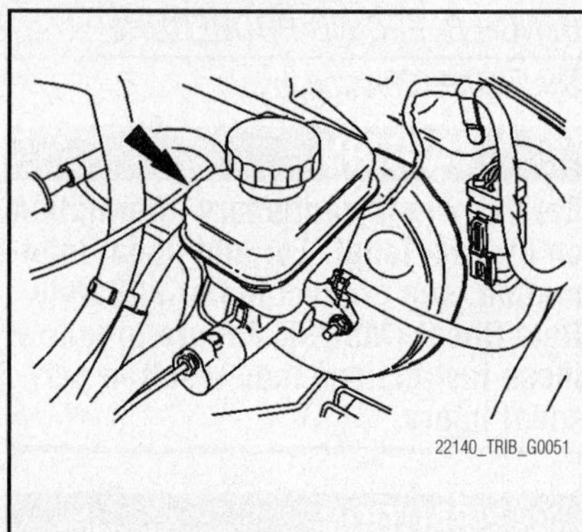

Fig. 28 Hydraulic clutch master cylinder

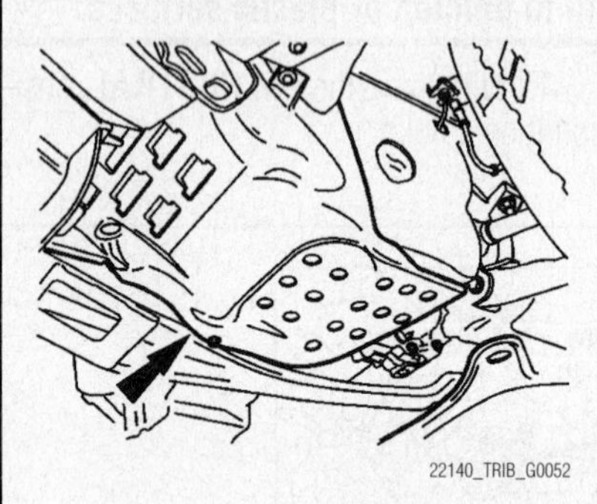

Fig. 29 Splash shield bolts and 1 pushpin to expose the clutch slave cylinder

2. Check the fluid level of the brake/clutch reservoir. Fill the reservoir with the specified fluid to the MAX mark

3. Remove the 7 splash shield bolts and 1 pushpin, then remove the splash shield.

4. Remove the bleeder screw cover and attach a rubber hose to the bleeder screw. Place the other end of the rubber hose into a clear container partially filled with the specified brake fluid.

5. Have an assistant depress and release the clutch pedal 5 to 7 times. Fully depress the clutch pedal to the floor and hold down.

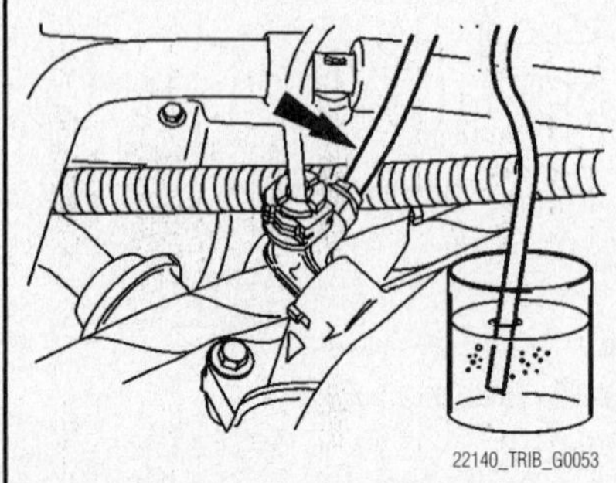

Fig. 30 Bleeder screw cover and attachment of a rubber hose to the bleeder screw.

6. With the clutch pedal depressed, loosen the bleeder screw until fluid and air escape the system. With the clutch being held to the floor, tighten the bleeder screw. Repeat Steps 5 and 6 until no air comes from the rubber hose.

7. Tighten the bleeder screw.

8. Check the fluid level of the reservoir. Fill the reservoir with the specified fluid to the MAX mark. Install the reservoir cap.

9. Depress and release the clutch pedal several times.

10. Test the clutch system for normal operation.

CLUTCH SLAVE CYLINDER

REMOVAL & INSTALLATION

**** CAUTION**

Carefully read cautionary information on product label. For additional information, see product Material Safety Data Sheet (MSDS). Failure to follow these instructions may result in personal injury.

**** CAUTION**

Do not spill brake fluid onto painted or plastic surfaces. If spilled, wipe up immediately. Brake fluid is harmful to painted or plastic surfaces.

1. Before servicing the vehicle, refer to the Precautions Section.

**** WARNING**

Brake fluid is harmful to painted and plastic surfaces. If brake fluid is spilled onto a painted or plastic surface, wash it immediately with water.

2. Remove the transaxle.

3. Disconnect the clutch slave cylinder-to-clutch hydraulic fluid tube adapter.

4. Remove and the clutch slave cylinder.

5. To install, reverse the removal procedure. Tighten the 3 clutch slave cylinder bolts to 9 ft. lbs. (12 Nm)

6. Bleed the air from the system.

CLUTCH HYDRAULIC SYSTEM BLEEDING

The following procedure is recommended for bleeding the clutch hydraulic system installed on the vehicle. It is recommended that the original clutch tube, with quick-connect fitting be replaced when servicing the hydraulic system, because air can be trapped in the quick-connect fitting and prevent complete bleeding of the system.

1. Before servicing the vehicle, refer to the Precautions Section.

2. Clean the dirt and grease from the dust cap.

3. Remove the cap and diaphragm and fill the reservoir ¾ of the way with approved brake fluid C6AZ-19542-AB or DOT 3 equivalent fluid (ESA-M6C25-A).

4. Loosen the bleeder screw cover from the slave cylinder and attach a hose to the screw.

5. Place the hose in a container and slowly pump the clutch pedal several times.

6. With the clutch pedal depressed, loosen the bleeder screw to release the fluid and air.

7. Remove the hose and tighten the bleeder screw.

8. Repeat this procedure until all the air is removed from the hydraulic system

FRONT HALFSHAFTS

REMOVAL & INSTALLATION

1. With the vehicle in **NEUTRAL**, position it on a hoist.

2. Remove the front tire and wheel.

3. Remove and discard the front wheel hub nut.

4. Remove the ABS wheel speed sensor bolt and position the sensor aside.

5. Remove the lower arm pinch bolt and nut from the lower arm.

6. Separate the lower arm from the front wheel knuckle.

7. Separate the front drive halfshaft from the front wheel hub using a suitable tool.

8. Remove the front drive halfshaft from the differential.

To install:

➡When seated correctly, the front drive half shaft bearing retainer cir-clip can be felt as it snaps into the differential side gear groove.

9. Position the front drive halfshaft so the splines line up with the differential side gear splines. Push the front drive halfshaft into the differential side gear.

10. Install the front drive halfshaft into the front wheel hub.

11. Position the lower arm into the front wheel knuckle.

12. Install the lower arm pinch bolt and nut. Tighten to 52 ft. lbs. (70 Nm).

13. Install the ABS wheel speed sensor and bolt. Tighten bolt to 80 inch lbs. (9 Nm).

WARNING

Do not tighten the front wheel hub nut with the vehicle on the ground. The nut must be tightened to specification before the vehicle is lowered onto the wheels. Wheel bearing damage will occur if the wheel bearing is loaded with the weight of the vehicle applied.

➡Apply the brake to keep the halfshaft from rotating.

14. Install a new front wheel hub nut and tighten to 222 ft. lbs. (300 Nm).
15. Install the front tire and wheel.
16. Check and fill the transaxle fluid as necessary.

INTERMEDIATE SHAFT

REMOVAL & INSTALLATION

See Figure 31.

1. Before servicing the vehicle, refer to the Precautions Section.

➡If removing the intermediate shaft in order to repair a separate component, it should only be removed as an assembly with the right front halfshaft.

2. Remove the right front halfshaft.
3. Remove the inner halfshaft bearing retainer nuts
4. Remove the intermediate shaft
5. To install, reverse the removal procedure. Apply a thin coat of grease to the splines of the intermediate shaft.
6. Verify the front axle lubricant level is to specifications.

REAR DRIVESHAFT

REMOVAL & INSTALLATION

See Figure 32.

1. Before servicing the vehicle, refer to the Precautions Section.

CAUTION

The normal operating temperature of the exhaust system is very high. Never attempt to remove any part of the system until it has cooled. Be especially careful when working around the catalytic converters. The temperature of the converter rises to a high level after only a few minutes of engine operation. Failure to follow these instructions may result in personal injury.

2. With the vehicle in **NEUTRAL**, position it on a hoist.
3. Remove the ground strap bolt.

WARNING

Do not reuse the CV-joint bolts and washers. Install new bolts and washers or damage to the vehicle may occur.

4. Remove and discard the 6 front driveshaft-to-transfer case bolts and washers. Index-mark the front driveshaft to the center bearing.

WARNING

Do not reuse the bolts and cap straps for the center U-joint. Install new bolts and cap straps or damage to the vehicle may occur.

➡There is a difference in the length of the head of the replacement cap strap bolts from the production bolts. The longer head pinion bolts can be used in either location.

5. Remove and discard the 4 universal joint cap strap bolts and 2 cap straps and remove the front driveshaft.

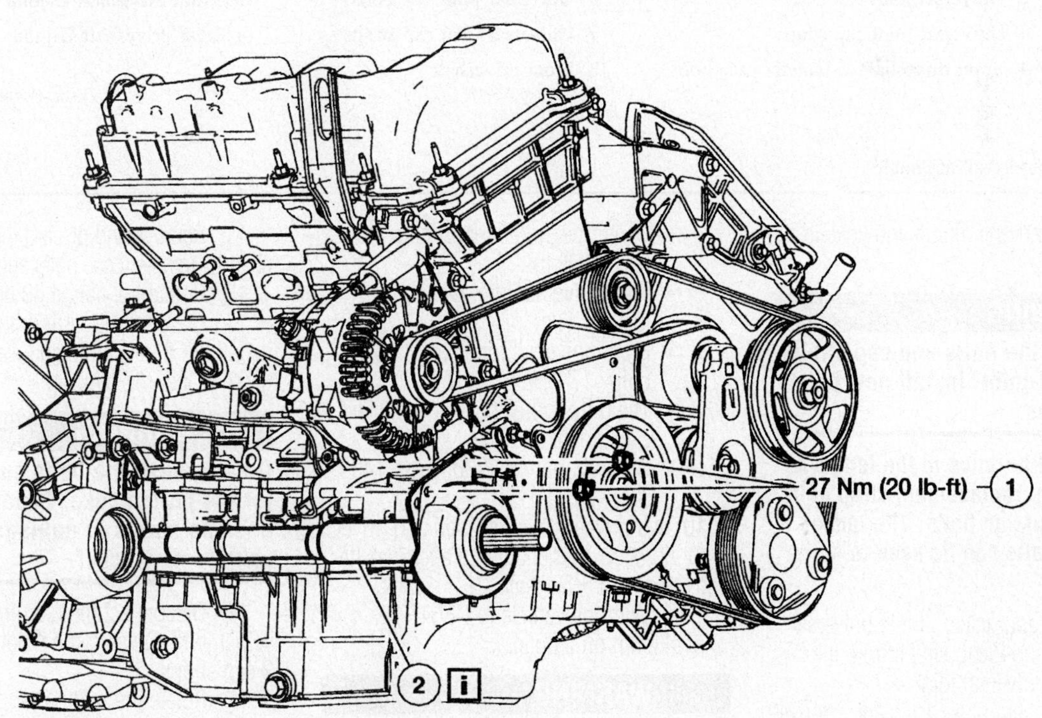

Fig. 31 Intermediate shaft

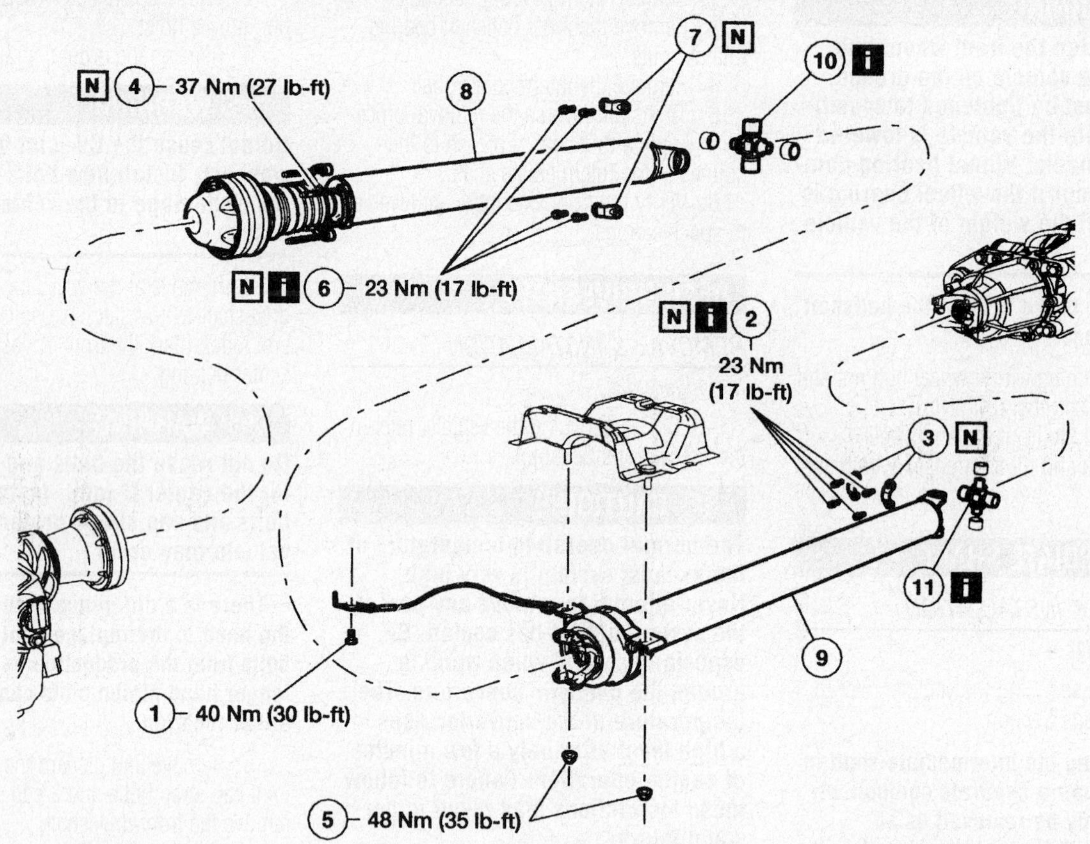

1 Ground strap bolt
2 Universal joint cap bolts
3 Universal joint cap straps
4 Front driveshaft-to-transfer case bolts
5 Center bearing support nuts
6 Universal joint cap bolts
7 Universal joint cap straps
8 Front driveshaft
9 Rear driveshaft
10 Front driveshaft U-joint
11 Rear driveshaft U-joint

67197-ESCA-G42

Fig. 32 Rear driveshaft assembly

6. Index-mark the pinion and yoke to the driveshaft.

WARNING

Do not reuse the bolts and cap straps for the rear U-joint. Install new bolts and cap straps.

➡There is a difference in the length of the head of the replacement strap bolts from the production bolts. The longer head pinion bolts can be used in either location.

7. Remove and discard the 4 universal joint cap bolts and 2 cap straps from the rear driveshaft universal joint.
8. With the help of an assistant, remove the center bearing support nuts and the driveshaft.

To install:

9. To install, reverse the removal procedure. Observe the following torques:

- Center bearing support nuts: 35 ft. lbs. (48 Nm)
- Rear universal joint cap bolts: 17 ft. lbs. (23 Nm)
- Front universal joint cap strap bolts: 17 ft. lbs. (23 Nm)
- The 6 front driveshaft-to-transfer case bolts: 27 ft. lbs. (37 Nm)
- Ground strap bolt: 30 ft. lbs. (40 Nm)

10. If a driveshaft is installed and driveshaft vibration is encountered after installation, index the driveshaft.
   a. With the vehicle in NEUTRAL, position it on a hoist.

WARNING

Do not reuse the CV-joint bolts and washers. Install new bolts and washers or damage to the vehicle may occur.

11. Remove and discard the 6 front driveshaft-to-transfer case bolts and washers.
12. Rotate the flange 60 degrees.
13. Connect the front driveshaft and install the 6 new bolts and washers. Tighten to 27 ft. lbs. (37 Nm).

WARNING

Do not reuse the bolts and cap straps for the pinion yoke. Install new bolts and cap straps or damage to the vehicle may occur.

14. Disconnect the rear driveshaft universal joint. Discard the 4 bolts and the 2 cap straps.
15. Rotate the rear pinion 180 degrees.
16. Connect the rear driveshaft and install 4 new bolts and 2 new cap straps. Tighten to 17 ft. lbs. (23 Nm).
17. Lower the vehicle and test drive.
18. Repeat the procedure if necessary.

ENGINE COOLING

COOLANT TEMPERATURE SENSOR

REMOVAL & INSTALLATION

2.3L Engine

See Figure 33.

➡This sensor is also referred to as the Cylinder Head Temperature (CHT) sensor

1. Disconnect the negative battery cable.
2. Detach the cylinder head temperature sensor cover and position aside.
3. Disconnect the cylinder head temperature sensor connector.
4. Remove cylinder head temperature sensor.
5. To install, reverse the removal procedure.
6. Tighten sensor to 9 ft. lbs. (12 Nm).

3.0L Engine

See Figure 34.

1. Partially drain the cooling system to a level below the sensor.
2. Disconnect the electrical connector.
3. Unclip and remove the Engine Coolant Temperature (ECT) sensor.

1
2
1. Engine coolant temperature (ECT) sensor electrical connector
2. ECT sensor
22086_ESCA_G0005

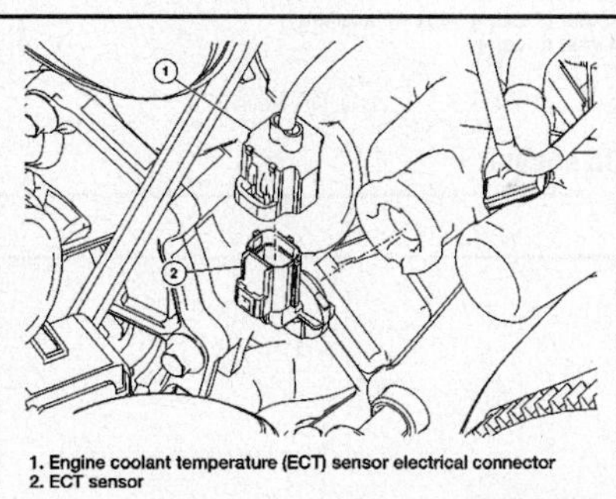

Fig. 34 Coolant temperature sensor and connector view—3.0L engine

4. Installation is the reverse of the removal procedure.

THERMOSTAT

REMOVAL & INSTALLATION

2.3L Engine

See Figure 35.

➡On the 2.3L engine, the thermostat and thermostat housing are serviced as an assembly.

1. Raise and safely support the vehicle.
2. Drain the cooling system.
3. Remove the accessory drive belt tensioner.
4. Disconnect the heater hose at the thermostat housing.
5. Disconnect the lower radiator hose at the thermostat housing.
6. Remove the 3 bolts, thermostat housing and gasket.
7. Clean and inspect the gasket, replace if necessary.

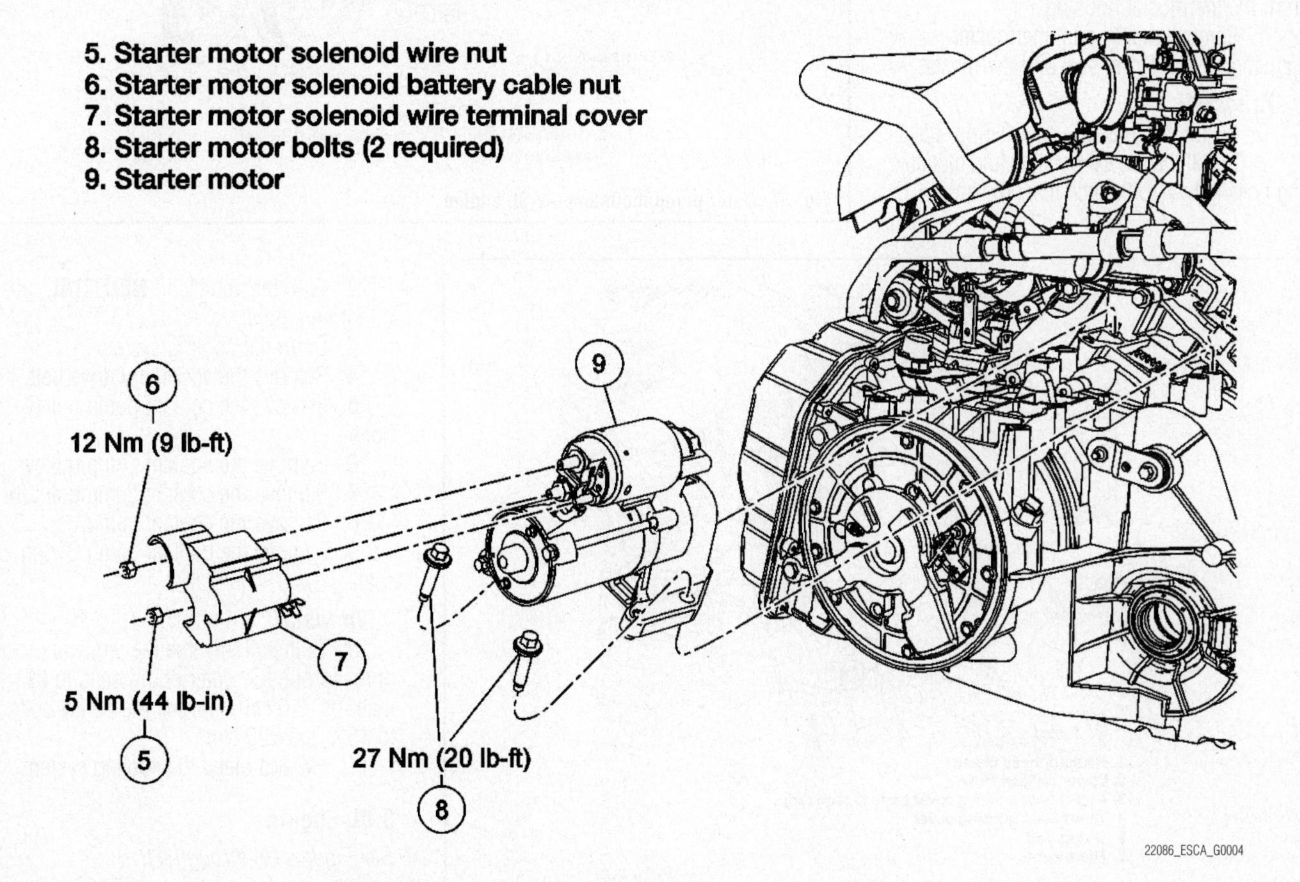

5. Starter motor solenoid wire nut
6. Starter motor solenoid battery cable nut
7. Starter motor solenoid wire terminal cover
8. Starter motor bolts (2 required)
9. Starter motor
6
12 Nm (9 lb-ft)
9
7
5 Nm (44 lb-in)
5
27 Nm (20 lb-ft)
8
22086_ESCA_G0004

Fig. 33 Cylinder head temperature sensor view—2.3L engine

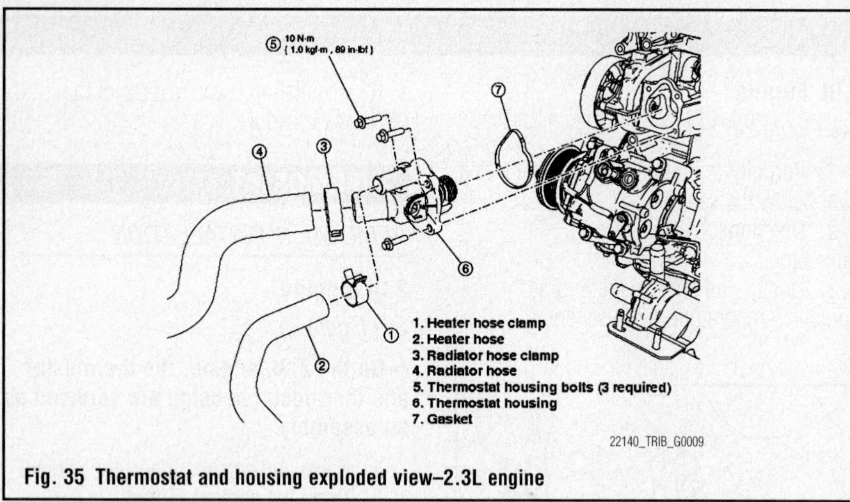

Fig. 35 Thermostat and housing exploded view–2.3L engine

8. To install, reverse the removal procedure. Tighten the thermostat housing bolts to 89 inch lbs. (10 Nm).

3.0L Engine

See Figure 36.

1. With the vehicle in **NEUTRAL**, position it on a hoist.
2. Drain the cooling system.
3. Remove the air cleaner outlet pipe.
4. Disconnect the lower radiator hose from the thermostat housing.
5. Remove the 3 bolts, thermostat housing cover, O-ring seal and thermostat.

To install:

6. Install thermostat a new O-ring seal.
7. Install thermostat housing and mounting bolts tighten to 89 inch lbs. (10 Nm).

➡To install, lubricate the thermostat housing O-ring seal with clean engine coolant

8. Reconnect lower radiator hose.
9. Install air cleaner outlet pipe.
10. Fill and bleed the cooling system.

WATER PUMP

REMOVAL & INSTALLATION

2.3L Engine

See Figure 37.

1. Before servicing the vehicle, refer to the Precautions Section.

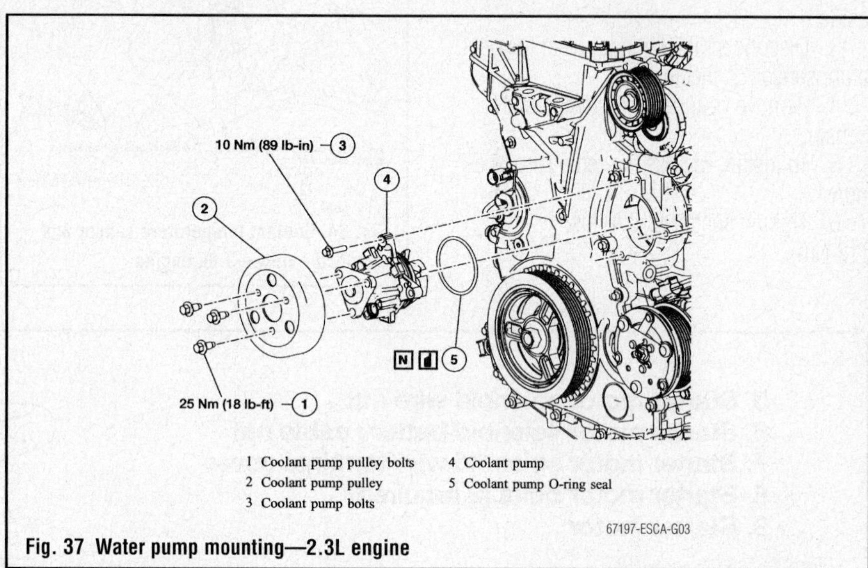

Fig. 37 Water pump mounting—2.3L engine

2. With the vehicle in **NEUTRAL**, position it on a hoist.
3. Drain the cooling system.
4. Remove the accessory drive belt.
5. Remove the coolant pump pulley bolts.
6. Remove the coolant pump pulley.
7. Remove the coolant pump bolts(3).
8. Remove the coolant pump.
9. Remove the coolant pump O-ring seal.

To install:

10. To install, reverse the removal procedure. Torque the water pump bolts to 89 inch lbs. (10 Nm). Torque the pulley bolts to 15 ft. lbs. (20 Nm).
11. Fill and bleed the cooling system.

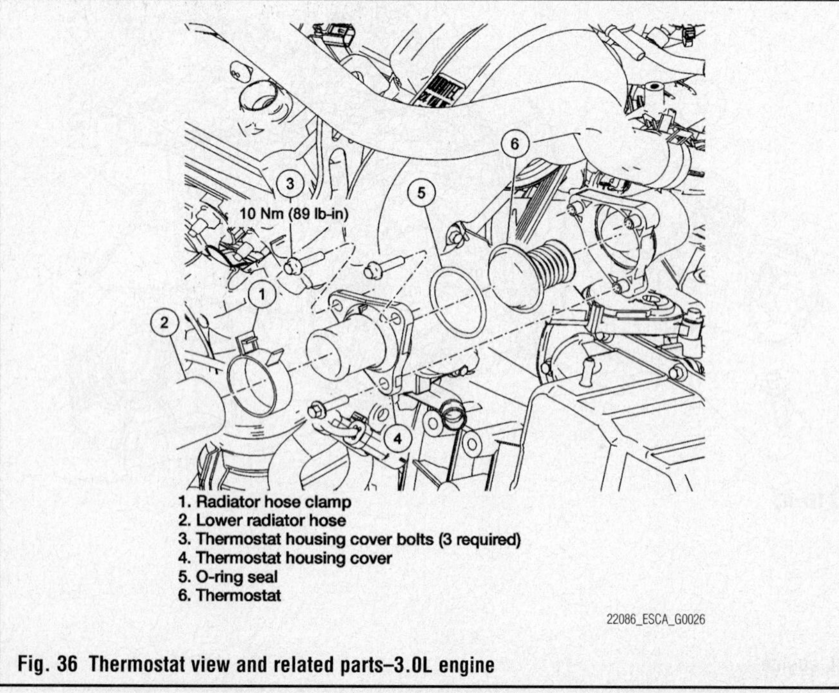

Fig. 36 Thermostat view and related parts–3.0L engine

3.0L Engine

See Figures 38 through 41.

1. With the vehicle in **NEUTRAL**, position it on a hoist.

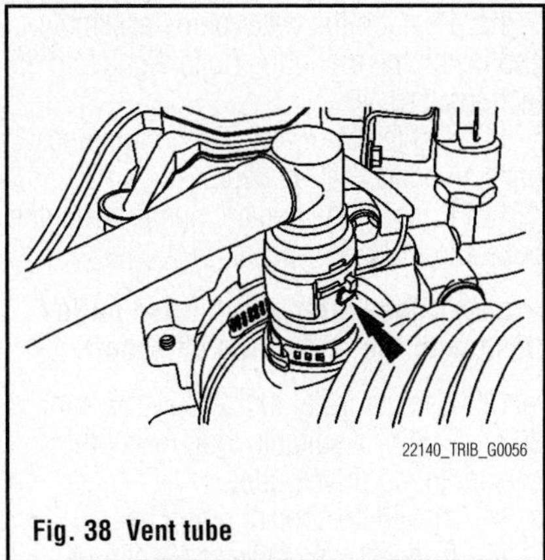

Fig. 38 Vent tube

2. Drain the cooling system.
3. Disconnect the crankcase vent tube and position it aside.
4. Remove the water pump belt.
5. Using Special tool 303—S455 and 303—009 or a suitable tool, remove the water pump drive pulley.
6. Disconnect the water pump-to-engine hose and position aside.
7. Remove the 3 mounting bolts from the water pump assembly.
8. Reposition the water pump-to-thermostat housing hose clamp and remove the coolant pump and hose as an assembly.

To install:

9. Connect the water pump-to-thermostat housing hose and reposition the clamp.

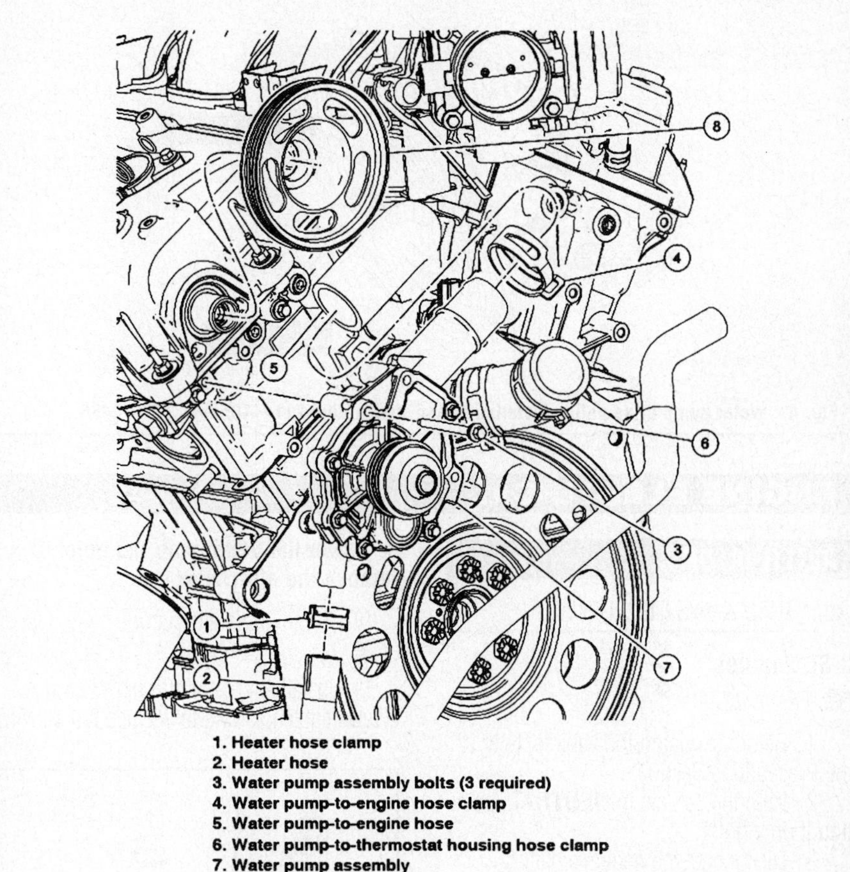

Fig. 40 Water pump—3.0L engine

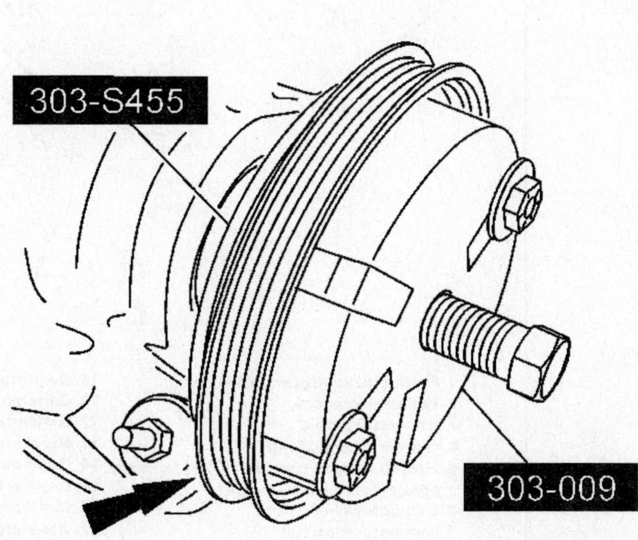

Fig. 39 Water pump drive pulley and puller 303—S455 and 303—009 shown

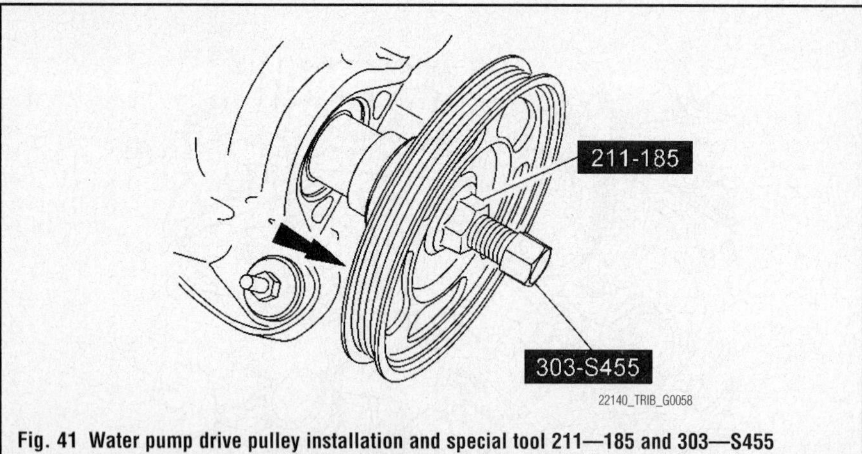

Fig. 41 Water pump drive pulley installation and special tool 211—185 and 303—S455

10. Position the water pump assembly and install the mounting bolts, tighten to 89 inch lbs. (10 Nm).
11. Tighten the water pump mounting bolts an additional 90 degrees.
12. Connect the coolant pump-to-engine hose and the heater hose.

➡Install the water pump drive pulley flush with the end of the camshaft.

13. Using special tool 211—185 and 303—S455or a suitable tool, install the coolant pump drive pulley.
14. Install the coolant pump belt.
15. Connect the crankcase vent tube.
16. Fill and bleed the cooling system.

ENGINE ELECTRICAL

CHARGING SYSTEM

ALTERNATOR

REMOVAL & INSTALLATION

2.3L Engines

See Figure 42.

1. Before servicing the vehicle, refer to the Precautions Section.
2. With the vehicle in **NEUTRAL**, position it on a hoist.
3. Disconnect the negative battery cable.
4. Remove the (5) bolts and the RH lower splash shield.
5. Rotate the front end accessory drive belt tensioner clockwise and position the accessory drive belt aside.
6. Press the locking tab to release the alternator lower air duct from the alternator and remove the lower air duct.
7. Remove the battery harness locators from the alternator shield.
8. Remove the heated oxygen sensor (HO2S) harness locator from the lower part of the alternator. shield.
9. Position the alternator B+ protective cover aside and remove the alternator B+ terminal nut.
10. Position the alternator B+ cable aside
11. Disconnect the alternator electrical connector.
12. Remove the alternator shield nuts.
13. Remove the pin-type retainer from the bottom of the alternator and the alternator shield.
14. Remove the alternator bolt.
15. Remove the alternator stud nuts.
16. Remove the alternator studs.
17. Remove the screws and the alternator upper air duct.

➡ Lower the vehicle on the hoist to remove the alternator.

18. Remove the alternator.

To install:

19. Install alternator and upper air duct screws tighten screws to 35 inch lbs. (4 Nm).
20. Install and tighten alternator studs to 18 ft. lbs. (25 Nm).
21. Install alternator stud nuts and tighten to 35 ft. lbs. (47 Nm).
22. Install alternator bolt and tighten to 35 ft. lbs. (47 Nm).

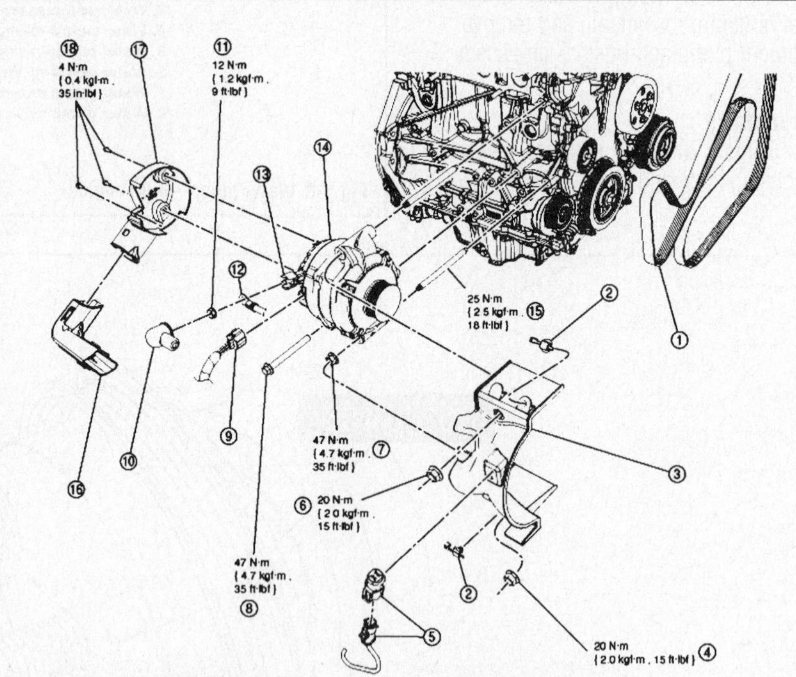

Fig. 42 Alternator and components exploded view—2.3L Engine

23. Install pin-type retainers to bottom of alternator and shield.
24. Install alternator shield nuts and tighten to 15 ft. lbs. (20 Nm).
25. Reconnect the alternator electrical connector.
26. Reposition the alternator B+ protective cover and tighten the alternator B+ terminal nut to 9 ft. lbs. (12 Nm).
27. Install the heated oxygen sensor (HO2S) harness locator to the lower part of the alternator shield.
28. Install the battery harness locators to the alternator shield.
29. Install the lower air duct.
30. Install drive belt
31. Install bolts and the RH lower splash shield. Tighten bolts to 71 inch lbs. (8 Nm).
32. Connect the negative battery cable.

3.0L Engine

See Figure 43.

WARNING

Do not allow any metal object to come in contact with the alternator housing and internal diode cooling fins. A short circuit may result and burn out the diodes. Failure to follow this instruction may result in component damage.

1. Disconnect the negative battery cable.
2. Remove the lower splash shield bolts and the pin-type retainers.

➡The LH lower splash shield must be removed before the RH lower splash shield.

3. Remove the lower splash shields.
4. Rotate the front end accessory drive tensioner counterclockwise and position the accessory drive belt aside

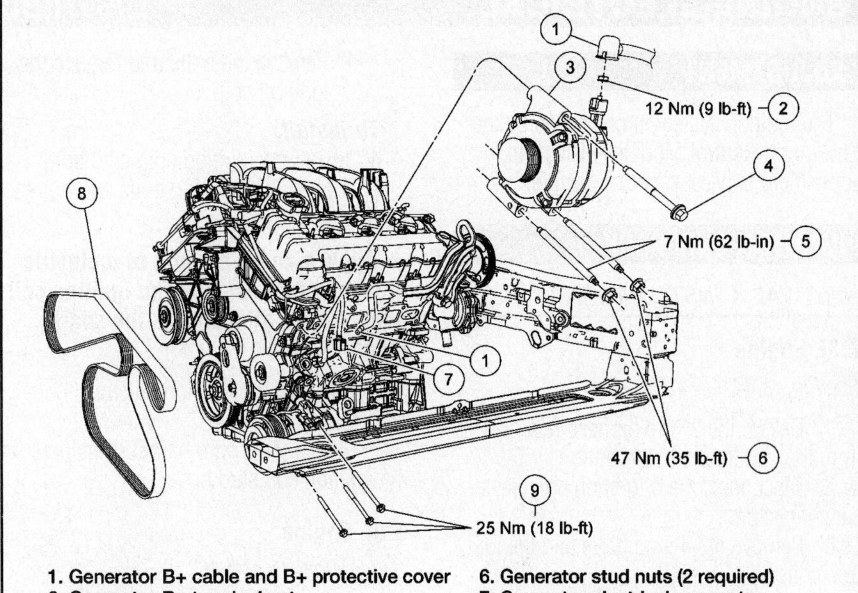

Fig. 43 Alternator mounting—3.0L engine

5. Disconnect the alternator electrical connector.
6. Position the alternator B+ protective cover aside and remove the alternator B+ terminal nut.
7. Remove the A/C compressor bolts. Use a tie-strap and position A/C compressor aside.
8. Loosen the alternator nuts.
9. Remove the lower alternator studs.
10. Remove the upper alternator bolt and the alternator.

To install:

11. Install alternator and upper bolt.
12. Tighten upper bolt to 35 ft. lbs. (47 Nm) with the vehicle on the ground.
13. Install the lower alternator studs and tighten to 62 inch lbs. (7 Nm).
14. Install and tighten the alternator nuts to 35 ft. lbs. (47 Nm).
15. Reposition and install A/C compressor bolts. Tighten to 18 ft. lbs. (25 Nm).
16. Reposition the alternator B+ protective cover and tighten terminal nut to 9 ft. lbs. (12 Nm).
17. Reconnect the alternator electrical connector.
18. Install accessory drive belt.
19. Install the splash shields and bolts and the 2 pin-type retainers.
20. Tighten the splash shield bolts to 71 inch lbs. (8 Nm).
21. Connect the negative battery cable.

ADJUSTMENT

The ignition system is controlled by the Powertrain Control Module (PCM). No adjustment is necessary or possible.

IGNITION COIL PACK

REMOVAL & INSTALLATION

2.3L Engine

See Figure 44.

1. Detach the 2 wire harness retainers from the ignition coil stud bolts.
2. Disconnect the 4 ignition coil electrical connectors.
3. Remove the 4 stud bolts and the ignition coils.

➡When removing the ignition coil-on-plugs, a slight twisting motion will break the seal and ease removal.

- Remove the bolts and remove the ignition coil.

To install:

4. Install the ignition coils and bolts.
5. Tighten the ignition coils and bolts to 71 inch lbs. (8 Nm).

➡ Apply a small amount of dielectric grease to the inside of the ignition coil boots before attaching to the spark plugs.

6. Connect the ignition electrical connectors.
7. Attach the 2 wire harness retainers to the ignition coil stud bolts.

3.0L Engine

See Figures 45 and 46.

1. Remove the upper intake manifold. .RH side only.
2. Disconnect the ignition coil electrical connectors.
3. Remove the bolts and the ignition coils.

➡ When removing the ignition coil-on-plugs, a slight twisting motion will break the seal and ease removal

To install:

4. Install the ignition coils and bolts.
5. Tighten the ignition coils and bolts to 62 inch lbs. (7 Nm).

➡ Apply a small amount of dielectric grease to the inside of the ignition coil boots before attaching to the spark plugs.

6. Connect the ignition coil electrical connectors.
7. Tighten the ignition coils electrical connectors to 11 ft. lbs. (15 Nm).
8. Install the upper intake manifold on the RH side.

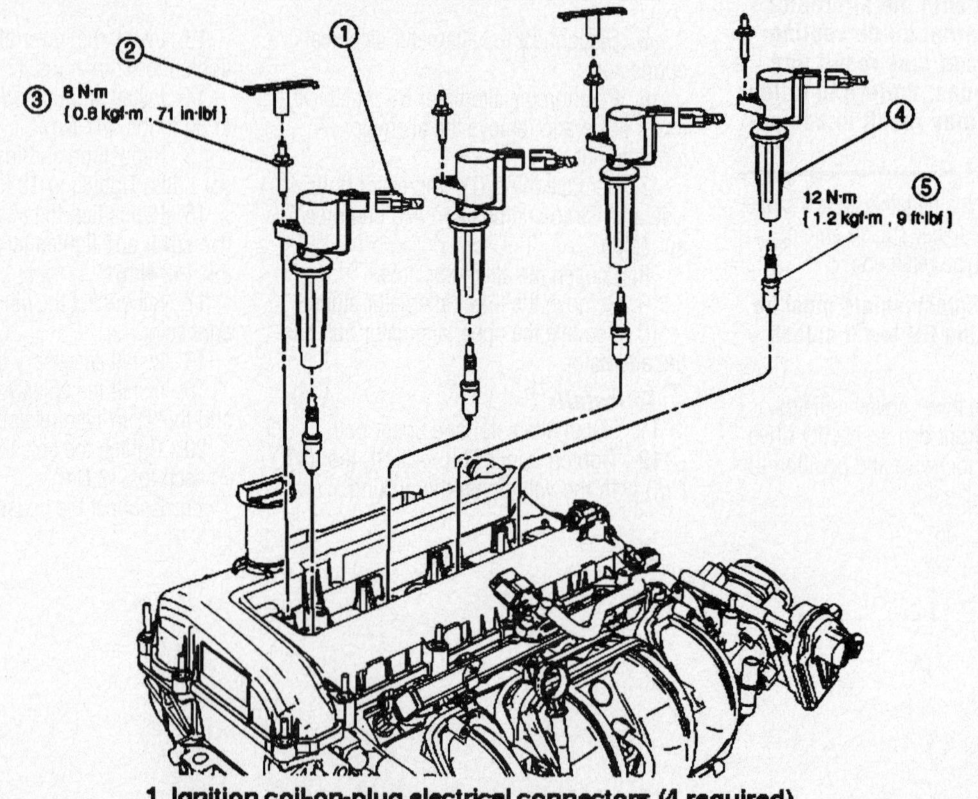

Fig. 44 Coil pack exploded view—2.3L engine

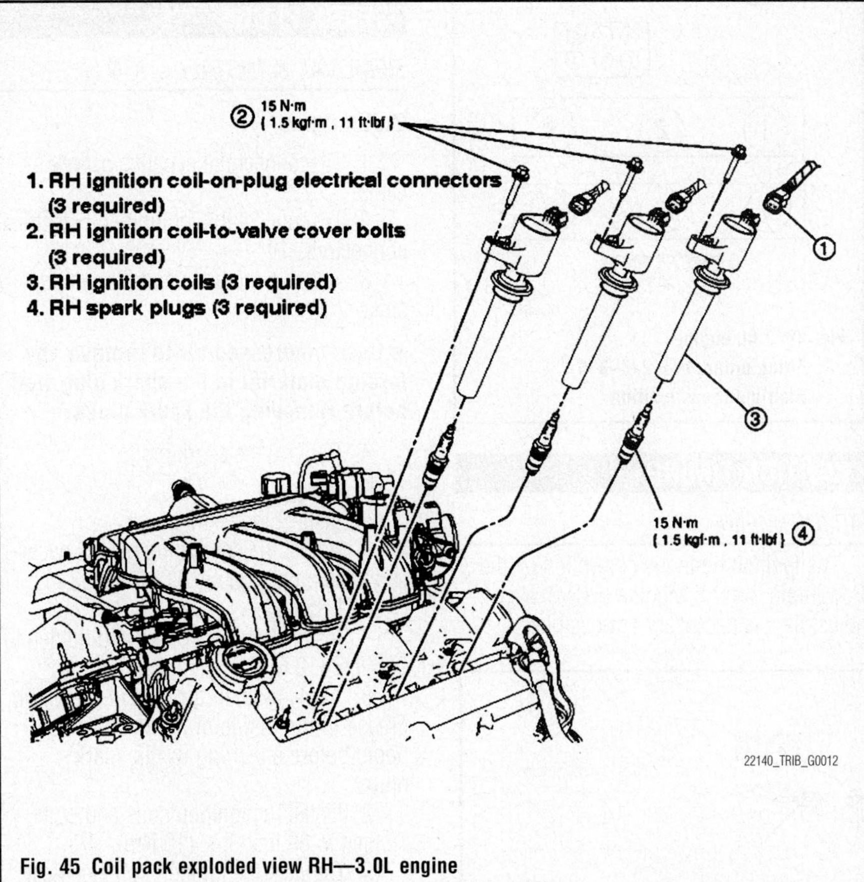

Fig. 45 Coil pack exploded view RH—3.0L engine

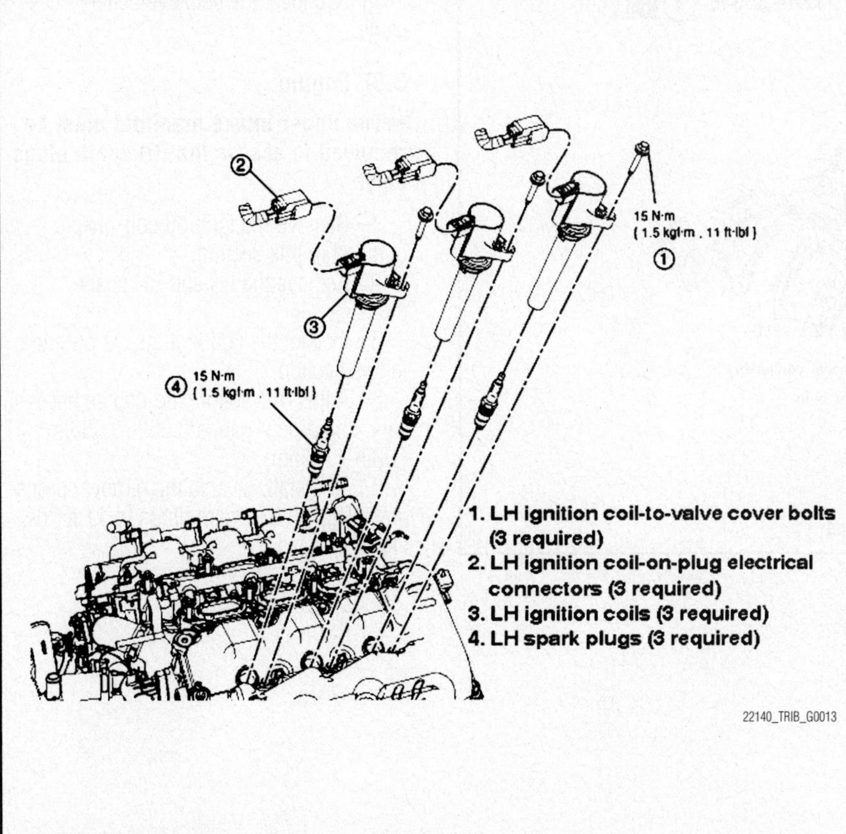

Fig. 46 Coil pack exploded view—3.0L engine

IGNITION COIL MODULE

REMOVAL & INSTALLATION

2.3L Engine

The 2.3L engine uses an ignition coil-on-plug type of ignition coil.

1. Disconnect the negative battery cable.
2. Disconnect the ignition coil electrical connectors.
3. Remove the bolts and the ignition coils.

To install:

4. Install the ignition coils. Tighten the bolts to 71 inch lbs. (8 Nm).
5. Apply a small amount of dielectric grease to the inside of the ignition coil boots before attaching to the spark plugs.

3.0L Engine—Left Side

See Figure 47.

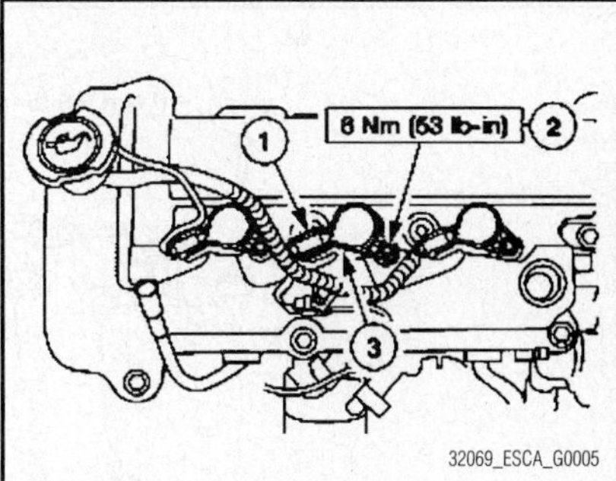

Fig. 47 Detach the electrical connector (1), remove the bolt (2) and the left side ignition coil-on-plug (3)—3.0L engine

➡The 3.0L engine uses an ignition coil-on-plug type of ignition coil.

1. Disconnect the negative battery cable.
2. Disconnect the ignition coil-on-plug electrical connector.
3. Remove the bolt, then remove the coil-on-plug.

To install:

4. Install the coil-on-plug. Tighten the coil-on-plug bolt to 11 ft. lbs. (15 Nm).
5. Attach the electrical connector.

➡Apply a light film of silicone brake caliper grease and dielectric compound to the interior of the spark plug boot prior to installation.

6. Connect the negative battery cable.

3.0L Engine— Right Side

➡The 3.0L engine uses and ignition coil-on-plug type of ignition coil.

1. Remove the upper intake manifold. Refer to the Engine Mechanical Section.
2. Disconnect the electrical connector.
3. Remove the bolt, then remove the coil-on-plug.

To install:

4. Install the coil-on-plug. Tighten the coil-on-plug bolt to 11 ft. lbs. (15 Nm).
5. Attach the electrical connector.

➡Apply a light film of silicone brake caliper grease and dielectric compound to the interior of the spark plug boot prior to installation.

6. Install the upper intake manifold.
7. Connect the negative battery cable.

FIRING ORDERS

See Figures 48 and 49.

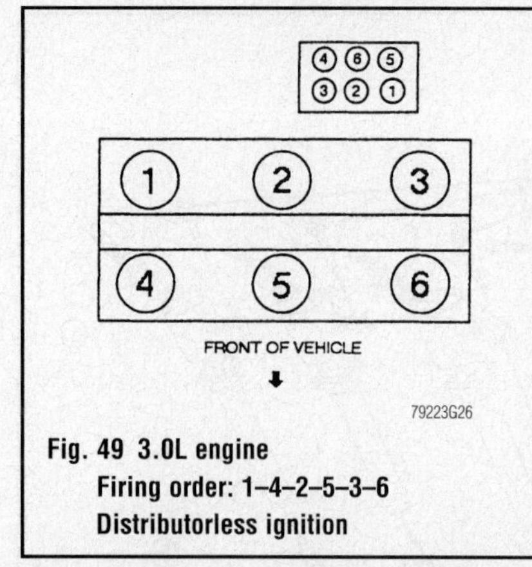

Fig. 49 3.0L engine
Firing order: 1-4-2-5-3-6
Distributorless ignition

IGNITION TIMING

ADJUSTMENT

The ignition timing is controlled by the Powertrain Control Module (PCM). No adjustment is necessary or possible.

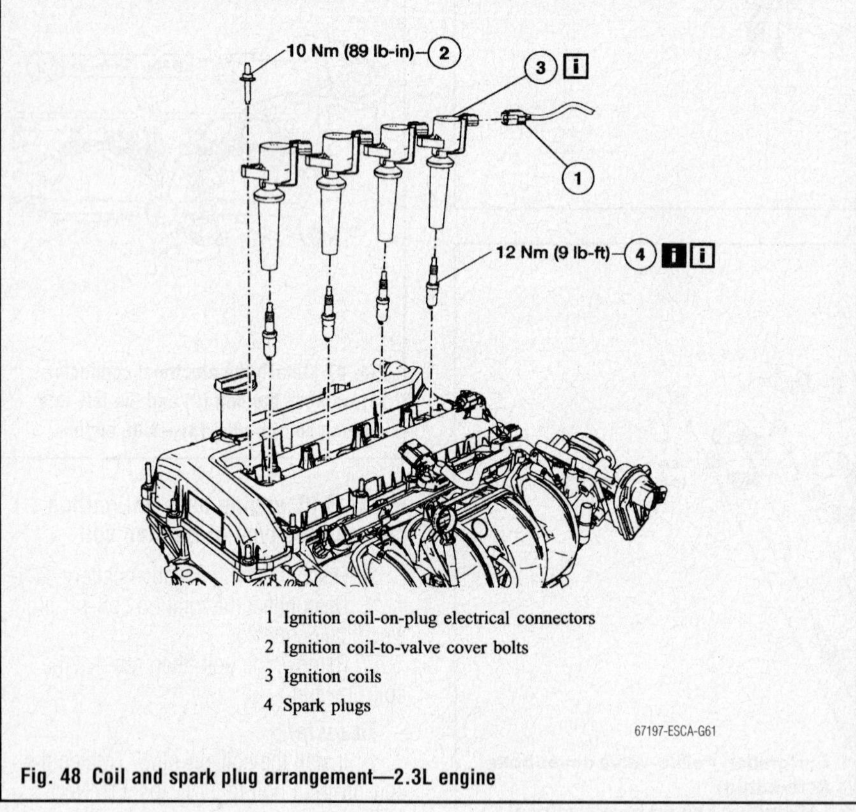

Fig. 48 Coil and spark plug arrangement—2.3L engine

SPARK PLUGS

REMOVAL & INSTALLATION

2.3L Engine

1. Disconnect the negative battery cable.
2. Disconnect the ignition coil electrical connectors.
3. Remove the bolts and the ignition coils.

➡Use compressed air to remove any foreign material in the spark plug well before removing the spark plugs.

4. Remove the spark plugs.

To install:

5. Inspect the spark plugs.
6. Adjust the spark plug gap as necessary. The proper gap is 0.049–0.053 in. (1.25–1.35mm).
7. Install the spark plugs and tighten to 9 ft. lbs. (12 Nm).
8. Apply a small amount of dielectric grease to the inside of the ignition coil boots before attaching to the spark plugs.
9. Install the ignition coils and bolts. Tighten to 89 inch lbs. (10 Nm).
10. Connect the ignition coil electrical connectors.
11. Connect the negative battery cable.

3.0L Engine

➡The upper intake manifold must be removed to access the RH spark plugs only.

1. Remove the ignition coil-on-plugs, as outlined in this section.
2. Remove the LH and RH spark plugs.
3. Inspect the spark plugs, as outlined in this section.
4. Adjust the spark plug gap as necessary. The proper gap is 0.052–0.056 in. (1.32–1.42mm).
5. To install, reverse the removal procedure. Tighten the spark plugs to 11 ft. lbs. (15 Nm).

STARTER

REMOVAL & INSTALLATION

2.3L Engine

1. With the vehicle in **NEUTRAL**, position it on a hoist.
2. Disconnect the negative battery cable.
3. Remove the starter solenoid wire nut.
4. Remove the starter solenoid battery cable nut and disconnect the starter motor solenoid terminal cover and cables.
5. Disconnect the wiring harness retainer and position aside the wiring harness.
6. Remove the ground strap nut and position aside the strap.
7. Remove the stud bolts and the starter motor.

To install:

8. Install starter motor and mounting stud bolts, tighten bolts to 26 ft. lbs. (35 Nm).
9. Reposition ground strap and tighten ground strap nut to 18 ft. lbs. (25 Nm).
10. Install battery cable and tighten nut to 9 ft. lbs (12 Nm).
11. Install solenoid wire and tighten nut to 44 inch lbs. (5 Nm).
12. Secure all retainers, wires and cover.
13. Connect the negative battery cable.

3.0L Engine

See Figure 50.

1. Disconnect the negative battery cable.
2. Remove the air cleaner.
3. Disconnect the transmission shift cable-to-manual lever.
4. Remove the transmission cable bracket bolts and detach the wire harness retainer and position aside the transmission cable and bracket.
5. Remove the starter motor solenoid wire nut.
6. Remove the starter motor solenoid battery cable nut and position aside the cables.
7. Remove the bolts and the starter motor.

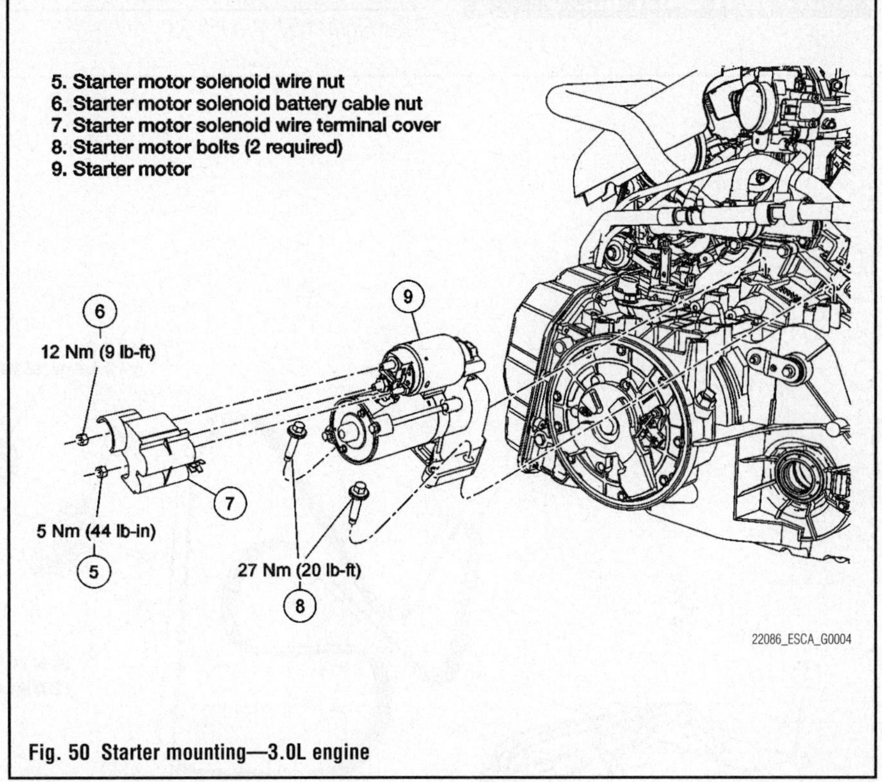

Fig. 50 Starter mounting—3.0L engine

To install:

8. Install starter motor and mounting bolts, tighten bolts to 20 ft. lbs. (27 Nm).
9. Reposition battery cable and tighten cable nut to 9 ft. lbs. (12 Nm).
10. Install solenoid wire and tighten nut to 44 inch lbs. (5 Nm).
11. Install transmission shift cable bracket, tighten mounting bolts to17 ft. lbs. (23 Nm).
12. Reconnect the transmission shift cable-to-manual lever.
13. Install the air cleaner.
14. Connect the negative battery cable.

ENGINE MECHANICAL

ACCESSORY DRIVE BELTS

ACCESSORY BELT ROUTING

See Figures 51 through 53.

Refer to the accompanying illustrations for accessory drive belt routing.

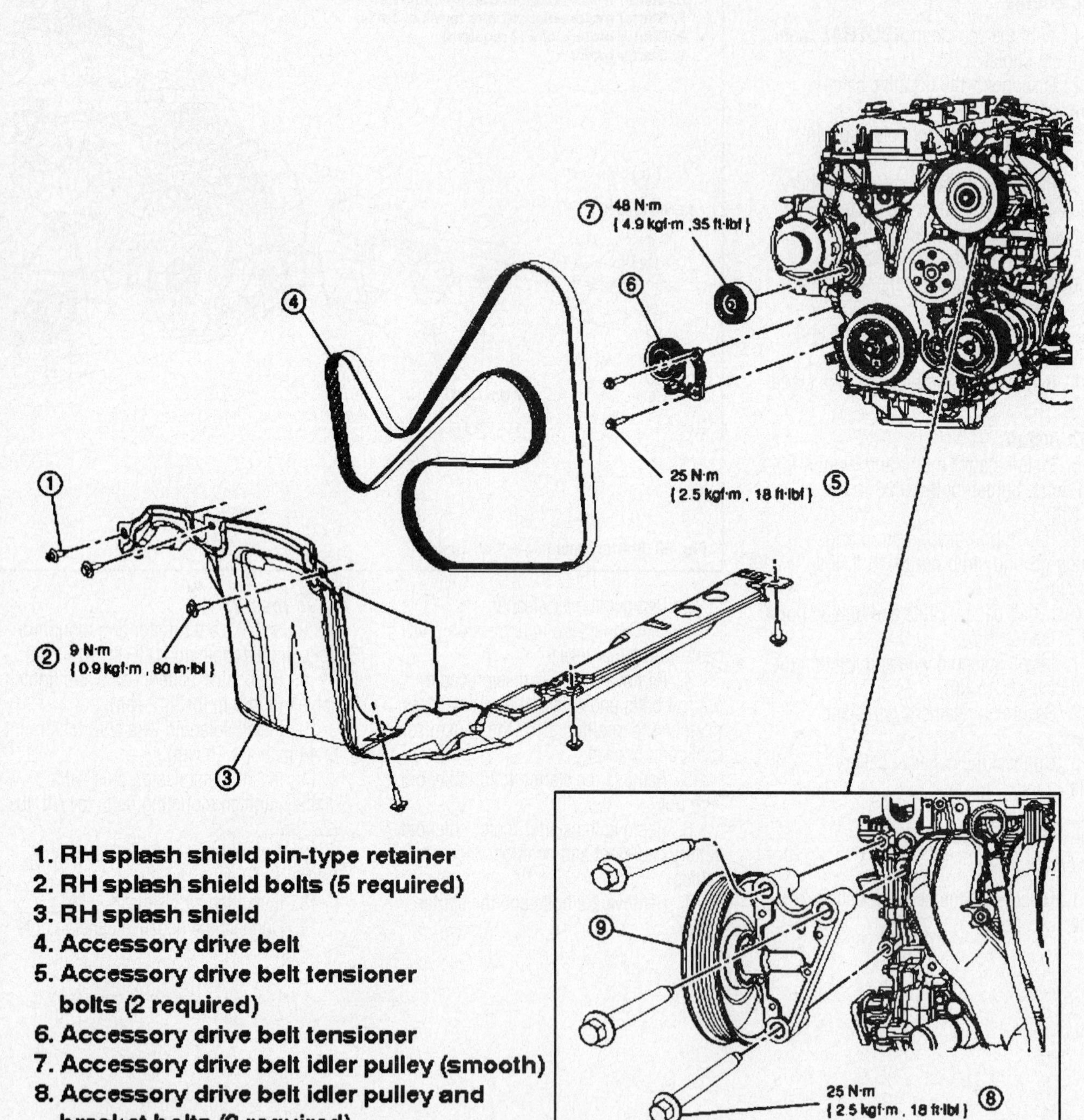

Fig. 51 Accessory drive belt routing and components exploded view s—2.3L engine

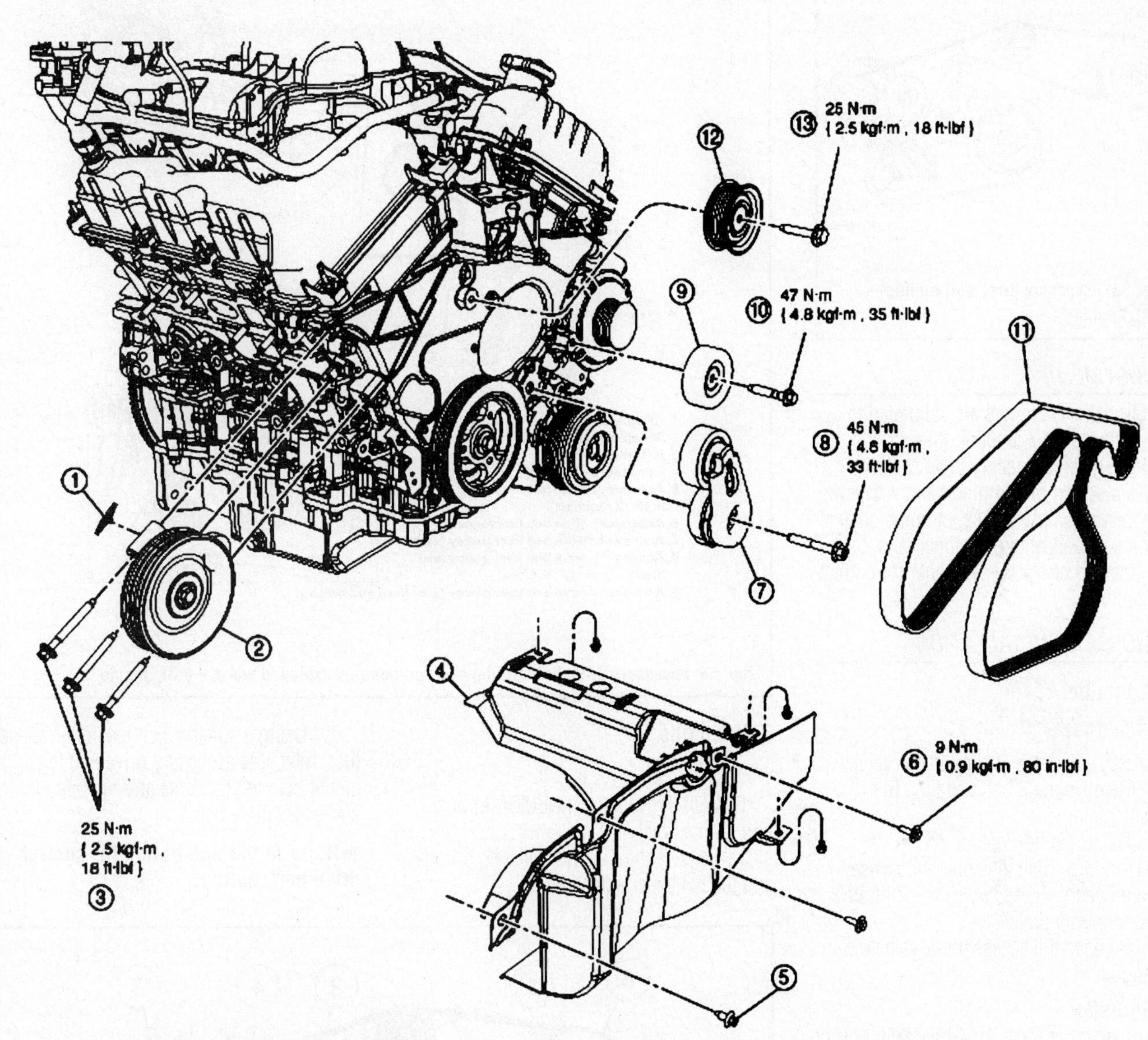

1. Wiring harness retainer
2. RH accessory drive belt idler pulley and bracket
3. RH accessory drive belt idler pulley and bracket bolts (3 required)
4. RH lower splash shield
5. RH lower splash shield pin-type retainer
6. RH lower splash shield bolt (5 required)
7. Accessory drive belt tensioner
8. Accessory drive belt tensioner bolt
9. LH accessory drive belt idler pulley
10. LH accessory drive belt idler pulley bolt
11. Accessory drive belt
12. Center accessory drive belt idler pulley
13. Center accessory drive belt idler pulley bolt

22140_TRIB_G0059

Fig. 52 Accessory drive belt routing and components exploded view—3.0L engine

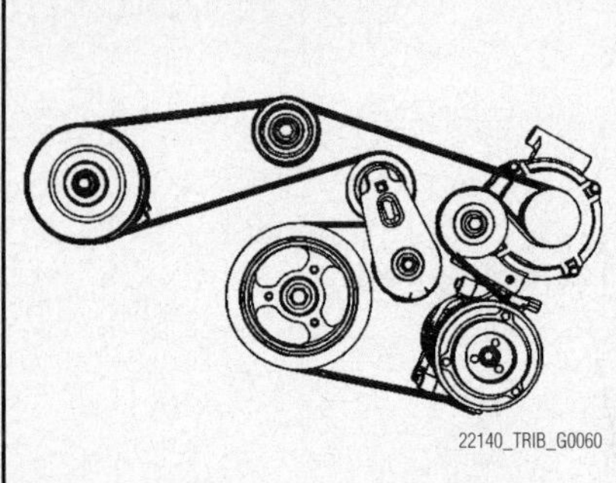

Fig. 53 Accessory drive belt routing—3.0L Engines

ADJUSTMENT

Automatic tensioners are calibrated to provide the correct amount of tension to the belt for a given accessory drive system. Unless a spring or damping band within the tensioner assembly breaks, or some other mechanical part of the tensioner fails, there is no need to check the tensioner for correct tension.

REMOVAL & INSTALLATION

2.3L Engine

See Figures 54 and 55.

1. Raise and safely support the vehicle..
2. Remove the 5 bolts and the RH splash shield.
3. Using the hex feature, rotate the accessory drive belt tensioner clockwise and remove the accessory drive belt from the water pump pulley.
4. Remove the accessory drive belt from the engine.

To install:

5. Install the accessory drive belt. Make sure it is routed correctly.
6. Install the RH splash shield and tighten the retaining bolts to 80 inch lbs. (9 Nm).

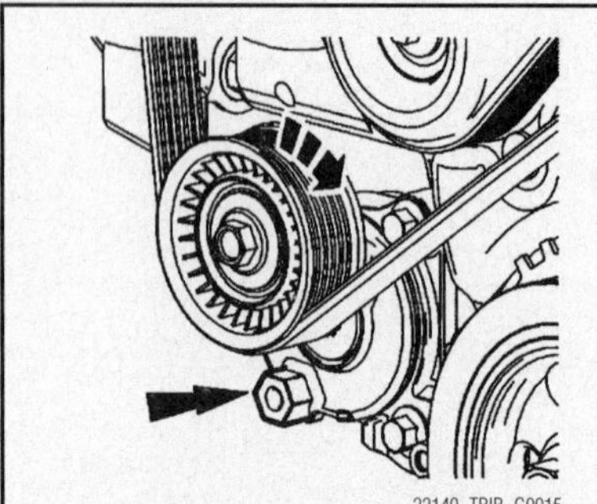

Fig. 54 Accessory drive belt removal and installation using the Hex to relieve tension—2.3L engine

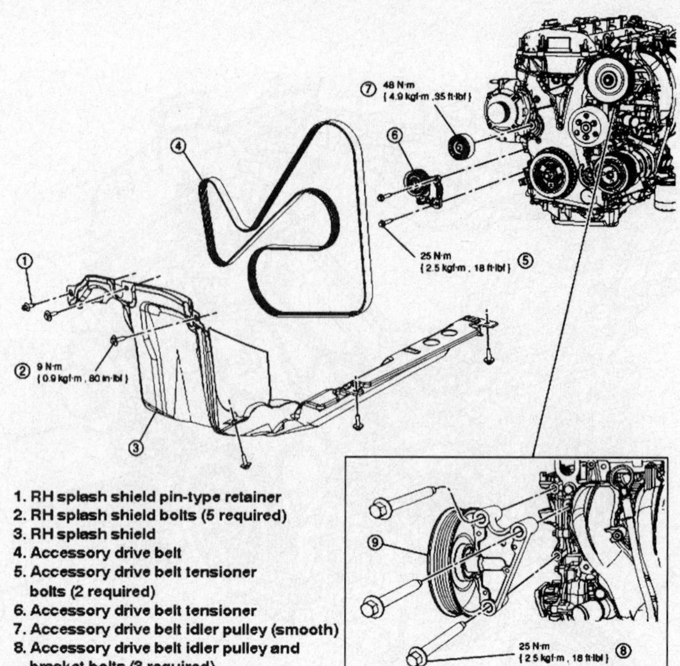

Fig. 55 Accessory drive belt routing and components exploded views—2.3L engine

3.0L Engine

See Figure 56.

1. With the vehicle in NEUTRAL, position it on a hoist.
2. Remove the pin-type retainer, 5 bolts and the RH lower splash shield.
3. Using a suitable belt tensioner release tool, rotate the accessory drive belt tensioner counterclockwise and remove the accessory drive belt.

➡**Refer to the illustration for correct drive belt routing.**

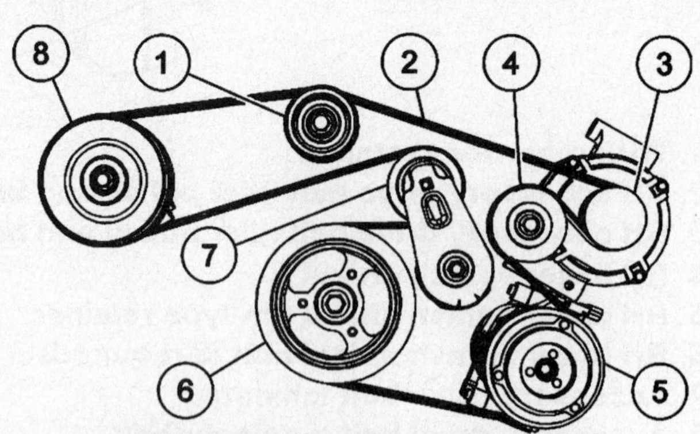

1. Center accessory drive belt idler pulley
2. Accessory drive belt
3. Generator pulley
4. LH accessory drive belt idler pulley
5. A/C clutch pulley
6. Crankshaft pulley
7. Accessory drive belt tensioner
8. RH accessory drive belt idler pulley and bracket

Fig. 56 Accessory drive belt routing— 3.0L engines with A/C and hydraulic power steering

4. Installation is the reverse of the removal procedure. Make sure the belt is properly routed.

Water Pump Belt

See Figures 57 and 58.

1. With the vehicle in **NEUTRAL**, position it on a hoist.
2. Cut and remove the coolant pump belt.

To install:

3. Install the coolant pump belt on the coolant pump pulley and position it on the camshaft pulley.

➡**This belt does not have any adjustments.**

⁂ WARNING

CAUTION: Do not use any screwdrivers, pliers or other metal objects that could cause damage to the belt or camshaft pulley while installing the belt.

4. Remove the pin-type retainer, 5 bolts and the RH lower splash shield.

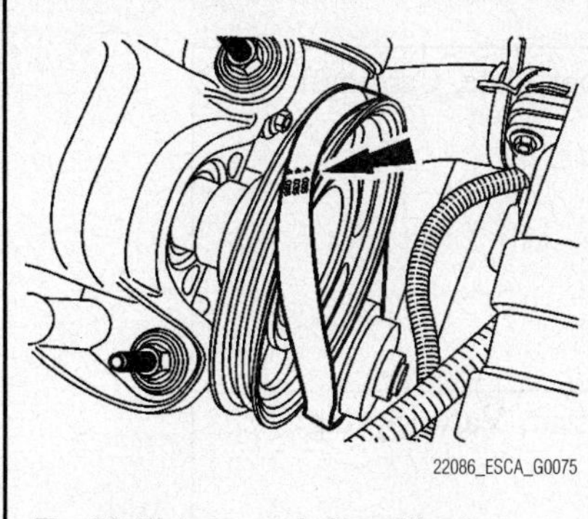

Fig. 58 Water pump belt position for installation

5. Rotate the crankshaft clockwise to seat the coolant pump belt on the camshaft pulley.

CAMSHAFT AND VALVE LIFTERS

REMOVAL & INSTALLATION

2.3L Engine

See Figures 59 through 63.

1. Before servicing the vehicle, refer to the Precautions Section.

⁂ WARNING

During engine repair procedures, cleanliness is extremely important. Any foreign material, including any material created while cleaning gasket surfaces that enters the oil passages, coolant passages or the oil pan can cause engine failure.

⁂ WARNING

The crankshaft, the crankshaft sprocket and the pulley are fitted together by friction, using diamond washers between the flange faces on each part. For that reason, the crankshaft sprocket is also unfastened if you loosen the pulley. Therefore, the engine must be retimed each time the damper is removed. Otherwise severe engine damage can occur.

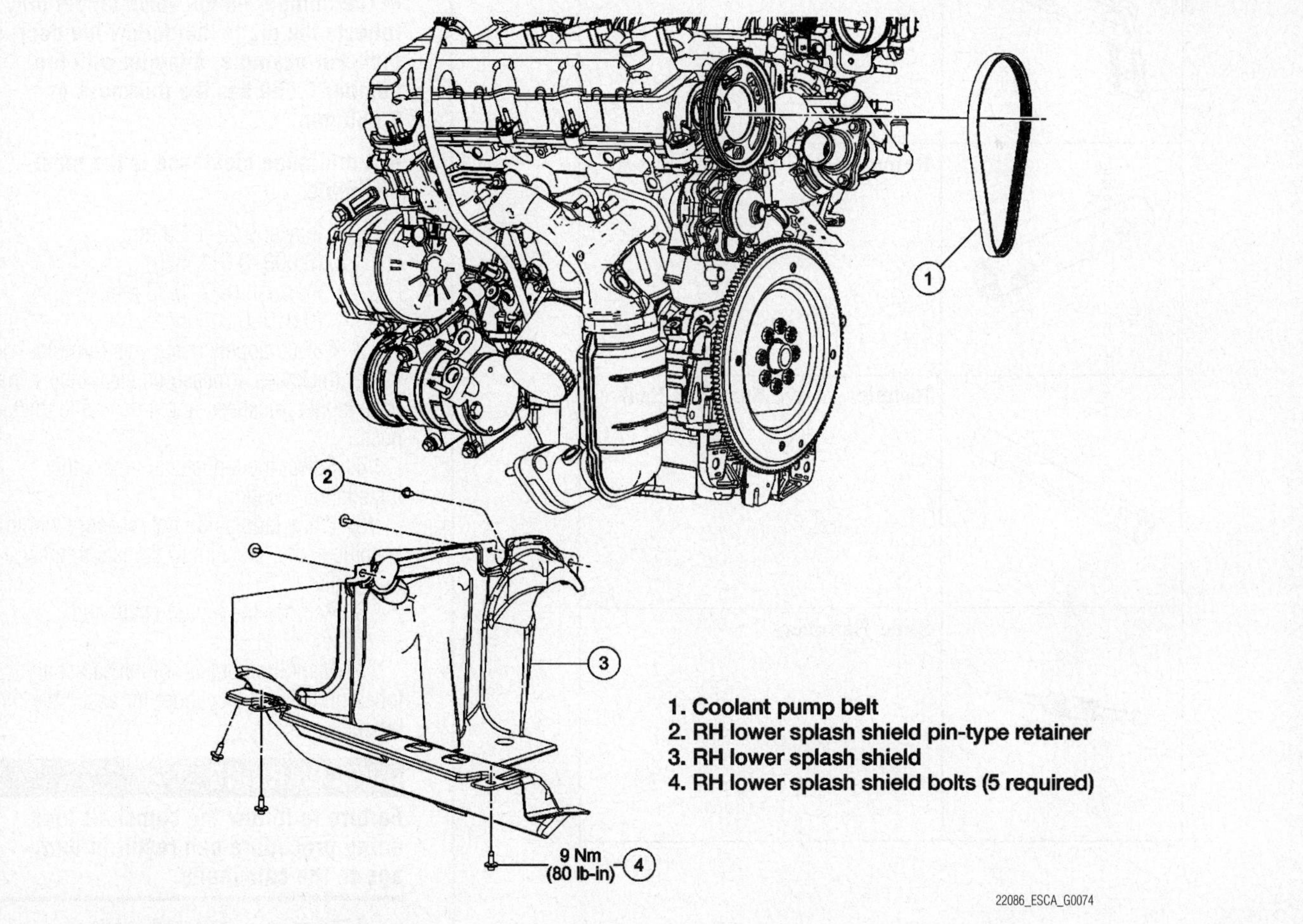

1. Coolant pump belt
2. RH lower splash shield pin-type retainer
3. RH lower splash shield
4. RH lower splash shield bolts (5 required)

9 Nm (80 lb-in)

22086_ESCA_G0074

Fig. 57 Water pump drive belt—3.0L engine

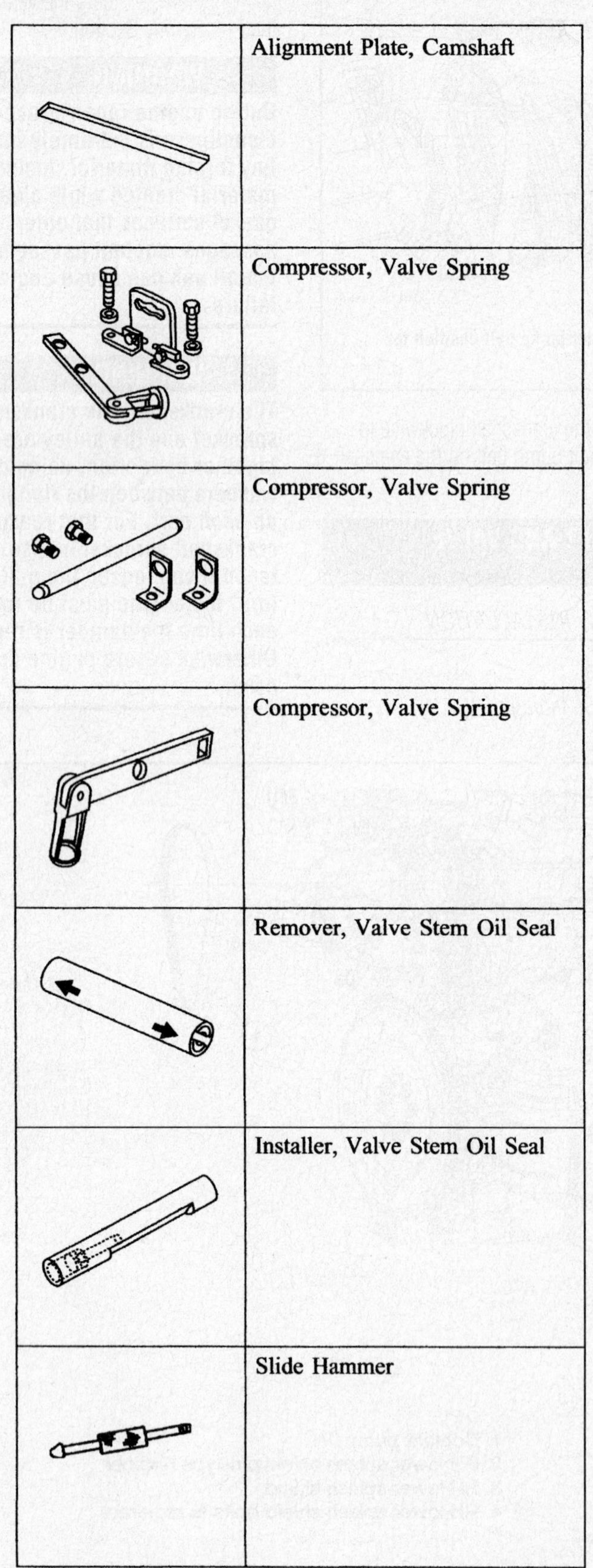

Alignment Plate, Camshaft
Compressor, Valve Spring
Compressor, Valve Spring
Compressor, Valve Spring
Remover, Valve Stem Oil Seal
Installer, Valve Stem Oil Seal
Slide Hammer
67197-ESCA-G15

Fig. 59 Tools necessary for camshaft and lifter removal—2.3L engine

WARNING

Do not rotate the camshafts unless instructed to in this procedure. Rotating the camshafts or crankshaft with timing components loosened or removed can cause serious damage to the valves or pistons.

➡**Valve tappets are select fit and the valve clearance must be checked before removing the tappets.**

WARNING

Turn the engine clockwise only, and only use the crankshaft bolt.

➡**Before removing the camshafts, measure the clearance of each valve at base circle, with the lobe pointed away from the tappet. Failure to measure all clearances prior to removing the camshafts will necessitate repeated removal and installation and wasted labor time.**

2. Use a feeler gauge to measure the clearance of each valve and record its location.

➡**The number on the valve tappet only reflects the digits that follow the decimal. For example, a tappet with the number 0.650 has the thickness of 3.650 mm.**

➡**A midrange clearance is the most desirable:**

- Intake: 0.22–0.28 mm (0.008–0.011 inch)
- Exhaust: 0.27–0.33 mm (0.010–0.013 inch)

3. Select tappets using this formula: tappet thickness = measured clearance + the base tappet thickness–most desirable thickness.

4. Select the tappets and mark the installation location.

5. If any tappets do not measure within specifications, install new tappets in these locations.

6. Remove the timing chain and sprockets.

7. Mark the position of the camshaft lobes on the No. 1 cylinder for assembly reference.

WARNING

Failure to follow the camshaft loosening procedure can result in damage to the camshafts.

8. Loosen the camshaft bearing bolts in the sequence shown, one turn at a

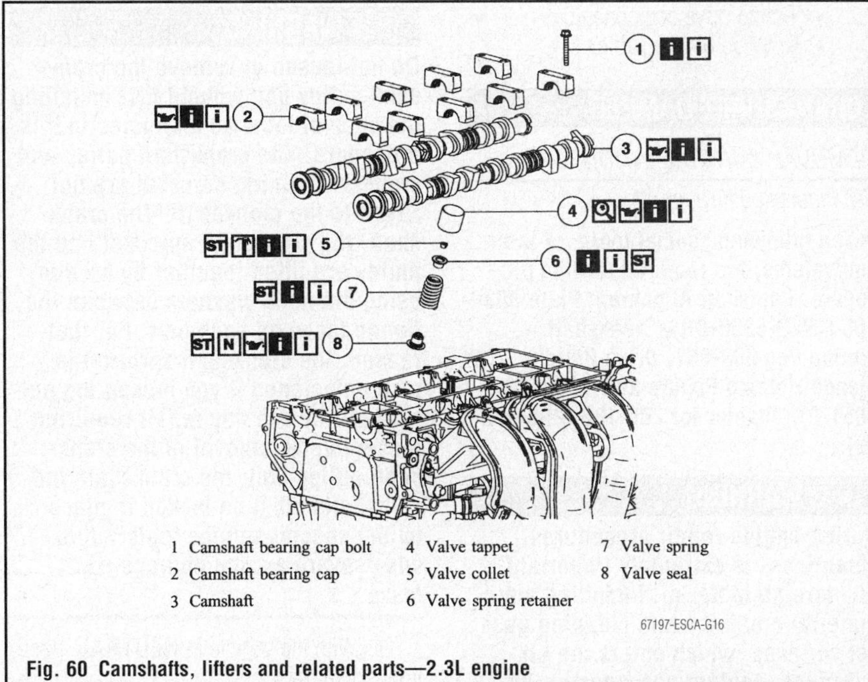

Fig. 60 Camshafts, lifters and related parts—2.3L engine

time. Repeat until all the tension is released.

9. Remove the camshaft bearing caps.

WARNING

If the camshafts and valve tappets are to be reused, mark the location of the valve tappets to make sure they are assembled in their original positions.

➡The number on the valve tappets only reflects the digits that follow the decimal. For example, a tappet with the number 0.650 has the thickness of 3.650 mm.

10. Remove the camshafts.
11. Valve tappets.
12. To install, reverse the removal procedure. Coat the valve tappets with clean engine oil and insert them.

WARNING

Install the camshafts with the alignment slots in the camshafts lined up so the Camshaft Alignment Plate can be installed without rotating the camshafts. Make sure the lobes on the No. 1 cylinder are in the same position as noted in the removal procedure. Rotating the camshafts when the timing chain is removed, or installing the camshafts 180 degrees out of position can cause severe damage to the valves and pistons.

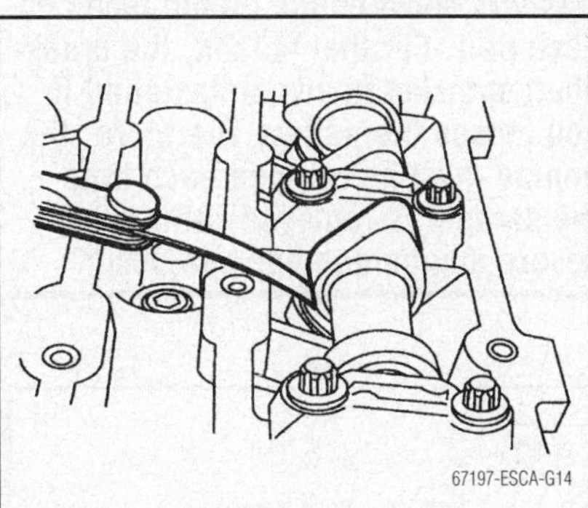

Fig. 62 Valve clearance check—2.3L engine

➡Lubricate the camshaft journals and bearing caps with clean engine oil.

13. Install the camshafts and bearing caps. Tighten the bolts in the sequence shown in three stages.
   a. Stage 1: Tighten the camshaft bearing bolt caps one turn at a time until tight.
   b. Stage 2: Tighten the bolts to 62 inch lbs. (7 Nm).
   c. Stage 3: Tighten the bolts to 12 ft. lbs. (16 Nm).

3.0L Engine—Left Side

See Figure 64.

1. Before servicing the vehicle, refer to the Precautions Section.
2. Remove or disconnect the following:
   - Negative battery cable
   - Water pump belt
   - Timing drive components
   - Camshaft oil seal
   - Camshaft oil seal retainer
   - Camshaft cap bolts by loosening them in sequence
   - Camshafts

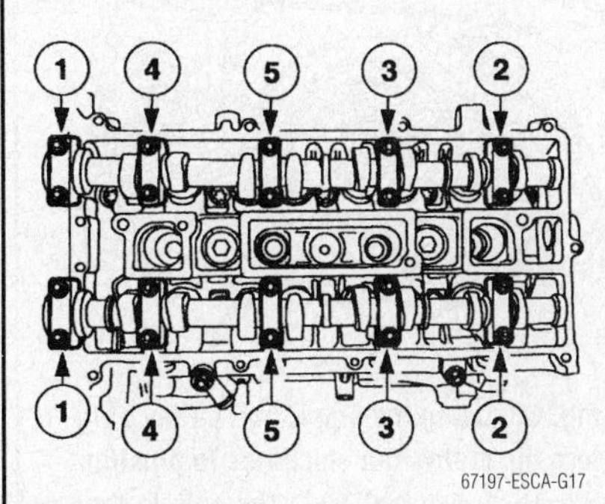

Fig. 61 Camshaft cap removal sequence—2.3L engine

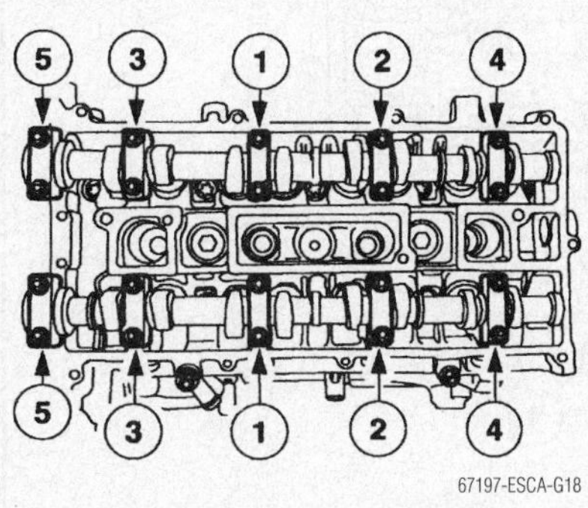

Fig. 63 Camshaft cap torque sequence—2.3L engine

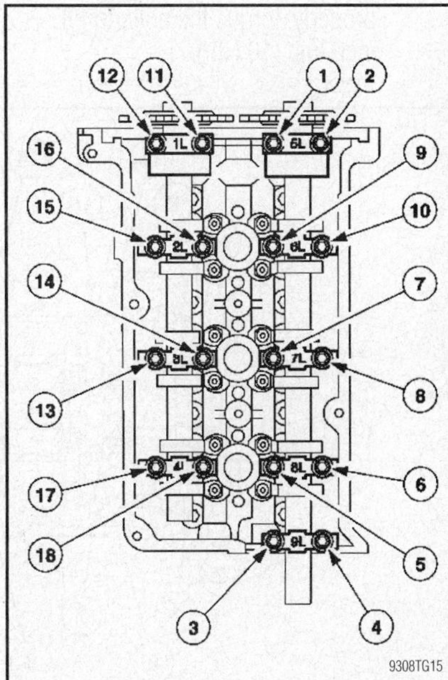

Fig. 64 Remove and install the left side camshaft bearing caps in sequence-3.0L engine

To install:

3. Install or connect the following:
   - Camshaft bearing caps in their original position
   - Align the camshafts
   - Bearing thrust caps and hand tighten the bolts. When aligned properly, torque the bolts to 89 inch lbs. (10 Nm).
   - Timing drive components
   - Camshaft oil seal retainer
   - Crankshaft oil seal
   - Water pump drive pulley
   - Water pump belt
   - Negative battery cable

3.0L Engine—Right Side

See Figure 65.

1. Before servicing the vehicle, refer to the Precautions Section.
2. Remove or disconnect the following:
   - Negative battery cable
   - Timing drive components
   - Camshaft cap bolts by loosening them in sequence
   - Camshafts caps
   - Camshafts

To install:

3. Install or connect the following:
   - Camshaft bearing caps in their original position
   - Align the camshafts
   - Bearing caps and hand tighten the bolts
   - Bearing thrust caps and hand tighten the bolts. When aligned properly, torque the bolts to 89 inch lbs. (10 Nm).
   - Timing drive components
   - Negative battery cable

CRANKSHAFT DAMPER

REMOVAL & INSTALLATION

See Figures 66 through 71.

➡**The following special tools, or their equivalents, are required for this procedure. Camshaft Alignment Plate 303-465 (T94P-6256-CH), Crankshaft Timing Peg 303-057, Drive Pinion Flange Holding Fixture 205-126 (T78P-4851-A), Adapter for 205-126 (205-072-02).**

✱✱ WARNING

During engine repair procedures, cleanliness is extremely important. Any foreign material, including any material created while cleaning gasket surfaces, which enters the oil passages, coolant passages or the oil pan can cause engine failure.

✱✱ WARNING

The crankshaft, the crankshaft sprocket and the pulley are fitted together by friction, using diamond washers between the flange faces on each part. For that reason, the crankshaft sprocket is also unfastened if you loosen the pulley. Therefore, the engine must be retimed each time the damper is removed. Otherwise severe engine damage can occur.

✱✱ WARNING

Do not loosen or remove the crankshaft pulley bolt without first installing the special tools as instructed in this procedure. The crankshaft pulley and crankshaft timing sprocket are not keyed to the crankshaft. The crankshaft, the crankshaft sprocket and the pulley are fitted together by friction, using diamond washers between the flange faces on each part. For that reason, the crankshaft sprocket is also unfastened if you loosen the pulley bolt. Before any repair requiring loosening or removal of the crankshaft pulley bolt, the crankshaft and camshafts must be locked in place by the special service tools, otherwise severe engine damage can occur.

1. With the vehicle in **NEUTRAL**, position it on a hoist
2. Remove the accessory drive belt.
3. Remove the valve cover, as outlined in this section.

✱✱ WARNING

Failure to position the No. 1 piston at top dead center (TDC) can result in damage to the engine. Turn the engine in the normal direction of rotation only.

4. Using the crankshaft pulley bolt, turn the crankshaft clockwise to position the No. 1 piston at TDC. The hole in the crankshaft pulley should be in the 6 o'clock position.

✱✱ WARNING

The special tool 303-465 is for camshaft alignment only. Using this

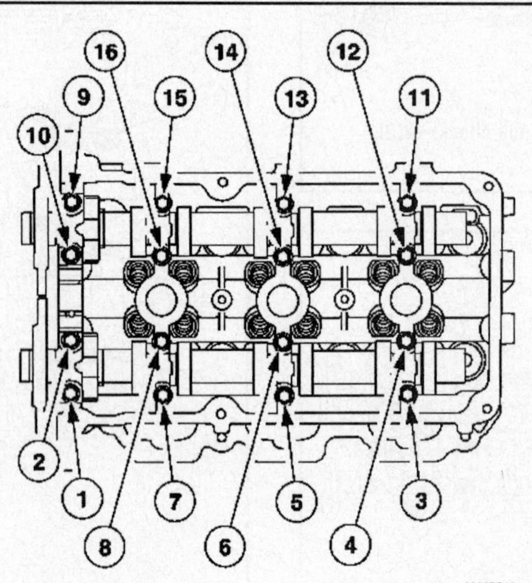

Fig. 65 Remove and install the right side camshaft bearing caps in sequence–3.0L engine

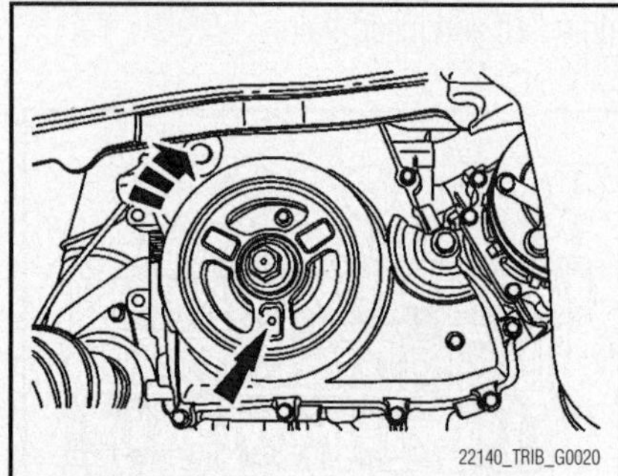

Fig. 66 Using the crankshaft pulley bolt, turn the crankshaft clockwise to position the No. 1 piston at TDC. The hole in the crankshaft pulley should be in the 6 o'clock position

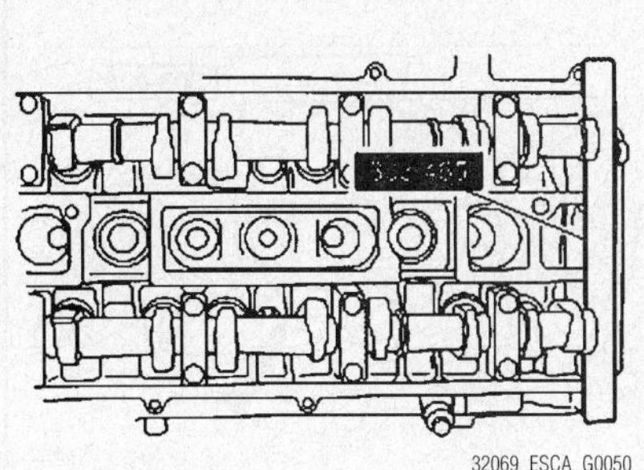

Fig. 67 Install the camshaft alignment plate special tool in the slots on the rear of both camshafts

tool to prevent engine rotation can result in engine damage.

➡The camshaft timing slots are offset. If the special tool cannot be installed, rotate the crankshaft one complete revolution clockwise to correctly position the camshafts.

5. Install Camshaft Alignment Plate 303-465, or equivalent special tool in the slots on the rear of both camshafts.
6. Remove the engine plug bolt.

➡The special tool will contact the crankshaft and prevent it from turning past TDC. However, the crankshaft can still be rotated in the counterclockwise direction. The crankshaft must remain at the TDC position during the crankshaft pulley removal and installation.

7. Install Crankshaft Timing Peg 303-057 or equivalent special tool.
8. Install Drive Pinion Flange Holding Fixture 205-126 (T78P-4851-A) and Adapter for 205-126 (205-072-02) or equivalent special tools.

**** WARNING**

Failure to hold the crankshaft pulley in place while loosening the bolt can result in damage to the engine.

**** WARNING**

If the crankshaft sprocket diamond washer comes off with the crankshaft pulley it must be installed back onto the crankshaft.

9. Remove the crankshaft pulley bolt and washer. Discard the bolt.
10. Remove the crankshaft pulley.

To install:

➡Do not reuse the crankshaft pulley bolt.

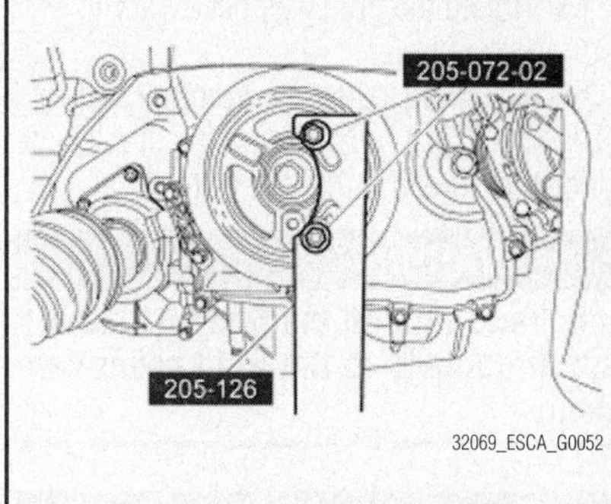

Fig. 69 Install Drive Pinion Flange Holding Fixture and Adapter special tools

➡Apply clean engine oil on the seal area before installing.

11. Position the crankshaft pulley onto the crankshaft with the hole in the pulley at the 6 o'clock position.
12. Install the crankshaft pulley and hand-tighten the bolt.

**** WARNING**

Only hand-tighten the bolt or damage can occur.

➡The following 2steps will correctly align the crankshaft pulley to the crankshaft.

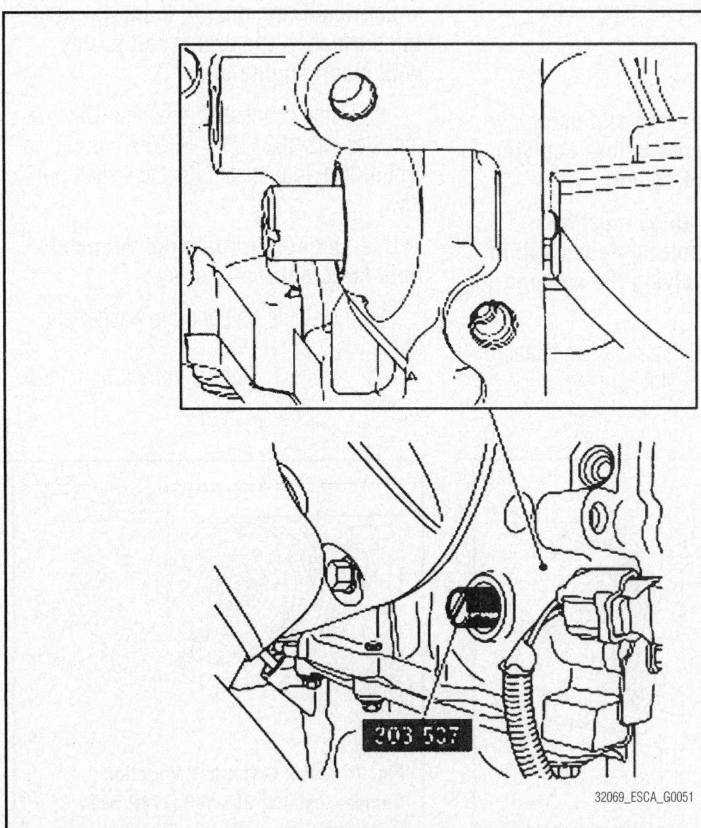

Fig. 68 Install Crankshaft Timing Peg 303-507 or equivalent special tool

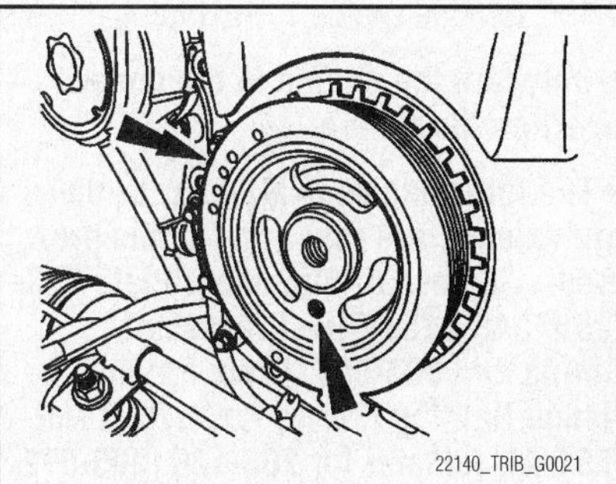

Fig. 70 Pulley positioned onto the crankshaft with the hole in the pulley at the 6 o'clock position

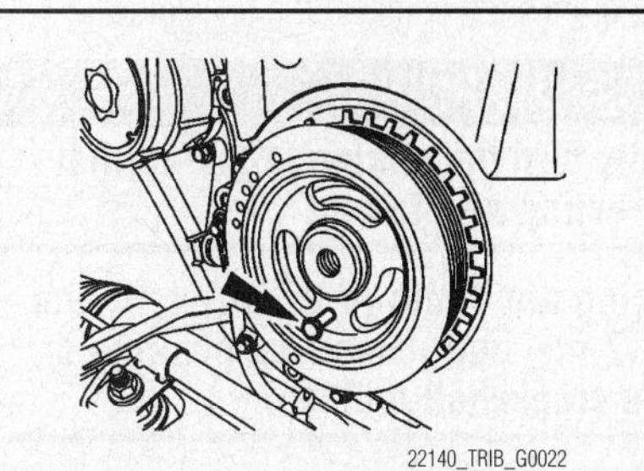

Fig. 71 Pulley positioned onto the crankshaft with the 6 mm (0.23 in) bolt in the pulley at the 6 o'clock position locking it at TDC

13. Rotate the pulley as necessary to align the bolt holes.

14. Install a standard 6-mm (0.23-in.) x 18-mm (0.7-in.) bolt through the crankshaft pulley and thread it into the front cover.

✻✻ WARNING

Only hand-tighten the 6 mm (0.23 in) bolt or damage to the front cover can occur.

✻✻ WARNING

Failure to hold the crankshaft pulley in place while tightening the bolt can cause damage to the engine front cover.

15. Using the special tools to hold the crankshaft pulley in place, tighten the crankshaft pulley bolt in 2 stages:

a. Stage 1: Tighten to 74 ft. lbs. (100 Nm).

b. Stage 2: Tighten an additional 90 degrees (1/4 turn).

16. Remove the 6-mm (0.23-in.) x 18-mm (0.7-in.) bolt.

17. Remove the special tools 303—507.

18. Remove the special tools 303—465.

➡Only turn the engine in the normal direction of rotation.

19. Turn the engine 1—3/4 turns.

➡Only turn the engine in the normal direction of rotation.

➡The following special tools, or their equivalents, are required for this procedure. Camshaft Alignment Plate 303-465 (T94P-6256-CH), Crankshaft Timing Peg 303-057, Drive Pinion Flange Holding Fixture 205-126 (T78P-4851-A), Adapter for 205-126 (205-072-02).

20. Install the special crankshaft timing peg tool.

21. Turn the crankshaft clockwise until the crankshaft contacts the special tool.

✻✻ WARNING

Only turn the engine in the normal direction of rotation.

Using the 6-mm (0.23-in.) x 18-mm (0.7-in.) bolt, check the position of the crankshaft pulley.

✻✻ WARNING

Only hand-tighten the bolt or damage to the front cover can occur.

22. If it is not possible to install the bolt, correct the engine timing.

23. Using the camshaft alignment plate special tool, check the position of the camshafts.

➡If it is not possible to install the special tool, correct the engine timing.

24. Remove the 6-mm (0.23-in.) x 18-mm (0.7-in.) bolt.

25. Remove the crankshaft timing peg 303-507 or equivalent special tool

26. Install the engine plug bolt and tighten to 15 ft. lbs. (20 Nm).

27. Install the accessory drive belt.

28. Install the valve cover.

3.0L Engine

See Figures 72 through 75.

1. With the vehicle in **NEUTRAL**, position it on a hoist.

2. Remove the accessory drive belt.

3. Remove the splash shield.

4. Install the Strap Wrench 303-D055 (D85L-6000-A), or equivalent special tool.

5. Remove the crankshaft pulley bolt and washer.

6. Using the Crankshaft Vibration Damper Remover 303-009 (T58P-6316-D) or equivalent special tool, remove the crankshaft pulley.

To install:

➡Clean the keyway and slot using metal surface cleaner before applying silicone gasket and sealer.

➡The crankshaft pulley must be installed and the bolt tightened within four minutes of applying the silicone gasket and sealer.

7. Apply silicone gasket and sealant to the end of the keyway slot.

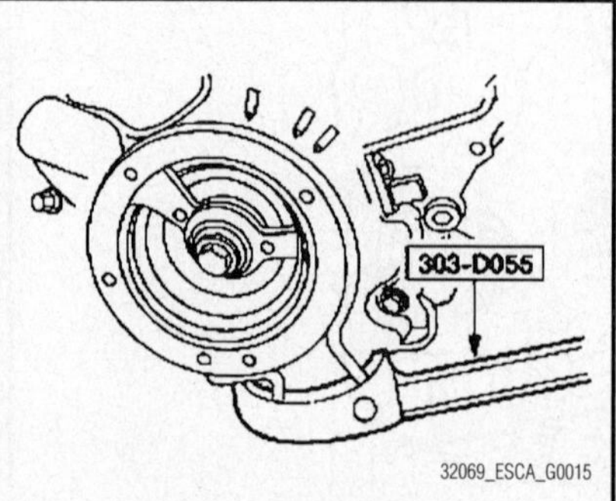

Fig. 72 Install the strap wrench to hold the crankshaft pulley stationary

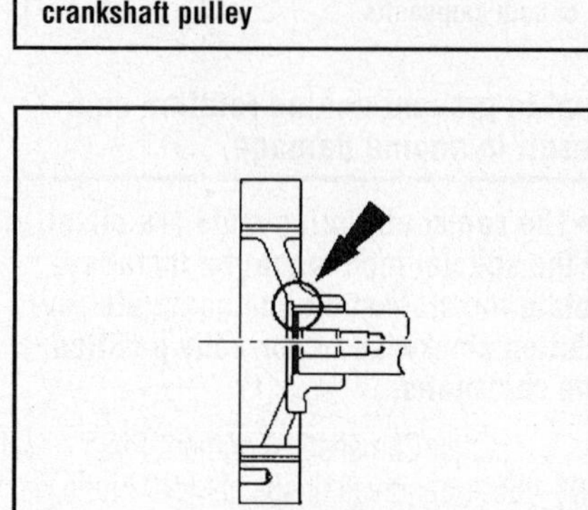

Fig. 73 Using the Crankshaft Vibration Damper Remover tool to remove the crankshaft pulley

32069_ESCA_G0017

Fig. 74 Apply silicone gasket and sealant to the end of the keyway slot

➡Lubricate the outside diameter sealing surface of the crankshaft pulley with clean engine oil.

8. Using Crankshaft Vibration Damper Installer 303-102 (T74P-6316-B) or equivalent special tool, install the crankshaft pulley.

➡Use an appropriate strap wrench to hold the crankshaft pulley.

9. Install the bolt and the washer and tighten in four stages.

a. Stage 1: Tighten to 89 ft. lbs. (120 Nm).

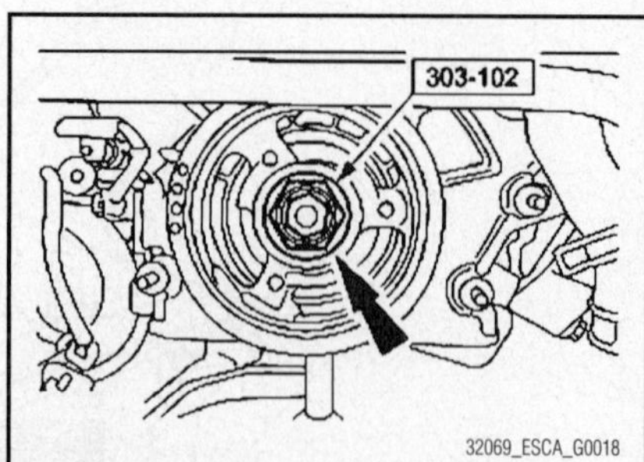

Fig. 75 Using Crankshaft Vibration Damper Installer 303-102 (T74P-6316-B) or equivalent special tool, install the crankshaft pulley

b. Stage 2: Loosen 360 degrees.
c. Stage 3: Tighten to 37 ft. lbs. (50 Nm).
d. Stage 4: Tighten an additional 90 degrees.
10. Install RH front inner splash shield. Tighten the retainers to 80 inch lbs. (9 Nm).
11. Install the accessory drive belt.

CRANKSHAFT FRONT SEAL

REMOVAL & INSTALLATION

2.3L Engine

See Figures 76 and 77.

1. Before servicing the vehicle, refer to the Precautions Section.

**** WARNING**

During engine repair procedures, cleanliness is extremely important. Any foreign material, including any material created while cleaning gasket surfaces that enters the oil passages, coolant passages or the oil pan can cause engine failure.

**** WARNING**

The crankshaft, the crankshaft sprocket and the pulley are fitted together by friction, using diamond washers between the flange faces on each part. For that reason, the crankshaft sprocket is also unfastened if you loosen the pulley. Therefore, the engine must be retimed each time the damper is removed. Otherwise severe engine damage can occur.

2. Remove the crankshaft pulley. Refer to crankshaft pulley removal and installation in this section.
3. Remove the front crankshaft seal using tool 303—409

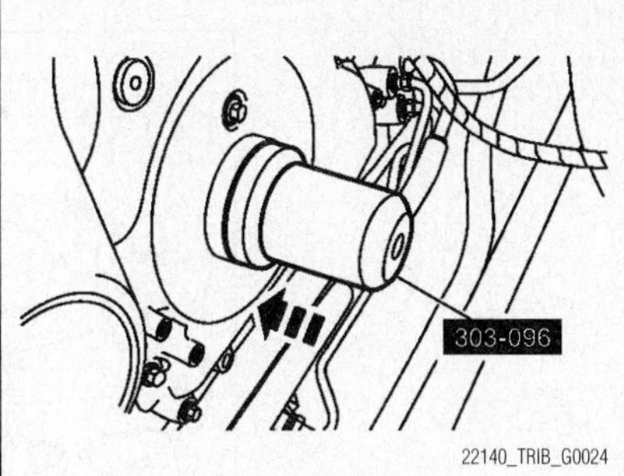

Fig. 77 Crankshaft seal installation tool 303—096

To install:

➡Lubricate the oil seal with clean engine oil.

4. Using the a seal driver 303–096, install the crankshaft front oil seal.

➡Do not reuse the crankshaft damper bolt.

5. Install the crankshaft pulley. Refer to crankshaft pulley removal and installation in this section.

3.0L Engine

See Figure 78.

1. Before servicing the vehicle, refer to the Precautions Section.
2. Remove or disconnect the following:
   - Negative battery cable
   - Crankshaft pulley
   - Front oil seal

To install:

3. Install or connect the following:
   - New front crankshaft oil seal
   - Crankshaft pulley. See the Torque Specifications Chart for the tightening torques.
   - Negative battery cable
4. Start the vehicle and check for leaks, repair if necessary.

CYLINDER HEAD

REMOVAL & INSTALLATION

2.3L Engine

See Figures 79 through 83.

1. Before servicing the vehicle, refer to the Precautions Section.

**** WARNING**

During engine repair procedures, cleanliness is extremely important. Any foreign material, including any material created while cleaning gasket surfaces that enters the oil passages, coolant passages or the oil pan can cause engine failure.

**** WARNING**

The crankshaft, the crankshaft sprocket and the pulley are fitted together by friction, using diamond washers between the flange faces on each part. For that reason, the crankshaft sprocket is also unfastened if you loosen the pulley. Therefore, the engine must be retimed each time the damper is removed. Otherwise severe engine damage can occur.

**** WARNING**

Do not loosen or remove the crankshaft pulley bolt without first installing the special tools as instructed in the timing chain section. The crankshaft pulley and the crankshaft timing sprocket are not keyed to the crankshaft.

2. With the vehicle in **NEUTRAL**, position it on a hoist.
3. Release the fuel system pressure.
4. Drain the engine cooling system.

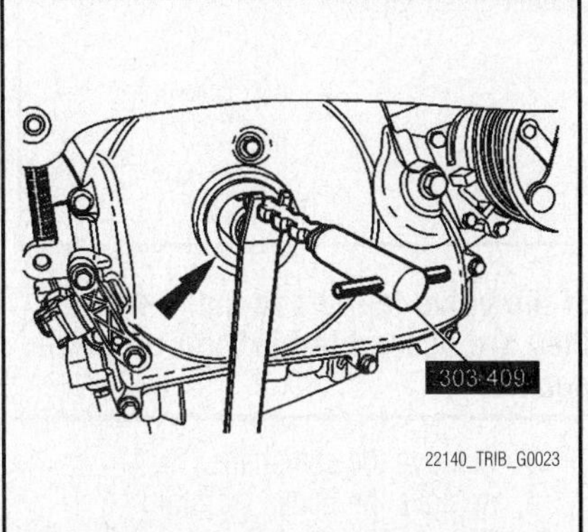

Fig. 76 Crankshaft seal removal tool

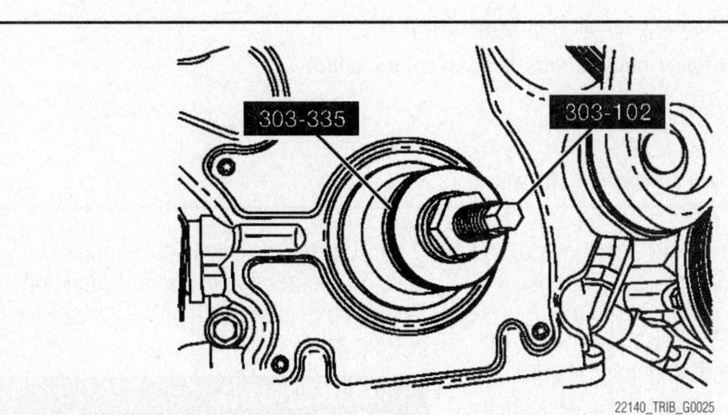

Fig. 78 Crankshaft seal installation too l303—335 and 303—102

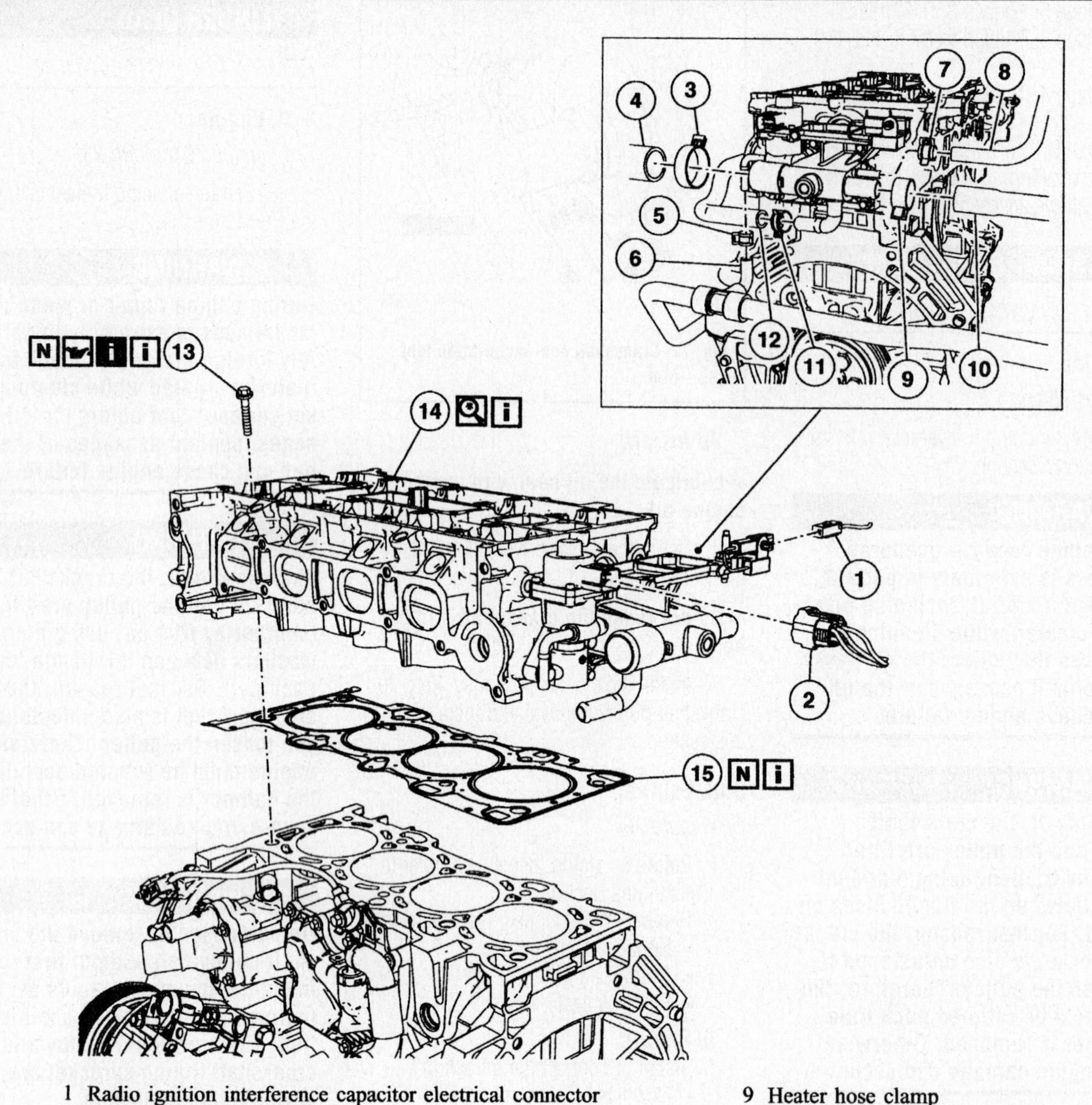

Fig. 79 Cylinder head removal—2.3L engine

5. Remove the timing drive components. For additional information, refer to Timing Drive Components in this section.

6. Mark the position of the camshaft lobes on the No. 1 cylinder for installation reference.

7. Loosen the camshaft bearing cap bolts, in sequence, one turn at a time until all tension is released from the camshaft bearing caps.

WARNING

If the camshafts and valve tappets are to be reused, mark the location of the valve tappets to make sure they are assembled in their original positions.

8. Remove the camshafts.

9. Remove the intake manifold.

10. Remove the catalytic converter/manifold.

11. Disconnect the radio ignition interference capacitor electrical connector
12. Disconnect the exhaust gas recirculation (EGR) valve electrical connector
13. Remove the upper radiator hose.
14. Remove the EGR coolant tube clamp.
15. Remove the EGR coolant hose.
16. Remove the engine coolant vent hose.
17. Remove the heater hose.
18. Remove the bypass hose.
19. Remove and discard the cylinder head bolts.
20. Remove the cylinder head.
21. Remove the cylinder head gasket.

To install:

22. Inspect the cylinder head for distortion.

✲✲ WARNING

Do not use metal scrapers, wire brushes, power abrasive discs or other abrasive means to clean the sealing surfaces. These tools cause scratches and gouges that make leak paths. Use a plastic scraping tool to remove all traces of the head gasket.

✲✲ WARNING

Observe all warnings or cautions and follow all application directions contained on the packaging of the silicone gasket remover and the metal surface prep.

➡If there is no residual gasket material present, metal surface prep can be used to clean and prepare the surfaces.

23. Clean the cylinder head-to-cylinder block mating surface of both the cylinder head and the cylinder block.
24. Remove any large deposits of silicone or gasket material with a plastic scraper.
25. Apply silicone gasket remover, following package directions, and allow to set for several minutes.
26. Remove the silicone gasket remover with a plastic scraper. A second application of silicone gasket remover may be required if residual traces of silicone or gasket material remain.
27. Apply metal surface prep, following package directions, to remove any traces of oil or coolant, and to prepare the surfaces to bond with the new gasket. Do not attempt to make the metal shiny. Some staining of the metal surfaces is normal.
28. Apply silicone gasket and sealant to the locations shown.
29. Install a new head gasket.

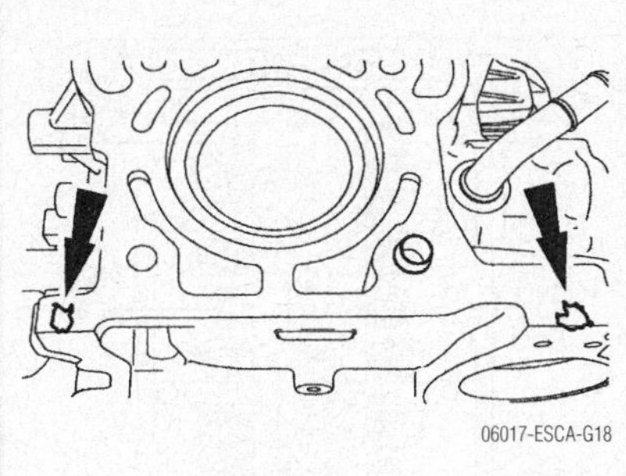

Fig. 80 Apply silicone gasket and sealant to the locations shown—2.3L engine

➡The cylinder head bolts are torque-to-yield and must not be reused. New cylinder head bolts must be installed.

➡Lubricate the bolts with clean engine oil prior to installation.

30. Install new cylinder head bolts. Tighten the bolts in the sequence shown in five stages.
    a. Tighten the bolts to 44 inch lbs. (5 Nm).
    b. Tighten the bolts to 11 ft. lbs. (15 Nm).
    c. Tighten the bolts to 33 ft. lbs. (45 Nm).
    d. Turn the bolts 90 degrees.
    e. Turn the bolts an additional 90 degrees.
31. Connect the upper radiator hose, coolant bypass hose, heater hose and coolant vent hose to the engine coolant outlet.
32. Attach the wire harness retainer to the coolant outlet stud.
33. Connect the EGR coolant hose to the EGR valve.
34. Connect the exhaust gas recirculation (EGR) valve electrical connector.
35. Install the catalytic converter and exhaust manifold.
36. Install the generator.

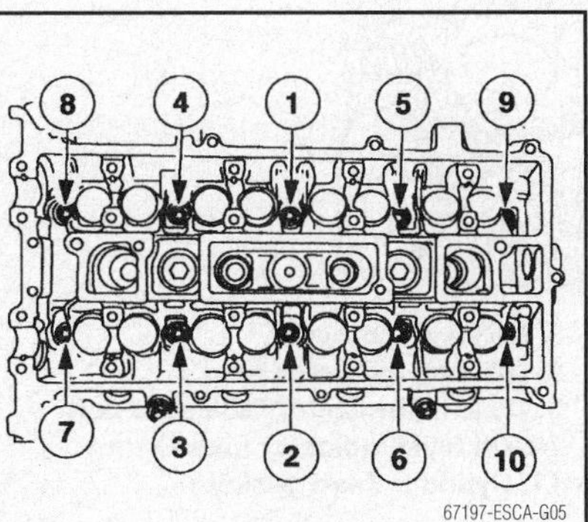

Fig. 81 Cylinder head bolt torque sequence—2.3L engine

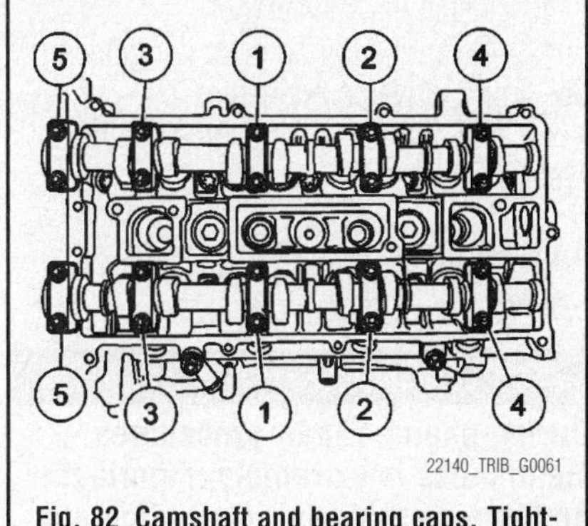

Fig. 82 Camshaft and bearing caps. Tightening sequence

37. Install the intake manifold.

➡Lubricate the valve tappets with clean engine oil.

38. Install the valve tappets in their original positions.
39. Install the camshaft and bearing caps. Tighten the bolts in the sequence shown in 3 stages.
    a. Tighten the camshaft bearing cap bolts, one turn at a time, until finger tight.
    b. Tighten the bolts to 62 inch lbs. (7 Nm).
    c. Tighten the bolts to 12 ft. lbs. (16 Nm).

✲✲ WARNING

Install the camshafts with the alignment notches in the camshafts lined up so the camshaft alignment plate can be installed. Make sure the lobes on the No. 1 cylinder are in the same position as noted in the removal procedure. Failure to follow this procedure can cause severe damage to the valves and pistons.

➡ Lubricate the camshaft journals and bearing caps with clean engine oil.

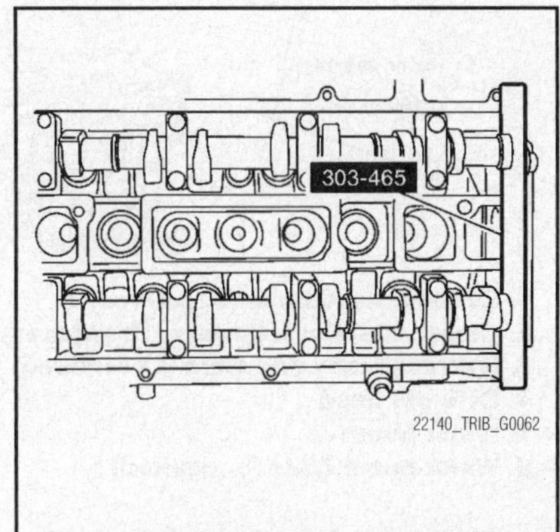

Fig. 83 Camshaft timing tool 303—465

40. Install the special tool.
41. Install the timing drive components. See timing drive components
42. Fill and bleed the engine cooling system

3.0L Engine

See Figures 84 through 88.

**** WARNING**

During engine repair procedures, cleanliness is extremely important. Any foreign material, (including any material created while cleaning gasket surfaces) that enters the oil passages, coolant passages or the oil pan, can cause engine failure.

The procedure for the left side cylinder head and right side are similar. Changes in the procedure will be noted for either side cylinder head.

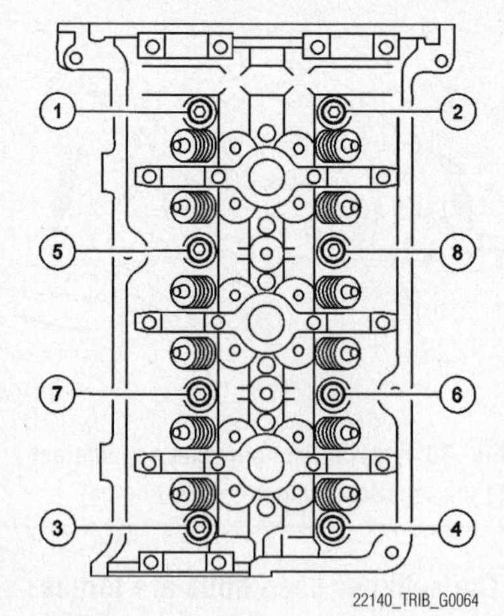

Fig. 85 Cylinder head bolt removal sequence— LH, 3.0L

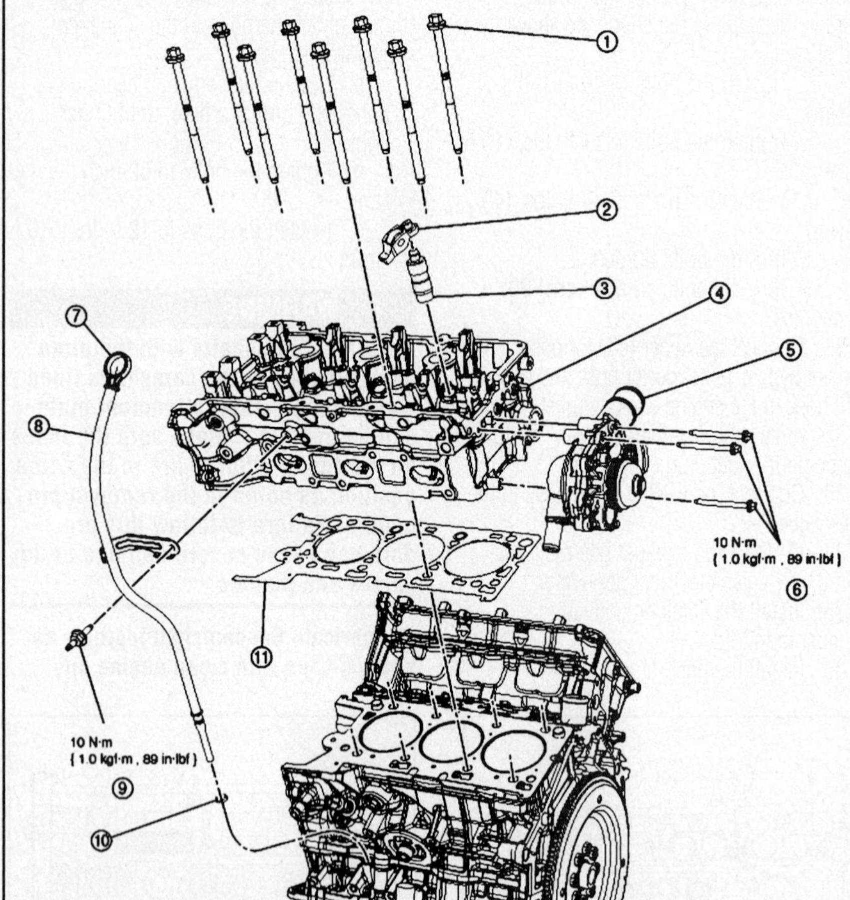

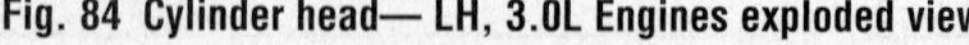
Fig. 84 Cylinder head— LH, 3.0L Engines exploded view

1. Before servicing the vehicle, refer to the Precautions Section.
2. Properly relieve the fuel system pressure.
3. Drain the cooling system.
4. With the vehicle in **NEUTRAL**, position it on a hoist.
5. Remove or disconnect the following:
   - Negative battery cable
   - Lower intake manifold
   - Coolant bypass tube

**** WARNING**

The hydraulic lash adjusters must be installed in their original positions.

   - Camshaft
   - Exhaust Gas Recirculation (EGR) tube, right side only
   - Exhaust manifold
   - oil level indicator and tube
   - Camshaft followers
   - Hydraulic lash adjusters and matchmark them for proper installation
   - Remove the 3 bolts and position the water pump aside.
   - Cylinder head bolts in sequence and discard them
6. Remove the cylinder head
7. Remove the cylinder head and support the cylinder head on a bench with the head gasket up.
8. Discard the gasket and bolts.
9. Inspect all areas of the deck face with a straightedge and feeler gauge. The cylinder head must not have depressions deeper than 0.001 in (0.0254 mm) across a 1.5 in.(38.1 mm) square area, or scratches more than 0.001 in (0.0254 mm).

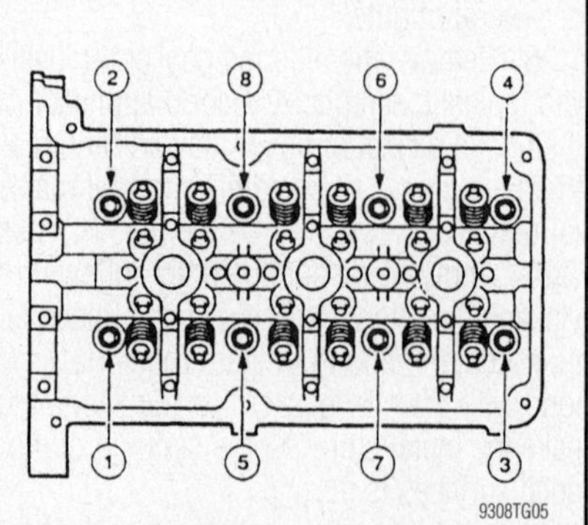

Fig. 86 Left side cylinder head bolt torque sequence 3.0L engine

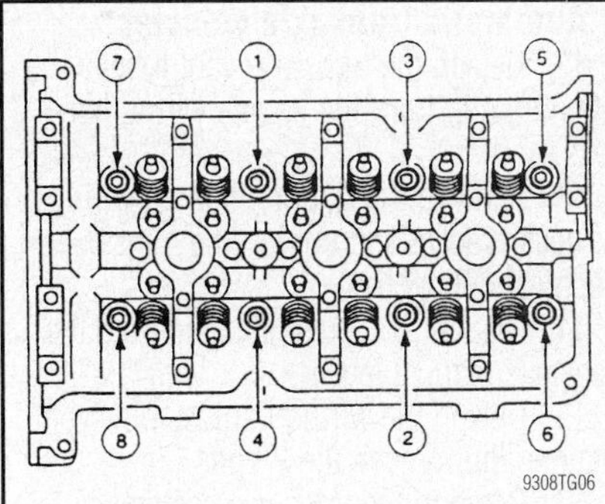

Fig. 87 Right side cylinder head bolt torque sequence 3.0L engine

WARNING

The straightedge used must be flat within 0.0051 mm (0.0002 in) per foot of tool length.

To install:

WARNING

Do not use metal scrapers, wire brushes, power abrasive discs or other abrasive means to clean the sealing surfaces. These tools cause scratches and gouges which make leak paths.

10. Position a new gasket and the cylinder head.
11. Install the bolts and tighten in 6 stages in the sequence shown.
    a. Tighten the bolts to 30 ft. lbs. (40 Nm).
    b. Turn the bolts an additional 90 degrees.
    c. Loosen 1 full turn.
    d. Tighten the bolts to 30 ft. lbs. (40 Nm).
    e. Turn the bolts 90 degrees.
    f. Turn the bolts an additional 90 degrees.
12. Lubricate the cylinder head bolt threads.

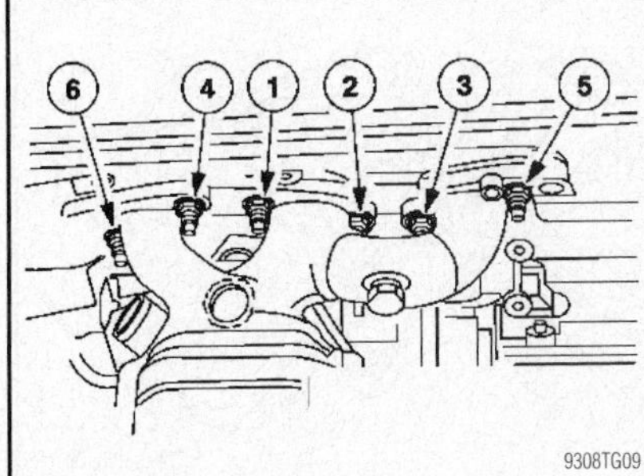

Fig. 88 Right side exhaust manifold bolt torque sequence–3.0L engine

13. Torque the cylinder head bolts in the proper sequence as follows:
    a. Tighten the bolts to 30 ft. lbs. (40 Nm).
    b. Turn the bolts an additional 90 degrees.
    c. Loosen 1 full turn.
    d. Tighten the bolts to 30 ft. lbs. (40 Nm).
    e. Turn the bolts 90 degrees.
    f. Turn the bolts an additional 90 degrees.
14. Install or connect the following:
    - Position the water pump and install the bolts
    - Tighten the bolts to 89 inch. lbs. (10 Nm).
    - Install the oil level indicator tube and stud bolt.
    - Install a new O-ring seal and lubricate with clean engine oil.
    - Hydraulic lash adjusters
    - Camshaft followers
    - Camshaft
    - Exhaust manifold. Torque the bolts in sequence to 15 ft. lbs. (20 Nm), right side only
    - EGR tube, right side only
    - Coolant bypass tube
    - Negative battery cable
15. Fill the coolant to the proper level.
16. Start the vehicle and check for leaks, repair if necessary.

ENGINE ASSEMBLY

REMOVAL & INSTALLATION

2.3L Engine

See Figures 89 through 116.

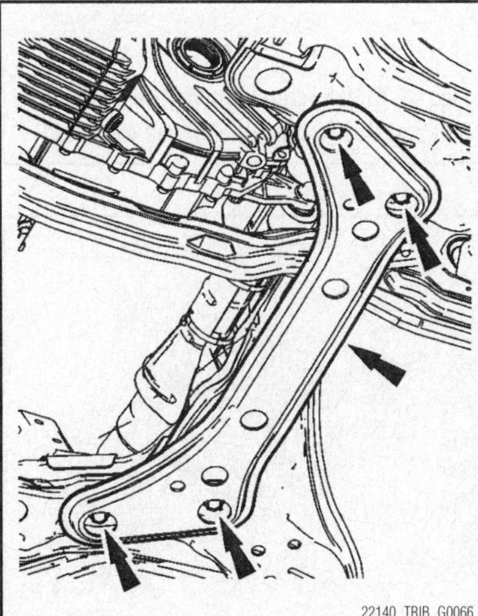

Fig. 89 Lateral support crossmember

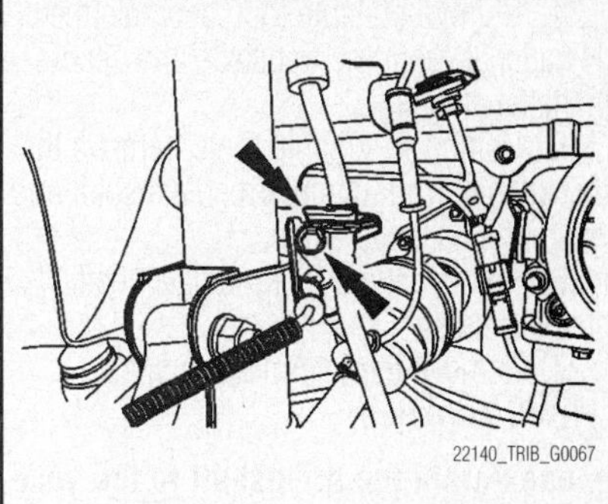

Fig. 90 Brake hose retainer and the ABS sensor retaining bolt on the RH strut.

1. With the vehicle in **NEUTRAL**, position it on a hoist.
2. Release the fuel system pressure.
3. Remove the engine air cleaner and air cleaner outlet pipe.
4. Remove the battery tray.
5. Drain the engine oil.
6. Remove the front wheels and tires.
7. Drain the cooling system.
8. Remove the exhaust flexible pipe.
9. Disconnect the heated oxygen sensor (HO2S) and the catalyst monitor sensor electrical connectors.
10. Remove the accessory drive belt and tensioner.
11. Press the 2 locking tabs (1 shown) to release the lower air duct from the upper air duct
12. Detach the wire retainers from the alternator shield and remove the nut and pin-type retainer, remove shield.
13. Remove the 4 bolts and the lateral support crossmember.
14. Remove the brake hose retainer and the ABS sensor retaining bolt from the LH strut.
15. Disconnect the LH stabilizer bar link.
16. Remove the LH tie rod end retaining nut.
17. Remove the LH lower control arm knuckle bolt.
18. Remove the brake hose retainer and the ABS sensor retaining bolt from the RH strut
19. Disconnect the RH stabilizer bar link.
20. Remove the RH tie rod end retaining nut.
21. Remove the RH lower control arm knuckle bolt.
22. Using Removal tool, disconnect the LH and RH tie rod end from the steering knuckle.
23. Separate the LH and RH lower control arms and disconnect the steering knuckle from the lower ball joint and position the steering knuckle aside.

24. Using a suitable tool, separate the LH halfshaft from the transaxle and secure the halfshaft aside.

25. Using the suitable tools, remove the RH halfshaft from the intermediate shaft and secure the halfshaft aside.

26. Remove the 2 intermediate shaft retaining nuts.

27. Remove the intermediate shaft

AWD vehicles:

➡Index-mark the driveshaft to the yoke for installation.

28. Remove the 6 bolts holding the driveshaft to the PTU and position aside with mechanic's wire.

All vehicles:

29. Remove the Power Distribution Box (PDB) cover.

30. Remove the nut and the cable from the PDB.

31. Disconnect the electrical connector from the PDB.

32. Remove the bolt and ground strap.

33. Disconnect the 34-pin electrical connector.

34. Detach the wiring harness retainers from the battery tray bracket and position the wiring harness aside

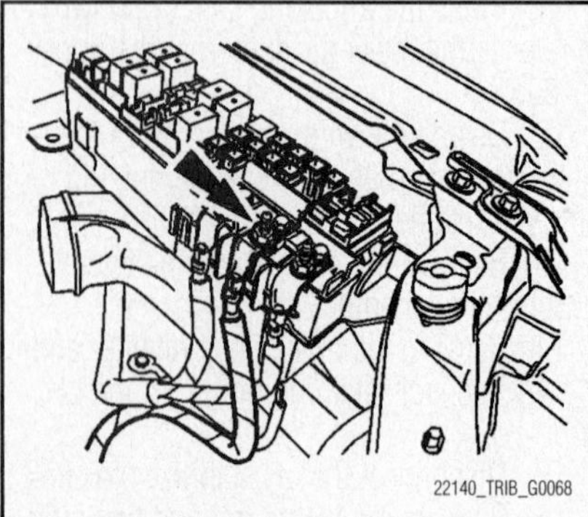

Fig. 91 Cable on the Power Distribution Box (PDB)

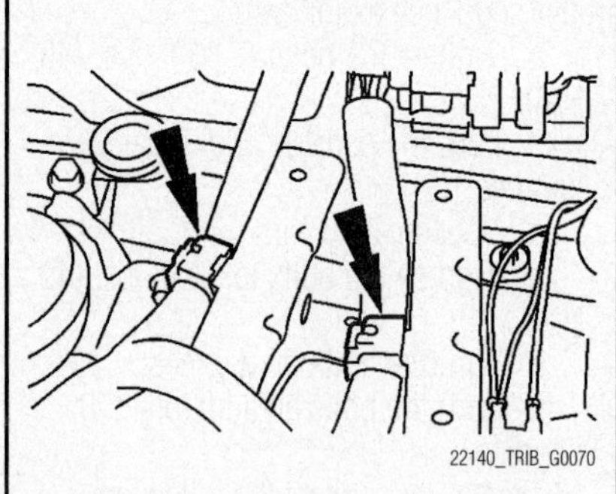

Fig. 93 Wiring harness retainers to the battery tray bracket

Manual transaxle vehicles:

35. Position the clutch hydraulic line aside.

36. Remove the clutch hydraulic line bracket-to-transaxle bolt.

37. Disconnect the clutch hydraulic line from the clutch slave cylinder.

38. Plug the hydraulic line.

39. Position the clutch hydraulic line aside.

40. Remove the 3 shift cable bracket bolts.

41. Position the bracket aside.

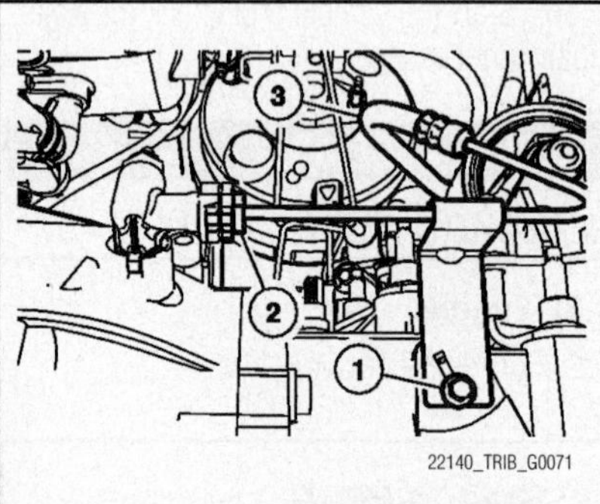

Fig. 94 Clutch, clutch bracket-to-transaxle bolt and line to the clutch slave cylinder.

Automatic transaxle vehicles:

42. Detach the transaxle vent tube retaining clip from the engine wiring harness.

43. Remove the nut holding the wiring harness bracket and unplug the transaxle electrical connector.

44. Disconnect the shift cable from the transaxle manual lever.

45. Detach the wiring harness pin-type retainer and remove the 2 bolts.

46. Position the transaxle cable and bracket aside.

47. Disconnect the transaxle range sensor electrical connector.

48. Disconnect the front transaxle fluid cooler tube.

49. Disconnect the rear transaxle fluid cooler tube.

50. Disconnect the transaxle fluid cooler tube.

a. Disconnect the Output Shaft Speed (OSS) sensor electrical connector (black).

b. Disconnect the Turbine Shaft Speed (TSS) sensor electrical connector (white).

c. Disconnect the wiring harness retainer from the transaxle case and position the harness aside.

d. Remove the transaxle fluid cooler retaining bracket bolt.

e. Position the fluid cooler tube aside.

51. Remove the bolt and the OSS sensor.

All vehicles:

52. Remove the wiring harness bracket nut and position aside

53. Disconnect the vehicle speed sensor (VSS) electrical connector and pin-type retainers

54. Disconnect the reversing lamp indicator switch.

55. Disconnect the shift cables.

56. If equipped, disconnect the block heater electrical connector.

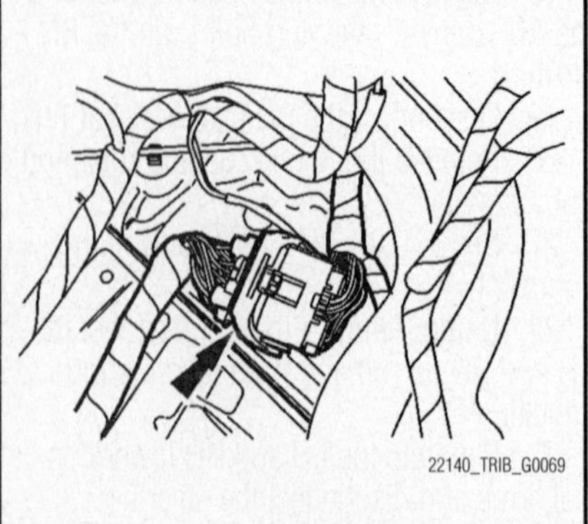

Fig. 92 34-pin electrical connector.

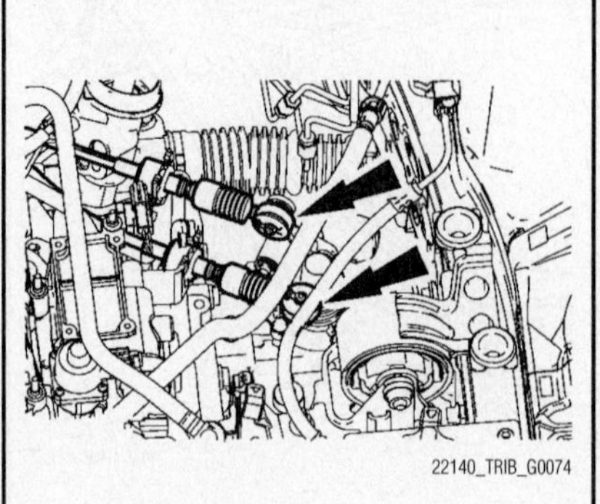

Fig. 95 Transmission shift cables.

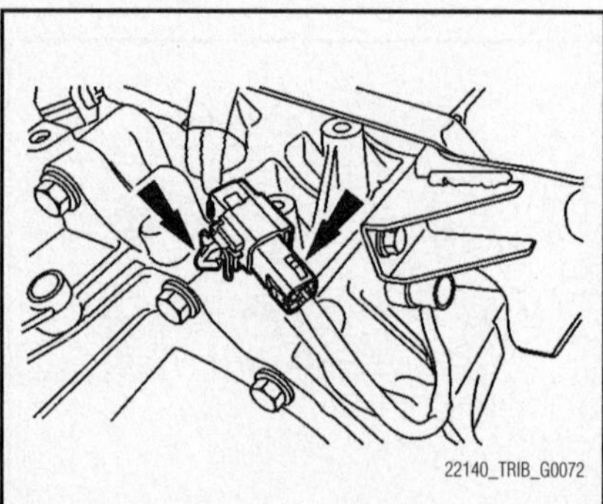

Fig. 96 Vehicle speed sensor (VSS) electrical connector

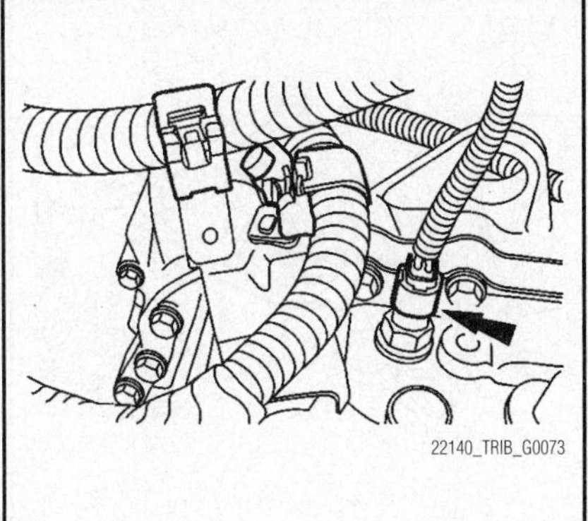
Fig. 97 Reverse lamp indicator switch

57. Detach all the block heater wiring harness retainers and position the wiring harness aside.
58. Disconnect the upper radiator hose.
59. Detach the heater hose support strap from the stud.
60. Disconnect the heater hoses from the heater core.
61. Remove the retainers and the accelerator cable snow shield
62. Disconnect the accelerator cable and speed control cable (if equipped).
63. Disconnect the accelerator and speed control cable (if equipped) from the throttle body.
64. Remove the bolts from the accelerator cable bracket.
65. Remove the nut from the accelerator control cable bracket.
66. Position the accelerator control cable and bracket assembly aside.

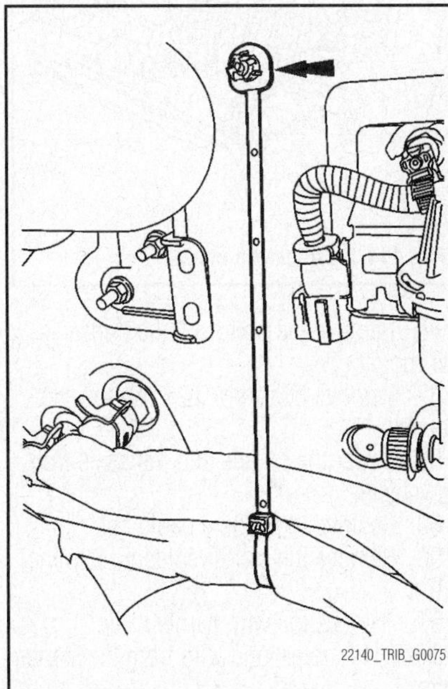
Fig. 98 Heater hose support strap and the stud

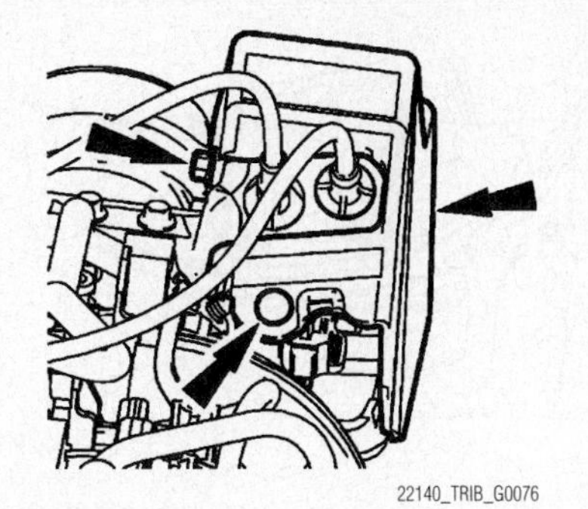
Fig. 99 The retainers and the accelerator cable snow shield

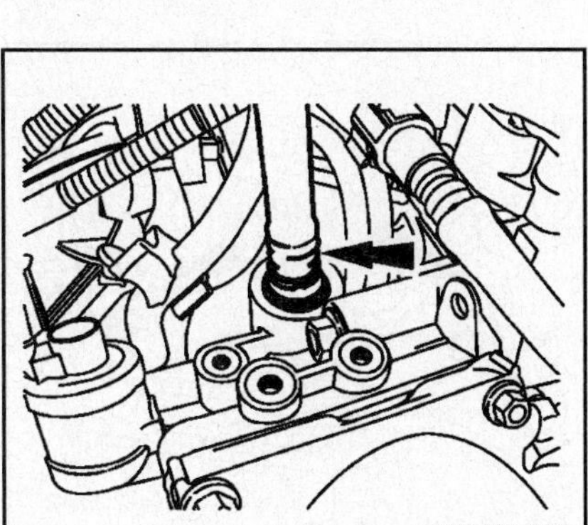
Fig. 100 The vacuum supply tube

Fig. 101 The fuel vapor tube

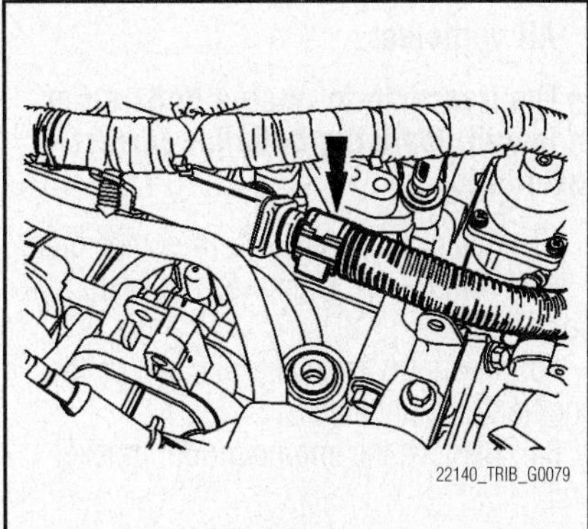
Fig. 102 The fuel supply tube

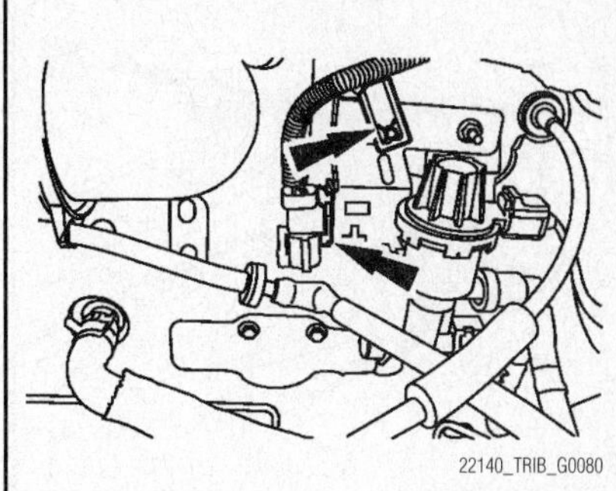
Fig. 103 The electrical connector retainers.

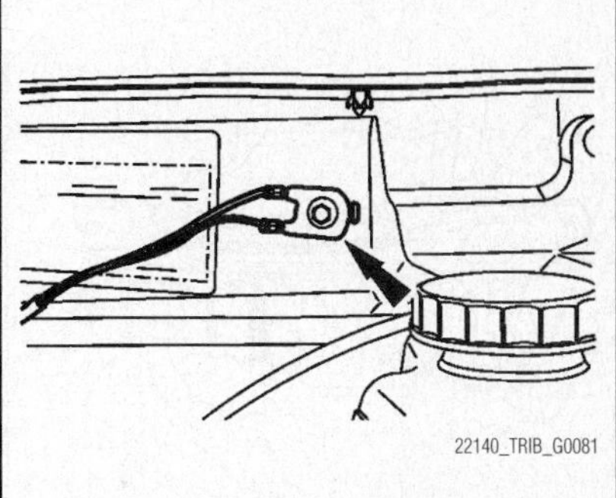
Fig. 104 The ground wire at the rear of the engine compartment

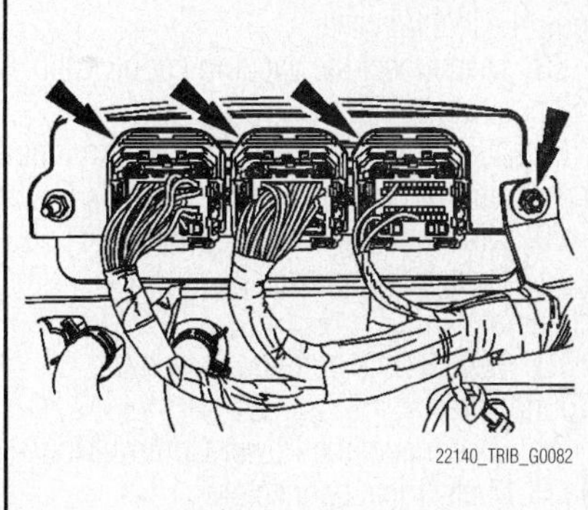
Fig. 105 PCM electrical connectors

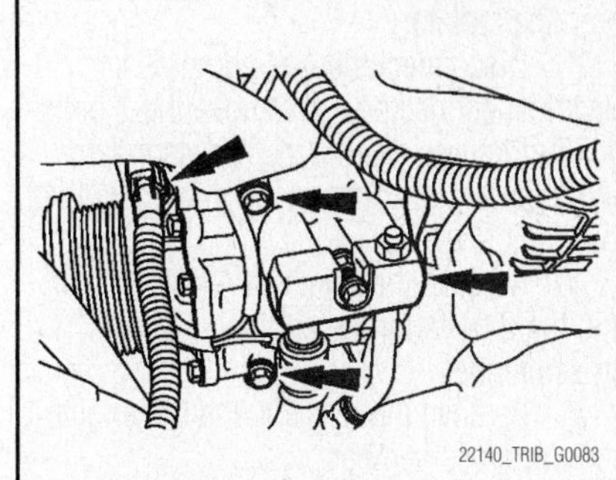
Fig. 106 A/C compressor electrical connector and the 3 bolts

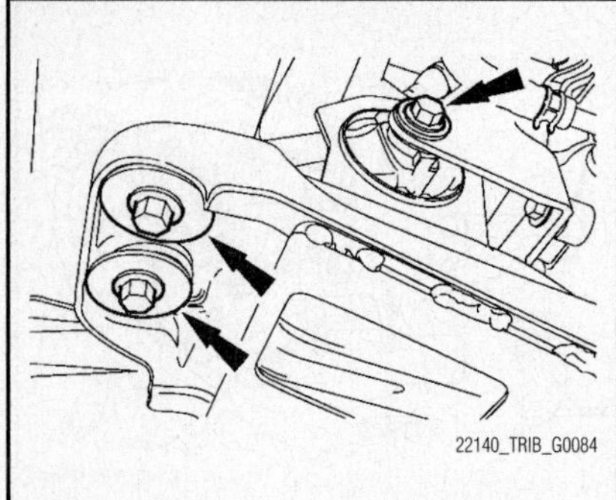

Fig. 107 Front roll restrictor bolt and the 2 bolts for the engine support crossmember

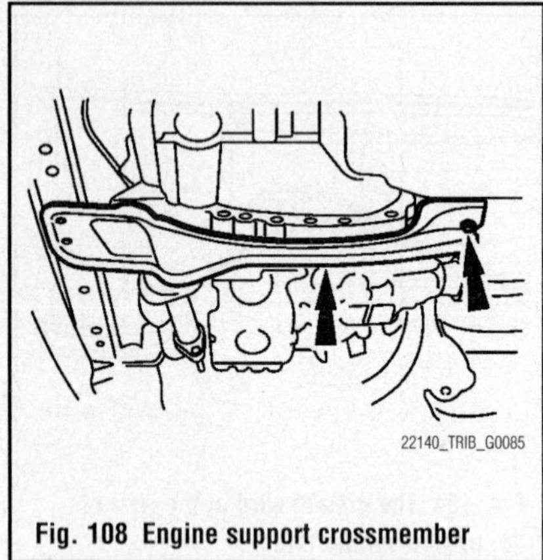

Fig. 108 Engine support crossmember

67. Remove the nut from the accelerator control cable bracket.
68. Disconnect the vacuum supply tube and position aside.
69. Disconnect the fuel vapor return tube and position aside.
70. Disconnect the fuel supply tube.
71. Detach the electrical connector retainers.
72. Remove the bolt and detach the ground wire.
73. Disconnect the Power Control Module PCM electrical connectors.
74. Remove the wiring harness retainer nut.
75. Disconnect the lower radiator hose from the radiator.
76. Disconnect the A/C compressor electrical connector and remove the 3 bolts.
77. Position the A/C compressor aside and support the compressor with a length of mechanic's wire.
78. Remove the front roll restrictor bolt and the 2 bolts for the engine support crossmember.
79. Remove the rear nut and the engine support crossmember.

Front wheel drive (FWD) vehicles:

80. Remove the 3 bolts and the damper.

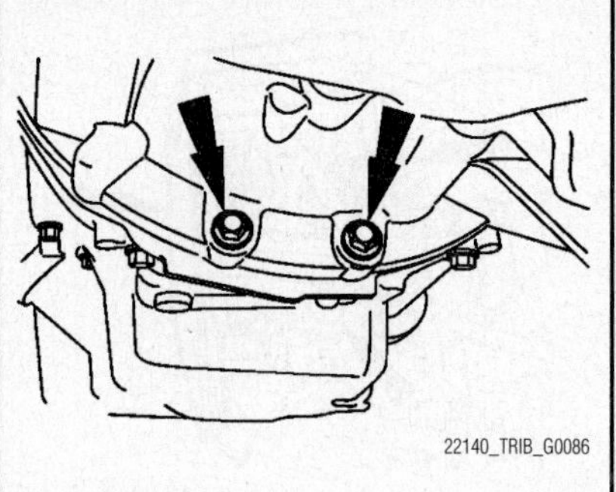

Fig. 109 Location of the 2 transaxle-to-engine bolts

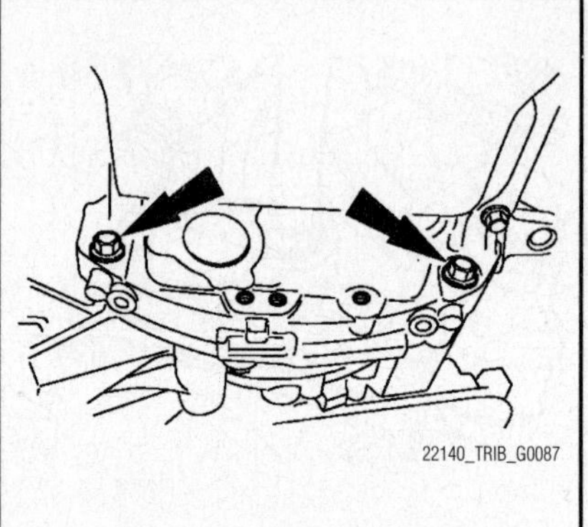

Fig. 110 2 engine-to-transaxle bolts

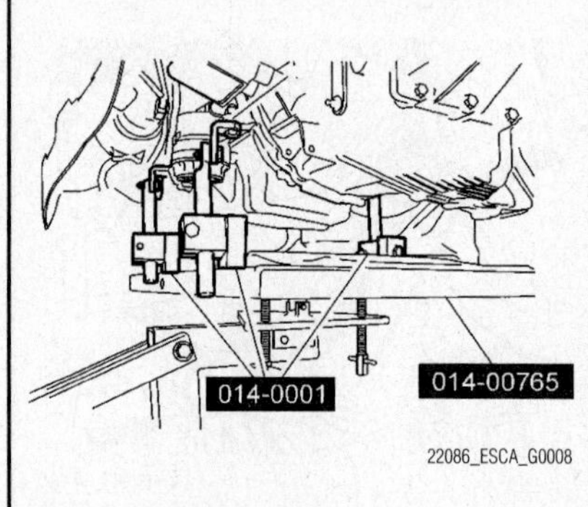

Fig. 111 Engine to lift table view

All vehicles:

➡The transaxle-to-engine bolts differ in length. Mark the bolts for correct installation.

81. Remove the transaxle-to-engine bolts.
82. Remove the engine-to-transaxle bolts
83. Using the special tools, secure the engine to the lift table
84. Remove the engine mount bracket bolt.
85. Remove the nuts and the engine mount bracket.

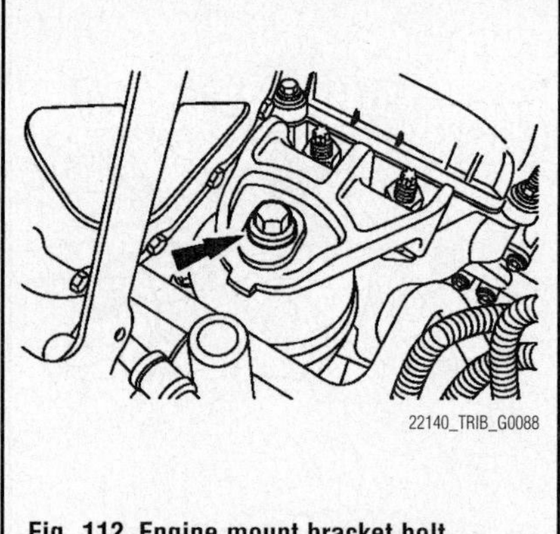

Fig. 112 Engine mount bracket bolt

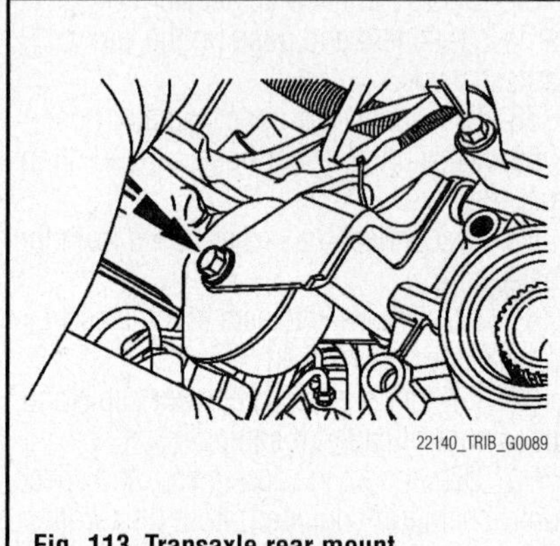

Fig. 113 Transaxle rear mount

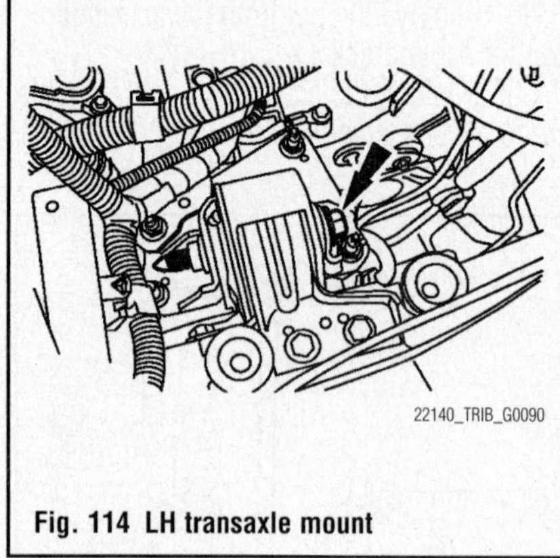

Fig. 114 LH transaxle mount

86. Remove the bolt from the transaxle rear mount.
87. Remove bolt from the LH transaxle mount.
88. Lower the engine and transaxle from the vehicle.
89. Remove the battery cable nut.
90. Remove the starter solenoid terminal nut.
91. Remove the wire harness clip retainer and the ground wire from the starter bolts.
92. Remove the 2 stud bolts and remove the starter.

AWD vehicles:

93. Remove the 2 lower catalytic converter bolts.
94. Remove the 6 bolts and the catalytic converter heat shield.
95. Remove and discard the 7 exhaust manifold nuts.
96. Remove the catalytic converter and discard the exhaust manifold gasket.
97. Remove and discard the 7 exhaust manifold studs.
98. Remove the 3 PTU bracket-to-engine bolts.
99. Remove the 2 PTU bracket-to-PTU bolts and remove the bracket.
100. Detach the PTU vent hose retainer.
101. Remove the transaxle-to-PTU bolt.
102. Remove the 3 PTU-to-transaxle bolts and the PTU.
103. Remove and discard the 4 torque converter nuts.

All vehicles:

104. Using the engine crane and spreader bar, remove the engine and transaxle from the lift table.

➡**The transaxle-to-engine bolts differ in length. Mark the bolts for correct installation.**

105. Remove the remaining 5 engine-to-transaxle bolts and separate the engine and transaxle.

To install:

106. Using the engine crane and spreader bar, position the engine and transaxle together. Install the 5 transaxle-to-engine bolts.
107. Using the engine crane and spreader bar, position the engine and transaxle onto the lift table.
108. Using the special tools, secure the engine to the lift table.
109. Install new torque converter nuts and tighten to 26 ft. lbs. (35 Nm).—Automatic Transaxles.

AWD vehicles:

110. Install the Power Transfer Unit (PTU) and the 3 PTU-to-transaxle bolts. Tighten bolts to 52 ft. lbs. (70 Nm).
111. Install the 1 transaxle-to-PTU bolt and tighten to 35 ft. lbs. (48 Nm).
112. Attach the PTU vent hose.
113. Install the PTU bracket and the 2 bolts and tighten to 33 ft. lbs. (45 Nm).
114. Install the 3 PTU bracket-to-engine bolts and tighten to 30 ft. lbs. (40 Nm).
115. Install 7 new exhaust manifold studs in the cylinder head and tighten to 13 ft. lbs. (17 Nm).

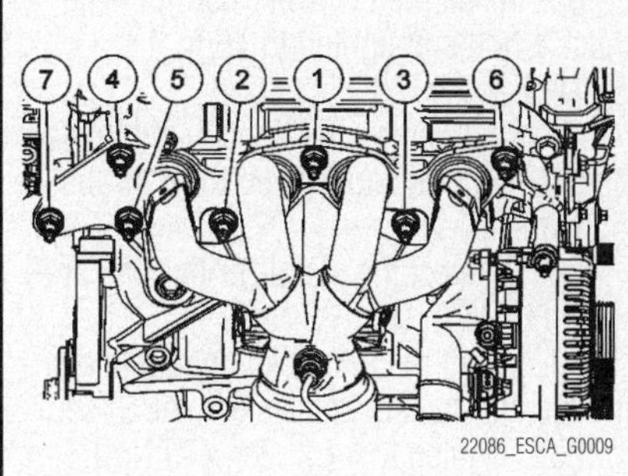

Fig. 115 Manifold tightening sequence—2008 2.3L engine

116. Position the catalytic converter and tighten the 7 new exhaust manifold nuts in 2 stages:
    a. Snug all nuts down evenly.
    b. Tighten nuts to 35 ft. lbs. (47 Nm).
117. Install the heat shield and the 6 bolts and tighten to 89 inch lbs. (10 Nm).
118. Install the 2 lower catalytic converter bolts and tighten to 18 ft. lbs. (25 Nm).

All vehicles:

119. Install the starter motor and tighten bolts to 26 ft. lbs. (35 Nm).
120. Install the starter motor harness connector.
121. Install the starter motor solenoid battery nut and tighten to 9 ft. lbs. (12 Nm).
122. Install the starter motor solenoid nut and tighten to 44 inch lbs. (5 Nm).
123. Install the wire harness clip retainer and the ground wire and nut to the starter bolts tighten to 18 ft. lbs. (25 Nm).
124. Raise the engine and transaxle into the vehicle.
125. Install the bolt in the LH transaxle mount and tighten to 76 ft. lbs. (23 Nm).
126. Install the bolt in the rear transaxle mount and tighten to 85 ft. lbs. (115 Nm).
127. Install the engine mount bracket bolt and tighten to 85 ft. lbs. (115 Nm).
128. Install the 2 engine-to-transaxle bolts and tighten to 35 ft. lbs. (48 Nm).
129. Install the 2 transaxle-to-engine bolts and tighten to 35 ft. lbs. (48 Nm).

FWD vehicles:

130. Install the damper and the 3 bolts. Tighten to 30 ft. lbs. (40 Nm).

All vehicles:

131. Install the engine support crossmember and new nut, tighten to 129 ft. lbs. (175 Nm).
132. Install the 2 bolts for the engine support crossmember and the front roll restrictor bolt.
133. Tighten the engine support crossmember bolts to 66 ft. lbs. (90 Nm).
134. Tighten the front roll restrictor bolt to 85 ft. lbs. (115 Nm).
135. Position the A/C compressor and install the bolts.
136. Tighten the bolts to 18 ft. lbs. (25 Nm).
137. Connect the A/C compressor electrical connector.
138. Connect the lower radiator hose to the radiator.
139. Connect the PCM electrical connectors.
140. Position the harness, install the nut and tighten to 53 inch lbs. (6 Nm).
141. Install the ground wire and bolt, tighten to 89 inch lbs. (10 Nm).
142. Attach the electrical connector retainers.
143. Connect the fuel supply tube quick connect coupling.
144. Connect the fuel vapor return tube and retainer.
145. Connect the vacuum supply tube.
146. Position the accelerator control cable and bracket and install the nut.
147. Tighten the bracket nut to 53 inch lbs. (6 Nm).
148. Install the accelerator cable and speed control cable (if equipped).
149. Connect the accelerator and speed control cable (if equipped) to the throttle body.
150. Install the accelerator cable bracket and bolts. Tighten to 89 inch lbs. (10 Nm).
151. Install the accelerator cable snow shield and retainers. Tighten to 35 inch lbs. (4 Nm).
152. Connect the heater hoses to the heater core.
153. Attach the heater hose support strap to the stud.
154. Connect the upper radiator hose.
155. If equipped, route the block heater wiring harness and attach all retainers.
156. Connect the block heater electrical connector.

Vehicles with automatic transaxles

157. Install the Output Shaft Speed (OSS) sensor tighten mounting bolt to 10 ft. lbs. (13 Nm).
158. Install the fluid cooler tube.
    a. Connect the OSS sensor.
    b. Connect the turbine shaft speed (TSS) sensor (white connector).
    c. Connect the wiring harness retainer to the transaxle case.
    d. Position the bracket and install the bolt. Tighten to 10 ft. lbs. (13 Nm).
    e. Install the fluid cooler tube fitting and tighten to 17 ft. lbs. (23 Nm).
159. Connect the rear transaxle fluid cooler tube and tighten to 17 ft. lbs. (23 Nm).

160. Connect the front transaxle fluid cooler tube and tighten to 17 ft. lbs. (23 Nm).

161. Connect the transaxle range sensor electrical connector.

162. Install the transaxle control cable, bracket and tighten bolts to 14 ft. lbs. (19 Nm).

163. Attach the wiring harness pin-type retainers.

164. Connect the shift cable to the transaxle manual lever.

165. Install the transaxle wiring harness bracket and nut and connect the transaxle electrical connector. Tighten bracket nut to 89 inch lbs. (10 Nm).

166. Attach the transaxle vent tube retaining clip to the engine wiring harness.

All vehicles

167. Install the shift cable bracket and tighten mounting bolts to 16 ft. lbs. (22 Nm).

168. Connect the shift cables.

169. Connect the reversing lamp indicator switch.

170. Connect the vehicle speed sensor (VSS) electrical connector and pin-type retainer.

171. Install the wiring harness bracket nut and tighten to 9 ft. lbs. (12 Nm).

Vehicles with manual transaxles

172. Connect the clutch hydraulic line to the clutch slave cylinder and tighten to 18 ft. lbs. (25 Nm).

173. Install the clutch hydraulic line bracket-to-transaxle bolt and tighten to 27 inch lbs. (3 Nm).

All vehicles

174. Attach the wiring harness retainers to the battery tray bracket.

175. Connect the 34-pin electrical connector.

176. Install the ground strap and bolt. Tighten to 89 inch lbs. (10 Nm).

177. Connect the electrical connector to the power distribution box (PDB).

178. Connect the cable to the PDB and install and tighten the nut to 9 ft. lbs. (12 Nm).

179. Install the PDB cover.

AWD vehicles:

180. Align index mark and position the driveshaft to the PTU and install the 6 bolts.

181. Tighten driveshaft mounting bolts to 27 ft. lbs. (37 Nm).

All vehicles:

182. Install the intermediate shaft.

183. Install the intermediate shaft retaining nuts and tighten to 20 ft. lbs. (27 Nm).

184. Install the LH half shaft in the transaxle and the RH half shaft to the intermediate shaft.

185. Install the ball joints in the knuckles.

186. Install the LH lower ball joint-to-knuckle bolt and tighten to 46 ft. lbs. (63 Nm).

187. Position the RH tie-rod end, install the retaining nut and tighten to 41 ft. lbs. (55 Nm).

188. Connect the stabilizer bar link and tighten to 46 ft. lbs. (63 Nm).

189. Install the RH brake hose retainer and the anti-lock brake system (ABS) sensor bolt. Tighten to 11 ft. lbs. (15 Nm).

190. Install the RH half shaft in the transaxle and the RH half shaft to the intermediate shaft.

191. Install the ball joints in the knuckles.

192. Install the RH lower ball joint-to-knuckle bolt and tighten to 46 ft. lbs. (63 Nm).

193. Position the RH tie-rod end, install the retaining nut and tighten to 41 ft. lbs. (55 Nm).

194. Connect the stabilizer bar link and tighten to 46 ft. lbs. (63 Nm).

195. Install the LH brake hose retainer and the ABS sensor bolt. Tighten to 11 ft. lbs. (15 Nm).

196. Install the lateral support crossmember and bolts. Tighten bolts to 85 ft. lbs. (115 Nm).

197. Install the alternator shield, the nut and the pin-type retainer.

198. Tighten to15 ft. lbs. (20 Nm).

199. Attach the wire retainers.

200. Install the lower air duct.

201. Install the accessory drive belt tensioner

202. Connect the heated oxygen sensor (HO2S) and the catalyst monitor sensor electrical connectors.

203. Install the exhaust flexible pipe.

204. Install the front wheels and tires.

205. Install the battery tray.

206. Install the engine air cleaner and air cleaner outlet pipe.

207. Fill the engine with clean engine oil.

208. Fill and bleed the cooling system.

209. Bleed the clutch system.

3.0L Engine

See Figures 116 through 154.

1. Disconnect the negative battery cable.
2. With the vehicle in **NEUTRAL**, position it on a hoist.
3. Release the fuel system pressure.
4. Remove the battery tray.
5. Remove the air cleaner outlet pipe and air cleaner.
6. Remove the front wheels and tires.
7. Remove the lower splash shields.

Fig. 116 Catalyst monitor sensor electrical connector and the 2 wiring harness retainers

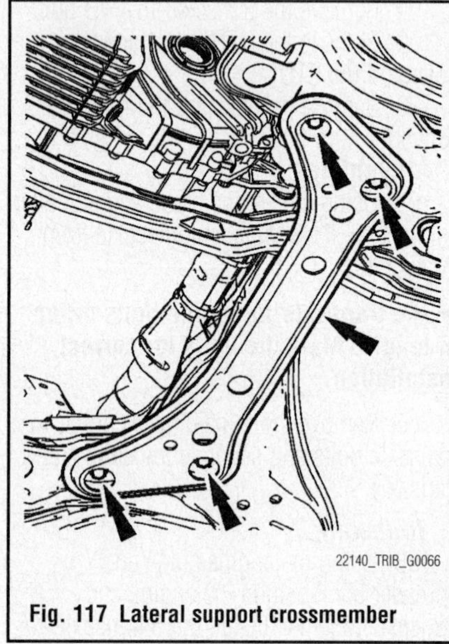

Fig. 117 Lateral support crossmember

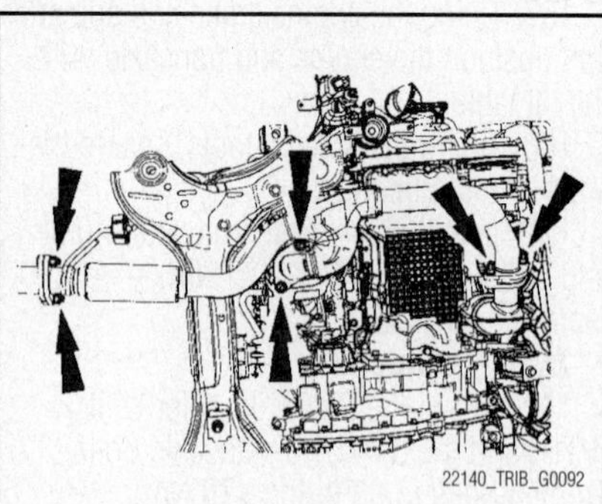

Fig. 118 6 exhaust Y-pipe-to-catalytic converter nuts

8. Disconnect the catalyst monitor sensor electrical connector and the 2 wiring harness retainers.
9. Remove the 4 bolts and the lateral support crossmember.
10. Remove and discard the 6 exhaust Y-pipe-to-catalytic converter nuts.
11. Detach the exhaust hanger and remove the exhaust Y-pipe.

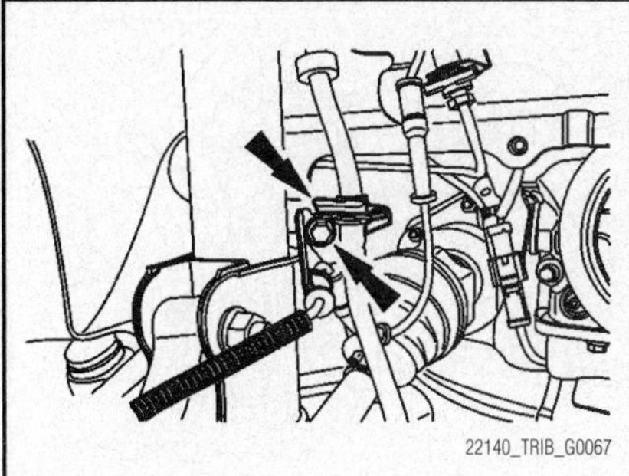

Fig. 119 Brake hose retainer and the ABS sensor retaining bolt on the RH strut

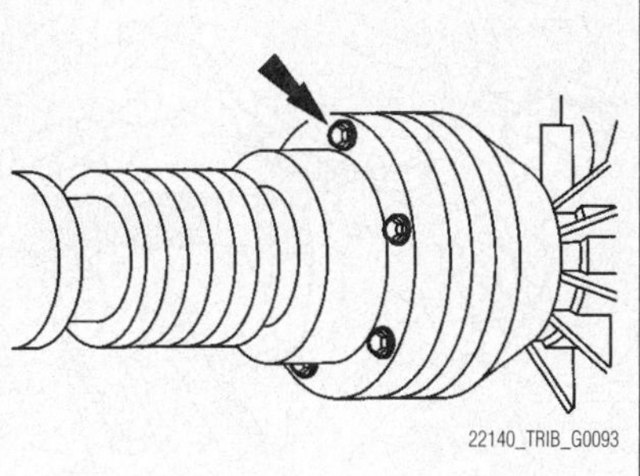

Fig. 120 6 front driveshaft-to-PTU bolts and washers

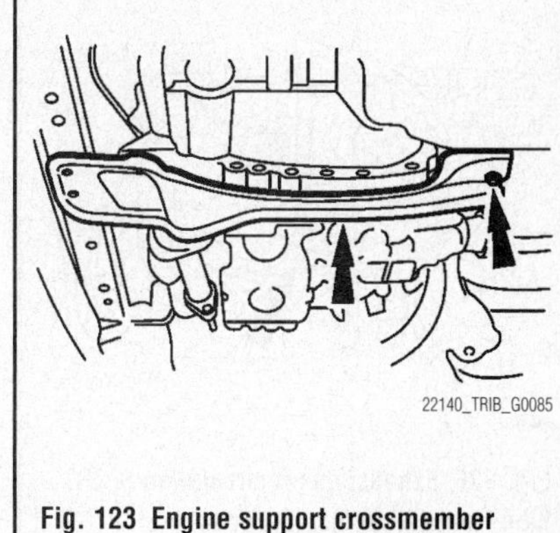

Fig. 123 Engine support crossmember

12. Remove the LH and RH brake hose retainers and ABS sensor wiring harness bolts.
13. Remove the LH and RH stabilizer link-to-strut nuts.
14. Remove the LH and RH tie-rod end nuts.
15. Using the suitable tool, separate the LH and RH tie-rod ends from the steering knuckles.
16. Remove the LH and RH lower ball joint pinch bolts and nuts.
17. Separate the steering knuckles from the lower ball joints.
18. Using a suitable tool, separate the LH halfshaft from the transaxle.
19. Support the halfshaft with a length of mechanic's wire.
20. Using the special tools, separate the RH halfshaft from the intermediate shaft.
21. Support the halfshaft with a length of mechanic's wire.
22. Remove the 2 intermediate shaft retaining nuts.
23. Remove the intermediate shaft.

All wheel drive (AWD) vehicles

WARNING

Do not reuse the driveshaft flange bolts and washers. Install new bolts and washers or damage to the vehicle may occur.

➡Index-mark the drive shaft flange and Power Transfer Unit (PTU) flange for installation.

24. Remove and discard the 6 front driveshaft-to-PTU bolts and washers.
25. Position the driveshaft aside and support with mechanic's wire.
26. Rotate the accessory drive belt tensioner counterclockwise and remove the accessory drive belt.
27. Drain the engine coolant.
28. Drain the engine oil and install the drain plug.

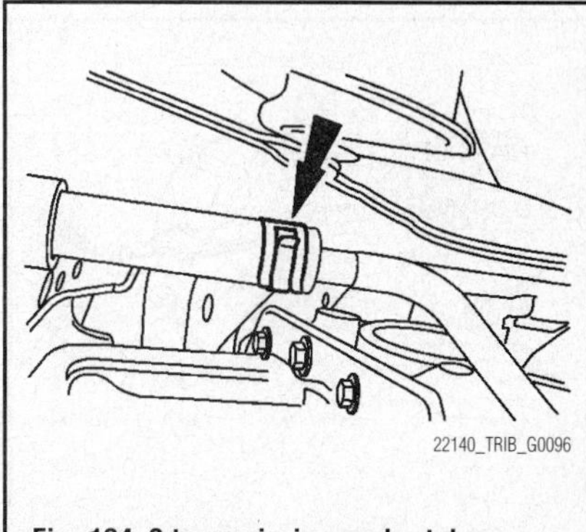

Fig. 124 2 transmission cooler tubes (1 shown)

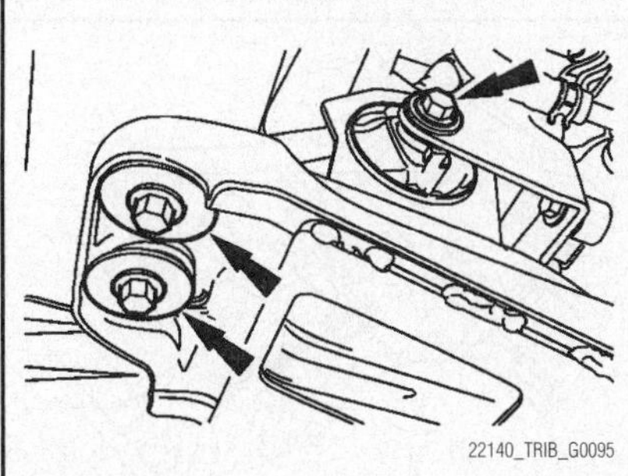

Fig. 121 A/C compressor bolt locations

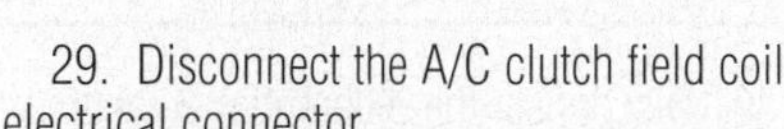

Fig. 122 Front roll restrictor bolt and the 2 bolts for the engine support cross brace

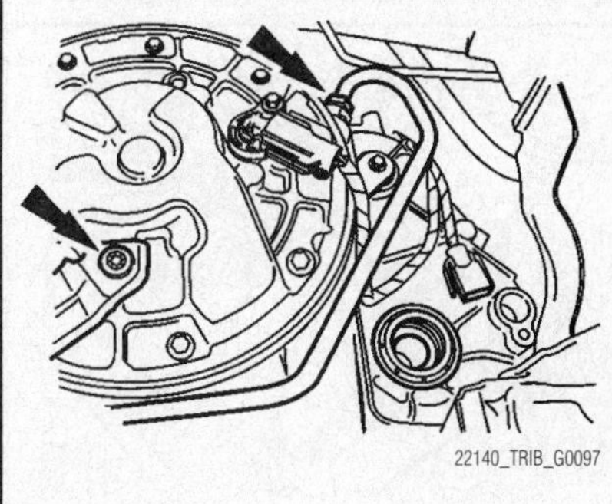

Fig. 125 Transaxle cooler tube, fitting bracket bolt and the transaxle cooler tube

29. Disconnect the A/C clutch field coil electrical connector.
30. Remove the 3 bolts and position the A/C compressor aside.
31. Remove the 3 accessory drive belt idler pulley assembly bolts.
32. Detach the wiring harness retainer and remove the idler pulley assembly.
33. Remove the front roll restrictor bolt and the 2 bolts for the engine support cross brace.
34. Remove the rear nut and the engine support crossmember.
35. Disconnect the 2 transmission cooler tubes.
36. Loosen the transaxle cooler tube fitting.
37. Remove the bracket bolt and the transaxle cooler tube.
38. Disconnect the EGR tube fitting from the converter.
39. Remove the 2 bolts and the accelerator cable snow shield.
40. Disconnect the accelerator cable and the speed control actuator cable.
41. Remove the 2 accelerator cable bracket bolts.
42. Remove the bolt and position the cables and brackets aside.

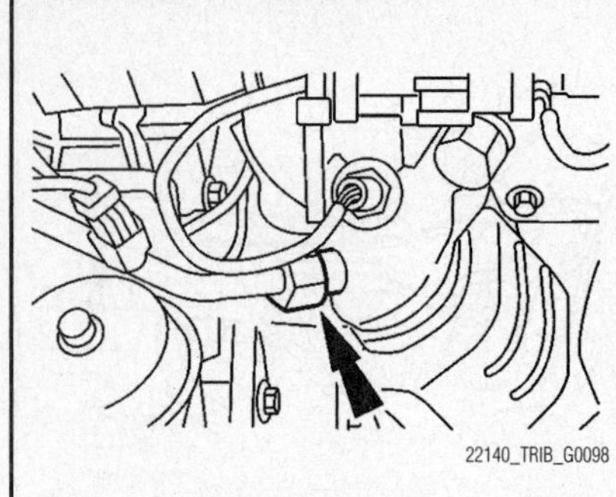

Fig. 126 Exhaust gas recirculation (EGR) tube fitting from the converter

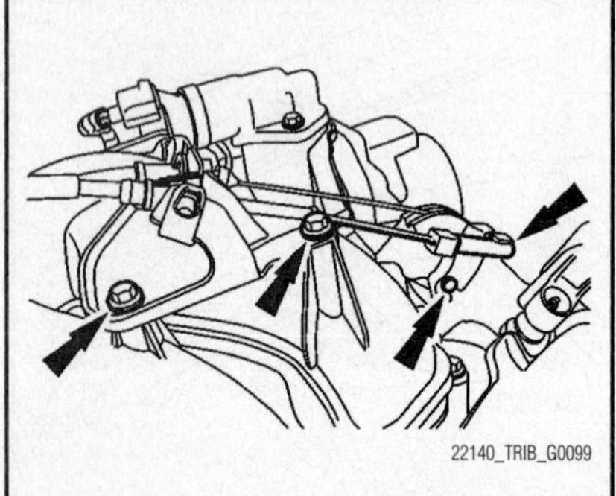

Fig. 127 Accelerator cable and the speed control actuator cable

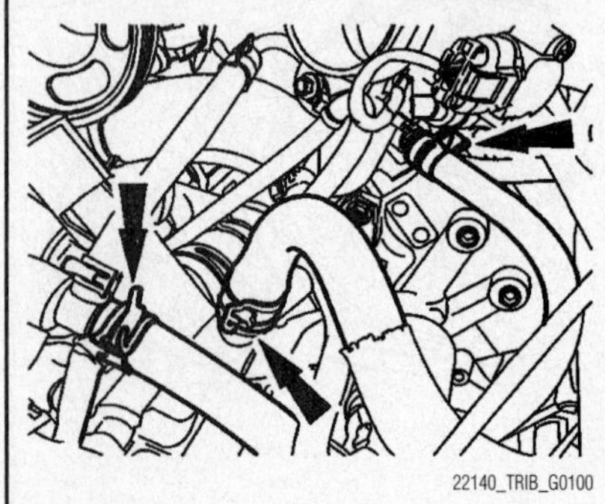
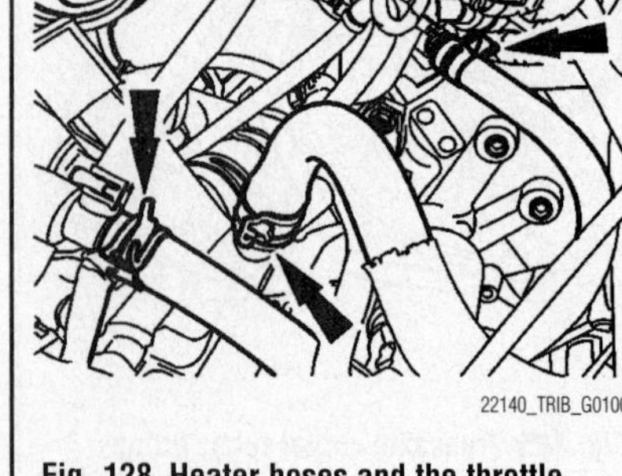

Fig. 128 Heater hoses and the throttle body coolant hose

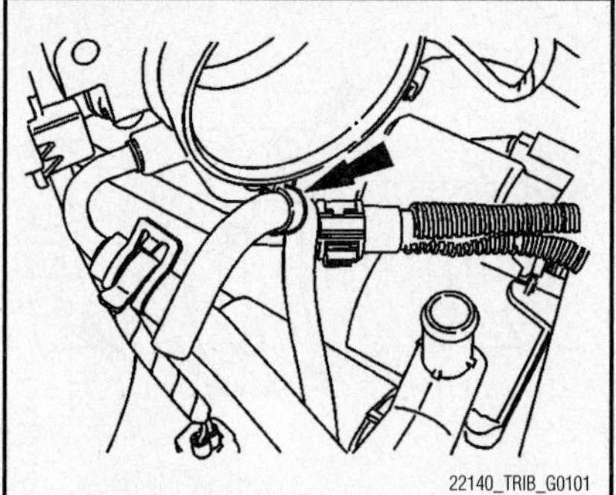

Fig. 129 Transaxle vent tube retainer at the throttle body

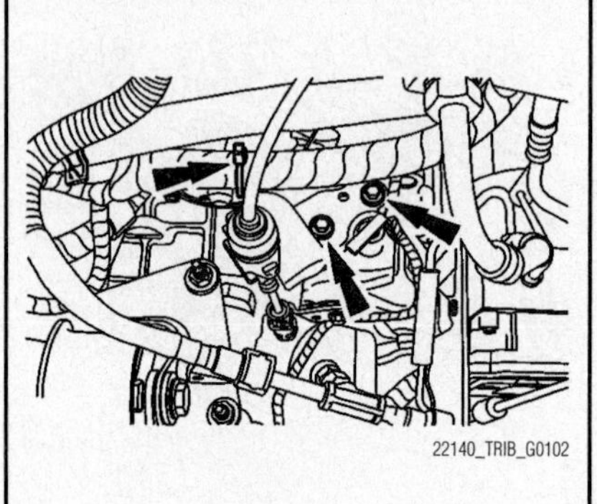

Fig. 130 Shift cable bracket

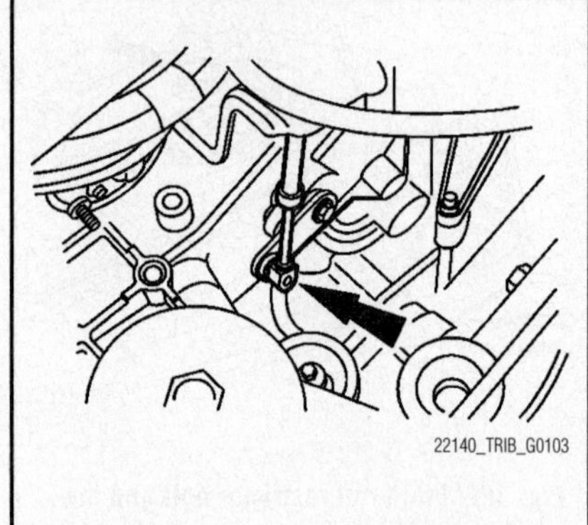

Fig. 131 Gearshift cable at the transaxle

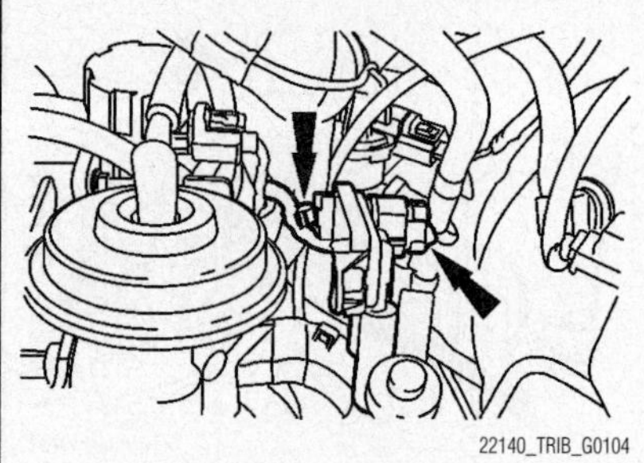

Fig. 132 Differential pressure feedback EGR sensor electrical connector and the pin-type retainer

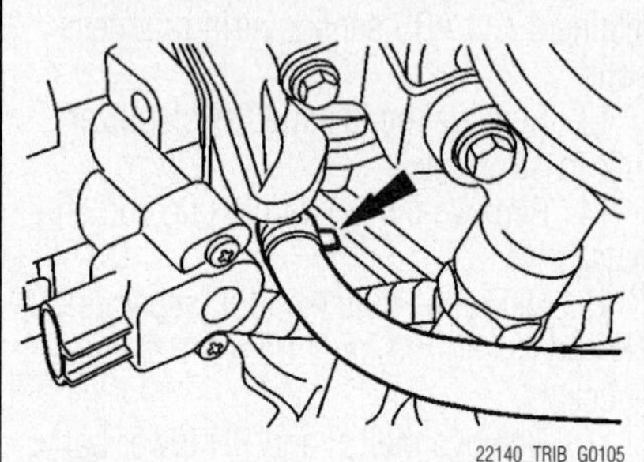

Fig. 133 Evaporative emissions (EVAP) canister purge valve tube at the intake manifold

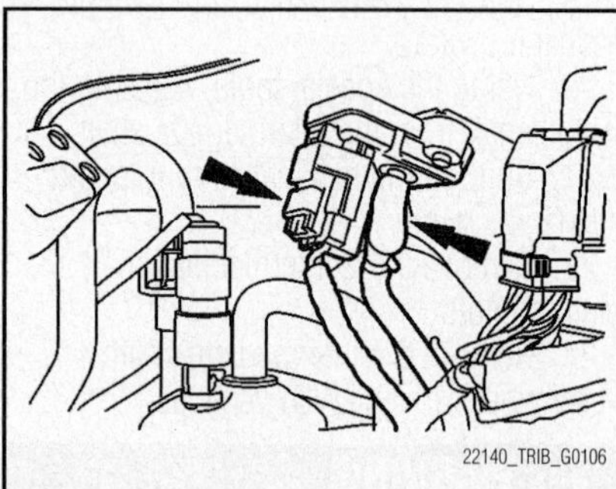

Fig. 134 Manifold absolute pressure (MAP) sensor vacuum tube and electrical connector

43. Disconnect the upper and lower radiator hoses.
44. Disconnect the heater hoses and the throttle body coolant hose.
45. Detach the transaxle vent tube retainer from the throttle body.
46. Disconnect the fuel supply tube quick connect coupling.
47. Disconnect the wire harness retainer from the shift cable bracket, remove the 2 remaining bolts, and position the cable and bracket aside.
48. Disconnect the gearshift cable from the transaxle
49. Disconnect the differential pressure feedback EGR sensor electrical connector and detach the pin-type retainer.
50. Disconnect the EGR tube fitting and remove the EGR tube and the differential pressure feedback EGR sensor as an assembly.
51. Disconnect the evaporative emissions (EVAP) canister purge valve tube from the intake manifold.
52. Disconnect the brake booster vacuum tube from the intake manifold.
53. Disconnect the manifold absolute pressure (MAP) sensor vacuum tube and electrical connector.
54. Disconnect the PCM electrical connectors and remove the nut.
55. Remove the bolt and detach the ground wire.
56. Remove the nut and disconnect the cable from the power distribution box.
57. Disconnect the electrical connector from the power distribution box.
58. Remove the bolt and disconnect the ground wire and the electrical connector.
59. Detach the wiring harness retainers from the battery tray bracket
60. Remove the nut and disconnect the wire from the battery cable.

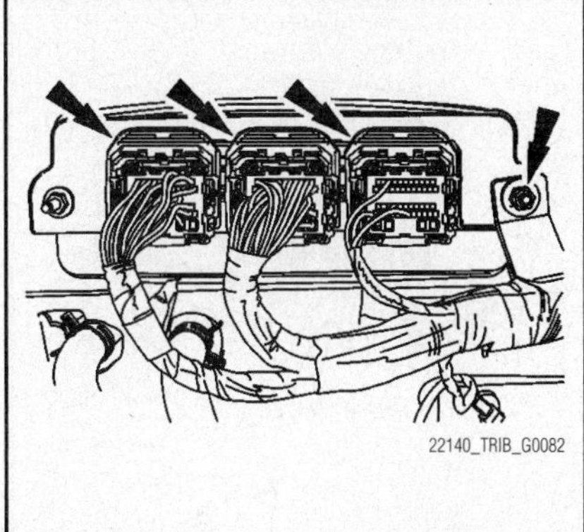

Fig. 135 PCM electrical connectors

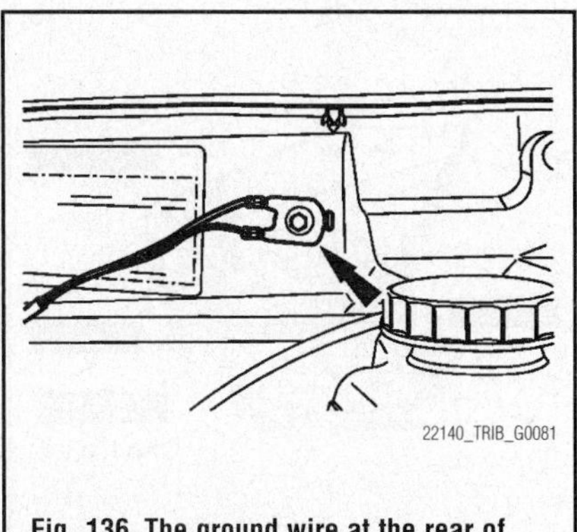

Fig. 136 The ground wire at the rear of the engine compartment

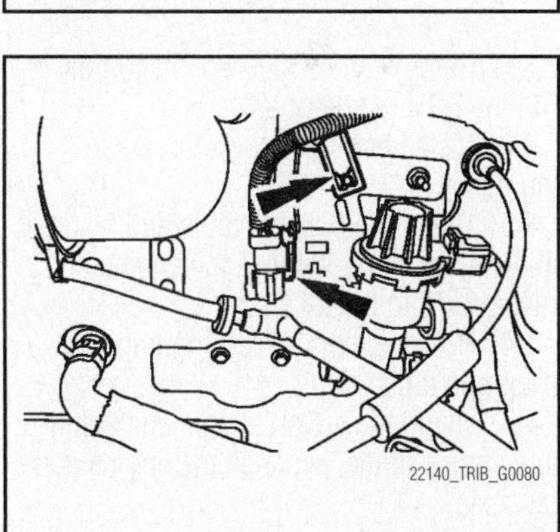

Fig. 137 The electrical connector retainers

61. Remove the nut and the ground wire from the engine mount stud.
62. If equipped, disconnect the engine block heater electrical connector.
63. Disconnect the Output Shaft Speed (OSS) sensor electrical connector and remove the bolt and the OSS sensor.
64. Remove the torque converter inspection cover.
65. Remove the 4 torque converter nuts.
66. Remove the 2 oil pan-to-transaxle bolts located next to the Torque converter access cover.
67. Remove the 2 nuts and the transaxle-to-engine stud.

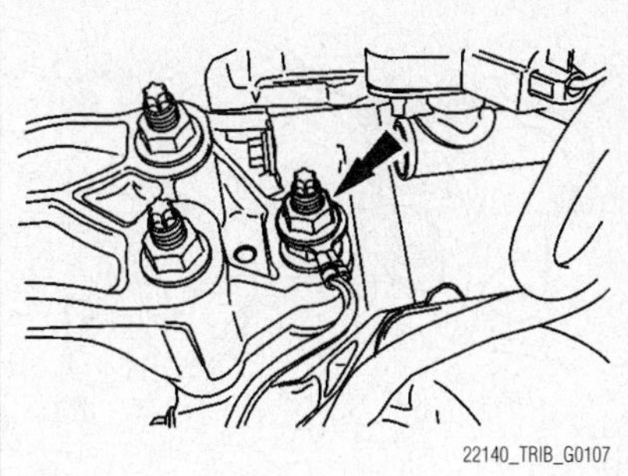

Fig. 138 Nut and the ground wire at the engine mount stud

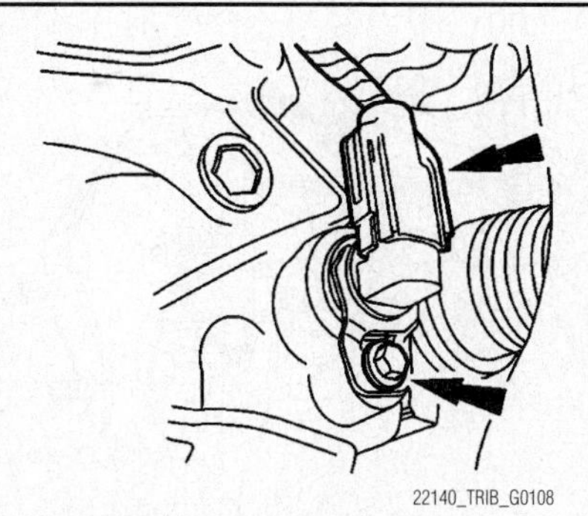

Fig. 139 Output Shaft Speed (OSS) sensor and electrical connector

Fig. 140 Torque converter access to the flexplate nuts

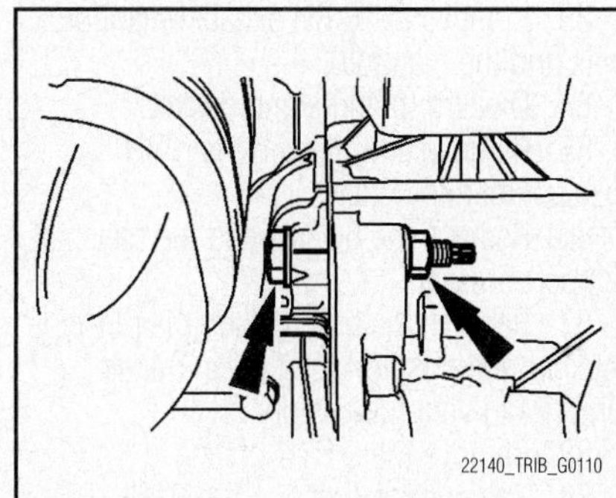

Fig. 141 2 nuts and transaxle-to-engine stud

✲✲ WARNING

Do not allow the engine oil pan to rest on the power train lift. Doing so may cause damage to the oil pan.

68. Using the suitable tools, secure the engine and transaxle to the powertrain lift.

➡The next 5 steps must be carried out with the vehicle raised and the powertrain lift in position.

69. Remove the RH transaxle support insulator bolt.

➡ The next 5 steps must be carried out with the vehicle raised and the powertrain lift in position.

70. Remove the bolt, nuts and the RH transaxle support insulator.
71. Remove the rear transaxle support bolt.
72. Remove the 3 engine support bracket nuts and the bolt.
73. Remove the engine support bracket.
74. Lower the powertrain from the vehicle.
75. Disconnect the LH Heated Exhaust Gas Oxygen (HEGO) sensor and LH catalyst monitor sensor electrical connectors.

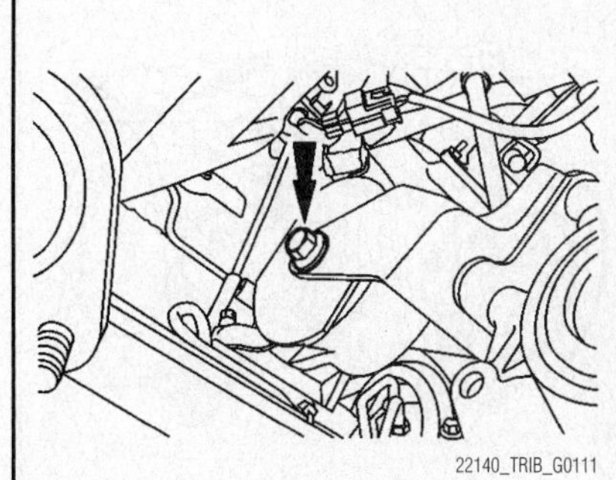

Fig. 142 RH transaxle support insulator bolt

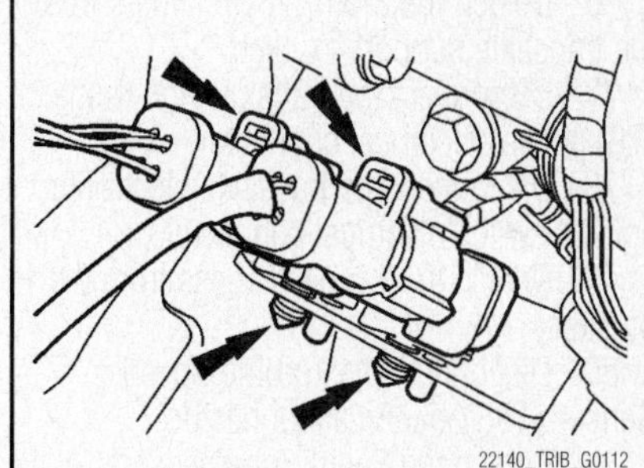

Fig. 143 LH heated exhaust gas oxygen (HEGO) sensor and LH catalyst monitor sensor electrical connectors

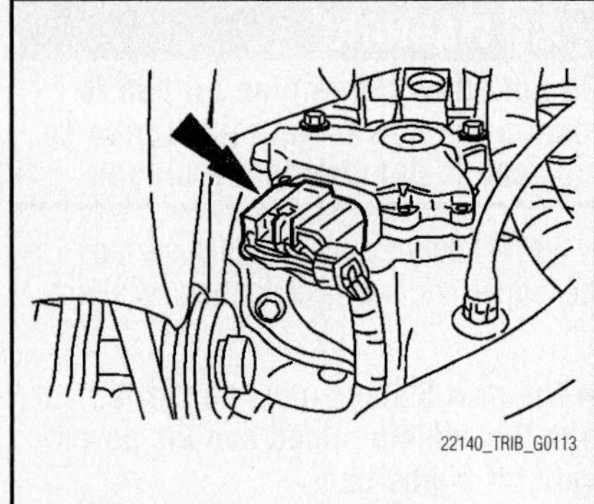

Fig. 144 Transmission Range (TR) sensor

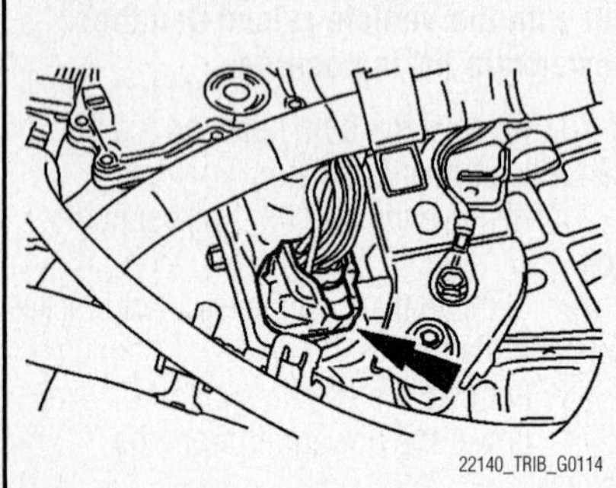

Fig. 145 Transaxle electronic control switch

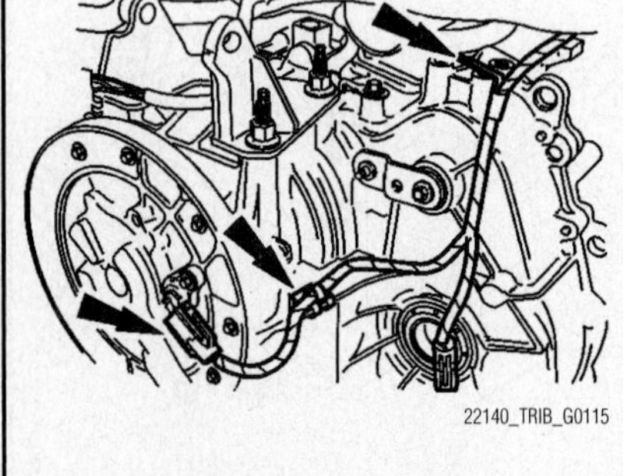

Fig. 146 TSS electrical sensor and 6 RH exhaust manifold nuts

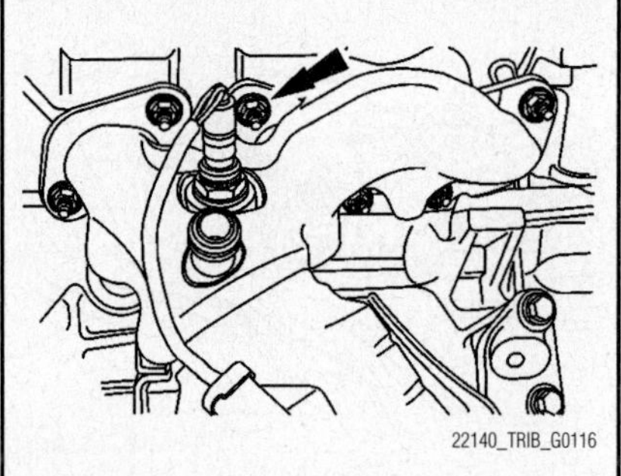

Fig. 147 RH HO2S sensor and the 6 RH exhaust manifold nuts

Fig. 148 6 bolts and the halfshaft support bracket

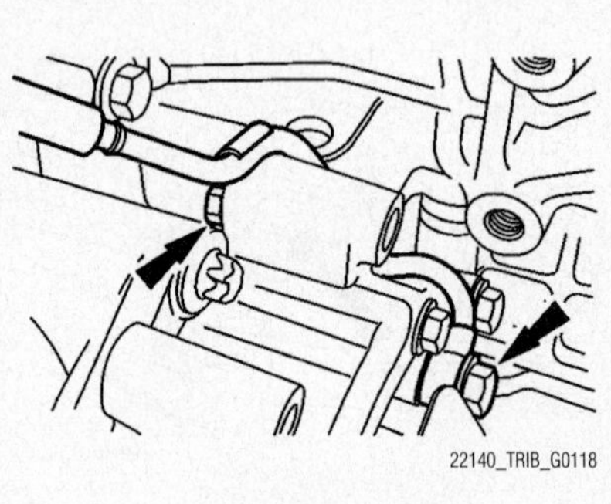

Fig. 149 Power Transfer Unit (PTU) vent tube

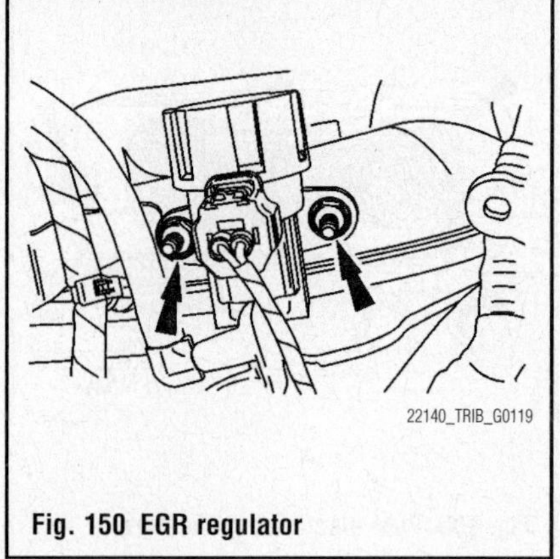

Fig. 150 EGR regulator

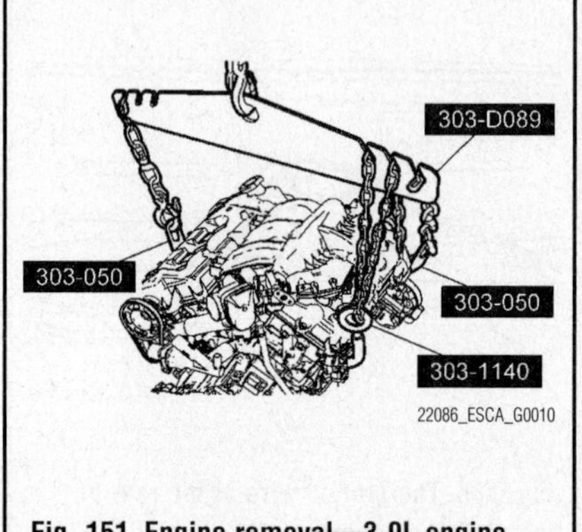

Fig. 151 Engine removal—3.0L engine

76. Detach the 3 pin-type retainers from the transaxle support bracket.
77. Disconnect the transmission range (TR) sensor electrical connector
78. Disconnect the transaxle electronic control switch electrical connector.
79. Detach the wiring harness from the bracket.
80. Disconnect the Turbine Speed Sensor (TSS) electrical connector.
81. Detach the 2 wiring harness retainer.

AWD vehicles:

82. Disconnect the RH HO2S electrical connector
83. Remove the 6 RH exhaust manifold nuts and the manifold.
84. Discard the nuts and gasket.
85. Remove and discard the 6 RH exhaust manifold studs.
86. Remove the 6 bolts and the halfshaft support bracket.
87. Remove the bolt, detach the pin-type retainer and position the Power Transfer Unit (PTU) vent tube aside.
88. Remove the 3 PTU bolts.
89. Remove the bolt and the PTU.

All vehicles:

90. Remove the nut and detach wiring harness retainer.
91. Remove the 2 stud bolts and position the EGR regulator aside.
92. Install engine lifting brackets to engine.
93. Using the suitable tools and a suitable engine crane, remove the engine and transaxle from the lift table.
94. Remove the 5 remaining transaxle-to-engine bolts.
95. Using the suitable tools and a suitable engine crane, separate the engine and transaxle.

To install:

➡ If the oil pan was removed during engine disassembly, it must be installed after the engine and transaxle are assembled and the transaxle-to-engine bolts are installed. Failure to follow this assembly sequence can result in engine oil leaks.

All vehicles:

96. Using the special tools, align the engine with the transaxle.
97. Install the 5 transaxle-to-engine bolts and tighten to 35 ft. lbs. (48 Nm).
98. Using the suitable tools, secure the engine and transaxle to the powertrain lift.
99. Install the EGR regulator and the 2 stud bolts. Tighten to 53 inch lbs. (6 Nm).

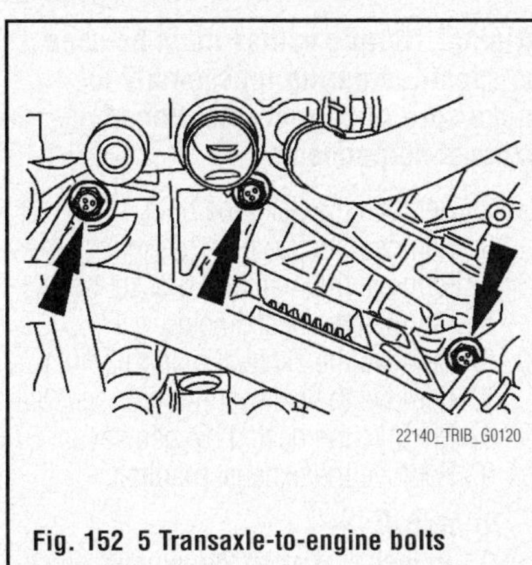
Fig. 152 5 Transaxle-to-engine bolts

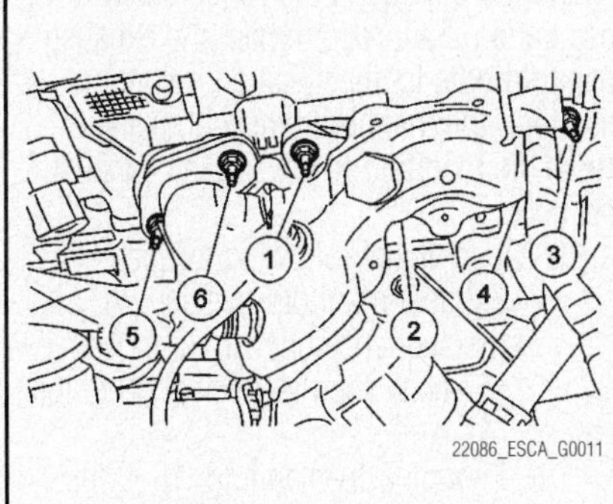

Fig. 153 Manifold tightening sequence—3.0L engine

100. Attach the wiring harness retainer and install the nut. Tighten to 53 inch lbs. (6 Nm).

AWD vehicles:

101. Position the Power Transfer Unit (PTU) and install the bolt. Tighten to 33 ft. lbs. (45 Nm).

102. Install the 3 PTU bolts and tighten to 33 ft. lbs. (45 Nm).

103. Install the PTU vent tube, the pin-type retainer and the bolt. Tighten to 20 ft. lbs. (14 Nm).

104. Install the halfshaft support bracket and the 6 bolts. Tighten bolts to 35 ft. lbs. (48 Nm).

105. Install 6 new RH exhaust manifold studs and tighten to 9 ft. lbs. (12 Nm).

106. Using a new gasket, install the RH exhaust manifold and 6 new nuts.

107. Tighten the manifold nuts to 15 ft. lbs. (20 Nm).

108. Connect the RH heated oxygen sensor (HO2S) electrical connector.

All vehicles:

109. Connect the turbine speed sensor (TSS) electrical connector.

110. Attach the wiring harness retainer.

111. Attach the transaxle control harness to the bracket.

112. Connect the transaxle wiring harness electronic control switch electrical connector.

113. Connect the transmission range (TR) sensor electrical connector

114. Detach the 3 pin-type retainers from the transaxle support bracket.

115. Disconnect the transmission range (TR) sensor electrical connector.

116. Attach the pin-type retainers to the transaxle support bracket and connect the HO2S and catalyst monitor sensor electrical connectors.

117. Position the powertrain into the vehicle

118. Install the engine support bracket, the bolt and the 3 nuts. Tighten the nuts to 56 ft. lbs. (41 Nm). And the bolts to 66 ft. lbs. (90 Nm).

119. Install the rear transaxle support through bolt and tighten to 76 ft. lbs. (103 Nm).

120. Position the RH transaxle support insulator and install the bolt and nuts. Tighten to 59 ft. lbs. (80 Nm).

121. Install the RH transaxle support insulator through bolt and tighten to 85 ft. lbs. (115 Nm).

122. Install the transaxle-to-engine stud and 2 nuts. Tighten to 35 ft. lbs. (48 Nm).

➡Clean and degrease all sealing surfaces with metal surface cleaner. The oil pan must be installed and the bolts tightened within 4 minutes of the sealant application.

123. Apply a 10 mm (0.39 in) dot of silicone gasket and sealant to the front cover-to-cylinder block sealing surface.

124. Install a new oil pan gasket.

125. Position the oil pan and gasket and loosely install the bolts and stud bolts.

126. Install the 2 oil pan-to-transaxle bolts. Tighten to 30 ft. lbs. (40 Nm).

127. Tighten the oil pan bolts in sequence to 18 ft. lbs. (25 Nm).

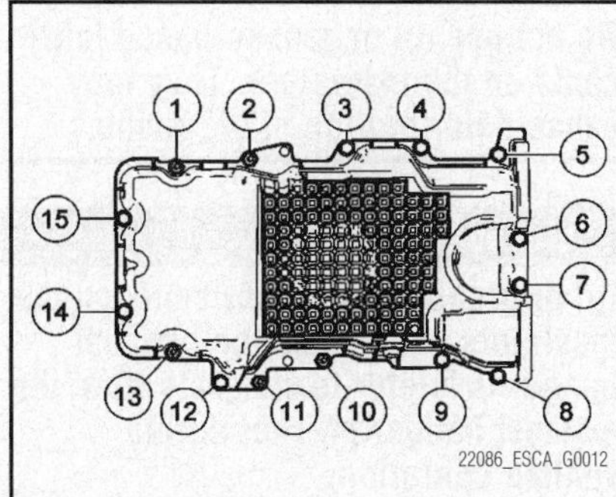

Fig. 154 Oil pan tightening sequence—3.0L engine

128. Remove the 2 oil pan-to-transaxle bolts located next to the Torque converter access cover.

129. Remove the 2 nuts and the transaxle-to-engine stud.

130. Install the torque converter inspection cover.

131. Install the output shaft speed (OSS) sensor and the bolt and connect the electrical connector. Tighten mounting bolt to 10 ft. lbs. (13 Nm).

132. If equipped, connect the engine block heater electrical connector.

133. Install the ground wire eyelet and nut to the engine mount stud. Tighten to 18 ft. lbs. (25 Nm).

134. Install the wire and nut to the battery cable. Tighten to 89 inch lbs. (10 Nm).

135. Attach the wiring harness retainers from the battery tray bracket.

136. Position the ground strap and the electrical connector and install the bolts. Tighten to 89 inch lbs. (10 Nm).

137. Connect the electrical connector to the power distribution box.

138. Install the cable and the nut. Tighten to 9 ft. lbs. (12 Nm).

139. Attach the ground wire and install the bolt. Tighten to 89 inch lbs. (10 Nm).

140. Position the wiring and install the nut. Tighten to 71 inch lbs. (8 Nm). Connect the PCM electrical connectors.

141. Connect the MAP sensor electrical connector and vacuum tube.

142. Connect the brake booster vacuum tube

143. Connect the evaporative emissions (EVAP) purge valve tube to the intake manifold.

144. Position the EGR tube and loosely install the fittings.

145. Tighten the EGR tube-to-EGR valve fitting to 30 ft. lbs. (40 Nm).

146. Connect the differential pressure feedback EGR sensor electrical connector and attach the pin-type retainer.

147. Position the shift cable and bracket in place, install the bolts and tighten to 17 ft. lbs. (23 Nm). Install the wiring harness retainer.

148. Connect the gearshift cable to the transaxle.

149. Attach the wiring harness pin-type retainer to the gearshift cable bracket.

150. Connect the fuel supply tube quick connect coupling at the fuel rail

151. Attach the transaxle vent tube retainer to the throttle body.

152. Connect the heater hoses and the throttle body coolant hose

153. Connect the upper and lower radiator hoses.

154. Position the accelerator and speed control cables and bracket. Install the nut and tighten to 89 inch lbs. (10 Nm).
155. Position the bracket and install and tighten the bolts to 89 inch lbs. (10 Nm).
156. Connect the accelerator cable and the speed control actuator cable
157. Position the accelerator cable snow shield, install and tighten the bolts to 89 inch lbs. (10 Nm).
158. Tighten the EGR tube-to-RH catalytic converter fitting to 30 ft. lbs. (40 Nm).
159. Install the transaxle cooler tube and the bracket bolt.
160. Tighten the bolt to 10 ft. lbs. (13 Nm), tighten the fitting to 17 ft. lbs. (23 Nm).
161. Connect the 2 transmission cooler tubes.
162. Install the cross brace and the new nut finger tight.
163. Install the 2 bolts for the cross brace and the bolt for the front roll restrictor.
 a. Tighten the 2 cross brace bolts to 66 ft. lbs. (90 Nm).
 b. Tighten the front roll restrictor bolt to 85 ft. lbs. (115 Nm).
164. Tighten the cross brace nut to 129 ft. lbs. (175 Nm).
165. Position the RH accessory drive belt idler pulley and bracket and attach the wiring harness retainer. Install the bolts and tighten to 18 ft. lbs. (25 Nm).
166. Install the A/C compressor and bolts, tighten to 18 ft. lbs. (25 Nm).
167. Connect the A/C clutch field coil electrical connector
168. Rotate the accessory drive belt tensioner counterclockwise and install the accessory drive belt.

AWD vehicles:

✲✲ WARNING

Do not reuse the driveshaft flange bolts and washers. Install new bolts and washers or damage to the vehicle may occur.

169. Align the index-marks made during removal and install 6 new front driveshaft-to-PTU bolts and washers. Tighten to 27 ft. lbs. (37 Nm).

All vehicles:

170. Install the intermediate shaft and the 2 nuts. Tighten to 20 ft. lbs. (27 Nm).
171. Install the RH halfshaft onto the intermediate shaft.
172. Install the LH halfshaft into the transaxle.
173. Attach the LH and RH lower ball joints to the steering knuckles and install the pinch bolts and nuts. Tighten to 46 ft. lbs (63 Nm).
174. Install the LH and RH tie-rod ends and nuts. Tighten to 41 ft. lbs. (55 Nm).
175. Install the LH and RH stabilizer link-to-strut nuts. Tighten to 46 ft. lbs (63 Nm).
176. Install the LH and RH brake hose retainers and ABS sensor wiring harness bolts. Tighten to 11 ft. lbs. (15 Nm).
177. Install the Y-pipe and the 2 exhaust flange-to-RH catalytic converter nuts. Tighten to 21 ft. lbs. (29 Nm).
178. Install the 6 exhaust flange-to-LH catalytic converter nuts. Tighten to 21 ft. lbs. (29 Nm).
179. Attach the Y-pipe to the rear catalytic converter.
180. Attach the exhaust hanger.
181. Install the 2 exhaust flange-to-rear catalytic converter nuts and tighten to 21 ft. lbs. (29 Nm).
182. Install the lateral support crossmember and the 4 bolts. Tighten to 85 ft. lbs. (115 Nm).
183. Connect the catalyst monitor sensor electrical connector and the 2 wiring harness retainers
184. Install the lower splash shields.
185. Install the front wheels and tires.
186. Install the air cleaner outlet pipe and air cleaner
187. Install the battery tray.
188. Fill the engine with clean engine oil.
189. Fill and bleed the cooling system.
190. Connect the negative battery cable.

EXHAUST MANIFOLD

REMOVAL & INSTALLATION

2.3L Engine

See Figures 155 through 157.

1. Before servicing the vehicle, refer to the Precautions Section.

✲✲ WARNING

Do not use oil or grease-based lubricants on the insulators. They may cause deterioration of the rubber.

✲✲ WARNING

Oil or grease-based lubricants on the insulators may cause the exhaust hanger insulator to separate from the exhaust hanger bracket during vehicle operation.

➡Exhaust fasteners are of a torque prevailing design. Use only new fasteners with the same part number as the original. Torque values must be used as specified during reassembly to make sure of correct retention of exhaust components.

2. Remove the flex pipe nuts.
3. Remove the flex pipe gasket.
4. Remove the manifold bracket bolts.
5. Remove the heat shield.
6. Remove the exhaust manifold nuts.
7. Remove the catalyst monitor sensor.
8. Remove the heated oxygen sensor.
9. Remove the exhaust manifold.

To install:

10. To install, reverse the removal procedure. Make sure to apply anti-seize lubricant to the threads of the sensors before installation. Failure to tighten the exhaust manifold nuts to specification before installing the manifold bracket bolts will cause the manifold to develop an exhaust gas leak.
11. Observe the following torques:
    - Exhaust manifold-to-head, in the sequence shown: 35 ft. lbs. (47 Nm), then recheck at 35 ft. lbs. (47 Nm)
    - Flex pipe-to-manifold: 18 ft. lbs. (25 Nm)
    - Heated oxygen sensor: 35 ft. lbs. (47 Nm)
    - Catalyst monitor sensor: 30 ft. lbs. (40 Nm)
12. Check the exhaust system for proper alignment.

3.0L Engine—Left Side

See Figure 158.

1. Before servicing the vehicle, refer to the Precautions Section.
2. With the vehicle in **NEUTRAL**, position it on a hoist. Remove or disconnect the following:
    - Negative battery cable
    - Exhaust flexible pipe
    - 2 catalytic converter bracket bolts
    - 6 heat shield bolts and the heat shield
    - Heated Oxygen (HO_2S) sensor and catalyst monitor
    - 7 catalytic converter and exhaust manifold nuts
3. Remove the catalytic converter and exhaust manifold from the vehicle
4. Discard the catalytic converter and exhaust manifold gasket.

➡ Clean and inspect the catalytic converter and exhaust manifold.

To install:

5. Clean the sealing surfaces of any old gasket material.

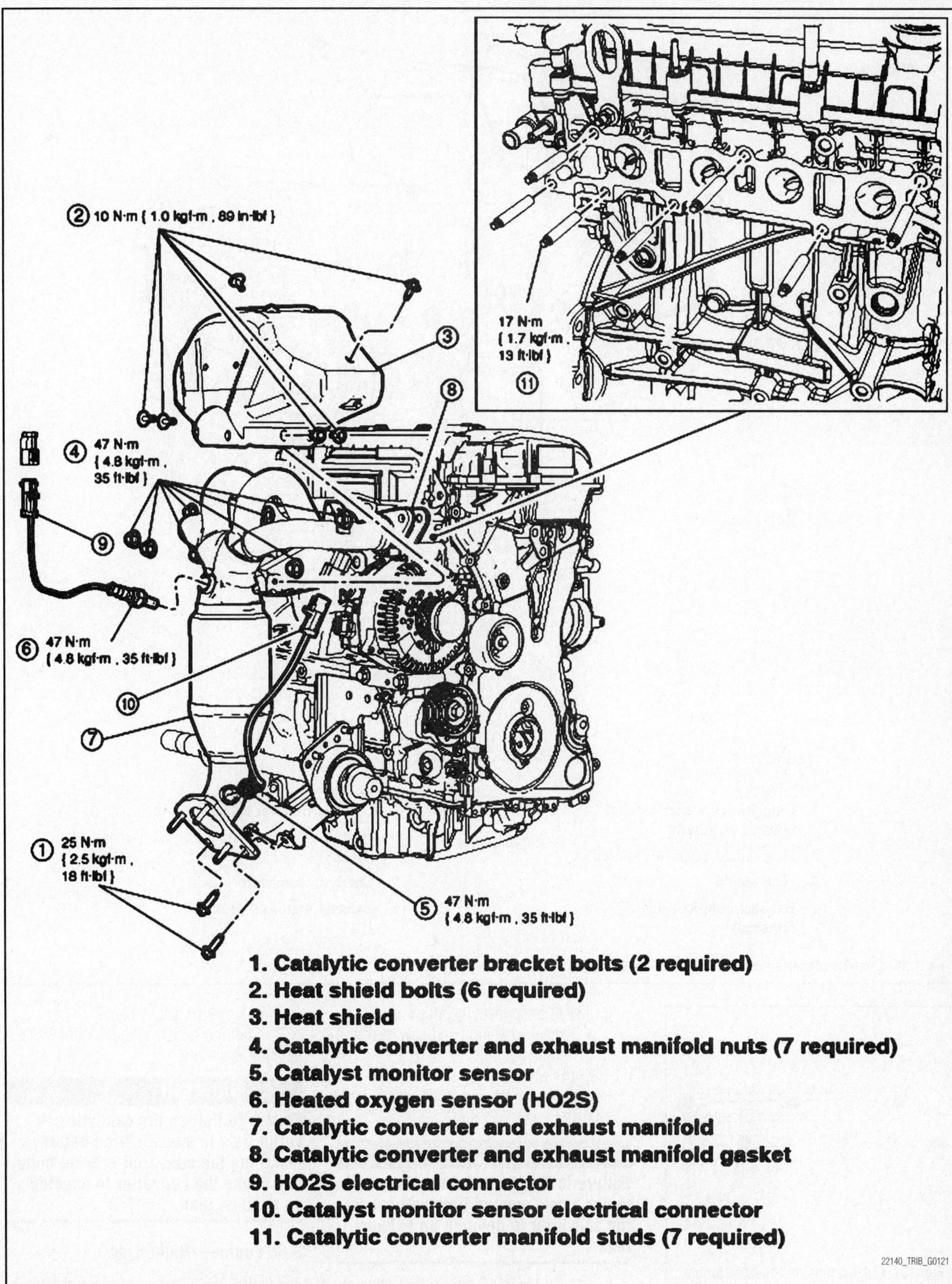

Fig. 155 Catalytic converter and exhaust manifold exploded view— 2.3L Engines

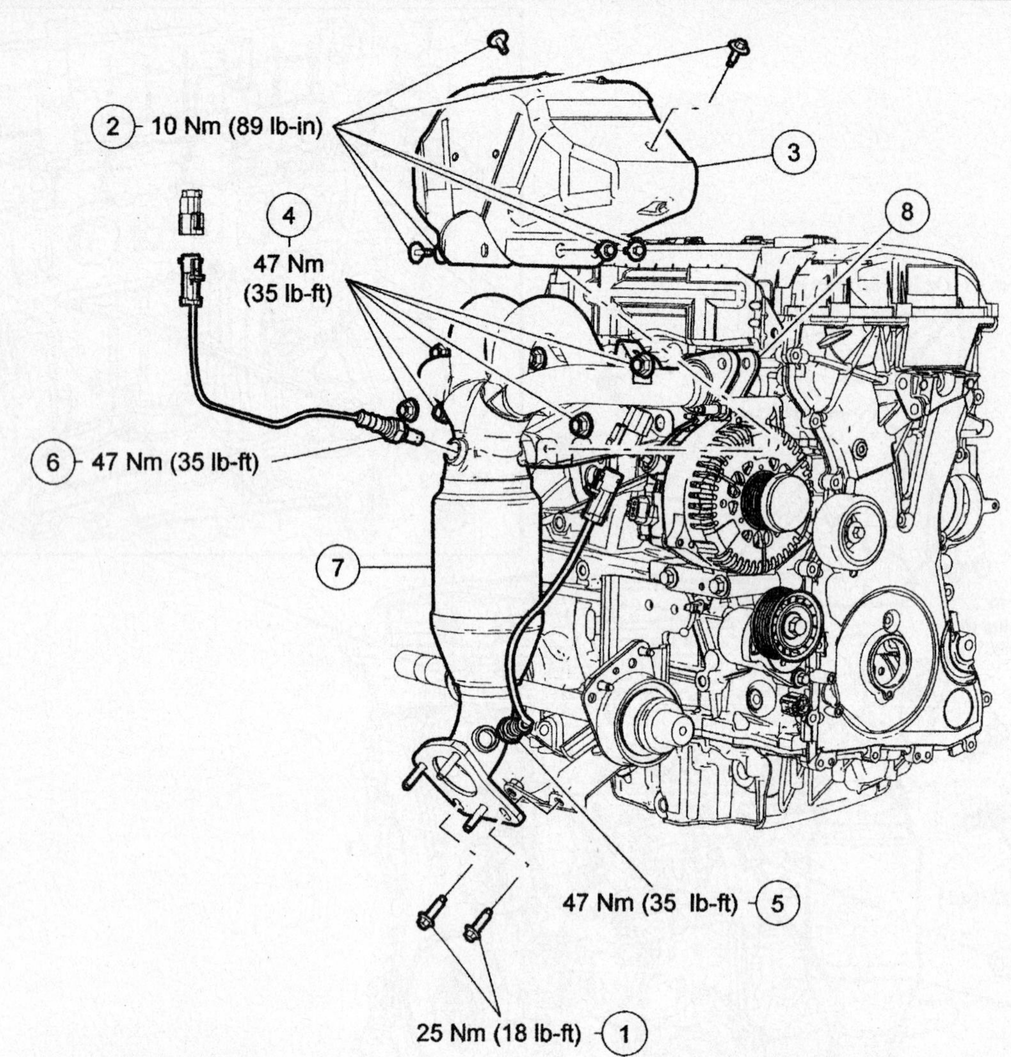

Fig. 156 Catalytic converter/exhaust manifold—2.3L engine

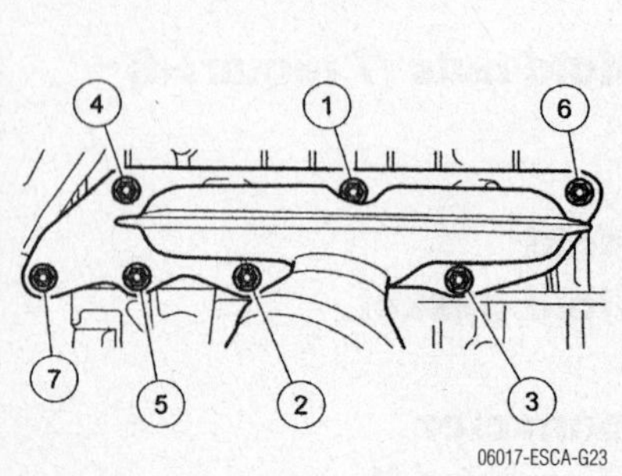

Fig. 157 Exhaust manifold-to-head torque sequence—2.3L engine

6. Install or connect the following:
   - Exhaust manifold and new gasket. Torque the bolts to 35 ft. lbs. (47 Nm).
   - Retighten the manifold bolts to 35 ft. lbs. (47 Nm).

✱✱ WARNING

Failure to tighten the catalytic converter nuts a second time will cause the converter to develop an exhaust leak.

   - Left side HO_2S sensor and catalyst monitor
   - Negative battery cable
7. Start the vehicle and check for leaks, repair if necessary.

✱✱ WARNING

Failure to tighten the catalytic converter nuts to specification before installing the converter bracket bolts will cause the converter to develop an exhaust leak.

3.0L Engine—Right Side

See Figure 159.

Fig. 158 Left side exhaust manifold bolt torque sequence–3.0L engine

1. Before servicing the vehicle, refer to the Precautions Section.
2. Remove or disconnect the following:
   - Negative battery cable
   - Exhaust Gas Recirculation (EGR) tube
   - Alternator
   - Right side Heated Oxygen (HO_2S) sensor
   - Right side exhaust manifold and discard the gasket

To install:

3. Clean the sealing surfaces of any old gasket material.
4. Install or connect the following:
   - Exhaust manifold and new gasket. Torque the bolts to 15 ft. lbs. (20 Nm).
   - Right side HO_2S sensor
   - Alternator
   - EGR tube
   - Negative battery cable

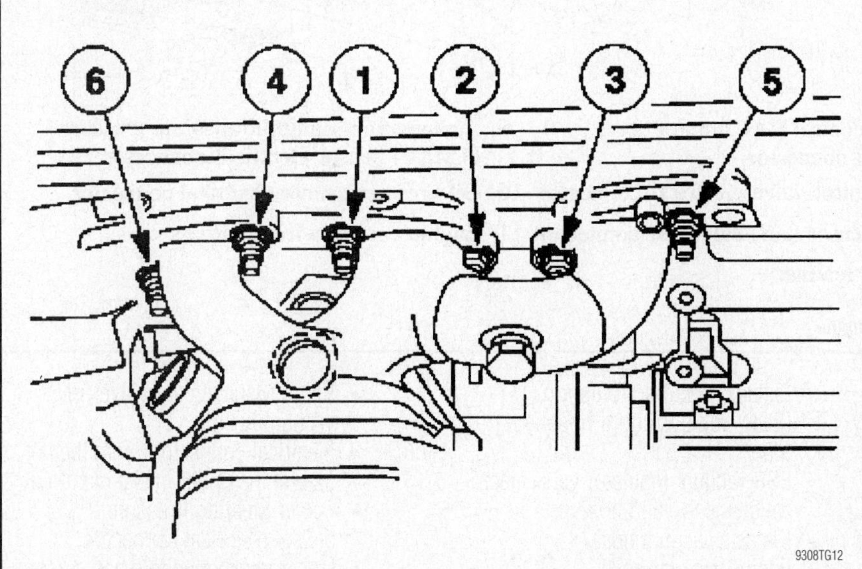

Fig. 159 Right side exhaust manifold bolt torque sequence–3.0L engine

5. Start the vehicle and check for leaks, repair if necessary.

INTAKE MANIFOLD

REMOVAL & INSTALLATION

2.3L Engine

See Figures 160 and 161.

1. Before servicing the vehicle, refer to the Precautions Section.
2. With the vehicle in NEUTRAL, position it on a hoist.
3. Remove the throttle body.
4. Remove the fuel rail.
5. Remove the oil level indicator tube.
6. Remove the vacuum tube.
7. Remove the vacuum supply hose.
8. Remove the fuel vapor return hose.
9. Remove the Idle Air Control (IAC) motor electrical connector.
10. Remove the swirl control valve electrical connector.
11. Remove the Knock Sensor (KS) electrical connector.
12. Remove the Temperature Manifold Absolute Pressure (TMAP) sensor electrical connector.
13. Remove the oil pressure sender electrical connector.
14. Remove the engine control wiring harness.
15. Remove the intake manifold bolts.

➡There are three different size bolts used. Mark the location of the bolts to make sure they are installed in the correct location.

16. Remove the bolts and position the intake manifold aside to access the crankcase vent hose clamp and the EGR tube.
17. Remove the crankcase vent hose.
18. Remove the exhaust gas recirculation (EGR) tube.
19. Remove the intake manifold.
20. Remove the intake manifold gasket.
21. To install, reverse the removal procedure. Torque the intake manifold bolts to 13 ft. lbs. (18 Nm).

3.0L Engine

Upper

See Figure 162.

1. Before servicing the vehicle, refer to the Precautions Section.
2. Properly relieve the fuel system pressure.
3. Drain the coolant system.
4. Remove or disconnect the following:
   - Negative battery cable

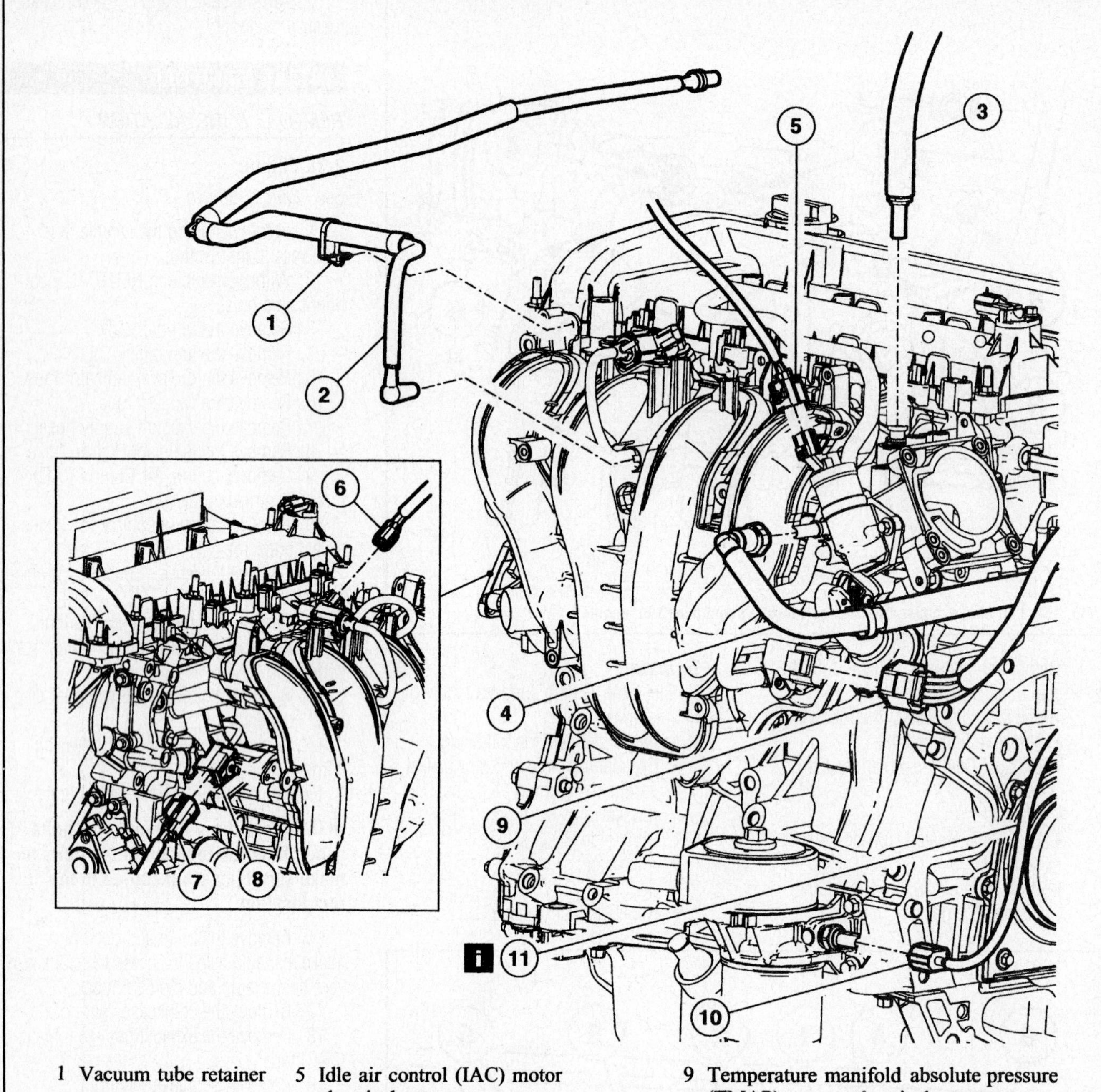

Fig. 160 Intake manifold and related parts—2.3L engine

- Air cleaner outlet tube
- Engine appearance cover
- Throttle cable
- Speed control cable, if equipped
- Throttle cable bracket
- Throttle Position (TP) sensor electrical connector
- Idle Air Control (IAC) valve electrical connector
- Exhaust Gas Recirculation (EGR) valve vacuum hose and tube
- EGR vacuum regulator valve electrical connector and hose
- Chassis vacuum hose
- Engine vacuum hose
- Positive Crankcase Ventilation (PCV) hose
- Vapor Management Valve (VMV) vacuum hose
- Electrical connectors from the left side of the upper intake manifold
- Power Steering Pressure (PSP) sensor electrical connector
- Upper intake manifold and discard the gasket

5. Clean the mating surfaces.

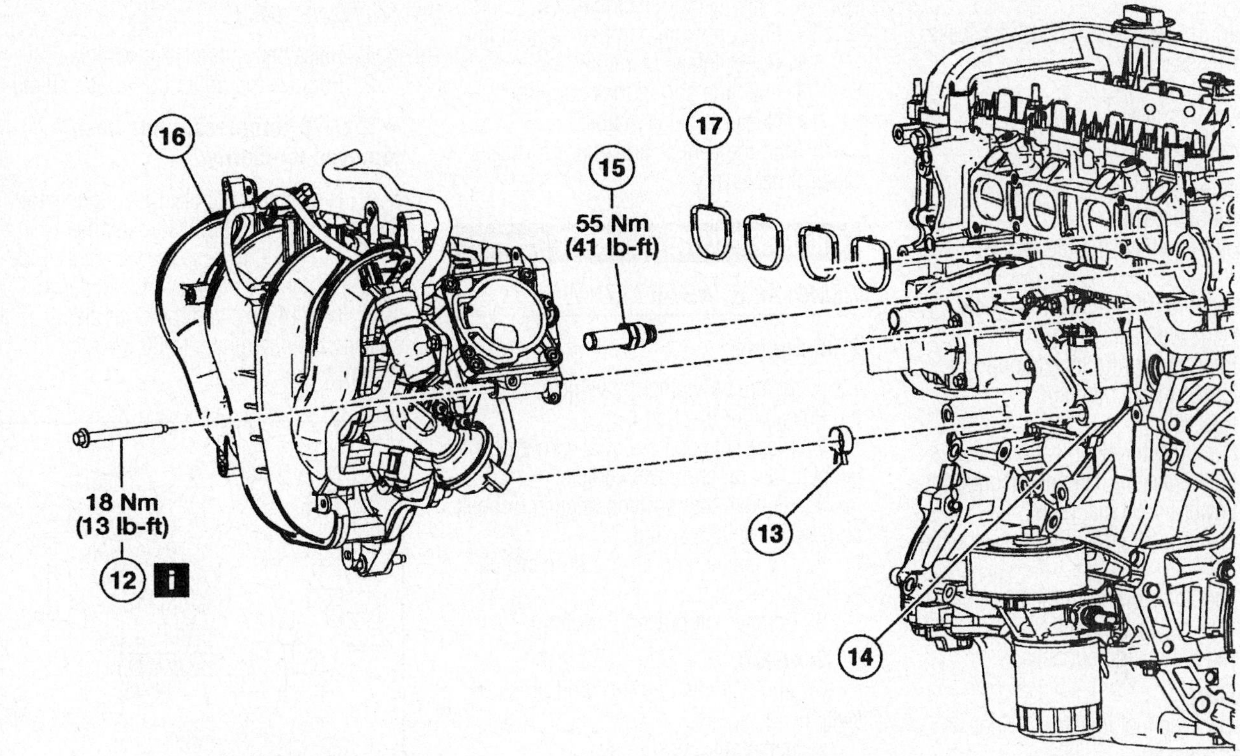

16
17
15
55 Nm
(41 lb-ft)
18 Nm
(13 lb-ft)
12
13
14
67197-ESCA-G07

12 Intake manifold bolts
13 Crankcase vent hose clamp
14 Crankcase vent hose (position aside)
15 Exhaust gas recirculation (EGR) tube
16 Intake manifold
17 Intake manifold gasket

Fig. 161 Intake manifold installation—2.3L engine

7
1
3
5
8
4
2
6
9308TG02

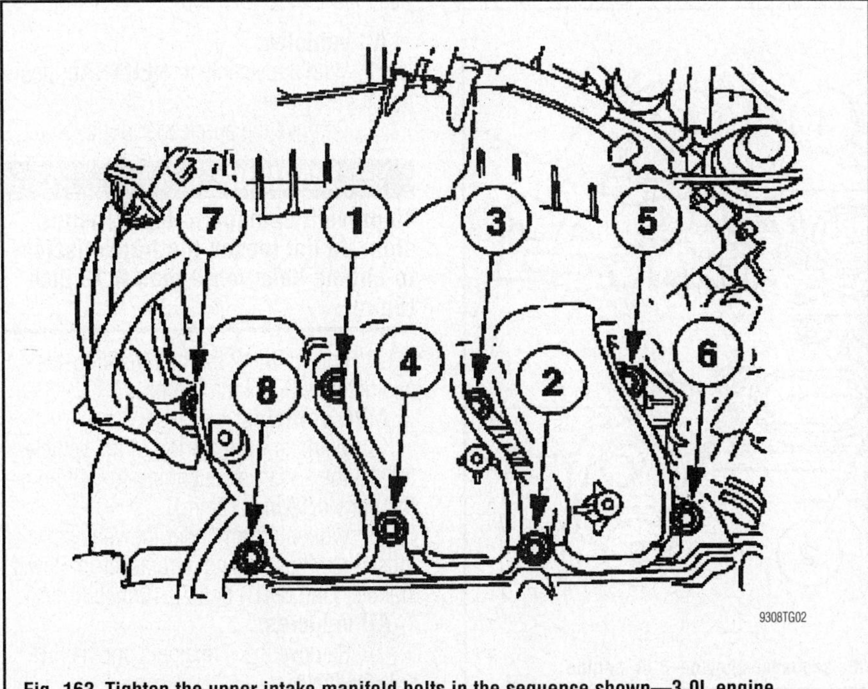

Fig. 162 Tighten the upper intake manifold bolts in the sequence shown—3.0L engine

To install:

6. Install or connect the following:
   - New gasket
   - Intake manifold. Torque the bolts, in sequence, to 89 inch lbs. (10 Nm).
   - PSP electrical connector
   - Electrical connectors on the left side of the upper intake manifold
   - VMV vacuum hose
   - Chassis, engine and PCV hoses
   - EGR valve vacuum regulator
   - EGR valve vacuum hose and tube. Torque the nut to 30 ft. lbs. (40 Nm).
   - TP sensor electrical connector
   - IAC valve electrical connector
   - Throttle cable and speed control cable, if equipped. Torque the bracket bolts to 89 inch lbs. (10 Nm).
   - Air cleaner outlet tube
   - Engine appearance cover. Torque the bolts to 53 inch lbs. (6 Nm).
   - Negative battery cable

7. Fill the coolant system to the proper level.
8. Start the vehicle and check for leaks, repair if necessary.

Lower

See Figure 163.

1. Before servicing the vehicle, refer to the Precautions Section.
2. Properly relieve the fuel system pressure.
3. Remove or disconnect the following:
   - Negative battery cable
   - Fuel line spring lock coupling
   - Upper intake manifold
   - Fuel rail
   - Fuel injector electrical connectors
   - Fuel pressure damper vacuum line
   - Lower intake manifold
   - Lower intake manifold from the fuel rail
   - Fuel injectors from the manifold and discard the gasket
4. Clean the mating surfaces.

To install:

5. Inspect the fuel injector O-rings and replace if necessary.
6. Install or connect the following:
   - Fuel injectors into the lower intake manifold
   - Fuel rail. Torque the bolts to 89 inch lbs. (10 Nm).
   - New gasket
   - Intake manifold. Torque the bolts, in sequence, to 89 inch lbs. (10 Nm).
   - Fuel rail electrical connectors
   - Fuel injector electrical connectors
   - Fuel pressure damper vacuum line
   - Upper intake manifold
   - Fuel line spring lock coupling
   - Negative battery cable
7. Start the vehicle and check for leaks, repair if necessary.

OIL PRESSURE SENSOR

REMOVAL & INSTALLATION

2.3L Engine

1. Before servicing the vehicle, refer to the Precautions Section.
2. Place vehicle in neutral with emergency brake on and raise vehicle.
3. Remove the pin-type retainer, 5 bolts and the RH splash shield.
4. Disconnect the oil pressure switch electrical connector.
5. Remove oil pressure switch.

To install:

6. Apply thread sealant to oil pressure switch.
7. Install oil pressure switch.
8. Tighten oil pressure switch to 11 ft. lbs (15 Nm).
9. Install oil pressure switch electrical connector.
10. Install splash shield, retainer and bolts.
11. Tighten shield bolts to 80 inch lbs. (9 Nm).
12. Lower vehicle.

3.0L Engine

See Figure 164.

1. Raise and support the vehicle.
2. Remove the right side splash shield.

➡**The A/C compressor has been removed for clarity.**

3. Disconnect the electrical connector, then remove the oil pressure sender.
4. To install, reverse the removal procedure. Tighten the oil pressure sender to 10 ft. lbs. (14 Nm) and the right side splash shield retainers to 80 inch lbs. (9 Nm).

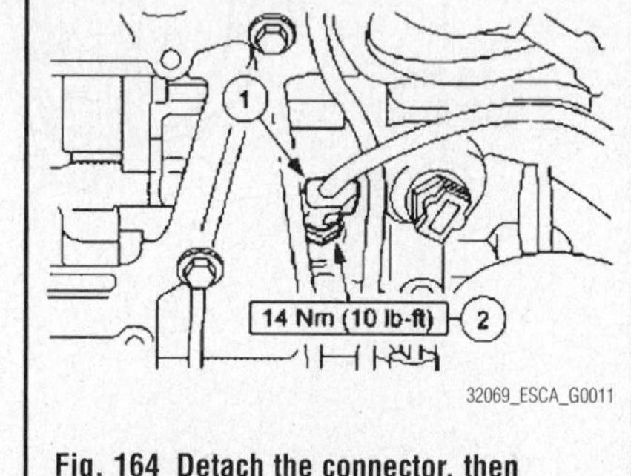

Fig. 164 Detach the connector, then remove the oil pressure sender. Note: A/C compressor removed for clarity

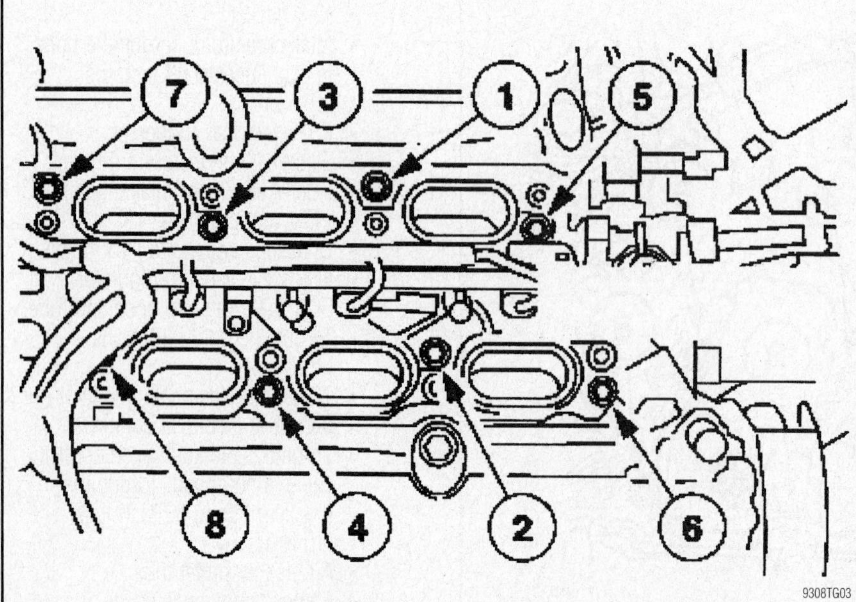

Fig. 163 Tighten the lower intake manifold bolts in the sequence shown—3.0L engine

OIL PAN

REMOVAL & INSTALLATION

2.3L Engine

See Figures 165 and 166.

All vehicles:

1. With the vehicle in **NEUTRAL**, position it on a hoist.
2. Remove the air cleaner outlet pipe.

⁂ WARNING

To prevent damage to the transmission, do not loosen the transmission-to-engine bolts more than 0.19 inch (5mm).

3. Loosen the 2 top bell housing-to-engine bolts 0.19 inch (5mm).

AWD vehicles:

4. Working from the top of the vehicle, loosen the 2 rear lower engine-to-bell housing bolts 0.19 inch (5mm).
5. Working from under the vehicle, loosen the 2 upper engine bracket-to-Power Transfer Unit (PTU) bolts 0.19 inch (5mm).

All vehicles:

6. Remove the 7 retainers and the LH splash shield.

7. Remove the oil level indicator and tube

8. Loosen the 2 front lower bell housing-to-engine bolts 0.19 inch (5mm).

FWD vehicles:

9. Loosen the 1 (manual transmission) and 2 (automatic transmission) rear lower engine-to-bell housing bolt 0.19 inch (5mm).

All vehicles:

10. Remove the 2 oil pan-to-bell housing bolts.

11. Remove the 2 bell housing-to-oil pan bolt.

12. Slide the transmission rearward 5 mm (0.19 in).

13. Drain the engine oil.

14. Remove the 4 engine front cover-to-oil pan bolts.

15. Remove the 13 bolts and the oil pan.

To install:

All vehicles:

16. Clean and inspect all mating surfaces.

➡If the oil pan is not secured within 10 minutes of sealant application, the sealant must be removed and the sealing area cleaned with metal surface cleaner. Allow to dry until there is no sign of wetness, or 10 minutes, whichever is longer. Failure to follow this procedure can cause future oil leakage.

17. Apply a 0.09 inch. (2.5mm) bead of silicone gasket and sealant to the oil pan-to-engine block and to the oil pan-to-engine front cover mating surface.

18. Position the oil pan onto the engine and install the oil pan bolts finger-tight.

⁂ WARNING

The engine front cover-to-oil pan bolts must be tightened first to align the front surface of the oil pan flush with the front surface of the engine block.

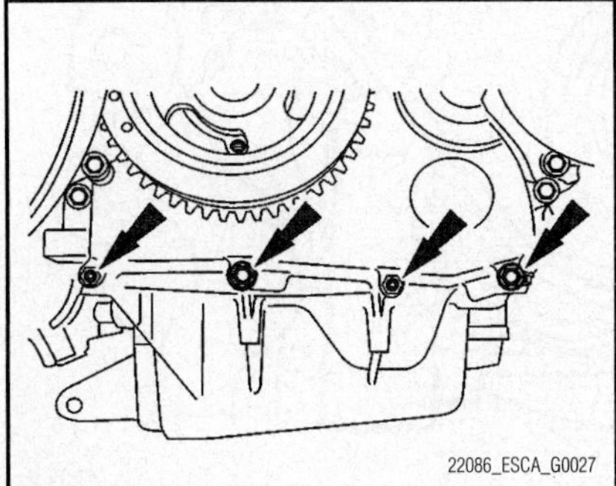

Fig. 165 Front cover-to-oil pan bolts shown

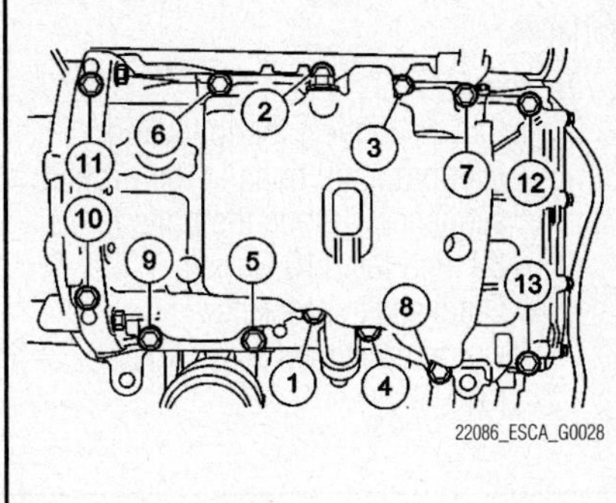

Fig. 166 Oil pan bolt tightening sequence

19. Install the 4 engine front cover-to-oil pan bolts and tighten to 89 inch lbs. (10 Nm).

20. Tighten the oil pan bolts in sequence to 18 ft. lbs. (25 Nm).

FWD vehicles:

21. Alternate tightening the 1 front and 1 rear lower bolts to slide the transmission and engine together. Tighten bolts to 35 ft. lbs. (48 Nm).

22. Tighten the remaining front lower bolt and rear lower bolt (automatic transmission) to 35 ft. lbs. (48 Nm).

AWD vehicles:

23. Alternate tightening the 1 upper engine-to-PTU bracket bolt and 1 front lower bolt to slide transmission and engine together.

24. Tighten the PTU bracket bolt to 33 ft. lbs. (45 Nm).

25. Tighten the front lower bolt to 33 ft. lbs. (45 Nm).

26. Tighten the remaining upper engine-to-PTU bracket bolt to 33 ft. lbs. (45 Nm).

27. Tighten the remaining front lower bolt to 33 ft. lbs. (45 Nm).

All vehicles:

28. Install the 2 bell housing-to-oil pan bolts to 33 ft. lbs. (45 Nm).

29. Install the 2 oil pan-to-bell housing bolts to 33 ft. lbs. (45 Nm).

30. Install the oil level indicator and tube.

31. Install the LH splash shield and the 7 retainers. Tighten to 80 inch lbs. (90 Nm).

AWD vehicles:

32. Working from the top of vehicle, tighten the 2 rear lower engine-to-bell housing bolts to 35 ft. lbs. (48 Nm).

All vehicles:

33. Tighten the 2 top bell housing-to-engine bolts to 35 ft. lbs. (48 Nm).

34. Install the air cleaner outlet pipe.

35. Fill the engine with clean engine oil.

36. Recheck for leaks.

3.0L Engine

See Figure 167.

1. Before servicing the vehicle, refer to the Precautions Section.
2. Drain the engine oil.
3. Remove or disconnect the following:
   - Negative battery cable
   - Flexible exhaust pipe
   - Downstream catalyst monitor sensor
   - Oil pan and gasket
4. Thoroughly clean the gasket mating surfaces.

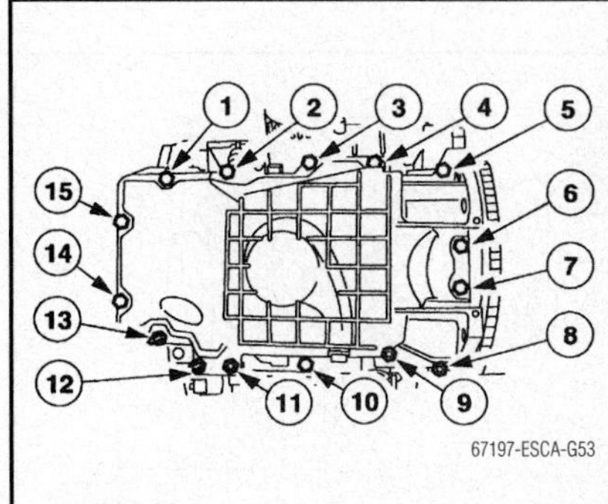

Fig. 167 Oil pan torque sequence—3.0L engine

To install:

5. Apply silicone sealer to the oil pan.
6. Install or connect the following:
   - New gasket on the oil pan
   - Oil pan. Torque the bolts in sequence to 19 ft. lbs. (26 Nm).
   - Flexible exhaust pipe
   - Downstream catalyst monitor sensor
   - Negative battery cable
7. Fill the engine with clean oil.
8. Start the vehicle and check for leaks, repair if necessary.

OIL PUMP

REMOVAL & INSTALLATION

2.3L Engine

See Figure 168.

1. Before servicing the vehicle, refer to the Precautions Section.
2. Remove the engine from the vehicle and mount it on an engine stand.
3. Remove the oil pan.
4. Remove the oil pump pickup tube and screen.
5. Remove the front cover and the timing chain.
6. Release the tension on the tensioner spring.

7. Remove the tensioner and the shoulder bolt.
8. Remove the guide.

➡**The oil pump chain sprocket must be held in place.**

9. Remove the oil pump chain and sprockets.
10. Remove the oil pump assembly and gasket.

To install:

11. Install the oil pump with a new gasket. Tighten the bolts in sequence as follows:
    a. Step 1: 89 inch lbs. (10 Nm).
    b. Step 2: 17 ft. lbs. (23 Nm).

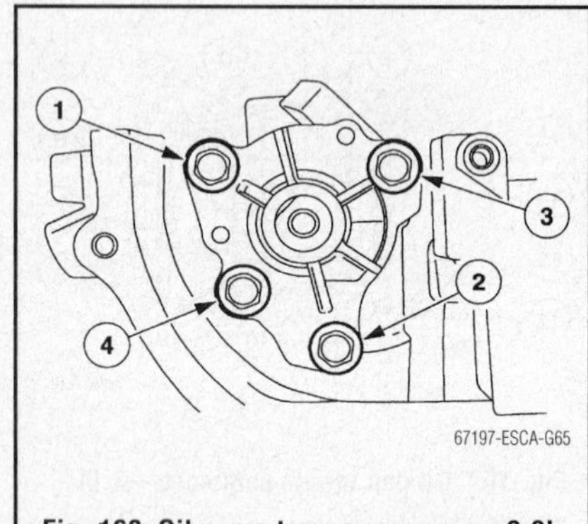

Fig. 168 Oil pump torque sequence—2.3L engine

12. Install the pump chain and sprockets. Tighten the pump sprocket bolt to 18 ft. lbs. (25 Nm).
13. Install the chain guide, tensioner, and shoulder bolt. Tighten the bolts to 89 inch lbs. (10 Nm).
14. Hook the tensioner spring around the shoulder bolt.
15. Install the oil pump pickup tube and screen with a new gasket. Tighten the bolts to 89 ft. lbs. (10 Nm).
16. Install the oil pan.
17. Install the timing chain and front cover.
18. Install the engine into the vehicle.

3.0L Engine

See Figures 169 and 170.

1. Before servicing the vehicle, refer to the Precautions Section.
2. Drain the engine oil.
3. Remove or disconnect the following:
    - Negative battery cable
    - Timing drive components
    - Oil pump screen cover and tube
    - Damper bolt and crankshaft sprockets
    - Oil pump bolts in the proper sequence
4. Thoroughly clean the gasket mating surfaces.

To install:

5. Install or connect the following:
    - Oil pump and bolts in the proper sequence. Torque the bolts to 89 inch lbs. (10 Nm).
    - Crankshaft sprockets
    - Oil pump screen cover and tube
    - Timing drive components
    - Negative battery cable
6. Refill the engine with clean oil.
7. Start the engine and check for leaks; repair if necessary.

MAIN BEARING TORQUE SEQUENCE

See Figure 171.

Fig. 169 Remove the oil pump bolts in the proper sequence–3.0L engine

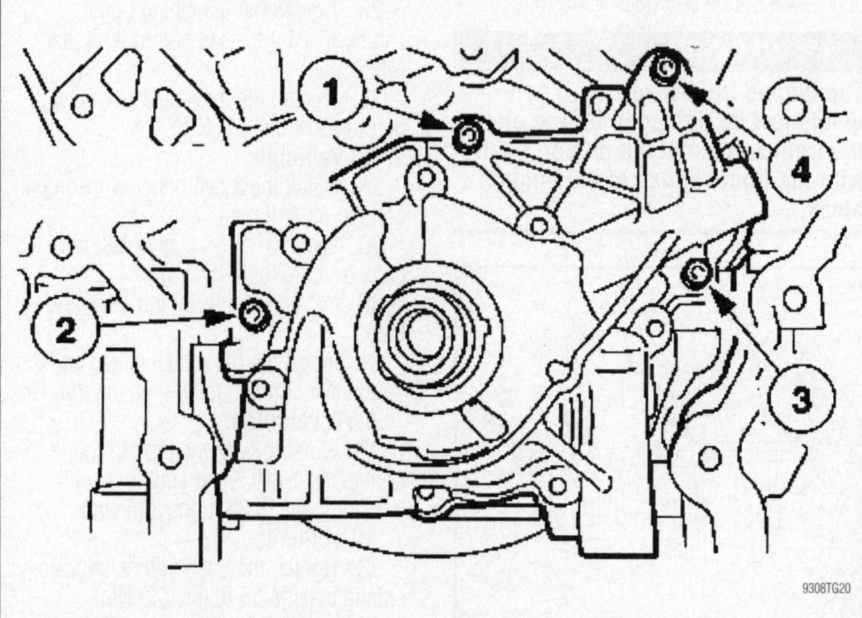

Fig. 170 Install the oil pump bolts in the proper sequence–3.0L engine

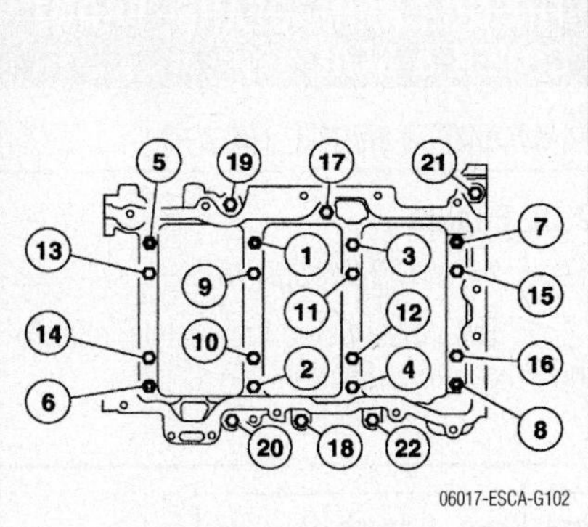

Fig. 171 3.0L engine main bearing (bed plate) torque sequence

PISTON AND RING

POSITIONING

See Figures 172 and 173.

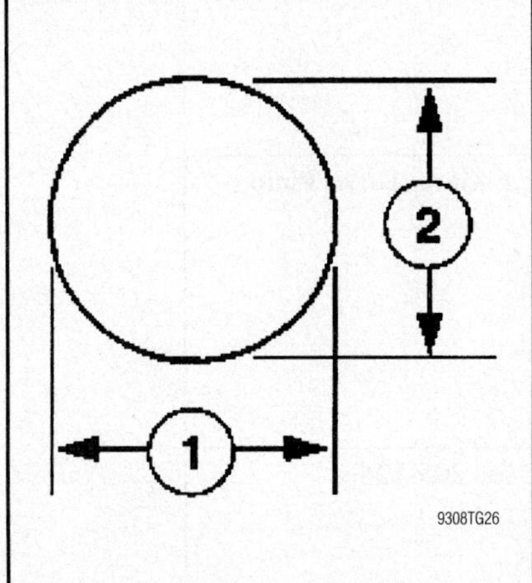

Fig. 172 2.3L engine—piston ring end-gap spacing

REAR MAIN SEAL

REMOVAL & INSTALLATION

2.3L Engine

See Figures 174 through 176.

1. Before servicing the vehicle, refer to the Precautions Section.
2. With the vehicle in NEUTRAL, position it on a hoist.
3. If equipped, remove the automatic transaxle.
4. If equipped, remove the manual transaxle and clutch.
5. Remove the flexplate or flywheel.
6. Remove the oil pan.
7. Remove the crankshaft rear oil seal with retainer plate

To install:

8. Using a seal installer, position the crankshaft rear oil seal with retainer plate onto the crankshaft.
9. Install the crankshaft rear oil seal with retainer plate. Tighten the bolts in the sequence shown to 89 inch lbs. (10 Nm).
10. Install the oil pan.

➡Special bolts are used for installation. Do not use standard bolts.

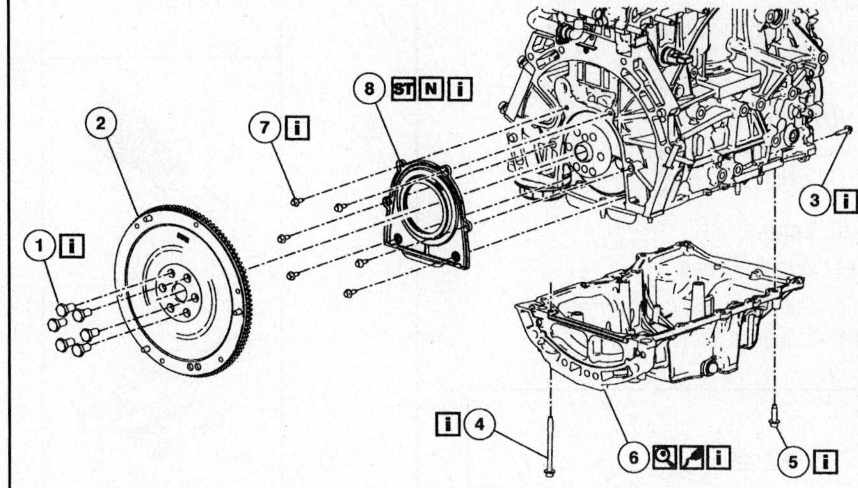

Fig. 174 Rear main seal and related parts—2.3L engine

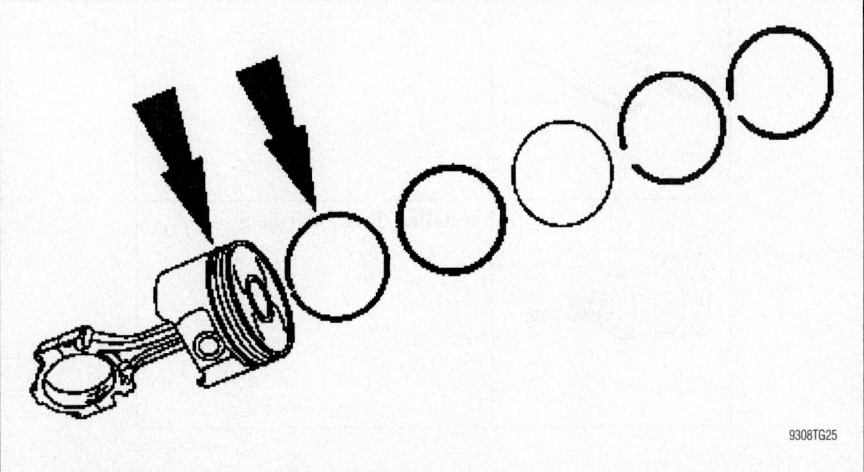

Fig. 173 3.0L engine—piston ring end-gap spacing

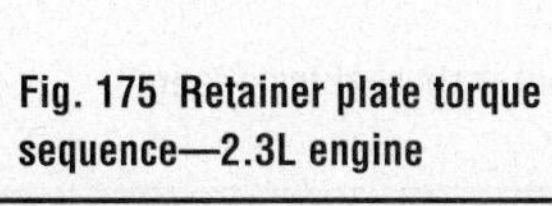

Fig. 175 Retainer plate torque sequence—2.3L engine

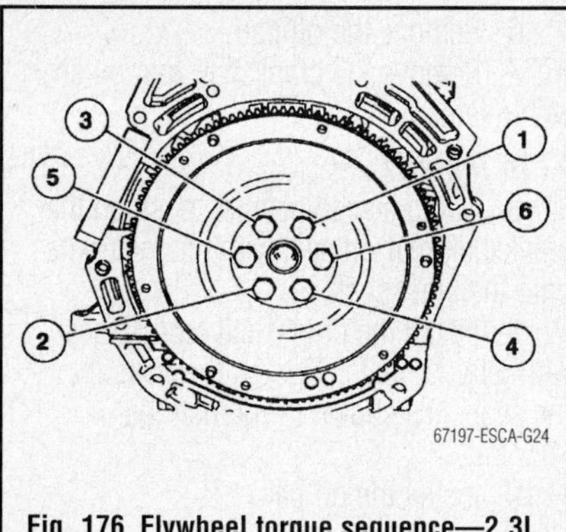

Fig. 176 Flywheel torque sequence—2.3L engine

11. Install the flywheel/flexplate. Tighten the bolts in the sequence shown in three stages.
    a. Stage 1: Tighten to 37 ft. lbs. (50 Nm).
    b. Stage 2: Tighten to 50 ft. lbs. (80 Nm).
    c. Stage 3: Tighten to 83 ft. lbs. (112 Nm).

3.0L Engine

See Figures 177 and 178.

1. Before servicing the vehicle, refer to the Precautions Section.

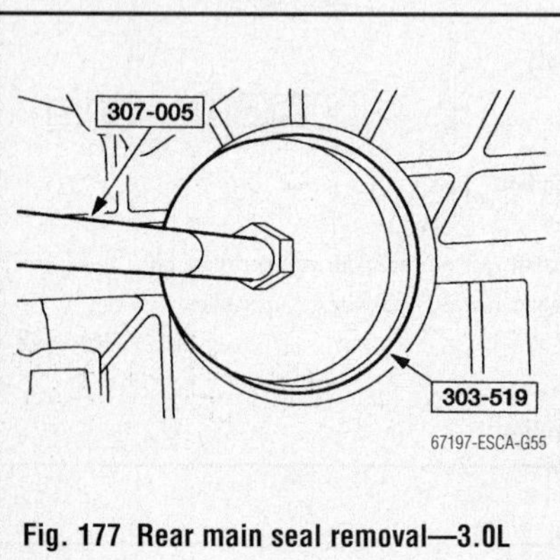

Fig. 177 Rear main seal removal—3.0L engine

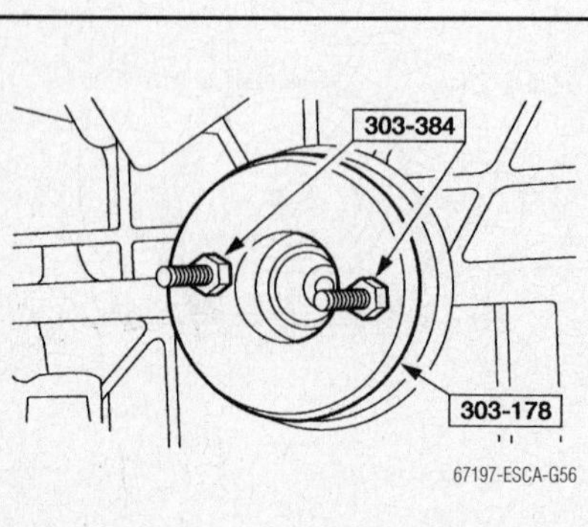

Fig. 178 Rear main seal installation—3.0L engine

2. Remove or disconnect the following:
   - Negative battery cable
   - Flexplate
   - Rear main oil seal

To install:

3. Coat the oil seal with clean engine oil.
4. Install or connect the following:
   - Crankshaft rear oil seal
   - Flywheel
   - Negative battery cable

TIMING CHAIN, GEARS, FRONT COVER & SEAL

REMOVAL & INSTALLATION

2.3L Engine

See Figures 179 through 187.

1. Before servicing the vehicle, refer to the Precautions Section.

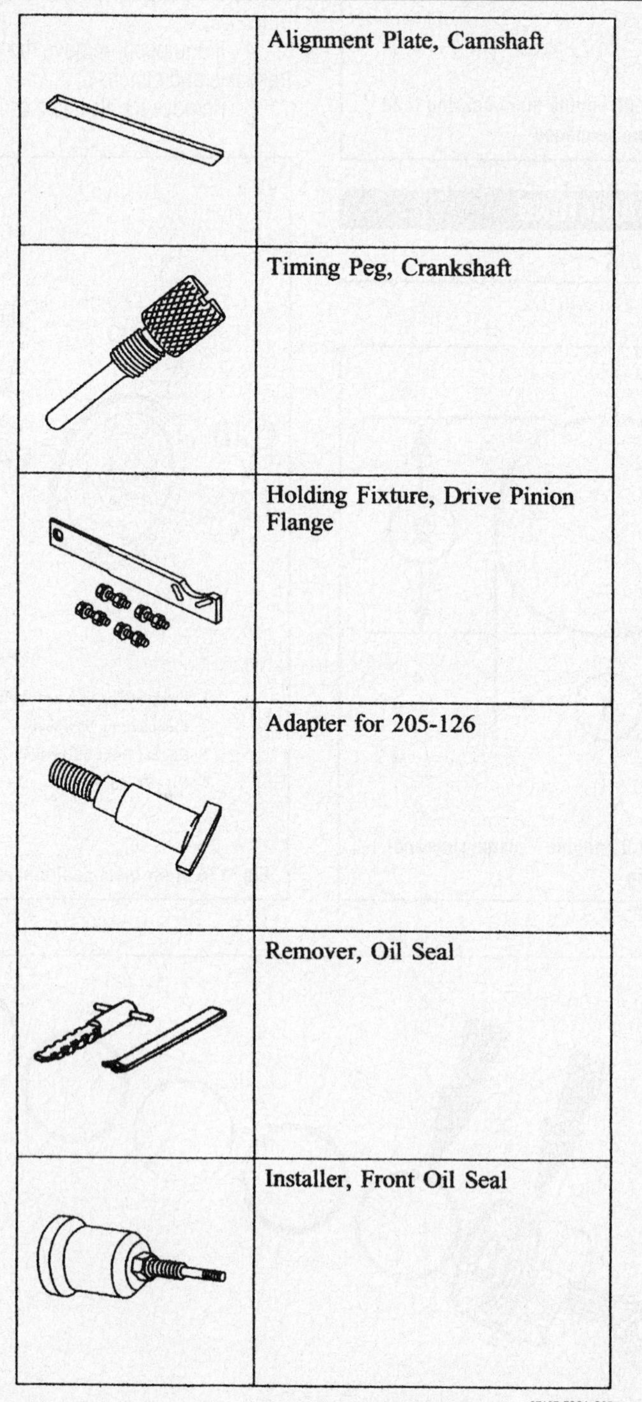

	Alignment Plate, Camshaft
	Timing Peg, Crankshaft
	Holding Fixture, Drive Pinion Flange
	Adapter for 205-126
	Remover, Oil Seal
	Installer, Front Oil Seal

67197-ESCA-G25

Fig. 179 Tools needed for timing chain and gears replacement—2.3L engine

✲✲ CAUTION

During engine repair procedures, cleanliness is extremely important. Any foreign material, including any material created while cleaning gasket surfaces that enters the oil passages, coolant passages or the oil pan can cause engine failure.

✲✲ CAUTION

The crankshaft, the crankshaft sprocket and the pulley are fitted together by friction, using diamond washers between the flange faces on each part. For that reason, the crankshaft sprocket is also unfastened if you loosen the pulley. Therefore, the engine must be retimed each time the damper is removed. Otherwise severe engine damage can occur.

2. With the vehicle in **NEUTRAL**, position it on a hoist.
3. Remove the accessory drive belt and idler pulleys.
4. Remove the engine mount.
5. Remove the valve cover.

✲✲ CAUTION

Failure to position the No. 1 piston at top dead center (TDC) can result in damage to the engine. Turn the engine in the normal direction of rotation only.

6. Using the crankshaft pulley bolt, turn the crankshaft clockwise to position the No. 1 piston at TDC.

✲✲ CAUTION

The special tool 303-465 is for camshaft alignment only. Using this tool to prevent engine rotation can result in engine damage.

➡The camshaft timing slots are offset. If the special tool cannot be installed, rotate the crankshaft one complete revolution clockwise to correctly position the camshafts.

7. Install special tool 303-465 in the slots on the rear of both camshafts.
8. Remove the engine plug bolt.

➡Only turn the engine in the normal direction of rotation.

➡Installing the special tool in this step will prevent the engine from being rotated in the clockwise direction.

9. Install special tool 303-507.
10. Install the special tools 205-126 and 205-072-02.

✲✲ CAUTION

Failure to hold the crankshaft pulley in place while loosening the bolt can result in damage to the engine.

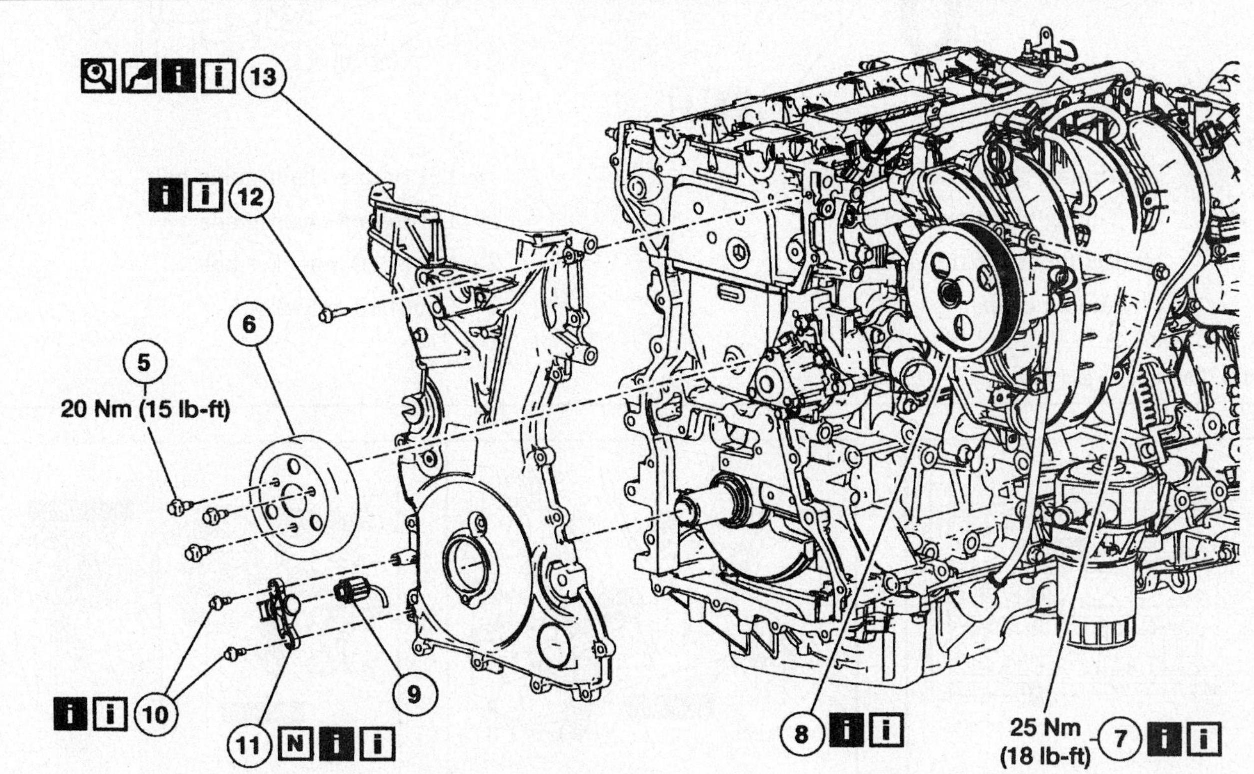

Fig. 180 Front cover and related parts—2.3L engine

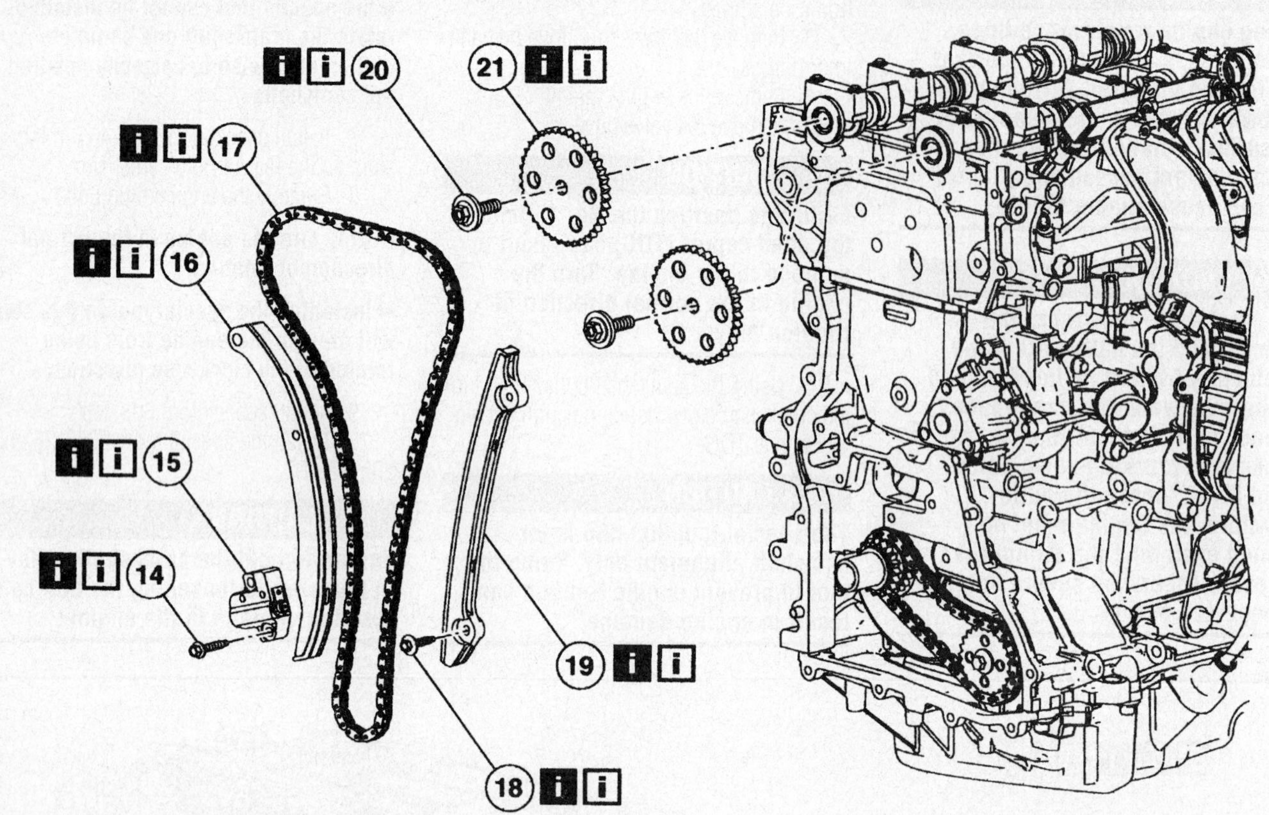

Fig. 181 Timing chain and related parts—2.3L engine

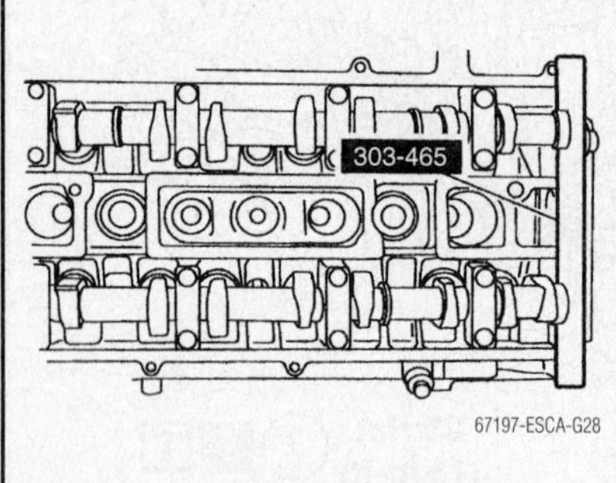

Fig. 182 Install special tool 303-465 in the slots on the rear of both camshafts—2.3L engine

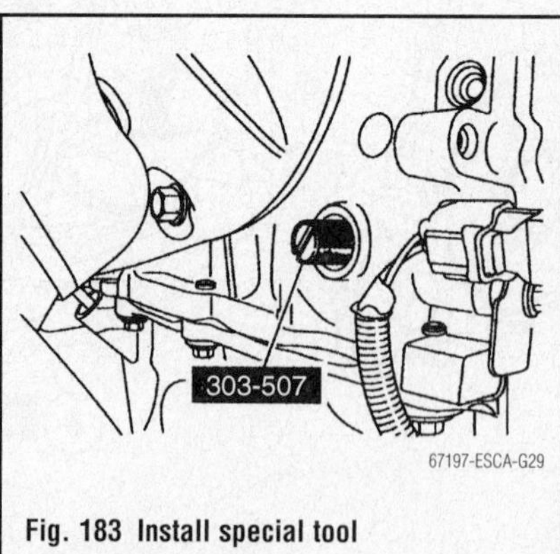

Fig. 183 Install special tool 303-507—2.3L engine

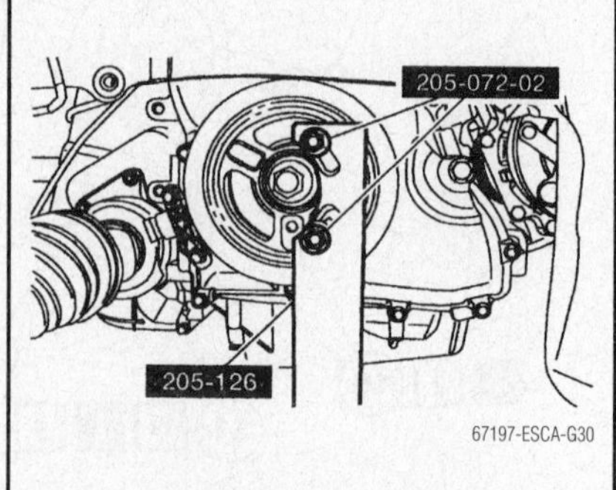

Fig. 184 Install the special tools 205-126 and 205-072-02—2.3L engine

11. Remove the crankshaft pulley bolt and washer.
12. Remove the crankshaft pulley.
13. Remove the crankshaft front seal.
14. Remove the coolant pump pulley.
15. Remove the power steering pump and position it aside.

➡The bolt under the power steering pressure tube will remain with the power steering pump.

16. Remove the CKP sensor.

➡Whenever the crankshaft position (CKP) sensor is removed, a new one must be installed, using the alignment jig supplied with the new part.

17. Remove the engine front cover bolts (there are 22).

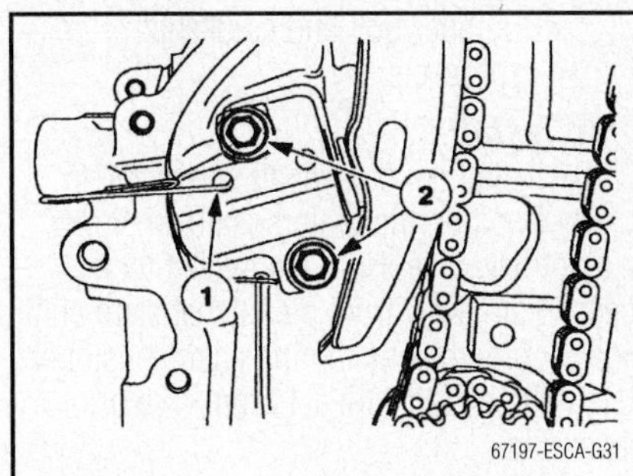

Fig. 185 Compress the timing chain tensioner, and insert a paper clip into the hole to retain the tensioner—2.3L engine

Fig. 186 Use the flats on the camshaft to prevent camshaft rotation—2.3L engine

18. Remove the engine front cover.
19. Remove the timing chain tensioner. Compress the timing chain tensioner, and insert a paper clip into the hole to retain the tensioner.
20. Remove the right timing chain guide.
21. Remove the timing chain.
22. Remove the left timing chain guide.
23. Remove the camshaft sprocket bolts.
24. Remove the camshaft sprockets.

⁂ CAUTION

Do not rely on the Camshaft Alignment Plate to prevent camshaft rotation. Damage to the tool or the camshaft can occur. Use the flats on the camshaft to prevent camshaft rotation.

To install:

25. Installation is the reverse of removal. Note the following:

⁂ CAUTION

Do not use metal scrapers, wire brushes, power abrasive disks or other abrasive means to clean sealing surfaces. These tools cause scratches and gouges which make leak paths.

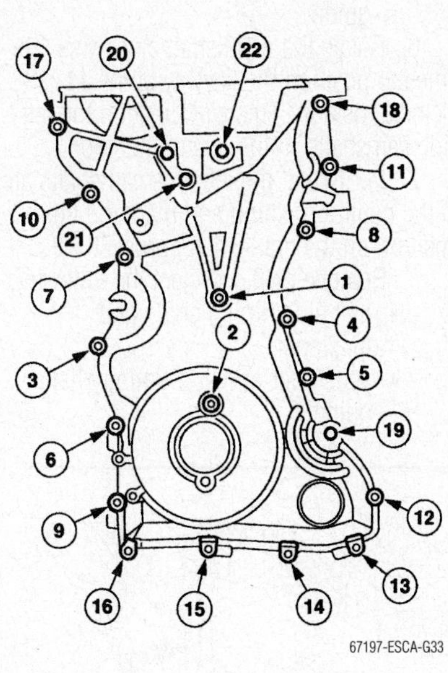

Fig. 187 Front cover bolt torque sequence—2.3L engine

26. Clean and inspect the mounting surfaces of the engine and the front cover.

➡The engine front cover must be installed and the bolts tightened within four minutes of applying the silicone gasket and sealant.

27. Apply a 2.5 mm bead of silicone gasket and sealant to the cylinder head and oil pan joint areas. Apply a 2.5 mm bead of silicone gasket and sealant to the front cover.
28. Install the engine front cover. Tighten the bolts in the sequence shown, to the following specifications:
    a. Tighten the 8 mm bolts to 89 inch lbs. (10 Nm).
    b. Tighten the 13 mm bolts to 35 ft. lbs. (48 Nm).
29. Position the power steering pump and install the bolts.

➡Remove the through-bolt from the special tool.

➡Lubricate the oil seal with clean engine oil.

30. Using a seal driver, install the crankshaft front oil seal.

➡Do not reuse the crankshaft damper bolt.

➡Apply clean engine oil on the seal area before installing.

31. Install the crankshaft pulley and hand-tighten the bolt.

⁂ CAUTION

Only hand-tighten the bolt or damage to the front cover can occur.

➡This step will correctly align the crankshaft pulley to the crankshaft.

32. Install a standard 6 mm x 18 mm bolt through the crankshaft pulley and thread it into the front cover. Rotate the pulley as necessary to align the bolt holes.

⁂ CAUTION

Failure to hold the crankshaft pulley in place while tightening the bolt can cause damage to the engine front cover.

33. Using the special tools to hold the crankshaft pulley in place, tighten the crankshaft pulley bolt in two stages:
    a. Stage 1: Tighten to 74 ft. lbs. (100 Nm).
    b. Stage 2: Tighten an additional 90 degrees (¼ turn).
34. Remove the 6 mm x 18 mm bolt.
35. Remove special tool 303–507.
36. Remove special tool 303–465.

➡Only turn the engine in the normal direction of rotation.

37. Turn the engine two complete revolutions.

➡Only turn the engine in the normal direction of rotation.

38. Turn the crankshaft until the No. 1 piston is at TDC.
39. Install special tool 303–507.

⁂ CAUTION

Only hand-tighten the bolt or damage to the front cover can occur.

40. Using the 6 mm x 18 mm bolt, check the position of the crankshaft pulley. If it is not possible to install the bolt, correct the engine timing.
41. Using special tool 303–465, check the position of the camshafts. If it is not possible to install the special tool, correct the engine timing.
42. Install the CKP sensor. Do not tighten the bolts at this time.
43. Adjust the CKP sensor alignment jig and tighten the bolts.
44. Remove the 6 mm x 18 mm bolt.
45. Install the engine plug bolt.

3.0L Engine

See Figures 188 through 199.

1. Before servicing the vehicle, refer to the Precautions Section.
2. Remove or disconnect the following:
   - Negative battery cable
   - Engine front cover

➡This pulse wheel is used in several different engines. Install the pulse wheel with the keyway in the slot stamped "30" or "30RFF" (orange in color).

   - Ignition pulse wheel and install the damper bolt
   - Spark plugs
3. Rotate the crankshaft clockwise to position the keyway at the 11 o'clock position and the camshafts in the correct positions. The No. 1 cylinder will be at Top Dead Center (TDC).
4. Rotate the crankshaft clockwise 120 degrees to the 3 o'clock position to locate the right side camshafts in the neutral position. Verify that the right camshafts are in the neutral position.
5. Remove or disconnect the following:
   - Right side timing chain and tensioner

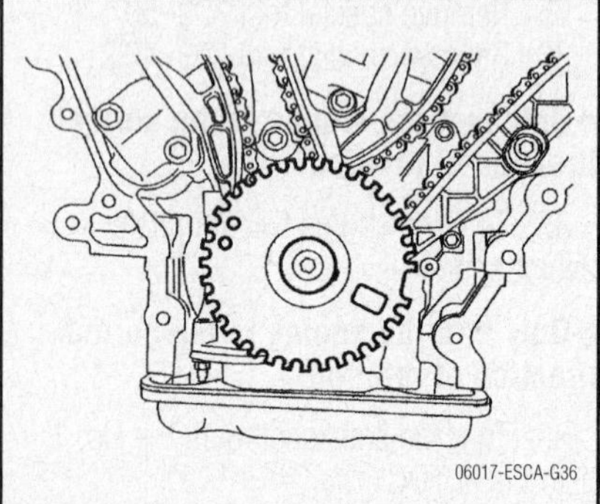

Fig. 188 Ignition pulse wheel—3.0L engine

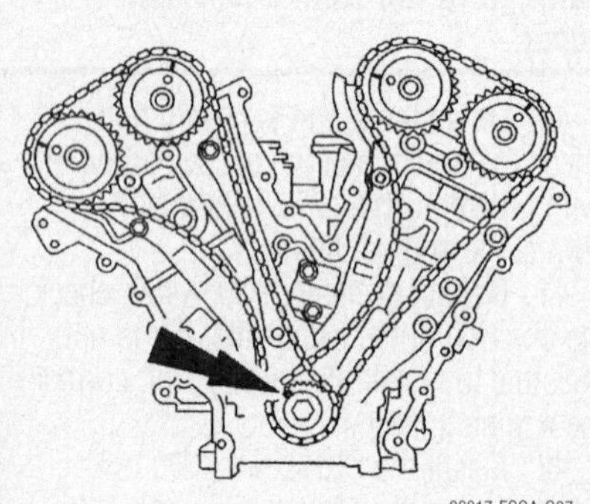

Fig. 189 Rotate the crankshaft clockwise to position the keyway at the 11 o'clock position—3.0L engine

   - Tensioner arm and timing chain guide
6. Rotate the crankshaft clockwise 1⅔ times to position the keyway at the 11 o'clock position. This will position the left side camshafts in the neutral position.
7. Verify that the left side camshafts are in the neutral position and mark the link position on the crankshaft sprocket.
8. Remove or disconnect the following:
   - Left side timing chain and tensioner
   - Tensioner arm and timing chain guide

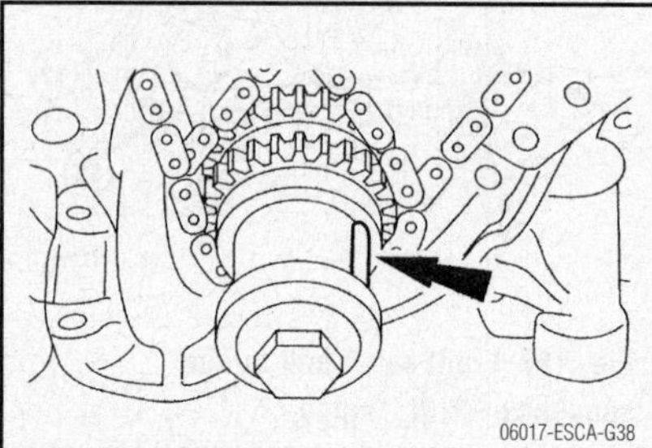

Fig. 190 Rotate the crankshaft clockwise 120 degrees to the 3 o'clock position to locate the right side camshafts in the neutral position—3.0L engine

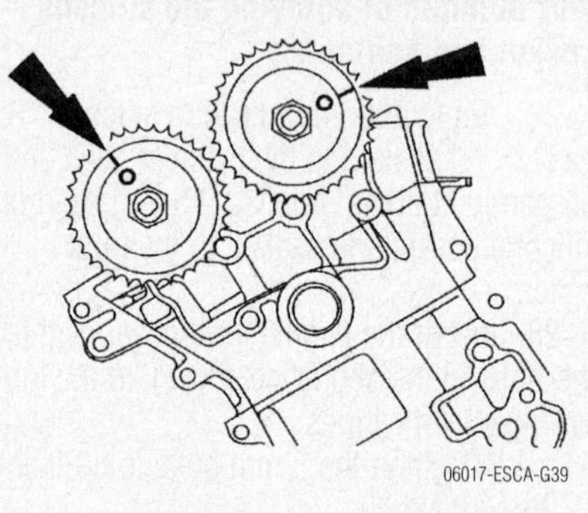

Fig. 191 Verify that the right camshafts are in the neutral position—3.0L engine

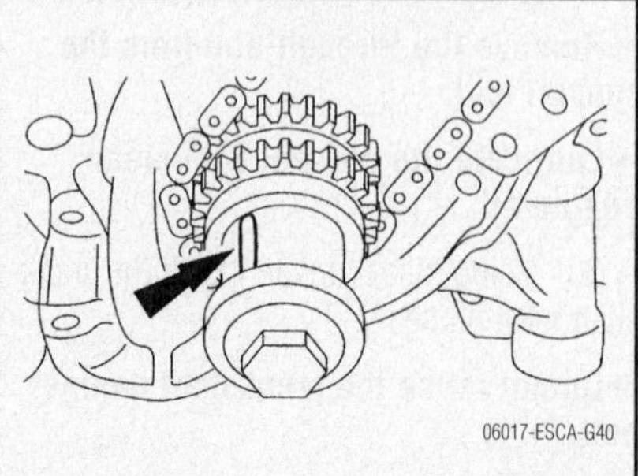

Fig. 192 Rotate the crankshaft clockwise 1⅔ times to position the keyway at the 11 o'clock position—3.0L engine

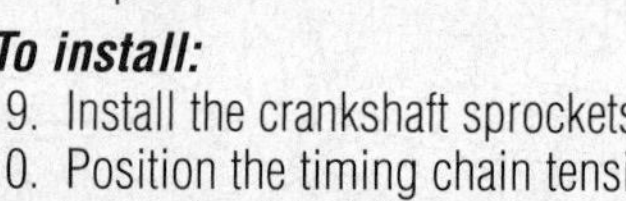

   - Damper bolt and crankshaft sprockets

To install:

9. Install the crankshaft sprockets.
10. Position the timing chain tensioner in a soft jaw vise. Hold the ratchet lock mechanism away from the ratchet stem and slowly compress the timing chain tensioner. Retain the piston with a 1.5mm wire or paper clip.
11. If the timing marks on the chain are not visible, use a permanent marker to mark the left and right side timing chains.

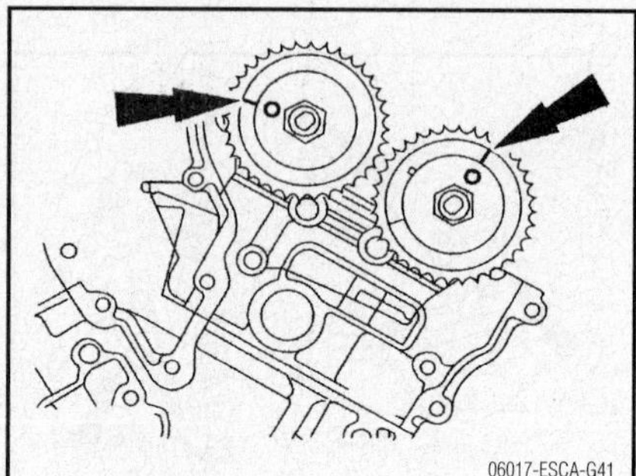

Fig. 193 Verify that the left side camshafts are in the neutral position—3.0L engine

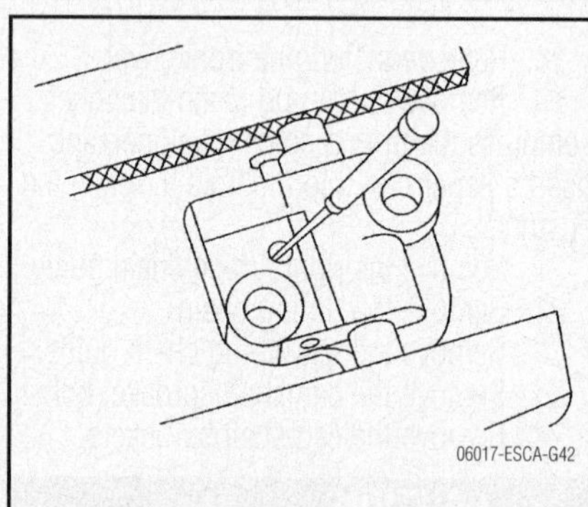

Fig. 194 Hold the ratchet lock mechanism away from the ratchet stem—3.0L engine

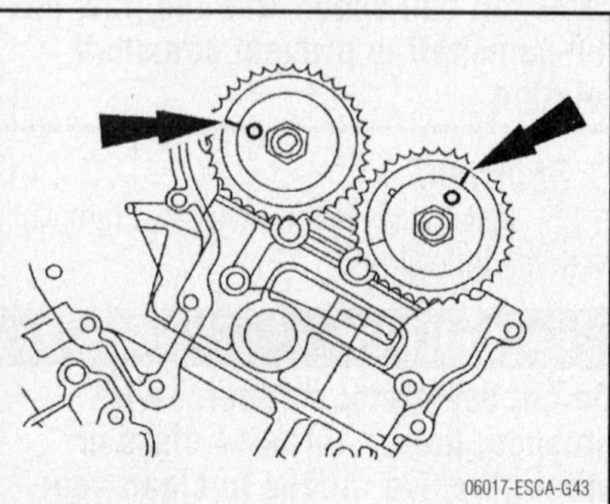

Fig. 195 Verify that the left camshafts are correctly positioned–3.0L engine

Mark the timing chains in the following sequence:

a. Mark any link to use as the crankshaft timing mark.

b. Count 29 links from the crankshaft timing mark and mark the link as the exhaust cam sprocket timing mark.

c. Continue counting to 42 and mark the link as the intake sprocket timing mark.

12. Verify that the left camshafts are correctly positioned.

13. Install the guide. Torque the bolts to 18 ft. lbs. (25 Nm).

14. Install the left side timing chain and align the chain in the following sequence:

a. Mark any link to use as the crankshaft timing mark.

b. Count 29 links from the crankshaft timing mark and mark the link as the exhaust cam sprocket timing mark.

c. Continue counting to 42 and mark the link as the intake sprocket timing mark

15. Install or connect the following:

- Left side timing chain and tensioner arm. Torque the bolts to 18 ft. lbs. (25 Nm).
- Crankshaft damper bolt and rotate the keyway to the 3 o'clock position.

16. Verify that the right side camshafts are properly positioned and install the right side timing chain and guide. Torque the bolts to 18 ft. lbs. (25 Nm).

17. Make certain that the timing chain aligns with the marks on the camshaft and crankshaft sprockets

CAUTION

Install the pulse wheel with the keyway in the slot stamped 20–25–34Y–30M (Color Blur).

18. Install or connect the following:

- Right side timing chain tensioner and arm. Torque the bolts to 18 ft. lbs. (25 Nm) and remove the damper bolt

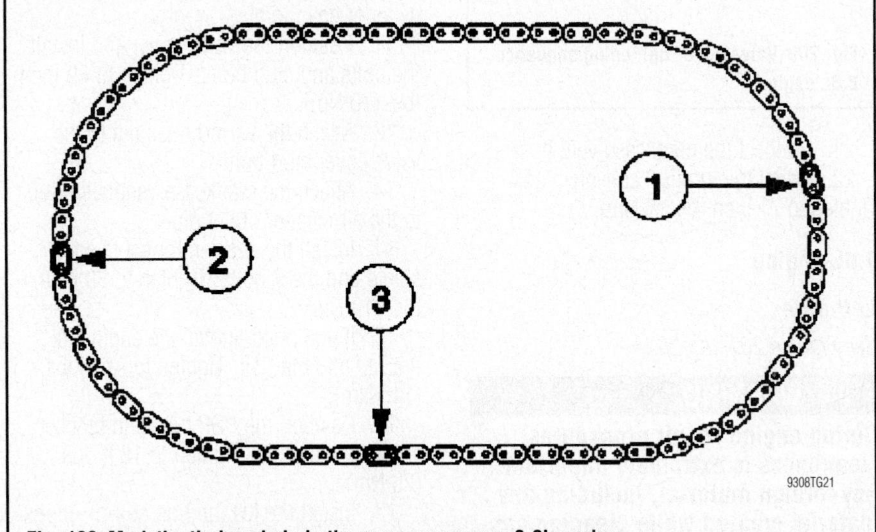

Fig. 196 Mark the timing chain in the proper sequence–3.0L engine

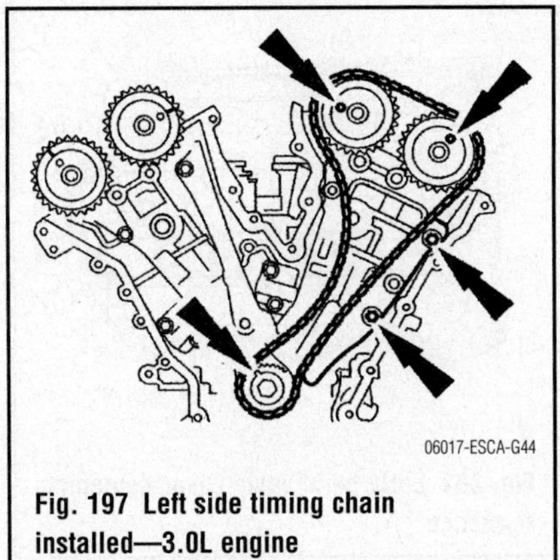

Fig. 197 Left side timing chain installed—3.0L engine

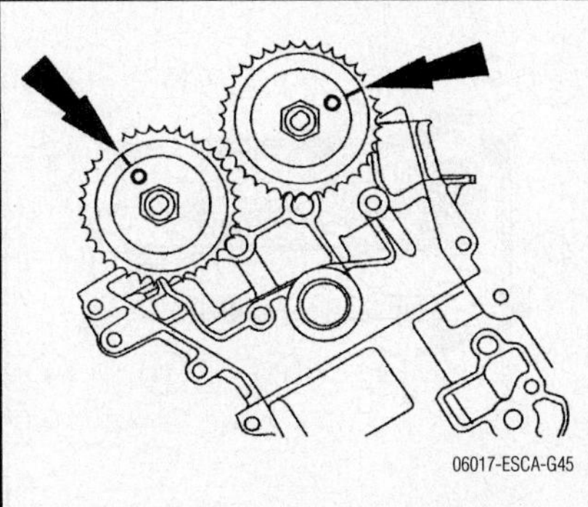

Fig. 198 Verify that the right camshafts are correctly positioned–3.0L engine

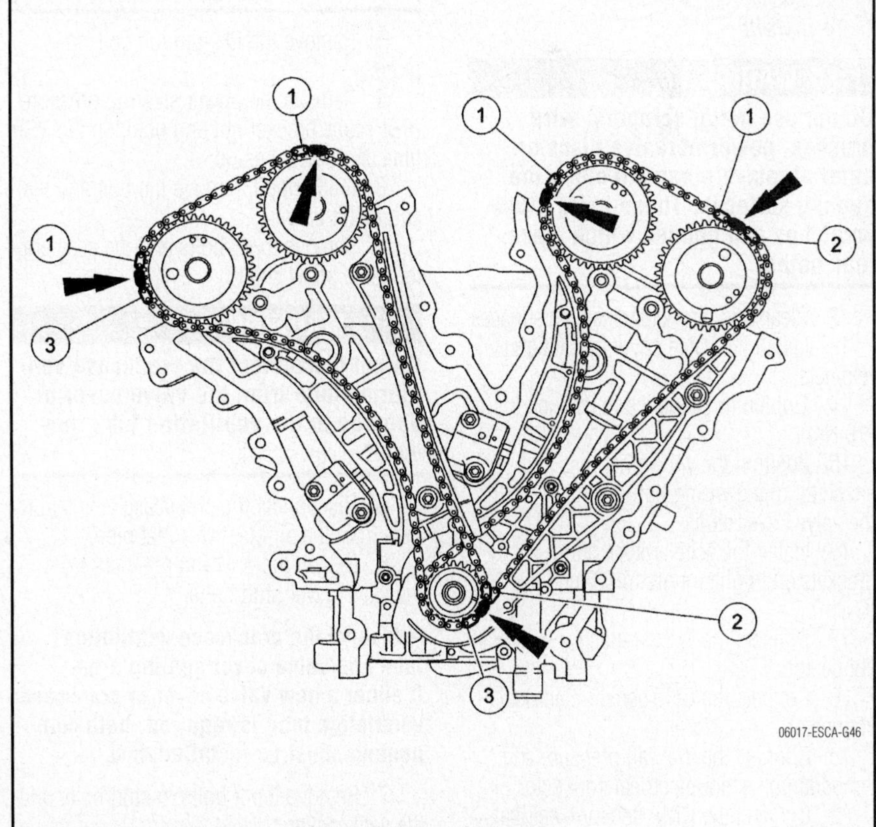

Fig. 199 Timing chains correctly installed—3.0L engine

- Ignition pulse wheel
- Spark plugs
- Engine front cover
- Negative battery cable

VALVE COVERS

REMOVAL & INSTALLATION

2.3L Engine

See Figure 200.

1. With the vehicle in NEUTRAL, position it on a hoist.
2. Remove the ignition coil-on-plugs.
3. Disconnect the crankcase vent hose.
4. Disconnect the Camshaft Position (CMP) sensor electrical connector.
5. Disconnect the fuel rail pressure and temperature sensor electrical connector.
6. Disconnect the Cylinder Head Temperature (CHT) sensor electrical connector.
7. Disconnect the radio capacitor electrical connector.
8. Remove the 2 accelerator cable bracket nuts.
9. Position the accelerator cable and brackets aside.
10. Detach all of the wiring harness retainers from the valve cover studs and position the harness aside.
11. Remove the 14 valve cover retainers, the valve cover and gasket.

To install:

✱✱ WARNING

Do not use metal scrapers, wire brushes, power abrasive discs or other abrasive means to clean the sealing surfaces. These tools cause scratches and gouges which make leak paths.

12. Clean and inspect the sealing surfaces
13. Install the valve cover, gasket and retainers.
14. Tighten in sequence to 89 inch lbs. (10 Nm).
15. Position the wiring harness and attach all of the wiring harness retainers to the valve cover studs.
16. Install the accelerator control cable brackets and tighten nuts to 53 inch lbs. (6 Nm).
17. Connect the radio capacitor electrical connector.
18. Connect the CHT sensor electrical connector.
19. Connect the fuel rail pressure and temperature sensor electrical connector
20. Connect the CMP sensor electrical connector.

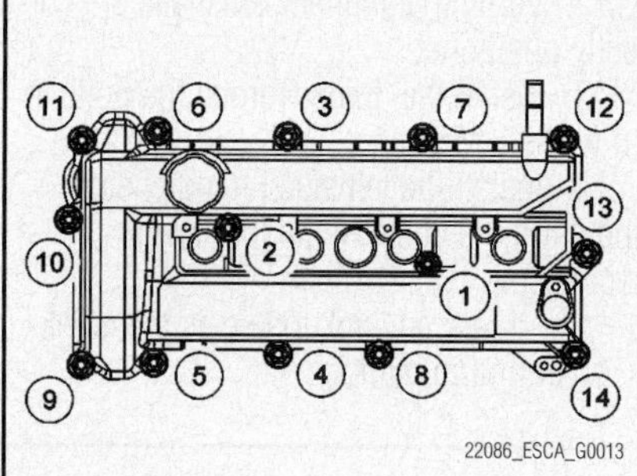

Fig. 200 Valve cover tightening sequence 2.3L engine

21. Connect the crankcase vent hose.
22. Install the ignition coil-on-plugs. Tighten to 71 inch lbs. (8 Nm).

3.0L Engine

Left Side

See Figures 201 and 202.

✱✱ WARNING

During engine repair procedures, cleanliness is extremely important. Any foreign material, including any material created while cleaning gasket surfaces that enters the oil passages, coolant passages or the oil pan, can cause engine failure.

1. Remove the LH ignition coil-on-plugs.
2. Remove the power steering pressure (PSP) tube bracket nut and position the PSP tube and bracket aside.
3. If equipped, remove the bolt and the engine lift bracket.
4. Remove the 2 bolts and the accelerator cable snow shield.

✱✱ WARNING

Do not disconnect the crankcase ventilation tube from the valve cover or damage to the ventilation tube may occur.

5. Disconnect the crankcase ventilation tube from the air cleaner outlet pipe.
6. Detach the 5 wiring retainers from the valve cover stud bolts.

➡Inspect the crankcase ventilation tube and valve cover sealing area. If either a new valve cover or crankcase ventilation tube is required, both components must be installed new.

7. Remove the 2 bolts, 9 stud bolts and the valve cover.
8. Remove and discard the gasket.

To install:

9. Clean the valve cover, cylinder head and front cover sealing surfaces with metal surface cleaner and install a new valve cover gasket.

➡The valve cover must be installed and the bolts and stud bolts tightened within 4 minutes of sealant application.

10. Apply a 19 inch (5mm) dot of silicone gasket sealant to the front cover-to-cylinder head joints.
11. Position the valve cover and install the bolts and stud bolts. Tighten to 89 inch lbs. (10 Nm).
12. Attach the wiring retainers to the valve cover stud bolts.
13. Attach the crankcase ventilation tube to the air cleaner outlet pipe
14. Install the accelerator cable snow shield and the 2 bolts. Tighten to 89 inch lbs. (10 Nm).
15. If equipped, install the engine lift bracket and the bolt. Tighten to 87 ft. lbs. (118 Nm).
16. Position the PSP tube and bracket and install the nut. Tighten to 18 ft. lbs. (25 Nm).
17. Install the LH ignition coil-on-plugs and tighten to 11 ft. lbs. (15 Nm).

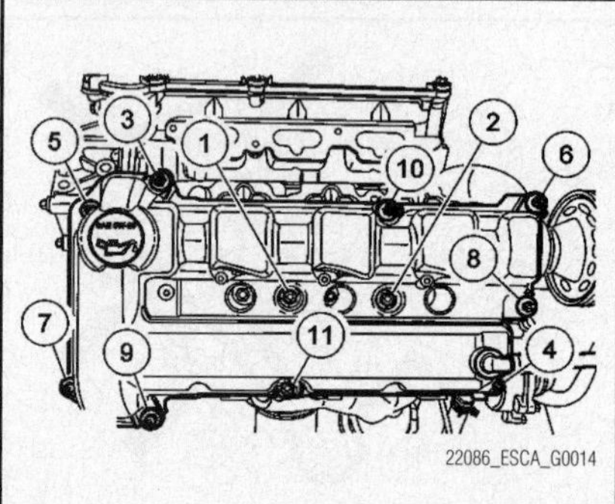

Fig. 201 Early build valve cover tightening sequence

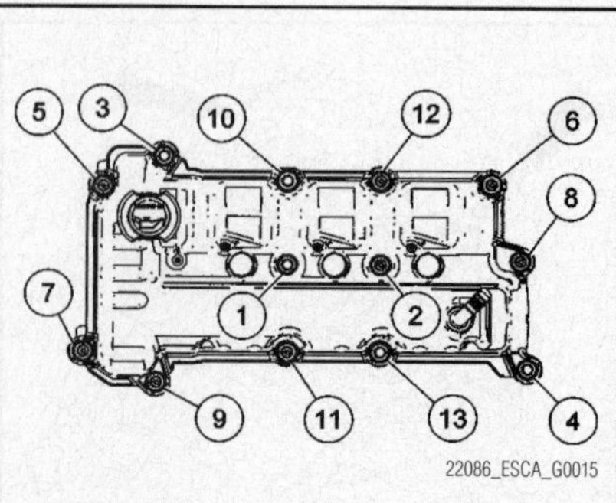

Fig. 202 Late build valve cover tightening sequence

Right Side

See Figures 203, 204.

1. Remove the upper intake manifold.
2. Remove the RH ignition coil-on-plugs.
3. Disconnect the positive crankcase ventilation (PCV) valve electrical connector.
4. Disconnect the PCV tube from the PCV valve and position it aside.
5. Disconnect the radio ignition interference capacitor electrical connector and detach the 3 wiring retainers from the stud bolts.
6. Remove the nut and the radio ignition interference capacitor.
7. Remove the 2 nuts and position the engine control harness aside.
8. Remove the bolt, 9 stud bolts and the valve cover.
9. Remove and discard the gasket.

To install:

10. Clean the valve cover, cylinder head and front cover sealing surfaces with metal surface cleaner and install a new valve cover gasket.

➡The valve cover must be installed and the bolts and stud bolts tightened within 4 minutes of sealant application. Apply a 19 inch. (5mm) dot of silicone gasket sealant to the front cover-to-cylinder head joints.

11. Position the valve cover and install the stud bolts.
12. Tighten in sequence to 89 inch lbs. (10 Nm).
13. Position the engine control harness and install the nuts. Tighten to 53 inch lbs. (6 Nm).
14. Install the radio ignition interference capacitor and the nut. Tighten to 53 inch lbs. (6 Nm).
15. Attach the wiring retainers to the stud bolts and connect the radio ignition interference capacitor electrical connector.
16. Connect the PCV tube to the PCV valve.
17. Connect the PCV valve electrical connector.
18. Install the RH ignition coil-on-plugs and tighten to 11 ft. lbs. (15 Nm).

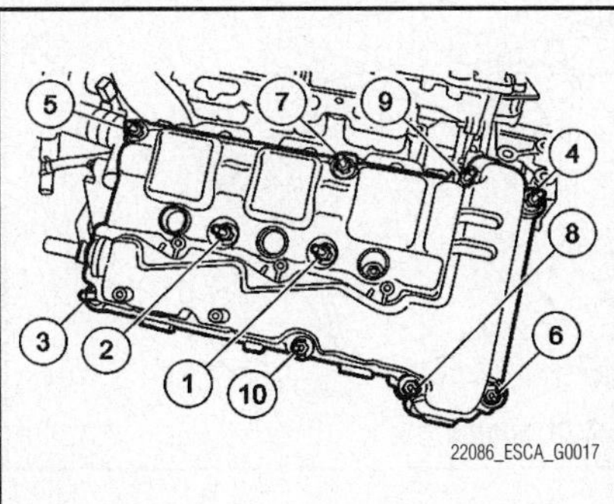

Fig. 204 Right valve cover tightening sequence 3.0L engine

1. Positive crankcase ventilation (PCV) valve electrical connector
2. PCV tube
3. Wiring retainer (3 required)
4. Valve cover stud bolt (10 required)
5. Valve cover
6. Valve cover gasket

22086_ESCA_G0016

Fig. 203 Valve cover right side view—3.0L engine

VALVE LASH

ADJUSTMENT

2.3L Engine

1. Before servicing the vehicle, refer to the Precautions Section.

➡Before removing the camshafts, measure the clearance of each valve at base circle, with the lobe pointed away from the tappet. Failure to measure all clearances prior to removing the camshafts will necessitate repeated removal and installation and wasted labor time.

2. Use a feeler gauge to measure the clearance of each valve and record its location.

➡The number on the valve tappet only reflects the digits that follow the decimal. For example, a tappet with the number 0.650 has the thickness of 3.650 mm.

3. A midrange clearance is the most desirable:
   - Intake: 0.22–0.28 mm (0.008–0.011 inch)
   - Exhaust: 0.27–0.33 mm (0.010–0.013 inch)
4. Select tappets using this formula: tappet thickness = measured clearance + the base tappet thickness–most desirable thickness.
5. Select the tappets and mark the installation location.
6. If any tappets do not measure within specifications, install new tappets in these locations.

3.0L Engine

1. Before servicing the vehicle, refer to the Precautions Section.
2. Remove or disconnect the following:
   - Negative battery cable
   - Camshaft followers
   - Hydraulic lash adjusters

➡Mark the position of the hydraulic lash adjusters to assure they are assembled in their original position

3. Inspect the adjusters for scoring or uneven wear in the bore and replace them as required.

To install:

4. Install or connect the following:
   - Hydraulic lash adjusters after lubricating them with clean engine oil
   - Camshaft followers
   - Negative battery cable

ENGINE PERFORMANCE & EMISSION CONTROL

CAMSHAFT POSITION (CMP) SENSOR

LOCATION

2.3L Engine

See Figure 205.

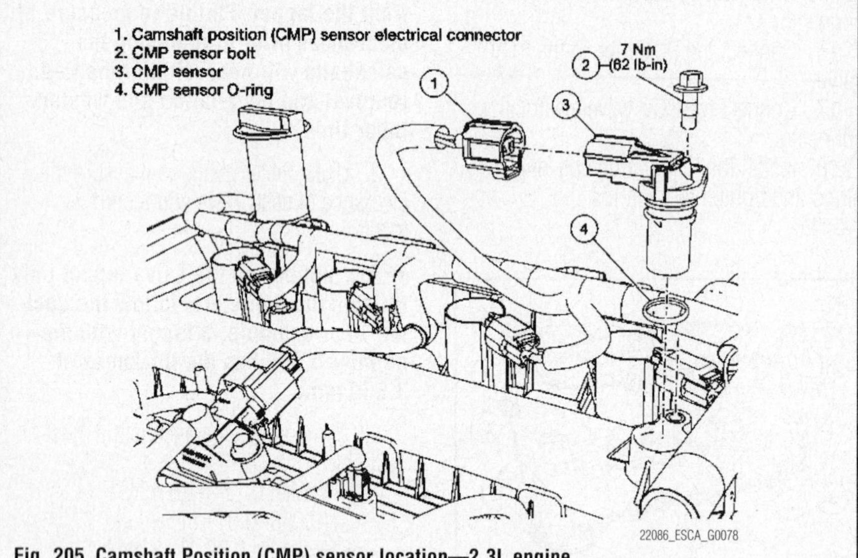

Fig. 205 Camshaft Position (CMP) sensor location—2.3L engine

The Camshaft Position (CMP) sensor is located on top the valve cover towards the front of the vehicle.

3.0L Engine

See Figure 206.

The Camshaft Position (CMP) sensor is located on left cylinder head just below the valve cover.

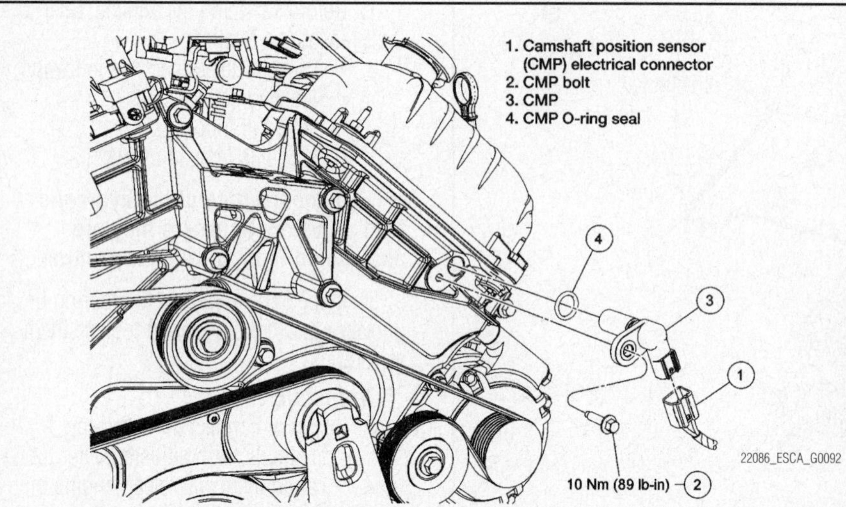

Fig. 206 Camshaft Position (CMP) sensor location—3.0L engine

REMOVAL & INSTALLATION

2.3L Engine

1. Disconnect the Camshaft Position (CMP) sensor electrical connector.
2. Remove the bolt and the CMP sensor.

To install:

➡**Lubricate the CMP sensor O-ring seal with clean engine oil.**

3. To install, reverse the removal procedure.
4. Tighten the mounting bolt to 62 inch lbs. (7 Nm)

3.0L Engine

1. Disconnect the Camshaft Position (CMP) sensor electrical connector.
2. Remove the bolt and the CMP sensor.

To install:

➡**Lubricate the CMP O-ring seal with clean engine oil.**

3. To install, reverse the removal procedure and tighten the mounting bolt to 89 inch lbs. (10 Nm).

CRANKSHAFT POSITION (CKP) SENSOR

LOCATION

2.3L Engine

See Figure 207.

The Crankshaft Position (CKP) sensor is located to the left of the crankshaft pulley.

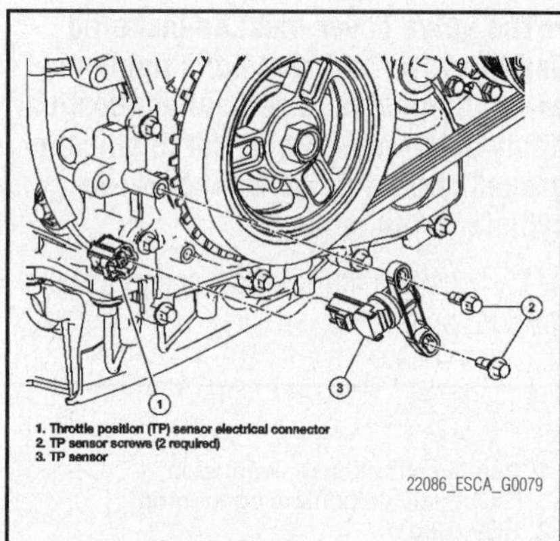

Fig. 207 Crankshaft Position (CKP) sensor location—2.3L engine

3.0L Engine

See Figure 208.

The Crankshaft Position (CKP) sensor is located just behind the crankshaft pulley on engine block.

REMOVAL & INSTALLATION

2.3L Engine

See Figures 209 through 213.

1. With the vehicle in **NEUTRAL**, position it on a hoist.
2. Remove the 5 bolts and the RH splash shield.
3. Remove the engine plug bolt.
4. Turn the crankshaft pulley bolt to position the number one cylinder at Top Dead Center (TDC) and install the special tool.

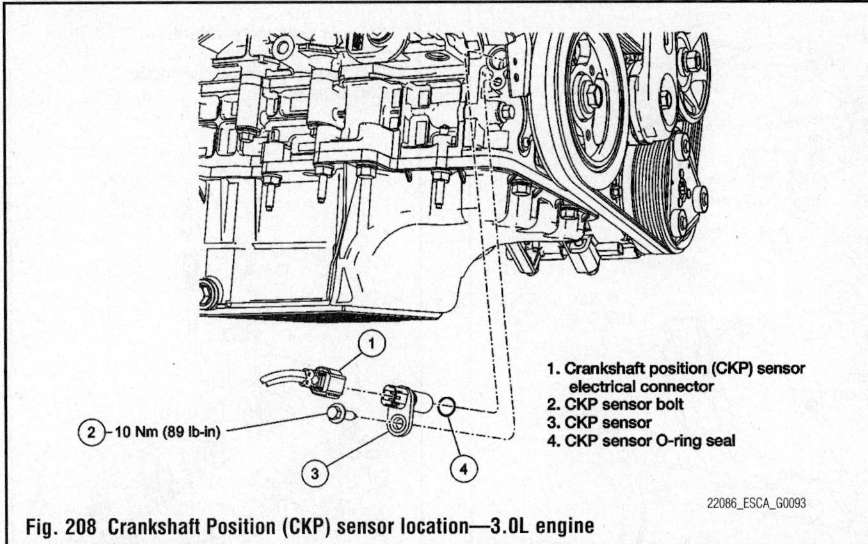

Fig. 208 Crankshaft Position (CKP) sensor location—3.0L engine

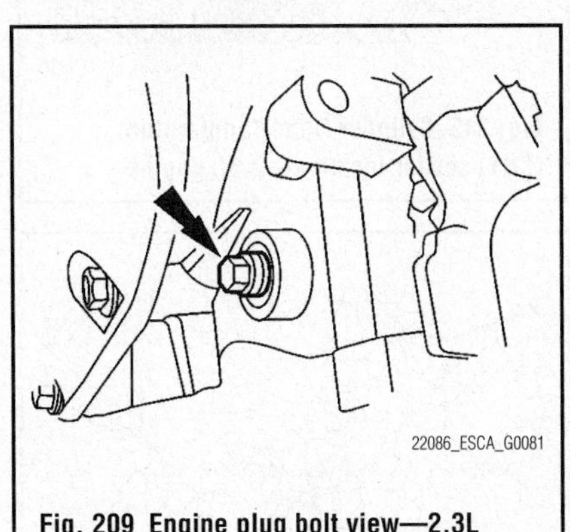

Fig. 209 Engine plug bolt view—2.3L engine

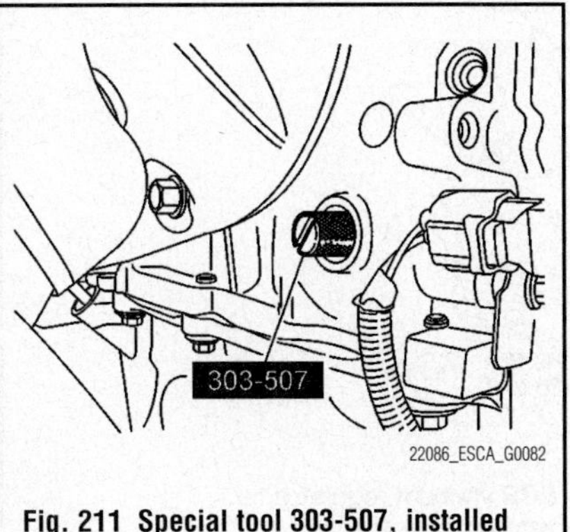

Fig. 211 Special tool 303-507, installed view

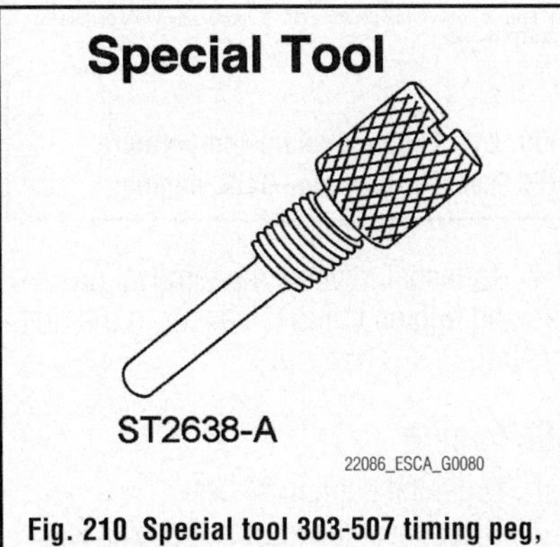

Fig. 210 Special tool 303-507 timing peg, crankshaft

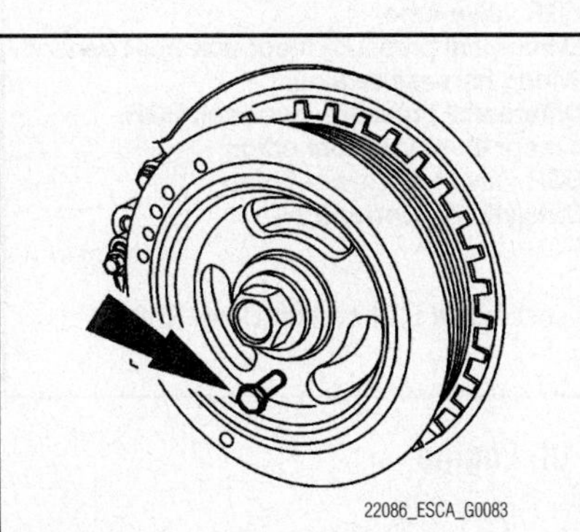

Fig. 212 Crankshaft pulley and bolt view—2.3L engine

5. Disconnect the Crankshaft Position (CKP) sensor electrical connector.
6. Remove the bolts and the CKP sensor.

To install:

7. Install a (6mm) 0.23 inch. x (18mm) 0.7 inch standard bolt in the crankshaft pulley.

**** WARNING**

Only hand-tighten the bolt or damage to the front cover can occur.

➡Whenever the CKP sensor is removed, a new one must be installed using the alignment tool supplied with the new part.

8. Install a new CKP sensor and the bolts. Do not tighten the bolts at this time.

➡The CKP sensor alignment tool is supplied with the new sensor and is not available separately.

Fig. 213 Crankshaft sensor and alignment tool shown—2.3L engine

9. Adjust the CKP sensor with the alignment tool and tighten mounting bolts to 62 inch lbs. (7 Nm).
10. Connect the CKP sensor electrical connector.
11. Remove the (6mm) 0.23 inch bolt from the crankshaft pulley.
12. Install the engine plug bolt and tighten to 15 ft. lbs. (20 Nm).
13. Install the RH splash shield and tighten the bolts to 80 inch lbs. (9 Nm).

3.0L Engine

1. With the vehicle in **NEUTRA**, position it on a hoist.
2. Remove the 5 bolts and the RH splash shield.
3. Disconnect the Crankshaft Position (CKP) sensor electrical connector.

➡Lubricate the CKP O-ring seal with clean engine oil.

4. Remove the bolt and the CKP sensor.
5. To install, reverse the removal procedure and tighten sensor mounting bolt to 89 inch lbs. (10 Nm).

DIFFERENTIAL PRESSURE FEEDBACK EGR (DPFE) SENSOR

LOCATION

3.0L Engine

See Figure 214.

The Differential Pressure Feedback EGR (DPFE) sensor is located to the rear of the engine and is above RH converter.

REMOVAL & INSTALLATION

1. Disconnect the Differential Pressure Feedback EGR (DPFE) sensor electrical connector and detach the wiring harness retainer.
2. Carefully detach the DPFE sensor from the EGR valve tube

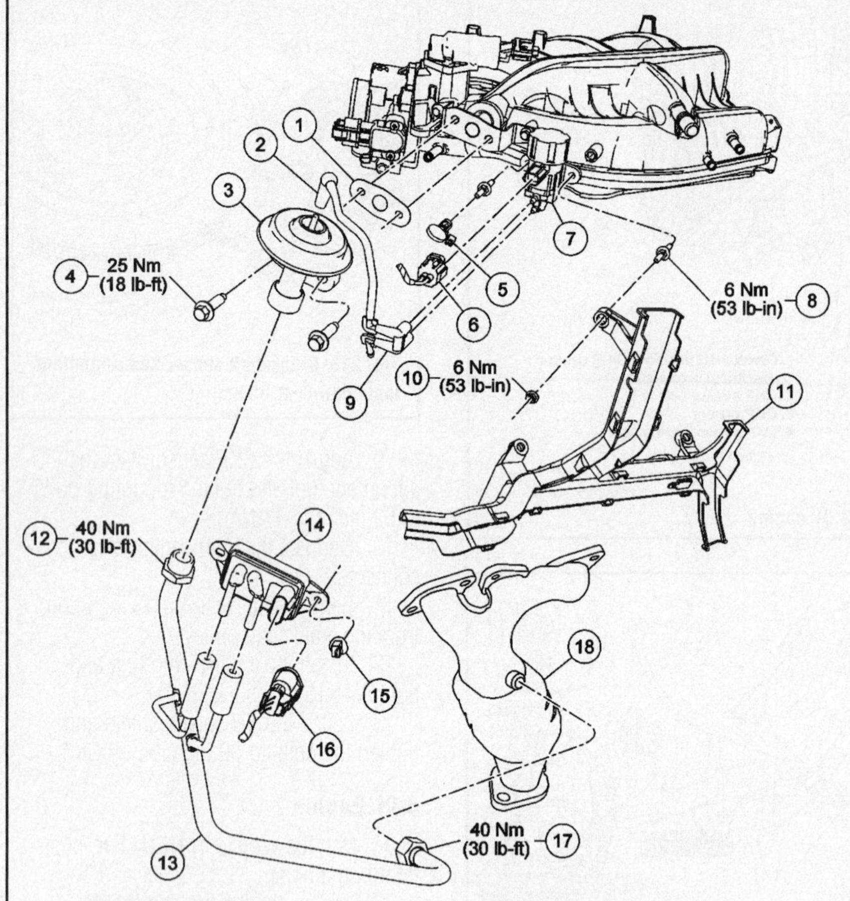

1. EGR valve gasket
2. Vacuum tube fitting
3. EGR valve
4. EGR valve bolt (2 required)
5. Wiring harness retainer
6. EGR vacuum regulator valve electrical connector
7. EGR vacuum regulator valve
8. EGR vacuum regulator valve stud bolt (2 required)
9. Vacuum tube fitting
10. EGR vacuum regulator valve stud bolt nut
11. Engine wiring harness bracket
12. EGR valve tube upper fitting
13. EGR valve tube
14. Differential pressure feedback EGR sensor
15. Wiring harness retainer
16. Differential pressure feedback EGR sensor electrical connector
17. EGR valve tube lower fitting
18. Catalytic converter (RH)

22086_ESCA_G0099

Fig. 214 Differential Pressure Feedback EGR (DPFE) sensor and EGR system view—3.0L engine

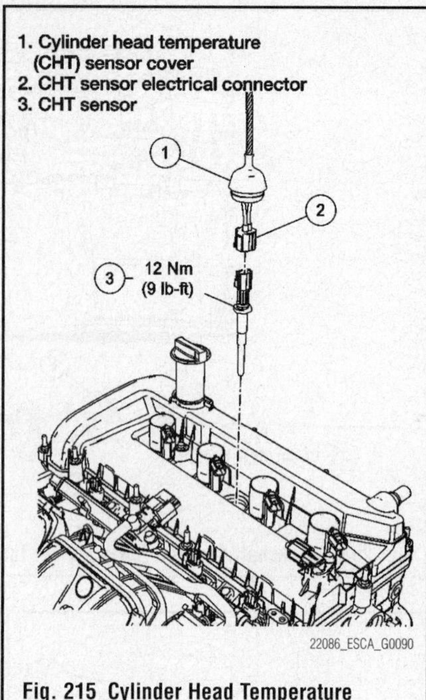

Fig. 215 Cylinder Head Temperature (CHT) sensor location—2.3L engine

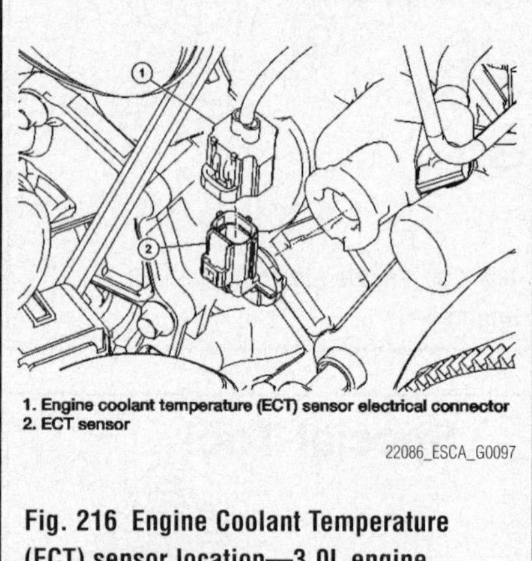

Fig. 216 Engine Coolant Temperature (ECT) sensor location—3.0L engine

3. To install, reverse the removal procedure.

ENGINE COOLANT TEMPERATURE (ECT) SENSOR

LOCATION

2.3L Engine

See Figure 215.

➡On 2.3L engine, this sensor is referred to as the Cylinder Head Temperature (CHT) sensor.

The Cylinder Head Temperature (CHT) sensor is located between the two center ignition coils.

3.0L Engine

See Figure 216.

The Engine Coolant Temperature (ECT) sensor is located on the front of left cylinder head, just under valve cover.

REMOVAL & INSTALLATION

2.3L Engine

1. Detach the Cylinder Head Temperature (CHT) sensor cover and position aside.
2. Disconnect the CHT sensor electrical connector.
3. Remove and discard the CHT sensor.
4. To install, reverse the removal procedure and tighten the CHT sensor to 9 ft. lbs. (12 Nm).

3.0L Engine

1. Drain the cooling system.
2. Disconnect the Engine Coolant Temperature (ECT) sensor electrical connector.
3. To remove the ECT sensor, pull up on locking tab and rotate the sensor clockwise.
4. To install, reverse the removal procedure.

FUEL RAIL TEMPERATURE PRESSURE (FRPT) SENSOR

LOCATION

2.3L Engine

See Figure 217.

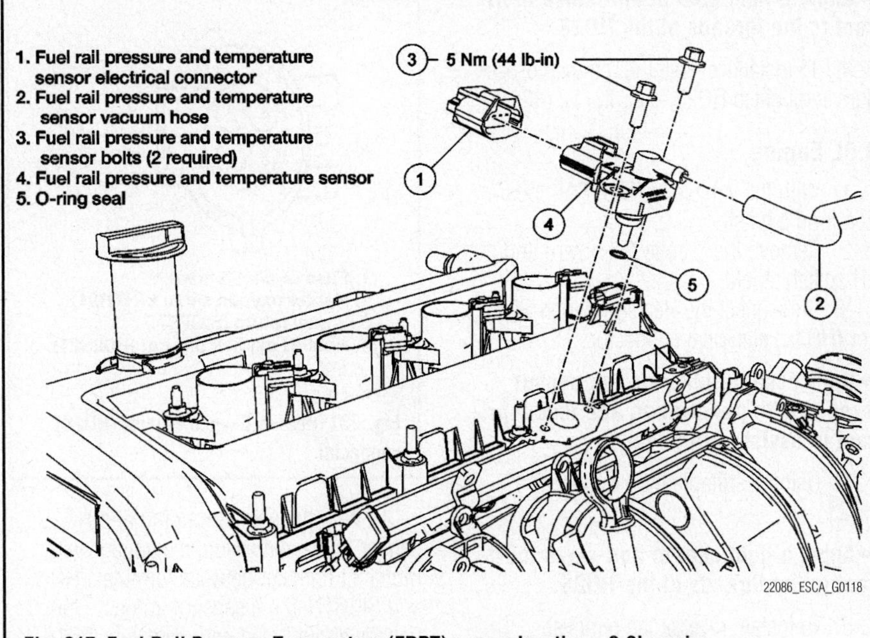

Fig. 217 Fuel Rail Pressure Temperature (FRPT) sensor location—2.3L engine

3.0L Engine

See Figure 218.

REMOVAL & INSTALLATION

2.3L Engine

⁂ CAUTION

Do not smoke or carry lighted tobacco or open flame of any type when working on or near any fuel-related components. Highly flammable mixtures are always present and may be ignited. Failure to follow these instructions may result in personal injury.

1. Release the fuel system pressure.
2. Disconnect the negative battery cable.
3. Disconnect the Fuel Rail Pressure Temperature (FRPT) sensor electrical connector and vacuum tube.
4. Remove the 2 bolts and the FRPT sensor.

To install:

➡Lubricate the FRPT sensor O-ring seal with clean engine oil.

5. To install, reverse the removal procedure.
6. Tighten the FRPT sensor mounting bolts to 44 inch lbs. (5 Nm).

Fig. 218 Fuel Rail Pressure Temperature (FRPT) sensor location—3.0L Engine

3.0L Engine

⁂ CAUTION

Do not smoke or carry lighted tobacco or open flame of any type when working on or near any fuel-related components. Highly flammable mixtures are always present and may be ignited. Failure to follow these instructions may result in personal injury.

1. Release the fuel system pressure.
2. Disconnect the negative battery cable.
3. Disconnect the Fuel Rail Pressure Temperature (FRPT) sensor electrical connector and vacuum tube.
4. Remove the 2 bolts and the FRPT sensor.

To install:

➡Lubricate the FRPT sensor O-ring seal with clean engine oil.

5. To install, reverse the removal procedure.
6. Tighten the FRPT sensor mounting bolts to 35 inch lbs. (4 Nm).

HEATED OXYGEN (HO2S) SENSOR

LOCATION

2.3L Engine

See Figure 219.

The Heated Oxygen Sensor (HO2S) is located just below exhaust manifold shield.

3.0L Engine

See Figure 220.

The Heated Oxygen Sensors (HO2S) are located at the top of the LH and RH converters.

REMOVAL & INSTALLATION

2.3L Engine

1. With the vehicle in NEUTRAL, position it on a hoist.
2. Disconnect the Heated Oxygen Sensor (HO2S) electrical connector.
3. Using a suitable tool, remove the HO2S.

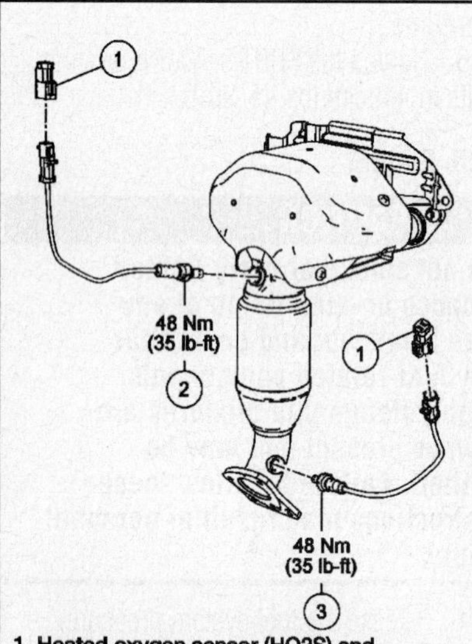

Fig. 219 Heated Oxygen Sensor (HO2S) location—2.3L engine

➡Apply a light coat of anti-seize lubricant to the threads of the HO2S.

4. To install, reverse the removal procedure and tighten HO2S to 35 ft. lbs. (48 Nm).

3.0L Engine

1. With the vehicle in NEUTRAL, position it on a hoist.
2. Remove the 7 bolts (5 shown) and the LH splash shield.
3. Disconnect the Heated Oxygen Sensor (HO2S) electrical connector.

➡If necessary, lubricate the sensor threads with penetrating and lock lubricant to assist in removal.

4. Using a suitable tool, remove the HO2S.

➡Apply a light coat of anti-seize lubricant to the threads of the HO2S.

5. To install, reverse the removal procedure and tighten HO2S to 35 ft. lbs. (38 Nm).

TESTING

See Figure 221.

1. Disconnect the Heated Oxygen Sensor (HO2S) connector. Measure the resistance between the HO2S heater and VPWR on sensor side of harness. The reading should be approximately 3—30 ohms. If resistance is not within specifications, the sensor may be faulty.

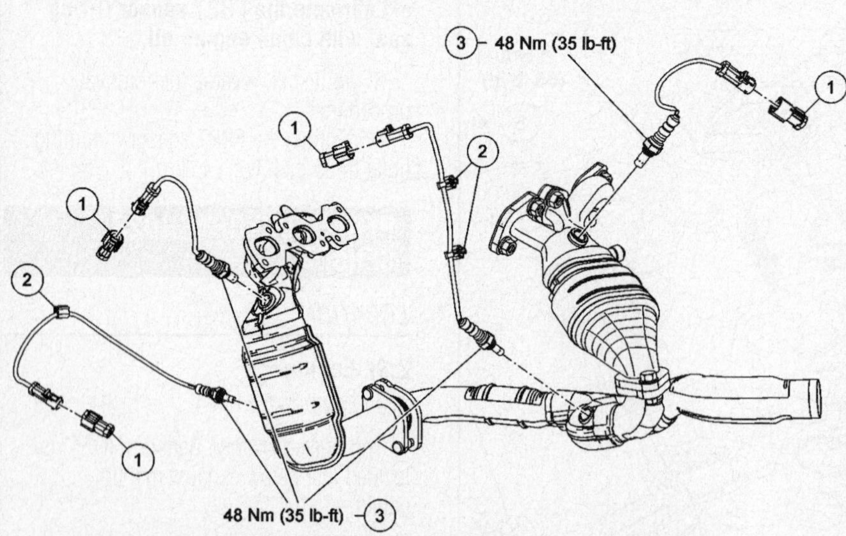

Fig. 220 Heated Oxygen Sensor (HO2S) locations—3.0L engine

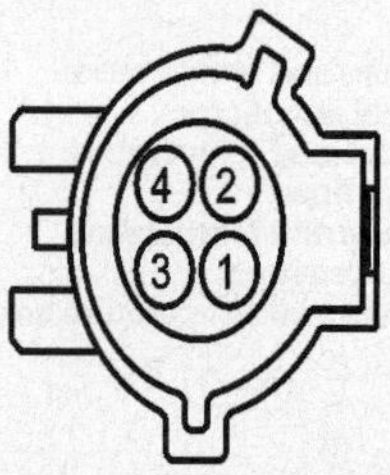

Fig. 221 Heated Oxygen Sensor (HO2S) connector

2. With the engine running and HO2S sensor in place measure the voltage with digital multimeter between terminals HO2S and SIGRTN at the sensor harness. The voltage should read approximately 0.01—1.0V. If the reading is, off or the voltage fluctuation is very slow suspect faulty sensor.
3. Check for unmetered air leaks at intake manifold gasket leaks, hoses connecting to the mass air flow (MAF) sensor assembly, PCV system. Fuel calculations can be affected by unmetered air leaks.

IDLE AIR CONTROL (IAC) VALVE

LOCATION

2.3L Engine

See Figure 222.

The Idle Air Control (IAC) valve is located just behind the throttle body on the intake manifold.

3.0L Engine

See Figure 223.

The Idle Air Control (IAC) valve is located just behind the throttle body on the top of the intake manifold.

REMOVAL & INSTALLATION

2.3L Engine

1. Remove the 3 screws, the pin-type retainer and the upper and lower snow shield.
2. Disconnect the Idle Air Control (IAC) valve electrical connector.
3. Remove the 2 bolts and the IAC valve.
4. To install, reverse the removal procedure and tighten IAC mounting bolts to 89 inch lbs. (10 Nm).

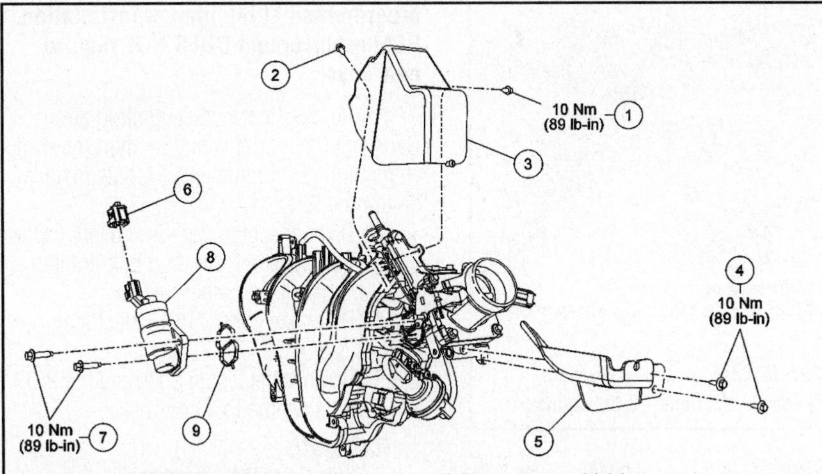

1. Upper snow shield-to-intake manifold screw
2. Upper snow shield-to-intake manifold pin-type retainer
3. Upper snow shield
4. Lower snow shield-to-intake manifold screws (2 required)
5. Lower snow shield
6. Idle air control (IAC) valve electrical connector
7. IAC valve bolts (2 required)
8. IAC valve
9. IAC valve gasket

22086_ESCA_G0087

Fig. 222 Idle Air Control (IAC) valve location—2.3L engine

1. Idle air control (IAC) valve electrical connector
2. IAC valve bolts (2 required)
3. IAC valve
4. IAC valve gasket

10 Nm (89 lb-in)

22086_ESCA_G0095

Fig. 223 Idle Air Control (IAC) valve location—3.0L engine

➡Inspect the gasket and install new as necessary.

3.0L Engine

1. Disconnect the Idle Air Control (IAC) valve electrical connector.
2. Remove the 2 bolts and the IAC valve and discard the gasket.
3. To install, reverse the removal procedure.
4. Clean and inspect all sealing surfaces. Install new gasket.
5. Tighten IAC mounting bolts to 89 inch lbs. (10 Nm).

INTAKE AIR TEMPERATURE (IAT) SENSOR

The Intake Air Temperature (IAT) sensor is incorporated in the Mass Air Flow assembly. Refer to the Mass Air Flow (MAF) sensor.

KNOCK SENSOR (KS)

LOCATION

See Figure 224.

The Knock Sensor (KS) is located behind the intake manifold to the rear of engine block.

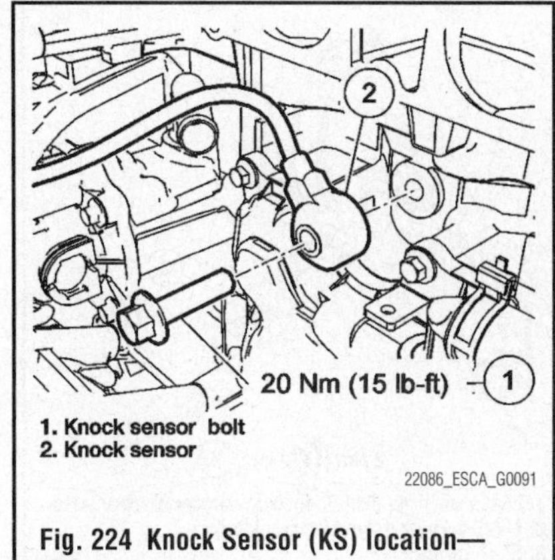

Fig. 224 Knock Sensor (KS) location—2.3L engine

REMOVAL & INSTALLATION

1. With the vehicle in **NEUTRA**, position it on a hoist.
2. Remove the intake manifold.
3. Remove the bolt and the Knock Sensor (KS).
4. To install, reverse the removal procedure and tighten the sensor mounting bolt to 15ft. lbs. (20 Nm).

MASS AIR FLOW (MAF) SENSOR

LOCATION

2.3L Engine

See Figure 225.

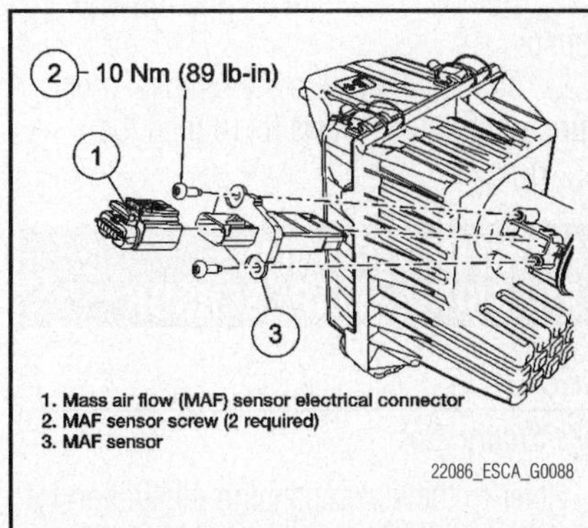

Fig. 225 Mass Air Flow (MAF) sensor location—2.3L engine

The Mass Air Flow (MAF) sensor is located on air supply tube at air filter housing.

3.0L Engine

See Figure 226.

The Mass Air Flow (MAF) sensor is located on air supply tube before throttle body.

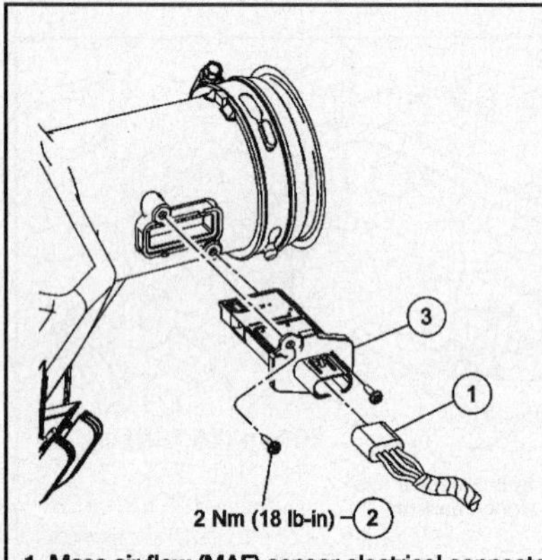

Fig. 226 Mass Air Flow (MAF) sensor location—3.0L engine

REMOVAL & INSTALLATION

2.3L Engine

1. Disconnect the Mass Air Flow (MAF) sensor electrical connector.
2. Remove the 2 screws and the MAF sensor.
3. To install, reverse the removal procedure and tighten MAF sensor mounting screws to 89 inch lbs. (10 Nm).

3.0L Engine

1. Disconnect the Mass Air Flow (MAF) sensor electrical connector.
2. Remove the 2 screws and the MAF sensor.
3. To install, reverse the removal procedure and tighten screws to 18 inch lbs. (2 Nm).

MANIFOLD ABSOLUTE PRESSURE (MAP) SENSOR

LOCATION

See Figure 227.

Refer to the accompanying illustration for Manifold Absolute Pressure (MAP) sensor location.

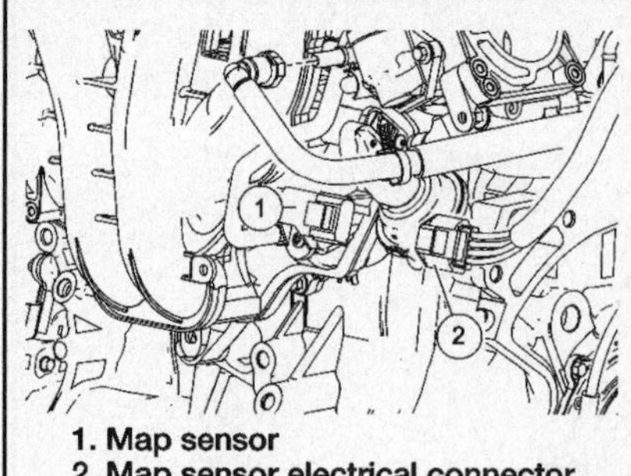

Fig. 227 Manifold Absolute Pressure (MAP) sensor locations—2.3L engine

REMOVAL & INSTALLATION

1. Disconnect the electrical connector from the Manifold Absolute Pressure (MAP) sensor.
2. Remove the MAP sensor mounting screws.
3. To install, reverse the removal procedure

POWERTRAIN CONTROL MODULE (PCM)

LOCATION

See Figure 228.

The Powertrain Control Module (PCM) is located behind the instrument panel (cowl), center to both driver and passenger sides (access from the engine compartment).

REMOVAL & INSTALLATION

➡Any Powertrain Control Module (PCM) replacement will require that ALL customer keys are available to be programmed at the time of installation. PCM replacement DOES NOT require new keys.

1. Retrieve the module configuration. Carry out the module configuration retrieval steps of the Programmable Module Installation procedure.
2. Disconnect the negative battery cable.
3. Remove the PCM stud bolt nut and position the wiring harness aside.
4. Disconnect the 3 PCM electrical connectors.
5. Remove the 2 stud bolts and the PCM
6. Remove the PCM cowl seal.

To install:

7. Install the PCM cowl seal.
8. Install the PCM and tighten the 2 stud bolts to 53 inch lbs. (6 Nm).
9. Connect the 3 PCM electrical connectors.
10. Position the wiring harness. Install and tighten the PCM stud bolt nut to 53 inch lbs. (6 Nm).
11. Restore the module configuration. Carry out the module configuration restore steps of the Programmable Module Installation procedure.
12. Reprogram the passive anti-theft system (PATS). Carry out the Key Programming Using Two Programmed Keys procedure.

THROTTLE POSITION SENSOR (TPS)

LOCATION

2.3L Engine

See Figure 229.

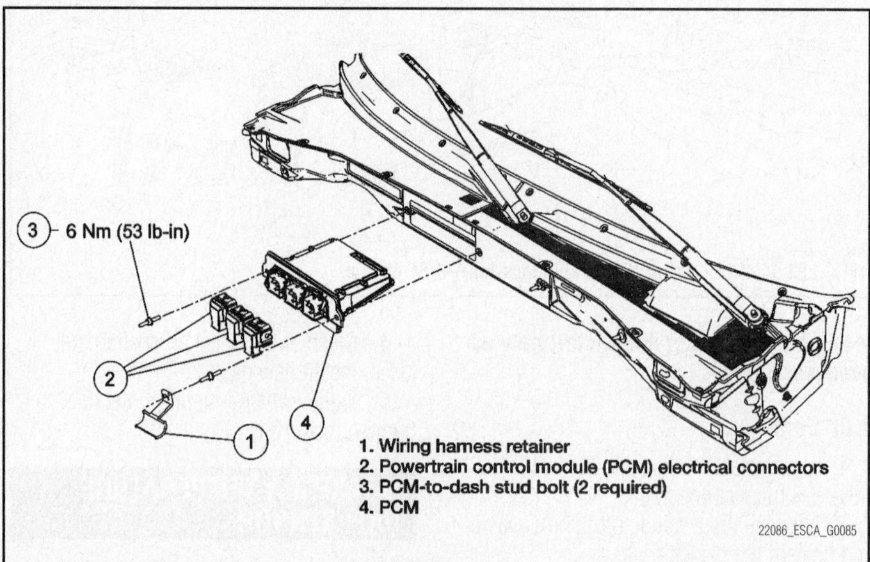

Fig. 228 Powertrain Control Module (PCM) view

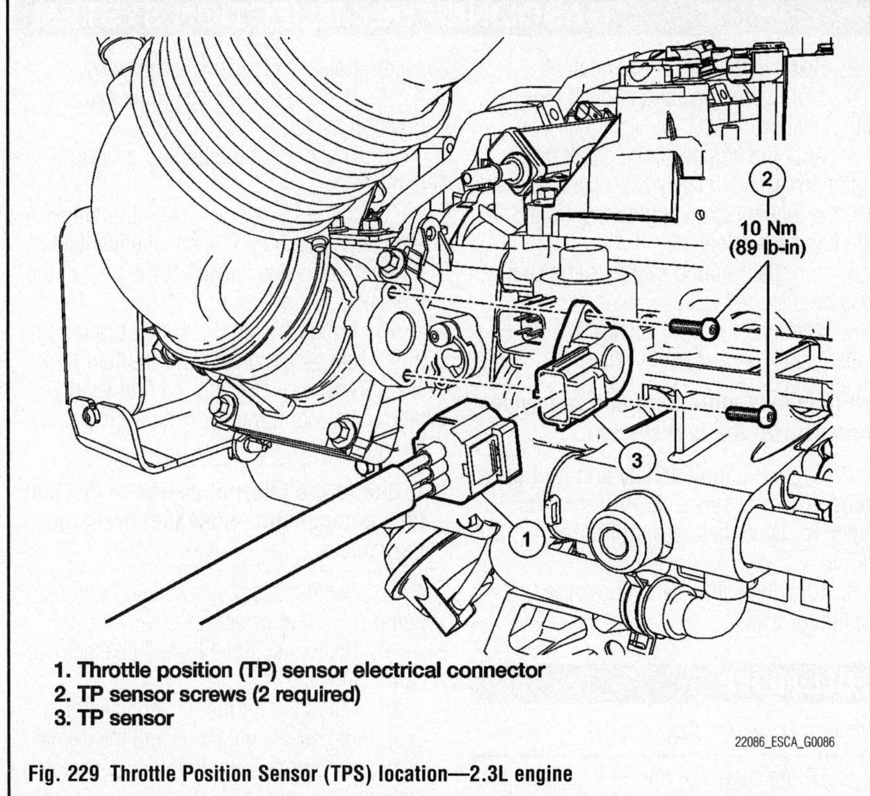

Fig. 229 Throttle Position Sensor (TPS) location—2.3L engine

The Throttle Position Sensor (TPS) is located to the right of throttle plate.

3.0L Engine

See Figure 230.

The Throttle Position Sensor (TPS) is located on the throttle body just behind the EGR valve tube.

REMOVAL & INSTALLATION

2.3L Engine

1. Disconnect the Throttle Position Sensor (TPS) electrical connector.
2. Remove the 2 screws and the TPS.
3. To install, reverse the removal procedure and tighten TPS screws to 89 inch lbs. (10 Nm).

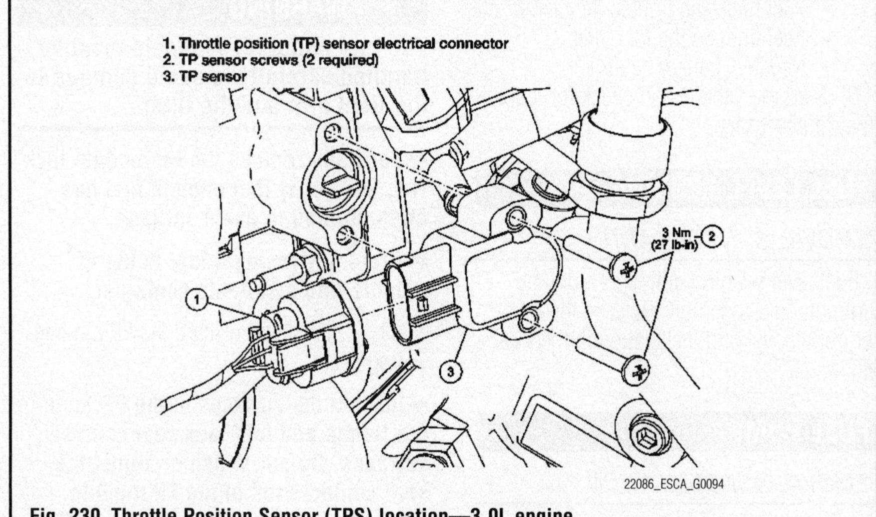

Fig. 230 Throttle Position Sensor (TPS) location—3.0L engine

3.0L Engine

1. Disconnect the Throttle Position Sensor (TPS) sensor electrical connector.
2. Remove the 2 screws and the TPS.
3. To install, reverse the removal procedure and tighten mounting screws to 27 inch lbs. (3 Nm).

VEHICLE SPEED SENSOR (VSS)

LOCATION

See Figure 231.

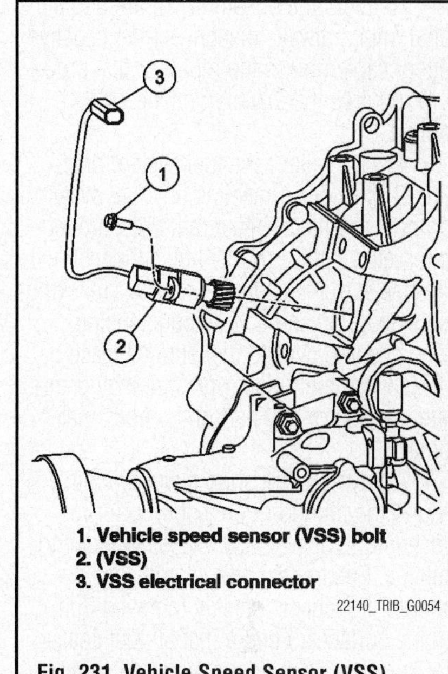

Fig. 231 Vehicle Speed Sensor (VSS) location

REMOVAL & INSTALLATION

1. Raise and support the vehicle on a hoist.
2. Disconnect the Vehicle Speed Sensor (VSS)/Output Shaft Speed (OSS) sensor electrical connector.
3. Remove the sensor bolt.
4. Remove the sensor.

To install:

➡When installing the sensor, lubricate the O-ring seal with clean transmission fluid.

5. To install, reverse the removal procedure and note the following:
   a. Tighten the sensor mounting bolt to 9 ft. lbs. (12 Nm).

FUEL SYSTEM SERVICE PRECAUTIONS

Safety is the most important factor when performing not only fuel system maintenance but any type of maintenance. Failure to conduct maintenance and repairs in a safe manner may result in serious personal injury or death. Maintenance and testing of the vehicle's fuel system components can be accomplished safely and effectively by adhering to the following rules and guidelines.

- To avoid the possibility of fire and personal injury, always disconnect the negative battery cable unless the repair or test procedure requires that battery voltage be applied.
- Always relieve the fuel system pressure prior to disconnecting any fuel system component (injector, fuel rail, pressure regulator, etc.), fitting or fuel line connection. Exercise extreme caution whenever relieving fuel system pressure to avoid exposing skin, face and eyes to fuel spray. Please be advised that fuel under pressure may penetrate the skin or any part of the body that it contacts.
- Always place a shop towel or cloth around the fitting or connection prior to loosening to absorb any excess fuel due to spillage. Ensure that all fuel spillage (should it occur) is quickly removed from engine surfaces. Ensure that all fuel soaked cloths or towels are deposited into a suitable waste container.
- Always keep a dry chemical (Class B) fire extinguisher near the work area.
- Do not allow fuel spray or fuel vapors to come into contact with a spark or open flame.
- Always use a back-up wrench when loosening and tightening fuel line connection fittings. This will prevent unnecessary stress and torsion to fuel line piping.
- Always replace worn fuel fitting O-rings with new. Do not substitute fuel hose or equivalent where fuel pipe is installed.

Before servicing the vehicle, make sure to also refer to the precautions in the beginning of this section as well.

RELIEVING FUEL SYSTEM PRESSURE

1. Before servicing the vehicle, refer to the Precautions Section.
2. Remove the fuel pump relay.
3. Start the engine and allow it to idle until it stalls.
4. After the engine stalls, crank the engine for approximately 5 seconds to make sure the fuel injection supply manifold pressure has been released.
5. Turn the ignition switch to the OFF position.
6. When fuel system service is complete, install the fuel pump relay.

➡It may take more than one key cycle to pressurize the fuel system.

7. Cycle the ignition key and wait three seconds to pressurize the fuel system. Check for leaks before starting the engine.
8. Start the vehicle and check the fuel system for leaks.

FUEL FILTER

REMOVAL & INSTALLATION

1. Before servicing the vehicle, refer to the Precautions Section.
2. Properly relieve the fuel system pressure.
3. Remove or disconnect the following:
   - Negative battery cable
   - Fuel line to the fuel filter
4. Loosen the clamp and remove the filter

To install:

5. Install or connect the following:
   - New clips to the fuel lines
   - Fuel filter and tighten the clamp
   - Fuel lines to the fuel filter
   - Negative battery cable
6. Start the vehicle and check for leaks, repair if necessary.

FUEL LEVEL SENDING UNIT

REMOVAL & INSTALLATION

Fuel level sending unit is part of fuel pump module assembly. Refer to the Fuel Pump Removal and Installation in the Fuel Section.

FUEL PUMP

REMOVAL & INSTALLATION

See Figure 232.

1. Before servicing the vehicle, refer to the Precautions in this section.
2. With the vehicle in **NEUTRAL**, position it on a hoist.
3. Release the fuel system pressure.
4. Release the fuel tank filler cap and position aside.
5. Disconnect the battery ground cable.
6. Insert the special tool into the fuel tank filler pipe until it opens the fuel tank level shutoff valve located at the inlet of the fuel tank.
7. Using the special tools, drain as much fuel as possible from the fuel tank and filler pipe, lowering the fuel level below the fuel pump (FP) mounting flange.

➡Due to the internal design of the fuel tank components, slow fuel drainage may occur.

8. Remove the 4 screws and the fuel pump (FP) module access cover.
9. Disconnect the FP module electrical connector.
10. Clean the FP module connection, couplings, mounting flange and the immediate surrounding area of any dirt or foreign material.
11. Disconnect the fuel supply tube and fuel vapor recirculation tube-to-FP module quick connect couplings.

➡Place absorbent toweling in the immediate surrounding area in case of fuel spills.

12. Using a suitable FP lock ring remover, rotate the lock ring counterclockwise and remove the FP module.

⁂ WARNING

The fuel pump (FP) module must be handled carefully to avoid damage to the float arm and the filter.

➡Carefully remove the FP module lock ring and verify that enough fuel has been drained to avoid spillage.

➡Drain any residual fuel in the FP module into a suitable container.

13. Remove and discard the FP module O-ring seal.

➡Inspect the surfaces of the FP module flange and fuel tank seal contact surfaces. Do not polish or adjust the seal contact area of the FP module flange or fuel tank. Install a new FP module or fuel tank if the seal contact area is bent, scratched or corroded.

14. To install, reverse the removal procedure.

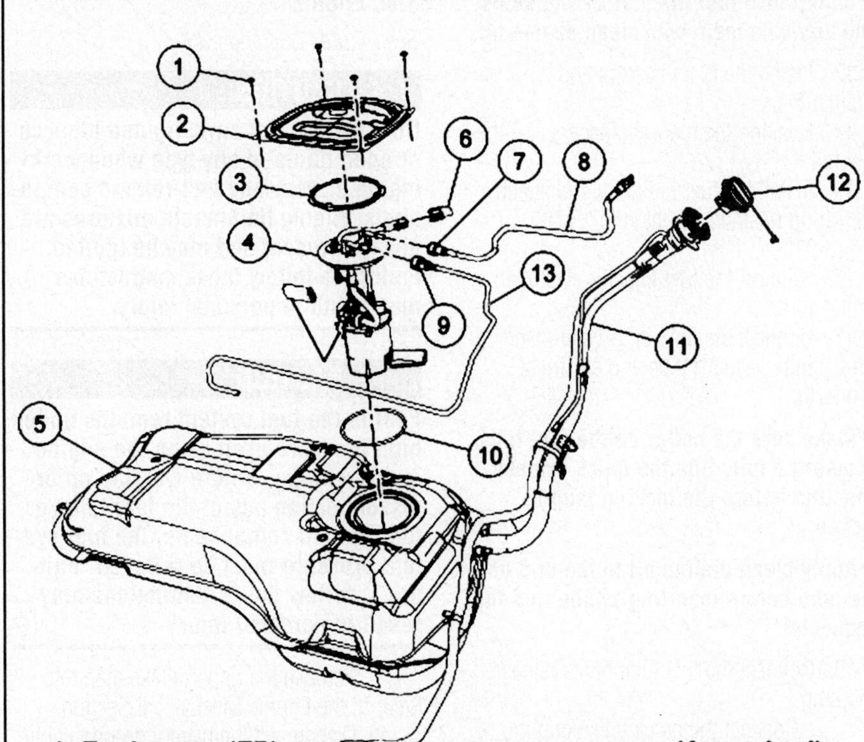

1. Fuel pump (FP) module access cover screws (4 required)
2. FP access cover
3. FP module lock ring
4. FP module
5. Fuel tank
6. FP module electrical connector
7. Fuel vapor tube-to-FP module quick connect coupling
8. Fuel vapor tube
9. Fuel supply tube assembly-to-fuel tank quick connect coupl
10. FP module O-ring
11. Fuel tank filler pipe
12. Fuel tank filler cap
13. Fuel supply tube

22140_TRIB_G0122

Fig. 232 Access to fuel pump module and exploded view of fuel pump and components

To install:

15. Turn the ignition key to the ON position to pressurize the fuel system.
16. Visually inspect the fuel system for leaks.

**** WARNING**

Make sure the fuel tube clicks into place when installing the tube. To make sure the tube is fully seated, pull on the tube.

➡Apply clean engine oil to the end of the tube before inserting the tube into the connector.

17. Install the fuel tube quick release coupling.

**** WARNING**

Inspect the surfaces of the fuel pump module flange and fuel tank O-ring contact surfaces. Do not polish or adjust the O-ring contact area of the fuel pump flange or fuel tank. Install a new fuel pump module or fuel tank if the O-ring contact area is bent, scratched or corroded.

**** WARNING**

Make sure to install a new fuel pump module O-ring and lock ring.

18. Lubricate and install a new fuel pump module O-ring seal upon installing the fuel pump module.
19. When installing the fuel pump module, make sure to align the locator tabs on the fuel tank mounting flange.

➡Be sure the aligning tabs of the fuel pump module unit are positioned in the slot before tightening the lock ring.

20. Holding the fuel pump module O-ring seal in place, rotate the lock ring clockwise until it stops against the retainer tabs.

➡Make sure the collar on the fuel tube is inserted fully into the quick release coupling before the locking tab is locked.

21. Connect the fuel supply quick connect coupling to the fuel supply manifold.
22. Press the fuel supply quick connect coupling locking tab into position.
23. Pull on the fitting to make sure it is fully engaged.

FUEL RAIL & INJECTORS

REMOVAL & INSTALLATION

2.3L Engine

See Figure 233.

**** CAUTION**

Do not smoke or carry lighted tobacco or open flame of any type when working on or near any fuel-related components. Highly flammable mixtures are always present and may be ignited. Failure to follow these instructions can result in personal injury.

**** CAUTION**

Fuel in the fuel system remains under high pressure even when the engine is not running. Before working on or disconnecting any of the fuel tubes or fuel system components, the fuel system pressure must be relieved. Failure to follow these instructions can result in personal injury.

1. Release the fuel pressure.
2. Disconnect the negative battery cable.

**** WARNING**

When reusing liquid or vapor tube connectors, make sure to use compressed air to remove any foreign material from the connector retaining clip area before separating the tube.

➡Carefully release the locking tabs to avoid breakage.

3. Disconnect the fuel supply tube quick connect coupling at the fuel rail.
4. Disconnect the fuel rail pressure and temperature vacuum tube and electrical connector.
5. Disconnect the 4 fuel injector electrical connectors.
6. Remove the nut and position the radio capacitor aside.
7. Detach the wire harness retainer from the valve cover stud bolt.
8. Detach the 2 pin-type wire harness retainers from the fuel rail.
9. Remove the 2 fuel rail stud bolts
10. Remove the fuel rail and injectors as an assembly and then remove the spacers.
11. Remove the 4 fuel injector retainer clips and the fuel injectors.
12. Remove and discard the 8 fuel injector O-ring seals.

To install:

⁂ WARNING

Use O-ring seals that are made of special fuel-resistant material. Use of ordinary O-rings can cause the fuel system to leak. Do not reuse the O-ring seals.

➡Install new fuel injector O-ring seals and lubricate them with clean engine oil.

13. Install the fuel injectors and the retainer clips.
14. Position the fuel rail spacers and the fuel rail.
15. Install the fuel rail bolts and attach the wiring retainers. Tighten to 23 Nm (17 lb-ft).
16. Connect the fuel injector electrical connectors.
17. Connect the fuel rail pressure and temperature vacuum tube and electrical connector.

➡Make sure the collar on the fuel tube is inserted fully into the quick release coupling before the locking tang is locked.

➡Apply clean engine oil to the end of the tube before inserting a tube into the connector.

18. Connect the fuel tube quick release coupling:
    a. Connect the quick lock coupling to the tube.
    b. Press the quick connect coupling locking tangs into position.
    c. Pull on the fitting to make sure it is fully engaged.
19. Connect the negative battery cable.

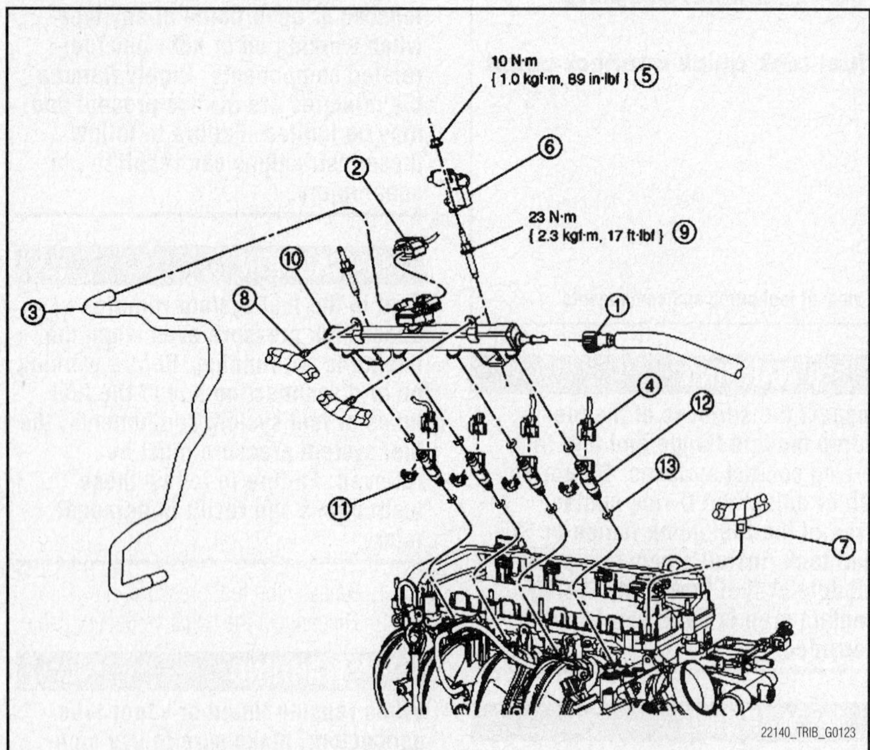

Fig. 233 Fuel rail and exploded view of components–2.3L engine

3.0L Engine

See Figures 234 and 235.

⁂ CAUTION

Do not smoke or carry lighted tobacco or open flame of any type when working on or near any fuel-related components. Highly flammable mixtures are always present and may be ignited. Failure to follow these instructions may result in personal injury.

⁂ CAUTION

Fuel in the fuel system remains under high pressure even when the engine is not running. Before working on or disconnecting any of the fuel lines or fuel system components, the fuel system pressure must be relieved. Failure to follow these instructions may result in personal injury.

1. Remove the upper intake manifold. Refer to the Engine Mechanical Section.
2. Disconnect the battery ground cable.
3. Disconnect the fuel line.
4. Disconnect the six fuel injector electrical connectors and release the wiring harness locators from the fuel injection supply manifold.
5. Remove the vacuum hose.
6. Remove the bolts and the fuel injection supply manifold.

To install:

⁂ WARNING

Use O-ring seals that are made of special fuel-resistant material. Use of ordinary O-rings can cause the fuel system to leak. Do not reuse the O-ring seals.

➡Lubricate the fuel injector O-ring seals with clean engine oil to aid installation.

7. Position the fuel rail and install the bolts.
8. Tighten the fuel injection supply manifold bolts to 89 inch lbs. (10 Nm).
9. Attach the fuel charging wiring harness retainers to the fuel rail.
10. Connect the 6 fuel injector electrical connectors.
11. Connect the fuel rail pressure and temperature sensor vacuum tube and electrical connector.
12. Connect the fuel supply tube quick connect coupling at the fuel rail.
13. Install the upper intake manifold.
14. Connect the battery ground cable.

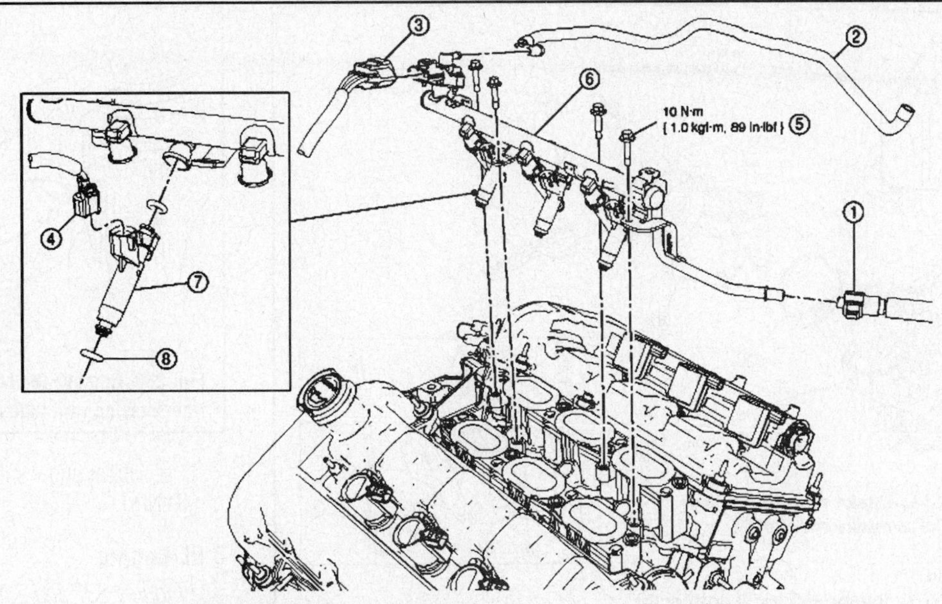

1. Fuel tube quick connect coupling
2. Fuel rail pressure and temperature sensor vacuum tube
3. Fuel rail pressure and temperature sensor electrical connector
4. Fuel injector electrical connector (6 required)
5. Fuel rail bolt (4 required)
6. Fuel rail
7. Fuel injectors (6 required)
8. Fuel injector O-ring seals (2 required per injector

22140_TRIB_G0124

Fig. 234 Fuel rail and exploded view of components–3.0L engine

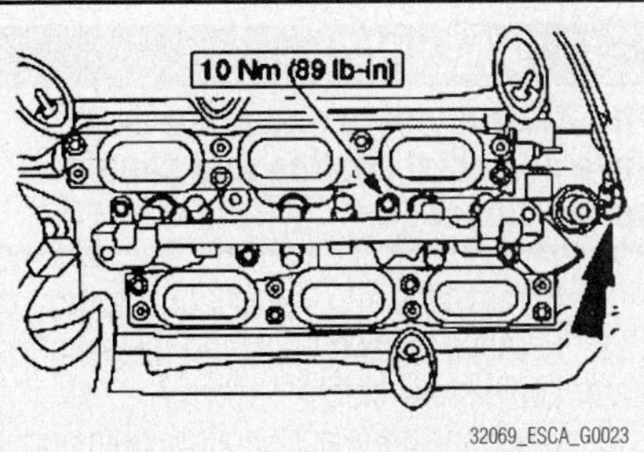

32069_ESCA_G0023

Fig. 235 Tighten the fuel injection supply manifold bolts to 89 inch lbs. (10 Nm)

FUEL TANK

REMOVAL & INSTALLATION

1. Disconnect the negative battery cable.
2. With the vehicle in **NEUTRAL**, position it on a hoist.
3. Drain the fuel tank.

➡All wheel drive (AWD) vehicles, remove the exhaust muffler and resonator and rear driveshaft.

4. Release the clamp and remove the fuel tank filler pipe hose from the fuel tank.
5. Position a suitable lifting device under the fuel tank.
6. Detach the 2 retainer clips from the LH fuel tank strap.
7. Remove the 2 bolts and position the 2 fuel tank straps aside
8. Partially lower the fuel tank enough to disconnect the fuel vapor tube assembly-to-fuel tank quick connect coupling.
9. Remove the fuel tank.

To install:

10. Install fuel tank.
11. Connect fuel vapor tube assembly to fuel tank before completely raising fuel tank.
12. support fuel tank.
13. Reposition fuel tank straps and tighten bolts to 41 ft. lbs. (55 Nm).
14. Remove fuel tank support.
15. Install retainer clips to LH fuel strap.
16. Install fuel tank filler pipe and tighten hose clamp.
17. Connect the negative battery cable.
18. Install exhaust and rear drive shaft if they were removed for (AWD) models.
19. Lower vehicle and refill fuel tank
20. Connect the negative battery cable.

IDLE SPEED

ADJUSTMENT

Idle speed is controlled by the ECM and not adjustments can be made.

THROTTLE BODY

REMOVAL & INSTALLATION

See Figure 236.

⁂ WARNING

Throttle body bore and plate area have a special coating and cannot be cleaned, or possible damage to the throttle body can occur.

1. Disconnect the negative battery cable.
2. Remove the air cleaner outlet pipe.
3. Remove the upper snow shield screw and pin-type retainer.
4. Remove the upper snow shield.
5. Detach the accelerator and speed control cables from the throttle body.
6. Disconnect the throttle position sensor electrical connector.
7. Remove the 4 bolts and the throttle body.
8. Inspect the throttle body gasket and install new as necessary.

To install:

9. To install, reverse the removal procedure. Note the following tightening specifications:
    a. Throttle body bolts: 89 inch lbs. (10 Nm).

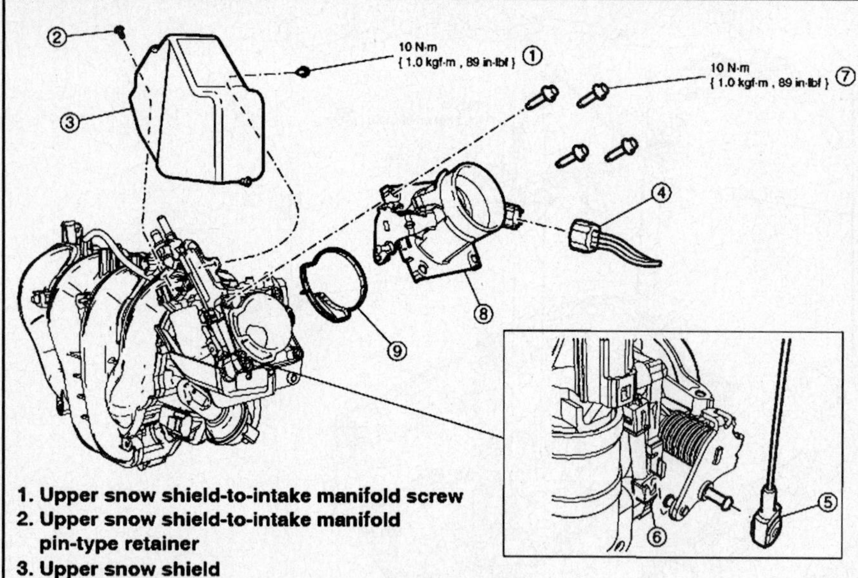

1. Upper snow shield-to-intake manifold screw
2. Upper snow shield-to-intake manifold pin-type retainer
3. Upper snow shield
4. Throttle position (TP) sensor electrical connector
5. Accelerator cable
6. Speed control cable
7. Throttle body-to-upper intake manifold bolts (4 required)
8. Throttle body
9. Gasket

22140_TRIB_G0125

Fig. 236 Throttle body and exploded view of components–2.3L engine

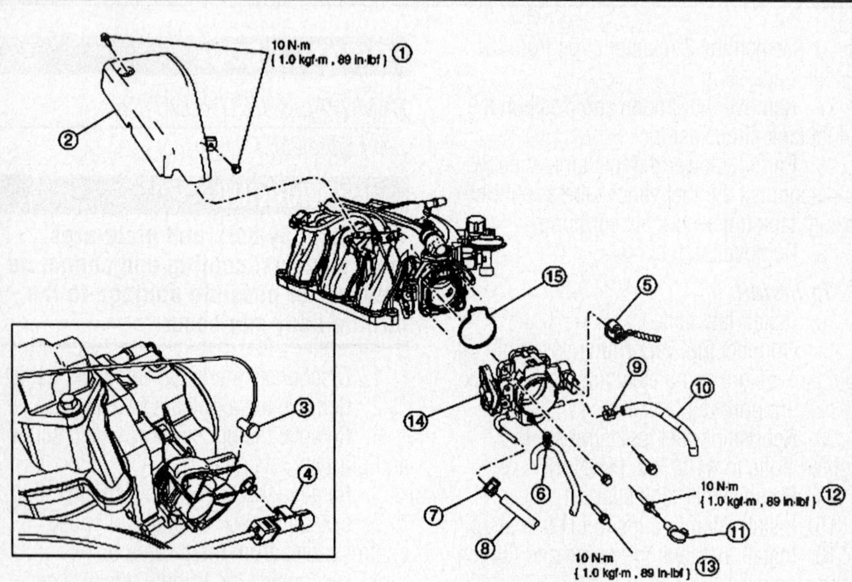

1. Snow shield bolts (2 required)
2. Snow shield
3. Accelerator cable
4. Speed control cable
5. Throttle position (TP) sensor electrical connector
6. Transaxle vent tube pin-type retainer
7. Clamp
8. Throttle body coolant hose
9. Clamp
10. Throttle body coolant hose
11. Wiring harness retainer
12. Throttle body-to-upper intake manifold stud bolt
13. Throttle body-to-upper intake manifold bolts (3 required)
14. Throttle body
15. Gasket

22140_TRIB_G0126

Fig. 237 Throttle body and exploded view of components—3.0L Engines

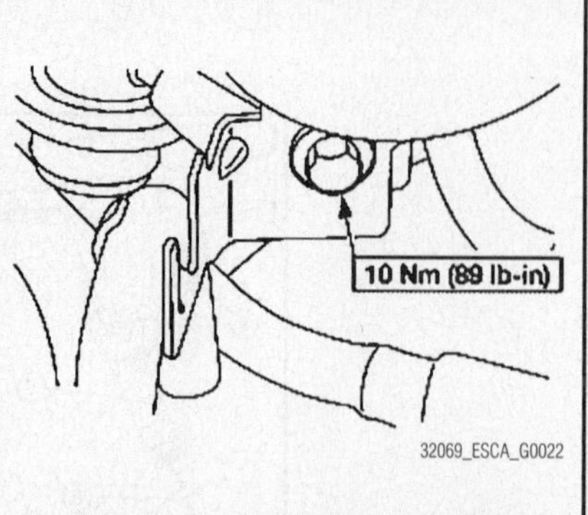

Fig. 238 Remove the bolt and position the transmission vent tube and bracket aside

b. Upper snow shield: 89 inch lbs. (10 Nm).

3.0L Engine

See Figures 237 and 238.

**** CAUTION**

Do not smoke or carry lighted tobacco or open flame of any type when working on or near any fuel-related components. Highly flammable mixtures are always present and may be ignited. Failure to follow these instructions may result in personal injury.

**** WARNING**

Throttle body bore and plate area have a special coating and should not be cleaned.

1. Remove the air cleaner outlet tube.
2. Remove the two bolts and the accelerator cable snow shield.
3. Disconnect the accelerator cable and the speed control cable (if equipped) from the throttle body lever.
4. Disconnect the Throttle Position (TP) sensor electrical connector.
5. Remove the bolt and position the transmission vent tube and bracket aside.
6. Disconnect and plug the 2 throttle body coolant hoses.
7. Remove the 3 bolts, the stud bolt and the throttle body.
8. Discard the gasket.

To install:

9. Installation is the reverse of the removal procedure, noting the following:
   - Install a new gasket
   - Tighten the throttle body retainers to 89 inch lbs. (10 Nm).
   - Tighten the transmission vent tube and bracket bolt to 89 inch lbs. (10 Nm).

HEATING & AIR CONDITIONING SYSTEM

BLOWER MOTOR

REMOVAL & INSTALLATION

1. Disconnect the blower motor electrical connector.
2. Release the 2 blower motor vent tube clips and pull the vent tube down until it is disengaged from the heater core and evaporator core housing.
3. The carpet below the blower motor must be slightly repositioned to remove the blower motor.
4. Rotate the blower motor counterclockwise to disengage it from the housing and remove the blower motor.
5. To install, reverse the removal procedure

HEATER CORE

REMOVAL & INSTALLATION

See Figures 239 and 240.

1. Before servicing the vehicle, refer to the Precautions Section.
2. Drain the engine coolant.
3. Recover the refrigerant.
4. Position the seats forward and remove the 2 floor console rear bolts
5. Position the seats rearward.
6. Remove the transmission selector lever trim ring.
7. Remove the floor console storage bin.
8. Remove the floor console finish panel.
9. Remove the 8 floor console bolts and remove the floor console.
10. Disarm the supplemental restraint system (SRS).
11. Remove the RH and LH A-pillar trim panels.
12. Remove the 4 pin-type retainers and the RH and LH front door opening scuff plates.
13. Remove the RH and LH lower A-pillar trim panels.
14. Remove the steering column opening cover.
15. Remove the RH and LH instrument panel side finish panels.
16. Disconnect the 2 electrical connectors at the LH side of the instrument panel.
17. Disconnect the main steering module electrical connector.
18. Remove the bolts and position aside the hood release handle and parking brake release handle.
19. Remove and slide the steering column intermediate shaft access cover and weather shield up the steering column intermediate shaft.

**** WARNING**

Do not allow the steering column shaft to rotate while the lower shaft is disconnected or damage to the clockspring can result. If there is evidence that the shaft has rotated, the clockspring must be removed and re-centered.

20. Remove and discard the steering column intermediate shaft-to-coupling bolt and slide the steering column intermediate shaft up.
21. Remove rivet and the Restraints Control Module (RCM) access cover.
22. Disconnect the LH RCM electrical connector.
23. For vehicles with automatic transmissions remove the selector lever cable from the selector lever assembly.

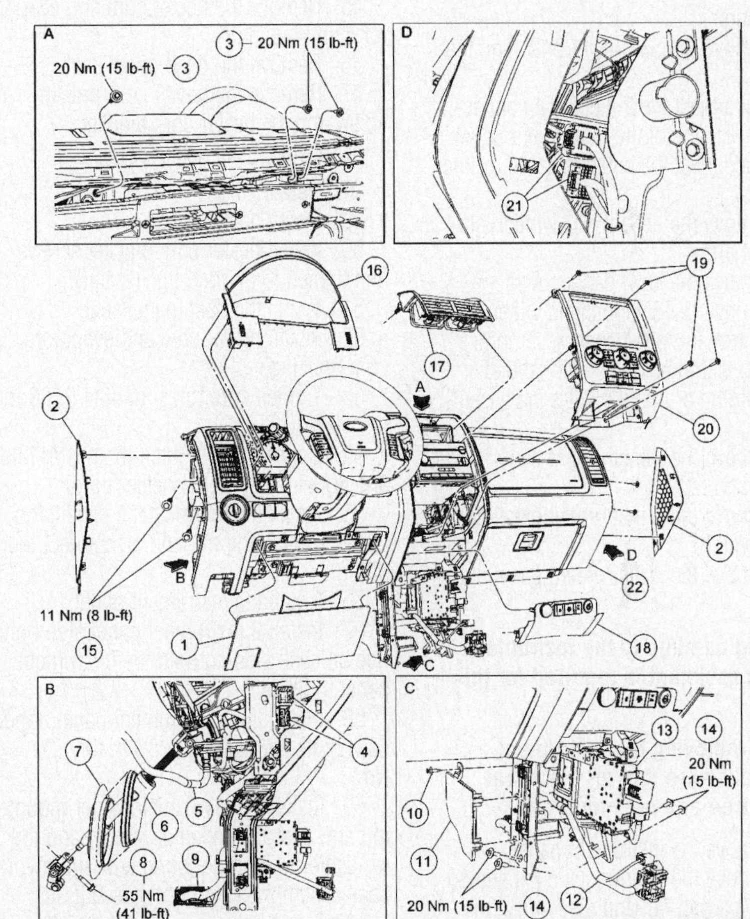

Fig. 239 Instrument panel exploded view

24. For vehicles with manual transmissions remove the shift cables from the shift lever assembly.
25. Disconnect the selector lever electrical connector and wiring harness pin-type retainers from the selector lever assembly.
26. If equipped, remove and position aside the Four Wheel Drive (4WD) control module and bracket from the selector lever assembly.
27. Remove the 4 selector lever assembly bolts and remove the selector lever assembly.
28. Disconnect the electrical connectors from the Smart Junction Box (SJB).
29. Disconnect the wiring harness pin-type retainers.
30. Remove the SJB lower bolts and the SJB.
31. Disconnect the 2 electrical connectors at the RH side of the instrument panel.
32. Disconnect the antenna cable in-line connector
33. Remove the LH and RH windshield wiper pivot arms.
34. Remove the cowl panel cover
35. Remove the 3 windshield wiper mounting arm and pivot shaft assembly bolts and position aside the windshield wiper mounting arm and pivot shaft assembly.
36. Remove the 3 instrument panel upper cowl bolts.
37. Remove the 4 instrument panel center brace bolts
38. Remove the 4 instrument panel side bolts.

➡**To avoid damage to the instrument panel, an assistant is required for this step.**

➡**Before removing the instrument panel, make sure that all electrical wiring is free and not hindered**

39. Remove the instrument panel.
40. Remove the Thermostatic Expansion Valve (TXV) fitting nut and disconnect the fitting.
41. Release the clamps and disconnect the heater inlet and outlet hoses from the heater core.
42. Remove the 6 heater core and evaporator core housing nuts.
43. Detach the heater core and evaporator core housing from the dash panel studs
44. Rotate the RH side of the heater core and evaporator core housing toward the rear of the vehicle while pulling the housing toward the RH door opening to detach it from the rear footwell duct.

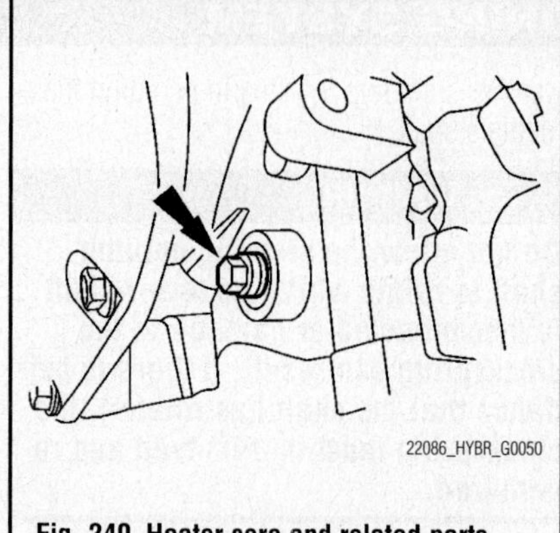

Fig. 240 Heater core and related parts

45. Remove the heater core and evaporator core housing
46. Remove the dash panel seal.
47. Remove the heater core bracket screw and the heater core bracket
48. Remove the heater core.

To install:

49. Install heater core
50. Install heater core bracket screws and tighten to 27 inch lbs. (3 Nm).
51. Install the dash panel seal.
52. Install heater core and evaporator core housing.
53. Tighten core housing nuts to 80 inch lbs. (9 Nm).
54. Connect and tighten the heater inlet and outlet hoses to the heater core.
55. Install the thermostatic expansion valve (TXV) fitting nut and disconnect the fitting.
56. Install the instrument panel.
57. Install 2 instrument panel side bolts, one on each side, to hold the instrument panel in place.
58. Install the 3 instrument panel upper cowl bolts and tighten to 15 ft. lbs. (20 Nm).
59. Install the windshield wiper mounting arm and pivot shaft assembly and the 3 windshield wiper mounting arm and pivot shaft assembly bolts tighten to 9 ft. lbs. (12 Nm).
60. Install the cowl panel cover.
61. Install the LH and RH windshield wiper pivot arms.
62. Install the instrument panel side bolts and tighten to 8 ft. lbs. (11 Nm).
63. Install the instrument panel center brace bolts and tighten to 15 ft. lbs. (20 Nm).
64. Connect the antenna cable in-line connector.
65. Connect the electrical connectors at the RH side of the instrument panel
66. Install the SJB, bolts and connect the wiring harness pin-type retainers.
67. Connect the electrical connectors to the SJB.
68. Install the selector lever assembly and the 4 selector assembly bolts and tighten to 18 ft. lbs (25 Nm).
69. If equipped, install the 4WD control module and bracket onto the selector lever assembly.
70. Connect the selector lever electrical connector.
71. Install the wiring harness pin-type retainers to the selector lever assembly.
72. Vehicles with automatic transmissions, install the selector lever cable to the selector lever assembly.
73. Vehicles with manual transmissions, install the shift cables to the shift lever assembly.
74. Connect the LH RCM electrical connector
75. Install the RCM access cover and rivet
76. Slide the steering column intermediate shaft down onto the coupling and install a new steering column intermediate shaft-to-coupling bolt, tighten to 42 ft. lbs. (55 Nm).
77. Install the steering column intermediate shaft access cover and weather shield.
78. Install the hood release and parking brake release handles and bolts.
79. Connect the main steering module electrical connector.
80. Connect the electrical connectors at the LH side of the instrument panel
81. Install the RH and LH instrument panel side finish panels
82. Install the steering column opening cover.
83. Install the RH and LH lower A-pillar trim panels.
84. Install the RH and LH A-pillar trim panels
85. Install the RH and LH front door opening scuff plates and the 4 pin-type retainers.
86. Rearm the SRS.
87. Install the 8 floor console bolts and the floor console.
88. Install the floor console finish panel.
89. Install the floor console storage bin.
90. Install the transmission selector lever trim ring.
91. Install and tighten the floor console rear bolts to 62 inch lbs. (7 Nm).
92. Evacuate, leak test and charge the refrigerant system.
93. Refill and bleed cooling system.

STEERING

POWER RACK & PINION STEERING GEAR

REMOVAL & INSTALLATION

See Figures 241 and 242.

1. Remove the front wheels and tires.
2. Turn the ignition key to the **OFF** position. Remove the ignition key.

WARNING

Do not allow the steering wheel to rotate while the intermediate shaft is disconnected or damage to the clockspring can result. If there is evidence that the shaft has rotated, the clockspring must be removed and re-centered

3. Remove and discard the steering column coupling-to-steering gear bolt and disconnect the coupling from the steering gear.
4. From the engine compartment, loosen the 2 steering gear bolts.
5. If equipped, remove the 3 pin-type retainers and the steering gear shield.
6. Remove and discard the 2 outer tie-rod end nuts.
7. Do not use a hammer to separate the tie-rod end from the wheel knuckle or damage to the wheel knuckle can result.
8. Using a suitable tool, separate the tie-rod ends from the wheel knuckles.
9. For AWD vehicles, remove the rear transaxle insulator through bolt.
10. For FWD vehicles with the 2.3L engine and an automatic transaxle, remove the 3 transmission damper bolts and the transmission damper.
11. Remove and discard the 2 steering gear bolts.

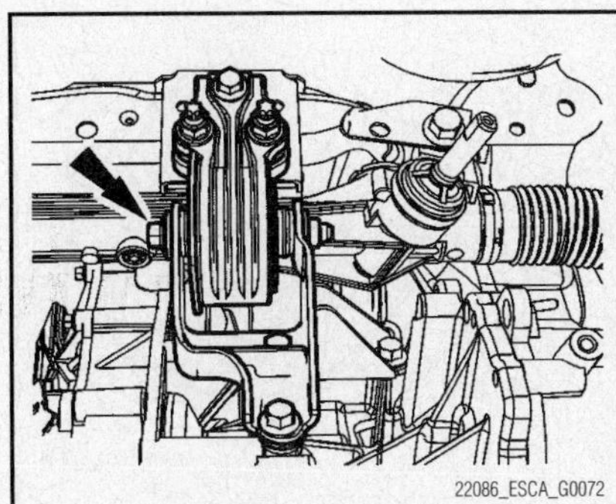

Fig. 241 Rear transaxle insulator and through bolt

➡For All Wheel Drive (AWD) vehicles, it is necessary to grasp the driveshaft by hand and apply slight downward pressure to obtain clearance for the removal of the steering gear.

12. Remove the steering gear from the LH side of the vehicle.

To install:

13. To install, reverse the removal procedure and note the following:
    a. Install a new steering column coupling-to-steering gear bolt and tighten to 41 ft. lbs. (55 Nm).
    b. Install and tighten the 2 steering gear bolts to 85 ft. lbs. (115 Nm).
    c. Install new outer tie-rod end nuts and tighten to 59 ft. lbs. (80 Nm).
    d. Tighten rear transaxle insulator through bolt to 66 ft. lbs. (90 Nm).
    e. Tighten transmission damper bolts to 30 ft. lbs. (40 Nm).
14. Check and, if necessary, align the front end.

POWER STEERING PUMP

➡Electronic Power Assist Steering (EPAS) System is used in the Tribute

The electronic power assist steering (EPAS) system provides power steering assist to the driver by replacing the conventional hydraulic valve system with an electric motor coupled to the steering shaft. The motor is controlled by the power steering control (PSC) module that senses the steering effort through the use of a torque sensor mounted between the steering column shaft and the steering gear. Steering assist is provided in proportion to the steering input effort and vehicle speed.

The EPAS system requires a 12-volt, hot at all times feed for system operation. The PSC module is activated when power is

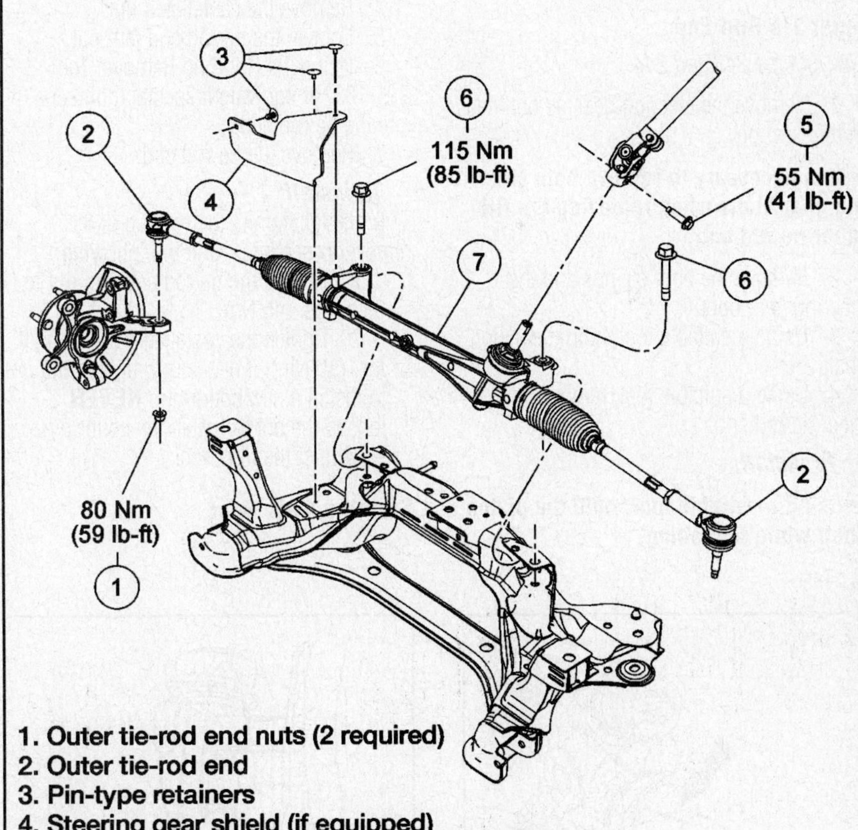

Fig. 242 Steering gear exploded view—2008 model

applied to the hardwired ignition/run input. After activation, the power steering control module monitors the high speed communications area network (HS-CAN) bus to determine if the vehicle is operating in a manner capable of supporting the electronic power steering system.

The vehicle speed, which is sent by the PCM over the HS-CAN bus, provides the necessary vehicle speed information to the electronic power steering system to determine the amount or level of power assist. As vehicle speed increases, the amount of power assist provided by the system is reduced to improve and enhance road feel at the steering wheel. If the vehicle speed is missing or out of range, the power steering control module defaults to a reduced level of assist. If the vehicle speed returns to the correct in-range values, the power steering control module adjusts the steering assist level accordingly.

STEERING LINKAGE

REMOVAL & INSTALLATION

Inner Tie Rod End

See Figures 243 and 244.

1. Remove the steering gear, as outlined in this section.

➡It is necessary to remove both steering gear boots when removing the RH inner tie rod end.

2. Remove the boot clamps and the steering gear boot.
3. Using a suitable tool, hold the piston shaft.
4. Using a suitable tool, remove the inner tie rod end.

To install:

➡Using a suitable tool, hold the piston shaft while tightening.

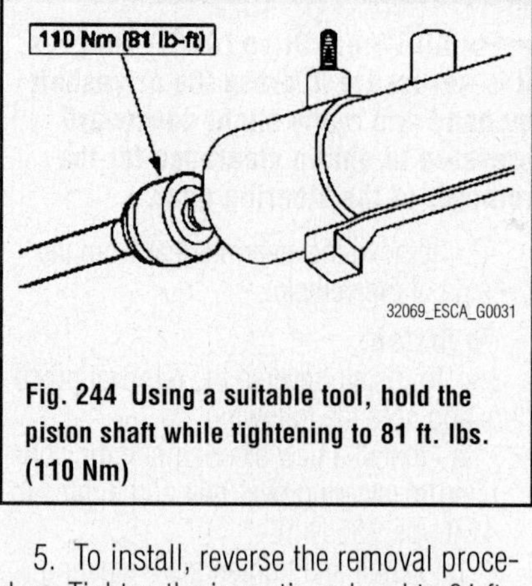

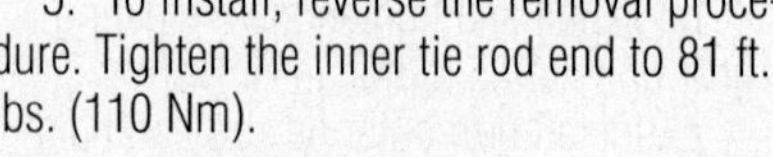

Fig. 244 Using a suitable tool, hold the piston shaft while tightening to 81 ft. lbs. (110 Nm)

5. To install, reverse the removal procedure. Tighten the inner tie rod end to 81 ft. lbs. (110 Nm).

Outer Tie Rod End

See Figures 245 through 247.

1. Turn the ignition switch to the **UNLOCK** position.
2. Remove the tire and wheel assembly.
3. Remove the cotter pin and discard.
4. Remove the castellated nut.
5. Loosen the tie rod end jam nut.
6. Using Tie Rod End Removal Tool 211-105, or equivalent special tool, separate the tie rod end.
7. Remove the tie rod end.

To install:

8. Installation is the reverse of the removal procedure. Note the following:
   a. Tighten the tie rod end jam nut to 33 ft. lbs. (45 Nm).
   b. Tighten the castellated nut to 37 ft. lbs. (50 Nm). If necessary, tighten the nut to install a new cotter pin. **NEVER** loosen the nut to install the cotter pin.
9. Adjust the front toe.

Steering Gear Boot

See Figures 248 and 249.

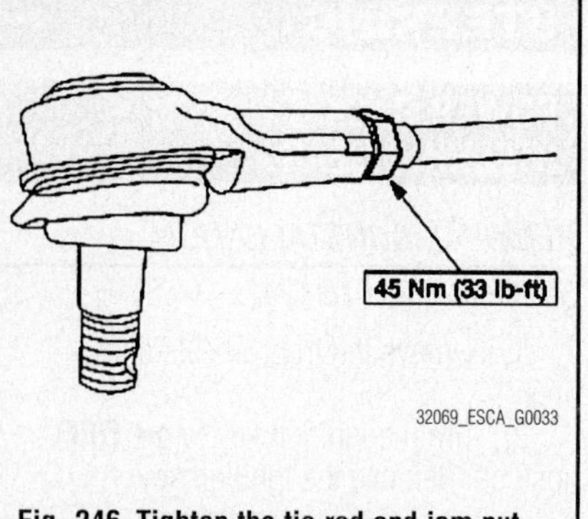

Fig. 246 Tighten the tie rod end jam nut to specifications

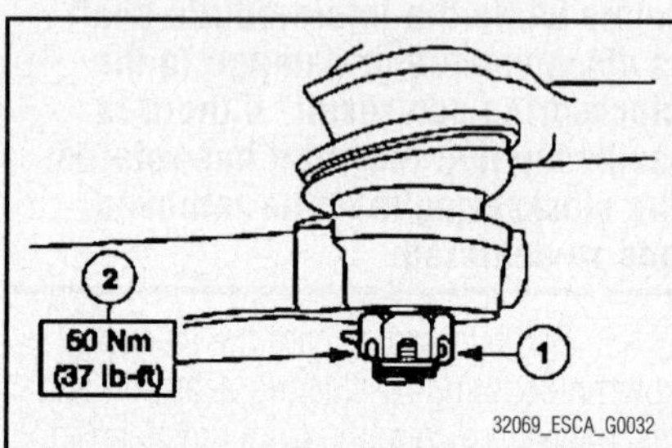

Fig. 247 After tightening the castellated nut (2) to specifications, install a new cotter pin (1). If necessary, tighten the nut further to install the pin. NEVER loosen the nut to install the cotter pin

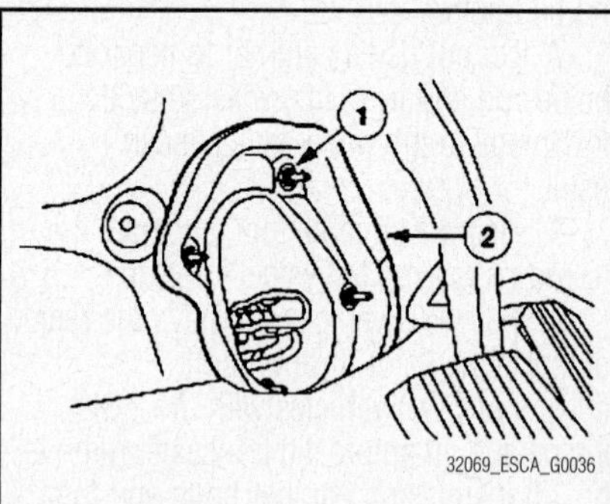

Fig. 248 Remove the nuts (1), then remove the steering gear boot (2)

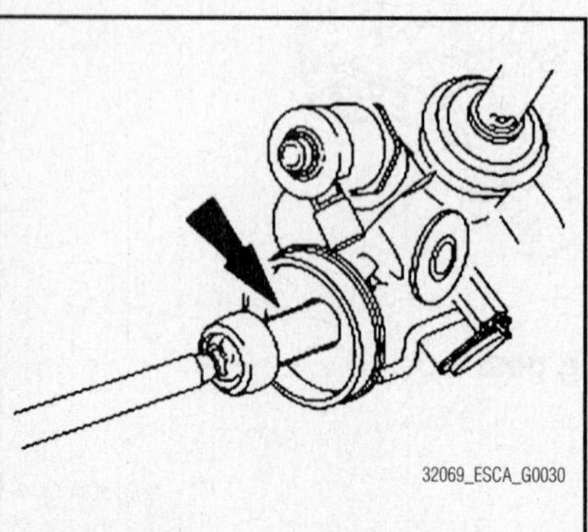

Fig. 243 Using a suitable tool, hold the piston shaft

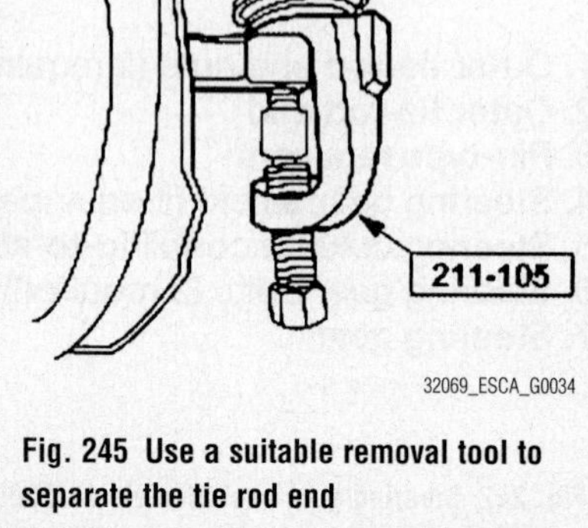

Fig. 245 Use a suitable removal tool to separate the tie rod end

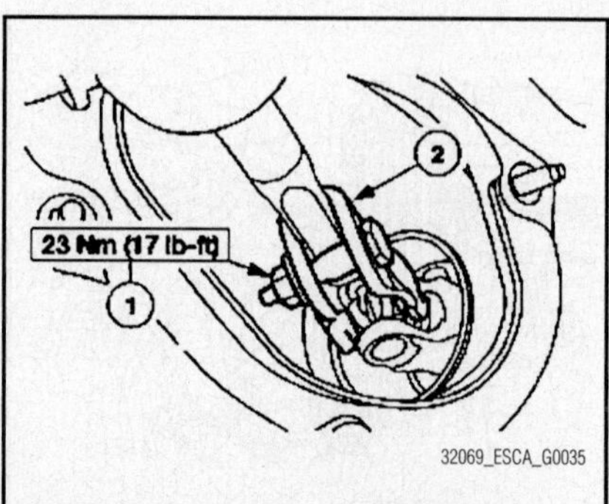

Fig. 249 Connect the steering column shaft to the flexible coupling (2), then install a new nut (1)

1. Position the steering gear boot cover aside.

➡**The nut cannot be reused. Install a new nut.**

2. Disconnect the steering column shaft from the flexible coupling, as follows:

a. Remove and discard the nut.

b. Disconnect the steering column shaft from the flexible coupling.

3. Remove the steering gear boot, as follows:

a. Remove the nuts.

b. Remove the steering gear boot.

4. To install, reverse the removal procedure. Tighten the flexible coupling-to-steering column shaft nut to 17 ft. lbs. (23 Nm).

SUSPENSION

FRONT SUSPENSION

COIL SPRING

REMOVAL & INSTALLATION

See Figure 250.

1. Before servicing the vehicle, refer to the Precautions Section.

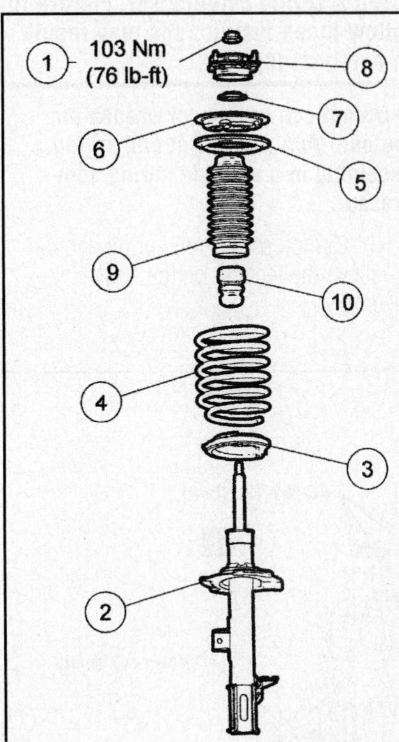

1 Strut piston rod-to-bushing nut
2 Strut (LH/RH)
3 Lower coil spring insulator
4 Coil spring
5 Upper coil spring insulator
6 Bearing plate
7 Bearing
8 Strut upper bushing
9 Dust boot
10 Bumper

06017-ESCA-G76

Fig. 250 Strut/spring disassembly

✲✲ CAUTION

Always wear safety goggles when using a spring compressor. Failure to follow these instructions may result in personal injury.

➡**Do not use an impact wrench on the nut.**

2. Mount the strut and spring assembly in a suitable spring compressor.
3. Compress the coil spring enough to relieve the tension on the strut assembly.
4. Remove the strut piston rod-to-bushing nut.
5. Remove the strut.
6. Remove the lower coil spring insulator.
7. Remove the coil spring.
8. Remove the upper coil spring insulator.

➡**During assembly, assemble the bearing plate to the strut so the arrow on the bearing plate points to the outboard side of the vehicle when the strut is installed.**

9. Remove the bearing plate.
10. Remove the bearing.
11. Remove the strut upper bushing.
12. Remove the dust boot and the bumper.
13. To assemble, reverse the disassembly procedure. Tighten the strut piston rod-to-bushing nut to 76 ft. lbs. (103 Nm).

CONTROL LINKS

REMOVAL & INSTALLATION

1. Before servicing the vehicle, refer to the Precautions Section.
2. Raise and support vehicle
3. Remove the wheel and tire.

➡**Use the hex holding feature to prevent the ball stud from turning while removing or installing the stabilizer bar link nut.**

4. Remove the upper stabilizer bar link nut.
5. Remove the lower stabilizer bar link nut.
6. Remove the stabilizer bar link.
7. Inspect the stabilizer bar link ball joints and boots for wear. If necessary, install new parts.

To install:

8. Install the stabilizer bar link.
9. Tighten the upper and lower stabilizer bar link nut to 41 ft. lbs. (55 Nm).
10. Install the wheel and tire.
11. Lower vehicle

LOWER BALL JOINT

REMOVAL & INSTALLATION

The lower ball joint is part of the lower control arm assembly.

LOWER CONTROL ARM

REMOVAL & INSTALLATION

See Figure 251.

1. Before servicing the vehicle, refer to the Precautions Section.
2. Remove the wheel.
3. Lift the lower arm with a floor jack until the vehicle starts to lift.
4. Record the ride height. It is measure from the center of the halfshaft to the fender lip.
5. Remove the floor jack.
6. Disconnect the ball joint from the knuckle.
7. Support the sub-frame and remove the lower arm.

To install:

8. Install the lower arm, with the bolts loose.
9. Connect the ball joint. Torque the bolt to 52 ft. lbs. (70 Nm).
10. Remove the support.
11. Position the jack under the ball joint and raise the arm to the previously recorded ride height.
12. Tighten the lower arm bolts. Horizontal 85 ft. lbs. (115 Nm); vertical 148 ft. lbs. (200 Nm).

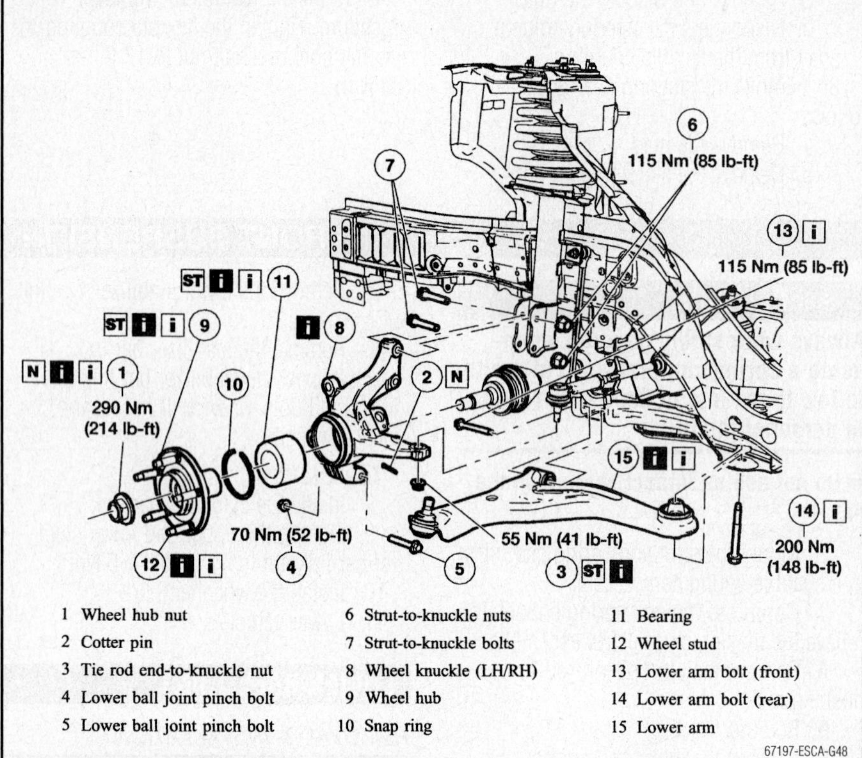

Fig. 251 Front lower control arm assembly

MACPHERSON STRUT

REMOVAL & INSTALLATION

See Figure 252.

1. Before servicing the vehicle, refer to the Precautions Section.

➡Make sure the steering wheel is in the unlocked position.

➡Use the hex holding feature to prevent the ball studs from turning while removing or installing the stabilizer bar link nuts.

2. Raise and support the vehicle.
3. Remove the brake jounce hose clip.
4. Remove the brake jounce hose. Pull the brake jounce hose downward slightly to remove the hose from the bracket.
5. Remove the ABS sensor harness bolt.
6. Remove the stabilizer bar link nut.
7. Remove the strut-to-knuckle nuts.
8. Remove the strut-to-knuckle bolts.
9. Remove the strut upper bushing nuts. Reference mark the strut mounting plate nuts.
10. Remove the strut and spring assembly.

✻✻ WARNING

Do not allow the axle shaft to move outboard. Over-extension of the tripod CV joint can result in separation of internal parts, causing failure of the axle shaft.

11. To install, reverse the removal procedure. See the illustration for the appropriate torque values.
12. Align the strut mounting plate nuts to the reference marks.
13. Check the front end alignment and adjust as necessary.

OVERHAUL

See Figure 253.

✻✻ CAUTION

Always wear safety goggles when using a spring compressor. Failure to follow these instructions may result in personal injury.

➡Do not use an impact wrench on the nut. Mount the strut and spring assembly in a suitable spring compressor.

1. Compress the coil spring enough to relieve the tension on the strut assembly.

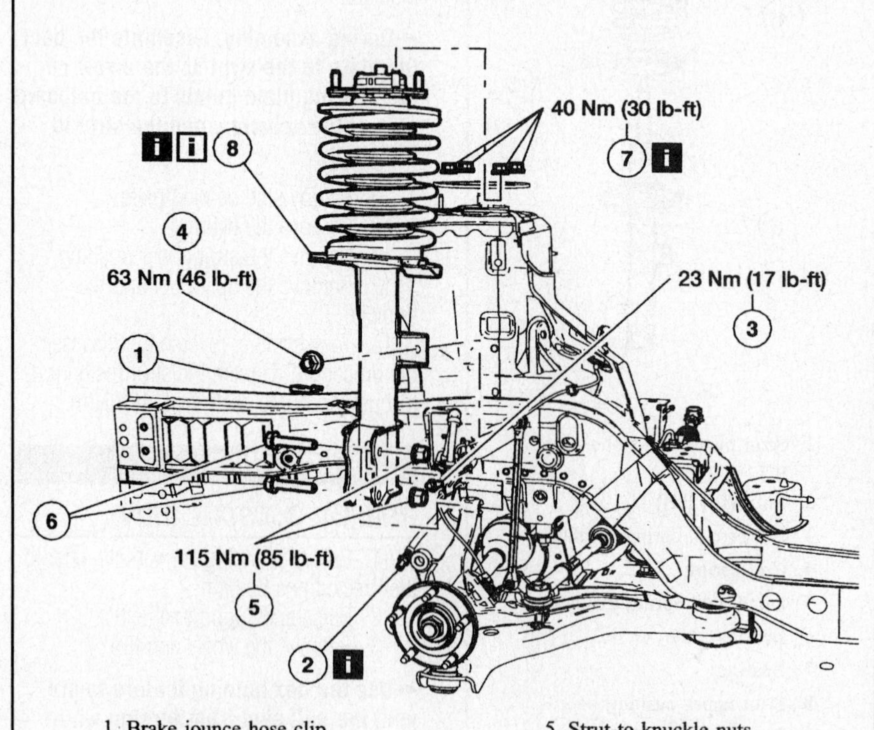

Fig. 252 Front strut installation

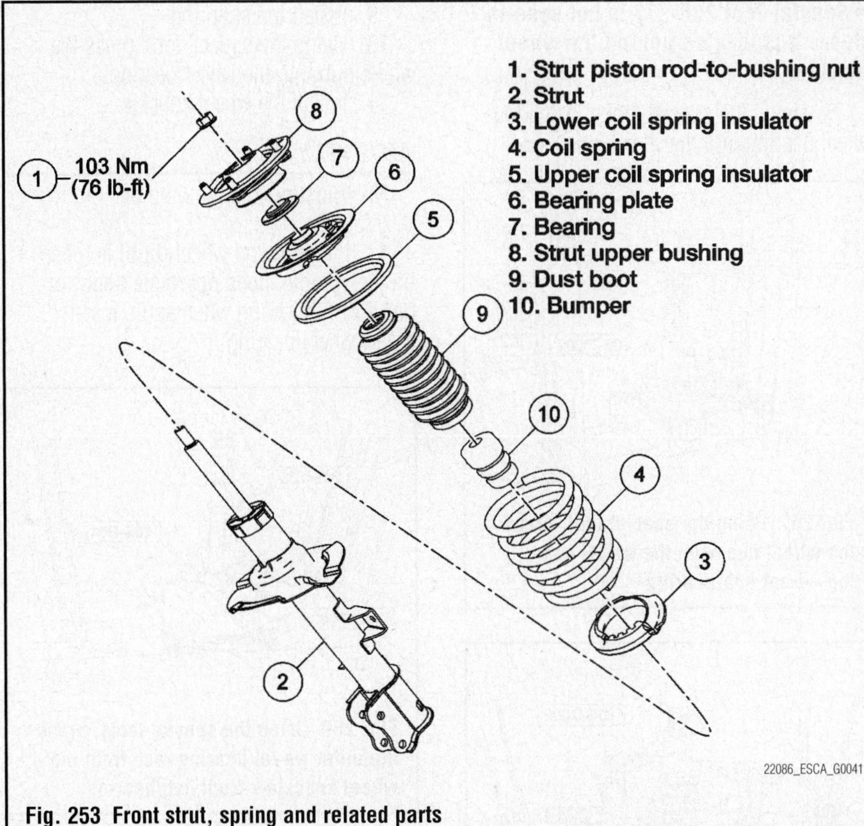

Fig. 253 Front strut, spring and related parts

2. Remove the strut piston rod-to-bushing nut.
3. Remove the strut.
4. Remove the lower coil spring insulator.
5. Remove the coil spring.
6. Remove the upper coil spring insulator.

➡During assembly, assemble the bearing plate to the strut so the arrow on the bearing plate points to the outboard side of the vehicle when the strut is installed.

7. Remove the bearing plate.
8. Remove the bearing.
9. Remove the strut upper bushing.
10. Remove the dust boot and the bumper.
11. To assemble, reverse the disassembly procedure.
12. Replace any parts that show excessive wear or defects.
13. Tighten the strut piston rod-to-bushing nut to 76 ft. lbs. (103 Nm).

STEERING KNUCKLE

REMOVAL & INSTALLATION

See Figures 254 and 255.

1. Before servicing the vehicle, refer to the Precautions Section.

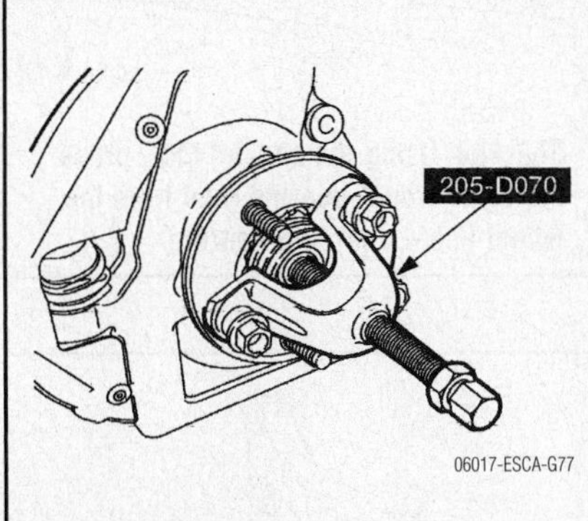

Fig. 254 Separate the outer CV-joint spindle from the wheel hub

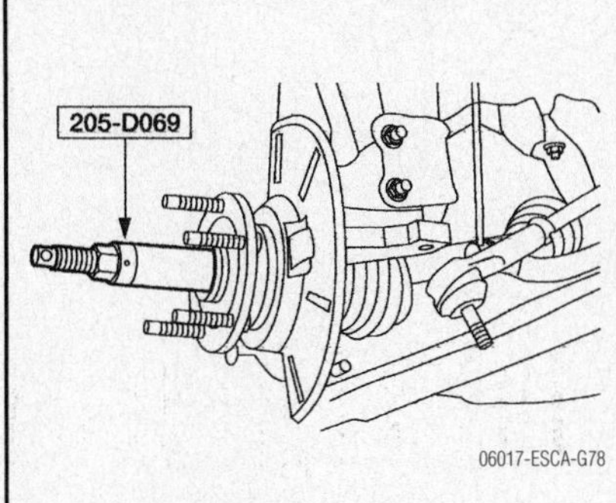

Fig. 255 Insert the halfshaft into the wheel hub

2. Remove the brake disc.
3. Remove and discard the wheel hub nut.
4. Separate the outer CV-joint spindle from the wheel hub.
5. Remove the cotter pin and the tie-rod end-to-knuckle nut.

⁂ WARNING

Do not use a hammer to separate the tie-rod end from the wheel knuckle or damage to the wheel knuckle can result. Do not damage the tie-rod end boot while installing the special tool.

6. Separate the tie-rod from the wheel knuckle.
7. Remove the lower ball joint pinch bolt nut and the pinch bolt.
8. Remove the anti-lock brake system (ABS) wheel speed sensor bolt and position the sensor aside.

Separate the lower ball joint from the wheel knuckle.

9. Remove the 2 strut-to-knuckle nuts, bolts and the wheel knuckle.

To install:

10. Position the wheel knuckle and install the 2 strut-to-knuckle bolts and nuts. Tighten to 85 ft. lbs. (115 Nm).
11. Position and align the ball joint stud into the wheel knuckle.
12. Install the lower ball joint pinch bolt and nut. Tighten to 52 ft. lbs. (70 Nm).
13. Install the ABS wheel speed sensor and the bolt. Tighten to 80 inch lbs. (9 Nm).
14. Position the tie rod-end into the wheel knuckle and install the tie-rod end-to-knuckle nut and a new cotter pin. Tighten to 41 ft. lbs. (55 Nm).
15. Insert the halfshaft into the wheel hub.
16. Install the wheel hub nut. Tighten to 221 ft. lbs. (300 Nm).
17. Install the brake disc.
18. Check and, if necessary, align the front end.

STABILIZER BAR

REMOVAL & INSTALLATION

1. Before servicing the vehicle, refer to the Precautions Section.
2. Remove the stabilizer bar bushing bracket bolts.

➡Use the hex holding feature to prevent the ball stud from turning while removing or installing the stabilizer link nut.

3. Remove the 2 lower stabilizer bar link nuts.

➡Access the stabilizer bar through the left wheel opening.

4. Remove the stabilizer bar.
5. To install, reverse the removal procedure. Observe the following torques:
   - Link nuts: 41 ft. lbs. (55 Nm)
   - Bushing bracket bolts: 52 ft. lbs. (70 Nm)

WHEEL HUB AND BEARING

REMOVAL & INSTALLATION

See Figures 256 through 261.

1. Before servicing the vehicle, refer to the Precautions Section.

➡If removing the wheel hub, the wheel bearing must be replaced.

2. Remove the wheel knuckle.
3. Using the special tool, press the wheel hub from the wheel bearing.

➡This step may not be necessary if the inner wheel bearing race remains in the wheel knuckle after removing the wheel hub.

4. Using the special tool, press the inner wheel bearing race from the wheel hub.
5. Remove the snapring.
6. Using the special tools, press the outer wheel bearing race from the wheel knuckle.

To install:

7. Position the wheel knuckle in a vise.

➡Special Tool 205-278 is not seen in place. It is located behind the wheel knuckle.

8. Using the special tools, install the wheel bearing into the wheel knuckle.

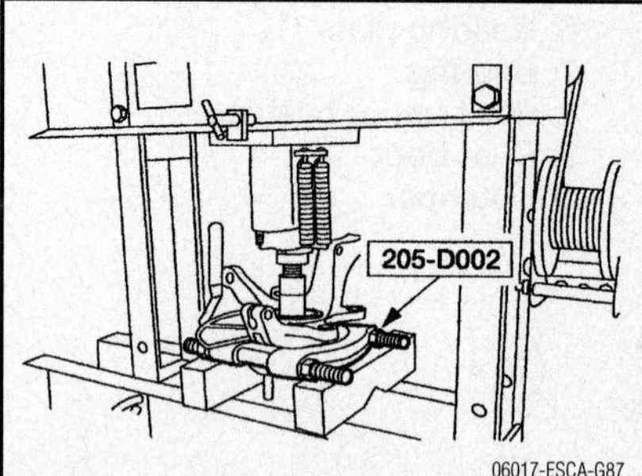

Fig. 257 Using the special tool, press the wheel hub from the wheel bearing—front hub/bearing

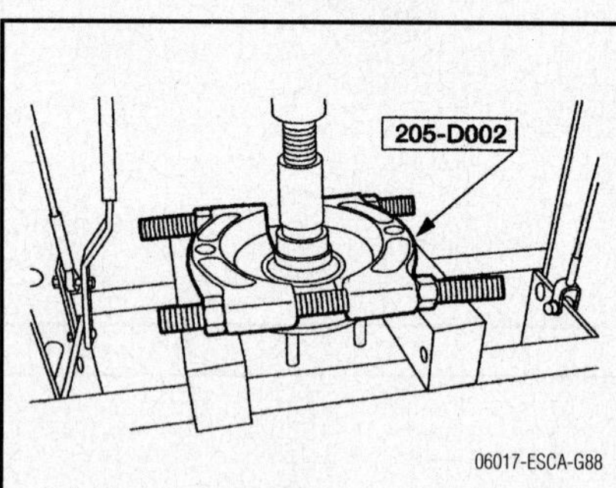

Fig. 258 Using the special tool, press the inner wheel bearing race from the wheel hub—front hub/bearing

9. Install the snapring.
10. Using the special tool, press the wheel hub into the wheel bearing.
11. Install the wheel knuckle.

ADJUSTMENT

No adjustment is required or possible.

1. If the tire and wheel (hub) is loose on the spindle, does not rotate freely, or has a rough feeling when spun, install a new wheel bearing.

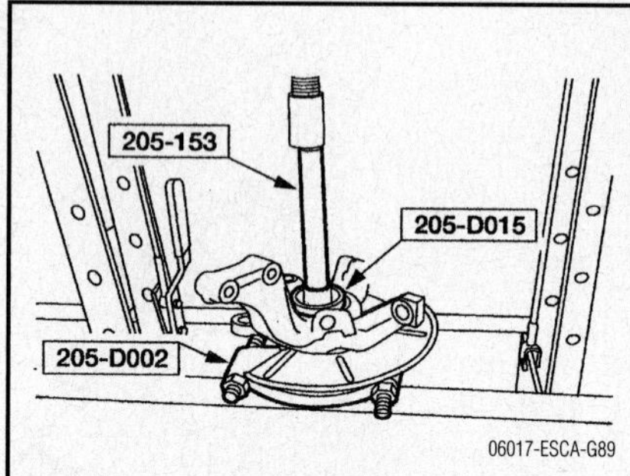

Fig. 259 Using the special tools, press the outer wheel bearing race from the wheel knuckle—front hub/bearing

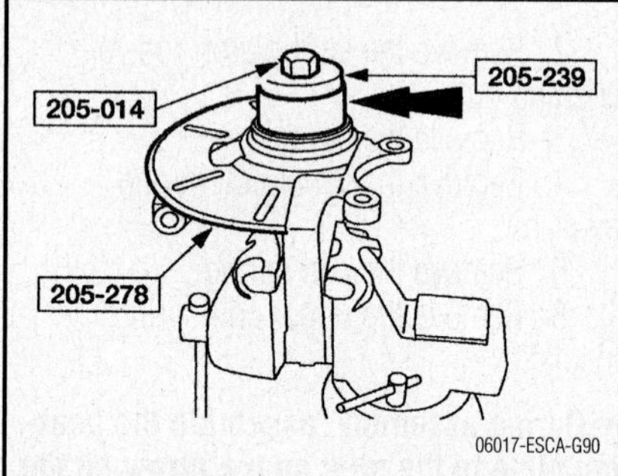

Fig. 260 Using the special tools, install the wheel bearing into the wheel knuckle—front hub/bearing

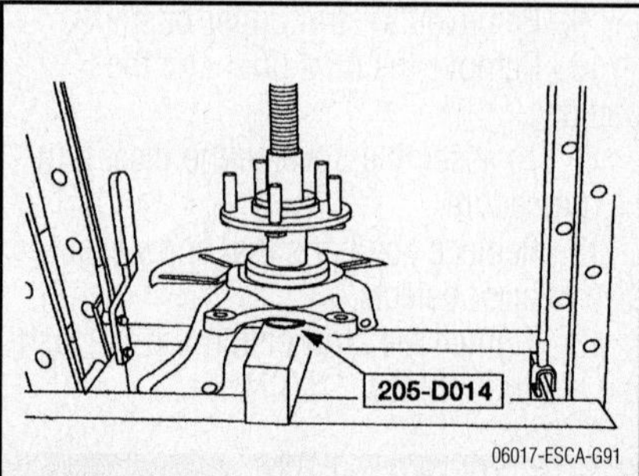

Fig. 261 Using the special tool, press the wheel hub into the wheel bearing—front hub/bearing

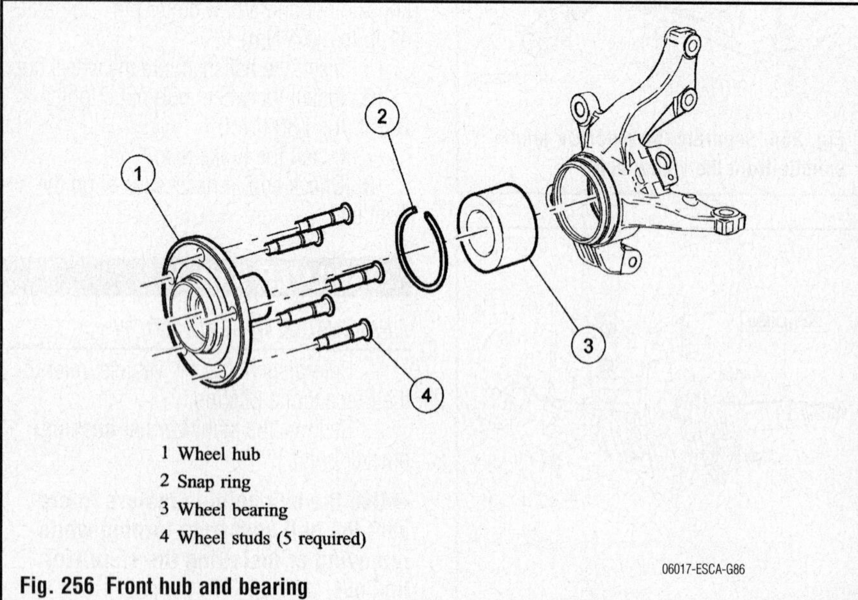

Fig. 256 Front hub and bearing

COIL SPRING

REMOVAL & INSTALLATION

1. Before servicing the vehicle, refer to the Precautions Section.
2. With the vehicle in **NEUTRAL**, position it on a hoist.
3. Remove the brake hose bracket-to-wheel knuckle bolt.
4. Vehicles with drum brakes, disconnect the brake line from the wheel cylinder.
5. Vehicles with disc brakes, remove the 2 brake caliper guide bolts and position the brake caliper aside. Support the caliper using mechanic's wire.
6. Support the wheel knuckle and remove the upper ball joint nut.
7. Remove the lower shock absorber nut, washer and bolt.
8. Remove the upper arm inner bolt and remove the upper arm.
9. Loosen the lower arm inner bolt.

➡Note the position of the coil spring insulator and coil spring for installation.

10. Carefully lower the wheel knuckle support.
11. Remove the coil spring.

To install:

12. Align the coil spring and coil spring insulator to the previously noted position.
13. Carefully raise the wheel knuckle support.
14. Position the upper control arm and install the upper arm inner bolt tighten to 85 ft. lbs. (115 Nm).
15. Install the lower shock absorber bolt, washer and nut tighten to 129 ft. lbs. (175 Nm).
16. Install the upper ball joint nut and remove the wheel knuckle support tighten the nut to 76 ft. lbs. (103 Nm).
17. Vehicles with disc brakes, position the brake caliper and install the 2 caliper guide bolts. Tighten bolts to 26 ft. lbs (35 Nm).
18. Vehicles with drum brakes, connect the brake line fitting to the wheel cylinder. Tighten the nut to 11 ft. lbs. (15 Nm).
19. Position the brake hose and install the brake hose bracket-to-wheel knuckle bolt. Tighten the nut to 13 ft. lbs (17 Nm).
20. Bleed the brake hydraulic system if line was removed.

CONTROL ARMS/LINKS

REMOVAL & INSTALLATION

Upper

1. Before servicing the vehicle, refer to the Precautions Section.
2. Remove the wheel and tire.

➡It may be necessary to hold the ball joint stud to keep it from turning while removing the nut.

3. Separate the upper arm from the wheel knuckle. Remove the upper ball joint nut.
4. Remove the upper arm inner bolt.
5. Remove the upper arm.
6. To install, reverse the removal procedure. Observe the following torques:
   - Ball joint nut: 76 ft. lbs. (103 Nm)
   - Lower arm bolts: 85 ft. lbs. (115 Nm)

Lower

See Figure 262.

1. Before servicing the vehicle, refer to the Precautions Section.
2. Remove or disconnect the following:
   - Negative battery cable
   - Lower ball joint from the knuckle while holding the ball joint stud from moving
   - Lower ball joint nut
   - Lower control arm
   - Lower control arm inner bolt

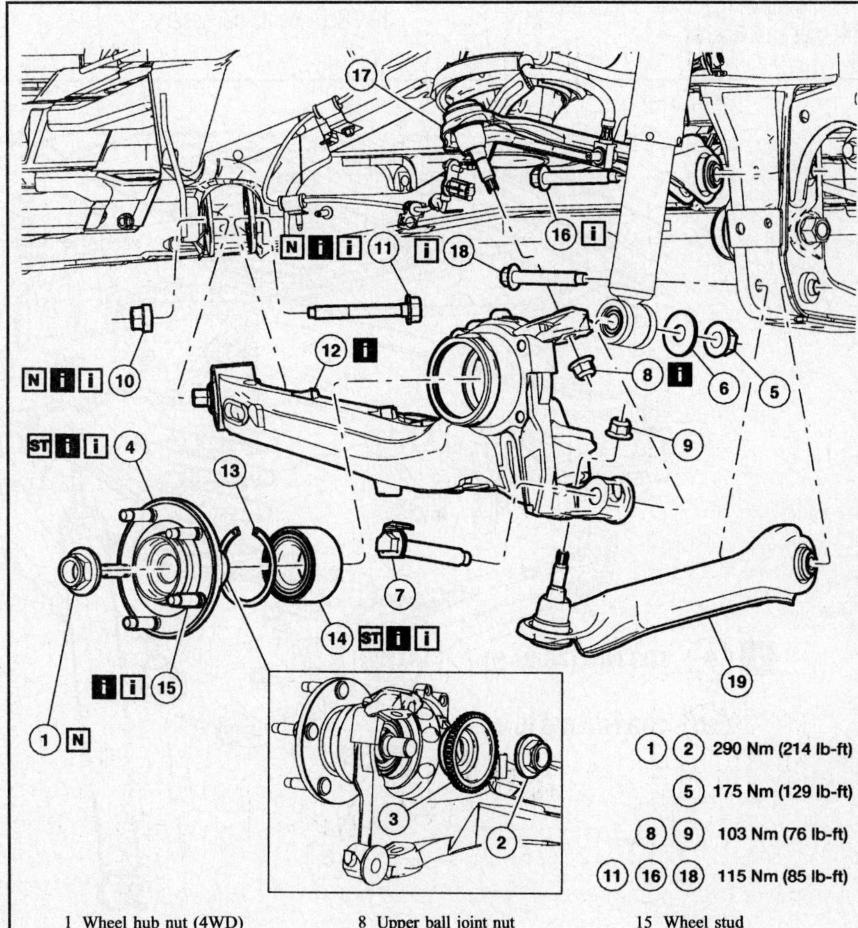

Fig. 262 Rear lower control arm and related parts

To install:

3. Install or connect the following:
   - Lower control arm inner bolt
   - Lower control arm. Torque the bolts to 85 ft. lbs. (115 Nm).
   - Lower ball joint nut
   - Lower ball joint the knuckle. Torque the ball joint nut to 76 ft. lbs. (103 Nm).
   - Rear wheel

SHOCK ABSORBER

REMOVAL & INSTALLATION

See Figure 263.

1. Remove the wheel and tire assemblies.
2. Remove the rear quarter trim panel. Remove the upper shock absorber nut, bushing and washer.
3. Remove the lower shock absorber nut, bolt and washer.
4. Remove the shock absorber and bushing.
5. To install, reverse the removal procedure. Torque the upper nut to 13 ft. lbs.; the lower nut to 129 ft. lbs. (175 Nm).

TESTING

1. Road test the vehicle.
2. On a smooth road see if any vibrations are present.
3. Use your hands in order to lift up and push down each corner of the vehicle 3 times.
4. Remove your hands from the vehicle.
5. Replace any shock that exceeds more than two bounces.
6. Raise vehicle for inspection
7. Inspect each shock absorber for external fluid leakage.
8. Inspect for deformation or damage.
9. Inspect bushings for wear or damage.
10. Replace as necessary.

WHEEL BEARINGS

REMOVAL & INSTALLATION

2WD Models

See Figures 264 through 269.

1. Before servicing the vehicle, refer to the Precautions Section.
2. Remove or disconnect the following:
   - Negative battery cable
   - Rear wheel
   - Rear brake drum
   - Wheel hub nut
   - Wheel hub
   - Inner wheel bearing race from the hub
   - Snapring
   - Wheel bearing outer race from the knuckle

To install:

3. Install or connect the following:

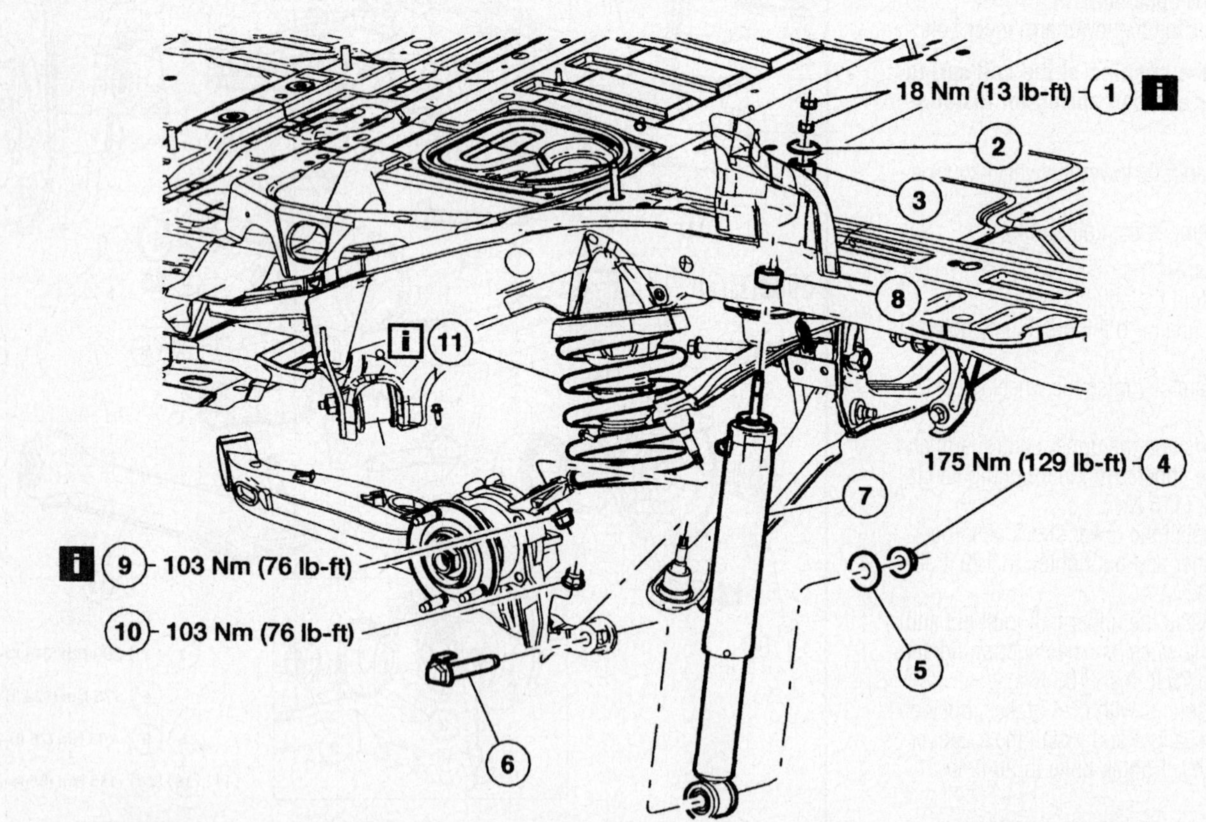

Fig. 263 Rear shock absorber and spring

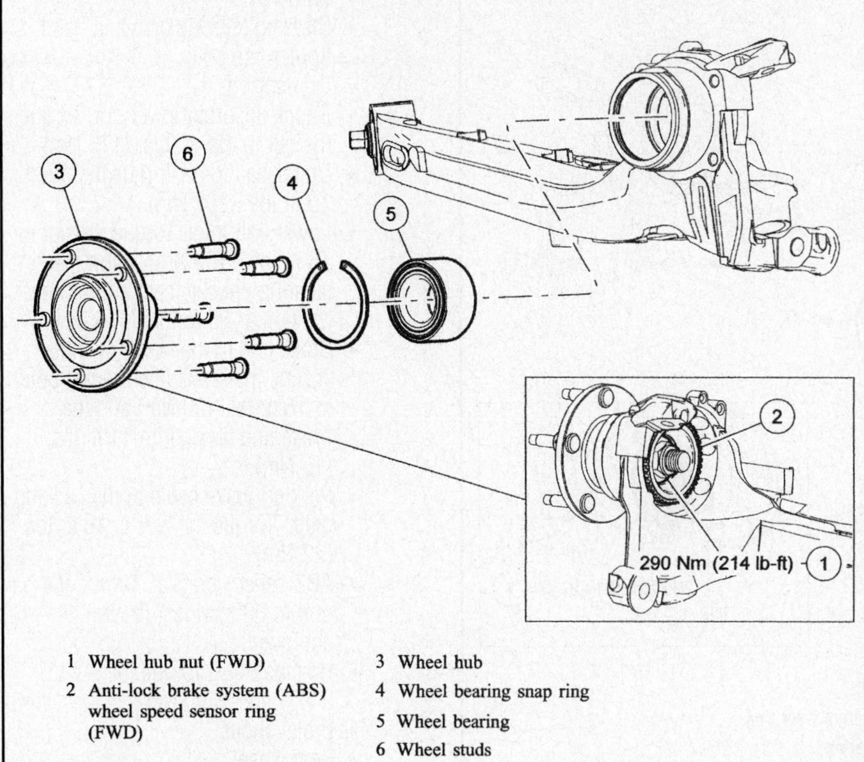

Fig. 264 Rear hub and bearing

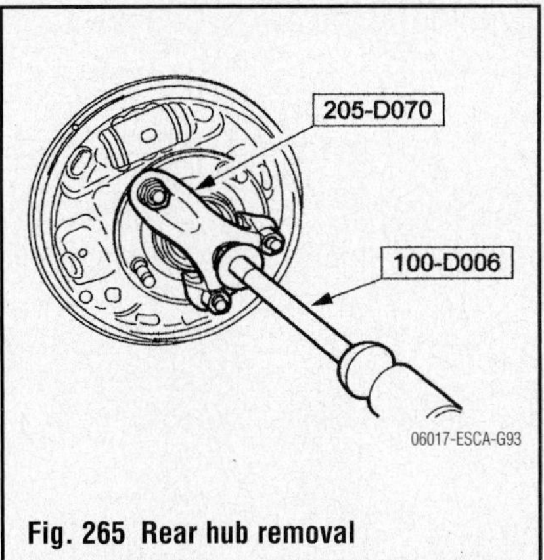

Fig. 265 Rear hub removal

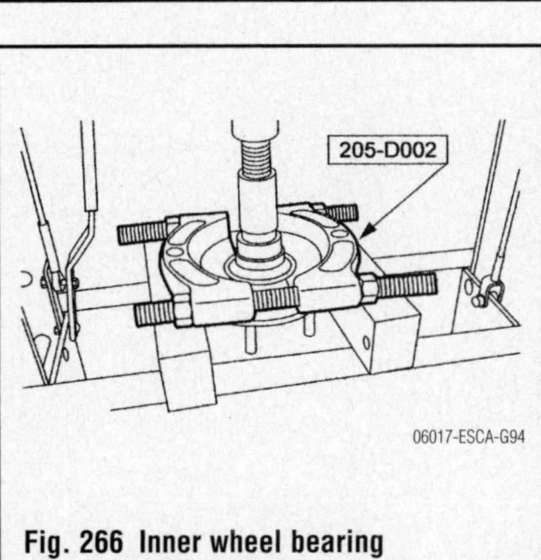

Fig. 266 Inner wheel bearing removal—rear hub/bearing

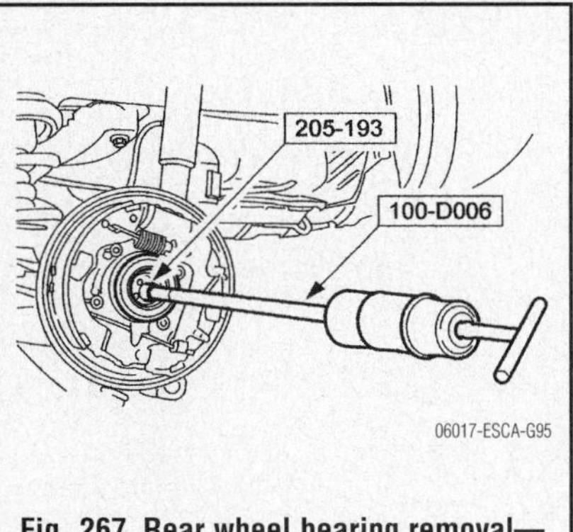

Fig. 267 Rear wheel bearing removal—2WD

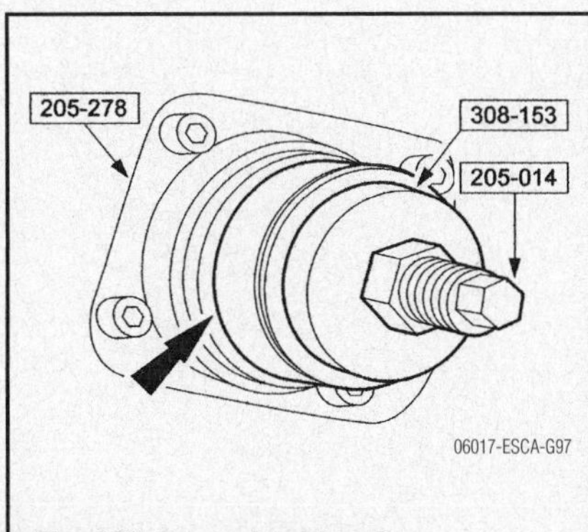

Fig. 268 Rear wheel bearing installation

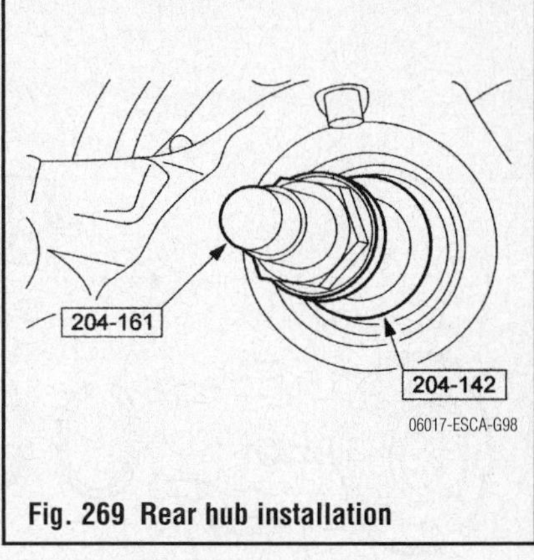

Fig. 269 Rear hub installation

- Wheel bearing in to the knuckle
- Snapring
- Wheel hub into the wheel bearing
- Wheel hub nut. Torque the nut to 214 ft. lbs. (290 Nm).
- Brake drum
- Rear wheel
- Negative battery cable

4WD Models

See Figure 270.

1. Before servicing the vehicle, refer to the Precautions Section.
2. Remove or disconnect the following:
   - Negative battery cable
   - Rear wheel
   - Rear brake shoes
   - Rear halfshaft nut and loosen the halfshaft from the hub
   - Wheel hub and place it in a vise
   - Inner wheel bearing race from the hub
   - Antilock Brake System (ABS) sensor bracket and move the sensor aside, if equipped
   - Parking brake cable from the steering knuckle
   - Brake line from the wheel cylinder and support the knuckle
   - Lower shock absorber nut
   - Lower ball joint by holding the ball joint stud
   - Upper ball joint
   - Coil spring while noting the location of the insulator
   - Steering knuckle cam
   - Steering knuckle
   - Snapring and press out the outer wheel bearing race from the knuckle

To install:

3. Install or connect the following:
   - New wheel bearing into the steering knuckle
   - Snapring to the knuckle

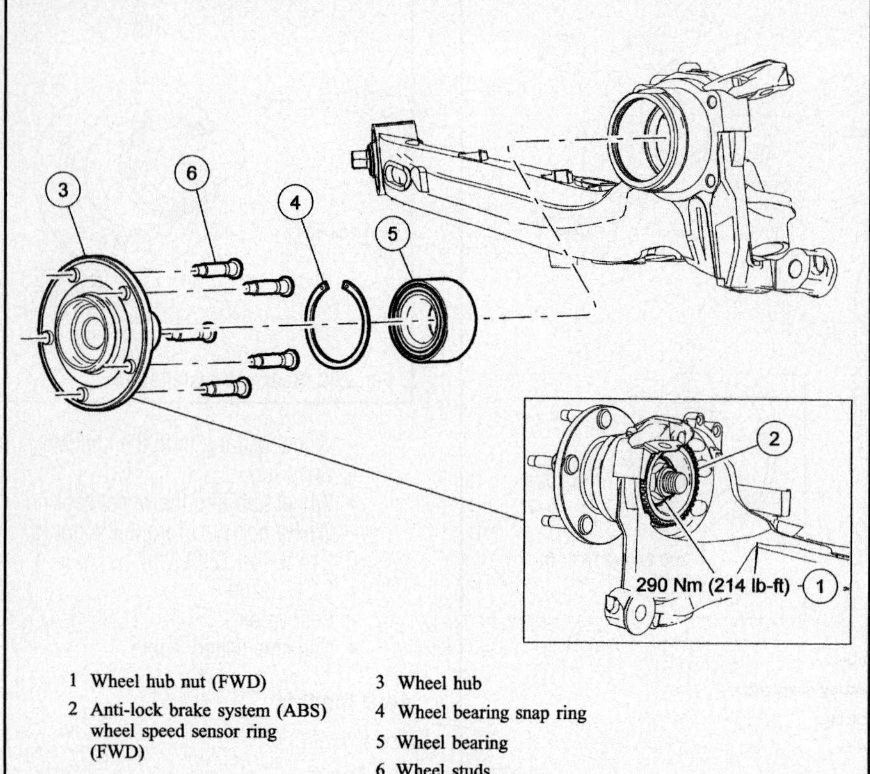

Fig. 270 Rear hub and bearing

- Wheel hub
- Steering knuckle cam and hand tighten the bolt
- Coil spring
- Shock absorber lower nut. Torque the nut to 129 ft. lbs. (175 Nm).
- Upper ball joint. Torque the nut to 76 ft. lbs. (103 Nm).
- Lower ball joint. Torque the nut to 76 ft. lbs. (103 Nm). Align the steering knuckle cam and torque the bolt to 85 ft. lbs. (115 Nm).
- Brake line to the wheel cylinder. Torque the brake line bracket bolt to 15 ft. lbs. (20 Nm) and the brake line fastener to 11 ft. lbs. (15 Nm).
- Parking brake cable to the backing plate. Torque the bolt to 16 ft. lbs. (22 Nm).
- ABS sensor bracket. Torque the bolt to 80 inch lbs. (9 Nm), if equipped
- Halfshaft nut. Torque the nut to 214 ft. lbs. (290 Nm).
- Brake shoes
- Rear wheel
- Negative battery cable

4. Fill and bleed the brake system.
5. Check and adjust the wheel alignment as needed.

7

MAZDA

Diagnostic Trouble Codes

DIAGNOSTIC TROUBLE CODES

OBD II VEHICLE APPLICATIONS

MAZDA

B-Series Truck
2007–2008
- 2.3L I4 .... VIN D

2007
- 3.0L V6 .... VIN U

2007–2008
- 4.0L V6 .... VIN E

CX-7
2007–2008
- 2.3L I4 .... VIN L (CA)
- 2.3L I4 .... VIN 3 (Fed)

CX-9
2007
- 3.5L V6 .... VIN C (CA)
- 3.5L V6 .... VIN Y (Fed)

2008
- 3.7L V6 .... VIN V (CA)
- 3.7L V6 .... VIN A (Fed)

Mazda3
2007–2008
- 2.0L I4 .... VIN G (CA)
- 2.0L I4 .... VIN F (Fed)

Mazda3 & MazdaSpeed3
2007–2008
- 2.3L I4 .... VIN 4 (CA)
- 2.3L I4 .... VIN 3 (Fed)
- 2.3L I4 .... VIN M (Turbo - CA)
- 2.3L I4 .... VIN L (Turbo – Fed.)

Mazda5
2007–2008
- 2.3L I4 .... VIN L (CA)
- 2.3L I4 .... VIN 3 (North America)

MazdaSpeed6
2007
- 2.3L I4 .... VIN C

Mazda6
2007–2008
- 2.3L I4 .... VIN C

2007–2008
- 3.0L V6 .... VIN D

Mazda6 Sport Wagon
2007
- 3.0L V6 .... VIN D

MX-5
2007–2008
- 2.0L I4 .... VIN F

RX-8
2007–2008
- 1.3L I4 .... VIN N (Std Power)
- 1.3L I4 .... VIN 3 (High Power)

Tribute
2008
- 2.3L I4 .... VIN Z

2008
- 3.0L V6 .... VIN 1

Gas Engine OBD II Trouble Code List (P0xxx Codes)

DTC	Trouble Code Title, Conditions & Possible Causes
DTC: P0010 **1T CCM, MIL: Yes** **Years:** 2007, 2008 **Models:** CX-9 **Engines:** All **Transmissions:** All	**Camshaft Position Actuator Circuit Malfunction** Engine running, and the PCM detected the Oil Control Valve (OCV) current was too high or too low (as calculated from system voltage). **Possible Causes:** • OCV solenoid control circuit is open or shorted to ground • OCV solenoid power circuit is open • OCV solenoid ground circuit is open • OCV is damaged or has failed • PCM has failed
DTC: P0011 **1T CCM, MIL: Yes** **Years:** 2007, 2008 **Models:** CX-7, CX-9, Mazda3, MazdaSpeed3, Mazda5, MazdaSpeed6, Mazda6, Mazda6 Sport Wagon, MX-5 **Engines:** All **Transmissions:** All	**Camshaft Position Timing Over-Advanced** Engine running, and the PCM detected the Actual valve timing was more than 10 degrees advanced for 1 second from the Target valve timing with the engine running at maximum valve timing retard. **Possible Causes:** • OCV valve is damaged or has failed • OCV spool valve is stuck in the advanced position • Variable valve timing actuator is stuck in the retard position • Oil runners between oil pressure switch and the OCV, or between the OCV and the VVT actuator are dirty or clogged
DTC: P0012 **1T CCM, MIL: Yes** **Years:** 2007, 2008 **Models:** CX-7, CX-9, Mazda3, MazdaSpeed3, Mazda5, MazdaSpeed6, Mazda6, Mazda6 Sport Wagon, MX-5 **Engines:** All **Transmissions:** All	**Camshaft Position Timing Over-Retarded** Engine running, and the PCM detected the Actual valve timing was more than 15 degrees retarded for 5 seconds from the Target valve timing with the Oil Control Valve system within feedback range. **Possible Causes:** • OCV valve is damaged or has failed • Low engine oil pressure condition • OCV spool valve is stuck in the retard position • Variable valve timing actuator is stuck in the advanced position • Timing belt is loose or improper valve timing due to a loose belt • PCM has failed
DTC: P0016 **2T CCM, MIL: Yes** **Years:** 2007, 2008 **Models:** CX-7, CX-9, Mazda3, MazdaSpeed3, Mazda5, MazdaSpeed6, Mazda6, Mazda6 Sport Wagon, MX-5 **Engines:** All **Transmissions:** All	**CKP-CMP Correlation (RH)** The PCM monitors the input pulses from the CKP sensor and CMP sensor. If the input pulse pick-up timing does not match each other, the PCM determines that the camshaft position does not coincide with the crankshaft position. **Possible Causes:** • Poor connection of connector • CMP sensor malfunction • CKP sensor malfunction • Damaged or foreign material on CKP or CMP sensor • Improper valve timing
DTC: P0018 **2T CCM, MIL: Yes** **Years:** 2007, 2008 **Models:** CX-9, Mazda6, Mazda6 Sport Wagon **Engines:** 3.0L, 3.5L, 3.7L **Transmissions:** All	**CKP-CMP Correlation (LH)** The PCM monitors the input pulses from the CKP sensor and CMP sensor. If the input pulse pick-up timing does not match each other, the PCM determines that the camshaft position does not coincide with the crankshaft position. **Possible Causes:** • Poor connection of connector • CMP sensor malfunction • CKP sensor malfunction • Damaged or foreign material on CKP or CMP sensor • Improper valve timing
DTC: P0020 **2T CCM, MIL: Yes** **Years:** 2007, 2008 **Models:** CX-9 **Engines:** All **Transmissions:** All	**CMP Actuator Circuit Open (LH)** The PCM monitors the OCV (LH) circuit to the PCM for high and low voltage. The test fails if the voltage exceeds or falls below a calibrated limit for a calibrated amount of time. **Possible Causes:** • OCV (LH) malfunction • Connector or terminal malfunction • Open circuit between OCV (LH) terminal A and main relay No.1 terminal D • Open circuit between OCV (LH) terminal B and PCM terminal 2BK • Short to ground circuit between OCV (LH) terminal B and PCM terminal 2BK • Short to power supply between OCV (LH) terminal B and PCM terminal 2BK

DTC	Trouble Code Title, Conditions & Possible Causes
DTC: P0021 **2T CCM, MIL: Yes** **Years:** 2007, 2008 **Models:** CX-9, Mazda6, Mazda6 Sport Wagon **Engines:** 3.0L, 3.5L, 3.7L **Transmissions:** All	**CMP Timing Over-Advanced (LH)** The PCM monitors the variable valve timing position for an over-advanced camshaft timing. The test fails when the camshaft timing exceeds a maximum calibrated value or remains in an advanced position. **Possible Causes:** • OCV (LH) malfunction • Spool valve in OCV (LH) is stuck in advanced position • Improper valve timing due to timing chain slippage
DTC: P0022 **2T CCM, MIL: Yes** **Years:** 2007, 2008 **Models:** CX-9, Mazda6, Mazda6 Sport Wagon **Engines:** 3.0L, 3.5L, 3.7L **Transmissions:** All	**CMP Timing Over-Retarded (LH)** The PCM monitors the variable valve timing position for over-retarded camshaft timing. The test fails when the camshaft timing exceeds a maximum calibrated value or remains in a retarded position. **Possible Causes:** • OCV (LH) malfunction • Low engine oil pressure • Spool valve in OCV (LH) is stuck in retard position • Improper valve timing due to timing chain slippage
DTC: P0030 **2T CCM, MIL: Yes** **Years:** 2007, 2008 **Models:** B-Series Truck, CX-7, Mazda3, MazdaSpeed3, Mazda5, MazdaSpeed6, Mazda6, MX-5, RX-8 **Engines:** All **Transmissions:** All	**HO2S-11 (Bank 1, Sensor 1) Heater Control Circuit Problem** The PCM monitors the front HO2S impedance when under the front HO2S heater control for 200 s. If the impedance is more than 44 ohms, the PCM determines that there is a front HO2S heater control circuit problem. **Possible Causes:** • Front HO2S heater malfunction • Connector or terminal damage • PCM has failed
DTC: P0031 **2T CCM, MIL: Yes** **Years:** 2007, 2008 **Models:** B-Series Truck, CX-7, Mazda3, MazdaSpeed3, Mazda5, MazdaSpeed6, Mazda6, Mazda6 Sport Wagon, MX-5, RX-8 **Engines:** All **Transmissions:** All	**HO2S-11 (Bank 1 Sensor 1) Heater Circuit Low Input** Engine running, and after the front HO2S Heater Control duty cycle signal was commanded "off", the PCM detected a low input condition on the heater control circuit. **Possible Causes:** • HO2S heater control circuit is shorted to chassis ground • HO2S heater is damaged or has failed • PCM has failed
DTC: P0032 **2T CCM, MIL: Yes** **Years:** B-Series Truck, CX-7, Mazda3, MazdaSpeed3, Mazda5, MazdaSpeed6, Mazda6, Mazda6 Sport Wagon, MX-5, RX-8 **Engines:** All **Transmissions:** All	**HO2S-11 (Bank 1 Sensor 1) Heater Circuit High Input** Engine running, and after the front HO2S Heater Control duty cycle signal was commanded "on", the PCM detected a high input condition on the heater control circuit. **Possible Causes:** • HO2S heater control circuit is shorted to system power (B+) • HO2S heater is damaged or has failed • PCM has failed
DTC: P0037 **2T CCM, MIL: Yes** **Years:** 2007, 2008 **Models:** B-Series Truck, CX-7, Mazda3, MazdaSpeed3, Mazda5, MazdaSpeed6, Mazda6, Mazda6 Sport Wagon, MX-5, RX-8 **Engines:** All **Transmissions:** All	**HO2S-12 (Bank 1 Sensor 2) Heater Circuit Low Input** Engine running, and after the rear HO2S Heater Control duty cycle signal was commanded "off", the PCM detected a low input condition on the heater control circuit. **Possible Causes:** • HO2S heater control circuit is shorted to chassis ground • HO2S heater is damaged or has failed • PCM has failed

DTC	Trouble Code Title, Conditions & Possible Causes
DTC: P0038 **2T CCM, MIL: Yes** **Years:** 2007, 2008 **Models:** B-Series Truck, CX-7, Mazda3, MazdaSpeed3, Mazda5, MazdaSpeed6, Mazda6, Mazda6 Sport Wagon, MX-5, RX-8 **Engines:** All **Transmissions:** All	**HO2S-12 (Bank 1 Sensor 2) Heater Circuit High Input** Engine running, and after the rear HO2S Heater Control duty cycle signal was commanded "on", the PCM detected a high input condition on the heater control circuit. **Possible Causes:** • HO2S heater control circuit is shorted to system power (B+) • HO2S heater is damaged or has failed • PCM has failed
DTC: P0040 **1T CCM, MIL: No** **Years:** 2007, 2008 **Models:** CX-9, Tribute **Engines:** All **Transmissions:** All	**Front HO2S (LH/RH) Signals Swapped** During KOER testing the front HO2S signal(s) response for a fuel shift to the correct engine bank was not indicated. **Possible Causes:** • Crossed HO2S harness connectors. • Crossed wiring at HO2S harness connectors. • Crossed wiring at PCM harness connector.
DTC: P0041 **1T CCM, MIL: No** **Years:** 2007, 2008 **Models:** CX-9, Tribute **Engines:** All **Transmissions:** All	**Rear HO2S (LH/RH) Signals Swapped** During KOER testing the rear HO2S signal(s) response for a fuel shift to the correct engine bank was not indicated. **Possible Causes:** • Crossed HO2S harness connectors. • Crossed wiring at HO2S harness connectors. • Crossed wiring at PCM harness connector.
DTC: P0043 **2T CCM, MIL: Yes** **Years:** 2007, 2008 **Models:** Mazda3, MazdaSpeed3, Mazda6, MazdaSpeed6 **Engines:** All **Transmissions:** All	**HO2S-12 (Bank 1 Sensor 2) Heater Circuit Low Input** Engine running, and after the rear HO2S Heater Control duty cycle signal was commanded "on", the PCM detected a high input condition on the heater control circuit. **Possible Causes:** • HO2S heater control circuit is shorted to system power (B+) • HO2S heater is damaged or has failed • PCM has failed
DTC: P0044 **2T CCM, MIL: Yes** **Years:** 2007, 2008 **Models:** Mazda3, MazdaSpeed3, Mazda6, MazdaSpeed6 **Engines:** All **Transmissions:** All	**HO2S-12 (Bank 1 Sensor 2) Heater Circuit High Input** Engine running, and after the rear HO2S Heater Control duty cycle signal was commanded "on", the PCM detected a high input condition on the heater control circuit. **Possible Causes:** • HO2S heater control circuit is shorted to system power (B+) • HO2S heater is damaged or has failed • PCM has failed
DTC: P0051 **2T CCM, MIL: Yes** **Years:** 2007, 2008 **Models:** B-Series Truck, Mazda6, MazdaSpeed6 Sport Wagon **Engines:** All **Transmissions:** All	**HO2S-21 (Bank 2 Sensor 1) Heater Circuit Low Input** Engine running, and after the rear HO2S Heater Control duty cycle signal was commanded "off", the PCM detected a low input condition on the heater control circuit. **Possible Causes:** • HO2S heater control circuit is shorted to chassis ground • HO2S heater is damaged or has failed • PCM has failed
DTC: P0052 **2T CCM, MIL: Yes** **Years:** 2007, 2008 **Models:** B-Series Truck, Mazda6, MazdaSpeed6 Sport Wagon **Engines:** All **Transmissions:** All	**HO2S-21 (Bank 2 Sensor 1) Heater Circuit High Input** Engine running, and after the rear HO2S Heater Control duty cycle signal was commanded "on", the PCM detected a high input condition on the heater control circuit. **Possible Causes:** • HO2S heater control circuit is shorted to system power (B+) • HO2S heater is damaged or has failed • PCM has failed

DTC	Trouble Code Title, Conditions & Possible Causes
DTC: P0053 **2T CCM, MIL: Yes** **Years:** 2007, 2008 **Models:** B-Series Truck, CX-9, Tribute **Engines:** All **Transmissions:** All	**Front HO2S (RH) Heater Resistance** Heater current requirements too low or high in the heated oxygen sensor (HO2S) heater control circuit. **Possible Causes:** • Open circuit between HO2S terminal C and main relay No.1 terminal E • Short to ground circuit between HO2S terminal A and body ground • HO2S circuits are shorted each other • Open circuit between front HO2S (RH) terminal A and PCM terminal 2J
DTC: P0054 **2T CCM, MIL: Yes** **Years:** 2007, 2008 **Models:** B-Series Truck, CX-9, Tribute **Engines:** All **Transmissions:** All	**Rear HO2S (RH) Heater Resistance** Heater current requirements too low or high in the heated oxygen sensor (HO2S) heater control circuit. **Possible Causes:** • Open circuit between HO2S terminal C and main relay No.1 terminal E • Short to ground circuit between HO2S terminal A and body ground • HO2S circuits are shorted each other • Open circuit between rear HO2S (RH) terminal A and PCM terminal 2AS
DTC: P0057 **2T CCM, MIL: Yes** **Years:** 2007, 2008 **Models:** Mazda6, Mazda6 Sport Wagon **Engines:** All **Transmissions:** All	**HO2S-22 (Bank 2, Sensor 2) Heater Circuit Low Input** Engine running, and after the rear HO2S Heater Control duty cycle signal was commanded "off", the PCM detected a low input condition on the heater control circuit. **Possible Causes:** • HO2S heater control circuit is shorted to chassis ground • HO2S heater is damaged or has failed • PCM has failed
DTC: P0058 **2T CCM, MIL: Yes** **Years:** 2007, 2008 **Models:** Mazda6, Mazda6 Sport Wagon **Engines:** All **Transmissions:** All	**HO2S-22 (Bank 2, Sensor 2) Heater Circuit High Input** Engine running, and after the rear HO2S Heater Control duty cycle signal was commanded "on", the PCM detected a high input condition on the heater control circuit. **Possible Causes:** • HO2S heater control circuit is shorted to system power (B+) • HO2S heater is damaged or has failed • PCM has failed
DTC: P0059 **2T CCM, MIL: Yes** **Years:** 2007, 2008 **Models:** B-Series Truck, CX-9, Tribute **Engines:** All **Transmissions:** All	**Front HO2S (LH) Heater Resistance** Heater current requirements too low or high in the heated oxygen sensor (HO2S) heater control circuit. **Possible Causes:** • Open circuit between HO2S terminal C and main relay No.1 terminal E • Short to ground circuit between HO2S terminal A and body ground • HO2S circuits are shorted each other • Open circuit between front HO2S (LH) terminal A and PCM terminal 2E
DTC: P0060 **2T CCM, MIL: Yes** **Years:** 2007, 2008 **Models:** B-Series Truck, CX-9, Tribute **Engines:** All **Transmissions:** All	**Rear HO2S (LH) Heater Resistance** Heater current requirements too low or high in the heated oxygen sensor (HO2S) heater control circuit. **Possible Causes:** • Open circuit between HO2S terminal C and main relay No.1 terminal E • Short to ground circuit between HO2S terminal A and body ground • HO2S circuits are shorted each other • Open circuit between rear HO2S (LH) terminal A and PCM terminal 2AO
DTC: P0068 **2T CCM, MIL: Yes** **Years:** 2007, 2008 **Models:** B-Series Truck, CX-9, Tribute **Engines:** All **Transmissions:** All	**PCM – Mass Air Flow (MAF/IAT) Sensor – Throttle Position (TP) Sensor** TP signal went below 0.24 volts with a load greater than 55 percent, or TP signal went above 2.44 volts with a load less than 30 percent. **Possible Causes:** • TP Sensor has failed • PCM has failed
DTC: P0069 **2T CCM, MIL: Yes** **Years:** 2007, 2008 **Models:** CX-7, Mazda3, MazdaSpeed3, Mazda5, Mazda6, MazdaSpeed6, MX-5, RX-8, **Engines:** All **Transmissions:** All	**Manifold Absolute Pressure/Atmospheric Pressure Correlation** PCM monitors differences between intake manifold vacuum and atmospheric pressure. If the difference is below 12 kPa {−90 mmHg, −3.5 inHg} or above 12 kPa {90 mmHg, 3.5 inHg}, the PCM determines that there is a MAP sensor performance problem. **Possible Causes:** • MAP Sensor has failed • Barometric Pressure Sensor failed • PCM has failed

DTC	Trouble Code Title, Conditions & Possible Causes
DTC: P0076 **2T CCM, MIL: No** **Years:** 2007, 2008 **Models:** RX-8, **Engines:** All **Transmissions:** All	**VDI Solenoid Valve Control Low Circuit** The PCM monitors the VDI solenoid valve control voltage when the PCM turns off the VDI solenoid valve. If the control voltage is low, the PCM determines that the VDI solenoid control circuit voltage is low. **Possible Causes:** • VDI solenoid valve malfunction • Connector or terminal malfunction • PCM has failed
DTC: P0077 **2T CCM, MIL: No** **Years:** 2007, 2008 **Models:** RX-8, **Engines:** All **Transmissions:** All	**VDI Solenoid Valve Control High Circuit** The PCM monitors the VDI solenoid valve control voltage when the PCM turns off the VDI solenoid valve. If the control voltage is high, the PCM determines that the VDI solenoid control circuit voltage is high. **Possible Causes:** • VDI solenoid valve malfunction • Connector or terminal malfunction • PCM has failed
DTC: P0089 **2T CCM, MIL: No** **Years:** 2007, 2008 **Models:** CX-7, **Engines:** All **Transmissions:** All	**Fuel Pressure Regulator Performance** If the fuel pressure average value measured by the PCM exceeds the specification when the camshaft is rotating at a specified rate, the PCM determines that there is a fuel pressure regulator performance problem. **Possible Causes:** • Spill valve control solenoid malfunction • Connector or terminal malfunction • PCM malfunction
DTC: P0091 **1T CCM, MIL: Yes** **Years:** 2007, 2008 **Models:** CX-7, **Engines:** All **Transmissions:** All	**Fuel Pressure Regulator Control Low Circuit** When the PCM turns the spill valve control solenoid valve off but the spill valve control solenoid valve control circuit voltage is low for 5 seconds, the PCM determines that the spill valve control solenoid valve control circuit has malfunction. **Possible Causes:** • Spill valve control solenoid valve malfunction • Open circuit in the wiring harness between high pressure fuel pump terminal A and PCM terminal 2F • Open circuit in the wiring harness between high pressure fuel pump terminal B and PCM terminal 2G • Short to ground in the wiring harness between high pressure fuel pump terminal A and PCM terminal 2F • Short to ground in the wiring harness between high pressure fuel pump terminal B and PCM terminal 2G • Connector or terminal malfunction • PCM malfunction
DTC: P0092 **1T CCM, MIL: Yes** **Years:** 2007, 2008 **Models:** CX-7, **Engines:** All **Transmissions:** All	**Fuel Pressure Regulator Control High Circuit** When the PCM turns the spill valve control solenoid valve off but the spill valve control solenoid valve control circuit voltage is high for 5 seconds, the PCM determines that the spill valve control solenoid valve control circuit has malfunction. **Possible Causes:** • Spill valve control solenoid valve malfunction • Short to power supply in the wiring harness between high pressure fuel pump terminal B and PCM terminal 2G • Connector or terminal malfunction • PCM malfunction
DTC: P0096 **2T CCM, MIL: Yes** **Years:** 2007, 2008 **Models:** CX-7 **Engines:** All **Transmissions:** All	**IAT No. 2 Circuit Performance Problem** If air intake temperature is lower than engine coolant temperature by −9.4 degrees F (−23 degrees C) for 1.2 seconds with the ignition switch ON, the PCM determines that there is a IAT No. 2 performance problem. **Possible Causes:** • IAT No. 2 sensor malfunction • ECT sensor malfunction • PCM has failed
DTC: P0097 **1T CCM, MIL: Yes** **Years:** 2007, 2008 **Models:** CX-7 **Engines:** All **Transmissions:** All	**IAT No. 2 Circuit Low Input** The PCM monitors the IAT No. 2 sensor at PCM terminal 4R. If the voltage at PCM terminal 4R is below 0.1 v, the PCM determines that the IAT NO. 2 sensor has malfunctioned. **Possible Causes:** • IAT No. 2 sensor malfunction • Short to the ground circuit between IAT No. 2 sensor and PCM terminal connection 4R • PCM has failed

DTC	Trouble Code Title, Conditions & Possible Causes
DTC: P0098 **1T CCM, MIL: Yes** **Years:** 2007, 2008 **Models:** CX-7 **Engines:** All **Transmissions:** All	**IAT No. 2 Circuit High Input** The PCM monitors the IAT No. 2 sensor at PCM terminal 4R. If the voltage at PCM terminal 4R is above 4.9 v, the PCM determines that the IAT NO. 2 sensor has malfunctioned. **Possible Causes:** • IAT No. 2 sensor malfunction • Short to the ground circuit between IAT No. 2 sensor and PCM terminal connection 4R • PCM has failed
DTC: P0100 **1T CCM, MIL: Yes** **Years:** 2007, 2008 **Models:** B-Series Truck **Engines:** All **Transmissions:** All	**MAF or VAF Sensor Circuit Malfunction** Key on or engine running and the PCM detected the MAF or VAF sensor signal was less than 0.20v, or that it was more than 4.90v at any time during the CCM test. **Possible Causes:** • MAF sensor signal circuit is open or shorted to ground • MAF sensor power circuit is open or the ground circuit is open • MAF sensor is damaged or has failed • PCM has failed • TSB SSP056 (2/02) contains a MAF sensor warranty ()
DTC: P0101 **2T CCM, MIL: Yes** **Years:** 2007, 2008 **Models:** B-Series Truck, CX-7, Mazda3, MazdaSpeed3, Mazda5, Mazda6, MazdaSpeed6, MazdaSpeed6 Sport Wagon, MX-5, RX-8, Tribute **Engines:** All **Transmissions:** All	**MAF Sensor inconsistent with TP Sensor** The PCM compares the actual input signal from MAF sensor with expected input signal from MAF sensor which PCM calculates by input voltage from TP sensor or engine speed. With engine running and throttle opening angle at 50 percent for 5 seconds, if MAF amount is less than 8.93 g/sec., PCM determines that detected MAF amount is too low. With engine running at 2,000 RPM for 5 seconds, if MAF amount is over 103 g/sec., PSM determines that detected MAF amount is too high. **Possible Causes:** • MAF sensor malfunction • TP sensor malfunction • Electric corrosion in MAF signal circuit or MAF return circuit • Voltage drops in MAF signal circuit • Voltage drops in ground circuit
DTC: P0102 **1T CCM, MIL: Yes** **Years:** 2007, 2008 **Models:** B-Series Truck, CX-7, CX-9, Mazda3, MazdaSpeed3, Mazda5, Mazda6, MazdaSpeed6, MX-5, RX-8, Tribute **Engines:** All **Transmissions:** All	**MAF Sensor Circuit Low Input** Key on or engine running and the PCM detected the MAF sensor signal was less than 0.36v at any time during the CCM test. **Possible Causes:** • MAF sensor signal circuit is open • MAF sensor signal is shorted to ground • MAF sensor power circuit is open • MAF sensor is damaged or has failed • PCM has failed
DTC: P0103 **1T CCM, MIL: Yes** **Years:** 2007, 2008 **Models:** B-Series Truck, CX-7, CX-9, Mazda3, MazdaSpeed3, Mazda5, Mazda6, MazdaSpeed6, MX-5, RX-8, Tribute **Engines:** All **Transmissions:** All	**MAF Sensor Circuit High Input** Key on or engine running and the PCM detected the MAF sensor signal was more than 4.97v at any time during the CCM test. **Possible Causes:** • MAF sensor ground circuit is open • MAF sensor signal is shorted to VREF or system power (B+) • MAF sensor is damaged or has failed • PCM has failed
DTC: P0104 **1T CCM, MIL: No** **Years:** 2007, 2008 **Models:** B-Series Truck, CX-9, Tribute **Engines:** All **Transmissions:** All	**MAF Circuit Intermittent/Erratic** A concern exists in the MAF sensor circuit, or the air tube containing the sensor, causing an incorrect air flow reading. **Possible Causes:** • Air leaks in the tube from the MAF to the throttle body • Open circuit wiring harness between MAF sensor terminal C and PCM terminal 1AJ • Open circuit wiring harness between MAF sensor terminal D and PCM terminal 1AN • Short to power supply in wiring harness between MAF sensor terminal C and PCM terminal 1AJ • Short to power supply in wiring harness between MAF sensor terminal D and PCM terminal 1AN • Short to ground circuit between MAF sensor terminal C and PCM terminal 1AJ • Short to ground circuit between MAF sensor terminal D and PCM terminal 1AN • Connector or terminal malfunction

DTC	Trouble Code Title, Conditions & Possible Causes
DTC: P0105 **2T CCM, MIL: Yes** **Years:** 2007, 2008 **Models:** B-Series Truck **Engines:** All **Transmissions:** All	**BARO Sensor Circuit Performance** Engine running, and the PCM detected the BARO sensor signal was less than 0.22v, or that it was more than 4.97v during the CCM test. **Possible Causes:** • BARO sensor signal circuit is open or shorted to ground • BARO sensor signal is shorted to VREF or system power • BARO sensor is damaged or has failed • PCM has failed
DTC: P0106 **2T CCM, MIL: Yes** **Years:** 2007, 2008 **Models:** B-Series Truck, Tribute **Engines:** All **Transmissions:** All	**BARO Sensor Circuit Performance** Engine running, and the PCM detected the BARO sensor signal was less than 0.22v, or that it was more than 4.97v during the CCM test. **Possible Causes:** • BARO sensor signal circuit is open or shorted to ground • BARO sensor signal is shorted to VREF or system power • BARO sensor is damaged or has failed • PCM has failed
DTC: P0107 **1T CCM, MIL: Yes** **Years:** 2007, 2008 **Models:** B-Series Truck, CX-7, Mazda3, MazdaSpeed3, Mazda5, Mazda6, MazdaSpeed6, MX-5, RX-8, Tribute **Engines:** All **Transmissions:** All	**BARO Sensor Circuit Low Input** Engine running, EGR Boost Sensor Solenoid commanded "off" so that BARO pressure is applied to the sensor), IAT sensor signal more than 50°F, and the PCM detected the BARO sensor signal indicated less than 0.21v during the CCM test. **Possible Causes:** • EGR Boost sensor signal circuit is shorted to ground • EGR boost sensor VREF circuit is open or shorted to ground • EGR Boost sensor is damaged or has failed • PCM has failed
DTC: P0108 **1T CCM, MIL: Yes** **Years:** 2007, 2008 **Models:** B-Series Truck, CX-7, Mazda3, MazdaSpeed3, Mazda5, Mazda6, MazdaSpeed6, MX-5, RX-8, Tribute **Engines:** All **Transmissions:** All	**BARO Sensor Circuit High Input** Engine running, EGR Boost Sensor Solenoid commanded "off" so that BARO pressure is applied to the sensor), IAT sensor signal more than 50°F, and the PCM detected the BARO sensor signal indicated more than 4.80v during the CCM test. **Possible Causes:** • EGR boost sensor signal circuit open from sensor to the PCM • EGR boost sensor ground circuit open from sensor to the PCM • EGR Boost sensor signal circuit is shorted to VREF or power • EGR Boost sensor is damaged or has failed • PCM has failed
DTC: P0109 **1T CCM, MIL: Yes** **Years:** 2007, 2008 **Models:** B-Series Truck, Tribute **Engines:** All **Transmissions:** All	**MAP Sensor Signal to PCM Intermittent Fail** MAP sensor signal to the PCM is failing intermittently. **Possible Causes:** • MAP sensor is damaged or has failed • PCM has failed
DTC: P0110 **1T CCM, MIL: Yes** **Years:** 2007, 2008 **Models:** B-Series Truck, Tribute **Engines:** All **Transmissions:** All	**Intake Air Temperature Sensor Circuit Malfunction** Key on or engine running, and the PCM detected the IAT sensor signal indicated less than 0.10v, or it detected the signal was more than 4.80v at any time during the CCM test period. **Possible Causes:** • IAT sensor signal circuit is open or shorted to ground • IAT sensor is damaged or has failed • PCM has failed
DTC: P0111 **2T CCM, MIL: Yes** **Years:** 2007, 2008 **Models:** B-Series Truck, CX-7, CX-9, Mazda3, MazdaSpeed3, Mazda5, Mazda6, MazdaSpeed6, MX-5, RX-8 **Engines:** All **Transmissions:** All	**Intake Air Temperature Sensor Circuit Range/Performance** Key on or engine running, and the PCM detected the IAT sensor signal was more than 104°F higher than the ECT sensor signal. **Possible Causes:** • IAT sensor signal circuit has high resistance • IAT sensor has drifted out of calibration • IAT sensor is damaged or has failed • PCM has failed

DTC	Trouble Code Title, Conditions & Possible Causes
DTC: P0112 **1T CCM, MIL: Yes** **Years:** 2007, 2008 **Models:** B-Series Truck, CX-7, CX-9, Mazda3, MazdaSpeed3, Mazda5, Mazda6, MazdaSpeed6, MX-5, RX-8, Tribute **Engines:** All **Transmissions:** All	**Intake Air Temperature Sensor Circuit Low Input** Key on or engine running, and the PCM detected the IAT sensor signal indicated less than 0.20v (a Scan Tool PID near 250°F) at any time during the CCM test period. **Possible Causes:** • IAT sensor signal circuit is shorted to ground • IAT sensor is damaged or has failed • PCM has failed
DTC: P0113 **1T CCM, MIL: Yes** **Years:** 2007, 2008 **Models:** B-Series Truck, CX-7, CX-9, Mazda3, MazdaSpeed3, Mazda5, Mazda6, MazdaSpeed6, MX-5, RX-8, Tribute **Engines:** All **Transmissions:** All	**Intake Air Temperature Sensor Circuit High Input** Key on or engine running, and the PCM detected the IAT sensor signal indicated more than 4.60v (a Scan Tool PID near −46°F) at any time during the CCM test period. **Possible Causes:** • IAT sensor signal circuit is open • IAT sensor signal circuit is shorted to VREF or system power • IAT sensor is damaged or has failed • PCM has failed
DTC: P0114 **1T CCM, MIL: Yes** **Years:** 2007, 2008 **Models:** B-Series Truck, CX-9, Tribute **Engines:** All **Transmissions:** All	**PCM - Intake Air Temperature (IAT) Sensor** IAT sensor signal erratic. **Possible Causes:** • IAT sensor signal circuit is open • IAT sensor signal circuit is shorted to VREF or system power • IAT sensor is damaged or has failed • PCM has failed
DTC: P0115 **1T CCM, MIL: Yes** **Years:** 2007, 2008 **Models:** B-Series Truck **Engines:** All **Transmissions:** All	**Intake Air Temperature Sensor Circuit Malfunction** Key on or engine running, and the PCM detected the IAT sensor signal indicated less than 0.10v, or it detected the signal was more than 4.80v at any time during the CCM test period. **Possible Causes:** • IAT sensor signal circuit is open or shorted to ground • IAT sensor is damaged or has failed • PCM has failed
DTC: P0116 **2T CCM, MIL: Yes** **Years:** 2007, 2008 **Models:** B-Series Truck, CX-7, CX-9, Mazda3, MazdaSpeed3, Mazda5, Mazda6, MazdaSpeed6, MX-5, RX-8, Tribute **Engines:** All **Transmissions:** All	**Engine Coolant Temperature Sensor Circuit Performance Problem** Key on or engine running, and the PCM detected the ECT sensor signal maximum value and minimum value is below 5.6 degrees C (4.2 degrees F), the PCM determines that ECT signal circuit has malfunctioned. **Possible Causes:** • ECT sensor malfunction • Poor connection at ECT sensor or PCM connector • PCM malfunction
DTC: P0117 **1T CCM, MIL: Yes** **Years:** 2007, 2008 **Models:** B-Series Truck, CX-7, CX-9, Mazda3, MazdaSpeed3, Mazda5, Mazda6, MazdaSpeed6, MX-5, RX-8, Tribute **Engines:** All **Transmissions:** All	**Engine Coolant Temperature Sensor Circuit Low Input** Key on or engine running, and the PCM detected the ECT sensor signal indicated less than 0.20v (a Scan Tool PID near 250°F) at any time during the CCM test period. **Possible Causes:** • ECT sensor signal circuit is shorted to ground • ECT sensor is damaged or has failed • PCM has failed

DTC	Trouble Code Title, Conditions & Possible Causes
DTC: P0118 **1T CCM, MIL: Yes** **Years:** 2007, 2008 **Models:** B-Series Truck, CX-7, CX-9, Mazda3, MazdaSpeed3, Mazda5, Mazda6, MazdaSpeed6, MX-5, RX-8, Tribute **Engines:** All **Transmissions:** All	**Engine Coolant Temperature Sensor Circuit High Input** Key on or engine running, and the PCM detected the ECT sensor signal indicated more than 4.60v (a Scan Tool PID near −46°F) at any time during the CCM test period. **Possible Causes:** • ECT sensor signal circuit is open • ECT sensor signal circuit is shorted to VREF or system power • ECT sensor is damaged or has failed • PCM has failed
DTC: P0119 **1T CCM, MIL: Yes** **Years:** 2007, 2008 **Models:** B-Series Truck, CX-9, Tribute **Engines:** All **Transmissions:** All	**PCM - Engine Coolant Temperature (ECT) Sensor/Coolant Head Temperature (CHT) Sensor** ECT / CHT sensor signal is erratic. **Possible Causes:** • ECT / CHT sensor signal circuit is open • ECT / CHT sensor signal circuit is shorted to VREF or system power • ECT / CHT sensor is damaged or has failed • PCM has failed
DTC: P0120 **1T CCM, MIL: Yes** **Years:** 2007, 2008 **Models:** B-Series Truck, Tribute **Engines:** All **Transmissions:** All	**Intake Air Temperature Sensor Circuit Malfunction** Key on or engine running, and the PCM detected the IAT sensor signal indicated less than 0.10v, or it detected the signal was more than 4.80v at any time during the CCM test period. **Possible Causes:** • IAT sensor signal circuit is open • IAT sensor signal circuit is shorted to ground • IAT sensor is damaged or has failed • PCM has failed
DTC: P0121 **2T CCM, MIL: Yes** **Years:** 2007, 2008 **Models:** B-Series Truck, Tribute **Engines:** All **Transmissions:** All	**TP Sensor In-Range Operating Circuit Malfunction** Vehicle driven at light engine load at over 20 MPH, and the PCM detected the TP sensor signal did not correlate when it was compared to the MAF sensor signal during the CCM test. **Possible Causes:** • TP sensor signal circuit is open to the PCM (intermittent fault) • TP sensor ground circuit is open • Throttle body is damaged or throttle linkage is bent or binding • TP sensor is damaged or has failed
DTC: P0122 **1T CCM, MIL: Yes** **Years:** 2007, 2008 **Models:** B-Series Truck, CX-7, CX-9, Mazda3, MazdaSpeed3, Mazda5, Mazda6, MazdaSpeed6, MX-5, RX-8, Tribute **Engines:** All **Transmissions:** All	**TP Sensor Circuit Low Input** Key on or engine running, and the PCM detected the TP sensor signal indicated less than 0.17v (a throttle opening of 3.43 percent) during the CCM test. **Possible Causes:** • TP sensor signal circuit is shorted to ground • TP sensor VREF circuit is open or shorted to ground • Throttle body is damaged • Throttle linkage is bent or binding (sticking) • TP sensor is damaged or has failed
DTC: P0123 **1T CCM, MIL: Yes** **Years:** 2007, 2008 **Models:** B-Series Truck, CX-7, CX-9, Mazda3, MazdaSpeed3, Mazda5, Mazda6, MazdaSpeed6, MX-5, RX-8, Tribute **Engines:** All **Transmissions:** All	**TP Sensor Circuit High Input** Key on or engine running, and the PCM detected the TP sensor signal indicated more than 4.60v (a throttle opening of 92.27 percent). **Note: This trouble code may set due to an intermittent fault.** **Possible Causes:** • TP sensor signal return circuit is open • TP sensor signal circuit is shorted to VREF or system power • TP sensor not mounted properly to the throttle body • TP sensor is damaged or has failed • PCM has failed

DTC	Trouble Code Title, Conditions & Possible Causes
DTC: P0124 **2T CCM, MIL: Yes** **Years:** 2007, 2008 **Models:** B-Series Truck, Tribute **Engines:** All **Transmissions:** All	**PCM Detects Intermittent Concern with TP Sensor** Key on or engine running, and the PCM detected the TP sensor signal indicated more than 4.60v (a throttle opening of 92.27 percent). **Note: This trouble code may set due to an intermittent fault.** **Possible Causes:** • TP sensor signal return circuit is open • TP sensor signal circuit is shorted to VREF or system power • TP sensor not mounted properly to the throttle body • TP sensor is damaged or has failed • PCM has failed
DTC: P0125 **2T CCM, MIL: Yes** **Years:** 2007, 2008 **Models:** B-Series Truck, CX-7, CX-9, Mazda3, MazdaSpeed3, Mazda5, Mazda6, MazdaSpeed6, MX-5, RX-8, Tribute **Engines:** All **Transmissions:** All	**Excessive Time To Enter Closed Loop** DTC P0115, P0117 and P0118 not set, cold startup requirement met (ECT input less than 140°F, IAT sensor more than 50°F at startup), engine runtime over 10 minutes, and the PCM detected the ECT sensor signal did not reach 78°F after the engine warm-up period. **Possible Causes:** • Inspect for low coolant level or an incorrect coolant mixture • Check the operation of the thermostat (it may be stuck open) • ECT sensor signal circuit has high resistance • ECT sensor has failed
DTC: P0126 **2T CCM, MIL: Yes** **Years:** 2007, 2008 **Models:** B-Series Truck, CX-7, CX-9, Mazda3, MazdaSpeed3, Mazda5, Mazda6, MazdaSpeed6, MX-5, RX-8, Tribute **Engines:** All **Transmissions:** All	**Thermostat Malfunction** Cold startup requirement met (IAT sensor more than 14°F and the difference between ECT and IAT sensor signals less than 43°F), engine runtime over 10 minutes, and the PCM detected the ECT sensor did not reach 160°F after the engine warm-up period. **Possible Causes:** • Inspect for low coolant level or an incorrect coolant mixture • Check the operation of the thermostat (it may be stuck open) • ECT sensor has drifted out of calibration or it has failed • PCM has failed
DTC: P0128 **1T CCM, MIL: Yes** **Years:** 2007, 2008 **Models:** B-Series Truck, CX-7, CX-9, Mazda3, MazdaSpeed3, Mazda5, Mazda6, MazdaSpeed6, MX-5, RX-8, Tribute **Engines:** All **Transmissions:** All	**Thermostat Malfunction** Cold startup requirement met (ECT input less than 95°F, IAT sensor more than 14°F and the difference between the ECT and IAT sensor signals less than 43°F), VSS input over 15 MPH and the MAF, IAT and ECT sensor signals all within normal range, the PCM detected the radiator heat ratio exceeded its threshold after a warm-up period. **Possible Causes:** • Check the operation of the thermostat (it may be stuck open) • ECT sensor has failed • PCM has failed
DTC: P0130 **2T OBD/O2S1, MIL: Yes** **Years:** 2007, 2008 **Models:** B-Series Truck, Mazda3, MazdaSpeed3, Mazda6, MazdaSpeed6, MX-5, RX-8 **Engines:** All **Transmissions:** All	**HO2S-11 (Bank 1 Sensor 1) Circuit Malfunction** Engine speed from 1500–3000 RPM while in closed loop, ECT sensor signal more than 14°F, engine load from 28–59 percent, vehicle speed over 3.5 MPH, and after the PCM monitored the HO2S inversion cycle period, lean to rich response and rich to lean response time under these conditions, it determined that the HO2S failed one of the tests. **Possible Causes:** • Air leaks in intake manifold, exhaust pipes or exhaust manifold • Fuel delivery system component has failed (i.e., a clogged fuel filter, dirty or restricted fuel injectors, low or high fuel pressure) • Base engine "mechanical" problem (low cylinder compression, incorrect camshaft timing, intake or exhaust manifold leaks) • HO2S is damaged or has failed • HO2S heater element has deteriorated or failed • PRC solenoid or Purge solenoid is damaged or has failed • PCM has failed
DTC: P0131 **2T CCM, MIL: Yes** **Years:** 2007, 2008 **Models:** B-Series Truck, CX-7, Mazda3, MazdaSpeed3, Mazda5, Mazda6, MazdaSpeed6, MX-5, RX-8 **Engines:** All **Transmissions:** All	**HO2S-11 (Bank 1 Sensor 1) Circuit Malfunction** Vehicle driven in closed loop at cruise speed and back to idle speed, and the PCM detected a negative voltage on the HO2S circuit. **Possible Causes:** • HO2S is water or fuel contaminated • HO2S is damaged or has failed • PCM has failed

DTC	Trouble Code Title, Conditions & Possible Causes
DTC: P0132 **2T CCM, MIL: Yes** **Years:** 2007, 2008 **Models:** B-Series Truck, CX-7, CX-9, Mazda3, MazdaSpeed3, Mazda5, Mazda6, MazdaSpeed6, MX-5, RX-8, Tribute **Engines:** All **Transmissions:** All	**HO2S-11 (Bank 1 Sensor 1) Circuit Malfunction** Vehicle driven in closed loop at cruise speed and back to idle speed, and the PCM detected a negative voltage on the HO2S circuit. **Possible Causes:** • HO2S is water or fuel contaminated • HO2S is damaged or has failed • PCM has failed
DTC: P0133 **2T OBD/O2S1, MIL: Yes** **Years:** 2007, 2008 **Models:** B-Series Truck, CX-7, CX-9, Mazda3, MazdaSpeed3, Mazda5, Mazda6, MazdaSpeed6, MX-5, RX-8, Tribute **Engines:** All **Transmissions:** All	**HO2S-11 (Bank 1 Sensor 1) Circuit Slow Response** Vehicle driven in closed loop at cruise speed and back to idle speed, and the PCM detected the frequency and the amplitude response rate of the front HO2S was less than a calibrated window in memory. **Possible Causes:** • Air leaks in the intake manifold or exhaust manifold or pipes • IAT sensor or MAF sensor has deteriorated (out of calibration) • HO2S signal circuit is open or shorted to ground (intermittent) • HO2S contaminated with wrong fuel, has deteriorated or failed • PCM has failed
DTC: P0134 **2T OBD/O2S1, MIL: Yes** **Years:** 2007, 2008 **Models:** B-Series Truck, CX-7, Mazda3, MazdaSpeed3, Mazda5, Mazda6, MazdaSpeed6, MX-5, RX-8 **Engines:** All **Transmissions:** All	**HO2S-11 (Bank 1 Sensor 1) Circuit No Activity Detected** Engine speed from 1500–3000 RPM while in closed loop, ECT sensor signal more than 176°F, and the PCM detected the front HO2S signal did not exceed 550 mv for over 54.2 seconds during the test. **Possible Causes:** • Air leaks in the intake manifold or exhaust manifold or pipes • IAT sensor or MAF sensor has deteriorated (out of calibration) • HO2S signal circuit is open or shorted to ground (intermittent) • HO2S contaminated with wrong fuel, has deteriorated or failed • PCM has failed
DTC: P0135 **2T OBD/O2S2, MIL: Yes** **Years:** 2007, 2008 **Models:** B-Series Truck, CX-9, Tribute **Engines:** All **Transmissions:** All	**HO2S-11 (Bank 1 Sensor 1) Heater Circuit Malfunction** Engine started, engine running, and the PCM detected the HO2S heater circuit was more than 11.5v with the heater turned "on", or it was less than 5.8v with the heater turned "off" for 331–361 seconds. **Possible Causes:** • HO2S heater control circuit is open or shorted to ground • HO2S heater control circuit is shorted to system power (B+) • HO2S heater power circuit open between the heater and PCM • HO2S heater is damaged or has failed • PCM has failed (it controls the heater with a duty cycle signal)
DTC: P0136 **2T OBD/O2S1, MIL: Yes** **Years:** 2007, 2008 **Models:** B-Series Truck **Engines:** All **Transmissions:** All	**HO2S-12 (Bank 1 Sensor 2) No Activity Detected** Engine running in closed loop for 5 minutes, and the PCM detected the middle HO2S signal was less than a calibrated value in memory. **Possible Causes:** • Air leaks in the intake manifold or exhaust manifold or pipes • IAT sensor or MAF sensor has deteriorated (out of calibration) • HO2S signal circuit is open or shorted to ground • HO2S contaminated with wrong fuel, has deteriorated or failed • PCM has failed
DTC: P0137 **2T CCM, MIL: Yes** **Years:** 2007, 2008 **Models:** B-Series Truck, CX-7, Mazda3, MazdaSpeed3, Mazda5, Mazda6, MazdaSpeed6, MX-5, RX-8 **Engines:** All **Transmissions:** All	**HO2S-12 (Bank 1 Sensor 2) Circuit Low Input** Vehicle driven in closed loop at cruise speed at low engine load for over 5 minutes, and the PCM detected the HO2S signal remained at less than 500 mv during the CCM test for 55 seconds. **Possible Causes:** • HO2S signal circuit is open or shorted to ground • Air leaks at the exhaust manifold or exhaust pipes • HO2S contaminated with wrong fuel, has deteriorated or failed • PCM has failed

DTC	Trouble Code Title, Conditions & Possible Causes
DTC: P0138 **2T CCM, MIL: Yes** **Years:** 2007, 2008 **Models:** B-Series Truck, CX-7, CX-9, Mazda3, MazdaSpeed3, Mazda5, Mazda6, MazdaSpeed6, MX-5, RX-8, Tribute **Engines:** All **Transmissions:** All	**HO2S-12 (Bank 1 Sensor 2) Circuit High Input** Vehicle driven in closed loop at cruise speed at low engine load for over 5 minutes, and the PCM detected the HO2S signal remained at more than 500 mv during the CCM test for 55 seconds. **Possible Causes:** • HO2S signal tracking (wet/oily) in connector causing a short between the signal circuit and heater power circuit • HO2S signal circuit is shorted to system power HO2S contaminated with wrong fuel, has deteriorated or failed • PCM has failed
DTC: P0139 **2T CCM, MIL: Yes** **Years:** 2007, 2008 **Models:** B-Series Truck, CX-7, CX-9, Mazda3, MazdaSpeed3, Mazda5, Mazda6, MazdaSpeed6, MX-5, RX-8 **Engines:** All **Transmissions:** All	**Middle/Rear HO2S Circuit Problem** The PCM monitors inversion cycle period, middle HO2S output voltage inclination. The PCM detects that the voltage inclinations are below threshold consecutive 5 times when following conditions are met, the PCM determines that circuit has malfunction. Under the following monitoring conditions, if 0.3 V or more is detected three times even if fuel cut is performed for 3 seconds or more, a circuit malfunction is determined. **Possible Causes:** • Middle HO2S circuit deterioration • Middle HO2S circuit malfunction • PCM has failed
DTC: P0140 **2T OBD/O2S1, MIL: Yes** **Years:** 2007, 2008 **Models:** B-Series Truck, CX-7, Mazda3, MazdaSpeed3, Mazda5, Mazda6, MazdaSpeed6, MX-5, RX-8 **Engines:** All **Transmissions:** All	**HO2S-12 (Bank 1 Sensor 2) Circuit No Activity Detected** Engine speed from 1500–3000 RPM while in closed loop, ECT sensor signal more than 176°F, and the PCM detected the HO2S signal voltage did not exceed 550 mv for over 54.2 seconds during the test. **Possible Causes:** • Air leaks in the intake manifold or exhaust manifold or pipes • IAT sensor or MAF sensor has deteriorated (out of calibration) • HO2S signal circuit is open or shorted to ground • HO2S contaminated with wrong fuel, has deteriorated or failed • TSB 0100700 (3/00) contains information for this code ()
DTC: P0141 **2T OBD/O2S, MIL: Yes** **Years:** 2007, 2008 **Models:** B-Series Truck, CX-9, Tribute **Engines:** All **Transmissions:** All	**HO2S-12 (Bank 1 Sensor 2) Heater Circuit Malfunction** Engine started, engine running, and the PCM detected the HO2S heater circuit indicated more than 11.5v with the heater turned "on", or it indicated less than 5.8v with the heater turned "off" for 331–361 seconds during the CCM test. **Possible Causes:** • HO2S heater control circuit is open or shorted to ground • HO2S heater control circuit is shorted to system power (B+) • HO2S heater power circuit open between the heater and PCM • HO2S heater is damaged or has failed • PCM has failed (it controls the heater with a duty cycle signal)
DTC: P0144 **2T CCM, MIL: Yes** **Years:** 2007, 2008 **Models:** Mazda3, MazdaSpeed3, Mazda6, MazdaSpeed6 **Engines:** All **Transmissions:** All	**HO2S-13 (Bank 1 Sensor 3) Circuit Malfunction** Vehicle driven in closed loop at cruise speed for 5 minutes, followed by a deceleration period, and the PCM detected the HO2S signal remained above 0.45v for 5 seconds under these conditions. **Possible Causes:** • HO2S signal circuit is shorted to system power (heater circuit) at the connector or somewhere in the wiring harness • HO2S contaminated due to wrong fuel, or it has deteriorated • PCM has failed
DTC: P0148 **2T OBD/O2S, MIL: Yes** **Years:** 2007, 2008 **Models:** CX-9, Tribute **Engines:** All **Transmissions:** All	**HO2S-13 (Bank 1 Sensor 3) Heater Circuit Malfunction** At least one bank lean at wide open throttle. **Possible Causes:** • HO2S heater control circuit is open or shorted to ground • HO2S heater control circuit is shorted to system power (B+) • HO2S heater power circuit open between the heater and PCM • HO2S heater is damaged or has failed • PCM has failed (it controls the heater with a duty cycle signal)
DTC: P0151 **2T CCM, MIL: Yes** **Years:** 2007, 2008 **Models:** B-Series Truck, Mazda6, Mazda6 Sport Wagon **Engines:** All **Transmissions:** All	**HO2S-21 (Bank 2 Sensor 1) Circuit Malfunction** Engine started, vehicle driven in closed loop at cruise speed and then back to idle speed, and the PCM detected a negative voltage on the HO2S circuit during the CCM test. **Possible Causes:** • HO2S is water or fuel contaminated • HO2S is damaged or has failed • PCM has failed

DTC	Trouble Code Title, Conditions & Possible Causes
DTC: P0152 **2T CCM, MIL: Yes** **Years:** 2007, 2008 **Models:** B-Series Truck, CX-9, Mazda6, Mazda6 Sport Wagon, Tribute **Engines:** All **Transmissions:** All	**HO2S-21 (Bank 2 Sensor 1) Circuit Malfunction** Engine started, vehicle driven in closed loop at cruise speed and then back to idle speed, and the PCM detected a negative voltage on the HO2S circuit during the CCM test. **Possible Causes:** • HO2S is water or fuel contaminated • HO2S is damaged or has failed • PCM has failed
DTC: P0153 **2T CCM, MIL: Yes** **Years:** 2007, 2008 **Models:** B-Series Truck, CX-9, Mazda6, Mazda6 Sport Wagon, Tribute **Engines:** All **Transmissions:** All	**HO2S-21 (Bank 2 Sensor 1) Circuit Slow Response** Vehicle driven in closed loop at cruise speed and back to idle speed, and the PCM detected the frequency and the amplitude response rate of the front HO2S was less than a calibrated window in memory. **Possible Causes:** • Air leaks in the intake manifold or exhaust manifold or pipes • IAT sensor or MAF sensor has deteriorated (out of calibration) • HO2S signal circuit is open or shorted to ground (intermittent) • HO2S contaminated with wrong fuel, has deteriorated or failed
DTC: P0154 **2T OBD/O2S1, MIL: Yes** **Years:** 2007, 2008 **Models:** B-Series Truck, Mazda6, Mazda6 Sport Wagon **Engines:** All **Transmissions:** All	**HO2S-21 (Bank 2 Sensor 1) Circuit No Activity Detected** Engine speed from 1500–3000 RPM while in closed loop, ECT sensor signal more than 176°F, and the PCM detected the front HO2S signal did not exceed 0.55v for over 54.2 seconds during the test. **Possible Causes:** • Air leaks in the intake manifold or exhaust manifold or pipes • IAT sensor or MAF sensor has deteriorated (out of calibration) • HO2S signal circuit is open or shorted to ground • HO2S contaminated with wrong fuel, has deteriorated or failed
DTC: P0155 **2T OBD/O2S, MIL: Yes** **Years:** 2007, 2008 **Models:** B-Series Truck, CX-9, Tribute **Engines:** All **Transmissions:** All	**HO2S-21 (B2 S1) Heater Circuit Conditions** Engine started, engine running, and the PCM detected the HO2S heater circuit indicated more than 11.5v with the heater turned "on", or it indicated less than 5.8v with the heater turned "off" for 331–361 seconds during the CCM test. **Possible Causes:** • HO2S heater control circuit is open or shorted to ground • HO2S heater control circuit is shorted to system power (B+) • HO2S heater power circuit open between the heater and PCM • HO2S heater is damaged or has failed • PCM has failed (it controls the heater with a duty cycle signal)
DTC: P0157 **2T OBD/O2S1, MIL: Yes** **Years:** 2007, 2008 **Models:** Mazda6, Mazda6 Sport Wagon **Engines:** All **Transmissions:** All	**HO2S-22 (Bank 2 Sensor 2) Circuit Low Input** Vehicle driven in closed loop at cruise speed at low engine load for over 5 minutes, and the PCM detected the HO2S signal remained at less than 500 mv during the CCM test for 55 seconds. **Possible Causes:** • HO2S signal circuit is open or shorted to ground • Air leaks at the exhaust manifold or exhaust pipes • HO2S contaminated with wrong fuel, has deteriorated or failed • PCM has failed
DTC: P0158 **2T OBD/O2S1, MIL: Yes** **Years:** 2007, 2008 **Models:** CX-9, Mazda6, Mazda6 Sport Wagon, Tribute **Engines:** All **Transmissions:** All	**HO2S-22 (Bank 2 Sensor 2) Circuit High Input** Engine started, vehicle driven at cruise at low engine load for over 5 minutes, and the PCM detected the rear HO2S signal remained too high, the signal was fixed from 350–550 mv, or the HO2S signal switch time was too long during the CCM test. **Possible Causes:** • Air leaks at the exhaust manifold or exhaust pipes • HO2S signal tracking (wet/oily) in connector causing a short between the signal circuit and heater power circuit • HO2S contaminated with wrong fuel, has deteriorated or failed • PCM has failed
DTC: P0159 **2T CCM MIL: Yes** **Years:** 2007, 2008 **Models:** B-Series Truck, CX-9, Mazda6, Mazda6 Sport Wagon **Engines:** All **Transmissions:** All	**Rear HO2S Circuit Problem** The PCM monitors inversion cycle period, rear HO2S output voltage inclination. The PCM detects that the voltage inclinations are below threshold consecutive 5 times when following conditions are met, the PCM determines that circuit has malfunction. Under the following monitoring conditions, if 0.3 V or more is detected three times even if fuel cut is performed for 3 seconds or more, a circuit malfunction is determined. **Possible Causes:** • Rear HO2S circuit deterioration • Rear HO2S circuit malfunction • PCM has failed

DTC	Trouble Code Title, Conditions & Possible Causes
DTC: P0160 **2T OBD/O2S1, MIL: Yes** **Years:** 2007, 2008 **Models:** Mazda6, Mazda6 Sport Wagon **Engines:** All **Transmissions:** All	**HO2S-22 (Bank 2 Sensor 2) No Activity Detected** Engine started, engine running in closed loop, and the PCM detected the HO2S signal remained under 500 mv, or with the ECT input over 154°F and engine speed over 1500 RPM for 2 minutes, it detected the HO2S-22 signal did not exceed 500 mv during the test. **Possible Causes:** • Air leaks in the intake manifold or exhaust manifold or pipes • IAT sensor or MAF sensor has deteriorated (out of calibration) • HO2S signal circuit is open or shorted to ground • HO2S contaminated with wrong fuel, has deteriorated or failed
DTC: P0161 **2T OBD/O2S, MIL: Yes** **Years:** 2007, 2008 **Models:** B-Series Truck, CX-9, Tribute **Engines:** All **Transmissions:** All	**HO2S-22 (Bank 2 Sensor 2) Heater Circuit Malfunction** Engine started, engine running, and the PCM detected the HO2S heater circuit indicated more than 11.5v with the heater turned "on", or it indicated less than 5.8v with the heater turned "off" for 331–361 seconds during the CCM test. **Possible Causes:** • HO2S heater control circuit is open or shorted to ground • HO2S heater control circuit is shorted to system power (B+) • HO2S heater power circuit open between the heater and PCM • HO2S heater is damaged or has failed • PCM has failed (it controls the heater with a duty cycle signal)
DTC: P0170 **2T CCM, MIL: Yes** **Years:** 2007, 2008 **Models:** B-Series Truck, Tribute **Engines:** All **Transmissions:** All	**Adaptive Fuel Trim Too Rich or Too Lean (Bank 1)** Engine running in closed loop at cruise speed for 2 minutes, and the PCM detected the A/F ratio remained richer or leaner than the fuel correction limit for 10 seconds during the Fuel System Monitor test. **Possible Causes:** • Air leaks after the MAF sensor, or air leaks in the PCV system • Air leaks at the EGR gasket, or at the EGR valve diaphragm • Base engine "mechanical" fault affecting one or more cylinders • Exhaust leaks before or near where the front HO2S is mounted • Fuel control sensor is out of calibration (i.e., ECT, IAT or MAP) • Fuel delivery system supplying too much or too little fuel during idle or cruise speed (weak fuel pump, weak pressure regulator) • Fuel Injectors (one or more) is dirty, restricted or leaking fuel • HO2S is contaminated, deteriorated or it has failed • Vehicle driven low on fuel or driven until it ran out of fuel
DTC: P0171 **2T CCM, MIL: Yes** **Years:** 2007, 2008 **Models:** B-Series Truck, CX-9, Mazda3, MazdaSpeed3, Mazda6, MazdaSpeed6, RX-8, Tribute **Engines:** All **Transmissions:** All	**Adaptive Fuel Trim Too Lean (Bank 1)** Engine running in closed loop at cruise speed for 2 minutes, and the PCM detected the A/F ratio remained leaner than the fuel correction limit in memory for 10 seconds during the Fuel System Monitor test. **Possible Causes:** • Air leaks after the MAF sensor, or in the PCV or EGR system • Base engine "mechanical" fault affecting one or more cylinders • Exhaust leaks before or near where the front HO2S is mounted • Fuel control sensor out of calibration (i.e., ECT, IAT or MAF) • Fuel delivery system supplying too much or too little fuel during idle or cruise speed (weak fuel pump, weak pressure regulator) • Fuel Injectors (one or more) is dirty, restricted or leaking fuel • HO2S is contaminated, deteriorated or it has failed • Vehicle driven low on fuel or driven until it ran out of fuel
DTC: P0172 **2T CCM, MIL: Yes** **Years:** 2007, 2008 **Models:** B-Series Truck, CX-9, Mazda3, MazdaSpeed3, Mazda6, MazdaSpeed6, RX-8, Tribute **Engines:** All **Transmissions:** All	**Adaptive Fuel Trim Too Rich (Bank 1)** Engine running in closed loop at cruise speed for 2 minutes, and the PCM detected the A/F ratio remained richer than the fuel correction limit in memory for 10 seconds during the Fuel System Monitor test. **Possible Causes:** • Base engine "mechanical" fault affecting one or more cylinders • EVAP system component has failed or canister fuel saturated • Exhaust leaks before or near where the front HO2S is mounted • Fuel control sensor out of calibration (i.e., IAT, MAP or MAF) • Fuel injectors (one or more) sticking or leaking fuel • Fuel delivery system supplying too much fuel during cruise or idle periods (e.g., faulty fuel pump, or faulty pressure regulator) • HO2S is contaminated, deteriorated or it has failed

DTC	Trouble Code Title, Conditions & Possible Causes
DTC: P0173 **2T CCM, MIL: Yes** **Years:** 2007, 2008 **Models:** B-Series Truck, Tribute **Engines:** All **Transmissions:** All	**Adaptive Fuel Trim Too Rich or Too Lean (Bank 2)** Engine running in closed loop at cruise speed for 2 minutes, and the PCM detected the A/F ratio remained leaner or leaner than the fuel correction limit for 10 seconds during the Fuel System Monitor test. **Possible Causes:** • Air leaks after the MAF sensor, or air leaks in the PCV system • Air leaks at the EGR gasket, or at the EGR valve diaphragm • Base engine "mechanical" fault affecting one or more cylinders • Exhaust leaks before or near where the front HO2S is mounted • Fuel control sensor is out of calibration (i.e., ECT, IAT or MAP) • Fuel delivery system supplying too much or too little fuel during idle or cruise speed (weak fuel pump, weak pressure regulator) • Fuel Injectors (one or more) is dirty, restricted or leaking fuel • HO2S is contaminated, deteriorated or it has failed • Vehicle driven low on fuel or driven until it ran out of fuel
DTC: P0174 **2T CCM, MIL: Yes** **Years:** 2007, 2008 **Models:** B-Series Truck, CX-9, Tribute **Engines:** All **Transmissions:** All	**Adaptive Fuel Trim Too Lean (Bank 2)** Engine running in closed loop at cruise speed for 2 minutes, and the PCM detected the A/F ratio remained leaner than the fuel correction limit in memory for 10 seconds during the Fuel System Monitor test. **Possible Causes:** • Vehicle driven low on fuel or until it ran out of fuel • One or more injectors restricted or pressure regulator has failed • Fuel delivery system supplying too much or too little fuel during cruise or idle periods (e.g., faulty fuel pump, or dirty fuel filter) • Air leaks after the MAF sensor, or air leaks in the PCV system • Air leaks at the EGR gasket, or at the EGR valve diaphragm • Exhaust leaks before or near where the front HO2S is mounted • HO2S is contaminated, deteriorated or it has failed • Fuel control sensor is out of calibration (i.e., ECT, IAT or MAP) • Base engine "mechanical" fault affecting one or more cylinders
DTC: P0175 **2T CCM, MIL: Yes** **Years:** 2007, 2008 **Models:** B-Series Truck, CX-9, Tribute **Engines:** All **Transmissions:** All	**Adaptive Fuel Trim Too Rich (Bank 2)** Engine running in closed loop at cruise speed for 2 minutes, and the PCM detected the A/F ratio remained richer than the fuel correction limit in memory for 10 seconds during the Fuel System Monitor test. **Possible Causes:** • One or more injectors leaking or pressure regulator is leaking • Fuel delivery system supplying too much fuel during cruise or idle periods (e.g., faulty fuel pump, or faulty pressure regulator) • Exhaust leaks before or near where the front HO2S is mounted • HO2S is contaminated, deteriorated or it has failed • EVAP system component has failed or canister fuel saturated • Fuel control sensor is out of calibration (i.e., ECT, IAT or MAP) • Base engine "mechanical" fault affecting one or more cylinders
DTC: P0180 **2T CCM, MIL: Yes** **Years:** 2008 **Models:** Tribute **Engines:** All **Transmissions:** All	**Flexible Fuel Sensor Circuit Malfunction** Key on or engine running and the PCM detected an unexpected voltage condition on the Flexible Fuel Sensor circuit during the test. **Possible Causes:** • Flexible fuel sensor signal circuit is open or shorted to ground • Flexible fuel sensor ground circuit is open • Flexible fuel sensor power circuit is open or shorted to ground • Flexible fuel sensor is damaged or has failed • PCM has failed
DTC: P0181 **2T CCM, MIL: Yes** **Years:** 2008 **Models:** Tribute **Engines:** All **Transmissions:** All	**Flexible Fuel Sensor Circuit Malfunction** Key on or engine running and the PCM detected an unexpected voltage condition on the Flexible Fuel Sensor circuit during the test. **Possible Causes:** • Flexible fuel sensor signal circuit is open or shorted to ground • Flexible fuel sensor ground circuit is open • Flexible fuel sensor power circuit is open or shorted to ground • Flexible fuel sensor is damaged or has failed • PCM has failed

DTC	Trouble Code Title, Conditions & Possible Causes
DTC: P0182 **2T CCM, MIL: Yes** **Years:** 2008 **Models:** Tribute **Engines:** All **Transmissions:** All	**Flexible Fuel Sensor Circuit Malfunction** Key on or engine running and the PCM detected an unexpected voltage condition on the Flexible Fuel Sensor circuit during the test. **Possible Causes:** • Flexible fuel sensor signal circuit is open or shorted to ground • Flexible fuel sensor ground circuit is open • Flexible fuel sensor power circuit is open or shorted to ground • Flexible fuel sensor is damaged or has failed • PCM has failed
DTC: P0183 **2T CCM, MIL: Yes** **Years:** 2008 **Models:** Tribute **Engines:** All **Transmissions:** All	**Flexible Fuel Sensor Circuit Malfunction** Key on or engine running and the PCM detected an unexpected voltage condition on the Flexible Fuel Sensor circuit during the test. **Possible Causes:** • Flexible fuel sensor signal circuit is open or shorted to ground • Flexible fuel sensor ground circuit is open • Flexible fuel sensor power circuit is open or shorted to ground • Flexible fuel sensor is damaged or has failed • PCM has failed
DTC: P0186 **2T CCM, MIL: Yes** **Years:** 2008 **Models:** Tribute **Engines:** All **Transmissions:** All	**Flexible Fuel Sensor Circuit Malfunction** Key on or engine running and the PCM detected an unexpected voltage condition on the Flexible Fuel Sensor circuit during the test. **Possible Causes:** • Flexible fuel sensor signal circuit is open or shorted to ground • Flexible fuel sensor ground circuit is open • Flexible fuel sensor power circuit is open or shorted to ground • Flexible fuel sensor is damaged or has failed • PCM has failed
DTC: P0187 **2T CCM, MIL: Yes** **Years:** 2008 **Models:** Tribute **Engines:** All **Transmissions:** All	**Flexible Fuel Sensor Circuit Malfunction** Key on or engine running and the PCM detected an unexpected voltage condition on the Flexible Fuel Sensor circuit during the test. **Possible Causes:** • Flexible fuel sensor signal circuit is open or shorted to ground • Flexible fuel sensor ground circuit is open • Flexible fuel sensor power circuit is open or shorted to ground • Flexible fuel sensor is damaged or has failed • PCM has failed
DTC: P0188 **2T CCM, MIL: Yes** **Years:** 2008 **Models:** Tribute **Engines:** All **Transmissions:** All	**Flexible Fuel Sensor Circuit Malfunction** Key on or engine running and the PCM detected an unexpected voltage condition on the Flexible Fuel Sensor circuit during the test. **Possible Causes:** • Flexible fuel sensor signal circuit is open or shorted to ground • Flexible fuel sensor ground circuit is open • Flexible fuel sensor power circuit is open or shorted to ground • Flexible fuel sensor is damaged or has failed • PCM has failed
DTC: P0190 **1T CCM, MIL: Yes** **Years:** 2008 **Models:** Tribute **Engines:** All **Transmissions:** All	**Flexible Fuel Sensor Circuit Malfunction** Key on or engine running and the PCM detected an unexpected voltage condition on the Flexible Fuel Sensor circuit during the test. **Possible Causes:** • Flexible fuel sensor signal circuit is open or shorted to ground • Flexible fuel sensor ground circuit is open • Flexible fuel sensor power circuit is open or shorted to ground • Flexible fuel sensor is damaged or has failed • PCM has failed
DTC: P0191 **1T CCM, MIL: Yes** **Years:** 2008 **Models:** Tribute **Engines:** All **Transmissions:** All	**Flexible Fuel Sensor Circuit Malfunction** Key on or engine running and the PCM detected an unexpected voltage condition on the Flexible Fuel Sensor circuit during the test. **Possible Causes:** • Flexible fuel sensor signal circuit is open or shorted to ground • Flexible fuel sensor ground circuit is open • Flexible fuel sensor power circuit is open or shorted to ground • Flexible fuel sensor is damaged or has failed • PCM has failed

DTC	Trouble Code Title, Conditions & Possible Causes
DTC: P0192 **1T CCM, MIL: Yes** **Years:** 2007, 2008 **Models:** CX-7, Tribute **Engines:** All **Transmissions:** All	**Flexible Fuel Sensor Circuit Malfunction** Key on or engine running and the PCM detected an unexpected voltage condition on the Flexible Fuel Sensor circuit during the test. **Possible Causes:** • Flexible fuel sensor signal circuit is open or shorted to ground • Flexible fuel sensor ground circuit is open • Flexible fuel sensor power circuit is open or shorted to ground • Flexible fuel sensor is damaged or has failed • PCM has failed
DTC: P0193 **1T CCM, MIL: Yes** **Years:** 2007, 2008 **Models:** CX-7, Tribute **Engines:** All **Transmissions:** All	**Flexible Fuel Sensor Circuit Malfunction** Key on or engine running and the PCM detected an unexpected voltage condition on the Flexible Fuel Sensor circuit during the test. **Possible Causes:** • Flexible fuel sensor signal circuit is open or shorted to ground • Flexible fuel sensor ground circuit is open • Flexible fuel sensor power circuit is open or shorted to ground • Flexible fuel sensor is damaged or has failed • PCM has failed
DTC: P0201 **1T CCM, MIL: Yes** **Years:** 2007, 2008 **Models:** B-Series Truck, CX-7, CX-9, Tribute **Engines:** All **Transmissions:** All	**Fuel Injector 1 Control Circuit Malfunction** Engine running and the PCM detected an unexpected voltage condition on the fuel injector control circuit during the CCM test. **Possible Causes:** • Fuel injector 1 control circuit is open or grounded • Fuel injector 1 power circuit open between injector and VPWR • Fuel injector 1 has failed • PCM has failed (i.e., the PCM driver for Injector 1 has failed)
DTC: P0202 **1T CCM, MIL: Yes** **Years:** 2007, 2008 **Models:** B-Series Truck, CX-7, CX-9, Tribute **Engines:** All **Transmissions:** All	**Fuel Injector 2 Control Circuit Malfunction** Engine running and the PCM detected an unexpected voltage condition on the fuel injector control circuit during the CCM test. **Possible Causes:** • Fuel injector 2 control circuit is open or grounded • Fuel injector 2 power circuit open between injector and VPWR • Fuel injector 2 has failed • PCM has failed (i.e., the PCM driver for Injector 2 has failed)
DTC: P0203 **1T CCM, MIL: Yes** **Years:** 2007, 2008 **Models:** B-Series Truck, CX-7, CX-9, Tribute **Engines:** All **Transmissions:** All	**Fuel Injector 3 Control Circuit Malfunction** Engine running and the PCM detected an unexpected voltage condition on the fuel injector control circuit during the CCM test. **Possible Causes:** • Fuel injector 3 control circuit is open or grounded • Fuel injector 3 power circuit open between injector and VPWR • Fuel injector 3 has failed • PCM has failed (i.e., the PCM driver for Injector 3 has failed)
DTC: P0204 **1T CCM, MIL: Yes** **Years:** 2007, 2008 **Models:** B-Series Truck, CX-7, CX-9, Tribute **Engines:** All **Transmissions:** All	**Fuel Injector 4 Control Circuit Malfunction** Engine running and the PCM detected an unexpected voltage condition on the fuel injector control circuit during the CCM test. **Possible Causes:** • Fuel injector 4 control circuit is open or grounded • Fuel injector 4 power circuit open between injector and VPWR • Fuel injector 4 has failed • PCM has failed (i.e., the PCM driver for Injector 4 has failed)
DTC: P0205 **1T CCM, MIL: Yes** **Years:** 2007, 2008 **Models:** B-Series Truck, CX-7, CX-9, Tribute **Engines:** 3.0L, 3.5L, 3.7L, 4.0L **Transmissions:** All	**Fuel Injector 5 Control Circuit Malfunction** Engine running and the PCM detected an unexpected voltage condition on the fuel injector control circuit during the CCM test. **Possible Causes:** • Fuel injector 5 control circuit is open or grounded • Fuel injector 5 power circuit open between injector and VPWR • Fuel injector 5 has failed • PCM has failed (i.e., the PCM driver for Injector 5 has failed)
DTC: P0206 **1T CCM, MIL: Yes** **Years:** 2007, 2008 **Models:** B-Series Truck, CX-7, CX-9, Tribute **Engines:** 3.0L, 3.5L, 3.7L, 4.0L **Transmissions:** All	**Fuel Injector 6 Control Circuit Malfunction** Engine running and the PCM detected an unexpected voltage condition on the fuel injector control circuit during the CCM test. **Possible Causes:** • Fuel injector 6 control circuit is open or grounded • Fuel injector 6 power circuit open between injector and VPWR • Fuel injector 6 has failed • PCM has failed (i.e., the PCM driver for Injector 6 has failed)

DTC	Trouble Code Title, Conditions & Possible Causes
DTC: P0217 **1T CCM, MIL: Yes** **Years:** 2008 **Models:** Tribute **Engines:** All **Transmissions:** All	**Engine Overheat Condition Sensed by ECT/CHT Sensor** The ECT sensor has detected an engine overheating condition. **Possible Causes:** • Low coolant level • Coolant leak • Defective thermostat
DTC: P0218 **1T CCM, MIL: Yes** **Years:** 2007, 2008 **Models:** B-Series Truck, Tribute **Engines:** All **Transmissions:** All	**Transaxle Fluid Over-Temperature Condition** TFT indicates transaxle over-temperature condition. **Possible Causes:** • Transmission fluid temperature exceeded 127°C • Damaged PCM
DTC: P0219 **1T CCM, MIL: Yes** **Years:** 2007, 2008 **Models:** B-Series Truck, Tribute **Engines:** All **Transmissions:** All	**PCM - Engine RPM Limiter, PCM Software** Indicates the vehicle has been operated in a manner, which caused the engine speed to exceed a calibration limit. The engine speed is continuously monitored and evaluated by the PCM. The DTC is set when the RPM exceeds the calibrated limit set with in the PCM. **Possible Causes:** • TP Sensor malfunction • Connector or Terminal Malfunction • Open circuit between throttle body terminal A and PCM terminal 3M • Open circuit between throttle body terminal E and PCM terminal 3N
DTC: P0220 **1T CCM, MIL: Yes** **Years:** 2007, 2008 **Models:** CX-9 **Engines:** All **Transmissions:** A/T	**Accelerator Pedal Position Signal Error** If the accelerator pedal position signal from the PCM, which is transmitted using the CAN communication, is invalid and the condition has continued for 500 ms or more. **Possible Causes:** • Accelerator pedal position sensor malfunction • PCM Malfunction • TCM Malfunction
DTC: P0221 **1T CCM, MIL: Yes** **Years:** 2007, 2008 **Models:** B-Series Truck, Tribute **Engines:** All **Transmissions:** All	**TP Sensor Circuit Malfunction** The TP sensor circuit registered an out of range in either the closed or wide open throttle positions. **Possible Causes:** • TP Sensor malfunction • Connector or Terminal Malfunction • TP sensor not seated properly or tightened down • Damaged PCM
DTC: P0222 **1T CCM, MIL: Yes** **Years:** 2007, 2008 **Models:** B-Series Truck, CX-7, CX-9, Mazda3, MazdaSpeed3, Mazda5, Mazda6, MazdaSpeed6, MX-5, RX-8, Tribute **Engines:** All **Transmissions:** All	**TP Sensor No. 2 Circuit High Input** Key on or engine running, and the PCM detects TP Sensor No. 2 voltage at PCM terminal 3J is below 0.2 v after ignition key to ON, PCM determines that TP circuit has a malfunction. **Possible Causes:** • TP Sensor malfunction • Connector or Terminal Malfunction • Open circuit between throttle body terminal A and PCM terminal 3M • Open circuit between throttle body terminal E and PCM terminal 3N
DTC: P0223 **1T CCM, MIL: Yes** **Years:** 2007, 2008 **Models:** B-Series Truck, CX-7, CX-9, Mazda3, MazdaSpeed3, Mazda5, Mazda6, MazdaSpeed6, MX-5, RX-8, Tribute **Engines:** All **Transmissions:** All	**TP Sensor No. 2 Circuit Low Input** Key on or engine running, and the PCM detects TP Sensor No. 2 voltage at PCM terminal 3J is above 4.85 v after ignition key to ON, PCM determines that TP circuit has a malfunction. **Possible Causes:** • TP Sensor malfunction • Connector or Terminal Malfunction • Open circuit between throttle body terminal D and PCM terminal 3K • Open circuit between throttle body terminal C and PCM terminal 3J
DTC: P0224 **1T CCM, MIL: Yes** **Years:** 2007, 2008 **Models:** B-Series Truck, Tribute **Engines:** All **Transmissions:** All	**TP Sensor Circuit Malfunction** PCM has registered an intermittent concern with TP sensor. **Possible Causes:** • TP Sensor malfunction • Connector or Terminal Malfunction

DTC	Trouble Code Title, Conditions & Possible Causes
DTC: P0230 **1T CCM, MIL: Yes** **Years:** 2007, 2008 **Models:** B-Series Truck, CX-9, Tribute **Engines:** All **Transmissions:** All	**Fuel Pump Primary or Secondary Circuit Malfunction** Key on or engine running and the PCM detected an unexpected voltage condition on the fuel pump primary circuit during the test. **Note: This trouble code may set due an intermittent fault!** **Possible Causes:** • Fuel pump primary circuit is open or shorted to ground • Fuel pump primary circuit is shorted to system power (VREF) • Fuel pump relay is damaged or has failed • PCM has failed
DTC: P0231 **1T CCM, MIL: Yes** **Years:** 2007, 2008 **Models:** B-Series Truck, CX-9, Tribute **Engines:** All **Transmissions:** All	**Fuel Pump Primary or Secondary Circuit Malfunction** Key on, then with the Fuel Pump commanded "on", the PCM did not detect system voltage (B+) on the fuel pump monitor circuit. **Possible Causes:** • Fuel pump feed circuit open between feed circuit and the pump • Fuel pump relay contacts "open" that provide B+ to the pump • FP PWR circuit open between relay and connection to the FPM • Fuel pump relay is damaged or has failed • PCM has failed (the engine will not start if the PCM circuit fails)
DTC: P0232 **1T CCM, MIL: Yes** **Years:** 2007, 2008 **Models:** B-Series Truck, CX-9, Tribute **Engines:** All **Transmissions:** All	**Fuel Pump Primary Secondary Circuit Malfunction** Key on, then with the Fuel Pump commanded "on", the PCM did not detect system voltage (B+) on the fuel pump monitor circuit. **Note: the Fuel Pump Driver Module (FPM) modulates the voltage to the fuel pump to achieve the correct fuel pressure. Power to the fuel pump is supplied by the power relay or FPDM power supply relay.** **Possible Causes (No Start Condition)** Inertia switch needs resetting or its internal contacts are open • Fuel pump circuit open between the FPM connection and the FP PWWR circuit • Fuel pump ground circuit is open or has high resistance • Fuel pump relay is damaged or has failed • PCM has failed (the engine will not start and run) • Possible Causes (Engine Starts Condition) • Fuel pump secondary circuit is shorted to system power (B+) • Fuel pump relay contacts closed all of the time • Fuel pump circuit open between the PCM connection and the FP PWR circuit • PCM has failed (the engine will start and run)
DTC: P0233 **1T CCM, MIL: Yes** **Years:** 2007, 2008 **Models:** B-Series Truck **Engines:** All **Transmissions:** All	**Fuel Pump Primary or Secondary Circuit Malfunction** Key on or engine running and the PCM detected an intermittent concern with the FPM circuit. **Note: This trouble code may set due an intermittent fault!** **Possible Causes:** • Fuel pump primary circuit is open or shorted to ground • Fuel pump primary circuit is shorted to system power (VREF) • Fuel pump relay is damaged or has failed • PCM has failed
DTC: P0234 **1T CCM, MIL: No** **Years:** 2007, 2008 **Models:** CX-7 **Engines:** All **Transmissions:** All	**Turbocharger Over-Boost Condition** If the manifold absolute pressure or charging efficiency are more than the specification (engine speed is 2,000 RPM or more) for the specified period of time, the PCM determines that the turbocharger is in an over-boost condition. **Possible Causes:** • Wastegate control solenoid valve malfunction • Vacuum hose looseness or damage • Improper installation of the vacuum hose
DTC: P0245 **1T CCM, MIL: Yes** **Years:** 2007, 2008 **Models:** CX-7 **Engines:** All **Transmissions:** All	**Turbocharger Wastegate Regulating Valve Control Low Circuit** The PCM monitors the turbocharger wastegate regulating valve control voltage when the PCM turns the turbocharger wastegate regulating valve on. If the control voltage is low, the PCM determines that the turbocharger wastegate regulating valve control circuit voltage is low. **Possible Causes:** • Turbo charger wastegate malfunctioning • Connector or terminal malfunction • Short to ground in wiring harness between turbo charger wastegate regulating valve terminal A and PCM terminal 4D • Short to ground in wiring harness between turbo charger wastegate regulating valve terminal A and PCM terminal 4E • PCM has failed

DTC	Trouble Code Title, Conditions & Possible Causes
DTC: P0246 **1T CCM, MIL: Yes** **Years:** 2007, 2008 **Models:** CX-7 **Engines:** All **Transmissions:** All	**Turbocharger Wastegate Regulating Valve Control High Circuit** The PCM monitors the turbocharger wastegate regulating valve control voltage when the PCM turns the turbocharger wastegate regulating valve on. If the control voltage is high, the PCM determines that the turbocharger wastegate regulating valve control circuit voltage is high. **Possible Causes:** • Turbo charger wastegate malfunctioning • Connector or terminal malfunction • Short to ground in wiring harness between turbo charger wastegate regulating valve terminal A and PCM terminal 4D • Short to ground in wiring harness between turbo charger wastegate regulating valve terminal A and PCM terminal 4E • PCM has failed
DTC: P0261 **1T CCM, MIL: Yes** **Years:** 2007, 2008 **Models:** B-Series Truck, Tribute **Engines:** All **Transmissions:** All	**Fuel Injector Cylinder 1 Low Circuit** A concern is detected in the fuel injector interface module in the primary (forward) fuel injector circuit. **Possible Causes:** • Fuel injector • Fuel Harness circuits: VPWR and Injector 1 • Damaged PCM
DTC: P0262 **1T CCM, MIL: Yes** **Years:** 2007, 2008 **Models:** B-Series Truck, Tribute **Engines:** All **Transmissions:** All	**Fuel Injector Cylinder 1 High Circuit** A concern is detected in the fuel injector interface module in the primary (forward) fuel injector circuit. **Possible Causes:** • Fuel injector • Fuel Harness circuits: VPWR and Injector 1 • Damaged PCM
DTC: P0264 **1T CCM, MIL: Yes** **Years:** 2007, 2008 **Models:** B-Series Truck, Tribute **Engines:** All **Transmissions:** All	**Fuel Injector Cylinder 2 Low Circuit** A concern is detected in the fuel injector interface module in the primary (forward) fuel injector circuit. **Possible Causes:** • Fuel injector • Fuel Harness circuits: VPWR and Injector 2 • Damaged PCM
DTC: P0265 **1T CCM, MIL: Yes** **Years:** 2007, 2008 **Models:** B-Series Truck, Tribute **Engines:** All **Transmissions:** All	**Fuel Injector Cylinder 2 High Circuit** A concern is detected in the fuel injector interface module in the primary (forward) fuel injector circuit. **Possible Causes:** • Fuel injector • Fuel Harness circuits: VPWR and Injector 2 • Damaged PCM
DTC: P0267 **1T CCM, MIL: Yes** **Years:** 2007, 2008 **Models:** B-Series Truck, Tribute **Engines:** All **Transmissions:** All	**Fuel Injector Cylinder 3 Low Circuit** A concern is detected in the fuel injector interface module in the primary (forward) fuel injector circuit. **Possible Causes:** • Fuel injector • Fuel Harness circuits: VPWR and Injector 3 • Damaged PCM
DTC: P0268 **1T CCM, MIL: Yes** **Years:** 2007, 2008 **Models:** B-Series Truck, Tribute **Engines:** All **Transmissions:** All	**Fuel Injector Cylinder 3 High Circuit** A concern is detected in the fuel injector interface module in the primary (forward) fuel injector circuit. **Possible Causes:** • Fuel injector • Fuel Harness circuits: VPWR and Injector 3 • Damaged PCM
DTC: P0270 **1T CCM, MIL: Yes** **Years:** 2007, 2008 **Models:** B-Series Truck, Tribute **Engines:** All **Transmissions:** All	**Fuel Injector Cylinder 4 Low Circuit** A concern is detected in the fuel injector interface module in the primary (forward) fuel injector circuit. **Possible Causes:** • Fuel injector • Fuel Harness circuits: VPWR and Injector 4 • Damaged PCM

DTC	Trouble Code Title, Conditions & Possible Causes
DTC: P0271 **1T CCM, MIL: Yes** **Years:** 2007, 2008 **Models:** B-Series Truck, Tribute **Engines:** All **Transmissions:** All	**Fuel Injector Cylinder 4 High Circuit** A concern is detected in the fuel injector interface module in the primary (forward) fuel injector circuit. **Possible Causes:** • Fuel injector • Fuel Harness circuits: VPWR and Injector 4 • Damaged PCM
DTC: P0273 **1T CCM, MIL: Yes** **Years:** 2007, 2008 **Models:** B-Series Truck, Tribute **Engines:** 3.0L, 4.0L **Transmissions:** All	**Fuel Injector Cylinder 5 Low Circuit** A concern is detected in the fuel injector interface module in the primary (forward) fuel injector circuit. **Possible Causes:** • Fuel injector • Fuel Harness circuits: VPWR and Injector 5 • Damaged PCM
DTC: P0274 **1T CCM, MIL: Yes** **Years:** 2007, 2008 **Models:** B-Series Truck, Tribute **Engines:** 3.0L, 4.0L **Transmissions:** All	**Fuel Injector Cylinder 5 High Circuit** A concern is detected in the fuel injector interface module in the primary (forward) fuel injector circuit. **Possible Causes:** • Fuel injector • Fuel Harness circuits: VPWR and Injector 5 • Damaged PCM
DTC: P0276 **1T CCM, MIL: Yes** **Years:** 2007, 2008 **Models:** B-Series Truck, Tribute **Engines:** 3.0L, 4.0L **Transmissions:** All	**Fuel Injector Cylinder 6 Low Circuit** A concern is detected in the fuel injector interface module in the primary (forward) fuel injector circuit. **Possible Causes:** • Fuel injector • Fuel Harness circuits: VPWR and Injector 6 • Damaged PCM
DTC: P0277 **1T CCM, MIL: Yes** **Years:** 2007, 2008 **Models:** B-Series Truck, Tribute **Engines:** 3.0L, 4.0L **Transmissions:** All	**Fuel Injector Cylinder 6 High Circuit** A concern is detected in the fuel injector interface module in the primary (forward) fuel injector circuit. **Possible Causes:** • Fuel injector • Fuel Harness circuits: VPWR and Injector 6 • Damaged PCM
DTC: P0297 **1T CCM, MIL: Yes** **Years:** 2007, 2008 **Models:** B-Series Truck, CX-9, Tribute **Engines:** All **Transmissions:** All	**Vehicle Overspeed Condition** The vehicle has been operated in a manner which caused the engine or vehicle to exceed a calibration limit. **Possible Causes:** • Wheel slippage (water, ice, mud and snow) • Excessive engine RPM in neutral • Vehicle drive at a high rate of speed
DTC: P0298 **1T CCM, MIL: Yes** **Years:** 2007, 2008 **Models:** B-Series Truck, Tribute **Engines:** All **Transmissions:** All	**Engine Oil Over Temperature Condition** Oil protection strategy in the PCM has been activated. PCM uses an oil algorithm to infer actual oil temperature. **Possible Causes:** • Overheating condition • Basic engine concerns • Engine cooling concerns • Open circuit or short in ECT harness • PCM has failed
DTC: P0300 **2T CCM, MIL: Yes** **Years:** 2007, 2008 **Models:** B-Series Truck, CX-7, CX-9, Mazda3, MazdaSpeed3, Mazda5, Mazda6, MazdaSpeed6, MX-5, RX-8, Tribute **Engines:** All **Transmissions:** All	**Random Misfire Detected** Engine started, vehicle speed over 3 MPH at 400–4000 RPM, and the PCM detected a random Misfire condition in two or more cylinders during the 200 or 1000-revolution Misfire test, or the PCM could not identify the misfiring cylinder due to a problem in the CMP sensor. **Note: If the misfire is severe, the MIL will flash on/off on the 1st trip.** **Possible Causes:** • Base engine mechanical fault affecting one or more cylinders • CMP sensor is damaged or failed (problem may be intermittent) • Fuel metering fault that affects more than one cylinder • Fuel pressure too low or too high, fuel supply contaminated • EVAP system problem or the EVAP canister is fuel saturated • EGR valve is stuck open or the PCV system has a vacuum leak • Ignition system fault that affects more than one cylinder • MAF sensor is contaminated (this can cause a lean condition)

DTC	Trouble Code Title, Conditions & Possible Causes
DTC: P0301 **2T CCM, MIL: Yes** **Years:** 2007, 2008 **Models:** B-Series Truck, CX-7, CX-9, Mazda3, MazdaSpeed3, Mazda5, Mazda6, MazdaSpeed6, MX-5, RX-8, Tribute **Engines:** All **Transmissions:** All	**Cylinder 1 Misfire Detected** Engine started, vehicle speed over 3 MPH at 400–4000 RPM, and the PCM detected a Misfire condition in a single cylinder during the 200 or 1000-revolution Misfire test under positive engine load conditions. **Note: If the misfire is severe, the MIL will flash on/off on the 1st trip.** **Possible Causes:** • Base engine mechanical fault affecting only Cylinder 1 • CKP sensor is damaged or failed (problem may be intermittent) • Fuel metering fault that affects only Cylinder 1 • Ignition system fault that affects only Cylinder 1
DTC: P0302 **2T CCM, MIL: Yes** **Years:** 2007, 2008 **Models:** B-Series Truck, CX-7, CX-9, Mazda3, MazdaSpeed3, Mazda5, Mazda6, MazdaSpeed6, MX-5, RX-8, Tribute **Engines:** All **Transmissions:** All	**Cylinder 2 Misfire Detected** Engine started, vehicle speed over 3 MPH at 400–4000 RPM, and the PCM detected a Misfire condition in a single cylinder during the 200 or 1000-revolution Misfire test under positive engine load conditions. **Note: If the misfire is severe, the MIL will flash on/off on the 1st trip.** **Possible Causes:** • Base engine mechanical fault affecting only Cylinder 2 • CKP sensor is damaged or failed (problem may be intermittent) • Fuel metering fault that affects only Cylinder 2 • Ignition system fault that affects only Cylinder 2
DTC: P0303 **2T CCM, MIL: Yes** **Years:** 2007, 2008 **Models:** B-Series Truck, CX-7, CX-9, Mazda3, MazdaSpeed3, Mazda5, Mazda6, MazdaSpeed6, MX-5, RX-8, Tribute **Engines:** All **Transmissions:** All	**Cylinder 3 Misfire Detected** Engine started, vehicle speed over 3 MPH at 400–4000 RPM, and the PCM detected a Misfire condition in a single cylinder during the 200 or 1000-revolution Misfire test under positive engine load conditions. **Note: If the misfire is severe, the MIL will flash on/off on the 1st trip.** **Possible Causes:** • Base engine mechanical fault affecting only Cylinder 3 • CKP sensor is damaged or failed (problem may be intermittent) • Fuel metering fault that affects only Cylinder 3 • Ignition system fault that affects only Cylinder 3
DTC: P0304 **2T CCM, MIL: Yes** **Years:** 2007, 2008 **Models:** B-Series Truck, CX-7, CX-9, Mazda3, MazdaSpeed3, Mazda5, Mazda6, MazdaSpeed6, MX-5, RX-8, Tribute **Engines:** All **Transmissions:** All	**Cylinder 4 Misfire Detected** Engine started, vehicle speed over 3 MPH at 400–4000 RPM, and the PCM detected a Misfire condition in a single cylinder during the 200 or 1000-revolution Misfire test under positive engine load conditions. **Note: If the misfire is severe, the MIL will flash on/off on the 1st trip.** **Possible Causes:** • Base engine mechanical fault affecting only Cylinder 4 • CKP sensor is damaged or failed (problem may be intermittent) • Fuel metering fault that affects only Cylinder 4 • Ignition system fault that affects only Cylinder 4
DTC: P0305 **2T CCM, MIL: Yes** **Years:** 2007, 2008 **Models:** B-Series Truck, CX-9, Mazda6, Mazda6 Sport Wagon, Tribute **Engines:** 3.0L, 3.5L, 3.7L, 4.0L **Transmissions:** All	**Cylinder 5 Misfire Detected** Engine started, vehicle speed over 3 MPH at 400–4000 RPM, and the PCM detected irregular CKP signals indicating a Misfire condition present in one cylinder during the 200 or 1000-revolution Misfire test. **Note: If the misfire is severe, the MIL will flash on/off on the 1st trip.** **Possible Causes:** • Base engine mechanical fault affecting only Cylinder 5 • CKP sensor is damaged or failed (problem may be intermittent) • Fuel metering fault that affects only Cylinder 5 • Ignition system fault that affects only Cylinder 5
DTC: P0306 **2T CCM, MIL: Yes** **Years:** 2007, 2008 **Models:** B-Series Truck, CX-9, Mazda6, Mazda6 Sport Wagon, Tribute **Engines:** 3.0L, 3.5L, 3.7L, 4.0L **Transmissions:** All	**Cylinder 6 Misfire Detected** Engine started, vehicle speed over 3 MPH at 400–4000 RPM, and the PCM detected irregular CKP signals indicating a Misfire condition present in one cylinder during the 200 or 1000-revolution Misfire test. **Note: If the misfire is severe, the MIL will flash on/off on the 1st trip.** **Possible Causes:** • Base engine mechanical fault affecting only Cylinder 6 • CKP sensor is damaged or failed (problem may be intermittent) • Fuel metering fault that affects only Cylinder 6 • Ignition system fault that affects only Cylinder 6

DTC	Trouble Code Title, Conditions & Possible Causes
DTC: P0315 **2T CCM, MIL: Yes** **Years:** 2007, 2008 **Models:** B-Series Truck, CX-9, Mazda6, MazdaSpeed6, MX-5, RX-8, Tribute **Engines:** All **Transmissions:** All	**Misfire Detected** Engine started, vehicle speed over 3 MPH at 400–4000 RPM, and the PCM detected irregular CKP signals indicating a Misfire condition present in one cylinder during the 200 or 1000-revolution Misfire test. **Note: If the misfire is severe, the MIL will flash on/off on the 1st trip!** **Possible Causes:** • Base engine mechanical fault • CKP sensor is damaged or failed (problem may be intermittent) • Fuel metering fault • Ignition system fault
DTC: P0316 **2T CCM, MIL: Yes** **Years:** 2007, 2008 **Models:** B-Series Truck, CX-9, Tribute **Engines:** All **Transmissions:** All	**Misfire Detected** Engine started, vehicle speed over 3 MPH at 400–4000 RPM, and the PCM detected irregular CKP signals indicating a Misfire condition present in one cylinder during the 200 or 1000-revolution Misfire test. **Note: If the misfire is severe, the MIL will flash on/off on the 1st trip!** **Possible Causes:** • Base engine mechanical fault • CKP sensor is damaged or failed (problem may be intermittent) • Fuel metering fault • Ignition system fault
DTC: P0320 **1T CCM, MIL: Yes** **Years:** 2007, 2008 **Models:** B-Series Truck, CX-9, Tribute **Engines:** All **Transmissions:** All	**Ignition Engine Speed (PIP) Signal Error** Engine running, and the PCM did not detect any engine speed (PIP) signals, or it detected that erratic speed signals were present. **Possible Causes:** • Engine speed or PIP signal circuit is open (an intermittent fault) • Engine speed signal circuit shorted to ground (intermittent) • Ignition system components arcing (ignition coils or wires) • Vehicle onboard transmitter interruption (i.e., 2-way radio)
DTC: P0321 **1T CCM, MIL: Yes** **Years:** 2007, 2008 **Models:** B-Series Truck, Tribute **Engines:** All **Transmissions:** All	**Ignition Engine Speed (PIP) Signal Range/Performance** Engine running, and the PCM did not detect any engine speed (PIP) signals, or it detected that erratic speed signals were present. **Possible Causes:** • Engine speed or PIP signal circuit is open (an intermittent fault) • Engine speed signal circuit shorted to ground (intermittent) • Ignition system components arcing (ignition coils or wires) • Vehicle onboard transmitter interruption (i.e., 2-way radio)
DTC: P0322 **1T CCM, MIL: Yes** **Years:** 2007, 2008 **Models:** B-Series Truck, Tribute **Engines:** All **Transmissions:** All	**Ignition Engine Speed (PIP) Signal Range/Performance** Engine running, and the PCM did not detect any engine speed (PIP) signals, or it detected that erratic speed signals were present. **Possible Causes:** • Engine speed or PIP signal circuit is open (an intermittent fault) • Engine speed signal circuit shorted to ground (intermittent) • Ignition system components arcing (ignition coils or wires) • Vehicle onboard transmitter interruption (i.e., 2-way radio)
DTC: P0323 **1T CCM, MIL: Yes** **Years:** 2007, 2008 **Models:** B-Series Truck, Tribute **Engines:** All **Transmissions:** All	**Ignition Engine Speed (PIP) Signal Range/Performance** Engine running, and the PCM did not detect any engine speed (PIP) signals, or it detected that erratic speed signals were present. **Possible Causes:** • Engine speed or PIP signal circuit is open (an intermittent fault) • Engine speed signal circuit shorted to ground (intermittent) • Ignition system components arcing (ignition coils or wires) • Vehicle onboard transmitter interruption (i.e., 2-way radio)
DTC: P0325 **1T CCM, MIL: Yes** **Years:** 2007, 2008 **Models:** B-Series Truck, CX-9, Tribute **Engines:** All **Transmissions:** All	**Knock Sensor Circuit Malfunction** Engine running at idle speed, and the PCM detected an unexpected voltage condition on the Knock Sensor circuit during the CCM test. **Note: Check the Knock Sensor installation torque and connection.** **Possible Causes:** • Knock sensor signal circuit is open • Knock sensor signal circuit is shorted to ground • Knock sensor is damaged or has failed • PCM is damaged

DTC	Trouble Code Title, Conditions & Possible Causes
DTC: P0326 **1T CCM, MIL: Yes** **Years:** 2007, 2008 **Models:** B-Series Truck, Tribute **Engines:** All **Transmissions:** All	**Knock Sensor Circuit Range/Performance** Engine running at idle speed, and the PCM detected the knock sensor (KS1) signals during engine acceleration and deceleration changes that were outside of a calibrated value stored in memory. **Possible Causes:** • Knock sensor signal circuit is open (intermittent fault) • Knock sensor signal circuit shorted to ground (intermittent fault) • Knock sensor is damaged or has failed • PCM is damaged
DTC: P0327 **1T CCM, MIL: Yes** **Years:** 2007, 2008 **Models:** B-Series Truck, CX-7, Mazda3, MazdaSpeed3, Mazda5, Mazda6, MazdaSpeed6, MX-5, RX-8 **Engines:** All **Transmissions:** All	**Knock Sensor Circuit Low Input** Key on or engine running and the PCM detected the knock sensor (KS1) signal indicated less than 1.25v during the CCM test. **Possible Causes:** • Knock sensor signal circuit is shorted to ground • Knock sensor is damaged or has failed • PCM is damaged
DTC: P0328 **1T CCM, MIL: Yes** **Years:** 2007, 2008 **Models:** B-Series Truck, CX-7, Mazda3, MazdaSpeed3, Mazda5, Mazda6, MazdaSpeed6, MX-5, RX-8 **Engines:** All **Transmissions:** All	**Knock Sensor Circuit High Input** Key on or engine running and the PCM detected the knock sensor (KS1) signal is intermittent during the CCM test. **Possible Causes:** • Knock sensor signal circuit is open • Knock sensor signal circuit is shorted to VREF or system power • Knock sensor is damaged or has failed • PCM is damaged
DTC: P0329 **1T CCM, MIL: Yes** **Years:** 2007, 2008 **Models:** B-Series Truck **Engines:** All **Transmissions:** All	**Knock Sensor Circuit Range/Performance** Key on or engine running and the PCM detected the knock sensor (KS1) signal indicated more than 3.75v during the CCM test. **Possible Causes:** • Knock sensor signal circuit is open • Knock sensor signal circuit is shorted to ground • Knock sensor is damaged or has failed • PCM is damaged
DTC: P0330 **1T CCM, MIL: Yes** **Years:** 2007, 2008 **Models:** CX-9 **Engines:** All **Transmissions:** All	**Knock Sensor Circuit Malfunction** Key on or engine running and the PCM detected the knock sensor (KS1) signal indicated a malfunction **Possible Causes:** • Knock sensor signal circuit is open • Knock sensor signal circuit is shorted to VREF or system power • Knock sensor is damaged or has failed • PCM is damaged
DTC: P0335 **1T CCM, MIL: Yes** **Years:** 2007, 2008 **Models:** CX-7, Mazda3, MazdaSpeed3, Mazda5, Mazda6, MazdaSpeed6, MX-5, RX-8 **Engines:** All **Transmissions:** All	**Crankshaft Position Sensor Circuit Malfunction** Engine cranking for 1.4 seconds, or engine running, and the PCM did not detect any CKP (NE) signals for 1.1 seconds during the test. **Possible Causes:** • CKP sensor signal is open or shorted to ground • CKP sensor signal is shorted to VREF or system power (B+) • CKP sensor is damaged or has failed • Trigger wheel or tone wheel is damaged • PCM has failed
DTC: P0336 **1T CCM, MIL: Yes** **Years:** 2007, 2008 **Models:** RX-8 **Engines:** All **Transmissions:** All	**Crankshaft Position (CKP) Sensor Circuit Malfunction** Engine running, MAF sensor signal indicating over 2.43 g/sec, and the PCM did not detect any CKP (NE) signals for 4.2 seconds. **Possible Causes:** • CKP sensor signal is open or shorted to ground • CKP sensor signal is shorted to VREF or system power (B+) • CKP sensor is damaged or has failed • Trigger wheel or tone wheel is damaged • PCM has failed

DTC	Trouble Code Title, Conditions & Possible Causes
DTC: P0340 **1T CCM, MIL: Yes** **Years:** 2007, 2008 **Models:** B-Series Truck, CX-7, CX-9, Mazda3, MazdaSpeed3, Mazda5, Mazda6, MazdaSpeed6, MX-5, Tribute **Engines:** All **Transmissions:** All	**Camshaft Position Sensor Circuit Malfunction** Engine running, MAF sensor signal over 4.2 g/sec, and the PCM did not detect any CMP (SGT) sensor signals for 4.2 seconds during the CCM test. **Possible Causes:** • CMP sensor signal is open or shorted to ground • CMP sensor signal is shorted to VREF or system power (B+) • CMP sensor is damaged, or the sensor signal shielding is open • PCM has failed
DTC: P0344 **1T CCM, MIL: Yes** **Years:** 2007, 2008 **Models:** CX-9, Tribute **Engines:** All **Transmissions:** All	**Camshaft Position Sensor Circuit Malfunction** Engine running, MAF sensor signal over 4.2 g/sec, and the PCM did not detect any CMP (SGT) sensor signals for 4.2 seconds during the CCM test. **Possible Causes:** • CMP sensor signal is open or shorted to ground • CMP sensor signal is shorted to VREF or system power (B+) • CMP sensor is damaged, or the sensor signal shielding is open • PCM has failed
DTC: P0345 **1T CCM, MIL: Yes** **Years:** 2007, 2008 **Models:** CX-9, Tribute **Engines:** All **Transmissions:** All	**Camshaft Position Sensor Circuit Malfunction** Engine running, MAF sensor signal over 4.2 g/sec, and the PCM did not detect any CMP (SGT) sensor signals for 4.2 seconds during the CCM test. **Possible Causes:** • CMP sensor signal is open or shorted to ground • CMP sensor signal is shorted to VREF or system power (B+) • CMP sensor is damaged, or the sensor signal shielding is open • PCM has failed
DTC: P0349 **1T CCM, MIL: Yes** **Years:** 2007, 2008 **Models:** CX-9 **Engines:** All **Transmissions:** All	**Camshaft Position Sensor Circuit Malfunction** Engine running, MAF sensor signal over 4.2 g/sec, and the PCM did not detect any CMP (SGT) sensor signals for 4.2 seconds during the CCM test. **Possible Causes:** • CMP sensor signal is open or shorted to ground • CMP sensor signal is shorted to VREF or system power (B+) • CMP sensor is damaged, or the sensor signal shielding is open • PCM has failed
DTC: P0350 **2T CCM, MIL: Yes** **Years:** 2007, 2008 **Models:** B-Series Truck, Tribute **Engines:** All **Transmissions:** All	**Ignition Coil Primary Circuit Malfunction** Engine running and the PCM detected an unexpected voltage condition on the ignition coil primary circuit during the CCM test. **Possible Causes:** • Ignition coil primary circuit is open or shorted to ground • Ignition coil primary circuit is shorted to system power (B+) • Ignition coil power circuit is open between coil and Start circuit • PCM has failed
DTC: P0350 **2T CCM, MIL: Yes** **Years:** 2007, 2008 **Models:** B-Series Truck, CX-9, Tribute **Engines:** All **Transmissions:** All	**Ignition Coil Primary Circuit Malfunction** Engine running and the PCM detected an unexpected voltage condition on the ignition coil primary circuit during the CCM test. **Possible Causes:** • Ignition coil primary circuit is open or shorted to ground • Ignition coil primary circuit is shorted to system power (B+) • Ignition coil power circuit is open between coil and Start circuit • PCM has failed
DTC: P0351 **2T CCM, MIL: Yes** **Years:** 2007, 2008 **Models:** B-Series Truck, CX-9, Tribute **Engines:** All **Transmissions:** All	**Ignition Coil 1 Primary Circuit Malfunction** Engine running and the PCM detected an unexpected voltage condition on the Ignition Coil 1 primary circuit during the CCM test. **Note: DTC P0351 is related to the ignition coil on cylinder 1.** **Possible Causes:** • Ignition coil primary circuit is open, shorted to ground or power • Ignition coil power circuit is open between coil and Start circuit • Ignition coil has failed (arcing between primary and secondary) • PCM has failed

DTC	Trouble Code Title, Conditions & Possible Causes
DTC: P0352 **2T CCM, MIL: Yes** **Years:** 2007, 2008 **Models:** B-Series Truck, CX-9, Tribute **Engines:** All **Transmissions:** All	**Ignition Coil 2 Primary Circuit Malfunction** Engine running and the PCM detected an unexpected voltage condition on the Ignition Coil 2 primary circuit during the CCM test. **Note: DTC P0352 is related to the ignition coil on cylinder 2.** **Possible Causes:** • Ignition coil primary circuit is open, shorted to ground or power • Ignition coil power circuit is open between coil and Start circuit • Ignition coil has failed (arcing between primary and secondary) • PCM has failed
DTC: P0353 **2T CCM, MIL: Yes** **Years:** 2007, 2008 **Models:** B-Series Truck, CX-9, Tribute **Engines:** All **Transmissions:** All	**Ignition Coil 3 Primary Circuit Malfunction** Engine running and the PCM detected an unexpected voltage condition on the Ignition Coil 3 primary circuit during the CCM test. **Note: DTC P0353 is related to the ignition coil on cylinder 3.** **Possible Causes:** • Ignition coil primary circuit is open, shorted to ground or power • Ignition coil power circuit is open between coil and Start circuit • Ignition coil has failed (arcing between primary and secondary) • PCM has failed
DTC: P0354 **2T CCM, MIL: Yes** **Years:** 2007, 2008 **Models:** B-Series Truck, CX-9, Tribute **Engines:** All **Transmissions:** All	**Ignition Coil 4 Primary Circuit Malfunction** Engine running and the PCM detected an unexpected voltage condition on the Ignition Coil 4 primary circuit during the CCM test. **Note: DTC P0354 is related to the ignition coil on cylinder 4.** **Possible Causes:** • Ignition coil primary circuit is open, shorted to ground or power • Ignition coil power circuit is open between coil and Start circuit • Ignition coil has failed (arcing between primary and secondary) • PCM has failed
DTC: P0355 **2T CCM, MIL: Yes** **Years:** 2007, 2008 **Models:** B-Series Truck, CX-9, Tribute **Engines:** All **Transmissions:** All	**Ignition Coil 5 Primary Circuit Malfunction** Engine running and the PCM detected an unexpected voltage condition on the Ignition Coil 5 primary circuit during the CCM test. **Note: DTC P0354 is related to the ignition coil on cylinder 5.** **Possible Causes:** • Ignition coil primary circuit is open, shorted to ground or power • Ignition coil power circuit is open between coil and Start circuit • Ignition coil has failed (arcing between primary and secondary) • PCM has failed
DTC: P0356 **2T CCM, MIL: Yes** **Years:** 2007, 2008 **Models:** B-Series Truck, CX-9, Tribute **Engines:** All **Transmissions:** All	**Ignition Coil 6 Primary Circuit Malfunction** Engine running and the PCM detected an unexpected voltage condition on the Ignition Coil 6 primary circuit during the CCM test. **Note: DTC P0354 is related to the ignition coil on cylinder 6.** **Possible Causes:** • Ignition coil primary circuit is open, shorted to ground or power • Ignition coil power circuit is open between coil and Start circuit • Ignition coil has failed (arcing between primary and secondary) • PCM has failed
DTC: P0400 **2T CCM, MIL: Yes** **Years:** 2007, 2008 **Models:** B-Series Truck, Tribute **Engines:** All **Transmissions:** All	**EGR System Flow Malfunction** Vehicle driven to a speed of over 7 MPH, then back to idle speed, ECT sensor signal more than 131°F, and the PCM detected little or no change in the EGR Differential Pressure sensor (DPFE) with the EGR vent commanded "on" and then "off" in the EGR Monitor test. **Possible Causes:** • EGR vent solenoid circuit open or shorted to ground • EGR vent solenoid power circuit is open • EGR vent solenoid is damaged or has failed • DPFE sensor signal circuit is open or shorted to ground • DPFE sensor is damaged or has failed • EGR valve assembly is leaking, sticking, damaged or has failed • EGR valve vacuum hose(s) lose, damaged or disconnected • Exhaust pipe to EGR valve is restricted or plugged • PCM has failed

DTC	Trouble Code Title, Conditions & Possible Causes
DTC: P0401 **2T CCM, MIL: Yes** **Years:** 2007, 2008 **Models:** B-Series Truck, CX-7, Mazda3, MazdaSpeed3, Mazda5, Mazda6, MazdaSpeed6, MX-5, Tribute **Engines:** All **Transmissions:** All	**EGR System Insufficient Flow Detected** DTC P0102, P0103, P0106, P0107, P0108, P0111, P0112, P0113, P0122, P0123, P0111, P0112, P0113, P0122, P0123, P1122, P1123, P1487, P1496, P1497, P1498 and P1499 not set, engine running in closed loop at 35–55 MPH for two minutes, and the PCM detected insufficient EGR gas flow during the EGR Monitor test. **Possible Causes:** • EGR boost solenoid control circuit is open or shorted to ground • EGR boost solenoid power circuit is open to the Main Relay • EGR boost solenoid is damaged or has failed • EGR boost sensor is damaged or has failed • EGR valve position sensor is damaged, stuck or has failed • EGR valve assembly is leaking, damaged or has failed • EGR valve vacuum hose(s) lose, damaged or disconnected • IAT, MAF, TP and VSS (sensor) has drifted out-of-range • PCM has failed
DTC: P0402 **2T CCM, MIL: Yes** **Years:** 2008 **Models:** Tribute **Engines:** All **Transmissions:** All	**EGR System Excessive Flow At Idle Speed Detected** DTC P0102, P0103, P0106, P0107, P0108, P0111, P0112, P0113, P0122, P0123, P0111, P0112, P0113, P0122, P0123, P1122, P1123, P1487, P1496, P1497, P1498 and P1499 not set, engine running in closed loop at 35–55 MPH, then back to idle speed and the PCM detected excessive EGR gas flow during the EGR Monitor test. **Possible Causes:** • EGR boost solenoid control circuit is open or shorted to ground • EGR boost solenoid power circuit is open to the Main Relay • EGR boost solenoid is damaged or has failed • EGR boost sensor is damaged or has failed • EGR valve position sensor is damaged, stuck or has failed • EGR valve assembly is leaking, damaged or has failed • EGR valve vacuum hose(s) lose, damaged or disconnected • IAT, MAF, TP and VSS (sensor) has drifted out-of-range • PCM has failed
DTC: P0403 **2T CCM, MIL: Yes** **Years:** 2007, 2008 **Models:** B-Series Truck, CX-7, Mazda3, MazdaSpeed3, Mazda5, Mazda6, MazdaSpeed6, MX-5, Tribute **Engines:** All **Transmissions:** All	**EGR Valve (Stepper Motor) Malfunction** PCM monitors input voltage from EGR valve. If voltage at PCM terminals 4E, 4H, 4K and/or 4N remain low or high, PCM determines that ERG valve circuit has a malfunction. **Possible Causes:** • EGR valve malfunction • Connector or terminal malfunction • Short to power circuit in wiring to EGR valve terminals and PCM terminals • Open circuit in wiring between EGR valve terminals and PCM terminals • PCM has failed
DTC: P0404 **2T CCM, MIL: Yes** **Years:** 2007, 2008 **Models:** B-Series Truck, Tribute **Engines:** All **Transmissions:** All	**EGR System Range/Performance** PCM has detected the EGR system operating out of expected parameters. **Possible Causes:** • EGR valve malfunction • Connector or terminal malfunction • Short to power circuit in wiring to EGR valve terminals and PCM terminals • Open circuit in wiring between EGR valve terminals and PCM terminals • PCM has failed
DTC: P0405 **2T CCM, MIL: Yes** **Years:** 2008 **Models:** Tribute **Engines:** All **Transmissions:** All	**EGR Valve Position Sensor Circuit Range/Performance** Engine speed from 1810–2190, TP angle from 3–13.4 percent, engine load 25–50 percent, and the PCM detected that the EGR solenoid accumulated on-time period exceeded a threshold with the EGR system enabled, or it detected an EGR vent solenoid problem existed during the test. **Possible Causes:** • EGR boost solenoid control circuit is open or shorted to ground • EGR boost solenoid power circuit is open to the Main Relay • EGR boost solenoid is damaged or has failed • EGR boost sensor is damaged or has failed • EGR valve position sensor is damaged, stuck or has failed • EGR valve assembly is leaking, damaged or has failed • EGR valve vacuum hose(s) lose, damaged or disconnected • Exhaust pipe to EGR boost sensor or solenoid leaking/plugged • PCM has failed

DTC	Trouble Code Title, Conditions & Possible Causes
DTC: P0406 **2T CCM, MIL: Yes** **Years:** 2008 **Models:** Tribute **Engines:** All **Transmissions:** All	**EGR Valve Position Sensor Circuit Range/Performance** Engine speed from 1810–2190, TP angle from 3–13.4 percent, engine load 25–50 percent, and the PCM detected that the EGR solenoid accumulated on-time period exceeded a threshold with the EGR system enabled, or it detected an EGR vent solenoid problem existed during the test. **Possible Causes:** • EGR boost solenoid control circuit is open or shorted to ground • EGR boost solenoid power circuit is open to the Main Relay • EGR boost solenoid is damaged or has failed • EGR boost sensor is damaged or has failed • EGR valve position sensor is damaged, stuck or has failed • EGR valve assembly is leaking, damaged or has failed • EGR valve vacuum hose(s) lose, damaged or disconnected • Exhaust pipe to EGR boost sensor or solenoid leaking/plugged • PCM has failed
DTC: P0410 **2T CCM, MIL: Yes** **Years:** 2007, 2008 **Models:** RX-8 **Engines:** All **Transmissions:** All	**Secondary Air Injection System Problem** The PCM monitors the front HO2S output current when the Secondary air injection system is operating. If the output current is less than the specification, the PCM determines that there is a Secondary air injection system problem. **Possible Causes:** • AIR system malfunction • Restrict or damaged in exhaust system • Restrict in three way catalytic converter • PCM malfunction
DTC: P0411 **2T CCM, MIL: Yes** **Years:** 2007, 2008 **Models:** RX-8 **Engines:** All **Transmissions:** All	**Secondary Air Injection System Incorrect Upstream Flow** The PCM monitors the front HO2S output current when the Secondary air injection system is operating. If the output current is less than the specification, the PCM determines that there is a Secondary air injection system problem. **Possible Causes:** • AIR system malfunction • Restrict or damaged in exhaust system • Restrict in three way catalytic converter • PCM malfunction
DTC: P0420 **2T OBD/CAT1, MIL: Yes** **Years:** 2007, 2008 **Models:** B-Series Truck, CX-9, Mazda6, MazdaSpeed6, RX-8 **Engines:** All **Transmissions:** All	**Catalyst Efficiency Below Normal (Bank 1)** Vehicle driven at a speed of 16–64 MPH at 1090–3090 RPM with the calculated engine load from 16–55 percent for 2–3 minutes, and the PCM detected the inversion ratio of the rear HO2S and front HO2S was less than a stored threshold during the Catalyst Monitor test. **Possible Causes:** • Air leaks at the exhaust manifold or in the exhaust pipes • Catalytic converter is contaminated, damaged or has failed • Front HO2S and/or the rear HO2S is loose in the mounting hole • Front HO2S older (aged) than the rear HO2S (HO2S is lazy) • Front HO2S or rear HO2S is contaminated with fuel or moisture
DTC: P0421 **2T OBD/CAT2, MIL: Yes** **Years:** 2007, 2008 **Models:** B-Series Truck, CX-7, Mazda3, MazdaSpeed3, Mazda5, Mazda6, MazdaSpeed6, MX-5, Tribute **Engines:** 2.3L VIN 2 **Transmissions:** All	**Catalyst Efficiency Below Normal (Bank 1)** Vehicle driven at a speed of 22–75 MPH at 1250–2094 RPM with the calculated engine load from 20–50 percent for 2–3 minutes, and the PCM detected the inversion ratio of the right bank rear HO2S and front HO2S was less than a stored threshold in the Catalyst Monitor test. **Possible Causes:** • Air leaks at the exhaust manifold or in the exhaust pipes • Catalytic converter is contaminated, damaged or has failed • Front HO2S and/or the rear HO2S is loose in the mounting hole • Front HO2S older (aged) than the rear HO2S (HO2S is lazy) • Front HO2S or rear HO2S is contaminated with fuel or moisture
DTC: P0430 **2T OBD/CAT3, MIL: Yes** **Years:** 2007, 2008 **Models:** B-Series Truck, CX-9 **Engines:** All **Transmissions:** All	**Catalyst Efficiency Below Normal (Bank 2)** Vehicle driven at a speed of 16–64 MPH at 1090–3090 RPM with the calculated engine load from 16–55 percent for 2–3 minutes, and the PCM detected the switch rate of the rear HO2S was close to the switch rate of the front HO2S-11 during the Catalyst Monitor test. **Possible Causes:** • Air leaks at the exhaust manifold or in the exhaust pipes • Catalytic converter is contaminated, damaged or has failed • Front HO2S and/or the rear HO2S is loose in the mounting hole • Front HO2S older (aged) than the rear HO2S (HO2S is lazy) • Front HO2S or rear HO2S is contaminated with fuel or moisture

DTC	Trouble Code Title, Conditions & Possible Causes
DTC: P0431 **2T OBD/CAT4, MIL: Yes** **Years:** 2007, 2008 **Models:** B-Series Truck, Mazda6, Mazda6 Sport Wagon, Tribute **Engines:** 2.3L VIN 2 **Transmissions:** All	**Warm-up Catalyst Efficiency Below Normal (Bank 2)** Vehicle driven at a speed of 22–75 MPH at 1250–2094 RPM with the calculated engine load from 20–50 percent for 2–3 minutes, and the PCM detected the inversion ratio of the right bank rear HO2S and front HO2S was less than a stored threshold in the Catalyst Monitor test. **Possible Causes:** • Air leaks at the exhaust manifold or in the exhaust pipes • Catalytic converter is contaminated, damaged or has failed • Front HO2S and/or the rear HO2S is loose in the mounting hole • Front HO2S older (aged) than the rear HO2S (HO2S is lazy) • Front HO2S or rear HO2S is contaminated with fuel or moisture
DTC: P0440 **2T CCM, MIL: Yes** **Years:** 2007, 2008 **Models:** B-Series Truck, Tribute **Engines:** All **Transmissions:** All	**EVAP System Malfunction** DTC P0443 not set, ECT sensor signal from 14–90°F and IAT sensor signal less than 14°F, engine running in closed loop at a cruise speed over 15 MPH, then with the purge solenoid commanded "on", the PCM detected a fault in the EVAP Monitor purge flow test. **Possible Causes:** • EVAP purge solenoid is damaged or has failed • EVAP charcoal canister is loaded with fuel or moisture • Vapor line between the purge solenoid and the intake manifold reservoir, or the vapor line between the EVAP purge solenoid, check valve and canister is leaking, restricted or disconnected • PCM has failed
DTC: P0441 **2T CCM, MIL: Yes** **Years:** 2007, 2008 **Models:** B-Series Truck, CX-7, Mazda3, MazdaSpeed3, Mazda5, Mazda6, MazdaSpeed6, MX-5, RX-8, Tribute **Engines:** All **Transmissions:** All	**EVAP System Malfunction** DTC P0443 not set, ECT sensor signal from 14–90°F and IAT sensor signal less than 14°F, engine running in closed loop at a cruise speed over 15 MPH, then with the purge solenoid commanded "on", the PCM detected the fuel control "feedback" signal was less than a value stored in memory with "purge" enabled during the EVAP test. **Possible Causes:** • EVAP purge solenoid is damaged or has failed • EVAP charcoal canister is loaded with fuel or moisture • Vapor line is damaged or restricted between the purge solenoid and intake manifold, or between check valve and vapor valve • PCM has failed
DTC: P0442 **2T CCM, MIL: Yes** **Years:** 2007, 2008 **Models:** B-Series Truck, CX-7, CX-9, Mazda3, MazdaSpeed3, Mazda5, Mazda6, MazdaSpeed6, MX-5, RX-8, Tribute **Engines:** All **Transmissions:** All	**EVAP System Small Leak (0.040") Detected** DTC P0443 not set, ECT sensor signal from 14–90°F, IAT sensor signal more than 14°F at startup, BARO sensor signal over 72 kPa, IAT sensor signal from 14–140°F and ECT sensor signal from 158–212°F during the EVAP leak test, fuel level from 15–85 percent, vehicle driven at 24–65 MPH at an engine speed of 1000–4000 RPM, engine load from 9–65 percent with the throttle opening from 3.1–31.6 percent, then with the purge and CDCV valves both closed, the PCM detected a leak as small as 0.040" in the system during the EVAP Leak Monitor test. **Note: The fuel tank target pressure for this test is plus (+) 127 kPa.** **Possible Causes:** • Fuel filler cap loose, cross-threaded, incorrect part or damaged • Purge solenoid valve is damaged or has failed • Charcoal canister is loaded with fuel or moisture • Canister drain cut valve (CDCV) is damaged or has failed • Rollover valve, catch tank valve or fuel tank damaged/ leaking • Fuel tank pressure sensor is damaged or has failed • ECT, IAT, MAF, VSS or TP sensor signals out-of-calibration • Vapor line(s) damaged or leaking between the purge solenoid and the intake manifold, EVAP purge solenoid valve and the canister, canister drain cut valve and the rollover valve • PCM has failed
DTC: P0443 **1T CCM, MIL: Yes** **Years:** 2007, 2008 **Models:** B-Series Truck, CX-7, CX-9, Mazda3, MazdaSpeed3, Mazda5, Mazda6, MazdaSpeed6, MX-5, RX-8, Tribute **Engines:** All **Transmissions:** All	**EVAP Canister Purge Solenoid Circuit Malfunction** Key on or engine running and the PCM detected an unexpected voltage condition on the EVAP Purge solenoid circuit during the test. **Note: This is a diagnostic support code only – not stored in the PCM.** **Possible Causes:** • Purge solenoid control circuit is open or shorted to ground • Purge solenoid control circuit is shorted to system power (B+) • Purge solenoid power circuit is open or shorted to ground • Purge solenoid is damaged or has failed • PCM has failed (the purge solenoid driver may be damaged)

DTC	Trouble Code Title, Conditions & Possible Causes
DTC: P0444 **1T CCM, MIL: Yes** **Years:** 2007, 2008 **Models:** B-Series Truck **Engines:** All **Transmissions:** All	**EVAP Purge Solenoid Circuit Malfunction** Key on or engine running and the PCM detected an unexpected voltage condition on the EVAP purge solenoid circuit during the test. **Note: This is a diagnostic support code only – not stored in the PCM!** **Possible Causes:** • VMV solenoid control circuit is open or shorted to ground • VMV solenoid control circuit is shorted to system power (B+) • VMV solenoid power circuit is open or shorted to ground • VMV solenoid is damaged or has failed • PCM has failed (the VMV solenoid driver may be damaged)
DTC: P0445 **1T CCM, MIL: Yes** **Years:** 2007, 2008 **Models:** B-Series Truck **Engines:** All **Transmissions:** All	**EVAP Purge Solenoid Circuit Malfunction** Key on or engine running and the PCM detected an unexpected voltage condition on the EVAP purge solenoid circuit during the test. **Note: This is a diagnostic support code only – not stored in the PCM!** **Possible Causes:** • VMV solenoid control circuit is open or shorted to ground • VMV solenoid control circuit is shorted to system power (B+) • VMV solenoid power circuit is open or shorted to ground • VMV solenoid is damaged or has failed • PCM has failed (the VMV solenoid driver may be damaged)
DTC: P0446 **1T CCM, MIL: Yes** **Years:** 2007, 2008 **Models:** B-Series Truck, CX-7, CX-9, Mazda3, MazdaSpeed3, Mazda5, Mazda6, MazdaSpeed6, MX-5, RX-8, Tribute **Engines:** All **Transmissions:** All	**EVAP System Vent Control Malfunction** DTC P0443 not set, ECT sensor signal from 14–90°F and IAT sensor signal more than 14°F at startup, engine speed at 1000–3000 RPM, VSS input from 25–63 MPH, calculated load at 9–70 percent, TP angle from 3–44 percent, then with the CDCV closed and then reopened, the PCM detected the change in fuel tank pressure was too small. **Possible Causes:** • Canister drain cut valve (CDCV) is damaged, sticking or failed • Tank pressure control valve (TPCV) is damaged or has failed • Charcoal canister is loaded with fuel or moisture • Air filter is severely restricted, or 2-way check valve is clogged • FTP sensor signal circuit is open, shorted to ground or to power • FTP sensor power circuit is open or shorted to ground • FTP sensor is damaged or has failed • BARO, ECT, IAT, MAF, VSS or TP sensor is out-of-calibration • Fuel tank level sensor is damaged or out-of-calibration • Fuel vapor line(s) kinked or blocked between the CDVC valve and intake manifold, or between the TPCV and the canister
DTC: P0450 **2T CCM, MIL: Yes** **Years:** 2007, 2008 **Models:** B-Series Truck **Engines:** All **Transmissions:** All	**EVAP Pressure Sensor Circuit Malfunction** Engine running, ECT sensor signal less than 176°F, and the PCM detected the Fuel Tank Pressure (FTP) sensor signal was less than 0.20v, or that it was more than 4.80v during the CCM test. **Possible Causes:** • FTP sensor signal circuit is open (4.80v reading) • FTP sensor signal circuit is shorted to ground (0.20v reading) • FTP sensor power circuit is open between sensor and the PCM • FTP sensor ground circuit is open between sensor and ground • FTP sensor signal circuit is shorted to VREF or system power • FTP sensor is damaged or has failed • PCM has failed
DTC: P0451 **2T CCM, MIL: Yes** **Years:** 2007, 2008 **Models:** B-Series Truck, CX-9, Tribute **Engines:** All **Transmissions:** All	**EVAP Pressure Sensor Circuit Range/Performance** Engine running, and the PCM detected the Fuel Tank Pressure (FTP) sensor changed more than 14" H2O within a 10 second period (indicating the FTP sensor circuit is noisy). **Possible Causes:** • FTP sensor signal circuit is open (intermittent fault) • FTP sensor signal circuit is shorted to ground (intermittent fault) • FTP sensor is damaged or has failed • PCM has failed

DTC	Trouble Code Title, Conditions & Possible Causes
DTC: P0452 **2T CCM, MIL: Yes** **Years:** 2007, 2008 **Models:** B-Series Truck, CX-9, Tribute **Engines:** All **Transmissions:** All	**EVAP Pressure Sensor Circuit Low Input** Engine running and the PCM detected the Fuel Tank Pressure (FTP) sensor signal indicated less than 0.22v during the CCM test. **Possible Causes:** • FTP sensor signal circuit is shorted to sensor ground • FTP sensor signal circuit is shorted to chassis ground • FTP sensor signal is shorted in the connector due to moisture • FTP sensor is damaged or has failed • PCM has failed
DTC: P0453 **2T CCM, MIL: Yes** **Years:** 2007, 2008 **Models:** B-Series Truck, CX-9, Tribute **Engines:** All **Transmissions:** All	**EVAP Pressure Sensor Circuit High Input** Engine running and the PCM detected the Fuel Tank Pressure (FTP) sensor signal indicated less than 0.22v during the CCM test. **Possible Causes:** • FTP sensor signal circuit is shorted to VREF or power (B+) • FTP sensor ground circuit is open • FTP sensor is damaged or has failed • PCM has failed
DTC: P0454 **2T CCM, MIL: Yes** **Years:** 2007, 2008 **Models:** B-Series Truck, CX-9, Tribute **Engines:** All **Transmissions:** All	**EVAP Pressure Sensor Circuit Range/Performance** Engine running and the PCM detected the Fuel Tank Pressure (FTP) sensor signal indicated less than 0.22v during the CCM test. **Possible Causes:** • FTP sensor signal circuit is open (intermittent fault) • FTP sensor signal circuit is shorted to ground (intermittent fault) • FTP sensor is damaged or has failed • PCM has failed
DTC: P0455 **2T CCM, MIL: Yes** **Years:** 2007, 2008 **Models:** B-Series Truck, CX-7, CX-9, Mazda3, MazdaSpeed3, Mazda5, Mazda6, MazdaSpeed6, MX-5, RX-8, Tribute **Engines:** All **Transmissions:** All	**EVAP System Large Leak (0.080") Detected** DTC P0443 not set, ECT sensor signal from 14–90°F, IAT sensor signal more than 14°F at startup, BARO sensor more than 72 kPa, fuel level from 15–85 percent, ECT sensor signal from 158–212°F and IAT sensor signal from 14–131°F during testing, vehicle driven at 24–65 MPH at an engine speed of 1000–4000 RPM, calculated engine load of 9–65 percent and a throttle angle from 3–31.6 percent, then with the CDCV (valve) closed, the PCM detected the fuel tank pressure was less than a threshold due to a blockage or a large leak (0.080") somewhere in the system during the EVAP Monitor test. **Note: The target fuel tank pressure for this test is minus (-) 0.99 kPa.** **Possible Causes:** • Fuel filler cap loose, cross-threaded, incorrect part or damaged • Purge solenoid valve is damaged or has failed • Charcoal canister is loaded with fuel or moisture, or has failed • Canister drain cut valve (CDCV) is damaged or leaking • Fuel tank pressure (FTP) sensor is damaged or has failed • Catch tank or rollover valve is damaged, or fuel tank is leaking • Vapor line(s) damaged or leaking between the purge solenoid and intake manifold, purge solenoid and canister, or between the pressure control valve, check valve and rollover valve • PCM has failed
DTC: P0456 **2T CCM, MIL: Yes** **Years:** 2007, 2008 **Models:** B-Series Truck, CX-7, CX-9, Mazda3, MazdaSpeed3, Mazda5, Mazda6, MazdaSpeed6, MX-5, RX-8, Tribute **Engines:** All **Transmissions:** All	**EVAP System Small Leak (0.020") Detected** DTC P0443 not set, ECT sensor signal from 14–90°F, IAT sensor signal more than 14°F at startup, engine speed at 1000–3000 RPM, VSS input from 25–63 MPH, calculated load at 9–70 percent, fuel level over 75 percent, TP angle from 3–44 percent, and the PCM detected less than 2.5" H2O bleed-up over a 15 second period during the EVAP leak test. **Note: The vapor generation limit is over 2.5" H2O in 120 seconds.** **Possible Causes:** • Fuel filler cap loose, cross-threaded, incorrect part or damaged • Small holes or cuts in any of the vapor hoses and/or tubes • CV solenoid stuck part-way open while it is commanded closed • EVAP system component seals leaking at the purge valve, FTP sensor, CV solenoid, fuel vapor control valve or vapor vent tube • Fuel vapor hose or tube connections loose at the component • PCM has failed
DTC: P0457 **2T CCM, MIL: Yes** **Years:** 2007, 2008 **Models:** B-Series Truck, CX-9, Tribute **Engines:** All **Transmissions:** All	**EVAP System Gross Leak Detected** Engine running, and immediately after a vehicle "refueling" event, the PCM detected it could not achieve any initial vacuum in the EVAP system with excessive vapor flow present (Gross EVAP leak). **Possible Causes:** • Fuel filler cap does not fit properly (it is the wrong part number) • Fuel filler cap is missing

DTC	Trouble Code Title, Conditions & Possible Causes
DTC: P0460 **2T CCM, MIL: Yes** **Years:** 2007, 2008 **Models:** B-Series Truck, CX-9, Tribute **Engines:** All **Transmissions:** All	**Fuel Level Indicator Signal Circuit Malfunction** Engine running, and the PCM detected the Fuel Level Indicator (FLI) signal indicated less than 0.1v, or that it indicated more than 3.48v during the CCM test period. **Possible Causes:** • FLI signal circuit is open, shorted to ground or to power (B+) • Fuel tank empty or overfull (FP module is stuck mechanically) • Wrong fuel gauge is installed, or instrument panel is damaged • Fuel gauge sender unit is damaged or has failed • PCM has failed
DTC: P0461 **2T CCM, MIL: Yes** **Years:** 2007, 2008 **Models:** B-Series Truck, CX-7, CX-9, Mazda3, MazdaSpeed3, Mazda5, Mazda6, MazdaSpeed6, MX-5, RX-8, Tribute **Engines:** All **Transmissions:** All	**Fuel Tank Level Sensor Circuit Range/Performance** Engine running and the PCM determined the fuel gauge sender unit signal was operating in too narrow a range during the CCM test. **Possible Causes:** • Fuel gauge sending unit is damaged or has failed • Fuel gauge sending unit signal is open or shorted to ground • Fuel gauge sending unit power circuit is open • Instrument cluster is damaged or has failed • PCM has failed
DTC: P0462 **2T CCM, MIL: Yes** **Years:** 2007, 2008 **Models:** B-Series Truck, CX-7, CX-9, Mazda3, MazdaSpeed3, Mazda5, Mazda6, MazdaSpeed6, MX-5, RX-8, Tribute **Engines:** All **Transmissions:** All	**Fuel Level Sensor Circuit Low Input** Engine running, system voltage from 11–16v, and the PCM detected the fuel level sensor signal was less than 0.08v during the CCM test. **Possible Causes:** • Fuel gauge sending unit signal is shorted to sensor ground • Fuel gauge sending unit signal is shorted to chassis ground • Fuel gauge sending unit is damaged or has failed • PCM has failed
DTC: P0463 **2T CCM, MIL: Yes** **Years:** 2007, 2008 **Models:** B-Series Truck, CX-7, CX-9, Mazda3, MazdaSpeed3, Mazda5, Mazda6, MazdaSpeed6, MX-5, RX-8, Tribute **Engines:** All **Transmissions:** All	**Fuel Level Sensor Circuit High Input** Engine running, system voltage from 11–16v, and the PCM detected the fuel level sensor signal was more than 4.92v during the CCM test. **Possible Causes:** • Fuel gauge sending unit signal shorted to VREF or power • Fuel gauge sending unit ground circuit is open • Fuel gauge sender signal is damaged or has failed • PCM has failed
DTC: P0464 **2T CCM, MIL: Yes** **Years:** 2007, 2008 **Models:** B-Series Truck, Tribute **Engines:** All **Transmissions:** All	**Fuel Tank Level Sensor Circuit Range/Performance** Engine running and the PCM determined the fuel gauge sender unit signal was operating in too narrow a range during the CCM test. **Possible Causes:** • Fuel gauge sending unit is damaged or has failed • Fuel gauge sending unit signal is open or shorted to ground • Fuel gauge sending unit power circuit is open • Instrument cluster is damaged or has failed • PCM has failed
DTC: P0480 **2T CCM, MIL: No** **Years:** 2007, 2008 **Models:** CX-7, CX-9, Mazda3, MazdaSpeed3, Mazda5, Mazda6, MazdaSpeed6, MX-5, RX-8, Tribute **Engines:** All **Transmissions:** All	**Condenser Fan Relay Control Circuit Malfunction** Key on or engine running and the PCM detected an unexpected high or low voltage condition on the Condenser Fan Relay control circuit during the CCM test. **Possible Causes:** • Condenser fan relay control circuit open or shorted to ground • Condenser fan relay control power circuit is open • Condenser fan relay is damaged or has failed • PCM has failed
DTC: P0481 **2T CCM, MIL: No** **Years:** 2007, 2008 **Models:** CX-7, Mazda6, MazdaSpeed6, MX-5, RX-8, Tribute **Engines:** All **Engines:** All **Transmissions:** All	**Cooling Fan Relay 1 and Condenser Fan Relay 1 Control Circuit Malfunction** Key on, and the PCM detected an unexpected high or low voltage on the Cooling Fan Relay 1 or Condenser Fan Relay 1 control circuit. **Possible Causes:** • Cooling or condenser fan relay 1 circuit open, shorted to ground • Cooling or condenser fan relay 1 control power circuit is open • Cooling or condenser fan relay 1 is damaged or has failed • PCM has failed

DTC	Trouble Code Title, Conditions & Possible Causes
DTC: P0482 **2T CCM, MIL: No** **Years:** 2007, 2008 **Models:** Mazda6, MazdaSpeed6, MX-5, Tribute **Engines:** All **Transmissions:** All	**Cooling Fan Relay and Condenser Fan Relay 2 Control Circuit Malfunction** Key on, and the PCM detected an unexpected high or low voltage on the Cooling Fan Relay or Condenser Fan Relay 2 control circuit. **Possible Causes:** • Cooling or condenser fan relay 2 circuit open, shorted to ground • Cooling or condenser fan relay 2 control power circuit is open • Cooling or condenser fan relay 1 is damaged or has failed • PCM has failed
DTC: P0483 **2T CCM, MIL: No** **Years:** 2008 **Models:** Tribute **Engines:** All **Transmissions:** All	**Cooling Fan Performance** Key on, and the PCM detected a concern with the cooling fan system. **Possible Causes:** • Cooling or condenser fan relay circuit open, shorted to ground • Cooling or condenser fan relay control power circuit is open • Cooling or condenser fan relay is damaged or has failed • PCM has failed
DTC: P0484 **2T CCM, MIL: No** **Years:** 2008 **Models:** Tribute **Engines:** All **Transmissions:** All	**Cooling Fan Control Primary Circuit Failure** Key on, and the PCM detected a concern with the cooling fan system. **Possible Causes:** • Cooling or condenser fan relay circuit open, shorted to ground • Cooling or condenser fan relay control power circuit is open • Cooling or condenser fan relay is damaged or has failed • PCM has failed
DTC: P0485 **2T CCM, MIL: No** **Years:** 2008 **Models:** Tribute **Engines:** All **Transmissions:** All	**Cooling Fan Control Primary Circuit Failure** Key on, and the PCM detected a concern with the cooling fan system. **Possible Causes:** • Cooling or condenser fan relay circuit open, shorted to ground • Cooling or condenser fan relay control power circuit is open • Cooling or condenser fan relay is damaged or has failed • PCM has failed
DTC: P0489 **2T CCM, MIL: No** **Years:** 2007, 2008 **Models:** B-Series Truck Tribute **Engines:** All **Transmissions:** All	**Exhaust Gas Recirculation Malfunction** DPFEGR sensor pressure electrical concern. **Possible Causes:** • EGR valve stuck open or closed • Harness open or shorted to power or ground • No power to EGR • PCM has failed
DTC: P0490 **2T CCM, MIL: No** **Years:** 2007, 2008 **Models:** B-Series Truck Tribute **Engines:** All **Transmissions:** All	**Exhaust Gas Recirculation Malfunction** DPFEGR sensor pressure electrical concern. **Possible Causes:** • EGR valve stuck open or closed • Harness open or shorted to power or ground • No power to EGR • PCM has failed
DTC: P0496 **1T CCM, MIL: Yes** **Years:** 2007, 2008 **Models:** B-Series Truck **Engines:** All **Transmissions:** All	**EVAP Canister Purge Solenoid Circuit Malfunction** Key on or engine running and the PCM detected an unexpected voltage condition on the EVAP Purge solenoid circuit during the test. **Note: This is a diagnostic support code only – not stored in the PCM.** **Possible Causes:** • Purge solenoid control circuit is open or shorted to ground • Purge solenoid control circuit is shorted to system power (B+) • Purge solenoid power circuit is open or shorted to ground • Purge solenoid is damaged or has failed • PCM has failed (the purge solenoid driver may be damaged)

DTC	Trouble Code Title, Conditions & Possible Causes
DTC: P0497 **1T CCM, MIL: Yes** **Years:** 2007, 2008 **Models:** B-Series Truck **Engines:** All **Transmissions:** All	**EVAP Canister Purge Solenoid Circuit Malfunction** Key on or engine running and the PCM detected an unexpected voltage condition on the EVAP Purge solenoid circuit during the test. **Note: This is a diagnostic support code only – not stored in the PCM.** **Possible Causes:** • Purge solenoid control circuit is open or shorted to ground • Purge solenoid control circuit is shorted to system power (B+) • Purge solenoid power circuit is open or shorted to ground • Purge solenoid is damaged or has failed • PCM has failed (the purge solenoid driver may be damaged)
DTC: P0498 **1T CCM, MIL: Yes** **Years:** 2007, 2008 **Models:** B-Series Truck, Tribute **Engines:** All **Transmissions:** All	**EVAP Canister Vent Solenoid Malfunction** This code sets when there is less than 2.5" H2O bleed-up over a 15 second period with the fuel tank level more than 75 percent full. The bleed-up and evaluation time vary as a function of fuel level. Vapor generation is more than 2.5" H2O over a 120-second period of time. **Possible Causes:** • Fuel filler cap loose, cross-threaded, incorrect part or damaged • EVAP purge solenoid valve is damaged or has failed • EVAP vent control solenoid is damaged or has failed • EVAP charcoal canister is loaded with fuel or moisture • Fuel tank pressure sensor damaged or has failed • Vapor line(s) damaged or leaking between the purge solenoid and the intake manifold, purge solenoid valve and the canister, fuel vapor control valve tube or fuel vapor vent valve assembly • PCM has failed
DTC: P0499 **1T CCM, MIL: Yes** **Years:** 2007, 2008 **Models:** B-Series Truck, Tribute **Engines:** All **Transmissions:** All	**EVAP Canister Vent Solenoid Malfunction** This code sets when there is less than 2.5" H2O bleed-up over a 15 second period with the fuel tank level more than 75 percent full. The bleed-up and evaluation time vary as a function of fuel level. Vapor generation is more than 2.5" H2O over a 120-second period of time. **Possible Causes:** • Fuel filler cap loose, cross-threaded, incorrect part or damaged • EVAP purge solenoid valve is damaged or has failed • EVAP vent control solenoid is damaged or has failed • EVAP charcoal canister is loaded with fuel or moisture • Fuel tank pressure sensor damaged or has failed • Vapor line(s) damaged or leaking between the purge solenoid and the intake manifold, purge solenoid valve and the canister, fuel vapor control valve tube or fuel vapor vent valve assembly • PCM has failed
DTC: P0500 **2T CCM, MIL: Yes** **Years:** 2007, 2008 **Models:** CX-7, Mazda3, MazdaSpeed3, Mazda5, Mazda6, MazdaSpeed6, MX-5, RX-8, Tribute **Engines:** All **Transmissions:** A/T	**Vehicle Speed Sensor Circuit Malfunction** Vehicle driven in Drive, 2nd or Low gear position, engine speed over RPM with the charging efficiency over 40 percent, and the PCM detected the VSS input indicated less than 2.34 MPH for 33 seconds. **Possible Causes:** • VSS signal circuit is open, shorted to ground or to power • VSS power circuit is open or shorted to ground • VSS is damaged or has failed • PCM has failed
DTC: P0501 **2T CCM, MIL: No** **Years:** 2008 **Models:** Tribute **Engines:** All **Transmissions:** A/T	**Vehicle Speed Sensor Circuit Malfunction** Vehicle driven in gear at more than 1000 RPM, ECT sensor signal more than 140°F, and the PCM detected an intermittent VSS signal during the CCM test. **Possible Causes:** • VSS positive (+) signal open or shorted to ground (intermittent) • VSS negative (-) signal open or shorted to ground (intermittent) • VSS is damaged or has failed • PCM has failed
DTC: P0502 **2T CCM, MIL: No** **Years:** 2008 **Models:** Tribute **Engines:** All **Transmissions:** A/T	**Vehicle Speed Sensor Circuit Malfunction** Vehicle driven in gear at more than 1000 RPM, ECT sensor signal more than 140°F, and the PCM detected an intermittent VSS signal during the CCM test. **Possible Causes:** • VSS positive (+) signal open or shorted to ground (intermittent) • VSS negative (-) signal open or shorted to ground (intermittent) • VSS is damaged or has failed • PCM has failed

DTC	Trouble Code Title, Conditions & Possible Causes
DTC: P0503 **2T CCM, MIL: Yes** **Years:** 2008 **Models:** Tribute **Engines:** All **Transmissions:** A/T	**Vehicle Speed Sensor Circuit Malfunction** Vehicle driven in gear at more than 1000 RPM, ECT sensor signal more than 140°F, and the PCM detected a poor or noisy VSS signal during the CCM test. **Possible Causes:** • Check for any devices that could cause RFI on the VSS circuit • VSS positive (+) signal open or shorted to ground (intermittent) • VSS negative (-) signal open or shorted to ground (intermittent) • VSS is damaged or has failed • PCM has failed
DTC: P0505 **2T CCM, MIL: Yes** **Years:** 2007, 2008 **Models:** B-Series Truck, CX-7, CX-9, Mazda3, MazdaSpeed3, Mazda5, Mazda6, MazdaSpeed6, MX-5, RX-8, Tribute **Engines:** All **Transmissions:** All	**Idle Speed Control System Malfunction** Key on or engine running, system voltage over 11 volts, and the PCM detected an unexpected voltage condition on the IAC valve control circuit, condition met for 14 seconds, or the PCM could not control the IAC valve in order to reach the desired engine speed. **Possible Causes:** • Throttle body is damaged, dirty or idle speed out of adjustment • High resistance between PCM and IAC valve control circuits • IAC valve circuits open, shorted to ground or shorted to power • IAC valve air inlet is plugged, dirty or restricted • IAC valve is damaged or has failed
DTC: P0506 **2T CCM, MIL: Yes** **Years:** 2007, 2008 **Models:** B-Series Truck, CX-7, CX-9, Mazda3, MazdaSpeed3, Mazda5, Mazda6, MazdaSpeed6, MX-5, RX-8, Tribute **Engines:** All **Transmissions:** All	**Idle Control System RPM Lower Than Expected** Engine running at idle, and the PCM detected the Actual idle speed was more than 100 RPM lower than the Target idle speed with the brake depressed and steering wheel straight ahead for 14 seconds. **Note: If the atmospheric pressure (BARO reading) is less than 72 kPa, or the IAT sensor signal is less than 14°F, the test is cancelled.** **Possible Causes:** • Air intake system leaks (in the intake manifold or PCV valve) • IAC valve power circuit is open between main relay and valve • High resistance between PCM and IAC valve control circuits • IAC valve circuits open, shorted to ground or shorted to power • IAC valve air inlet is plugged, dirty or restricted • IAC valve is damaged or has failed • Throttle body is damaged or dirty (it may need to be cleaned)
DTC: P0507 **2T CCM, MIL: Yes** **Years:** 2007, 2008 **Models:** B-Series Truck, CX-7, CX-9, Mazda3, MazdaSpeed3, Mazda5, Mazda6, MazdaSpeed6, MX-5, RX-8, Tribute **Engines:** All **Transmissions:** All	**Idle Control System RPM Higher Than Expected** Engine running at idle, and the PCM detected the Actual idle speed was more than 200 RPM higher than the Target idle speed with the brake depressed and steering wheel straight ahead for 14 seconds. **Note: If the atmospheric pressure (BARO reading) is less than 72 kPa, or the IAT sensor signal is less than 14°F, the test is cancelled.** **Possible Causes:** • Air intake system leaks (in the intake manifold or PCV valve) • IAC valve power circuit is open between main relay and valve • High resistance between PCM and IAC valve control circuits • IAC valve circuits open, shorted to ground or shorted to power • IAC valve air inlet is plugged, dirty or restricted • IAC valve is damaged or has failed • Throttle body is damaged or dirty (it may need to be cleaned)
DTC: P050A **2T CCM, MIL: Yes** **Years:** 2007, 2008 **Models:** CX-7, Mazda3, MazdaSpeed3, Mazda5, Mazda6, MazdaSpeed6, MX-5, RX-8 **Engines:** All **Transmissions:** All	**Cold Start Idle Air Control System Performance** Actual idle speed is lower than expected by 100 RPM for 8.4 seconds when target idle speed is above 0 RPM at cold start or ignition retard value is above 8.5°CA. **Note: If the atmospheric pressure (BARO reading) is less than 72 kPa, or the IAT sensor signal is less than 14°F, the test is cancelled.** **Possible Causes:** • The air into the fuel line by exchange or re-installation • Electronic throttle control system malfunction • Throttle valve stuck or blockage • Air suction in intake air system • PCM malfunction

DTC	Trouble Code Title, Conditions & Possible Causes
DTC: P050B **2T CCM, MIL: Yes** **Years:** 2007, 2008 **Models:** CX-7, Mazda3, MazdaSpeed3, Mazda5, Mazda6, MazdaSpeed6, MX-5 **Engines:** All **Transmissions:** All	**Cold Start Ignition Timing Performance** The PCM monitors actual ignition timing using the CKP sensor while electronic spark advance control fast idle correction operating. If the ignition timing is out of specified range, the PCM determines that the ignition timing at cold condition has performance problem. **Possible Causes:** • The air into the fuel line by exchange or re-installation • Electronic throttle control system malfunction • Throttle valve stuck or blockage • Air suction in intake air system • PCM malfunction
DTC: P050E **2T CCM, MIL: Yes** **Years:** 2007, 2008 **Models:** CX-9 **Engines:** All **Transmissions:** All	**Cold Start Engine Exhaust Temperature Out of Range** The PCM calculates the actual catalyst warm up temperature during a cold start. The PCM then compares the actual temperature to the expected catalyst temperature model. The difference between the actual and expected temperatures is a ratio. When this ratio exceeds the calibrated value this DTC is set and the malfunction indicator lamp (MIL) illuminates. **Possible Causes:** • Intake air restriction • Exhaust restriction • Mechanical concern with the engine • Throttle body malfunction • Vacuum leakage • PCM malfunction
DTC: P0510 **2T CCM, MIL: Yes** **Years:** 2007, 2008 **Models:** B-Series Truck, Tribute **Engines:** All **Transmissions:** All	**Closed Throttle Position Switch Circuit Malfunction** Engine running at idle speed, and the PCM detected the CTP switch signal was less than 2.0v for over 33 seconds, or the CTP switch signal was more than 8.0v for 33 seconds. **Note: This test is stopped if the TP sensor signal is less than 0.9v.** **Possible Causes:** • Closed throttle switch signal circuit is open or shorted to ground • Closed throttle switch signal circuit is shorted to system power • Closed throttle switch is damaged or has failed • PCM has failed
DTC: P0511 **2T CCM, MIL: Yes** **Years:** 2007, 2008 **Models:** B-Series Truck, Tribute **Engines:** All **Transmissions:** All	**PCM - Idle Air Control (IAC) Valve Assembly** PCM detects electrical load failure on IAC output circuit **Possible Causes:** • IAC output circuit is open or shorted to ground • IAC output circuit is shorted to system power • IAC valve assembly is damaged or has failed • PCM has failed
DTC: P0512 **2T CCM, MIL: Yes** **Years:** 2008 **Models:** Tribute **Engines:** All **Transmissions:** All	**Starter Request Circuit** Indicates one touch integrated starting system voltage circuit to the starter relay has a short to voltage. **Possible Causes:** • Battery • Starter motor
DTC: P0518 **2T CCM, MIL: Yes** **Years:** 2007, 2008 **Models:** B-Series Truck **Engines:** All **Transmissions:** All	**PCM - Idle Air Control (IAC) Valve Assembly** PCM detects an intermittent electrical failure on IAC output circuit. **Possible Causes:** • IAC output circuit is open or shorted to ground • IAC output circuit is shorted to system power • IAC valve assembly is damaged or has failed • PCM has failed
DTC: P0519 **2T CCM, MIL: Yes** **Years:** 2007, 2008 **Models:** B-Series Truck **Engines:** All **Transmissions:** All	**PCM - Idle Air Control (IAC) Valve Assembly** IAC operating out of range. **Possible Causes:** • IAC output circuit is open or shorted to ground • IAC output circuit is shorted to system power • IAC valve assembly is damaged or has failed • PCM has failed

DTC	Trouble Code Title, Conditions & Possible Causes
DTC: P0525 **2T CCM, MIL: Yes** **Years:** 2007, 2008 **Models:** B-Series Truck, Tribute **Engines:** All **Transmissions:** All	**Cruise Control Servo Performance** Cruise control servo operating out of expected range. **Possible Causes:** • Cruise control switch • Cruise control servo • Circuitry • PCM has failed
DTC: P0530 **2T CCM, MIL: Yes** **Years:** 2007, 2008 **Models:** B-Series Truck **Engines:** All **Transmissions:** All	**PCM - Air Conditioning Pressure (ACP) Sensor Performance** ACP sensor is operating out of expected range. **Possible Causes:** • ACP sensor is open or shorted to ground • ACP sensor is shorted to system power • ACP sensor is damaged or has failed • PCM has failed
DTC: P0531 **2T CCM, MIL: Yes** **Years:** 2007, 2008 **Models:** B-Series Truck **Engines:** All **Transmissions:** All	**PCM - Air Conditioning Pressure (ACP) Sensor Performance** ACP sensor is operating out of expected range. **Possible Causes:** • ACP sensor is open or shorted to ground • ACP sensor is shorted to system power • ACP sensor is damaged or has failed • PCM has failed
DTC: P0532 **2T CCM, MIL: Yes** **Years:** 2007, 2008 **Models:** B-Series Truck **Engines:** All **Transmissions:** All	**PCM - Air Conditioning Pressure (ACP) Sensor** ACP sensor voltage is above calibrated limit. **Possible Causes:** • ACP sensor is open or shorted to ground • ACP sensor is shorted to system power • ACP sensor is damaged or has failed • PCM has failed
DTC: P0533 **2T CCM, MIL: Yes** **Years:** 2007, 2008 **Models:** B-Series Truck **Engines:** All **Transmissions:** All	**PCM - Air Conditioning Pressure (ACP) sensor** ACP sensor voltage is below calibrated limit. **Possible Causes:** • ACP sensor is open or shorted to ground • ACP sensor is shorted to system power • ACP sensor is damaged or has failed • PCM has failed
DTC: P0534 **2T CCM, MIL: Yes** **Years:** 2007, 2008 **Models:** B-Series Truck **Engines:** All **Transmissions:** All	**PCM - Air Conditioning (A/C) system** A/C compressor clutch cycling too frequently. **Possible Causes:** • A/C system is low on Freon • A/C system has too much Freon • A/C compressor clutch is damaged or failed • PCM has failed
DTC: P053A **2T CCM, MIL: Yes** **Years:** 2007, 2008 **Models:** CX-9 **Engines:** All **Transmissions:** All	**PCV Valve Heater Control Circuit Performance** The PCM detects a PCV valve heater circuit malfunction. **Possible Causes:** • Open circuit between PCV valve heater control terminal A and main relay No.1 • Open circuit between PCV valve heater fitting terminal A and main relay No.1 • Short to ground circuit between PCV valve heater control terminal B and PCM terminal 2BL • Short to ground circuit between PCV valve heater fitting terminal B and PCM terminal 2AU • Short to power circuit between PCV valve heater control terminal B and PCM terminal 2BL • Short to power circuit between PCV valve heater fitting terminal B and PCM terminal 2AU • Open circuit between PCV valve heater control terminal B and PCM terminal 2BL • Open circuit between PCV valve heater fitting B and PCM terminal 2AU • Connector or terminal malfunction • Damaged PCV heater assembly

DTC	Trouble Code Title, Conditions & Possible Causes
DTC: P0550 **2T CCM, MIL: Yes** **Years:** 2007, 2008 **Models:** B-Series Truck, CX-7, Mazda6, Mazda6 Sport Wagon, MazdaSpeed6, MX-5 **Engines:** All **Transmissions:** All	**Power Steering Pressure Switch Circuit Malfunction** Engine running, ECT sensor signal more than 14°F, VSS more than 37.4 MPH, and the PCM detected the PSP switch signal low signal (switch "on" signal), condition met for over one minute. **Note: This is a normally open (N.O.) type of pressure switch.** **Possible Causes:** • Power steering switch signal circuit shorted to sensor ground • Power steering switch signal circuit shorted to chassis ground • Power steering switch is damaged or has failed • Engine Speed System (power steering function) is damaged • PCM has failed
DTC: P0551 **2T CCM, MIL: Yes** **Years:** 2007, 2008 **Models:** B-Series Truck **Engines:** All **Transmissions:** All	**Power Steering Pressure Switch Circuit Malfunction** Engine running, ECT sensor signal more than 14°F, VSS more than 37.4 MPH, and the PCM detected the PSP switch signal low signal (switch "on" signal), condition met for over one minute. **Note: This is a normally open (N.O.) type of pressure switch.** **Possible Causes:** • Power steering switch signal circuit shorted to sensor ground • Power steering switch signal circuit shorted to chassis ground • Power steering switch is damaged or has failed • Engine Speed System (power steering function) is damaged • PCM has failed
DTC: P0562 **2T CCM, MIL: Yes** **Years:** 2007, 2008 **Models:** RX-8 **Engines:** All **Transmissions:** All	**Cruise Control Switch Circuit Malfunction** The PCM monitors the cruise control switch signal at the PCM terminal 3P. If the PCM detects that any one of the following switches (MAIN, CANCEL, SET/COAST, RESUME/ACCEL) remains on for 2 min., the PCM determines that the cruise control switch has a malfunction. **Possible Causes:** • Cruise control switch malfunction • Connector or terminal malfunction • Short to power circuit in wiring from cruise control terminal B and PCM terminal 3P. • Short to ground circuit in wiring from cruise control terminal B and PCM terminal 3P • PCM has failed
DTC: P0564 **2T CCM, MIL: Yes** **Years:** 2007, 2008 **Models:** CX-7, Mazda3, MazdaSpeed3, Mazda5, Mazda6, MazdaSpeed6, MX-5, RX-8, Tribute **Engines:** All **Transmissions:** All	**Cruise Control Switch Circuit Malfunction** The PCM monitors the cruise control switch signal at the PCM terminal 3P. If the PCM detects that any one of the following switches (MAIN, CANCEL, SET/COAST, RESUME/ACCEL) remains on for 2 min., the PCM determines that the cruise control switch has a malfunction. **Possible Causes:** • Cruise control switch malfunction • Connector or terminal malfunction • Short to power circuit in wiring from cruise control terminal B and PCM terminal 3P. • Short to ground circuit in wiring from cruise control terminal B and PCM terminal 3P • PCM has failed
DTC: P0565 **2T CCM, MIL: Yes** **Years:** 2008 **Models:** Tribute **Engines:** All **Transmissions:** All	**Cruise Control Switch Circuit Malfunction** PCM detects a concern with the cruise control ON circuit. **Possible Causes:** • Cruise control switch malfunction • Connector or terminal malfunction • Short to power circuit in wiring from cruise control terminal B and PCM terminal 3P. • Short to ground circuit in wiring from cruise control terminal B and PCM terminal 3P • PCM has failed
DTC: P0566 **2T CCM, MIL: Yes** **Years:** 2008 **Models:** Tribute **Engines:** All **Transmissions:** All	**Cruise Control Switch Circuit Malfunction** PCM detects a concern with the cruise control OFF circuit. **Possible Causes:** • Cruise control switch malfunction • Connector or terminal malfunction • Short to power circuit in wiring from cruise control terminal B and PCM terminal 3P. • Short to ground circuit in wiring from cruise control terminal B and PCM terminal 3P • PCM has failed

DTC	Trouble Code Title, Conditions & Possible Causes
DTC: P0567 **2T CCM, MIL: Yes** **Years:** 2008 **Models:** Tribute **Engines:** All **Transmissions:** All	**Cruise Control Switch Circuit Malfunction** PCM detects a concern with the cruise control RESUME circuit. **Possible Causes:** • Cruise control switch malfunction • Connector or terminal malfunction • Short to power circuit in wiring from cruise control terminal B and PCM terminal 3P. • Short to ground circuit in wiring from cruise control terminal B and PCM terminal 3P • PCM has failed
DTC: P0568 **2T CCM, MIL: Yes** **Years:** 2008 **Models:** Tribute **Engines:** All **Transmissions:** All	**Cruise Control Switch Circuit Malfunction** PCM detects a concern with the cruise control SET circuit. **Possible Causes:** • Cruise control switch malfunction • Connector or terminal malfunction • Short to power circuit in wiring from cruise control terminal B and PCM terminal 3P. • Short to ground circuit in wiring from cruise control terminal B and PCM terminal 3P • PCM has failed
DTC: P0569 **2T CCM, MIL: Yes** **Years:** 2008 **Models:** Tribute **Engines:** All **Transmissions:** All	**Cruise Control Switch Circuit Malfunction** PCM detects a concern with the cruise control COAST circuit. **Possible Causes:** • Cruise control switch malfunction • Connector or terminal malfunction • Short to power circuit in wiring from cruise control terminal B and PCM terminal 3P. • Short to ground circuit in wiring from cruise control terminal B and PCM terminal 3P • PCM has failed
DTC: P0570 **2T CCM, MIL: Yes** **Years:** 2008 **Models:** Tribute **Engines:** All **Transmissions:** All	**Cruise Control Switch Circuit Malfunction** PCM detects a concern with the cruise control ACCEL circuit. **Possible Causes:** • Cruise control switch malfunction • Connector or terminal malfunction • Short to power circuit in wiring from cruise control terminal B and PCM terminal 3P. • Short to ground circuit in wiring from cruise control terminal B and PCM terminal 3P • PCM has failed
DTC: P0571 **2T CCM, MIL: Yes** **Years:** 2007, 2008 **Models:** CX-7, Mazda3, MazdaSpeed3, Mazda5, Mazda6, MazdaSpeed6, MX-5, RX-8 **Engines:** All **Transmissions:** All	**Cruise Control Switch Circuit Malfunction** The PCM monitors the cruise control switch signal at the PCM terminal 3P. If the PCM detects that any one of the following switches (MAIN, CANCEL, SET/COAST, RESUME/ACCEL) remains on for 2 min., the PCM determines that the cruise control switch has a malfunction. **Possible Causes:** • Cruise control switch malfunction • Connector or terminal malfunction • Short to power circuit in wiring from cruise control terminal B and PCM terminal 3P. • Short to ground circuit in wiring from cruise control terminal B and PCM terminal 3P • PCM has failed
DTC: P0574 **2T CCM, MIL: Yes** **Years:** 2008 **Models:** Tribute **Engines:** All **Transmissions:** All	**Cruise Control Switch Circuit Malfunction** Vehicle speed above set parameter. **Possible Causes:** • Cruise control switch malfunction • Connector or terminal malfunction • PCM has failed
DTC: P0575 **2T CCM, MIL: Yes** **Years:** 2008 **Models:** Tribute **Engines:** All **Transmissions:** All	**Cruise Control Input Circuit Performance** Cruise control input circuit is out of range. **Possible Causes:** • Cruise control switch malfunction • Connector or terminal malfunction • PCM has failed

DTC	Trouble Code Title, Conditions & Possible Causes
DTC: P0576 **2T CCM, MIL: Yes** **Years:** 2008 **Models:** Tribute **Engines:** All **Transmissions:** All	**Cruise Control Input Circuit Low Voltage** Cruise control input circuit voltage is below minimum parameter. **Possible Causes:** • Cruise control switch malfunction • Connector or terminal malfunction • PCM has failed
DTC: P0577 **2T CCM, MIL: Yes** **Years:** 2008 **Models:** Tribute **Engines:** All **Transmissions:** All	**Cruise Control Input Circuit Low Voltage** Cruise control input circuit voltage is above minimum parameter. **Possible Causes:** • Cruise control switch malfunction • Connector or terminal malfunction • PCM has failed
DTC: P0578 **2T CCM, MIL: Yes** **Years:** 2008 **Models:** Tribute **Engines:** All **Transmissions:** All	**PCM – Vehicle Speed Signal (VSS) Malfunction** The PCM monitors the Vehicle speed signal (VSS) at the PCM terminal. If the PCM detects that the Turbine Shaft Speed (TSS) sensor or the Output Shaft Sensor (OSS) remains on for 2 min., the PCM determines that the VSS has a malfunction. **Possible Causes:** • VSS malfunction • Connector or terminal malfunction • Short to power circuit in wiring from VSS and PCM terminal. • Short to ground circuit in wiring from VSS and PCM terminal. • PCM has failed
DTC: P0579 **2T CCM, MIL: Yes** **Years:** 2007, 2008 **Models:** CX-9, Tribute **Engines:** All **Transmissions:** All	**PCM – Vehicle Speed Signal (VSS) Malfunction** The PCM monitors the Vehicle speed signal (VSS) at the PCM terminal. If the PCM detects that the Turbine Shaft Speed (TSS) sensor or the Output Shaft Sensor (OSS) remains on for 2 min., the PCM determines that the VSS has a malfunction. **Possible Causes:** • VSS malfunction • Connector or terminal malfunction • Short to power circuit in wiring from VSS and PCM terminal. • Short to ground circuit in wiring from VSS and PCM terminal. • PCM has failed
DTC: P0580 **2T CCM, MIL: Yes** **Years:** 2008 **Models:** Tribute **Engines:** All **Transmissions:** All	**PCM – Vehicle Speed Signal (VSS) Malfunction** The PCM monitors the Vehicle speed signal (VSS) at the PCM terminal. If the PCM detects that the Turbine Shaft Speed (TSS) sensor or the Output Shaft Sensor (OSS) remains on for 2 min., the PCM determines that the VSS has a malfunction. **Possible Causes:** • VSS malfunction • Connector or terminal malfunction • Short to power circuit in wiring from VSS and PCM terminal. • Short to ground circuit in wiring from VSS and PCM terminal. • PCM has failed
DTC: P0581 **2T CCM, MIL: Yes** **Years:** 2007, 2008 **Models:** CX-9, Tribute **Engines:** All **Transmissions:** All	**PCM – Vehicle Speed Signal (VSS) Malfunction** The PCM monitors the Vehicle speed signal (VSS) at the PCM terminal. If the PCM detects that the Turbine Shaft Speed (TSS) sensor or the Output Shaft Sensor (OSS) remains on for 2 min., the PCM determines that the VSS has a malfunction. **Possible Causes:** • VSS malfunction • Connector or terminal malfunction • Short to power circuit in wiring from VSS and PCM terminal. • Short to ground circuit in wiring from VSS and PCM terminal. • PCM has failed
DTC: P0594 **2T CCM, MIL: Yes** **Years:** 2008 **Models:** Tribute **Engines:** All **Transmissions:** All	**PCM – Vehicle Speed Signal (VSS) Malfunction** The PCM monitors the Vehicle speed signal (VSS) at the PCM terminal. If the PCM detects that the Turbine Shaft Speed (TSS) sensor or the Output Shaft Sensor (OSS) remains on for 2 min., the PCM determines that the VSS has a malfunction. **Possible Causes:** • VSS malfunction • Connector or terminal malfunction • Short to power circuit in wiring from VSS and PCM terminal. • Short to ground circuit in wiring from VSS and PCM terminal. • PCM has failed

DTC	Trouble Code Title, Conditions & Possible Causes
DTC: P0595 **2T CCM, MIL: Yes** **Years:** 2008 **Models:** Tribute **Engines:** All **Transmissions:** All	**PCM – Vehicle Speed Signal (VSS) Malfunction** The PCM monitors the Vehicle speed signal (VSS) at the PCM terminal. If the PCM detects that the Turbine Shaft Speed (TSS) sensor or the Output Shaft Sensor (OSS) remains on for 2 min., the PCM determines that the VSS has a malfunction. **Possible Causes:** • VSS malfunction • Connector or terminal malfunction • Short to power circuit in wiring from VSS and PCM terminal. • Short to ground circuit in wiring from VSS and PCM terminal. • PCM has failed
DTC: P0596 **2T CCM, MIL: Yes** **Years:** 2008 **Models:** Tribute **Engines:** All **Transmissions:** All	**PCM – Vehicle Speed Signal (VSS) Malfunction** The PCM monitors the Vehicle speed signal (VSS) at the PCM terminal. If the PCM detects that the Turbine Shaft Speed (TSS) sensor or the Output Shaft Sensor (OSS) remains on for 2 min., the PCM determines that the VSS has a malfunction. **Possible Causes:** • VSS malfunction • Connector or terminal malfunction • Short to power circuit in wiring from VSS and PCM terminal. • Short to ground circuit in wiring from VSS and PCM terminal. • PCM has failed
DTC: P0600 **1T CCM, MIL: Yes** **Years:** 2007, 2008 **Models:** B-Series Truck, CX-9, Mazda5, Tribute **Engines:** All **Transmissions:** All	**Internal Control Module Memory Check Sum Error** PCM internal ROM malfunction **Possible Causes:** • Reprogramming has not been completed properly • PCM has failed
DTC: P0601 **1T CCM, MIL: Yes** **Years:** 2007, 2008 **Models:** B-Series Truck, CX-7, CX-9, Mazda3, MazdaSpeed3, Mazda5, Mazda6, MazdaSpeed6, MX-5, RX-8, Tribute **Engines:** All **Transmissions:** All	**Internal Control Module Memory Check Sum Error** PCM internal ROM malfunction **Possible Causes:** • Reprogramming has not been completed properly • PCM has failed
DTC: P0602 **1T CCM, MIL: Yes** **Years:** 2007, 2008 **Models:** B-Series Truck, CX-7, CX-9, Mazda3, MazdaSpeed3, Mazda5, Mazda6, MazdaSpeed6, MX-5, RX-8, Tribute **Engines:** All **Transmissions:** All	**PCM Programming Error** No configuration data in PCM **Possible Causes:** • Reprogramming has not been completed properly • PCM has failed
DTC: P0603 **1T CCM, MIL: No** **Years:** 2007, 2008 **Models:** B-Series Truck, CX-7, CX-9, Mazda6, Mazda6 Sport Wagon, MazdaSpeed6, MX-5, RX-8, Tribute **Engines:** All **Transmissions:** All	**PCM Keep Alive Memory Test Error** Key on, and the PCM detected an interruption to the Keep Alive Memory (KAM) circuit during the initial key "on" sequence. **Possible Causes:** • KAM circuit is open (it may be an intermittent problem) • KAM circuit is shorted to ground • Battery terminals have high resistance due to corrosion or dirt • PCM has failed

DTC	Trouble Code Title, Conditions & Possible Causes
DTC: P0604 **1T CCM, MIL: Yes** **Years:** 2007, 2008 **Models:** B-Series Truck, CX-7, CX-9, Mazda3, MazdaSpeed3, Mazda5, Mazda6, MazdaSpeed6, MX-5, RX-8, Tribute **Engines:** All **Transmissions:** All	**PCM RAM Error** PCM RAM Malfunction **Possible Causes:** • Reprogramming has not been completed properly • PCM has failed
DTC: P0605 **1T CCM, MIL: No** **Years:** 2007, 2008 **Models:** B-Series Truck, CX-9, Tribute **Engines:** All **Transmissions:** All	**PCM Read Only Memory Test Error** Key on, and the PCM detected a Read Only Memory test error during its initial Self-Test key "on" sequence. **Possible Causes:** • PCM has failed (an internal fault) • PCM must be replaced when this code is set
DTC: P0606 **1T CCM, MIL: Yes** **Years:** 2007, 2008 **Models:** B-Series Truck, CX-7, CX-9, Mazda3, MazdaSpeed3, Mazda5, Mazda6, Mazda6 Sport Wagon, MazdaSpeed6, MX-5, Tribute **Engines:** All **Transmissions:** All	**ECM/PCM Processor** PCM internal malfunction. **Possible Causes:** • PCM internal CPU malfunction
DTC: P0607 **1T CCM, MIL: Yes** **Years:** 2007, 2008 **Models:** B-Series Truck, CX-7, CX-9, Mazda3, MazdaSpeed3, Mazda5, Mazda6, Mazda6 Sport Wagon, MazdaSpeed6, Tribute **Engines:** All **Transmissions:** All	**Control Module vehicle operations error** PCM data configuration error **Possible Causes:** • PCM internal malfunction
DTC: P060A **1T CCM, MIL: Yes** **Years:** 2007, 2008 **Models:** CX-9 **Engines:** All **Transmissions:** All	**Internal Control Module Monitoring Processor Performance** Indicates an error occurred in the PCM. **Possible Causes:** • Software incompatibility issue • PCM malfunction
DTC: P060B **1T CCM, MIL: Yes** **Years:** 2007, 2008 **Models:** CX-9 **Engines:** All **Transmissions:** All	**Internal Control Module A/D Processing Performance** Indicates an error occurred in the PCM. **Possible Causes:** • PCM malfunction
DTC: P060C **1T CCM, MIL: Yes** **Years:** 2007, 2008 **Models:** CX-9 **Engines:** All **Transmissions:** All	**Internal Control Module Main Processor Performance** Indicates an error occurred in the PCM. **Possible Causes:** • Software incompatibility issue • PCM malfunction

DTC	Trouble Code Title, Conditions & Possible Causes
DTC: P0610 **1T CCM, MIL: Yes** **Years:** 2007, 2008 **Models:** B-Series Truck, CX-7, CX-9, Mazda3, MazdaSpeed3, Mazda5, Mazda6, MazdaSpeed6, MX-5, RX-8, Tribute **Engines:** All **Transmissions:** All	**Control Module Performance** PCM internal CPU malfunction. **Possible Causes:** • PCM internal malfunction • Configuration has not been properly completed.
DTC: P061B **1T CCM, MIL: Yes** **Years:** 2007, 2008 **Models:** CX-9 **Engines:** All **Transmissions:** All	**Internal Control Module Torque Calculation Performance** Indicates that a calculation error occurred in the PCM. **Note: This DTC is an informational DTC and may be set in combination with a number of other DTCs which are causing the FMEM. Diagnose other DTCs first.** **Possible Causes:** • Connector or terminal malfunction
DTC: P061C **1T CCM, MIL: Yes** **Years:** 2007, 2008 **Models:** CX-9 **Engines:** All **Transmissions:** All	**Internal Control Module Engine RPM Performance** Indicates that a calculation error occurred in the PCM. **Possible Causes:** • CKP sensor circuit is open or short • CKP sensor circuit intermittent • CKP sensor malfunction • CMP sensor circuit is open or short • CMP sensor circuit intermittent • CMP sensor malfunction • PCM malfunction
DTC: P061D **1T CCM, MIL: Yes** **Years:** 2007, 2008 **Models:** CX-9 **Engines:** All **Transmissions:** All	**Internal Control Module Engine Air Mass Performance** Indicates an error occurred in the PCM. **Possible Causes:** • Software incompatibility issue • PCM malfunction
DTC: P061F **1T CCM, MIL: Yes** **Years:** 2007, 2008 **Models:** CX-9 **Engines:** All **Transmissions:** All	**Internal Control Module Throttle Actuator Controller Performance** Indicates that a calculation error occurred in the PCM. **Note: This DTC is an informational DTC and may be set in combination with a number of other DTCs which are causing the FMEM. Diagnose other DTCs first.** **Possible Causes:** • PCM malfunction
DTC: P0620 **1T CCM, MIL: Yes** **Years:** 2007, 2008 **Models:** CX-9, Tribute **Engines:** All **Transmissions:** All	**Regulator / Generator malfunction** PCM internal CPU malfunction. **Possible Causes:** • PCM internal malfunction • Configuration has not been properly completed.
DTC: P0622 **1T CCM, MIL: Yes** **Years:** 2008 **Models:** Tribute **Engines:** All **Transmissions:** All	**Control Module performance** PCM internal CPU malfunction. **Possible Causes:** • PCM internal malfunction • Configuration has not been properly completed

DTC	Trouble Code Title, Conditions & Possible Causes
DTC: P0625 **1T CCM, MIL: Yes** **Years:** 2007, 2008 **Models:** CX-9, Tribute **Engines:** All **Transmissions:** All	**Generator Field Terminal Low Circuit** The PCM monitors generator load from the generator/regulator in the form of frequency. The concern indicates the input is lower than the load should be in normal operation. The load input could be low when no generator output exists. **Possible Causes:** • Drive belt damaged • Generator malfunction • Short to ground between generator terminal RC and PCM terminal 2S • Short to ground between generator terminal LI and PCM terminal 2AG • Open circuit between terminal B and battery positive terminal • Low system voltage • Connector or terminal malfunction
DTC: P0626 **1T CCM, MIL: Yes** **Years:** 2007, 2008 **Models:** CX-9, Tribute **Engines:** All **Transmissions:** All	**Generator Field Terminal High Circuit** The PCM monitors generator load from the generator/regulator in the form of frequency. The concern indicates the input is higher than the load should be in normal operation. The load input could be high when a battery short to ground exists. **Possible Causes:** • Drive belt damaged • Short to power supply between generator terminal RC and PCM terminal 2S • Short to power supply between generator terminal LI and PCM terminal 2AG • Open circuit in wiring harness between generator terminal RC and PCM terminal 2S • Connector or terminal malfunction
DTC: P0630 **1T CCM, MIL: Yes** **Years:** 2007, 2008 **Models:** B-Series Truck, Tribute **Engines:** All **Transmissions:** All	**VIN Not Programmed or Incompatible – ECM/PCM** There is a PCM programming error. **Note: The IGN START/RUN and ground circuits, or the B+ and VPWR circuits may be reversed in the harness connector. Refer to the Wiring Diagrams Manual for schematic and connector information.** **Possible Causes:** • Harness circuits: B+, IGN START/RUN, INJPWRM, ISP-R, PCMRC, VPWR and GND • PCM power relay • PCM malfunction
DTC: P0632 **1T CCM, MIL: Yes** **Years:** 2007, 2008 **Models:** B-Series Truck, Tribute **Engines:** All **Transmissions:** All	**Odometer Not Programmed – ECM/PCM** There is a PCM programming error. **Note: The IGN START/RUN and ground circuits, or the B+ and VPWR circuits may be reversed in the harness connector. Refer to the Wiring Diagrams Manual for schematic and connector information.** **Possible Causes:** • Harness circuits: B+, IGN START/RUN, INJPWRM, ISP-R, PCMRC, VPWR and GND • PCM power relay • PCM malfunction
DTC: P0633 **1T CCM, MIL: Yes** **Years:** 2007, 2008 **Models:** B-Series Truck, Tribute **Engines:** All **Transmissions:** All	**Immobilizer Key Not Programmed – ECM/PCM** There is a PCM programming error. **Note: The IGN START/RUN and ground circuits, or the B+ and VPWR circuits may be reversed in the harness connector. Refer to the Wiring Diagrams Manual for schematic and connector information.** **Possible Causes:** • Harness circuits: B+, IGN START/RUN, INJPWRM, ISP-R, PCMRC, VPWR and GND • PCM power relay • PCM malfunction
DTC: P0638 **1T CCM, MIL: Yes** **Years:** 2007, 2008 **Models:** CX-7, Mazda3, MazdaSpeed3, Mazda5, Mazda6, MazdaSpeed6, MX-5, RX-8, Tribute **Engines:** All **Transmissions:** All	**Throttle Actuator Control Range Performance** If the PCM detects that actual throttle angle opening is smaller or larger than the target throttle opening angle, the PCM determines that the throttle actuator control system has a malfunction **Possible Causes:** • PCM internal malfunction • Throttle Body malfunction
DTC: P0639 **1T CCM, MIL: Yes** **Years:** 2008 **Models:** Tribute **Engines:** All **Transmissions:** All	**Throttle Actuator Control Range Performance** If the PCM detects that actual throttle angle opening is smaller or larger than the target throttle opening angle, the PCM determines that the throttle actuator control system has a malfunction **Possible Causes:** • PCM internal malfunction • Throttle Body malfunction

DTC	Trouble Code Title, Conditions & Possible Causes
DTC: P0642 **1T CCM, MIL: Yes** **Years:** 2007, 2008 **Models:** CX-9 **Engines:** All **Transmissions:** All	**Sensor Reference Voltage Low Circuit** This code indicates the reference voltage circuit is lower than reference voltage minimum. **Possible Causes:** • Short to ground between fuel tank pressure sensor terminal C and PCM terminal 1U • Fuel tank pressure sensor malfunction • Connector or terminal malfunction
DTC: P0643 **1T CCM, MIL: Yes** **Years:** 2007, 2008 **Models:** CX-9 **Engines:** All **Transmissions:** All	**Sensor Reference Voltage High Circuit** This code indicates the reference voltage circuit is higher than reference voltage maximum. **Possible Causes:** • Short to power supply between fuel tank pressure sensor terminal C and PCM terminal 1U • Fuel tank pressure sensor malfunction • Connector or terminal malfunction
DTC: P0645 **1T CCM, MIL: Yes** **Years:** 2007, 2008 **Models:** B-Series Truck, CX-9, Tribute **Engines:** All **Transmissions:** All	**A/C Clutch Relay (ACCR) Malfunction** If the PCM detects excessive current draw when PCM grounds the circuit or voltage is not detected on the ACCR circuit when it is not grounded by the PCM, the PCM determines that the ACCR has a malfunction **Possible Causes:** • PCM internal malfunction • ACCR malfunction
DTC: P0646 **1T CCM, MIL: Yes** **Years:** 2007, 2008 **Models:** B-Series Truck, Tribute **Engines:** All **Transmissions:** All	**A/C Clutch Relay (ACCR) Control Low Circuit** If the PCM detects excessive current draw when PCM grounds the circuit or voltage is not detected on the ACCR circuit when it is not grounded by the PCM, the PCM determines that the ACCR has a malfunction **Possible Causes:** • Damaged A/C Demand Switch • Damaged A/C clutch relay • A/C On during self-test • PCM internal malfunction • ACCR malfunction
DTC: P0647 **1T CCM, MIL: Yes** **Years:** 2007, 2008 **Models:** B-Series Truck, Tribute **Engines:** All **Transmissions:** All	**A/C Clutch Relay (ACCR) Control High Circuit** If the PCM detects excessive current draw when PCM grounds the circuit or voltage is not detected on the ACCR circuit when it is not grounded by the PCM, the PCM determines that the ACCR has a malfunction **Possible Causes:** • Damaged A/C Demand Switch • Damaged A/C clutch relay • A/C On during self-test • PCM internal malfunction • ACCR malfunction
DTC: P064D **2T CCM, MIL: Yes** **Years:** 2008 **Models:** Mazda5 **Engines:** All **Transmissions:** All	**Internal Control Module Front HO2S Process Performance** The front HO2S IC integrated in PCM converts to voltage value for fuel control and for diagnosis based on the front HO2S signal current and sends to CPU (integrated in PCM). The CPU switches the voltage value for the fuel control and for the diagnosis in the fuel control once during one drive cycle, and compares the voltage values. If the difference of compared value is below threshold, the PCM determines that the signal circuit between CPU and front HO2S control IC has malfunction. **Possible Causes:** • PCM internal malfunction (between CPU and front HO2S control IC communication line)
DTC: P0660 **1T CCM, MIL: Yes** **Years:** 2008 **Models:** Tribute **Engines:** All **Transmissions:** All	**Variable Inertia Charging System Circuit Malfunction** Key on or engine running and the PCM detected an unexpected voltage condition on the Variable Inertia Charging System Solenoid (VICS) circuit during the CCM test period. **Possible Causes:** • VICS control circuit is open or shorted to ground • VICS power circuit is open (check the power from the relay) • VICS valve is damaged or has failed • PCM has failed

DTC	Trouble Code Title, Conditions & Possible Causes
DTC: P0661 **2T CCM, MIL: No** **Years:** 2007, 2008 **Models:** Mazda3, MazdaSpeed3, Mazda5, Mazda6, MazdaSpeed6, MX-5, RX-8, Tribute **Engines:** All **Transmissions:** All	**Variable Intake-Air System (VIS) Control Solenoid Valve Circuit Low Input** PCM monitors the VIS control solenoid valve signal at PCM terminal 4R. If PCM turns VIS control solenoid valve OFF but voltage at PCM terminal 4R remains low, PCM determines that VIS control solenoid valve circuit has a malfunction. **Possible Causes:** • VIS control valve malfunction • Connector or terminal malfunction • Open circuit between VIS and PSM • PCM has failed
DTC: P0662 **2T CCM, MIL: No** **Years:** 2007, 2008 **Models:** Mazda3, MazdaSpeed3, Mazda5, Mazda6, MazdaSpeed6, MX-5, RX-8, Tribute **Engines:** All **Transmissions:** All	**Variable Intake-Air System (VIS) Control Solenoid Valve Circuit High Input** PCM monitors the VIS control solenoid valve signal at PCM terminal 4R. If PCM turns VIS control solenoid valve OFF but voltage at PCM terminal 4R remains high, PCM determines that VIS control solenoid valve circuit has a malfunction. **Possible Causes:** • VIS control valve malfunction • Connector or terminal malfunction • Open circuit between VIS and PSM • PCM has failed
DTC: P0663 **2T CCM, MIL: No** **Years:** 2008 **Models:** Tribute **Engines:** All **Transmissions:** All	**Variable Intake-Air System (VIS) Control Solenoid Valve Open Circuit** PCM monitors the VIS control solenoid valve signal at PCM terminal 4R. If PCM turns VIS control solenoid valve OFF but voltage at PCM terminal 4R remains high, PCM determines that VIS control solenoid valve circuit has a malfunction. **Possible Causes:** • VIS control valve malfunction • Connector or terminal malfunction • Open circuit between VIS and PSM • PCM has failed
DTC: P0664 **2T CCM, MIL: No** **Years:** 2008 **Models:** Tribute **Engines:** All **Transmissions:** All	**Variable Intake-Air System (VIS) Control Solenoid Valve Circuit Low Input** PCM monitors the VIS control solenoid valve signal at PCM terminal 4R. If PCM turns VIS control solenoid valve OFF but voltage at PCM terminal 4R remains high, PCM determines that VIS control solenoid valve circuit has a malfunction. **Possible Causes:** • VIS control valve malfunction • Connector or terminal malfunction • Open circuit between VIS and PSM • PCM has failed
DTC: P0665 **2T CCM, MIL: No** **Years:** 2008 **Models:** Tribute **Engines:** All **Transmissions:** All	**Variable Intake-Air System (VIS) Control Solenoid Valve Circuit High Input** PCM monitors the VIS control solenoid valve signal at PCM terminal 4R. If PCM turns VIS control solenoid valve OFF but voltage at PCM terminal 4R remains high, PCM determines that VIS control solenoid valve circuit has a malfunction. **Possible Causes:** • VIS control valve malfunction • Connector or terminal malfunction • Open circuit between VIS and PSM • PCM has failed
DTC: P0668 **2T CCM, MIL: No** **Years:** 2008 **Models:** B-Series Truck, Tribute **Engines:** All **Transmissions:** All	**PCM/ECM/TCM Internal Temperature Sensor Low Circuit** PCM programming error. **Note: The IGN START/RUN and ground circuits, or the B+ and VPWR circuits may be reversed in the harness connector. Refer to the Wiring Diagrams Manual for schematic and connector information.** **Possible Causes:** • Harness circuits: B+, IGN START/RUN, INJPWRM, ISP-R, PCMRC, VPWR and GND • PCM power relay • PCM malfunction
DTC: P0669 **2T CCM, MIL: No** **Years:** 2008 **Models:** B-Series Truck, Tribute **Engines:** All **Transmissions:** All	**PCM/ECM/TCM Internal Temperature Sensor High Circuit** PCM programming error. **Note: The IGN START/RUN and ground circuits, or the B+ and VPWR circuits may be reversed in the harness connector. Refer to the Wiring Diagrams Manual for schematic and connector information.** **Possible Causes:** • Harness circuits: B+, IGN START/RUN, INJPWRM, ISP-R, PCMRC, VPWR and GND • PCM power relay • PCM malfunction

DTC	Trouble Code Title, Conditions & Possible Causes
DTC: P0685 **2T CCM, MIL: No** **Years:** 2008 **Models:** CX-9, Tribute **Engines:** All **Transmissions:** All	**PCM Power Relay Control** The PCM power relay circuit is open. **Possible Causes:** • Harness circuits: open or short to ground • PCM power relay • PCM malfunction
DTC: P0689 **2T CCM, MIL: No** **Years:** 2008 **Models:** CX-9, Tribute **Engines:** All **Transmissions:** All	**PCM Power Relay Control** The PCM power relay sense circuit is low. **Possible Causes:** • Harness circuits: open or short to ground • PCM power relay • PCM malfunction
DTC: P0690 **2T CCM, MIL: No** **Years:** 2008 **Models:** CX-9, Tribute **Engines:** All **Transmissions:** All	**PCM Power Relay Control** The PCM power relay sense circuit is high. **Possible Causes:** • Harness circuits: open or short to ground • PCM power relay • PCM malfunction
DTC: P0703 **2T CCM, MIL: No** **Years:** 2007, 2008 **Models:** B-Series Truck, CX-7, CX-9, Mazda3, MazdaSpeed3, Mazda5, Mazda6, MazdaSpeed6, MX-5, RX-8, Tribute **Engines:** All **Transmissions:** A/T	**Brake On/Off Switch Circuit Malfunction (Self-Test)** KOEO Self-Test: Key on, and the PCM did not detect any change in the Brake Switch status after the brake pedal was pressed and released during the self-test. KOER Self-Test: and the PCM did not detect any change in the Brake Switch status after the brake pedal was pressed and released during the self-test. **Possible Causes:** • Brake switch signal circuit is open or shorted ground • Brake switch power circuit (B+) open between switch and PCM • Brake switch is damaged, misadjusted or installed improperly • PCM has failed
DTC: P0704 **2T CCM, MIL: No** **Years:** 2007, 2008 **Models:** B-Series Truck, Mazda3, MazdaSpeed3, Mazda5, Mazda6, MazdaSpeed6, MX-5, RX-8, Tribute **Engines:** All **Transmissions:** M/T	**Clutch Pedal Switch Circuit Malfunction** Engine started, the after the vehicle was accelerated from 0–16 MPH, the PCM detected the Clutch Pedal Position (CPP) switch input did not change status correctly under these conditions in the CCM test. **Possible Causes:** • Clutch switch signal circuit is open or shorted ground • Clutch switch power circuit (B+) open between switch and PCM • Clutch switch is damaged, misadjusted or installed improperly • PCM has failed
DTC: P0705 **2T CCM, MIL: No** **Years:** 2007, 2008 **Models:** B-Series Truck, RX-8, Tribute **Engines:** All **Transmissions:** A/T	**Clutch Pedal Switch Circuit Malfunction** Engine started, the after the vehicle was accelerated from 0–16 MPH, the PCM detected the Clutch Pedal Position (CPP) switch input did not change status correctly under these conditions in the CCM test. **Possible Causes:** • Clutch switch signal circuit is open or shorted ground • Clutch switch power circuit (B+) open between switch and PCM • Clutch switch is damaged, misadjusted or installed improperly • PCM has failed
DTC: P0706 **2T CCM, MIL: Yes** **Years:** 2007, 2008 **Models:** B-Series Truck, CX-7, CX-9, Mazda3, MazdaSpeed3, Mazda5, Mazda6, MazdaSpeed6, RX-8, Tribute **Engines:** All **Transmissions:** A/T	**Transmission Range Switch Circuit Malfunction** Key on or engine running, and the PCM did not detect any TR switch inputs, or it detected 2 switch inputs simultaneously during the test. **Possible Causes:** • TR signal 'P' circuit open between the switch and the PCM (0v) • TR signal 'D' circuit open between the switch and the PCM (0v) • TR signal 'S' circuit open between the switch and the PCM (0v) • TR signal 'L' circuit open between the switch and the PCM (0v) • TR switch or its connector is damaged, or the switch has failed • PCM has failed

DTC	Trouble Code Title, Conditions & Possible Causes
DTC: P0707 **2T CCM, MIL: Yes** **Years:** 2007, 2008 **Models:** CX-7, CX-9, Mazda3, MazdaSpeed3, Mazda5, Mazda6, MazdaSpeed6, MX-5, RX-8, Tribute **Engines:** 2.0L VIN C **Transmissions:** A/T	**Transmission Range Switch Range/Performance** Engine running, and the PCM detected multiple T/R switch inputs, or the TR Switch signal did not change with the vehicle moving. **Possible Causes:** • TR switch signal circuit is shorted to another switch signal • TR switch signal circuit is open (problem may be intermittent) • TR switch is damaged, out of adjustment or it has failed • PCM has failed
DTC: P0708 **2T CCM, MIL: Yes** **Years:** 2007, 2008 **Models:** B-Series Truck, CX-7, CX-9, Mazda3, MazdaSpeed3, Mazda5, Mazda6, MazdaSpeed6, MX-5, RX-8, Tribute **Engines:** All **Transmissions:** A/T	**Transmission Range Switch Circuit Malfunction** Key on or engine running and the PCM detected an unexpected high or low voltage condition on the TR switch circuit during the test. **Possible Causes:** • TR switch signal circuit is open between the switch and PCM • TR switch signal circuit is shorted between switch and PCM • TR switch signal circuit is shorted to VREF or system power • TR switch ground circuit is open between switch and ground • TR switch or its connector is damaged, or the switch has failed • PCM has failed
DTC: P0709 **2T CCM, MIL: Yes** **Years:** 2007, 2008 **Models:** B-Series Truck, Tribute **Engines:** All **Transmissions:** A/T	**Transmission Range Switch Circuit Malfunction** Key on or engine running and the PCM detected an unexpected high or low voltage condition on the TR switch circuit during the test. **Possible Causes:** • TR switch signal circuit is open between the switch and PCM • TR switch signal circuit is shorted between switch and PCM • TR switch signal circuit is shorted to VREF or system power • TR switch ground circuit is open between switch and ground • TR switch or its connector is damaged, or the switch has failed • PCM has failed
DTC: P0710 **1T CCM, MIL: Yes** **Years:** 2007, 2008 **Models:** B-Series Truck, Tribute **Engines:** All **Transmissions:** A/T	**Transmission Fluid Temperature Sensor Circuit Malfunction** Engine started, vehicle driven to a speed of over 12 MPH, and the PCM detected the TFT sensor signal was less than 0.10v, or it was more than 4.90v, condition met for 100 seconds. **Possible Causes:** • TFT sensor signal circuit is open or shorted to ground • TFT sensor ground circuit is open between sensor and PCM • TFT sensor signal circuit shorted to VREF or system power • TFT sensor is damaged or has failed • PCM has failed
DTC: P0711 **2T CCM, MIL: Yes** **Years:** 2007, 2008 **Models:** B-Series Truck, CX-7, CX-9, Mazda3, MazdaSpeed3, Mazda5, Mazda6, MazdaSpeed6, MX-5, RX-8, Tribute **Engines:** All **Transmissions:** A/T	**Transmission Fluid Temperature Sensor Circuit Range/Performance** DTC P0710 not set, engine started, vehicle driven to a speed of over 37 MPH for 430 seconds, and the PCM detected the TFT sensor signal was less than 0.09v, or it was more than 4.99v in the test. **Possible Causes:** • TFT sensor signal circuit is open between sensor and the PCM • TFT sensor signal circuit is shorted to ground • TFT sensor is damaged or has failed • PCM has failed
DTC: P0712 **1T CCM, MIL: No** **Years:** 2007, 2008 **Models:** B-Series Truck, CX-7, CX-9, Mazda3, MazdaSpeed3, Mazda5, Mazda6, MazdaSpeed6, MX-5, RX-8, Tribute **Engines:** All **Transmissions:** A/T	**Transmission Fluid Temperature Sensor Low Input (High Temperature)** DTC P0710 not set, engine started, vehicle driven to a speed of over 37 MPH for 430 seconds, and the PCM detected the TFT sensor indicated less than 0.20v (Scan Tool reads more than 315°F) during the CCM rationality test. **Possible Causes:** • TFT sensor signal circuit has a short toe ground condition • TFT sensor is damaged or has failed (it may be shorted) • PCM has failed

DTC	Trouble Code Title, Conditions & Possible Causes
DTC: P0713 **1T CCM, MIL: No** **Years:** 2007, 2008 **Models:** B-Series Truck, CX-7, CX-9, Mazda3, MazdaSpeed3, Mazda5, Mazda6, MazdaSpeed6, MX-5, RX-8, Tribute **Engines:** All **Transmissions:** A/T	**Transmission Fluid Temperature Sensor High Input (Low Temperature)** DTC P0710 not set, engine started, vehicle driven to a speed of over 37 MPH for 430 seconds, and the PCM detected the TFT sensor indicated more than 4.96v (Scan Tool reads less than −40°F) during the CCM rationality test. **Possible Causes:** • TFT sensor signal circuit is open between sensor and the PCM • TFT sensor signal circuit has a short to power condition • TFT sensor is damaged or has failed (it may be open) • PCM has failed
DTC: P0714 **1T CCM, MIL: Yes** **Years:** 2007, 2008 **Models:** B-Series Truck, Tribute **Engines:** All **Transmissions:** A/T	**Transmission Fluid Temperature Sensor Circuit Malfunction** DTC P0710 not set, engine started, vehicle driven to a speed of over 37 MPH for 430 seconds, and the PCM detected the TFT sensor indicated more than 4.96v (Scan Tool reads less than −40°F) during the CCM rationality test. **Possible Causes:** • TFT sensor signal circuit is open between sensor and the PCM • TFT sensor signal circuit has a short to power condition • TFT sensor is damaged or has failed (it may be open) • PCM has failed
DTC: P0715 **1T CCM, MIL: Yes** **Years:** 2007, 2008 **Models:** B-Series Truck, Mazda3, MazdaSpeed3, Mazda5, Mazda6, MazdaSpeed6, RX-8, Tribute **Engines:** All **Transmissions:** A/T	**Input Shaft or Turbine Shaft Sensor Circuit Malfunction** DTC P0710 not set, engine started, vehicle driven to a speed of over 25 MPH for 430 seconds, and the PCM detected the ISS/TSS signal dropped out for more than 1 second during the CCM test. **Possible Causes:** • ISS/TSS (+) signal circuit is open (an intermittent fault) • ISS/TSS (+) signal circuit shorted to ground (intermittent fault) • ISS/TSS (-) signal circuit is open (an intermittent fault) • ISS/TSS (-) signal circuit shorted to ground (intermittent fault) • ISS/TSS is damaged or has failed • PCM has failed
DTC: P0716 **1T CCM, MIL: Yes** **Years:** 2007, 2008 **Models:** B-Series Truck, Tribute **Engines:** All **Transmissions:** A/T	**Input Shaft or Turbine Shaft Sensor Circuit Malfunction** DTC P0710 not set, engine started, vehicle driven to a speed of over 25 MPH for 430 seconds, and the PCM detected the ISS/TSS signal dropped out for more than 1 second during the CCM test. **Possible Causes:** • ISS/TSS (+) signal circuit is open (an intermittent fault) • ISS/TSS (+) signal circuit shorted to ground (intermittent fault) • ISS/TSS (-) signal circuit is open (an intermittent fault) • ISS/TSS (-) signal circuit shorted to ground (intermittent fault) • ISS/TSS is damaged or has failed • PCM has failed
DTC: P0717 **1T CCM, MIL: Yes** **Years:** 2007, 2008 **Models:** B-Series Truck, CX-7, CX-9, Mazda6, MazdaSpeed6, MX-5, RX-8, Tribute **Engines:** All **Transmissions:** A/T	**Input Shaft or Turbine Shaft Sensor Circuit Malfunction** DTC P0710 not set, engine started, vehicle driven to a speed of over 25 MPH for 430 seconds, and the PCM detected the ISS/TSS signal dropped out for more than 1 second during the CCM test. **Possible Causes:** • ISS/TSS (+) signal circuit is open (an intermittent fault) • ISS/TSS (+) signal circuit shorted to ground (intermittent fault) • ISS/TSS (-) signal circuit is open (an intermittent fault) • ISS/TSS (-) signal circuit shorted to ground (intermittent fault) • ISS/TSS is damaged or has failed • PCM has failed
DTC: P0718 **1T CCM, MIL: Yes** **Years:** 2007, 2008 **Models:** B-Series Truck, Tribute **Engines:** All **Transmissions:** A/T	**Input Shaft or Turbine Shaft Sensor Circuit Malfunction** DTC P0710 not set, engine started, vehicle driven to a speed of over 25 MPH for 430 seconds, and the PCM detected the ISS/TSS signal dropped out for more than 1 second during the CCM test. **Possible Causes:** • ISS/TSS (+) signal circuit is open (an intermittent fault) • ISS/TSS (+) signal circuit shorted to ground (intermittent fault) • ISS/TSS (-) signal circuit is open (an intermittent fault) • ISS/TSS (-) signal circuit shorted to ground (intermittent fault) • ISS/TSS is damaged or has failed • PCM has failed

<table>
<tr><th>DTC</th><th>Trouble Code Title, Conditions & Possible Causes</th></tr>
<tr><td>DTC: P0719
1T CCM, MIL: Yes
Years: 2007, 2008
Models: B-Series Truck, Tribute
Engines: All
Transmissions: A/T</td><td>Brake Pedal Position (BPP) Switch Malfunction
The brake switch circuit has low voltage.
Possible Causes:
• Open or short in BPP circuit
• Open or short in stoplamp circuits
• Malfunction in module(s) connected to BPP circuit
• Damaged brake switch
• Misadjusted brake switch
• Damaged PCM</td></tr>
<tr><td>DTC: P0720
1T CCM, MIL: Yes
Years: 2007, 2008
Models: B-Series Truck, Mazda3, MazdaSpeed3, Mazda5, Mazda6, MazdaSpeed6, RX-8, Tribute
Engines: All
Transmissions: A/T</td><td>Output Shaft Sensor Circuit Malfunction
Engine started, vehicle driven to a speed of over 37 MPH, and the PCM detected the OSS signal had dropped out (the signal was lost) for more than 85 seconds during the CCM rationality test.
Possible Causes:
• OSS (+) signal circuit is open (fault may be intermittent)
• OSS (+) signal circuit shorted to ground (an intermittent fault)
• OSS (-) signal circuit is open (fault may be intermittent)
• OSS (-) signal circuit shorted to ground (an intermittent fault)
• OSS is damaged or has failed
• PCM has failed</td></tr>
<tr><td>DTC: P0721
1T CCM, MIL: Yes
Years: 2007, 2008
Models: B-Series Truck, Tribute
Engines: All
Transmissions: A/T</td><td>Output Shaft Sensor Circuit Signal Erratic
Engine started, vehicle driven to a speed of over 12 MPH, and the PCM detected the OSS signal was erratic during the CCM test.
Possible Causes:
• OSS (+) circuit is open or shorted to ground (noisy circuit)
• OSS (-) circuit is open or shorted to ground (noisy circuit)
• OSS reluctor contains metal chips or is damaged
• OSS is damaged or has failed
• PCM has failed</td></tr>
<tr><td>DTC: P0722
1T CCM, MIL: Yes
Years: 2007, 2008
Models: B-Series Truck, CX-7, CX-9, Mazda6, MazdaSpeed6, MX-5, RX-8, Tribute
Engines: All
Transmissions: A/T</td><td>Output Shaft Sensor Circuit Signal Intermittent
Engine started, vehicle driven to a speed of over 12 MPH, and the PCM detected the OSS signal was lost (dropped out) during the test.
Note: This code can set if the OSS signal drops "in" and "out".
Possible Causes:
• OSS (+) circuit is open or shorted to ground (intermittent)
• OSS (-) circuit is open or shorted to ground (intermittent)
• OSS is damaged or has failed
• PCM has failed</td></tr>
<tr><td>DTC: P0723
1T CCM, MIL: Yes
Years: 2007, 2008
Models: B-Series Truck, Tribute
Engines: All
Transmissions: A/T</td><td>Output Shaft Sensor Circuit Signal Intermittent
Engine started, vehicle driven to a speed of over 12 MPH, and the PCM detected the OSS signal was lost (dropped out) during the test.
Note: This code can set if the OSS signal drops "in" and "out".
Possible Causes:
• OSS (+) circuit is open or shorted to ground (intermittent)
• OSS (-) circuit is open or shorted to ground (intermittent)
• OSS is damaged or has failed
• PCM has failed</td></tr>
<tr><td>DTC: P0724
2T CCM, MIL: No
Years: 2008
Models: Tribute
Engines: All
Transmissions: A/T</td><td>Brake On/Off Switch Circuit Malfunction (Self-Test)
KOEO Self-Test: Key on, and the PCM did not detect any change in the Brake Switch status after the brake pedal was pressed and released during the self-test.
KOER Self-Test: and the PCM did not detect any change in the Brake Switch status after the brake pedal was pressed and released during the self-test.
Possible Causes:
• Brake switch signal circuit is open or shorted ground
• Brake switch power circuit (B+) open between switch and PCM
• Brake switch is damaged, misadjusted or installed improperly
• PCM has failed</td></tr>
</table>

DTC	Trouble Code Title, Conditions & Possible Causes
DTC: P0725 **1T CCM, MIL: Yes** **Years:** 2008 **Models:** Tribute **Engines:** All **Transmissions:** A/T	**Engine Speed Signal Circuit Malfunction** Engine started, vehicle driven to a speed of over 37 MPH, and the PCM detected the Engine Speed signal indicated less than 300 RPM for more than 10 seconds during the CCM rationality test. **Possible Causes:** • CMP sensor signal erratic or missing (an intermittent fault) • PCM to TCM circuit is open or shorted to ground • TCM has failed • PCM has failed
DTC: P0726 **1T CCM, MIL: Yes** **Years:** 2008 **Models:** Tribute **Engines:** All **Transmissions:** A/T	**Engine Speed Signal Circuit Malfunction** Engine started, vehicle driven to a speed of over 37 MPH, and the PCM detected the Engine Speed signal indicated less than 300 RPM for more than 10 seconds during the CCM rationality test. **Possible Causes:** • CMP sensor signal erratic or missing (an intermittent fault) • PCM to TCM circuit is open or shorted to ground • TCM has failed • PCM has failed
DTC: P0727 **1T CCM, MIL: Yes** **Years:** 2007, 2008 **Models:** CX-9, Tribute **Engines:** All **Transmissions:** A/T	**Engine Speed Signal Circuit Malfunction** Engine started, vehicle driven to a speed of over 37 MPH, and the PCM detected the Engine Speed signal indicated less than 300 RPM for more than 10 seconds during the CCM rationality test. **Possible Causes:** • CMP sensor signal erratic or missing (an intermittent fault) • PCM to TCM circuit is open or shorted to ground • TCM has failed • PCM has failed
DTC: P0728 **1T CCM, MIL: Yes** **Years:** 2008 **Models:** Tribute **Engines:** All **Transmissions:** A/T	**Engine Speed Signal Circuit Malfunction** Engine started, vehicle driven to a speed of over 37 MPH, and the PCM detected the Engine Speed signal indicated less than 300 RPM for more than 10 seconds during the CCM rationality test. **Possible Causes:** • CMP sensor signal erratic or missing (an intermittent fault) • PCM to TCM circuit is open or shorted to ground • TCM has failed • PCM has failed
DTC: P0729 **1T CCM, MIL: Yes** **Years:** 2007, 2008 **Models:** CX-7, CX-9, Mazda6, Mazda6 Sport Wagon, MazdaSpeed6, **Engines:** All **Transmissions:** A/T	**Incorrect Gear Ratio Detected** With engine running, vehicle speed signal normal, driving in 6th gear with the accelerator opening angle at 10 percent or more, the PCM detected an incorrect gear ratio. **Possible Causes:** • ATF level low • Deteriorated ATF • Shift solenoid C stuck • Shift solenoid E stuck • Line pressure control solenoid stuck • Line pressure low • B1 brake slipping • C2 clutch slipping • Control valve stuck • Oil pump malfunction • TCM malfunction
DTC: P0730 **1T CCM, MIL: Yes** **Years:** 2007, 2008 **Models:** CX-7, CX-9, Mazda6, Mazda6 Sport Wagon,MazdaSpeed6, Tribute **Engines:** All **Transmissions:** A/T	**Engine Speed Signal Circuit Malfunction** Engine started, vehicle driven to a speed of over 37 MPH, and the PCM detected the Engine Speed signal indicated less than 300 RPM for more than 10 seconds during the CCM rationality test. **Possible Causes:** • CMP sensor signal erratic or missing (an intermittent fault) • PCM to TCM circuit is open or shorted to ground • PCM or the TCM has failed

DTC	Trouble Code Title, Conditions & Possible Causes
DTC: P0731 **2T CCM, MIL: Yes** **Years:** 2007, 2008 **Models:** B-Series Truck, CX-7, CX-9, Mazda3, MazdaSpeed3, Mazda5, Mazda6, MazdaSpeed6, MX-5, RX-8, Tribute **Engines:** All **Transmissions:** A/T	**Transmission 1st Gear Ratio Incorrect** DTC P0500, P0710, P0715, P0755 and P0760 not set, engine started, then driven in Drive in 1st Gear (1GR) to a speed from 4–37 MPH, engine speed over 500 RPM, TSS signal more than 75 RPM, TP angle over 6.25 percent, ATF sensor signal more than 68°F, brake switch indicating "off", and the PCM detected the ratio of the Turbine speed to the Vehicle speed was less than a preset value stored in memory. **Possible Causes:** • ATF level is too low, or the ATF is badly deteriorated • A/T line pressure is low, or the oil pump has failed • SSA, SSB, SSC or the PCS is stuck (mechanical fault) • A/T 1–2 clutch is slipping (mechanical fault) • A/T 1–2 shift valve is stuck (mechanical fault) • A/T Pressure regulator or the pressure modifier valve is stuck • A/T solenoid reducing valve is stuck (mechanical fault) • PCM has failed
DTC: P0732 **2T CCM, MIL: Yes** **Years:** 2007, 2008 **Models:** B-Series Truck, CX-7, CX-9, Mazda3, MazdaSpeed3, Mazda5, Mazda6, MazdaSpeed6, MX-5, RX-8, Tribute **Engines:** All **Transmissions:** A/T	**Transmission 2nd Gear Ratio Incorrect** DTC P0500, P0710, P0715, P0755 and P0760 not set, engine started, then driven in Drive in 2nd Gear (2GR) to a speed of 66 MPH (± 6 MPH), engine speed over 500 RPM, throttle opening over 6.25 percent, TSS signal more than 75 RPM, ATF sensor signal more than 68°F, brake switch indicating "off", and the PCM detected the ratio of the Turbine speed to the Vehicle speed was less than a preset value stored in memory. **Possible Causes:** • ATF level is too low, or the ATF is badly deteriorated • A/T line pressure is low, or the oil pump has failed • SSA, SSB, SSC or the PCS is stuck (mechanical fault) • A/T 2–3 clutch is slipping (mechanical fault) • A/T 1–2 or the 2–3 shift valve is stuck (mechanical fault) • A/T Pressure regulator or the pressure modifier valve is stuck • A/T solenoid reducing valve is stuck (mechanical fault) • PCM has failed
DTC: P0733 **2T CCM, MIL: Yes** **Years:** 2007, 2008 **Models:** B-Series Truck, CX-7, CX-9, Mazda3, MazdaSpeed3, Mazda5, Mazda6, MazdaSpeed6, MX-5, RX-8, Tribute **Engines:** All **Transmissions:** A/T	**Transmission 3rd Gear Ratio Incorrect** DTC P0500, P0710, P0715, P0755 and P0760 not set, engine started, then driven in Drive in 3rd Gear (3GR) to a speed of 19–30 MPH, engine speed over 500 RPM, TSS signal more than 75 RPM, ATF sensor signal more than 68°F, brake switch indicating "off", and the PCM detected the ratio of the Turbine speed to the Vehicle speed was less than a preset value stored in memory. **Possible Causes:** • ATF level is too low, or the ATF is badly deteriorated • A/T line pressure is low, or the oil pump has failed • SSA, SSB, SSC or the PCS is stuck (mechanical fault) • A/T 3–4 clutch is slipping (mechanical fault) • A/T 2–3 or the 3–4 shift valve is stuck (mechanical fault) • A/T Pressure regulator or the pressure modifier valve is stuck • A/T solenoid reducing valve is stuck (mechanical fault) • PCM has failed
DTC: P0734 **2T CCM, MIL: Yes** **Years:** 2007, 2008 **Models:** B-Series Truck, CX-7, CX-9, Mazda3, MazdaSpeed3, Mazda5, Mazda6, MazdaSpeed6, MX-5, RX-8, Tribute **Engines:** All **Transmissions:** A/T	**Transmission 4th Gear Ratio Incorrect** DTC P0500, P0710, P0715, P0755 and P0760 not set, engine started, then driven in Drive in 4th Gear (4GR) to a speed of 47–64 MPH, engine speed over 500 RPM, TSS signal more than 75 RPM, ATF sensor signal more than 68°F, brake switch indicating "off", and the PCM detected the ratio of the Turbine speed to the Vehicle speed was less than a preset value stored in memory. **Possible Causes:** • ATF level is too low, or the ATF is badly deteriorated • A/T line pressure is low, or the oil pump has failed • SSA, SSB, SSC or the PCS is stuck (mechanical fault) • A/T 2–4 brake bank or 3–4 clutch is slipping (mechanical fault) • A/T 1–2, 2–3 or 3–4 shift valve is stuck (mechanical fault) • A/T Pressure regulator, pressure modifier or solenoid reducing valve is stuck • PCM has failed

DTC	Trouble Code Title, Conditions & Possible Causes
DTC: P0735 **2T CCM, MIL: Yes** **Years:** 2007, 2008 **Models:** B-Series Truck, CX-9, Mazda6, Mazda6 Sport Wagon, MazdaSpeed6, Tribute **Engines:** All **Transmissions:** A/T	**Transmission 5th Gear Ratio Incorrect** DTC P0500, P0710, P0715, P0755 and P0760 not set, engine started, then driven in Drive in 5th Gear (5GR) to a speed of 47–64 MPH, engine speed over 500 RPM, TSS signal more than 75 RPM, ATF sensor signal more than 68°F, brake switch indicating "off", and the PCM detected the ratio of the Turbine speed to the Vehicle speed was less than a preset value stored in memory. **Possible Causes:** • ATF level is too low, or the ATF is badly deteriorated • A/T line pressure is low, or the oil pump has failed • SSA, SSB, SSC or the PCS is stuck (mechanical fault) • A/T 2–4 brake bank or 3–4 clutch is slipping (mechanical fault) • A/T 1–2, 2–3 or 3–4 shift valve is stuck (mechanical fault) • A/T Pressure regulator, pressure modifier or solenoid reducing valve is stuck • PCM has failed
DTC: P0736 **1T CCM, MIL: No** **Years:** 2007, 2008 **Models:** CX-7, CX-9, Mazda6, Mazda6 Sport Wagon, MazdaSpeed6, **Engines:** All **Transmissions:** A/T	**Incorrect Gear Ratio Detected** With engine running, vehicle speed signal normal, driving in reverse with the accelerator opening angle at 10 percent or more, the PCM detected an incorrect gear ratio. **Possible Causes:** • ATF level low • Deteriorated ATF • Shift solenoid C stuck • Shift solenoid E stuck • Line pressure control solenoid stuck • Line pressure low • B1 brake slipping • C2 clutch slipping • Control valve stuck • Oil pump malfunction • TCM malfunction
DTC: P0740 **2T CCM, MIL: Yes** **Years:** 2007, 2008 **Models:** B-Series Truck, RX-8, Tribute **Engines:** All **Transmissions:** A/T	**Torque Converter Clutch System Malfunction** DTC P0500, P0710, P0715, P0755 and P0760 not set, engine started, then driven in Drive in 4th Gear (4GR) to a speed of 47–64 MPH with the throttle open, engine speed over 500 RPM, ATF sensor signal more than 68°F, Turbine speed over 75 RPM, brake switch indicating "off", TCC system operating, O/D OFF Switch is "off", and the PCM detected the difference between the engine speed and the Turbine speed was more than a preset value stored in memory. **Possible Causes:** • ATF level is too low, or the ATF is badly deteriorated • A/T line pressure is low, or the oil pump has failed • A/T control valve has failed (it may be stuck) • TCC solenoid valve or pressure control valve is stuck • TCC in the torque converter is slipping (mechanical fault) • TCC shift valve or converter relief valve is stuck • Pressure modifier or the pressure regulator valve is stuck • Solenoid reducing valve is stuck (mechanical fault) • PCM has failed
DTC: P0741 **2T CCM, MIL: Yes** **Years:** 2007, 2008 **Models:** B-Series Truck, Mazda3, MazdaSpeed3, Mazda5, Mazda6, MazdaSpeed6, Tribute **Engines:** All **Transmissions:** All	**Torque Converter Clutch Performance** DTC P0500, P0705, P0706, P0715, P0720, P1740, P1742, P1751, P1752, P1756, P1757, P1771 and P1772 not set, engine started, vehicle driven in Drive at 60 MPH (± 3 MPH) for 20 seconds, engine speed over 600 RPM, ECT sensor signal more than 113°F, brake pedal switch indicating "off", throttle angle over 8 percent, and the PCM detected the difference between the engine speed and Turbine speed was more than a predetermined value. **Possible Causes:** • ATF level is too low, or the ATF is badly deteriorated • A/T line pressure is low, or the oil pump has failed • A/T control valve is stuck, or it is damaged (a mechanical fault) • TCC solenoid is stuck "off", or is damaged (a mechanical fault) • TCC system has failed • TCM has failed

DTC	Trouble Code Title, Conditions & Possible Causes
DTC: P0742 **2T CCM, MIL: Yes** **Years:** 2007, 2008 **Models:** B-Series Truck, Mazda3, MazdaSpeed3, Mazda5, Mazda6, MazdaSpeed6, Tribute **Engines:** All **Transmissions:** All	**Torque Converter Clutch Performance** DTC P0705, P0706, P0715, P0720, P1740, P1742, P1751, P1752, P1756, P1757, P1771 and P1772 not set, engine started, vehicle driven in Drive at 60 MPH (± 3 MPH) for 20 seconds, engine speed over 600 RPM, ECT sensor signal more than 113°F, brake pedal switch indicating "off", throttle angle over 8 percent, and the PCM detected the difference between the engine speed and Turbine speed was more than a predetermined value. **Possible Causes:** • ATF level is too low, or the ATF is badly deteriorated • A/T line pressure is low, or the oil pump has failed • A/T control valve is stuck, or it is damaged (a mechanical fault) • TCC solenoid is stuck "on", or is damaged (a mechanical fault) • TCC system has failed • TCM has failed
DTC: P0743 **1T CCM, MIL: Yes** **Years:** 2007, 2008 **Models:** B-Series Truck, RX-8, Tribute **Engines:** All **Transmissions:** A/T	**Torque Converter Clutch Solenoid Circuit Malfunction** Engine running and the PCM detected an unexpected voltage condition on the Torque Converter Clutch (TCC) solenoid control circuit during the CCM test. **Possible Causes:** • TCC control circuit is open (continuous high signal) • TCC control circuit is shorted to ground (continuous low signal) • TCC control circuit is shorted to system power (B+) • TCC solenoid is damaged or has failed • PCM has failed
DTC: P0744 **1T CCM, MIL: Yes** **Years:** 2007, 2008 **Models:** B-Series Truck, Mazda5, Mazda6, MazdaSpeed6, Tribute **Engines:** All **Transmissions:** A/T	**Torque Converter Clutch Solenoid Circuit Malfunction** Engine running and the PCM detected an unexpected voltage condition on the Torque Converter Clutch (TCC) solenoid control circuit during the CCM test. **Possible Causes:** • TCC control circuit is open (continuous high signal) • TCC control circuit is shorted to ground (continuous low signal) • TCC control circuit is shorted to system power (B+) • TCC solenoid is damaged or has failed • PCM has failed
DTC: P0745 **1T CCM, MIL: No** **Years:** 2007, 2008 **Models:** Mazda3, MazdaSpeed3, Mazda5, Mazda6, MazdaSpeed6, Tribute **Engines:** All **Transmissions:** A/T	**Electronic Pressure Control Solenoid Circuit Malfunction** Engine running vehicle driven to a speed of over 37 MPH, and the PCM detected an unexpected voltage condition on the Electronic Pressure Control (EPC) solenoid control circuit during the CCM test. **Possible Causes:** • PCS control circuit is open (continuous high signal) • PCS control circuit is shorted to ground (continuous low signal) • PCS control circuit is shorted to system power (B+) • PCS is damaged or has failed • PCM has failed
DTC: P0746 **1T CCM, MIL: No** **Years:** 2007, 2008 **Models:** B-Series Truck, Tribute **Engines:** All **Transmissions:** A/T	**Electronic Pressure Control Solenoid Circuit Malfunction** Engine running vehicle driven to a speed of over 37 MPH, and the PCM detected an unexpected voltage condition on the Electronic Pressure Control (EPC) solenoid control circuit during the CCM test. **Possible Causes:** • PCS control circuit is open (continuous high signal) • PCS control circuit is shorted to ground (continuous low signal) • PCS control circuit is shorted to system power (B+) • PCS is damaged or has failed • PCM has failed
DTC: P0747 **1T CCM, MIL: No** **Years:** 2007, 2008 **Models:** B-Series Truck, Tribute **Engines:** All **Transmissions:** A/T	**Electronic Pressure Control Solenoid Circuit Malfunction** Engine running vehicle driven to a speed of over 37 MPH, and the PCM detected an unexpected voltage condition on the Electronic Pressure Control (EPC) solenoid control circuit during the CCM test. **Possible Causes:** • PCS control circuit is open (continuous high signal) • PCS control circuit is shorted to ground (continuous low signal) • PCS control circuit is shorted to system power (B+) • PCS is damaged or has failed • PCM has failed

DTC	Trouble Code Title, Conditions & Possible Causes
DTC: P0748 **1T CCM, MIL: No** **Years:** 2007, 2008 **Models:** B-Series Truck, RX-8, Tribute **Engines:** All **Transmissions:** A/T	**Electronic Pressure Control Solenoid Circuit Malfunction** Engine running vehicle driven to a speed of over 37 MPH, and the PCM detected an unexpected voltage condition on the Electronic Pressure Control (EPC) solenoid control circuit during the CCM test. **Possible Causes:** • PCS control circuit is open (continuous high signal) • PCS control circuit is shorted to ground (continuous low signal) • PCS control circuit is shorted to system power (B+) • PCS is damaged or has failed • PCM has failed
DTC: P0749 **1T CCM, MIL: No** **Years:** 2007, 2008 **Models:** B-Series Truck, Tribute **Engines:** All **Transmissions:** A/T	**Electronic Pressure Control Solenoid Circuit Malfunction** Engine running vehicle driven to a speed of over 37 MPH, and the PCM detected an unexpected voltage condition on the Electronic Pressure Control (EPC) solenoid control circuit during the CCM test. **Possible Causes:** • PCS control circuit is open (continuous high signal) • PCS control circuit is shorted to ground (continuous low signal) • PCS control circuit is shorted to system power (B+) • PCS is damaged or has failed • PCM has failed
DTC: P0750 **1T CCM, MIL: Yes** **Years:** 2007, 2008 **Models:** B-Series Truck, Tribute **Engines:** All **Transmissions:** A/T	**A/T Shift Solenoid 'A' Circuit Malfunction** Engine started, and the PCM detected an unexpected voltage condition on the Shift Solenoid 'A' (SSA) circuit during the CCM test. **Possible Causes:** • SSA control circuit is open (continuous high signal) • SSA control circuit is shorted to ground (continuous low signal) • SSA control circuit is shorted to system power (B+) • SSA is damaged or has failed • PCM has failed
DTC: P0751 **2T CCM, MIL: Yes** **Years:** 2007, 2008 **Models:** B-Series Truck, Mazda3, MazdaSpeed3, Mazda5, Mazda6, MazdaSpeed6, MX-5, RX-8, Tribute **Engines:** All **Transmissions:** A/T	**A/T Shift Solenoid 'A' Performance (Mechanical)** DTC P0705, P0706, P0715, P0720, P1740, P1742, P1751, P1752, P1756, P1757, P1771 and P1772 not set, engine started, vehicle driven in Drive at 60 MPH (± 3 MPH) for 20 seconds, engine speed over 600 RPM, ECT sensor signal more than 113°F, brake pedal switch indicating "off", throttle angle over 8 percent, and the PCM detected the difference between the engine speed and Turbine speed was more than a predetermined value. **Possible Causes:** • ATF level is too low, or the ATF is badly deteriorated • A/T line pressure is low, or the oil pump has failed • A/T control valve is stuck, or it is damaged (a mechanical fault) • SSA is stuck "off", or is damaged (a mechanical fault) • Transmission is damaged or has failed • TCM has failed
DTC: P0752 **2T CCM, MIL: Yes** **Years:** 2007, 2008 **Models:** B-Series Truck, Mazda3, MazdaSpeed3, Mazda5, Mazda6, MazdaSpeed6, MX-5, RX-8, Tribute **Engines:** All **Transmissions:** A/T	**A/T Shift Solenoid 'A' Performance (Mechanical)** DTC P0705, P0706, P0715, P0720, P1740, P1742, P1751, P1752, P1756, P1757, P1771 and P1772 not set, engine started, vehicle driven in Drive at 60 MPH (± 3 MPH) for 20 seconds, engine speed over 600 RPM, ECT sensor signal more than 113°F, brake pedal switch indicating "off", throttle angle over 8 percent, and the PCM detected the difference between the engine speed and Turbine speed was more than a predetermined value. **Possible Causes:** • ATF level is too low, or the ATF is badly deteriorated • A/T line pressure is low, or the oil pump has failed • A/T control valve is stuck, or it is damaged (a mechanical fault) • SSA is stuck "on", or is damaged (a mechanical fault) • Transmission is damaged or has failed • TCM has failed

DTC	Trouble Code Title, Conditions & Possible Causes
DTC: P0753 **1T CCM, MIL: Yes** **Years:** 2007, 2008 **Models:** B-Series Truck, Mazda3, MazdaSpeed3, Mazda5, Mazda6, MazdaSpeed6, RX-8, Tribute **Engines:** All **Transmissions:** A/T	**A/T Shift Solenoid 'A' Circuit Malfunction** Engine started, then driven to a speed of over 37 MPH, and the PCM detected an unexpected voltage condition on the Shift Solenoid 'A' (SSB) control circuit during the CCM test. **Possible Causes:** • SSB control circuit is open (continuous high signal) • SSB control circuit is shorted to ground (continuous low signal) • SSB control circuit is shorted to system power (B+) • SSB is damaged or has failed • PCM has failed
DTC: P0754 **1T CCM, MIL: Yes** **Years:** 2007, 2008 **Models:** B-Series Truck, Tribute **Engines:** All **Transmissions:** A/T	**A/T Shift Solenoid 'A' Circuit Malfunction** Engine started, and the PCM detected an unexpected voltage condition on the Shift Solenoid 'A' (SSA) circuit during the CCM test. **Possible Causes:** • SSA control circuit is open (continuous high signal) • SSA control circuit is shorted to ground (continuous low signal) • SSA control circuit is shorted to system power (B+) • SSA is damaged or has failed • PCM has failed
DTC: P0755 **1T CCM, MIL: Yes** **Years:** 2007, 2008 **Models:** B-Series Truck, Tribute **Engines:** All **Transmissions:** A/T	**A/T Shift Solenoid 'B' Circuit Malfunction** Engine started, and the PCM detected an unexpected voltage condition on the Shift Solenoid 'B' (SSA) circuit during the CCM test. **Possible Causes:** • SSA control circuit is open (continuous high signal) • SSA control circuit is shorted to ground (continuous low signal) • SSA control circuit is shorted to system power (B+) • SSA is damaged or has failed • PCM has failed
DTC: P0756 **2T CCM, MIL: Yes** **Years:** 2007, 2008 **Models:** B-Series Truck, Mazda3, MazdaSpeed3, Mazda5, Mazda6, MazdaSpeed6, MX-5, RX-8, Tribute **Engines:** All **Transmissions:** A/T	**A/T Shift Solenoid 'B' Performance (Mechanical)** DTC P0705, P0706, P0715, P0720, P1740, P1742, P1751, P1752, P1756, P1757, P1771 and P1772 not set, engine started, vehicle driven in Drive at 60 MPH (± 3 MPH) for 20 seconds, engine speed over 600 RPM, ECT sensor signal more than 113°F, brake pedal switch indicating "off", throttle angle over 8 percent, and the PCM detected the difference between the engine speed and Turbine speed was more than a predetermined value. **Possible Causes:** • ATF level is too low, or the ATF is badly deteriorated • A/T line pressure is low, or the oil pump has failed • A/T control valve is stuck, or it is damaged (a mechanical fault) • SSB is stuck "off", or is damaged (a mechanical fault) • Transmission is damaged or has failed • TCM has failed
DTC: P0757 **2T CCM, MIL: Yes** **Years:** 2007, 2008 **Models:** B-Series Truck, Mazda3, MazdaSpeed3, Mazda5, Mazda6, MazdaSpeed6, MX-5, RX-8, Tribute **Engines:** All **Transmissions:** A/T	**A/T Shift Solenoid 'B' Performance (Mechanical)** DTC P0705, P0706, P0715, P0720, P1740, P1742, P1751, P1752, P1756, P1757, P1771 and P1772 not set, engine started, vehicle driven in Drive at 60 MPH (± 3 MPH) for 20 seconds, engine speed over 600 RPM, ECT sensor signal more than 113°F, brake pedal switch indicating "off", throttle angle over 8 percent, and the PCM detected the difference between the engine speed and Turbine speed was more than a predetermined value. **Possible Causes:** • ATF level is too low, or the ATF is badly deteriorated • A/T line pressure is low, or the oil pump has failed • A/T control valve is stuck, or it is damaged (a mechanical fault) • SSB is stuck "on", or is damaged (a mechanical fault) • Transmission is damaged or has failed • TCM has failed
DTC: P0758 **1T CCM, MIL: Yes** **Years:** 2007, 2008 **Models:** B-Series Truck, Mazda3, MazdaSpeed3, Mazda5, Mazda6, MazdaSpeed6, RX-8, Tribute **Engines:** All **Transmissions:** A/T	**A/T Shift Solenoid 'B' Circuit Malfunction** Engine started, and the PCM detected an unexpected voltage condition on the Shift Solenoid 'B' (SSB) control circuit during the CCM test. **Possible Causes:** • SSB control circuit is open (continuous high signal) • SSB control circuit is shorted to ground (continuous low signal) • SSB control circuit is shorted to system power (B+) • SSB is damaged or has failed • PCM has failed

DTC	Trouble Code Title, Conditions & Possible Causes
DTC: P0759 **1T CCM, MIL: Yes** **Years:** 2007, 2008 **Models:** B-Series Truck, Tribute **Engines:** All **Transmissions:** A/T	**A/T Shift Solenoid 'B' Circuit Malfunction** Engine started, and the PCM detected an unexpected voltage condition on the Shift Solenoid 'B' (SSA) circuit during the CCM test. **Possible Causes:** • SSA control circuit is open (continuous high signal) • SSA control circuit is shorted to ground (continuous low signal) • SSA control circuit is shorted to system power (B+) • SSA is damaged or has failed • PCM has failed
DTC: P0760 **1T CCM, MIL: Yes** **Years:** 2007, 2008 **Models:** B-Series Truck **Engines:** All **Transmissions:** A/T	**A/T Shift Solenoid 'C' Circuit Malfunction** Engine started, and the PCM detected an unexpected voltage condition on the Shift Solenoid 'C' (SSA) circuit during the CCM test. **Possible Causes:** • SSA control circuit is open (continuous high signal) • SSA control circuit is shorted to ground (continuous low signal) • SSA control circuit is shorted to system power (B+) • SSA is damaged or has failed • PCM has failed
DTC: P0761 **1T CCM, MIL: Yes** **Years:** 2007, 2008 **Models:** B-Series Truck, Mazda3, MazdaSpeed3, Mazda5, Mazda6, MazdaSpeed6, MX-5, RX-8 **Engines:** All **Transmissions:** A/T	**A/T Shift Solenoid 'C' Performance (Mechanical)** 1st Gear Test DTC P0733 and P0734 not set, engine started, and then driven in Drive in 1st Gear (1GR), Turbine speed at 4988 RPM (± 225), ATF temperature at 68°F or higher, engine speed over 450 RPM, differential gear case output speed over 35 MPH, and the PCM detected the revolution ratio of the forward clutch drum revolution to the differential gear case revolution was less than 2.157. **Possible Causes:** • ATF level is too low, or the ATF is badly deteriorated • A/T SSC is stuck "off", or it is damaged (a mechanical fault) • A/T control valve is stuck, or it is damaged (a mechanical fault) • PCM has failed
DTC: P0762 **2T CCM, MIL: Yes** **Years:** 2007, 2008 **Models:** B-Series Truck, Mazda3, MazdaSpeed3, Mazda5, Mazda6, MazdaSpeed6, MX-5, RX-8 **Engines:** All **Transmissions:** A/T	**A/T Shift Solenoid 'C' Performance (Mechanical)** 3rd Gear Test DTC P0731 and P0732 not set, engine started, and then driven in Drive in 3rd Gear (3GR), Turbine speed at 4988 RPM (± 225), ATF temperature at 68°F or higher, engine speed over 450 RPM, differential gear case output speed over 35 MPH, and the PCM detected the revolution ratio of the forward clutch drum revolution to the differential gear case revolution was less than 0.863, or it was more than 1.249 during the CCM rationality test. 4th Gear Test With the conditions listed above for the 3rd Gear Test met, the PCM detected the revolution ratio of the forward clutch drum revolution to differential gear case revolution was less than 0.6, or it was 1.249 or higher while driving in 4th gear (4GR) in the CCM rationality test. **Possible Causes:** • ATF level is too low, or the ATF is badly deteriorated • A/T SSC is stuck "on", or it is damaged (a mechanical fault) • A/T control valve stuck "on", or it is damaged (mechanical fault) • PCM has failed
DTC: P0763 **1T CCM, MIL: Yes** **Years:** 2007, 2008 **Models:** B-Series Truck, Mazda3, MazdaSpeed3, Mazda5, Mazda6, MazdaSpeed6, RX-8 **Engines:** All **Transmissions:** A/T	**A/T Shift Solenoid 'C' Circuit Malfunction** Engine started, then driven in Drive in 4th Gear and the PCM detected an unexpected voltage condition on the Shift Solenoid 'C' (SSC) circuit during the CCM test. **Possible Causes:** • SSC control circuit is open (continuous high signal) • SSC control circuit is shorted to ground (continuous low signal) • SSC control circuit is shorted to system power (B+) • SSC is damaged or has failed • PCM has failed

DTC	Trouble Code Title, Conditions & Possible Causes
DTC: P0764 **1T CCM, MIL: Yes** **Years:** 2007, 2008 **Models:** B-Series Truck **Engines:** All **Transmissions:** A/T	**A/T Shift Solenoid 'C' Circuit Malfunction** Engine started, then driven in Drive in 4th Gear and the PCM detected an unexpected voltage condition on the Shift Solenoid 'C' (SSC) circuit during the CCM test. **Possible Causes:** • SSC control circuit is open (continuous high signal) • SSC control circuit is shorted to ground (continuous low signal) • SSC control circuit is shorted to system power (B+) • SSC is damaged or has failed • PCM has failed
DTC: P0765 **1T CCM, MIL: Yes** **Years:** 2007, 2008 **Models:** B-Series Truck **Engines:** All **Transmissions:** A/T	**A/T Shift Solenoid 'D' Circuit Malfunction** Engine started, then driven in Drive in 4th Gear and the PCM detected an unexpected voltage condition on the Shift Solenoid 'D' (SSC) circuit during the CCM test. **Possible Causes:** • SSC control circuit is open (continuous high signal) • SSC control circuit is shorted to ground (continuous low signal) • SSC control circuit is shorted to system power (B+) • SSC is damaged or has failed • PCM has failed
DTC: P0766 **2T CCM, MIL: Yes** **Years:** 2007, 2008 **Models:** B-Series Truck, Mazda3, MazdaSpeed3, Mazda5, Mazda6, MazdaSpeed6, MX-5, RX-8 **Engines:** All **Transmissions:** A/T	**A/T Shift Solenoid 'D' Performance (Mechanical)** 4th Gear Test DTC P0731, P0132 and P0733 not set, engine started, then driven in Drive in 4th Gear (4GR), Turbine speed at 4988 RPM (± 225), ATF temperature over 68°F, engine speed over 450 RPM, differential gear case output speed over 35 MPH, and the PCM detected the revolution ratio of the forward clutch drum revolution to the differential gear case revolution was less than 0.60, or it was 1.249 or higher during the CCM rationality test. **Possible Causes:** • ATF level is too low, or the ATF is badly deteriorated • A/T SSD is stuck "off", or it is damaged (a mechanical fault) • A/T control valve is stuck, or it is damaged (a mechanical fault) • PCM has failed
DTC: P0767 **2T CCM, MIL: Yes** **Years:** 2007, 2008 **Models:** B-Series Truck, Mazda3, MazdaSpeed3, Mazda5, Mazda6, MazdaSpeed6, RX-8 **Engines:** All **Transmissions:** A/T	**A/T Shift Solenoid 'D' Performance (Mechanical)** 3rd Gear Test DTC P0731, P0132, P0734 and P0741 not set, engine started, then driven in Drive in 3rd Gear (3GR) Turbine speed at 4988 RPM (± 225), engine speed over 450 RPM, differential gear case output speed over 35 MPH, ATF temperature over 68°F, and the PCM detected the revolution ratio of the forward clutch drum revolution to the differential gear case revolution was less than 0.863, or it was 1.249 or higher during the CCM rationality test. **Possible Causes:** • ATF level is too low, or the ATF is badly deteriorated • A/T SSD is stuck "on", or it is damaged (a mechanical fault) • A/T control valve is stuck, or it is damaged (a mechanical fault) • PCM has failed
DTC: P0768 **1T CCM, MIL: Yes** **Years:** 2007, 2008 **Models:** B-Series Truck, Mazda3, MazdaSpeed3, Mazda5, Mazda6, MazdaSpeed6, RX-8 **Engines:** All **Transmissions:** A/T	**A/T Shift Solenoid 'D' Circuit Malfunction** Engine started, then driven in Drive in 4th Gear, and the PCM detected an unexpected voltage condition on the Shift Solenoid 'D' (SSD) control circuit during the CCM test. **Possible Causes:** • SSD control circuit is open (continuous high signal) • SSD control circuit is shorted to ground (continuous low signal) • SSD control circuit is shorted to system power (B+) • SSD is damaged or has failed • PCM has failed
DTC: P0769 **1T CCM, MIL: Yes** **Years:** 2007, 2008 **Models:** B-Series Truck **Engines:** All **Transmissions:** A/T	**A/T Shift Solenoid 'D' Circuit Malfunction** Engine started, then driven in Drive in 4th Gear, and the PCM detected an unexpected voltage condition on the Shift Solenoid 'D' (SSD) control circuit during the CCM test. **Possible Causes:** • SSD control circuit is open (continuous high signal) • SSD control circuit is shorted to ground (continuous low signal) • SSD control circuit is shorted to system power (B+) • SSD is damaged or has failed • PCM has failed

DTC	Trouble Code Title, Conditions & Possible Causes
DTC: P0771 **2T CCM, MIL: Yes** **Years:** 2007, 2008 **Models:** Mazda3, MazdaSpeed3, Mazda5, Mazda6, Mazda6 Sport Wagon, MazdaSpeed6 **Engines:** All **Transmissions:** A/T	**A/T Shift Solenoid 'E' Performance (Mechanical)** 4th Gear Test DTC P0731, P0132 and P0733 not set, engine started, then driven in Drive in 4th Gear (4GR) at 37–62 MPH, Turbine speed at 4988 RPM (± 225), engine speed over 450 RPM, TCC operating in normal or power mode, SSA commanded to over 99 percent, ATF temperature at 68°F or higher, and the PCM detected the difference between the engine speed and turbine speed was more than 100 RPM. **Possible Causes:** • ATF level is too low, or the ATF is badly deteriorated • A/T SSE is stuck "off", or it is damaged (a mechanical fault) • A/T control valve is stuck, or it is damaged (a mechanical fault) • PCM has failed
DTC: P0772 **2T CCM, MIL: Yes** **Years:** 2007, 2008 **Models:** Mazda3, MazdaSpeed3, Mazda5, Mazda6, Mazda6 Sport Wagon, MazdaSpeed6 **Engines:** All **Transmissions:** A/T	**A/T Shift Solenoid 'E' Performance (Mechanical)** 4th Gear Test DTC P0734 not set, engine started, then driven in Drive in 4th Gear (4GR) at a speed of less than 43 MPH, Turbine speed at 4988 RPM (± 225), engine speed over 450 RPM, ATF temperature over 68°F, TCC not enabled, throttle angle over 6.25 percent for 10 seconds, or throttle angle closed for 10 seconds, or throttle angle from 3.125–6.25 percent for at least 3 seconds, and the PCM detected the difference between the engine speed and turbine speed was less than 50 RPM. **Possible Causes:** • ATF level is too low, or the ATF is badly deteriorated • A/T SSE is stuck "on", or it is damaged (a mechanical fault) • A/T control valve is stuck, or it is damaged (a mechanical fault) • PCM has failed
DTC: P0773 **1T CCM, MIL: Yes** **Years:** 2007, 2008 **Models:** Mazda3, MazdaSpeed3, Mazda5, Mazda6, Mazda6 Sport Wagon, MazdaSpeed6 **Engines:** All **Transmissions:** A/T	**A/T Shift Solenoid 'E' Circuit Malfunction** Engine started, vehicle driven in Drive (4th Gear) with the TCC system engaged, and the PCM detected an unexpected voltage condition on the SSE circuit during the CCM test. **Possible Causes:** • SSE control circuit is open (continuous high signal) • SSE control circuit is shorted to ground (continuous low signal) • SSE control circuit is shorted to system power (B+) • SSE is damaged or has failed • PCM has failed
DTC: P0775 **1T CCM, MIL: Yes** **Years:** 2007, 2008 **Models:** Tribute **Engines:** All **Transmissions:** A/T	**A/T Shift Solenoid 'E' Circuit Malfunction** Engine started, vehicle driven in Drive (4th[h] Gear) with the TCC system engaged, and the PCM detected an unexpected voltage condition on the SSE circuit during the CCM test. **Possible Causes:** • SSE control circuit is open (continuous high signal) • SSE control circuit is shorted to ground (continuous low signal) • SSE control circuit is shorted to system power (B+) • SSE is damaged or has failed • PCM has failed
DTC: P0776 **1T CCM, MIL: Yes** **Years:** 2007, 2008 **Models:** Tribute **Engines:** All **Transmissions:** A/T	**A/T Shift Solenoid 'E' Circuit Malfunction** Engine started, vehicle driven in Drive (4th[h] Gear) with the TCC system engaged, and the PCM detected an unexpected voltage condition on the SSE circuit during the CCM test. **Possible Causes:** • SSE control circuit is open (continuous high signal) • SSE control circuit is shorted to ground (continuous low signal) • SSE control circuit is shorted to system power (B+) • SSE is damaged or has failed • PCM has failed
DTC: P0777 **1T CCM, MIL: Yes** **Years:** 2007, 2008 **Models:** Mazda5, Mazda6, Mazda6 Sport Wagon, MazdaSpeed6, Tribute **Engines:** All **Transmissions:** A/T	**A/T Shift Solenoid 'E' Circuit Malfunction** Engine started, vehicle driven in Drive (4th[h] Gear) with the TCC system engaged, and the PCM detected an unexpected voltage condition on the SSE circuit during the CCM test. **Possible Causes:** • SSE control circuit is open (continuous high signal) • SSE control circuit is shorted to ground (continuous low signal) • SSE control circuit is shorted to system power (B+) • SSE is damaged or has failed • PCM has failed

DTC	Trouble Code Title, Conditions & Possible Causes
DTC: P0778 **1T CCM, MIL: Yes** **Years:** 2007, 2008 **Models:** Mazda5, Mazda6, Mazda6 Sport Wagon, MazdaSpeed6, Tribute **Engines:** All **Transmissions:** A/T	**A/T Shift Solenoid 'E' Circuit Malfunction** Engine started, vehicle driven in Drive (4th[h] Gear) with the TCC system engaged, and the PCM detected an unexpected voltage condition on the SSE circuit during the CCM test. **Possible Causes:** • SSE control circuit is open (continuous high signal) • SSE control circuit is shorted to ground (continuous low signal) • SSE control circuit is shorted to system power (B+) • SSE is damaged or has failed • PCM has failed
DTC: P0779 **1T CCM, MIL: Yes** **Years:** 2007, 2008 **Models:** Tribute **Engines:** All **Transmissions:** A/T	**A/T Shift Solenoid 'E' Circuit Malfunction** Engine started, vehicle driven in Drive (4th[h] Gear) with the TCC system engaged, and the PCM detected an unexpected voltage condition on the SSE circuit during the CCM test. **Possible Causes:** • SSE control circuit is open (continuous high signal) • SSE control circuit is shorted to ground (continuous low signal) • SSE control circuit is shorted to system power (B+) • SSE is damaged or has failed • PCM has failed
DTC: P0780 **1T CCM, MIL: Yes** **Years:** 2007, 2008 **Models:** CX-7, CX-9, Mazda6, Mazda6 Sport Wagon, MazdaSpeed6 **Engines:** All **Transmissions:** A/T	**Valve Control Solenoid Circuit Malfunction** Engine started, vehicle driven in Drive with counter drive gear revolution speed at 300RPM ore more, the PCM detected an irregular shift control. **Possible Causes:** • Shift solenoid A stuck • Shift solenoid B stuck • Shift solenoid C stuck • Shift solenoid D stuck • Shift solenoid E stuck • Shift solenoid F stuck • Line pressure control solenoid stuck • Control valve stuck • TCM malfunction
DTC: P0781 **2T CCM, MIL: Yes** **Years:** 2007, 2008 **Models:** MX-5, RX-8 **Engines:** All **Transmissions:** A/T	**1–2 Shift Valve Malfunction** The TCM detects a 1–2 shift valve malfunction. **Possible Causes:** • ATF level low • Deteriorated ATF • Line pressure low • Stuck 1–2 shift valve • Control valve body malfunction • TCM malfunction
DTC: P0791 **2T CCM, MIL: Yes** **Years:** 2007, 2008 **Models:** B-Series Truck, Mazda5, Mazda6, Mazda6 Sport Wagon, MazdaSpeed6 **Engines:** All **Transmissions:** A/T	**Input Shaft Speed (ISS) Sensor** PCM detected ISS sensor signal failure. **Possible Causes:** • Input Shaft Speed Sensor circuit is open • ISS circuit is shorted to ground • ISS circuit is shorted to system power (B+) • ISS Sensor is damaged or has failed • PCM has failed
DTC: P0794 **2T CCM, MIL: Yes** **Years:** 2007, 2008 **Models:** B-Series Truck **Engines:** All **Transmissions:** A/T	**Input Shaft Speed (ISS) Sensor** PCM detected an intermittent failure with the ISS sensor signal. **Possible Causes:** • Input Shaft Speed Sensor circuit is open • ISS circuit is shorted to ground • ISS circuit is shorted to system power (B+) • ISS Sensor is damaged or has failed • PCM has failed

DTC	Trouble Code Title, Conditions & Possible Causes
DTC: P0841 **2T CCM, MIL: Yes** **Years:** 2007, 2008 **Models:** Mazda3 **Engines:** All **Transmissions:** A/T	**Neutral switch input circuit problem** The PCM monitors changes in input voltage from the neutral switch. If the PCM does not detect PCM terminal 1S voltage changes while running vehicle with vehicle speed above 30 km/h {19 MPH} and clutch pedal turns press and depress 10 times repeatedly, the PCM determines that the neutral switch circuit has malfunction. **Possible Causes:** • Neutral switch malfunction • Poor connection of neutral switch connector or PCM connector • Short to ground in wiring harness between neutral switch terminal A and PCM terminal 1S • Open circuit in wiring harness between ground and neutral switch terminal B • PCM has failed
DTC: P0850 **2T CCM, MIL: Yes** **Years:** 2007, 2008 **Models:** B-Series Truck, CX-7, CX-9, Mazda3, MazdaSpeed3, Mazda5, Mazda6, MazdaSpeed6, MX-5, RX-8, Tribute **Engines:** All **Transmissions:** A/T	**Neutral switch input circuit problem** The PCM monitors changes in input voltage from the neutral switch. If the PCM does not detect PCM terminal 1S voltage changes while running vehicle with vehicle speed above 30 km/h {19 MPH} and clutch pedal turns press and depress 10 times repeatedly, the PCM determines that the neutral switch circuit has malfunction. **Possible Causes:** • Neutral switch malfunction • Poor connection of neutral switch connector or PCM connector • Short to ground in wiring harness between neutral switch terminal A and PCM terminal 1S • Open circuit in wiring harness between ground and neutral switch terminal B • PCM has failed

Gas Engine OBD II Trouble Code List (P1xxx Codes)

DTC	Trouble Code Title, Conditions & Possible Causes
DTC: P1000 **1T CCM, MIL: Yes** **Years:** 2007, 2008 **Models:** B-Series Truck, CX-7, CX-9, Mazda3, MazdaSpeed3, Mazda5, Mazda6, MazdaSpeed6, MX-5, RX-8, Tribute **Engines:** All **Transmissions:** All	**OBD II Monitor Testing Not Completed** Key on or engine running, and the PCM detected that DTC P1000 was set because one of the OBD II Main Monitors did not complete the I/M Readiness Test sequence. The recommended next step is to drive vehicle through the complete Mazda Drive Cycle pattern. **Note: This trouble code is deleted if the MIL is activated.** **Possible Causes:** • A PCM Reset step was performed with an OBD II Scan Tool • Battery keep alive power (KAPWR) was removed to the PCM • One or more OBD II Monitors did not complete during an official OBD II Drive Cycle
DTC: P1001 **1T CCM, MIL: No** **Years:** 2007, 2008 **Models:** B-Series Truck, CX-7, CX-9, Mazda3, MazdaSpeed3, Mazda5, Mazda6, MazdaSpeed6, MX-5, RX-8, Tribute **Engines:** All **Transmissions:** All	**Data Link Connector Circuit Malfunction** Engine running and the PCM detected an unexpected voltage condition on the SCP communication circuit to the PCM. **Note: This code indicates the Scan Tool could not "talk" to the PCM.** **Possible Causes:** • DLC connector or the related terminals/pins are damaged • DLC Bus (+) circuit is open or shorted to ground • DLC Bus (+) circuit is shorted at an associated module • DLC ground circuit is open or has a high resistance condition • PCM power relay is damaged or has failed (no power to DLC) • Incorrect Self-Test procedure followed, or idle speed to low • PCM has failed (e.g., the PCM VREF may be out-of-range)
DTC: P1260 **1T CCM, MIL: Yes** **Years:** 2007, 2008 **Models:** Mazda3 **Engines:** All **Transmissions:** All	**Anti-Theft System Signal Detected, Engine Disabled** Key on, and the PCM received a signal from the Anti-Theft System that a theft condition had occurred. The theft indicator on the dash will flash rapidly or remain on "solid" with the ignition switch in the "on" position. The engine may "start and stall", or may not crank if the vehicle is equipped with the PATS starter "disable" feature. **Possible Causes:** • Previous theft condition has occurred • Anti-Theft System is damaged or has failed

DTC	Trouble Code Title, Conditions & Possible Causes
DTC: P1487 **1T CCM, MIL: Yes** **Years:** 2007, 2008 **Models:** B-Series Truck, CX-7, CX-9, Mazda3, MazdaSpeed3, Mazda5, Mazda6, MazdaSpeed6, MX-5, RX-8, Tribute **Engines:** All **Transmissions:** All	**EGR Boost Sensor Solenoid Valve Circuit Malfunction** Key on, and the PCM detected an unexpected voltage condition on the EGR Boost Sensor Check Solenoid valve during the CCM test. **Note: This is a diagnostic support code - the PCM monitors this circuit once per key cycle, but does not set a pending code.** **Possible Causes:** • EGR boost sensor solenoid circuit is open or shorted to ground • EGR boost sensor solenoid circuit is shorted to system power • EGR boost sensor solenoid circuit is open or shorted to ground • EGR boost sensor solenoid is damaged or has failed • PCM has failed
DTC: P1562 **1T CCM, MIL: Yes** **Years:** 2007, 2008 **Models:** Mazda6 **Engines:** All **Transmissions:** All	**PCM +BB Voltage Low** Key on, engine cranking, and the PCM detected the +BB circuit voltage was too low (less than 2.5v) during the CCM test period. **Note: The +BB circuit is the backup or direct voltage circuit.** **Possible Causes:** • +BB circuit is open • +BB circuit is shorted to ground (check for a blown fuse) • +BB circuit has high resistance (check the battery connections) • PCM has failed
DTC: P1602 **1T PCM, MIL: No** **Years:** 2007, 2008 **Models:** B-Series Truck, CX-7, CX-9, Mazda3, MazdaSpeed3, Mazda5, Mazda6, MazdaSpeed6, MX-5, RX-8, Tribute **Engines:** All **Transmissions:** All	**PCM Communication Line to TCM Error** Key on, and the PCM detected the command transmission to the Immobilizer unit was too long, or it did not receive any response. **Possible Causes:** • Key (the transponder) is damaged or has failed • PCM communication line to Immobilizer is open, shorted to ground or shorted to system power • Transponder "coil" is damaged or has failed • PCM or the Immobilizer unit has failed
DTC: P1603 **1T PCM, MIL: No** **Years:** 2007, 2008 **Models:** B-Series Truck, CX-7, CX-9, Mazda3, MazdaSpeed3, Mazda5, Mazda6, MazdaSpeed6, MX-5, RX-8, Tribute **Engines:** All **Transmissions:** All	**Code Word Not Registered In The PCM** Key on, and the PCM detected the Immobilizer procedure "code word" was not performed after the PCM was replaced. **Possible Causes:** • Check for any related Technical Service Bulletins • Perform the Immobilizer "code word" reprogramming procedure
DTC: P1604 **1T PM, MIL: No** **Years:** 2007, 2008 **Models:** B-Series Truck, CX-7, CX-9, Mazda3, MazdaSpeed3, Mazda5, Mazda6, MazdaSpeed6, MX-5, RX-8, Tribute **Engines:** All **Transmissions:** All	**Key Identification Numbers Not Registered In The PCM** Key on, and the PCM detected the Key Identification (ID) numbers were not registered in the PCM was it was replaced. **Possible Causes:** • Check for any related Technical Service Bulletins • Perform the Immobilizer "Key ID" reprogramming procedure
DTC: P1608 **1T PCM, MIL: No** **Years:** 2007, 2008 **Models:** B-Series Truck, CX-7, CX-9, Mazda3, MazdaSpeed3, Mazda5, Mazda6, MazdaSpeed6, MX-5, RX-8, Tribute **Engines:** All **Transmissions:** All	**PCM Output Device Circuit Malfunction** Key on or engine running and the PCM detected an unexpected voltage condition on one or more of the output device circuits. The PCM cannot diagnose the devices properly under these conditions. **Possible Causes:** • Output device (relay or solenoid) control circuit(s) shorted to ground between the device and the PCM • If the output device circuits are okay, the PCM has failed

DTC	Trouble Code Title, Conditions & Possible Causes
DTC: P1621 **1T PCM, MIL: No** **Years:** 2007, 2008 **Models:** B-Series Truck, CX-7, CX-9, Mazda3, MazdaSpeed3, Mazda5, Mazda6, MazdaSpeed6, MX-5, RX-8, Tribute **Engines:** All **Transmissions:** All	**Key Identification Number Not Registered In The PCM** Engine cranking and the PCM detected the "code word" in the PCM and in the Immobilizer did not match. **Possible Causes:** • Immobilizer problem during transformation of the "code word" • PCM problem during transformation of the "code word" • Immobilizer Unit has failed • PCM has failed
DTC: P1622 **1T PCM, MIL: No** **Years:** 2007, 2008 **Models:** B-Series Truck, CX-7, CX-9, Mazda3, MazdaSpeed3, Mazda5, Mazda6, MazdaSpeed6, MX-5, RX-8, Tribute **Engines:** All **Transmissions:** All	**Key Identification Number Does Not Match In The PCM** Engine cranking, and the PCM detected the "code word" in the PCM and the Immobilizer did not match after the Immobilizer Unit was replaced. **Possible Causes:** • An "unregistered" key was used during Step 3 of the Immobilizer Unit replacement procedure • A problem occurred during transformation of the Key ID number stored in the PCM
DTC: P1623 **1T PCM, MIL: No** **Years:** 2007, 2008 **Models:** B-Series Truck, CX-7, CX-9, Mazda3, MazdaSpeed3, Mazda5, Mazda6, MazdaSpeed6, MX-5, RX-8, Tribute **Engines:** All **Transmissions:** All	**Code Word or Key ID Number Read/Write Error in the PCM** Engine cranking, and the PCM detected a "code word" or Key ID read/write error during the key on initialization period. **Possible Causes:** • Perform a PCM "reset" procedure - if the same trouble code resets, the PCM has failed • Reprogram the Immobilizer Unit after the PCM is replaced
DTC: P1624 **1T PCM, MIL: No** **Years:** 2007, 2008 **Models:** B-Series Truck, CX-7, CX-9, Mazda3, MazdaSpeed3, Mazda5, Mazda6, MazdaSpeed6, MX-5, RX-8, Tribute **Engines:** All **Transmissions:** All	**Immobilizer Communication Error** Engine cranking, engine does not start at least three (3) times, and the PCM detected a communication error occurred three (3) time while attempting to communicate with the Immobilizer Unit. **Possible Causes:** • Perform a PCM "reset" procedure • Attempt to start the engine three (3) times • If the code resets, check for other codes related to a No Start • Perform engine diagnostics to determine the cause of No Start

Gas Engine OBD II Trouble Code List (P2xxx Codes)

DTC	Trouble Code Title, Conditions & Possible Causes
DTC: P2005 **2T CCM, MIL: Yes** **Years:** 2007, 2008 **Models:** B-Series Truck, CX-7, CX-9, Mazda3, MazdaSpeed3, Mazda5, Mazda6, MazdaSpeed6, MX-5, RX-8, Tribute **Engines:** All **Transmissions:** All	**Variable Tumble Control System (VTCS) Shutter Valve Stuck Open** PCM monitors mass VTCS shutter valve position using VTCS position sensor. If PCM turns VTCS solenoid valve ON but the VTCS position still remains open (VTCS shutter valve switch output: approx. 5.0V), PCM determines VTCS shutter valve has been stuck open. **Possible Causes:** • VTCS shutter valve malfunction (stuck open) • Misconnecting or pull out the vacuum hose • Variable tumble control valve malfunction • PCM has failed
DTC: P2006 **2T CCM, MIL: Yes** **Years:** 2007, 2008 **Models:** B-Series Truck, CX-7, CX-9, Mazda3, MazdaSpeed3, Mazda5, Mazda6, MazdaSpeed6, MX-5, RX-8, Tribute **Engines:** All **Transmissions:** All	**Variable Tumble Control System (VTCS) Shutter Valve Stuck Closed** PCM monitors mass VTCS shutter valve position using VTCS position sensor. If PCM turns VTCS solenoid valve ON but the VTCS position still remains closed (VTCS shutter valve switch output: approx. 5.0V), PCM determines VTCS shutter valve has been stuck closed. **Possible Causes:** • VTCS shutter valve malfunction (stuck closed) • Misconnecting or pull out the vacuum hose • Variable tumble control valve malfunction • PCM has failed

DTC	Trouble Code Title, Conditions & Possible Causes
DTC: P2009 **2T CCM, MIL: Yes** **Years:** 2007, 2008 **Models:** B-Series Truck, CX-7, CX-9, Mazda3, MazdaSpeed3, Mazda5, Mazda6, MazdaSpeed6, MX-5, RX-8, Tribute **Engines:** All **Transmissions:** All	**Variable Tumble Control Solenoid Valve Circuit Input Low** PCM monitors Variable Tumble Control Solenoid Valve signal at PCM terminal 4T. If PCM turns variable tumble control solenoid valve OFF but voltage at PCM terminal 4T remains low, PCM determines VTCS solenoid valve circuit has a malfunction. **Possible Causes:** • Poor connection of connectors for PCM and/or VTCS • Open circuit or short between VTCS terminals and PCM terminals • Variable tumble control valve malfunction • PCM has failed
DTC: P2010 **2T CCM, MIL: Yes** **Years:** 2007, 2008 **Models:** B-Series Truck, CX-7, CX-9, Mazda3, MazdaSpeed3, Mazda5, Mazda6, MazdaSpeed6, MX-5, RX-8, Tribute **Engines:** All **Transmissions:** All	**Variable Tumble Control Solenoid Valve Circuit Input High** PCM monitors Variable Tumble Control Solenoid Valve signal at PCM terminal 4T. If PCM turns variable tumble control solenoid valve OFF but voltage at PCM terminal 4T remains high, PCM determines VTCS solenoid valve circuit has a malfunction. **Possible Causes:** • Poor connection of connectors for PCM and/or VTCS • Open circuit or short between VTCS terminals and PCM terminals • Variable tumble control valve malfunction • PCM has failed
DTC: P2088 **2T CCM, MIL: Yes** **Years:** 2007, 2008 **Models:** B-Series Truck, CX-7, CX-9, Mazda3, MazdaSpeed3, Mazda5, Mazda6, MazdaSpeed6, MX-5, RX-8, Tribute **Engines:** All **Transmissions:** All	**CMP Actuator Circuit Low** PCM monitors OCV voltage. If PCM detects OCV voltage (calculated from OCV) is below the threshold voltage (calculated from battery positive voltage), PCM determines that OCV circuit has a malfunction. **Possible Causes:** • Poor connection of connectors for PCM and/or OCV • Open circuit or short between OCV terminals and PCM terminals • OCV malfunction • PCM has failed
DTC: P2089 **2T CCM, MIL: Yes** **Years:** 2007, 2008 **Models:** B-Series Truck, CX-7, CX-9, Mazda3, MazdaSpeed3, Mazda5, Mazda6, MazdaSpeed6, MX-5, RX-8, Tribute **Engines:** All **Transmissions:** All	**CMP Actuator Circuit High** PCM monitors OCV voltage. If PCM detects OCV voltage (calculated from OCV) is above the threshold voltage (calculated from battery positive voltage), PCM determines that OCV circuit has a malfunction. **Possible Causes:** • Poor connection of connectors for PCM and/or OCV • Open circuit or short between OCV terminals and PCM terminals • OCV malfunction • PCM has failed
DTC: P2096 **2T CCM, MIL: Yes** **Years:** 2007, 2008 **Models:** Mazda3, Mazda6 **Engines:** All **Transmissions:** All	**Target A/F Feedback System Too Lean (Right Bank)** The PCM monitors the target A/F fuel trim when under the target A/F feedback control. If the fuel trim is more than the specification, the PCM determines that the target A/F feedback system too lean. **Possible Causes:** • Leaking exhaust gas • HO2S (RR, LR) malfunction • MAF malfunction • PCM has failed
DTC: P2097 **2T CCM, MIL: Yes** **Years:** 2007, 2008 **Models:** B-Series Truck, CX-7, CX-9, Mazda3, MazdaSpeed3, Mazda5, Mazda6, MazdaSpeed6, MX-5, RX-8, Tribute **Engines:** All **Transmissions:** All	**Target A/F feedback System Too Rich (Right Bank)** The PCM monitors the target A/F fuel trim when under the target A/F feedback control. If the fuel trim is more than the specification, the PCM determines that the target A/F feedback system too rich. **Possible Causes:** • Leaking exhaust gas • HO2S (RR, LR) malfunction • MAF malfunction • PCM has failed

DTC	Trouble Code Title, Conditions & Possible Causes
DTC: P2100 **1T CCM, MIL: Yes** **Years:** 2007, 2008 **Models:** B-Series Truck, CX-7, CX-9, Mazda3, MazdaSpeed3, Mazda5, Mazda6, MazdaSpeed6, MX-5, RX-8, Tribute **Engines:** All **Transmissions:** All	**Throttle Actuator Circuit Open** PCM monitors electronic throttle valve motor current. If PCM detects that the electronic throttle valve motor current is below the threshold current, PCM determines that the electronic throttle valve motor current has a malfunction. **Possible Causes:** • Poor connection of connectors for PCM and/or OCV • Open circuit or short between OCV terminals and PCM terminals • OCV malfunction • PCM has failed
DTC: P2101 **1T CCM, MIL: Yes** **Years:** 2007, 2008 **Models:** B-Series Truck, CX-7, CX-9, Mazda3, MazdaSpeed3, Mazda5, Mazda6, MazdaSpeed6, MX-5, RX-8, Tribute **Engines:** All **Transmissions:** All	**Throttle Actuator Circuit Range/Performance** If the PCM detects any of the following conditions, the PCM determines that the electronic throttle valve motor current has a malfunction: Default throttle angle that PCM memorized and the throttle angle with ET control relay OFF is not much; Voltage from ET relay is too high or too low; PCM detects a big voltage difference between the ET control relay and from tha main relay; or PCM internal malfunction. **Possible Causes:** • ET control relay and related circuit malfunction • Main relay and related circuit malfunction • PCM has failed
DTC: P2102 **1T CCM, MIL: Yes** **Years:** 2007, 2008 **Models:** B-Series Truck, CX-7, CX-9, Mazda3, MazdaSpeed3, Mazda5, Mazda6, MazdaSpeed6, MX-5, RX-8, Tribute **Engines:** All **Transmissions:** All	**Throttle Actuator Circuit Low Input** The PCM monitors the throttle actuator circuit current. If PCM detects throttle actuator circuit current is excessively low, the PCM determines the throttle actuator circuit has a malfunction. **Possible Causes:** • Open circuits between throttle body terminals and PCM terminals • Short circuit between throttle body terminals and PCM terminals • Poor connection of throttle body or PCM connector • Throttle Valve motor malfunction • PCM has failed
DTC: P2103 **1T CCM, MIL: Yes** **Years:** 2007, 2008 **Models:** B-Series Truck, CX-7, CX-9, Mazda3, MazdaSpeed3, Mazda5, Mazda6, MazdaSpeed6, MX-5, RX-8, Tribute **Engines:** All **Transmissions:** All	**Throttle Actuator Circuit High Input** The PCM monitors the throttle actuator circuit current. If PCM detects throttle actuator circuit current is excessively high, the PCM determines the throttle actuator circuit has a malfunction. **Possible Causes:** • Open circuits between throttle body terminals and PCM terminals • Short circuits between throttle body terminals and PCM terminals • Poor connection of throttle body or PCM connector • Throttle Valve motor malfunction • PCM has failed
DTC: P2107 **1T CCM, MIL: Yes** **Years:** 2007, 2008 **Models:** B-Series Truck, CX-7, CX-9, Mazda3, MazdaSpeed3, Mazda5, Mazda6, MazdaSpeed6, MX-5, RX-8, Tribute **Engines:** All **Transmissions:** All	**Throttle Actuator Control Module Processor** If the PCM detects that either the electronic control module has malfunctioned or the target throttle opening angle is more than the actual throttle opening angle, the PCM determines that the throttle actuator control module has a malfunction. **Possible Causes:** • PCM has failed
DTC: P2108 **1T CCM, MIL: Yes** **Years:** 2007, 2008 **Models:** B-Series Truck, CX-7, CX-9, Mazda3, MazdaSpeed3, Mazda5, Mazda6, MazdaSpeed6, MX-5, RX-8, Tribute **Engines:** All **Transmissions:** All	**Throttle Actuator Control Module Performance** If the PCM detects any of the following conditions, the PCM determines that the throttle actuator control system has a malfunction: TP sensor power supply voltage below 4.4V; TP sensor No. 1 output voltage is below 0.20V or above 4.85V; TP sensor No. 2 output voltage is below 0.20V or above 4.85V; PCM internal circuit for TP sensor No. 1 input circuit malfunction; or the wrong communication between main CPU and throttle control system CPU in PCM internal. **Possible Causes:** • Open circuits between throttle body terminals and PCM terminals • Short circuits between throttle body terminals and PCM terminals • Poor connection of throttle body or PCM connector • TP sensor No. 1 or TP sensor No. 2 malfunction • PCM has failed

DTC	Trouble Code Title, Conditions & Possible Causes
DTC: P2109 **1T CCM, MIL: Yes** **Years:** 2007, 2008 **Models:** Mazda3 **Engines:** All **Transmissions:** All	**TP Sensor Minimum Stop Range/Performance Problem** The PCM monitors the minimum TP when the closed TP learning is completed. If the TP is less than 11.5 percent or more than 24.3 percent, the PCM determines that there is a TP sensor minimum stop range/performance problem. **Possible Causes:** • Drive-by-wire control system malfunction • Throttle actuator malfunction • Throttle valve malfunction • PCM has failed
DTC: P2112 **1T CCM, MIL: Yes** **Years:** 2007, 2008 **Models:** B-Series Truck, CX-7, CX-9, Mazda3, MazdaSpeed3, Mazda5, Mazda6, MazdaSpeed6, MX-5, RX-8, Tribute **Engines:** All **Transmissions:** All	**TP Sensor Minimum Stop Range/Performance Problem** The PCM monitors the minimum TP when the closed TP learning is completed. If the TP is less than 11.5 percent or more than 24.3 percent, the PCM determines that there is a TP sensor minimum stop range/performance problem. **Possible Causes:** • Drive-by-wire control system malfunction • Throttle actuator malfunction • Throttle valve malfunction • PCM has failed
DTC: P2119 **1T CCM, MIL: Yes** **Years:** 2007, 2008 **Models:** B-Series Truck, CX-7, CX-9, Mazda3, MazdaSpeed3, Mazda5, Mazda6, MazdaSpeed6, MX-5, RX-8, Tribute **Engines:** All **Transmissions:** All	**TP Sensor Minimum Stop Range/Performance Problem** The PCM monitors the minimum TP when the closed TP learning is completed. If the TP is less than 11.5 percent or more than 24.3 percent, the PCM determines that there is a TP sensor minimum stop range/performance problem. **Possible Causes:** • Drive-by-wire control system malfunction • Throttle actuator malfunction • Throttle valve malfunction • PCM has failed
DTC: P2122 **1T CCM, MIL: Yes** **Years:** 2007, 2008 **Models:** B-Series Truck, CX-7, CX-9, Mazda3, MazdaSpeed3, Mazda5, Mazda6, MazdaSpeed6, MX-5, RX-8, Tribute **Engines:** All **Transmissions:** All	**APP Sensor No. 1 Circuit Low Input** The PCM monitors the input voltage from the APP sensor No.1 when the engine is running. If the input voltage is less than 0.35 V, the PCM determines that the APP sensor No.1 circuit input voltage is low. **Possible Causes:** • Open circuits between APP sensor terminals and PCM terminals • Short circuits between APP sensor terminals and PCM terminals • Poor connection of APP sensor or PCM connector • APP sensor No. 1 malfunction • PCM has failed
DTC: P2123 **1T CCM, MIL: Yes** **Years:** 2007, 2008 **Models:** B-Series Truck, CX-7, CX-9, Mazda3, MazdaSpeed3, Mazda5, Mazda6, MazdaSpeed6, MX-5, RX-8, Tribute **Engines:** All **Transmissions:** All	**APP Sensor No. 1 Circuit High Input** The PCM monitors the input voltage from the APP sensor No.1 when the engine is running. If the input voltage is more than 4.8 V, the PCM determines that the APP sensor No.1 circuit input voltage is high. **Possible Causes:** • Open circuits between APP sensor terminals and PCM terminals • Short circuits between APP sensor terminals and PCM terminals • Poor connection of APP sensor or PCM connector • APP sensor No. 1 malfunction • PCM has failed
DTC: P2126 **1T CCM, MIL: Yes** **Years:** 2007, 2008 **Models:** Mazda3 **Engines:** All **Transmissions:** All	**Throttle Actuator Control Module Performance** If the PCM detects any of the following conditions, the PCM determines that the throttle actuator control system has a malfunction: TP sensor power supply voltage below 4.4 V; TP sensor No. 1 output voltage is below 0.20 V or above 4.85 V; TP sensor No. 2 output voltage is below 0.20 V or above 4.85 V; PCM internal circuit for TP sensor No. 1 input circuit malfunction; or the wrong communication between main CPU and throttle control system CPU in PCM internal. **Possible Causes:** • Open circuits between throttle body terminals and PCM terminals • Short circuits between throttle body terminals and PCM terminals • Poor connection of throttle body or PCM connector • TP sensor No. 1 or TP sensor No. 2 malfunction • PCM has failed

DTC	Trouble Code Title, Conditions & Possible Causes
DTC: P2127 **1T CCM, MIL: Yes** **Years:** 2007, 2008 **Models:** B-Series Truck, CX-7, CX-9, Mazda3, MazdaSpeed3, Mazda5, Mazda6, MazdaSpeed6, MX-5, RX-8, Tribute **Engines:** All **Transmissions:** All	**APP Sensor No. 2 Circuit Low Input** The PCM monitors the input voltage from the APP sensor No.1 when the engine is running. If the input voltage is less than 0.35 V, the PCM determines that the APP sensor No.1 circuit input voltage is low. **Possible Causes:** • Open circuits between APP sensor terminals and PCM terminals • Short circuits between APP sensor terminals and PCM terminals • Poor connection of APP sensor or PCM connector • APP sensor No. 1 malfunction • PCM has failed
DTC: P2128 **1T CCM, MIL: Yes** **Years:** 2007, 2008 **Models:** B-Series Truck, CX-7, CX-9, Mazda3, MazdaSpeed3, Mazda5, Mazda6, MazdaSpeed6, MX-5, RX-8, Tribute **Engines:** All **Transmissions:** All	**APP Sensor No. 2 Circuit High Input** The PCM monitors the input voltage from the APP sensor No.1 when the engine is running. If the input voltage is more than 4.8 V, the PCM determines that the APP sensor No.1 circuit input voltage is high. **Possible Causes:** • Open circuits between APP sensor terminals and PCM terminals • Short circuits between APP sensor terminals and PCM terminals • Poor connection of APP sensor or PCM connector • APP sensor No. 1 malfunction • PCM has failed
DTC: P2135 **1T CCM, MIL: Yes** **Years:** 2007, 2008 **Models:** B-Series Truck, CX-7, CX-9, Mazda3, MazdaSpeed3, Mazda5, Mazda6, MazdaSpeed6, MX-5, RX-8, Tribute **Engines:** All **Transmissions:** All	**TP Sensor No. 1/No. 2 Voltage Correlation Problem** The PCM compares the input voltage from TP sensor No.1 with the input voltage from TP sensor No.2 when the engine is running. If the difference is more than the specification, the PCM determines that there is a TP sensor No.1/No.2 voltage correlation problem. **Possible Causes:** • TP sensor No. 1 malfunction • TP sensor No. 2 malfunction • Poor connection of throttle body or PCM connector • PCM has failed
DTC: P2138 **1T CCM, MIL: Yes** **Years:** 2007, 2008 **Models:** B-Series Truck, CX-7, CX-9, Mazda3, MazdaSpeed3, Mazda5, Mazda6, MazdaSpeed6, MX-5, RX-8, Tribute **Engines:** All **Transmissions:** All	**TP Sensor No. 3/No. 4 Voltage Correlation Problem** The PCM compares the input voltage from TP sensor No.3 with the input voltage from TP sensor No.4 when the engine is running. If the difference is more than the specification, the PCM determines that there is a TP sensor No.3/No.4 voltage correlation problem. **Possible Causes:** • TP sensor No. 3 malfunction • TP sensor No. 4 malfunction • Poor connection of throttle body or PCM connector • PCM has failed
DTC: P2177 **1T CCM, MIL: Yes** **Years:** 2007, 2008 **Models:** Mazda6, MX-5 **Engines:** All **Transmissions:** All	**Fuel System Too Lean At Off Idle (Right Bank)** PCM monitors short term fuel trim (SHRTFT), long term fuel trim (LONGFT) during closed loop fuel control at off-idle. If the LONGFT and the sum total of these fuel trims exceed preprogrammed criteria. PCM determines that fuel system is too lean at off-idle. **Possible Causes:** • Misfire • Exhaust system leak • Fuel pump malfunction • Fuel filter clogged • PCM has failed
DTC: P2178 **1T CCM, MIL: Yes** **Years:** 2007, 2008 **Models:** Mazda6, MX-5 **Engines:** All **Transmissions:** All	**Fuel System Too Rich At Off Idle (Right Bank)** PCM monitors short term fuel trim (SHRTFT), long term fuel trim (LONGFT) during closed loop fuel control at off-idle. If the LONGFT and the sum total of these fuel trims exceed preprogrammed criteria. PCM determines that fuel system is too rich at off-idle. **Possible Causes:** • Misfire • Exhaust system leak • Fuel pump malfunction • Fuel filter clogged • PCM has failed

DTC	Trouble Code Title, Conditions & Possible Causes
DTC: P2187 **1T CCM, MIL: Yes** **Years:** 2007, 2008 **Models:** Mazda6, MX-5 **Engines:** All **Transmissions:** All	**Fuel System Too Lean At Off Idle (Right Bank)** PCM monitors short term fuel trim (SHRTFT), long term fuel trim (LONGFT) during closed loop fuel control at off-idle. If the LONGFT and the sum total of these fuel trims exceed preprogrammed criteria. PCM determines that fuel system is too lean at off-idle. **Possible Causes:** • Misfire • Exhaust system leak • Fuel pump malfunction • Fuel filter clogged • PCM has failed
DTC: P2188 **1T CCM, MIL: Yes** **Years:** 2007, 2008 **Models:** Mazda6, MX-5 **Engines:** All **Transmissions:** All	**Fuel System Too Lean At Off Idle (Left Bank)** PCM monitors short term fuel trim (SHRTFT), long term fuel trim (LONGFT) during closed loop fuel control at off-idle. If the LONGFT and the sum total of these fuel trims exceed preprogrammed criteria. PCM determines that fuel system is too lean at off-idle. **Possible Causes:** • Misfire • Exhaust system leak • Fuel pump malfunction • Fuel filter clogged • PCM has failed
DTC: P2195 **1T CCM, MIL: Yes** **Years:** 2007, 2008 **Models:** Mazda6, MX-5 **Engines:** All **Transmissions:** All	**HO2S (RF) Signal Stuck Lean** The PCM monitors the HO2S (RF, LF) output voltage when the following conditions are met. If output voltage is less than 0.45 V for 25.6 s, the PCM determines that the HO2S (RF, LF) signal remains lean. **Possible Causes:** • HO2S (RF, LF) malfunction • Exhaust gas leaking • MAF sensor malfunction • ECT sensor malfunction • PCM has failed
DTC: P2196 **1T CCM, MIL: Yes** **Years:** 2007, 2008 **Models:** Mazda6 **Engines:** All **Transmissions:** All	**HO2S (RF) Signal Stuck Rich** The PCM monitors the HO2S (RF, LF) output voltage when the following conditions are met. If output voltage is more than 0.45 V for 25.6 s, the PCM determines that the HO2S (RF, LF) signal remains rich. **Possible Causes:** • HO2S (RF, LF) malfunction • Exhaust gas leaking • MAF sensor malfunction • ECT sensor malfunction • PCM has failed
DTC: P2228 **1T CCM, MIL: Yes** **Years:** 2007, 2008 **Models:** B-Series Truck, CX-7, CX-9, Mazda3, MazdaSpeed3, Mazda5, Mazda6, MazdaSpeed6, MX-5, RX-8, Tribute **Engines:** All **Transmissions:** All	**EGR Boost Sensor Solenoid Valve Circuit Low Input** The PCM monitors the EGR boost sensor solenoid valve control signal. If PCM turns the EGR boost sensor solenoid valve off but voltage still remains low, the PCM determines that the EGR boost sensor solenoid valve circuit has malfunction. **Possible Causes:** • EGR boost sensor solenoid malfunction • Connector or terminal malfunction • PCM has failed
DTC: P2229 **1T CCM, MIL: Yes** **Years:** 2007, 2008 **Models:** B-Series Truck, CX-7, CX-9, Mazda3, MazdaSpeed3, Mazda5, Mazda6, MazdaSpeed6, MX-5, RX-8, Tribute **Engines:** All **Transmissions:** All	**EGR Boost Sensor Solenoid Valve Circuit High Input** The PCM monitors the EGR boost sensor solenoid valve control signal. If PCM turns the EGR boost sensor solenoid valve off but voltage still remains high, the PCM determines that the EGR boost sensor solenoid valve circuit has malfunction. **Possible Causes:** • EGR boost sensor solenoid malfunction • Connector or terminal malfunction • PCM has failed

DTC	Trouble Code Title, Conditions & Possible Causes
DTC: P2270 **1T CCM, MIL: Yes** **Years:** 2007, 2008 **Models:** B-Series Truck, CX-7, CX-9, Mazda3, MazdaSpeed3, Mazda5, Mazda6, MazdaSpeed6, MX-5, RX-8, Tribute **Engines:** All **Transmissions:** All	**Rear HO2S Signal Stuck Lean** The PCM monitors the input voltage from the rear HO2S when the following conditions are met. If the input voltage is less than 0.4 V for 40 seconds, the PCM determines that the rear HO2S signal remains lean. **Possible Causes:** • Front or Rear HO2S malfunction • Front or Rear HO2S heater malfunction • Fuel pressure malfunction • Fuel injector malfunction • PCM has failed
DTC: P2271 **1T CCM, MIL: Yes** **Years:** 2007, 2008 **Models:** B-Series Truck, CX-7, CX-9, Mazda3, MazdaSpeed3, Mazda5, Mazda6, MazdaSpeed6, MX-5, RX-8, Tribute **Engines:** All **Transmissions:** All	**Rear HO2S Signal Stuck Rich** The PCM monitors the input voltage from the rear HO2S when the following conditions are met. If the input voltage is more than 0.85 V for 40 seconds, the PCM determines that the rear HO2S signal remains lean. **Possible Causes:** • Front or Rear HO2S malfunction • Front or Rear HO2S heater malfunction • Fuel pressure malfunction • Fuel injector malfunction • PCM has failed
DTC: P2272 **1T CCM, MIL: Yes** **Years:** 2007, 2008 **Models:** Mazda6 **Engines:** All **Transmissions:** All	**Rear HO2S Signal Stuck Lean** The PCM monitors the input voltage from the rear HO2S when the following conditions are met. If the input voltage is less than 0.4 V for 40 seconds, the PCM determines that the rear HO2S signal remains lean. **Possible Causes:** • Front or Rear HO2S malfunction • Front or Rear HO2S heater malfunction • Fuel pressure malfunction • Fuel injector malfunction • PCM has failed
DTC: P2274 **1T CCM, MIL: Yes** **Years:** 2007, 2008 **Models:** B-Series Truck, CX-7, CX-9, Mazda3, MazdaSpeed3, Mazda5, Mazda6, MazdaSpeed6, MX-5, RX-8, Tribute **Engines:** All **Transmissions:** All	**Rear HO2S Signal Stuck Lean** The PCM monitors the input voltage from the rear HO2S when the following conditions are met. If the input voltage is less than 0.4 V for 40 seconds, the PCM determines that the rear HO2S signal remains lean. **Possible Causes:** • Front or Rear HO2S malfunction • Front or Rear HO2S heater malfunction • Fuel pressure malfunction • Fuel injector malfunction • PCM has failed
DTC: P2275 **1T CCM, MIL: Yes** **Years:** 2007, 2008 **Models:** B-Series Truck, CX-7, CX-9, Mazda3, MazdaSpeed3, Mazda5, Mazda6, MazdaSpeed6, MX-5, RX-8, Tribute **Engines:** All **Transmissions:** All	**Rear HO2S Signal Stuck Rich** The PCM monitors the input voltage from the rear HO2S when the following conditions are met. If the input voltage is more than 0.85 V for 40 seconds, the PCM determines that the rear HO2S signal remains lean. **Possible Causes:** • Front or Rear HO2S malfunction • Front or Rear HO2S heater malfunction • Fuel pressure malfunction • Fuel injector malfunction • PCM has failed
DTC: P2401 **1T CCM, MIL: Yes** **Years:** 2007, 2008 **Models:** B-Series Truck, CX-7, CX-9, Mazda3, MazdaSpeed3, Mazda5, Mazda6, MazdaSpeed6, MX-5, RX-8, Tribute **Engines:** All **Transmissions:** All	**EVAP System Leak Detection Pump Motor Circuit Low** The PCM monitors pump load current (EVAP line pressure), while evaporative leak monitor is operating. If the pump load current is lower than specified, the PCM determines EVAP system leak detection pump motor circuit has a malfunction. **Possible Causes:** • EVAP system leak detection pump malfunction • Poor connection of EVAP system or PCM connector • PCM has failed

DTC	Trouble Code Title, Conditions & Possible Causes
DTC: P2402 **1T CCM, MIL: Yes** **Years:** 2007, 2008 **Models:** B-Series Truck, CX-7, CX-9, Mazda3, MazdaSpeed3, Mazda5, Mazda6, MazdaSpeed6, MX-5, RX-8, Tribute **Engines:** All **Transmissions:** All	**EVAP System Leak Detection Pump Motor Circuit Low** The PCM monitors pump load current (EVAP line pressure), while evaporative leak monitor is operating. If the pump load current is higher than specified, the PCM determines EVAP system leak detection pump motor circuit has a malfunction. **Possible Causes:** • EVAP system leak detection pump malfunction • Poor connection of EVAP system or PCM connector • PCM has failed
DTC: P2404 **1T CCM, MIL: Yes** **Years:** 2007, 2008 **Models:** B-Series Truck, CX-7, CX-9, Mazda3, MazdaSpeed3, Mazda5, Mazda6, MazdaSpeed6, MX-5, RX-8, Tribute **Engines:** All **Transmissions:** All	**EVAP System Leak Detection Pump Sense Circuit Problem** The PCM monitors pump load current (EVAP line pressure), while evaporative leak monitor is operating. After obtaining the reference current value, if the time in which the pump load current reaches the reference current value is less than the specification, the PCM determines air filter has a malfunction. **Possible Causes:** • Air filter clogging • EVAP hose bending • PCM has failed
DTC: P2405 **1T CCM, MIL: Yes** **Years:** 2007, 2008 **Models:** Mazda6, MX-5 **Engines:** All **Transmissions:** All	**EVAP System Leak Detection Pump Sense Circuit Low Input** The PCM monitors pump load current (EVAP line pressure), while evaporative leak monitor is operating. If the current is lower than the specification while the PCM obtains the reference current value, the PCM determines EVAP system leak detection pump orifice has a malfunction. **Possible Causes:** • EVAP system leak detection orifice has fallen off • EVAP system leak detection pump motor malfunction • PCM has failed
DTC: P2407 **1T CCM, MIL: Yes** **Years:** 2007, 2008 **Models:** Mazda3. Mazda6, MX-5 **Engines:** All **Transmissions:** All	**EVAP System Leak Detection Pump Sense Circuit Intermittent** The PCM monitors pump load current (EVAP line pressure), while evaporative leak monitor is operating. When either of the following is detected 6 times or more successively, the PCM determines EVAP system leak detection pump heater has a malfunction. **Possible Causes:** • EVAP system leak detection pump heater malfunction • PCM has failed
DTC: P2502 **1T CCM, MIL: Yes** **Years:** 2007, 2008 **Models:** Mazda6, MX-5 **Engines:** All **Transmissions:** All	**Charging System Voltage Problem** PCM judges' generator output voltage is above 17 V or battery voltage is below 11 V during engine running. **Possible Causes:** • Open circuits between battery terminals • Battery is malfunctioning • Poor connection of PCM connectors • Generator malfunction • PCM has failed
DTC: P2503 **1T CCM, MIL: Yes** **Years:** 2007, 2008 **Models:** Mazda6, MX-5 **Engines:** All **Transmissions:** All	**Charging System Voltage Low** PCM needs more than 20 A from generator, and judges' generator output voltage to be below 8.5 V during engine running. **Possible Causes:** • Open circuits between battery terminals • Battery is malfunctioning • Poor connection of PCM connectors • Generator malfunction • PCM has failed
DTC: P2504 **1T CCM, MIL: Yes** **Years:** 2007, 2008 **Models:** B-Series Truck, CX-7, CX-9, Mazda3, MazdaSpeed3, Mazda5, Mazda6, MazdaSpeed6, MX-5, RX-8, Tribute **Engines:** All **Transmissions:** All	**Charging System Voltage High** PCM judges generator output voltage is above 18.5 V or battery voltage is above 16.0 V during engine running. **Possible Causes:** • Short to power circuit between generator and PCM • Generator malfunction • PCM has failed

DTC	Trouble Code Title, Conditions & Possible Causes
DTC: P2507 **1T CCM, MIL: Yes** **Years:** 2007, 2008 **Models:** B-Series Truck, CX-7, CX-9, Mazda3, MazdaSpeed3, Mazda5, Mazda6, MazdaSpeed6, MX-5, RX-8, Tribute **Engines:** All **Transmissions:** All	**PCM B+ Voltage Low** The PCM monitors the voltage of back-up battery positive terminal at PCM terminal 4AG. If the PCM detected battery positive terminal voltage below 2.5 V for 2 seconds, the PCM determines that the backup voltage circuit has malfunction. **Possible Causes:** • Melt down fuse • Poor connection of PCM • PCM has failed
DTC: P2510 **1T CCM, MIL: Yes** **Years:** 2007, 2008 **Models:** MX-5 **Engines:** All **Transmissions:** All	**Timer Error In PCM** PCM internal timer is damaged. **Possible Causes:** • PCM internal timer is damaged
DTC: P2676 **1T CCM, MIL: Yes** **Years:** 2007, 2008 **Models:** Mazda6 **Engines:** All **Transmissions:** All	**Variable Air Duct (VAD) Control Solenoid Valve Circuit Low Input** The PCM monitors VAD solenoid valve control signal at PCM terminal 68. If the PCM turns VAD solenoid valve off but voltage at PCM terminal 68 still remains low, the PCM determines that VAD solenoid valve circuit has malfunction. **Possible Causes:** • Open circuits between main relay and VAD terminals • Short circuits between VAD terminals and PCM terminals • Poor connection of PCM connector • VAD solenoid valve malfunction • PCM has failed
DTC: P2677 **1T CCM, MIL: Yes** **Years:** 2007, 2008 **Models:** Mazda6 **Engines:** All **Transmissions:** All	**Variable Air Duct (VAD) Control Solenoid Valve Circuit High Input** The PCM monitors VAD solenoid valve control signal at PCM terminal 68. If the PCM turns VAD solenoid valve off but voltage at PCM terminal 68 still remains high, the PCM determines that VAD solenoid valve circuit has malfunction. **Possible Causes:** • Open circuits between main relay and VAD terminals • Short circuits between VAD terminals and PCM terminals • Poor connection of PCM connector • VAD solenoid valve malfunction • PCM has failed

Gas Engine OBD II Trouble Code List (U1xxx Codes)

DTC	Trouble Code Title, Conditions & Possible Causes
DTC: U0073 **1T CCM, MIL: No** **Years:** 2007, 2008 **Models:** All **Engines:** All **Transmissions:** A/T	**Data Circuit Malfunction** Key on, and the PCM detected the SCP was invalid, or that it was missing data from the A/C System during the CCM test. **Possible Causes:** • SCP data bus to other controller(s) is open or shorted to ground • A/C System controller is damaged or has failed • PCM has failed
DTC: U0101 **1T CCM, MIL: No** **Years:** 2007, 2008 **Models:** All **Engines:** All **Transmissions:** A/T	**Data Circuit Malfunction** Key on, and the PCM detected the SCP was invalid, or that it was missing data from the A/C System during the CCM test. **Possible Causes:** • SCP data bus to other controller(s) is open or shorted to ground • A/C System controller is damaged or has failed • PCM has failed
DTC: U0121 **1T CCM, MIL: No** **Years:** 2007, 2008 **Models:** All **Engines:** All **Transmissions:** A/T	**Data Circuit Malfunction** Key on, and the PCM detected the SCP was invalid, or that it was missing data from the A/C System during the CCM test. **Possible Causes:** • SCP data bus to other controller(s) is open or shorted to ground • A/C System controller is damaged or has failed • PCM has failed

DTC	Trouble Code Title, Conditions & Possible Causes
DTC: U0155 **1T CCM, MIL: No** **Years:** 2007, 2008 **Models:** All **Engines:** All **Transmissions:** A/T	**Data Circuit Malfunction** Key on, and the PCM detected the SCP was invalid, or that it was missing data from the A/C System during the CCM test. **Possible Causes:** • SCP data bus to other controller(s) is open or shorted to ground • A/C System controller is damaged or has failed • PCM has failed
DTC: U0167 **1T CCM, MIL: No** **Years:** 2007, 2008 **Models:** All **Engines:** All **Transmissions:** A/T	**Data Circuit Malfunction** Key on, and the PCM detected the SCP was invalid, or that it was missing data from the A/C System during the CCM test. **Possible Causes:** • SCP data bus to other controller(s) is open or shorted to ground • A/C System controller is damaged or has failed • PCM has failed
DTC: U0302 **1T CCM, MIL: No** **Years:** 2007, 2008 **Models:** All **Engines:** All **Transmissions:** A/T	**Data Circuit Malfunction** Key on, and the PCM detected the SCP was invalid, or that it was missing data from the A/C System during the CCM test. **Possible Causes:** • SCP data bus to other controller(s) is open or shorted to ground • A/C System controller is damaged or has failed • PCM has failed
DTC: U1020 **1T CCM, MIL: No** **Years:** 2007, 2008 **Models:** All **Engines:** All **Transmissions:** A/T	**Data Circuit Malfunction** Key on, and the PCM detected the SCP was invalid, or that it was missing data from the A/C System during the CCM test. **Possible Causes:** • SCP data bus to other controller(s) is open or shorted to ground • A/C System controller is damaged or has failed • PCM has failed
DTC: U1039 **1T CCM, MIL: No** **Years:** 2007, 2008 **Models:** All **Engines:** All **Transmissions:** A/T	**Data Circuit Malfunction** Key on, and the PCM detected the SCP was invalid, or that it was missing data from the Vehicle Speed System during the CCM test. **Possible Causes:** • SCP data bus to other controller(s) is open or shorted to ground • Vehicle Speed System controller is damaged or has failed • PCM has failed
DTC: U1040 **1T CCM, MIL: No** **Years:** 2007, 2008 **Models:** All **Engines:** All **Transmissions:** A/T	**Data Circuit Malfunction** Key on, and the PCM detected the SCP was invalid, or that it was missing data from the Vehicle Speed System during the CCM test. **Possible Causes:** • SCP data bus to other controller(s) is open or shorted to ground • Vehicle Speed System controller is damaged or has failed • PCM has failed
DTC: U1051 **1T CCM, MIL: No** **Years:** 2007, 2008 **Models:** All **Engines:** All **Transmissions:** A/T	**Data Circuit Malfunction** Key on, and the PCM detected the SCP was invalid, or that it was missing data from the Antilock Brake System during the CCM test. **Possible Causes:** • SCP data bus to other controller(s) is open or shorted to ground • ABS System controller is damaged or has failed • PCM has failed
DTC: U1131 **1T CCM, MIL: No** **Years:** 2007, 2008 **Models:** All **Engines:** All **Transmissions:** A/T	**Data Circuit Malfunction** Key on, and the PCM detected the SCP was invalid, or that it was missing data from the Fuel System during the CCM test. **Possible Causes:** • SCP data bus to other controller(s) is open or shorted to ground • Fuel System controller is damaged or has failed • PCM has failed

DTC	Trouble Code Title, Conditions & Possible Causes
DTC: U1147 **1T CCM, MIL: No** **Years:** 2007, 2008 **Models:** All **Engines:** All **Transmissions:** A/T	**Data Circuit Malfunction** Key on, and the PCM detected the SCP was invalid, or that it was missing data from the Vehicle Security System during the CCM test. **Possible Causes:** • SCP data bus to other controller(s) is open or shorted to ground • Vehicle Security System controller is damaged or has failed • PCM has failed
DTC: U1451 **1T CCM, MIL: No** **Years:** 2007, 2008 **Models:** All **Engines:** All **Transmissions:** A/T	**Data Circuit Malfunction** Key on, and the PCM detected the SCP was invalid, or that it was missing data from the Anti-Theft System during the CCM test. **Possible Causes:** • SCP data bus to other controller(s) is open or shorted to ground • Anti-Theft System controller is damaged or has failed • PCM has failed
DTC: U1262 **1T CCM, MIL: No** **Years:** 2007, 2008 **Models:** All **Engines:** All **Transmissions:** A/T	**Data Circuit Malfunction** Key on, and the PCM detected a problem in the SCP communications data bus circuit during the CCM test. **Note: Perform a network communications test of the whole system.** **Possible Causes:** • SCP data bus to other controller(s) is open or shorted to ground • Anti-Theft System controller is damaged or has failed • PCM has failed
DTC: U1900 **1T CCM, MIL: No** **Years:** 2007, 2008 **Models:** All **Engines:** All **Transmissions:** A/T	**Data Circuit Malfunction** Key on, and the PCM detected the SCP was invalid, or that it was missing data from the Exterior Environment System in the CCM test. **Possible Causes:** • SCP data bus to other controller(s) is open or shorted to ground • Vehicle Security System controller is damaged or has failed • PCM has failed
DTC: U2013 **1T CCM, MIL: No** **Years:** 2007, 2008 **Models:** All **Engines:** All **Transmissions:** A/T	**Data Circuit Malfunction** Key on, and the PCM detected the SCP was invalid, or that it was missing data from the Exterior Environment System in the CCM test. **Possible Causes:** • SCP data bus to other controller(s) is open or shorted to ground • Vehicle Security System controller is damaged or has failed • PCM has failed
DTC: U2023 **1T CCM, MIL: No** **Years:** 2007, 2008 **Models:** All **Engines:** All **Transmissions:** A/T	**Data Circuit Malfunction** Key on, and the PCM detected the SCP was invalid, or that it was missing data from the Exterior Environment System in the CCM test. **Possible Causes:** • SCP data bus to other controller(s) is open or shorted to ground • Vehicle Security System controller is damaged or has failed • PCM has failed
DTC: U2050 **1T CCM, MIL: No** **Years:** 2007, 2008 **Models:** All **Engines:** All **Transmissions:** A/T	**Data Circuit Malfunction** Key on, and the PCM detected the SCP was invalid, or that it was missing data from the Exterior Environment System in the CCM test. **Possible Causes:** • SCP data bus to other controller(s) is open or shorted to ground • Vehicle Security System controller is damaged or has failed • PCM has failed
DTC: U2051 **1T CCM, MIL: No** **Years:** 2007, 2008 **Models:** All **Engines:** All **Transmissions:** A/T	**Data Circuit Malfunction** Key on, and the PCM detected the SCP was invalid, or that it was missing data from the Exterior Environment System in the CCM test. **Possible Causes:** • SCP data bus to other controller(s) is open or shorted to ground • Vehicle Security System controller is damaged or has failed • PCM has failed
DTC: U2243 **1T CCM, MIL: No** **Years:** 2007, 2008 **Models:** All **Engines:** All **Transmissions:** A/T	**Data Circuit Malfunction** Key on, and the PCM detected the SCP was invalid, or that it was missing data from the Exterior Environment System in the CCM test. **Possible Causes:** • SCP data bus to other controller(s) is open or shorted to ground • Vehicle Security System controller is damaged or has failed • PCM has failed

MITSUBISHI

Eclipse • Galant • Lancer • Lancer Evolution • Spyder

8

SPECIFICATIONS AND MAINTENANCE CHARTS

ENGINE AND VEHICLE IDENTIFICATION

Engine						
Code ①	Liters (cc)	Cu. In.	Cyl.	Fuel Sys.	Engine Type	Eng. Mfg.
4B11/U	2.0 (1,998)	121.9	4	MPFI	DOHC	Mitsubishi
4B11/V	2.0 (1,998)	121.9	4	MPFI-Turbo	DOHC	Mitsubishi
4G63/C	2.0 (1,997)	121.9	4	MPFI-Turbo	DOHC	Mitsubishi
4G94/E	2.0 (1,999)	122.0	4	MPFI	SOHC	Mitsubishi
4G69/F	2.4 (2,378)	145.1	4	MPFI	SOHC	Mitsubishi
6G75/S	3.8 (3,828)	233.6	6	MPFI	SOHC	Mitsubishi
6G75/T	3.8 (3,828)	233.6	6	MPFI	SOHC	Mitsubishi

Model Year	
Code ②	Year
6	2006
7	2007
8	2008

MPFI: Multi-Point Fuel Injection

DOHC: Double Overhead Camshafts

① Engine Code/8th digit of VIN

② 10th digit of VIN

22140_MITS_C0001

GENERAL ENGINE SPECIFICATIONS

All measurements are given in inches.

Year	Model	Engine Displacement Liters	Engine Series ID/VIN	Net Horsepower @ rpm	Net Torque @ rpm (ft. lbs.)	Bore x Stroke (in.)	Compression Ratio	Oil Pressure @ rpm
2006	Eclipse	2.4	4G69/F	162@5750	162@4000	3.43 x 3.94	9.5:1	43-100@3500
	Eclipse	3.8	6G75/T	263@5750	260@4500	3.74 x 3.54	10.5:1	43-100@3500
	Galant	2.4	4G69/F	162@5750	162@4000	3.43 x 3.94	9.5:1	43-100@3500
	Galant	3.8	6G75/S	230@5250	250@4000	3.74 x 3.54	10.0:1	43-100@3500
	Lancer	2.0	4G94/E	120@5500	130@4250	3.21 x 3.77	9.5:1	43-100@3500
	Lancer	2.4	4G69/F	162@5750	162@4000	3.43 x 3.94	9.5:1	43-100@3500
	Lancer Evolution	2.0	4G63/C	286@6500	289@3500	3.35 x 3.46	8.8:1	43-100@3500
	Lancer Sportback	2.4	4G69/F	160@5750	162@4000	3.43 x 3.94	9.5:1	43-100@3500
2007	Eclipse/Spyder	2.4	4G69/F	162@5750	162@4000	3.43 x 3.94	9.5:1	43-100@3500
	Eclipse/Spyder	3.8	6G75/T	263@5750	260@4500	3.74 x 3.54	10.5:1	43-100@3500
	Galant	2.4	4G69/F	160@5500	157@4000	3.43 x 3.94	9.5:1	43-100@3500
	Galant	3.8	6G75/S	230@5250	250@4000	3.74 x 3.54	10.0:1	43-100@3500
	Galant	3.8	6G75/T	258@5750	258@4500	3.74 x 3.54	10.5:1	43-100@3500
	Lancer	2.0	4G94/E	120@5500	130@4250	3.21 x 3.77	9.5:1	43-100@3500
2008	Eclipse/Spyder	2.4	4G69/F	162@5750	162@4000	3.43 x 3.94	9.5:1	43-100@3500
	Eclipse/Spyder	3.8	6G75/T	263@5750	260@4500	3.74 x 3.54	10.5:1	43-100@3500
	Galant	2.4	4G69/F	160@5500	157@4000	3.43 x 3.94	9.5:1	43-100@3500
	Galant	3.8	6G75/T	258@5750	258@4500	3.74 x 3.54	10.5:1	43-100@3500
	Lancer	2.0	4B11/U	152@6000	146@4250	3.39 x 3.39	10.0:1	43-100@3500
	Lancer Evolution	2.0	4B11/V	291@6500	300@4400	3.39 x 3.39	10.0:1	43-100@3500

22140_MITS_C0002

GASOLINE ENGINE TUNE-UP SPECIFICATIONS

Year	Engine Displacement Liters	Engine Series ID/VIN	Spark Plug Gap (in.)	Ignition Timing (deg.) MT	Ignition Timing (deg.) AT	Fuel Pump (psi)	Idle Speed (rpm) MT	Idle Speed (rpm) AT	Valve Clearance (in.) In.	Valve Clearance (in.) Ex.
2006	2.0	4G63/C	0.020-0.024	①	—	33	②	—	HYD	HYD
	2.0	4G94/E	0.039-0.043	①	①	47	②	②	HYD	HYD
	2.4	4G69/F	0.028-0.031	①	①	47	②	②	0.0080	0.0120
	3.8	6G75/S	0.028-0.031	—	①	47	—	②	HYD	HYD
	3.8	6G75/T	0.028-0.031	①	①	47	②	②	0.0080	③
2007	2.0	4G94/E	0.039-0.043	①	①	47	②	②	HYD	HYD
	2.4	4G69/F	0.028-0.031	①	①	47	②	②	0.0080	0.0120
	3.8	6G75/S	0.028-0.031	—	①	47	—	②	HYD	HYD
	3.8	6G75/T	0.028-0.031	①	①	47	②	②	0.0080	③
2008	2.0	4B11/U	0.028-0.031	①	①	47	②	②	④	⑤
	2.0	4B11/V	0.020-0.023	①	—	38	②	—	④	⑤
	2.4	4G69/F	0.028-0.031	①	①	47	②	②	0.0080	0.0120
	3.8	6G75/T	0.028-0.031	①	①	47	②	②	0.0080	③

NOTE: The Vehicle Emission Control Information label reflects specification changes made during production.
Follow the figures on the label if they differ from those in this chart.

HYD: Hydraulic

① Ignition timing is preset and cannot be adjusted

② Idle speed is maintained by the Electronic Control Module (ECM)

③ Valve clearance check and adjustment is unnecessary for the exhaust side as an auto lash adjuster is installed.

④ Intake valve clearance: 0.008 +/- 0.0012 inch

⑤ Exhaust valve clearance: 0.012 +/- 0.0012 inch

22140_MITS_C0003

CAPACITIES

Year	Model	Engine Displacement Liters	Engine Series ID/VIN	Engine Oil with Filter (qts.)	Transmission (pts.) Manual	Transmission (pts.) Auto. ①	Transfer Case (pts.)	Drive Axle Front (pts.)	Drive Axle Rear (pts.)	Fuel Tank (gal.)	Cooling System (qts.)
2006	Eclipse	2.4	4G69/F	4.5	4.6	16.2	—	—	—	17.7	9.2-9.3
	Eclipse	3.8	6G75/T	4.5	4.6	17.8	—	—	—	17.7	8.5-8.6
	Galant	2.4	4G69/F	4.5	—	16.2	—	—	—	17.7	8.5
	Galant	3.8	6G75/S	4.8	—	17.8	—	—	—	17.7	9.2
	Lancer	2.0	4G94/E	4.0	4.6	16.2	—	—	—	13.2	6.3
	Lancer	2.4	4G69/F	4.5	4.6	16.2	—	—	—	13.2	7.4
	Lancer Evolution	2.0	4G63/C	②	③	—	1.3	—	1.9	14.0	6.3
	Lancer Sportback	2.4	4G69/F	4.5	—	16.2	—	—	—	13.2	7.4
2007	Eclipse/Spyder	2.4	4G69/F	4.5	4.6	16.2	—	—	—	17.7	9.2-9.3
	Eclipse/Spyder	3.8	6G75/T	4.5	4.6	17.8	—	—	—	17.7	8.5-8.6
	Galant	2.4	4G69/F	4.5	—	16.2	—	—	—	17.7	8.5
	Galant	3.8	6G75/S	4.8	—	17.8	—	—	—	17.7	9.2
	Galant	3.8	6G75/T	4.8	—	17.8	—	—	—	17.7	9.2
	Lancer	2.0	4G94/E	4.5	4.6	16.2	—	—	—	13.2	7.4
2008	Eclipse/Spyder	2.4	4G69/F	4.5	4.6	16.2	—	—	—	17.7	9.2-9.3
	Eclipse/Spyder	3.8	6G75/T	4.5	4.6	17.8	—	—	—	17.7	8.5-8.6
	Galant	2.4	4G69/F	4.5	—	16.2	—	—	—	17.7	8.5
	Galant	3.8	6G75/T	4.8	—	17.8	—	—	—	17.7	9.2
	Lancer	2.0	4B11/U	4.5	5.2	16.4	—	—	—	15.5	7.4
	Lancer Evolution	2.0	4B11/V	④	⑤	—	1.9	15.0	1.2	14.5	7.9

NOTE: All capacities are approximate. Add fluid gradually and check to be sure a proper fluid level is obtained.

① Drain and refill

② Oil pan (4.8 qts.), Oil filter (0.32 qt.), Oil cooler (0.52 qt.)

③ W5M51 (6.0 pts.), W6MAA (4.6 pts.)

④ Oil pan (5.07 qts.), Oil cooler (0.52 qt.), Oil filter (0.32 qt.)

⑤ W5M6A (5.3 pts.), W6DGA - Twin Clutch Sport Shift Transaxle (TC-SST) (16.5 pts.)

22140_MITS_C0004

FLUID SPECIFICATIONS

Year	Model	Engine Displacement Liters	Engine Series ID/VIN	Engine Oil	Manual Trans.	Auto. Trans.	Drive Axle	Transfer Case	Power Steering Fluid	Brake Master Cylinder	Cooling System
2006	Eclipse	2.4	4G69/F	5W-20	①	②	—	—	③	④	⑤
	Eclipse	3.8	6G75/T	5W-20	①	②	—	—	③	④	⑤
	Galant	2.4	4G69/F	5W-20	—	②	—	—	③	④	⑤
	Galant	3.8	6G75/S	5W-20	—	②	—	—	③	④	⑤
	Lancer	2.0	4G94/E	5W-20	① or ⑥	⑦	—	—	③	④	⑤
	Lancer	2.4	4G69/F	5W-20	① or ⑥	⑦	—	—	③	④	⑤
	Lancer Evolution	2.0	4G63/C	10W-30	⑧	—	⑨	⑩	③	④	⑤
	Lancer Sportback	2.4	4G69/F	5W-20	—	②	—	—	③	④	⑤
2007	Eclipse/Spyder	2.4	4G69/F	5W-20	①	②	—	—	③	④	⑤
	Eclipse/Spyder	3.8	6G75/T	5W-20	①	②	—	—	③	④	⑤
	Galant	2.4	4G69/F	5W-20	—	②	—	—	③	④	⑤
	Galant	3.8	6G75/S	5W-20	—	②	—	—	③	④	⑤
	Galant	3.8	6G75/T	5W-20	—	②	—	—	③	④	⑤
	Lancer	2.0	4G94/E	5W-20	① or ⑥	⑦	—	—	③	④	⑤
2008	Eclipse/Spyder	2.4	4G69/F	5W-20	①	②	—	—	③	④	⑤
	Eclipse/Spyder	3.8	6G75/T	5W-20	①	②	—	—	③	④	⑤
	Galant	2.4	4G69/F	5W-20	—	②	—	—	③	④	⑤
	Galant	3.8	6G75/T	5W-20	—	②	—	—	③	④	⑤
	Lancer	2.0	4B11/U	5W-20	①	⑪	—	—	③	④	⑤
	Lancer Evolution	2.0	4B11/V	5W-30	⑫	—	⑨	⑩	③	④	⑤

DOT: Department Of Transportation

① DiaQueen NEW MULTI GEAR OIL (GL-3), SAE 75W-80

② DIAMOND ATF SP-3 or equivalent

③ Genuine MITSUBISHI power steering fluid

④ DOT 3, DOT 4, or equivalent

⑤ Long life antifreeze/coolant or an equivalent

⑥ Gear oil API classification GL-4, SAE 75W-85W, 75W-90

⑦ MITSUBISHI genuine ATF SP-3

⑧ W5M51: Use DiaQueen NEW MULTI GEAR OIL (GL-3), SAE 75W-80
or Gear oil API classification GL-4, SAE 75W-85/75W-90
W6MAA: Use DiaQueen super multi-gear 75W-85W (GL-4) or exact equivalent

⑨ Active Center Differential (ACD) fluid: DIAMOND ATF SP-3
Rear differential gear oil: MITSUBISHI Limited Slip Differential Oil (LSD) or equivalent

⑩ MITSUBISHI Limited Slip Differential Oil (LSD) or equivalent

⑪ DiaQueen CVTF-J1

⑫ W5M6A: Use DiaQueen NEW MULTI GEAR OIL (GL-3), SAE 75W-80
W6DGA - Twin Clutch Sport Shift Transaxle (TC-SST): Use DiaQueen SSTF-1

22140_MITS_C0005

VALVE SPECIFICATIONS

Year	Engine Displacement Liters	Engine Series ID/VIN	Seat Angle (deg.)	Face Angle (deg.)	Spring Test Pressure (lbs. @ in.)	Spring Installed Height (in.)	Stem-to-Guide Clearance (in.) Intake	Stem-to-Guide Clearance (in.) Exhaust	Stem Diameter (in.) Intake	Stem Diameter (in.) Exhaust
2006	2.0	4G63/C	NA	45.0-45.5	63.0 @1.57	1.57	0.0008-0.0020	0.0020-0.0035	0.236	0.236
	2.0	4G94/E	NA	43.5-44.0	53.0 @1.74	1.74	0.0008-0.0019	0.0014-0.0024	0.240	0.240
	2.4	4G69/F	NA	43.5-44.0	60.0 @1.74	1.74	0.0008-0.0016	0.0016-0.0024	0.240	0.240
	3.8	6G75/S	NA	43.5-44.0	60.0 @1.74	1.74	0.0008-0.0019	0.0014-0.0024	0.240	0.240
	3.8	6G75/T	NA	43.5-44.0	①	1.74	0.0008-0.0019	0.0016-0.0023	0.240	0.240
2007	2.0	4G94/E	NA	43.5-44.0	53.0 @1.74	1.74	0.0008-0.0019	0.0014-0.0024	0.240	0.240
	2.4	4G69/F	NA	43.5-44.0	60.0 @1.74	1.74	0.0008-0.0019	0.0014-0.0024	0.240	0.240
	3.8	6G75/S	NA	43.5-44.0	60.0 @1.74	1.74	0.0008-0.0019	0.0014-0.0024	0.240	0.240
	3.8	6G75/T	NA	43.5-44.0	①	1.74	0.0008-0.0019	0.0016-0.0023	0.240	0.240
2008	2.0	4B11/U	②	②	NA	③	0.0008-0.0019	0.0012-0.0021	NA	NA
	2.0	4B11/V	②	②	NA	④	0.0008-0.0019	0.0012-0.0022	NA	NA
	2.4	4G69/F	NA	43.5-44.0	60.0 @1.74	1.74	0.0008-0.0019	0.0014-0.0024	0.240	0.240
	3.8	6G75/T	NA	43.5-44.0	①	1.74	0.0008-0.0019	0.0016-0.0023	0.240	0.240

NA: Not Available

① Intake: 59.0 lbs. @ 1.74 in.
Exhaust: 53.0 @ 1.74 in.

② Valve seat contact width - Intake: 0.046 - 0.058 in.
Exhaust: 0.053 - 0.065 in.

③ Free height of valve spring: 1.868 inches

④ Free height of valve spring: 2.03 inches

22140_MITS_C0006

CAMSHAFT AND BEARING SPECIFICATIONS CHART

All measurements are given in inches.

Year	Engine Displ. Liters	Engine Series ID/VIN	Journal Dia.	Brg. Oil Clearance	Shaft End-play	Runout	Journal Bore	Lobe Height Intake	Lobe Height Exhaust
2006	2.0	4G63/C	NA	NA	NA	NA	NA	1.400-1.420	1.380 1.400
	2.0	4G94/E	NA	NA	NA	NA	NA	1.470-1.490	1.460-1.480
	2.4	4G69/F	1.800	NA	NA	NA	NA	①	1.471-1.491
	3.8	6G75/S	1.800	NA	NA	NA	NA	1.465-1.485	1.443-1.462
	3.8	6G75/T	1.800	NA	NA	NA	NA	②	1.471-1.491
2007	2.0	4G94/E	NA	NA	NA	NA	NA	1.470-1.490	1.460-1.480
	2.4	4G69/F	1.800	NA	NA	NA	NA	③	1.471-1.491
	3.8	6G75/S	1.800	NA	NA	NA	NA	1.465-1.485	1.443-1.462
	3.8	6G75/T	1.800	NA	NA	NA	NA	②	1.471-1.491
2008	2.0	4B11/U	NA	0.0000-0.0013	NA	NA	NA	1.683-1.703	1.752-1.772
	2.0	4B11/V	NA	0.0014-0.0028	NA	NA	NA	1.720-1.740	1.750-1.770
	2.4	4G69/F	1.800	NA	NA	NA	NA	③	1.471-1.491
	3.8	6G75/T	1.800	NA	NA	NA	NA	②	1.471-1.491

NA: Not Available

① Intake Low speed cam A and B: 1.456-1.475 in.
Intake High speed cam: 1.445-1.465 in.

② Intake Low speed cam A: 1.301-1.321 in.
Intake Low speed cam B: 1.451-1.4705 in.
Intake High speed cam: 1.445-1.465 in.

③ Intake Low speed cam A: 1.335-1.355 in.
Intake Low speed cam B: 1.456-1.475 in.
Intake High speed cam: 1.445-1.465 in.

22140_MITS_C0007

CRANKSHAFT AND CONNECTING ROD SPECIFICATIONS

All measurements are given in inches.

Year	Engine Displacement Liters	Engine Series ID/VIN	Crankshaft Main Brg. Journal Dia.	Crankshaft Main Brg. Oil Clearance	Crankshaft Shaft End-play	Crankshaft Thrust on No.	Connecting Rod Journal Diameter	Connecting Rod Oil Clearance	Connecting Rod Side Clearance
2006	2.0	4G63/C	2.240	0.0008-0.0016	0.0020-0.0098	3	1.770	0.0012-0.0020	0.0039-0.0098
	2.0	4G94/E	1.970	0.0008-0.0012	0.0020-0.0098	3	1.850	0.0008-0.0016	0.0039-0.0098
	2.4	4G69/F	2.240	0.0008-0.0015	0.0020-0.0090	3	1.770	0.0008-0.0019	0.0040-0.0090
	3.8	6G75/S	2.520	①	0.0020-0.0090	3	2.165	0.0008-0.0019	0.0030-0.0090
	3.8	6G75/T	2.520	②	0.0020-0.0090	3	2.165	0.0008-0.0015	0.0030-0.0090
2007	2.0	4G94/E	1.970	0.0008-0.0012	0.0020-0.0098	3	1.850	0.0008-0.0016	0.0039-0.0098
	2.4	4G69/F	2.240	③	0.0020-0.0090	3	1.770	④	0.0040-0.0090
	3.8	6G75/S	2.520	①	0.0020-0.0090	3	2.165	0.0008-0.0019	0.0030-0.0090
	3.8	6G75/T	2.520	⑤	0.0020-0.0090	3	2.165	0.0008-0.0015	0.0040-0.0090
2008	2.0	4B11/U	NA	0.0005-0.0012	0.0020-0.0010	3	NA	0.0007-0.0018	0.0040-0.0010
	2.0	4B11/V	NA	⑥	0.0020-0.0100	3	NA	0.0015-0.0027	0.0040-0.0010
	2.4	4G69/F	2.240	③	0.0020-0.0090	3	1.770	④	0.0040-0.0090
	3.8	6G75/T	2.520	⑤	0.0020-0.0090	3	2.165	0.0008-0.0015	0.0040-0.0090

NA: Not Available

① Number 1 and 4: 0.0007-0.0014 in.
Number 2 and 3: 0.0009-0.0017 in.

② Number 1 and 4: 0.0008-0.0012 in.
Number 2 and 3: 0.0012-0.0016 in.

③ Number 1, 2, 4, 5: 0.0008-0.0012 in.
Number 3: 0.0012-0.0016 in.

④ ID Mark or color - I or Yellow: 0.0007-0.0089 in.
II or None: 0.0006-0.0019 in.
III or Blue: 0.0007-0.0018 in.

⑤ Number 1 and 4: 0.0008-0.0012 in.
Number 2 and 3: 0.0008-0.0015 in.

⑥ Number 1, 2, 4, 5: 0.0015-0.0026 in.
Number 3: 0.0020-0.0030 in.

22140_MITS_C0008

PISTON AND RING SPECIFICATIONS

All measurements are given in inches.

	Engine	Engine		Ring Gap			Ring Side Clearance		
	Displ.	Series	Piston	Top	Bottom	Oil	Top	Bottom	Oil
Year	Liters	ID/VIN	Clearance	Compression	Compression	Control	Compression	Compression	Control
2006	2.0	4G63/C	0.0008-0.0016	0.0079-0.0118	0.0128-0.0197	0.0039-0.0157	0.0012-0.0028	0.0008-0.0024	NA
	2.0	4G94/E	0.0008-0.0012	0.0059-0.0118	0.0157-0.0217	0.0039-0.0138	0.0016-0.0023	0.0008-0.0024	NA
	2.4	4G69/F	0.0008-0.0015	0.0060-0.0120	0.0110-0.0170	0.0040-0.0160	0.0012-0.0028	0.0008-0.0023	NA
	3.8	6G75/S	0.0008-0.0016	0.0100-0.0160	0.0140-0.0190	0.0030 0.0140	0.0120 0.0270	0.0008-0.0023	NA
	3.8	6G75/T	0.0012-0.0019	0.0100-0.0160	0.0140-0.0190	0.0030 0.0140	0.0120 0.0270	0.0008-0.0023	NA
2007	2.0	4G94/E	0.0008-0.0012	0.0059-0.0118	0.0157-0.0217	0.0039-0.0138	0.0016-0.0023	0.0008-0.0024	NA
	2.4	4G69/F	0.0008-0.0015	0.0060-0.0120	0.0110-0.0170	0.0030-0.0150	0.0012-0.0028	0.0008-0.0024	NA
	3.8	6G75/S	0.0008-0.0016	0.0100-0.0160	0.0140-0.0190	0.0030 0.0140	0.0120 0.0270	0.0008-0.0023	NA
	3.8	6G75/T	0.0012-0.0019	0.0100-0.0160	0.0140-0.0190	0.0040-0.0140	0.0012-0.0027	0.0008-0.0023	NA
2008	2.0	4B11/U	NA	0.0060-0.0110	0.0100-0.0160	0.0040-0.0140	0.0010-0.0030	0.0010-0.0030	NA
	2.0	4B11/V	NA	0.0070-0.0110	0.0110-0.0170	0.0040-0.0130	0.0010-0.0020	0.0010-0.0020	NA
	2.4	4G69/F	0.0008-0.0015	0.0060-0.0120	0.0110-0.0170	0.0030-0.0150	0.0012-0.0028	0.0008-0.0024	NA
	3.8	6G75/T	0.0012-0.0019	0.0100-0.0160	0.0140-0.0190	0.0040-0.0140	0.0012-0.0027	0.0008-0.0023	NA

NA: Not Available

22140_MITS_C0009

TORQUE SPECIFICATIONS

All readings in ft. lbs.

Year	Engine Displacement Liters	Engine Series ID/VIN	Cylinder Head Bolts	Main Bearing Bolts	Rod Bearing Bolts	Crankshaft Damper Bolts	Flywheel Bolts	Manifold Intake	Manifold Exhaust	Spark Plugs	Oil Pan Drain Plug
2006	2.0	4G63/C	①	②	③	123	95-101	④	⑤	15-22	25-33
	2.0	4G94/E	⑥	②	③	130-138	69-75	⑦	⑧	15-22	25-33
	2.4	4G69/F	①	②	③	123	95-101	⑨	⑩	15-22	25-33
	3.8	6G75/S	⑪	51-57	⑫	134-140	53-56	⑬	⑭	15-22	25-33
	3.8	6G75/T	⑪	51-57	⑫	134-140	53-56	⑮	⑭	15-22	25-33
2007	2.0	4G94/E	⑥	②	③	130-138	69-75	⑦	⑧	15-22	25-33
	2.4	4G69/F	①	②	③	123	95-101	⑨	⑩	15-22	25-33
	3.8	6G75/S	⑪	51-57	⑫	134-140	53-56	⑬	⑭	15-22	25-33
	3.8	6G75/T	⑪	51-57	⑫	134-140	53-56	⑮	⑭	15-22	25-33
2008	2.0	4B11/U	⑯	⑰	⑱	155	⑲	⑳	㉑	15-22	25-33
	2.0	4B11/V	⑯	㉒	⑱	㉓	⑲	13-21	㉔	12-14	25-33
	2.4	4G69/F	①	②	③	123	95-101	⑨	⑩	15-22	25-33
	3.8	6G75/T	⑪	51-57	⑫	134-140	53-56	⑮	⑭	15-22	25-33

NOTE: Dip main bearing bolts and crankshaft damper bolt in clean engine oil prior to tightening.

① Step 1: Tighten all bolts to 57-59 ft. lbs.
Step 2: Loosen all bolts to 0 ft. lbs.
Step 3: Tighten all bolts to 14-16 ft. lbs.
Step 4: Plus 90 degrees
Step 4: Plus another 90 degrees

② Step 1: 17-19 ft. lbs.
Step 2: Plus 90 degrees

③ Step 1: 14-16 ft. lbs.
Step 2: Plus 90-94 degrees

④ M8 bolts (14-16 ft. lbs.), M10 bolts (22-30 ft. lbs.)
Engine hanger bolt 22-30 ft. lbs.
Stay bolt 21-25 ft. lbs.

⑤ M8 bolts (20-28 ft. lbs.), M10 bolts (35-47 ft. lbs.)
Heat protector bolt 15-19 ft. lbs.
Stay bolt 21-25 ft. lbs.

⑥ Step 1: 52-58 ft. lbs.
Step 2: 14-16 ft. lbs.
Step 3: Plus 90 degrees
Step 4: Plus another 90 degrees

⑦ Intake manifold bolts 12-16 ft. lbs.
Stay bolt 21-25 ft. lbs.

⑧ M8 bolts (12-14 ft. lbs.), M10 bolts (19-23 ft. lbs.)
Cover bolt 107-122 inch lbs.

⑨ Intake manifold bolt 17-19 ft. lbs.
Intake manifold nut 14-16 ft. lbs.
Stay bolt 21-25 ft. lbs.

⑩ Exhaust manifold nut 33-39 ft. lbs.
Cover bolt 116-132 inch lbs.

⑪ Step 1: Tighten all bolts to 76-84 ft. lbs.
Step 2: Loosen all bolts to 0 ft. lbs.
Step 3: Tighten all bolts to 76-84 ft. lbs.

⑫ Step 1: 19-21 ft. lbs.
Step 2: Plus 90-94 degrees

⑬ Intake manifold bolt 18-24 ft. lbs.
Stay bolts: M8 (12-14 ft. lbs.), M10 (23-31 ft. lbs.)

⑭ Exhaust manifold nut 29-37 ft. lbs.
Stay bolts: M8 (12-16 ft. lbs.), M10 (28-38 ft. lbs.)
Stay bolts: M12 (48-63 ft. lbs.)
Engine hanger bolt 22-30 ft. lbs.

⑮ Intake manifold bolts 15-17 ft. lbs.
Stay bolts: M8 (15-17 ft. lbs.), M10 (31-39 ft. lbs.)

⑯ Step 1: 25-27 ft. lbs.
Step 2: Plus 90 degrees
Step 3: Plus another 90 degrees

⑰ Step 1: 19-21 ft. lbs.
Step 2: Plus 45 degrees

⑱ Step 1: 44 inch lbs.
Step 2: 15 ft. lbs.
Step 3: Plus 90 degrees

⑲ Step 1: 30 ft. lbs.
Step 2: 96 ft. lbs.

⑳ Step 1: 2-28 inch lbs.
Step 2: 14-16 ft. lbs.

㉑ Exhaust manifold nut 33-39 ft. lbs.
Upper and lower covers: 116-132 inch lbs.

㉒ Step 1: M8 bolts (72-106 inch lbs.)
and M10 bolts (27-29 ft. lbs.)
Step 2: Plus 90 degrees

㉓ Step 1: 184 ft. lbs.
Step 2: Loosen to 0 ft. lbs.
Step 3: 81 ft. lbs.
Step 4: Plus 60 degrees

㉔ Exhaust manifold nut 33-39 ft. lbs.
Cover bolt: 72-106 inch lbs.

22140_MITS_C0010

WHEEL ALIGNMENT

Year	Model		Caster Range (+/-Deg.)	Caster Preferred Setting (Deg.)	Camber Range (+/-Deg.)	Camber Preferred Setting (Deg.)	Toe-in (Deg.)
2006	Eclipse	Front	0.30	+3.00	0.30	0.00	0 +/- 0.12
		Rear	—	—	0.30	-0.50	0.12 +/- 0.12
	Galant	Front	0.30	+3.00	0.30	0.00	0 +/- 0.12
		Rear	—	—	0.30	-0.50	0.12 +/- 0.12
	Lancer ①	Front	0.30	+2.55	0.30	0.00	0.04 +/- 0.08
		Rear	—	—	0.30	-0.40	0.12 +/- 0.08
	Lancer ②	Front	0.30	+2.45	0.30	0.05	0.04 +/- 0.08
		Rear	—	—	0.30	-0.40	0.12 +/- 0.08
	Lancer ③	Front	0.30	+2.55	0.30	-0.05	0.04 +/- 0.08
		Rear	—	—	0.30	-0.40	0.12 +/- 0.08
	Lancer Evolution	Front	0.30	+3.55	0.30	④	0 +/- 0.08
		Rear	—	—	0.30	-1.00	0.12 +/- 0.07
	Lancer Sportback ②	Front	0.30	+2.35	0.30	0.05	0.04 +/- 0.07
		Rear	—	—	0.30	-0.40	0.12 +/- 0.08
	Lancer Sportback ③	Front	0.30	+2.45	0.30	-0.05	0.04 +/- 0.07
		Rear	—	—	0.30	-0.40	0.12 +/- 0.08
2007	Eclipse/Spyder	Front	0.30	+3.00	0.30	0.00	0 +/- 0.12
		Rear	—	—	0.30	-0.50	0.12 +/- 0.12
	Galant	Front	0.30	+3.00	0.30	0.00	0 +/- 0.12
		Rear	—	—	0.30	-0.50	0.12 +/- 0.12
	Lancer	Front	0.30	+2.40	0.30	-0.50	0.04 +/- 0.09
		Rear	—	—	0.30	-0.55	0.12 +/- 0.08
2008	Eclipse/Spyder	Front	0.30	+3.00	0.30	0.00	0 +/- 0.12
		Rear	—	—	0.30	-0.50	0.12 +/- 0.12
	Galant	Front	0.30	+3.00	0.30	0.00	0 +/- 0.12
		Rear	—	—	0.30	-0.50	0.12 +/- 0.12
	Lancer	Front	0.30	+2.40	0.30	-0.50	0.04 +/- 0.09
		Rear	—	—	0.30	-0.55	0.12 +/- 0.08
	Lancer Evolution	Front	0.30	+4.25	0.30	-1.00	0 +/- 0.07
		Rear	—	—	0.30	-1.00	0.12 +/- 0.08

① With 14 in. wheels
② With 15 in. wheels
③ With 16 in. wheels
④ Select from 2 options: -1.00 +/- 0.30 or -2.00 +/- 0.30

22140_MITS_C0011

TIRE, WHEEL AND BALL JOINT SPECIFICATIONS

Year	Model	OEM Tires		Tire Pressures (psi)		Wheel	Ball Joint	Lug Nut
		Standard	Optional	Front	Rear	Size	Inspection	Torque (ft. lbs.)
2006	Eclipse	P225/50R17	P235/45R18	①	①	7.5JJ x 17 8.0JJ x 18	②	65-80
	Galant	P215/60R16	P235/55R17	①	①	6.5JJ x 16 7.0JJ x 17	②	65-80
	Lancer DE	P185/65R14	—	①	①	5.5JJ x 14	③	65-80
	Lancer ES, OZ-Rally	P195/60R15	—	①	①	6.0JJ x 15	③	65-80
	Lancer Ralliart	P205/50R16	—	①	①	6.0JJ x 16	③	65-80
	Lancer Evolution	P235/45R17	—	①	①	8.0JJ x 17	④	65-80
	Lancer Sportback LS, Ralliart	P195/60R15	P205/60R16	①	①	6.0JJ x 15 6.0JJ x 16	③	65-80
2007	Eclipse/Spyder	P225/50R17	P235/45R18	①	①	7.5JJ x 17 8.0JJ x 18	⑤	65-80
	Galant	P215/60R16 P235/45R18	P215/55R17	①	①	⑥	②	65-80
	Lancer DE, ES	P205/60R16	—	①	①	6.5JJ x 16	⑤	65-80
	Lancer GTS	P215/45R18	—	①	①	7.0JJ x 18	⑤	65-80
2008	Eclipse/Spyder	P225/50R17	P235/45R18	①	①	7.5JJ x 17 8.0JJ x 18	⑤	65-80
	Galant	P215/60R16 P235/45R18	P215/55R17	①	①	⑥	②	65-80
	Lancer DE, ES	P205/60R16	—	①	①	6.5JJ x 16	⑤	65-80
	Lancer GTS	P215/45R18	—	①	①	7.0JJ x 18	⑤	65-80
	Lancer Evolution	P245/40R18	—	①	①	8.5JJ x 18	⑦	65-80

OEM: Original Equipment Manufacturer

PSI: Pounds Per Square Inch

① Refer to placard on vehicle for proper inflation pressure.

② Replace the ball joint if too loose or if rotating torque exceeds specification: 31-61 inch lbs.

③ Replace the ball joint if too loose or if rotating torque exceeds specification: 0-35 inch lbs.

④ Replace the ball joint if too loose or if rotating torque exceeds specification: 4-30 inch lbs.

⑤ Replace the ball joint if too loose or if rotating torque exceeds specification: 19-36 inch lbs.

⑥ 2.4L model: 6.5JJ x 16 and 7.0JJ x 17. 3.8L model: 8.0JJ x 18

⑦ Replace the ball joint if too loose or if rotating torque exceeds specification: 29 inch lbs. or less

22140_MITS_C0012

BRAKE SPECIFICATIONS

All measurements in inches unless noted

			Brake Disc			Brake Drum Diameter				Brake Caliper	
Year	Model		Original Thickness	Minimum Thickness	Maximum Runout	Original Inside Diameter	Max. Wear Limit	Maximum Machine Diameter	Minimum Lining Thickness	Bracket Bolts (ft. lbs.)	Mounting Bolts (ft. lbs.)
2006	Eclipse	F	1.020	0.960	0.0039	—	—	—	0.080	25-31	67-81
		R	0.390	0.330	0.0016	—	—	—	0.080	28-36	42-48
	Galant	F	1.020	0.960	0.0039	—	—	—	0.080	25-31	67-81
		R	0.390	0.330	0.0016	—	—	—	0.080	28-36	42-48
	Lancer	F	1.020	0.960	0.0006	—	—	—	0.080	①	67-81
		R	0.390	0.330	0.0015	—	—	—	0.080	28-36	42-48
	Lancer	F	1.020	0.960	0.0006	—	—	—	0.080	①	67-81
		R	—	—	—	7.990	8.070	8.070	0.040	—	—
	Lancer Evolution	F	1.260	1.170	0.0012	—	—	—	0.080	NA	73-87
		R	0.870	0.800	0.0012	—	—	—	0.080	NA	36-44
	Lancer Sportback	F	1.020	0.960	0.0015	—	—	—	0.080	25-31	67-81
		R	0.390	0.330	0.0015	—	—	—	0.080	28-36	42-48
2007	Eclipse/Spyder	F	1.020	0.960	0.0039	—	—	—	0.080	25-31	67-81
		R	0.390	0.330	0.0016	—	—	—	0.080	28-36	42-48
	Galant	F	1.020	0.960	0.0039	—	—	—	0.080	25-31	67-81
		R	0.390	0.330	0.0016	—	—	—	0.080	28-36	42-48
	Lancer	F	1.020	0.960	0.0024	—	—	—	0.080	28-36	67-81
		R	0.390	0.330	0.0032	—	—	—	0.080	28-36	42-48
	Lancer	F	1.020	0.960	0.0024	—	—	—	0.080	28-36	67-81
		R	—	—	—	7.990	8.070	8.070	0.040	—	—
2008	Eclipse/Spyder	F	1.020	0.960	0.0039	—	—	—	0.080	25-31	67-81
		R	0.390	0.330	0.0016	—	—	—	0.080	28-36	42-48
	Galant	F	1.020	0.960	0.0039	—	—	—	0.080	25-31	67-81
		R	0.390	0.330	0.0016	—	—	—	0.080	28-36	42-48
	Lancer	F	1.020	0.960	0.0024	—	—	—	0.080	28-36	67-81
		R	0.390	0.330	0.0032	—	—	—	0.080	28-36	42-48
	Lancer	F	1.020	0.960	0.0024	—	—	—	0.080	28-36	67-81
		R	—	—	—	7.990	8.070	8.070	0.040	—	—
	Lancer Evolution	F	1.260	1.180	0.0024	—	—	—	0.080	NA	73-87
		R	0.870	0.790	0.0032	—	—	—	0.080	NA	52-66

NA: Not Available

F: Front

R: Rear

① Main slide pin (2.0L): 59-65 ft. lbs.
Sub slide pin (2.0L): 34-40 ft. lbs.
Lock pin bolt (2.4L): 25-31 ft. lbs.

22140_MITS_C0013

SCHEDULED MAINTENANCE INTERVALS

MITSUBISHI—ECLIPSE/SPYDER, GALANT, LANCER & LANCER EVOLUTION

		VEHICLE MILEAGE INTERVAL (x1000)														
TO BE SERVICED	TYPE OF	7.5	15	22.5	30	37.5	45	52.5	60	67.5	75	82.5	90	97.5	105	120
Accessory drive belts	S/I				✓				✓				✓			✓
Air cleaner element (engine)	R				✓				✓				✓			✓
Air conditioner system	S/I	Inspect the system operation annually														
Automatic transaxle fluid	S/I				✓				✓				✓			✓
Ball joint and steering linkage seals	S/I				✓				✓				✓			✓
Brake lines, hoses, and connections	S/I		✓		✓		✓		✓		✓		✓		✓	✓
Brake pads, calipers, & rotors	S/I		✓		✓		✓		✓		✓		✓		✓	✓
Cooling system hoses and coolant level	S/I				✓				✓				✓			✓
Driveshafts and CV-boots	S/I		✓		✓		✓		✓		✓		✓		✓	✓
Engine coolant	R								✓				✓			✓
Engine oil and filter	R	✓	✓	✓	✓	✓	✓	✓	✓	✓	✓	✓	✓	✓	✓	✓
Exhaust pipe connections, muffler, and suspension bolts	S/I				✓				✓				✓			✓
Evaporative emission control system (except canister)	S/I								✓							✓
Fuel hoses	S/I				✓				✓				✓			✓
Fuel System (tank, connections, gas cap)	S/I								✓							✓
Manual transaxle oil	S/I				✓				✓				✓			✓
Rear drum brake linings and wheel cylinders	S/I				✓				✓				✓			✓
Spark plugs (standard)	R				✓				✓				✓			✓
Spark plugs (Iridium coated)	R	Every 84 months or 105,000 miles (under normal usage)														
Spark plugs (Platinum	R								✓							✓
Suspension system	S/I				✓				✓				✓			✓
Timing Belt	R														✓	
Tires (rotate)	S/I	✓	✓	✓	✓	✓	✓	✓	✓	✓	✓	✓	✓	✓	✓	✓
Valve clearance	S/I				✓				✓				✓			✓

R: Replace S/I: Service or Inspect

FREQUENT OPERATION MAINTENANCE (SEVERE SERVICE)

If a vehicle is operated under any of the following conditions it is considered severe service:

- Extremely dusty areas.

- 50% or more of the vehicle operation is in 90°F (32°C) or higher temperatures, or constant operation in temperatures below 32°F (0°C).

- Prolonged idling (vehicle operation in stop and go traffic).

- Frequent short running periods (engine does not warm to normal operating temperatures).

- Police, taxi, delivery usage, or trailer towing usage.

Air cleaner filter: service or inspect every 15,000 miles.

Automatic transaxle fluid & filter: check every 15,000 miles, replace every 30,000 miles.

Brake pads, calipers & rotors: service or inspect every 7,500 miles.

Manual transaxle oil change every 30,000 miles

Oil & oil filter: change every 3,750 miles.

Rear brake drums & linings: service or inspect every 15,000 miles.

Spark plugs: service or inspect every 15,000 miles (standard plugs only)

Suspension system inspect for looseness and damage every 7,500 miles

22140_MITS_C0014

PRECAUTIONS

Before servicing any vehicle, please be sure to read all of the following precautions, which deal with personal safety, prevention of component damage, and important points to take into consideration when servicing a motor vehicle:

- Never open, service or drain the radiator or cooling system when the engine is hot; serious burns can occur from the steam and hot coolant.
- Observe all applicable safety precautions when working around fuel. Whenever servicing the fuel system, always work in a well-ventilated area. Do not allow fuel spray or vapors to come in contact with a spark, open flame, or excessive heat (a hot drop light, for example). Keep a dry chemical fire extinguisher near the work area. Always keep fuel in a container specifically designed for fuel storage; also, always properly seal fuel containers to avoid the possibility of fire or explosion. Refer to the additional fuel system precautions later in this section.
- Fuel injection systems often remain pressurized, even after the engine has been turned **OFF**. The fuel system pressure must be relieved before disconnecting any fuel lines. Failure to do so may result in fire and/or personal injury.
- Brake fluid often contains polyglycol ethers and polyglycols. Avoid contact with the eyes and wash your hands thoroughly after handling brake fluid. If you do get brake fluid in your eyes, flush your eyes with clean, running water for 15 minutes. If eye irritation persists, or if you have taken brake fluid internally, IMMEDIATELY seek medical assistance.
- The EPA warns that prolonged contact with used engine oil may cause a number of skin disorders, including cancer. You should make every effort to minimize your exposure to used engine oil. Protective gloves should be worn when changing oil. Wash your hands and any other exposed skin areas as soon as possible after exposure to used engine oil. Soap and water, or waterless hand cleaner should be used.
- All new vehicles are now equipped with an air bag system, often referred to as a Supplemental Restraint System (SRS) or Supplemental Inflatable Restraint (SIR) system. The system must be disabled before performing service on or around system components, steering column, instrument panel components, wiring and sensors. Failure to follow safety and disabling procedures could result in accidental air bag deployment, possible personal injury and unnecessary system repairs.
- Always wear safety goggles when working with, or around, the air bag system. When carrying a non-deployed air bag, be sure the bag and trim cover are pointed away from your body. When placing a non-deployed air bag on a work surface, always face the bag and trim cover upward, away from the surface. This will reduce the motion of the module if it is accidentally deployed. Refer to the additional air bag system precautions later in this section.
- Clean, high quality brake fluid from a sealed container is essential to the safe and proper operation of the brake system. You should always buy the correct type of brake fluid for your vehicle. If the brake fluid becomes contaminated, completely flush the system with new fluid. Never reuse any brake fluid. Any brake fluid that is removed from the system should be discarded. Also, do not allow any brake fluid to come in contact with a painted surface; it will damage the paint.
- Never operate the engine without the proper amount and type of engine oil; doing so WILL result in severe engine damage.
- Timing belt maintenance is extremely important. Many models utilize an interference-type, non-freewheeling engine. If the timing belt breaks, the valves in the cylinder head may strike the pistons, causing potentially serious (also time-consuming and expensive) engine damage. Refer to the maintenance interval charts for the recommended replacement interval for the timing belt, and to the timing belt section for belt replacement and inspection.
- Disconnecting the negative battery cable on some vehicles may interfere with the functions of the on-board computer system(s) and may require the computer to undergo a relearning process once the negative battery cable is reconnected.
- When servicing drum brakes, only disassemble and assemble one side at a time, leaving the remaining side intact for reference.
- Only an MVAC-trained, EPA-certified automotive technician should service the air conditioning system or its components.

BRAKES

ANTI-LOCK BRAKE SYSTEM (ABS)

GENERAL INFORMATION

The Anti-Lock Brake System (ABS) controls the hydraulic brake pressure of all four wheels during sudden braking and braking on hazardous road surfaces, preventing the wheels from locking. The ABS provides the following benefits:

- Enables steering around obstacles with a greater degree of certainty during panic braking
- Enables stopping during panic braking while allowing stability and control, even on curves
- If a malfunction occurs in the ABS, the system will operate as a normal brake (fail safe mode). A diagnostic function and a fail-safe system have been included for serviceability.

BRAKES — BLEEDING THE BRAKE SYSTEM

BLEEDING PROCEDURE

When any part of the hydraulic system has been disconnected for repair or replacement, air may get into the lines and cause spongy pedal action (because air can be compressed and brake fluid cannot). To correct this condition, it is necessary to bleed the hydraulic system so to be sure all air is purged.

When bleeding the brake system, bleed one brake cylinder at a time, beginning at the cylinder with the longest hydraulic line (farthest from the master cylinder) first. ALWAYS keep the master cylinder reservoir filled with brake fluid during the bleeding operation. Never use brake fluid that has been drained from the hydraulic system, no matter how clean it is.

The primary and secondary hydraulic brake systems are separate and are bled independently. During the bleeding operation, do not allow the reservoir to run dry. Keep the master cylinder reservoir filled with brake fluid.

1. Clean all dirt from around the master cylinder fill cap, remove the cap and fill the master cylinder with brake fluid until the level is within 1/4 inch (6mm) of the top edge of the reservoir.
2. Clean the bleeder screws at all 4 wheels. The bleeder screws are located on the back of the brake backing plate (drum brakes) and on the top of the brake calipers (disc brakes).
3. Attach a length of rubber hose over the bleeder screw and place the other end of the hose in a glass jar, submerged in brake fluid.
4. Open the bleeder screw 1/2–3/4 turn. Have an assistant slowly depress the brake pedal.

✱✱ CAUTION

Brake fluid contains polyglycol ethers and polyglycols. Avoid contact with the eyes and wash your hands thoroughly after handling brake fluid. If you do get brake fluid in your eyes, flush your eyes with clean, running water for 15 minutes. If eye irritation persists, or if you have taken brake fluid internally, IMMEDIATELY seek medical assistance.

5. Close the bleeder screw and tell your assistant to allow the brake pedal to return slowly. Continue this process to purge all air from the system.
6. When bubbles cease to appear at the end of the bleeder hose, close the bleeder screw and remove the hose. Tighten the bleeder screw to the proper torque.
7. Check the master cylinder fluid level and add fluid accordingly. Do this after bleeding each wheel.
8. Repeat the bleeding operation at the remaining 3 wheels, ending with the one closet to the master cylinder.
9. Fill the master cylinder reservoir to the proper level.

BLEEDING THE ABS SYSTEM

There are no special procedures for bleeding the ABS system. Refer to the conventional bleeding procedures.

BRAKES — FRONT DISC BRAKES

✱✱ CAUTION

Dust and dirt accumulating on brake parts during normal use may contain asbestos fibers from production or aftermarket brake linings. Breathing excessive concentrations of asbestos fibers can cause serious bodily harm. Exercise care when servicing brake parts. Do not sand or grind brake lining unless equipment used is designed to contain the dust residue. Do not clean brake parts with compressed air or by dry brushing. Cleaning should be done by dampening the brake components with a fine mist of water, then wiping the brake components clean with a dampened cloth. Dispose of cloth and all residue containing asbestos fibers in an impermeable container with the appropriate label. Follow practices prescribed by the Occupational Safety and Health Administration (OSHA) and the Environmental Protection Agency (EPA) for the handling, processing, and disposing of dust or debris that may contain asbestos fibers.

BRAKE CALIPER

REMOVAL & INSTALLATION

1. As required, partially drain the master cylinder.
2. Raise and support the vehicle safely.
3. Remove the tire and wheel assembly.
4. Disconnect and plug the brake hose connection. Discard the gasket.
5. Remove the caliper retaining bolts. Remove the caliper from the vehicle.

To install

6. Position the caliper to its mounting on the vehicle.
7. Install the caliper mounting bolts and tighten to:
   a. Except Lancer Evolution: 67–81 ft. lbs. (91–110 Nm).
   b. Lancer Evolution: 73–87 ft. lbs. (99–118 Nm).
8. Connect the brake line to the caliper with new gaskets. Torque the brake line union bolt to 18–22 ft. lbs. (25–30 Nm).
9. Bleed the brake system.
10. Install the wheel and tire.
11. Before attempting to move the vehicle, pump the brake pedal to seat the pads against the rotors. Make sure the vehicle has a firm brake pedal. Check the level of the brake fluid and add fluid if necessary.

DISC BRAKE PADS

REMOVAL & INSTALLATION

1. Before servicing the vehicle, refer to the Precautions Section.
2. Drain some of the brake fluid from the master cylinder reservoir.
3. Remove the front wheels.
4. Remove the caliper guide and lock pins and lift the caliper assembly from the caliper support.

➡On some vehicles, the caliper can be flipped up by leaving the upper pin in place and using it as a pivot point.

5. Remove the brake pads, spring clip, and shims.

To install:

6. Compress pistons back into the caliper bore.
7. Lubricate slide points and install the brake pads, shims, and spring clips onto the caliper support.

8. Install the caliper over the brake pads.

9. Lubricate and install the caliper guide and lock pins in their original positions.

10. Install the wheels.

11. Before attempting to move the vehicle, pump the brake pedal to seat the pads against the rotors. Make sure the vehicle has a firm brake pedal. Check the level of the brake fluid and add fluid if necessary.

BRAKES

REAR DISC BRAKES

⁂ CAUTION

Dust and dirt accumulating on brake parts during normal use may contain asbestos fibers from production or aftermarket brake linings. Breathing excessive concentrations of asbestos fibers can cause serious bodily harm. Exercise care when servicing brake parts. Do not sand or grind brake lining unless equipment used is designed to contain the dust residue. Do not clean brake parts with compressed air or by dry brushing. Cleaning should be done by dampening the brake components with a fine mist of water, then wiping the brake components clean with a dampened cloth. Dispose of cloth and all residue containing asbestos fibers in an impermeable container with the appropriate label. Follow practices prescribed by the Occupational Safety and Health Administration (OSHA) and the Environmental Protection Agency (EPA) for the handling, processing, and disposing of dust or debris that may contain asbestos fibers.

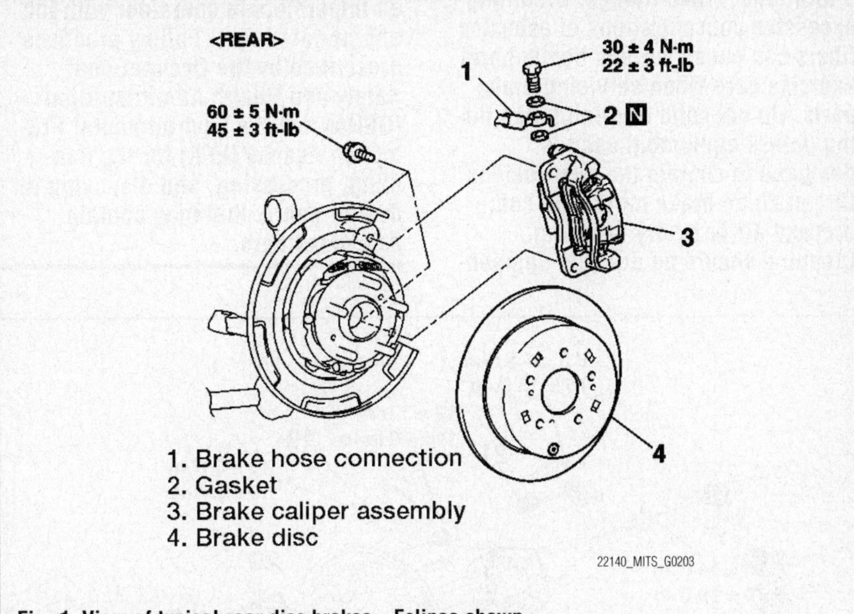

Fig. 1 View of typical rear disc brakes—Eclipse shown

BRAKE CALIPER

REMOVAL & INSTALLATION

See Figure 1.

1. Before servicing the vehicle, refer to the Precautions Section.

2. As required, partially drain the master cylinder.

3. Remove or disconnect the following:
   - Wheels
   - Brake hose from the caliper
   - Caliper guide and lock pins and lift the caliper assembly from the caliper support

To install

4. Install or connect the following:
   - Caliper onto the caliper support
   - Guide pin and lock pin and tighten to specification
   - Brake hose or banjo bolt with new washers

5. Bleed the brake system.
6. Install the wheels.

DISC BRAKE PADS

REMOVAL & INSTALLATION

See Figure 1.

1. Before servicing the vehicle, refer to the Precautions Section.

2. Remove or disconnect the following:
   - Rear wheels
   - Lower caliper mounting bolt and rotate the caliper upward
   - Pads from the caliper support
   - Pad retainers, if necessary

To install:

3. Install or connect the following:
   - Pad retainers, if removed
   - Pads onto the pad retainers

4. Compress the caliper piston using a C-clamp.

5. Rotate the caliper downward and install the mounting bolt.

6. Install the wheel.

7. Pump the brake pedal until the brake pads are seated and a firm pedal is achieved before attempting to move the vehicle.

⁂ CAUTION

Do not move the vehicle until a firm pedal is obtained.

8. Road test the vehicle to check for proper brake operation.

⁂ CAUTION

Dust and dirt accumulating on brake parts during normal use may contain asbestos fibers from production or aftermarket brake linings. Breathing excessive concentrations of asbestos fibers can cause serious bodily harm. Exercise care when servicing brake parts. Do not sand or grind brake lining unless equipment used is designed to contain the dust residue. Do not clean brake parts with compressed air or by dry brushing. Cleaning should be done by dampening the brake components with a fine mist of water, then wiping the brake components clean with a dampened cloth. Dispose of cloth and all residue containing asbestos fibers in an impermeable container with the appropriate label. Follow practices prescribed by the Occupational Safety and Health Administration (OSHA) and the Environmental Protection Agency (EPA) for the handling, processing, and disposing of dust or debris that may contain asbestos fibers.

BRAKE DRUM

REMOVAL & INSTALLATION

Lancer

See Figure 2.

⁂ WARNING

Do not disassemble the rear wheel hub assembly. The magnetic encoder collects metallic particles easily, because it is magnetized. When the rear wheel hub assembly is removed/installed, make sure that

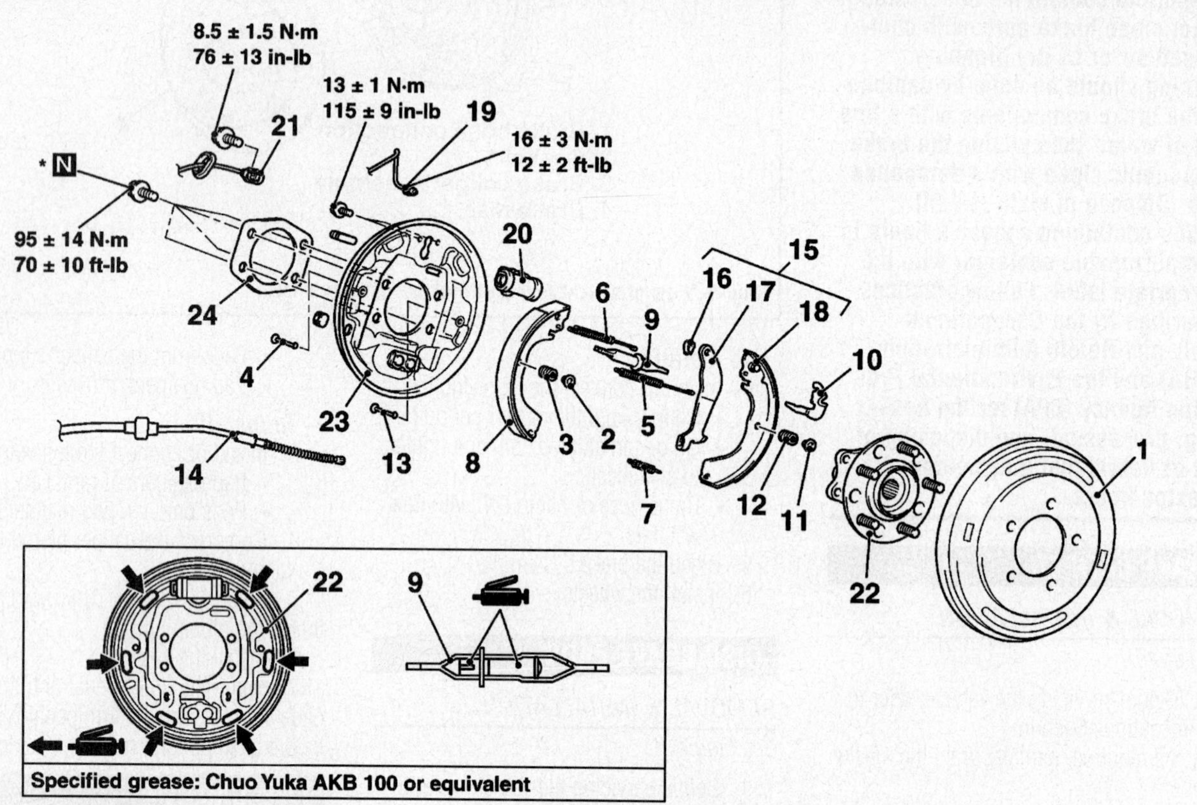

1. Rear brake drum
2. Shoe hold down cup
3. Shoe hold down spring
4. Shoe hold down pin
5. Shoe-to-lever spring
6. Shoe-to-shoe spring
7. Retainer spring
8. Shoe and lining assembly
9. Auto adjuster assembly
10. Adjuster lever
11. Shoe hold down cup
12. Shoe hold down spring
13. Shoe hold down pin
14. Parking brake rear cable removal
15. Shoe and lever assembly
16. Retainer
17. Parking brake lever
18. Shoe and lining assembly
19. Brake pipe connection
20. Wheel cylinder assembly
21. Rear wheel speed sensor (Vehicles with ABS)
22. Rear hub assembly
23. Backing plate
24. Spacer

22140_MITS_G0204

Fig. 2 Exploded view of rear drum brakes—Lancer

the magnetic encoder (integrated with inner oil seal) does not contact with surrounding parts to avoid damage (vehicles with ABS).

**** WARNING**

When removing and installing the rear wheel speed sensor, make sure that its pole piece at the end does not make contact with surrounding parts in order to avoid damage (vehicles with ABS).

**** CAUTION**

Frequent inhalation of brake pad dust, regardless of material composition, could be hazardous to your health. Avoid breathing dust particles. Never use an air hose or brush to clean brake assemblies.

1. Before servicing the vehicle, refer to the Precautions Section.
2. Check to ensure that the parking brake is fully released.
3. Raise and safely support the vehicle.
4. Remove the tire and wheel assembly.
5. Remove the brake drum.

To install:

6. If installing a new brake drum, use denatured alcohol or an equivalent approved brake cleaner and a clean shop towel to remove the protective coating from the friction surface of the drum.
7. Install the brake drum.
8. Install the tire and wheel assembly.
9. Apply the brakes approximately 3 times in order to seat and center the brake shoes within the drum.
10. Lower the vehicle.

BRAKE SHOES

REMOVAL & INSTALLATION

See Figure 2.

**** CAUTION**

Frequent inhalation of brake pad dust, regardless of material composition, could be hazardous to your health. Avoid breathing dust particles. Never use an air hose or brush to clean brake assemblies.

1. Before servicing the vehicle, refer to the Precautions Section.
2. Check to ensure that the parking brake is fully released.
3. Raise and safely support the vehicle.
4. Remove the tire and wheel assembly.
5. Remove the brake drum.
6. Remove the shoe-to-shoe spring.
7. Remove the shoe-to-lever spring and adjuster assembly.
8. Remove the shoe hold-down clips and the brake shoes.
9. Remove the parking brake cable from the rear shoes by spreading the horseshoe clip apart.

To install

10. Connect the parking brake cable to the brake assembly.
11. Clean the threaded portions of adjuster sleeve and push rod female. Coat the threads of the adjuster assembly with grease. To shorten the clevises, turn the adjuster bolt.
12. Hook the shoe adjuster lever, then install it to the brake shoe.
13. Install the adjuster assembly and upper return spring.

**** WARNING**

Be careful not to damage the wheel cylinder dust covers.

14. Install the lower return spring.
15. Apply brake cylinder grease, or equivalent rubber grease, to the sliding surfaces and brake shoe ends and opposite edges of the shoes.

➡Be careful not to get grease on the brake linings.

16. Install the brake shoes onto the backing plate.
17. Install the shoe hold down pins and the shoe hold down springs.
18. Install the rear brake drum.
19. Install the tire and wheel assembly.
20. Depress the brake pedal several times to set the self-adjusting brake.
21. Adjust the parking brake, as necessary.

ADJUSTMENT

The rear drum brakes are automatically adjusted while driving the vehicle. The brakes are also adjusted each time the parking brake is applied. Manual brake adjustment is only required after the brake shoes or hardware have been replaced, or the adjuster has been replaced.

1. Remove the brake drum as described in this section.
2. Remove any excessive dust and dirt present on the brakes using the appropriate methods.
3. Using a brake adjustment gauge, measure the inside diameter of the brake drum.
4. Adjust the brake shoes to the same diameter as the drum by placing the brake adjustment gauge on the shoes and turn the adjusting star wheel.
5. Install the brake drum as described in this section.

PARKING BRAKE CABLES

ADJUSTMENT

See Figure 3.

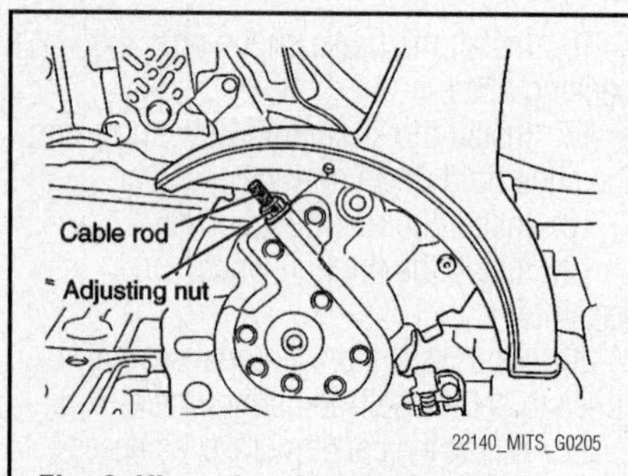

Fig. 3 View of parking brake adjustment nut

1. Before servicing the vehicle, refer to the Precautions Section.
2. Pull the parking brake lever with a force of 44 lbs. (196 N) and count the number of notches. Standard value is: 3–5 notches.
3. If the parking brake lever is not the standard value, adjust in the following manner:
   a. Remove the inner compartment mat of the floor console.
   b. Loosen the adjusting nut at the end of the cable rod, freeing the parking brake.
   c. With the engine idling, forcefully depress the brake pedal 5–6 times and confirm that the pedal stroke stops changing. If the pedal stroke stops changing, the automatic-adjustment mechanism is functioning normally, and the clearance between the shoe and the drum is correct.
   d. After adjusting the parking brake lever stroke, safely raise and support the rear of the vehicle and with the parking brake lever in the released position, turn the rear wheels to confirm that there is no brake drag.

➡Be careful that the parking brake lever notch number is within the standard range. If the notch number is too low, rear brake dragging may result.

PARKING BRAKE SHOES

REMOVAL & INSTALLATION

Except Lancer—With Rear Drum Brakes

See Figures 4 and 5.

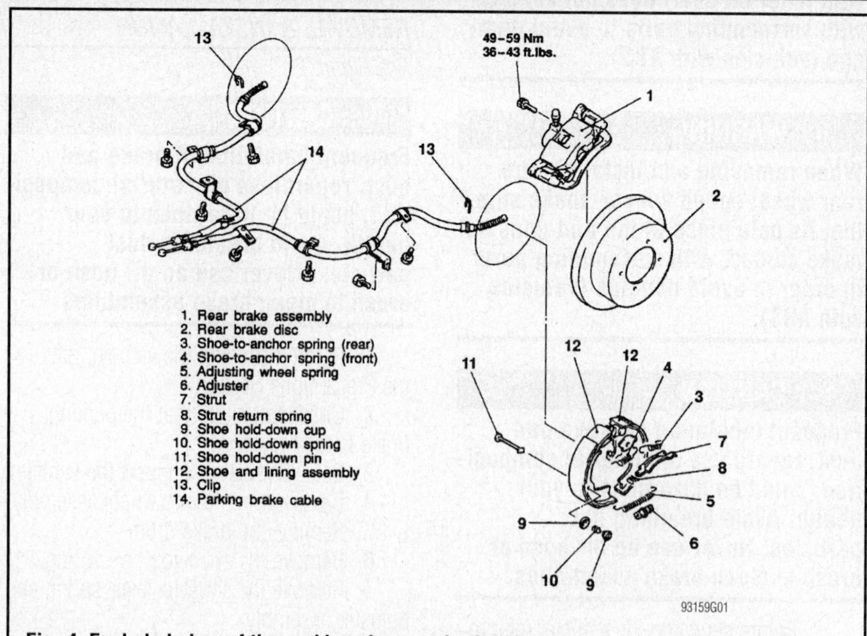

Fig. 4 Exploded view of the parking shoes and related components—with disc brakes

1. Before servicing the vehicle, refer to the Precautions Section.
2. Raise and safely support the vehicle.
3. Remove the caliper assembly.
4. Remove the rear brake rotor.

➡When servicing, only dissemble and assemble one side at a time, leaving the remaining side intact for reference.

5. Remove the front and rear shoe-to-anchor springs.
6. Remove the adjusting wheel spring and the adjuster.
7. Remove the strut and the strut return spring.
8. Remove the shoe hold-down cup, spring, and pin.
9. Remove the shoe and lining assembly.
10. Unfasten the clips and the retaining bolts.
11. Remove the parking brake cable.

To install:

12. Installation is the reverse of the removal procedure.

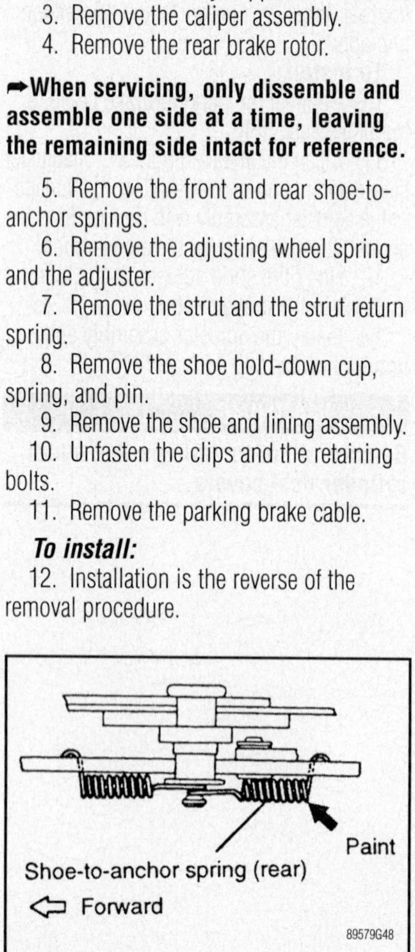

Paint
Shoe-to-anchor spring (rear)
⇦ Forward
89579G48

Fig. 5 Shoe-to-anchor spring installation

13. Install the adjuster so the shoe adjusting bolt of the left hand wheel is attached toward the front of the vehicle and the shoe adjusting bolt of the right hand wheel is toward the rear of the vehicle.
14. The load on the respective shoe-to-anchor springs is different, so the spring in the figure has been painted, as shown in the accompanying figure.

Lancer—With Rear Drum Brakes

The rear drum brake shoes serve as the parking brakes. Refer to the procedures under Rear Drum Brakes, Brake Shoes, Removal & Installation.

ADJUSTMENT

1. Before servicing the vehicle, refer to the Precautions Section.
2. Remove the floor console, release the lever and back off the cable adjuster locknut at the base of the lever.
3. Raise and safely support the vehicle.
4. Remove the wheel.
5. Remove the hole plug in the brake rotor.
6. Remove the brake caliper and hang it out of the way with wire. Do not disconnect the fluid line.
7. Use a suitable prybar to pry up on the self-adjuster wheel until the rotor will not turn.
8. Return the adjuster 5 notches in the opposite direction. Make sure the rotor turns freely with a slight drag.
9. Install the caliper and check brake operation.

CHASSIS ELECTRICAL AIR BAG (SUPPLEMENTAL RESTRAINT SYSTEM)

GENERAL INFORMATION

CAUTION

This vehicle is equipped with an air bag system. The system must be disarmed before performing service on, or around, system components, the steering column, instrument panel components, wiring, and sensors. Failure to follow the safety precautions and the disarming procedure could result in accidental air bag deployment, possible injury, and unnecessary system repairs.

SERVICE PRECAUTIONS

Disconnect and isolate the battery negative cable before beginning any airbag system component diagnosis, testing, removal, or installation procedures. Allow system capacitor to discharge for 3 minutes before beginning any component service. This will disable the airbag system. Failure to disable the airbag system may result in accidental airbag deployment, personal injury, or death.

Do not place an intact undeployed airbag face down on a solid surface. The airbag will propel into the air if accidentally deployed and may result in personal injury or death.

When carrying or handling an undeployed airbag, the trim side (face) of the airbag should be pointing towards the body to minimize possibility of injury if accidental deployment occurs. Failure to do this may result in personal injury or death.

Replace airbag system components with OEM replacement parts. Substitute parts may appear interchangeable, but internal differences may result in inferior occupant protection. Failure to do so may result in occupant personal injury or death.

Wear safety glasses, rubber gloves, and long sleeved clothing when cleaning powder residue from vehicle after an airbag deployment. Powder residue emitted from a deployed airbag can cause skin irritation. Flush affected area with cool water if irritation is experienced. If nasal or throat irritation is experienced, exit the vehicle for fresh air until the irritation ceases. If irritation continues, see a physician.

Do not use a replacement airbag that is not in the original packaging. This may result in improper deployment, personal injury, or death.

The factory installed fasteners, screws and bolts used to fasten airbag components have a special coating and are specifically designed for the airbag system. Do not use substitute fasteners. Use only original equipment fasteners listed in the parts catalog when fastener replacement is required.

During, and following, any child restraint anchor service, due to impact event or vehicle repair, carefully inspect all mounting hardware, tether straps, and anchors for proper installation, operation, or damage. If a child restraint anchor is found damaged in any way, the anchor must be replaced. Failure to do this may result in personal injury or death.

Deployed and non-deployed airbags may or may not have live pyrotechnic material within the airbag inflator.

Do not dispose of driver/passenger/curtain airbags or seat belt tensioners unless you are sure of complete deployment. Refer to the Hazardous Substance Control System for proper disposal.

Dispose of deployed airbags and tensioners consistent with state, provincial, local, and federal regulations.

After any airbag component testing or service, do not connect the battery negative cable. Personal injury or death may result if the system test is not performed first.

If the vehicle is equipped with the Occupant Classification System (OCS), do not connect the battery negative cable before performing the OCS Verification Test using the scan tool and the appropriate diagnostic information. Personal injury or death may result if the system test is not performed properly.

Never replace both the Occupant Restraint Controller (ORC) and the Occupant Classification Module (OCM) at the same time. If both require replacement, replace one, then perform the Airbag System test before replacing the other.

Both the ORC and the OCM store Occupant Classification System (OCS) calibration data, which they transfer to one another when one of them is replaced. If both are replaced at the same time, an irreversible fault will be set in both modules and the OCS may malfunction and cause personal injury or death.

If equipped with OCS, the Seat Weight Sensor is a sensitive, calibrated unit and must be handled carefully. Do not drop or handle roughly. If dropped or damaged, replace with another sensor. Failure to do so may result in occupant injury or death.

If equipped with OCS, the front passenger seat must be handled carefully as well. When removing the seat, be careful when it setting on the floor not to drop it. If dropped, the sensor may be inoperative, could result in occupant injury, or possibly death.

If equipped with OCS, when the passenger front seat is on the floor, no one should sit in the front passenger seat. This uneven force may damage the sensing ability of the seat weight sensors. If sat on and damaged, the sensor may be inoperative, could result in occupant injury, or possibly death.

Several precautions must be observed when handling the inflator module to avoid accidental deployment and possible personal injury.

- Never carry the inflator module by the wires or connector on the underside of the module
- When carrying a live inflator module, hold it securely with both hands, and ensure that the bag and trim cover are pointed away
- Place the inflator module on a bench or other surface with the bag and trim cover facing up
- With the inflator module on the bench, never place anything on or close to the module which may be thrown in the event of an accidental deployment

Before servicing the vehicle, make sure to refer to the precautions in the beginning of this section as well.

DISARMING THE SYSTEM

1. Before servicing the vehicle, refer to the Precautions Section.
2. Remove the ignition key from the vehicle.
3. Disconnect the negative battery cable and isolate it from accidental reconnection. Insulate the cable end with high-quality electrical tape or a similar non-conductive wrapping.
4. Wait at least 1 minute for the system capacitor to discharge before performing any service. The air bag system is designed to retain enough voltage to deploy the air bag for a short period of time after the battery has been disconnected.

ARMING THE SYSTEM

1. Before servicing the vehicle, refer to the Precautions Section.
2. Reconnect the negative battery cable.
3. To confirm proper system operation, turn the ignition switch to the **ON** position. The SRS indicator light will be lit for at least 7 seconds and then go off.

DRIVETRAIN

AUTOMATIC TRANSAXLE ASSEMBLY

REMOVAL & INSTALLATION

Eclipse

1. Before servicing the vehicle, refer to the Precautions Section.
2. Remove or disconnect the following:
   - Battery, battery tray, and support
   - Air intake hoses
   - Auto-cruise actuator and bracket, if equipped with cruise control
   - Charcoal canister and bracket
   - Shift and select cables from the transaxle
   - Back-up light switch and the vehicle speed sensor connectors
   - Dipstick and tube assembly
   - Starter assembly
   - Park/neutral switch
   - Oil temperature sensor
   - Kick down servo switch
   - Solenoid valve
   - Pulse generator
   - Speedometer connections
3. Attach an engine support fixture to the engine and remove the transaxle mounting bolts.
4. Remove or disconnect the following:
   - Rear roll stopper bracket mounting bolts
   - Transaxle mounting bracket mounting nuts
5. Raise the vehicle and remove the engine undercovers.
6. Remove or disconnect the following:
   - Front exhaust pipe
   - Axle shafts
   - Bell housing cover and the right-hand center member stay (support)
   - Center member
   - Drive plate connecting bolts
7. Place a transmission jack under the transaxle and remove the transaxle mounting bolt.
8. Lower the transaxle.

To install:

9. Raise the transaxle into position and install the transaxle mounting bracket. Torque the through-bolt to 51 ft. lbs. (69 Nm).
10. Install or connect the following:
    - Transaxle assembly mounting bolt. Torque the bolt to 22–25 ft. lbs. (29–34 Nm)
    - Drive plate connecting bolts. Torque the bolts to 33–38 ft. lbs. (45–52 Nm)
    - Center member assembly and the right-hand stay
    - Bell housing cover and the slave cylinder
    - Axle shafts.
    - Front exhaust pipe
    - Engine undercovers and lower the vehicle
    - Transaxle mounting bracket mounting nuts
    - Rear roll stopper bracket mounting bolts
    - Transaxle assembly mounting bolts. Torque the bolts to 35 ft. lbs. (48 Nm)
11. Remove the engine support fixture.
12. Install or connect the following:
    - Park/neutral switch
    - Oil temperature sensor
    - Kick down servo switch
    - Solenoid valve
    - Pulse generator
    - Speedometer connections
    - Starter assembly
    - Dipstick and tube assembly
    - Vehicle speed sensor and the back-up light connectors
    - Cruise control actuator if removed
    - Battery tray support and the tray
    - Charcoal canister bracket and the canister
    - Air duct and the air cleaner assembly
13. Refill the transaxle with the proper amount and type of fluid.

Galant

4-Speed Transaxle

1. Before servicing the vehicle, refer to the Precautions Section.
2. Disconnect the negative battery cable.
3. Drain the engine coolant.
4. Drain the transaxle fluid.
5. Remove the left side undercover.
6. Remove the air cleaner assembly.
7. Remove the front exhaust pipe.
8. Remove the PCM.
9. Remove the battery and the battery tray.
10. Remove the transaxle control cable adjusting nut. Remove the transaxle control cable.
11. Disconnect the transaxle range switch connection. Disconnect the A/T control solenoid valve assembly connector. Disconnect the input shaft speed sensor connector.
12. Disconnect the output shaft speed sensor connector. Disconnect the output shaft speed sensor connector.
13. Remove the front tire and wheel assembly. Remove the cotter pin. Remove the locknut. Remove the wheel speed sensor, sensor bracket, and brake hose clamp.
14. Remove the stabilizer link connection (strut side). Remove the tie rod end self-locking nut. Separate the tie rod end from the steering knuckle.
15. Remove the self-locking nut for the lower ball joint connection.
16. Using tools MB990242, MB990244, MB991354, and MB990767, remove the left side driveshaft, circlip, right side driveshaft, and circlip.
17. Disconnect the transaxle cooler lines.
18. Remove the starter.
19. Remove the transaxle upper retaining bolts. Remove the bell housing cover.
20. Remove the torque converter and drive plate retaining bolts.
21. Remove the center member assembly.
22. Remove the rear roll stopper bracket. Remove the transaxle mounting bracket assembly.
23. Remove the transaxle mounting stopper. Remove the transaxle mounting body side bracket.
24. Install engine support tool MB991895. Set the tool to the front fender rear mounting bolts and the upper radiator support insulator mounting bolts, which are located in the engine compartment.
25. Use tools MB991527 (engine lifting fixture) and tool MB991454 (chain) to hold the engine and transaxle assembly in place.
26. Remove the lower transaxle retaining bolts. Remove the transaxle assembly from the vehicle.

To install:

27. Engage the torque converter into the transaxle, securely. Raise the transaxle into position and install the lower transaxle retaining bolts. Install the transaxle mounting stopper.
28. Install the driveshaft. Be sure to properly install the driveshaft washer.

➡Before securely tightening the driveshaft nuts, make sure there is no load on the wheel bearings. Otherwise the wheel bearing will be damaged. Using tool MB990767, to hold the assembly in place, torque the retaining bolt to 146–188 ft. lbs. (197–155 Nm).

29. Place the selector lever and the manual control lever in the neutral "N" position. Position the cable stud into the manual control lever slot and loosely install the nut. Gently push the transaxle control cable into the manual control lever slot until the cable is taut. Tighten the nut to 90–124 inch lbs. (10–14 Nm).

30. Continue the installation in the reverse order of the removal procedure.

31. Check and adjust fluid levels, as required.

32. Check and adjust the front alignment.

33. To initialize the PCM, turn the ignition switch ON, then OFF, and keep it in the OFF position for at least 10 seconds.

5-Speed Transaxle

1. Before servicing the vehicle, refer to the Precautions Section.
2. Disconnect the negative battery cable.
3. Drain the transaxle fluid.
4. Remove the side undercover. Remove the air cleaner assembly. Remove the engine cover assembly.
5. Remove the PCM.
6. Remove the battery and the battery tray.
7. Disconnect the front exhaust pipe.
8. Remove the starter.
9. Remove the transaxle control cable adjusting nut. Remove the transaxle control cable. Inhibitor switch sensor connector.
10. Disconnect the A/T control solenoid valve assembly connector. Disconnect the input shaft speed sensor connector. Disconnect the output shaft speed sensor connector.
11. Remove the front tire and wheel assembly. Remove the cotter pin. Remove the locknut. Remove the wheel speed sensor, sensor bracket, and brake hose clamp.
12. Remove the stabilizer link connection (strut side). Remove the tie rod end self-locking nut. Separate the tie rod end from the steering knuckle.
13. Remove the self-locking nut for the lower ball joint connection. Remove the lower arm ball joint connection.
14. Using tools MB990242, MB990244, MB991354, and MB990767, remove the left side driveshaft, circlip, right side driveshaft, and circlip.
15. Disconnect the transaxle cooler lines.
16. Remove the transaxle upper retaining bolts. Remove the bell housing cover.
17. Remove the torque converter and drive plate retaining bolts.
18. Remove the center member assembly.
19. Remove the engine oil pan to transaxle retaining bolts.
20. Remove the rear roll stopper bracket. Remove the transaxle mounting bracket assembly.
21. Remove the transaxle mounting stopper. Remove the transaxle mounting body side bracket.
22. Remove the engine hanger.
23. Remove the intake manifold rear plenum stay and engine hanger.
24. Install special tool MB992012 and MB992013 to cylinder head.
25. Install engine support tool MB991895, to hold the engine and transaxle in position. Set the tool to the front fender rear mounting bolts and the upper radiator support insulator mounting bolts, which are located in the engine compartment.
26. Use tool MB991454 to hold the engine and transaxle assembly in place.
27. Remove the lower transaxle retaining bolts. Remove the transaxle assembly from the vehicle.

To install:

28. Engage the torque converter into the transaxle, securely. Raise the transaxle into position and install the lower transaxle retaining bolts. Install the transaxle mounting stopper.

29. Install the driveshaft. Be sure to properly install the driveshaft washer.

➡Before securely tightening the driveshaft nuts, make sure there is no load on the wheel bearings. Otherwise the wheel bearing will be damaged. Using tool MB990767, to hold the assembly in place, torque the retaining bolt to 146–188 ft. lbs. (197–155 Nm).

30. Place the selector lever and the manual control lever in the neutral "N" position. Position the cable stud into the manual control lever slot and loosely install the nut. Gently push the transaxle control cable into the manual control lever slot until the cable is taut. Tighten the nut to 90–124 inch lbs. (10–14 Nm).

31. Continue the installation in the reverse order of the removal procedure.

32. Check and adjust fluid levels, as required.

33. Check and adjust the front alignment.

34. To initialize the PCM, turn the ignition switch ON, then OFF, and keep it in the OFF position for at least 10 seconds.

Lancer

1. Before servicing the vehicle, refer to the Precautions Section.
2. Disconnect the negative battery cable.
3. Drain the engine coolant. Drain the transaxle fluid.
4. Remove the left side undercover.
5. Remove the air cleaner assembly.
6. Remove the front exhaust pipe.
7. Remove the battery and the battery tray.
8. Remove the transaxle control cable adjusting nut. Remove the transaxle control cable.
9. Disconnect the transaxle range switch connection.
10. Disconnect the A/T control solenoid valve assembly connector. Disconnect the input shaft speed sensor connector.
11. Disconnect the output shaft speed sensor connector. Disconnect the output shaft speed sensor connector.
12. Remove the front tire and wheel assembly. Remove the cotter pin. Remove the locknut. Remove the wheel speed sensor, sensor bracket, and brake hose clamp.
13. Remove the stabilizer link connection (strut side) Remove the tie rod end self-locking nut. Separate the tie rod end from the steering knuckle.
14. Remove the self-locking nut for the lower ball joint connection.
15. Using tools MB990242, MB990244, MB991354, and MB990767, remove the left side driveshaft, circlip, right side driveshaft, and circlip.
16. Disconnect the transaxle cooler lines.
17. Remove the starter.
18. Remove the transaxle upper retaining bolts.
19. Remove the bell housing cover. Remove the torque converter and drive plate retaining bolts.
20. Remove the center member assembly.
21. Remove the rear roll stopper bracket. Remove the transaxle mounting bracket assembly.
22. Remove the transaxle mounting stopper. Remove the transaxle mounting body side bracket.
23. Install engine support tool MB991895. Set the tool to the front fender rear mounting bolts and the upper radiator support insulator mounting bolts, which are located in the engine compartment.
24. Use tools MB991527 (engine lifting fixture) and tool MB991454 (chain) to hold the engine and transaxle assembly in place.
25. Remove the lower transaxle retaining bolts. Remove the transaxle assembly from the vehicle.

To install:

26. Engage the torque converter into the transaxle, securely. Raise the transaxle into position and install the lower transaxle retaining bolts. Install the transaxle mounting stopper.

27. Install the driveshaft. Be sure to properly install the driveshaft washer.

➡Before securely tightening the driveshaft nuts, make sure there is no load on the wheel bearings. Otherwise the wheel bearing will be damaged. Using tool MB990767, to hold the assembly in place, torque the retaining bolt to 146–188 ft. lbs. (197–155 Nm).

28. Place the selector lever and the manual control lever in the neutral "N" position. Position the cable stud into the manual control lever slot and loosely install the nut. Gently push the transaxle control cable into the manual control lever slot until the cable is taut. Tighten the nut to 90–124 inch lbs. (10–14 Nm).

29. Continue the installation in the reverse order of the removal procedure.

30. Check and adjust fluid levels, as required.

31. Check and adjust the front alignment.

MANUAL TRANSAXLE ASSEMBLY

REMOVAL & INSTALLATION

Eclipse

1. Before servicing the vehicle, refer to the Precautions Section.
2. Remove or disconnect the following:
   - Negative battery cable and positive battery cable
   - Battery and the air intake hoses
   - Battery tray and support
   - Auto-cruise actuator and bracket, if equipped with cruise control
   - Charcoal canister and bracket
   - Shift and select cables from the transaxle
   - Back-up light switch and the Vehicle Speed Sensor (VSS) connectors
   - Starter assembly
   - Engine support fixture to the engine and remove the transaxle mounting bolts
3. Remove or disconnect the following:
   - Rear roll stopper bracket mounting bolts
   - Transaxle mounting bracket mounting nuts
   - Engine undercovers
   - Axle shafts
   - Slave cylinder from the bell housing without disconnecting the fluid line. Position it out of the way.
   - Bell housing cover and the right-hand center member stay (support)
   - Center member
4. Place a transmission jack under the transaxle and remove the transaxle mounting bolt.
5. Remove the transaxle mounting and lower the transaxle.

To install:

6. Raise the transaxle into position and install the transaxle mounting. Torque the through-bolt to 50 ft. lbs. (69 Nm).
7. Install or connect the following:
   - Transaxle assembly mounting bolt. Torque the bolt to 22–25 ft. lbs. (30–34 Nm)
   - Center member assembly and the right-hand stay
   - Bell housing cover and the slave cylinder
   - Axle shafts. Be sure to install the washer in the proper direction
   - Engine undercovers and lower the vehicle
   - Transaxle mounting bracket mounting nuts
   - Rear roll stopper bracket mounting bolts
   - Transaxle assembly mounting bolts. Torque the mounting bolts to 35 ft. lbs. (48 Nm)
8. Remove the engine support fixture.
9. Install or connect the following:
   - Starter assembly
   - VSS and the back-up light connectors
   - Cruise control actuator if removed
   - Battery tray support and the tray
   - Charcoal canister bracket and the canister
   - Air duct and the air cleaner assembly

Galant

1. Before servicing the vehicle, refer to the Precautions Section.
2. Remove or disconnect the following:
   - Negative battery cable
   - Air cleaner and intake hoses
   - Cotter pins and clips securing the select and shift cables and remove the cable ends from the transaxle
   - Air compressor, if equipped with Active Electronic Control Suspension (Active-ECS)
   - Back-up light switch harness and position aside
   - Speedometer electrical connector from the transaxle assembly
   - Starter motor and position aside
3. Support the engine assembly.
4. Remove or disconnect the following:
   - Rear roll stopper mounting bracket
   - Transaxle mount bracket
   - Upper transaxle mounting bolts
   - Front wheel assemblies
   - Right-hand undercover
   - Cotter pin and disconnect the tie rod end, from the steering knuckle
   - Stabilizer bar link from the damper fork
   - Damper fork from the lateral lower control arm
   - Lateral lower arm and the compression arm lower ball joints from the steering knuckle
   - Halfshafts from the transaxle, and secure aside
   - Clutch release cylinder without disconnecting the hydraulic line and secure aside
   - Cover from the transaxle bell housing
   - Engine front roll stopper through-bolt
   - Crossmember and the triangular right-hand stay
5. Support the transaxle with a transmission jack and remove the transaxle lower coupling bolt.

➡The coupling bolt threads from the engine side into the transaxle and is located just above the halfshaft opening.

6. Slide the transaxle rearward and carefully lower it from the vehicle.

To install:

7. Install or connect the following:
   - Transaxle to the engine and install the mounting bolts. Tighten to 35 ft. lbs. (48 Nm)
   - The transaxle lower coupling bolt and tighten to 22–25 ft. lbs. (30–34 Nm)
   - Cover to the transaxle bell housing and tighten the mounting bolts to 84 inch lbs. (9 Nm)
   - Crossmember and tighten the front mounting bolts to 65 ft. lbs. (88 Nm) and the rear bolt to 54 ft. lbs. (73 Nm).
   - The front engine roll stopper through-bolt and lightly tighten. Once the full weight of the engine is on the mounts, tighten the bolt to 42 ft. lbs. (57 Nm)

- Triangular stay bracket and tighten the mounting bolts to 65 ft. lbs. (88 Nm)
- Clutch release cylinder
- Halfshafts, using new circlips on the axle ends
- Tie rod and ball joints to the steering knuckle. Tighten the ball joint self-locking nuts to 48 ft. lbs. (65 Nm). Tighten the tie rod end nut to 21 ft. lbs. (28 Nm) and secure with a new cotter pin
- Damper fork to the lower control arm and tighten the through-bolt to 65 ft. lbs. (88 Nm)
- Stabilizer link to the damper fork and tighten the self-locking nut to 29 ft. lbs. (39 Nm)
- Undercover
- Wheels and lower vehicle
- Transaxle mount bracket to the transaxle and tighten the mounting nuts to 32 ft. lbs. (43 Nm)
- Rear roll stopper mounting bracket

8. Remove the engine support. Tighten the transaxle mount through-bolt to 51 ft. lbs. (69 Nm) and tighten the front engine roll stopper through-bolt.
9. Install or connect the following:
   - Upper transaxle mounting bolts and tighten to 35 ft. lbs. (48 Nm)
   - Starter motor
   - Back-up light switch and the speedometer connector
   - Air compressor, if equipped with Active Electronic Control Suspension (Active-ECS)
   - Select and shift cables and install new cotter pins
   - Air cleaner and the air intake hose
   - Negative battery cable
10. Check the transaxle for proper operation.

Lancer

1. Before servicing the vehicle, refer to the Precautions Section.
2. Drain the transaxle fluid.
3. Remove or disconnect the following:
   - Negative battery cable
   - Engine undercover
   - Evaporative canister
   - Positive battery cable, battery, and battery tray
   - Shifter cables
   - Back-up light switch and Vehicle Speed Sensor (VSS) connector
   - Starter motor
   - Clutch hose
   - Upper engine-to-transaxle bolts
   - Transaxle mount
   - Transaxle mount stopper
4. Install a suitable engine support assembly, then raise and safely support the vehicle.
5. Remove or disconnect the following:
   - Stabilizer bar
   - Wheel Speed Sensor (WSS) connector, if equipped with Anti-lock Brakes (ABS)
   - Brake hose clamp
   - Tie rod end
   - Lower control arm
   - Center member
   - Halfshafts by inserting a prybar between the transaxle case and the driveshaft and prying the shaft from the transaxle. Do not pull on the driveshaft.
   - Bell housing lower cover
   - Transaxle to engine bolts and lower the transaxle from the vehicle

To install:

6. Install or connect the following:
   - Transaxle to the engine and install the lower mounting bolts
   - Bell housing cover

➡When installing the halfshafts, use new circlips on the axle ends.

7. Install or connect the following:
   - Halfshafts into the transaxle
   - Center member
   - Lower control arm
   - Tie rod end
   - Brake hose clamp
   - WSS connector, if equipped
   - Stabilizer bar
8. Lower the vehicle, then remove the engine support.
9. Install or connect the following:
   - Transaxle mount bracket. Tighten the nuts to 35 ft. lbs. (47 Nm)
   - Transaxle mount stopper. Tighten the nuts to 61 ft. lbs. (82 Nm)
   - Transaxle mount
   - Upper transaxle-to-engine bolts and torque to 36 ft. lbs. (48 Nm)
   - Clutch line
   - Starter motor
   - VSS connector
   - Back-up light switch connector
   - Shifter cables, adjust
   - Evaporative canister
   - Battery and battery tray
   - Engine undercover
   - Positive and negative battery cables
10. Fill the transaxle with the proper amount and type of fluid.
11. Bleed the clutch, check and adjust the front wheel alignment, then check the transaxle for proper operation.

Lancer Evolution

5-Speed Transaxle

See Figure 6.

1. Before servicing the vehicle, refer to the Precautions Section.
2. Disconnect the negative battery cable.
3. Remove the transfer case.
4. Remove the starter.
5. Remove the air cleaner bracket.
6. Remove the rear roll rod assembly.
7. Disconnect the transaxle harness clamp, backup light switch connector and the VSS connector.
8. Remove the clutch release cylinder and clutch oil pipe. Remove the snap pin. Remove the cable bracket and cable assembly (transaxle side). Remove the rear roll mount bracket.
9. While supporting the engine and transaxle with a floor jack, remove the transaxle mounting insulator assembly.
10. Remove the upper transaxle assembly retaining bolts. Remove the upper transaxle mounting insulator assembly. Remove the transaxle mounting insulator stopper. Remove the transaxle mounting insulator.
11. Install engine support tool MB991895. Set the tool to the front fender rear mounting bolts and the upper radiator support insulator mounting bolts, which are located in the engine compartment.
12. Use tool MB991454 to hold the engine and transaxle assembly in place.
13. Remove the cover from the clutch release service hole in the clutch housing.

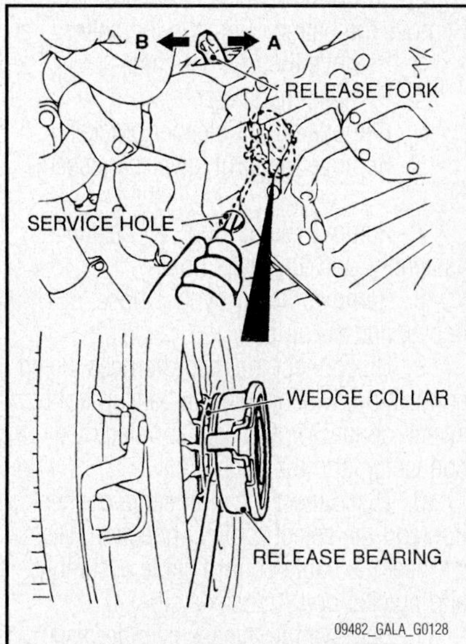

Fig. 6 Clutch release bearing separation location points—Lancer Evolution

➡If it is hard to turn the suitable tool, to pry off the release bearing, remove the tool and repeat the procedure below, after pushing the release fork fully in direction "A" 2–3 times. Forcibly prying can cause the release bearing to be damaged.

14. While pushing the release fork by hand in direction "A" (in the illustration), insert a suitable tool between the release bearing and the wedge collar.

➡Be sure to push the release fork in direction "A" before inserting the suitable tool.

15. Separate the release bearing from the wedge collar by prying the suitable tool at a 90° angle.

➡The release fork is forced to move fully in direction "B" (in the illustration) by the return spring as soon as it is separated from the wedge collar.

16. Remove the transaxle assembly lower retaining bolts. Remove the transaxle from the vehicle.

To install:

17. Raise the transaxle into position and install the lower transaxle retaining bolts.
18. Continue the installation in the reverse order of the removal procedure.
19. Check and adjust fluid levels, as required.

6-Speed Transaxle

See Figure 7.

1. Before servicing the vehicle, refer to the Precautions Section.
2. Disconnect the battery cables. Remove the battery. Remove the battery tray.
3. Remove the transfer case.
4. Remove the starter.
5. Remove the air cleaner bracket.
6. Remove the strut tower bar assembly.
7. Remove the air duct, air cleaner assembly, and air intake hose.
8. Remove the air bypass hose, air hoses, and air pipe.
9. Disconnect the main harness clamp connection. Disconnect the backup light switch connector, VSS sensor electrical connector, and the snap pin.
10. Disconnect the shift cable connection and the select cable connection. Disconnect the cable control cable assembly and bracket bolt (transaxle side).
11. Remove the release cylinder and clutch oil pipe.
12. Remove the transaxle upper retaining bolts.
13. Remove the harness clamp and clamp mounting bolt.
14. Using a floor jack support the engine and transaxle assembly. Remove the transaxle mounting insulator assembly.
15. Remove the transaxle mounting insulator stopper. Remove the transaxle mounting insulator.
16. Install engine support tool MB991895. Set the tool to the front fender rear mounting bolts and the upper radiator support insulator mounting bolts, which are located in the engine compartment.
17. Use tool MB991454 to hold the engine and transaxle assembly in place.
18. Remove the cover from the clutch release service hole in the clutch housing.

➡If it is hard to turn the suitable tool, to pry off the release bearing, remove the tool and repeat the procedure below, after pushing the release fork fully in direction "A" 2–3 times. Forcibly prying can cause the release bearing to be damaged.

19. While pushing the release fork by hand in direction "A" (in the illustration), insert a suitable tool between the release bearing and the wedge collar.

➡Be sure to push the release fork in direction "A" before inserting the suitable tool.

20. Separate the release bearing from the wedge collar by prying the suitable tool at a 90° angle.

➡The release fork is forced to move fully in direction "B" (in the illustration) by the return spring as soon as it is separated from the wedge collar.

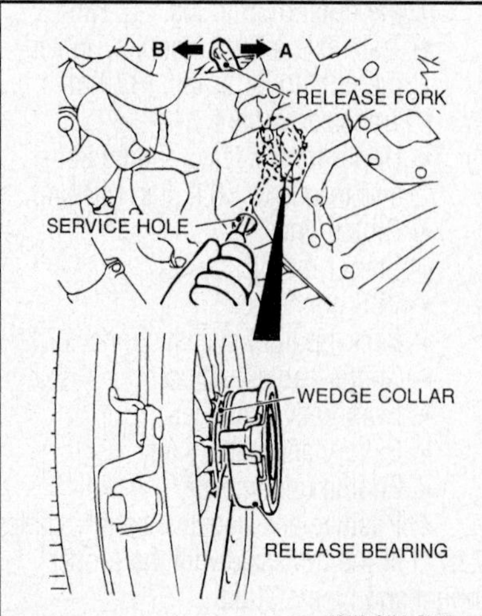

Fig. 7 Clutch release bearing separation location points—Lancer Evolution

21. Remove the transaxle assembly lower retaining bolts. Remove the transaxle from the vehicle.

To install:

22. Raise the transaxle into position and install the lower transaxle retaining bolts.
23. Continue the installation in the reverse order of the removal procedure.
24. Check and adjust fluid levels, as required.

CLUTCH DRIVEN DISC & PRESSURE PLATE

REMOVAL & INSTALLATION

CAUTION

The clutch driven disc may contain asbestos, which has been determined to be a cancer causing agent. Never clean clutch surfaces with compressed air! Avoid inhaling any dust from any clutch surface! When cleaning clutch surfaces, use a commercially available brake cleaning fluid.

1. Before servicing the vehicle, refer to the Precautions Section.
2. Disconnect the negative battery cable.
3. Raise and safely support the vehicle.
4. Remove the transaxle assembly from the vehicle.
5. Remove the pressure plate attaching bolts, pressure plate and clutch disc. If the pressure plate is to be reused, loosen the bolts in a diagonal pattern, 1 or 2 turns at a time. This will prevent warping the clutch cover assembly.
6. Remove the return clip and the pressure plate release bearing. Do not use solvent to clean the bearing.
7. Inspect the clutch release fork and fulcrum for damage or wear. If necessary, remove the release fork and unthread the fulcrum from the transaxle.
8. Carefully inspect the condition of the clutch components and replace any worn or damaged parts.

To install:

9. Inspect the flywheel for heat damage or cracks. Resurface or replace the flywheel as required.
10. Install the fulcrum and tighten to 25 ft. lbs. (35 Nm).
11. Install the release fork.
12. Apply a coating of multi-purpose grease to the point of contact with the fulcrum and the point of contact with the release bearing.

13. Apply a coating of multi-purpose grease to the end of the release cylinder pushrod and the pushrod hole in the release fork.

14. Apply multi-purpose grease to the clutch release bearing. Pack the bearing inner surface and the groove with grease. Do not apply grease to the resin portion of the bearing.

15. Place the bearing in position and install the return clip.

16. Using the proper alignment tool, install the clutch disc to the flywheel.

17. Install the pressure plate assembly.

18. Install the retainer bolts and tighten a little at a time, in a diagonal sequence. Tighten them to a final torque of 16 ft. lbs. (22 Nm). Remove the aligning tool.

19. Install the transaxle assembly.

20. Check for proper clutch operation.

ADJUSTMENTS

The clutch system is hydraulic and requires no adjustment.

CLUTCH MASTER CYLINDER

REMOVAL & INSTALLATION

1. Before servicing the vehicle, refer to the Precautions Section.

2. Disconnect the negative battery cable.

3. Remove necessary under hood components in order to gain access to the clutch master cylinder.

4. Place a suitable drain pan under the vehicle to catch the fluid once the line is disconnected, or place a rag or shop towel under the fluid line of the master cylinder.

5. Loosen the line at the cylinder and allow the fluid to drain.

**** WARNING**

Clean, high quality brake fluid is essential to the safe and proper operation of the brake system. You should always buy the highest quality brake fluid that is available. If the brake fluid becomes contaminated, drain and flush the system, then refill the master cylinder with new fluid. Never reuse any brake fluid. Any brake fluid that is removed from the system should be discarded. Also, do not allow any brake fluid to come in contact with a painted surface; it will damage the paint.

6. Remove the clevis pin retainer at the clutch pedal and remove the washer and clevis pin.

7. Remove the 2 nuts and pull the cylinder from the firewall. A seal should be between the mounting flange and firewall. This seal should be replaced.

8. The installation is the reverse of the removal procedure.

9. Lubricate all pivot points with grease.

10. Bleed the system at the slave cylinder using DOT 3 brake fluid and check the adjustment of the clutch pedal.

CLUTCH HYDRAULIC SYSTEM BLEEDING

Bleeding air from the hydraulic clutch system is necessary whenever any part of the system has been disconnected or the fluid level (in the reservoir) has been allowed to fall so low that air has been drawn into the master cylinder.

**** WARNING**

NEVER use fluid that has been bled from a clutch system to fill the master cylinder reservoir, as it may be aerated, contain excessive moisture and/or be contaminated in some other way.

1. Before servicing the vehicle, refer to the Precautions Section.

2. Fill the clutch master cylinder reservoir with new hydraulic clutch fluid.

3. Attach a hose to the bleeder on the clutch actuator and submerge the other end of the hose in a container of hydraulic clutch fluid.

4. Have an assistant slowly depress and hold the clutch pedal.

5. Loosen the bleeder to purge air.

6. Tighten the bleeder.

7. Repeat the above 3 steps until all air is completely purged from the system.

8. Refill the clutch master cylinder reservoir.

TRANSFER CASE ASSEMBLY

REMOVAL & INSTALLATION

Lancer Evolution

See Figure 8.

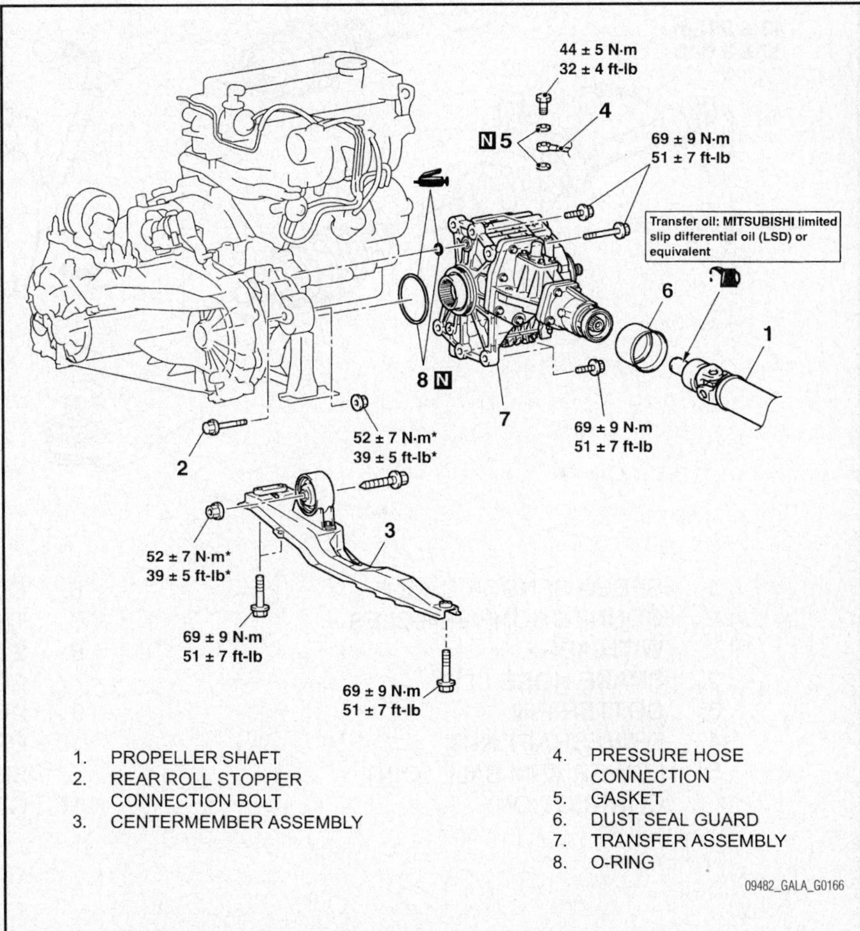

Fig. 8 Transfer case and related components—Lancer Evolution

1. Before servicing the vehicle, refer to the Precautions Section.
2. Disconnect the negative battery cable.
3. Remove the undercover assembly.
4. Drain the transaxle fluid.
5. Drain the transfer case fluid.
6. Drain the engine coolant.
7. Remove the front axle crossmember assembly.
8. Remove the front exhaust pipe.
9. Remove the battery and the battery tray.
10. Remove the air cleaner and air intake hose assembly.
11. Remove the strut tower bar assembly.
12. Remove the air hose, air by-pass hose, and air by-pass valve.
13. Remove the radiator.
14. Remove the output shaft. Remove the driveshaft.
15. Remove the rear roll stopper connection bolt.
16. Remove the crossmember assembly.
17. Disconnect the pressure hose connection and discard the gasket.
18. Remove the dust seal guard.
19. Remove the transfer case retaining bolts. Remove the transfer case from the vehicle. Discard the O-ring.

To install:

20. Position the transfer case on a suitable holding fixture. Install the transfer case to its mounting in the vehicle.
21. Continue the installation in the reverse order of the removal procedure.
22. Be sure to check and adjust all fluid levels, as necessary.

HALFSHAFTS

REMOVAL & INSTALLATION

Eclipse

See Figure 9.

1. Before servicing the vehicle, refer to the Precautions Section.
2. Raise and support the vehicle safely.
3. If equipped with ABS, disconnect the speed sensor connection. Remove the brake hose clip.

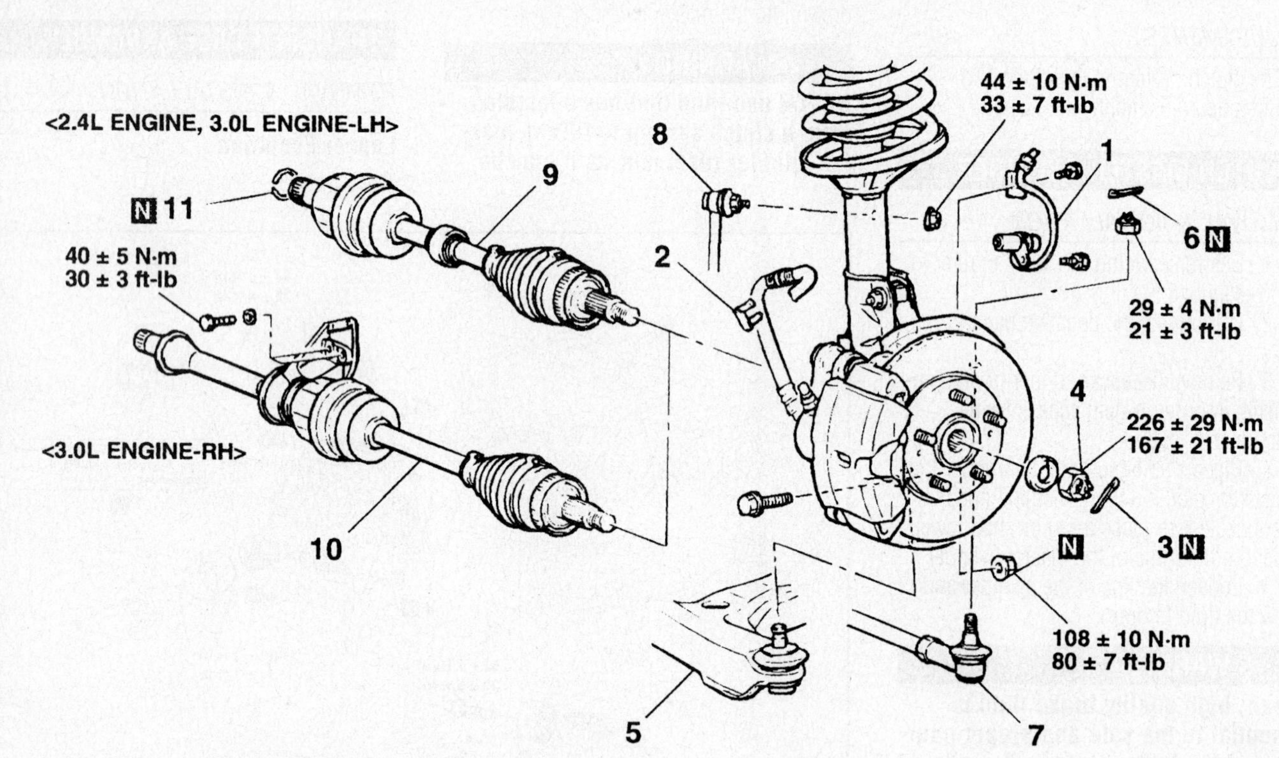

Fig. 9 Halfshaft and related components—Eclipse

4. Remove the cotter pin. Install tool MB990767 to the hub and remove the halfshaft nut. Remove the washer.

5. Remove the lower ball joint cotter pin.

➡Do not remove the nut from the ball joint. Loosen it and use special tool MB991897 to avoid possible damage to the ball joint threads. Hang the special tool in place with wire or string to prevent it from falling.

6. Install the special tool. Turn the bolt and knob as necessary to make the jaws of the tool parallel. Tighten the bolt by hand and confirm that the jaws are still parallel.

➡When adjusting the jaws in parallel, make sure the knob is in the vertical (upward) position.

7. Tighten the bolt with a wrench to disconnect the lower arm ball joint connection.

8. Remove the tie rod end cotter pin. Install tool MB990767 to the hub and remove the halfshaft nut. Remove the washer.

➡Do not remove the nut from the tie rod end. Loosen it and use special tool MB991897 to avoid possible damage to the ball joint threads. Hang the special tool in place with wire or string to prevent it from falling.

9. Install the special tool. Turn the bolt and knob as necessary to make the jaws of the tool parallel. Tighten the bolt by hand and confirm that the jaws are still parallel.

➡When adjusting the jaws in parallel, make sure the knob is in the vertical (upward) position.

10. Tighten the bolt with a wrench to disconnect the tie rod end.

11. Disconnect the stabilizer link connection.

➡Do not damage the ABS rotor attached to the BJ outer race, on vehicle equipped with ABS.

12. Use special tools MB991354, MB990242 and MB990767 to push the halfshaft out from the hub.

➡Do not pull on the halfshaft, doing so will damage the TJ. Be sure to use a prybar. Do not insert the prybar so deep as to damage the oil seal.

13. Insert a prybar between the transaxle case and the halfshaft to remove the halfshaft.

14. If the inner shaft and transaxle are tightly joined, tap the center bearing bracket with a plastic hammer to remove the halfshaft and inner shaft from the transaxle.

To install:

15. Installation is the reverse of the removal procedure.

16. Check and adjust the front end alignment, as necessary.

Galant

See Figure 10.

1. Before servicing the vehicle, refer to the Precautions Section.

2. Raise and support the vehicle safely.

3. Remove the front undercover. Remove the side undercover.

4. Drain the transaxle fluid.

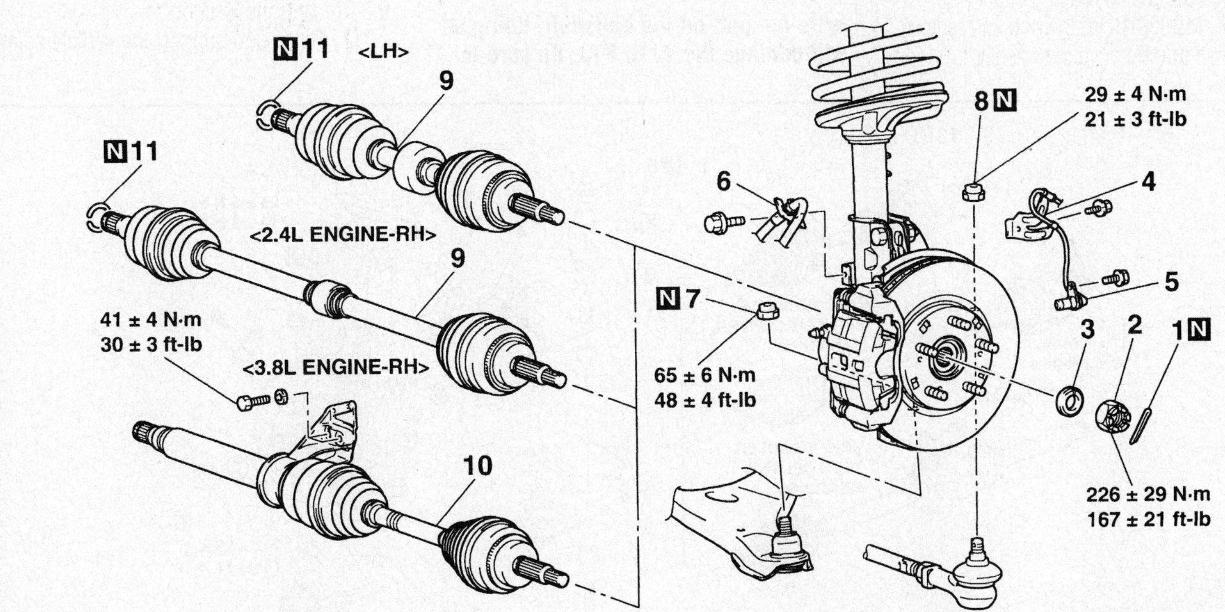

Fig. 10 Halfshaft and related components—Galant

5. On vehicles equipped with the 3.8L engine, disconnect the front exhaust pipe if working on the right side halfshaft.

6. Disconnect the speed sensor connection. Remove the wheel speed sensor. Remove the brake hose clip.

7. Remove the cotter pin. Install tool MB990767 to the hub and remove the halfshaft nut. Remove the washer.

8. Remove the lower ball joint cotter pin.

➡Do not remove the nut from the ball joint. Loosen it and use special tool MB991897 to avoid possible damage to the ball joint threads. Hang the special tool in place with wire or string to prevent it from falling.

9. Install the special tool. Turn the bolt and knob as necessary to make the jaws of the tool parallel. Tighten the bolt by hand and confirm that the jaws are still parallel.

➡When adjusting the jaws in parallel, make sure the knob is in the vertical (upward) position.

10. Tighten the bolt with a wrench to disconnect the lower arm ball joint connection.

11. Remove the tie rod end cotter pin. Install tool MB990767 to the hub and remove the halfshaft nut. Remove the washer.

➡Do not remove the nut from the tie rod end. Loosen it and use special tool MB991897 to avoid possible damage to the ball joint threads. Hang the special tool in place with wire or string to prevent it from falling.

12. Install the special tool. Turn the bolt and knob as necessary to make the jaws of the tool parallel. Tighten the bolt by hand and confirm that the jaws are still parallel.

➡When adjusting the jaws in parallel, make sure the knob is in the vertical (upward) position.

13. Tighten the bolt with a wrench to disconnect the tie rod end.

14. Disconnect the stabilizer link connection.

➡Do not strike the ABS rotor attached to the BJ or EBJ outer race, of the halfshaft against other parts when removing the halfshaft as damage to the rotors will result.

15. Use special tools MB991354, MB990242, and MB990767 to push the halfshaft out from the hub.

➡Do not pull on the halfshaft, doing so will damage the TJ or PTJ. Be sure to use a prybar. Do not insert the prybar so deep as to damage the oil seal.

16. Remove the halfshaft from the hub by pulling the bottom of the brake disc towards you.

17. Insert a prybar between the transaxle case and the halfshaft, and then pry and remove the halfshaft from the transaxle.

➡Insert a prybar, taking care not to damage the protrusion of the transaxle case when removing the halfshaft (left side).

18. If the inner shaft and transaxle are tightly joined, tap the center bearing bracket with a plastic hammer to remove the halfshaft and inner shaft from the transaxle.

To install:

19. Installation is the reverse of the removal procedure.

20. Check and adjust the front end alignment, as necessary.

Lancer

See Figure 11.

1. Before servicing the vehicle, refer to the Precautions Section.

2. Raise and support the vehicle safely.

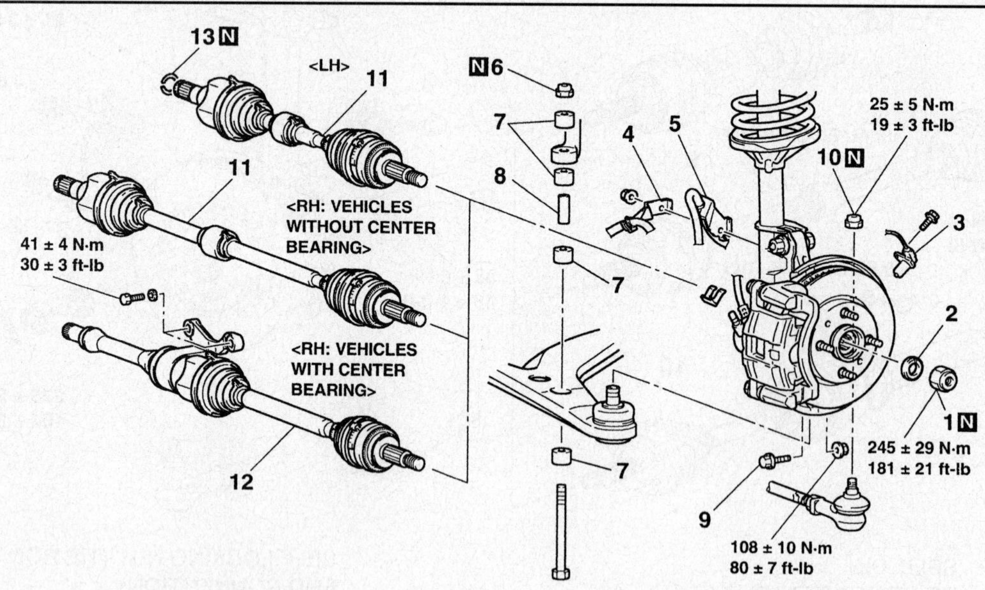

Fig. 11 Halfshaft and related components—Lancer

3. If equipped with ABS, disconnect the speed sensor connection. Remove the ABS sensor and bracket. Remove the brake hose clip.

4. Remove the stabilizer bar locknut, rubber insulator, and collar.

5. Remove the cotter pin. Install tool MB990767 to the hub and remove the halfshaft nut. Remove the washer.

6. Remove the lower ball joint cotter pin.

➡Do not remove the nut from the ball joint. Loosen it and use special tool MB991897 to avoid possible damage to the ball joint threads. Hang the special tool in place with wire or string to prevent it from falling.

7. Install the special tool. Turn the bolt and knob as necessary to make the jaws of the tool parallel. Tighten the bolt by hand and confirm that the jaws are still parallel.

➡When adjusting the jaws in parallel, make sure the knob is in the vertical (upward) position.

8. Tighten the bolt with a wrench to disconnect the lower arm ball joint connection.

9. Remove the tie rod end cotter pin. Install tool MB990767 to the hub and remove the halfshaft nut. Remove the washer.

➡Do not remove the nut from the tie rod end. Loosen it and use special tool MB991897 to avoid possible damage to the ball joint threads. Hang the special tool in place with wire or string to prevent it from falling.

10. Install the special tool. Turn the bolt and knob as necessary to make the jaws of the tool parallel. Tighten the bolt by hand and confirm that the jaws are still parallel.

➡When adjusting the jaws in parallel, make sure the knob is in the vertical (upward) position.

11. Tighten the bolt with a wrench to disconnect the tie rod end.

➡Do not damage the ABS rotor attached to the BJ outer race, on vehicle equipped with ABS.

12. Use special tools MB990241 and MB990767 (vehicles without center bearing) or tools MB991354, MB990242 and MB990244 (vehicles equipped with center bearing) to push the halfshaft out from the hub and knuckle.

13. Remove the halfshaft from the hub by pulling the bottom of the brake disc toward you, and then remove the hub retaining bolts.

➡Do not pull on the halfshaft, doing so will damage the TJ or ETJ. Be sure to use a prybar. Do not insert the prybar so deep as to damage the oil seal.

14. Insert a prybar between the transaxle case and the halfshaft, and then pry the halfshaft from the transaxle.

15. If the inner shaft and transaxle are tightly joined, tap the center bearing bracket with a plastic hammer to remove the halfshaft and inner shaft from the transaxle.

To install:

16. Installation is the reverse of the removal procedure.

17. Check and adjust the front end alignment, as necessary.

Lancer Evolution

See Figure 12.

1. Before servicing the vehicle, refer to the Precautions Section.

2. Raise and support the vehicle safely.

3. Remove the undercover. Remove the side cover. Drain the transaxle fluid. Drain the transfer case.

4. Remove the cotter pin. Install tool MB990767 to the hub and remove the driveshaft nut. Remove the washer.

5. Disconnect the speed sensor connection. Remove the ABS sensor and bracket. Remove the brake hose clip.

6. Remove the stabilizer bar locknut.

7. Remove the lower ball joint cotter pin.

➡Do not remove the nut from the ball joint. Loosen it and use special tool MB991897 to avoid possible damage to the ball joint threads. Hang the special tool in place with wire or string to prevent it from falling.

8. Install the special tool. Turn the bolt and knob as necessary to make the jaws of the tool parallel. Tighten the bolt by hand and confirm that the jaws are still parallel.

➡When adjusting the jaws in parallel, make sure the knob is in the vertical (upward) position.

9. Tighten the bolt with a wrench to disconnect the lower arm ball joint connection.

10. Remove the tie rod end cotter pin. Install tool MB990767 to the hub and remove the halfshaft nut. Remove the washer.

➡Do not remove the nut from the tie rod end. Loosen it and use special tool MB991897 to avoid possible damage to the ball joint threads. Hang the special tool in place with wire or string to prevent it from falling.

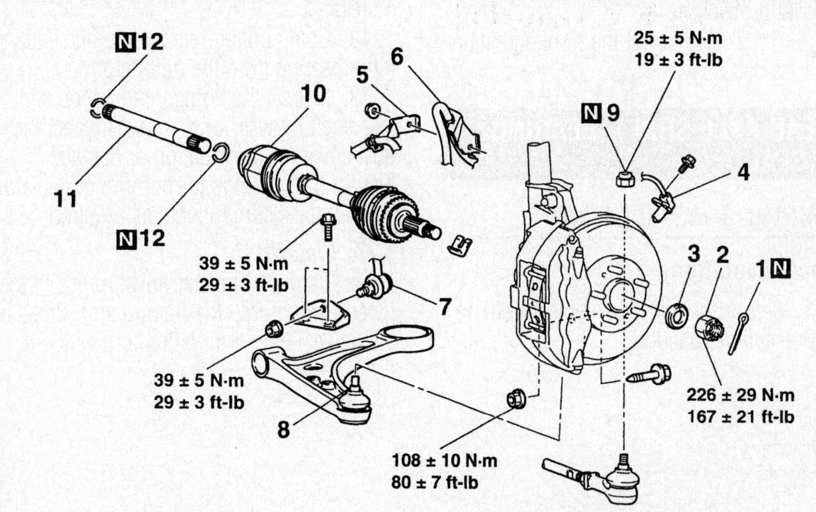

Fig. 12 Halfshaft and related components—Evolution

11. Install the special tool. Turn the bolt and knob as necessary to make the jaws of the tool parallel. Tighten the bolt by hand and confirm that the jaws are still parallel.

➡When adjusting the jaws in parallel, make sure the knob is in the vertical (upward) position.

12. Tighten the bolt with a wrench to disconnect the tie rod end.

➡Do not strike the ABS rotor attached to the EBJ outer race, of the halfshaft against other parts when removing the halfshaft as damage to the rotors will result.

13. Use special tools MB990241 (MB990242 and MB990244), MB991354 and MB990767 to push the halfshaft out from the hub.
14. Remove the halfshaft from the hub by pulling the bottom of the brake disc toward you, and then remove the hub retaining bolts.

➡Do not pull on the halfshaft, doing so will damage the TJ. Be sure to use a prybar. Do not insert the prybar so deep as to damage the oil seal.

15. Insert a prybar between the transaxle case and the halfshaft, and then pry and remove the driveshaft from the transaxle.

To install:

16. Installation is the reverse of the removal procedure.
17. Check and adjust the front end alignment, as necessary.

REAR AXLE SHAFT, BEARING & SEAL

REMOVAL & INSTALLATION

Lancer Evolution

1. Before servicing the vehicle, refer to the Precautions Section.
2. Disconnect the negative battery cable.
3. Raise and support the vehicle safely.
4. Remove the tire and wheel assembly from the vehicle.
5. If equipped with ABS, remove the rear wheel speed sensor.

➡Be cautious to ensure that the tip of the pole piece on the rear speed sensor does not come in contact with other parts during removal. Sensor damage could occur.

6. Remove the rear caliper and support assembly out of the way. Remove the brake disc.
7. Remove the driveshaft and companion flange installation bolts, nuts and washers. Move the end of shaft slightly to access the self-locking nut.
8. Using axle holding tool MB990211-01 or equivalent, secure the rear axle shaft in position, then remove the self-locking nut.
9. Using puller and adapter MB990211-01 and MB990241-01 or equivalents, remove the rear axle shaft from the trailing arm.
10. If equipped with ABS, remove the rear rotor from the axle assembly using collar and press. The rotor is a press fit.
11. Remove the outer bearing and dust cover concurrently from the axle shaft using a press.
12. Using puller, remove the oil seal and inner bearing from the trailing arm.
13. Inspect the companion flange and axle shaft for wear or damage. Inspect the dust cover for deformation or damage. Inspect the bearings for burning or declaration. Replace components as required.

To install:

14. Using the proper driver, press fit the inner bearing onto the trailing arm. Press fit the oil seal onto the trailing arm with the depression in the oil seal facing upward, and until it contacts the shoulder on the inner arm.

➡When tapping the oil seal in, use a plastic hammer to lightly tap the top and circumference of the seal installation tool, press fitting gradually and evenly.

15. Press fit the dust covers onto the axle until it contacts the axle shaft shoulder. Install the innermost cover so the depression is facing upward.

➡When tapping the oil seal in, use a plastic hammer to lightly tap the top and circumference of the seal installation tool, press fitting gradually and evenly.

16. Apply multi-purpose grease around the entire circumference of the inner side of the outer bearing seal lip. Press fit the outer bearing to the axle shaft so that the bearing seal lip surface is facing towards the axle shaft flange.
17. Press fit the rear rotor to the axle shaft with the rear rotor groove surface towards the axle shaft flange.
18. Install the rear axle shaft to the trailing arm temporarily. Install the companion flange to the rear axle shaft, then install a new self-locking nut.
19. While holding the rear axle shaft in position using holding fixture tool MB990767-01 or equivalent, tighten a new self-locking nut to 159 ft. lbs. (220 Nm).
20. Install the drive shaft nuts, washers and bolts. Tighten to 40–47 ft. lbs. (55–65 Nm).
21. Install the rear brake disc, caliper assembly and parking brake.
22. Install the tire and wheel assembly and lower the vehicle. Check the parking brake stroke and adjust as required.
23. Before moving the vehicle, pump the brakes until a firm pedal is achieved.

ENGINE COOLING

COOLANT TEMPERATURE SENSOR

REMOVAL & INSTALLATION

See Figure 13.

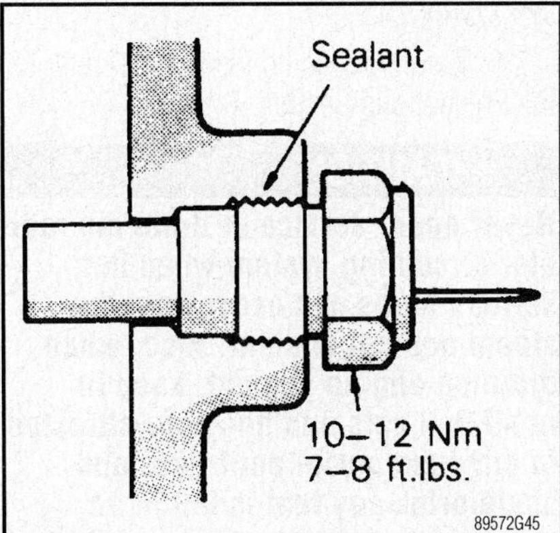

Fig. 13 Before installing the sending unit, coat the threads with a suitable sealant

1. Before servicing the vehicle, refer to the Precautions Section.
2. Disconnect the negative battery cable.
3. Position a suitable drain pan under the radiator.
4. Drain the engine coolant a level below the coolant temperature sending unit.
5. Disconnect the sending unit wiring harness, then remove the coolant temperature sending unit from the engine.

To install:

6. Coat the sending unit threads with a suitable thread sealant.
7. Install the coolant temperature sensor and tighten to 7–8 ft. lbs. (10–12 Nm).
8. Attach the electrical harness connector to the sending unit.
9. Fill the cooling system to the proper level.
10. Connect the negative battery cable.
11. Run the engine to test for leaks.

THERMOSTAT

REMOVAL & INSTALLATION

Eclipse & Galant

2.4L Engine

See Figures 14 and 15.

1. Before servicing the vehicle, refer to the Precautions Section.

**** CAUTION**

Never open, service or drain the radiator or cooling system when hot; serious burns can occur from the steam and hot coolant. Also, when draining engine coolant, keep in mind that cats and dogs are attracted to ethylene glycol antifreeze and could drink any that is left in an uncovered container or in puddles on the ground. This will prove fatal in sufficient quantities. Always drain coolant into a sealable container. Coolant should be reused unless it is contaminated or is several years old.

2. Drain the engine coolant.
3. Remove the ECM (M/T) or the PCM (A/T).
4. Remove the air cleaner housing cover and air intake hose.
5. Remove the battery and battery tray.
6. Remove the harness connection.
7. Remove the harness bracket.
8. Remove the radiator lower hose connection.
9. Remove the water inlet fitting.
10. Remove the thermostat.

To install:

11. Install the thermostat so that the jiggle valve is facing straight up. Be careful not to fold or scratch the rubber ring.
12. Install the water inlet fitting.
13. Install the radiator lower hose connection.
14. Install the harness bracket.
15. Install the harness connection.
16. Install the battery and battery tray.
17. Install the air cleaner housing cover and air intake hose.
18. Install the ECM (M/T) or the PCM (A/T).
19. Connect the negative battery cable, run the vehicle until the thermostat opens and fill the radiator completely.
20. Once the vehicle has cooled, recheck the coolant level.

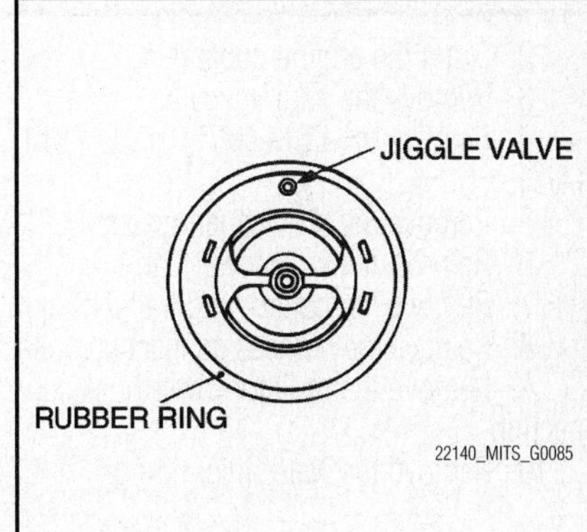

Fig. 15 Install the thermostat so that the jiggle valve is facing straight up

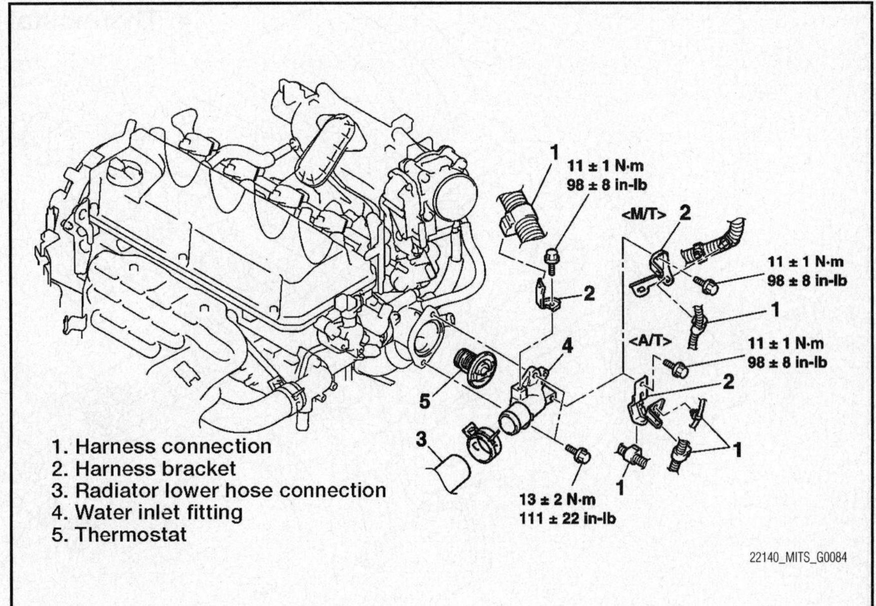

Fig. 14 Exploded view of the thermostat and related parts—2.4L engine

3.8L Engine

See Figures 16 and 17.

1. Before servicing the vehicle, refer to the Precautions Section.

✲✲ CAUTION

Never open, service or drain the radiator or cooling system when hot; serious burns can occur from the steam and hot coolant. Also, when draining engine coolant, keep in mind that cats and dogs are attracted to ethylene glycol antifreeze and could drink any that is left in an uncovered container or in puddles on the ground. This will prove fatal in sufficient quantities. Always drain coolant into a sealable container. Coolant should be reused unless it is contaminated or is several years old.

2. Drain the engine coolant.
3. Remove the engine cover.
4. Remove the ECM (M/T) or the PCM (A/T).
5. Remove the air cleaner assembly.
6. Remove the strut tower bar.
7. Remove the battery and battery tray.
8. Remove the harness connection bolts.
9. Remove the radiator lower hose connection.
10. Remove the water inlet fitting.

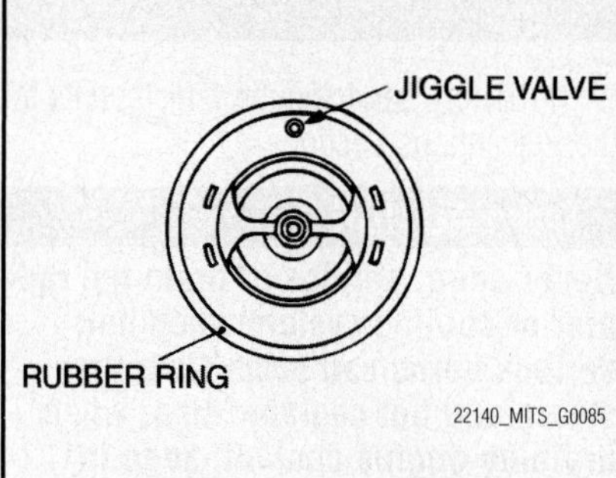

Fig. 17 Install the thermostat so that the jiggle valve is facing straight up

11. Remove the thermostat.

To install:

12. Install the thermostat so that the jiggle valve is facing straight up. Be careful not to fold or scratch the rubber ring.
13. Install the water inlet fitting.
14. Install the radiator lower hose connection.
15. Install the harness bracket.
16. Install the harness connection.
17. Install the battery and battery tray.
18. Install the strut tower bar.
19. Install the air cleaner housing cover and air intake hose.
20. Install the ECM (M/T) or the PCM (A/T).
21. Install the engine cover.
22. Connect the negative battery cable, run the vehicle until the thermostat opens and fill the radiator completely.
23. Once the vehicle has cooled, recheck the coolant level.

Lancer

2006–07 2.0L Engine

See Figure 18.

1. Before servicing the vehicle, refer to the Precautions Section.

✲✲ CAUTION

Never open, service or drain the radiator or cooling system when hot; serious burns can occur from the steam and hot coolant. Also, when draining engine coolant, keep in mind that cats and dogs are attracted to ethylene glycol antifreeze and could drink any that is left in an uncovered container or in puddles on the ground. This will prove fatal in sufficient quantities. Always drain coolant into a sealable container. Coolant should be reused unless it is contaminated or is several years old.

2. Disconnect the negative battery cable.
3. Drain the cooling system.
4. Remove any necessary components to access the thermostat.

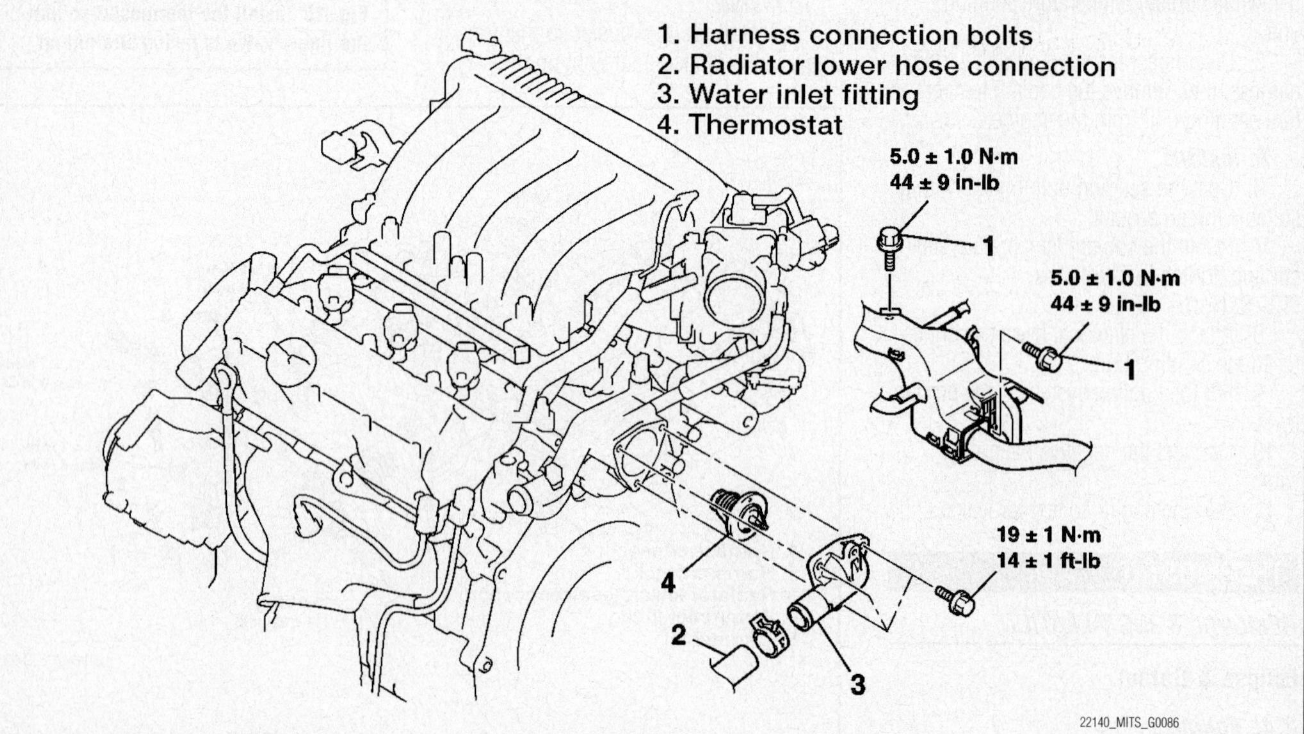

Fig. 16 Exploded view of the thermostat and related parts—3.8L engine

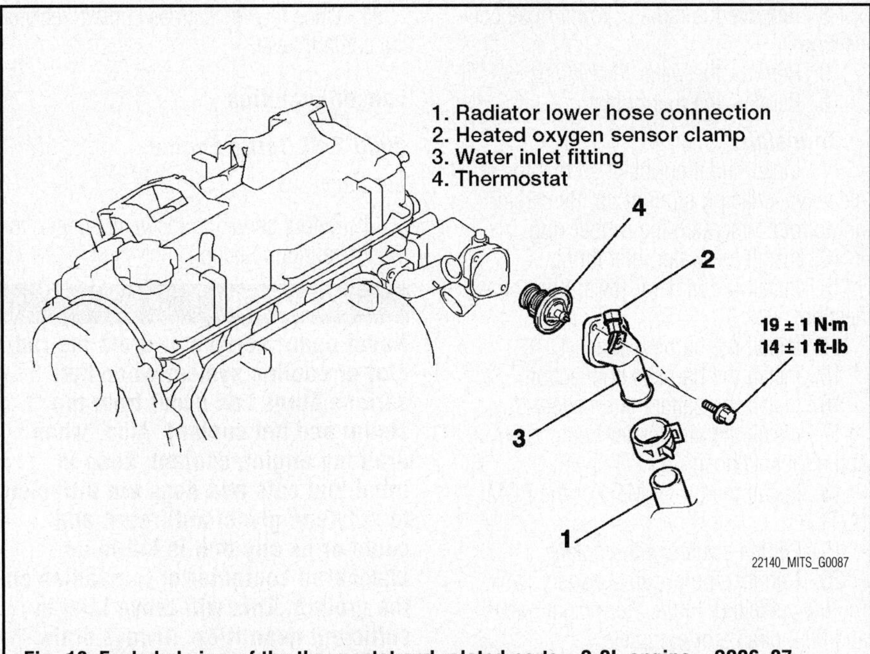

Fig. 18 Exploded view of the thermostat and related parts—2.0L engine—2006-07

5. Remove the thermostat housing retaining bolts.

6. Lift the housing from the engine.

7. Remove the thermostat taking note of its original position in the housing.

To install:

➡In order to prevent leakage, make sure both mating surfaces are clean and free of any old gasket material.

8. Install the thermostat so its flange seats tightly in the machined groove in the intake manifold or thermostat case. Refer to its location prior to removal. Align the relief valve with the alignment mark on the thermostat housing.

9. Use a new gasket or O-ring and reinstall the thermostat housing. Tighten the housing mounting bolts

10. Fill the system with coolant.

11. Install the removed air intake plumbing.

12. Connect the negative battery cable, run the vehicle until the thermostat opens and fill the radiator completely.

13. Once the vehicle has cooled, recheck the coolant level.

2008 2.0L Engine

See Figures 19 and 20.

1. Before servicing the vehicle, refer to the Precautions Section.

⁂ CAUTION

Never open, service or drain the radiator or cooling system when hot; serious burns can occur from the steam and hot coolant. Also, when draining engine coolant, keep in mind that cats and dogs are attracted to ethylene glycol antifreeze and could drink any that is left in an uncovered container or in puddles on the ground. This will prove fatal in sufficient quantities. Always drain coolant into a sealable container. Coolant should be reused unless it is contaminated or is several years old.

2. Disconnect the negative battery cable.

3. Drain the cooling system.

4. Remove the engine upper cover.

5. Remove the air cleaner assembly.

6. Remove the radiator lower hose connection.

7. Remove the control wiring harness clamp connection (CVT).

8. Remove the cooling water inlet hose fitting.

9. Remove the thermostat.

To install:

10. Install the thermostat so that the jiggle valve is facing straight up. Be careful not to fold or scratch the rubber ring.

11. Install the cooling water inlet hose fitting.

12. Install the control wiring harness clamp connection (CVT).

13. Install the radiator lower hose connection.

14. Install the air cleaner assembly.

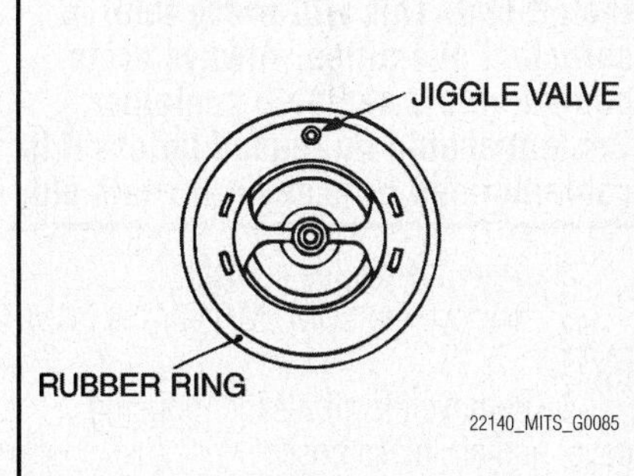

Fig. 20 Install the thermostat so that the jiggle valve is facing straight up

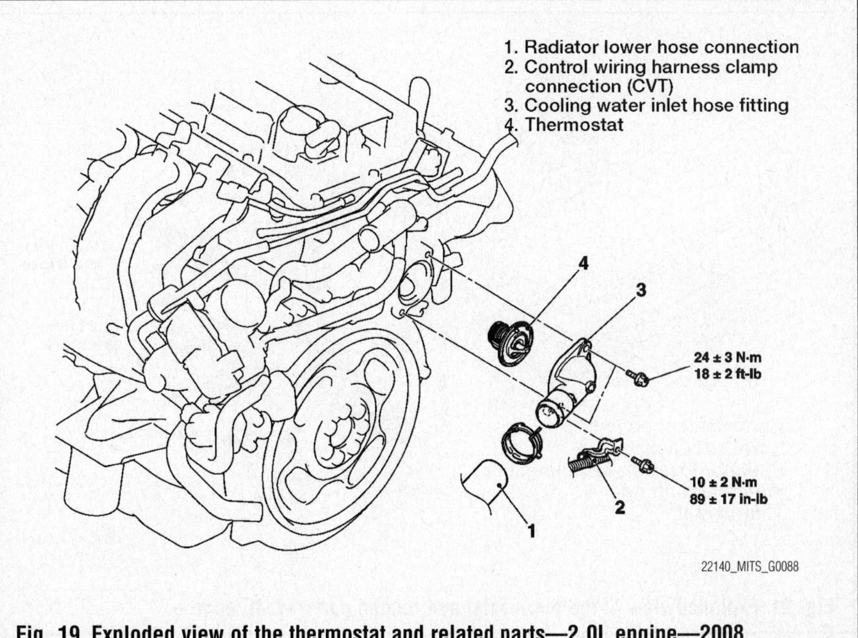

Fig. 19 Exploded view of the thermostat and related parts—2.0L engine—2008

15. Install the engine upper cover.
16. Fill the system with coolant.
17. Connect the negative battery cable, run the vehicle until the thermostat opens and fill the radiator completely.
18. Once the vehicle has cooled, recheck the coolant level.

2006 2.4L Engine

See Figures 21 and 22.

1. Before servicing the vehicle, refer to the Precautions Section.

⁂ CAUTION

Never open, service or drain the radiator or cooling system when hot; serious burns can occur from the steam and hot coolant. Also, when draining engine coolant, keep in mind that cats and dogs are attracted to ethylene glycol antifreeze and could drink any that is left in an uncovered container or in puddles on the ground. This will prove fatal in sufficient quantities. Always drain coolant into a sealable container. Coolant should be reused unless it is contaminated or is several years old.

2. Drain the engine coolant.
3. Remove the ECM (M/T) or the PCM (A/T).
4. Remove the air cleaner housing cover and air intake hose.
5. Remove the battery and battery tray.
6. Remove the harness connection.
7. Remove the harness bracket.
8. Remove the radiator lower hose connection.
9. Remove the water inlet fitting.
10. Remove the thermostat.

To install:

11. Install the thermostat so that the jiggle valve is facing straight up. Be careful not to fold or scratch the rubber ring.
12. Install the water inlet fitting.
13. Install the radiator lower hose connection.
14. Install the harness bracket.
15. Install the harness connection.
16. Install the battery and battery tray.
17. Install the air cleaner housing cover and air intake hose.
18. Install the ECM (M/T) or the PCM (A/T).
19. Fill the system with coolant.
20. Connect the negative battery cable, run the vehicle until the thermostat opens and fill the radiator completely.

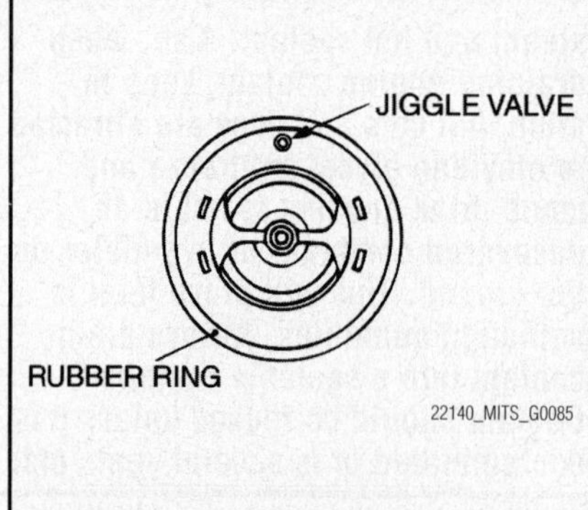

Fig. 22 Install the thermostat so that the jiggle valve is facing straight up

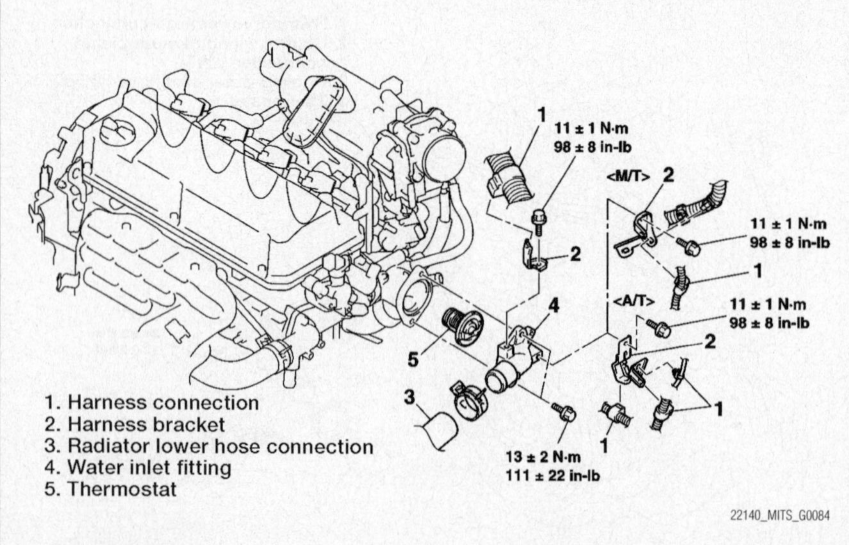

Fig. 21 Exploded view of the thermostat and related parts—2.4L engine

21. Once the vehicle has cooled, recheck the coolant level.

Lancer Evolution

2006 2.0L Turbo Engine

See Figure 23.

1. Before servicing the vehicle, refer to the Precautions Section.

⁂ CAUTION

Never open, service or drain the radiator or cooling system when hot; serious burns can occur from the steam and hot coolant. Also, when draining engine coolant, keep in mind that cats and dogs are attracted to ethylene glycol antifreeze and could drink any that is left in an uncovered container or in puddles on the ground. This will prove fatal in sufficient quantities. Always drain coolant into a sealable container. Coolant should be reused unless it is contaminated or is several years old.

2. Remove the engine undercover.
3. Drain the engine coolant.
4. Remove the intake air duct from the air cleaner assembly.
5. Remove air hose E, air pipe C, and air hose D of the charge air cooler.
6. Remove the accelerator cable connection.
7. Remove the control wiring harness connection.
8. Remove the vacuum hose and pipe assembly.
9. Remove the radiator upper hose connection.
10. Remove the wiring harness clamp.
11. Remove the harness bracket.
12. Remove the water outlet fitting.
13. Remove the thermostat.

To install:

➡Make sure there is no oil adhering to the rubber ring of the thermostat. In addition, be careful not to fold over or scratch the rubber ring when inserting the thermostat. If the rubber ring is damaged, replace the thermostat.

14. Install the thermostat being careful not to fold over or scratch the rubber ring.
15. Install the water outlet fitting.
16. Install the harness bracket.
17. Install the wiring harness clamp.
18. Install the radiator upper hose connection.
19. Install the vacuum hose and pipe assembly.

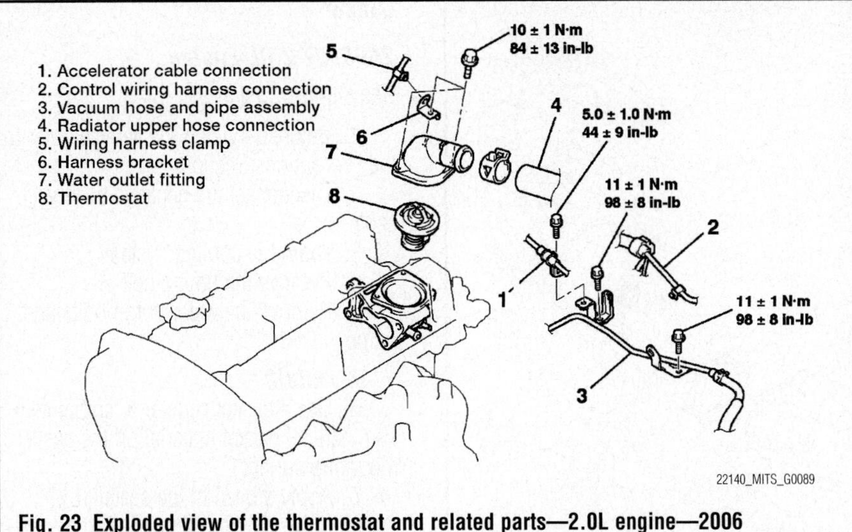

Fig. 23 Exploded view of the thermostat and related parts—2.0L engine—2006

20. Install the control wiring harness connection.
21. Install the accelerator cable connection.
22. Install air hose E, air pipe C, and air hose D of the charge air cooler.
23. Install the intake air duct and air cleaner assembly.
24. Fill the system with coolant.
25. Install the engine undercover.
26. Connect the negative battery cable, run the vehicle until the thermostat opens and fill the radiator completely.
27. Once the vehicle has cooled, recheck the coolant level.

2008 2.0L Turbo Engine

See Figures 24 and 25.

1. Before servicing the vehicle, refer to the Precautions Section.

⁂ CAUTION

Never open, service or drain the radiator or cooling system when hot; serious burns can occur from the steam and hot coolant. Also, when draining engine coolant, keep in mind that cats and dogs are attracted to ethylene glycol antifreeze and could drink any that is left in an uncovered container or in puddles on the ground. This will prove fatal in sufficient quantities. Always drain coolant into a sealable container. Coolant should be reused unless it is contaminated or is several years old.

2. Remove the engine undercover.
3. Drain the engine coolant.
4. Remove the engine upper cover.
5. Remove the air cleaner intake hose from the air cleaner assembly.
6. Remove the radiator lower hose connection.
7. Remove the control wiring harness clamp connection.
8. Remove the harness bracket.
9. Remove the cooling water inlet hose fitting.
10. Remove the thermostat.

To install:

11. Install the thermostat so that the jiggle valve is facing straight up. Be careful not to fold or scratch the rubber ring.
12. Install the cooling water inlet hose fitting.
13. Install the harness bracket.
14. Install the control wiring harness clamp connection.

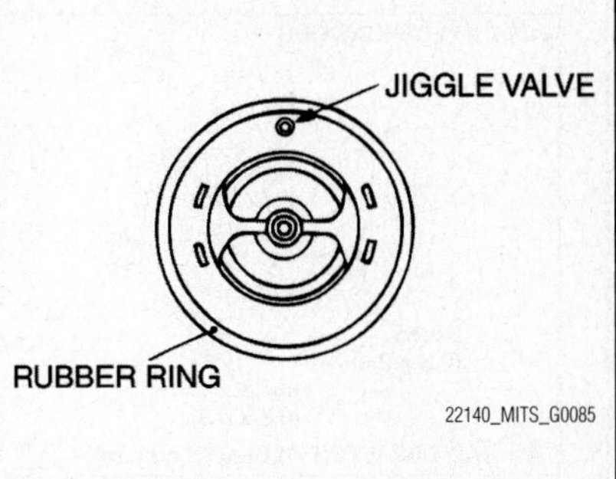

Fig. 25 Install the thermostat so that the jiggle valve is facing straight up

15. Install the radiator lower hose connection.
16. Install the air cleaner intake hose.
17. Install the engine upper cover.
18. Fill the system with coolant.
19. Install the engine undercover.
20. Connect the negative battery cable, run the vehicle until the thermostat opens and fill the radiator completely.
21. Once the vehicle has cooled, recheck the coolant level.

WATER PUMP

REMOVAL & INSTALLATION

Eclipse & Galant

2.4L Engine

See Figure 26.

1. Before servicing the vehicle, refer to the Precautions Section.

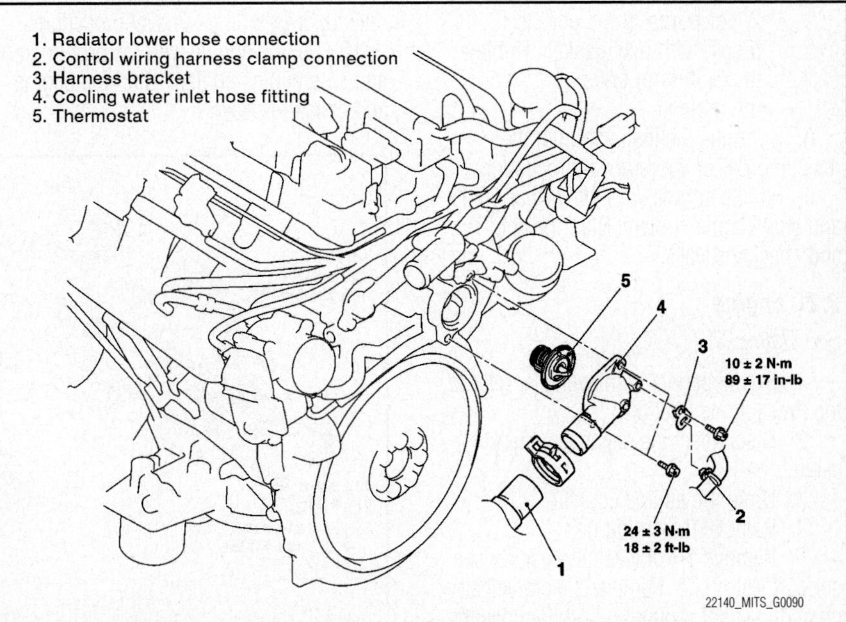

Fig. 24 Exploded view of the thermostat and related parts—2.0L engine—2008

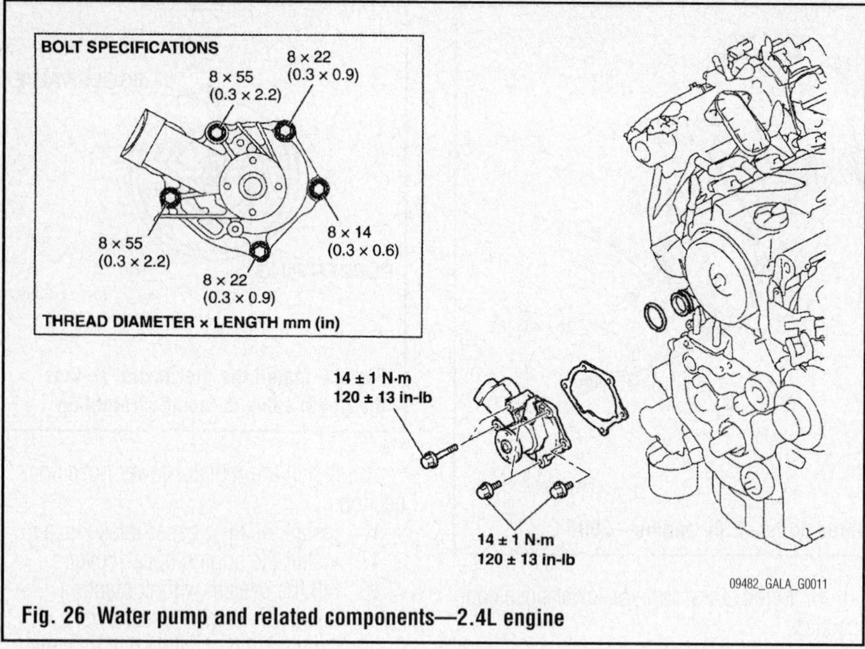

Fig. 26 Water pump and related components—2.4L engine

2. Disconnect the negative battery cable.
3. Drain the engine coolant.
4. Remove the timing belt.
5. Remove the water pump retaining bolts.
6. Remove the water pump from the engine. Discard the water pump gasket and O-ring.

To install:

7. Install or connect the following:
   - New O-ring on the water inlet pipe. Coat the O-ring with water or coolant. Do not allow oil or other grease to contact the O-ring.
   - Water pump to the engine block, with new gasket. Tighten the mounting bolts as illustrated.
8. Continue the installation in the reverse order of the removal procedure.
9. Fill the engine with the proper grade and type engine coolant. Start the engine and check for leaks.

3.8L Engine

See Figure 27.

1. Before servicing the vehicle, refer to the Precautions Section.
2. Disconnect the negative battery cable.
3. Drain the engine coolant.
4. Remove the timing belt.
5. Remove the crankshaft position sensor connector clip. Remove the crankshaft position sensor mounting bolt. Remove the sensor.
6. Remove the water pump retaining bolts.
7. Remove the water pump from the engine. Discard the water pump gasket and O-ring.

To install:

8. Install or connect the following:
   - New O-ring on the water inlet pipe. Coat the O-ring with water or coolant. Do not allow oil or other grease to contact the O-ring.
   - Water pump to the engine block, with new gasket. Tighten the mounting bolts as in the illustration.
9. Continue the installation in the reverse order of the removal procedure.
10. Fill the engine with the proper grade and type engine coolant. Start the engine and check for leaks.

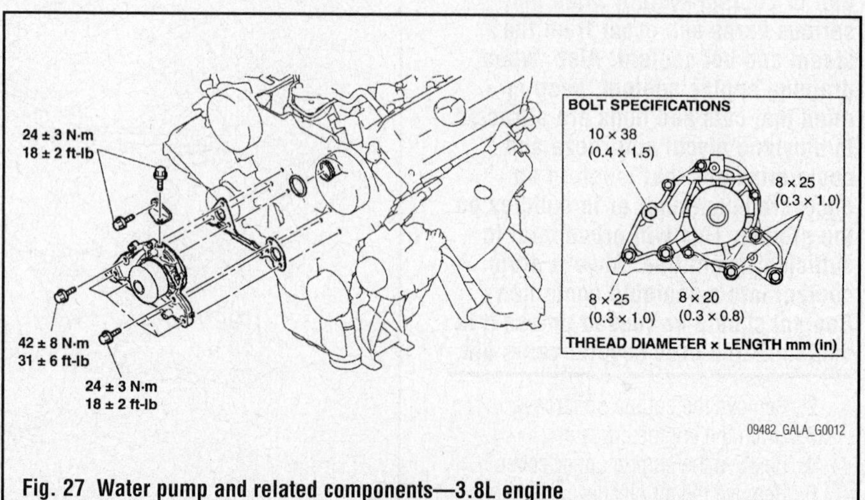

Fig. 27 Water pump and related components—3.8L engine

Lancer

2006–07 2.0L Engine

See Figures 28 and 29.

1. Before servicing the vehicle, refer to the Precautions Section.
2. Disconnect the negative battery cable.
3. Drain the cooling system.
4. Remove the timing belt.
5. Remove the water pump bolts and pump.

To install:

6. Use a gasket scraper to completely eliminate all gasket material on the gasket mounting surface.
7. Apply a bead of the sealant 3M® AAD Part No. 8672, 3M® AAD Part No. 8679/8678, or equivalent.
8. Within 15 minutes after the sealant is applied, install the water pump. Do not apply the sealant in an area more than the required.
9. Install the water pump and tighten the mounting bolt to 15–19 ft. lbs. (20–26 Nm).

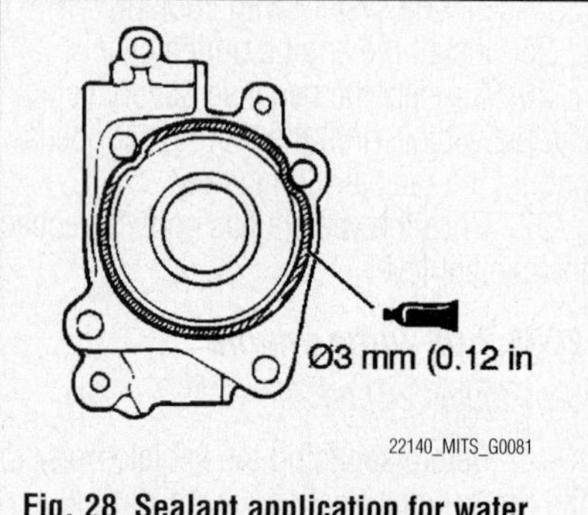

Fig. 28 Sealant application for water pump installation—Lancer 2.0L engine—2006-07

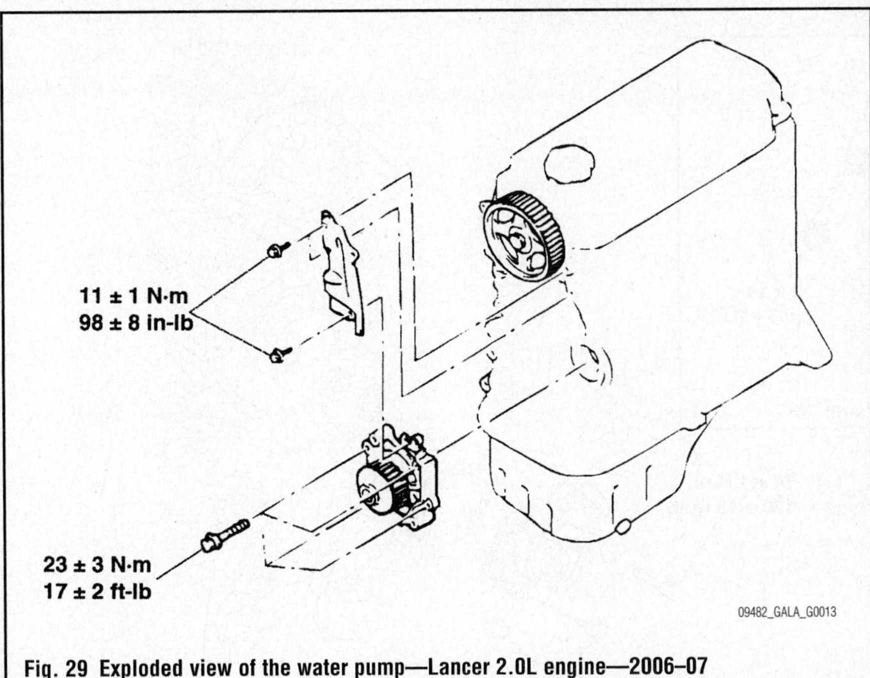

Fig. 29 Exploded view of the water pump—Lancer 2.0L engine—2006–07

10. Install the timing belt.
11. Refill the cooling system and connect the negative battery cable.

2008 2.0L Engine

See Figure 30.

1. Before servicing the vehicle, refer to the Precautions Section.
2. Disconnect the negative battery cable.
3. Drain the engine coolant.
4. Remove the accessory drive belt. Refer to Accessory Drive Belts, Removal & Installation.
5. Remove the water pump pulley.
6. Remove the water pump inlet pipe mounting nuts.
7. Remove the water pump.

To install:

8. Completely remove the water pump gasket material from the mating surfaces.
9. Install new gaskets and the water pump into position. Tighten the bolts according to the illustration.
10. Install the water pump inlet pipe mounting nuts.
11. Install the water pump pulley.
12. Install the accessory drive belt. Refer to Accessory Drive Belts, Removal & Installation.
13. Refill the cooling system and connect the negative battery cable.

2006 2.4L Engine

See Figure 31.

1. Before servicing the vehicle, refer to the Precautions Section.
2. Disconnect the negative battery cable.
3. Drain the engine coolant.
4. Remove the timing belt.
5. Remove the water pump retaining bolts.
6. Remove the water pump from the engine. Discard the water pump gasket and O-ring.

To install:

7. Install or connect the following:
   - New O-ring on the water inlet pipe. Coat the O-ring with water or coolant. Do not allow oil or other grease to contact the O-ring.
   - Water pump to the engine block, with new gasket. Tighten the mounting bolts as illustrated.
8. Continue the installation in the reverse order of the removal procedure.

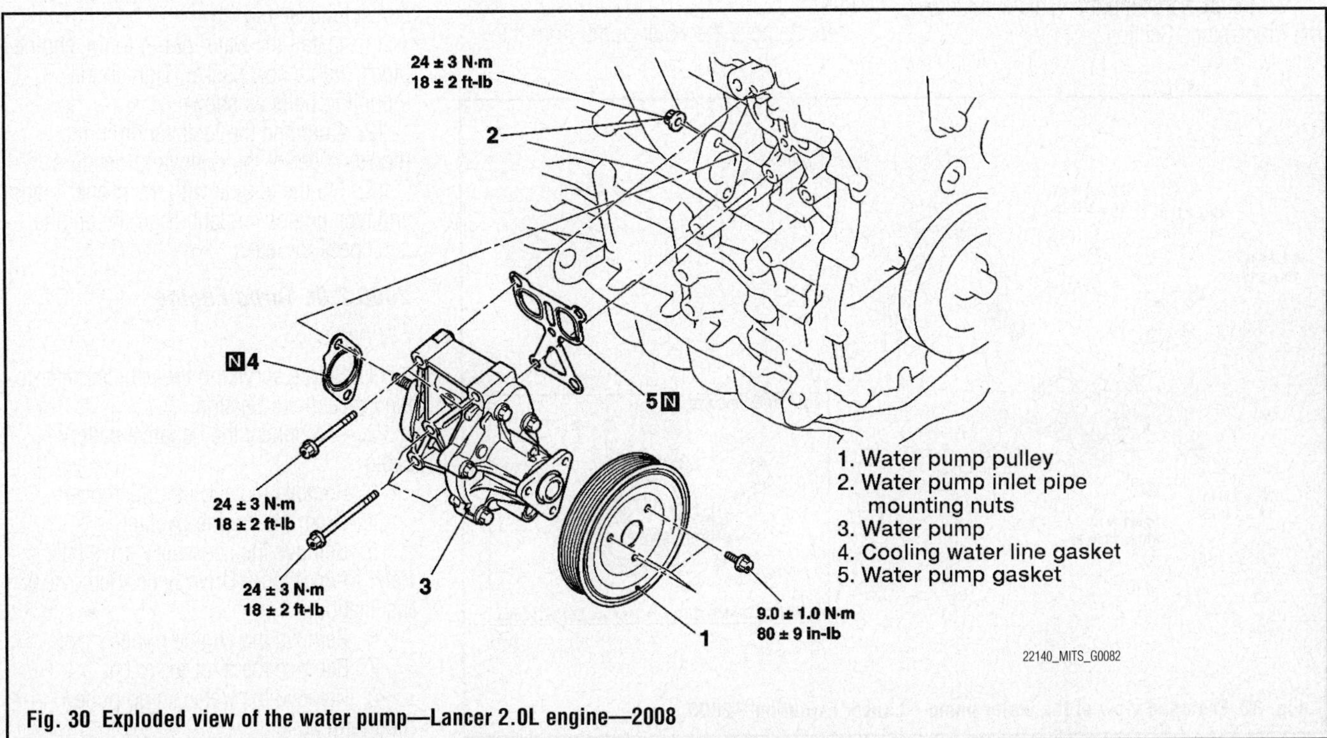

Fig. 30 Exploded view of the water pump—Lancer 2.0L engine—2008

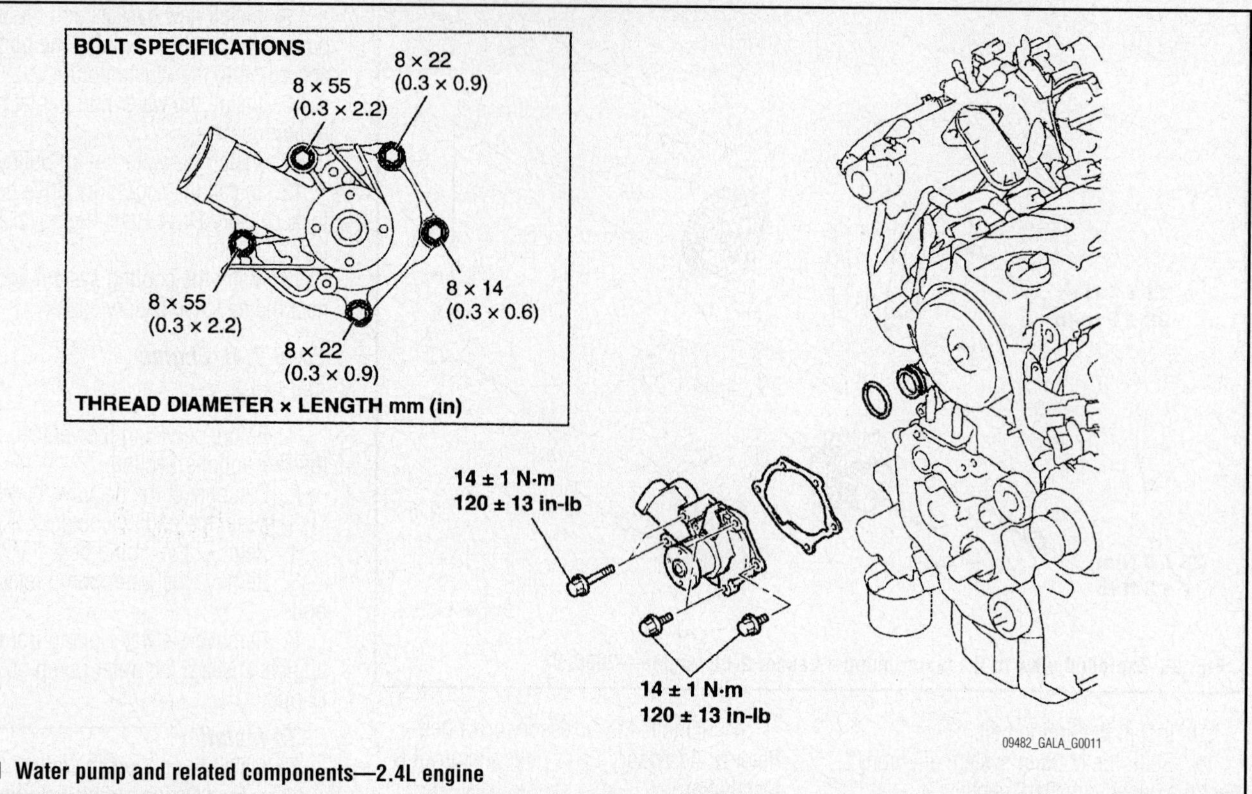

Fig. 31 Water pump and related components—2.4L engine

9. Fill the engine with the proper grade and type engine coolant. Start the engine and check for leaks.

Lancer Evolution

2006 2.0L Turbo Engine

See Figure 32.

1. Before servicing the vehicle, refer to the Precautions Section.
2. Disconnect the negative battery cable.
3. Remove the engine undercover.
4. Drain the engine coolant.
5. Remove the timing belt tension adjuster.
6. Remove the alternator brace.
7. Remove the water pump retaining bolts.
8. Remove the water pump from the engine. Discard the water pump gasket and O-ring.

To install:

9. Completely remove any gasket material from the mating surfaces.
10. Install a new O-ring on the water inlet pipe. Coat the O-ring with water or coolant. Do not allow oil or other grease to contact the O-ring.
11. Install the water pump to the engine block, with a new gasket. Tighten the mounting bolts as shown.
12. Continue the installation in the reverse order of the removal procedure.
13. Fill the engine with the proper grade and type engine coolant. Start the engine and check for leaks.

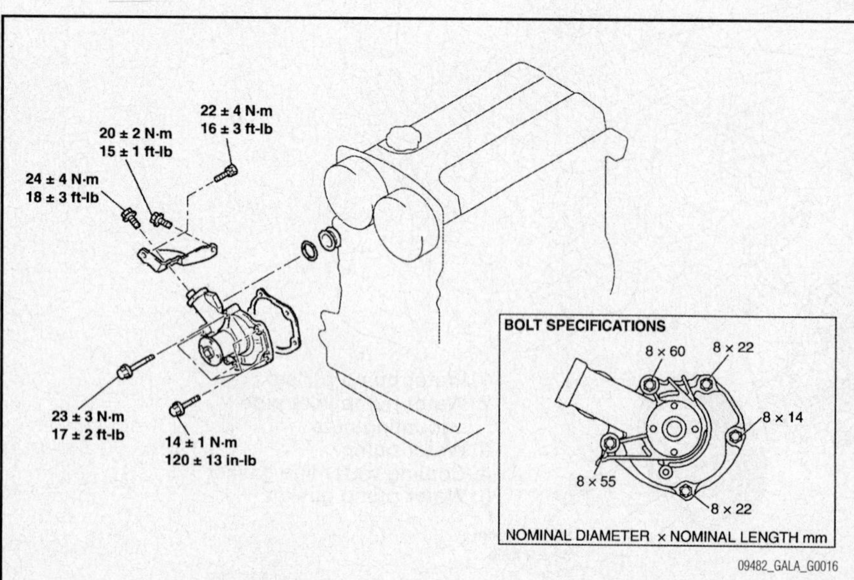

Fig. 32 Exploded view of the water pump—Lancer Evolution—2006

2008 2.0L Turbo Engine

See Figure 33.

1. Before servicing the vehicle, refer to the Precautions Section.
2. Disconnect the negative battery cable.
3. Remove the engine undercover.
4. Drain the engine coolant.
5. Remove the accessory drive belt. Refer to Accessory Drive Belts, Removal & Installation.
6. Remove the engine upper cover.
7. Remove the strut tower bar.
8. Remove the water pump pulley mounting bolts.

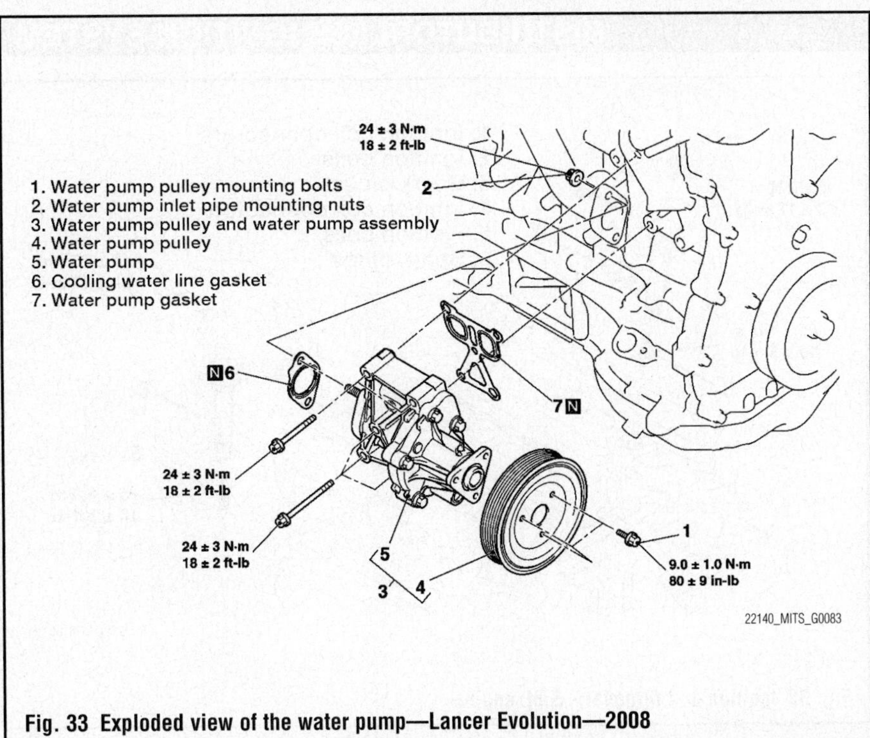

Fig. 33 Exploded view of the water pump—Lancer Evolution—2008

9. Remove the water pump inlet pipe mounting nuts.
10. Remove the water pump pulley.
11. Remove the water pump.

To install:

12. Completely remove the cooling water line gasket and the water pump gasket material.
13. Install the water pump and tighten the bolts as illustrated.
14. Install the water pump pulley.
15. Install the water pump inlet pipe mounting nuts.
16. Install the water pump pulley mounting bolts.
17. Install the strut tower bar.
18. Install the engine upper cover.
19. Install the accessory drive belt. Refer to Accessory Drive Belts, Removal & Installation.
20. Fill the engine with the proper grade and type engine coolant.
21. Install the engine undercover.
22. Connect the negative battery cable.
23. Start the engine and check for leaks.

ENGINE ELECTRICAL

CHARGING SYSTEM

ALTERNATOR

REMOVAL & INSTALLATION

2.0L & 2.4L Engines

1. Before servicing the vehicle, refer to the Precautions Section.
2. Remove or disconnect the following:
   - The negative battery cable
   - The left side cover panel under the vehicle
   - The accessory drive belt. Refer to Accessory Drive Belts, Removal & Installation
   - The regulator connector from the alternator
   - The nut retaining the battery cable to the alternator
   - The battery cable from the alternator
   - The nut retaining the battery cable to the alternator
   - The regulator connector from the alternator
   - The nut retaining the wire harness to the alternator, and remove the harness from the alternator
   - The nut for the pivot bolt on the rear of the alternator
   - The pivot bolt from the alternator
   - The alternator adjusting bolt
   - The alternator from the vehicle

To install:

3. Position the alternator on the lower mounting fixture and install the lower mounting bolt and nut. Tighten nut just enough to allow for movement of the alternator.
4. Install the alternator upper bracket/brace and connect the alternator electrical harness.
5. Install the accessory drive belt. Refer to Accessory Drive Belts, Removal & Installation.
6. Install the left side cover panel under the vehicle as required.
7. Connect the negative battery cable and check for proper operation.

3.8L Engine

1. Disconnect the negative battery cable.
2. Remove the necessary components in order to gain access to the alternator assembly.
3. Disconnect the alternator electrical connections.
4. Remove the alternator mounting bolts.
5. Remove the alternator from the vehicle.

➡On some vehicles it may be easier to remove the alternator from underneath of the vehicle.

To install:

6. Position the alternator assembly to its mounting.
7. Install the retaining bolts.
8. Connect the electrical connectors.
9. Continue the installation in the reverse order of the removal procedure.

ADJUSTMENT

These engines are equipped with a Distributorless Ignition System (DIS). No adjustment is necessary.

IGNITION COIL

REMOVAL & INSTALLATION

Eclipse & Galant

2.4L Engine

See Figure 34.

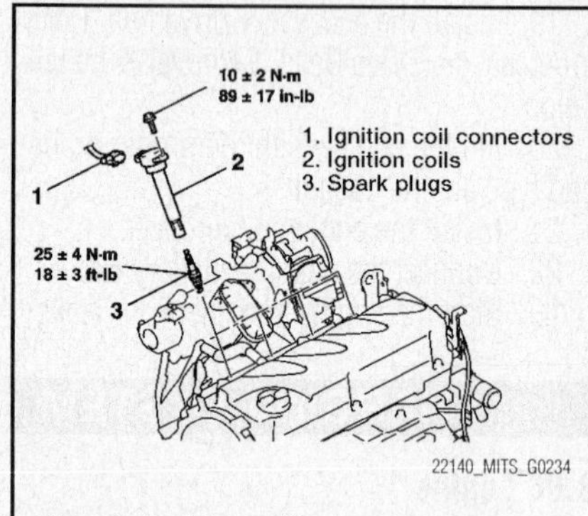

Fig. 34 Ignition coil removal—2.4L engine

1. Before servicing the vehicle, refer to the Precautions Section.
2. Disconnect the negative battery cable.
3. Remove the air cleaner resonator.
4. Remove the ignition coil connectors.
5. Remove the ignition coil retaining bolt.
6. Remove the ignition coils.

To install:

7. Installation is the reverse of the removal procedure.
8. Tighten the retaining bolts as illustrated.

3.8L Engine

See Figure 35.

1. Before servicing the vehicle, refer to the Precautions Section.
2. Disconnect the negative battery cable.
3. Remove the engine cover.
4. Left bank ignition coil removal:
   a. Remove the ignition coil connectors.
   b. Remove the ignition coil retaining bolt.
   c. Remove the ignition coils.
5. Right bank ignition coil removal:
   a. Remove the intake manifold plenum.

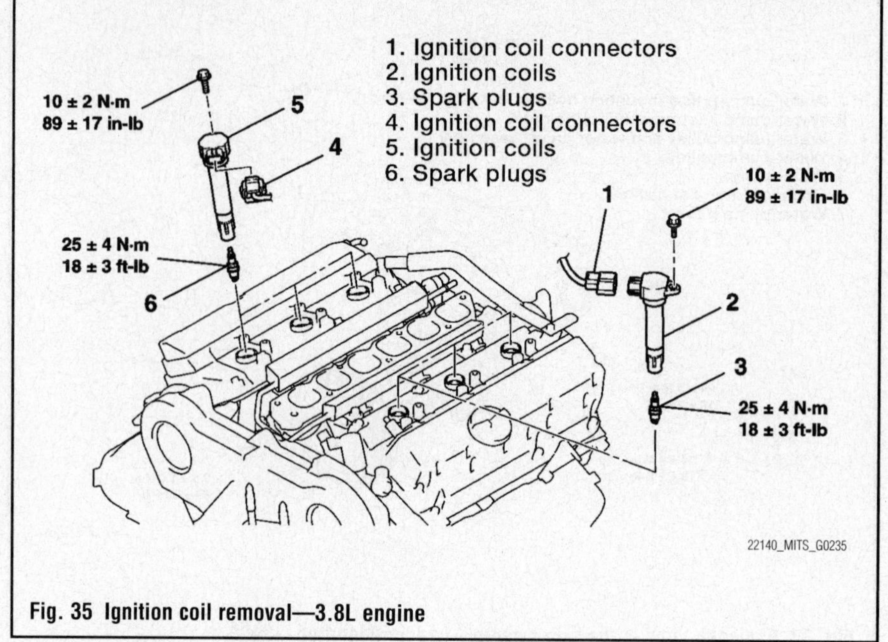

Fig. 35 Ignition coil removal—3.8L engine

   b. Remove the ignition coil connectors.
   c. Remove the ignition coil retaining bolt.
   d. Remove the ignition coils.

To install:

6. Installation is the reverse of the removal procedure.
7. Tighten the retaining bolts as illustrated.

Lancer

2006–07 2.0L Engine

See Figure 36.

1. Before servicing the vehicle, refer to the Precautions Section.
2. Disconnect the negative battery cable.
3. Remove the engine cover.
4. Remove the ignition coil connectors.

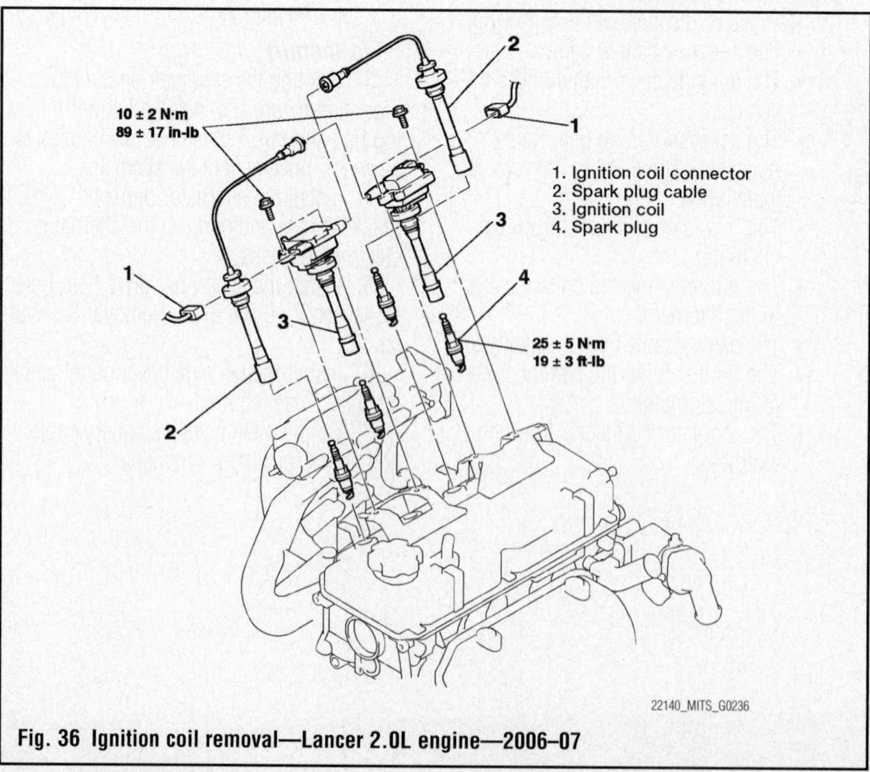

Fig. 36 Ignition coil removal—Lancer 2.0L engine—2006–07

5. Remove the spark plug cable where applicable (see illustration).
6. Remove the ignition coil retaining bolt.
7. Remove the ignition coil.

To install:

8. Installation is the reverse of the removal procedure.
9. Tighten the retaining bolts to 72–106 inch lbs. (8–12 Nm).

2008 2.0L Engine

See Figure 37.

1. Before servicing the vehicle, refer to the Precautions Section.
2. Disconnect the negative battery cable.
3. Remove the cylinder head cover center cover.
4. Remove the ignition coil connector.
5. Remove the ignition coil retaining bolt.
6. Remove the ignition coil.

To install:

7. Installation is the reverse of the removal procedure.
8. Tighten the retaining bolts as illustrated.

Lancer Evolution

2008 2.0L Turbo Engine

See Figure 38.

1. Before servicing the vehicle, refer to the Precautions Section.
2. Disconnect the negative battery cable.
3. Remove the engine upper cover.

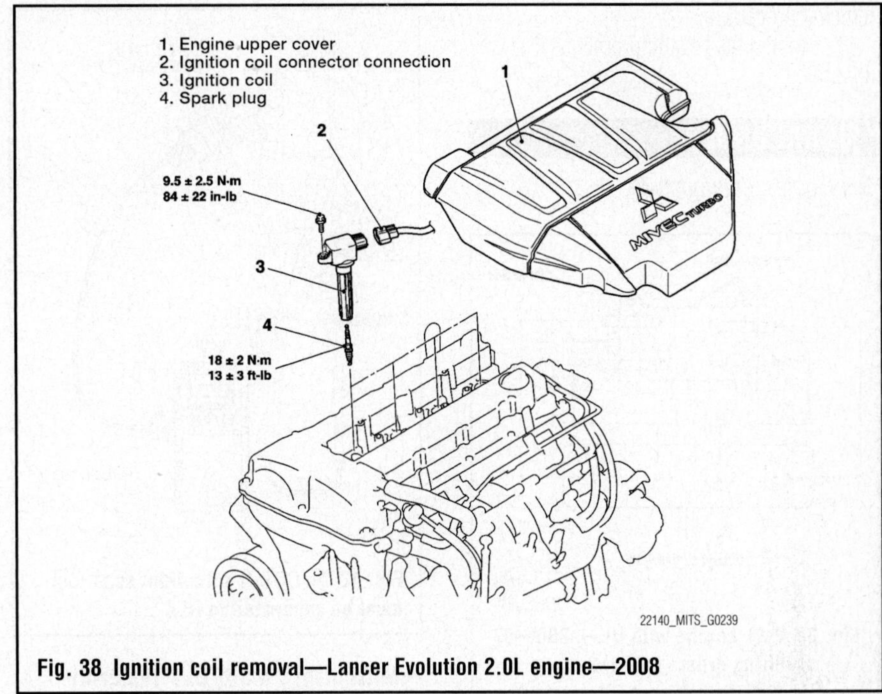

Fig. 38 Ignition coil removal—Lancer Evolution 2.0L engine—2008

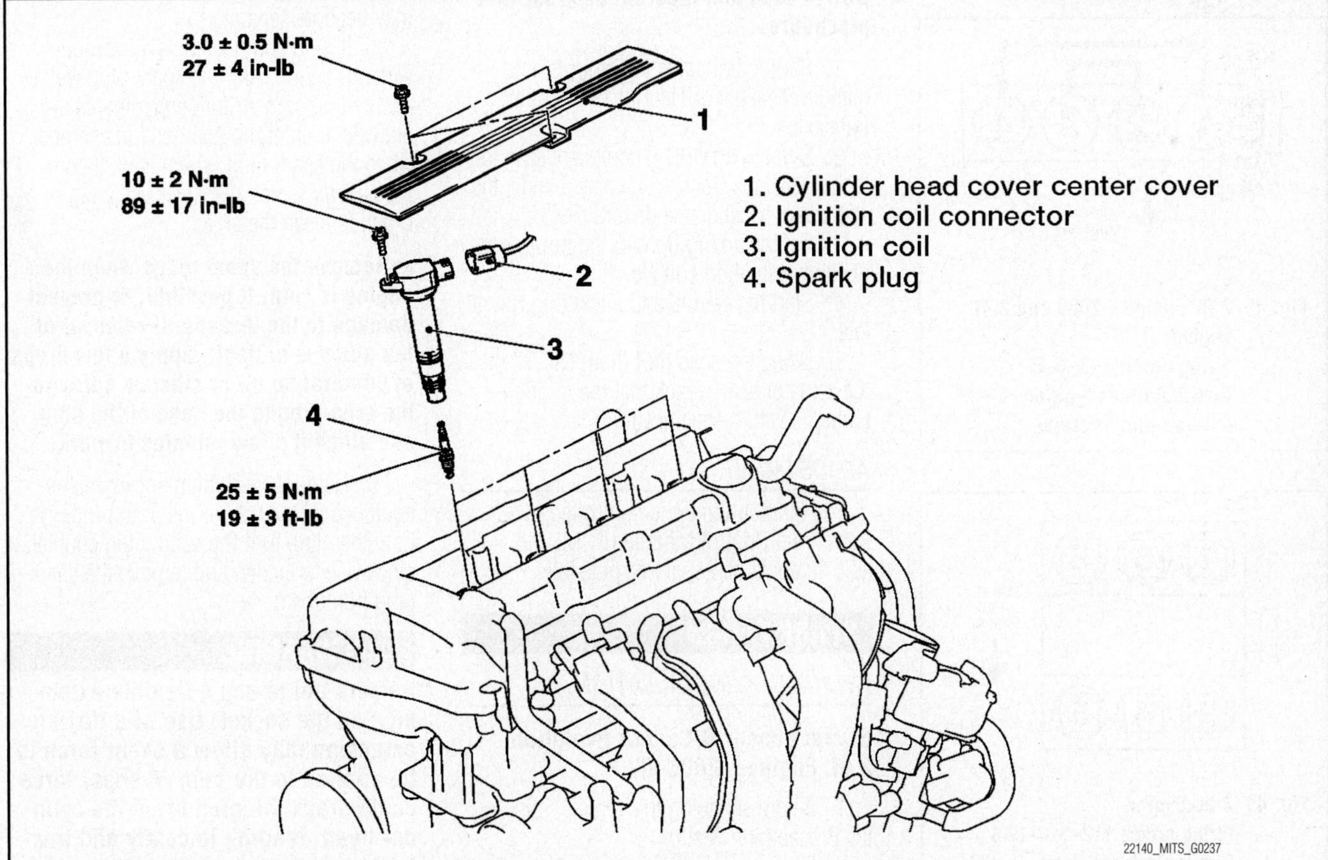

3.0 ± 0.5 N·m
27 ± 4 in-lb
1
10 ± 2 N·m
89 ± 17 in-lb
2
3
4
25 ± 5 N·m
19 ± 3 ft-lb

1. Cylinder head cover center cover
2. Ignition coil connector
3. Ignition coil
4. Spark plug

22140_MITS_G0237

Fig. 37 Ignition coil removal—Lancer 2.0L engine—2008

4. Remove the ignition coil connectors.
5. Remove the ignition coil retaining bolt.
6. Remove the ignition coil.

To install:

7. Installation is the reverse of the removal procedure.
8. Tighten the retaining bolts as illustrated.

FIRING ORDER

See Figures 39 through 41.

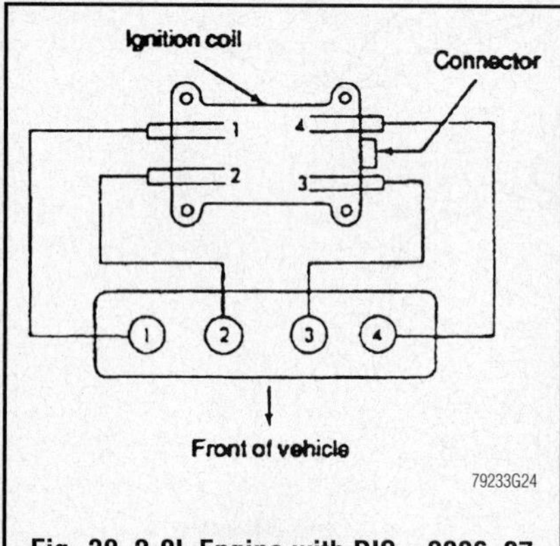

Fig. 39 2.0L Engine with DIS—2006–07 Firing order: 1–3–4–2 Distributorless ignition system

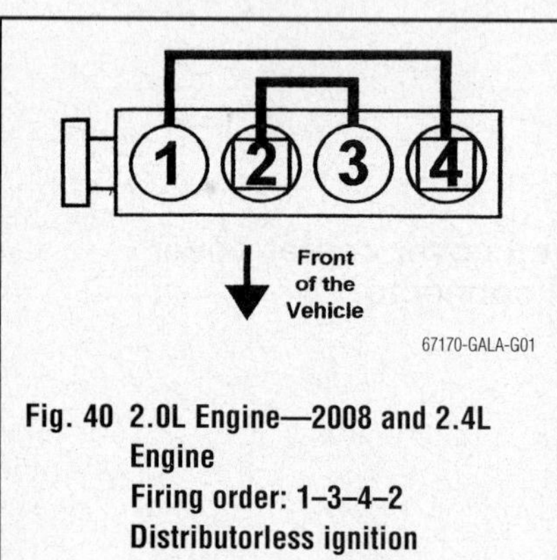

Fig. 40 2.0L Engine—2008 and 2.4L Engine Firing order: 1–3–4–2 Distributorless ignition (coil-on-plug) system

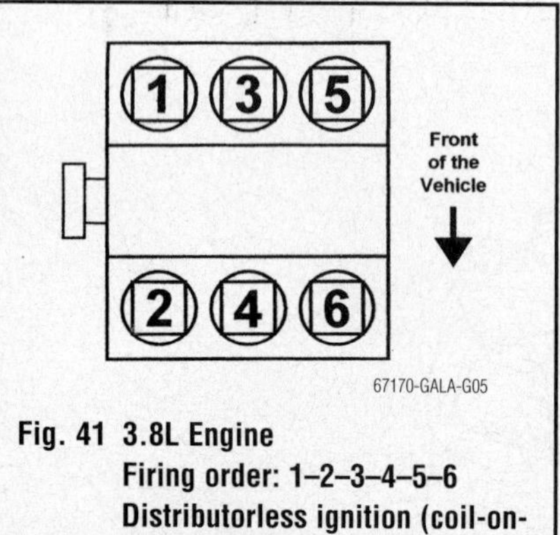

Fig. 41 3.8L Engine Firing order: 1–2–3–4–5–6 Distributorless ignition (coil-on-plug) system

IGNITION TIMING

INSPECTION

See Figure 42.

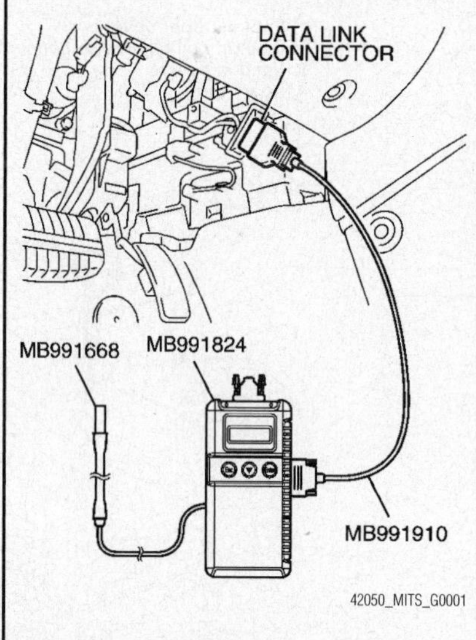

Fig. 42 An OBD-II compliant scan tool must be connected to DLC

➡Mitsubishi's Multi-Use Tester-III (MUT III) scan tool or an equivalent OBD-II scan tool must be used for this procedure.

1. Set the transaxle in Park for automatic transaxles or Neutral for manual transaxles.
2. Connect an OBD-II compliant scan tool to the Data Link Connector, located under the instrument panel on the driver's side.
3. Set the timing light to the power supply line of ignition coil No. 1.
4. Start the engine and allow it to run at idle.
5. Using the scan tool (Item No. 17 of the actuator test), check that the ignition timing is within specification.

ADJUSTMENT

The ignition timing is controlled by the Powertrain Control Module (PCM). No adjustment is necessary or possible.

SPARK PLUGS

REMOVAL & INSTALLATION

Except Lancer & Lancer Evolution 2.0L Engine—2006–07

1. Before servicing the vehicle, refer to the Precautions Section.
2. Remove the engine cover (as necessary).
3. Disconnect the ignition coil connector.
4. Remove the ignition coil. Refer to Ignition Coil Pack, Removal & Installation.
5. Use a spark plug socket and wrench to remove the spark plugs.

✱✱ WARNING

Be careful that no contaminates enter through the spark plug holes.

➡Check the electrode gap on the spark plugs before installation.

6. To install, reverse the removal procedure. Tighten the spark plugs to 15–22 ft. lbs. (20–30 Nm).

Lancer & Lancer Evolution 2.0L Engine—2006–07

1. Before servicing the vehicle, refer to the Precautions Section.
2. Disconnect the negative battery cable, and if the vehicle has been run recently, allow the engine to cool.
3. If equipped, remove the center cover.
4. Carefully twist the spark plug wire boot to loosen it, then pull upward and remove the boot from the plug. Be sure to pull on the boot and not on the wire, otherwise the connector located inside the boot may become separated.
5. Using compressed air, blow any water or debris from the spark plug well to assure that no harmful contaminants are allowed to enter the combustion chamber when the spark plug is removed. If compressed air is not available, use a rag or a brush to clean the area.

➡Remove the spark plugs when the engine is cold, if possible, to prevent damage to the threads. If removal of the plugs is difficult, apply a few drops of penetrating oil or silicone spray to the area around the base of the plug, and allow it a few minutes to work.

6. Using a spark plug socket that is equipped with a rubber insert to properly hold the plug, turn the spark plug counterclockwise to loosen and remove the spark plug from the bore.

✱✱ WARNING

Be sure not to use a flexible extension on the socket. Use of a flexible extension may allow a shear force to be applied to the plug. A shear force could break the plug off in the cylinder head, leading to costly and frustrating repairs.

To install:

7. Inspect the spark plug boot for tears or damage. If a damaged boot is found, the spark plug wire must be replaced.

8. Using a wire feeler gauge, check and adjust the spark plug gap. When using a gauge, the proper size should pass between the electrodes with a slight drag. The next larger size should not be able to pass while the next smaller size should pass freely.

9. Carefully thread the plug into the bore by hand. If resistance is felt before the plug is almost completely threaded, back the plug out and begin threading again. In small, hard to reach areas, an old spark plug wire and boot could be used as a threading tool. The boot will hold the plug while you twist the end of the wire and the wire is supple enough to twist before it would allow the plug to cross thread.

⁂ WARNING

Do not use the spark plug socket to thread the plugs. Always carefully thread the plug by hand or using an old plug wire to prevent the possibility of cross threading and damaging the cylinder head bore.

10. Carefully tighten the spark plug. If the plug you are installing is equipped with a crush washer, seat the plug, then tighten about 1/4 turn to crush the washer. If you are installing a tapered seat plug, tighten the plug to specifications provided by the vehicle or plug manufacturer.

11. Apply a small amount of silicone dielectric compound to the end of the spark plug lead or inside the spark plug boot to prevent sticking, then install the boot to the spark plug and push until it clicks into place. The click may be felt or heard, then gently pull back on the boot to assure proper contact.

ENGINE ELECTRICAL

STARTING SYSTEM

STARTER

REMOVAL & INSTALLATION

1. Before servicing the vehicle, refer to the Precautions Section.

2. Remove the negative battery cable and wait at least 3 minutes.

3. Raise and safely support the vehicle.

4. Remove the engine undercover.

5. If necessary, detach the speedometer cable connector at the transaxle end.

6. Detach the starter motor electrical connections.

7. Remove the starter motor mounting bolts and remove the starter.

To install:

8. Installation is the reverse of removal. Tighten starter motor bolts to 20–25 ft. lbs. (27–34 Nm).

ENGINE MECHANICAL

ACCESSORY DRIVE BELTS

ACCESSORY BELT ROUTING

See Figures 43 through 45.

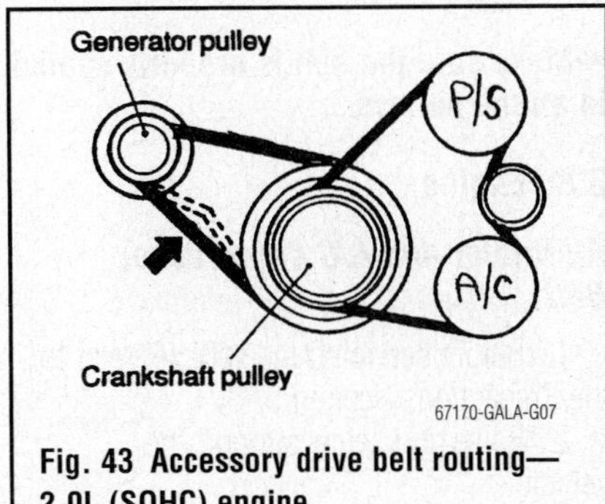

Fig. 43 Accessory drive belt routing—2.0L (SOHC) engine

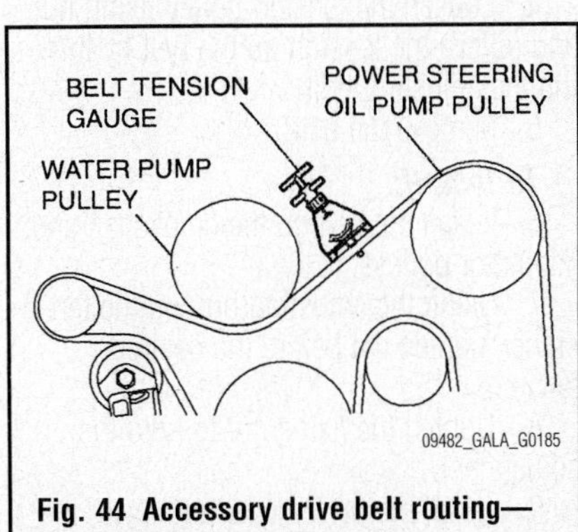

Fig. 44 Accessory drive belt routing—2.0L (DOHC) and 2.4L engines

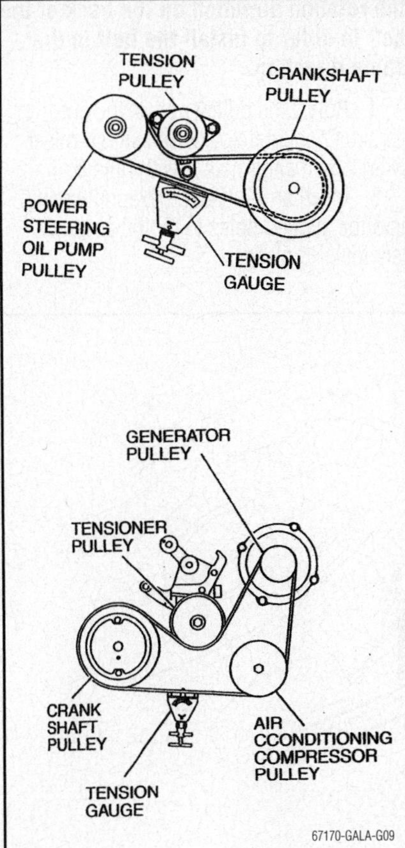

Fig. 45 Accessory drive belt routing — 3.8L engine

ADJUSTMENT

Excessive belt tension will cause damage to the alternator and water pump pulley bearings, while, on the other hand, loose belt tension will produce slip and premature wear on the belt. Therefore, be sure to adjust the belt tension to the proper level.

If the engine is not equipped with an auto-tensioner, loosen the adjusting bolt or fixing bolt locknut on the alternator, alternator bracket or tension pulley. Then, move the alternator or turn the adjusting bolt to adjust belt tension. Once the desired value is reached, secure the bolt or locknut and recheck tension.

REMOVAL & INSTALLATION

2.0L (SOHC) Engine

Alternator Belt

1. Before servicing the vehicle, refer to the Precautions Section.

2. Loosen the alternator support nut.

3. Loosen the adjuster lock bolt.

4. Rotate the adjuster bolt counter clockwise to release the tension on the belt.

5. Remove the belt.

To install:

6. Install the belt on the pulleys.

7. Rotate the adjuster bolt clockwise until the proper tension is reached.
8. Tighten the adjuster lock bolt and the alternator support nut.

Power Steering Belt

1. Before servicing the vehicle, refer to the Precautions Section.
2. Remove the alternator belt.
3. Loosen the power steering pump adjusting bolts.
4. Remove the power steering pump fixed bolt on the rear of the bracket.
5. Rotate the pump toward the engine and remove the belt.

To install:

6. Install the belt on the pulleys.
7. Rotate the pump until the proper tension is reached.
8. Tighten the adjusting bolts on the pump.
9. Tighten the fixed bolt on the rear of the bracket.
10. Install the alternator belt.

A/C Compressor Belt

1. Before servicing the vehicle, refer to the Precautions Section.
2. Loosen the tension pulley and remove the belt.
3. The installation is the reverse of the removal.

2.0L (DOHC) & 2.4L Engines

See Figures 46 and 47.

1. Before servicing the vehicle, refer to the Precautions Section.
2. Remove the engine undercover.

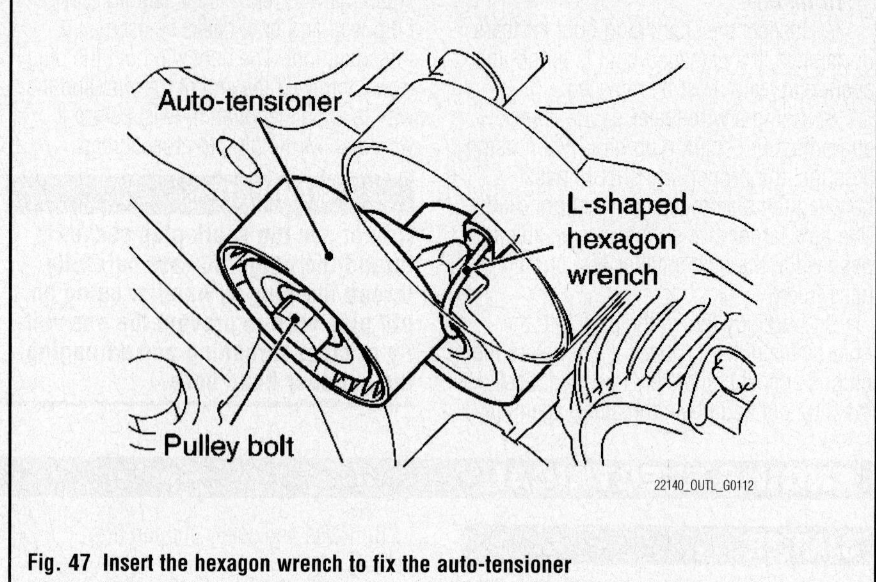

Fig. 47 Insert the hexagon wrench to fix the auto-tensioner

3. Remove the radiator condenser tank assembly mounting bolt, and move the radiator condenser tank assembly to a place where it does not interfere with the drive belt removal and installation.

➡To reuse the accessory drive belt, use chalk to draw an arrow indicating the rotation direction on the back of the belt in order to install the belt in the same direction.

4. Rotate the pulley bolt of the auto-tensioner counterclockwise with an offset wrench and align hole A with hole B.
5. Insert an L-shaped hexagon wrench into the aligned holes to fix the auto-tensioner in place.

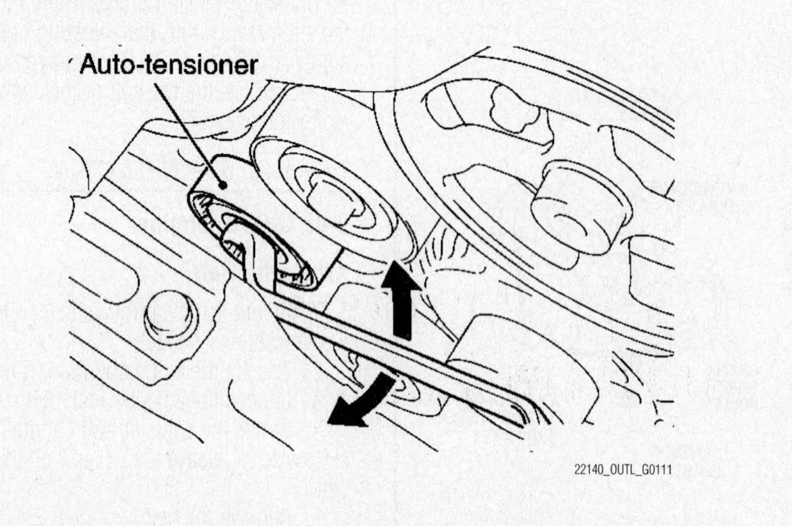

Fig. 46 Rotate the pulley bolt of the auto-tensioner counterclockwise with an offset wrench

6. Remove the accessory drive belt from the vehicle.

To install:

7. Install the drive belt to each pulley.
8. Set an offset wrench to the pulley bolt of the auto-tensioner. Then, rotate the auto-tensioner counter-clockwise and remove the L-shaped hexagon wrench fixing the auto-tensioner.
9. Apply tension to the drive belt while slowly turning the auto-tensioner clockwise.

➡Make sure the belt is properly seated in all the pulleys.

3.8L Engine

Alternator and A/C Compressor Belt

1. Before servicing the vehicle, refer to the Precautions Section.
2. Raise and safely support the vehicle.
3. Remove the front undercover.
4. Loosen the tension pulley fixing nut and relieve the tension on the belt by turning the adjusting bolt.
5. Remove the belt.

To install:

6. Install the belt on the crankshaft and alternator pulleys.
7. Using the adjusting bolt on the tensioner, tighten the belt to the desired tension.
8. Tighten the fixing nut to hold the adjustment.
9. Install the undercover and lower the vehicle.

Power Steering Belt

1. Before servicing the vehicle, refer to the Precautions Section.
2. Raise and safely support the vehicle.
3. Remove the front undercover.
4. Remove the alternator and A/C compressor belt.
5. Lower the vehicle and remove the cruise control pump link assembly.
6. Place the power steering hose under the oil reservoir.
7. Loosen the tension pulley fixing bolts and remove the power steering pump drive belt.

To install:

8. Install the power steering pump drive belt.
9. Insert an extension bar or equivalent into the opening at the end of the tension pulley bracket and pivot the pulley to apply tension to the belt.
10. Tighten the fixing bolts.
11. Raise the vehicle and install the alternator and A/C compressor belt.
12. Install the undercover and lower the vehicle.

CAMSHAFT AND VALVE LIFTERS

REMOVAL & INSTALLATION

Eclipse & Galant

2.4L Engine

See Figures 48 through 51.

1. Before servicing the vehicle, refer to the Precautions Section.
2. Disconnect the negative battery cable.
3. If equipped with manual transaxle, remove the ECM. If equipped with automatic transaxle, remove the PCM.

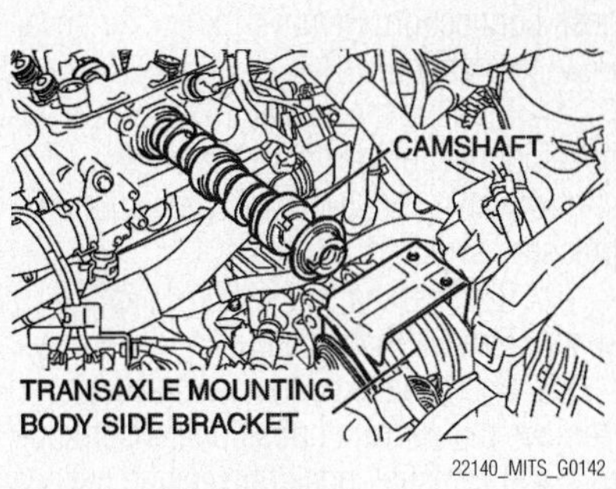

Fig. 48 Camshaft and transaxle mounting side bracket location—2.4L engine

4. Remove the air cleaner assembly.
5. Remove the battery.
6. Remove the ignition coils.
7. Remove the timing belt. Refer to Timing Belt and Sprockets, Removal & Installation.

1. Control wiring harness connection
2. Rocker cover PCV hose
3. Rocker cover breather hose
4. Engine oil control valve
5. O-ring
6. Engine oil pressure switch
7. Rocker cover assembly
8. Rocker cover gasket
9. Spark plug guide oil seals
10. Accumulator assembly

22140_MITS_G0143

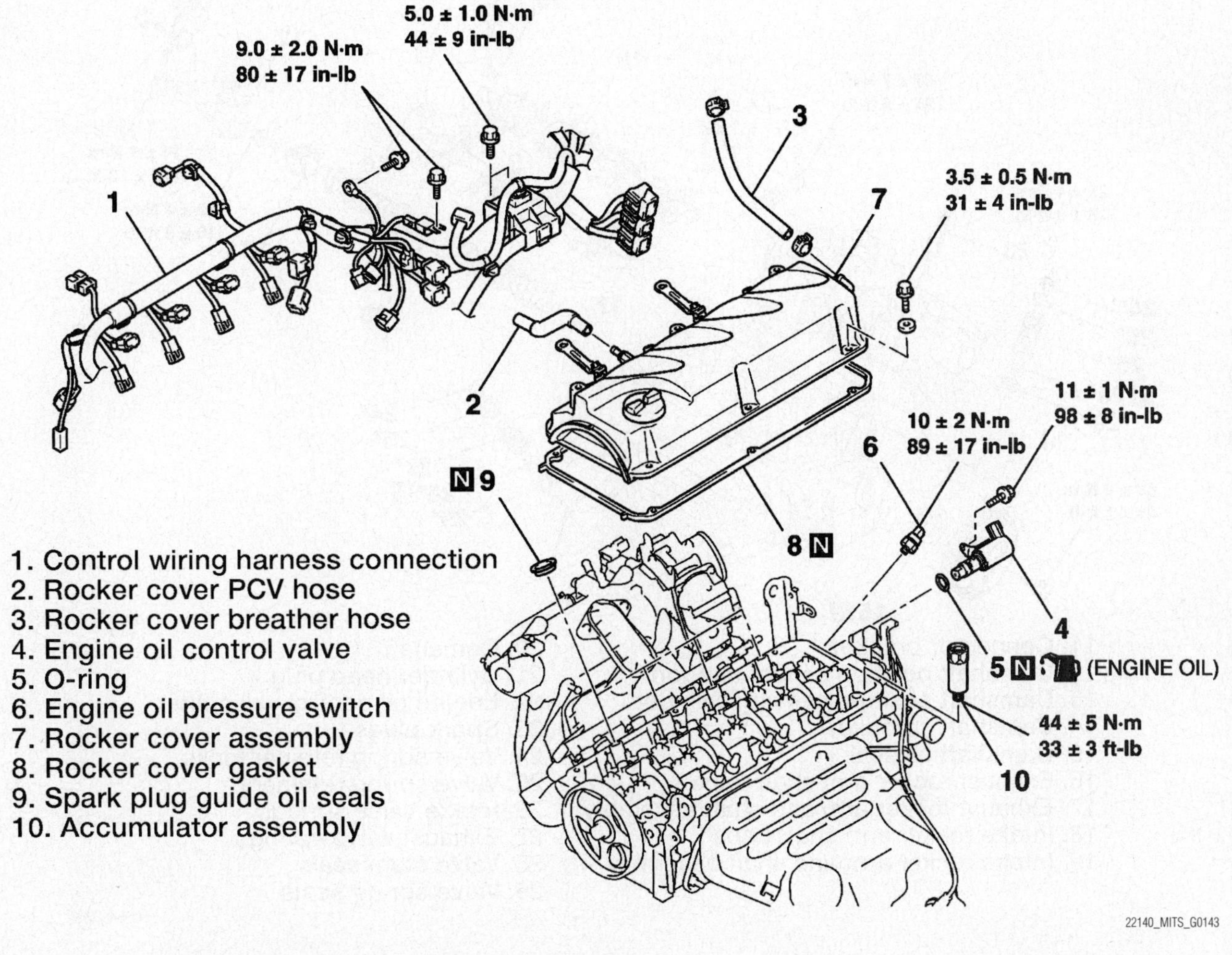

Fig. 49 Remove the rocker cover assembly and engine electrical connections—2.4L engine

8. Disconnect the control wiring harness connection. Remove the rocker cover PCV connection.

9. Remove the rocker cover breather hose connection. Remove the engine oil control valve and O-ring. Remove the oil pressure switch.

10. Remove the spark plug guide oil seals. Remove the accumulator assembly.

11. Remove the connector bracket. Remove the camshaft position sensor support and camshaft position sensing cylinder.

12. Use tool MB998719 to hold the camshaft sprocket in place. Remove the camshaft sprocket retaining bolt. Remove the camshaft sprocket. Remove the oil seal.

13. Remove the rocker cover retaining bolts. Remove the rocker cover. Discard the gasket.

14. Remove the exhaust rocker arm shaft caps. Remove the exhaust rocker arm shaft assembly.

15. Remove the intake rocker arm shaft caps. Remove the intake rocker arm shaft assembly.

16. Raise the transaxle until the camshaft and transaxle mounting side bracket do not touch when removing the camshaft.

17. Remove the camshaft from its mounting on the engine.

To install:

18. Lubricate the camshaft with clean engine oil. Install the camshaft. Set the dowel pin of the camshaft in the position as shown in the illustration.

19. Install the rocker arm assemblies.

20. Continue the installation in the reverse order of the removal procedure.

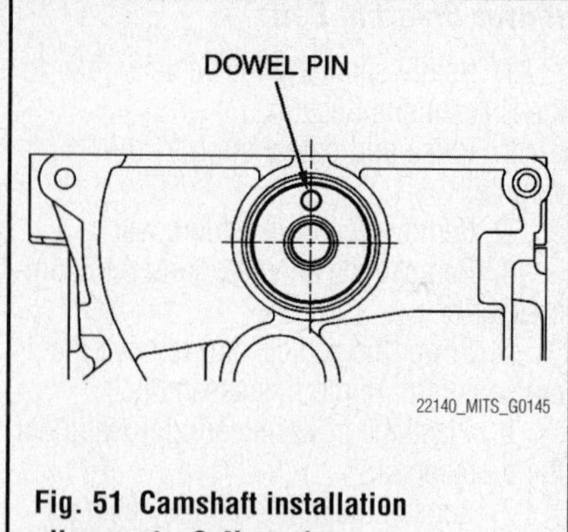

Fig. 51 Camshaft installation alignment—2.4L engine

21. Adjust the valves as required.

22. To initialize the ECM, manual transaxle equipped vehicles, or the PCM, automatic transaxle equipped vehicles,

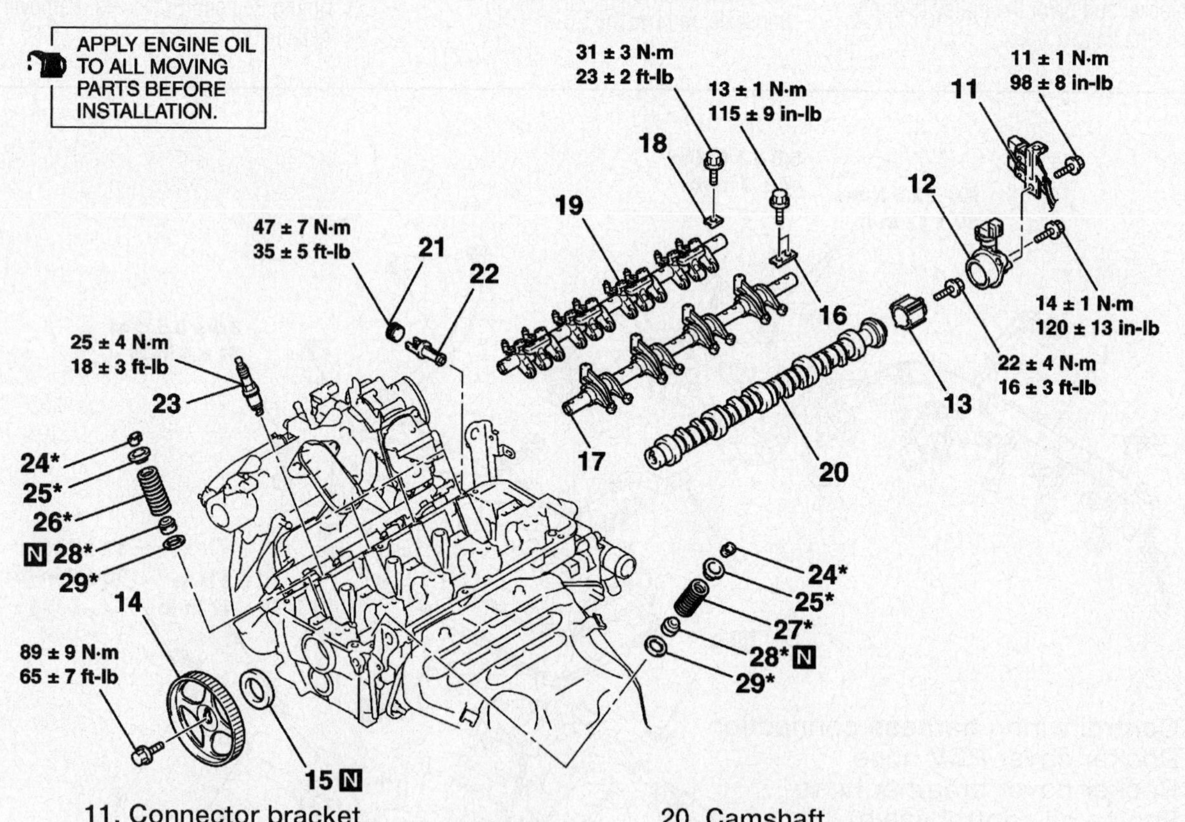

Fig. 50 Camshaft and related components—2.4L engine

turn the ignition switch ON then OFF and keep it in the OFF position for at least 10 seconds.

3.8L Engine

See Figures 52 through 54.

1. Before servicing the vehicle, refer to the Precautions Section.
2. Disconnect the negative battery cable.
3. Drain the cooling system.
4. Remove the timing belt. Refer to Timing Belt and Sprockets, Removal & Installation.
5. On the left side, remove the thermostat housing assembly. Remove the PCV valve hose connection, ignition coil connectors and ignition coils. Disconnect the engine control wiring harness clamp and harness bracket.
6. On the right side, remove the intake manifold plenum, breather hose connection, heater hose and water hose connection. Remove the ignition coil connectors and coils.

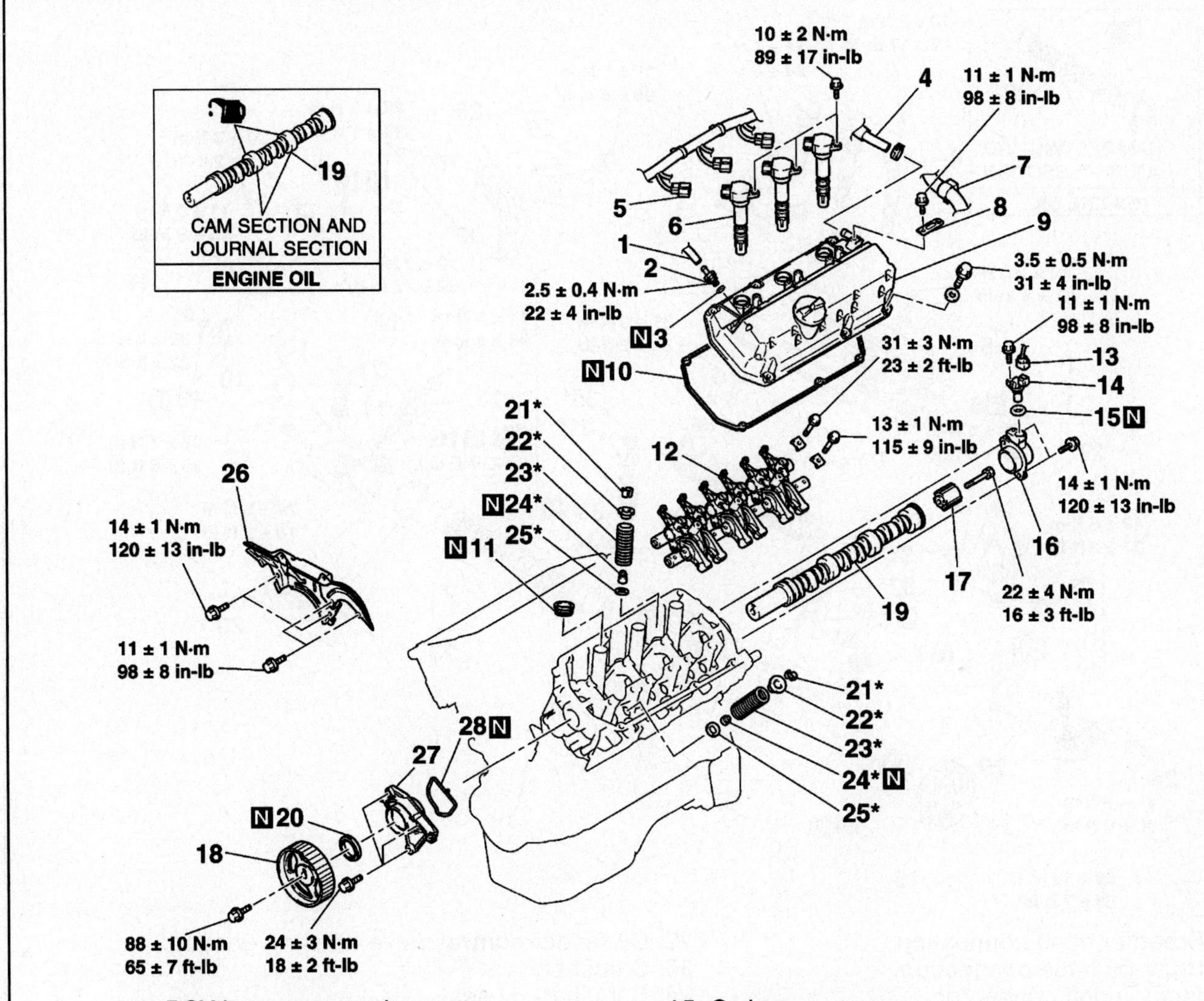

1. PCV hose connection
2. PCV valve
3. O-ring
4. Blow-by hose connection
5. Ignition coil connector
6. Ignition coil
7. Engine control wiring harness clamp
8. Harness bracket
9. Rocker cover
10. Rocker cover gasket
11. Spark plug guide oil seal
12. Rocker arm and shaft assembly
13. Camshaft position sensor connector
14. Camshaft position sensor
15. O-ring
16. Camshaft position sensor support
17. Camshaft position sensing cylinder
18. Camshaft sprocket
19. Camshaft
20. Camshaft oil seal
21. Valve spring retainer lock
22. Valve spring retainer
23. Valve spring
24. Valve stem seal
25. Valve spring seat
26. Timing belt rear center cover
27. Camshaft oil seal case
28. Camshaft oil seal case gasket

22140_MITS_G0146

Fig. 52 Camshaft and related components—3.8L engine (left bank)

7. Remove the rocker arm cover retaining bolts. Remove the rocker arm cover. Discard the gasket.

8. Install tool MD998443 on the rocker arms, so that the lash adjusters will not fall out.

9. Loosen the rocker arm and shaft assembly mounting bolts. Remove the rocker arm and shaft assembly with the bolts still attached.

➡**Do not disassemble the rocker arm and shaft assembly.**

10. On the left side, disconnect the camshaft position sensor and remove the sensor support. Remove the camshaft position sensing cylinder. Remove the camshaft sprocket. Remove the camshaft and camshaft oil seal.

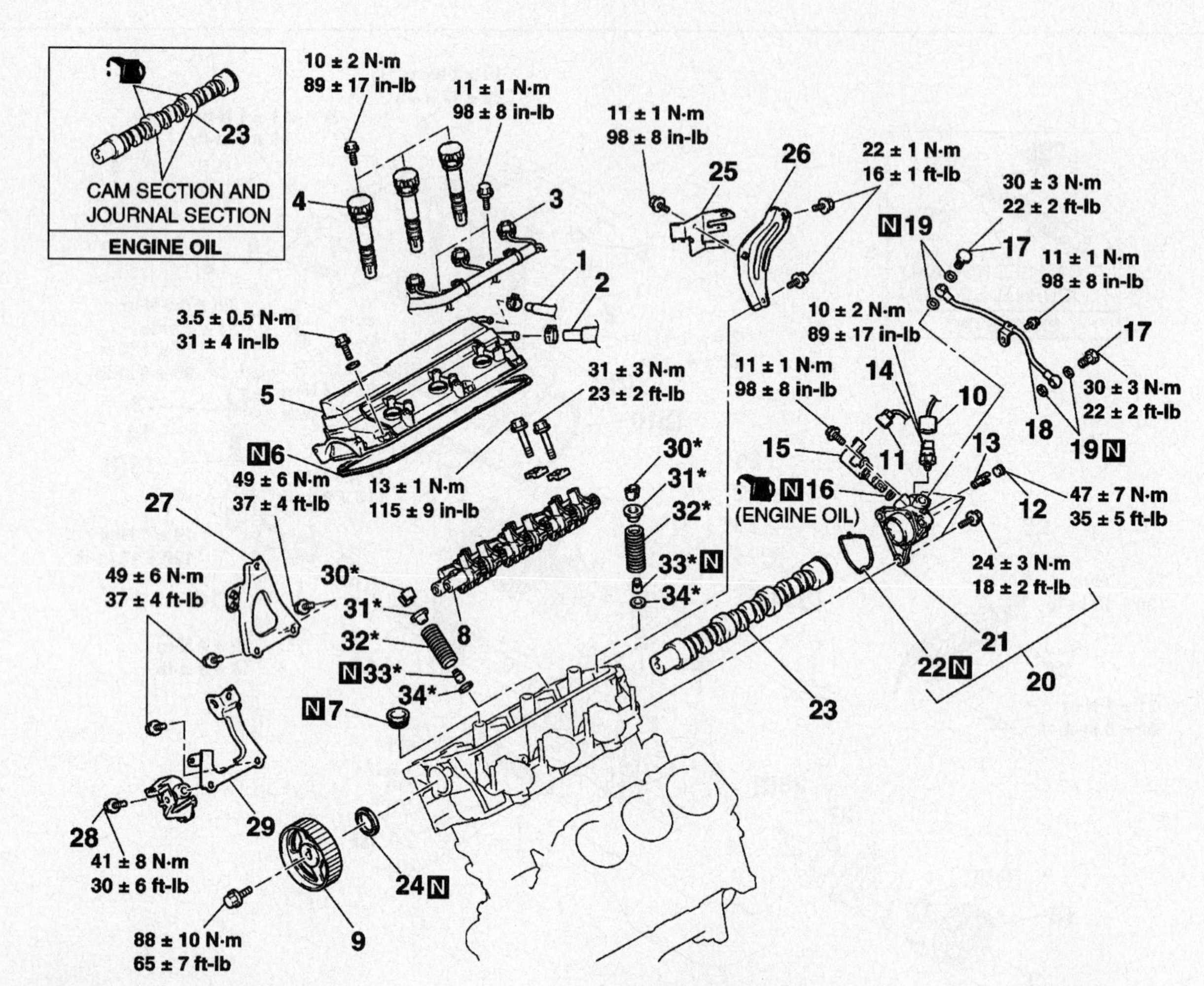

Fig. 53 Camshaft and related components—3.8L engine (right bank)

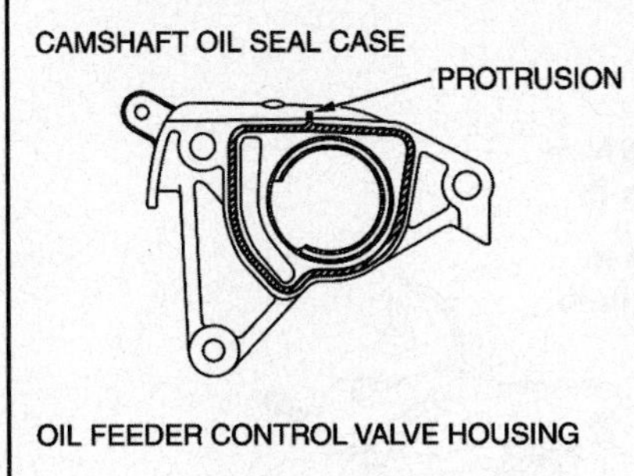

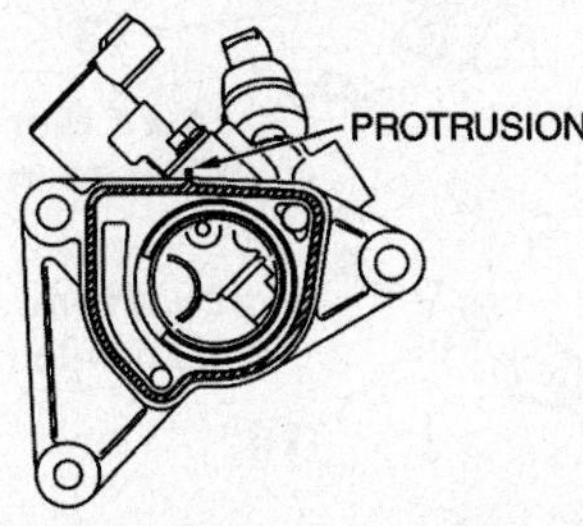

22140_MITS_G0148

Fig. 54 Install the camshaft oil seal case gasket and oil feeder control valve housing gasket with their protrusions in the direction shown—3.8L engine (right bank)

11. On the right side, disconnect the engine oil control valve connector and engine oil pressure switch connector. Remove the oil feed control valve right housing assembly. Remove the camshaft sprocket.

12. Remove the camshaft and camshaft oil seal.

To install:

13. Install the camshaft oil seal case gasket and oil feeder control valve housing gasket with their protrusions in the direction shown.

**** WARNING**

Be careful that no foreign material gets into the oil passages. Make sure the mating surfaces are clean.

14. Install the camshaft oil seal case and oil feeder control valve housing to the cylinder head.

15. Tighten the mounting bolt to: 16–20 ft. lbs. (21–27 Nm).

16. Install a gasket to one of oil feeder control valve pipes and tighten the eye bolt by hand.

17. Install a gasket to the other oil feeder control valve pipe and tighten the eye bolt to the specified torque. Tightening torque: 20–24 ft. lbs. (27–33 Nm).

18. Tighten the eye bolt which is temporarily tightened to the specified torque. Tightening torque: 20–24 ft. lbs. (27–33 Nm).

19. Tighten the oil feeder control valve pipe mounting bolt to the specified torque. Tightening torque: 90–106 inch lbs. (10–12 Nm).

➡Do not re-use the O-ring for the engine oil control valve.

20. Before installing a new O-ring on the engine oil control valve, wind adhesive tape around the oil passages cut-out area of the engine oil control valve to prevent damage. If the O-ring is damaged, it may cause an oil leak.

21. Apply a small amount of engine oil to the O-ring and install it to the oil control valve.

22. Assemble the engine oil control valve to the cylinder head.

23. Tighten the engine oil control valve mounting bolt to the specified torque. Tightening torque: 90–106 inch lbs. (10–12 Nm).

24. Install the intake and exhaust side rocker arms and shaft assemblies. Refer to Rocker Arms/Shafts, Removal & Installation.

25. Continue the installation in the reverse order of the removal procedure.

26. Adjust the valves as required.

27. To initialize the ECM, manual transaxle equipped vehicles, or the PCM, automatic transaxle equipped vehicles, turn the ignition switch ON then OFF and keep it in the OFF position for at least 10 seconds.

Lancer

2006–07 2.0L Engine

See Figures 55 and 56.

1. Before servicing the vehicle, refer to the Precautions Section.
2. Disconnect the negative battery cable.
3. Remove the air cleaner assembly.
4. Remove the ignition coils.
5. Remove the breather hose.
6. Remove the Positive Crankcase Ventilation (PCV) hose.
7. Remove the rocker arm (valve) cover and gasket.
8. Remove the spark plug guide.
9. Remove the timing belt front upper cover.
10. Remove the camshaft sprocket, as follows:
    a. Secure the camshaft sprocket and timing belt with wire ties to prevent them from slipping out of place.
    b. While holding the sprocket from turning with special tools MB990767 and MD998719, remove the camshaft sprocket bolt and sprocket.

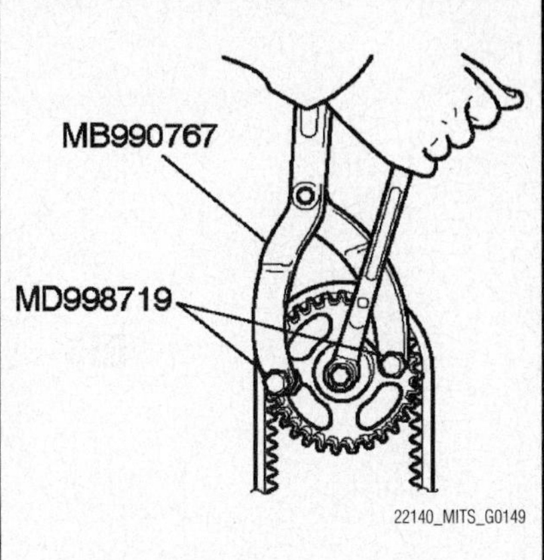

22140_MITS_G0149

Fig. 55 Use special tools MB990767 and MD998719 to remove/install the camshaft sprocket bolt and sprocket—Lancer 2.0L engine—2006–07

11. Remove the intake and exhaust rocker arm and shaft assemblies. Loosen both rocker arm assemblies gradually and evenly and remove the rocker shafts from the vehicle. Do NOT disassemble.
12. Remove the Camshaft Position (CMP) sensor connector.
13. Remove the CMP sensor support.
14. Remove the CMP sensor sensing cylinder.
15. Remove the camshaft.

➡Do not turn the crankshaft after the camshaft sprocket is removed.

To install:

16. Place the camshaft into position.
17. Install the CMP sensor sensing cylinder, support and connector.
18. Install the rocker arm and shaft assemblies. Refer to Rocker Arms/Shafts, Removal & Installation.
19. Apply engine oil to the camshaft oil seal lip. Use special tool MD998713 to install the camshaft oil seal.
20. Install the camshaft sprocket. Tighten the bolt to 58–72 ft. lbs. (78–98 Nm).
21. Install the timing belt front upper cover.
22. Install the spark plug guide.
23. Install the rocker arm cover, with a new gasket. Tighten the bolts to 23–31 inch lbs. (3–4 Nm).
24. Install the accelerator cable clamp.
25. Install the PCV and breather hoses.
26. Install the negative battery cable.
27. Start the engine and check for proper operation.

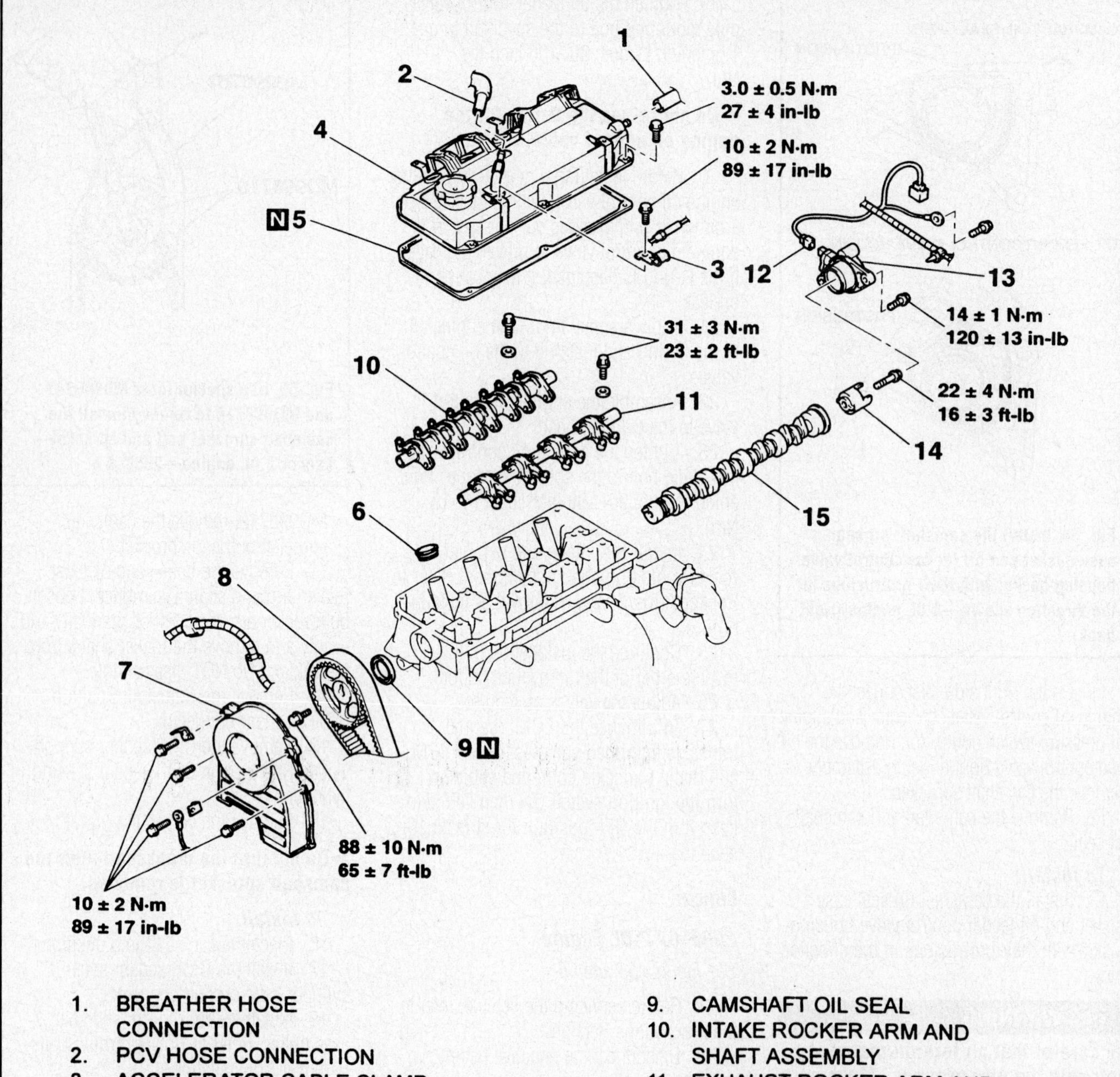

Fig. 56 Camshaft and related components—Lancer 2.0L engine—2006–07

2008 2.0L Engine

See Figures 57 through 64.

1. Before servicing the vehicle, refer to the Precautions Section.
2. Disconnect the negative battery cable.
3. Remove the engine undercover front A, B and side cover (RH).
4. Remove the air cleaner assembly.
5. Remove the ignition coils.
6. Remove the engine upper cover.
7. Remove the breather hose connection and the PCV hose connection.
8. Remove the control wiring harness connection.
9. Remove the rocker cover assembly mounting bolts in the order shown and remove the rocker cover assembly and gasket.

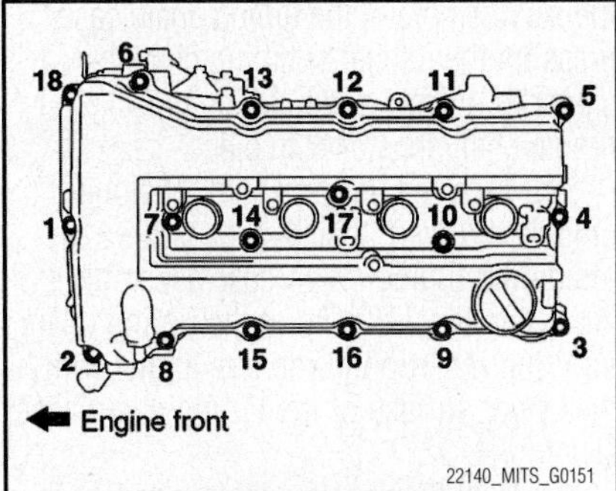

Fig. 57 Remove the rocker cover assembly mounting bolts in the order shown—Lancer 2.0L engine—2008

WARNING

Turn the crankshaft in a clockwise direction only.

10. Turn the crankshaft clockwise so that the camshaft sprocket timing marks become horizontal to the cylinder head upper surface, and set the cylinder number 1 to the Top Dead Center (TDC) of compression.
11. Check that the crankshaft pulley timing mark is in the 0° position of the ignition timing indicator of the timing chain case assembly.
12. Put paint marks on both the camshaft sprocket and timing chain at the position of camshaft sprocket timing chain mating mark (circular hole).
13. Remove the timing chain upper guide.
14. Remove the service hole bolt.
15. Insert a flat-tipped screwdriver through the service hole of the timing chain case, press up the timing chain tensioner ratchet to unlock, and keep the timing chain tensioner in this position.

➡Lightly press down the tail end of the flat-tipped screwdriver to press up the tip of the flat-tipped screwdriver inserted into the timing chain tensioner in order to unlock it.

16. With the timing chain tensioner unlocked, insert special tool MB992103 inside the timing chain case assembly along the tension side of the timing chain until the insertion guide line aligns with the upper surface of the timing chain case assembly.
17. With the special tool inserted up to the insertion guide line, press the special tool against the intake side camshaft sprocket and spread and hold the timing chain tension side guide.
18. Remove the flat-tipped precision screwdriver unlocking the timing chain tensioner.

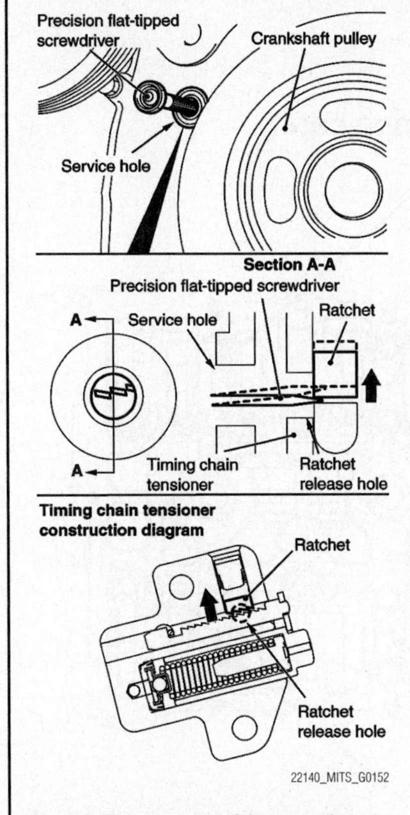

Fig. 58 Using a screwdriver to release the timing chain tensioner mechanism—Lancer 2.0L engine—2008

WARNING

The timing chain may snag on other parts. After removing the tension from the timing chain, never rotate the crankshaft.

19. With the timing chain tension side guide spread, hook the special tool over the hexagon part of the camshaft on the exhaust side, and turn the camshaft clockwise to apply slack to the timing chain between the camshaft sprockets.
20. Remove the mounting bolts of front camshaft bearing cap in the order shown in the figure and remove the front camshaft bearing cap assembly.

WARNING

When the camshaft bearing cap mounting bolts are loosened all at once, the mounting bolts jump out by spring force and the threads are damaged. Always loosen the mounting bolts in 4–5 steps.

21. Loosen the mounting bolts of the camshaft bearing caps in the order shown in the figure in 4–5 steps. Remove the camshaft bearing caps.

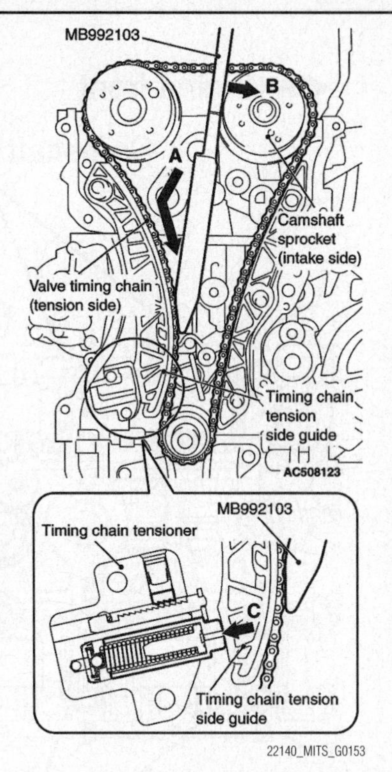

Fig. 59 With the timing chain tensioner unlocked, insert special tool MB992103 inside the timing chain case assembly along the tension side of the timing chain—Lancer 2.0L engine—2008

22. Slightly raise the transaxle side of the camshaft and camshaft sprocket assembly (exhaust side) by using the slack of the timing chain, and remove from the cam bearing.
23. Remove the timing chain from the camshaft and camshaft sprocket assembly (exhaust side) toward the timing chain case assembly, and remove the camshaft and camshaft sprocket assembly (exhaust side) toward the transaxle.
24. Remove special tool MB992103 inserted into the timing chain case assembly.

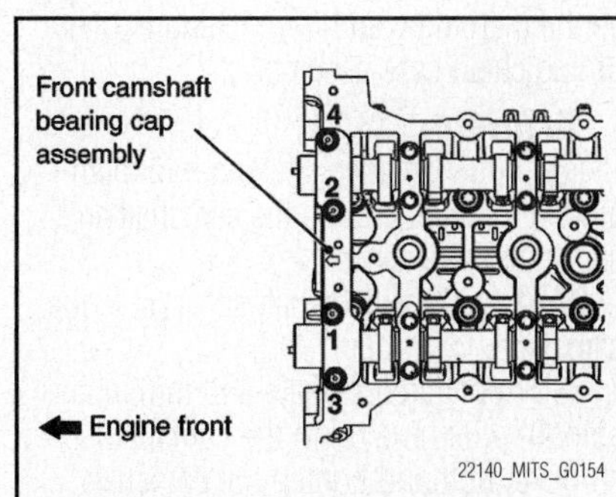

Fig. 60 Remove the mounting bolts of front camshaft bearing cap in the order shown—Lancer 2.0L engine—2008

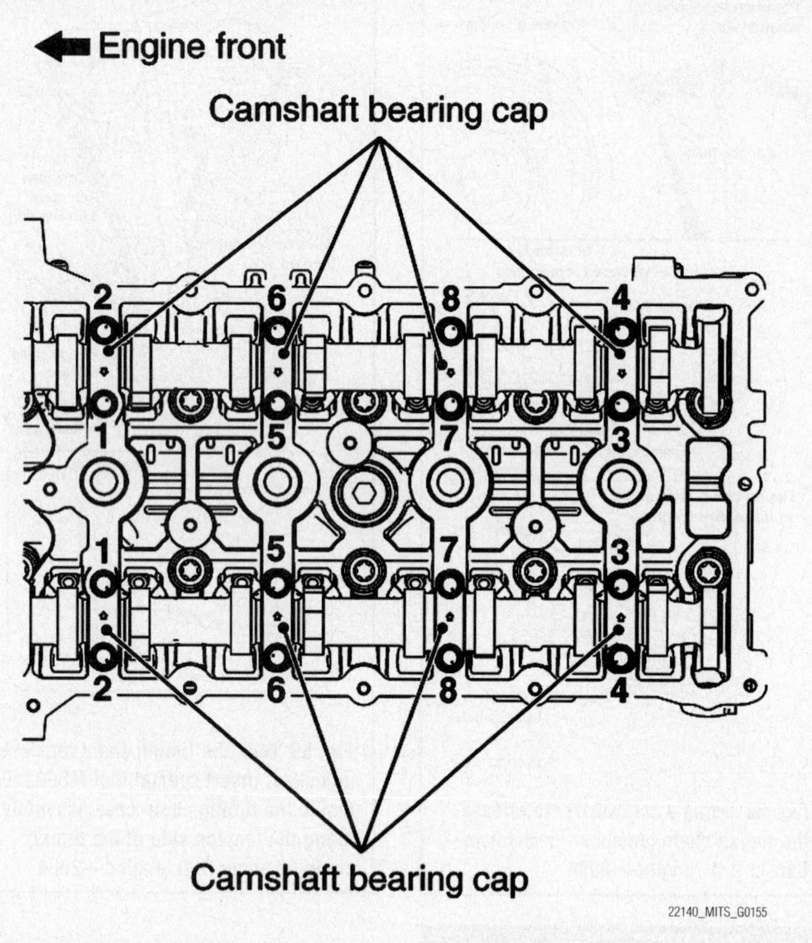

Fig. 61 Loosen the mounting bolts of the camshaft bearing caps in the order shown—Lancer 2.0L engine—2008

WARNING

The timing chain may snag on other parts. After removing the camshaft and camshaft sprocket assembly, never rotate the crankshaft.

25. After removing the camshaft and camshaft sprocket assembly (exhaust side), hang up the timing chain with a rope to prevent the timing chain from falling into the timing chain case assembly.

To install:

26. Apply an adequate and minimum amount of engine oil to the camshaft and camshaft sprocket.

27. Install the camshaft sprocket to the camshaft.

28. Apply an adequate and minimum amount of engine oil to the camshaft sprocket bolt and tighten the camshaft sprocket bolts to the specified torque. Tightening torque: 41–47 ft. lbs. (54–64 Nm).

29. Align the intake side paint mark of the timing chain, which was made at removal, with the paint mark of the intake side camshaft sprocket and install the camshaft sprocket to the timing chain.

30. Install the camshaft and camshaft sprocket assembly (intake side) to the cylinder head.

31. Install the camshaft bearing caps to the cylinder heads.

➡Because the thrust camshaft bearing cap and other camshaft bearing caps are the same in shape, check the bearing cap number and additionally its symbol to identify the intake and exhaust sides for correct installation.

32. Tighten each camshaft bearing cap mounting bolt to the specified torque in the order shown in the figure in 2–3 steps. Tightening torque: 99–115 inch lbs. (11–13 Nm).

33. In the same manner as removal, insert the flat-tipped screwdriver through the service hole of the timing chain case, press up the ratchet of timing chain tensioner to unlock, and hold the unlocked timing chain tensioner in place.

34. With the timing chain tensioner unlocked, insert special tool MB992103 inside the timing chain case assembly along the tension side of the timing chain until the insertion guide line aligns with the upper surface of the timing chain case assembly.

35. With the special tool inserted up to the insertion guide line, press the special tool against the intake side camshaft sprocket and spread and hold the timing chain tension side guide.

36. Remove the flat-tipped screwdriver unlocking the timing chain tensioner.

37. Pull up the camshaft and camshaft sprocket assembly (exhaust side) mounting area of the timing chain to provide allowance for easy installation of the camshaft and camshaft sprocket assembly (exhaust side) to the timing chain.

➡When installing the camshaft and camshaft sprocket assembly (exhaust side), be careful not to let the camshaft bearing which is installed to the front cam bearing deviate from its position.

38. Align the exhaust side paint mark of the timing chain, which was made during removal, with the paint mark of the exhaust side camshaft sprocket and install the timing chain to the camshaft sprocket.

39. Install the camshaft and camshaft sprocket assembly (exhaust side) to the cylinder head.

40. Remove the special tool inserted into the timing chain case assembly inside.

WARNING

When the mounting bolts are tightened with the front camshaft bearing cap tilted, the front camshaft bearing cap is damaged. Install the front camshaft bearing cap properly to the cylinder head and camshaft.

41. Install the front camshaft bearing cap to the cylinder head and temporarily tighten the camshaft bearing front cap to the specified torque in the order of the figure (1). Tightening torque: 11–15 ft. lbs. (14–20 Nm).

42. Tighten the front camshaft bearing cap again to the specified torque in the order of the figure (2). Tightening torque: 22–24 ft. lbs. (28–32 Nm).

43. After the front camshaft bearing cap installation, check that the paint markings of

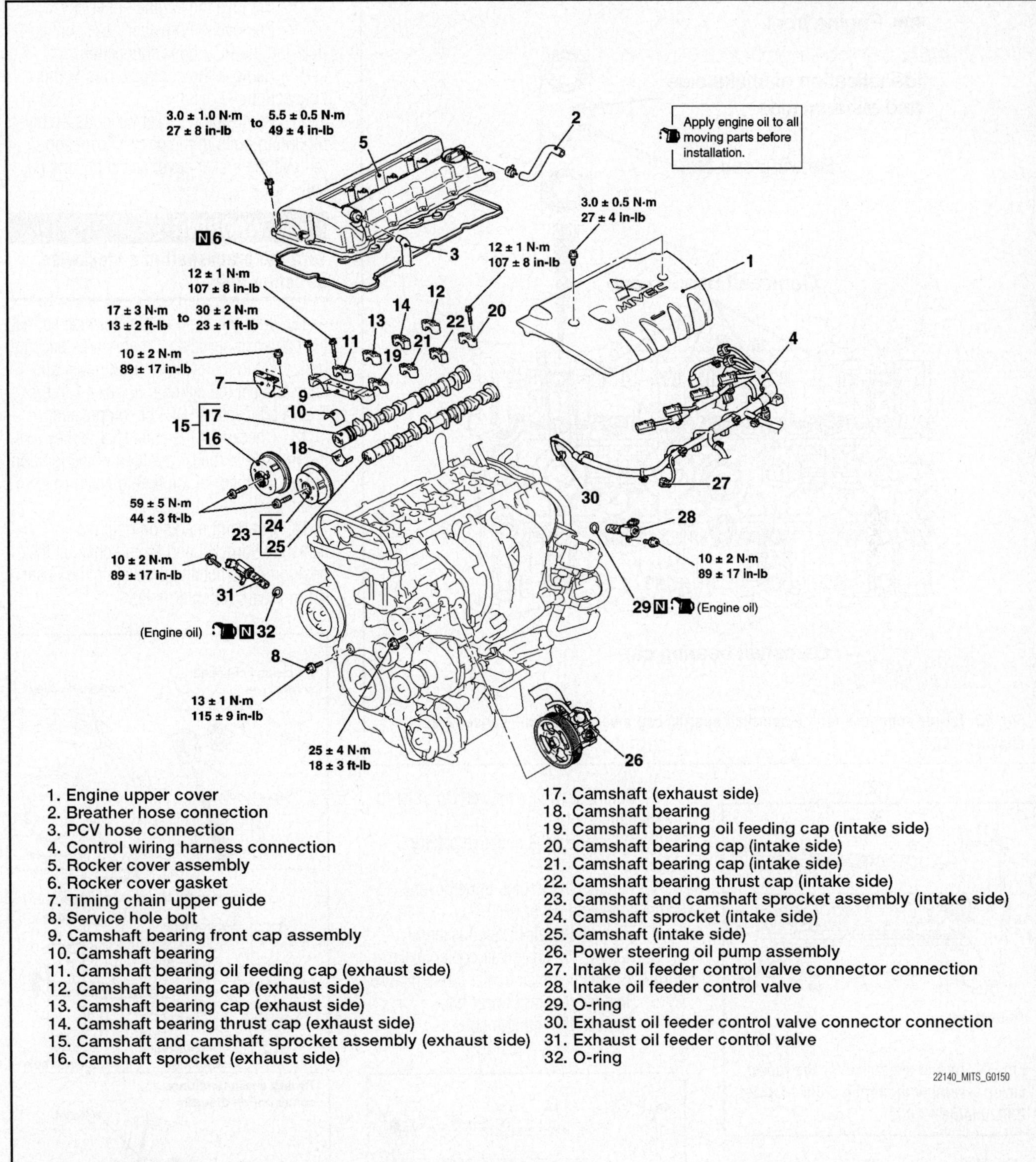

Fig. 62 Camshaft and related components—Lancer 2.0L engine—2008

the camshaft sprocket and the timing chain and the timing mark of the crankshaft pulley and the 0° position of ignition timing indicator are aligned properly.

44. Wipe off any sealant on the mating surface of the rocker cover assembly and the cylinder head and timing chain case assembly, and degrease the surface where the sealant is applied with white gasoline or an equivalent degreaser.

45. Apply sealant to the joint between the cylinder head and timing chain case assembly and install the rocker cover assembly to the cylinder head. Specified sealant: Three Bond® 1217G or equivalent.

➡**Install the rocker cover assembly within 3 minutes after the application of sealant.**

46. Tighten the rocker cover assembly mounting bolts to the specified torque in the order shown. Tightening torque: 19–35 inch lbs. (2–4 Nm).

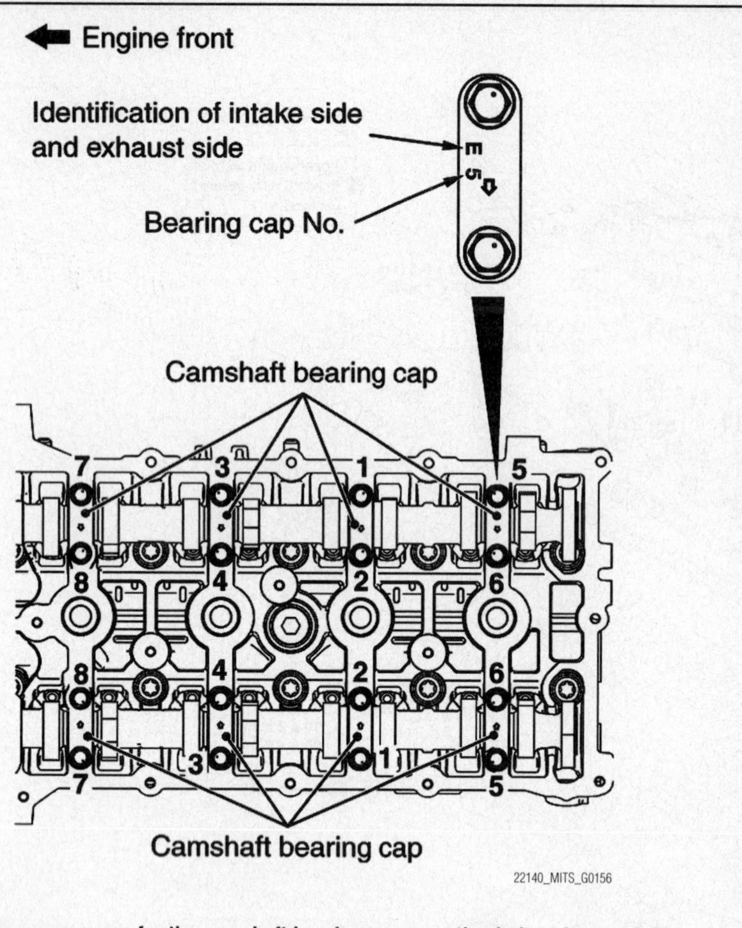

Fig. 63 Torque sequence for the camshaft bearing cap mounting bolts—Lancer 2.0L engine—2008

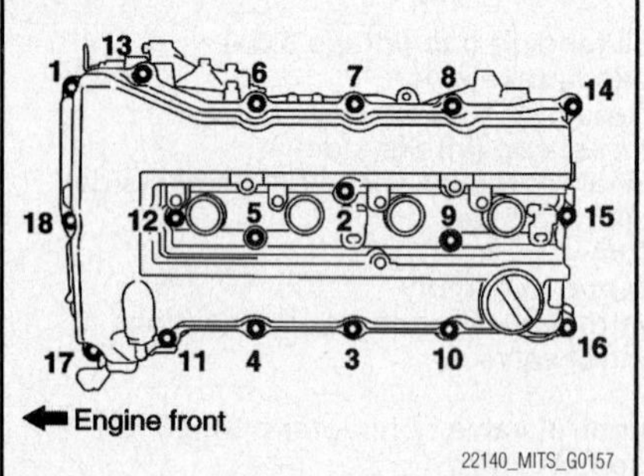

Fig. 64 Torque sequence for the rocker cover assembly mounting bolts—Lancer 2.0L engine—2008

47. Tighten the rocker cover assembly mounting bolts again to the specified torque in the order shown. Tightening torque: 45–53 inch lbs. (5–6 Nm).

48. Installation continues in the reverse order of the removal procedure.

Lancer Evolution

2008 2.0L Turbo Engine

See Figures 65 through 72.

1. Before servicing the vehicle, refer to the Precautions Section.
2. Disconnect the negative battery cable.
3. Remove the engine undercover. Remove the side cover (RH).
4. Remove the air cleaner assembly.
5. Remove the charge air cooler intake hose A, B and charge air cooler intake pipe A.
6. Remove the strut tower bar.
7. Remove the ignition coils.

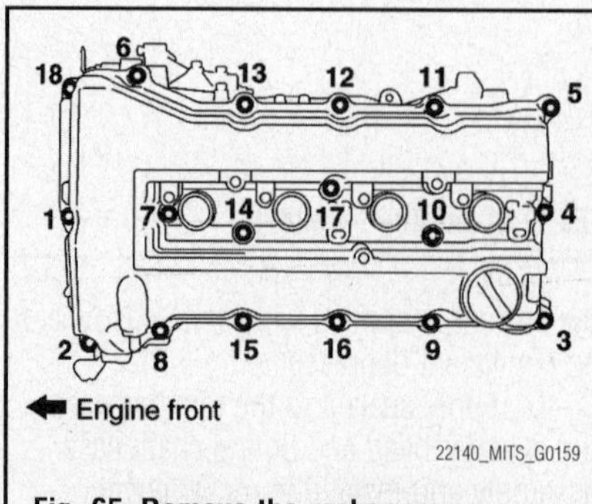

Fig. 65 Remove the rocker cover assembly mounting bolts in the order shown—Lancer Evolution—2008

8. Remove the engine upper cover.
9. Remove the breather hose connection and the PCV hose connection.
10. Remove the control wiring harness connection.
11. Remove the rocker cover assembly mounting bolts in the order shown and remove the rocker cover assembly and gasket.

WARNING

Turn the crankshaft in a clockwise direction only.

12. Turn the crankshaft clockwise so that the camshaft sprocket timing marks become horizontal to the cylinder head upper surface, and set the cylinder number 1 to the Top Dead Center (TDC) of compression.
13. Check that the crankshaft pulley timing mark is in the 0° position of the ignition timing indicator of the timing chain case assembly.
14. Put paint marks on both the camshaft sprocket and timing chain at the position of camshaft sprocket timing chain mating mark (circular hole).

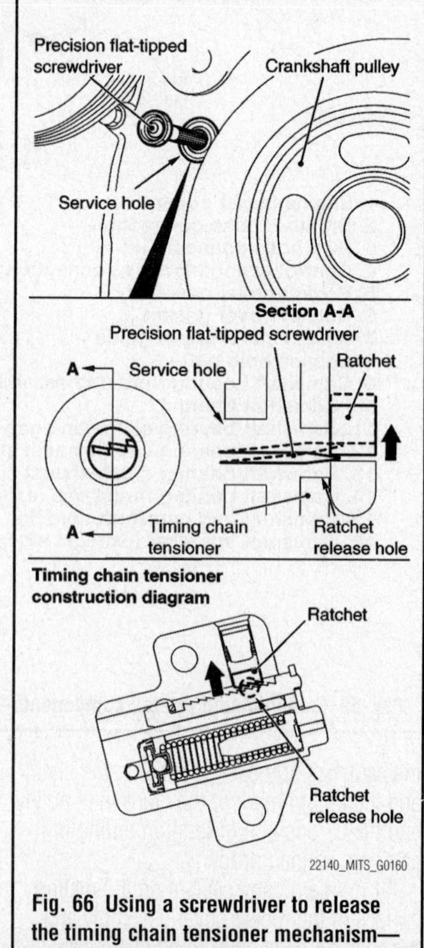

Fig. 66 Using a screwdriver to release the timing chain tensioner mechanism—Lancer Evolution—2008

15. Remove the timing chain upper guide.
16. Remove the service hole bolt.
17. Insert a flat-tipped screwdriver through the service hole of the timing chain case, press up the timing chain tensioner ratchet to unlock, and keep the timing chain tensioner in this position.

➡Lightly press down the tail end of the flat-tipped screwdriver to press up the tip of the flat-tipped screwdriver inserted into the timing chain tensioner in order to unlock it.

18. With the timing chain tensioner unlocked, insert special tool MB992103 inside the timing chain case assembly along the tension side of the timing chain until the insertion guide line aligns with the upper surface of the timing chain case assembly.
19. With the special tool inserted up to the insertion guide line, press the special tool against the intake side camshaft sprocket and spread and hold the timing chain tension side guide.

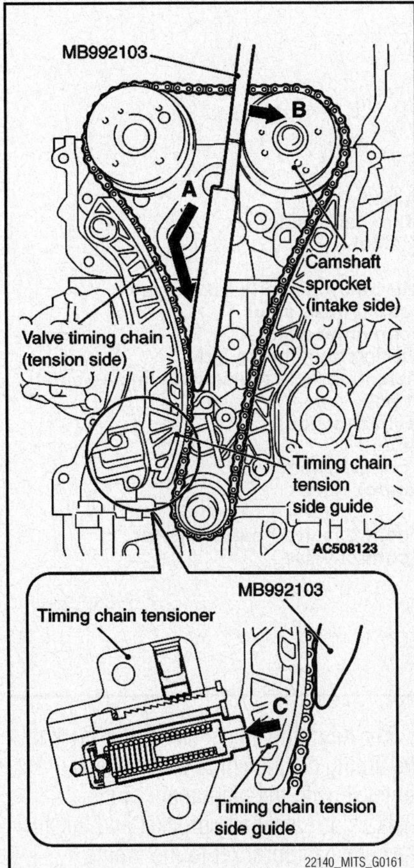

Fig. 67 With the timing chain tensioner unlocked, insert special tool MB992103 inside the timing chain case assembly along the tension side of the timing chain—Lancer Evolution—2008

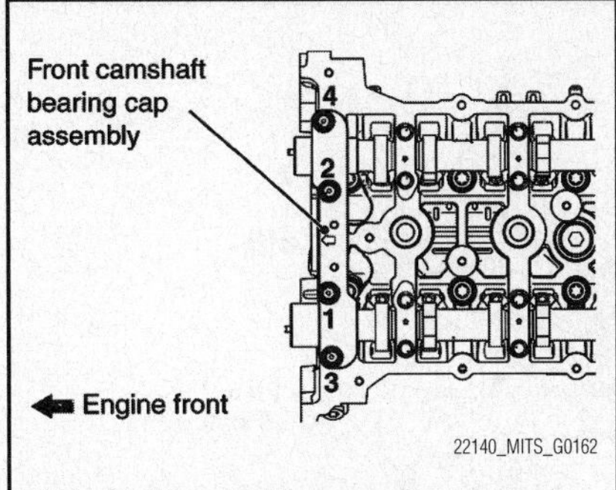

Fig. 68 Remove the mounting bolts of front camshaft bearing cap in the order shown—Lancer Evolution—2008

20. Remove the flat-tipped precision screwdriver unlocking the timing chain tensioner.

⁂ WARNING

The timing chain may snag on other parts. After removing the tension from the timing chain, never rotate the crankshaft.

21. With the timing chain tension side guide spread, hook the special tool over the hexagon part of the camshaft on the exhaust side, and turn the camshaft clockwise to apply slack to the timing chain between the camshaft sprockets.

22. Remove the mounting bolts of front camshaft bearing cap in the order shown in the figure and remove the front camshaft bearing cap assembly.

⁂ WARNING

When the camshaft bearing cap mounting bolts are loosened all at once, the mounting bolts jump out by spring force and the threads are damaged. Always loosen the mounting bolts in 4–5 steps.

23. Loosen the mounting bolts of the camshaft bearing caps in the order shown in the figure in 4–5 steps. Remove the camshaft bearing caps.
24. Slightly raise the transaxle side of the camshaft and camshaft sprocket assembly (exhaust side) by using the slack of the timing chain, and remove from the cam bearing.
25. Remove the timing chain from the camshaft and camshaft sprocket assembly (exhaust side) toward the timing chain case assembly, and remove the camshaft and camshaft sprocket assembly (exhaust side) toward the transaxle.
26. Remove special tool MB992103 inserted into the timing chain case assembly.

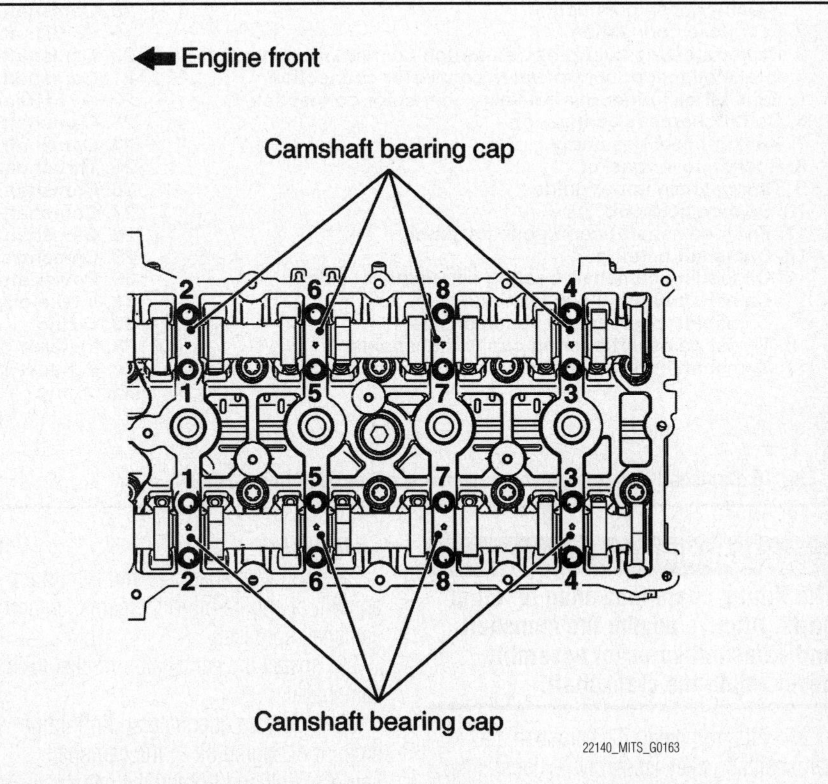

Fig. 69 Loosen the mounting bolts of the camshaft bearing caps in the order shown—Lancer Evolution—2008

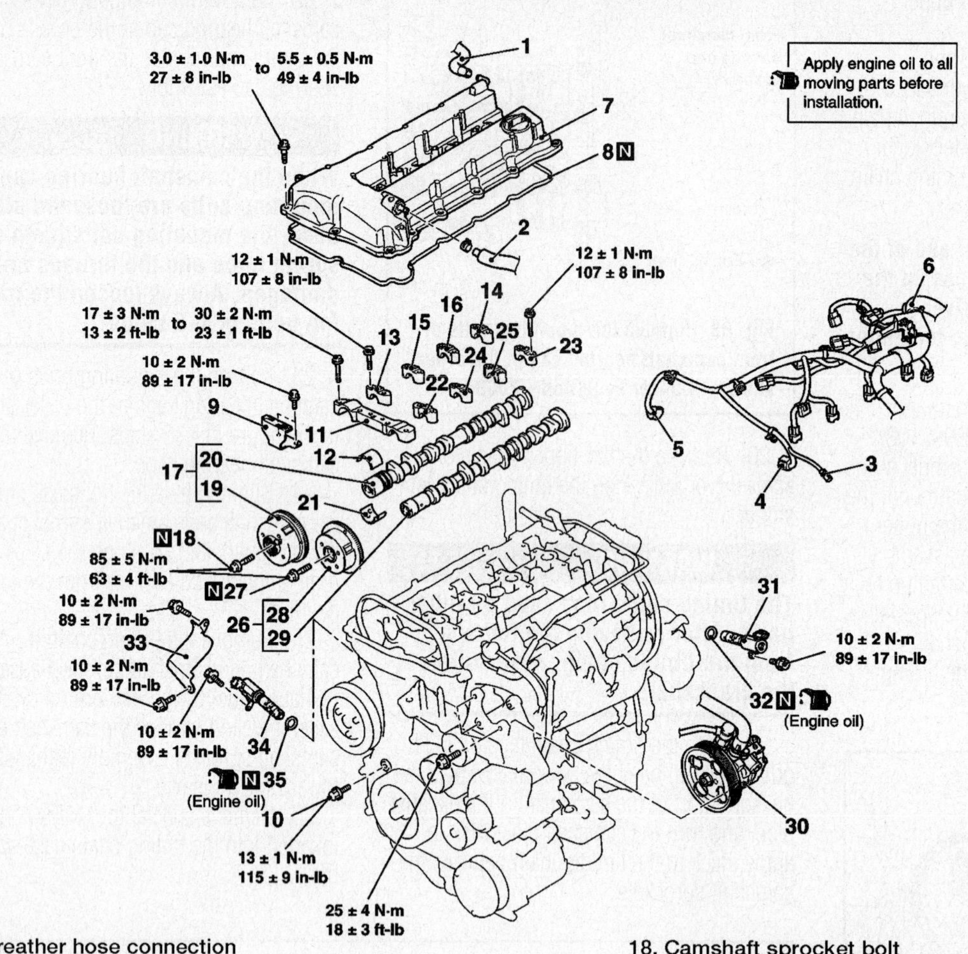

1. Breather hose connection
2. PCV hose connection
3. Power steering fluid pressure switch connector connection
4. Intake oil feeder control valve connector connection
5. Exhaust oil feeder control valve connector connection
6. Control harness connection
7. Rocker cover assembly
8. Rocker cover gasket
9. Timing chain upper guide
10. Service hole bolt
11. Front camshaft bearing cap assembly
12. Camshaft bearing
13. Oil feeding camshaft bearing cap (exhaust side)
14. Camshaft bearing cap (exhaust side)
15. Camshaft bearing cap (exhaust side)
16. Thrust camshaft bearing cap (exhaust side)
17. Camshaft and camshaft sprocket assembly (exhaust side)
18. Camshaft sprocket bolt
19. Camshaft sprocket (exhaust side)
20. Camshaft (exhaust side)
21. Camshaft bearing
22. Oil feeding camshaft bearing cap (intake side)
23. Camshaft bearing cap (intake side)
24. Camshaft bearing cap (intake side)
25. Thrust camshaft bearing cap (intake side)
26. Camshaft and camshaft sprocket assembly (intake side)
27. Camshaft sprocket bolt
28. Camshaft sprocket (intake side)
29. Camshaft (intake side)
30. Power steering oil pump assembly
31. Intake oil feeder control valve
32. O-ring
33. Exhaust oil feeder control valve heat protector
34. Exhaust oil feeder control valve
35. O-ring

22140_MITS_G0158

Fig. 70 Camshaft and related components—Lancer Evolution—2008

⁂ WARNING

The timing chain may snag on other parts. After removing the camshaft and camshaft sprocket assembly, never rotate the crankshaft.

27. After removing the camshaft and camshaft sprocket assembly (exhaust side), hang up the timing chain with a rope to prevent the timing chain from falling into the timing chain case assembly.

To install:

28. Apply an adequate and minimum amount of engine oil to the camshaft and camshaft sprocket.

29. Install the camshaft sprocket to the camshaft.

30. Apply an adequate and minimum amount of engine oil to the camshaft sprocket bolt and tighten the camshaft sprocket bolts to the specified torque. Tightening torque: 41–47 ft. lbs. (54–64 Nm).

31. Align the intake side paint mark of the timing chain, which was made at removal, with the paint mark of the intake side camshaft sprocket and install the camshaft sprocket to the timing chain.

32. Install the camshaft and camshaft sprocket assembly (intake side) to the cylinder head.

33. Install the camshaft bearing caps to the cylinder heads.

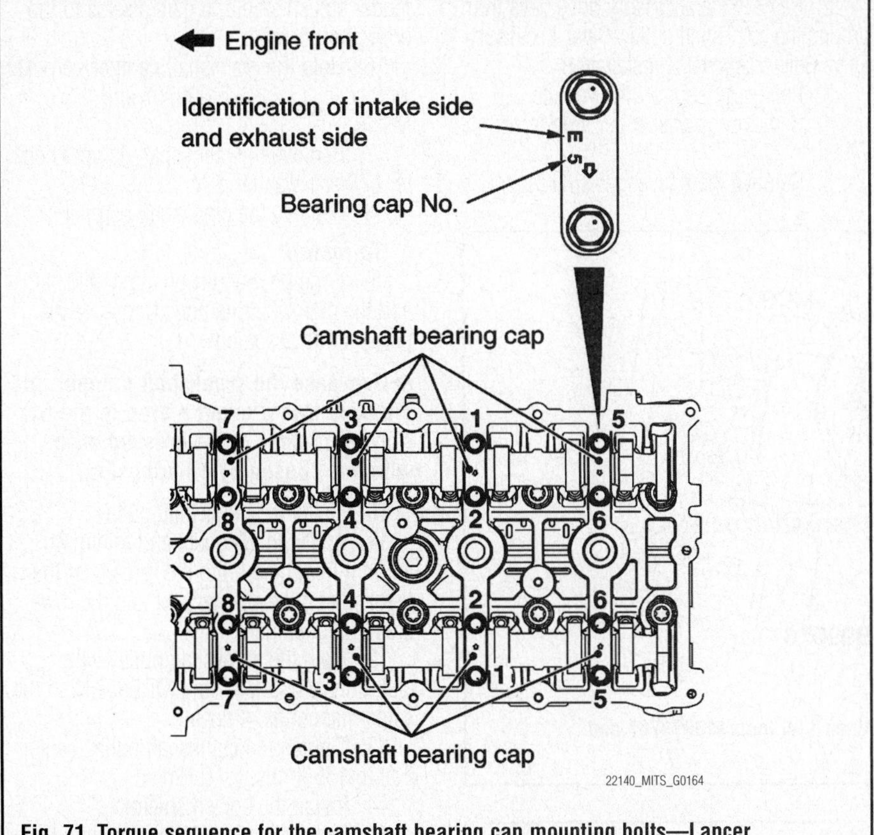

Fig. 71 Torque sequence for the camshaft bearing cap mounting bolts—Lancer Evolution—2008

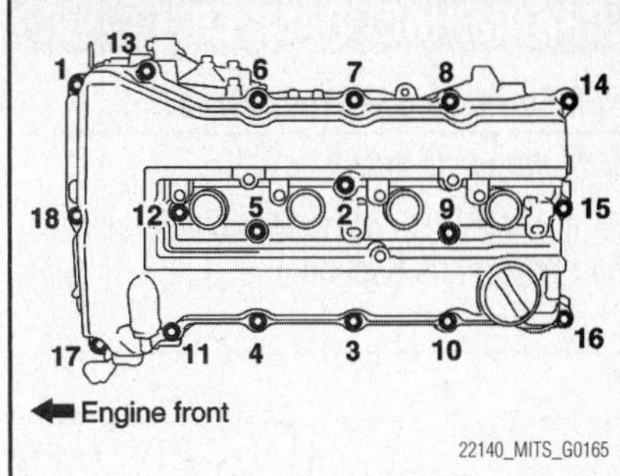

Fig. 72 Torque sequence for the rocker cover assembly mounting bolts—Lancer Evolution—2008

➡Because the thrust camshaft bearing cap and other camshaft bearing caps are the same in shape, check the bearing cap number and additionally its symbol to identify the intake and exhaust sides for correct installation.

34. Tighten each camshaft bearing cap mounting bolt to the specified torque in the order shown in the figure in 2–3 steps. Tightening torque: 99–115 inch lbs. (11–13 Nm).
35. In the same manner as removal, insert the flat-tipped screwdriver through the service hole of the timing chain case, press up the ratchet of timing chain tensioner to unlock, and hold the unlocked timing chain tensioner in place.
36. With the timing chain tensioner unlocked, insert special tool MB992103 inside the timing chain case assembly along the tension side of the timing chain until the insertion guide line aligns with the upper surface of the timing chain case assembly.
37. With the special tool inserted up to the insertion guide line, press the special tool against the intake side camshaft sprocket and spread and hold the timing chain tension side guide.
38. Remove the flat-tipped screwdriver unlocking the timing chain tensioner.
39. Pull up the camshaft and camshaft sprocket assembly (exhaust side) mounting area of the timing chain to provide allowance for easy installation of the camshaft and camshaft sprocket assembly (exhaust side) to the timing chain.

➡When installing the camshaft and camshaft sprocket assembly (exhaust side), be careful not to let the camshaft bearing which is installed to the front cam bearing deviate from its position.

40. Align the exhaust side paint mark of the timing chain, which was made during removal, with the paint mark of the exhaust side camshaft sprocket and install the timing chain to the camshaft sprocket.
41. Install the camshaft and camshaft sprocket assembly (exhaust side) to the cylinder head.
42. Remove the special tool inserted into the timing chain case assembly inside.

✲✲ WARNING

When the mounting bolts are tightened with the front camshaft bearing cap tilted, the front camshaft bearing cap is damaged. Install the front camshaft bearing cap properly to the cylinder head and camshaft.

43. Install the front camshaft bearing cap to the cylinder head and temporarily tighten the camshaft bearing front cap to the specified torque in the order of the figure (1). Tightening torque: 11–15 ft. lbs. (14–20 Nm).
44. Tighten the front camshaft bearing cap again to the specified torque in the order of the figure (2). Tightening torque: 22–24 ft. lbs. (28–32 Nm).
45. After the front camshaft bearing cap installation, check that the paint markings of the camshaft sprocket and the timing chain and the timing mark of the crankshaft pulley and the 0° position of ignition timing indicator are aligned properly.
46. Wipe off any sealant on the mating surface of the rocker cover assembly and the cylinder head and timing chain case assembly, and degrease the surface where the sealant is applied with white gasoline or an equivalent degreaser.
47. Apply sealant to the joint between the cylinder head and timing chain case assembly and install the rocker cover assembly to the cylinder head. Specified sealant: Three Bond® 1217G or equivalent.

➡Install the rocker cover assembly within 3 minutes after the application of sealant.

48. Tighten the rocker cover assembly mounting bolts to the specified torque in the order shown. Tightening torque: 19–35 inch lbs. (2–4 Nm).
49. Tighten the rocker cover assembly mounting bolts again to the specified torque in the order shown. Tightening torque: 45–53 inch lbs. (5–6 Nm).
50. Installation continues in the reverse order of the removal procedure.

CAMSHAFT BEARING REPLACEMENT

Check each bearing for damage. If the bearing surface is excessively damaged, replace the cylinder head assembly or camshaft bearing cap, as necessary.

CRANKSHAFT DAMPER

REMOVAL & INSTALLATION

See Figures 73 and 74.

1. Before servicing the vehicle, refer to the Precautions Section.
2. Remove the accessory drive belts from around the crankshaft pulley. Refer Accessory Drive Belts, Removal & Installation.
3. Raise and support the vehicle.
4. Remove the passenger side front wheel.
5. Remove the passenger side inner fender splash shield to gain access to the crankshaft damper.
6. Hold the crankshaft damper sprocket with special tools MB990767 and MD998719.
7. Remove the crankshaft damper center bolt and washer.
8. Remove the crankshaft damper.

To install:

9. Wipe off any dirt on the crankshaft and the crankshaft damper. Degrease the parts before assembly.

➡Degrease the crankshaft damper and crankshaft to prevent a drop in the friction coefficient of the pressed area, which is caused by oil adhesion.

10. Install the crankshaft pulley.
11. Apply an adequate and minimum amount of engine oil to the threads of the crankshaft pulley center bolt and the lower area of the flange.
12. Hold the crankshaft pulley with special tools MB990767 and MD998719 in the same manner as removal.
13. Tighten the crankshaft pulley center bolt to 155 ft. lbs. (210 Nm).
14. Install the splash shield.
15. Install the wheel, then carefully lower the vehicle.
16. Install the accessory drive belts. Refer Accessory Drive Belts, Removal & Installation.
17. Connect the negative battery cable.

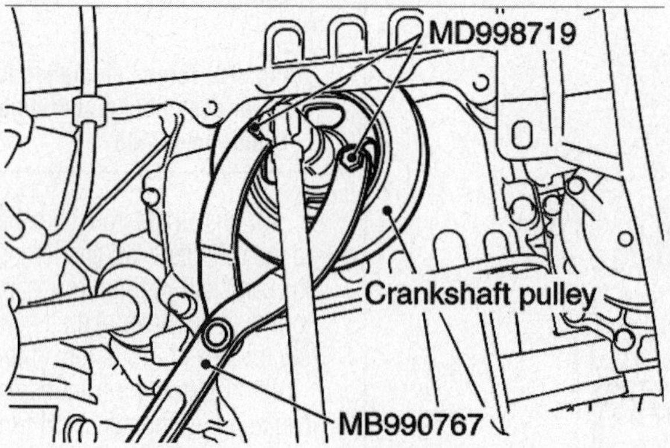

Fig. 73 Hold the crankshaft damper sprocket with special tools MB990767 and MD998719 for removal

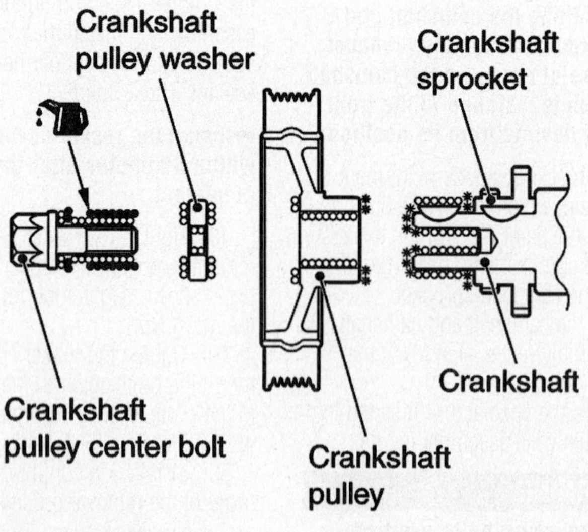

Fig. 74 Wipe off any dirt on the crankshaft and the crankshaft damper and degrease the parts before assembly

CRANKSHAFT FRONT SEAL

REMOVAL & INSTALLATION

Eclipse & Galant

2.4L Engine

See Figures 75 and 76.

1. Before servicing the vehicle, refer to the Precautions Section.
2. Disconnect the negative battery cable.
3. Remove the timing belt. Refer to Timing Belt and Sprockets, Removal & Installation.
4. Remove the crankshaft sprocket.
5. Carefully pry the oil seal out of the front case. Be careful not to damage the oil seal bore or the crankshaft sealing surface.

To install:

6. Apply clean engine oil to the oil seal lip. Using a seal driver, install the oil seal.
7. Install the crankshaft sprocket.
8. Install the timing belt. Refer to Timing Belt and Sprockets, Removal & Installation.
9. Install the negative battery cable.

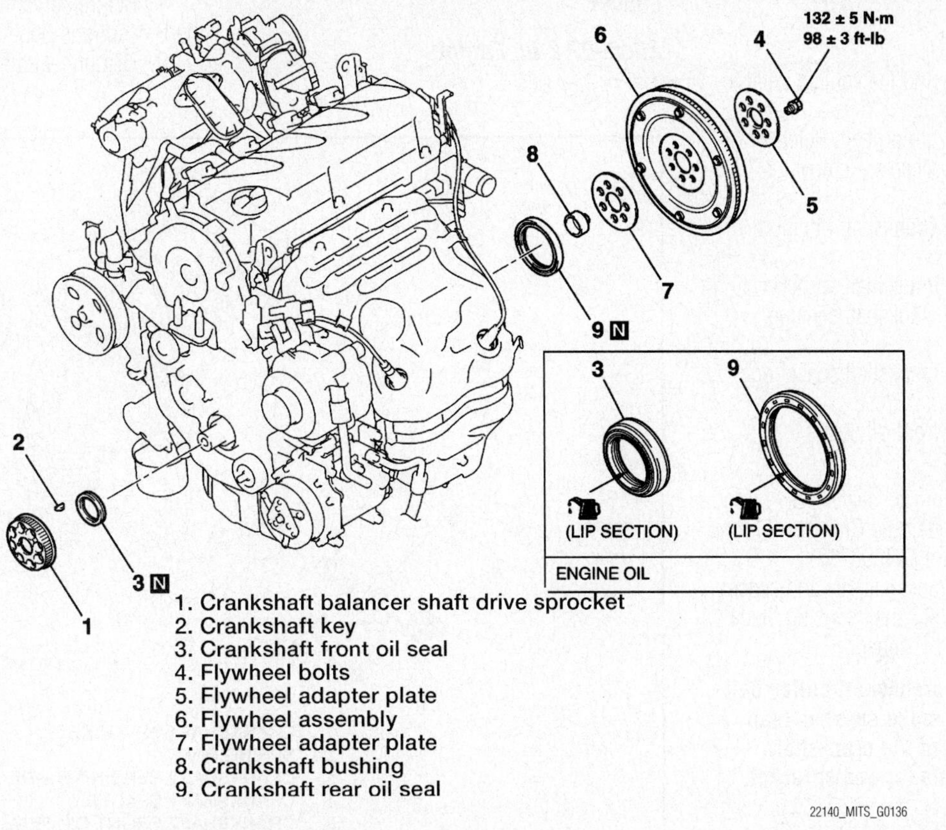

Fig. 75 Expanded view of the crankshaft oil seals (front and rear)—2.4L engine (M/T)

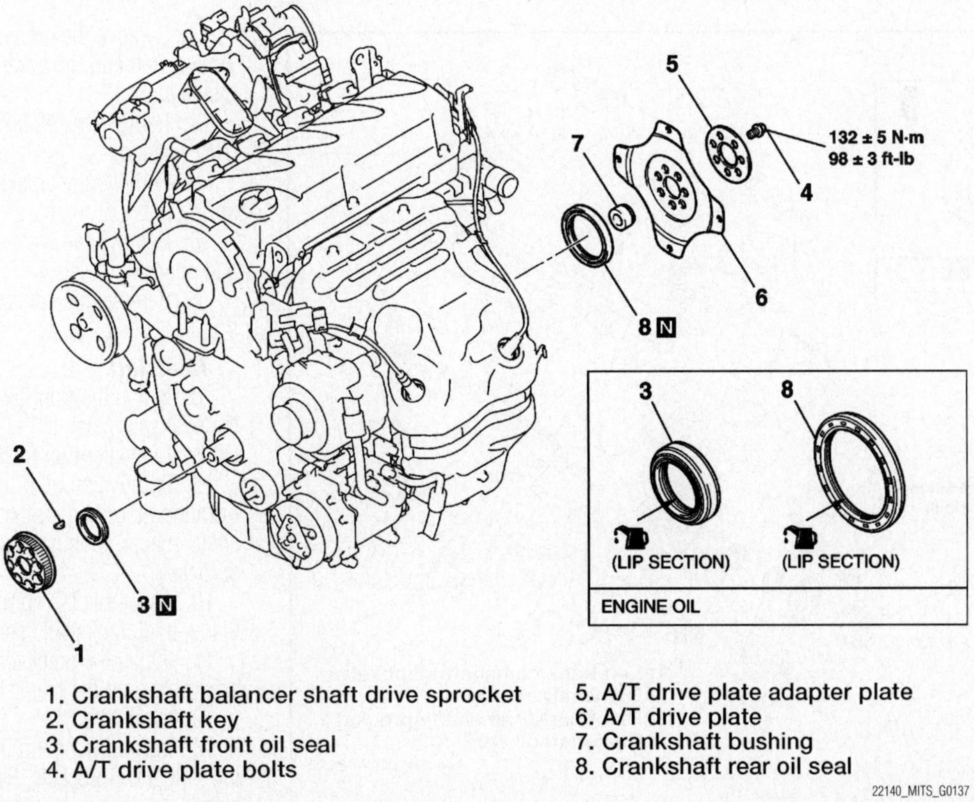

Fig. 76 Expanded view of the crankshaft oil seals (front and rear)—2.4L engine (A/T)

3.8L Engine

See Figure 77.

1. Before servicing the vehicle, refer to the Precautions Section.
2. Remove the timing belt. Refer to Timing Belt and Sprockets, Removal & Installation.
3. Remove the Crankshaft Position (CKP) sensor.
4. Remove the crankshaft sprocket.
5. Remove the crankshaft sensing blade.
6. Remove the crankshaft spacer and key.
7. Remove the front oil seal.

To install:

8. Install the front oil seal. Apply oil to the seal and install using a Crankshaft Front Oil Seal Installer tool MD998717.
9. Install the crankshaft key and spacer.
10. Install the crankshaft sensing blade.
11. Install the CKP sensor.

➡To be sure the crankshaft pulley bolt does not loosen, make sure to clean the mating areas of the crankshaft, spacer, sensing blade, and sprocket.

12. Install the crankshaft sprocket.
13. Install the timing belt. Refer to Timing Belt and Sprockets, Removal & Installation.

ENGINE OIL
8.5 ± 0.5 N·m
76 ± 4 in-lb
1. Left bank camshaft sprocket
2. Camshaft oil seal
3. Right bank camshaft sprocket
4. Camshaft oil seal
22140_MITS_G0138

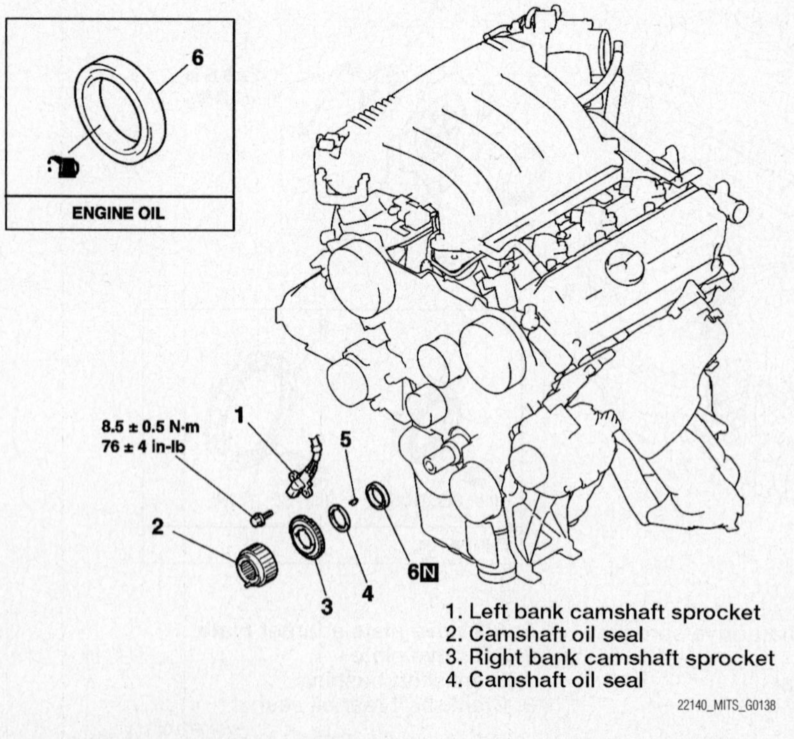

Fig. 77 Expanded view of the crankshaft oil front seal—3.8L engine

Lancer

2006–07 2.0L Engine

See Figure 78.

1. Before servicing the vehicle, refer to the Precautions Section.
2. Disconnect the negative battery cable.

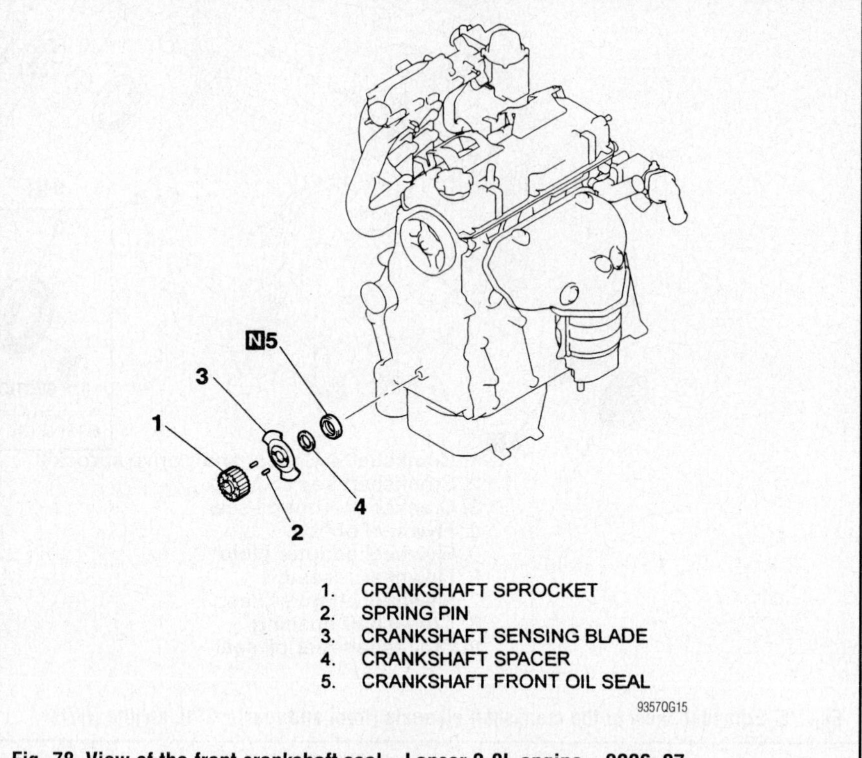

Fig. 78 View of the front crankshaft seal—Lancer 2.0L engine—2006–07

3. Remove the timing belt. Refer to Timing Belt and Sprockets, Removal & Installation.
4. Remove the Crankshaft Position (CKP) sensor.
5. Remove the crankshaft sprocket and spring pin.
6. Remove the crankshaft sensing blade and spacer.
7. Remove the crankshaft front oil seal.

To install:

8. Install the front oil seal. Apply oil to the seal and install using a Crankshaft Front Oil Seal Installer tool MD998717.
9. Remove all oil or other lubricants from the mounting surfaces on the crankshaft, spacer, sensing blade, and crankshaft sprocket.
10. Assemble the spring pin, sensing blade, and crankshaft spacer together.
11. Install the crankshaft sprocket assembly onto the crankshaft.
12. Install the CKP.
13. Install the timing belt. Refer to Timing Belt and Sprockets, Removal & Installation.
14. Connect the negative battery cable.

2008 2.0L Engine

See Figure 79.

1. Before servicing the vehicle, refer to the Precautions Section.
2. Disconnect the negative battery cable.
3. Remove the timing chain. Refer to Timing Chain and Sprockets, Removal & Installation.
5. Carefully pry the oil seal out of the front case. Be careful not to damage the oil seal bore or the crankshaft sealing surface.

To install:

6. Apply a small amount of engine oil to the entire inner diameter of the oil seal lip.
7. Using special tool MB991448, press in the crankshaft front oil seal up to the chamfered surface of timing chain case.
8. Install the crankshaft sprocket.
9. Install the timing chain. Refer to Timing Chain and Sprockets, Removal & Installation.
10. Connect the negative battery cable.

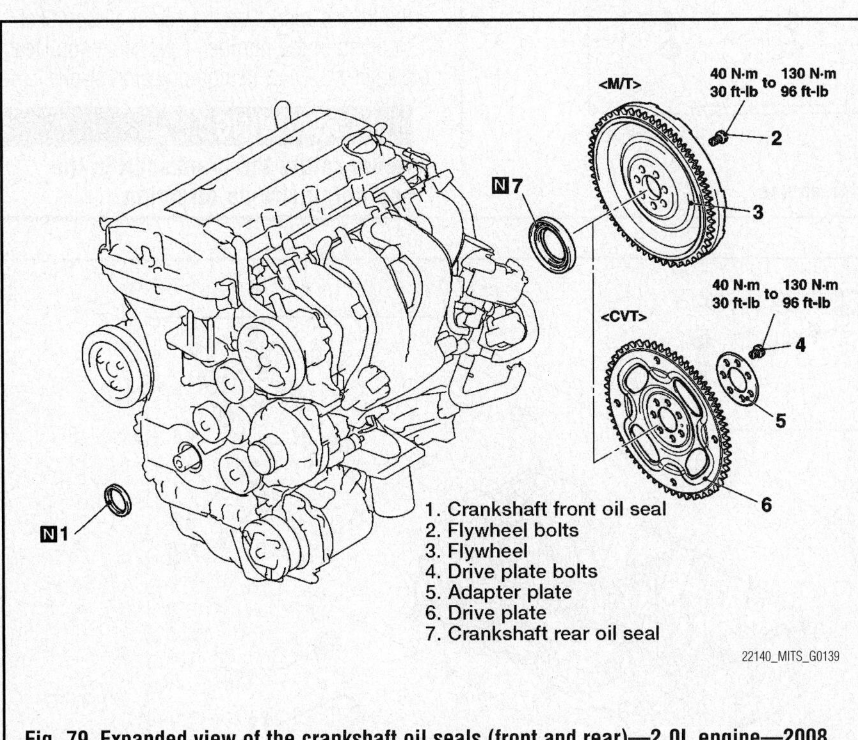

Fig. 79 Expanded view of the crankshaft oil seals (front and rear)—2.0L engine—2008

4. Remove the crankshaft sprocket.

To install:

5. Apply oil to the seal and install using a Crankshaft Front Oil Seal Installer tool MB991448, press in the crankshaft front oil seal up to the chamfered surface of timing chain case.
6. Install the crankshaft sprocket.
7. Install the timing chain. Refer to Timing Chain and Sprockets, Removal & Installation.
8. Connect the negative battery cable.

Lancer Evolution

2008 2.0L Turbo Engine

See Figure 80.

1. Before servicing the vehicle, refer to the Precautions Section.
2. Disconnect the negative battery cable.
3. Remove the timing chain. Refer to Timing Chain and Sprockets, Removal & Installation.
4. Remove the crankshaft sprocket.

CYLINDER HEAD

REMOVAL & INSTALLATION

Eclipse & Galant

2.4L Engine

See Figures 81 through 83.

1. Before servicing the vehicle, refer to the Precautions Section.
2. Relieve the fuel system pressure.
3. Disconnect the negative battery cable.
4. Drain the cooling system.
5. Remove the ECM, if the vehicle is equipped with a manual transaxle. Remove the PCM, if the vehicle is equipped with an automatic transaxle.
6. Remove the air cleaner assembly.
7. Remove the battery and battery tray.
8. Disconnect the control wiring harness, radiator hose lower clamp, and the water hose clamp.
9. Disconnect the EVAP hose connection. Disconnect the brake booster vacuum hose, pressure hose clamp, and knock sensor connector.

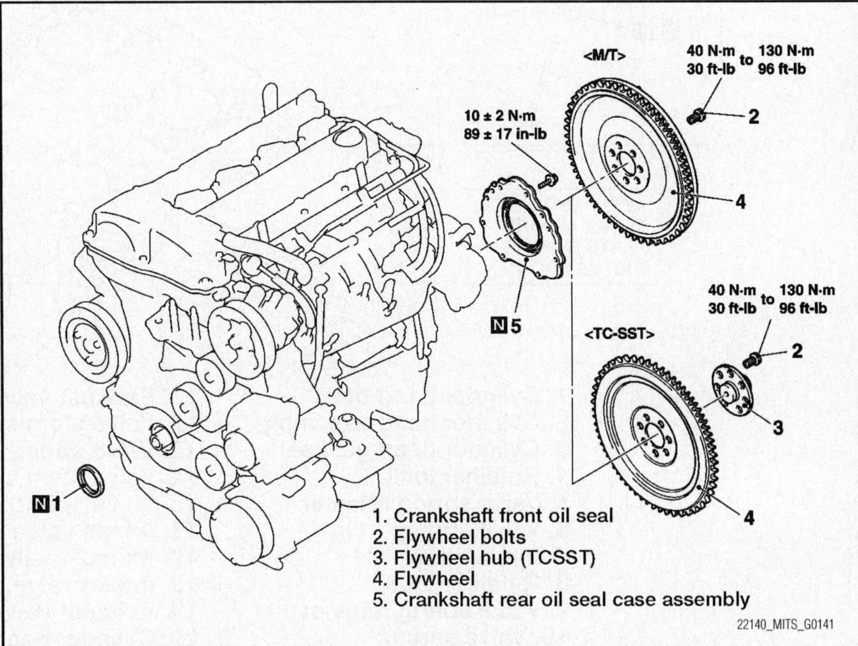

Fig. 80 Expanded view of the crankshaft oil seals (front and rear)—Lancer Evolution—2008

ENGINE FRONT
3
5
10
8
2
1
7
9
6
4
09482_GALA_G0036

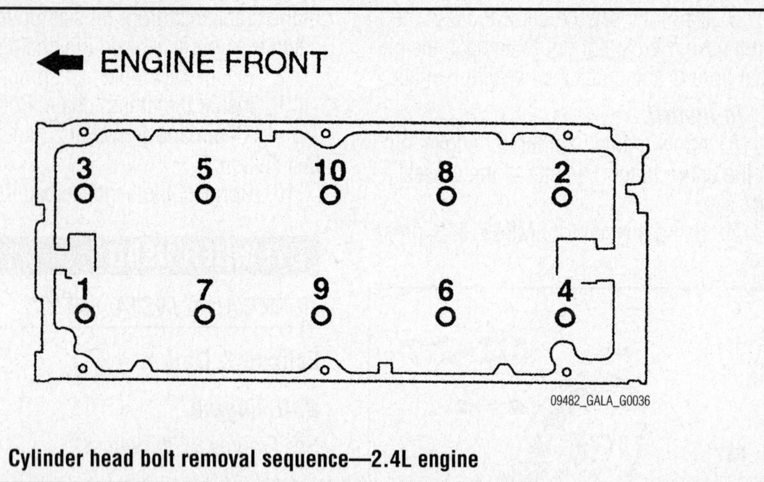

Fig. 81 Cylinder head bolt removal sequence—2.4L engine

10. Remove the engine oil dipstick and dipstick guide.
11. Remove the intake manifold stay.
12. Remove the exhaust manifold.
13. Remove the timing belt upper cover.
14. Remove the engine front mounting bracket.
15. Turn the crankshaft clockwise, align the timing marks on the camshaft sprocket to position the number 1 piston to Top Dead Center (TDC) of its compression stroke.

WARNING

Never rotate the crankshaft in the counterclockwise direction.

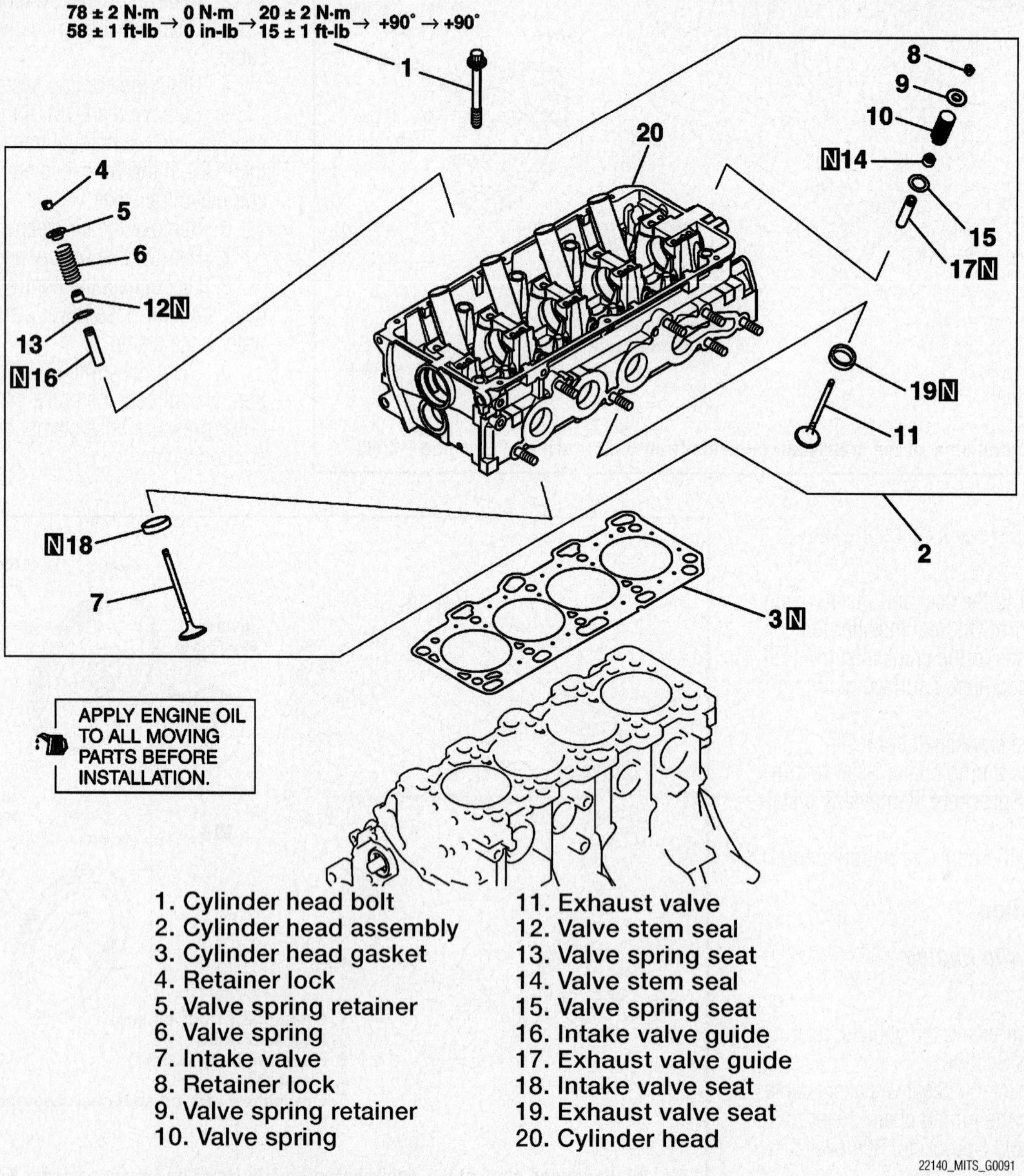

78 ± 2 N·m → 0 N·m → 20 ± 2 N·m → +90° → +90°
58 ± 1 ft-lb → 0 in-lb → 15 ± 1 ft-lb
APPLY ENGINE OIL TO ALL MOVING PARTS BEFORE INSTALLATION.
1. Cylinder head bolt
2. Cylinder head assembly
3. Cylinder head gasket
4. Retainer lock
5. Valve spring retainer
6. Valve spring
7. Intake valve
8. Retainer lock
9. Valve spring retainer
10. Valve spring
11. Exhaust valve
12. Valve stem seal
13. Valve spring seat
14. Valve stem seal
15. Valve spring seat
16. Intake valve guide
17. Exhaust valve guide
18. Intake valve seat
19. Exhaust valve seat
20. Cylinder head
22140_MITS_G0091

Fig. 82 Expanded view of cylinder head components—2.4L engine

16. Remove the timing belt undercover rubber plug, using tool MD998738. Screw the tool in until it contacts the timing belt tensioner arm.

17. Secure the camshaft and valve timing belt with wire to prevent slippage between the camshaft sprocket and valve timing belt.

18. Remove the camshaft sprocket. Do not turn the crankshaft after the camshaft sprocket is removed.

19. Remove the upper radiator hose connection, water cooler hose connection (automatic transaxle), and water hose connection. Disconnect the high pressure hose connection.

20. Remove the valve cover.

21. Using special tool MB991654, loosen the cylinder head bolts. Loosen each bolt evenly, little by little, in 2–3 steps and in the proper bolt removal sequence.

22. Remove the cylinder head from the engine. Discard the cylinder head gasket.

➡If the cylinder head bolts cannot be pulled out due to the washer being trapped in the valve spring, raise the bolt slightly, then remove it while holding it with a magnet.

To install:

23. Thoroughly clean the mating surfaces of the head and block.

24. Place a new head gasket on the cylinder block. Position the gasket on the block so that the identification mark "381" is at the top surface and on the exhaust side.

25. Inspect the cylinder head bolt length prior to installation. If the length exceeds 3.91 inches (99.4mm), the bolt must be replaced. Apply a small amount of engine oil to the thread section and washer of the bolt.

26. Position the cylinder head on the engine. Install and torque the retaining bolts to specification and in the proper sequence.

a. Step 1: 57–59 ft. lbs. (76–80 Nm).

b. Step 2: Loosen all bolts fully in the reverse order of tightening.

c. Step 3: Retighten the loosened bolts to the specified torque in the tightening sequence shown. Tightening torque: 14–16 ft. lbs. (18–22 Nm).

d. Step 4: Make a paint mark across each bolt head and cylinder head. Tighten the cylinder head bolts 90° in the specified order.

e. Step 5: Tighten the bolts another 90° degrees in the same order and check that the paint marks on the cylinder head bolt are aligned with the paint marks on the cylinder head.

ENGINE FRONT

8 6 1 3 9
10 4 2 5 7

09482_GALA_G0037

Fig. 83 Cylinder head bolt torque sequence—2.4L engine

⁂ WARNING

If the bolt is turned less than 90°, proper fastening performance may not be achieved. Be careful to turn each bolt exactly 90°. If the bolt is over-tightened, loosen the bolt completely and then retighten it by repeating the tightening procedure from step 1.

27. Torque the camshaft bolt to 58—72 ft. lbs. (80—98 Nm).

28. Continue the installation in the reverse order of the removal procedure.

29. Fill the engine with the proper grade and type engine oil.

30. Fill the cooling system with the proper grade and type engine coolant.

31. Initialize the ECM, if equipped with manual transaxle or the PCM, if equipped with automatic transaxle by turning the ignition switch ON then OFF. Keep it off for at least 10 seconds.

32. Start the engine and check for leaks.

3.8L Engine

See Figures 84 through 87.

1. Before servicing the vehicle, refer to the Precautions Section.

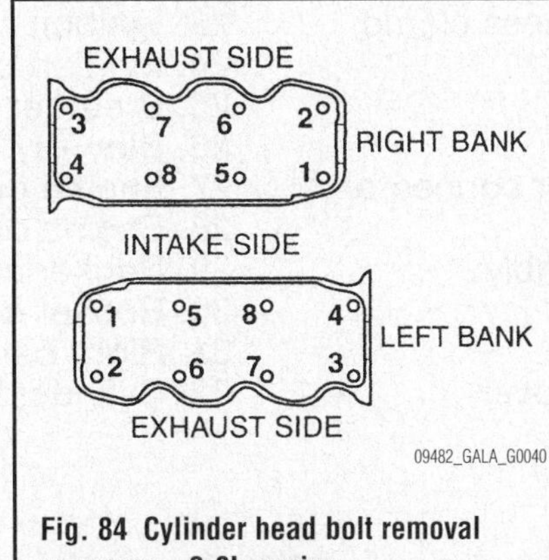

Fig. 84 Cylinder head bolt removal sequence—3.8L engine

2. Relieve the fuel system pressure.

3. Disconnect the negative battery cable.

4. Drain the cooling system.

5. Remove the intake manifold. Remove the exhaust manifolds.

6. Remove the timing belt.

7. Remove the thermostat housing. Disconnect the PCV hose connection. Remove the PCV valve.

8. Remove the ignition coils.

9. Disconnect the engine control wiring harness clamp. Remove the harness bracket.

10. Remove the rocker arm covers and gaskets. Remove the camshaft position sensor connector.

11. Remove the engine oil control valve connector. Remove the grounding bolt. Remove the engine oil dipstick assembly.

12. Remove the camshaft sprockets, using tool MB990767 or equivalent. Remove the timing belt rear center cover.

13. Remove the power steering oil pump assembly and position it to the side. Remove the power steering oil pump bracket bolt. Remove the engine oil control valve connector.

14. Loosen all cylinder head bolts in 2–3 steps and in the proper bolt removal sequence. Remove the cylinder head from the engine. Discard the cylinder head gasket.

To install:

15. Thoroughly clean the sealing surfaces of the head and block.

16. Place a new head gasket on the cylinder block making sure the identification mark on the cylinder head gasket is in the front top (upward) location. Do not use sealer on the gasket.

17. Carefully install the cylinder head on the block. Be sure the head bolt washers are installed with the beveled side facing upward. Torque the cylinder head bolts to specification in 2–3 passes. Tightening torque:

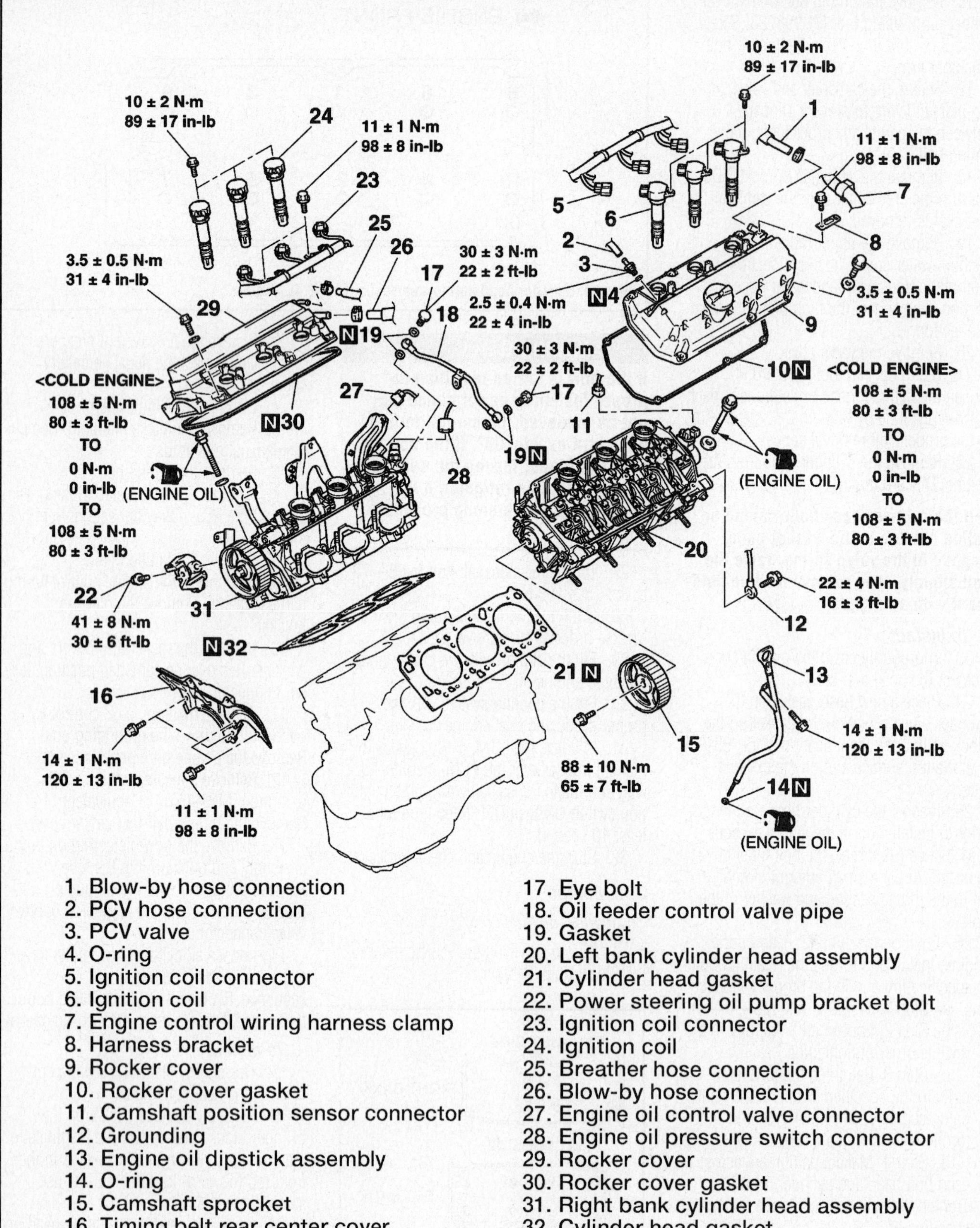

Fig. 85 Expanded view of cylinder head and related components—3.8L engine

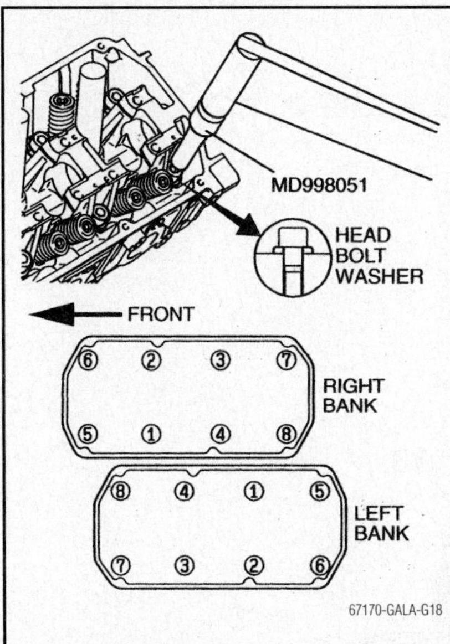

Fig. 86 Cylinder head bolt torque sequence—3.8L engine

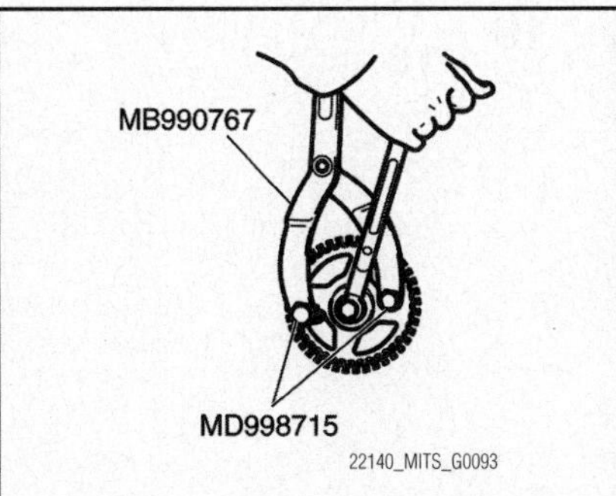

Fig. 87 Use special tools MD998715 and MB990767 to install the camshaft sprocket—3.8L engine

a. Step 1: 77–83 ft. lbs. (103–113 Nm).
b. Step 2: 0 inch lbs. (0 Nm).
c. Step 3: 77–83 ft. lbs. (103–113 Nm).

18. Use special tools MD998715 and MB990767 in the same way as during removal to install the camshaft sprocket.
19. Tighten the camshaft sprocket mounting bolt to 58–72 ft. lbs. (78–98 Nm).
20. Torque the camshaft bolt to 58—72 ft. lbs. (80—98 Nm).
21. Continue the installation in the reverse order of the removal procedure.
22. Fill the engine with the proper grade and type engine oil.
23. Fill the cooling system with the proper grade and type engine coolant.
24. Initialize the ECM, if equipped with manual transaxle or the PCM, if equipped with automatic transaxle by turning the ignition switch ON then OFF. Keep it off for at least 10 seconds.
25. Start the engine and check for leaks.

Lancer

2006–07 2.0L Engine

See Figures 88 through 91.

1. Before servicing the vehicle, refer to the Precautions Section.
2. Relieve the fuel system pressure.
3. Disconnect the negative battery cable.
4. Drain the cooling system.
5. Remove the engine undercover. Remove the air cleaner assembly.
6. Remove the exhaust manifold. Remove the water hose and pipe.
7. Disconnect the accelerator cable connection.
8. Disconnect the air conditioning compressor connector, power steering oil pressure switch connector, and the crankshaft position sensor connector.
9. Disconnect the manifold absolute pressure sensor connector, EVAP solenoid connector, and EGR solenoid valve connector.
10. Disconnect the ignition coil connector, injector connector, and throttle position sensor connector.
11. Disconnect the idle air control motor connector, engine coolant temperature sensor connector, and the camshaft position sensor connector.
12. Disconnect the knock sensor connector, engine coolant temperature gauge unit connector, and the heated oxygen sensor connector.
13. Disconnect the brake booster vacuum hose. Disconnect the high pressure hose connection. Remove the timing belt upper cover.
14. Turn the crankshaft clockwise, align the timing marks on the camshaft sprocket to position number 1 piston to Top Dead Center (TDC) of its compression stroke.

⁂ WARNING

Never rotate the crankshaft in the counterclockwise direction.

15. Remove the timing belt undercover rubber plug, using tool MD998738. Screw the tool in until it contacts the timing belt tensioner arm.
16. Secure the camshaft and valve timing belt, with wire to prevent slippage between the camshaft sprocket and valve timing belt. Remove the camshaft sprocket.

➡Do not turn the crankshaft after the camshaft sprocket is removed.

17. Remove the intake manifold stay connecting bolts.

➡Be careful not to damage or deform the plug guides when removing the cylinder head bolts. Plug guides cannot be replaced separately.

18. Loosen all cylinder head bolts in 2–3 steps and in the proper bolt removal sequence.
19. Remove the cylinder head from the engine. Discard the cylinder head gasket.

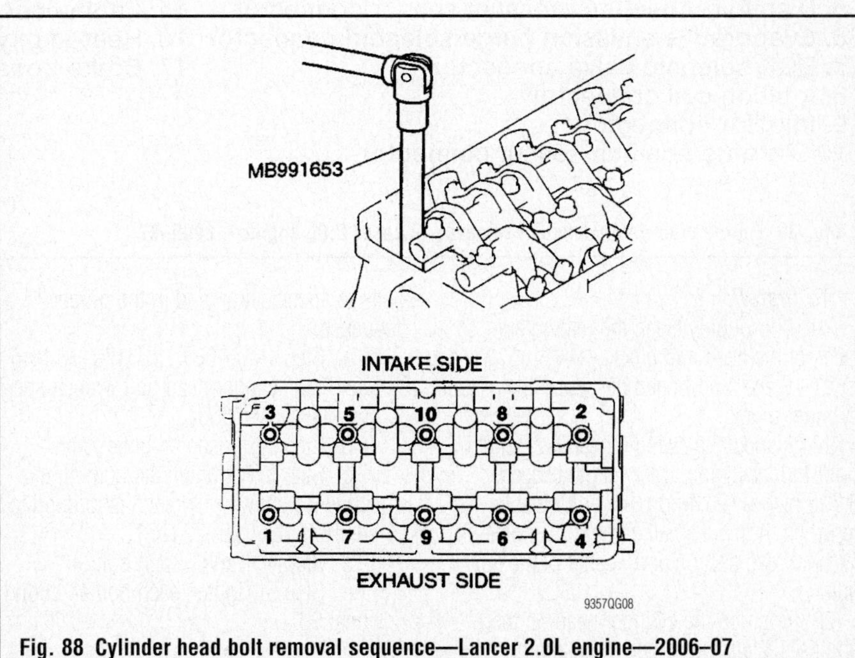

Fig. 88 Cylinder head bolt removal sequence—Lancer 2.0L engine—2006-07

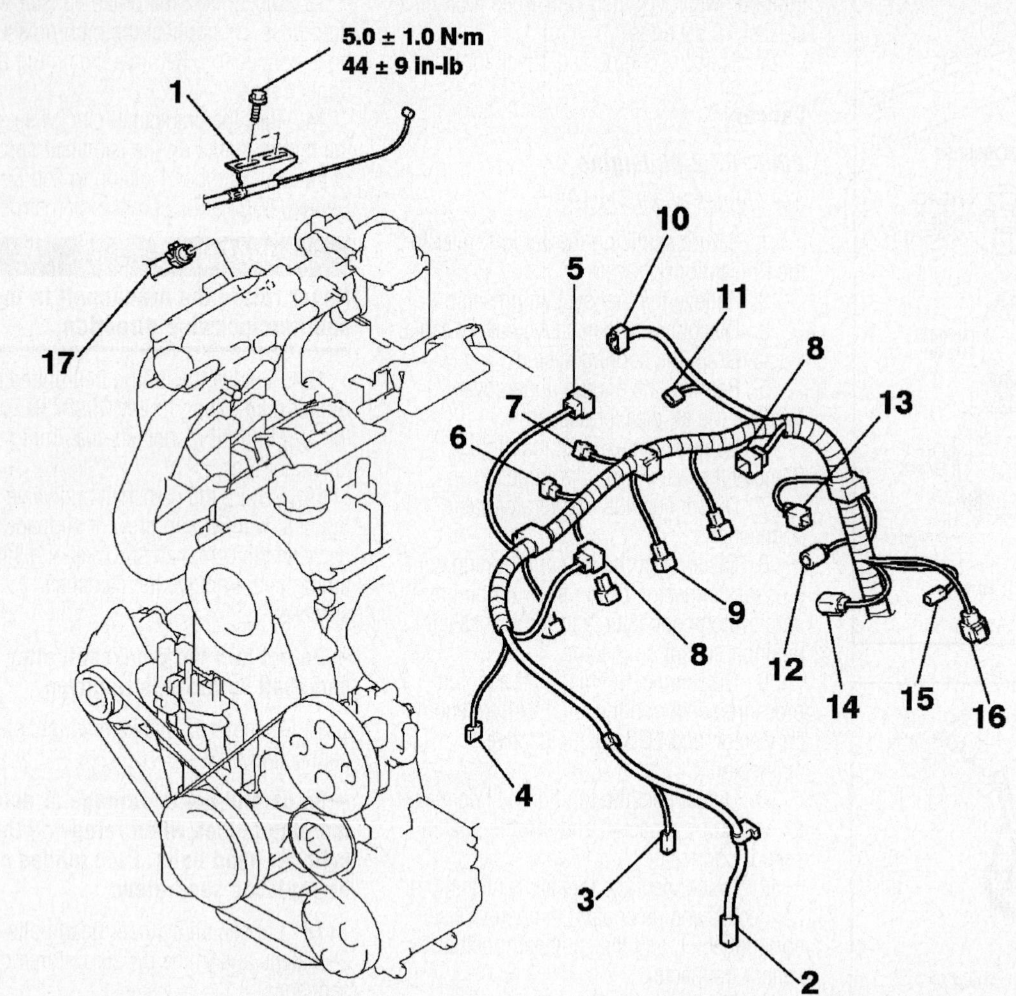

1. Accelerator cable connection
2. A/C compressor connector
3. Power steering oil pressure switch connector
4. Crankshaft position sensor connector
5. Manifold Absolute pressure sensor connector
6. Evaporative emission purge solenoid connector
7. EGR solenoid valve connector
8. Ignition coil connector
9. Injector connector
10. Throttle position sensor connector
11. Idle air control motor connector
12. Engine coolant temperature sensor connector
13. Camshaft position sensor connector
14. Knock sensor connector
15. Engine coolant temperature gauge unit
16. Heated oxygen sensor connector
17. Brake booster vacuum hose connection

22140_MITS_G0094

Fig. 89 Engine electrical connector removal—Lancer 2.0L engine—2006-07

To install:

20. Thoroughly clean the mating surfaces of the head and block.

21. Place a new head gasket on the cylinder block.

22. Inspect the cylinder head bolt length prior to installation. If the length exceeds 3.795 inches (96.4mm), the bolt must be replaced. Apply a small amount of engine oil to the thread section and washer of the bolt.

23. Position the cylinder head on the engine. Install and torque the retaining bolts to specification and in the proper sequence.

a. Step 1: 52–58 ft. lbs. (70–78 Nm) using the cylinder head bolt wrench special tool MB991653.

b. Step 2: Loosen all bolts fully.

c. Step 3: Retighten the bolts in the sequence shown until each is torqued to 14–16 ft. lbs. (18–22 Nm).

d. Step 4: Tighten each bolt 90°.

e. Step 5: Tighten each bolt an additional 90°.

➡If the bolts are tightened by an angle of less than 90°, they may not hold the cylinder head with sufficient strength. If the bolts are tightened by an angle exceeding 90°, completely remove them and repeat the installation procedure.

24. Continue the installation in the reverse order of the removal procedure.

25. Fill the engine with the proper grade and type engine oil.

26. Fill the cooling system with the proper grade and type engine coolant.

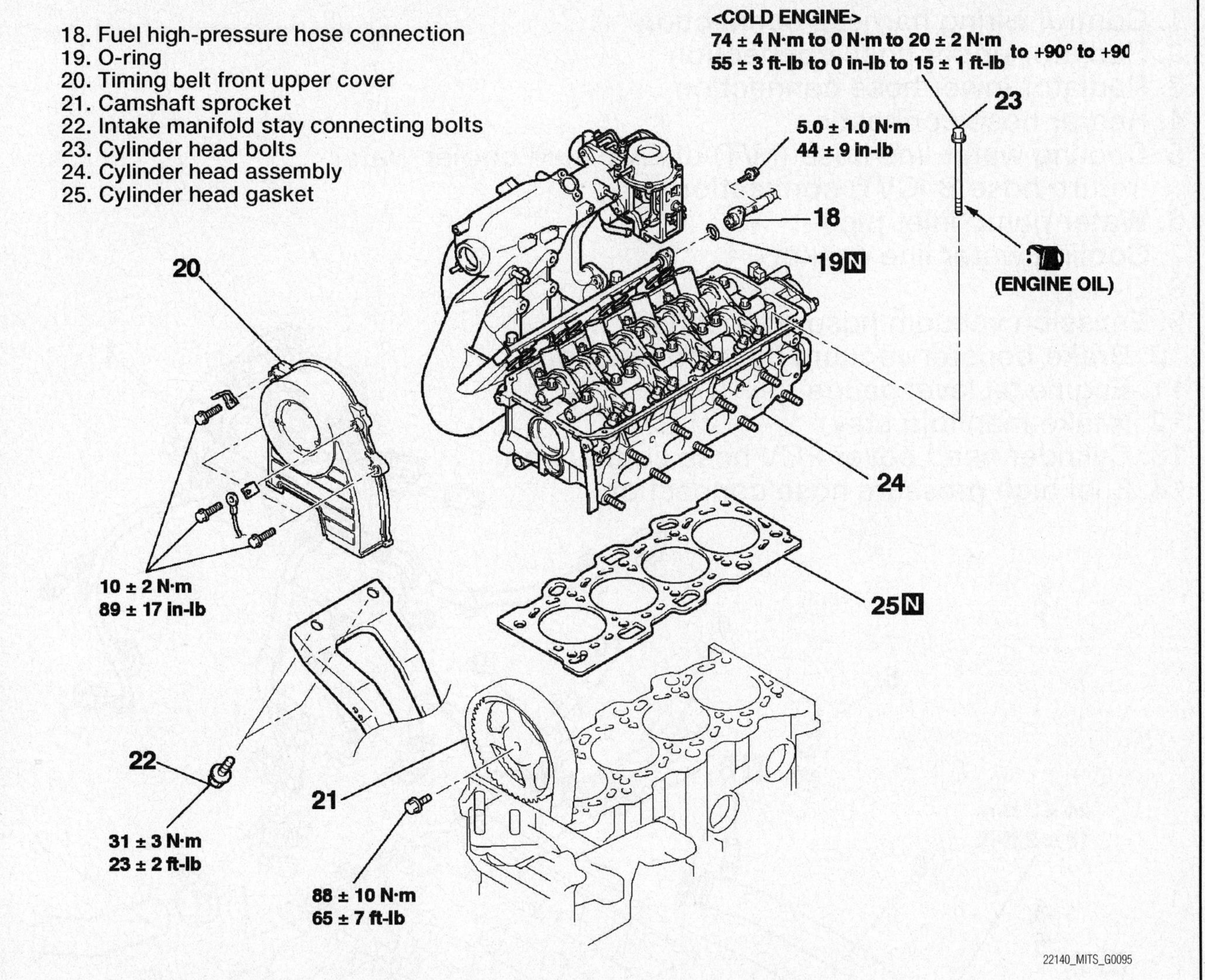

18. Fuel high-pressure hose connection
19. O-ring
20. Timing belt front upper cover
21. Camshaft sprocket
22. Intake manifold stay connecting bolts
23. Cylinder head bolts
24. Cylinder head assembly
25. Cylinder head gasket
<COLD ENGINE>
74 ± 4 N·m to 0 N·m to 20 ± 2 N·m
55 ± 3 ft-lb to 0 in-lb to 15 ± 1 ft-lb
to +90° to +90
23
5.0 ± 1.0 N·m
44 ± 9 in-lb
18
19N
(ENGINE OIL)
20
24
10 ± 2 N·m
89 ± 17 in-lb
25N
22
21
31 ± 3 N·m
23 ± 2 ft-lb
88 ± 10 N·m
65 ± 7 ft-lb
22140_MITS_G0095

Fig. 90 Expanded view of cylinder head and related components—Lancer 2.0L engine—2006–07

MB991653
INTAKE SIDE
8
6
1
3
9
10
4
2
5
7
EXHAUST SIDE
9357QG10

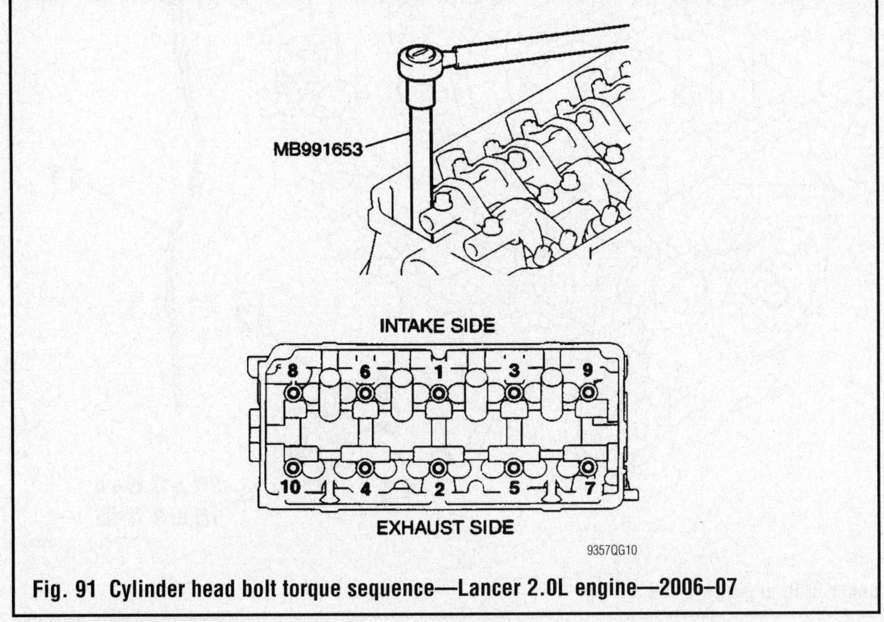

Fig. 91 Cylinder head bolt torque sequence—Lancer 2.0L engine—2006–07

27. Start the engine and check for leaks.

2008 2.0L Engine

See Figures 92 through 99.

1. Before servicing the vehicle, refer to the Precautions Section.
2. Relieve the fuel system pressure.
3. Disconnect the negative battery cable.
4. Remove the engine room undercover and side cover (RH).
5. Drain the cooling system.
6. Remove the air cleaner assembly.
7. Remove the ignition coils.
8. Remove the exhaust manifold.
9. Remove the throttle body assembly.
10. Remove the EGR valve and EGR valve stay (California Vehicles).
11. Remove the water pump.
12. Remove the control wiring harness connection.

1. Control wiring harness connection
2. Radiator upper hose connection
3. Radiator lower hose connection
4. Heater hose connection
5. Cooling water line hose (M/T) or CVT fluid cooler water return hose B (CVT) connection
6. Water pump inlet pipe
7. Cooling water line gasket
8. O-ring
9. Emission vacuum hose connection
10. Brake booster vacuum hose connection
11. Engine oil level gauge
12. Intake manifold stay
13. Cylinder head cover PCV hose connection
14. Fuel high pressure hose connection

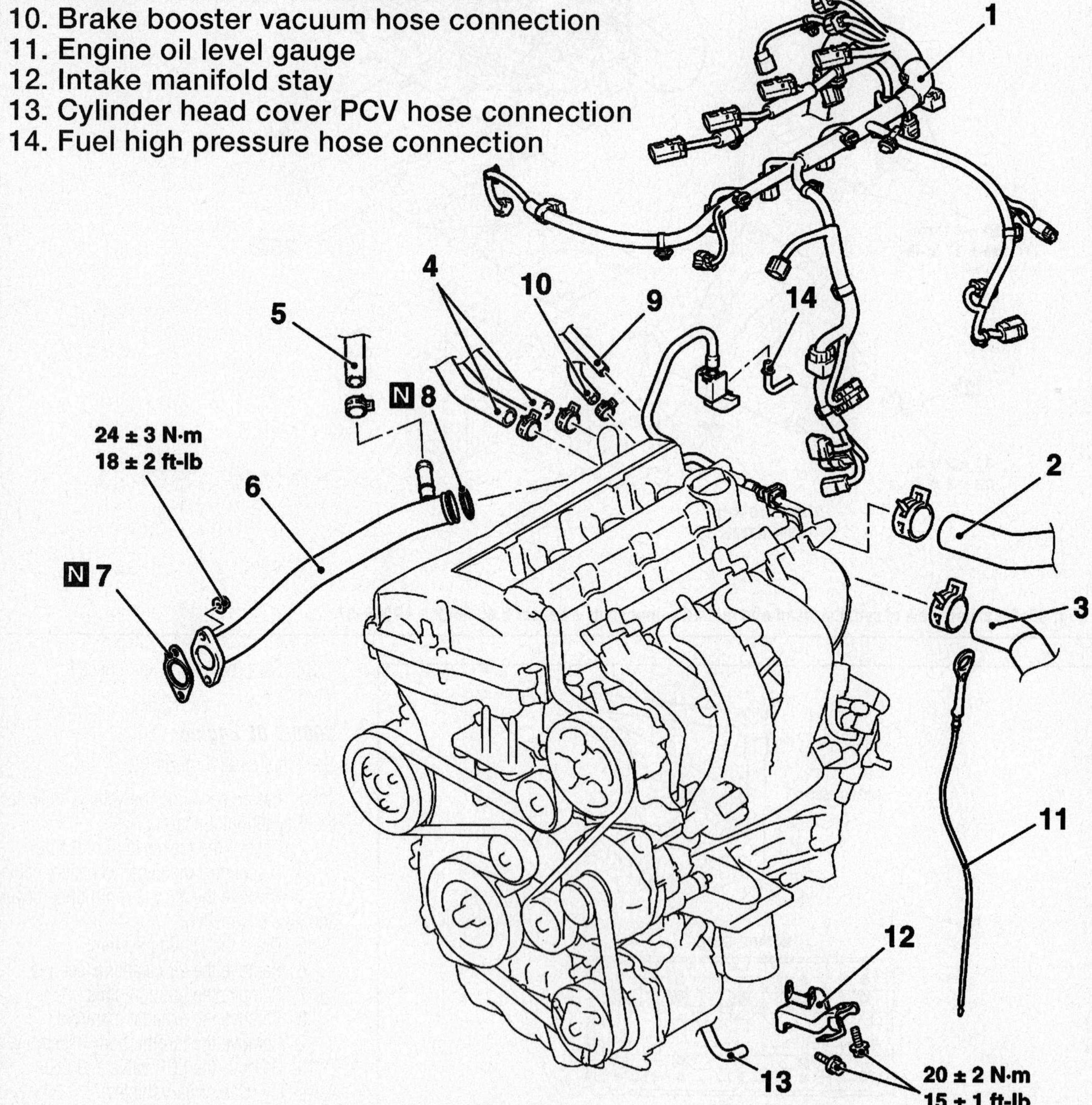

Fig. 92 Engine electrical connector removal—Lancer 2.0L engine—2008

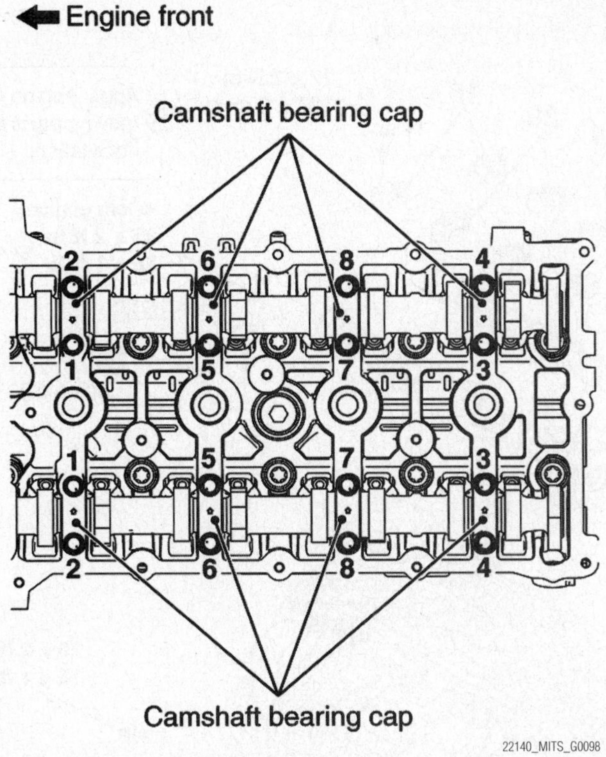

Engine front

Camshaft bearing cap

Camshaft bearing cap

22140_MITS_G0098

Fig. 93 Loosen the mounting bolts of the camshaft bearing caps in the order shown—Lancer 2.0L engine—2008

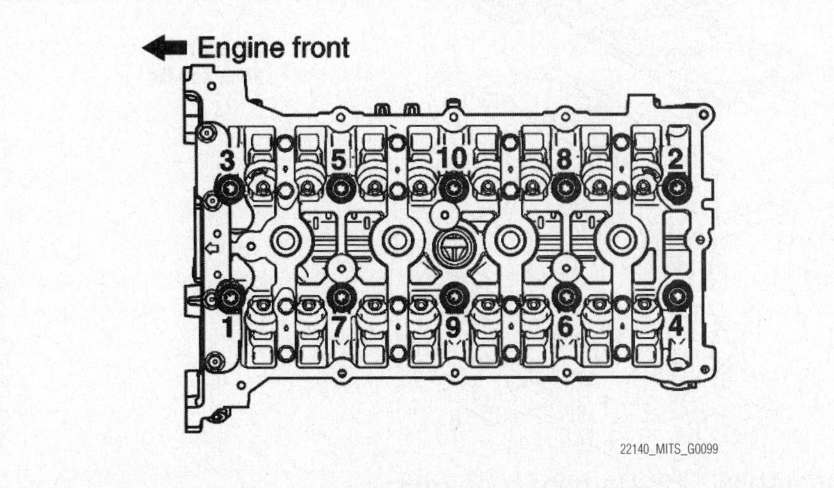

Fig. 94 Loosen and remove the cylinder head bolts in 2–3 steps in the order shown—Lancer 2.0L engine—2008

13. Remove the radiator upper and lower hose connection.
14. Remove the heater hose connection.
15. Remove the cooling water line hose (M/T) or CVT fluid cooler water return hose B (CVT) connection.
16. Remove the water pump inlet pipe.
17. Remove the cooling water line gasket and O-ring.
18. Remove the emission vacuum hose connection and brake booster vacuum hose connection.
19. Remove the engine oil level gauge.
20. Remove the intake manifold stay.
21. Remove the cylinder head cover PCV hose connection.
22. Remove the fuel high pressure hose connection.
23. Remove the valve timing chain.

WARNING

When the camshaft bearing cap mounting bolts are loosened all at once, the mounting bolts jump out by the spring force and the threads are damaged. Always loosen the mounting bolts in 4–5 steps.

24. Loosen the mounting bolts of the camshaft bearing caps in the order of number shown in the figure in 4–5 steps, and remove the camshaft bearing caps.
25. Remove the camshaft and camshaft sprocket assembly.
26. Remove the camshaft bearing.
27. Remove the intake manifold stay B (except for California).
28. Temporarily install the engine oil pan which was removed at the valve timing chain removal.
29. Place a garage jack against the engine oil pan with a piece of wood in between to support the engine and transaxle assembly.
30. Remove special tool MB991928 or MB991895 which was installed for supporting the engine and transaxle assembly when the valve timing chain was removed.
31. Loosen and remove the cylinder head bolts in 2–3 steps in the order of number shown in the figure.
32. Remove the cylinder head assembly and the gasket.

To install:

WARNING

Do not allow any foreign materials get into the coolant passages, oil passages, or cylinder.

33. Remove the sealant and grease on the top surface of cylinder block and on the bottom surface of the cylinder head. Then, use a quick-drying degreasing agent (white gasoline) to degrease the sealant application surface.
34. Apply sealant to the top surface of cylinder block as shown in the figure. Specified sealant: Three Bond® 1217G.
35. Within 3 minutes after the sealant application, install the cylinder head gasket to the cylinder block.

➡When the cylinder gasket is installed to the cylinder block, check that the sealant is securely applied to the bead line of the cylinder head gasket.

36. Apply the sealant to the top surface of cylinder head gasket as shown in the figure. Specified sealant: Three Bond® 1217G.

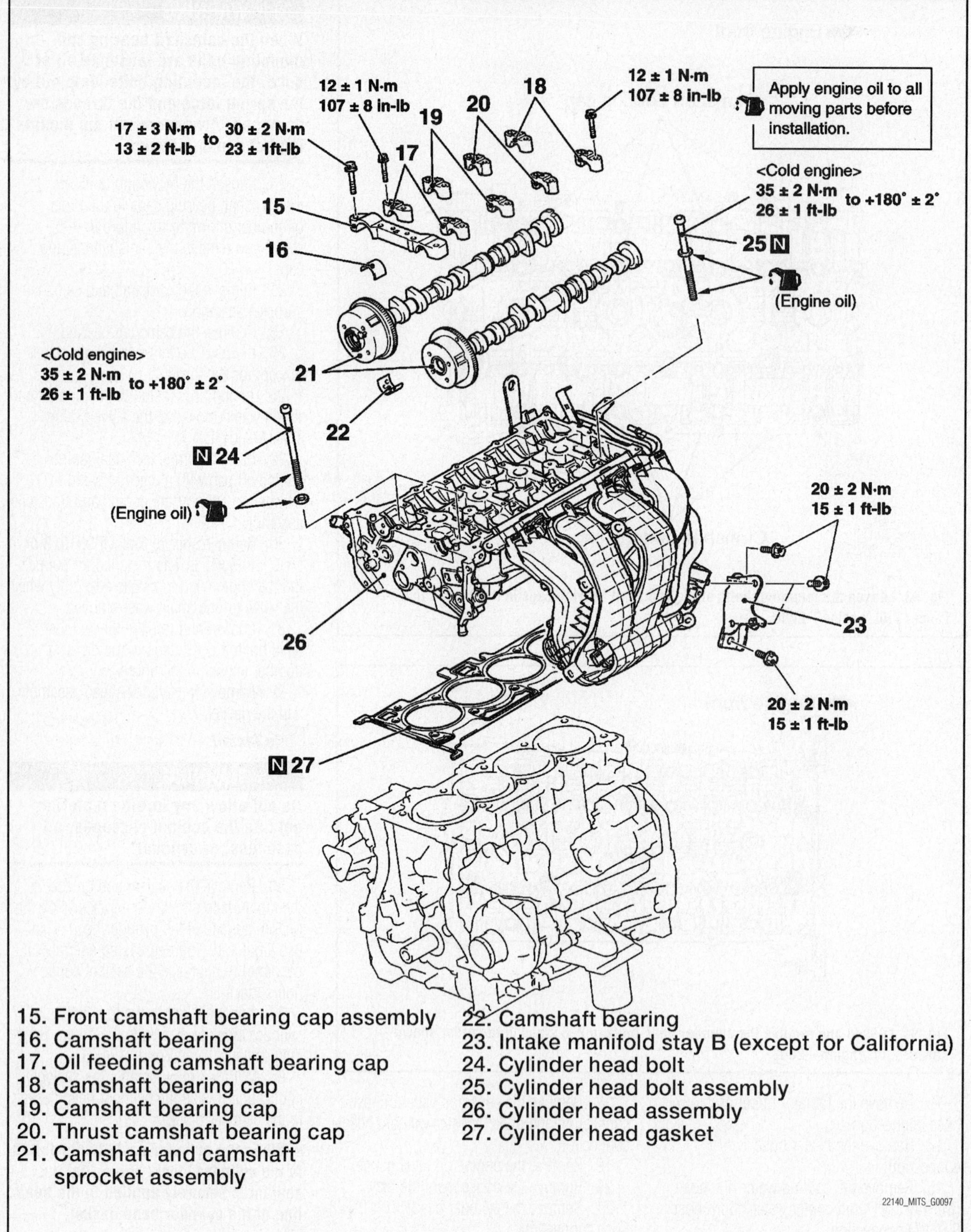

Fig. 95 Expanded view of cylinder head and related components—Lancer 2.0L engine—2008

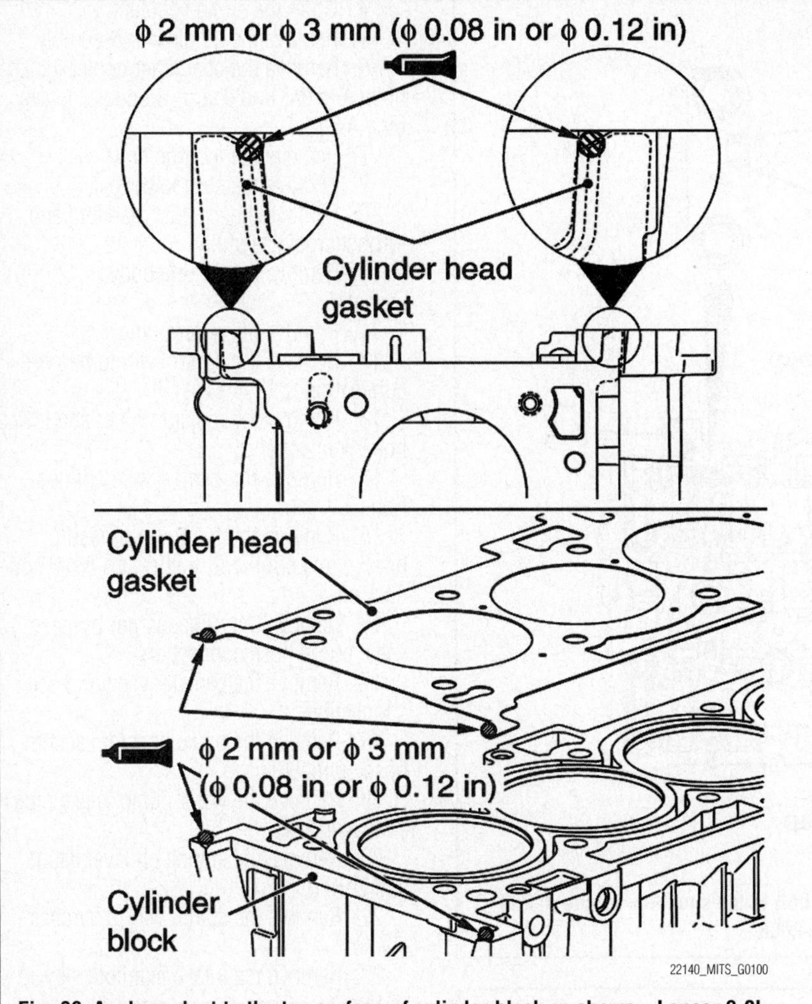

Fig. 96 Apply sealant to the top surface of cylinder block as shown—Lancer 2.0L engine—2008

WARNING

Within 2 hours after the cylinder head assembly installation, do not apply oil or water to the sealant application area or start the engine.

37. Within 3 minutes after the sealant application, install the cylinder head assembly.
38. Replace cylinder head bolts with a new ones.
39. For the 2 bolts of the timing chain side, the washer can be removed from the bolt. Install the washer, with its sag facing upward, to the bolts.
40. Apply a small amount of engine oil to the cylinder head bolt threads and the washers.
41. Tighten the bolts by the following procedure (plastic region angular tightening method).
   a. Step 1: tighten the bolts to 25–27 ft. lbs. (33–37 Nm) in the order shown in the figure.

WARNING

The bolt is not tightened sufficiently if the tightening angle is less than a 180° angle. If the tightening angle exceeds the standard specification, remove the bolt and repeat the installation steps.

   b. Step 2: Put a paint mark on the cylinder head bolt head and cylinder head, tighten to 178–182° in the order shown in the figure, and check that the paint mark on the cylinder head bolt head aligns with the paint mark on the cylinder head.
42. Install special tool MB991928 or MB991895 which was installed for supporting the engine and transaxle assembly when the valve timing chain was removed.
43. Remove the garage jack which supports the engine and transaxle assembly.
44. Remove the engine oil pan installed temporarily.

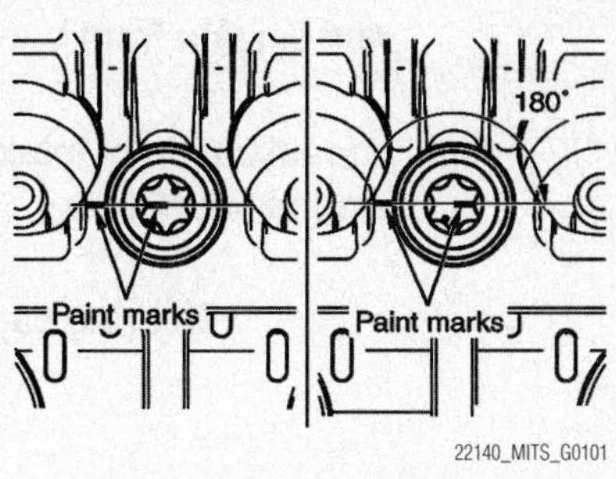

Fig. 97 Check that the paint mark on the cylinder head bolt head aligns with the paint mark on the cylinder head when tightening—Lancer 2.0L engine—2008

WARNING

Be careful not to drop the camshaft bearing. When installing the camshaft and camshaft sprocket assembly (exhaust side), be careful not to let the camshaft bearing, which is installed to the front cam bearing, deviate from its position.

45. Install the camshaft bearing caps to the cylinder heads.

➡Because the thrust camshaft bearing cap and the other camshaft bearing caps are the same in shape, check the bearing cap number and additionally its symbol to identify the intake and exhaust sides for correct installation.

46. Tighten each camshaft bearing cap mounting bolt to the specified torque in 2–3 passes according to the figure shown. Tightening torque: 99–115 inch lbs. (11–13 Nm).

WARNING

When the mounting bolts are tightened with the front camshaft bearing cap tilted, the front camshaft bearing cap is damaged. Install the front camshaft bearing cap properly to the cylinder head and camshaft.

47. Install the front camshaft bearing cap to the cylinder head and temporarily tighten the front camshaft bearing cap to the specified torque in the order shown (1). Tightening torque: 11–15 ft. lbs. (14–20 Nm).
48. Tighten the front camshaft bearing cap again to the specified torque in the order shown (2). Tightening torque: 22–24 ft. lbs. (28–32 Nm).
49. Installation continues in the reverse of the removal procedure.

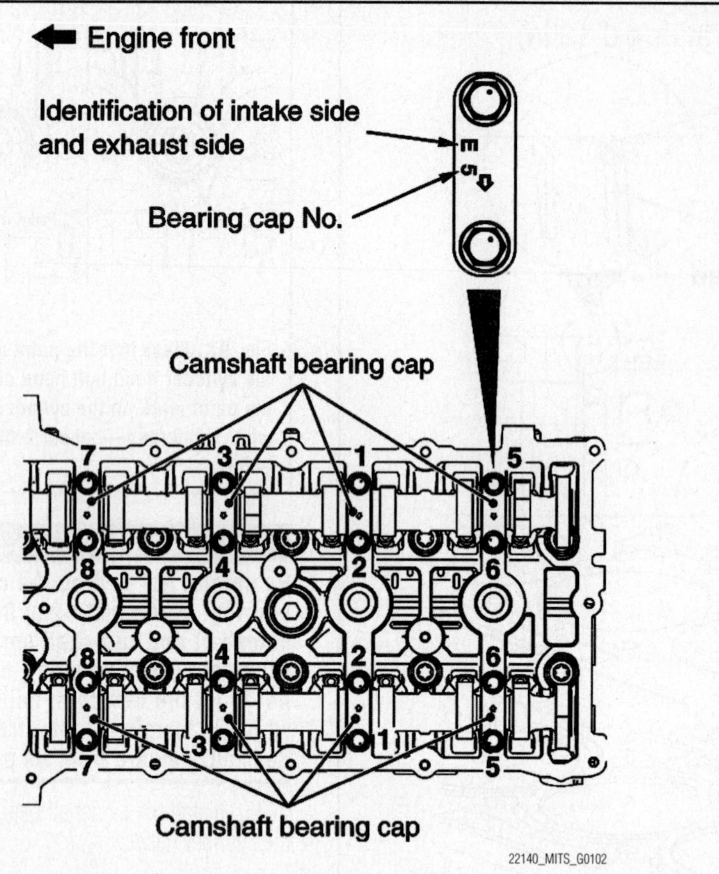

Fig. 98 Tighten each camshaft bearing cap mounting bolt to the specified torque in 2–3 passes in the sequence shown—Lancer 2.0L engine—2008

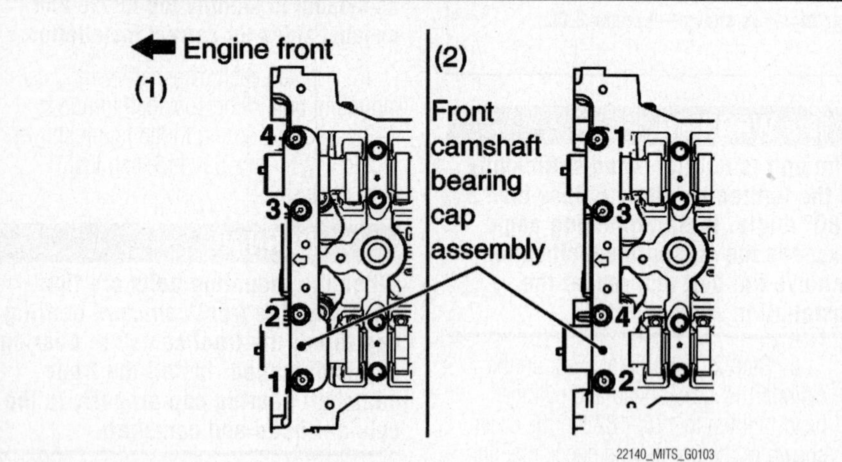

Fig. 99 Install the front camshaft bearing cap to the cylinder head and tighten the front camshaft bearing cap to the specified torque in the order shown—Lancer 2.0L engine—2008

50. Run the engine and check for fluid leaks and proper operation.

Lancer Evolution

2008 2.0L Turbo Engine

See Figures 100 through 107.

1. Before servicing the vehicle, refer to the Precautions Section.
2. Relieve the fuel system pressure. Disconnect the negative battery cable. Drain the cooling system.
3. Remove the engine undercover. Remove the side cover.
4. Drain the engine coolant.
5. Remove the air cleaner assembly.
6. Remove the charge air cooler intake hose A and B and charge air cooler intake pipe A.
7. Remove the ignition coils.
8. Remove the strut tower bar.
9. Remove the exhaust manifold and turbocharger assembly.
10. Remove the throttle body assembly.
11. Remove the water pump.
12. Remove the control wiring harness connection.
13. Remove the radiator upper and lower hose connections.
14. Remove the heater hose connection.
15. Remove the fuel high pressure hose connection and fuel return hose connection.
16. Remove the turbocharger by-pass valve purge hose connection.
17. Remove the canister vacuum hose connection.
18. Remove the brake booster vacuum hose connection.
19. Remove the water pump intake pipe and O-ring.
20. Remove the engine oil level gauge and O-ring.
21. Remove the starter wiring harness clamp.
22. Remove the intake manifold stay (front).
23. Remove the starter wiring harness clamp.
24. Remove the intake manifold stay (rear).
25. Remove the valve timing chain.

✲✲ WARNING

When the camshaft bearing cap mounting bolts are loosened all at once, the mounting bolts jump out by the spring force and the threads are damaged. Always loosen the mounting bolts in 4–5 steps.

26. Loosen the mounting bolts of the camshaft bearing caps in the order of number shown in the figure in 4–5 steps, and remove the camshaft bearing caps.
27. Remove the camshaft and camshaft sprocket assembly.
28. Remove the camshaft bearing.
29. Remove the intake manifold stay B (except for California).
30. Temporarily install the engine oil pan which was removed at the valve timing chain removal.

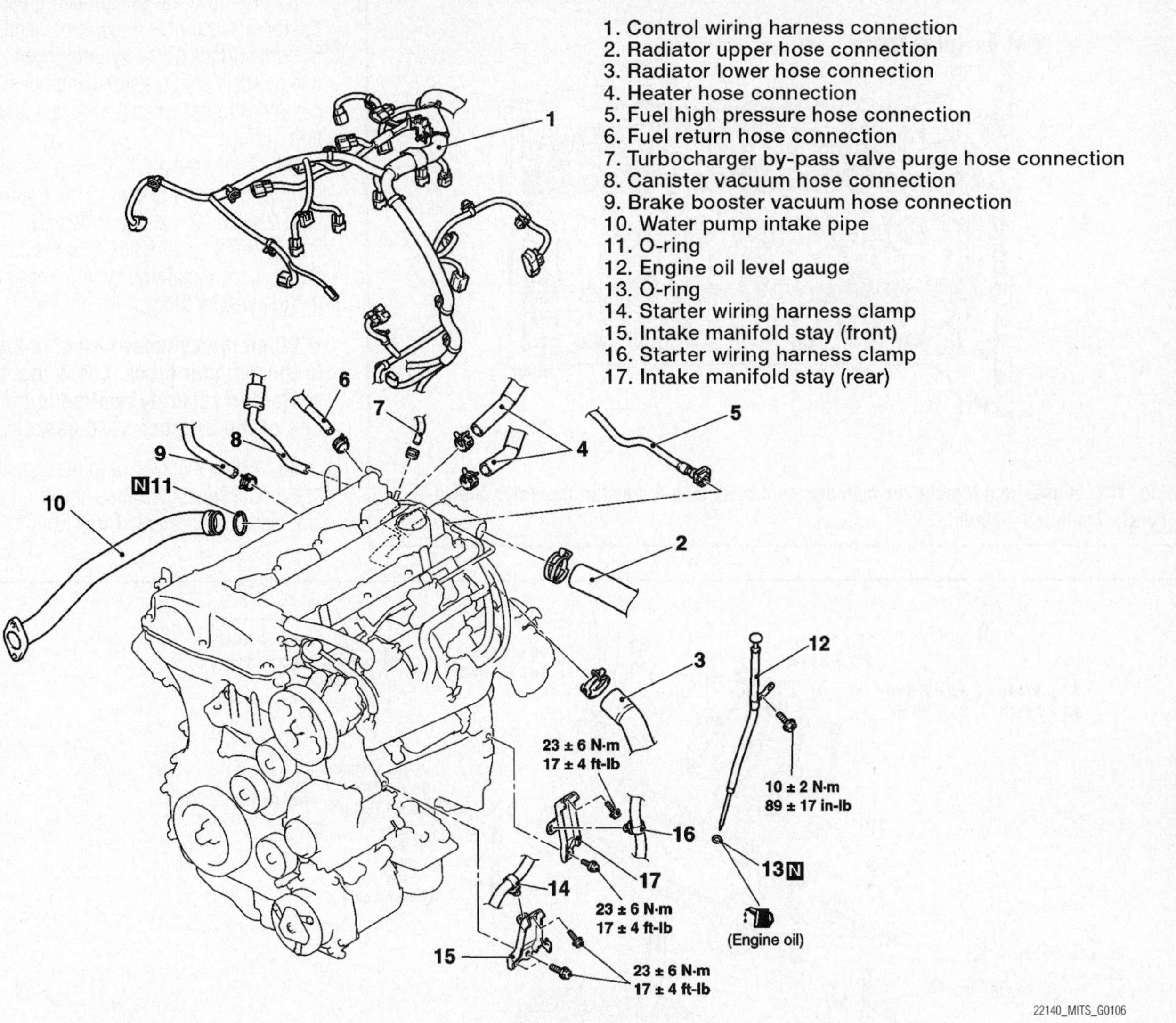

1. Control wiring harness connection
2. Radiator upper hose connection
3. Radiator lower hose connection
4. Heater hose connection
5. Fuel high pressure hose connection
6. Fuel return hose connection
7. Turbocharger by-pass valve purge hose connection
8. Canister vacuum hose connection
9. Brake booster vacuum hose connection
10. Water pump intake pipe
11. O-ring
12. Engine oil level gauge
13. O-ring
14. Starter wiring harness clamp
15. Intake manifold stay (front)
16. Starter wiring harness clamp
17. Intake manifold stay (rear)
23 ± 6 N·m
17 ± 4 ft-lb
10 ± 2 N·m
89 ± 17 in-lb
(Engine oil)
22140_MITS_G0106

Fig. 100 Engine component and connection removal—Lancer Evolution—2008

Engine front
Camshaft bearing cap
Camshaft bearing cap
22140_MITS_G0108

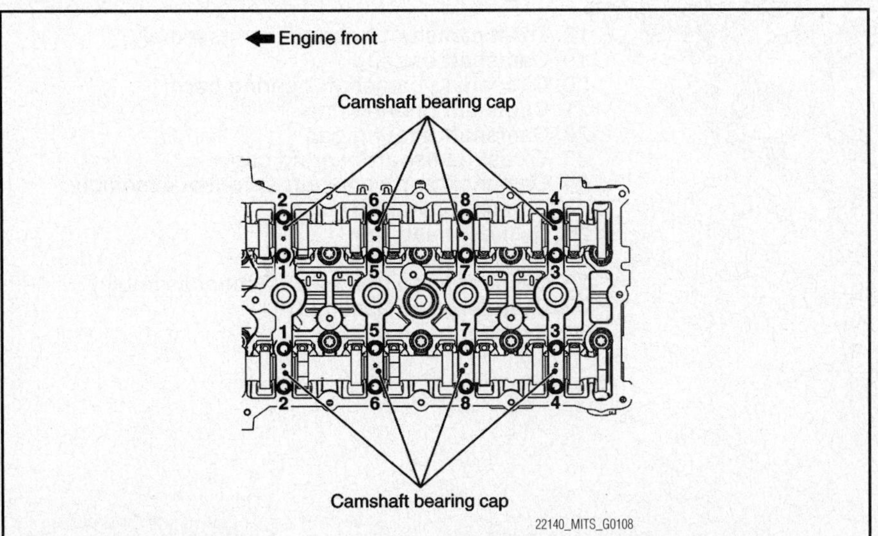

Fig. 101 Loosen the mounting bolts of the camshaft bearing caps in the order shown—Lancer Evolution—2008

31. Place a garage jack against the engine oil pan with a piece of wood in between to support the engine and transaxle assembly.

32. Remove special tool MB991928 or MB991895 which was installed for supporting the engine and transaxle assembly when the valve timing chain was removed.

33. Loosen and remove the cylinder head bolts in 2–3 steps in the order of number shown in the figure.

34. Remove the cylinder head assembly and the gasket.

To install:

WARNING

Do not allow any foreign materials get into the coolant passages, oil passages, or cylinder.

Engine front
3
5
10
8
2
1
7
9
6
4
22140_MITS_G0109

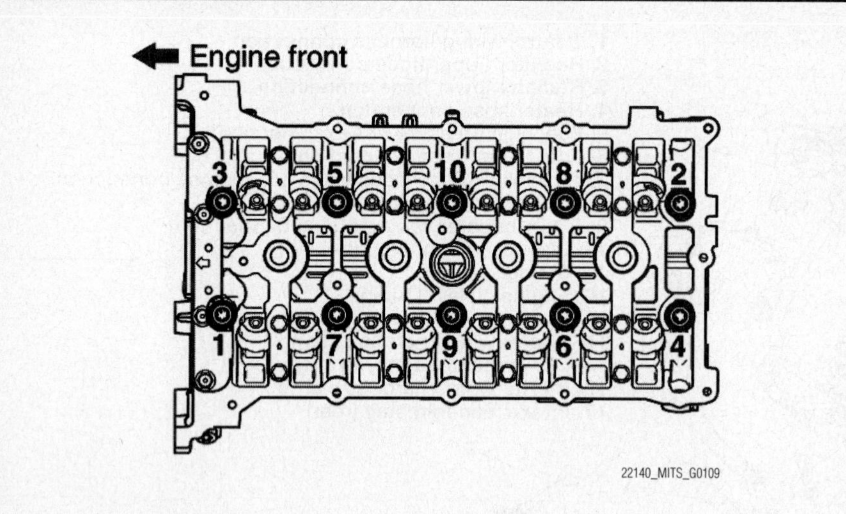

Fig. 102 Loosen and remove the cylinder head bolts in 2–3 steps in the order shown—Lancer Evolution—2008

35. Remove the sealant and grease on the top surface of cylinder block and on the bottom surface of the cylinder head. Then, use a quick-drying degreasing agent (white gasoline) to degrease the sealant application surface.

36. Apply sealant to the top surface of cylinder block as shown in the figure. Specified sealant: Three Bond® 1217G.

37. Within 3 minutes after the sealant application, install the cylinder head gasket to the cylinder block.

➡When the cylinder gasket is installed to the cylinder block, check that the sealant is securely applied to the bead line of the cylinder head gasket.

38. Apply the sealant to the top surface of cylinder head gasket as shown in the figure. Specified sealant: Three Bond® 1217G.

12 ± 1 N·m
107 ± 8 in-lb
17 ± 3 N·m
13 ± 2 ft-lb
to
30 ± 2 N·m
23 ± 1ft-lb
12 ± 1 N·m
107 ± 8 in-lb
Apply engine oil to all moving parts before installation.
<Cold engine>
35 ± 2 N·m
26 ± 1 ft-lb
to +90° to +90°
(Engine oil)
<Cold engine>
35 ± 2 N·m
26 ± 1 ft-lb
to +90° to +90°
(Engine oil)
18. Front camshaft bearing cap assembly
19. Camshaft bearing
20. Oil feeding camshaft bearing cap
21. Camshaft bearing cap
22. Camshaft bearing cap
23. Thrust camshaft bearing cap
24. Camshaft and camshaft sprocket assembly
25. Camshaft bearing
26. Cylinder head bolt
27. Cylinder head bolt washer
28. Cylinder head bolt and washer assembly
29. Cylinder head assembly
30. Cylinder head gasket
22140_MITS_G0107

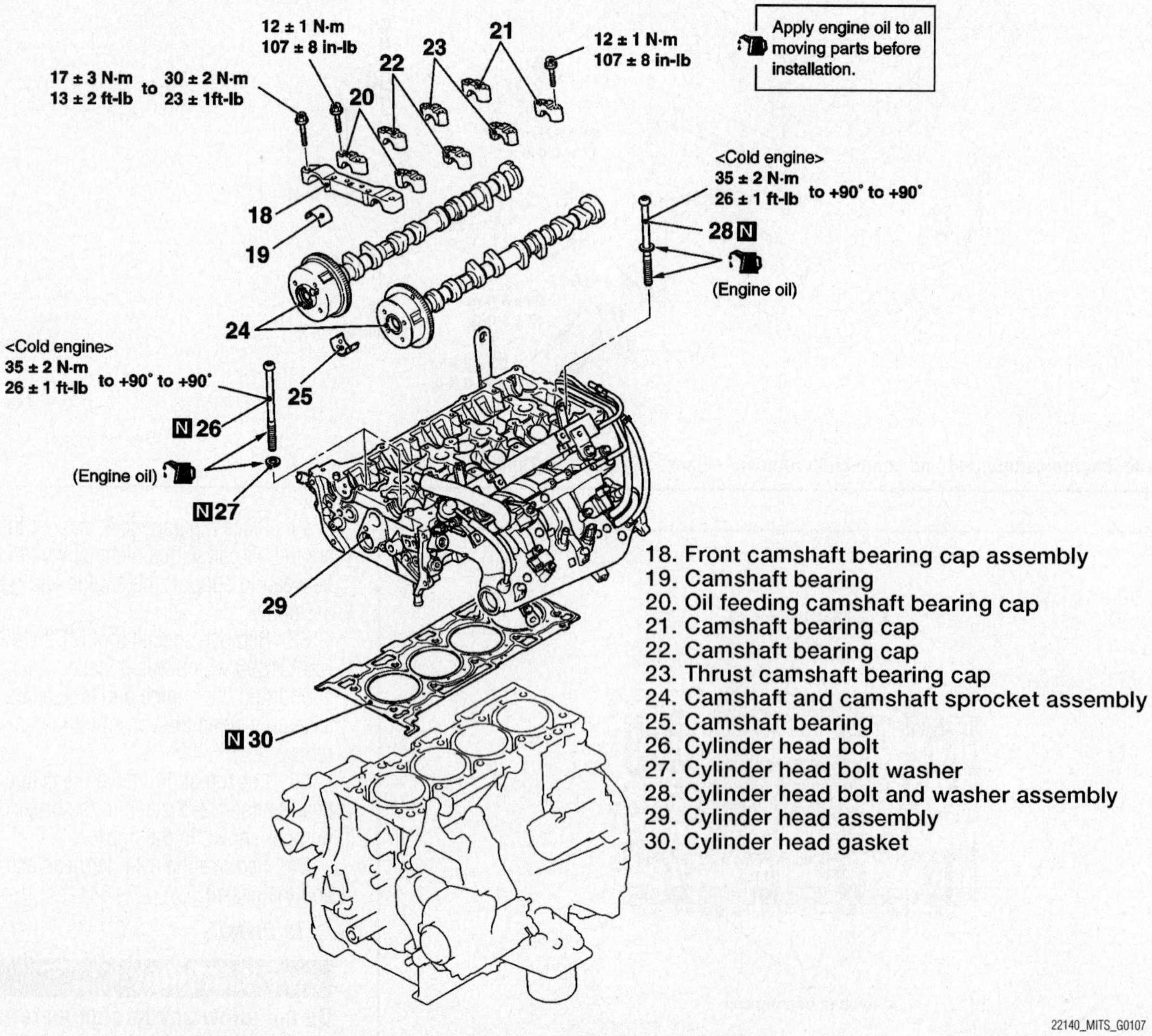

Fig. 103 Expanded view of cylinder head and related components—Lancer Evolution—2008

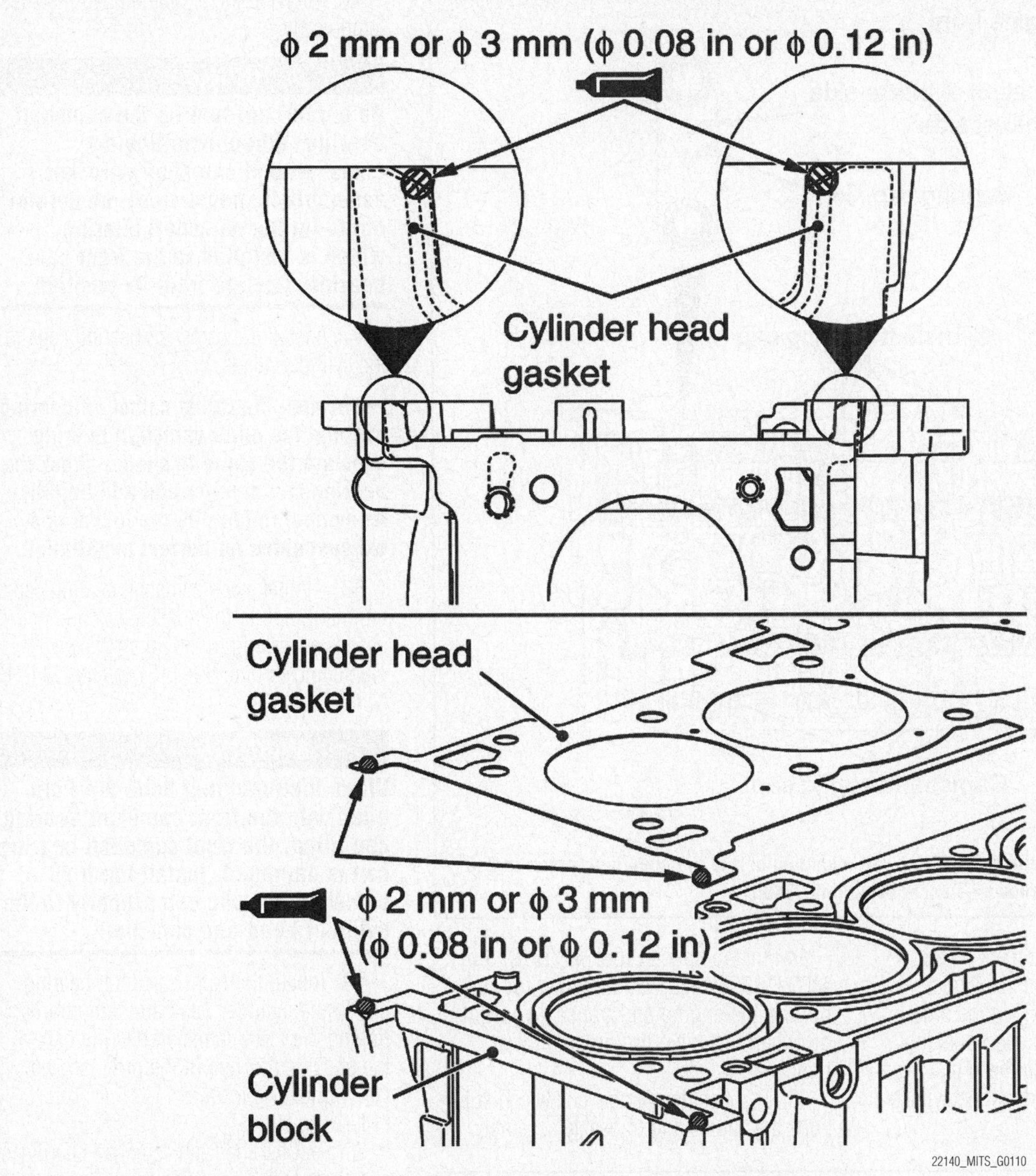

Fig. 104 Apply sealant to the top surface of cylinder block as shown—Lancer Evolution—2008

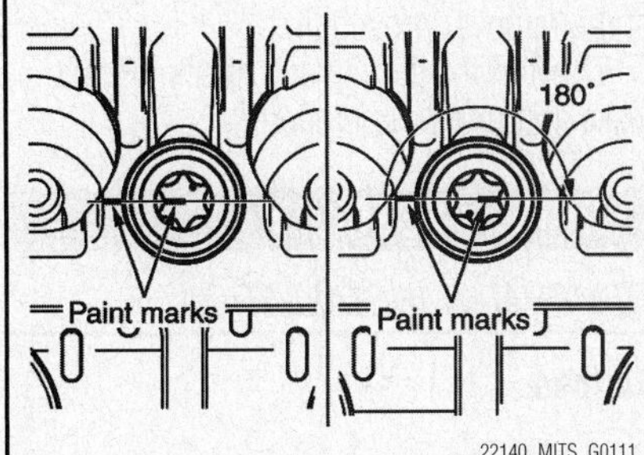

Fig. 105 Check that the paint mark on the cylinder head bolt head aligns with the paint mark on the cylinder head when tightening—Lancer Evolution—2008

WARNING

Within 2 hours after the cylinder head assembly installation, do not apply oil or water to the sealant application area or start the engine.

39. Within 3 minutes after the sealant application, install the cylinder head assembly.
40. Replace cylinder head bolts with a new ones.
41. For the 2 bolts of the timing chain side, the washer can be removed from the bolt. Install the washer, with its sag facing upward, to the bolts.
42. Apply a small amount of engine oil to the cylinder head bolt threads and the washers.
43. Tighten the bolts by the following procedure (plastic region angular tightening method).
    a. Step 1: tighten the bolts to 25–27 ft. lbs. (33–37 Nm) in the order shown in the figure.

WARNING

The bolt is not tightened sufficiently if the tightening angle is less than a 180° angle. If the tightening angle exceeds the standard specification, remove the bolt and repeat the installation steps.

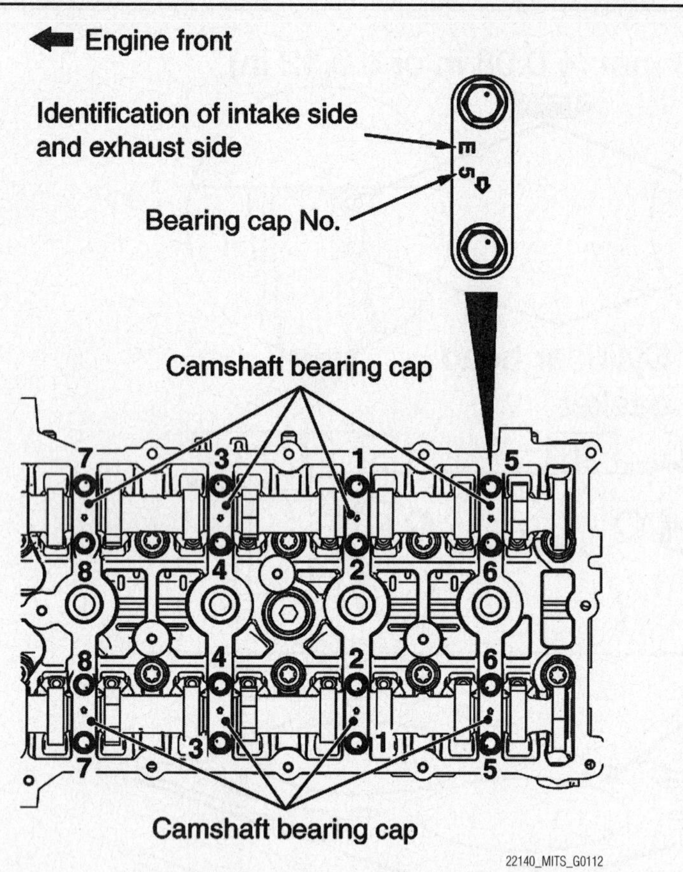

Fig. 106 Tighten each camshaft bearing cap mounting bolt to the specified torque in 2–3 passes in the sequence shown—Lancer Evolution—2008

b. Step 2: Put a paint mark on the cylinder head bolt head and cylinder head, tighten to 178–182° in the order shown in the figure, and check that the paint mark on the cylinder head bolt head aligns with the paint mark on the cylinder head.

44. Install special tool MB991928 or MB991895 which was installed for supporting the engine and transaxle assembly when the valve timing chain was removed.

45. Remove the garage jack which supports the engine and transaxle assembly.

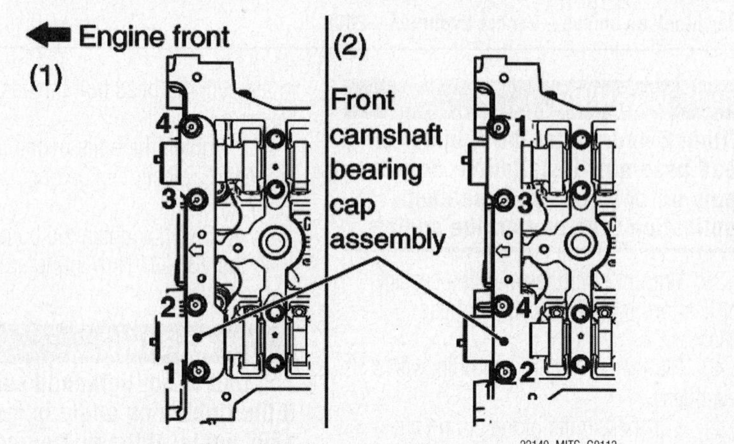

Fig. 107 Install the front camshaft bearing cap to the cylinder head and tighten the front camshaft bearing cap to the specified torque in the order shown—Lancer Evolution—2008

46. Remove the engine oil pan installed temporarily.

WARNING

Be careful not to drop the camshaft bearing. When installing the camshaft and camshaft sprocket assembly (exhaust side), be careful not to let the camshaft bearing, which is installed to the front cam bearing, deviate from its position.

47. Install the camshaft bearing caps to the cylinder heads.

➡Because the thrust camshaft bearing cap and the other camshaft bearing caps are the same in shape, check the bearing cap number and additionally its symbol to identify the intake and exhaust sides for correct installation.

48. Tighten each camshaft bearing cap mounting bolt to the specified torque in 2–3 passes according to the figure shown. Tightening torque: 99–115 inch lbs. (11–13 Nm).

WARNING

When the mounting bolts are tightened with the front camshaft bearing cap tilted, the front camshaft bearing cap is damaged. Install the front camshaft bearing cap properly to the cylinder head and camshaft.

49. Install the front camshaft bearing cap to the cylinder head and temporarily tighten the front camshaft bearing cap to the specified torque in the order shown (1). Tightening torque: 11–15 ft. lbs. (14–20 Nm).

50. Tighten the front camshaft bearing cap again to the specified torque in the order shown (2). Tightening torque: 22–24 ft. lbs. (28–32 Nm).

51. Installation continues in the reverse of the removal procedure.

52. Run the engine and check for fluid leaks and proper operation.

ENGINE ASSEMBLY

REMOVAL & INSTALLATION

Eclipse

2.4L Engine

See Figures 108 and 109.

1. Before servicing the vehicle, refer to the Precautions Section.

2. Relieve the fuel system pressure. Refer to Relieving Fuel System Pressure.

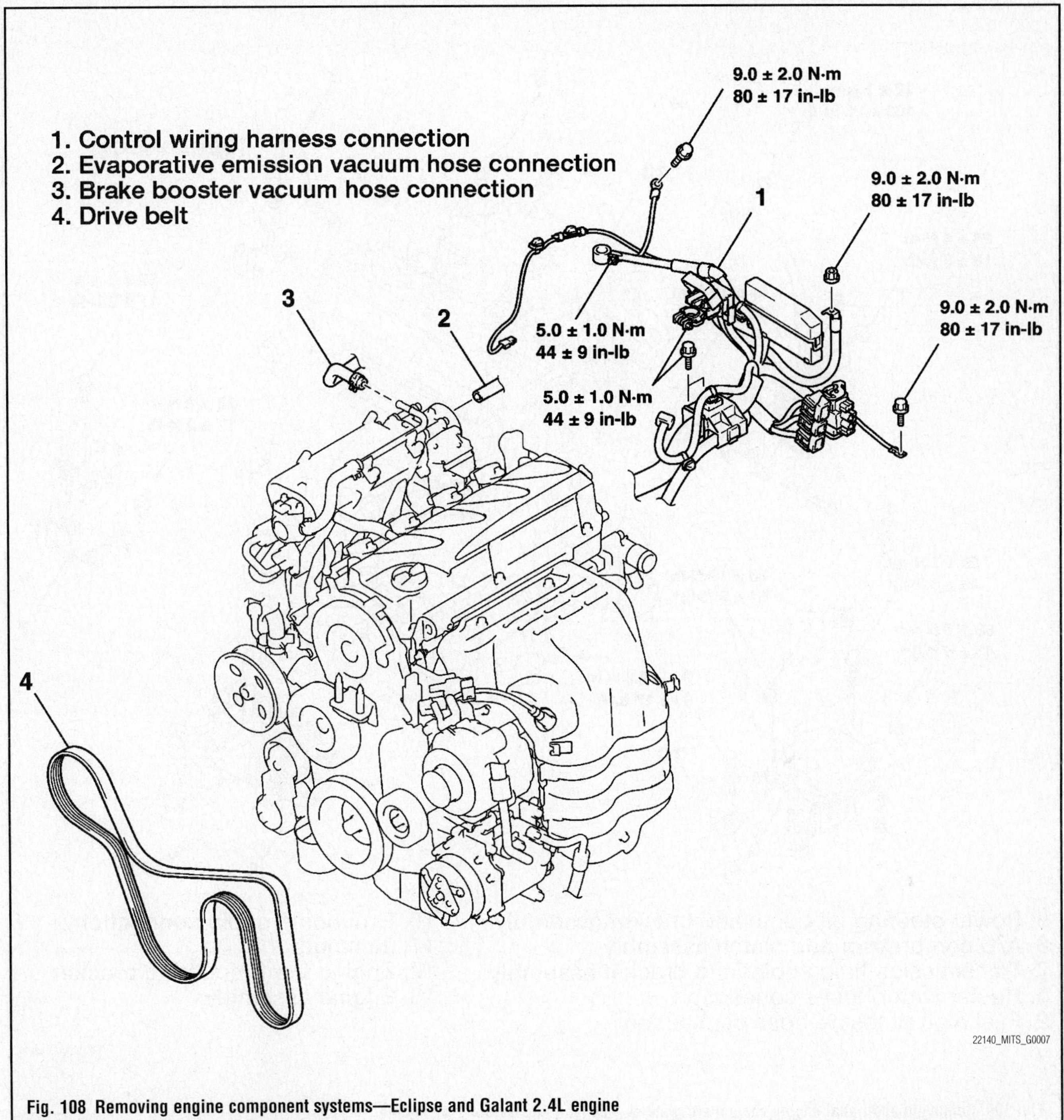

Fig. 108 Removing engine component systems—Eclipse and Galant 2.4L engine

CAUTION

The fuel injection system remains under pressure after the engine has been OFF. Properly relieve fuel pressure before disconnecting any fuel lines. Failure to do so may result in fire or personal injury.

3. Drain the engine oil.
4. Remove or disconnect the following:
   - Negative battery cable
   - Hood
   - Intake air duct
5. Drain the engine coolant.
6. Remove or disconnect the following:
   - Hoses and remove the radiator
   - Engine undercover
7. Attach an engine lifting fixture to the engine and remove the transaxle assembly.
8. Disconnect the following connectors:
   - Air conditioning compressor
   - Power steering pressure switch
   - Heated Oxygen (HO_2S) sensor
   - Engine Coolant Temperature (ECT) gauge sender
   - ECT sensor
   - Manifold Absolute Pressure (MAP) sensor
   - Intake Air Temperature (IAT) sensor
9. Remove or disconnect the following:
   - Power steering pump from the bracket and position the pump out of the way

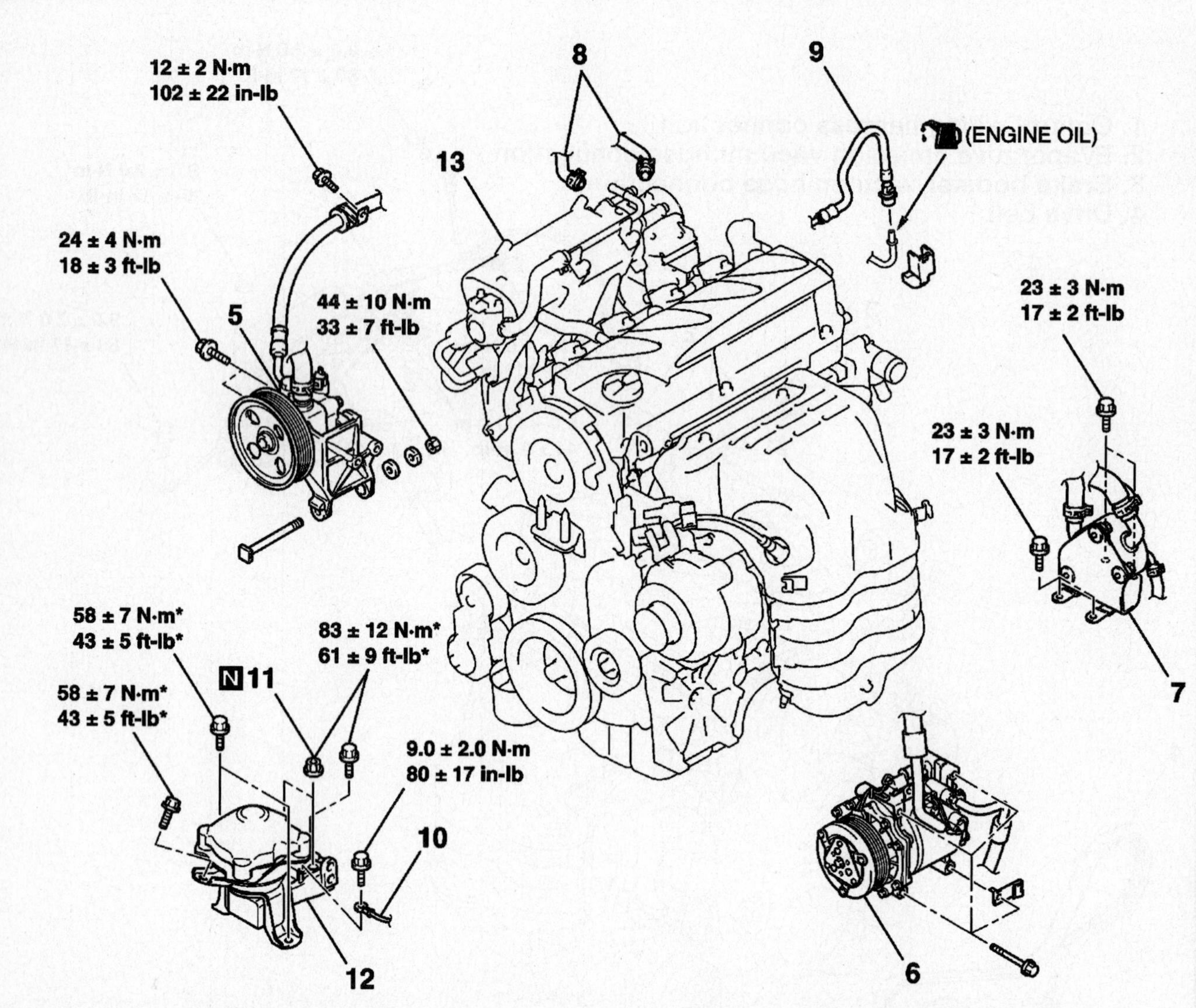

Fig. 109 Removing additional engine component systems—Eclipse and Galant 2.4L engine

- Air conditioning compressor from the bracket and position it out of the way. Do not disconnect the hoses
- Accelerator cable from the throttle body and mounting bracket

10. Disconnect the following connectors:

- Idle Air Control (IAC)
- Knock Sensor (KS)
- Oil pressure switch
- Throttle Position (TP) sensor
- Condenser
- Injectors
- Ignition coil(s)
- Camshaft Position (CMP) sensor
- Crankshaft Position (CKP) sensor
- Engine control wiring harness

11. Remove or disconnect the following:

- Heater hoses from the engine
- Fuel lines from the fuel supply rail
- Purge air hose and the brake booster vacuum hose
- Front exhaust pipe from the manifold

12. Place a floor jack against the oil pan with a piece of wood in between to protect the oil pan.

13. Raise the engine with the jack and remove the engine support fixture.

14. Install a chain hoist to the top of the engine.

15. Remove the engine mount bracket.

16. Lift the engine up and slowly remove it from the engine compartment.

To install:

17. Slowly lower the engine assembly into the vehicle.
18. Position the floor jack under the oil pan with a piece of wood in between. Use the floor jack to adjust the height of the engine while installing the engine mount bracket.
19. Remove the chain hoist and install the engine support fixture.
20. Install or connect the following:
    - Front exhaust pipe to the manifold
    - Brake booster vacuum hose
    - New O-ring on the high pressure fuel line. Apply a small amount of clean engine oil to the O-ring and connect the fuel lines to the fuel supply rail
21. Connect the following connectors:
    - IAC
    - KS
    - Oil pressure switch
    - TP sensor
    - Condenser
    - Injectors
    - Ignition coil(s)
    - CMP sensor
    - CKP sensor
    - Engine control wiring harness
22. Install or connect the following:
    - Accelerator cable, adjust
    - Air conditioning compressor and the power steering pump in their brackets
    - IAT sensor, MAP sensor, ECT sensor and gauge sender, HO2S sensor, power steering pressure switch, and the air conditioning compressor harness connectors
    - Radiator and hoses
    - Transaxle and remove the engine support fixture
    - Engine undercover
    - Intake air duct
    - Hood
    - Negative battery cable
23. Refill the engine oil to the correct level.
24. Refill the cooling system to the correct level.
25. Start the engine and check for leaks.

3.8L Engine

See Figures 110 and 111.

1. Before servicing the vehicle, refer to the Precautions Section.

➡When the engine assembly replacement is performed, use scan tool MB991958 to initialize the learning values needed for the engine to run properly.

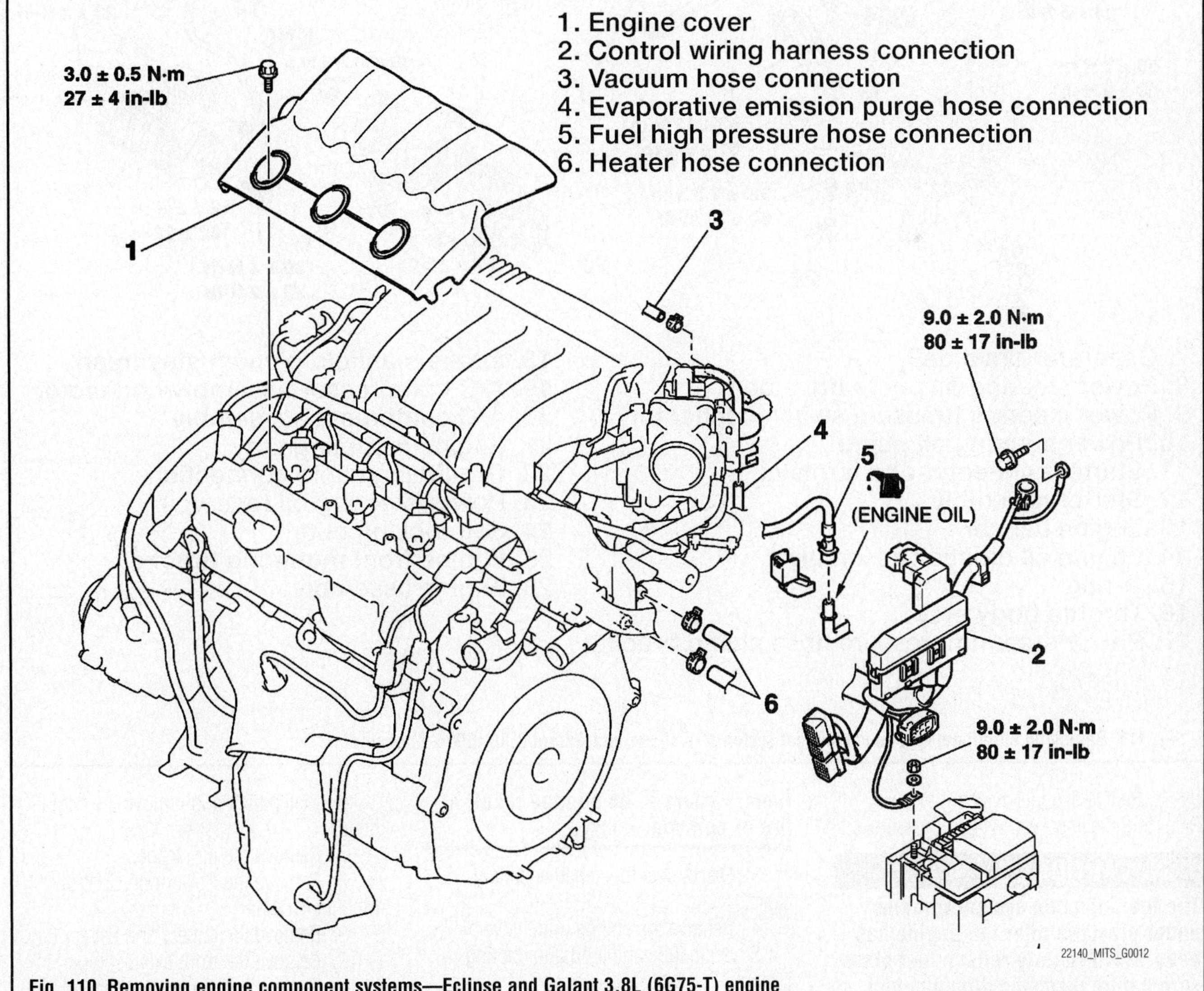

Fig. 110 Removing engine component systems—Eclipse and Galant 3.8L (6G75-T) engine

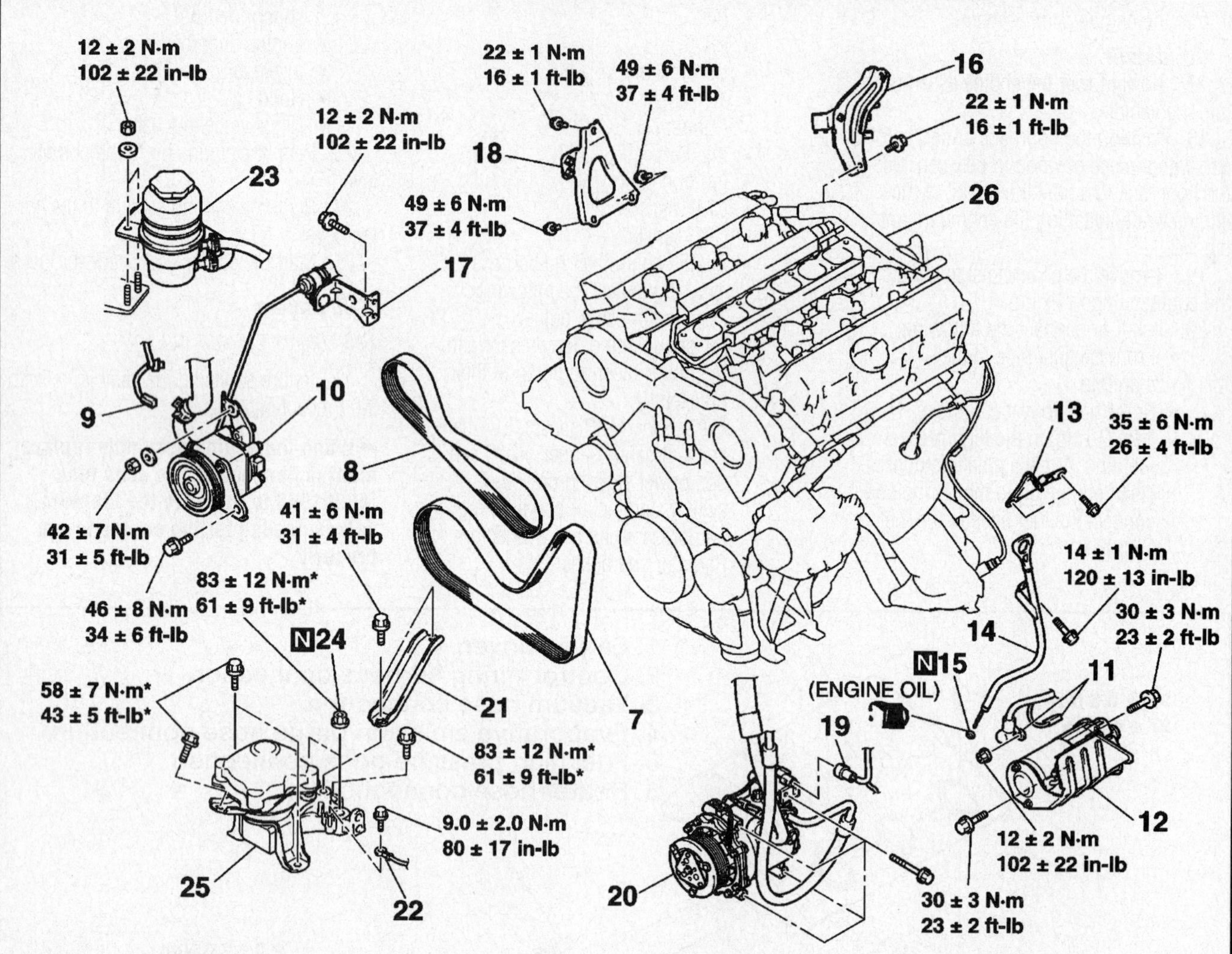

7. Generator drive belt
8. Power steering oil pump drive belt
9. Power steering pressure switch connector
10. Power steering oil pump
11. Starter connector and terminal
12. Starter assembly
13. Engine hanger
14. Engine oil dipstick assembly
15. O-ring
16. Throttle body stay
17. Power steering pressure hose clamp bracket
18. Intake manifold plenum stay (rear)
19. A/C compressor assembly connector
20. A/C compressor assembly
21. Engine mounting stay
22. Grounding cable connection
23. Power steering oil reservoir
24. Self-locking nuts
25. Engine front mounting bracket
26. Engine assembly

22140_MITS_G0013

Fig. 111 Removing additional engine component systems—Eclipse and Galant 3.8L (6G75-T) engine

2. Relieve the fuel system pressure. Refer to Relieving Fuel System Pressure.

⁂ CAUTION

The fuel injection system remains under pressure after the engine has been OFF. Properly relieve fuel pressure before disconnecting any fuel lines. Failure to do so may result in fire or personal injury.

3. Disconnect the negative battery cable.
4. Remove the engine undercover.
5. Drain the engine oil and cooling system.
6. Matchmark and remove the engine hood.
7. Remove the air cleaner.
8. Remove the Powertrain Control Module (PCM).
9. Remove the battery and battery tray.
10. Remove the front exhaust pipe.
11. Remove the strut tower bar.

12. Remove the engine cover.
13. Disconnect the control wiring harness.
14. Disconnect the vacuum hose.
15. Disconnect the evaporative emission purge hose.
16. Disconnect the fuel high pressure hose.
17. Disconnect the heater hose.
18. Remove the drive shaft assembly.
19. Remove the right exhaust manifold.
20. Remove the radiator.
21. Remove the intake manifold plenum.
22. Remove the reservoir assembly.
23. Remove the alternator and power steering pump drive belts.
24. Disconnect the air conditioning compressor electrical connector.
25. Remove the air conditioning compressor from its mounting and position and set it to the side.

➡Do not disconnect the refrigerant lines.

26. Disconnect the power steering pressure switch connector.
27. Remove the power steering pump and position it to the side.

➡Do not disconnect the fluid lines.

28. Remove the power steering pressure hose clamp and bracket.
29. Remove the engine mount stay.
30. Remove the ground cable.
31. Remove the locknuts.
32. Support the engine with a floor jack.
33. Install the engine lifting fixture.
34. Raise the engine slightly to take the tension off the engine mount.
35. Loosen the engine mount retaining nuts and bolts then remove the engine mount.
36. After checking that all cables, hoses, and wiring harness connectors are disconnected from the engine, remove the engine from the vehicle.

To install:

37. Lower the engine into position and install the engine mount nuts and bolts. Tighten the nuts to 18–20 ft. lbs. (25–27 Nm), and the bolts to 33 ft. lbs. (44 Nm).
38. Raise the engine slightly to take the tension off the engine mount.
39. Remove the engine lifting fixture.
40. Install the locknuts.
41. Install the ground cable.
42. Install the engine mount stay.
43. Install the power steering pressure hose clamp and bracket.
44. Install the power steering pump and position it to the side.
45. Connect the power steering pressure switch connector.
46. Install the air conditioning compressor on its mounting and position.
47. Connect the air conditioning compressor electrical connector.
48. Install the alternator and power steering pump drive belts.
49. Install the reservoir assembly.
50. Connect the heater hose connection.
51. Connect the fuel pipe retainer, and main pipe connector.
52. Install the ground wire retaining screw and ground wire.
53. Install the front wiring harness and control wiring harness.
54. Connect the purge hose connection.
55. Connect the vacuum hose connection.
56. Install the EVAP solenoid.
57. Connect the EVAP hose.
58. Connect the EVAP solenoid connector.
59. Install the transaxle.
60. Install the right side exhaust manifold.
61. Install the radiator assembly.
62. Install the battery and battery tray.
63. Install the PCM.
64. Install the engine hood using the matchmarks made during removal.
65. Install the air cleaner.
66. Fill the engine oil to the proper level.
67. Fill the cooling system to the proper level.
68. Install the engine undercover.
69. Connect the negative battery cable.
70. Turn the ignition **ON** and then **OFF**, and keep it **OFF** for at least 10 seconds.
71. Connect a scan tool and complete the vehicle initialization procedure:

➡To complete the initialization procedure the following tools are needed: MB991958 scan tool, MB991824 VCI, MB991827 MUT III USB cable, MB991910 MUT III Main harness "A".

a. Connect the scan tool to the data link connector. To prevent damage to the scan tool be sure that the ignition switch is in the **LOCK** position before connecting the scan tool.
b. Turn the ignition switch to the **ON** position.
c. Select "check mode" from the menu screen.
d. Select "erase memory" from the menu screen.
e. Initialize the learning value.
f. After initialization complete the idle learning procedure.

➡This procedure must be performed when the PCM is replaced, or when the learning value is initialized, as the idling is not stabilized because the learning value in the MFI engine is not completed.

g. Start the engine. Allow the coolant temperature to reach 176°F (80°C) or more.
h. Stop the engine and place the ignition switch in the **LOCK** position.
i. After 10 seconds, restart the engine.
j. For 10 minutes carry out the idling procedure below to confirm that the engine has normal idling.
- Position the transaxle selector lever in the **P** range
- The engine fan is not to be operated
- The engine coolant temperature should be 176°F (80°C) or more

➡If the engine stalls during idling, check the throttle valve of the throttle body for dirt. Correct and perform the procedure again.

72. Be sure to check and adjust all fluid levels, as required.
73. Check fuel system for leaks.

Galant

2.4L Engine

See Figures 112 through 115.

1. Before servicing the vehicle, refer to the Precautions Section.
2. Relieve the fuel system pressure. Refer to Relieving Fuel System Pressure.

CAUTION

The fuel injection system remains under pressure after the engine has been OFF. Properly relieve fuel pressure before disconnecting any fuel lines. Failure to do so may result in fire or personal injury.

3. Disconnect the negative battery cable.
4. Remove the engine undercover.
5. Drain the engine oil and engine coolant.
6. Drain the transaxle fluid.
7. Matchmark and remove the engine hood.

FRONT END STRUCTURE BAR
22140_MITS_G0005

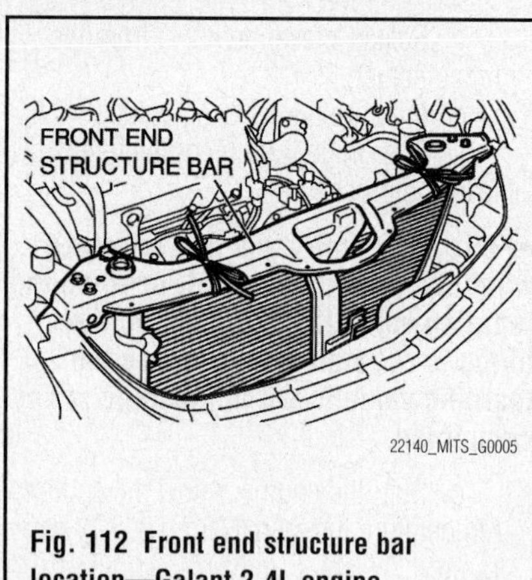

Fig. 112 Front end structure bar location—Galant 2.4L engine

MB991895
MB991527
MB991454
22140_MITS_G0006

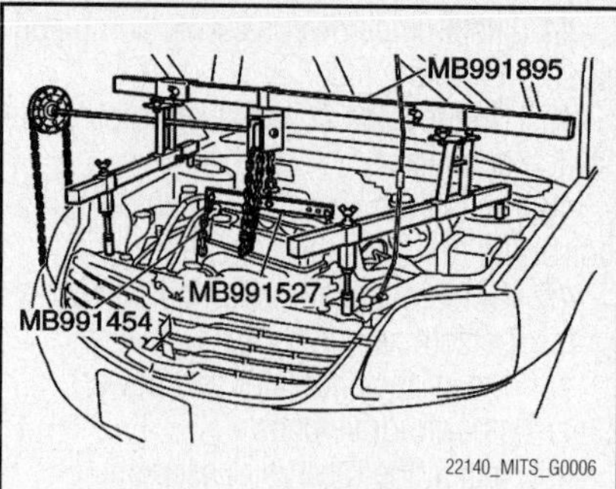

Fig. 113 These special tools are needed to support the engine while the transaxle is removed—Galant 2.4L engine

8. Remove the battery and battery tray.
9. Remove the engine air cleaner.
10. Remove the ECM/PCM.
11. Remove the upper and lower radiator hoses.
12. Remove the radiator.
13. Remove the front exhaust pipes.
14. Disconnect the engine electrical wiring harness.
15. Disconnect the EVAP hose connection.
16. Disconnect the brake booster vacuum hose connection.
17. Remove the drive belt.

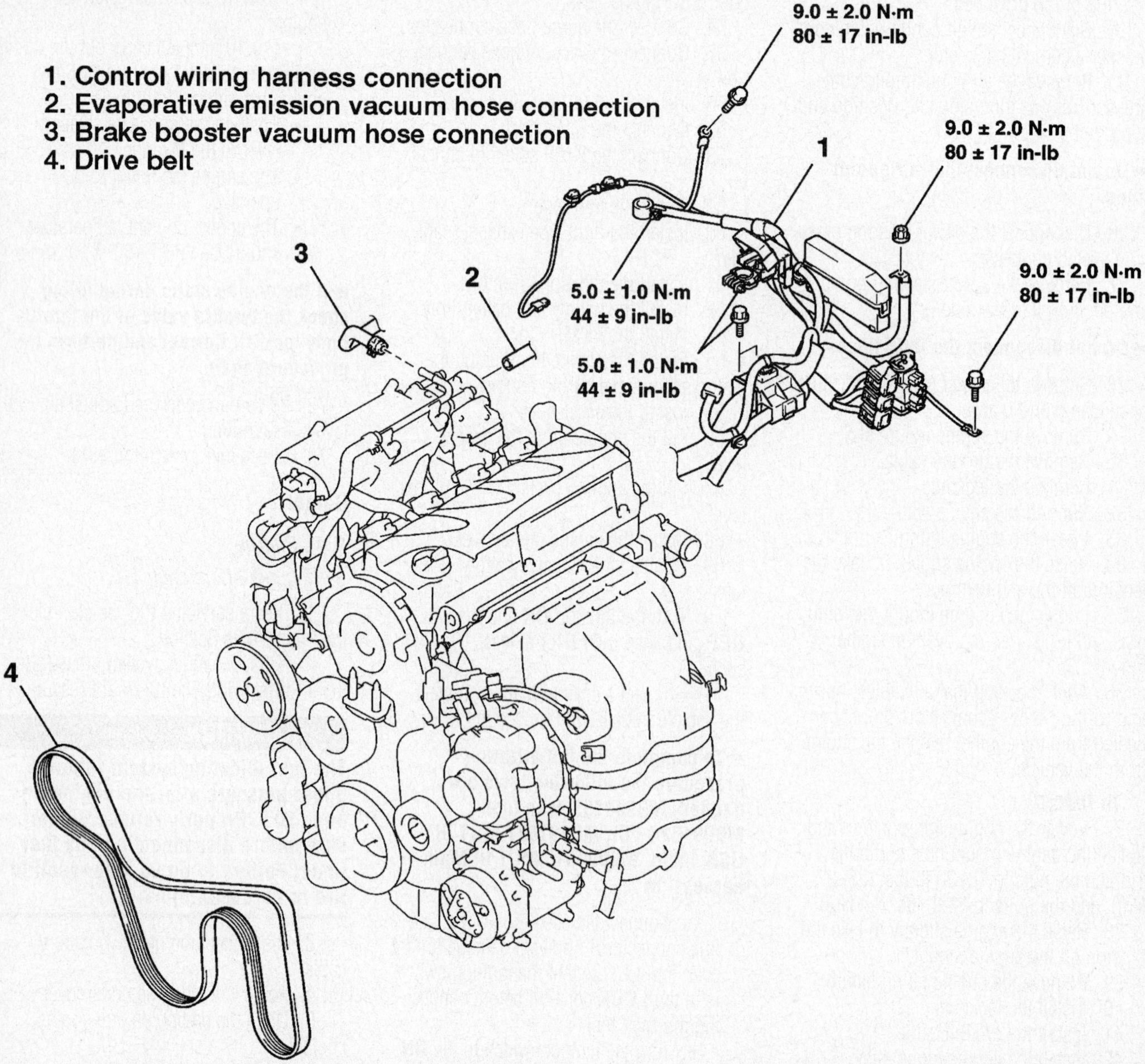

1. Control wiring harness connection
2. Evaporative emission vacuum hose connection
3. Brake booster vacuum hose connection
4. Drive belt
9.0 ± 2.0 N·m
80 ± 17 in-lb
1
9.0 ± 2.0 N·m
80 ± 17 in-lb
9.0 ± 2.0 N·m
80 ± 17 in-lb
3
2
5.0 ± 1.0 N·m
44 ± 9 in-lb
5.0 ± 1.0 N·m
44 ± 9 in-lb
4
22140_MITS_G0007

Fig. 114 Removing engine component systems—Eclipse and Galant 2.4L engine

18. Remove the power steering pump and bracket assembly. Lay the power steering pump to the side. Do not disconnect the fluid lines from the pump.

19. Remove the air conditioning compressor. Lay the air conditioning compressor to the side. Do not disconnect the pressure lines from the pump.

20. With the hose installed, remove the automatic transaxle fluid cooler and bracket assembly from the transaxle case front stopper bracket.

21. Remove the heater hoses.

22. Remove the fuel high pressure hose connection.

23. Secure the engine front end structure bar tool in a location that will not interfere with the engine removal.

24. Remove the transaxle assembly from the vehicle.

25. Remove the grounding cable connection and power steering fluid reservoir.

➡Support the engine with a garage jack when removing the special tools MB991895, MB991454, and MB991527. These special tools were needed and installed to remove the transaxle from the vehicle.

26. Hold the engine assembly in place with the engine lifting device. Properly position a garage jack against the engine oil pan with a piece of wood in between so that the weight of the engine assembly is no longer

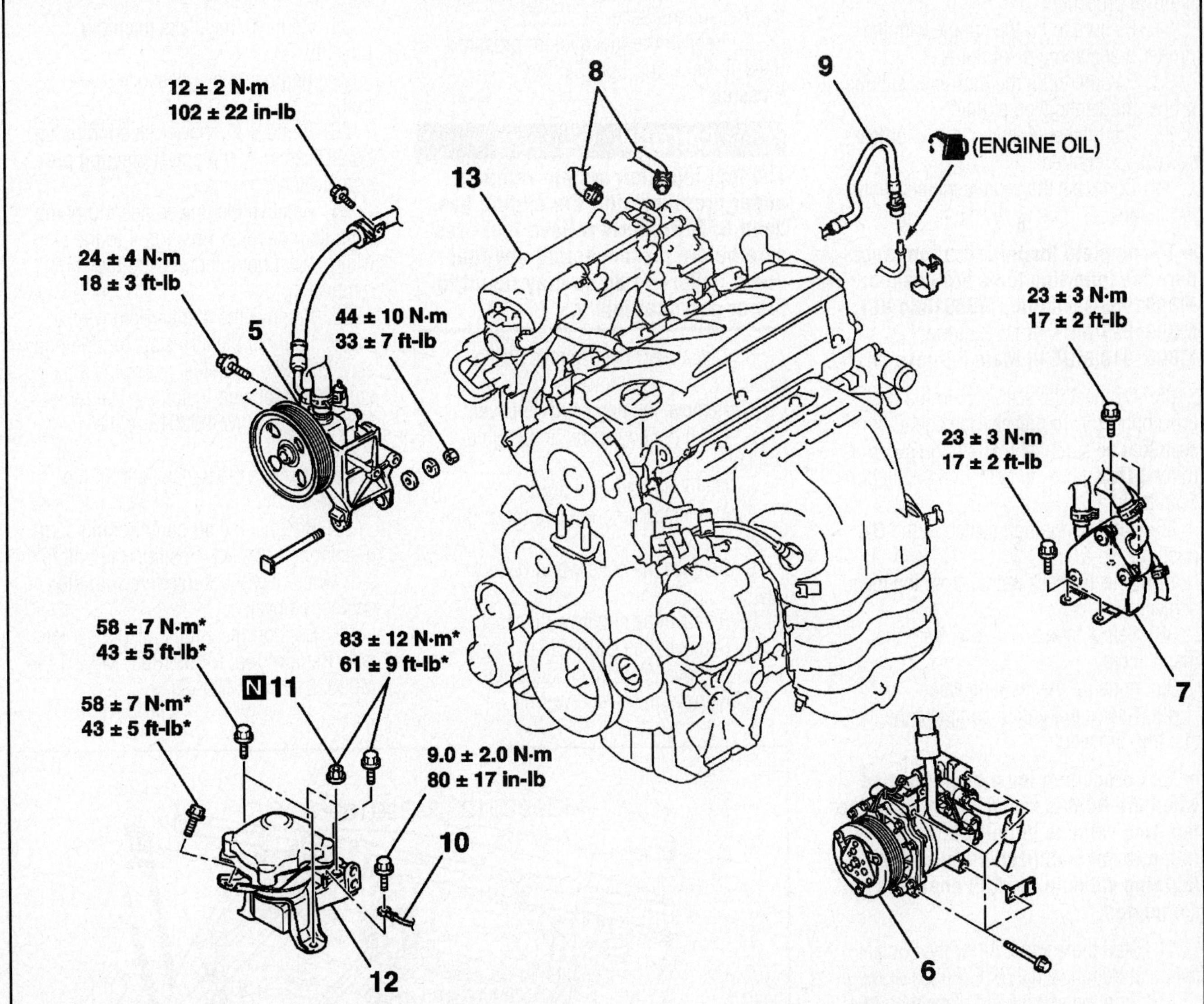

Fig. 115 Removing additional engine component systems—Eclipse and Galant 2.4L engine

being applied to the front mounting bracket.

27. Loosen the engine front mounting bracket mounting nuts and bolts. Remove the front engine mounting bracket.

28. After checking that all cables, hoses, and wiring are disconnected from the engine, lift the engine hoist slowly and remove the engine from the vehicle.

To install:

29. Slowly lower the engine assembly into the vehicle.

30. The installation is the reverse of the removal procedure.

31. Be sure to fill the engine with the proper grade and type of fluids.

32. Be sure to fill the transaxle with the proper grade and type of fluid.

33. Start the engine and check for leaks, correct as required.

34. Complete the vehicle initialization procedure.

➡To complete the initialization procedure the following tools are needed. MB991958 scan tool, MB991824 VCI, MB991827 MUT III USB cable, MB991910 MUT III Main harness "A."

35. Connect the scan tool to the data link connector. To prevent damage to the scan tool be sure that the ignition switch is in the **LOCK** position before connecting the scan tool.

36. Turn the ignition switch to the **ON** position.

37. Select "check mode" from the menu screen.

38. Select "erase memory" from the menu screen.

39. Initialize the learning value.

40. After initialization complete the idle learning procedure.

➡This procedure must be performed when the PCM is replaced, or when the learning value is initialized, as the idling is not stabilized because the learning value in the MFI engine is not completed.

41. Start the engine. Allow the coolant temperature to reach 176°F (80°C) or more.

42. Stop the engine and place the ignition switch in the **LOCK** position.

43. After 10 seconds, restart the engine.

44. For 10 minutes carry out the idling procedure below to confirm that the engine has normal idling.

- Position the transaxle selector lever in the "P" range
- The engine fan is not to be operated
- The engine coolant temperature should be 176°F (80°C) or more

➡If the engine stalls during idling, check the throttle valve of the throttle body for dirt. Correct and perform the procedure again.

45. Road test the vehicle.

46. Check the front end alignment and correct as required.

47. Reprogram the radio stations and security codes, as necessary.

3.8L Engine

See Figures 116 through 121.

1. Before servicing the vehicle, refer to the Precautions Section.

2. Relieve the fuel system pressure. Refer to Relieving Fuel System Pressure.

**** CAUTION**

The fuel injection system remains under pressure after the engine has been OFF. Properly relieve fuel pressure before disconnecting any fuel lines. Failure to do so may result in fire or personal injury.

3. Disconnect the negative battery cable.

4. Remove the engine undercover.

5. Drain the engine oil and engine coolant.

6. Drain the transaxle fluid.

7. Matchmark and remove the engine hood.

8. Remove the battery and battery tray.

9. Remove the engine air cleaner.

10. Remove the ECM/PCM.

11. Remove the front exhaust pipe.

12. Remove the strut tower bar.

13. Remove the radiator grille.

14. Remove the top engine cover.

15. Disconnect the control wiring harness connection.

16. Disconnect the EVAP hose connection.

17. Disconnect the high pressure fuel hose connection.

18. Disconnect and remove the top and bottom radiator hoses. Remove the radiator.

19. Remove the driveshaft.

20. Remove the exhaust manifold right bank.

21. Remove the intake manifold plenum.

22. Remove the accessory drive belt.

23. Remove the power steering pump belt. Disconnect the power steering pressure switch connector.

24. Remove the power steering pump from its mounting. Position it to the side where it will not interfere with the engine removal.

25. Remove the starter. Remove the intake manifold plenum stay, throttle body stay, engine oil dipstick assembly and engine hanger, and install special tools MB992012 and MB992013 to the engine.

26. Remove the transaxle from the vehicle.

27. Remove the air conditioning compressor and bracket. Position the unit to the side where it will not interfere with the engine removal.

28. Support the engine with a garage jack. Remove tool MB991895, MB991454, MB992012 and MB992013.

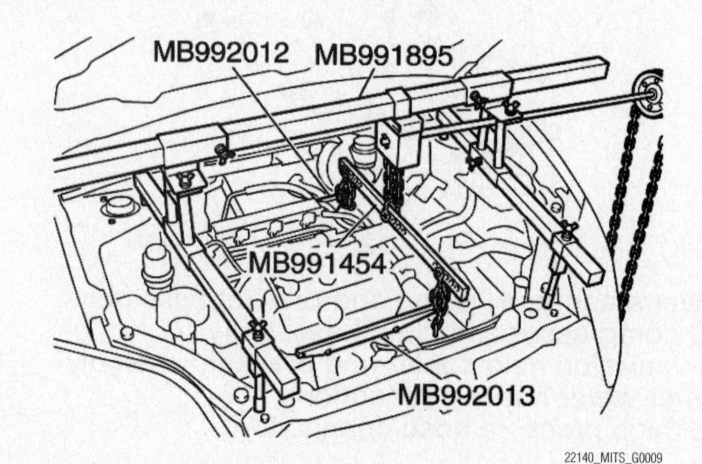

Fig. 116 These special tools are needed to support the engine while the transaxle is removed—Galant 3.8L engine

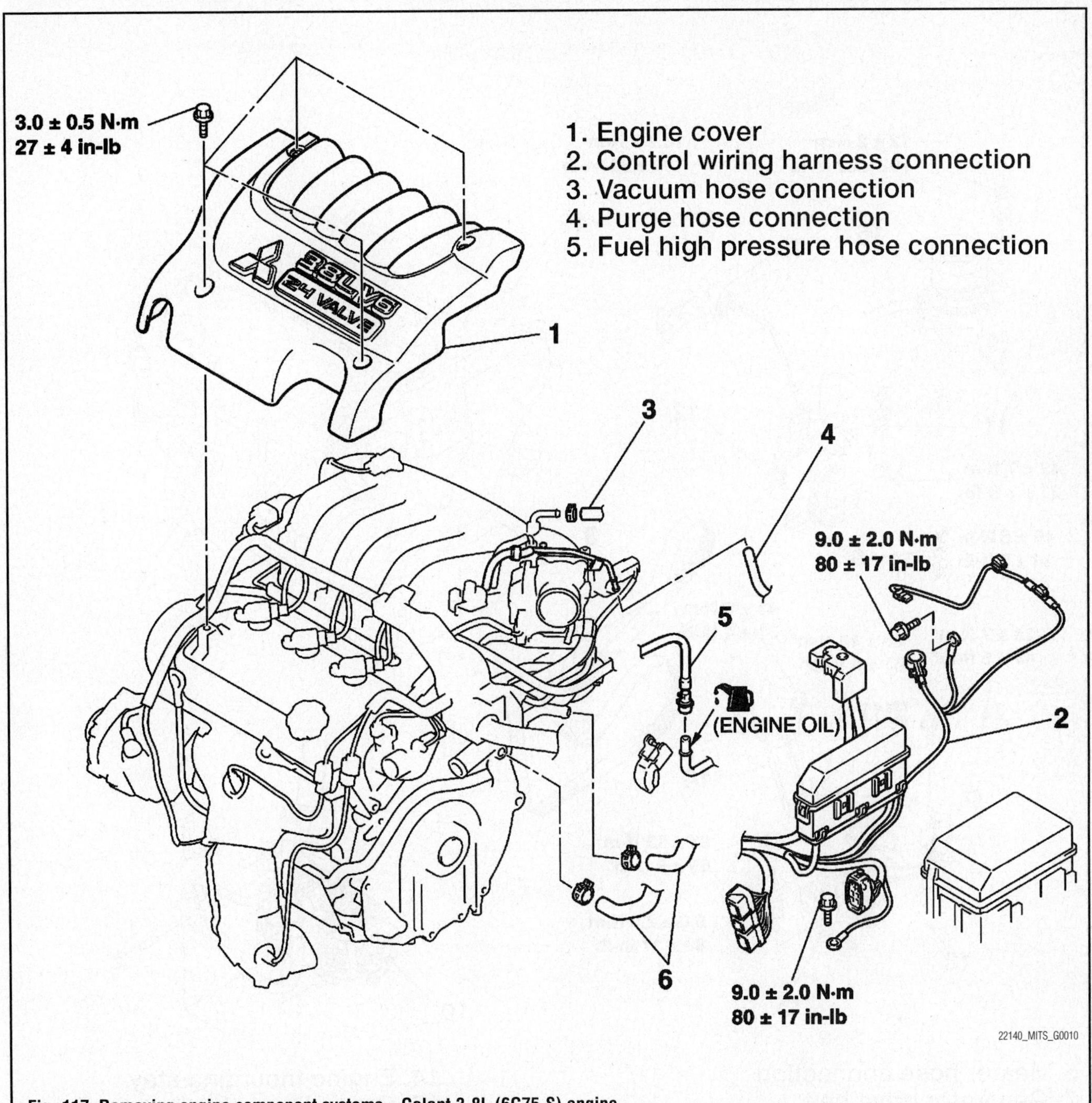

Fig. 117 Removing engine component systems—Galant 3.8L (6G75-S) engine

➡These special tools were needed and installed to remove the transaxle from the vehicle.

29. Hold the engine assembly in place with the engine lifting device. Properly position a garage jack against the engine oil pan with a piece of wood in between so that the weight of the engine assembly is no longer being applied to the front mounting bracket.

30. Loosen the engine front mounting bracket mounting nuts and bolts. Remove the front engine mounting bracket.

31. After checking that all cables, hoses and wiring are disconnected from the engine, lift the engine hoist slowly and remove the engine from the vehicle.

To install:

32. Slowly lower the engine assembly into the vehicle.

33. Place a garage jack against the engine oil pan, with a piece of wood in between. Install the engine mount while adjusting the position of the engine.

34. Support the engine with a garage jack. Remove the chain block (engine lifting device).

35. Install engine hanger tool MB991895 to the strut mounting nuts and the front end structure bar assembling bolts.

36. Continue the installation in the reverse order of the removal procedure.

37. Be sure to fill the engine with the proper grade and type of fluids.

38. Be sure to fill the transaxle with the proper grade and type of fluid.

39. Start the engine and check for leaks, correct as required.

40. Complete the vehicle initialization procedure.

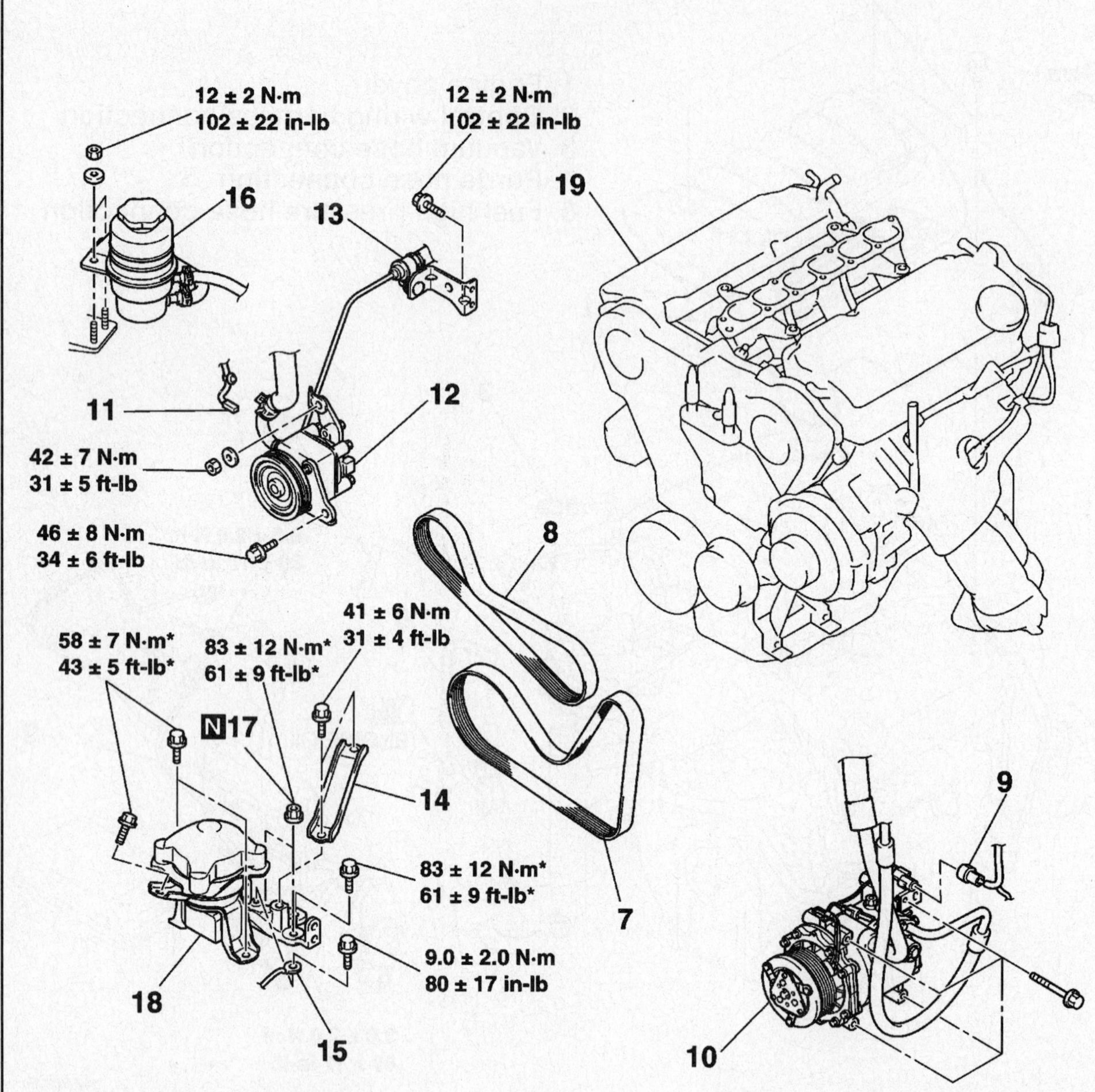

6. Heater hose connection
7. Generator drive belt
8. Power steering oil pump drive belt
9. A/C compressor assembly connector
10. A/C compressor assembly
11. Power steering pressure switch connector
12. Power steering oil pump
13. Power steering pressure hose clamp bracket
14. Engine mounting stay
15. Grounding cable connection
16. Power steering oil reservoir
17. Jam nuts
18. Engine front mounting bracket
19. Engine assembly

22140_MITS_G0011

Fig. 118 Removing additional engine component systems—Galant 3.8L (6G75-S) engine

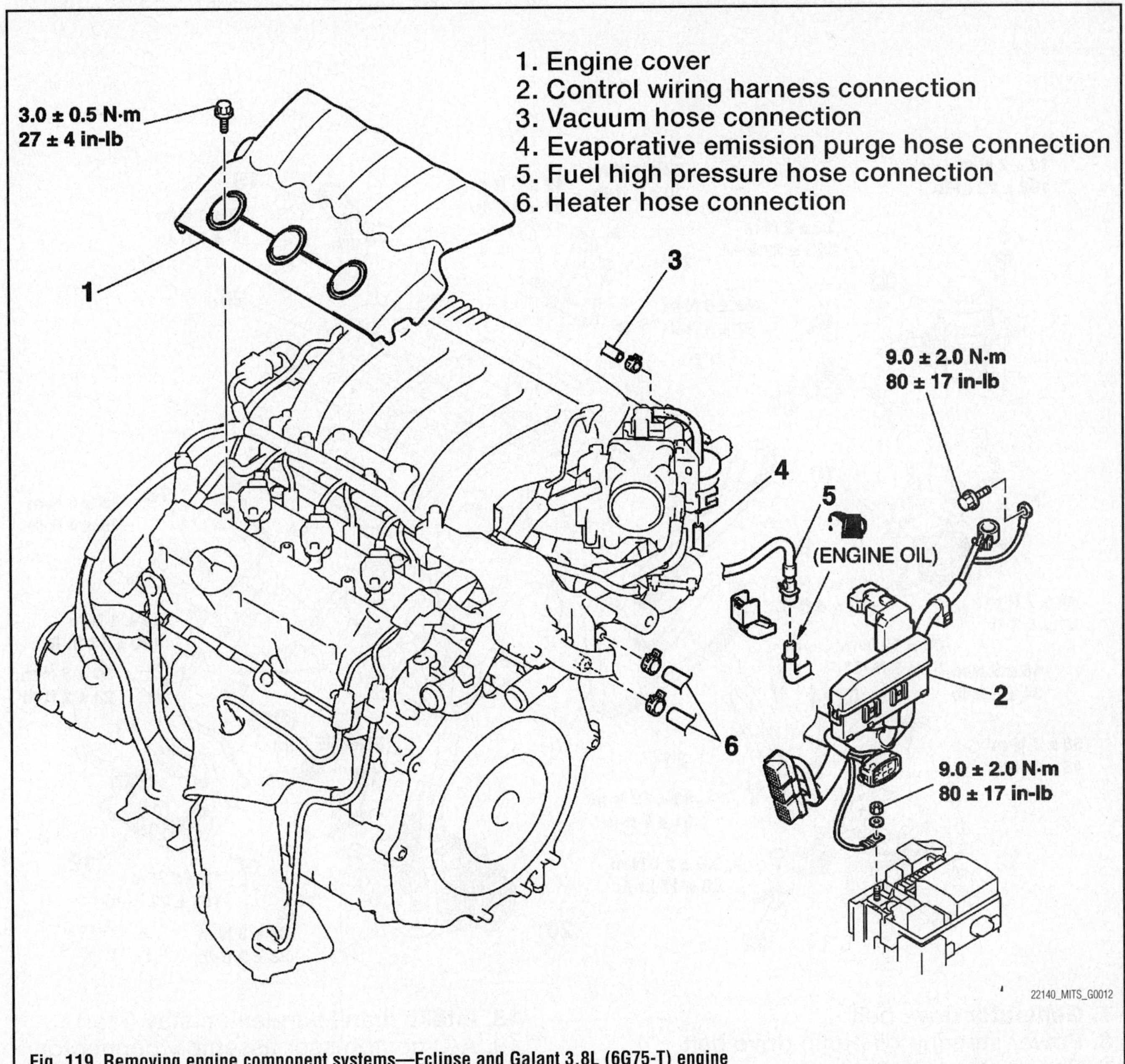

Fig. 119 Removing engine component systems—Eclipse and Galant 3.8L (6G75-T) engine

➡To complete the initialization procedure the following tools are needed. MB991958 scan tool, MB991824 VCI, MB991827 MUT III USB cable, MB991910 MUT III Main harness "A."

41. Connect the scan tool to the data link connector. To prevent damage to the scan tool be sure that the ignition switch is in the **LOCK** position before connecting the scan tool.
42. Turn the ignition switch to the **ON** position.
43. Select "check mode" from the menu screen.
44. Select "erase memory" from the menu screen.
45. Initialize the learning value.
46. After initialization complete the idle learning procedure.

➡This procedure must be performed when the PCM is replaced, or when the learning value is initialized, as the idling is not stabilized because the learning value in the MFI engine is not completed.

47. Start the engine. Allow the coolant temperature to reach 176°F (80°C) or more.
48. Stop the engine and place the ignition switch in the **LOCK** position.
49. After 10 seconds, restart the engine.
50. For 10 minutes carry out the idling procedure below to confirm that the engine has normal idling.
    - Position the transaxle selector lever in the "P" range
    - The engine fan is not to be operated
    - The engine coolant temperature should be 176°F (80°C) or more

➡If the engine stalls during idling, check the throttle valve of the throttle body for dirt. Correct and perform the procedure again.

51. Road test the vehicle.
52. Check the front end alignment and correct as required.
53. Reprogram the radio stations and security codes, as necessary.

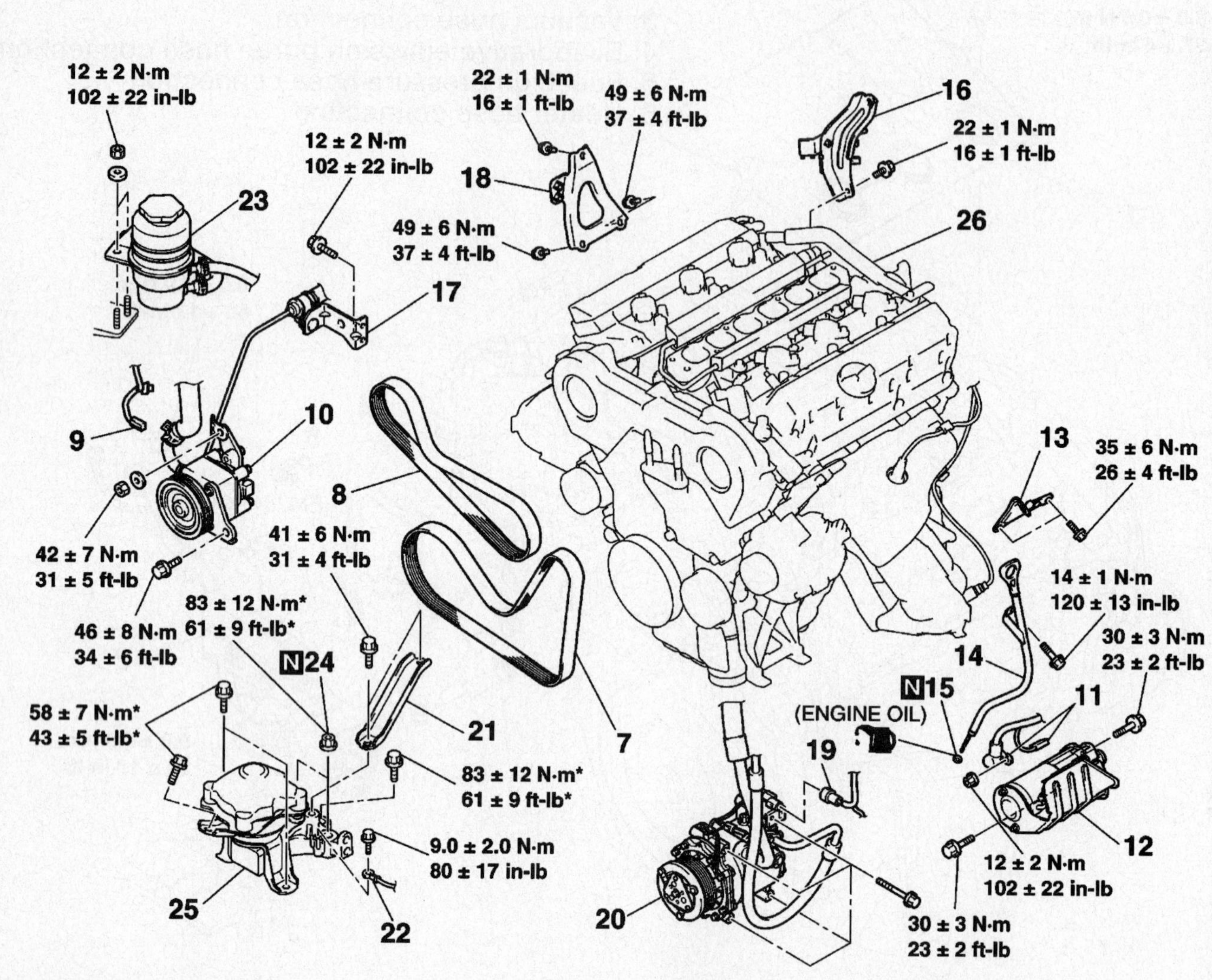

7. Generator drive belt
8. Power steering oil pump drive belt
9. Power steering pressure switch connector
10. Power steering oil pump
11. Starter connector and terminal
12. Starter assembly
13. Engine hanger
14. Engine oil dipstick assembly
15. O-ring
16. Throttle body stay
17. Power steering pressure hose clamp bracket
18. Intake manifold plenum stay (rear)
19. A/C compressor assembly connector
20. A/C compressor assembly
21. Engine mounting stay
22. Grounding cable connection
23. Power steering oil reservoir
24. Self-locking nuts
25. Engine front mounting bracket
26. Engine assembly

22140_MITS_G0013

Fig. 120 Removing additional engine component systems—Eclipse and Galant 3.8L (6G75-T) engine

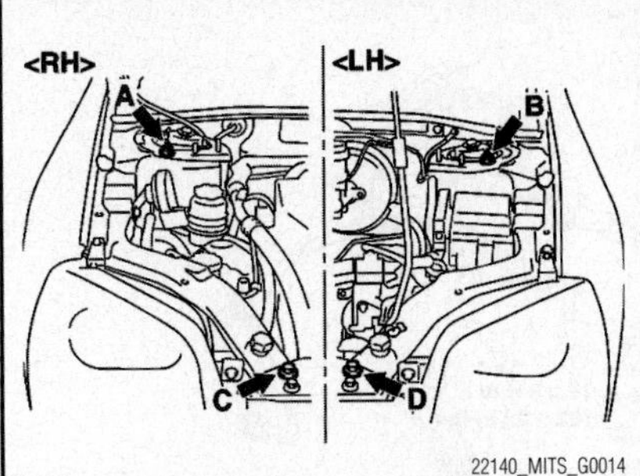

Fig. 121 Install engine hanger tool MB991895 to the strut mounting nuts (A, B) and the front end structure bar assembling bolts (C, D)–Galant 3.8L engine

Lancer

2006–07 2.0L Engine

See Figures 122 through 124.

1. Before servicing the vehicle, refer to the Precautions Section.

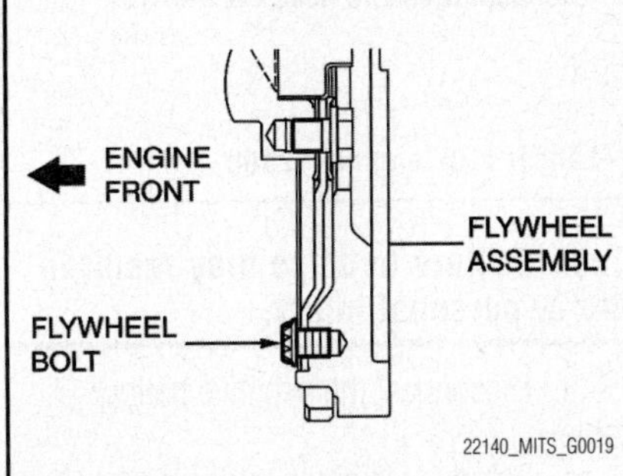

Fig. 122 Do NOT remove the flywheel bolt indicated by the arrow in the accompanying illustration. Removal of this bolt will cause the flywheel to be out of balance—Lancer 2.0L engine—2006–07

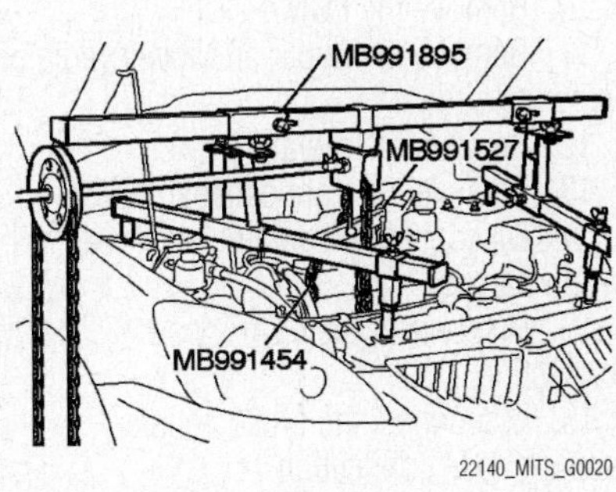

Fig. 123 These special tools are needed to support the engine while the transaxle is removed—Lancer 2.0L engine—2006–07

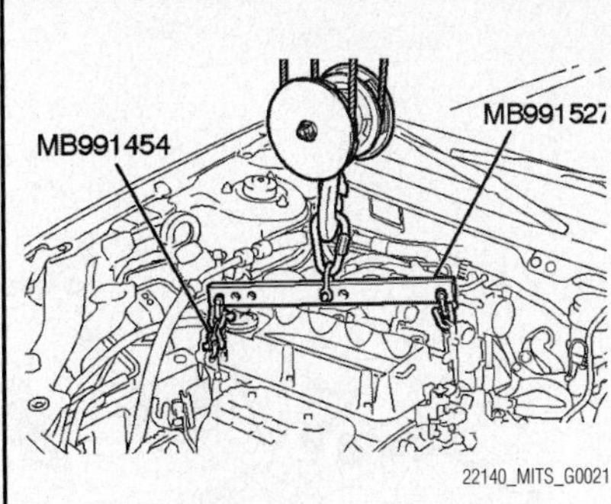

Fig. 124 View of the special tools needed to support the engine during removal—Lancer 2.0L engine—2006–07

2. Relieve the fuel system pressure. Refer to Relieving Fuel System Pressure.

CAUTION

The fuel injection system remains under pressure after the engine has been OFF. Properly relieve fuel pressure before disconnecting any fuel lines. Failure to do so may result in fire or personal injury.

3. Disconnect the negative battery cable.
4. Remove the engine undercover.
5. Drain the engine oil and engine coolant.
6. Drain the transaxle fluid.
7. Matchmark and remove the engine hood.
8. Remove the air cleaner assembly and all adjoining air intake duct work.
9. Remove or disconnect the following:
   - Radiator
   - Front exhaust pipe
   - Battery and battery tray
   - Accelerator cable
10. Detach the electrical connectors from the following components:
    - Air Conditioning (A/C) compressor
    - Power steering oil pressure switch
    - Crank angle sensor
    - Manifold differential pressure sensor
    - Evaporative emission (EVAP) purge solenoid
    - Exhaust Gas Recirculation (EGR) solenoid valve
    - Ignition coil
    - Fuel injectors
    - Throttle Position (TP) sensor
    - Idle Air Control (IAC) motor
    - Engine Coolant Temperature (ECT) sensor
    - Camshaft Position (CMP) sensor
    - Knock Sensor (KS)
    - ECT gauge unit
    - Heated Oxygen (HO_2S) sensor
    - Starter and alternator
    - Oil pressure switch
11. Remove or disconnect the following:
    - Brake booster vacuum hose
    - Power steering pump and A/C compressor drive belt
    - Power steering pump and brace. Position the assembly aside, but do NOT disconnect the fluid line.
    - A/C compressor, but do NOT disconnect the lines

➡Matchmark the installed position of the radiator hoses before disconnecting them.

- Upper and lower radiator hoses
- Heater and purge hoses
- Fuel lines. Discard the O-rings
- Transaxle assembly

WARNING

Do NOT remove the flywheel bolt indicated by the arrow in the accompanying illustration. Removal of this bolt will cause the flywheel to be out of balance.

12. Remove the engine mount insulator and bracket as follows:
    a. Support the engine with a suitable floor jack.
    b. Remove the special tools that were installed for transaxle removal.
    c. Support the engine with special tools MB991454 and MB991527 attached to a chain block or engine hoist.
    d. Place a jack under the oil pan with a block of wood in between to protect the pan. Jack up the engine to take the weight off the engine mount insulator and bracket, then remove the insulator and bracket.
13. Make sure that all cables, hoses and harnesses are disconnected from the engine, then use the engine hoist to slowly lift the engine up and out of the engine compartment.

To install:

14. Installation is the reverse of the removal procedure.
15. Fill the coolant system and engine crankcase with the proper amount and type of fluids.

16. Connect the negative battery cable.
17. Connect the scan tool to the data link connector. To prevent damage to the scan tool be sure that the ignition switch is in the **LOCK** position before connecting the scan tool.
18. Turn the ignition switch to the **ON** position.
19. Select "check mode" from the menu screen.
20. Select "erase memory" from the menu screen.
21. Initialize the learning value.
22. Start the engine. Allow the coolant temperature to reach 176°F (80°C) or more.
23. Stop the engine and place the ignition switch in the **LOCK** position.
24. After 10 seconds, restart the engine.
25. For 10 minutes carry out the idling procedure below to confirm that the engine has normal idling.
   - Position the transaxle selector lever in the "P" range
   - The engine fan is not to be operated
   - The engine coolant temperature should be 176°F (80°C) or more.

➡If the engine stalls during idling, check the throttle valve of the throttle body for dirt. Correct and perform the procedure again.

26. Road test the vehicle.
27. Check the front end alignment and correct as required.
28. Reprogram the radio stations and security codes, as necessary.

2008 2.0L Engine

See Figures 125 through 127.

1. Before servicing the vehicle, refer to the Precautions Section.

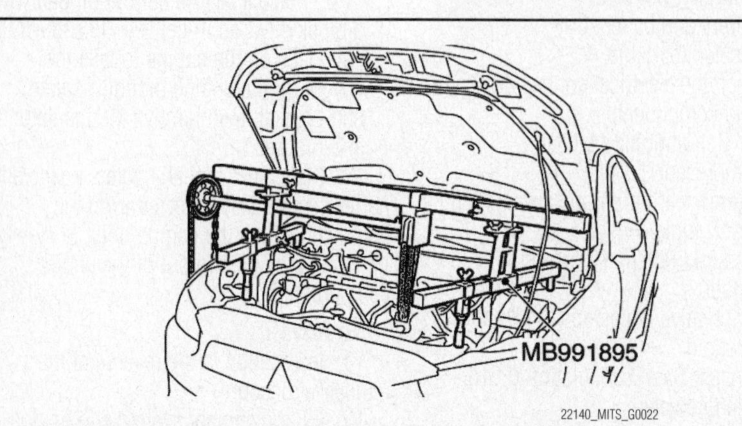

Fig. 125 These special tools are needed to support the engine while the transaxle is removed—Lancer 2.0L engine—2008

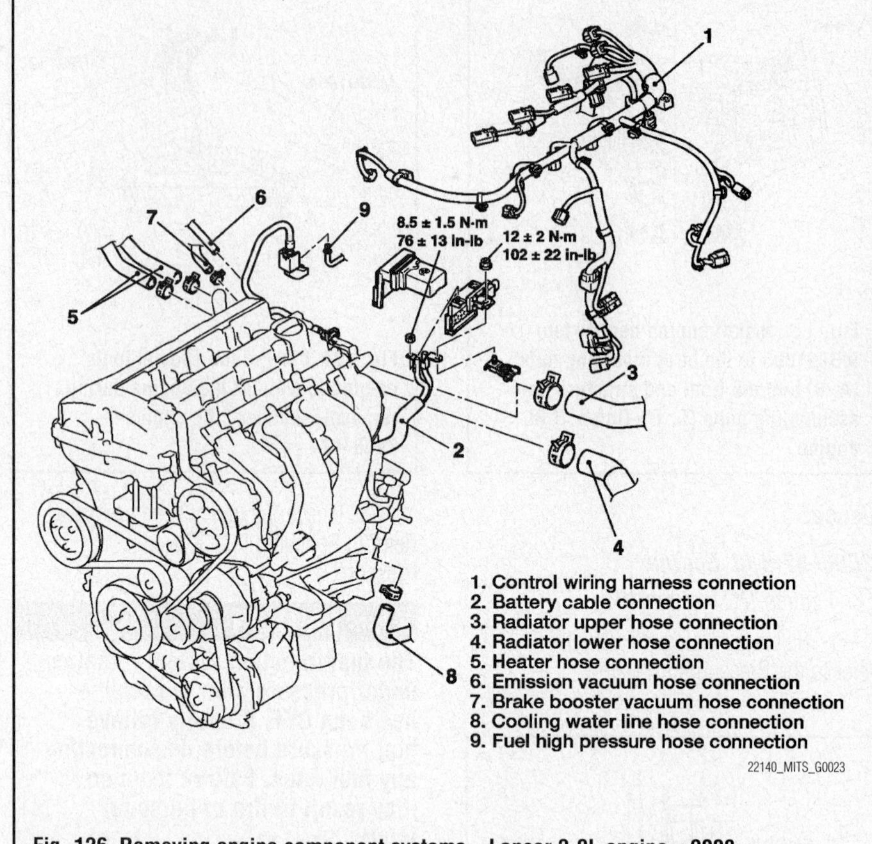

Fig. 126 Removing engine component systems—Lancer 2.0L engine—2008

➡When the engine assembly is replaced, use scan tool MB991958 to initialize the learning values for engine performance.

2. Relieve the fuel system pressure. Refer to Relieving Fuel System Pressure.

⁂ CAUTION

The fuel injection system remains under pressure after the engine has been OFF. Properly relieve fuel pressure before disconnecting any fuel lines. Failure to do so may result in fire or personal injury.

3. Disconnect the negative battery cable.
4. Remove the engine undercover.
5. Drain the engine oil and engine coolant.
6. Drain the transaxle fluid.
7. Matchmark and remove the engine hood.
8. Remove the battery and battery tray.
9. Remove the engine air cleaner.
10. Remove the ECM/PCM.
11. Remove the upper and lower radiator hoses.
12. Remove the radiator.
13. Remove the front exhaust pipes.
14. Disconnect the engine electrical wiring harness.
15. Disconnect the EVAP hose connection.
16. Disconnect the brake booster vacuum hose connection.
17. Remove the drive belt.
18. Remove the power steering pump and bracket assembly. Lay the power steering pump to the side. Do not disconnect the fluid lines from the pump.

19. Remove the air conditioning compressor. Lay the air conditioning compressor to the side. Do not disconnect the pressure lines from the pump.

20. For automatic transaxle, remove the transaxle fluid cooler and bracket assembly from the transaxle case front stopper bracket.

21. Remove the heater hoses.

22. Remove the fuel high pressure hose connection.

23. Remove the transaxle assembly from the vehicle.

24. Remove the grounding cable connection and power steering fluid reservoir.

➡**Support the engine with a garage jack when removing the special tools MB991895, MB991454, and MB991527. These special tools were needed and installed to remove the transaxle from the vehicle.**

25. Hold the engine assembly in place with the engine lifting device. Properly position a garage jack against the engine oil pan with a piece of wood in between so that the weight of the engine assembly is no longer being applied to the front mounting bracket.

26. Loosen the engine front mounting bracket mounting nuts and bolts. Remove the front engine mounting bracket.

27. After checking that all cables, hoses, and wiring are disconnected from the engine, lift the engine hoist slowly and remove the engine from the vehicle.

To install:

28. Slowly lower the engine assembly into the vehicle.

29. The installation is the reverse of the removal procedure.

30. Be sure to fill the engine with the proper grade and type of fluids.

31. Be sure to fill the transaxle with the proper grade and type of fluid.

32. Start the engine and check for leaks, correct as required.

33. Complete the vehicle initialization procedure.

➡**To complete the initialization procedure the following tools are needed. MB991958 scan tool, MB991824 VCI, MB991827 MUT III USB cable, MB991910 MUT III Main harness "A."**

34. Connect the scan tool to the data link connector. To prevent damage to the scan tool be sure that the ignition switch is in the **LOCK** position before connecting the scan tool.

35. Turn the ignition switch to the **ON** position.

36. Select "check mode" from the menu screen.

37. Select "erase memory" from the menu screen.

38. Initialize the learning value.

39. After initialization complete the idle learning procedure.

➡**This procedure must be performed when the PCM is replaced, or when the learning value is initialized, as the idling is not stabilized because the learning value in the MFI engine is not completed.**

40. Start the engine. Allow the coolant temperature to reach 176°F (80°C) or more.

41. Stop the engine and place the ignition switch in the **LOCK** position.

42. After 10 seconds, restart the engine.

43. For 10 minutes carry out the idling procedure below to confirm that the engine has normal idling.

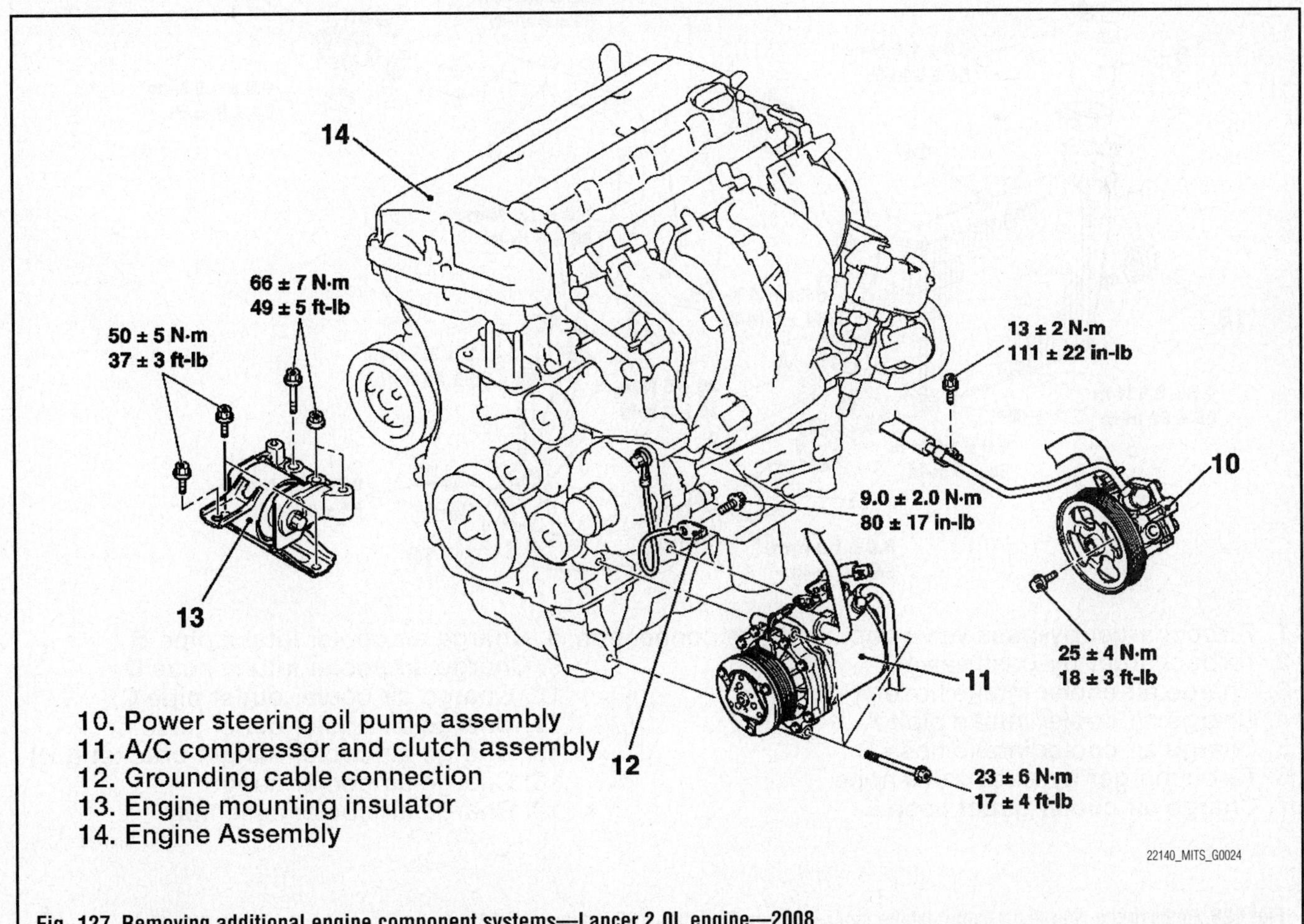

Fig. 127 Removing additional engine component systems—Lancer 2.0L engine—2008

- Position the transaxle selector lever in the "P" range
- The engine fan is not to be operated
- The engine coolant temperature should be 176°F (80°C) or more

➡If the engine stalls during idling, check the throttle valve of the throttle body for dirt. Correct and perform the procedure again.

44. Road test the vehicle.
45. Check the front end alignment and correct as required.
46. Reprogram the radio stations and security codes, as necessary.

Lancer Evolution

2008 2.0L Turbo Engine

See Figures 128 through 132.

1. Before servicing the vehicle, refer to the Precautions Section.

➡When the engine assembly is replaced, use scan tool MB991958 to initialize the learning values for engine performance.

2. Relieve the fuel system pressure. Refer to Relieving Fuel System Pressure.

⁂ CAUTION

The fuel injection system remains under pressure after the engine has been OFF. Properly relieve fuel pressure before disconnecting any fuel lines. Failure to do so may result in fire or personal injury.

3. Disconnect the negative battery cable.
4. Remove the engine undercover.
5. Drain the engine oil and engine coolant.
6. Drain the transaxle fluid.

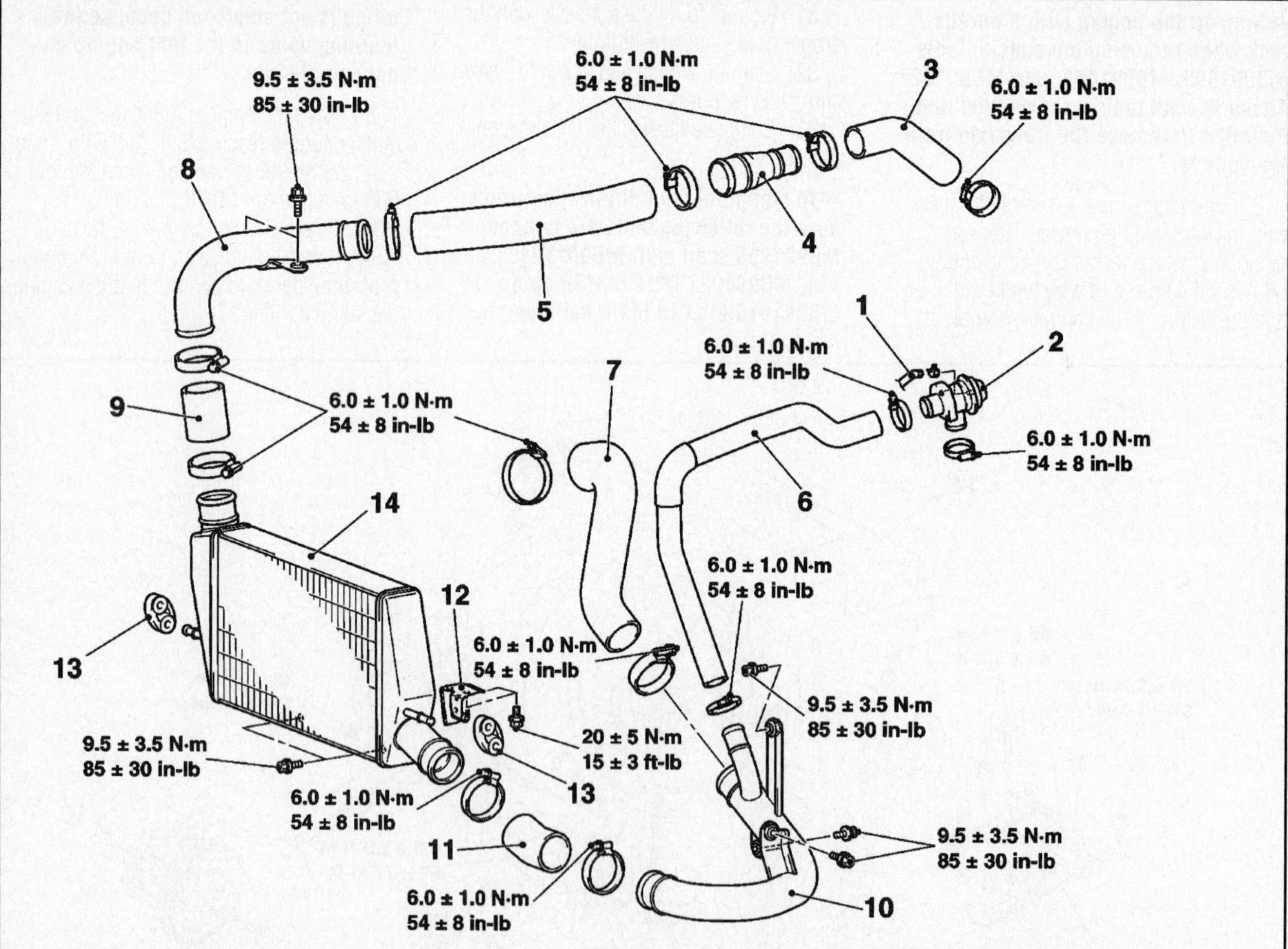

Fig. 128 Remove the charge air cooler intake pipes—Lancer Evolution 2.0L engine—2008

7. Drain the transfer case.
8. Matchmark and remove the engine hood.
9. Remove the battery and battery tray.
10. Remove the strut tower bar.
11. Remove the charge air cooler intake pipes.
12. Remove the upper and lower radiator hoses. Remove the radiator.
13. Remove the front axle crossmember bar.
14. Remove the front exhaust pipe. Remove the turbocharger air outlet fitting.
15. Disconnect the ignition coil connectors.
16. Disconnect the Heated Oxygen (HO_2S) sensor connector.
17. Disconnect the Crankshaft Position (CKP) sensor connector.
18. Disconnect the manifold differential pressure sensor connector.
19. Disconnect the fuel pressure solenoid connector.
20. Disconnect the Knock Sensor (KS) connector.
21. Disconnect the evaporative emission purge solenoid connector.
22. Disconnect the Throttle Position (TP) sensor connector.
23. Disconnect the Idle Air Control (IAC) motor connector.
24. Disconnect the fuel injector connectors.
25. Disconnect the Camshaft Position (CMP) sensor connector.
26. Disconnect the Engine Coolant Temperature (ECT) gauge unit connector.
27. Disconnect the ECT sensor connector.
28. Remove the control wiring harness and transaxle wiring harness combination.
29. Disconnect the ground cable connection.
30. Disconnect the alternator wiring connectors.
31. Disconnect the engine oil pressure switch connector.
32. Remove the turbocharger wastegate actuator bolts.
33. Remove the accessory drive belt.
34. Disconnect the brake booster vacuum hose.
35. Disconnect the purge hose connection.
36. Disconnect the power steering oil pressure switch connector.
37. Remove the power steering oil pump heat protector.
38. Remove the power steering oil pump, mounting bracket, and reservoir assembly. Do not disconnect hoses from pump.
39. Disconnect the A/C compressor connector.
40. Remove the A/C compressor and clutch assembly with hoses intact to one side and tie back out of the way of engine removal.
41. Disconnect the engine oil cooler feed and return hose connections.
42. Disconnect the heater water hose connections.
43. Disconnect the fuel return line and high-pressure hose connections. Discard the O-rings.
44. Remove the transaxle and transfer case assemblies from the engine.
45. Remove the engine front roll stopper bracket and engine front mounting bracket through-bolts as follows:
   a. Support the engine with a suitable floor jack.
   b. Remove the special tools that were installed for transaxle removal.
   c. Support the engine with special tool MB991454 attached to a chain block or engine hoist.

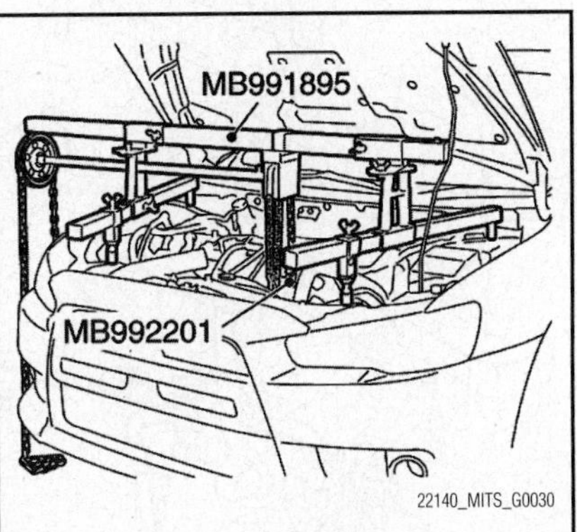

Fig. 129 These special tools are needed to support the engine while the transaxle is removed—Lancer Evolution 2.0L engine—2008

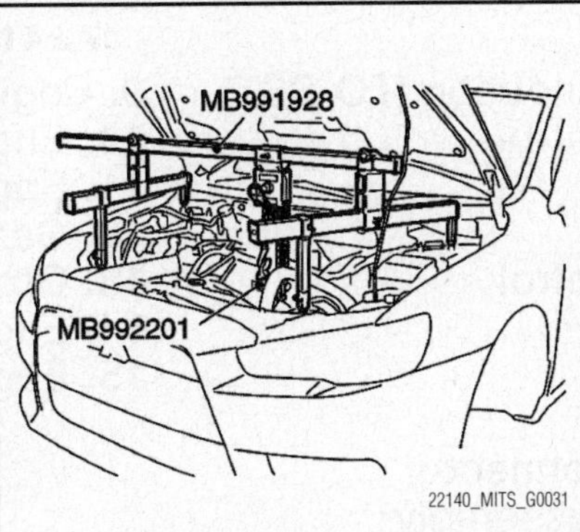

Fig. 130 Or these special tools can be used to support the engine while the transaxle is removed—Lancer Evolution 2.0L engine—2008

   d. Place a jack under the oil pan with a block of wood in between to protect the pan. Jack up the engine to take the weight off the engine mount insulator and bracket, then remove the front roll stopper bracket and engine front mounting bracket through-bolts.
46. Make sure that all cables, hoses, and harnesses are disconnected from the engine, then use the engine hoist to slowly lift the engine up and out of the engine compartment.

To install:

47. Install the engine and secure in position. The engine mount through-bolts should not be tightened until the full weight of the engine is on the mounts. Tighten the engine front roll stopper bracket through-bolt to 34–44 ft. lbs. (45–59 Nm). Tighten the engine front mounting bracket bolt to 66–80 ft. lbs. (88–108 Nm).
48. Continue installation in the reverse of the removal procedure.
49. Be sure to fill the engine with the proper grade and type of fluids.
50. Be sure to fill the transaxle with the proper grade and type of fluid.
51. Start the engine and check for leaks, correct as required.
52. Complete the vehicle initialization procedure.

➡To complete the initialization procedure the following tools are needed. MB991958 scan tool, MB991824 VCI, MB991827 MUT III USB cable, MB991910 MUT III Main harness "A."

53. Connect the scan tool to the data link connector. To prevent damage to the scan tool be sure that the ignition switch is in the **LOCK** position before connecting the scan tool.

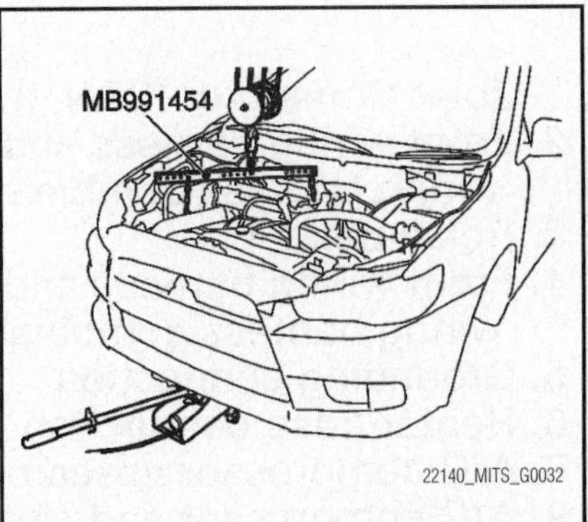

Fig. 131 Support the engine with special tool MB991454 attached to a chain block or engine hoist—Lancer Evolution 2.0L engine—2008

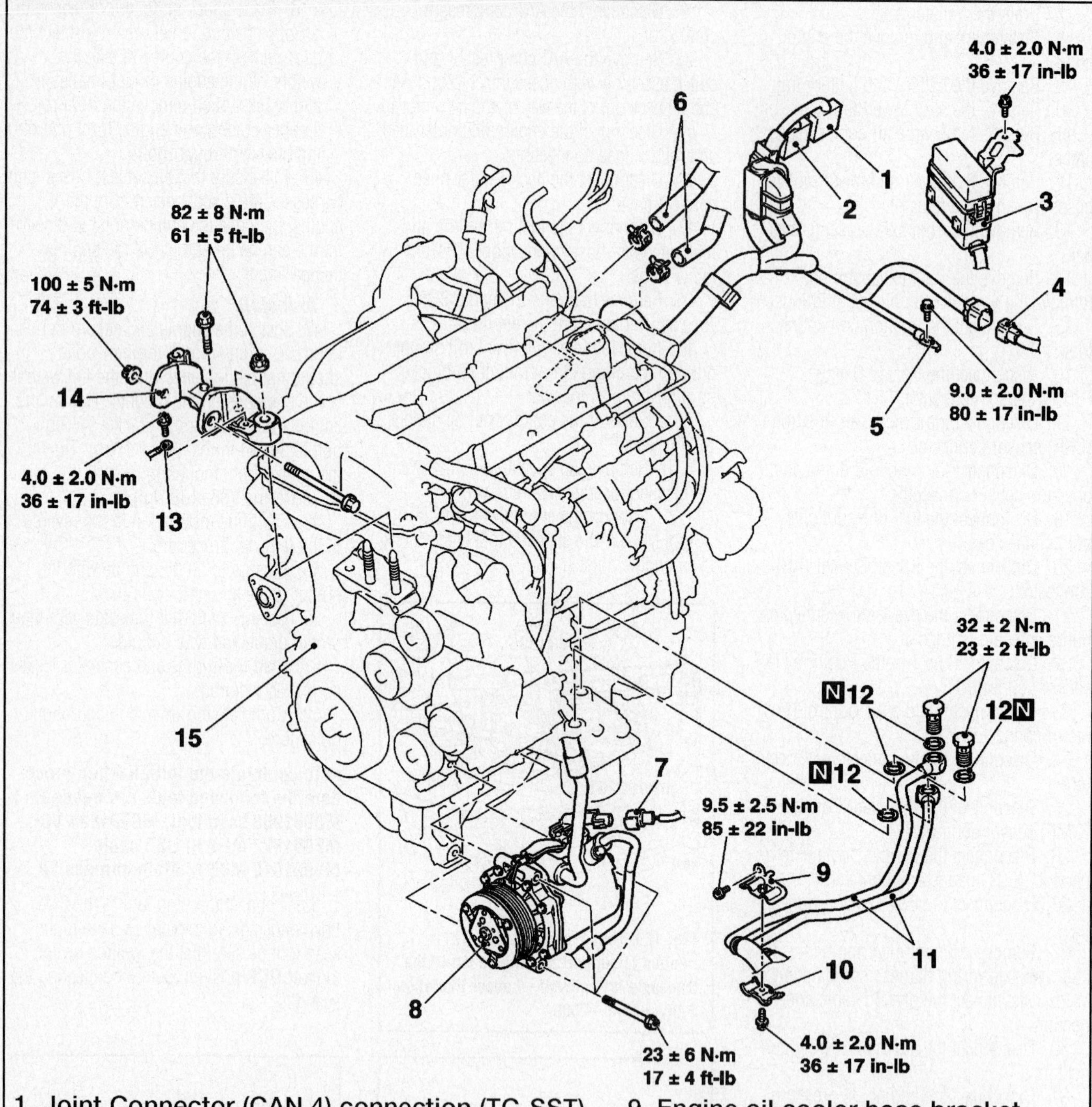

1. Joint Connector (CAN 4) connection (TC-SST)
2. Front wiring harness and control wiring harness combination
3. Relay box
4. Front wiring harness and control wiring harness combination
5. Grounding connection
6. Heater hose connection
7. A/C compressor assembly connector
8. A/C compressor and clutch assembly
9. Engine oil cooler hose bracket
10. Engine oil cooler line clamp
11. Engine oil cooler hose connection
12. Gasket
13. Grounding cable connection
14. Engine mounting bracket
15. Engine assembly

22140_MITS_G0033

Fig. 132 Removing engine component systems—Lancer Evolution 2.0L engine—2008

54. Turn the ignition switch to the **ON** position.

55. Select "check mode" from the menu screen.

56. Select "erase memory" from the menu screen.

57. Initialize the learning value.

58. After initialization, complete the idle learning procedure.

➡**This procedure must be performed when the PCM is replaced, or when the learning value is initialized, as the idling is not stabilized because the learning value in the MFI engine is not completed.**

59. Start the engine. Allow the coolant temperature to reach 176°F (80°C) or more.

60. Stop the engine and place the ignition switch in the **LOCK** position.

61. After 10 seconds, restart the engine.

62. For 10 minutes carry out the idling procedure below to confirm that the engine has normal idling.

- Position the transaxle selector lever in the "P" range
- The engine fan is not to be operated
- The engine coolant temperature should be 176°F (80°C) or more

➡**If the engine stalls during idling, check the throttle valve of the throttle body for dirt. Correct and perform the procedure again.**

63. Road test the vehicle.

64. Check the front end alignment and correct as required.

65. Reprogram the radio stations and security codes, as necessary.

EXHAUST MANIFOLD

REMOVAL & INSTALLATION

Eclipse & Galant

2.4L Engine

See Figure 133.

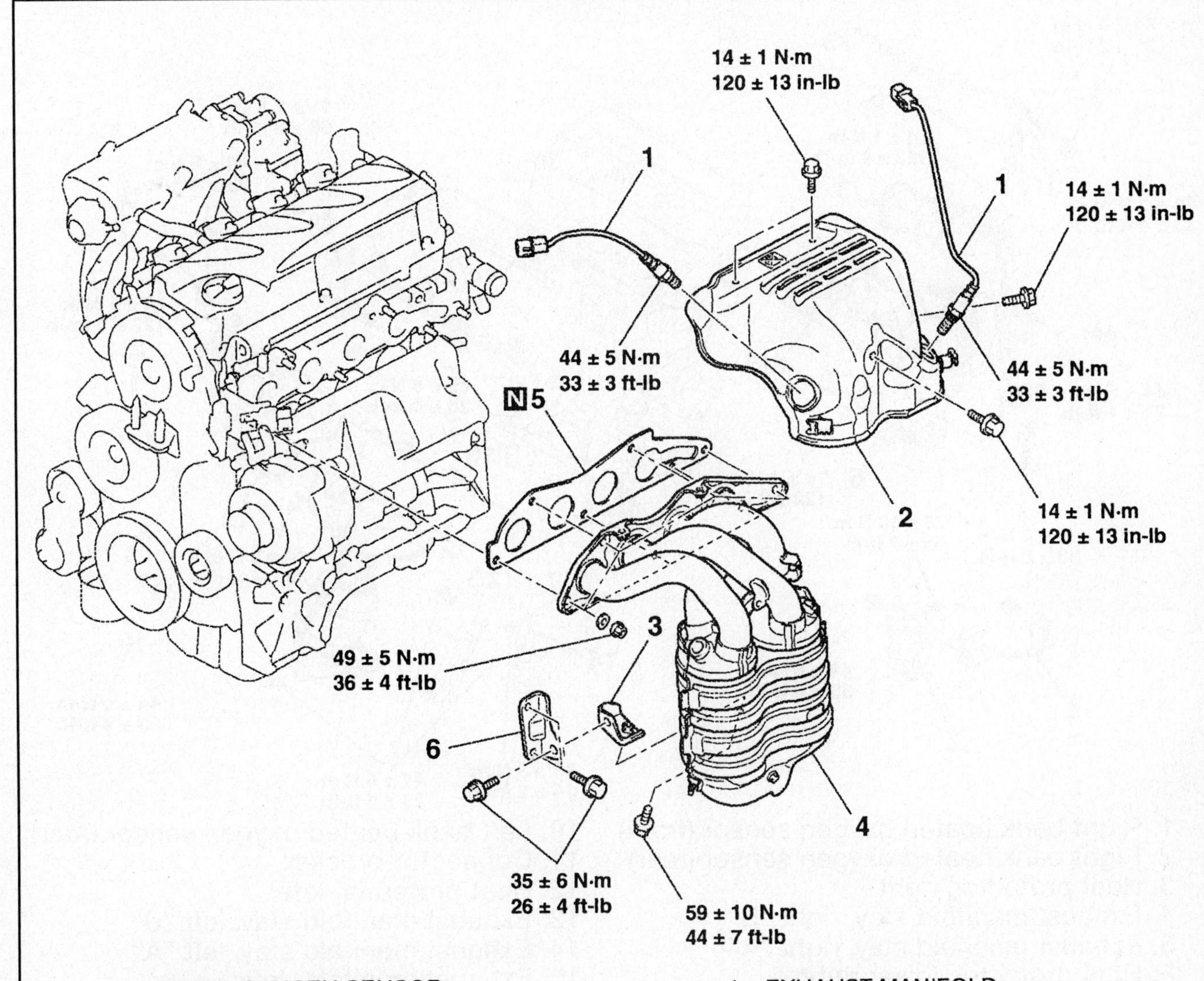

Fig. 133 Exhaust manifold and related components—2.4L engine

1. Before servicing the vehicle, refer to the Precautions Section.
2. Disconnect the negative battery cable.
3. Disconnect the heated oxygen sensor electrical connection. Remove the heated oxygen sensor.
4. Remove the exhaust manifold cover retaining screws. Remove the cover.
5. Remove the exhaust manifold bracket. Remove the lower heat shield.
6. Remove the exhaust manifold retaining bolts.
7. Remove the exhaust manifold from the engine.
8. Discard the old gasket.

To install:

9. Position a new exhaust manifold gasket to the mating surface of the cylinder.
10. Install the exhaust manifold to the engine. Torque all bolts to specification according to the illustration.
11. Continue the installation in the reverse order of the removal procedure.

3.8L Engine

See Figure 134.

1. Before servicing the vehicle, refer to the Precautions Section.
2. Disconnect the negative battery cable.
3. Remove the engine undercover. Remove the air cleaner intake duct assembly.
4. Remove the right side exhaust manifold:
    a. Remove the right bank heated oxygen sensor (front).
    b. Remove the right bank heated oxygen sensor (rear).
    c. Remove the heat protector (right) and the exhaust manifold stay (right "B").
    d. Remove the exhaust manifold stay (right "A").
    e. Remove the heat protector, lower right.
    f. Remove the right side exhaust manifold and discard the gasket.

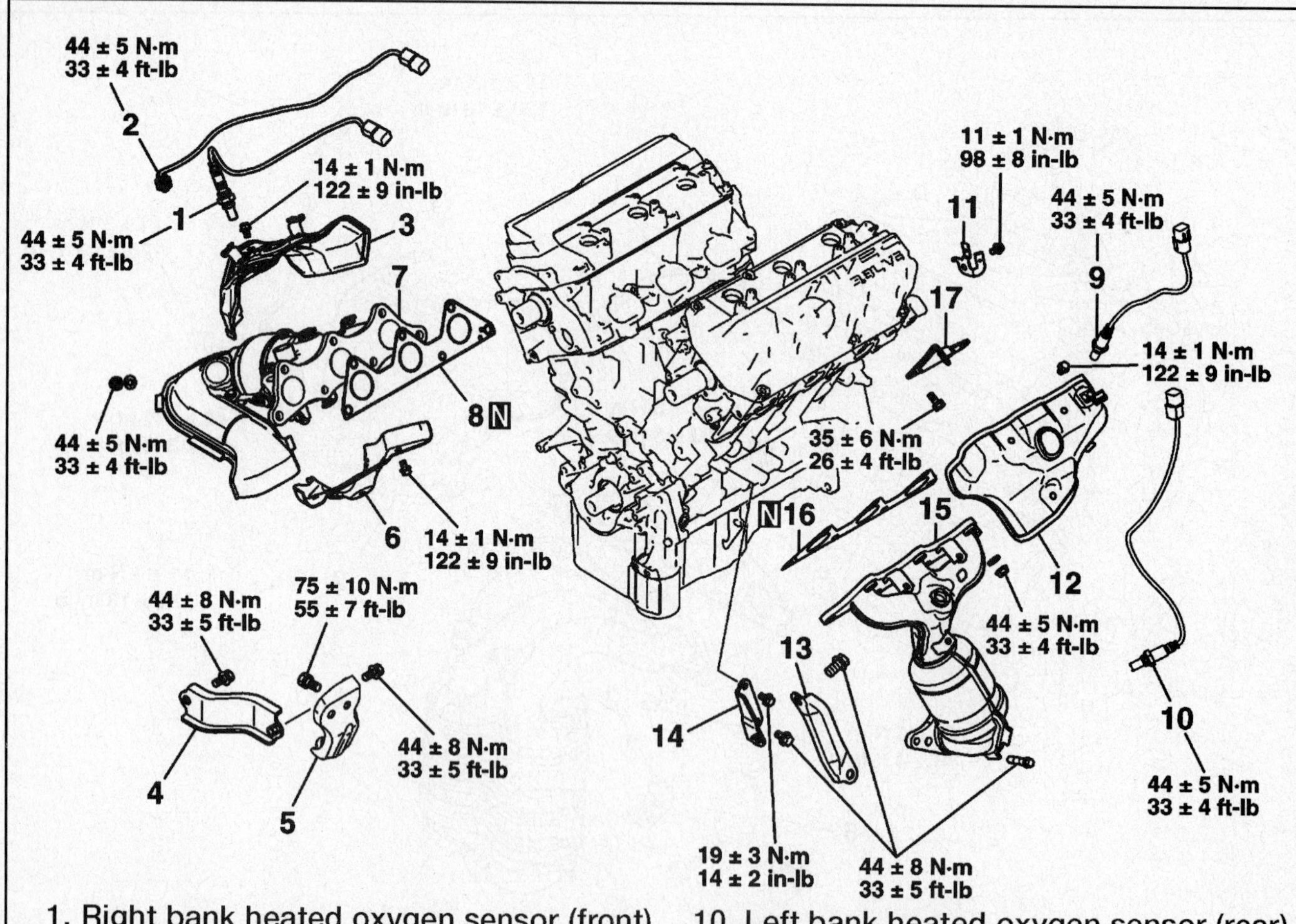

Fig. 134 Exhaust manifolds and related components—3.8L engine

5. Remove the left side exhaust manifold:
   a. Remove the left bank heated oxygen sensor (front).
   b. Remove the left bank heated oxygen sensor (rear).
   c. Remove the connector bracket and the heat protector (left).
   d. Remove the exhaust manifold stay (left "B").
   e. Remove the left side exhaust manifold and discard the gasket.

To install:

6. Position new exhaust manifold gaskets to the mating surfaces on the left and right sides.
7. Install the exhaust manifolds to the engine. Torque all bolts to specification according to the illustration.
8. Continue the installation in the reverse order of the removal procedure.
9. Start the engine and check for exhaust leaks.

Lancer

2006–07 2.0L Engine

See Figure 135.

1. Before servicing the vehicle, refer to the Precautions Section.
2. Remove or disconnect the following:
   - Negative battery cable
   - Engine undercover
   - Pressure hose clamp bolt

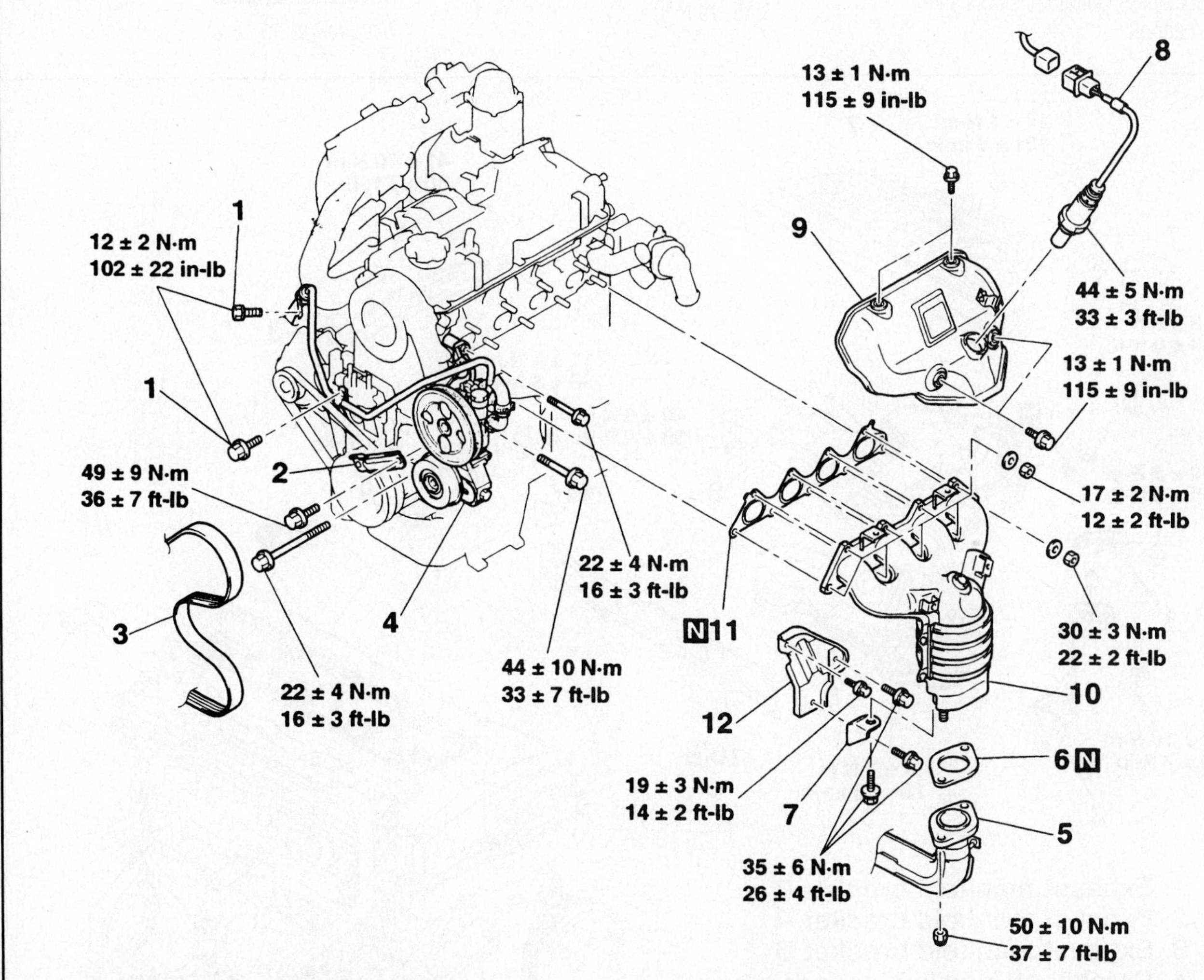

Fig. 135 Exhaust manifold and related components—2.0L engine—2006-07

- Power steering pump bracket stay bolt
- Drive belt
- Power steering pump and bracket assembly and position it aside. Do NOT disconnect the fluid lines.
- Front exhaust pipe and gasket from the manifold. Discard the gasket.
- Exhaust manifold bracket "B" (see illustration)
- Heated Oxygen (HO_2S) sensor
- Heat shield
- Exhaust manifold retainers and manifold
- Exhaust manifold gasket, and discard
- Exhaust manifold bracket "A" (see illustration)

To install:

3. Clean all gasket material from the mating surfaces.
4. Install or connect the following:
   - Exhaust manifold bracket "A"
   - New gasket and exhaust manifold. Tighten the upper retainers to 20–24 ft. lbs. (27–33 Nm) and the lower retainers to 10–14 ft. lbs. (15–19 Nm).
   - Heat shield
   - HO_2S
   - Exhaust manifold bracket "B"
   - Front exhaust pipe to the exhaust manifold, using a new gasket
   - Power steering pump assembly
   - Drive belt
   - Power steering pump bracket stay bolt
   - Pressure hose clamp bolt
   - Engine undercover
   - Negative battery cable

2008 2.0L Engine

See Figures 136 and 137.

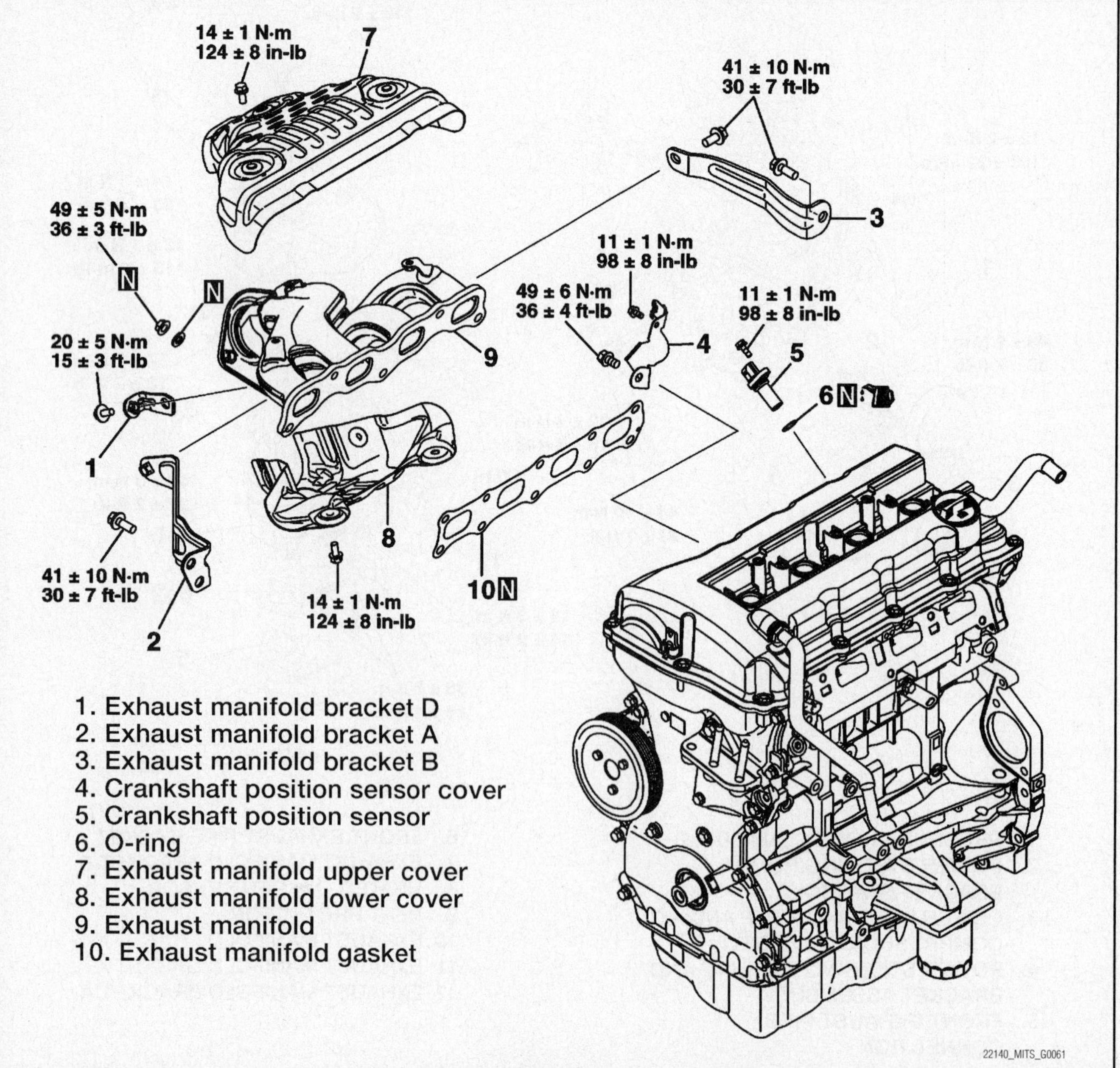

Fig. 136 Exhaust manifold and related components—2.0L engine (except California models)—2008

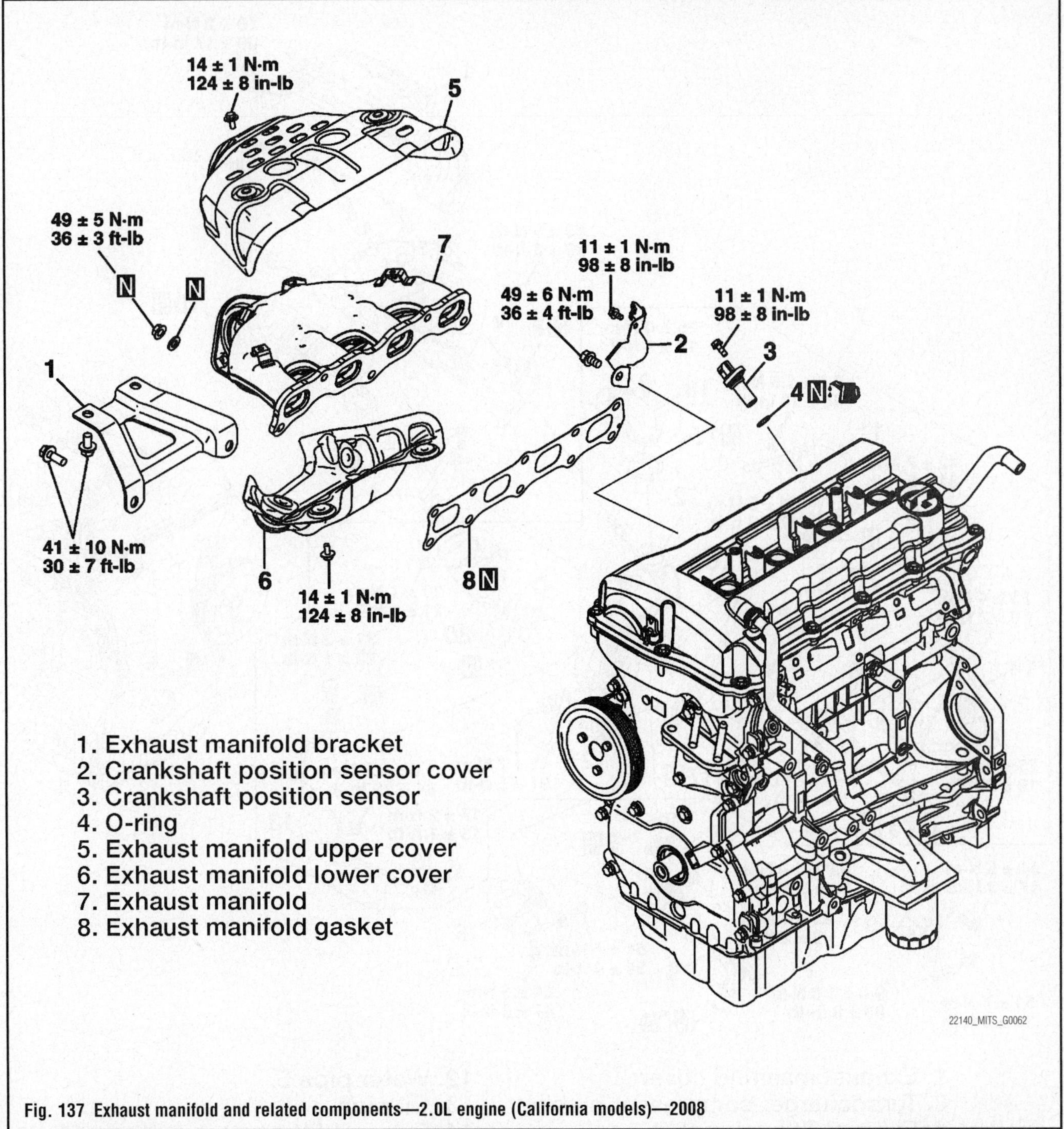

Fig. 137 Exhaust manifold and related components—2.0L engine (California models)—2008

1. Before servicing the vehicle, refer to the Precautions Section.
2. Disconnect the negative battery cable.
3. Remove the air cleaner assembly.
4. Remove the exhaust manifold bracket D—except California models.
5. Remove the exhaust manifold bracket A—except California models.
6. Remove the exhaust manifold bracket B—except California models.
7. Remove the crankshaft position sensor cover.
8. Remove the crankshaft position sensor and O-ring.
9. Remove the exhaust manifold upper cover and lower cover.
10. Remove the exhaust manifold and discard the gasket.

To install:

11. Position a new exhaust manifold gasket to the mating surface of the cylinder.

➡The exhaust manifold gasket, washers and nuts must not be reused.

12. Install the exhaust manifold to the engine. Torque all exhaust manifold nuts to: 33–39 ft. lbs. (44–54 Nm).
13. Continue the installation in the reverse order of the removal procedure. Tighten the all bolts to specification according to the illustration.

Lancer Evolution

2008 2.0L Turbo Engine

See Figure 138.

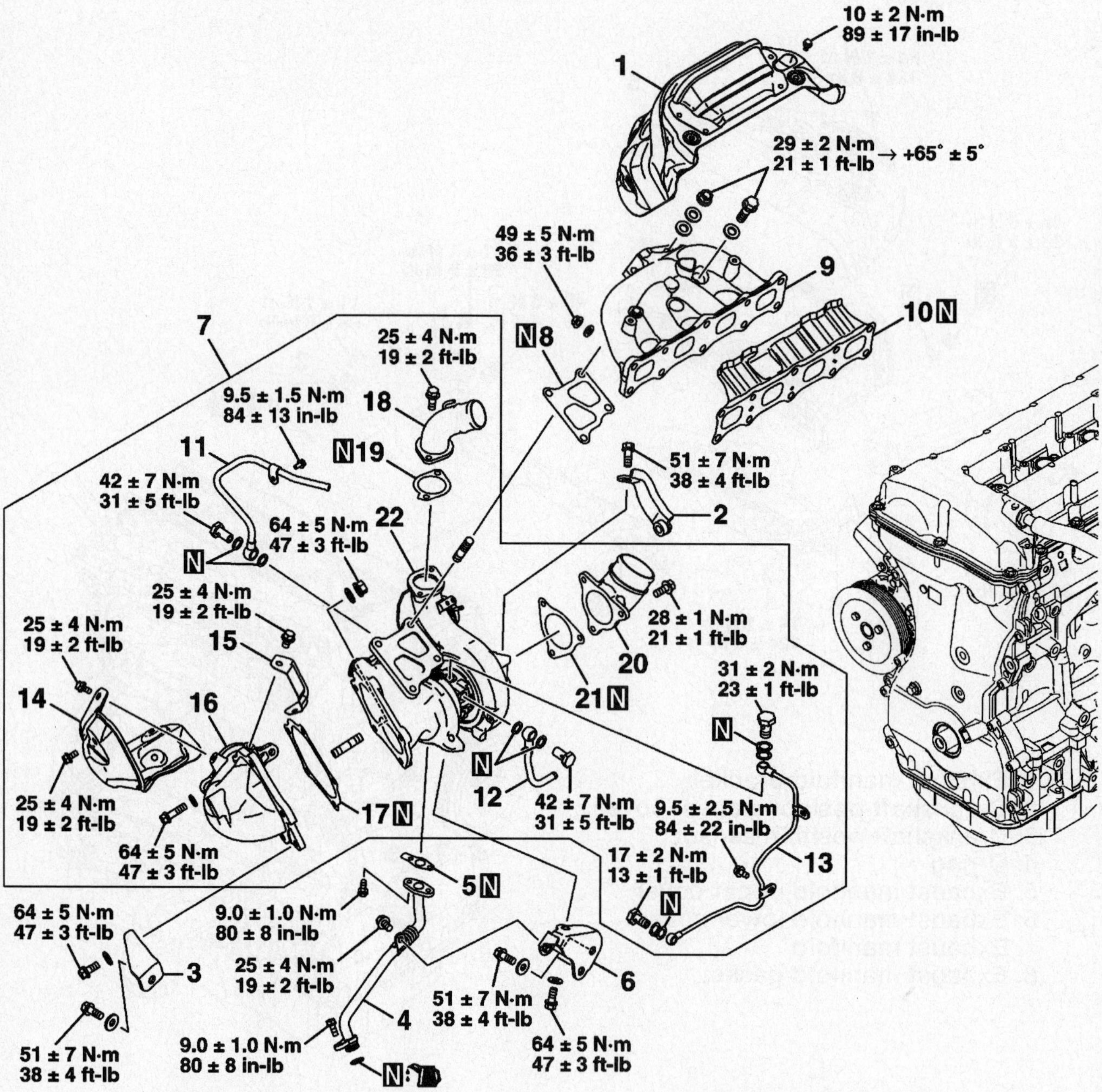

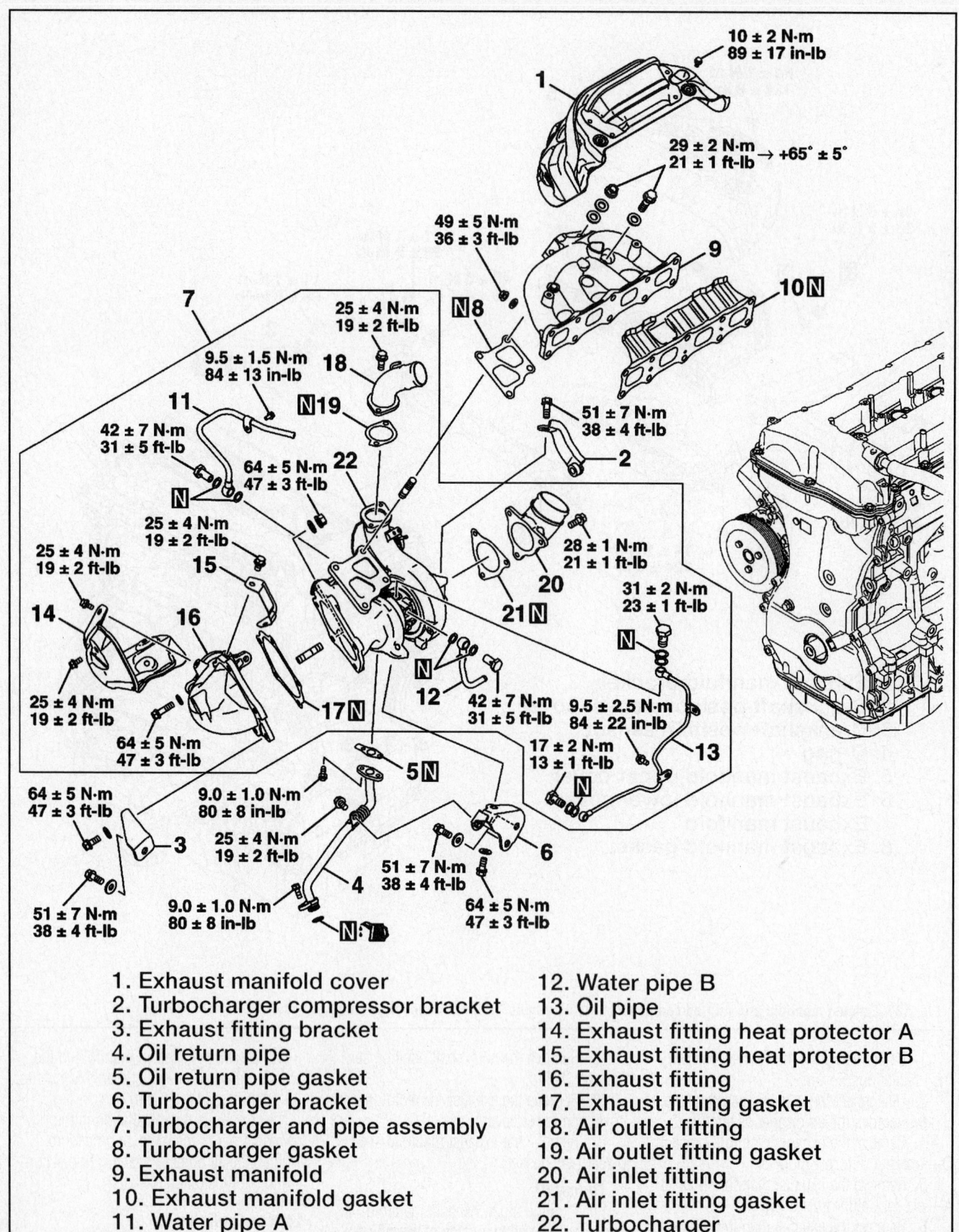

1. Exhaust manifold cover
2. Turbocharger compressor bracket
3. Exhaust fitting bracket
4. Oil return pipe
5. Oil return pipe gasket
6. Turbocharger bracket
7. Turbocharger and pipe assembly
8. Turbocharger gasket
9. Exhaust manifold
10. Exhaust manifold gasket
11. Water pipe A
12. Water pipe B
13. Oil pipe
14. Exhaust fitting heat protector A
15. Exhaust fitting heat protector B
16. Exhaust fitting
17. Exhaust fitting gasket
18. Air outlet fitting
19. Air outlet fitting gasket
20. Air inlet fitting
21. Air inlet fitting gasket
22. Turbocharger

22140_MITS_G0065

Fig. 138 Exhaust manifold and related components—Lancer Evolution 2.0L engine—2008

1. Before servicing the vehicle, refer to the Precautions Section.
2. Disconnect the negative battery cable.
3. Remove the air cleaner assembly.
4. Remove the exhaust manifold cover.
5. Remove the turbocharger compressor bracket.
6. Remove the exhaust fitting bracket.
7. Remove the oil return pipe and gasket.
8. Remove the turbocharger bracket, turbocharger, pipe assembly, and gasket.
9. Remove the exhaust manifold and discard the gasket.

To install:

10. Clean all gasket material from the mating surfaces.
11. Install a new gasket and the exhaust manifold. Torque the manifold nuts to: 33–39 ft. lbs. (44–54 Nm).
12. Install the turbocharger assembly. Refer to Turbocharger, Removal & Installation.
13. Install the exhaust manifold cover.
14. Install the air cleaner assembly.
15. Connect the negative battery cable.
16. Start the engine and check for exhaust leaks.

FLYWHEEL

REMOVAL & INSTALLATION

1. Before servicing the vehicle, refer to the Precautions Section.
2. Disconnect the negative battery cable.
3. Remove the transaxle. Refer to Automatic Transaxle Assembly or Manuel Transaxle Assembly, Removal & Installation.
4. If equipped with a manual transaxle, remove the clutch disc and pressure plate.
5. Mark the position of the flywheel/driveplate on the crankshaft and remove the retaining bolts.
6. On automatic transaxle equipped models, remove the driveplate adapter.
7. Remove the flywheel/driveplate from the engine.

To install:

8. Coat the threads of the driveplate/flywheel retaining bolts with thread locking compound.
9. Position the driveplate/flywheel on the crankshaft flange.
10. On automatic transaxle equipped models, install the driveplate adapter.
11. Install and tighten the bolts, in an alternating star pattern, to the following specifications:
    a. 2.0L engine (2006—07): 70–76 ft. lbs. (93–103 Nm).
    b. 2.0L engine (2008): 30–96 ft. lbs. (40–130 Nm).
    c. 2.4L engine: 95–101 ft. lbs. (127–137 Nm).
    d. 3.8L engine: 53–55 ft. lbs. (73–75 Nm).
12. If equipped with a manual transaxle, install the clutch and pressure plate.
13. Install the transaxle. Refer to Automatic Transaxle Assembly or Manuel Transaxle Assembly, Removal & Installation.
14. Connect the negative battery cable.

INTAKE MANIFOLD

REMOVAL & INSTALLATION

Eclipse & Galant

2.4L Engine

See Figures 139 and 140.

1. Before servicing the vehicle, refer to the Precautions Section.
2. Relieve the fuel system pressure. Disconnect the negative battery cable.
3. Remove the air cleaner assembly. Drain the cooling system. Remove the battery.

➡Do not loosen the retaining screws for the resin cover of the throttle body assembly. If these screws are loosened, the sensor incorporated in the resin cover becomes misaligned and the throttle body will not function properly.

4. Disconnect the accelerator cable connection, throttle position sensor connector and water return hose connection
5. Remove the throttle body bracket. Remove the throttle body retaining bolts. Remove the throttle body. Discard the gasket.
6. Remove the delivery pipe and injector assembly.
7. Disconnect the EVAP purge solenoid valve connector, the evaporative emission purge hose connection and the EGR valve connector.
8. Disconnect the MAP sensor connector, brake booster vacuum hose connection, pressure hose clamp, and knock sensor connector bracket.
9. Disconnect the harness clamp bracket and harness clamp. Remove the engine oil dipstick and tube assembly.
10. Disconnect the lower radiator hose. Remove the thermostat case assembly, resonator, and gasket.
11. Remove the EGR valve and gasket. Remove the harness clamp bracket, EVAP purge solenoid valve, and brake booster vacuum pipe.
12. Remove the brake booster vacuum hose and MAP sensor.
13. Remove the intake manifold bracket.
14. Remove the intake manifold retaining bolts. Remove the intake manifold from the engine. Discard the gasket.

To install:

15. Be sure to use a new intake manifold gasket. Position the gasket on the engine.
16. Install the intake manifold. Torque the retaining bolts to specification.
17. When installing the thermostat case assembly, be sure to apply a bead of sealant, part number 8672 or equivalent, to the cylinder head mating surface of the case assembly.

➡Be sure to assemble the components with 15 minutes of applying the sealant to the surfaces.

18. Continue the installation in the reverse order of the removal procedure.

➡Be sure to install the throttle body gasket in the proper direction. If this gasket is not installed correctly, poor engine idling may result.

19. Complete the vehicle initialization procedure.

➡To complete the initialization procedure the following tools are needed: MB991958 scan tool, MB991824 VCI, MB991827 MUT III USB cable, MB991910 MUT III Main harness "A."

20. Connect the scan tool to the data link connector. To prevent damage to the scan tool, be sure that the ignition switch is in the **LOC** position before connecting the scan tool.
21. Turn the ignition switch to the **ON** position.
22. Select "check mode" from the menu screen.
23. Select "erase memory" from the menu screen.
24. Initialize the learning value.
25. After initialization, complete the idle learning procedure.

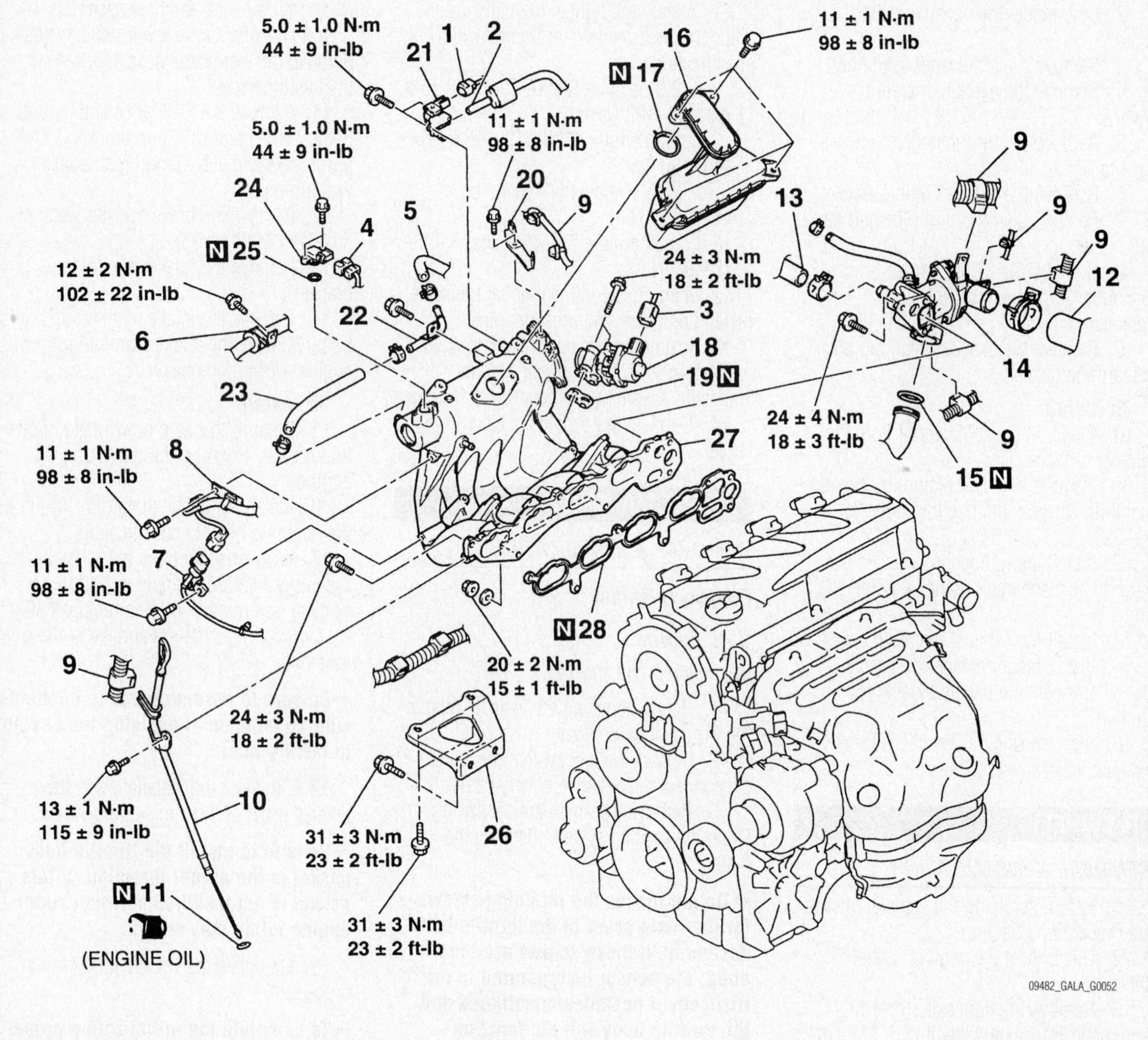

Fig. 139 Intake manifold and related components—2.4L engine

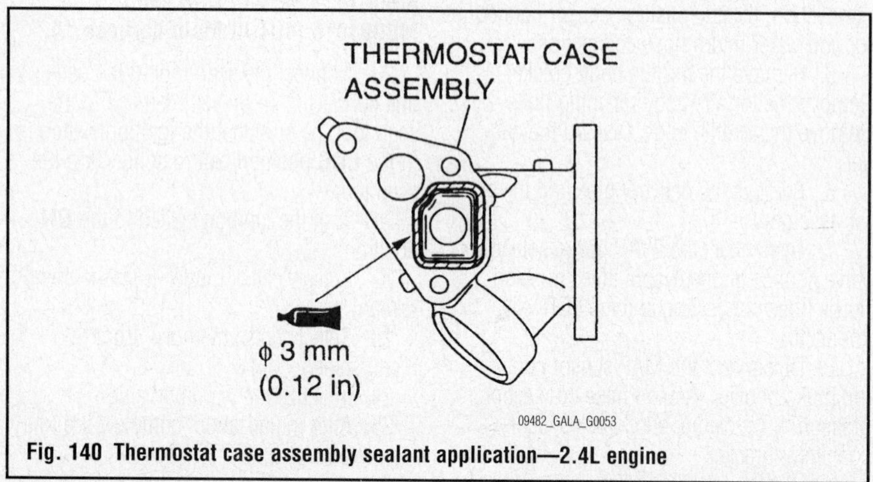

Fig. 140 Thermostat case assembly sealant application—2.4L engine

➡This procedure must be performed when the PCM is replaced, or when the learning value is initialized, as the idling is not stabilized because the learning value in the MFI engine is not completed.

26. Start the engine. Allow the coolant temperature to reach 176°F (80°C) or more.
27. Stop the engine and place the ignition switch in the **LOCK** position.
28. After 10 seconds, restart the engine.
29. For 10 minutes, carry out the idling procedure below to confirm that the engine has normal idling.
   - Position the transaxle selector lever in the "P" range

- The engine fan is not to be operated
- The engine coolant temperature should be 176°F (80°C) or more

➡If the engine stalls during idling, check the throttle valve of the throttle body for dirt. Correct and perform the procedure again.

3.8L Engine

See Figure 141.

1. Before servicing the vehicle, refer to the Precautions Section.
2. Relieve the fuel system pressure.
3. Disconnect the negative battery cable.

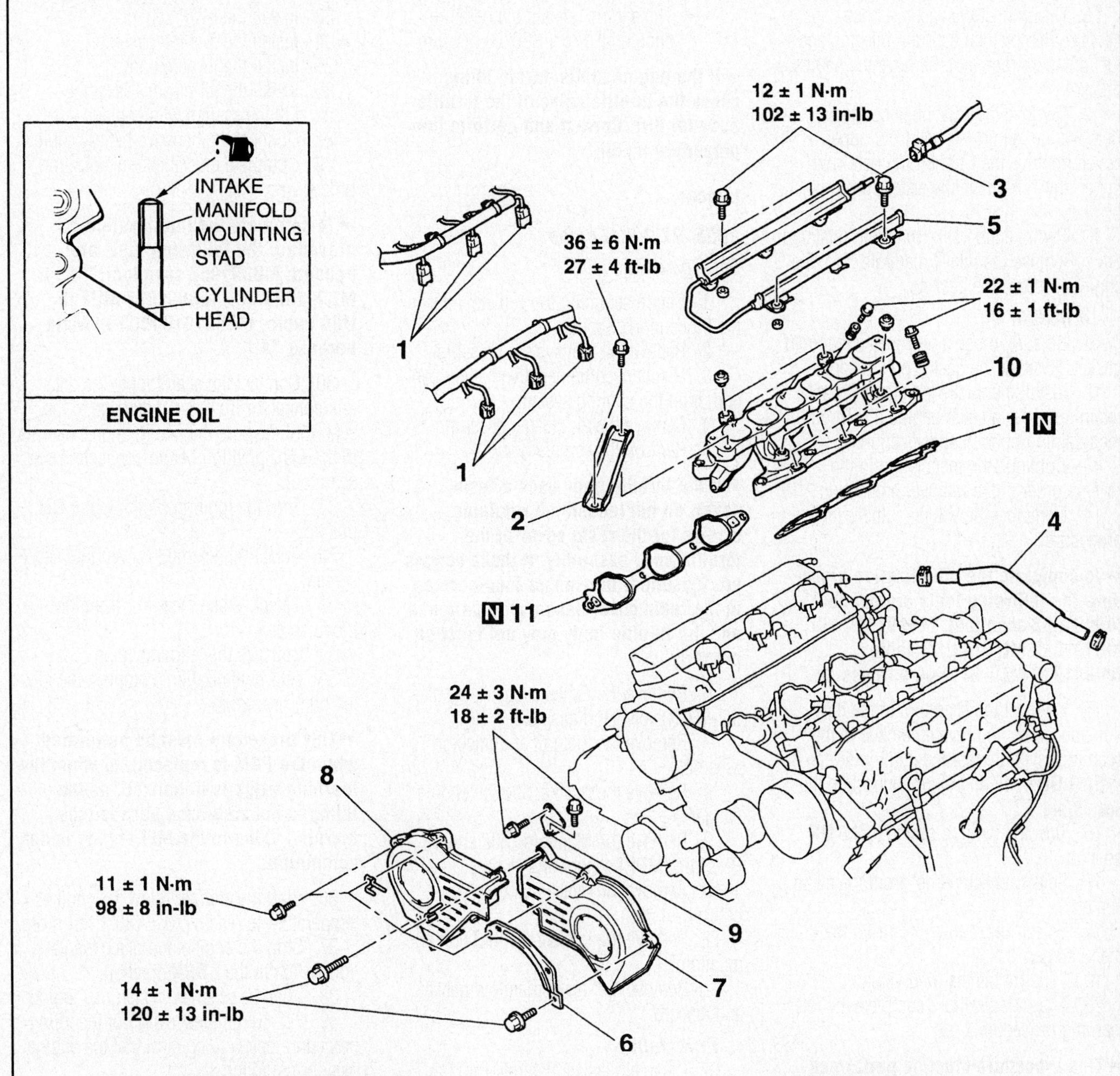

Fig. 141 Intake manifold and related components—3.8L engine

4. Remove the air cleaner assembly.
5. Drain the cooling system.
6. Remove the intake manifold plenum assembly.
7. Disconnect the injector connector wiring assembly, high pressure hose connection, O-ring, and fuel return hose.
8. Disconnect the vacuum hose connection. Remove the fuel rail, injector, and fuel pressure regulator. Remove the harness bracket.
9. Remove the PCV hose connection. Remove the left timing belt front upper cover. Remove the right timing belt front upper cover. Remove the water pump bracket.
10. Remove the intake manifold retaining bolts. Remove the intake manifold from the engine. Discard the gaskets.

To install:

11. Be sure to use a new intake manifold gasket. Position the gasket on the engine.
12. Install the intake manifold. Coat the retaining bolts in clean engine oil. Torque the retaining bolts to specification.
13. Continue the installation in the reverse order of the removal procedure.
14. Complete the vehicle initialization procedure.

➡To complete the initialization procedure the following tools are needed: MB991958 scan tool, MB991824 VCI, MB991827 MUT III USB cable, MB991910 MUT III Main harness "A."

15. Connect the scan tool to the data link connector. To prevent damage to the scan tool, be sure that the ignition switch is in the **LOC** position before connecting the scan tool.
16. Turn the ignition switch to the **ON** position.
17. Select "check mode" from the menu screen.
18. Select "erase memory" from the menu screen.
19. Initialize the learning value.
20. After initialization, complete the idle learning procedure.

➡This procedure must be performed when the PCM is replaced, or when the learning value is initialized, as the idling is not stabilized because the learning value in the MFI engine is not completed.

21. Start the engine. Allow the coolant temperature to reach 176°F (80°C) or more.
22. Stop the engine and place the ignition switch in the **LOCK** position.
23. After 10 seconds, restart the engine.
24. For 10 minutes, carry out the idling procedure below to confirm that the engine has normal idling.
- Position the transaxle selector lever in the "P" range
- The engine fan is not to be operated
- The engine coolant temperature should be 176°F (80°C) or more

➡If the engine stalls during idling, check the throttle valve of the throttle body for dirt. Correct and perform the procedure again.

Lancer

2006–07 2.0L Engine

See Figure 142.

1. Before servicing the vehicle, refer to the Precautions Section.
2. Relieve the fuel system pressure.
3. Disconnect the negative battery cable and drain the cooling system.
4. Remove the air cleaner assembly.
5. Remove the throttle body.

➡If the throttle body uses a resin cover, do not loosen the retaining screws for the resin cover of the throttle body assembly. If these screws are loosened, the sensor incorporated in the resin cover becomes misaligned and the throttle body may not function properly.

6. Remove the Exhaust Gas Recirculation (EGR) valve and gasket.
7. Remove the fuel rail and injector assembly.
8. Remove the bracket and the engine hanger.
9. Remove the Manifold Differential Pressure (MDP) sensor and O-ring.
10. Remove the intake manifold stay bracket.
11. Remove the intake manifold bolt and manifold.
12. Remove the intake manifold gasket and discard.

To install:

13. Clean all gasket material from the cylinder head intake mounting surface and intake manifold assembly.
14. Install the intake manifold, using a new gasket. Torque the manifold in a crisscross pattern, starting from the inside and working outwards to 12–16 ft. lbs. (16–22 Nm).
15. Install the intake manifold stay.
16. Install the throttle body stay.
17. Install the engine hanger.
18. Install the EGR valve and gasket.
19. Install the vacuum hose and pipe assembly.
20. Connect the brake booster vacuum hose connection.
21. Connect the vacuum pipe.
22. Connect the auto cruise vacuum hose connection.
23. Install the MDP sensor.
24. Install the fuel rail assembly.
25. Install the throttle body.
26. Install the air cleaner assembly.
27. Fill the system with coolant.
28. Connect the negative battery cable.
29. Complete the vehicle initialization procedure.

➡To complete the initialization procedure the following tools are needed: MB991958 scan tool, MB991824 VCI, MB991827 MUT III USB cable, MB991910 MUT III Main harness "A."

30. Connect the scan tool to the data link connector. To prevent damage to the scan tool, be sure that the ignition switch is in the **LOC** position before connecting the scan tool.
31. Turn the ignition switch to the **ON** position.
32. Select "check mode" from the menu screen.
33. Select "erase memory" from the menu screen.
34. Initialize the learning value.
35. After initialization, complete the idle learning procedure.

➡This procedure must be performed when the PCM is replaced, or when the learning value is initialized, as the idling is not stabilized because the learning value in the MFI engine is not completed.

36. Start the engine. Allow the coolant temperature to reach 176°F (80°C) or more.
37. Stop the engine and place the ignition switch in the **LOCK** position.
38. After 10 seconds, restart the engine.
39. For 10 minutes, carry out the idling procedure below to confirm that the engine has normal idling.
- Position the transaxle selector lever in the "P" range
- The engine fan is not to be operated
- The engine coolant temperature should be 176°F (80°C) or more

➡If the engine stalls during idling, check the throttle valve of the throttle body for dirt. Correct and perform the procedure again.

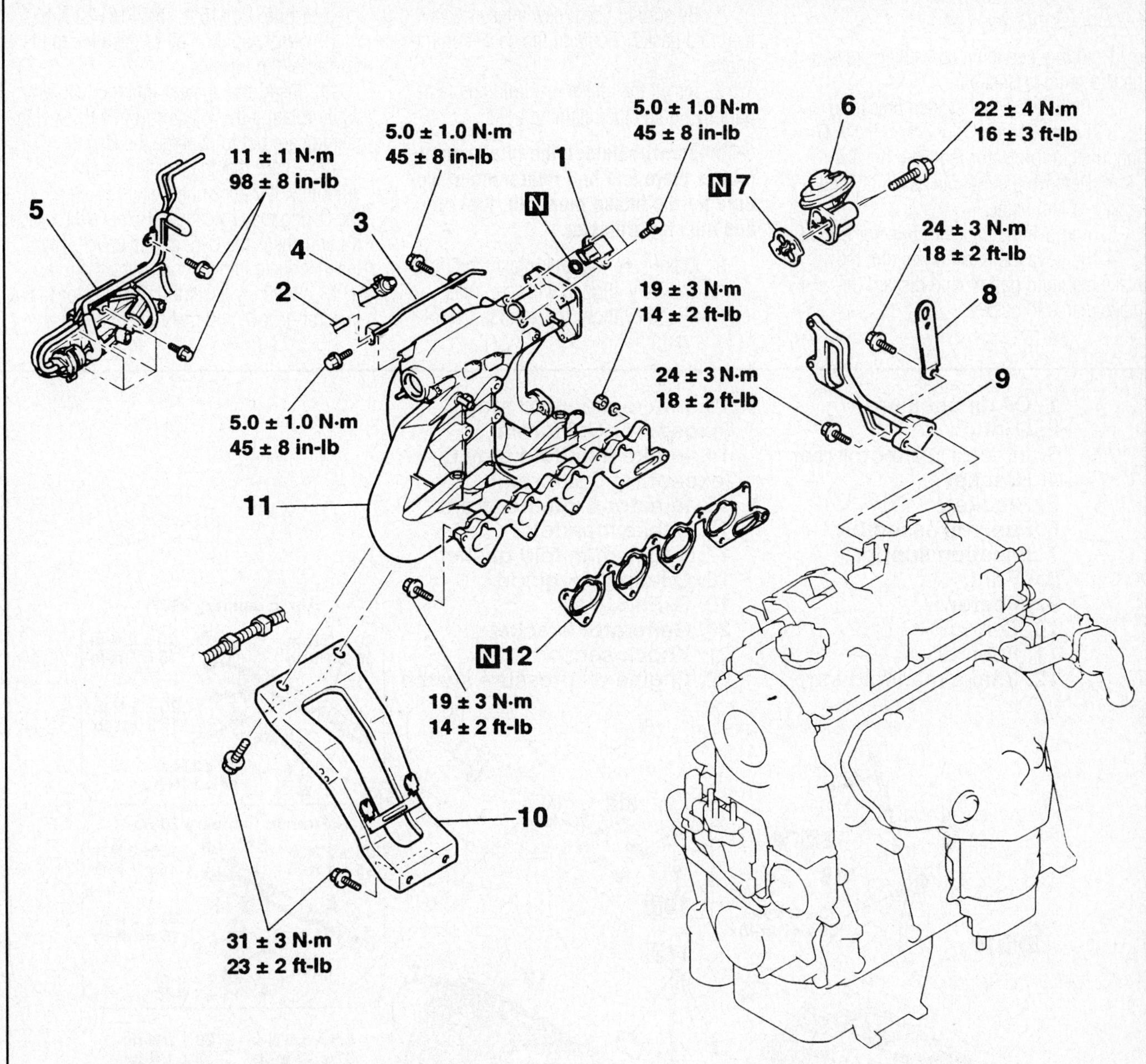

REMOVAL STEPS

1. MANIFOLD DIFFERENTIAL PRESSURE SENSOR
2. AUTO CRUISE VACUUM HOSE CONNECTION
3. VACUUM PIPE
4. BRAKE BOOSTER VACUUM HOSE CONNECTION
5. VACUUM HOSE AND PIPE ASSEMBLY

REMOVAL STEPS (Continued)

6. EGR VALVE
7. EGR VALVE GASKET
8. ENGINE HANGER
9. THROTTLE BODY STAY
10. INTAKE MANIFOLD STAY
11. INTAKE MANIFOLD
12. INTAKE MANIFOLD GASKET

9357QG13

Fig. 142 Intake manifold and related components—2.0L

2008 2.0L Engine

See Figures 143 and 144.

1. Before servicing the vehicle, refer to the Precautions Section.
2. Relieve the fuel system pressure.
3. Remove the oil dipstick rod with O-ring, injector protector rear, and bracket.
4. Remove the fuel rail assembly and intake manifold stay.
5. Remove the injector protector front.
6. Remove the intake manifold, the intake manifold gasket, and discard the intake manifold gasket.

To install:

7. Be sure to use a new intake manifold gasket. Position the gasket on the engine.
8. Install the intake manifold and temporarily tighten the retaining bolts.

➡Temporarily tighten the intake manifold as there is a bolt tightening procedure for the intake manifold, fuel rail, and injector protector.

9. Make sure that the intake manifold stay is securely aligned with the intake manifold and cylinder block boss before tightening it to the specified torque. Specified torque: 14–16 ft. lbs. (18–22 Nm).
10. Apply spindle oil or gasoline to the O-ring of the injector.
11. Insert the injector into the fuel rail while rotating the injector from side to side, taking care not to damage the O-ring.
12. Check that the injector rotates smoothly. If it does not rotate smoothly, the O-ring may be caught. Remove the injector and check the O-ring for damage. Then, insert it again into the fuel rail and check.
13. Securely assemble the injector to the injector groove and fuel rail collar.

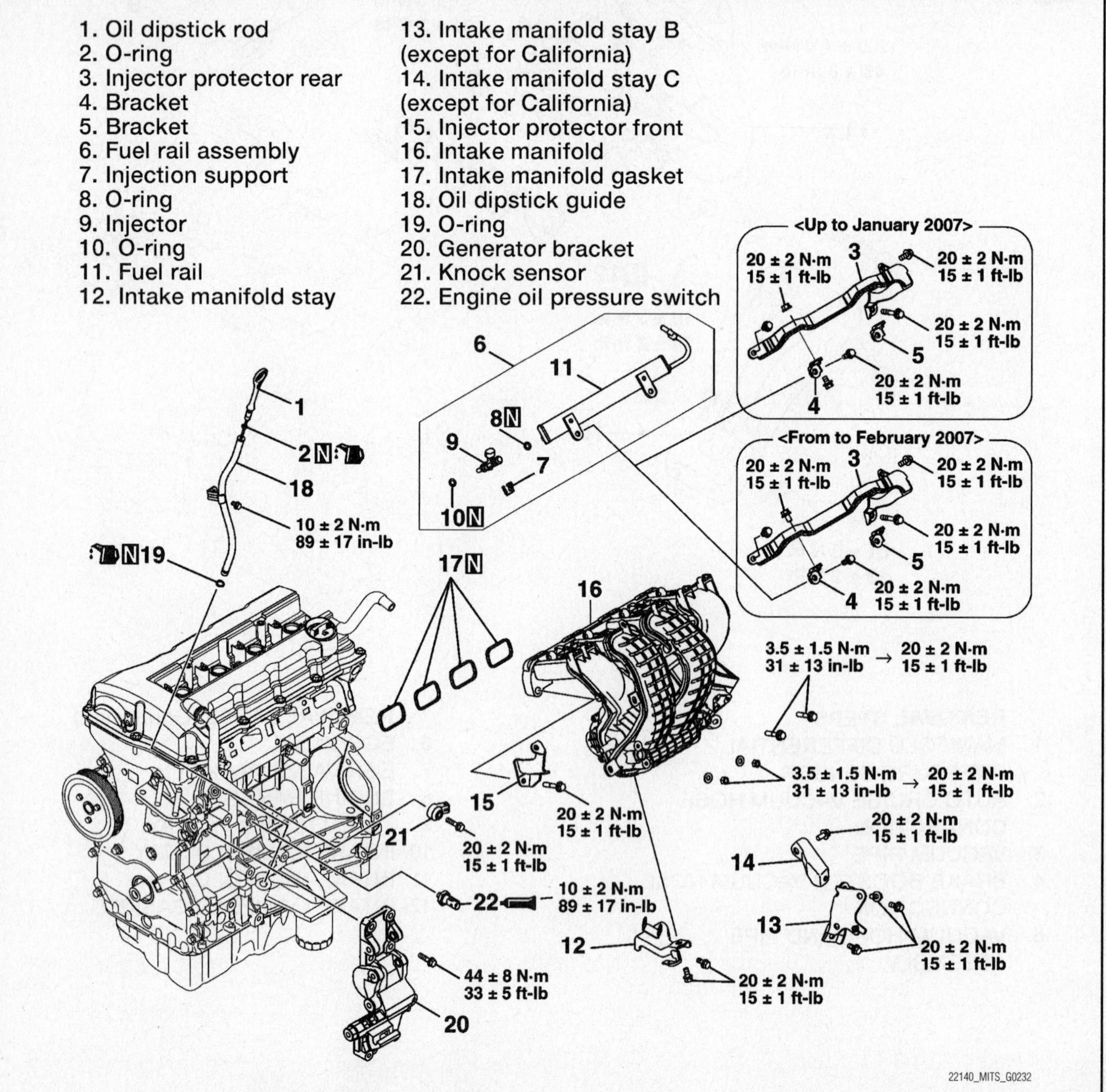

Fig. 143 Intake manifold and related components—Lancer—2008

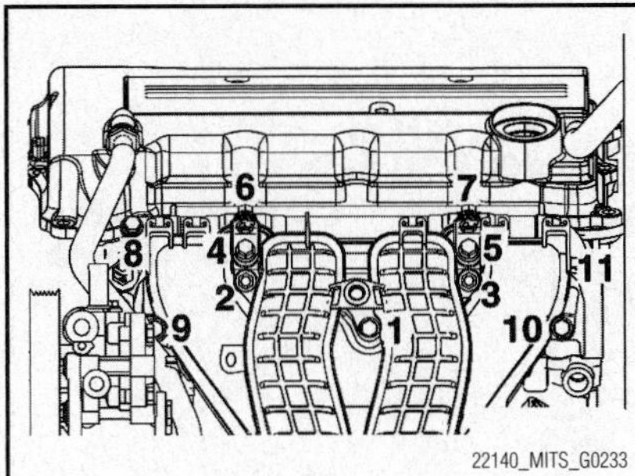

Fig. 144 Torque sequence for intake manifold and related components—Lancer—2008

14. Install the fuel rail assembly, bracket, and injector protector on the cylinder head.

15. Tighten the mounting bolts to the specified torque together with the temporarily tightened intake manifold mounting bolts in the order shown in the illustration. Specified torque: 14–16 ft. lbs. (18–22 Nm).

16. Continue the installation in the reverse order of the removal procedure.

Lancer Evolution

2008 2.0L Turbo Engine

See Figures 145 and 146.

1. Before servicing the vehicle, refer to the Precautions Section.
2. Relieve the fuel system pressure.
3. Remove the oil dipstick rod with O-ring, fuel return pipe, and fuel hose.
4. Remove the vacuum hose, fuel pressure regulator with O-ring, fuel rail assembly, and insulator.
5. Remove the injection support, injector, and fuel rail.
6. Remove the front and rear intake manifold stays.
7. Remove the intake manifold, the intake manifold gasket, and discard the intake manifold gasket.

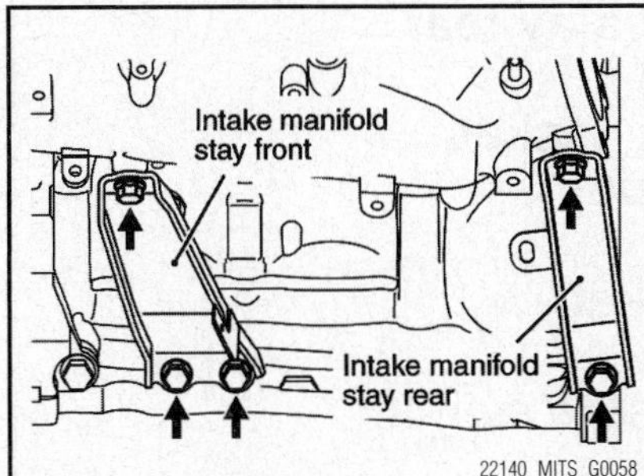

Fig. 145 Remove the front and rear intake manifold stays—Lancer Evolution—2008

To install:

8. Be sure to use a new intake manifold gasket. Position the gasket on the engine.
9. Install the intake manifold. Coat the retaining bolts in clean engine oil. Torque the retaining bolts to specification.
10. Continue the installation in the reverse order of the removal procedure.
11. Complete the vehicle initialization procedure.

➡To complete the initialization procedure the following tools are needed: MB991958 scan tool, MB991824 VCI, MB991827 MUT III USB cable, MB991910 MUT III Main harness "A."

12. Connect the scan tool to the data link connector. To prevent damage to the scan tool, be sure that the ignition switch is in the **LOC** position before connecting the scan tool.
13. Turn the ignition switch to the **ON** position.
14. Select "check mode" from the menu screen.
15. Select "erase memory" from the menu screen.
16. Initialize the learning value.
17. After initialization, complete the idle learning procedure.

➡This procedure must be performed when the PCM is replaced, or when the learning value is initialized, as the idling is not stabilized because the learning value in the MFI engine is not completed.

18. Start the engine. Allow the coolant temperature to reach 176°F (80°C) or more.
19. Stop the engine and place the ignition switch in the **LOCK** position.
20. After 10 seconds, restart the engine.
21. For 10 minutes, carry out the idling procedure below to confirm that the engine has normal idling.
    - Position the transaxle selector lever in the "P" range
    - The engine fan is not to be operated
    - The engine coolant temperature should be 176°F (80°C) or more

➡If the engine stalls during idling, check the throttle valve of the throttle body for dirt. Correct and perform the procedure again.

OIL PAN

REMOVAL & INSTALLATION

Eclipse & Galant

2.4L Engine

See Figures 147 and 148.

1. Before servicing the vehicle, refer to the Precautions Section.
2. Remove the negative battery cable.
3. Drain the engine oil.
4. Remove the engine undercover. Remove the front exhaust pipe.
5. Remove the torque converter housing front lower cover.
6. Remove the oil pan retaining bolts. Remove the oil pan. Discard the gasket.

➡Use tool MD998727 to remove the oil pan. Tap the tool into range B between the cylinder block and the engine oil pan, and then slide the tool sideway. Do not position the tool in area A of the engine oil pan as this may cause deformation of the front case because the front case is made of aluminum.

To install:

7. Clean the sealing surface on the oil pan and engine block. Apply a continuous bead of sealant to the oil pan.
8. Install the oil pan. Torque the bolts to specification illustrated.
9. Continue the installation in the reverse order of the removal procedure.

3.8L Engine

See Figures 149 through 153.

1. Before servicing the vehicle, refer to the Precautions Section.
2. Remove the negative battery cable.
3. Remove the engine undercover. Remove the starter. Remove the engine oil dipstick and guide.
4. If removing the upper oil pan, remove the front exhaust pipe. If equipped with an automatic transaxle, remove the cover and connecting bolt.
5. Remove the lower engine oil pan retaining bolts. Remove the engine lower oil pan. Discard the gasket.
6. Remove the upper oil pan retaining bolts. Remove the upper oil pan. Discard the gasket.

➡If the vehicle is equipped with a manual transaxle, align the recessed area in the flywheel with the location shown in the illustration. Mark the flywheel. Turn the crankshaft so that the alignment mark is positioned as shown in the illustration.

To install:

7. Clean the sealing surface on the oil pan and engine block. Apply a continuous bead of sealant to the oil pan.
8. Install the upper oil pan. Torque the bolts to specification.

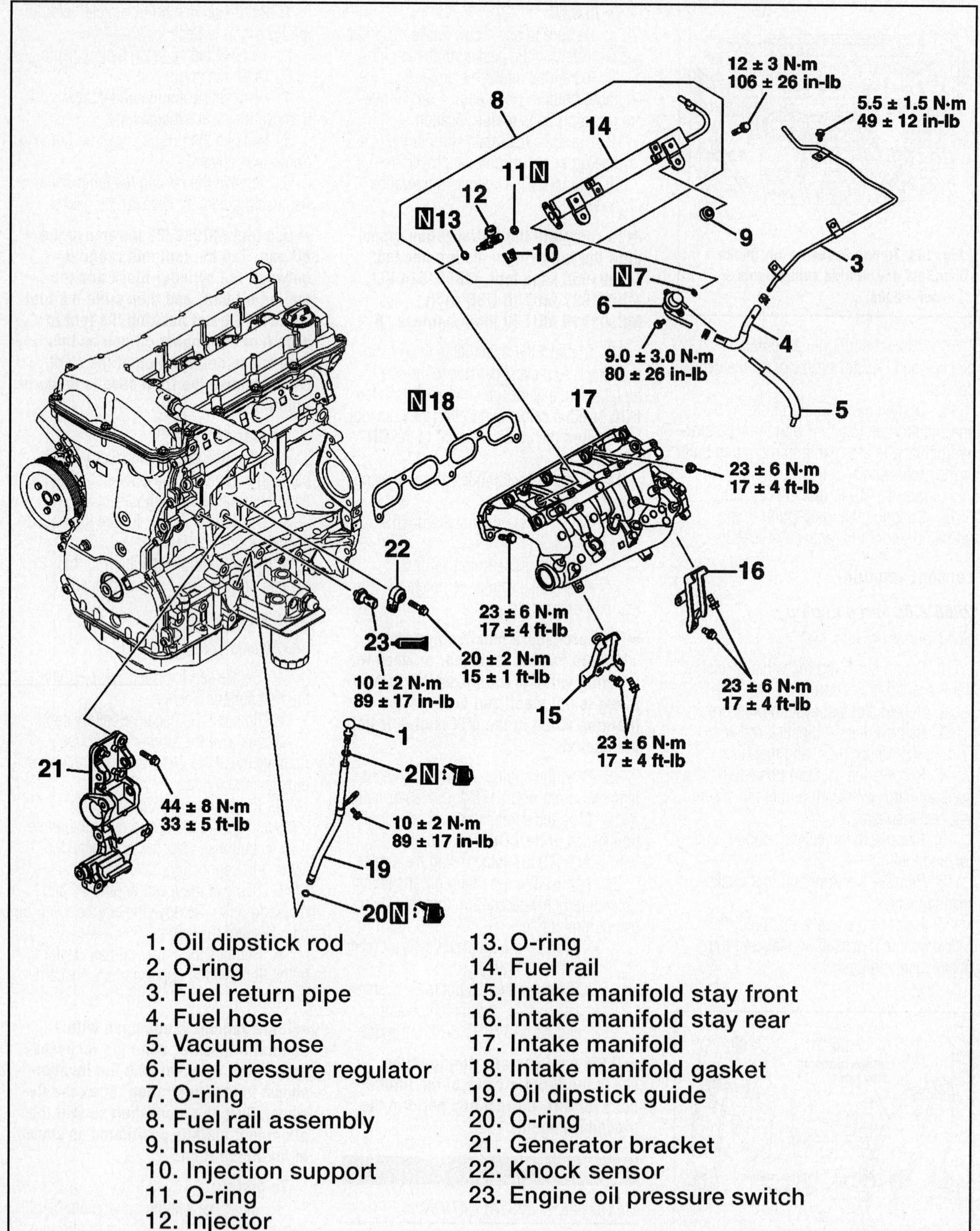

Fig. 146 Intake manifold and related components—Lancer Evolution—2008

B
A
MD998727
MD998727
09482_GALA_G0085

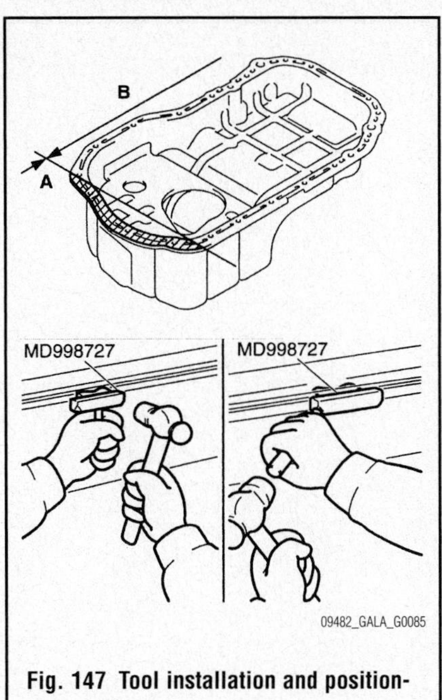

Fig. 147 Tool installation and positioning—2.4L engine

➡The bolt holes for bolts 13 and 14 are cut away on the transaxle side. Be careful not to insert these bolts at an angle.

9. Install the lower oil pan. Torque the bolts to specification.

49 ± 3 N·m
36 ± 2 ft-lb
09482_GALA_G0091

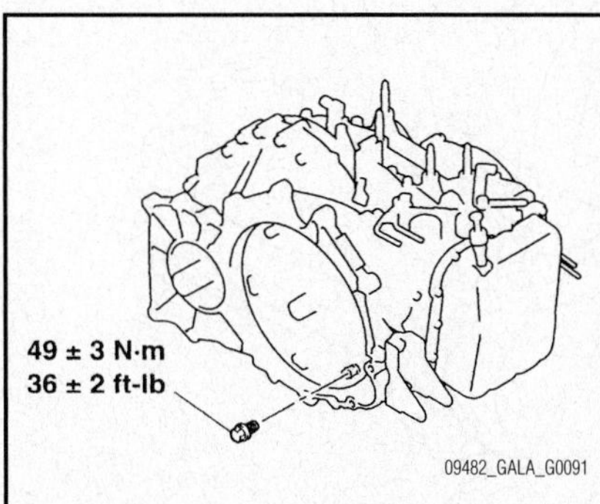

Fig. 149 Automatic transaxle torque converter connecting bolt location—3.8L engine

FLYWHEEL
FLYWHEEL
09482_GALA_G0092

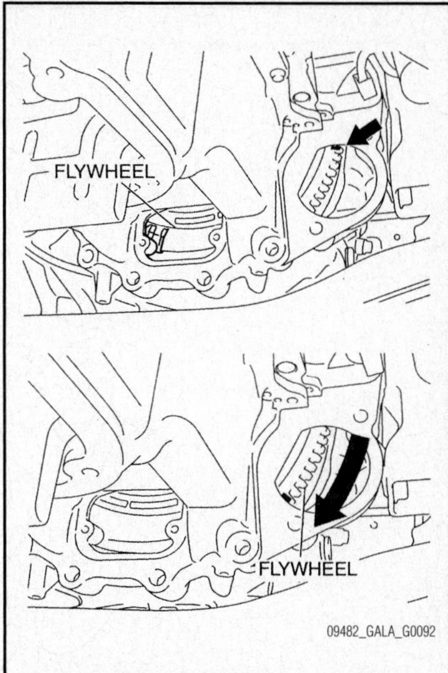

Fig. 150 Manual transaxle alignment location—3.8L engine

4
N 3
39 ± 5 N·m
29 ± 3 ft-lb
2
9.0 ± 3.0 N·m
80 ± 26 in-lb
9.0 ± 3.0 N·m
80 ± 26 in-lb
4 mm
(0.16 in)
GROOVE
PORTION
BOLT HOLE
PORTION
26 ± 5 N·m
19 ± 4 ft-lb
1
9.0 ± 1.0 N·m
80 ± 9 in-lb

1. TORQUE CONVERTER HOUSING FRONT LOWER COVER
2. ENGINE OIL PAN DRAIN PLUG
3. ENGINE OIL PAN DRAIN PLUG GASKET
4. ENGINE OIL PAN

09482_GALA_G0086

Fig. 148 Oil pan and related components—2.4L engine

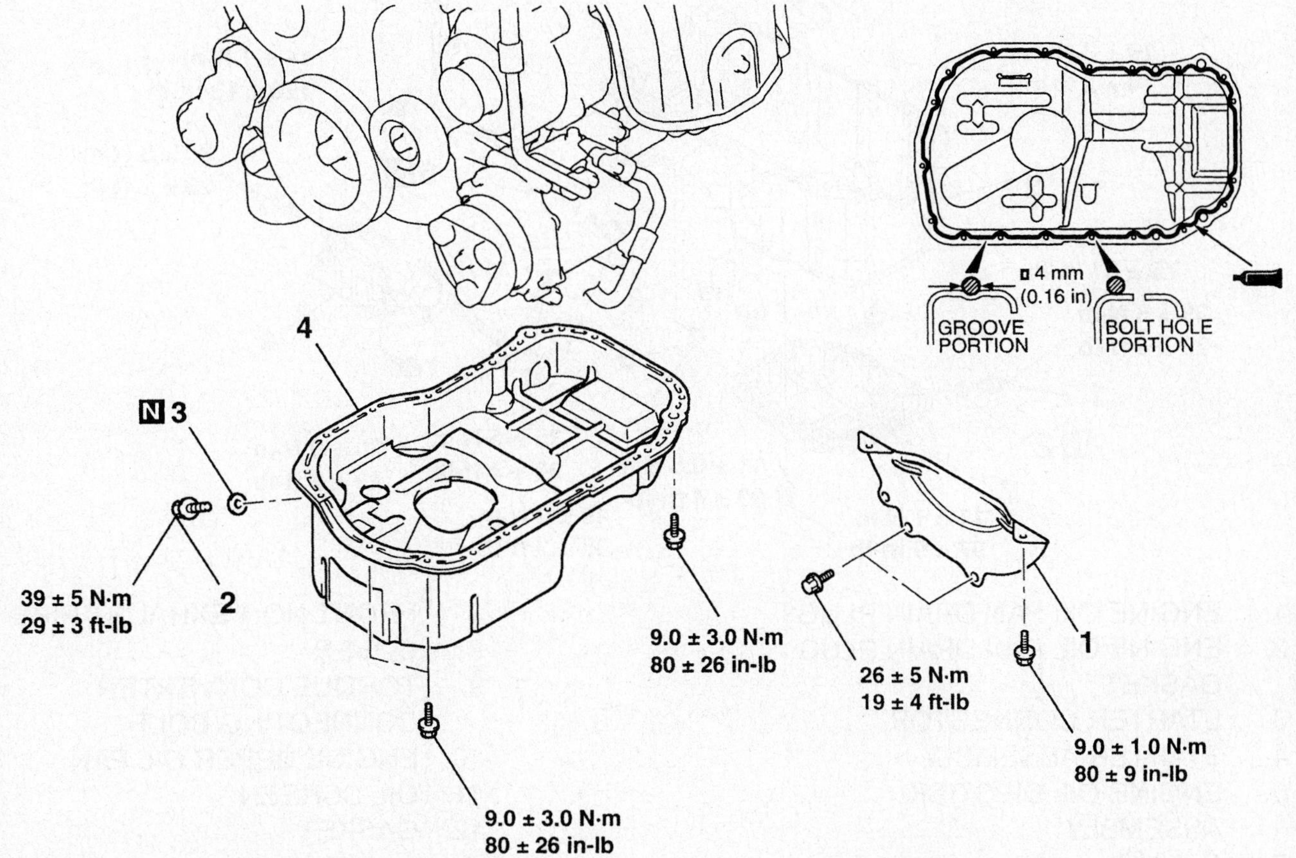

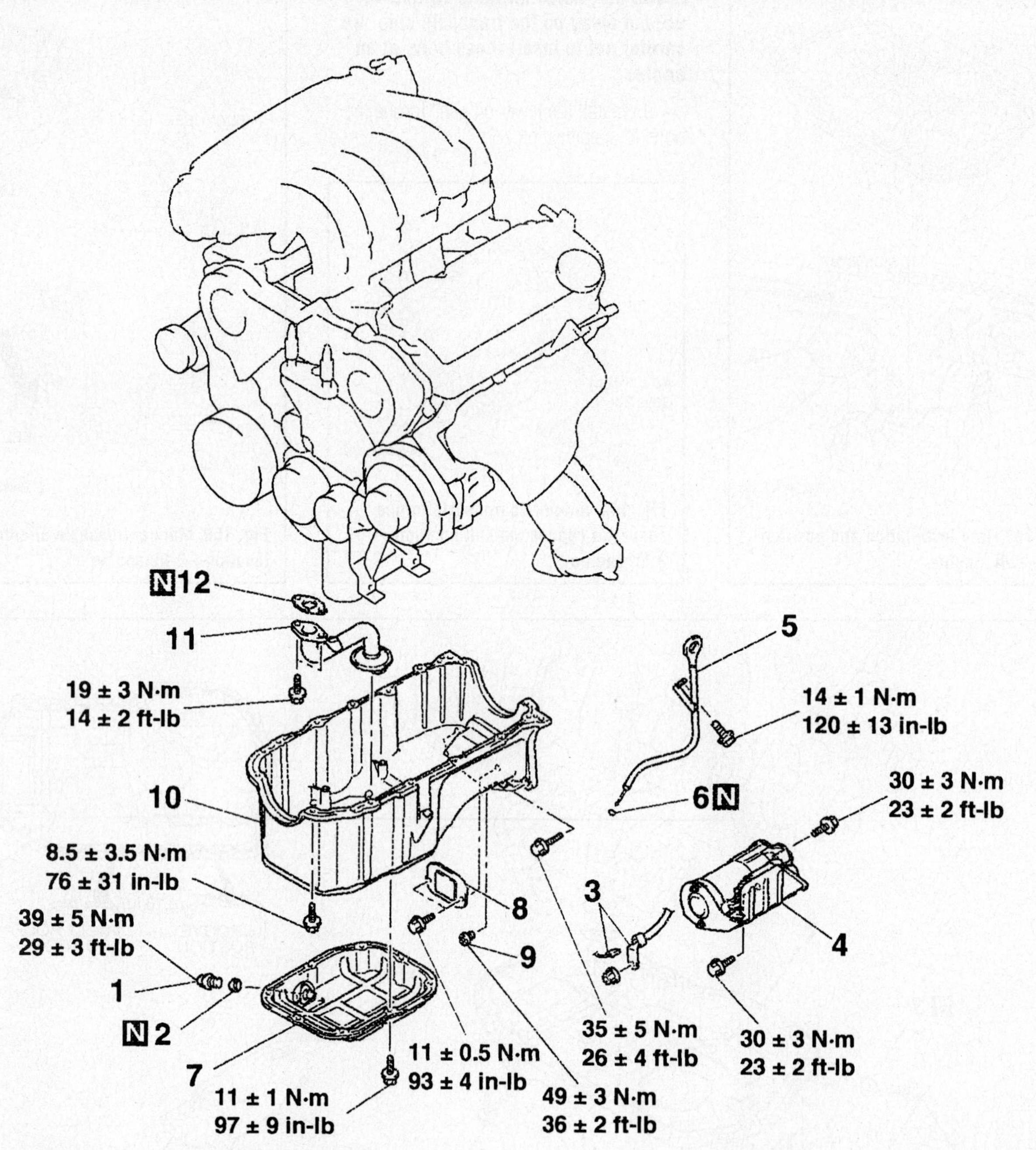

1. ENGINE OIL PAN DRAIN PLUG
2. ENGINE OIL PAN DRAIN PLUG GASKET
3. STARTER CONNECTOR
4. STARTER ASSEMBLY
5. ENGINE OIL DIPSTICK ASSEMBLY
6. O-RING
7. ENGINE LOWER OIL PAN

- FRONT NO.1 EXHAUST PIPE

8. COVER
9. TORQUE CONVERTER CONNECTING BOLT
10. ENGINE UPPER OIL PAN
11. OIL SCREEN
12. GASKET

09482_GALA_G0093

Fig. 151 Oil pan and related components—3.8L engine

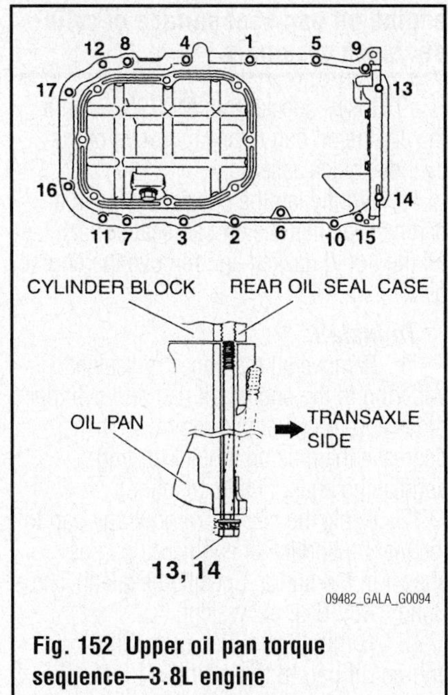

Fig. 152 Upper oil pan torque sequence—3.8L engine

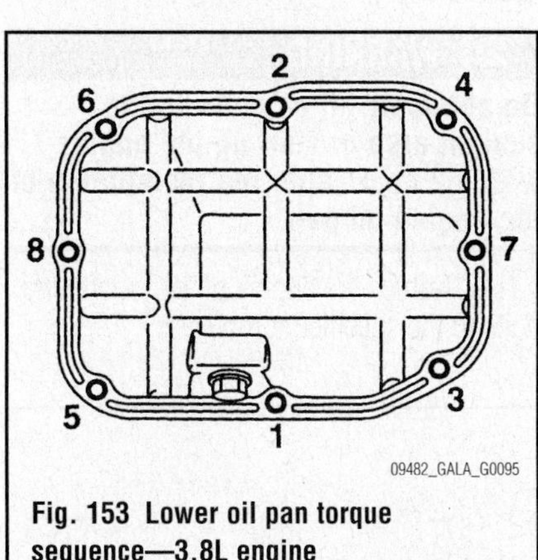

Fig. 153 Lower oil pan torque sequence—3.8L engine

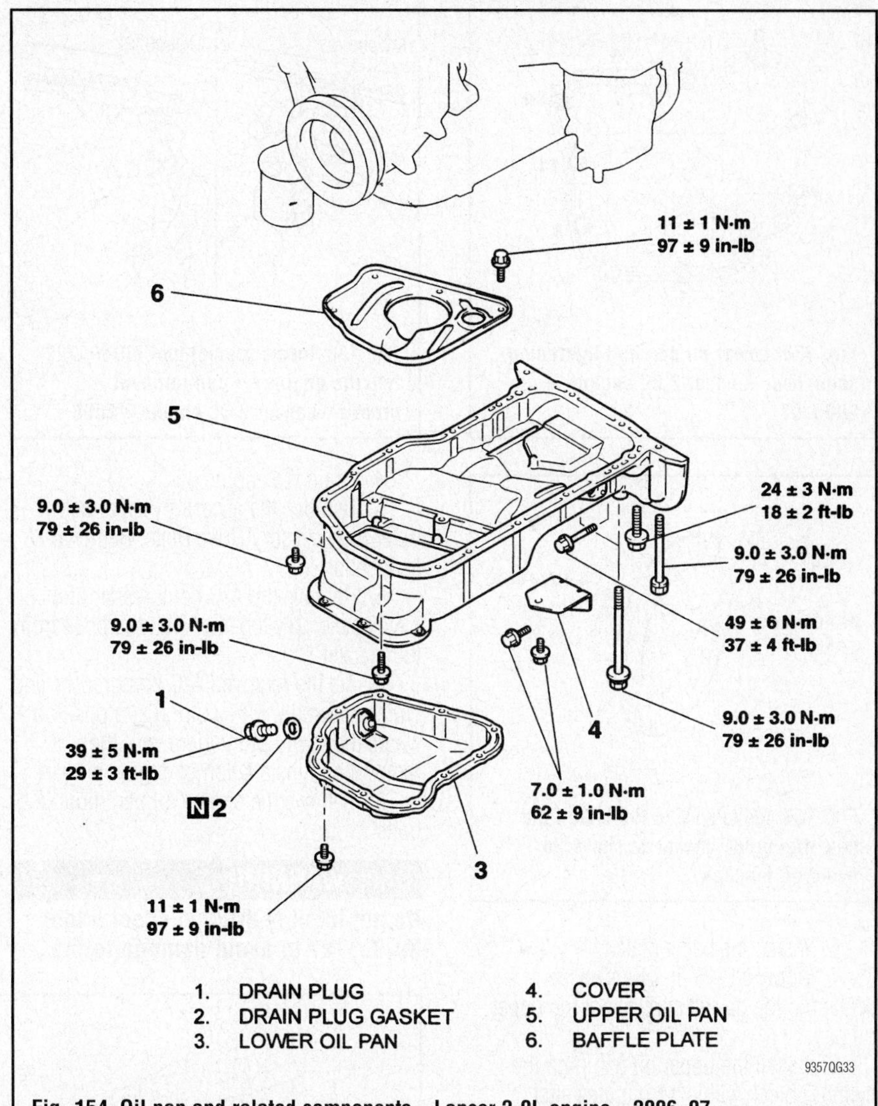

Fig. 154 Oil pan and related components—Lancer 2.0L engine—2006-07

10. Continue the installation in the reverse order of the removal procedure.

Lancer

2006-07 2.0L Engine

See Figures 154 through 157.

1. Before servicing the vehicle, refer to the Precautions Section.
2. Disconnect the negative battery cable.
3. Drain the engine oil.
4. Remove or disconnect the following:
   - Engine undercover
   - Front exhaust pipe
   - Lower oil pan bolts and lower pan
   - Cover
   - Upper oil pan bolt and upper pan
   - Baffle plate

To install:

5. Clean all gasket surfaces of the cylinder block and the upper and lower oil pan.

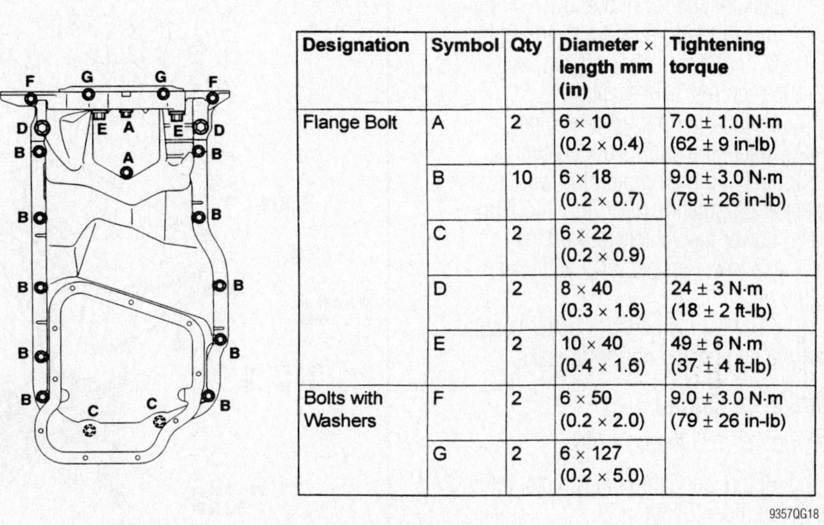

Designation	Symbol	Qty	Diameter × length mm (in)	Tightening torque
Flange Bolt	A	2	6 × 10 (0.2 × 0.4)	7.0 ± 1.0 N·m (62 ± 9 in-lb)
	B	10	6 × 18 (0.2 × 0.7)	9.0 ± 3.0 N·m (79 ± 26 in-lb)
	C	2	6 × 22 (0.2 × 0.9)	
	D	2	8 × 40 (0.3 × 1.6)	24 ± 3 N·m (18 ± 2 ft-lb)
	E	2	10 × 40 (0.4 × 1.6)	49 ± 6 N·m (37 ± 4 ft-lb)
Bolts with Washers	F	2	6 × 50 (0.2 × 2.0)	9.0 ± 3.0 N·m (79 ± 26 in-lb)
	G	2	6 × 127 (0.2 × 5.0)	

9357QG18

Fig. 155 Upper oil pan bolt location and torque sequence—Lancer 2.0L engine—2006-07

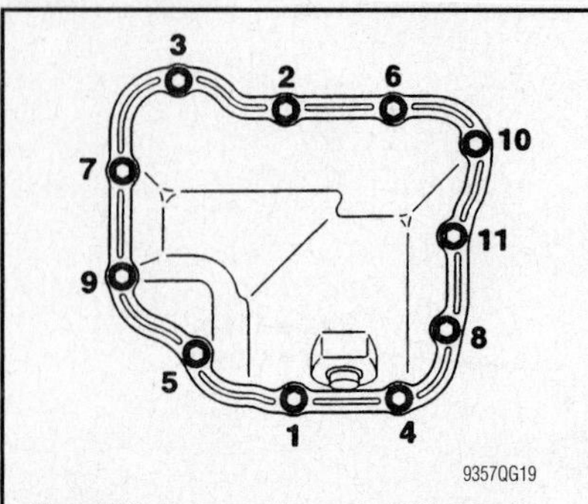

Fig. 156 Lower oil pan bolt tightening sequence—Lancer 2.0L engine—2006–07

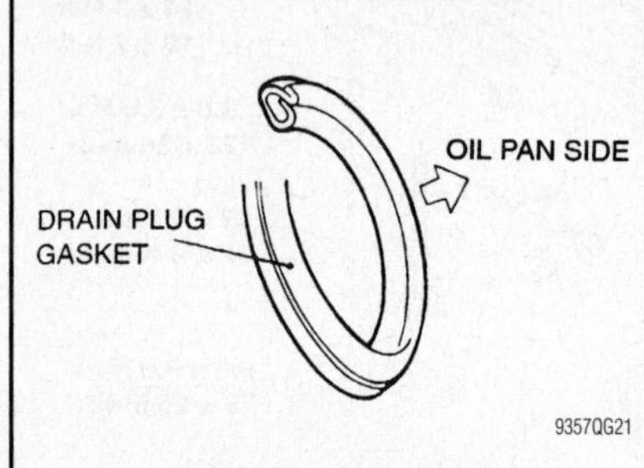

Fig. 157 Make sure to the install the new drain plug gasket as shown to avoid oil leakage

6. Install the baffle plate.
7. Apply a 0.16 in. (4mm) bead of sealant to the gasket surfaces of the upper oil pan.
8. Install the upper oil pan onto the cylinder block within 15 minutes after applying sealant. Tighten the bolts as shown in the accompanying figure.
9. Apply 0.16 in. (4mm) bead of sealant to the gasket surfaces of the lower oil pan.
10. Install the lower oil pan and tighten the bolts, in the sequence shown, to 88–106 inch lbs. (10–12 Nm).
11. Install the front exhaust pipe.
12. Install the engine undercover.
13. Install the oil drain plug with a new gasket and tighten to 29 ft. lbs. (40 Nm).
14. Lower the vehicle and fill the crankcase to the proper level with clean engine oil.
15. Connect the negative battery cable. Start the engine and check for leaks.

2008 2.0L Engine

See Figures 158 through 160.

1. Before servicing the vehicle, refer to the Precautions Section.
2. Remove the negative battery cable.
3. Remove the engine undercover and side cover (RH).

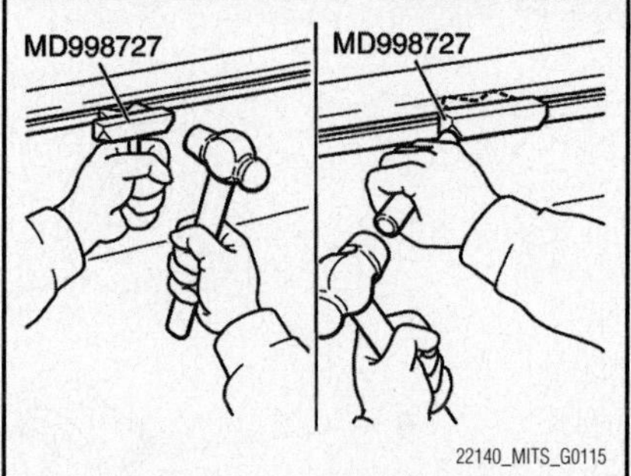

Fig. 158 Insert special tool MD998727 into the engine oil pan removal groove—Lancer 2.0L engine—2008

4. Drain the engine oil.
5. Remove the accessory drive belt. Refer to Accessory Drive Belts, Removal & Installation.
6. Remove the A/C compressor and clutch assembly together with the hose from the bracket.
7. Tie the removed A/C compressor and clutch assembly with a string at a position where they will not interfere with the removal and installation of engine oil pan.
8. Remove the engine oil pan mounting bolts.

⁂ WARNING

Do not forcibly drive in special tool MD998727 to avoid damage to the engine oil pan seal surface of cylinder block assembly.

9. Insert special tool MD998727 into the engine oil pan removal groove of the cylinder block assembly.
10. Lightly tap the special tool with a hammer to slide the oil pan seal surface, cut off the liquid gasket, and remove the engine oil pan.

To install:

11. Remove all the traces of sealant adhering to the engine oil pan and cylinder block assembly using a remover. Then, degrease them using a quick-drying degreasing agent (white gasoline).
12. Apply the sealant without any gap to the mating surface of engine oil pan as shown in the figure. Specified sealant: Three Bond® 1217G or equivalent.
13. Within 3 minutes, install the engine oil pan to the cylinder block assembly.

⁂ WARNING

Do not apply oil or water to the sealant area or start up the engine within 2 hours after the installation of the engine oil pan.

14. Tighten the engine oil pan mounting bolts to the specified torque.

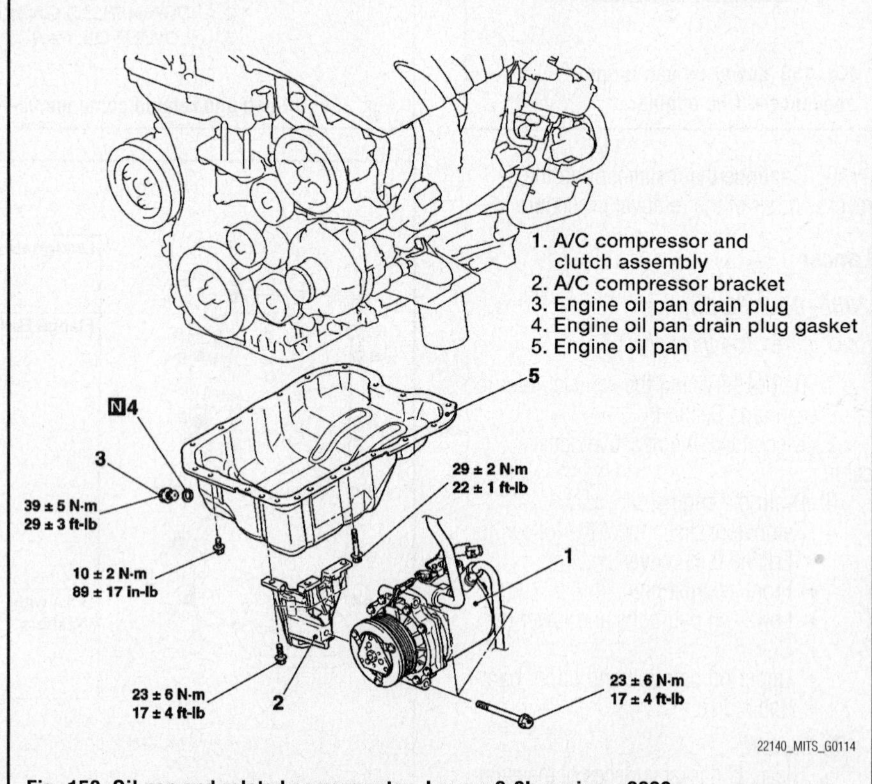

Fig. 159 Oil pan and related components—Lancer 2.0L engine—2008

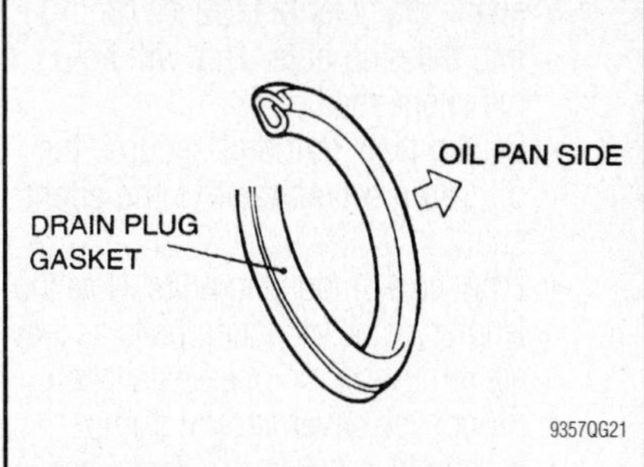

Fig. 160 Make sure to the install the new drain plug gasket as shown to avoid oil leakage

a. M6 bolts: 72–106 inch lbs. (8–12 Nm).

b. M8 bolts: 21–23 ft. lbs. (27–31 Nm).

15. Install the engine undercover.
16. Install the oil drain plug with a new gasket and tighten to 29 ft. lbs. (40 Nm).
17. Lower the vehicle and fill the crankcase to the proper level with clean engine oil.
18. Connect the negative battery cable. Start the engine and check for leaks.

Lancer Evolution

2008 2.0L Turbo Engine

See Figures 161 through 164.

1. Before servicing the vehicle, refer to the Precautions Section.
2. Remove the negative battery cable.
3. Remove the engine undercover and side cover (RH).
4. Drain the engine oil.
5. Remove the accessory drive belt. Refer to Accessory Drive Belts, Removal & Installation.
6. Remove the A/C compressor and clutch assembly together with the hose from the bracket.

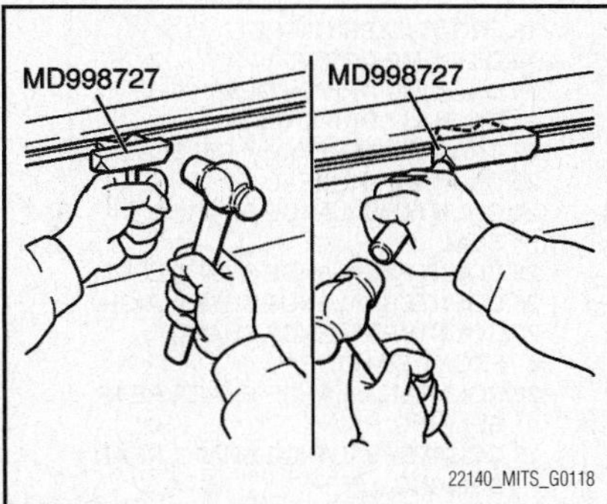

Fig. 161 Insert special tool MD998727 into the engine oil pan removal groove—Lancer Evolution—2008

7. Tie the removed A/C compressor and clutch assembly with a string at a position where they will not interfere with the removal and installation of engine oil pan.
8. Remove the engine oil pan mounting bolts.

✲✲ WARNING

Do not forcibly drive in special tool MD998727 to avoid damage to the engine oil pan seal surface of cylinder block assembly.

9. Insert special tool MD998727 into the engine oil pan removal groove of the cylinder block assembly.
10. Lightly tap the special tool with a hammer to slide the oil pan seal surface, cut off the liquid gasket, and remove the engine oil pan.

To install:

11. Remove all the traces of sealant adhering to the engine oil pan and cylinder block assembly using a remover. Then, degrease them using a quick-drying degreasing agent (white gasoline).
12. Apply the sealant without any gap to the mating surface of engine oil pan as shown in the figure. Specified sealant: Three Bond® 1217G or equivalent.
13. Within 3 minutes, install the engine oil pan to the cylinder block assembly.

✲✲ WARNING

Do not apply oil or water to the sealant area or start up the engine within 2 hours after the installation of the engine oil pan.

14. Tighten the engine oil pan mounting bolts to the specified torque.

a. M6 bolts: 72–106 inch lbs. (8–12 Nm).

b. M8 bolts: 21–23 ft. lbs. (27–31 Nm).

15. Install the engine undercover.
16. Install the oil drain plug with a new gasket and tighten to 29 ft. lbs. (40 Nm).
17. Tighten A/C compressor and clutch assembly mounting bolts to the specified torque in the order shown in the illustration. Tightening torque: 13–21 ft. lbs. (17–29 Nm).

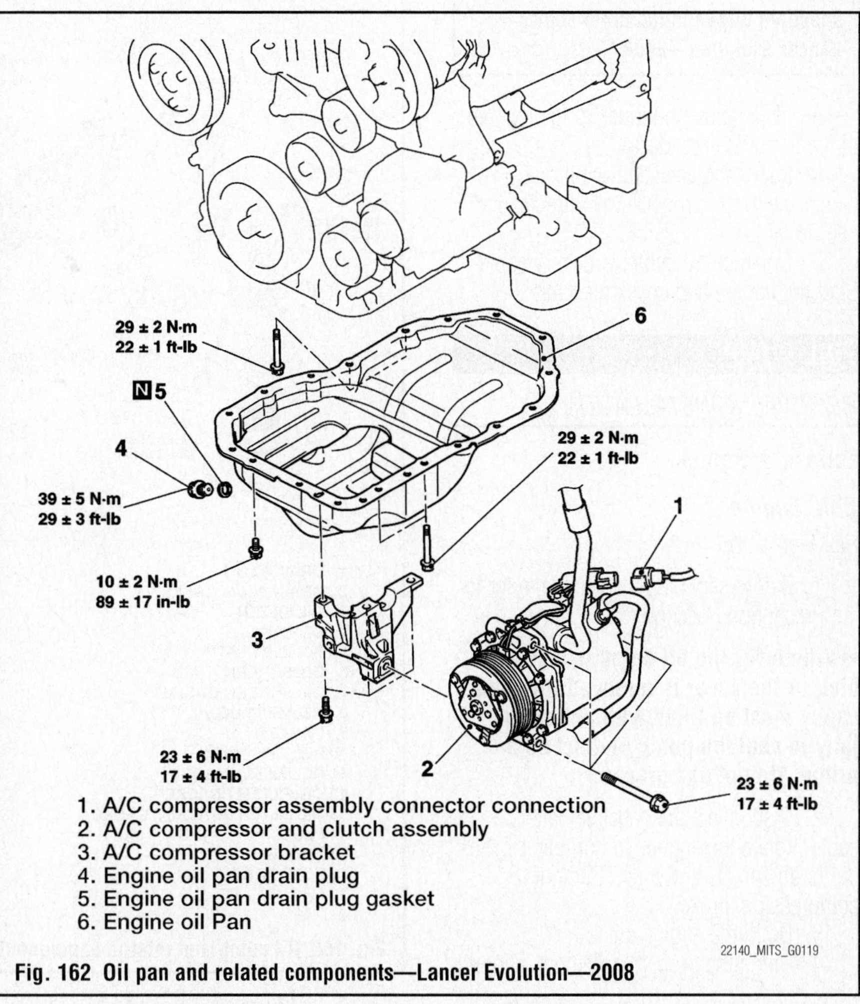

Fig. 162 Oil pan and related components—Lancer Evolution—2008

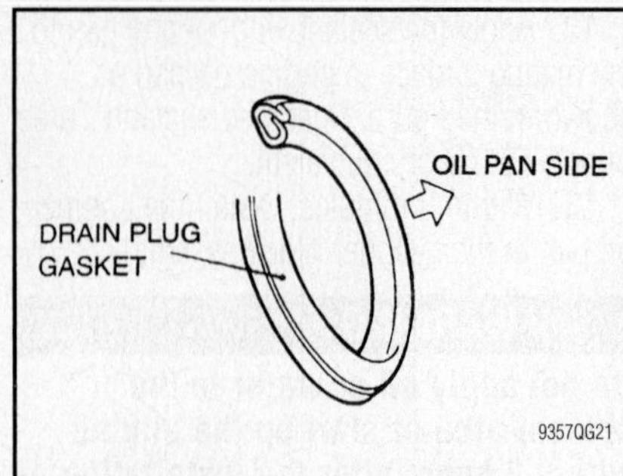

Fig. 163 Make sure to the install the new drain plug gasket as shown to avoid oil leakage

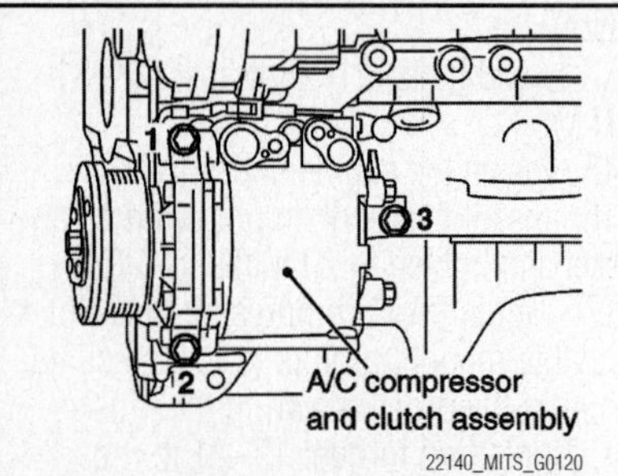

Fig. 164 Tighten A/C compressor and clutch assembly mounting bolts to the specified torque in the order shown—Lancer Evolution—2008

18. Installation continues in the reverse of the removal procedure.
19. Lower the vehicle and fill the crankcase to the proper level with clean engine oil.
20. Connect the negative battery cable. Start the engine and check for leaks.

OIL PUMP

REMOVAL & INSTALLATION

Eclipse & Galant

2.4L Engine

See Figure 165.

1. Before servicing the vehicle, refer to the Precautions Section.

➡Whenever the oil pump is disassembled or the cover is removed, the gear cavity must be filled with petroleum jelly to seal the pump and act as a prime. Do not use grease.

2. Disconnect the negative battery cable. Rotate the engine so number 1 cylinder is on Top Dead Center (TDC) of its compression stroke.
3. Drain the engine oil.
4. Using the proper equipment, support the weight of the engine. Remove the front engine mount bracket and accessory drive belts.
5. Remove or disconnect the following:
   - Timing belt upper and lower covers
   - Timing belt and crankshaft sprocket
   - Electrical connector from the oil pressure sending unit
   - Oil pressure sensor
   - Oil filter and the oil filter bracket
   - Oil pan, oil screen and gasket
6. Using special tool MD998162, remove the plug cap in the engine front cover.
7. Remove or disconnect the following:
   - Plug on the side of the engine block. Insert a steel rod with a shank diameter of 0.32 in. (8mm) into the plug hole. This will hold the silent shaft
   - Driven gear bolt that secures the oil pump driven gear to the silent shaft
   - Front cover mounting bolts. Note the lengths of the mounting bolts as they are removed for proper installation
   - Front case cover and oil pump assembly. If necessary, the silent shaft can come out with the cover assembly
   - Oil pump cover, located on the back of the engine front cover. Remove the oil pump drive and driven gears
8. After disassembling the oil pump, clean all components and remove any gasket material from mating surfaces

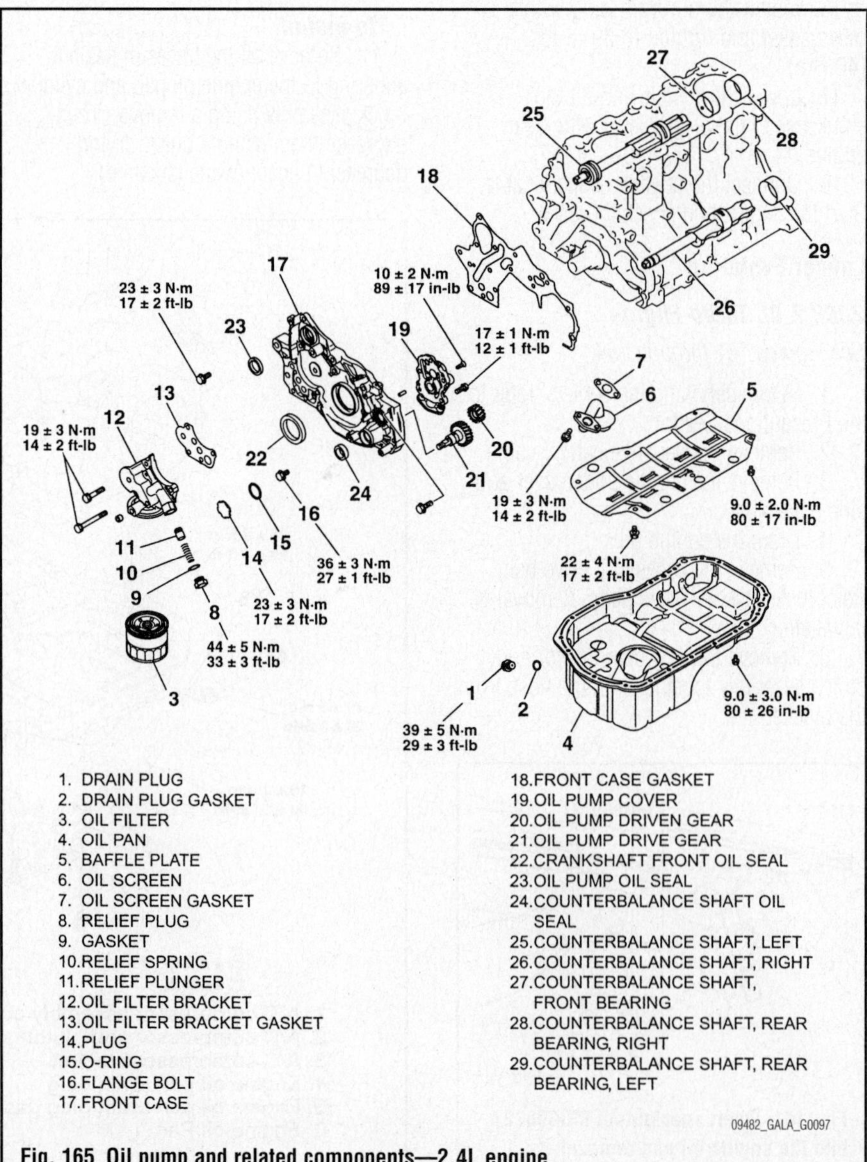

Fig. 165 Oil pump and related components—2.4L engine

9. Assemble the oil pump gears into the front case and rotate it to ensure smooth rotation and no looseness. Be sure there is no ridge wear on the contact surface between the front case and the gear surface of the oil pump front cover.

To install

10. Align the timing mark on the oil pump drive gear with that on the driven gear and install them into the engine front case. Apply engine oil to the gears.
11. Install the oil pump cover and tighten the retainer bolts to 13 ft. lbs. (18 Nm) on Eclipse models and 17 ft. lbs. (24 Nm) on Galant models.
12. Using the appropriate driver, install a new crankshaft seal into the front case.
13. Position new front case gasket in place. Set seal guide tool MD998285 on the front end of the crankshaft to protect the seal from damage. Apply a thin coat of oil to the outer circumference of the seal pilot tool.
14. Install the front case assembly through a new front case gasket and temporarily tighten the flange bolts.
15. Mount the oil filter on the bracket with new oil filter bracket gasket in place. Install the bolts with washers and tighten to 14 ft. lbs. (19 Nm).
16. Insert a Phillips screwdriver into the hole in the left side of the engine block to lock the silent shaft in place.
17. Install or connect the following:
    - Oil pump drive gear onto the left silent shaft. Tighten the driven gear bolt to 27 ft. lbs. (37 Nm).
    - New O-ring to the groove in the front case and install the plug cap. Tighten the cap to 17 ft. lbs. (24 Nm).
    - Oil screen in position with new gasket in place
18. Clean both mating surfaces of the oil pan and the cylinder block. Apply sealant in the groove in the oil pan flange.

➡After applying sealant to the oil pan, do not exceed 15 minutes before installing the oil pan.

19. Install or connect the following:
    - Oil pan to the engine and secure with the retainers. Tighten bolts to 60 inch lbs. (7 Nm).
    - Oil pressure gauge unit and the oil pressure switch
    - Electrical harness connector
    - Oil cooler. Oil cooler bolt to 31 ft. lbs. (43 Nm).
20. Refill the crankcase with the correct amount and type of oil. Install new oil filter.
21. Connect the negative battery cable and start the engine.
22. Verify there is correct oil pressure. Inspect for leaks.

3.8L Engine

1. Before servicing the vehicle, refer to the Precautions Section.
2. Disconnect the negative battery cable.
3. Remove the timing belt.
4. Drain the engine oil.
5. Remove or disconnect the following:
    - Splash shield from the wheel well, as needed
    - Oil filter adapter
    - Lower and upper oil pans
    - Lower baffle, oil pump pick-up, and upper baffle
    - Oil pump case mounting bolts and the oil pump case
    - Oil pump gear cover
6. Make matchmarks on the oil pump rotors before removing them.
7. Remove the crankshaft seal from the oil pump case.

To install:

8. Install a new crankshaft seal in the oil pump cover.
9. Apply engine oil to the rotors, then align the matchmarks and install the rotors in the oil pump case.
10. Install the rotor cover and tighten the bolts to 84 inch lbs. (10 Nm).
11. Apply a 0.113 in. (3mm) bead of sealant to the back of the oil pump case. Install the case on the engine and tighten the bolts to 10 ft. lbs. (13 Nm).
12. Install the upper baffle plate and oil pump pick-up using a new gasket.
13. Tighten the baffle bolts to 84 inch lbs. (10 Nm) and the pick-up bolts to 13 ft. lbs. (18 Nm).
14. Lower the baffle in the upper oil pan. Tighten the bolts to 96 inch lbs. (11 Nm).
15. Install the oil pans.
16. Install the oil filter adapter using a new gasket. Tighten the larger bolt to 30 ft. lbs. (41 Nm) and the smaller bolt to 17 ft. lbs. (23 Nm).
17. Install the timing belt and remaining the components from the removal procedure.
18. Fill the engine with the correct amount of oil.
19. Connect the negative battery cable.
20. Start the engine and check for leaks.

Lancer

2006–07 2.0L Engine

See Figure 166.

1. Before servicing the vehicle, refer to the Precautions Section.
2. Drain the engine oil.
3. Remove or disconnect the following:
    - Negative battery cable
    - Oil pressure switch
    - Oil filter
    - Drain plug and gasket. Discard the gasket.
    - Cover
    - Upper oil pan. Remove the 5 inch (127mm) bolt, which is closest to the flywheel/flexplate first, then, the other bolts.
    - Baffle plate
    - Lower oil pan. Remove the 5 inch (127mm) bolt, which is closest to the flywheel/flexplate first, then, the other bolts.
    - Oil screen and gasket
    - Relief plug and spring
    - Oil seal
    - Front case
    - O-ring
    - Oil pump case cover

➡Matchmark the installed position of the pump rotors before removing them.

- Outer and inner oil pump rotors

To install:

4. Install or connect the following:
    - Inner and outer rotors, making sure the alignment marks are matched up
    - Oil pump case cover
    - O-ring

➡After installation or the front case, wait at least one hour before filling the crankcase with oil or starting the engine.

- Front case. Apply a 0.12 inch (3mm) bead of sealant, then tighten the case bolts to 124 inch lbs. (14 Nm)
- Oil seal
- Relief plunger, spring and plug
- Oil screen gasket and screen
- Lower oil pan. Refer to the oil pan procedure for sealant application and torque specifications.
- Baffle plate
- Upper oil pan. Refer to the oil pan procedure for sealant application and torque specifications.
- Cover
- Drain plug with a new gasket. Tighten the plug to 29 ft. lbs. (35 Nm).
- Oil filter
- Oil pressure switch

5. Fill the engine with the correct amount and type of oil.
6. Connect the negative battery cable.
7. Start the engine and check for leaks.

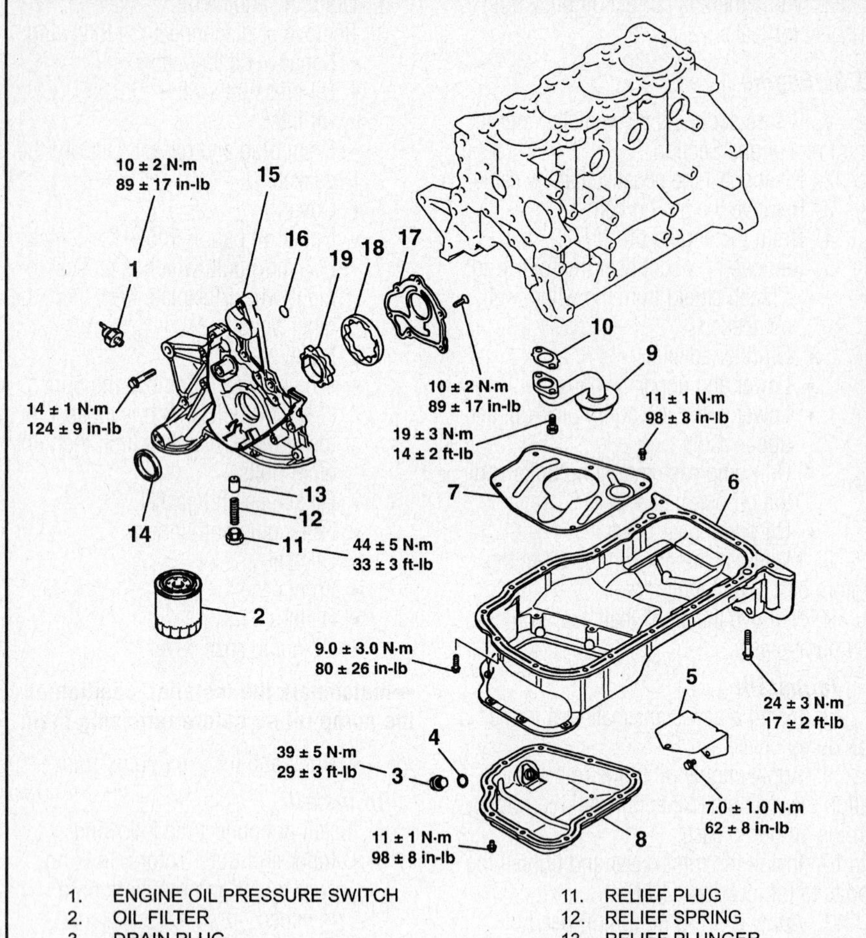

Fig. 166 Oil pump and related components—Lancer 2.0L engine—2006-07

2008 2.0L Engine

See Figures 167 and 168.

1. Before servicing the vehicle, refer to the Precautions Section.

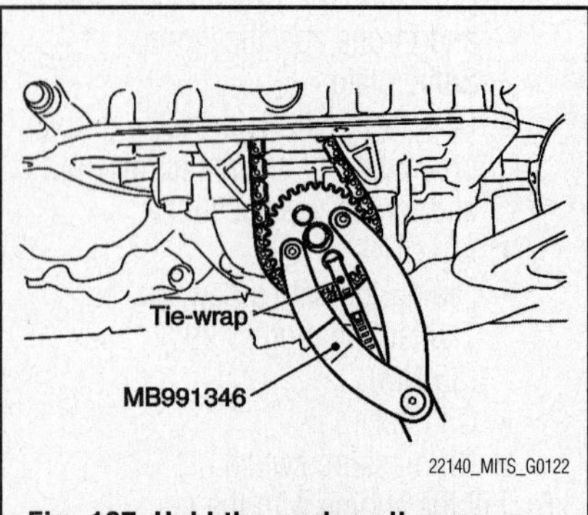

Fig. 167 Hold the engine oil pump sprocket with special tool MB991346 for removal—2.0L engine—2008

2. Disconnect the negative battery cable.
3. Remove the engine oil pan. Refer to Oil Pan, Removal & Installation.
4. Secure the engine oil pump sprocket and the engine oil pump drive chain with tie-wrap to prevent slippage between the engine oil pump sprocket and the engine oil pump drive chain.
5. Hold the engine oil pump sprocket with special tool MB991346.
6. Remove the engine oil pump sprocket with the engine oil pump drive chain attached.

To install:

7. Installation is the reverse of the removal procedure.
8. Tighten the engine oil pump sprocket center bolt to 16–18 ft. lbs. (21–25 Nm).
9. Tighten the engine oil pump assembly attaching bolts to 19–21 ft. lbs. (24–28 Nm).
10. Fill the engine with the correct amount and type of oil.
11. Connect the negative battery cable.
12. Start the engine and check for leaks.

Lancer Evolution

2008 2.0L Turbo Engine

See Figures 169 and 170.

1. Before servicing the vehicle, refer to the Precautions Section.
2. Disconnect the negative battery cable.
3. Remove the engine oil pan. Refer to Oil Pan, Removal & Installation.
4. Secure the engine oil pump sprocket and the engine oil pump drive chain with

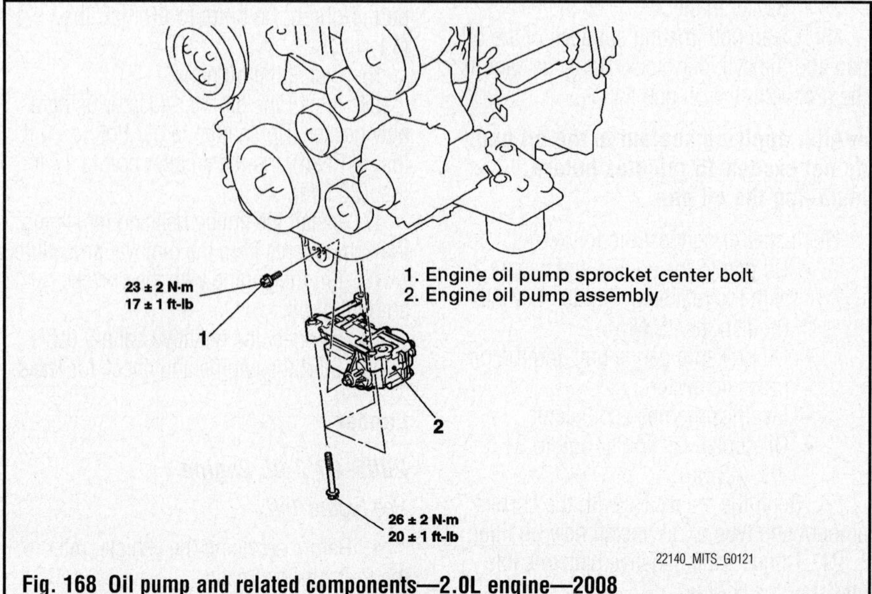

Fig. 168 Oil pump and related components—2.0L engine—2008

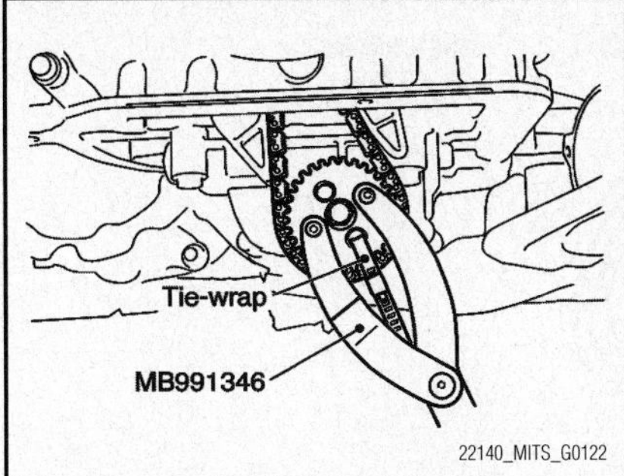

Fig. 169 Hold the engine oil pump sprocket with special tool MB991346 for removal—2.0L engine—2008

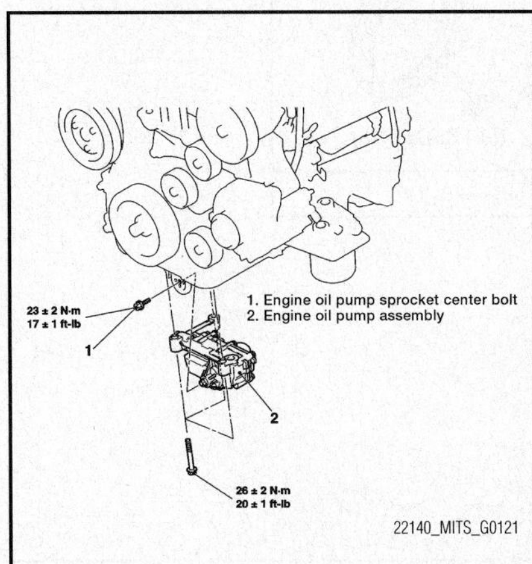

Fig. 170 Oil pump and related components—2.0L engine—2008

tie-wrap to prevent slippage between the engine oil pump sprocket and the engine oil pump drive chain.

5. Hold the engine oil pump sprocket with special tool MB991346.
6. Remove the engine oil pump sprocket with the engine oil pump drive chain attached.

To install:

7. Installation is the reverse of the removal procedure.
8. Tighten the engine oil pump sprocket center bolt to 16–18 ft. lbs. (21–25 Nm).
9. Tighten the engine oil pump assembly attaching bolts to 19–21 ft. lbs. (24–28 Nm).
10. Fill the engine with the correct amount and type of oil.
11. Connect the negative battery cable.
12. Start the engine and check for leaks.

PISTON AND RING

POSITIONING

See Figures 171 through 177.

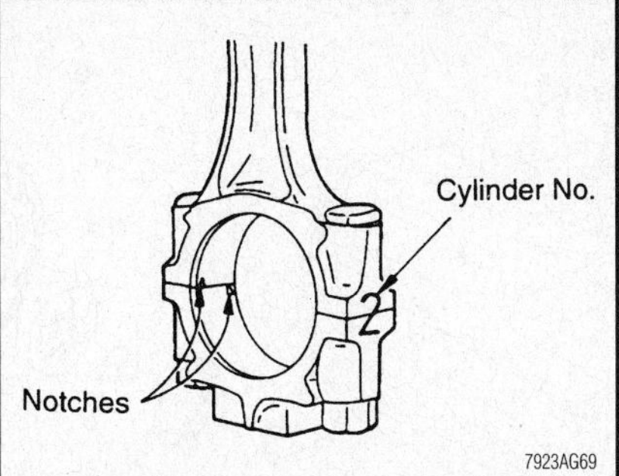

Fig. 171 Before removing the caps from the connecting rods, be sure to matchmark them as shown

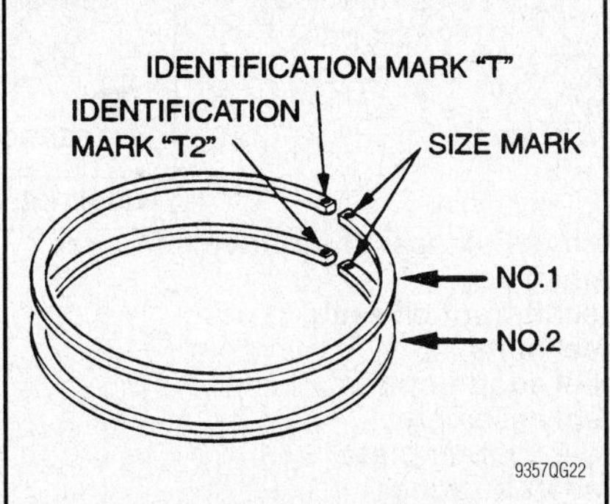

Fig. 172 Compression ring identification mark—2.0L SOHC engine

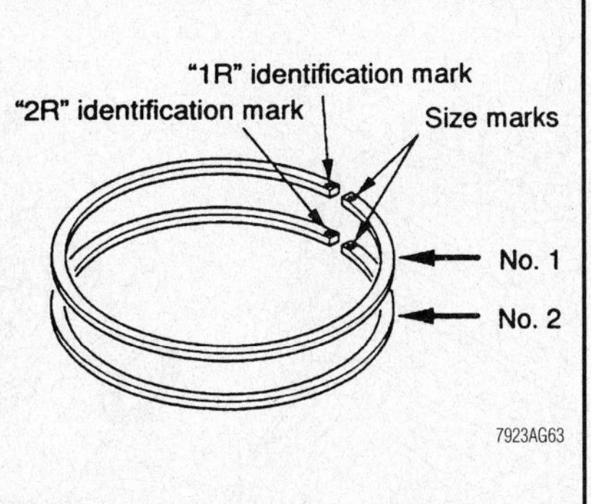

Fig. 173 Compression ring identification mark—2.0L DOHC engine

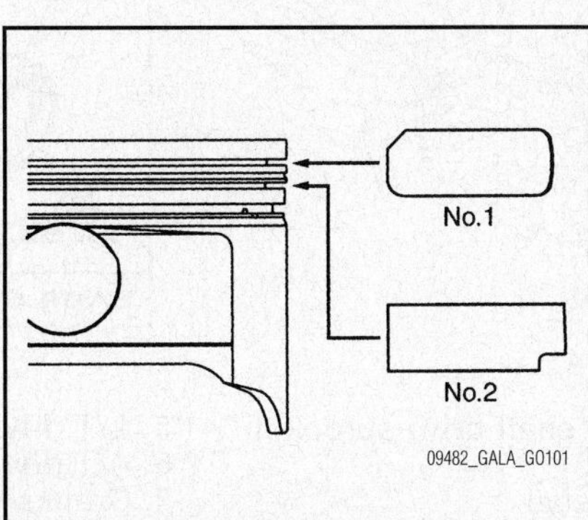

Fig. 174 Compression ring identification mark—2.4L engine

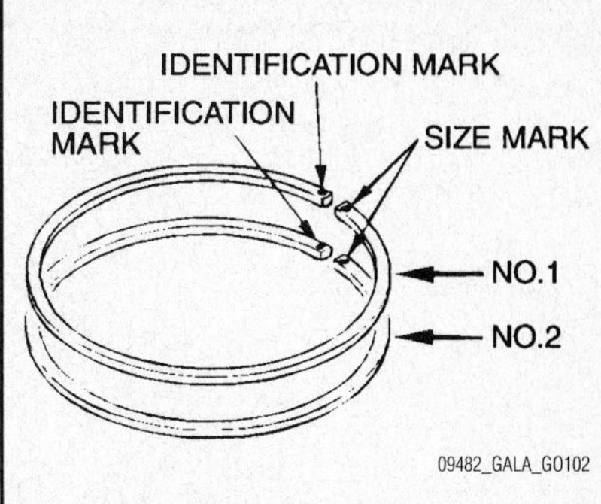

Fig. 175 Compression ring identification mark—3.8L engine

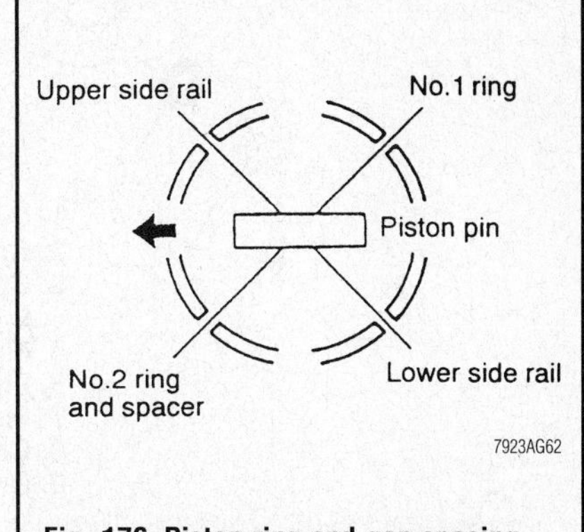

Fig. 176 Piston ring end-gap spacing

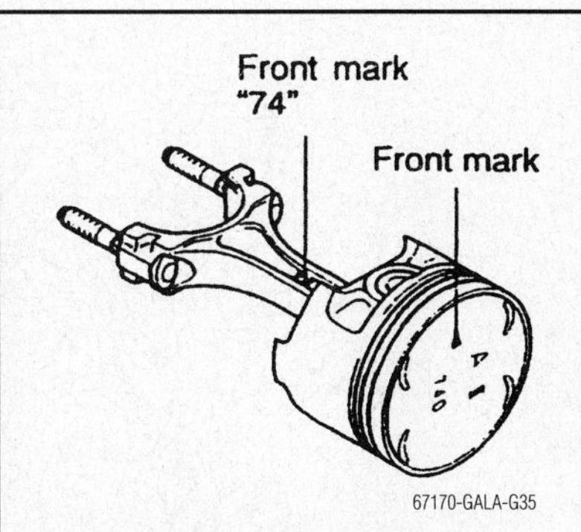

Fig. 177 Piston and connecting rod assembly positioning—3.8L engine

REAR MAIN SEAL

REMOVAL & INSTALLATION

Eclipse & Galant

2.4L Engine

See Figures 178 and 179.

1. Before servicing the vehicle, refer to the Precautions Section.
2. Disconnect the negative battery cable.
3. Remove the transaxle assembly. Refer to Automatic Transaxle Assembly or Manuel Transaxle Assembly, Removal & Installation.

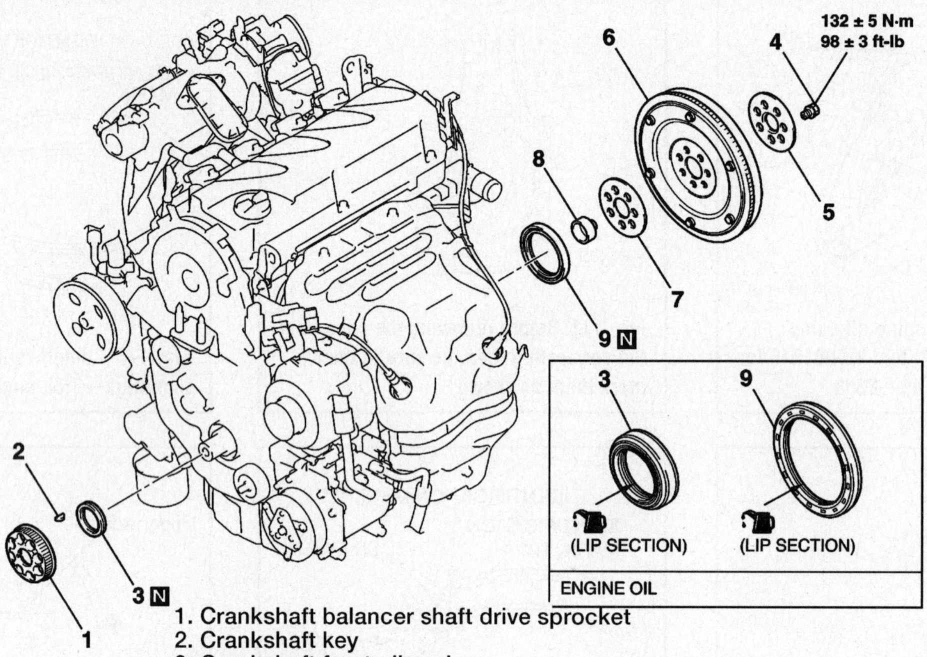

1. Crankshaft balancer shaft drive sprocket
2. Crankshaft key
3. Crankshaft front oil seal
4. Flywheel bolts
5. Flywheel adapter plate
6. Flywheel assembly
7. Flywheel adapter plate
8. Crankshaft bushing
9. Crankshaft rear oil seal

22140_MITS_G0136

Fig. 178 Expanded view of the crankshaft oil seals (front and rear)—2.4L engine (M/T)

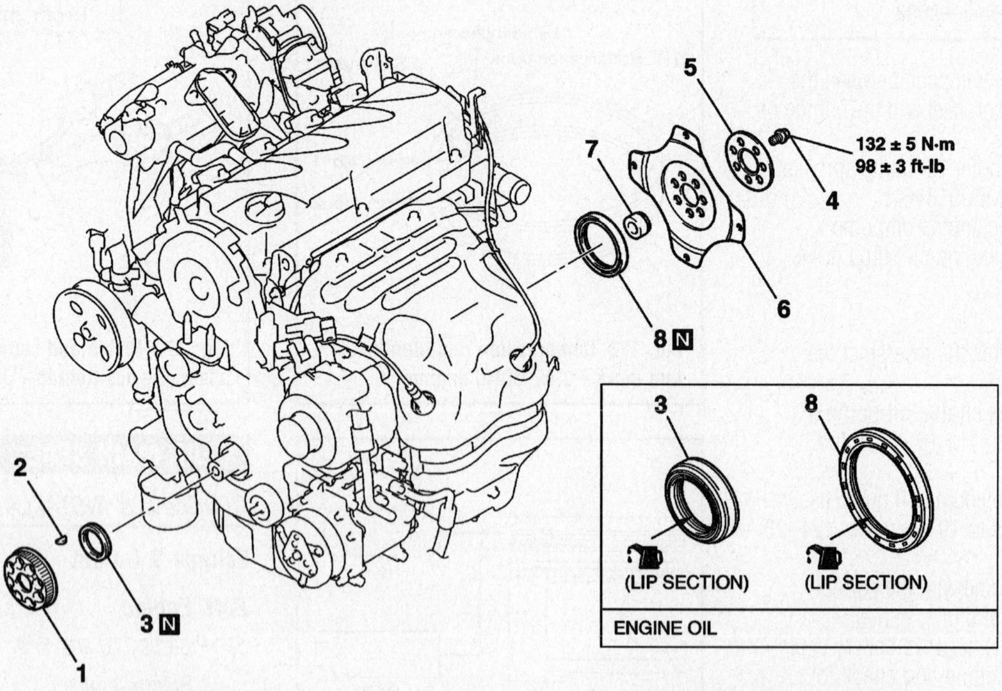

1. Crankshaft balancer shaft drive sprocket
2. Crankshaft key
3. Crankshaft front oil seal
4. A/T drive plate bolts
5. A/T drive plate adapter plate
6. A/T drive plate
7. Crankshaft bushing
8. Crankshaft rear oil seal

22140_MITS_G0137

Fig. 179 Expanded view of the crankshaft oil seals (front and rear)—2.4L engine (A/T)

4. Remove the flywheel. Refer to Flywheel, Removal & Installation.
5. Remove the crankshaft bushing.
6. Remove the rear main seal. Be careful not to damage the oil seal bore or the crankshaft sealing surface.

To install:

7. Apply clean engine oil to the oil seal lip. Using a seal driver, install the oil seal.
8. Use special tools MB990938 and MD998776 to press-fit the oil seal.
9. Installation continues in the reverse of the removal procedure.

3.8L Engine

See Figure 180.

1. Before servicing the vehicle, refer to the Precautions Section.
2. Disconnect the negative battery cable.
3. Remove the transaxle assembly. Refer to Automatic Transaxle Assembly or Manuel Transaxle Assembly, Removal & Installation.
4. Remove the flywheel. Refer to Flywheel, Removal & Installation.
5. Remove the rear oil seal.

To install:

6. Apply a small amount of engine oil to the entire circumference of the rear oil seal lip.
7. Use special tool MD998718 to tap in the oil seal.
8. Installation continues in the reverse of the removal procedure.

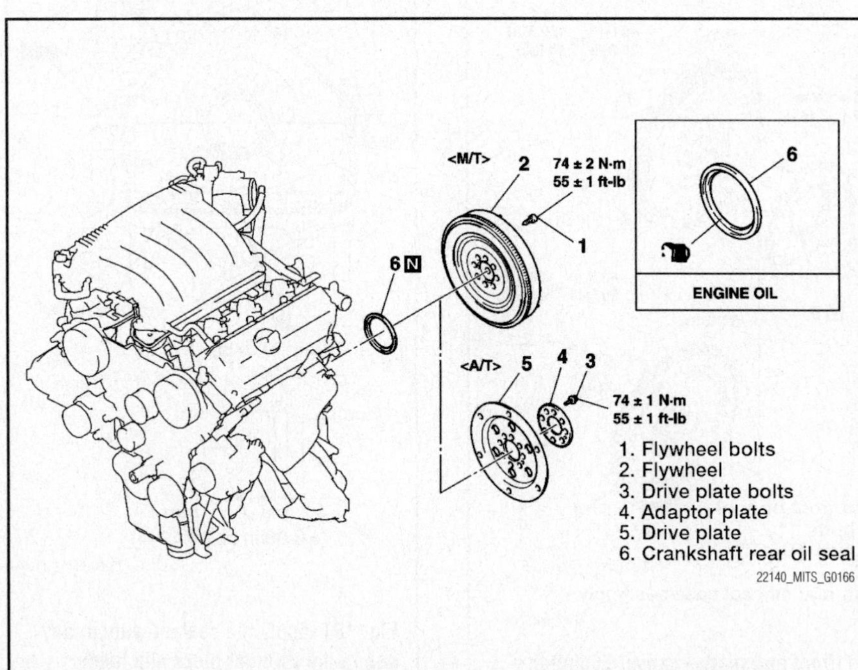

Fig. 180 Expanded view of the crankshaft oil rear seal—3.8L engine (M/T and A/T)

Lancer

2006–07 2.0L Engine

See Figure 181.

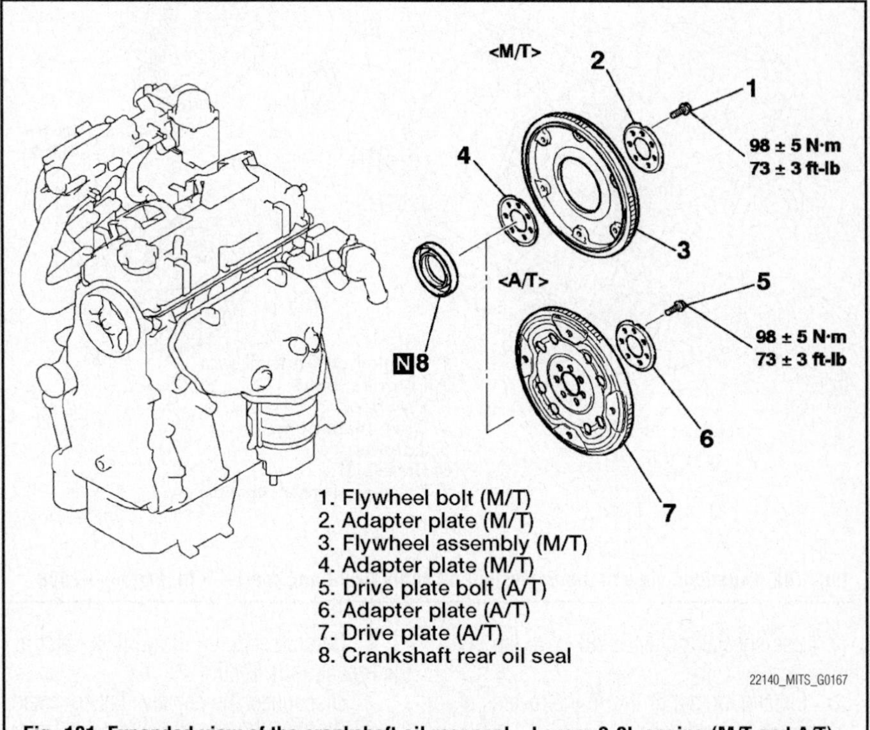

Fig. 181 Expanded view of the crankshaft oil rear seal—Lancer 2.0L engine (M/T and A/T)

1. Before servicing the vehicle, refer to the Precautions Section.
2. Disconnect the negative battery cable.
3. Remove the transaxle assembly. Refer to Automatic Transaxle Assembly or Manuel Transaxle Assembly, Removal & Installation.
4. Remove the flywheel. Refer to Flywheel, Removal & Installation.
5. Remove the crankshaft rear oil seal.

To install:

6. Apply a small amount of engine oil to the entire circumference of the rear oil seal lip.
7. Use special tool MD998718 to tap in the oil seal.
8. Installation continues in the reverse of the removal procedure.

2008 2.0L Engine

See Figure 182.

1. Before servicing the vehicle, refer to the Precautions Section.
2. Disconnect the negative battery cable.
3. Remove the transaxle assembly. Refer to Automatic Transaxle Assembly or Manuel Transaxle Assembly, Removal & Installation.
4. Remove the flywheel. Refer to Flywheel, Removal & Installation.
5. Remove the crankshaft rear oil seal.

To install:

6. Apply a small amount of engine oil to the entire circumference of the rear oil seal lip.

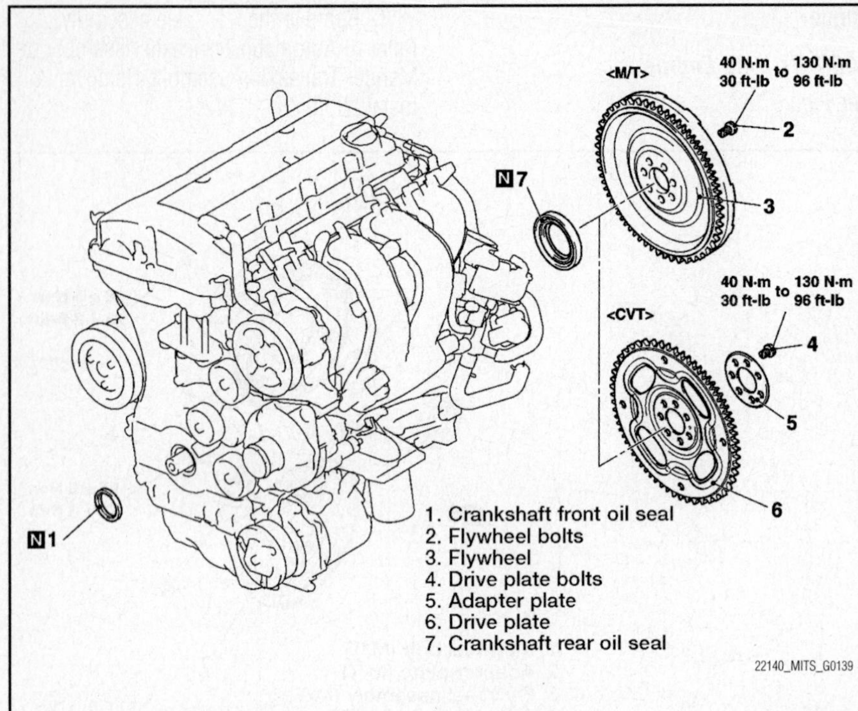

Fig. 182 Expanded view of the crankshaft oil seals (front and rear)—2.0L engine—2008

7. Use special tool MD998718 to tap in the oil seal.
8. Installation continues in the reverse of the removal procedure.

Lancer Evolution

2008 2.0L Turbo Engine

See Figures 183 and 184.

1. Before servicing the vehicle, refer to the Precautions Section.
2. Disconnect the negative battery cable.
3. Remove the transfer assembly. Refer to Transfer Case Assembly, Removal & Installation.
4. Remove the transaxle assembly. Refer to Manual Transaxle Assembly, Removal & Installation.

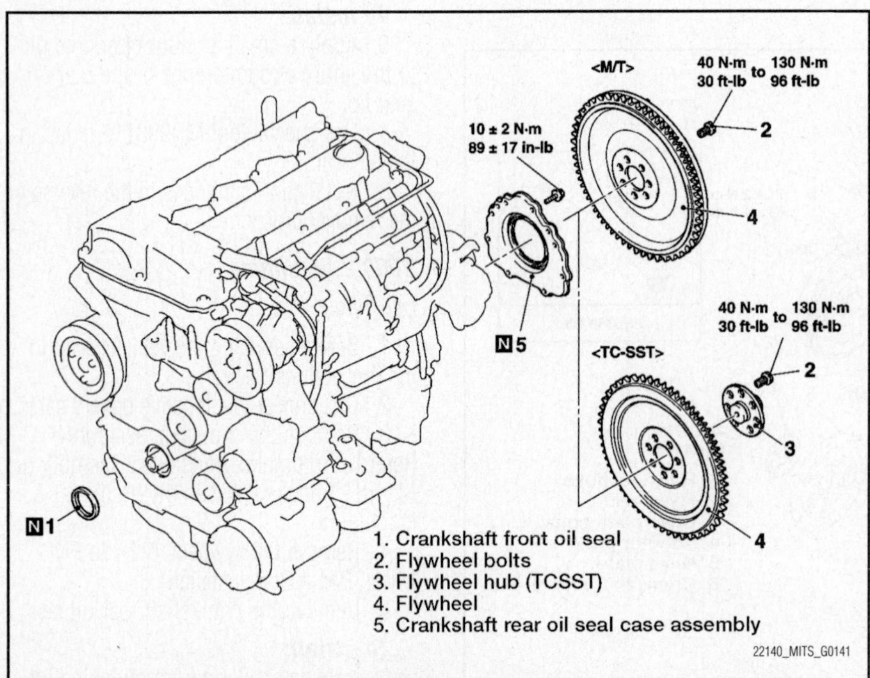

Fig. 183 Expanded view of the crankshaft oil seals (front and rear)—Lancer Evolution—2008

5. Remove the flywheel. Refer to Flywheel, Removal & Installation.
6. Remove the crankshaft rear main seal case assembly.

To install:

7. Remove all the traces of sealant adhering to the cylinder block and ladder frame using a remover. Degrease using quick-drying degreasing agent (white gasoline).
8. Apply the sealant without any gap to the cylinder block and ladder frame as shown. Within 3 minutes, install the crankshaft rear oil seal case assembly. Specified sealant: Three Bond® 1227D or equivalent.
9. Apply a small amount of engine oil to the entire inner diameter of the oil seal lip, and install the crankshaft rear oil seal case assembly.

✲✲ WARNING

Do not apply oil or water to the sealant-applied area or start the engine within 2 hours after the installation of the crankshaft rear oil seal.

10. Tighten the crankshaft rear oil seal case assembly mounting bolts to the specified torque. Tightening torque: 72– 106 inch lbs. (8–12 Nm).

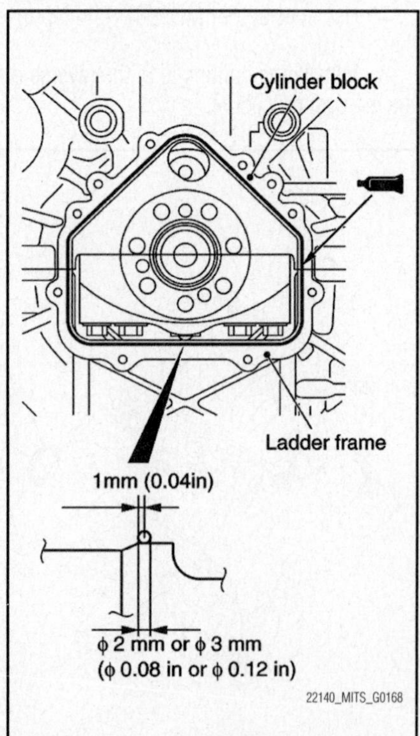

Fig. 184 Apply the sealant without any gap to the cylinder block and ladder frame—Lancer Evolution—2008

11. Installation continues in the reverse of the removal procedure.

ROCKER ARMS/SHAFTS

REMOVAL & INSTALLATION

Eclipse & Galant

2.4L ENGINE

See Figures 185 and 186.

1. Before servicing the vehicle, refer to the Precautions Section.
2. Disconnect the negative battery cable.
3. If equipped with a manual transaxle, remove the ECM. If equipped with an automatic transaxle, remove the PCM.
4. Remove the air cleaner assembly.
5. Remove the battery.
6. Remove the ignition coils.
7. Remove the timing belt upper cover.
8. Disconnect the control wiring harness connection.
9. Remove the rocker cover PCV connection.
10. Remove the rocker cover breather hose connection.
11. Remove the engine oil control valve and O-ring. Remove the oil pressure switch.

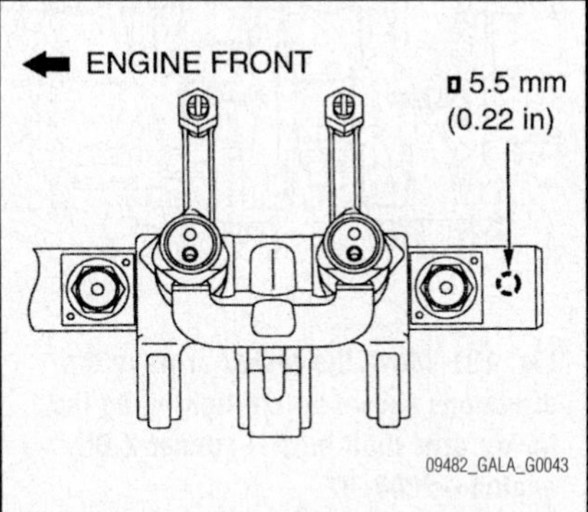

Fig. 185 Intake rocker arm shaft installation locating point—2.4L engine

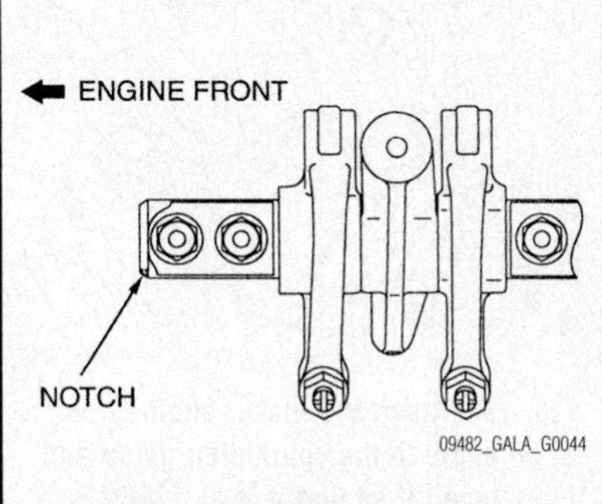

Fig. 186 Exhaust rocker arm shaft installation locating point—2.4L engine

12. Remove the rocker cover retaining bolts. Remove the rocker cover. Discard the gasket.
13. Remove the spark plug guide oil seals. Remove the accumulator assembly.
14. Remove the exhaust rocker arm shaft caps. Remove the exhaust rocker arm shaft assembly.
15. Remove the intake rocker arm shaft caps. Remove the intake rocker arm shaft assembly.

➡**Do not disassemble the exhaust or intake rocker arm assembly.**

To install:

16. Position the intake rocker arm assembly so that its 0.22 inch hole faces toward the cylinder head.
17. Install the rocker shaft mounting bolts. Torque to 21—25 ft. lbs. (29—34 Nm).
18. Position the exhaust rocker arm assembly so that its notch is positioned as shown in the illustration.
19. Install the rocker shaft mounting bolts. Torque to 106—124 inch lbs. (12—14 Nm).
20. Continue the installation in the reverse order of the removal procedure.
21. To initialize the ECM, manual transaxle equipped vehicles, or the PCM, automatic transaxle equipped vehicles, turn the ignition switch ON then OFF and keep it in the OFF position for at least 10 seconds.

3.8L Engine

See Figure 187.

1. Before servicing the vehicle, refer to the Precautions Section.
2. Disconnect the negative battery cable.
3. Drain the cooling system.
4. On the left side:
   a. Remove the thermostat housing assembly.
   b. Remove the PCV valve hose connection, ignition coil connector, and ignition coil.
   c. Disconnect the engine control wiring harness clamp and harness bracket.
5. On the right side:
   a. Remove the intake manifold plenum, breather hose connection, heater hose, and water hose connection.
   b. Remove the ignition coil connector and coil.
6. Remove the rocker arm cover retaining bolts. Remove the rocker arm cover. Discard the gasket.
7. Install tool MD998443 on the rocker arms, so that the lash adjusters will not fall out.

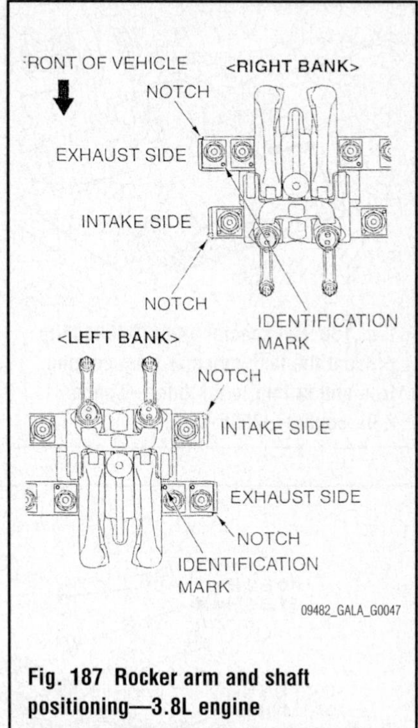

Fig. 187 Rocker arm and shaft positioning—3.8L engine

8. Loosen the rocker arm and shaft assembly mounting bolts. Remove the rocker arm and shaft assembly with the bolts still attached.

➡**Do not disassemble the rocker arm and shaft assembly.**

To install:

9. Install the rocker arm, shaft, and lash adjuster assembly.
10. Check that the notches in the rocker shaft are facing the direction, shown in the illustration. Install the rocker shaft cap with its identification mark as shown in the illustration.
11. Tighten the intake side rocker arm and shaft assembly mounting bolts to 21—25 ft. lbs. (28—34 Nm).
12. Tighten the exhaust side rocker arm and shaft assembly mounting bolts to 106—124 inch lbs. (12—14 Nm).
13. Remove special tool MD998443.
14. Continue the installation in the reverse order of the removal procedure. Be sure to use new valve cover gaskets.

Lancer

2006–07 2.0L Engine

See Figures 188 through 192.

1. Before servicing the vehicle, refer to the Precautions Section.
2. Disconnect the negative battery cable.
3. Remove the air cleaner assembly.
4. Remove the ignition coil.

Fig. 188 Set special tool MD998443 to prevent the lash adjuster from coming free and falling to the floor—Lancer 2.0L engine—2006–07

5. Disconnect the breather hose connection, PCV valve connection, and accelerator cable clamp.

6. Remove the rocker arm cover retaining bolts. Remove the rocker arm cover. Discard the gasket.

7. Set special tool MD998443 to prevent the lash adjuster from coming free and falling to the floor.

8. Loosen the rocker arm and shaft assembly mounting bolts. Remove the rocker arm and shaft assembly with the bolts still attached.

➡Do not disassemble the rocker arm and shaft assembly.

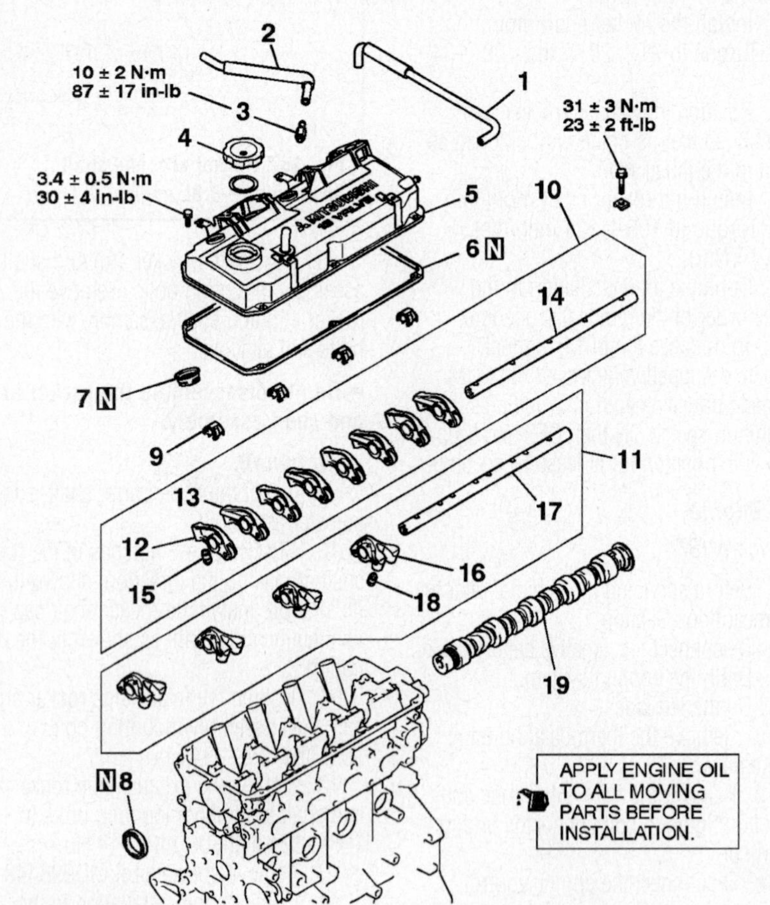

Fig. 189 Exploded view of rocker arms and camshaft—Lancer 2.0L engine—2006–07

To install:

9. Install the rocker arm shafts, place the end with the notched side toward the timing belt side as shown.

10. Move the rocker arms in the directions shown in the illustration before tightening the rocker arm shaft bolts.

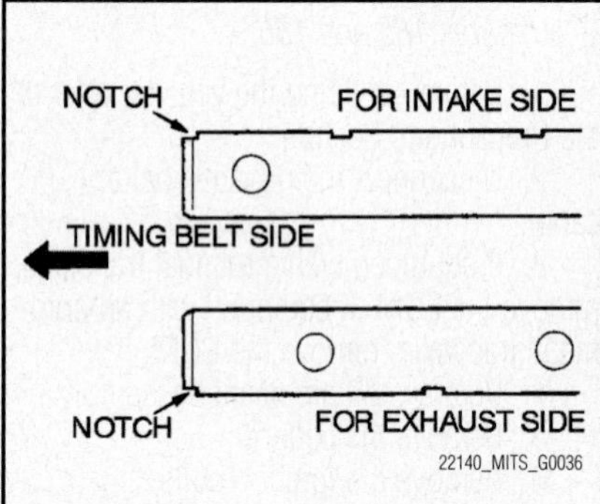

Fig. 190 Install the rocker arm shafts, place the end with the notched side toward the timing belt side—Lancer 2.0L engine—2006–07

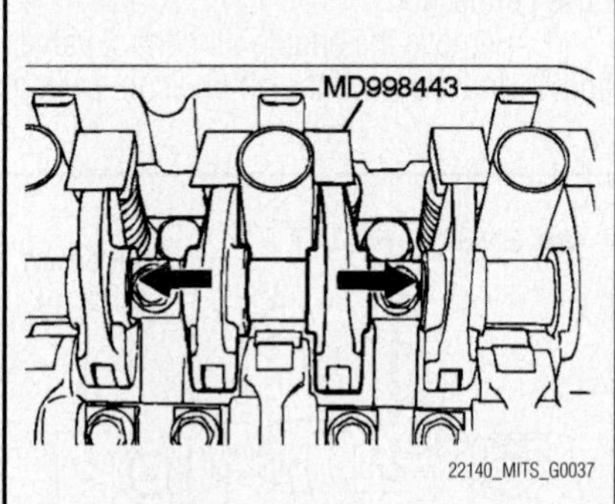

Fig. 191 Move the rocker arms in the directions shown before tightening the rocker arm shaft bolts—Lancer 2.0L engine—2006–07

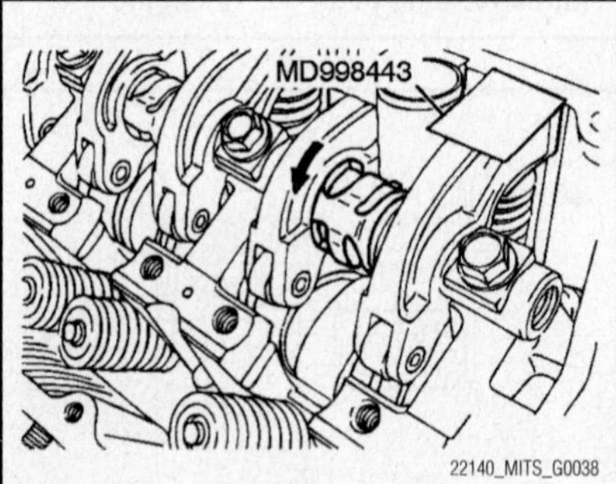

Fig. 192 Insert the rocker shaft spring at an angle to the spark plug guide and then install it so that it is at a right angle to the guide—Lancer 2.0L engine—2006–07

11. Insert the rocker shaft spring at an angle to the spark plug guide and then install it so that it is at a right angle to the guide.

12. Remove the special tool from the rocker arms.

13. Installation continues in the reverse of the removal procedure.

14. Torque the following to specification:

a. Positive Crankcase Ventilation (PCV) valve bolt to: 72 –106 inch lbs. (8–12 Nm).

b. Rocker arm shaft bolts to: 21–25 ft. lbs. (28–34 Nm).

c. Rocker cover bolt 26–34 inch lbs. (3–4 Nm).

2008 2.0L Engine

These engines are not equipped with rocker arms. The camshafts act directly on the valves through the valve tappets.

Lancer Evolution

2008 2.0L Turbo Engine

These engines are not equipped with rocker arms. The camshafts act directly on the valves through the valve tappets.

TURBOCHARGER

REMOVAL & INSTALLATION

Lancer Evolution

See Figures 193 through 196.

1. Before servicing the vehicle, refer to the Precautions Section.

2. Remove the engine room undercover and side cover.

3. Drain the engine oil and coolant.

4. Remove the engine upper cover and the ignition coil assemblies.

5. Remove the charge air cooler intake hose A and charge air cooler intake pipe A.

6. Remove the air cleaner assembly.

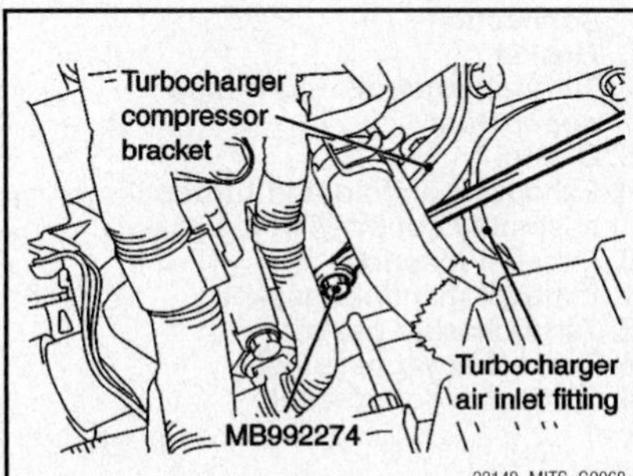

Fig. 193 Use special MB992274 to remove the turbocharger compressor bracket mounting bolt—2.0L engine

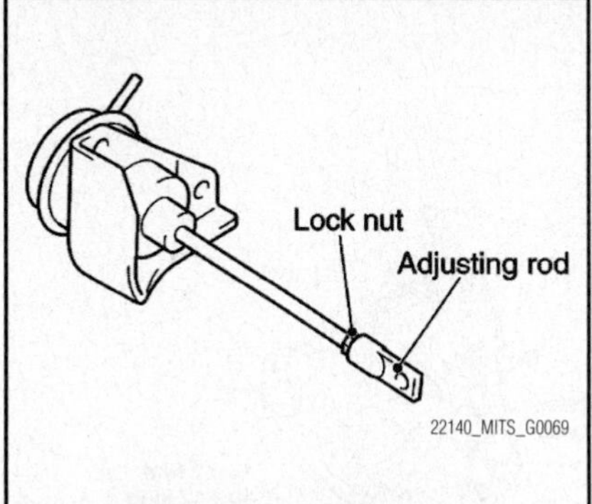

Fig. 194 View of turbocharger waste gate actuator—2.0L engine

7. Remove the front exhaust pipe.

8. Remove the strut tower bar.

9. Remove the cowl top panel.

10. Remove the dash panel heat protector.

11. Remove the exhaust manifold cover.

12. Remove the emission vacuum control hose connection.

13. Remove the turbocharger air outlet fitting and gasket.

14. Remove the turbocharger compressor bracket.

➡Use special MB992274 to remove the turbocharger compressor bracket mounting bolt (cylinder block side).

15. Remove the turbocharger air inlet fitting and gasket.

16. Remove the turbocharger water feed hose, water feed pipe, and gasket.

17. Remove the turbocharger water return hose.

18. Remove the turbocharger exhaust outlet fitting bracket.

19. Remove the transfer heat protector and the drive shaft heat protector.

20. Remove the steering gear and linkage heat protector, the turbocharger protectors A and B.

21. Remove the turbocharger pin and the waste gate actuator.

➡Do not loosen the locking nuts and adjusting rod of the waste gate actuator.

22. Remove the turbocharger exhaust outlet fitting and gasket.

23. Remove the turbocharger bracket and turbocharger assembly coupling bolt.

24. Remove the turbocharger oil feed tube connection and gasket.

⁂ WARNING

Take care not to allow foreign objects to get into the oil passage hole of the turbocharger assembly after the turbocharger oil feed tube is removed.

25. Remove the turbocharger oil return tube connection and O-ring.

26. Remove the exhaust manifold and turbocharger assembly coupling bolt and nut.

27. Remove the exhaust manifold and gasket.

28. Remove the turbocharger assembly and gasket.

To install:

29. Clean all gasket material from the mating surfaces.

30. Install a new gasket and exhaust manifold. Tighten the retainers to 32–40 ft. lbs. (44–54 Nm).

31. Install the turbocharger assembly. Be sure to clean the oil and water pipe fittings, inside of eye bolts and individual pipes of any clogs. Clean or use compressed air to remove any carbon matter stuck in the oil passage of the turbocharger. Refill new engine oil at the oil feed pipe fitting hole.

32. Install the exhaust fitting assembly, along with a new gasket.

33. Install the exhaust fitting bracket.

34. Install the air outlet fitting and gasket.

35. Connect the vacuum hose connections.

36. Connect the turbocharger oil feed and oil return pipes, along with new gaskets. Tighten the oil feed pipe fastener to 10–14 ft. lbs. (15–19 Nm). Tighten the oil return pipe fasteners to 71–89 inch lbs. (8–10 Nm).

37. Connect the turbocharger water feed pipe and return hoses, along with new gaskets. Tighten the pipe fasteners to 26–36 ft. lbs. (35–49 Nm).

38. Install the turbocharger heat protector.

39. Install the front Heated Oxygen (HO_2S) sensor.

40. Install the exhaust manifold cover.

41. Install the front exhaust pipe to the exhaust manifold.

42. Install the air pipes and hoses to the charge air cooler assembly.

43. Install the air intake hose to the air cleaner assembly.

44. Install the front undercover and secure with the retaining clips.

45. Refill the cooling system and the engine oil with the correct amount and type of fluid.

46. Connect the negative battery cable.

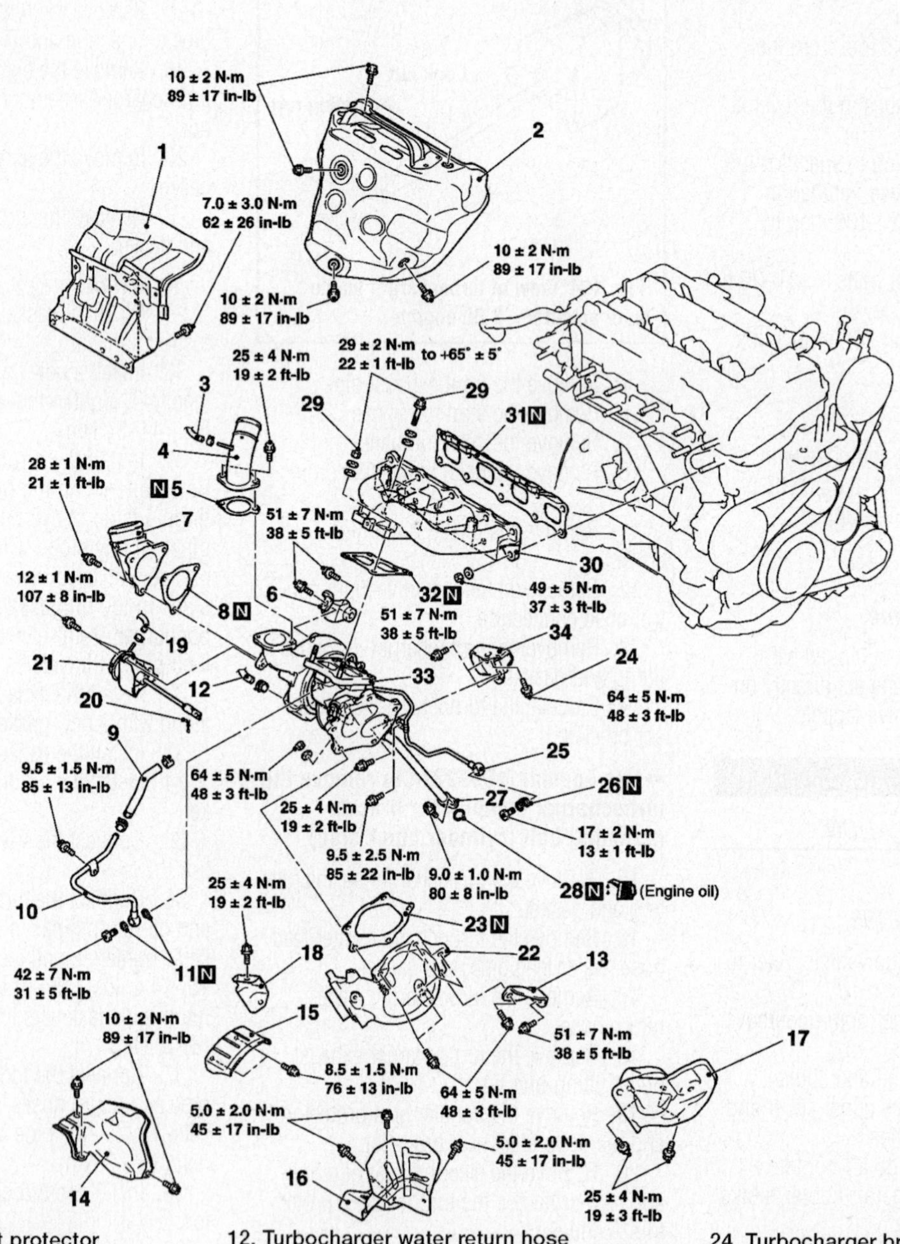

Fig. 195 Turbocharger assembly and related components—2.0L engine

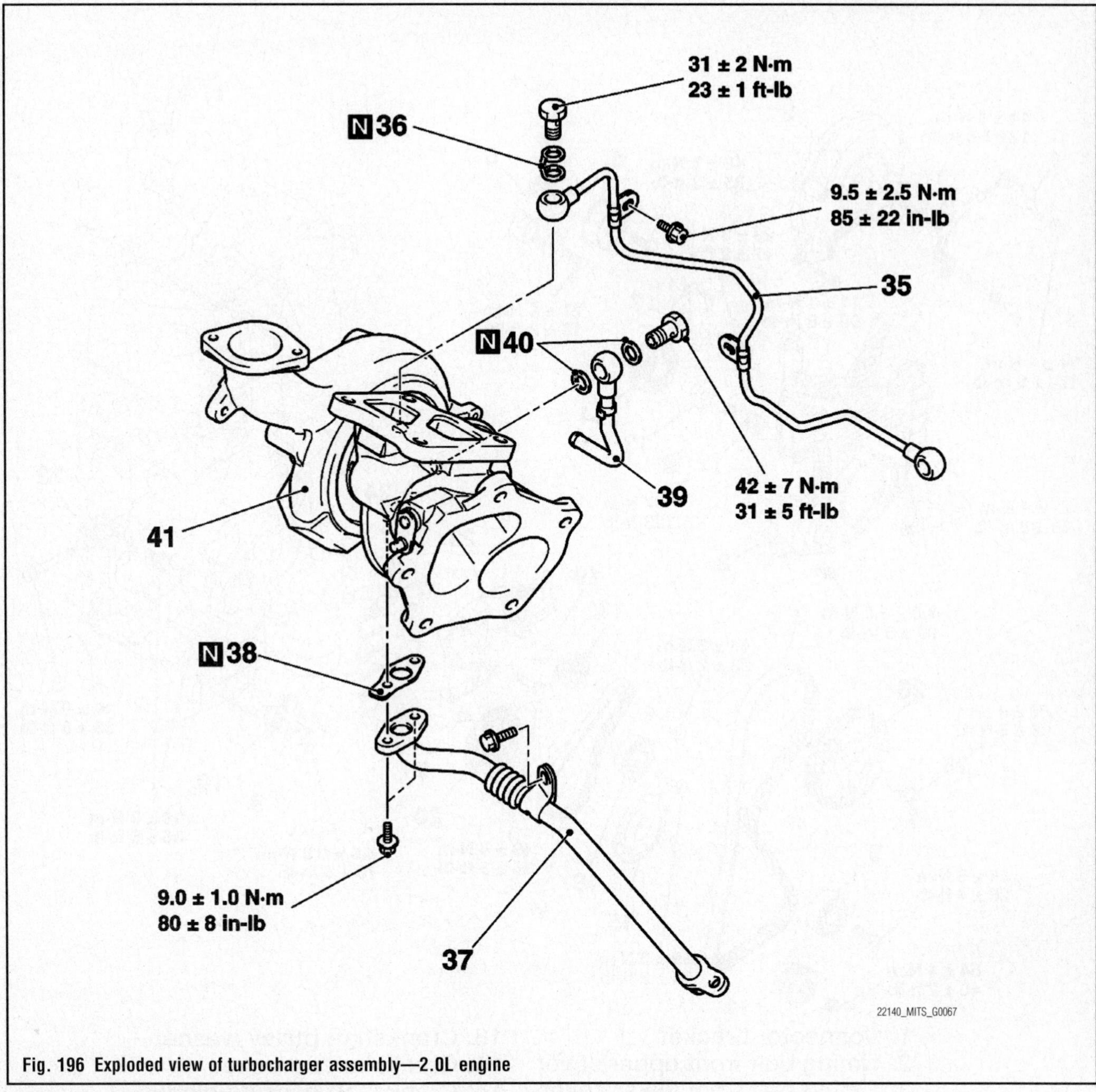

Fig. 196 Exploded view of turbocharger assembly—2.0L engine

TIMING BELT FRONT COVER

REMOVAL & INSTALLATION

Eclipse & Galant

2.4L Engine

See Figure 197.

1. Before servicing the vehicle, refer to the Precautions Section.
2. Disconnect the negative battery cable.
3. Remove the engine undercover.
4. Remove the crankshaft damper pulley. Refer to Crankshaft Damper, Removal & Installation.
5. Disconnect the control wiring harness connector, battery wiring harness connector, and connector bracket to engine mounting insulator.
6. Remove the harness bracket.
7. Remove the front upper timing belt cover.
8. Remove the water pump pulley.
9. Remove the idler pulley.
10. Remove the auto tensioner.
11. Remove the front lower timing belt cover.

To install:

12. Install the front lower timing belt cover.
13. Install the auto tensioner.
14. Install the idler pulley.
15. Install the water pump pulley.
16. Install the front upper timing belt cover.
17. Tighten all bolts to specification as illustrated.
18. Install the harness bracket.
19. Connect the control wiring harness connector, battery wiring harness connector, and connector bracket to engine mounting insulator.
20. Install the crankshaft damper pulley. Refer to Crankshaft Damper, Removal & Installation.
21. Install the engine undercover.
22. Connect the negative battery cable.

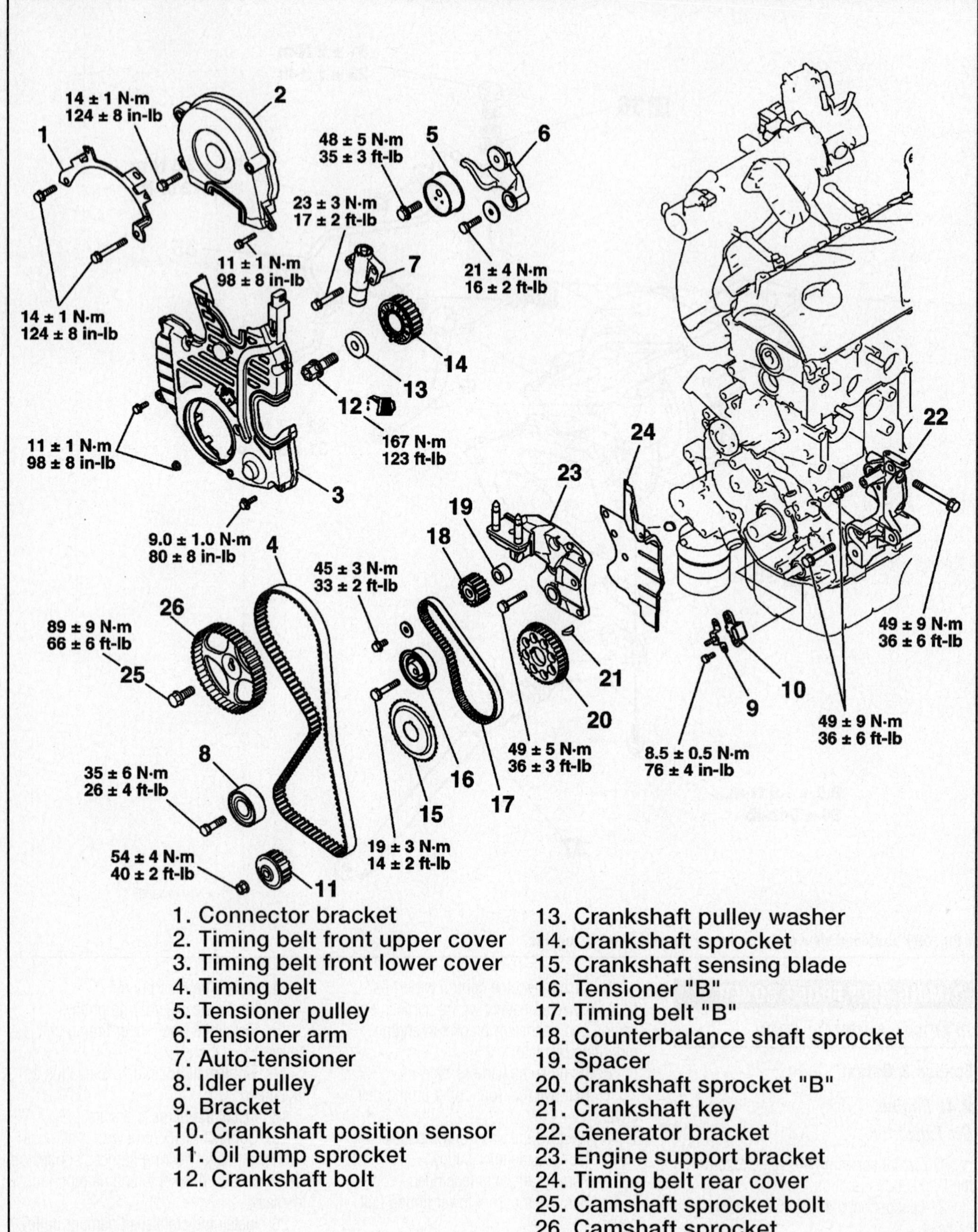

Fig. 197 Exploded view of timing belt and related components—2.4L engine

3.8L Engine

See Figure 198.

1. Before servicing the vehicle, refer to the Precautions Section.
2. Disconnect the negative battery cable.
3. Remove the engine undercover.
4. Remove the engine room cover.
5. Remove the engine side cover.
6. Remove the alternator.
7. Remove the power steering fluid pump drive belt.
8. Remove the crankshaft damper pulley. Refer to Crankshaft Damper, Removal & Installation.
9. Disconnect the control wiring harness and injector wiring connector.
10. Disconnect the knock sensor connector and crankshaft position sensor connector.
11. Remove the connector bracket. Disconnect the oil pressure switch connector and engine oil control valve connector.
12. Remove the engine mounting stay and the connector bracket.

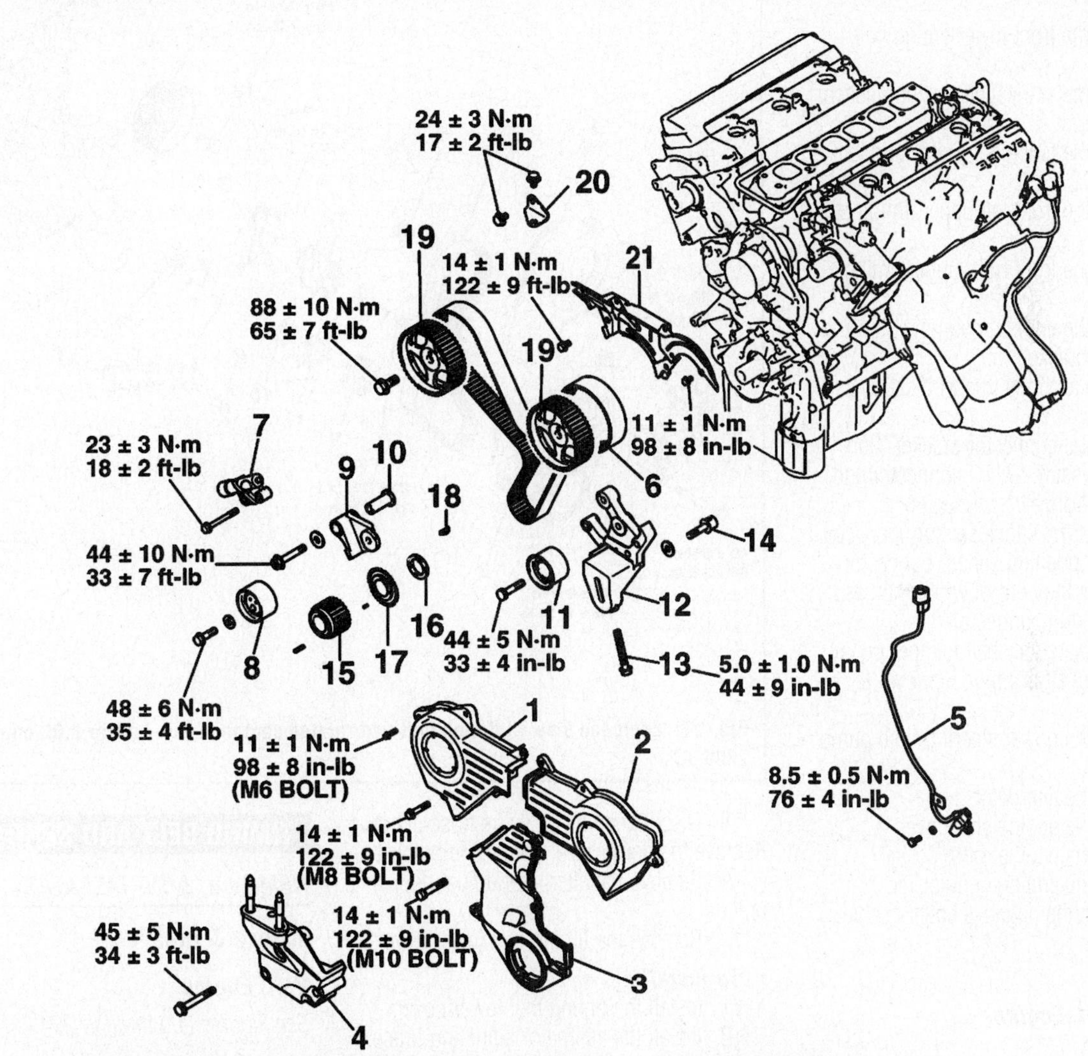

Fig. 198 Exploded view of timing belt and related components—3.8L engine

13. Remove the front right upper timing belt cover.
14. Remove the front left upper timing belt cover.
15. Remove the tensioner pulley and tensioner bracket.
16. Remove the crankshaft position sensor harness clamp.
17. Remove the front lower timing belt cover.

To install:

18. Install the front lower timing belt cover.
19. Install the crankshaft position sensor harness clamp.
20. Install the tensioner pulley and tensioner bracket.
21. Install the front left upper timing belt cover.
22. Install the front right upper timing belt cover.
23. Install the engine mounting stay and the connector bracket.
24. Tighten all bolts to specification as illustrated.
25. Install the connector bracket. Connect the oil pressure switch connector and engine oil control valve connector.
26. Connect the knock sensor connector and crankshaft position sensor connector.
27. Connect the control wiring harness and injector wiring connector.
28. Install the crankshaft damper pulley. Refer to Crankshaft Damper, Removal & Installation.
29. Install the power steering fluid pump drive belt.
30. Install the alternator.
31. Install the engine side cover.
32. Install the engine room cover.
33. Install the engine undercover.
34. Connect the negative battery cable.

Lancer

2006–07 2.0L Engine

See Figure 199.

1. Before servicing the vehicle, refer to the Precautions Section.
2. Remove the left side engine undercover.
3. Disconnect the negative battery cable.
4. Remove the engine mounting insulator.
5. Remove the crankshaft damper pulley. Refer to Crankshaft Damper, Removal & Installation.
6. Disconnect the crankshaft position sensor connection.
7. Remove the timing belt front upper cover.

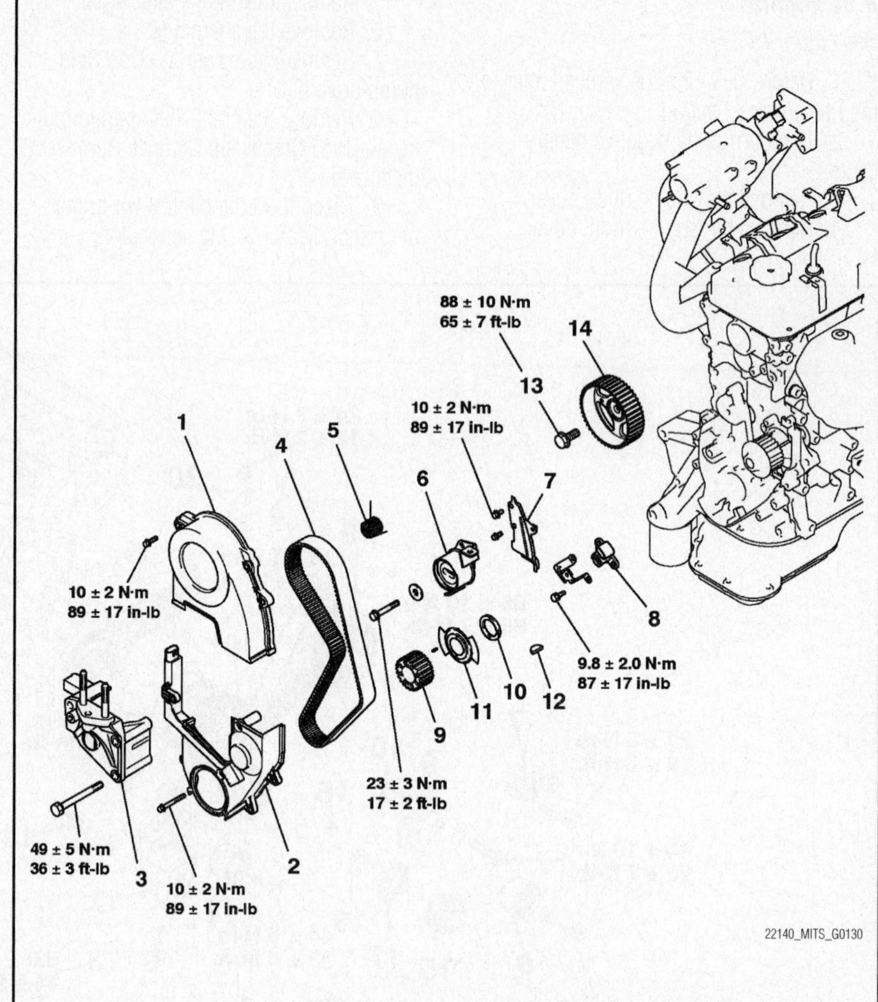

Fig. 199 Exploded view of timing belt and related components—Lancer 2.0L engine—2006–07

8. Disconnect the pressure hose clamp. Remove the power steering oil line bracket.
9. Remove the alternator adjusting brace.
10. Remove the timing belt lower cover.

To install:

11. Install the timing belt lower cover.
12. Install the alternator adjusting brace.
13. Connect the pressure hose clamp. Install the power steering oil line bracket.
14. Install the timing belt front upper cover.
15. Tighten all bolts to specification as illustrated.
16. Connect the crankshaft position sensor connection.
17. Install the crankshaft damper pulley. Refer to Crankshaft Damper, Removal & Installation.
18. Install the engine mounting insulator.
19. Connect the negative battery cable.
20. Install the left side engine undercover.

TIMING BELT AND SPROCKETS

REMOVAL & INSTALLATION

Eclipse & Galant

2.4L Engine

See Figures 200 through 204.

1. Before servicing the vehicle, refer to the Precautions Section.
2. Disconnect the negative battery cable.
3. Remove the timing belt front cover(s). Refer to Timing Belt Front Cover, Removal & Installation.
4. Turn the crankshaft clockwise and align each timing mark to set the number 1 piston to Top Dead Center (TDC) of its compression stroke.
5. Remove the timing belt undercover rubber plug and then install special tool MD998738. Screw the special tool until it contacts the timing belt tensioner arm.

✲✲ WARNING

The special tool must be screwed in gradually at the rate of a 30° turn per second. If it is turned in all at once, the timing belt tensioner adjuster rod will not easily retract and the tool may bend.

6. Gradually screw in the special tool and then align the timing belt tensioner adjuster rod set hole "A" with the timing belt tensioner adjuster cylinder set hole "B".

7. Insert a wire or pin in the set holes to lock the assembly in place. After removing the special tool, loosen the timing belt tensioner pulley mounting bolts and remove the timing belt.

➡If the belt is being reused be sure to mark the direction of rotation (clockwise) on the belt.

8. Remove the timing belt tensioner pulley, tensioner arm, and adjuster.

9. Remove the timing belt idler pulley.

10. Remove the timing belt lower cover bracket.

11. Remove the crankshaft position sensor.

12. Remove the crankshaft pulley center bolt and washer. Remove the drive sprocket. Remove the crankshaft angle sensing blade.

13. Remove the balancer timing belt tensioner.

14. Remove the timing belt "B" (balancer belt) from its mounting. Remove the sprocket.

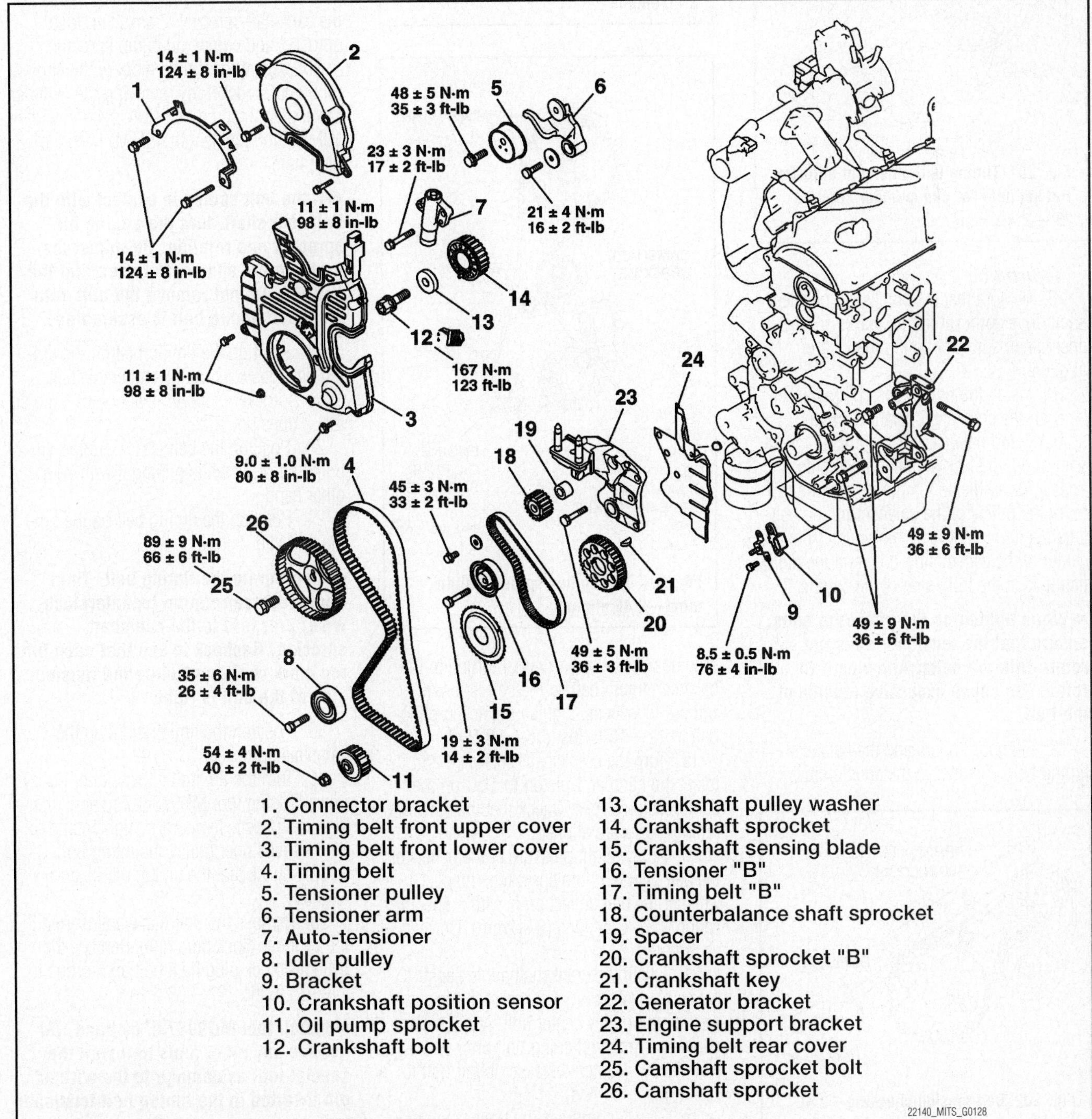

Fig. 200 Exploded view of timing belt and related components—2.4L engine

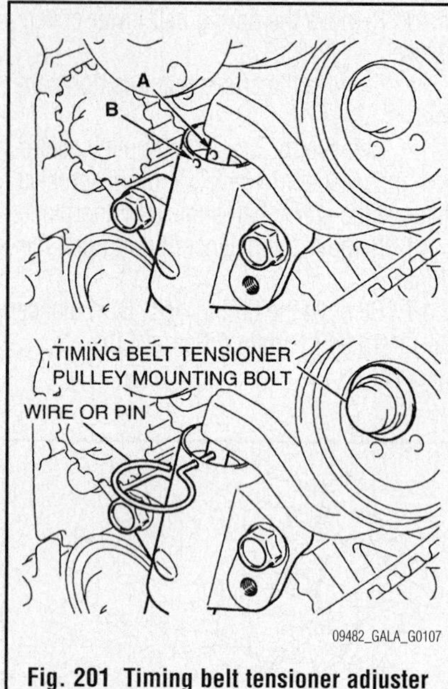

Fig. 201 Timing belt tensioner adjuster rod set hole "A" and cylinder set hole "B"—2.4L engine

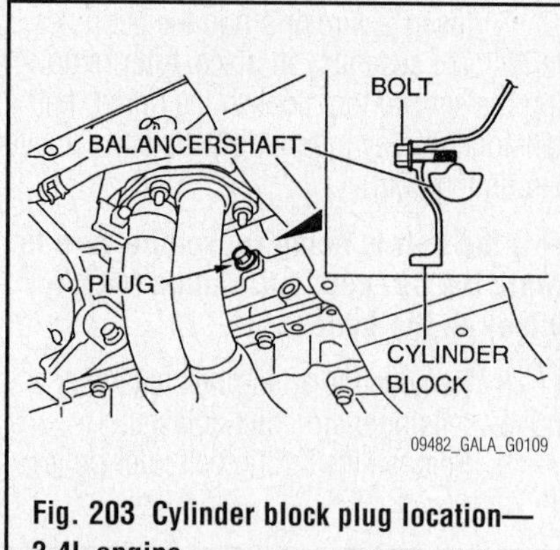

Fig. 203 Cylinder block plug location—2.4L engine

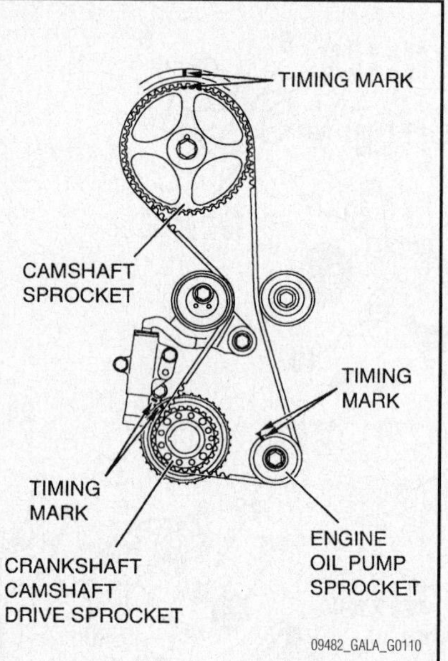

Fig. 204 Timing belt sprocket alignment—2.4L engine

To install:

15. Be sure that the crankshaft balancer shaft drive sprocket timing marks and balancer shaft sprocket timing marks are aligned.
16. Install the timing belt "B" (balancer belt) on the crankshaft balancer drive sprocket and balancer shaft sprocket. There should be no slack on the tension side.
17. Assemble an temporarily fix the center of the pulley of the balancer timing belt tensioner so that it is at the top left from the center of the assembling bolt, and the pulley flange is at the front side of the engine.

➡When tightening the mounting bolts, ensure that the tensioner does not rotate with the bolts. Allowing it to rotate can cause excessive tension of the belt.

18. Lift your fingers and the balancer timing belt tensioner in the counterclockwise position. Apply minimal torque to the balancer timing belt so the belt is tense without looseness. Tighten the assembling bolt to 12—16 ft. lbs. (16—22 Nm).
19. Turn the crankshaft clockwise 2 turns to set the number 1 piston to TDC on the compression stroke. Check that the sprocket timing marks are aligned.
20. Apply slight pressure at the center of the belt between both sprockets. Inspect whether the belt deflection is within specification 0.31–0.47 inch (8–12mm). Correct as required.
21. Install the crankshaft angle sensing blade, crankshaft drive sprocket, pulley washer and pulley center bolt. Apply clean engine oil to the retaining bolt prior to installation. Tighten the assembling bolt to 123 ft. lbs. (167 Nm).
22. Slowly compress the timing belt tensioner adjuster rod using a press or vise. Align the set hole "A" of the rod with set hole "B" of the timing belt tensioner adjuster cylinder.

➡Do not compress the assembly too fast as damage to the rod may occur.

23. Insert a wire or pin in the aligned set holes to lock the assembly in place.
24. Install the timing belt tensioner adjuster to the engine and tighten the mounting bolt to 15—19 ft. lbs. (20—26 Nm).
25. Temporarily tighten the timing belt tensioner pulley. Align the timing marks on the camshaft sprocket, crankshaft drive sprocket and engine oil pump sprocket.
26. Adjust the timing mark of the engine oil pump sprocket, by removing the cylinder block plug. Insert a bolt (M6, section width 10MM, nominal length 45MM) from the plug hole.

➡If the bolt comes in contact with the balancer shaft, turn the engine oil sprocket one rotation. Re-adjust the timing mark and check to see that the bolt fits. Do not remove the bolt until the valve timing belt is assembled.

27. Position the timing belt on the timing belt tensioner pulley and crankshaft driver sprocket, support it with your hand so that it does not slide.
28. Position the belt on the engine oil pump sprocket while pulling it with your other hand.
29. Position the timing belt on the timing belt idler pulley.

➡Incorporate the timing belt. Then apply reverse rotation (counterclockwise) pressure to the camshaft sprocket. Recheck to see that each timing mark is aligned while the tension side of the belt is right.

30. Position the timing mark on the camshaft sprocket.
31. Turn the timing belt tensioner pulley upward using tool MD998767 to apply tension to the belt. Temporarily tighten and fix the belt tensioner pulley mounting bolt.
32. Check that the timing marks are aligned.
33. Remove the bolt that was inserted in the cylinder block plug hole. Replace the cylinder block plug hole bolt and torque to 21—25 ft. lbs. (27—33 Nm).

➡Install tool MD998738 by hand. Do not use any other tools to install the special tool as damage to the wire or pin inserted in the timing belt tensioner may occur.

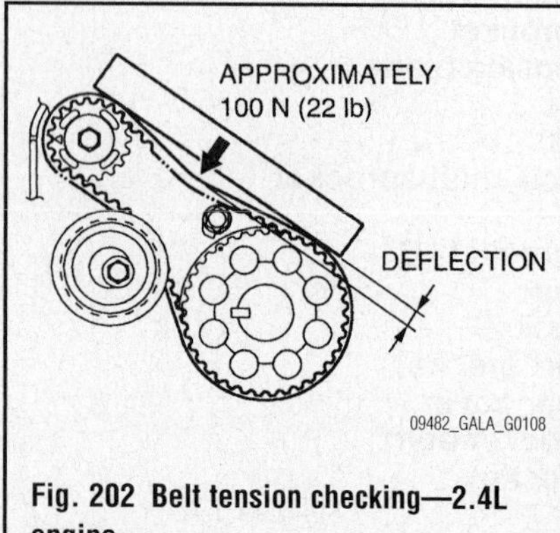

Fig. 202 Belt tension checking—2.4L engine

34. Gradually screw the tool into position so that the wire or pin inserted in the timing belt tensioner adjuster moves slightly.

35. Turn the crankshaft in the clockwise direction; align each timing mark to set number one piston to TDC on the compression stroke.

36. Loosen the timing belt tensioner pulley mounting bolt. With tool MD998767 and a torque wrench, apply tension torque 31 inch lbs. (42 Nm) to the timing belt. Tighten the timing belt tensioner pulley mounting bolt to 33—39 ft. lbs. (43—53 Nm).

37. Remove the wire or pin inserted in the timing belt tensioner adjuster. Remove tool MD998738. Install the rubber plug of the timing belt undercover.

38. Rotate the crankshaft clockwise 2 revolutions and leave it for about 15 minutes.

39. Insert the wire or pin, previously removed, and ensure that it can be pulled out with a light load. When the wire or pin can be pulled out, appropriate tension is applied on the timing belt.

40. If the wire or pin cannot be pulled out easily, repeat the above step until it can.

➡Always check the tightening torque of the crankshaft pulley center bolt when turning the crankshaft pulley center bolt counterclockwise. Retighten if it is loose.

41. Check again that the timing marks on the sprockets are aligned.

42. Continue the installation in the reverse order of the removal procedure.

3.8L Engine

See Figures 205 and 206.

1. Before servicing the vehicle, refer to the Precautions Section.
2. Disconnect the negative battery cable.
3. Remove the timing belt front cover(s). Refer to Timing Belt Front Cover, Removal & Installation.
4. Remove the engine support bracket.
5. Turn the crankshaft clockwise to align each timing mark and set the number 1 cylinder to Top Dead Center (TDC) on the compression stroke.

➡If the belt is going to be reused, mark the direction of rotation.

6. Remove the timing belt.
7. Remove the auto tensioner, tensioner pulley, and tensioner arm.

⁂ WARNING

Water or oil on the belt shortens its life drastically, so the removed timing belt, sprocket, and tensioner must be kept free from oil and water. These parts should not be washed or immersed in solvent. Replace parts if contaminated. If there is oil or water on any part, check the front case oil seal, camshaft oil seal, and water pump for leaks.

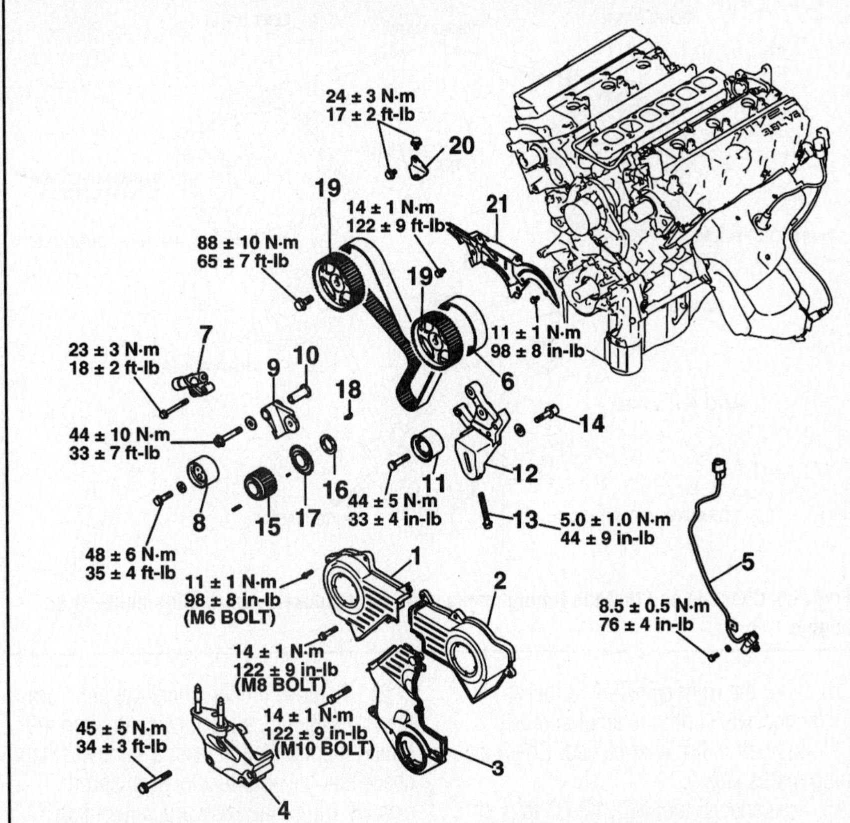

1. Timing belt front upper cover, right
2. Timing belt front upper cover, left
3. Timing belt front lower cover
4. Engine support bracket, right
5. Crankshaft position sensor
6. Timing belt
7. Auto-tensioner
8. Tensioner pulley
9. Tensioner arm
10. Shaft
11. Idler pulley
12. Tensioner bracket
13. Adjusting bolt
14. Adjusting stud
15. Crankshaft sprocket
16. Crankshaft spacer
17. Crankshaft sensing blade
18. Key
19. Camshaft sprocket
20. Bracket
21. Timing belt rear cover

22140_MITS_G0129

Fig. 205 Exploded view of timing belt and related components—3.8L engine

8. While holding the camshaft sprocket with special tools MB990767 and MD998715, loosen the camshaft sprocket bolt. Remove the camshaft sprocket.
9. Remove the crankshaft sprocket.

To install:

10. Install the crankshaft sprocket. Ensure the keyway and key are in position.
11. Install the camshaft sprocket. Tighten the bolt to 58–72 ft. lbs. (78–98 Nm).
12. Align the timing marks on the camshaft sprockets with those on the rocker cover and the timing mark on the crankshaft sprocket with that on the engine block.

⁂ CAUTION

The right side camshaft sprocket can turn easily due to the spring force applied. Be careful of hands and fingers when turning the camshaft as they could get caught in an uncomfortable position.

13. Install the timing belt first on the crankshaft sprocket, than on the idler pulley, then on the left camshaft sprocket, then on the water pump pulley, then on the right camshaft sprocket, and finally on the tensioner pulley.

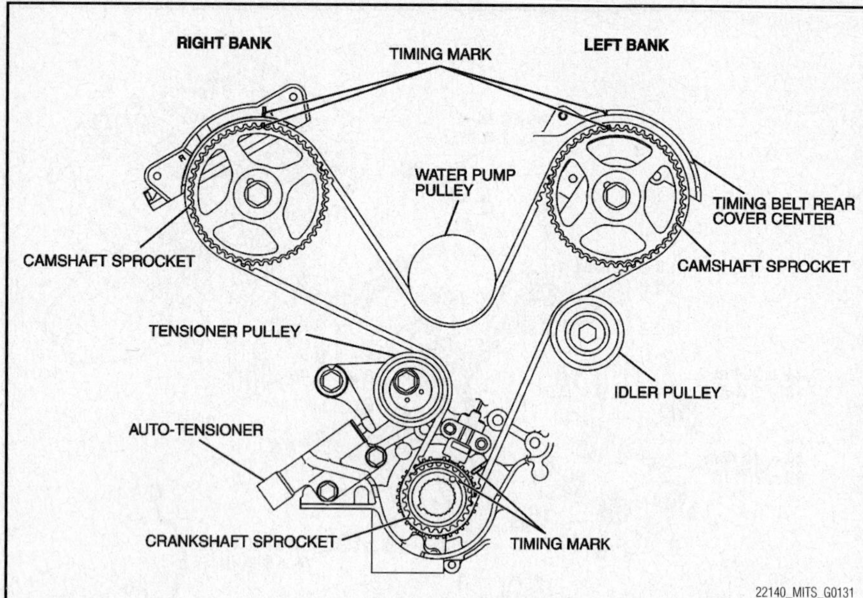

Fig. 206 Check to see that the timing marks of all the sprockets are in alignment—3.8L engine

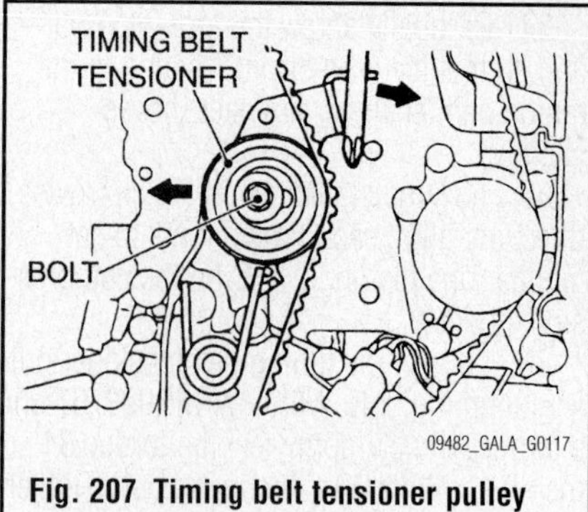

Fig. 207 Timing belt tensioner pulley and belt removal—Lancer 2.0L engine—2006–07

14. Turn the right camshaft sprocket counterclockwise until the tension side of the timing belt is firmly stretched. Check all timing marks, again.
15. Use special tool MD998767 to push the tensioner pulley into the timing belt, and then temporarily tighten the center bolt.
16. Use special tool MD998769 to turn the crankshaft 1⁄4 turn counterclockwise, then turn it again clockwise until the timing marks are aligned.
17. Holding the auto tensioner, with your hand, press the end of the pushrod against a metal surface with a force of about 32 lbs. Measure how far in the push rod is pushed. Specification should be 0.189–0.236 inch (4.8–6.0mm). If not within specification, replace the auto tensioner assembly.
18. Position the auto tensioner perpendicular in a vise. If the tensioner has a plug at the base, be sure to use a washer to protect the plug.
19. Slowly compress the pushrod of the auto tensioner until the pin hole "A" is aligned the pin hole "B" in the cylinder. Insert a pin into both holes once they are aligned. Install the auto tensioner to the engine.

➡When tightening the center bolt, be careful that the tensioner pulley does not turn with the bolt.

20. Loosen the center bolt of the tensioner pulley. Use tool MD998767 and a torque wrench and apply tension torque to the timing belt in a downward motion. Tighten the center bolt to 32—40 ft. lbs. (42—54 Nm).
21. Remove the tensioner set pin. Turn the crankshaft clockwise twice to align the timing marks. Wait at least 5 minutes, then check that the auto tensioner pushrod extends within the standard value range "A" of 0.19—0.24 inch. If not repeat the operation again.
22. Check again that the timing marks of the sprockets are aligned.
23. Continue the installation in the reverse order of the removal procedure.

Lancer

2006–07 2.0L Engine

See Figures 207 through 211.

1. Before servicing the vehicle, refer to the Precautions Section.
2. Disconnect the negative battery cable.
3. Remove the timing belt front cover(s). Refer to Timing Belt Front Cover, Removal & Installation.
4. Remove the power steering pump bracket stay.
5. Remove the flange. Remove the engine support bracket.
6. Turn the crankshaft clockwise to align each timing mark and set the number 1 cylinder to Top Dead Center (TDC) on the compression stroke. If the belt is going to be reused, mark the direction of rotation.
7. Loosen the tension pulley retaining bolt. Position a suitable tool against the tensioner pulley and pry it fully back in the direction of the arrow in the illustration. Temporarily tighten the tensioner pulley bolt.
8. Remove the timing belt.
9. Remove the auto tensioner, tensioner pulley, and tensioner arm.

⁂ WARNING

Water or oil on the belt shortens its life drastically, so the removed timing belt, sprocket, and tensioner must be kept free from oil and water. These parts should not be washed or immersed in solvent. Replace parts if contaminated. If there is oil or water on any part, check the front case oil seal, camshaft oil seal, and water pump for leaks.

10. Use special tools MB990767 and MD998719 to prevent the camshaft sprocket from turning, and then loosen the camshaft sprocket bolt. Remove the camshaft sprocket.
11. Remove the crankshaft sprocket.

To install:

12. Use special tools MB990767 and MD998719 to prevent the camshaft sprocket from turning, and then tighten the camshaft sprocket bolt. Tightening torque: 58–72 ft. lbs. (78–98 Nm).
13. Install the crankshaft sprocket. Ensure the keyway and key are in position.

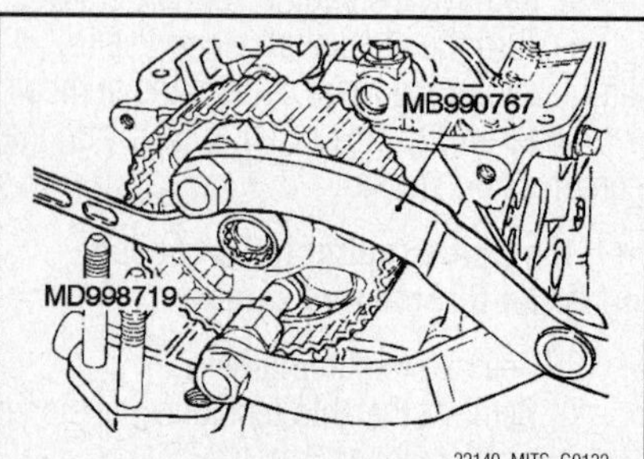

Fig. 208 Use special tools MB990767 and MD998719 to prevent the camshaft sprocket from turning—Lancer 2.0L engine—2006–07

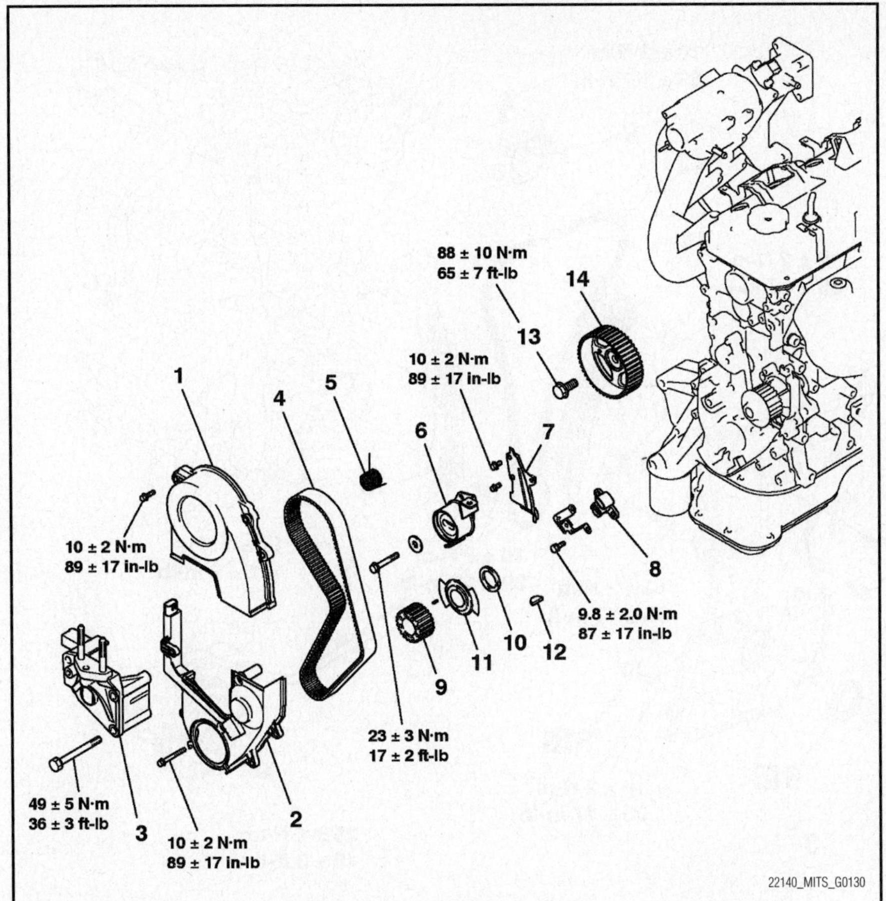

Fig. 209 Exploded view of timing belt and related components—Lancer 2.0L engine—2006–07

14. With the timing belt tensioner pulley bolt loosened, use a suitable tool and pry the tensioner pulley as close to the engine mount as possible. Temporarily tighten the tensioner bolt.

15. Align each of the crankshaft and camshaft sprocket timing marks.

16. Install the timing belt in the following order: crankshaft sprocket, then water pump sprocket, then camshaft sprocket, and finally tensioner pulley.

➡After installing the belt, try to rotate the camshaft sprocket in the reverse direction. Recheck to be sure that the belt is fully tensioned and that each timing mark is in the proper direction.

17. Initially loosen the fixing bolt of the tensioner pulley fixed to the engine mount side by ¼–½ turn. Use the force of the tensioner spring to apply tension to the belt.

➡As the purpose of this procedure is to apply the proper amount of tension to the tension side of the timing belt by using the cam driving torque, turn the crankshaft only by the amount given below. Do not turn the crankshaft in the opposite direction (counterclockwise).

18. Turn the crankshaft in the clockwise direction for 2 rotations, and recheck to be sure that the timing marks on each sprocket are aligned.

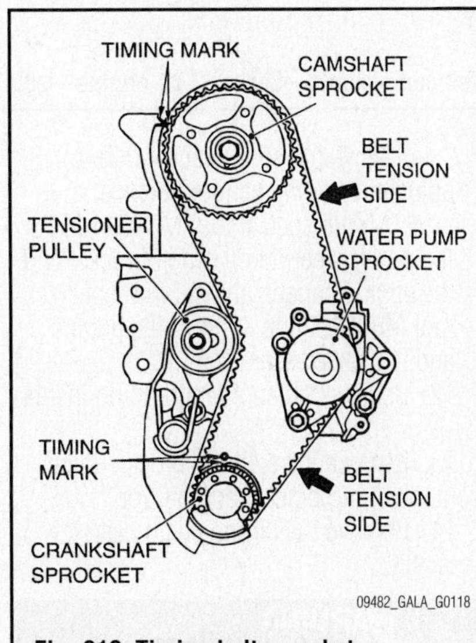

Fig. 210 Timing belt sprocket alignment—Lancer 2.0L engine—2006–07

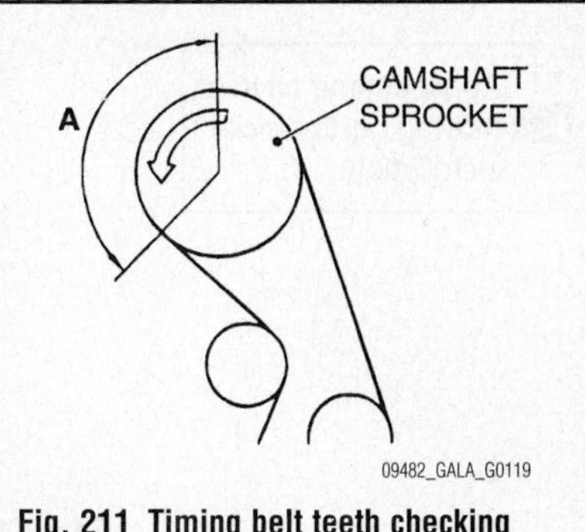

Fig. 211 Timing belt teeth checking location—Lancer 2.0L engine—2006–07

19. After checking to be sure that no belt teeth in the section marked "A" in the illustration are lifted up and that the teeth in each sprocket are engaged, secure the tensioner pulley.

20. Install the flange.

21. Continue the installation in the reverse order of the removal procedure.

2008 2.0L Engine

Refer to Timing Chain and Sprockets, Removal & Installation.

Lancer Evolution

2008 2.0L Turbo Engine

Refer to Timing Chain and Sprockets, Removal & Installation.

TIMING BELT REAR COVER

REMOVAL & INSTALLATION

On models with a rear timing belt cover, the timing belt and sprockets must be removed to access the cover. Refer to Timing Belt and Sprocket, Removal & Installation.

TIMING CHAIN COVER AND SEAL

REMOVAL & INSTALLATION

Lancer

2008 2.0L Engine

See Figures 212 through 214.

1. Before servicing the vehicle, refer to the Precautions Section.
2. Remove the front engine undercover and side covers (RH).
3. Drain the engine oil.
4. Remove the rocker cover assembly.
5. Remove the oil pan.

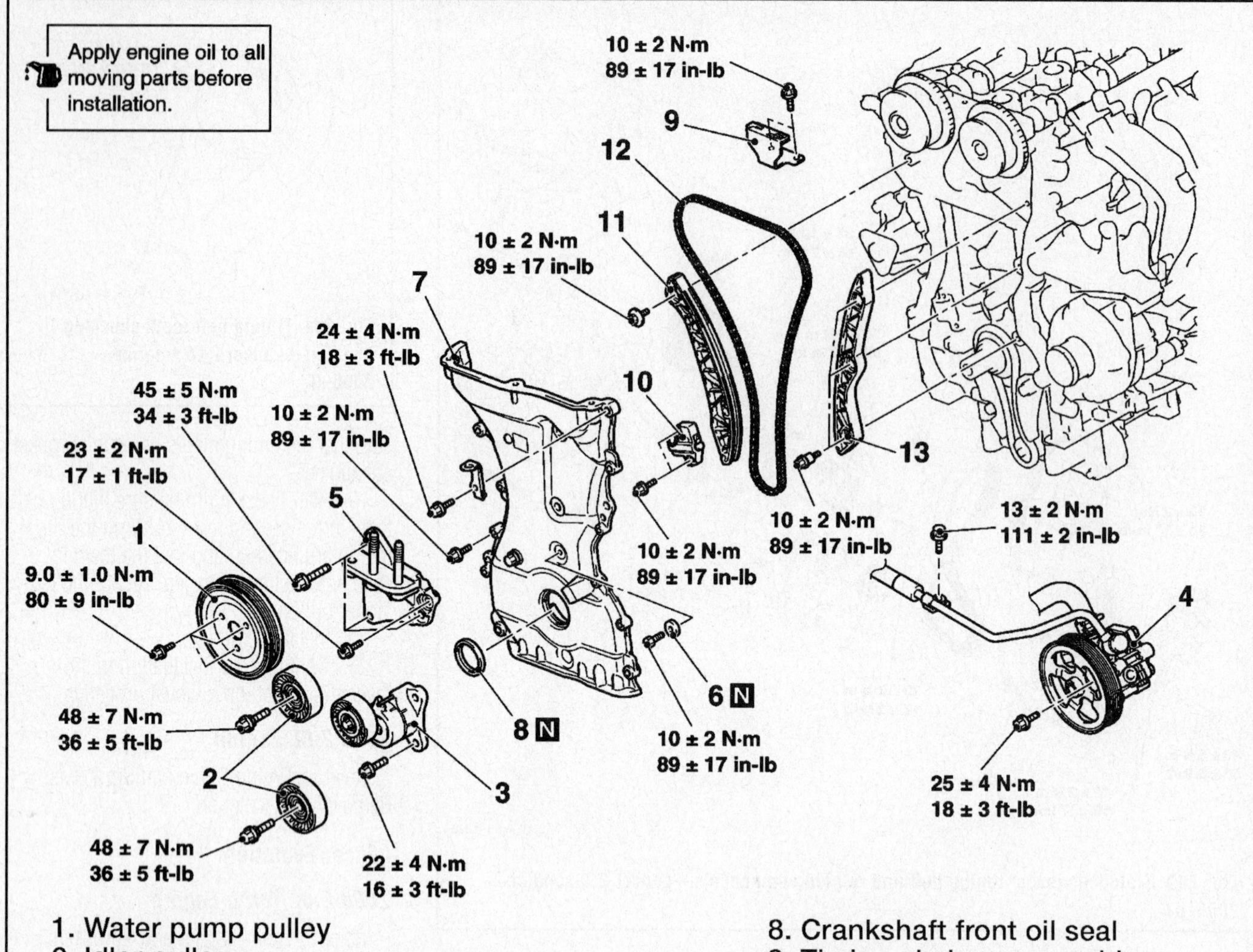

1. Water pump pulley
2. Idler pulley
3. Autotensioner
4. Power steering oil pump assembly
5. Cylinder block engine front mounting bracket
6. Gasket
7. Timing chain case assembly
8. Crankshaft front oil seal
9. Timing chain upper guide
10. Timing chain tensioner
11. Timing chain tension side guide
12. Timing chain
13. Timing chain loose side guide

Fig. 212 Exploded view of timing chain and related components—Lancer 2.0L engine—2008

6. Remove the crankshaft damper pulley. Refer to Crankshaft Damper, Removal & Installation.
7. Remove the water pump pulley.
8. Remove the idler pulley.
9. Remove the autotensioner.
10. Remove the power steering oil pump assembly.
11. Install a special tool, engine hanger MB991928, for holding the engine and transaxle assembly.
    a. Assemble the engine hanger. Set the following parts on the base hanger.
    - Slide bracket (HI)
    - Foot x 4 (standard) (MB991932)
    - Joint x 2 (90) (MB991930)

    b. Set the foot of the special tool to the strut mounts and front bumper brace.
    c. Move the slide bracket (HI) to adjust the engine hanger balance.
    d. Mount special tool MB991454 to the power steering oil pump bracket and the engine hanger, and set it to special tool MB991928 to support the engine and transaxle assembly.
12. Remove the engine mounting insulator.
13. Remove the cylinder block engine front mounting bracket and gasket.
14. Remove the timing chain case assembly.

✱✱ WARNING

The adhesive strength of the sealant on the timing chain case assembly may be so strong that the boss may be damaged by peeling it off. Do not peel it off forcibly.

a. After removing the timing chain case assembly mounting bolts, slightly pry the boss of the timing chain case assembly using a flat-tipped screwdriver
b. Remove the timing chain case assembly from the cylinder head and cylinder block.
c. If the sealant cannot be peeled off easily, insert a wooden hammer shank into the timing chain case assembly, pry slightly, and remove the timing chain case assembly from the cylinder head and cylinder block.

15. Remove the crankshaft front oil seal.

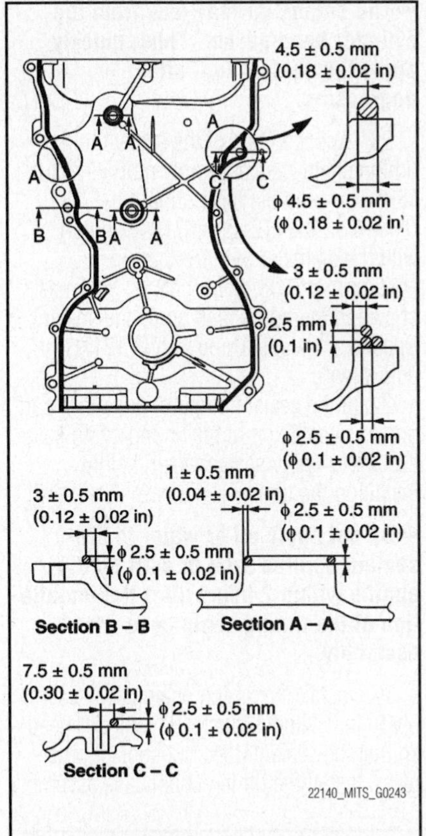

Fig. 213 Apply the sealant as shown to the timing chain case assembly mating surface—Lancer and Lancer Evolution—2008

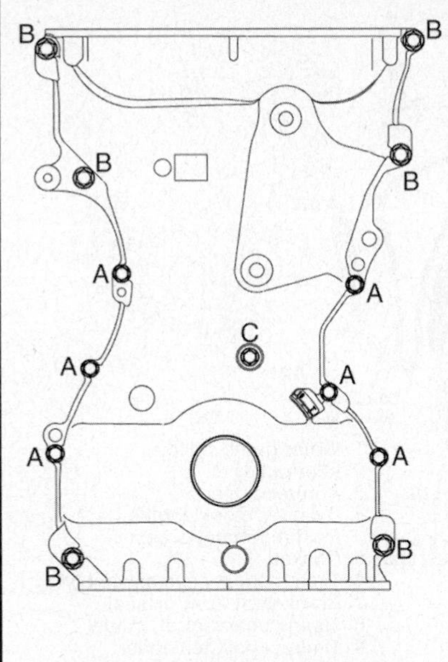

Bolt (symbol)	Thread diameter x Length mm	Tightening torque
Flange bolt (A)	M6 × 25	10 ± 2 N·m (89 ± 17 in-lb)
Flange bolt (B)	M8 × 28	24 ± 4 N·m (18 ± 2 ft-lb)
Bolt (C)	M6 × 25	10 ± 2 N·m (89 ± 17 in-lb)

22140_MITS_G0244

Fig. 214 Tighten the timing chain case assembly mounting bolts to the specified torque shown—Lancer and Lancer Evolution—2008

To install:

16. Apply a small amount of engine oil to the entire inner diameter of the crankshaft front oil seal lip.

17. Using special tool MB991448, press in the crankshaft front oil seal up to the chamfered surface of timing chain case.

➡Be sure to remove the sealant remaining in the mounting hole, O-ring groove, and the gap between parts. Do not touch the degreased area once cleaned.

18. Remove all the traces of sealant adhering to the timing chain case assembly installation surfaces of the case assembly, cylinder block, and cylinder head. Then, degrease the surfaces with a quick-drying degreasing agent (white gasoline).

19. Remove all the sealant adhering to the gasket between the cylinder head and cylinder block (3-surface aligned part). Then, degrease the surfaces with the quick-drying degreasing agent (white gasoline).

➡The engine oil may leak from the cylinder head gasket. Thus, quickly apply the sealant to it after degreasing.

20. Apply a bead of the sealant to the timing chain case assembly mounting surface. The bead diameter should be 0.08–0.12 inch (2–3mm). Overlap part "A" with the diameter of 0.16–0.20 inch (4–5mm) or 0.08–0.12 inch (2–3mm) as shown in the figure, and apply the sealant. Specified sealant: Three Bond® 1217G or equivalent.

21. If the sealant contacts any other part during installation of the timing chain case assembly, apply sealant again before installing the timing chain case assembly.

➡Do not apply oil or water to the sealant applied area or start up the engine within 2 hours after the installation of the timing chain case assembly.

22. Install the timing chain case assembly to the cylinder block and cylinder head so that the sealant does not contact other parts. Install the timing chain case assembly within 3 minutes after the application of sealant.

23. Insert the bolts to the timing chain case assembly as shown, and tighten them to the specified torque.

Lancer Evolution

2008 2.0L Turbo Engine

See Figures 215 through 217.

1. Before servicing the vehicle, refer to the Precautions Section.
2. Remove the front engine undercover and side covers (RH).
3. Remove the strut tower bar.
4. Drain the engine oil.
5. Remove the rocker cover assembly.
6. Remove the oil pan.
7. Remove the power steering oil pump assembly.
8. Remove the headlight support panel cover.
9. Remove the engine mounting bracket.
10. Remove the crankshaft damper pulley. Refer to Crankshaft Damper, Removal & Installation.
11. Remove the water pump pulley.
12. Remove the idler pulley.
13. Remove the autotensioner.
14. Install a special tool, engine hanger MB991928, for holding the engine and transaxle assembly.
    a. Assemble the engine hanger. Set the following parts on the base hanger.
        - Slide bracket (HI)
        - Foot x 4 (standard) (MB991932)
        - Joint x 2 (90) (MB991930)
    b. Set the foot of the special tool to the strut mounts and front bumper brace.
    c. Move the slide bracket (HI) to adjust the engine hanger balance.
    d. Mount special tool MB991454 to the power steering oil pump bracket and the engine hanger, and set it to special tool MB991928 to support the engine and transaxle assembly.
15. Remove the engine mounting insulator.
16. Remove the cylinder block engine front mounting bracket and gasket.
17. Remove the timing chain case assembly.

⁂ WARNING

The adhesive strength of the sealant on the timing chain case assembly may be so strong that the boss may be damaged by peeling it off. Do not peel it off forcibly.

a. After removing the timing chain case assembly mounting bolts, slightly pry the boss of the timing chain case assembly using a flat-tipped screwdriver
b. Remove the timing chain case assembly from the cylinder head and cylinder block.

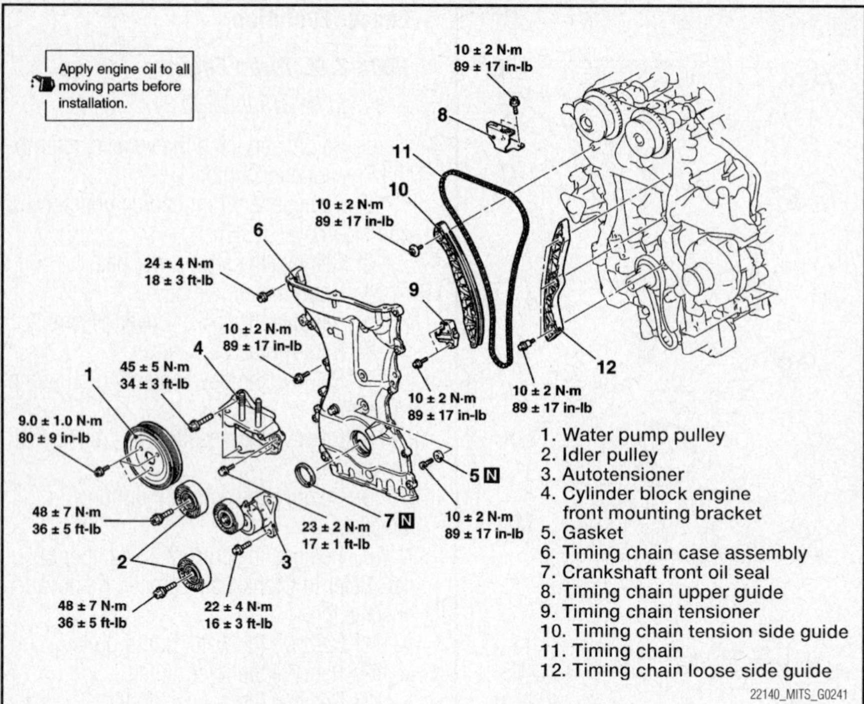

Fig. 215 Exploded view of timing chain and related components—Lancer Evolution 2.0L engine—2008

c. If the sealant cannot be peeled off easily, insert a wooden hammer shank into the timing chain case assembly, pry slightly, and remove the timing chain case assembly from the cylinder head and cylinder block.

18. Remove the crankshaft front oil seal.

To install:

19. Apply a small amount of engine oil to the entire inner diameter of the crankshaft front oil seal lip.

20. Using special tool MB991448, press in the crankshaft front oil seal up to the chamfered surface of timing chain case.

➡Be sure to remove the sealant remaining in the mounting hole, O-ring groove, and the gap between parts. Do not touch the degreased area once cleaned.

21. Remove all the traces of sealant adhering to the timing chain case assembly installation surfaces of the case assembly, cylinder block, and cylinder head. Then, degrease the surfaces with a quick-drying degreasing agent (white gasoline).

22. Remove all the sealant adhering to the gasket between the cylinder head and cylinder block (3-surface aligned part). Then, degrease the surfaces with the quick-drying degreasing agent (white gasoline).

➡The engine oil may leak from the cylinder head gasket. Thus, quickly apply the sealant to it after degreasing.

23. Apply a bead of the sealant to the timing chain case assembly mounting surface. The bead diameter should be 0.08–0.12 inch (2–3mm). Overlap part "A" with the diameter of 0.16–0.20 inch (4–5mm) or 0.08–0.12 inch (2–3mm) as shown in the figure, and apply the sealant. Specified sealant: Three Bond® 1217G or equivalent.

24. If the sealant contacts any other part during installation of the timing chain case assembly, apply sealant again before installing the timing chain case assembly.

➡Do not apply oil or water to the sealant applied area or start up the engine within 2 hours after the installation of the timing chain case assembly.

25. Install the timing chain case assembly to the cylinder block and cylinder head so that the sealant does not contact other parts. Install the timing chain case assembly

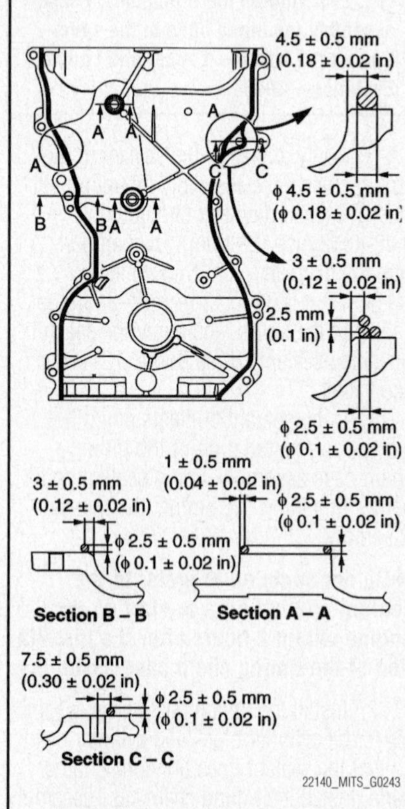

Fig. 216 Apply the sealant as shown to the timing chain case assembly mating surface—Lancer and Lancer Evolution—2008

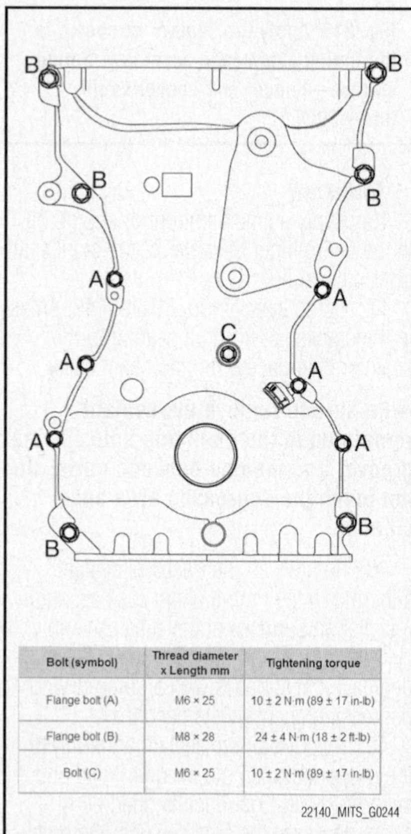

Bolt (symbol)	Thread diameter x Length mm	Tightening torque
Flange bolt (A)	M6 × 25	10 ± 2 N·m (89 ± 17 in-lb)
Flange bolt (B)	M8 × 28	24 ± 4 N·m (18 ± 2 ft-lb)
Bolt (C)	M6 × 25	10 ± 2 N·m (89 ± 17 in-lb)

Fig. 217 Tighten the timing chain case assembly mounting bolts to the specified torque shown—Lancer and Lancer Evolution—2008

within 3 minutes after the application of sealant.

26. Insert the bolts to the timing chain case assembly as shown, and tighten them to the specified torque.

TIMING CHAIN AND SPROCKETS

REMOVAL & INSTALLATION

Lancer and Lancer Evolution

2008 2.0L Engine

See Figures 218 through 220.

1. Before servicing the vehicle, refer to the Precautions Section.
2. Remove the timing chain cover and crankshaft oil seal. Refer to Timing Chain Cover and Seal, Removal & Installation.
3. Remove the timing chain upper guide (1).
4. Insert a flat blade screwdriver into the release hole of the timing chain tensioner to release the latch.
5. Push the tensioner lever by hand and push in the plunger of the timing chain tensioner until it hits the bottom. Then, insert a hard wire or hexagonal bar wrench 0.05 inch (1.5mm) into the fixing hole of the plunger.
6. Remove the timing chain tensioner (2).
7. Remove the timing chain lever (4).
8. Remove the timing chain (5).
9. Remove the chain oil jet (6).
10. Remove the exhaust Variable Valve Timing (VVT) sprocket assembly.
    a. Hold the hexagonal portion of the exhaust camshaft with a wrench and loosen the exhaust VVT sprocket bolt.
    b. Remove the exhaust VVT sprocket assembly.
11. Remove the intake VVT sprocket assembly.
    a. Hold the hexagonal portion of the intake camshaft with a wrench and loosen the intake VVT sprocket bolt.
    b. Remove the intake VVT sprocket assembly.

To install:

12. Assemble the intake VVT sprocket assembly in the following procedure.
    a. Make sure that the knock pin of the inlet camshaft assembly is positioned facing straight upward.
    b. Apply an appropriate and minimum amount of engine oil to the circumference of the tip of the intake VVT sprocket assembly and the entire circumference of the area into which the intake VVT sprocket assembly is inserted.
    c. Slowly insert the intake VVT sprocket assembly into the normal position of the inlet camshaft assembly with its knock pin hole facing straight upward.
    d. Install the VVT sprocket.
    e. Make sure that the VVT sprocket is securely inserted into the bottom and that the VVT sprocket does not rotate with the hexagonal portion of the camshaft secured with a wrench.
    f. Hold the hexagonal portion of the camshaft with a wrench and tighten the intake VVT sprocket bolt to 60–66 ft. lbs. (80–90 Nm).
13. Assemble the exhaust VVT sprocket assembly in the following procedure.
    a. Make sure that the knock pin of the exhaust camshaft assembly is positioned facing straight upward.
    b. Apply an appropriate and minimum amount of engine oil to the circumference of the tip of the exhaust VVT sprocket assembly and the entire circumference of the area into which the exhaust VVT sprocket assembly is inserted.
    c. Slowly insert the exhaust VVT sprocket assembly into the normal position of the exhaust camshaft assembly with its knock pin hole facing straight upward.
    d. Install the VVT sprocket.

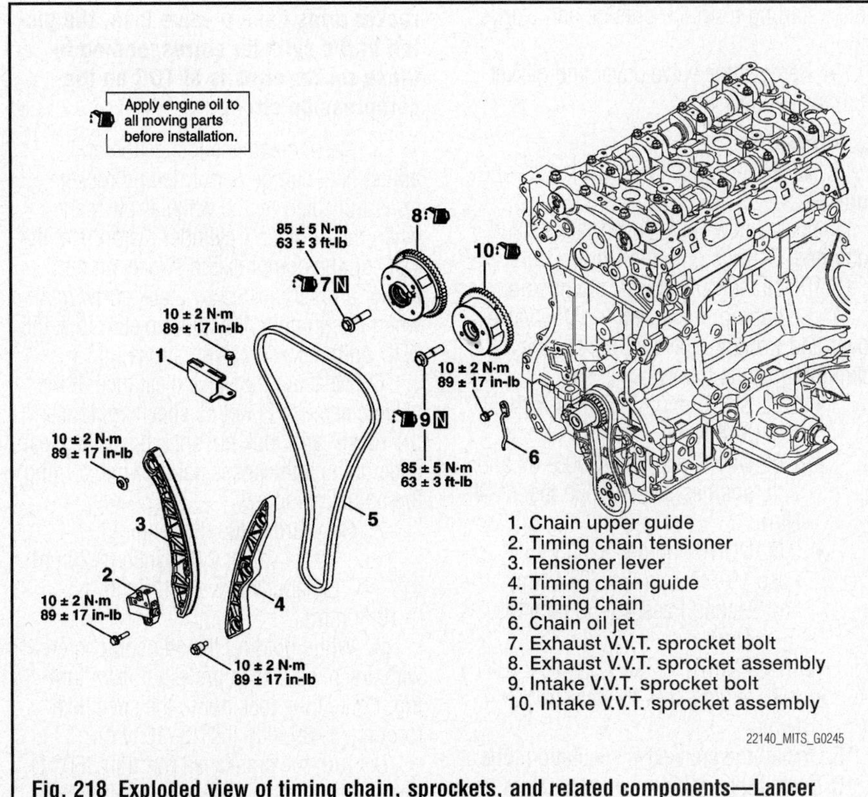

Fig. 218 Exploded view of timing chain, sprockets, and related components—Lancer and Lancer Evolution—2008

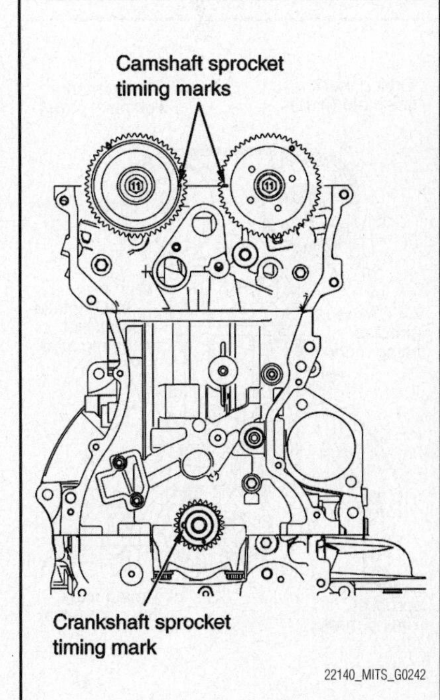

Fig. 219 Set the timing marks of the camshaft sprockets and the crankshaft sprocket as shown—Lancer and Lancer Evolution—2008

e. Make sure that the VVT sprocket is securely inserted into the bottom and that the VVT sprocket does not rotate with the hexagonal portion of the camshaft secured with a wrench.

f. Hold the hexagonal portion of the camshaft with a wrench and tighten the camshaft sprocket bolt to 60–66 ft. lbs. (80–90 Nm).

14. Align the timing marks of the VVT sprockets and the crankshaft sprocket key as illustrated.

15. Align the link plate (Orange) with the timing mark of the exhaust VVT sprocket and loop the timing chain.

16. Align the link plate (Blue) with the timing mark of the intake VVT sprocket to loop the timing chain. Rotate the intake VVT sprocket by 1 or 2 teeth to align with the timing mark.

17. Align the timing mark of the crankshaft sprocket with the link plate (Blue) to loop the timing chain. Because of the timing chain slacks, hold it to prevent the timing mark from coming off the link plate (Blue).

18. Make sure that the timing mark of each sprocket is aligned with the link plate (blue) of the timing chain at all 3 locations.

19. Install the timing chain guide.

20. Install the timing chain tensioner on the cylinder block and tighten it to the specified torque. Specified torque: 72–106 inch lbs. (8–12 Nm).

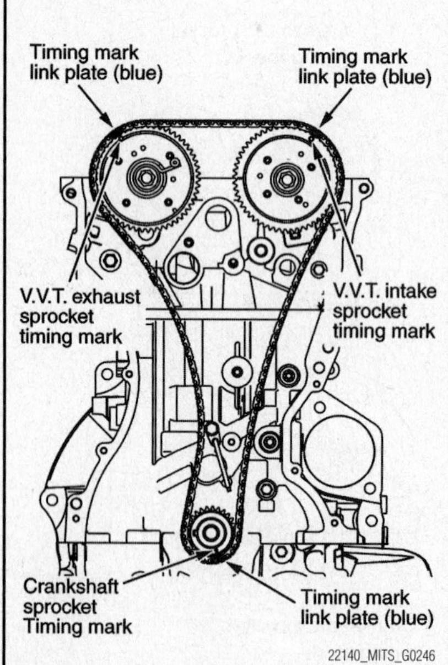

Fig. 220 Make sure that the timing mark of each sprocket is aligned with the link plate (blue) of the timing chain at all 3 locations—Lancer and Lancer Evolution—2008

21. Remove the hard wire or hexagonal bar wrench 0.05 inch (1.5mm) from the timing chain tensioner. This enables the plunger of the timing chain tensioner to push the tensioner lever to keep the timing chain tight.

22. Install the timing chain upper guide.

23. Install the timing chain cover and crankshaft oil seal. Refer to Timing Chain Cover and Seal, Removal & Installation.

24. Run the engine and check for proper operation and for any oil leakage.

VALVE COVERS

REMOVAL & INSTALLATION

1. Before servicing the vehicle, refer to the Precautions Section.

2. Disconnect the negative battery cable.

3. If necessary, remove the air intake hose.

4. If necessary, remove the throttle cable from the cable routing clips.

5. If equipped, remove the breather hose from the valve cover.

6. Remove the crankcase ventilation tube from the valve cover.

7. On the 2.0L (DOHC) engines, remove the ignition wire cover.

8. Label and remove the ignition wires and separators.

9. Remove the valve cover retaining bolts, starting from the outside and working in.

10. Remove the valve cover and gasket from the cylinder head.

To install:

11. Clean the valve cover gasket sealing surfaces.

12. Install new valve cover gaskets and if equipped, O-rings onto the valve covers.

13. Place the valve cover into position.

14. Beginning in the center of the valve cover and working outward, tighten the retaining bolts as follows:

- 2.0L SOHC engine: 26–34 inch lbs. (3–4 Nm)
- 2.0L DOHC engine—2006–07 and 2.4L engines: 27–35 inch lbs. (3–4 Nm)
- 2.0L DOHC engine—2008: First Pass 19–35 inch lbs. (2–4 Nm) and Second Pass 45–53 inch lbs. (5–6 Nm)

19–(2–3 Nm)

- 3.8L engine: 27–35 inch lbs. (3–4 Nm)

15. Install the crankcase ventilation tube.

16. Install the ignition wire separators and the ignition wires.

17. On the 2.0L (DOHC) engines, install the ignition wire cover.

18. Install the air intake hose.

19. Connect the negative battery cable.

20. Run the engine and check for leaks and proper operation.

VALVE LASH

ADJUSTMENT

Eclipse & Galant

2.4L Engine

See Figure 221.

1. Before servicing the vehicle, refer to the Precautions Section.

➡Before adjusting the valves, check that the engine coolant temperature is between 176—203°F (80–95°C and that all lights and accessories are OFF.

2. Remove the ignition coils. Remove the rocker arm cover.

3. Turn the crankshaft clockwise until the notch on the pulley is lined up with the "T" mark on the timing indicator.

4. Move the rocker arms on the number 1 and number 4 cylinders up and down by hand to determine which cylinder has its piston at the Top Dead Center (TDC) position on the compression stroke.

➡If both intake and exhaust valve rocker arms have a valve lash, the piston in the cylinder corresponding to these rocker arms is at TDC on the compression stroke.

5. Valve clearance inspection and adjustment can be performed on rocker arms indicated by the white arrow mark when the number 1 cylinder piston is at the TDC on the compression stroke, and on rocker arms indicated by black arrow mark when the number 4 cylinder piston is at the TDC on the compression stroke.

6. Measure the valve clearance. If the valve clearance is not as specified, loosen the rocker arm lock nut and adjust the clearance using a thickness gauge while turning the adjusting screw.

7. Standard value (hot engine):

a. Intake valve: 0.008 inch (0.20mm).

b. Exhaust valve: 0.012 inch (0.30mm).

8. While holding the adjusting screw with a screwdriver to prevent it from turning, tighten the lock nut to the specified torque: 71–89 inch lbs. (8–10 Nm).

9. Turn the crankshaft through 360° to line up the notch on the crankshaft pulley with the "T" mark on the timing indicator.

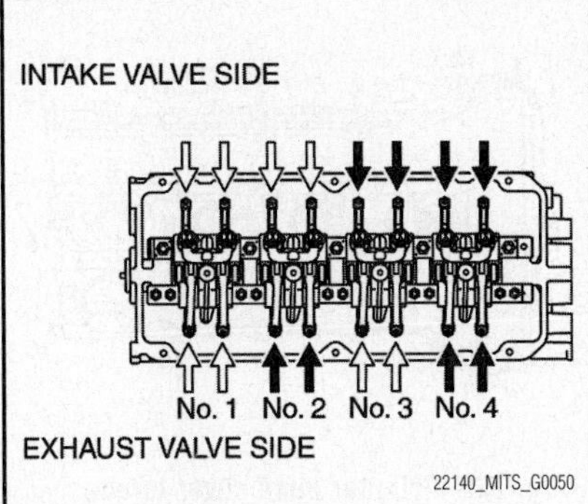

Fig. 221 Valve clearance inspection and adjustment can be performed on rocker arms indicated by the white arrow mark when the number 1 cylinder piston is at the TDC on the compression stroke, and on rocker arms indicated by black arrow mark when the number 4 cylinder piston is at the TDC on the compression stroke—2.4L engine

10. Repeat the steps on other valves for clearance adjustment.
11. Install the rocker arm cover.
12. Install the ignition coils.

3.8L Engine

See Figure 222.

1. Before servicing the vehicle, refer to the Precautions Section.

➡Perform the valve clearance check and adjustment when the engine is cold.

2. Remove all of the ignition coils.
3. Remove the rocker arm covers.
4. Turn the crankshaft clockwise until the notch on the pulley is lined up with "T" mark on the timing indicator.
5. Move the rocker arms on the number 1 and number 4 cylinders up and down by hand to determine which cylinder has its piston at the Top Dead Center (TDC) on the compression stroke.

➡If both intake and exhaust valve rocker arms have a valve lash, the piston in the cylinder corresponding to these rocker arms is at TDC on the compression stroke.

6. Valve clearance inspection and adjustment can be performed on rocker arms indicated by white arrow mark when the number 1 cylinder piston is at the TDC on the compression stroke, and on rocker arms indicated by black arrow mark when the number 4 cylinder piston is at the TDC on the compression stroke.

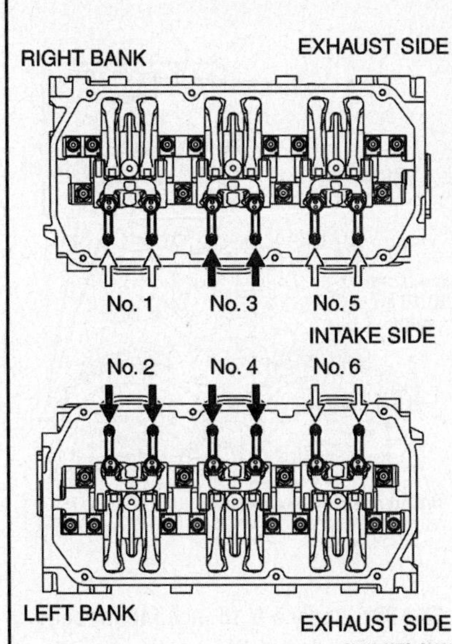

Fig. 222 Valve clearance inspection and adjustment can be performed on rocker arms indicated by the white arrow mark when the number 1 cylinder piston is at the TDC on the compression stroke, and on rocker arms indicated by black arrow mark when the number 4 cylinder piston is at the TDC on the compression stroke—3.8L engine

7. Measure the valve clearance for the intake side. If the valve clearance is not as specified, loosen the rocker arm lock nut and adjust the clearance using a thickness gauge while turning the adjusting screw. Standard value (cold engine): 0.004 inch (0.10mm).

➡Valve clearance check and adjustment is unnecessary for the exhaust side due to auto lash adjusters installed.

8. While holding the adjusting screw with a screwdriver to prevent it from turning, tighten the lock nut to the specified torque: 71–89 inch lbs. (8–10 Nm).
9. Turn the crankshaft through 360° to line up the notch on the crankshaft pulley with the "T" mark on the timing indicator.
10. Repeat the steps on other valves for clearance adjustment.
11. Install the rocker arm covers.
12. Install the ignition coils.

Lancer

2.0L Engine

These engines use hydraulic valve lash adjusters. Valve lash adjustments are not necessary or possible on these engines.

Lancer Evolution

2008 2.0L Turbo Engine

See Figures 223 through 228.

1. Before servicing the vehicle, refer to the Precautions Section.

➡Perform the valve clearance check and adjustment when the engine is cold.

2. Remove all of the ignition coils.
3. Remove the cylinder head cover.

⁂ WARNING

Always turn the crankshaft in a clockwise direction.

4. Turn the crankshaft clockwise, and align the timing mark on the exhaust camshaft sprocket against the upper face of the cylinder head as shown in the illustration.

➡When the marks are aligned, number 1 cylinder will move to the compression stroke at Top Dead Center (TDC).

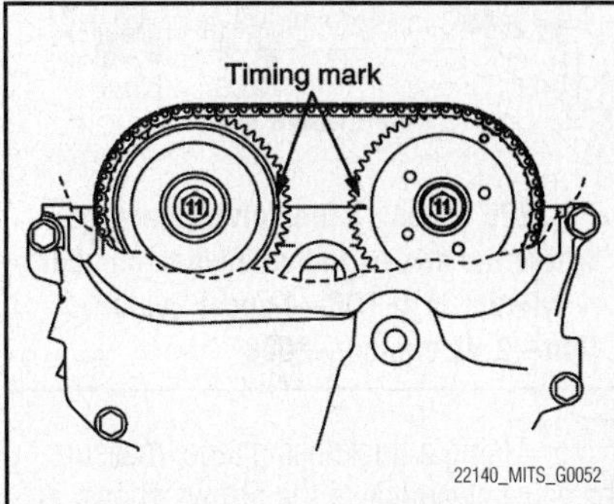

Fig. 223 Turn the crankshaft clockwise, and align the timing mark on the exhaust camshaft sprocket against the upper face of the cylinder head—Lancer Evolution—2.0L engine—2008

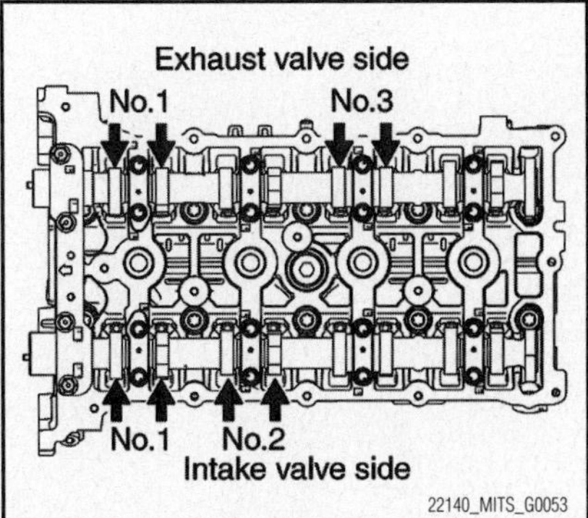

Fig. 224 Measure the valve clearance where the arrows indicate when number 1 cylinder is at TDC—Lancer Evolution—2.0L engine—2008

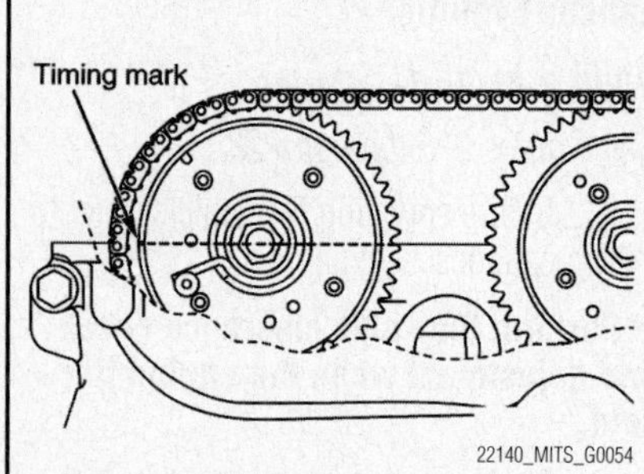

Fig. 225 Turn the crankshaft clockwise 360 degrees and put the timing mark on the exhaust camshaft sprocket in the position shown—Lancer Evolution—2.0L engine—2008

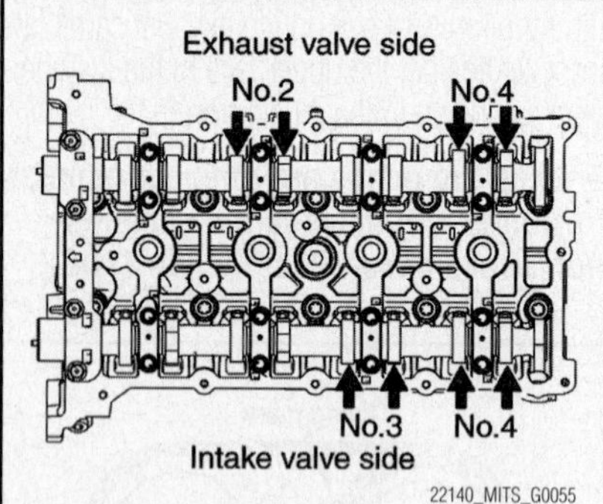

Fig. 226 Measure the valve clearance where the arrows indicate when number 4 cylinder is at TDC—Lancer Evolution—2.0L engine—2008

5. Using a thickness gauge, measure the valve clearance at the arrows shown in the illustration.

6. Standard values:

a. Intake valve: 0.0068– 0.0092 inch (0.17–0.23mm).

b. Exhaust valve: 0.0108– 0.0132 inch (0.27–0.33mm).

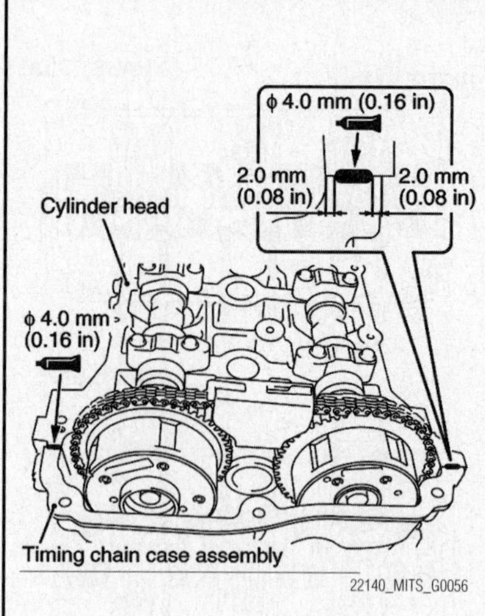

Fig. 227 Apply a 0.16 inch (4mm) bead of liquid gasket as illustrated—Lancer Evolution—2.0L engine—2008

7. Turn the crankshaft clockwise 360° and put the timing mark on the exhaust camshaft sprocket in the position shown.

➡**When the timing marks are aligned, number 4 cylinder will be in the compression stroke at TDC.**

8. Using a thickness gauge, measure the valve clearance at the arrows shown in the illustration.

9. If the valve clearance deviates from the standard value, remove the camshaft and the valve tappet. Refer to Camshaft and Valve Lifters, Removal & Installation.

10. Replace the valve tappet with one of the correct thickness.

11. Install the valve tappet selected and install the camshaft. Refer to Camshaft and Valve Lifters, Removal & Installation.

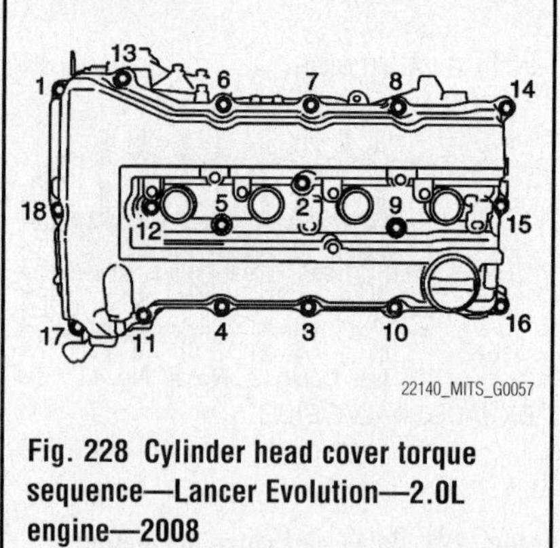

Fig. 228 Cylinder head cover torque sequence—Lancer Evolution—2.0L engine—2008

12. After installing the timing chain, measure the valve clearance using the above procedures. Confirm the clearance is within the standard value.

➡**Completely remove all the old gasket material, which might be remaining among the components.**

13. After completely removing the liquid gasket adhering on the timing chain case, cylinder block, and cylinder head, degrease them with white gasoline.

➡**The cylinder head cover should be installed within 3 minutes of applying liquid gasket.**

14. Apply a 0.16 inch (4mm) bead of liquid gasket as illustrated. Specified sealant: THREE BOND 1217G or exact equivalent.

15. Install the cylinder head cover and tighten the bolts, in the order shown, using the following procedure:

a. Step 1: Tighten the bolts to 18–36 inch lbs. (2–4 Nm).

b. Step 2: Tighten the bolts to 40–58 inch lbs. (5–6 Nm).

16. Install the ignition coils.

ENGINE PERFORMANCE & EMISSION CONTROL

ACCELERATOR PEDAL POSITION (APP) SENSOR

LOCATION

See Figure 229.

The Accelerator Pedal Position (APP) sensor is located inside the vehicle. It is part of the accelerator pedal assembly.

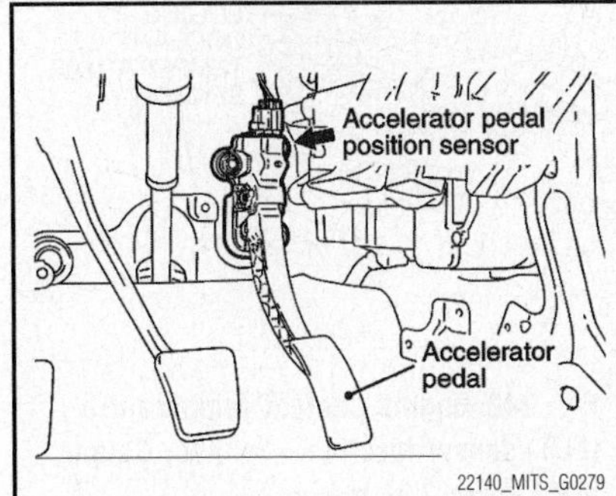

Fig. 229 Accelerator Pedal Position (APP) sensor location

REMOVAL & INSTALLATION

1. Before servicing the vehicle, refer to the Precautions Section.
2. Disconnect the ground cable from the battery.
3. Disconnect the Accelerator Pedal Position (APP) sensor connector.
4. Remove the nut securing accelerator pedal assembly.

To install:

5. Install the accelerator pedal assembly.
6. Tighten the pedal mounting nuts to 111 inch lbs. (13 Nm).
7. Connect the electrical connector at the sensor.

CAMSHAFT POSITION (CMP) SENSOR

LOCATION

See Figures 230 through 239.

REMOVAL & INSTALLATION

1. Before servicing the vehicle, refer to the Precautions Section.
2. Disconnect the negative battery cable.
3. Disconnect the connector from the CMP sensor.
4. Remove the bolt that retains the CMP sensor.
5. Remove the CMP sensor.

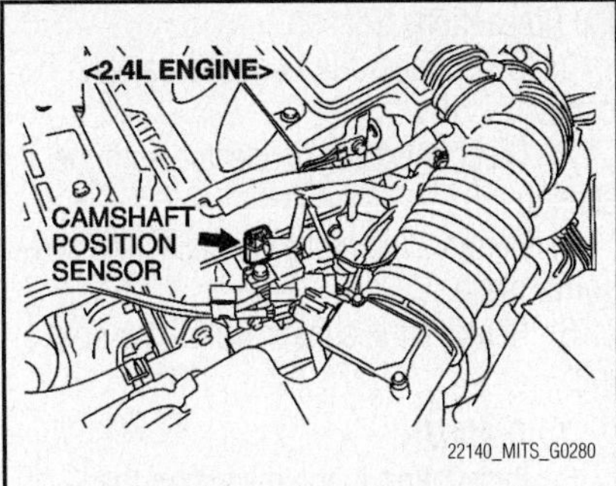

Fig. 230 Camshaft Position (CMP) sensor location—Eclipse 2.4L engine

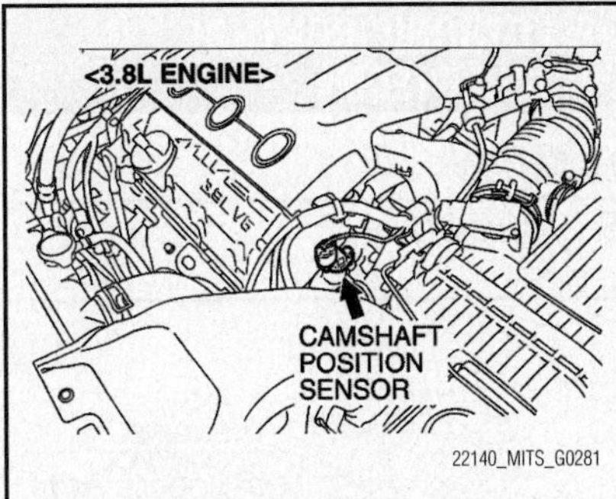

Fig. 231 Camshaft Position (CMP) sensor location—Eclipse 3.8L engine

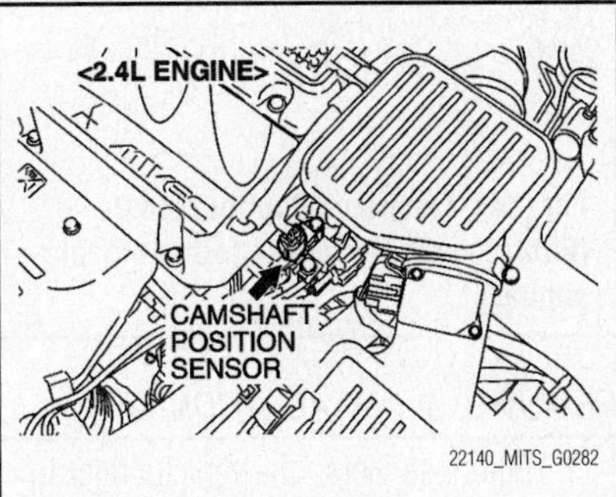

Fig. 232 Camshaft Position (CMP) sensor location—Galant 2.4L engine

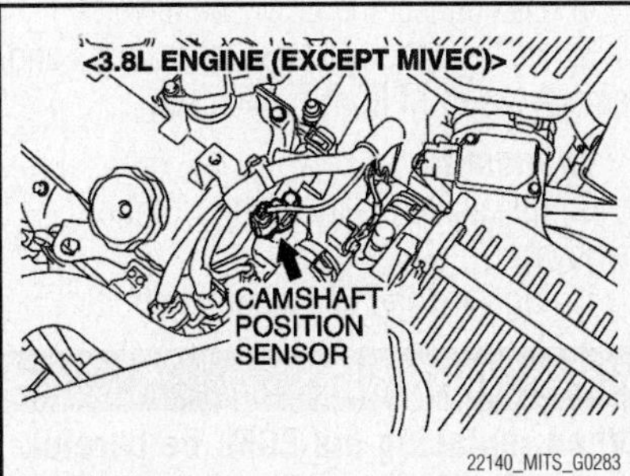

Fig. 233 Camshaft Position (CMP) sensor location—Galant 3.8L (except MIVEC) engine

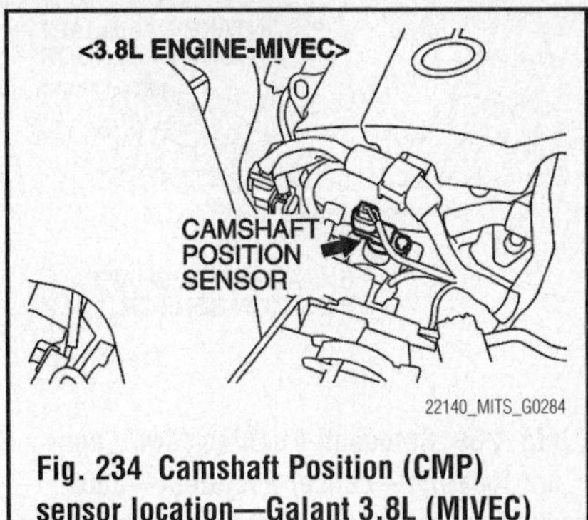

Fig. 234 Camshaft Position (CMP) sensor location—Galant 3.8L (MIVEC) engine

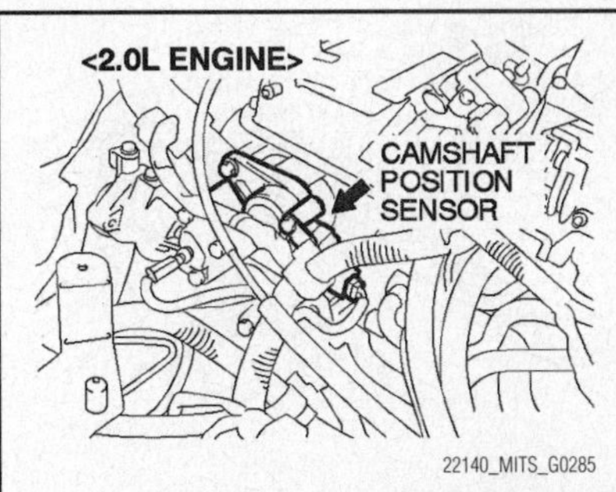

Fig. 235 Camshaft Position (CMP) sensor location—Lancer 2.0L engine—2006-07

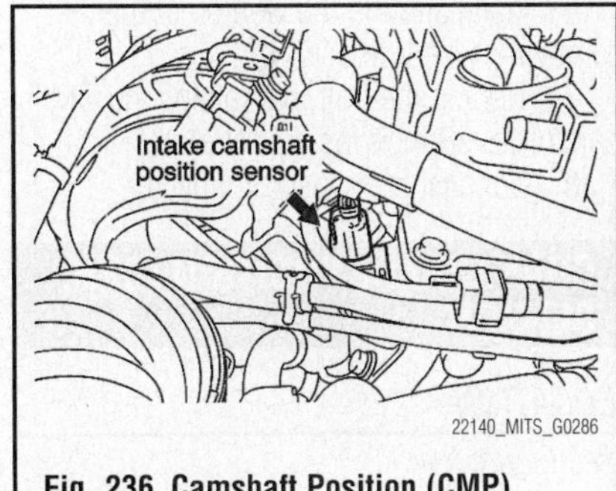

Fig. 236 Camshaft Position (CMP) sensor location—Lancer 2.0L engine—2008

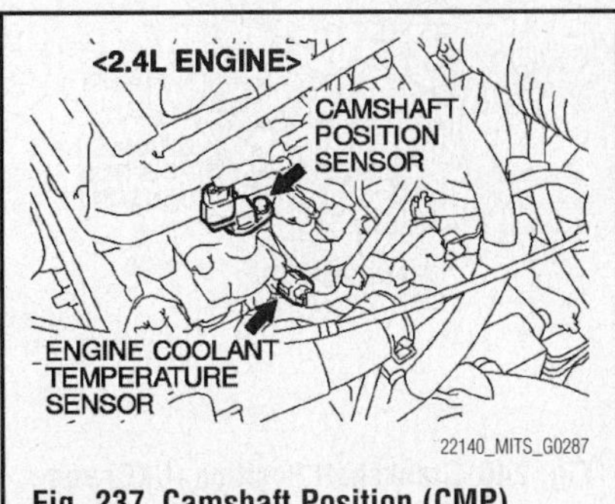

Fig. 237 Camshaft Position (CMP) sensor location—Lancer 2.4L engine—2006

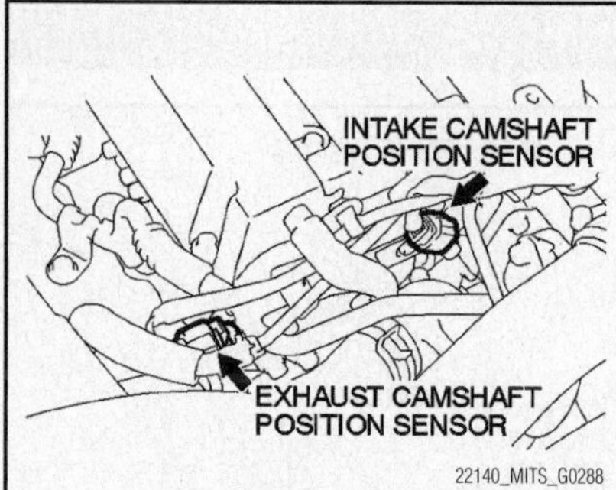

Fig. 238 Camshaft Position (CMP) sensor location—Lancer Evolution—2006

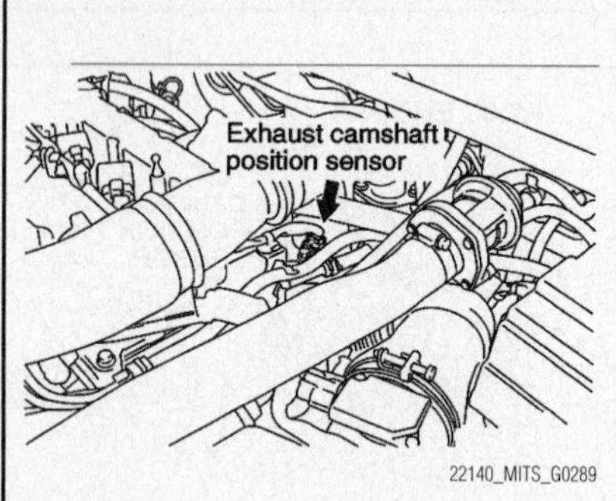

Fig. 239 Camshaft Position (CMP) sensor location—Lancer Evolution—2008

To install:

6. Installation is the reverse of the removal procedure.
7. Tighten the bolt that retains the CMP sensor to: 98 inch lbs. (11 Nm).
8. Connect the sensor connector.

CRANKSHAFT POSITION (CKP) SENSOR

LOCATION

See Figure 240.

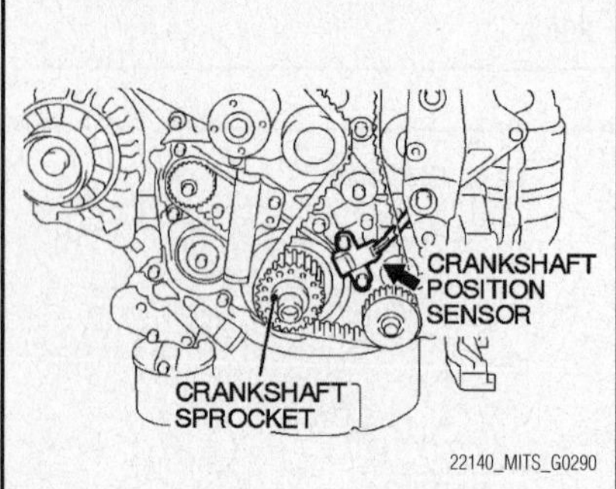

Fig. 240 Crankshaft Position (CKP) sensor location—Lancer Evolution (other models similar)

REMOVAL & INSTALLATION

1. Before servicing the vehicle, refer to the Precautions Section.
2. Disconnect the negative battery cable.
3. Disconnect the connector from the sensor.
4. Remove the bolt that retains the sensor in place.
5. Remove the sensor from its mounting.

To install:

6. Installation is the reverse of the removal procedure.
7. Tighten the sensor retaining bolt to: 78 inch lbs. (9 Nm).

ELECTRONIC CONTROL MODULE (ECM)

LOCATION

See Figure 241.

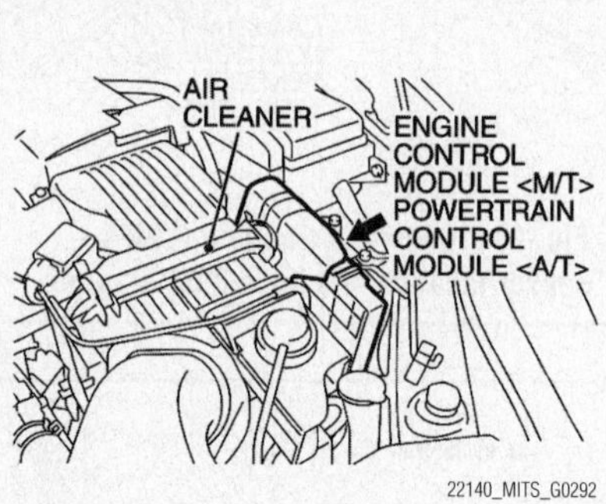

Fig. 241 Electronic Control Module (ECM) location—Eclipse (other models similar)

REMOVAL & INSTALLATION

1. Before servicing the vehicle, refer to the Precautions Section.
2. Turn ignition switch off.
3. Disconnect the negative battery cable from the battery.
4. Disconnect the ECM connectors.
5. Remove the ECM mounting bolts and remove the ECM from the vehicle.

To install:

6. Installation is the reverse of the removal.
7. Tighten the ECM mounting bolts.

**** WARNING**

When replacing the ECM, be careful to use the right part number, as damage to the injection system could occur.

ENGINE COOLANT TEMPERATURE (ECT) SENSOR

LOCATION

See Figures 242 through 247.

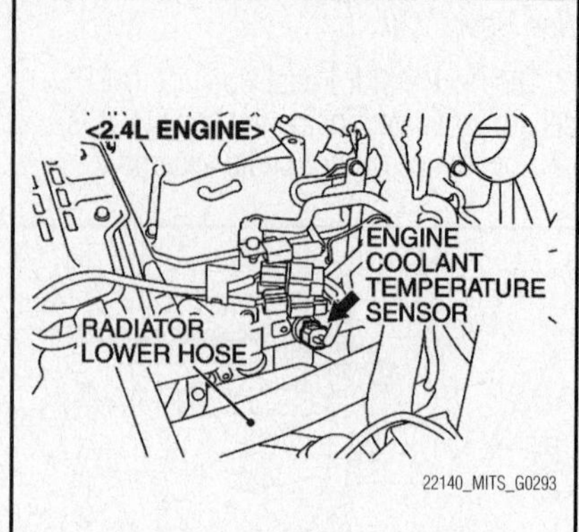

Fig. 242 Engine Coolant Temperature (ECT) sensor location—Eclipse, Galant, and Lancer 2.4L engine

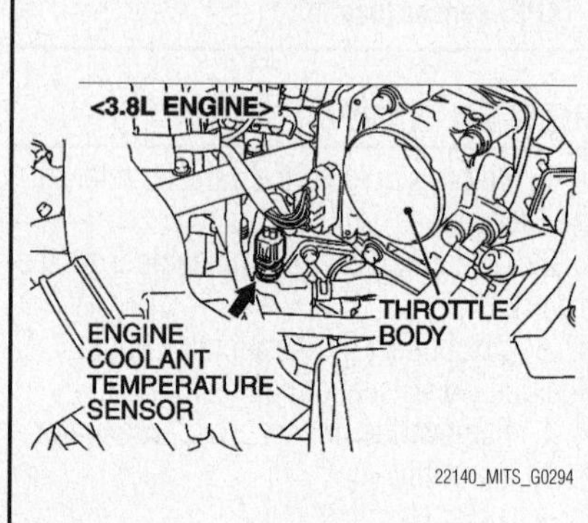

Fig. 243 Engine Coolant Temperature (ECT) sensor location—Eclipse and Galant 3.8L engine

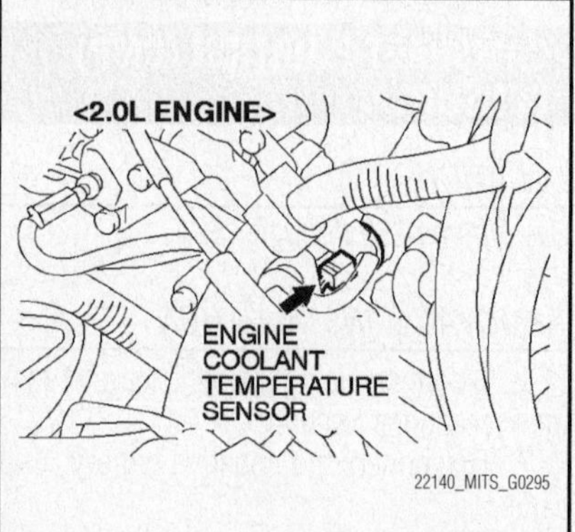

Fig. 244 Engine Coolant Temperature (ECT) sensor location—Lancer 2.0L engine—2006-07

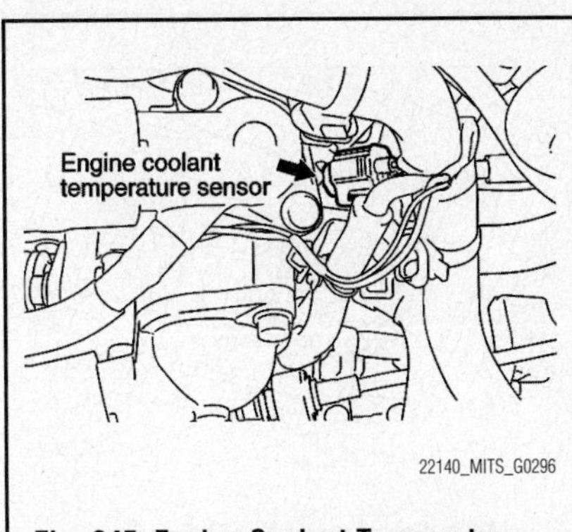

Fig. 245 Engine Coolant Temperature (ECT) sensor location—Lancer—2008

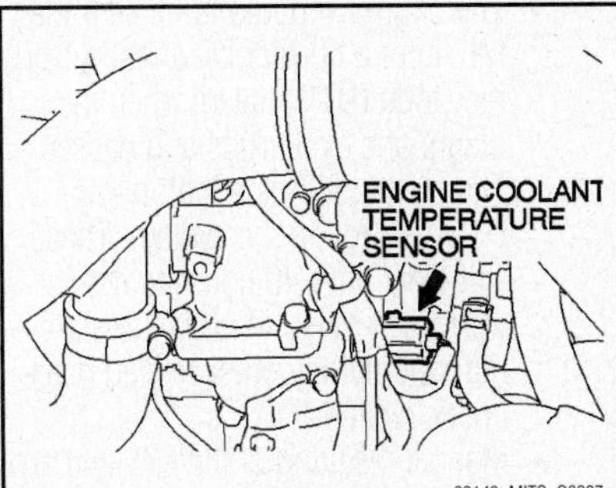

Fig. 246 Engine Coolant Temperature (ECT) sensor location—Lancer Evolution—2006

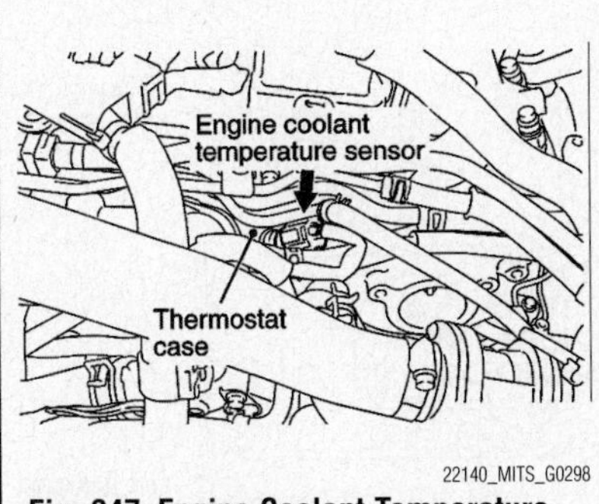

Fig. 247 Engine Coolant Temperature (ECT) sensor location—Lancer Evolution—2008

REMOVAL & INSTALLATION

1. Before servicing the vehicle, refer to the Precautions Section.
2. Drain the coolant to a level below the bottom of the sensor.
3. Disconnect the ground cable from the battery and then remove the sensor connector.
4. Remove the coolant temperature sensor.

To install:

5. Coat the threads of the sensor with a suitable sealant and thread into the housing.
6. Tighten the sensor to 22 ft. lbs. (30 Nm).
7. Refill the cooling system to the proper level.
8. Attach the electrical connector to the sensor securely.
9. Connect the negative battery cable.

HEATED OXYGEN (HO2S) SENSOR

LOCATION

The sensors are located in the exhaust system. On some vehicles, one sensor is located up at the exhaust manifold(s) and the other sensor is located down at the catalytic converter.

REMOVAL & INSTALLATION

**** CAUTION**

The temperature of the exhaust system is extremely high after the engine has been run. To prevent personal injury, allow the exhaust system to cool before removing the sensor from the exhaust system.

1. Before servicing the vehicle, refer to the Precautions Section.
2. Disconnect the negative battery cable.
3. Raise and safely support the vehicle, as needed.
4. Detach the electrical connector from the oxygen sensor.
5. Using socket MD998770, or equivalent oxygen sensor socket, remove the heated oxygen sensor.

To install:

6. If installing the old oxygen sensor, coat the threads with anti-seize compound. New sensors are already coated. Take care not to contaminate the oxygen sensor probe with the anti-seize compound.
7. Install the oxygen sensor. Using the correct tool, tighten the sensor to 33 ft. lbs. (45 Nm).
8. Attach the wiring to the sensor.
9. Connect the negative battery cable.

INTAKE AIR TEMPERATURE (IAT) SENSOR

LOCATION

The Intake Air Temperature (IAT) sensor is mounted in the intake air hose of the air cleaner assembly.

REMOVAL & INSTALLATION

1. Before servicing the vehicle, refer to the Precautions Section.
2. Disconnect the negative battery cable.
3. Disconnect the connector from the sensor.
4. Remove the sensor retaining screws.
5. Remove the sensor from its mounting.

To install:

6. Installation is the reverse of the removal procedure.
7. Handle the sensor assembly carefully, protecting it from impact, extremes of temperature and/or exposure to shop chemicals.

KNOCK SENSOR (KS)

LOCATION

See Figures 248 through 253.

REMOVAL & INSTALLATION

1. Before servicing the vehicle, refer to the Precautions Section.
2. Disconnect the negative battery cable.
3. Disconnect the sensor connector.
4. Remove the sensor from its mounting.

To install:

5. Installation is the reverse of the removal procedure.
6. Tighten the sensor to 12–17 ft. lbs. (16–24 Nm).

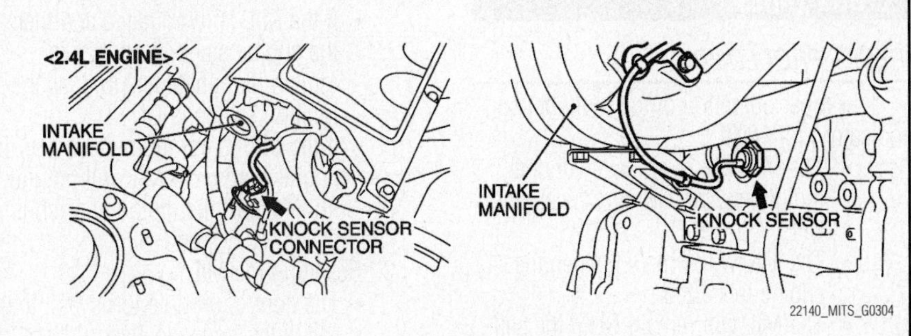

Fig. 248 Location of Knock Sensor (KS)—Eclipse, Galant, and Lancer 2.4L engine

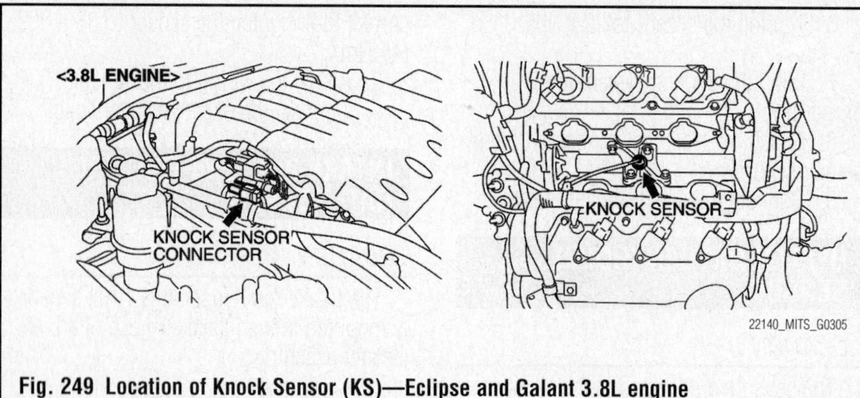

Fig. 249 Location of Knock Sensor (KS)—Eclipse and Galant 3.8L engine

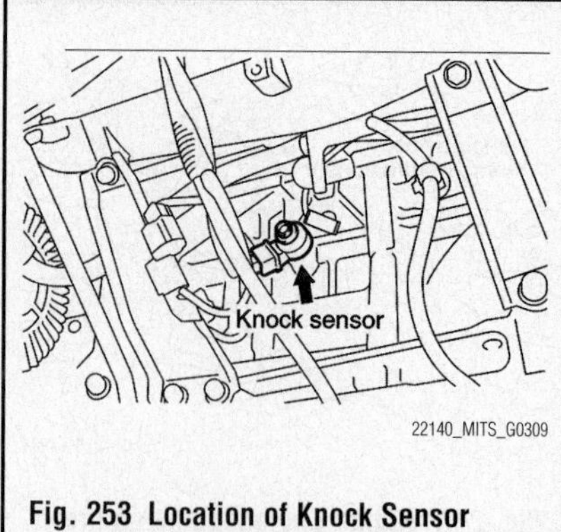

Fig. 253 Location of Knock Sensor (KS)—Lancer Evolution—2008

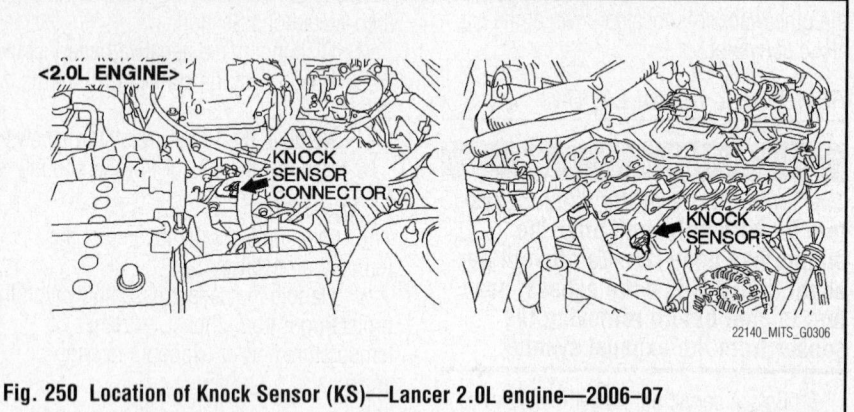

Fig. 250 Location of Knock Sensor (KS)—Lancer 2.0L engine—2006-07

- The control module turns OFF the MIL after a current Diagnostic Trouble Code (DTC) clears when the diagnostic cycle runs and passes
- There may still be a history of DTC's stored in the system. These will clear after 40 consecutive warm-up cycles, if no failures are reported by any other related diagnostic system
- Manual resetting of the MIL and any DTC stored in the system, requires the use of an OBD2 scan tool connected to the Data Link Connector

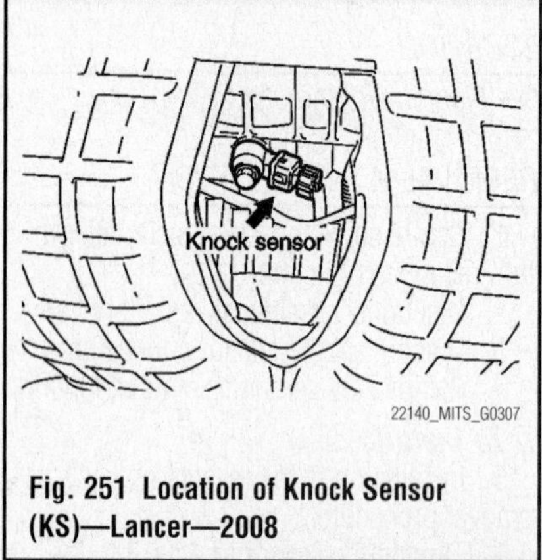

Fig. 251 Location of Knock Sensor (KS)—Lancer—2008

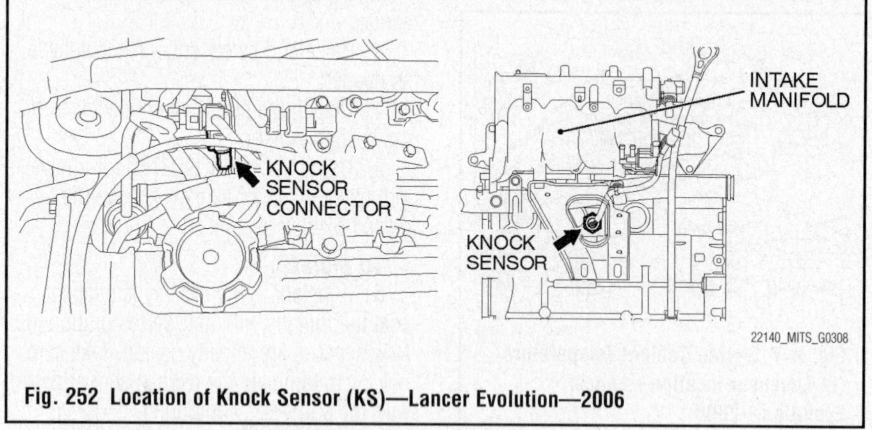

Fig. 252 Location of Knock Sensor (KS)—Lancer Evolution—2006

MALFUNCTION INDICATOR LIGHT (MIL)

RESET PROCEDURES

1. Proper operation of the Malfunction Indicator Light (MIL):
   - The MIL will illuminate with the ignition switch ON and the engine OFF
   - The MIL will turn OFF when the engine is started
   - The MIL will remain ON if the self-diagnostic system has detected a malfunction
   - The MIL may turn OFF if the malfunction is no longer present
   - If the MIL is illuminated and then the engine stalls, the MIL will remain illuminated as long as the ignition switch is ON
   - If the MIL is not illuminated and the engine stalls, the MIL will not illuminate until the ignition switch is cycled OFF, then ON
2. Resetting the MIL:
   - The control module turns OFF the MIL after 3 consecutive ignition cycles that the diagnostic system runs and does not fail

(DLC) for communication with the vehicle. Follow the instructions of the scan tool for both retrieval and resetting of DTC's.

➡If the error symptoms causing the MIL to illuminate have been corrected, the MIL will return to normal operation.

MASS AIR FLOW (MAF) SENSOR

LOCATION

See Figure 254.

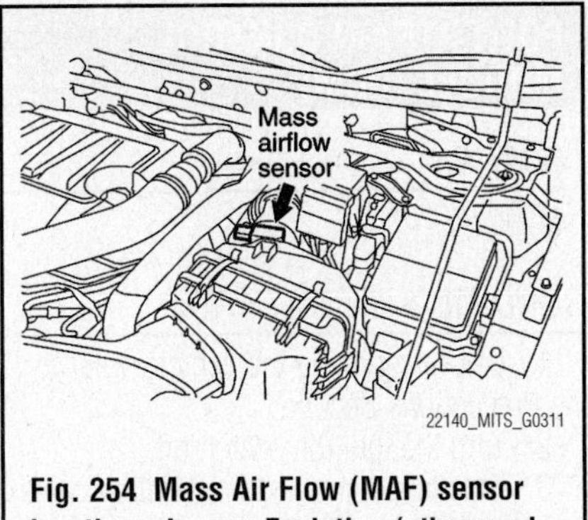

Fig. 254 Mass Air Flow (MAF) sensor location—Lancer Evolution (other models similar)

The Mass Air Flow (MAF) sensor is mounted in the intake air hose of the air cleaner assembly.

REMOVAL & INSTALLATION

1. Before servicing the vehicle, refer to the Precautions Section.
2. Disconnect the negative battery cable.
3. Disconnect the connector from the sensor.
4. Remove the air cleaner and air intake assembly, as required.
5. Remove the sensor from its mounting.

To install:

6. Installation is the reverse of the removal procedure.
7. Handle the sensor assembly carefully, protecting it from impact, extremes of temperature, and exposure to shop chemicals.

MANIFOLD ABSOLUTE PRESSURE (MAP) SENSOR

LOCATION

See Figures 255 through 260.

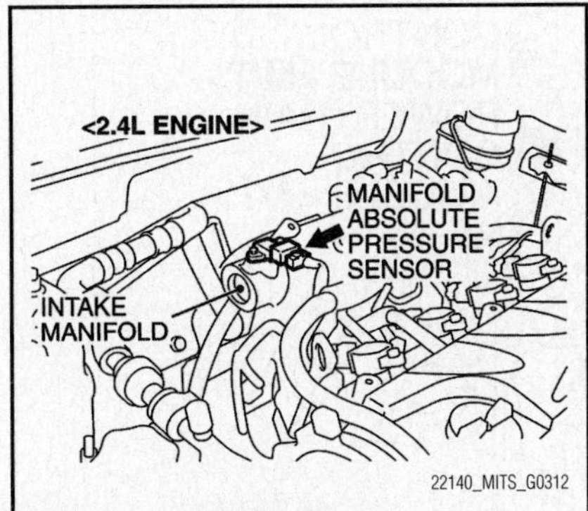

Fig. 255 Manifold Absolute Pressure Sensor location—Eclipse, Galant, and Lancer 2.4L engines

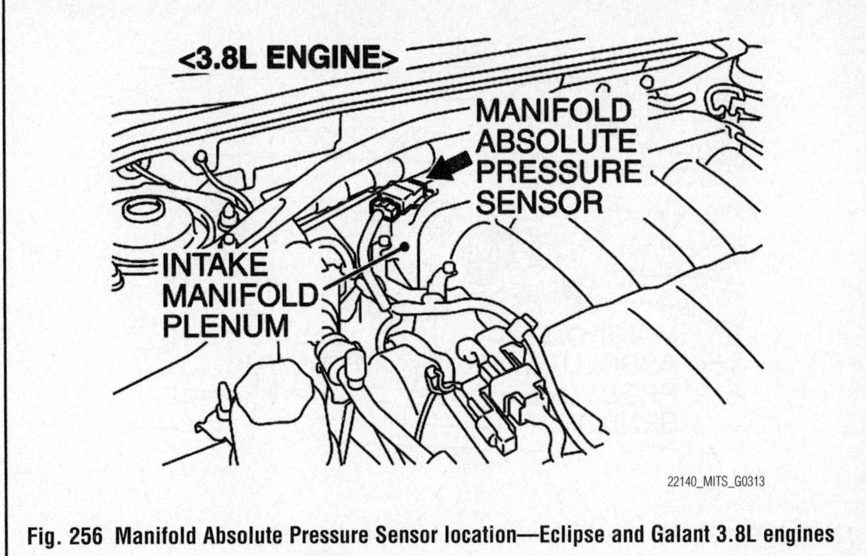

Fig. 256 Manifold Absolute Pressure Sensor location—Eclipse and Galant 3.8L engines

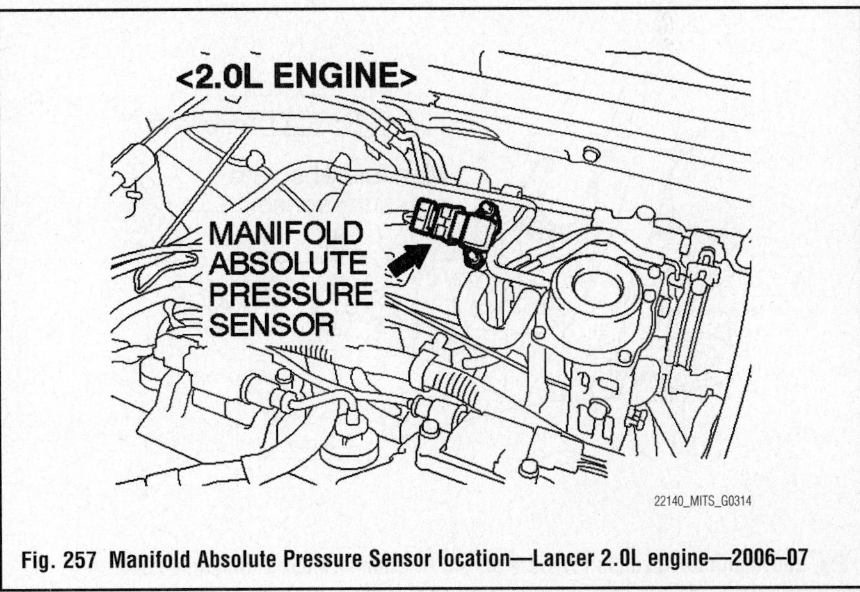

Fig. 257 Manifold Absolute Pressure Sensor location—Lancer 2.0L engine—2006-07

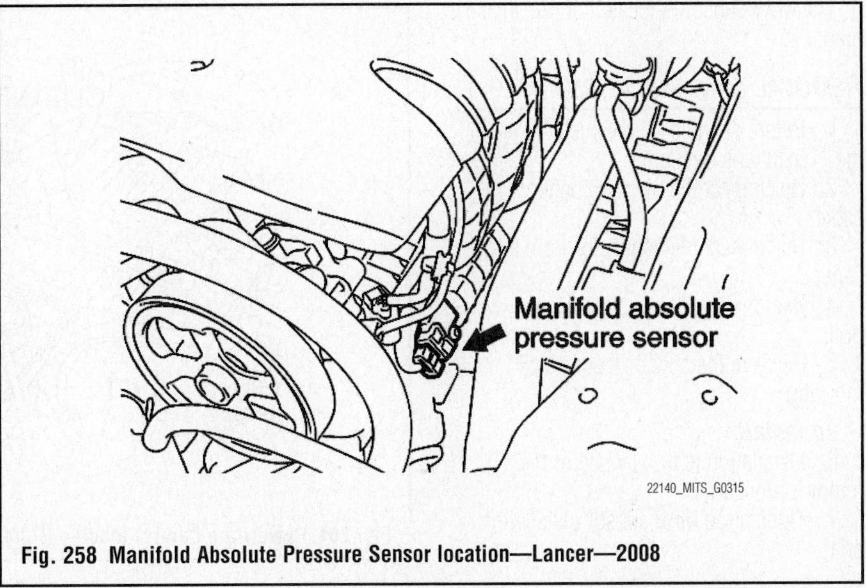

Fig. 258 Manifold Absolute Pressure Sensor location—Lancer—2008

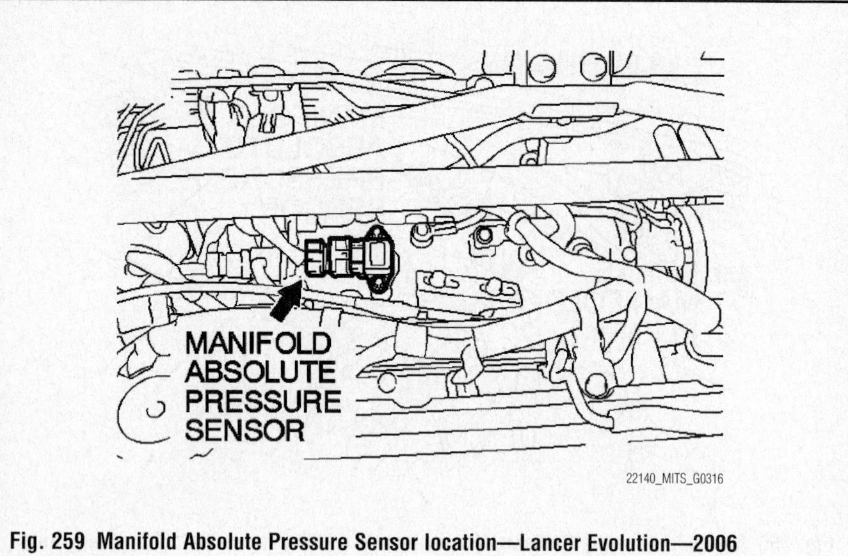

Fig. 259 Manifold Absolute Pressure Sensor location—Lancer Evolution—2006

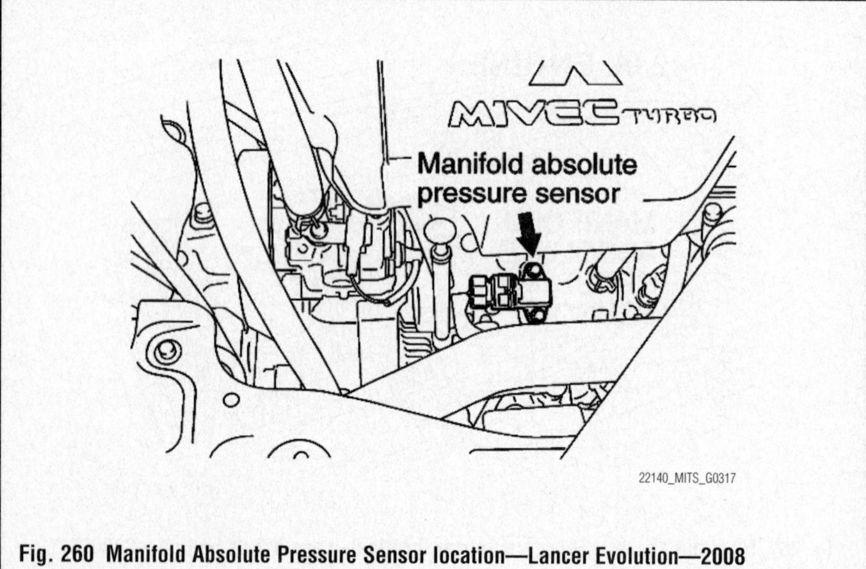

Fig. 260 Manifold Absolute Pressure Sensor location—Lancer Evolution—2008

The MAP sensor is located on the intake manifold.

REMOVAL & INSTALLATION

1. Before servicing the vehicle, refer to the Precautions Section.
2. Disconnect the negative battery cable.
3. Disconnect the connector from the sensor.
4. Remove the sensor retaining bolt.
5. Remove the sensor from its mounting.

To install:

6. Installation is the reverse of the removal procedure.
7. Tighten the MAP sensor installation bolt.

POWERTRAIN CONTROL MODULE (PCM)

LOCATION

See Figure 261.

REMOVAL & INSTALLATION

1. Before servicing the vehicle, refer to the Precautions Section.
2. Turn the ignition switch off.
3. Disconnect the negative battery cable from the battery.
4. Disconnect the PCM connector(s).
5. Remove the PCM mounting bolts and remove the PCM from the air cleaner assembly.

To install:

6. Installation is the reverse of the removal.
7. Tighten the PCM mounting bolts to: 86–104 inch lbs. (10–12 Nm).

THROTTLE POSITION SENSOR (TPS)

LOCATION

The Throttle Position Sensor (TPS) is mounted on the throttle body and is incorporated into the throttle body assembly.

REMOVAL & INSTALLATION

The throttle position sensor is an integral part of the throttle body.

1. Before servicing the vehicle, refer to the Precautions Section.
2. Properly relieve the fuel system pressure.
3. Drain the engine coolant.

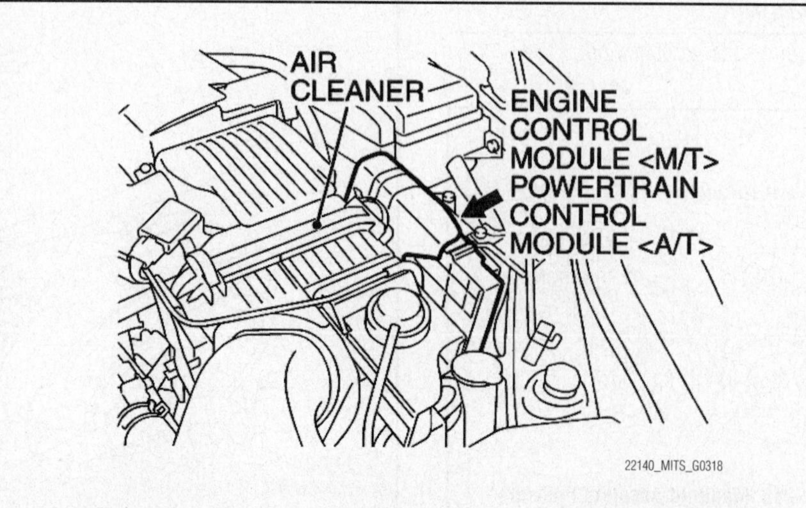

Fig. 261 Powertrain Control Module (PCM) location—Eclipse (other models similar)

4. Remove the air intake hose.
5. Remove the battery.
6. Disconnect the throttle position sensor connector.
7. Disconnect the water hose connection.
8. Remove the throttle body retaining bolts.
9. Remove the throttle body from the engine.
10. Discard the gasket.

To install:

***** WARNING**

Do not loosen the retaining screws for the resin cover of the throttle body assembly. If the screws are loosened, the sensor incorporated in the resin cover becomes misaligned and the throttle body may not work properly.

11. Align the recess on the intake manifold plenum with the projection of the throttle body gasket.
12. Install the gasket. Install the throttle body to the engine and tighten the retaining bolts to 18–24 ft. lbs. (24–32 Nm).

➡Poor idling may result if the throttle body gasket is not installed properly.

13. Continue the installation in the reverse order of the removal procedure.
14. Connect the negative battery cable.
15. Turn the ignition **ON** and then **OFF**, and keep it off for at least 10 seconds.
16. Complete the vehicle initialization procedure.

➡To complete the initialization procedure the following tools are needed. MB991958 scan tool, MB991824 VCI, MB991827 MUT III USB cable, MB991910 MUT III Main harness "A."

a. Connect the scan tool to the Data Link Connector (DLC). To prevent damage to the scan tool, be sure that the ignition switch is in the **LOCK** position before connecting the scan tool.
b. Turn the ignition switch to the **ON** position.
c. Select "check mode" from the menu screen.
d. Select "erase memory" from the menu screen.
e. Initialize the learning value.
f. After initialization, complete the idle learning procedure.

➡This procedure must be performed when the PCM is replaced, or when the learning value is initialized, as the idling is not stabilized when the learning value in the MFI engine is not completed.

g. Start the engine. Allow the coolant temperature to reach 176°F (80°C) or more.
h. Stop the engine and place the ignition switch in the **LOCK** position.
i. After 10 seconds, restart the engine.
j. Carry out the following idling procedure for 10 minutes to confirm that the engine has normal idling:

- Position the transaxle selector lever in the **P** or neutral range
- The engine fan is not to be operated
- The engine coolant temperature should be 176°F (80°C) or more

➡If the engine stalls during idling, check the throttle valve of the throttle body for dirt. Correct and perform the procedure again.

VARIABLE CAMSHAFT TIMING OIL CONTROL SOLENOID

LOCATION

See Figure 262.

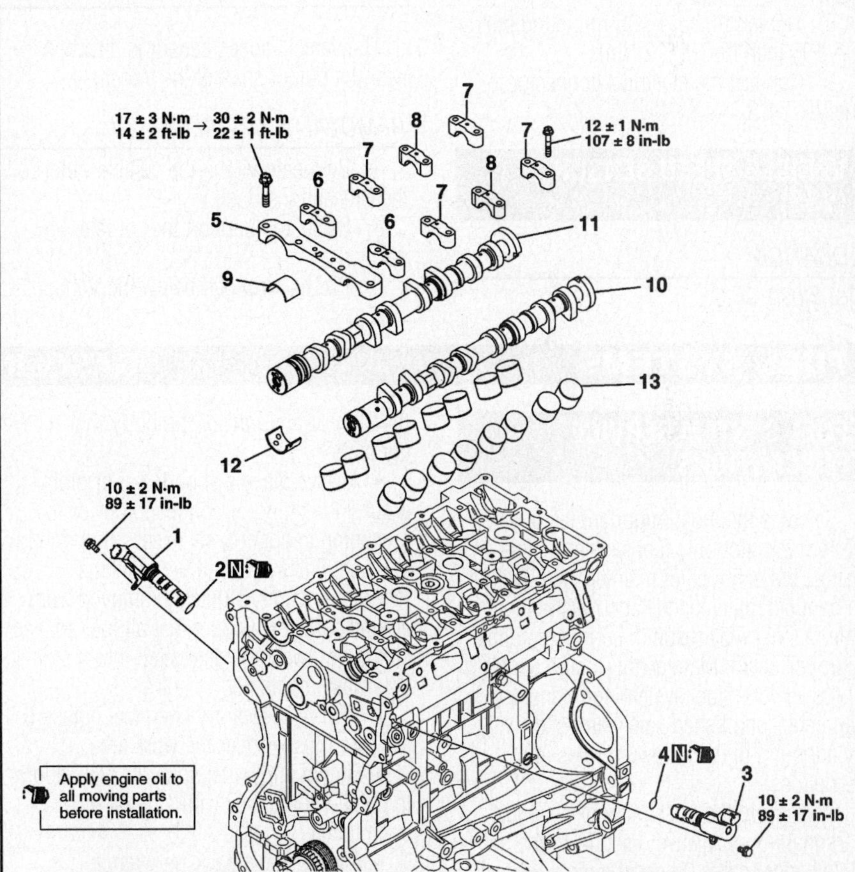

Fig. 262 Location of the variable camshaft timing oil control solenoid (1), (3), also called the engine Oil Control Valve (OCV)

REMOVAL & INSTALLATION

1. Before servicing the vehicle, refer to the Precautions Section.
2. Disconnect the ground cable from the battery.
3. Disconnect the connector from Variable Camshaft Timing Oil Control Solenoid (VCTOCS) on the right-hand bank and/or left-hand bank.
4. Remove the bolt retaining the VCTOCS.
5. Remove the VCTOCS.

To install:

➡**Always use a new gasket/O-ring.**

6. Apply a small amount of engine oil to the new O-ring of the VCTOCS.
7. Install the VCTOCS on the cylinder head.
8. Tighten the VCTOCS mounting bolt to 72–106 inch lbs. (8–12 Nm).
9. Connect the electrical connector to the VCTOCS.

VEHICLE SPEED SENSOR (VSS)

LOCATION

See Figure 263.

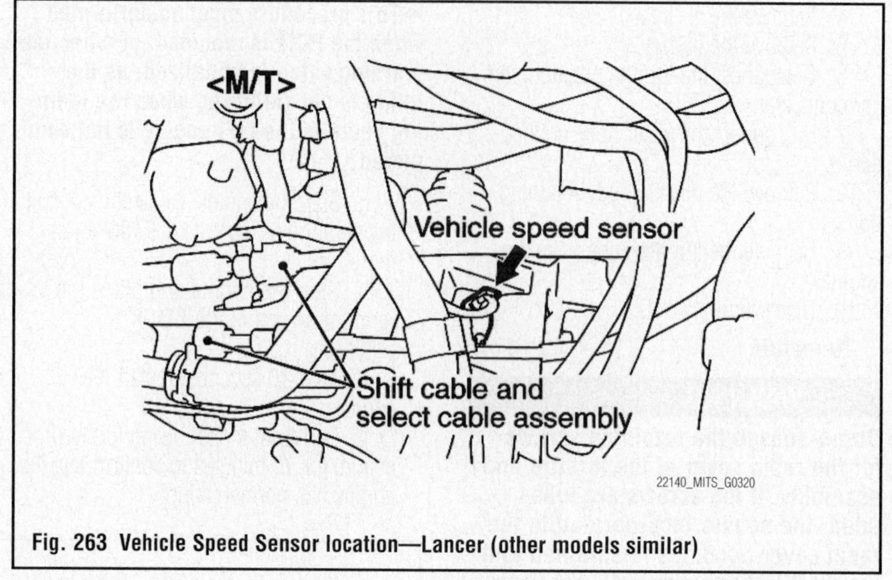

Fig. 263 Vehicle Speed Sensor location—Lancer (other models similar)

The vehicle speed sensor is attached near the output shaft of the transaxle.

REMOVAL & INSTALLATION

1. Before servicing the vehicle, refer to the Precautions Section.
2. Raise and support the vehicle safely.
3. Place a drip pan below the Vehicle Speed Sensor (VSS) to catch any spilled fluid when it is removed.
4. Disconnect the VSS connector.
5. Remove the sensor from its mounting.

To install:

6. Install the sensor and tighten the attaching bolt to 97 inch lbs. (11 Nm).
7. Replace any lost transaxle fluid.
8. Connect the sensor electrical connector.

FUEL SYSTEMS — GASOLINE FUEL INJECTION SYSTEM

FUEL SYSTEM SERVICE PRECAUTIONS

Safety is the most important factor when performing, not only fuel system maintenance, but any type of maintenance. Failure to conduct maintenance and repairs in a safe manner may result in serious personal injury or death. Maintenance and testing of the vehicle's fuel system components can be accomplished safely and effectively by adhering to the following rules and guidelines.

- To avoid the possibility of fire and personal injury, always disconnect the negative battery cable unless the repair or test procedure requires that battery voltage be applied.
- Always relieve the fuel system pressure prior to disconnecting any fuel system component (injector, fuel rail, pressure regulator, etc.), fitting, or fuel line connection. Exercise extreme caution whenever relieving fuel system pressure to avoid exposing skin, face, and eyes to fuel spray. Please be advised that fuel under pressure may penetrate the skin or any part of the body that it contacts.
- Always place a shop towel or cloth around the fitting or connection prior to loosening to absorb any excess fuel due to spillage. Ensure that all fuel spillage (should it occur) is quickly removed from engine surfaces. Ensure that all fuel soaked cloths or towels are deposited into a suitable waste container.
- Always keep a dry chemical (Class B) fire extinguisher near the work area.
- Do not allow fuel spray or fuel vapors to come into contact with a spark or an open flame.
- Always use a back-up wrench when loosening and tightening fuel line connection fittings. This will prevent unnecessary stress and torsion to fuel line piping.
- Always replace worn fuel fitting O-rings with new. Do not substitute fuel hose or equivalent where fuel pipe is installed.

Before servicing the vehicle, make sure to refer to the precautions in the beginning of this section as well.

RELIEVING FUEL SYSTEM PRESSURE

1. Before servicing the vehicle, refer to the Precautions Section.
2. Remove the rear seat cushion assembly.
3. Remove the fuel pump module access hole cover.
4. Disconnect the fuel pump module connector.
5. After starting the engine and letting it run until it stops, turn the ignition switch to the **OFF** position.
6. Crank the engine for 2 seconds or more.
7. If the engine does not start, turn the ignition switch to the **OFF** position.
8. If the engine starts, let it run until it stalls and turn the ignition switch to the **OFF** position.
9. Connect the fuel pump module connector.
10. Install the fuel pump module access hole cover.
11. Install the rear seat cushion assembly.

FUEL FILTER

REMOVAL & INSTALLATION

The fuel delivery system integrates the fuel filter with the in-tank fuel pump. To service this filter, remove the fuel pump. Refer to Fuel Pump, Removal & Installation.

FUEL PUMP

REMOVAL & INSTALLATION

See Figure 264.

✲✲ CAUTION

Observe all applicable safety precautions when working around fuel. Whenever servicing the fuel system, always work in a well ventilated area. Do not allow fuel spray or vapors to come in contact with a spark or open flame. Keep a dry chemical fire extinguisher near the work area. Always keep fuel in a container specifically designed for fuel

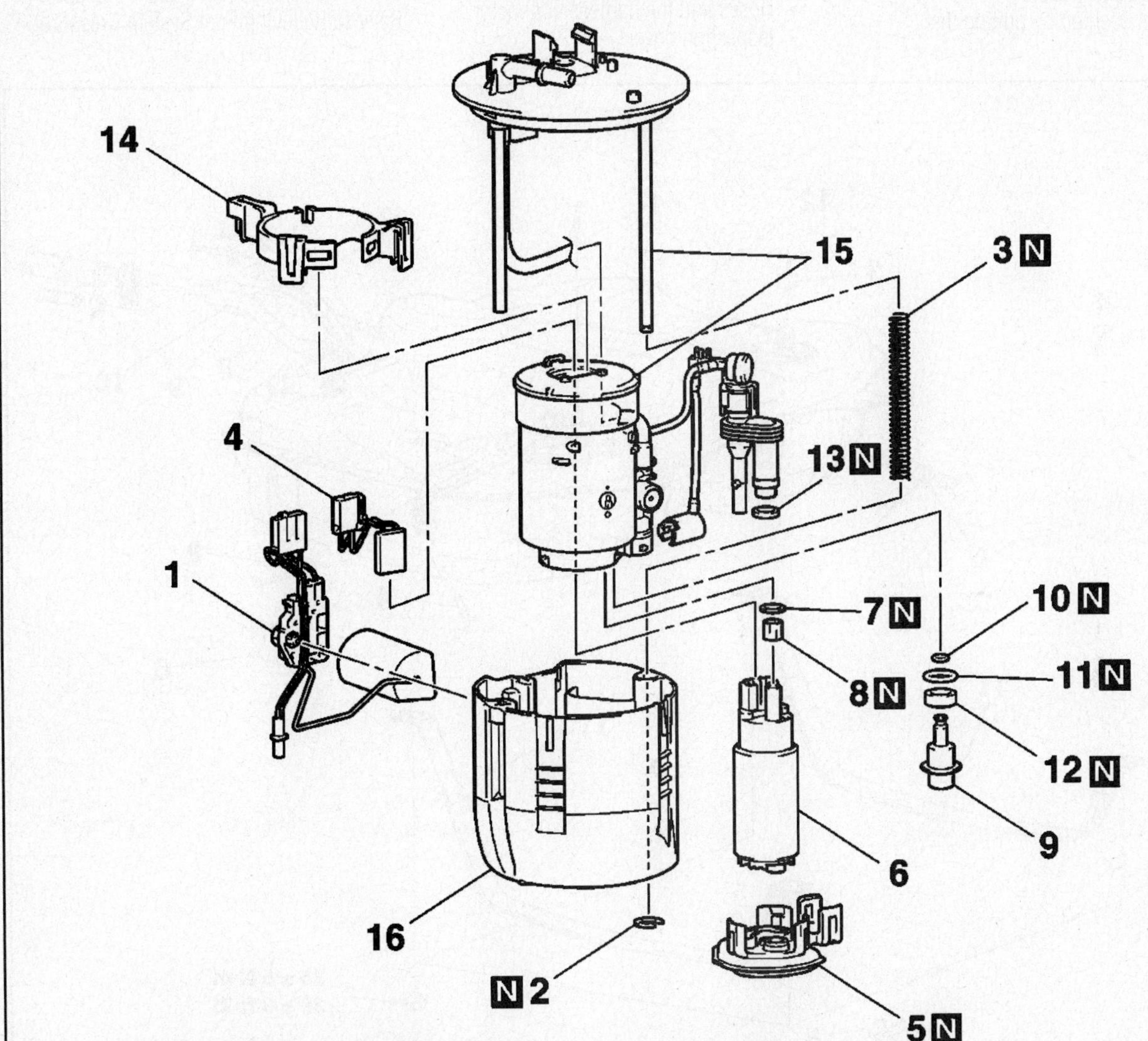

Fig. 264 Exploded view of the fuel pump module

storage; also, always properly seal fuel containers to avoid the possibility of fire or explosion.

1. Before servicing the vehicle, refer to the Precautions Section.
2. Relieve the fuel system pressure. Refer to Relieving Fuel System Pressure.
3. Remove or disconnect the following:
   - Negative battery cable
   - Rear seat cushion by pulling the seat stopper near the floor and lifting the cushion up
   - Inspection cover on the right side of the vehicle
   - Harness connector and the fuel lines
   - Fuel pump assemble from the tank.

To install:

4. Install or connect the following:
   - Fuel pump in the tank
   - Hoses and the harness connector
   - Inspection cover
   - Rear seat
   - Negative battery cable

FUEL TANK

REMOVAL & INSTALLATION

See Figures 265 through 267.

1. Before servicing the vehicle, refer to the Precautions Section.
2. Relieve the fuel system pressure. Refer to Relieving Fuel System Pressure.

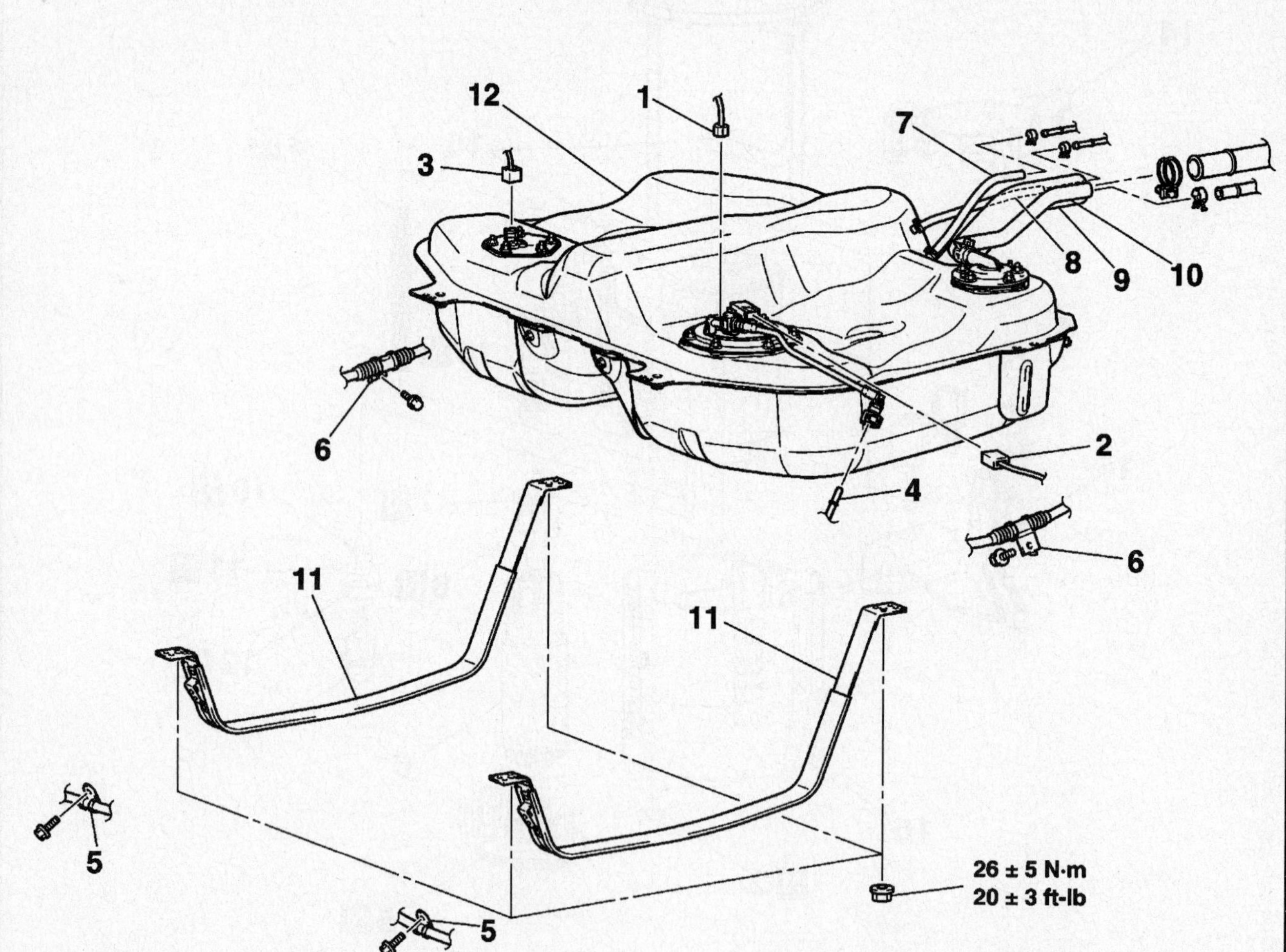

Fig. 265 Fuel tank band and related components—Eclipse and Galant

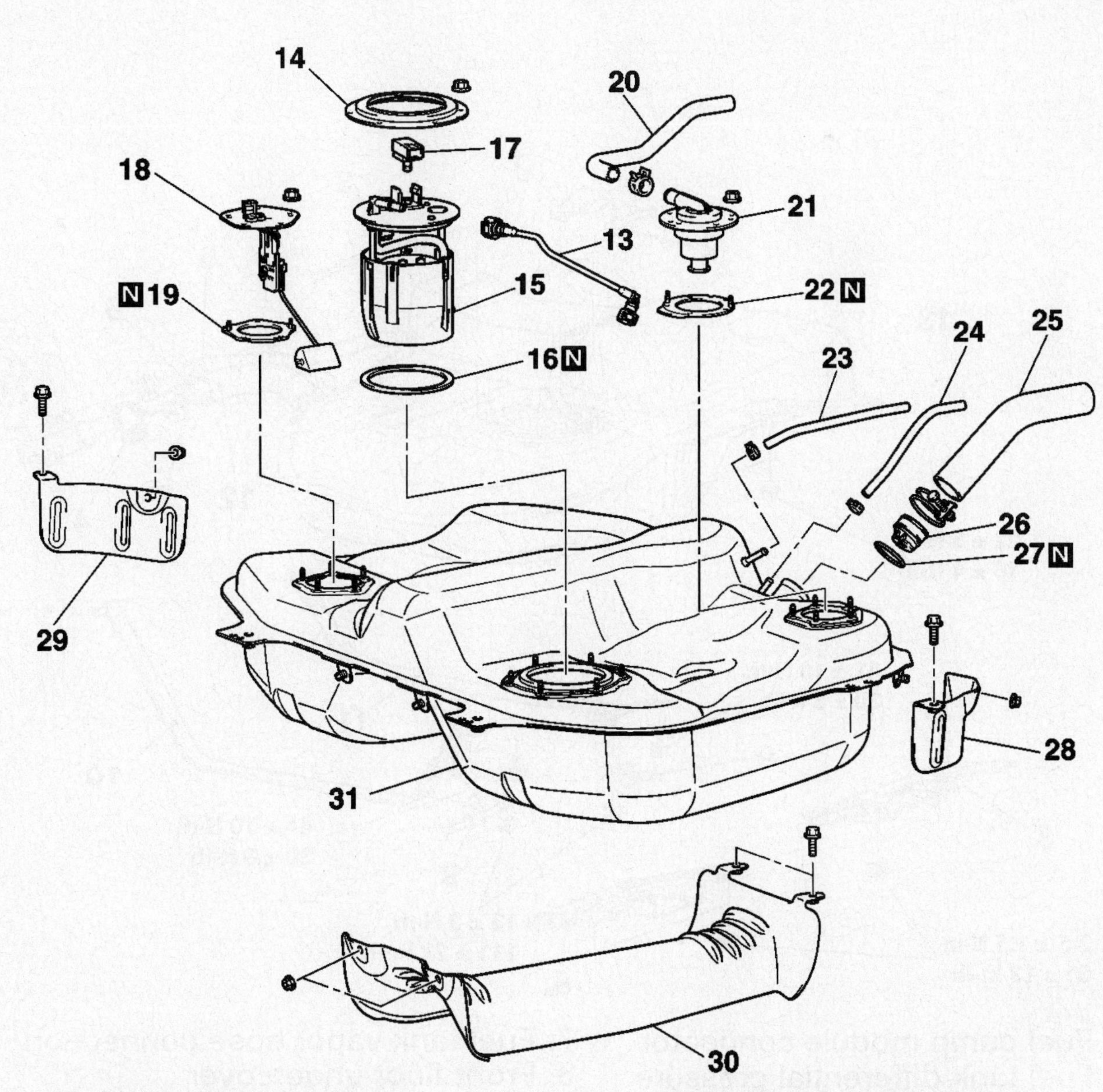

13. Fuel highpressure hose
14. Plate
15. Fuel pump module
16. Packing
17. Fuel tank differential pressure sensor
18. Fuel level sensor (sub)
19. Packing
20. Fuel tank vapor hose A
21. Fuel tank leveling valve assembly
22. Packing
23. Fuel tank vapor hose C
24. Fuel tank vapor hose B
25. Fuel filler hose
26. Fuel tank shut-off valve
27. O-ring
28. Fuel tank protector (A)
29. Fuel tank protector (B)
30. Fuel tank center protector
31. Fuel tank

22140_MITS_G0253

Fig. 266 Fuel tank, fuel pump module, and related components—Eclipse and Galant

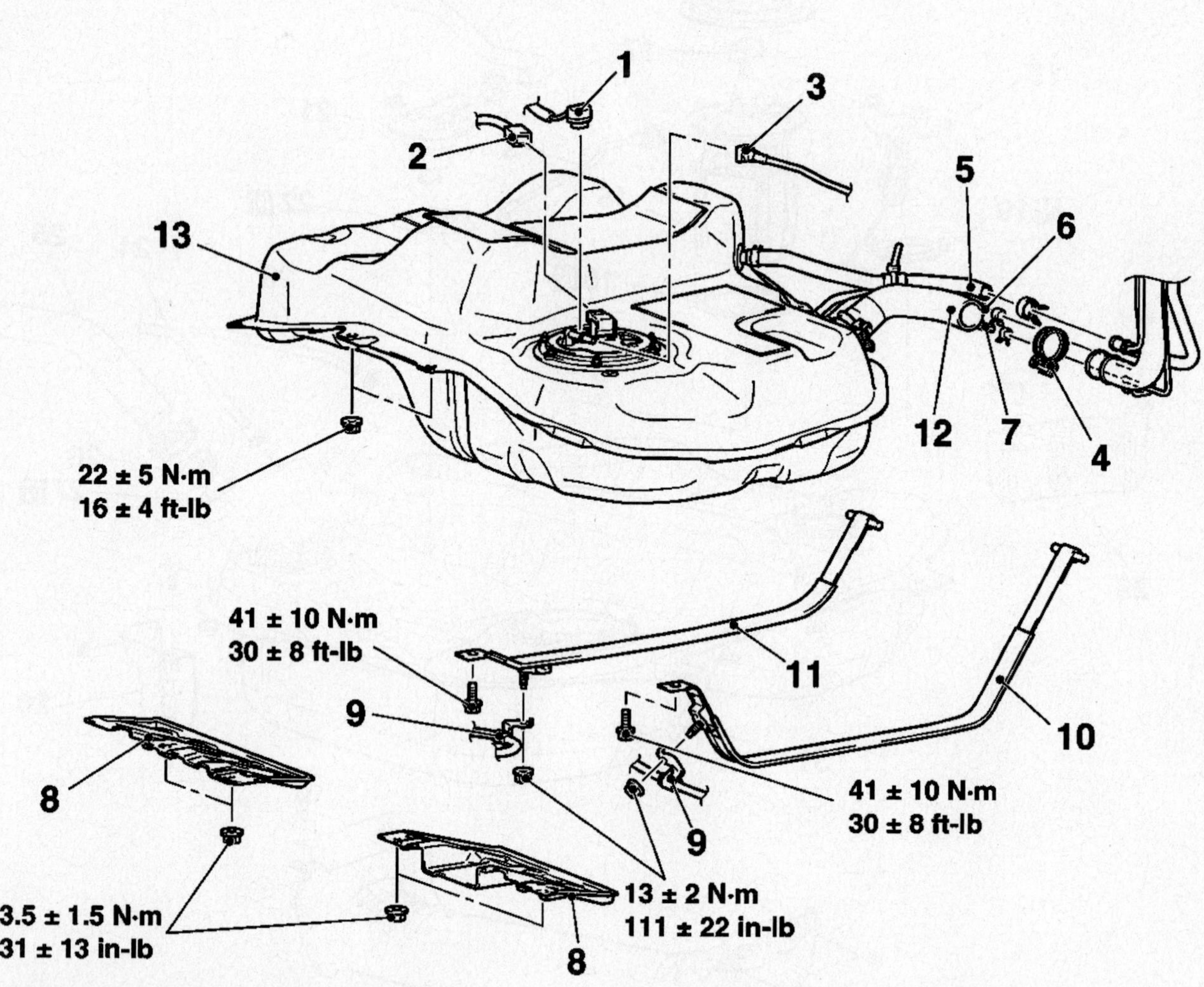

1. Fuel pump module connector
2. Fuel tank differential pressure sensor connector connection
3. Fuel main pipe connection
4. Hose clamp
5. Fuel leveling hose connection
6. Fuel tank vapor hose connection
7. Fuel tank vapor hose connection
8. Front floor undercover
9. Parking brake rear cable clamp connection
10. Fuel tank band LH
11. Fuel tank band RH
12. Fuel filler hose connection
13. Fuel tank assembly

22140_MITS_G0255

Fig. 267 Fuel tank bands and related components—Lancer and Lancer Evolution

CAUTION

The fuel injection system remains under pressure even after the engine has been turned OFF. Properly relieve fuel pressure before disconnecting any fuel lines. Failure to do so may result in fire or personal injury. Do not allow fuel spray or fuel vapors to come in contact with a spark or an open flame. Keep a dry chemical fire extinguisher nearby. Never store fuel in an open container due to risk of fire or explosion.

3. Drain the fuel from the tank.
4. Remove the negative battery cable.
5. Remove the rear seat cushion assembly.
6. Remove the fuel pump module hole cover.
7. Disconnect the fuel pump module connector and fuel tank differential pressure sensor connector connection.
8. Raise and safely support the vehicle.
9. Remove the center exhaust pipe.
10. Remove the propeller shaft on AWD vehicles.
11. Remove the fuel tank protector, if equipped.
12. Disconnect the parking brake cable clamp.
13. Disconnect the fuel tank vapor hose connection.
14. Disconnect the fuel filler hose connection.
15. Disconnect the fuel high-pressure hose connection.

CAUTION

As there will be some pressure remaining in the fuel pipe line, cover it with a shop towel to prevent fuel from spraying out.

16. Remove the fuel tank band.
    a. Support the fuel tank with a transaxle jack.
    b. Remove the front securing nut of the fuel tank band. Tilt the fuel tank assembly forward and lower it gradually to remove it. Remove the fuel tank band.
17. Carefully lower the fuel tank assembly from the vehicle.

To install:

18. Install the fuel tank bands.
    a. Raise the fuel tank assembly carefully with a transaxle jack.
    b. Ensure that the fuel tank assembly does not interfere with surrounding parts. Install the fuel tank band and tighten to 15–23 ft. lbs. (21–31 Nm).
19. Connect the fuel high-pressure hose connection.
    a. After installing, slightly pull the fuel high-pressure hose and ensure a good connection.
20. Connect the fuel filler hose connection.
21. Connect the fuel tank vapor hose connection.
22. Connect the parking brake cable clamp.
23. Install the fuel tank protector, if equipped.
24. Install the propeller shaft on AWD vehicles
25. Install the center exhaust pipe.
26. Connect the fuel level sensor connector.
27. Connect the fuel tank differential pressure sensor connector.
28. Connect the fuel pump module connector.
29. Install the service hole cover.
30. Install the rear seat cushion assembly.
31. Fill the fuel tank.
32. Start the engine and check for leaks.

FUEL RAIL & INJECTORS

REMOVAL & INSTALLATION

Eclipse & Galant

2.4L Engine

1. Before servicing the vehicle, refer to the Precautions Section.
2. Relieve the fuel system pressure.
3. Disconnect the negative battery cable.
4. Remove the air cleaner cover and air intake hose assembly.
5. Disconnect the PCV valve hose connection, ignition coil connectors, EGR valve connector and fuel injector connectors.
6. Disconnect the throttle position sensor connector, manifold absolute pressure sensor connector, EVAP emission purge solenoid connector and knock sensor connector.
7. Disconnect the power steering pressure switch connector.
8. Remove the rocker cover bracket bolts.
9. Disconnect the high pressure hose connection. Remove the fuel rail, insulators, grommets, and injectors. Discard the O-rings.

To install:

10. Installation is the reverse of the removal procedure. Be sure to use new O-rings and gaskets as required.
11. Connect the scan tool to the data link connector. To prevent damage to the scan tool be sure that the ignition switch is in the LOCK position before connecting the scan tool.
12. Turn the ignition switch to the ON position.
13. Select "check mode" from the menu screen.
14. Select "erase memory" from the menu screen.
15. Initialize the learning value.
16. Start the engine and check for leaks, correct as required.

3.8L Engine

See Figure 268.

1. Before servicing the vehicle, refer to the Precautions Section.
2. Relieve the fuel system pressure.
3. Disconnect the negative battery cable.
4. Remove the intake manifold plenum assembly.
5. Disconnect the fuel injector connectors.
6. Remove the control wiring harness bracket mounting bolts.
7. Remove the engine mounting stay.
8. Disconnect the high pressure fuel hose connection. Remove the fuel rail and fuel injector assembly.
9. Remove the insulators, O-rings, and fuel injectors. Discard the gaskets and O-rings.

To install:

10. Installation is the reverse of the removal procedure. Be sure to use new O-rings and gaskets as required.
11. Connect the scan tool to the data link connector. To prevent damage to the scan tool be sure that the ignition switch is in the LOCK position before connecting the scan tool.
12. Turn the ignition switch to the ON position.
13. Select "check mode" from the menu screen.
14. Select "erase memory" from the menu screen.
15. Initialize the learning value.
16. Start the engine and check for leaks, correct as required.

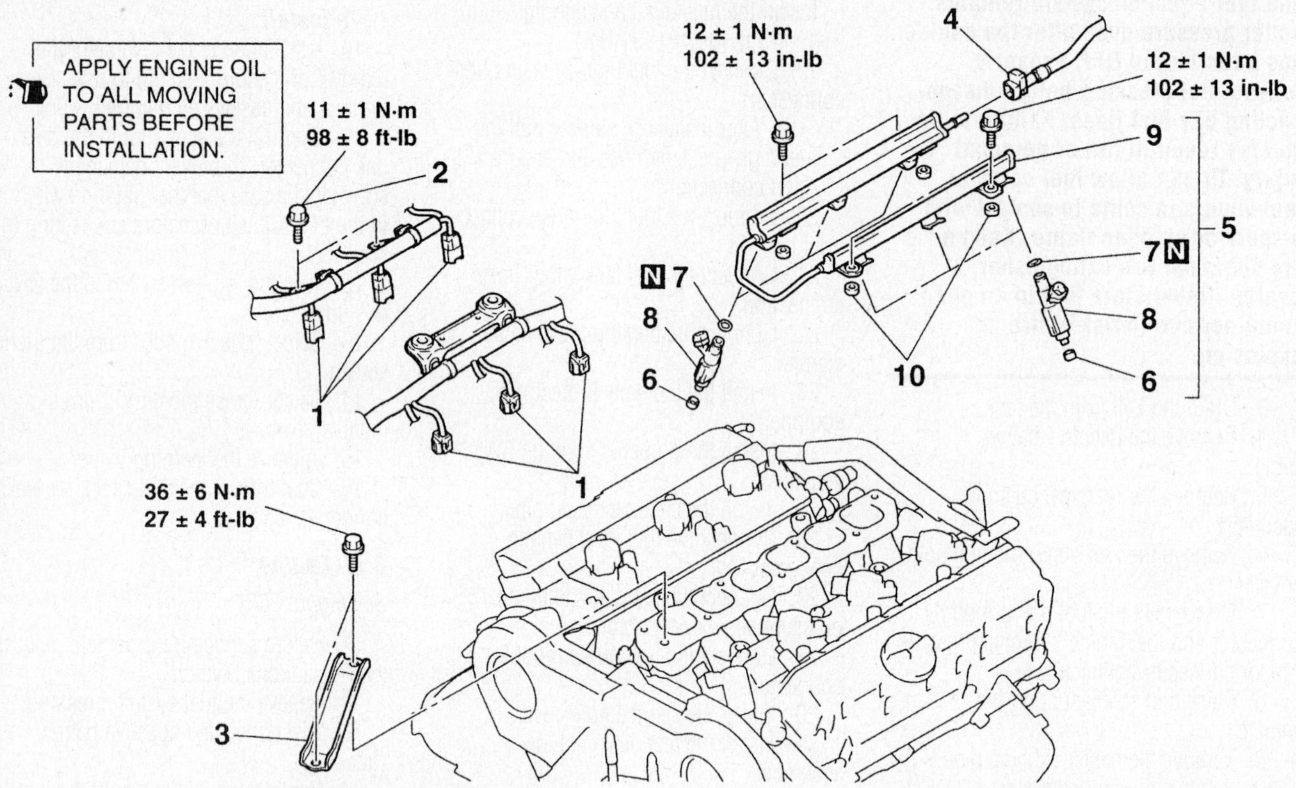

APPLY ENGINE OIL TO ALL MOVING PARTS BEFORE INSTALLATION.

1. FUEL INJECTOR CONNECTORS
2. CONTROL WIRING HARNESS BRACKET MOUNTING BOLTS
3. ENGINE MOUNTING STAY
4. FUEL HIGH-PRESSURE HOSE CONNECTION (FUEL RAIL SIDE)
5. FUEL RAIL AND FUEL INJECTOR ASSEMBLY
6. INSULATORS
7. O-RINGS
8. FUEL INJECTORS
9. FUEL RAIL
10. INSULATORS

09482_GALA_G0126

Fig. 268 Fuel injectors and related components—3.8L engine

Lancer

2006–07 2.0L Engine

See Figure 269.

1. Before servicing the vehicle, refer to the Precautions Section.
2. Relieve the fuel system pressure.
3. Disconnect the negative battery cable.
4. Wrap the connection with a shop towel and disconnect the high pressure fuel line at the fuel rail.

CAUTION

Observe all applicable safety precautions when working around fuel. Whenever servicing the fuel system, always work in a well ventilated area. Do not allow fuel spray or vapors to come in contact with a spark or open flame. Keep a dry chemical fire extinguisher near the work area. Always keep fuel in a container specifically designed for fuel storage; also, always properly seal fuel containers to avoid the possibility of fire or explosion.

5. Remove or disconnect the following:
   - Positive Crankcase Ventilation (PCV) hose
   - Exhaust Gas Recirculation (EGR) solenoid valve connector
   - Manifold Differential Pressure (MDP) sensor connector
   - Purge control solenoid valve connector
   - Throttle Position (TP) sensor connector
   - Idle Air Control (IAC) motor connector
   - Electrical connector from each injector, label for reference
   - High pressure fuel hose
   - Fuel return hose
   - Vacuum hose(s)
   - Fuel pressure regulator
   - Fuel hose
   - Fuel return pipe
   - Bolt(s) holding the fuel rail to the manifold. Carefully lift the rail up and remove it with the injectors attached. Take great care not to drop an injector. Place the rail and injectors in a safe location on the workbench. Protect the tips of the injectors from dirt and/or impact
   - Injector insulators from the intake manifold, discard. The insulators are not reusable

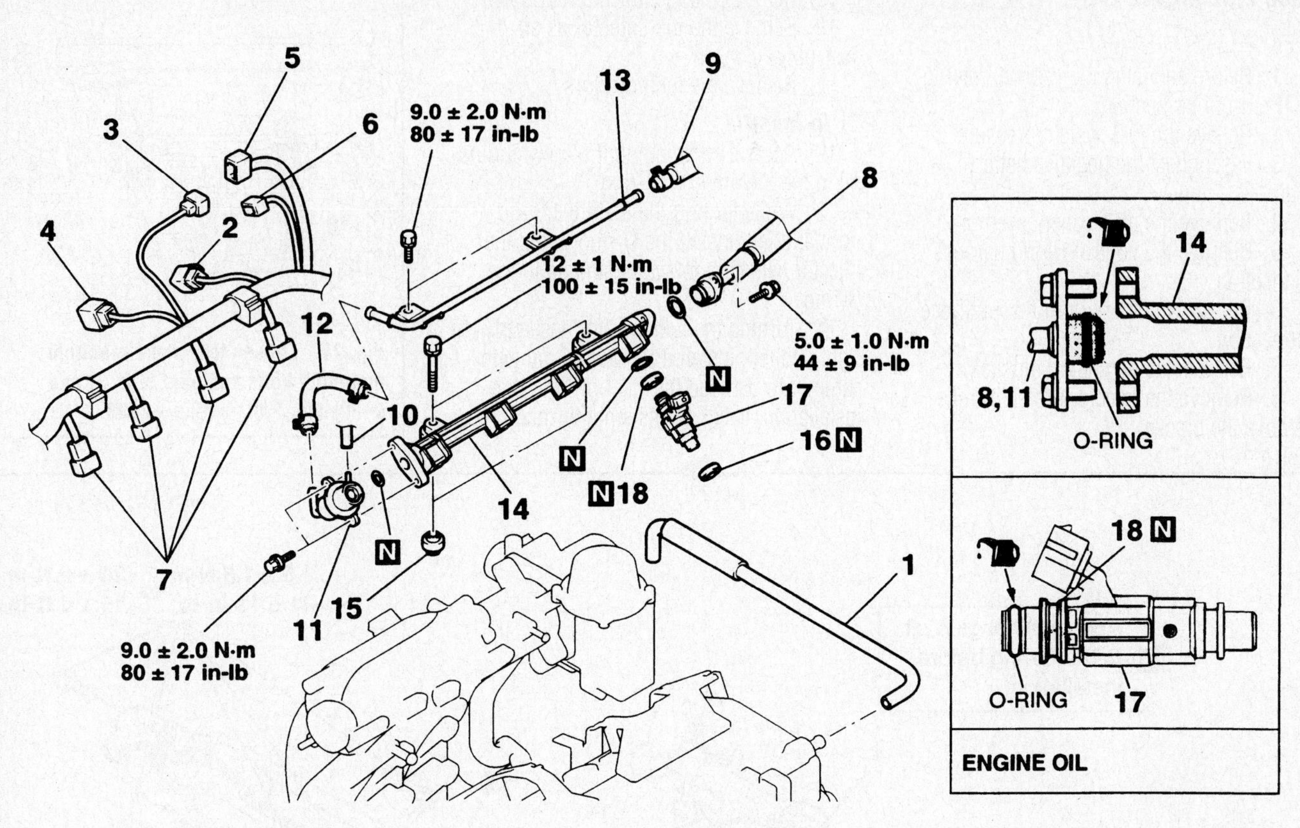

1. PCV HOSE CONNECTION
2. EGR SOLENOID VALVE CONNECTOR
3. MANIFOLD DIFFERENTIAL PRESSURESENSOR CONNECTOR
4. PURGE CONTROL SOLENOID VALVE CONNECTOR
5. THROTTLE POSITION SENSOR CONNECTOR
6. IDLE AIR CONTROL MOTOR CONNECTOR
7. INJECTOR CONNECTOR

9357QG25

Fig. 269 Fuel injectors and related components—Lancer 2.0L engine—2006–07

- Injectors from the fuel rail by pulling gently in a straight outward motion. Make certain the grommet and O-ring come off with the injector

To install:

6. Install a new insulator in each injector port in the manifold.
7. Remove the old grommet and O-ring from each injector. Install a new grommet and O-ring; coat the O-ring lightly with clean, thin oil.
8. If the fuel pressure regulator was removed, replace the O-ring with a new one and coat it lightly with clean, thin oil. Insert the regulator straight into the rail, then check that it can be rotated freely. If it does not rotate smoothly, remove it and inspect the O-ring for deformation or jamming. When properly installed, align the mounting holes and tighten the retaining bolts to 84 inch lbs. (9 Nm). This procedure must be followed even if the fuel rail was not removed.
9. Install or connect the following:
   - Injector into the fuel rail, constantly turning the injector left and right during installation. When fully installed, the injector should still turn freely in the rail. If it does not, remove the injector and inspect the O-ring for deformation or damage
   - Fuel rail and injectors to the engine. Make certain that each injector fits correctly into its port and that the rubber insulators for the fuel rail mounts are in position
   - Fuel return pipe and fuel hose
   - Fuel return hose to the fuel rail
   - High pressure fuel line to the fuel rail
   - Fuel injector connectors
   - IAC motor connector
   - TP sensor connector
   - Purge control solenoid connector
   - MDP sensor connector
   - EGR solenoid valve connector
   - PCV hose connector
   - Negative battery cable
10. Pressurize the fuel system and inspect all connections for leaks.

2008 2.0L Engine

See Figures 270 and 271.

1. Before servicing the vehicle, refer to the Precautions Section.
2. Relieve the fuel system pressure.
3. Disconnect the negative battery cable.
4. Remove the air cleaner assembly.
5. Remove the control wiring harness connection.
6. Remove the fuel high-pressure hose connection.
7. Remove the engine oil dipstick.
8. Remove the injector protector rear. Remove the bracket.
9. Remove the fuel injector support.
10. Remove the fuel injector as an assembly.
11. Remove the fuel injectors.

To install:

12. Apply a small amount of new engine oil to the fuel injector O-ring.
13. While turning the fuel injector to right and left, install the O-ring to the fuel injector with care to avoid damage to the O-ring.
14. Turning the fuel injector assembly to right and left, install it to the fuel rail with care not to damage the O-ring. After the installation, check for its smooth rotation.

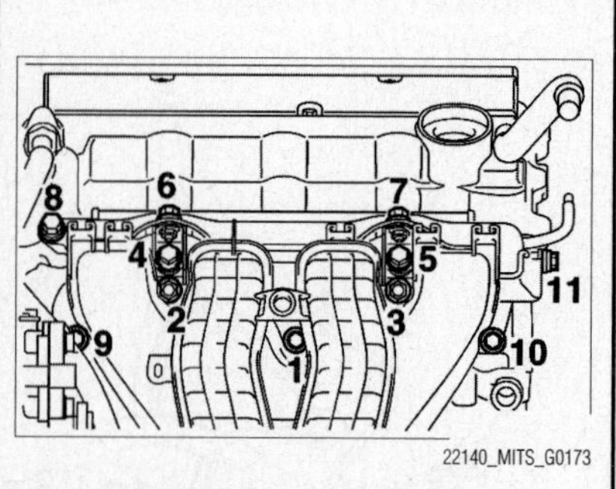

Fig. 271 Loosen the intake manifold mounting bolts and nuts 1, 2, 3, and 9—Lancer 2.0L engine—2008

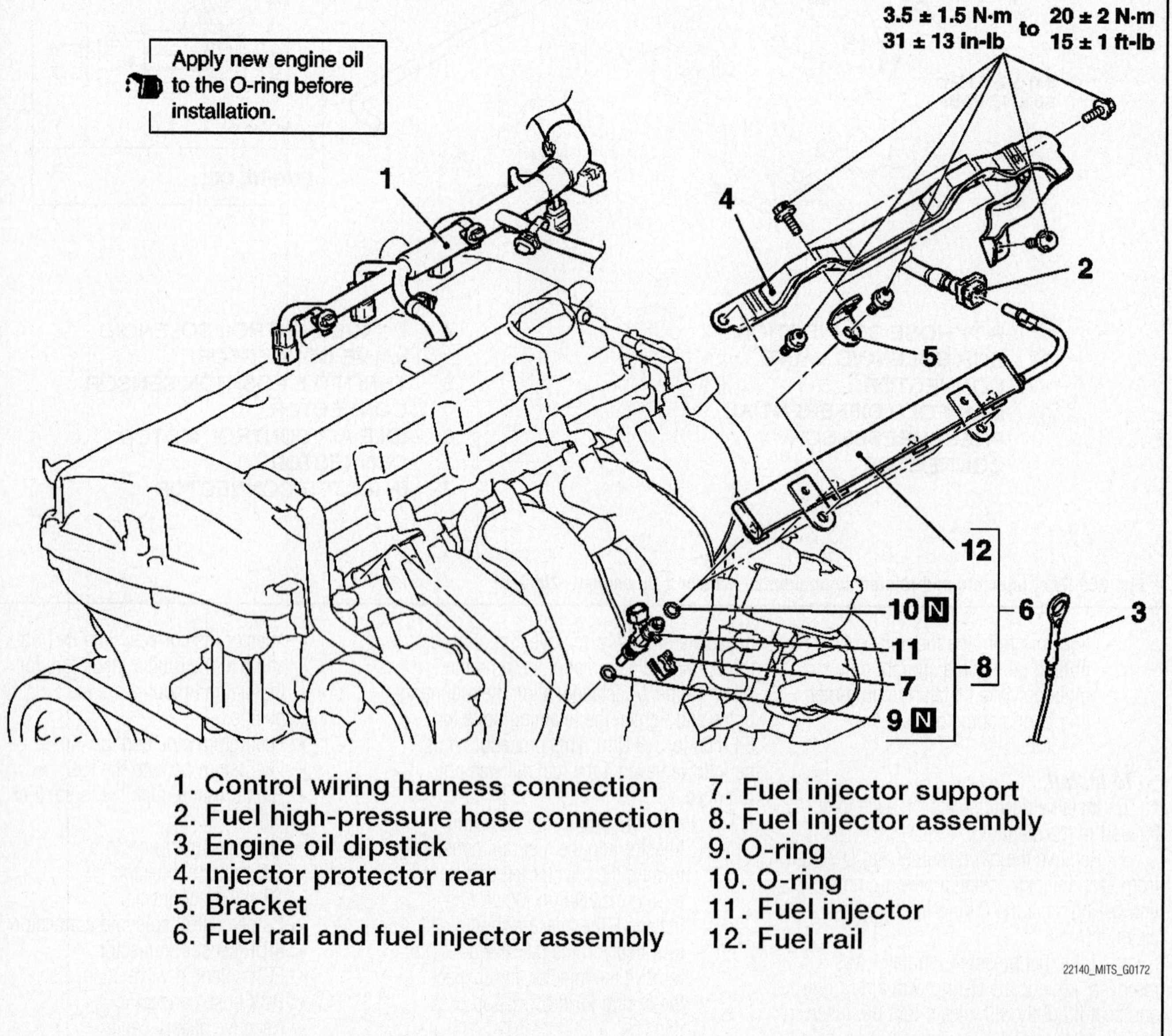

3.5 ± 1.5 N·m to 20 ± 2 N·m
31 ± 13 in-lb to 15 ± 1 ft-lb

Apply new engine oil to the O-ring before installation.

1. Control wiring harness connection
2. Fuel high-pressure hose connection
3. Engine oil dipstick
4. Injector protector rear
5. Bracket
6. Fuel rail and fuel injector assembly
7. Fuel injector support
8. Fuel injector assembly
9. O-ring
10. O-ring
11. Fuel injector
12. Fuel rail

22140_MITS_G0172

Fig. 270 Fuel injectors and related components—Lancer 2.0L engine—2008

At this time, check that the projection of the fuel injector assembly is in the center.

15. If the rotation is not smooth, the O-ring may be caught. Remove the fuel injector assembly and check the O-ring for damage. Re-insert it into the fuel rail and check for its smooth rotation.

16. Apply a small amount of new engine oil to the O-ring at the end of fuel injector assembly.

**** WARNING**

When installing the fuel rail and fuel injector assembly to the intake manifold, pay attention to avoid damage to the O-ring at the end of the fuel injector assembly.

17. Install the fuel rail and fuel injector assembly to the intake manifold.

18. Install the bracket and injector protector rear.

19. Loosen the intake manifold mounting bolts and nuts. (Bolts and nuts 1, 2, 3, and 9 shown in the figure).

20. Remove the EGR valve (vehicles for California).

21. Loosen the intake manifold stay B and intake manifold coupling bolts (except for California).

22. Loosen the EGR valve support and intake manifold coupling bolts (vehicles for California).

23. Loosen the intake manifold stay mounting bolts.

24. Temporarily tighten the mounting bolts of the fuel rail, and the mounting bolts and nuts of the bracket, injector protector rear, and intake manifold to the specified torque of 18–44 inch lbs. (2–5 Nm) in the order of number shown in the figure.

25. Temporarily tighten the mounting bolts of the fuel rail, and then mounting bolts and nuts of the bracket, injector protector rear, and intake manifold to the specified torque of 14–16 ft. lbs. (18–22 Nm) in the order of number shown in the figure.

26. Tighten the intake manifold stay to the specified torque. Tightening torque: 14–16 ft. lbs. (18–22 Nm).

27. Tighten the intake manifold stay B and intake manifold coupling bolts to the specified torque (except for California). Tightening torque: 14–16 ft. lbs. (18–22 Nm).

28. Tighten the EGR valve support and intake manifold coupling bolts to the specified torque (vehicles for California). Tightening torque: 14–16 ft. lbs. (18–22 Nm).

29. Install the EGR valve (vehicles for California).

30. Installation continues in the reverse of the removal procedure.

Lancer Evolution

2008 2.0L Turbo Engine

See Figure 272.

1. Before servicing the vehicle, refer to the Precautions Section.
2. Relieve the fuel system pressure.
3. Disconnect the negative battery cable.
4. Remove the engine upper cover.
5. Remove the charge air cooler intake hose A, B and charge air cooler intake pipe A.
6. Remove the air cleaner assembly.
7. Remove the control wiring harness connection.
8. Remove the fuel return hose connection.
9. Remove the PCV hose.
10. Remove the emission control equipment hose connection.
11. Remove the fuel high-pressure hose connection.
12. Remove the engine upper cover bracket rear.
13. Remove the fuel injector hose.
14. Remove the fuel injector return pipe.
15. Remove the MFI fuel rail pressure regulator.
16. Remove the O-ring.
17. Remove the fuel rail and fuel injector assembly.
18. Remove the fuel rail insulator.
19. Remove the fuel injector support.
20. Remove the fuel injector assembly.
21. Remove the fuel injector.

To install:

22. Apply a small amount of new engine oil to the fuel injector O-ring.
23. While turning the fuel injector to right and left, install the O-ring to the fuel injector with care to avoid damage to the O-ring.
24. Turning the fuel injector assembly to right and left, install it to the fuel rail with care not to damage the O-ring. After the installation, check for its smooth rotation. At this time, check that the projection of the fuel injector assembly is in the center.
25. If the rotation is not smooth, the O-ring may be caught. Remove the fuel injector assembly and check the O-ring for damage. Re-insert it into the fuel rail and check for its smooth rotation.
26. Apply a small amount of new engine oil to the O-ring at the end of fuel injector assembly.

**** WARNING**

When installing the fuel rail and fuel injector assembly to the intake manifold, pay attention to avoid damage to the O-ring at the end of the fuel injector assembly.

27. Install the fuel rail and fuel injector assembly to the intake manifold.
28. Installation continues in the reverse of the removal procedure.

IDLE SPEED

ADJUSTMENT

Idle speed is maintained by the Powertrain Control Module (PCM). No adjustment is necessary or possible.

THROTTLE BODY

REMOVAL & INSTALLATION

See Figures 273 through 276.

➡When the throttle body assembly replacement is performed, use scan tool MB991958 to initialize the learning value. Do not loosen the fixing screws for the resin cover of throttle body assembly. If the screws are loosened, the sensor incorporated in the resin cover becomes misaligned and the throttle body may not work normally.

1. Before servicing the vehicle, refer to the Precautions Section.
2. Relieve the fuel system pressure.
3. Drain the engine cooling system.
4. Remove the engine air cleaner assembly.
5. Matchmark the location of the adjuster bolt on the accelerator cable mounting flange. This will assure that the cable is installed in its original location. Remove the throttle cable adjusting bolt and disconnect the cable from the lever on the throttle body. Position cable aside.
6. Remove the connection for the breather hose and the air intake hose from the throttle body and position aside.
7. Tag and disconnect the necessary vacuum hoses.
8. Label and detach the electrical connectors at the throttle body, as necessary.
9. Disconnect the water and water by-pass hoses at the base of the throttle body.

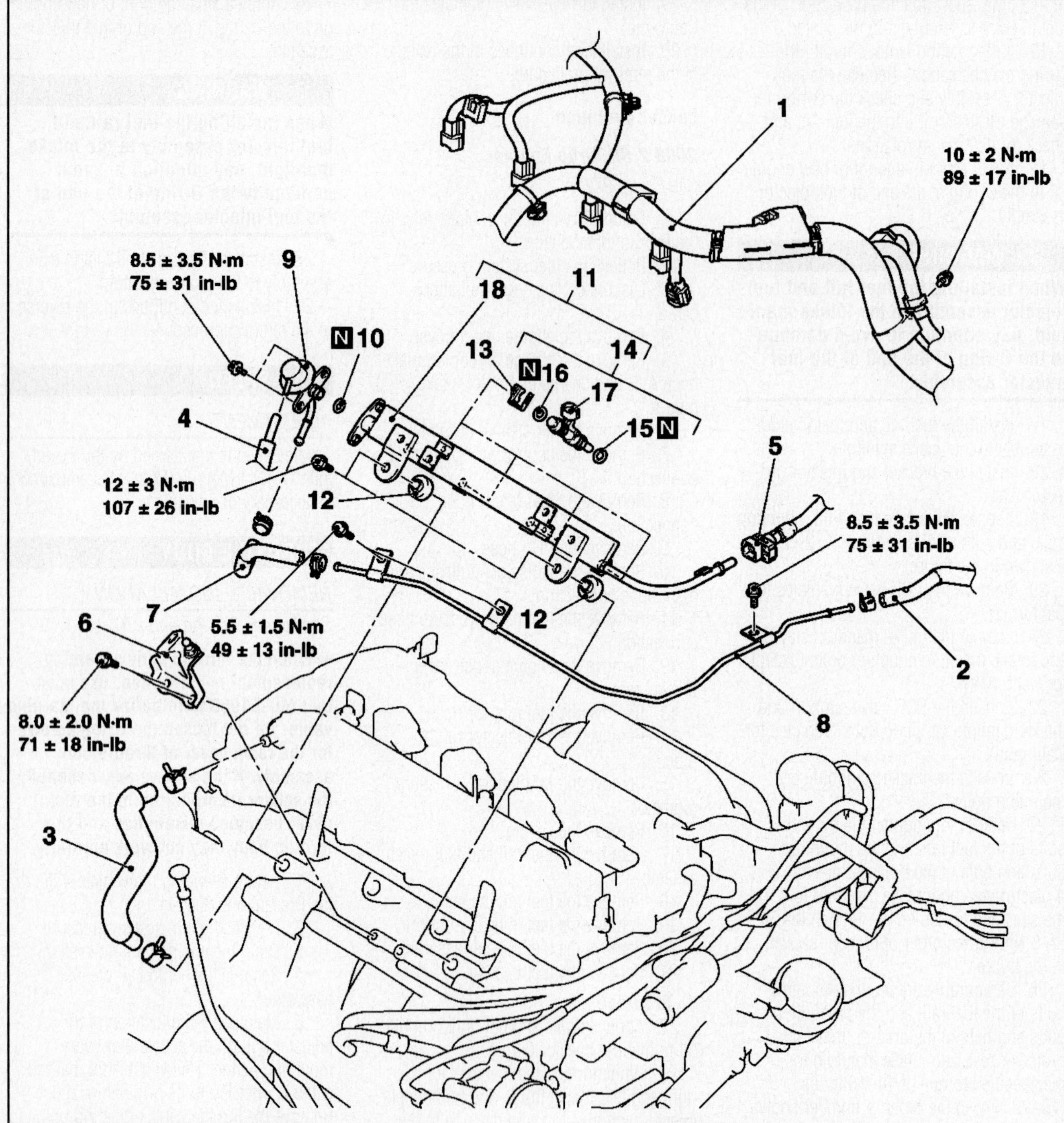

1. Control wiring harness connection
2. Fuel return hose connection
3. PCV hose
4. Emission control equipment hose connection
5. Fuel high-pressure hose connection
6. Engine upper cover bracket rear
7. Fuel injector hose
8. Fuel injector return pipe
9. MFI fuel rail pressure regulator
10. O-ring
11. Fuel rail and fuel injector assembly
12. Fuel rail insulator
13. Fuel injector support
14. Fuel injector assembly
15. O-ring
16. O-ring
17. Fuel injector
18. Fuel rail

22140_MITS_G0174

Fig. 272 Fuel injectors and related components—Lancer Evolution—2008

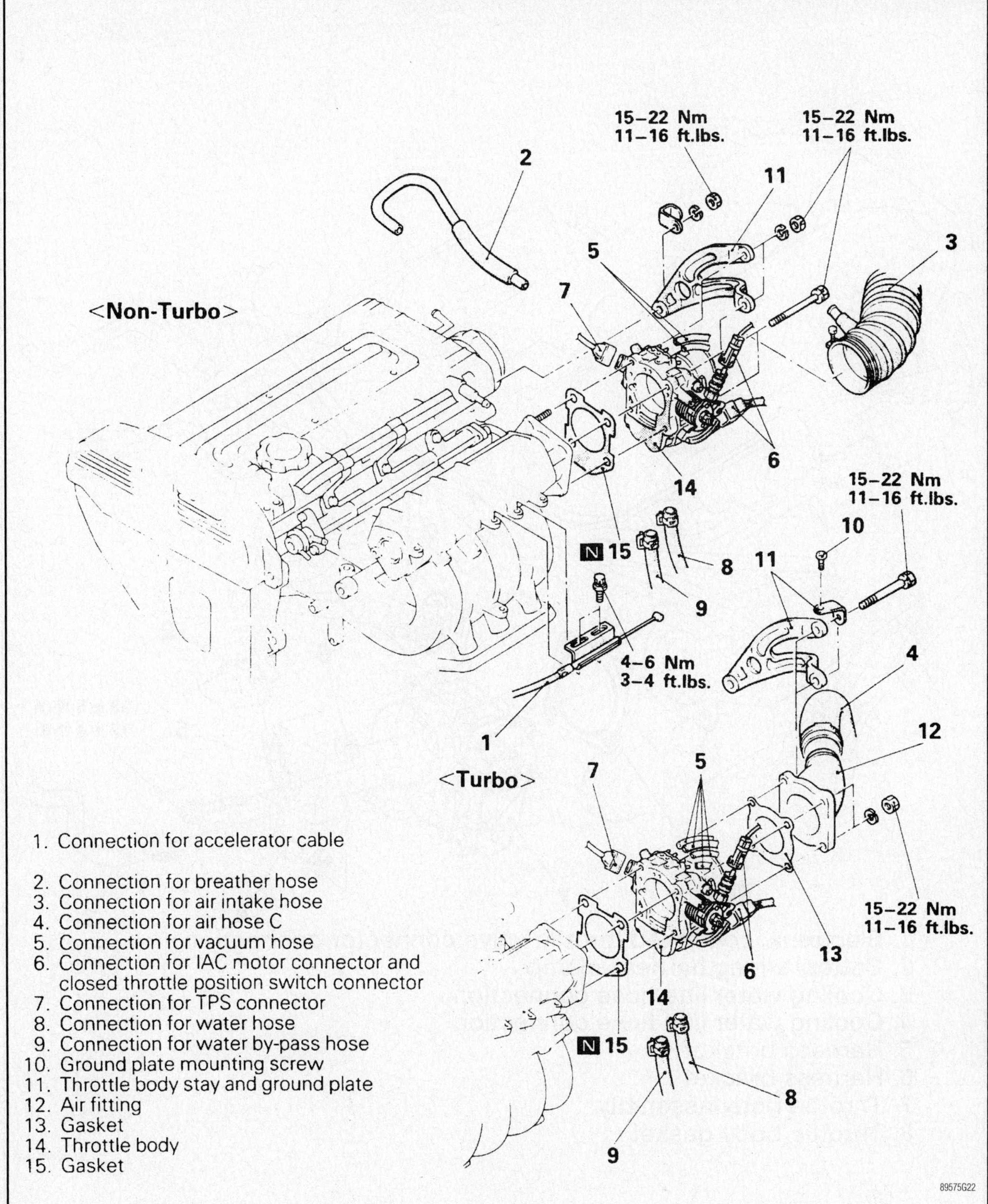

Fig. 273 Exploded view of the throttle body and related components—Lancer 2.0L engine

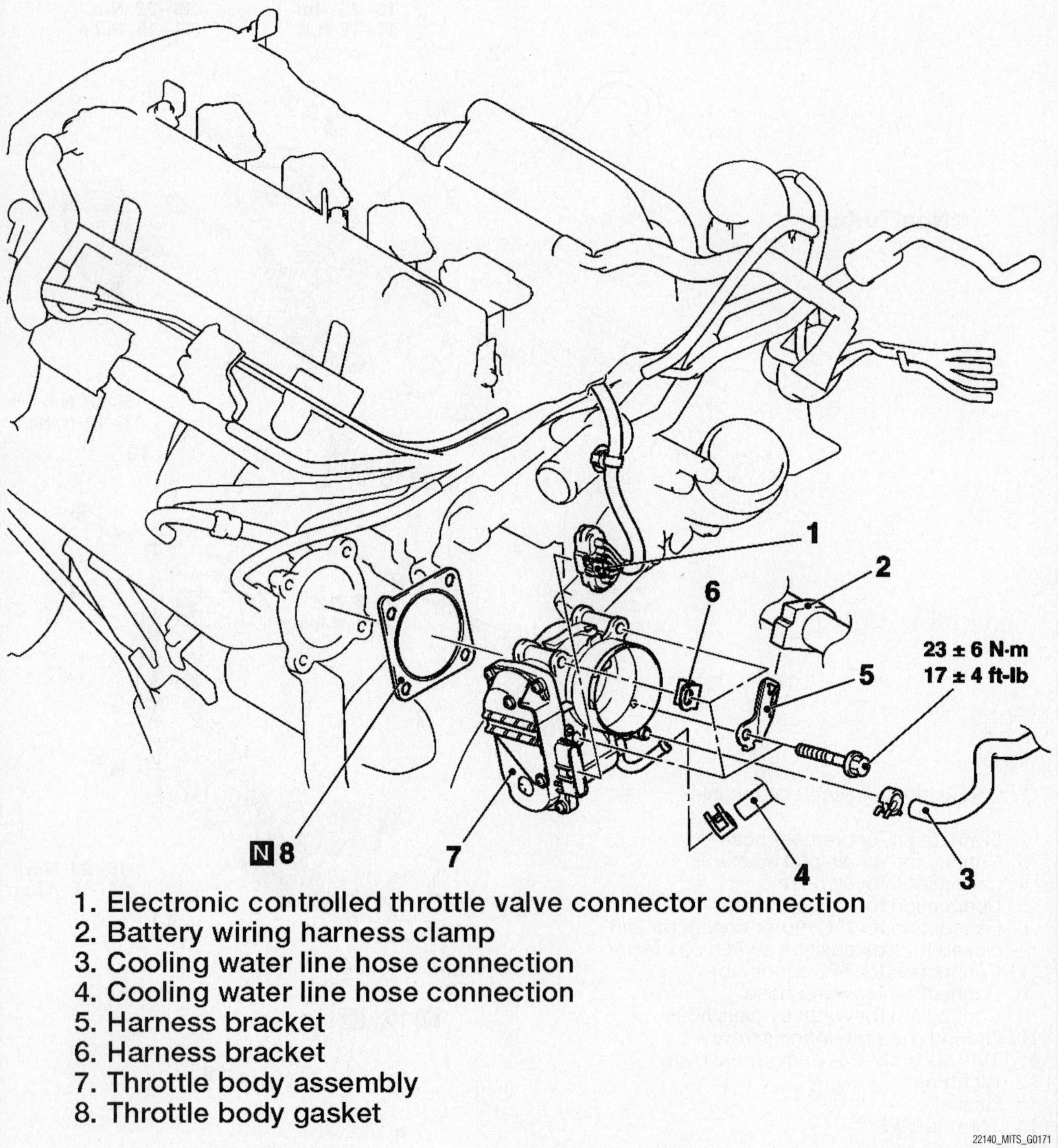

Fig. 274 Exploded view of the throttle body and related components—Lancer Evolution 2.0L engine

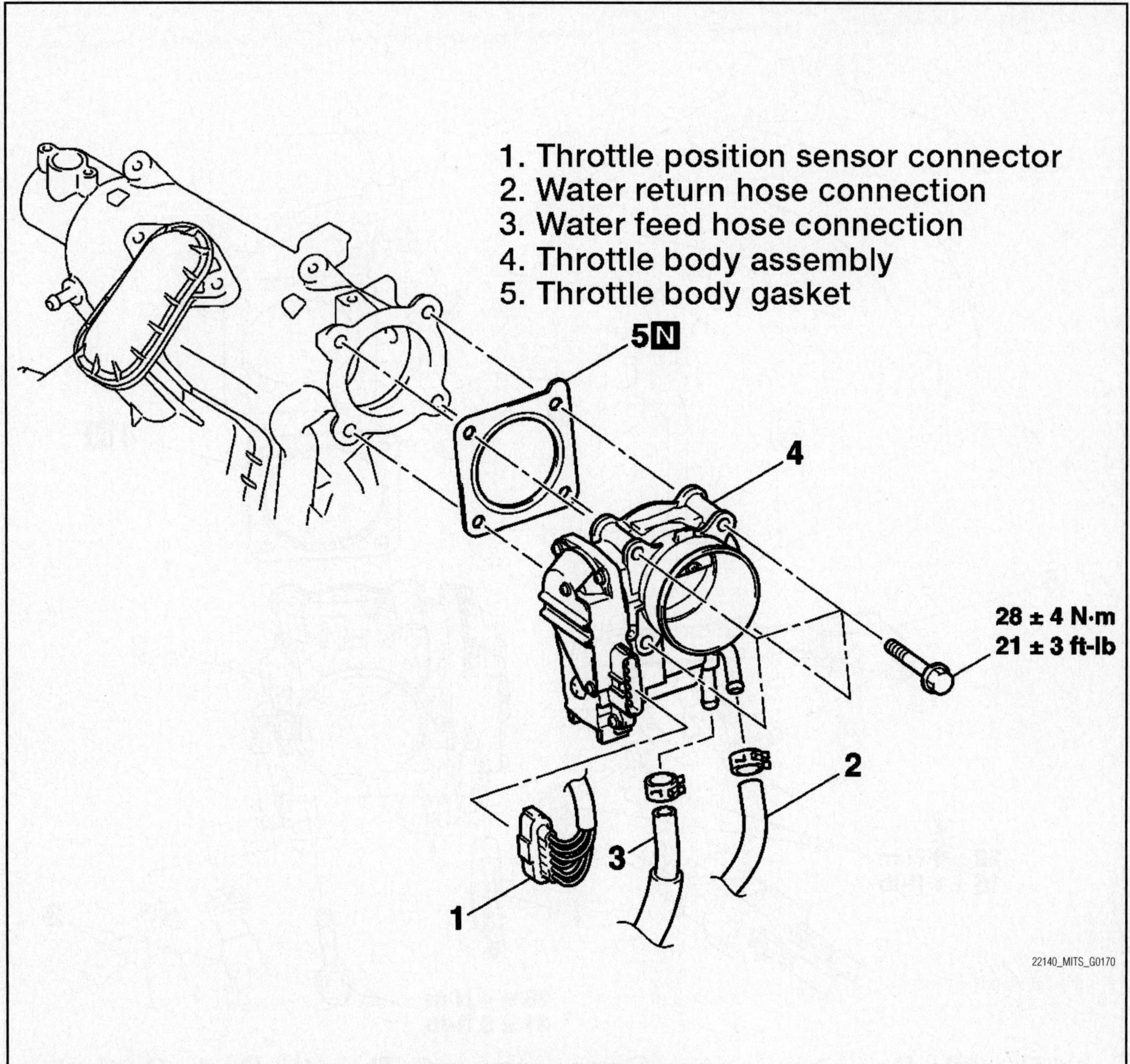

Fig. 275 Exploded view of the throttle body and related components—2.4L engine

10. If equipped, unfasten the ground plate mounting screws, then remove the throttle body stay and ground plate from the engine.

11. Remove the air fitting and gasket.

12. Unfasten the throttle body mounting bolts, then remove the throttle body from the engine. Remove and discard the gasket.

To install:

13. Clean all old gasket material from the both throttle body mounting surfaces. Install new gasket onto the intake manifold plenum mounting surface.

➡Poor idling quality and poor performance may be experienced if the gasket is installed incorrectly.

14. Install the throttle body to the intake manifold plenum and tighten the mounting bolts.

15. Install the air fitting, if equipped, making sure new gasket is in place.

16. If equipped, install the throttle body stay and ground plate. Secure with retainers tightened to 11–16 ft. lbs. (15–22 Nm).

17. Install the ground plate mounting screw.

18. Connect the water hoses to the throttle body. Install new hose clamps if required.

19. Attach the electrical and vacuum connectors to the throttle body, as tagged during removal.

20. Connect the accelerator cable to the throttle body and install the adjusting bolt in original position. Check adjustment of cable.

21. Install the air intake and breather hoses.

22. If removed, install the battery and connect the positive cable.

23. Connect the negative battery cable.

24. Refill the cooling system.

25. Run the engine and check for leaks and proper operation.

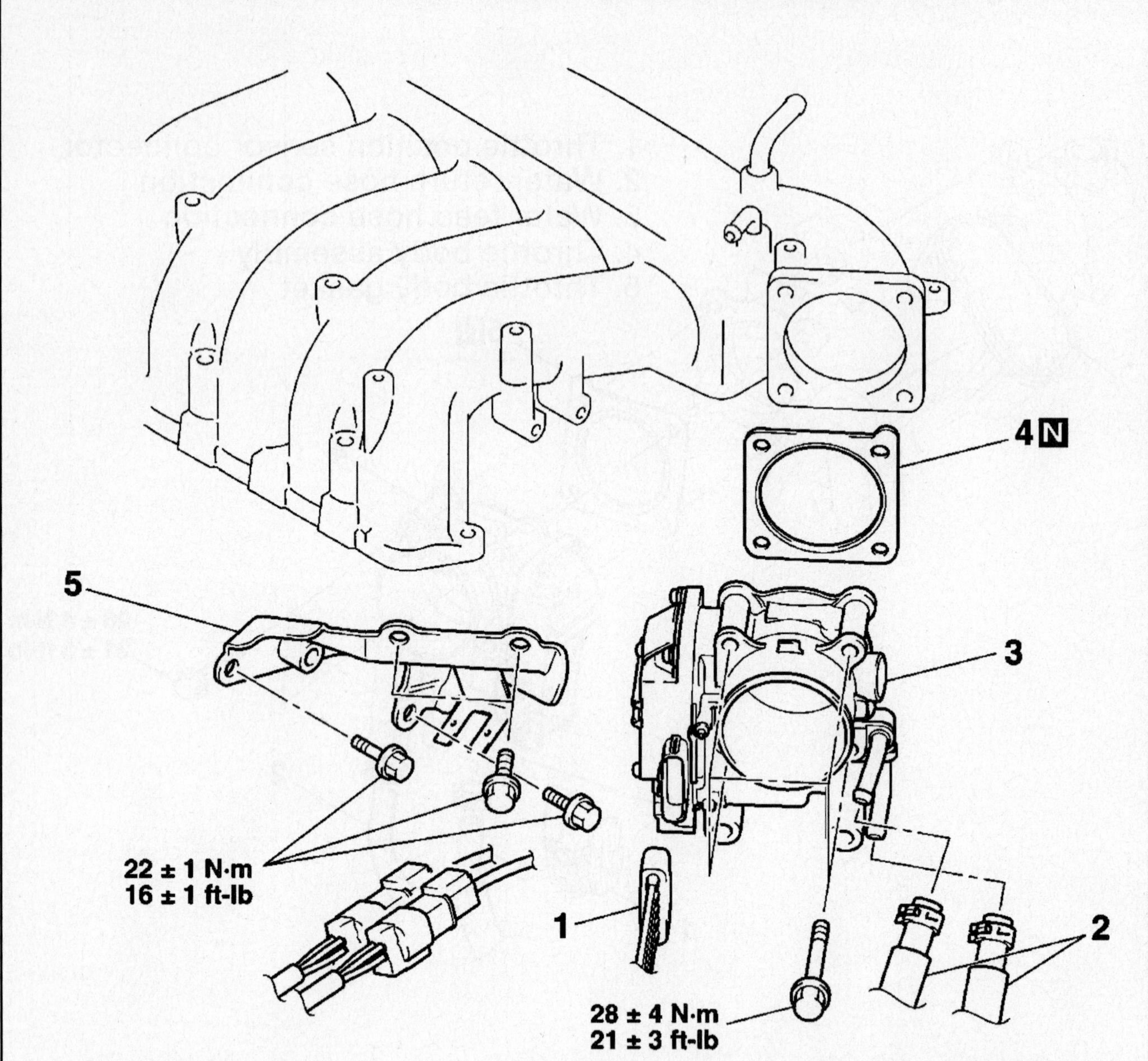

Fig. 276 Exploded view of the throttle body and related components—3.8L engine

HEATING & AIR CONDITIONING SYSTEM

BLOWER MOTOR

REMOVAL & INSTALLATION

Eclipse

See Figure 277.

1. Before servicing the vehicle, refer to the Precautions Section.
2. Disconnect the negative battery cable.
3. Disable the SIR system. Refer to Air Bag (Supplemental Restraint System), disarming the system.
4. Remove the stopper, then lower the glove compartment assembly.

**** CAUTION**

Some models may be equipped with a Supplemental Restraint System (SRS), which uses an air bag. Whenever working near any of the SRS components, such as the impact sensors, the air bag module, steering column, and instrument panel, disable the SRS.

5. If necessary, remove the blower resistor.
6. For vehicles equipped with A/C, unfasten the retainers, then remove the automatic compressor ECM from the vehicle.
7. Remove the retaining bolts, then remove the blower motor and fan from the vehicle.
8. If necessary, to remove the entire blower unit, perform the following:
   a. Remove the instrument panel.
   b. Remove the retaining clip by using a Philips-head screwdriver to push the pin (in the center of the clip) inward about 0.08 inch (2mm), then pull the clip out to remove it.

**** WARNING**

Do not push the pin in more than necessary as the grommet may be damaged, or the pin may fall in.

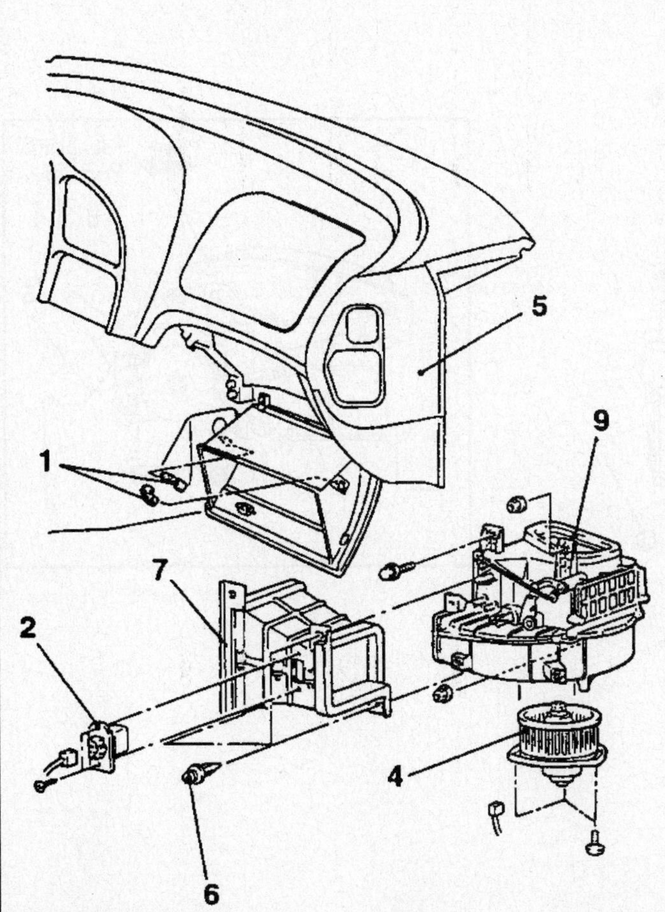

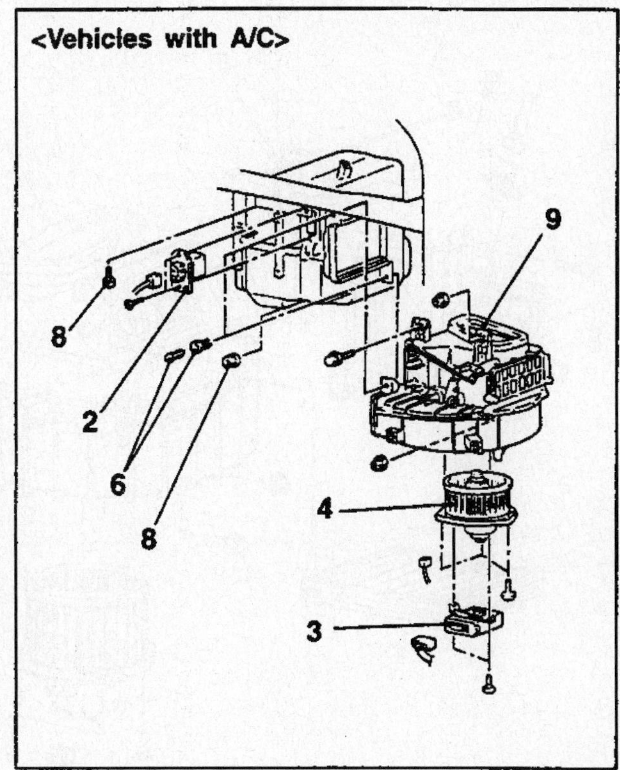

Resistor removal steps
1. Stopper
2. Resistor

Blower fan and motor removal steps
3. Automatic compressor-ECM <Vehicles with A/C for non-turbo>
4. Blower fan and motor

Blower unit removal steps
5. Instrument panel
6. Clip
7. Joint duct <Vehicles without A/C>
8. Cooling unit installation bolts and nuts <Vehicles with A/C>
9. Blower unit assembly

89576G06

Fig. 277 Exploded view of the blower motor unit and related components—Eclipse

9. For vehicles without A/C, remove the joint duct.

10. If equipped with A/C, unfasten the cooling unit bolts and nuts.

11. Remove the blower unit assembly from the vehicle.

To install:

12. If removed, position the blower unit in the vehicle and perform the following:

a. If equipped with A/C, install the cooling unit bolts and nuts.

b. For vehicles not equipped with A/C, install the joint duct.

c. Install the retaining clip by inserting with the pin pulled out, then pushing the pin inward until the head is flush with the grommet.

d. Install the instrument panel.

13. Position the blower fan and motor in the vehicle, and secure with the retainer(s).

14. For vehicles with A/C, install the automatic compressor ECM.

15. If removed, install the resistor.

16. Place the glove compartment in the proper closed position, then install the stopper.

17. Connect the negative battery cable.

Galant

See Figure 278.

1. Before servicing the vehicle, refer to the Precautions Section.

2. Disable the SIR system. Refer to Air Bag (Supplemental Restraint System), disarming the system.

3. Disconnect the negative battery cable.

4. Remove the three instrument panel undercover mounting screws and remove the cover.

5. If equipped with A/C, unplug and remove the compressor module.

6. Detach the electrical connector from the fan motor.

7. Remove the three small bolts holding the motor to the housing and remove the motor and fan.

To install:

8. Check the inside of the case carefully; any debris can snag the fan and cause noise or poor airflow.

9. Install the blower motor, in the blower case and secure with the three mounting bolts.

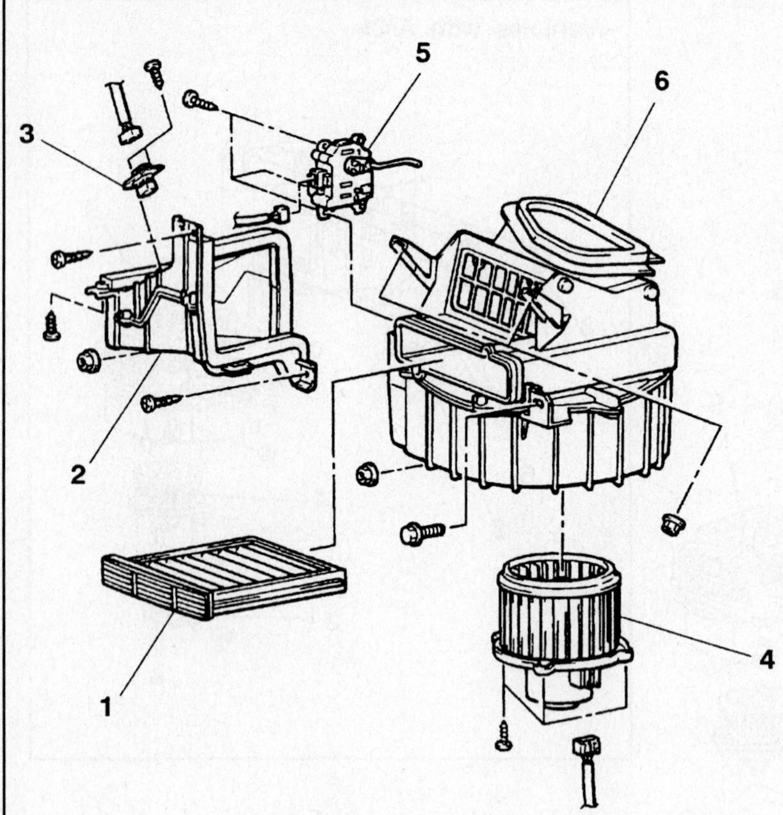

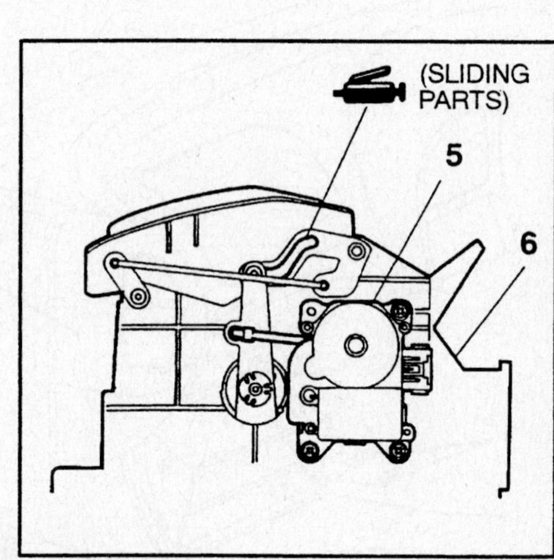

1. AIR PURIFIER ASSEMBLY
2. JOINT DUCT
3. RESISTOR
4. BLOWER FAN AND MOTOR
5. INSIDE/OUTSIDE AIR CHANGEOVER DAMPER MOTOR
6. BLOWER ASSEMBLY

93112GF3

Fig. 278 Blower motor assembly and related components—Galant

10. Attach the blower motor electrical connector.

11. Install the compressor module, if removed.

12. Install the undercover, taking care to insure it is in place and all the fasteners are secure.

13. Connect the negative battery cable.

Lancer

See Figure 279.

1. Before servicing the vehicle, refer to the Precautions Section.

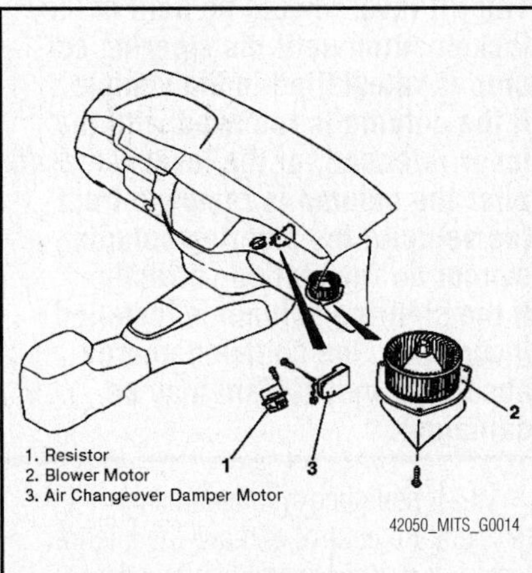

Fig. 279 Exploded view of the blower motor assembly—Lancer

2. Disable the SIR system. Refer to Air Bag (Supplemental Restraint System), disarming the system.

CAUTION

Wait at least 1 minute before working on the vehicle. The air bag system is designed to retain enough voltage to deploy the air bag for a short period of time even after the battery has been disconnected.

3. Remove the glove box.

4. Unbolt and remove the blower motor.

5. Installation is the reverse order of removal.

HEATER CORE

REMOVAL & INSTALLATION

Eclipse

See Figure 280.

1. Before servicing the vehicle, refer to the Precautions Section.

2. Disable the SIR system. Refer to Air Bag (Supplemental Restraint System), disarming the system.

CAUTION

Wait for 1 minute after disconnecting the negative battery cable before working inside the vehicle. The air bag system is set to deploy for a short period of time after the battery is disconnected.

3. If not already done, disconnect the negative battery cable.

4. Drain the cooling system.

5. Discharge the air conditioning system.

6. Be sure the front tires are in the straight ahead position.

7. Remove the center console box assembly, lid assembly (except Eclipse RS), door mirror control switch and harness, accessory socket harness, ashtray, top switch and harness (Spyder), shift lever panel assembly and boot (manual transaxle) and garnish (automatic transaxle)

8. Remove the front and rear floor console retaining screws. Remove the front and rear consoles.

9. Remove the driver's air bag module. Remove the steering wheel nut. Remove the steering wheel.

10. Remove the auto-cruise control switch. Remove the instrument panel undercover switch. Remove the lower column cover. Remove the upper column cover.

11. Remove the clockspring and column switch assembly.

12. If equipped with automatic transaxle remove the cover and key interlock cable.

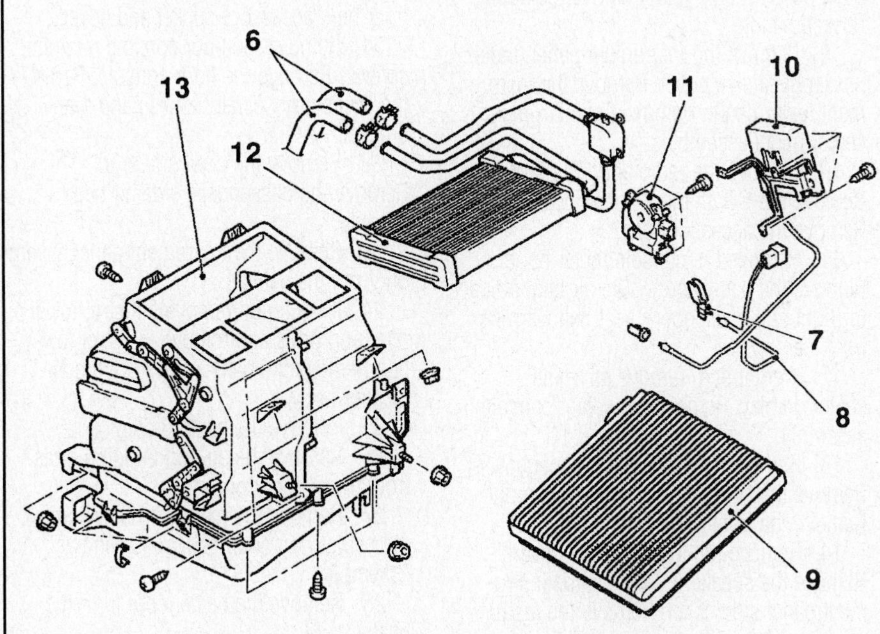

Fig. 280 Heater core and related components—Eclipse

13. Remove the steering shaft assembly. Remove the steering cover assembly.

⁂ WARNING

The tilt lever should be held in the lock position until the steering column is reinstalled in the vehicle. If the column is removed with the lever released, or the lever released after the column is removed from the vehicle, the steering column cannot be reinstalled correctly. If the steering column is installed incorrectly, the collision energy absorbing mechanism may be damaged.

14. Remove the front pillar trim panel. Remove the instrument panel side cover. Remove the fog lamp switch assembly.

15. Remove the hood release handle. Remove the meter bezel assembly. Remove the center panel assembly.

16. Remove the center air outlet. Remove the radio assembly. Remove the heater control assembly.

17. Remove the center hood. Remove the multi center display. Remove the center speaker. Remove the inner and outer glove box and light switch. Remove the glove box striker.

18. Remove the passenger's side air bag module. Remove the instrument panel assembly.

19. Remove the instrument panel reinforcement and both knee absorbers.

20. Remove the auto-cruise control ECU and the immobilizer ECU.

21. Remove the instrument panel center reinforcement. Remove the joint duct.

22. Remove the strut tower bar assembly.

23. Remove the air intake hose. Remove the air conditioning pipe connection, expansion valve, heater hose, and thermistor sensor.

24. Remove the evaporator. Remove the automatic compressor controller.

25. Remove the heater core.

To install:

26. Installation is the reverse of the removal procedure.

⁂ WARNING

The tilt lever should be held in the lock position until the steering column is reinstalled in the vehicle. If the column is removed with the lever released, or the lever released after the column is removed from the vehicle, the steering column cannot be reinstalled correctly. If the steering column is installed incorrectly, the collision energy absorbing mechanism may be damaged.

27. Be sure to fill the cooling system with the proper grade and type coolant.

28. Recharge the air conditioning system.

Galant

See Figure 281.

1. Before servicing the vehicle, refer to the Precautions Section.

2. Disable the SIR system. Refer to Air Bag (Supplemental Restraint System), disarming the system.

⁂ CAUTION

Wait for 1 minute after disconnecting the negative battery cable before working inside the vehicle. The air bag system is set to deploy for a short period of time after the battery is disconnected.

3. If not already done, disconnect the negative battery cable.

4. Drain the cooling system.

5. Discharge the air conditioning system.

6. Be sure the front tires are in the straight ahead position.

7. Remove the front seat. Remove the glovebox assembly.

8. Remove the instrument panel parcel box and glovebox lock. Remove the hood release handle.

9. Remove the instrument panel under passenger's side cover. Remove the instrument lower panel. Remove the instrument center panel assembly.

10. Remove the radio and CD player. Remove the CD changer, if equipped. Remove the accessory box.

11. Remove the console meter hood. Remove the passenger's side air bag indicator light and passenger's seat belt warning light.

12. If equipped, remove the multi center display. Remove the center console assembly.

13. Remove the instrument panel cover. Remove the combination meter assembly. Remove the front pillar trim.

14. Remove the front speaker covers. Remove the speakers. Remove the instrument panel side cover. Remove the instrument panel side air outlet.

15. Remove the driver's side air bag module. Remove the steering wheel. Remove the steering column cover. Remove the clockspring connector and clockspring and column switch assembly.

16. Disconnect the electrical connections for the ignition switch. Remove the key interlock cable.

17. Remove the steering shaft pad. Remove the steering column retaining bolts. Remove the steering column assembly. Pinch the steering column shaft clip with a pliers, and pull up the shaft to disengage the steering column assembly.

⁂ WARNING

The tilt lever should be held in the lock position until the steering column is reinstalled in the vehicle. If the column is removed with the lever released, or the lever released after the column is removed from the vehicle, the steering column cannot be reinstalled correctly. If the steering column is installed incorrectly, the collision energy absorbing mechanism may be damaged.

18. If equipped with automatic temperature control, remove the interior temperature sensor and photo sensor. Remove the instrument panel front end garnish.

19. Remove the gearshift lever handle, heated seat switch, accessory socket, socket cover, front plate, and front box.

20. Remove the lid lock lever, hinge, lid, liner box, accessory socket and cover.

21. Remove the floor console retaining screws. Remove the floor console. Remove the accessory socket harness and rear bracket.

22. Remove the cowl side trim. Remove the passenger's side air bag module.

23. Remove the instrument panel assembly from the vehicle.

24. If equipped remove the strut tower retaining bolts and remove the strut tower bar. Remove the battery. Remove the air cleaner body.

25. Remove the heater hoses.

26. Remove the air conditioning lines from the evaporator.

27. Remove the rear heat duct. Remove the heater unit and deck crossmember assembly.

28. Remove the heater core from the heater case.

To install:

29. Installation is the reverse of the removal procedure.

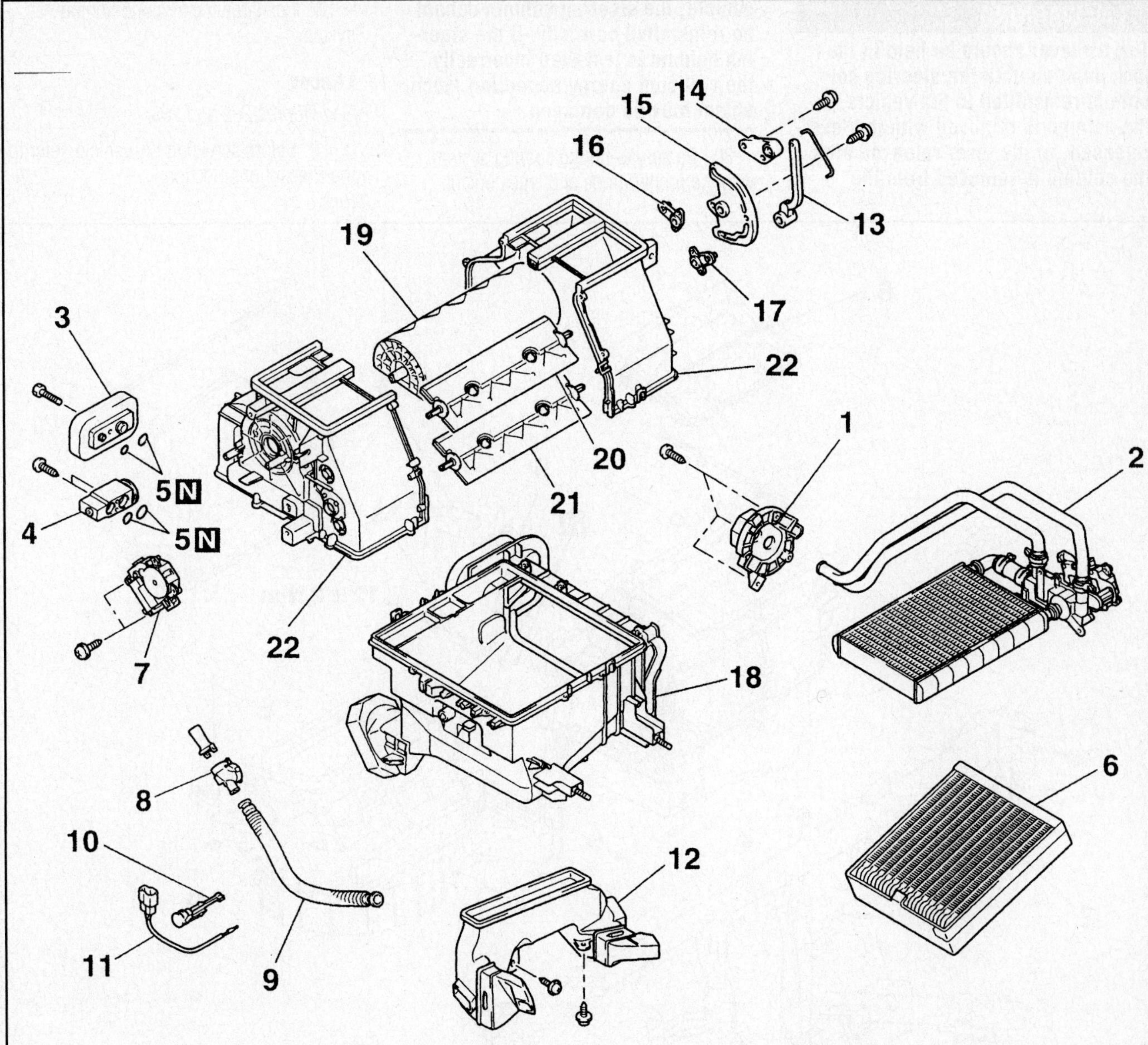

1. PACKING
2. HEATER CORE ASSEMBLY
3. EXPANSION VALVE JOINT
4. EXPANSION VALVE
5. O-RING
6. EVAPORATOR
7. MODE SELECTION DAMPER CONTROL MOTOR AND POTENTIOMETER
8. ASPIRATOR
9. ASPIRATOR HOSE
10. AIR THERMO SENSOR CLIP
11. AIR THERMO SENSOR
12. FOOT DUCT
13. LEVER A
14. LEVER B
15. LEVER C
16. LEVER D
17. LEVER E
18. HEATER CASE LOWER
19. MODE SELECTION DAMPER
20. MAX A/C DAMPER
21. AIR MIXING DAMPER
22. HEATER CASE UPPER

09482_GALA_G0024

Fig. 281 Heater core and related components—Galant

WARNING

The tilt lever should be held in the lock position until the steering column is reinstalled in the vehicle. If the column is removed with the lever released, or the lever released after the column is removed from the vehicle, the steering column cannot be reinstalled correctly. If the steering column is installed incorrectly, the collision energy absorbing mechanism may be damaged.

30. Be sure to fill the cooling system with the proper grade and type coolant.

31. Recharge the air conditioning system.

Lancer

See Figures 282 and 283.

1. Before servicing the vehicle, refer to the Precautions Section.

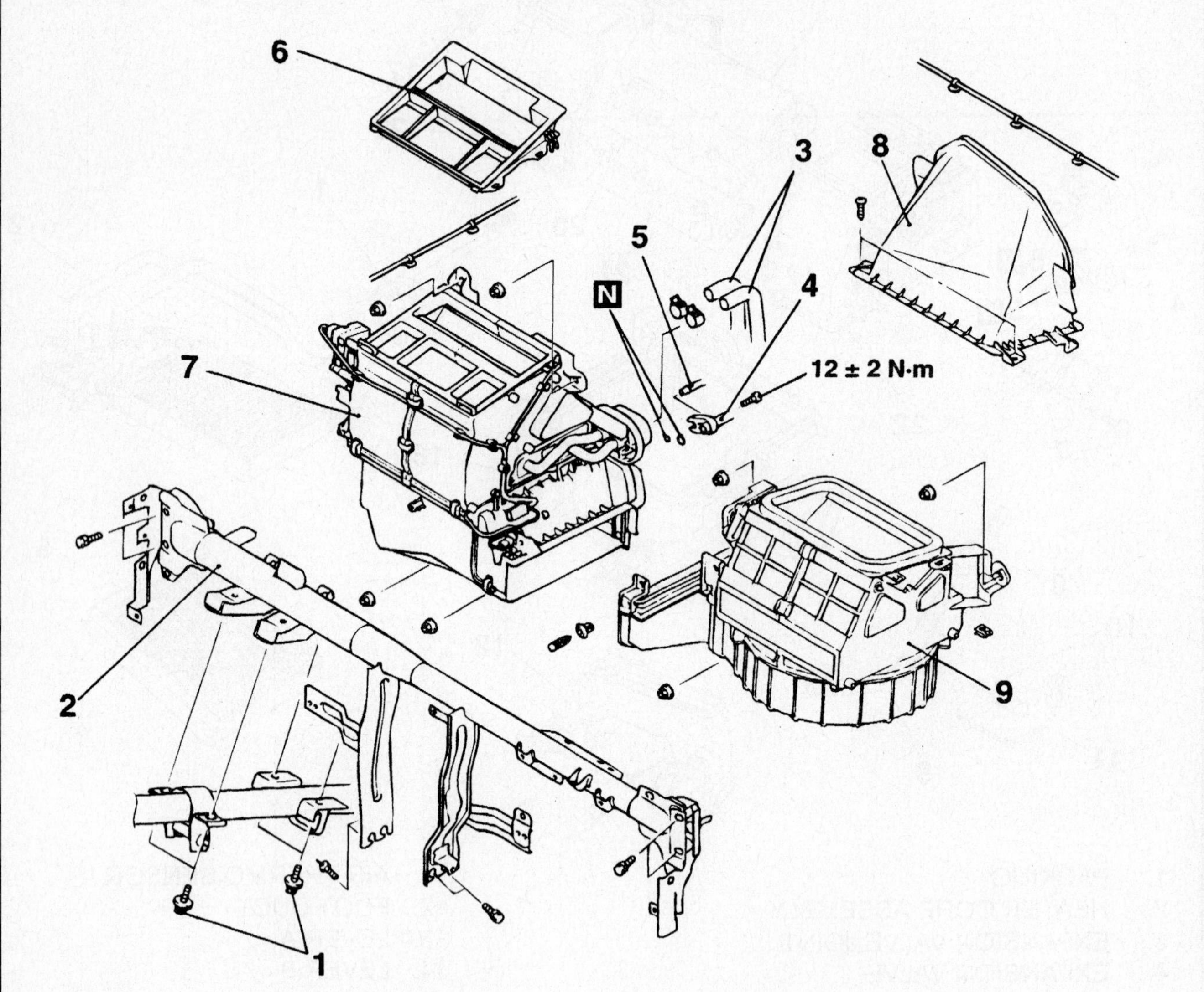

1. STEERING SHAFT ATTACHMENT BOLT
2. FRONT DECK CROSSMEMBER
3. HEATER HOSE CONNECTION
4. FLEXIBLE SUCTION HOSE CONNECTION
5. LIQUID PIPE B CONNECTION
6. CENTER DUCT
7. HEATER UNIT
8. INTAKE DUCT
9. BLOWER ASSEMBLY

42356-GALA-G01

Fig. 282 Exploded view of the heater core and related components—Lancer

2. Disable the SIR system. Refer to Air Bag (Supplemental Restraint System), disarming the system.

CAUTION

Wait for 1 minute after disconnecting the negative battery cable before working inside the vehicle. The air bag system is set to deploy for a short period of time after the battery is disconnected.

3. If not already done, disconnect the negative battery cable.
4. Drain the cooling system.
5. Discharge and recover the air conditioning system refrigerant.
6. Remove or disconnect the following:
   - Instrument panel
   - Front seat assembly
   - Front console assembly
   - Front floor carpet
   - Steering shaft attachment bolt
   - Front deck crossmember
   - Heater hoses

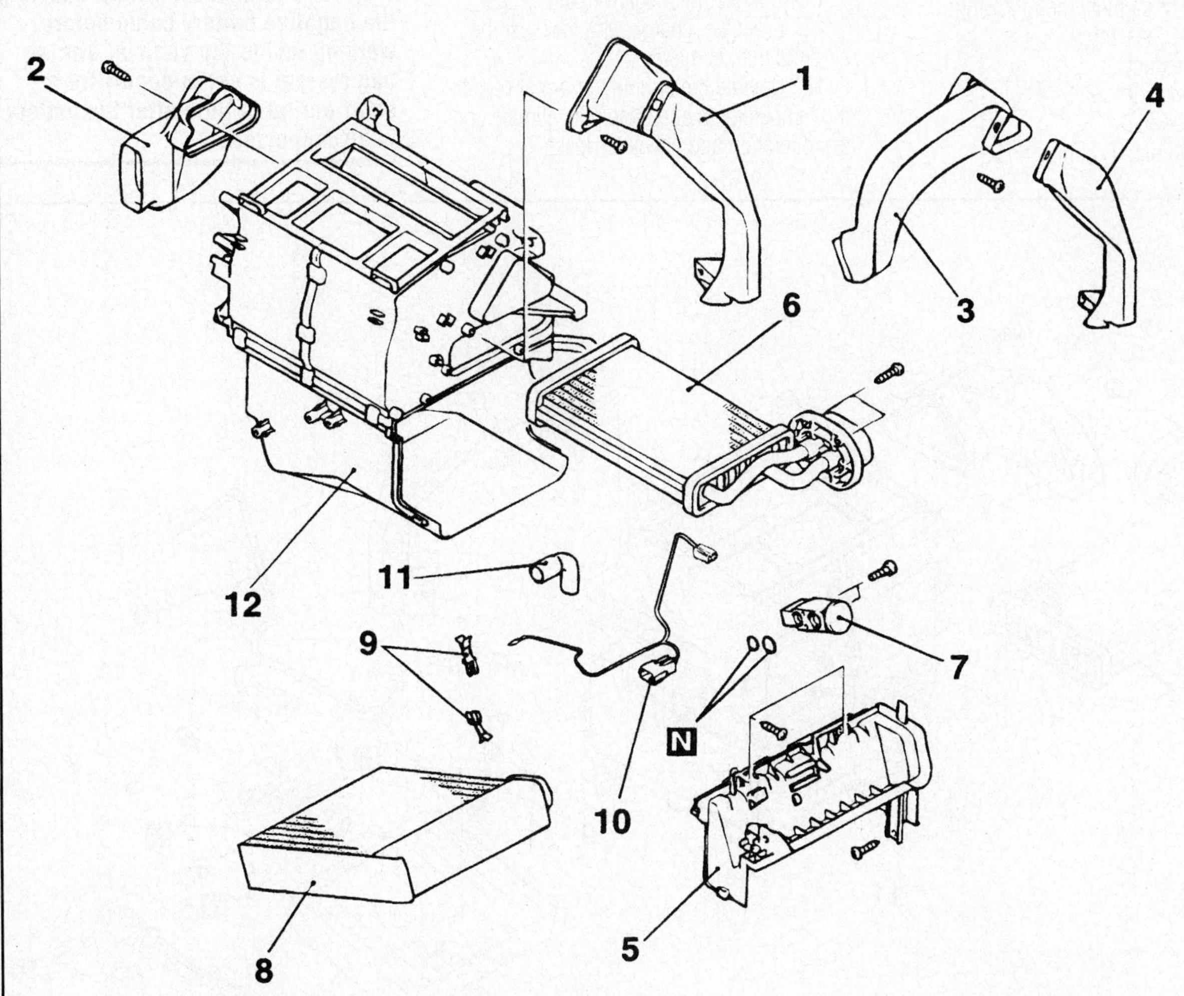

Fig. 283 Disassembly of the heater unit—Lancer

- Flexible suction hose
- Liquid pipe B connection
- Center duct
- Heater unit
- Intake duct
- Blower assembly

7. Disassemble the heater unit and remove the heater core.

To install:

8. Assemble the heater unit.
9. Install or connect the following:
   - Blower assembly
   - Intake duct
   - Heater unit
   - Center duct
   - Liquid pipe B connection
   - Flexible suction hose
   - Heater hoses
   - Front deck crossmember
   - Steering shaft attachment bolt
   - Front floor carpet
   - Front console assembly
   - Front seat assembly
   - Instrument panel
10. Be sure to fill the cooling system with the proper grade and type coolant.
11. Connect the negative battery cable.
12. Evacuate, charge, and leak test the air conditioning system.
13. Operate the engine to normal operating temperatures and check the climate control operation and check for leaks.

Lancer Evolution

See Figures 284 and 285.

1. Before servicing the vehicle, refer to the Precautions Section.
2. Disable the SIR system. Refer to Air Bag (Supplemental Restraint System), disarming the system.

CAUTION

Wait for 1 minute after disconnecting the negative battery cable before working inside the vehicle. The air bag system is set to deploy for a short period of time after the battery is disconnected.

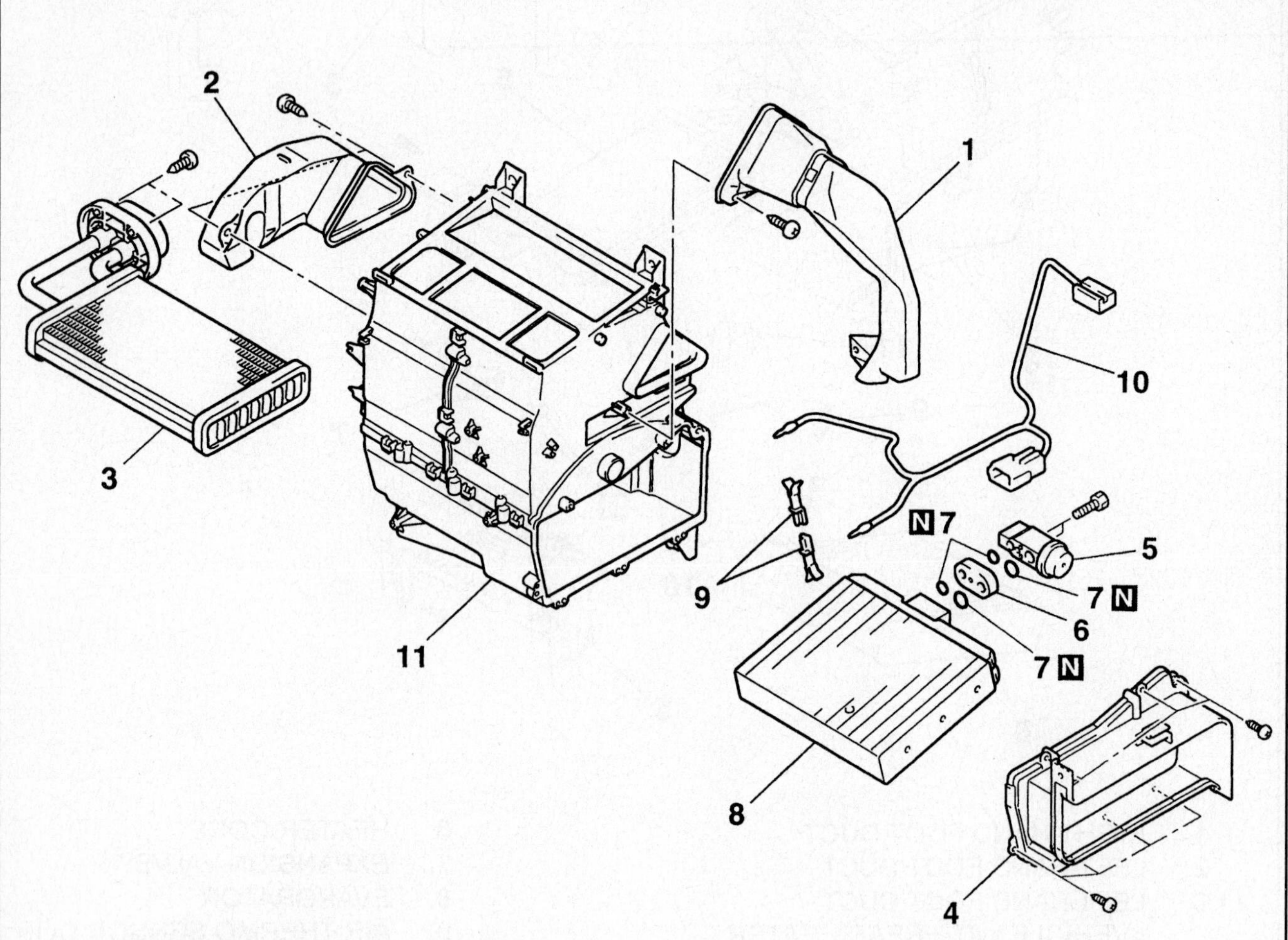

Fig. 284 Heater core and related components—Lancer Evolution

3. If not already done, disconnect the negative battery cable.

4. Drain the cooling system.

5. Discharge the air conditioning system.

6. Be sure the front tires are in the straight ahead position.

7. Remove the front seat.

8. Remove the column cover. Remove the meter bezel. Remove the combination meter.

9. Remove the instrument panel ornament. Remove the undercover, switch panel, switch holder, and lower frame.

10. Remove the heater control knob, center panel, and heater control assembly.

11. Remove the radio and CD player.

12. Remove the center air outlet panel and the center lower case, if equipped.

13. If equipped with a CD changer, remove the center lower panel.

14. Remove the stopper, glovebox, and harness cover. Remove the instrument panel side cover.

15. Remove the passenger's side air bag module mounting bolt and air bag module.

16. Remove the steering column cover.

17. Remove the driver's side air bag module.

18. Remove the steering wheel.

19. Remove the steering column clockspring and column switch assembly. Remove the shaft cover.

20. Remove the steering column shaft assembly. Remove the cover assembly.

WARNING

The tilt lever should be held in the lock position until the steering column is reinstalled in the vehicle. If the column is removed with the lever released, or the lever released after the column is removed from the vehicle, the steering column cannot be reinstalled correctly. If the steering column is installed incorrectly, the collision energy absorbing mechanism may be damaged.

21. Remove the instrument panel assembly.

22. Remove the console side cover, shift lever knob and front floor console assembly.

23. Remove the front floor console panel. If equipped with manual transaxle remove the shift lever cover.

24. Remove the shift lever bezel, lighter and ashtray.

25. Remove the console retaining screws. Remove the console from the

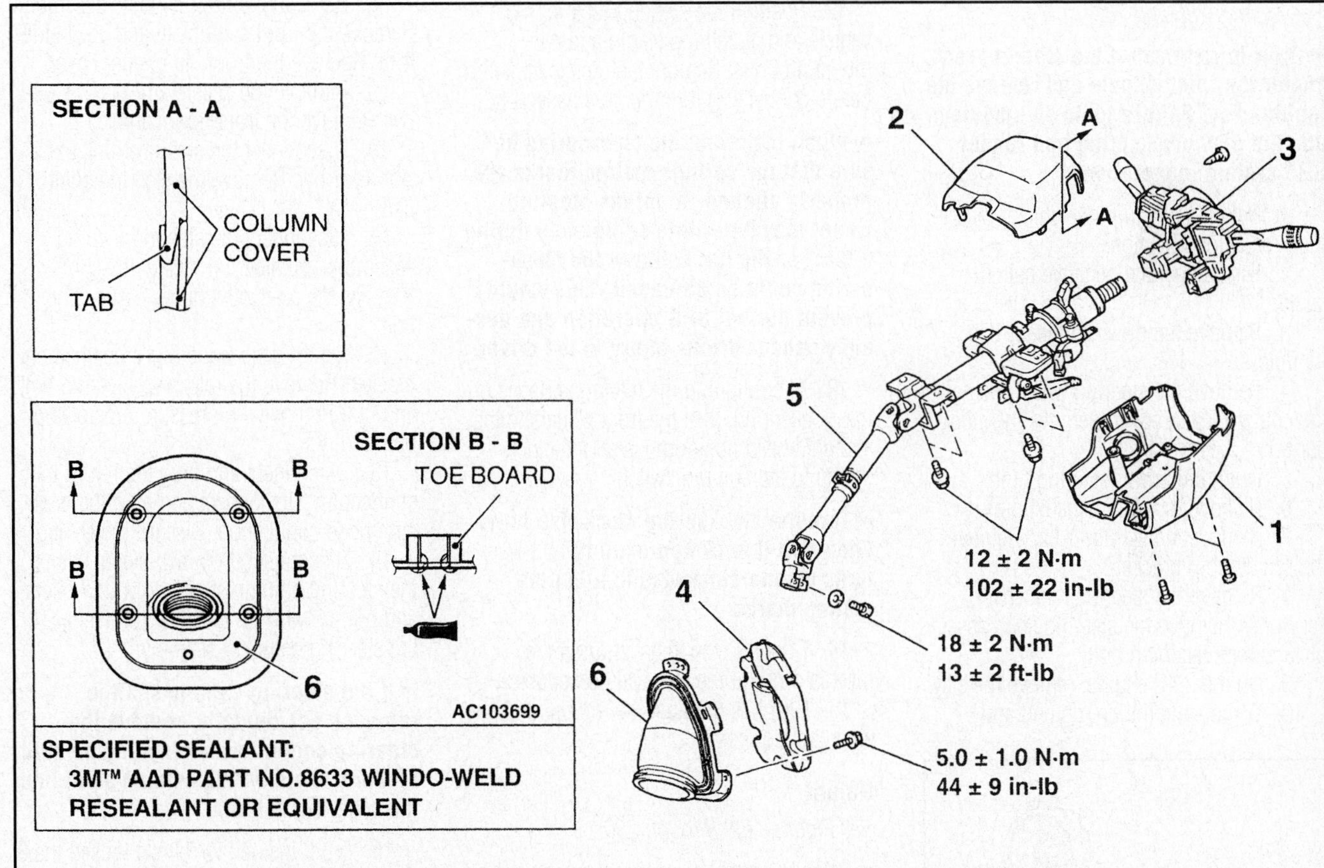

Fig. 285 Steering column installation—Lancer Evolution

vehicle. Remove the electrical harness and bracket assembly, as required.

26. Remove the front deck crossmember.
27. Remove the heater hoses.
28. Remove the air conditioning lines.
29. Remove the center duct.
30. Remove the heater unit, intake duct, and blower assembly from the vehicle.
31. Remove the heater core.

To install:

32. Installation is the reverse of the removal procedure.

➡The tilt lever should be held in the lock position until the steering column is reinstalled in the vehicle. If the column is removed with the lever released, or the lever released after the column is removed from the vehicle, the steering column cannot be reinstalled correctly. If the steering column is installed incorrectly, the collision energy absorbing mechanism may be damaged.

33. If the steering column shaft was removed accidentally, remove the steering column assembly and be sure to insert the steering column shaft using the alignment illustration.
34. Be sure to fill the cooling system with the proper grade and type coolant.
35. Recharge the air conditioning system.

STEERING

POWER RACK & PINION STEERING GEAR

REMOVAL & INSTALLATION

Eclipse

See Figures 286 and 287.

➡Prior to removal of the steering rack, center the front wheels and remove the ignition key. Failure to do so may damage the SRS clockspring and render SRS system inoperative.

1. Before servicing the vehicle, refer to the Precautions Section.
2. Disconnect the negative battery cable.
3. Remove the driver's side air bag module.
4. Remove the steering wheel. Remove the column lower cover. Remove the clockspring.
5. Drain the power steering fluid.
6. Remove the front exhaust pipe.
7. Remove the stabilizer bar. Remove the roll stopper bracket.
8. Remove the steering rack cover assembly. Remove the steering shaft assembly and rack retaining bolt.
9. On the 2.4L engine, remove the stay.
10. Disconnect the tie rod end and knuckle connection. Disconnect and plug the pressure tube connection.
11. Remove the power steering rack retaining clamp. Remove the assembly from the vehicle.

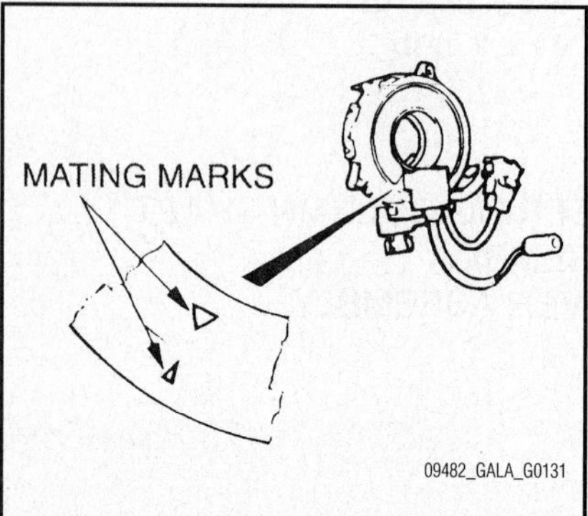

Fig. 286 Clockspring mating marks—Eclipse

To install:

12. Position the steering rack in the vehicle and install the clamp and the mounting bolts. Be sure the rack is centered before connecting it to the joint assembly.

➡When installing the clockspring be sure that the springs mating marks are properly aligned. If not the steering wheel may not rotate completely during a turn, or the flat cable in the clockspring could be damaged. This would prevent normal SRS operation and possibly cause serious injury to the driver.

13. After aligning the mating surfaces of the clockspring, turn the front wheels to the straight ahead position. Install the clockspring to the column switch.

➡Turn the clockspring clockwise fully. Then turn it back approximately 3 ¾ turns counterclockwise to align the mating marks.

14. Continue the installation in the reverse order of the removal procedure.
15. Check and adjust the front end alignment, as required.

Galant

See Figures 288 through 290.

➡Prior to removal of the steering rack, center the front wheels and remove the ignition key. Failure to do so may damage the SRS clockspring and render SRS system inoperative.

1. Before servicing the vehicle, refer to the Precautions Section.
2. Disconnect the negative battery cable.
3. Drain the power steering fluid.
4. Remove the front undercover.
5. Remove the center crossmember.
6. Remove the lower arm assembly.
7. Remove the driver's side air bag module. Remove the steering wheel. Remove the column lower cover. Remove the clockspring.
8. Remove the floor console assembly. Remove the front scuff plate and cowl side trim. Remove the trunk lid opener cover.
9. Remove the accelerator pedal stopper. Remove the front floor carpet.
10. Disconnect the stabilizer link and stabilizer bar. Remove the steering shaft pad.
11. Disconnect the steering column assembly and steering rack connection. Remove the rear roll stopper connecting bolt.
12. Remove the front axle crossmember bracket. Remove the return tube clamp bolt and nut. Remove the pressure tube clamp bolt and nut.
13. Disconnect and plug the return tube connection. Disconnect and plug the pressure hose connection. Discard the O-ring.
14. Disconnect the pinch clamp. Pinch the steering column shaft clip with a pliers, and pull upward on the shaft to disengage the steering column assembly.

➡If the steering column shaft is removed accidentally, remove the steering column assembly and be sure to insert the steering column shaft into the steering column.

15. From inside the vehicle loosen the 3 clips from the body panel, near the steering column to floor hole.
16. Use a transmission jack to hold the crossmember, and then remove the crossmember mounting nuts and bolts. Lower the crossmember with the rear roll stopper, then the stabilizer bar and steering rack.
17. Remove the steering column dash panel cover.
18. Remove the steering rack linkage protector, on the 3.8L engine.

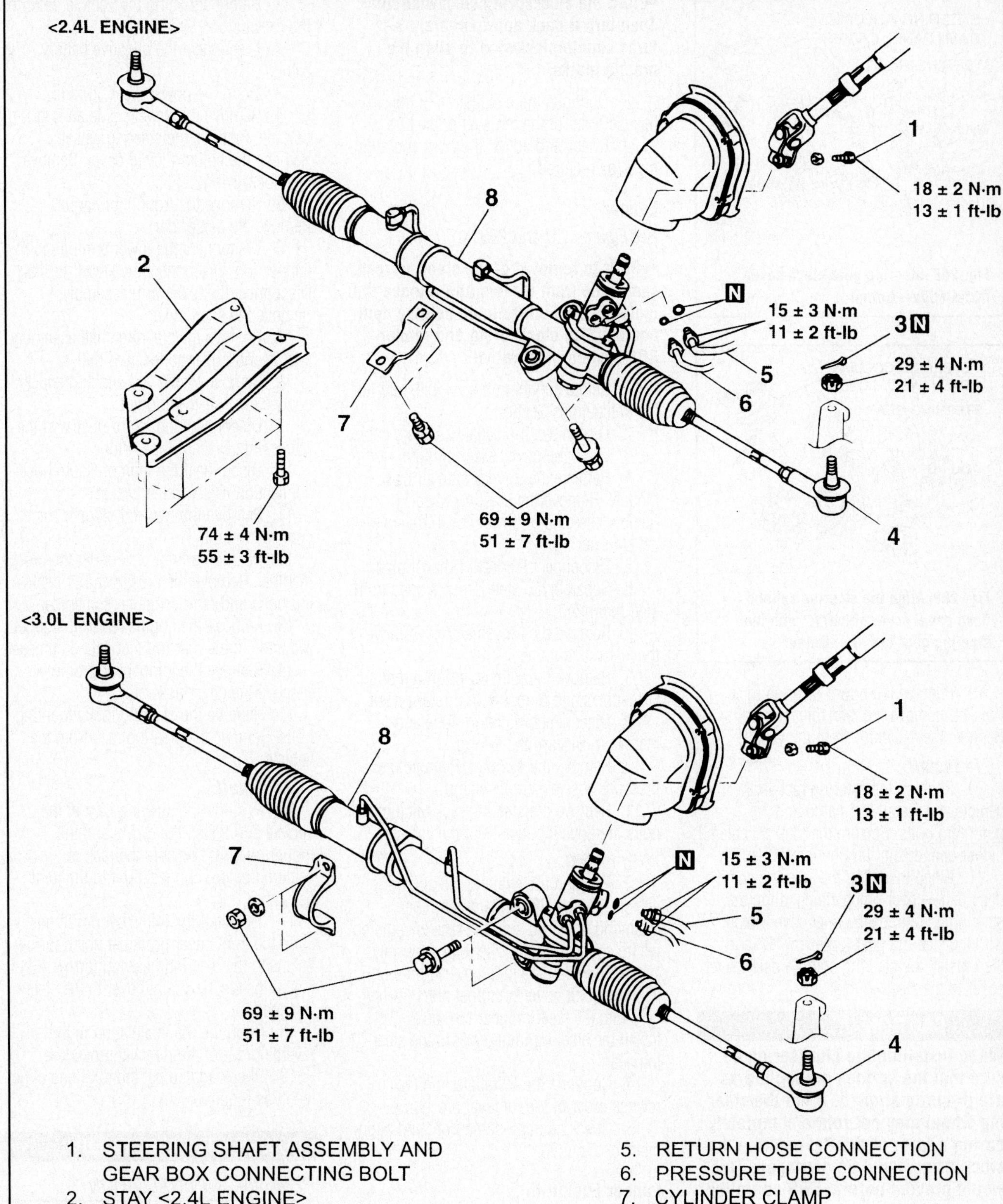

Fig. 287 Power steering rack and related components—Eclipse

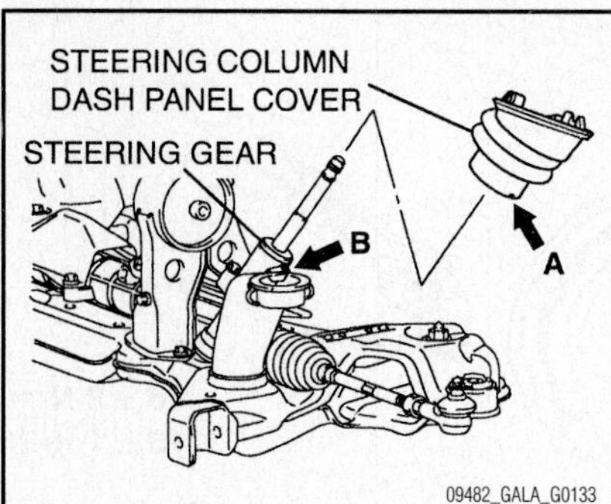

Fig. 288 Steering gear pinch cover installation—Galant

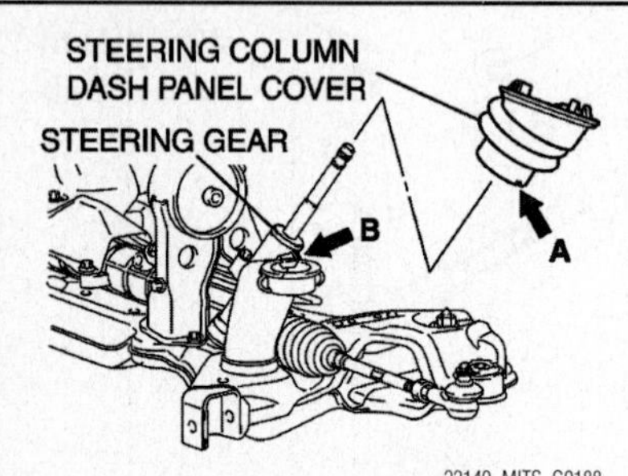

Fig. 289 Align the steering column dash panel cover notch "A" with the steering gear lug "B"—Galant

19. Remove the power steering rack bracket. Remove the rack retaining bolts. Remove the assembly from the vehicle.

To install:

20. Position the steering rack in the vehicle and install the clamp and the mounting bolts. Be sure the rack is centered before connecting it to the joint assembly.
21. When installing the steering column dash panel cover pinch clamp, align the steering column dash panel cover notch "A" with the steering gear lug arrow "B" and then install the steering column dash panel cover to the steering gear.

CAUTION

When installing the clockspring be sure that the springs mating marks are properly aligned. If not the steering wheel may not rotate completely during a turn, or the flat cable in the clockspring could be damaged. This would prevent normal SRS operation and possibly cause serious injury to the driver.

22. After aligning the mating surfaces of the clockspring, turn the front wheels to the straight ahead position. Install the clockspring to the column switch.

➡Turn the clockspring clockwise fully. Then turn it back approximately 3 ¾ turns counterclockwise to align the mating marks.

23. Continue the installation in the reverse order of the removal procedure.
24. Check and adjust the front end alignment, as required.

Lancer

See Figures 291 and 292.

➡Prior to removal of the steering rack, center the front wheels and remove the ignition key. Failure to do so may damage the SRS clockspring and render SRS system inoperative.

1. Before servicing the vehicle, refer to the Precautions Section.
2. Disconnect the negative battery cable.
3. Drain the power steering fluid.
4. Remove the driver's side air bag module. Remove the steering wheel. Remove the column lower cover. Remove the clockspring.
5. Disconnect the front exhaust pipe.
6. Separate the steering rack ends from their mounting.
7. Remove the crossmember. Remove the rear roll stopper.
8. Remove the joint cover grommet. Disconnect and plug the fluid return hose.
9. Disconnect the return tube and remove it. Discard the O-ring.
10. Remove the eye bolt. Remove the pressure hose assembly clamp.
11. Remove the steering rack retaining bolts. Remove the rack from the vehicle.

To install:

12. Position the steering rack in the vehicle and install the clamp and the mounting bolts. Be sure the rack is centered before connecting it to the joint assembly.
13. To install the joint cover grommet, align the joint cover grommet notch (arrow "A") with the steering gear lug (arrow "B"). Install the steering joint cover to the steering gear.
14. Continue the installation in the reverse order of the removal procedure.
15. Check and adjust the front end alignment, as required.

Lancer Evolution

See Figures 293 and 294.

➡Prior to removal of the steering rack, center the front wheels and remove the ignition key. Failure to do so may damage the SRS clockspring and render SRS system inoperative.

1. Before servicing the vehicle, refer to the Precautions Section.
2. Disconnect the negative battery cable.
3. Drain the power steering fluid.
4. Remove the driver's side air bag module. Remove the steering wheel. Remove the column lower cover. Remove the clockspring.
5. Remove the front floor carpet. Remove the center bar.
6. Disconnect the lower arm ball joint connection. Disconnect the stabilizer link. Disconnect the lower arm assembly. Remove the shaft cover.
7. Remove the steering shaft assembly and steering gear connecting bolt.
8. Remove the tie rod end and knuckle connection self-locking nut.
9. Disconnect the return hose and the return line. Discard the O-ring.
10. Disconnect the high pressure hose connection. Discard the gasket.
11. Remove the rear roll stopper rod connecting bolt.
12. Position a floor jack under the crossmember. Remove the crossmember retaining bolts and remove the crossmember.
13. Remove the engine rear roll stopper rod and engine rear roll stopper rod bracket.
14. Remove the joint cover grommet. Remove the cover assembly.
15. Remove the steering rack mounting bolts. Remove the steering rack from the vehicle.

To install:

16. Position the steering rack in the vehicle and install the clamp and the mounting bolts. Be sure the rack is centered before connecting it to the joint assembly.
17. To install the joint cover grommet, align the joint cover grommet notch (arrow "A") with the steering gear lug (arrow "B"). Install the steering joint cover to the steering gear.
18. Continue the installation in the reverse order of the removal procedure.
19. Check and adjust the front end alignment, as required.

POWER STEERING PUMP

REMOVAL & INSTALLATION

Eclipse

1. Before servicing the vehicle, refer to the Precautions Section.
2. Disconnect the battery negative cable.
3. Drain the power steering fluid into a suitable container.

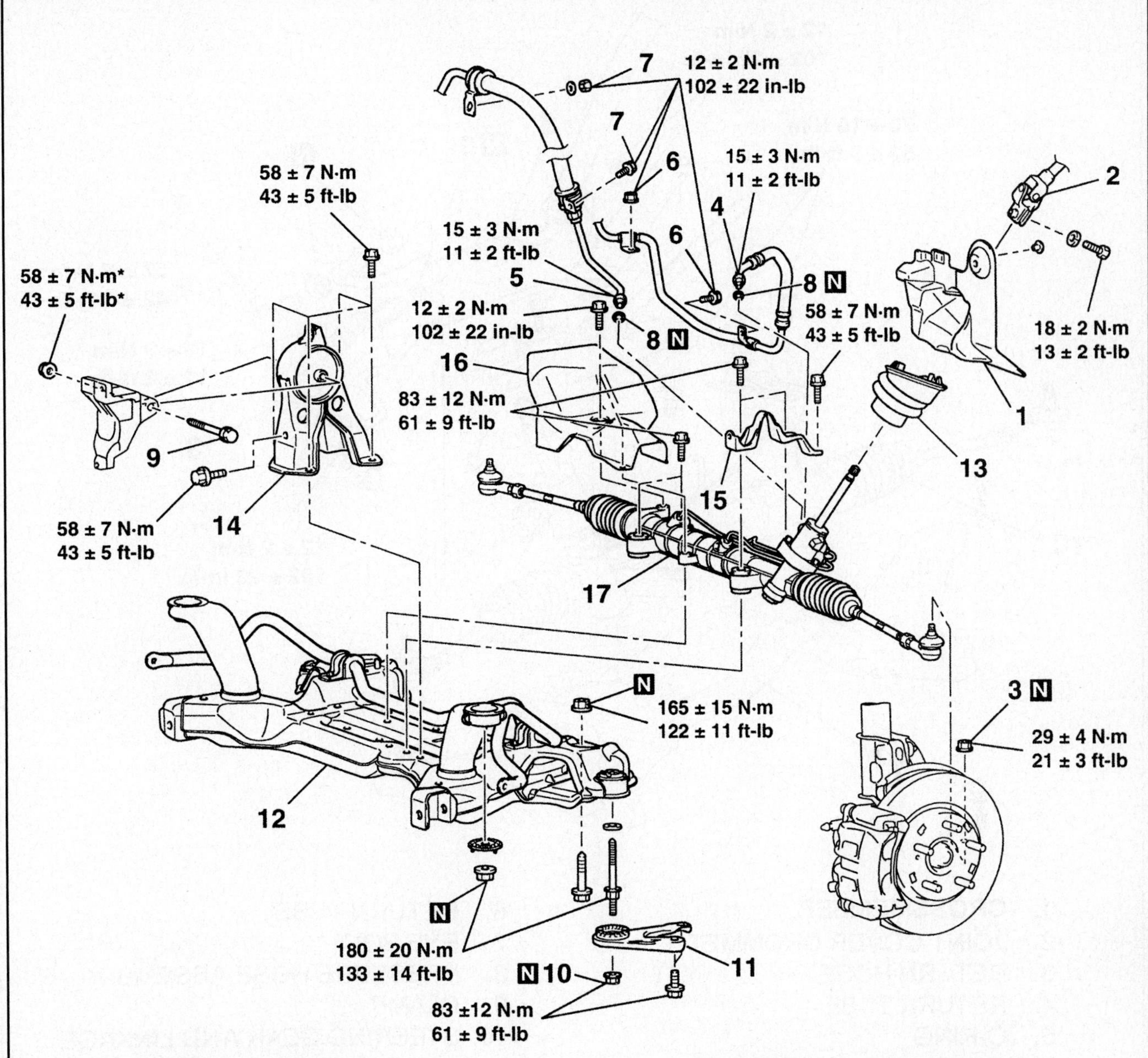

1. STEERING SHAFT PAD
2. STEERING COLUMN SHAFT ASSEMBLY AND STEERING GEAR CONNECTION
3. JAM NUT (TIE ROD END AND KNUCKLE CONNECTION)
4. RETURN TUBE CONNECTION
5. PRESSURE HOSE CONNECTION
6. RETURN TUBE CLAMP
7. PRESSURE HOSE CLAMP
8. O-RING
9. REAR ROLL STOPPER CONNECTING BOLT
10. JAM NUT
11. FRONT AXLE CROSSMEMBER STAY
12. CROSSMEMBER ASSEMBLY
13. STEERING COLUMN DASH PANEL COVER
14. REAR ROLL STOPPER
15. POWER STEERING GEAR BRACKET
16. STEERING GEAR AND LINKAGE PROTECTOR <3.8L ENGINE>
17. POWER STEERING GEAR AND LINKAGE

09482_GALA_G0135

Fig. 290 Power steering rack and related components—Galant

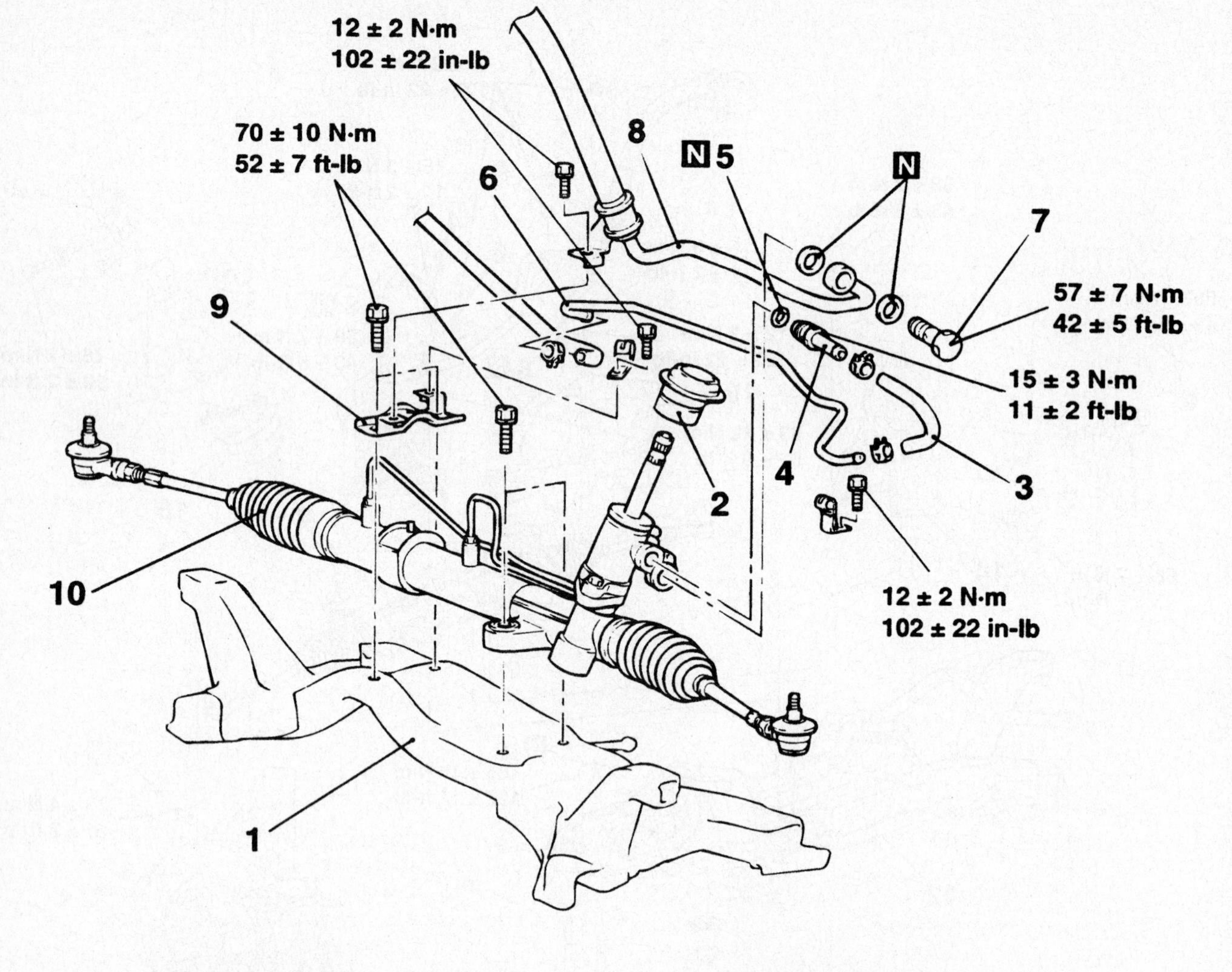

1. CROSSMEMBER
2. JOINT COVER GROMMET
3. RETURN HOSE
4. RETURN TUBE
5. O-RING
6. RETURN TUBE
7. EYE BOLT
8. PRESSURE HOSE ASSEMBLY
9. CLAMP
10. STEERING GEAR AND LINKAGE

93570G28

Fig. 291 Power steering rack and related components—Lancer

4. Remove the power steering drive belt.
5. Disconnect the supply and return lines.
6. Remove the pump mounting bolts.
7. Remove the power steering pump mounting bracket.
8. Remove the power steering pump.

To install:

9. Installation is the reverse order of removal.
   a. Tighten the power steering pump bracket bolt to 36 ft. lbs. (49 Nm).
   b. Tighten the power steering pump mounting bolts to 21 ft. lbs. (29 Nm).
10. Bleed the power steering system.
11. Check the tension of the power steering pump belt.

Galant

2.4l Engine

See Figure 295.

1. Before servicing the vehicle, refer to the Precautions Section.
2. Disconnect the battery negative cable.
3. Loosen and remove the power steering pump drive belt.
4. Remove the pressure switch connector from the side of the pump.

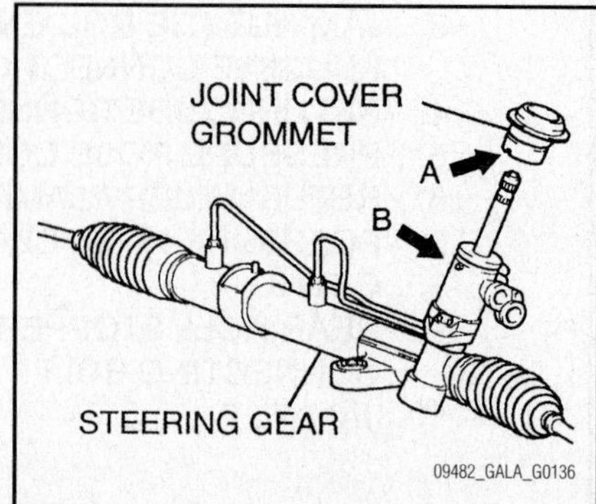

Fig. 292 Joint cover installation—Lancer and Lancer Evolution

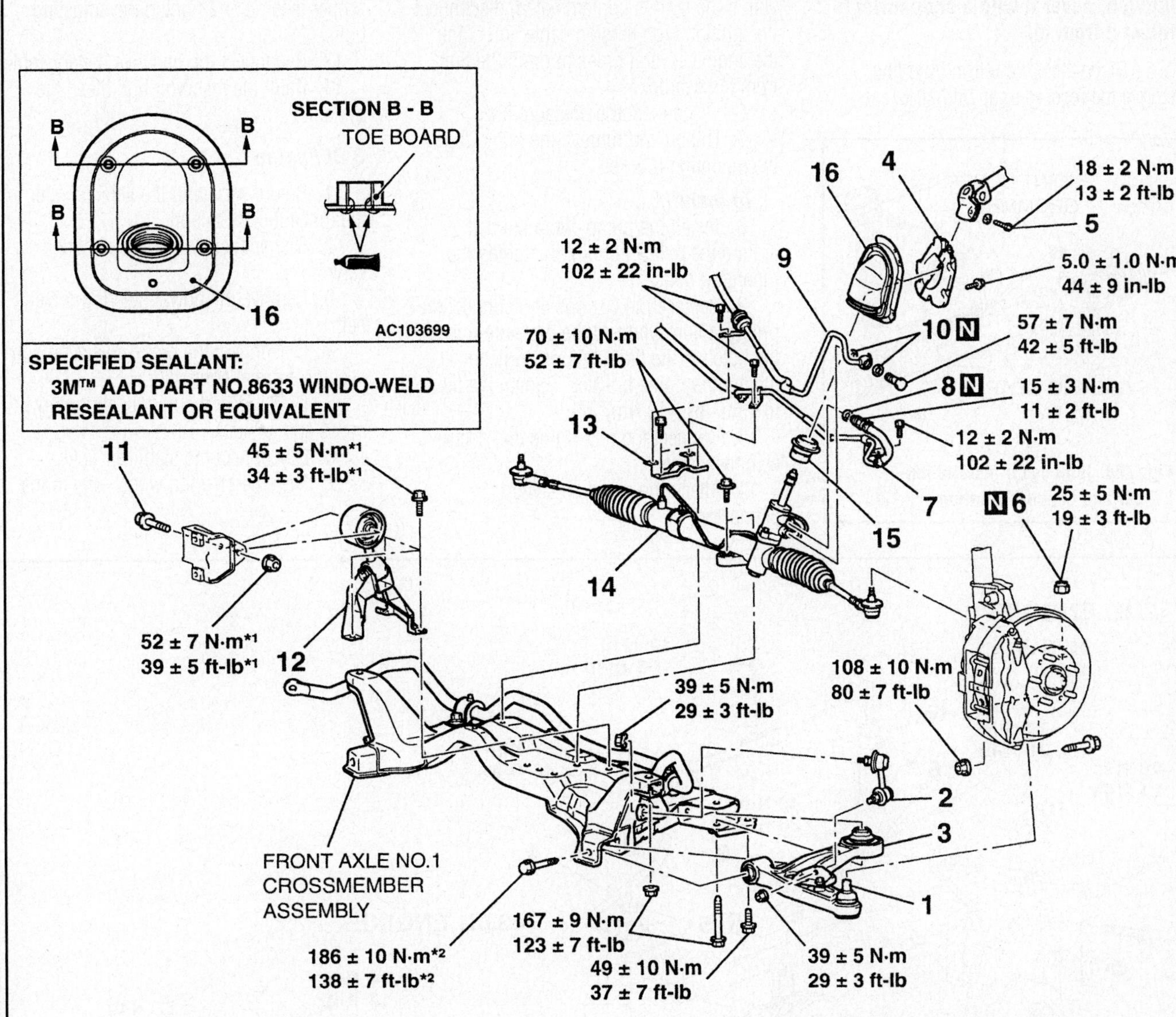

1. LOWER ARM BALL JOINT CONNECTION
2. STABILIZER LINK
3. LOWER ARM ASSEMBLY
4. SHAFT COVER
5. STEERING COLUMN SHAFT ASSEMBLY AND STEERING GEAR CONNECTING BOLT
6. SELF LOCKING NUT (TIE ROD END AND KNUCKLE CONNECTION)
7. RETURN HOSE AND RETURN TUBE
8. O-RING
9. PRESSURE HOSE CONNECTION
10. GASKET
11. ENGINE REAR ROLL STOPPER ROD CONNECTING BOLT
12. ENGINE REAR ROLL STOPPER ROD AND ENGINE REAR ROLL STOPPER ROD BRACKET
13. CLAMP
14. STEERING GEAR
15. JOINT COVER GROMMET
16. COVER ASSEMBLY

09482_GALA_G0137

Fig. 293 Power steering rack and related components—Lancer Evolution

➡If the alternator is located under the oil pump, cover it with a shop towel to protect it from oil.

5. Disconnect the return fluid line. Remove the reservoir cap and allow the return line to drain the fluid from the reservoir. If the fluid is contaminated, disconnect the ignition high tension cable and crank the engine several times to drain the fluid from the gearbox.

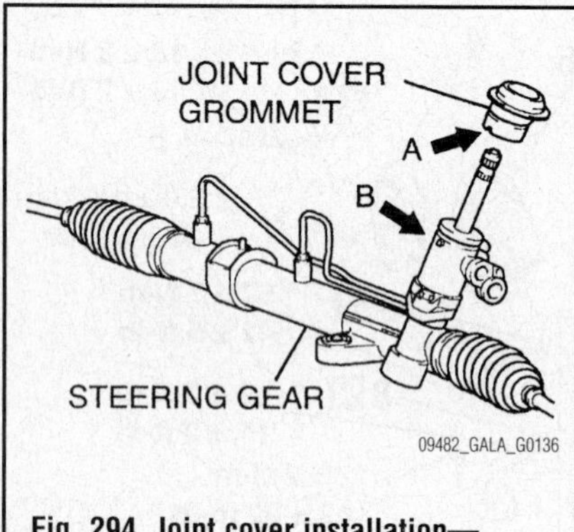

Fig. 294 Joint cover installation—Lancer and Lancer Evolution

6. Disconnect the pressure line.
7. Unbolt and remove the pump from the mounting bracket.

To install:

8. Install the pump, wrap the belt around the pulley and lightly tighten the mounting bolts.
9. Replace the O-rings and connect the pressure line. Connect the pressure line so the notch in the fitting aligns and contacts the pump's guide bracket. Tighten the fitting to 13 ft. lbs. (18 Nm).
10. Connect the return line and secure with the clamp.
11. Attach the pressure switch connector.
12. Adjust the power steering belt for proper tension and tighten the adjusting bolts.
13. Reconnect the negative battery cable.
14. Refill the reservoir and bleed the system.

3.8l Engine

1. Before servicing the vehicle, refer to the Precautions Section.
2. Disconnect the negative battery cable.
3. Remove the power steering drive belt.
4. Disconnect the pressure switch connector electrical connector.
5. Disconnect the supply and return hoses and gaskets.
6. Disconnect the stabilizer links.
7. Turn the steering wheel fully to the left.

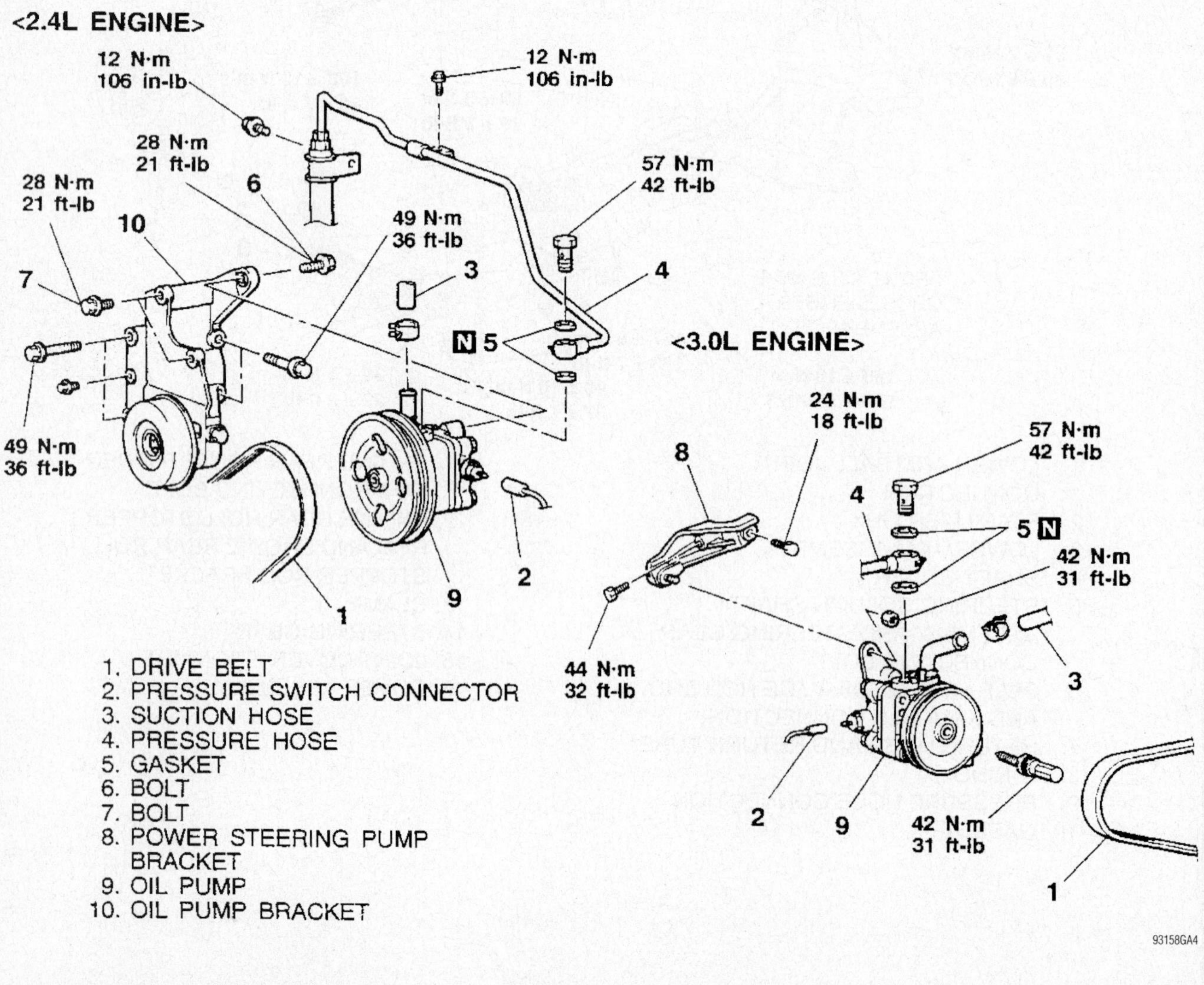

<2.4L ENGINE>

<3.0L ENGINE>

1. DRIVE BELT
2. PRESSURE SWITCH CONNECTOR
3. SUCTION HOSE
4. PRESSURE HOSE
5. GASKET
6. BOLT
7. BOLT
8. POWER STEERING PUMP BRACKET
9. OIL PUMP
10. OIL PUMP BRACKET

93158GA4

Fig. 295 Exploded view of the power steering pump—2.4L engine

8. Remove the power steering pump mounting bolts.
9. Remove the power steering pump.

To install:

10. Installation is the reverse order of removal.
11. Torque the power steering pump mounting bolts to 36 ft. lbs. (49 Nm).

Lancer & Lancer Evolution

See Figure 296.

1. Before servicing the vehicle, refer to the Precautions Section.
2. Disconnect the negative battery cable.
3. Remove accessory drive belt.
4. Remove the engine cover.
5. Remove the radiator condenser tank.
6. Disconnect the supply and return hoses and gaskets.
7. Remove the power steering pump mounting bolts.
8. Remove the power steering pump.

To install:

9. Installation is the reverse order of removal.
10. Torque the power steering pump mounting bolts according to the illustration.

BLEEDING

1. Raise the vehicle and support safely.
2. Manually turn the oil pump pulley a few times.
3. Turn the steering wheel all the way to the left and to the right 5 or 6 times.
4. Disconnect the ignition high tension cable and, while operating the starter motor intermittently, turn the steering wheel all the way to the left and right 5–6 times for 15–20 seconds. During bleeding, make sure the fluid in the reservoir never falls below the lower position of the filter. If bleeding is attempted with the engine running, the air will be absorbed in the fluid. Bleed only while cranking.
5. Connect ignition high tension cable, start engine and allow to idle.
6. Turn the steering wheel left and right until there are no air bubbles in the reservoir. Confirm that the fluid is not milky and the level is up to the specified position on the gauge. Confirm that there is very little change in the fluid level when the steering wheel is turned.

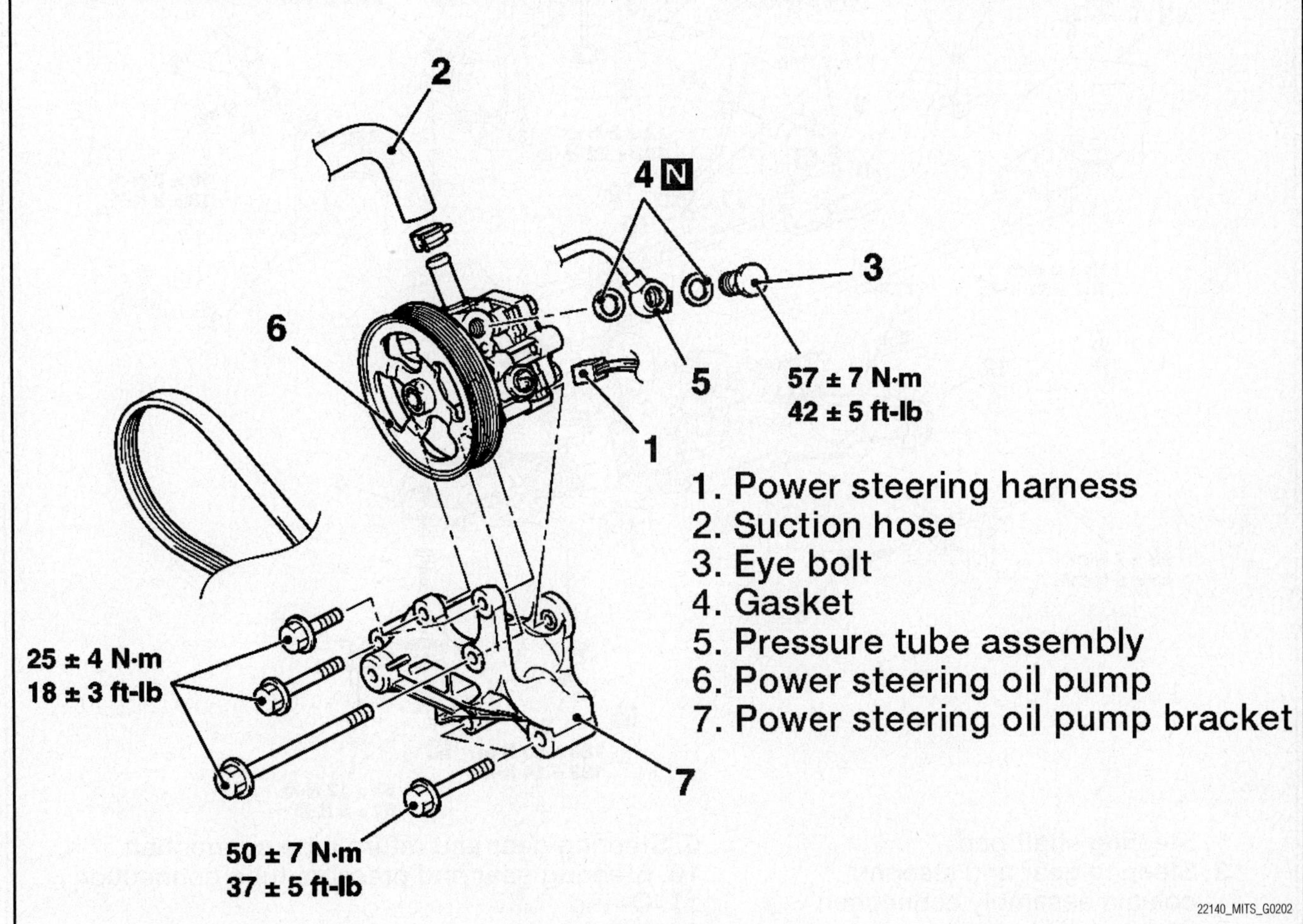

Fig. 296 Exploded view of the power steering pump—Lancer and Lancer Evolution

CONTROL LINKS

REMOVAL & INSTALLATION

See Figures 297 through 299.

1. Before servicing the vehicle, refer to the Precautions Section.
2. Raise and safely support the vehicle.
3. Remove the front wheel and tire from the front hub.
4. Remove the stabilizer bar control link by removing the upper and lower mounting nuts.
5. Remove the stabilizer bar control link.

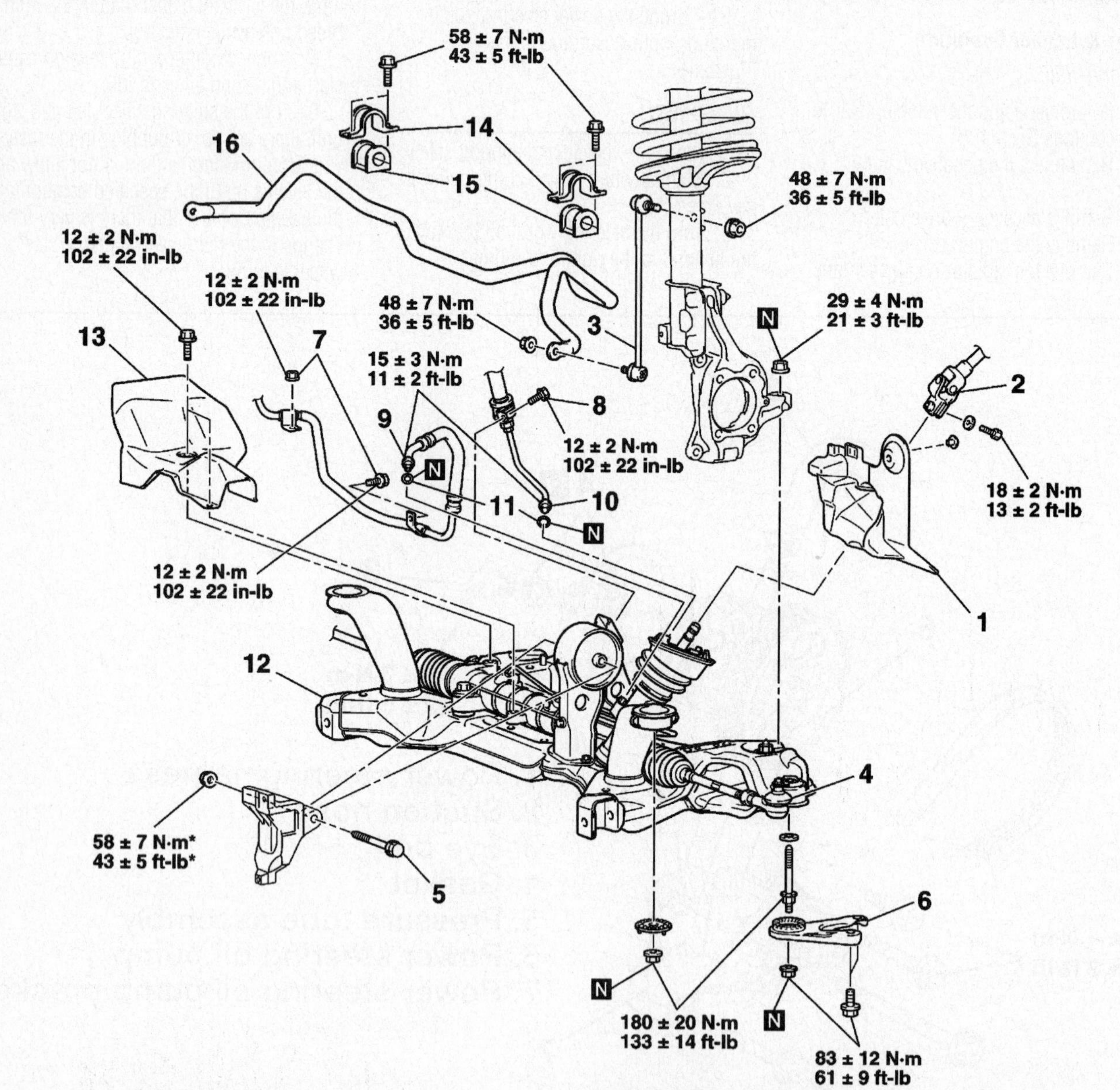

Fig. 297 Exploded view of the stabilizer bar and related components—Eclipse & Galant

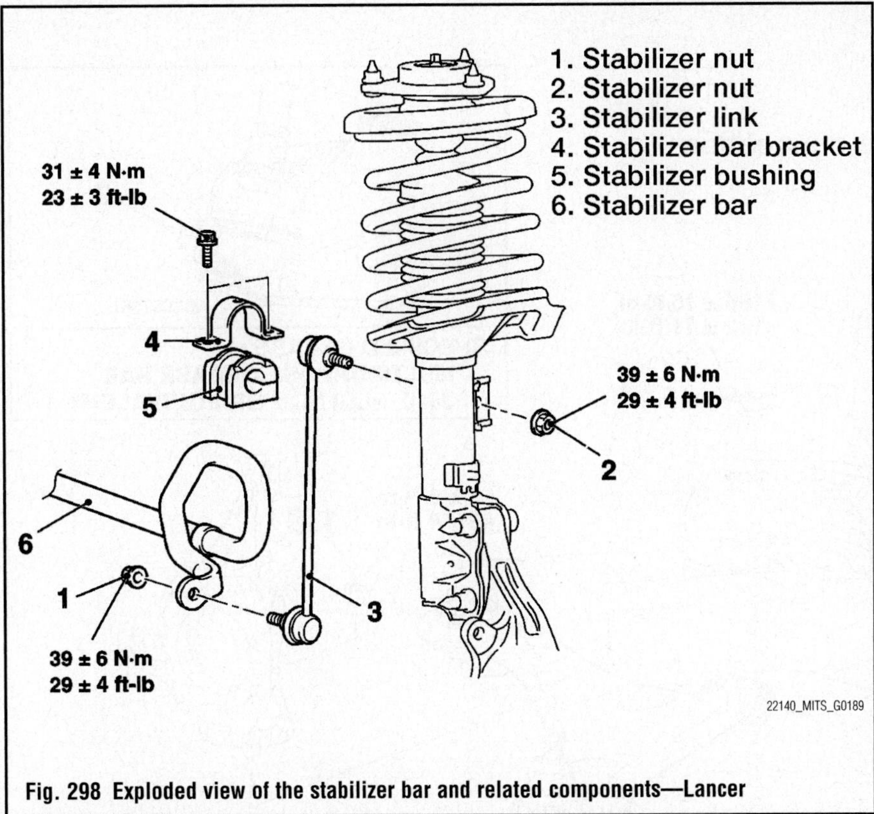

Fig. 298 Exploded view of the stabilizer bar and related components—Lancer

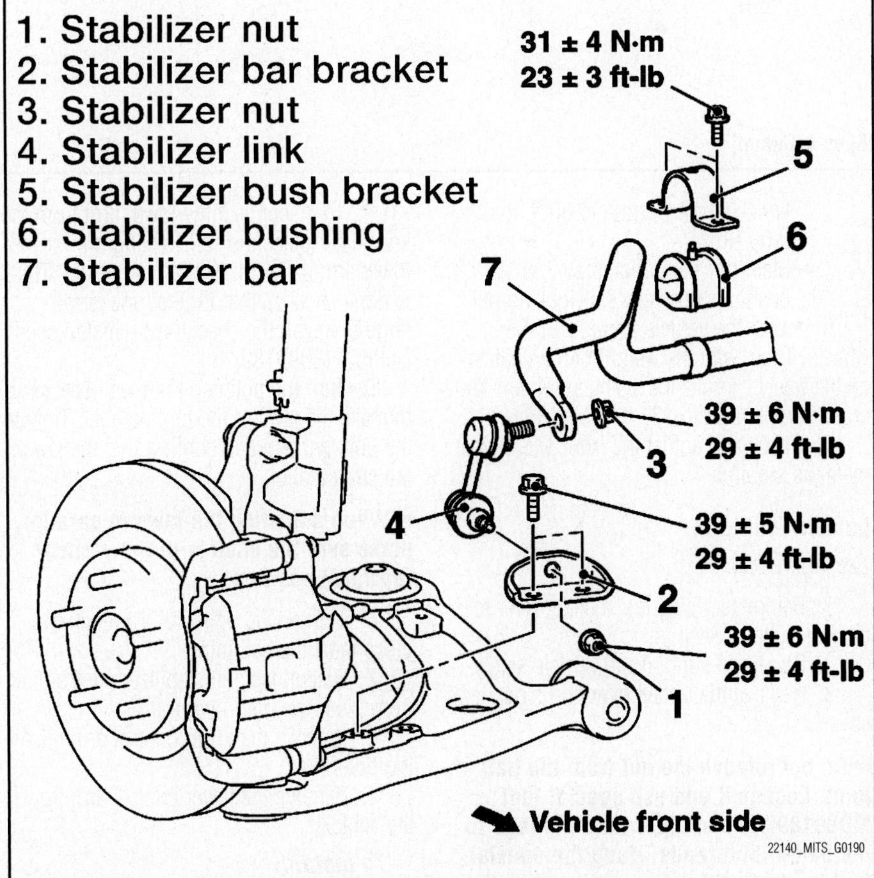

Fig. 299 Exploded view of the stabilizer bar and related components—Lancer Evolution

To install:

6. Install the stabilizer bar control link.
7. Tighten the stabilizer bar control link upper and lower mounting nuts to specification as illustrated per vehicle.
8. Install the front wheels.

LOWER BALL JOINT

REMOVAL & INSTALLATION

The lower ball joint is an integral part of the lower control arm assembly and cannot be serviced separately. A worn or damaged ball joint requires replacement of lower control arm assembly. Refer to Lower Control Arm, Removal & Installation.

LOWER CONTROL ARM

REMOVAL & INSTALLATION

Eclipse & Galant

See Figure 300.

1. Before servicing the vehicle, refer to the Precautions Section.
2. Raise and support the vehicle safely.

➡Do not remove the nut from the ball joint. Loosen it and use special tool MB991897 to avoid possible damage to the ball joint threads. Hang the special tool in place with wire or string to prevent it from falling.

3. To disconnect the lower arm from the knuckle, replace the self-locking nut for the lower arm ball joint with a regular nut. This is done because the original one is too large to install the special tool. Install special tool MB991897.
4. Turn the bolt and knob as necessary to make the jaws of the tool parallel. Tighten the bolt by hand and confirm that the jaws are still parallel.

➡When adjusting the jaws in parallel, make sure the knob is in the vertical (upward) position.

5. Tighten the bolt with a wrench to disconnect the ball joint.
6. Remove the lower control arm retaining bolts.
7. Remove the lower control arm from the vehicle.

To install:

8. Installation is the reverse of the removal procedure.
9. Tighten the bolts as illustrated in the figure.
10. Check and adjust the front end alignment, as required.

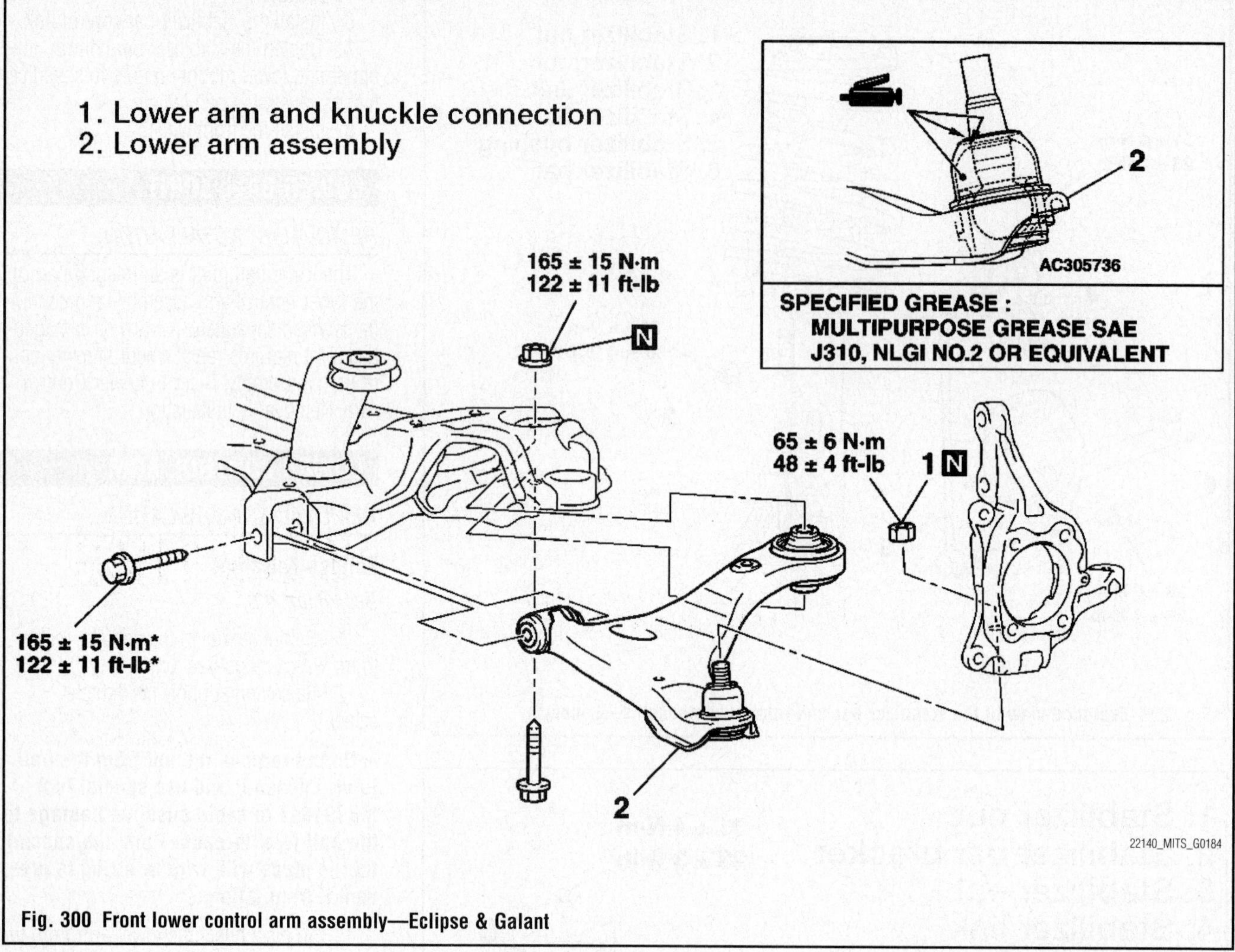

Fig. 300 Front lower control arm assembly—Eclipse & Galant

Lancer

See Figure 301.

1. Before servicing the vehicle, refer to the Precautions Section.
2. Raise the vehicle and support safely.
3. Remove the wheel and tire assembly.
4. Remove the stabilizer bar self-locking nut, rubber bushings, stabilizer bar, and collar. Discard the nut.
5. Remove the lower control arm-to-knuckle bolt and nut.
6. Lift the transaxle with a jack, then remove the front control arm-to-crossmember bolt.
7. Remove the lower control arm.

To install:

8. Install or connect the following:
   - Lower control arm into the vehicle
   - Lower control arm-to-crossmember bolts. Torque the bottom bolt to 123 ft. lbs. (167 Nm) and the side bolt snug until the vehicle is lowered.
   - Lower control arm-to-steering knuckle bolt. Torque to 80 ft. lbs. (108 Nm).
   - Stabilizer bar collar, stabilizer bar, bushings and new self-locking nut.
9. Lower the vehicle and install the wheels. Then, with the weight of the vehicle on the wheels, torque the side control arm-to-crossmember bolt to 137 ft. lbs. (186 Nm).
10. Check and adjust the front end alignment, as required.

Lancer Evolution

See Figure 302.

1. Before servicing the vehicle, refer to the Precautions Section.
2. Raise and support the vehicle safely.
3. Remove the undercover and side cover.

➡Do not remove the nut from the ball joint. Loosen it and use special tool MB991897 to avoid possible damage to the ball joint threads. Hang the special tool in place with wire or string to prevent it from falling.

4. To disconnect the lower arm from the knuckle, replace the self-locking nut for the lower arm ball joint with a regular nut. This is done because the original one is too large to install the special tool. Install special tool MB991897.
5. Turn the bolt and knob as necessary to make the jaws of the tool parallel. Tighten the bolt by hand and confirm that the jaws are still parallel.

➡When adjusting the jaws in parallel, make sure the knob is in the vertical (upward) position.

6. Tighten the bolt with a wrench to disconnect the ball joint.
7. Disconnect the stabilizer control link from the lower control arm.
8. Remove the lower control arm retaining bolts.
9. Remove the lower control arm from the vehicle.

To install:

10. Installation is the reverse of the removal procedure.

110 ± 11 N·m*1
81 ± 8 ft-lb
*2
*2
110 ± 11 N·m*1
81 ± 8 ft-lb
1*2
71 ± 10 N·m
53 ± 7 ft-lb
2
*2
22140_MITS_G0186

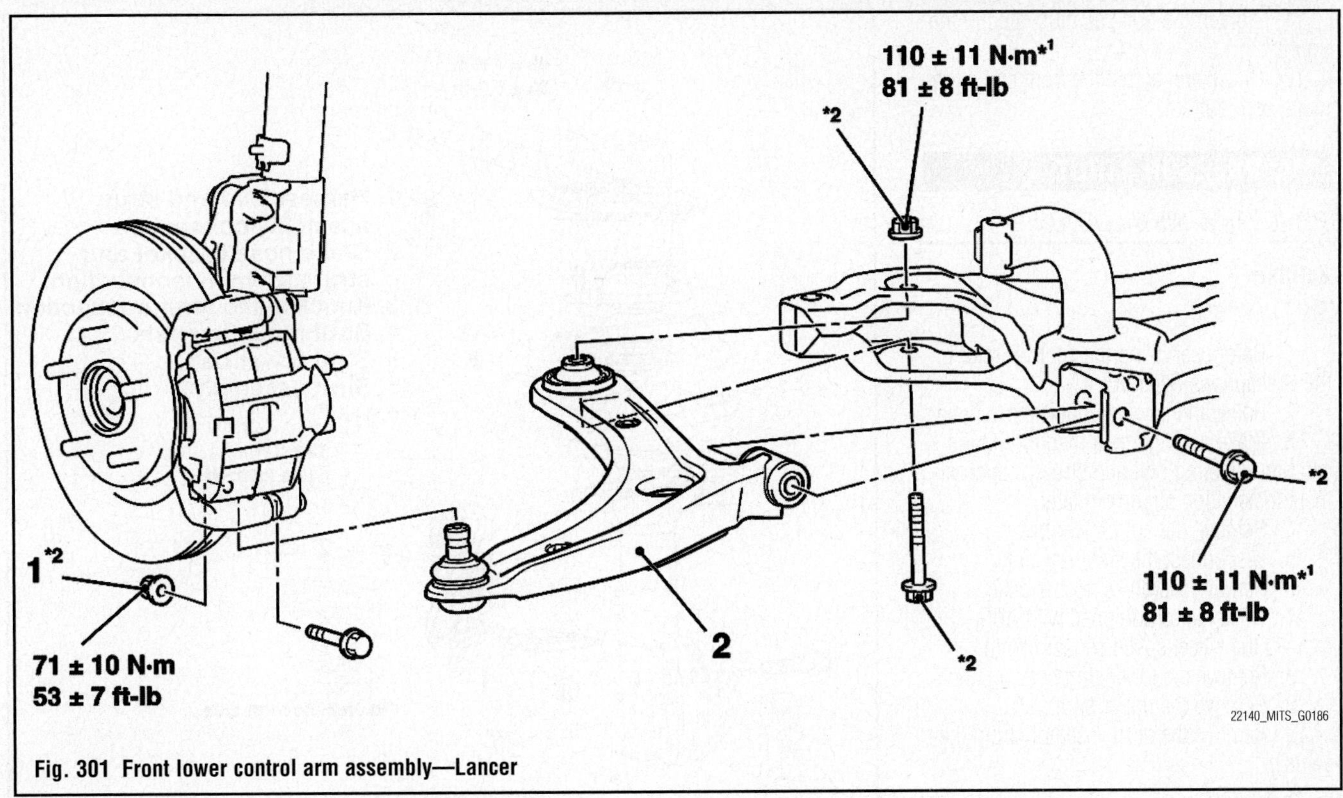

Fig. 301 Front lower control arm assembly—Lancer

1. Stabilizer link bracket and lower arm connection
2. Lower arm assembly
2
Multipurpose grease SAE J310, NLGI No.2 or equivalent
39 ± 5 N·m
29 ± 4 ft-lb
1
110 ± 11 N·m
81 ± 8 ft-lb
*2
*1, 2
110 ± 11 N·m
81 ± 8 ft-lb
*2
71 ± 10 N·m
53 ± 7 ft-lb
2
*2
Vehicle front side
22140_MITS_G0185

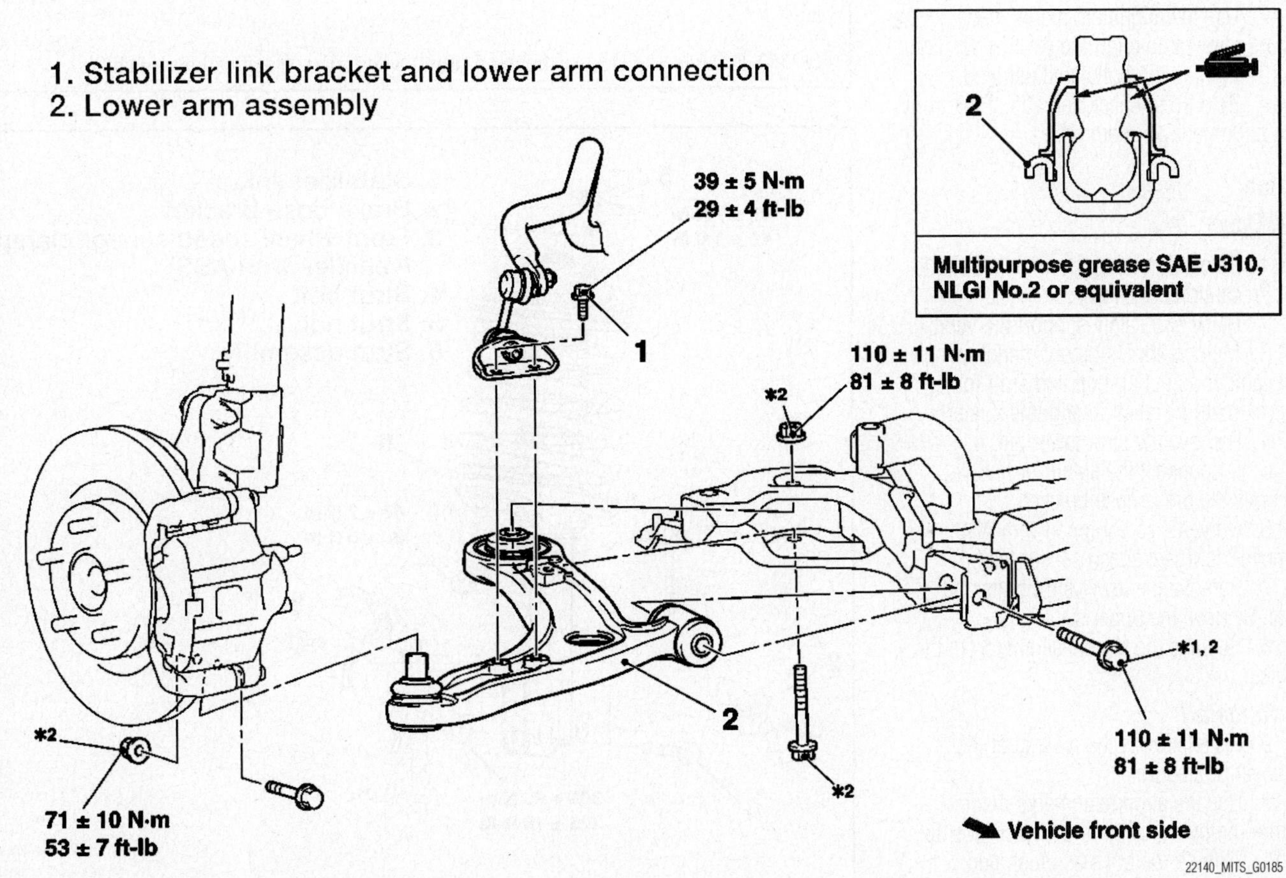

Fig. 302 Front lower control arm assembly—Lancer Evolution

11. Tighten the bolts as illustrated in the figure.
12. Check and adjust the front end alignment, as required.

MACPHERSON STRUT

REMOVAL & INSTALLATION

Eclipse

See Figure 303.

1. Before servicing the vehicle, refer to the Precautions Section.
2. Raise and safely support the vehicle.
3. Make an alignment marking on the camber adjusting bolt and strut for approximate installation alignment later.
4. Remove the strut tower bar.
5. Disconnect the stabilizer link.
6. Remove the brake hose bracket.
7. On vehicles equipped with ABS, remove the wheel speed sensor clamp.
8. Remove the lower strut bolts and nuts.
9. Remove the upper strut nuts.
10. Remove the strut assembly from the vehicle.

To install:

11. Installation is the reverse of the removal procedure.
12. Use the alignment marks made before removal to align the strut in position.
13. Tighten the bolts as illustrated.
14. Be sure to check and adjust the front end alignment, as required.

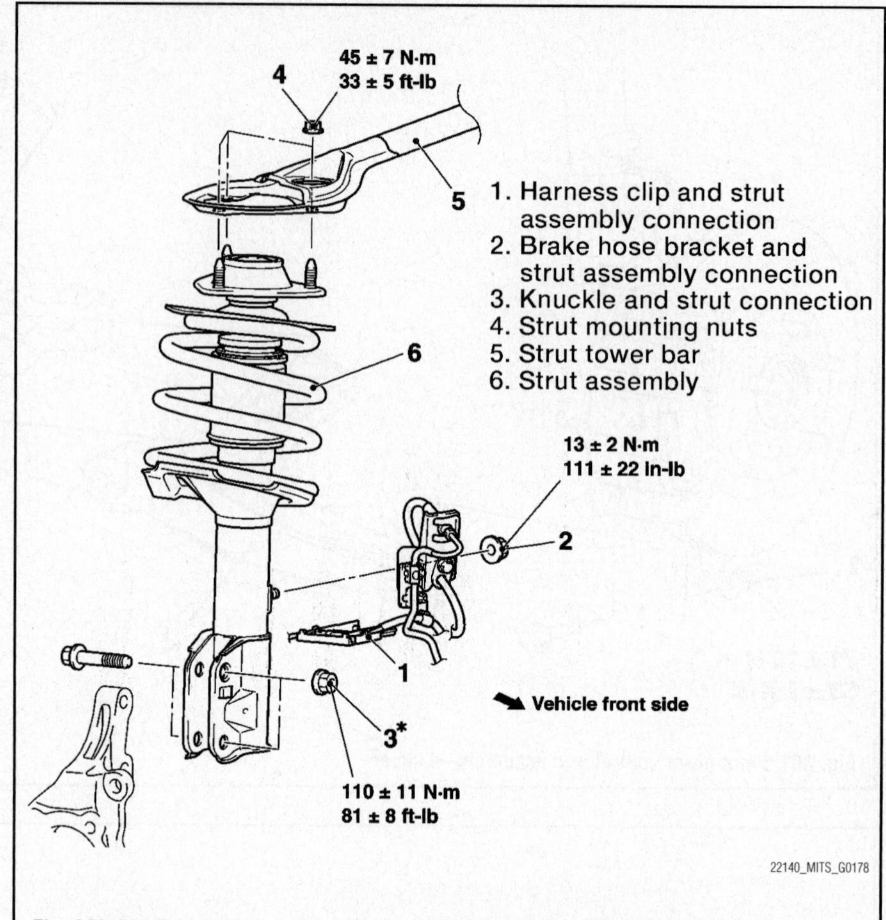

Fig. 303 MacPherson strut suspension (front) with torque specifications—Eclipse

Galant

See Figure 304.

1. Before servicing the vehicle, refer to the Precautions Section.
2. Raise and safely support the vehicle.
3. Matchmark alignment markings on the camber adjusting bolt and strut for approximate installation alignment later.
4. Remove the strut tower bar.
5. Disconnect the stabilizer link. Remove the brake hose bracket.
6. On vehicles equipped with ABS, remove the wheel speed sensor clamp.
7. Remove the lower strut bolts and nuts. Remove the upper strut nuts.
8. Remove the strut assembly from the vehicle.

To install:

9. Installation is the reverse of the removal procedure.
10. Use the alignment marks made before removal to align the strut in position.
11. Tighten the bolts as illustrated.
12. Be sure to check and adjust the front end alignment, as required.

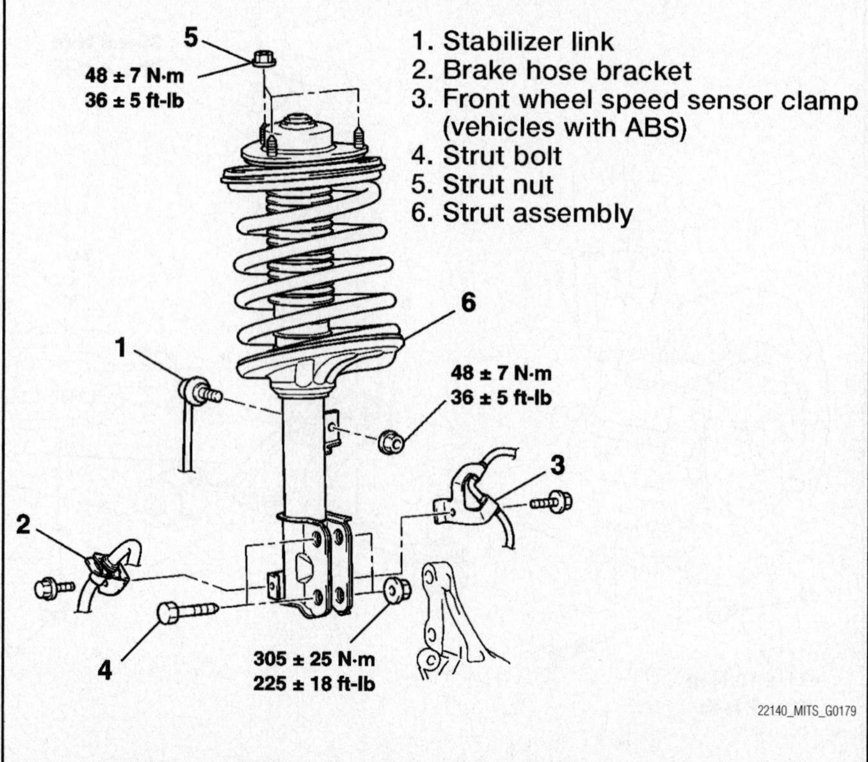

Fig. 304 MacPherson strut suspension (front) with torque specifications—Galant

Lancer

See Figure 305.

1. Before servicing the vehicle, refer to the Precautions Section.
2. Raise and safely support the vehicle.
3. Matchmark alignment markings on the camber adjusting bolt and strut for approximate installation alignment later.
4. Remove the brake hose bracket.
5. On vehicles equipped with ABS, remove the wheel speed sensor clamp.
6. Remove the lower strut bolts and nuts from the knuckle connection. Remove the upper strut nuts.
7. Remove the strut assembly from the vehicle.

To install:

8. Installation is the reverse of the removal procedure.
9. Be sure to check and adjust the front end alignment, as required.

Lancer Evolution

See Figure 306.

1. Before servicing the vehicle, refer to the Precautions Section.
2. Raise and safely support the vehicle.
3. Matchmark alignment markings on the camber adjusting bolt and strut for approximate installation alignment later.
4. Remove the brake hose bracket.

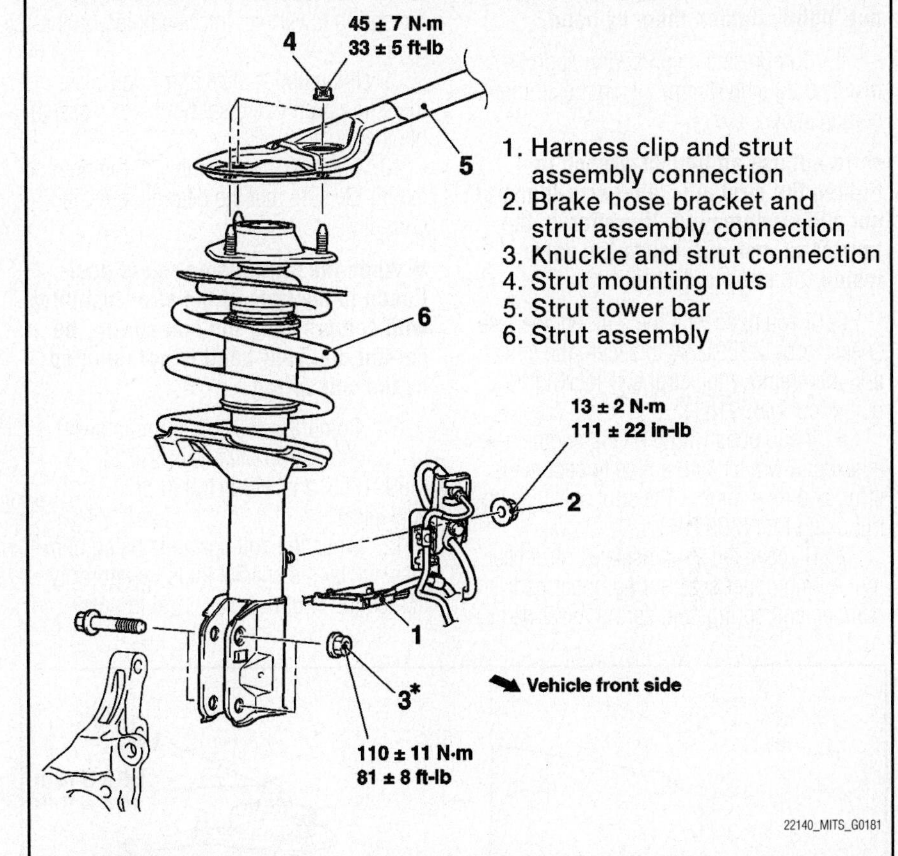

Fig. 306 MacPherson strut suspension (front) with torque specifications—Lancer Evolution

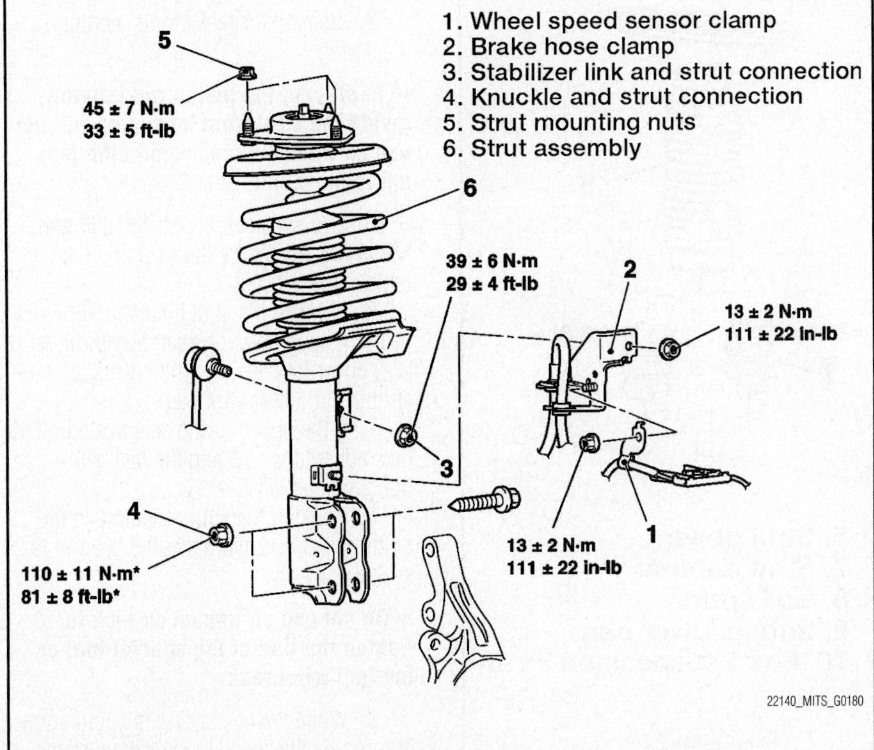

Fig. 305 MacPherson strut suspension (front) with torque specifications—Lancer

5. Remove the wheel speed sensor harness bracket.
6. Remove the lower strut bolts and nuts from the knuckle connection. Remove the upper strut nuts.
7. Remove the strut assembly from the vehicle.

To install:

8. Installation is the reverse of the removal procedure.
9. Be sure to check and adjust the front end alignment, as required.

OVERHAUL

Eclipse & Galant

See Figure 307.

1. Before servicing the vehicle, refer to the Precautions Section.
2. Remove the strut from the vehicle.
3. Position the strut assembly using fixture tools MB991793, MB991794, MB991795, and MB9910830 or equivalent.

➡Position the strut assembly in the disassembly fixture using the bolt and nut that were removed from the

vehicle. When installing the bolt and nut, lightly tighten them by hand.

4. Compress the coil spring approximately 0.20 inch (5mm), using the spring compressor.

➡Do not use an impact wrench to tighten the strut nut, otherwise the strut nut will be damaged. Vibration of the impact wrench will cause the valve inside the strut to drop out.

5. If equipped with the 2.4L engine, use special tool MB991682 to secure the strut, and then remove the strut self-locking nut using tool MB991681.

6. If equipped with the 3.8L engine, use a hexagon wrench and a pipe to secure the strut, and then remove the strut self-locking nut using tool MB991681.

7. Remove the strut insulator, strut bearing, spring upper seat, spring upper pad, damper, coil spring, and spring lower pad.

8. To assemble, engage the 3 lugs of the spring lower pad into the holes on the strut.

9. If the upper plate and lower plate of the strut have been disassembled, reassemble them.

10. Position the assembly in the holding fixture. Be sure that the bearing is seated correctly.

➡When the strut piston rod is positioned to the hole of the strut insulator with compressing the coil spring, be careful that your hand is not jammed by the coil spring.

11. Compress the coil spring slowly using tools MB991793, MB991794, MB991795, and MB9910830 or equivalent.

12. While the coil spring is being compressed, by the special tools, temporarily tighten the strut nut (self-locking nut).

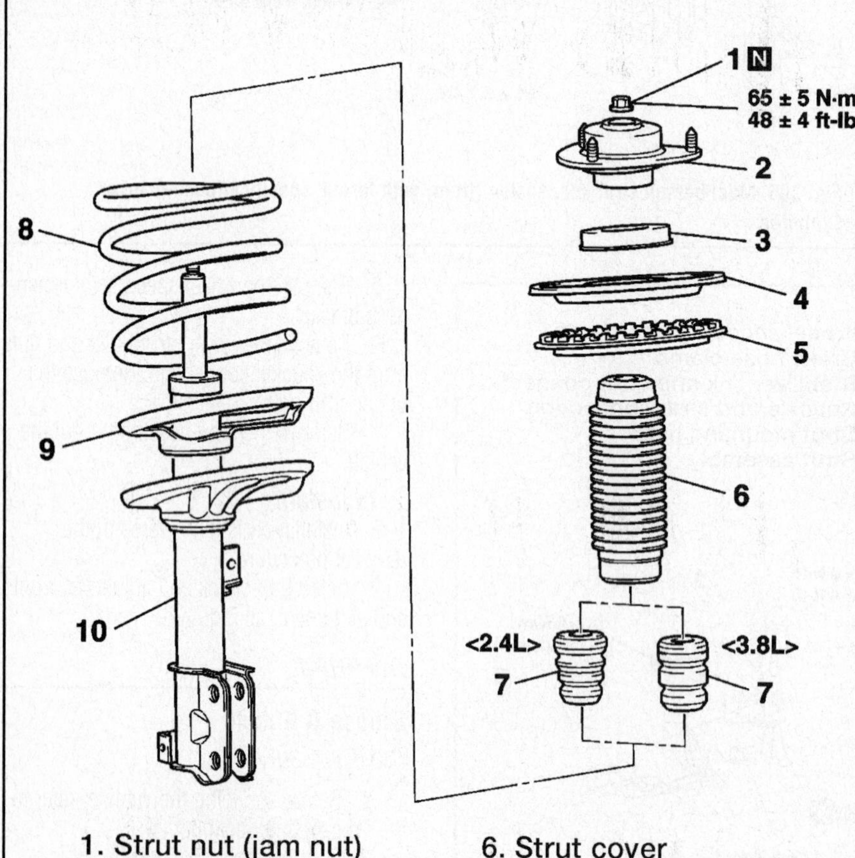

Fig. 307 Expanded view of MacPherson strut—Eclipse

➡Do not use an impact wrench to tighten the strut nut, otherwise the strut nut will be damaged. Vibration of the impact wrench can cause the valve inside the strut to drop out.

13. If equipped with the 2.4L engine, use special tools MB991681 and MB991682 to tighten the strut nut (self-locking nut) to 44–52 ft. lbs. (60–70 Nm).

14. If equipped with the 3.8L engine, use special tool MB991681, a hexagon wrench and a pipe to tighten the strut (self-locking nut) to 44–52 ft. lbs. (60–70 Nm).

15. Install the strut to the vehicle.

Lancer & Lancer Evolution

See Figure 308.

1. Before servicing the vehicle, refer to the Precautions Section.

2. Remove the strut from the vehicle.

3. Position the strut assembly using fixture tools MB991238 and MB9911237, or equivalent.

➡Do not tighten the special tool bolt too tight or the tool will break. Install the tools evenly and so that the maximum length will be attained within the installation range. Do not use an impact wrench to tighten the bolt of special tool MB991237, otherwise the tool may be damaged.

4. Using the special tools, compress the spring.

➡To prevent the piston rod jam nut inside the strut from loosening, do not use an impact wrench when the jam nut is loosened.

5. Use special tools MB991681 and MB991682 to secure the strut and then remove the jam nut.

6. Remove the strut insulator assembly, bearing, upper bearing seat, bump rubber or dust cover (depending on the vehicle), coil spring, and lower strut pad.

7. If the upper spring seat peeled off the pad, adhere the seat and the pad with double-sided tape.

8. Position the strut assembly in the holding fixture. Ensure that the bearing is seated correctly.

➡Do not use an impact wrench to tighten the bolt of the special tool or the tool will break.

9. While the coil spring is being compressed by the special tools, temporarily tighten the jam nut.

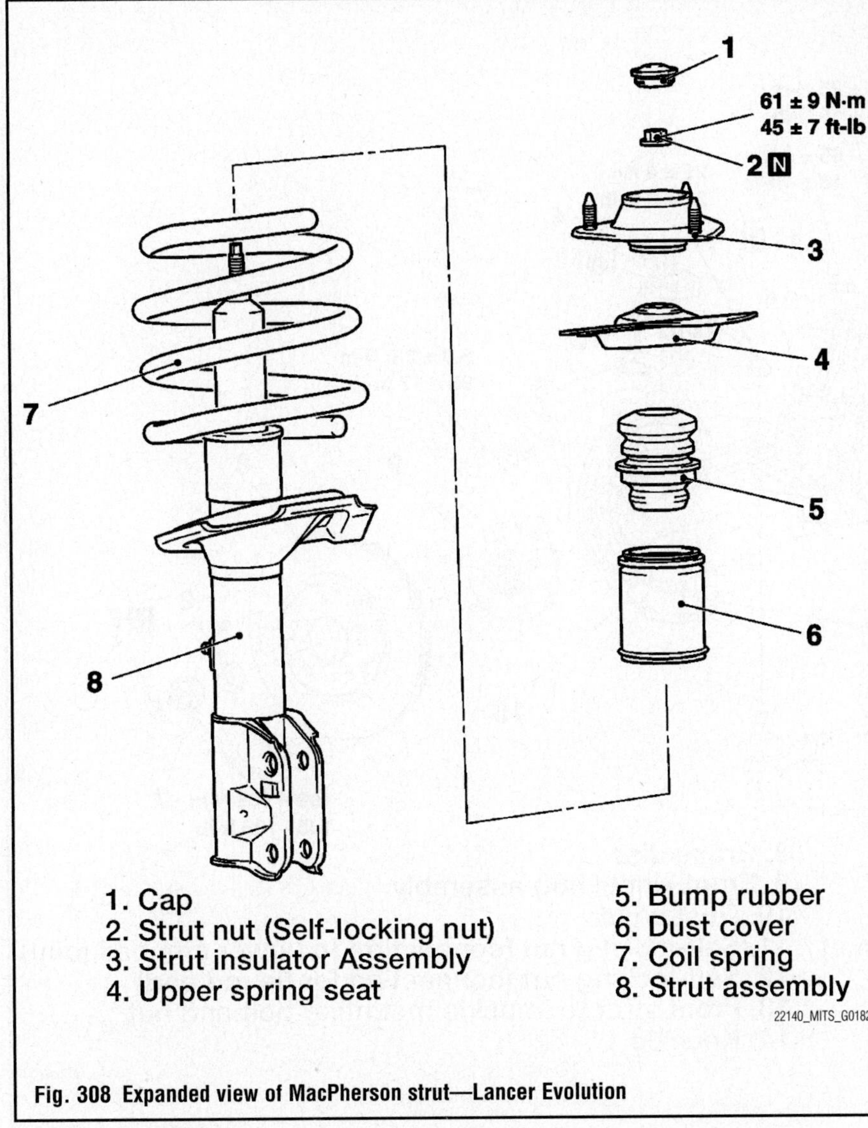

Fig. 308 Expanded view of MacPherson strut—Lancer Evolution

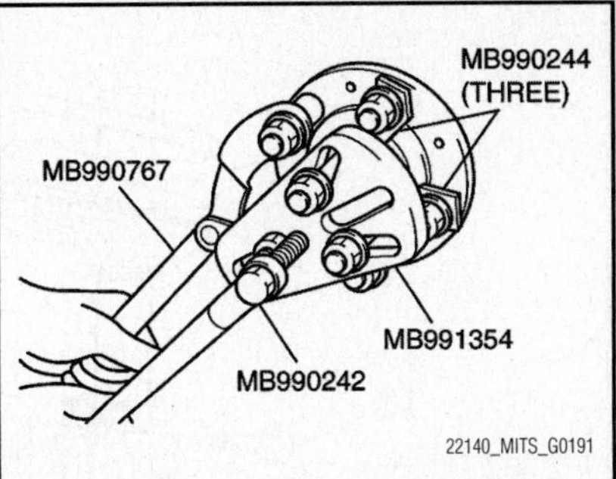

Fig. 309 Use special tools MB990242, MB990244, MB991354, and MB990767 to push out the driveshaft from the hub and knuckle

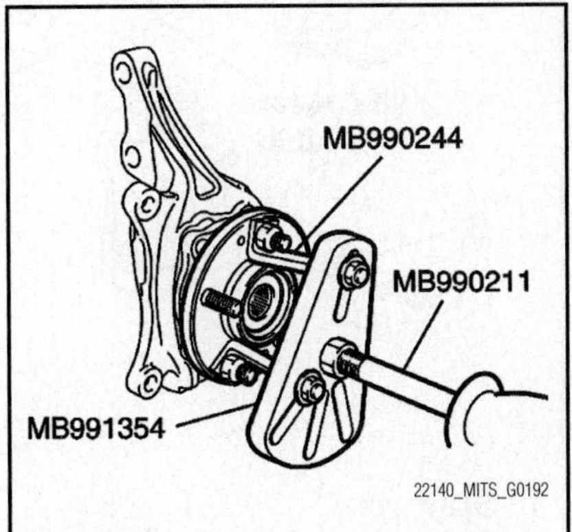

Fig. 310 Use special tools MB990244, MB991354, and MB990211 to pull out the front wheel hub from the steering knuckle

10. Align the hole in the strut assembly lower spring seat with the hole in the upper spring seat. You can use a rod to facilitate the alignment process.
11. Align both ends of the coil spring with the grooves in the spring seat, and then loosen the special tools.

➡Do not use an impact wrench to tighten the jam nut, otherwise the jam nut will not be tightened securely.

12. Use special tools MB991681 and MB991682 and tighten the jam nut to 44–52 ft. lbs. (60–70 Nm).
13. Install the strut in the vehicle.

STEERING KNUCKLE

REMOVAL & INSTALLATION

See Figures 309 through 311.

1. Before servicing the vehicle, refer to the Precautions Section.
2. Raise and safely support the vehicle.
3. Remove the front wheels and tires.
4. Remove the brake rotor. Refer to Front Disc Brakes, Rotor, Removal & Installation.
5. Remove the lower ball joint. Refer to Lower Ball Joint, Removal & Installation.
6. Remove the wheel speed sensor, the strut lower mounting bolt, and the lower arm mounting bolt from the steering knuckle.
7. Remove the front wheel hub mounting bolts while pushing out the driveshaft by hand.
8. If it is difficult to push out the driveshaft by hand, use special tools MB990242, MB990244, MB991354, and MB990767 to push out the driveshaft from the hub and knuckle.
9. If the front wheel hub is seized, remove the knuckle together with front wheel hub and fix them in a vise.
10. Hang the driveshaft on the vehicle body with a rope.
11. Use special tools MB990244, MB991354, and MB990211 to pull out the front wheel hub from the steering knuckle.

To install:

12. Install the hub and knuckle assembly.
13. Install the wheel speed sensor, the strut lower mounting bolt, and the lower arm mounting bolt to the steering knuckle.
14. Tighten the bolts according to the illustration.
15. Install the lower ball joint. Refer to Lower Ball Joint, Removal & Installation.
16. Install the brake rotor. Refer to Front Disc Brakes, Rotor, Removal & Installation.
17. Install the front wheels and tires.
18. Check alignment and adjust as necessary.

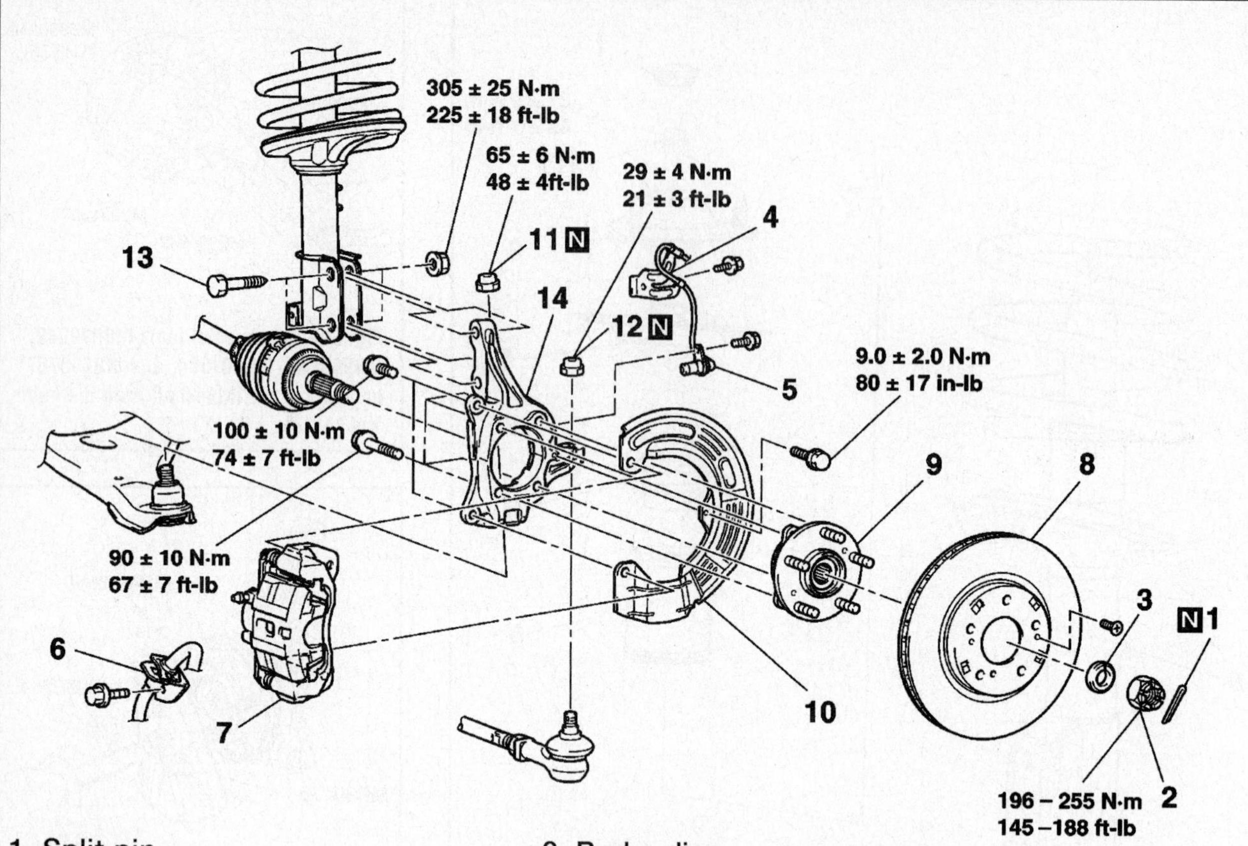

1. Split pin
2. Driveshaft nut
3. Washer
4. Front wheel speed sensor bracket
5. Front wheel speed sensor
6. Brake hose bracket
7. Caliper assembly
8. Brake disc
9. Front wheel hub assembly
10. Dust cover
11. Self-locking nut (connection for lower arm ball joint)
12. Self-locking nut (connection for tie rod end)
13. Front strut to knuckle mounting bolt and nut
14. Knuckle

22140_MITS_G0193

Fig. 311 View of wheel hub, steering knuckle, and related components

STABILIZER BAR

REMOVAL & INSTALLATION

Eclipse & Galant

See Figures 312 and 313.

➡Prior to removal of the stabilizer bar, center the front wheels and remove the ignition key. Failure to do so may damage the SRS clockspring and render the SRS system inoperative.

1. Before servicing the vehicle, refer to the Precautions Section.
2. Disconnect the negative battery cable.
3. Drain the power steering fluid.
4. Remove the front undercover.
5. Remove the center crossmember.
6. Remove the lower arm assembly.
7. Remove the driver's side air bag module.
8. Remove the steering wheel. Remove the column lower cover. Remove the clockspring.
9. Remove the floor console assembly. Remove the front scuff plate and cowl side trim. Remove the trunk lid opener cover.
10. Remove the accelerator pedal stopper. Remove the front floor carpet.
11. Disconnect the upper stabilizer link.
12. Remove the steering shaft pad.
13. Disconnect the steering column assembly and steering rack connection. Remove the rear roll stopper connecting bolt.
14. Remove the front axle crossmember bracket. Remove the return tube clamp bolt and nut. Remove the pressure tube clamp bolt and nut.

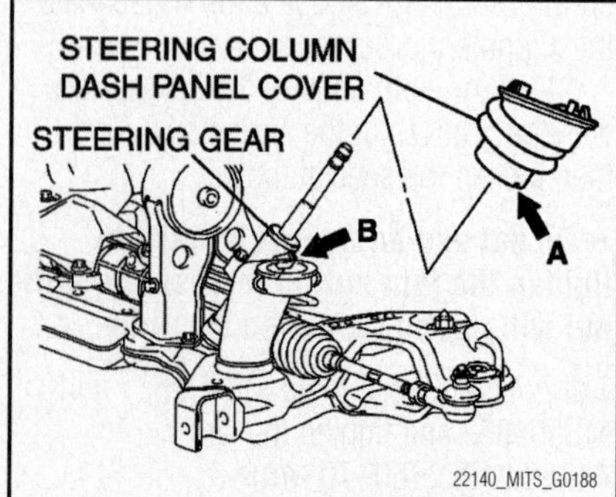

Fig. 312 Align the steering column dash panel cover notch "A" with the steering gear lug "B"—Eclipse & Galant

15. Disconnect and plug the return tube connection. Disconnect and plug the pressure hose connection. Discard the O-ring.

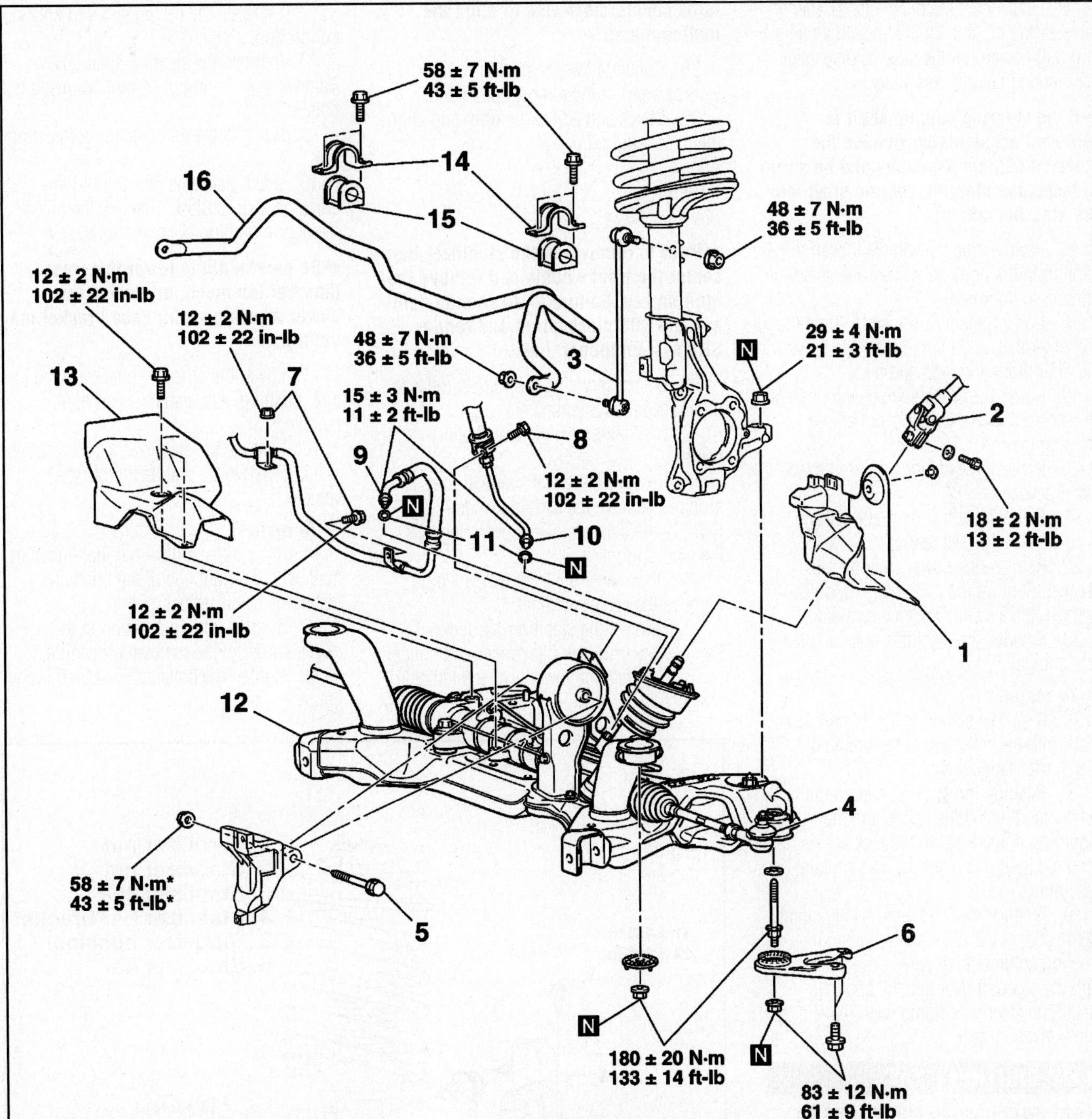

1. Steering shaft pad
2. Steering gear and steering column assembly connection
3. Stabilizer link
4. Tie rod end and knuckle connection
5. Rear roll stopper connection bolt
6. Front axle crossmember stay
7. Return tube clamp nut and bolt
8. Pressure tube clamp bolt
9. Steering gear and return tube connection
10. Steering gear and pressure tube connection
11. O-ring
12. Front axle crossmember, rear roll stopper and steering gear assembly
13. Steering gear and linkage protector
14. Stabilizer bracket
15. Stabilizer bushing
16. Stabilizer bar

22140_MITS_G0187

Fig. 313 Exploded view of the stabilizer bar and related components—Eclipse & Galant

16. Disconnect the pinch clamp. Pinch the steering column shaft clip with a pliers and pull upward on the shaft to disengage the steering column assembly.

➡If the steering column shaft is removed accidentally, remove the steering column assembly and be sure to insert the steering column shaft into the steering column.

17. From inside the vehicle loosen the 3 clips from the body panel near the steering column to floor hole.
18. Use a transmission jack to hold the crossmember, and then remove the crossmember mounting nuts and bolts.
19. Lower the crossmember with the rear roll stopper, then the stabilizer bar and steering rack.
20. Remove the steering column dash panel cover.
21. Remove the steering rack linkage protector, on the 3.8L engine.
22. Remove the power steering rack bracket. Remove the rack retaining bolts. Remove the assembly from the vehicle.
23. Remove the stabilizer bracket, bushing, and bar.

To install:

24. Align the stabilizer bar identification mark with the right end of the bushing. Install the stabilizer bar.
25. Position the steering rack in the vehicle and install the clamp and the mounting bolts. Be sure the rack is centered before connecting it to the joint assembly.
26. When installing the steering column dash panel cover pinch clamp, align the steering column dash panel cover notch "A" with the steering gear lug "B" and then install the steering column dash panel cover to the steering gear.

⁂ CAUTION

When installing the clockspring be sure that the springs mating marks are properly aligned. If not the steering wheel may not rotate completely during a turn, or the flat cable in the clockspring could be damaged. This would prevent normal SRS operation and possibly cause serious injury to the driver.

27. After aligning the mating surfaces of the clockspring, turn the front wheels to the straight ahead position. Install the clockspring to the column switch.

➡Turn the clockspring clockwise fully. Then turn it back approximately 3 ¾ turns counterclockwise to align the mating marks.

28. Continue the installation in the reverse order of the removal procedure.
29. Check and adjust the front end alignment, as required.

Lancer

See Figure 314.

➡Prior to removal of the stabilizer bar, center the front wheels and remove the ignition key. Failure to do so may damage the SRS clockspring and render SRS system inoperative.

1. Before servicing the vehicle, refer to the Precautions Section.
2. Disconnect the negative battery cable.
3. Remove the driver's side air bag module. Remove the steering wheel. Remove the column lower cover. Remove the clockspring.
4. Disconnect the front exhaust pipe. Remove the center member.
5. Remove the stabilizer bar locknut. Remove the stabilizer bar rubber and collar.
6. Separate the lower arm and knuckle connection
7. Separate the tie rod end and knuckle connection.
8. Remove the steering shaft cover. Remove the steering gear joint connecting bolt.
9. Remove the rear stopper connecting bolt.
10. Use a transaxle jack to hold the crossmember in place. Remove the crossmember mounting bolts and nuts.

➡Be careful not to lower the crossmember too much, otherwise the power steering return hose bracket may deform.

11. Lower the crossmember until the fixture, bushing, and stabilizer can be removed.
12. Remove the stabilizer retaining bolts.
13. Remove the stabilizer from the vehicle.

To install:

14. Align the stabilizer bar identification mark with the right end of the bushing. Install the stabilizer bar.
15. Continue the installation in the reverse order of the removal procedure.
16. Tighten the bolts as illustrated in the figure.

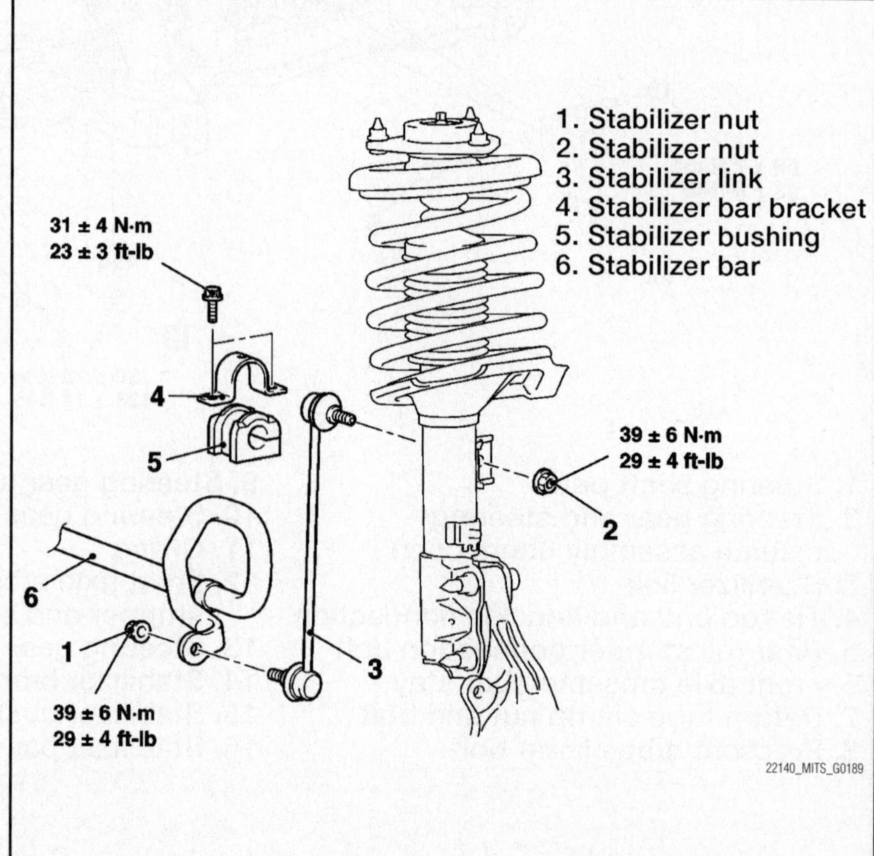

Fig. 314 Exploded view of the stabilizer bar and related components—Lancer

17. Check and adjust the front end alignment, as required.

Lancer Evolution

See Figure 315.

➡Prior to removal of the stabilizer bar, center the front wheels and remove the ignition key. Failure to do so may damage the SRS clockspring and render SRS system inoperative.

1. Before servicing the vehicle, refer to the Precautions Section.
2. Disconnect the negative battery cable.
3. Remove the driver's side air bag module. Remove the steering wheel. Remove the column lower cover. Remove the clockspring.
4. Disconnect the front exhaust pipe. Remove the center member.
5. Remove the stabilizer bar locknut. Remove the stabilizer bar rubber and collar.
6. Separate the lower arm and knuckle connection
7. Separate the tie rod end and knuckle connection.
8. Remove the steering shaft cover. Remove the steering gear joint connecting bolt.
9. Remove the rear stopper connecting bolt.
10. Use a transaxle jack to hold the crossmember in place. Remove the crossmember mounting bolts and nuts.

➡Be careful not to lower the crossmember too much, otherwise the power steering return hose bracket may deform.

11. Lower the crossmember until the fixture, bushing and stabilizer can be removed.
12. Remove the stabilizer retaining bolts. Remove the stabilizer from the vehicle.

To install:

13. Align the stabilizer bar identification mark with the right end of the bushing. Install the stabilizer bar.
14. Continue the installation in the reverse order of the removal procedure.
15. Tighten the bolts as illustrated in the figure.
16. Check and adjust the front end alignment, as required.

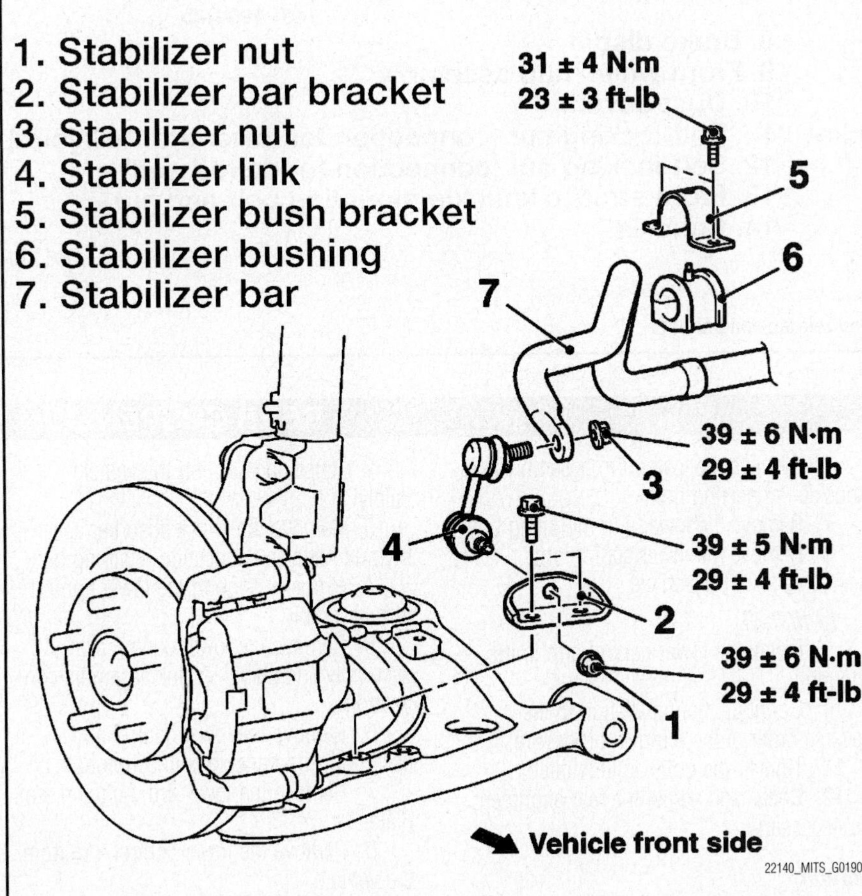

Fig. 315 Exploded view of the stabilizer bar and related components—Lancer Evolution

WHEEL HUB AND BEARING

REMOVAL & INSTALLATION

See Figures 316 and 317.

1. Before servicing the vehicle, refer to the Precautions Section.
2. Remove the steering knuckle. Refer to Steering Knuckle, Removal & Installation.
3. Use special tools MB990244, MB991354, and MB990211 to pull out the front wheel hub from the steering knuckle.

To install:

4. Install the hub bearing into the knuckle assembly.
5. Install the knuckle. Refer to Steering Knuckle, Removal & Installation.
6. Tighten the bolts according to the illustration.
7. Install the lower ball joint. Refer to Lower Ball Joint, Removal & Installation.
8. Install the brake rotor. Refer to Front Disc Brakes, Rotor, Removal & Installation.
9. Install the front wheels and tires.
10. Check alignment and adjust as necessary.

ADJUSTMENT

The front wheel bearings are not adjustable. If the bearings are noisy or become loose, they must be replaced.

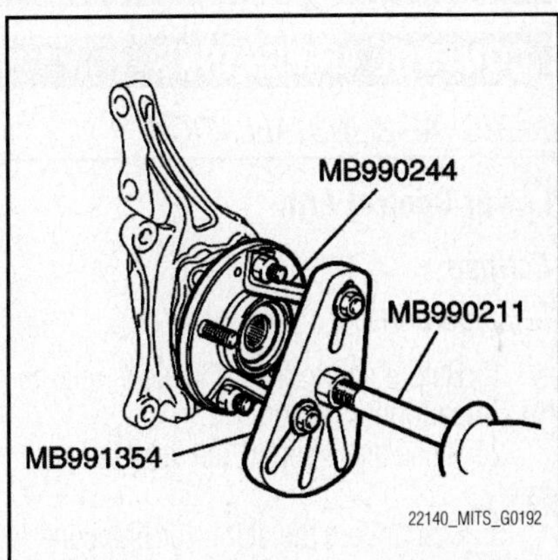

Fig. 316 Use special tools MB990244, MB991354, and MB990211 to pull out the front wheel hub from the steering knuckle

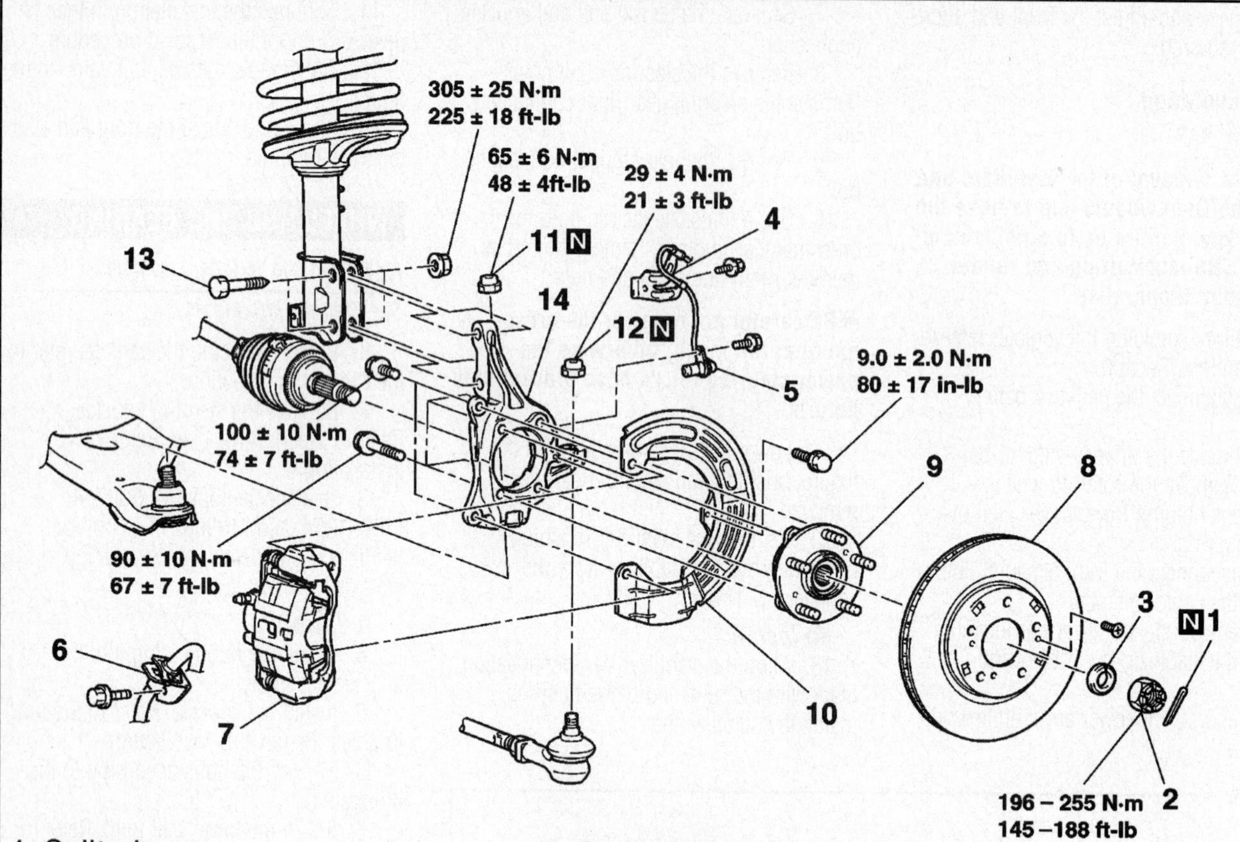

Fig. 317 View of wheel hub, steering knuckle, and related components

CONTROL ARMS/LINKS

REMOVAL & INSTALLATION

Lower Control Arm

Eclipse

See Figure 318.

1. Before servicing the vehicle, refer to the Precautions Section.
2. Raise and support the vehicle safely.
3. Disconnect the stabilizer link connection.
4. Disconnect the wheel speed sensor mounting bolts.
5. Properly support the lower control arm assembly.
6. Remove the lower arm assembly and knuckle connecting bolt.
7. Remove the lower arm mounting bolt.
8. Remove the lower control arm assembly from the vehicle.

To install:

9. Install the lower control arm to its mounting.
10. Continue the installation in the reverse order of the removal procedure.
11. Tighten the bolts as illustrated.
12. Check and adjust the rear alignment, as necessary.

Galant

See Figure 319.

1. Before servicing the vehicle, refer to the Precautions Section.
2. Raise and support the vehicle safely.
3. Remove the shock absorber assembly and knuckle connection retaining bolt.
4. Properly support the lower control arm assembly.
5. Remove the lower control arm assembly and stabilizer link assembly connection.
6. Remove the lower control arm assembly and knuckle connection bolt.
7. Remove the lower arm bolt and arm plate.
8. Remove the lower control arm from the vehicle.

To install:

9. Install the lower control arm to its mounting.

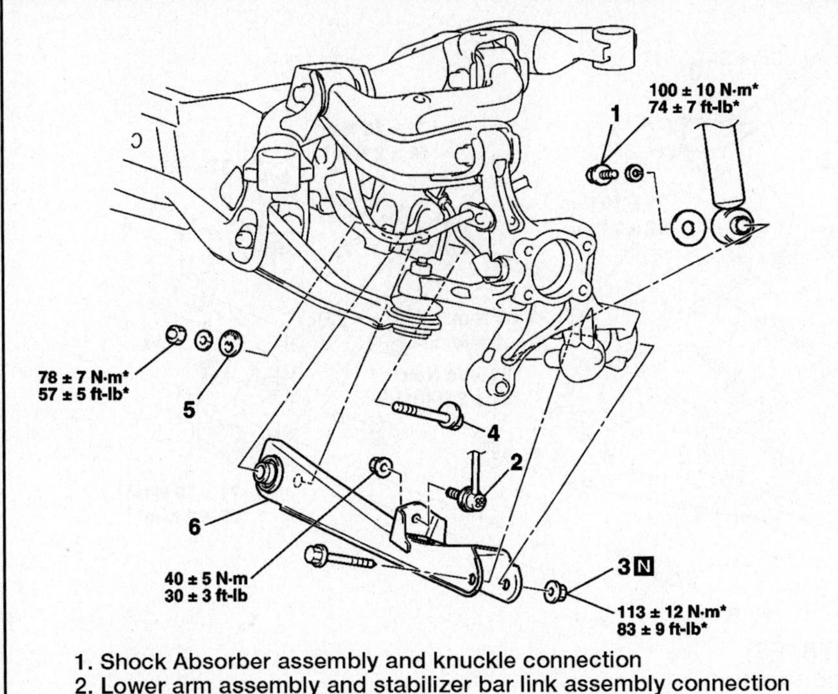

Fig. 318 View of lower control arm assembly—Eclipse

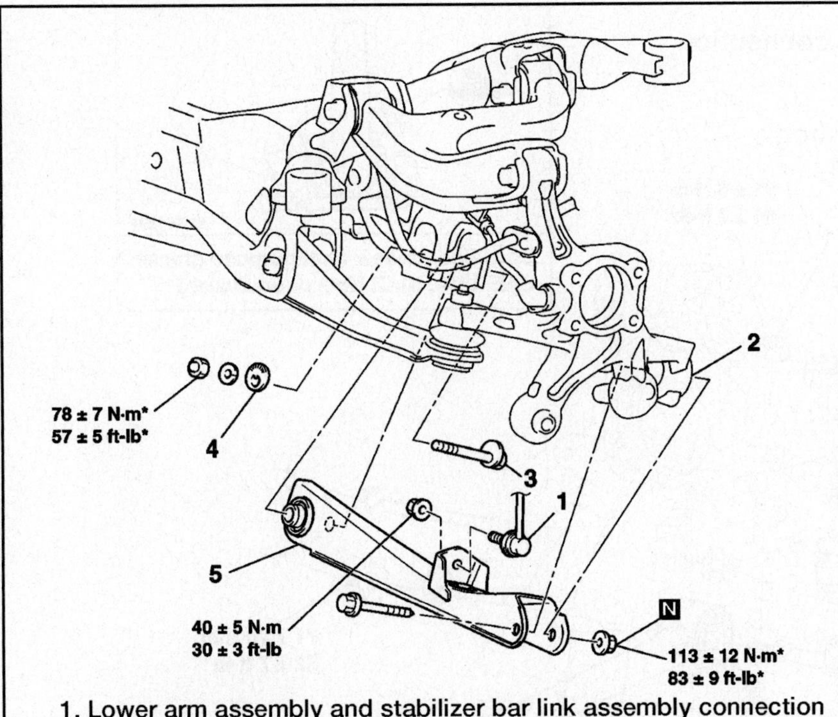

Fig. 319 View of lower control arm assembly—Galant

10. Continue the installation in the reverse order of the removal procedure.
11. Tighten the bolts as illustrated.
12. Check and adjust the rear alignment, as necessary.

Lancer

See Figure 320.

1. Before servicing the vehicle, refer to the Precautions Section.
2. Raise and support the vehicle safely.
3. Support the lower arm assembly, using a floor jack.
4. After making a mating mark on the toe-in or camber adjusting bolt, remove the control link or lower arm.
5. Remove the lower arm and trailing arm bolt and nut.
6. Remove the lower shock absorber retaining bolt and nut.
7. Remove the lower control arm from the vehicle.

To install:

8. Install the lower control arm to its mounting.
9. Continue the installation in the reverse order of the removal procedure.
10. Tighten the bolts as illustrated.
11. Check and adjust the rear alignment, as necessary.

Lancer Evolution

See Figure 321.

1. Before servicing the vehicle, refer to the Precautions Section.
2. Raise and support the vehicle safely.
3. After making a mating mark on the toe-in or camber adjusting bolt, remove the control link or lower arm.
4. Properly support the lower control arm assembly.
5. Remove the lower control arm assembly mounting bolt.
6. Remove the lower control arm assembly.

To install:

7. Install the lower control arm to its mounting.
8. Continue the installation in the reverse order of the removal procedure.
9. Tighten the bolts as illustrated.
10. Check and adjust the rear alignment, as necessary.

Upper Control Arm

Eclipse

See Figure 322.

1. Before servicing the vehicle, refer to the Precautions Section.

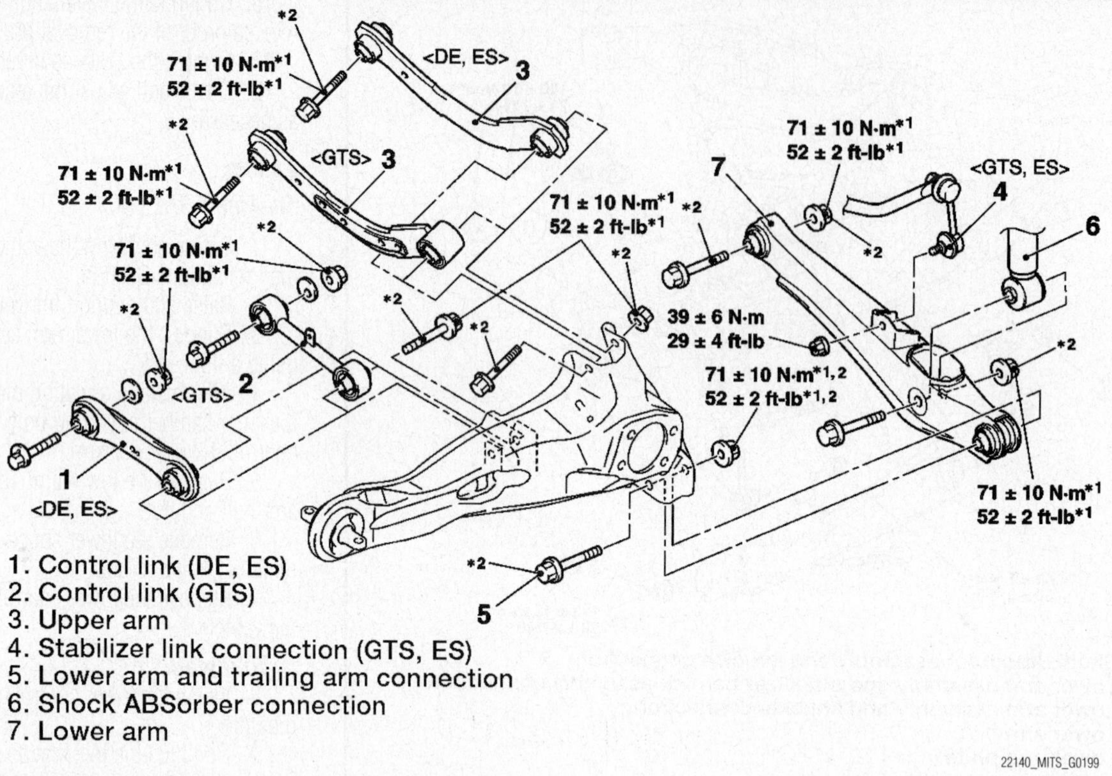
*2
71 ± 10 N·m*1
52 ± 2 ft-lb*1
<DE, ES> 3
<GTS> 3
7
<GTS, ES>
4
6
39 ± 6 N·m
29 ± 4 ft-lb
71 ± 10 N·m*1, 2
52 ± 2 ft-lb*1, 2
<GTS> 2
1
<DE, ES>
5
1. Control link (DE, ES)
2. Control link (GTS)
3. Upper arm
4. Stabilizer link connection (GTS, ES)
5. Lower arm and trailing arm connection
6. Shock ABSorber connection
7. Lower arm
22140_MITS_G0199

Fig. 320 View of lower and upper control arm assemblies—Lancer

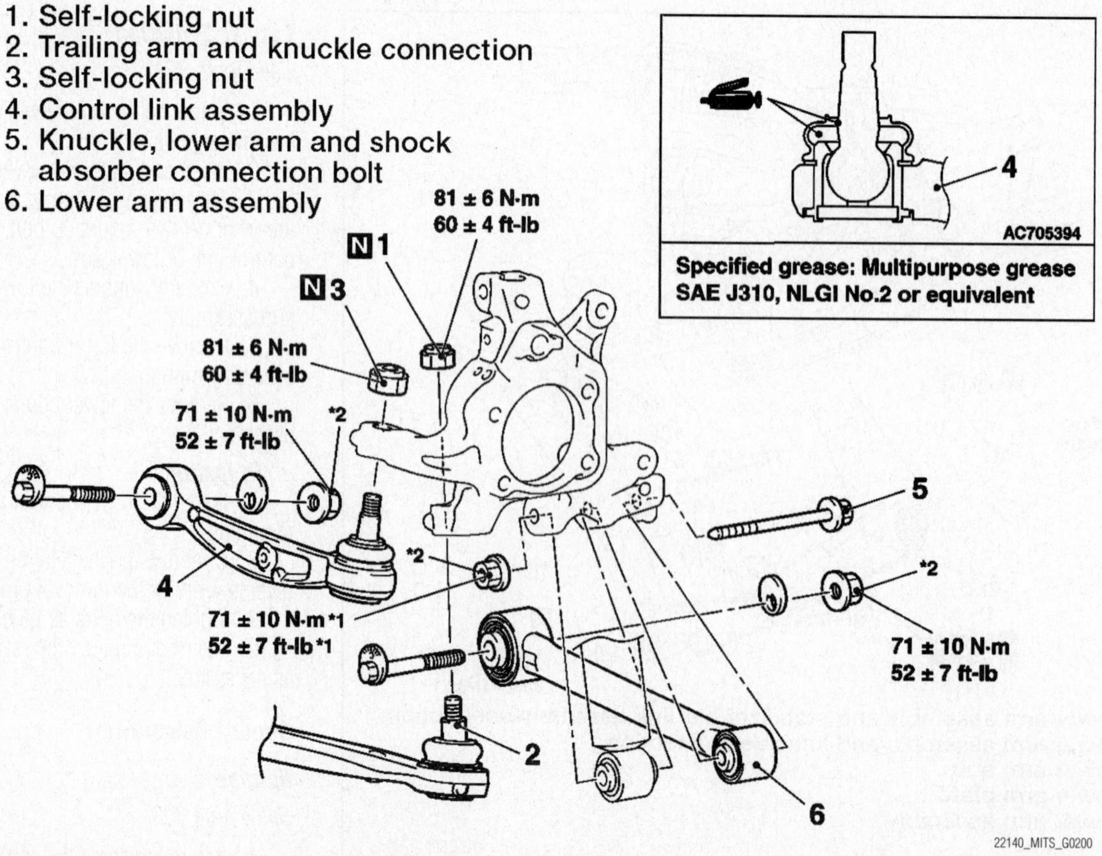
1. Self-locking nut
2. Trailing arm and knuckle connection
3. Self-locking nut
4. Control link assembly
5. Knuckle, lower arm and shock absorber connection bolt
6. Lower arm assembly
81 ± 6 N·m
60 ± 4 ft-lb
N 1
N 3
71 ± 10 N·m
52 ± 7 ft-lb
*2
4
71 ± 10 N·m*1
52 ± 7 ft-lb*1
2
5
6
AC705394
Specified grease: Multipurpose grease SAE J310, NLGI No.2 or equivalent
22140_MITS_G0200

Fig. 321 View of lower and upper control arm assemblies—Lancer Evolution

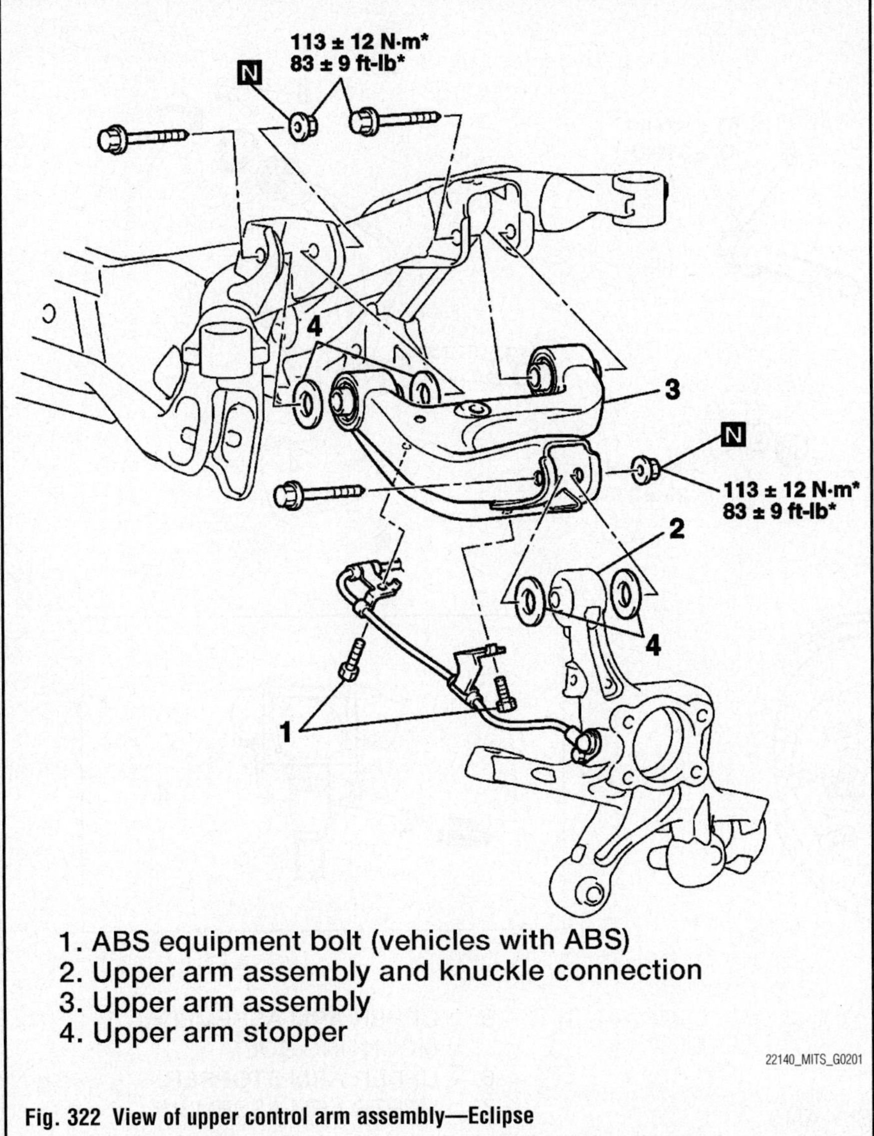

Fig. 322 View of upper control arm assembly—Eclipse

2. Raise and support the vehicle safely.
3. Position a floor jack to properly support the knuckle before removing the upper control arm.
4. Remove the upper control arm and knuckle connecting bolt.
5. Remove the upper control arm assembly mounting bolts.
6. Remove the upper control arm bracket assembly.
7. Remove the upper control arm bracket.
8. Remove the upper control arm from the vehicle.

To install:

9. Install the upper control arm to its mounting.
10. Continue the installation in the reverse order of the removal procedure.
11. Tighten the bolts as illustrated.
12. Check and adjust the rear alignment, as necessary.

Galant

1. Before servicing the vehicle, refer to the Precautions Section.
2. Raise and support the vehicle safely.
3. If equipped with ABS, remove the wheel speed sensor retaining bolts.
4. Position a floor jack to properly support the knuckle before removing the upper control arm.
5. Remove the upper arm assembly and knuckle retaining bolt and nut.
6. Remove the upper arm assembly retaining bolts and nuts.
7. Remove the upper control arm from the vehicle.

To install:

8. Install the upper control arm to its mounting.
9. Continue the installation in the reverse order of the removal procedure. Note the following torques:
   a. Upper arm assembly mounting bolts: 29 ft. lbs. (39 Nm).
   b. Upper arm to knuckle connecting bolt: 72 ft. lbs. (98 Nm).
10. Check and adjust the rear alignment, as necessary.

Lancer

1. Before servicing the vehicle, refer to the Precautions Section.
2. Raise and support the vehicle safely.
3. Support the lower arm assembly, using a floor jack.
4. After making a mating mark on the toe-in or camber adjusting bolt, remove the control link or lower arm.
5. Remove the upper control arm retaining bolts. Remove the upper control arm from the vehicle.

To install:

6. Install the upper control arm to its mounting.
7. Continue the installation in the reverse order of the removal procedure.
8. Check and adjust the rear alignment, as necessary.

Lancer Evolution

See Figure 323.

1. Remove the fuel filler cap, when removing the left side upper arm assembly.
2. Raise and support the vehicle safely.
3. Remove the fuel protector and bolt, when removing the left side upper arm assembly.
4. Position a floor jack to properly support the rotor and hub assembly before removing the upper control

➡Do not remove the nut from the ball joint. Loosen it and use special tool MB991897 to avoid possible damage to the ball joint threads. Hang the special tool in place with wire or string to prevent it from falling.

5. Install the special tool. Turn the bolt and knob as necessary to make the jaws of the tool parallel. Tighten the bolt by hand and confirm that the jaws are still parallel.

➡When adjusting the jaws in parallel, make sure the knob is in the vertical (upward) position.

6. Tighten the bolt with a wrench to disconnect the upper arm assembly and the knuckle.
7. Remove the upper arm assembly mounting bolts. Remove the upper arm stopper.
8. Remove the upper control arm from the vehicle.

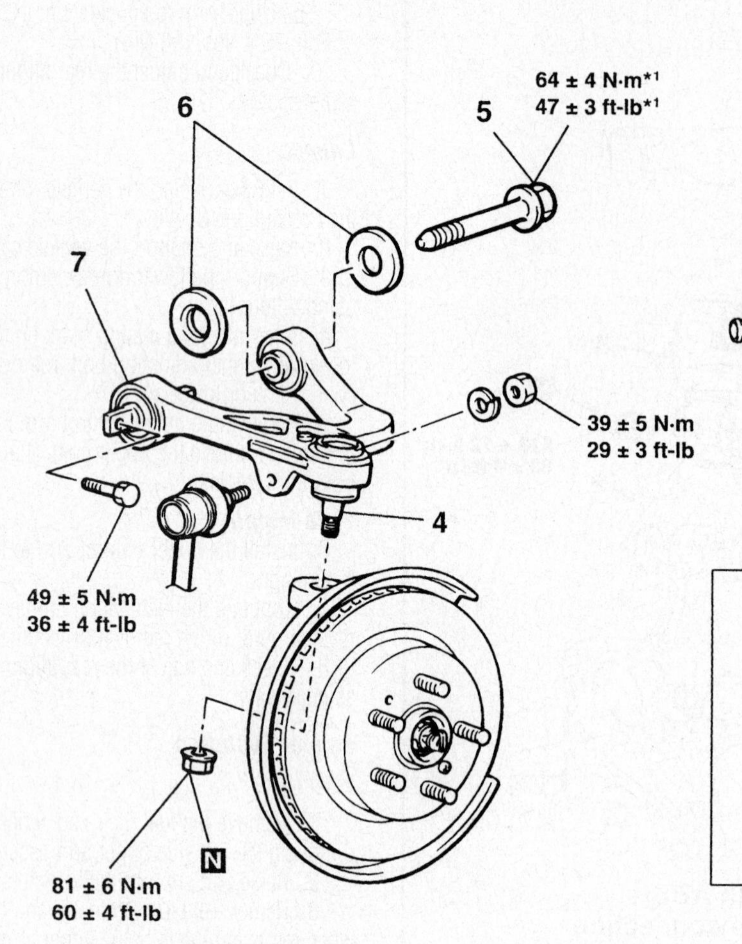

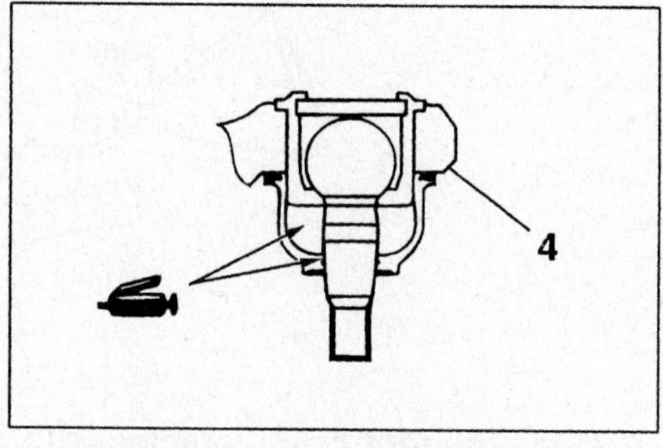

1. FUEL FILLER CAP
2. PROTECTOR
3. BOLT
4. UPPER ARM ASSEMBLY AND KNUCKLE CONNECTION
5. UPPER ARM ASSEMBLY MOUNTING BOLT
6. UPPER ARM STOPPER
7. UPPER ARM ASSEMBLY

09482_GALA_G0155

Fig. 323 Rear upper control arm and related components—Evolution

To install:

9. Install the upper control arm to its mounting.
10. Continue the installation in the reverse order of the removal procedure.
11. Tighten the bolts as illustrated.
12. Check and adjust the rear alignment, as necessary.

MACPHERSON STRUTS

REMOVAL & INSTALLATION

Eclipse

See Figure 324.

1. Before servicing the vehicle, refer to the Precautions Section.
2. Remove the rear trunk trim panel, on Eclipse.
3. On Eclipse Spyder (convertible) make sure the top is in the fully closed position.
4. Raise and support the vehicle safely.
5. Remove the drain turf mounting hose, so that the cap can be seen from the passenger compartment.
6. Remove the strut/shock absorber cap. Remove the mounting nuts.
7. Remove the lower shock absorber retaining bolt.
8. Remove the shock absorber assembly from the vehicle.

To install:

9. Installation is the reverse of the removal procedure.
10. Tighten the bolts according to the illustration.

Galant

See Figure 325.

1. Before servicing the vehicle, refer to the Precautions Section.
2. Remove the rear sub woofer speaker.

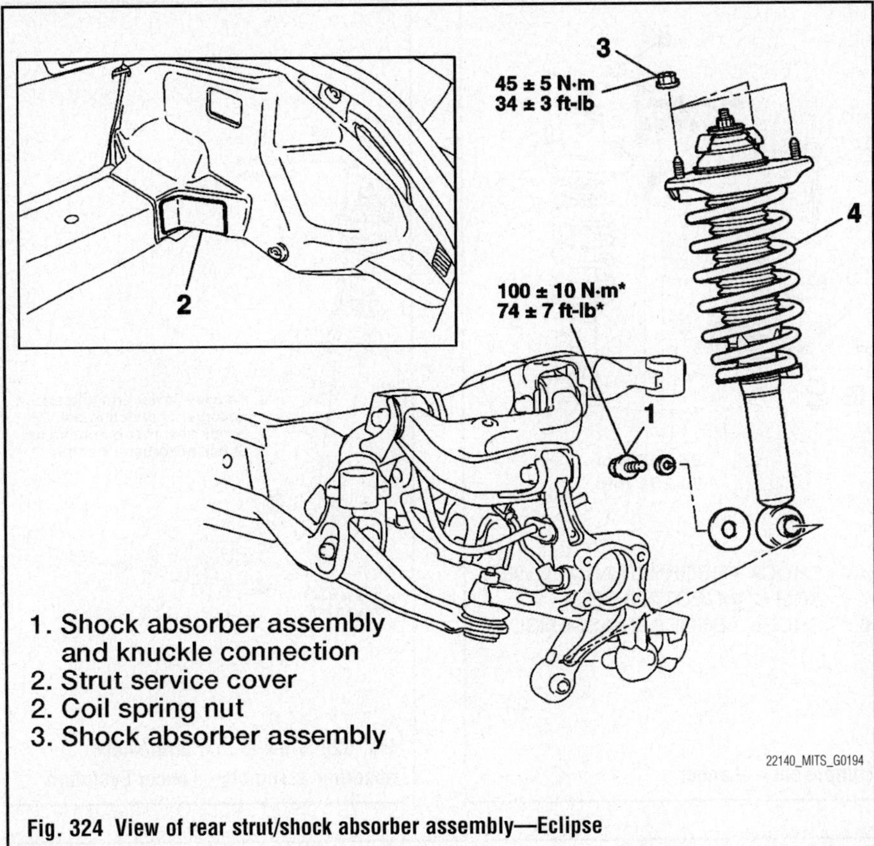

Fig. 324 View of rear strut/shock absorber assembly—Eclipse

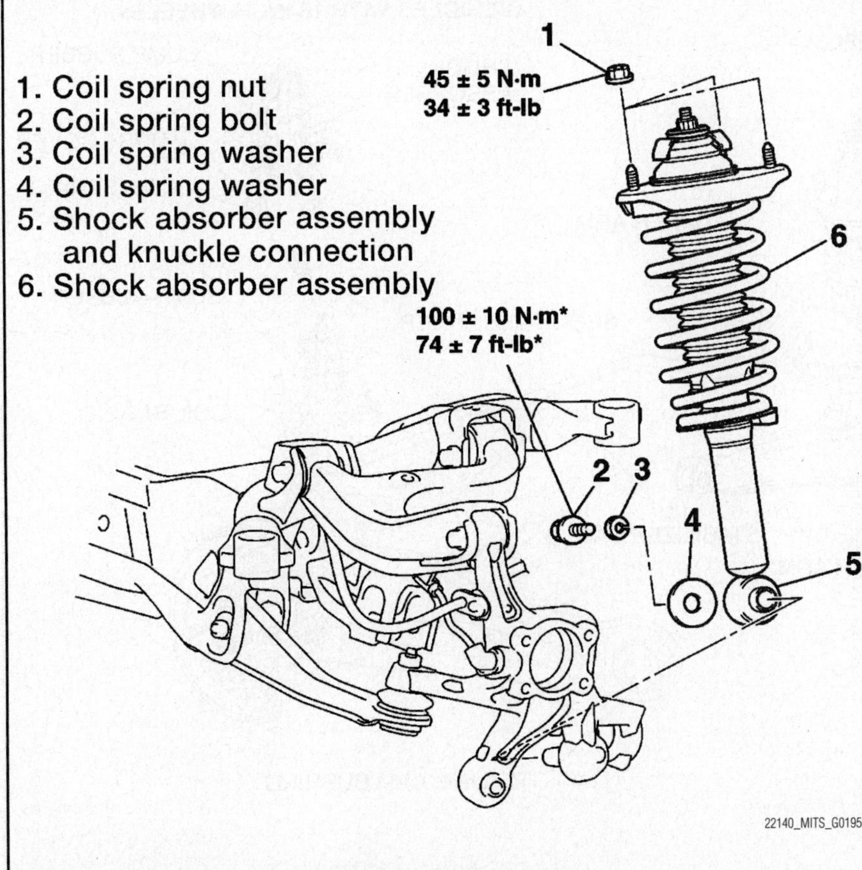

Fig. 325 View of rear strut/shock absorber assembly—Galant

3. Remove the trunk trim (front side).
4. Raise and support the vehicle safely.
5. Remove the shock absorber assembly and knuckle retaining bolt.
6. Remove the shock absorber service cover.
7. Remove the upper shock absorber retaining nuts.
8. Remove the strut/shock absorber assembly from the vehicle.

To install:

9. Installation is the reverse of the removal procedure.
10. Tighten the bolts according to the illustration.

Lancer

See Figures 326 and 327.

1. Before servicing the vehicle, refer to the Precautions Section.
2. Remove the stabilizer link connection.
3. Support the lower control arm with a jack.
4. Remove or disconnect the following:
   - Lower control arm and trailing arm bolt
   - Upper shock absorber mounting nut
   - Shock absorber-to-lower control arm attaching bolt
   - Shock absorber/strut assembly

To install:

5. Position the shock absorber into the vehicle. Install the spring seat stepped section so that it points toward the rear side of the vehicle.
6. Install or connect the following:
   - Shock absorber-to-lower control arm bolt and nut. Tighten the nut to 70 ft. lbs. (95 Nm)
   - Upper shock mounting nut and tighten to 32 ft. lbs. (44 Nm)
   - Lower control arm and trailing arm
   - Stabilizer link. Tighten the self-locking nuts so that the end of the stabilizer line bolt protrudes 0.24–0.31 inch (6–8mm)

Lancer Evolution

See Figures 328 and 329.

1. Before servicing the vehicle, refer to the Precautions Section.
2. Raise and support the vehicle safely.
3. Remove the upper shock absorber retaining nuts.

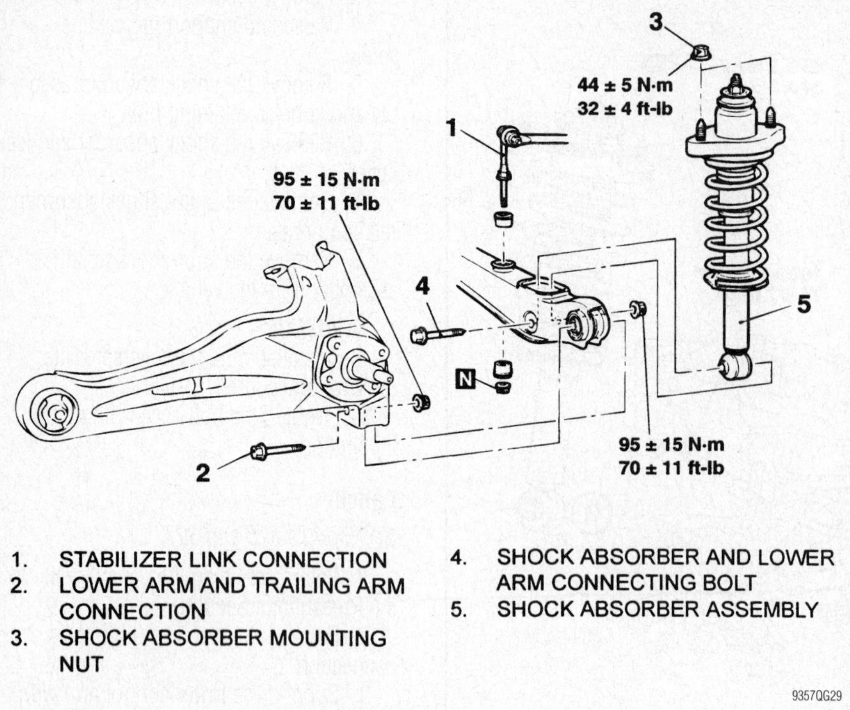
3
44 ± 5 N·m
32 ± 4 ft-lb
1
95 ± 15 N·m
70 ± 11 ft-lb
4
5
N
2
95 ± 15 N·m
70 ± 11 ft-lb
1. STABILIZER LINK CONNECTION
2. LOWER ARM AND TRAILING ARM CONNECTION
3. SHOCK ABSORBER MOUNTING NUT
4. SHOCK ABSORBER AND LOWER ARM CONNECTING BOLT
5. SHOCK ABSORBER ASSEMBLY
9357QG29

Fig. 326 Rear shock absorber/strut and related components—Lancer

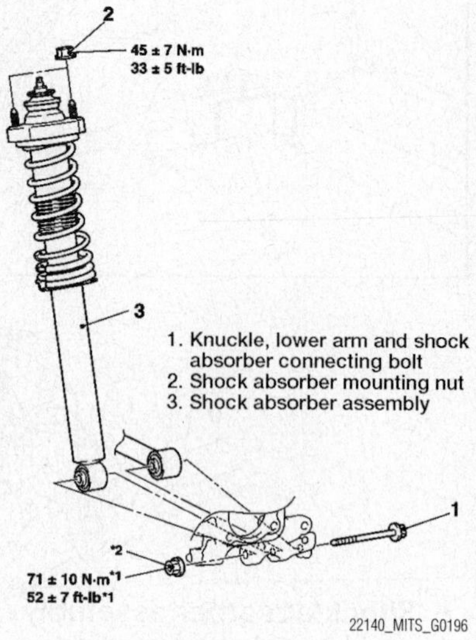
2
45 ± 7 N·m
33 ± 5 ft-lb
3
1. Knuckle, lower arm and shock absorber connecting bolt
2. Shock absorber mounting nut
3. Shock absorber assembly
1
*2
71 ± 10 N·m*1
52 ± 7 ft-lb*1
22140_MITS_G0196

Fig. 328 View of rear strut/shock absorber assembly—Lancer Evolution

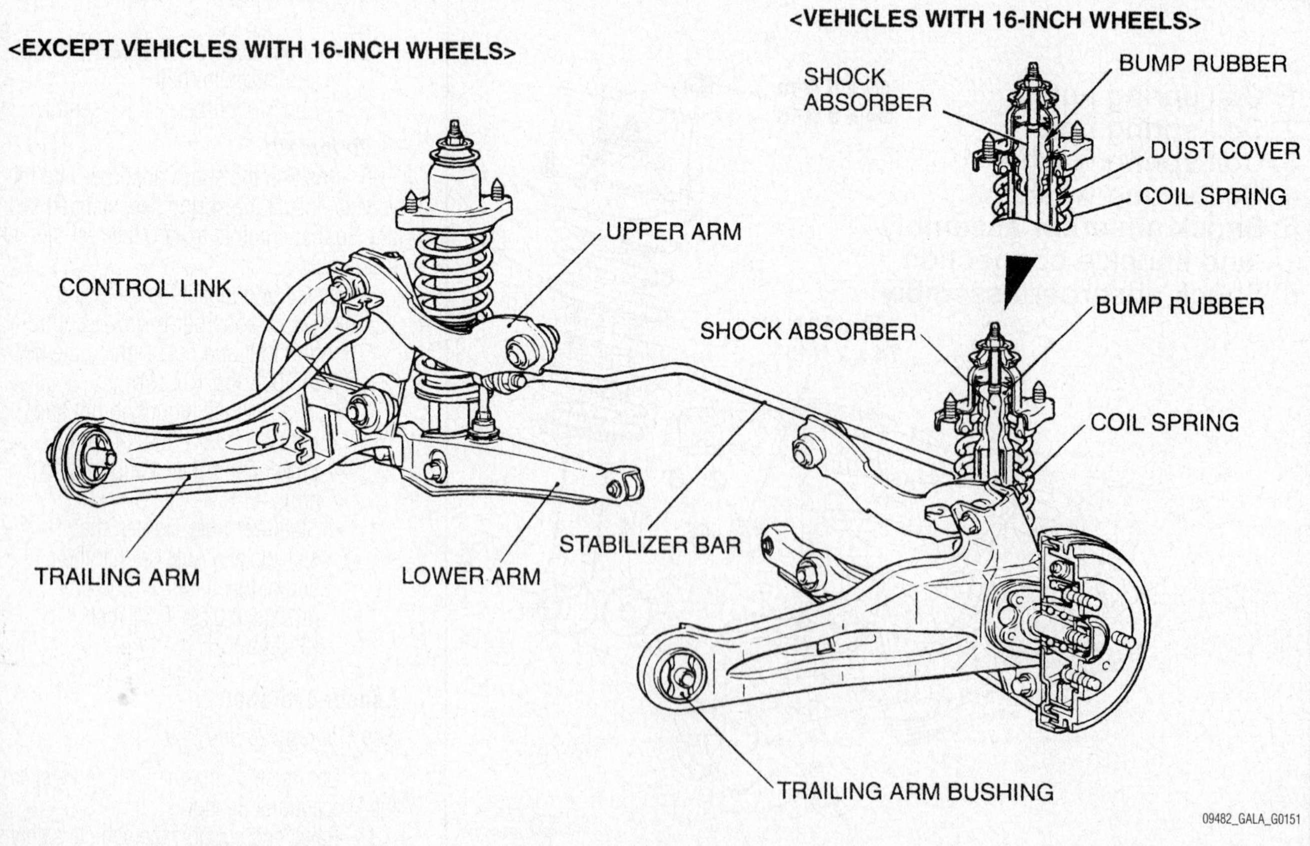
<EXCEPT VEHICLES WITH 16-INCH WHEELS>
<VEHICLES WITH 16-INCH WHEELS>
SHOCK ABSORBER
BUMP RUBBER
DUST COVER
COIL SPRING
UPPER ARM
CONTROL LINK
SHOCK ABSORBER
BUMP RUBBER
COIL SPRING
STABILIZER BAR
TRAILING ARM
LOWER ARM
TRAILING ARM BUSHING
09482_GALA_G0151

Fig. 327 Rear suspension and related components—Lancer

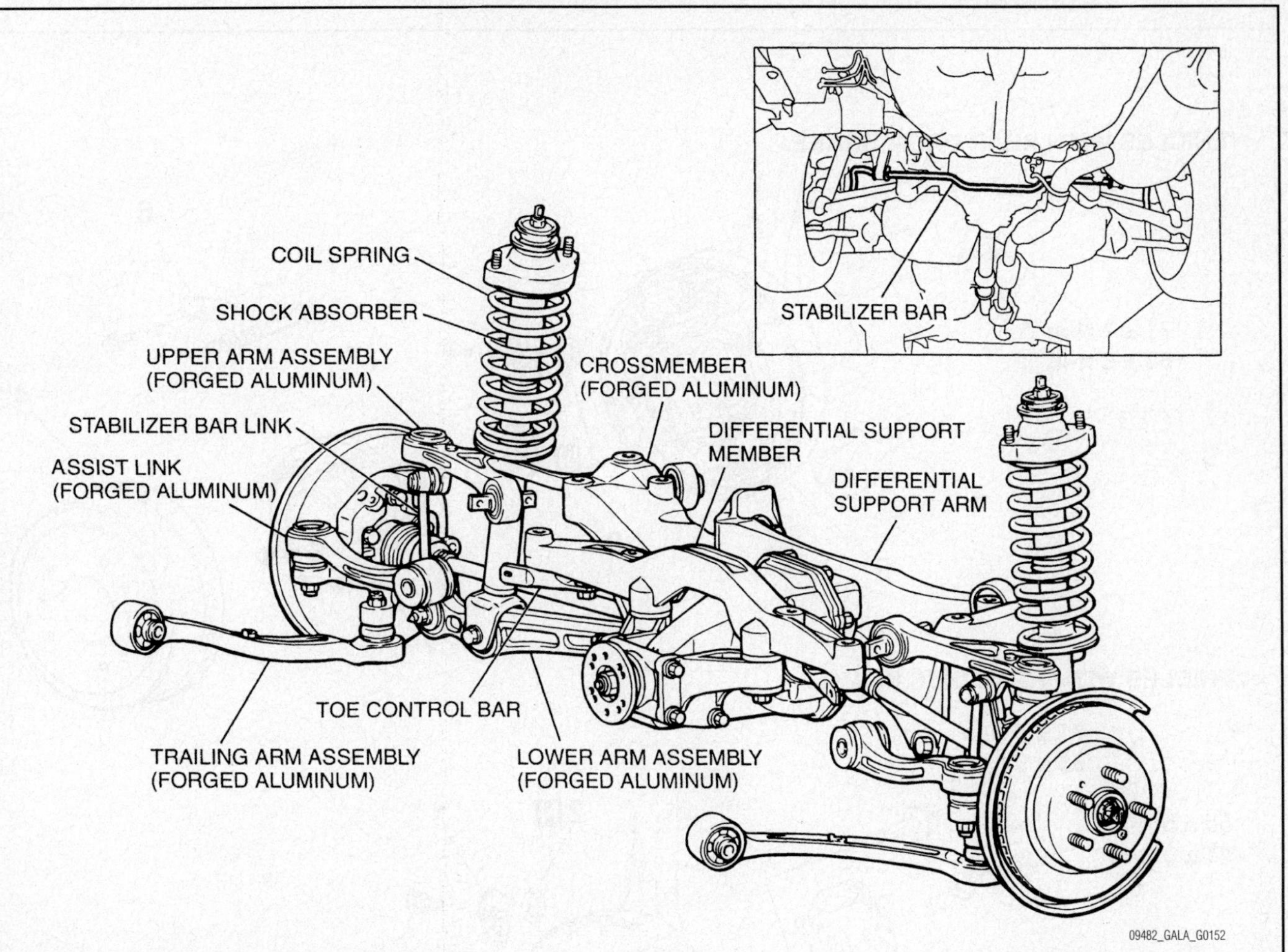

Fig. 329 Rear suspension and related components—Lancer Evolution

4. Remove the lower shock absorber retaining nuts.
5. Remove the shock absorber from the vehicle.

To install:

6. Installation is the reverse of the removal procedure.
7. Tighten the bolts according to the illustration.

OVERHAUL

See Figure 330.

1. Before servicing the vehicle, refer to the Precautions Section.
2. Use a suitable coil spring compressor and compress the coil spring.

➡An air tool should not be used to tighten the spring compressor.

3. While holding the piston rod, remove the self-locking nut.
4. Remove the upper bracket assembly and spring pad.
5. Remove the collar, upper bushing, cup assembly, bump rubber, and dust cover.

1. Self-locking nut
2. Washer
3. Upper bushing A
4. Upper bracket assembly
5. Upper spring pad
6. Upper bushing B
7. Collar
8. Cup assembly
9. Dust cover
10. Bump rubber
11. Coil spring
12. Shock absorber assembly

89578G52

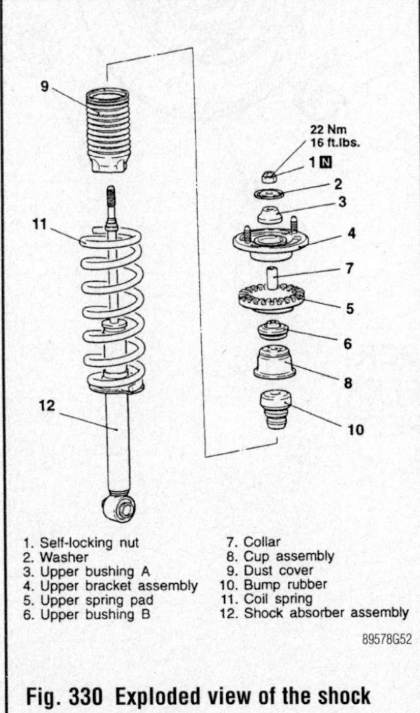

Fig. 330 Exploded view of the shock and strut assembly

6. Remove the coil spring from the strut.

To install:

7. Align the end of the coil spring with the stepped part of the spring seat and install the compressed coil spring on the strut.
8. Install the dust cover, bump rubber, cup assembly, upper bushing, collar, upper spring pad, and bracket assembly on the strut.
9. Install the upper bushing and washer on the piston rod.
10. Install a new self-locking nut on the piston rod. Temporarily tighten the nut.
11. Carefully remove the spring compressor from the spring. Tighten the self-locking nut to 16 ft. lbs. (25 Nm).

WHEEL HUB AND BEARING

REMOVAL & INSTALLATION

Eclipse

See Figure 331.

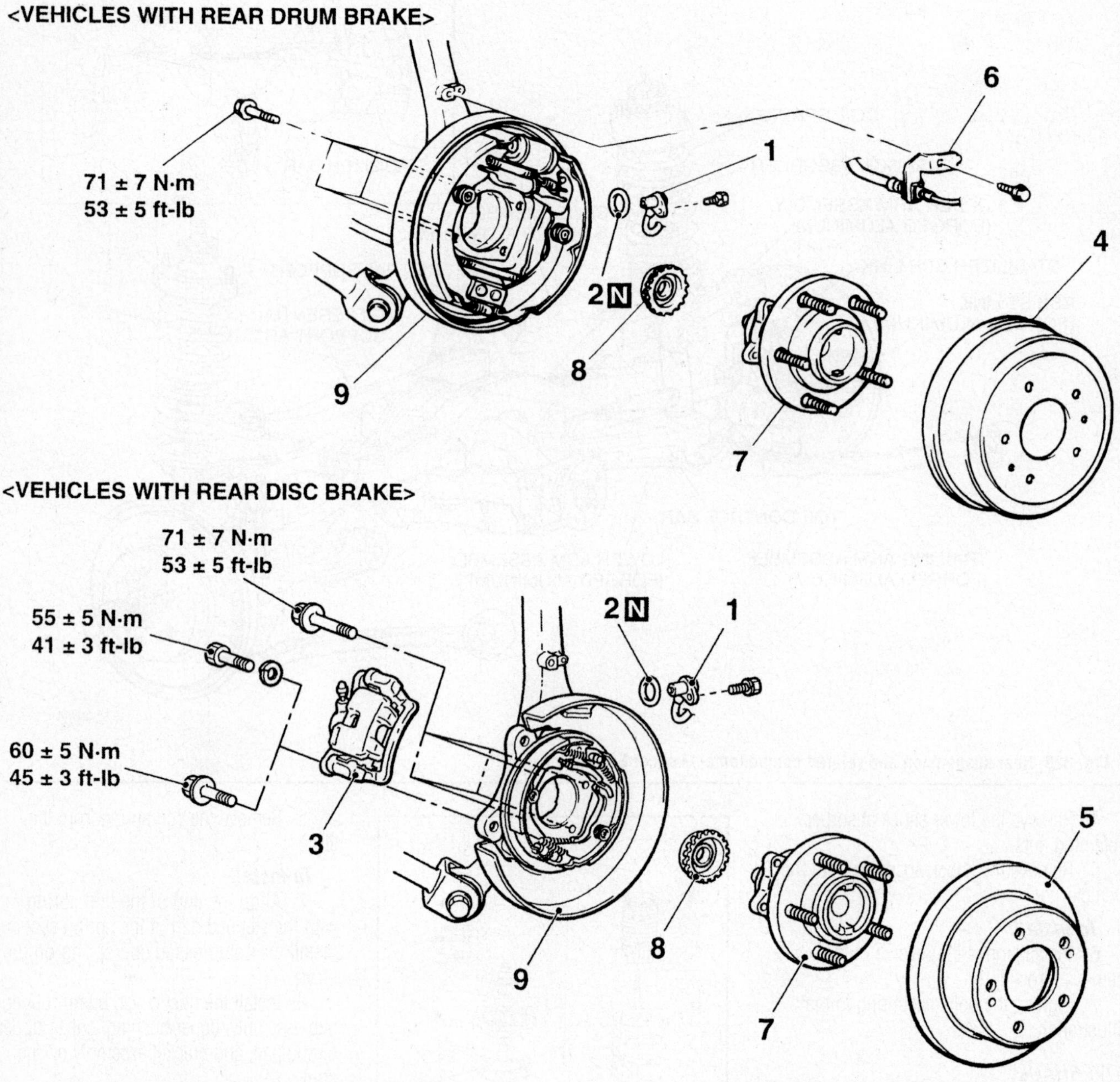

<VEHICLES WITH REAR DRUM BRAKE>

71 ± 7 N·m
53 ± 5 ft-lb

<VEHICLES WITH REAR DISC BRAKE>

71 ± 7 N·m
53 ± 5 ft-lb

55 ± 5 N·m
41 ± 3 ft-lb

60 ± 5 N·m
45 ± 3 ft-lb

1. REAR WHEEL SPEED SENSOR <VEHICLES WITH ABS> (REFER TO GROUP 35B, WHEEL SPEED SENSOR P.35B-56.)
2. O-RING
3. CALIPER ASSEMBLY
4. BRAKE DRUM
5. BRAKE DISC
6. BRAKE HOSE INSTALLATION BRACKET
7. REAR HUB ASSEMBLY
8. ABS ROTOR <VEHICLES WITH ABS>
9. BACKING PLATE

09482_GALA_G0156

Fig. 331 Rear axle hub and related components—Eclipse

➡**The hub and bearing assembly is serviced as a unit.**

1. Before servicing the vehicle, refer to the Precautions Section.
2. Disconnect the negative battery cable.
3. Raise and safely support the vehicle.
4. Remove or disconnect the following:
   - Wheel and tire assembly
   - Rear wheel speed sensor if equipped with Anti-Lock Brake (ABS)
   - Remove the caliper assembly and rotor. Suspend the caliper out of the way with wire.
5. Remove the hub mounting bolts from behind the backing plate and remove the hub.

➡**The rotor for the ABS must be removed and installed using a press.**

To install:

6. Press the rotor (ABS) to the hub.
7. Install or connect the following:
   - Hub and tighten the mounting bolts to 54–65 ft. lbs. (74–88 Nm)
   - Parking brake shoes if equipped
   - Rotor and caliper
   - Speed sensor if equipped
   - Wheel and tire assembly
8. Lower the vehicle to the floor.
9. Connect the negative battery cable.

Galant

See Figure 332.

➡**The hub and bearing assembly is serviced as a unit.**

1. Before servicing the vehicle, refer to the Precautions Section.
2. Disconnect the negative battery cable.
3. Raise and safely support the vehicle.
4. Remove the appropriate wheel assembly.
5. If equipped with Anti-Lock Brake (ABS), remove the Vehicle Speed Sensor (VSS).
6. Remove the caliper assembly and rotor. Suspend the caliper out of the way with wire.
7. From the back of the knuckle, remove the 4 bolts securing the hub to the knuckle.
8. Remove the hub and bearing assembly from the knuckle.

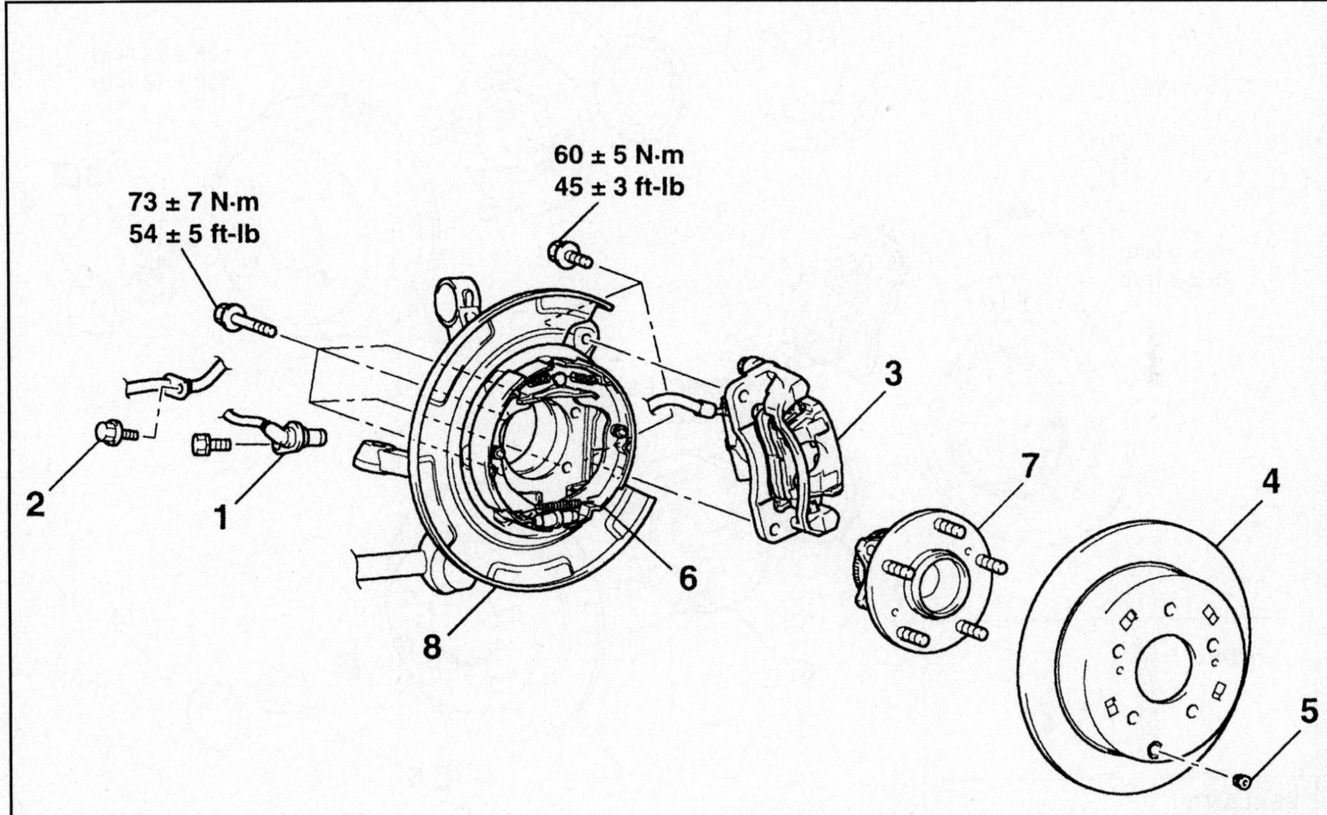

Fig. 332 Rear axle hub and related components—Galant

➡The hub assembly is not serviceable and should not be disassembled.

9. If replacing the hub, use special socket MB991248 and a press, to remove the wheel sensor rotor from the hub.

To install:

10. Press the wheel sensor rotor onto the hub.
11. Install or connect the following:
    - Hub to the knuckle and tighten the mounting bolts to 54–65 ft. lbs. (74–88 Nm)
    - Brake drum on the hub
    - VSS, if equipped with ABS
    - Wheel assembly and lower the vehicle

Lancer

See Figure 333.

➡The hub and bearing assembly is serviced as a unit.

1. Before servicing the vehicle, refer to the Precautions Section.
2. Disconnect the negative battery cable.
3. If equipped with Anti-Lock Brake (ABS), remove the wheel speed sensor.
4. Raise and safely support the vehicle.
5. Remove or disconnect the following:
    - Rear wheel

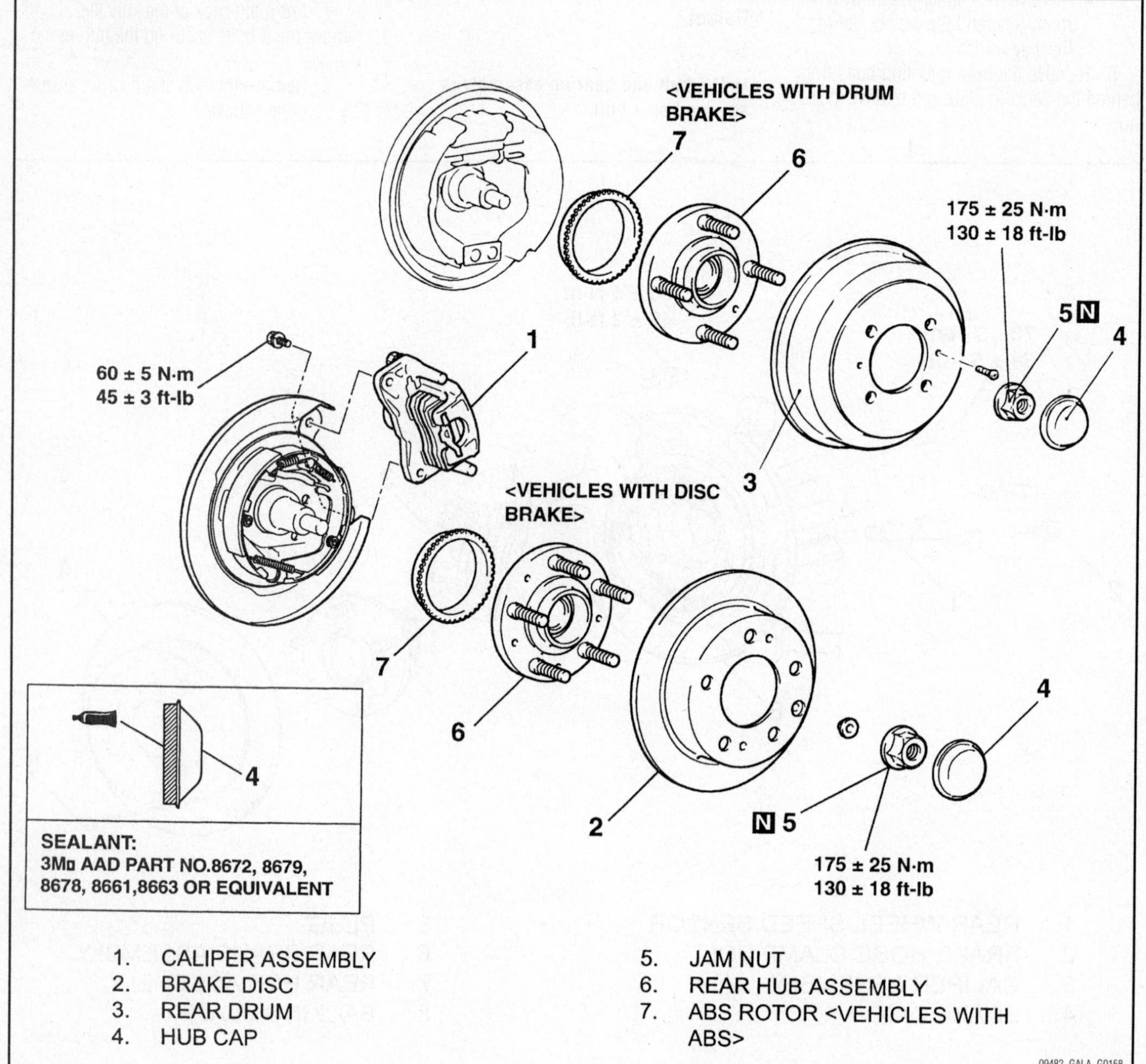

Fig. 333 Rear axle hub and related components—Lancer

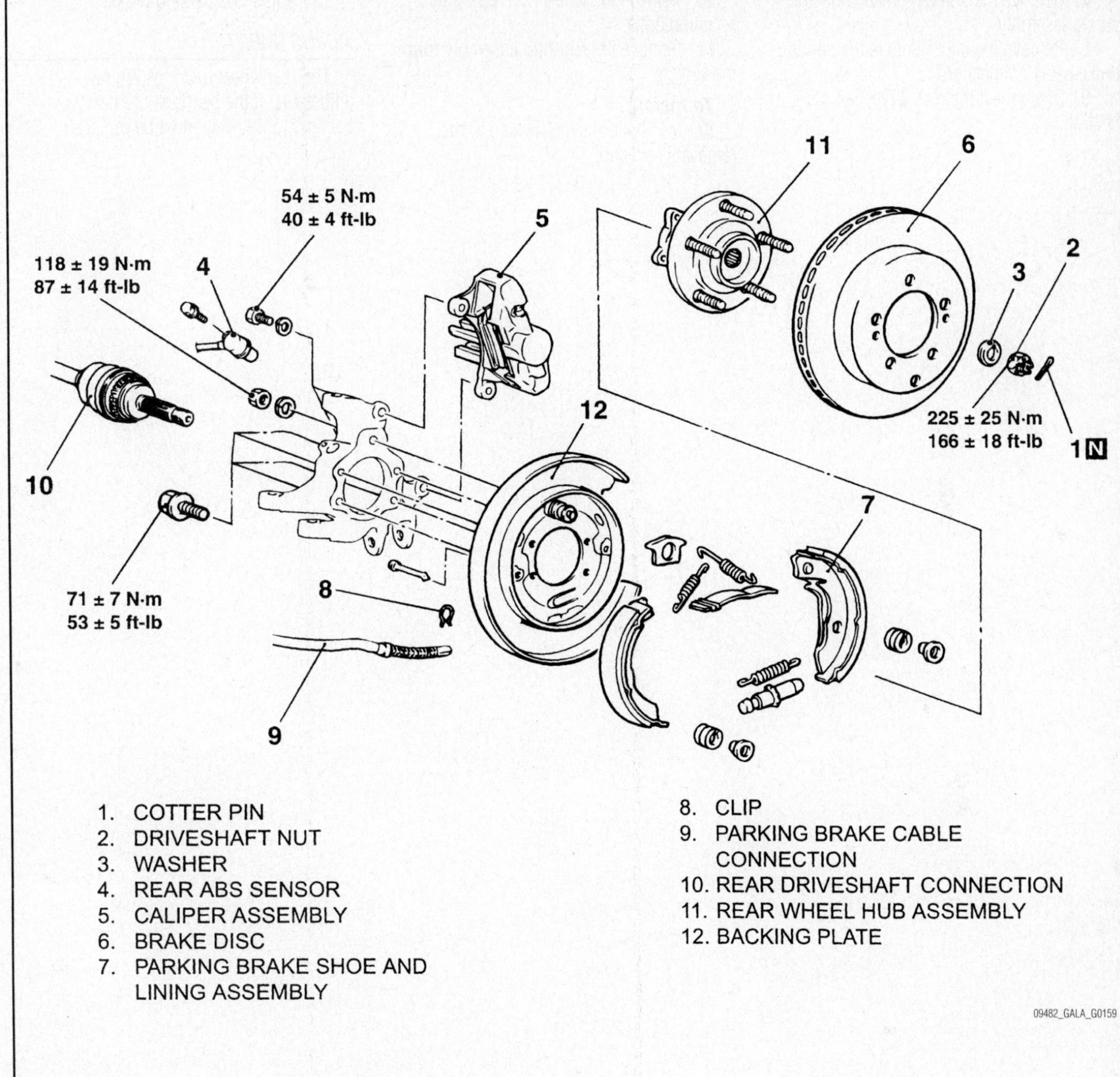

Fig. 334 Rear axle hub and related components—Lancer Evolution

- Caliper and brake disc or brake drum
- Dust cap and flange nut
- Rear hub assembly

To install:

6. Install or connect the following:
   - Rear hub assembly using a new flange nut. Torque the flange nut to 130 ft. lbs. (180 Nm).
   - Dust cap
   - Wheel speed sensor if removed. The air gap should be 0.012–0.035 in. (0.3–0.9mm).
   - Brake disc and caliper, or brake drum
   - Rear wheel assembly and lower the vehicle to the floor

Lancer Evolution

See Figure 334.

➡The hub and bearing assembly is serviced as a unit.

1. Before servicing the vehicle, refer to the Precautions Section.
2. Disconnect the negative battery cable.
3. Raise and support the vehicle safely. Remove the tire and wheel assembly.
4. Remove the cotter pin. Remove the driveshaft nut and washer.
5. Remove the rear ABS sensor.
6. Remove the caliper. Position it to the side with wire. Do not disconnect the brake hose.

7. Remove the parking brake shoe and lining assembly.

8. Remove the clip. Remove the parking brake cable connection.

9. Disconnect the rear driveshaft connection.

10. Remove the wheel hub assembly retaining bolts.

11. Remove the rear hub assembly from the vehicle.

To install:

12. Installation is the reverse of the removal procedure.

13. Adjust the parking brake.

ADJUSTMENT

The rear wheel bearings are not adjustable. If the bearings are noisy or become loose, they must be replaced.

MITSUBISHI

Endeavor

9

SPECIFICATIONS AND MAINTENANCE CHARTS

ENGINE AND VEHICLE IDENTIFICATION CHART

Engine						
Code	Liters (cc)	Cu. In.	Cyl.	Fuel Sys.	Engine Type	Eng. Mfg.
6G75/S	3.8 (3828)	233.6	6	MFI	SOHC	Mitsubishi

Model Year	
Code	Year
7	2007
8	2008

MFI: Multi-port Fuel Injection

22140_ENDE_C0001

GENERAL ENGINE SPECIFICATIONS

Year	Engine Displacement Liters	Engine ID/VIN	Net Horsepower @ rpm	Net Torque @ rpm (ft. lbs.)	Bore x Stroke (in.)	Com-pression Ratio	Oil Pressure @ rpm
2007	3.8	6G75/S	225@5000	255@3750	3.74x3.54	10.0:1	①
2008	3.8	6G75/S	225@5000	255@3750	3.74x3.54	10.0:1	①

① 4.2 or more psi @ idle
43-100 @ 3500 rpm

22140_ENDE_C0002

ENGINE TUNE-UP SPECIFICATIONS

			Spark Plugs	Ignition Timing (deg.)		Fuel	Idle Speed (rpm)		Valve Clearance	
Year	Engine Displacement Liters	Engine ID/VIN	Gap (in.)	MT	AT	Pump (psi)	MT	AT	In.	Ex.
2007	3.8	6G75/S	0.028-0.031	—	①	47	—	②	HYD	HYD
2008	3.8	6G75/S	0.028-0.031	—	①	47	—	②	HYD	HYD

HYD: Hydraulic

① Base ignition timing: 2-8 degrees BTDC
Actual ignition timing: 10 degrees BTDC

② 580-780 rpm's

22140_ENDE_C0003

CAPACITIES

Year	Model	Engine Displacement Liters	Engine ID/VIN	Engine Oil with Filter (qts.)	Transmission (pts.) 5-Spd	Transmission (pts.) Auto.	Transfer Case (pts.)	Drive Axle Front (pts.)	Drive Axle Rear (pts.)	Fuel Tank (gal.)	Cooling System (qts.)
2007	Endeavor	3.8	6G75/S	4.5	—	①	1.12	—	2.4	21.4	②
2008	Endeavor	3.8	6G75/S	4.5	—	①	1.12	—	2.4	21.4	②

① FWD: 17.8 pts.
AWD: 18.6 pts.

② FWD & AWD without tow kit: 9.5 qts.
AWD with tow kit: 10.1 qts.

22140_ENDE_C0004

FLUID SPECIFICATIONS

Year	Model	Engine Displ. Liters (VIN)	Engine Oil	Man. Trans.	Auto. Trans.	Drive Axle Front	Drive Axle Rear	Transfer Case	Power Steering Fluid	Brake Master Cylinder	Cooling System
2007	Endeavor	6G75/S	5W-20	—	ATF SPIII	—	80W-90	SAE 90	Mitsubishi	DOT 3-4	Long Life
2008	Endeavor	6G75/S	5W-20	—	ATF SPIII	—	80W-90	SAE 90	Mitsubishi	DOT 3-4	Long Life

DOT: Department Of Transpotation

22140_ENDE_C0014

VALVE SPECIFICATIONS

Year	Engine Displacement Liters	Engine ID/VIN	Seat Angle (deg.)	Face Angle (deg.)	Spring Test Pressure (lbs. @ in.)	Spring Installed Height (in.)	Stem-to-Guide Clearance (in.) Intake	Stem-to-Guide Clearance (in.) Exhaust	Stem Diameter (in.) Intake	Stem Diameter (in.) Exhaust
2007	3.8	6G75/S	NS	43.5-44	60@1.74	1.740	0.0008-0.0019	0.0016-0.0023	0.240	0.240
2008	3.8	6G75/S	NS	43.5-44	60@1.74	1.740	0.0008-0.0019	0.0014-0.0024	0.240	0.240

NS - Not specified by manufacturer

22140_ENDE_C0005

CAMSHAFT SPECIFICATIONS CHART

All measurements are given in inches.

Year	Engine Displ. Liters	Engine VIN	Journal Dia.	Brg. Oil Clearance	Shaft End-play	Runout	Lobe Height	
							Intake	Exhaust
2007	3.8	6G75/S	1.8000	NS	NS	NS	①	②
2008	3.8	6G75/S	1.8000	NS	NS	NS	①	②

NS - Not specified by manufacturer

① Standard value: 1.472
Minimum value: 1.452

② Standard value: 1.462
Minimum value: 1.443

22140_ENDE_C0006

CRANKSHAFT AND CONNECTING ROD SPECIFICATIONS

All measurements are given in inches.

Year	Engine Displacement Liters	Engine ID/VIN	Crankshaft				Connecting Rod		
			Main Brg. Journal Dia.	Main Brg. Oil Clearance	Shaft End-play	Thrust on No.	Journal Diameter	Oil Clearance	Side Clearance
2007	3.8	6G75/S	2.2500	①	0.0020-0.0090	NS	1.9700	0.0008-0.0015	0.0030-0.0090
2008	3.8	6G75/S	2.2500	②	0.0020-0.0090	NS	1.9700	0.0008-0.0015	0.0030-0.0090

NS - Not specified by manufacturer

① Nos. 1 & 4: 0.0008-0.0012 inch
Nos. 2 & 3: 0.0012-0.0016 inch

② Nos. 1 & 4: 0.0007-0.0017 inch
Nos. 2 & 3: 0.0009-0.0017 inch

22140_ENDE_C0007

PISTON AND RING SPECIFICATIONS

All measurements are given in inches.

Year	Engine Displacement Liters	Engine ID/VIN	Piston Clearance	Ring Gap			Ring Side Clearance		
				Top Compression	Bottom Compression	Oil Control	Top Compression	Bottom Compression	Oil Control
2007	3.8	6G75/S	0.0008-0.0015	0.0100-0.0170	0.0140-0.0190	0.0030-0.0140	0.0012-0.0027	0.0008-0.0023	Snug
2008	3.8	6G75/S	0.0008-0.0015	0.0100-0.0160	0.0140-0.0200	0.0030-0.0140	0.0012-0.0027	0.0008-0.0023	Snug

22140_ENDE_C0008

TORQUE SPECIFICATIONS

All readings in ft. lbs.

Year	Engine Displacement Liters	Engine ID/VIN	Cylinder Head Bolts	Main Bearing Bolts	Rod Bearing Bolts	Crankshaft Damper Bolts	Flywheel Bolts	Manifold Intake	Manifold Exhaust	Spark Plugs	Oil Pan Drain Plug
2007	3.8	6G75/S	①	②	③	136	54	④	⑤	⑥	⑦
2008	3.8	6G75/S	①	②	③	136	54	④	⑤	⑥	⑦

① Step 1: 76-84 ft. lbs.
Step 2: Loosen completely
Step 3: 76-84 ft. lbs.

② 51-57 ft. lbs.

③ 20 ft. lbs., plus an additional 90-94 degrees

④ Step 1: 45-71 inch lbs.
Step 2: 15-17 ft. lbs.

⑤ Exhaust manifold nut: 29-37 ft. lbs.
Exhaust manifold stay bolt (M8): 12-16 ft. lbs.
Exhaust manifold stay bolt (M10): 28-38 ft. lbs.
Exhaust manifold stay bolt (M12): 49-63 ft. lbs.

⑥ 14-22 ft. lbs.

⑦ 25-33 ft. lbs.

22140_ENDE_C0009

WHEEL ALIGNMENT

Year	Model		Caster Range (+/-Deg.)	Caster Preferred Setting (Deg.)	Camber Range (+/-Deg.)	Camber Preferred Setting (Deg.)	Toe-in (in.)
2007	Endeavor	F	0.30	+3.00	0.30	0	0.0+/-0.12
		R	—	—	0.30	-0.50	0.12+/-0.12
2008	Endeavor	F	0.30	+3.00	0.30	0	0.0+/-0.12
		R	—	—	0.30	-0.50	0.12+/-0.12

22140_ENDE_C0010

TIRE, WHEEL AND BALL JOINT SPECIFICATIONS

Year	Model	OEM Tires Standard	OEM Tires Optional	Tire Pressures (psi) Front	Tire Pressures (psi) Rear	Wheel Size	Ball Joint Inspection	Lug Nut (ft. lbs.)
2007	Endeavor	P235/65R17	None	29	29	7-JJ	U: 7-30 in. ① L: 0.010 in.	②
2008	Endeavor	P235/65R17	None	29	29	7-JJ	U: 7-30 in. ① L: 0.010 in.	②

OEM: Original Equipment Manufacturer
PSI: Pounds Per Square Inch
STD: Standard
OPT: Optional

① Torque required in inch lbs. to rotate ball joint when removed from the knuckle

② 66-80 ft. lbs.

22140_ENDE_C0011

BRAKE SPECIFICATIONS

All measurements in inches unless noted

			Brake Disc			Brake Drum Diameter			Minimum Lining Thickness		Brake Caliper	
Year	Model		Original Thickness	Minimum Thickness	Maximum Runout	Original Inside Diameter	Max. Wear Limit	Maximum Machine Diameter	Front	Rear	Bracket Bolts (ft. lbs.)	Mounting Bolts (ft. lbs.)
2007	Endeavor	F	1.020	0.960	0.0012	—	—	—	0.080	—	74	34
		R	0.390	0.330	0.0031	—	—	—	—	0.080	45	32
2008	Endeavor	F	1.020	0.960	0.0012	—	—	—	0.080	—	74	34
		R	0.390	0.330	0.0031	—	—	—	—	0.080	45	32

22140_ENDE_C0012

SCHEDULED MAINTENANCE INTERVALS

2007-08 Mitsubishi—Endeavor

TO BE SERVICED	TYPE OF SERVICE	VEHICLE MILEAGE INTERVAL (x1000)												
		7.5	15	22.5	30	37.5	45	52.5	60	67.5	75	82.5	90	97.5
Air cleaner filter	R				✓				✓				✓	
Automatic transaxle & transfer case oil (AWD)	S/I		✓		✓		✓		✓		✓		✓	
Automatic transaxle & transfer case oil (AWD)	R				✓				✓				✓	
Ball joints & steering linkage seals	S/I				✓				✓				✓	
Brake hoses	S/I		✓		✓		✓		✓		✓		✓	
Disc brake pads & rotors	S/I		✓		✓		✓		✓		✓		✓	
Drive belt(s)	S/I				✓				✓				✓	
Drive shaft boots	S/I		✓		✓		✓		✓		✓		✓	
Engine coolant	R	Initially at 60,000, then every 30,000												
Engine coolant hoses	S/I				✓				✓				✓	
Engine oil & filter	R	✓	✓	✓	✓	✓	✓	✓	✓	✓	✓	✓	✓	✓
EVAP canister	S/I	Every 100,000												
EVAP system (except EVAP canister)	S/I								✓					
Exhaust system	S/I				✓				✓				✓	
Front & rear axle	S/I				✓				✓				✓	
Fuel hoses	S/I				✓				✓				✓	
Fuel system	S/I								✓					
Ignition cables	R								✓					
PCV system	S/I	✓	✓	✓	✓	✓	✓	✓	✓	✓	✓	✓	✓	✓
Rear differential	S/I				✓				✓				✓	
Spark plugs	R	Every 105,000												
Suspension system	S/I				✓				✓				✓	
Timing belt	R	Every 105,000												
Tires (rotate)	R	✓	✓	✓	✓	✓	✓	✓	✓	✓	✓	✓	✓	✓
Transfer case oil	R								✓					
Transfer case oil	S/I				✓				✓				✓	

R: Replace S/I: Service or Inspect

FREQUENT OPERATION MAINTENANCE (SEVERE SERVICE)

If a vehicle is operated under any of the following conditions it is considered severe service:

- Extremely dusty areas.

- 50% or more of the vehicle operation is in 32°C (90°F) or higher temperatures, or constant operation in temperatures below 0°C (32°F).

- Prolonged idling (vehicle operation in stop and go traffic).

- Frequent short running periods (engine does not warm to normal operating temperatures).

- Police, taxi, delivery usage or trailer towing usage.

Air cleaner filter: service or inspect every 15,000 miles.

Front disc brake pads: service or inspect every 7500 miles.

Oil & oil filter: replace every 3750 miles.

PCV system: service or inspect every 60,000 miles.

Suspension system: service or inspect every 7500 miles.

Transmission fluid: check every 15,000 miles. Replace every 30,000 miles.

22140_ENDE_C0013

PRECAUTIONS

Before servicing any vehicle, please be sure to read all of the following precautions, which deal with personal safety, prevention of component damage, and important points to take into consideration when servicing a motor vehicle:

- Never open, service or drain the radiator or cooling system when the engine is hot; serious burns can occur from the steam and hot coolant.
- Observe all applicable safety precautions when working around fuel. Whenever servicing the fuel system, always work in a well-ventilated area. Do not allow fuel spray or vapors to come in contact with a spark, open flame, or excessive heat (a hot drop light, for example). Keep a dry chemical fire extinguisher near the work area. Always keep fuel in a container specifically designed for fuel storage; also, always properly seal fuel containers to avoid the possibility of fire or explosion. Refer to the additional fuel system precautions later in this section.
- Fuel injection systems often remain pressurized, even after the engine has been turned **OFF**. The fuel system pressure must be relieved before disconnecting any fuel lines. Failure to do so may result in fire and/or personal injury.
- Brake fluid often contains polyglycol ethers and polyglycols. Avoid contact with the eyes and wash your hands thoroughly after handling brake fluid. If you do get brake fluid in your eyes, flush your eyes with clean, running water for 15 minutes. If eye irritation persists, or if you have taken brake fluid internally, IMMEDIATELY seek medical assistance.
- The EPA warns that prolonged contact with used engine oil may cause a number of skin disorders, including cancer. You should make every effort to minimize your exposure to used engine oil. Protective gloves should be worn when changing oil. Wash your hands and any other exposed skin areas as soon as possible after exposure to used engine oil. Soap and water, or waterless hand cleaner should be used.
- All new vehicles are now equipped with an air bag system, often referred to as a Supplemental Restraint System (SRS) or Supplemental Inflatable Restraint (SIR) system. The system must be disabled before performing service on or around system components, steering column, instrument panel components, wiring and sensors. Failure to follow safety and disabling procedures could result in accidental air bag deployment, possible personal injury and unnecessary system repairs.
- Always wear safety goggles when working with, or around, the air bag system. When carrying a non-deployed air bag, be sure the bag and trim cover are pointed away from your body. When placing a non-deployed air bag on a work surface, always face the bag and trim cover upward, away from the surface. This will reduce the motion of the module if it is accidentally deployed. Refer to the additional air bag system precautions later in this section.
- Clean, high quality brake fluid from a sealed container is essential to the safe and proper operation of the brake system. You should always buy the correct type of brake fluid for your vehicle. If the brake fluid becomes contaminated, completely flush the system with new fluid. Never reuse any brake fluid. Any brake fluid that is removed from the system should be discarded. Also, do not allow any brake fluid to come in contact with a painted surface; it will damage the paint.
- Never operate the engine without the proper amount and type of engine oil; doing so WILL result in severe engine damage.
- Timing belt maintenance is extremely important. Many models utilize an interference-type, non-freewheeling engine. If the timing belt breaks, the valves in the cylinder head may strike the pistons, causing potentially serious (also time-consuming and expensive) engine damage. Refer to the maintenance interval charts for the recommended replacement interval for the timing belt, and to the timing belt section for belt replacement and inspection.
- Disconnecting the negative battery cable on some vehicles may interfere with the functions of the on-board computer system(s) and may require the computer to undergo a relearning process once the negative battery cable is reconnected.
- When servicing drum brakes, only disassemble and assemble one side at a time, leaving the remaining side intact for reference.
- Only an MVAC-trained, EPA-certified automotive technician should service the air conditioning system or its components.

BRAKES

ANTI-LOCK BRAKE SYSTEM (ABS)

GENERAL INFORMATION

The ABS ensures directional stability and control during hard braking. This ABS uses a 4-sensor 4-channel system that controls all four wheels independently of each other. The EBD (Electronic Brake-force Distribution system) control has been installed to provide the ideal braking force for the rear wheels.

In ABS, electronic control is used so the rear wheel brake hydraulic pressure during braking is regulated by rear wheel control solenoid valves in accordance with the vehicle's rate of deceleration. The front and rear wheel slippages are calculated from the signals received from the various wheel sensors. EBD control is a control system which provides a high level of control for both vehicle braking force and vehicle stability. The system has the following features:

- Because the system provides the optimum rear wheel braking force regardless of vehicle load and the condition of the road surface, the system reduces the required pedal depression force, especially when the vehicle is heavily loaded.
- Because the duty placed on the front brakes is reduced, the increases in pad temperature can be controlled during front brake application to improve pad wear.
- Control valves such as the proportioning valve are not required.

Below some of the characteristics of the Anti-lock Brake System (ABS) are described.

System Check Sound: When starting the engine, a thudding sound can sometimes be heard coming from the engine compartment. This is a normal sound during the ABS self-check.

ABS Operation Sounds and Sensations: During normal operation, the ABS makes several sounds that may seem unusual at first:

- A whining sound is caused by the ABS hydraulic unit motor.
- When pressure is applied to the brake pedal, the pulsation of the pedal causes a scraping sound.
- When the brakes are applied firmly, the ABS operates, rapidly applying and releasing the brakes many times per second. This repeated application and release of braking forces can cause the suspension to make a thumping sound and the tires to squeak.

Long Stopping Distances on Loose Road Surfaces: When braking on loose surfaces like snow-covered or gravel roads, the stopping distance can be longer for an ABS-equipped vehicle than the stopping distance for a vehicle with a conventional brake system.

Shock at Starting Check: This may be felt when the brake pedal is lightly pressed while driving at a low speed. This is a normal characteristic because the ABS system operation check is carried out when vehicle speed is 8 km/h (5 mph) or less.

WHEEL SPEED SENSORS

REMOVAL & INSTALLATION

Front

See Figure 1.

1. Before servicing the vehicle, refer to the precautions.
2. Raise and safely support the vehicle.
3. Remove the front wheel and tire assemblies.
4. Remove the front mud guard and splash shield.
5. Remove the bolts and the front ABS speed sensor.

To install:

6. Install the bolts and the front ABS speed sensor.
7. Install the front mud guard and splash shield.
8. Install the front wheel and tire assemblies.
9. Lower the vehicle.

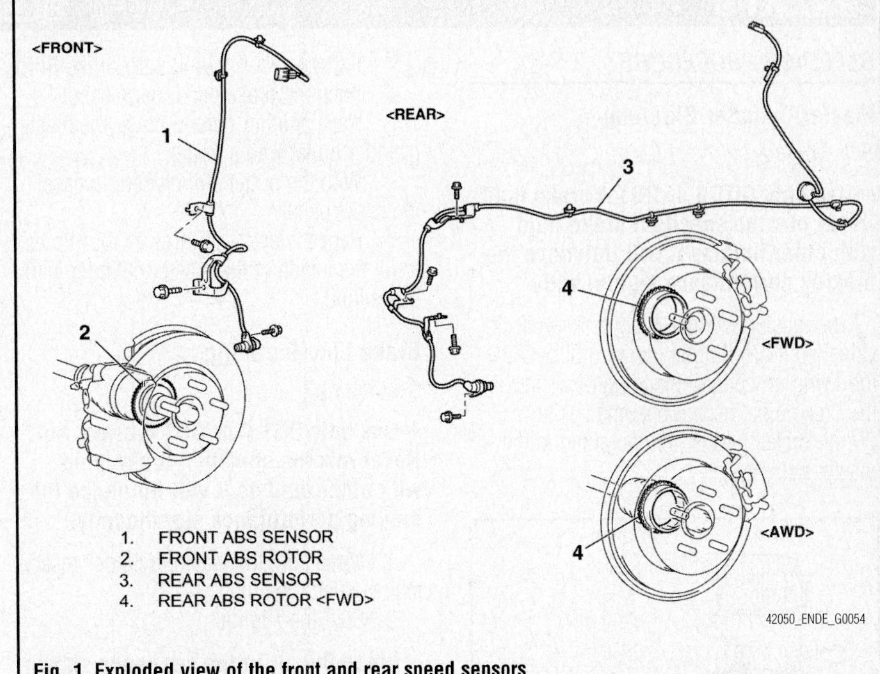

Fig. 1 Exploded view of the front and rear speed sensors

Rear

See Figure 1.

1. Before servicing the vehicle, refer to the precautions.
2. Raise and safely support the vehicle.
3. Remove the rear wheel and tire assemblies.
4. Remove the luggage floor rear board.
5. Remove the tonneau cover.
6. Remove the luggage floor front board.
7. Remove the parcel strap hook.
8. Remove the luggage floor side board.
9. Remove the rear end trim.
10. Remove the rear seat assembly.
11. Remove the scuff plate cover and rear scuff plate.
12. Remove the rear seat belt lower anchor bolt.
13. Remove the sash guide cover.
14. Remove the rear seat belt shoulder anchor bolt.
15. Remove the seat belt cover.
16. Remove the liftgate door opening trim.
17. Remove the rear door opening trim.
18. Remove the lower quarter trim.
19. Remove the bolts and the rear ABS sensor.

To install:

20. Install the bolts and the rear ABS sensor
21. Install the lower quarter trim.
22. Install the rear door opening trim.
23. Install the liftgate door opening trim.
24. Install the seat belt cover.
25. Install the rear seat belt shoulder anchor bolt.
26. Install the sash guide cover.
27. Install the rear seat belt lower anchor bolt.
28. Install the scuff plate cover and rear scuff plate.
29. Install the rear seat assembly.
30. Install the rear end trim.
31. Install the luggage floor side board.
32. Install the parcel strap hook.
33. Install the luggage floor front board.
34. Install the tonneau cover.
35. Install the luggage floor rear board.
36. Install the rear wheel and tire assemblies.
37. Lower the vehicle.

WHEEL SPEED SENSOR RINGS (TOOTHED RINGS)

REMOVAL & INSTALLATION

The front ABS rotor and the rear ABS rotor on AWD models are integrated with the BJ assembly of the Halfshaft and cannot be disassembled. The rear ABS sensor rotor on FWD models is integrated with the rear hub assembly and cannot be disassembled.

BRAKES — BLEEDING THE BRAKE SYSTEM

BLEEDING PROCEDURE

Master Cylinder Bleeding

See Figure 2.

➡Use only DOT 3 or DOT 4 brake fluid. Never mix the specified brake fluid with other fluid as it will influence the braking performance significantly.

The master cylinder used has no check valve, so if bleeding is carried out by the following procedure, bleeding of air from the brake pipeline will become easier. (When brake fluid is not contained in the master cylinder).

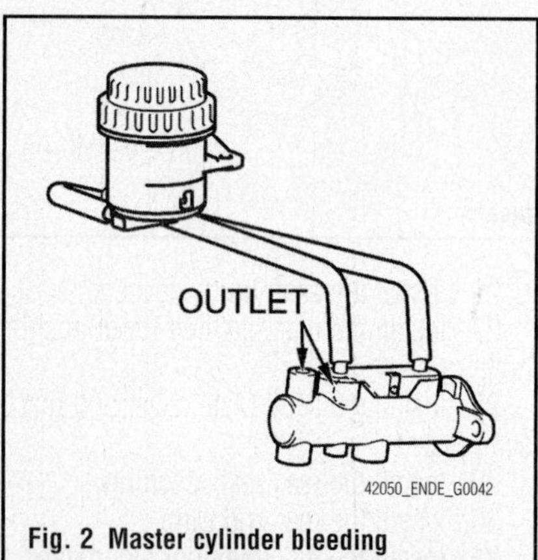

Fig. 2 Master cylinder bleeding

1. Fill the reserve tank with brake fluid.
2. Keep the brake pedal depressed.
3. Have another person cover the master cylinder outlet with a finger.
4. With the outlet still closed, release the brake pedal.
5. Repeat steps 2–4 three or four times to fill the inside of the master cylinder with brake fluid.

Brake Line Bleeding

See Figure 3.

➡Use only DOT 3 or DOT 4 brake fluid. Never mix the specified brake fluid with other fluid as it will influence the braking performance significantly.

1. Make sure the master cylinder reserve tank is filled the fluid.
2. Start the engine.

➡Bleed the air in the sequence shown in the figure.

3. Raise and safely support the vehicle.
4. Connect a piece of vinyl tubing to the brake caliper.
5. Have an assistant depress the brake pedal several times, then loosen the bleeder plug with the pedal held down.
6. When fluid stops coming out, tighten the bleeder plug, then release the brake pedal.

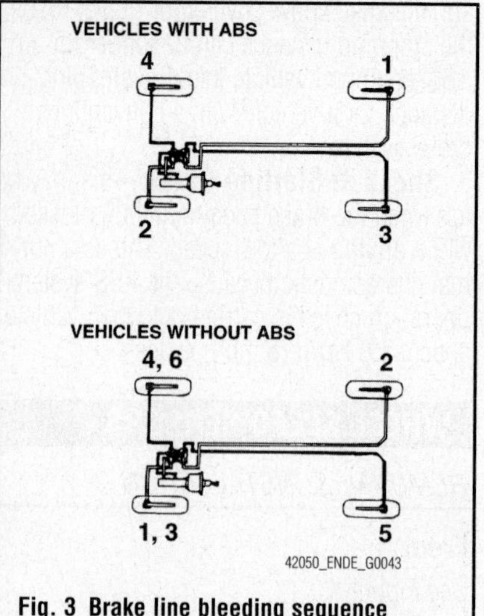

Fig. 3 Brake line bleeding sequence

7. Repeat steps 2. and 3. until all the air in the fluid has been bled out.
8. Tighten the brake bleeder plug to 74 inch lbs. (8.3 Nm).
9. Repeat the above steps to bleed the air out of the brake line for each wheel.

Bleeding the ABS System

There are no special procedures for bleeding the ABS system. Refer to the conventional bleeding procedures.

BRAKES — FRONT DISC BRAKES

⁂ CAUTION

Dust and dirt accumulating on brake parts during normal use may contain asbestos fibers from production or aftermarket brake linings. Breathing excessive concentrations of asbestos fibers can cause serious bodily harm. Exercise care when servicing brake parts. Do not sand or grind brake lining unless equipment used is designed to contain the dust residue. Do not clean brake parts with compressed air or by dry brushing. Cleaning should be done by dampening the brake components with a fine mist of water, then wiping the brake components clean with a dampened cloth. Dispose of cloth and all residue containing asbestos fibers in an impermeable container with the appropriate label. Follow practices prescribed by the Occupational Safety and Health Administration (OSHA) and the Environmental Protection Agency (EPA) for the handling, processing, and disposing of dust or debris that may contain asbestos fibers.

BRAKE CALIPER

REMOVAL & INSTALLATION

See Figure 4.

1. Before servicing the vehicle, refer to the Precautions Section.
2. Raise and safely support the vehicle.
3. Remove the wheel and tire assembly.
4. Drain the brake fluid.
5. Remove the brake hose connection, if you are replacing the caliper.
6. Remove the caliper mounting bolts.
7. Remove the caliper.

To install:

8. Position the caliper over the rotor so the caliper engages the adapter correctly.
9. Install the caliper. Tighten the front caliper mounting bolts to 74 ft. lbs. (100 Nm).

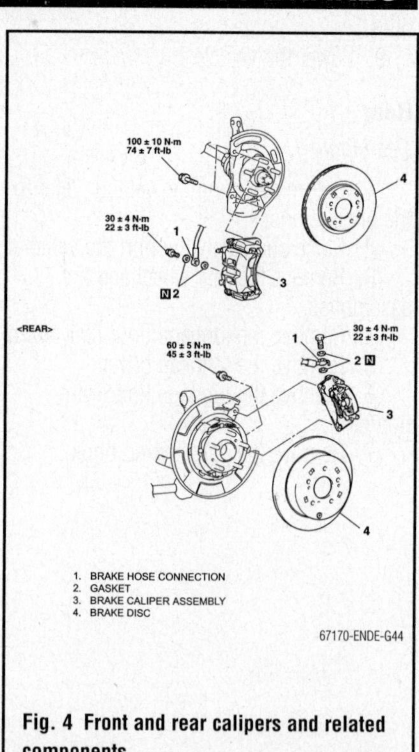

Fig. 4 Front and rear calipers and related components

10. Install the brake hose, if removed. Tighten banjo bolt to 22 ft. lbs. (30 Nm).
11. Install the wheel and tire assembly.
12. Fill and bleed the brake system.
13. Lower the vehicle.

DISC BRAKE PADS

REMOVAL & INSTALLATION

See Figure 5.

1. Before servicing the vehicle, refer to the Precautions Section.
2. Remove ½ of the brake fluid from the master cylinder.
3. Raise and safely support the vehicle.
4. Remove the wheel and tire assembly.
5. Remove the lower caliper guide pin bolt.
6. Remove the caliper from the caliper support but do NOT disconnect the fluid line. Suspend the caliper from the suspension with a piece of wire.
7. Remove the disc brake pads, shims, and the clips from the caliper support.

To install:

8. Clean the exposed portion of the caliper piston, then press the piston back into the caliper bore using the old inner brake pad and a C-clamp.
9. Install the disc brake pads, shims, and the clips. make sure the shims and clips are properly positioned.
10. Install the caliper over the rotor so the caliper engages the adapter correctly.
11. Install the mounting pin(s).
12. Install the wheel and tire assembly and lower the vehicle.
13. Apply the brake pedal several times until a firm pedal is obtained. Check the fluid level in the master cylinder and add fluid, as necessary.

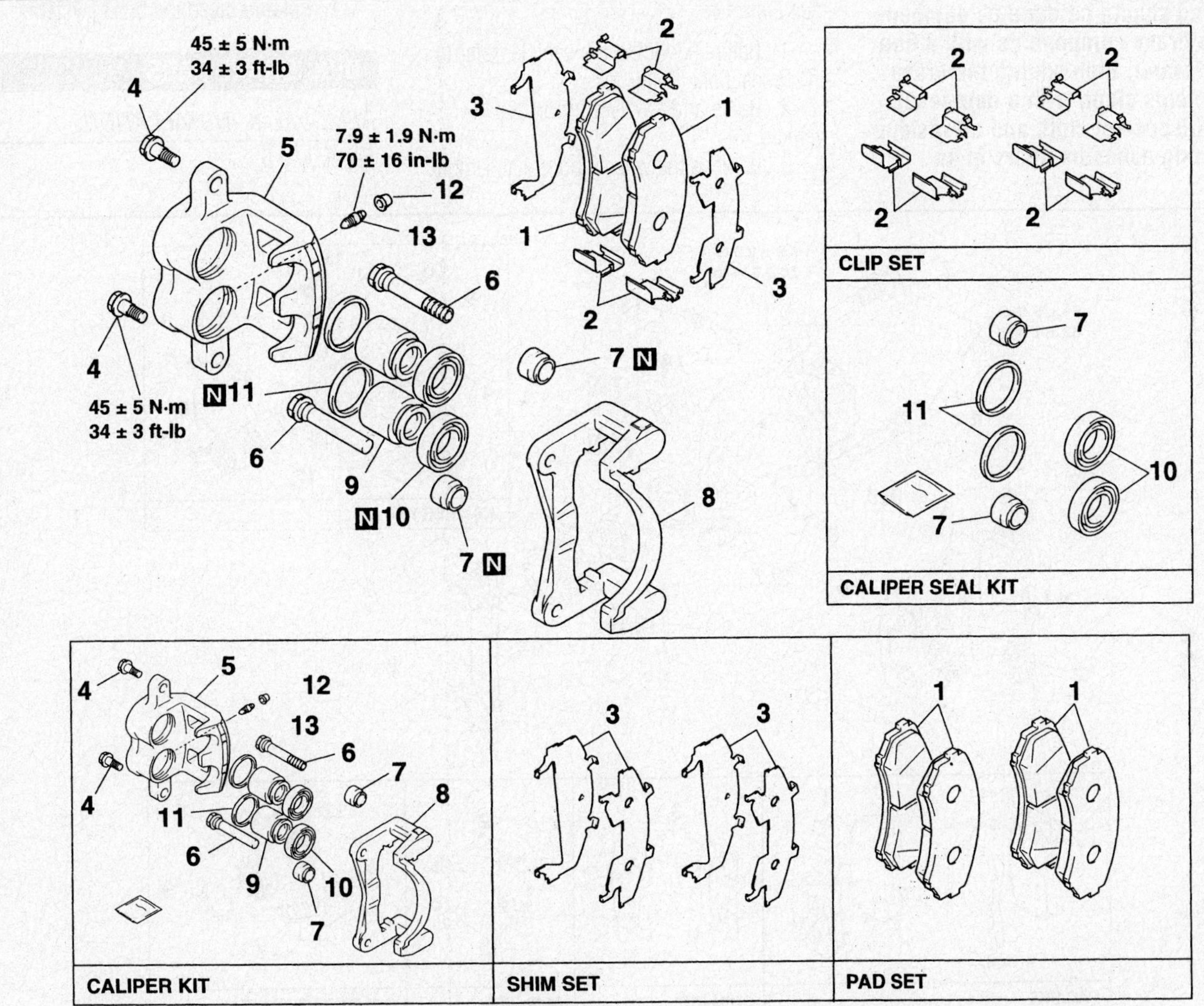

1. PAD (AND WEAR INDICATOR) ASSEMBLY
2. CLIP
3. SHIM
4. FRONT BRAKE BOLT
5. CALIPER BODY
6. FRONT BRAKE PIN
7. BOOT
8. CALIPER SUPPORT
9. CALIPER PISTON
10. PISTON BOOT
11. PISTON SEAL
12. CALIPER BLEEDER CAP
13. CALIPER BLEEDER

\67170-ENDE-G45

Fig. 5 Front brake pads and related components

✲✲ CAUTION

Dust and dirt accumulating on brake parts during normal use may contain asbestos fibers from production or aftermarket brake linings. Breathing excessive concentrations of asbestos fibers can cause serious bodily harm. Exercise care when servicing brake parts. Do not sand or grind brake lining unless equipment used is designed to contain the dust residue. Do not clean brake parts with compressed air or by dry brushing. Cleaning should be done by dampening the brake components with a fine mist of water, then wiping the brake components clean with a dampened cloth. Dispose of cloth and all residue containing asbestos fibers in an impermeable container with the appropriate label. Follow practices prescribed by the Occupational Safety and Health Administration (OSHA) and the Environmental Protection Agency (EPA) for the handling, processing, and disposing of dust or debris that may contain asbestos fibers.

BRAKE CALIPER

REMOVAL & INSTALLATION

See Figure 4.

1. Before servicing the vehicle, refer to the Precautions Section.
2. Raise and safely support the vehicle.
3. Remove the wheel and tire assembly.
4. Remove the brake hose connection, if you are replacing the caliper.
5. Remove the caliper mounting bolts.
6. Remove the caliper.

To install:

7. Position the caliper over the rotor so the caliper engages the adapter correctly
8. Install the caliper. tighten the rear caliper mounting bolts to 45 ft. lbs. (60 Nm).
9. Install the brake hose. Tighten to 22 ft. lbs. (30 Nm).
10. Install the wheel and tire assembly.
11. Fill and bleed the brake system.

DISC BRAKE PADS

REMOVAL & INSTALLATION

See Figure 6.

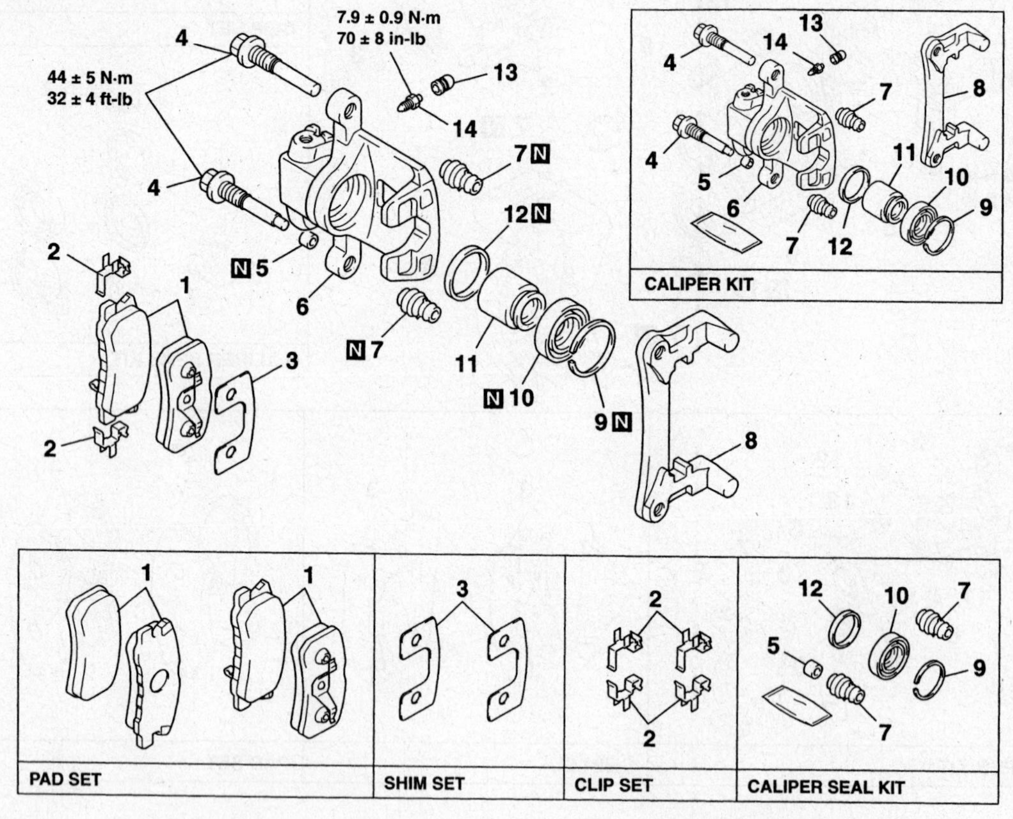

1. PAD (AND WEAR INDICATOR) ASSEMBLY
2. CLIP
3. SHIM
4. REAR BRAKE PIN
5. REAR BRAKE BUSHING
6. CALIPER BODY
7. PIN BOOT
8. CALIPER SUPPORT
9. BOOT RING
10. PISTON BOOT
11. CALIPER PISTON
12. PISTON SEAL
13. REAR BRAKE CAP
14. CALIPER BLEEDER

67170-ENDE-G46

Fig. 6 Rear brake pads and related components

1. Before servicing the vehicle, refer to the Precautions Section.
2. Remove ½ of the brake fluid from the master cylinder.
3. Raise and safely support the vehicle.
4. Remove the wheel and tire assembly.
5. Remove the lower caliper guide pin bolt.
6. Remove the caliper from the caliper support.
7. Remove the disc brake pads, shims, and the clips from the caliper support.

To install:

8. Clean the exposed portion of the caliper piston, then press the piston back into the caliper bore using the old inner brake pad and a C-clamp.
9. Install the disc brake pads, shims, and the clips. make sure the shims and clips are properly positioned.
10. Install the caliper over the rotor so the caliper engages the adapter correctly.
11. Install the mounting pin(s).
12. Install the wheel and tire assembly and lower the vehicle.
13. Apply the brake pedal several times until a firm pedal is obtained. Check the fluid level in the master cylinder and add fluid, as necessary.

BRAKES

PARKING BRAKE

PARKING BRAKE CABLES

ADJUSTMENT

Parking Brake Pedal Stroke Check and Adjustment

See Figures 7 through 10.

1. Depress the parking brake pedal with a force of 44–55 lbs. (196–245 N) and count the number of notches. The standard value is 1 notch
2. If the parking brake pedal stroke is not within the standard value, adjust as follows:
   a. Release the parking brake.
   b. Remove the console inner box tray and plate, and then loosen the adjusting nut to move it to the cable rod end so that the cable will be free.
   c. Remove the rear wheels, and then remove the adjustment hole plug on the brake disc.
   d. Use a flat-tip screwdriver to turn the adjuster in the direction of the arrow (the direction which expands the shoe) so that the disc will not rotate by hand. Return the adjuster five notches in the direction opposite to the direction of the arrow.

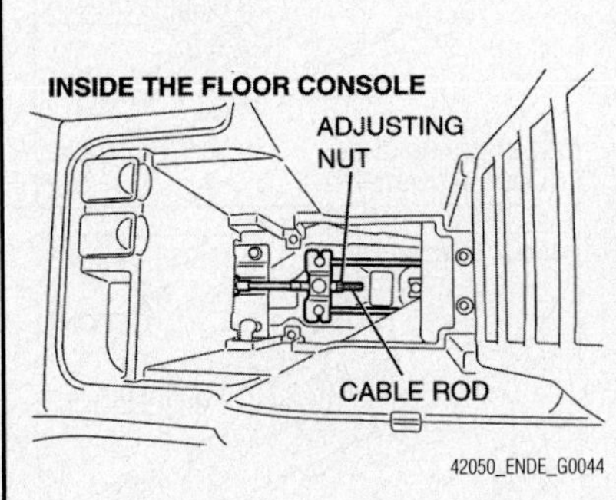

Fig. 7 Remove the console inner box tray and plate, and then loosen the adjusting nut to move it to the cable rod end so that the cable will be free

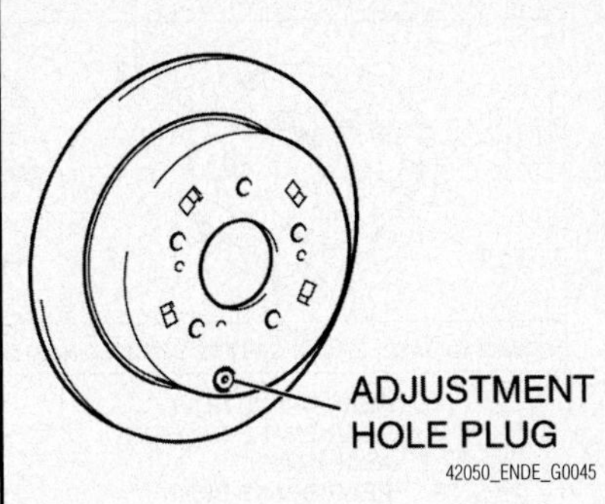

Fig. 8 Remove the rear wheels, and then remove the adjustment hole plug on the brake disc

   e. Install the rear wheels, and then tighten the wheel nuts to 66–80 ft. lbs. (88–108 Nm).

WARNING

Make sure that the parking brake pedal stroke is within the standard value. If the stroke is too short, it may cause the brakes to drag.

   f. Turn the adjusting nut to adjust the parking brake pedal stroke to the standard value. After adjustment, check that the adjust nut and the cable rod is not loose.
   g. Release the parking brake and turn the rear wheels to check that the rear brakes are not dragging.

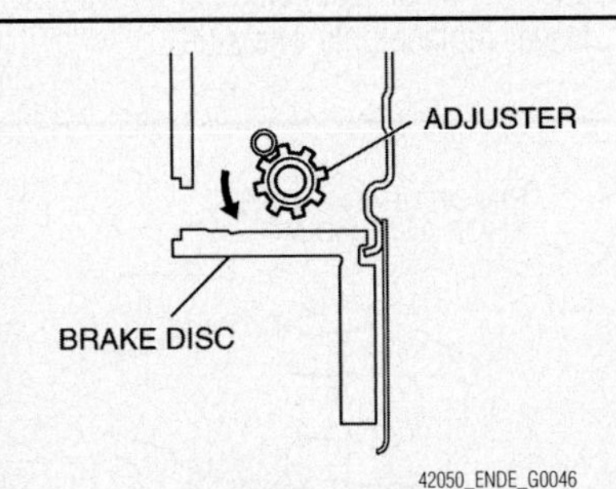

Fig. 9 Use a flat-tip screwdriver to turn the adjuster in the direction of the arrow (the direction which expands the shoe) so that the disc will not rotate by hand. Return the adjuster five notches in the direction opposite to the direction of the arrow

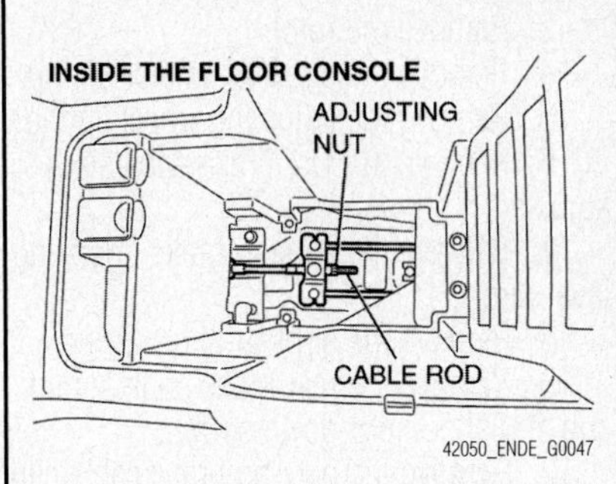

Fig. 10 Turn the adjusting nut to adjust the parking brake pedal stroke to the standard value. After adjustment, check that the adjust nut and the cable rod is not loose

3. If either of the parking brake cables is replaced, adjust the parking brake pedal stroke as described previously, depress the parking brake pedal 10 times with approximately 135 lbs. (600 N) to eliminate the initial slack of the cable. Then adjust the parking brake pedal stroke as outlined above again.

Parking Brake Lining Seating Procedure

Perform lining seating in a place with good visibility, and pay careful attention to safety. Perform lining seating by the following procedure when replacing the parking brake shoe assemblies or the rear brake discs, or when brake performance is insufficient.

1. Adjust the parking brake pedal stroke to the standard value.
2. Depress the parking brake pedal with a force of 23–34 lbs. (100–150 N).
3. Drive the vehicle at a constant speed of 22–31 mph (35—50 km/h) for about 328 ft. (100 meters).

4. Release the parking brake and let the brakes cool for five to ten minutes.
5. Repeat the procedure four to five times.

PARKING BRAKE SHOES

REMOVAL & INSTALLATION

See Figures 11 through 15.

1. Raise and safely remove the vehicle.
2. Unbolt and remove the caliper, but do not disconnect the fluid line. Suspend the caliper with a piece of wire. Do NOT let it hang by its brake line.
3. Remove the rotor.
4. Remove the shoe-to-anchor spring.
5. Remove the adjusting wheel spring.
6. Remove the rear brake shoe slack adjuster.
7. Remove the parking brake operating lever strut.
8. Remove the strut-to-shoe spring.
9. Remove the rear brake shoe spring cup and shoe hold-down spring.
10. Remove the parking brake cable clip.
11. Remove the rear parking brake cable connection.
12. Remove the rear brake shoe assembly.
13. Use a flat-tipped screwdriver or a similar tool to open up the rear brake chamber retainer joint. Then remove the rear brake chamber retainer.
14. Remove the parking brake operating lever.
15. If wheel hub, backing plate and shoe-hold down pin removal are necessary, perform the following:
    a. Remove the driveshaft nut (AWD models) and rear ABS sensor.
    b. Remove the rear wheel hub assembly.
    c. Remove the backing plate.
    d. Remove the shoe hold-down pin.
16. Inspect the parking brake lining, as follows:
    a. Measure the thickness of the brake lining at several places. The standard value is 2.8 mm (0.11 inch) and the minimum limit is 1.0 mm (0.04 inch)
    b. If the thickness of the brake lining is below the limit, replace the shoe assemblies on both sides of the vehicle. Never replace only one side.

To install:

17. Installation is the reverse of the removal procedure, noting the following:

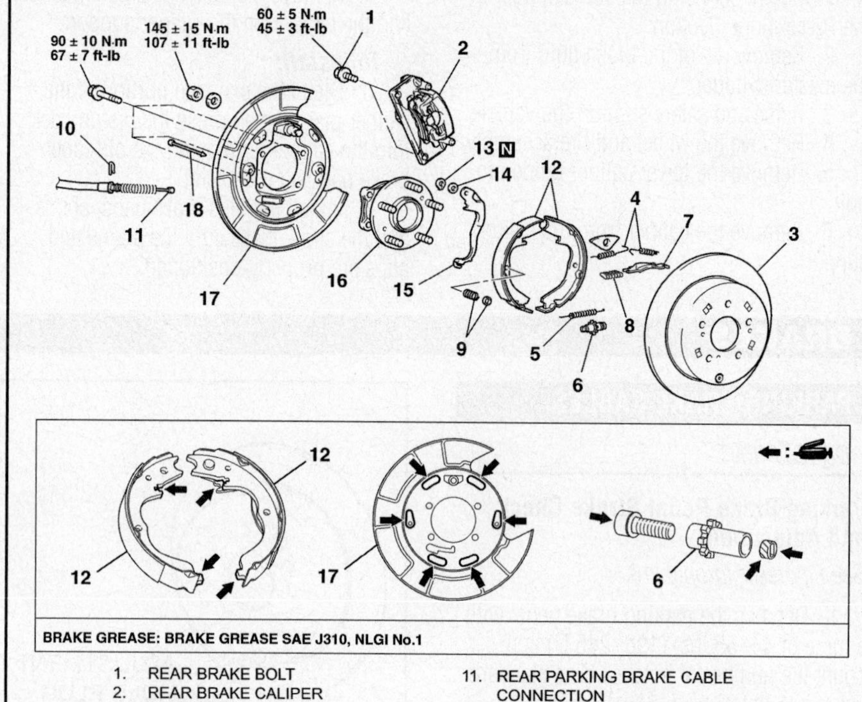

1. REAR BRAKE BOLT
2. REAR BRAKE CALIPER ASSEMBLY
3. REAR BRAKE DISC
4. SHOE-TO-ANCHOR SPRING
5. ADJUSTING WHEEL SPRING
6. REAR BRAKE SHOE SLACK ADJUSTER
7. PARKING BRAKE OPERATING LEVER STRUT
8. STRUT-TO- SHOE SPRING
9. REAR BRAKE SHOE SPRING CUP AND SHOE HOLD-DOWN SPRING
10. PARKING BRAKE CABLE CLIP
11. REAR PARKING BRAKE CABLE CONNECTION
12. REAR BRAKE SHOE ASSEMBLY
13. REAR BRAKE CHAMBER RETAINER
14. REAR BRAKE WASHER
15. PARKING BRAKE OPERATING LEVER

- DRIVE SHAFT NUT <AWD>
- REAR ABS SENSOR <VEHICLES WITH ABS>

16. REAR WHEEL HUB ASSEMBLY
17. BACKING PLATE
18. SHOE HOLD-DOWN PIN

42050_ENDE_G0048

Fig. 12 Exploded view of the parking brake shoes (linings) and related components

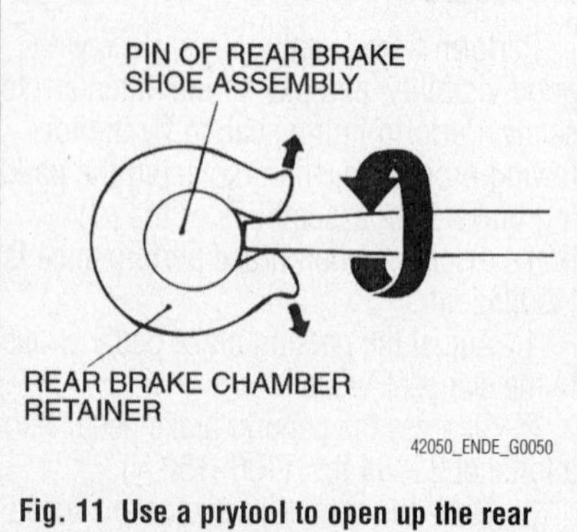

Fig. 11 Use a prytool to open up the rear brake chamber retainer joint in order to remove the chamber retainer

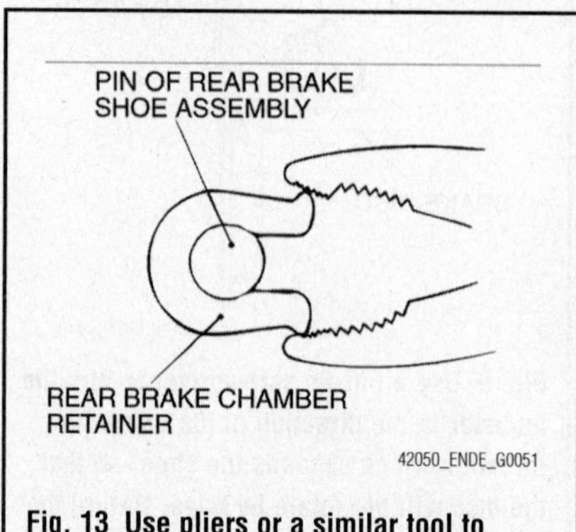

Fig. 13 Use pliers or a similar tool to close the rear brake chamber retainer end onto the pin

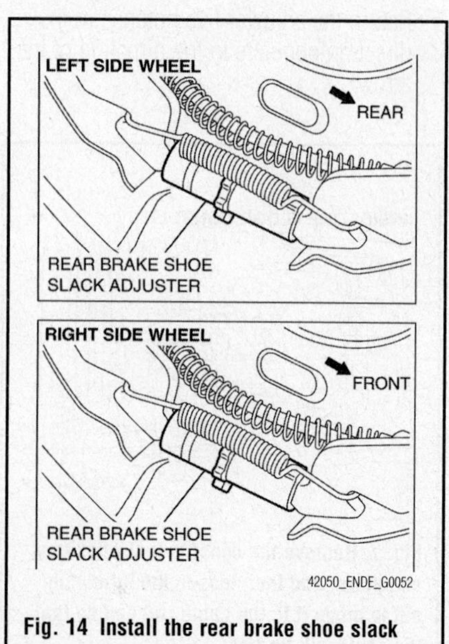

Fig. 14 Install the rear brake shoe slack adjuster as shown

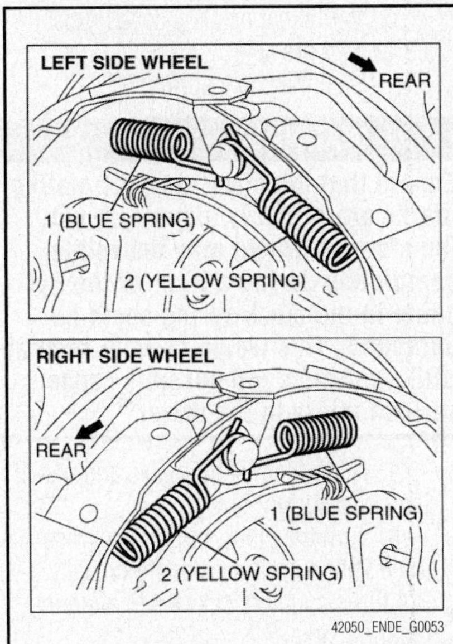

Fig. 15 Install the shoe-to-anchor springs in the order shown

a. Refer to the tightening specifications shown on the accompanying illustration.

b. Use pliers or a similar tool to close the rear brake chamber retainer end onto the pin.

c. Install the rear brake shoe slack adjuster as shown in the accompanying illustration.

⁂ WARNING

The front and rear shoe-to-anchor springs are not interchangeable, so the blue spring must be installed at the front side and the yellow spring must be installed at the rear side.

d. Install the shoe-to-anchor springs in the order shown in the accompanying illustration.

18. Perform the parking brake pedal stroke check and adjustment, and the parking brake seating procedure.

ADJUSTMENT

Parking Brake Pedal Stroke Check and Adjustment

See Figures 7 through 10.

1. Depress the parking brake pedal with a force of 44–55 lbs. (196–245 N) and count the number of notches. The standard value is 1 notch

2. If the parking brake pedal stroke is not within the standard value, adjust as follows:

a. Release the parking brake.

b. Remove the console inner box tray and plate, and then loosen the adjusting nut to move it to the cable rod end so that the cable will be free.

c. Remove the rear wheels, and then remove the adjustment hole plug on the brake disc.

d. Use a flat-tip screwdriver to turn the adjuster in the direction of the arrow (the direction which expands the shoe) so that the disc will not rotate by hand. Return the adjuster five notches in the direction opposite to the direction of the arrow.

e. Install the rear wheels, and then tighten the wheel nuts to 66–80 ft. lbs. (88–108 Nm).

⁂ WARNING

Make sure that the parking brake pedal stroke is within the standard value. If the stroke is too short, it may cause the brakes to drag.

f. Turn the adjusting nut to adjust the parking brake pedal stroke to the standard value. After adjustment, check that the adjust nut and the cable rod is not loose.

g. Release the parking brake and turn the rear wheels to check that the rear brakes are not dragging.

3. If either of the parking brake cables is replaced, adjust the parking brake pedal stroke as described previously, depress the parking brake pedal 10 times with approximately 135 lbs. (600 N) to eliminate the initial slack of the cable. Then adjust the parking brake pedal stroke as outlined above again.

Parking Brake Lining Seating Procedure

Perform lining seating in a place with good visibility, and pay careful attention to safety. Perform lining seating by the following procedure when replacing the parking brake shoe assemblies or the rear brake discs, or when brake performance is insufficient.

1. Adjust the parking brake pedal stroke to the standard value.

2. Depress the parking brake pedal with a force of 23–34 lbs. (100–150 N).

3. Drive the vehicle at a constant speed of 22–31 mph (35—50 km/h) for about 328 ft. (100 meters).

4. Release the parking brake and let the brakes cool for five to ten minutes.

5. Repeat the procedure in steps 2. to 4. four to five times.

CHASSIS ELECTRICAL

AIR BAG (SUPPLEMENTAL RESTRAINT SYSTEM)

GENERAL INFORMATION

⁂ CAUTION

This vehicle is equipped with an air bag system. The system must be disarmed before performing service on, or around, system components, the steering column, instrument panel components, wiring and sensors. Failure to follow the safety precautions and the disarming procedure could result in accidental air bag deployment, possible injury and unnecessary system repairs.

SERVICE PRECAUTIONS

Disconnect and isolate the battery negative cable before beginning any airbag system component diagnosis, testing, removal, or installation procedures. Allow system capacitor to discharge for two minutes before beginning any component service. This will disable the airbag system. Failure to disable the airbag system may result in accidental airbag deployment, personal injury, or death.

Do not place an intact undeployed airbag face down on a solid surface. The airbag will propel into the air if accidentally deployed and may result in personal injury or death.

When carrying or handling an undeployed airbag, the trim side (face) of the airbag should be pointing towards the body to minimize possibility of injury if accidental deployment occurs. Failure to do this may result in personal injury or death.

Replace airbag system components with OEM replacement parts. Substitute parts may appear interchangeable, but internal differences may result in inferior occupant protection. Failure to do so may result in occupant personal injury or death.

Wear safety glasses, rubber gloves, and long sleeved clothing when cleaning powder residue from vehicle after an airbag deployment. Powder residue emitted from a deployed airbag can cause skin irritation. Flush affected area with cool water if irritation is experienced. If nasal or throat irritation is experienced, exit the vehicle for fresh air until the irritation ceases. If irritation continues, see a physician.

Do not use a replacement airbag that is not in the original packaging. This may result in improper deployment, personal injury, or death.

The factory installed fasteners, screws and bolts used to fasten airbag components have a special coating and are specifically designed for the airbag system. Do not use substitute fasteners. Use only original equipment fasteners listed in the parts catalog when fastener replacement is required.

During, and following, any child restraint anchor service, due to impact event or vehicle repair, carefully inspect all mounting hardware, tether straps, and anchors for proper installation, operation, or damage. If a child restraint anchor is found damaged in any way, the anchor must be replaced. Failure to do this may result in personal injury or death.

Deployed and non-deployed airbags may or may not have live pyrotechnic material within the airbag inflator.

Do not dispose of driver/passenger/curtain airbags or seat belt tensioners unless you are sure of complete deployment. Refer to the Hazardous Substance Control System for proper disposal.

Dispose of deployed airbags and tensioners consistent with state, provincial, local, and federal regulations.

After any airbag component testing or service, do not connect the battery negative cable. Personal injury or death may result if the system test is not performed first.

DISARMING THE SYSTEM

To avoid personal injury when working on vehicles equipped with an air bag, the negative battery cable must be disconnected and at least 90 seconds must elapse before working on the system. Failure to do so may result in deployment of the air bag. You should also wrap or isolate the negative battery cable with electrical or other non-conductive tape.

ARMING THE SYSTEM

To arm the system after service is completed, connect the negative battery cable.

CLOCKSPRING CENTERING

WARNING

Ensure that the clock spring's mating marks are properly aligned. If not, the steering wheel may not rotate completely during a turn, or the flat cable in the clock spring could be damaged, This would prevent normal SRS operation and possibly cause serious injury to the driver.

1. Align the mating marks and install the clock spring.
   a. Turn the clock spring clockwise fully. Then turn it back approximately 3-3/4 turns counterclockwise to align the mating marks.
   b. Turn the front wheels to the straight-ahead position. Then install the clock spring to the column switch.

WARNING

Ensure that the steering wheel sensor's mating marks are properly aligned. If not, the steering wheel sensor could be damaged.

2. Align the mating marks and install the steering wheel sensor.
   a. Turn the steering wheel sensor to the arrow direction shown to align the mating marks.
   b. Align the mating marks on the clock spring and the steering wheel sensor, and install the steering wheel sensor to the column switch assembly.
   c. Connect the steering wheel sensor connector.
3. Install the column switch, clock spring and steering wheel assembly.
4. Connect a scan tool and calibrate the steering wheel sensor.

AIR BAG MODULE

REMOVAL & INSTALLATION

See Figure 16.

1. Disconnect the negative battery cable.

➡**Wait at least 90 seconds after disconnecting the cable from the negative (-) battery terminal to prevent airbag activation.**

2. Remove the steering column cover.
3. Disconnect the connectors for the horn and air bag module.
4. Remove the air bag module assembly.
5. Remove the steering wheel dynamic damper.
6. Disconnect steering wheel remote control harness
7. Remove the steering wheel.
8. Remove the lower column cover.
9. Remove the clock spring.

To install:

10. Perform a pre-installation inspection.
    a. Connect the negative battery cable.
    b. Connect a scan tool and read the DTC codes.
    c. Verify that no DTC codes except 21 and 24 are set.
11. Install the clock spring.

WARNING

Ensure that the clock spring's mating marks are properly aligned. If not, the steering wheel may not rotate completely during a turn, or the flat cable in the clock spring could be damaged, This would prevent normal SRS operation and possibly cause serious injury to the driver.

a. Align the mating marks and install the clock spring.
b. Turn the clock spring clockwise fully. Then turn it back approximately 3-3/4 turns counterclockwise to align the mating marks.
c. Turn the front wheels to the straight-ahead position. Then install the clock spring to the column switch.

WARNING

Ensure that the steering wheel sensor's mating marks are properly aligned. If not, the steering wheel sensor could be damaged.

12. Align the mating marks and install the steering wheel sensor.
    a. Turn the steering wheel sensor to the arrow direction shown to align the mating marks.
    b. Align the mating marks on the clock spring and the steering wheel sensor, and install the steering wheel sensor to the column switch assembly.
    c. Connect the steering wheel sensor connector.
13. Install the lower column cover.
14. Install the steering wheel assembly.

WARNING

When installing the steering wheel and air bag module, ensure that the harness of the clock spring does not become caught or tangled.

a. Before installing the steering wheel and air bag module, turn the vehicle's front wheels to the straight-ahead position and align the mating marks of the clock spring.
b. After securing the steering wheel, turn the steering wheel all the way in both directions to confirm that the steering wheel rotation is normal.

15. Connect the steering wheel remote control harness.
16. Install the steering wheel dynamic damper.

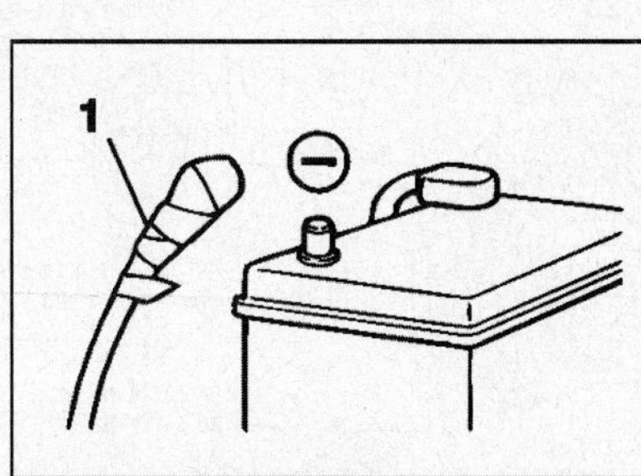

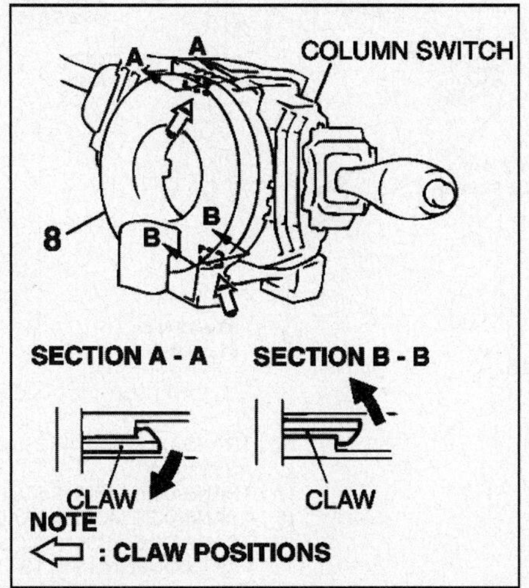

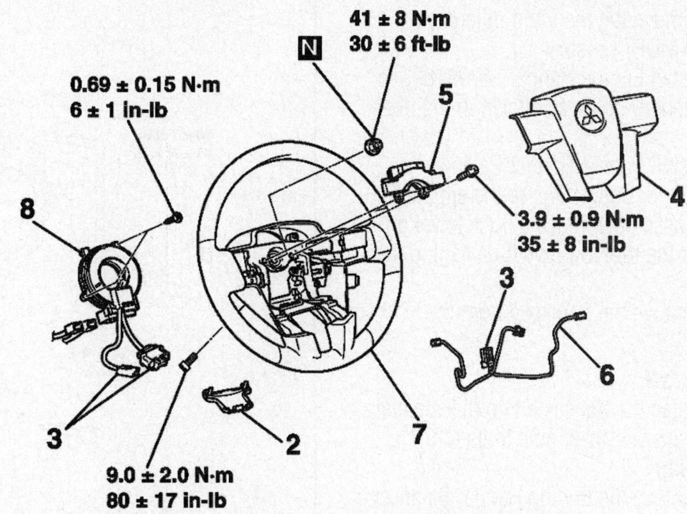

Fig. 16 Exploded view of the air bag and clock spring

17. Install the air bag module assembly.
    a. Connect the connectors for the horn, air bag module, and steering wheel remote control harness.
18. Install the cover.
19. Connect the negative battery cable.
20. Perform the post-installation inspection.
    a. Turn the ignition switch to **ON** position.
    b. If the "SRS" warning light illuminates for approximately seven seconds the SRS system is functioning properly.
    c. If not perform diagnostics to determine the problem.

DRIVETRAIN

AUTOMATIC TRANSAXLE ASSEMBLY

REMOVAL & INSTALLATION

See Figures 17 through 19.

1. Before servicing the vehicle, refer to the Precautions Section.
2. Remove the engine under cover.
3. Drain the transaxle fluid.
4. On AWD vehicles, drain the engine coolant.
5. Remove the hood.
6. Remove the negative battery cable.
7. Remove the air cleaner assembly.
8. Remove the powertrain control module (PCM).
9. Remove the battery and battery tray.
10. Remove the intake manifold rear plenum stay, EGR pipe and intake manifold plenum.
11. Remove the upper radiator hose, engine oil dipstick and engine hanger.
12. Remove the front exhaust pipe from the 2 exhaust manifolds, then disconnect it from the intermediate pipe/catalytic converter (make certain to retain the bolts and nuts for reassembly).
13. Remove the adjusting nut.
14. Remove the transaxle control cable connection.
15. Remove the inhibitor switch connector.
16. Remove the A/T control solenoid valve connector.
17. Remove the input and output shaft speed sensor connectors.
18. Remove the driveshaft (2WD) or driveshaft and output shaft assembly (AWD).
19. Remove the transfer case assembly.
20. Remove the starter.
21. Remove the transaxle oil cooler hose.
22. Remove the transaxle assembly upper part coupling bolts.
23. Remove the cover.
24. Remove the engine oil pan and transaxle assembly coupling bolts.
25. Remove the drive plate bolts. use Mitsubishi tool MD998781, or equivalent, to hold the driveplate in position to remove the bolts.
26. Remove the center member assembly.
27. Remove the rear roll stopper bracket.
28. Remove the crossmember plate, left side.
29. Remove the air cleaner bracket.
30. Remove the transaxle mount bracket assembly.

31. Remove the transaxle mount stopper.

32. Remove the manifold differential pressure (MDP) sensor.

33. Install Engine Hanger MB991895 to the front fender bolts, as shown in the illustration.

34. Install Engine Hanger Balancer MB991454, or equivalent, to the engine/transaxle assembly. Place a thick towel on the end of the tool to prevent damaging the firewall.

35. Remove the transaxle from the vehicle.

To install:

36. Install the transaxle into the vehicle. Tighten transaxle-to-engine bolts to 65 ft. lbs. (68 Nm)

37. Remove the Engine Hanger Balancer MB991454, or equivalent, from the engine/transaxle assembly.

38. Remove Engine Hanger MB991895 from the front fender bolts, as shown in the illustration.

39. Install the manifold differential pressure (MDP) sensor.

40. Install the transaxle mount stopper and tighten through bolts to 43 ft. lbs. (58 Nm).

41. Install the transaxle mount bracket assembly.

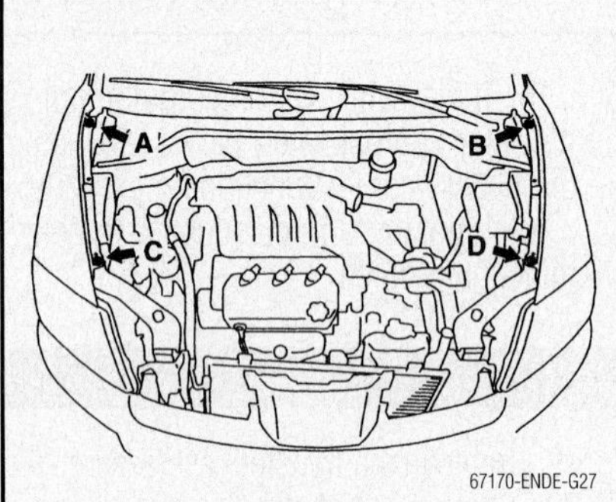

Fig. 17 Front fender assembling bolts for special tool installation

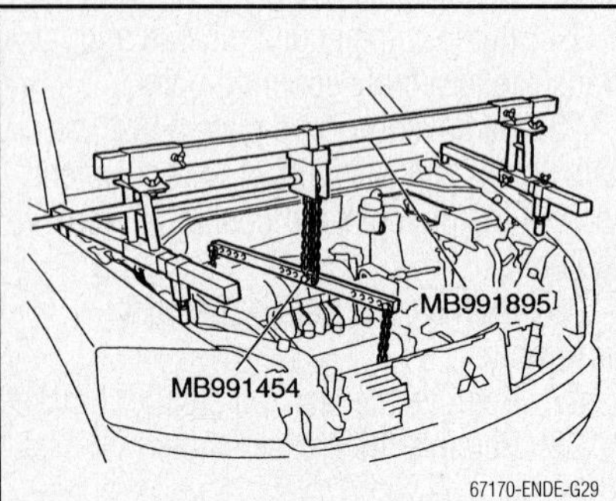

Fig. 18 View of the engine hangers installed on the engine

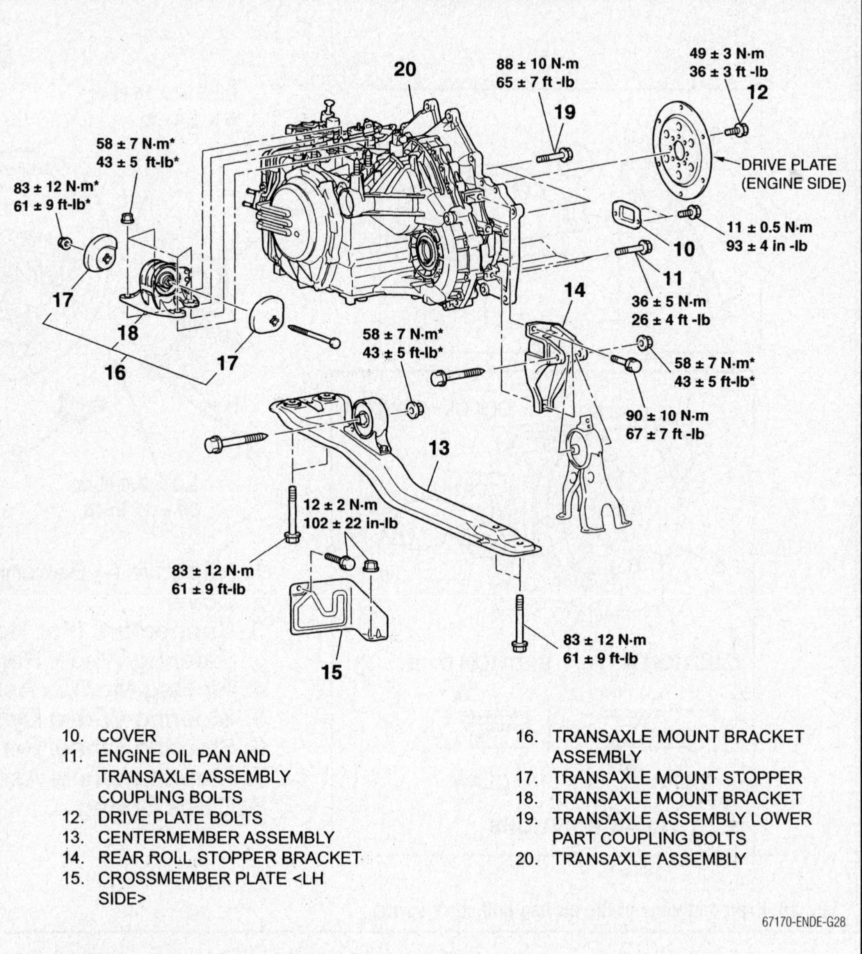

Fig. 19 Automatic transaxle and related components

42. Install the air cleaner bracket.

43. Install the crossmember plate, left side and tighten through bolts to 43 ft. lbs. (58 Nm).

44. Install the rear roll stopper bracket and tighten through bolts to 43 ft. lbs. (58 Nm).

45. Install the center member assembly. position to Install the bolts.

46. Install the drive plate bolts. Use Mitsubishi tool MD998781, or equivalent, to hold the driveplate in place while tightening bolts to 36 ft. lbs. (49 Nm).

47. Install the engine oil pan and transaxle assembly coupling bolts.

48. Install the cover.

49. Install the transaxle assembly upper part coupling bolts.

50. Install the transaxle oil cooler hose.

51. Install the starter.

52. Install the transfer case assembly.

53. Install the driveshaft (2WD) or driveshaft and output shaft assembly (AWD).

54. Install the input and output shaft speed sensor connectors.

55. Install the A/T control solenoid valve connector.

56. Install the inhibitor switch connector.

57. Install the transaxle control cable connection.

58. Install the adjusting nut. pipe/catalytic converter (make certain to retain the bolts and nuts for reassembly).

59. Install the front exhaust pipe from the 2 exhaust manifolds, then disconnect it from the intermediate

60. Install the upper radiator hose, engine oil dipstick and engine hanger.

61. Install the intake manifold rear plenum stay, EGR pipe and intake manifold plenum.

62. Install the battery and battery tray.

63. Install the powertrain control module (PCM).

64. Install the air cleaner assembly.

65. Install the negative battery cable.

66. Install the hood.

67. On AWD vehicles, drain the engine coolant.

68. Refill the transaxle and transfer case with oil.

69. Install the engine under cover.

70. Start the vehicle and check for leaks.

TRANSFER CASE ASSEMBLY

REMOVAL & INSTALLATION

See Figure 20.

1. Before servicing the vehicle, refer to the Precautions Section.
2. Drain the transaxle and transfer case fluid. Drain the engine coolant.
3. Remove the engine under cover.
4. Remove the front exhaust pipe.
5. Remove the front propeller shaft.
6. Remove the driveshaft and output shaft.
7. Remove the right exhaust manifold stay "b".
8. Remove the air guide.
9. Remove the water feed tube assembly and gasket.
10. Remove the water return hose and gasket.
11. Remove the retaining bolts and detach the transfer case assembly from the transaxle. lower the transfer case between the engine and crossmember. remove and discard the o-rings.
12. As necessary, remove the water feed hose, air cleaner assembly, water return hose, battery and battery tray.

To install:

13. Install the water return hose.
14. Install the battery tray and battery.
15. Install the air cleaner assembly.
16. Install the water feed hose.
17. Install the new o-rings onto the transfer case.
18. Install the transfer case, by maneuvering it up between the engine and crossmember. tighten the retaining bolts to 51 ft. lbs. (69 Nm).
19. Install the water return hose gasket and hose.
20. Install the water feed tube gasket and tube.
21. Install the air guide.
22. Install the right exhaust manifold stay "b".
23. Install the driveshaft and output shaft.
24. Install the front propeller shaft.
25. Install the front exhaust pipe.
26. Install the engine under cover.
27. Fill the engine cooling system. Fill the transaxle and transfer case with the proper grade and type fluid.
28. Connect the negative battery cable. Turn the ignition **ON** and then **OFF**, and keep it off for at least ten seconds.

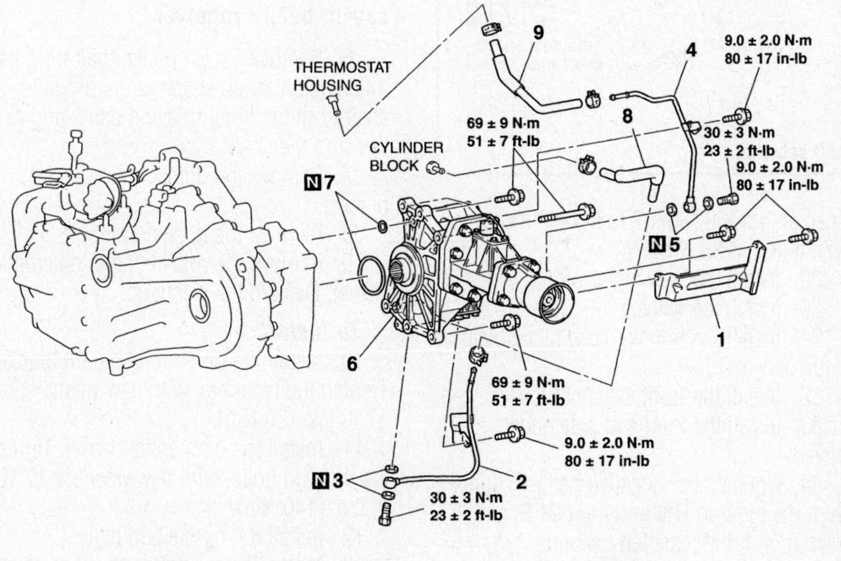

1. AIR GUIDE
2. WATER FEED TUBE ASSEMBLY
3. GASKET
4. WATER RETURN HOSE
5. GASKET
6. TRANSFER
7. O-RING
8. WATER FEED HOSE
9. WATER RETURN HOSE

67170-ENDE-G30

Fig. 20 Transfer case and related components

FRONT HALFSHAFT

REMOVAL & INSTALLATION

See Figures 21 and 22.

➡If the vehicle is equipped with ABS, do not strike the ABS rotors installed to the BJ outer race of the halfshaft against other parts when removing or installing the halfshaft. Otherwise the ABS rotors will be damaged.

1. Before servicing the vehicle, refer to the Precautions Section.
2. Drain the transaxle and transfer case fluid, as necessary.
3. Remove the negative battery cable.
4. Remove the front and side under covers.
5. Remove the front exhaust pipe.
6. Remove the transfer case heat shield, if equipped.
7. Remove the wheels.
8. Remove the cotter pin.
9. Remove the halfshaft nut and washer.
10. Remove the speed sensor, if equipped with abs.
11. Remove the brake hose bracket.
12. Remove the self locking nut and separate the lower ball joint.
13. Remove the loosen the tie rod end nut and separate the tie rod end. remove the nut.
14. Remove the halfshaft or halfshaft and inner shaft from the hub using Mitsubishi Special Tools MB990242, MB990244, MB991354 and MB990767 to push the halfshaft or halfshaft and inner shaft from the hub.
15. Remove the halfshaft from the hub by pulling the bottom of the brake rotor toward you.

⁂ WARNING

Never pull on the halfshaft or you risk damaging the joints. Only use a prybar to remove the halfshaft from the transaxle.

16. Insert a prybar between the transaxle case and halfshaft, then pry the halfshaft from the transaxle. Make sure the spline of the halfshaft does not damage the oil seal.
17. If you have trouble removing the inner shaft from the transaxle, hit the bracket assembly lightly with a plastic hammer and remove the inner shaft.

➡Do not apply pressure to the wheel bearing while the halfshafts are removed.

18. On AWD vehicles, use Mitsubishi Special tool MB991721 to remove the output shaft from the transaxle. Make sure the splined part of the output shaft does not damage the oil seal.

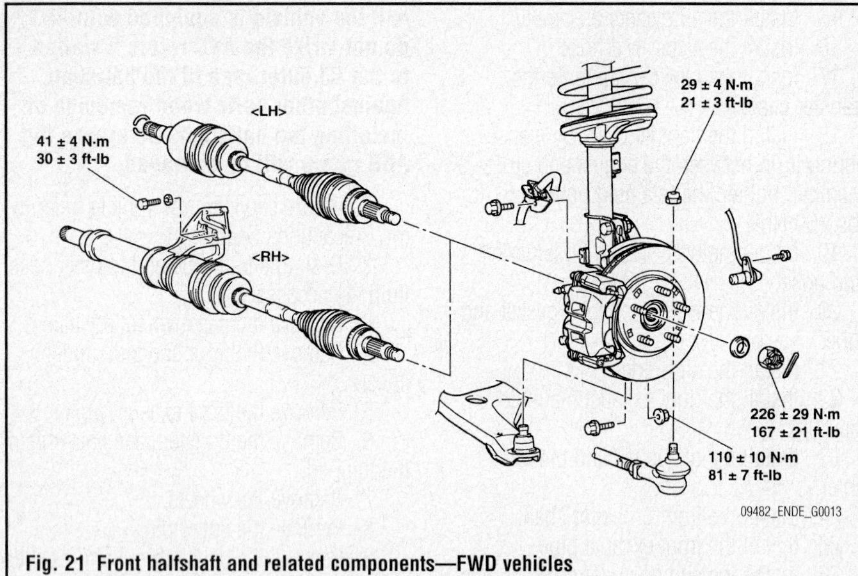

Fig. 21 Front halfshaft and related components—FWD vehicles

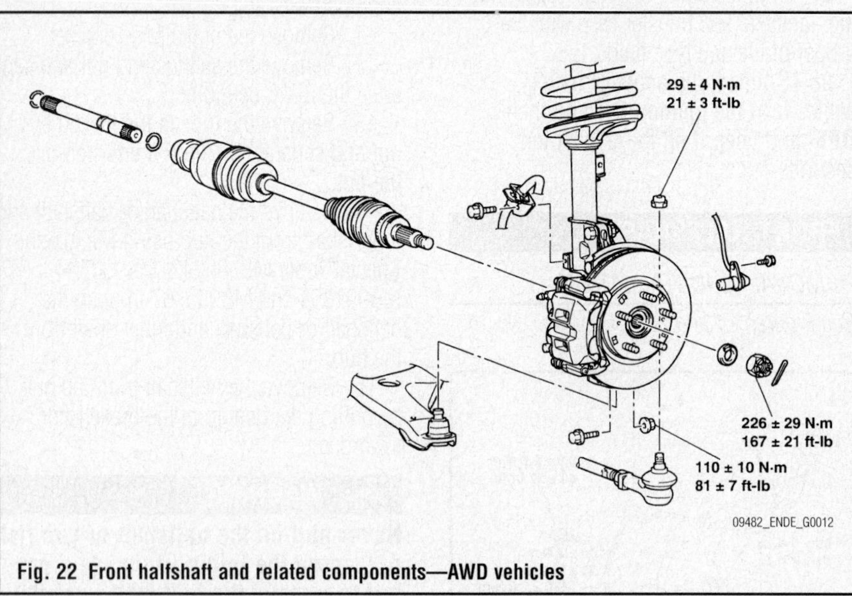

Fig. 22 Front halfshaft and related components—AWD vehicles

19. Remove the circlips from the ends of the shafts.

To install:

20. Install the new circlips.
21. Install the output shaft, AWD vehicles.
22. Install the halfshaft or halfshaft and inner shaft assembly, as necessary.
23. Install the tie rod end and tighten the nut to 21 ft. lbs. (29 Nm).
24. Install the lower ball joint and tighten the nuts to 81 ft. lbs. (110 Nm).
25. Install the brake hose bracket.
26. Install the speed sensor, if equipped with abs.
27. Install the new halfshaft washer, with the beveled edge facing out.
28. Install the halfshaft nut. with no load on the wheel bearings, use Mitsubishi special tool MB990767 to tighten the nut to 174 ft. lbs. (236 Nm).
29. Install the new cotter pin.
30. Install the wheels.
31. Install the transfer case heat shield, if equipped.
32. Install the front exhaust pipe.
33. Install the front and side under covers.
34. Connect the negative battery cable. Turn the ignition **ON** and then **OFF**, and keep it off for at least ten seconds.
35. Fill the transaxle and transfer case (if equipped) with the proper grade and type fluid

CV-BOOT INSPECTION

1. Raise and safely support the vehicle securely on jackstands.
2. Turn the steering wheel to full right lock.
3. Inspecting from behind the halfshaft, examine the driver's side CV boot while slowly turning the wheel.
4. Inspect the valleys between the folds. If cracks are apparent, replace the boot.
5. Also inspect the circumference where the boot covers the edge of the hub, as cracks can appear there too. If large cracks (large enough to fit a finger nail in) are apparent, replace the boot.
6. Repeat for the inner boot.
7. Turn the steering wheel to full left lock and examine the passenger side boots in the same manner.

REAR DIFFERENTIAL CARRIER ASSEMBLY

REMOVAL & INSTALLATION

See Figure 23.

1. Before servicing the vehicle, refer to the Precautions Section.
2. Drain the differential gear oil. Raise and support the vehicle safely.
3. Remove the center exhaust pipe.
4. Remove the rear driveshaft.
5. Remove the stabilizer bar bushing.

➡Matchmark the installed relation of the propeller shaft to the differential carrier before removal.

6. Remove the propeller shaft from the differential carrier. suspend the propeller shaft from the body to avoid damaging or bending the shaft.
7. Remove the differential mount bracket.
8. Remove the hose and nipple.
9. Remove the retainers and differential carrier. Discard the washers.

To install:

10. Install the differential mount bracket. Tighten the retainers, with new washers, to 67 ft. lbs. (90 Nm).
11. Install the differential carrier. Tighten the through bolts, with new washers, to 104 ft. lbs. (140 Nm).
12. Install the nipple and hose.
13. Install the propeller shaft, aligning the matchmarks made during removal. Tighten bolts to 24 ft. lbs. (32 Nm).
14. Fill the differential with the proper grade and type gear oil.
15. Check and adjust the rear wheel alignment, as necessary.

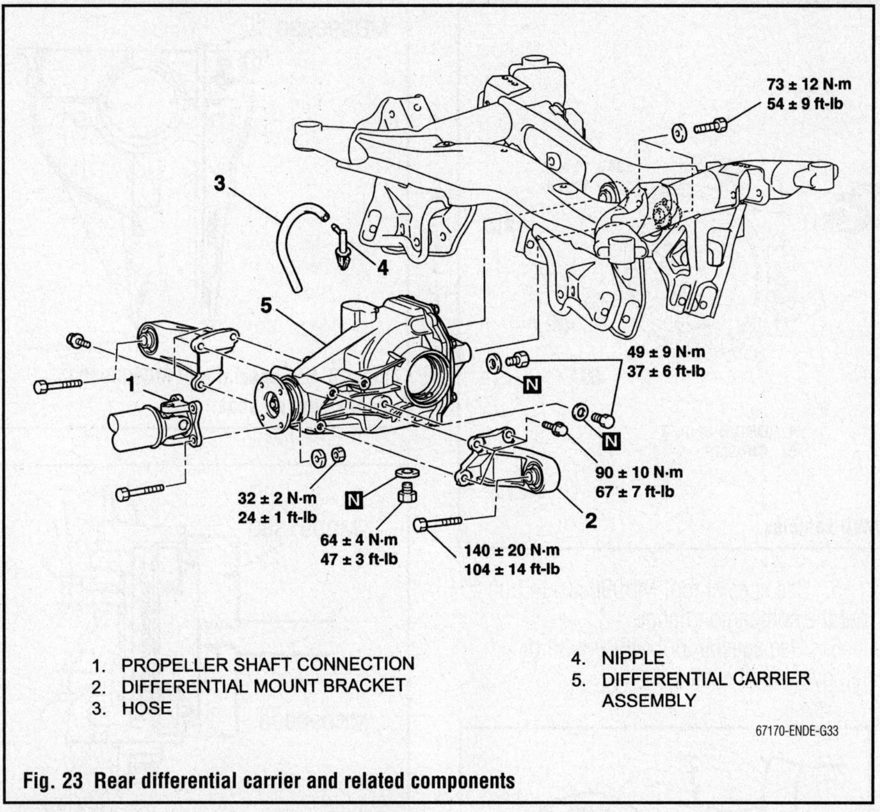

Fig. 23 Rear differential carrier and related components

REAR DRIVESHAFT

REMOVAL & INSTALLATION

See Figure 24.

1. Raise and safely support the vehicle securely on jackstands.
2. Make mating marks on the differential companion flange and the propeller shaft assembly.

⁂ WARNING

Be careful not to bend the joint assembly when removing the propeller shaft because this may cause damage to the joint boot.

3. Remove the self locking nut.
4. Remove the insulator and spacer.
5. Insert a rag so as to avoid boot damage, and remove the propeller shaft assembly in a straight and level manner.

To install:

⁂ WARNING

Do not damage the oil seal lips on the transfer case. Be careful not to bend the joint portion when removing the propeller shaft, because this will damage the joint boot.

6. Remove oil and grease from the threads of the mounting bolts and nuts before tightening, or they will loosen.
7. If reusing the propeller shaft, align the mating marks of differential companion flange and propeller shaft assembly to install. Tighten bolts to 24 ft. lbs. (32 Nm).
8. Lower the vehicle.

REAR HALFSHAFT

REMOVAL & INSTALLATION

See Figure 25.

1. Before servicing the vehicle, refer to the Precautions Section.
2. Remove the undercover.
3. Remove the wheels.
4. Remove the clevis pin.
5. Remove the driveshaft nut, using and end yoke holder to hold the hub.
6. Remove the washer.
7. Remove the rear wheel speed sensor, if equipped with abs.
8. Remove the lower control arm, shock absorber, trailing arm and toe control arm connection.
9. Remove the driveshaft. use Mitsubishi special tools MB990242, MB990244, MB991354 and MB990767 to push the driveshaft from the hub.

⁂ WARNING

Never pull on the driveshaft or you risk damaging the joints. Only use a prybar to remove the driveshaft from the transaxle.

10. Use a slide hammer to remove the driveshaft from the differential carrier. Make sure the spline of the driveshaft does not damage the oil seal.
11. Remove the circlip

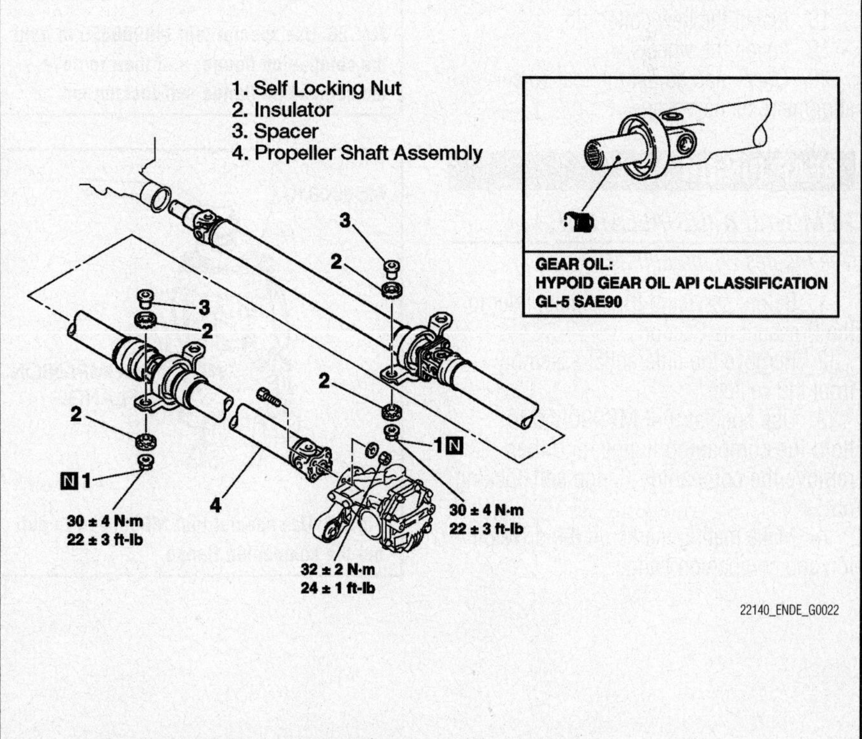

Fig. 24 Exploded view of the driveshaft assembly

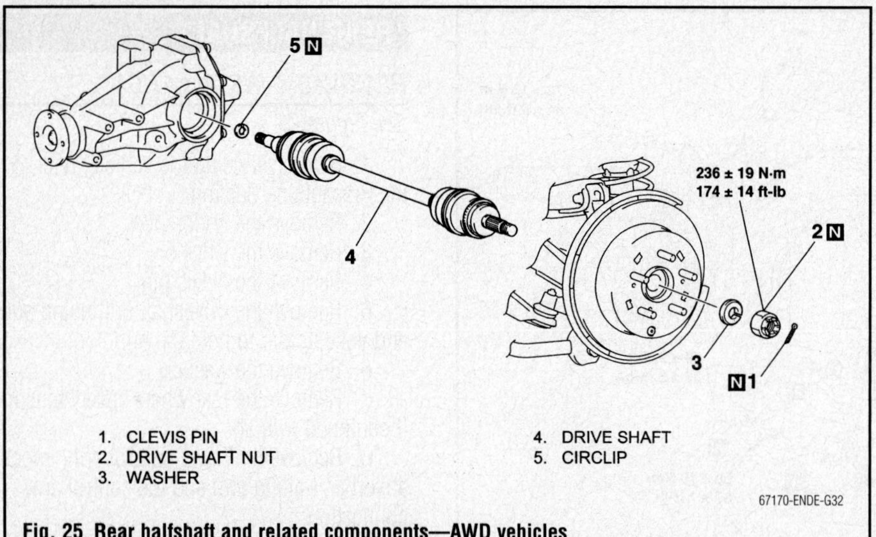

Fig. 25 Rear halfshaft and related components—AWD vehicles

To install:

12. Install the new circlip on the driveshaft.
13. Install the driveshaft into the differential carrier and hub.
14. Install the lower control arm, shock absorber, trailing arm and toe control arm connection.
15. Install the rear wheel speed sensor, if equipped with abs.
16. Install the new driveshaft washer, with the beveled edge facing out.
17. Install the driveshaft nut. with no load on the wheel bearings, use Mitsubishi special tool MB990767 to tighten the nut to 174 ft. lbs. (236 Nm).
18. Install the new cotter pin.
19. Install the wheels.
20. Check and adjust the rear wheel alignment, as necessary.

REAR PINION SEAL

REMOVAL & INSTALLATION

See Figures 26 through 29.

1. Before servicing the vehicle, refer to the Precautions Section.
2. Remove the differential assembly from the vehicle.
3. Use special tool MB990850 to hold the companion flange, and then remove the companion flange self-locking nut.
4. Make mating marks on the drive pinion and companion flange.
5. Use special tool MB990810 to pull out the companion flange.
6. Use special tool MB990939 to remove the oil seal.

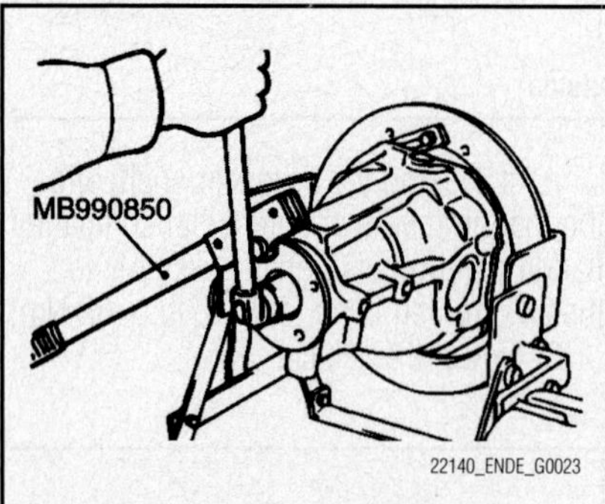

Fig. 26 Use special tool MB990850 to hold the companion flange, and then remove the companion flange self-locking nut

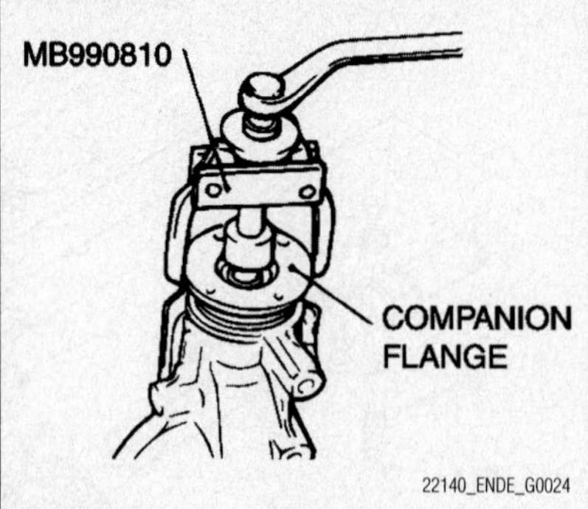

Fig. 27 Use special tool MB990810 to pull out the companion flange

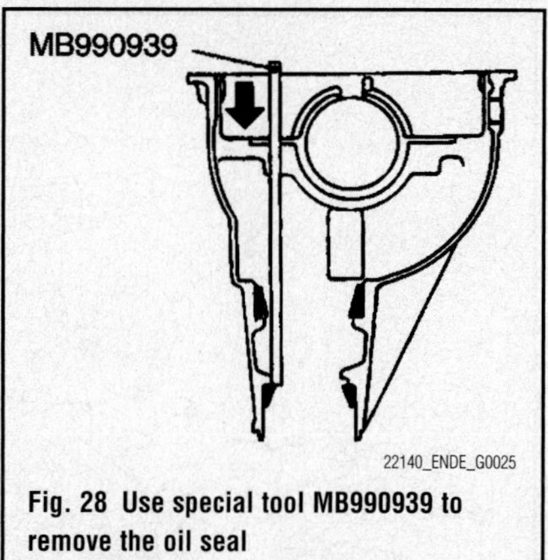

Fig. 28 Use special tool MB990939 to remove the oil seal

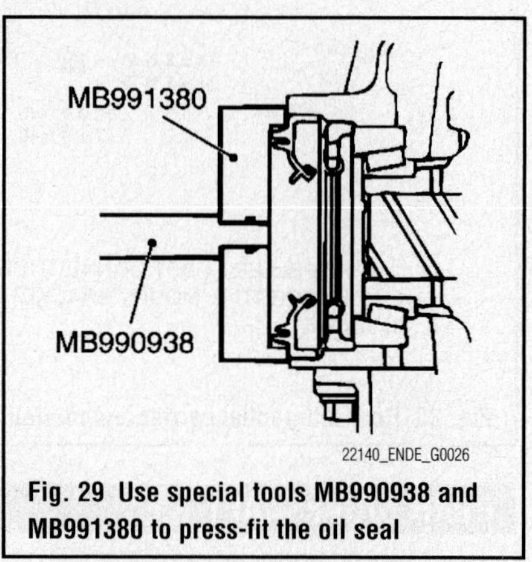

Fig. 29 Use special tools MB990938 and MB991380 to press-fit the oil seal

To install:

7. Use special tools MB990938 and MB991380 to press-fit the oil seal.
8. Adjust the drive pinion turning torque by the following procedures:
    a. Install the drive pinion assembly and companion flange with the mating marks properly aligned. Tighten the companion flange self-locking nut to 137 ft. lbs. (186 Nm) while holding the companion flange with special tool MB990850.
    b. Use special tools MB990685 and MB990326 to measure the drive pinion turning torque (with drive pinion oil seal) to verify that the drive pinion turning torque is 12.22–15.66 inch lbs. (1.38–1.77 Nm).
    c. If the turning torque is not within the standard value, check the tightening torque of the companion flange self-locking nut, and the installation of the oil seal.
9. Once proper turning torque has been achieved, install the differential assembly in the vehicle.

ENGINE COOLING

ENGINE FAN

REMOVAL & INSTALLATION

See Figures 30 and 31.

1. Before servicing the vehicle, refer to the Precautions Section.
2. Drain the engine coolant.
3. Remove the air cleaner assembly.
4. Remove the radiator grille.
5. Remove the rubber hose.
6. Matchmark the installed position of the hose and clamp, then disconnect the upper radiator hose.
7. Detach the fan motor connector.
8. Remove the shroud assembly.
9. Remove the radiator fan.
10. Remove the radiator fan motor (integrated with the fan control module).
11. Remove the condenser fan.
12. Remove the condenser fan motor.

To install:

13. Install the condenser fan.
14. Install the radiator fan motor.
15. Install the radiator fan.
16. Install the shroud assembly.
17. Connect the fan motor connector.
18. Connect the upper radiator hose.
    a. During installation of the upper radiator hose, insert the hose as far as the projection of the water inlet fitting. Align the matchmarks on the hose and hose clamp, then connect the hose.
19. Install the rubber hose.
    a. Insert each hose as far as the projection of the water inlet fitting. Make sure to align the matchmarks before connecting the hoses.
20. Install the radiator grille.
21. Install the air cleaner assembly.
22. Fill the engine with coolant.
23. Start the engine and check for proper fan operation.

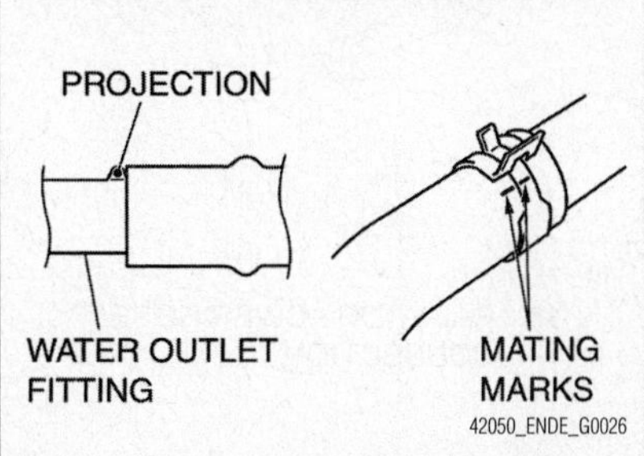

Fig. 31 Insert each hose as far as the projection of the water inlet fitting. Make sure to align the matchmarks before connecting the hoses

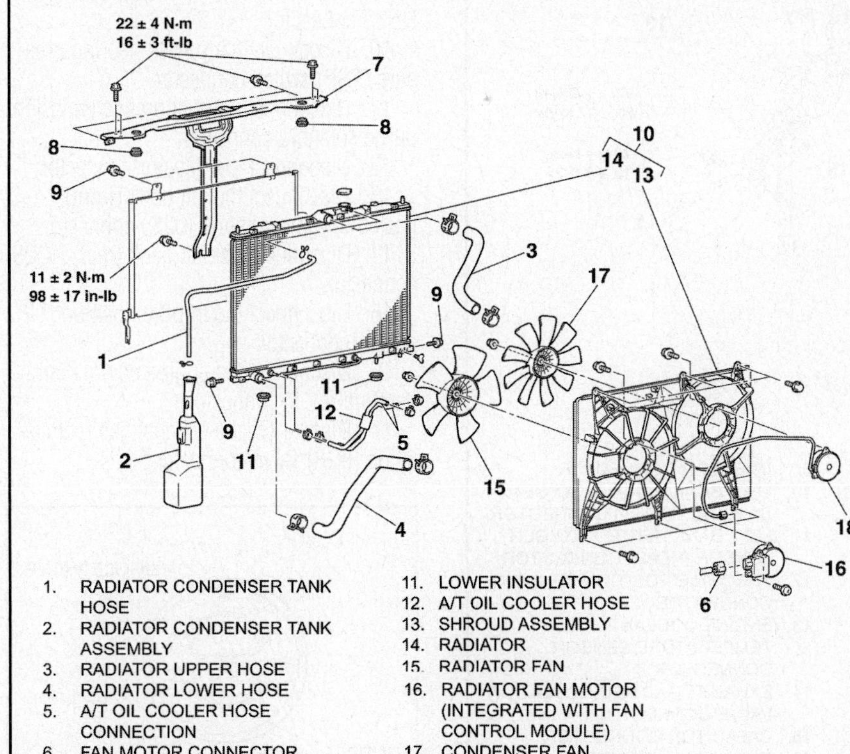

Fig. 30 Exploded view of the radiator, fan and related components

RADIATOR

REMOVAL & INSTALLATION

See Figures 30 and 31.

1. Before servicing the vehicle, refer to the Precautions Section.
2. Drain the engine coolant.
3. Remove the air cleaner assembly.
4. Remove the radiator grille.
5. Disconnect the radiator condenser tank hose.
6. Remove the radiator condenser tank assembly.
7. Matchmark the installed positions of the hoses and hose clamps, then disconnect the upper and lower radiator hoses.
8. Detach the A/T oil cooler hose connection. After disconnecting the hose from the radiator, plug the hose and radiator nipple to prevent foreign debris from entering the cooling system.
9. Disconnect the fan motor connector.
10. Remove the hood latch.
11. Remove the front bumper clips.
12. Remove the front end upper bar.
13. Remove the upper insulator.
14. Remove the condenser bolts.
15. Remove the radiator assembly.
16. Remove the lower insulator.
17. Remove the A/T oil cooler hose.
18. Remove the shroud assembly.

To install:

19. Install the shroud assembly.
20. Install the A/T oil cooler hose.
21. Install the lower insulator.
22. Install the radiator assembly.
23. Install the condenser bolts.
24. Install the upper insulator.
25. Install the front end upper bar.
26. Install the front bumper clips.
27. Install the hood latch.
28. Connect the fan motor connector.
29. Connect the A/T oil cooler hose connection.
30. Connect the upper and lower radiator hoses.
    a. During installation of the upper and lower radiator hoses, insert the hoses as far as the projection of the water inlet fitting. Align the matchmarks on the hoses and hose clamps, then connect the hoses.

31. Install the radiator condenser tank assembly.
32. Disconnect the radiator condenser tank hose.
33. Install the radiator grille.
34. Install the air cleaner assembly.
35. Fill the engine with coolant.
36. Start the engine and check for proper fan operation.

THERMOSTAT

REMOVAL & INSTALLATION

See Figures 32 through 34 and 31.

1. Before servicing the vehicle, refer to the Precautions Section.
2. Drain the cooling system.
3. Remove the air cleaner.
4. Disconnect the manifold differential pressure (MDP) sensor connector.
5. Disconnect the knock sensor (KS) connector.
6. Disconnect the crankshaft position (CKP) sensor connector.

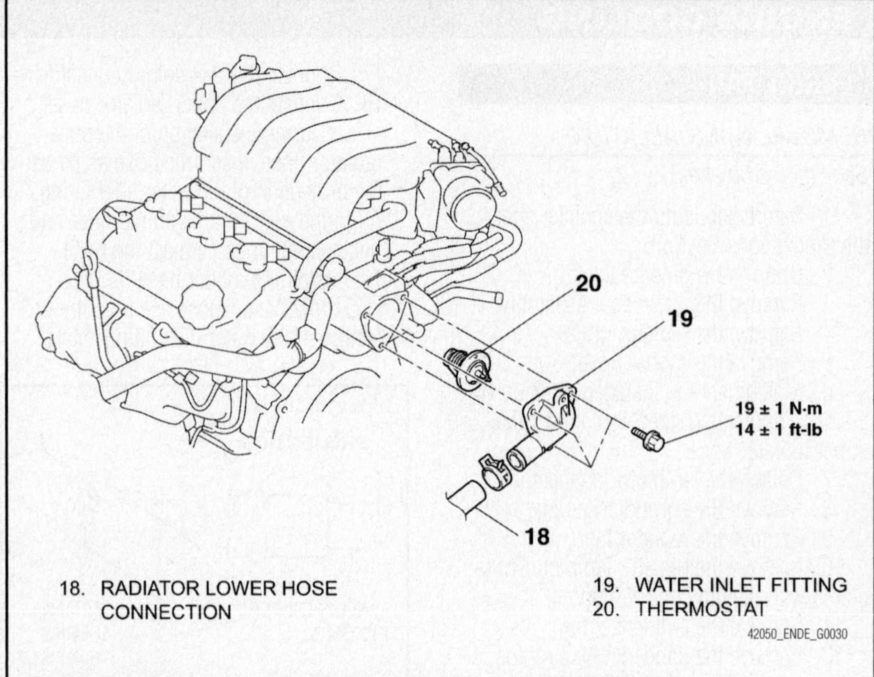

Fig. 33 Exploded view of the thermostat, water inlet and radiator hose

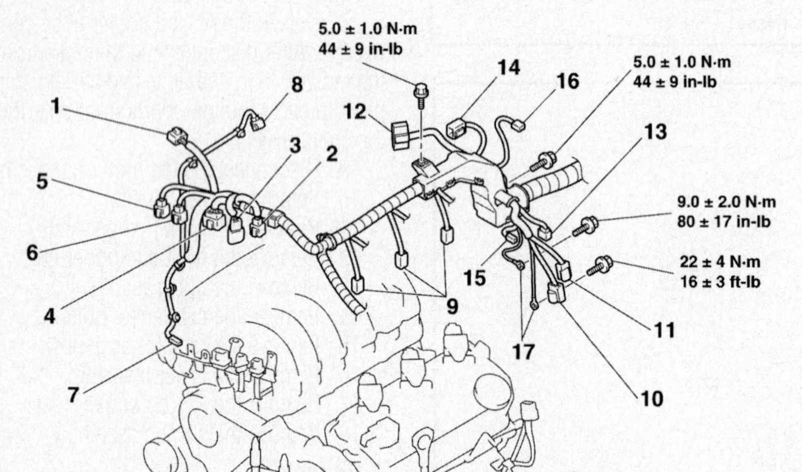

Fig. 32 Exploded view of the components you need to remove to access the thermostat

7. Disconnect the control wiring harness and wiring harness combination connector.
8. Disconnect the right bank (rear) heated oxygen sensor (HO2S) connector.
9. Disconnect the right bank (front) HO2S connector.
10. Disconnect the power steering pressure (PSP) switch connector.
11. Disconnect the evaporative emission purge solenoid connector.
12. Disconnect the injector connector.
13. Disconnect the left bank (front) heated oxygen sensor (HO2S) connector.
14. Disconnect the left bank (rear) HO2S connector.
15. Disconnect the throttle position (TP) sensor connector.
16. Disconnect the engine coolant temperature (ECT) connector.
17. Disconnect the exhaust gas recirculation (EGR) valve connector.

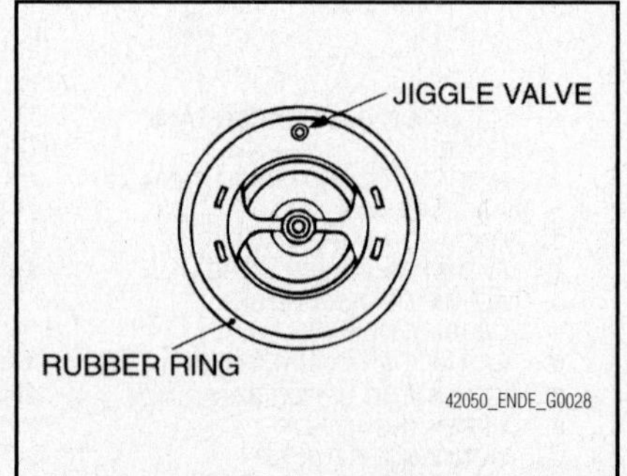

Fig. 34 The thermostat's jiggle valve must be facing up, at the 12 o'clock position during installation

18. Disconnect the capacitor connector.
19. Disconnect the output shaft speed sensor connector.
20. Disconnect the ground strap.
21. Matchmark the installed position of the hose and clamp, then disconnect the radiator lower hose connection.
22. Remove the water inlet fitting, then remove the thermostat.
23. Thoroughly clean the gasket mating surfaces.

To install:

24. Install the thermostat so the jiggle valve is facing straight up. Be careful not to fold or otherwise damage the rubber ring.

➡Make absolutely sure that no oil adheres to the rubber ring of the thermostat. Also do not fold or scratch the rubber ring during installation.

25. Install the water inlet fitting. Tighten the bolts to 14 ft. lbs. (19 Nm).
26. Install the lower radiator hose. Insert the hose as far as the projection of the water inlet fitting. Align the matchmarks on the hose and hose clamp, then connect the hose.
27. Connect the ground strap.
28. Connect the output shaft speed sensor connector.
29. Connect the capacitor connector.
30. Connect the exhaust gas recirculation (EGR) valve connector.
31. Connect the engine coolant temperature (ECT) connector.
32. Connect the throttle position (TP) sensor connector.
33. Connect the left bank (rear) HO2S connector.
34. Connect the left bank (front) heated oxygen sensor (HO2S) connector.
35. Connect the injector connector.
36. Connect the evaporative emission purge solenoid connector.
37. Connect the power steering pressure (PSP) switch connector.
38. Connect the right bank (front) HO2S connector.
39. Connect the right bank (rear) heated oxygen sensor (HO2S) connector.
40. Connect the control wiring harness and wiring harness combination connector.
41. Connect the crankshaft position (CKP) sensor connector.
42. Connect the knock sensor (KS) connector.
43. Connect the manifold differential pressure (MDP) sensor connector.
44. Install the air cleaner.
45. Fill the engine with coolant.
46. Start the engine and check for proper fan operation.

WATER PUMP

REMOVAL & INSTALLATION

See Figure 35.

1. Before servicing the vehicle, refer to the Precautions Section.
2. If necessary, properly release the fuel pressure.
3. Drain the cooling system.
4. Disconnect the negative battery cable.

CAUTION

Wait at least 90 seconds after the negative battery cable is disconnected to prevent possible deployment of the air bag.

5. Remove the timing belt.
6. Remove the crankshaft position (CKP) sensor.
7. Remove the water pump bracket.
8. Remove the retaining bolts, water pump assembly, gasket and o-ring.

To install:

9. Clean and dry the mating surfaces of the block and water pump
10. Install the water pump assembly, using a new gasket and O-ring. Torque the water pump bolts to the specifications shown in the accompanying figure.
11. Install the water outlet fitting bracket.
12. Install the CKP sensor.
13. Install the timing belt.
14. Refill the radiator with coolant. This cooling system has a self-bleeding thermostat, so system bleeding is not required.
15. Run the vehicle until the thermostat opens and fill the overflow tank. Check for leaks.
16. Once the vehicle has cooled, recheck the coolant level.

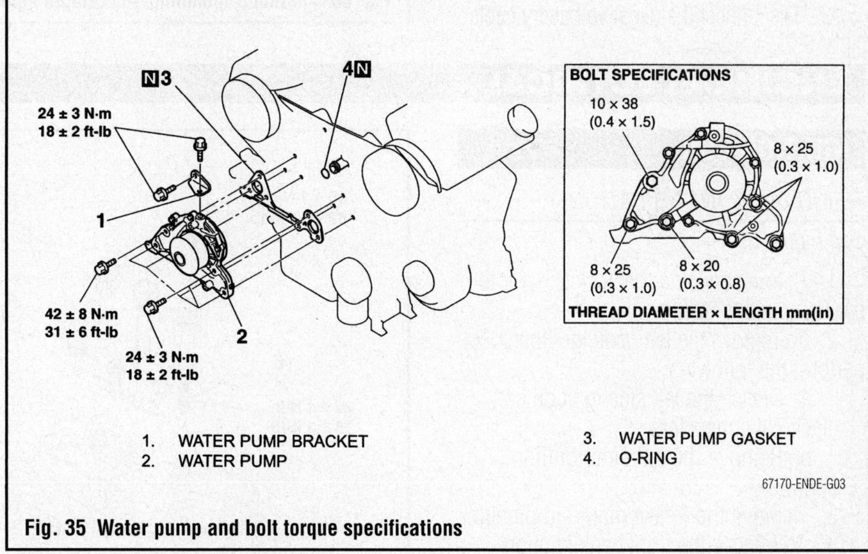

Fig. 35 Water pump and bolt torque specifications

ENGINE ELECTRICAL — CHARGING SYSTEM

ALTERNATOR

REMOVAL & INSTALLATION

See Figure 36.

1. Before servicing the vehicle, refer to the Precautions Section.
2. Disconnect the negative battery cable.
3. Remove the undercover.
4. Remove the side under cover.
5. Remove the drive belt.
6. Remove the alternator connector.
7. Disconnect the A/C compressor connector.
8. Remove the A/C compressor and position aside. do not disconnect the refrigerant lines.
9. Remove the alternator.

To install:

10. Install the alternator. Torque the bolts to 36 ft. lbs. (49 Nm).
11. Install the A/C compressor.
12. Install the A/C compressor connector.
13. Connect the alternator connector. Torque the nut to 102 inch lbs. (12 Nm).
14. Install the drive belt.
15. Install the side under cover.
16. Install the undercover.
17. Disconnect the negative battery cable.

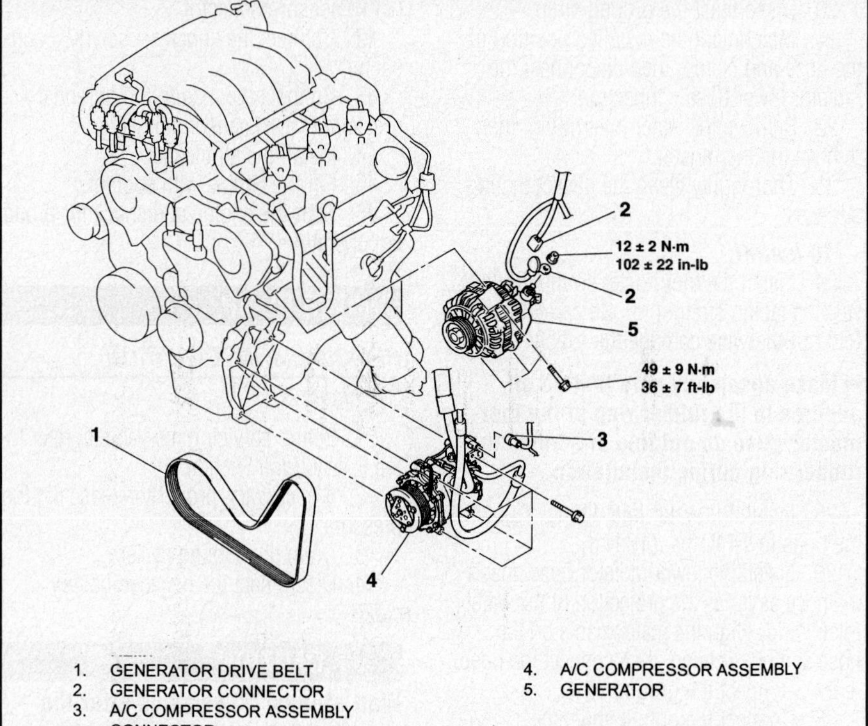

Fig. 36 Alternator mounting and related components

ENGINE ELECTRICAL — IGNITION SYSTEM

IGNITION COIL

REMOVAL & INSTALLATION

See Figure 37.

1. Before servicing the vehicle, refer to the Precautions Section.
2. To remove the left bank ignition coils, perform the following:
   a. Detach the left side ignition coil electrical connectors.
   b. Remove the left side ignition coils.
3. Remove the intake manifold plenum.
4. To remove the right bank ignition coils, perform the following:
   a. Detach the right side ignition coil electrical connectors.
   b. Remove the right side ignition coils.

To install:

5. To install the left bank ignition coils, perform the following:
   a. Install the left side ignition coils. Tighten screws to 89 inch lbs. (10 Nm).
   b. Connect the left side ignition coil electrical connectors.

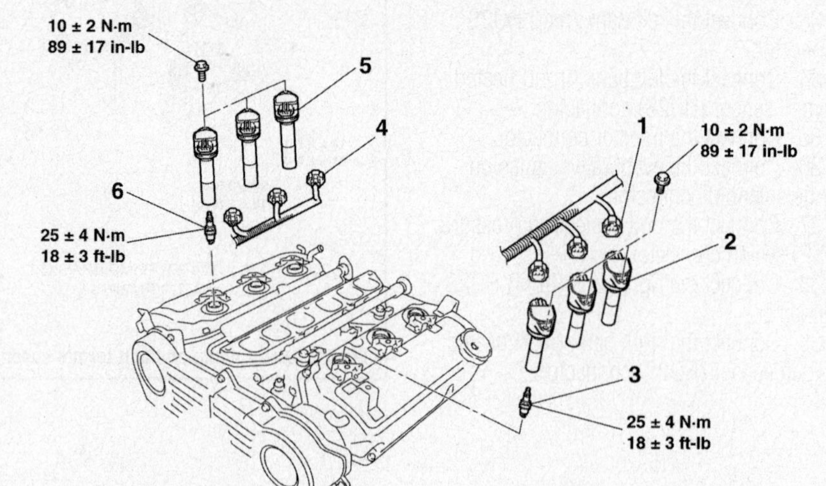

Fig. 37 Exploded view of the ignition coils and related components

6. To install the right bank ignition coils, perform the following:

a. Install the right side ignition coils. Tighten screws to 89 inch lbs. (10 Nm).

b. Connect the right side ignition coil electrical connectors.

c. Install the intake manifold plenum.

IGNITION TIMING

INSPECTION

1. Before servicing the vehicle, refer to the Precautions Section.

Before attempting to adjust the ignition timing, be sure of the following:

- The engine should be at normal operating temperature.
- The lights and all accessories should be OFF.
- The transaxle should be in **P** or **N**.

2. Connect scan tool MB991958 to the data link connector
3. Set up the timing light.
4. Start the engine and run at idle.
5. Verify that the idle speed is about 680 rpm.
6. Select scan tool MB991958 actuator test "item number 17".
7. Check that basic timing is within standard, it should be 2–8°BTDC.
8. If the base timing is out of specification:

a. Check to see if the distributor is aligned properly

b. Check to see if the timing belt cover and Crankshaft Position (CKP) sensor installation is conditions.

c. Crankshaft sensing blade conditions.

9. Press the clear key on the scan tool, select forced drive stop mode and cancel the actuator test.

CAUTION

If the actuator test is not canceled, the forced drive will continue for 27 minutes. Driving in this state could lead to engine failure.

10. Check that the actual ignition timing is approximately 10°BTDC.

➡Keep in mind that the ignition timing may fluctuate by as much as +/-7°BTDC even under normal operation conditions. It is also further advanced by 5–10°BTDC at higher altitudes.

ADJUSTMENT

The ignition timing is controlled by the Powertrain Control Module (PCM) and is not adjustable. The PCM determines the timing based on input from the crankshaft position sensor.

SPARK PLUGS

REMOVAL & INSTALLATION

See Figures 38 and 39.

Endeavor models use iridium spark plugs. Be careful not to damage the iridium tips of the plugs. Do not adjust the spark plug gap. Spark plugs must spark properly to assure proper engine performance and reduce exhaust emission level. Therefore, they should be replaced periodically with new ones.

1. Disconnect the negative battery cable.
2. Remove the ignition coil(s).
3. Remove the spark plug(s).
4. Check, but do not adjust, the plugs to make sure they have the proper gap.
5. Once the spark plugs is removed, you can test them, as follows:

a. Remove the spark plug and connect to the ignition coil.

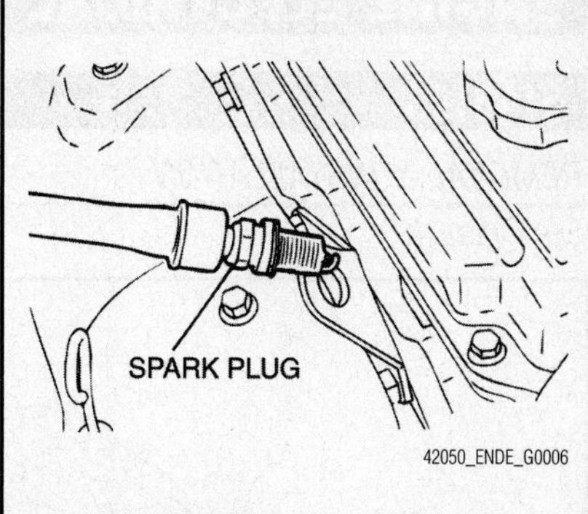

Fig. 38 Remove the spark plug and connect to the ignition coil

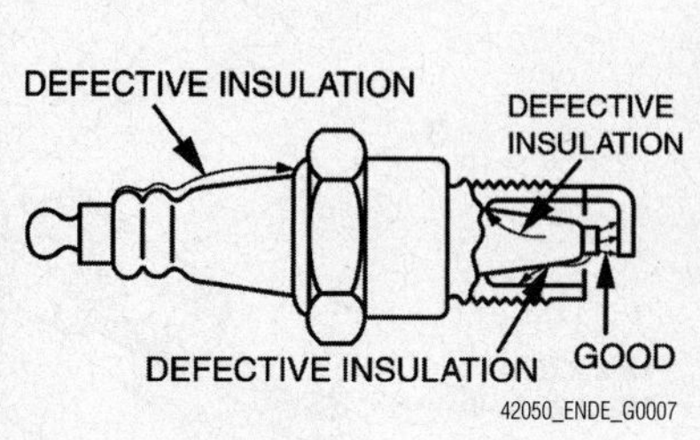

Fig. 39 Ground the spark plug outer electrode (body), and crank the engine. Check that there is an electrical discharge between the electrodes

b. Ground the spark plug outer electrode (body), and crank the engine. Check that there is an electrical discharge between the electrodes. If not, replace the plug.

To install:

6. Install the spark plugs and tighten to 14–22 ft. lbs. (21–29 Nm).
7. Install the ignition coil(s).
8. Connect the negative battery cable. Turn the ignition **ON** and then **OFF**, and keep it off for at least ten seconds.

ENGINE ELECTRICAL

STARTER

REMOVAL & INSTALLATION

See Figure 40.

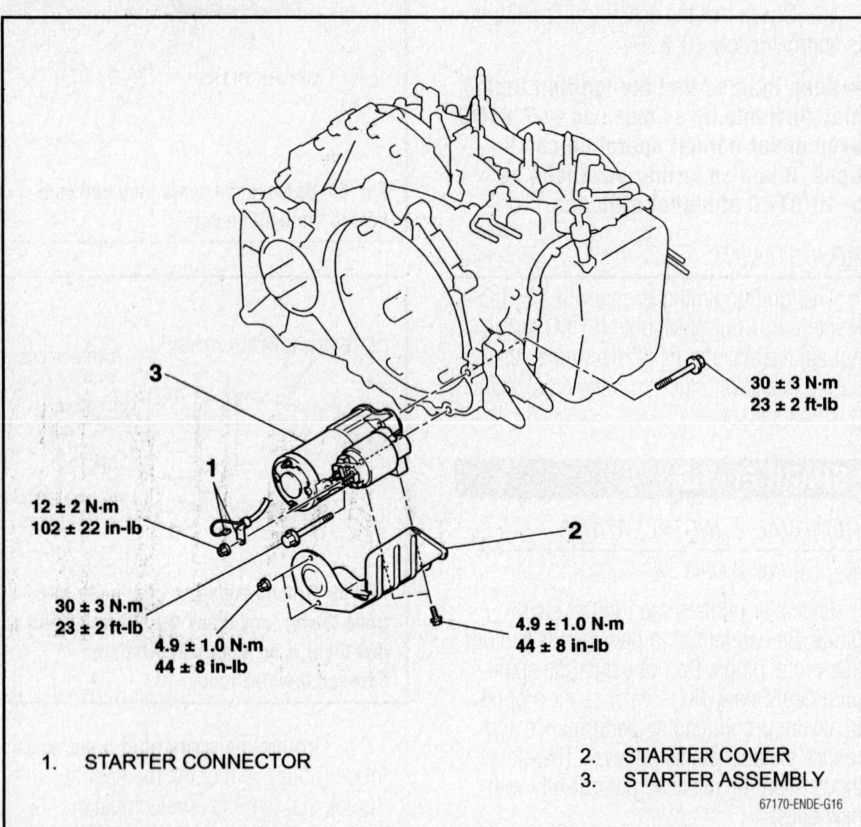

Fig. 40 Starter motor mounting

1. Before servicing the vehicle, refer to the precautions section.
2. Disconnect the negative battery cable.
3. Remove the engine under cover, if equipped.
4. Remove the wires.
5. Remove the starter cover.
6. Remove the starter motor.

To install:

7. Install the starter motor and cover. Torque the starter mounting bolts to 20–25 ft. lbs. (26–33 Nm) and the cover retainers to 36–42 inch lbs. (3.9–5.9 Nm).
8. Install the wires. Tighten the terminal nut to 80–124 inch lbs. (10–14 Nm).
9. Install the engine under cover, if equipped.
10. Disconnect the negative battery cable.

SOLENOID OR RELAY REPLACEMENT

1. Before servicing the vehicle, refer to the precautions section.
2. Disconnect the negative battery cable.
3. Remove the starter.
4. Remove the solenoid attaching screws.
5. Remove the solenoid.

To install:

6. Install the solenoid.
7. Install the solenoid attaching screws.
8. Install the starter.
9. Connect the negative battery cable.

ENGINE MECHANICAL

ACCESSORY DRIVE BELTS

ACCESSORY BELT ROUTING

See Figure 41.

Refer to the accompanying illustration for accessory drive belt routing.

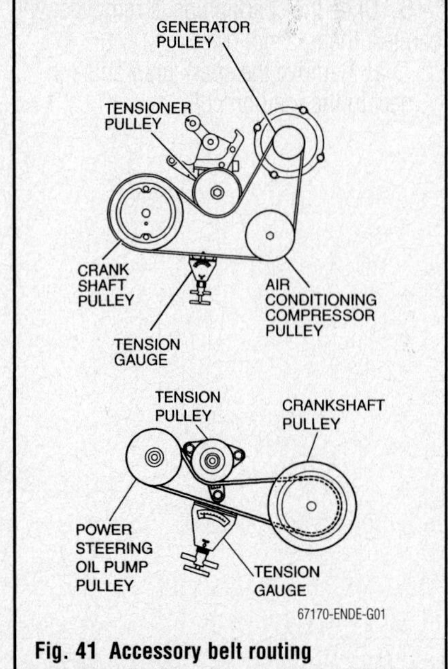

Fig. 41 Accessory belt routing

INSPECTION

Inspect the drive belt for signs of glazing or cracking. A glazed belt will be perfectly smooth from slippage, while a good belt will have a slight texture of fabric visible. Cracks will usually start at the inner edge of the belt and run outward. All worn or damaged drive belts should be replaced immediately.

Alternator-A/C Compressor Belt

There are 3 methods to check the belt tension: using a scan tool, using a belt tension gauge or the measuring the belt deflection. Each method is outlined below.

Belt Tension Gauge Method

See Figure 42.

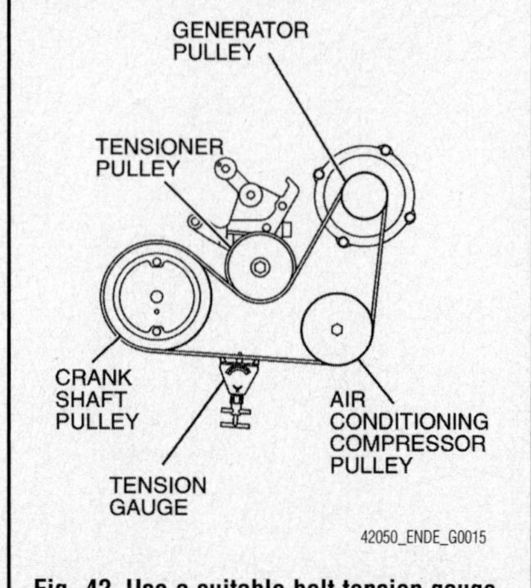

Fig. 42 Use a suitable belt tension gauge to check that the belt tension is within specifications

Use a suitable belt tension gauge to check that the belt tension is within 110–154 lbs. (490–686 N). If not, adjust the belt.

Belt Deflection Method

See Figure 43.

Apply approximately 22 lbs. (98 N) of force to the middle of the drive belt between the pulleys (at the place indicated by the arrow) and check that the amount of deflection in within 0.33–0.42 in. (8.4–10.7mm). If not, adjust the belt.

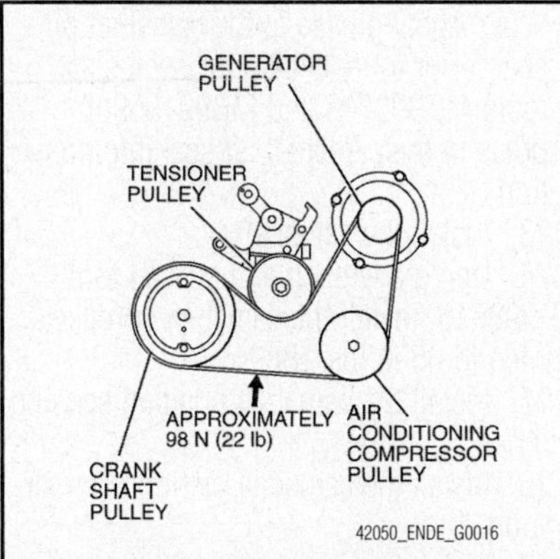

Fig. 43 Apply approximately 22 lbs. (98 N) of force to the middle of the drive belt between the pulleys (at the place indicated by the arrow) and check that the amount of deflection in within specifications

Power Steering Belt

Belt Tension Gauge Method

See Figures 44 and 45.

Use a suitable belt tension gauge to check that the belt tension is within the specifications shown on the accompanying table.

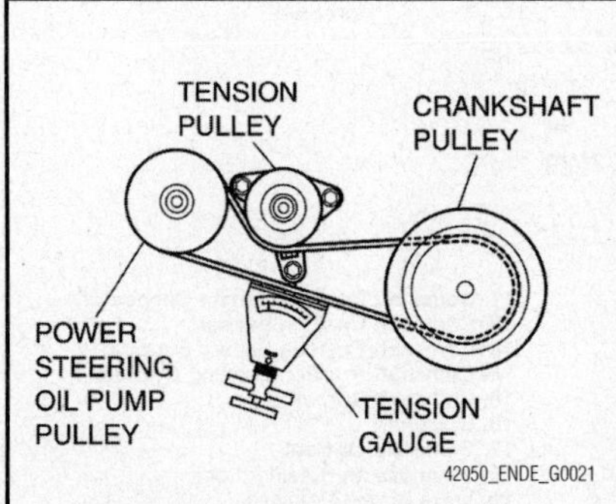

Fig. 44 Use a suitable belt tension gauge to check that the belt tension is within specifications

ITEM	WHEN CHECKED	DURING ADJUSTMENT	DURING REPLACEMENT
Tension N (lb)	294 – 490 (66 – 110)	343 – 441 (77 – 99)	490 – 686 (110 – 154)

42050_ENDE_G0022

Fig. 45 Power steering belt tension specifications

Belt Deflection Method

See Figures 46 and 47.

Apply approximately 22 lbs. (98 N) of force to the middle of the drive belt between the pulleys (at the place indicated by the arrow) and check that the amount of deflection in within the specifications shown on the accompanying table. If not, adjust the belt.

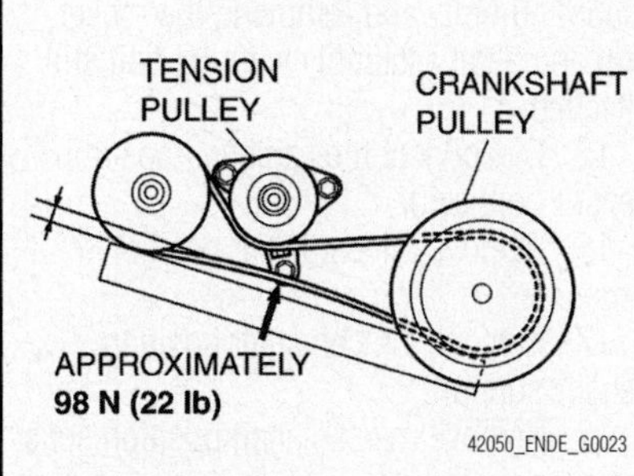

Fig. 46 Apply approximately 22 lbs. (98 N) of force to the middle of the drive belt between the pulleys (arrow) and check that the amount of deflection in within specifications

ITEM	WHEN CHECKED	DURING ADJUSTMENT	DURING REPLACEMENT
Deflection (Reference value) mm (in)	12.3 – 16.2 (0.48 – 0.64)	13.2 – 15.1 (0.52 – 0.59)	9.6 – 12.3 (0.38 – 0.48)

42050_ENDE_G0024

Fig. 47 Power steering belt deflection specifications

ADJUSTMENT

Alternator–A/C Compressor Belt

See Figures 48 and 49.

1. Loosen the tensioner pulley fixing nut.
2. With the tensioner pulley fixing nut temporarily tightened to 7–15 ft. lbs. (11–19 Nm), set the belt tension or defection amount to the standard value (shown in the accompanying table) using the adjusting bolt.

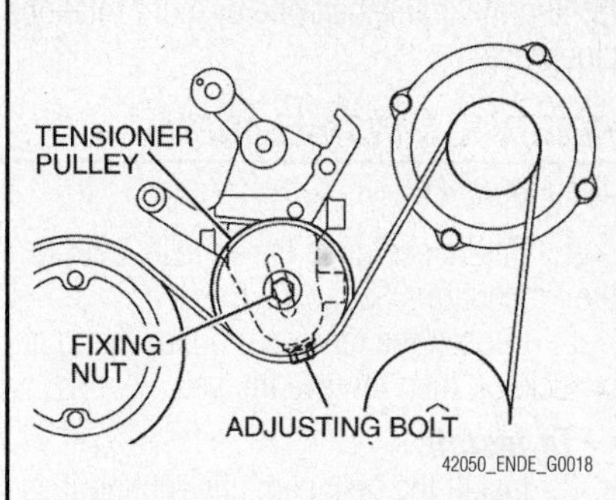

Fig. 48 Location of the tensioner pulley fixing nut and adjusting bolt

ITEM	Part No.	DURING ADJUSTMENT	DURING REPLACEMENT
Vibration frequency Hz	MD368275	140 – 152	168 – 188
	MN158101, MN187016	150 – 163	180 – 202
Tension N (lb)		539 – 637 (121 – 143)	785 – 981 (176 – 221)
Deflection (Reference value) mm (in)		8.9 – 10.1 (0.35 – 0.40)	6.2 – 7.5 (0.24 – 0.30)

42050_ENDE_G0017

Fig. 49 Belt tension specification table

➡Because the frequency depends on the belt material, confirm Part No. shown on the reverse of the belt.

3. After the proper specification is achieved, tighten the tension pulley fixing nut to 29–43 ft. lbs. (39–59 Nm).

Power Steering Belt

See Figure 50.

1. Loosen the tensioner pulley lock nut.
2. Adjust the belt tension to the standard value by turning the adjusting bolt. The tension will increase when turning the adjusting bolt clockwise, and decrease when turning counterclockwise.
3. Tighten the lock nut to 29–43 ft. lbs. (40–58 Nm).
4. Tighten the adjusting bolt to 35–53 inch lbs. (4–6 Nm).
5. Check the belt deflection amount and tension, and readjust if necessary after

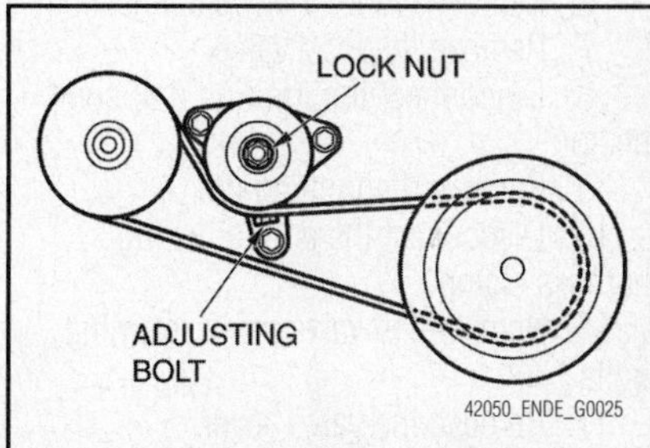

Fig. 50 Location of the power steering belt tensioner lock nut and adjusting bolt

turning the crankshaft one or more rotations clockwise

REMOVAL & INSTALLATION

See Figure 41.

1. Before servicing the vehicle, refer to the Precautions Section.
2. Loosen the tensioner pulley fixing nut or locknut, then remove the belt.

To install:

3. Install the belt, carefully routing it as shown in the accompanying illustration.
4. Adjust the belt tension and tighten the fixing bolt or lock nut and adjusting bolt as specified in the adjustment procedures.

CAMSHAFT AND VALVE LIFTERS

INSPECTION

1. Remove the camshaft from the engine.
2. Check the camshaft bearing journals for damage and binding.
3. If the journals are binding, check the cylinder head for damage.
4. Check the cylinder head for clogged oil holes.
5. As required, check the tooth surface of the distributor drive gear teeth of the camshaft. Replace the camshaft if wear is evident.
6. Check the camshaft surface for abnormal wear and damage. Replace the camshaft, as required.
7. Measure the cam height and replace the camshaft if not within specification.

REMOVAL & INSTALLATION

Left Bank

See Figures 51 through 53.

1. Before servicing the vehicle, refer to the Precautions Section.
2. Drain the cooling system.
3. Remove the timing belt.
4. Remove the thermostat housing.
5. Remove the blow-by hose connection.
6. Remove the PCV valve connection.
7. Remove the PCV valve.
8. Disconnect the ignition coil connector.
9. Remove the ignition coil.
10. Disconnect the control wiring harness clamp.
11. Remove the valve cover retaining bolts.
12. Remove the valve cover.
13. Remove the spark plug guide oil seal.

➡Install auto lash adjuster retainers SST MD998443 on the rocker arms.

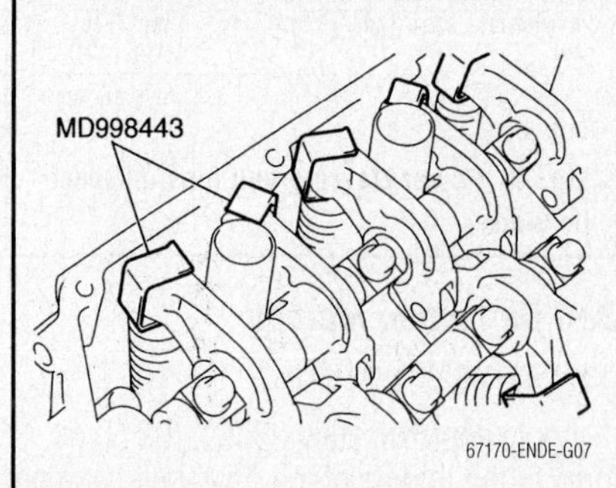

Fig. 51 Install the special tool to prevent the lash adjusters from falling to the floor during rocker arm removal

14. Remove the intake and exhaust rocker arms and shafts by loosening the mounting bolts and removing the rocker arm and shaft assembly with the bolt still attached.
15. Disconnect the camshaft position sensor connector.
16. Remove the camshaft position sensor.
17. Remove the camshaft position sensor support.
18. Remove the camshaft position sensing cylinder.
19. Using a wrench and special tool MD998715, remove the camshaft sprocket.
20. Remove the camshaft.
21. Remove the camshaft oil seal.
    a. Make a notch in the oil seal lip section with a knife.
    b. Cover the end of a flat-tipped screwdriver with a shop towel and insert into the notched section of the oil seal, and pry out the oil seal to remove it.

To install:

22. Install the camshaft oil seal.
    a. Apply grease to the camshaft oil seal, prior to installation. Use special tools MD998713 and MB991559 to press fit the camshaft oil seal into position.
23. Install the camshaft.
24. Using a wrench and special tool MD998715, install the camshaft sprocket. Tighten to 65 ft. lbs. (88 Nm).
25. Install the camshaft position sensing cylinder.
26. Install the camshaft position sensor support.
27. Install the camshaft position sensor. Tighten to 120 inch lbs. (14 Nm).
28. Connect the camshaft position sensor connector.

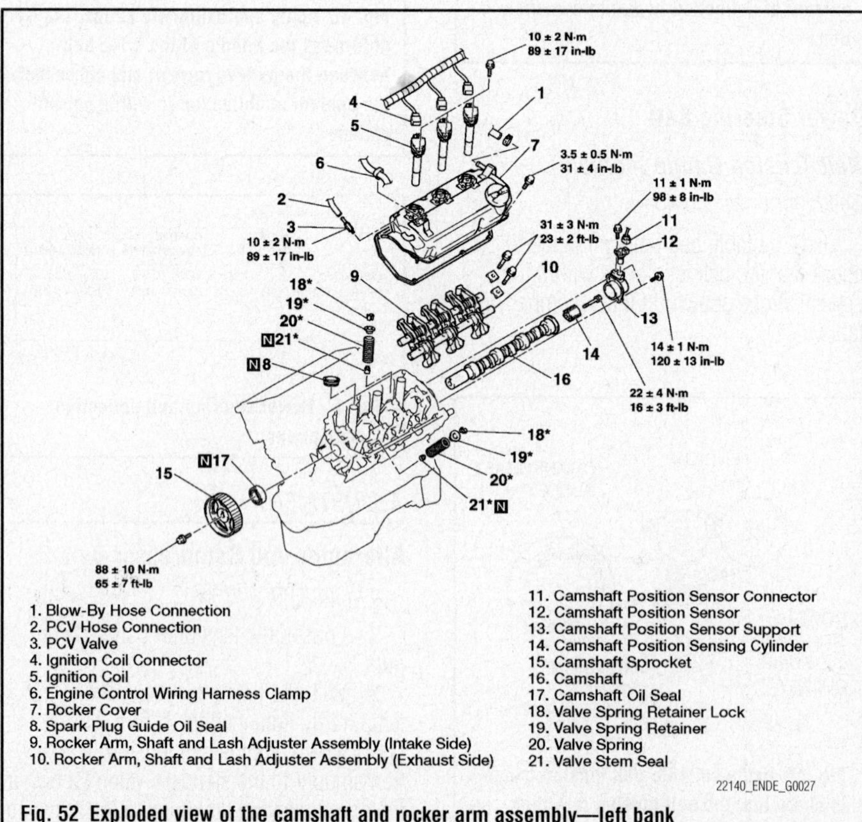

Fig. 52 Exploded view of the camshaft and rocker arm assembly—left bank

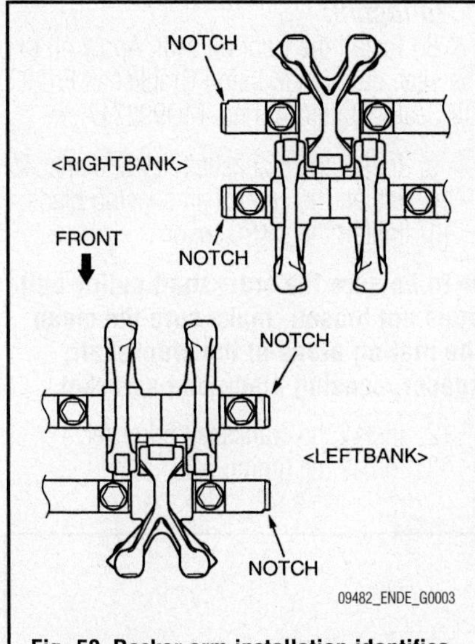

Fig. 53 Rocker arm installation identification points

29. Install the intake and exhaust rocker arms and shafts. Tighten bolts to 23 ft. lbs. (31 Nm).

➡Remove auto lash adjuster retainers SST MD998443 from the rocker arms.

30. Install the spark plug guide oil seal.
31. Install the valve cover.
32. Install the valve cover retaining bolts. Tighten to 31 inch lbs. (4 Nm).
33. Disconnect the control wiring harness clamp.
34. Install the ignition coil.
35. Disconnect the ignition coil connector.
36. Install the PCV valve.
37. Install the PCV valve connection.
38. Install the blow-by hose connection.
39. Install the thermostat housing.
40. Install the timing belt.
41. Fill the cooling system.
42. Run the engine and check for leaks.

Right Bank

See Figures 51, 54 and 53.

1. Before servicing the vehicle, refer to the Precautions Section.
2. Drain the cooling system.
3. Remove the intake manifold plenum assembly.
4. Remove the timing belt.
5. Remove the thermostat housing.
6. Remove the breather hose connection.
7. Remove the blow-by hose connection.
8. Remove the ignition coil connector.
9. Remove the ignition coil.
10. Remove the control wiring harness clamp.
11. Remove the valve cover retaining bolts.
12. Remove the valve cover.
13. Remove the spark plug guide oil seal.

➡Remove the auto lash adjuster retainers SST MD998443 on the rocker arms.

14. Remove the intake and exhaust rocker arms and shafts by loosening the mounting bolts and removing the rocker arm and shaft assembly with the bolt still attached.
15. Using a wrench and special tool MD998715, remove the camshaft sprocket.
16. Remove the EGR pipe "B", gasket and valve support.
17. Remove the thrust case and O-ring.
18. Remove the camshaft.
    a. Make a notch in the oil seal lip section with a knife.
    b. Cover the end of a flat-tipped screwdriver with a shop towel and insert into the notched section of the oil seal, and pry out the oil seal to remove it.

To install:

19. Install the camshaft oil seal.
    a. Apply grease to the camshaft oil seal, prior to installation. Use special tools MD998713 and MB991559 to press fit the camshaft oil seal into position.
20. Install the camshaft.
21. Install the thrust case and O-ring. Tighten bolts to 14 ft. lbs. (19 Nm).
22. Install the EGR pipe "B", gasket and valve support. Tighten to 120 inch lbs. (14 Nm).
23. Using a wrench and special tool MD998715, install the camshaft sprocket.
24. Install the intake and exhaust rocker arms and shafts. Tighten bolts to 23 ft. lbs. (31 Nm).

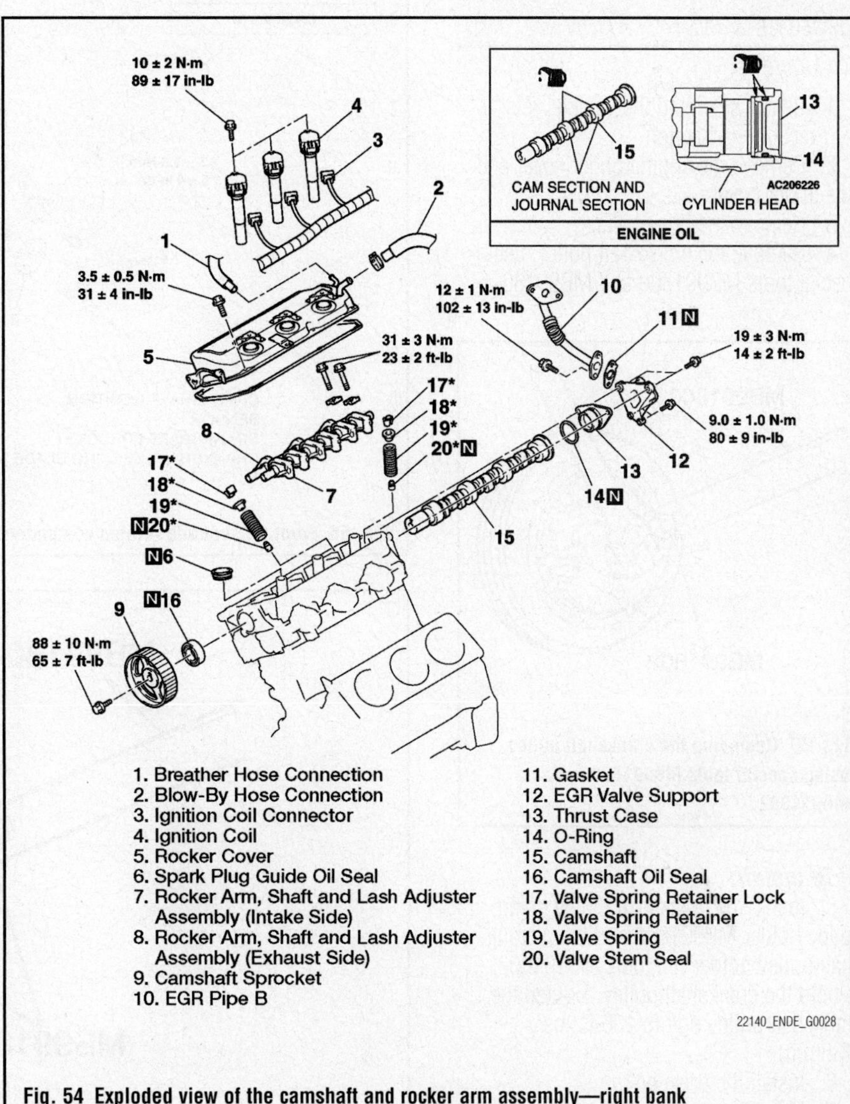

Fig. 54 Exploded view of the camshaft and rocker arm assembly—right bank

➡**Remove the auto lash adjuster retainers SST MD998443 on the rocker arms.**

25. Install the spark plug guide oil seal.
26. Install the valve cover.
27. Install the valve cover retaining bolts. Tighten to 31 inch lbs. (4 Nm).
28. Install the control wiring harness clamp.
29. Install the ignition coil.
30. Install the ignition coil connector.
31. Install the blow-by hose connection.
32. Install the breather hose connection.
33. Install the thermostat housing.
34. Install the timing belt.
35. Install the intake manifold plenum assembly.
36. Fill the cooling system.
37. Run the engine and check for leaks, correct as required.

CRANKSHAFT DAMPER

REMOVAL & INSTALLATION

See Figure 55.

1. Before servicing the vehicle, refer to the Precautions Section.
2. Remove the engine under cover and side under cover.
3. Remove the drive belts.
4. Remove the crankshaft pulley, using special tools MB991800 and MB991802.

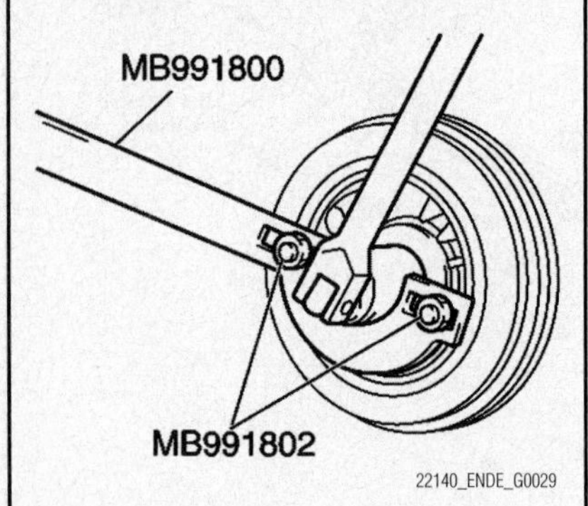

Fig. 55 Removing the crankshaft pulley, using special tools MB991800 and MB991802

To install:

5. Install the crankshaft pulley using pulley holder MB991800 and the 2 crankshaft pulley holder pin tools MB991802, to hold the crankshaft pulley. Tighten the crankshaft pulley bolt to 136 ft. lbs. (185 nm).
6. Install the drive belts.
7. Install the side and under cover.

CRANKSHAFT FRONT SEAL

REMOVAL & INSTALLATION

See Figures 56 through 58.

1. Before servicing the vehicle, refer to the Precautions Section.
2. Remove the timing belt.
3. Remove the crankshaft position (CKP) sensor.
4. Remove the crankshaft sprocket.
5. Remove the crankshaft sensing blade.
6. Remove the crankshaft spacer and key.
7. Remove the front oil seal.

To install:

8. Install the front oil seal. Apply oil to the seal, and install using Crankshaft Front Oil Seal Installer tool no. MD998717.
9. Install the crankshaft key and spacer.
10. Install the crankshaft sensing blade.
11. Install the CKP sensor.

➡**To be sure the crankshaft pulley bolt does not loosen, make sure the clean the mating areas of the crankshaft, spacer, sensing blade and sprocket.**

12. Install the crankshaft sprocket.
13. Install the timing belt.

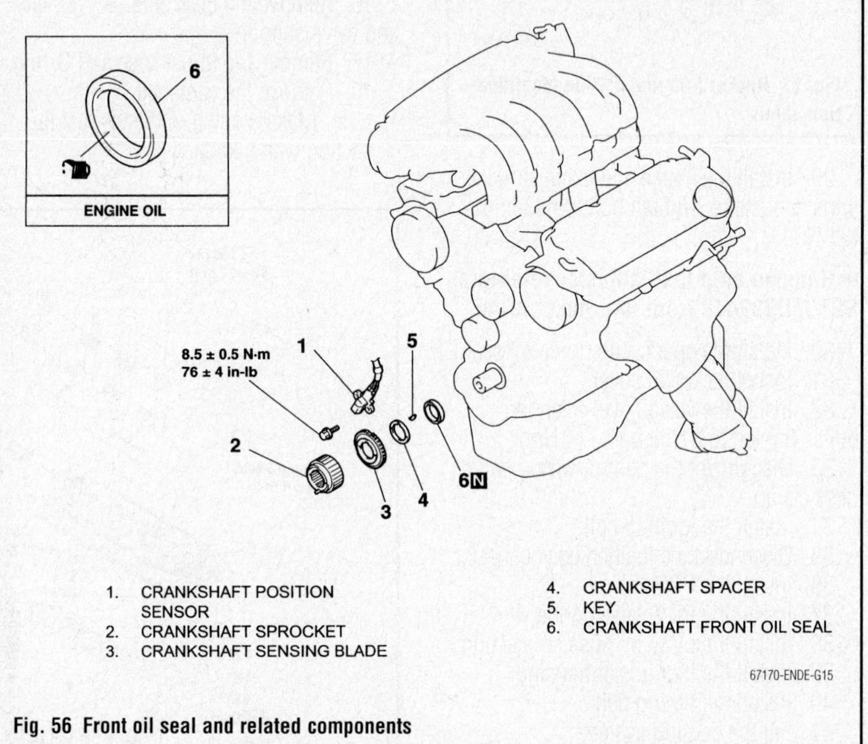

Fig. 56 Front oil seal and related components

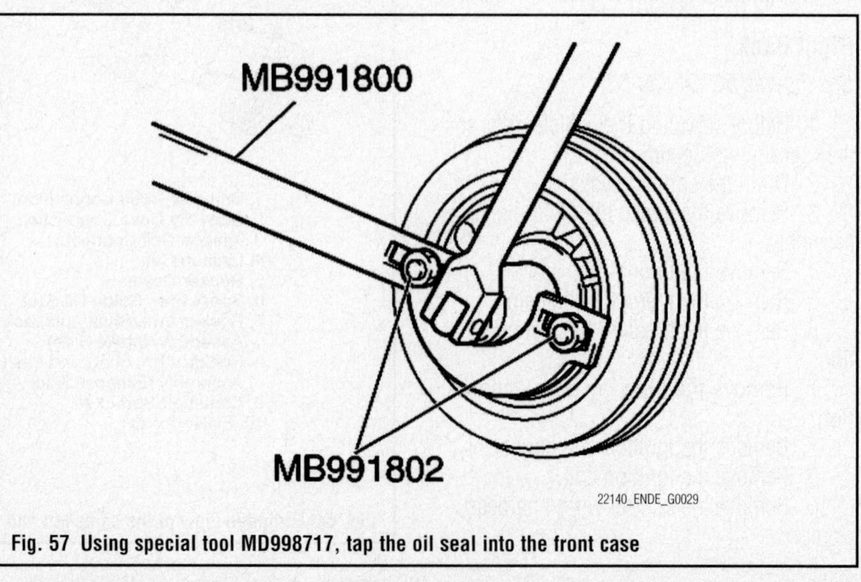

Fig. 57 Using special tool MD998717, tap the oil seal into the front case

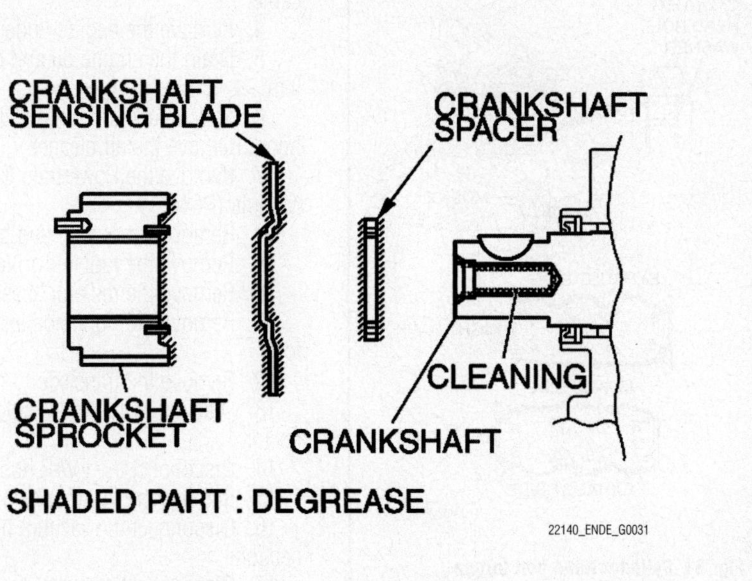

Fig. 58 To prevent the crankshaft pulley mounting bolt from loosening, degrease or clean the crankshaft, the crankshaft spacer, the crankshaft sensing blade and the crankshaft at the shown positions

CYLINDER HEAD

REMOVAL & INSTALLATION

See Figures 59 through 61.

1. Before servicing the vehicle, refer to the Precautions Section.

**** CAUTION**

The fuel injection system remains under pressure after the engine has been OFF. Properly relieve fuel pressure before disconnecting any fuel lines. Failure to do so may result in fire or personal injury.

2. Relieve fuel system pressure.
3. Drain the cooling system.
4. Disconnect the negative battery cable.

➡Wait at least 90 seconds after disconnecting the cable from the negative (-) battery terminal to prevent airbag activation.

5. Remove the intake manifold.
6. Remove the exhaust manifold.
7. Remove the timing belt.
8. Remove the thermostat housing.
9. Remove the alternator.
10. Remove the blow-by hose from the left and right valve covers.
11. Remove the positive crankcase ventilation (PCV) hose from the left and right valve covers.
12. Remove the spark plug wires; tag before disconnecting.
13. Remove the ignition coil connectors and ignition coils.

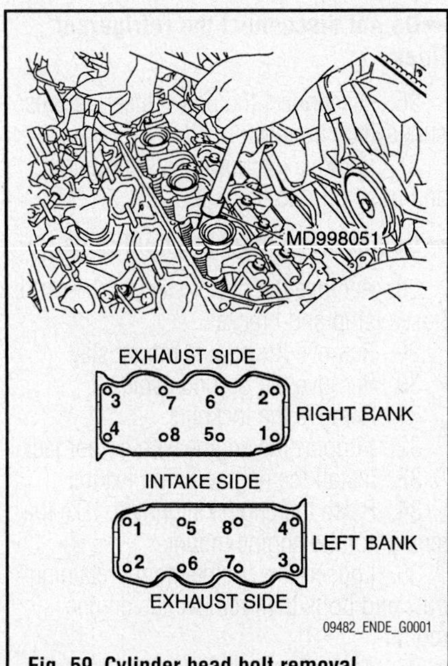

Fig. 59 Cylinder head bolt removal sequence

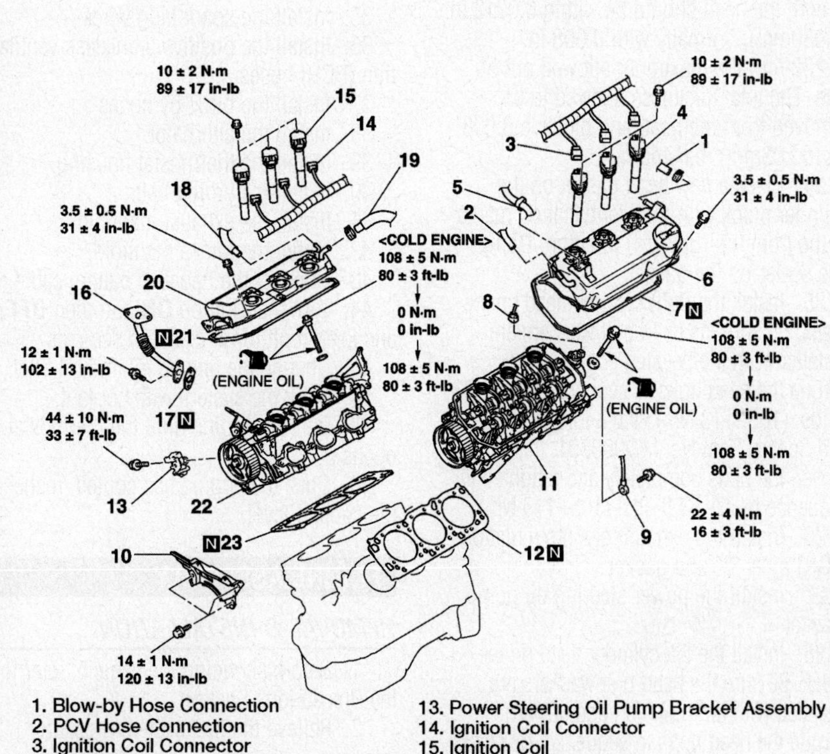

1. Blow-by Hose Connection
2. PCV Hose Connection
3. Ignition Coil Connector
4. Ignition Coil
5. Engine Control Wiring Harness Clamp
6. Rocker Cover
7. Rocker Cover Gasket
8. Camshaft Position Sensor Connector
9. Grounding
10. Timing Belt Rear Center Cover
11. Left Bank Cylinder Head Assembly
12. Cylinder Head Gasket
13. Power Steering Oil Pump Bracket Assembly
14. Ignition Coil Connector
15. Ignition Coil
16. EGR Pipe B
17. Gasket
18. Breather Hose Connection
19. Blow-by Hose Connection
20. Rocker Cover
21. Rocker Cover Gasket
22. Right Bank Cylinder Head Assembly
23. Cylinder Head Gasket

22140_ENDE_G0032

Fig. 60 Exploded view of the cylinder head assemblies

14. Remove the engine control wiring harness clamp.
15. Remove the rocker covers and gaskets.
16. Disconnect the camshaft position (CMP) sensor connector.
17. Remove the ground strap.
18. Remove the timing belt rear center cover.
19. Loosen the left cylinder head mounting bolts in 3 steps, in the sequence shown. Lift off the left cylinder head assembly and remove the head gasket.
20. Remove the power steering oil pump bracket.
21. Remove the Exhaust Gas Recirculation (EGR) pipe B and gasket.
22. Loosen the right cylinder head mounting bolts in 3 steps, in the sequence shown. Lift off the right cylinder head assembly and remove the head gasket.

To install:

23. Thoroughly clean and dry the mating surfaces of the head and block. Check the cylinder head for cracks, damage or engine coolant leakage. Remove scale, sealing compound and carbon. Clean oil passages thoroughly. Check the head for flatness. End to end, the head should be within 0.0012 in. (0.030mm), normally with 0.008 in. (0.203mm) the maximum allowed out of true. The total thickness allowed to be removed from the head and block is 0.008 in. (0.203mm) maximum.
24. Place a new head gasket on the cylinder block with the identification marks in the front top (upward) position. Do not use sealer on the gasket.
25. Install the right cylinder head on the block. Be sure the head bolt washers are installed with the beveled edge upward. Torque the head bolts in sequence, to 77–83 ft. lbs. (105–113 Nm) with a torque wrench and Special Tool No. MD998501, then loosen the bolts completely and retighten in sequence to 77–83 ft. lbs. (105–113 Nm).
26. Install the exhaust gas recirculation (EGR) pipe "B" and gasket
27. Install the power steering oil pump bracket
28. Install the left cylinder head on the block. Be sure the head bolt washers are installed with the beveled edge upward. Torque the head bolts in sequence, to 77–83 ft. lbs. (105–113 Nm) with a torque wrench and Special Tool No. MD998501, then loosen the bolts completely and retighten in sequence to 77–83 ft. lbs. (105–113 Nm).
29. Install the timing belt rear center cover.
30. Install the ground strap.
31. Install the CMP sensor connector.
32. Install the rocker covers with new gaskets.

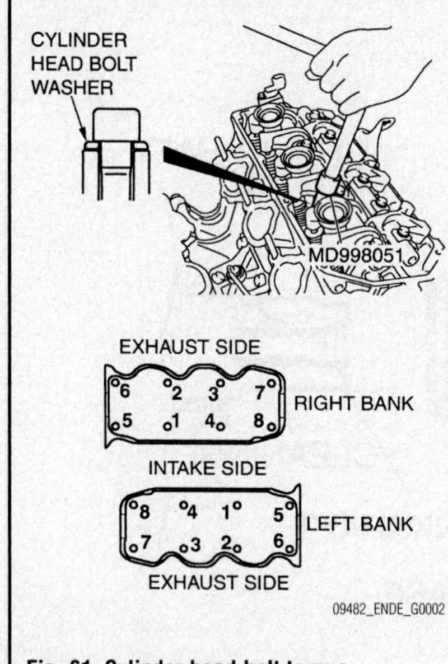

Fig. 61 Cylinder head bolt torque sequence

33. Install the engine control wiring harness clamp.
34. Install the ignition coils and their connectors.
35. Install the spark plug wires.
36. Install the positive crankcase ventilation (PCV) hoses.
37. Install the blow-by hoses .
38. Install the alternator.
39. Install the thermostat housing.
40. Install the timing belt.
41. Install the exhaust manifold.
42. Install the intake manifold.
43. Connect the negative battery cable.
44. Turn the ignition **ON** and then **OFF**, and keep it off for at least ten seconds.
45. Change the engine oil and oil filter.
46. Refill the system with coolant.
47. Run the engine until the thermostat opens.
48. Once the engine has cooled, recheck the coolant level.

ENGINE ASSEMBLY

REMOVAL & INSTALLATION

1. Before servicing the vehicle, refer to the Precautions Section.
2. Relieve the fuel system pressure.

CAUTION

The fuel injection system remains under pressure after the engine has beenOFF. Properly relieve fuel pressure before disconnecting any fuel lines. Failure to do so may result in fire or personal injury.

3. Disconnect the negative battery cable.
4. Remove the engine under cover.
5. Drain the engine oil and cooling system.
6. Matchmark and remove the engine hood. Remove the air cleaner.
7. Remove the Powertrain Control Module (PCM).
8. Remove the battery and battery tray.
9. Remove the radiator grille.
10. Remove the radiator assembly.
11. Remove the right side exhaust manifold.
12. Remove the transaxle.
13. Disconnect the EVAP solenoid connector.
14. Disconnect the EVAP hose.
15. Remove the EVAP solenoid.
16. Disconnect the vacuum hose connection.
17. Disconnect the purge hose connection.
18. Remove the front wiring harness and control wiring harness.
19. Remove the ground wire retaining screw and ground wire.
20. Disconnect the fuel pipe retainer, and main pipe connector.
21. Disconnect the heater hose connection.
22. Remove the reservoir assembly.
23. Remove the alternator and power steering pump drive belts.
24. Disconnect the air conditioning compressor electrical connector.
25. Remove the air conditioning compressor from its mounting and position and set it to the side.

➡Do not disconnect the refrigerant lines.

26. Disconnect the power steering pressure switch connector.
27. Remove the power steering pump and position it to the side.

➡Do not disconnect the fluid lines.

28. Remove the power steering pressure hose clamp and bracket.
29. Remove the engine mount stay.
30. Remove the ground cable.
31. Remove the locknuts.
32. Support the engine with a floor jack.
33. Install the engine lifting fixture.
34. Raise the engine slightly to take the tension of the engine mount.
35. Loosen the engine mount retaining nuts and bolts then remove the engine mount.
36. After checking that all cables, hoses and wiring harness connectors are

disconnected from the engine, remove the engine from the vehicle.

To install:

37. Lower the engine into position and install the engine mount nuts and bolts. Tighten the nuts to 18–20 ft. lbs. (25–27 Nm), and the bolts to tighten to 33 ft. lbs. (44 Nm).
38. Raise the engine slightly to take the tension of the engine mount.
39. Remove the engine lifting fixture.
40. Install the locknuts.
41. Install the ground cable.
42. Install the engine mount stay.
43. Install the power steering pressure hose clamp and bracket.
44. Install the power steering pump and position it to the side.
45. Connect the power steering pressure switch connector.
46. Install the air conditioning compressor on its mounting and position.
47. Connect the air conditioning compressor electrical connector.
48. Install the alternator and power steering pump drive belts.
49. Install the reservoir assembly.
50. Connect the heater hose connection.
51. Connect the fuel pipe retainer, and main pipe connector.
52. Install the ground wire retaining screw and ground wire.
53. Install the front wiring harness and control wiring harness.
54. Connect the purge hose connection.
55. Connect the vacuum hose connection.
56. Install the EVAP solenoid.
57. Connect the EVAP hose.
58. Connect the EVAP solenoid connector.
59. Install the transaxle.
60. Install the right side exhaust manifold.
61. Install the radiator assembly.
62. Install the radiator grille.
63. Install the battery and battery tray.
64. Install the Powertrain Control Module (PCM).
65. Matchmark and Install the engine hood. Install the air cleaner.
66. Drain the engine oil and cooling system.
67. Install the engine under cover.
68. Connect the negative battery cable.
69. Turn the ignition **ON** and then **OFF**, and keep it off for at least ten seconds.
70. Connect a scan tool and complete the vehicle initialization procedure.

➡To complete the initialization procedure the following tools are needed. MB991958 scan tool, MB991824 VCI, MB991827 MUT III USB cable, MB991910 MUT III Main harness "A".

a. Connect the scan tool to the data link connector. To prevent damage to the scan tool be sure that the ignition switch is in the **LOCK** position before connecting the scan tool.
b. Turn the ignition switch to the **ON** position.
c. Select "check mode" from the menu screen.
d. Select "erase memory" from the menu screen.
e. Initialize the learning value.
f. After initialization complete the idle learning procedure.

➡This procedure must be performed when the PCM is replaced, or when the learning value is initialized, as the idling is not stabilized because the learning value in the MFI engine is not completed.

g. Start the engine. Allow the coolant temperature to reach 176 degrees or more.
h. Stop the engine and place the ignition switch in the **LOCK** position.
i. After ten seconds, restart the engine.
j. For ten minutes carry out the idling procedure below to confirm that the engine has normal idling:
   - Position the transaxle selector lever in the **P** range
   - The engine fan is not to be operated
   - The engine coolant temperature should be 176 degrees or more

➡If the engine stalls during idling, check the throttle valve of the throttle body for dirt. Correct and perform the procedure again.

71. Be sure to check and adjust all fluid levels, as required.
72. Check fuel system for leaks.
73. Connect the negative battery cable.

EXHAUST MANIFOLD

REMOVAL & INSTALLATION

Left Bank

See Figure 62.

1. Before servicing the vehicle, refer to the Precautions Section.
2. Remove the engine under cover
3. Remove the air duct.
4. Remove the left Heated Oxygen Sensor (HO2S) connectors and sensors.
5. Remove the engine oil dipstick, guide and O-ring.
6. Remove the heat shield.
7. Remove the front exhaust pipe from the manifold. Remove and discard the gasket.
8. Remove the exhaust manifold stay (left B).
9. Remove the left side exhaust manifold and gasket.
10. Remove the exhaust manifold stay (left A).
11. Clean the gasket mounting surfaces. Inspect the manifolds for cracks, flatness and/or damage.

To install:

12. Install the exhaust manifold stay (left A).
13. Install the new gasket and left side exhaust manifold. Torque the nuts to specification.
14. Install the exhaust manifold stay (left B).
15. Install the exhaust pipe to the manifold with a new gasket. Torque the bolts to 31–43 ft. lbs. (40–58 Nm).
16. Install the left side heat shield.
17. Install the oil dipstick, guide and new O-ring.
18. Install the left HO2S and connectors.
19. Install the air duct.
20. Install the undercover.
21. Start the engine and check for exhaust leaks.

Right Bank

See Figure 63.

1. Before servicing the vehicle, refer to the Precautions Section.
2. Drain the transfer case fluid and the engine coolant.
3. Remove the air cleaner assembly.
4. Remove the battery and battery tray.
5. Remove the undercover.
6. Remove the front exhaust pipe from the right exhaust manifold.
7. Remove the propeller shaft, if AWD.
8. Remove the manifold differential pressure (MDP) sensor connector.
9. Remove the knock sensor (KS) connector.
10. Remove the crankshaft position (CKP) sensor connector.
11. Remove the control wiring harness and injector wiring harness combination connector.
12. Remove the right heated oxygen sensor (HO2S) connectors.

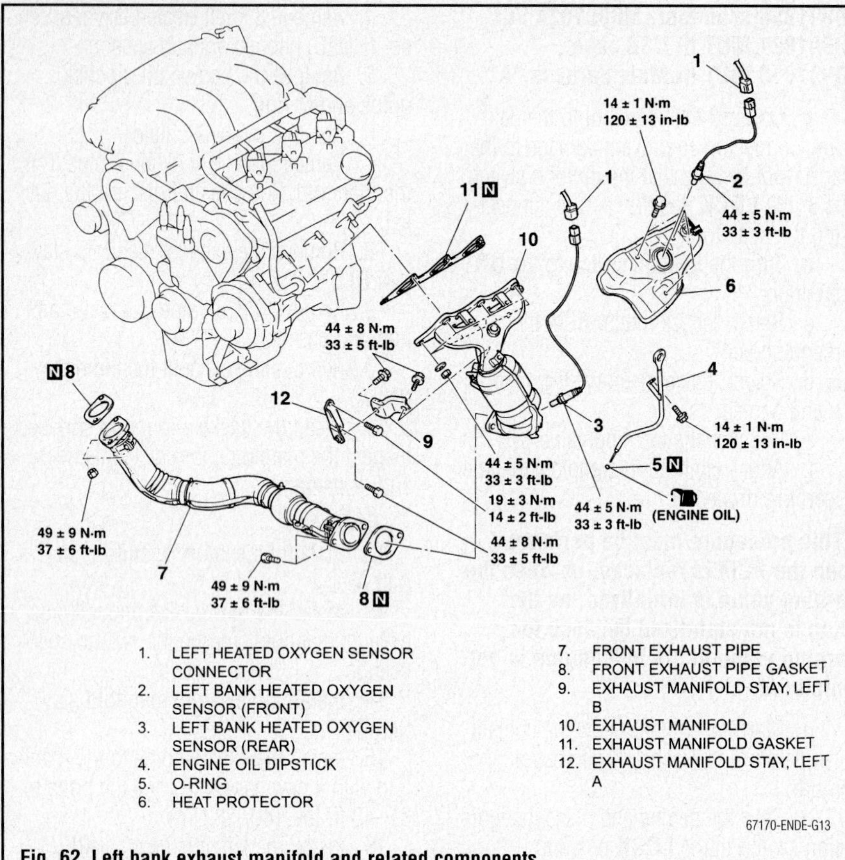

Fig. 62 Left bank exhaust manifold and related components

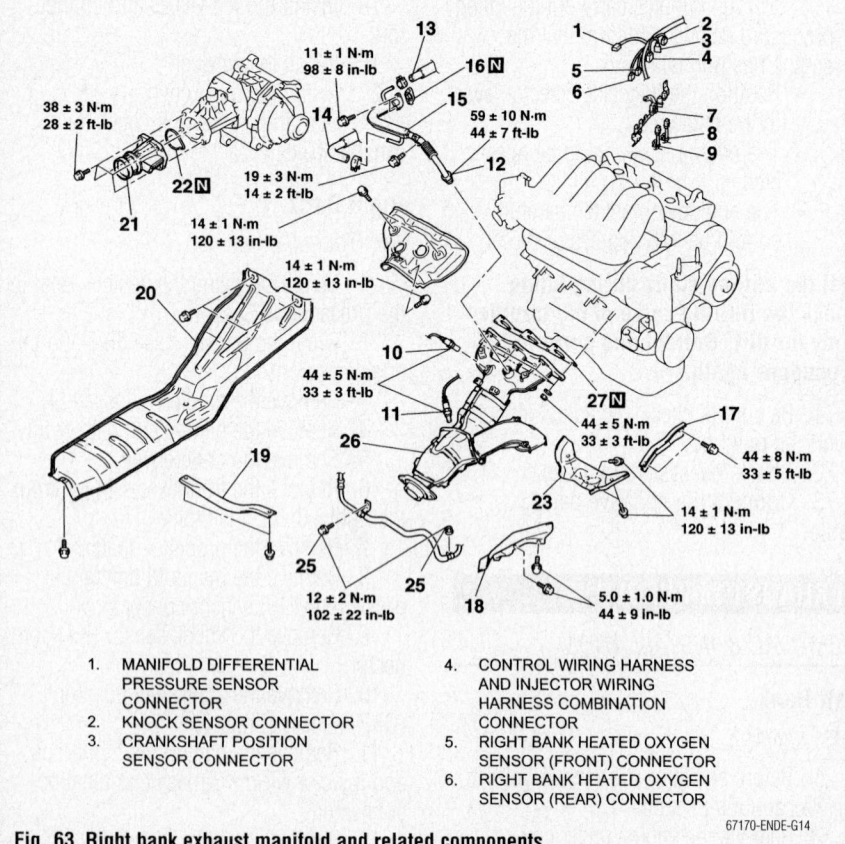

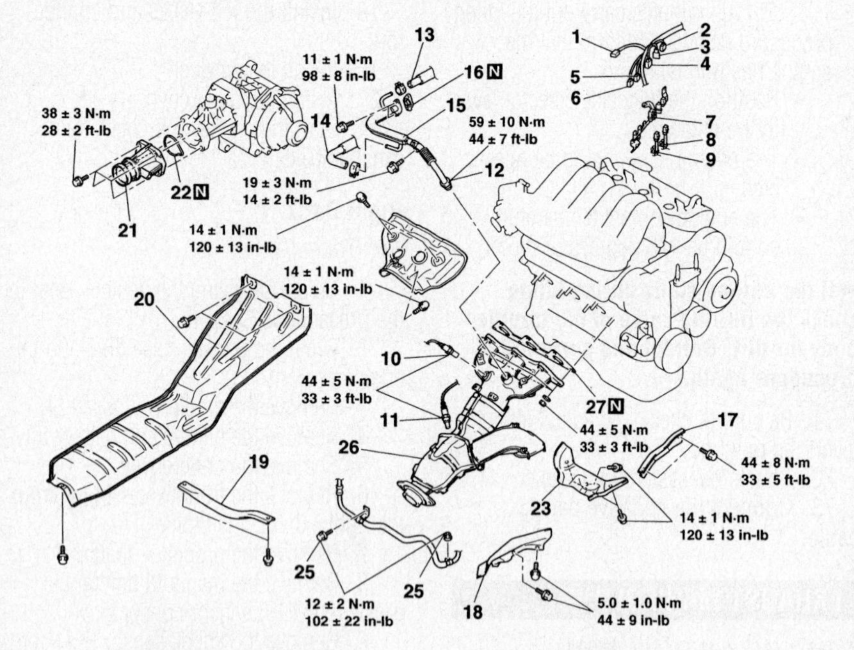

1. MANIFOLD DIFFERENTIAL PRESSURE SENSOR CONNECTOR
2. KNOCK SENSOR CONNECTOR
3. CRANKSHAFT POSITION SENSOR CONNECTOR
4. CONTROL WIRING HARNESS AND INJECTOR WIRING HARNESS COMBINATION CONNECTOR
5. RIGHT BANK HEATED OXYGEN SENSOR (FRONT) CONNECTOR
6. RIGHT BANK HEATED OXYGEN SENSOR (REAR) CONNECTOR

67170-ENDE-G14

Fig. 63 Right bank exhaust manifold and related components

13. Remove the connector bracket.
14. Remove the right ho2s sensor connector clamps and sensors.
15. Remove the exhaust gas recirculation (EGR) pipe a connection.
16. Remove the water hoses.
17. Remove the EGR pipe a and gasket.
18. Remove the exhaust manifold stay (right b).
19. Remove the steering gear and linkage protector.
20. Remove the front floor backbone brace.
21. Remove the front heat protector panel.
22. Remove the transfer extension housing and o-ring, if AWD.
23. Remove the lower and upper heat shields.
24. Remove the power steering return pipe clamp connecting bolt and nut, if AWD.
25. Remove the right side exhaust manifold and gasket.
26. Clean the gasket mounting surfaces. Inspect the manifolds for cracks, flatness and/or damage.

To install:

27. Install the new gasket and right side exhaust manifold. torque the nuts to specification.
28. Install the power steering return pipe clamp connecting bolt and nut, if AWD.
29. Install the lower and upper heat shields.
30. Install the transfer extension housing and o-ring, if AWD.
31. Install the front heat protector panel.
32. Install the front floor backbone brace.
33. Install the steering gear and linkage protector.
34. Install the exhaust manifold stay (right b).
35. Install the EGR pipe a and gasket.
36. Install the water hoses.
37. Install the EGR pipe a connection.
38. Install the right ho2s sensors and connector clamps.
39. Install the connector bracket.
40. Install the right ho2s connectors.
41. Install the control wiring harness and injector wiring harness combination connector.
42. Install the CKP sensor connector.
43. Install the knock sensor (KS) connector.
44. Install the manifold differential pressure (MDP) sensor connector.

45. Install the propeller shaft, if AWD.
46. Install the front exhaust pipe to the exhaust manifold.
47. Install the undercover.
48. Install the battery and battery tray.
49. Install the air cleaner assembly.
50. Fill the transfer case with fluid. Fill the engine cooling system.
51. Start the engine and check for exhaust leaks.

FLEXPLATE

REMOVAL & INSTALLATION

See Figures 64 and 65.

1. Before servicing the vehicle, refer to the Precautions Section.
2. Remove the transaxle assembly.
3. Remove the transfer case, if AWD.
4. Remove the driveplate and adapter plate, matchmark for reassembly. Use Mitsubishi tool (md998781) to hold the driveplate in position to remove the bolts.

To install:

5. Install the drive plate and adapter. Use Mitsubishi tool (md998781) to hold the driveplate in position while tightening the bolts to 55 ft. lbs. (75 Nm).
6. Install the transfer case, if AWD.
7. Install the transaxle and related components as necessary.

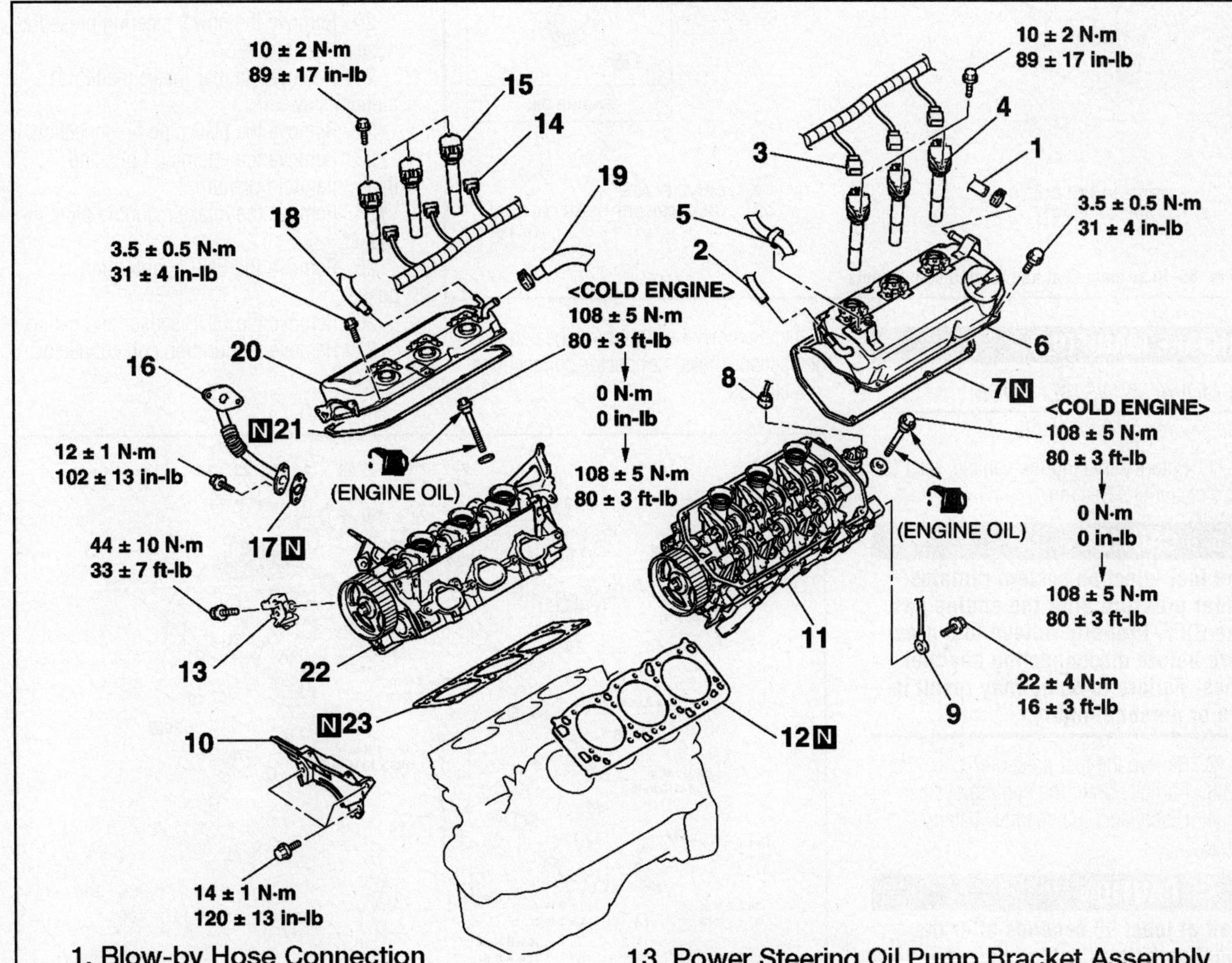

1. Blow-by Hose Connection
2. PCV Hose Connection
3. Ignition Coil Connector
4. Ignition Coil
5. Engine Control Wiring Harness Clamp
6. Rocker Cover
7. Rocker Cover Gasket
8. Camshaft Position Sensor Connector
9. Grounding
10. Timing Belt Rear Center Cover
11. Left Bank Cylinder Head Assembly
12. Cylinder Head Gasket
13. Power Steering Oil Pump Bracket Assembly
14. Ignition Coil Connector
15. Ignition Coil
16. EGR Pipe B
17. Gasket
18. Breather Hose Connection
19. Blow-by Hose Connection
20. Rocker Cover
21. Rocker Cover Gasket
22. Right Bank Cylinder Head Assembly
23. Cylinder Head Gasket

22140_ENDE_G0032

Fig. 64 Use special tool MD998781 to secure the drive plate and remove the drive plate bolts

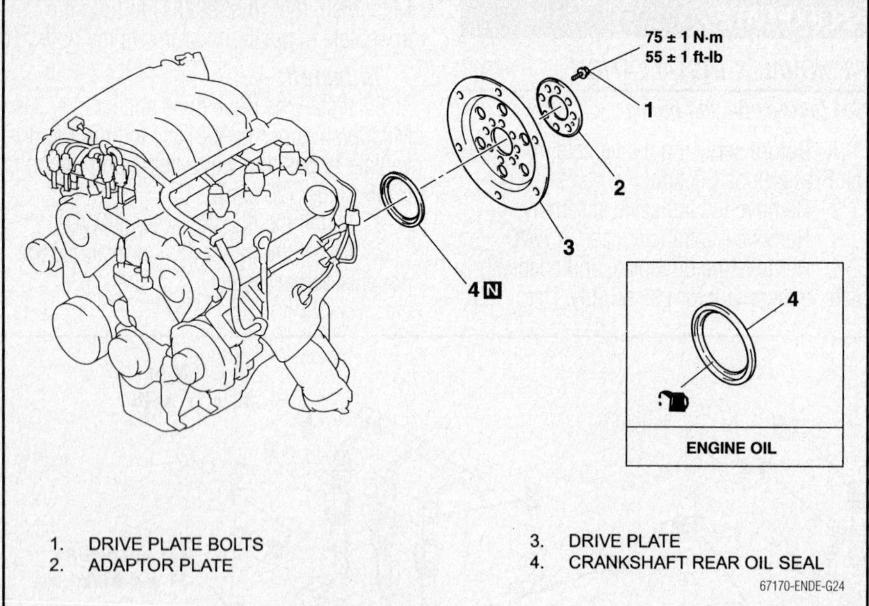

Fig. 65 Rear main seal and related components

INTAKE MANIFOLD

REMOVAL & INSTALLATION

See Figures 66 through 69.

1. Before servicing the vehicle, refer to the Precautions Section.

✱✱ CAUTION

The fuel injection system remains under pressure after the engine has beenOFF. Properly relieve fuel pressure before disconnecting any fuel lines. Failure to do so may result in fire or personal injury.

2. Relieve the fuel pressure.
3. Partially drain the cooling system.
4. Disconnect the negative battery cable.

✱✱ CAUTION

Wait at least 90 seconds after the negative battery cable is disconnected to prevent possible deployment of the air bag.

5. Remove the air cleaner assembly.
6. Remove the throttle body.
7. Remove the manifold differential pressure (MDP) sensor connector.
8. Remove the knock sensor (KS) connector.
9. Remove the crankshaft position (CKP) sensor connector.
10. Remove the control wiring harness and injector wiring harness combination connector.
11. Remove the right bank heated oxygen sensor (HO_2S) connector connections and clamps.
12. Remove the connector bracket.
13. Remove the evaporative emission (EVAP) purge solenoid connector.
14. Remove the purge hose.
15. Remove the purge hose connection.
16. Remove the EVAP purge solenoid.
17. Remove the front intake manifold plenum stay.
18. Remove the exhaust gas recirculation (EGR) pipe a clamp.
19. Remove the power steering pressure hose clamp.
20. Remove the power steering pressure hose clamp bracket.
21. Remove the rear intake manifold plenum stay.
22. Remove the EGR pipe b connection.
23. Remove the retaining bolts and intake manifold plenum.
24. Remove the intake manifold plenum gasket.
25. Remove the EGR adapter and gasket.
26. Remove the MDP sensor and o-ring.
27. Remove the ignition coil connector and coil.

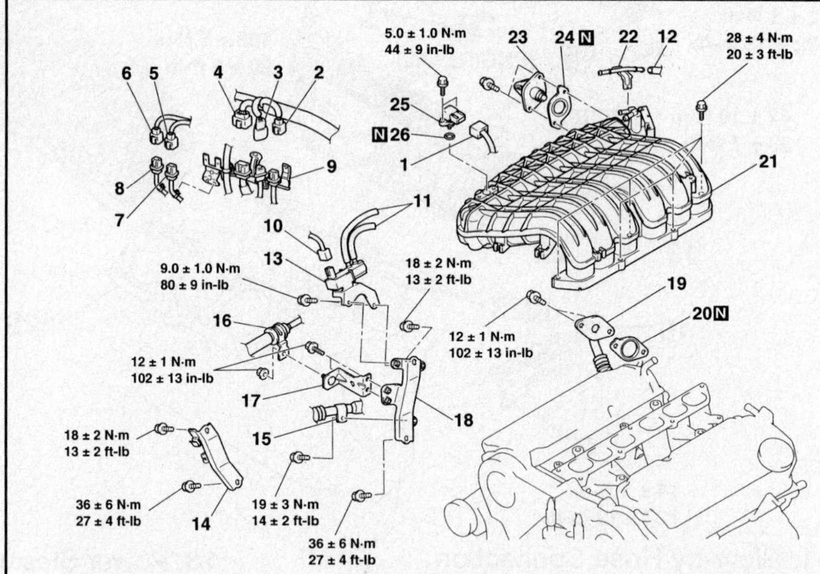

Fig. 66 Intake manifold plenum and related components

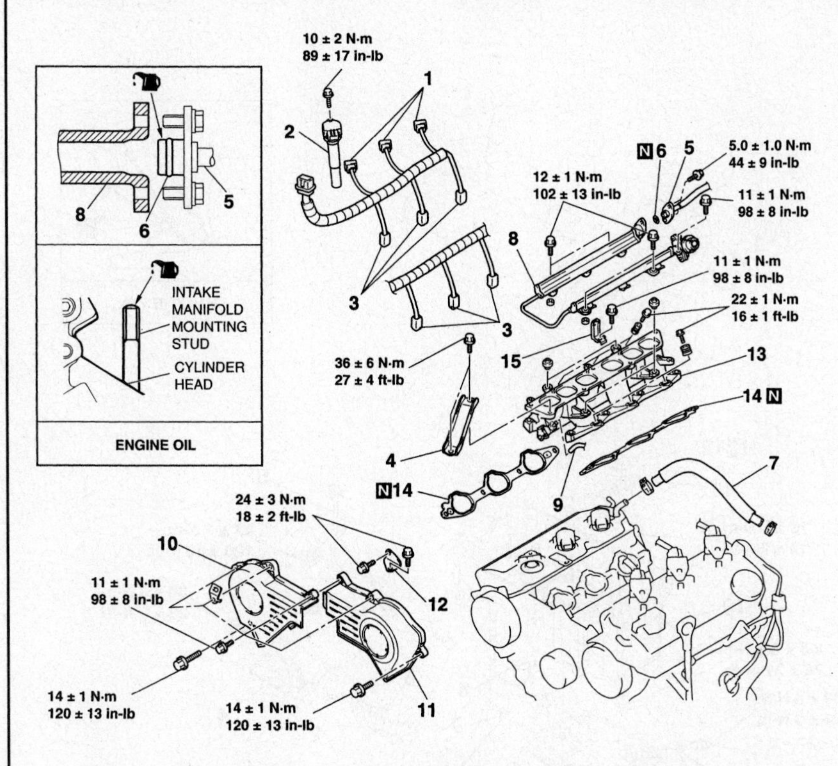

Fig. 67 Intake manifold and related components

28. Remove the injector connector.
29. Remove the engine mount stay.
30. Remove the high pressure fuel hose and o-ring.
31. Remove the blow-by hose.
32. Remove the fuel rail and injectors. also remove the damper assembly, if equipped.
33. Remove the positive crankcase ventilation (PCV) hose.
34. Remove the right and left front upper timing belt cover.
35. Remove the water pump bracket.
36. Remove the intake manifold retainers, manifold and gasket. thoroughly clean and dry the mating surfaces of the manifold and heads.

To install:

37. Install the new intake manifold gasket. Make sure the gaskets are installed with the protrusions as shown in the illustration.
38. Install the intake manifold.
39. Coat the intake manifold retaining studs with clean engine oil. Install the intake manifold bolts and tighten as follows:

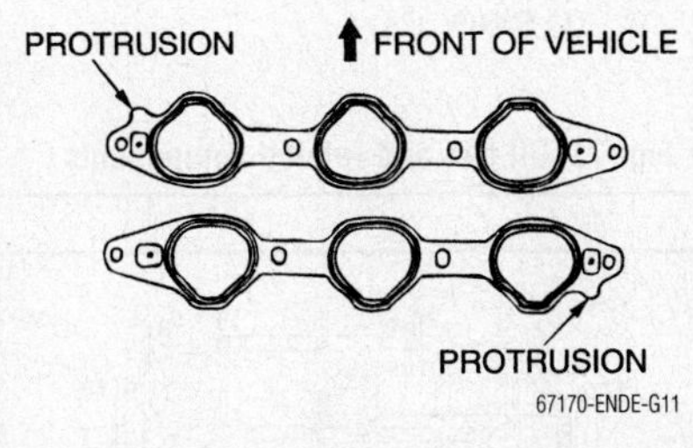

Fig. 68 Intake manifold gasket positioning

a. 1st step: Right bank nuts to 45–71 inch lbs. (5–8 Nm).
b. 2nd step: Left bank nuts to 15–17 ft. lbs. (21–23 Nm).
c. 3rd step: Right bank nuts to 15–17 ft. lbs. (21–23 Nm).
d. 4th step: Left bank nuts to 15–17 ft. lbs. (21–23 Nm).
e. 5th step: Right bank nuts to 15–17 ft. lbs. (21–23 Nm).

40. Install the water pump bracket.
41. Install the right and left front upper timing belt cover.

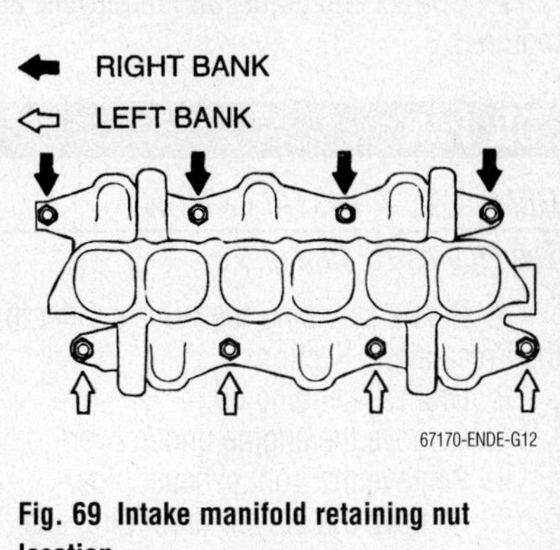

Fig. 69 Intake manifold retaining nut location

42. Install the PCV hose.
43. Install the fuel rail and injectors. also remove the damper assembly, if equipped.
44. Install the blow-by hose.
45. Install the high pressure fuel hose with a new o-ring.
46. Install the engine mount stay.
47. Install the injector connectors.
48. Install the ignition coil connectors and coils.
49. Install the MDP sensor and o-ring.
50. Install the EGR adapter and gasket.
51. Install the new intake manifold plenum gasket.
52. Install the intake manifold plenum. tighten the plenum mounting bolts to 20 ft. lbs. (28 Nm).
53. Install the EGR pipe b connection.
54. Install the rear intake manifold plenum stay.
55. Install the power steering pressure hose clamp bracket.
56. Install the power steering pressure hose clamp.
57. Install the EGR pipe a clamp.
58. Install the front intake manifold plenum stay.
59. Install the EVAP purge solenoid, hose and connector.
60. Install the connector bracket.
61. Install the right bank ho2s clamps and connectors.
62. Install the control wiring harness and injector wiring harness combination connector.
63. Install the CKP sensor connector.
64. Install the KS connector.
65. Install the MDP sensor connector.
66. Install the throttle body.
67. Install the air cleaner assembly.
68. Refill the radiator with coolant.
69. Connect the negative battery cable.
70. Turn the ignition **ON**and then **OFF**, and keep it off for at least ten seconds.
71. Run the engine and check for fuel leaks, correct as required.

72. Correct and adjust all fluid levels, as required.

OIL PAN

REMOVAL & INSTALLATION

See Figures 70 through 73.

1. Before servicing the vehicle, refer to the Precautions Section.
2. Drain the engine oil.
3. Remove the engine undercover.
4. Remove the front exhaust pipe.
5. Remove the oil pan drain plug and gasket.
6. Remove the starter motor.
7. Remove the engine oil dipstick and o-ring.
8. Remove the lower oil pan. if necessary, use a block of wood and hammer to carefully dislodge the lower oil pan.
9. Remove the cover.
10. Remove the 2 lower torque converter connecting bolts.
11. Remove the upper oil pan. screw the m10 bolts holding the oil pan to the transaxle assembly into the bolt hole shown in the illustration to remove the pan.

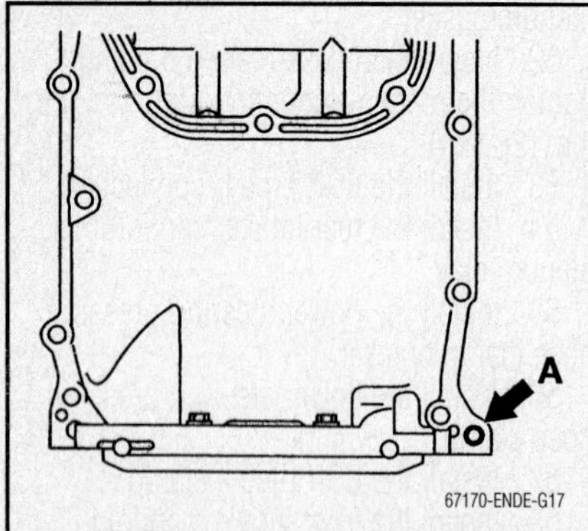

Fig. 70 Oil pan removal bolt installation location

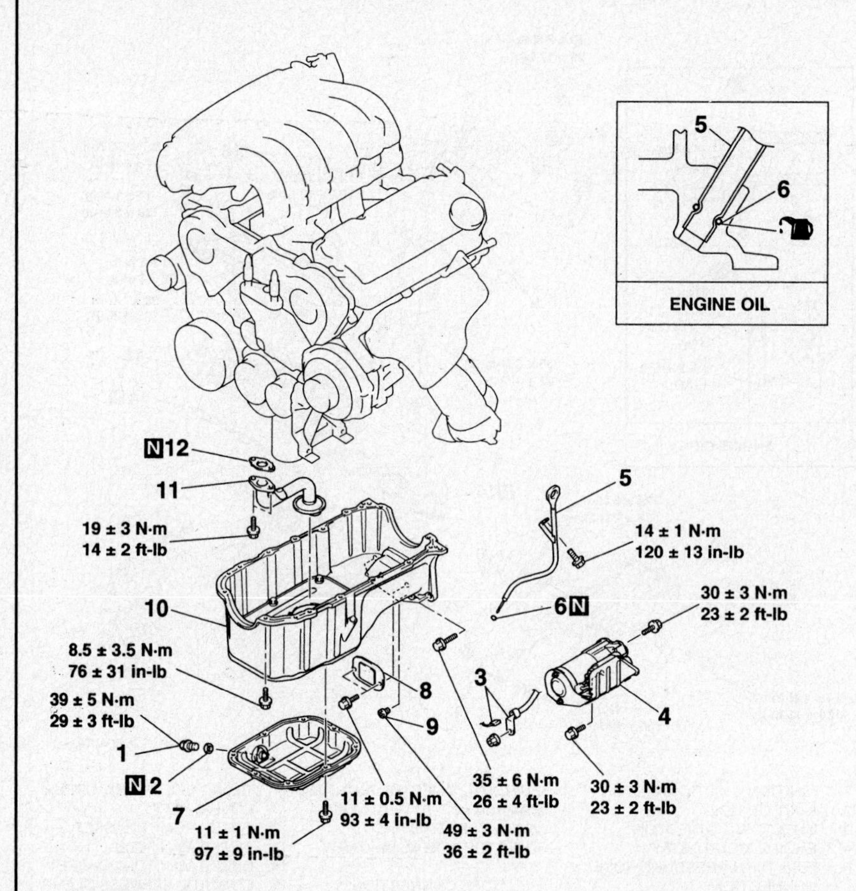

Fig. 71 Oil pan and related components

12. Remove the oil screen.
13. Remove the oil pan gasket.

To install:

14. Before installing, thoroughly clean the oil pan and cylinder block mating surfaces.
15. Apply liquid gasket around the surface of the oil pan.

➡Assemble the oil pan to the cylinder block within 30 minutes after applying the liquid gasket.

16. Install the oil screen. Torque the bolts to 12–16 ft. lbs. (16–22 Nm).
17. Install the upper oil pan. torque the bolts to specification and, in the proper sequence. the bolt holes for bolts 13 and 14 are cut away on the transaxle side. be sure you do not insert the bolts at an angle.

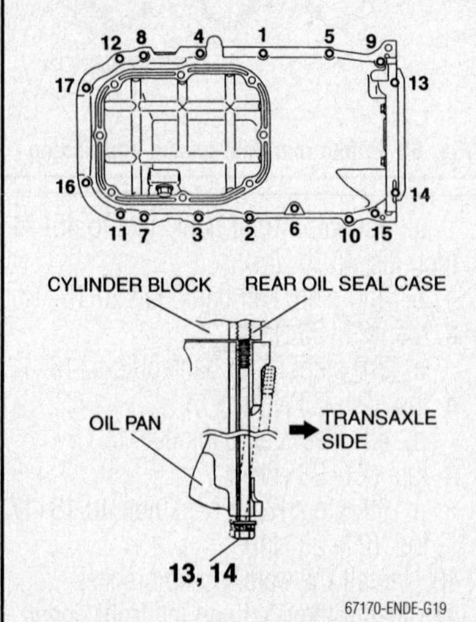

Fig. 72 Upper oil pan bolt torque sequence

18. Install the two lower torque converter connecting bolts. tighten the bolts to 36 ft. lbs. (49 Nm).
19. Install the cover.
20. Install the lower oil pan. torque the bolts to specification and in the proper sequence.
21. Install the engine oil dipstick and o-ring.
22. Install the starter motor.
23. Install the oil pan drain plug with a new gasket. tighten to specification.
24. Install the front exhaust pipe.
25. Install the engine undercover.
26. Fill the crankcase with oil.
27. Start the engine and check for leaks.

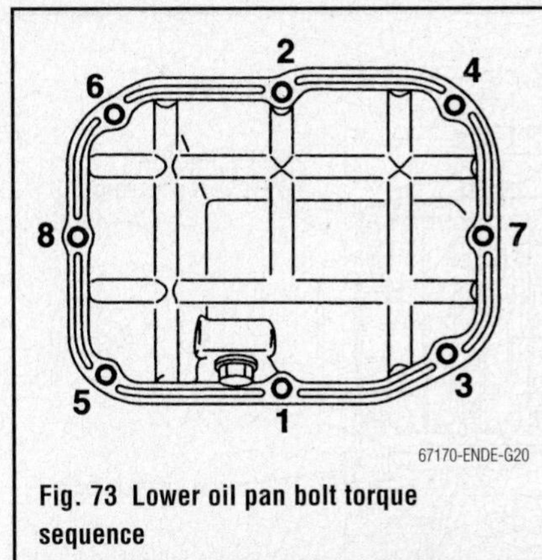

Fig. 73 Lower oil pan bolt torque sequence

OIL PUMP

REMOVAL & INSTALLATION

See Figures 74 and 75.

1. Before servicing the vehicle, refer to the Precautions Section.
2. Drain the engine oil.
3. Remove the timing belt.
4. Remove the oil pressure switch.
5. Remove the oil dipstick.
6. Remove the oil pans from the engine.
7. Remove the oil baffle and screen.
8. Remove the oil pump mounting bolts and the pump from the front of the engine.

➡Note the position of each oil pump case retaining bolts to facilitate installation. The bolts are of different length.

To install:

9. Clean the gasket mounting surfaces of the pump and engine block.
10. Prime the pump by pouring fresh oil into the inlet and turning the rotors or by packing pump with petroleum jelly. Using a new gasket, install the oil pump on the engine and tighten all bolts to 10 ft. lbs. (14 Nm).
11. Clean the gasket mounting surfaces of the pump and engine block.
12. Apply a 0.1 inch diameter bead of sealant (part number MD970389 or equivalent) to the oil pump case.

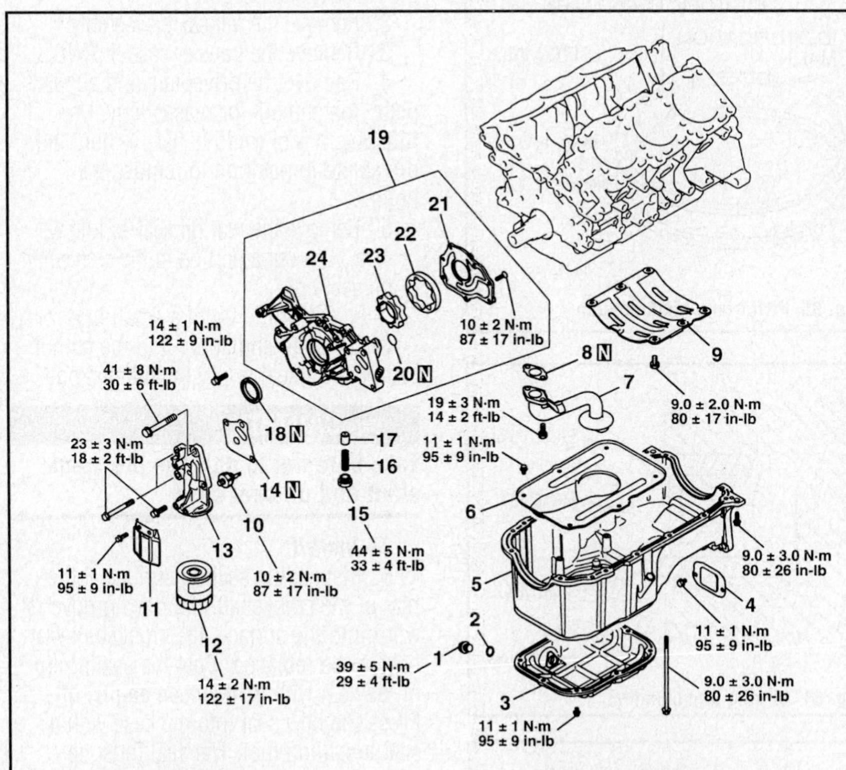

Fig. 74 Oil pump and related components

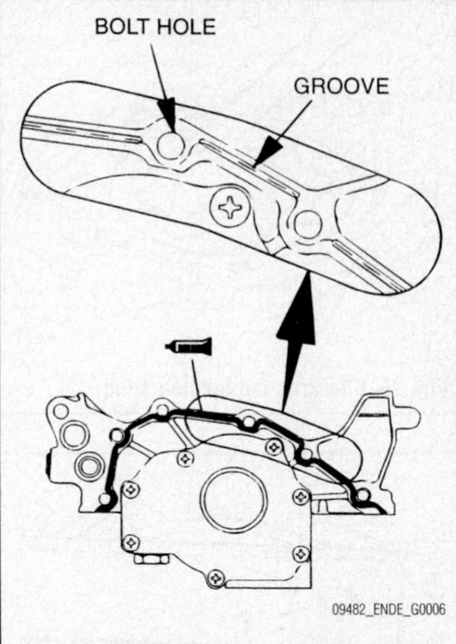
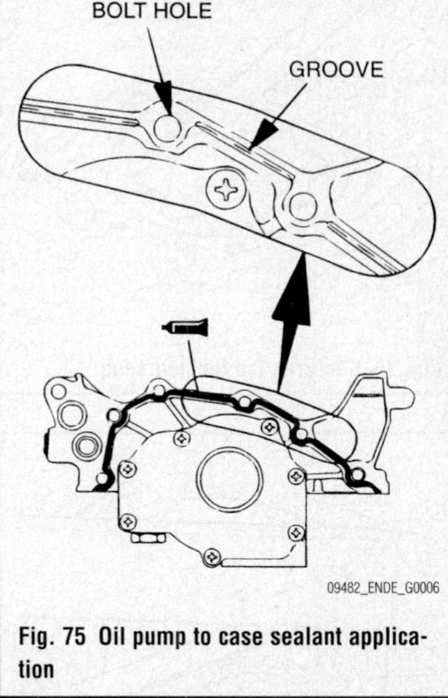

Fig. 75 Oil pump to case sealant application

➡Apply sealant as indicated by the broken line in the illustration. The grooves must be traced and the bolt holes must be surrounded with a bead of sealant.

13. Install the oil pump case assembly to the front of the cylinder block.

➡Be sure to install the oil pump case quickly, while the sealant is wet. There is a fifteen minute working window.

14. Torque the oil pump case mounting bolts to 113–131 inch lbs. (13–15 Nm).

➡After installation keep the sealed area free from oil and coolant for at least one hour.

15. Clean out the oil pick-up or replace as required. Replace the oil pick-up gasket ring and install the pick-up to the pump.
16. Install the oil filter and the bracket. torque the bolts to 17 ft. lbs. (23 Nm).
17. Install the oil baffle and screen. torque the bolts to 13 ft. lbs. (18 Nm).
18. Install the oil pans.
19. Install the oil pressure switch. torque the switch to 87 inch lbs. (9.8 Nm).
20. Install the timing belt.
21. Install the dipstick.
22. Refill the engine with the proper amount of oil.
23. Start the engine and check for proper oil pressure.
24. Check for and correct leaks.

INSPECTION

See Figures 76 through 78.

1. Check the tip clearance by inserting a feeler gauge between the inner and outer

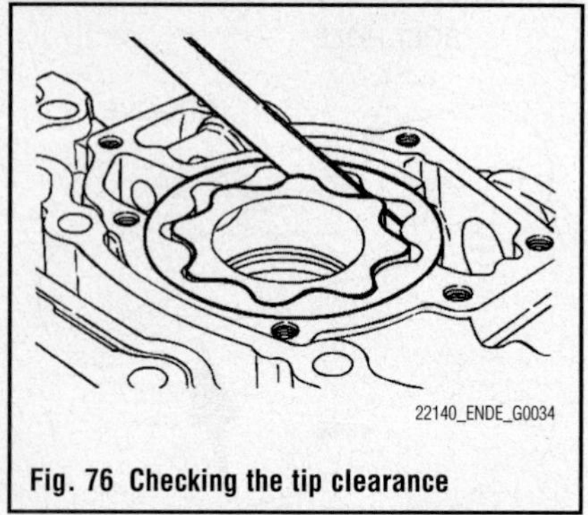
Fig. 76 Checking the tip clearance

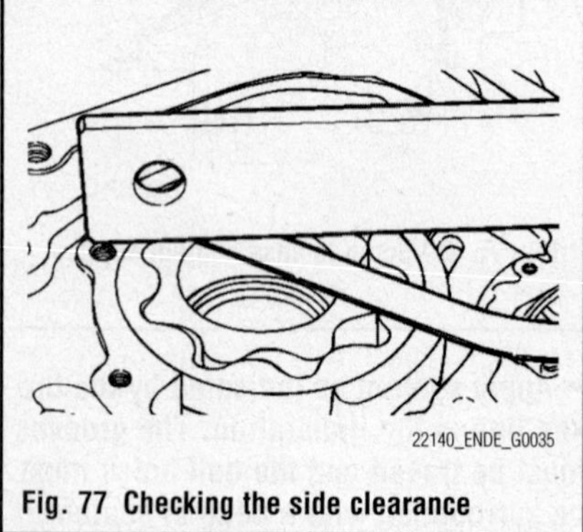
Fig. 77 Checking the side clearance

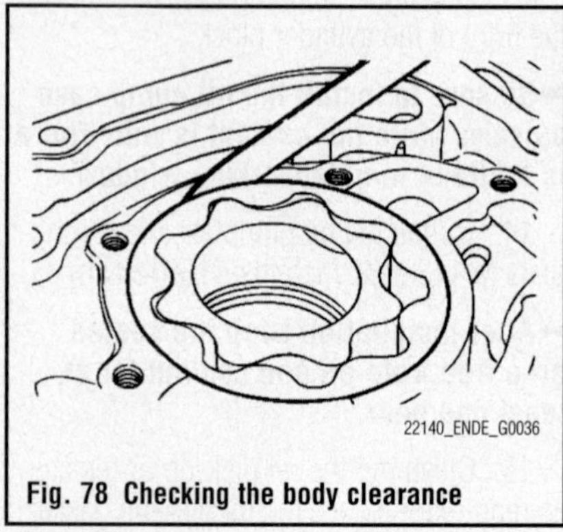
Fig. 78 Checking the body clearance

gears at the point of the tip. Clearance should be 0.003–0.007 inch (0.06–0.18 mm).

2. Check the side clearance by laying a straight edge across the gears and inserting a feeler gauge between the strait edge and the gears. Clearance should be 0.002–0.003 inch (0.04–0.10 mm).
3. Check the body clearance by inserting a feeler gauge between the outer gear and the body. Clearance should be 0.004–0.007 inch (0.10–0.18 mm) with a limit of 0.013 inch (0.35 mm).
4. If any clearance is out of specification, replace the oil pump.

MAIN BEARING TORQUE SEQUENCE

See Figure 79.

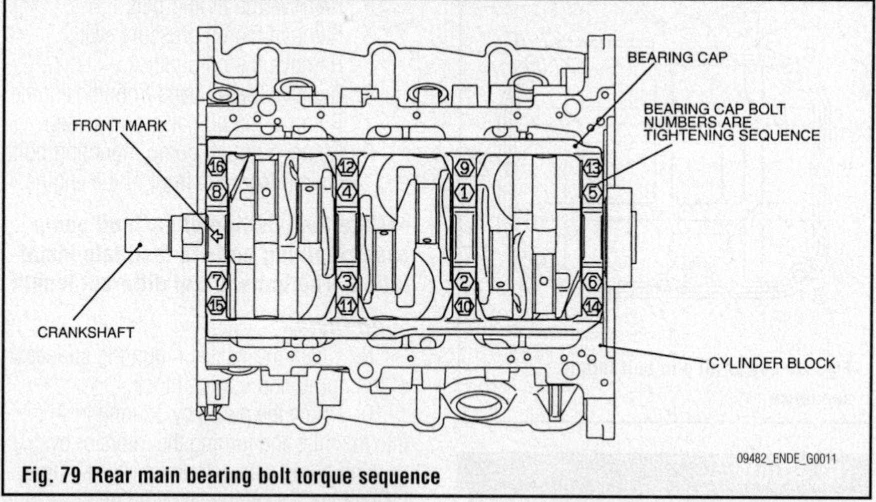

Fig. 79 Rear main bearing bolt torque sequence

PISTON AND RING

POSITIONING

See Figures 80 through 82.

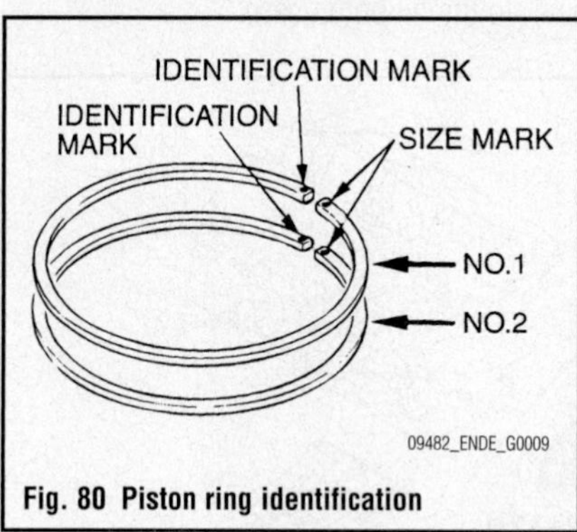

Fig. 80 Piston ring identification

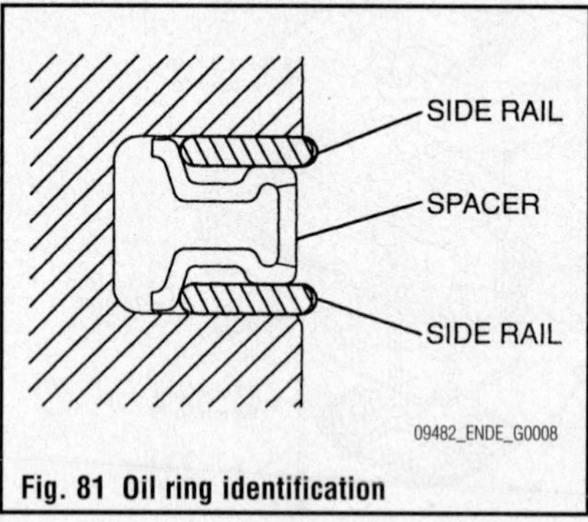

Fig. 81 Oil ring identification

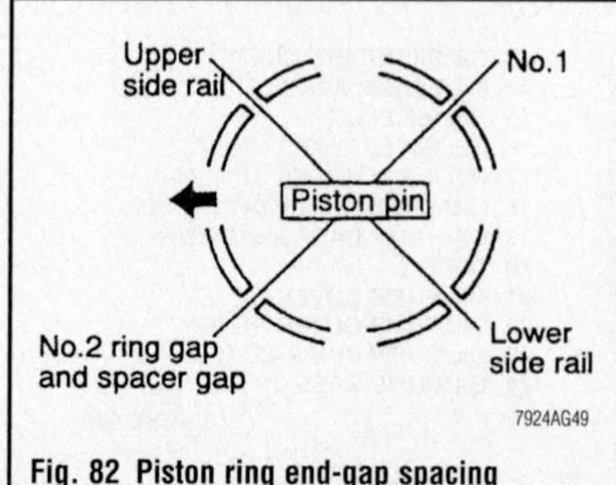

Fig. 82 Piston ring end-gap spacing

REAR MAIN SEAL

REMOVAL & INSTALLATION

See Figure 60.

1. Before servicing the vehicle, refer to the Precautions Section.
2. Remove the transaxle assembly.
3. Remove the transfer case, if AWD.
4. Remove the driveplate and adapter plate, matchmark for reassembly. Use Mitsubishi tool (md998781) to hold the driveplate in position to remove the bolts.
5. Remove the rear oil seal as follows:
   a. Cut out a portion in the crankshaft oil seal lip.
   b. Cover the tip of a small prytool with a cloth and apply it to the cutout in the oil seal to pry the oil seal out.

**** CAUTION**

Take care not to damage the crankshaft and oil seal case.

To install:

6. Inspect the sealing surface at the rear of the crankshaft. If a deep groove is worn into the surface, the crankshaft will have to be replaced. Coat the sealing lip of the seal with fresh, clean engine oil. Press the new seal into the case with a seal installing tool. The seal must be pressed in squarely until it bottoms in the case. It is necessary to use the proper tool (MD998718-01) to fit the seal into place.
7. Install the drive plate and adapter. Use Mitsubishi tool (md998781) to hold the driveplate in position while tightening the bolts to 55 ft. lbs. (75 Nm).
8. Install the transfer case, if AWD.
9. Install the transaxle and related components as necessary.

ROCKER ARMS/SHAFTS

REMOVAL & INSTALLATION

Left Bank

See Figures 51 through 53.

1. Before servicing the vehicle, refer to the Precautions Section.
2. Drain the cooling system.
3. Remove the timing belt.
4. Remove the thermostat housing.
5. Remove the blow-by hose connection.
6. Remove the PCV valve connection.
7. Remove the PCV valve.
8. Disconnect the ignition coil connector.
9. Remove the ignition coil.
10. Disconnect the control wiring harness clamp.
11. Remove the valve cover retaining bolts.
12. Remove the valve cover.
13. Remove the spark plug guide oil seal.

➡Install auto lash adjuster retainers SST MD998443 on the rocker arms.

14. Remove the intake and exhaust rocker arms and shafts by loosening the mounting bolts and removing the rocker arm and shaft assembly with the bolt still attached.

To install:

➡Lubricate the valve train components with clean engine oil.

15. Bleed and install the lash adjusters in their original bores in the cylinder head.
16. Install the intake and exhaust rocker arms and shafts.
    a. Check that notches in the each rocker shaft are facing the direction shown.
    b. Tighten bolts to 23 ft. lbs. (31 Nm).

➡Remove auto lash adjuster retainers SST MD998443 from the rocker arms.

17. Install the spark plug guide oil seal.
18. Install the valve cover.
19. Install the valve cover retaining bolts. Tighten to 31 inch lbs. (4 Nm).
20. Disconnect the control wiring harness clamp.
21. Install the ignition coil.
22. Disconnect the ignition coil connector.
23. Install the PCV valve.
24. Install the PCV valve connection.
25. Install the blow-by hose connection.
26. Install the thermostat housing.
27. Install the timing belt.
28. Fill the cooling system.
29. Run the engine and check for leaks.

Right Bank

See Figures 51, 53 and 54.

1. Before servicing the vehicle, refer to the Precautions Section.
2. Drain the cooling system.
3. Remove the intake manifold plenum assembly.
4. Remove the timing belt.
5. Remove the thermostat housing.
6. Remove the breather hose connection.
7. Remove the blow-by hose connection.
8. Remove the ignition coil connector.
9. Remove the ignition coil.
10. Remove the control wiring harness clamp.
11. Remove the valve cover retaining bolts.
12. Remove the valve cover.
13. Remove the spark plug guide oil seal.

➡Install the auto lash adjuster retainers SST MD998443 on the rocker arms.

14. Remove the intake and exhaust rocker arms and shafts by loosening the mounting bolts and removing the rocker arm and shaft assembly with the bolt still attached.

To install:

➡Lubricate the valve train components with clean engine oil.

15. Bleed and install the lash adjusters in their original bores in the cylinder head.
16. Install the intake and exhaust rocker arms and shafts.
    a. Check that notches in the each rocker shaft are facing the direction shown.
    b. Tighten bolts to 23 ft. lbs. (31 Nm).

➡Remove the auto lash adjuster retainers SST MD998443 on the rocker arms.

17. Install the spark plug guide oil seal.
18. Install the valve cover.
19. Install the valve cover retaining bolts. Tighten to 31 inch lbs. (4 Nm).
20. Install the control wiring harness clamp.
21. Install the ignition coil.
22. Install the ignition coil connector.
23. Install the blow-by hose connection.
24. Install the breather hose connection.
25. Install the thermostat housing.
26. Install the timing belt.
27. Install the intake manifold plenum assembly.
28. Fill the cooling system.
29. Run the engine and check for leaks, correct as required.

TIMING BELT FRONT COVER

REMOVAL & INSTALLATION

See Figure 84.

1. Before servicing the vehicle, refer to the Precautions Section.
2. Remove the engine under cover and side under cover.
3. Remove the drive belts.
4. Remove the crankshaft pulley, using special tools MB991800 and MB991802.
5. Remove the manifold differential pressure (MDP) sensor connector.
6. Remove the knock sensor (KS) connector.
7. Remove the crankshaft position (CKP) sensor connector.
8. Remove the control wiring harness and injector wiring harness combination connector.
9. Remove the right bank heated oxygen sensor (ho2s) connectors.
10. Remove the connector bracket.
11. Remove the engine mount stay.
12. Remove the right and left timing belt upper covers.
13. Remove the tensioner pulley.
14. Remove the tensioner bracket.
15. Remove the CKP sensor harness clamp.
16. Remove the lower timing belt cover.

To install:

17. Install the lower timing belt cover.
18. Install the CKP sensor harness clamp.
19. Install the tensioner bracket.
20. Install the tensioner pulley.
21. Install the left and right upper timing belt covers.
22. Install the engine mount stay.
23. Install the connector bracket.
24. Install the HO2S connectors.
25. Install the wiring harness and injector connector.
26. Install the CKP sensor connector.
27. Install the KS connector.
28. Install the MDP sensor connector.
29. Install the crankshaft pulley. Using pulley holder MB991800 and the 2 crankshaft pulley holder pin tools MD991802 to hold the crankshaft pulley, tighten the crankshaft pulley bolt to 136 ft. Lbs. (185 nm).
30. Install the drive belts.
31. Install the side and under covers.

TIMING BELT AND SPROCKETS

REMOVAL & INSTALLATION

See Figures 83 through 85.

1. Before servicing the vehicle, refer to the Precautions Section.
2. Remove the engine under cover and side under cover.
3. Remove the drive belts.
4. Remove the crankshaft pulley, using special tools MB991800 and MB991802.
5. Remove the manifold differential pressure (MDP) sensor connector.
6. Remove the knock sensor (KS) connector.
7. Remove the crankshaft position (CKP) sensor connector.
8. Remove the control wiring harness and injector wiring harness combination connector.
9. Remove the right bank heated oxygen sensor (ho2s) connectors.
10. Remove the connector bracket.
11. Remove the engine mount stay.
12. Remove the right and left timing belt upper covers.
13. Remove the tensioner pulley.
14. Remove the tensioner bracket.
15. Remove the CKP sensor harness clamp.
16. Remove the lower timing belt cover.
17. Remove the engine mount.
18. Remove the engine support bracket.
19. Turn the crankshaft clockwise to align the timing marks and set the No. 1 cylinder at Top Dead Center (TDC). If you are reusing the timing belt, mark the flat side of the belt with an arrow showing the clockwise direction.
20. Loosen the center bolt of the tension pulley, and then remove the timing belt.
21. Remove the auto tensioner.
22. Remove the tensioner pulley.
23. Remove the tensioner arm.
24. Remove the shaft.

To install:

25. Install the idler pulley.
26. Install the shaft.
27. Install the tensioner arm assembly.
28. Install the tension pulley.
29. Press the end of the auto-tensioner inward with 72–145 ft. lbs. (98–196 nm) of force and measure the distance that the pushrod is pushed in. If the standard distance is not 0.04 in. (1mm), replace the auto-tensioner.
30. Position the auto tensioner in a soft-jawed vise and slowly compress the pushrod until the pushrod and housing holes align; then, install a setting pin to secure the auto-tensioner in the retracted position.

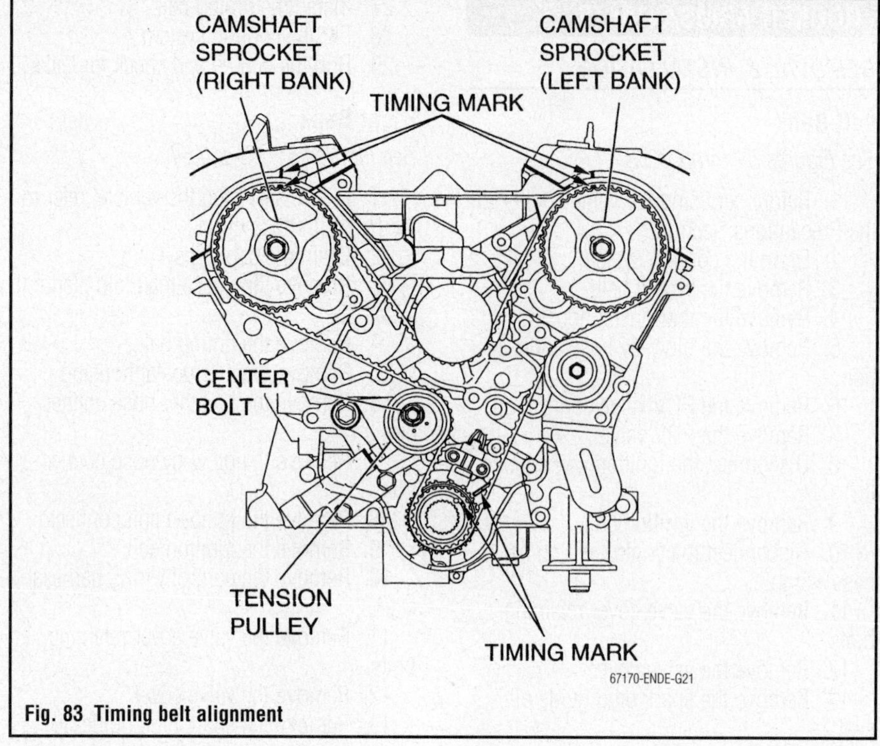

Fig. 83 Timing belt alignment

➡**If you are installing a new auto-tensioner, the pin will already be inserted into the pin holes of the new tensioner.**

31. Install the auto-tensioner.
32. Align the camshaft and crankshaft TDC timing marks.
33. Install the timing belt (noting its rotational direction) so that there is no deflection between the sprockets and pulleys in the following manner:
34. Install the crankshaft sprocket.
35. Install the idler pulley.
36. Install the left camshaft sprocket.
37. Install the water pump pulley.
38. Install the right camshaft sprocket.
39. Install the tension pulley.
40. Turn the camshaft sprocket counterclockwise until the tension side of the timing belt is firmly stretched, then, recheck the timing marks.
41. Using the tension pulley socket wrench tool MD998767, or equivalent, push the tensioner pulley into the timing belt and secure the center bolt.
42. Using the crankshaft pulley spacer tool MD998769, or equivalent, rotate the crankshaft ¼turn counterclockwise, then, turn it again clockwise to align the timing marks.
43. Loosen the timing belt tensioner center bolt. Using the tension pulley socket wrench tool MD998767, or equivalent, and a torque wrench, apply 39 inch lbs. (4.4Nm) pressure on the timing belt. torque the tensioner pulley center bolt to 35 ft. lbs. (48 Nm).
44. Remove the setting pin from the auto-tensioner.
45. Rotate the crankshaft 2 complete revolutions and realign the timing marks. then, wait for 5 minutes until the auto-tensioner pushrod extends to its standard value. If the standard value is not 0.19–0.24 in. (4.8–6.0 mm), repeat the adjustment procedure. If the standard value is still not achieved, replace the auto-tensioner.
46. Install the engine support bracket.
47. Install the engine mount.
48. Install the lower timing belt cover.
49. Install the CKP sensor harness clamp.
50. Install the tensioner bracket.
51. Install the tensioner pulley.
52. Install the left and right upper timing belt covers.
53. Install the engine mount stay.
54. Install the connector bracket.
55. Install the HO2S connectors.
56. Install the wiring harness and injector connector.
57. Install the CKP sensor connector.
58. Install the KS connector.
59. Install the MDP sensor connector.
60. Install the crankshaft pulley. Using pulley holder MB991800 and the 2 crankshaft pulley holder pin tools MD991802 to hold the crankshaft pulley, tighten the

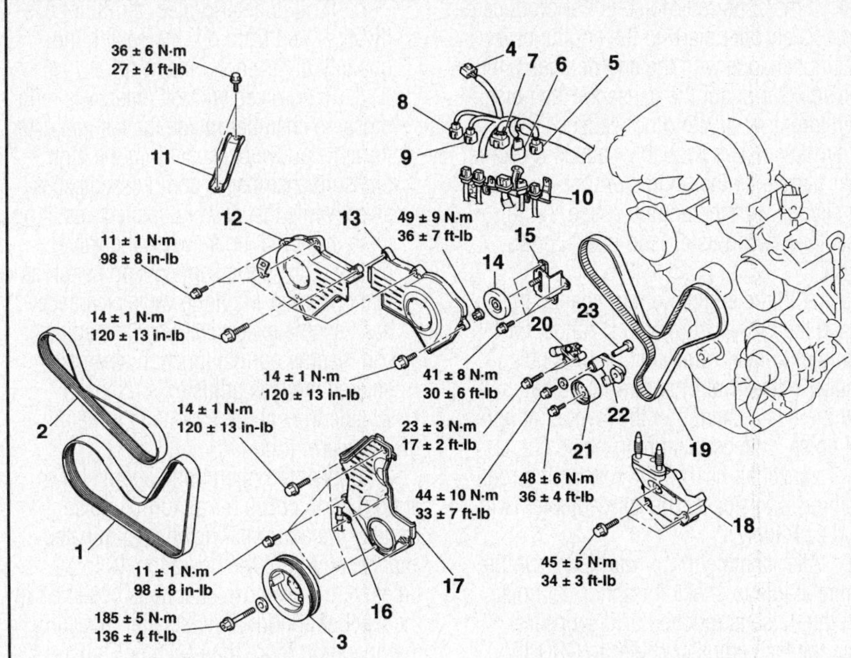

1. GENERATOR DRIVE BELT
2. POWER STEERING OIL PUMP DRIVE BELT
3. CRANKSHAFT PULLEY
4. MANIFOLD DIFFERENTIAL PRESSURE SENSOR CONNECTOR
5. KNOCK SENSOR CONNECTOR
6. CRANKSHAFT POSITION SENSOR CONNECTOR
7. CONTROL WIRING HARNESS AND INJECTOR WIRING HARNESS COMBINATION CONNECTOR
8. RIGHT BANK HEATED OXYGEN SENSOR (REAR) CONNECTOR
9. RIGHT BANK HEATED OXYGEN SENSOR (FRONT) CONNECTOR
10. CONNECTOR BRACKET
11. ENGINE MOUNT STAY
12. TIMING BELT FRONT UPPER COVER, RIGHT
13. TIMING BELT FRONT UPPER COVER, LEFT
14. TENSIONER PULLEY
15. TENSIONER BRACKET
16. CRANKSHAFT POSITION SENSOR HARNESS CLAMP
17. TIMING BELT LOWER COVER

67170-ENDE-G22

Fig. 84 Timing belt and related components

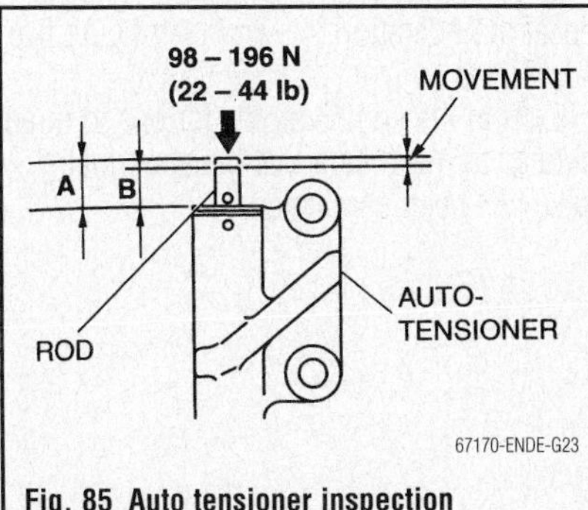

Fig. 85 Auto tensioner inspection

crankshaft pulley bolt to 136 ft. Lbs. (185 nm).

61. Install the drive belts.
62. Install the side under cover.
63. Install the undercover.
64. Connect the negative battery cable.
65. Turn the ignition **ON** and then **OFF**, and keep it off for at least ten seconds.
66. Refill the cooling system.

VALVE COVERS

REMOVAL & INSTALLATION

Left Bank

See Figure 52.

1. Before servicing the vehicle, refer to the Precautions Section.
2. Drain the cooling system.
3. Remove the timing belt.
4. Remove the thermostat housing.
5. Remove the blow-by hose connection.
6. Remove the PCV valve connection.
7. Remove the PCV valve.
8. Disconnect the ignition coil connector.
9. Remove the ignition coil.
10. Disconnect the control wiring harness clamp.
11. Remove the valve cover retaining bolts.
12. Remove the valve cover.

To install:

13. Install the valve cover.
14. Install the valve cover retaining bolts. Tighten to 31 inch lbs. (4 Nm).
15. Disconnect the control wiring harness clamp.
16. Install the ignition coil.
17. Disconnect the ignition coil connector.
18. Install the PCV valve.
19. Install the PCV valve connection.
20. Install the blow-by hose connection.
21. Install the thermostat housing.
22. Install the timing belt.
23. Fill the cooling system.
24. Run the engine and check for leaks.

Right Bank

See Figure 54.

1. Before servicing the vehicle, refer to the Precautions Section.
2. Drain the cooling system.
3. Remove the intake manifold plenum assembly.
4. Remove the timing belt.
5. Remove the thermostat housing.
6. Remove the breather hose connection.
7. Remove the blow-by hose connection.
8. Remove the ignition coil connector.
9. Remove the ignition coil.
10. Remove the control wiring harness clamp.
11. Remove the valve cover retaining bolts.
12. Remove the valve cover.

To install:

13. Install the valve cover.
14. Install the valve cover retaining bolts. Tighten to 31 inch lbs. (4 Nm).
15. Install the control wiring harness clamp.
16. Install the ignition coil.
17. Install the ignition coil connector.
18. Install the blow-by hose connection.
19. Install the breather hose connection.
20. Install the thermostat housing.
21. Install the timing belt.
22. Install the intake manifold plenum assembly.
23. Fill the cooling system.
24. Run the engine and check for leaks, correct as required.

VALVE LASH

ADJUSTMENT

Lash Adjuster Check

1. The engine used in the Endeavor uses hydraulic lash adjusters that do not

require adjustment. However, if a valve tap (abnormal noise) is noticed, perform the following procedure to attempt to fix the problem.

a. If an abnormal noise (chattering noise) suspected to be caused by malfunction of the lash adjuster is produced immediately after starting the engine and does not disappear, perform the following check.

b. Parking the vehicle on a grade for a long time may decrease oil in the lash adjuster, causing air to enter the high pressure chamber when starting the engine. After parking for many hours, oil may run out from the oil passage and take time before oil is supplied to the lash adjuster, causing air to enter the high pressure chamber. Abnormal noise can be eliminated by bleeding the lash adjuster system.

c. An abnormal noise due to malfunction of the lash adjuster is produced immediately after starting the engine and changes with the engine speed, irrespective of the engine load. If, the abnormal noise is not produced immediately after starting the engine or does not change with the engine speed, or it changes with the engine load, the lash adjuster is not the cause for the abnormal noise.

d. When the lash adjuster is malfunctioning, the abnormal noise is rarely eliminated by continuing the warming-up of the engine at idle speed.

e. The abnormal noise may disappear only when seizure is caused by oil sludge in the engine whose oil is not maintained properly.

2. Start the engine.

3. Check if the abnormal noise produced immediately after starting the engine, changes with the change in the engine speed.

4. If the abnormal noise is not produced immediately after starting the engine or it does not change with the engine speed, the lash adjuster is not the cause for the noise. Therefore, investigate other causes. The abnormal noise is probably caused by some other parts than the engine proper if it does not change with the engine speed. (In this case, the lash adjuster is in good condition.)

5. With the engine idling, change the engine load (shift from N to D range, for example) to make sure that there is no change in the level of abnormal noise.

If there is a change in the level of abnormal noise, suspect a tapping noise due to worn crankshaft bearing or connecting rod bearing (In this case, the lash adjuster is in good condition.).

6. After completion of warm-up, run the engine at idle to check for abnormal noise.

If the noise is reduced or disappears, clean the lash adjuster (Refer to GROUP 11B, Engine Overhaul - Rocker Arms and Camshaft - Inspection). As it is suspected that the noise is due to seizure of the lash adjuster. If there is no change in the level of the abnormal noise, proceed to step 5.

7. Run the engine to bleed the lash adjuster system.

8. If the abnormal noise does not disappear after air bleeding operation, clean the lash adjuster.

Bleeding Lash Adjuster

See Figure 86.

1. Check engine oil and add or change oil if required.

a. If the engine oil level is low, air is sucked from the oil screen, causing air to enter the oil passage.

b. If the engine oil level is higher than specification, oil may be stirred by the crankshaft, causing oil to be mixed with a large quantity of air.

c. If oil is deteriorated, air is not easily separated from oil, increasing the quantity of air contained in oil.

d. If air mixed with oil enters the high pressure chamber inside the lash adjuster from the above causes, air in the high pressure chamber is compressed excessively while the valve is opened, resulting in an abnormal noise when the valve closes. This is the same phenomenon as that observed when the valve clearance has become excessive. The lash adjuster can resume normal function when air entered the lash adjuster is removed.

2. Idle the engine for one to three minutes to warm it up.

3. Repeat the operation pattern shown, at no load to check for abnormal noise. Usually the abnormal noise is eliminated after repetition of the operation 10 to 30 times. If, however, no change is observed in the level of abnormal noise after repeating the operation more than 30 times, suspect that the abnormal noise is due to some other factors.

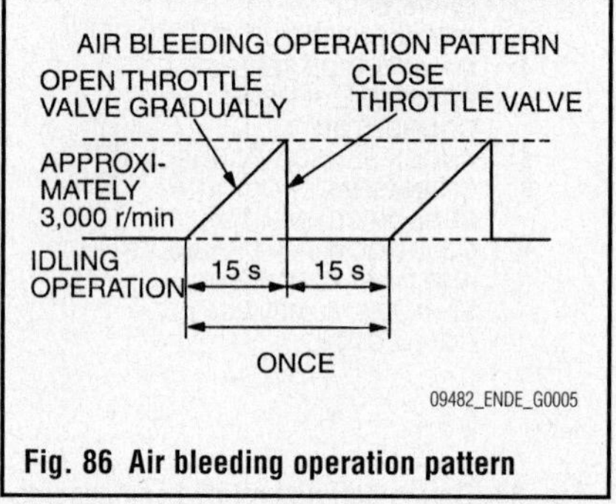

Fig. 86 Air bleeding operation pattern

4. After elimination of abnormal noise, repeat the operation shown in left figure five more times.

5. Run the engine at idle for one to three minutes to make sure that the abnormal noise has been eliminated.

ENGINE PERFORMANCE & EMISSION CONTROL

COMPONENT LOCATIONS

See Figure 87.

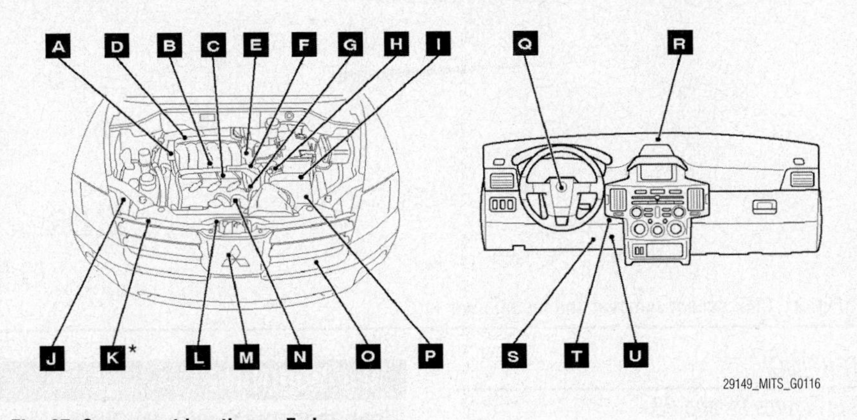

Fig. 87 Component locations—Endeavor

ACCELERATOR PEDAL POSITION (APP) SENSOR

LOCATION

See Figure 88.

Refer to the accompanying illustration for Accelerator Pedal Position (APP) sensor location.

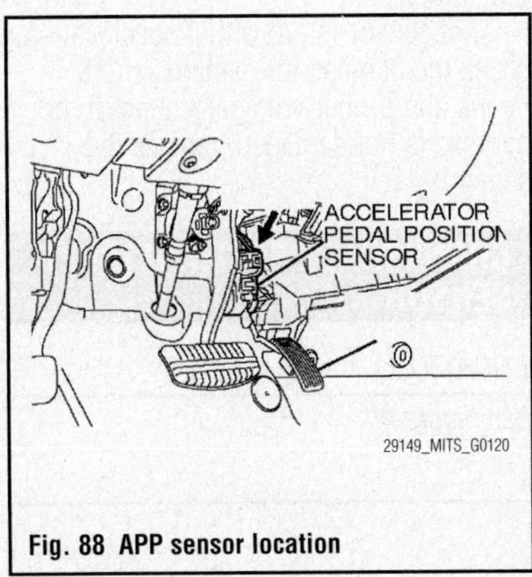

Fig. 88 APP sensor location

OPERATION

A 5 volt power supply is applied on the Accelerator Pedal Position (APP) Sensor (main) power terminal from the PCM. The ground terminal is grounded with PCM. When the accelerator pedal is moved from the idle position to the fully opened position, the resistance between the accelerator pedal position sensor (main) output terminal and ground terminal will increase according to the depression.

REMOVAL & INSTALLATION

The accelerator pedal position sensor is an integral part of the accelerator pedal assembly. It should not be serviced unnecessarily; it has been precisely adjusted by the manufacturer.

WARNING

Never loosen the screw fixing the accelerator pedal assembly resin cover. If the screw is loosened, the sensor position which exists inside the resin cover is off and the accelerator pedal position sensor does not work normally. Do not remove the accelerator pedal pad. If the pad is removed and installed, excessive force may damage accelerator pedal position sensor.

1. Disconnect the sensor electrical connector.
2. Remove the accelerator pedal.

To install:

3. Install the accelerator pedal assembly. Tighten the pedal mounting nuts to 111 inch lbs. (13 Nm).
4. Connect the electrical connector at the sensor.

TESTING

1. Connect a scan tool to the data link connector.
2. Set the scan tool to the data reading mode the sensor.
3. Output voltage should be between 735 and 1,335 millivolts when foot is released from accelerator pedal.
4. Output voltage should be 4,000 millivolts or more when accelerator pedal is fully depressed.
5. If sensor is not within specification, check the harness for damage, short/opens and proper voltage/ground. If the harness is not damaged, replace the sensor.

BAROMETRIC PRESSURE (BARO) SENSOR

LOCATION

See Figure 89.

Refer to the accompanying illustration for Barometric Pressure (BARO) sensor location.

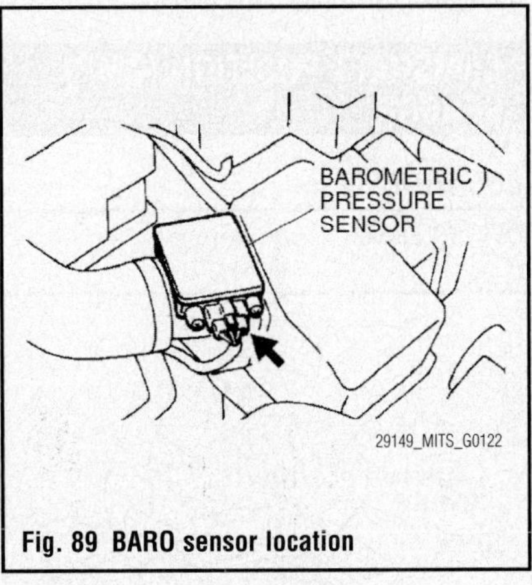

Fig. 89 BARO sensor location

OPERATION

A 5 volt reference and ground is supplied to the Barometric Pressure Sensor from the PCM. A sensor voltage that is proportional to the atmospheric pressure is sent to the PCM.

REMOVAL & INSTALLATION

1. Disconnect the sensor electrical connector.
2. Remove the barometric sensor from the air intake duct.

To install:

3. Install the barometric sensor on the air intake duct.
4. Connect the electrical connector at the sensor.

TESTING

1. Connect a scan tool to the data link connector.
2. Set the scan tool to the data reading mode the sensor.
3. Sensor data should read as follows:
   - When altitude is 0 foot (0 m), 29.8 in. Hg (101 kPa).

- When altitude is 1,969 feet (600 m), 28.1 in. Hg (95 kPa).
- When altitude is 3,937 feet (1,200 m), 26.0 in. Hg (88 kPa).
- When altitude is 5,906 feet (1,800 m), 23.9 in. Hg (81 kPa).

4. Start the engine.
5. Sensor data should read as follows:
   - When the engine is idling, 4.7 - 10.6 in. Hg (16 - 36 kPa).
   - When the engine is suddenly revved, manifold absolute pressure varies.
6. If sensor is not within specification, check the harness for damage, short/opens and proper voltage/ground. If the harness is not damaged, replace the sensor.

CAMSHAFT POSITION (CMP) SENSOR

LOCATION

See Figure 90.

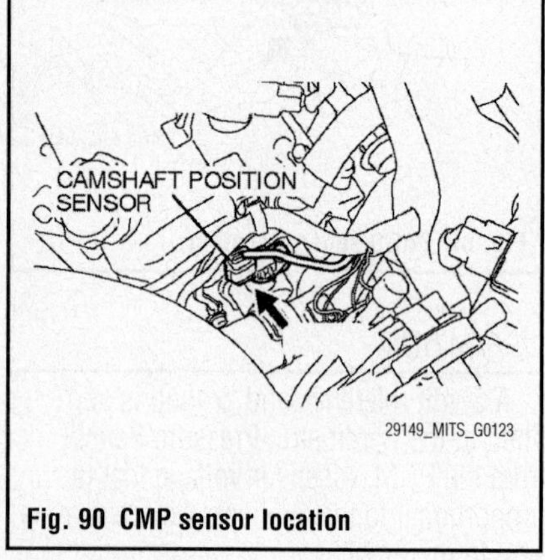

Fig. 90 CMP sensor location

Refer to the accompanying illustration for Camshaft Position (CMP) sensor location.

OPERATION

A 5 volt voltage is applied on the Camshaft Position (CMP) Sensor output terminal from the PCM. The sensor generates a pulse signal when the output terminal is opened and grounded.

REMOVAL & INSTALLATION

See Figure 91.

1. Disconnect the sensor connector.
2. Remove the retaining bolt, then remove the sensor and the O-ring.

To install:

3. Install the O-ring, sensor and retaining bolt. Tighten to 98 inch lbs. (11 Nm).
4. Connect the sensor connector.

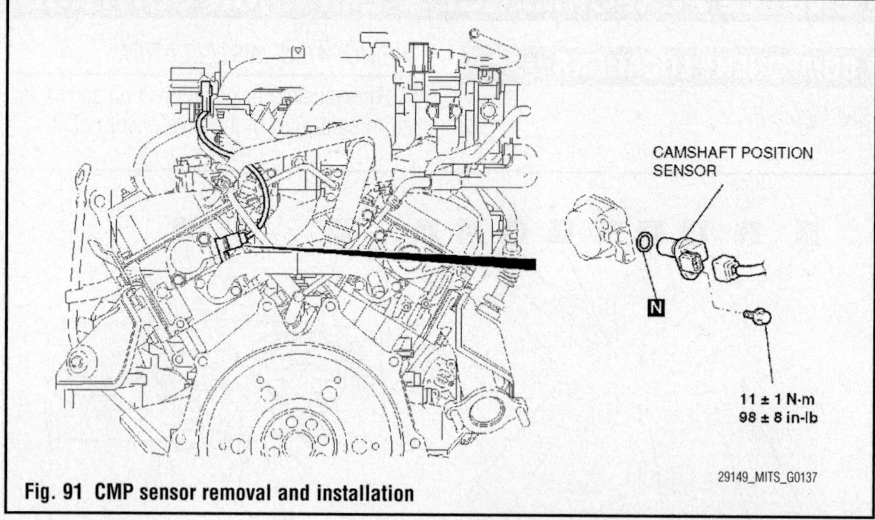

Fig. 91 CMP sensor removal and installation

TESTING

See Figures 92 and 93.

1. Disconnect the camshaft position sensor connector, and connect test harness special tool (MB991709) between the separated connectors (All terminals should be connected).
2. Connect the oscilloscope probe to the camshaft position sensor side connector terminal No. 2.

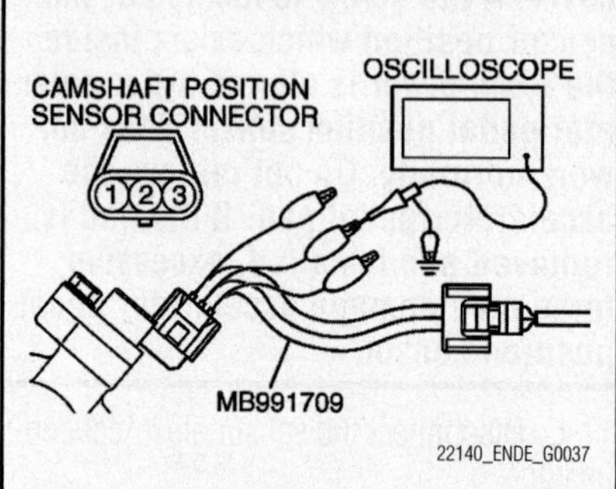

Fig. 92 Connect test harness special tool (MB991709) between the separated connectors

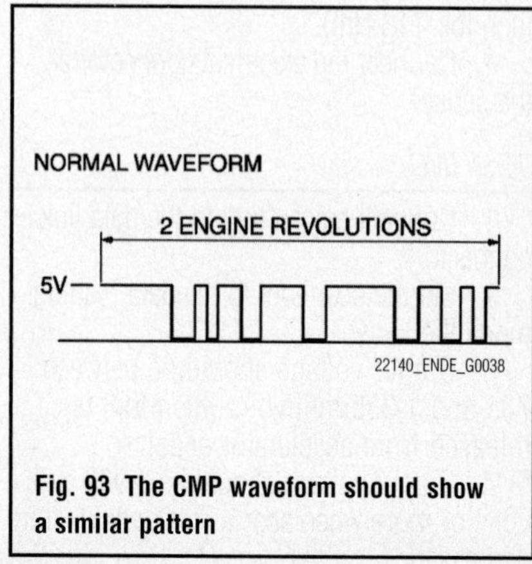

Fig. 93 The CMP waveform should show a similar pattern

⁂ WARNING

When measuring with the PCM side connector, disconnect all PCM connectors. Connect the check harness special tool (MB991923) between the separated connectors. Then connect the oscilloscope probe to the check harness connector terminal No. 104.

3. Start the engine and run at idle.
4. Check the waveform. The waveform should show a pattern similar to the illustration.
5. If sensor is not within specification, check the harness for damage, short/opens and proper voltage/ground. If the harness is not damaged, replace the sensor.

CRANKSHAFT POSITION (CKP) SENSOR

LOCATION

See Figure 94.

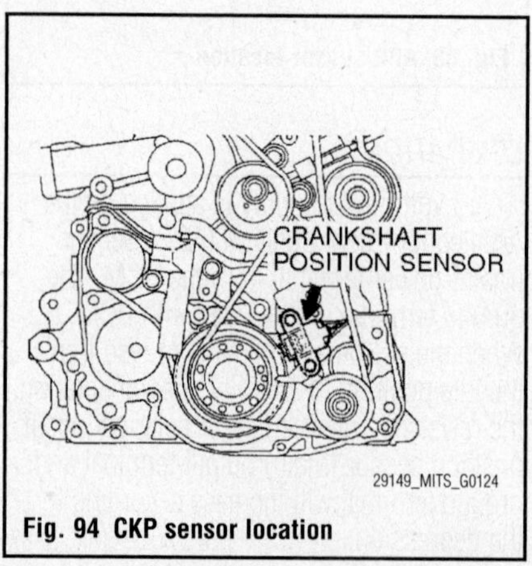

Fig. 94 CKP sensor location

Refer to the accompanying illustration for Crankshaft Position (CKP) sensor location.

OPERATION

A 5 volt voltage is applied on the Crankshaft Position (CKP) Sensor output terminal from the PCM. The sensor generates a pulse signal when the output terminal is opened and grounded.

REMOVAL & INSTALLATION

See Figure 95.

1. Detach the crankshaft position sensor electrical connector.
2. Remove the retaining bolt, then remove the sensor.

To install:

3. Position the sensor and install the retaining bolt. Tighten to 78 inch lbs. (9 Nm).
4. Connect the crankshaft position sensor electrical connector.

TESTING

See Figures 96 and 97.

1. Connect scan tool to the data link connector.
2. Start the engine and run at idle.
3. Set scan tool to the data reading mode for the sensor.
4. The tachometer and engine speed indicated on the scan tool should match.
5. If this is not the case, proceed as follows:
   a. Disconnect the crankshaft position sensor connector, and connect the test harness special tool (MD998478) between the separated connectors.
   b. Connect the oscilloscope probe to crankshaft position sensor connector terminal No. 2 (black clip of special tool).

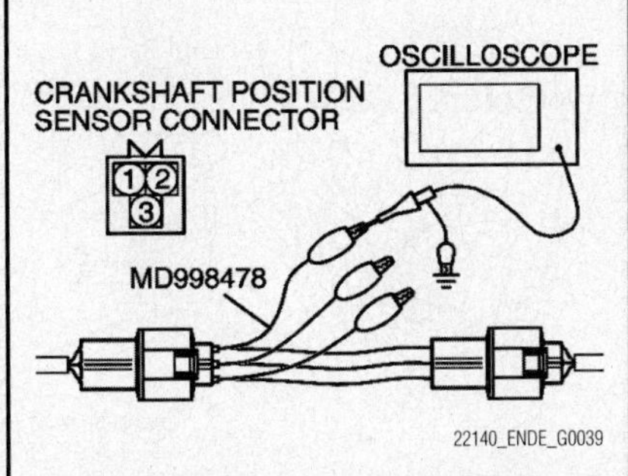

Fig. 96 Connect test harness special tool (MB991923) between the separated connectors

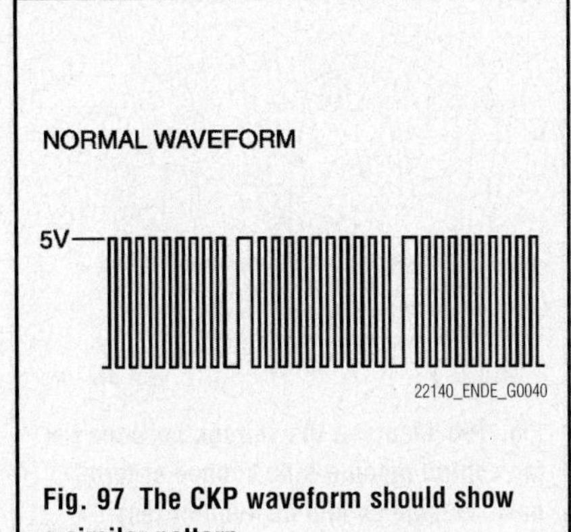

Fig. 97 The CKP waveform should show a similar pattern

⁂ WARNING

When measuring with the PCM side connector, disconnect all PCM connectors. Connect the check harness special tool (MB991923) between the separated connectors. Then connect the oscilloscope probe to the check harness connector terminal No. 103.

   c. Start the engine and run at idle.
   d. Check the waveform. The waveform should show a pattern similar to the illustration.
6. If sensor is not within specification, check the harness for damage, short/opens and proper voltage/ground. If the harness is not damaged, replace the sensor.

ELECTRIC FAN SWITCH

LOCATION

See Figure 98.

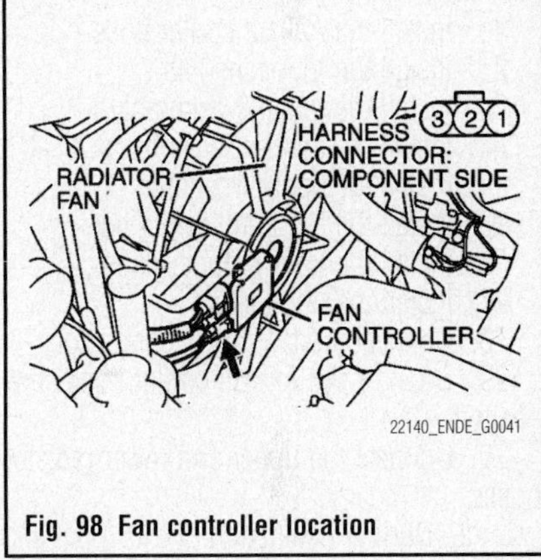

Fig. 98 Fan controller location

Refer to the accompanying illustration for fan controller location.

OPERATION

The PCM sends a duty signal to the fan controller according to engine coolant temperature, vehicle speed, and the condition of the A/C switch. (The closer the average voltage at the terminal comes to 5 volts, the higher the fan speed becomes.)

REMOVAL & INSTALLATION

1. Before servicing the vehicle, refer to the Precautions Section.
2. Drain the engine coolant.
3. Remove the air cleaner assembly.
4. Remove the radiator grille.
5. Disconnect the radiator condenser tank hose.
6. Remove the radiator condenser tank assembly.

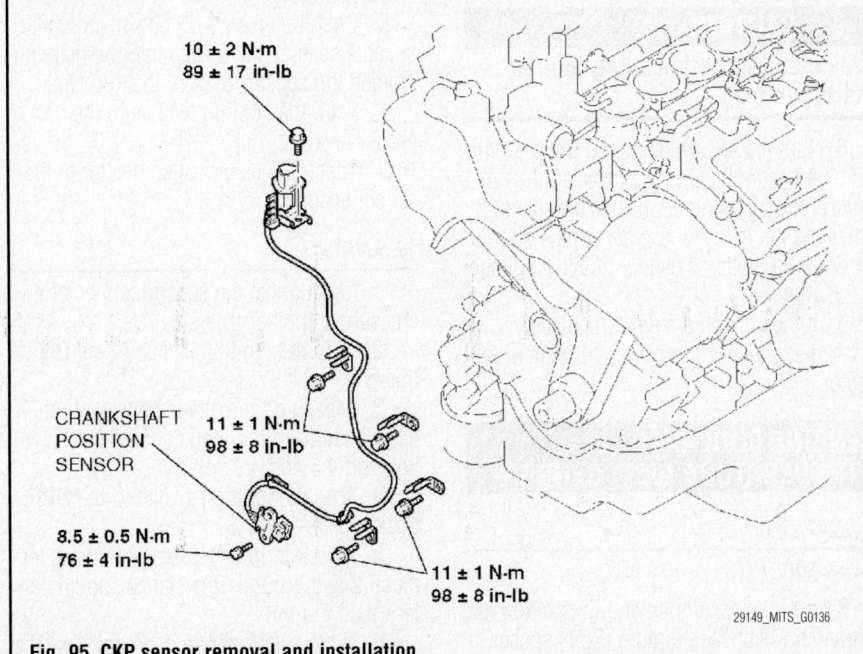

Fig. 95 CKP sensor removal and installation

7. Matchmark the installed positions of the hoses and hose clamps, then disconnect the upper and lower radiator hoses.

8. Detach the A/T oil cooler hose connection. After disconnecting the hose from the radiator, plug the hose and radiator nipple to prevent foreign debris from entering the cooling system.

9. Disconnect the fan motor connector.

10. Remove the hood latch.

11. Remove the front bumper clips.

12. Remove the front end upper bar.

13. Remove the upper insulator.

14. Remove the condenser bolts.

15. Remove the radiator assembly.

16. Remove the lower insulator.

17. Remove the A/T oil cooler hose.

18. Remove the shroud and fan assembly.

To install:

19. Install the shroud and fan assembly.

20. Install the A/T oil cooler hose.

21. Install the lower insulator.

22. Install the radiator assembly.

23. Install the condenser bolts.

24. Install the upper insulator.

25. Install the front end upper bar.

26. Install the front bumper clips.

27. Install the hood latch.

28. Connect the fan motor connector.

29. Connect the A/T oil cooler hose connection.

30. Connect the upper and lower radiator hoses.

a. During installation of the upper and lower radiator hoses, insert the hoses as far as the projection of the water inlet fitting. Align the matchmarks on the hoses and hose clamps, then connect the hoses.

31. Install the radiator condenser tank assembly.

32. Disconnect the radiator condenser tank hose.

33. Install the radiator grille.

34. Install the air cleaner assembly.

35. Fill the engine with coolant.

36. Start the engine and check for proper fan operation.

TESTING

See Figures 99 and 100.

1. Remove the fan control module connector.

2. Turn the ignition switch to the **ON**, and measure the voltage between the harness-side connector terminals. Voltage should be battery positive voltage

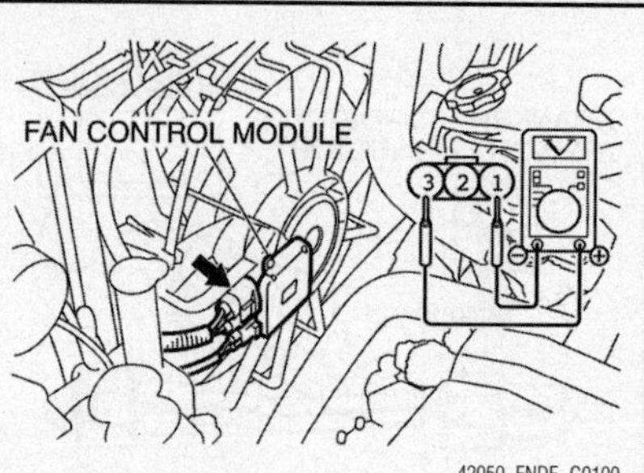

Fig. 99 Turn the ignition switch to the "ON" position, and measure the voltage between the harness-side connector terminals

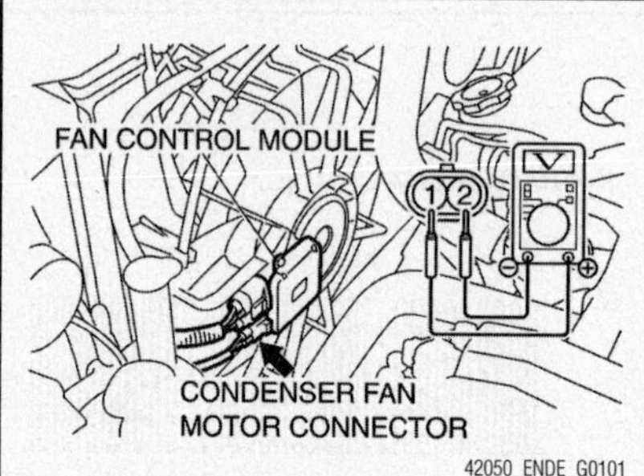

Fig. 100 Measure the voltage between the fan control module-side connector terminals. Voltage should be 1volt or less

3. Connect the fan control module connector, and disconnect the condenser fan motor connector.

4. Ensure that the A/C switch is off, and start the engine and run it at idle.

5. Measure the voltage between the fan control module-side connector terminals. Voltage should be 1volt or less.

⁂ CAUTION

Stay clear of the fan when the fan starts running.

6. Turn the A/C switch to the **ON** position.

7. Measure the voltage between the fan control module-side connector terminals while the fan is running. The voltage should be 8.2 ± 2.6 volts or battery positive voltage ± 2.6 volts.

8. If the voltage does not repeatedly change as indicated, replace the radiator fan motor.

ENGINE COOLANT TEMPERATURE (ECT) SENSOR

LOCATION

See Figure 101.

Refer to the accompanying illustration for Engine Coolant Temperature (ECT) sensor location.

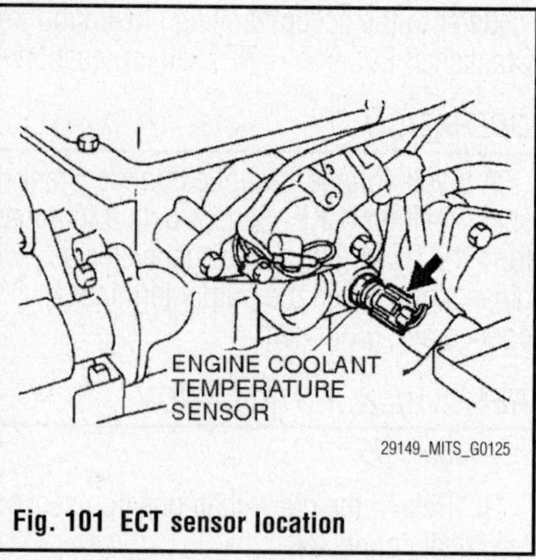

Fig. 101 ECT sensor location

OPERATION

A 5 volt voltage is applied to the Engine Coolant Temperature (ECT) sensor output terminal from the PCM via the resistor in the PCM. The ground terminal is grounded with PCM. The engine coolant temperature sensor is a negative temperature coefficient type of resistor. It has the characteristic that when the engine coolant temperature rises the resistance decreases. The engine coolant temperature sensor output voltage increases when the resistance increases and decreases when the resistance decreases.

REMOVAL & INSTALLATION

1. Drain the engine coolant to a level below the intake manifold.

2. Unplug the ECT sensor electrical connector

3. Use a deep socket and an extension to reach the ECT sensor, remove the ECT sensor from the thermostat housing.

To install:

4. Coat the threads of the sensor with a suitable sealant and thread into the housing. Tighten the sensor to 22 ft. lbs. (30 Nm).

5. Refill the cooling system to the proper level.

6. Attach the electrical connector to the sensor securely.

TESTING

1. Disconnect the sensor connector and measure at the harness side.

2. Turn the ignition switch to the **ON** position.

3. Measure the voltage between terminal 1 and ground. Voltage should be between 4.5 and 4.9 volts.

4. Turn the ignition switch to the **OFF** position.

5. Check for the resistance between terminal 2 and ground. Resistance should be less than 2 ohm.

6. If any check above does not meet the specifications, check connectors and wiring

between the sensor and PCM. If ok, replace PCM.

7. If all checks above meet the specifications, measure the resistance between the sensor side connector terminals 1 and 2. There should be resistance of 50–72 kilohms.

➡**Check that the circuit is not open circuit.**

8. If no continuity, replace the sensor.
9. If sensor has continuity, remove the sensor.
10. With the temperature sensing portion of sensor immersed in hot water, check resistance.
11. Resistance should be as follows:
    - 5.1–6.5 Kilohms at 32°F
    - 2.1–2.7 Kilohms at 68°F
    - 0.9–1.3 Kilohms at 68°F
    - 0.26–0.36 Kilohms 176°F
12. If sensor is not within specification, check the harness for damage, short/opens and proper voltage/ground. If the harness is not damaged, replace the sensor.

FUEL LEVEL SENDING UNIT

LOCATION

See Figure 102.

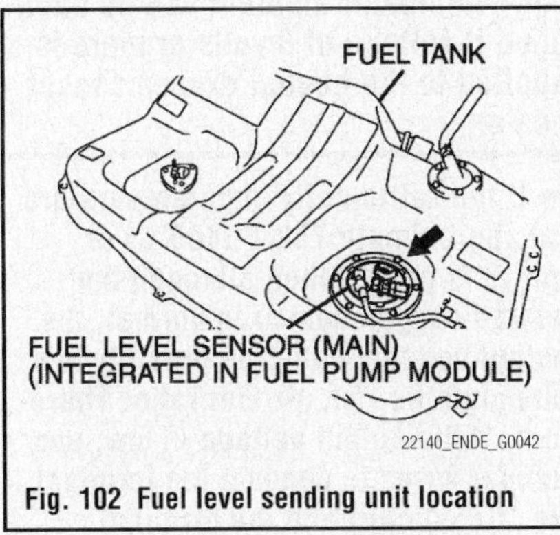

Fig. 102 Fuel level sending unit location

Refer to the accompanying illustration for fuel level sending unit location.

OPERATION

The fuel level sending unit consists of a variable resistor that changes resistance based on the fuel level in the tank. The PCM monitors the voltage across the sender resistance in order to determine the fuel level.

REMOVAL & INSTALLATION

1. Disconnect the negative battery cable.

⁂ CAUTION

The fuel injection system remains under pressure even after the engine has been turned OFF. Properly relieve fuel pressure before disconnecting any fuel lines. Failure to do so may result in fire or personal injury. Do not allow fuel spray or fuel vapors to come in contact with a spark or open flame. Keep a dry chemical fire extinguisher nearby. Never store fuel in an open container due to risk of fire or explosion.

2. Remove the left rear seat cushion and tie out of the way to the head restraint.
3. Push on the floor mat to find the notches, then cut the carpet along the notches to access the service hole cover. Remove the service hole cover.
4. Remove the fuel pump module connector.
5. Remove the fuel tank differential pressure sensor connector.
6. Remove the fuel high-pressure hose.
7. Remove the mounting nuts and plate and fuel pump module assembly.

⁂ WARNING

When removing the fuel pump module from the tank, be careful not to damage the module unit and float.

To install:

8. Install the fuel pump assembly into the fuel tank. torque the nuts to 24 inch lbs. (2.5 Nm)..
9. Install the fuel lines.
10. Install the fuel tank differential pressure sensor connector.
11. Install the fuel pump module connector.
12. Install the fuel pump cover. torque the bolts to 14 inch lbs. (1.5 Nm)..
13. Install the rear floor carpeting..
14. Connect the negative battery cable. Turn the ignition **ON** and then **OFF**, and keep it off for at least ten seconds.
15. Start the vehicle, check for leaks and proper operation.

TESTING

Resistance Test

See Figures 103 and 104.

1. Remove the fuel level sensor.
2. Check the resistances between fuel level sensor main terminals 1 and 2 as well as sub terminals 1 and 2 of the fuel pump module, respectively. Compare with the accompanying chart.
3. Check that resistance value changes smoothly when the float moves slowly between points.
4. If all checks are correct, go to fuel unit height check. If any check is not

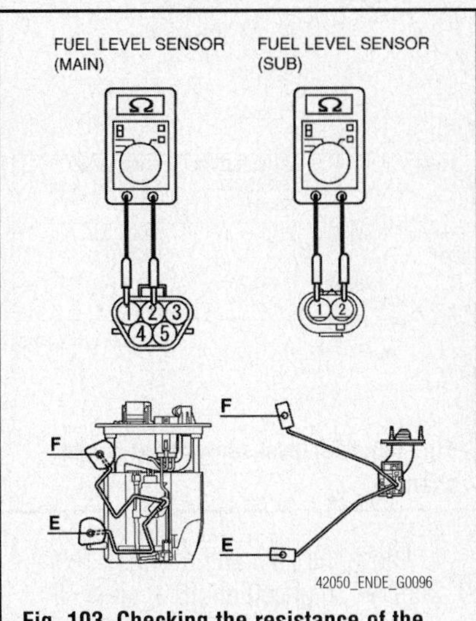

Fig. 103 Checking the resistance of the fuel level sensor

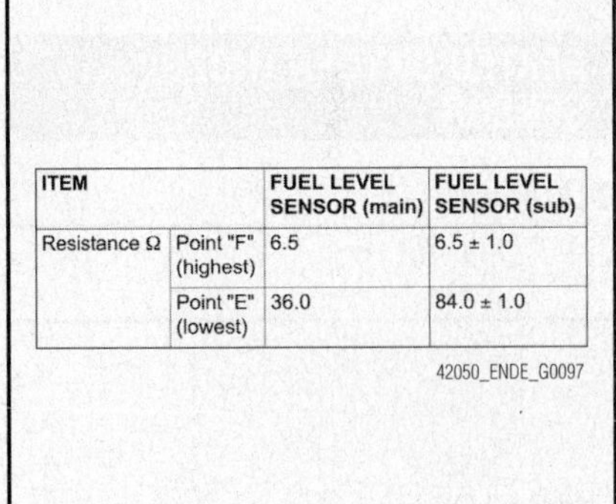

ITEM		FUEL LEVEL SENSOR (main)	FUEL LEVEL SENSOR (sub)
Resistance Ω	Point "F" (highest)	6.5	6.5 ± 1.0
	Point "E" (lowest)	36.0	84.0 ± 1.0

42050_ENDE_G0097

Fig. 104 Fuel level sensor resistance chart

correct, replace the fuel pump module or fuel level sensor .

Float Height Check

See Figures 105 and 106.

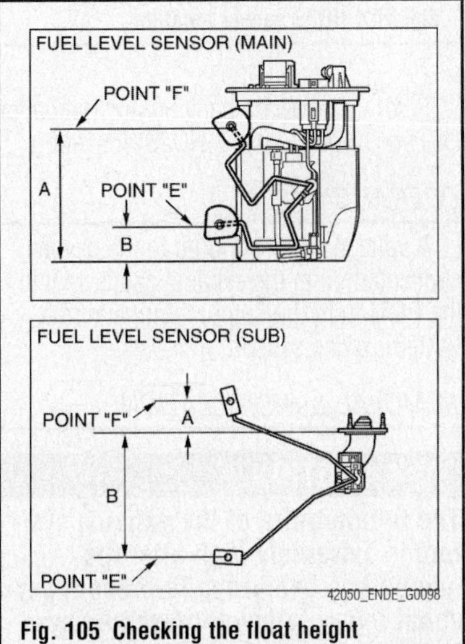

Fig. 105 Checking the float height

ITEM		FUEL LEVEL SENSOR (main)	FUEL LEVEL SENSOR (sub)
Height mm (in)	A at point "F"	34.7 (1.4)	30.5 ± 3.0 (1.2 ± 0.1)
	B at point "E"	149.3 (5.9)	177.7 ± 3.0 (7.0 ± 0.1)

42050_ENDE_G0099

Fig. 106 Fuel level sensor float height chart

1. Move the float and measure height A at point "F" (highest) and B at point "E" (lowest). Compare with the accompanying chart.
2. If any check is not correct, replace the fuel pump module or fuel level sensor .

HEATED OXYGEN (HO2S) SENSOR

LOCATION

See Figure 107.

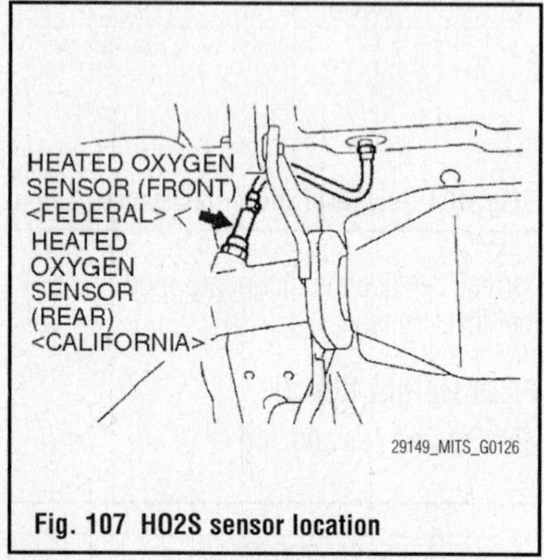

Fig. 107 HO2S sensor location

Refer to the accompanying illustration for Heated Oxygen Sensor (HO2S) location.

OPERATION

A voltage corresponding to the oxygen concentration in the exhaust gas is sent to the PCM from the output terminal of the heated oxygen sensor.

REMOVAL & INSTALLATION

CAUTION

The temperature of the exhaust system is extremely high after the engine has been run. To prevent personal injury, allow the exhaust system to cool completely before removing sensor from the exhaust system.

1. Raise and safely support the vehicle.
2. Disconnect the sensor electrical connector.
3. Using oxygen sensor socket, remove the oxygen sensor.

To install:

4. If installing old oxygen sensor, coat the threads with anti-seize compound. New sensors are already coated. Take care not to contaminate the oxygen sensor probe with the anti-seize compound.
5. Install the oxygen sensor into the exhaust manifold. Tighten the sensor, using the correct tool, to 33 ft. lbs. (45 Nm)
6. Connect the sensor electrical connector.
7. Lower the vehicle.

TESTING

Using a Scan Tool

1. Connect a scan tool to the data link connector.
2. Start the engine and run at idle.
3. Set scan tool to the data reading mode for the heated oxygen sensor.
4. Warm up the engine.
5. After increasing the output voltage 0.5 volt or more by the engine revving, confirm that the output voltage reduces to 0.2 volt or less within 6 seconds when the engine is allowed to return to idle.
6. If sensor is not within specification, check the harness for damage, short/opens and proper voltage/ground. If the harness is not damaged, replace the sensor.

Using Special Tool

See Figure 108.

Front

1. Disconnect the heated oxygen sensor connector and connect special tool MD998464 to the connector on the heated oxygen sensor side.
2. Make sure that there is continuity [4.5–8.0 ohms at 68°F (20°C)] between terminal No. 1 (red clip) and terminal No. 3 (blue clip) on the heated oxygen sensor connector.
3. If there is no continuity, replace the heated oxygen sensor.
4. Warm up the engine until engine coolant is 176°F (80°C) or higher.
5. Rev the engine for 5 minutes or more with the engine speed of 2,000 rpm.
6. Connect a digital voltage meter between terminal No. 2 (black clip) and terminal No. 4 (white clip).

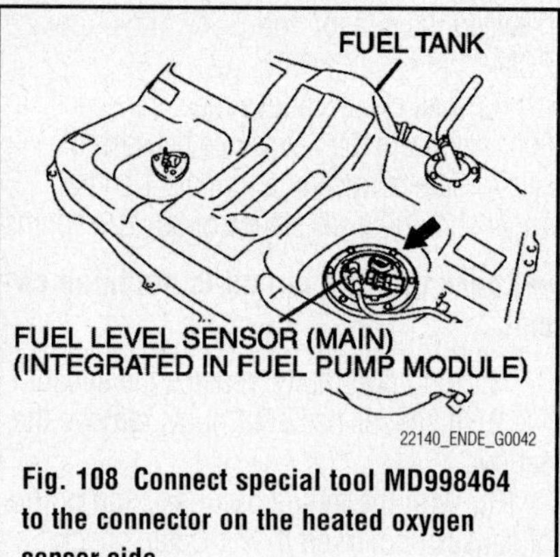

Fig. 108 Connect special tool MD998464 to the connector on the heated oxygen sensor side

7. While repeatedly revving the engine, measure the heated oxygen sensor output voltage.
8. Oxygen sensor output voltage should be 0.6–1.0 volts.
9. If you make the air/fuel ratio rich by revving the engine repeatedly, a normal heated oxygen sensor will output a voltage of 0.6–1.0 volt.

WARNING

Be very careful when connecting the jumper wire; incorrect connection can damage the heated oxygen sensor. The heater element will be damaged if voltage of 8 volts or more is applied to the heated oxygen sensor heater.

➡ If the sufficiently high temperature [of approximate 752°F (400°C) or more] is not reached although the heated oxygen sensor is normal, the output voltage would be possibly low although the rich air/fuel ratio. Therefore, if the output voltage is low, use jumper wires to connect the terminal No. 1 (red clip) and the terminal No. 3 (blue clip) of the heated oxygen sensor with a (+) terminal and (-) terminal of 8 volts power supply respectively, then check again.

10. If sensor is not within specification, check the harness for damage, short/opens and proper voltage/ground. If the harness is not damaged, replace the sensor.

Rear

1. Disconnect the heated oxygen sensor connector and connect special tool MD998464 to the connector on the heated oxygen sensor side.
2. Make sure that there is continuity [11–18 ohms at 68° F (20°C)] between

terminal No. 1 (red clip) and terminal No. 3 (blue clip) on the heated oxygen sensor connector.

3. If there is no continuity, replace the heated oxygen sensor.

4. Warm up the engine until engine coolant is 176°F (80° C) or higher.

5. Drive at 31mph (50 km/h) or more for 10 minutes.

6. Connect a digital voltage meter between terminal No. 2 (black clip) and terminal No. 4 (white clip).

7. Measure the output voltage of the heated oxygen sensor under the following driving conditions:

- Transaxle in drive position **2ND**
- Drive with wide open throttle
- Engine: 3,500 rpm or more

8. Oxygen sensor output voltage should be 0.6–1.0 volts.

9. If you make the air/fuel ratio rich by revving the engine repeatedly, a normal heated oxygen sensor will output a voltage of 0.6–1.0 volt.

➡ If the sufficiently high temperature [of approximate 400°C (752°F) or more] is not reached although the heated oxygen sensor is normal, the output voltage would be possibly low although the rich air/fuel ratio.

10. If sensor is not within specification, check the harness for damage, short/opens and proper voltage/ground. If the harness is not damaged, replace the sensor.

INTAKE AIR TEMPERATURE (IAT) SENSOR

LOCATION

See Figure 109.

Refer to the accompanying illustration for Intake Air Temperature (IAT) sensor location.

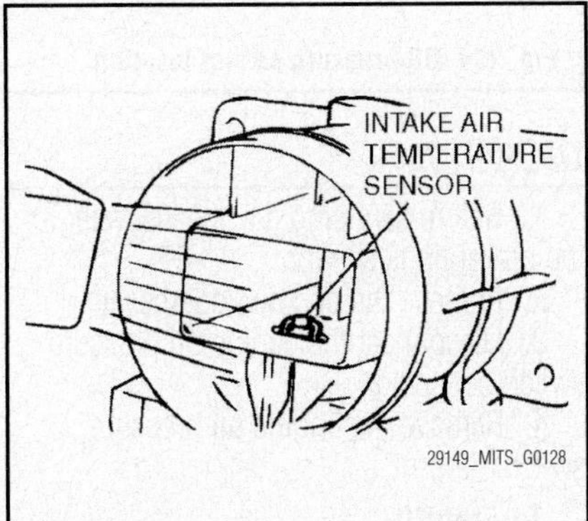

Fig. 109 Intake Air Temperature (IAT) Sensor

OPERATION

Approximately 5 volts are applied to the Intake Air Temperature (IAT) Sensor output terminal from the PCM via the resistor in the PCM. The ground terminal is grounded with PCM. The sensor is a negative temperature coefficient type of resistor. When the intake air temperature rises, the resistance decreases. The sensor output voltage increases when the resistance increases and decreases when the resistance decreases.

REMOVAL & INSTALLATION

1. Disconnect the sensor electrical connector.

2. Remove the sensor from the air intake duct.

To install:

3. Install the sensor on the air intake duct.

4. Connect the electrical connector at the sensor.

TESTING

1. Connect a scan tool to the data link connector.

2. Set the scan tool to the data reading mode for the sensor.

3. The intake air temperature and temperature shown on the scan tool should match.

4. If not, disconnect the intake air temperature sensor connector.

5. Measure the resistance between intake air Temperature sensor side connector terminal No. 1 and No. resistance should be 0.30–20 kilohms.

6. If sensor is not within specification, check the harness for damage, short/opens and proper voltage/ground. If the harness is not damaged, replace the sensor.

KNOCK SENSOR (KS)

LOCATION

See Figure 110.

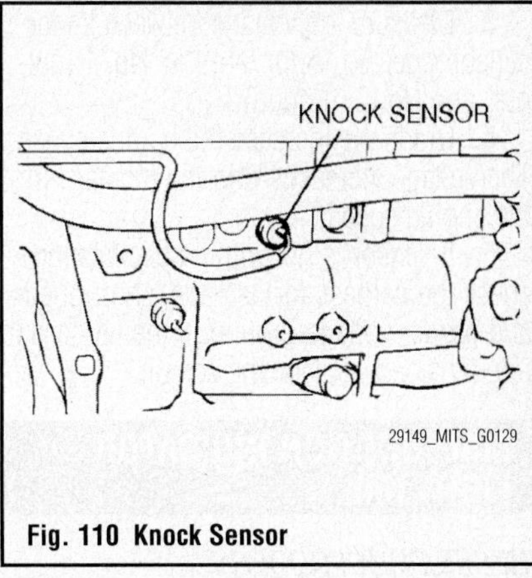

Fig. 110 Knock Sensor

Refer to the accompanying illustration for Knock Sensor (KS) location.

OPERATION

The Knock Sensor (KS) converts the vibration of the cylinder block into a voltage and outputs it. If there is a malfunction of the knock sensor, the voltage output will not change. The PCM checks whether the voltage output changes.

REMOVAL & INSTALLATION

See Figure 111.

1. Disconnect the electrical connector at the sensor.

2. Remove the sensor from the engine block.

To install:

3. Install the knock sensor in the opening in the engine block and tighten the sensor retainer to 17 ft. lbs. (23 Nm).

4. Attach the electrical connector to the sensor.

TESTING

1. Disconnect the knock sensor connector.

2. Start the engine and run at idle.

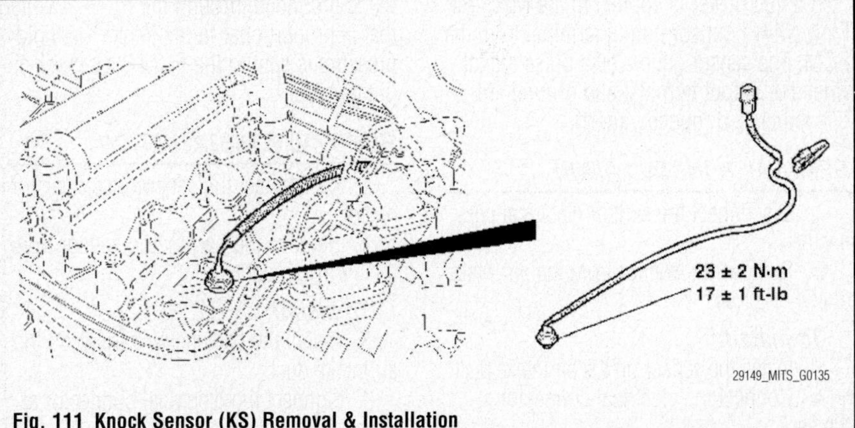

Fig. 111 Knock Sensor (KS) Removal & Installation

3. Measure the voltage between knock sensor side connector terminal No. 1 (output) and No. 2 (ground).
4. Gradually increase the engine speed. The voltage increases with the increase in the engine speed.
5. If sensor is not within specification, check the harness for damage, short/opens and proper voltage/ground. If the harness is not damaged, replace the sensor.

MALFUNCTION INDICATOR LIGHT (MIL)

RESET PROCEDURES

1. Connect a scan tool to the diagnostic connector.
2. Clear DTCs.
3. The MIL should turn **OFF**.

MASS AIR FLOW (MAF) SENSOR

LOCATION

See Figure 112.

Refer to the accompanying illustration for Mass Air Flow (MAF) sensor location.

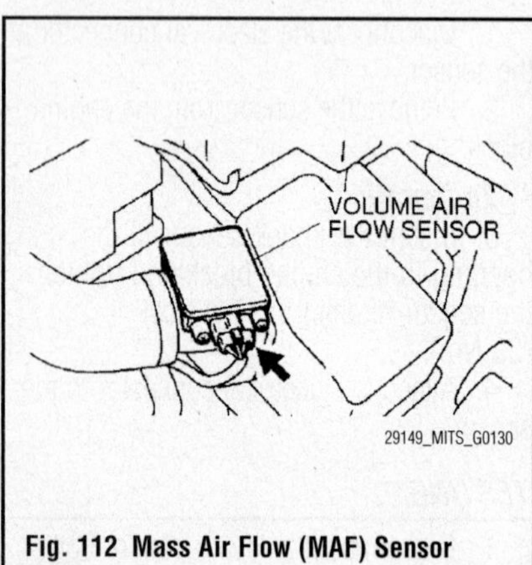

Fig. 112 Mass Air Flow (MAF) Sensor

OPERATION

A 5 volt power is applied to the Mass Air Flow (MAF) Sensor output terminal from the PCM. The sensor generates a pulse signal when the output terminal and ground are opened/closed (opened/short).

REMOVAL & INSTALLATION

1. Disconnect the sensor electrical connector.
2. Remove the sensor from the air intake duct.

To install:

3. Install the sensor on the air intake duct.
4. Connect the electrical connector at the sensor.

TESTING

1. Connect a suitable scan tool to read data.
2. Start the engine and run at idle. Warm up the engine to normal operating temperature: 176° to 203°F.
3. The standard value during idling should be between 1,360 and 1,650 millivolts. When the engine is revved, the mass airflow rate should increase according to the increase in engine speed.
4. If sensor is not within specification, check the harness for damage, short/opens and proper voltage/ground. If the harness is not damaged, replace the sensor.

MANIFOLD ABSOLUTE PRESSURE (MAP) SENSOR

LOCATION

See Figure 113.

Refer to the accompanying illustration for Manifold Absolute Pressure (MAP) sensor location.

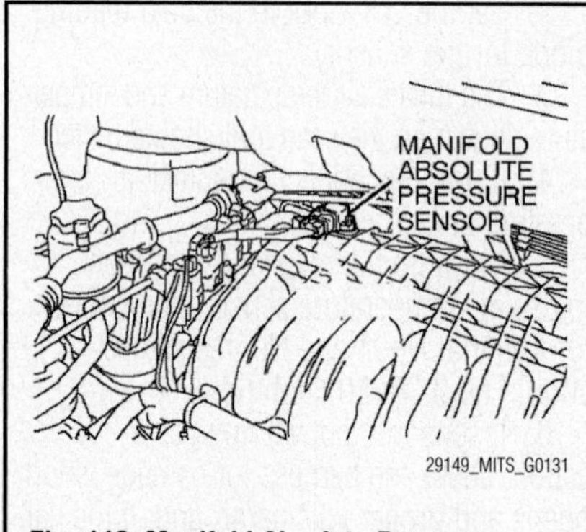

Fig. 113 Manifold Absolute Pressure Sensor

OPERATION

A 5 volt voltage is supplied to the Manifold Absolute Pressure (MAP) Sensor power terminal from the PCM. The ground terminal is grounded through the PCM. A voltage that is proportional to the intake manifold pressure is sent to the PCM the sensor output terminal.

REMOVAL & INSTALLATION

1. Disconnect the sensor electrical connector.
2. Remove the barometric sensor from the air intake duct.

To install:

3. Install the barometric sensor on the air intake duct.
4. Connect the electrical connector at the sensor.

TESTING

1. Connect a scan tool to the data link connector.
2. Set the scan tool to the data reading mode the sensor.
3. Sensor data should read as follows:
   - When altitude is 0 foot (0 m), 29.8 in. Hg (101 kPa).
   - When altitude is 1,969 feet (600 m), 28.1 in. Hg (95 kPa).
   - When altitude is 3,937 feet (1,200 m), 26.0 in. Hg (88 kPa).
   - When altitude is 5,906 feet (1,800 m), 23.9 in. Hg (81 kPa).
4. Start the engine.
5. Sensor data should read as follows:
   - When the engine is idling, 4.7–10.6 in. Hg (16 - 36 kPa).
   - When the engine is suddenly revved, manifold absolute pressure varies.
6. If sensor is not within specification, check the harness for damage, short/opens and proper voltage/ground. If the harness is not damaged, replace the sensor.

OIL PRESSURE SENSOR

LOCATION

See Figure 114.

Refer to the accompanying illustration for oil pressure sensor location.

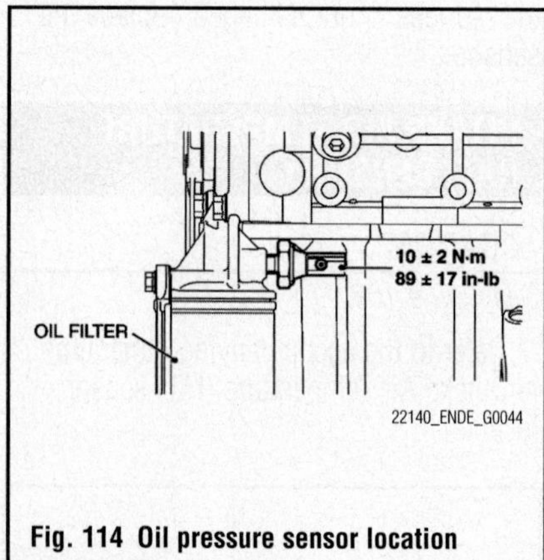

Fig. 114 Oil pressure sensor location

OPERATION

1. Before servicing the vehicle, refer to the precautions section.
2. Remove the engine undercover.
3. Disconnect the engine oil pressure switch connector.
4. Remove the engine oil pressure switch.

To install:

5. Apply Three bond 1215 or equivalent sealant to 2 or 3 threads of the sensor.

6. Install the sensor and tighten to 89 inch lbs. (10 Nm).

WARNING

Do not start the engine for 2 hour or more after the engine oil pressure switch installation.

7. Connect the engine oil pressure switch connector.
8. Install the engine undercover.

TESTING

1. Disconnect the sender electrical connection.
2. Connect one lead of a ohmmeter to the sender terminal and the other lead to the sender body.
3. With the engine off the resistance should be low.
4. Start the engine. The resistance should increase as the engine speed increases.
5. If sensor is not within specification, replace the sensor.

POWERTRAIN CONTROL MODULE (PCM)

LOCATION

See Figure 115.

Refer to the accompanying illustration for Powertrain Control Module (PCM) location.

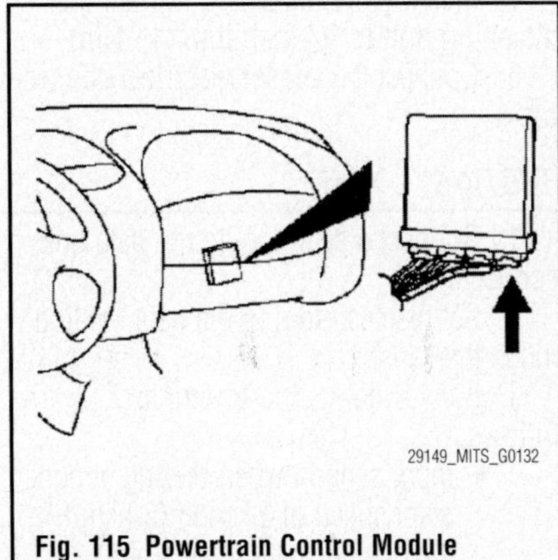

Fig. 115 Powertrain Control Module

OPERATION

The powertrain has electronic controls to reduce exhaust emissions while maintaining excellent driveability and fuel economy. The Powertrain Control Module (PCM) is the control center of this system. The PCM monitors numerous engine and vehicle functions. The PCM constantly monitors the information from various sensors and other inputs, and controls the systems that affect vehicle performance and emissions.

The PCM also performs the diagnostic tests on various parts of the system. The PCM can recognize operational problems and alert the driver via the Malfunction Indicator Lamp (MIL). When the PCM detects a malfunction, the PCM stores a Diagnostic Trouble Code (DTC). The problem area is identified by the particular DTC that is set. The control module supplies a buffered voltage to various sensors and switches. Review the components and wiring diagrams in order to determine which systems are controlled by the PCM.

REMOVAL & INSTALLATION

1. Turn the ignition switch to the **OFF** position.
2. If the ECU is mounted under the dash, remove the left and/or right side panel from the center console, or remove the under dash panel.
3. Remove the bolts holding the ECU to the mounting bracket.
4. Disconnect the wiring harness from the ECU and remove ECU from the vehicle.

To install:

5. Connect the electrical harness to the ECU. Make certain the multi–pin connector is firmly and squarely seated to the ECU.
6. Install the ECU in the mounting bracket and secure in position.
7. If necessary, install the side panels to the center console or dash panel.
8. Connect the negative battery cable.

TESTING

1. Connect a scan tool to the data link connector.
2. When the power of Vehicle Communications Interface (VCI) is turned **ON**, the indicator lamp of the VCI illuminates in green.
3. If the indicator lamp of the VCI does not illuminate in green, use the scan tool to perform CAN bus line diagnosis.
4. Measure the battery positive voltage during cranking. Voltage should be 8 volts or more.
5. If voltage is less than 8 volts, check the battery.
6. Measure the power supply voltage at data link connector between terminals 16–ground. Voltage should be battery positive voltage.
7. If voltage is less, repair the open circuit between battery and data link connector terminal No. 16.
8. Check for the continuity between data link connector terminals 4–5 and ground. Resistance should be less than 2 ohms.
9. If resistance is not as specified, repair the open circuit or harness damage between the data link connector and ground.

THROTTLE POSITION SENSOR (TPS)

LOCATION

See Figure 116.

Refer to the accompanying illustration for Throttle Position Sensor (TPS) location.

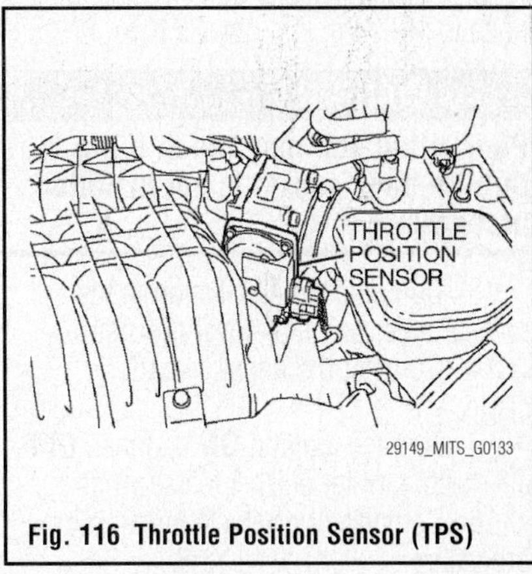

Fig. 116 Throttle Position Sensor (TPS)

OPERATION

A 5 volt power supply is applied to the Throttle Position Sensor (TPS) power terminal from the PCM. The ground terminal is grounded with PCM. When the throttle valve shaft is turned from the idle position to the fully opened position, the resistance between the TPS output terminal and the ground terminal would increase according to the rotation.

REMOVAL & INSTALLATION

The throttle position sensor is a integral part of the throttle body.

1. Before servicing the vehicle, refer to the Precautions Section.
2. Properly relieve the fuel system pressure.
3. Drain the engine coolant.
4. Remove the air intake hose.
5. Remove the battery.
6. Disconnect the throttle position sensor connector.
7. Disconnect the water hose connection.
8. Remove the throttle body retaining bolts.
9. Remove the throttle body from the engine.
10. Discard the gasket.

To install:

⁂ WARNING

Do not loosen the retaining screws for the resin cover of the throttle body assembly. If the screws are loosened, the sensor incorporated in the resin cover becomes misaligned and the throttle body cannot work properly.

11. Align the recess on the intake manifold plenum with the projection of the throttle body gasket.
12. Install the gasket. Install the throttle body to the engine and tighten the retaining bolts to 18–24 ft. lbs. (24–32 Nm).

⁂ WARNING

Poor idling etc. may result if the throttle body gasket is not installed properly.

13. Continue the installation in the reverse order of the removal procedure.
14. Connect the negative battery cable.
15. Turn the ignition **ON** and then **OFF**, and keep it off for at least ten seconds.
16. Complete the vehicle initialization procedure.

➡To complete the initialization procedure the following tools are needed. MB991958 scan tool, MB991824 VCI, MB991827 MUT III USB cable, MB991910 MUT III Main harness "A".

a. Connect the scan tool to the data link connector. To prevent damage to the scan tool be sure that the ignition switch is in the **LOCK** position before connecting the scan tool.
b. Turn the ignition switch to the **ON** position.
c. Select "check mode" from the menu screen.
d. Select "erase memory" from the menu screen.
e. Initialize the learning value.
f. After initialization complete the idle learning procedure.

➡This procedure must be performed when the PCM is replaced, or when the learning value is initialized, as the idling is not stabilized because the learning value in the MFI engine is not completed.

g. Start the engine. Allow the coolant temperature to reach 176 degrees or more.
h. Stop the engine and place the ignition switch in the **LOCK** position.
i. After ten seconds, restart the engine.
j. For ten minutes carry out the idling procedure below to confirm that the engine has normal idling:
- Position the transaxle selector lever in the **P** range
- The engine fan is not to be operated
- The engine coolant temperature should be 176 degrees or more

➡If the engine stalls during idling, check the throttle valve of the throttle body for dirt. Correct and perform the procedure again.

TESTING

1. Connect a scan tool to the data link connector.
2. Set the scan tool to the data reading mode the sensor.
3. Disconnect the intake air hose at the throttle body.
4. Disconnect the TPS connector.
5. Use test harness special tool (MB991658) to connect only terminals No. 3, No. 4, No. 5, and No. 6.
6. Set scan tool to the data reading mode for the sensor.
7. For the main sensor, output voltage should measure between 300 and 700 millivolts when the throttle valve is fully closed with your finger. Output voltage should measure 4.0 volts or more when the throttle valve is fully open with your finger.
8. For the sub sensor, output voltage should measure between 2,200 and 2,800 millivolts when the throttle valve is fully closed with your finger. Output voltage should measure 4,000 millivolts or more when the throttle valve is fully open with your finger.
9. If sensor is not within specification, check the harness for damage, short/opens and proper voltage/ground. If the harness is not damaged, replace the sensor.

VEHICLE SPEED SENSOR (VSS)

LOCATION

See Figure 117.

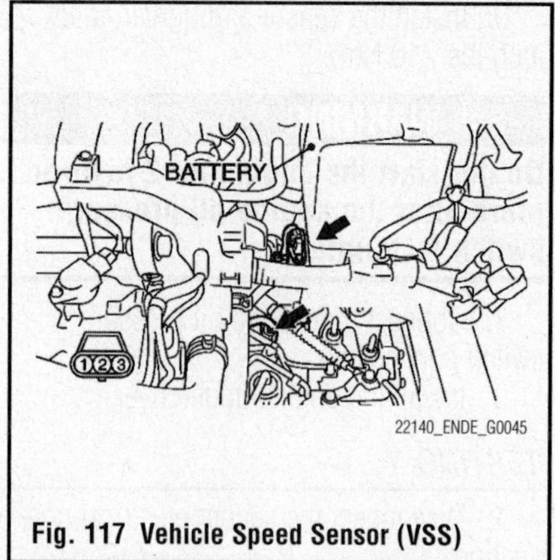

Fig. 117 Vehicle Speed Sensor (VSS)

Refer to the accompanying illustration for Vehicle Speed Sensor (VSS) location.

OPERATION

A 5 volt voltage is applied to the Vehicle Speed Sensor (VSS) output terminal from the PCM. The sensor generates a pulse signal when the output terminal is opened and grounded.

REMOVAL & INSTALLATION

1. Before servicing the vehicle, refer to the precautions section.
2. Disconnect the sensor electrical connector.
3. Remove the sensor attaching bolt/screw.
4. Remove the sensor.

To install:

5. Install the sensor and tighten the attaching bolt to 97 inch lbs. (11 Nm).
6. Connect the sensor electrical connector.

TESTING

1. Connect a scan tool to the data link connector.
2. Set the scan tool to the data reading mode the sensor.
3. Sensor data should read as follows:
- Input sensor: when driving at constant speed of 31 mph (50 km/h), the display should be "1,400 - 1,700 rpm" in 3rd gear.

4. If sensor is not within specification, check the harness for damage, short/opens and proper voltage/ground. If the harness is not damaged, replace the sensor.

FUEL SYSTEM SERVICE PRECAUTIONS

Safety is the most important factor when performing not only fuel system maintenance but any type of maintenance. Failure to conduct maintenance and repairs in a safe manner may result in serious personal injury or death. Maintenance and testing of the vehicle's fuel system components can be accomplished safely and effectively by adhering to the following rules and guidelines.

- To avoid the possibility of fire and personal injury, always disconnect the negative battery cable unless the repair or test procedure requires that battery voltage be applied.
- Always relieve the fuel system pressure prior to disconnecting any fuel system component (injector, fuel rail, pressure regulator, etc.), fitting or fuel line connection. Exercise extreme caution whenever relieving fuel system pressure to avoid exposing skin, face and eyes to fuel spray. Please be advised that fuel under pressure may penetrate the skin or any part of the body that it contacts.
- Always place a shop towel or cloth around the fitting or connection prior to loosening to absorb any excess fuel due to spillage. Ensure that all fuel spillage (should it occur) is quickly removed from engine surfaces. Ensure that all fuel soaked cloths or towels are deposited into a suitable waste container.
- Always keep a dry chemical (Class B) fire extinguisher near the work area.
- Do not allow fuel spray or fuel vapors to come into contact with a spark or open flame.
- Always use a back-up wrench when loosening and tightening fuel line connection fittings. This will prevent unnecessary stress and torsion to fuel line piping.
- Always replace worn fuel fitting O-rings with new. Do not substitute fuel hose or equivalent where fuel pipe is installed.

Before servicing the vehicle, make sure to also refer to the general precautions section.

RELIEVING FUEL SYSTEM PRESSURE

See Figure 118.

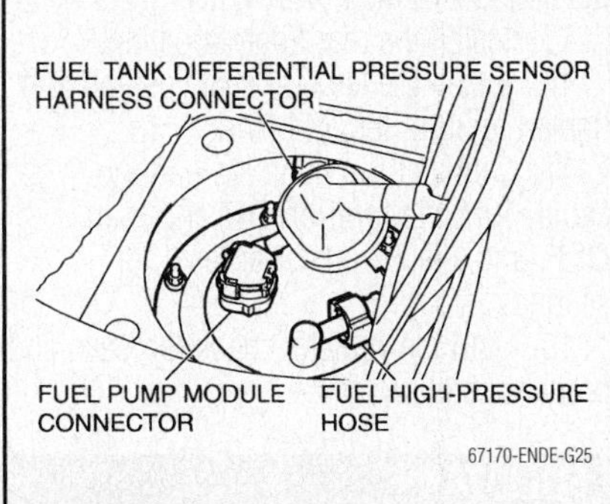

Fig. 118 Fuel pump connector location

**** CAUTION**

The fuel system is under constant pressure, even with the engine off. This pressure must be relieved before disconnecting any fuel system component, fitting or fuel line connection. Failure to do so may result in personal injury.

1. Turn the ignition switch to **LOCK**.
2. Remove the left rear seat cushion mounting bolts, and lift the seat cushion to tie it to the head restraint.
3. Push on the floor mat to find the notches, then cut the carpet along the notches to access the service hole cover. Remove the service hole cover.
4. Detach the fuel pump module connector.
5. Start the engine and let it run out of fuel.
6. Attach the fuel pump module connector.
7. Install the service hole cover, floor mat and left rear seat cushion.

FUEL FILTER

REMOVAL & INSTALLATION

The fuel filter used on this vehicle is an integral part of the fuel pump module and is not normally serviced separately.

FUEL PUMP

REMOVAL & INSTALLATION

See Figures 119 and 120.

1. Before servicing the vehicle, refer to the Precautions Section.
2. Relieve the fuel system pressure.
3. Disconnect the negative battery cable.

**** CAUTION**

The fuel injection system remains under pressure even after the engine has been turned OFF. Properly relieve fuel pressure before disconnecting any fuel lines. Failure to do so may result in fire or personal injury. Do not allow fuel spray or fuel vapors to come in contact with a spark or open flame. Keep a dry chemical fire extinguisher nearby. Never store fuel in an open container due to risk of fire or explosion.

4. Remove the left rear seat cushion and tie out of the way to the head restraint.
5. Push on the floor mat to find the notches, then cut the carpet along the notches to access the service hole cover. Remove the service hole cover.
6. Disconnect the fuel pump module connector.
7. Remove the fuel tank differential pressure sensor connector.

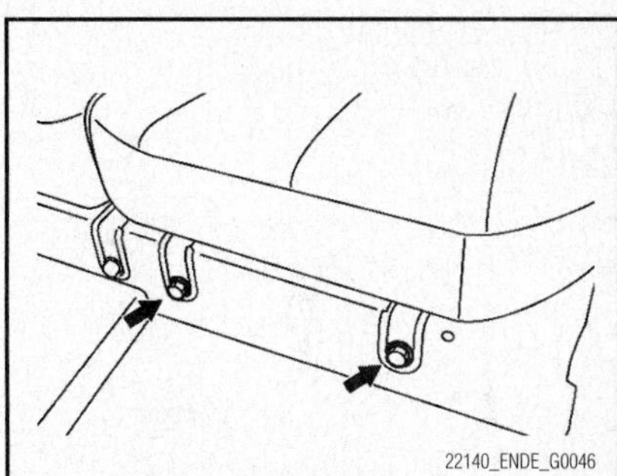

Fig. 119 Remove the left rear seat cushion and tie out of the way to the head restraint.

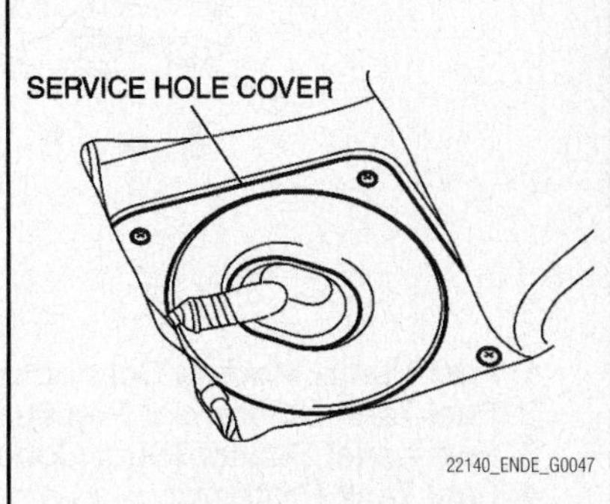

Fig. 120 Push on the floor mat to find the notches, then cut the carpet along the notches to access the service hole cover. Remove the service hole cover

8. Remove the fuel high-pressure hose.
9. Remove the mounting nuts and plate and fuel pump module assembly.

WARNING

When removing the fuel pump module from the tank, be careful not to damage the module unit and float.

To install:

10. Install the fuel pump assembly into the fuel tank. torque the nuts to 24 inch lbs. (2.5 Nm).
11. Install the fuel lines.
12. Install the fuel tank differential pressure sensor connector.
13. Install the fuel pump module connector.
14. Install the fuel pump cover. torque the bolts to 14 inch lbs. (1.5 Nm).
15. Install the rear floor carpeting.
16. Install the left rear seat cushion and tighten to 33 ft. lbs. (44 Nm).
17. Connect the negative battery cable. Turn the ignition **ON** and then **OFF**, and keep it off for at least ten seconds.
18. Start the vehicle, check for leaks and proper operation.

FUEL TANK

REMOVAL & INSTALLATION

See Figures 121 and 122.

1. Before servicing the vehicle, refer to the Precautions Section.
2. Properly relieve the fuel system pressure.

CAUTION

The fuel injection system remains under pressure even after the engine has been turned OFF. Properly relieve fuel pressure before disconnecting any fuel lines. Failure to do so may result in fire or personal injury. Do not allow fuel spray or fuel vapors to come in contact with a spark or open flame. Keep a dry chemical fire extinguisher nearby. Never store fuel in an open container due to risk of fire or explosion.

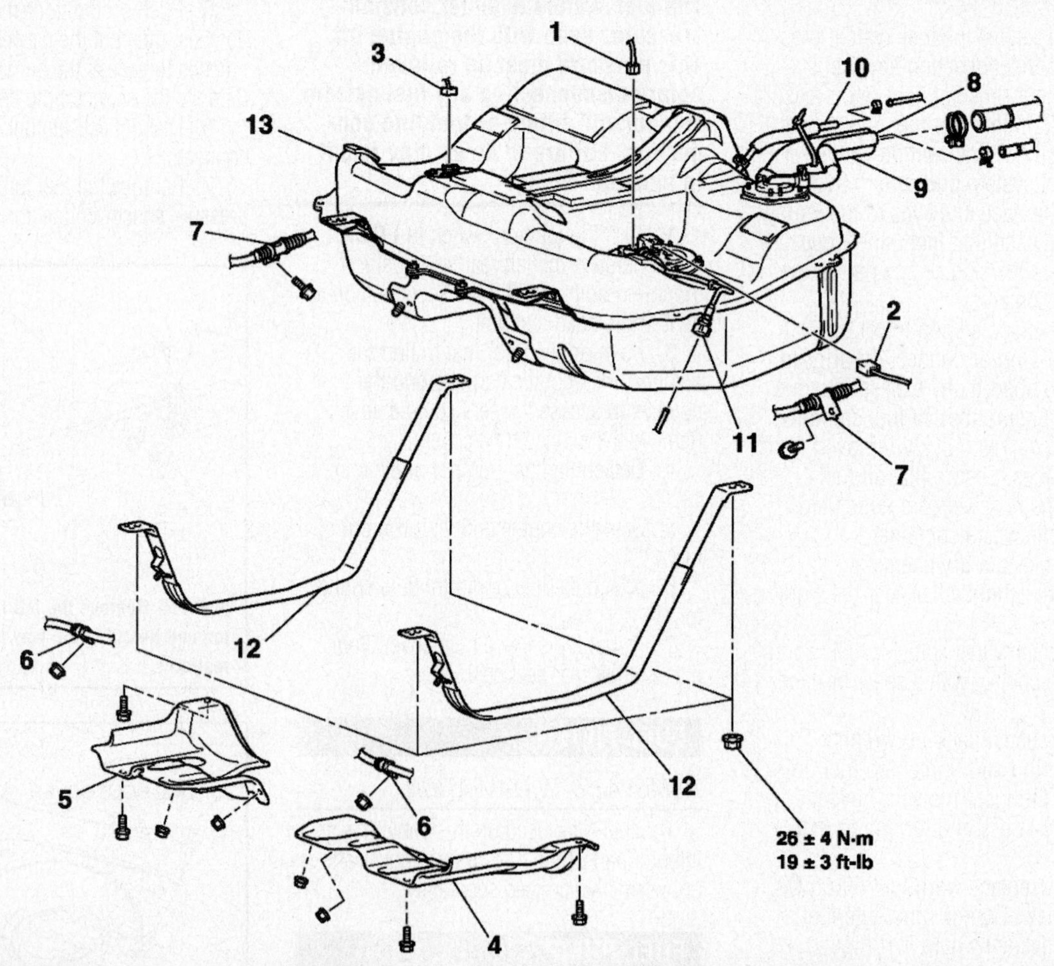

Fig. 121 Exploded view of the fuel tank attaching straps

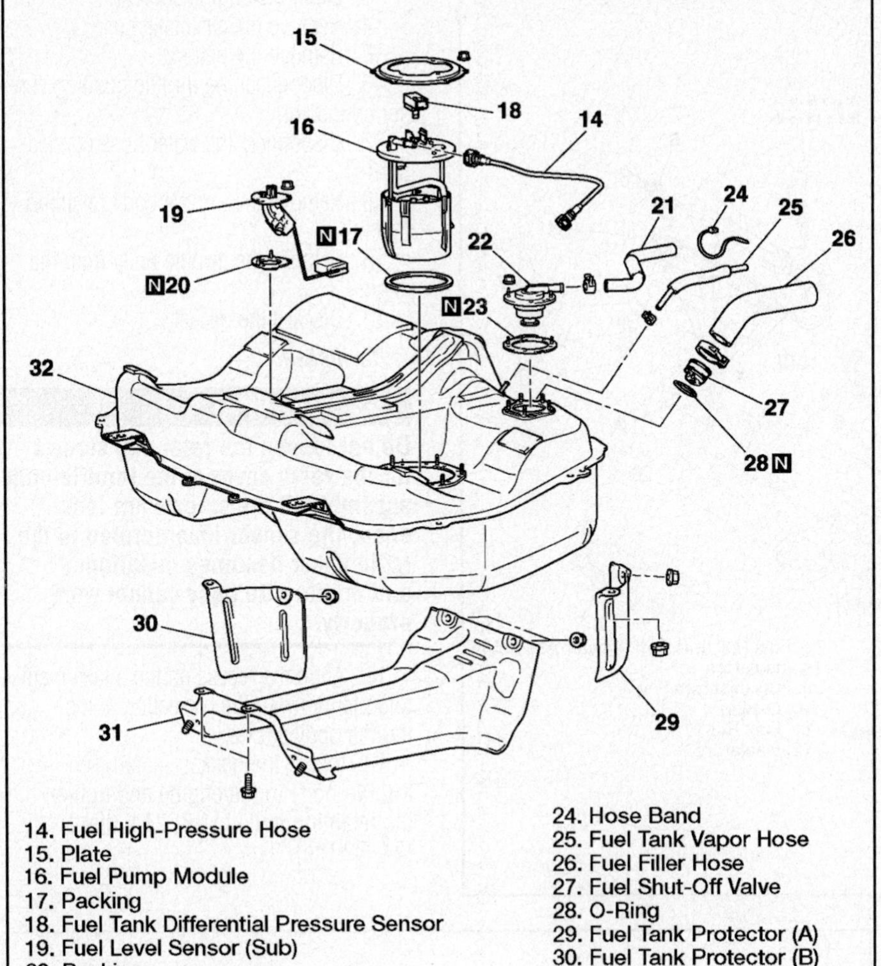

Fig. 122 Exploded view of the fuel tank hoses and components

3. Drain the fuel from the tank.
4. Remove the left rear seat cushion and tie out of the way to the head restraint.
5. Push on the floor mat to find the notches, then cut the carpet along the notches to access the service hole cover. Remove the service hole cover.
6. Disconnect the fuel pump module connector.
7. Disconnect the fuel tank differential pressure sensor connector.
8. Disconnect the fuel level sensor connector.
9. Raise and safely support the vehicle securely on jackstands.
10. Remove the center exhaust pipe.
11. Remove the propeller shaft on AWD vehicles
12. Remove the fuel tank protector.
13. Disconnect the parking brake cable clamp.
14. Disconnect the fuel tank vapor hose connection.
15. Disconnect the fuel filler hose connection.
16. Disconnect the fuel high-pressure hose connection.

**** WARNING**

As there will be some pressure remaining in the fuel pipe line, cover it with a shop towel to prevent fuel from spraying out.

17. Remove the fuel tank band.
    a. Support the fuel tank with a transaxle jack.
    b. On FWD vehicles, remove the fuel tank band and lower the fuel tank assembly gradually to remove it.
    c. On AWD vehicles, remove the front securing nut of the fuel tank band. Tilt the fuel tank assembly forward and lower it gradually to remove it. Remove the fuel tank band.
18. Carefully lower the fuel tank assembly from the vehicle.

To install:

19. Install the fuel tank bands
    a. Raise the fuel tank assembly carefully with a transaxle jack.
    b. Ensure that the fuel tank assembly does not interfere with surrounding parts. Then install the fuel tank band and tighten to 19 ft. lbs. (26 Nm).
20. Connect the fuel high-pressure hose connection.
    a. After installing, slightly pull the fuel high-pressure hose and ensure that there is no disengaged fuel high-pressure hose. Also confirm that there is approximately 0.12 inch (3 mm) play.
21. Connect the fuel filler hose connection.
22. Connect the fuel tank vapor hose connection.
23. Connect the parking brake cable clamp.
24. Install the fuel tank protector.
25. Install the propeller shaft on AWD vehicles
26. Install the center exhaust pipe.
27. Raise and safely support the vehicle securely on jackstands.
28. Connect the fuel level sensor connector.
29. Connect the fuel tank differential pressure sensor connector.
30. Connect the fuel pump module connector.
31. Install the service hole cover.
32. Install the left rear seat cushion.
33. Fill the fuel tank.
34. Start the engine and check for leaks.

FUEL RAIL & INJECTORS

REMOVAL & INSTALLATION

See Figure 123.

1. Before servicing the vehicle, refer to the Precautions Section.
2. Properly relieve the fuel system pressure.
3. Remove the intake manifold plenum.
4. Remove the ignition coil connectors and coils.
5. Remove the injector connectors.
6. Remove the high pressure fuel hose and o-ring.
7. Remove the blow-by hose connection.
8. Remove the fuel rail, injector and damper assembly, as necessary.
9. Remove the insulators from the fuel rail.

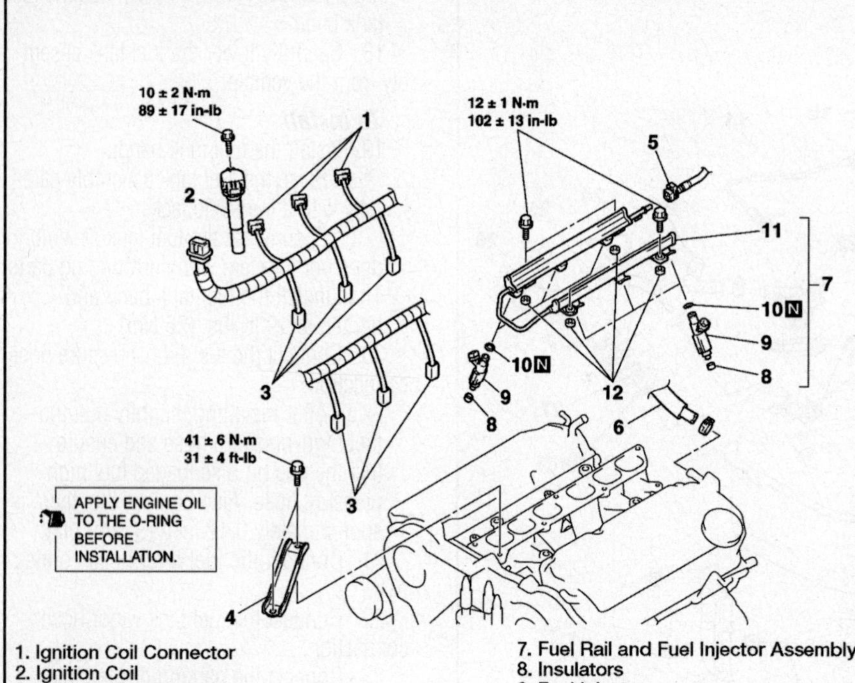

Fig. 123 Fuel rail and related components

To install:

10. Install the fuel rail and insulators. Tighten the bolts to 102 inch lbs. (12 Nm).
11. Install the injector connectors.
12. Install the blow-by hose connection.
13. Install the high pressure fuel hose and o-ring. torque the bolts to 44 inch lbs. (5.0 Nm). .
14. Install the ignition coils and connectors.
15. Install the intake manifold plenum.

IDLE SPEED

ADJUSTMENT

Idle speed is maintained by the Powertrain Control Module (PCM). No adjustment is possible.

THROTTLE BODY

REMOVAL & INSTALLATION

See Figure 124.

The throttle position sensor is a integral part of the throttle body.

1. Before servicing the vehicle, refer to the Precautions Section.
2. Properly relieve the fuel system pressure.
3. Drain the engine coolant.
4. Remove the air intake hose.
5. Remove the battery.
6. Disconnect the throttle position sensor connector.
7. Disconnect the water hose connection.
8. Remove the throttle body retaining bolts.
9. Remove the throttle body from the engine.
10. Discard the gasket.

To install:

WARNING

Do not loosen the retaining screws for the resin cover of the throttle body assembly. If the screws are loosened, the sensor incorporated in the resin cover becomes misaligned and the throttle body cannot work properly.

11. Align the recess on the intake manifold plenum with the projection of the throttle body gasket.
12. Install the gasket. Install the throttle body to the engine and tighten the retaining bolts to 18–24 ft. lbs. (24–32 Nm).

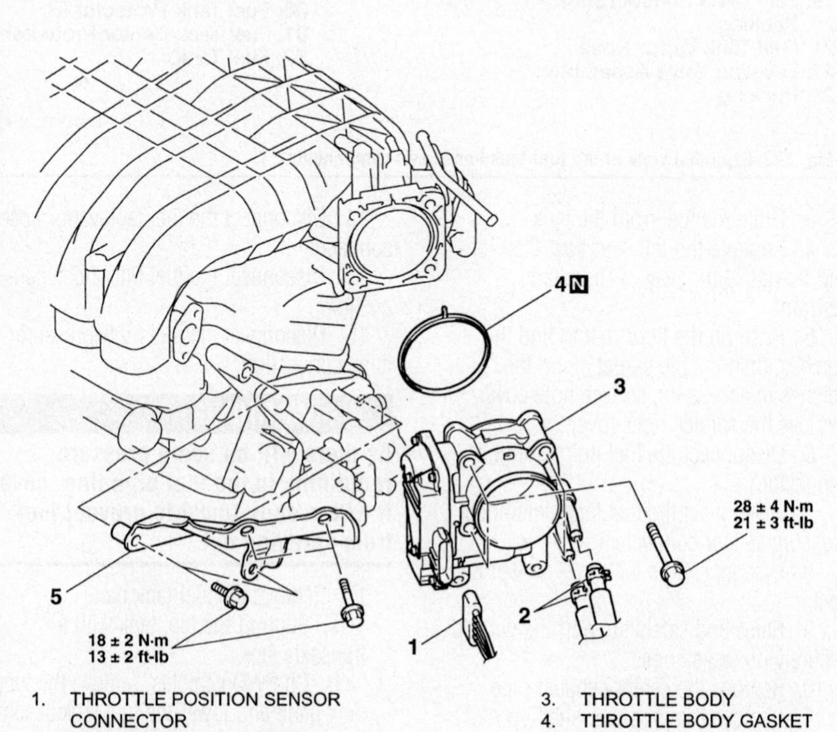

Fig. 124 Throttle body and related components

⁂ WARNING

Poor idling etc. may result if the throttle body gasket is not installed properly.

13. Continue the installation in the reverse order of the removal procedure.
14. Connect the negative battery cable.
15. Turn the ignition **ON** and then **OFF**, and keep it off for at least ten seconds.
16. Complete the vehicle initialization procedure.

➡To complete the initialization procedure the following tools are needed. MB991958 scan tool, MB991824 VCI, MB991827 MUT III USB cable, MB991910 MUT III Main harness "A".

a. Connect the scan tool to the data link connector. To prevent damage to the scan tool be sure that the ignition switch is in the **LOCK** position before connecting the scan tool.
b. Turn the ignition switch to the **ON** position.
c. Select "check mode" from the menu screen.
d. Select "erase memory" from the menu screen.
e. Initialize the learning value.
f. After initialization complete the idle learning procedure.

➡This procedure must be performed when the PCM is replaced, or when the learning value is initialized, as the idling is not stabilized because the learning value in the MFI engine is not completed.

g. Start the engine. Allow the coolant temperature to reach 176 degrees or more.
h. Stop the engine and place the ignition switch in the **LOCK** position.
i. After ten seconds, restart the engine.
j. For ten minutes carry out the idling procedure below to confirm that the engine has normal idling:
- Position the transaxle selector lever in the **P** range
- The engine fan is not to be operated
- The engine coolant temperature should be 176 degrees or more

➡If the engine stalls during idling, check the throttle valve of the throttle body for dirt. Correct and perform the procedure again.

HEATING & AIR CONDITIONING SYSTEM

BLOWER MOTOR

REMOVAL & INSTALLATION

Front

See Figure 125.

1. Before servicing the vehicle, refer to the Precautions Section.
2. Disconnect the blower motor connector.
3. Remove the blower motor attaching screws.
4. Carefully lower the blower motor from the heater unit assembly.

To install:

5. Installation is the reverse of the removal procedure.

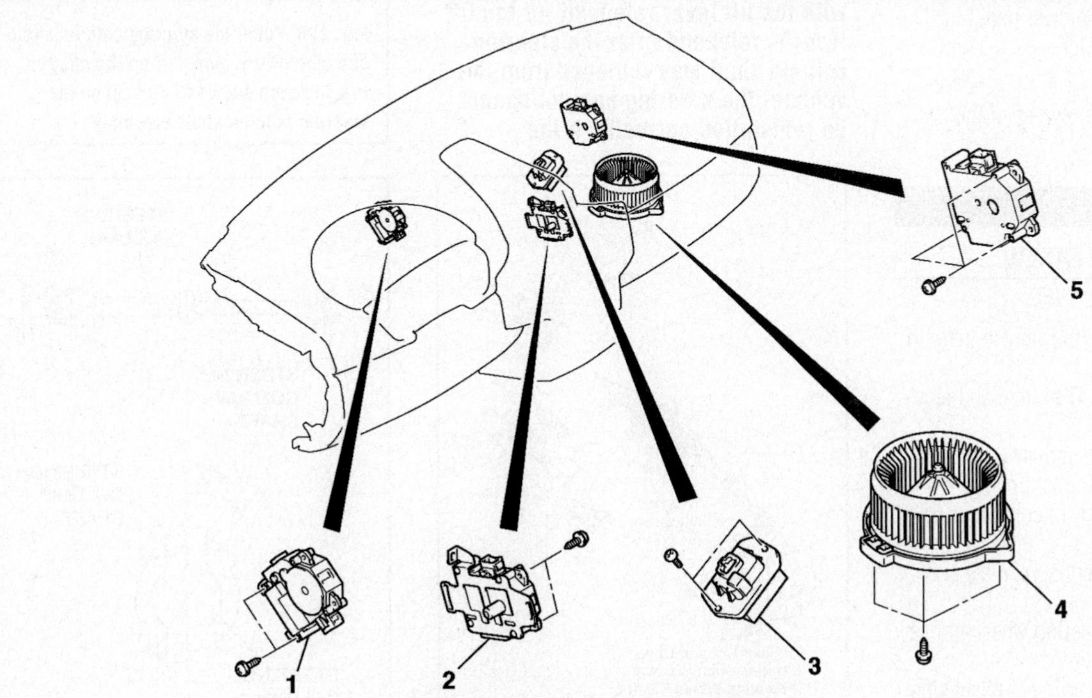

Fig. 125 Blower motor and actuator locations

Rear

See Figure 126.

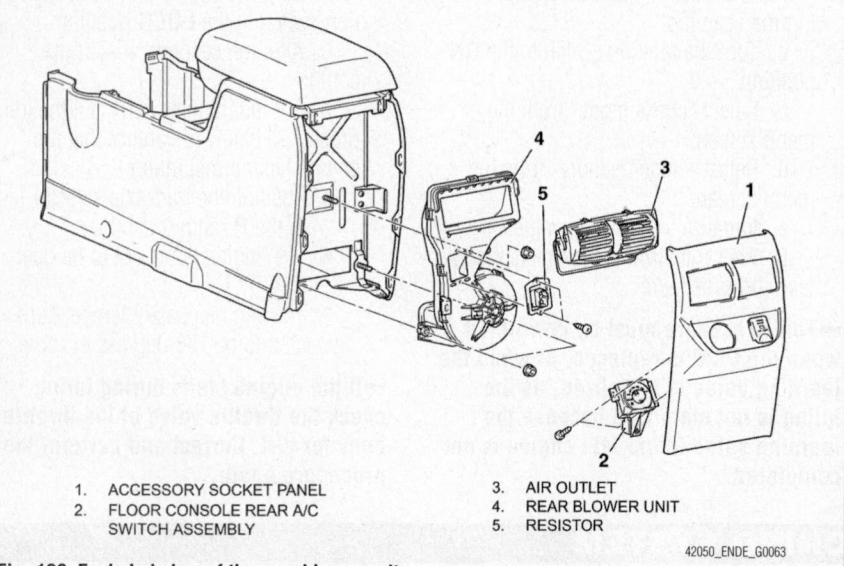

1. ACCESSORY SOCKET PANEL
2. FLOOR CONSOLE REAR A/C SWITCH ASSEMBLY
3. AIR OUTLET
4. REAR BLOWER UNIT
5. RESISTOR

42050_ENDE_G0063

Fig. 126 Exploded view of the rear blower unit

1. Before servicing the vehicle, refer to the Precautions Section.
2. Remove the accessory socket panel.
3. Remove the rear A/C floor console switch assembly.
4. Remove the air outlet.
5. Remove the rear blower unit.
6. Remove the resistor.

To install:

7. Installation is the reverse of the removal procedure.

HEATER CORE

REMOVAL & INSTALLATION

See Figures 127 through 130.

1. Before servicing the vehicle, refer to the Precautions Section.
2. Discharge the A/C system and recycle the refrigerant.
3. Drain the engine coolant.
4. Remove the instrument panel.
5. Remove the steering column as follows:
    a. Remove the front scuff plate and cowl side trim.
    b. Remove the steering wheel and air bag module assembly.
    c. Remove the instrument panel side cover and under cover.
    d. Remove the floor console side cover.
    e. Remove the front floor carpet
    f. Remove the lower and upper column covers.
    g. Remove the steering wheel sensor, clock spring and column switch.
    h. Disconnect the key interlock cable.
    i. Remove the steering shaft pad.

✲✲ WARNING

The tilt lever should be held in the lock position until the steering column shaft is installed to the vehicle. If the steering column is removed with the tilt lever released, or the tilt lever is released after the steering column shaft was removed from the vehicle, the steering column cannot be reinstalled correctly. If the steering column is installed incorrectly, the collision energy absorbing mechanism may be damaged.

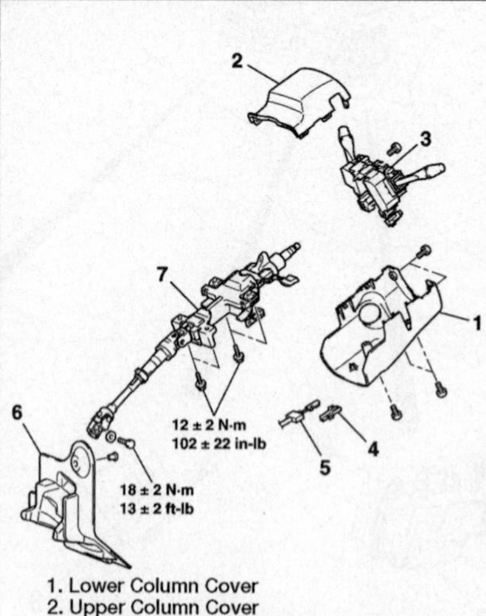

1. Lower Column Cover
2. Upper Column Cover
3. Steering Wheel Sensor, Clock Spring and Column Switch Assembly
4. Cover
5. Key Interlock Cable
6. Steering Shaft Pad
7. Steering Column Shaft Assembly

22140_ENDE_G0056

Fig. 127 Exploded view of the steering column assembly

    j. Ensure that the tilt lever is in the lock position, and remove the steering column mounting bolts.
    k. Pinch the steering column shaft clip with pliers, and pull up the shaft in the direction shown to disengage the steering column shaft assembly.

➡If the steering column shaft is removed accidentally, remove the steering column assembly and be sure to insert the steering column shaft into the steering column as shown.

6. Disconnect the heater hose.

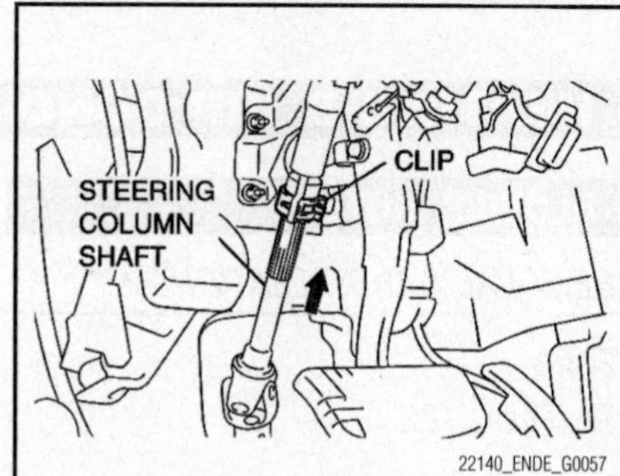

Fig. 128 Pinch the steering column shaft clip with pliers, and pull up the shaft in the direction shown to disengage the steering column shaft assembly

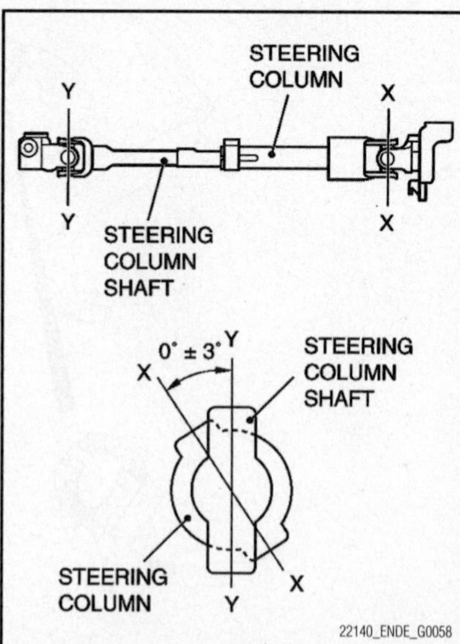

Fig. 129 If the steering column shaft is removed accidentally, remove the steering column assembly and be sure to insert the steering column shaft into the steering column as shown

7. Disconnect the flexible suction hose, then plug the hose and nipple to prevent contaminants from entering the system. Remove and discard the O-ring.

8. Disconnect the flexible discharge hose, then plug the hose and nipple to prevent contaminants from entering the system. Remove and discard the O-ring.

9. Remove the junction block.

10. Remove the deck crossmember cowl top stay.

11. Remove the deck crossmember assembly.

a. Disconnect the following connectors to gain access to the front deck crossmember:

- Audio amplifier
- Instrument panel wiring harness and front door wiring harness (RH) combination
- Capacitor
- Instrument panel wiring harness and front wiring harness combination
- SRS-ECU
- Instrument panel wiring harness and front door wiring harness (LH) combination
- Instrument panel wiring harness and floor wiring harness combination
- Instrument panel wiring harness and front wiring harness combination
- Front wiring harness and floor wiring harness combination
- Instrument panel wiring harness and floor wiring harness combination
- Instrument panel wiring harness and console wiring harness combination
- Floor wiring harness and junction block combination
- Roof wiring harness and junction block combination
- Front wiring harness and junction block combination
- Front wiring harness and junction block combination
- Front wiring harness and junction block combination
- Ignition switch
- Key reminder switch

12. Remove the heater unit assembly.

13. Disassemble the heater unit as follows:

a. Remove the foot light bracket.

b. Remove the air mixing damper control motor and potentiometer.

c. Remove the antenna cable.

d. Remove the console duct c.

e. Remove the deck crossmember.

f. Remove the joint bracket.

g. Remove the outside/inside air selection damper control motor.

h. Remove the front blower motor.

i. Remove the power transistor.

j. Remove the blower case.

k. Remove the console duct a.

l. Remove the console duct b.

m. Remove the aspirator hose.

n. Remove the aspirator.

o. Remove the mode selection damper control motor and potentiometer.

p. Remove the joint.

q. Remove the expansion valve.

r. Remove the o-ring.

s. Remove the evaporator.

t. Remove the heater pipe.

u. Remove the flow rate control valve.

v. Remove the heater core.

To install:

14. Assemble the heater unit in the reverse of the disassembly procedure.

15. Install the heater unit assembly in the vehicle.

16. Install the deck crossmember assembly.

17. Connect the previously disconnected connectors.

18. Install the deck crossmember cowl top stay.

19. Install the junction block.

20. Lubricate and install a new O-ring. Connect the flexible discharge hose and tighten to 43 inch lbs. (5 Nm).

21. Lubricate and install a new O-ring. Connect the flexible suction hose and tighten to 43 inch lbs. (5 Nm).

22. Connect the heater hose.

23. Install the steering column as follows:

24. Install the instrument panel.

25. Fill the engine with coolant.

26. Evacuate and charge the A/C system.

27. Check for leaks.

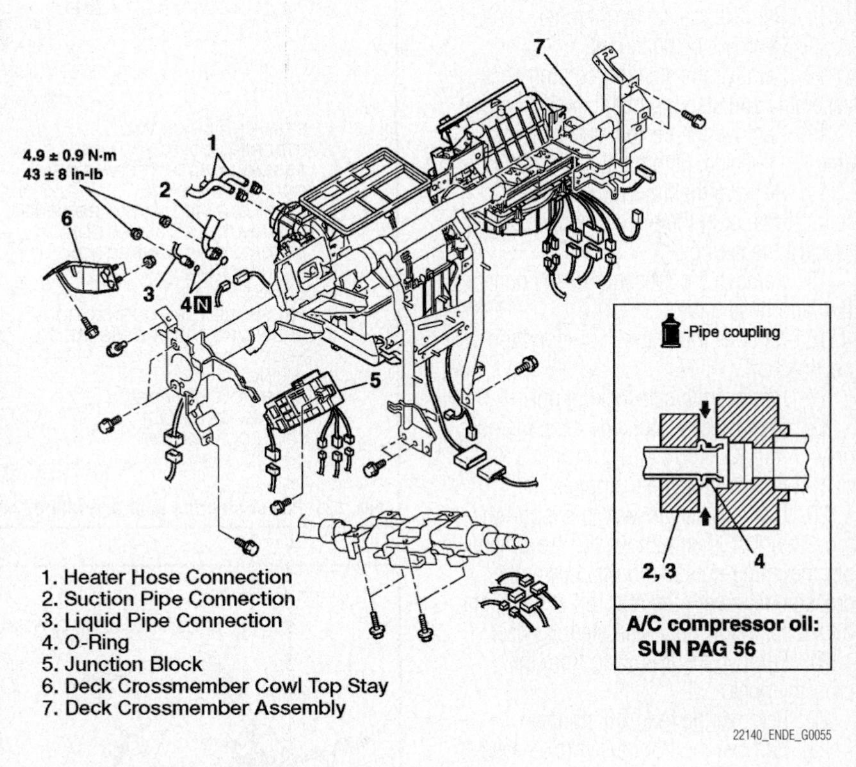

Fig. 130 Heater unit assembly

STEERING

POWER RACK & PINION STEERING GEAR

REMOVAL & INSTALLATION

See Figures 131 through 136.

1. Before servicing the vehicle, refer to the precautions.
2. Disconnect the negative battery cable.

➡Wait at least 90 seconds after disconnecting the cable from the negative (-) battery terminal to prevent airbag activation.

3. Drain the power steering system.
4. Remove the steering wheel.
5. Remove the clock spring and put aside in a safe place.
6. Remove the front scuff plate and cowl side trim.
7. Remove the front floor left side console cover.
8. Remove the front floor carpet.
9. Remove the stabilizer link and lower control arm assembly.
10. Remove the center member.
11. Remove the shaft pad.
12. Remove the steering column shaft assembly and steering gear connecting bolt.
13. Remove the tie rod end from the steering knuckle, using a suitable puller.
14. Remove the fluid return hose and return tube. plug the hoses. remove and discard the o-ring.
15. Remove the pressure hose connection and hose gasket.
16. Remove the rear roll stopper connecting bolt.
17. Remove the self-locking nut.
18. Remove the front axle crossmember stay.
19. Remove the plate stopper.
20. Use a transaxle jack to support the crossmember, then remove the crossmember mounting nuts and bolts. Lower the crossmember with the rear roll stopper, stabilizer bar, return tube and steering gear.
21. Remove the following from the crossmember.
22. Remove the rear roll stopper.
23. Remove the front axle no. 1 crossmember.
24. Remove the power steering gear bracket.
25. Remove the heat protector.
26. Remove the steering gear mounting gear side bracket.
27. Remove the steering gear.
28. Remove the steering gear joint cover.

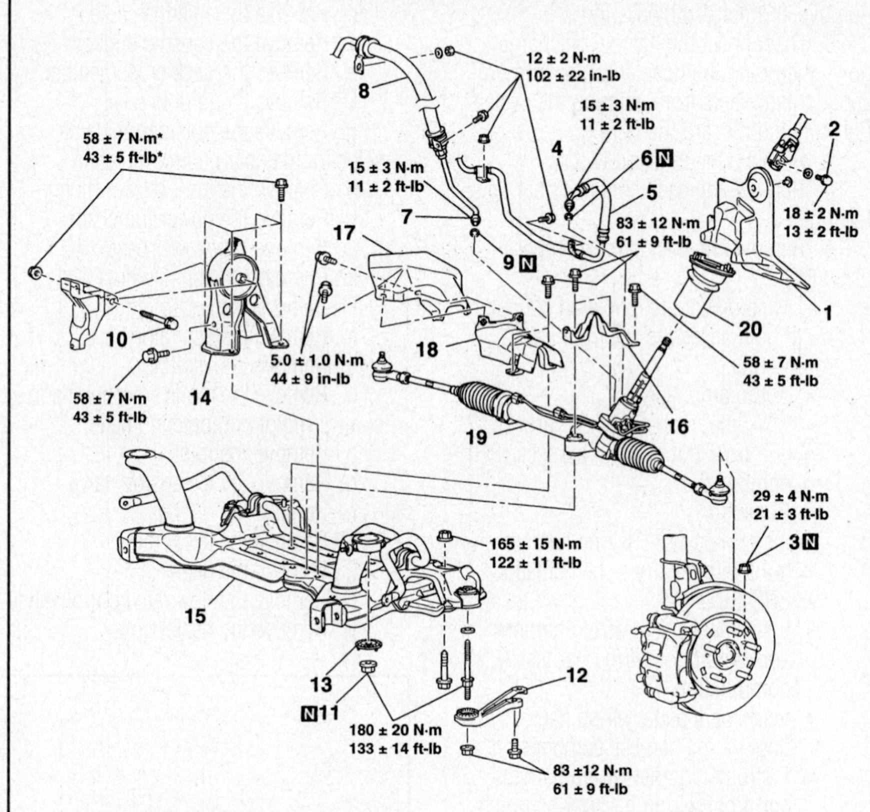

Fig. 131 Power steering gear and related components

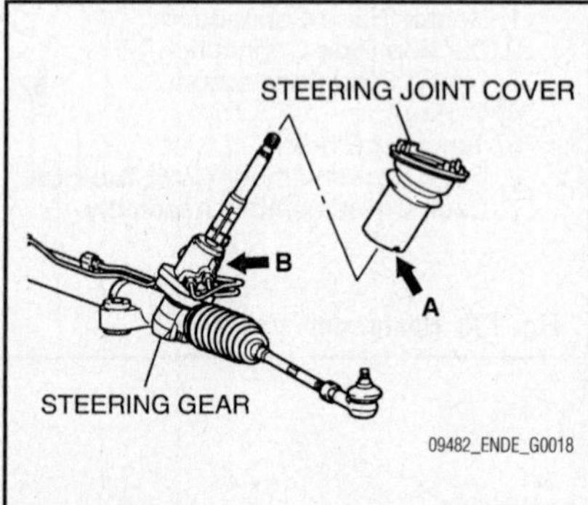

Fig. 132 Steering joint cover notch alignment

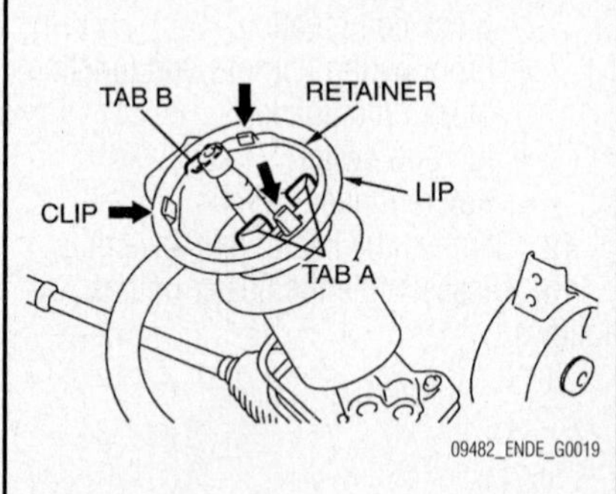

Fig. 133 Steering joint cover tab "A" and tab "B" location

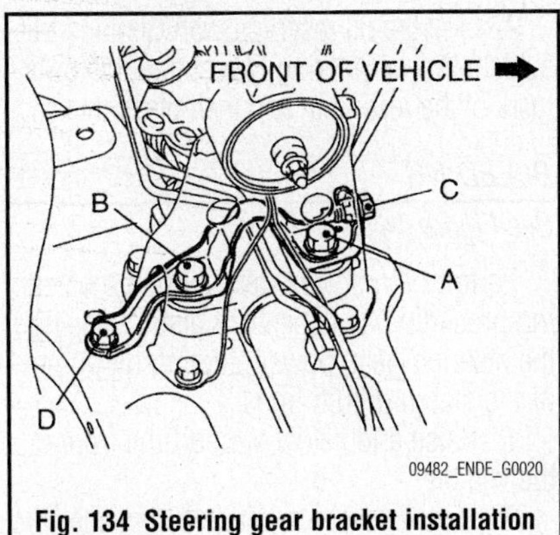

Fig. 134 Steering gear bracket installation points

To install:

29. Align the steering joint cover notch with the steering gear lug. Install the steering joint cover to the steering gear.

30. From inside the vehicle, pull tab "A" and then tab "B" to secure the three clips to the body panel.

➡When securing the joint cover to the body panel, be careful that the cover seal lip does not move backwards.

31. After installing the joint cover, check that the joint cover rubber is not disengaged from the retainer. If there is any doubt,

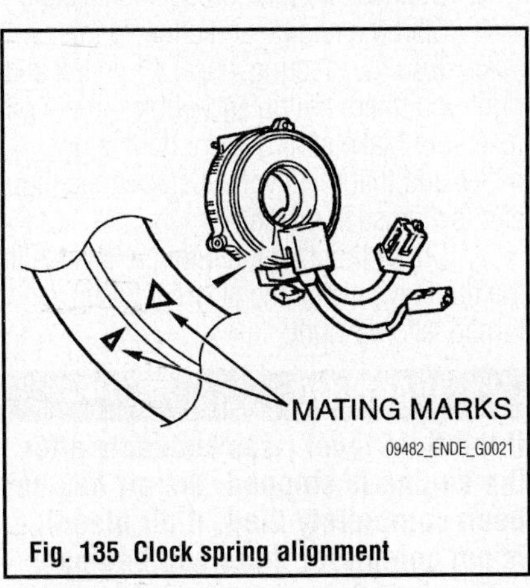

Fig. 135 Clock spring alignment

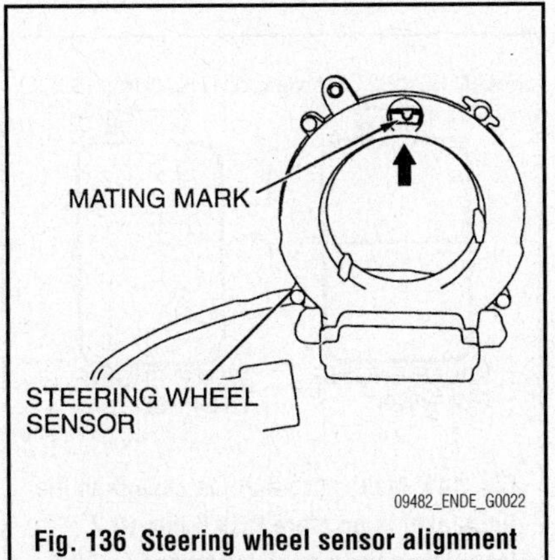

Fig. 136 Steering wheel sensor alignment

release the clips from the body and start the procedure again.

32. Tighten the power steering gear bracket bolts in the order shown in the illustration.

33. Continue the installation in the reverse order of the removal procedure.

➡When installing the clock spring, ensure that the mating marks are properly aligned. If not the steering wheel may not rotate completely during a turn, or the flat cable in the clock spring could be damaged. This would prevent normal SRS operation and possibly cause serious injury to the driver.

34. To align the mating marks of the clock spring, turn the clock spring counterclockwise fully. Then turn it back approximately 3 3/4 turns counterclockwise to align the mating marks.

35. Turn the front wheels to the straight ahead position. Then install the clock spring to the column switch.

➡If the vehicle is equipped with Active Skid Control (ASC), ensure that the steering wheel sensor's mating marks are properly aligned. If not, the steering wheel sensor could be damaged.

36. To align the mating marks, turn the steering wheel sensor clockwise fully. Then turn it back approximately 2 ¾ turns counterclockwise to align the mating marks. Align the mating marks on the clock spring and the steering wheel sensor, and install the steering wheel sensor to the column switch assembly.

37. Connect the steering wheel sensor connector.

38. Connect the negative battery cable. Turn the ignition **ON** and then **OFF**, and keep it off for at least ten seconds.

39. Refill the reservoir and bleed the system.

40. Check and adjust the front end alignment, as required.

POWER STEERING PUMP

REMOVAL & INSTALLATION

See Figures 137 through 140.

1. Raise and safely support the front of the vehicle.

2. Remove the front and side under covers.

3. Drain the power steering fluid, as follows:

a. Disconnect the return hose connection, and then connect a vinyl hose to the return hose, and drain the fluid into a container.

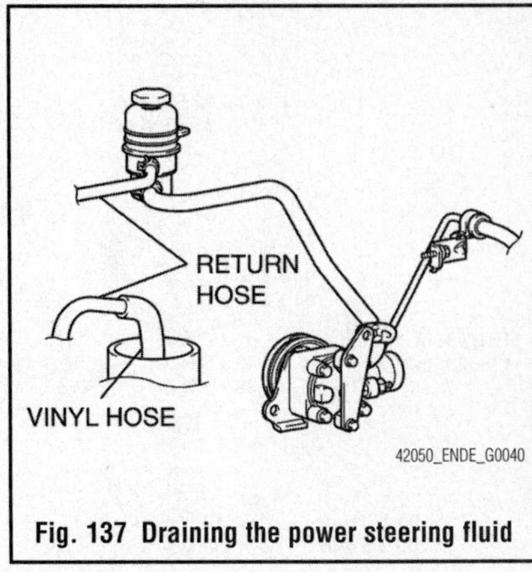

Fig. 137 Draining the power steering fluid

b. Disconnect the ignition coil connectors.

c. While operating the starter motor intermittently, turn the steering wheel all the way to the left and right several times to drain all of the fluid.

d. Connect the return hose securely, and then secure with the clip.

4. Remove the power steering pump drive belt, by loosen the tensioner pulley fixing nut or locknut, then remove the belt.

5. Remove the belt tensioner pulley and bracket.

6. Detach the pressure switch connector.

7. Disconnect the power steering pressure hose and gasket.

8. Disconnect the power steering suction hose.

9. Remove the power steering pump bracket.

10. Remove the power steering pump assembly.

➡Before removing the power steering pressure switch, wipe the power steering pump clean to prevent foreign debris from entering the pump.

11. If necessary, remove the power steering pressure switch, as follows:

a. Detach the harness connector from the male terminal of the pressure switch.

b. Use a socket wrench to loosen and remove the switch.

To install:

⁂ WARNING

Do not touch the terminal part of the switch during installation as this may damage the switch.

12. If removed, install the power steering switch, as follows:

a. Insert the new switch and tighten hand-tight.

b. Use a socket to tighten the pressure switch to 14.5 ft. lbs. (19.6 Nm).

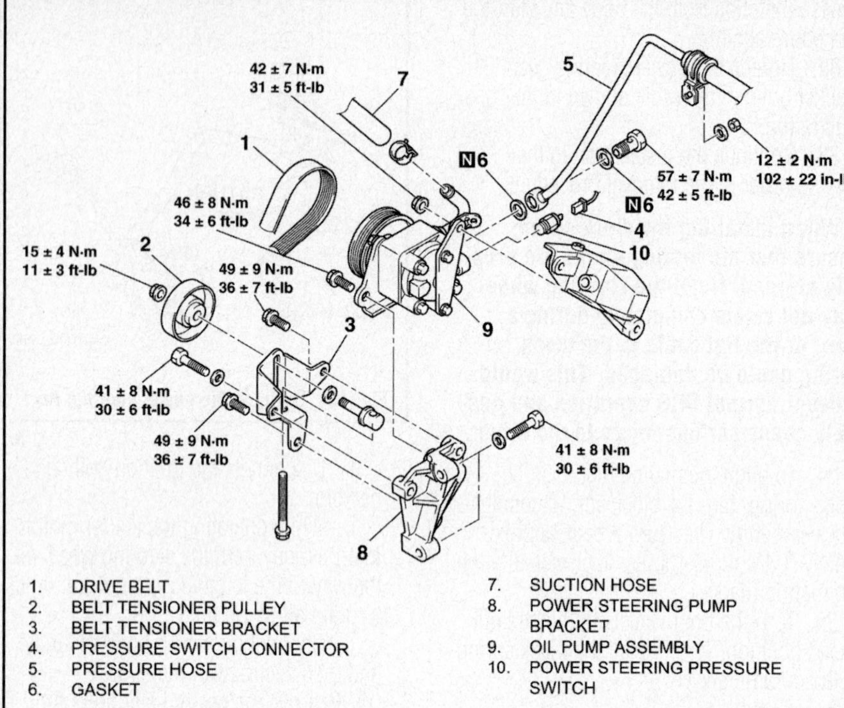

Fig. 138 Exploded view of the power steering pump and fastener tightening specifications

c. Attach the harness connector to the pressure switch.

13. Install the power steering pump and tighten the retainers to the specifications shown in the accompanying illustration.

14. Install the power steering pump bracket and tighten the retainers to the specifications shown in the accompanying illustration.

15. Install the suction hose. Make sure the marks on the hose and nipple are positioned as shown in the accompanying illustration.

16. Install the power steering pressure hose and gasket. Tighten the eye bolt to 42 ft. lbs. (57 Nm).

17. Install the belt tensioner bracket and tighten the retainers to the specifications shown in the accompanying illustration.

18. Install the belt tensioner pulley. Tighten the bolt to 11 ft. lbs. (15 Nm).

19. Install the power steering pump belt, carefully routing it as shown in the accompanying illustration.

20. Adjust the belt tension and tighten the fixing bolt or lock nut and adjusting bolt as specified in the adjustment procedures.

21. Install the front and side under covers.

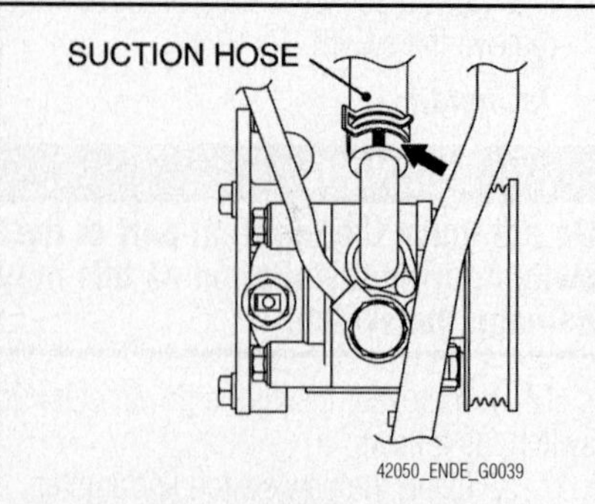

Fig. 139 When installing the suction hose, make sure the marks are aligned as shown

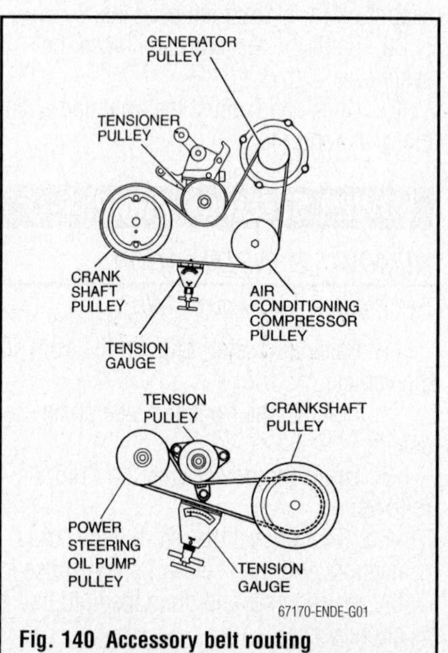

Fig. 140 Accessory belt routing

22. Fill the oil reservoir with Genuine Mitsubishi Power Steering Fluid up to the lower mark of the reservoir, and then bleed the air.

BLEEDING

See Figure 141.

Perform the power steering pump bleeding procedure as necessary after replacing the steering gear, power steering oil pump or the steering fluid lines.

1. Raise and safely support the front of the vehicle.
2. Disconnect the ignition coil connectors.

**** WARNING**

Perform air bleeding only while cranking the engine. Do not perform air bleeding while the engine is running. If you do so, air in the fluid will be increased and air bleeding will become more difficult. During air bleeding, refill the steering fluid so that the level never falls below the lower mark on the dipstick.

3. Turn the steering wheel all the way to the left and right five or six times while using the starter motor to crank the engine intermittently several times (for 15 to 20 seconds).
4. Connect the ignition coil connectors.
5. Start the engine and allow to idle.
6. Turn the steering wheel to the left and right until there are no air bubbles in the oil reservoir. Make sure that the fluid is not milky, and that the level is between the high and low dipstick marks.
7. Check that there is minimal change in the fluid level when the steering wheel is turned left and right.

**** WARNING**

If the fluid level rises suddenly after the engine is stopped, the air has not been completely bled. If air bleeding is not complete, there will be abnormal noises from the pump and the

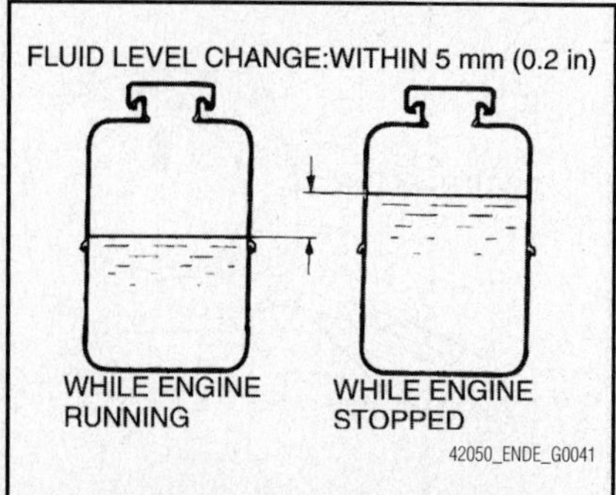

Fig. 141 Make sure that the change in the fluid level is no more than 5 mm (0.2 inch) when the engine is stopped

flow-control valve, and this condition could reduce the life of the power steering components.

8. Make sure that the change in the fluid level is no more than 5 mm (0.2 inch) when the engine is stopped.

9. If the change of the fluid level is 5 mm (0.2 inch) or more, the air has not been completely bled from the system. The air bleeding procedure must be repeated.

STEERING LINKAGE

REMOVAL & INSTALLATION

Tie Rod Ends

See Figure 142.

1. Before servicing the vehicle, refer to the precautions.
2. Raise and safely support the vehicle securely on jackstands.

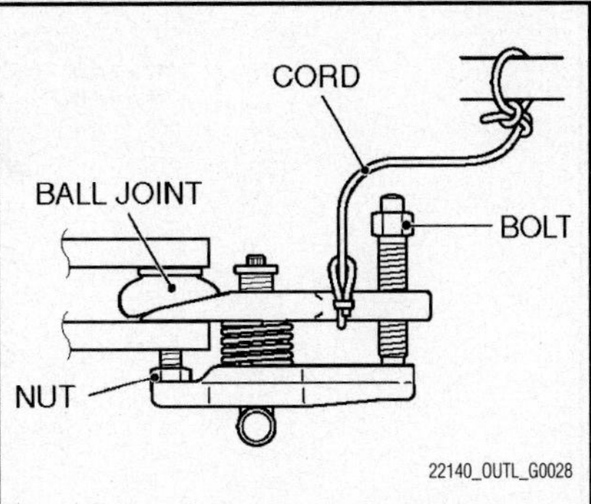

Fig. 142 Using special tool MB991897 tie rod ball joint separator

3. Remove the wheels.
4. Disconnect tie rod end.
   a. Remove the cotter pin and nut.
   b. Using special tool special tool MB991897 or a tie rod ball joint separator, disconnect tie rod end from the steering knuckle.
5. Matchmark the tie rod end to the tie rod.
6. Loosen the tie rod end locking nut.
7. Carefully remove the tie rod end, counting the number of revolutions (or threads) for installation reference.

To install:

8. Install the tie rod end.
   a. Carefully install the tie rod end, counting the number of revolutions until it reaches the reference from the removal.
   b. Tighten the tie rod end locking nut.
   c. Connect the tie rod end to the steering knuckle and tighten the nut to 19 ft. lbs. (25 Nm).
   d. Install a new cotter pin.
9. Install the wheels.
10. Lower the vehicle.
11. Check the toe-in measurement.

SUSPENSION

FRONT SUSPENSION

LOWER BALL JOINT

REMOVAL & INSTALLATION

The lower ball joint is a integral part of the lower control arm and is not serviced separately.

LOWER CONTROL ARM

REMOVAL AND & INSTALLATION

See Figure 143.

1. Before servicing the vehicle, refer to the Precautions Section.
2. Raise and support the vehicle safely.
3. Remove the wheel.
4. Remove the lower control arm mounting bolts and nuts.
5. Remove the lower ball joint from knuckle.
6. Remove the lower control arm.

To install:

7. Install the lower control arm.
8. Install the lower control arm retainers. Tighten the knuckle nut and bolt to 81 ft. lbs. (110 Nm) and the other retainers hand-tight only at this time.
9. Install the wheel.
10. Once the weight of the vehicle is resting on the suspension, torque the lower control arm mounting bolts and nuts 122 ft. lbs. (165 Nm).
11. Check and adjust the front wheel alignment.

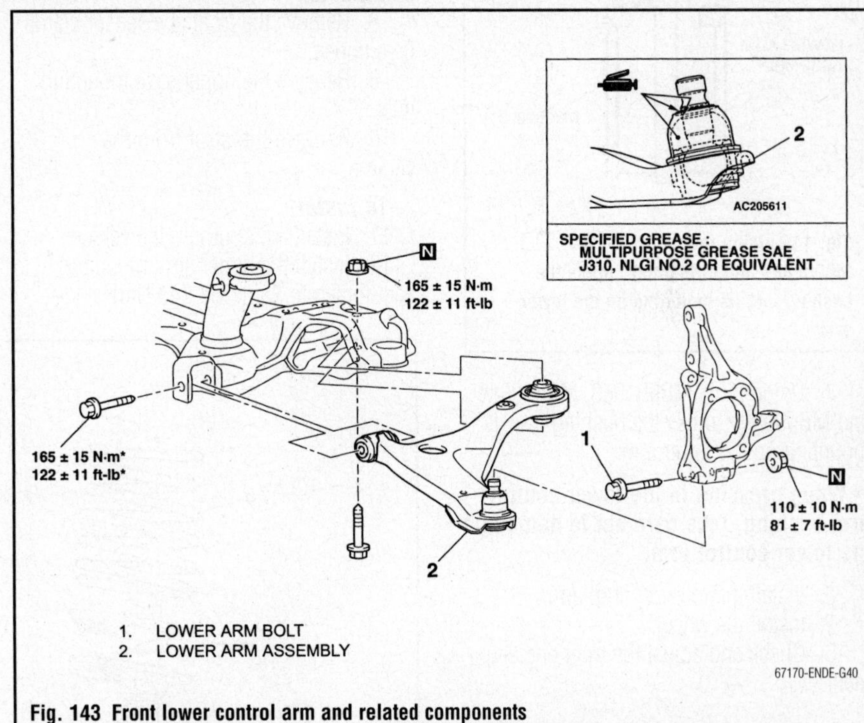

Fig. 143 Front lower control arm and related components

CONTROL ARM BUSHING REPLACEMENT

See Figures 144 and 145.

1. Before servicing the vehicle, refer to the Precautions Section.
2. Raise and support the vehicle safely.
3. Remove the wheel.
4. Remove the lower control arm and place in a vise.
5. Using tools MB991963 and MB990890 remove the bushing.

To install:

6. Position the bushing with the larger end facing the front of the vehicle.

➡Coat the bushing with a soap solution and take care not to twist.

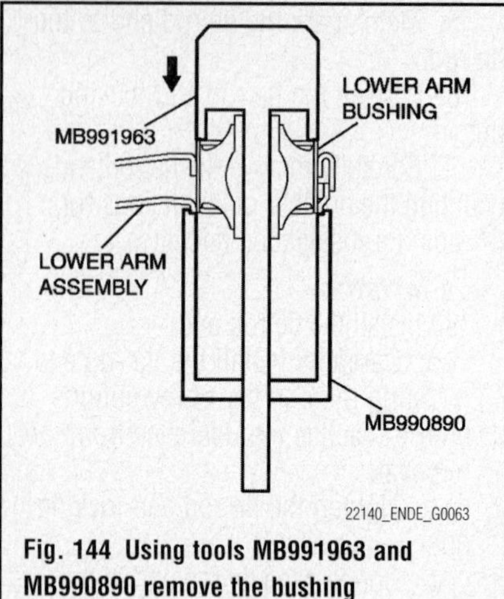

Fig. 144 Using tools MB991963 and MB990890 remove the bushing

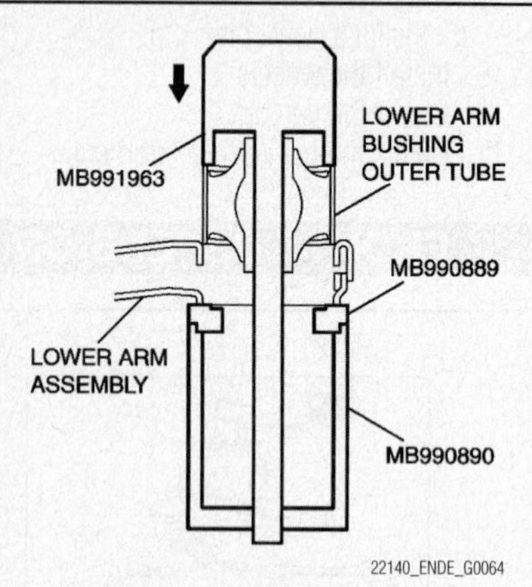

Fig. 145 Using tools MB991963, MB990889 and MB990890, press the bushing into its mounting on the lower arm

7. Using tools MB991963, MB990889 and MB990890, press the bushing into its mounting on the lower arm.

➡When pressing in the lower control arm bushing, take care not to damage the lower control arm.

8. Install the lower control arm.
9. Install the wheel.
10. Check and adjust the front end alignment.

MACPHERSON STRUT

REMOVAL & INSTALLATION

See Figure 146.

1. Before servicing the vehicle, refer to the Precautions Section.
2. Remove the windshield wiper arm assemblies and front deck garnish.
3. Raise and support the vehicle safely.
4. Remove the tire and wheel.
5. Remove the stabilizer link.

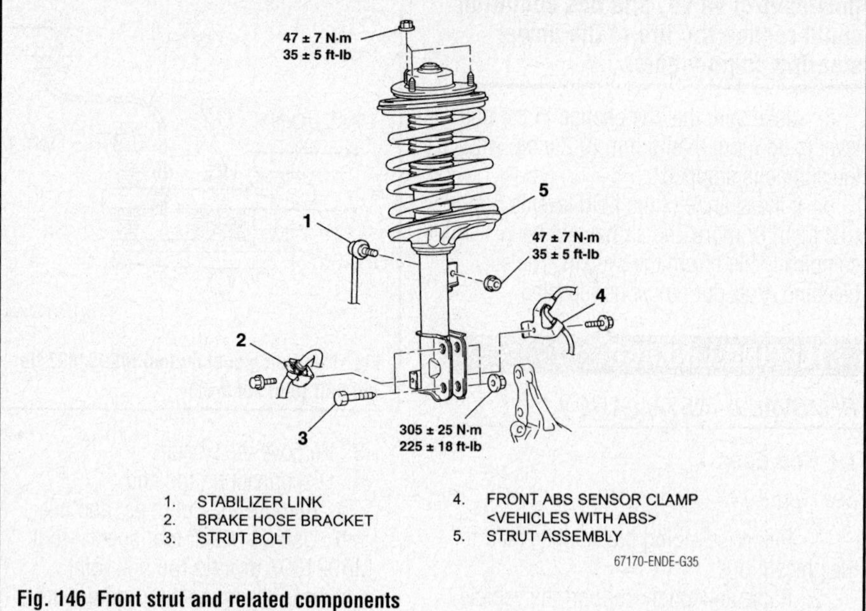

Fig. 146 Front strut and related components

6. Remove the brake hose bracket.
7. Remove the lower strut mounting bolt and nut.
8. Remove the front abs sensor clamp, if equipped.
9. Remove the upper strut mounting nuts.
10. Remove the strut from the vehicle.

To install:

11. Install the strut into the vehicle.
12. Install the upper strut mounting nuts and tighten to 35 ft. lbs. (47 Nm).
13. Install the front abs sensor clamp, if equipped.
14. Install the lower strut mounting bolt and nut. torque the nut to 225 ft. lbs. (305 Nm).
15. Install the brake hose bracket.
16. Install the stabilizer link.
17. Install the tire and wheel.
18. Install the front deck garnish and wiper arms.

OVERHAUL

See Figure 147.

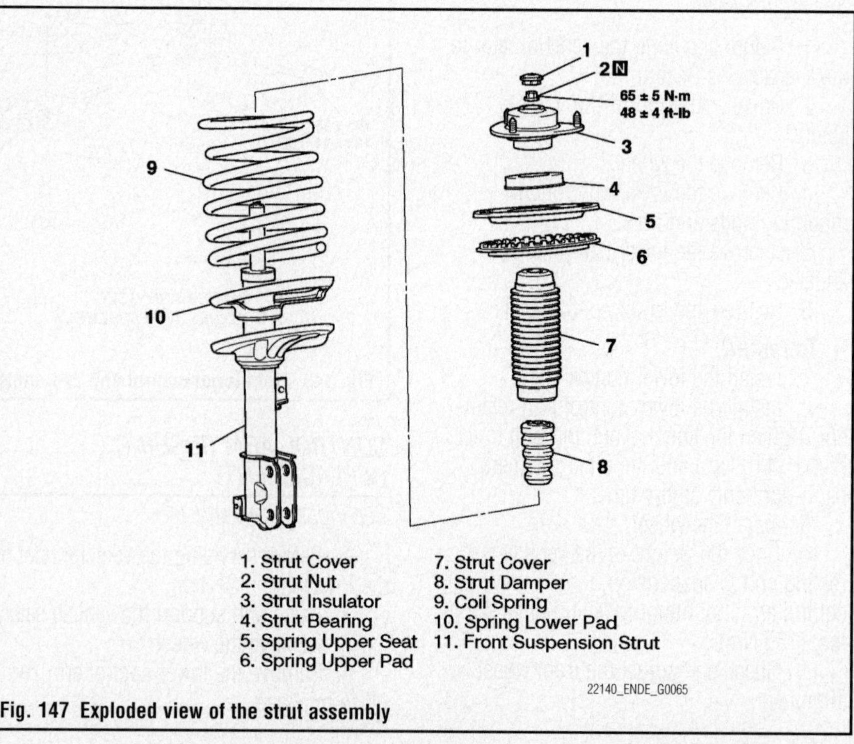

Fig. 147 Exploded view of the strut assembly

1. Before servicing the vehicle, refer to the Precautions Section.
2. Remove the strut.
3. Remove the strut cover.
4. Compress the coil spring until there is a clearance on both ends
5. Remove the strut center nut.
6. Remove the insulator.
7. Remove the bearing.
8. Remove the spring upper seat.
9. Remove the spring upper pad.
10. Remove the strut cover.
11. Remove the strut damper.
12. Remove the coil spring.
13. Remove the spring lower pad.
14. Remove the strut.

To Assemble:

15. Assembly is the reverse of disassembly. Torque the center nut to 48 ft. lbs. (65 Nm).

STABILIZER BAR

REMOVAL & INSTALLATION

See Figures 148 through 150.

➡To avoid personal injury when working on vehicles equipped with an air bag, the negative battery cable must be disconnected and at least 60 seconds must elapse before working on the system. Failure to do so may result in deployment of the air bag. You should also wrap or isolate the negative battery cable with electrical or other non-conductive tape.

1. Before servicing the vehicle, refer to the precautions.
2. Disconnect the negative battery cable.

➡Wait at least 90 seconds after disconnecting the cable from the negative (-) battery terminal to prevent airbag activation.

3. Raise and safely support the vehicle securely on jackstands.
4. Remove the wheels.
5. Remove the front under cover.
6. Drain the power steering fluid.
7. Remove the center member.
8. Remove the steering wheel.
9. Remove the floor console side cover.
10. Remove the front scruff plate and cowl side trim.
11. Pull back the floor carpeting.
12. Remove the stabilizer nut.
13. Disconnect the tie rod end front the knuckle.
14. Disconnect the lower arm assembly from the knuckle.
15. Disconnect the steering gear from the steering column shaft.
    a. Remove the steering gear and steering column shaft connecting bolt.
    b. Pinch the steering column shaft clip with pliers, and pull up the shaft towards the direction shown to disengage the steering column shaft from the steering gear.

⁂ WARNING

If the steering column shaft is removed accidentally, remove the steering column assembly and be sure to insert the steering column shaft into the steering column as shown in the figure.

16. Disconnect the steering gear from the return tube.
17. Remove the return tube clamp.
18. Remove the pressure hose clamp.
19. Disconnect the steering gear from the pressure tube.
20. Remove the rear roll stopper connection bolt.
21. Remove the self-locking nut.
22. Remove the front axle crossmember stay.
23. Remove the front axle No.1 crossmember, rear roll stopper and steering gear assembly.

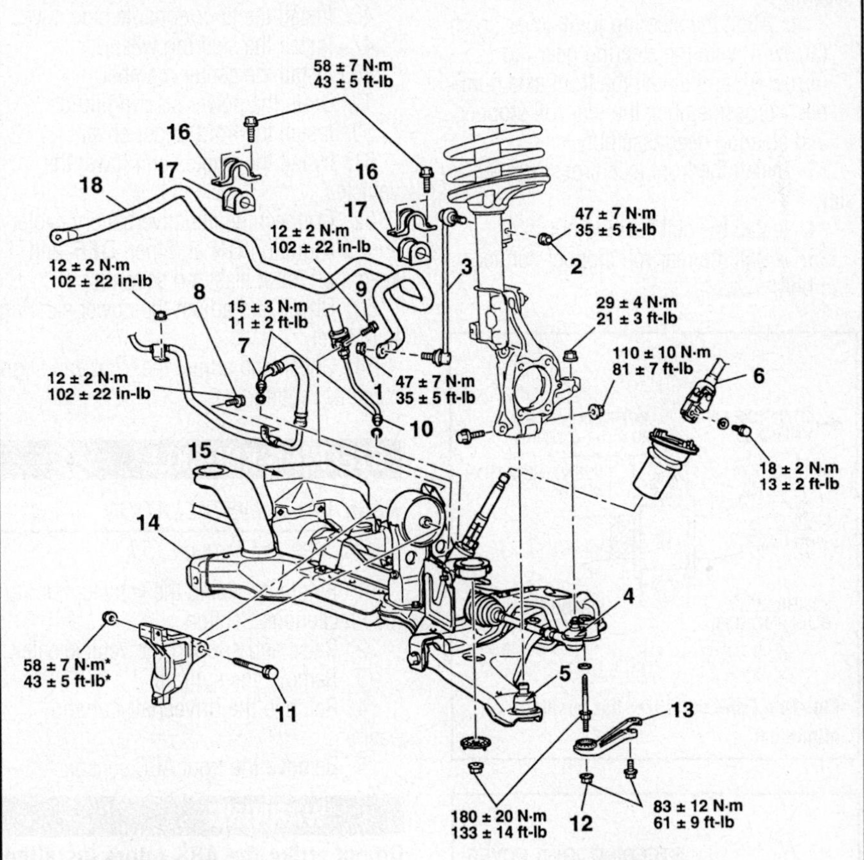

1. STABILIZER NUT
2. STABILIZER NUT
3. STABILIZER LINK
4. TIE ROD END AND KNUCKLE CONNECTION
5. LOWER ARM ASSEMBLY AND KNUCKLE CONNECTION
6. STEERING GEAR AND STEERING COLUMN SHAFT CONNECTION
7. STEERING GEAR AND RETURN TUBE CONNECTION
8. RETURN TUBE CLAMP
9. PRESSURE HOSE CLAMP
10. STEERING GEAR AND PRESSURE TUBE CONNECTION
11. REAR ROLL STOPPER CONNECTION BOLT
12. SELF-LOCKING NUT
13. FRONT AXLE CROSSMEMBER STAY
14. FRONT AXLE NO.1 CROSSMEMBER, REAR ROLL STOPPER AND STEERING GEAR ASSEMBLY
15. STEERING GEAR AND LINKAGE PROTECTOR
16. STABILIZER BRACKET
17. STABILIZER BUSHING
18. STABILIZER BAR

09482_ENDE_G0025

Fig. 148 Front stabilizer bar and related components

24. Remove the steering gear and linkage protector.
25. Remove the stabilizer bracket.
26. Remove the stabilizer bushing.
27. Remove the stabilizer bar.

To install:

28. Install the stabilizer bar.
    a. Align the stabilizer bar identification mark with the right end of the bushing.
29. Install the stabilizer bushing.
30. Install the stabilizer bracket. Tighten to 43 ft. lbs. (58 Nm).
31. Install the steering gear and linkage protector.
32. Install the front axle No.1 crossmember, rear roll stopper and steering gear assembly.
    a. Align the steering joint cover notch (arrow A) with the steering gear lug (arrow B), and install the front axle number 1 crossmember, the rear roll stopper and steering gear assembly.
33. Install the front axle crossmember stay.
34. Install the self-locking nut.
35. Install the rear roll stopper connection bolt.

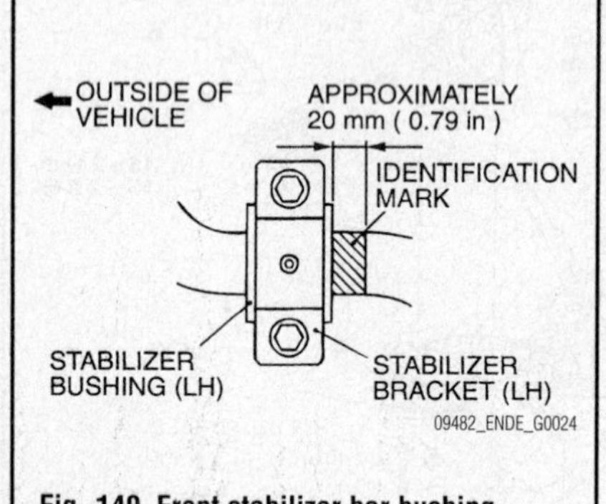

Fig. 149 Front stabilizer bar bushing alignment

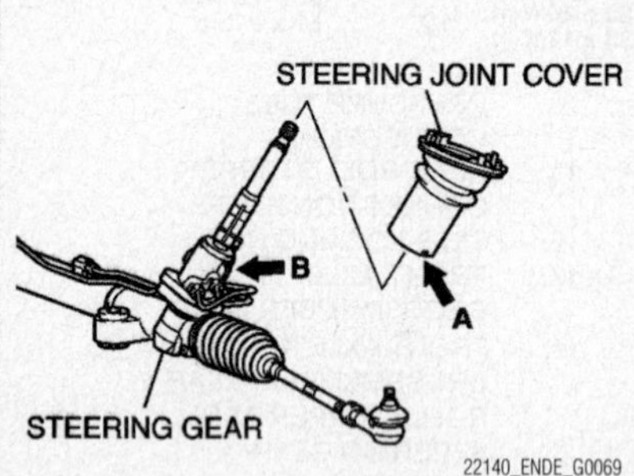

Fig. 150 Align the steering joint cover notch (arrow A) with the steering gear lug (arrow B), and install the front axle number 1 crossmember, the rear roll stopper and steering gear assembly

36. Connect the steering gear from the pressure tube.
37. Install the pressure hose clamp.
38. Install the return tube clamp.
39. Connect the steering gear from the return tube.
40. Connect the steering gear from the steering column shaft.
41. Connect the lower arm assembly from the knuckle.
42. Connect the tie rod end front the knuckle.
43. Install the stabilizer link nut. Tighten to 35 ft. lbs. (47 Nm).
44. Pull back the floor carpeting.
45. Install the front scruff plate and cowl side trim.
46. Install the floor console side cover.
47. Install the steering wheel.
48. Install the center member.
49. Drain the power steering fluid.
50. Install the front under cover.
51. Install the wheels and lower the vehicle.
52. Connect the negative battery cable. Turn the ignition **ON** and then **OFF**, and keep it off for at least ten seconds.
53. Check and adjust the power steering fluid level,
54. Check and adjust the front end alignment, as required.

STEERING KNUCKLE

REMOVAL & INSTALLATION

See Figures 151 through 153.

1. Before servicing the vehicle, refer to the Precautions Section.
2. Raise and support the vehicle safely.
3. Remove the split pin.
4. Remove the driveshaft nut and washer.
5. Remove the front ABS sensor.

**** WARNING**

Do not strike the ABS rotors installed to the BJ outer race of driveshaft against other parts when removing or installing the driveshaft. Otherwise the ABS rotors will be damaged. Be careful not to strike the pole piece at the tip of the front ABS sensor with tools during servicing work.

6. Remove the brake hose bracket.
7. Remove the caliper assembly.
    a. Remove the caliper assembly with brake hose.
    b. Retain the removed caliper assembly with a wire to prevent from falling.
8. Remove the brake disc.
    a. If the brake disc is seized, install a M8 × 1.25-mm bolts in the provided holes, and remove the disc by tightening the bolts evenly and gradually.
9. Remove the front wheel hub assembly.
    a. Use special tools MB990242, MB990244, MB991354 and MB990767 to push out the driveshaft from the hub and knuckle.
    b. If the front wheel hub is seized, remove the knuckle together with front wheel hub and fix them with a vise.
    c. Hang the driveshaft on the vehicle body with a rope.
    d. Use special tools MB990244, MB991354 and MB990211 to pull out the front wheel hub from the knuckle.
10. Remove the dust cover.
11. Remove the self locking nut for lower arm ball joint.
12. Remove the self locking nut for tie rod end.
13. Remove the front strut to hub and knuckle mounting bolt and nut.
14. Remove the knuckle.

To install:

15. Install the knuckle.
16. Install the front strut to hub and knuckle mounting bolt and nut.

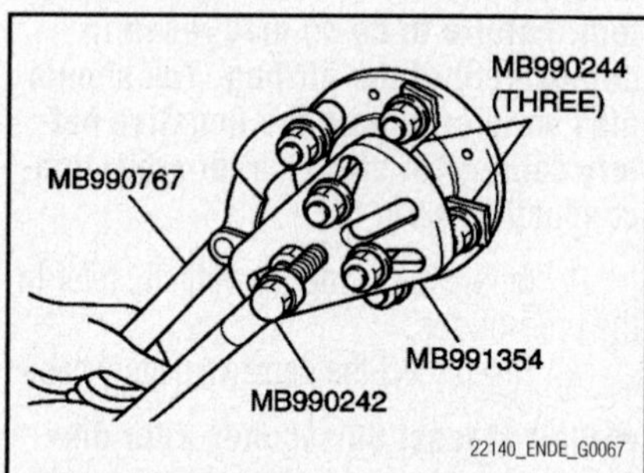

Fig. 151 Use special tools MB990242, MB990244, MB991354 and MB990767 to push out the driveshaft from the hub and knuckle

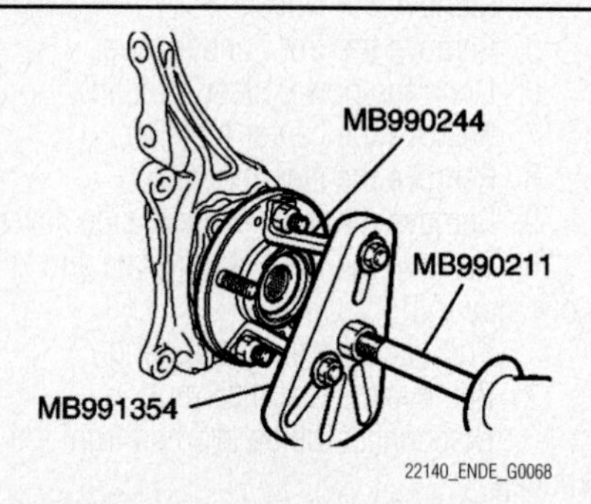

Fig. 152 Use special tools MB990244, MB991354 and MB990211 to pull out the front wheel hub from the knuckle

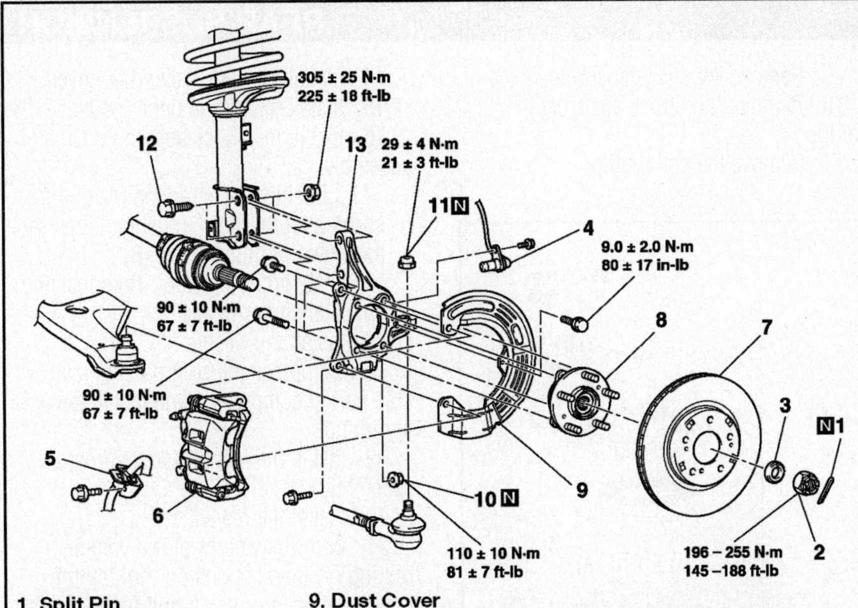

1. Split Pin
2. Driveshaft Nut
3. Washer
4. Front ABS Sensor
5. Brake Hose Bracket
6. Caliper Assembly
7. Brake Disc
8. Front Wheel Hub Assembly
9. Dust Cover
10. Self Locking Nut (Connection For Lower Arm Ball Joint)
11. Self Locking Nut (Connection For Tie Rod End)
12. Front Strut To Hub and Knuckle Mounting Bolt and Nut
13. Knuckle

22140_ENDE_G0066

Fig. 153 Exploded view of the front steering knuckle

17. Install the self locking nut for tie rod end.
18. Install the self locking nut for lower arm ball joint.
19. Install the dust cover.
20. Install the front wheel hub assembly.
21. Install the brake disc.
22. Install the caliper assembly.
23. Install the brake hose bracket.
24. Install the front ABS sensor.
25. Install the driveshaft nut and washer.
    a. Before securely tightening the driveshaft nuts, make sure there is no load on the wheel bearings. Otherwise the wheel bearings will be damaged.
    b. Be sure to install the driveshaft washer in the specified direction.
    c. Using special tool MB990767, tighten the driveshaft nut. At this time, tighten the nut to 145–188 ft. lbs. (196–255 Nm) torque so that the pin hole may align with split pin.
    d. If the pin hole does not align with the pin, tighten the driveshaft nut to less than 188 ft. lbs. (255 Nm) and find the nearest hole, then fit the split pin.
26. Install the split pin.

WHEEL HUB AND BEARING

REMOVAL & INSTALLATION

See Figures 151 through 153.

1. Before servicing the vehicle, refer to the Precautions Section.
2. Raise and support the vehicle safely.
3. Remove the split pin.
4. Remove the driveshaft nut and washer.
5. Remove the front ABS sensor.

✱✱ WARNING

Do not strike the ABS rotors installed to the BJ outer race of driveshaft against other parts when removing or installing the driveshaft. Otherwise the ABS rotors will be damaged. Be careful not to strike the pole piece at the tip of the front ABS sensor with tools during servicing work.

6. Remove the brake hose bracket.
7. Remove the caliper assembly.
    a. Remove the caliper assembly with brake hose.
    b. Retain the removed caliper assembly with a wire to prevent from falling.
8. Remove the brake disc.
    a. If the brake disc is seized, install a M8 × 1.25-mm bolts in the provided holes, and remove the disc by tightening the bolts evenly and gradually.
9. Remove the front wheel hub assembly.
    a. Use special tools MB990242, MB990244, MB991354 and MB990767 to push out the driveshaft from the hub and knuckle.
    b. If the front wheel hub is seized, remove the knuckle together with front wheel hub and fix them with a vise.
    c. Hang the driveshaft on the vehicle body with a rope.
    d. Use special tools MB990244, MB991354 and MB990211 to pull out the front wheel hub from the knuckle.
10. Remove the dust cover.
11. Remove the self locking nut for lower arm ball joint.
12. Remove the self locking nut for tie rod end.
13. Remove the front strut to hub and knuckle mounting bolt and nut.
14. Remove the steering knuckle.
15. Remove the hub from the steering knuckle.

To install:

16. Install the hub on the steering knuckle. Tighten the attaching bolts to 67 ft. lbs. (90 Nm).
17. Install the knuckle.
18. Install the front strut to hub and knuckle mounting bolt and nut.
19. Install the self locking nut for tie rod end.
20. Install the self locking nut for lower arm ball joint.
21. Install the dust cover.
22. Install the front wheel hub assembly.
23. Install the brake disc.
24. Install the caliper assembly.
25. Install the brake hose bracket.
26. Install the front ABS sensor.
27. Install the driveshaft nut and washer.
    a. Before securely tightening the driveshaft nuts, make sure there is no load on the wheel bearings. Otherwise the wheel bearings will be damaged.
    b. Be sure to install the driveshaft washer in the specified direction.
    c. Using special tool MB990767, tighten the driveshaft nut. At this time, tighten the nut to 145–188 ft. lbs. (196–255 Nm) torque so that the pin hole may align with split pin.
    d. If the pin hole does not align with the pin, tighten the driveshaft nut to less than 188 ft. lbs. (255 Nm) and find the nearest hole, then fit the split pin.
28. Install the split pin.

ADJUSTMENT

The wheel bearings are not adjustable. If the bearings are noisy or become loose, they must be replaced.

COIL SPRING

REMOVAL & INSTALLATION

See Figures 154 and 155.

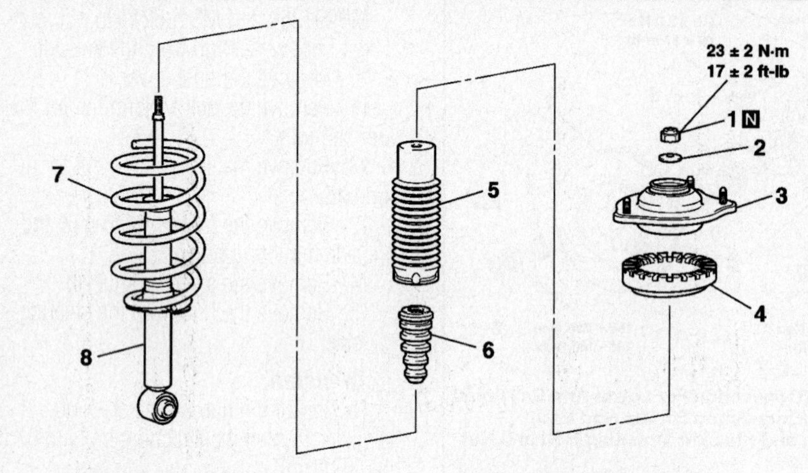

Fig. 154 Exploded view of the rear shock and coil spring

1. Before servicing the vehicle, refer to the Precautions Section.
2. Raise and support the vehicle safely.
3. Remove the rear wheel.
4. Remove the rear shock absorber assembly.
5. Compress the spring using a suitable spring compressor and remove the coil spring nut.
6. Remove the washer.
7. Remove the shock absorber insulator assembly.
8. Remove the spring upper pad.
9. Remove the shock absorber cover.
10. Remove the shock absorber damper.
11. Remove the coil spring.

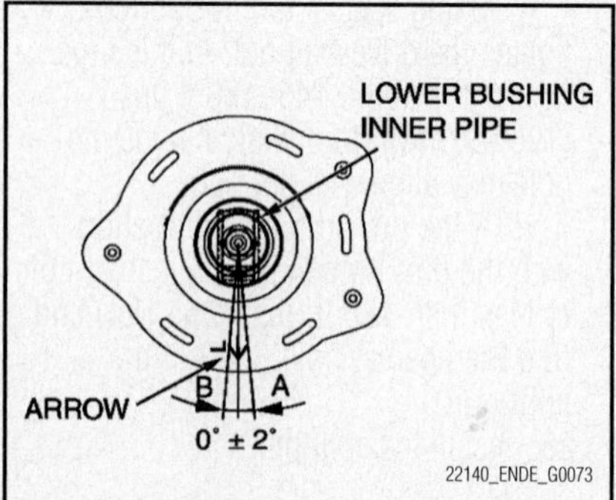

Fig. 155 Position a center line (A) of the shock absorber lower bushing inner pipe as shown from the arrow (B) on the shock absorber insulator

To install:

12. Install the coil spring.
13. Install the shock absorber damper.
14. Install the shock absorber cover.
15. Install the spring upper pad.
16. Install the shock absorber insulator assembly.
    a. Position a center line (A) of the shock absorber lower bushing inner pipe as shown from the arrow (B) on the shock absorber insulator. Then install the shock absorber insulator.
17. Install the washer.
18. Compress the spring using a suitable spring compressor and install the coil spring nut.
19. Install the rear shock absorber assembly.
20. Install the rear wheel.
21. With the weight of the vehicle resting on the suspension, tighten the lower shock-to-knuckle bolt to 74 ft. lbs. (100 Nm).

CONTROL ARMS/LINKS

REMOVAL & INSTALLATION

Lower

See Figure 156.

1. Before servicing the vehicle, refer to the Precautions Section.
2. Raise and support the vehicle safely.

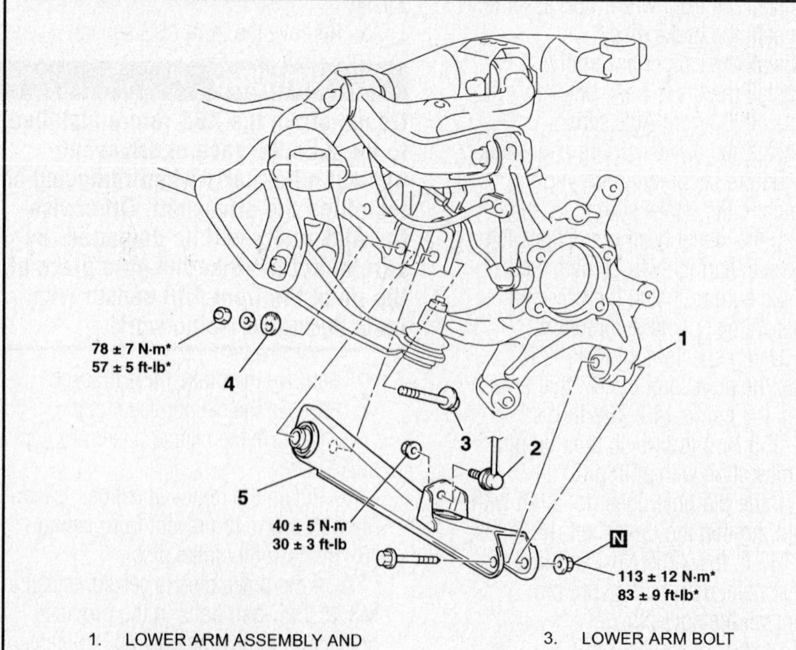

Fig. 156 Rear lower control arm and related components

3. Remove the rear wheel.
4. Remove the lower control arm from the knuckle.
5. Remove the retaining nut, then separate the lower control arm from the stabilizer bar link.

➡Matchmark the crossmember and the plate before removing the lower control arm bolt.

6. Remove the lower control arm bolt, plate and lower control arm assembly.

To install:

7. Install the lower control arm.
8. Install the lower control arm plate.
9. Install the lower control arm bolt, hand-tight only at this time.
10. Install the lower control arm to the stabilizer bar link. Tighten the retaining to 30 ft. lbs. (40 Nm).
11. Install the lower control arm to the knuckle. Use a new nut and secure hand-tight.
12. Install the rear wheel.
13. Once the weight of the vehicle is resting on the suspension, torque the lower control arm mounting bolt to 57 ft. lbs. (78 Nm), and the lower control arm to knuckle nut to 83 ft. lbs. (113 Nm).
14. Check and adjust the rear wheel alignment.

Upper

See Figure 157.

1. Before servicing the vehicle, refer to the Precautions Section.
2. Raise and support the vehicle safely.
3. Remove the wheel assembly.
4. Remove the upper control arm from the knuckle.
5. Remove the ABS sensor clamp bolts, if equipped.
6. Remove the upper control arm.
7. Remove the upper arm stopper.

To install:

8. Install the upper arm stoppers.
9. Install the upper arm into the vehicle and hand-tighten the retainers.
10. Install the upper control arm-to-knuckle bolt and nut and tighten hand-tight.
11. Install the ABS sensor clamp bolts.
12. Install the wheel assembly.
13. Once the weight of the vehicle is on the suspension, tighten the control arm-to-knuckle nut and the upper control arm mounting nuts to 83 ft. lbs. (113 Nm).
14. Check and adjust the rear wheel alignment.

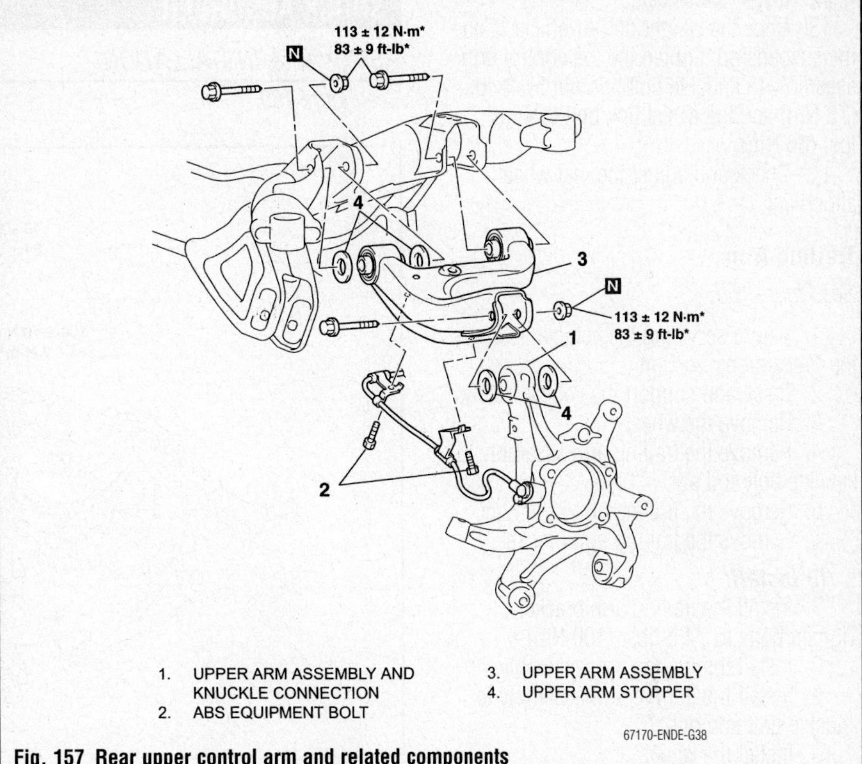

Fig. 157 Rear upper control arm and related components

Toe Control

See Figure 158.

1. Before servicing the vehicle, refer to the Precautions Section.
2. Raise and support the vehicle safely.
3. Remove the wheel.
4. Remove the toe control arm assembly to knuckle bolt and nut.
5. Remove the assist link bolt.
6. Remove the assist link plate.
7. Remove the toe control arm assembly.

To install:

8. Install the toe control arm assembly.
9. Install the assist link plate.
10. Install the assist link bolt.
11. Install the toe control arm assembly to knuckle bolt and nut.

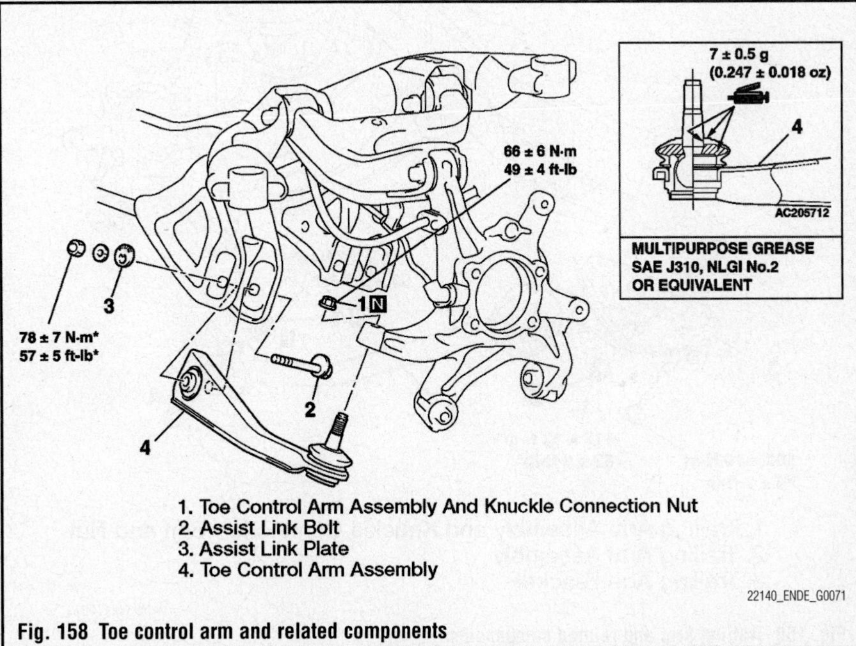

Fig. 158 Toe control arm and related components

12. Install the wheel.
13. Once the weight of the vehicle is on the suspension, tighten the toe control arm assembly to knuckle bolt and nut 57 ft. lbs. (78 Nm) and the assist link bolt to 49 ft. lbs. (66 Nm).
14. Check and adjust the rear wheel alignment.

Trailing Arm

See Figure 159.

1. Before servicing the vehicle, refer to the Precautions Section.
2. Raise and support the vehicle safely.
3. Remove the wheel.
4. Remove the trailing arm assembly to knuckle bolt and nut.
5. Remove the trailing arm assembly.
6. Remove the trailing arm bracket.

To install:

7. Install the trailing arm bracket. Tighten bolts to 74 ft. lbs. (100 Nm).
8. Install the trailing arm assembly.
9. Install the trailing arm assembly to knuckle bolt and nut.
10. Install the wheel.
11. Once the weight of the vehicle is on the suspension, tighten the trailing arm assembly to knuckle bolt and nut 83 ft. lbs. (113 Nm).
12. Check and adjust the rear wheel alignment.

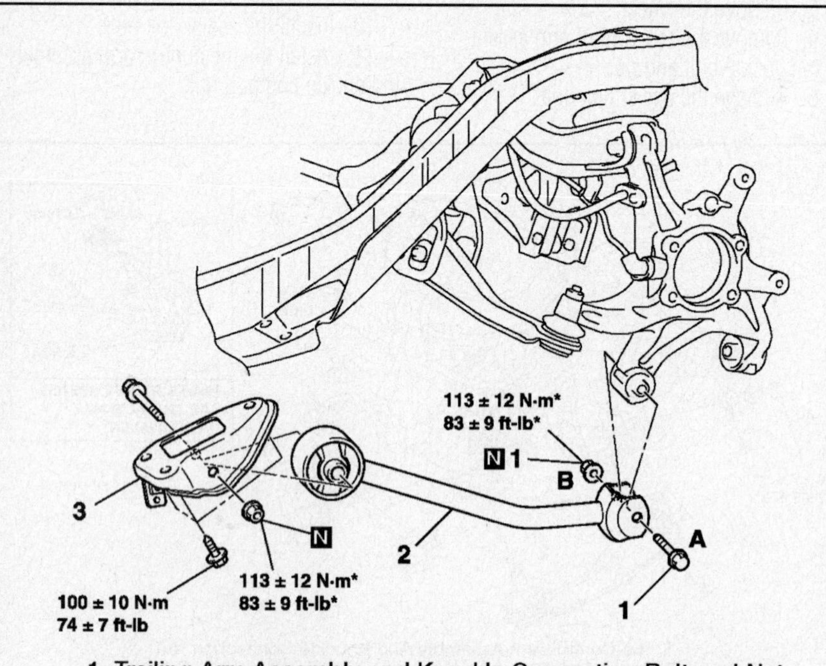

Fig. 159 Trailing arm and related components

SHOCK ABSORBER

REMOVAL & INSTALLATION

See Figure 160.

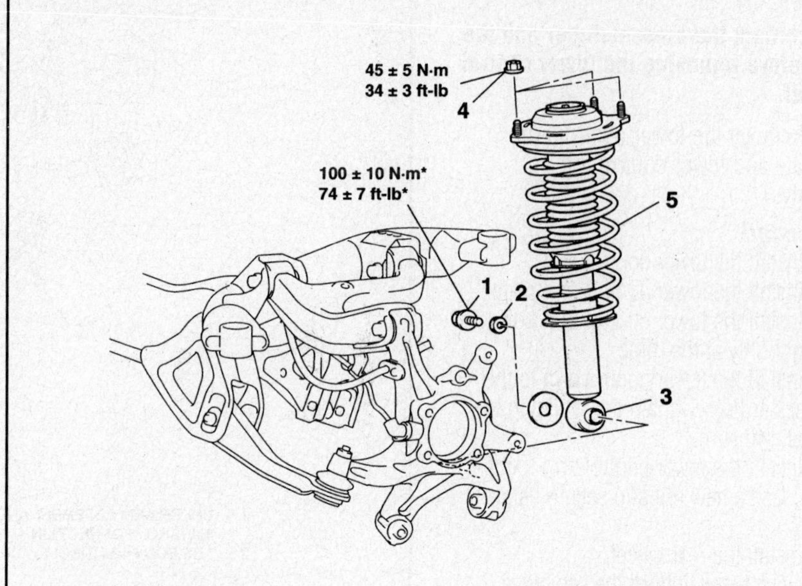

Fig. 160 Rear shock absorber and related components

1. Before servicing the vehicle, refer to the Precautions Section.
2. Remove the luggage floor rear board.
3. Remove the tonneau cover.
4. Remove the luggage floor front board.
5. Remove the parcel strap hook.
6. Remove the luggage floor side board.
7. Remove the rear end trim.
8. Remove the luggage floor carpet bracket.
9. Raise and support the vehicle safely.
10. Remove the wheel and tire.
11. Remove the lower shock absorber bolt and washer, and separate the shock from the knuckle.
12. Remove the upper shock mounting nuts.
13. Remove the shock absorber from the vehicle.

To install:

14. Install the shock absorber. Tighten the upper nuts to 34 ft. lbs. (45 Nm) and the lower shock-to-knuckle bolt hand-tight.
15. Install the wheel and tire
16. Install the luggage floor carpet bracket
17. Install the rear end trim
18. Install the luggage floor side board

19. Install the parcel strap hook
20. Install the luggage floor front board
21. Install the tonneau cover
22. Install the luggage floor rear board
23. With the weight of the vehicle resting on the suspension, tighten the lower shock-to-knuckle bolt to 74 ft. lbs. (100 Nm).

TESTING

1. Check the rubber parts for cracks and wear.
2. Check the shock absorber for malfunctions, oil leakage, or abnormal noise.
3. If shock absorber does not perform properly, replace the shock absorber.

STABILIZER BAR

REMOVAL & INSTALLATION

See Figure 161.

1. Before servicing the vehicle, refer to the Precautions Section.
2. Raise and support the vehicle safely.
3. Remove the tire and wheel assembly.
4. Remove the stabilizer link retaining bolts.
5. Remove the stabilizer link assembly.
6. Remove the stabilizer bar bracket retaining bolts.
7. Remove the bushing.
8. Remove the stabilizer from the vehicle.

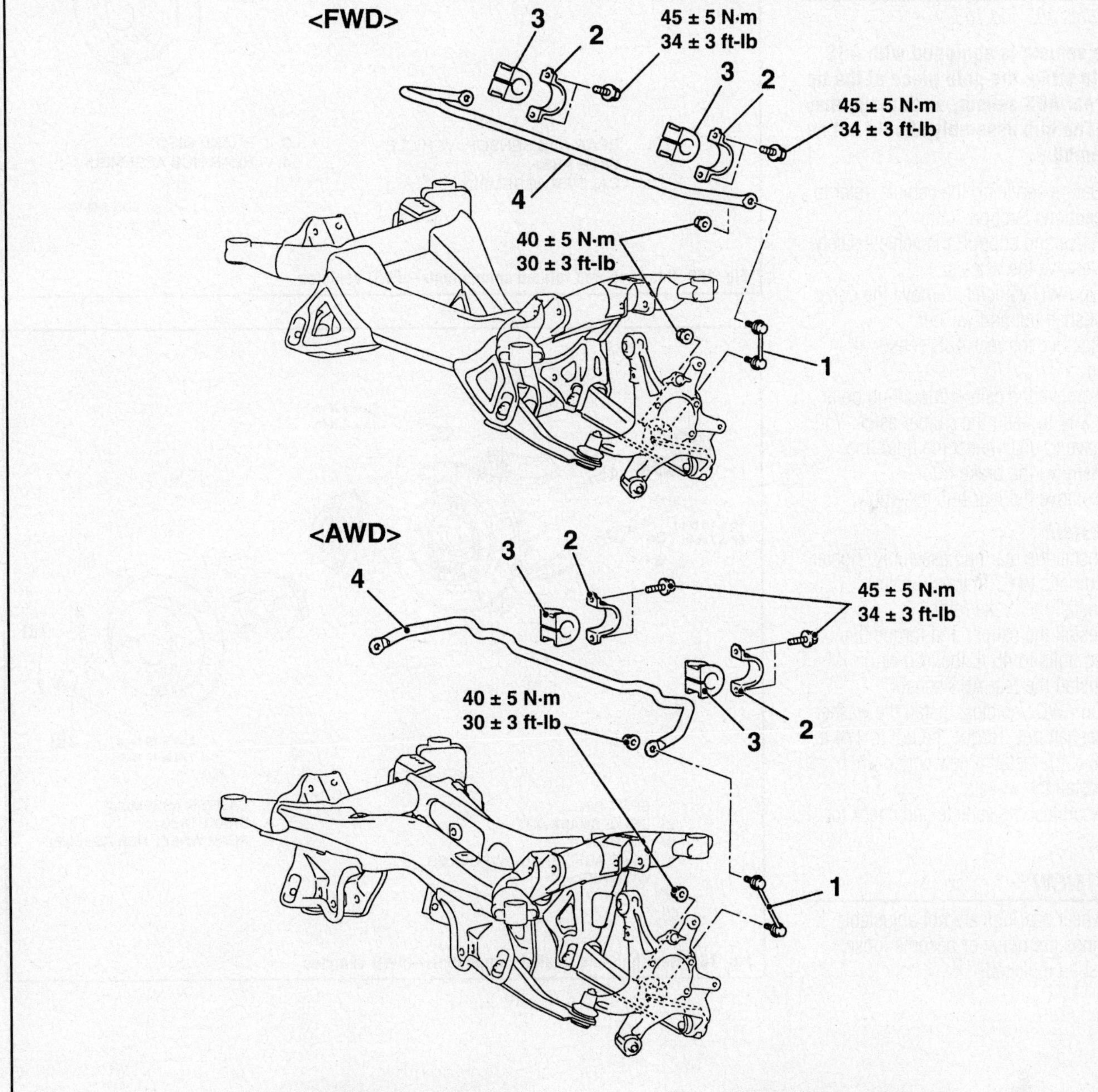

Fig. 161 Rear stabilizer bar and related components

To install:

9. Install the stabilizer from the vehicle.
10. Install the bushing.
11. Install the stabilizer bar bracket retaining bolts. Tighten to 34 ft. lbs. (45 Nm).
12. Install the stabilizer link assembly.
13. Install the stabilizer link retaining bolts. Tighten to 30 ft. lbs. (40 Nm).
14. Install the tire and wheel assembly.
15. Lower the vehicle.

WHEEL HUB AND BEARING

REMOVAL & INSTALLATION

See Figures 162 and 163.

➡If the vehicle is equipped with ABS, do not to strike the pole piece at the tip of the rear ABS sensor, as damage may result. The hub assembly should not be disassembled.

1. Before servicing the vehicle, refer to the Precautions Section.
2. Raise and support the vehicle safely.
3. Remove the wheels.
4. On AWD vehicles, remove the cotter pin, driveshaft nut and washer.
5. Remove the rear ABS sensor, if equipped.
6. Remove the caliper mounting bolts and use wire to hang the caliper aside. You do not have to disconnect the fluid line.
7. Remove the brake rotor.
8. Remove the rear hub assembly.

To install:

9. Install the rear hub assembly. Tighten mounting bolts to 67 ft. lbs. (90 Nm).
10. Install the brake rotor.
11. Install the caliper and torque the mounting bolts to 45 ft. lbs. (60 Nm).
12. Install the rear ABS sensor.
13. On AWD vehicles, install the washer and driveshaft nut. Torque the nut to 174 ft. lbs. (236 Nm). Install a new cotter pin.
14. Install the wheels.
15. Road test the vehicle and check for leaks.

ADJUSTMENT

The wheel bearings are not adjustable. If the bearings are noisy or become loose, they must be replaced.

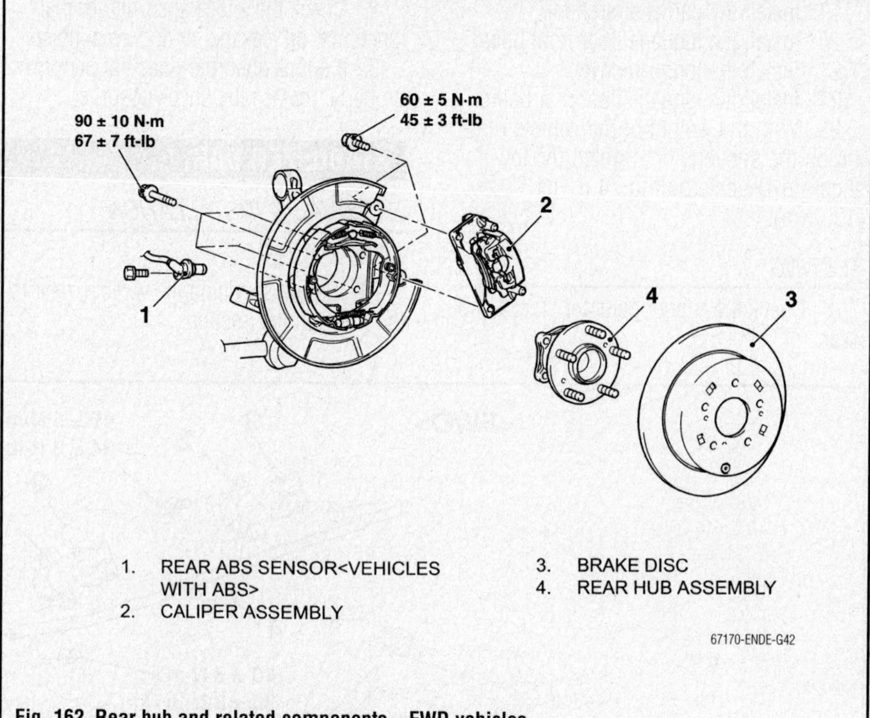

Fig. 162 Rear hub and related components—FWD vehicles

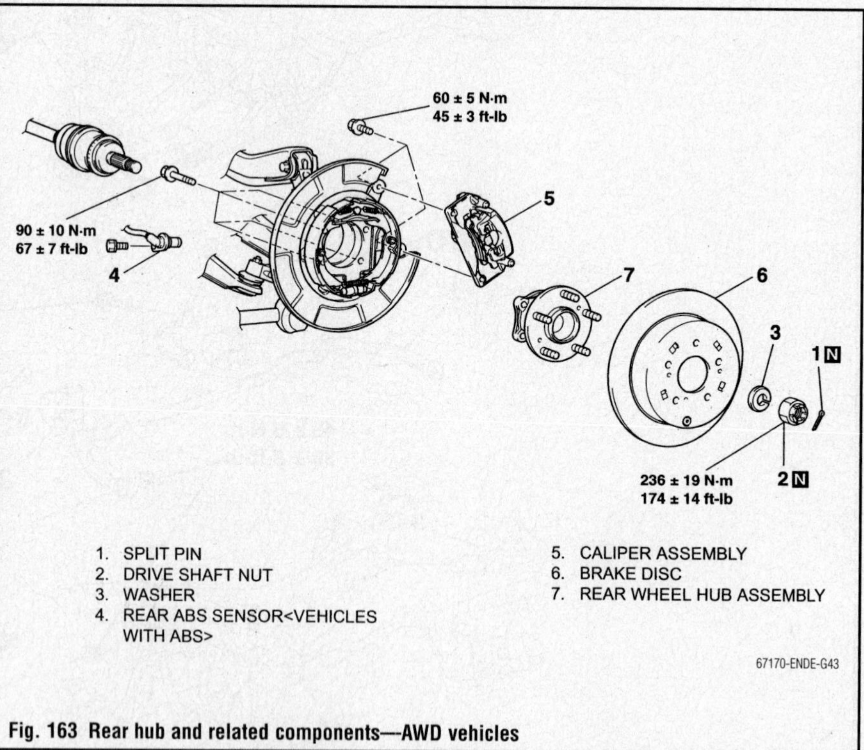

Fig. 163 Rear hub and related components—AWD vehicles

MITSUBISHI

Outlander

10

SPECIFICATIONS AND MAINTENANCE CHARTS

ENGINE AND VEHICLE IDENTIFICATION

Engine						
Code ①	Liters (cc)	Cu. In	Cyl.	Fuel Sys.	Type	Eng. Mfg.
4B12/W	2.4 (2359)	144	4	MFI	DOHC	Mitsubishi
4G69/F	2.4 (2378)	143	4	MFI	SOHC	Mitsubishi
6B31/X	3.0 (2998)	183	6	MFI	SOHC	Mitsubishi

Model Year	
Code ②	Year
7	2007
8	2008

MFI: Multiport fuel injection

SOHC: Single overhead camshaft

① Engine ID / 8th digit of the VIN

② 10th digit of the VIN

22140_OUTL_C0001

GENERAL ENGINE SPECIFICATIONS

Year	Model	Engine Displacement Liters	Engine ID/VIN	Net Horsepower @ rpm	Net Torque @ rpm (ft. lbs.)	Bore x Stroke (in.)	Com-pression Ratio	Oil Pressure @ rpm
2007	Outlander	3.0	6B31/X	220@6250	204@4000	3.45x3.54	9.5:1	①
2008	Outlander	2.4	4B12/W	168@6000	167@4100	3.43x3.78	10.5:1	①
	Outlander	3.0	6B31/X	220@6250	204@4000	3.45x3.54	9.5:1	①

① 4.2 psi or more at curb idle speed

22140_OUTL_C0002

ENGINE TUNE-UP SPECIFICATIONS

Year	Engine Displacement Liters	Engine ID/VIN	Spark Plugs Gap (in.)	Ignition Timing (deg.)		Fuel Pump (psi)	Idle Speed (rpm)		Valve Clearance	
				MT	AT		MT	AT	In.	Ex.
2007	3.0	4G69/F	0.028-0.031	—	2-8B	47	500-700	500-700	0.004	HYD
2008	2.4	4B12/W	0.028-0.031	—	2-8B	47	600-800	600-800	0.007-0.009	0.011-0.013
	3.0	6B31/X	0.028-0.031	—	2-8B	47	500-700	500-700	0.004	HYD

NOTE: The Vehicle Emission Control Information label often reflects specification changes made during production.

The label figures must be used if they differ from those in this chart.

B: Before top dead center

① Engine cold: 0.004

Engine hot: 0.008

② Engine cold: 0.008

Engine hot: 0.012

22140_OUTL_C0003

CAPACITIES

Year	Model	Engine Displacement Liters	Engine ID/VIN	Engine Oil with Filter	Transmission (pts.) 5-Spd	Transmission (pts.) Auto.	Transfer Case (pts.)	Drive Axle Front (pts.)	Drive Axle Rear (pts.)	Fuel Tank (gal.)	Cooling System (qts.)
2007	Outlander	3.0	6B31/X	4.5	—	③	1.12	—	1.06	16.6	④
2008	Outlander	2.4	4B12/W	4.8	—	③	1.12	—	1.06	16.6	④
	Outlander	3.0	6B31/X	4.5	—	③	1.12	—	1.06	16.6	④

NOTE: All capacities are approximate. Add fluid gradually and ensure a proper fluid level is obtained.

① FWD transaxle: 4.6 pts.
AWD tranaxle: 4.9 pts.

② FWD transaxle: 16.2 pts.
AWD tranaxle: 17.2 pts.

③ Automatic transaxle: 17.4 pts.
CVT tranaxle: 16.4 pts.

④ Automatic transaxle: 8.7 qts.
CVT tranaxle: 8.2 qts.

22140_OUTL_C0004

FLUID SPECIFICATIONS

Year	Model	Engine Displacemen Liters	Engine ID/VIN	Engine Oil	Auto. Trans.	Manual Trans.	Drive Axle	Transfer Case	Power Steering Fluid	Brake Master Cylinder
2007	Outlander	3.0	6B31/X	5W-20	①	—	85W-90	85W-90	Genuine Mitsubishi	DOT 3 or 4
2008	Outlander	2.4	4B12/W	5W-20	①	—	85W-90	85W-90	Genuine Mitsubishi	DOT 3 or 4
	Outlander	3.0	6B31/X	5W-20	①	—	85W-90	85W-90	Genuine Mitsubishi	DOT 3 or 4

DOT: Department Of Transpotation

① Automatic transaxle: DIA Queen J2
CVT transaxle: DIA Queen CVTF-J1

22140_OUTL_C0014

VALVE SPECIFICATIONS

Year	Engine Displacement Liters	Engine ID/VIN	Seat Angle (deg.)	Face Angle (deg.)	Spring Test Pressure (lbs. @ in.)	Spring Installed Height (in.)	Stem-to-Guide Clearance (in.) Intake	Stem-to-Guide Clearance (in.) Exhaust	Stem Diameter (in.) Intake	Stem Diameter (in.) Exhaust
2007	3.0	6B31/X	NA	43.5-44	55.13@1.940	1.9400	0.0008-0.0018	0.0014-0.0024	0.240	0.240
2008	2.4	4B12/W	NA	NA	NA	NA	0.0008-0.0019	0.0012-0.0021	NA	NA
	3.0	6B31/X	NA	43.5-44	55.13@1.940	1.9400	0.0008-0.0018	0.0014-0.0024	0.240	0.240

NA - Not Available

22140_OUTL_C0007

CAMSHAFT SPECIFICATIONS CHART

All measurements are given in inches.

Year	Engine Displacement Liters	Engine VIN	Journal Dia.	Brg. Oil Clearance	Shaft End-play	Runout	Lobe Height Intake	Lobe Height Exhaust
2007	3.0	6B31/X	1.8000	NS	NS	NS	③	④
2008	2.4	4B12/W	1.8000	NS	NS	NS	1.736	1.772
	3.0	6B31/X	1.8000	NS	NS	NS	③	④

NS - Not specified by manufacturer

① Standard value low speed cam: 1.475

Standard value high speed cam: 1.465

② Standard value: 1.491

③ Standard value low speed cam : 1.468

Standard value high speed cam: 1.426

④ Standard value: 1.490

22140_OUTL_C0013

CRANKSHAFT AND CONNECTING ROD SPECIFICATIONS

All measurements are given in inches.

Year	Engine Displacement Liters	Engine ID/VIN	Crankshaft Main Brg. Journal Dia.	Crankshaft Main Brg. Oil Clearance	Crankshaft Shaft End-play	Crankshaft Thrust on No.	Connecting Rod Journal Diameter	Connecting Rod Oil Clearance	Connecting Rod Side Clearance
2007	3.0	6B31/X	2.700	①	0.0020-0.0090	3	NA	0.0005-0.0015	0.0040-0.0090
2008	2.4	4B12/W	NA	0.0005-0.0012	0.0020-0.0010	3	NA	0.0007-0.0018	0.0040-0.0100
	3.0	6B31/X	2.700	①	0.0020-0.0090	3	NA	0.0005-0.0015	0.0040-0.0090

NA - Not Available

① Journal 1 and 4: 0.0008-0.0014 in..

Journal 2 and 3: 0.0010-0.0017 in..

22140_OUTL_C0005

PISTON AND RING SPECIFICATIONS

All measurements are given in inches.

Year	Engine Displacement Liters	Engine ID/VIN	Piston Clearance	Ring Gap Top Compression	Ring Gap Bottom Compression	Ring Gap Oil Control	Ring Side Clearance Top Compression	Ring Side Clearance Bottom Compression	Ring Side Clearance Oil Control
2007	3.0	6B31/X	N/A	0.0080-0.0120	0.0120-0.0180	0.0040-0.0230	0.0016-0.0031	0.0012-0.0027	NA
2008	2.4	4B12/W	N/A	0.0060-0.0110	0.0100-0.0160	0.0040-0.0140	0.0010-0.0030	0.0010-0.0030	NA
	3.0	6B31/X	N/A	0.0080-0.0120	0.0120-0.0180	0.0040-0.0230	0.0016-0.0031	0.0012-0.0027	NA

NA - Not Available

22140_OUTL_C0006

TORQUE SPECIFICATIONS

All readings in ft. lbs.

Year	Engine Displacement Liters	Engine ID/VIN	Cylinder Head Bolts	Main Bearing Bolts	Rod Bearing Bolts	Crankshaft Damper Bolts	Flywheel Bolts	Manifold Intake	Manifold Exhaust	Spark Plugs	Oil Pan Drain Plug
2007	3.0	6B31/X	⑤	18 ②	15 ②	⑥	56	16	36	13	29
2008	2.4	4B12/W	⑦	20 ⑧	14 ②	155	96	15	36	19	29
	3.0	6B31/X	⑤	18 ②	15 ②	⑥	56	16	36	13	29

① Step 1: Tighten all bolts to 58 ft. lbs.
Step 2: Loosen all bolts to 0 ft. lbs.
Step 3: Tighten all bolts to 15 ft. lbs.
Step 4: Tighten all bolts 90 degrees.
Step 5: Tighten all bolts an additional 90 degrees.

② Torque to specification plus an additional 90 degrees.

③ Bolt: 18 ft. lbs.
Nut: 15 ft. lbs.

④ Cover bolt, except California: 124 inch lbs.
Cover bolt, California: 22 ft. lbs.
Exhaust bracket bolt, California: 24 ft. lbs.
Nut: 36 ft. lbs.

⑤ Step 1: Tighten all bolts to 33 ft. lbs.
Step 2: Tighten all bolts 150 degrees.

⑥ Step 1: Tighten bolt to 148 ft. lbs.
Step 2: Loosen bolt to 0 ft. lbs.
Step 3: Tighten bolt to 81 ft. lbs.
Step 4: Tighten bolt 60 degrees.

⑦ Step 1: Tighten all bolts to 26 ft. lbs.
Step 2: Tighten all bolts 90 degrees.
Step 3: Tighten all bolts an additional 90 degrees.

⑧ Torque to specification plus an additional 45 degrees.

22140_OUTL_C0008

WHEEL ALIGNMENT

Year	Model		Caster Range (+/-Deg.)	Caster Preferred Setting (Deg.)	Camber Range (+/-Deg.)	Camber Preferred Setting (Deg.)	Toe-in (in.)
2007	Outlander	F	0.30	+2.35	0.30	0.20	0.04 +/- 0.09
		R	—	—	0.30	-0.25	0.12 +/- 0.08
2008	Outlander	F	0.30	+2.35	0.30	0.20	0.04 +/- 0.09
		R	—	—	0.30	-0.25	0.12 +/- 0.08

22140_OUTL_C0009

TIRE, WHEEL AND BALL JOINT SPECIFICATIONS

Year	Model	OEM Tires		Tire Pressures (psi)		Wheel	Ball Joint	Lug Nut
		Standard	Optional	Front	Rear	Size	Inspection ①	(ft. lbs.)
2007	Outlander	P215/70R16	P225/55R18	③	③	④	U: 4-26 in. L: 0-35 in.	66-80
2008	Outlander	P215/70R16	P225/55R18	③	③	④	U: 4-26 in. L: 0-35 in.	66-80

OEM: Original Equipment Manufacturer

PSI: Pounds Per Square Inch. If specification differs from the one located on driver's door, use the driver's door specification

L: Lower

U: Upper

① Torque required in inch lbs. to rotate ball joint when removed from the knuckle

② Standard: 6-JJ. Optional: 6.5JJ

③ Refer to the label attached to the center pillar on the Driver's side of the vehicle

④ Standard: 6.5JJ. Optional: 7JJ

22140_OUTL_C0010

BRAKE SPECIFICATIONS

All measurements in inches unless noted

Year	Model		Brake Disc			Brake Drum Diameter			Minimum Lining Thickness		Brake Caliper	
			Original Thickness	Minimum Thickness	Maximum Runout	Original Inside Diameter	Max. Wear Limit	Maximum Machine Diameter	Front	Rear	Bracket Bolts (ft. lbs.)	Mounting Bolts (ft. lbs.)
2007	Outlander	F	1.020	0.960	0.0024	—	—	—	0.080	—	55	74
		R	0.390	0.330	0.0032	—	—	—	—	0.080	33	33
2008	Outlander	F	1.020	0.960	0.0024	—	—	—	0.080	—	55	74
		R	0.390	0.330	0.0032	—	—	—	—	0.080	33	33

NA: Not Available

F: Front

R: Rear

22140_OUTL_C0011

SCHEDULED MAINTENANCE INTERVALS
Mitsubishi—Outlander

TO BE SERVICED	TYPE OF SERVICE	VEHICLE MILEAGE INTERVAL (x1000)													
		7.5	15	22.5	30	37.5	45	52.5	60	67.5	75	82.5	90	97.5	102.5
Engine oil & filter	R	✓	✓	✓	✓	✓	✓	✓	✓	✓	✓	✓	✓	✓	✓
Air cleaner element	R				✓				✓				✓		
Automatic transaxle fluid	S/I		✓		✓		✓		✓		✓		✓		
Manual transaxle fluid	S/I		✓		✓		✓		✓		✓		✓		
Brake hoses	S/I		✓		✓		✓		✓		✓		✓		
Disc brake pads	S/I		✓		✓		✓		✓		✓		✓		
Driveshaft boots	S/I		✓		✓		✓		✓		✓		✓		
Valve clearance	S/I				✓				✓				✓		
Engine coolant	R								✓				✓		
Spark plugs (standard)	R								✓				✓		
Spark plugs (platinum)	R								✓				✓		
Spark plugs (iridium)	R														✓
Ball joints & steering linkage seals	S/I				✓				✓				✓		
Drive belt(s)	S/I				✓				✓				✓		
Exhaust system	S/I				✓				✓				✓		
Fuel hoses	S/I				✓				✓				✓		
Transfer case fluid	S/I				✓				✓				✓		
Transfer case fluid	R								✓						
Rear drum brake linings & rear wheel cylinders	S/I				✓				✓				✓		
Ignition cables	R								✓						
Timing belt(s)	R								✓						
EVAP system (except canister)	S/I								✓						
Fuel system (tank, pipe line, connection & fuel tank filler tube cap)	S/I								✓						
Tires (rotate)	S/I	✓	✓	✓	✓	✓	✓	✓	✓	✓	✓	✓	✓	✓	✓

R: Replace S/I: Service or Inspect

FREQUENT OPERATION MAINTENANCE (SEVERE SERVICE)

If a vehicle is operated under any of the following conditions it is considered severe service:

- Extremely dusty areas.

- 50% or more of the vehicle operation is in 32°C (90°F) or higher temperatures, or constant operation in temperatures below 0°C (32°F).

- Prolonged idling (vehicle operation in stop and go traffic).

- Frequent short running periods (engine does not warm to normal operating temperatures).

- Police, taxi, delivery usage or trailer towing usage.

Oil & oil filter: change every 3750 miles.

Disc brake pads: service or inspect every 6000 miles.

Rear drum brake linings and rear wheel cylinders: service or inspect every 15,000 miles

Air filter element: service or inspect every 15,000 miles.

Automatic transaxle fluid & filter: replace every 15,000 miles.

22140_OUTL_C0012

PRECAUTIONS

Before servicing any vehicle, please be sure to read all of the following precautions, which deal with personal safety, prevention of component damage, and important points to take into consideration when servicing a motor vehicle:

- Never open, service or drain the radiator or cooling system when the engine is hot; serious burns can occur from the steam and hot coolant.
- Observe all applicable safety precautions when working around fuel. Whenever servicing the fuel system, always work in a well-ventilated area. Do not allow fuel spray or vapors to come in contact with a spark, open flame, or excessive heat (a hot drop light, for example). Keep a dry chemical fire extinguisher near the work area. Always keep fuel in a container specifically designed for fuel storage; also, always properly seal fuel containers to avoid the possibility of fire or explosion. Refer to the additional fuel system precautions later in this section.
- Fuel injection systems often remain pressurized, even after the engine has been turned **OFF**. The fuel system pressure must be relieved before disconnecting any fuel lines. Failure to do so may result in fire and/or personal injury.
- Brake fluid often contains polyglycol ethers and polyglycols. Avoid contact with the eyes and wash your hands thoroughly after handling brake fluid. If you do get brake fluid in your eyes, flush your eyes with clean, running water for 15 minutes. If eye irritation persists, or if you have taken brake fluid internally, IMMEDIATELY seek medical assistance.
- The EPA warns that prolonged contact with used engine oil may cause a number of skin disorders, including cancer. You should make every effort to minimize your exposure to used engine oil. Protective gloves should be worn when changing oil. Wash your hands and any other exposed skin areas as soon as possible after exposure to used engine oil. Soap and water, or waterless hand cleaner should be used.
- All new vehicles are now equipped with an air bag system, often referred to as a Supplemental Restraint System (SRS) or Supplemental Inflatable Restraint (SIR) system. The system must be disabled before performing service on or around system components, steering column, instrument panel components, wiring and sensors. Failure to follow safety and disabling procedures could result in accidental air bag deployment, possible personal injury and unnecessary system repairs.
- Always wear safety goggles when working with, or around, the air bag system. When carrying a non-deployed air bag, be sure the bag and trim cover are pointed away from your body. When placing a non-deployed air bag on a work surface, always face the bag and trim cover upward, away from the surface. This will reduce the motion of the module if it is accidentally deployed. Refer to the additional air bag system precautions later in this section.
- Clean, high quality brake fluid from a sealed container is essential to the safe and proper operation of the brake system. You should always buy the correct type of brake fluid for your vehicle. If the brake fluid becomes contaminated, completely flush the system with new fluid. Never reuse any brake fluid. Any brake fluid that is removed from the system should be discarded. Also, do not allow any brake fluid to come in contact with a painted surface; it will damage the paint.
- Never operate the engine without the proper amount and type of engine oil; doing so WILL result in severe engine damage.
- Timing belt maintenance is extremely important. Many models utilize an interference-type, non-freewheeling engine. If the timing belt breaks, the valves in the cylinder head may strike the pistons, causing potentially serious (also time-consuming and expensive) engine damage. Refer to the maintenance interval charts for the recommended replacement interval for the timing belt, and to the timing belt section for belt replacement and inspection.
- Disconnecting the negative battery cable on some vehicles may interfere with the functions of the on-board computer system(s) and may require the computer to undergo a relearning process once the negative battery cable is reconnected.
- When servicing drum brakes, only disassemble and assemble one side at a time, leaving the remaining side intact for reference.
- Only an MVAC-trained, EPA-certified automotive technician should service the air conditioning system or its components.

BRAKES

ANTI-LOCK BRAKE SYSTEM (ABS)

GENERAL INFORMATION

The Anti-lock Brake System (ABS) electronically controls the brake fluid pressure of all 4 wheels to ensure stability and operability (driveability) in the direction of the vehicle upon quick braking. This ABS features a 4 sensor, 3 channel system with left/right independent control of the front wheels and integrated control of the rear wheels (select low control*).

➡*Select low control: Control system that compares the speeds of the right and left wheels and performs the same fluid pressure control on both wheels according to the speed of the wheel that is likely to be locked

In ABS, electronic control method is used whereby the rear wheel brake hydraulic pressure during braking is regulated by rear wheel control solenoid valves in accordance with the vehicle's rate of deceleration and the front and rear wheel slippage which are calculated from the signals received from the various wheel sensors. EBD control is a control system which provides a high level of control for both vehicle braking force and vehicle stability. The system has the following features:

- EBD (Electronic Brake-force Distribution system) control has been added to provide the ideal braking force for the rear wheels.
- Fail-safe function which ensures that safety is maintained
- Diagnostic function which provides improved serviceability
- Because the system provides the optimum rear wheel braking force regardless of the vehicle laden condition and the condition of the road surface, the system reduces the required pedal depression force, particularly when the vehicle is heavily laden or driving on road surfaces with high frictional coefficients.
- Because the duty placed on the front brakes has been reduced, the increases in pad temperature can be controlled during front brakes applying to improve the wear resistance characteristics of the pad.

WHEEL SPEED SENSORS

REMOVAL & INSTALLATION

Front

See Figure 1.

1. Before servicing the vehicle, refer to the precautions.
2. Remove the air cleaner assembly.
3. Raise and safely support the vehicle.
4. Remove the front wheel and tire assemblies.

⁂ WARNING

The wheel speed sensor collects any metallic particles easily, because it is magnetized. Make sure that the sensor does not collect any metallic particles. Check that there is no trouble prior to reassembling it.

5. Remove the bolt from the front wheel speed sensor and knuckle connection.
6. Disconnect the front wheel speed sensor connector.
7. Remove the front wheel speed sensor grommet
8. Remove the front wheel speed sensor.

To install:

9. Install the front wheel speed sensor. Tighten to 76 inch lbs. (9 Nm).
10. Install the front wheel speed sensor grommet
11. Connect the front wheel speed sensor connector.
12. Install the bolt from the front wheel speed sensor and knuckle connection. Tighten to 98 inch lbs. (11 Nm).

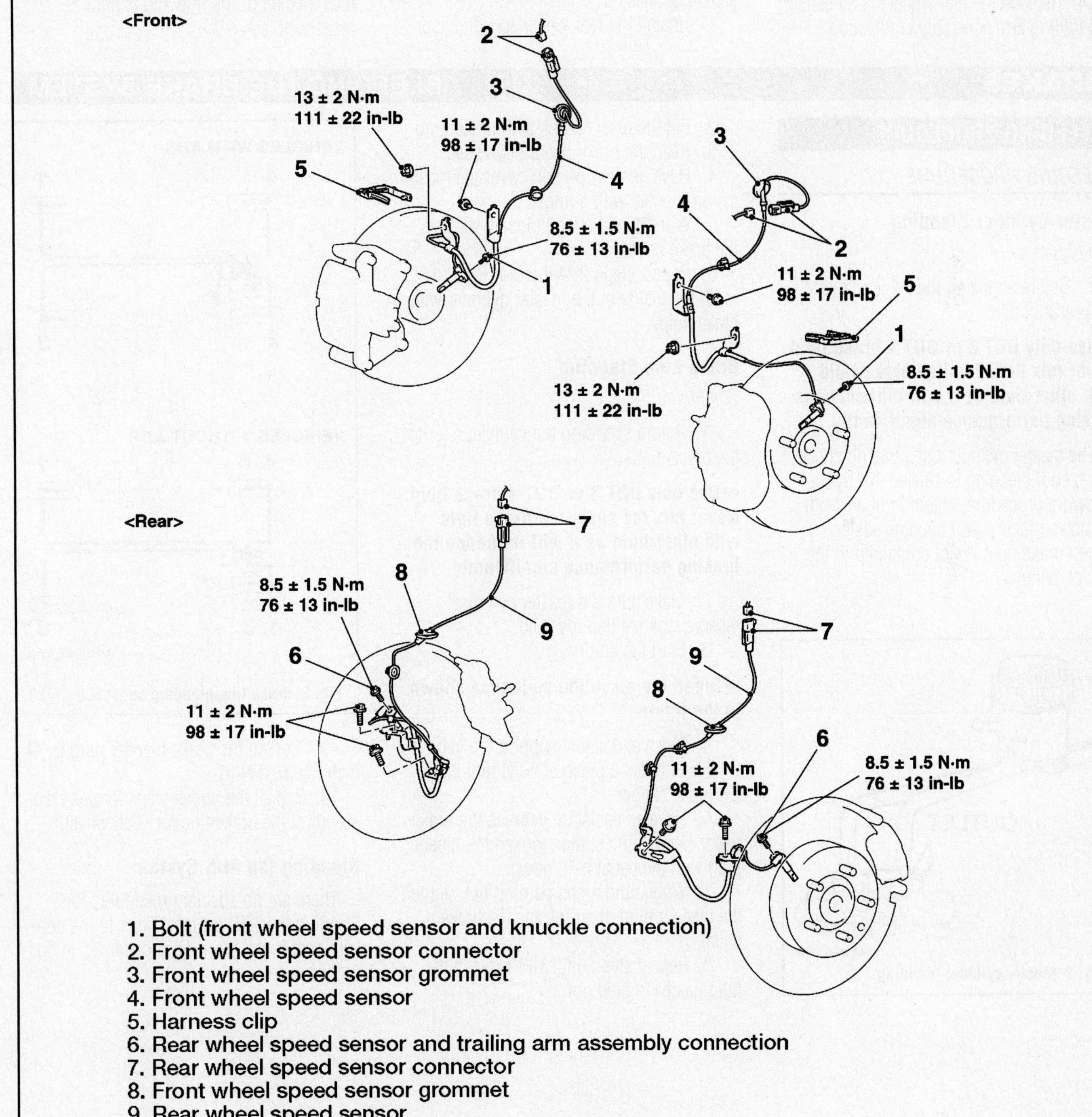

Fig. 1 Exploded view of the front and rear speed sensors

13. Install the front wheel and tire assemblies.
14. Lower the vehicle.
15. Install the air cleaner assembly.

Rear

See Figure 1.

1. Before servicing the vehicle, refer to the precautions.
2. Raise and safely support the vehicle.
3. Remove the rear wheel and tire assemblies.
4. Remove the rear wheel speed sensor and trailing arm assembly connection.
5. Disconnect the rear wheel speed sensor connector.
6. Remove the rear wheel speed sensor grommet.
7. Remove the rear wheel speed sensor.

To install:

8. Install the rear wheel speed sensor. Tighten to 76 inch lbs. (9 Nm).
9. Install the rear wheel speed sensor grommet.
10. Connect the rear wheel speed sensor connector.
11. Install the rear wheel speed sensor and trailing arm assembly connection. Tighten to 98 inch lbs. (11 Nm).
12. Install the rear wheel and tire assemblies.
13. Lower the vehicle.

WHEEL SPEED SENSOR RINGS (TOOTHED RINGS)

REMOVAL & INSTALLATION

The ABS wheel speed sensor ring is an integral part of either the BJ assembly of the halfshaft or the hub and cannot be disassembled.

BRAKES

BLEEDING THE BRAKE SYSTEM

BLEEDING PROCEDURE

BLEEDING PROCEDURE

Master Cylinder Bleeding

See Figure 2.

1. Before servicing the vehicle, refer to the precautions.

➡Use only DOT 3 or DOT 4 brake fluid. Never mix the specified brake fluid with other fluid as it will influence the braking performance significantly.

The master cylinder used has no check valve, so if bleeding is carried out by the following procedure, bleeding of air from the brake pipeline will become easier. (When brake fluid is not contained in the master cylinder).

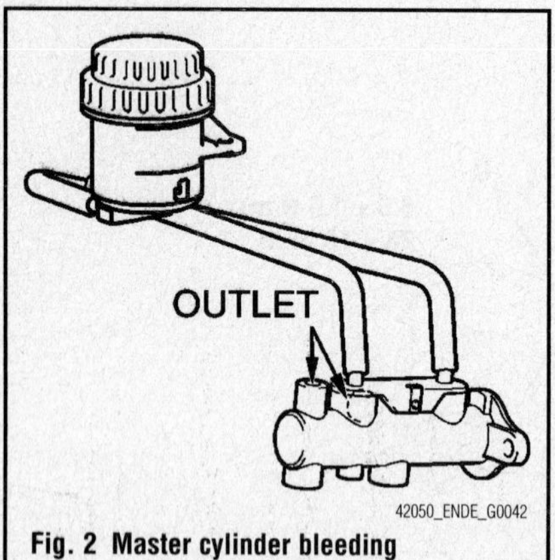

Fig. 2 Master cylinder bleeding

2. Fill the reserve tank with brake fluid.
3. Keep the brake pedal depressed.
4. Have another person cover the master cylinder outlet with a finger.
5. With the outlet still closed, release the brake pedal.
6. Repeat steps 2–4 three or four times to fill the inside of the master cylinder with brake fluid.

Brake Line Bleeding

See Figure 3.

1. Before servicing the vehicle, refer to the precautions.

➡Use only DOT 3 or DOT 4 brake fluid. Never mix the specified brake fluid with other fluid as it will influence the braking performance significantly.

2. Make sure the master cylinder reserve tank is filled the fluid.
3. Start the engine.

➡Bleed the air in the sequence shown in the figure.

4. Raise and safely support the vehicle.
5. Connect a piece of vinyl tubing to the brake caliper.
6. Have an assistant depress the brake pedal several times, then loosen the bleeder plug with the pedal held down.
7. When fluid stops coming out, tighten the bleeder plug, then release the brake pedal.
8. Repeat steps until all the air in the fluid has been bled out.

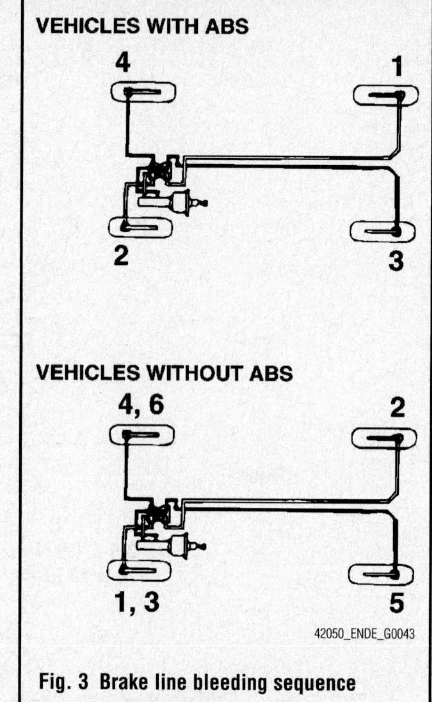

Fig. 3 Brake line bleeding sequence

9. Tighten the brake bleeder plug to 74 inch lbs. (8.3 Nm).
10. Repeat the above steps to bleed the air out of the brake line for each wheel.

Bleeding the ABS System

There are no special procedures for bleeding the ABS system. Refer to the conventional brake bleeding procedures in this section.

CAUTION

Dust and dirt accumulating on brake parts during normal use may contain asbestos fibers from production or aftermarket brake linings. Breathing excessive concentrations of asbestos fibers can cause serious bodily harm. Exercise care when servicing brake parts. Do not sand or grind brake lining unless equipment used is designed to contain the dust residue. Do not clean brake parts with compressed air or by dry brushing. Cleaning should be done by dampening the brake components with a fine mist of water, then wiping the brake components clean with a dampened cloth. Dispose of cloth and all residue containing asbestos fibers in an impermeable container with the appropriate label. Follow practices prescribed by the Occupational Safety and Health Administration (OSHA) and the Environmental Protection Agency (EPA) for the handling, processing, and disposing of dust or debris that may contain asbestos fibers.

BRAKE CALIPER

REMOVAL & INSTALLATION

See Figures 4 through 6.

1. Before servicing the vehicle, refer to the precautions.
2. Drain brake fluid from the master cylinder.
3. Raise and support the vehicle safely.
4. Remove the wheels.
5. Disconnect the brake hose. Discard the gaskets.
6. Remove the caliper mounting bolts.
7. Remove the caliper assembly.

To install:

8. Install the brake caliper into position. Tighten bolt to 74 ft. lbs. (100 Nm).
9. Connect the brake hose with new gaskets. Tighten banjo bolt to 22 ft. lbs. (30 Nm).
10. Bleed the brake system.
11. Install the wheels and lower the vehicle.
12. Depress the brake pedal several times to seat the brake pads against the rotors.

DISC BRAKE PADS

REMOVAL & INSTALLATION

See Figures 4 through 6.

1. Before servicing the vehicle, refer to the precautions.
2. Raise and support the vehicle safely.
3. Remove some of the brake fluid from the master cylinder reservoir.
4. Remove the wheels.
5. Remove the caliper guide and lock pins and lift the caliper assembly from the caliper support.

➡On some vehicles, the front caliper can be flipped up by leaving the upper pin in place and using it as a pivot point.

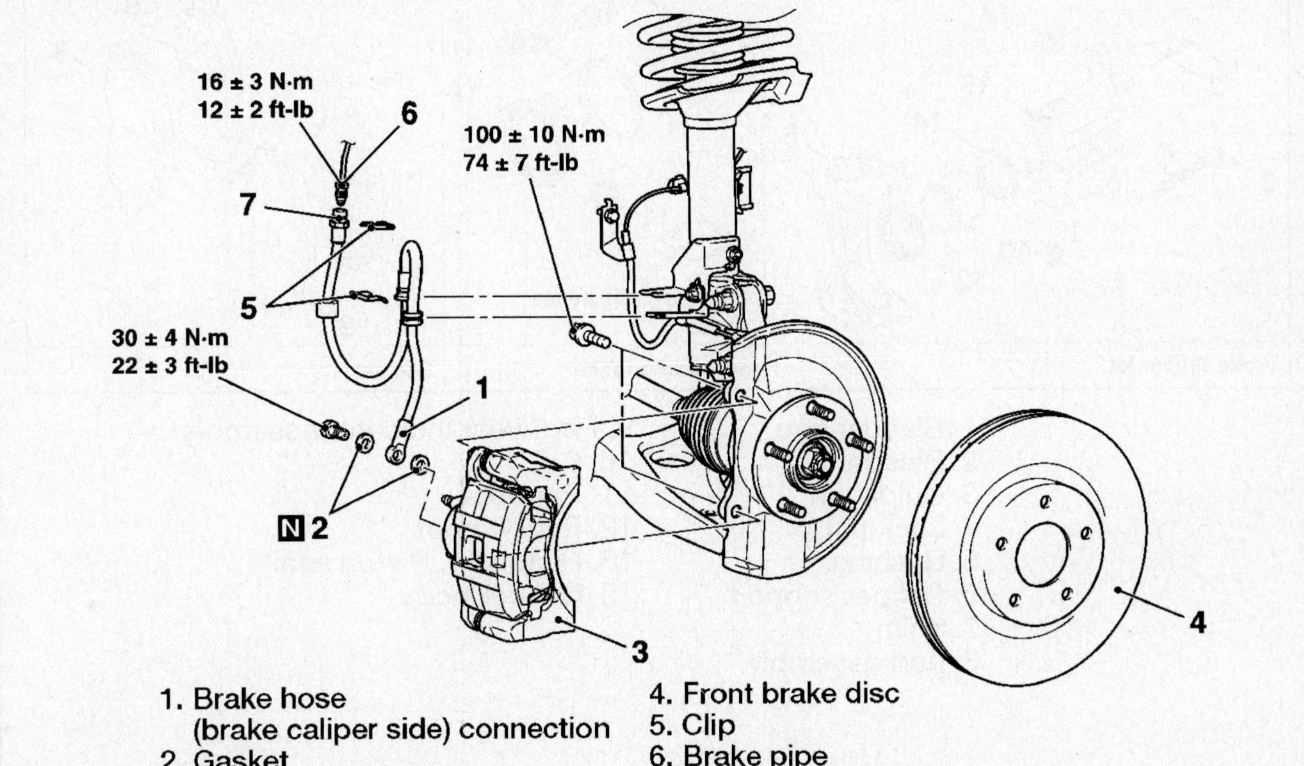

Fig. 4 Front brake caliper and related components

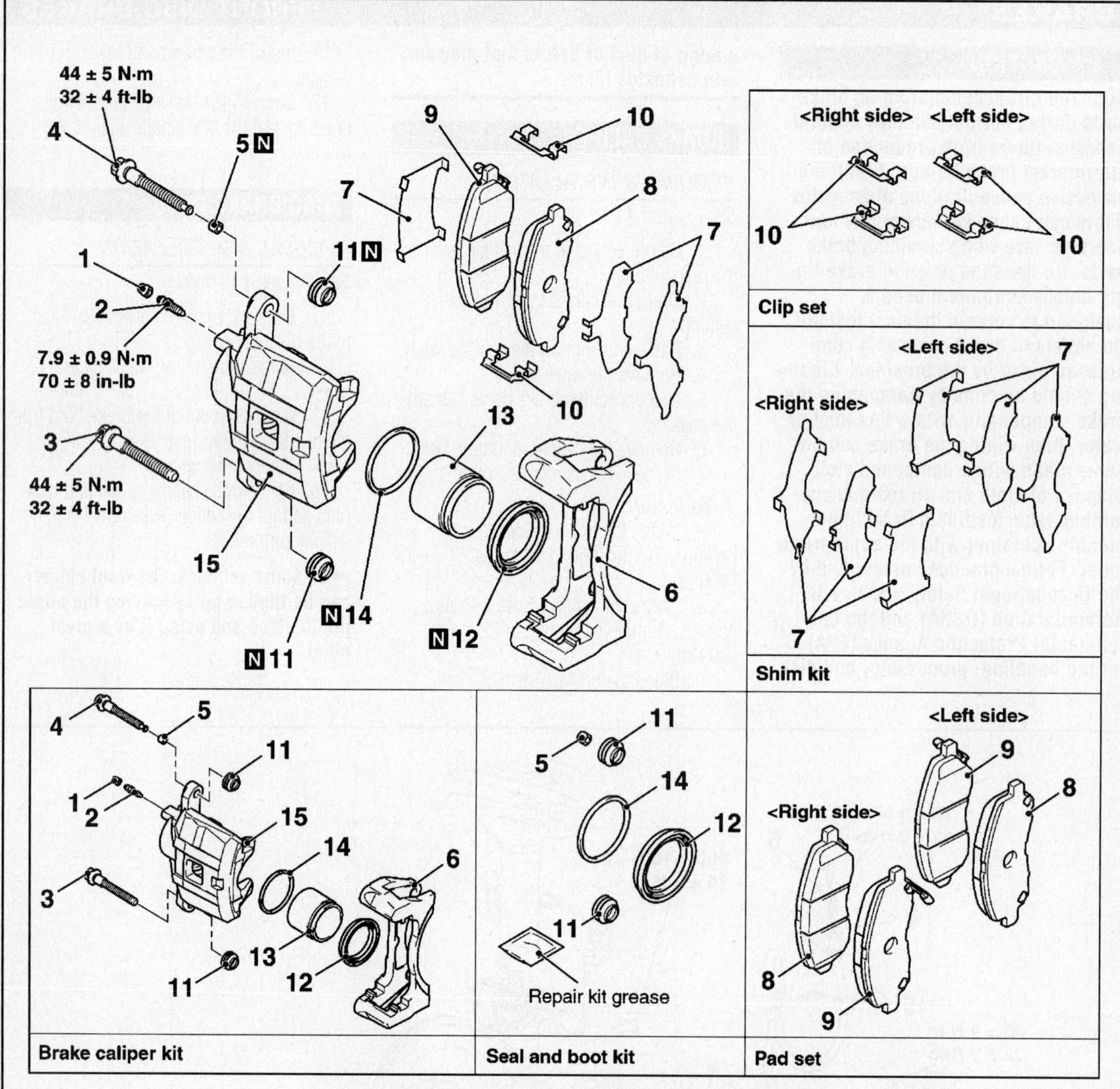

1. Bleeder cap
2. Bleeder
3. Guide pin
4. Lock pin
5. Bushing
6. Caliper support
7. Shim
8. Pad assembly
9. Pad/wear indicator assembly
10. Clip
11. Pin boot
12. Piston boot
13. Piston 14. Piston seal
15. Caliper body

22140_OUTL_G0069

Fig. 5 Exploded view of the front brake caliper

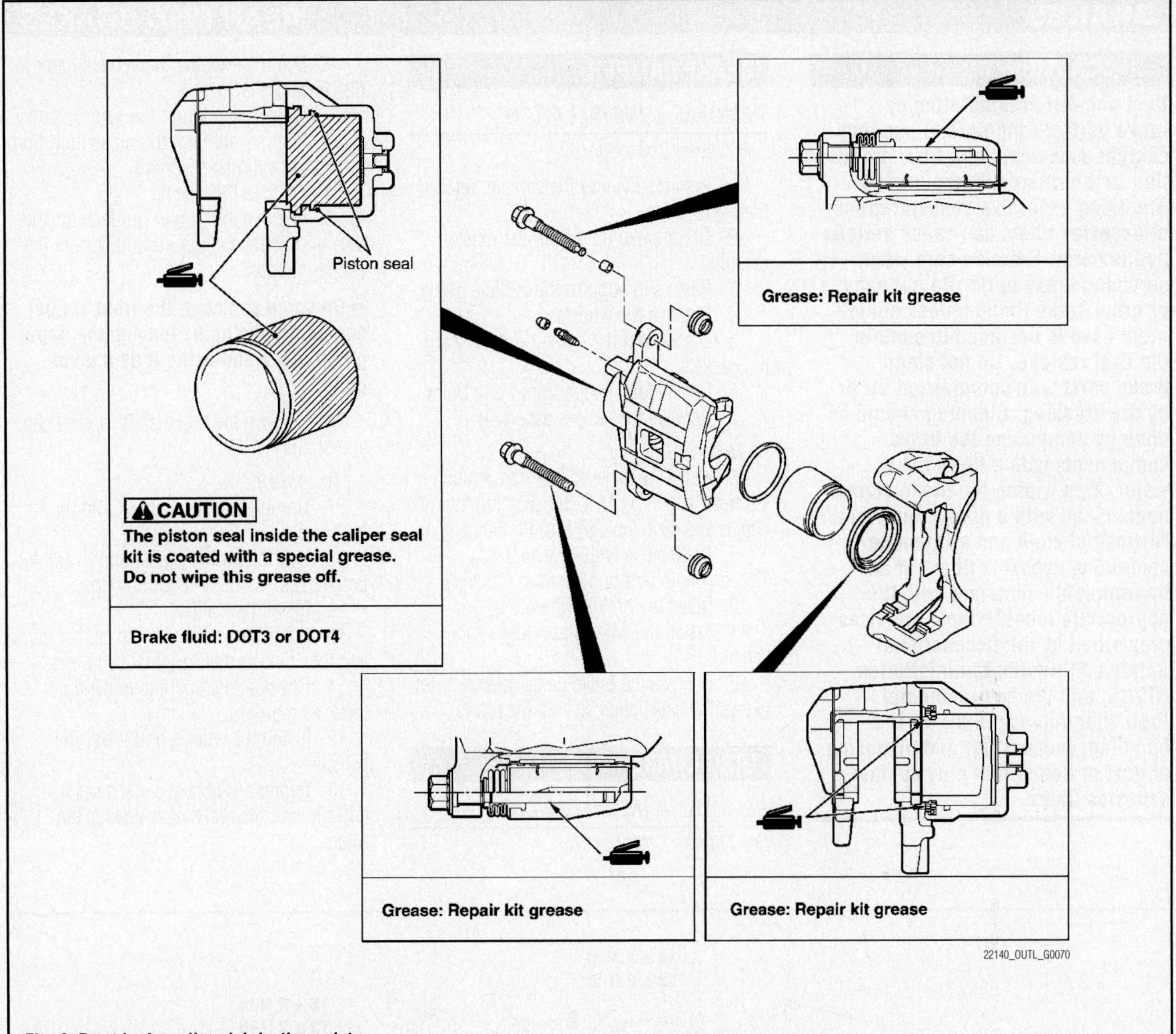
Piston seal
CAUTION
The piston seal inside the caliper seal kit is coated with a special grease. Do not wipe this grease off.
Brake fluid: DOT3 or DOT4
Grease: Repair kit grease
Grease: Repair kit grease
Grease: Repair kit grease
22140_OUTL_G0070

Fig. 6 Front brake caliper lubrication points

6. Remove the brake pads, spring clip and shims.

To install:

7. Compress pistons back into the caliper bore.
8. Lubricate slide points and install the brake pads, shims and spring clips.
9. Install the caliper.
10. Lubricate and install the caliper guide and lock pins in their original positions.
11. Check and adjust the brake fluid level, as required.
12. Install the wheels and lower the vehicle.
13. Depress the brake pedal several times to seat the brake pads against the rotors.

⁂ CAUTION

Dust and dirt accumulating on brake parts during normal use may contain asbestos fibers from production or aftermarket brake linings. Breathing excessive concentrations of asbestos fibers can cause serious bodily harm. Exercise care when servicing brake parts. Do not sand or grind brake lining unless equipment used is designed to contain the dust residue. Do not clean brake parts with compressed air or by dry brushing. Cleaning should be done by dampening the brake components with a fine mist of water, then wiping the brake components clean with a dampened cloth. Dispose of cloth and all residue containing asbestos fibers in an impermeable container with the appropriate label. Follow practices prescribed by the Occupational Safety and Health Administration (OSHA) and the Environmental Protection Agency (EPA) for the handling, processing, and disposing of dust or debris that may contain asbestos fibers.

BRAKE CALIPER

REMOVAL & INSTALLATION

See Figures 7 through 9.

1. Before servicing the vehicle, refer to the precautions.
2. Drain brake fluid from the master cylinder.
3. Raise and support the vehicle safely.
4. Remove the wheels.
5. Disconnect the brake hose. Discard the gaskets.
6. Remove the caliper mounting bolts
7. Remove the caliper assembly.

To install:

8. Install the brake caliper into position. Tighten the bolts to 41 ft. lbs. (55 Nm) for 2.4L engine and 43 ft. lbs. (58 Nm) for 3.0L engine.
9. Brake hose with new gaskets. Tighten banjo bolt to 12 ft. lbs. (18 Nm).
10. Bleed the brake system.
11. Install the wheels and lower the vehicle.
12. Depress the brake pedal several times to seat the brake pads against the rotors.

DISC BRAKE PADS

REMOVAL & INSTALLATION

See Figures 7 through 9.

1. Before servicing the vehicle, refer to the precautions.
2. Raise and support the vehicle safely.
3. Remove some of the brake fluid from the master cylinder reservoir.
4. Remove the wheels.
5. Remove the caliper guide and lock pins and lift the caliper assembly from the caliper support.

➡On some vehicles, the front caliper can be flipped up by leaving the upper pin in place and using it as a pivot point.

6. Remove the brake pads, spring clip and shims.

To install:

7. Compress pistons back into the caliper bore.
8. Lubricate slide points and install the brake pads, shims and spring clips.
9. Install the caliper.
10. Lubricate and install the caliper guide and lock pins in their original positions.
11. Check and adjust the brake fluid level, as required.
12. Install the wheels and lower the vehicle.
13. Depress the brake pedal several times to seat the brake pads against the rotors.

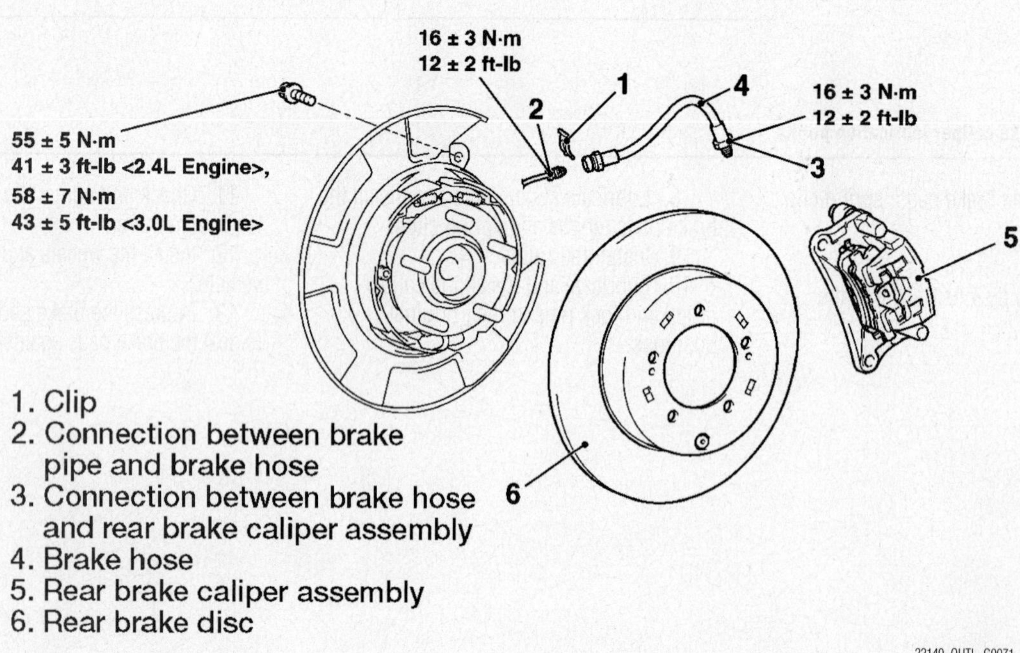

Fig. 7 Rear brake caliper and related components

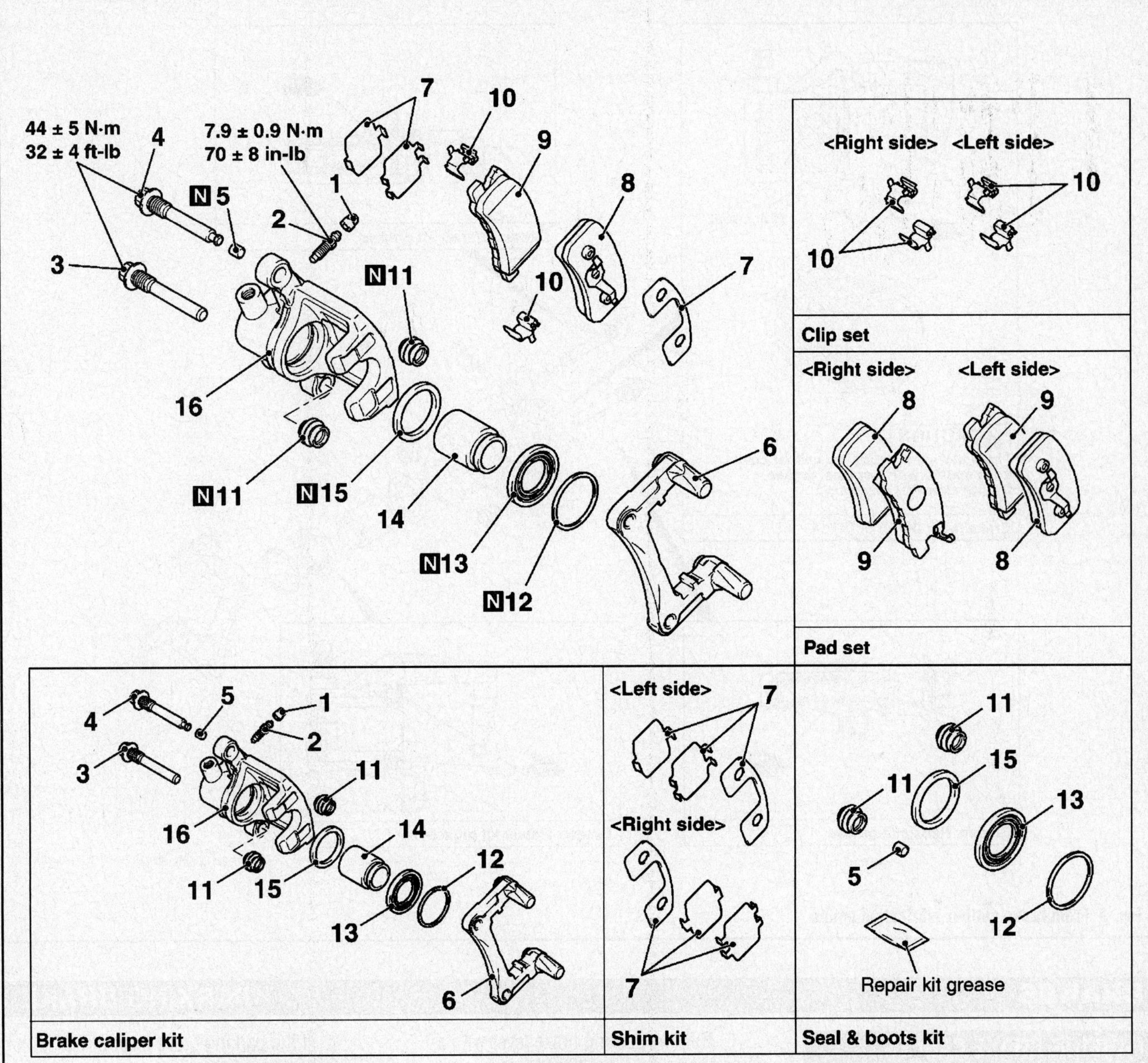

Fig. 8 Exploded view of the front brake caliper

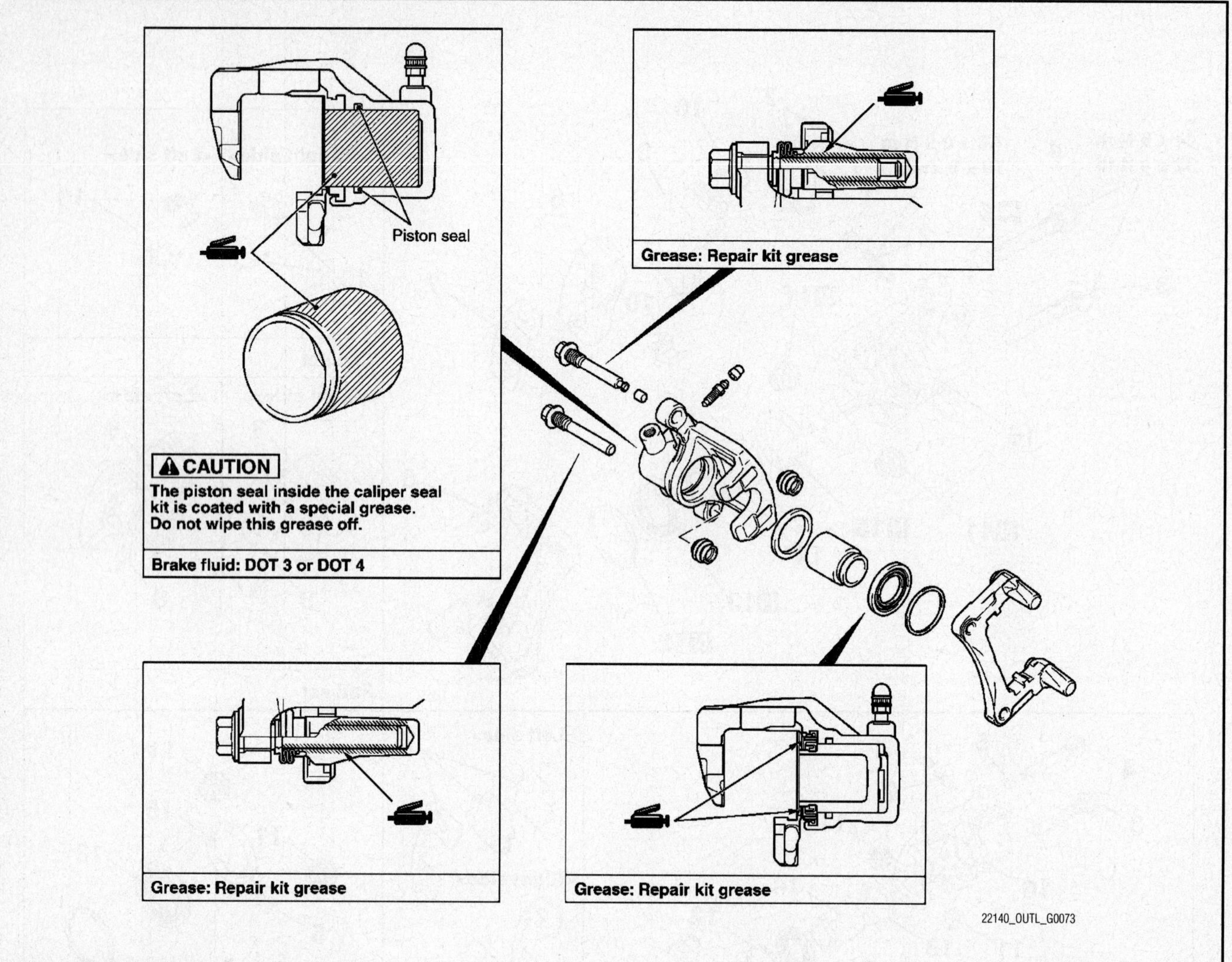

Piston seal

CAUTION

The piston seal inside the caliper seal kit is coated with a special grease. Do not wipe this grease off.

Brake fluid: DOT 3 or DOT 4

Grease: Repair kit grease

Grease: Repair kit grease

Grease: Repair kit grease

22140_OUTL_G0073

Fig. 9 Front brake caliper lubrication points

BRAKES

PARKING BRAKE

PARKING BRAKE CABLES

ADJUSTMENT

See Figures 10 and 11.

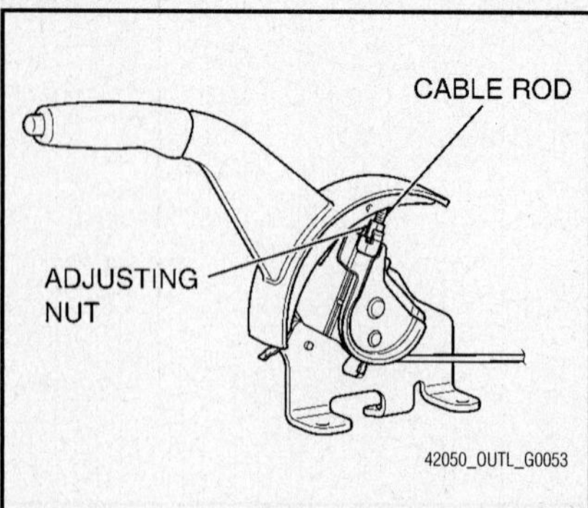

Fig. 10 Loosen the adjusting nut to move it to the cable rod end so that the cable will be free

1. Pull the parking brake lever with a force of about 45 lbs. (200 N) and count the number of notches. The proper range is 4–5 notches.

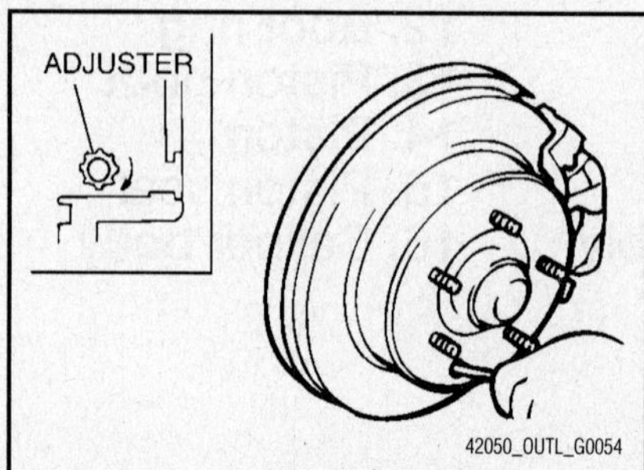

Fig. 11 Insert a flat-tip screwdriver to turn the adjuster to the arrow direction (to expand the shoe) until the parking brake shoe makes contact and the disc can no longer be turned

2. If the parking brake lever stroke is not the standard value, adjust as outlined in the following steps.
3. Remove the floor console assembly.
4. Loosen the adjusting nut to move it to the cable rod end so that the cable will be free.
5. Remove the rear brake adjusting hole plug. Then insert a flat-tip screwdriver to turn the adjuster to the arrow direction (to expand the shoe) until the parking brake shoe makes contact and the disc can no longer be turned. Back off the adjuster to the opposite direction by five notches.
6. If the parking brake lever stroke is below the standard value and the braking is too firm, the rear brakes may drag.
7. Adjust the parking brake lever stroke to the standard value by turning the adjusting nut. After the adjustment, ensure that

there is no free play between the adjusting nut and the parking brake lever.

8. After adjusting the parking brake lever stroke, raise and safely support the rear end of the vehicle, then release the parking brake and turn the rear wheels to make sure the rear brakes are not dragging.

PARKING BRAKE SHOES

REMOVAL & INSTALLATION

See Figures 12 through 14.

1. Raise and safely remove the vehicle.
2. Release the parking brake lever.
3. Unbolt and remove the caliper, but do not disconnect the fluid line. Suspend the caliper with a piece of wire. Do NOT let it hang by its brake line.
4. Remove the rotor.
5. Remove the shoe-to-anchor springs.
6. Remove the adjusting wheel spring
7. Remove the shoe guide plate.
8. Remove the adjusting wheel spring.
9. Remove the adjuster assembly.
10. Remove the strut.
11. Remove the strut-to-shoe spring.
12. Remove the shoe hold-down cup.
13. Remove the shoe hold-down spring.
14. Remove the shoe hold-down pin.
15. Remove the shoe and lining assembly.
16. Remove the clip.
17. If wheel hub, backing plate and shoe-hold down pin removal are necessary, perform the following:
    a. Remove the rear wheel hub assembly.
    b. Remove the backing plate.
    c. Remove the shoe hold-down pin.

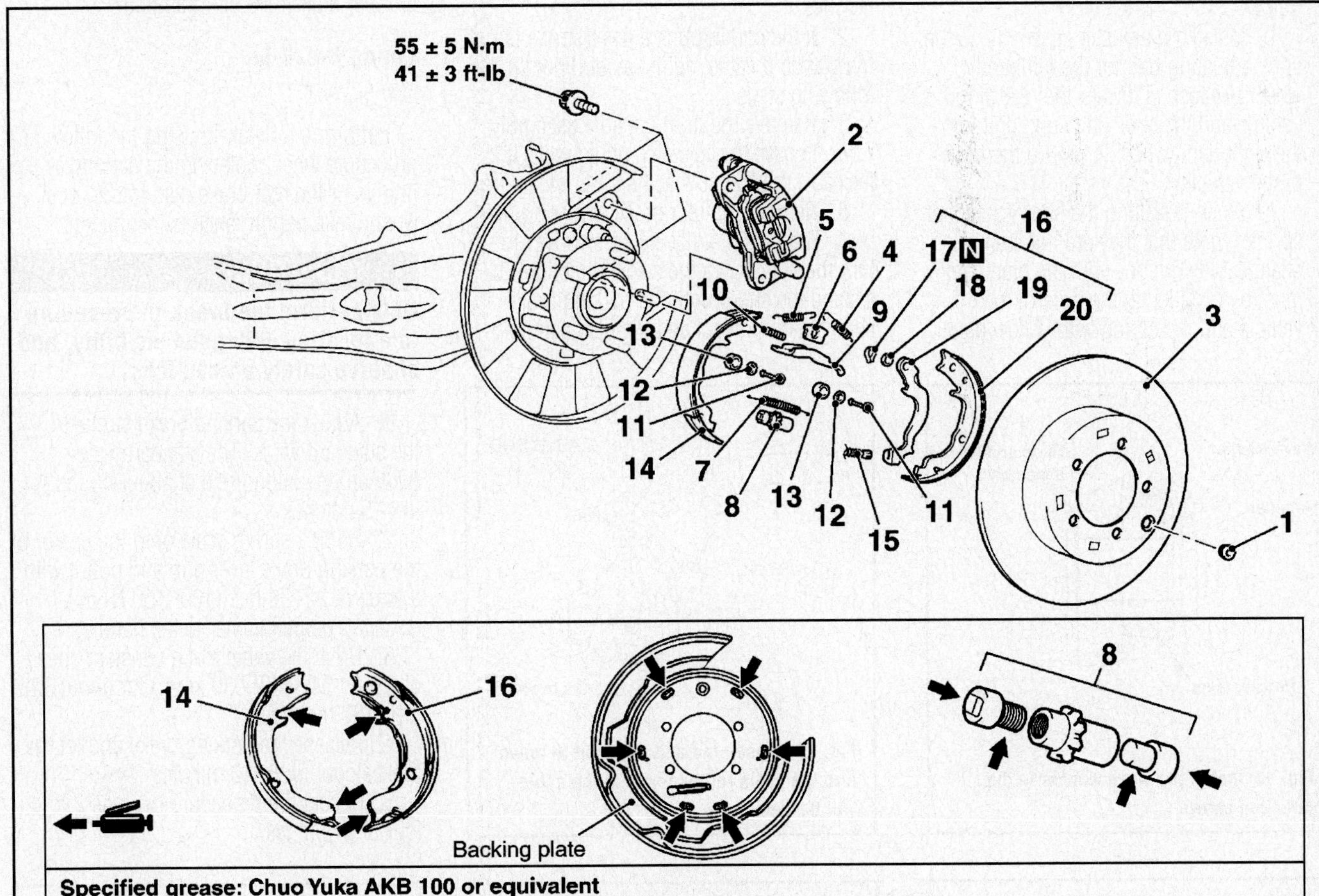

Fig. 12 Exploded view of the parking brake shoes (linings) and related components

18. Inspect the parking brake lining.
   a. Measure the thickness of the brake lining at several places. The standard value is 2.8 mm (0.11 inch) and the minimum limit is 1.0 mm (0.04 inch)
   b. If the thickness of the brake lining is below the limit, replace the shoe assemblies on both sides of the vehicle. Never replace only one side.

To install:

19. Installation is the reverse of the removal procedure, noting the following:
   a. Refer to the tightening specifications shown on the accompanying illustration.
   b. Install the adjuster assembly so the shoe adjusting bolt for the left hand wheel is attached toward the rear of the vehicle and the shoe adjusting bolt for the right hand wheel is toward the front of the vehicle.
   c. When installing the shoe-to-anchor springs, note that they are not interchangeable. The one with the blue paint must be installed at the front of the vehicle and the spring with yellow paint must be installed at the rear of the vehicle.
20. Perform the parking brake pedal stroke check and adjustment, and the parking brake seating procedure.

ADJUSTMENT

Parking Brake Lever Stroke Check And Adjustment

See Figures 15 and 16.

1. Pull the parking brake lever with a force of about 45 lbs. (200 N) and count the number of notches. The proper range is 4–5 notches.
2. If the parking brake lever stroke is not the standard value, adjust as outlined in the following steps.
3. Remove the floor console assembly.
4. Loosen the adjusting nut to move it to the cable rod end so that the cable will be free.
5. Remove the rear brake adjusting hole plug. Then insert a flat-tip screwdriver to turn the adjuster to the arrow direction (to expand the shoe) until the parking brake shoe makes contact and the disc can no

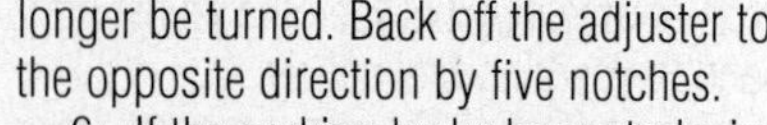

longer be turned. Back off the adjuster to the opposite direction by five notches.
6. If the parking brake lever stroke is below the standard value and the braking is too firm, the rear brakes may drag.
7. Adjust the parking brake lever stroke to the standard value by turning the adjusting nut. After the adjustment, ensure that there is no free play between the adjusting nut and the parking brake lever.
8. After adjusting the parking brake lever stroke, raise and safely support the rear end of the vehicle, then release the parking brake and turn the rear wheels to make sure the rear brakes are not dragging.

Lining Break-In

See Figure 17.

Perform the break-in using the following procedure when replacing the parking brake linings or the rear brake disc rotors, or when brake performance is insufficient.

CAUTION

Only perform the break-in procedure in a location with good visibility, and observe safety precautions.

1. Adjust the parking brake stroke to the standard value. The standard value (with an operation force of about 45 lbs.) is 4–5 notches.
2. Hook a spring scale onto the center of the parking brake lever grip and pull it with a force of 22–34 lbs. (100–150 N) in a direction perpendicular to the handle.
3. Drive the vehicle at a constant speed of 22–31 MPH (35–50 km/h) for about 328 feet (100 meters).
4. Release the parking brake and let the brakes cool for 5–10 minutes.
5. Repeat the procedure in steps 2 to 4, four or five times.

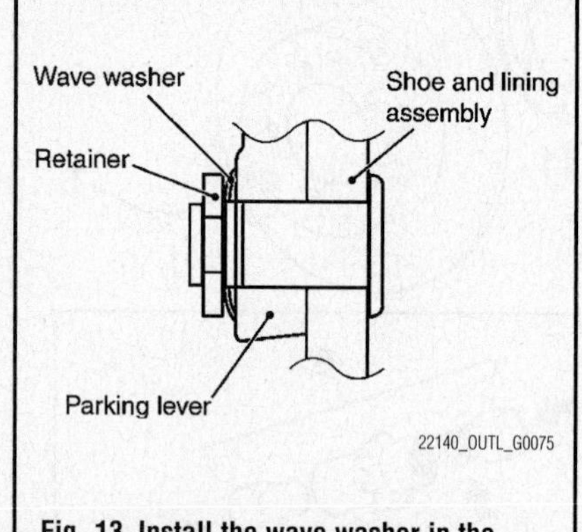

Fig. 13 Install the wave washer in the direction shown

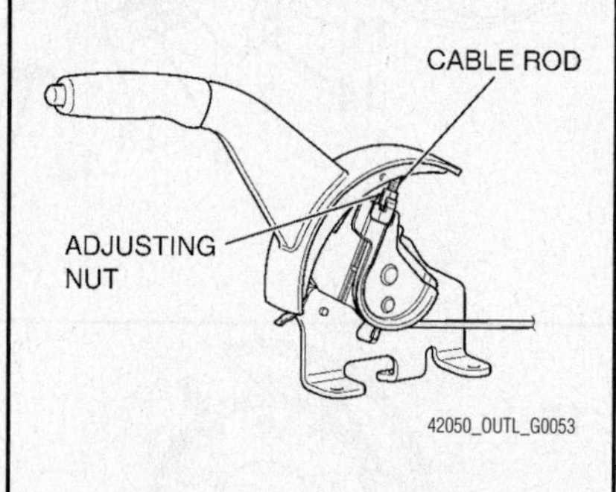

Fig. 15 Loosen the adjusting nut to move it to the cable rod end so that the cable will be free

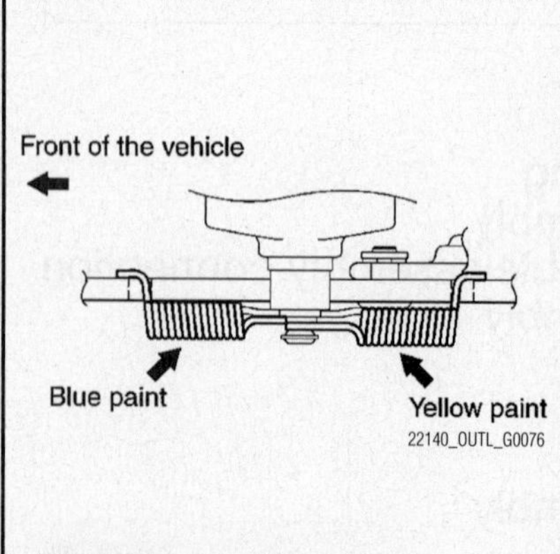

Fig. 14 Shoe-to-anchor spring installation—left rear wheel shown, right rear wheel is symmetrical

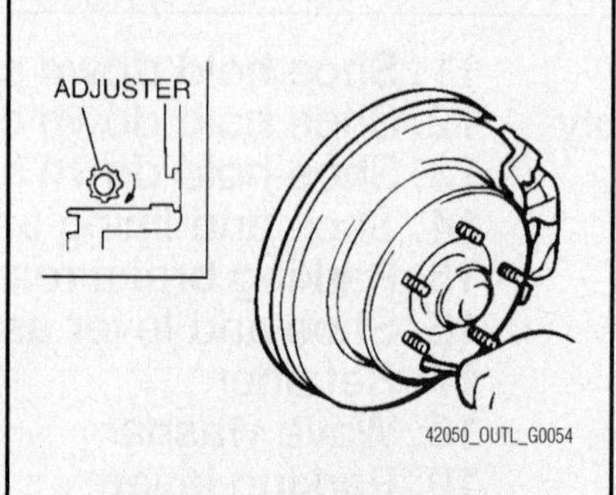

Fig. 16 Insert a flat-tip screwdriver to turn the adjuster to the arrow direction (to expand the shoe) until the parking brake shoe makes contact and the disc can no longer be turned

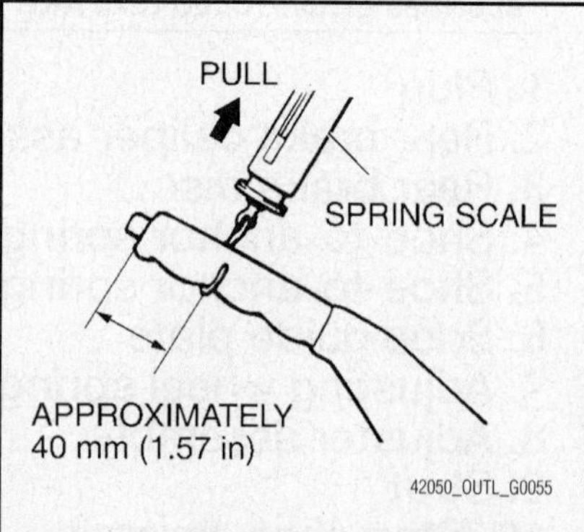

Fig. 17 Hook a spring scale onto the center of the parking brake lever grip and pull it with a force of 22–34 lbs. (100–150 N) in a direction perpendicular to the handle

GENERAL INFORMATION

CAUTION

Some vehicles are equipped with an air bag system. The system must be disarmed before performing service on, or around, system components, the steering column, instrument panel components, wiring and sensors. Failure to follow the safety precautions and the disarming procedure could result in accidental air bag deployment, possible injury and unnecessary system repairs.

SERVICE PRECAUTIONS

Disconnect and isolate the battery negative cable before beginning any airbag system component diagnosis, testing, removal, or installation procedures. Allow system capacitor to discharge for two minutes before beginning any component service. This will disable the airbag system. Failure to disable the airbag system may result in accidental airbag deployment, personal injury, or death.

Do not place an intact undeployed airbag face down on a solid surface. The airbag will propel into the air if accidentally deployed and may result in personal injury or death.

When carrying or handling an undeployed airbag, the trim side (face) of the airbag should be pointing towards the body to minimize possibility of injury if accidental deployment occurs. Failure to do this may result in personal injury or death.

Replace airbag system components with OEM replacement parts. Substitute parts may appear interchangeable, but internal differences may result in inferior occupant protection. Failure to do so may result in occupant personal injury or death.

Wear safety glasses, rubber gloves, and long sleeved clothing when cleaning powder residue from vehicle after an airbag deployment. Powder residue emitted from a deployed airbag can cause skin irritation. Flush affected area with cool water if irritation is experienced. If nasal or throat irritation is experienced, exit the vehicle for fresh air until the irritation ceases. If irritation continues, see a physician.

Do not use a replacement airbag that is not in the original packaging. This may result in improper deployment, personal injury, or death.

The factory installed fasteners, screws and bolts used to fasten airbag components have a special coating and are specifically designed for the airbag system. Do not use substitute fasteners. Use only original equipment fasteners listed in the parts catalog when fastener replacement is required.

During, and following, any child restraint anchor service, due to impact event or vehicle repair, carefully inspect all mounting hardware, tether straps, and anchors for proper installation, operation, or damage. If a child restraint anchor is found damaged in any way, the anchor must be replaced. Failure to do this may result in personal injury or death.

Deployed and non-deployed airbags may or may not have live pyrotechnic material within the airbag inflator.

Do not dispose of driver/passenger/curtain airbags or seat belt tensioners unless you are sure of complete deployment. Refer to the Hazardous Substance Control System for proper disposal.

Dispose of deployed airbags and tensioners consistent with state, provincial, local, and federal regulations.

After any airbag component testing or service, do not connect the battery negative cable. Personal injury or death may result if the system test is not performed first.

If the vehicle is equipped with the Occupant Classification System (OCS), do not connect the battery negative cable before performing the OCS Verification Test using the scan tool and the appropriate diagnostic information. Personal injury or death may result if the system test is not performed properly.

Never replace both the Occupant Restraint Controller (ORC) and the Occupant Classification Module (OCM) at the same time. If both require replacement, replace one, then perform the Airbag System test before replacing the other.

Both the ORC and the OCM store Occupant Classification System (OCS) calibration data, which they transfer to one another when one of them is replaced. If both are replaced at the same time, an irreversible fault will be set in both modules and the OCS may malfunction and cause personal injury or death.

If equipped with OCS, the Seat Weight Sensor is a sensitive, calibrated unit and must be handled carefully. Do not drop or handle roughly. If dropped or damaged, replace with another sensor. Failure to do so may result in occupant injury or death.

If equipped with OCS, the front passenger seat must be handled carefully as well. When removing the seat, be careful when setting on floor not to drop. If dropped, the sensor may be inoperative, could result in occupant injury, or possibly death.

If equipped with OCS, when the passenger front seat is on the floor, no one should sit in the front passenger seat. This uneven force may damage the sensing ability of the seat weight sensors. If sat on and damaged, the sensor may be inoperative, could result in occupant injury, or possibly death.

DISARMING THE SYSTEM

See Figure 18.

1. Before servicing the vehicle, refer to the precautions.
2. Position the front wheels in the straight-ahead position and place the key in the **LOCK** position. Remove the key from the ignition lock cylinder.
3. Disconnect the negative battery cable and insulate the cable end with high-quality electrical tape or similar non-conductive wrapping.
4. Wait at least 60 seconds minute before working on the vehicle. The air bag system is designed to retain enough voltage to deploy the air bag for a short period of time after the battery has been disconnected.

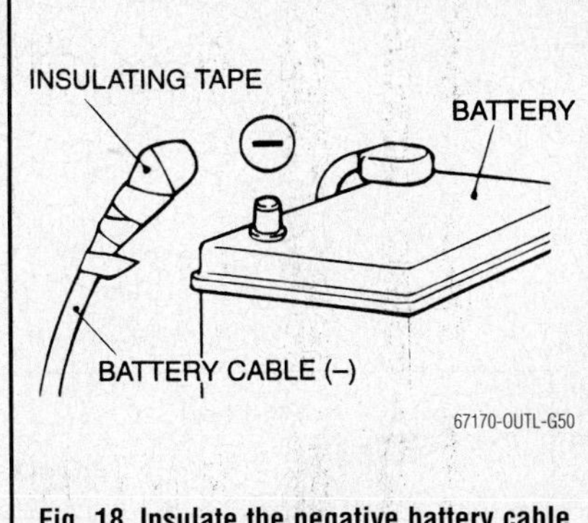

Fig. 18 Insulate the negative battery cable to prevent accidental deployment of the air bag

ARMING THE SYSTEM

1. Connect the negative battery cable, turn the ignition switch to the **ON** position and check the Supplemental Restraint System (SRS) warning light for proper operation.

AIRBAG MODULE

REMOVAL & INSTALLATION

See Figure 19.

1. Disconnect the negative battery cable.

CAUTION

Wait at least 60 seconds after disconnecting the cable from the negative (-) battery terminal to prevent airbag activation.

2. Position the front wheels in a straight ahead direction.

WARNING

SRS-ECU adopts the rollover specification that the curtain air bag and seat belt pre-tensioner operate at the occurrence of rollover. Therefore, do not tilt the vehicle to the right and left with the ignition ON or tilt the SRS-ECU to the right and left with the ignition ON and the harness installed.

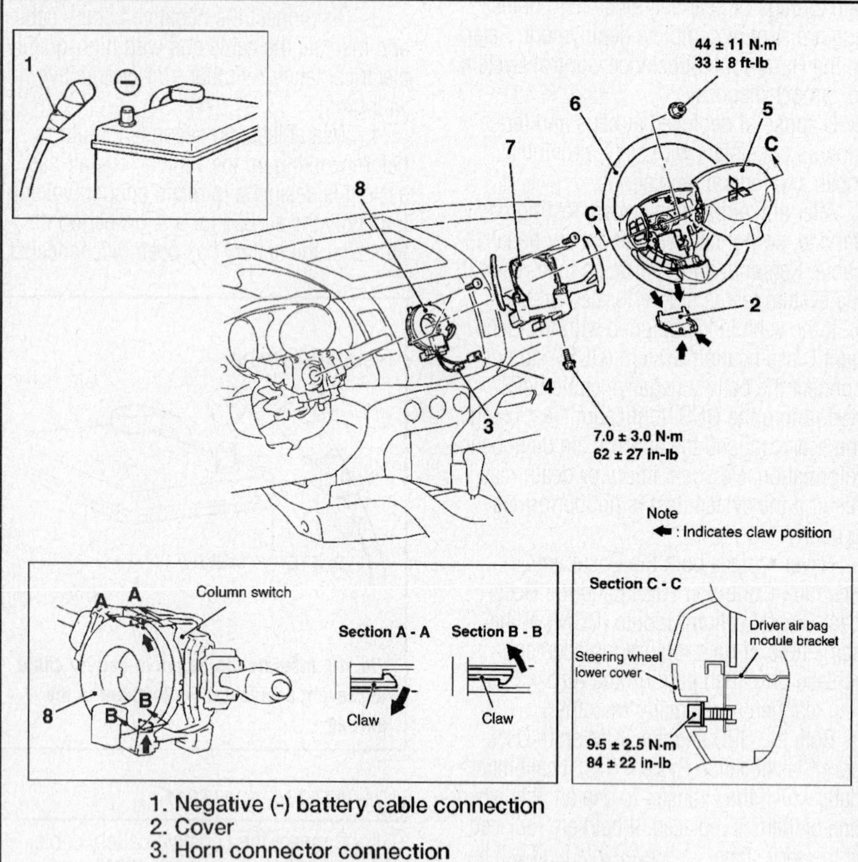

Fig. 19 Exploded view of the driver's side air bag module

3. Remove the cover.
4. Disconnect the horn connector.
5. Disconnect the driver's air bag module connector.
6. Remove the driver's air bag module.
7. Remove the steering wheel.
8. Remove the lower and upper column covers.
9. Remove the paddle shift assembly, if equipped.
10. Remove the clock spring.

To install:

WARNING

If the center of the clock spring is not correctly aligned, the steering wheel may not be turned fully or the cable inside the clock spring may be broken, causing the SRS air bag to be inoperative or operated incorrectly.

11. Align the mating marks of the clock spring.
    a. Turn the clock spring clockwise fully.
    b. Turn the clock spring counterclockwise approximately three and 3/4 turns to align the mating marks.
    c. Install the clock spring to the column switch.

WARNING

Always align three mating marks of the steering wheel sensor simultaneously as shown in the figure. If these mating marks are not aligned correctly, the steering wheel sensor may be damaged.

12. Align three mating marks of the steering wheel sensor simultaneously as shown in the figure.
    a. Check the window for inspecting the neutral position of the steering wheel sensor. If the mating marks cannot be seen from the window, align the mating marks as shown in the figure.
    b. Install the steering wheel sensor to the column switch assembly, maintaining the neutral position correctly.

➡A new steering wheel sensor has a pin for preventing the rotation of (fixing) the steering wheel sensor. After installing the column switch assembly, remove this pin.

    c. Install the column switch assembly to the vehicle, maintaining the neutral position correctly.

13. Install the paddle shift assembly, if equipped.
14. Install the lower and upper column covers.
15. Install the steering wheel.
16. Connect the driver's air bag module connector.
17. Install the driver's air bag module.
18. Connect the horn connector.
19. Install the cover.
20. After installation is complete, use a scan tool to perform the following:
    a. Update the neutral position stored in the steering wheel sensor.
    b. Reset the calibrated value of the steering angle stored in ASC-ECU.

DRIVETRAIN

AUTOMATIC TRANSAXLE ASSEMBLY

REMOVAL & INSTALLATION

See Figures 20 and 21.

1. Before servicing the vehicle, refer to the precautions.
2. Disconnect the negative battery cable.
3. Remove the engine undercovers.
4. Drain the transaxle fluid.
5. Remove the air cleaner bracket.
6. Remove the battery and battery tray.
7. Remove the ECM.
8. Remove the wiper arm assembly and front deck garnish.
9. Remove the strut tower bar.
10. Remove the halfshafts.
11. Remove the transmission fluid filler tube.
12. Disconnect the A/T control solenoid valve connector.
13. Disconnect the input shaft speed sensor connector.
14. Disconnect the transmission range switch connector.
15. Disconnect the output shaft speed sensor connector.
16. Disconnect the transaxle control cable.
17. Disconnect the battery ground and harness clamp.
18. Remove the upper transaxle coupling bolt.
19. Remove the starter assembly.
20. On AWD, remove the transfer case.
21. Disconnect the transmission fluid cooler hose.
22. Disconnect the water return hose "A", Water return tube, Water return hose "B".
23. Remove the transmission fluid warmer.
24. Remove the center member and body bolts.
25. Remove the center member and crossmember bolts.
26. Remove the front roll stopper and center member assembly.
27. Remove the flange nut.
28. Remove the rear roll stopper and bracket.
29. Remove the oil pan cover
30. Remove the torque converter and flywheel coupling bolts.
    a. Remove the coupling bolts while turning the crankshaft.
    b. Fully push the torque converter into the transaxle side so that it does not remain on the engine side.
31. Remove the intake manifold plenum.
32. Raise the engine and transaxle assembly to the position where the engine weight is not applied to the transaxle mounting insulator.
33. Remove the transaxle mounting bracket.
34. Remove the transaxle mounting insulator stopper.
35. Remove the transaxle mounting insulator.
36. Install the engine holding assembly.

⁂ WARNING

The engine hanger plate (special tool: MB992208) should be secured by tightening bolts with the engine hanger plate to 16 inch lbs. (22 Nm). If the other bolts are used, the engine assembly may fall down when it is raised.

    a. Remove the intake manifold plenum bracket, rear on the right bank and the engine hanger on the left bank, and then install the engine hanger plate (Special tool: MB992208) to the place.
    b. When engine hanger (Special tool: MB991928) is used, assemble the engine hanger (Special tool: MB991928), set the feet of the special tool and adjust the engine hanger balance by sliding the slide bracket. Set the chains of the engine hanger (Special tool: MB991527) and the engine hanger balancer (Special tool: MB991454) to support the engine and transaxle assembly. Remove the garage jack and then remove the transaxle assembly upper part coupling bolts that have been loosened previously.
    c. When using an engine mechanical hanger (Special tool: MZ203830 or MZ203831), set the foot of the engine mechanical hanger (Special tool: MZ203830 or MZ203831), slide the front foot of the engine mechanical hanger (Special tool: MZ203830 or MZ203831) to balance the engine hanger. Place a rag between the engine mechanical hanger (Special tool: MZ203830 or MZ203831) and the windshield to prevent the special tool from interfering with the windshield. Set the chains of the engine hanger (Special tool: MB991527) and the engine hanger balancer (Special tool: MB991454) to support the engine and transaxle assembly. Remove the garage jack and then remove the transaxle assembly upper part coupling bolts that have been loosened previously.

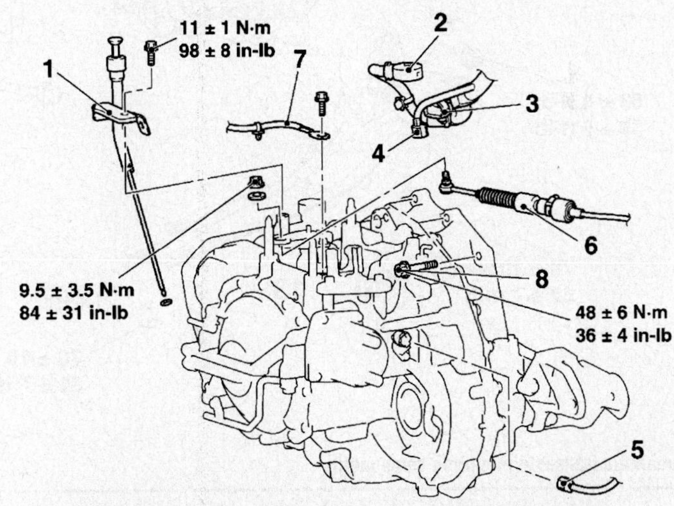

Fig. 20 Automatic transaxle exterior components

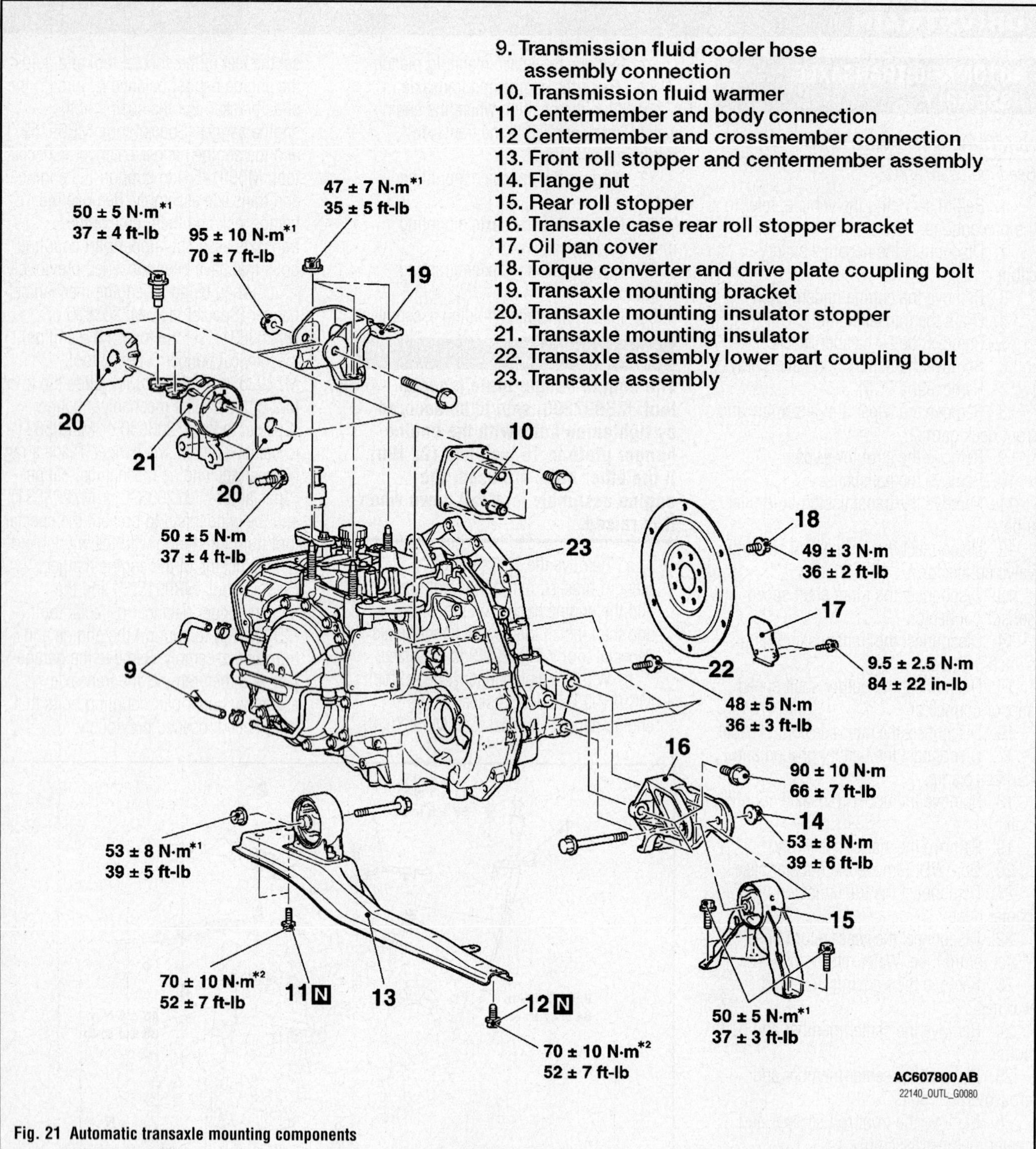

Fig. 21 Automatic transaxle mounting components

37. Remove the transaxle assembly lower part coupling bolt.

38. Remove the transaxle assembly.

To install:

39. Installation is the reverse of the removal procedure, noting the following:

- Transaxle-to-engine lower mounting bolts: 34–40 ft. lbs. (46–54 Nm)
- Flywheel bolts: 34–40 ft. lbs. (46–54 Nm)
- Transaxle mount bracket nuts: 36 ft. lbs. (49 Nm)
- Transaxle mount stopper nuts: 56–66 ft. lbs. (75–89 Nm)
- Transaxle-to-engine upper mounting bolts: 34–40 ft. lbs. (46–54 Nm)

40. Fill the transaxle using the proper grade and type transaxle fluid.

41. If AWD, fill the transfer case to the correct level.

42. Start the engine and check for leaks.

CVT TRANSAXLE ASSEMBLY

REMOVAL & INSTALLATION

See Figures 22 through 24.

1. Before servicing the vehicle, refer to the precautions.

2. Disconnect the negative battery cable.
3. Remove the engine undercovers.
4. Drain the transaxle fluid.
5. Remove the air cleaner bracket.
6. Remove the battery and battery tray.
7. Remove the ECM.
8. Remove the wiper arm assembly and front deck garnish.
9. Remove the strut tower bar.
10. Remove the transaxle control cable connection.
11. Remove the transaxle control bracket.
12. Disconnect the secondary pulley rotation sensor harness connector
13. Disconnect the crank angle sensor harness connector.
14. Disconnect the battery ground
15. Disconnect the CVT fluid cooler hose.
16. Disconnect the CVT assembly connector.
17. Disconnect the primary pulley rotation sensor connector.
18. Disconnect the transmission range switch connector.
19. Remove the oil filler tube.
20. Remove the water tube.
21. Remove the starter mounting bolt.
22. Remove the upper transaxle assembly bolt.
23. Remove the cover.
24. Remove the torque converter and flywheel bolts.
25. Remove the centre member.
26. On AWD vehicles, remove the transfer case.
27. Remove the rear rolling stopper.
28. Remove the transaxle case rear roll stopper bracket .
29. Remove the transaxle mounting bracket
30. Remove the transaxle mounting insulator stopper
31. Remove the transaxle mounting insulator
32. Install the engine holding assembly
33. Remove the transaxle assembly lower part coupling bolt
34. Remove the transaxle assembly

To install:

35. Installation is the reverse of the removal procedure, noting the following:
    - Upper transaxle bolt: 36 ft. lbs. (49Nm)
    - Flywheel bolts: 36 ft. lbs. (49Nm)
    - Centre member: 52 ft. lbs. (71Nm)
    - Rear roll stopper: 39 ft. lbs. (90 Nm)
    - Transaxle mounting insulator: 70 ft. lbs. (90 Nm)

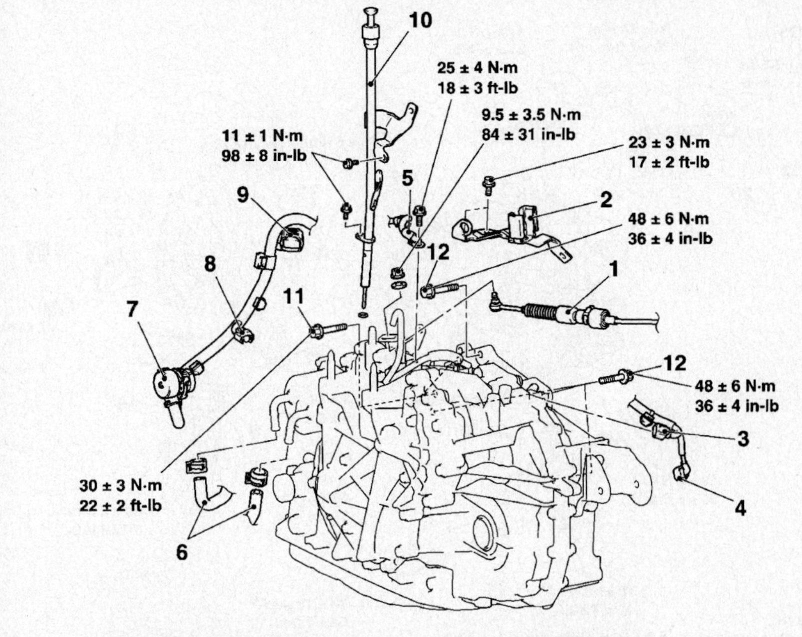

Fig. 22 CVT transaxle exterior components

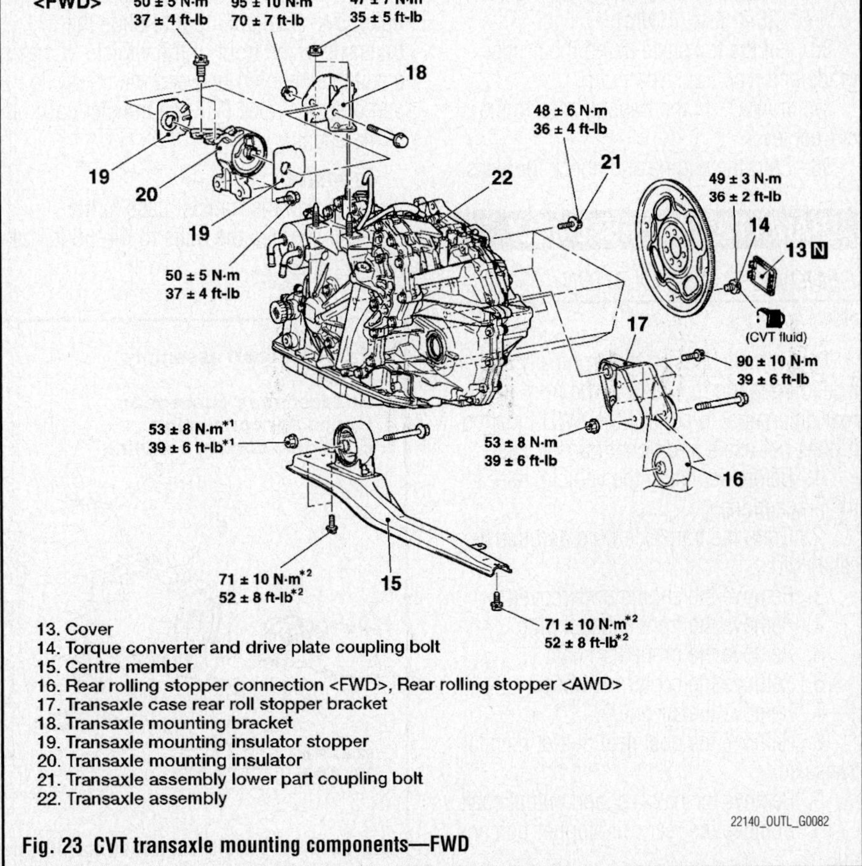

Fig. 23 CVT transaxle mounting components—FWD

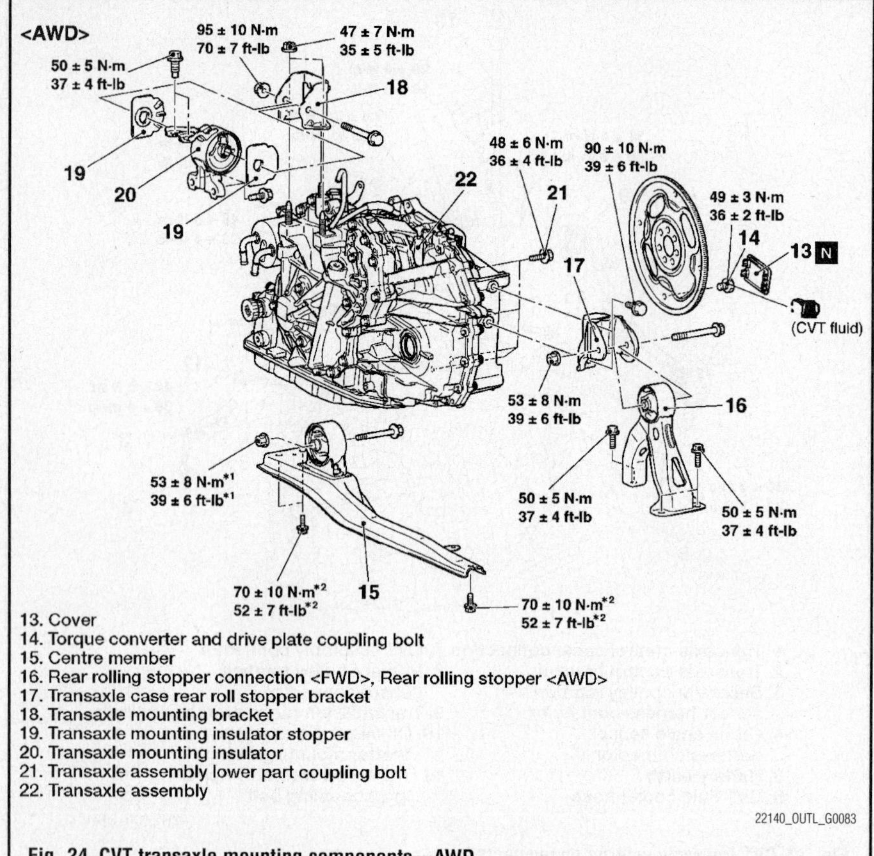

Fig. 24 CVT transaxle mounting components—AWD

- Transaxle assembly coupling bolt: 36 ft. lbs. (49Nm)

36. Fill the transaxle using the proper grade and type transaxle fluid.
37. If AWD, fill the transfer case to the correct level.
38. Start the engine and check for leaks.

TRANSFER CASE ASSEMBLY

REMOVAL & INSTALLATION

See Figure 25.

This vehicle uses an electronically controlled coupling mounted to the front of the rear differential to control the AWD function. It does not use a traditional transfer case.

1. Before servicing the vehicle, refer to the precautions.
2. Drain the transaxle fluid and transfer case fluid.
3. Remove the engine undercover(s).
4. Remove the front exhaust pipe.
5. Remove the propeller shaft.
6. Remove the center member.
7. Remove the air guide.
8. Remove the dust seal guard, manual transaxle.
9. Remove the halfshaft and output shaft.
10. Remove the rear roll stopper bolt and nut.
11. Remove the transfer case mounting bolts. Use a suitable tool to slide the transaxle to the front of the vehicle to make a suitable opening between the transaxle and crossmember. Pull the transfer case out of the opening.

To install:

12. Install the transfer case to the transaxle. Torque the bolts to 44–58 ft. lbs. (60–78 Nm).
13. Install the rear roll stopper bolt and nut hand-tight. After the full weight of the vehicle is on the ground, torque the nut to 34–44 ft. lbs. (45–59 Nm).
14. Install the halfshaft and output shaft.
15. Install the air guide.
16. Install the dust seal guard, manual transaxle.
17. Install the center member.
18. Install the propeller shaft.
19. Install the front exhaust pipe.
20. Install the engine undercover(s).
21. Fill the transaxle and the transfer case to the correct level.

FRONT HALFSHAFT

REMOVAL & INSTALLATION

See Figures 26 and 27.

⁂ WARNING

The wheel speed sensor/toothed ring collect metallic particles easily, because it is magnetized. Make sure that the sensor does not collect metallic particles. Check that there is not any trouble prior to reassembling it. When removing and installing the driveshaft assembly, make sure that the sensor and ring (integrated with the inner oil seal) do not contact with surrounding parts to avoid damage. When removing and installing the front wheel speed sensor, make sure that the sensor head at the end does not contact with surrounding parts to avoid damage.

1. Before servicing the vehicle, refer to the Precautions Section.
2. Raise the vehicle and support it safely.

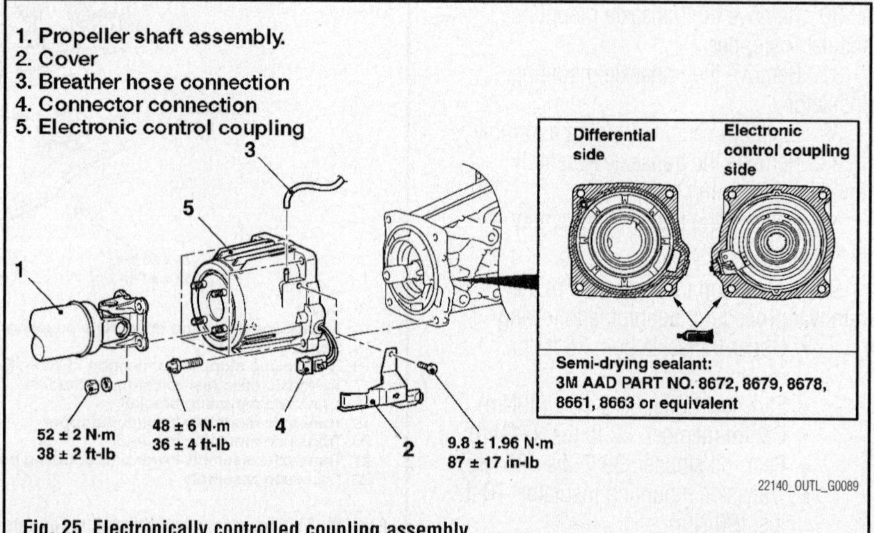

Fig. 25 Electronically controlled coupling assembly

3. Drain the transaxle fluid.
4. Drain the transfer case, if equipped with AWD.
5. Remove the wheel and tire assembly.
6. Remove the halfshaft cotter pin, nut and washer.
7. Remove the front wheel speed sensor and harness bracket.
8. Disconnect the front height sensor from the lower arm.
9. Remove the stabilizer link.
10. Remove the lower control arm ball joint and tie rod end from the steering knuckle.
11. Remove the halfshaft or halfshaft and inner shaft as follows:
   a. Use Mitsubishi Special tool Nos. MB990241, MB991354 and MB990767, or suitable puller, to push the halfshaft or halfshaft and inner shaft assembly from the hub.
   b. Remove the halfshaft from the hub by pulling the bottom of the rotor toward you.

WARNING

When pulling the halfshaft from the transaxle, be careful that the spline of the halfshaft does not damage the oil seal.

   c. Insert a pry bar between the transaxle case and halfshaft, then pry the halfshaft out of the transaxle.
   d. If the inner shaft is difficult to remove, tap the bracket assembly lightly with a plastic hammer, then remove the inner shaft.
   e. Cover the halfshaft opening in the transaxle case to prevent foreign debris from entering.

➡Do not pull on the shaft; doing so damages the inboard joint.

12. For AWD vehicles, Use Mitsubishi Special tool No. MB991721 to remove the output shaft.

To install:

13. Replace the circlip on the ends of the halfshafts.
14. If AWD, install the output shaft.

➡When installing the output shaft, halfshaft or halfshaft and inner shaft assembly, make sure the splines do not damage the oil seal.

15. Insert the halfshaft or halfshaft and inner shaft into the transaxle. Be sure it is fully seated.
16. Push out on the knuckle assembly and install the halfshaft through the hub.

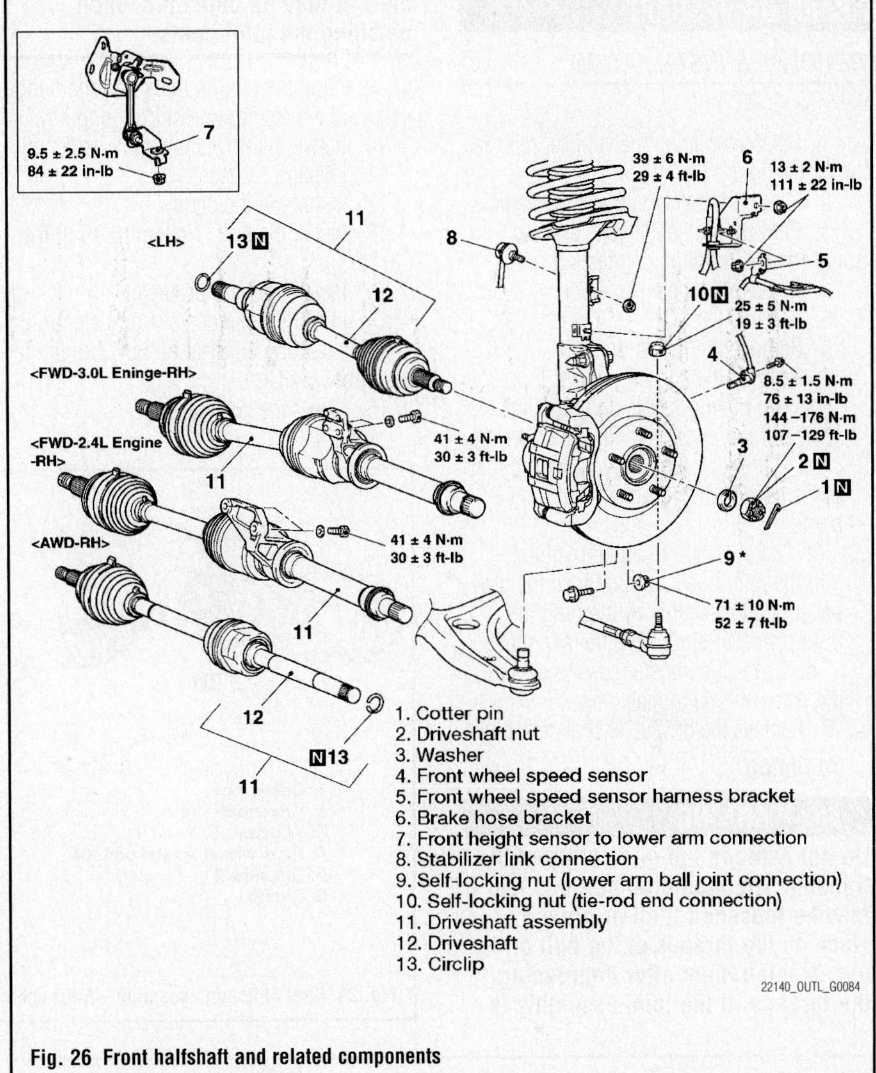

Fig. 26 Front halfshaft and related components

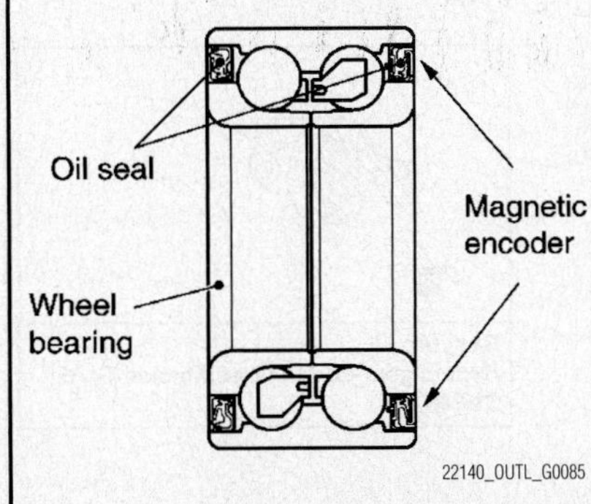

Fig. 27 The wheel speed sensor assembly is an integral part of the wheel bearing

17. Install the self-locking nut. Tighten the nut so the protruding length of the stabilizer link is 0.35–0.39 inches (9.0–9.8mm).
18. Connect the tie rod end to the steering knuckle. Torque the retaining nut to 21 ft. lbs. (29 Nm) and secure with a new cotter pin.
19. Connect the ball joint to the steering knuckle. Torque the new retaining nut to 43–52 ft. lbs. (60–72 Nm) and secure with a new cotter pin.
20. Install the stabilizer link and rubber insulator.
21. Connect the front height sensor from the lower arm.
22. Install the ABS sensor harness bracket and ABS sensor.
23. Install the washer so the chamfered edge faces outward. Install the halfshaft nut and tighten to 107–129 ft. lbs. (144–176 Nm). Secure with a new cotter pin.
24. Install the wheel and lower the vehicle to the floor.
25. Fill the transaxle with fluid.
26. On AWD models, fill the transfer case.
27. Test drive the vehicle and check for leaks.
28. Check and adjust the alignment, as required.

REAR DRIVESHAFT

REMOVAL & INSTALLATION

See Figure 28.

1. Before servicing the vehicle, refer to the precautions.
2. Raise and safely support the vehicle.
3. Remove the flange yoke and electronic control coupling connecting nut.
4. Remove the hear protector.
5. Remove the bolt.
6. Remove the insulator.
7. Remove the spacer.
   a. Put mating marks on the flange yoke and the electronic control coupling.
   b. If the joint assembly is bent, it may be damaged when pinching the joint boots.
   c. Insert a rag or similar materials into the joint boots, and remove the propeller shaft assembly by aligning the front propeller shaft with the rear shaft.
   d. Cover the transfer case to prevent the entry of foreign materials.
8. Remove the driveshaft shaft assembly.

To install:

⁂ WARNING

Do not damage the oil seal lip of the transfer. The mounting bolt and nut may be loosened if oil or grease is stuck on the threads of the bolt and nut. Tighten them after degreasing the threads. If the joint assembly is bent, it may be damaged when pinching the joint boots.

9. Align the mating marks on the flange yoke and the electronic control coupling.
10. Install the driveshaft shaft assembly.
11. Install the spacer.
12. Install the insulator.
13. Install the bolt. Tighten to 30 ft. lbs. (40 Nm).
14. Install the hear protector.
15. Install the flange yoke and electronic control coupling connecting nut. Tighten to 40 ft. lbs. (54 Nm).
16. Lower the vehicle.

REAR HALFSHAFT

REMOVAL & INSTALLATION

See Figure 29.

➡If the vehicle is equipped with ABS, do not strike the ABS rotors installed to the outer BL outer race of the halfshaft against other parts when removing or installing the halfshaft. Otherwise the ABS rotors will be damaged.

1. Before servicing the vehicle, refer to the precautions.
2. Raise and safely support the vehicle.
3. Drain the gear oil.

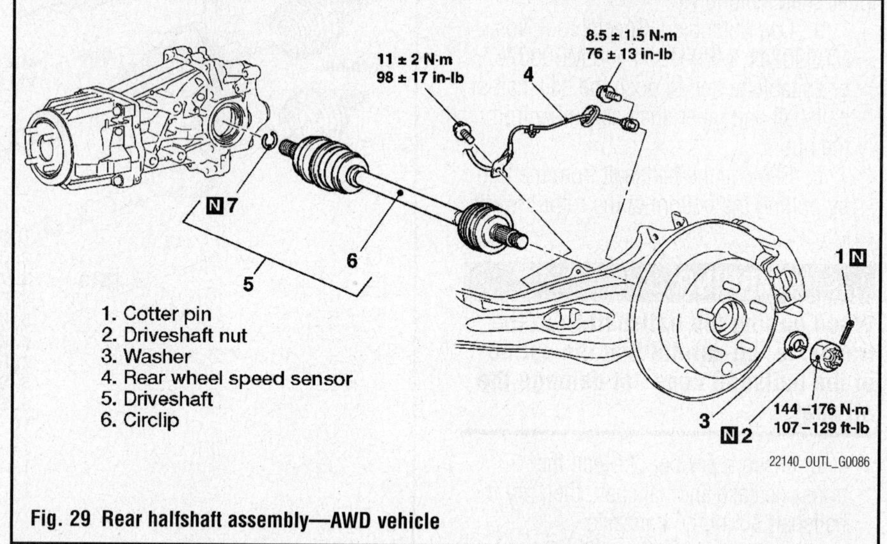

Fig. 29 Rear halfshaft assembly—AWD vehicle

4. Remove the tire and wheel assembly.
5. Remove the halfshaft nut and washer.

➡Prevent the hub assembly from turning by using a tool such as MB990767 and remove the halfshaft nut and washer.

⁂ WARNING

Be careful not to strike the pole piece at the tip of the rear wheel speed sensor, as damage may result.

6. Disconnect the lower arm from the trailing arm, the shock absorber and the stabilizer link.
7. Disconnect the upper arm from the trailing arm.
8. Remove the rear wheel speed sensor.
9. Remove the halfshaft and circlip.

To install:

10. Replace the circlip on the end of the halfshaft.
11. Install the halfshaft.
12. Install the rear wheel speed sensor.
13. Connect the upper arm to the trailing arm.

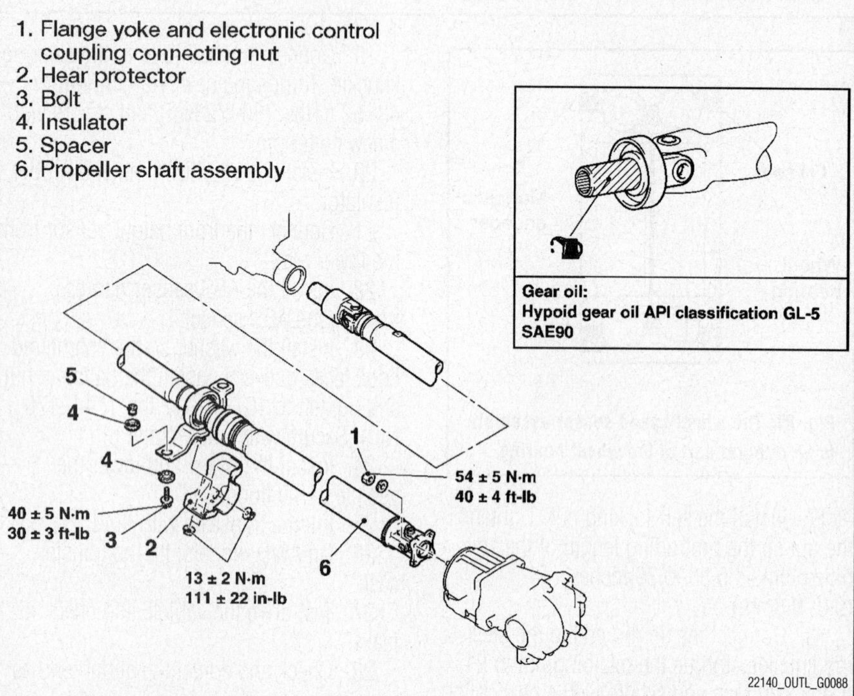

1. Flange yoke and electronic control coupling connecting nut
2. Hear protector
3. Bolt
4. Insulator
5. Spacer
6. Propeller shaft assembly

Gear oil:
Hypoid gear oil API classification GL-5 SAE90

5, 4, 4, 1, 3, 2, 6
40 ± 5 N·m 30 ± 3 ft-lb
13 ± 2 N·m 111 ± 22 in-lb
54 ± 5 N·m 40 ± 4 ft-lb
22140_OUTL_G0088

Fig. 28 Driveshaft assembly

14. Connect the lower arm to the trailing arm, the shock absorber and the stabilizer link.

15. Install the halfshaft nut and washer. Tighten to 107–129 ft. lbs. (144–176 Nm).

16. Install the tire and wheel assembly.

17. Fill the differential with oil.

18. Test drive the vehicle and check for leaks.

19. Check and adjust the rear alignment, as required.

REAR PINION SEAL

REMOVAL & INSTALLATION

See Figure 30.

1. Raise and safely support the vehicle securely on jackstands.

2. Remove the halfshaft from the differential carrier.

3. Remove the differential carrier oil seal.

To install:

4. Using the special tools MB990938 and MB991115, press fit a new oil seal.

5. Apply multi-purpose grease to the oil seal lip and driveshaft oil seal seating area.

6. Replace the halfshaft circlip with a new one, and install the halfshaft to the differential carrier.

7. Lower the vehicle.

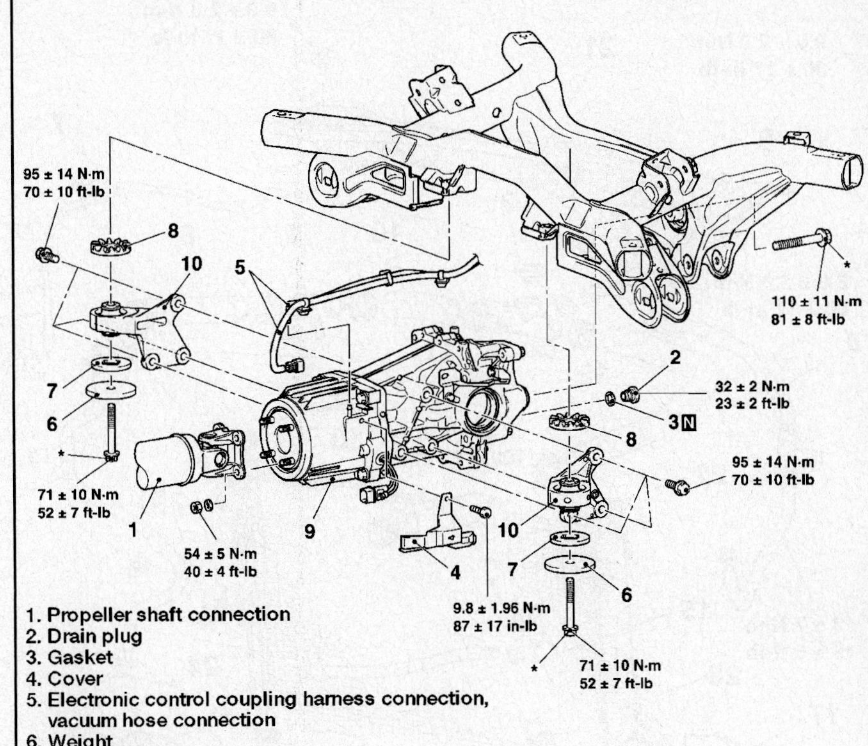

Fig. 30 Differential carrier assembly

ENGINE COOLING

ENGINE FAN

REMOVAL & INSTALLATION

See Figures 31 through 33.

1. Before servicing the vehicle, refer to the precautions.

2. Disconnect the negative battery cable.

⁂ CAUTION

Wait at least 90 seconds after the negative battery cable is disconnected to prevent possible deployment of the air bag.

3. Remove the transaxle fluid cooler hose.

4. Remove the water feed hose.

5. Remove the radiator condenser tank hose.

6. Make mating marks on the radiator hoses and the hose clamps.

7. Remove the radiator hoses.

8. Remove the water pipes.

9. Remove the radiator cap.

10. Disconnect the radiator fan motor connector.

11. Disconnect the condenser fan motor connector.

12. Remove the fan, fan motor and fan shroud.

13. Remove the radiator fan.

14. Remove the radiator fan motor.

15. Remove the condenser fan.

16. Remove the condenser fan motor.

17. Remove the fan shroud.

To install:

18. Install the fan shroud.

19. Install the condenser fan motor.

20. Install the condenser fan.

21. Install the radiator fan motor.

22. Install the radiator fan.

23. Install the fan, fan motor and fan shroud.

24. Connect the condenser fan motor connector.

25. Connect the radiator fan motor connector.

26. Install the radiator cap.

27. Install the water pipes.

28. Install the radiator hoses.

29. Make mating marks on the radiator hoses and the hose clamps.

30. Install the radiator condenser tank hose.

31. Install the water feed hose.

32. Install the transaxle fluid cooler hose.

33. Connect the negative battery cable.

34. Start the engine and check for proper fan operation.

RADIATOR

REMOVAL & INSTALLATION

See Figures 31 through 33.

1. Before servicing the vehicle, refer to the precautions.

2. Disconnect the negative battery cable.

⁂ CAUTION

Wait at least 90 seconds after the negative battery cable is disconnected to prevent possible deployment of the air bag.

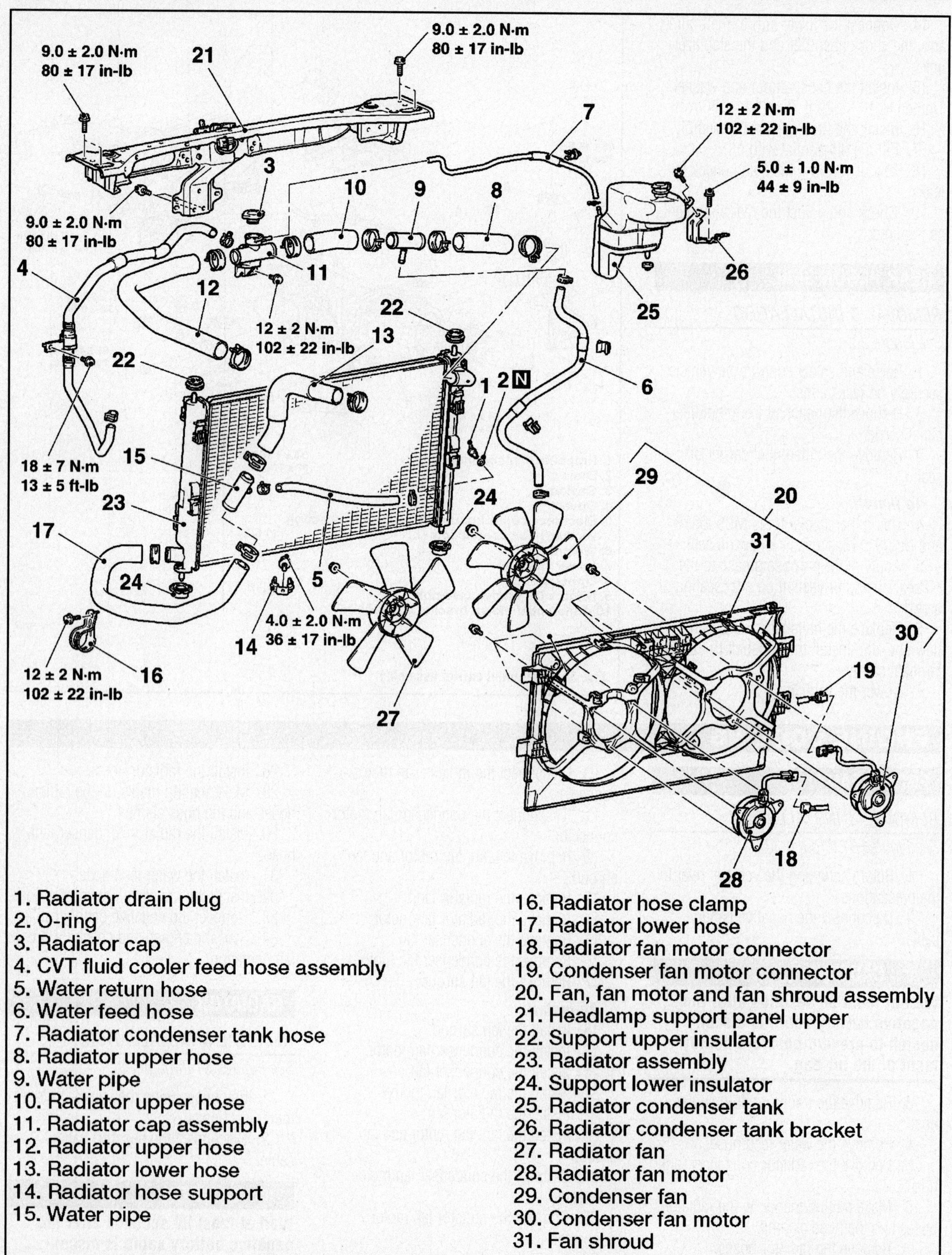

Fig. 31 Exploded view of the radiator assembly—2.4L engine

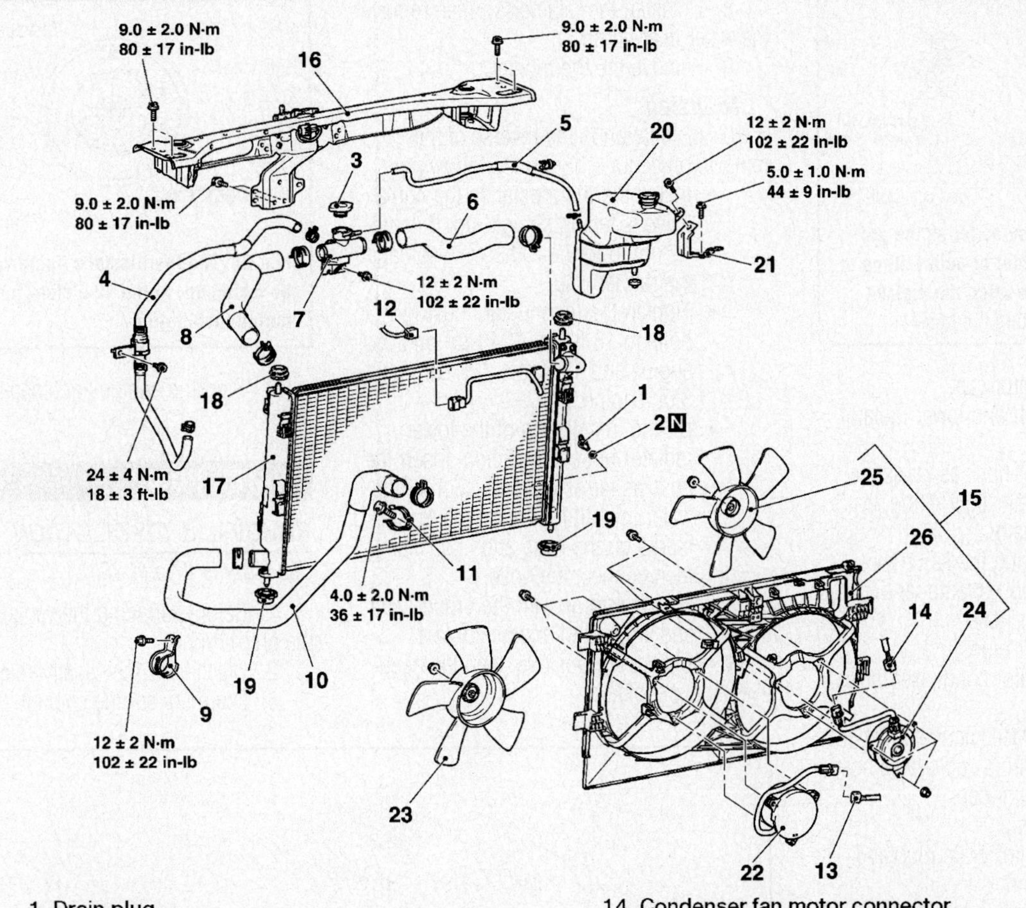

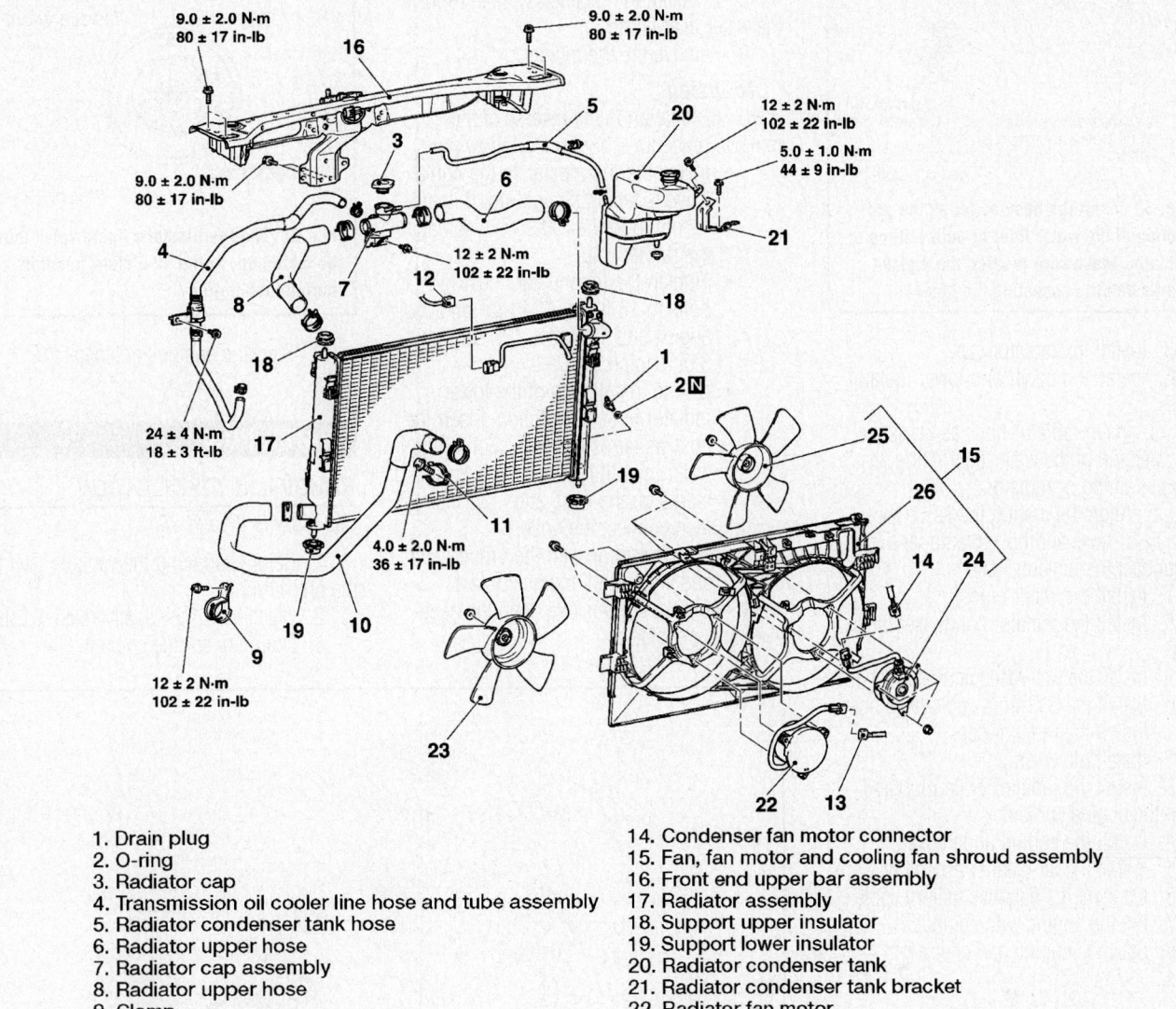

1. Drain plug
2. O-ring
3. Radiator cap
4. Transmission oil cooler line hose and tube assembly
5. Radiator condenser tank hose
6. Radiator upper hose
7. Radiator cap assembly
8. Radiator upper hose
9. Clamp
10. Radiator lower hose
11. Radiator hose support
12. Radiator sensor connector
13. Radiator fan motor connector
14. Condenser fan motor connector
15. Fan, fan motor and cooling fan shroud assembly
16. Front end upper bar assembly
17. Radiator assembly
18. Support upper insulator
19. Support lower insulator
20. Radiator condenser tank
21. Radiator condenser tank bracket
22. Radiator fan motor
23. Fan
24. Condenser fan motor
25. Fan
26. Cooling fan shroud

22140_OUTL_G0093

Fig. 32 Exploded view of the radiator assembly—3.0L engine

3. Remove the air cleaner assembly.
4. Remove the engine undercover.
5. Remove the radiator drain plug and drain the engine coolant.
6. Remove the O-ring.
7. Remove the radiator cap.
8. Remove the transaxle fluid cooler hose.
9. Remove the water feed and return hoses.
10. Remove the radiator condenser tank hose.
11. Make mating marks on the radiator hoses and the hose clamps.
12. Remove the radiator hoses.
13. Remove the water pipes.
14. Remove the radiator cap.
15. Remove the radiator hose support.
16. Disconnect the radiator fan motor connector.
17. Disconnect the condenser fan motor connector.
18. Remove the fan, fan motor and fan shroud assembly.
19. Remove the radiator grille, headlamp support upper panel cover.
20. Remove the front impact sensor.
21. Remove the hood lock release cable, hood switch.
22. Remove the upper headlamp support panel.
23. Remove the support upper insulator.
24. Remove the radiator assembly.

To install:

25. Install the radiator assembly.
26. Install the support upper insulator.
27. Install the upper headlamp support panel.
28. Install the hood lock release cable, hood switch.
29. Install the front impact sensor.
30. Install the radiator grille, headlamp support upper panel cover.
31. Install the fan, fan motor and fan shroud assembly.
32. Connect the condenser fan motor connector.
33. Connect the radiator fan motor connector.
34. Install the radiator hose support.

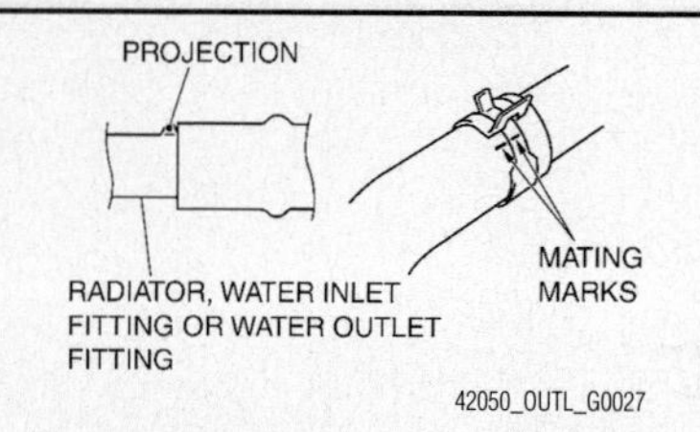

Fig. 33 Insert the hose as far as the projection of the water inlet or outlet fitting or radiator. Make sure to align the matchmarks before connecting the hose

35. Install the radiator cap.
36. Install the upper and lower radiator hoses.
    a. Insert radiator hose as far as the projection of the water inlet fitting, water outlet fitting or radiator.
    b. Align the mating marks on the radiator hose and hose clamp, and then connect the radiator hose.
37. Install the water pipes.
38. Install the radiator condenser tank hose.
39. Install the water feed and return hoses.
40. Install the CVT fluid cooler feed hose.
41. Install the radiator cap.
42. Install the O-ring.
43. Install the radiator drain plug and drain the engine coolant.
44. Install the engine undercover.
45. Install the air cleaner assembly.
46. Connect the negative battery cable.
47. Fill the engine with coolant.
48. Start the engine and check for leaks.

THERMOSTAT

REMOVAL & INSTALLATION

See Figures 34 through 36.

1. Disconnect the negative battery cable, then the positive battery cable.

**** CAUTION**

Wait at least 90 seconds after the negative battery cable is disconnected to prevent possible deployment of the air bag.

2. Drain the engine coolant.
3. Remove the air cleaner assembly.
4. Remove the battery and battery tray, as necessary.
5. Matchmark the installed positions of the hose and hose clamp, then disconnect the lower radiator hose from the water inlet fitting.
6. Detach the control hose electrical connection.
7. Remove the control wiring harness connection bracket, as necessary.
8. Unfasten the retainer(s), then remove the water inlet fitting.
9. Remove the thermostat.

To install:

10. Installation is the reverse of the removal procedure, noting the following:
    - Install the thermostat so the jiggle valve is facing straight up. Be careful not to fold or otherwise damage the rubber ring.
    - Tighten the thermostat housing bolts to 18 ft. lbs. (24 Nm) on 2.4L engine and 17 ft. lbs. (23 Nm) on 3.0L engine
    - During installation of the lower radiator hose connection, insert the hose as far as the projection of the water inlet fitting. Align the matchmarks on the hose and hose clamp, then connect the hose.
    - Fill the engine with the proper type and amount of engine coolant.
11. Connect the positive, then the negative battery cable.

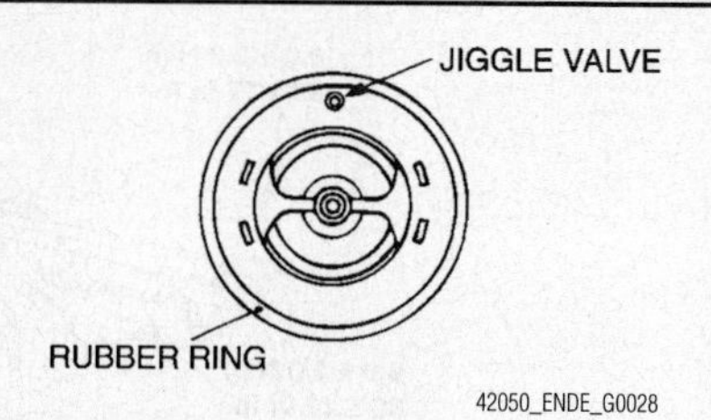

Fig. 36 The thermostat's jiggle valve must be facing up, at the 12 o'clock position during installation

12. Start the engine and check for leaks.

WATER PUMP

REMOVAL & INSTALLATION

See Figures 37 and 38.

1. Before servicing the vehicle, refer to the precautions.
2. Disconnect the negative battery cable.
3. Drain the engine coolant.

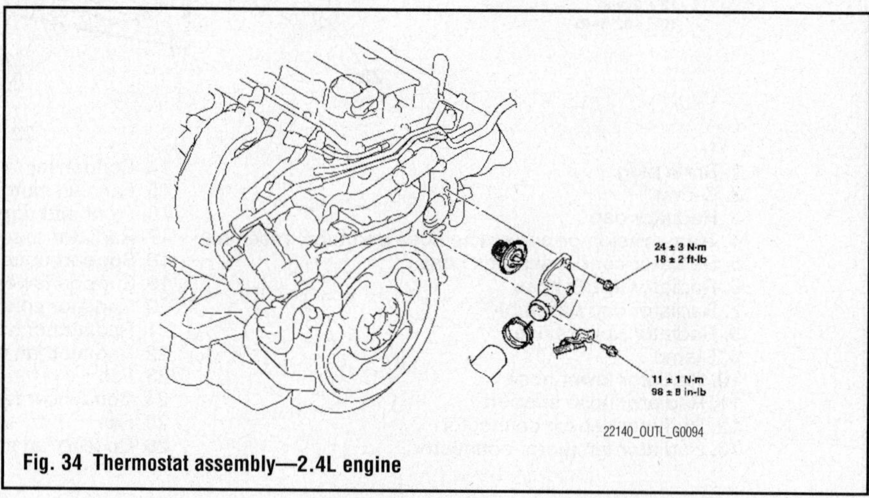

Fig. 34 Thermostat assembly—2.4L engine

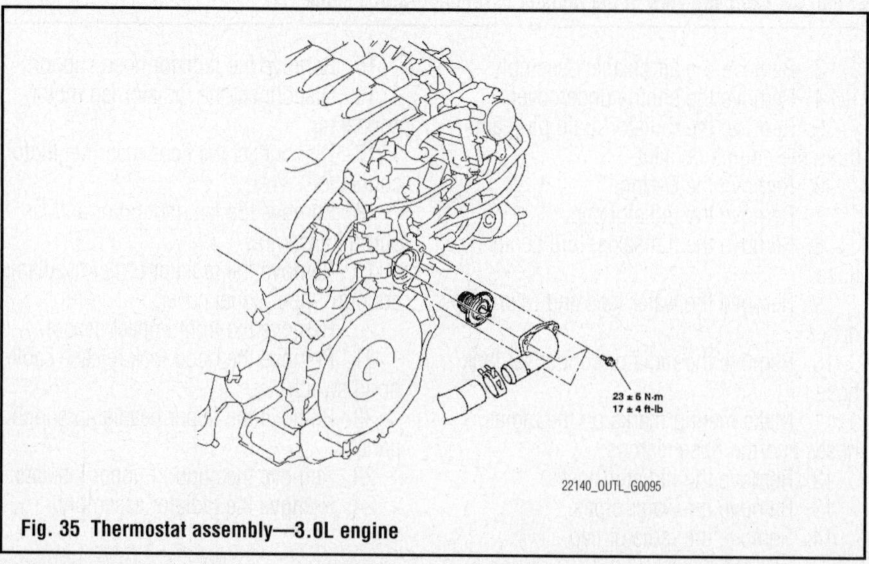

Fig. 35 Thermostat assembly—3.0L engine

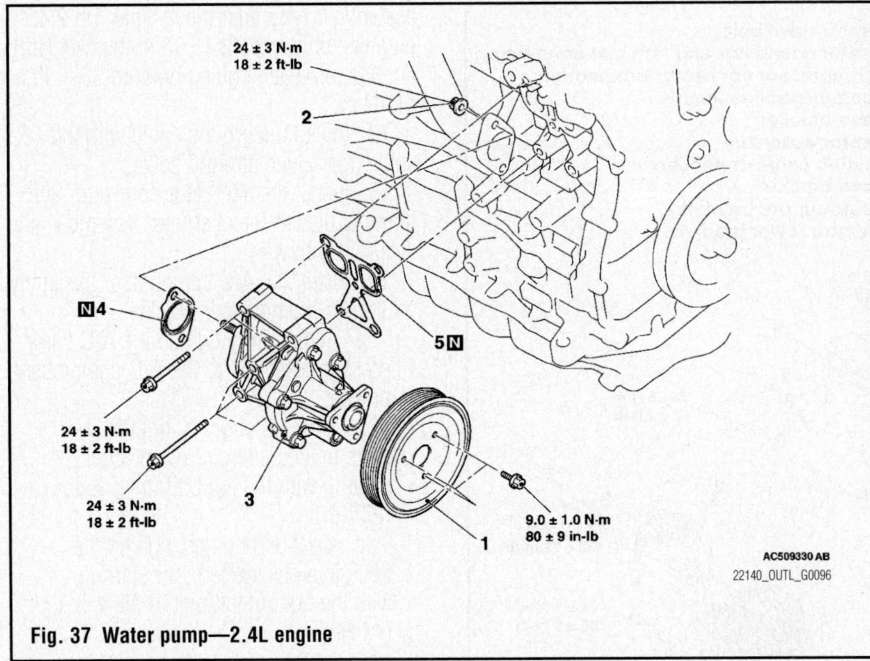

Fig. 37 Water pump—2.4L engine

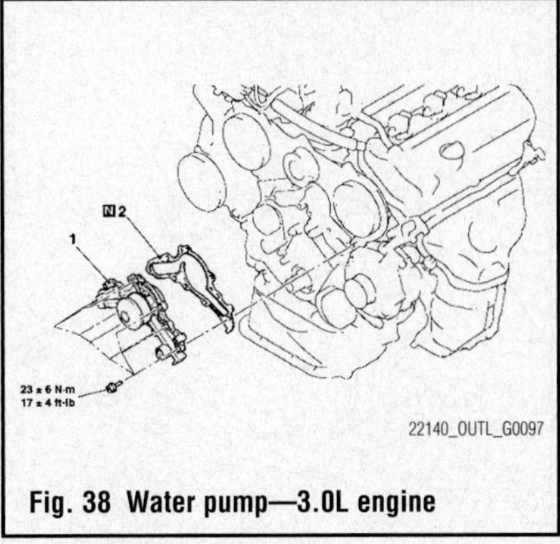

Fig. 38 Water pump—3.0L engine

4. On 2.4L engines, remove the accessory drive belt(s).
5. On 3.0L engines, remove the timing belt.
6. Remove the water pump mounting bolts.
7. Remove the water pump, gasket.

To install:

8. Install the water pump to the engine block, with new gasket.
9. Tighten the mounting bolts to 18 ft. lbs. (24 Nm) on 2.4L engine and 17 ft. lbs. (23 Nm) on 3.0L engine.
10. On 2.4L engines, install the accessory drive belt(s).
11. On 3.0L engines, install the timing belt.
12. Refill the engine with coolant.
13. Connect the negative battery cable.
14. Start the engine and check for leaks.

ENGINE ELECTRICAL — CHARGING SYSTEM

ALTERNATOR

REMOVAL & INSTALLATION

See Figures 39 and 40.

1. Before servicing the vehicle, refer to the precautions.
2. Disconnect the negative battery cable.
3. Remove the engine undercover.
4. Remove the drive belt and idler pulley.
5. Remove the A/C compressor connector and connector clamp.
6. Remove the A/C compressor assembly from the A/C compressor bracket with the hose still attached.
7. Place the removed A/C compressor assembly where it will not be a hindrance when removing and installing the alternator assembly, and secure it with wire.
8. Remove the alternator connector and terminal.
9. Remove the ground cable.
10. Move the A/C compressor assembly to one side, and then remove the alternator assembly from under the vehicle.

To install:

11. Move the A/C compressor assembly to one side, and then install the alternator

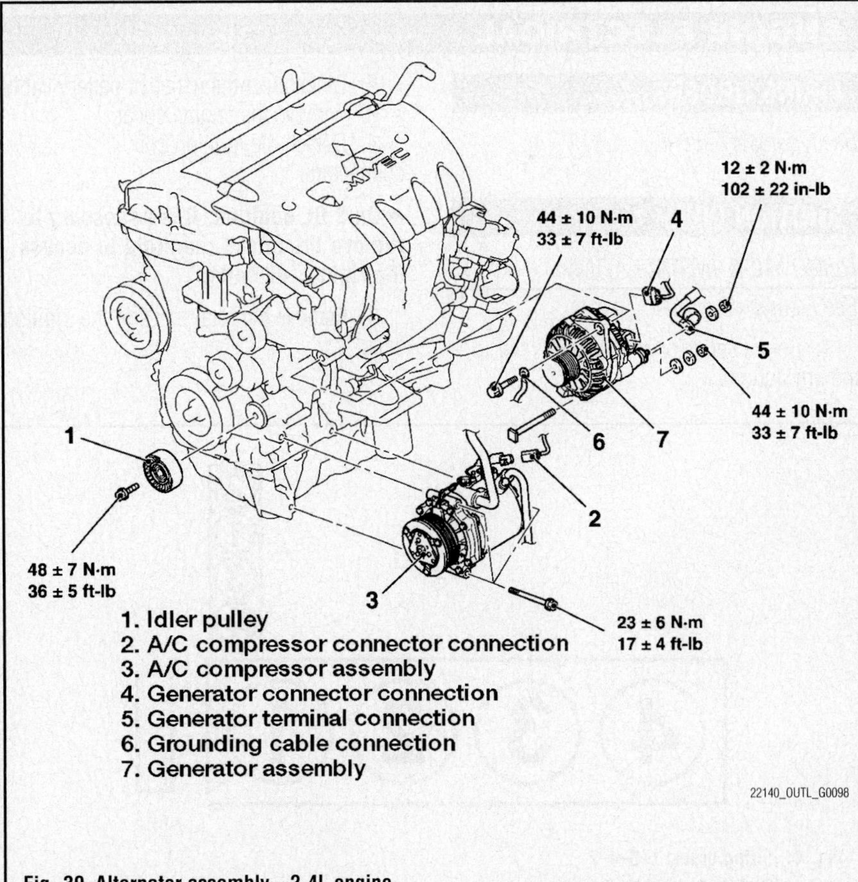

Fig. 39 Alternator assembly—2.4L engine

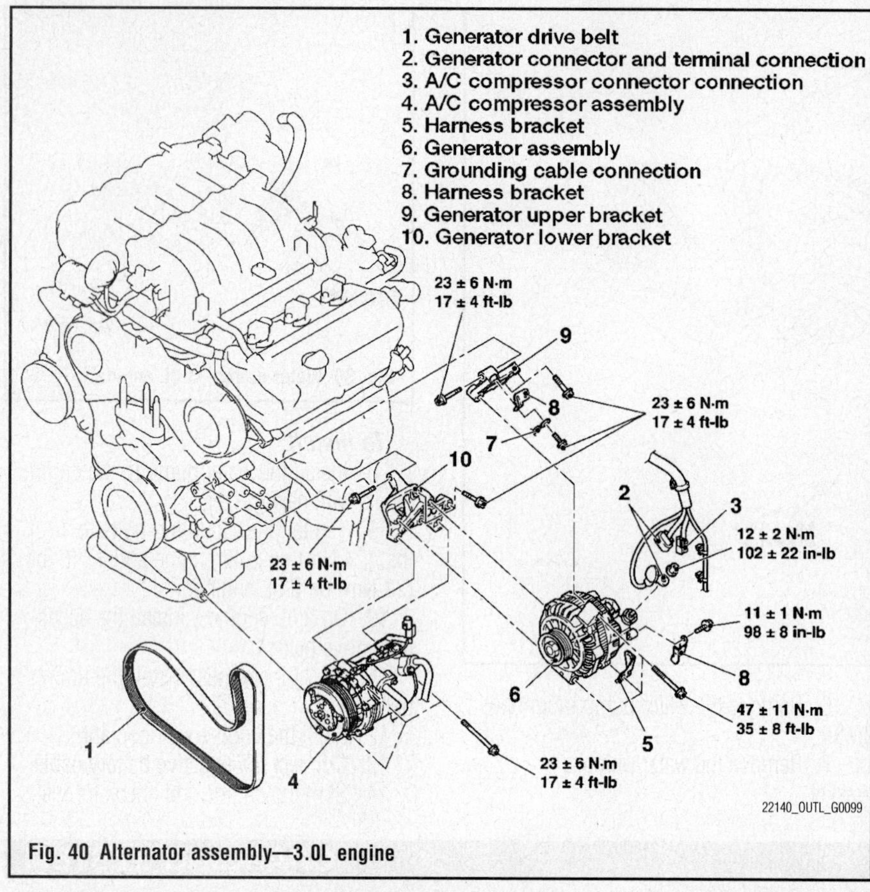

Fig. 40 Alternator assembly—3.0L engine

assembly from under the vehicle. On 2.4L engines, tighten bolts to 33 ft. lbs. (44 Nm). On 3.0L engines, tighten bolts to 35 ft. lbs. (47 Nm).

12. Install the ground cable with the cable positioned straight down.
13. Install the alternator connector and terminal. Install the alternator assembly, and secure it with wire.
14. Install the A/C compressor assembly on the A/C compressor bracket.
    a. Temporarily tighten the A/C compressor assembly to the A/C compressor bracket.
    b. On 2.4L engines, tighten the bolts in a counterclockwise order starting with the top bolt to 17 ft. lbs. (23 Nm).
    c. On 3.0L engines, tighten the bolts in a clockwise order starting with the bottom left bolt to 35 ft. lbs. (47 Nm).
15. Install the A/C compressor connector and connector clamp.
16. Install the drive belt and idler pulley.
17. Install the engine undercover.
18. Disconnect the negative battery cable.

ENGINE ELECTRICAL

IGNITION SYSTEM

FIRING ORDERS

See Figures 41 and 42.

IGNITION COIL

REMOVAL & INSTALLATION

See Figures 43 and 44.

1. Before servicing the vehicle, refer to the precautions.
2. Disconnect the negative battery cable.
3. Remove the engine cover.
4. Detach the ignition coil connectors.

➡On 3.0L engines, it is necessary to remove the intake manifold to access the right bank coils.

5. Remove the retainers and the ignition coils.

To install:

6. Install the ignition coils. Tighten the retainers to 89 inch lbs. (10 Nm) on 2.4L engine and 84 inch lbs. (10 Nm) on 3.0L engine.
7. Attach the ignition coil connectors.
8. Install the engine cover.
9. Connect the negative battery cable.

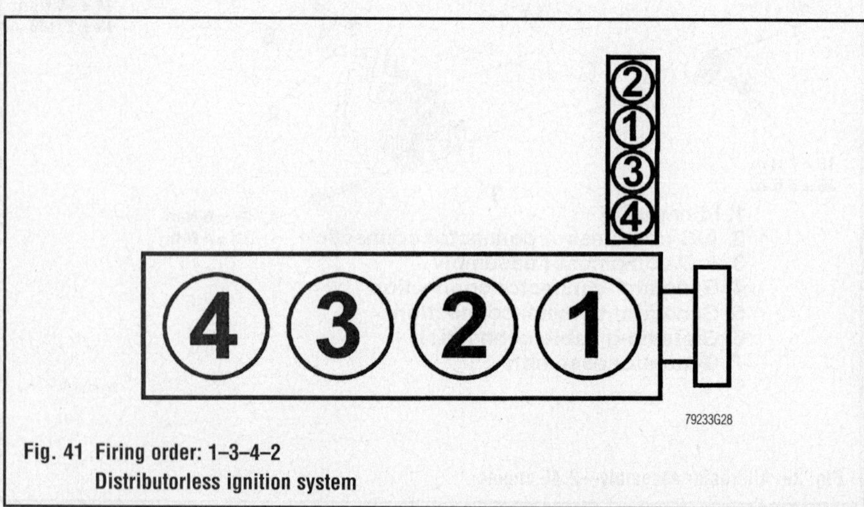

Fig. 41 Firing order: 1–3–4–2 Distributorless ignition system

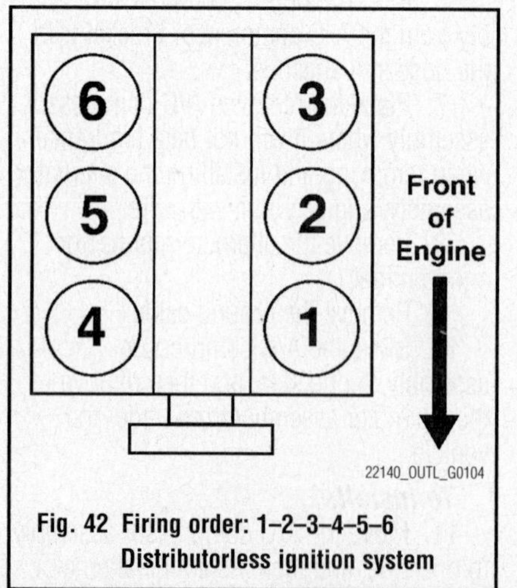

Fig. 42 Firing order: 1–2–3–4–5–6 Distributorless ignition system

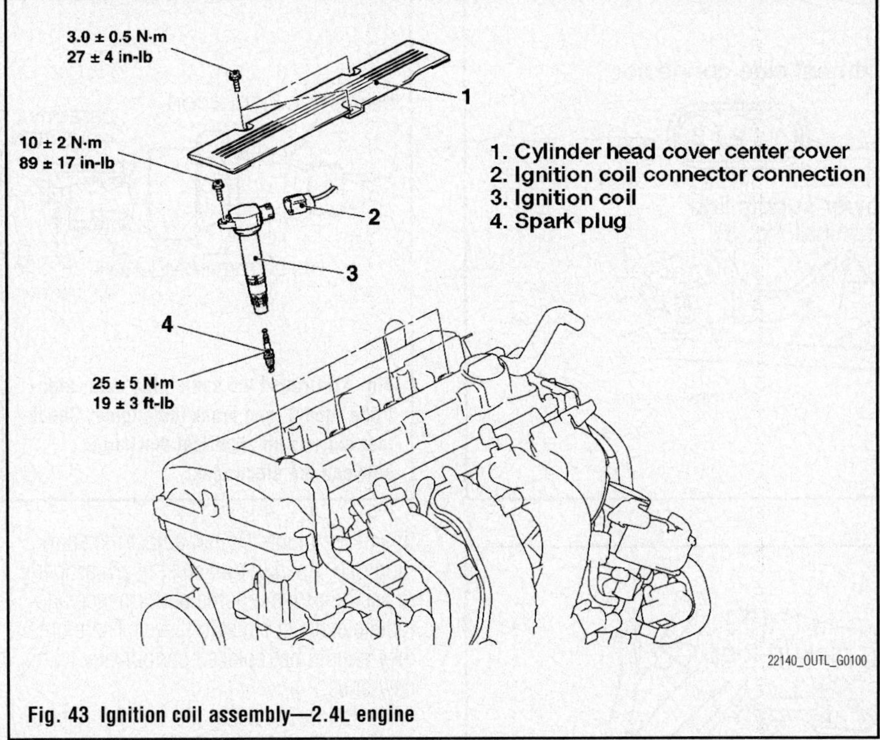

Fig. 43 Ignition coil assembly—2.4L engine

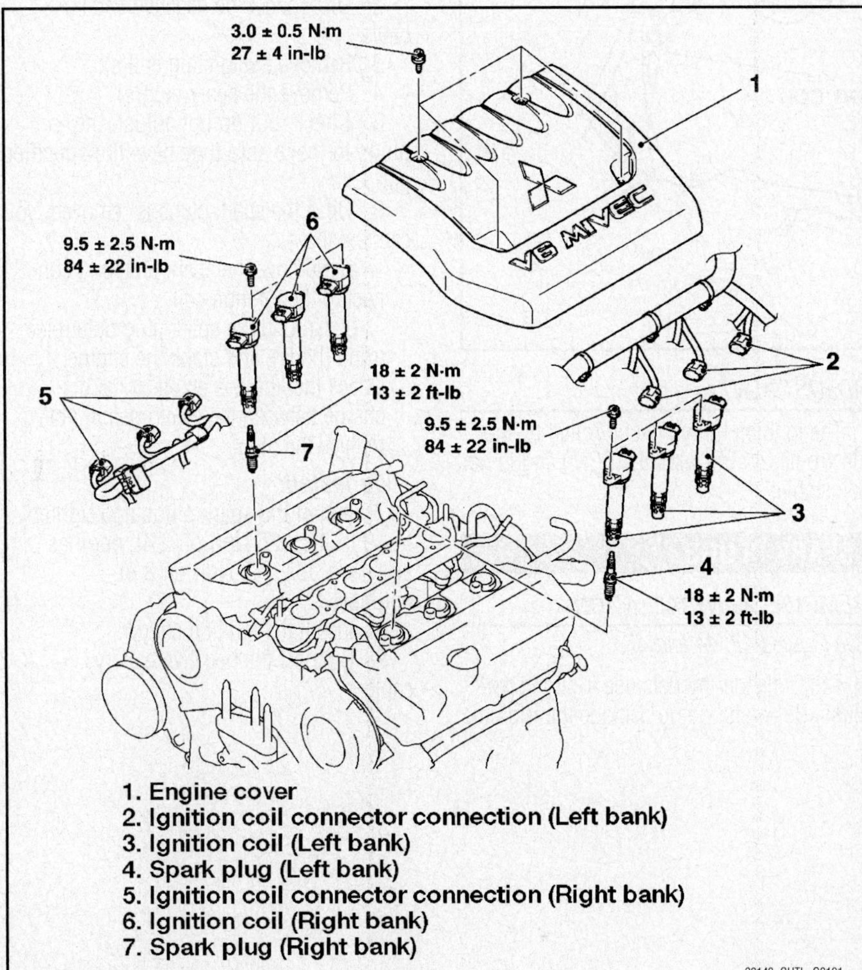

Fig. 44 Ignition coil assembly—3.0L engine

IGNITION TIMING

INSPECTION

See Figures 45 and 46.

This procedure requires the use of the following special tools, or their equivalents:

- MB991958: Scan Tool (MUT-III Sub Assembly)
- MB991824: V.C.I.
- MB991827: MUT-III USB Cable
- MB991910: MUT-III Main Harness A

1. Before starting the check, make sure that the vehicle is in the following condition:
   a. Engine coolant temperature: 80–95°C (176–203°F)
   b. Lights and all accessories: OFF
   c. Transaxle: Neutral (or P range on vehicles with A/T)

➡On Canadian vehicles, the headlight, taillight, etc. remain lit even when the light switch is in the "OFF" position. This is not a problem for checks and adjustment.

⁂ WARNING

To prevent damage to scan tool MB991958, always turn the ignition switch to the OFF position before connecting or disconnecting scan tool MB991958.

2. Connect the scan tool to the data link connector.
3. Set the timing light to the power supply line (terminal No. 1) of the No. 1 ignition coil.

➡The power supply line is looped and also longer than the other ones.

4. Start the engine and allow it to idle.
5. Check that the idle speed is within specification.
6. Select scan tool MB991958 actuator test "item number 17".
7. Check that base ignition timing is within the standard value: 2–8°BTDC (Before Top Dead Center).
8. If the base ignition timing is not within the standard value, check the following:
   a. Diagnostic output
   b. Timing belt cover and crankshaft position sensor installation conditions
   c. Crankshaft sensing blade condition

⁂ WARNING

If the actuator test is not canceled, the forced drive will continue for 27 minutes. Driving in this state could lead to engine failure.

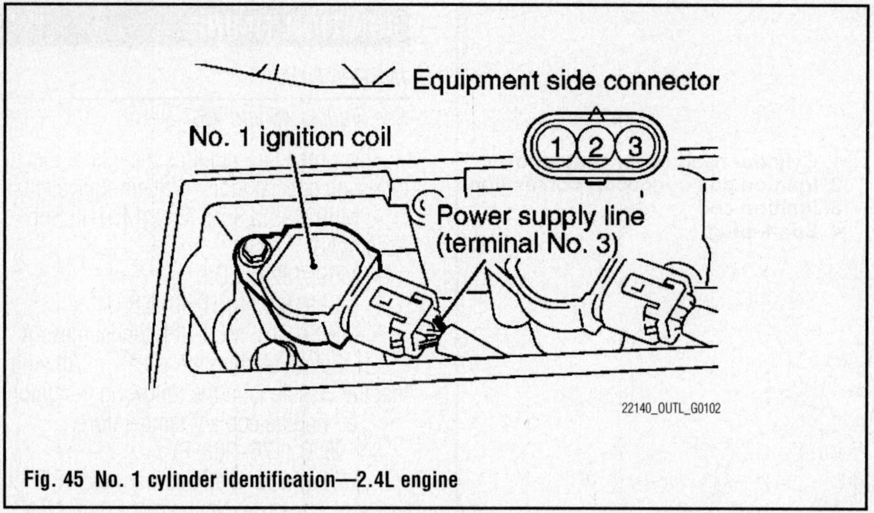

Fig. 45 No. 1 cylinder identification—2.4L engine

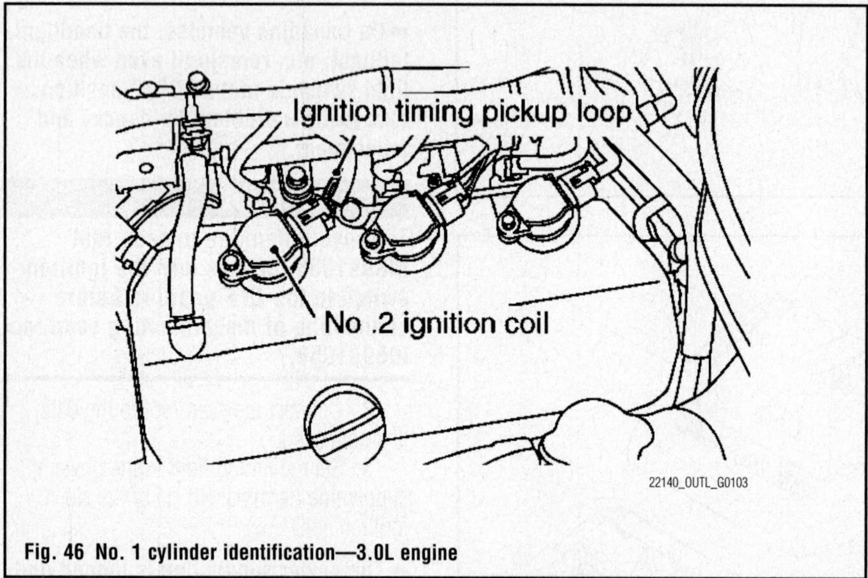

Fig. 46 No. 1 cylinder identification—3.0L engine

9. Press the clear key on scan tool MB991958 (select forced drive stop mode), and cancel the actuator test.
10. Check that the actual ignition timing is at the standard value: about 10°BTDC.

➡Ignition timing fluctuates approximately +/- 7°Before Top Dead Center, even under normal operating conditions.

➡At higher altitude, ignition timing is automatically further advanced by about 5–10°BTDC.

ADJUSTMENT

The ignition timing is controlled by the Electronic Control Module (ECM) and is not adjustable.

SPARK PLUGS

REMOVAL & INSTALLATION

See Figures 43, 44 and 47.

The Outlander models use iridium spark plugs. Be careful not to damage the iridium tips of the plugs. Do not adjust the spark plug gap. Spark plugs must spark properly to assure proper engine performance and reduce exhaust emission level. Therefore, they should be replaced periodically with new ones.

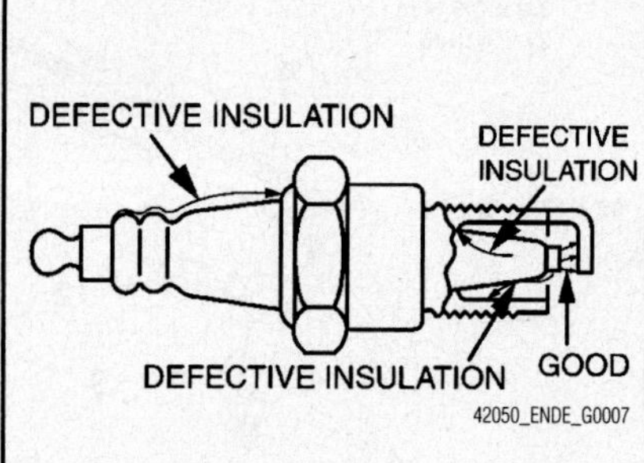

Fig. 47 Ground the spark plug outer electrode (body), and crank the engine. Check that there is an electrical discharge between the electrodes

1. Before servicing the vehicle, refer to the precautions.
2. Disconnect the negative battery cable.
3. Remove the ignition coil(s).
4. Remove the spark plug(s).
5. Check, but do not adjust, the plugs to make sure they have the specified gap.
6. Once the spark plugs is removed, you can test them.
   a. Remove the spark plug and connect to the ignition coil.
   b. Ground the spark plug outer electrode (body), and crank the engine. Check that there is an electrical discharge between the electrodes. If not, replace the plug.

To install:

7. Install the spark plugs and tighten to 19 ft. lbs. (25 Nm) on 2.4L engines and 13 ft. lbs. (18 Nm) on 3.0L engines.
8. Install the ignition coil(s).
9. Connect the negative battery cable.

STARTER

REMOVAL & INSTALLATION

2.4L Engine

See Figure 48.

1. Before servicing the vehicle, refer to the precautions.
2. Disconnect the negative battery cable.
3. Remove the air cleaner duct and intake hose.
4. Remove the battery and battery tray.
5. Remove the engine undercover.
6. Disconnect the emission vacuum hose and brake booster vacuum hose.
7. Remove the throttle body support.
8. Disconnect the starter connector and terminal.
9. Remove the starter assembly.
   a. Slide the starter assembly, and disconnect the starter connector and terminal.
   b. Remove the starter assembly from the lower front of the engine.

To install:

10. Install the starter assembly.
    a. Install the starter assembly from the lower front of the engine.
    b. Connect the starter connector and terminal and tighten to 93 inch lbs. (11 Nm).
    c. Slide the starter assembly into place. Tighten starter mounting bolts to 27 ft. lbs. (35 Nm).
11. Install the throttle body support.
12. Connect the emission vacuum hose and brake booster vacuum hose.
13. Install the engine undercover.
14. Install the battery and battery tray.
15. Install the air cleaner duct and intake hose.
16. Connect the negative battery cable.

3.0L Engine

See Figure 49.

1. Before servicing the vehicle, refer to the precautions.
2. Disconnect the negative battery cable.
3. Remove the engine undercover.
4. Disconnect the starter connector and terminal.
5. Remove the starter assembly.
   a. Disconnect the starter connector and terminal.
   b. Remove the starter cover.
   c. Remove the starter assembly from under the vehicle..

To install:

6. Install the starter assembly.
   a. Install the starter assembly from under the vehicle.
   b. Connect the starter connector and terminal and tighten to 93 inch lbs. (11 Nm).
   c. Remove the starter cover. Tighten to 44 inch lbs. (5 Nm).
   d. Slide the starter assembly into place. Tighten starter mounting bolts to 27 ft. lbs. (35 Nm).
7. Install the engine undercover.
8. Connect the negative battery cable.

SOLENOID OR RELAY REPLACEMENT

See Figure 50.

1. Remove the starter from the vehicle.

➡Do not clamp the yoke assembly with a vise.

2. Disconnect the lead from the M terminal of the magnetic switch.
3. Remove the screw and bracket from the solenoid.
4. Remove the solenoid from the starter.

To install:

5. Installation is the reverse of removal.
6. Tighten the solenoid screws to 51 inch lbs. (6 Nm).

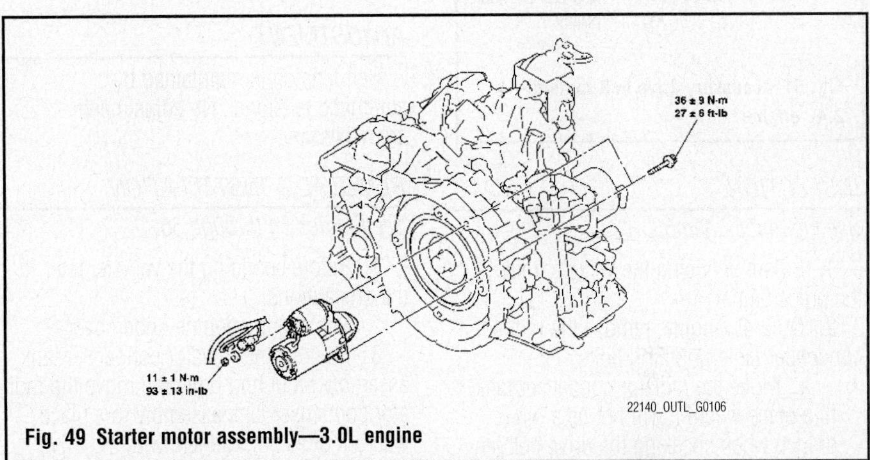

Fig. 49 Starter motor assembly—3.0L engine

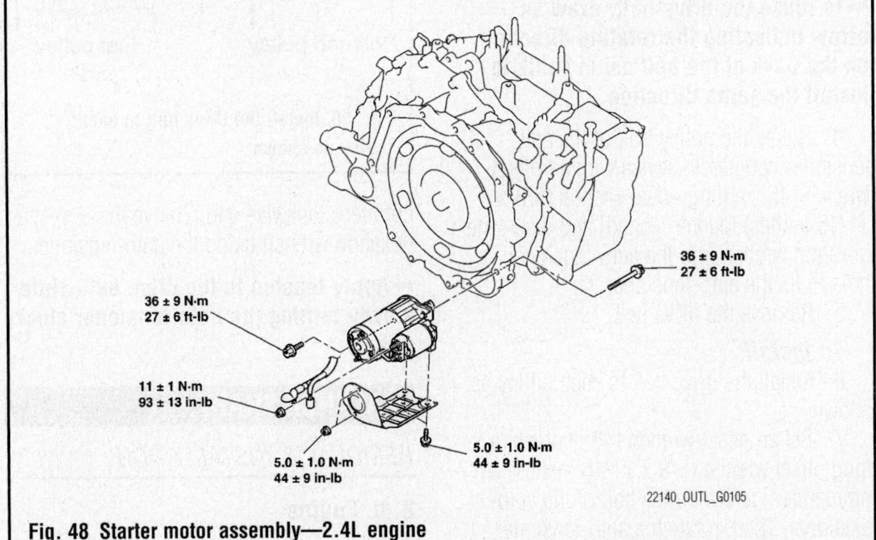

Fig. 48 Starter motor assembly—2.4L engine

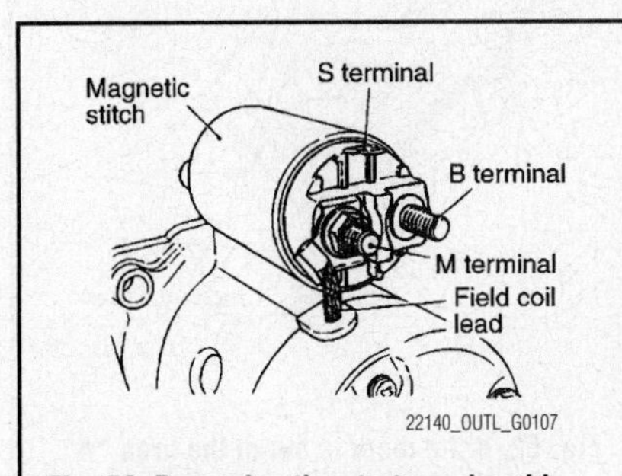

Fig. 50 Removing the starter solenoid

ENGINE MECHANICAL

ACCESSORY DRIVE BELTS

ACCESSORY BELT ROUTING

See Figure 51.

Refer to the accompanying illustration for accessory drive belt routing.

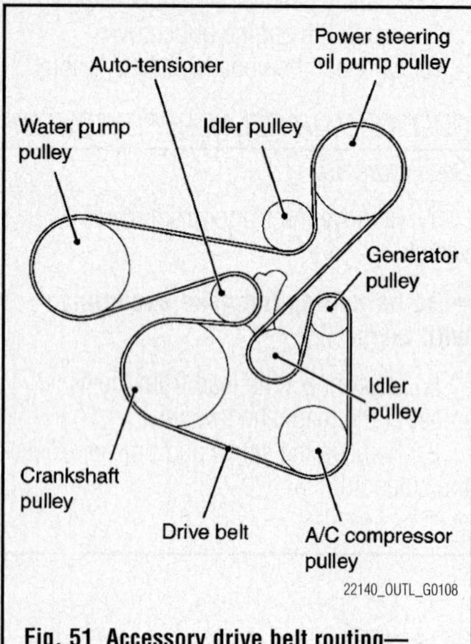

Fig. 51 Accessory drive belt routing—2.4L engine

INSPECTION

See Figures 52 and 53.

1. Before servicing the vehicle, refer to the precautions.
2. On 2.4L engine, remove the radiator condenser tank mounting bolts.
   a. Move the radiator condenser tank to a place where it will not be a hindrance when checking the drive belt tension.

➡Check the drive belt tension after turning the crankshaft clockwise one turn or more.

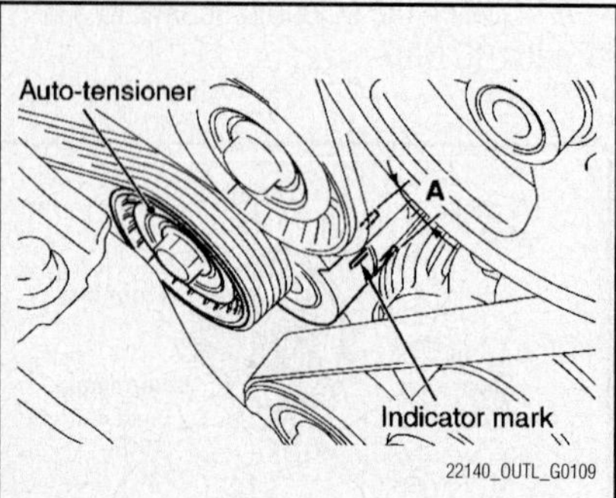

Fig. 52 If the mark is out of the area "A", replace the drive belt—2.4L engine

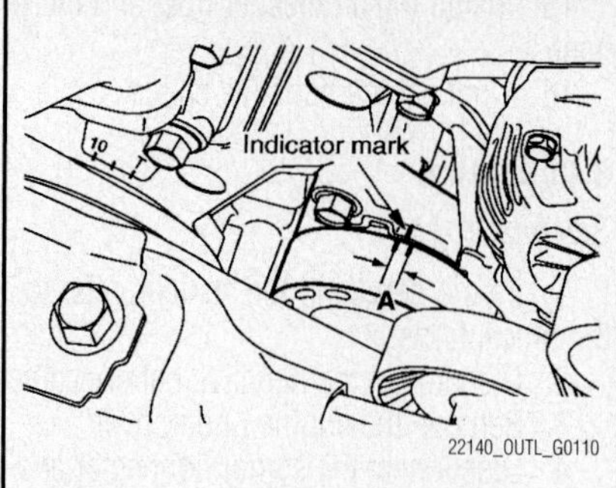

Fig. 53 If the mark is out of the area "A", replace the drive belt —3.0L engine

3. Make sure that the indicator mark on the auto-tensioner is within the area marked with "A".
4. If the mark is out of the area A, replace the drive belt.
5. On 2.4L engine, tighten the radiator condenser tank mounting bolts to 102 inch lbs. (12 Nm).

ADJUSTMENT

Belt tension is maintained by a automatic tensioner. No adjustments are necessary.

REMOVAL & INSTALLATION

See Figures 54 through 56.

1. Before servicing the vehicle, refer to the precautions.
2. Remove the engine undercover.
3. Remove the radiator condenser tank assembly mounting bolt, and move the radiator condenser tank assembly to a place where it does not interfere with the drive belt removal and installation.

➡To reuse the drive belt, draw an arrow indicating the rotating direction on the back of the belt using chalk to install the same direction.

4. Rotate the pulley bolt of the auto-tensioner counterclockwise with an offset wrench [45°, a long offset wrench (5/8 x 11/16 inches) recommended] and insert the hexagon wrench into the auto-tensioner hole to fix the auto-tensioner.
5. Remove the drive belt.

To install:

6. Install the drive belt to each pulley as shown.
7. Set an offset wrench [45 degrees, a long offset wrench (5/8 x 11/16 inches) recommended] to the pulley bolt of the auto-tensioner. Then, rotate the auto-tensioner

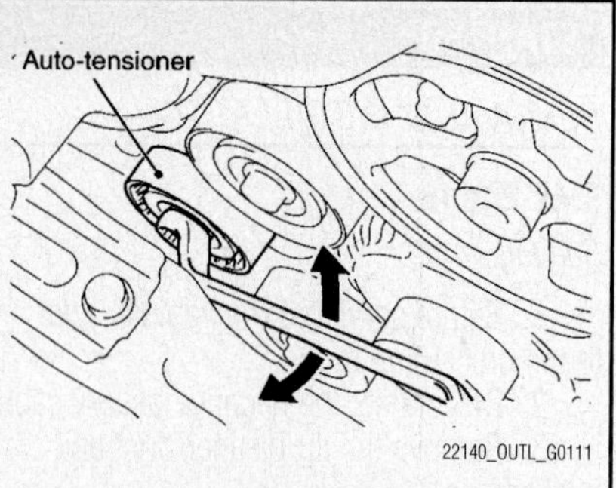

Fig. 54 Rotate the pulley bolt of the auto-tensioner counterclockwise with an offset wrench

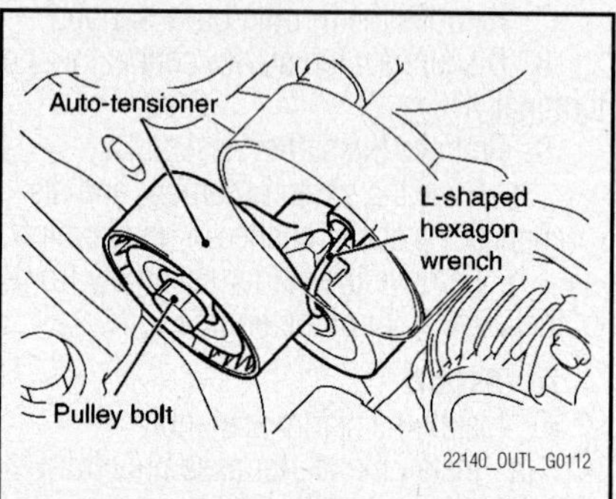

Fig. 55 Insert the hexagon wrench into the auto-tensioner hole to fix the auto-tensioner

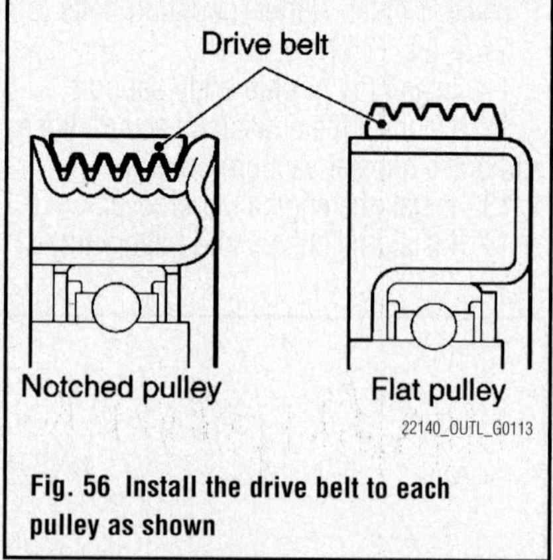

Fig. 56 Install the drive belt to each pulley as shown

counter-clockwise and remove the L-shaped hexagon wrench fixing the auto-tensioner.

➡Apply tension to the drive belt while slowly turning the auto-tensioner clockwise.

BALANCE SHAFT

REMOVAL & INSTALLATION

2.4L Engine

See Figures 57 through 60.

1. Before servicing the vehicle, refer to the precautions.
2. Remove the timing chain.
3. Remove the balance shaft timing chain tensioner.

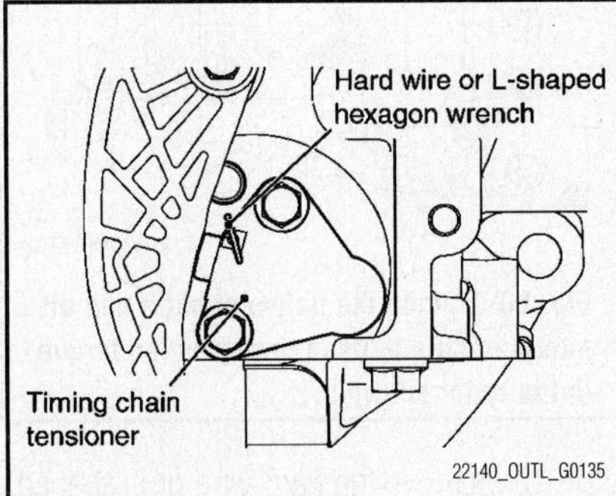

Fig. 57 Compress the plunger of the timing chain tensioner and insert hard wire or L-shaped hexagon wrench to fix the plunger of the timing chain tensioner

a. Securely install the plunger of the timing chain tensioner. Otherwise, it may pop out.

b. Press the balancer timing chain against the timing chain tensioner, compress the plunger of the timing chain tensioner and insert hard wire (piano wire, etc.) or L-shaped hexagon wrench 0.05 inch (1.5 mm) to fix the plunger of the timing chain tensioner.

c. Remove the balance shaft timing chain tensioner.

4. Remove the balance shaft timing chain guides.
5. Remove the balance shaft and oil pump module.
6. Remove the balance shaft timing chain.

To install:

7. Install the balance shaft and oil pump module.

a. When installing the new balancer shaft and oil pump module, apply oil to the oil pump in the balancer shaft and oil pump module and the balancer shaft bearing as follows:

b. Clean the inside of the removed engine oil pan, and put the balancer shaft and oil pump module into the engine oil pan with its oil inlet port facing up.

c. Pour new engine oil until two-thirds of the balancer shaft and oil pump module is soaked.

d. Fill the engine oil (approximately 3.05 cu. in. [50 cm3]) into the balancer shaft and oil pump module from the oil inlet port.

e. Turn the balancer shaft sprocket of the balancer shaft and oil pump module clockwise four rotations or more to apply the engine oil to the entire area of the oil pump and the balancer shaft bearing.

Apply engine oil to all moving parts before installation.

10 ± 2 N·m
89 ± 17 in-lb

10 ± 2 N·m
89 ± 17 in-lb

10 ± 2 N·m
89 ± 17 in-lb

(Engine oil)

20 N·m 15 ft-lb to 44 N·m 33 ft-lb to 0 N·m to 20 N·m 15 ft-lb to +135°

1. Timing chain tensioner
2. Balancer timing chain guide
3. Balancer timing chain guide
4. Balancer shaft and oil pump module
5. Balancer timing chain
6. Crankshaft sprocket

22140_OUTL_G0136

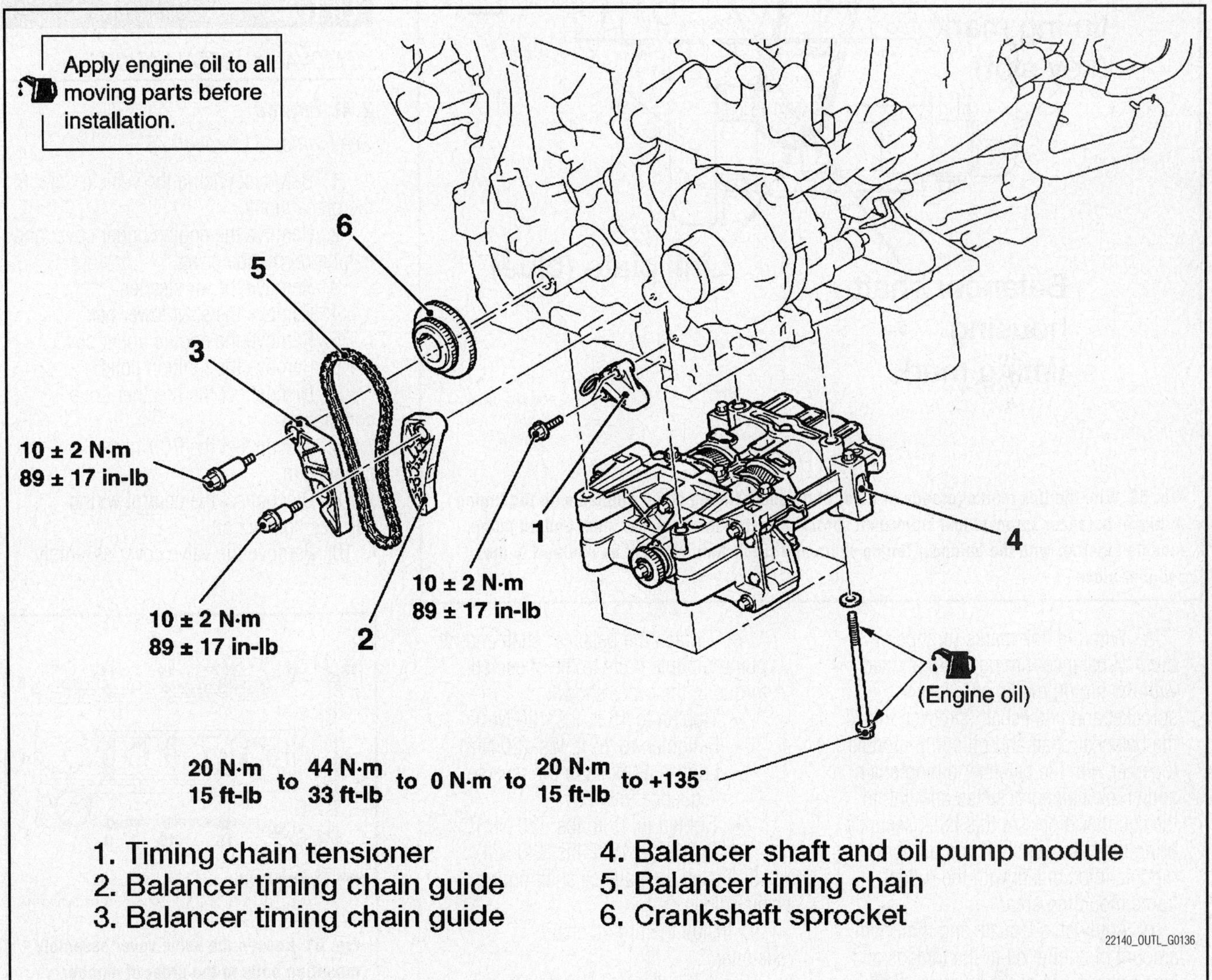

Fig. 58 Balance shaft timing chain and module

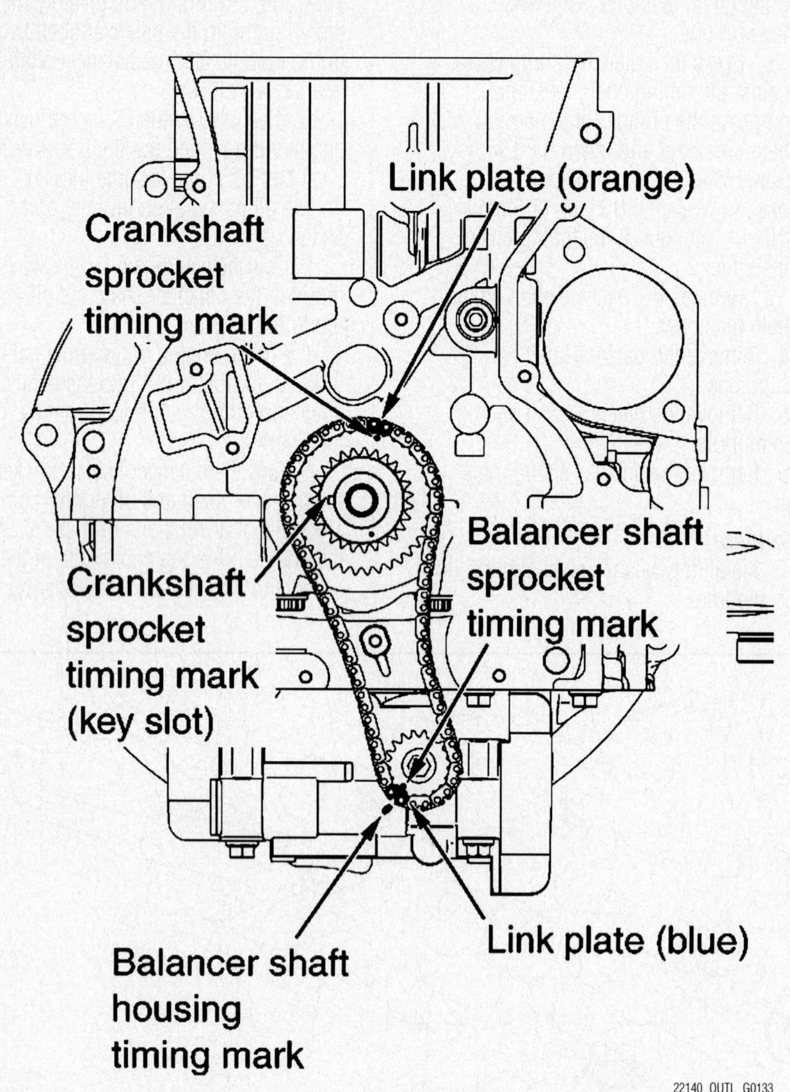

Fig. 59 With the link marks (orange or blue) of balancer timing chain aligned with the timing marks of balancer sprocket and crankshaft sprocket, install the balancer shaft and oil pump module together with the balancer timing chain and crankshaft sprocket as one unit to the cylinder block

f. With the link marks (orange or blue) of balancer timing chain aligned with the timing marks of balancer sprocket and crankshaft sprocket, install the balancer shaft and oil pump module together with the balancer timing chain and crankshaft sprocket as one unit to the cylinder block. At this time, securely bring the balancer shaft and oil pump module into contact with the rudder frame mounting area.

g. Apply an adequate and minimum amount of engine oil to the threads and bearing surfaces of the balancer shaft and oil pump module bolts.

h. Tighten the balancer shaft and oil pump module bolts to the specified torque in the order shown:

- Tighten to 15 ft. lbs. (20 Nm)
- Retighten to 15 ft. lbs. (20 Nm)
- Loosen all bolts in the reverse sequence fully
- Tighten to 15 ft. lbs. (20 Nm)
- Tighten to 15 ft. lbs. (20 Nm)

8. Install the balance shaft timing chain guides.
9. Install the timing chain tensioner.

a. Install the timing chain tensioner to the cylinder block.

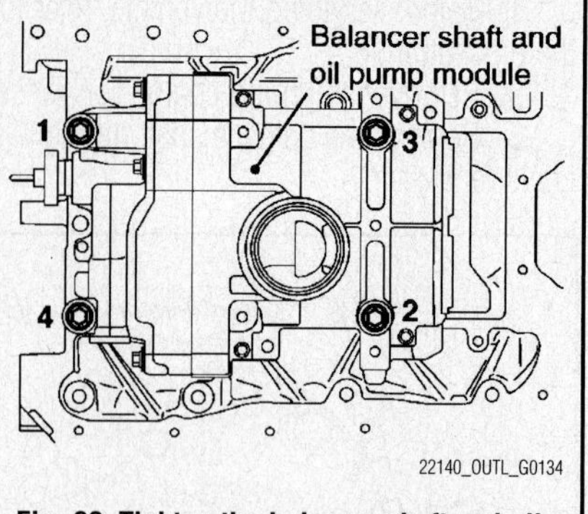

Fig. 60 Tighten the balancer shaft and oil pump module bolts to the specified torque in the order shown

b. Remove the hard wire or L-shaped hexagon wrench fixing the plunger of the timing chain tensioner to apply tension to the balancer timing chain.

10. Install the timing chain.

CAMSHAFT AND VALVE LIFTERS

REMOVAL & INSTALLATION

2.4L Engine

See Figures 61 through 78.

1. Before servicing the vehicle, refer to the precautions.
2. Remove the engine under cover and engine room side cover.
3. Remove the air cleaner.
4. Remove the strut tower bar.
5. Remove the engine upper cover.
6. Remove the ignition coils.
7. Disconnect the breather hose connection.
8. Disconnect the PCV hose connection.
9. Disconnect the control wiring harness connection.
10. Remove the valve cover assembly.

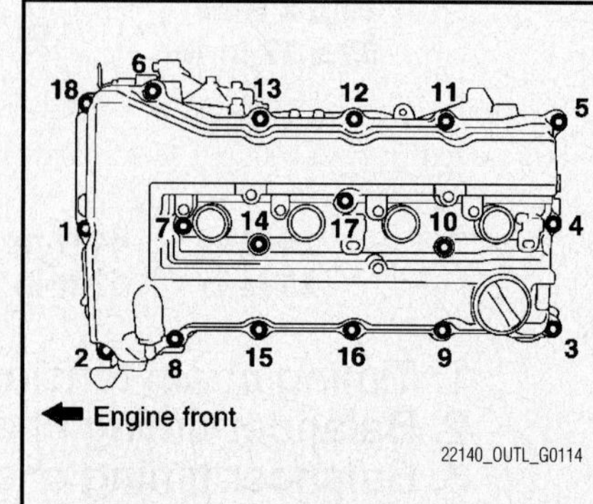

Fig. 61 Loosen the valve cover assembly mounting bolts in the order of number shown

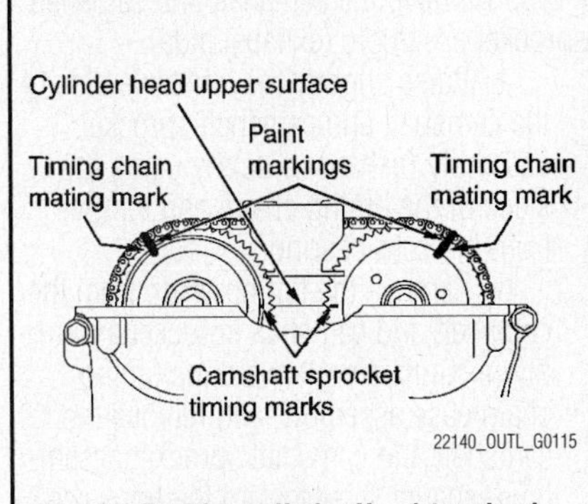

Fig. 62 Position cylinder No. 1 top dead center on the compression stroke

a. Loosen the valve cover assembly mounting bolts in the order of number shown, and remove the valve cover assembly.

b. Remove the valve cover gasket.

11. Position cylinder No. 1 top dead center on the compression stroke.

a. Turn the crankshaft clockwise so that the camshaft sprocket timing marks become horizontal to the cylinder head upper surface, and set cylinder No. 1 to Top Dead Center (TDC) on the compression stroke. At this time, check that the crankshaft pulley timing mark is in the 0-degree position of the ignition timing indicator of the timing chain case assembly.

b. Put paint marks on both the camshaft sprocket and timing chain at the position of camshaft sprocket timing chain mating mark (circular hole).

12. Remove the timing chain upper guide.

13. Remove the service hole bolt.

14. Remove the camshaft and camshaft sprocket assembly (exhaust side).

a. Insert a precision flat-tipped screwdriver through the service hole of the timing chain case, press up the timing chain tensioner ratchet to unlock, and keep the timing chain tensioner in that state.

b. Lightly press down the tail end of the precision flat-tipped screwdriver to press up the tip of the precision flat-tipped screwdriver inserted to the timing chain tensioner to unlock.

c. When inserting special tool MB992103 into the timing chain case assembly inside, pay attention to the position of the timing chain to avoid damage to the timing chain and timing chain tension side guide. Do not insert the special tool beyond its insertion guideline.

d. If unlocking the timing chain tensioner is insufficient, the special tool cannot be inserted to the insertion guideline. Do not insert the special tool forcibly, follow steps again to unlock the timing chain tensioner and insert the special tool.

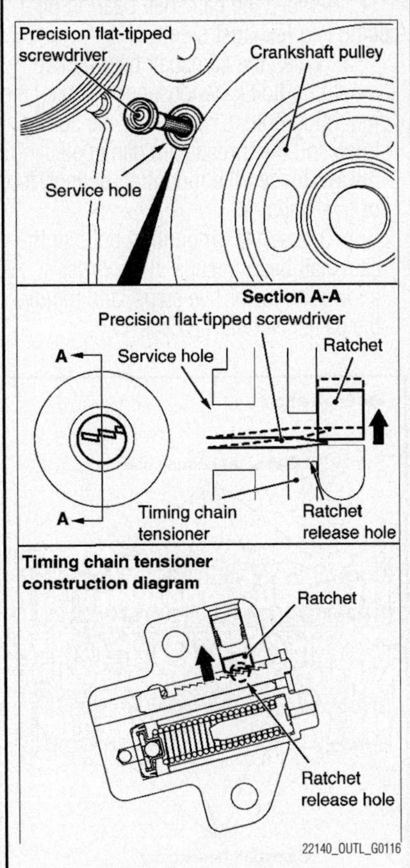

Fig. 63 Insert a precision flat-tipped screwdriver through the service hole of the timing chain case, press up the timing chain tensioner ratchet to unlock

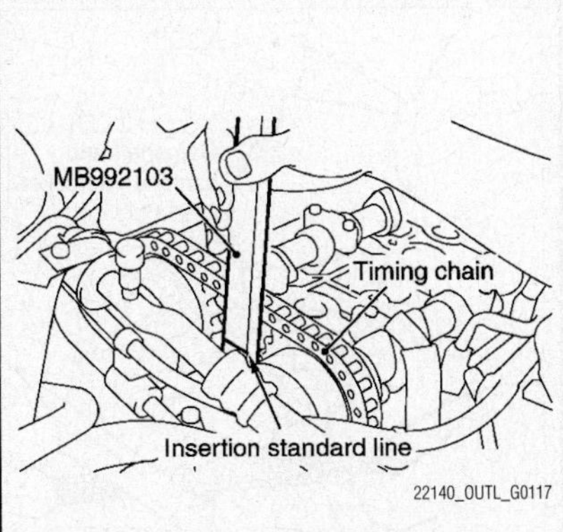

Fig. 64 When inserting special tool MB992103 into the timing chain case assembly inside, pay attention to the position of the timing chain to avoid damage to the timing chain and timing chain tension side guide

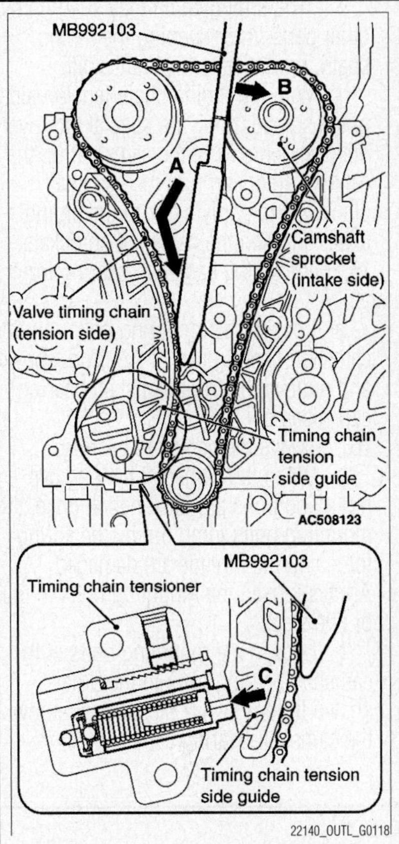

Fig. 65 With the timing chain tensioner unlocked, insert special tool MB992103 inside the timing chain case assembly along the tension side of the timing chain until the insertion guide line aligns with the upper surface of the timing chain case assembly

e. With the timing chain tensioner unlocked, insert special tool MB992103 inside the timing chain case assembly along the tension side of the timing chain until the insertion guide line aligns with the upper surface of the timing chain case assembly.

f. With the timing chain tensioner unlocked, insert the special tool along the tension side of the timing chain, according to the special tool top shape. The special tool can be inserted smoothly to the position where the special tool insertion guide line aligns with the timing chain case assembly top surface, and the spread timing chain tension side guide can be held.

g. With the special tool inserted up to the insertion guide line, press the special tool against the intake side camshaft sprocket and spread and hold the timing chain tension side guide.

h. Remove the flat-tipped precision screwdriver unlocking the timing chain tensioner.

i. The timing chain may snag on by other parts. After sagging the timing chain, never rotate the crankshaft.

j. With the timing chain tension side guide spread, hook the special tool over the hexagon part of the camshaft on the exhaust side, and turn the camshaft clockwise to apply slack to the timing chain between the camshaft sprockets.

15. Remove the camshaft bearing front caps.

a. Loosen the mounting bolts of front camshaft bearing cap in the order shown, and remove the front camshaft bearing cap assembly.

16. Remove the camshaft bearing.

a. When the camshaft bearing cap mounting bolts are loosened at once, the mounting bolts jump out by the spring force and the threads are damaged. Always loosen the mounting bolts in four or five steps.

b. Loosen the mounting bolts of the camshaft bearing caps in the order shown in four or five steps, and remove the camshaft bearing caps.

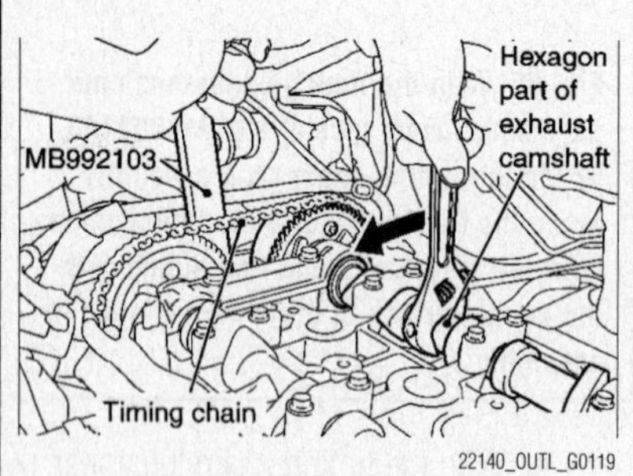

Fig. 66 With the timing chain tension side guide spread, hook the special tool over the hexagon part of the camshaft on the exhaust side, and turn the camshaft clockwise to apply slack to the timing chain between the camshaft sprockets

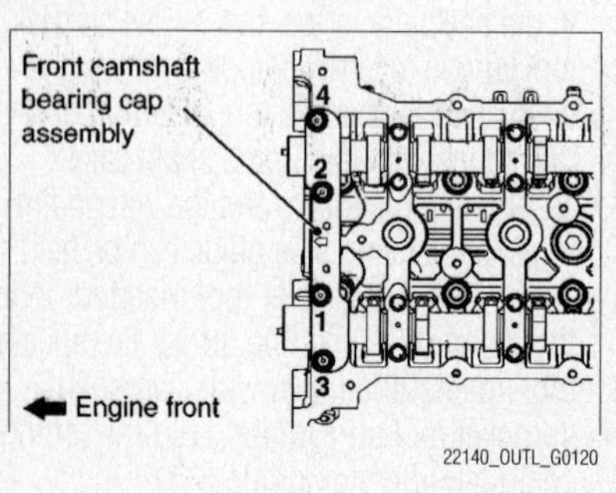

Fig. 67 Loosen the mounting bolts of front camshaft bearing cap in the order shown

17. Remove the camshaft bearing oil feeding cap (exhaust side).

a. When the camshaft bearing cap mounting bolts are loosened at once, the mounting bolts jump out by the spring force and the threads are damaged. Always loosen the mounting bolts in four or five steps.

b. Loosen the mounting bolts of the camshaft bearing caps in the order shown in four or five steps, and remove the camshaft bearing caps.

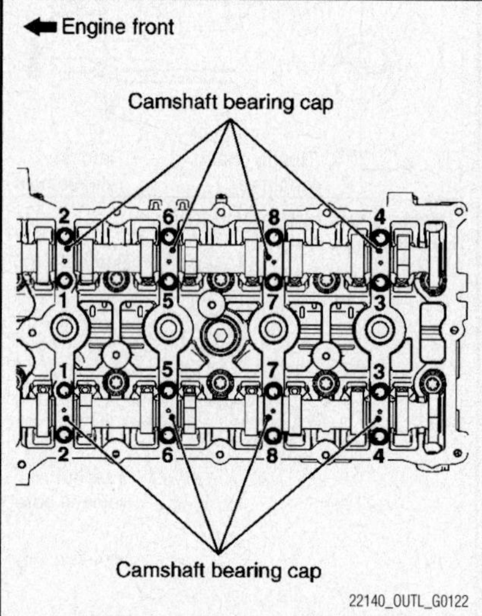

Fig. 68 Loosen the mounting bolts of the camshaft bearing caps in the order shown in four or five steps

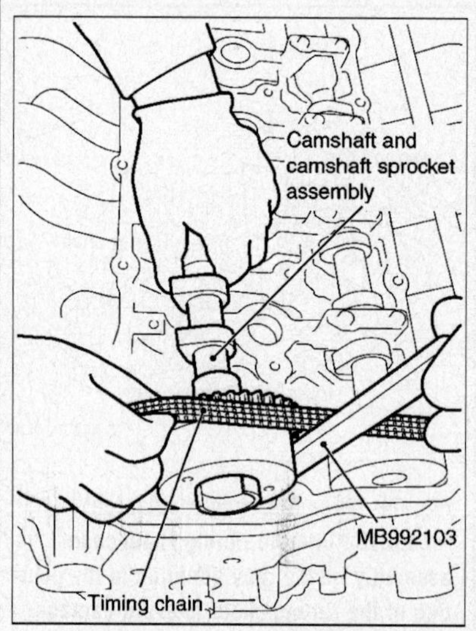

Fig. 69 Remove the timing chain from the camshaft and camshaft sprocket assembly (exhaust side) toward the timing chain case assembly, and remove the camshaft and camshaft sprocket assembly (exhaust side) toward the transaxle

18. Remove the camshaft and camshaft sprocket assembly (exhaust side).

a. Raise slightly the transaxle side of the camshaft and camshaft sprocket assembly (exhaust side) by using the slack of the timing chain, and remove from the cam bearing.

b. Remove the timing chain from the camshaft and camshaft sprocket assembly (exhaust side) toward the timing chain case assembly, and remove the camshaft and camshaft sprocket assembly (exhaust side) toward the transaxle.

c. Remove special tool MB992103 inserted into the timing chain case assembly.

d. The timing chain may snag on other parts. After removing the camshaft and camshaft sprocket assembly, never rotate the crankshaft.

e. After removing the camshaft and camshaft sprocket assembly (exhaust side), hang up the timing chain with a rope to prevent the timing chain from falling into the timing chain case assembly.

19. Remove the camshaft sprocket (exhaust side).

a. Hold the flats of the camshaft with an adjustable wrench. Loosen the camshaft sprocket mounting bolts and remove the camshaft sprocket from the camshaft.

20. Remove the camshaft (exhaust side).

21. Remove the camshaft bearing oil feeding cap (intake side).

22. Remove the camshaft bearing cap (intake side).

23. Remove the camshaft sprocket (intake side).

24. Remove the camshaft (intake side).

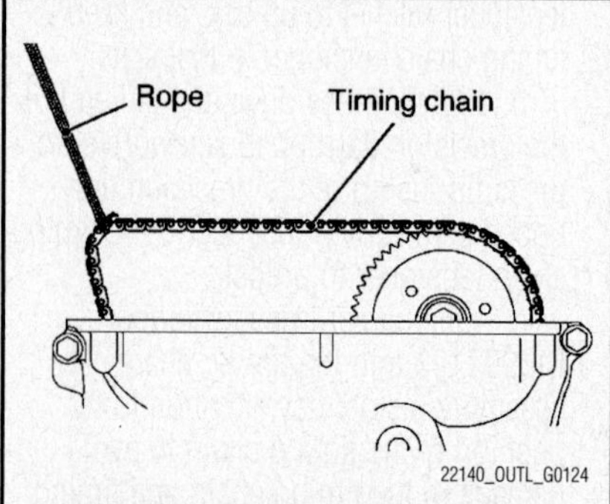

Fig. 70 After removing the camshaft and camshaft sprocket assembly (exhaust side), hang up the timing chain with a rope to prevent the timing chain from falling into the timing chain case assembly

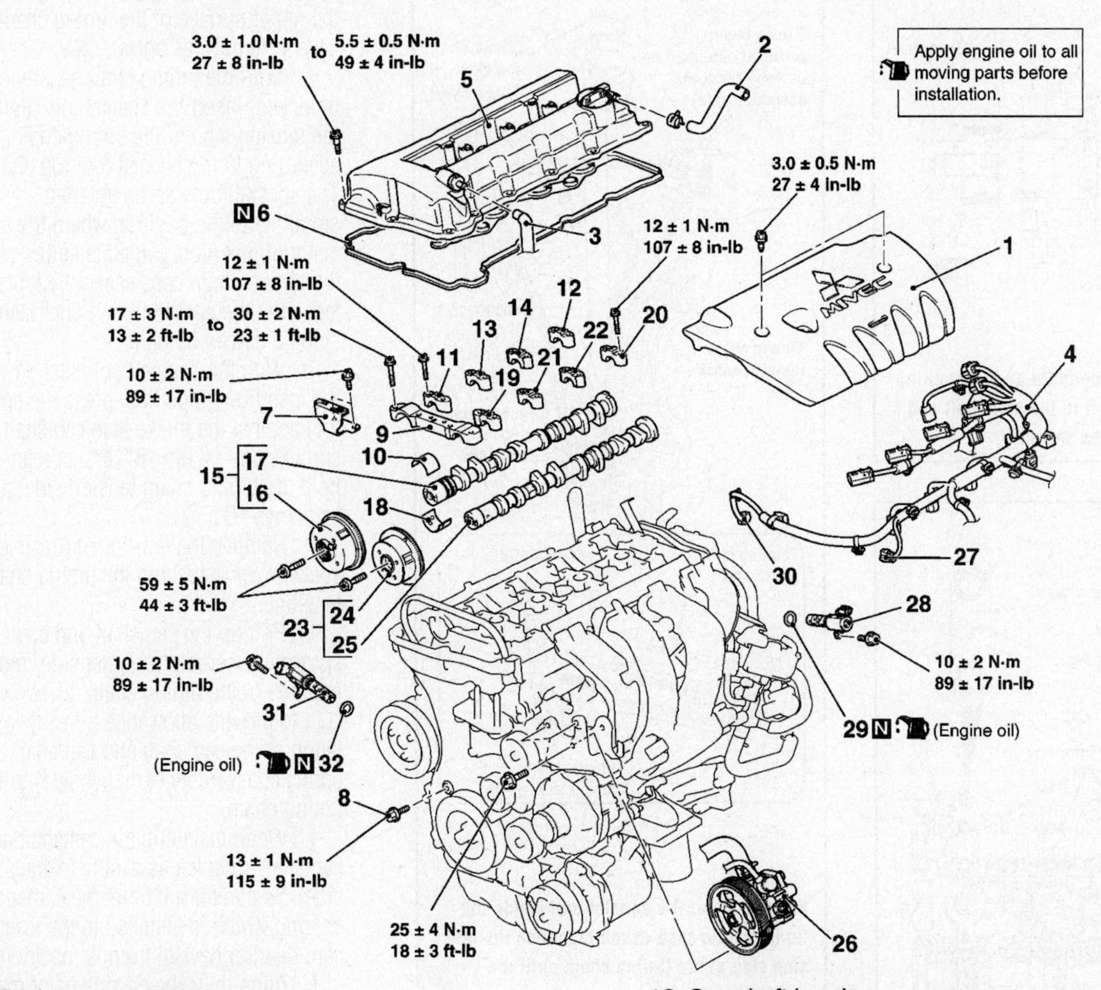

1. Engine upper cover
2. Breather hose connection
3. PCV hose connection
4. Control wiring harness connection
5. Rocker cover assembly
6. Rocker cover gasket
7. Timing chain upper guide
8. Service hole bolt
9. Camshaft bearing front cap assembly
10. Camshaft bearing
11. Camshaft bearing oil feeding cap (exhaust side)
12. Camshaft bearing cap (exhaust side)
13. Camshaft bearing cap (exhaust side)
14. Camshaft bearing thrust cap (exhaust side)
15. Camshaft and camshaft sprocket assembly (exhaust side)
16. Camshaft sprocket (exhaust side)
17. Camshaft (exhaust side)
18. Camshaft bearing
19. Camshaft bearing oil feeding cap (intake side)
20. Camshaft bearing cap (intake side)
21. Camshaft bearing cap (intake side)
22. Camshaft bearing thrust cap (intake side)
23. Camshaft and camshaft sprocket assembly (intake side)
24. Camshaft sprocket (intake side)
25. Camshaft (intake side)
26. Power steering oil pump assembly
27. Intake oil feeder control valve connector connection
28. Intake oil feeder control valve
29. O-ring
30. Exhaust oil feeder control valve connector connection
31. Exhaust oil feeder control valve
32. O-ring

22140_OUTL_G0132

Fig. 71 Exploded view of the camshaft assemblies—2.4L engine

To install:

25. Install the camshaft sprocket (intake side).
   a. Apply an adequate and minimum amount of engine oil to the camshaft and camshaft sprocket as shown.
   b. Install the camshaft sprocket to the camshaft.
   c. Apply an adequate and minimum amount of engine oil to the camshaft sprocket bolt. Tighten the camshaft sprocket bolts to 44 ft. lbs (59 Nm).
26. Install the camshaft (intake side).
   a. Align the intake side paint mark of the timing chain which was put at removal with the paint mark of the intake side camshaft sprocket, and install the camshaft sprocket to the timing chain.
   b. Install the camshaft and camshaft sprocket assembly (intake side) to the cylinder head.
27. Install the camshaft bearing caps to the cylinder head (intake side).
   a. Because the thrust camshaft bearing cap and camshaft bearing cap are the same in shape, check the bearing cap number and additionally its symbol to

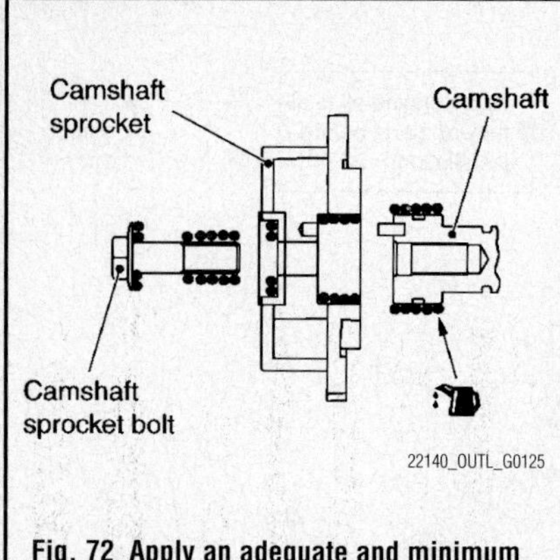

Fig. 72 Apply an adequate and minimum amount of engine oil to the camshaft and camshaft sprocket as shown

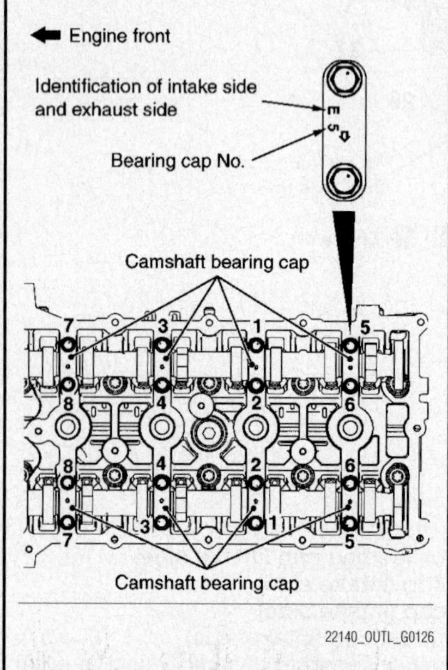

Fig. 73 Tighten each camshaft bearing cap mounting bolt in the order shown, using two or three steps

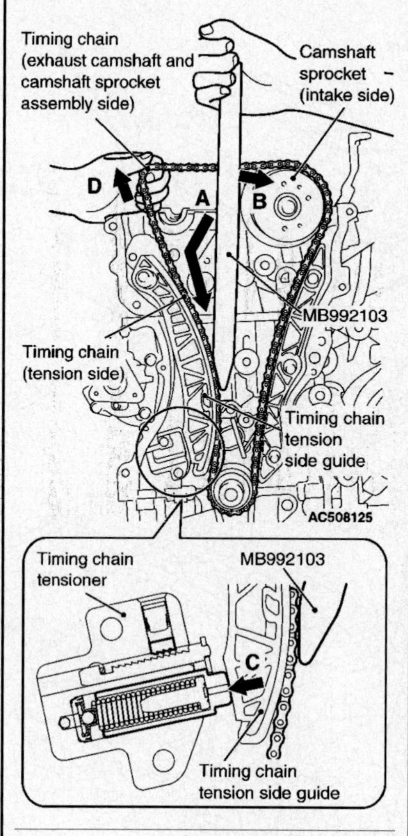

Fig. 74 Insert the special tool inside the timing chain case assembly along the tension side of the timing chain until the insertion guide line aligns with the upper surface of the timing chain case assembly "A", press the special tool against the intake side camshaft sprocket, "B", and spread and hold the timing chain tension side guide "C". Pull up the camshaft and camshaft sprocket assembly (exhaust side) mounting area of the timing chain "D"

identify the intake and exhaust sides for correct installation.

b. Tighten each camshaft bearing cap mounting bolt to 107 inch lbs. 912 Nm) in the order shown, using two or three steps.

28. Install the camshaft (exhaust side).

a. Insert the precision flat-tipped screwdriver through the service hole of the timing chain case, press up the ratchet of timing chain tensioner to unlock, and hold the unlocked timing chain tensioner.

b. Lightly press down the tail end of the precision flat-tipped screwdriver to press up the tip of the precision flat-tipped screwdriver inserted to the timing chain tensioner to unlock.

c. When inserting special tool MB992103 into the timing chain case assembly, pay attention to the position of the timing chain to avoid damage to the timing chain and timing chain tension side guide. Do not insert the special tool beyond its insertion guideline.

d. If unlocking the timing chain tensioner is insufficient, the special tool cannot be inserted to the insertion guideline. Do not insert the special tool forcibly, follow Step 2 again to unlock the timing chain tensioner and insert the special tool.

e. With the timing chain tensioner unlocked, insert special tool MB992103 inside the timing chain case assembly along the tension side of the timing chain until the insertion guide line aligns with the upper surface of the timing chain case assembly, as shown "A".

f. With the timing chain tensioner unlocked, insert the special tool along the tension side of the timing chain, according to the special tool top shape. The special tool can be inserted smoothly to the position where the special tool insertion guideline aligns with the timing chain case assembly top surface, and the spread timing chain tension side guide can be hold.

g. With the special tool inserted up to the insertion guide line, press the special tool against the intake side camshaft sprocket, as shown "B", and spread and hold the timing chain tension side guide, as shown "C".

h. Remove the flat-tipped precision screwdriver unlocking the timing chain tensioner.

i. Pull up the camshaft and camshaft sprocket assembly (exhaust side) mounting area of the timing chain, as shown "D", to provide allowance for easy installation of the camshaft and camshaft sprocket assembly (exhaust side) to the timing chain.

j. When installing the camshaft and camshaft sprocket assembly (exhaust side), be careful not to let the camshaft bearing which is installed to the front cam bearing deviate from its position.

k. Align the exhaust side paint mark of the timing chain which was put at removal with the paint mark of the exhaust side camshaft sprocket, and install the timing chain to the camshaft sprocket.

l. Install the camshaft and camshaft sprocket assembly (exhaust side) to the cylinder head.

m. Remove the special tool inserted into the timing chain case assembly inside.

29. Install the camshaft bearing caps to the cylinder head (exhaust side).

a. Because the thrust camshaft bearing cap and camshaft bearing cap are the same in shape, check the bearing cap number and additionally its symbol to identify the intake and exhaust sides for correct installation.

b. Tighten each camshaft bearing cap mounting bolt to 107 inch lbs. 912 Nm) in the order shown, using two or three steps.

30. Install the camshaft bearing front caps.

a. When the mounting bolts are tightened with the front camshaft bearing cap

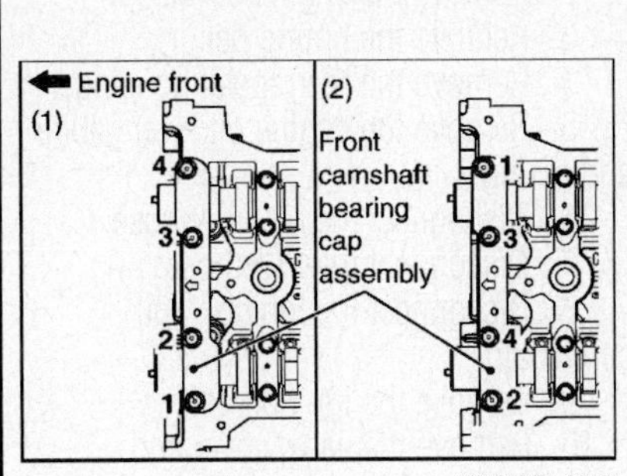

Fig. 75 Temporarily tighten the camshaft bearing front cap in the order shown (1). Then tighten again in the order (2)

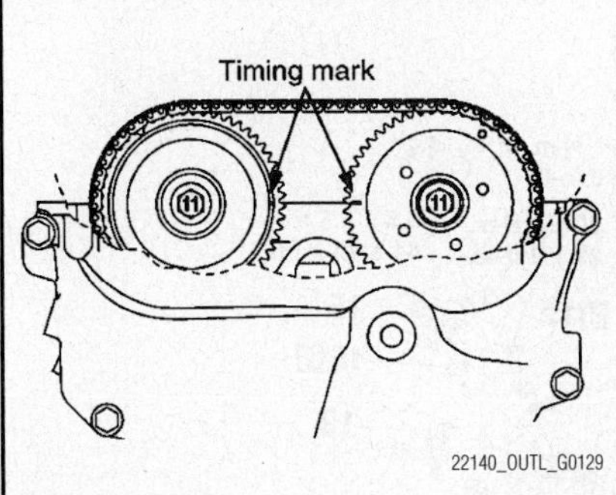

Fig. 76 Timing sprockets with No. 1 cylinder at TDC

tilted, the front camshaft bearing cap is damaged. Install the front camshaft bearing cap properly to the cylinder head and camshaft.

b. Install the front camshaft bearing cap to the cylinder head, and temporarily tighten the camshaft bearing front cap to 13 ft. lbs. (17 Nm) in the order shown of the figure (1). Then tighten again in the order of the figure (2).

c. After the front camshaft bearing cap installation, check that the paint markings of the camshaft sprocket and the timing chain and the timing mark of the crankshaft pulley and the "T" mark position of ignition timing indicator are aligned respectively.

31. Install the service hole bolt.
32. Install the timing chain upper guide.
33. Adjust the valve clearance.

a. Perform the valve clearance check and adjustment at the engine cold state.

b. If the engine has been turned, turn the crankshaft clockwise, and align the timing mark on the exhaust camshaft sprocket against the upper face of the cylinder head. The No.1 cylinder should now be at TDC on the compression stroke.

c. Using a thickness gauge, measure the valve clearance of cylinders No. 1 and 2 on the intake side and cylinders No. 1 and 3 on the exhaust side. If not within specification, make note for the valve clearance adjustment.

d. Turn the crankshaft clockwise 360 degrees, and put the timing mark on the exhaust camshaft sprocket as illustrated. The No.1 cylinder should now be at TDC on the compression stroke.

e. Check the valve clearance of cylinders No. 3 and 4 on the intake side and cylinders No. 2 and 4 on the exhaust side.

f. If the valve clearance is not within specification, remove the camshaft and the valve tappet.

g. Using a micrometer, measure the thickness of the removed valve tappet.

h. Calculate the thickness of the newly installed valve tappet through the following equation:

- A: thickness of newly installed valve tappet
- B: thickness of removed valve tappet
- C: measured valve clearance

i. Calculate the thickness of the valve tappet as follows:

- Intake valve: A = B + [C - 0.20 mm (0.008 inch)]
- Exhaust valve: A = B + [C - 0.30 mm (0.012 inch)]

j. Install the valve tappet selected and put the camshaft in position.

k. After installing the timing chain, measure the valve clearance again.

34. Install the valve cover assembly

a. Wipe off the sealant on the mating surface of the valve cover assembly and the cylinder head and timing chain case assembly, and degrease the surface where the sealant is applied by white gasoline or the like.

b. Apply Three bond 1217G or equivalent sealant to the joint between the cylinder head and timing chain case assembly as shown and install the valve cover assembly to the cylinder head.

➡Install the valve cover assembly within 3 minutes after the application of sealant.

c. Tighten the valve cover assembly mounting bolts to 27 inch lbs. (3 Nm) in the order shown. Then tighten again to specification in the order shown.

35. Connect the control wiring harness connection.
36. Connect the PCV hose connection.
37. Connect the breather hose connection.
38. Install the ignition coils.

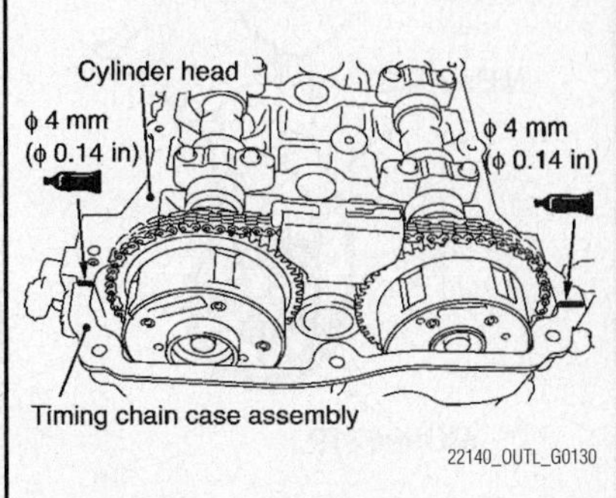

Fig. 77 Apply Three bond 1217G or equivalent sealant to the joint between the cylinder head and timing chain case assembly as shown

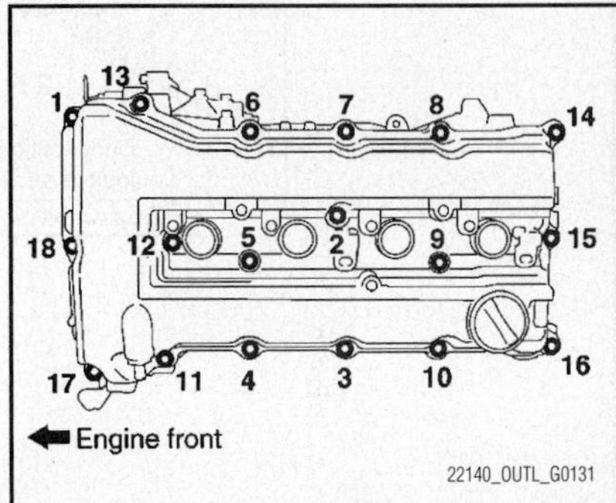

Fig. 78 Tighten the valve cover assembly mounting bolts in the order shown

39. Install the engine upper cover.
40. Install the strut tower bar .
41. Install the air cleaner.
42. Install the engine under cover and engine room side cover.

3.0L Engine

Left Bank

See Figures 79 through 86.

1. Before servicing the vehicle, refer to the precautions.

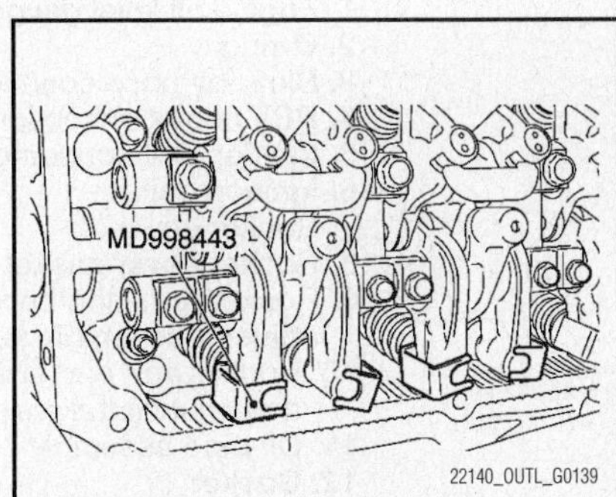

Fig. 79 Install special tool MD998443 as shown so that the lash adjusters will not fall out

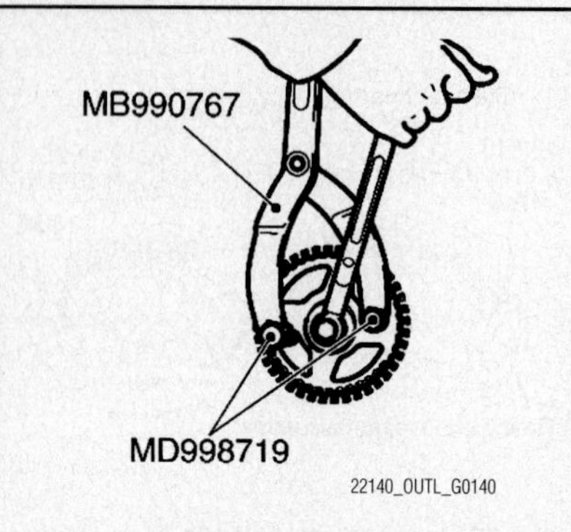

Fig. 80 Using special tools MB990767 and MD998719, remove the camshaft sprocket

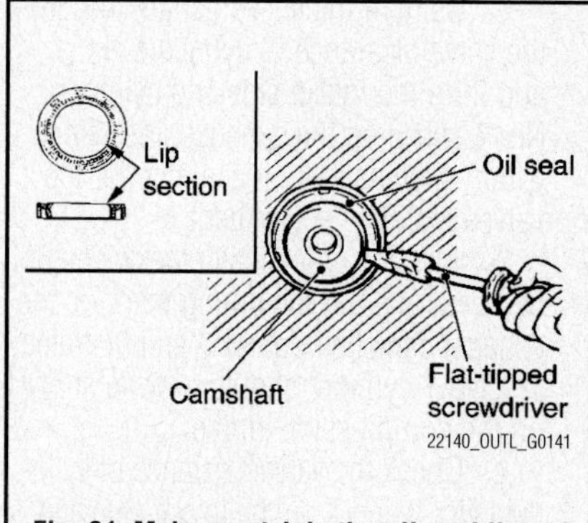

Fig. 81 Make a notch in the oil seal lip section with a knife

2. Remove the engine cover.
3. Remove the timing belt.
4. Remove the thermostat housing.
5. Remove the engine oil level gauge and o-ring.
6. Disconnect the blow-by hose.
7. Disconnect the PCV hose.
8. Disconnect the ignition coil connector.
9. Remove the ignition coil.
10. Remove the valve cover and gasket.
11. Remove the rocker arm, shaft and the lash adjuster assembly (exhaust side).

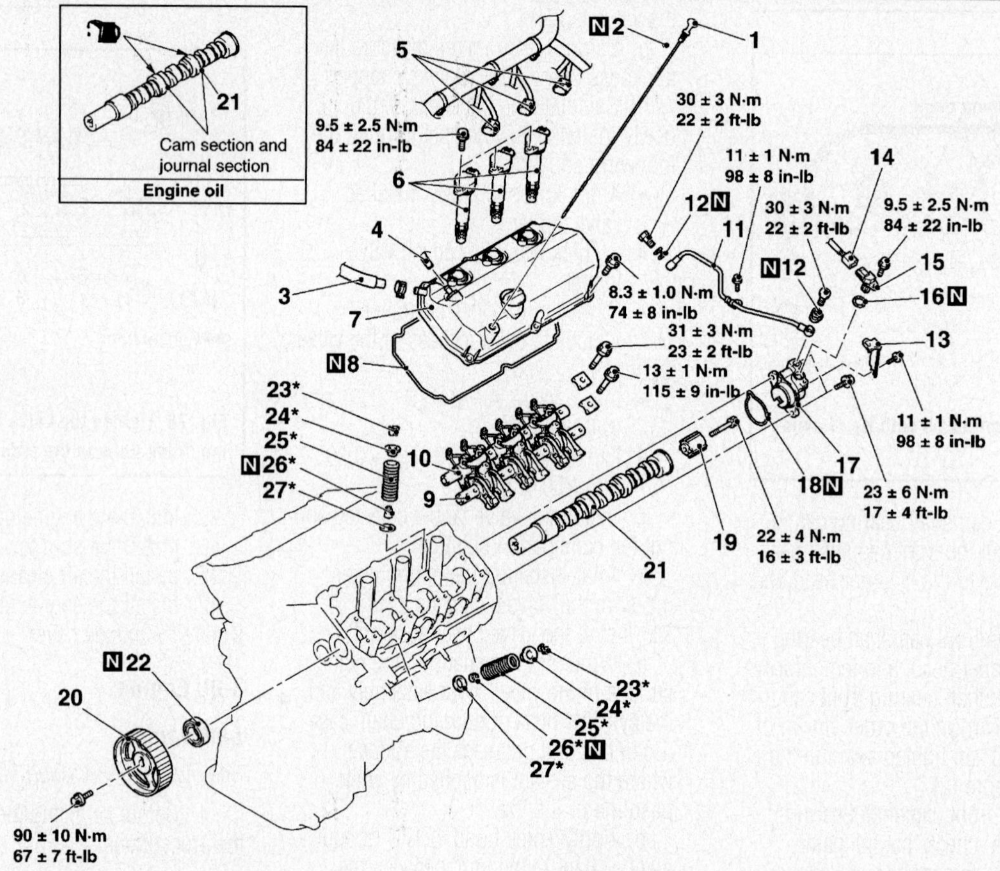

1. Engine oil level gauge
2. O-ring
3. Blow-by hose connection
4. PCV hose connection
5. Ignition coil connector
6. Ignition coil
7. Rocker cover
8. Rocker cover gasket
9. Rocker arm, shaft and lash adjuster assembly (exhaust side)
10. Rocker arm and shaft assembly (intake side)
11. Oil pipe assembly
12. Gasket
13. Harness bracket
14. Camshaft position sensor connector
15. Camshaft position sensor
16. O-ring
17. Camshaft position sensor support
18. Camshaft position sensor support gasket
19. Camshaft position sensing cylinder
20. Camshaft sprocket
21. Camshaft
22. Camshaft oil seal
23. Valve spring retainer lock
24. Valve spring retainer
25. Valve spring
26. Valve stem seal
27. Valve spring seat

22140_OUTL_G0137

Fig. 82 Exploded view of the left bank camshaft assembly —3.0L engine

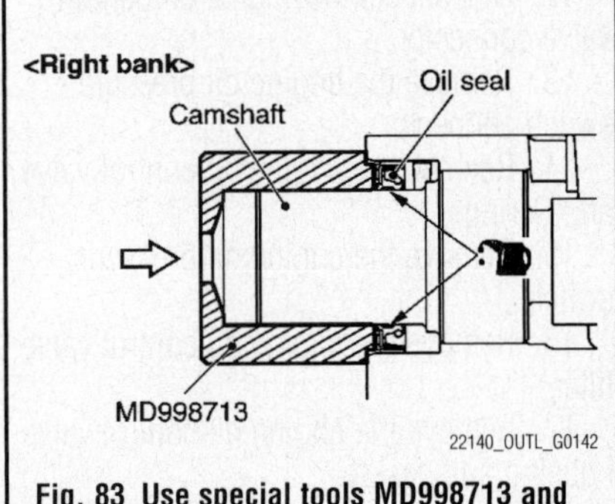

Fig. 83 Use special tools MD998713 and MD998777 to press-fit the camshaft oil seal

a. Install special tool MD998443 as shown so that the lash adjusters will not fall out.

WARNING

Never disassemble the rocker arm and shaft assembly.

b. Loosen the rocker arm and shaft assembly mounting bolt, and then remove the rocker arm and shaft assembly with the bolt still attached.

12. Remove the rocker arm and shaft assembly (intake side).
13. Remove the oil pipe assembly and gasket.
14. Remove the harness bracket.
15. Disconnect the camshaft position sensor connector.
16. Remove the camshaft position sensor and O-ring.
17. Remove the camshaft position sensor support and gasket.
18. Remove the camshaft position sensing cylinder.
19. Using special tools MB990767 and MD998719, remove the camshaft sprocket.
20. Remove the camshaft.
21. Remove the camshaft oil seal.

a. Make a notch in the oil seal lip section with a knife.

b. Be careful not to damage the camshaft and the cylinder head.

c. Cover the end of a flat-tipped screwdriver with a shop towel and insert into the notched section of the oil seal, and pry out the oil seal to remove it.

To install:

22. Install the camshaft oil seal.

a. Apply engine oil to the camshaft oil seal lip.

b. Use special tools MD998713 and MD998777 to press-fit the camshaft oil seal.

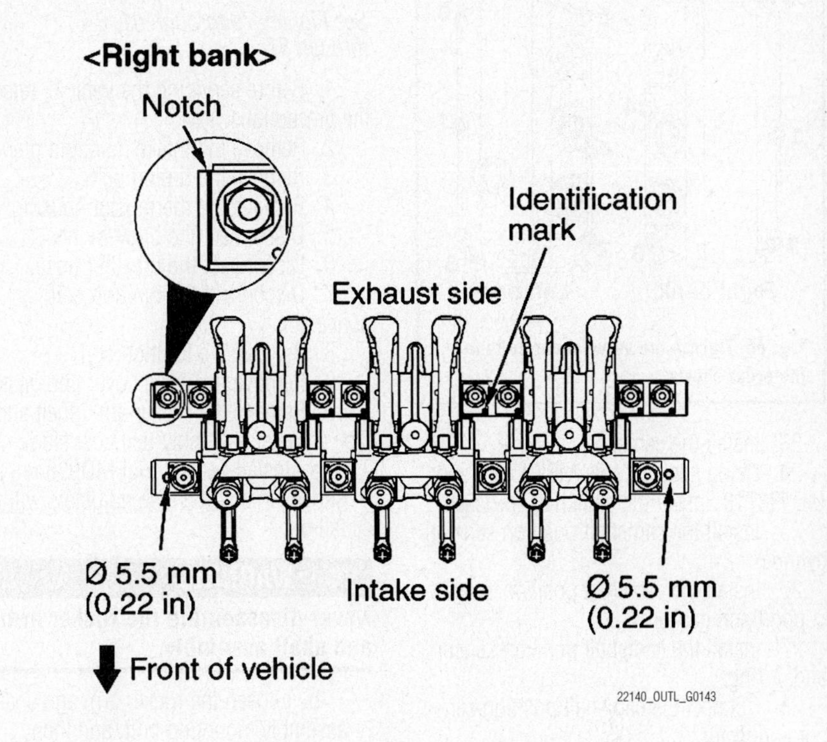

Fig. 84 Install the intake side rocker arm and shaft assembly so that the holes of rocker arm shaft face the cylinder head side

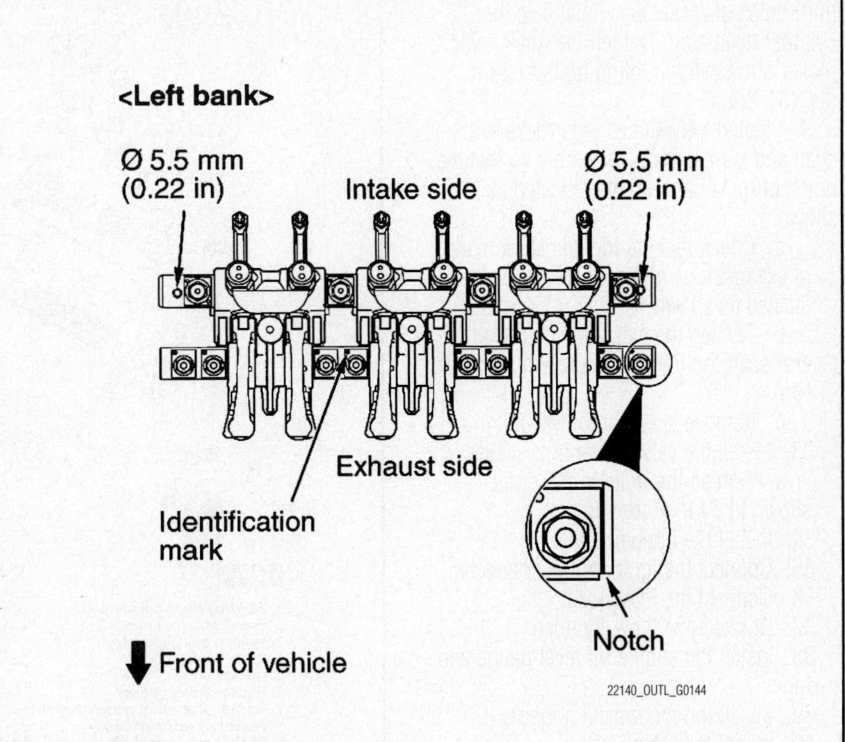

Fig. 85 Install the exhaust side rocker arm, shaft and lash adjuster assembly so that the notch of rocker arm shaft is located as shown

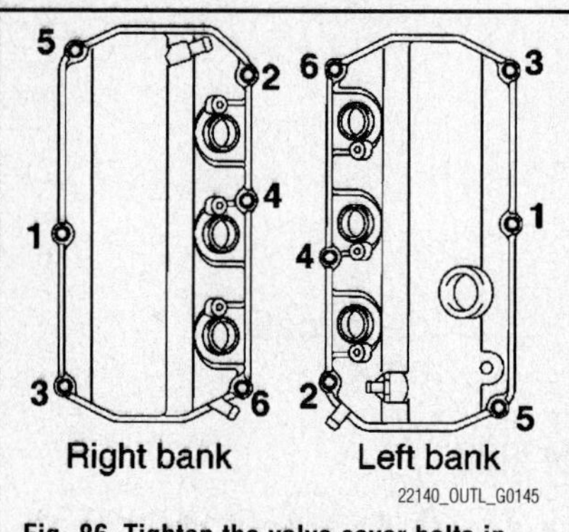

Fig. 86 Tighten the valve cover bolts in the order shown

23. Install the camshaft.
24. Using special tools MB990767 and MD998719, install the camshaft sprocket.
25. Install the camshaft position sensing cylinder.
26. Install the camshaft position sensor support and gasket.
27. Install the camshaft position sensor and O-ring.
28. Connect the camshaft position sensor connector.
29. Install the harness bracket.
30. Install the oil pipe assembly and gasket.
31. Install the intake side rocker arm and shaft assembly so that the 0.22 inch (5.5 mm) holes of rocker arm shaft face the cylinder head side. Tighten the intake side rocker arm shaft mounting bolts to 23 ft. lbs. (31 Nm).
32. Install the exhaust side rocker arm, shaft and lash adjuster assembly so that the notch of rocker arm shaft is located as shown.
    a. Check that the identification mark of exhaust side rocker shaft cap is located as shown.
    b. Tighten the exhaust side rocker arm shaft mounting bolts to 9 ft. lbs. (13 Nm).
    c. Remove special tool MD998443.
33. Install the valve cover and gasket.
    a. Tighten the bolts in the order shown to 74 inch lbs. (8 Nm).
34. Install the ignition coil.
35. Connect the ignition coil connector.
36. Connect the PCV hose.
37. Connect the blow-by hose.
38. Install the engine oil level gauge and o-ring.
39. Install the thermostat housing.
40. Install the timing belt.
41. Install the engine cover.
42. Change the engine oil.
43. Start the engine and check for leaks.

Right Bank

See Figures 79 through 81, 83 through 87.

1. Before servicing the vehicle, refer to the precautions.
2. Remove the intake manifold plenum.
3. Remove the timing belt.
4. Remove the thermostat housing.
5. Disconnect the blow-by hose.
6. Disconnect the breather hose.
7. Disconnect the ignition coil connector.
8. Remove the ignition coil.
9. Remove the valve cover and gasket.
10. Remove the rocker arm, shaft and the lash adjuster assembly (exhaust side).
    a. Install special tool MD998443 as shown so that the lash adjusters will not fall out.

WARNING

Never disassemble the rocker arm and shaft assembly.

    b. Loosen the rocker arm and shaft assembly mounting bolt, and then remove the rocker arm and shaft assembly with the bolt still attached.
11. Remove the rocker arm and shaft assembly (intake side).
12. Disconnect the engine oil control valve connector.
13. Remove the engine oil pressure switch connector.
14. Remove the engine oil control valve and O-ring.
15. Remove the engine oil pressure switch.
16. Remove the engine oil control valve filter.
17. Remove the engine oil control valve housing and gasket.
18. Remove the engine oil control valve housing gasket
19. Using special tools MB990767 and MD998719, remove the camshaft sprocket.
20. Remove the engine control module.
21. Remove the air cleaner bracket.
22. Remove the camshaft.
23. Remove the camshaft oil seal.
    a. Make a notch in the oil seal lip section with a knife.
    b. Be careful not to damage the camshaft and the cylinder head.
    c. Cover the end of a flat-tipped screwdriver with a shop towel and insert into the notched section of the oil seal, and pry out the oil seal to remove it.

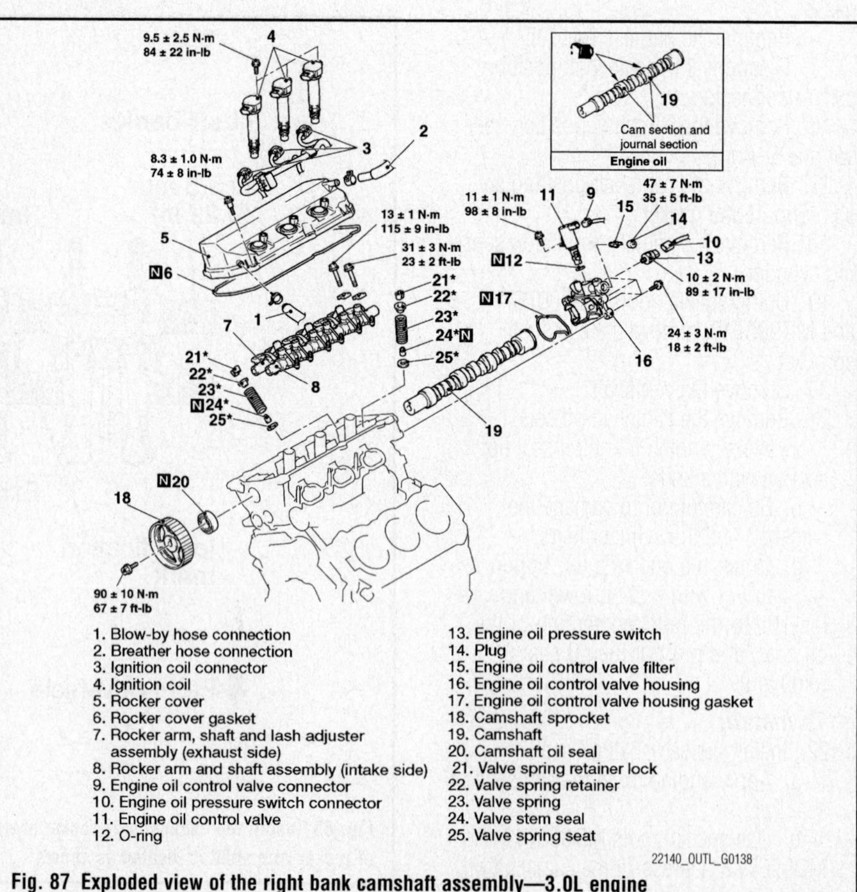

Fig. 87 Exploded view of the right bank camshaft assembly—3.0L engine

To install:

24. Install the camshaft oil seal.
   a. Apply engine oil to the camshaft oil seal lip.
   b. Use special tools MD998713 and MD998777 to press-fit the camshaft oil seal
25. Install the camshaft.
26. Install the air cleaner bracket.
27. Install the engine control module.
28. Using special tools MB990767 and MD998719, install the camshaft sprocket.
29. Install the engine oil control valve housing gasket
30. Install the engine oil control valve housing and gasket.
31. Install the engine oil control valve filter.
32. Install the engine oil pressure switch.
   a. Apply Three bond 1215 or equivalent sealant to the thread of the engine oil pressure switch.
   b. Tighten to 89 inch lbs. (10 Nm).
33. Install the engine oil control valve and O-ring.

WARNING

Never re-use the O-ring.

   c. Before installing O-ring, wind sealing tape around the oil passages cut-out area of engine oil control valve, to prevent damage. If the O-ring is damaged, it can cause an oil leak.
   d. Apply a small amount of engine oil to the O-ring and then install it to the engine oil control valve.
   e. Install the engine oil control valve to the cylinder head and tighten to 98 inch lbs. (11 Nm).
34. Install the engine oil pressure switch connector.
35. Disconnect the engine oil control valve connector.
36. Install the intake side rocker arm and shaft assembly so that the 0.22 inch (5.5 mm) holes of rocker arm shaft face the cylinder head side. Tighten the intake side rocker arm shaft mounting bolts to 23 ft. lbs. (31 Nm).
37. Install the exhaust side rocker arm, shaft and lash adjuster assembly so that the notch of rocker arm shaft is located as shown.
   a. Check that the identification mark of exhaust side rocker shaft cap is located as shown.
   b. Tighten the exhaust side rocker arm shaft mounting bolts to 9 ft. lbs. (13 Nm).
   c. Remove special tool MD998443.
38. Install the valve cover and gasket.
   a. Tighten the valve cover bolts in the order shown to 74 inch lbs. (8 Nm).
39. Install the ignition coil.
40. Disconnect the ignition coil connector.
41. Disconnect the breather hose.
42. Disconnect the blow-by hose.
43. Install the thermostat housing.
44. Install the timing belt.
45. Install the intake manifold plenum.
46. Change the engine oil.
47. Start the engine and check for leaks.

INSPECTION

1. Before servicing the vehicle, refer to the precautions.
2. Remove the camshaft from the engine.
3. Check the camshaft bearing journals for damage and binding.
4. If the journals are binding, check the cylinder head for damage.
5. Check the cylinder head for clogged oil holes.
6. Check the camshaft surface for abnormal wear and damage. Replace the camshaft, as required.
7. Measure the cam height and replace the camshaft if not within specification.

CRANKSHAFT DAMPER

REMOVAL & INSTALLATION

2.4L Engine

See Figures 88 and 89.

1. Before servicing the vehicle, refer to the precautions.
2. Remove the radiator condenser tank assembly.
3. Remove the drive belt.
4. Remove the crankshaft pulley center bolt.

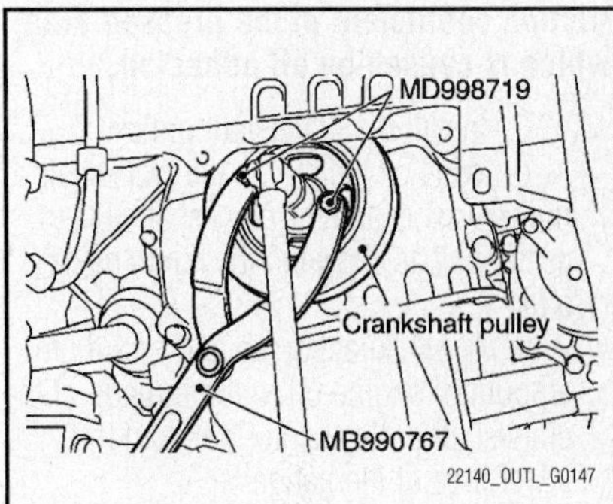

Fig. 88 Hold the crankshaft pulley with special tools MB990767 and MD998719

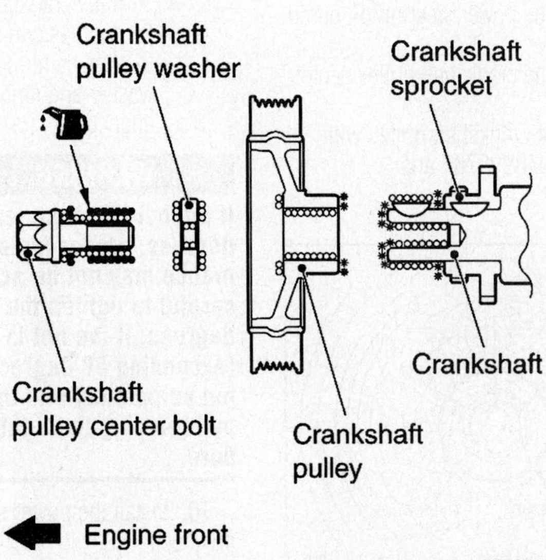

Fig. 89 Clean and lubricate the crankshaft and pulley bolt as shown

a. Hold the crankshaft pulley with special tools MB990767 and MD998719.

5. Remove the crankshaft pulley washer.

6. Remove the crankshaft pulley.

To install:

7. Install the crankshaft pulley.

a. Wipe off the dirt on the crankshaft and the crankshaft pulley as shown in the figure using a rag.

b. Wipe off the dirt on the crankshaft sprocket, the crankshaft and the crankshaft pulley as shown in the figure using a rag, and then degrease them.

➡Degrease them to prevent drop in the friction coefficient of the pressed area, which is caused by oil adhesion.

c. Install the crankshaft pulley.

d. Wipe off the dirt on the crankshaft pulley washer and the crankshaft pulley center bolt as shown in the figure using a rag.

e. Apply an adequate and minimum amount of engine oil to the threads of the crankshaft pulley center bolt and the lower area of the flange.

f. Hold the crankshaft pulley with special tools MB990767 and MD998719.

g. Tighten the crankshaft pulley center bolt to 155 ft. lbs. (210 Nm).

8. Install the drive belt.

9. Install the radiator condenser tank assembly.

3.0L Engine

See Figures 88 and 90.

1. Remove the alternator drive belt.

2. Remove the power steering tensioner pulley bracket.

3. Remove the power steering oil pump drive belt.

4. Remove the crankshaft pulley center bolt.

a. Hold the crankshaft pulley with special tools MB990767 and MD998719.

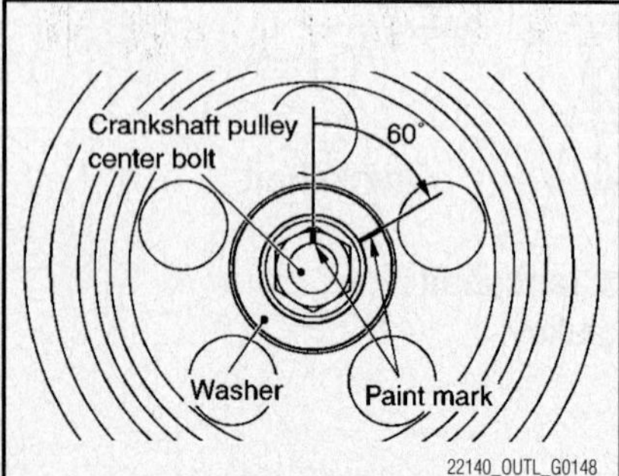

Fig. 90 Correct location of paint marks for crankshaft pulley bolt and washer

b. Provide one punch mark on the head of the crankshaft bolt each time the bolt is removed. Replace the bolt that already has three punch marks.

5. Remove the crankshaft pulley washer.

6. Remove the crankshaft pulley.

To install:

7. Install the crankshaft pulley.

8. Install the crankshaft pulley washer.

9. Install the crankshaft pulley center bolt.

a. Clean the bolt hole in crankshaft bolt and crankshaft pulley's seating surface.

b. Degrease the cleaned seating surface of the front flange and crankshaft pulley.

c. Install the front flange and crankshaft pulley.

d. Apply oil to the threads of crankshaft bolt and the outer surface of washer.

e. Use special tools MB990767 and MD998719 to install the crankshaft pulley.

f. Tighten the bolt as follows:

- Tighten to 148 ft. lbs. (200 Nm)
- Loosen the crankshaft bolt fully
- Tighten to 81 ft. lbs. (110 Nm)
- Make a paint mark on the crankshaft bolt.
- Make a paint mark on the bolt end at a position 60 degrees from the paint mark made on the washer in the direction of tightening the crankshaft bolt
- Turn the crankshaft bolt another 60 degrees and make sure that the paint marks on the washer and crankshaft bolt are aligned.

WARNING

If the nut is turned less than 60 degrees, proper fastening performance may not be achieved. Be careful to tighten the nut exactly 60 degrees. If the nut is over tightened (exceeding 60 degrees), loosen the nut completely and then retighten it by repeating the tightening procedure.

10. Install the power steering oil pump drive belt.

11. Install the power steering tensioner pulley bracket.

12. Install the alternator drive belt.

CRANKSHAFT FRONT SEAL

REMOVAL & INSTALLATION

See Figures 91 and 92.

1. Before servicing the vehicle, refer to the precautions.

2. On 3.0L engines, remove the timing belt.

3. Remove the crankshaft pulley.

4. Carefully pry the oil seal out of the front case.

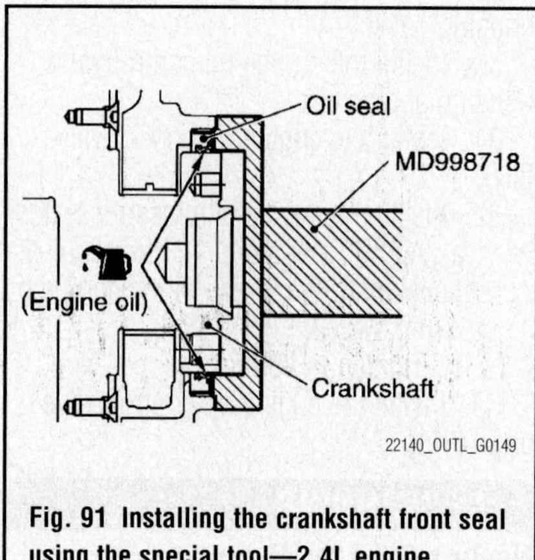

Fig. 91 Installing the crankshaft front seal using the special tool—2.4L engine

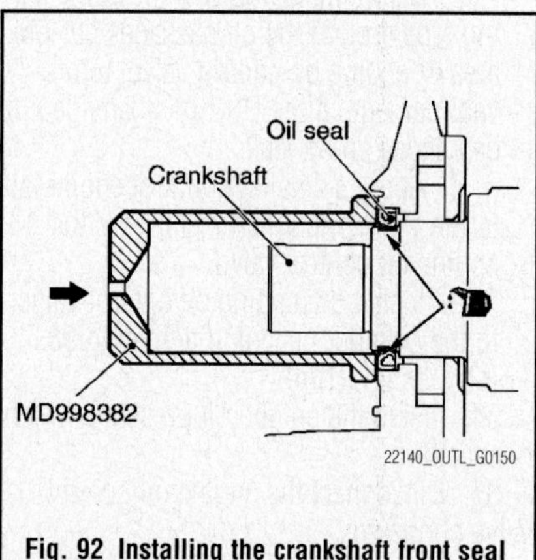

Fig. 92 Installing the crankshaft front seal using the special tool—3.0L engine

WARNING

Be careful not to damage the oil seal bore or the crankshaft sealing surface.

To install:

5. Install the crankshaft seal.

a. Apply a small amount of engine oil to the entire inner diameter of the oil seal lip.

b. Using special tool MD998718, press in the crankshaft rear oil seal up to the cylinder block assembly end surface.

6. Install the crankshaft pulley.
7. On 3.0L engines, install the timing belt.

CYLINDER HEAD

REMOVAL & INSTALLATION

2.4L Engine

See Figures 93 through 100.

1. Before servicing the vehicle, refer to the precautions.
2. Relieve fuel system pressure.
3. Remove the engine under cover and engine side cover.
4. Drain the engine coolant.
5. Remove the air cleaner assembly.
6. Remove the ignition coil.
7. Remove the strut tower bar.

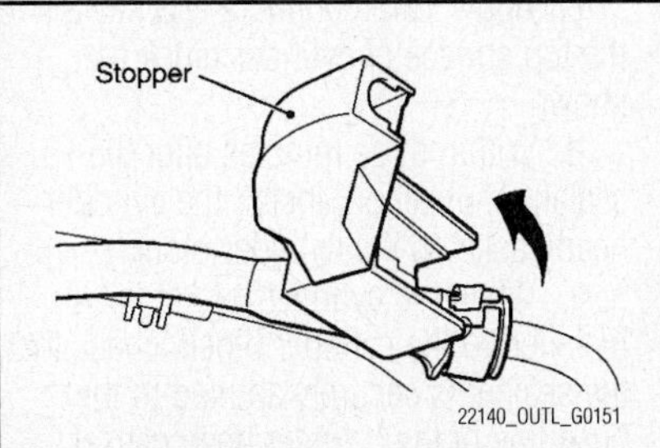

Fig. 93 Remove the stopper of the fuel high-pressure hose

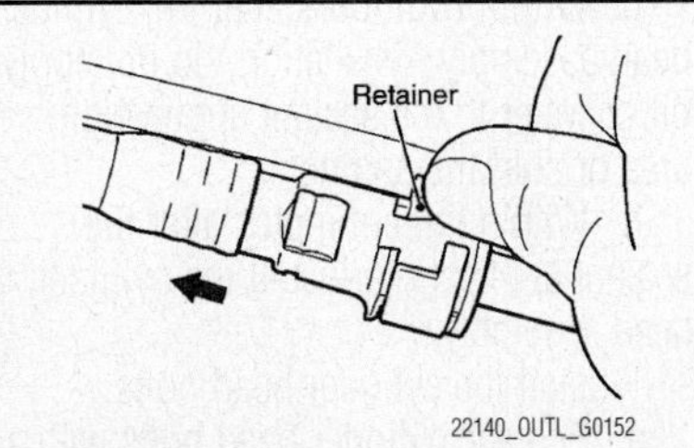

Fig. 94 Raise the retainer of the fuel high-pressure hose and pull out the fuel high-pressure hose in the direction shown

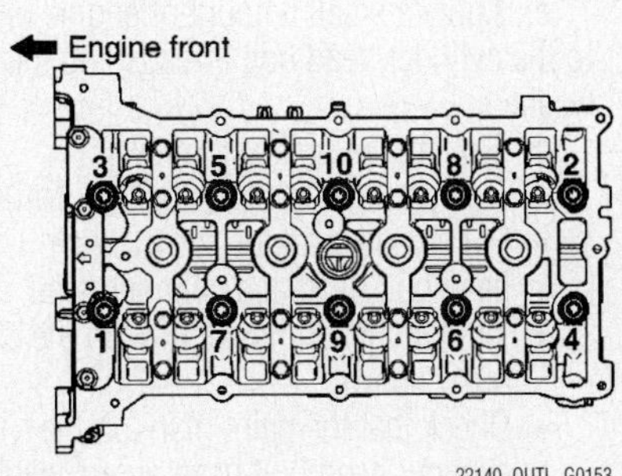

Fig. 95 Loosen and remove the bolts in two or three steps in the order shown

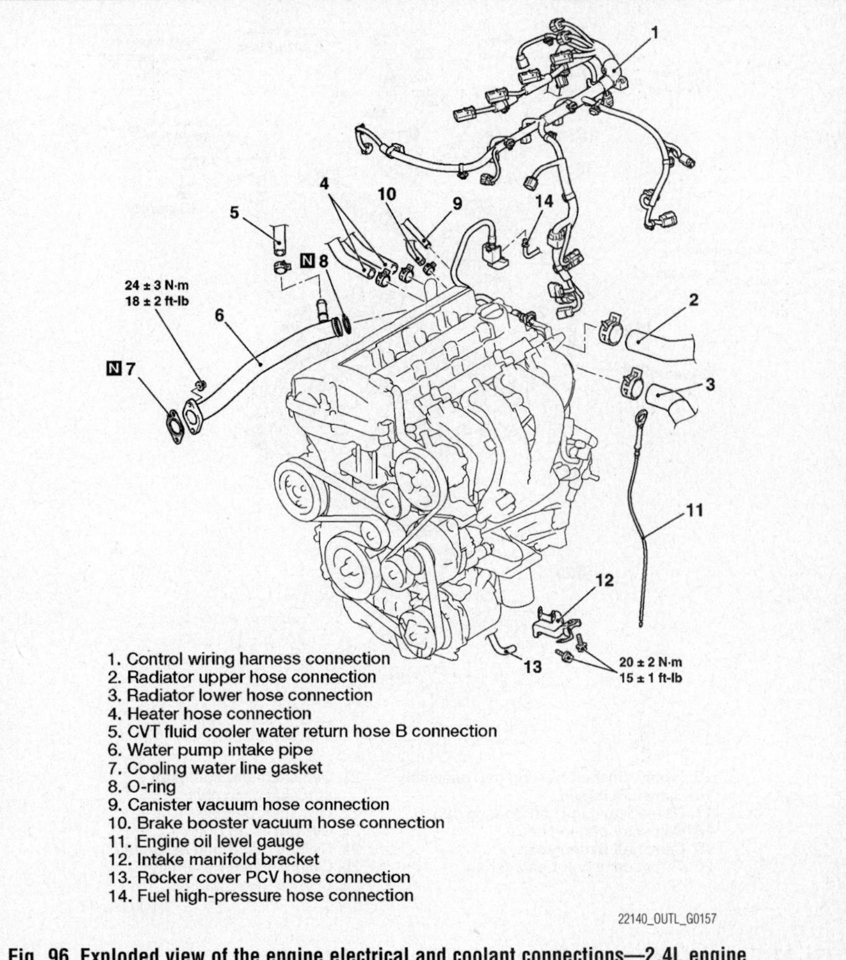

Fig. 96 Exploded view of the engine electrical and coolant connections—2.4L engine

8. Remove the exhaust manifold. Refer to "Engine Mechanical Components, Exhaust Manifold, Removal & Installation."
9. Remove the throttle body assembly.
10. Remove the EGR valve and EGR valve bracket.
11. Remove the water pump.
12. Disconnect the control wiring harness.
13. Disconnect the upper and lower radiator hoses
14. Disconnect the heater hose.
15. Disconnect the CVT fluid cooler water return hose.
16. Disconnect the water pump intake pipe.
17. Remove the cooling water line gasket.
18. Remove the o-ring.
19. Disconnect the canister vacuum hose.
20. Disconnect the brake booster vacuum hose.
21. Disconnect the engine oil level gauge.
22. Remove the intake manifold bracket
23. Disconnect the valve cover PCV hose.
24. Disconnect the fuel high-pressure hose.
    a. Remove the stopper of the fuel high-pressure hose.
    b. Raise the retainer of the fuel high-pressure hose and pull out the fuel high-pressure hose in the direction shown.
    c. If the retainer is released, install it securely after removing the fuel high-pressure hose.
25. Remove the timing chain. Refer to "Engine Mechanical Components, Timing Chain and Sprockets, Removal & Installation."
26. Remove the camshaft and camshaft sprocket. Refer to "Engine Mechanical Components, Camshaft and Valve Lifters, Removal & Installation."
27. Remove the camshaft bearings.
28. Remove the cylinder head bolts.
    a. Loosen and remove the bolts in two or three steps in the order shown.
29. Remove the cylinder head and gasket.

To install:

30. Install the cylinder head gasket and cylinder head.
    a. Do not allow any foreign materials get into the coolant passages, oil passages and cylinder.

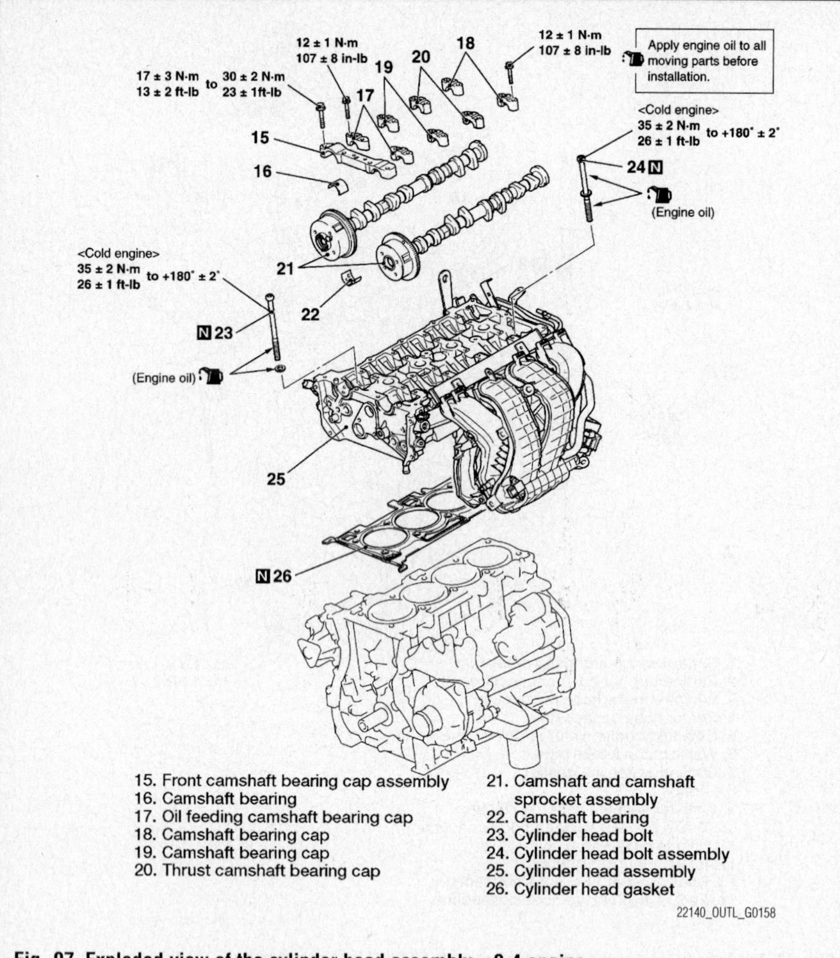

Fig. 97 Exploded view of the cylinder head assembly—2.4 engine

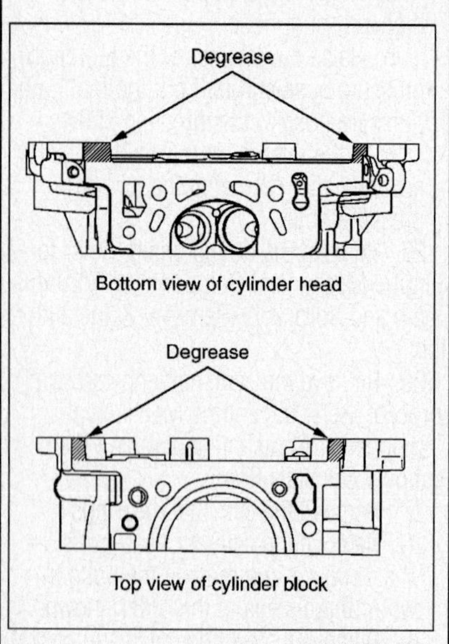

Fig. 98 Remove the sealant and grease on the top surface of cylinder block and on the bottom surface of the cylinder head. Then degrease the sealant application surface

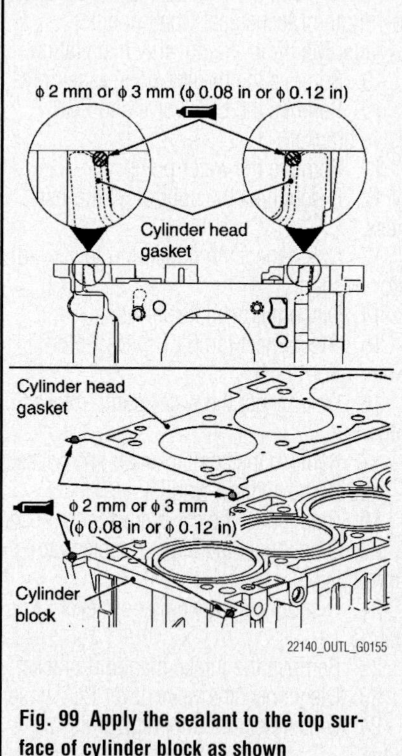

Fig. 99 Apply the sealant to the top surface of cylinder block as shown

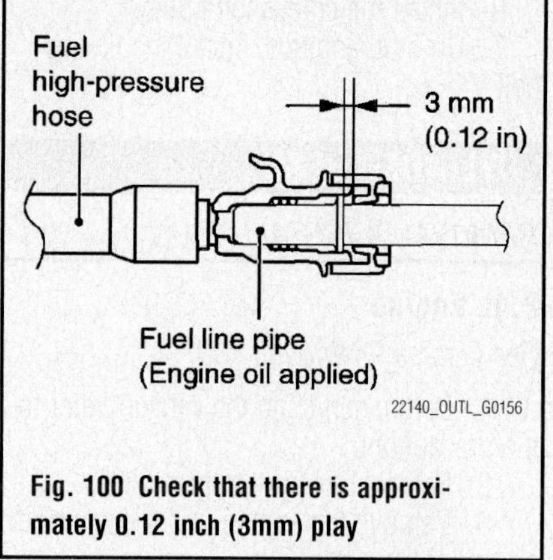

Fig. 100 Check that there is approximately 0.12 inch (3mm) play

b. Remove the sealant and grease on the top surface of cylinder block and on the bottom surface of the cylinder head. Then degrease the sealant application surface.

c. Apply Three bond 1217G sealant to the top surface of cylinder block as shown.

d. Within three minutes after the sealant application, install the cylinder head gasket to the cylinder block.

e. When the cylinder gasket is installed to the cylinder block, check that the sealant is securely applied to the bead line of the cylinder head gasket.

f. Apply Three bond 1217G sealant to the top surface of cylinder head gasket as shown.

g. Within two hours after the cylinder head assembly installation, do not apply oil or water to the sealant application area or start the engine.

h. Within three minutes after the sealant application, install the cylinder head assembly.

31. Install the cylinder head bolts.

a. Replace cylinder head bolts with a new ones.

b. For two bolts of the timing chain side, the washer can be removed from the bolt. Install the washer, with its sag facing upward, to the bolts.

c. Apply a small amount of engine oil to the cylinder head bolt threads and the washers.

d. Tighten the bolts as follows:

- Tighten bolts to 26 ft. lbs. (35 Nm).
- Put a paint mark on the cylinder head bolt head and cylinder head
- Tighten an additional 180° in the order shown
- Check that the paint mark on the cylinder head bolt head aligns with the paint mark on the cylinder head.

WARNING

The bolt is not tightened sufficiently if the tightening angle is less than a 180°. If the tightening angle exceeds the standard specification, remove the bolt and repeat the installation steps.

32. Install the camshaft bearings, camshaft and camshaft sprocket.
33. Install the timing chain.
34. Connect the fuel high-pressure hose.
    a. After connecting the fuel high-pressure hose, slightly pull it in the pull-out direction to check that it is installed firmly. In addition, check that there is approximately 0.12 inch (3mm)play. After the check, install the stopper securely.
    b. Apply a small amount of engine oil to the fuel line pipe, and install the fuel high-pressure hose.
35. Connect the valve cover PCV hose.
36. Install the intake manifold bracket
37. Connect the engine oil level gauge.
38. Connect the brake booster vacuum hose.
39. Connect the canister vacuum hose.
40. Install the o-ring.
41. Install the cooling water line gasket.
42. Connect the water pump intake pipe.
43. Connect the CVT fluid cooler water return hose.
44. Connect the heater hose.
45. Connect the upper and lower radiator hoses
46. Connect the control wiring harness.
47. Install the water pump.
48. Install the EGR valve and EGR valve bracket.
49. Install the throttle body assembly.
50. Install the exhaust manifold.
51. Install the strut tower bar.
52. Install the ignition coil.
53. Install the air cleaner assembly.
54. Fill the engine with coolant.
55. Install the engine under cover and engine side cover.
56. Start the engine and check for leaks.

3.0L Engine

See Figures 101 through 105.

1. Before servicing the vehicle, refer to the precautions.
2. Remove the intake manifold. Refer to "Engine Mechanical Components, Exhaust Manifold, Removal & Installation."
3. Remove the exhaust manifold. Refer to "Engine Mechanical Components, Intake Manifold, Removal & Installation."
4. Remove the timing belt. Refer to "Engine Mechanical Components, Timing Belt and Sprockets, Removal & Installation."

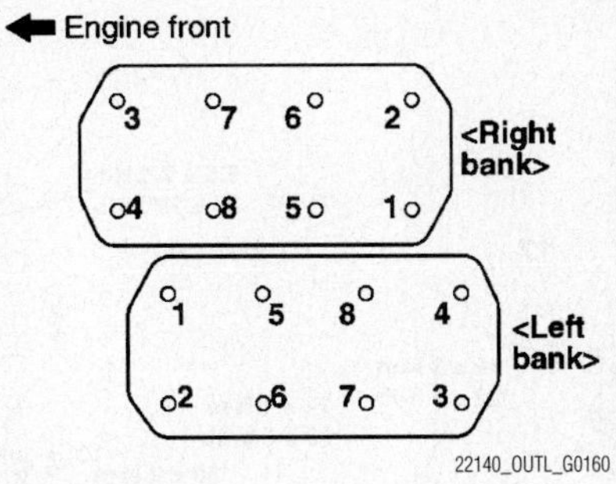

Fig. 101 Loosen the bolts in two or three steps in the order shown

5. Remove the thermostat housing.
6. Remove the alternator.
7. Disconnect the blow-by hose.
8. On the left bank, disconnect the PCV hose.
9. On the left bank, remove the engine oil dipstick.
10. On the right bank, disconnect the breather hose.
11. On the right bank, disconnect the engine oil control valve connector.
12. On the right bank, disconnect the engine oil pressure switch connector.
13. On the right bank, disconnect the solenoid valve connector.
14. Disconnect the ignition coil connector.
15. Remove the ignition coil.
16. Remove the valve cover and gasket.
17. Remove the oil pipe and gasket.
18. Disconnect the camshaft position sensor connector.
19. Remove the camshaft sprocket. Refer to "Engine Mechanical Components, Timing Belt and Sprockets, Removal & Installation."
20. Remove the idler pulley
21. Remove the rear timing belt cover.
22. Remove the cylinder head and gasket assembly.
    a. Loosen the bolts in two or three steps in the order shown.

To install:

23. Install the cylinder head and gasket assembly.
    a. Be careful that no foreign material gets into the cylinder, coolant passages

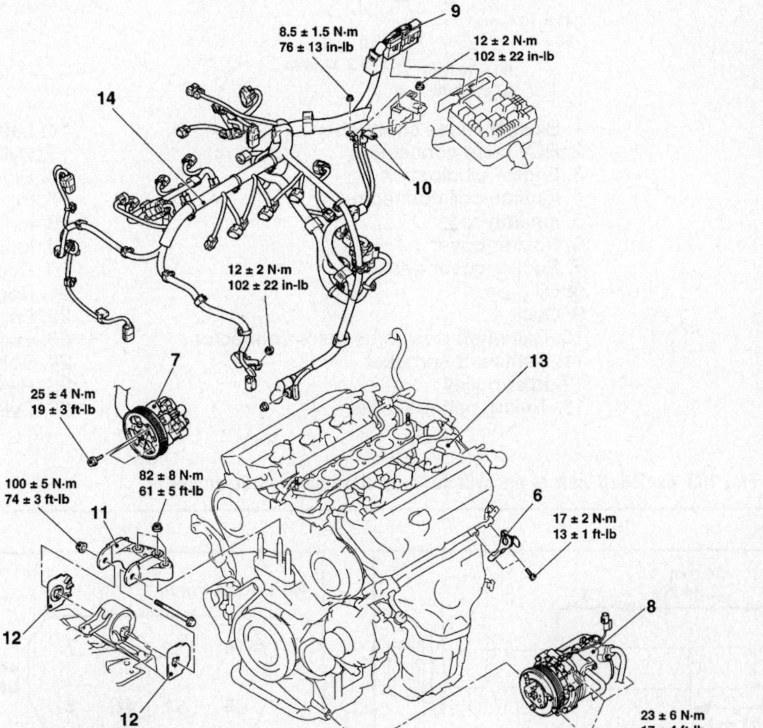

6. Engine hanger
7. Power steering oil pump assembly
8. A/C compressor assembly
9. Control wiring harness (relay box side)
10. Control wiring harness (battery side)
11. Engine mounting bracket
12. Engine mounting insulator stopper
13. Engine assembly
14. Control wiring harness

22140_OUTL_G0165

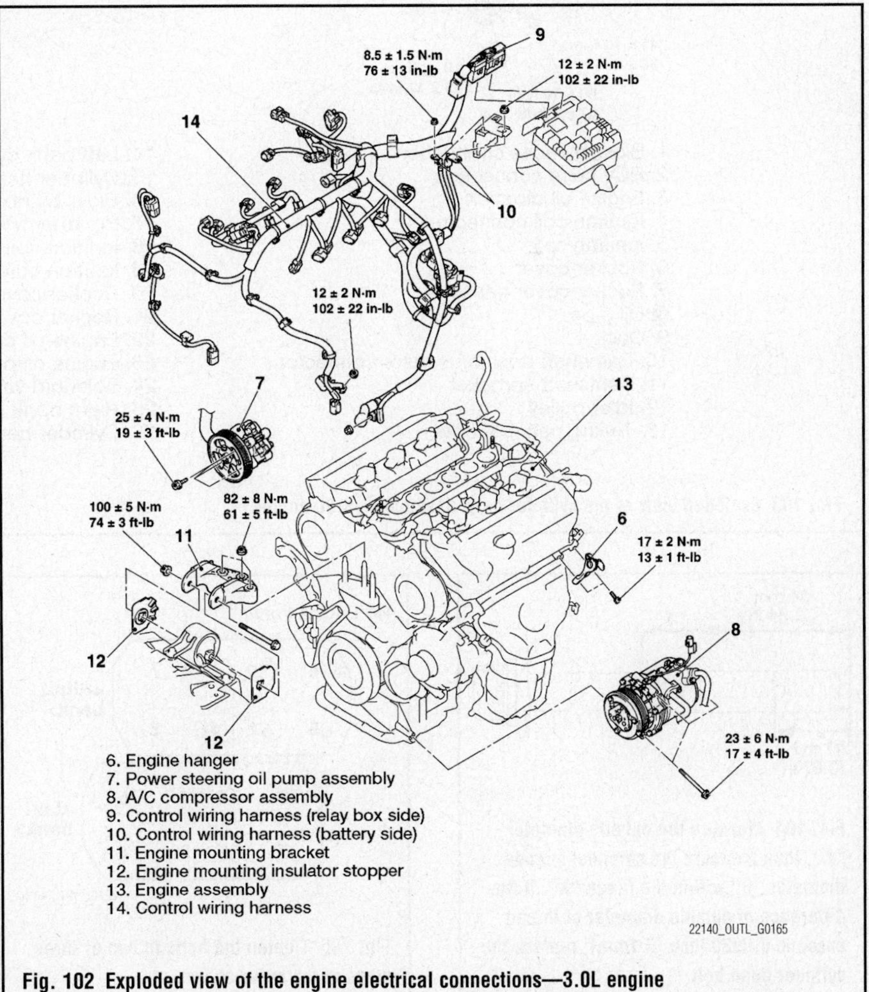

Fig. 102 Exploded view of the engine electrical connections—3.0L engine

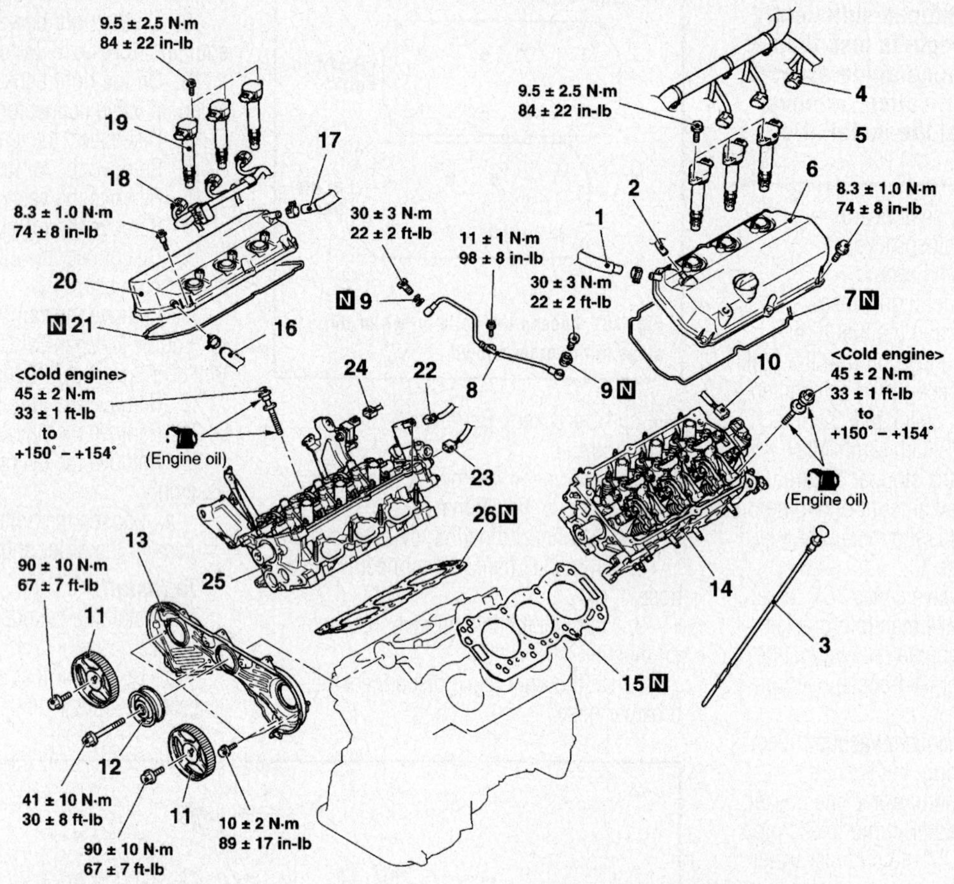

1. Blow-by hose connection
2. PCV hose connection
3. Engine oil dipstick
4. Ignition coil connector
5. Ignition coil
6. Rocker cover
7. Rocker cover gasket
8. Oil pipe
9. Gasket
10. Camshaft position sensor connector
11. Camshaft sprocket
12. Idler pulley
13. Timing belt rear cover
14. Left bank cylinder head assembly
15. Cylinder head gasket
16. Blow-by hose connection
17. Breather hose connection
18. Ignition coil connector
19. Ignition coil
20. Rocker cover
21. Rocker cover gasket
22. Engine oil control valve connector
23. Engine oil pressure switch connector
24. Solenoid valve connector
25. Right bank cylinder head assembly
26. Cylinder head gasket

22140_OUTL_G0159

Fig. 103 Exploded view of the cylinder head assembly—3.0L engine

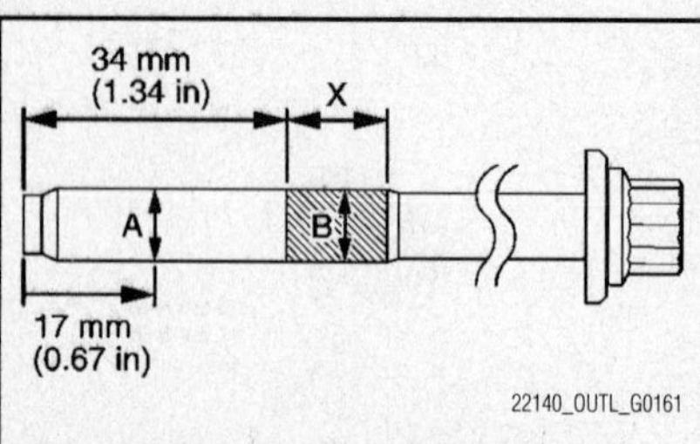

Fig. 104 Measure the outside diameter "A", then measure the smallest outside diameter "B" within the range "X". If the difference of outside diameter of thread exceeds 0.0039 inch (0.1mm), replace the cylinder head bolt

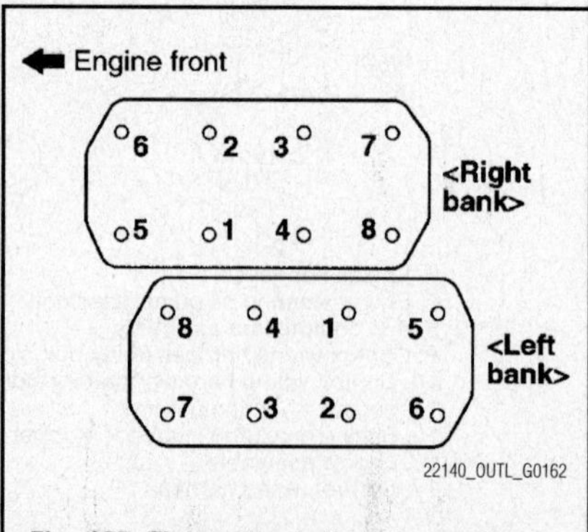

Fig. 105 Tighten the bolts in two or three steps in the order shown

or oil passages. Engine damage may result.

b. Use a scraper to clean the gasket surface of the cylinder head assembly.

c. Check the cylinder head bolt as follows:

- Measure the outside diameter "A"
- Measure the smallest outside diameter "B" within the range "X" shown
- If the difference of outside diameter of thread exceeds 0.0039 inch (0.1mm), replace the cylinder head bolt.

d. Tighten the cylinder head bolts as follows:

- Tighten in the order shown in two or three steps to 33 ft. lbs. (45 Nm).
- Tighten an additional 150–154°

✱✱ WARNING

If the bolt is turned less than 150–154°, proper fastening performance may not be achieved. Be sure to turn the bolt exactly 150 to 154 degrees. If the bolts is over tightened, loosen the bolt completely and then retighten it by repeating the tightening procedure.

24. Install the rear timing belt cover.
25. Install the idler pulley
26. Install the camshaft sprocket.
27. Connect the camshaft position sensor connector.
28. Install the oil pipe and gasket.
29. Install the valve cover and gasket.
30. Install the ignition coil.
31. Connect the ignition coil connector.
32. On the right bank, connect the solenoid valve connector.
33. On the right bank, connect the engine oil pressure switch connector.
34. On the right bank, connect the engine oil control valve connector.
35. On the right bank, connect the breather hose.
36. On the left bank, Install the engine oil dipstick.
37. On the left bank, connect the PCV hose.
38. Connect the blow-by hose.
39. Install the alternator.
40. Install the thermostat housing.
41. Install the timing belt.
42. Install the exhaust manifold.
43. Install the intake manifold.
44. Fill the engine with coolant.
45. Change the engine oil.
46. Start the engine and check for leaks.

ENGINE ASSEMBLY

REMOVAL & INSTALLATION

2.4L Engine

See Figures 93, 94, 96, 100 and 106.

1. Disconnect the negative battery cable.
2. Remove the hood.
3. Relieve the fuel line pressure.
4. Remove the engine under cover and engine side cover.
5. Drain the engine coolant.
6. Drain the engine oil.
7. Drain the transaxle fluid.

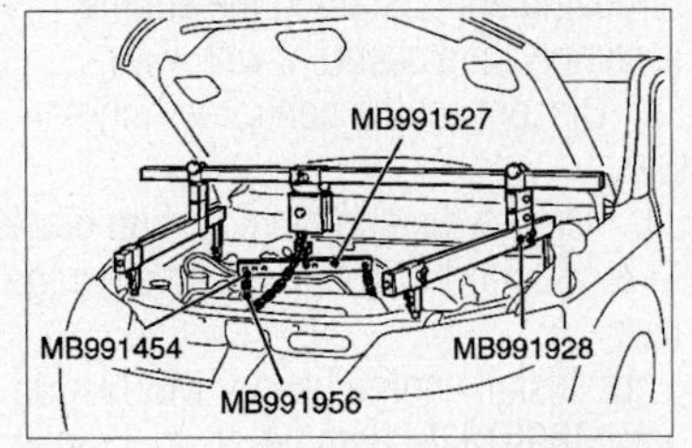

Fig. 106 Remove special tools engine hanger (MB991928 or MB991895) which was installed for supporting the engine assembly when the transaxle assembly was removed

8. Remove the engine upper cover.
9. Remove the air cleaner.
10. Remove the battery and battery tray.
11. Remove the engine-ECU.
12. Remove the drive shaft.
13. Remove the transfer case.
14. Remove the strut tower bar.
15. Remove the exhaust manifold
16. Remove the throttle body.
17. Disconnect the control wiring harness.
18. Disconnect the upper and lower radiator hose
19. Disconnect the heater hose.
20. Disconnect the emission vacuum hose.
21. Disconnect the brake booster vacuum hose.
22. Disconnect the cooling water line hose.
23. Disconnect the fuel high-pressure hose.
24. Remove the drive belt.
25. Remove the transaxle assembly.
26. Remove the intake manifold bracket
27. Remove the power steering oil pump.
    a. Remove the power steering oil pump assembly with hose on it.
    b. Tie the removed power steering oil pump with wire at a position where it will not interfere with the removal and installation of engine assembly.
28. Remove the A/C compressor and clutch.
    a. Remove the A/C compressor and clutch assembly together with the hose from the bracket.
    b. Tie the removed A/C compressor and clutch assembly with a string at a position where it will not interfere with the removal and installation of engine assembly.
29. Disconnect the ground cable.
30. Install the engine mounting bracket.

✱✱ WARNING

When supporting the engine and transaxle assembly with a garage jack, be careful not to deform the engine oil pan.

    a. Place a garage jack against the engine oil pan with a piece of wood in between to support the engine assembly.
    b. Remove special tools engine hanger which were installed for supporting the engine assembly when the transaxle assembly was removed.
    c. Operate a garage jack so that the engine weight is not applied to the engine mounting insulator, and remove the engine mounting bracket.
31. Remove the engine assembly.
    a. After checking that all cables, hoses and wiring harness connectors and so on are disconnected from the engine, lift the engine assembly slowly with the chain block to remove the engine assembly upward from the engine compartment.

To install:

32. Install the engine assembly.
33. Install the engine mounting bracket.
34. Connect the ground cable.
35. Install the A/C compressor and clutch.
36. Install the power steering oil pump.
37. Install the intake manifold bracket
38. Install the transaxle assembly.
39. Install the drive belt.
40. Connect the fuel high-pressure hose. Check that there is approximately 0.12 inch (3mm) play as shown.
41. Connect the cooling water line hose.
42. Connect the brake booster vacuum hose.
43. Connect the emission vacuum hose.
44. Connect the heater hose.
45. Connect the upper and lower radiator hose
46. Connect the control wiring harness.
47. Install the throttle body.
48. Install the exhaust manifold
49. Install the strut tower bar.
50. Install the transfer case.
51. Install the drive shaft.
52. Install the engine-ECU.
53. Install the battery and battery tray.
54. Install the air cleaner.
55. Install the engine upper cover.
56. Fill the transaxle fluid.
57. Fill the engine oil.
58. Fill the engine coolant.
59. Install the engine under cover and engine side cover.
60. Install the hood.

61. Connect the negative battery cable.
62. Use a scan tool to initialize the learning value in the MFI.
63. Start the engine and check for leaks.

3.0L Engine

See Figures 93, 94, 100, 102 and 107.

1. Disconnect the negative battery cable.
2. Remove the engine under cover and engine side cover.
3. Relieve the fuel line pressure.
4. Drain the engine coolant.
5. Drain the engine oil.
6. Drain the transaxle fluid.
7. Drain the transfer case oil.
8. Remove the hood.
9. Remove the upper engine cover.
10. Remove the air cleaner.
11. Remove the engine-ECU.
12. Remove the battery and battery tray.
13. Remove the front exhaust pipes.
14. Remove the strut tower bar.
15. Remove the halfshafts.
16. Remove the driveshaft.
17. Disconnect the power steering pressure hose assembly and return tube at the reservoir.
18. Remove the rear roll stopper.
19. Remove the transfer case.
20. Remove the starter.
21. Disconnect the upper and lower radiator hoses.
22. Remove the alternator drive belt.
23. Remove the power steering oil pump drive belt.
24. Disconnect the ground cable.
25. Disconnect the fuel high-pressure hose
26. Disconnect the heater hose.
27. Remove the intake manifold plenum.
28. Remove the right bank exhaust manifold.
29. Remove the transaxle assembly.
30. Install the engine hanger.
31. Remove the power steering oil pump.
    a. Remove the power steering oil pump from the engine with the hose attached.
    b. Place the removed power steering oil pump in a place where it will not be a hindrance when removing and installing the engine assembly, and secure it with wire.
32. Remove the A/C compressor.
    a. Remove the compressor from the compressor bracket with the hose still attached.
    b. Place the removed A/C compressor where it will not be a hindrance when removing and installing the engine assembly, and secure it with wire.
33. Disconnect the control wiring harnesses.
34. Remove the engine mounting bracket.
    a. Support the engine with a garage jack.
    b. Install engine hanger MB991895 and MB991928. Remove special tool MB991895 and MB991928.
    c. When removing the transaxle assembly, remove the special tool MB992208 (Right bank) that supported the engine assembly.
    d. Mount the special tool MB991454 to the engine right hanger and special tool MB992208 (Left bank), and support the engine assembly using the chain block or others.
    e. Place a garage jack against the engine oil pan with a piece of wood in between so that the weight of the engine is no longer being applied to the engine mounting bracket.
    f. Loosen the engine mount mounting nuts and bolt, and remove the engine mounting bracket.
35. Remove the engine mounting insulator stopper.
36. Remove the engine assembly.
    a. After checking that all cables, hoses and wiring harness connectors and so on are disconnected from the engine, lift the chain block slowly to remove the engine assembly upward from the engine compartment.

To install:

37. Install the engine assembly.
    a. Install the engine assembly, being careful not to pinch the cables, hoses or wiring harness connectors.
    b. Support the engine assembly with a garage jack.
38. Install the engine mounting insulator stopper.

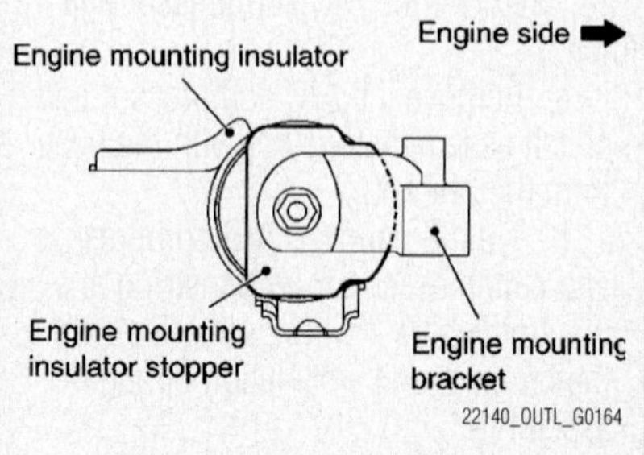

Fig. 107 Mount the engine mounting insulator stopper as shown

    a. Mount the engine mounting insulator stopper to be positioned as shown, then mount the engine mounting bracket.
39. Install the engine mounting bracket.
40. Connect the control wiring harnesses.
41. Install the A/C compressor.
42. Install the power steering oil pump.
43. Install the engine hanger.
44. Install the transaxle assembly.
45. Install the right bank exhaust manifold.
46. Install the intake manifold plenum.
47. Connect the heater hose.
48. Connect the fuel high-pressure hose
    a. After connecting the fuel high-pressure hose, slightly pull it to ensure that it is installed securely. Also confirm that there is a play approximately 0.12 inch (3 mm). Then install the stopper securely.
    b. Apply a small amount of engine oil to the fuel line pipe and then install the fuel high-pressure hose.
49. Connect the ground cable.
50. Install the power steering oil pump drive belt.
51. Install the alternator drive belt.
52. Connect the upper and lower radiator hoses.
53. Install the starter.
54. Install the transfer case.
55. Install the rear roll stopper.
56. Connect the power steering pressure hose assembly and return tube at the reservoir.
57. Install the driveshaft.
58. Install the halfshafts.
59. Install the strut tower bar.
60. Install the front exhaust pipes.
61. Install the battery and battery tray.
62. Install the engine-ECU.
63. Install the air cleaner.
64. Install the upper engine cover.
65. Install the hood.
66. Fill the transfer case oil.
67. Fill the transaxle fluid.
68. Fill the engine oil.
69. Fill the engine coolant.
70. Install the engine under cover and engine side cover.
71. Connect the negative battery cable.
72. Disconnect the upper and lower radiator hose.

EXHAUST MANIFOLD

REMOVAL & INSTALLATION

2.4L Engine

See Figures 108 through 112.

1. Before servicing the vehicle, refer to the precautions.

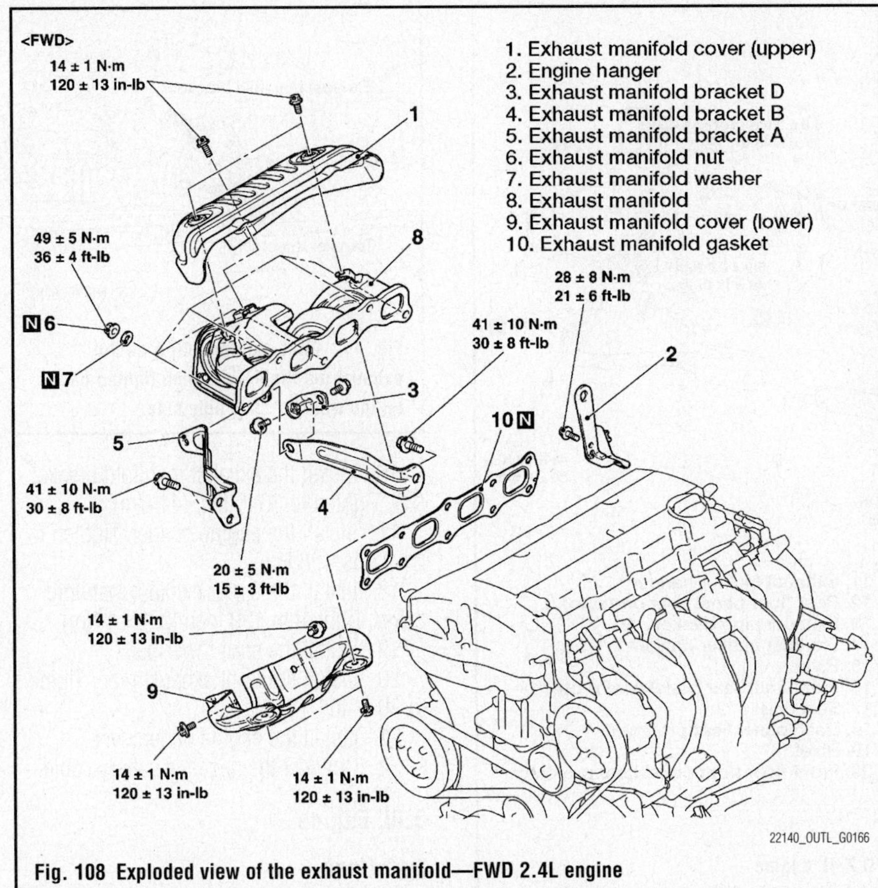

Fig. 108 Exploded view of the exhaust manifold—FWD 2.4L engine

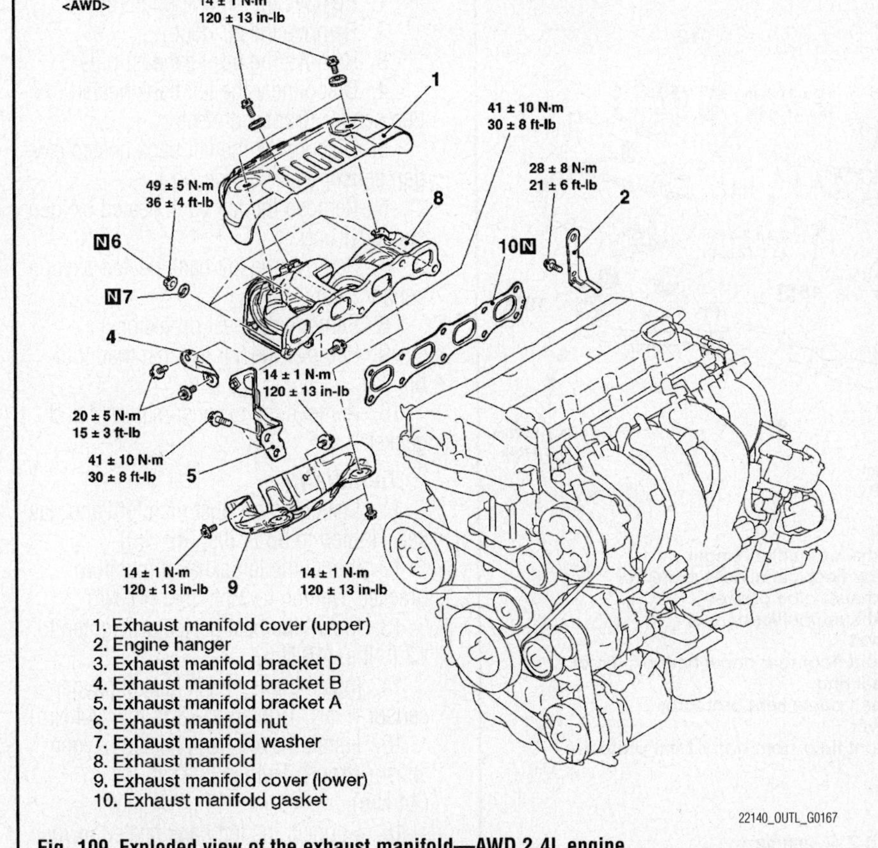

Fig. 109 Exploded view of the exhaust manifold—AWD 2.4L engine

2. Disconnect the negative battery cable.
3. Remove the engine undercover.
4. Remove the front exhaust pipe.
5. Remove the strut tower bar.
6. Remove the upper exhaust manifold cover.
7. Remove the engine hanger.
8. Remove the exhaust manifold bracket (D).
9. Remove the exhaust manifold bracket (B).
10. Remove the exhaust manifold bracket (A).
11. Remove the exhaust manifold.
  a. Remove the mounting bolts of the exhaust manifold cover (lower) and move them to the position where they will not interfere with the loosening of the exhaust manifold nuts.
  b. Loosen the exhaust manifold nuts, and remove the exhaust manifold nuts and the exhaust manifold washers.
  c. Remove the exhaust manifold and the exhaust manifold cover (lower) as a set.
  exhaust manifold washer
12. Remove the exhaust manifold gasket.

To install:

13. Install the exhaust manifold and gasket.
  a. Install the exhaust manifold cover (lower) to the exhaust manifold (with mounting bolts removed), install the exhaust manifold and the exhaust manifold cover (lower) to the engine as a set.
  b. Move the exhaust manifold cover (lower) to the position where it will not interfere with the installation of the exhaust manifold nuts.
  c. Install the new exhaust manifold nuts and exhaust manifold washers, and tighten the exhaust manifold nuts to 36 ft. lbs. (49 Nm).
  d. Install the exhaust manifold cover (lower) to the exhaust manifold, and tighten the exhaust manifold cover (lower) mounting bolts to 10 ft. lbs. (14 Nm).
14. On AWD vehicles, install the exhaust manifold bracket (B).
  a. Make sure that the exhaust manifold bracket is closely contacted with the exhaust manifold and transfer assembly, and then temporarily tighten the bolts.
  b. First, tighten bolt A on the exhaust manifold side to 15 ft. lbs. (20 Nm).
  c. Then, tighten bolt B on the transfer assembly side to 15 ft. lbs. (20 Nm).
15. Install the exhaust manifold bracket (A). Tighten to 30 ft. lbs. (41 Nm).

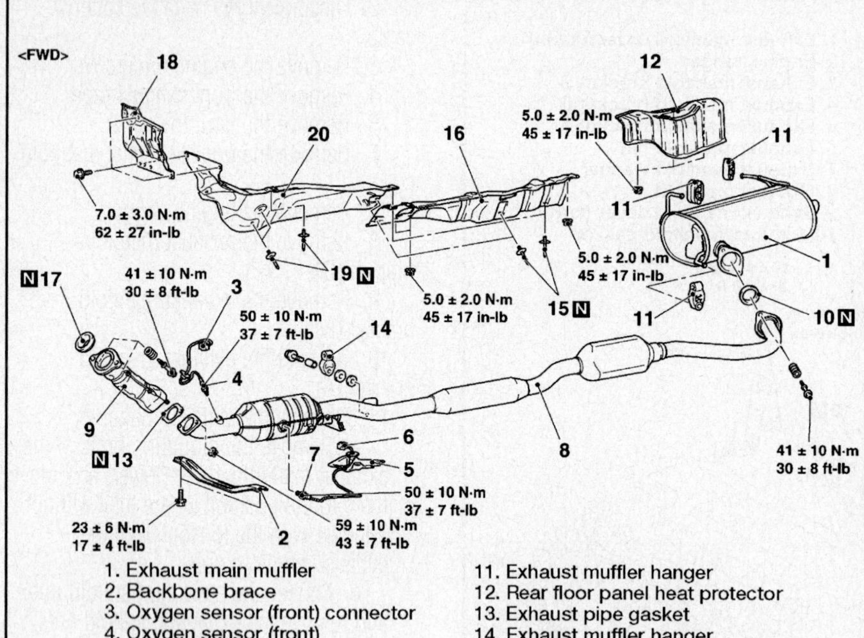

Fig. 110 Exploded view of the exhaust system—FWD 2.4L engine

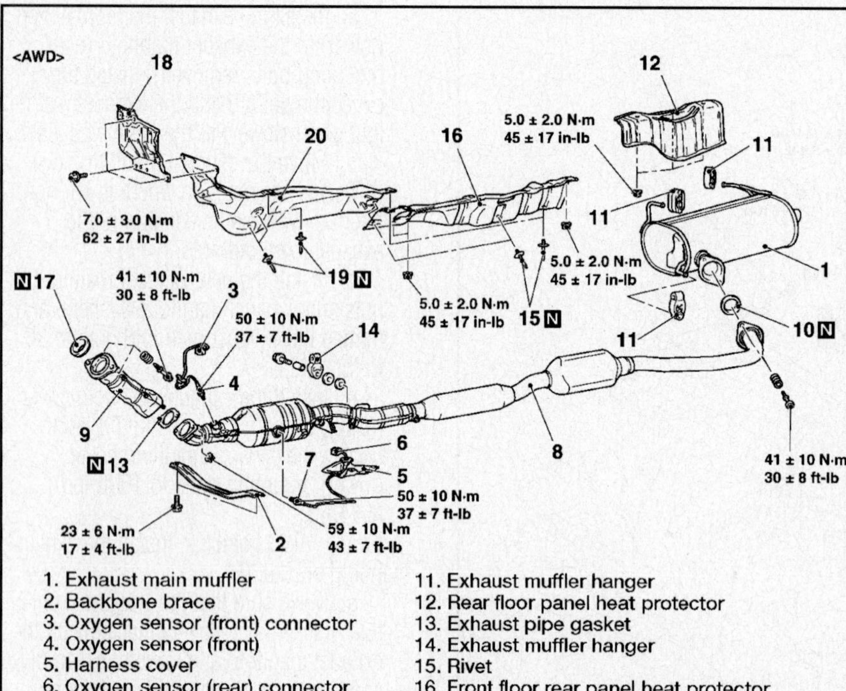

Fig. 111 Exploded view of the exhaust system—AWD 2.4L engine

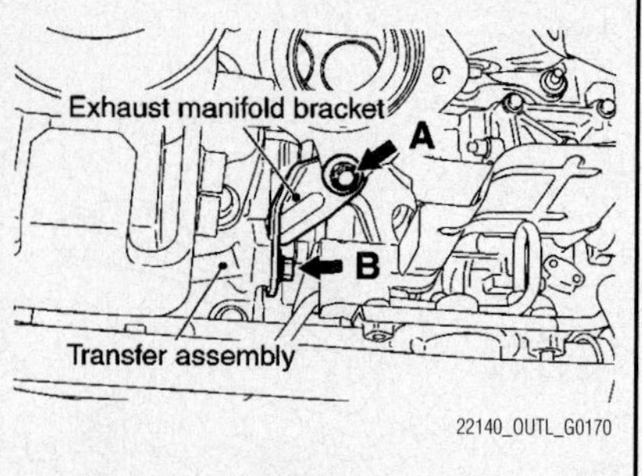

Fig. 112 First, tighten bolt A on the exhaust manifold side, then tighten bolt B on the transfer assembly side

16. Install the exhaust manifold bracket (D). Tighten to 30 ft. lbs. (41 Nm).
17. Install the engine hanger. Tighten to 21 ft. lbs. (28 Nm).
18. Install the upper exhaust manifold cover. Tighten to 120 inch lbs. (14 Nm).
19. Install the strut tower bar.
20. Install the front exhaust pipe. Tighten to 30 ft. lbs. (41 Nm).
21. Install the engine undercover.
22. Connect the negative battery cable.

3.0L Engine

Left Bank

See Figures 113 through 115.

1. Remove the engine under cover.
2. Remove the air duct.
3. Remove the front exhaust pipe.
4. Disconnect the left bank heated oxygen sensor (front) connector.
5. Disconnect the left bank heated oxygen sensor (rear) connector.
6. Remove the left bank heated oxygen sensor (front).
7. Remove the left bank heated oxygen sensor (rear).
8. Remove the heat protector.
9. Remove the left exhaust manifold bracket.
10. Remove the exhaust manifold and gasket.

To install:

11. Install the exhaust manifold and gasket. Tighten to 36 ft. lbs. (49 Nm).
12. Install the left exhaust manifold bracket. Tighten to 35 ft. lbs. (47 Nm).
13. Install the heat protector. Tighten to 12 ft. lbs. (16 Nm).
14. Install the left bank heated oxygen sensor (rear). Tighten to 33 ft. lbs. (44 Nm).
15. Install the left bank heated oxygen sensor (front). Tighten to 33 ft. lbs. (44 Nm).
16. Connect the left bank heated oxygen sensor (rear) connector.

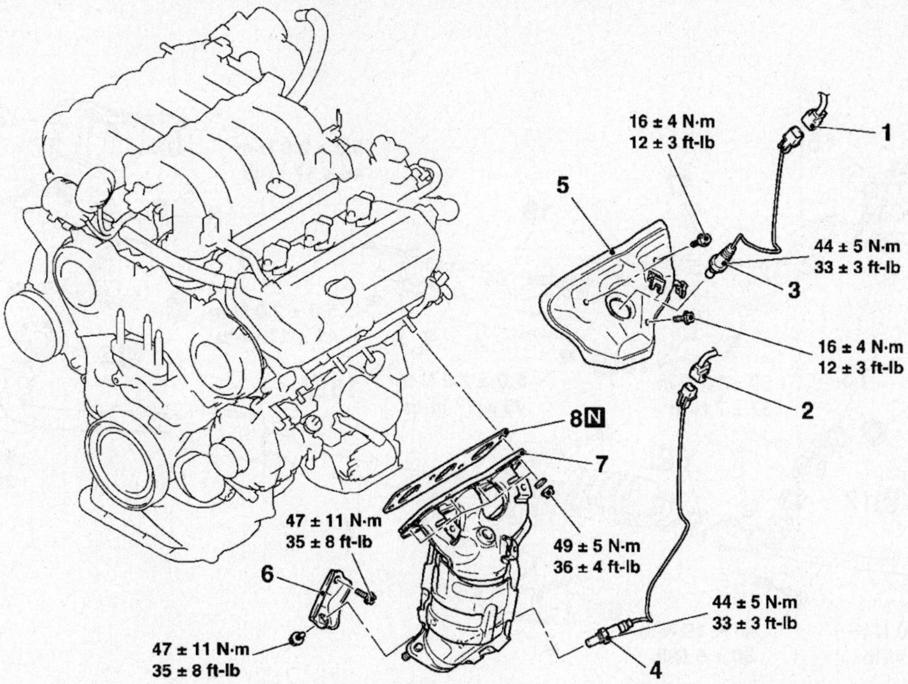

1. Left bank heated oxygen sensor (Front) connector
2. Left bank heated oxygen sensor (Rear) connector
3. Left bank heated oxygen sensor (Front)
4. Left bank heated oxygen sensor (Rear)
5. Heat protector
6. Left exhaust manifold bracket
7. Exhaust manifold
8. Exhaust manifold gasket

22140_OUTL_G0171

Fig. 113 Exploded view of the left bank exhaust manifold—3.0L engine

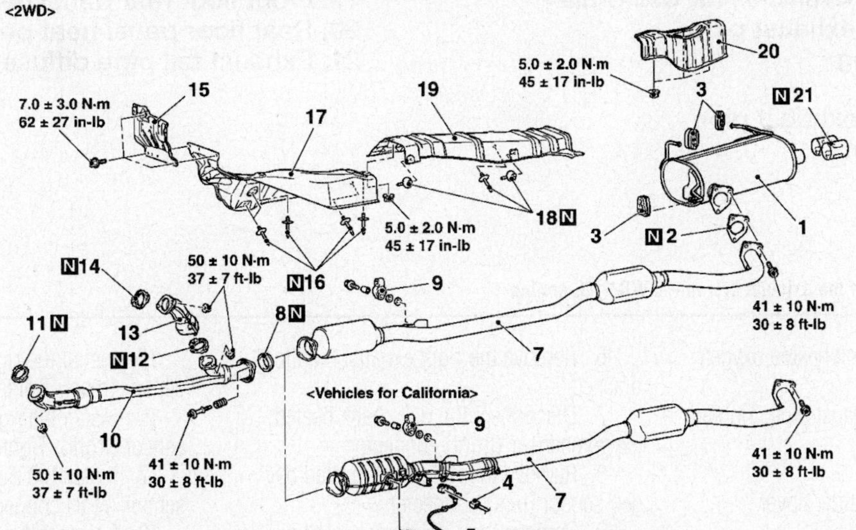

<2WD>

1. Exhaust main muffler
2. Gasket
3. Hanger
4. Center exhaust pipe heated oxygen sensor connector <Vehicles for California>
5. Harness cover <Vehicles for California>
6. Center exhaust pipe heated oxygen sensor <Vehicles for California>
7. Center exhaust pipe
8. Seal ring
9. Hanger
10. Front exhaust pipe
11. Gasket
12. Gasket
13. Front exhaust pipe RH
14. Gasket
15. Dash panel heat protector
16. Rivet
17. Front floor front panel heat protector
18. Rivet
19. Front floor rear panel heat protector
20. Rear floor panel heat protector
21. Exhaust tail pipe diffuser

22140_OUTL_G0173

Fig. 114 Exploded view of the exhaust system—FWD 3.0L engine

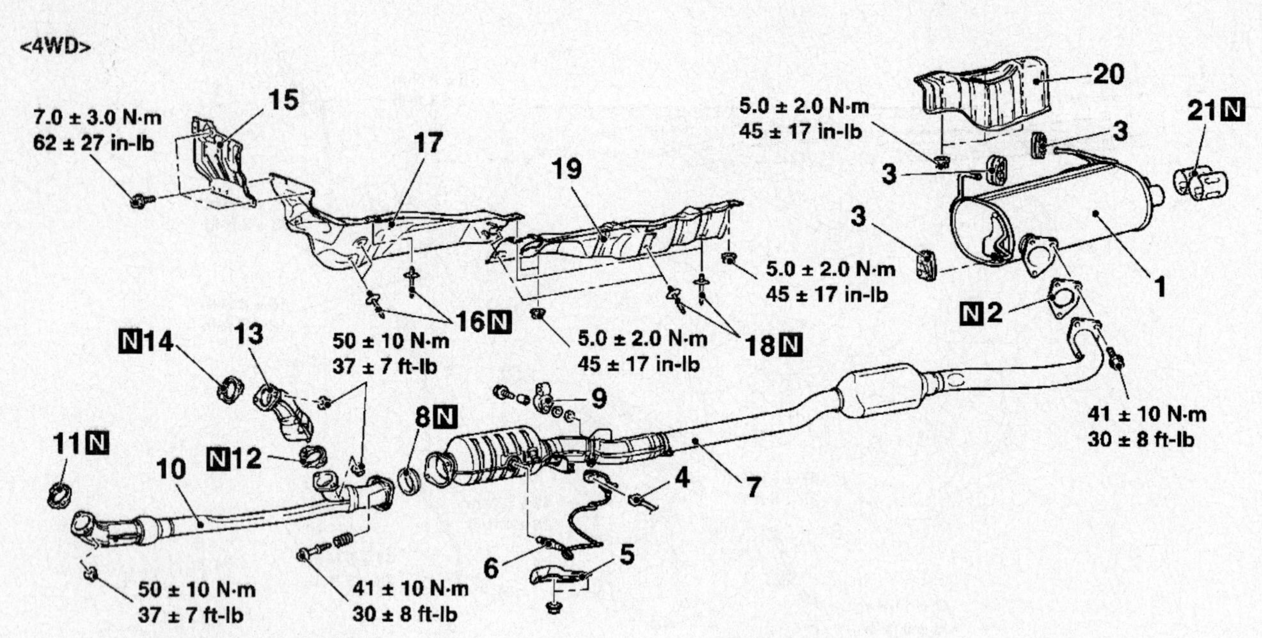

1. Exhaust main muffler
2. Gasket
3. Hanger
4. Center exhaust pipe heated oxygen sensor connector <Vehicles for California>
5. Harness cover <Vehicles for California>
6. Center exhaust pipe heated oxygen sensor <Vehicles for California>
7. Center exhaust pipe
8. Seal ring
9. Hanger
10. Front exhaust pipe
11. Gasket
12. Gasket
13. Front exhaust pipe RH
14. Gasket
15. Dash panel heat protector
16. Rivet
17. Front floor front panel heat protector
18. Rivet
19. Front floor rear panel heat protector
20. Rear floor panel heat protector
21. Exhaust tail pipe diffuser

22140_OUTL_G0174

Fig. 115 Exploded view of the exhaust system—AWD 3.0L engine

17. Connect the left bank heated oxygen sensor (front) connector.
18. Install the front exhaust pipe. Tighten to 30 ft. lbs. (41 Nm).
19. Install the air duct.
20. Install the engine under cover.

Right Bank

See Figures 113 through 115.

1. Remove the air cleaner cover and air intake duct.
2. Remove the battery.
3. Remove the engine under cover.
4. Remove the front exhaust pipe, center exhaust pipe and front floor front panel heat protector.
5. Remove the strut tower bar.
6. Remove the right exhaust manifold bracket.
7. Disconnect the right bank heated oxygen sensor (front) connector.
8. Remove the right bank heated oxygen sensor (rear) connector.
9. Remove the right bank heated oxygen sensor (front).
10. Remove the right bank heated oxygen sensor (rear).
11. Remove the heat protector.
12. Remove the exhaust manifold and gasket.

To install:

13. Install the exhaust manifold and gasket. Tighten to 36 ft. lbs. (49 Nm).
14. Install the heat protector. Tighten to 12 ft. lbs. (16 Nm).
15. Install the right bank heated oxygen sensor (rear). Tighten to 33 ft. lbs. (44 Nm).
16. Install the right bank heated oxygen sensor (front). Tighten to 33 ft. lbs. (44 Nm).
17. Install the right bank heated oxygen sensor (rear) connector.
18. Connect the right bank heated oxygen sensor (front) connector.
19. Install the right exhaust manifold bracket. Tighten to 35 ft. lbs. (47 Nm).
20. Install the strut tower bar.
21. Install the front exhaust pipe and center exhaust pipe. Tighten to 37 ft. lbs. (50 Nm).
22. Install the floor panel heat protector.
23. Install the engine under cover.
24. Install the battery.
25. Install the air cleaner cover and air intake duct.

FLYWHEEL

REMOVAL & INSTALLATION

See Figures 116 through 118.

1. Remove the transaxle.
2. Use special tool MB991883 on 2.4L engine and MD998781 on 3.0L engine to secure the flywheel and adapter plate .
3. Remove the flywheel bolts, then remove the flywheel.

To install:

4. Remove the sealant, the engine oil, and other adhering materials from the flywheel and adapter plate installation face, the crankshaft screw hole and flywheel bolts.
5. Install the flywheel and adapter plate to the crankshaft.
6. Use special tool MB991883 on 2.4L engine and MD998781 on 3.0L engine to secure the flywheel and adapter plate in the same manner as removal.

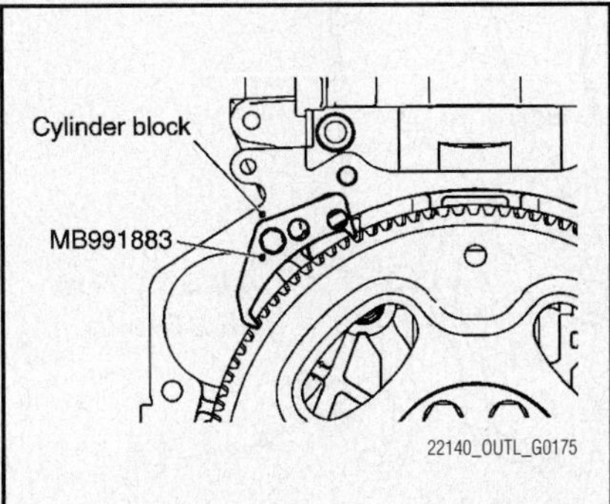

Fig. 116 Hold the flywheel using special tool MB991883—2.4L engine

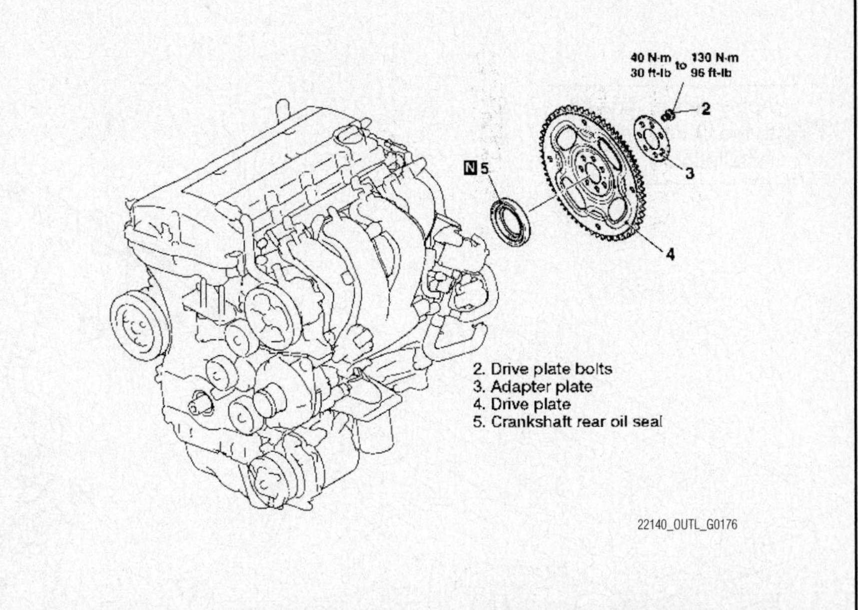

Fig. 118 Exploded view of flywheel assembly—2.4L engine shown, 3.0L similar

7. Apply a small amount of engine oil to the screw holes of the crankshaft and the bearing surface of the flywheel bolts and the adapter plate bolts.
8. Apply Three bond 1324 or equivalent sealant to the thread of the flywheel bolts and the flywheel bolts.
9. On 2.4L engines, tighten flywheel bolts as follows:
   a. Tighten flywheel bolts to temporary torque 30 ft. lbs. (40 Nm) in a star pattern.
   b. Tighten flywheel bolts to 96 ft. lbs. (130 Nm) in a star pattern.
10. On 3.0L engines, tighten flywheel bolts to 56 ft. lbs. (76 Nm) in a star pattern.
11. Install the transaxle.

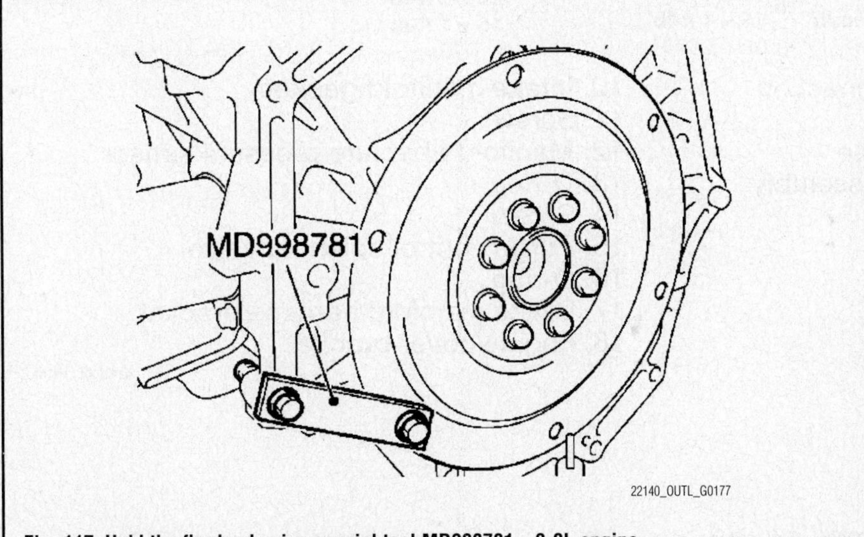

Fig. 117 Hold the flywheel using special tool MD998781—3.0L engine

INTAKE MANIFOLD

REMOVAL & INSTALLATION

See Figures 119 and 120.

1. Before servicing the vehicle, refer to the precautions.
2. Drain the engine coolant.
3. Remove the upper engine cover.
4. Remove the drive belt.
5. Remove the air cleaner intake hose and air cleaner assembly.
6. Remove the throttle body assembly.
7. Remove the fuel injectors.
8. Remove the EGR valve and EGR valve bracket.
9. Disconnect the control wiring harness.
10. Disconnect the emission vacuum hose.
11. Disconnect the brake booster vacuum hose
12. Remove the power steering oil pump.
    a. With the hose installed, remove the power steering oil pump assembly from the bracket.
    b. Tie the removed power steering oil pump assembly with wire at a position where it will not interfere with the removal and installation of the intake manifold.
13. Disconnect the valve cover PCV hose.
14. Remove the engine oil level gauge.
15. Remove the intake manifold bracket.
16. Remove the front injector protector.
17. Remove the intake manifold and gasket.

To install:

18. Install the intake manifold and gasket.

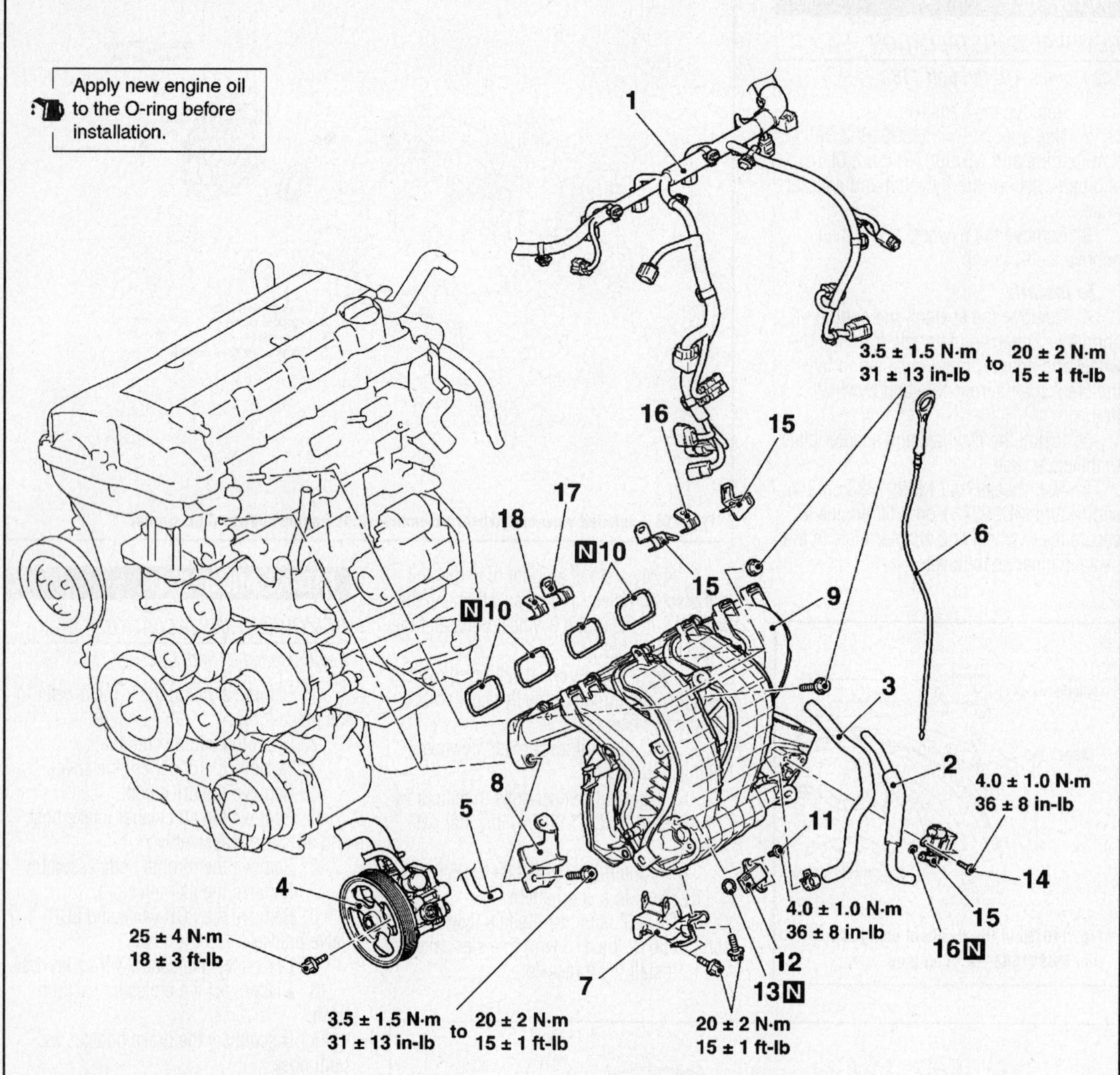

1. Control wiring harness connection
2. Emission vacuum hose
3. Brake booster vacuum hose
4. Power steering oil pump assembly
5. Rocker cover PCV hose
6. Engine oil level gauge
7. Intake manifold bracket
8. Injector protector front
9. Intake manifold assembly
10. Intake manifold gasket
11. Screw
12. Manifold absolute pressure sensor
13. O-ring
14. Screw
15. Purge control solenoid valve
16. O-ring
17. Intake manifold harness bracket
18. Engine cover bracket

Fig. 119 Exploded view of the intake manifold assembly—2.4L engine

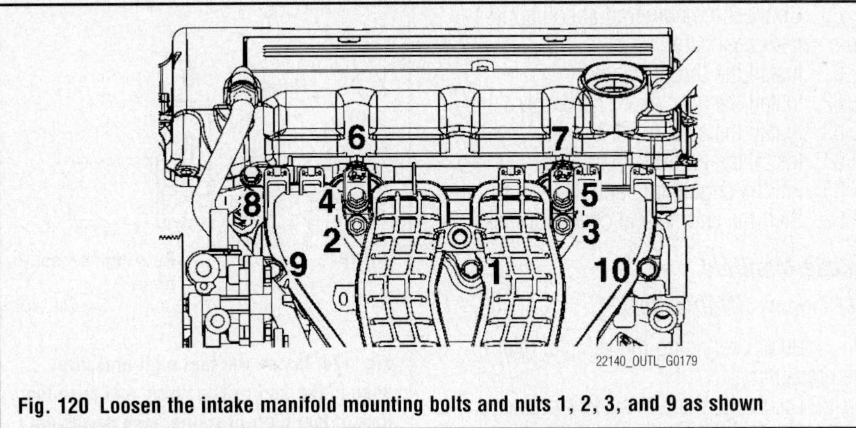

Fig. 120 Loosen the intake manifold mounting bolts and nuts 1, 2, 3, and 9 as shown

a. Install the intake manifold assembly and the injector protector front, and tighten mounting bolts and nuts temporarily.

➡**The tightening of the fuel delivery pipe, the intake manifold assembly and the injector protector front and rear has the specified order. Temporarily tighten the intake manifold assembly and injector protector front mounting bolts and nuts.**

b. Loosen the intake manifold mounting bolts and nuts 1, 2, 3, and 9 as shown.

c. Remove the EGR valve.

d. Loosen the EGR valve support and intake manifold coupling bolts.

e. Loosen the intake manifold bracket mounting bolts.

f. Temporarily tighten the mounting bolts and nuts of the intake manifold, bracket, fuel rail and injector protector rear to 31 inch lbs. (4 Nm) in the order shown.

g. Tighten the mounting bolts and nuts of the intake manifold, bracket, fuel rail and injector protector rear to 15 ft. lbs. (20 Nm) in the order shown.

h. Tighten the intake manifold bracket mounting bolts to 15 ft. lbs. (20 Nm).

i. Tighten the EGR valve support and intake manifold coupling bolts to 15 ft. lbs. (20 Nm).

19. Install the EGR valve to 18 ft. lbs. (24 Nm).
20. Install the engine oil level gauge.
21. Connect the valve cover PCV hose.
22. Install the power steering oil pump.
23. Connect the brake booster vacuum hose
24. Connect the emission vacuum hose.
25. Connect the control wiring harness.
26. Install the EGR valve and EGR valve bracket.
27. Install the fuel injectors.
28. Install the throttle body assembly.
29. Install the air cleaner intake hose and air cleaner assembly.
30. Install the drive belt.
31. Install the upper engine cover.
32. Fill the engine coolant.

3.0L Engine

Intake Manifold Plenum

See Figure 121.

1. Before servicing the vehicle, refer to the precautions.
2. Drain the engine coolant
3. Remove the engine cover.
4. Remove the strut tower bar.
5. Remove the air cleaner assembly.
6. Remove the throttle body.
7. Disconnect the manifold absolute pressure sensor connector.
8. Remove the manifold absolute pressure sensor.
9. Remove the o-ring.
10. Disconnect the variable intake air control solenoid valve connector.
11. Disconnect the right bank heated oxygen sensor connectors.
12. Remove the right bank heated oxygen sensors.
13. Disconnect the purge hose.
14. Disconnect the vacuum hose.
15. Remove the vacuum tank and bracket.
16. Remove the variable intake air control solenoid valve.
17. Disconnect the left bank knock sensor connector
18. Remove the left bank knock sensor.
19. Disconnect the right bank knock sensor connector
20. Remove the right bank knock sensor.
21. Disconnect the control wiring harness and injector wiring harness combination connector.

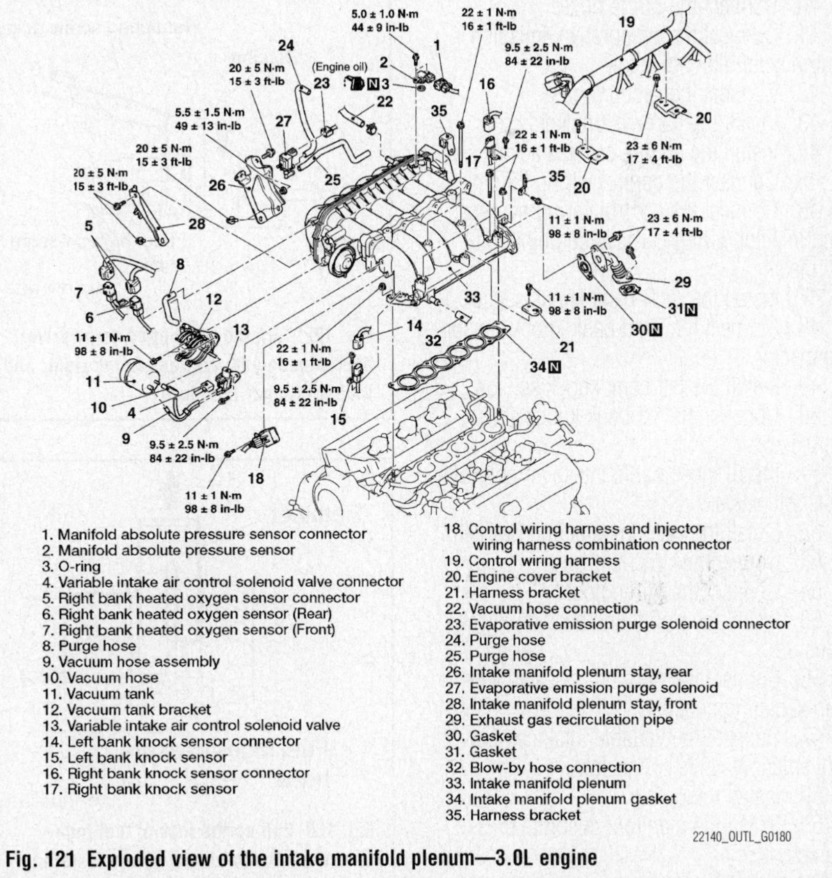

Fig. 121 Exploded view of the intake manifold plenum—3.0L engine

22. Disconnect the control wiring harness.
23. Remove the engine cover bracket.
24. Remove the harness bracket.
25. Disconnect the vacuum hose.
26. Disconnect the evaporative emission purge solenoid connector
27. Disconnect the purge hose.
28. Remove the intake manifold plenum rear bracket.
29. Remove the evaporative emission purge solenoid.
30. Remove the intake manifold plenum front bracket.
31. Remove the exhaust gas recirculation pipe and gasket.
32. Disconnect the blow-by hose.
33. Remove the intake manifold plenum and gasket.

To install:

34. Install the intake manifold plenum and gasket.
35. Connect the blow-by hose.
36. Install the exhaust gas recirculation pipe and gasket.
37. Install the intake manifold plenum front bracket.
38. Install the evaporative emission purge solenoid.
39. Install the intake manifold plenum rear bracket.
40. Connect the purge hose.
41. Connect the evaporative emission purge solenoid connector
42. Connect the vacuum hose.
43. Install the harness bracket.
44. Install the engine cover bracket.
45. Connect the control wiring harness.
46. Connect the control wiring harness and injector wiring harness combination connector.
47. Install the right bank knock sensor.
48. Connect the right bank knock sensor connector
49. Install the left bank knock sensor.
50. Connect the left bank knock sensor connector
51. Install the variable intake air control solenoid valve.
52. Install the vacuum tank and bracket.
53. Connect the vacuum hose.
54. Connect the purge hose.
55. Install the right bank heated oxygen sensors.
56. Connect the right bank heated oxygen sensor connectors.
57. Connect the variable intake air control solenoid valve connector.
58. Install the o-ring.
59. Install the manifold absolute pressure sensor.
60. Connect the manifold absolute pressure sensor connector.
61. Install the throttle body.
62. Install the air cleaner assembly.
63. Install the strut tower bar.
64. Install the engine cover.
65. Fill the engine coolant.
66. Start the engine and check for leaks.

Intake Manifold

See Figures 121 through 124.

1. Before servicing the vehicle, refer to the precautions.
2. Discharge the fuel system pressure.
3. Remove the intake manifold plenum.
4. Disconnect the injector connector.
5. Disconnect the fuel high-pressure hose.

✲✲ WARNING

So not kink the fuel high-pressure hose as it is made of plastic.

a. Insert a flat-tipped screwdriver [width 0.24 inch (6 mm), thickness 0.04 inch (1 mm)] to the retainer.

b. Turn the flat-tipped screwdriver approximately 90 degrees to the right, and lift the retainer to unlock.

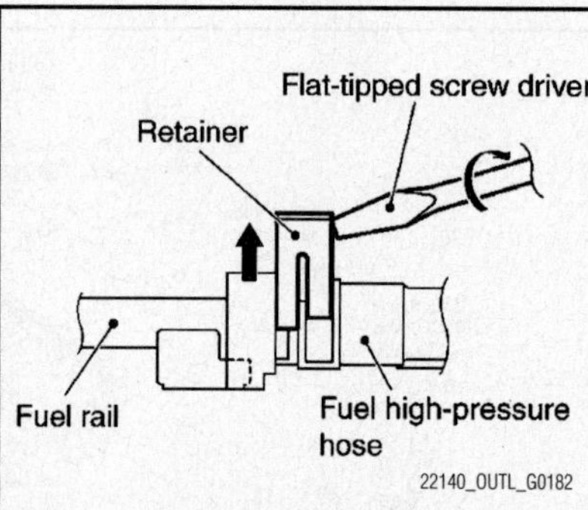

Fig. 122 Turn the flat-tipped screwdriver approximately 90 degrees to the right, and lift the retainer to unlock

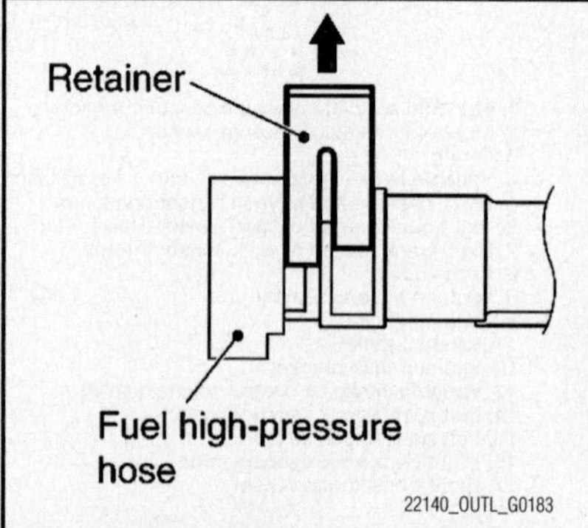

Fig. 123 Pull up the lock of fuel high-pressure hose to unlock before installing

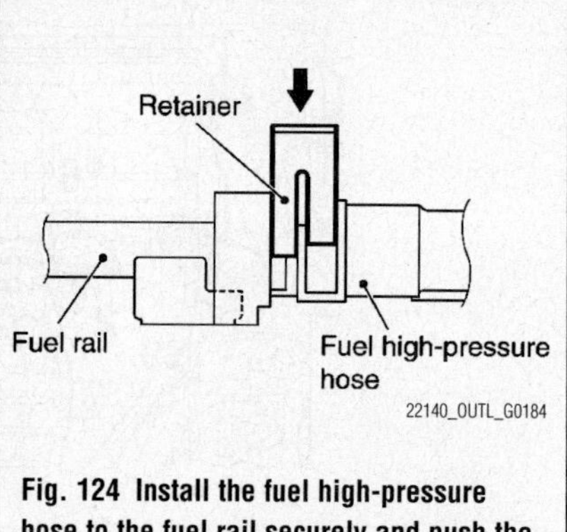

Fig. 124 Install the fuel high-pressure hose to the fuel rail securely and push the lock of fuel high-pressure hose downward and lock thoroughly

6. Remove the fuel rail and fuel injector assemblies.

✲✲ WARNING

Do not drop the fuel injectors.

➡Remove the fuel rail with the fuel injectors still attached.

7. Disconnect the harness bracket.
8. Remove the intake manifold and gasket.

To install:

9. Install the intake manifold and gasket.
   a. Coat the intake manifold mounting studs with engine oil.
   b. Tighten the intake manifold mounting nuts in the following order:
      - Left bank nuts to 58 inch lbs. (7 Nm)
      - Right bank nuts to 16 ft. lbs. (22 Nm)
      - Left bank nuts to 16 ft. lbs. (22 Nm)
      - Right bank nuts to 16 ft. lbs. (22 Nm)
      - Left bank nuts to 16 ft. lbs. (22 Nm)
10. Connect the harness bracket.
11. Install the fuel rail and fuel injector assemblies.
12. Connect the fuel high-pressure hose.
   a. Pull up the lock of fuel high-pressure hose to unlock before installing.
   b. Install the fuel high-pressure hose to the fuel rail securely and push the lock of fuel high-pressure hose downward and lock thoroughly.
   c. After installing, slightly pull the fuel high-pressure hose and ensure that there is no disengaged fuel high-pressure hose. Also confirm that there is approximately 0.04 inch (1 mm) play at this time.
13. Connect the injector connector.
14. Install the intake manifold plenum.

OIL PAN

REMOVAL & INSTALLATION

2.4L Engine

See Figures 125 through 128.

1. Before servicing the vehicle, refer to the precautions.
2. Remove the engine under cover and engine side cover.
3. Drain the engine oil.
4. Remove the drive belt.
5. Remove the A/C compressor and clutch assembly
   a. Remove the A/C compressor and clutch assembly together with the hose from the bracket.
   b. Tie the removed A/C compressor and clutch assembly with a string at a position where they will not interfere with the removal and installation of engine oil pan.
6. Remove the A/C compressor bracket.
7. Remove the engine oil pan.
   a. Remove the engine oil pan mounting bolts.

WARNING

Do not forcibly drive in special tool MD998727 to avoid damage to the engine oil pan seal surface of cylinder block assembly.

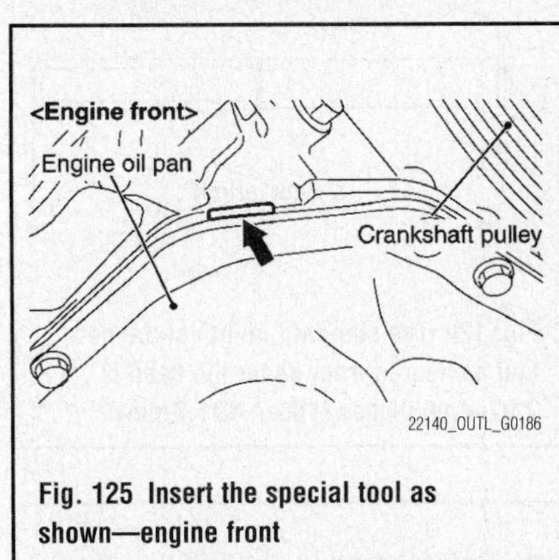

Fig. 125 Insert the special tool as shown—engine front

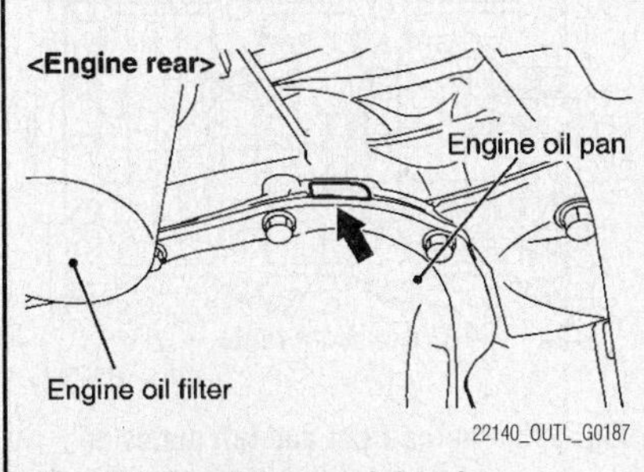

Fig. 126 Insert the special tool as shown—engine rear

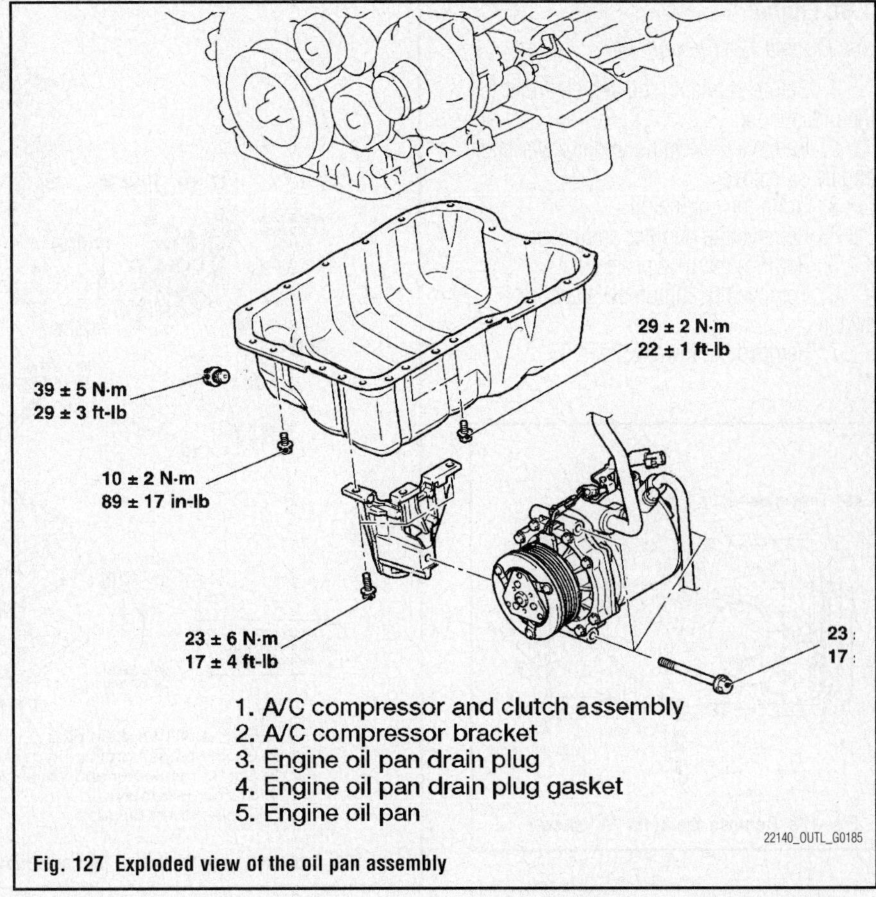

Fig. 127 Exploded view of the oil pan assembly

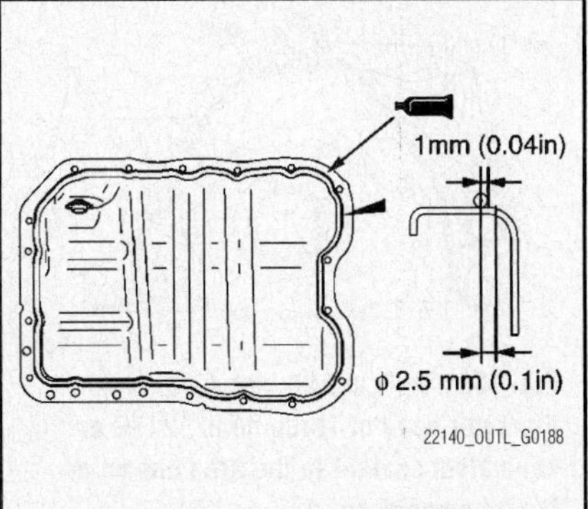

Fig. 128 Apply the sealant without any gap to the mating surface of engine oil pan as shown

   b. Insert special tool MD998727 from the engine oil pan removal groove of the cylinder block assembly.
   c. Lightly tap the special tool with a hammer to slide the oil pan seal surface, cut off the liquid gasket, and remove the engine oil pan.

To install:

8. Install the engine oil pan.
   a. Remove all the traces of sealant adhering to the engine oil pan and cylinder block assembly using a remover or others. Then, degrease them using white gasoline.
   b. Apply the sealant without any gap to the mating surface of engine oil pan as shown in the figure. Within three minutes, install the engine oil pan to the cylinder block assembly.
   Specified sealant: Three bond 1217G or equivalent
   c. Do not apply oil or water to the sealant-applied area or start up the engine within 2 hours after the installation of the engine oil pan.
   d. Tighten the M6 89 inch lbs. (10 Nm) and the M8 to 22 ft. lbs. (29 Nm).
9. Install the A/C compressor bracket.
10. Install the A/C compressor and clutch assembly. Tighten bolts to 17 ft. lbs. (23 Nm) in a counter clockwise direction from the top.
11. Install the drive belt.
12. Replace the engine oil pan drain plug gasket with a new one. Install the new gasket with the chamfer toward the engine oil pan.
13. Fill the engine with oil.
14. Install the engine under cover and engine side cover.
15. Start the engine and check for leaks.

3.0L Engine

See Figures 129 through 140.

1. Before servicing the vehicle, refer to the precautions.
2. Remove the engine under cover and engine side cover.
3. Drain the engine oil.
4. Remove the oil filter and cover.
5. Remove the heat protector.
6. Remove the engine oil pressure switch.
7. Remove the lower oil pan.

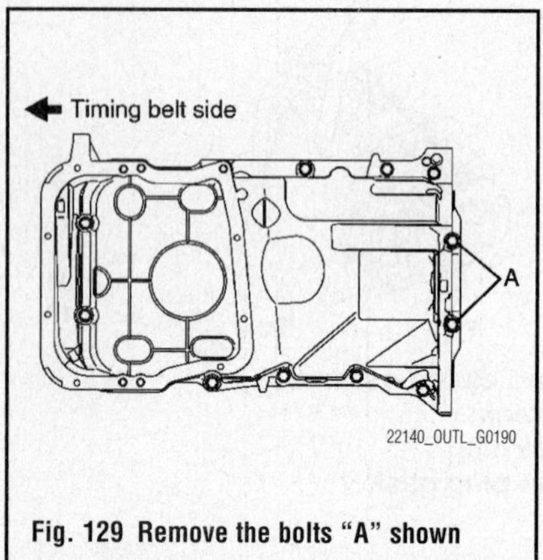

Fig. 129 Remove the bolts "A" shown

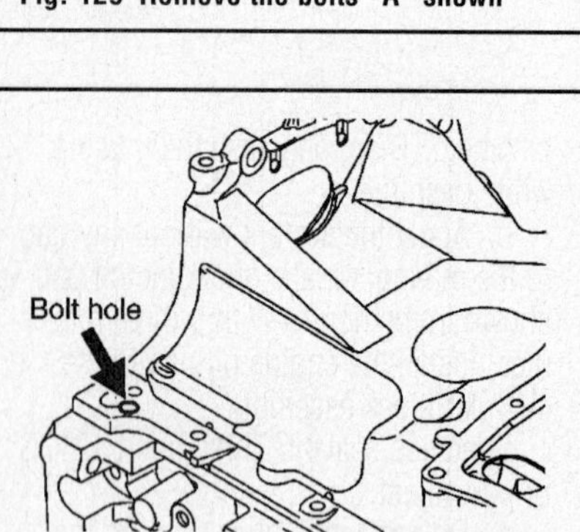

Fig. 130 Thread the M10 x 1.25 pitch bolt into the illustrated bolt hole to remove the oil pan upper

✱✱ WARNING

To prevent the sealed area from being damaged, carefully insert the special tool.

a. Insert the special tool MD998727, into the groove. Strike and slide it and then cut the liquid gasket.

8. Remove the oil pan cover.
9. Remove the upper oil pan.

a. Remove the bolts A shown in the illustration first.

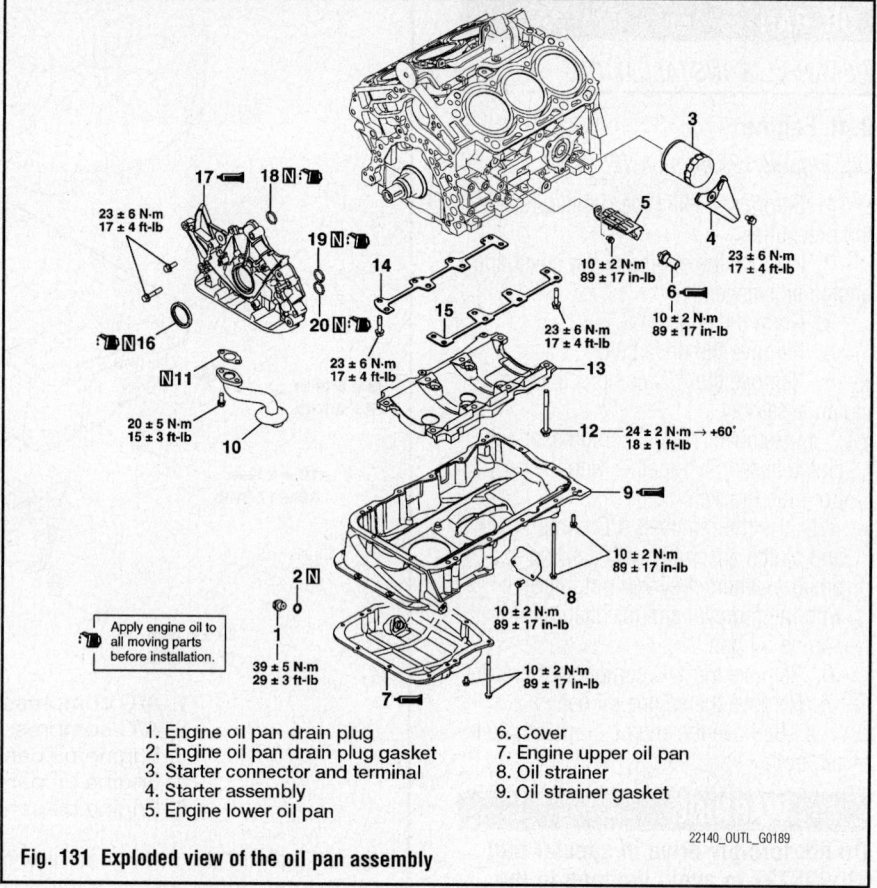

b. Remove all other bolts.

1. Engine oil pan drain plug
2. Engine oil pan drain plug gasket
3. Starter connector and terminal
4. Starter assembly
5. Engine lower oil pan
6. Cover
7. Engine upper oil pan
8. Oil strainer
9. Oil strainer gasket

22140_OUTL_G0189

Fig. 131 Exploded view of the oil pan assembly

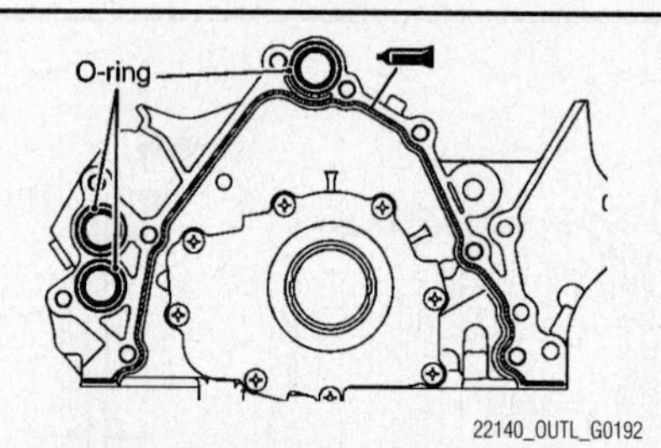

Fig. 132 Apply a 0.08 inch (2.0 mm) diameter bead of Three bond 1217G or equivalent sealant to the area shown on the oil pump case

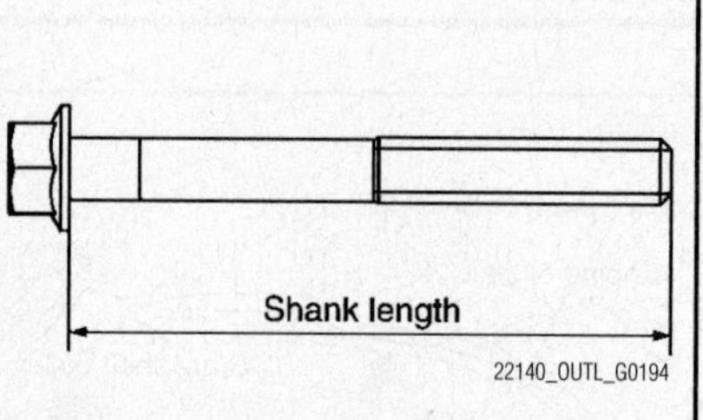

Fig. 134 The standard length of the new bolt measured from under the head is 3.97–4.00 inches (100.7–101.7 mm).

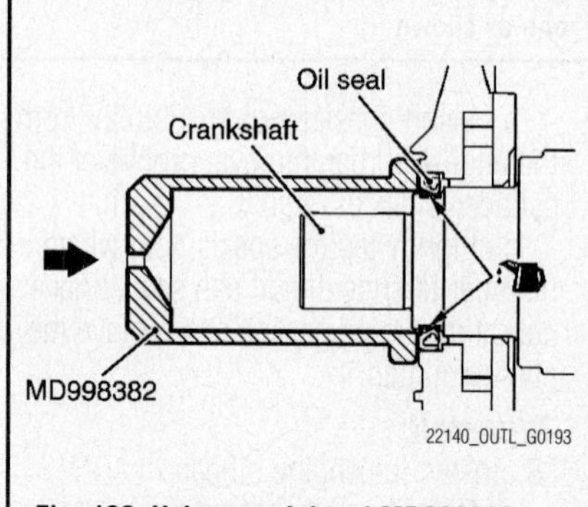

Fig. 133 Using special tool MD998382, press-fit the oil seal into the oil pump case

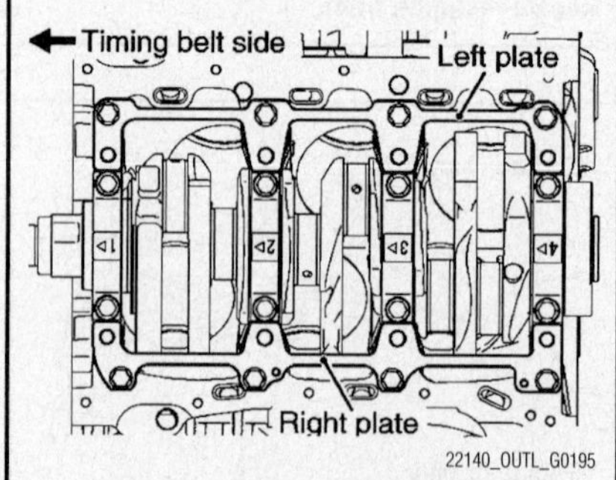

Fig. 135 Put the right and left plates on the crankshaft bearing, temporarily tighten bolts

✱✱ WARNING

Do not use a scraper or special tool to remove the oil pan upper.

c. Thread the M10 × 1.25 pitch bolt into the illustrated bolt hole to remove the oil pan upper.

10. Remove the oil screen and gasket.
11. Remove the bearing cap bolt.
12. Remove the beam.
13. Remove the right and left plate.
14. Remove the crankshaft front oil seal.
15. Remove the oil pump case.
16. Remove the o-rings.

To install:

17. Install the o-rings.
18. Install the oil pump case.

✱✱ WARNING

Carefully work so that the rests of the liquid gasket cannot enter to the oil passage or water passage.

a. Remove the remaining liquid gasket on the oil pump case and the cylinder block.

b. Degrease and clean the plane where the liquid gasket is applied and the both sides of the chamfered areas where the liquid gasket collects.

c. After degreasing and cleaning the plane and the both sides, always spray compressed air into the oil passage and the water passage. Check that the oil passage and the water passage are free of foreign materials such as liquid gasket.

d. As shown in the illustration, apply a 0.08 inch (2.0 mm) diameter bead of Three bond 1217G or equivalent sealant to the area shown on the oil pump case.

e. After checking that the o-ring is installed at three places, install the oil pump case to the cylinder block.

f. Tighten the oil pump case to 17 ft. lbs. (23 Nm).

19. Install the crankshaft front oil seal.
20. Install the right and left plate, and the beam.

a. Before installing the bearing cap bolt, check the bolt screw head is not damaged. If the screw head extremely gets damaged, replace it with a new bolt. The standard length of the new bolt measured from under the head is 3.97–4.00 inches (100.7–101.7 mm).

✱✱ WARNING

When the beam is removed and installed, never loosen the tightening bolts of the crankshaft bearing cap.

b. Put the right and left plates on the crankshaft bearing, temporarily tighten bolts.

c. Put the beam on them, Temporarily tighten bolt.

d. Tighten the beam bolts to 18 ft. lbs. (24 Nm) in the sequence shown.

✱✱ WARNING

When the tightening angle is smaller than the specified tightening angle, the appropriate tightening capacity cannot be secured. When the tightening angle is larger than the specified tightening angle, remove the bolt to start from the beginning again according to the procedure.

✱✱ WARNING

Using a torque angle meter, tighten the beam bolt (M9) 60 degrees in the specified order.

e. Tighten the plate bolts to 17 ft. lbs. (23 Nm) in the sequence shown.

21. Install the bearing cap bolt.
22. Install the oil screen and gasket.
23. Install the upper oil pan.

a. Clean the gasket surfaces of the oil pan upper and oil pan lower.

✱✱ WARNING

When installing the upper oil pan, be sure not to expel the sealant from the oil pan flange at portion A.

b. Apply a 0.10 inch (2.5 mm) diameter bead of Three bond 1217G or equivalent sealant to the oil pan.

c. Be sure to install the oil pan quickly while the sealant is wet. Install the oil pan within 15 minutes after applying liquid gasket.

d. Then wait at least one hour. Never start the engine or let engine oil or coolant touch the adhesion surface during that time.

e. Tighten the upper oil pan bolts to 89 ft. lbs. (10 Nm) in the sequence shown.

f. After installation, keep the sealed area away from the oil and coolant for approximately one hour.

24. Install the oil pan cover.
25. Install the lower oil pan.

a. Clean the gasket surfaces of the cylinder block and oil pan lower.

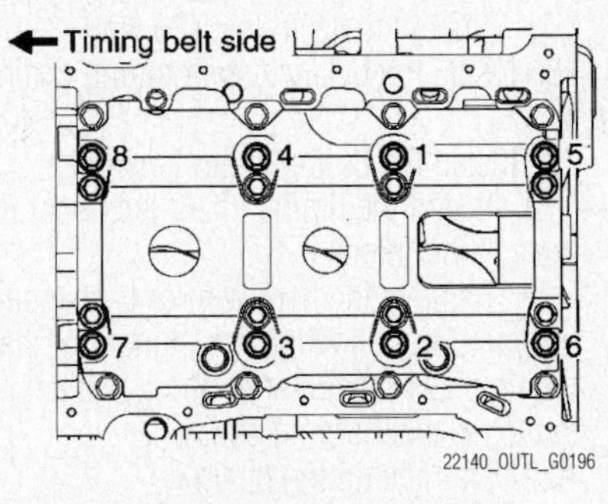

Fig. 136 Tighten the beam bolts in the sequence shown

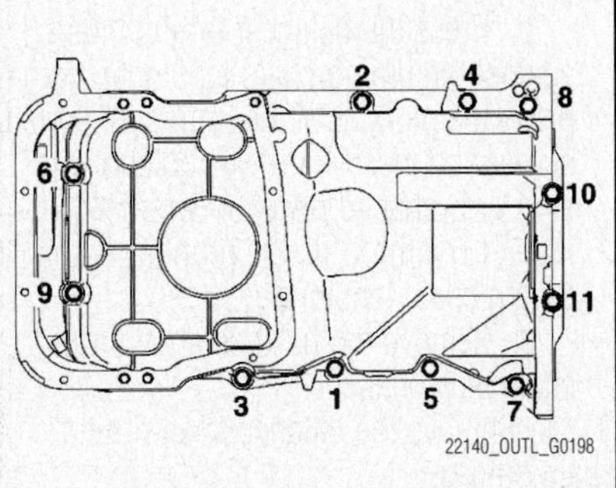

Fig. 138 Tighten the upper oil pan bolts in the sequence shown

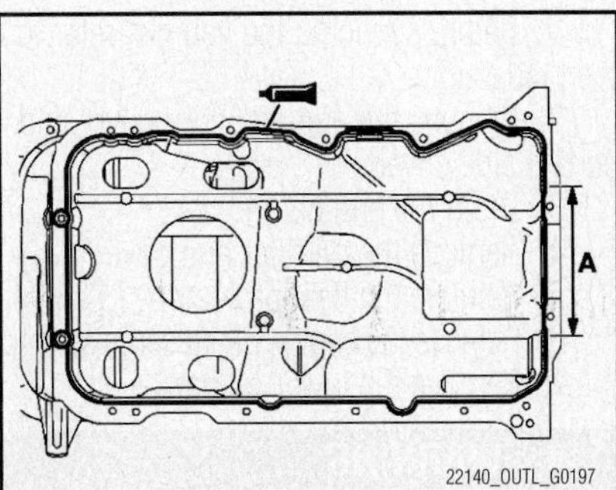

Fig. 137 Apply a 0.10 inch (2.5 mm) diameter bead of Three bond 1217G or equivalent sealant to the oil pan

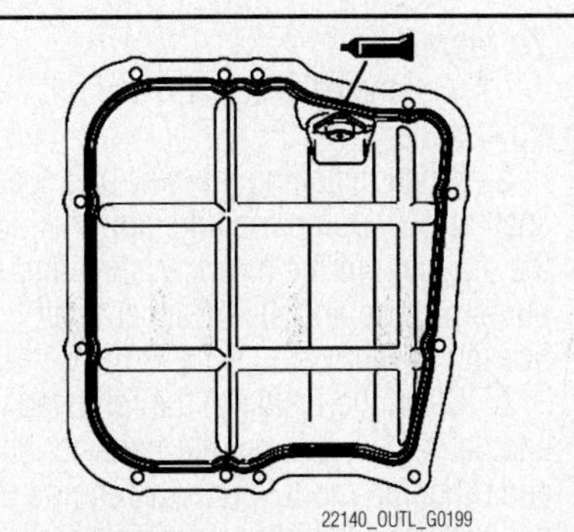

Fig. 139 Apply a 0.10 inch (2.5 mm) diameter bead of Three bond 1217G or equivalent sealant to the oil pan

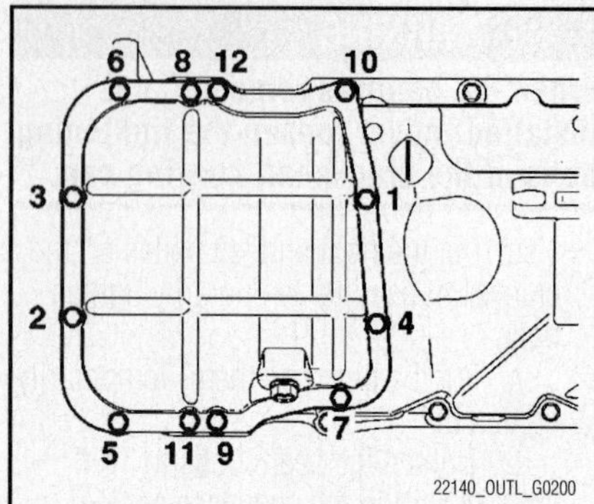

Fig. 140 Tighten the upper oil pan bolts in the sequence shown

b. Apply a 2.5 ± 0.5 mm (0.10 ± 0.01 inch) diameter bead of sealant (Three bond 1217G or exact equivalent) to the oil pan. Be sure to install the oil pan quickly while the sealant is wet.

**** WARNING**

Install the oil pan within 15 minutes after applying liquid gasket. Then wait at least one hour. Never start the engine or let engine oil or coolant touch the adhesion surface during that time.

c. Tighten the upper oil pan bolts to 89 inch lbs. (10 Nm) in the sequence shown.

d. After installation, keep the sealed area away from the oil and coolant for approximately one hour.

26. Install the engine oil pressure switch.

a. Completely remove existing sealant from the oil pressure switch and the switch mounting hole on the oil pump case.

b. Apply Three bond 1215, Three bond 1212D or equivalent sealant to the threaded part of the oil pressure switch.

c. Install the oil pressure switch to the oil pump case and tighten to 89 inch lbs. (10 Nm).

27. Install the heat protector.

28. Install the oil filter and cover.

a. Check the right and left ribs of the oil filter cover touch the outer circumference of the oil filter boss.

b. 1.Clean the filter mounting surface on the front case.

c. Apply engine oil to the O-ring of the oil filter.

**** WARNING**

If the filter is tightened by hand only, it will be insufficiently torqued, resulting in oil leaks.

d. Screw the oil filter in and tighten the oil filter to the specified torque using a commercially available special tool MB991396 or general tool from where the O-ring has come into contact with the oil filter mounting surface.

e. Tighten MD332687, MD365876 to 13 ft. lbs. (16Nm) <approx. 1 turn>. Tighten MD360935 124 inch lbs. (14 Nm) <approx. 3/4 of a turn>.

29. Fill the engine with oil. Install the drain plug using a new gasket. Install the gasket with the chamfer toward the oil pan.

30. Install the engine under cover and engine side cover.

31. Start the engine and check for leaks.

OIL PUMP

REMOVAL & INSTALLATION

2.4L Engine

See Figures 57 through 60.

1. Before servicing the vehicle, refer to the precautions.

2. Remove the timing chain.

3. Remove the balance shaft timing chain tensioner.

a. Securely install the plunger of the timing chain tensioner. Otherwise, it may pop out.

b. Press the balancer timing chain against the timing chain tensioner, compress the plunger of the timing chain tensioner and insert hard wire (piano wire, etc.) or L-shaped hexagon wrench 0.05 inch (1.5 mm) to fix the plunger of the timing chain tensioner.

c. Remove the balance shaft timing chain tensioner.

4. Remove the balance shaft timing chain guides.

5. Remove the balance shaft and oil pump module.

6. Remove the balance shaft timing chain.

To install:

7. Install the balance shaft and oil pump module.

a. When installing the new balancer shaft and oil pump module, apply oil to the oil pump in the balancer shaft and oil pump module and the balancer shaft bearing as follows:

b. Clean the inside of the removed engine oil pan, and put the balancer shaft and oil pump module into the engine oil pan with its oil inlet port facing up.

c. Pour new engine oil until two-thirds of the balancer shaft and oil pump module is soaked.

d. Fill the engine oil (approximately 3.05 cu. in. [50 cm3]) into the balancer shaft and oil pump module from the oil inlet port.

e. Turn the balancer shaft sprocket of the balancer shaft and oil pump module clockwise four rotations or more to apply the engine oil to the entire area of the oil pump and the balancer shaft bearing.

f. With the link marks (orange or blue) of balancer timing chain aligned with the timing marks of balancer sprocket and crankshaft sprocket, install the balancer shaft and oil pump module together with the balancer timing chain and crankshaft sprocket as one unit to the cylinder block. At this time, securely bring the balancer shaft and oil pump module into contact with the rudder frame mounting area.

g. Apply an adequate and minimum amount of engine oil to the threads and bearing surfaces of the balancer shaft and oil pump module bolts.

h. Tighten the balancer shaft and oil pump module bolts to the specified torque in the order shown:

- Tighten to 15 ft. lbs. (20 Nm)
- Retighten to 15 ft. lbs. (20 Nm)
- Loosen all bolts in the reverse sequence fully
- Tighten to 15 ft. lbs. (20 Nm)
- Tighten to 15 ft. lbs. (20 Nm)

8. Install the balance shaft timing chain guides.

9. Install the timing chain tensioner.

a. Install the timing chain tensioner to the cylinder block.

b. Remove the hard wire or L-shaped hexagon wrench fixing the plunger of the timing chain tensioner to apply tension to the balancer timing chain.

10. Install the timing chain.

3.0L Engine

See Figures 129 through 140.

1. Before servicing the vehicle, refer to the precautions.

2. Remove the engine under cover and engine side cover.

3. Drain the engine oil.

4. Remove the oil filter and cover.

5. Remove the heat protector.

6. Remove the engine oil pressure switch.

7. Remove the lower oil pan.

**** WARNING**

To prevent the sealed area from being damaged, carefully insert the special tool.

a. Insert the special tool MD998727, into the groove. Strike and slide it and then cut the liquid gasket.

8. Remove the oil pan cover.
9. Remove the upper oil pan.

a. Remove the bolts A shown in the illustration first.

b. Remove all other bolts.

WARNING

Do not use a scraper or special tool to remove the oil pan upper.

c. Thread the M10 × 1.25 pitch bolt into the illustrated bolt hole to remove the oil pan upper.

10. Remove the oil screen and gasket.
11. Remove the bearing cap bolt.
12. Remove the beam.
13. Remove the right and left plate.
14. Remove the crankshaft front oil seal.
15. Remove the oil pump case.
16. Remove the o-rings.

To install:

17. Install the o-rings.
18. Install the oil pump case.

WARNING

Carefully work so that the rests of the liquid gasket cannot enter to the oil passage or water passage.

a. Remove the remaining liquid gasket on the oil pump case and the cylinder block.

b. Degrease and clean the plane where the liquid gasket is applied and the both sides of the chamfered areas where the liquid gasket collects.

c. After degreasing and cleaning the plane and the both sides, always spray compressed air into the oil passage and the water passage. Check that the oil passage and the water passage are free of foreign materials such as liquid gasket.

d. As shown in the illustration, apply a 0.08 inch (2.0 mm) diameter bead of Three bond 1217G or equivalent sealant to the area shown on the oil pump case.

e. After checking that the o-ring is installed at three places, install the oil pump case to the cylinder block.

f. Tighten the oil pump case to 17 ft. lbs. (23 Nm).

19. Install the crankshaft front oil seal.
20. Install the right and left plate, and the beam.

a. Before installing the bearing cap bolt, check the bolt screw head is not damaged. If the screw head extremely gets damaged, replace it with a new bolt. The standard length of the new bolt measured from under the head is 3.97–4.00 inches (100.7–101.7 mm).

WARNING

When the beam is removed and installed, never loosen the tightening bolts of the crankshaft bearing cap.

b. Put the right and left plates on the crankshaft bearing, temporarily tighten bolts.

c. Put the beam on them, Temporarily tighten bolt.

d. Tighten the beam bolts to 18 ft. lbs. (24 Nm) in the sequence shown.

WARNING

When the tightening angle is smaller than the specified tightening angle, the appropriate tightening capacity cannot be secured. When the tightening angle is larger than the specified tightening angle, remove the bolt to start from the beginning again according to the procedure.

WARNING

Using a torque angle meter, tighten the beam bolt (M9) 60 degrees in the specified order.

e. Tighten the plate bolts to 17 ft. lbs. (23 Nm) in the sequence shown.

21. Install the bearing cap bolt.
22. Install the oil screen and gasket.
23. Install the upper oil pan.

a. Clean the gasket surfaces of the oil pan upper and oil pan lower.

WARNING

When installing the upper oil pan, be sure not to expel the sealant from the oil pan flange at portion A.

b. Apply a 0.10 inch (2.5 mm) diameter bead of Three bond 1217G or equivalent sealant to the oil pan.

c. Be sure to install the oil pan quickly while the sealant is wet. Install the oil pan within 15 minutes after applying liquid gasket.

d. Then wait at least one hour. Never start the engine or let engine oil or coolant touch the adhesion surface during that time.

e. Tighten the upper oil pan bolts to 89 ft. lbs. (10 Nm) in the sequence shown.

f. After installation, keep the sealed area away from the oil and coolant for approximately one hour.

24. Install the oil pan cover.
25. Install the lower oil pan.

a. Clean the gasket surfaces of the cylinder block and oil pan lower.

b. Apply a 2.5 ± 0.5 mm (0.10 ± 0.01 inch) diameter bead of sealant (Three bond 1217G or exact equivalent) to the oil pan. Be sure to install the oil pan quickly while the sealant is wet.

WARNING

Install the oil pan within 15 minutes after applying liquid gasket. Then wait at least one hour. Never start the engine or let engine oil or coolant touch the adhesion surface during that time.

c. Tighten the upper oil pan bolts to 89 inch lbs. (10 Nm) in the sequence shown.

d. After installation, keep the sealed area away from the oil and coolant for approximately one hour.

26. Install the engine oil pressure switch.

a. Completely remove existing sealant from the oil pressure switch and the switch mounting hole on the oil pump case.

b. Apply Three bond 1215, Three bond 1212D or equivalent sealant to the threaded part of the oil pressure switch.

c. Install the oil pressure switch to the oil pump case and tighten to 89 inch lbs. (10 Nm).

27. Install the heat protector.
28. Install the oil filter and cover.

a. Check the right and left ribs of the oil filter cover touch the outer circumference of the oil filter boss.

b. Clean the filter mounting surface on the front case.

c. Apply engine oil to the O-ring of the oil filter.

WARNING

If the filter is tightened by hand only, it will be insufficiently torqued, resulting in oil leaks.

d. Screw the oil filter in and tighten the oil filter to the specified torque using a commercially available special tool MB991396 or general tool from where the O-ring has come into contact with the oil filter mounting surface.

e. Tighten MD332687, MD365876 to 13 ft. lbs. (16Nm) <approx. 1 turn>. Tighten MD360935 124 inch lbs. (14 Nm) <approx. 3/4 of a turn>.

29. Fill the engine with oil. Install the drain plug using a new gasket. Install the gasket with the chamfer toward the oil pan.

30. Install the engine under cover and engine side cover.
31. Start the engine and check for leaks.

MAIN BEARING TORQUE SEQUENCE

See Figures 141 and 142.

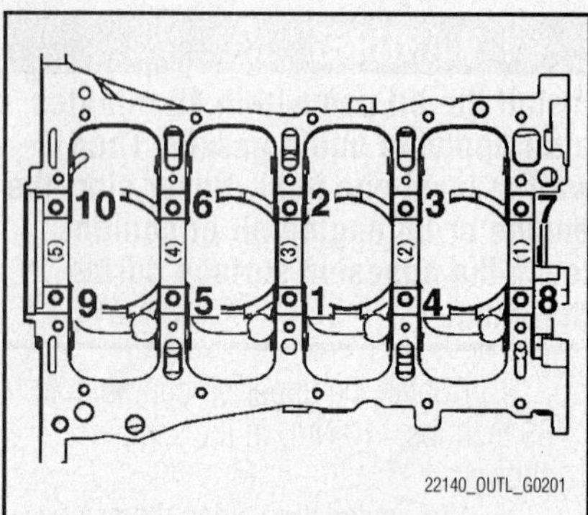

Fig. 141 Main bearing torque sequence—2.4L engine

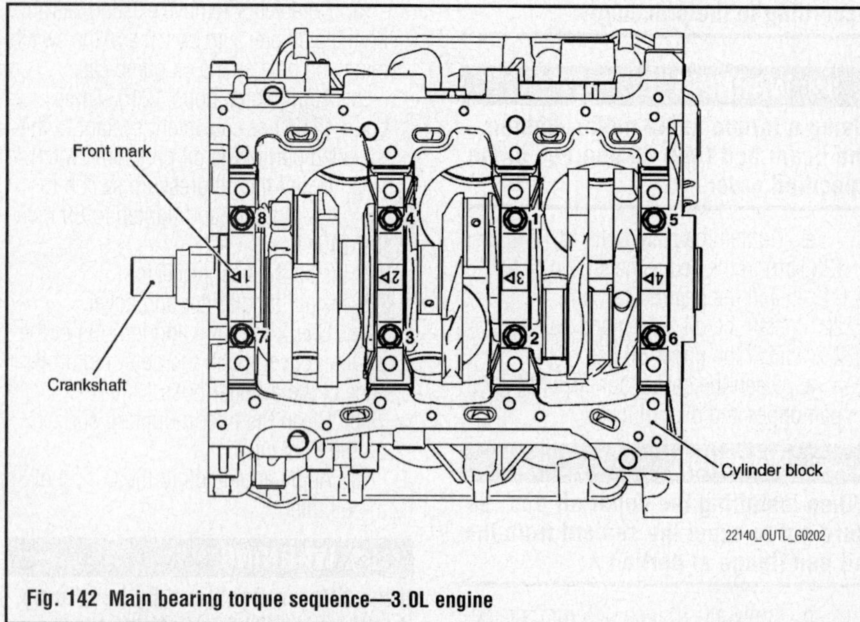

Fig. 142 Main bearing torque sequence—3.0L engine

PISTON AND RING

POSITIONING

See Figures 143 through 145.

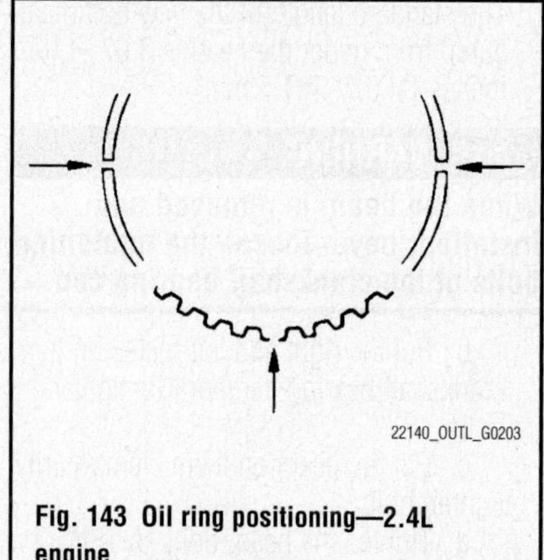

Fig. 143 Oil ring positioning—2.4L engine

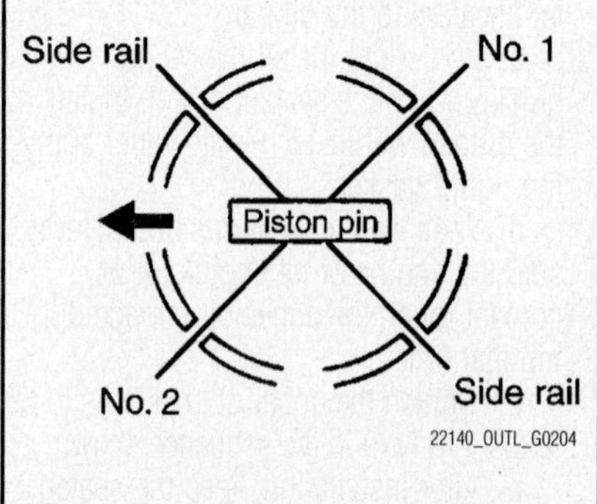

Fig. 144 Piston ring positioning—2.4L engine

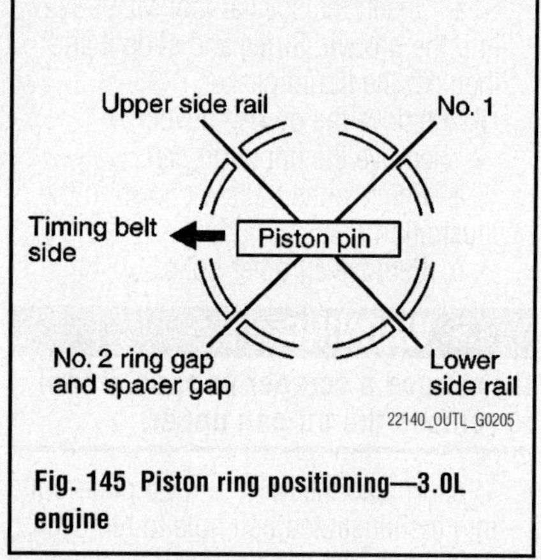

Fig. 145 Piston ring positioning—3.0L engine

REAR MAIN SEAL

REMOVAL & INSTALLATION

See Figures 146 and 147.

1. Before servicing the vehicle, refer to the precautions.
2. Remove the transaxle from the vehicle.
3. Remove the flywheel bolts. Use special tool MB991883 for 2.4L engine or MD998781 for 3.0L engine to hold the flywheel while loosening the bolts.
   a. Remove the flywheel adapter plate and flywheel.
4. Carefully pry the seal out of the oil seal case without damaging the sealing surface of the crankshaft.

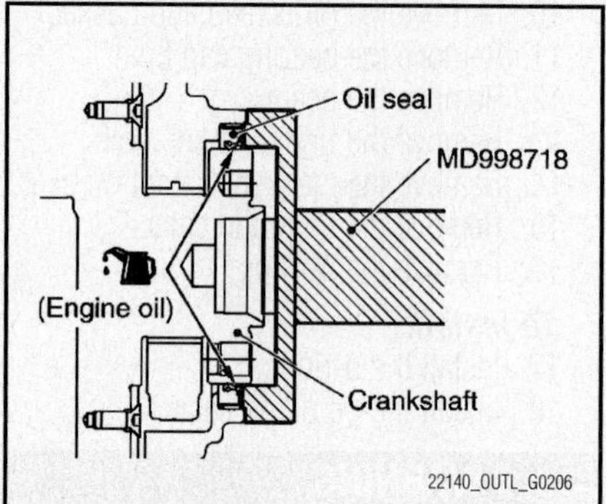

Fig. 146 Using special tool MD998718, press in the crankshaft rear oil seal up to the cylinder block assembly end surface—2.4L engine

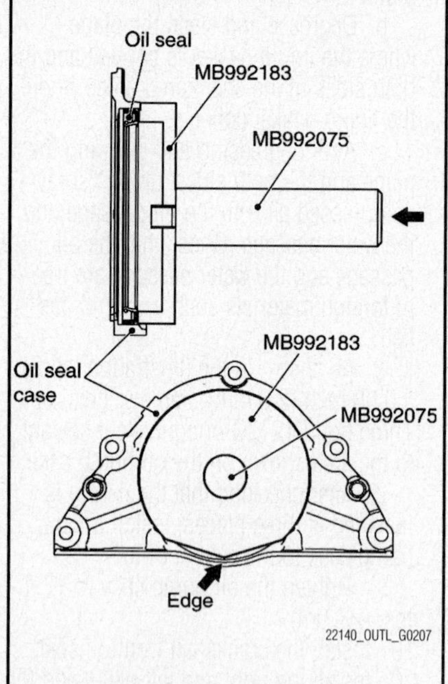

Fig. 147 Using special tool MB992075 and MB992183, press in the crankshaft rear oil seal up to the cylinder block assembly end surface—3.0L engine

To install:

5. Install a new rear main seal.
   a. Apply a small amount of engine oil to the entire inner diameter of the oil seal lip.
   b. Using special tool MD998718, press in the crankshaft rear oil seal up to the cylinder block assembly end surface.
6. Install the flywheel and flywheel adapter plate.
   a. Remove the sealant, the engine oil, and other adhering materials from the drive plate and adapter plate installation face, the crankshaft screw hole and drive plate bolts.
   b. Install the drive plate and adapter plate to the crankshaft.
   c. Use special tool MB991883 for 2.4L engine or MD998781 for 3.0L engine to hold the drive plate and adapter plate in the same manner as removal.
   d. Apply a small amount of engine oil to the screw holes of the crankshaft and the bearing surface of the drive plate bolts and the adapter plate bolts.
   e. Apply Three bond 1324 or equivalent sealant to the thread of the drive plate bolts and the flywheel bolts.
   f. Tighten drive plate bolts to temporary torque 30 ft. lbs. (40 Nm) in a star pattern.
   g. Tighten drive plate bolts to 96 ft. lbs. (130 Nm) in a star pattern.
7. Install the transaxle assembly.

ROCKER ARMS/SHAFTS

REMOVAL & INSTALLATION

3.0L Engine (Only)

Left Bank

See Figures 79, 82, 84, 85, 148.

1. Before servicing the vehicle, refer to the precautions.

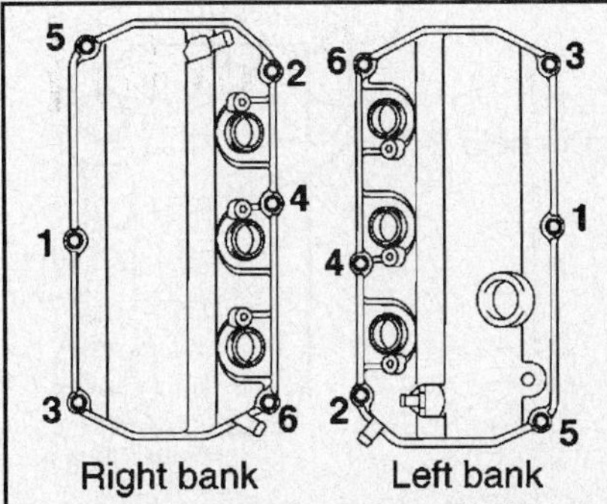

Fig. 148 Tighten the valve cover bolts in the order shown

2. Remove the engine cover.
3. Remove the timing belt.
4. Remove the thermostat housing.
5. Remove the engine oil level gauge and o-ring.
6. Disconnect the blow-by hose.
7. Disconnect the PCV hose.
8. Disconnect the ignition coil connector.
9. Remove the ignition coil.
10. Remove the valve cover and gasket.
11. Remove the rocker arm, shaft and the lash adjuster assembly (exhaust side).
   a. Install special tool MD998443 as shown so that the lash adjusters will not fall out.

✻✻ WARNING

Never disassemble the rocker arm and shaft assembly.

   b. Loosen the rocker arm and shaft assembly mounting bolt, and then remove the rocker arm and shaft assembly with the bolt still attached.
12. Remove the rocker arm and shaft assembly (intake side).

To install:

13. Install the intake side rocker arm and shaft assembly so that the 0.22 inch (5.5 mm) holes of rocker arm shaft face the cylinder head side. Tighten the intake side rocker arm shaft mounting bolts to 23 ft. lbs. (31 Nm).
14. Install the exhaust side rocker arm, shaft and lash adjuster assembly so that the notch of rocker arm shaft is located as shown.
   a. Check that the identification mark of exhaust side rocker shaft cap is located as shown.
   b. Tighten the exhaust side rocker arm shaft mounting bolts to 9 ft. lbs. (13 Nm).
   c. Remove special tool MD998443.
15. Install the valve cover and gasket.
   a. Tighten the bolts in the order shown to 74 inch lbs. (8 Nm).
16. Install the ignition coil.
17. Connect the ignition coil connector.
18. Connect the PCV hose.
19. Connect the blow-by hose.
20. Install the engine oil level gauge and o-ring.
21. Install the thermostat housing.
22. Install the timing belt.
23. Install the engine cover.
24. Change the engine oil.
25. Start the engine and check for leaks.

Right Bank

See Figures 79, 87, 84, 85 and 86, 148, 169.

1. Before servicing the vehicle, refer to the precautions.
2. Remove the intake manifold plenum.
3. Remove the timing belt.
4. Remove the thermostat housing.
5. Disconnect the blow-by hose.
6. Disconnect the breather hose.
7. Disconnect the ignition coil connector.
8. Remove the ignition coil.
9. Remove the valve cover and gasket.
10. Remove the rocker arm, shaft and the lash adjuster assembly (exhaust side).
   a. Install special tool MD998443 as shown so that the lash adjusters will not fall out.

✻✻ WARNING

Never disassemble the rocker arm and shaft assembly.

   b. Loosen the rocker arm and shaft assembly mounting bolt, and then remove the rocker arm and shaft assembly with the bolt still attached.

To install:

11. Install the intake side rocker arm and shaft assembly so that the 0.22 inch (5.5 mm) holes of rocker arm shaft face the cylinder head side. Tighten the intake side rocker arm shaft mounting bolts to 23 ft. lbs. (31 Nm).
12. Install the exhaust side rocker arm, shaft and lash adjuster assembly so that the notch of rocker arm shaft is located as shown.
   a. Check that the identification mark of exhaust side rocker shaft cap is located as shown.
   b. Tighten the exhaust side rocker arm shaft mounting bolts to 9 ft. lbs. (13 Nm).
   c. Remove special tool MD998443.
13. Install the valve cover and gasket
   a. Tighten the bolts in the order shown to 74 inch lbs. (8 Nm).
14. Install the ignition coil.
15. Disconnect the ignition coil connector.
16. Disconnect the breather hose.
17. Disconnect the blow-by hose.
18. Install the thermostat housing.
19. Install the timing belt.
20. Install the intake manifold plenum.
21. Change the engine oil.
22. Start the engine and check for leaks.

TIMING BELT AND SPROCKETS

REMOVAL & INSTALLATION

See Figures 149 through 155.

1. Before servicing the vehicle, refer to the precautions.

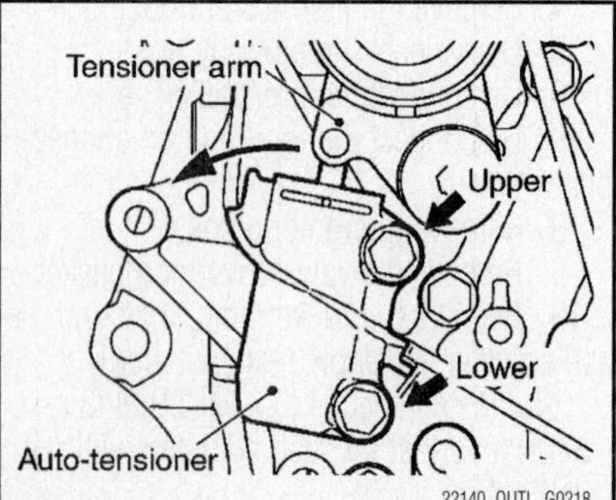

Fig. 149 Loosen the lower tightening bolt of the auto-tensioner slowly and slide the auto-tensioner slightly. Remove the rod from the tensioner arm. Remove the lower tightening bolt of the auto-tensioner

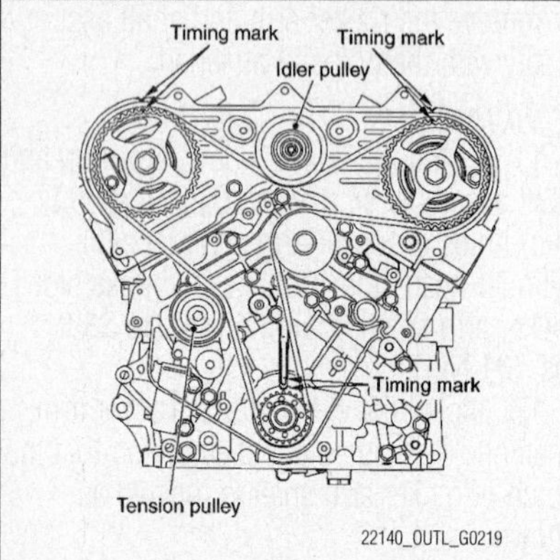

Fig. 150 Loosen the lower tightening bolt of the auto-tensioner slowly and slide the auto-tensioner slightly. Remove the rod from the tensioner arm. Remove the lower tightening bolt of the auto-tensioner

2. Remove the engine cover.
3. Remove the engine under cover and side cover.
4. Remove the drive belt auto-tensioner.
5. Remove the alternator drive belt.

WARNING

When the alternator drive belt is reused, draw an arrow indicating the rotating direction on the back of the belt using chalk to install the same direction.

a. Turn the drive belt auto-tensioner to counterclockwise, and insert the L-shaped hexagon wrench to the auto-tensioner hole in order to fix the auto-tensioner.

b. Remove the alternator drive belt.

c. Remove the drive belt auto-tensioner.

6. Remove the power steering tensioner pulley bracket.
7. Remove the power steering oil pump drive belt.
8. Remove the crankshaft pulley.

a. Use special tools MB990767 and MD998719 to remove the crankshaft pulley from the crankshaft.

9. Remove the timing belt front upper covers.
10. Remove the timing belt lower cover
11. Remove the engine mounting bracket.
12. Remove the engine support bracket.
13. Remove the front flange.
14. Remove the timing belt auto-tensioner

a. Remove the upper tightening bolt of the auto-tensioner.

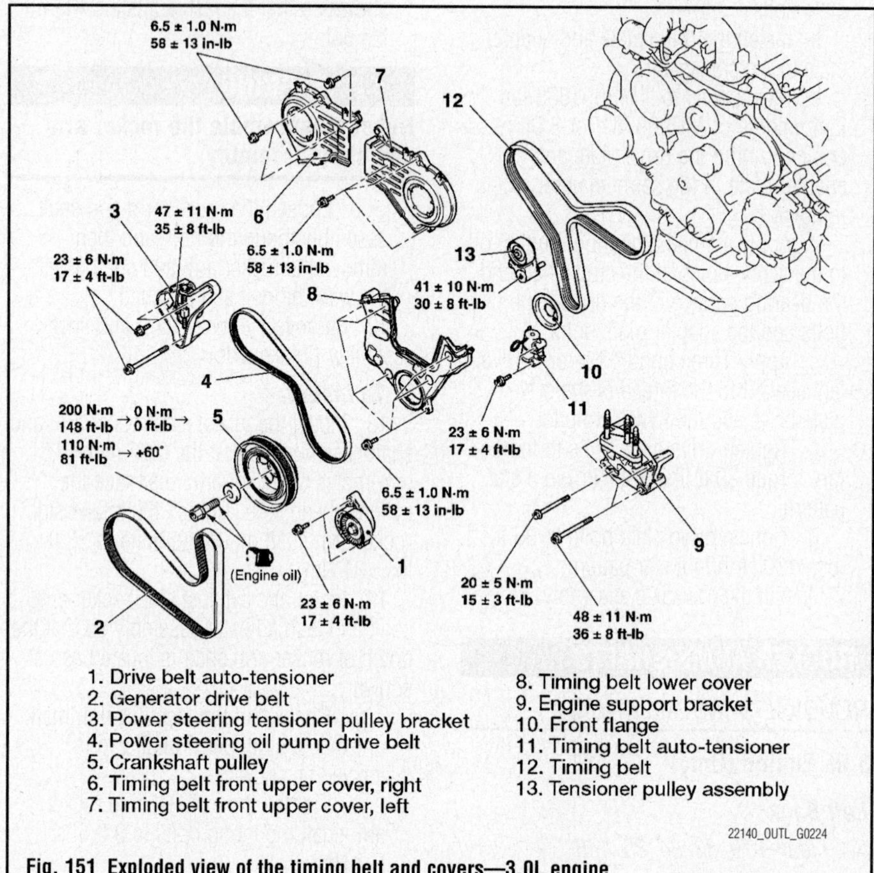

Fig. 151 Exploded view of the timing belt and covers—3.0L engine

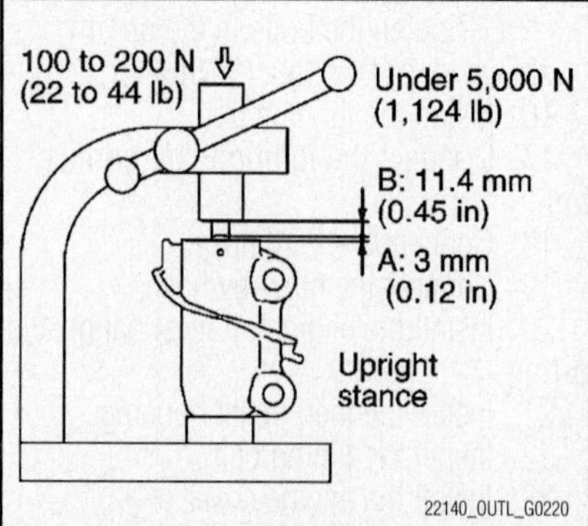

Fig. 152 Set the auto-tensioner as shown and press the rod slowly down to the lowest point "A"

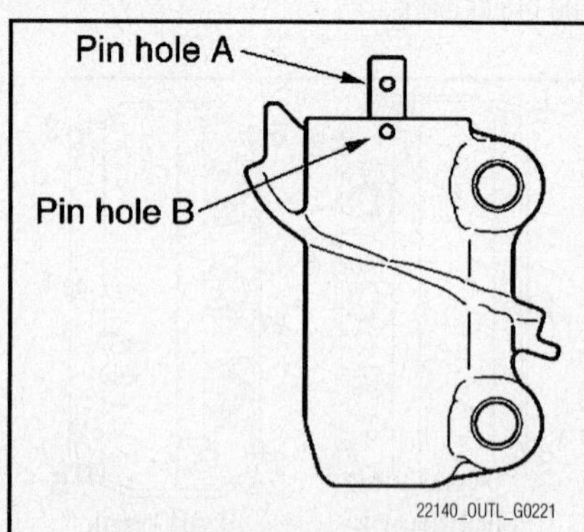

Fig. 153 Slowly close the vice to force the rod in until the hole (A) of the rod is lined up with set hole (B) of the cylinder

⁂ WARNING

The auto-tensioner rotates centering on the flange bolt due to the rod thrust, so please make sure your finger is not trapped.

b. Loosen the lower tightening bolt of the auto-tensioner slowly and slide the auto-tensioner slightly. Remove the rod from the tensioner arm.

c. Remove the lower tightening bolt of the auto-tensioner.

15. Remove the timing belt.

⁂ WARNING

Never turn the crankshaft counterclockwise.

a. Turn the crankshaft clockwise to align each timing mark and to set the number 1 cylinder to compression top dead center.

b. If the timing belt is to be reused, chalk an arrow on the flat side of the belt, indicating the clockwise direction.

c. Loosen the center bolt of the tensioner pulley, then remove the timing belt.

16. Remove the tensioner pulley.

To install:

⁂ WARNING

Always bleed the auto-tensioner of air before installing the auto-tensioner.

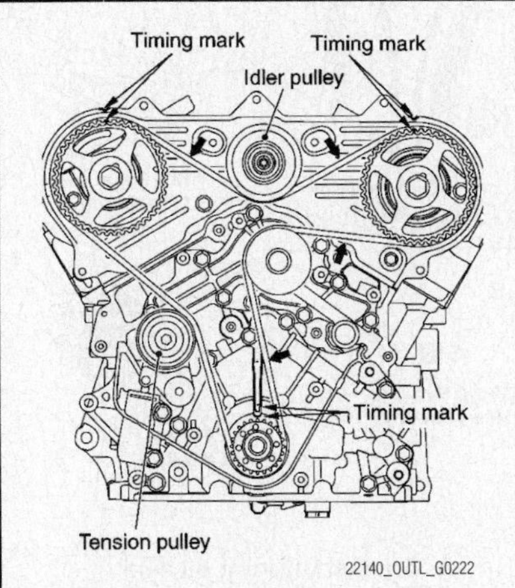

Fig. 154 Align the timing marks on the camshaft sprockets with those on the timing belt rear cover and the timing mark on the crankshaft sprocket with that on the engine block as shown

17. Bleed the auto-tensioner.

a. Always use the vertical press and put the auto-tensioner vertically.

⁂ WARNING

Do not apply the load of 1,124 pound (5,000 N) or more to the rod. Do not press the rod beyond Dimension "A" shown.

b. Set the auto-tensioner as shown.

c. Press the rod slowly down to the lowest point "A" shown.

d. Repeat the procedure 2 three times.

e. While the rod is projected at the point "B" shown in the illustration, push the rod with 22–44 pound (100–200 N). Check the enough stiffness. If the stiffness is not enough, replace the auto-tensioner.

f. Press down the rod slowly. Put the pin through the hole and secure it.

18. Insert the pin into the rod of the auto-tensioner under the following procedures.

a. Put the auto-tensioner vertically to the vertical press not to be in the sideway direction.

b. Slowly close the vice to force the rod in until the hole (A) of the rod is lined up with set hole (B) of the cylinder.

c. Insert a pin into the set holes.

d. Remove the auto-tensioner from the vice.

19. Install the timing belt and auto-tensioner.

a. Align the timing marks on the camshaft sprockets with those on the timing belt rear cover and the timing mark on the crankshaft sprocket with that on the engine block as shown.

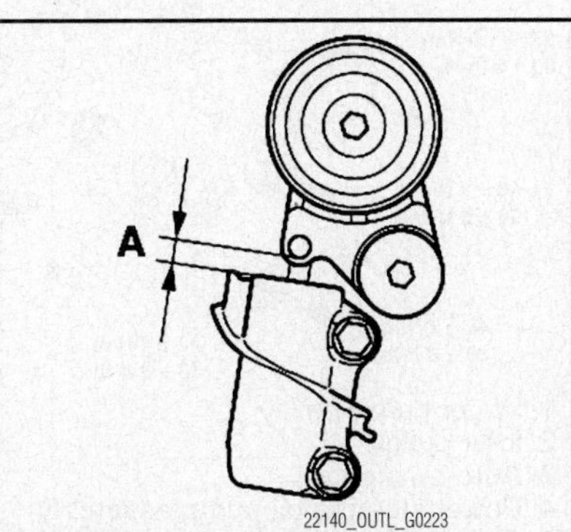

Fig. 155 Wait for at least five minutes, then check that the auto-tensioner pushrod extends within 0.36–0.52 inch (9.1–13.4 mm)

⁂ WARNING

The camshaft sprocket (right bank) can turn easily due to the spring force applied, so be careful not to get your fingers caught.

b. Install the timing belt in the following order so that there is no deflection in the timing belt between each sprocket and pulley.

- Crankshaft sprocket
- Water pump pulley
- Camshaft sprocket (Left bank)
- Idler pulley
- Camshaft sprocket (Right bank)
- Tensioner pulley

c. Turn the camshaft sprocket (Right bank) counterclockwise until the tension side of the timing belt is firmly stretched. Check all the timing marks again.

d. Use special tool MD998716 to turn the crankshaft 1/4 turn counterclockwise, then turn it again clockwise until the timing marks are aligned.

e. Remove the setting pin that has been inserted into the auto-tensioner.

f. Turn the crankshaft clockwise twice to align the timing marks.

g. Wait for at least five minutes, then check that the auto-tensioner pushrod extends within 0.36–0.52 inch (9.–13.4 mm).

h. If not, repeat the operation.

i. Check again that the timing marks of the sprockets are aligned.

20. Install the front flange.

21. Install the engine support bracket.

22. Install the engine mounting bracket.

23. Install the timing belt lower cover. Tighten to 58 inch lbs. (7 Nm).

24. Install the timing belt front upper covers. Tighten to 58 inch lbs. (7 Nm).

25. Install the crankshaft pulley.

a. Use special tools MB990767 and MD998719 to install the crankshaft pulley.

26. Install the power steering oil pump drive belt.

27. Install the power steering tensioner pulley bracket.

28. Install the alternator drive belt.

29. Install the drive belt auto-tensioner.

30. Install the engine under cover and side cover.

31. Install the engine cover.

TIMING CHAIN COVER AND SEAL

REMOVAL & INSTALLATION

See Figures 156 through 161.

1. Before servicing the vehicle, refer to the precautions.
2. Remove the engine room under cover and engine side.

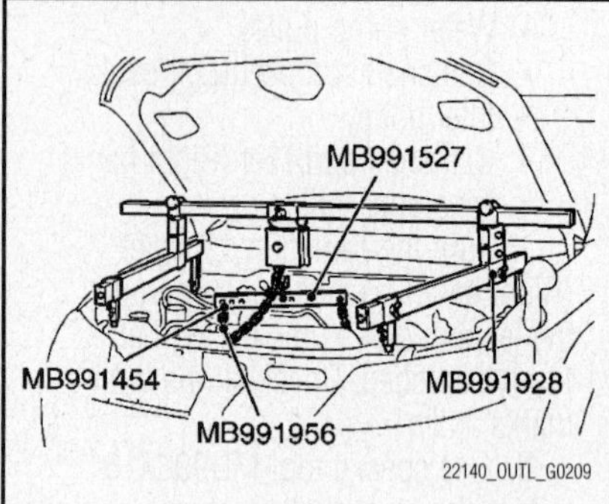

Fig. 156 Install a special tool for holding the engine and transaxle assembly

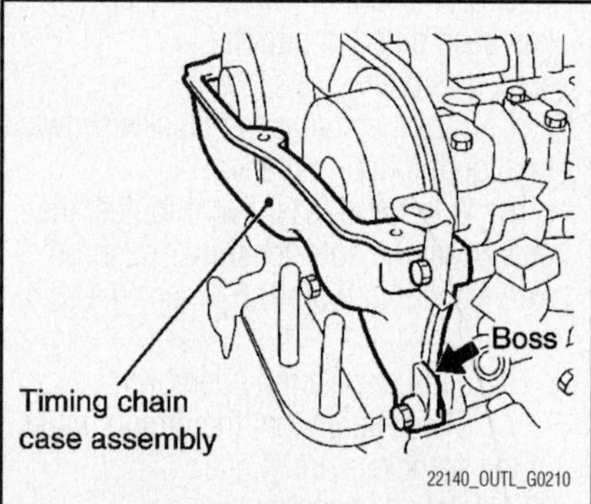

Fig. 157 Slightly pry the boss of the timing chain case assembly using a flat-tipped screwdriver (-), and remove the timing chain case assembly

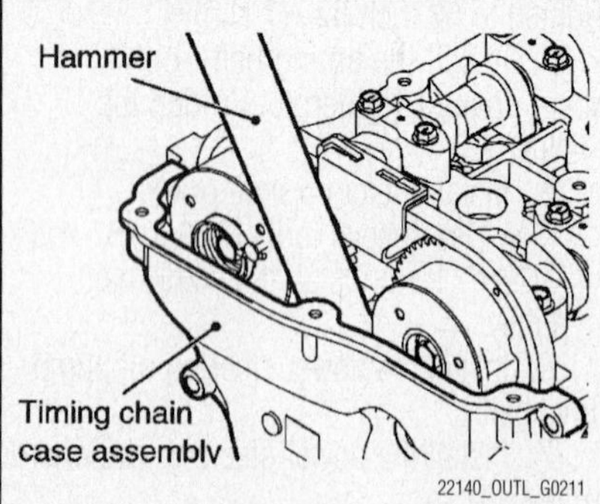

Fig. 158 If the sealant cannot be peeled off easily, insert a wooden hammer shank into the timing chain case assembly inside as shown in the figure, pry slightly, and remove the timing chain case assembly

3. Drain the engine oil.
4. Remove the valve cover.
5. Remove the engine oil pan.
6. Remove the crankshaft pulley.
   a. When removing the crankshaft pulley, slightly loosen the water pump pulley mounting bolts before removal of the drive belt.
7. Remove the water pump pulley.
8. Remove the idler pulley.
9. Remove the auto-tensioner.
10. Remove the power steering oil pump.
    a. With the hose installed, remove the power steering oil pump assembly from the bracket.
    b. Tie the removed power steering oil pump assembly with a string at a position where it will not interfere with the removal and installation of valve timing chain.
11. Install the engine and transaxle assembly holding fixture.
    a. If special tool engine hanger MB991928 is used:
    b. Set the foot of the special tools (MB991930, MB991932 and MB991933) as shown.
    c. Slide the slide bracket (HI) to adjust the engine hanger balance.
    d. Install special tool MB991956 to the cylinder head, and set special tool MB991527 and the chains of special tool MB991454 to the engine assembly to hold the engine and transaxle assembly.
    e. If special tool engine hanger MB991895 is used:
    f. Place a rag between special tool MB991895 and the windshield to prevent the special tool MB991895 from interfering with the windshield.
    g. Set the foot of special tool MB991895 as shown.
    h. Slide the foot to adjust the engine hanger balance.
    i. Install special tool MB991956 to the cylinder head, and set special tool MB991527 and the chains of special tool MB991454 to the engine assembly to hold the engine and transaxle assembly.
12. Remove the engine mounting insulator.
13. Remove the cylinder block engine front mounting bracket
14. Remove the gasket.
15. Remove the timing chain case.
    a. If the adhesive strength of sealant on the timing chain case assembly is so strong that the boss may be damaged by peeling off, do not peel it off forcibly.
    b. After removing the timing chain case assembly mounting bolts, slightly

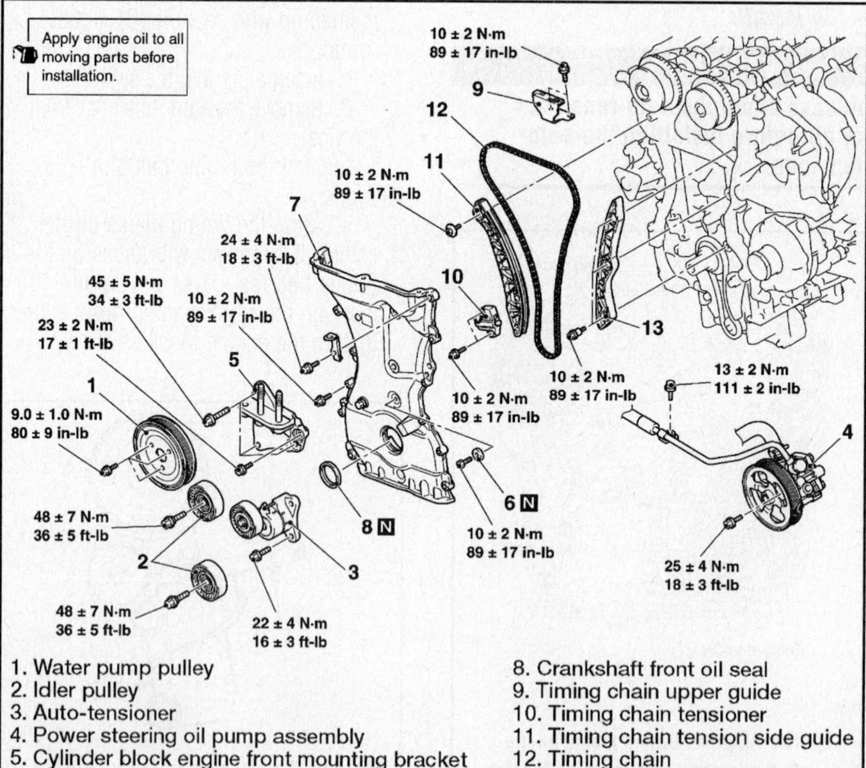

Fig. 159 Exploded view of timing chain and cover assembly—2.4L engine

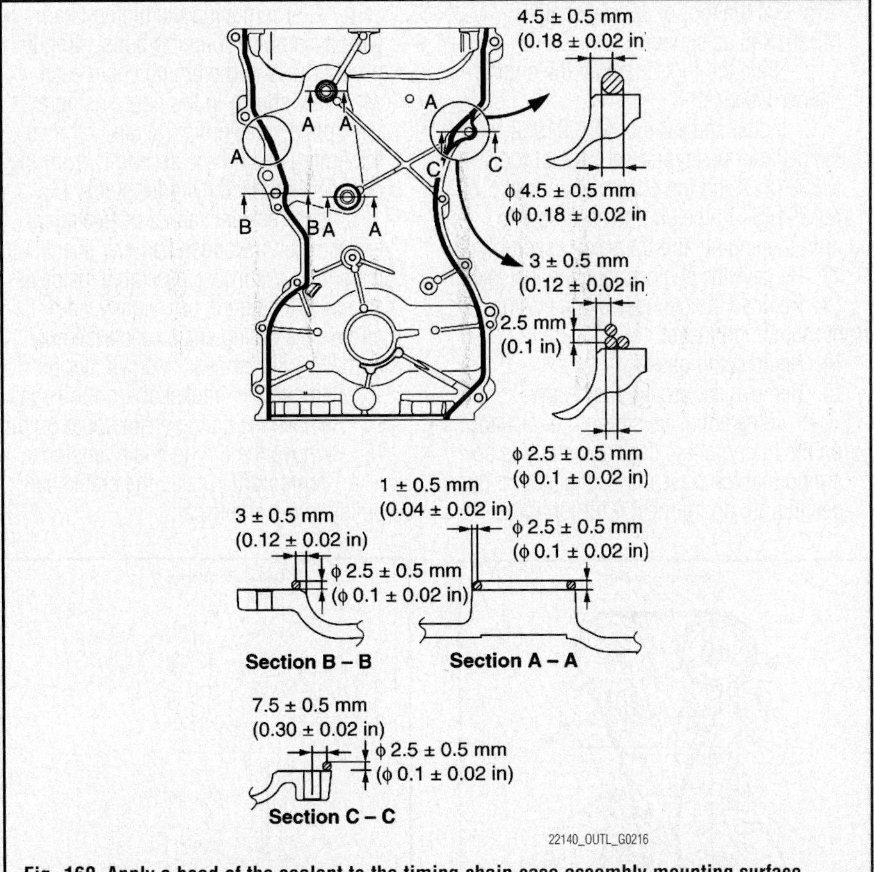

Fig. 160 Apply a bead of the sealant to the timing chain case assembly mounting surface as shown

pry the boss of the timing chain case assembly shown in the figure using a flat-tipped screwdriver (-), and remove the timing chain case assembly from the cylinder head and cylinder block.

c. If the sealant cannot be peeled off easily, insert a wooden hammer shank into the timing chain case assembly inside as shown in the figure, pry slightly, and remove the timing chain case assembly from the cylinder head and cylinder block.

16. Remove the crankshaft front oil seal.

To install:

17. Install the crankshaft front oil seal.

a. Apply a small amount of engine oil to the entire inner diameter of the crankshaft front oil seal lip.

WARNING

When installing the crankshaft front oil seal, be careful to avoid damage to the crankshaft front oil seal.

b. Using a seal installer, press in the crankshaft front oil seal up to the chamfered surface of timing chain case.

18. Install the timing chain case.

WARNING

Be sure to remove the sealant inside the mounting holes and the O-ring grooves.

After degreasing with white gasoline or the like, check that there is no oil on the surface where the sealant is applied. After degreasing, never touch the degreased area with fingers.

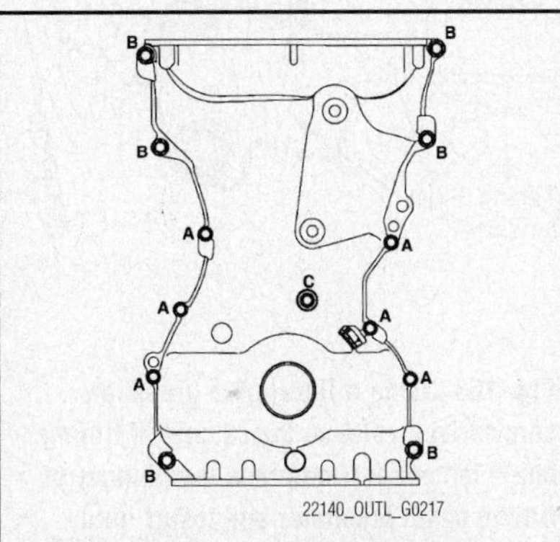

Fig. 161 Insert the bolts to the timing chain case assembly as shown, and tighten them to the specified torque

a. Remove sealant from the timing chain case assembly and the timing chain case assembly mounting surface of the cylinder block and the cylinder head, and degrease the surface where the sealant is applied by white gasoline or the like.

b. Remove all the sealant adhering to the gasket between the cylinder head and cylinder block (three-surface aligned part). Then, degrease the surfaces with the white gasoline.

c. As for the three-surface aligned part that is indicated in Step 2 above, the engine oil oozes from the cylinder head gasket. Thus, quickly apply the sealant to it after degreasing.

d. Apply a bead of Three bond 1217G or equivalent sealant to the timing chain case assembly mounting surface as shown. The bead diameter should be 0.1 ± 0.02 inch (2.5 ± 0.5 mm). Overlap the part "A" with the diameter of 0.18 ± 0.02 inch (4.5 ± 0.5 mm) or 0.1 ± 0.02 inch (2.5 ± 0.5 mm) as shown, and apply the sealant.

WARNING

If the sealant contacts any other part during installation of the timing chain case assembly, apply sealant again before installing the timing chain case assembly. Do not apply oil or water to the sealant-applied area or start up the engine within 2 hours after the installation of the timing chain case assembly.

e. Install the timing chain case assembly to the cylinder block and cylinder head so that the sealant does not contact other parts.

➡**Install the timing chain case assembly within 3 minutes after the application of sealant.**

f. Insert the bolts to the timing chain case assembly as shown, and tighten them as follows:

- Flange bolt (A)–M6 × 25–Tighten to 89 inch lbs. (10 Nm)
- Flange bolt (B)–M8 × 28–Tighten to 18 ft. lbs. (24 Nm)
- Bolt (C)–M6 × 25–Tighten to 89 inch lbs. (10 Nm)

19. Install the gasket.
20. Install the cylinder block engine front mounting bracket
21. Install the engine mounting insulator.
22. Install the engine and transaxle assembly holding fixture.
23. Install the power steering oil pump.
24. Install the auto-tensioner.

25. Install the idler pulley.
26. Install the water pump pulley.
    a. Temporarily tighten the water pump pulley mounting bolts.
    b. Then, tighten them to 80 inch lbs. (9 Nm) after the installation of drive belt.
27. Install the crankshaft pulley.
28. Install the engine oil pan.
29. Install the valve cover.
30. Drain the engine oil.
31. Install the engine room under cover and engine side.

TIMING CHAIN AND SPROCKETS

REMOVAL & INSTALLATION

See Figures 156 through 165.

1. Before servicing the vehicle, refer to the precautions.
2. Remove the engine room under cover and engine side.
3. Drain the engine oil.
4. Remove the valve cover.
5. Remove the engine oil pan.
6. Remove the crankshaft pulley.
    a. When removing the crankshaft pulley, slightly loosen the water pump pulley mounting bolts before removal of the drive belt.
7. Remove the water pump pulley.
8. Remove the idler pulley.
9. Remove the auto-tensioner.
10. Remove the power steering oil pump.
    a. With the hose installed, remove the power steering oil pump assembly from the bracket.
    b. Tie the removed power steering oil pump assembly with a string at a position where it will not interfere with the removal and installation of valve timing chain.
11. Install the engine and transaxle assembly holding fixture.
    a. If special tool engine hanger MB991928 is used:
    b. Set the foot of the special tools (MB991930, MB991932 and MB991933) as shown.
    c. Slide the slide bracket (HI) to adjust the engine hanger balance.
    d. Install special tool MB991956 to the cylinder head, and set special tool MB991527 and the chains of special tool MB991454 to the engine assembly to hold the engine and transaxle assembly.
    e. If special tool engine hanger MB991895 is used:
    f. Place a rag between special tool MB991895 and the windshield to prevent the special tool MB991895 from interfering with the windshield.
    g. Set the foot of special tool MB991895 as shown.
    h. Slide the foot to adjust the engine hanger balance.
    i. Install special tool MB991956 to the cylinder head, and set special tool MB991527 and the chains of special tool MB991454 to the engine assembly to hold the engine and transaxle assembly.
12. Remove the engine mounting insulator.
13. Remove the cylinder block engine front mounting bracket
14. Remove the gasket.
15. Remove the timing chain case.
    a. If the adhesive strength of sealant on the timing chain case assembly is so strong that the boss may be damaged by peeling off, do not peel it off forcibly.
    b. After removing the timing chain case assembly mounting bolts, slightly pry the boss of the timing chain case assembly shown in the figure using a flat-tipped screwdriver (-), and remove the timing chain case assembly from the cylinder head and cylinder block.
    c. If the sealant cannot be peeled off easily, insert a wooden hammer shank into the timing chain case assembly inside as shown in the figure, pry slightly, and remove the timing chain case assembly from the cylinder head and cylinder block.
16. Remove the crankshaft front oil seal.
17. Remove the timing chain upper guide.
18. Remove the timing chain tensioner.
    a. Temporarily install the crankshaft pulley to the crankshaft.

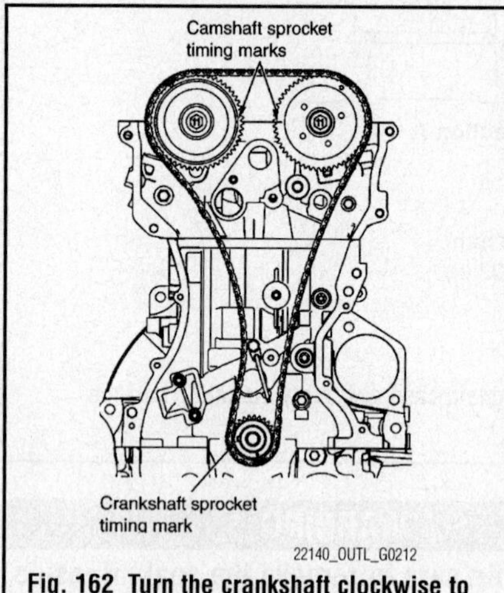

Fig. 162 Turn the crankshaft clockwise to align the sprocket timing marks as shown in the figure and set the cylinder No. 1 to the top dead center of compression stroke

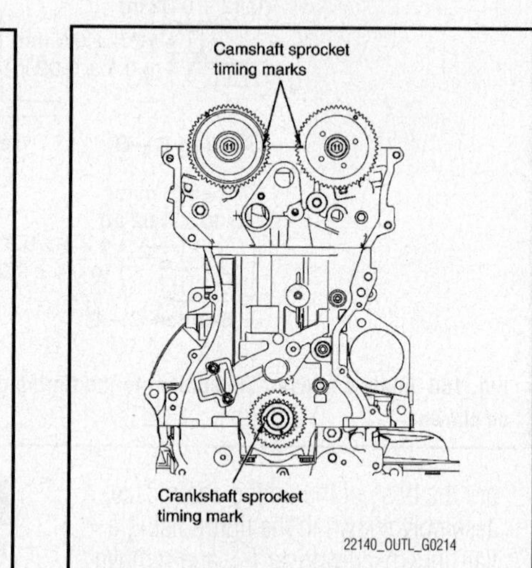

Fig. 164 Set the timing marks of the camshaft sprockets and the crankshaft sprocket as shown

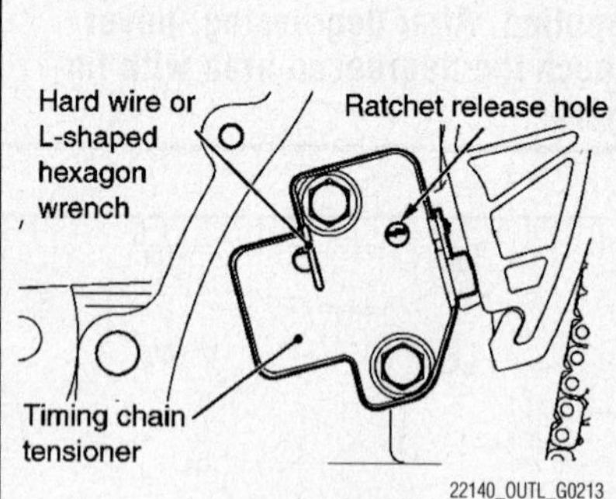

Fig. 163 Using a flat-tipped precision screwdriver, release the ratchet of timing chain tensioner. Compress the plunger of timing chain tensioner and insert hard wire (such as piano wire) or an L-shaped hexagon wrench to hold the plunger of the timing chain tensioner

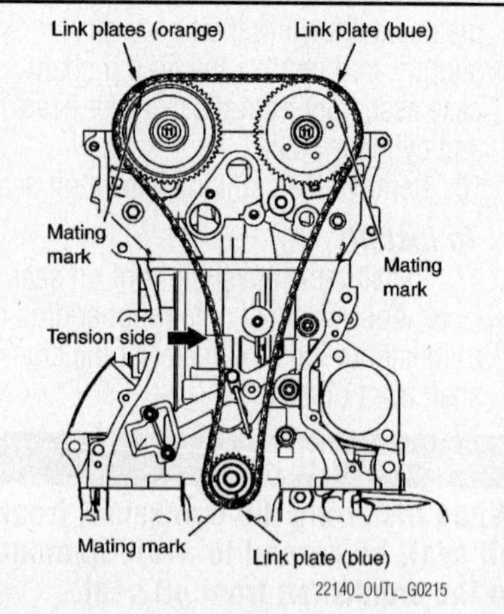

Fig. 165 Align each sprocket timing chain mating mark with the link plate (orange or blue) of timing chain to avoid slack of the timing chain tension side, and install the timing chain to the sprockets

b. Turn the crankshaft clockwise.

c. Turn the crankshaft clockwise to align the sprocket timing marks as shown in the figure and set the cylinder No. 1 to the top dead center of compression stroke.

d. At this time, it is not necessary that the link plate (orange or blue) of the timing chain always aligns with each sprocket timing mark.

e. Remove the crankshaft pulley installed temporarily.

f. Using a flat-tipped precision screwdriver, release the ratchet of timing chain tensioner.

g. Compress the plunger of timing chain tensioner and insert hard wire (such as piano wire) or the L-shaped hexagon wrench (1.5 mm[0.05 inch]) to fix the plunger of the timing chain tensioner.

h. Remove the timing chain tensioner.

19. Remove the timing chain tension side guide.

20. Remove the timing chain.

21. Remove the timing chain loose side guide.

To install:

22. Install the timing chain loose side guide.

23. Install the timing chain.

a. Set the timing marks of the camshaft sprockets and the crankshaft sprocket as shown in the figure.

b. Align each sprocket timing chain mating mark with the link plate (orange or blue) of timing chain to avoid slack of the timing chain tension side, and install the timing chain to the sprockets.

24. Install the timing chain tension side guide.

25. Install the timing chain tensioner.

a. Check that the sprocket timing chain mating marks align with the link plates (orange or blue) of the timing chain, and install the timing chain tensioner to the cylinder block.

b. Remove the hard wire or L-shaped hexagon wrench fixing the plunger of the timing chain tensioner to apply tension to the timing chain.

26. Install the timing chain upper guide.

27. Install the crankshaft front oil seal.

a. Apply a small amount of engine oil to the entire inner diameter of the crankshaft front oil seal lip.

✱✱ WARNING

When installing the crankshaft front oil seal, be careful to avoid damage to the crankshaft front oil seal.

b. Using a seal installer, press in the crankshaft front oil seal up to the chamfered surface of timing chain case.

28. Install the timing chain case.

✱✱ WARNING

Be sure to remove the sealant inside the mounting holes and the O-ring grooves. After degreasing with white gasoline or the like, check that there is no oil on the surface where the sealant is applied. After degreasing, never touch the degreased area with fingers.

a. Remove sealant from the timing chain case assembly and the timing chain case assembly mounting surface of the cylinder block and the cylinder head, and degrease the surface where the sealant is applied by white gasoline or the like.

b. Remove all the sealant adhering to the gasket between the cylinder head and cylinder block (three-surface aligned part). Then, degrease the surfaces with the white gasoline.

c. As for the three-surface aligned part that is indicated in Step 2 above, the engine oil oozes from the cylinder head gasket. Thus, quickly apply the sealant to it after degreasing.

d. Apply a bead of Three bond 1217G or equivalent sealant to the timing chain case assembly mounting surface as shown. The bead diameter should be 0.1 ± 0.02 inch (2.5 ± 0.5 mm). Overlap the part "A" with the diameter of 0.18 ± 0.02 inch (4.5 ± 0.5 mm) or 0.1 ± 0.02 inch (2.5 ± 0.5 mm) as shown, and apply the sealant.

✱✱ WARNING

If the sealant contacts any other part during installation of the timing chain case assembly, apply sealant again before installing the timing chain case assembly. Do not apply oil or water to the sealant-applied area or start up the engine within 2 hours after the installation of the timing chain case assembly.

e. Install the timing chain case assembly to the cylinder block and cylinder head so that the sealant does not contact other parts.

➡Install the timing chain case assembly within 3 minutes after the application of sealant.

f. Insert the bolts to the timing chain case assembly as shown, and tighten them as follows:

- Flange bolt (A)–M6 × 25–Tighten to 89 inch lbs. (10 Nm)
- Flange bolt (B)–M8 × 28–Tighten to 18 ft. lbs. (24 Nm)
- Bolt (C)–M6 × 25–Tighten to 89 inch lbs. (10 Nm)

29. Install the gasket.

30. Install the cylinder block engine front mounting bracket

31. Install the engine mounting insulator.

32. Install the engine and transaxle assembly holding fixture.

33. Install the power steering oil pump.

34. Install the auto-tensioner.

35. Install the idler pulley.

36. Install the water pump pulley.

a. Temporarily tighten the water pump pulley mounting bolts.

b. Then, tighten them to 80 inch lbs. (9 Nm) after the installation of drive belt.

37. Install the crankshaft pulley.

38. Install the engine oil pan.

39. Install the valve cover.

40. Drain the engine oil.

41. Install the engine room under cover and engine side.

VALVE COVERS

REMOVAL & INSTALLATION

2.4L Engine

See Figures 166 through 168.

1. Before servicing the vehicle, refer to the precautions.

2. Remove the engine under cover and engine room side cover.

3. Remove the air cleaner.

4. Remove the strut tower bar .

5. Remove the engine upper cover.

6. Remove the ignition coils.

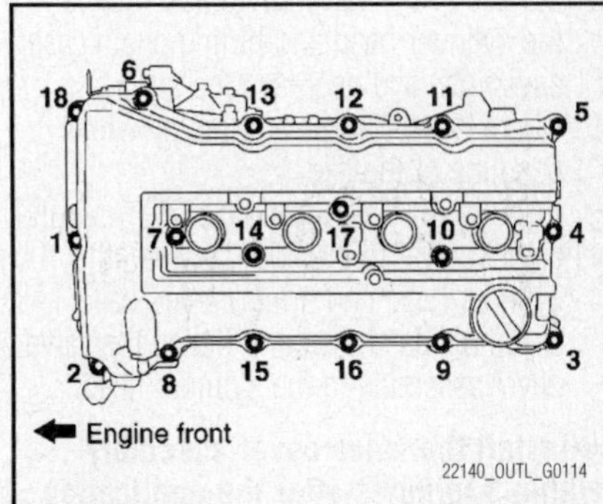

Fig. 166 Loosen the valve cover assembly mounting bolts in the order of number shown

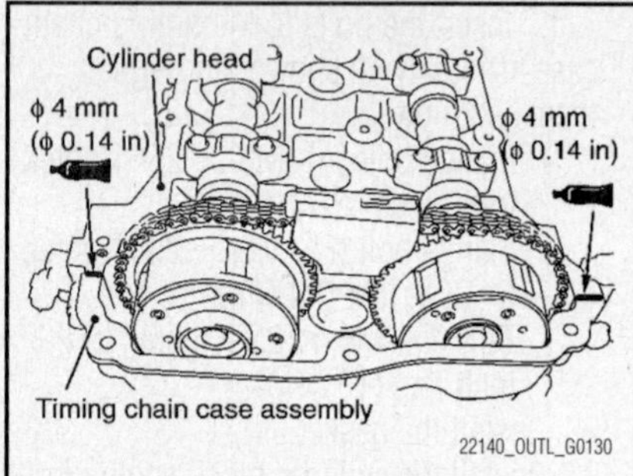

Fig. 167 Apply Three bond 1217G or equivalent sealant to the joint between the cylinder head and timing chain case assembly as shown

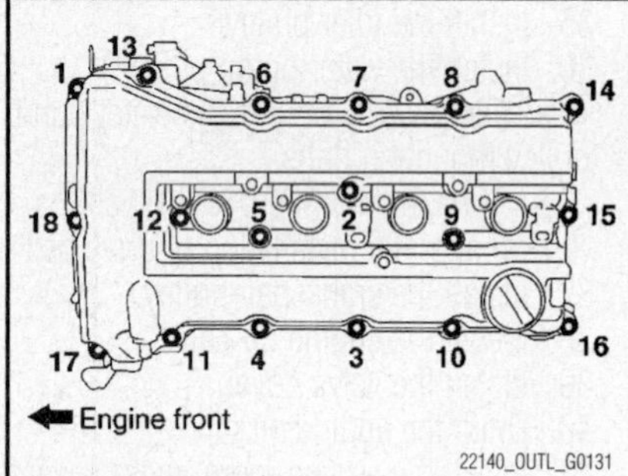

Fig. 168 Tighten the valve cover assembly mounting bolts in the order shown

7. Disconnect the breather hose connection.
8. Disconnect the PCV hose connection.
9. Disconnect the control wiring harness connection.
10. Remove the valve cover assembly
    a. Loosen the valve cover assembly mounting bolts in the order of number shown, and remove the valve cover assembly.
    b. Remove the valve cover gasket.

To install:

11. Install the valve cover assembly
    a. Wipe off the sealant on the mating surface of the valve cover assembly and the cylinder head and timing chain case assembly, and degrease the surface where the sealant is applied by white gasoline or the like.
    b. Apply Three bond 1217G or equivalent sealant to the joint between the cylinder head and timing chain case assembly as shown and install the valve cover assembly to the cylinder head.

➡Install the valve cover assembly within 3 minutes after the application of sealant.

    c. Tighten the valve cover assembly mounting bolts to 27 inch lbs. (3 Nm) in the order shown. Then tighten again to specification in the order shown.
12. Connect the control wiring harness connection.
13. Connect the PCV hose connection.
14. Connect the breather hose connection.
15. Install the ignition coils.
16. Install the engine upper cover.
17. Install the strut tower bar .
18. Install the air cleaner.
19. Install the engine under cover and engine room side cover.

3.0L Engine

Left Bank

See Figure 169.

1. Before servicing the vehicle, refer to the precautions.
2. Remove the engine cover.
3. Remove the timing belt.
4. Remove the thermostat housing.
5. Remove the engine oil level gauge and o-ring.

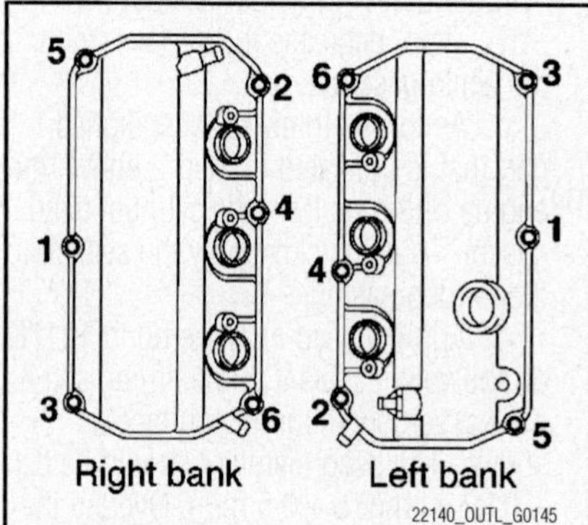

Fig. 169 Tighten the valve cover bolts in the order shown

6. Disconnect the blow-by hose.
7. Disconnect the PCV hose.
8. Disconnect the ignition coil connector.
9. Remove the ignition coil.
10. Remove the valve cover and gasket.

To install:

11. Install the valve cover and gasket.
    a. Tighten the bolts in the order shown to 74 inch lbs. (8 Nm).
12. Install the ignition coil.
13. Connect the ignition coil connector.
14. Connect the PCV hose.
15. Connect the blow-by hose.
16. Install the engine oil level gauge and o-ring.
17. Install the thermostat housing.
18. Install the timing belt.
19. Install the engine cover.
20. Change the engine oil.
21. Start the engine and check for leaks.

Right Bank

See Figure 86, 148, 169.

1. Before servicing the vehicle, refer to the precautions.
2. Remove the intake manifold plenum.
3. Remove the timing belt.
4. Remove the thermostat housing.
5. Disconnect the blow-by hose.
6. Disconnect the breather hose.
7. Disconnect the ignition coil connector.
8. Remove the ignition coil.
9. Remove the valve cover and gasket.

To install:

10. Install the valve cover and gasket.
    a. Tighten the bolts in the order shown to 74 inch lbs. (8 Nm).
11. Install the ignition coil.
12. Disconnect the ignition coil connector.
13. Disconnect the breather hose.
14. Disconnect the blow-by hose.
15. Install the thermostat housing.
16. Install the timing belt.
17. Install the intake manifold plenum.
18. Change the engine oil.
19. Start the engine and check for leaks.

VALVE LASH

ADJUSTMENT

2.4L Engine

See Figure 170.

1. Before servicing the vehicle, refer to the precautions.
2. Remove the valve cover.

➡Perform the valve clearance check and adjustment at the engine cold state.

3. If the engine has been turned, turn the crankshaft clockwise, and align the timing mark on the exhaust camshaft sprocket

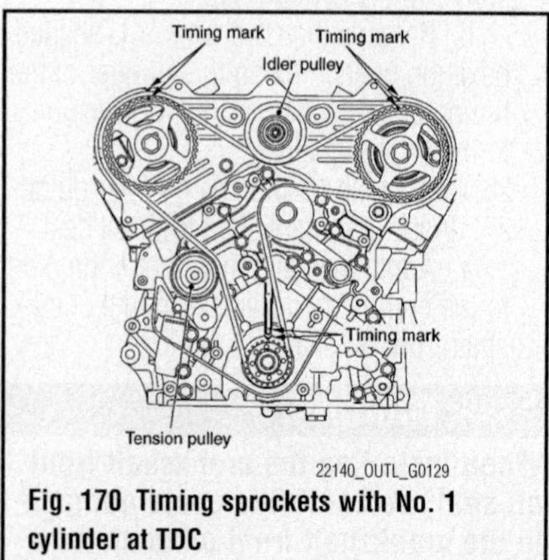

Fig. 170 Timing sprockets with No. 1 cylinder at TDC

against the upper face of the cylinder head. The No.1 cylinder should now be at TDC on the compression stroke.

4. Using a thickness gauge, measure the valve clearance of cylinders No. 1 and 2 on the intake side and cylinders No. 1 and 3 on the exhaust side. If not within specification, make note for the valve clearance adjustment.

5. Turn the crankshaft clockwise 360 degrees, and put the timing mark on the exhaust camshaft sprocket as illustrated. The No.1 cylinder should now be at TDC on the compression stroke.

6. Check the valve clearance of cylinders No. 3 and 4 on the intake side and cylinders No. 2 and 4 on the exhaust side.

7. If the valve clearance is not within specification, remove the camshaft and the valve tappet.

8. Using a micrometer, measure the thickness of the removed valve tappet.

9. Calculate the thickness of the newly installed valve tappet through the following equation.

- A: thickness of newly installed valve tappet
- B: thickness of removed valve tappet
- C: measured valve clearance

10. Calculate the thickness of the valve tappet as follows:

- Intake valve: A = B + [C - 0.20 mm (0.008 inch)]
- Exhaust valve: A = B + [C - 0.30 mm (0.012 inch)]

11. Install the valve tappet selected and put the camshaft in position.

12. After installing the timing chain, measure the valve clearance again.

13. Install the valve cover.

ENGINE PERFORMANCE & EMISSION CONTROLS

ACCELERATOR PEDAL POSITION (APP) SENSOR

LOCATION

See Figure 171.

Refer to the accompanying illustration for Accelerator Pedal Position (APP) sensor location.

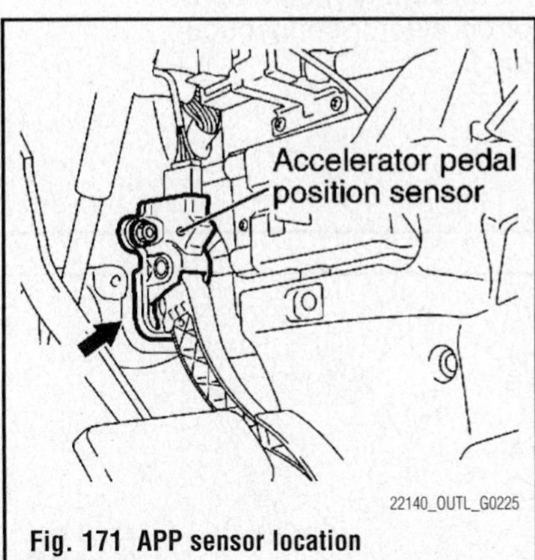

Fig. 171 APP sensor location

OPERATION

A 5 volt power supply is applied on the Accelerator Pedal Position (APP) Sensor (main) power terminal from the PCM. The ground terminal is grounded with PCM. When the accelerator pedal is moved from the idle position to the fully opened position, the resistance between the accelerator pedal position sensor (main) output terminal and ground terminal will increase according to the depression.

REMOVAL & INSTALLATION

CAUTION

The accelerator pedal position sensor should not be moved unnecessarily; it has been precisely adjusted by the manufacturer.

1. Before servicing the vehicle, refer to the precautions.

2. Remove the bottom cover assembly from under the steering wheel.

3. Detach the electrical connector at the sensor.

4. Remove the accelerator pedal assembly from the vehicle.

To install:

5. Installation is the reverse of the removal procedure.

6. Tighten accelerator pedal nuts to 111 inch lbs. (13 Nm)

TESTING

1. Connect a scan tool to the data link connector.

2. Turn the ignition switch **ON**.

3. Check output voltage for the accelerator pedal position sensor.

4. Voltage should be between 0.9–1.1 volts when the pedal is released and 4.0 volts and higher when the accelerator is fully depressed.

5. If voltage is not as specified, replace the sensor.

BAROMETRIC PRESSURE (BARO) SENSOR

LOCATION

See Figure 172.

Refer to the accompanying illustration for Barometric Pressure (BARO) sensor location.

OPERATION

A 5 volt reference and ground is supplied to the Barometric Pressure Sensor from the ECM/PCM. A sensor voltage that is proportional to the atmospheric pressure is sent to the ECM/PCM.

REMOVAL & INSTALLATION

1. Before servicing the vehicle, refer to the precautions.

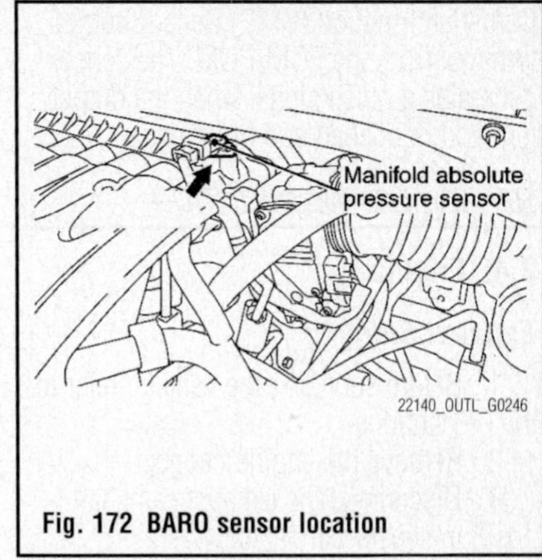

Fig. 172 BARO sensor location

2. Disconnect the negative battery cable.

3. Replacing the sensor requires disconnecting the electrical connector, then carefully removing the lid of the air filter housing.

➡Handle the sensor assembly carefully, protecting it from impact, extremes of temperature and/or exposure to shop chemicals.

To install:

4. Installation is the reverse of the removal procedure.

TESTING

1. Connect a scan tool to the data link connector.

2. Turn the ignition switch **ON**.

- When altitude is 0 foot, pressure should be 29.8 in. Hg (101 kPa).
- When altitude is 1,969 feet, pressure should be 28.1 in. Hg (95 kPa).
- When altitude is 3,937 feet, pressure should be 26.1 in. Hg (88 kPa).
- When altitude is 5,906 feet, pressure should be 23.9 in. Hg (81 kPa).

3. Start the engine.

4. When the engine is idling, pressure should be 4.7 - 10.6 in. Hg (16 - 36 kPa).

5. When the engine is suddenly revved, manifold absolute pressure varies.

6. Turn the ignition switch to the **OFF** position.
7. If not within specifications, replace the Mass Air Flow (MAF) Sensor.

CAMSHAFT POSITION (CMP) SENSOR

LOCATION

See Figures 173 and 174.

Refer to the accompanying illustrations for Camshaft Position (CMP) sensor location.

OPERATION

A 5 volt voltage is applied on the Camshaft Position (CMP) Sensor output terminal from the ECM/PCM. The sensor generates a pulse signal when the output terminal is opened and grounded.

REMOVAL & INSTALLATION

2.4L Engine

Exhaust Side

1. Before servicing the vehicle, refer to the precautions.
2. Remove the engine hanger.
3. Disconnect the exhaust camshaft position sensor connector.
4. Remove the exhaust camshaft position sensor.
5. Remove the O-ring.

To install:

6. Lubricate the O-ring with engine oil and install on the sensor.
7. Install the exhaust camshaft position sensor. Tighten to 98 inch lbs. (11 Nm).
8. Connect the exhaust camshaft position sensor connector.
9. Install the engine hanger.

Intake Side

1. Before servicing the vehicle, refer to the precautions.
2. Remove the air cleaner cover and air cleaner intake hose.
3. Disconnect the control harness clamp.
4. Remove the emission vacuum hose and pipe assembly mounting bolt.
5. Disconnect the intake camshaft position sensor connector.
6. Remove the intake camshaft position sensor.
7. Remove the O-ring.

To install:

8. Lubricate the O-ring with engine oil and install on the sensor.
9. Install the exhaust camshaft position sensor. Tighten to 98 inch lbs. (11 Nm).

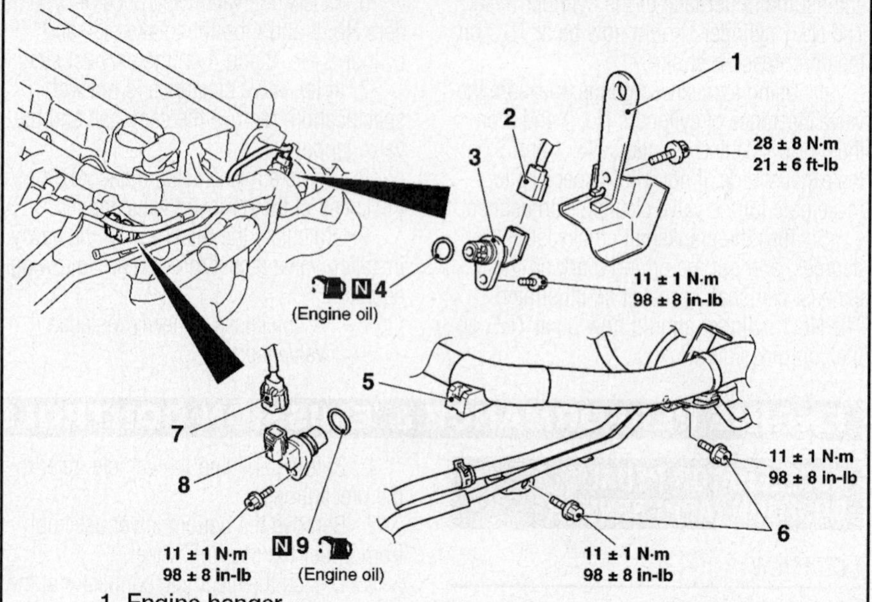

Fig. 173 CMP sensor location—2.4L engine

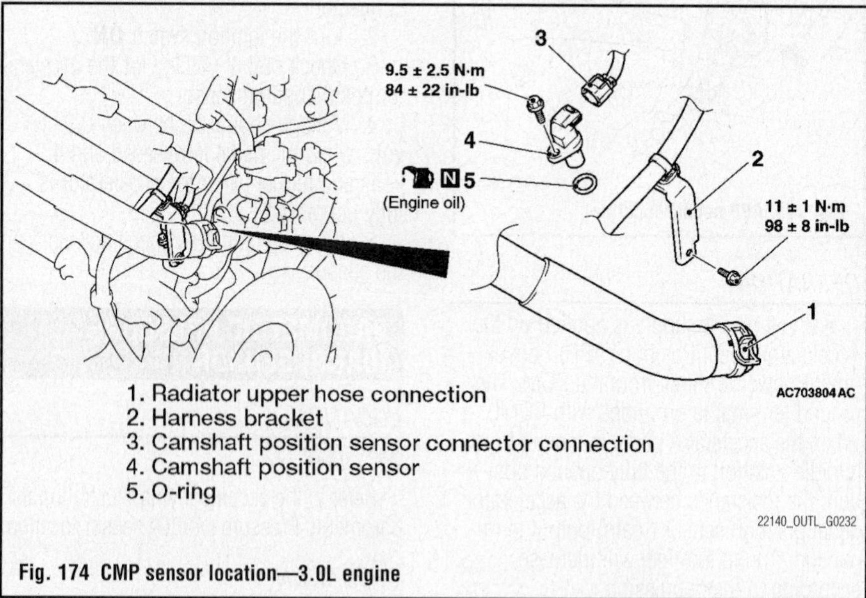

Fig. 174 CMP sensor location—3.0L engine

10. Connect the exhaust camshaft position sensor connector.
11. Install the emission vacuum hose and pipe assembly mounting bolt.
12. Connect the control harness clamp.
13. Install the air cleaner cover and air cleaner intake hose.

3.0L Engine

1. Before servicing the vehicle, refer to the precautions.
2. Drain the engine coolant.
3. Remove the battery and battery tray.
4. Disconnect the radiator upper hose.
5. Remove the harness bracket.

6. Disconnect the camshaft position sensor connector.
7. Remove the camshaft position sensor.
8. Remove the O-ring.

To install:

9. Lubricate the O-ring with engine oil and install on the sensor.
10. Install the exhaust camshaft position sensor. Tighten to 84 inch lbs. (10 Nm).
11. Connect the exhaust camshaft position sensor connector.
12. Install the harness bracket.
13. Connect the radiator upper hose.
14. Install the battery and battery tray.
15. Fill the engine with coolant.

TESTING

See Figures 175 and 176.

1. Disconnect the intake camshaft position sensor connector and connect test harness special tool (MB991709) between the separated connectors. (All terminals should be connected.)
2. Connect the oscilloscope probe to the intake camshaft position sensor side connector terminal No. 3.
3. Start the engine and run at idle.

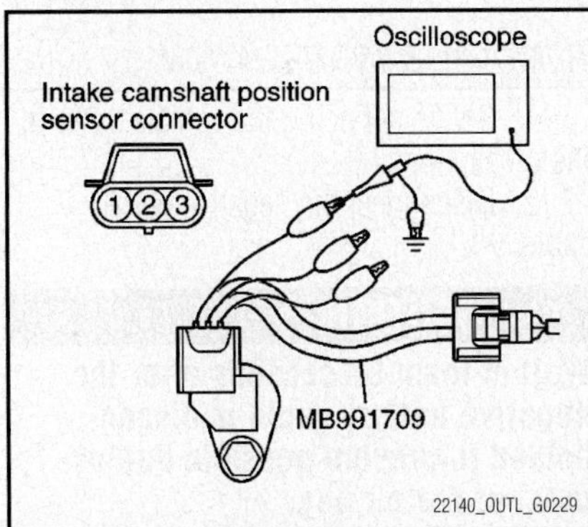

Fig. 175 Disconnect the intake camshaft position sensor connector and connect test harness special tool (MB991709) between the separated connectors

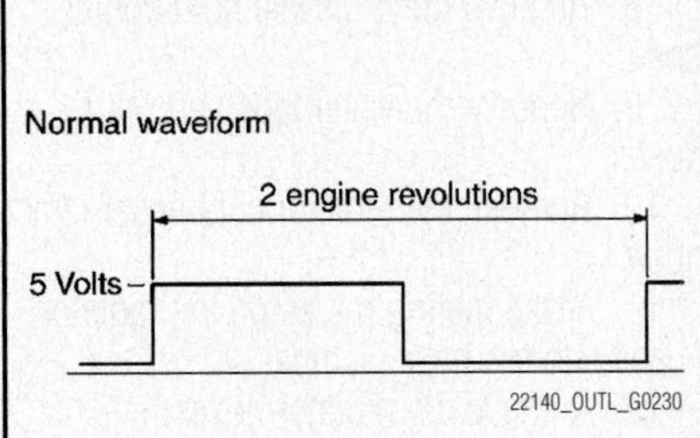

Fig. 176 The oscilloscope waveform should show a pattern similar to the illustration

4. Check the waveform.
5. The waveform should show a pattern similar to the illustration.
6. Turn the ignition switch to the **OFF** position.

CRANKSHAFT POSITION (CKP) SENSOR

LOCATION

See Figures 177 and 178.

Refer to the accompanying illustrations for Crankshaft Position (CKP) sensor location.

OPERATION

A 5 volt voltage is applied on the Crankshaft Position (CKP) Sensor output terminal from the ECM/PCM. The sensor generates a pulse signal when the output terminal is opened and grounded.

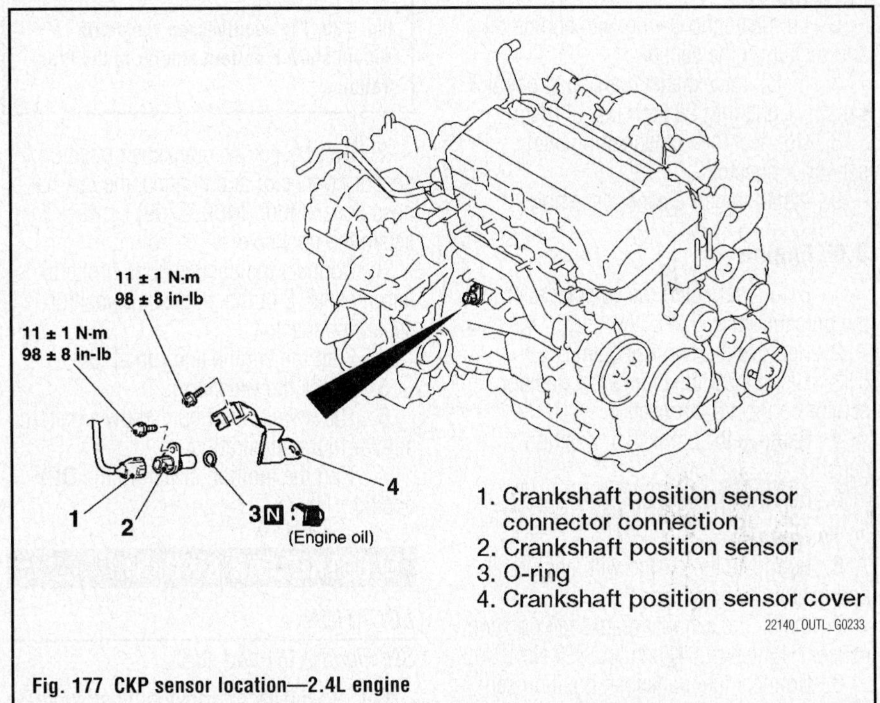

Fig. 177 CKP sensor location—2.4L engine

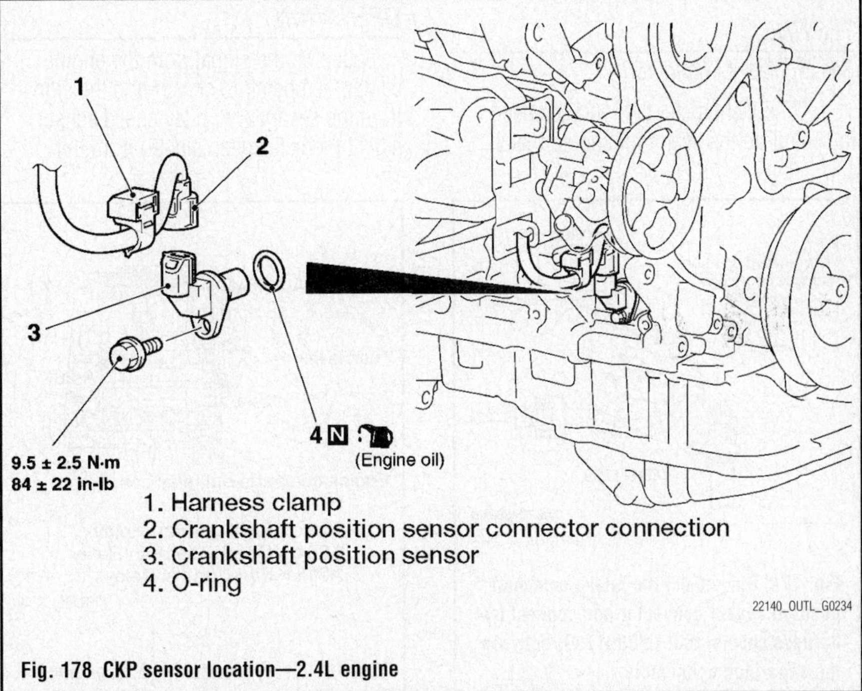

Fig. 178 CKP sensor location—2.4L engine

REMOVAL & INSTALLATION

2.4L Engine

1. Before servicing the vehicle, refer to the precautions.
2. Remove the air cleaner assembly.
3. Disconnect the crankshaft position sensor connector.
4. Remove the crankshaft position sensor.
5. Remove the o-ring.

To install:

6. Lubricate the O-ring with engine oil and install on the sensor.
7. Install the exhaust crankshaft position sensor. Tighten to 98 inch lbs. (11 Nm)
8. Connect the crankshaft position sensor connector.
9. Install the air cleaner assembly.

3.0L Engine

1. Before servicing the vehicle, refer to the precautions.
2. Remove the harness clamp.
3. Disconnect the crankshaft position sensor connector connection
4. Remove the crankshaft position sensor.
5. Remove the o-ring.

To install:

6. Lubricate the O-ring with engine oil and install on the sensor.
7. Install the exhaust crankshaft position sensor. Tighten to 84 inch lbs. (10 Nm).
8. Connect the crankshaft position sensor connector.
9. Install the harness clamp.

TESTING

See Figures 179 and 180.

1. The tachometer and engine speed indicated on the scan tool should match.

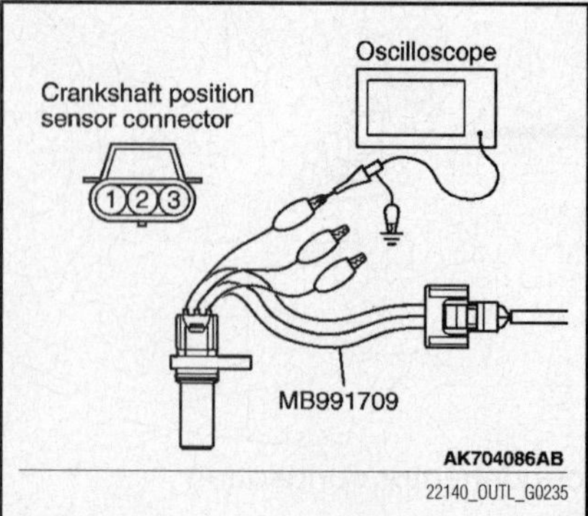

Fig. 179 Disconnect the intake camshaft position sensor connector and connect test harness special tool (MB991709) between the separated connectors

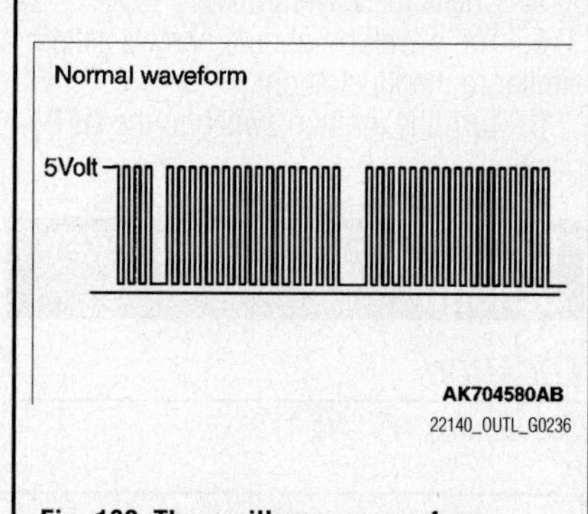

Fig. 180 The oscilloscope waveform should show a pattern similar to the illustration

2. Disconnect the crankshaft position sensor connector and connect the test harness special tool (MB991709) between the separated connectors.
3. Connect the oscilloscope probe to terminal No. 3 of the crankshaft position sensor connector.
4. Start the engine and run at idle.
5. Check the waveform.
6. The waveform should show a pattern similar to the illustration.
7. Turn the ignition switch to the **OFF** position.

ELECTRIC FAN SWITCH

LOCATION

See Figures 181 and 182.

Refer to the accompanying illustrations for fan relay locations.

OPERATION

Based on the signal from the engine coolant temperature sensor and the signal from the secondary pulley speed sensor signal (vehicle speed signal) from the transaxle control module, and the A/C switch signal from the A/C-ECU, the engine control module sends the radiator fan and condenser fan control signal to the ETACS-ECU via CAN bus line.

Based on the radiator fan and condenser fan control signal from the engine control module, the ETACS-ECU turns ON/OFF the radiator fan relay, condenser fan relay, and the fan control relay to control the rotation of the radiator fan motor and condenser fan motor.

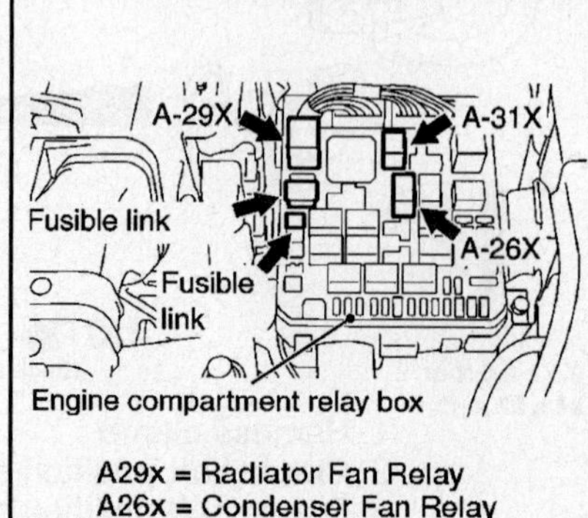

Fig. 181 Relay locations—2.4L engine

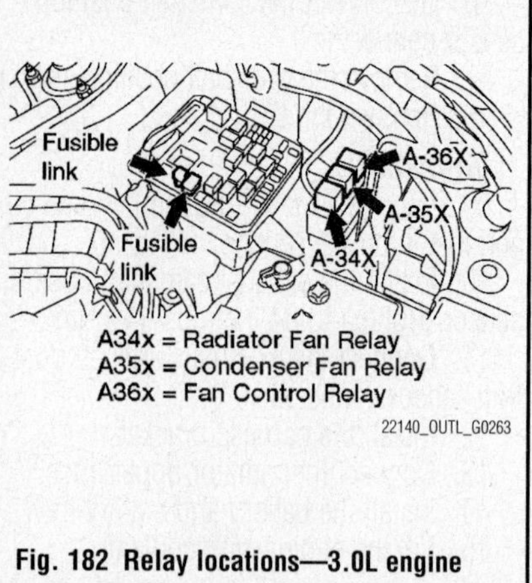

Fig. 182 Relay locations—3.0L engine

REMOVAL & INSTALLATION

1. Before servicing the vehicle, refer to the precautions.
2. Disconnect the negative battery cable.

⁂ CAUTION

Wait at least 90 seconds after the negative battery cable is disconnected to prevent possible deployment of the air bag.

3. Remove the air cleaner assembly.
4. Remove the engine undercover.
5. Remove the radiator drain plug and drain the engine coolant.
6. Remove the O-ring.
7. Remove the radiator cap.
8. Remove the transaxle fluid cooler hose.
9. Remove the water feed and return hoses.
10. Remove the radiator condenser tank hose.
11. Make mating marks on the radiator hoses and the hose clamps.
12. Remove the radiator hoses.
13. Remove the water pipes.
14. Remove the radiator cap.
15. Remove the radiator hose support.
16. Disconnect the radiator fan motor connector.

17. Disconnect the condenser fan motor connector.
18. Remove the fan, fan motor and fan shroud assembly.

To install:

19. Install the fan, fan motor and fan shroud assembly.
20. Connect the condenser fan motor connector.
21. Connect the radiator fan motor connector.
22. Install the radiator hose support.
23. Install the radiator cap.
24. Install the upper and lower radiator hoses.
    a. Insert radiator hose as far as the projection of the water inlet fitting, water outlet fitting or radiator.
    b. Align the mating marks on the radiator hose and hose clamp, and then connect the radiator hose.
25. Install the water pipes.
26. Install the radiator condenser tank hose.
27. Install the water feed and return hoses.
28. Install the CVT fluid cooler feed hose.
29. Install the radiator cap.
30. Install the O-ring.
31. Install the radiator drain plug and drain the engine coolant.
32. Install the engine undercover.
33. Install the air cleaner assembly.
34. Connect the negative battery cable.
35. Fill the engine with coolant.
36. Start the engine and check for leaks.

TESTING

1. Disconnect the fan motor connector.
2. Connect the fan motor-side connector terminal 1 to battery positive and terminal number 2 to ground.
3. Check that the fan motor runs and there is no abnormal sound emitted from the fan motor at this time.
4. If the fan motor runs properly, replace the fan control relay, radiator fan relay and/or condenser fan relay.

ELECTRONIC CONTROL MODULE (ECM)

LOCATION

See Figure 183.

Refer to the accompanying illustration for Electronic Control Module (ECM) location.

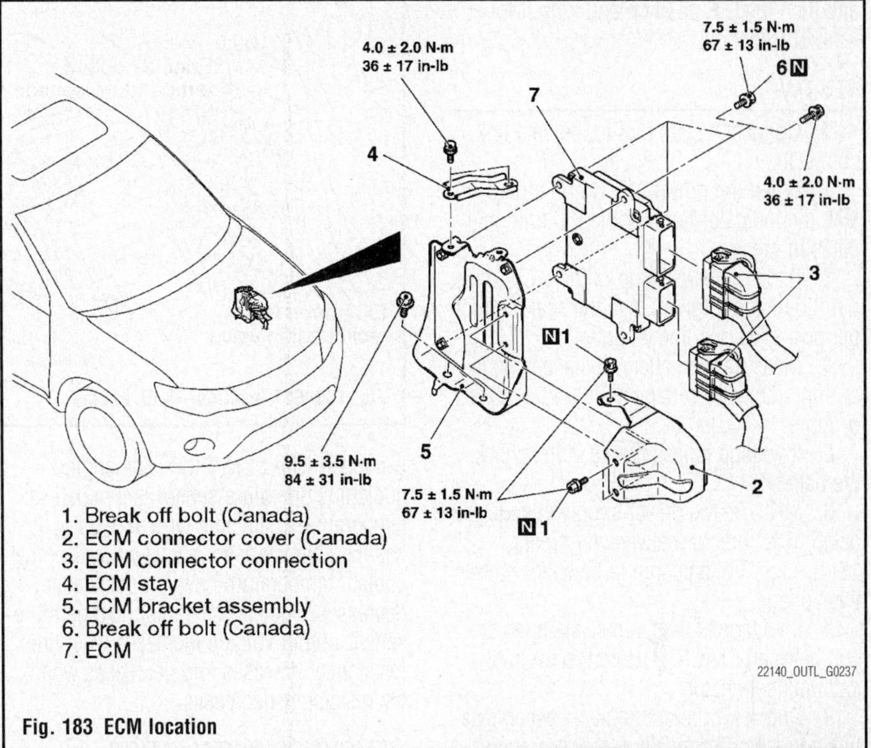

Fig. 183 ECM location

OPERATION

The powertrain has electronic controls to reduce exhaust emissions while maintaining excellent driveability and fuel economy. The Engine Control Module (ECM) is the control center of this system. The ECM monitors numerous engine and vehicle functions. The ECM constantly monitors the information from various sensors and other inputs, and controls the systems that affect vehicle performance and emissions.

The ECM also performs the diagnostic tests on various parts of the system. The ECM can recognize operational problems and alert the driver via the Malfunction Indicator Lamp (MIL). When the ECM detects a malfunction, the ECM stores a Diagnostic Trouble Code (DTC). The problem area is identified by the particular DTC that is set. The control module supplies a buffered voltage to various sensors and switches. Review the components and wiring diagrams in order to determine which systems are controlled by the ECM.

REMOVAL & INSTALLATION

**** WARNING**

When the ECM is replaced, do not replace the immobilizer ECU (WCM) or KOS-ECU simultaneously. When multiple ECUs are to be replaced, always replace only one ECU at a time and complete necessary ID registration using a scan tool.

**** WARNING**

After the ECM replacement, idling speed may be unstable because the MFI engine learning is not completed. To make it stable, let the system learn the idling.

**** WARNING**

After the ECM replacement, register a key code using a scan tool.

**** WARNING**

To replace the ECM, use an ECM for which the vehicle identification number entry and coding are completed. If DTC P0630 vehicle identification number-related or DTC P1676 variant coding-related is set in the ECM, replace it with an ECM to which the same vehicle identification number has been entered and the same coding has been made as for the relevant vehicle.

1. Remove the air cleaner assembly.
2. Disconnect the ECM connector.
3. Remove the ECM bracket.
4. Remove the ECM.

To install:

5. Installation is the reverse of removal.
6. Perform the initialization procedure. Turn the ignition switch to the **ON** position

and then to **OFF** position and hold it for at least 10 seconds.

TESTING

1. Connect a scan tool to the data link connector.
2. When the power of V.C.I. is turned **ON**, the indicator lamp of the V.C.I. illuminates in green.
3. If the indicator lamp of the V.C.I. does not illuminate in green, use the scan tool to perform CAN bus line diagnosis.
4. Measure the battery positive voltage during cranking. Voltage should be 8 volts or more.
5. If voltage is less than 8 volts, check the battery.
6. Measure the power supply voltage at data link connector between terminals 16–ground. Voltage should be battery positive voltage.
7. If voltage is less, repair the open circuit between ETACS-ECU connector and data link connector.
8. Check for the continuity between data link connector terminals 4–5 and ground. Resistance should be less than 2 ohms.
9. If resistance is not as specified, repair the open circuit or harness damage between the data link connector and ground.

ENGINE COOLANT TEMPERATURE (ECT) SENSOR

LOCATION

See Figures 184 and 185.

Refer to the accompanying illustrations for Engine Coolant Temperature (ECT) sensor location.

OPERATION

A 5 volt voltage is applied to the Engine Coolant Temperature (ECT) sensor output terminal from the ECM/PCM via the resistor in the ECM/PCM. The ground terminal is grounded with ECM/PCM. The engine coolant temperature sensor is a negative temperature coefficient type of resistor. It has the characteristic that when the engine coolant temperature rises the resistance decreases. The engine coolant temperature sensor output voltage increases when the resistance increases and decreases when the resistance decreases.

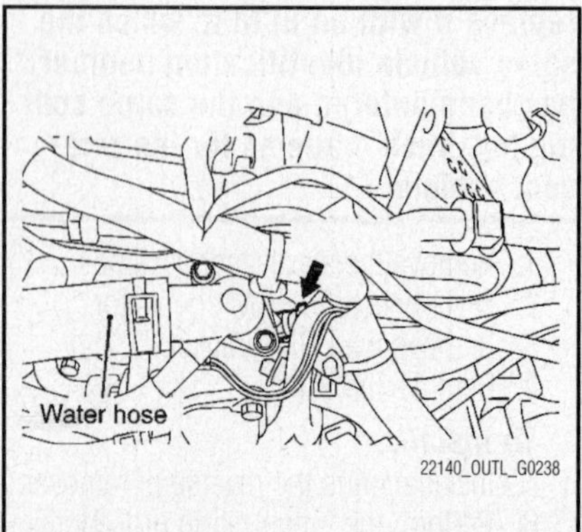

Fig. 184 ECT location—2.4L engine

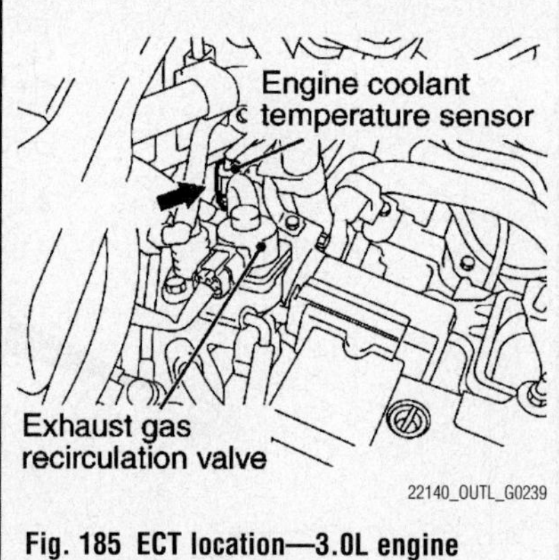

Fig. 185 ECT location—3.0L engine

REMOVAL & INSTALLATION

1. Before servicing the vehicle, refer to the precautions.
2. Drain the engine coolant to a level below the intake manifold.
3. Unplug the sensor wiring harness.
4. Unplug the ECT sensor electrical connector
5. Unthread and remove the sensor from the engine.

Use a deep socket and an extension to reach the ECT sensor, remove the ECT sensor from the thermostat housing

To install:

6. Coat the threads of the sensor with a suitable sealant and thread into the housing. Tighten to 22 ft. lbs. (30 Nm).
7. Refill the cooling system to the proper level.
8. Connect the electrical connector to the sensor.

TESTING

See Figure 186.

1. With the temperature sensing portion of sensor immersed in hot water, check resistance.
2. Standard values are as follows:
   - 5.1–6.5 Kilohms at 32°F
   - 2.1–2.7 Kilohms at 68°F
   - 0.9–1.3 Kilohms [at 40°C (68°F)
   - 0.26–0.36 Kilohms [at 80°C (176°F)
3. If not within specifications, replace the sensor.

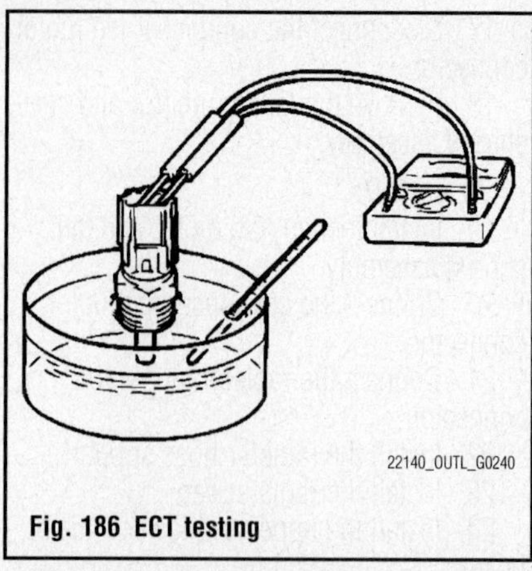

Fig. 186 ECT testing

FUEL LEVEL SENDING UNIT

LOCATION

See Figure 187.

Refer to the accompanying illustration for fuel level sending unit location.

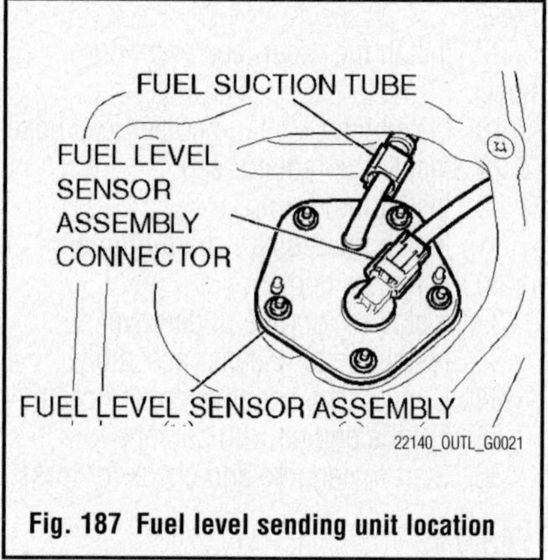

Fig. 187 Fuel level sending unit location

OPERATION

The fuel level sending unit consists of a variable resistor that changes resistance based on the fuel level in the tank. The ECM monitors the voltage across the sender resistance in order to determine the fuel level.

REMOVAL & INSTALLATION

See Figure 188.

1. Before servicing the vehicle, refer to the precautions.
2. Disconnect the negative battery cable.
3. Remove the rear seats.
4. Remove hole cover and remove the fuel pump module and fuel level sensor.

To install:

5. Installation is the reverse of the removal procedure.
6. Connect the negative battery cable.

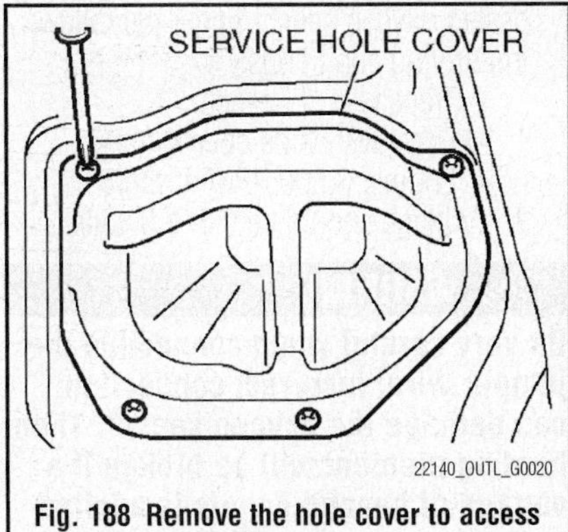

Fig. 188 Remove the hole cover to access the fuel level sending unit.

TESTING

Resistance Test

FWD Models

See Figure 189.

1. Remove the fuel pump module.
2. Remove the fuel level sensor.
3. Check that resistance value between the fuel gauge terminal and ground terminal is at the standard value when the fuel level sensor float is between point "F" (highest) and point "E" (lowest). Standard value:
   - Point "F": 12–14 ohms
   - Point "E": 119–121 ohms
4. Check that resistance value changes smoothly when the float moves slowly between point "F" (highest) and point "E" (lowest).
5. If all checks are correct, go to fuel level sensor float height check. If any check is not correct, replace the fuel level sensor.

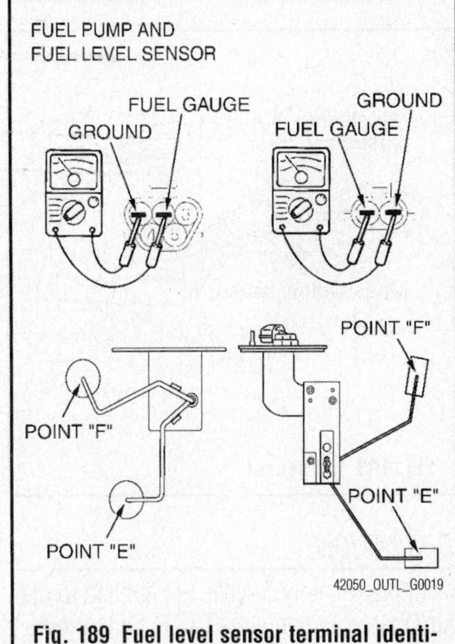

Fig. 189 Fuel level sensor terminal identification

AWD Models

See Figure 189.

1. Remove the fuel pump module.
2. Remove the fuel level sensor.
3. Check that resistance value between the fuel gauge terminal and ground terminal is at the standard value when the fuel level sensor float is between point "F" (highest) and point "E" (lowest). Standard value:
   - Point "F" (Main): 5.5–7.5 ohms
   - Point "F" (Sub): 5.5–7.5 ohms
   - Point "E" (Main): 43.7–45.7 ohms
   - Point "E" (Sub): 74.3–76.3 ohms
4. Check that resistance value changes smoothly when the float moves slowly between point "F" (highest) and point "E" (lowest).
5. If all checks are correct, go to fuel level sensor float height check. If any check is not correct, replace the fuel level sensor.

Float Height Check

FWD Models

See Figure 190.

1. Remove the fuel pump module.
2. Remove the fuel level sensor.
3. Move the float and measure height A at point "F" (highest) and B at point "E" (lowest) with the float arm touching stopper.
   - Point "F" (Height "A"): 7.9 in. (201.6 mm)
   - Point "E" (Height "B"): 0.7 in. (18 mm)

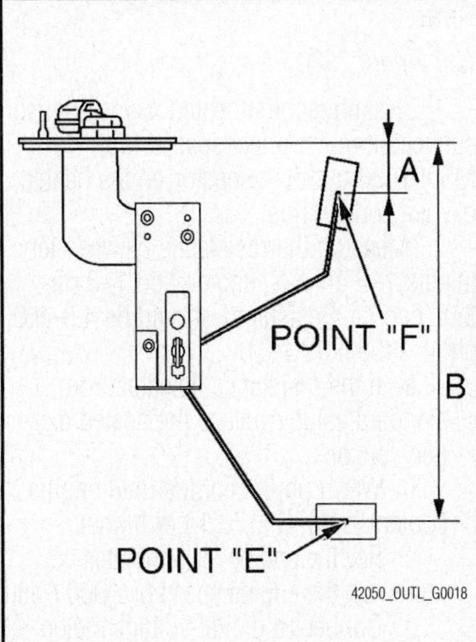

Fig. 190 Float height check—FWD models

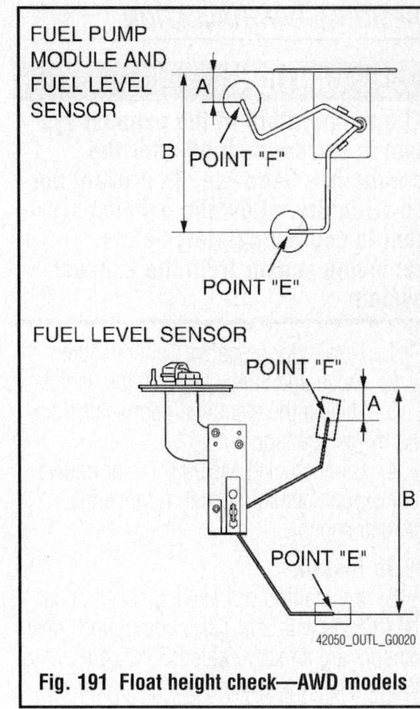

Fig. 191 Float height check—AWD models

4. Adjust the float arm to the standard value, then go to the thermistor check.

AWD Models

See Figure 191.

1. Remove the fuel pump module.
2. Remove the fuel level sensor.
3. Move the float and measure height A at point "F" (highest) and B at point "E" (lowest) with the float arm touching stopper. Standard value:
   a. Fuel pump module & fuel level sensor:
   - Point "F" (Height "A") (Main): 6.5 in. (166.2 mm)
   - Point "F" (Height "A") (Sub): 1.8–2.0 in. (45.8–51.8 mm)
   - Point "E" (Height "B") (Main): 1.5 in. (38.8 mm)
   - Point "E" (Height "B") (Sub): 9.2–9.4 in. (235–241 mm)
4. Adjust the float arm to the standard value, then go to the thermistor check.

HEATED OXYGEN (HO2S) SENSOR

OPERATION

A voltage corresponding to the Heated Oxygen (HO2S) Sensor concentration in the exhaust gas is sent to the ECM/PCM from the output terminal of the sensor.

The sensor is grounded with the ECM/PCM.

REMOVAL & INSTALLATION

✻✻ CAUTION

The temperature of the exhaust system is extremely high after the engine has been run. To prevent personal injury, allow the exhaust system to cool completely before removing sensor from the exhaust system.

1. Detach the negative battery cable.
2. Raise and safely support the vehicle.
3. Detach the electrical connector from the oxygen sensor.
4. Using socket MD998770, or equivalent oxygen sensor socket, remove the oxygen sensor.

To install:

5. If installing old oxygen sensor, coat the threads with anti-seize compound. New sensors are already coated. Take care not to contaminate the oxygen sensor probe with the anti-seize compound.
6. Install the oxygen sensor into the exhaust manifold. Tighten the sensor, using the correct tool, to 37 ft. lbs. (50 Nm)
7. Attach the wiring to the sensor.
8. Carefully lower the vehicle.
9. Connect the negative battery cable.

TESTING

Front

See Figure 192.

1. Disconnect the heated oxygen sensor connector and connect special tool MB991658 to the connector on the heated oxygen sensor side.
2. Measure the resistance between terminals 1–2 on 2.4L engine and 1–3 on 3.0L engine. Resistance should be 4.5–8.0 ohms at 68°F (20°C)

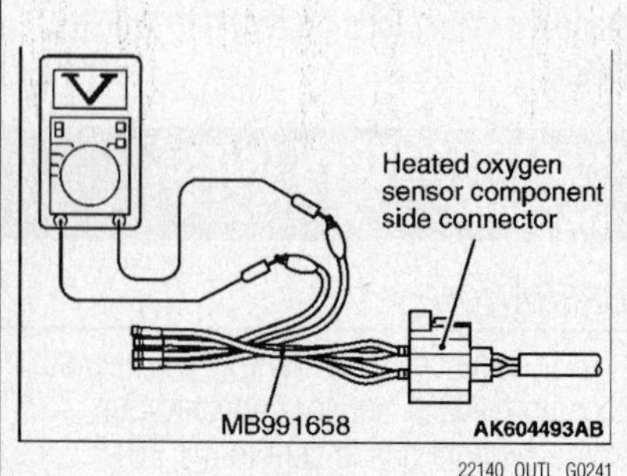

Fig. 192 Connect special tool MB991658 to the connector on the heated oxygen sensor side

a. If the resistance deviates from standard value, replace the heated oxygen sensor.

b. Warm up the engine until engine coolant is 80°C (176°F) or higher.

c. Rev the engine for 5 minutes or more with the engine speed of 2,000 r/min.

d. Connect a digital voltage meter between terminal 3–4 on 2.4L engine and 2–4 on 3.0L engine

e. While repeatedly revving the engine, measure the heated oxygen sensor output voltage. Voltage should be 0.6–1.0 volts

✻✻ WARNING

Be very careful when connecting the jumper wire; incorrect connection can damage the oxygen sensor. Then heating element will be broken if a voltage of beyond 8 volts is applied to the oxygen sensor heater.

➡If the temperature of sensing area does not reach the high temperature [of approximately 752°F (400°C) or more] even though the oxygen sensor is normal, the output voltage would be possibly low in spite of the rich air-fuel ratio. Therefore, if the output voltage is low, use a jumper wire to connect terminal No. 1 and terminal No. 2 of the oxygen sensor with a (+) terminal and (-) terminal of and 8 volt power supply respectively, then check again.

3. If the output voltage is not within the standard value, replace the heated oxygen sensor.

Rear

See Figure 192.

1. Disconnect the heated oxygen sensor connector and connect special tool MB991658 to the connector on the heated oxygen sensor side.
2. Measure the resistance between terminals 1–2 on 2.4L engine and 1–3 on 3.0L engine. Resistance should be 4.5–8.0 ohms at 68°F (20°C)

a. If the resistance deviates from standard value, replace the heated oxygen sensor.

b. Warm up the engine until engine coolant is 80°C (176°F) or higher.

c. Rev the engine for 5 minutes or more with the engine speed of 2,000 r/min.

d. Connect a digital voltage meter between terminal 1–2 on 2.4L engine and 1–3 on 3.0L engine.

e. Measure the output voltage of the heated oxygen sensor under the following driving conditions.

- Transaxle: "L" range
- Drive with wide open throttle
- Engine: 3,500 RPM or more

3. Voltage should be 0.6–1.0 volts

✻✻ WARNING

Be very careful when connecting the jumper wire; incorrect connection can damage the oxygen sensor. Then heating element will be broken if a voltage of beyond 8 volts is applied to the oxygen sensor heater.

➡If the temperature of sensing area does not reach the high temperature of approximately 752°F (400°C) or more even though the oxygen sensor is normal, the output voltage would be possibly low in spite of the rich air-fuel ratio. Therefore, if the output voltage is low, use a jumper wire to connect terminal No. 1 and terminal No. 2 of the oxygen sensor with a (+) terminal and (-) terminal of and 8 volt power supply respectively, then check again.

4. If the output voltage is not within the standard value, replace the heated oxygen sensor.

INTAKE AIR TEMPERATURE (IAT) SENSOR

LOCATION

See Figure 193.

Refer to the accompanying illustration for Intake Air Temperature (IAT) sensor location.

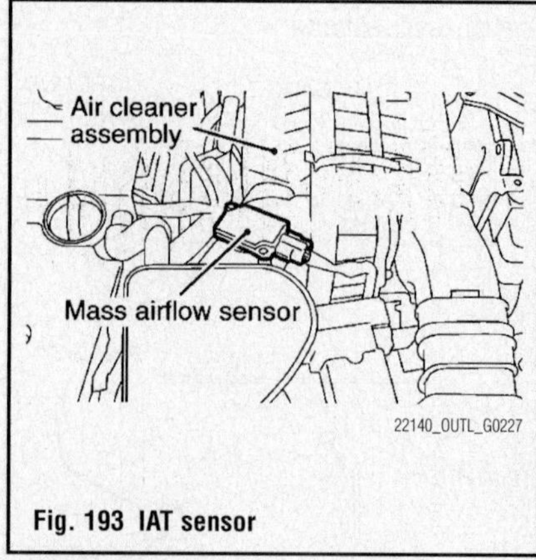

Fig. 193 IAT sensor

OPERATION

Approximately 5–volts are applied to the Air Charge Temperature (ACT) Sensor output terminal from the ECM/PCM via the resistor in the ECM/PCM. The ground terminal is grounded with ECM/PCM. The sensor is a negative temperature coefficient type of resistor. When the intake air temperature rises, the resistance decreases. The sensor output voltage increases when the resistance increases and decreases when the resistance decreases.

REMOVAL & INSTALLATION

1. Before servicing the vehicle, refer to the precautions.
2. Replacing the sensor requires disconnecting the electrical connector, then carefully removing the lid of the air filter housing.

➡Handle the sensor assembly carefully, protecting it from impact, extremes of temperature and/or exposure to shop chemicals.

To install:

3. Installation is the reverse of the removal procedure.

TESTING

See Figure 194.

1. Disconnect the sensor connector.
2. Measure the resistance between the sensor side connector terminals 1–4.
3. Standard values are as follows:
   - 5.3–6.7 Kilohms at 32°F
   - 2.3–3.0 Kilohms at 68°F
   - 1.0–1.5 Kilohms at 68°F
   - 0.30–0.42 Kilohms at 176°F
4. Measure resistance while heating the sensor using warm, dry air. As the temperature rises, the resistance value should drop. If within specifications, replace the ECM or PCM.

➡Check that the circuit is not Open circuit.

5. If no continuity, replace the Mass Air Flow (MAF) Sensor.

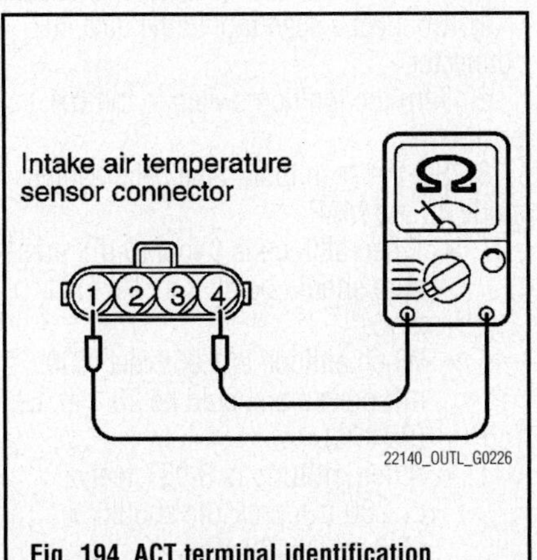

Fig. 194 ACT terminal identification

KNOCK SENSOR (KS)

LOCATION

See Figures 195 and 196.

Refer to the accompanying illustrations for Knock Sensor (KS) location.

OPERATION

The Knock Sensor (KS) converts the vibration of the cylinder block into a voltage and outputs it. If there is a malfunction of the knock sensor, the voltage output will not change. The PCM checks whether the voltage output changes.

REMOVAL & INSTALLATION

1. Before servicing the vehicle, refer to the precautions.

⁂ WARNING

Do not drop or hit the knock sensor against other components. Internal damage may result, and the knock sensor will need to be replaced.

2. Remove the intake manifold assembly.
3. Disconnect the knock sensor connector.
4. Remove the knock sensor.

To install:

5. Install the knock sensor. Tighten to 15 ft. lbs. (20 Nm).
6. Connect the knock sensor connector.
7. Install the intake manifold assembly.
8. When the knock sensor replacement is performed, use a scan tool to initialize the learning value.

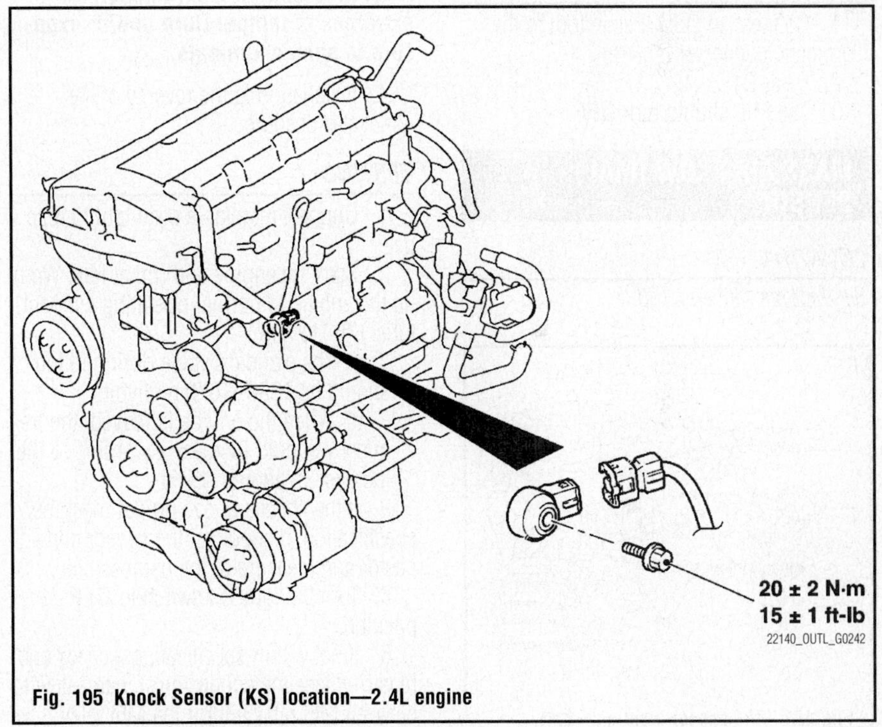

Fig. 195 Knock Sensor (KS) location—2.4L engine

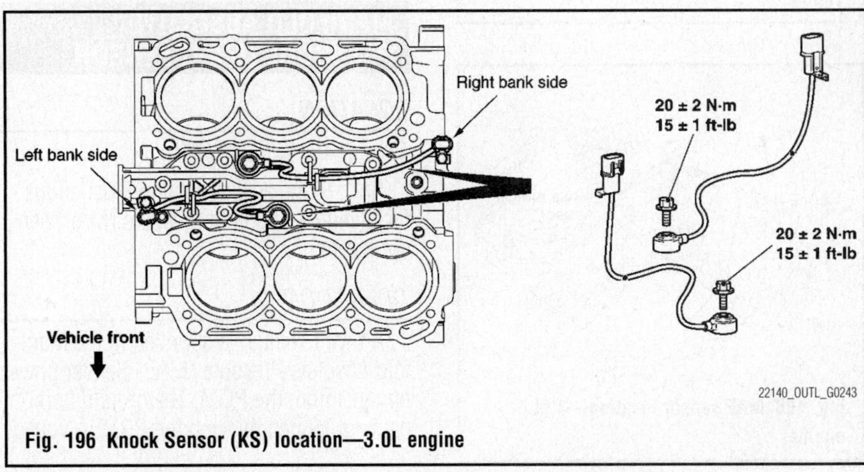

Fig. 196 Knock Sensor (KS) location—3.0L engine

TESTING

1. Disconnect the knock sensor connector.
2. Start the engine and run at idle.
3. Measure the voltage between knock sensor side connector terminals 1–2.
4. Gradually increase the engine speed. The voltage increases with the increase in the engine speed.
5. Turn the ignition switch to the **OFF** position.
6. If within specifications, check connectors and wiring between the sensor and ECM/PCM. If ok, replace ECM/PCM.
7. If not within specifications, replace the knock sensor.

MALFUNCTION INDICATOR LIGHT (MIL)

RESET PROCEDURES

1. Connect an OBD II scan tool to the diagnostic connector.
2. Clear DTCs.
3. The MIL should turn **OFF**.

MASS AIR FLOW (MAF) SENSOR

LOCATION

See Figures 197 and 198.

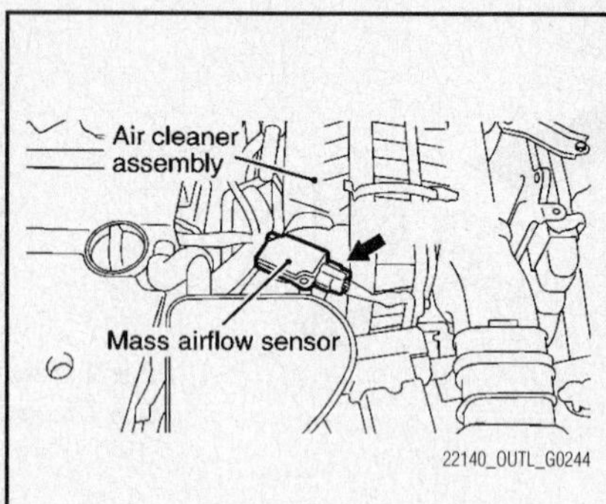

Fig. 197 MAF sensor location—2.4L engine

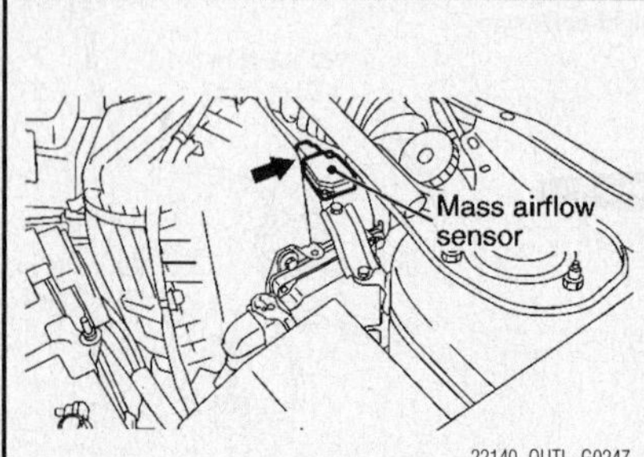

Fig. 198 MAF sensor location—3.0L engine

Refer to the accompanying illustrations for Mass Air Flow (MAF) sensor location.

OPERATION

A 5-volt power is applied to the Mass Air Flow (MAF) Sensor output terminal from the ECM/PCM. The sensor generates a pulse signal when the output terminal and ground are opened/closed (opened/short).

REMOVAL & INSTALLATION

1. Before servicing the vehicle, refer to the precautions.
2. Disconnect the negative battery cable.
3. Replacing the sensor requires disconnecting the electrical connector, then carefully removing the lid of the air filter housing.

➡Handle the sensor assembly carefully, protecting it from impact, extremes of temperature and/or exposure to shop chemicals.

4. Installation is the reverse of the removal procedure.

TESTING

1. Connect a suitable scan tool to read data.
2. Start the engine and run at idle. Warm up the engine to normal operating temperature: 176° to 205°F.
   a. The standard value during idling should be 1350–1670 millivolts.
   b. When the engine is revved, the frequency should increase according to the increase in engine speed.
3. If the checks above do not meet the specification, disconnect the sensor connector and measure at the harness side.
4. Turn the ignition switch to **OFF** position.
5. If not within specifications, check and repair/replace the connector(s), harness wire between MFI relay and/or the MFI relay.

MANIFOLD ABSOLUTE PRESSURE (MAP) SENSOR

LOCATION

See Figures 199 and 200.

Refer to the accompanying illustrations for Manifold Absolute Pressure (MAP) sensor location.

OPERATION

A 5 volt voltage is supplied to the Manifold Absolute Pressure (MAP) Sensor power terminal from the PCM. The ground terminal is grounded through the PCM. A voltage that is proportional to the intake manifold pressure is sent to the PCM the sensor output terminal.

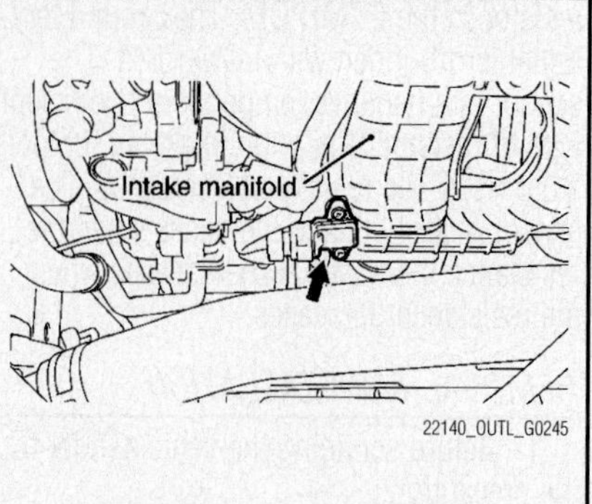

Fig. 199 MAP sensor location—2.4L engine

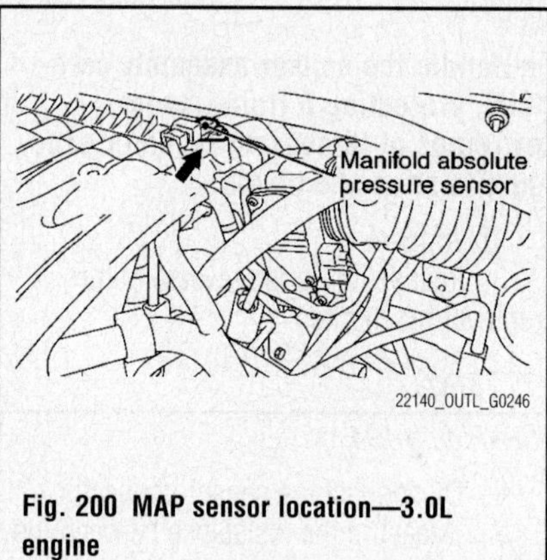

Fig. 200 MAP sensor location—3.0L engine

REMOVAL & INSTALLATION

1. Before servicing the vehicle, refer to the precautions.
2. Disconnect the negative battery cable.
3. Detach the electrical connector at the sensor.
4. Remove the sensor.
5. Installation is the reverse of the removal procedure.

TESTING

1. Connect a scan tool to the data link connector.
2. Turn the ignition switch to the **ON** position.
3. Set the scan tool to the data reading mode for the MAP.
   - When altitude is 0 foot (0 m), pressure should be 29.8 in. Hg (101 kPa).
   - When altitude is 1,969 feet (600 m), pressure should be 28.1 in. Hg (95 kPa).
   - When altitude is 3,937 feet (1,200 m), pressure should be 26.0 in. Hg (88 kPa).

- When altitude is 5,906 feet (1,800 m), pressure should be 23.9 in. Hg (81 kPa).

4. Start the engine.
   - When the engine is idling, pressure should be 4.7 - 10.6 in. Hg (16 - 36 kPa).
   - When the engine is suddenly revved, pressure should vary.

OIL PRESSURE SENSOR

LOCATION

See Figures 201 and 202.

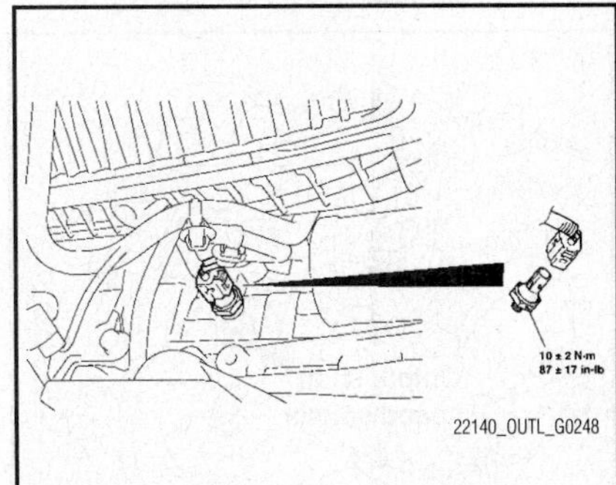

Fig. 201 Oil pressure sensor location—2.4L engine

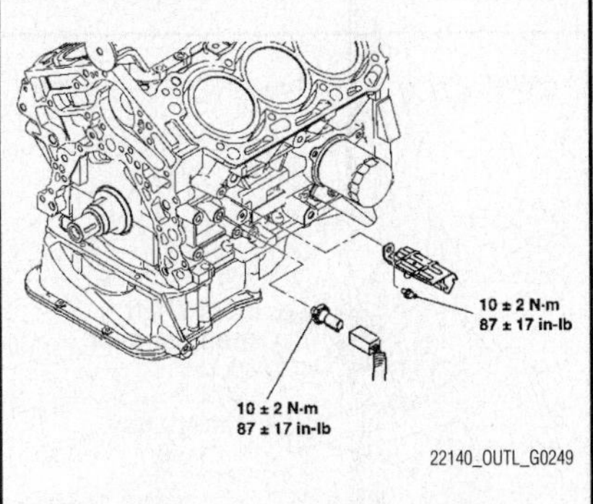

Fig. 202 Oil pressure sensor location—3.0L engine

Refer to the accompanying illustrations for oil pressure sensor location.

OPERATION

The oil pressure sensor reads engine oil pressure and illuminates a dash mounted light when pressure drops below a predetermined pressure.

REMOVAL & INSTALLATION

1. Before servicing the vehicle, refer to the precautions section.
2. Remove the engine undercover.
3. On 2.4L engine, remove the intake manifold bracket.
4. On 2.4L engine, remove the throttle body mounting bolts and put the throttle body aside so as not to interfere the oil pressure switch connector disconnection.
5. Disconnect the engine oil pressure switch connector.
6. Remove the engine oil pressure switch.

To install:

7. Apply Three bond 1215 or equivalent sealant to 2 or 3 threads of the sensor.
8. Install the sensor and tighten to 87 inch lbs. (10 Nm).

✻✻ WARNING

Do not start the engine for 2 hour or more after the engine oil pressure switch installation.

9. Connect the engine oil pressure switch connector.
10. On 2.4L engine, install the throttle body mounting bolts.
11. On 2.4L engine, install the intake manifold bracket.
12. Install the engine undercover.

TESTING

1. Disconnect the sender electrical connection.
2. Connect one lead of a ohmmeter to the sender terminal and the other lead to the sender body.
3. With the engine off the resistance should be low.
4. Start the engine. The resistance should increase as the engine speed increases.
5. If specification is not as stated, replace the sender.

THROTTLE POSITION SENSOR (TPS)

LOCATION

See Figures 203 and 204.

Refer to the accompanying illustrations for Throttle Position Sensor (TPS) location.

OPERATION

A 5 volt power supply is applied to the Throttle Position Sensor (TPS) power terminal from the ECM/PCM. The ground terminal is grounded with ECM/PCM. When the throttle valve shaft is turned from the idle position to the fully opened position, the resistance between the TPS output terminal and the ground terminal would increase according to the rotation.

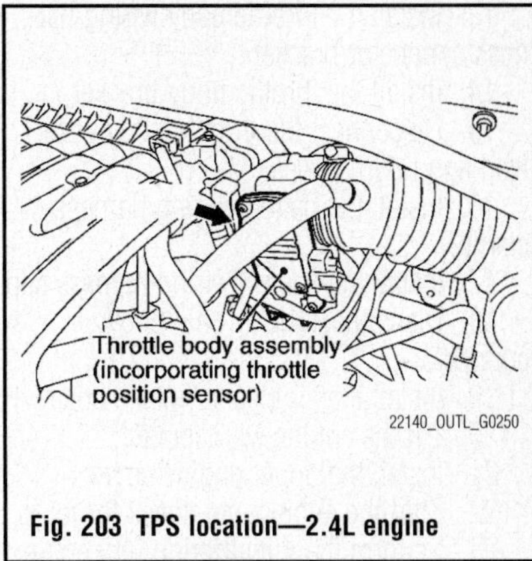

Fig. 203 TPS location—2.4L engine

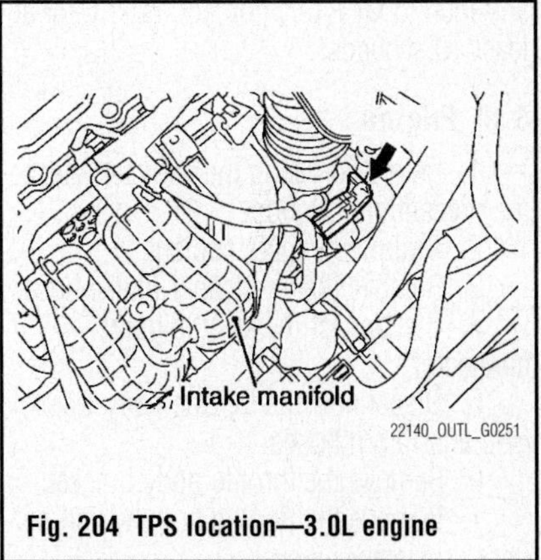

Fig. 204 TPS location—3.0L engine

REMOVAL & INSTALLATION

The throttle position sensor is an integral part of the throttle body assembly and cannot be serviced separately.

2.4L Engine

1. Before servicing the vehicle, refer to the precautions section.
2. Remove the upper engine cover.
3. Drain the engine coolant.
4. Remove the air cleaner intake hose.
5. Disconnect the throttle body connector.
6. Remove the control wiring harness clip.
7. Remove the battery wiring harness clamp.
8. Disconnect the throttle body water feed and return hoses.
9. Remove the throttle body bracket.
10. Remove the throttle body wiring harness connector bracket
11. Remove the throttle body and gasket.

To install:

12. Install the throttle body and gasket. Tighten to 88 inch lbs. (10 Nm).
    a. Fit the throttle body gasket in the intake manifold groove securely without twisting or damage.

13. Install the throttle body wiring harness connector bracket
14. Install the throttle body bracket.
15. Disconnect the throttle body water feed and return hoses.
16. Install the battery wiring harness clamp.
17. Install the control wiring harness clip.
18. Disconnect the throttle body connector.
19. Install the air cleaner intake hose.
20. Fill the engine with coolant.
21. Install the upper engine cover.
22. Start the engine and check for leaks.
23. Perform the initialization procedure. Turn the ignition switch to the **ON** position and then to **OFF** position and hold it for at least 10 seconds.

3.0L Engine

1. Before servicing the vehicle, refer to the precautions section.
2. Drain the engine coolant.
3. Remove the air cleaner intake hose.
4. Disconnect the throttle body connector.
5. Disconnect the throttle body water feed and return hoses.
6. Remove the throttle body bracket.
7. Remove the throttle body wiring harness connector bracket
8. Remove the throttle body and gasket.

To install:

9. Install the throttle body and gasket. Tighten to 15 ft. lbs. (20 Nm).
10. Install the throttle body wiring harness connector bracket
11. Install the throttle body bracket.
12. Disconnect the throttle body water feed and return hoses.
13. Disconnect the throttle body connector.
14. Install the air cleaner intake hose.
15. Fill the engine with coolant.
16. Start the engine and check for leaks.
17. Perform the initialization procedure. Turn the ignition switch to the **ON** position and then to **OFF** position and hold it for at least 10 seconds.

TESTING

Main

1. Turn the ignition switch to the **ON** position.
2. Disconnect the intake air hose at the throttle body.
3. Disconnect the of the throttle position sensor connector.
4. Use test harness (MB991658) to connect only terminals 3–4–5–6.
5. Set a scan tool to the data reading mode TPS.
6. Output voltage should be between 0.3–0.7 volts when the throttle valve is fully closed with your finger.
7. Output voltage should be 4.0 volts or more when the throttle valve is fully open with your finger.
8. If sensor is not within specification, replace the sensor.
9. Turn the ignition switch to the **OFF** position.

Sub

1. Turn the ignition switch to the **ON** position.
2. Disconnect the intake air hose at the throttle body.
3. Disconnect the of the throttle position sensor connector.
4. Use test harness (MB991658) to connect only terminals 3–4–5–6.
5. Set a scan tool to the data reading mode TPS.
6. Output voltage should be between 0.3–0.7 volts when the throttle valve is fully open with your finger.
7. Output voltage should be 4.0 volts or more when the throttle valve is fully closed with your finger.
8. If sensor is not within specification, replace the sensor.
9. Turn the ignition switch to the **OFF** position.

VEHICLE SPEED SENSOR (VSS)

LOCATION

See Figures 205 through 208.

Refer to the accompanying illustrations for sensor location.

OPERATION

A 5 volt voltage is applied to the Vehicle Speed Sensor (VSS) output terminal from the ECM/PCM. The sensor generates a pulse signal when the output terminal is opened and grounded.

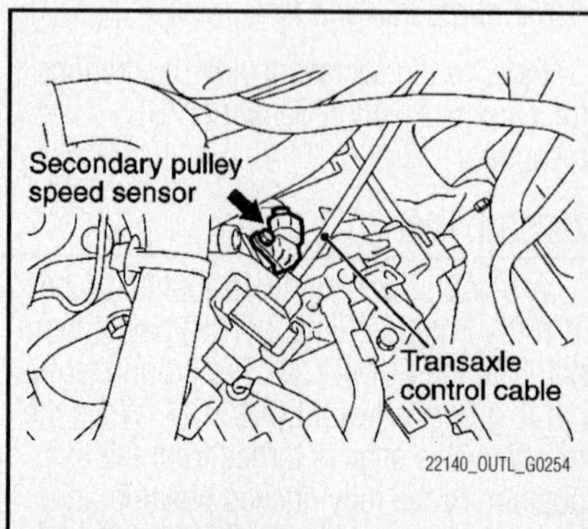

Fig. 205 Secondary pulley speed sensor—CVT transmission

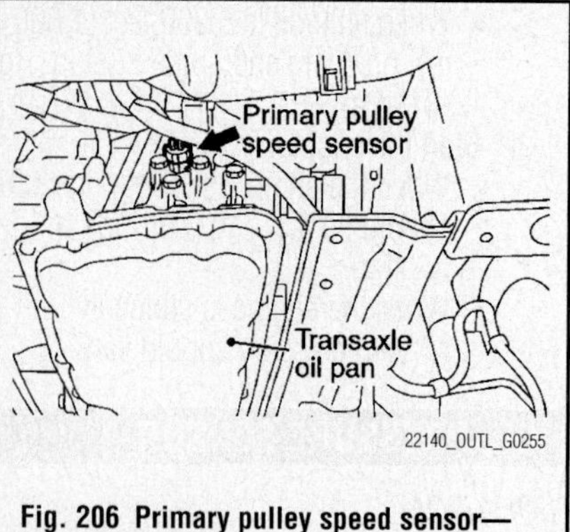

Fig. 206 Primary pulley speed sensor—CVT transmission

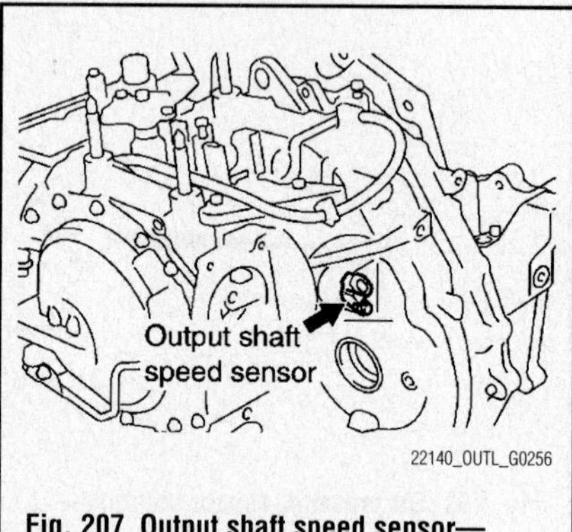

Fig. 207 Output shaft speed sensor—automatic transmission

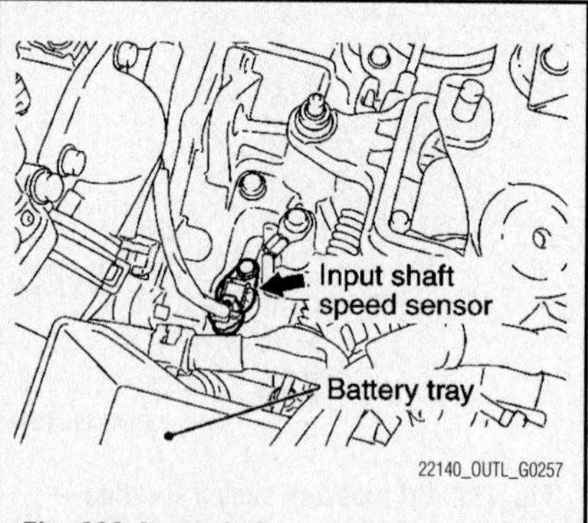

Fig. 208 Input shaft speed sensor—automatic transmission

REMOVAL & INSTALLATION

1. Before servicing the vehicle, refer to the precautions section.
2. Disconnect the sensor electrical connector.
3. Remove the sensor attaching bolt/screw.
4. Remove the sensor.

To install:

5. Installation is the reverse of removal.

TESTING

Automatic Transaxle

1. With the connector connected, connect the oscilloscope probe to each terminal of TCM.

2. With the transmission in **D** range and the vehicle being driven at a constant speed of approximately 19 MPH (30 km /h)

3. Check that the voltage between TCM connector C-37 terminal 48 and C-38 terminal 13 or C-38 terminal 26 is 0 V.

Check the frequency of TCM connector C-37 terminal 37 is constant at 588 Hz.

CVT Transaxle

1. With the connector connected, connect the oscilloscope probe to each terminal of TCM.

2. With the transmission in **D** range and the vehicle being driven at a constant speed of approximately 12 MPH (20 km /h), check the wave pattern cycle is constant and voltage is between 0 and 4.5 volts.

- Terminals 36–26: secondary pulley rotation sensor
- Terminals 37–26: primary pulley rotation sensor

FUEL — GASOLINE FUEL INJECTION SYSTEM

FUEL SYSTEM SERVICE PRECAUTIONS

Safety is the most important factor when performing not only fuel system maintenance but any type of maintenance. Failure to conduct maintenance and repairs in a safe manner may result in serious personal injury or death. Maintenance and testing of the vehicle's fuel system components can be accomplished safely and effectively by adhering to the following rules and guidelines.

- To avoid the possibility of fire and personal injury, always disconnect the negative battery cable unless the repair or test procedure requires that battery voltage be applied.
- Always relieve the fuel system pressure prior to disconnecting any fuel system component (injector, fuel rail, pressure regulator, etc.), fitting or fuel line connection. Exercise extreme caution whenever relieving fuel system pressure to avoid exposing skin, face and eyes to fuel spray. Please be advised that fuel under pressure may penetrate the skin or any part of the body that it contacts.
- Always place a shop towel or cloth around the fitting or connection prior to loosening to absorb any excess fuel due to spillage. Ensure that all fuel spillage (should it occur) is quickly removed from engine surfaces. Ensure that all fuel soaked cloths or towels are deposited into a suitable waste container.
- Always keep a dry chemical (Class B) fire extinguisher near the work area.
- Do not allow fuel spray or fuel vapors to come into contact with a spark or open flame.
- Always use a back-up wrench when loosening and tightening fuel line connection fittings. This will prevent unnecessary stress and torsion to fuel line piping.
- Always replace worn fuel fitting O-rings with new. Do not substitute fuel hose or equivalent where fuel pipe is installed.

Before servicing the vehicle, make sure to also refer to the precautions in the beginning of this section as well.

RELIEVING FUEL SYSTEM PRESSURE

See Figure 209.

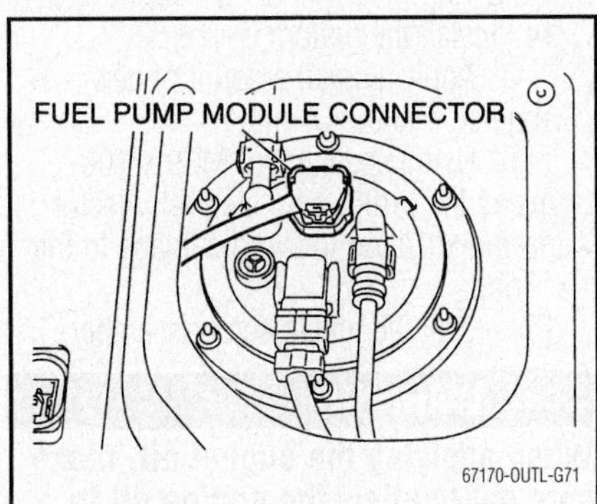

Fig. 209 Fuel pump module connector location

1. Before servicing the vehicle, refer to the precautions.
2. Turn the ignition to the **OFF** position.
3. Loosen the fuel filler cap to release fuel tank pressure.
4. Remove the rear seat assembly, then remove the protector and disconnect the fuel pump module connector.
5. Start the vehicle and allow it to run until it stalls from lack of fuel. Turn the key to the **OFF** position.
6. Disconnect the negative battery cable, then reconnect the fuel pump connector. Install the rear seat assembly.

FUEL FILTER

REMOVAL & INSTALLATION

The fuel filter is an integral part of the fuel pump module and is not normally serviced.

FUEL PUMP

REMOVAL & INSTALLATION

See Figure 210.

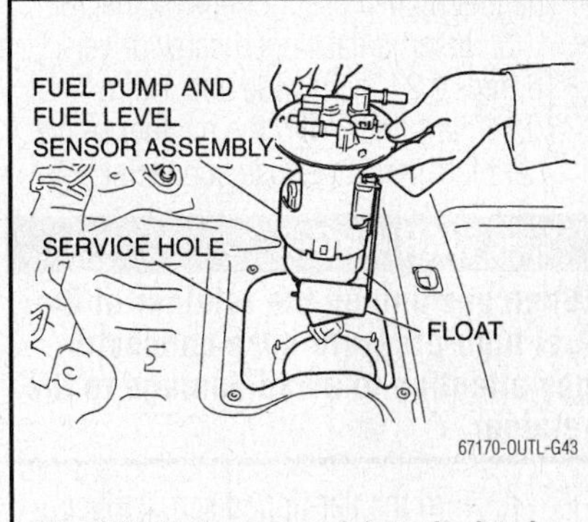

Fig. 210 Fuel pump module and related components

1. Before servicing the vehicle, refer to the precautions.
2. Remove the harness electrical connector(s).
3. Properly relieve the fuel system pressure.
4. Remove the rear seat cushion.
5. Remove the retainer screws and service hole cover.
6. Remove the fuel lines.
7. Remove the fuel pump module mounting nuts and plate.

WARNING

Be careful not to damage the module unit and float, when removing the fuel pump module from the fuel tank.

8. Remove the fuel pump module (FWD) or Fuel pump and fuel level sensor (AWD) from the service hole.
9. Disassemble the fuel pump module as necessary.

To install:

10. Installation is the reverse of the removal procedure.

11. Replace the fuel pump module gasket with a new one.
12. Torque the fuel pump module place nuts to 22 inch lbs. (2.5 Nm).

FUEL RAIL & INJECTORS

REMOVAL & INSTALLATION

2.4L Engine

See Figure 211.

1. Disconnect the fuel pump connector.
2. Remove the engine upper cover.
3. Remove the air cleaner intake hose.
4. Disconnect the control wiring harness.
5. Disconnect the fuel high-pressure hose.
   a. Follow the steps below to unlock the fuel high-pressure hose connector.
   b. Insert a flat-tipped screwdriver [6 mm (0.24 inch) wide and 1 mm (0.04inch) thick] into the retainer of the fuel high-pressure hose connector.

***** WARNING**

When pushing up the retainer of the fuel high-pressure hose connector, pay attention to avoid damage to the retainer.

   c. Turn the flat-tipped screwdriver inserted into the retainer by 90° to push up the retainer and unlock the fuel high-pressure hose connector.
   d. Disconnect the fuel high-pressure hose.

6. Remove the engine oil dipstick.
7. Remove the rear injector protector.
8. Remove the bracket
9. Remove the fuel rail and fuel injector assembly

***** WARNING**

Do not drop the fuel injector.

   a. Remove the fuel rail with the fuel injectors attached to it.

10. Remove the fuel injector support.
11. Remove the fuel injector assembly.
12. Remove the O-ring.

To install:

13. Install the top O-ring
   a. Apply a small amount of new engine oil to the O-ring.
   b. While turning the fuel injector to right and left, install the O-ring to the fuel injector with care to avoid damage to the O-ring.
14. Install the bottom O-ring
   a. Apply a small amount of new engine oil to the O-ring.
   b. Using special tool MB992106, install the o-ring onto the fuel injector paying attention to avoid damage to the O-ring.
15. Install the fuel injector assembly

***** WARNING**

When applying the engine oil, make sure not to allow the engine oil to enter the fuel rail inside.

   a. Apply a small amount of new engine oil to the O-ring.
   b. Turning the fuel injector assembly to right and left, install it to the fuel rail with care not to damage the O-ring. After the installation, check for its smooth rotation. At this time, check that the projection of the fuel injector assembly is in the centre.
   c. If the rotation is not smooth, the O-ring may be caught. Remove the fuel injector assembly and check the O-ring for damage. After this, re-insert it to the fuel rail and check for its smooth rotation.

16. Install the fuel injector support.
   a. Install the fuel injector support to the fuel injector groove and fuel rail flange, and fix the fuel injector assembly and fuel rail.
17. Install the fuel rail and fuel injector assembly.
   a. Apply a small amount of new engine oil to the O-ring at the end of fuel injector assembly.

***** WARNING**

When installing the fuel rail and fuel injector assembly to the intake manifold, pay attention to avoid damage to the O-ring at the end of the fuel injector assembly.

   b. Install the fuel rail and fuel injector assembly to the intake manifold.
   c. Install the bracket and injector protector rear.
   d. Loosen the intake manifold mounting bolts and nuts (Bolts and nuts 1, 2, 3, and 9 shown in the figure).
   e. Remove the EGR valve.
   f. Loosen the EGR valve support and intake manifold coupling bolts.
   g. Loosen the intake manifold stay mounting bolts.
   h. Temporarily tighten the mounting bolts and nuts of the inlet manifold, bracket, fuel rail and injector protector rear 31 inch lbs. (4 Nm) in the order of number shown in the figure.
   i. Tighten the mounting bolts and nuts of the inlet manifold, bracket, fuel rail and injector protector rear to 15 ft. lbs. (20 Nm) in the order of number shown in the figure again.
   j. Tighten the inlet manifold stay mounting bolts to 15 ft. lbs. (20 Nm).
   k. Tighten the EGR valve support and intake manifold coupling bolts to 15 ft. lbs. (20 Nm).
   l. Install the EGR valve.

18. Install the bracket

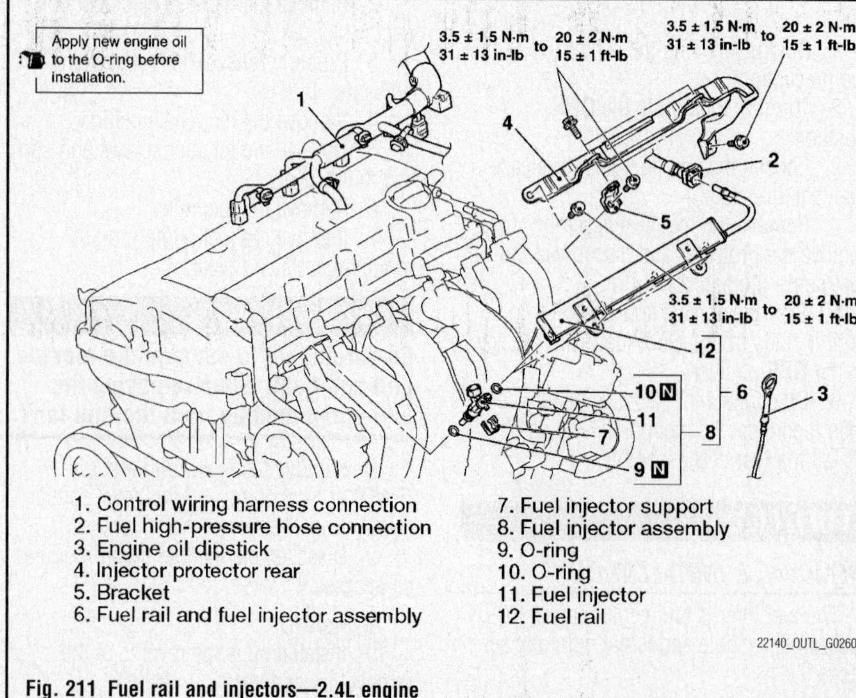

Fig. 211 Fuel rail and injectors—2.4L engine

19. Install the rear injector protector.
20. Install the engine oil dipstick.
21. Connect the fuel high-pressure hose.
 a. Pull up the retainer of fuel high-pressure hose to unlock before installing.
 b. Securely insert the fuel rail stopper into the fuel high-pressure hose connector groove to install the fuel high-pressure hose to the fuel rail.

**** WARNING**

When pushing in the retainer of the fuel high-pressure hose connector, pay attention to avoid damage to the retainer.

 c. After the installation of the fuel high-pressure hose, slightly pull the fuel high-pressure hose to check that it is connected securely. At this time, also check that there is approximately 0.04inch (1 mm) play.
 d. Push in the retainer of the fuel high-pressure hose connector to lock the fuel high-pressure hose and fuel rail.
22. Connect the control wiring harness.
23. Install the air cleaner intake hose.
24. Install the engine upper cover.

3.0L Engine

See Figure 212.

1. Disconnect the fuel pump connector.
2. Remove the intake manifold plenum.
3. Disconnect the fuel injector connectors.
4. Disconnect the fuel high pressure hose.
 a. Follow the steps below to unlock the fuel high-pressure hose connector.
 b. Insert a flat-tipped screwdriver [6 mm (0.24 inch) wide and 1 mm (0.04inch) thick] into the retainer of the fuel high-pressure hose connector.

**** WARNING**

When pushing up the retainer of the fuel high-pressure hose connector, pay attention to avoid damage to the retainer.

 c. Turn the flat-tipped screwdriver inserted into the retainer by 90 degrees to push up the retainer and unlock the fuel high-pressure hose connector.
 d. Disconnect the fuel high-pressure hose.
5. Remove the insulator.
6. Remove the fuel injector support.
7. Remove the fuel injector assembly.

**** WARNING**

Do not drop the fuel injector.

 a. Remove the fuel rail with the fuel injectors attached to it.
8. Remove the O-rings.
9. Repeat the process for the other bank.

To install:

10. Install the top O-ring
 a. Apply a small amount of new engine oil to the O-ring.
 b. While turning the fuel injector to right and left, install the O-ring to the fuel injector with care to avoid damage to the O-ring.
11. Install the bottom O-ring
 a. Apply a small amount of new engine oil to the O-ring.
 b. Using special tool MB992106, install the o-ring onto the fuel injector paying attention to avoid damage to the O-ring.
 c. Install the fuel rail with the fuel injectors attached to it.
12. Install the fuel injector assembly.

**** WARNING**

When applying the engine oil, make sure not to allow the engine oil to enter the fuel rail inside.

 a. Apply a small amount of new engine oil to the O-ring.
 b. Turning the fuel injector assembly to right and left, install it to the fuel rail with care not to damage the O-ring. After the installation, check for its smooth rotation. At this time, check that the projection of the fuel injector assembly is in the centre.
 c. If the rotation is not smooth, the O-ring may be caught. Remove the fuel injector assembly and check the O-ring for damage. After this, re-insert it to the fuel rail and check for its smooth rotation.
13. Install the fuel injector support.
 a. Install the fuel injector support to the fuel injector groove and fuel rail flange, and fix the fuel injector assembly and fuel rail.
14. Install the insulator.
15. Install the fuel rail and injector assembly.

**** WARNING**

When applying the engine oil, make sure not to allow the engine oil to enter the intake manifold inside.

 a. Apply a small amount of new engine oil to the O-ring at the end of fuel injector assembly.

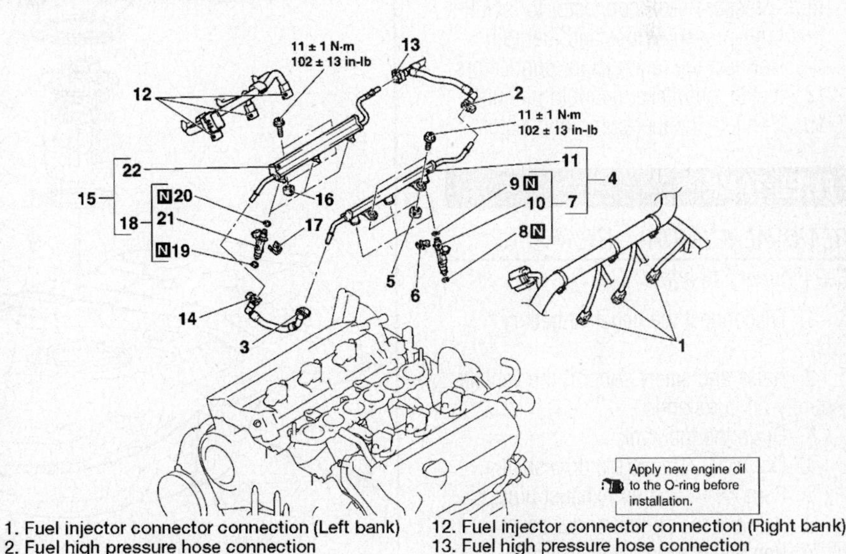

Fig. 212 Fuel rail and injectors—3.0L engine

⁂ WARNING

When installing the fuel rail and fuel injector assembly to the intake manifold, pay attention to avoid damage to the O-ring at the end of the fuel injector assembly.

b. Install the fuel rail and fuel injector assembly to the intake manifold.

c. Tighten the fuel rail mounting bolts to 102 inch lbs. (11 Nm).

16. Connect the fuel high pressure hose.

a. Pull up the retainer of fuel high-pressure hose to unlock before installing.

b. Securely insert the fuel rail stopper into the fuel high-pressure hose connector groove to install the fuel high-pressure hose to the fuel rail.

⁂ WARNING

When pushing in the retainer of the fuel high-pressure hose connector, pay attention to avoid damage to the retainer.

c. After the installation of the fuel high-pressure hose, slightly pull the fuel high-pressure hose to check that it is connected securely. At this time, also check that there is approximately 0.04inch (1 mm) play.

d. Push in the retainer of the fuel high-pressure hose connector to lock the fuel high-pressure hose and fuel rail.

17. Connect the fuel injector connectors.

18. Install the intake manifold plenum.

19. Connect the fuel pump connector.

FUEL TANK

REMOVAL & INSTALLATION

See Figures 213 and 214.

1. Disconnect the negative battery cable.
2. Raise and safely support the vehicle securely on jackstands.
3. Drain the fuel tank.
4. On AWD, remove the driveshaft.
5. Remove the center exhaust pipe.
6. Remove the engine under cover.
7. Remove the rear seat assembly.
8. Disconnect the fuel pump module connector.
9. Disconnect the fuel tank differential pressure sensor connector.
10. Disconnect the fuel main pipe.
11. Disconnect the fuel filler hose.
12. Disconnect the fuel leveling hose.
13. Disconnect the fuel tank vapor hoses.

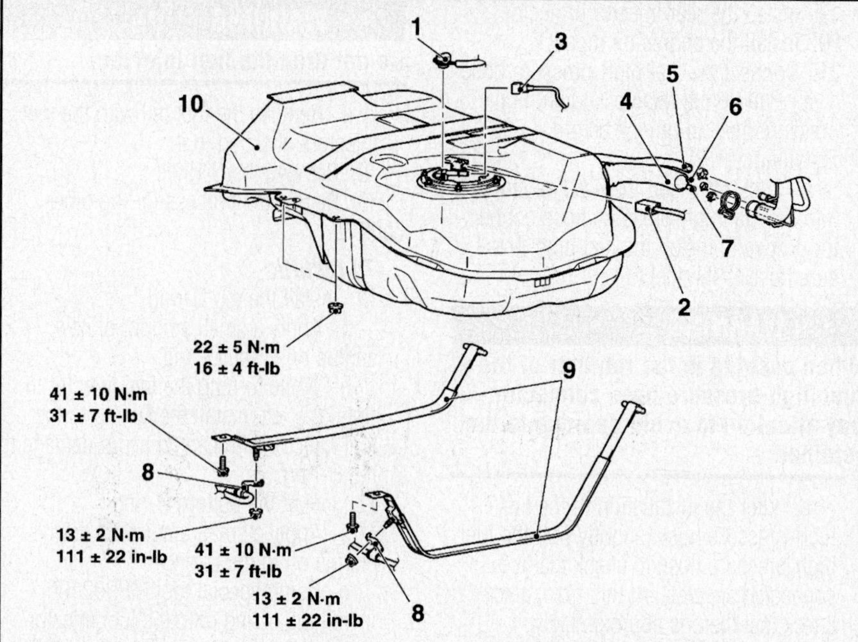

Fig. 213 Fuel tank attaching components

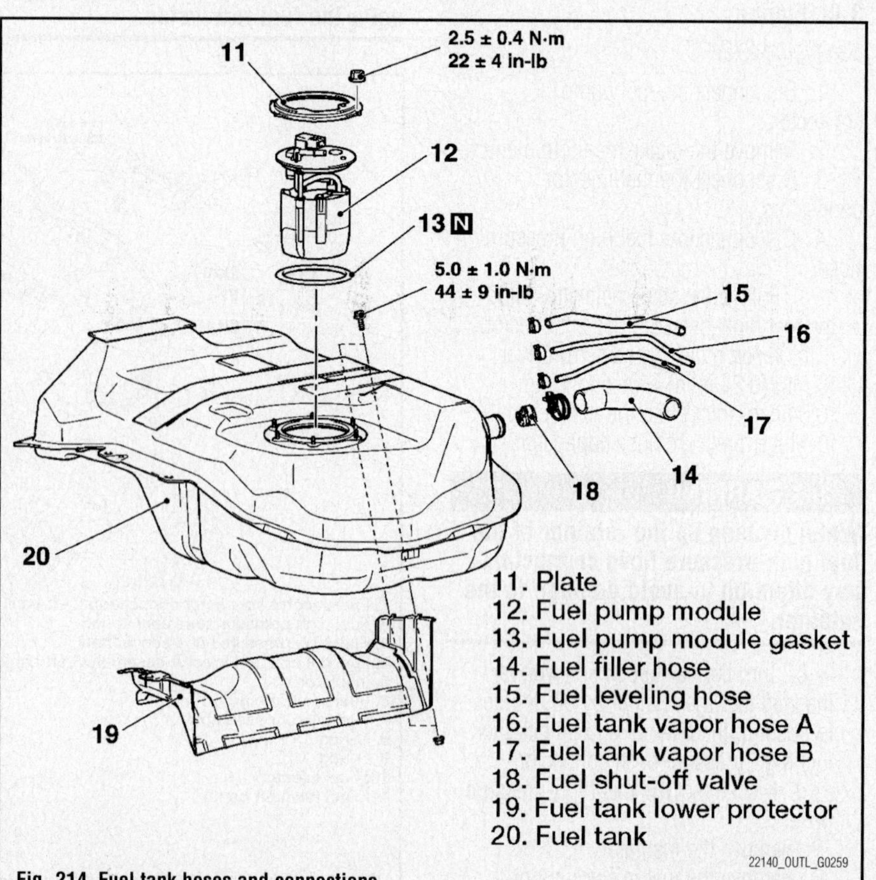

Fig. 214 Fuel tank hoses and connections

14. Disconnect the parking brake rear cable clamp.
15. Hold the fuel tank assembly with a transaxle jack, and remove the connecting bolts of fuel tank band and the connecting nuts of fuel tank assembly.
16. Remove the fuel tank band.
17. On AWD, remove the rear deferential support member mounting bolt.
    a. Remove the mounting bolts of rear differential support member and tilt the differential carrier assembly.
    b. Hold the fuel tank assembly with the transaxle jack, and remove the connecting bolts of fuel tank band and the connecting nuts of fuel tank assembly.
    c. Remove the fuel tank assembly in the tilt direction, paying attention not to bump it against the rear differential carrier.
18. Slowly lower the fuel tank assembly from the vehicle.

To install:

19. Slowly raise the fuel tank assembly into the vehicle.
20. Install the fuel tank band and tighten to 31 ft. lbs. (41 Nm).
21. On AWD, install the rear deferential support member mounting bolt.
22. Connect the parking brake rear cable clamp.
23. Connect the fuel tank vapor hoses.
24. Connect the fuel leveling hose.
25. Connect the fuel filler hose.
26. Connect the fuel main pipe.
27. Connect the fuel tank differential pressure sensor connector.
28. Connect the fuel pump module connector.
29. Install the rear seat assembly.
30. Install the engine under cover.
31. Install the center exhaust pipe.
32. On AWD, install the driveshaft.
33. Fill the fuel tank.
34. Lower the vehicle.
35. Connect the negative battery cable.
36. Check for leaks.

IDLE SPEED

ADJUSTMENT

Idle speed is maintained by the Powertrain Control Module (PCM). No adjustment is necessary or possible.

THROTTLE BODY

REMOVAL & INSTALLATION

2.4L Engine

See Figure 215.

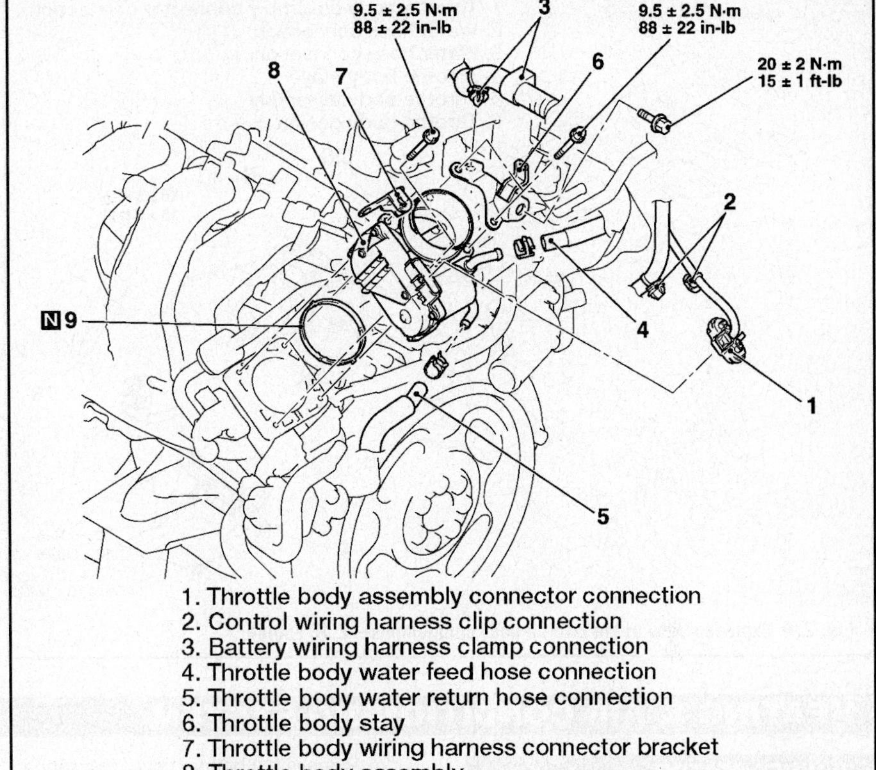

Fig. 215 Exploded view of the throttle body components—2.4L engine

The throttle position sensor is an integral part of the throttle body assembly and cannot be serviced separately.

1. Before servicing the vehicle, refer to the precautions section.
2. Remove the upper engine cover.
3. Drain the engine coolant.
4. Remove the air cleaner intake hose.
5. Disconnect the throttle body connector.
6. Remove the control wiring harness clip.
7. Remove the battery wiring harness clamp.
8. Disconnect the throttle body water feed and return hoses.
9. Remove the throttle body bracket.
10. Remove the throttle body wiring harness connector bracket
11. Remove the throttle body and gasket.

To install:

12. Install the throttle body and gasket. Tighten to 88 inch lbs. (10 Nm).
    a. Fit the throttle body gasket in the intake manifold groove securely without twisting or damage.
13. Install the throttle body wiring harness connector bracket
14. Install the throttle body bracket.
15. Disconnect the throttle body water feed and return hoses.
16. Install the battery wiring harness clamp.
17. Install the control wiring harness clip.
18. Disconnect the throttle body connector.
19. Install the air cleaner intake hose.
20. Fill the engine with coolant.
21. Install the upper engine cover.
22. Start the engine and check for leaks.
23. Perform the initialization procedure. Turn the ignition switch to the **ON** position and then to **OFF** position and hold it for at least 10 seconds.

3.0L Engine

See Figure 216.

1. Before servicing the vehicle, refer to the precautions section.
2. Drain the engine coolant.
3. Remove the air cleaner intake hose.
4. Disconnect the throttle body connector.
5. Disconnect the throttle body water feed and return hoses.

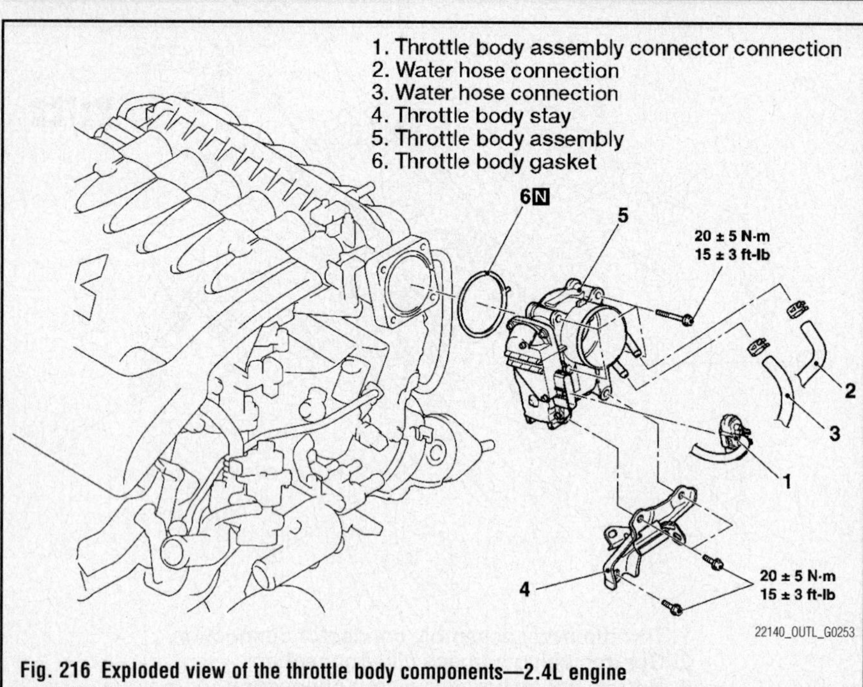

Fig. 216 Exploded view of the throttle body components—2.4L engine

6. Remove the throttle body bracket.
7. Remove the throttle body wiring harness connector bracket
8. Remove the throttle body and gasket.

To install:

9. Install the throttle body and gasket. Tighten to 15 ft. lbs. (20 Nm).
10. Install the throttle body wiring harness connector bracket
11. Install the throttle body bracket.
12. Disconnect the throttle body water feed and return hoses.
13. Disconnect the throttle body connector.
14. Install the air cleaner intake hose.
15. Fill the engine with coolant.
16. Start the engine and check for leaks.
17. Perform the initialization procedure. Turn the ignition switch to the **ON** position and then to **OFF** position and hold it for at least 10 seconds.

HEATING & AIR CONDITIONING SYSTEM

BLOWER MOTOR

REMOVAL & INSTALLATION

See Figure 217.

1. Before servicing the vehicle, refer to the precautions.
2. Remove the bottom cover (passenger side).
3. Disconnect the blower hose.
4. Remove the blower motor

To install:

5. Installation is the reverse of the removal procedure

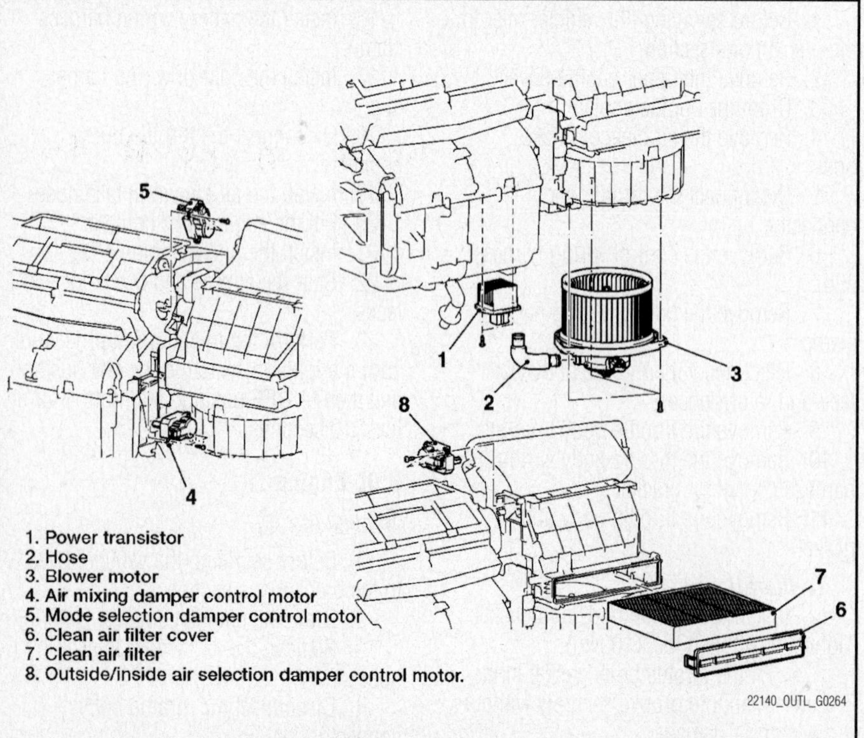

1. Power transistor
2. Hose
3. Blower motor
4. Air mixing damper control motor
5. Mode selection damper control motor
6. Clean air filter cover
7. Clean air filter
8. Outside/inside air selection damper control motor.

22140_OUTL_G0264

Fig. 217 Exploded view of the blower motor and related components

HEATER CORE

REMOVAL & INSTALLATION

See Figures 218 through 220.

1. Drain the engine coolant.
2. Discharge and recycle the refrigerant.
3. Remove the front seat assembly.
4. Remove the rear heater duct
   a. Remove the inner front scuff plate and cowl side trim.
   b. Remove the selector lever, selector lever panel, glove box, lower panel, console side cover and center console.
   c. Remove the rear heater ducts.
5. Remove the instrument panel.
6. Disconnect the wiring harness and clamps.
7. Disconnect the heater hose.
8. Remove the heat protector.
9. Disconnect the suction pipe.

⁂ WARNING

Use the plug which is not breathable because A/C compressor oil or receiver are highly hygroscopic.

10. Disconnect the liquid pipe.
11. Remove the O-ring.
12. Disconnect the drain hose.
13. Remove the front deck crossmember.
14. Remove the heater unit assembly.
15. Remove the heater core from the heater unit.

To install:

16. Install the heater core into the heater unit.
17. Install the heater unit assembly. Tighten fasteners to 98 inch lbs. (11 Nm).
18. Install the front deck crossmember.
19. Connect the drain hose.
20. Coat the O-ring with compressor oil and install.
21. Connect the liquid pipe. Tighten to 48 inch lbs. (5 Nm).

WARNING

Use the plug which is not breathable because A/C compressor oil or receiver are highly hygroscopic.

22. Connect the suction pipe.
23. Install the heat protector.
24. Connect the heater hose.
25. Connect the wiring harness and clamps.
26. Install the instrument panel.
    a. Install the rear heater ducts. console.
    b. Install the selector lever, selector lever panel, glove box, lower panel, console side cover and center
    c. Install the inner front scuff plate and cowl side trim.
27. Install the front seat assembly.
28. Evacuate and charge the refrigerant system.
29. Fill the engine with coolant.
30. Check for leaks.

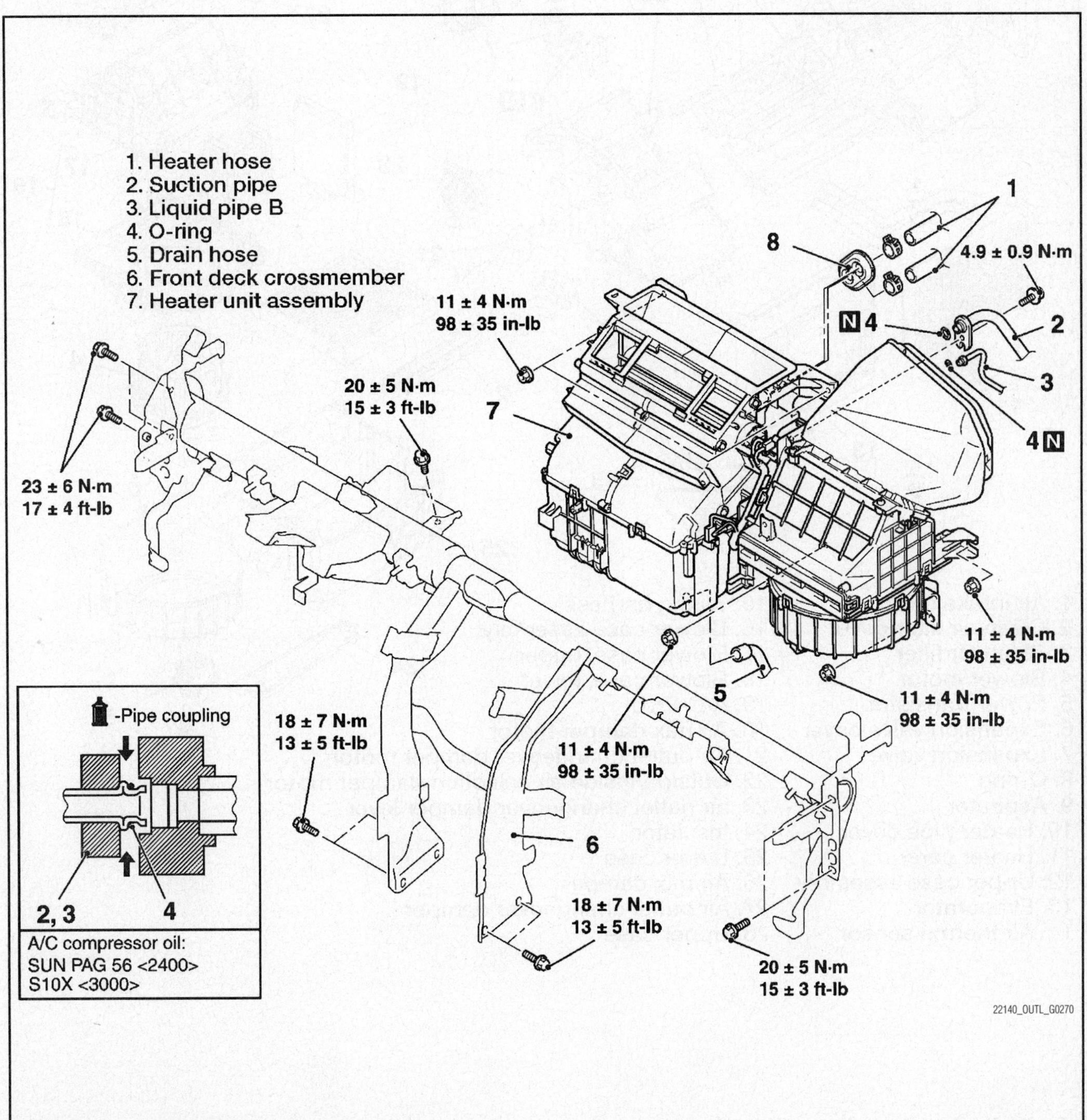

Fig. 218 Heater unit and related components

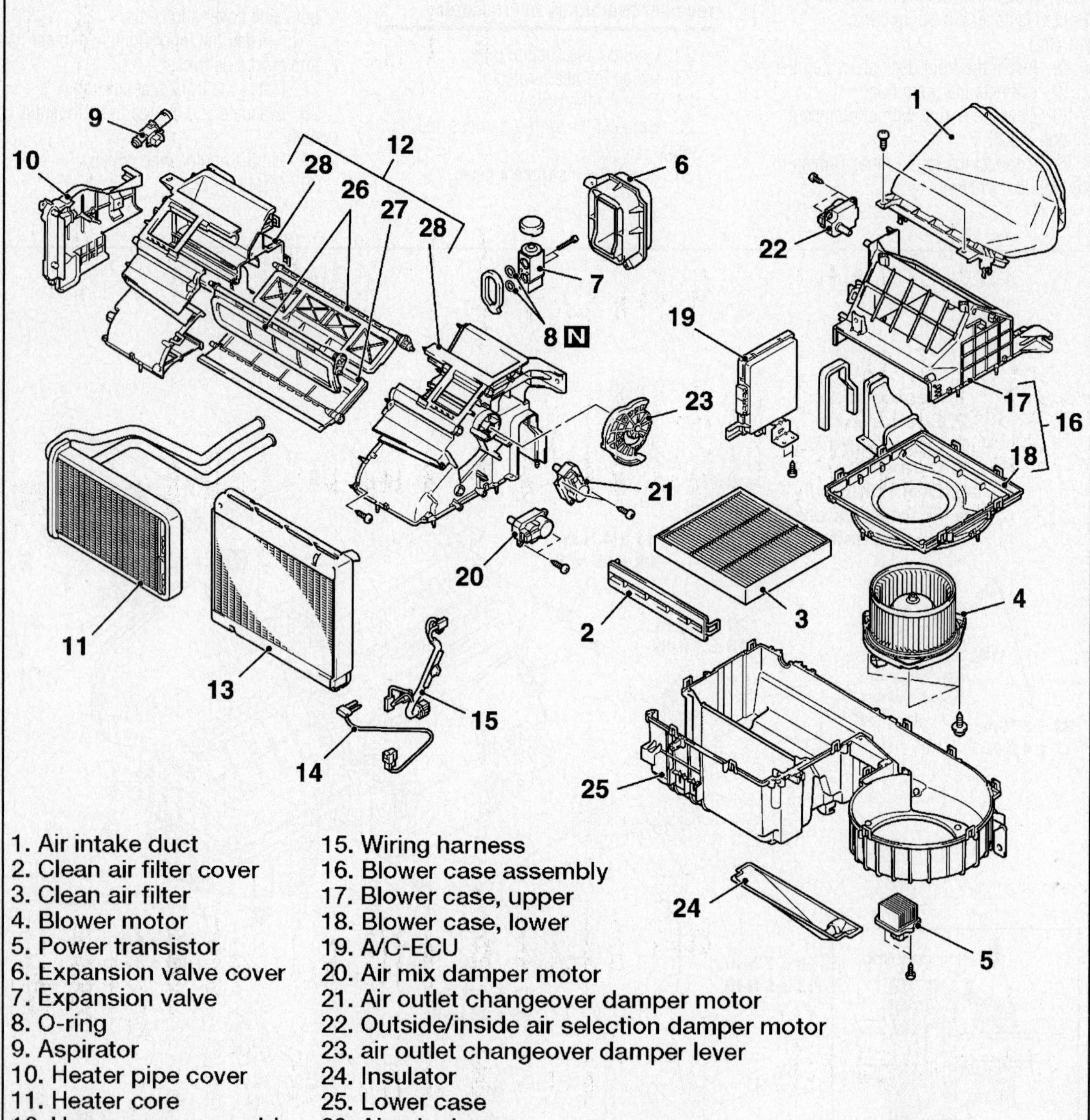

Fig. 219 Exploded view of the heater unit internal components

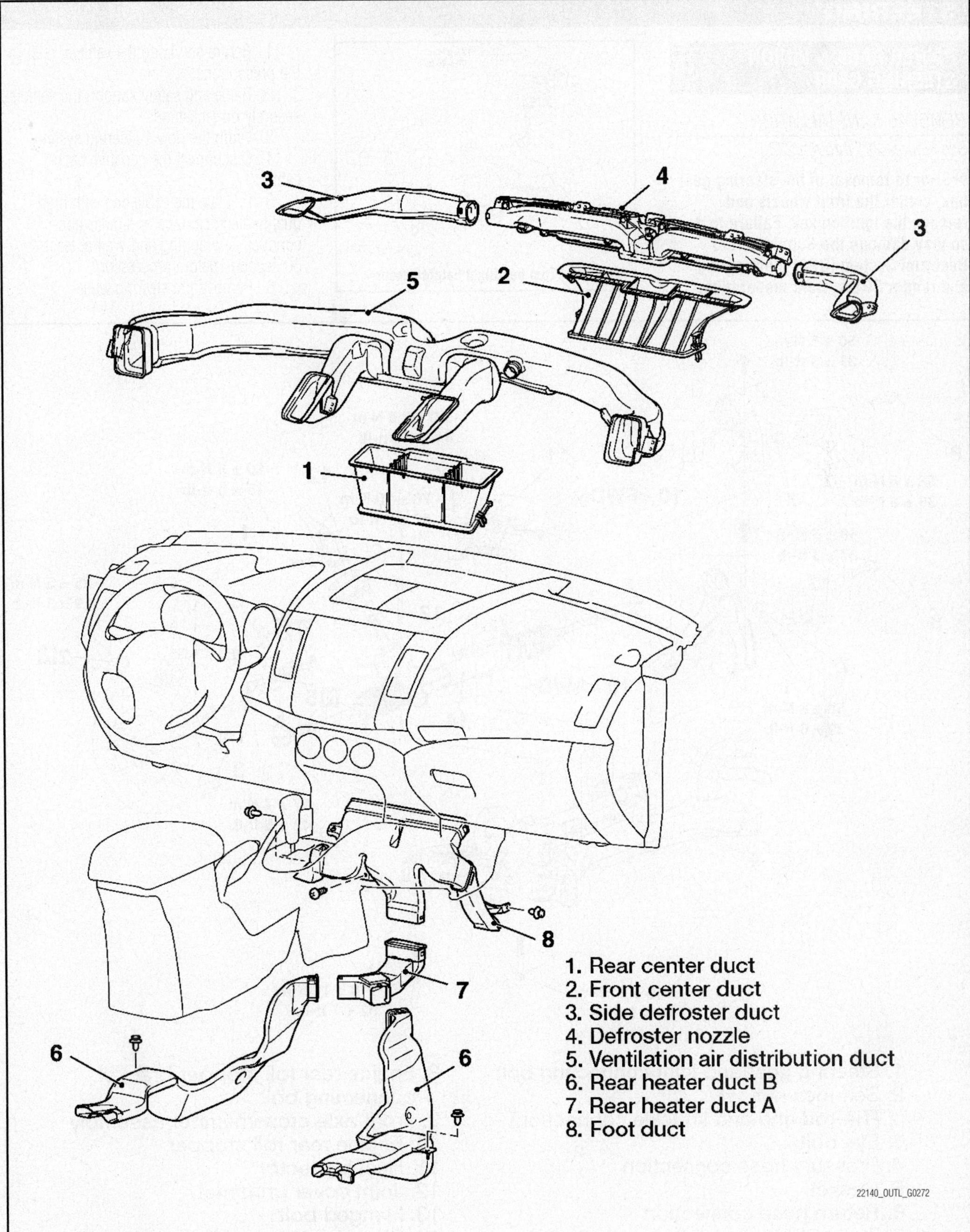

Fig. 220 Ventilation and air distribution ducts

STEERING

POWER RACK & PINION STEERING GEAR

REMOVAL & INSTALLATION

See Figures 221 through 223.

➡Prior to removal of the steering gear box, center the front wheels and remove the ignition key. Failure to do so may damage the Supplemental Restraint System (SRS) clock spring and render SRS system inoperative.

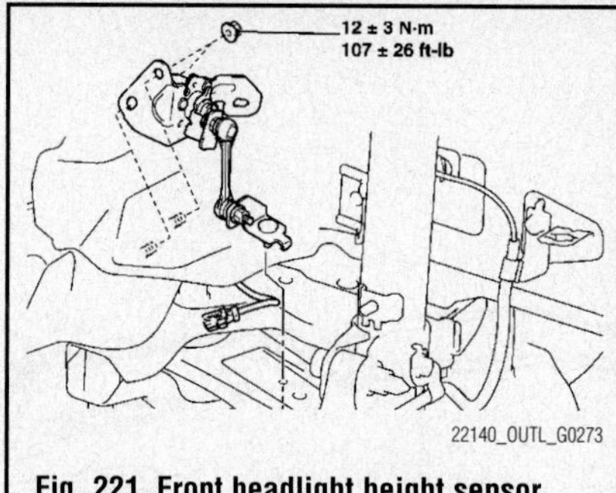

Fig. 221 Front headlight height sensor

1. Before servicing the vehicle, refer to the precautions.
2. Raise and safely support the vehicle securely on jackstands.
3. Drain the power steering system.
4. Disconnect the negative battery cable.
5. Insulate the cable end with high-quality electrical tape or similar non-conductive wrapping and Wait at least 60 seconds before proceeding.
6. Remove the steering wheel.

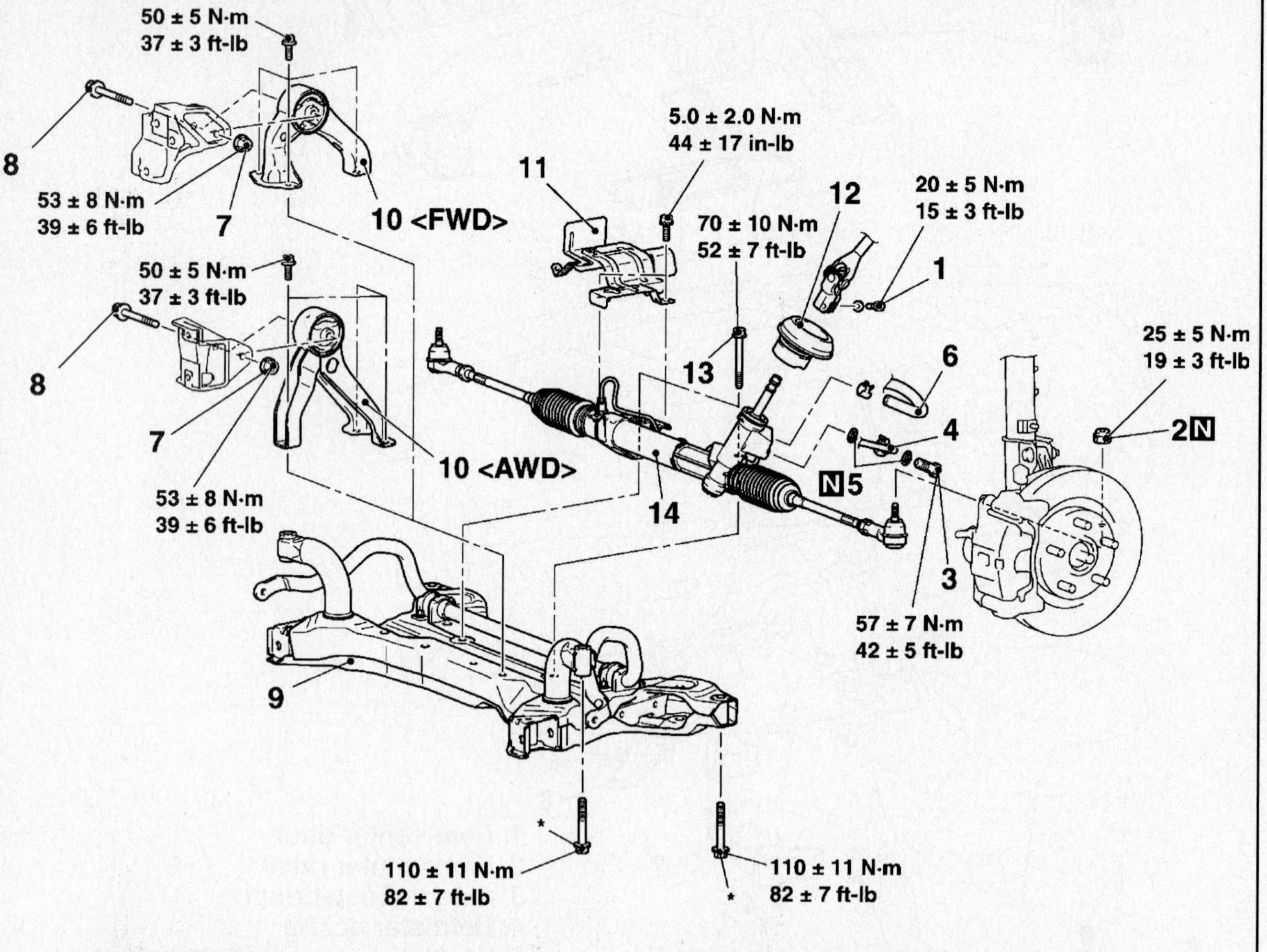

1. Steering gear and joint connecting bolt
2. Self-lock nut (Tie-rod end and knuckle connection)
3. Eye bolt
4. Pressure hose connection
5. Gasket
6. Return hose connection
7. Flanged nut
8. Engine rear roll stopper bracket connecting bolt
9. Front axle crossmember assembly
10. Engine rear roll stopper
11. Heat protector
12. Joint cover grommet
13. Flanged bolt
14. Steering gear and linkage

22140_OUTL_G0274

Fig. 222 Rack and pinion steering components—2.4L engine

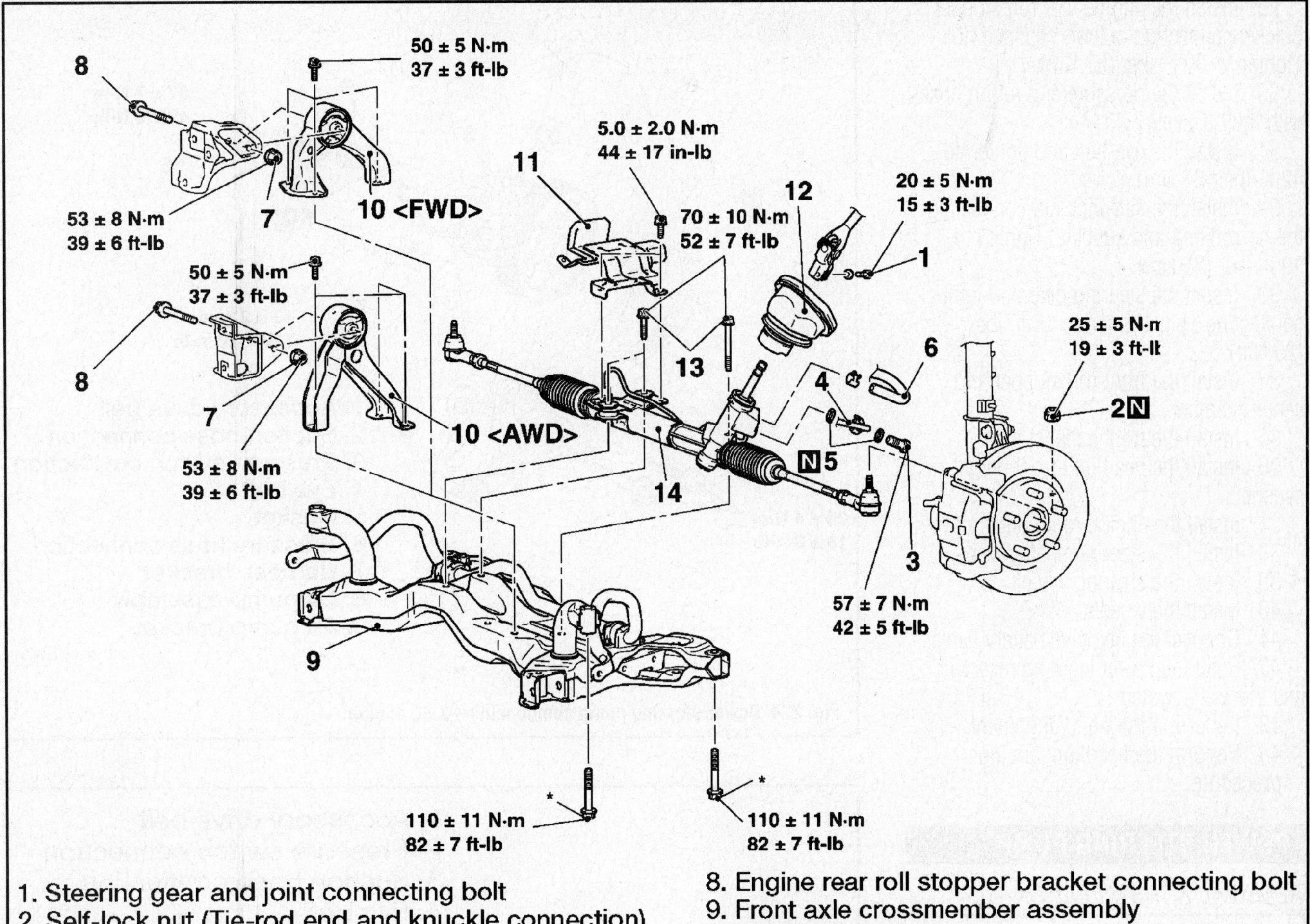

Fig. 223 Rack and pinion steering components—3.0L engine

7. Remove the steering shaft cover.
8. Remove the engine room under covers.
9. Remove the headlight front height sensor.
10. Remove the front exhaust pipe.
11. Remove the front roll stopper and center member.
12. Remove the steering gear and joint connecting bolt.
13. Remove the self-lock nut connecting the tie-rod end and knuckle.

⁂ WARNING

Loosen the self-lock nut from the ball joint. Do not remove.

a. Install the ball joint remover (Special tool MB991897 or MB992011).

b. Turn the bolt and knob to make the special tool jaws parallel, then hand-tighten the bolt. After tightening, check that the jaws are still parallel.

c. Unscrew the bolt to disconnect the ball joint.

14. Remove the eye bolt.
15. Remove the pressure and return hoses with their gaskets.

connection

16. Remove the flanged nut.
17. Remove the engine rear roll stopper bracket connecting bolt.
18. Remove the front axle crossmember.

a. Jack up and support the crossmember, and remove it.

b. Check the hoses and harnesses for roughness, and then remove the front axle No.1 crossmember assembly with the rear roll stopper and the steering gear and linkage installed.

19. Remove the engine rear roll stopper.
20. Remove the heat protector.
21. Remove the joint cover grommet
22. Remove the flanged bolt.
23. Remove the steering gear and linkage

To install:

24. Install the steering gear and linkage. Tighten the flanged bolt to 52 ft. lbs. (70 Nm).
25. Install the joint cover grommet.

a. Install the joint cover grommet to the steering gear and linkage by aligning the mating marks.

26. Install the heat protector. Tighten to 44 inch lbs. (5 Nm).
27. Install the engine rear roll stopper. Tighten to 37 ft. lbs. (50 Nm).
28. Install the front axle crossmember. Tighten to 82 ft. lbs. (110 Nm).

29. Install the engine rear roll stopper bracket connecting bolt and flanged nut. Tighten to 39 ft. lbs. (53 Nm).
30. Install the pressure and return hoses with their gaskets.
31. Install the eye bolt and tighten to 42 ft. lbs. (57 Nm).
32. Install the self-lock nut connecting the tie-rod end and knuckle. Tighten to 19 ft. lbs. (25 Nm).
33. Install the steering gear and joint connecting bolt. Tighten to 15 ft. lbs. (20 Nm).
34. Install the front roll stopper and center member.
35. Install the front exhaust pipe.
36. Install the headlight front height sensor.
37. Install the engine room under covers.
38. Install the steering shaft cover.
39. Install the steering wheel.
40. Install the wheels.
41. Connect the negative battery cable.
42. Refill the power steering reservoir and bleed the system.
43. Perform a front end alignment.
44. Perform the headlight aiming procedure.

POWER STEERING PUMP

REMOVAL & INSTALLATION

See Figures 224 and 225.

1. Before servicing the vehicle, refer to the precautions.
2. Drain the power steering fluid.
3. Remove the right engine room side cover.
4. Remove the accessory drive belt.
5. Remove the eye bolt and gasket.
6. Disconnect the suction and pressure hoses.
7. Remove the pressure switch.
8. Remove the harness bracket
9. Remove the oil pump assembly.

To install:

10. Install the oil pump assembly. Tighten bolts to 19 ft. lbs. (25 Nm). On 3.0L engines, tighten two additional bolts to 37 ft. lbs. (50 Nm) as shown.
11. Install the harness bracket
12. Install the pressure switch.
13. Connect the suction and pressure hoses.
14. Install the eye bolt and gasket. Tighten to 42 ft. lbs. (57 Nm).
15. Install the accessory drive belt.
16. Install the right engine room side cover.
17. Fill and bleed the power steering system.

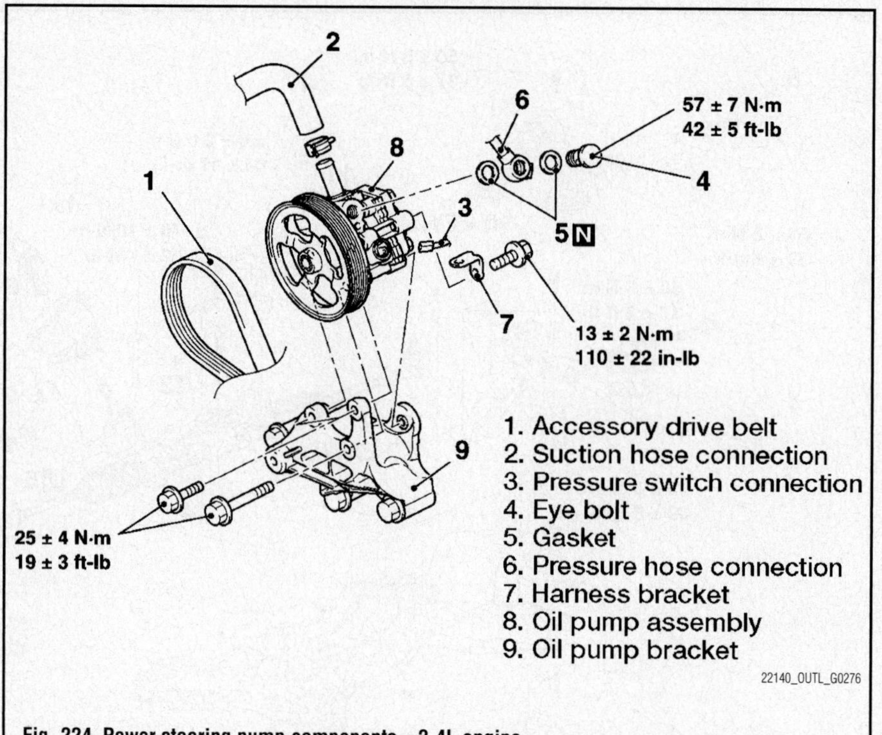

Fig. 224 Power steering pump components—2.4L engine

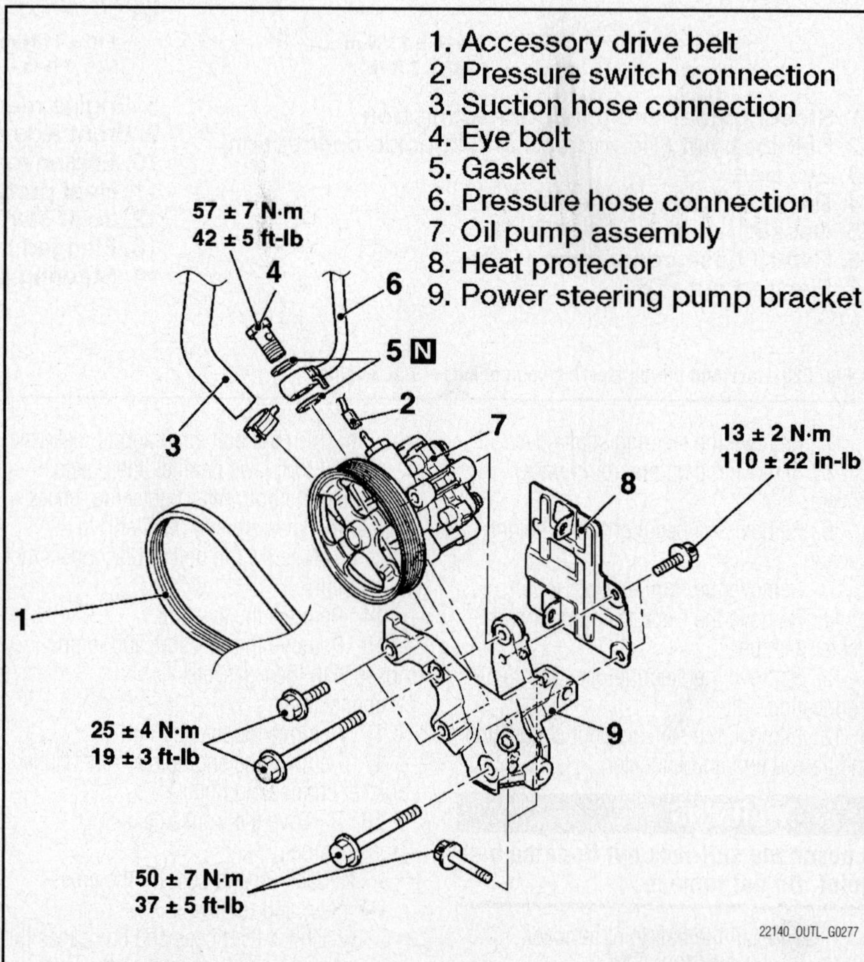

Fig. 225 Power steering pump components—3.0L engine

BLEEDING

1. Raise and support the front wheels.
2. Disconnect the all of the injector connectors.

WARNING

Perform air bleeding only while cranking the engine. If air bleeding is performed while the engine is running, air could enter the fluid. During air bleeding, refill the steering fluid supply so that the level never falls below the lower mark on the dipstick.

3. Turn the steering wheel all the way to the left and right five or six times while using the starter motor to crank the engine intermittently several times (for 15 to 20 seconds).
4. Connect the all of the injector connectors.
5. Start the engine (idling).
6. Turn the steering wheel to the left and right until there are no air bubbles in the oil reservoir.
7. Confirm that the fluid is not milky, and that the level is between the high and low dipstick marks.
8. Confirm that there is very little change in the fluid level when the steering wheel is turned left and right.

WARNING

If the fluid level rises suddenly after the engine is stopped, the air has not been completely bled. If air bleeding is not complete, there will be abnormal noises from the pump and the flow-control valve, and this condition could cause reduce the life of the power steering components.

9. Confirm that the change in the fluid level is no more than 5 mm (0.2 inch) when the engine is stopped and when it is running.
10. If the change of the fluid level is 5 mm (0.2 inch) or more, the air has not been completely bled from the system. The air bleeding procedure must be repeated.
11. Use a scan tool to check if a DTC is set. If the diagnosis code is set, erase it.

STEERING LINKAGE

REMOVAL & INSTALLATION

Tie Rods

See Figure 226.

1. Before servicing the vehicle, refer to the precautions.
2. Raise and safely support the vehicle securely on jackstands.
3. Remove the wheels.
4. Disconnect tie rod end.
   a. Remove the cotter pin and nut.
   b. Using special tool special tool MB991897 or a tie rod ball joint separator, disconnect tie rod end from the steering knuckle.
5. Matchmark the tie rod end to the tie rod.

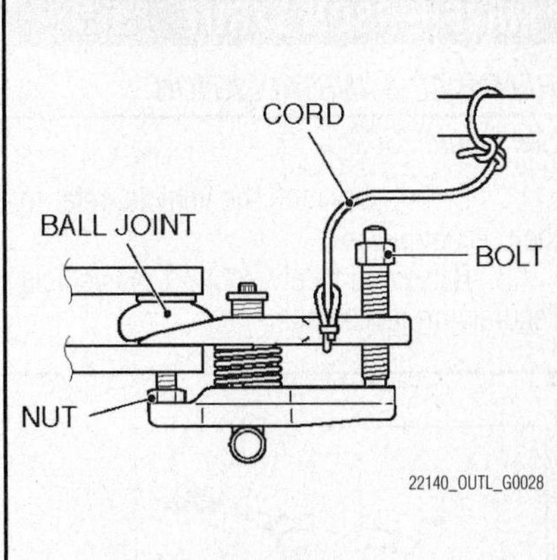

Fig. 226 Using special tool MB991897 tie rod ball joint separator

6. Loosen the tie rod end locking nut.
7. Carefully remove the tie rod end, counting the number of revolutions (or threads) for installation reference.

To install:

8. Install the tie rod end.
   a. Carefully install the tie rod end, counting the number of revolutions until it reaches the reference from the removal.
   b. Tighten the tie rod end locking nut.
   c. Connect the tie rod end to the steering knuckle and tighten the nut to 19 ft. lbs. (25 Nm).
   d. Install a new cotter pin.
9. Install the wheels.
10. Lower the vehicle.
11. Check the toe-in measurement.

SUSPENSION

LOWER CONTROL ARM

REMOVAL & INSTALLATION

See Figures 227 and 231.

1. Before servicing the vehicle, refer to the precautions.
2. Raise and safely support the vehicle securely on jackstands.
3. Remove the headlight front suspension height sensor.
4. Remove the lower arm ball joint nut (self-locking nut).
5. Remove the lower arm to chassis bolts.
6. Remove the lower arm assembly

To install:

7. Using your fingers, press the dust cover to check for a crack or damage.
8. Install the lower arm assembly.
9. Install the lower arm to chassis bolts. Tighten to 81 ft. lbs. (110 Nm).
10. Install the lower arm ball joint nut (self-locking nut). Tighten to 53 ft. lbs. (71 Nm).
11. Install the headlight front suspension height sensor.
12. Lower the vehicle.
13. Check the wheel alignment.
14. Check the headlight aim.

CONTROL ARM BUSHING REPLACEMENT

See Figures 228 and 229.

Use the following special tools to remove the bushing:

- MB992119 (Arm bushing remover and installer)
- MB990979 (Ring)
- MB990890 (Rear suspension bushing base)

Use the following special tools to press-fit the bushing.

- MB992119 (Arm bushing remover and installer)
- MB990979 (Ring)
- MB990890 (Rear suspension bushing base)

1. Press-fit the bushing so that the bushing protrusion is in the direction shown in the figure.
2. Press-fit the bushing until the special tool contacts with the lower arm assembly.

MACPHERSON STRUT

REMOVAL & INSTALLATION

See Figure 230.

1. Before servicing the vehicle, refer to the precautions.
2. Raise and safely support the vehicle securely on jackstands.
3. Remove the headlight front suspension height sensor.
4. Remove the wheel speed sensor clamp (strut side).
5. Remove the brake hose clamp (strut side).
6. Disconnect the stabilizer link and strut.
7. Disconnect the knuckle and strut connection (self-locking nut).
8. Remove the stabilizer link mounting nuts.
9. Remove the strut assembly.

To install:

10. Install the strut assembly.
11. Install the stabilizer link mounting nuts.
12. Connect the knuckle and strut connection (self-locking nut).
13. Connect the stabilizer link and strut.
14. Install the brake hose clamp (strut side).
15. Install the wheel speed sensor clamp (strut side).
16. Install the headlight front suspension height sensor.
17. Raise and safely support the vehicle securely on jackstands.
18. Install the headlight front suspension height sensor.
19. Lower the vehicle.
20. Check the wheel alignment.
21. Check the headlight aim.

Specified grease: Multipurpose grease SAE J310, NLGI No.2 or equivalent

110 ± 10 N·m 81 ± 8 ft-lb

71 ± 10 N·m 53 ± 7 ft-lb

22140_OUTL_G0282

Fig. 227 Lower control arm assembly

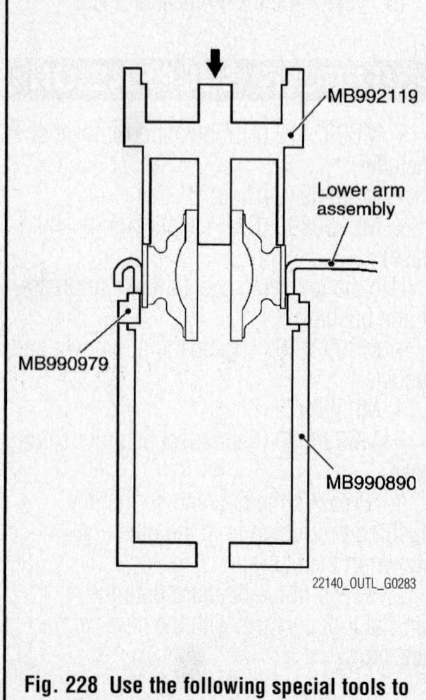
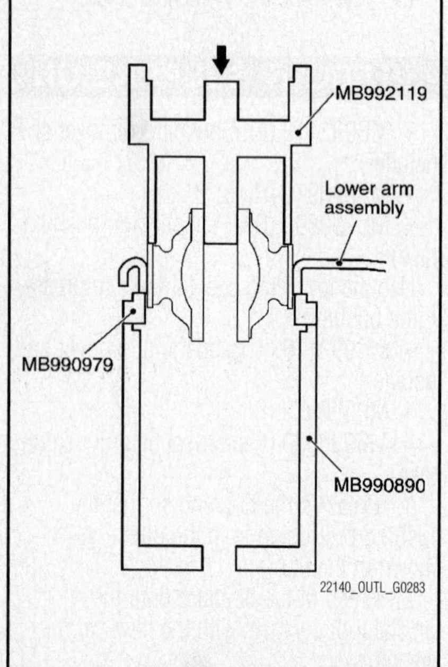

Fig. 228 Use the following special tools to remove the bushing

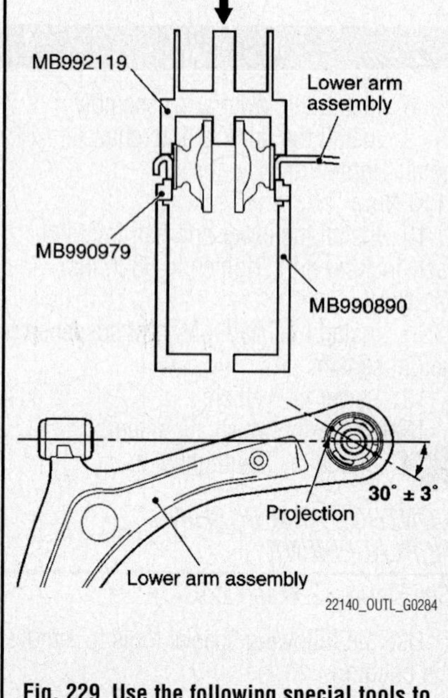

Fig. 229 Use the following special tools to press-fit the bushing

OVERHAUL

See Figure 231.

1. Before servicing the vehicle, refer to the precautions.
2. Remove the strut assembly from the vehicle.
3. Compress the coil spring using a spring compressor until the spring just comes away from one of the seats.
4. Remove or disconnect the following:
   - Center nut from the strut. Do NOT use an impact wrench to remove the center nut.
   - Strut insulator assembly
   - Bearing
   - Upper spring seat and upper spring pad
   - Rubber bumper
   - Coil spring
   - Lower spring pad
   - Strut assembly

To install:

5. Install or connect the following:
   - Strut assembly
   - Lower spring pad
   - Compressed spring on the strut assembly
   - Rubber bumper
   - Upper spring pad and spring seat
   - Bearing
   - Strut insulator
   - Nut and tighten it to 38–52 ft. lbs. (50–70 Nm)
   - Strut to the vehicle

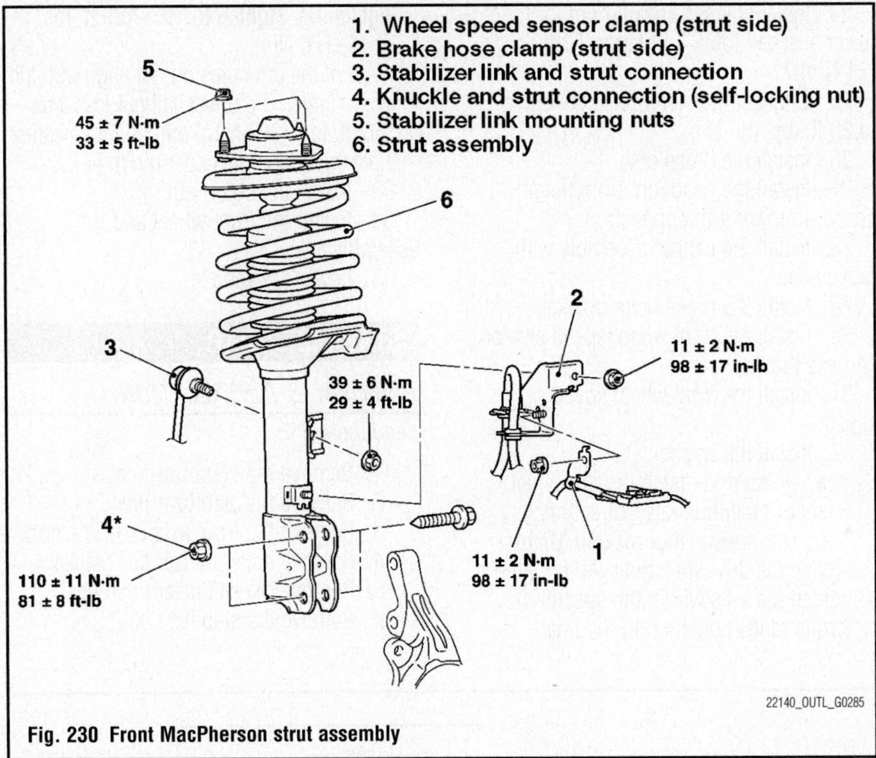

Fig. 230 Front MacPherson strut assembly

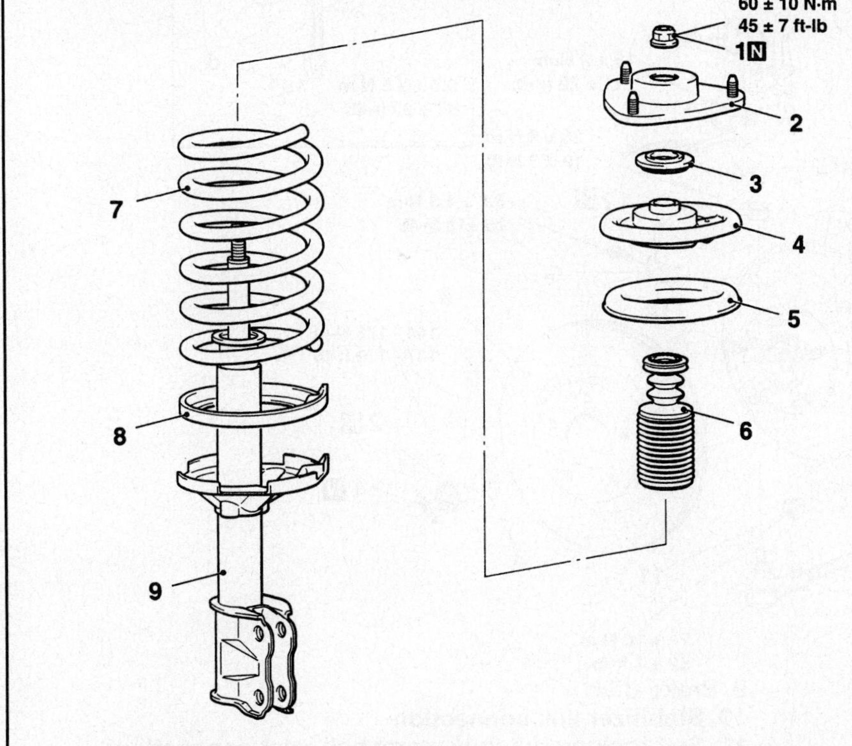

Fig. 231 Front strut assembly exploded view

STEERING KNUCKLE

REMOVAL & INSTALLATION

See Figure 232.

➡If the vehicle is equipped with ABS, do not strike the ABS rotors installed to the outer BL outer race of the halfshaft against other parts when removing or installing the halfshaft. Otherwise the ABS rotors will be damaged.

1. Before servicing the vehicle, refer to the precautions.
2. Raise and safely support the vehicle.
3. Remove the front wheel and tire assembly.
4. Remove the cotter pin.
5. Remove the driveshaft nut.

✱✱ WARNING

Do not apply the vehicle weight on the front wheel hub assembly with the driveshaft nut loosened. Otherwise, the wheel bearing may be broken.

   a. Use special tool MB990767 to hold the hub when removing the driveshaft nut.

6. Remove the washer.
7. Remove the front wheel speed sensor.
8. Remove the front wheel speed sensor harness bracket.
9. Remove the brake hose bracket.
10. Remove the caliper assembly with brake hose.
11. Secure the removed caliper assembly with a wire or other similar material at a position where it will not interfere with the removal and installation of the hub knuckle assembly.
12. Remove the headlight front height sensor to lower arm connection.
13. Remove the brake disc.
14. Disconnect the stabilizer link.
15. Remove the self-locking nut for the lower arm ball joint.
16. Remove the self-locking nut for the tie-rod end.

✱✱ WARNING

Loosen the self-locking nut (tie-rod end connection) from the ball joint, but do not remove here. Use the special tool. To prevent the special tool from dropping off, suspend it with a cord.

   a. Install special tool MB991897 or MB992011 as shown in the figure.

b. Turn the bolt and knob to make the special tool jaws parallel, then hand-tighten the bolt. After tightening, check that the jaws are still parallel.

c. Unscrew the bolt to disconnect the ball joint.

17. Remove the driveshaft.

a. If the driveshaft is seized, use special tools MB990242 and MB990244, MB991354 and MB990767 to push the driveshaft out from the hub.

18. Remove the hub knuckle assembly and strut mounting bolt and nuts.

19. Remove the hub knuckle assembly.

To install:

20. Install the hub knuckle assembly.

21. Install the hub knuckle assembly and strut mounting bolt and nuts. Tighten to 81 ft. lbs. (110 Nm).

22. Install the driveshaft.

23. Install the self-locking nut for the tie-rod end. Tighten to19 ft. lbs. (25 Nm).

24. Install the self-locking nut for the lower arm ball joint. Tighten to 53 ft. lbs. (71 Nm).

25. Connect the stabilizer link. Tighten to 29 ft. lbs. (39 Nm).

26. Install the brake disc.

27. Install the headlight front height sensor to lower arm connection.

28. Install the caliper assembly with brake hose.

29. Install the brake hose bracket.

30. Install the front wheel speed sensor harness bracket.

31. Install the front wheel speed sensor.

32. Install the washer.

a. Be sure to install the driveshaft washer in the illustrated direction.

b. Use special tool MB990767 to tighten the driveshaft nuts. At this time, tighten the nuts within the specified torque range considering the final tightening. Tighten to 107–129 ft. lbs. (144–176 Nm).

c. If the pin holes do not align with the pins, tighten the driveshaft nut less than 147 ft. lbs. (200 Nm)] and find the nearest hole then bend the cotter pin to fit it.

33. Install the cotter pin.

34. Install the front wheel and tire assembly.

35. Lower the vehicle.

STABILIZER BAR

REMOVAL & INSTALLATION

See Figure 233.

1. Remove the stabilizer nuts.
2. Remove the stabilizer link.
3. Remove the front axle crossmember.
4. Remove the stabilizer bar bracket.
5. Remove the stabilizer bushing.
6. Remove the stabilizer bar.

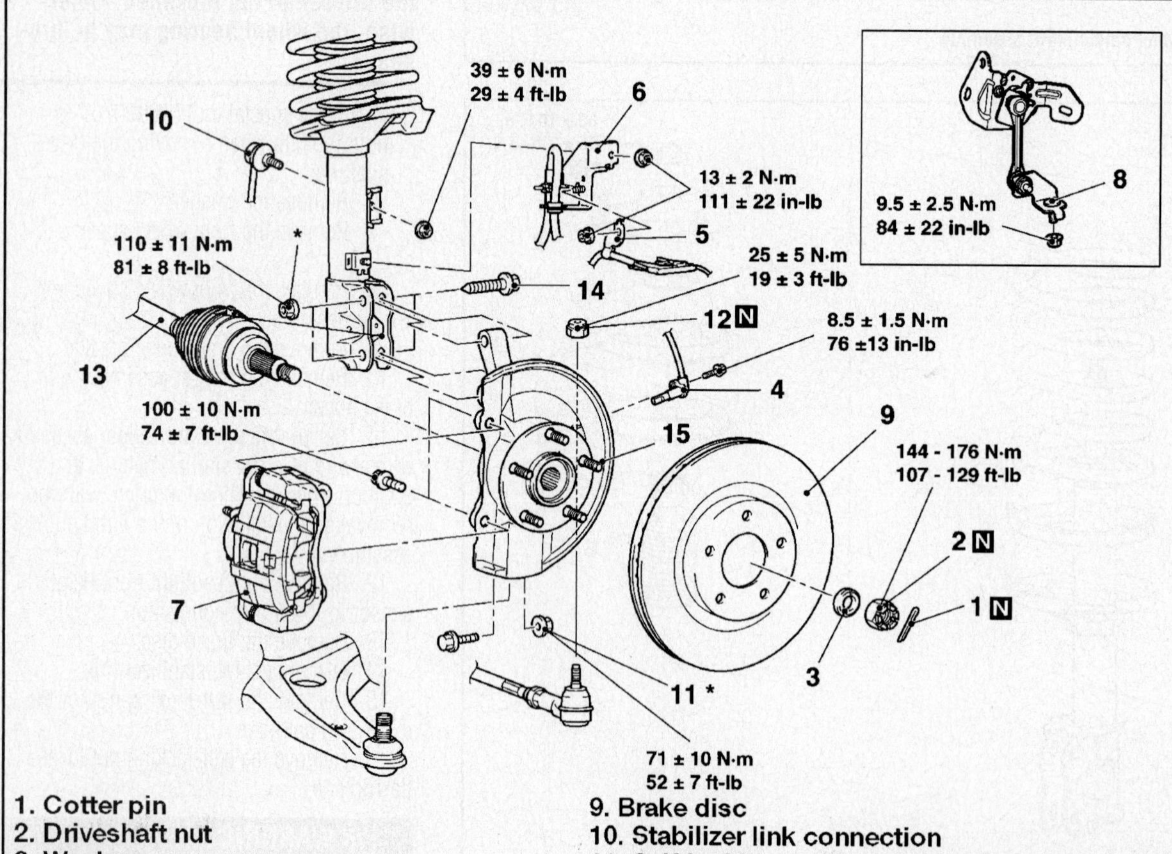

Fig. 232 Steering knuckle assembly

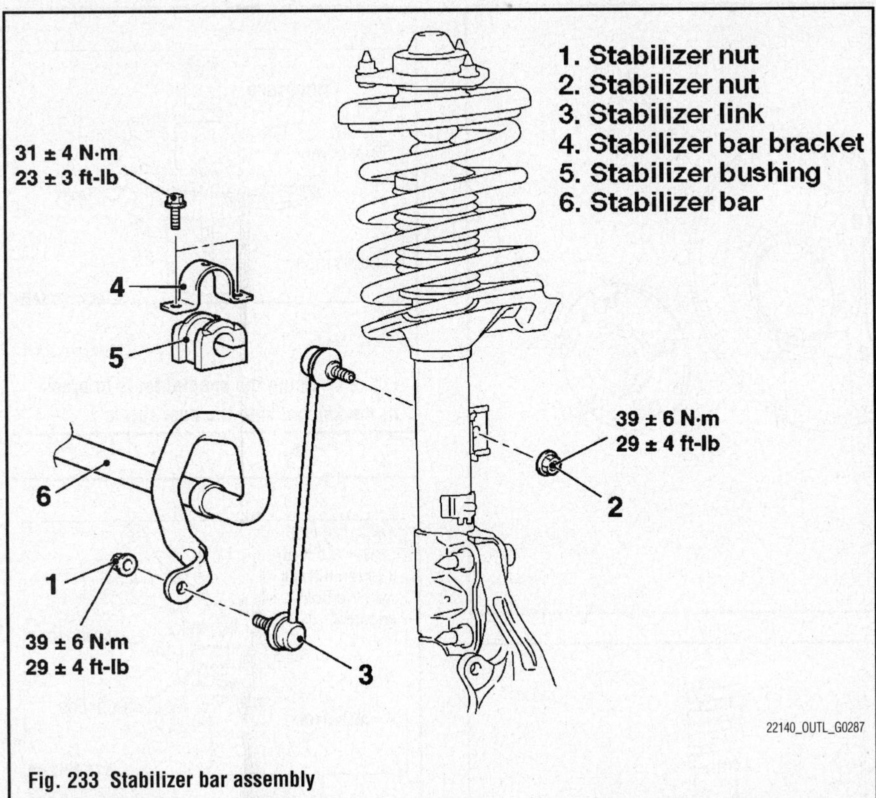

Fig. 233 Stabilizer bar assembly

To install:

7. Install the stabilizer bar.
8. Install the stabilizer bushing.
9. Install the stabilizer bar bracket.

Tighten bushing bracket bolts to 23 ft. lbs. (31 Nm).

10. Install the front axle crossmember.
11. Install the stabilizer link.
12. Install the stabilizer link nuts.

Tighten to 29 ft. lbs. (39 Nm).

WHEEL HUB AND BEARING

REMOVAL & INSTALLATION

See Figures 234 through 243.

1. Raise and safely support the vehicle securely on jackstands.
2. Remove the front wheel.
3. Remove the hub.
   a. Replace special tool MB992250 with a guide of special MB991355 as shown in the figure.
   b. Insert special tool MB992250 in the knuckle and tighten it with a bolt and nut.

➡Set the spacers of special tool MB992250 as shown in the figure.

   c. Use special tools MB991000, MB991017, MB991355 and MB992250 to remove the hub.
4. Remove the dust shield.
5. Remove the snap ring.

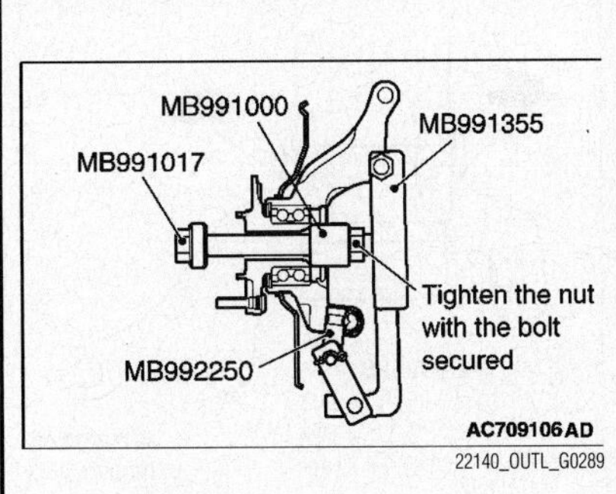

Fig. 234 Using the special tools to remove the hub

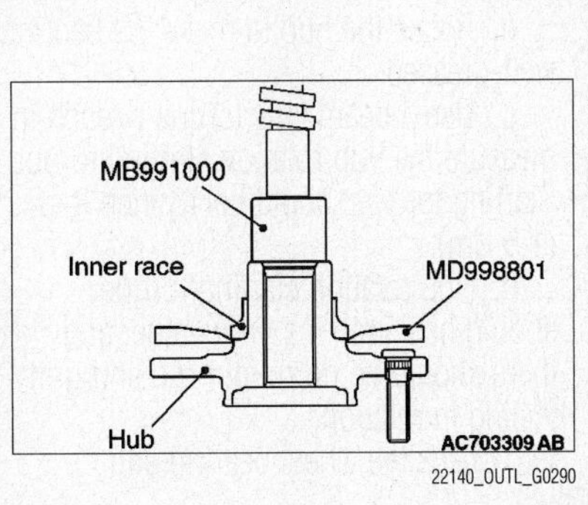

Fig. 235 Using the special tool to remove the wheel bearing inner race from the hub

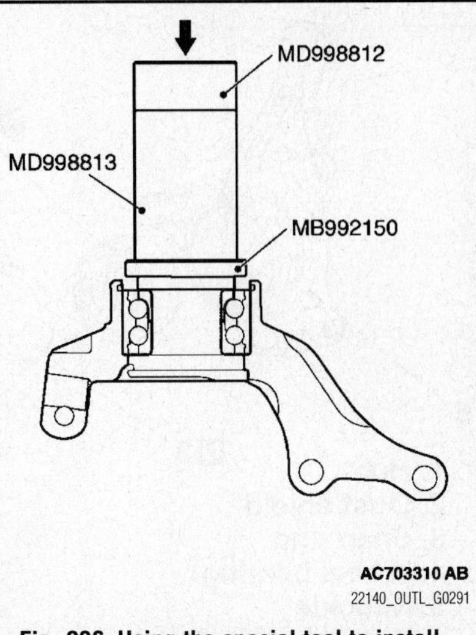

Fig. 236 Using the special tool to install the wheel bearing inner race from the hub

6. Remove the wheel bearing.
   a. Use special tools MD998801 and MB991000 to remove the wheel bearing inner race (outside) from the hub.
   b. Use special tools MB992150, MD998812 and MD998813 to assemble the inner race (outside) removed from the hub to the wheel bearing.
   c. Use special tools MB990935 and MB990938 to remove the wheel bearing.

To install:

7. Install the wheel bearing.

⁂ WARNING

The magnetic encoder for wheel speed sensor is installed in the wheel bearing.

   a. Install the wheel bearing so that the encoder is positioned in the direction shown in the figure.
   b. When press-fit the wheel bearing, push the outer race. After press-fit the wheel bearing, wipe off the extra grease in order not to remain on the magnetic encoder.
   c. Remove grease and foreign material cleanly from the inside of knuckle bore.
   d. Apply Molykote BR2 Plus thinly and evenly to the inside of knuckle as shown in the figure.
   e. Use special tools MB990883 and MB 990890 to press-fit the wheel bearing.
   f. Remove excessive grease seeped out between knuckle and wheel bearing outer race after press-fitting the wheel bearing.
8. Install the dust shield.
   a. Use special tools MB991388 and MB991576 to press-fit the knuckle into the dust shield.

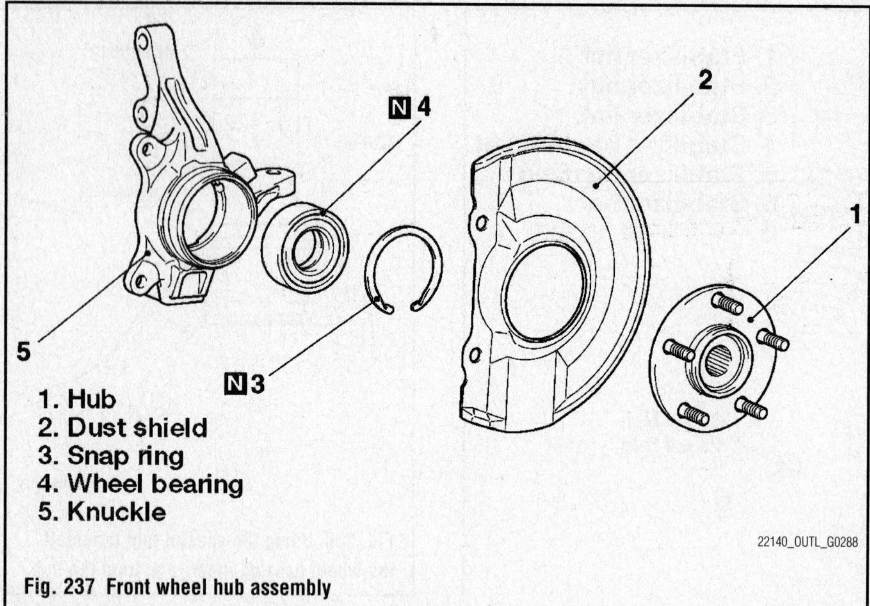

Fig. 237 Front wheel hub assembly

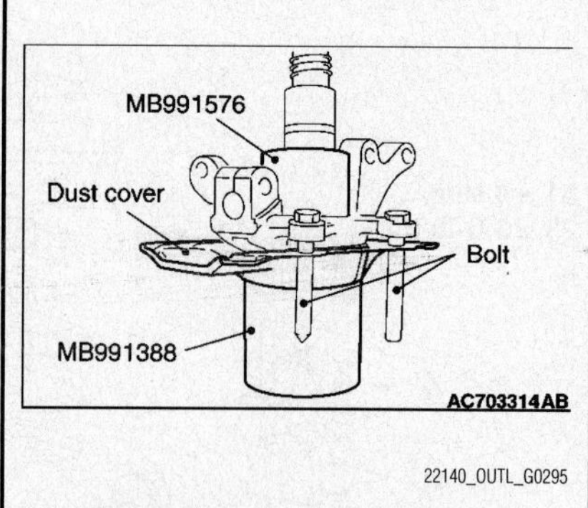

Fig. 241 Using the special tools to press-fit the knuckle into the dust shield

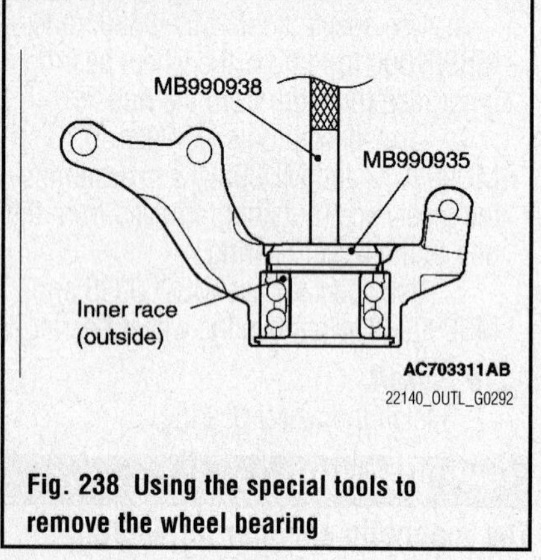

Fig. 238 Using the special tools to remove the wheel bearing

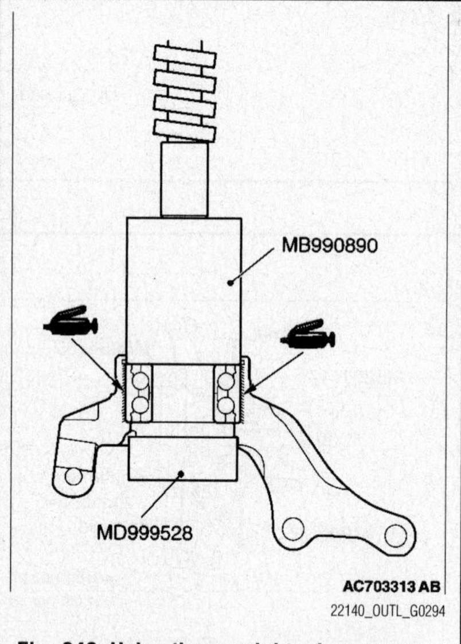

Fig. 240 Using the special tools to press-fit the wheel bearing

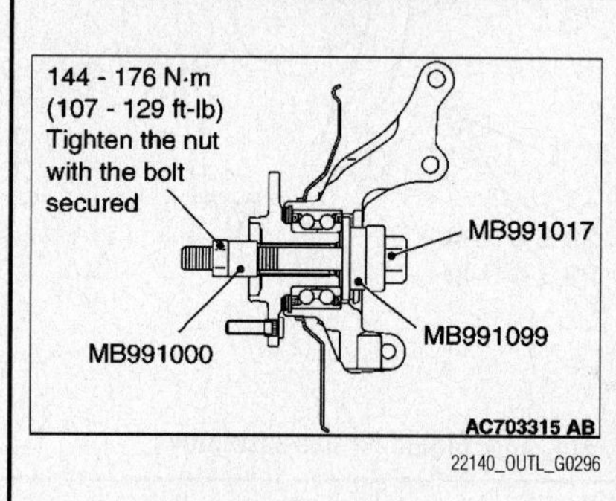

Fig. 242 Using the special tools as shown in the figure, tighten the nut to the specified torque, and press-fit the hub into the knuckle

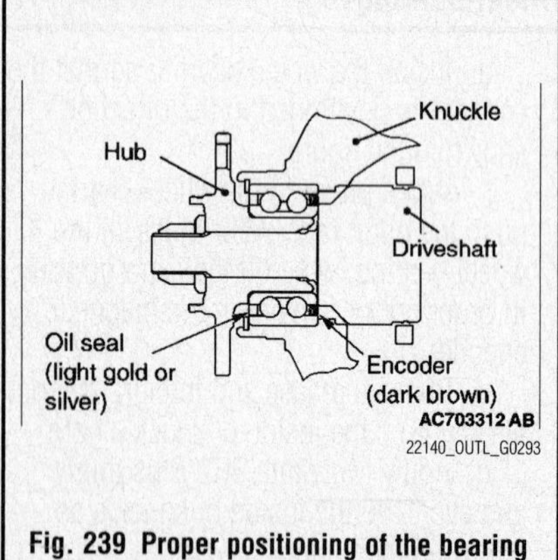

Fig. 239 Proper positioning of the bearing

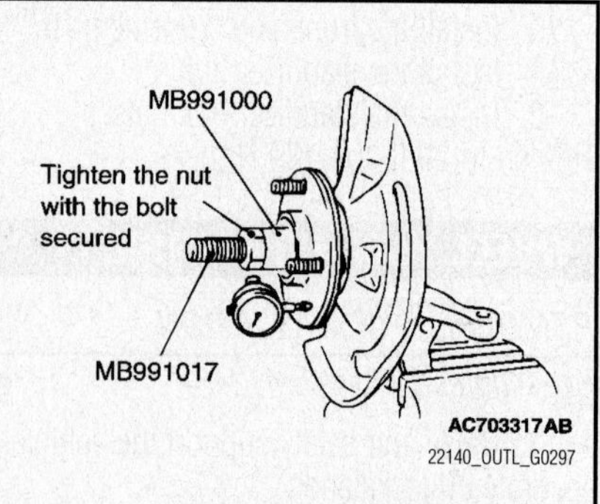

Fig. 243 Using the special tools to measure to determine whether the wheel bearing end play is within the specified limit

➡Use the bolts (M12) to align the caliper mounting holes.

9. Install the hub.
10. Check the hub rotation starting torque.
    a. Set special tools MB991000 and MB991017 as shown in the figure, tighten the nut to 107–129 ft. lbs. (144–176 Nm), and press-fit the hub into the knuckle.
    b. Rotate the hub to make the bearing well-greased.
    c. Use a beam type torque wrench to measure the hub rotation starting torque. Starting torque should be 13 inch lbs. (1.5 Nm).
    d. Hub rotation starting torque should be within the limit value, and there should be no roughness and gritty feeling in rotation.
11. Check the wheel bearing end play.
    a. Use special tools MB991000 and MB991017 to measure to determine whether the wheel bearing end play is 0.002 inch (0.05 mm).
    b. If the end play is not within the limit range while the nut is tightened to specified torque, the bearing, hub and/or knuckle have probably not been installed correctly. Replace the bearing and re-install.

ADJUSTMENT

The front wheel bearings are not adjustable. If the bearings are noisy or become loose, they must be replaced.

COIL SPRING

REMOVAL & INSTALLATION

2.4L Engine

See Figures 244 and 245.

1. Remove the coil over shock assembly from the vehicle.
2. Remove the self-locking nut.

⁂ WARNING

To hold the coil spring securely, install the spring compressor evenly, and so that the space between both arms of the special tool will be maximum within the installation range.

   a. The locking nut for the piston rod inside the shock absorber may be loose. Do not use an impact wrench to loosen the self-locking nut.
   b. While holding the piston rod, remove the self-locking nut.
3. Remove the washer.
4. Remove the bushing "B".
5. Remove the collar.
6. Remove the spring upper bracket assembly.
7. Remove the spring upper pad.
8. Remove the bushing "A".
9. Remove the plate.
10. Remove the bump rubber.
11. Remove the coil spring.

To install:

12. Install a spring compressor to compress the coil spring, and install it to the lower spring pad.
13. Align the end of the coil spring with the shock absorber as shown in the figure.
14. Assemble the coil over shock in the reverse of the removal procedure.
15. Counterhold the piston rod of the shock absorber, and tighten the self-locking nut to 19ft. lbs. (25 Nm).
16. Install the coil over shock assembly on the vehicle.

3.0L Engine

See Figure 246.

1. Before servicing the vehicle, refer to the precautions.
2. Raise and safely support the vehicle.
3. Disconnect the rear height sensor to control link connection.
4. Disconnect the stabilizer link.
5. Disconnect the lower arm from the trailing arm.
   a. While jacking-up the lower arm with the garage jack, remove the mounting bolts.
6. Disconnect the shock absorber from the lower arm.
7. Remove the coil spring.
8. Remove the coil spring upper and lower pads.

To install:

➡The suspension components should not be fully tightened until the vehicle's weight is resting on its wheels.

9. Install the coil spring upper and lower pads.
10. Install the coil spring.
11. Connect the shock absorber to the lower arm.
   a. While jacking-up the lower arm with the garage jack, Install the mounting bolts.
12. Connect the lower arm and trailing arm.
13. Connect the stabilizer link.
14. Temporarily tighten the suspension bolts.
15. Connect the rear height sensor to control link connection.
16. Lower the vehicle.
17. Finish tightening the bolts as specified in the illustration.
18. Check the wheel alignment.
19. Check the headlight aiming.

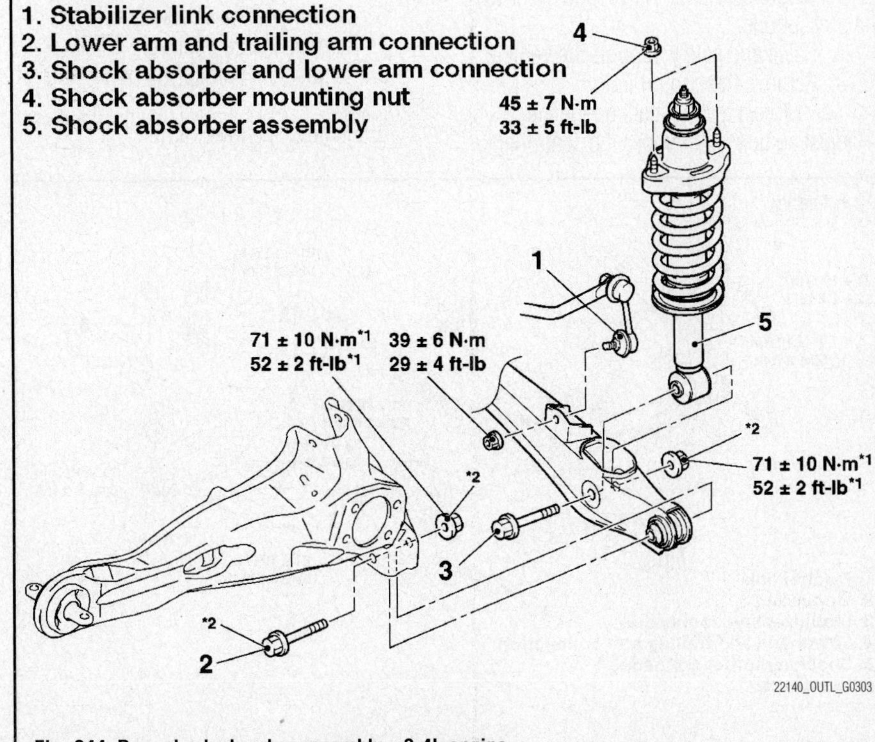

Fig. 244 Rear shock absorber assembly—2.4L engine

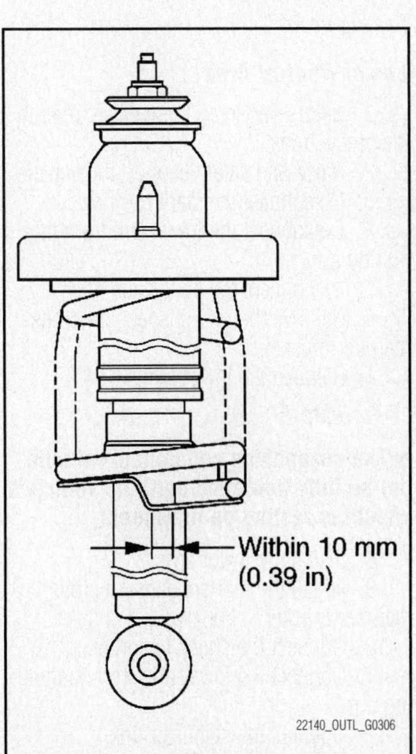

Fig. 245 Align the end of the coil spring with the shock absorber as shown

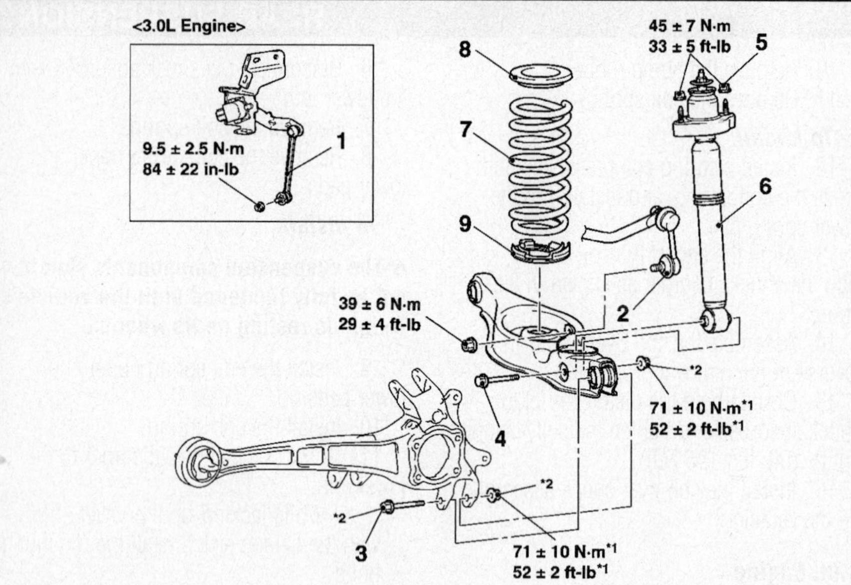

Fig. 246 Rear shock absorber assembly—3.0L engine

CONTROL ARMS/LINKS

REMOVAL & INSTALLATION

2.4L Engine

See Figure 247.

Lower Control Arm

1. Before servicing the vehicle, refer to the precautions.
2. Raise and safely support the vehicle.
3. Disconnect the stabilizer link.
4. Disconnect the lower arm from the trailing arm.
5. Disconnect the shock absorber.
6. Remove the rear suspension cross-member bracket.
7. Remove the lower arm.

To install:

➡The suspension components should not be fully tightened until the vehicle's weight is resting on its wheels.

8. Install the lower arm.
9. Install the rear suspension cross-member bracket.
10. Connect the shock absorber.
11. Connect the lower arm from the trailing arm.
12. Connect the stabilizer link.
13. Temporarily tighten the suspension bolts.
14. Lower the vehicle.
15. Finish tightening the bolts as specified in the illustration.
16. Check the wheel alignment.
17. Check the headlight aiming.

Upper Control Arm

1. Before servicing the vehicle, refer to the precautions.
2. Raise and safely support the vehicle.
3. Remove the control link.
   a. Make a mating mark on the toe adjusting bolt, and remove the control link.
4. Disconnect the fuel tank vapor hose.
5. Remove the upper arm.

To install:

➡The suspension components should not be fully tightened until the vehicle's weight is resting on its wheels.

6. Install the upper arm.
   a. Install the upper arm so that the hole faces the body side.
7. Connect the fuel tank vapor hose.
8. Install the control link.
   a. Align the mating mark on the toe adjusting bolt.
9. Temporarily tighten the suspension bolts.
10. Lower the vehicle.
11. Finish tightening the bolts as specified in the illustration.
12. Check the wheel alignment.
13. Check the headlight aiming.

3.0L Engine

See Figure 248.

Lower Control Arm

1. Before servicing the vehicle, refer to the precautions.
2. Raise and safely support the vehicle.
3. Disconnect the rear height sensor to control link.
4. Disconnect the stabilizer link.
5. Disconnect the lower arm and trailing arm connection.
6. Disconnect the shock absorber.
7. Remove the coil spring.
8. Remove the coil spring upper and lower pads.
9. Remove the lower arm.

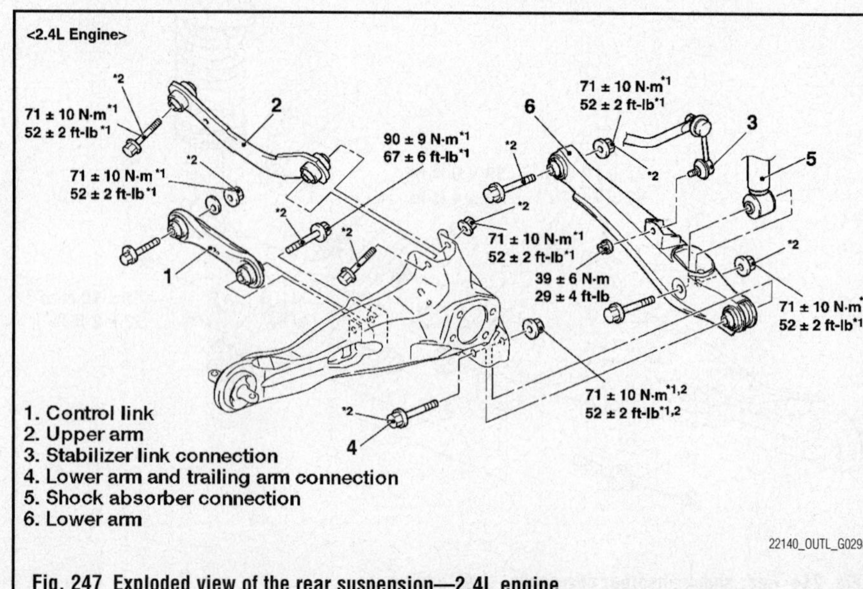

Fig. 247 Exploded view of the rear suspension—2.4L engine

To install:

➡The suspension components should not be fully tightened until the vehicle's weight is resting on its wheels.

10. Install the lower arm.
11. Install the coil spring upper and lower pads.
12. Install the coil spring.
13. Connect the shock absorber.
14. Connect the lower arm and trailing arm connection.
15. Connect the stabilizer link.
16. Connect the rear height sensor to control link.
17. Temporarily tighten the suspension bolts.
18. Lower the vehicle.
19. Finish tightening the bolts as specified in the illustration.
20. Check the wheel alignment.
21. Check the headlight aiming.

Upper Control Arm

1. Before servicing the vehicle, refer to the precautions.
2. Raise and safely support the vehicle.
3. Disconnect the rear height sensor to control link.
4. Remove the control link.
   a. Make a mating mark on the toe adjusting bolt, and remove the control link.
5. Remove the upper arm.

To install:

➡The suspension components should not be fully tightened until the vehicle's weight is resting on its wheels.

6. Install the upper arm.
   a. Install the upper arm so that the hole faces the body side.

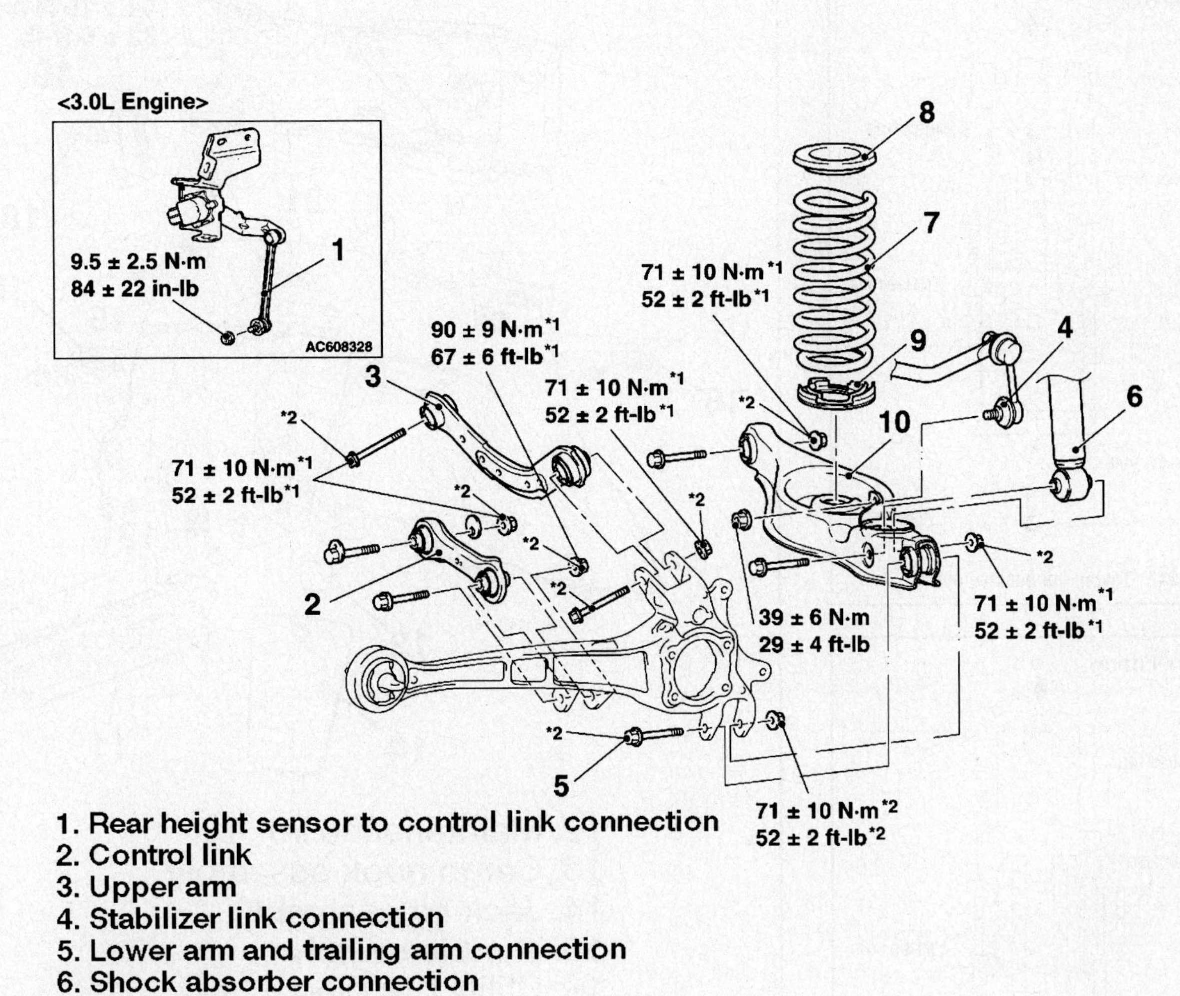

Fig. 248 Exploded view of the rear suspension—3.0L engine

7. Install the control link.
   a. Align the mating mark on the toe adjusting bolt.
8. Connect the rear height sensor to control link.
9. Temporarily tighten the suspension bolts.
10. Lower the vehicle.
11. Finish tightening the bolts as specified in the illustration.
12. Check the wheel alignment.
13. Check the headlight aiming.

CONTROL ARM BUSHING REPLACEMENT

See Figures 249 and 250.

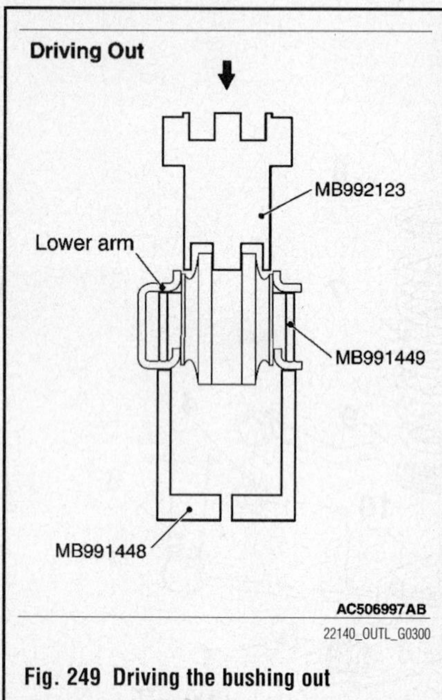

Fig. 249 Driving the bushing out

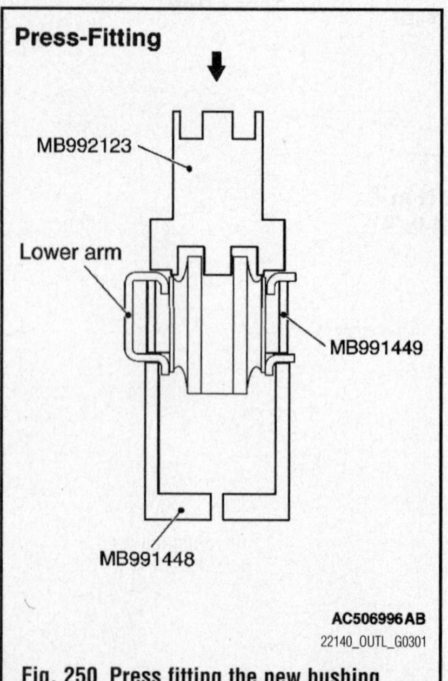

Fig. 250 Press fitting the new bushing

Use the following special tools to remove and install the bushing:

- MB992123: Arm Bushing Remover and Installer
- MB991448: Bushing Remover and Installer Base
- MB991449: Bushing Remover and Installer Supporter

✲✲ WARNING

As the bushing has different outer diameters at both ends, be careful not to confuse the removal direction with the press-fit direction.

Use the special tools MB992123, MB991448 and MB991449 to remove and press-fit the lower arm bushing.

SHOCK ABSORBER

REMOVAL & INSTALLATION

2.4L Engine

See Figures 251 and 252.

1. Before servicing the vehicle, refer to the precautions.
2. Raise and safely support the vehicle.
3. Remove the quarter trim.

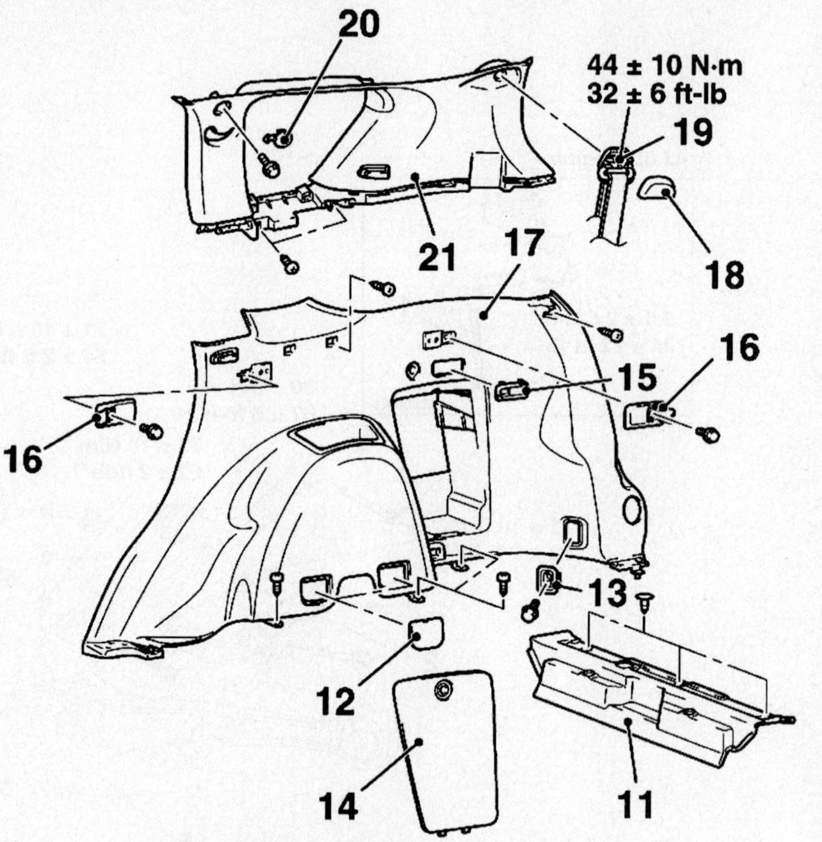

44 ± 10 N·m
32 ± 6 ft-lb

12. Maintenance lid
13. Cargo hook assembly
14. Jack lid assembly
15. Tumble switch
16. Utility bar plug
17. Quarter trim, lower
18. Sash guide cover
19. Third seat belt mounting bolt
20. Quarter trim upper cap
21. Quarter trim, upper

22140_OUTL_G0302

Fig. 251 Rear quarter trim components

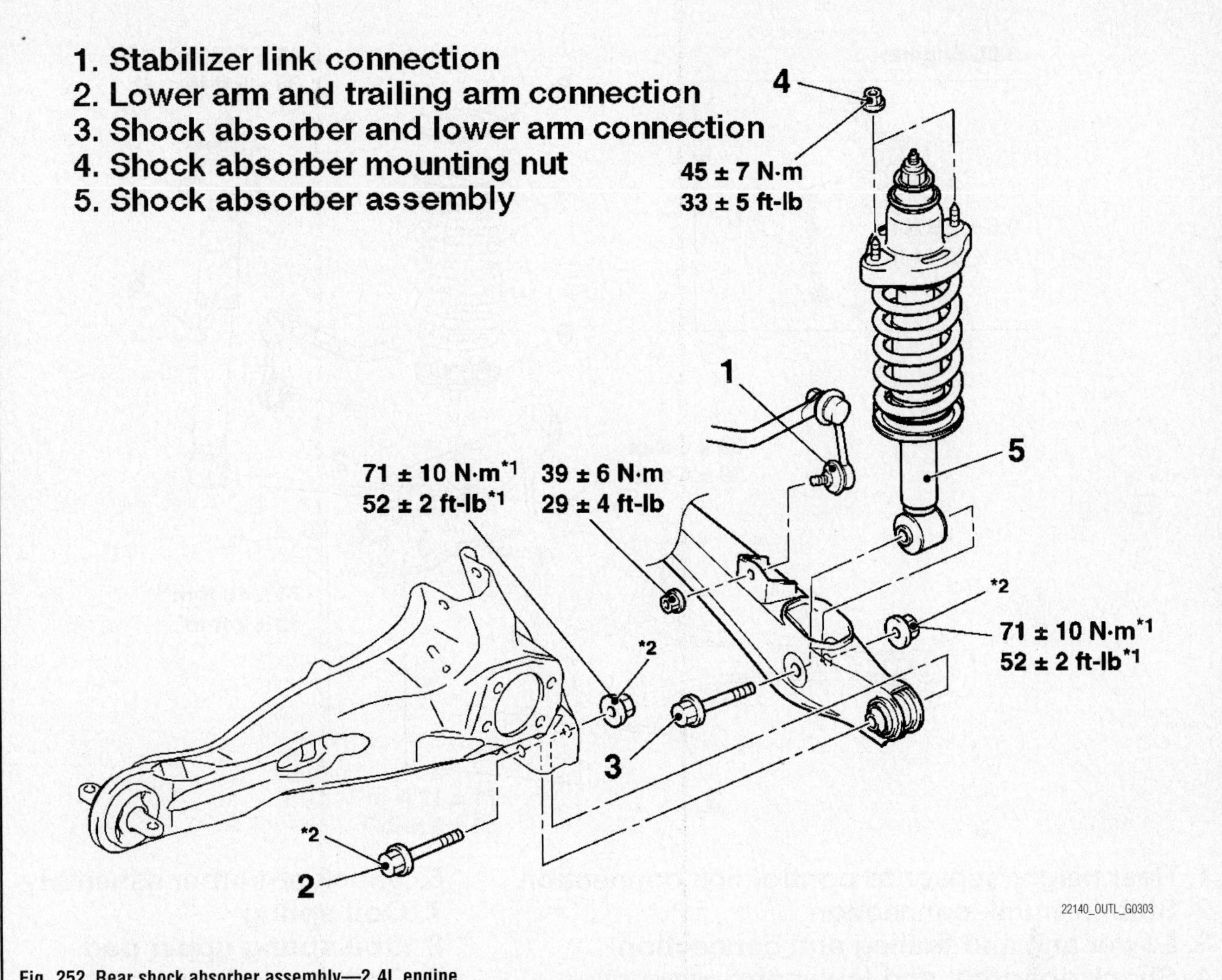

Fig. 252 Rear shock absorber assembly—2.4L engine

a. Remove the maintenance lid.
b. Remove the cargo hook.
c. Remove the jack lid.
d. Remove the tumble switch.
e. Remove the utility bar plug.
f. Remove the lower quarter trim
g. Remove the sash guide cover.
h. Remove the third seat belt mounting bolt.
i. Remove the quarter trim upper cap.
j. Remove the upper quarter trim.

4. Disconnect the stabilizer link.
5. Disconnect the lower arm from the trailing arm.
   a. While jacking-up the lower arm with the garage jack, remove the mounting bolts.
6. Disconnect the shock absorber from the lower arm.
7. Remove the maintenance lid, and then remove the shock absorber mounting nut.
8. Remove the shock absorber assembly.

To install:

➡The suspension components should not be fully tightened until the vehicle's weight is resting on its wheels.

9. Install the shock absorber assembly.
   a. Install the shock absorber assembly so that the coil spring end faces the rear of the vehicle.
10. Install the maintenance lid, and then Install the shock absorber mounting nut.
11. Connect the shock absorber to the lower arm.
   a. While jacking-up the lower arm with the garage jack, Install the mounting bolts.
12. Connect the lower arm and trailing arm.
13. Connect the stabilizer link.
14. Install the quarter trim.
15. Temporarily tighten the suspension bolts.
16. Lower the vehicle.
17. Finish tightening the bolts as specified in the illustration.
18. Check the wheel alignment.
19. Check the headlight aiming.

3.0L Engine

See Figure 253.

1. Before servicing the vehicle, refer to the precautions.
2. Raise and safely support the vehicle.
3. Disconnect the rear height sensor to control link connection.
4. Disconnect the stabilizer link.
5. Disconnect the lower arm from the trailing arm.
   a. While jacking-up the lower arm with the garage jack, remove the mounting bolts.
6. Disconnect the shock absorber from the lower arm.
7. Remove the shock absorber mounting nut.
8. Remove the shock absorber assembly.

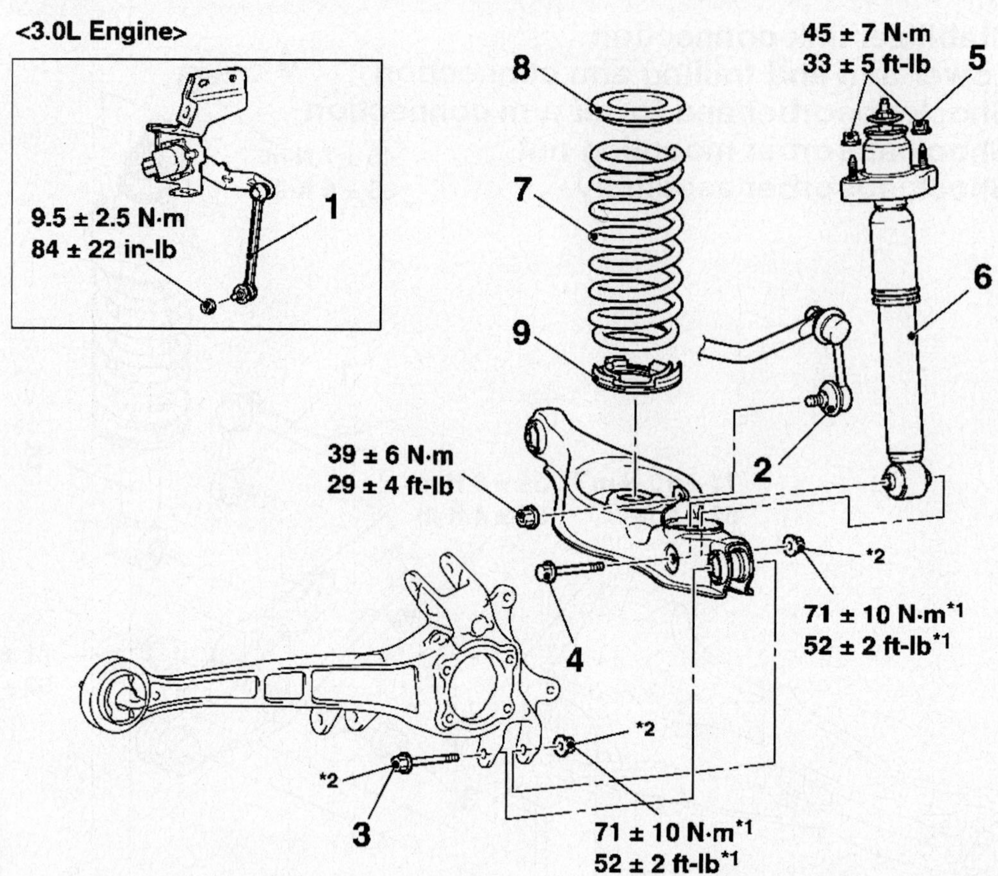

Fig. 253 Rear shock absorber assembly—3.0L engine

To install:

➡The suspension components should not be fully tightened until the vehicle's weight is resting on its wheels.

9. Install the shock absorber assembly.
   a. Install the shock absorber assembly so that the coil spring end faces the rear of the vehicle.
10. Install the shock absorber mounting nut.
11. Connect the shock absorber to the lower arm.
    a. While jacking-up the lower arm with the garage jack, Install the mounting bolts.
12. Connect the lower arm and trailing arm.
13. Connect the stabilizer link.
14. Temporarily tighten the suspension bolts.
15. Connect the rear height sensor to control link connection.
16. Lower the vehicle.
17. Finish tightening the bolts as specified in the illustration.
18. Check the wheel alignment.
19. Check the headlight aiming.

TESTING

1. Check the rubber parts for cracks and wear.
2. Check the shock absorber for malfunctions, oil leakage, or abnormal noise.
3. If shock absorber does not perform properly, replace the shock absorber.

STABILIZER BAR

REMOVAL & INSTALLATION

See Figures 254 through 257.

1. Before servicing the vehicle, refer to the precautions.
2. Raise and support the vehicle safely.
3. Disconnect the stabilizer link.
4. Remove the stabilizer bracket.
5. Remove the bushing.
6. Remove the rear differential carrier assembly.
7. Remove the stabilizer bar.

To install:

8. On 2.4L engine vehicles, position the identification mark of the stabilizer bar at the left side of the vehicle as shown in the

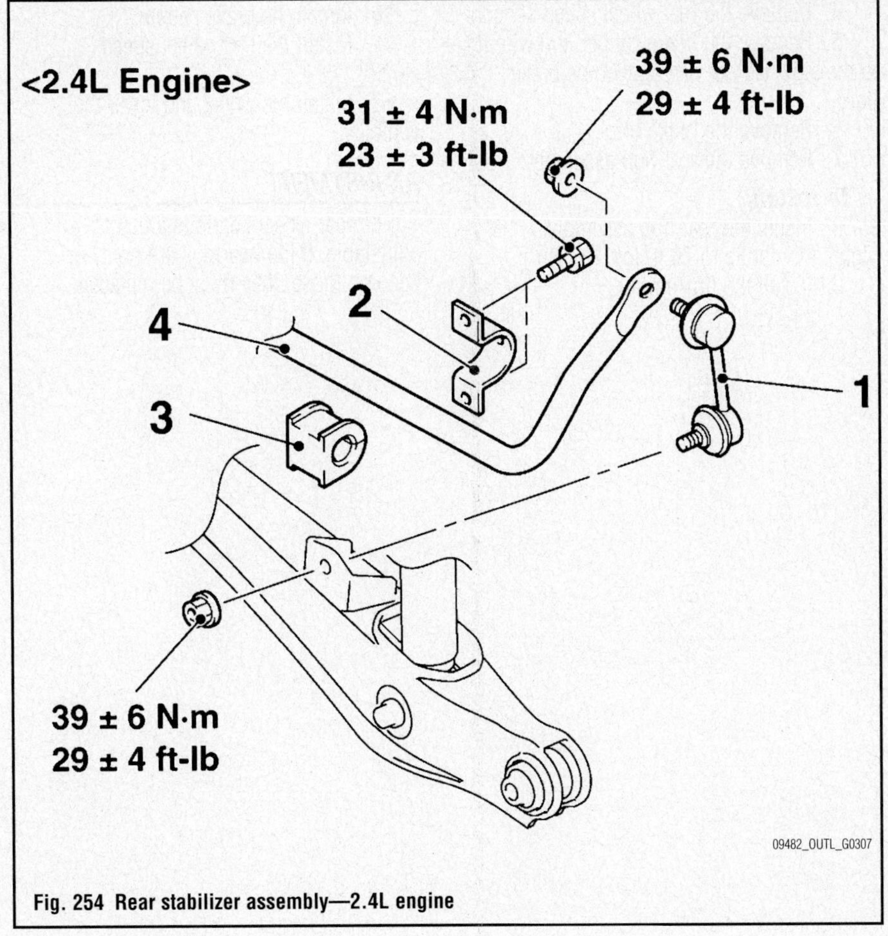

Fig. 254 Rear stabilizer assembly—2.4L engine

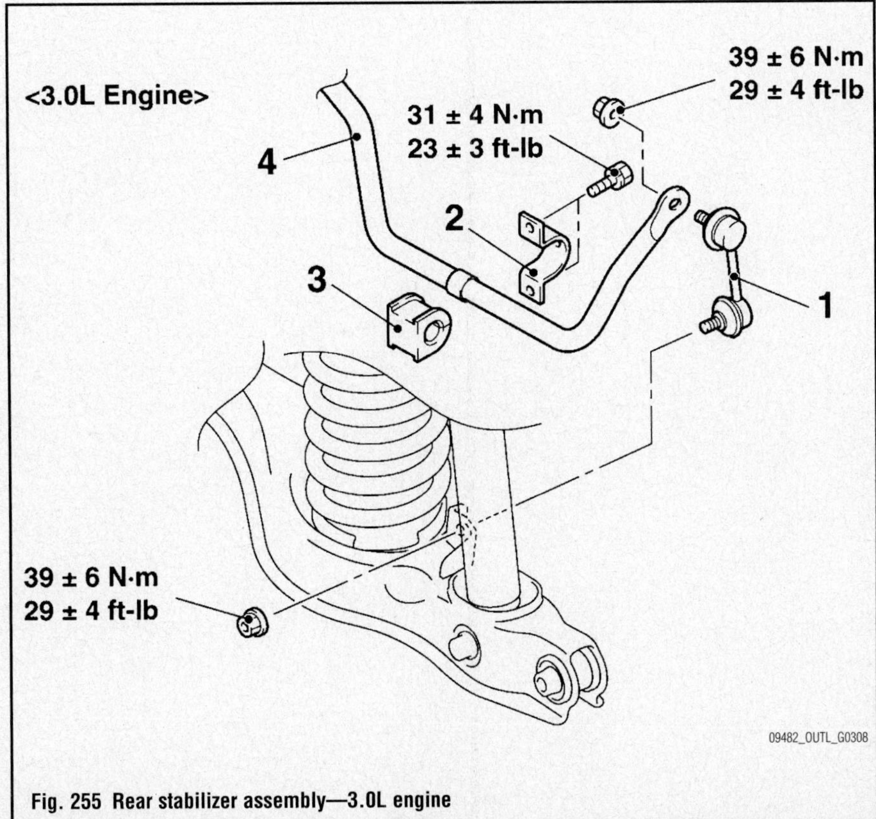

Fig. 255 Rear stabilizer assembly—3.0L engine

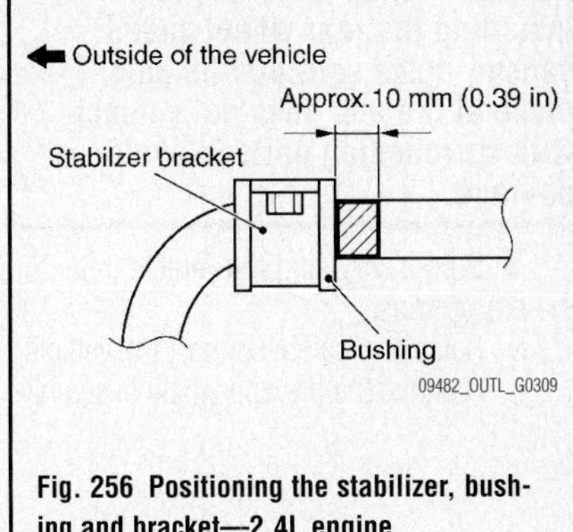

Fig. 256 Positioning the stabilizer, bushing and bracket—2.4L engine

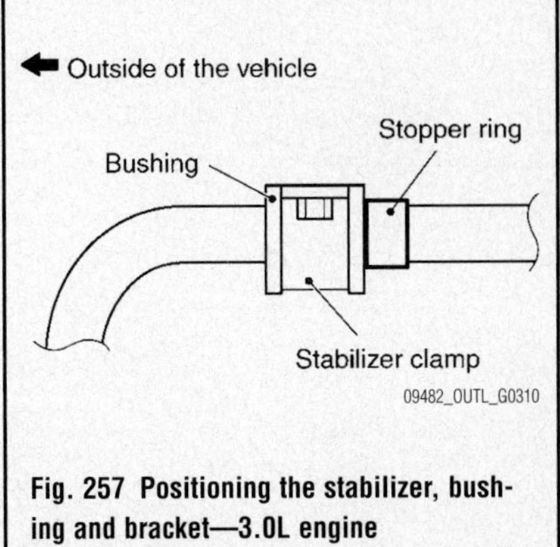

Fig. 257 Positioning the stabilizer, bushing and bracket—3.0L engine

figure, and tighten the stabilizer bracket mounting nut.

9. On 3.0L engine vehicles, install the stabilizer bracket as shown in the figure, and tighten the stabilizer bracket mounting nut.

10. Install the rear differential carrier assembly.

11. Lower the vehicle.

12. Check the wheel alignment.

13. Check the headlight aiming.

WHEEL HUB AND BEARING (SEALED UNIT)

REMOVAL & INSTALLATION

✱✱ WARNING

The magnetic encoder collects metallic particles easily, because it is magnetized. Make sure that the magnetic encoder should not collect metallic particles. Check that there is no trouble prior to reassembling it. When the rear wheel hub assembly is removed/installed, make sure that the magnetic encoder (integrated with inner oil seal) does not contact with surrounding parts to avoid

damage. When removing and installing the rear wheel speed sensor, make sure that its pole piece at the end does not contact with surrounding parts to avoid damage.

1. Before servicing the vehicle, refer to the precautions.
2. Raise and safely support the vehicle.
3. Remove the tire and wheel assembly.
4. Remove the rear wheel speed sensor.
5. Remove the brake caliper and wire it to the side. Do not disconnect the brake line.
6. Remove the brake rotor.
7. Remove the rear hub assembly.

To install:

8. Install the rear hub assembly. Tighten the bolts to 70 ft. lbs. (95 Nm).
9. Install the brake rotor.
10. Install the brake caliper.
11. Install the rear wheel speed sensor.
12. Install the wheel and lower the vehicle.

ADJUSTMENT

The rear wheel bearings are not adjustable. If the bearings are noisy or become loose, they must be replaced.

MITSUBISHI

Raider

11

SPECIFICATIONS AND MAINTENANCE CHARTS

ENGINE AND VEHICLE IDENTIFICATION

Engine							Model Year	
Code ①	Liters (cc)	Cu. In.	Cyl.	Fuel Sys.	Engine Type	Eng. Mfg.	Code ②	Year
K	3.7 (3701)	226	6	SEFI	SOHC	Chrysler	6	2006
N	4.7 (4701)	287	8	SEFI	SOHC	Chrysler	7	2007
P ③	4.7 (4701)	287	8	SEFI	SOHC	Chrysler	8	2008

① 8th position of VIN

② 10th position of VIN

③ Flex Fuel

22140_RAID_C0001

GENERAL ENGINE SPECIFICATIONS

Year	Engine Displacement Liters	Engine VIN	Net Horsepower @ rpm	Net Torque @ rpm (ft. lbs.)	Bore x Stroke (in.)	Com-pression Ratio	Oil Pressure @ rpm
2006	3.7	K	211@5200	236@4000	3.66x3.40	9.6:1	25-110@3000
	4.7	N	230@4600	290@3600	3.66x3.40	9.0:1	35-105@3000
2007	3.7	K	211@5200	236@4000	3.66x3.40	9.6:1	25-110@3000
	4.7	N	235@4800	295@3200	3.66x3.40	9.0:1	35-105@3000
	4.7	P	235@4800	295@3200	3.66x3.40	9.0:1	35-105@3000
2008	3.7	K	211@5200	236@4000	3.66x3.40	9.6:1	25-110@3000

22140_RAID_C0002

GASOLINE ENGINE TUNE-UP SPECIFICATIONS

Year	Engine Dispacement Liters	Engine VIN	Spark Plug Gap (in.)	Ignition Timing (deg.)	Fuel Pump (psi)	Idle Speed (rpm)	Valve Clearance	
							Intake	Exhaust
2006	3.7	K	0.043	②	56-60	②	HYD	HYD
	4.7	N	①	②	56-60	②	HYD	HYD
2007	3.7	K	0.043	②	56-60	②	HYD	HYD
	4.7	N	①	②	56-60	②	HYD	HYD
	4.7	P	①	②	56-60	②	HYD	HYD
2008	3.7	K	0.043	②	56-60	②	HYD	HYD

NOTE: The Vehicle Emission Control Information (VECI) label often reflects specification changes made during production. The label figures must be used if they differ from those in this chart.

HYD: Hydraulic

① Intake: 0.040 Exhaust: 0.050

② Ignition timing is controlled by the PCM and is not adjustable.

③ Idle speed is controlled by the PCM and is not adjustable

22140_RAID_C0003

CAPACITIES

Year	Engine Displ. Liters	Engine VIN	Oil with Filter (qts.)	Transmission (pts.) Manual	Transmission (pts.) Auto.	Transfer Case (pts.)	Drive Axle Front (pts.)	Drive Axle Rear (pts.)	Fuel Tank (gal.)	Cooling System (qts.)
2006	3.7	K	5.0	4.65	①	②	3.5	③	22.0	13.3
	4.7	N	6.0	4.65	①	②	3.5	③	22.0	13.3
2007	3.7	K	5.0	4.65	①	②	3.5	③	22.0	13.3
	4.7	N	6.0	4.65	①	②	3.5	③	22.0	13.3
	4.7	P	6.0	4.65	①	②	3.5	③	22.0	13.3
2008	3.7	K	5.0	4.65	①	②	3.5	③	22.0	13.3

NOTE: All capacities are approximate. Add fluid gradually and check to be sure a proper fluid level is obtained.

① 42RLE: 8.0 pts.
545RFE 2wd: 11.0 pts.
545RFE 4wd: 13.0 pts.

② NV233: 2.5 pts.
NV244: 2.85 pts.

③ The following values include 0.25 pt. of friction modifier for LSD axles.
8.25 in. axle: 4.4 pts.
9.25 in. axle: 4.9 pts.

22140_RAID_C0004

FLUID SPECIFICATIONS

Year	Model	Engine Displacement Liters	Engine VIN	Engine Oil	Auto. Trans.	Manual Trans.	Drive Axle	Transfer Case	Power Steering Fluid	Brake Master Cylinder
2006	Raider	3.7	K	5W-30	ATF+4	ATF+4	①	ATF+4	ATF+4	DOT 3 or 4
		4.7	N	5W-30	ATF+4	ATF+4	①	ATF+4	ATF+4	DOT 3 or 4
2007	Raider	3.7	K	5W-30	ATF+4	ATF+4	①	ATF+4	ATF+4	DOT 3 or 4
		4.7	N	5W-30	ATF+4	ATF+4	①	ATF+4	ATF+4	DOT 3 or 4
		4.7	P	5W-30	ATF+4	ATF+4	①	ATF+4	ATF+4	DOT 3 or 4
2008	Raider	3.7	K	5W-30	ATF+4	ATF+4	①	ATF+4	ATF+4	DOT 3 or 4

DOT: Department Of Transpotation

① Front: 75W-90
Rear: 75W-140 Synthetic (LSD add MS-10111 limited slip additive)

22140_RAID_C0014

VALVE SPECIFICATIONS

Year	Engine Displ. Liters	Engine VIN	Seat Angle (deg.)	Face Angle (deg.)	Spring Test Pressure (lbs. @ in.)	Spring Installed Height (in.)	Stem-to-Guide Clearance (in.) Intake	Stem-to-Guide Clearance (in.) Exhaust	Stem Diameter (in.) Intake	Stem Diameter (in.) Exhaust
2006	3.7	K	44.5-45	45-45.5	①	②	0.0008-0.0028	0.0019-0.0039	0.2729-0.2739	0.2717-0.2728
	4.7	N	44.5-45	45-45.5	174.5-195.6 @1.137	1.579	0.0008-0.0028	0.0019-0.0039	0.2729-0.2739	0.2717-0.2728
2007	3.7	K	44.5-45	45-45.5	①	②	0.0008-0.0028	0.0019-0.0039	0.2729-0.2739	0.2717-0.2728
	4.7	N	44.5-45	45-45.5	174.5-195.6 @1.137	1.579	0.0008-0.0028	0.0019-0.0039	0.2729-0.2739	0.2717-0.2728
	4.7	P	44.5-45	45-45.5	174.5-195.6 @1.137	1.579	0.0008-0.0028	0.0019-0.0039	0.2729-0.2739	0.2717-0.2728
2008	3.7	K	44.5-45	45-45.5	①	②	0.0008-0.0028	0.0019-0.0039	0.2729-0.2739	0.2717-0.2728

① Intake: 213-234 lbs @1.107 in.

Exhaust: 197-215 lbs @1.067 in.

② Intake: 1.5795 in.

Exhaust: 1.54 in.

22140_RAID_C0005

CAMSHAFT AND BEARING SPECIFICATIONS CHART

All measurements are given in inches.

Year	Engine Displ. Liters	Engine ID/VIN	Journal Dia.	Brg. Oil Clearance	Shaft End-play	Runout	Journal Bore	Lobe Height Intake	Lobe Height Exhaust
2006	3.7	K	1.0227-1.0235	0.0010-0.0026	0.0030-0.0079	0.0008	1.0245-1.0252	NA	NA
	4.7	N	1.0227-1.0235	0.0010-0.0026	0.0030-0.0079	0.0008	1.0245-1.0252	NA	NA
2007	3.7	K	1.0227-1.0235	0.0010-0.0026	0.0030-0.0079	0.0008	1.0245-1.0252	NA	NA
	4.7	N	1.0227-1.0235	0.0010-0.0026	0.0030-0.0079	0.0008	1.0245-1.0252	NA	NA
	4.7	P	1.0227-1.0235	0.0010-0.0026	0.0030-0.0079	0.0008	1.0245-1.0252	NA	NA
2008	3.7	K	1.0227-1.0235	0.0010-0.0026	0.0030-0.0079	0.0008	1.0245-1.0252	NA	NA

NA: Not Available

22140_RAID_C0006

CRANKSHAFT AND CONNECTING ROD SPECIFICATIONS

All measurements are given in inches.

	Engine		Crankshaft				Connecting Rod		
Year	Displ. Liters	Engine VIN	Main Brg. Journal Dia.	Main Brg. Oil Clearance	Shaft End-play	Thrust on No.	Journal Diameter	Oil Clearance	Side Clearance
2006	3.7	K	2.4996-2.5005	0.0008-0.0018	0.0021-0.0112	2	2.2792-2.2798	0.0002-0.0017	0.0040-0.0138
	4.7	N	2.4996-2.5005	0.0002-0.0013	0.0021-0.0112	2	2.0076-2.0082	0.0006-0.0022	0.0040-0.0138
2007	3.7	K	2.4996-2.5005	0.0008-0.0018	0.0021-0.0112	2	2.2792-2.2798	0.0002-0.0017	0.0040-0.0138
	4.7	N	2.4996-2.5005	0.0002-0.0013	0.0021-0.0112	2	2.0076-2.0082	0.0006-0.0022	0.0040-0.0138
	4.7	P	2.4996-2.5005	0.0002-0.0013	0.0021-0.0112	2	2.0076-2.0082	0.0006-0.0022	0.0040-0.0138
2008	3.7	K	2.4996-2.5005	0.0008-0.0018	0.0021-0.0112	2	2.2792-2.2798	0.0002-0.0017	0.0040-0.0138

22140_RAID_C0007

PISTON AND RING SPECIFICATIONS

All measurements are given in inches.

	Engine			Ring Gap			Ring Side Clearance		
Year	Displ. Liters	Engine VIN	Piston Clearance	Top Comp.	Bottom Comp.	Oil Control	Top Comp.	Bottom Comp.	Oil Control
2006	3.7	K	0.0014	0.0079-0.0142	0.0146-0.0249	0.0099-0.0300	0.0020-0.0037	0.0016-0.0031	0.0007-0.0091
	4.7	N	0.0014	0.0079-0.0142	0.0146-0.0249	0.0099-0.0300	0.0020-0.0037	0.0016-0.0031	0.0007-0.0091
2007	3.7	K	0.0014	0.0079-0.0142	0.0146-0.0249	0.0099-0.0300	0.0020-0.0037	0.0016-0.0031	0.0007-0.0091
	4.7	N	0.0014	0.0079-0.0142	0.0146-0.0249	0.0099-0.0300	0.0020-0.0037	0.0016-0.0031	0.0007-0.0091
	4.7	P	0.0014	0.0079-0.0142	0.0146-0.0249	0.0099-0.0300	0.0020-0.0037	0.0016-0.0031	0.0007-0.0091
2008	3.7	K	0.0014	0.0079-0.0142	0.0146-0.0249	0.0099-0.0300	0.0020-0.0037	0.0016-0.0031	0.0007-0.0091

22140_RAID_C0008

TORQUE SPECIFICATIONS

All readings in ft. lbs.

Year	Engine Displ. Liters	Engine VIN	Cylinder Head Bolts	Main Bearing Bolts	Rod Bearing Bolts	Crankshaft Damper Bolts	Flywheel Bolts	Manifold Intake	Manifold Exhaust	Spark Plugs	Oil Pan Drain Plug
2006	3.7	K	①	②	④	130	70	9	18	20	25
	4.7	N	①	③	④	130	45	⑤	18	20	25
2007	3.7	K	①	②	④	130	70	9	18	20	25
	4.7	N	①	③	④	130	45	⑤	18	20	25
	4.7	P	①	③	④	130	45	⑤	18	20	25
2008	3.7	K	①	②	④	130	70	9	18	20	25

① See text

② Bed plate bolt sequence. Refer to illustration

Step 1: Hand tighten bolts 1D,1G and 1F until the bedplate contacts the block.

Step 2: Tighten bolts 1A - 1J to 54 N·m (40 ft. lbs.)

Step 3: Tighten bolts 1 - 8 to 7 N·m (5 ft. lbs.)

Step 4: Turn bolts 1 - 8 an additional 90°.

Step 5: Tighten bolts A - E 27 N·m (20 ft. lbs.).

③ Bed plate bolt sequence. Refer to illustration

Step 1: Bolts A-L to 40 ft. lbs.

Step 2: Bolts 1-10 25 inch lbs.

Step 3: Bolts 1-10 plus 90 degrees

Step 4: Bolts A1-A6 20 ft. lbs.

④ 20 ft. lbs. plus 90 degrees

⑤ See illustration in text section

Step 1: 1-4 to 72 inch lbs. in 12 inch lb. Increments

Step 2: bolts 5-12: 72 inch lbs.

Step 3: Check that all bolts are at 72 inch lbs.

Step 4: All bolts, in sequence, to 12 ft. lbs.

Step 5: Check that all bolts are at 12 ft. lbs.

22140_RAID_C0009

WHEEL ALIGNMENT

Year	Model			Caster Range (+/-Deg.)	Caster Preferred Setting (Deg.)	Camber Range (+/-Deg.)	Camber Preferred Setting (Deg.)	Toe-in (Deg.)
2006	Raider	F	Left	0.05	+3.50	0.50	-0.25	0.20+/-0.10
			Right	0.05	+3.70	0.50	-0.25	0.20+/-0.10
		R		—	—	0.35	-0.10	0.30+/-0.35
2007	Raider	F	Left	0.05	+3.50	0.50	-0.25	0.20+/-0.10
			Right	0.05	+3.70	0.50	-0.25	0.20+/-0.10
		R		—	—	0.35	-0.10	0.30+/-0.35
2008	Raider	F	Left	0.05	+3.50	0.50	-0.25	0.20+/-0.10
			Right	0.05	+3.70	0.50	-0.25	0.20+/-0.10
		R		—	—	0.35	-0.10	0.30+/-0.35

22140_RAID_C0011

TIRE, WHEEL AND BALL JOINT SPECIFICATIONS

Year	Model	OEM Tires Standard	OEM Tires Optional	Tire Pressures (psi) Front	Tire Pressures (psi) Rear	Wheel Size	Ball Joint Inspection	Lug Nut Torque (ft. lbs.)
2006	LS	P245/70R16	P255/65R16	①	①	NA	0.020 in.	135
			P265/70R16	①	①	NA		
	DuroCross 2WD	P265/70R16	-	①	①	NA		
	DuroCross 4WD	LT265/70R16	-	①	①	NA		
	XLS	P265/65R17		①	①	NA		
2007	LS	P245/70R16	P255/65R16	①	①	NA	0.020 in.	135
			P265/70R16	①	①	NA		
	DuroCross 2WD	P265/70R16	-	①	①	NA		
	DuroCross 4WD	LT265/70R16	-	①	①	NA		
	SE	P265/65R17		①	①	NA		
2008	LS	P245/70R16	P255/65R16	①	①	NA	0.020 in.	135
			P265/70R16	①	①	NA		

OEM: Original Equipment Manufacturer

PSI: Pounds Per Square Inch

① See the tire placard on the vehicle

22140_RAID_C0010

BRAKE SPECIFICATIONS

All measurements in inches unless noted

Year		Brake Disc Original Thickness	Brake Disc Minimum Thickness	Brake Disc Maximum Run-out	Brake Drum Original Inside Diameter	Brake Drum Max. Wear Limit	Brake Drum Maximum Machine Diameter	Minimum Lining Thickness Front	Minimum Lining Thickness Rear	Brake Caliper Bracket Bolts (ft. lbs.)	Brake Caliper Mounting Bolts (ft. lbs.)
2006	F	1.100	1.039	0.0010	-	-	-	②	-	130	24
	R	-	-	-	11.50	①	11.693	-	③	-	-
2007	F	1.100	1.039	0.0010	-	-	-	②	-	130	24
	R	-	-	-	11.50	①	11.693	-	③	-	-
2008	F	1.100	1.039	0.0010	-	-	-	②	-	130	24
	R	-	-	-	11.50	①	11.693	-	③	-	-

F: Front

R: Rear

NA: Not Available

① Maximum allowable drum diameter, either from wear or machining, is stamped on the drum.

② Riveted brake pads: 0.0625 in.
Bonded brake pads: 0.1875 in.

③ Riveted brake shoes: 0.031 in.
Bonded brake shoes: 0.0625 in.

22140_RAID_C0012

SCHEDULED MAINTENANCE INTERVALS

Mistubishi Raider

TO BE SERVICED	SERVICE	VEHICLE MILEAGE INTERVAL (x1000)													
		6	12	18	24	30	36	42	48	54	60	66	72	78	84
Engine oil & filter	R	✓	✓	✓	✓	✓	✓	✓	✓	✓	✓	✓	✓	✓	✓
Tires	Rotate	✓	✓	✓	✓	✓	✓	✓	✓	✓	✓	✓	✓	✓	✓
Brake linings	S/I		✓		✓		✓		✓		✓		✓		✓
CV-Joints	S/I		✓		✓				✓				✓		
Exhaust system	S/I		✓		✓				✓				✓		
Front suspension	S/I		✓		✓				✓				✓		
Air cleaner element	S/I					✓					✓				
Transfer case fluid level	I					✓					✓				
Spark plugs	R					✓					✓				
PCV valve	S/I										✓				
Engine coolant	R	Replace every 60 months or 102,000 miles, whichever comes first													
Automatic trans fluid	R	Replace every 120 months or 120,000 miles, whichever comes first													
Accessory drive belt	R	Replace every 120 months or 120,000 miles, whichever comes first													

R: Replace S/I: Service or Inspect L: Lubricate Adj: Adjust

FREQUENT OPERATION MAINTENANCE (SEVERE SERVICE)

If a vehicle is operated under any of the following conditions it is considered severe service:

- Extremely dusty areas.

- 50%or more of the vehicle operation is in 32°C (90°F) or higher temperatures, or constant operation in temperatures below 0°C (32°F).

- Prolonged idling (vehicle operation in stop and go traffic.

- Frequent short running periods (engine does not warm to normal operating temperatures).

- Police, taxi, delivery usage or trailer towing usage.

Air filter (dusty or off road conditions): inspect every 12,000 miles.

Front and rear axle fluid: Inspect every 18,000 miles.

Automatic transmission fluid & filter: change every 60,000 miles.

Transfer case fluid: change every 60,000 miles.

Manual transmission fluid: change every 60,000 miles.

22140_RAID_C0013

PRECAUTIONS

Before servicing any vehicle, please be sure to read all of the following precautions, which deal with personal safety, prevention of component damage, and important points to take into consideration when servicing a motor vehicle:

- Never open, service or drain the radiator or cooling system when the engine is hot; serious burns can occur from the steam and hot coolant.
- Observe all applicable safety precautions when working around fuel. Whenever servicing the fuel system, always work in a well-ventilated area. Do not allow fuel spray or vapors to come in contact with a spark, open flame, or excessive heat (a hot drop light, for example). Keep a dry chemical fire extinguisher near the work area. Always keep fuel in a container specifically designed for fuel storage; also, always properly seal fuel containers to avoid the possibility of fire or explosion. Refer to the additional fuel system precautions later in this section.
- Fuel injection systems often remain pressurized, even after the engine has been turned **OFF**. The fuel system pressure must be relieved before disconnecting any fuel lines. Failure to do so may result in fire and/or personal injury.
- Brake fluid often contains polyglycol ethers and polyglycols. Avoid contact with the eyes and wash your hands thoroughly after handling brake fluid. If you do get brake fluid in your eyes, flush your eyes with clean, running water for 15 minutes. If eye irritation persists, or if you have taken brake fluid internally, IMMEDIATELY seek medical assistance.
- The EPA warns that prolonged contact with used engine oil may cause a number of skin disorders, including cancer. You should make every effort to minimize your exposure to used engine oil. Protective gloves should be worn when changing oil. Wash your hands and any other exposed skin areas as soon as possible after exposure to used engine oil. Soap and water, or waterless hand cleaner should be used.
- All new vehicles are now equipped with an air bag system, often referred to as a Supplemental Restraint System (SRS) or Supplemental Inflatable Restraint (SIR) system. The system must be disabled before performing service on or around system components, steering column, instrument panel components, wiring and sensors. Failure to follow safety and disabling procedures could result in accidental air bag deployment, possible personal injury and unnecessary system repairs.
- Always wear safety goggles when working with, or around, the air bag system. When carrying a non-deployed air bag, be sure the bag and trim cover are pointed away from your body. When placing a non-deployed air bag on a work surface, always face the bag and trim cover upward, away from the surface. This will reduce the motion of the module if it is accidentally deployed. Refer to the additional air bag system precautions later in this section.
- Clean, high quality brake fluid from a sealed container is essential to the safe and proper operation of the brake system. You should always buy the correct type of brake fluid for your vehicle. If the brake fluid becomes contaminated, completely flush the system with new fluid. Never reuse any brake fluid. Any brake fluid that is removed from the system should be discarded. Also, do not allow any brake fluid to come in contact with a painted surface; it will damage the paint.
- Never operate the engine without the proper amount and type of engine oil; doing so WILL result in severe engine damage.
- Timing belt maintenance is extremely important. Many models utilize an interference-type, non-freewheeling engine. If the timing belt breaks, the valves in the cylinder head may strike the pistons, causing potentially serious (also time-consuming and expensive) engine damage. Refer to the maintenance interval charts for the recommended replacement interval for the timing belt, and to the timing belt section for belt replacement and inspection.
- Disconnecting the negative battery cable on some vehicles may interfere with the functions of the on-board computer system(s) and may require the computer to undergo a relearning process once the negative battery cable is reconnected.
- When servicing drum brakes, only disassemble and assemble one side at a time, leaving the remaining side intact for reference.
- Only an MVAC-trained, EPA-certified automotive technician should service the air conditioning system or its components.

BRAKES

ANTI-LOCK BRAKE SYSTEM (ABS)

GENERAL INFORMATION

This vehicle uses an antilock brake system. If the Antilock Brake Module (ABM) senses impending rear wheel lock-up, it will energize the isolation solenoid. This prevents a further increase of driver induced brake pressure to the rear wheels. If this initial action is not enough to prevent rear wheel lock-up, the ABM will momentarily energize a dump solenoid. This opens the dump valve to vent a small amount of isolated rear brake pressure to an accumulator. The action of fluid moving to the accumulator reduces the isolated brake pressure at the wheel cylinders. The dump (pressure venting) cycle is limited to very short time periods (milliseconds). The ABM will pulse the dump valve until rear wheel deceleration reaches the desired slip rate programmed into the ABM. The system will switch to normal braking once wheel locking tendencies are no longer present.

As part of the anti-lock brake system, this vehicle is equipped with Electronic stability control (ESC). ESC is a computerized technology that improves the safety of a vehicle's handling by detecting and preventing skids. When ESC detects loss of steering control, ESC automatically applies individual brakes to help "steer" the vehicle where the driver wants to go. Braking is automatically applied to individual wheels, such as the outer front wheel to counter oversteer, or the inner rear wheel to counter under steer. Some ESC systems also reduce engine power until control is regained.

ESC compares the driver's intended direction (by measuring steering angle) to the vehicle's actual direction (by measuring lateral acceleration, vehicle rotation (yaw) and individual road wheel speeds). If the vehicle is not going where the driver is steering, ESC then brakes individual front or rear wheels and/or reduces excess engine power as needed to help correct under steer (plowing) or oversteer (fishtailing).

ESC incorporates yaw rate control into the anti-lock braking system (ABS). Yaw is rotation around the vertical axis; i.e. spinning left or right. Anti-lock brakes enable ESC to brake individual wheels. The ESC used on this vehicle also incorporates a Traction Control System (TCS), which senses drive-wheel slip under acceleration and individually brakes the slipping wheel or wheels and/or reduces excess engine power until control is regained.

WHEEL SPEED SENSORS

REMOVAL & INSTALLATION

Front

See Figure 1.

1. Before servicing the vehicle, refer to the precautions.
2. Raise and safely support the vehicle securely on jackstands.
3. Remove the wheel.
4. Remove the brake caliper.
5. Remove the rotor.

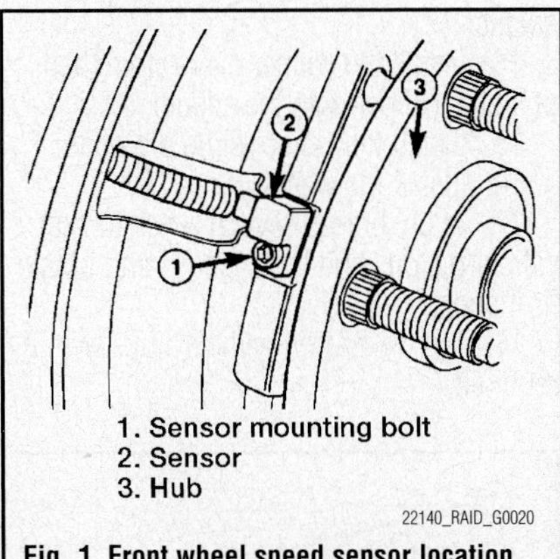

Fig. 1 Front wheel speed sensor location

6. Remove the sensor attaching bolts.
7. Disconnect the wire and remove the sensor from the vehicle.

To install:

8. Install the sensor from the vehicle. Tighten the bolts to 190 inch. lbs. (21 Nm).
9. Connect the sensor wire.
10. Install the sensor attaching bolts.
11. Install the rotor.
12. Install the brake caliper.
13. Install the wheel.
14. Lower the vehicle.

Rear

See Figure 2.

1. Before servicing the vehicle, refer to the precautions.
2. Raise and safely support the vehicle securely on jackstands.
3. Remove brake line mounting nut and remove the brake line from the sensor stud.
4. Remove the park brake cable and bracket from the sensor stud.
5. Disconnect the electrical connector.
6. Remove mounting stud from the sensor.
7. Remove sensor from differential housing.

To install:

8. Install O–ring on sensor if removed.
9. Install sensor on the differential housing. Tighten to 18 ft. lbs. (24 Nm).

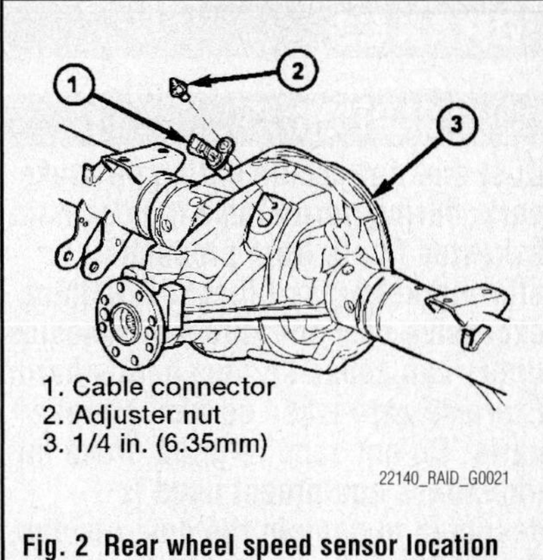

Fig. 2 Rear wheel speed sensor location

10. Connect harness to sensor. Be sure the seal is securely in place between the sensor and wiring connector.
11. Install the park brake cable and bracket from the sensor bolt.
12. Install brake line mounting nut and Install the brake line from the sensor bolt.
13. Lower the vehicle.

WHEEL SPEED SENSOR RINGS (TOOTHED RINGS)

REMOVAL & INSTALLATION

The ABS wheel speed sensor ring is an integral part of either the wheel hub (front) or the differential (rear) and cannot be disassembled.

BRAKES

BLEEDING THE BRAKE SYSTEM

BLEEDING PROCEDURE

Except ABS

Use Mitsubishi brake fluid, or an equivalent quality fluid meeting SAE J1703-F and DOT 3 standards only. Use fresh, clean fluid from a sealed container at all times.

1. Before servicing the vehicle, refer to the precautions.
2. Remove reservoir filler caps and fill reservoir.
3. If calipers were overhauled, open all caliper bleed screws. Then close each bleed screw as fluid starts to drip from it. Top off master cylinder reservoir once more before proceeding.
4. Attach one end of bleed hose to bleed screw and insert opposite end in glass container (2) partially filled with brake fluid. Be sure end of bleed hose is immersed in fluid.

➡Bleed procedure should be in this order (1) Right rear (2) Left rear (3) Right front (4) Left front.

5. Open up bleeder, then have a helper press down the brake pedal. Once the pedal is down close the bleeder. Repeat bleeding until fluid stream is clear and free of bubbles. Then move to the next wheel.
6. Before moving the vehicle verify the pedal is firm and not mushy.
7. Top off the brake fluid and install the reservoir cap.

ABS

ABS system bleeding requires conventional bleeding methods plus use of a scan tool. The procedure involves performing a base brake bleeding, followed by use of the scan tool to cycle and bleed the Hydraulic Control Unit (HCU) pump and solenoids. A second base brake bleeding procedure is then required to remove any air remaining in the system.

1. Perform base brake bleeding.
2. Connect the scan tool to the data link connector.
3. Select "Anti-lock Brakes" followed by "Miscellaneous", then "Bleed Brakes". Follow the instructions displayed until the unit displays "Test Complete", then disconnect the scan tool and proceed.
4. Perform a base brake bleeding a second time.
5. Top up the master cylinder.

✻✻ CAUTION

Dust and dirt accumulating on brake parts during normal use may contain asbestos fibers from production or aftermarket brake linings. Breathing excessive concentrations of asbestos fibers can cause serious bodily harm. Exercise care when servicing brake parts. Do not sand or grind brake lining unless equipment used is designed to contain the dust residue. Do not clean brake parts with compressed air or by dry brushing. Cleaning should be done by dampening the brake components with a fine mist of water, then wiping the brake components clean with a dampened cloth. Dispose of cloth and all residue containing asbestos fibers in an impermeable container with the appropriate label. Follow practices prescribed by the Occupational Safety and Health Administration (OSHA) and the Environmental Protection Agency (EPA) for the handling, processing, and disposing of dust or debris that may contain asbestos fibers.

BRAKE CALIPER

REMOVAL & INSTALLATION

See Figure 3.

1. Before servicing the vehicle, refer to the precautions.

✻✻ CAUTION

Never allow the disc brake caliper to hang from the brake hose. Damage to the brake hose with result. Provide a suitable support to hang the caliper securely.

2. Remove the wheel.
3. Compress the disc brake caliper.
4. Remove the banjo bolt and discard the copper washers.
5. Remove the caliper slide pin bolts.
6. Remove the disc brake caliper from the caliper adapter.
7. Remove the caliper slide pins from the adapter.

To install:

✻✻ WARNING

Petroleum based grease should not be used on any of the rubber components of the caliper, Use only Non-Petroleum based grease.

➡Clean slide pin bores thoroughly to remove any old grease.

➡Use grease packets included with the kit or Dow Corning-807T grease.

8. Thoroughly coat the new slide pins on all working surfaces.
9. Install the boot onto the slide pin and then insert into the adapter.
10. Push the pin all the way into the adapter and carefully expel the trapped air by gently pushing on the boot near the slide pin head.

➡Install a new copper washers on the banjo bolt when installing

11. Install the disc brake caliper to the brake caliper adapter.
12. Install the banjo bolt with new copper washers to the caliper. Tighten to 21 ft. lbs. (28 Nm).
13. Install the caliper slide pin bolts. Tighten to 24 ft. lbs. (32 Nm).
14. Bleed the brake system.
15. Install the wheel.

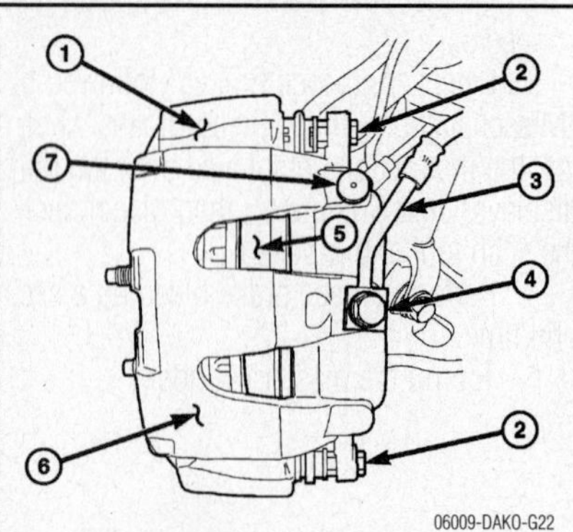

Fig. 3 (1) Caliper mounting adaptor, (2) Caliper mounting bolts, (3) Brake line, (4) Banjo bolt, (5) Pad, (6) Caliper

DISC BRAKE PADS

REMOVAL & INSTALLATION

See Figures 4 and 5.

1. Before servicing the vehicle, refer to the precautions.
2. Remove the wheel.
3. Compress the caliper.
4. Remove the caliper slide pin bolts.
5. Remove the caliper from the caliper adapter.

➡Do not allow brake hose to support caliper assembly.

6. Remove the inboard brake pad from the caliper adapter.
7. Remove the outboard brake pad from the caliper adapter.
8. Remove the anti-rattle clips from the pad.
9. Bottom pistons in caliper bore with C-clamp. Place an old brake shoe between a C-clamp and caliper piston.
10. Clean the caliper mounting adapter.
11. Install new anti-rattle clips to the brake pads.
12. Install the inboard brake pad in the adapter.
13. Install the outboard brake pad in the adapter.
14. Install the caliper over rotor. Then, push the caliper onto the adapter.
15. Install the caliper slide pin bolts.
16. Install the wheel.
17. Apply brakes several times to seat caliper pistons and brake shoes and obtain firm pedal.
18. Top off master cylinder fluid level if required.

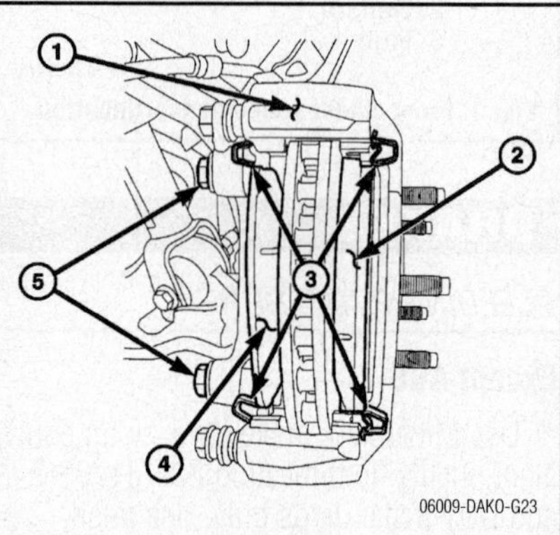

Fig. 4 (1) Caliper adapter, (2) Outboard pad, (3) Anti-rattle clips, (4) Inboard pad, (5) Caliper mounting bolts

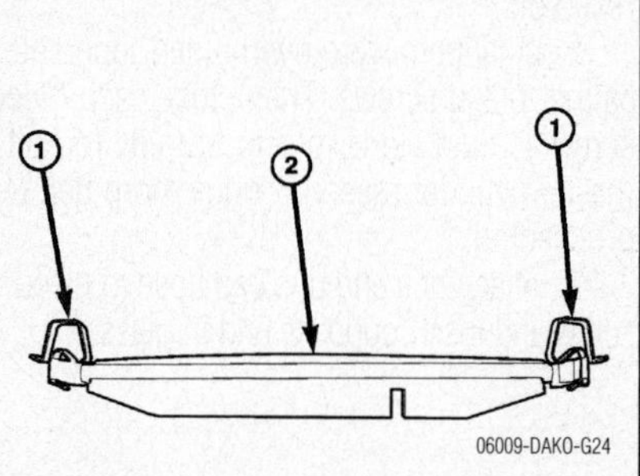

Fig. 5 Anti-rattle clips (1) installed on the pad (2)

CAUTION

Dust and dirt accumulating on brake parts during normal use may contain asbestos fibers from production or aftermarket brake linings. Breathing excessive concentrations of asbestos fibers can cause serious bodily harm. Exercise care when servicing brake parts. Do not sand or grind brake lining unless equipment used is designed to contain the dust residue. Do not clean brake parts with compressed air or by dry brushing. Cleaning should be done by dampening the brake components with a fine mist of water, then wiping the brake components clean with a dampened cloth. Dispose of cloth and all residue containing asbestos fibers in an impermeable container with the appropriate label. Follow practices prescribed by the Occupational Safety and Health Administration (OSHA) and the Environmental Protection Agency (EPA) for the handling, processing, and disposing of dust or debris that may contain asbestos fibers.

BRAKE DRUM

REMOVAL & INSTALLATION

1. Before servicing the vehicle, refer to the precautions.
2. Remove the axle shaft nuts, washers and cones. If the cones do not readily release, rap the axle shaft sharply in the center.
3. Remove the axle shaft.
4. Remove the outer hub nut.
5. Straighten the lockwasher tab and remove it along with the inner nut and bearing.
6. Carefully remove the drum.

To install:

7. Position the drum on the axle housing.
8. Install the bearing and inner nut. While rotating the wheel and tire, tighten the adjusting nut until a slight drag is felt.
9. Back off the adjusting nut ⅙ turn so that the wheel rotates freely without excessive end-play.
10. Install the lock ring s and nut. Place a new gasket on the hub and install the axle shaft, cones, lockwashers and nuts.
11. Install the wheel and tire.
12. Road-test the vehicle.

BRAKE SHOES

REMOVAL & INSTALLATION

See Figures 6 and 7.

1. Before servicing the vehicle, refer to the precautions.
2. Remove the wheel.
3. Remove the clip nuts securing brake drum to wheel studs.
4. Remove the drum. If drum is difficult to remove, remove rear plug from access hole in support plate. Back-off self adjusting by inserting a thin screwdriver into access hole and push lever away from adjuster star wheel. Then insert an adjuster tool into brake adjusting hole rotate adjuster star wheel to retract brake shoes.
5. Vacuum brake components to remove brake lining dust.
6. Remove shoe return spring with brake spring pliers tool.
7. Remove adjuster spring and lever. Disengage lever from spring by sliding lever forward to clear pivot and work lever out from under spring.
8. Disengage and remove shoe return spring from brake shoes.
9. Remove the brake shoe hold down clips.
10. Remove the rear brake shoe from support plate.
11. Remove the front brake shoe from support plate.
12. Remove the park brake lever from the brake shoe.

To install:

13. Clean and inspect individual brake components.

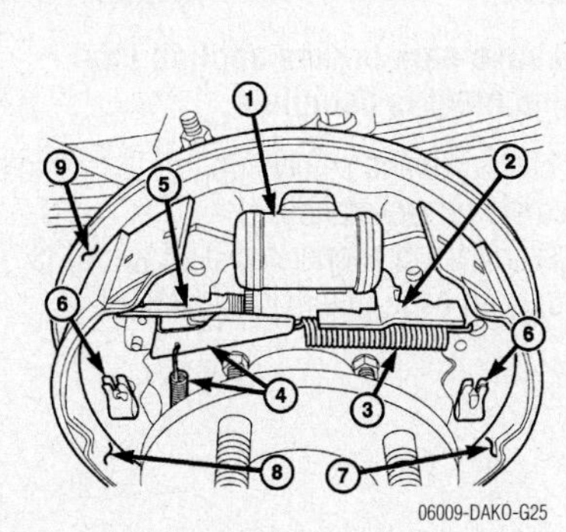

Fig. 6 Rear brake parts (1) wheel cylinder, (2) parking brake lever, (3) return spring, (4 & 5) adjuster spring and lever, (6) hold-down clips, (7 & 8) brake shoes, (9) backing plate

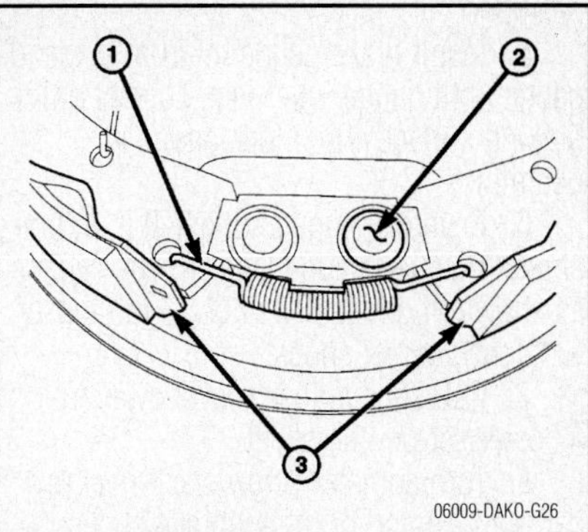

Fig. 7 Lower return spring (1), cam (2) and brake shoes, (3)

14. Lubricate where the brake shoe contacts the support plate with high temperature grease or Lubriplate®.
15. Lubricate adjuster screw socket, nut, button and screw thread surfaces with grease or Lubriplate®.
16. Install parking brake lever to the rear shoe and install the hold down clip.
17. Install the adjuster strut onto the shoes and park brake lever.
18. Install the front shoe on support plate, and install the hold down clip.
19. Install the adjuster spring and lever in the slot in the adjuster strut.
20. Install the lower return spring to the shoes.
21. Verify adjuster operation. Pull both shoes outward to move the adjuster lever to rotate the star wheel. Be sure adjuster lever properly engages star wheel teeth.
22. Adjust brake shoes to drum with a brake gauge.
23. Install wheel and tire assembly.

ADJUSTMENT

The rear drum brakes are equipped with a self-adjusting mechanism. Under normal circumstances, the only time adjustment is required is when the shoes are replaced, removed for access to other parts, or when one or both drums are replaced. Adjustment can be made with a standard brake gauge or with adjusting tool. Adjustment is performed with the complete brake assembly installed on the backing plate.

Adjustment with a Brake Gauge

See Figures 8 and 9.

1. Before servicing the vehicle, refer to the precautions.
2. Be sure parking brakes are fully released.
3. Raise rear of vehicle and remove wheels and brake drums.

4. Verify that left and right automatic adjuster levers and cables are properly connected.

5. Insert brake gauge in drum. Expand gauge until gauge inner legs contact drum braking surface. Then lock gauge in position.

6. Reverse gauge and install it on brake shoes. Position gauge legs at shoe centers as shown. If gauge does not fit (too loose/ too tight), adjust shoes.

7. Pull shoe adjuster lever away from adjuster screw star wheel.

8. Turn adjuster screw star wheel (by hand) to expand or retract brake shoes. Continue adjustment until gauge outside legs are light drag-fit on shoes.

9. Install brake drums and wheels and lower vehicle.

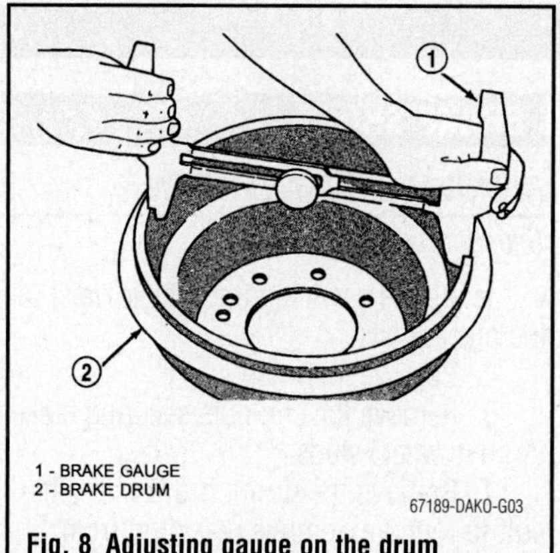

Fig. 8 Adjusting gauge on the drum

10. Drive vehicle and make one forward stop followed by one reverse stop. Repeat procedure 8-10 times to operate automatic adjusters and equalize adjustment.

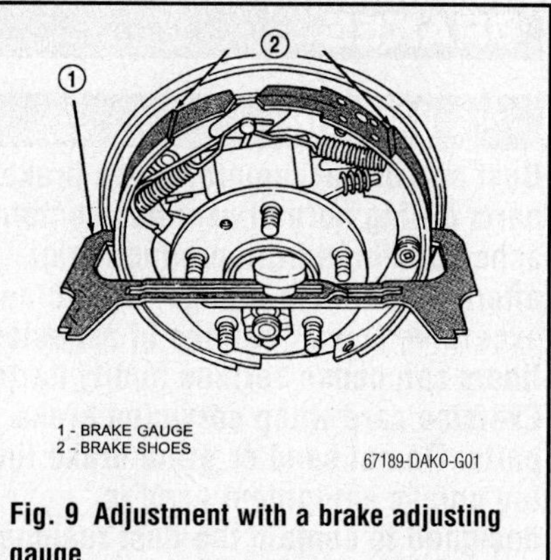

Fig. 9 Adjustment with a brake adjusting gauge

➡Bring vehicle to complete standstill at each stop. Incomplete, rolling stops will not activate automatic adjusters.

PARKING BRAKE CABLES

ADJUSTMENT

See Figure 10.

Tensioner adjustment is only necessary when the tensioner, or a cable has been replaced or disconnected for service. When adjustment is necessary, perform adjustment only as described in the following procedure. This is necessary to avoid faulty park brake operation.

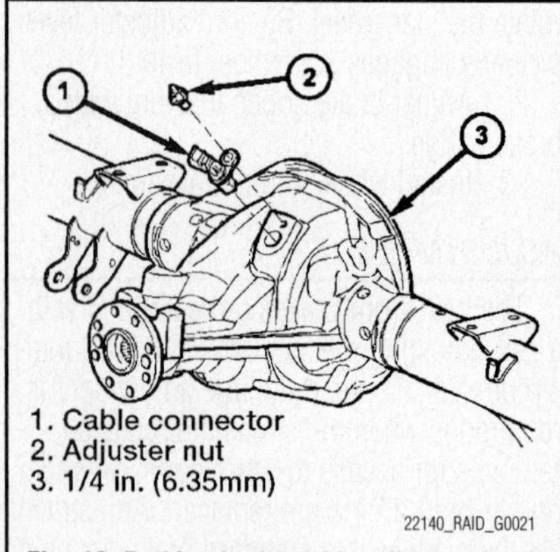

Fig. 10 Parking brake cable adjustment

1. Before servicing the vehicle, refer to the precautions.

2. Raise and safely support the vehicle securely on jackstands.

3. Back off cable tensioner adjusting nut at equalizer to create slack in cables.

4. Remove rear wheel/tire assemblies. Then remove brake drums.

5. Verify brakes are in good condition and operating properly.

6. Verify park brake cables operate freely and are not binding, or seized.

7. Check rear brake shoe adjustment with standard brake gauge.

8. Install drums and verify that drums rotate freely without drag.

9. Reinstall wheel/tire assemblies after brake shoe adjustment is complete.

10. Lower vehicle enough for access to park brake foot pedal. Then fully apply park brakes.

➡Leave park brakes applied until adjustment is complete.

11. Raise and safely support the vehicle securely on jackstands.

12. Mark tensioner rod 0.25 in. (6.35 mm) from edge of tensioner bracket.

13. Tighten adjusting nut at equalizer until mark on tensioner rod moves into alignment with tensioner bracket.

⁂ WARNING

Do not loosen, or tighten the tensioner adjusting nut for any reason after completing adjustment.

14. Lower vehicle until rear wheels are 6-8 in. (15-20 cm) off shop floor.

15. Release park brake foot pedal and verify that rear wheels rotate freely without drag. Then lower vehicle.

PARKING BRAKE SHOES

REMOVAL & INSTALLATION

The rear brake shoes serve as the parking brakes. Refer to Rear Drum Brakes, Brake Shoes, Removal & Installation.

ADJUSTMENT

The rear brake shoes serve as the parking brakes. Refer to Rear Drum Brakes, Brake Shoes, Adjustment.

GENERAL INFORMATION

CAUTION

Some vehicles are equipped with an air bag system. The system must be disarmed before performing service on, or around, system components, the steering column, instrument panel components, wiring and sensors. Failure to follow the safety precautions and the disarming procedure could result in accidental air bag deployment, possible injury and unnecessary system repairs.

SERVICE PRECAUTIONS

Disconnect and isolate the battery negative cable before beginning any airbag system component diagnosis, testing, removal, or installation procedures. Allow system capacitor to discharge for two minutes before beginning any component service. This will disable the airbag system. Failure to disable the airbag system may result in accidental airbag deployment, personal injury, or death.

Do not place an intact undeployed airbag face down on a solid surface. The airbag will propel into the air if accidentally deployed and may result in personal injury or death.

When carrying or handling an undeployed airbag, the trim side (face) of the airbag should be pointing towards the body to minimize possibility of injury if accidental deployment occurs. Failure to do this may result in personal injury or death.

Replace airbag system components with OEM replacement parts. Substitute parts may appear interchangeable, but internal differences may result in inferior occupant protection. Failure to do so may result in occupant personal injury or death.

Wear safety glasses, rubber gloves, and long sleeved clothing when cleaning powder residue from vehicle after an airbag deployment. Powder residue emitted from a deployed airbag can cause skin irritation. Flush affected area with cool water if irritation is experienced. If nasal or throat irritation is experienced, exit the vehicle for fresh air until the irritation ceases. If irritation continues, see a physician.

Do not use a replacement airbag that is not in the original packaging. This may result in improper deployment, personal injury, or death.

The factory installed fasteners, screws and bolts used to fasten airbag components have a special coating and are specifically designed for the airbag system. Do not use substitute fasteners. Use only original equipment fasteners listed in the parts catalog when fastener replacement is required.

During, and following, any child restraint anchor service, due to impact event or vehicle repair, carefully inspect all mounting hardware, tether straps, and anchors for proper installation, operation, or damage. If a child restraint anchor is found damaged in any way, the anchor must be replaced. Failure to do this may result in personal injury or death.

Deployed and non-deployed airbags may or may not have live pyrotechnic material within the airbag inflator.

Do not dispose of driver/passenger/curtain airbags or seat belt tensioners unless you are sure of complete deployment. Refer to the Hazardous Substance Control System for proper disposal.

Dispose of deployed airbags and tensioners consistent with state, provincial, local, and federal regulations.

After any airbag component testing or service, do not connect the battery negative cable. Personal injury or death may result if the system test is not performed first.

If the vehicle is equipped with the Occupant Classification System (OCS), do not connect the battery negative cable before performing the OCS Verification Test using the scan tool and the appropriate diagnostic information. Personal injury or death may result if the system test is not performed properly.

Never replace both the Occupant Restraint Controller (ORC) and the Occupant Classification Module (OCM) at the same time. If both require replacement, replace one, then perform the Airbag System test before replacing the other.

Both the ORC and the OCM store Occupant Classification System (OCS) calibration data, which they transfer to one another when one of them is replaced. If both are replaced at the same time, an irreversible fault will be set in both modules and the OCS may malfunction and cause personal injury or death.

If equipped with OCS, the Seat Weight Sensor is a sensitive, calibrated unit and must be handled carefully. Do not drop or handle roughly. If dropped or damaged, replace with another sensor. Failure to do so may result in occupant injury or death.

If equipped with OCS, the front passenger seat must be handled carefully as well. When removing the seat, be careful when setting on floor not to drop. If dropped, the sensor may be inoperative, could result in occupant injury, or possibly death.

If equipped with OCS, when the passenger front seat is on the floor, no one should sit in the front passenger seat. This uneven force may damage the sensing ability of the seat weight sensors. If sat on and damaged, the sensor may be inoperative, could result in occupant injury, or possibly death.

DISARMING THE SYSTEM

Disconnect and isolate the negative battery cable. Wait 2 minutes for the system capacitor to discharge before performing any service.

ARMING THE SYSTEM

When repairs are completed, connect the negative battery cable.

CLOCKSPRING CENTERING

See Figure 11.

The clockspring is mounted on the steering column behind the steering wheel. Its purpose is to maintain a continuous electrical circuit between the wiring harness and the driver's side air bag module. This assembly consists of a flat, ribbon-like electrically conductive tape that winds and unwinds with the steering wheel rotation.

Service replacement clocksprings are shipped pre-centered and with a molded plastic locking pin that snaps into a receptacle on the rotor and is engaged between two tabs on the upper surface of the rotor case. The locking pin secures the centered clockspring rotor to the clockspring case during shipment, but the locking pin must be removed from the clockspring after it is installed on the steering column. This locking pin should not be removed until the clockspring has been installed on the steering column. If the locking pin is removed before the clockspring is installed on a steering column, the clockspring centering procedure must be performed.

➡The clockspring cannot be repaired. If the clockspring is faulty, damaged, or if the driver airbag has been deployed, the clockspring must be replaced.

1. Before servicing the vehicle, refer to the precautions.
2. Disconnect the negative battery cable.

➡Wait at least 2 minutes after disconnecting the cable from the negative (-) battery terminal to prevent airbag activation.

➡Before starting this procedure, be certain to turn the steering wheel until the front wheels are in the straight-ahead position.

3. Place the front wheels in the straight-ahead position.
4. Remove the clockspring from the steering column.
5. Rotate the clockspring rotor clockwise to the end of its travel. Do not apply excessive torque.
6. From the end of the clockwise travel, rotate the rotor about two and one-half turns counterclockwise.
7. The engagement dowel and yellow rubber boot should end up at the bottom, and the arrows on the clockspring rotor and case should be in alignment. The clockspring is now centered.
8. The front wheels should still be in the straight-ahead position. Reinstall the clockspring onto the steering column.

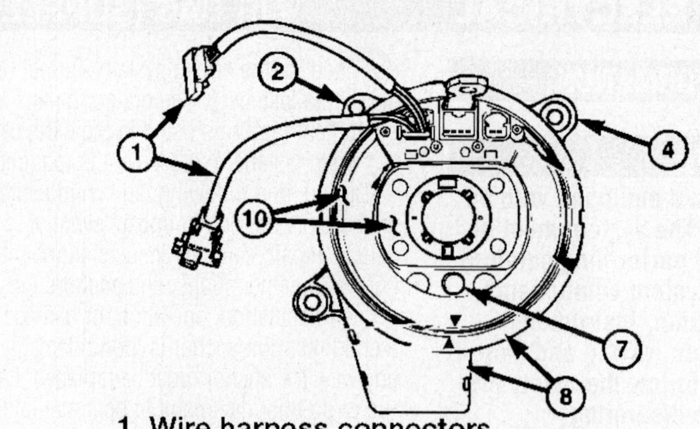

Fig. 11 Clockspring alignment marks

DRIVETRAIN

AUTOMATIC TRANSMISSION ASSEMBLY

REMOVAL & INSTALLATION

42RLE

See Figures 12 through 15.

1. Before servicing the vehicle, refer to the precautions.
2. Mark the driveshaft and axle companion flanges for assembly alignment.
3. Disconnect the negative battery cable.
4. Remove the skid plates, if equipped.
5. Remove the rear driveshaft.
6. Remove the front driveshaft, if equipped.
7. Disconnect the input and output speed sensors.

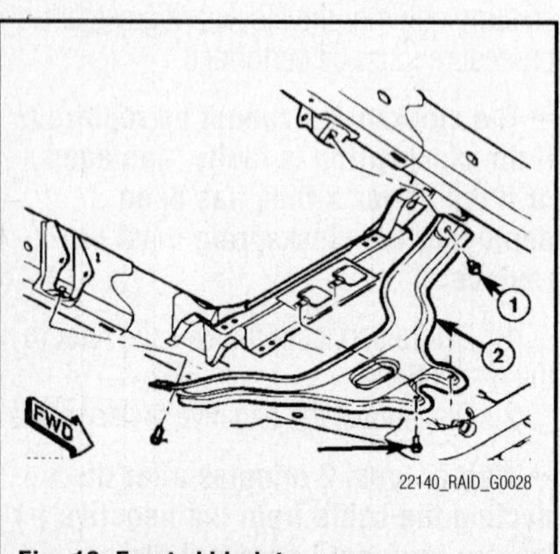

Fig. 12 Front skid plate

8. Remove the transfer case shift motor and mode sensor assembly.
9. Disconnect the variable line pressure connector from the transmission, if equipped.

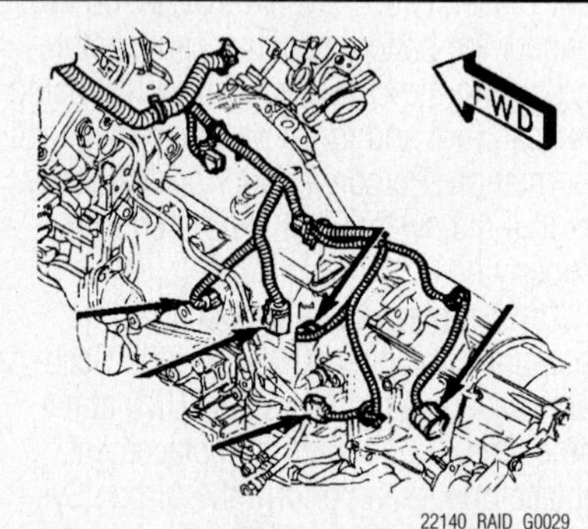

Fig. 13 Transmission electrical harness connectors—42RLE

Fig. 14 Transmission shift linkage—42RLE

10. Remove the transmission range sensor.
11. Disconnect the wires from the solenoid/pressure switch assembly.
12. Remove the bolts holding the exhaust crossover pipe to the pre-catalytic converter pipe flanges.
13. Remove the bolts holding the exhaust crossover pipe to the catalytic converter flange.
14. Remove the starter motor.
15. Remove the engine to transmission collar.
16. Rotate the crankshaft in a clockwise direction until converter bolts are accessible. Remove the bolts one at a time.
17. Remove the transmission vent hose .
18. Remove the transfer case, if equipped.
19. Support the rear of engine with a suitable jack.
20. Raise the transmission slightly with a suitable jack to relieve the load on the crossmember and supports.
21. Remove the bolts securing rear support and cushion to transmission and crossmember.
22. Remove the bolts attaching the crossmember to the frame and remove the crossmember.
23. Disconnect the transmission fluid cooler lines
24. Remove all remaining converter housing bolts.
25. Carefully work the transmission assembly rearward off engine block dowels.

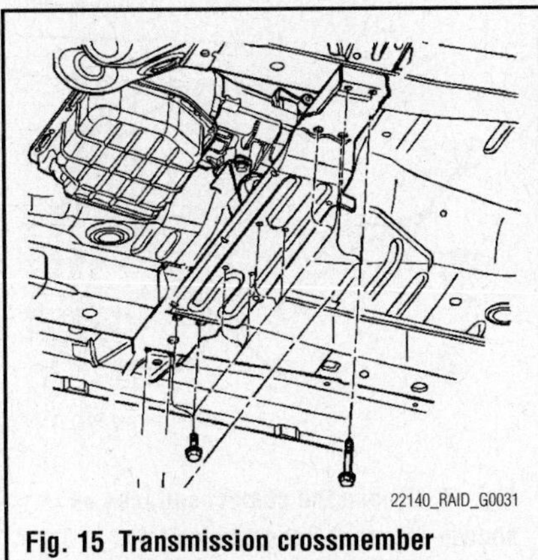

Fig. 15 Transmission crossmember

➡**Hold the tighten converter in place during transmission removal.**

26. Lower the transmission assembly and remove from the vehicle.

To install:

➡**If a replacement transmission is being installed, transfer any components necessary, such as the manual shift lever and shift cable bracket, from the original transmission onto the replacement transmission.**

27. Raise the transmission and align the tighten converter with the drive plate and transmission converter housing with the engine block.
28. Move the transmission forward and align the converter housing with engine block dowels.
29. Carefully work the transmission forward and over engine block dowels until converter hub is seated in crankshaft.

➡**Verify that no wires, or the transmission vent hose, have become trapped between the engine block and the transmission.**

30. Install two mounting bolts to attach the transmission to the engine.
31. Install the remaining tighten converter housing to engine bolts. Tighten to 50 ft. lbs. (68 nm).
32. Install the transfer case, if equipped.
33. Install the rear transmission crossmember. Tighten the crossmember to frame bolts to 50 ft. lbs. (68 nm).
34. Install the rear support to transmission. Tighten the bolts to 35 ft. lbs. (47 nm).
35. Lower the transmission assembly onto the crossmember and install bolts attaching transmission mount to crossmember. Tighten clevis bracket to crossmember bolts to 35 ft. lbs. (47 Nm). Tighten the clevis bracket to rear support bolt to 50 ft. lbs. (68 Nm)
36. Install the gearshift cable to support bracket and transmission manual lever.
37. Connect the input and output speed sensor and the transmission range sensor.
38. Connect the variable line pressure connector, if equipped.
39. Connect the wires to the solenoid/pressure switch assembly.
40. Install the tighten converter-to-driveplate bolts. tighten to 65 ft. lbs. (88 nm).
41. Install the starter motor and cooler line bracket.
42. Install the cooler lines to the transmission.
43. Install the transmission fill tube.
44. Install the exhaust components.
45. Install the rear driveshaft.
46. Install the front driveshaft, if equipped.
47. Install the skid plates, if equipped.
48. Connect the negative battery cable.
49. Adjust gearshift cable if necessary.
50. Refill the transmission with fluid to the correct level.

45RFE and 545RFE

See Figures 12, 15 and 16.

1. Before servicing the vehicle, refer to the precautions.
2. Disconnect the negative battery cable.
3. Remove the skid plates, if equipped.
4. Mark the driveshaft and axle companion flanges for assembly alignment.
5. Remove the rear driveshaft.
6. Remove the front driveshaft, if equipped.
7. Remove the engine to transmission collar.
8. Remove the exhaust support bracket from the rear of the transmission.
9. Remove the any necessary exhaust components required for clearance.
10. Remove the starter motor.
11. Rotate the crankshaft in clockwise direction until converter bolts are accessible. Then remove bolts one at a time.
12. Disconnect the output speed sensor connector.
13. Disconnect the input speed sensor connector.
14. Remove the transmission solenoid/TRS assembly connector.
15. Disconnect the line pressure sensor connector.
16. Remove the gearshift cable from transmission manual valve lever.
17. Remove the transmission fluid cooler lines.
18. Remove the transmission vent hose from the transmission.

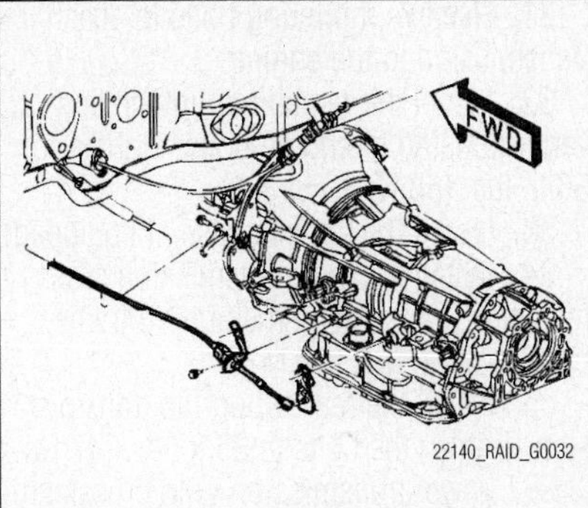

Fig. 16 Transmission shift linkage—45RFE and 545RFE

19. Support rear of engine with safety stand or jack.
20. Raise the transmission slightly with service jack to relieve load on crossmember and supports.
21. Remove the bolts securing rear support and cushion to transmission and crossmember.
22. Remove the bolts attaching crossmember to frame and remove crossmember.
23. Remove the transfer case, if equipped.
24. Remove the all remaining converter housing bolts.
25. Carefully work transmission and tighten converter assembly rearward off engine block dowels.
26. Hold tighten converter in place during transmission removal.
27. Lower transmission and remove the transmission assembly.

To install:

➡**If a replacement transmission is being installed, transfer any components necessary, such as the manual shift lever and shift cable bracket, from the original transmission onto the replacement transmission.**

28. Raise the transmission and align the tighten converter with the drive plate and transmission converter housing with the engine block.
29. Move transmission forward and align the converter housing with engine block dowels.
30. Carefully work transmission forward and over engine block dowels until converter hub is seated in crankshaft.

CAUTION

Verify that no wires, or the transmission vent hose, have become trapped between the engine block and the transmission.

31. Use two mounting bolts to attach the transmission to the engine.
32. Install the remaining tighten converter housing to engine bolts. Tighten to 50 ft. lbs. (68 Nm).
33. Install the transfer case, if equipped.
34. Install the rear transmission crossmember. Tighten crossmember to frame bolts to 50 ft. lbs. (68 Nm).
35. Install the rear support to transmission. Tighten the bolts to 35 ft. lbs. (47 Nm).
36. Lower transmission onto crossmember and install bolts attaching transmission mount to crossmember. Tighten clevis bracket to crossmember bolts to 35 ft. lbs. (47 Nm). Tighten the clevis bracket to rear support bolt to 50 ft. lbs. (68 Nm).
37. Install the gearshift cable to transmission.
38. Connect the wires to solenoid and pressure switch assembly connector, input and output speed sensors, and line pressure sensor.

➡Be sure the transmission harnesses are properly routed.

39. Install the tighten converter-to-driveplate bolts. Tighten to 22 ft. lbs. (31 Nm).
40. Install the starter motor and cooler line bracket.
41. Install the cooler lines to transmission.
42. Install the transmission fill tube.
43. Install the exhaust components.
44. Install the engine collar onto the transmission and the engine. Tighten the bolts to 40 ft. lbs. (54 Nm).
45. Install the rear driveshaft.
46. Install the front driveshaft, if equipped.
47. Install the skid plates, if equipped.
48. Connect the negative battery cable.
49. Adjust the gearshift cable as necessary.
50. Refill the transmission with fluid to the correct level.

MANUAL TRANSMISSION ASSEMBLY

REMOVAL & INSTALLATION

See Figures 17 and 18.

1. Before servicing the vehicle, refer to the precautions.
2. Shift transmission into Neutral.
3. Drain the transmission fluid.
4. Remove the negative battery cable.
5. Remove the shift knob.
6. Remove the shift lever boot.
7. Remove the shift lever extension.
8. Remove the lower shift lever boot assembly from the floor pan.
9. Remove the shift tower from the transmission.
10. Remove the skid plate, if equipped with 4wd.
11. Matchmark the driveshaft companion flanges for installation reference.
12. Remove the driveshafts.
13. Remove the y-pipe from the exhaust manifolds.
14. Remove the backup light switch connector.
15. Remove the clutch slave cylinder splash shield, if equipped.
16. Remove the clutch slave cylinder.
17. Support the transmission with a suitable jack.
18. Remove the transfer case if equipped.
19. Remove the starter.
20. Remove the transmission dust shield.
21. Remove the rear crossmember.
22. Remove the bolts/nuts from the rear transmission mount.
23. Remove the transmission harness wires from clips on transmission shift cover.
24. Lower the transmission slightly and remove transmission to engine bolts.

CAUTION

Do not remove structural dust cover from engine block.

25. Slide transmission rearward until input shaft clears clutch disc.
26. Lower the transmission and remove from the vehicle.

To install:

27. Clean the transmission front housing mounting surface
28. Apply a light coat of high temperature bearing grease or equivalent to contact surfaces of following components
29. Install the release fork.
30. Install the ball stud.
31. Install the release bearing slide surface.

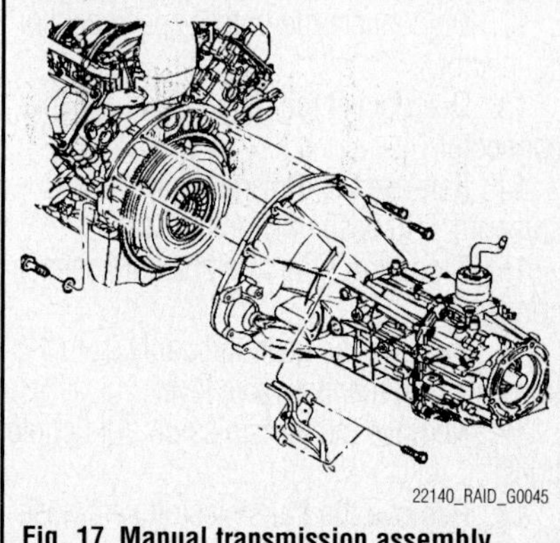

Fig. 17 Manual transmission assembly

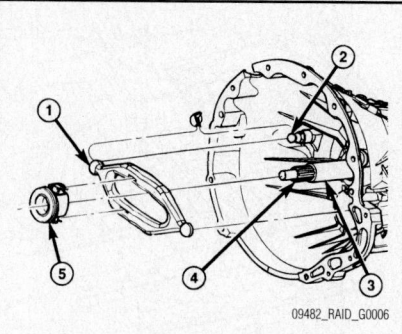

Fig. 18 Grease the contact surfaces as shown—Manual transmission

32. Install the input shaft splines.
33. Install the release bearing bore.
34. Install the propeller shaft slip yoke.
35. Support the transmission with a suitable jack.
36. Raise and align the transmission input shaft with clutch disc, then slide transmission into place.
37. Verify the front housing is fully seated.
38. Install or connect the following:.
39. Install the transmission to engine bolts without washers and tighten to 30 ft. lbs. (41 Nm).
40. Install the transmission bolts with washers and tighten to 50 ft. lbs. (68 Nm).
41. Install the dust shield and tighten bolts to 40 inch lbs. (4.5 Nm).
42. Install the rear crossmember and tighten nuts to 75 ft. lbs. (102 Nm).
43. Install the transmission rear mount bolts and tighten to 50 ft. lbs. (68 Nm).
44. Install the transmission harnesses to clips on shift cover .
45. Install the slave cylinder .
46. Install the transfer case, if equipped.
47. Install the driveshafts with reference marks aligned.
48. Install the y-pipe to the exhaust manifolds.
49. Install the shift tower, tighten bolts to 88 inch lbs. (10 Nm).
50. Install the lower shift boot.
51. Install the floor console.
52. Install the shift lever extension.
53. Install the upper shift boot.
54. Install the Negative battery cable.
55. Refill the transmission with fluid to the correct level.

CLUTCH DRIVEN DISC & PRESSURE PLATE

REMOVAL & INSTALLATION

See Figures 19 and 20.

1. Before servicing the vehicle, refer to the precautions.

2. Remove the transmission and clutch housing as assembly.

➡**If pressure plate is being removed for access to another component, mark position of pressure plate cover on flywheel with small punch marks.**

3. Loosen pressure plate cover bolts evenly and in rotation to relieve spring tension. Loosen bolts a few threads at a time to avoid warping cover.

4. Remove cover bolts, pressure plate and clutch disc.

To install:

➡**Clean flywheel surface with solvent. Scuff sand the surface with 120/180 grit emery cloth to remove minor scratches and glazing.**

5. Check new clutch disc for runout and free operation on input shaft splines.

6. Lubricate crankshaft pilot bearing with a NLGI—2 rated grease.

7. Position clutch disc with pressure plate on the flywheel.

8. Insert alignment tool or spare input shaft through clutch disc and into pilot bearing.

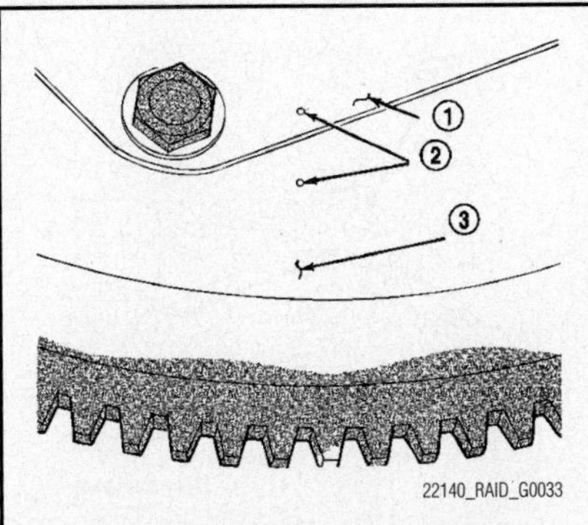

Fig. 19 Mark position of pressure plate cover on flywheel with small punch marks

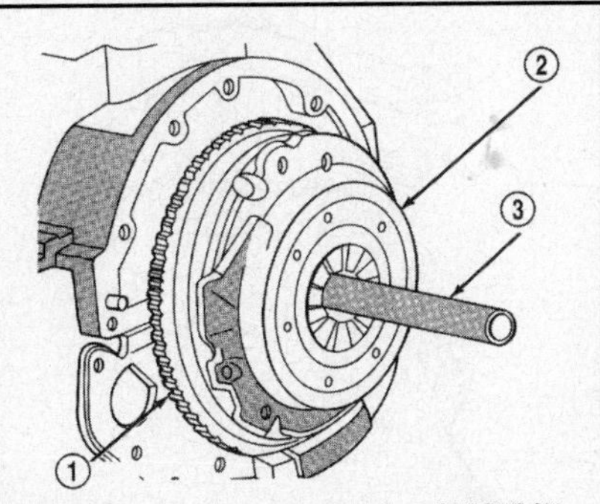

Fig. 20 Clutch installation. (1) flywheel, (2) pressure plate, (3) alignment tool

9. Verify that the disc hub is positioned correctly. The raised portion of the hub faces away from the flywheel.

10. Install the cover bolts finger tight.

11. Tighten cover bolts evenly (and in rotation) a few threads at a time. Cover bolts must be tightened evenly and to specified tighten to avoid distorting cover.

12. Tighten the cover bolts as follows:

a. 5⁄16 in. bolts to 17 ft. lbs. (23 Nm).

b. 3⁄8 in. bolts to 30 ft. lbs. (41 Nm).

13. Apply light coat of high temperature bearing grease to splines of transmission input shaft and to release bearing slide surface of front bearing retainer.

**** WARNING**

Do not over-lubricate shaft splines. This could result in grease contamination of disc.

14. Install the transmission assembly.

CLUTCH MASTER CYLINDER

REMOVAL & INSTALLATION

See Figure 21.

**** WARNING**

The hydraulic linkage is serviced as a complete assembly only. The individual components must not be overhauled or serviced separately.

1. Raise and safely support the vehicle securely on jackstands.

2. Remove slave cylinder nuts and remove cylinder from transmission.

**** WARNING**

Do not disconnect slave cylinder quick disconnect. If disconnected hydraulic linkage must be replaced.

3. Remove clip holding clutch cylinder push rod to clutch pedal .

4. Slide clutch cylinder push rod off clutch pedal pin .

5. Disconnect clutch pedal position switch connector from wiring harness.

6. Remove hydraulic fluid line clip from the lower dash panel flange.

7. Remove clutch cylinder nuts.

8. Ensure clutch reservoir cap is tight.

9. Remove clutch reservoir nuts.

10. Pull clutch reservoir and cylinder from dash panel.

11. Remove hydraulic linkage components from vehicle as an assembly.

To install:

12. Ensure clutch reservoir cap is tight

13. Install clutch reservoir in dash panel and tighten nuts to 40 inch lbs. (5 Nm).

14. Install clutch cylinder in dash panel and tighten nuts to 40 ft. lbs. (54 Nm).

15. Connect clutch pedal position switch connector to wiring harness.

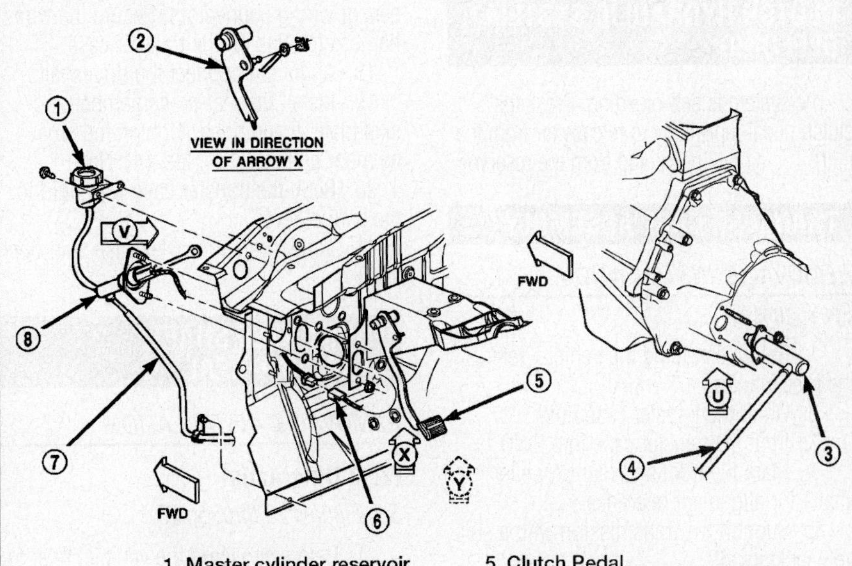

Fig. 21 The hydraulic linkage is serviced as a complete assembly only. The individual components must not be overhauled or serviced separately

16. Install hydraulic fluid line clip into the hole in the lower dash panel.

17. Apply light coating of grease to the inner diameter of the clutch cylinder push rod and outer diameter of the clutch pedal pin.

18. Install clutch cylinder push rod on clutch pedal pin and install retaining clip.

19. Install slave cylinder in transmission and ensure rod is securely engaged in release lever.

➡If new clutch linkage is being installed, do not remove plastic shipping strap from the slave cylinder push rod. The strap will break on its own upon first clutch application.

20. Install slave cylinder nuts and tighten to 17 ft. lbs. (23 Nm).

21. Verify that fluid line from master cylinder to slave cylinder is properly routed.

OVERHAUL

The hydraulic linkage is serviced as a complete assembly only. The individual components must not be overhauled or serviced separately.

CLUTCH SLAVE CYLINDER

REMOVAL & INSTALLATION

The hydraulic linkage is serviced as a complete assembly only. The individual components must not be overhauled or serviced separately. Refer to Clutch, Master Cylinder, Removal and Installation.

CLUTCH HYDRAULIC SYSTEM BLEEDING

The system is self-bleeding. Press the clutch pedal repeatedly to release air from the fluid. The air will be vented from the reservoir.

TRANSFER CASE ASSEMBLY

REMOVAL & INSTALLATION

See Figure 22.

1. Before servicing the vehicle, refer to the precautions.

2. Drain the transfer case fluid.

3. Shift the transfer case into 2WD.

4. Mark the front and rear propeller shafts for alignment reference.

5. Support the transmission with a suitable jack stand.

6. Remove the rear crossmember and skid plate, if equipped.

7. Disconnect the front and rear driveshafts at the transfer case.

8. Disconnect the transfer case shift motor and mode sensor wire connectors.

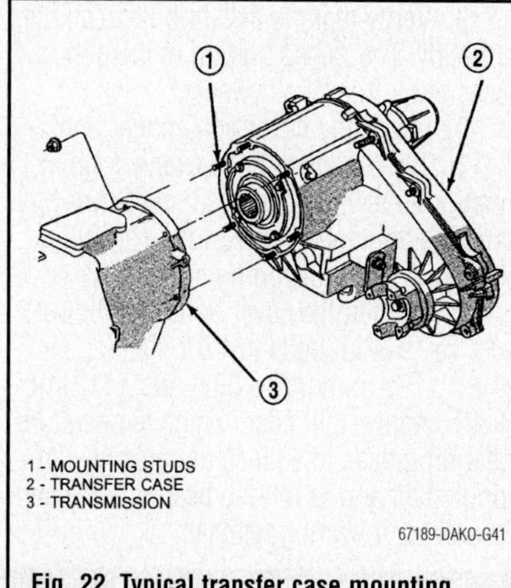

Fig. 22 Typical transfer case mounting

9. Disconnect the transfer case vent hose.

10. Support the transfer case with a suitable transmission jack and secure the transfer case to the jack with chains.

11. Remove the nuts attaching transfer case to the transmission.

12. Pull the transfer case and jack rearward to disengage the transfer case.

13. Remove the transfer case assembly.

To install:

14. Align the transfer case and transmission shafts and install the transfer case onto the transmission.

15. Install and the transfer case attaching nuts. Tighten to 20–25 ft. lbs. (27–34 Nm).

16. Connect the vent hose.

17. Connect the shift motor and mode sensor wiring connectors. Secure the wire harness to clips on the transfer case.

18. Align and connect the driveshafts.

19. Install the rear crossmember and skid plate, if equipped. Tighten the crossmember bolts to 30 ft. lbs. (41 Nm).

20. Refill the transfer case with fluid to the correct level.

21. Verify proper transfer case shift operation.

FRONT AXLE SHAFT, BEARING & SEAL

REMOVAL & INSTALLATION

Front Differential

See Figures 23 through 25.

1. Before servicing the vehicle, refer to the precautions.

2. With vehicle in neutral, position vehicle on hoist.

3. Remove the skid plate, if equipped.

4. Remove skid plate support crossmember, if necessary.

5. Remove hub/bearing wheel speed sensor .and dust shield from knuckles .

6. Remove both halfshafts.

7. Mark the propeller shaft, transfer case, and pinion companion flange for installation reference.

8. Remove the front propeller shaft.

9. Remove the axle vent tube.

10. Position and secure a jack under differential housing.

11. Remove bolts from the axle tube .

12. Remove bolts from axle differential housing .

13. Remove bolts from axle pinion nose bracket .

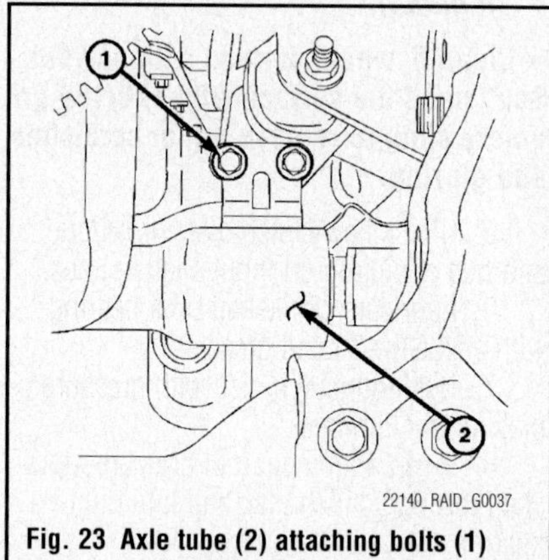

Fig. 23 Axle tube (2) attaching bolts (1)

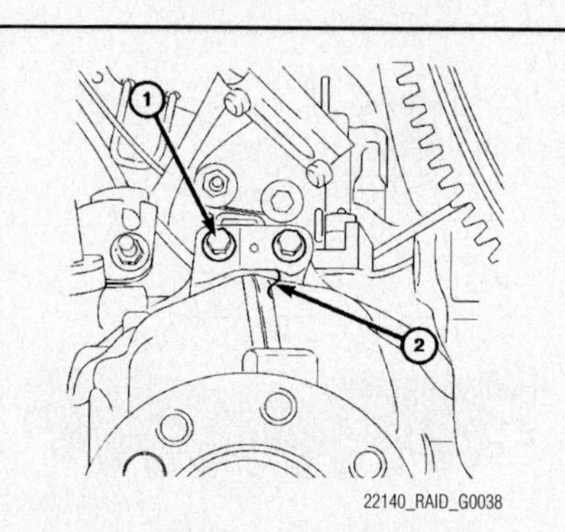

Fig. 24 Axle differential housing (2) attaching bolts (1)

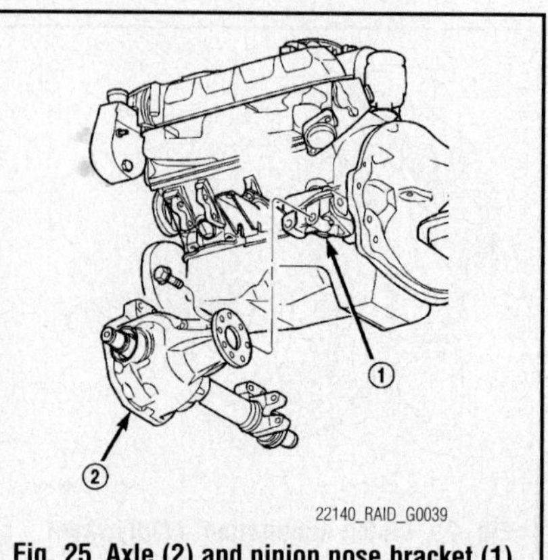

Fig. 25 Axle (2) and pinion nose bracket (1)

14. Lower jack and remove axle from vehicle.

To install:

15. Raise the axle into position.
16. Loosely install the bolts and nuts to hold the axle pinion nose bracket .
17. Loosely install bolts from axle differential housing .
18. Loosely install bolts from the axle tube.
19. Tighten all bolts to 85 ft. lbs. (115 Nm).
20. Install the axle vent tube.
21. Align the reference marks on the propeller shaft, transfer case, and pinion companion flange.
22. Install propeller shaft.
23. Install halfshafts.
24. Install hub/bearing wheel speed sensor and dust shield on knuckles .
25. Install skid plate support crossmember, if necessary.
26. Install skid plate, if necessary.
27. Check differential lubricant level and add lubricant, if necessary.

Axle Bearings

See Figure 26.

1. Before servicing the vehicle, refer to the precautions.
2. Remove the axle shaft.
3. Remove the axle shaft seal.
4. Install the axle shaft bearing remover C-4660-A, or equivalent, in the bearing. Then tighten the nut to spread the remover in the bearing.
5. Install the bearing remove cup, bearing and nut. Then tighten the nut to draw the bearing out.
6. Inspect the axle shaft tube bore for roughness and burrs. Remove as necessary.

To install:

7. Wipe the axle shaft tube bore clean.
8. Install the axle shaft bearing with installer 5063 and handle C-4171, or equivalent.

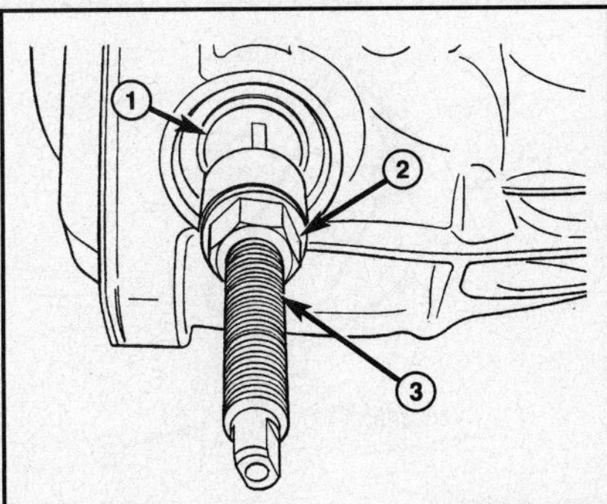

Fig. 26 Removing the axle shaft bearing (1) using remover C-4660-A (3) and nut (2)

9. Install a new axle shaft seal with installer 8402 and handle C-4171, or equivalent.
10. Install the axle shaft and halfshaft.

Axle Shaft Seal

See Figure 27.

1. Before servicing the vehicle, refer to the precautions.
2. Remove the axle shaft.
3. Remove the axle shaft seal with a small prybar.

To install:

4. Wipe the axle shaft tube bore clean.
5. Install a new axle shaft seal with installer 8402 and handle C-4171.
6. Install the axle shaft and halfshaft.

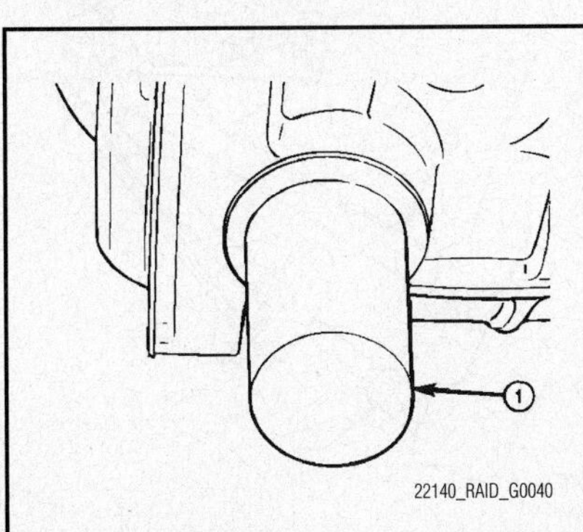

Fig. 27 Install a new axle shaft seal with Installer 8402 or equivalent

FRONT HALFSHAFTS

REMOVAL & INSTALLATION

See Figures 28 and 29.

1. Before servicing the vehicle, refer to the precautions.
2. Remove the front wheel.
3. Remove the skid plate, if equipped.
4. Remove the hub nut and washer.
5. Remove the brake caliper and rotor.
6. Remove the ABS wheel speed sensor if equipped.
7. Remove the hub bearing bolts from the knuckle.
8. Remove the hub bearing and brake shield from the knuckle.
9. Support the halfshaft at the CV joint housings.
10. Position two pry bars behind the inner CV housing and disengage the CV joint from the axle.
11. Remove the halfshaft from the vehicle.

To install:

12. Apply a light coating of wheel bearing grease on the axle splines.

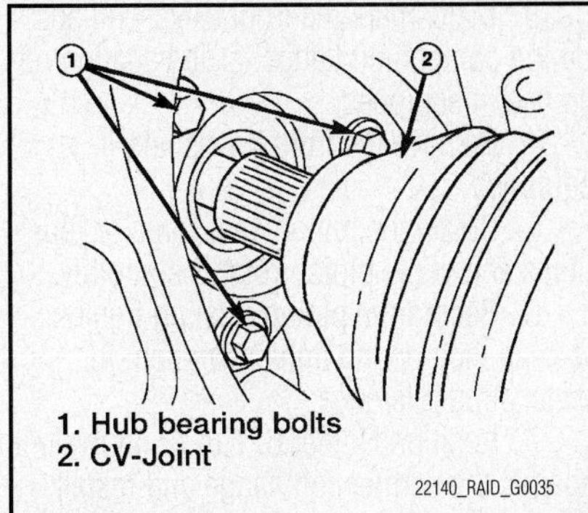

Fig. 28 Hub bearing bolts on the steering knuckle

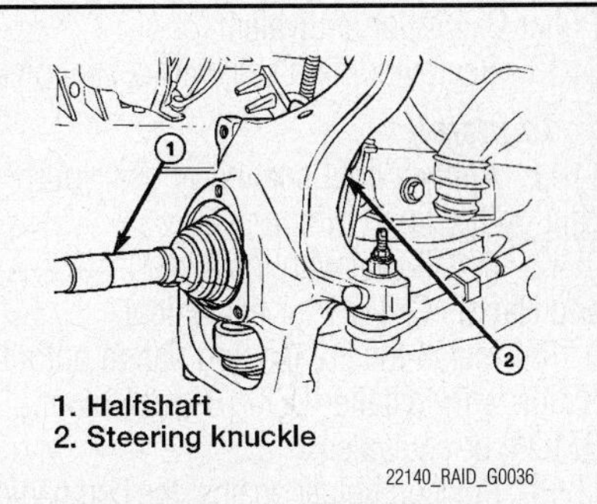

Fig. 29 Remove the halfshaft from the vehicle through the hole in the steering knuckle

13. Insert the halfshaft stub through the steering knuckle and onto the axle. Verify the shaft snapring engages with the groove on the inside of the joint housing.
14. Clean the hub bearing bore and hub bearing mating surface of all foreign materials. Apply a light coating of grease to all mating surfaces.
15. Install the hub bearing onto the axle halfshaft and steering knuckle.
16. Install the hub bearing bolts and tighten to 120 ft. lbs. (163 Nm).
17. Install the ABS wheel speed sensor, if equipped.
18. Install brake rotor and caliper.
19. Apply the brakes and tighten hub nut to 185 ft. lbs. (251 Nm).
20. Install the skid plate, if equipped.
21. Install the wheel and tire assembly.

FRONT PINION SEAL

REMOVAL & INSTALLATION

See Figures 30 through 33.

1. Before servicing the vehicle, refer to the precautions.
2. Remove both halfshafts.

3. Matchmark the front driveshaft and pinion companion flange for installation reference, if equipped.
4. Remove the front driveshaft, if equipped
5. Rotate the pinion gear three or four times and verify pinion rotates smoothly.
6. Record the pinion rotating tighten with an inch pound tighten wrench, for installation reference.
7. Position Holder 6719A or equivalent against the companion flange and install four bolts and washers into the threaded holes and tighten the bolts.
8. Remove the pinion nut.
9. Remove the companion flange with Puller C-452, or equivalent.
10. Remove pinion seal with a seal pick.

To install:

11. Apply a light coating of gear lubricant on the lip of pinion seal.
12. Install seal with Installer C-3972-A and Handle C-4171, or equivalent.
13. Install the companion flange onto the pinion with Installer C-3718 and Holder 6719A, or equivalent.
14. Position holder against the companion flange and install four bolts and washers into the threaded holes. Tighten the bolt and washer so that the holder is held to the flange.

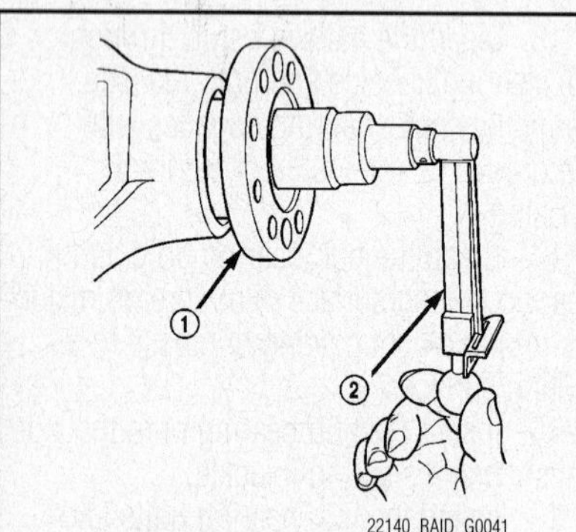

Fig. 30 Record the pinion (1) rotating torque with an inch pound torque wrench (2)

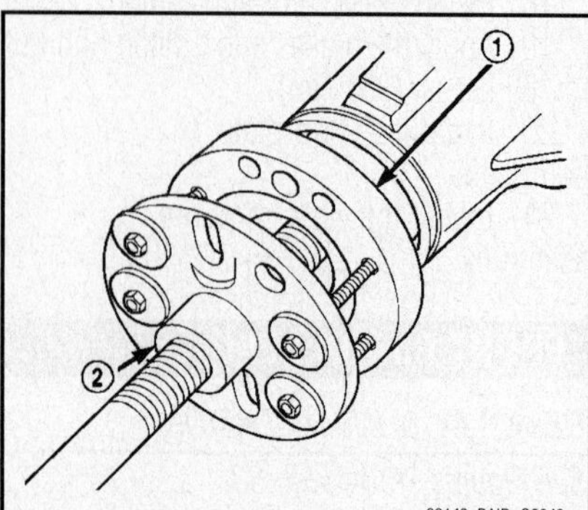

Fig. 31 Remove the companion flange (1) with Puller C-452, or equivalent (2)

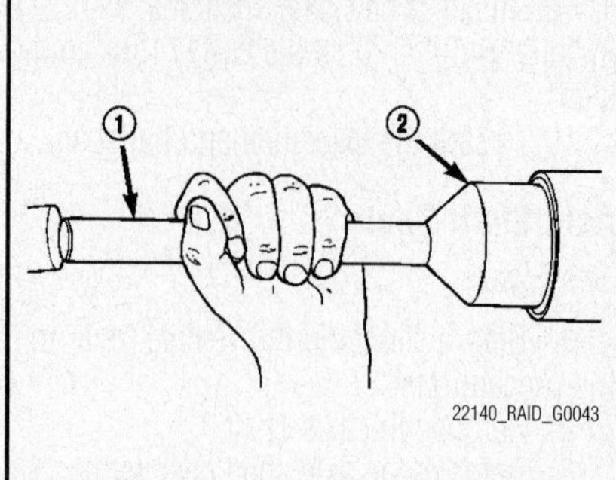

Fig. 32 Install seal with Installer C-3972-A (2) and Handle C-4171 (1), or equivalent

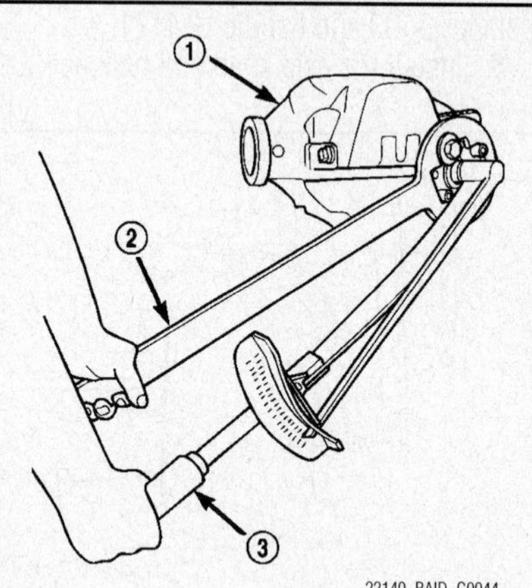

Fig. 33 Position holder (2) against the companion flange. Tighten the differential (1) bolt and washer using a torque wrench (3) to obtain the rotating torque referenced during removal

15. Install a new pinion nut onto the pinion shaft and tighten the pinion nut until there is zero bearing end-play.

WARNING

Do not exceed the minimum tightening tighten when installing the companion flange at this point. Damage to the collapsible spacer or bearings may result.

16. Tighten the nut to 200 ft. lbs. (271 Nm).

➡Never loosen pinion nut to decrease pinion bearing rotating tighten and never exceed specified preload tighten. If preload tighten or rotating tighten is exceeded a new collapsible spacer must be installed.

17. Record the pinion rotating tighten using a tighten wrench. The rotating tighten should be equal to the reading recorded during removal plus 5 inch lbs. (0.56 Nm).
18. If the rotating tighten is low, tighten the pinion nut in 5 ft. lbs. (6.8 Nm) increments until the proper rotating tighten is achieved.

➡If the maximum tightening tighten is reached prior to reaching the required rotating tighten, the collapsible spacer may have been damaged. Replace the collapsible spacer.

19. Install the driveshaft with the reference marks aligned.
20. Install halfshafts.

REAR AXLE HOUSING

REMOVAL & INSTALLATION

See Figure 34.

1. Before servicing the vehicle, refer to the precautions.
2. Place the transmission in neutral.
3. Raise and safely support the vehicle securely on jackstands.
4. Position a lift under axle and secure axle to lift.
5. Remove the wheels and rear brake components.
6. Remove ABS sensor from the differential housing.
7. Disconnect the brake hose at the axle junction block.
8. Disconnect the vent hose from the axle shaft tube.
9. Mark the propeller shaft and companion flange for installation alignment reference.
10. Remove propeller shaft.
11. Remove shock absorbers axle mounting bolts.
12. Remove stabilizer bar retainer clamp bolts and retainer clamps from the axle.
13. Remove spring clamps nuts and spring plates.
14. Remove the axle from the vehicle.

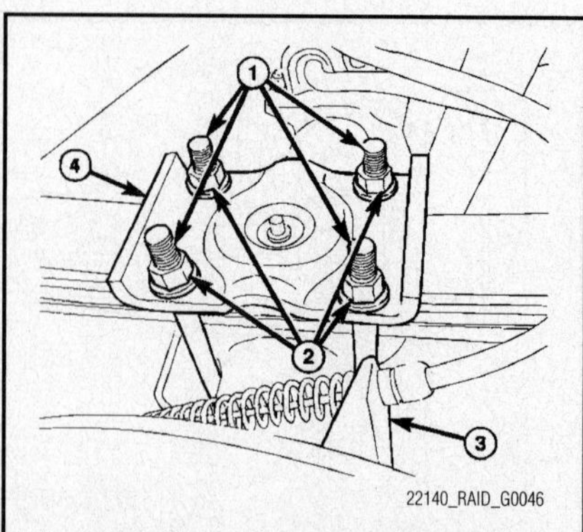

Fig. 34 Spring clamp (1), nuts (2), emergency brake cable (3) and spring plate (4)

To install:

15. Raise axle with lifting device and align to the leaf spring centering bolts.
16. Install spring clamps and spring plates. Tighten the U-bolt nuts to 110 ft. lbs. (149 Nm).
17. Install shock absorbers and tighten nuts to 75 ft. lbs. (102 Nm).
18. Install ABS sensor into the differential housing.
19. Install rear brake and park brake components.
20. Connect brake hose to axle junction block.
21. Install stabilizer bar and center it with equal spacing on both sides. Tighten retainer clamp bolts to 45 ft. lbs. (61 Nm).
22. Install axle vent hose.
23. Install propeller shaft with reference marks aligned. Tighten bolts to 80 ft. lbs. (108 Nm).
24. Add gear lubricant, if necessary.

REAR AXLE SHAFT, BEARING & SEAL

REMOVAL & INSTALLATION

Axle Shaft

See Figures 35 and 36.

1. Before servicing the vehicle, refer to the precautions.
2. Remove the rear brake components.
3. Remove the differential cover and drain the fluid.
4. Rotate the differential case to access the pinion mate shaft lock screw. Remove the lock screw and pinion mate shaft from the differential case.
5. Push the axle shaft inward then remove the C-clip from the axle shaft.
6. Remove the axle shaft.

To install:

7. Lubricate the bearing bore and seal lip with gear lubricant. Insert the axle shaft through the seal, bearing and engage it into side gear splines.

➡Use care to prevent the shaft splines from damaging the axle shaft seal.

8. Insert the C-clip in end of axle shaft. Push the axle shaft outward to seat the C-clip in side gear.
9. Insert the pinion shaft into the differential case, through the thrust washers and differential pinions.
10. Align the hole in the pinion mate shaft with the hole in the differential case and install the lock screw with Loctite® on the threads. Tighten the lock screw to 8 ft. lbs. (11 Nm).
11. Install the differential cover and tighten the bolts in a criss-cross pattern to 30 ft. lbs. (41 Nm).
12. Fill the differential with gear lubricant to the bottom of the fill plug hole.
13. Install the fill hole plug.
14. Install the rear brake components.

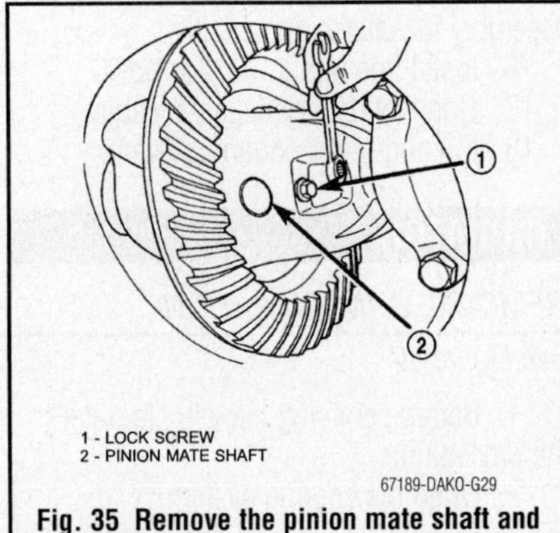

Fig. 35 Remove the pinion mate shaft and lock screw

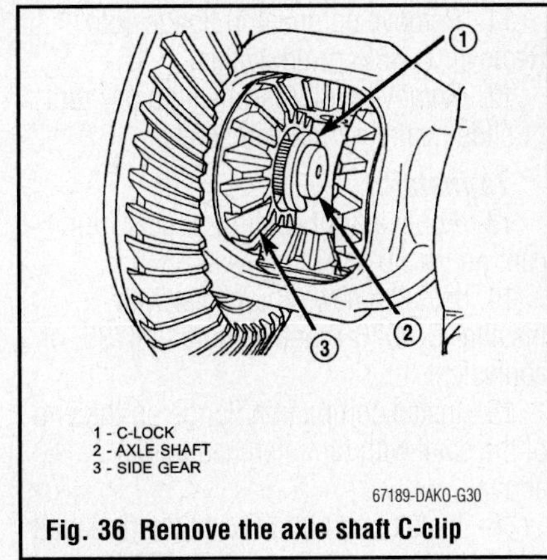

Fig. 36 Remove the axle shaft C-clip

Axle Shaft Seal

See Figure 37.

1. Before servicing the vehicle, refer to the precautions.
2. Remove the axle shaft.
3. Remove the axle shaft seal from the end of the axle tube with a seal pick.

To install:

4. Wipe the axle tube bore clean. Remove any old sealer or burrs from the tube.
5. Coat the lip of the new seal with axle lubricant for protection prior to installing the axle shaft.
6. Install the new axle seal with Installer C-4198 and Handle C-4171, or equivalent. When the tool contacts the axle tube, the seal is installed to the correct depth.
7. Install the axle shaft.

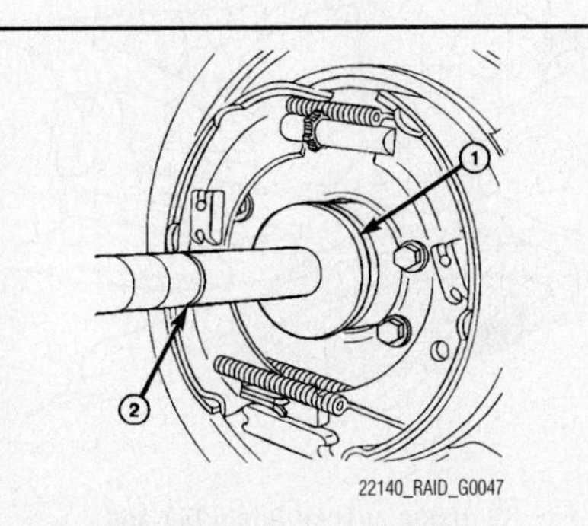

Fig. 37 Install the new axle seal with Installer C-4198 (1) and Handle C-4171 (2), or equivalent

Axle Bearings

See Figure 38.

1. Before servicing the vehicle, refer to the precautions.
2. Remove the axle shaft.
3. Remove the axle shaft seal from the axle tube with a seal pick.

➡The seal and bearing can be removed at the same time with the bearing removal tool.

4. Remove the axle shaft bearing with Bearing Removal Tool Set 6310 and Adapter Foot 6310-9, or equivalent.

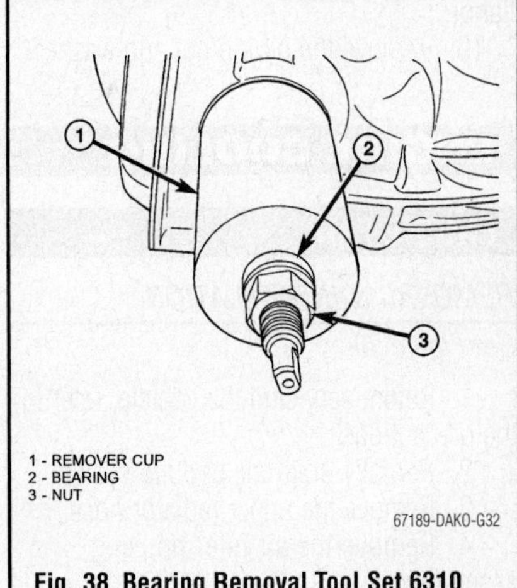

Fig. 38 Bearing Removal Tool Set 6310

To install:

5. Wipe the axle tube bore clean. Remove any old sealer or burrs from the tube.
6. Install the axle shaft bearing with Installer C-4198 and Handle C-4171, or equivalent.

➡Install the bearing with part number against the installer.

7. Install a new axle seal.
8. Install the axle shaft

REAR PINION SEAL

REMOVAL & INSTALLATION

See Figures 30 through 33.

1. Before servicing the vehicle, refer to the precautions.
2. Matchmark the universal joint, companion flange and pinion shaft for installation reference.
3. Disconnect the rear driveshaft.
4. Remove companion flange bolts and secure the shaft in an upright position to prevent damage to the rear universal joint.
5. Remove brake drums to prevent any drag.
6. Rotate companion flange three or four times and verify flange rotates smoothly.
7. Measure rotating torque of the pinion with an inch pound torque wrench and record the reading for installation reference.
8. Install bolts into two of the threaded holes in the companion flange 180° apart.
9. Position Holder 6719 or equivalent against the companion flange and install a bolt and washer into one of the remaining threaded holes. Tighten the bolts so the Holder 6719 or equivalent is held to the flange.
10. Remove the pinion nut and washer.
11. Remove companion flange with Remover C-452, or equivalent.
12. Remove pinion seal with a pry tool or slide hammer mounted screw.

To install:

13. Apply a light coating of gear lubricant on the lip of pinion seal.
14. Install a new pinion seal with Installer C-4076-B and Handle C-4735, or equivalent.
15. Install companion flange on the end of the shaft with the reference marks aligned.
16. Install bolts into two of the threaded holes in the companion flange 180°apart.
17. Position Holder 6719, or equivalent, against the companion flange and install a bolt and washer into one of the remaining threaded holes. Tighten the bolts so Holder 6719 is held to the flange.
18. Install companion flange on pinion shaft with Installer C-3718 and Holder 6719, or equivalent.
19. Install the pinion washer and a new pinion nut. The convex side of the washer must face outward.

➡Do not exceed the minimum tightening tighten when installing the companion flange retaining nut at this point. Damage to collapsible spacer or bearings may result.

20. Hold companion flange with Holder 6719 and tighten the pinion nut to 210 ft. lbs. (285 Nm). Rotate pinion several revolutions to ensure the bearing rollers are seated.
21. Rotate the pinion flange with an inch pound tighten wrench. Rotating tighten should be equal to the reading recorded during removal plus 5 inch lbs. (0.56 Nm).

➡Never loosen pinion nut to decrease pinion bearing rotating tighten and never exceed specified preload tighten. If rotating tighten is exceeded, a new collapsible spacer must be installed.

22. If rotating tighten is too low, use Holder 6719 to hold the companion flange and tighten pinion nut in 5 ft. lbs. (6.8 Nm) increments until proper rotating tighten is achieved.

➡The seal replacement is unacceptable if final pinion nut tighten is less than 210 ft. lbs. (285 Nm).

➡The bearing rotating tighten should be constant during a complete revolution of the pinion. If the rotating tighten varies, this indicates a binding condition.

23. Install driveshaft with the installation reference marks aligned.
24. Tighten companion flange bolts to 80 ft. lbs. (108 Nm).
25. Install the rear brake components.
26. Check the differential housing lubricant level.

ENGINE COOLING

ENGINE FAN

REMOVAL & INSTALLATION

See Figure 39.

1. Before servicing the vehicle, refer to the precautions.
2. Partially drain the cooling system.
3. Remove the upper radiator hose.
4. Remove the air filter housing assembly.
5. Using special tool 6958 and adapter 8346 or similar, remove the fan/viscous fan drive assembly from the water pump. Do not remove the fan/viscous fan drive assembly from the vehicle at this time.
6. Position the fan/fan drive assembly in the radiator shroud.
7. Remove the two shroud mounting screws.
8. Remove the radiator shroud and fan drive assembly.
9. Place the viscous fan drive in a horizontal position.
10. Remove the four bolts securing the fan blade assembly to viscous fan drive.

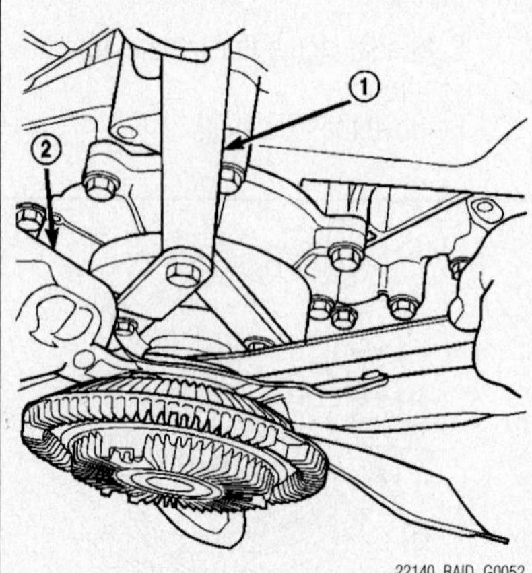

Fig. 39 Using special tool 6958 and adapter 8346 (1) or similar, remove the fan/viscous fan drive assembly (2)

To install:

11. Install fan blade assembly to viscous fan drive. Tighten bolts to 17 ft. lbs. (23 Nm).
12. Position fan blade/viscous fan drive assembly into the radiator shroud.
13. Install the radiator shroud and fan drive assembly into the vehicle.
14. Install fan shroud retaining screws. Tighten screws to 50 inch lbs. (6 Nm)
15. Install fan blade/viscous fan drive assembly to water pump shaft .
16. Install the upper radiator hose.
17. Connect battery negative cable.
18. Fill and bleed cooling system.

RADIATOR

REMOVAL & INSTALLATION

See Figure 40.

1. Before servicing the vehicle, refer to the precautions.
2. Drain the cooling system.
3. Disconnect the negative battery cable.

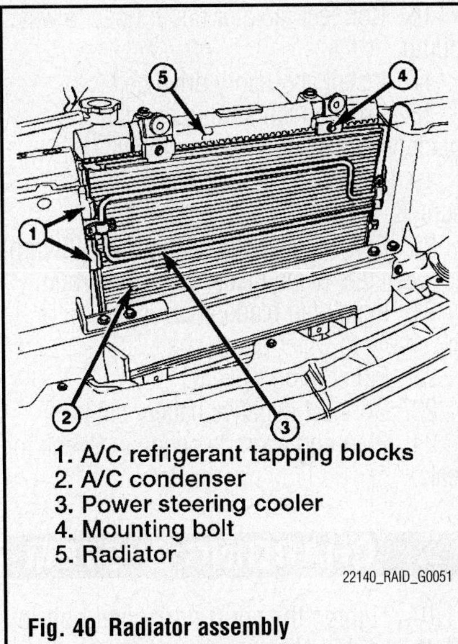

Fig. 40 Radiator assembly

4. Remove the pushpins and upper condenser/radiator seal.
5. Disconnect the upper radiator hose.
6. Disconnect the power steering hoses from power steering fluid/transmission cooler.
7. Disconnect the overflow tube.
8. Remove the radiator fan shroud from the radiator and position over the radiator fan.
9. Raise and safely support the vehicle securely on jackstands.
10. Disconnect the power steering cooler lines.
11. Remove the lower radiator hose.
12. Lower the vehicle.
13. Remove the upper radiator mounting bolts.
14. Remove the radiator.

To install:

15. Install the radiator.
16. Install the upper radiator mounting bolts. Tighten radiator support bolts to 200 inch lbs. (23 Nm).
17. Lower the vehicle.
18. Install the lower radiator hose.
19. Connect the power steering cooler lines.
20. Raise and safely support the vehicle securely on jackstands.
21. Install the radiator fan shroud to the radiator and position over the radiator fan.
22. Connect the overflow tube.
23. Connect the power steering hoses to power steering fluid/transmission cooler.
24. Connect the upper radiator hose.
25. Install the pushpins and upper condenser/radiator seal.
26. Connect the negative battery cable.
27. Fill and bleed the cooling system.

THERMOSTAT

REMOVAL & INSTALLATION

See Figure 41.

1. Before servicing the vehicle, refer to the precautions.
2. Drain the cooling system.
3. Raise the vehicle.
4. Remove the splash shield.
5. Remove the lower radiator hose at the thermostat housing.
6. Remove the thermostat housing mounting bolts and remove housing and thermostat.

To install:

7. Clean mating areas of timing chain cover and thermostat housing.
8. Install thermostat (spring side down) into recessed machined groove on timing chain cover .
9. Position thermostat housing on timing chain cover.
10. Install two housing-to-timing chain cover bolts. Tighten bolts to 112 inch lbs. (13 Nm).

⁂ WARNING

Housing must be tightened evenly and thermostat must be centered into recessed groove in timing chain cover. If not, it may result in a cracked housing, damaged timing chain cover threads or coolant leaks.

11. Install lower radiator hose on thermostat housing.
12. Install splash shield.
13. Lower vehicle.
14. Fill cooling system.
15. Connect negative battery cable to battery.
16. Start and warm the engine. Check for leaks.

WATER PUMP

REMOVAL & INSTALLATION

See Figure 42.

1. Before servicing the vehicle, refer to the precautions.
2. Disconnect negative battery cable.
3. Drain cooling system.
4. Remove fan/viscous fan drive assembly from water pump. Do not attempt to remove fan/viscous fan drive assembly from vehicle at this time.
5. If water pump is being replaced, do not unbolt fan blade assembly from thermal viscous fan drive.
6. Remove two fan shroud-to-radiator screws.
7. Remove fan shroud and fan blade/viscous fan drive assembly from vehicle.
8. After removing fan blade/viscous fan drive assembly, do not place thermal

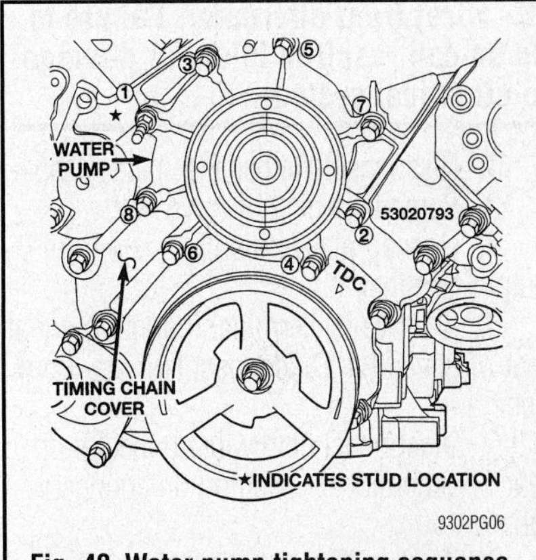

Fig. 42 Water pump tightening sequence

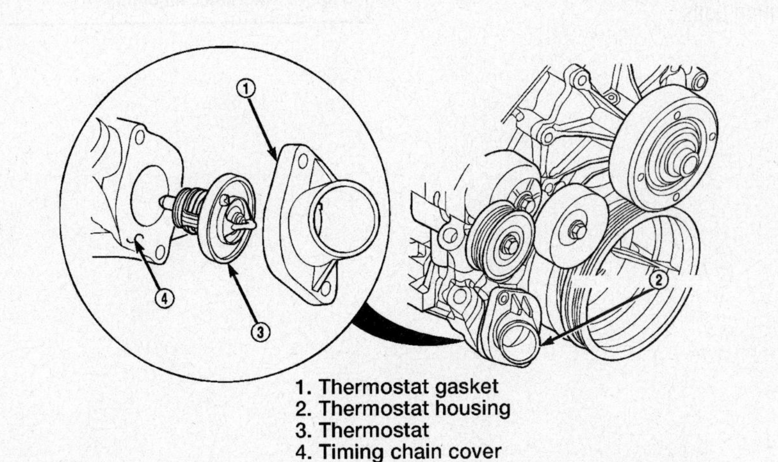

Fig. 41 Thermostat assembly

viscous fan drive in horizontal position. If stored horizontally, silicone fluid in viscous fan drive could drain into its bearing assembly and contaminate lubricant.

9. Remove accessory drive belt from the water pump pulley.

10. Remove upper radiator hose clamp and remove upper hose at water pump.

11. Remove seven water pump mounting bolts and one stud bolt.

⁂ WARNING

Do not pry water pump at timing chain case/cover. The machined surfaces may be damaged resulting in leaks.

12. Remove water pump and gasket. Discard gasket.

To install:

13. Clean gasket mating surfaces.

14. Using a new gasket, position water pump and install mounting bolts as shown. Tighten water pump mounting bolts to 43 ft. lbs. (58 Nm).

15. Spin water pump to be sure that pump impeller does not rub against timing chain case/cover.

16. Connect radiator upper hose to water pump.

17. Install accessory drive belt.

18. Position fan shroud and fan blade/viscous fan drive assembly.

19. Install two fan shroud-to-radiator screws.

20. Be sure of at least 1.0 inches (25 mm) between tips of fan blades and fan shroud.

21. Install fan blade/viscous fan drive assembly to water pump shaft.

22. Fill cooling system.

23. Connect negative battery cable.

24. Start and warm the engine. Check for leaks.

ENGINE ELECTRICAL

CHARGING SYSTEM

ALTERNATOR

REMOVAL & INSTALLATION

See Figure 43.

⁂ WARNING

Disconnect negative battery cable before removing battery output wire (B+ wire) from alternator. Failure to do so can result in injury or damage to electrical system.

1. Disconnect the negative battery cable.
2. Remove generator drive belt.
3. Unsnap plastic insulator cap from B+ output terminal.
4. Remove B+ terminal mounting nut at rear of generator. Disconnect terminal from generator.
5. Disconnect field wire connector at rear of generator by pushing on connector tab.
6. Remove 1 rear vertical generator mounting bolt.
7. Remove 2 front horizontal generator mounting bolts.
8. Remove generator from vehicle.

To install:

9. Position generator to engine and install 2 horizontal bolts and 1 vertical bolt.

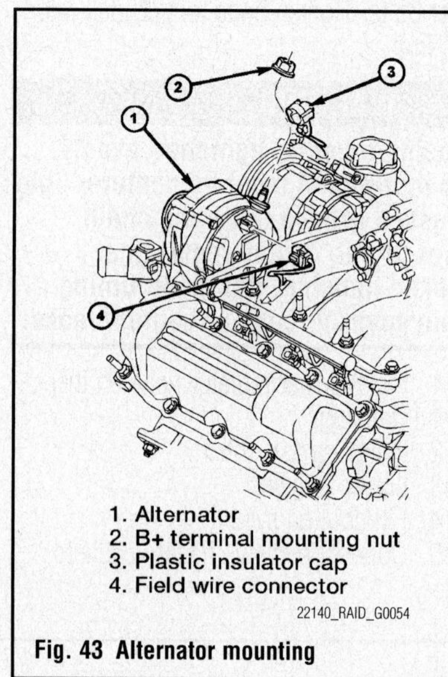

1. Alternator
2. B+ terminal mounting nut
3. Plastic insulator cap
4. Field wire connector

22140_RAID_G0054

Fig. 43 Alternator mounting

10. Tighten the short horizontal bolt to 55 ft. lbs. (74 Nm) and the vertical bolt and long horizontal bolt to 40 ft. lbs. (55 Nm).

11. Snap field wire connector into rear of generator.

12. Install B+ terminal eyelet to generator output stud.

⁂ WARNING

Never force a belt over a pulley rim using a screwdriver. The synthetic fiber of the belt can be damaged.

⁂ WARNING

When installing a serpentine accessory drive belt, the belt MUST be routed correctly. The water pump may be rotating in the wrong direction if the belt is installed incorrectly, causing the engine to overheat.

13. Install generator drive belt.

14. Connect the negative battery cable.

FIRING ORDER

3.7L Engine

Engine firing order is 1–6–5–4–3–2. The No.1 cylinder is located on the left bank.

4.7L Engine

Engine firing order is 1–8–4–3–6–5–7–2. The No.1 cylinder is located on the left bank.

IGNITION COIL PACK

REMOVAL & INSTALLATION

See Figures 44 and 45.

1. Before servicing the vehicle, refer to the precautions.
2. Certain coils may require removal of the throttle body air intake tube or intake box for access.
3. Disconnect the negative battery cable.
4. Detach the electrical connector from the coil by pushing downward on the release lock on top of the connector and pulling the connector from the coil.
5. Clean the area at the base of each coil with compressed air.
6. Remove the coil mounting nut(s). Pull the coil up with a slight twisting action and remove it from the vehicle.

To install:

7. Smear the coil O–ring with silicone grease.
8. Install the coil by pushing down with a slight twisting action.
9. Install the coil mounting nut(s). Tighten the mounting nut to 70 inch lbs. (8 Nm).

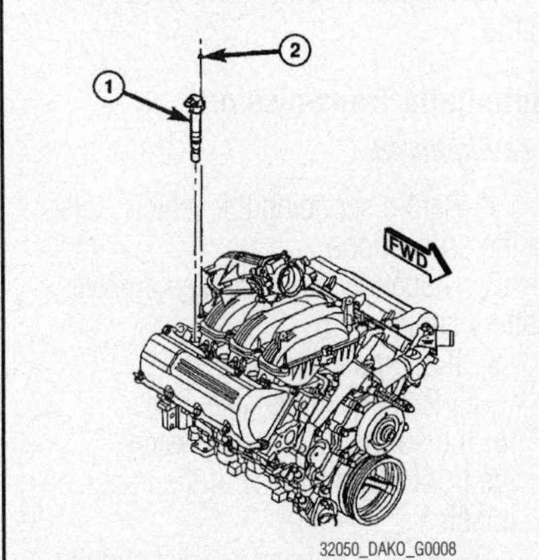

Fig. 44 3.7L V6 Coil Location (1) Ignition Coil (2) Coil Mounting Nut

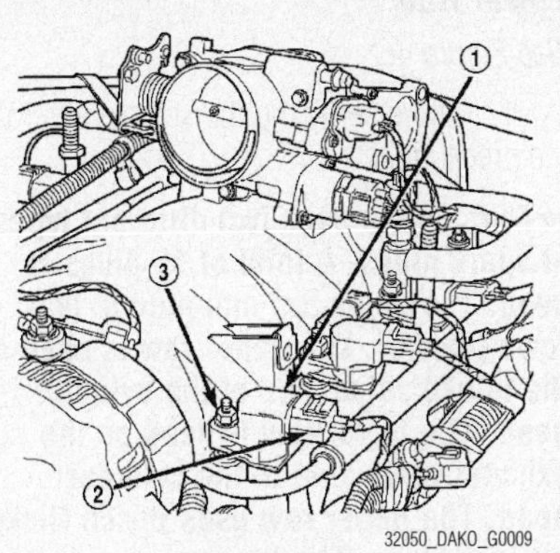

Fig. 45 4.7L V8 Coil Location (1) Ignition Coil (2) Coil Electrical Connector (3) Coil Mounting Stud/Nut

10. Connect the electrical connector.
11. Connect the negative battery cable.
12. Install the throttle body air intake tube or intake box if removed.

IGNITION TIMING

ADJUSTMENT

The ignition timing is controlled by the Engine Control Module and is not adjustable.

SPARK PLUGS

REMOVAL & INSTALLATION

3.7L Engine

1. Before servicing the vehicle, refer to the precautions.
2. Remove air filter tubing at throttle body.
3. Use compressed air around coil base at cylinder head and into cylinder head opening. This will help prevent foreign material from entering combustion chamber.
4. Remove ignition coil.
5. Remove spark plug from cylinder head.
6. Inspect spark plug condition.

To install:

⁂ WARNING

Do not attempt to clean any of the spark plugs. Replace only.

7. Start the spark plug into the cylinder head by hand to avoid cross threading.
8. Tighten spark plugs. Tighten to 20 ft. lbs. (30 Nm).
9. Before installing ignition coil(s), check condition of coil o-ring and replace as necessary. To aid in coil installation, apply silicone to coil o-ring.
10. Install ignition coil(s).
11. Install air filter tubing at throttle body.

4.7L Engine

Upper Row

See Figure 46.

1. Before servicing the vehicle, refer to the precautions.

➡This engine uses two different types of spark plugs. A total of 16 plugs is used. The plugs are mounted in two rows (banks). The upper row is used on the intake valve side of the cylinder head. The lower row is used on the exhaust valve side of the cylinder head. The upper row uses Bosch Nickel Yttrium plugs. The lower row uses Bosch Iridium plugs.

⁂ WARNING

Do not interchange these plugs.

2. Remove air filter tubing at throttle body.
3. Use compressed air around coil base at cylinder head and into cylinder head opening. This will help prevent foreign material from entering combustion chamber.

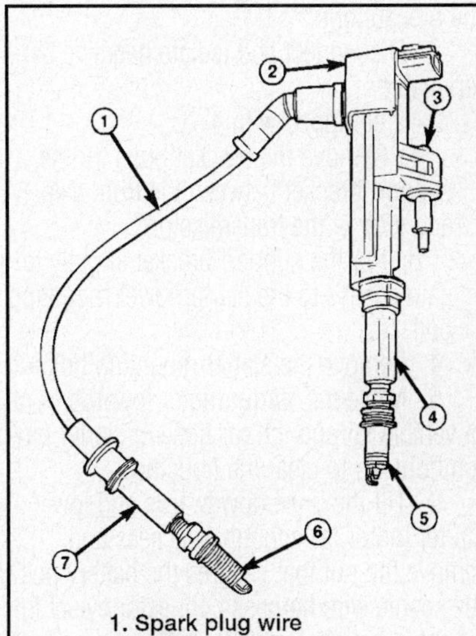

Fig. 46 Dual spark plug assembly

➡Due to tight clearances between upper row of plugs and cylinder head, a conventional deep, thick-wall spark plug socket will not fit. Use a deep, thin-wall spark plug socket for plug removal and installation.

4. Remove ignition coil.
5. Remove spark plug from cylinder head.
6. Inspect spark plug condition.

To install:

⁂ WARNING

Do not attempt to clean any of the spark plugs. Replace only.

7. Start the spark plug into the cylinder head by hand to avoid cross threading.
8. Tighten spark plugs. Tighten to 20 ft. lbs. (30 Nm).
9. Before installing ignition coil(s), check condition of coil o-ring and replace as necessary. To aid in coil installation, apply silicone to coil o-ring.
10. Install ignition coil(s).
11. Install air filter tubing at throttle body.

Lower Row

See Figure 46.

1. Before servicing the vehicle, refer to the precautions.

➡This engine uses two different types of spark plugs. A total of 16 plugs is used. The plugs are mounted in two rows (banks). The upper row is used on the intake valve side of the cylinder head. The lower row is used on the exhaust valve side of the cylinder head. The upper row uses Bosch Nickel Yttrium plugs. The lower row uses Bosch Iridium plugs.

⁂ WARNING

Do not interchange these plugs.

2. Remove air filter tubing at throttle body.
3. Use compressed air around coil base at cylinder head and into cylinder head opening. This will help prevent foreign material from entering combustion chamber.
4. Remove the spark plug wire using a twisting motion.

⁂ WARNING

Do not pull on the wire itself, rather on the boot. coil.

5. Remove spark plug from cylinder head.
6. Inspect spark plug condition.

To install:

⁂ WARNING

Do not attempt to clean any of the spark plugs. Replace only.

7. Start the spark plug into the cylinder head by hand to avoid cross threading.
8. Tighten spark plugs. Tighten to 20 ft. lbs. (30 Nm).
9. Connect the spark plug wire.
10. Install air filter tubing at throttle body.

ENGINE ELECTRICAL

STARTING SYSTEM

STARTER

REMOVAL & INSTALLATION

Manual Transmission

See Figure 47.

1. Before servicing the vehicle, refer to the precautions.
2. Disconnect and isolate negative battery cable.
3. If equipped with 4WD:
   a. Remove the bracket bolts for the support bracket between the front axle and side of the transmission.
   b. Pry the support bracket slightly to gain access to the starter lower mounting bolt.
4. Remove the starter mounting bolts.
5. Move the starter motor towards front of vehicle far enough for nose of starter pinion housing to clear the housing.
6. Tilt the nose downwards and lower starter motor far enough to access and remove the nut that secures the battery positive cable wire harness connector eyelet to solenoid battery terminal stud.
7. Remove the positive cable wire harness connector eyelet from solenoid battery terminal stud.
8. Disconnect the battery positive cable wire harness connector from the solenoid terminal connector receptacle.

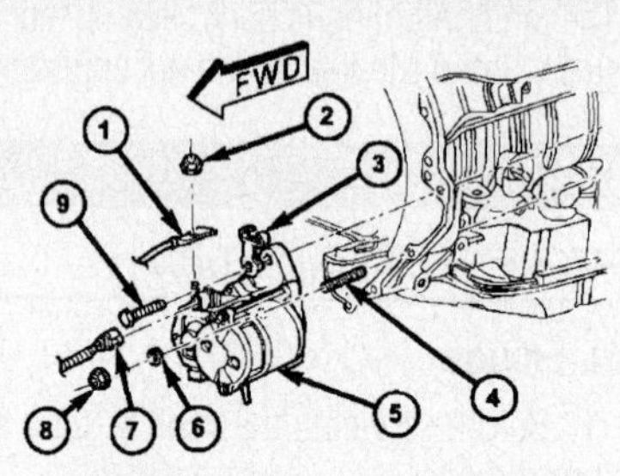

1. Wire harness connector eyelet
2. Nut
3. Wire harness clamp
4. Stud
5. Starter motor
6. Washer
7. Positive cable wire harness
8. Nut
9. Bolt

22140_RAID_G0056

Fig. 47 Starter motor assembly—manual transmission

9. Remove the starter motor.

To install:

10. Position starter motor to transmission housing.
11. Connect the battery cable solenoid terminal wire harness connector to connector receptacle on starter solenoid.

⁂ CAUTION

Always support the starter motor during this process. Do not let the starter motor hang from the wire harness.

12. Install the battery cable eyelet terminal onto solenoid B (+) terminal stud.
13. Install the nut securing battery cable eyelet terminal to starter solenoid B (+) terminal stud. Tighten nut to 120 inch lbs. (13.6 Nm).
14. Position the starter motor over stud on transmission housing.
15. Loosely install the washers, bolt, and nut to starter. Tighten bolt and nut to 50 ft. lbs. (67.8 Nm).
16. Connect the negative battery cable.

Automatic Transmission

See Figure 48.

1. Before servicing the vehicle, refer to the precautions.
2. Disconnect and isolate negative battery cable.
3. If equipped with 4WD:
   a. Remove the bracket bolts for the support bracket between the front axle and side of the transmission.
   b. Pry the support bracket slightly to gain access to the starter lower mounting bolt.

4. Remove mounting bolts (rearward facing) securing starter motor to the transmission housing.

⁂ CAUTION

Always support starter motor during this process. Do not let starter motor hang from wire harness.

5. Lower starter motor from front of transmission housing far enough to access and remove nut securing battery positive cable eyelet terminal to the starter solenoid B (+) terminal stud.

6. Remove the battery cable eyelet terminal from solenoid B (+) terminal stud.

7. Disconnect the battery cable solenoid terminal wire harness connector from receptacle on starter solenoid.

8. Remove the starter motor.

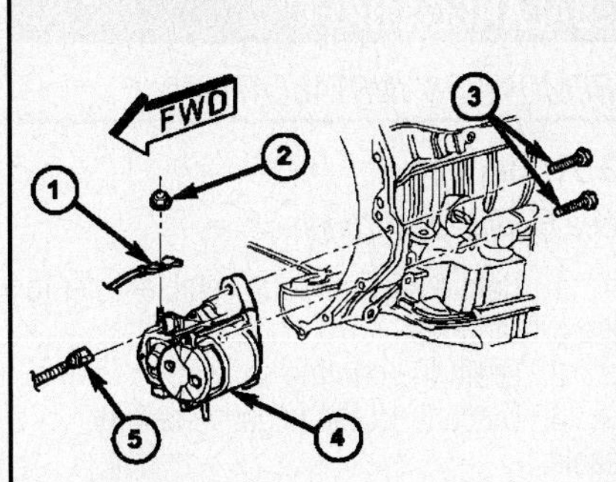

1. Wire harness connector eyelet
2. Nut
3. Mounting bolts
4. Starter motor
5. Positive cable wire harness

22140_RAID_G0057

Fig. 48 Starter motor assembly—automatic transmission

To install:

9. Position the starter motor to transmission housing.

10. Connect battery cable solenoid terminal wire harness connector to connector receptacle on starter solenoid.

11. Install the battery cable eyelet terminal onto solenoid B (+) terminal stud.

12. Install and tighten nut securing battery cable eyelet terminal to starter solenoid B (+) terminal stud. Tighten nut to 120 inch lbs. (13.6 Nm).

13. Position starter motor to transmission housing and loosely install two bolts/washers.

14. Tighten bolts to 50 ft. lbs. (67.8 Nm).

15. Connect negative battery cable.

ENGINE MECHANICAL

ACCESSORY DRIVE BELTS

ACCESSORY BELT ROUTING

See Figure 49.

Refer to the accompanying illustration for belt routing.

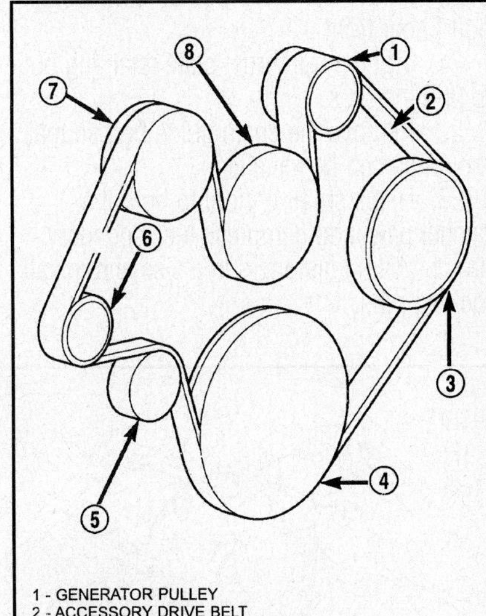

1 - GENERATOR PULLEY
2 - ACCESSORY DRIVE BELT
3 - POWER STEERING PUMP PULLEY
4 - CRANKSHAFT PULLEY
5 - IDLER PULLEY
6 - TENSIONER
7 - A/C COMPRESSOR PULLEY
8 - WATER PUMP PULLEY

67189-DAKO-G51

Fig. 49 Accessory drive belt routing 3.7L and 4.7L engine

INSPECTION

See Figure 50.

Inspect the drive belt for signs of glazing or cracking. A glazed belt will be perfectly

ACCESSORY DRIVE BELT DIAGNOSIS CHART

CONDITION	POSSIBLE CAUSES	CORRECTION
RIB CHUNKING (One or more ribs has separated from belt body)	1. Foreign objects imbedded in pulley grooves.	1. Remove foreign objects from pulley grooves. Replace belt.
	2. Installation damage	2. Replace belt
RIB OR BELT WEAR	1. Pulley misaligned	1. Align pulley(s)
	2. Abrasive environment	2. Clean pulley(s). Replace belt if necessary
	3. Rusted pulley(s)	3. Clean rust from pulley(s)
	4. Sharp or jagged pulley groove tips	4. Replace pulley. Inspect belt.
	5. Belt rubber deteriorated	5. Replace belt
BELT SLIPS	1. Belt slipping because of insufficient tension	1. Inspect/Replace tensioner if necessary
	2. Belt or pulley exposed to substance that has reduced friction (belt dressing, oil, ethylene glycol)	2. Replace belt and clean pulleys
	3. Driven component bearing failure (seizure)	3. Replace faulty component or bearing
	4. Belt glazed or hardened from heat and excessive slippage	4. Replace belt.
LONGITUDE BELT CRACKING	1. Belt has mistracked from pulley groove	1. Replace belt
	2. Pulley groove tip has worn away rubber to tensile member	2. Replace belt
"GROOVE JUMPING"	1. Incorrect belt tension	1. Inspect/Replace tensioner if necessary
(Belt does not maintain correct position on pulley)	2. Pulley(s) not within design tolerance	2. Replace pulley(s)
	3. Foreign object(s) in grooves	3. Remove foreign objects from grooves
	4. Pulley misalignment	4. Align component
	5. Belt cordline is broken	5. Replace belt
BELT BROKEN	1. Incorrect belt tension	1. Replace Inspect/Replace tensioner if necessary
(Note: Identify and correct problem before new belt is installed)	2. Tensile member damaged during belt installation	2. Replace belt
	3. Severe misalignment	3. Align pulley(s)
	4. Bracket, pulley, or bearing failure	4. Replace defective component and belt
NOISE	1. Incorrect belt tension	1. Inspect/Replace tensioner if necessary
(Objectionable squeal, squeak, or rumble is heard or felt while drive belt is in operation)	2. Bearing noise	2. Locate and repair
	3. Belt misalignment	3. Align belt/pulley(s)
	4. Belt to pulley mismatch	4. Install correct belt
	5. Driven component induced vibration	5. Locate defective driven component and repair
TENSION SHEETING FABRIC FAILURE	1. Tension sheeting contacting stationary object	1. Correct rubbing condition
(Woven fabric on outside, circumference of belt has cracked or separated from body of belt)	2. Excessive heat causing woven fabric to age	2. Replace belt
	3. Tension sheeting splice has fractured	3. Replace belt
CORD EDGE FAILURE	1. Incorrect belt tension	1. Inspect/Replace tensioner if necessary
(Tensile member exposed at edges of belt or separated from belt body)	2. Belt contacting stationary object	2. Replace belt
	3. Pulley(s) out of tolerance	3. Replace pulley
	4. Insufficient adhesion between tensile member and rubber matrix	4. Replace belt

22140_RAID_G0018A

Fig. 50 Accessory drive belt diagnosis chart

smooth from slippage, while a good belt will have a slight texture of fabric visible. Cracks will usually start at the inner edge of the belt and run outward. All worn or damaged drive belts should be replaced immediately.

ADJUSTMENT

See Figure 51.

Belt tension is not adjustable. Belt adjustment is maintained by an automatic (spring load) belt tensioner.

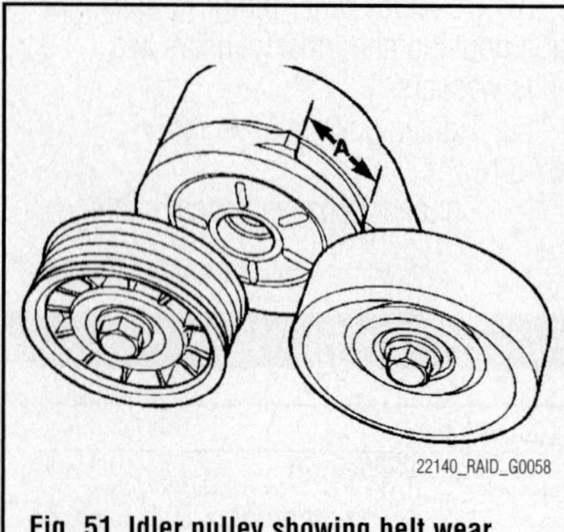

Fig. 51 Idler pulley showing belt wear indicator

REMOVAL & INSTALLATION

See Figures 49 and 51.

1. Disconnect negative battery cable.
2. Rotate the belt tensioner until it contacts its stops.
3. Remove the belt, then slowly release the tensioner

To install:

4. Check condition of all pulleys.

⁂ WARNING

When installing the serpentine accessory drive belt, the belt MUST be routed correctly. If not, the engine may overheat due to the water pump rotating in the wrong direction.

5. Install new belt . Route the belt around all pulleys except the idler pulley
6. Rotate the tensioner arm until it contacts it's stop position.
7. Route the belt around the idler and slowly let the tensioner rotate into the belt. Make sure the belt is seated onto all pulleys.
8. With the drive belt installed, inspect the belt wear indicator.
9. On 4.7L Engines only, the gap between the tang and the housing stop must not exceed .94 inches (24 mm). If the measurement exceeds this specification replace the serpentine accessory drive belt.

BALANCE SHAFT

REMOVAL & INSTALLATION

3.7L Engines

See Figures 52 through 58.

1. Before servicing the vehicle, refer to the precautions.
2. Drain the cooling system.
3. Disconnect the negative battery cable.
4. Remove the cylinder head covers.
5. Remove the radiator fan.
6. Rotate the crankshaft so that the crankshaft timing mark aligns with the Top Dead Center (TDC) mark on the front cover, and the **V6** marks on the camshaft sprockets are at 12 O'clock.
7. Remove the power steering pump.
8. Remove the access plugs from the cylinder heads.
9. Remove the oil fill housing.

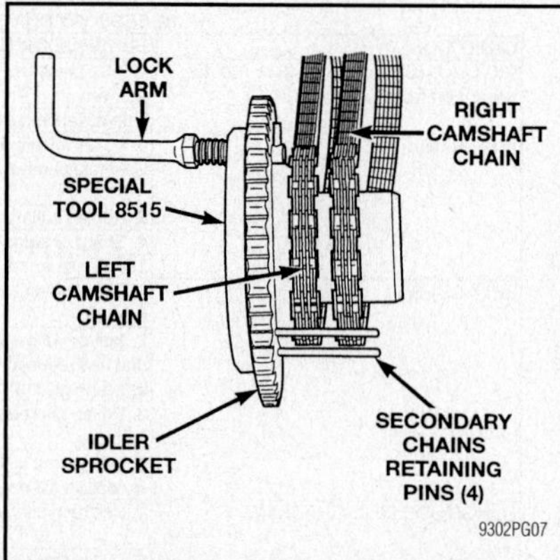

Fig. 52 Use the Timing Chain Locking tool to lock the timing chains on the idler gear—3.7L engine

10. Remove the crankshaft damper.
11. Compress the primary timing chain tensioner and install a lock pin.
12. Remove the secondary timing chain tensioners.
13. Hold the left camshaft with adjustable pliers and remove the sprocket and chain. Rotate the **left** camshaft 5 degrees **clockwise** to the neutral position.
14. Hold the right camshaft with adjustable pliers and remove the camshaft sprocket. Rotate the **right** camshaft 45 degrees **counterclockwise** to the neutral position.
15. Remove the primary and secondary timing chain and sprockets.
16. Remove the balance shaft from engine using special tool 8641 (balance shaft remover/installer tool) or equivalent.

To install:

17. Coat balance shaft bearing journals with clean engine oil.
18. Install balance shaft into engine using special tool 8641 (balance shaft remover/installer tool).
19. Install balance shaft thrust plate retaining bolt finger tight. Do not tighten at this time.
20. Position the right side of the thrust plate with the right chain guide bolt, install bolt finger tight.
21. Tighten the thrust plate retaining bolt to 250 inch lbs.
22. Remove the chain guide bolt so that the guide can be installed.
23. Use a small prytool to hold the ratchet pawl and compress the secondary timing chain tensioners in a vise and install locking pins.

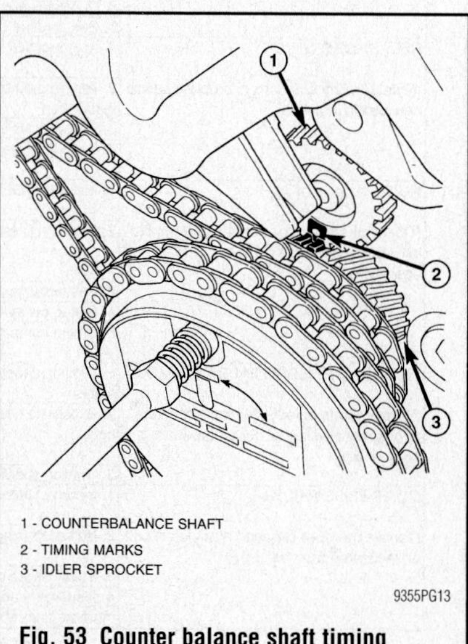

Fig. 53 Counter balance shaft timing marks—3.7L

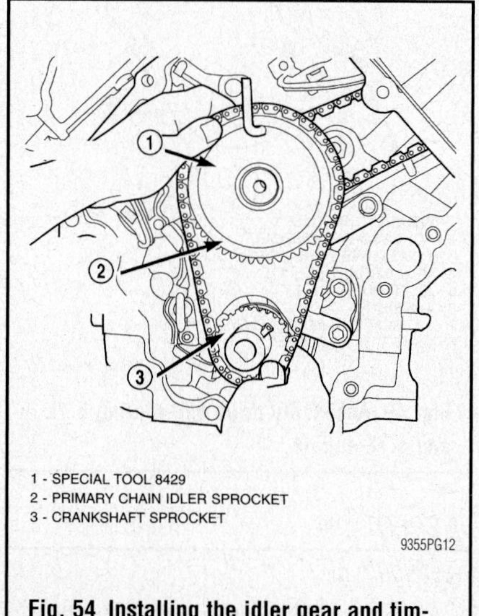

Fig. 54 Installing the idler gear and timing chains—3.7L

➡**The black bolts fasten the guide to the engine block and the silver bolts fasten the guide to the cylinder head.**

24. Install the secondary timing chain guides. Tighten the bolts to 21 ft. lbs. (28 Nm).
25. Install the secondary timing chains to the idler sprocket so that the double plated links on each chain are visible through the slots in the primary idler sprocket
26. Lock the secondary timing chains to the idler sprocket with Timing Chain Locking tool as shown.
27. Align the primary chain double plated links with the idler sprocket timing mark and the single plated link with the crankshaft sprocket timing mark.
28. Install the primary chain and sprockets. Tighten the idler sprocket bolt to 25 ft. lbs. (34 Nm).
29. Align the secondary chain single plated links with the timing marks on the secondary sprockets. Align the dot at the **L** mark on the left sprocket with the plated link on the left chain and the dot at the **R** mark on the right sprocket with the plated link on the right chain.
30. Rotate the camshafts back from the neutral position and install the camshaft sprockets.
31. Remove the secondary chain locking tool.
32. Remove the primary and secondary timing chain tensioner locking pins.
33. Hold the camshaft sprockets with a spanner wrench and tighten the retaining bolts to 90 ft. lbs. (122 Nm).
34. Install the front cover. Tighten the bolts, in sequence, to 40 ft. lbs. (54 Nm).

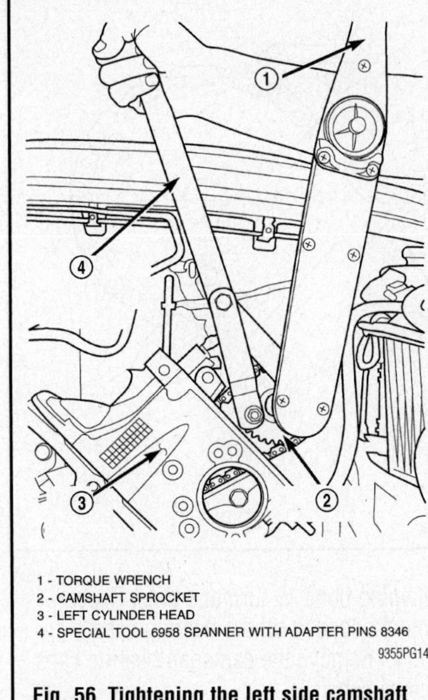

Fig. 56 Tightening the left side camshaft sprocket—3.7L

35. Install the front crankshaft seal.
36. Install the cylinder head access plugs.
37. Install the A/C compressor.
38. Install the alternator.
39. Install the accessory drive belt tensioner. tighten the bolt to 40 ft. lbs. (54 nm).
40. Install the oil fill housing.
41. Install the crankshaft damper. tighten the bolt to 130 ft. lbs. (175 nm).
42. Install the power steering pump.
43. Install the lower radiator hose.
44. Install the heater hoses.

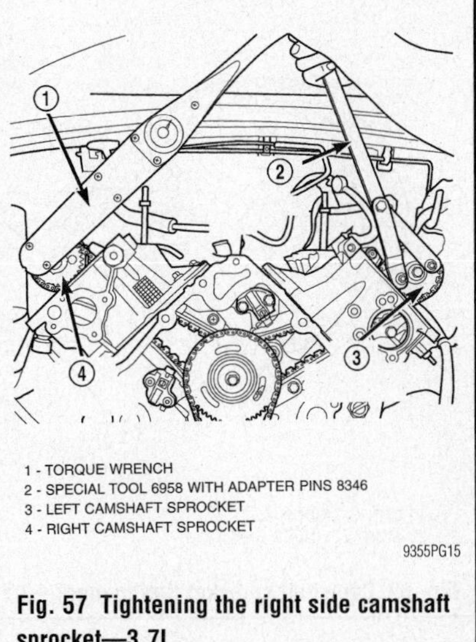

Fig. 57 Tightening the right side camshaft sprocket—3.7L

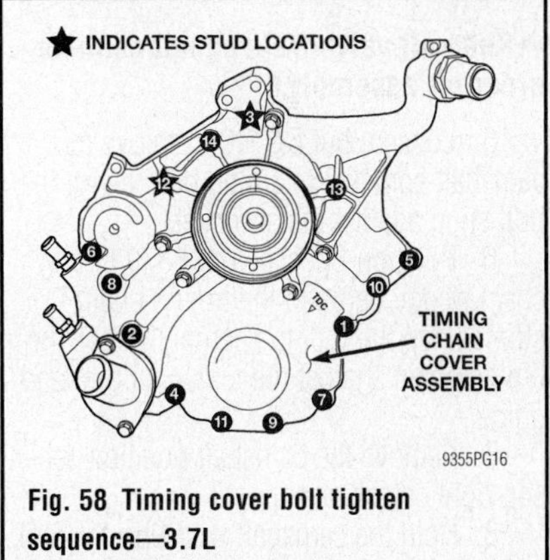

Fig. 58 Timing cover bolt tighten sequence—3.7L

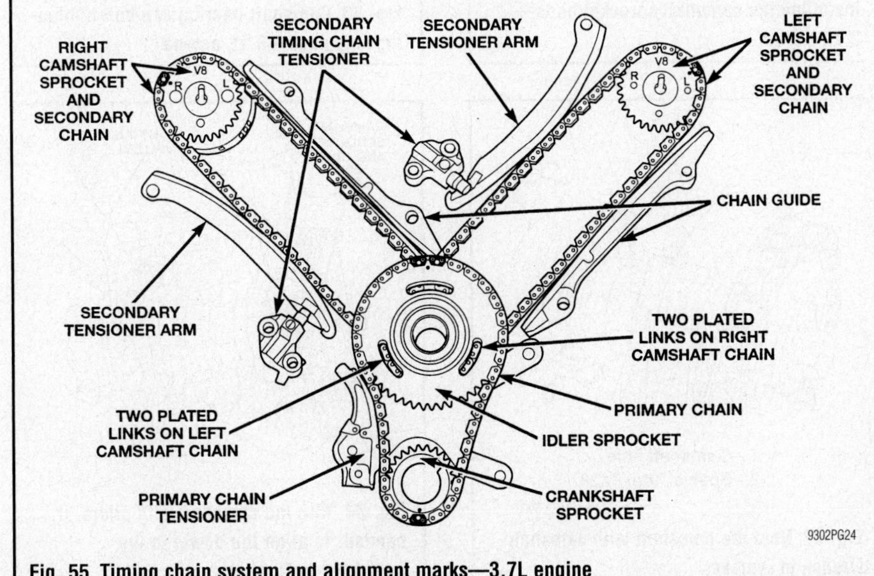

Fig. 55 Timing chain system and alignment marks—3.7L engine

45. Install the accessory drive belt.
46. Install the engine cooling fan and shroud.
47. Install the camshaft position (CMP) sensor.
48. Install the valve covers.
49. Connect the negative battery cable.
50. Fill and bleed the cooling system.
51. Start the engine, check for leaks and repair if necessary.

CAMSHAFT AND VALVE LIFTERS

REMOVAL & INSTALLATION

3.7L Engines

See Figures 59 through 64.

1. Before servicing the vehicle, refer to the precautions.
2. Disconnect the negative battery cable.
3. Remove the cylinder head cover.
4. Set the No. 1 cylinder to Top Dead Center (TDC). The camshaft sprocket

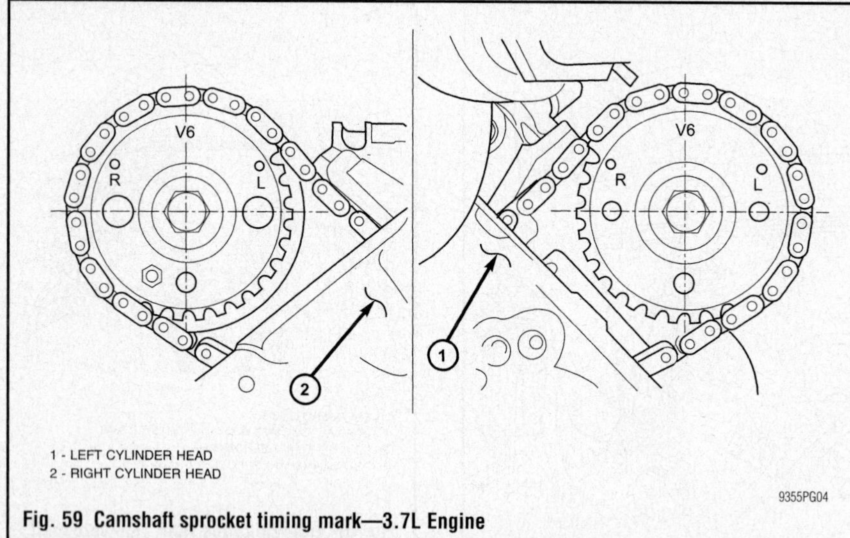

Fig. 59 Camshaft sprocket timing mark—3.7L Engine

alignment mark should be at the 12 o'clock position.

➡**Keep all valve train components in order for assembly.**

5. Loosen, but DO NOT remove the camshaft sprocket retaining bolt. Leave the bolt snug against the sprocket.

6. Position Special Tool 8379 timing chain wedge between the timing chain strands, tap the tool to securely wedge the timing chain against the tensioner arm and guide.

7. Remove the camshaft position sensor, right camshaft only.

8. Hold the camshaft with Special Tool 8428 Camshaft Wrench, while removing the camshaft sprocket bolt and sprocket.

9. Using Special Tool 8428 Camshaft Wrench, gently allow the camshaft to rotate 5°clockwise until the camshaft is in the neutral position (no valve load).

10. Starting at the outside working inward, loosen the camshaft bearing cap retaining bolts ½ turn at a time. Repeat until the load is off the bearing caps.

11. Remove the camshaft bearing caps and camshaft.

➡**When the camshaft is removed the rocker arms may slide downward. Mark the rocker arms before removing camshaft.**

To install:

12. Lubricate the camshaft journals with clean engine oil

13. Position the camshaft into the cylinder head.

14. Install the camshaft bearing caps and hand tighten.

➡**Caps should be installed so that the stamped arrows on the caps point toward the front of the engine.**

15. Tighten the bearing cap bolts in ½ turn increments, in sequence, to 100 inch lbs. (11 Nm).

16. Position the camshaft sprocket into the timing chain aligning the alignment mark between the two marked chain links (Two links marked during removal).

17. Using Tool 8428 Camshaft Wrench, rotate the camshaft until the camshaft sprocket dowel is aligned with the slot in the camshaft sprocket. Install the sprocket onto the camshaft.

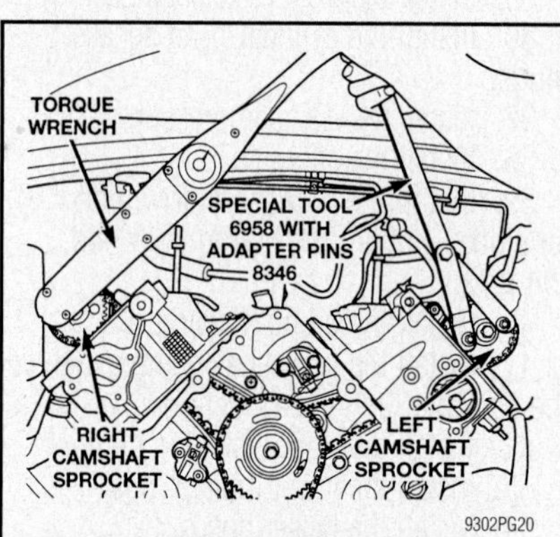

Fig. 61 Hold the left camshaft sprocket with a spanner wrench while removing or installing the camshaft sprocket bolts—3.7L engine

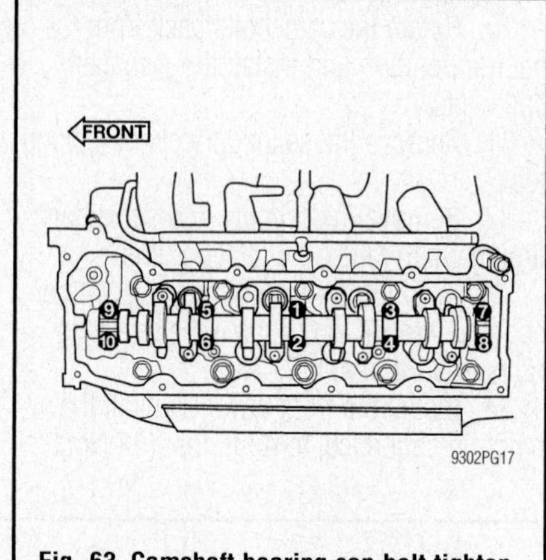

Fig. 63 Camshaft bearing cap bolt tightening sequence—3.7L engine

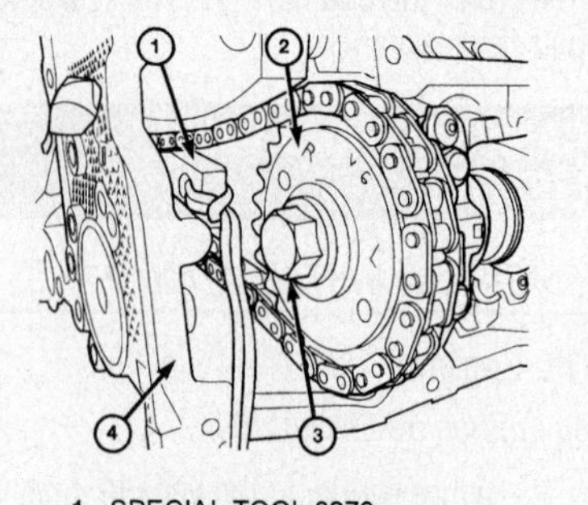

Fig. 60 Chain tensioner retaining wedge installation

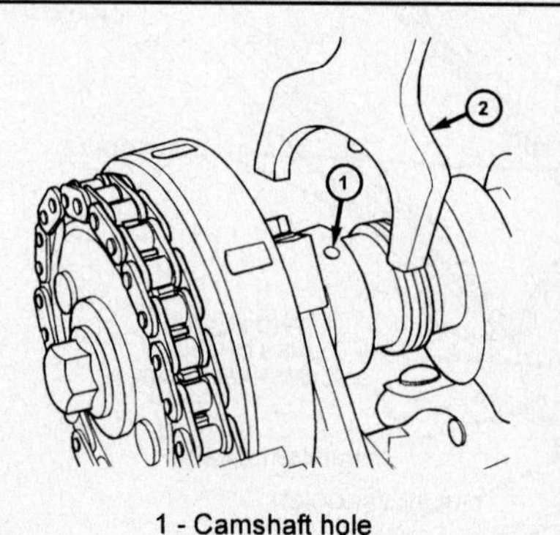

Fig. 62 Hold the camshaft with camshaft wrench to remove

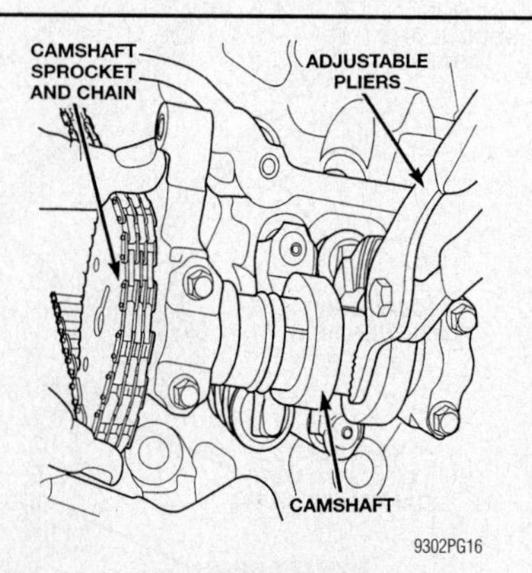

Fig. 64 Turn the camshaft with pliers, if needed, to align the dowel in the sprocket—3.7L engine

CAUTION

Remove any excess oil from the camshaft sprocket bolt. Failure to do so can cause bolt over-tighten.

18. Remove excess oil from bolt, then install the camshaft sprocket retaining bolt and hand tighten.
19. Remove Special Tool 8379 timing chain wedge.
20. Using Special Tool 6958 spanner wrench with adapter pins 8346, tighten the camshaft sprocket retaining bolt to 90 ft. lbs. (122 Nm).
21. Install or connect the following:
    - Camshaft position sensor, right side only
    - Cylinder head cover
22. Start the engine and check for leaks.

4.7L Engine

See Figures 65 through 68.

WARNING

When removing the cam sprocket, timing chains or camshaft, Failure to use special tool 9867 will result in hydraulic tensioner ratchet over extension, Requiring timing chain cover removal to re-set the tensioner ratchet.

1. Before servicing the vehicle, refer to the precautions.
2. Disconnect the negative battery cable.
3. Remove the cylinder head covers.
4. Set engine to TDC cylinder No. 1, camshaft sprocket V8 marks at the 12 o'clock position.
5. Mark one link on the secondary timing chain on both sides of the V8 mark on the camshaft sprocket to aid in installation.

WARNING

Do not hold or pry on the camshaft target wheel for any reason, Severe damage will occur to the target wheel. A damaged target wheel could cause a vehicle no start condition.

6. Loosen but DO NOT remove the camshaft sprocket retaining bolt. Leave bolt snug against sprocket.

➡The timing chain tensioners must be secured prior to removing the camshaft sprockets. Failure to secure tensioners will allow the tensioners to extend, requiring timing chain cover removal in order to reset tensioners.

WARNING

Do not force wedge past the narrowest point between the chain strands. Damage to the tensioners may occur.

7. Position Special Tool 9867 timing chain wedge between the timing chain strands. Tap the tool to securely wedge the timing chain against the tensioner arm and guide.
8. Remove the camshaft position sensor.

➡When gripping the camshaft, place the pliers on the tube portion of the camshaft only. Do not grip the lobes or the sprocket areas.

9. Hold the camshaft with adjustable pliers while removing the camshaft sprocket bolt and sprocket.
10. Using the pliers, gently allow the camshaft to rotate 45°counter-clockwise until the camshaft is in the neutral position (no valve load).

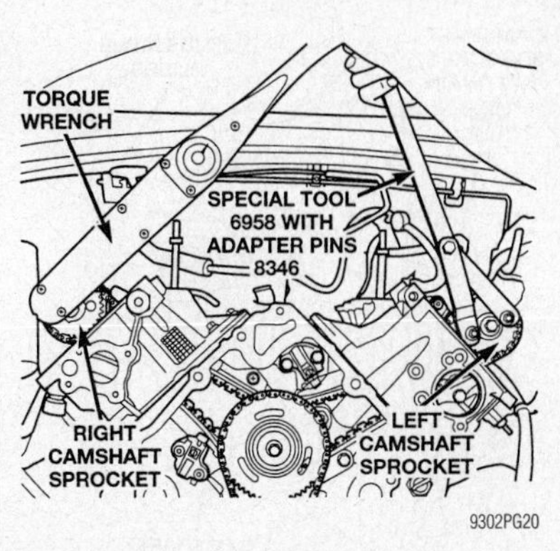

Fig. 66 Hold the camshaft sprocket with a spanner wrench while removing or installing the camshaft sprocket bolts—4.7L engine

11. Starting at the outside working inward, loosen the camshaft bearing cap retaining bolts ½ turn at a time. Repeat until all load is off the bearing caps.

WARNING

Do not stamp or strike the camshaft bearing caps. Severe damage will occur to the bearing caps.

12. When the camshaft is removed the rocker arms may slide downward, mark the rocker arms before removing camshaft.
13. Remove the camshaft bearing caps and the camshaft.

To install:

14. Lubricate camshaft journals with clean engine oil.
15. Position the right side camshaft so that the camshaft sprocket dowel is near the 10 o'clock position, This will place the camshaft at the neutral position easing the installation of the camshaft bearing caps.
16. Position the camshaft into the cylinder head.
17. Install the camshaft bearing caps, hand tighten the retaining bolts.

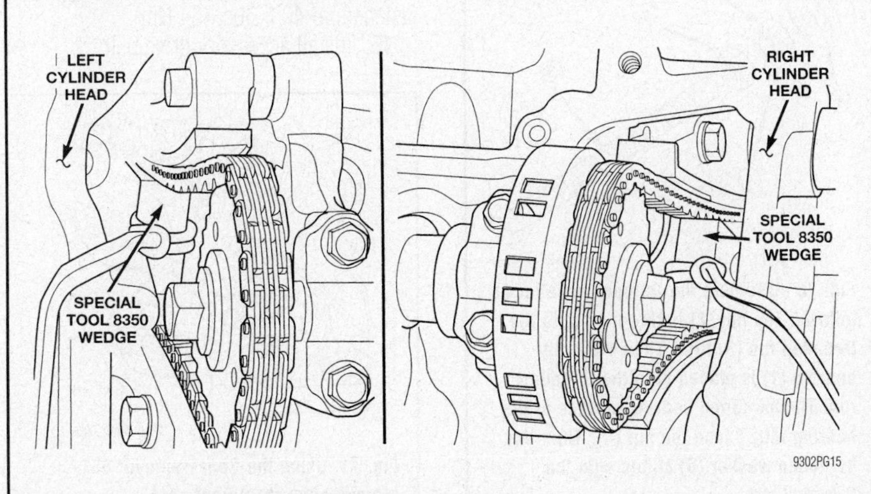

Fig. 65 Chain Tensioner Retaining Wedges—4.7L engine

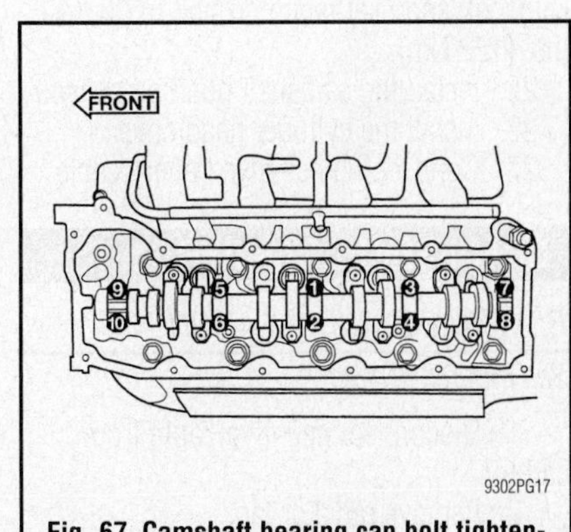

Fig. 67 Camshaft bearing cap bolt tightening sequence—4.7L engine

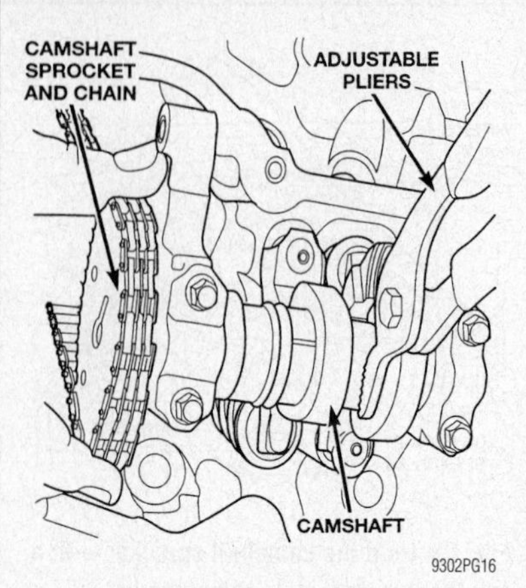

Fig. 68 Turn the camshaft with pliers, if needed, to align the dowel in the sprocket—4.7L engine

18. Working in ½ turn increments, tighten the bearing cap retaining bolts starting with the middle cap working outward.
19. Torque the camshaft bearing cap retaining bolts to 100 inch lbs. (11 Nm).
20. Position the camshaft drive gear into the timing chain aligning the V8 mark between the two marked chain links (Two links marked during removal).

**** WARNING**

When gripping the camshaft, place the pliers on the tube portion of the camshaft only. Do not grip the lobes or the sprocket areas.

21. Using the adjustable pliers, rotate the camshaft until the camshaft sprocket dowel is aligned with the slot in the camshaft sprocket . Install the sprocket onto the camshaft.
22. Remove excess oil from camshaft sprocket bolt, then install the camshaft sprocket retaining bolt and hand tighten.
23. Remove timing chain wedge special tool 9867.
24. Using Special Tool 6958 spanner wrench with adapter pins 8346 , torque the camshaft sprocket retaining bolt to 90 ft. lbs. (122 Nm).
25. Install the camshaft position sensor.
26. Install the cylinder head cover
27. Connect the negative battery cable.

CRANKSHAFT DAMPER

REMOVAL & INSTALLATION

See Figures 69 and 70.

1. Disconnect negative cable from battery.
2. Remove radiator fan.
3. Remove accessory drive belt,

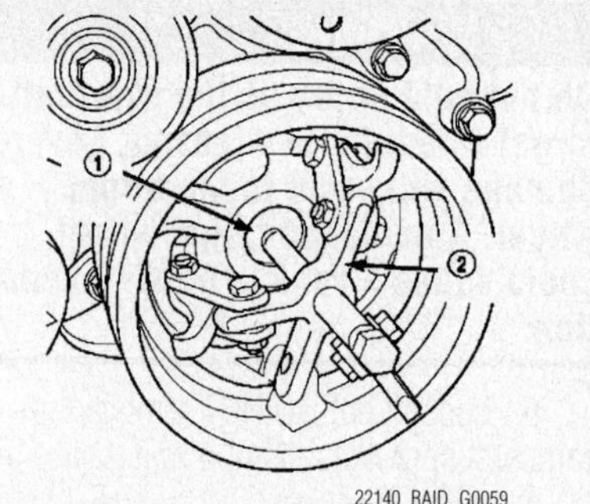

Fig. 69 Remove damper using the Crankshaft Insert 8513 (1) and Three Jaw Puller 1026 (2)

4. Remove crankshaft damper bolt.
5. Remove damper using Mitsubishi special tools 8513 insert and 1026 three jaw puller.

To install:

6. To prevent severe damage to the Crankshaft, Damper or Damper Installer 8512-A, thoroughly clean the damper bore and the crankshaft nose before installing Damper.
7. Align vibration damper slot with key in crankshaft. Slide damper onto crankshaft slightly.

**** CAUTION**

Damper Installer 8512-A, is assembled in a specific sequence. Failure to assemble this tool in this sequence can result in tool failure and severe damage to either the tool or the crankshaft.

8. Assemble Damper Installer 8512-A as follows: The nut is threaded onto the threaded rod first. Then the roller bearing is placed onto the threaded rod (The hardened bearing surface of the bearing MUST face the nut). Then the hardened washer slides onto the threaded rod . Once assembled coat the threaded rod's threads with Mitsubishi Nickel Anti-Seize or equivalent.
9. Using the Damper Installer 8512A, press the damper onto the crankshaft.
10. Install and tighten vibration damper bolt to 130 ft. lbs. (175 Nm).
11. Install the radiator fan.
12. Install the accessory drive belt.
13. Connect the negative cable to battery.

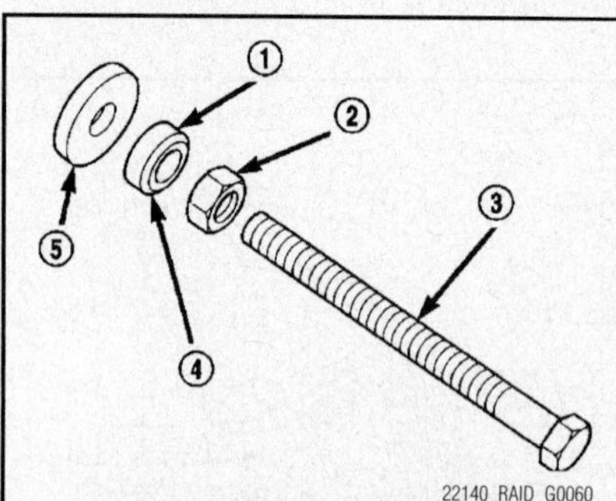

Fig. 70 Assemble the Damper Installer as follows: The nut (2) is threaded onto the threaded rod (3) first. Then the roller bearing (1) is placed onto the threaded rod. The hardened bearing surface of the bearing MUST face the nut (4). Then the hardened washer (5) slides onto the threaded rod

CRANKSHAFT FRONT SEAL

REMOVAL & INSTALLATION

See Figures 71 and 72.

1. Disconnect negative cable from battery.
2. Remove accessory drive belt.
3. Remove A/C compressor mounting fasteners and set aside.
4. Drain cooling system.
5. Remove upper radiator hose.
6. Disconnect electrical connector for fan mounted inside radiator shroud.
7. Remove radiator cooling fan
8. Remove crankshaft damper bolt.
9. Remove damper using Mitsubishi special tools 8513 and tool 1026 three jaw puller.
10. Using the Seal Remover 8511, remove crankshaft front seal.

To install:

11. Using Mitsubishi special tool 8348 and 8512, install crankshaft front seal.
12. Install vibration damper.
13. Install radiator cooling fan and shroud.
14. Install upper radiator hose.
15. Install A/C compressor and tighten fasteners to 40 ft. lbs. (54 Nm).
16. Install accessory drive belt.

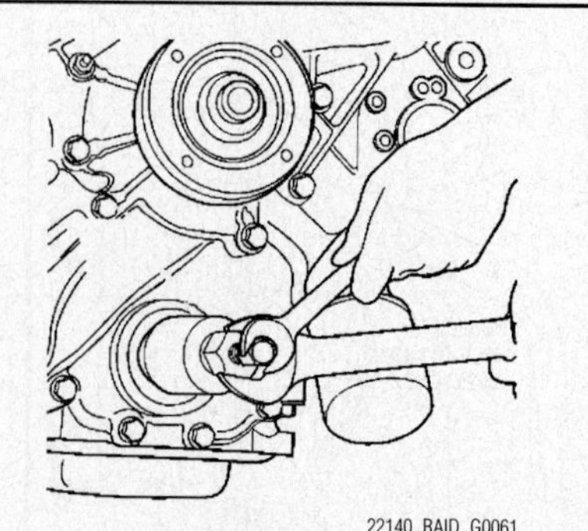

Fig. 71 Using the Seal Remover 8511, remove crankshaft front seal

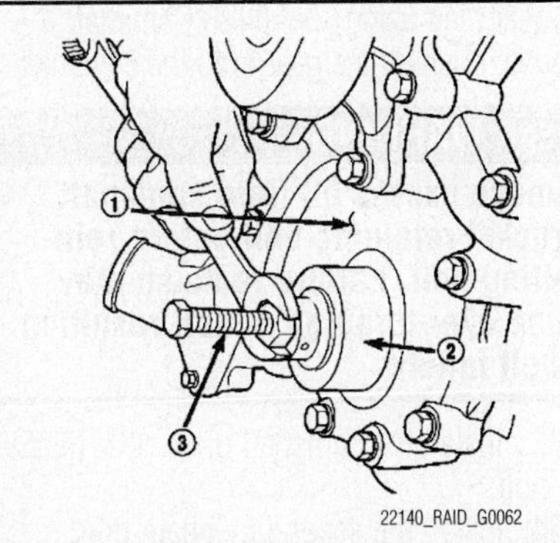

Fig. 72 Using the Seal Installer 8348 (2) and Damper Installer 8512A (3), install the crankshaft front seal

17. Refill cooling system.
18. Connect negative cable to battery.

CYLINDER HEAD

REMOVAL & INSTALLATION

3.7L Engine

See Figures 73 through 78.

1. Before servicing the vehicle, refer to the precautions.
2. Disconnect the negative battery cable.
3. Raise and safely support the vehicle securely on jackstands.
4. Disconnect the exhaust pipe at the exhaust manifold.
5. Drain the engine coolant.
6. Lower the vehicle.
7. Remove the intake manifold.
8. On the left side, remove the master cylinder and booster assembly.

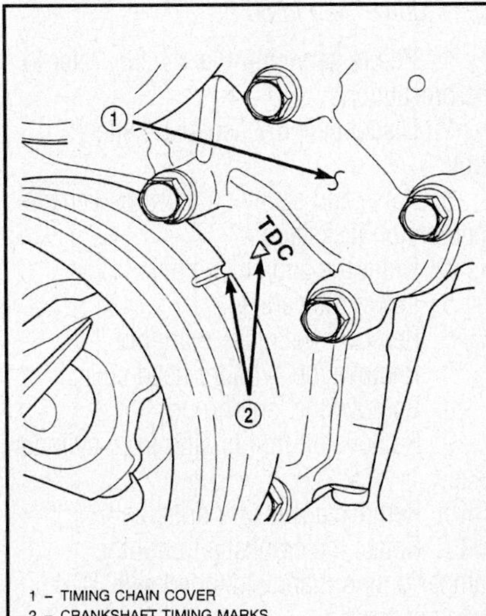

Fig. 73 Rotate the crankshaft until the damper timing mark is aligned with TDC indicator mark

9. Remove the cylinder head cover.
10. Remove the radiator fan.
11. Remove oil fill housing from cylinder head.
12. Remove accessory drive belt.
13. On the left side, remove the power steering pump and set it aside.
14. Rotate the crankshaft until the damper timing mark is aligned with TDC indicator mark.
15. Verify the V6 mark on the camshaft sprocket is at the 12 o'clock position. Rotate the crankshaft one turn if necessary.
16. Remove the crankshaft damper.
17. Remove the timing chain cover.
18. Lock the secondary timing chains to the idler sprocket using Special Tool 8429 Timing Chain Holding Fixture.

➡Mark the secondary timing chain prior to removal to aid in installation.

19. Mark the secondary timing chain, one link on each side of the V6 mark on the camshaft drive gear.
20. Remove the secondary chain tensioner.
21. Remove the cylinder head access plug.

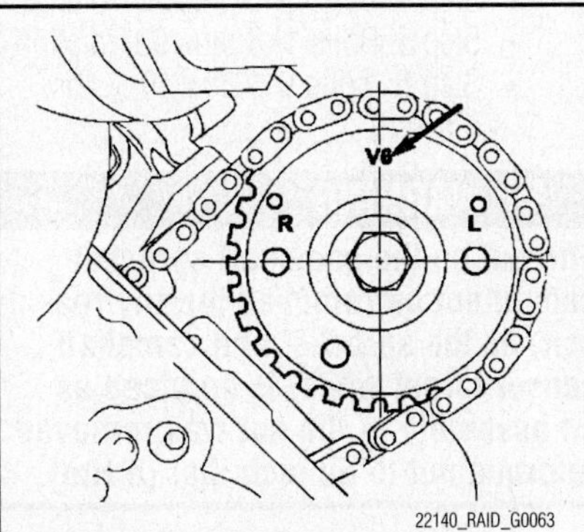

Fig. 74 Verify the V6 mark on the camshaft sprocket is at the 12 o'clock position. Rotate the crankshaft one turn if necessary

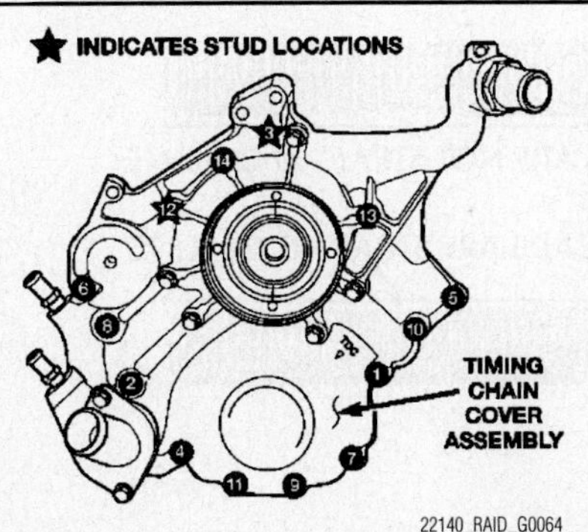

Fig. 75 Timing cover bolt locations—3.7L engine

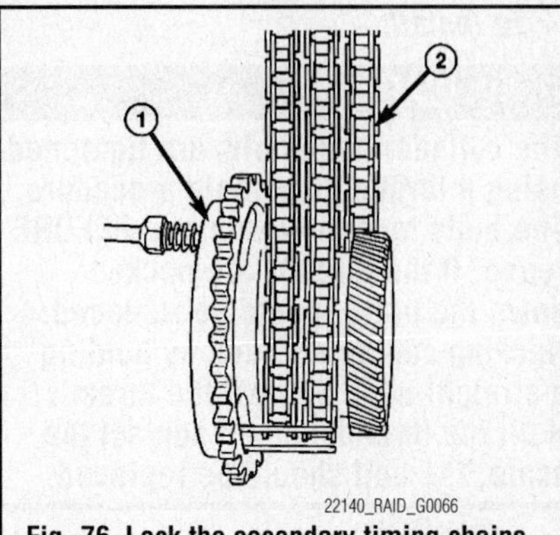

Fig. 76 Lock the secondary timing chains (2) to the idler sprocket using Secondary Camshaft Chain Holder 8429 (1)—3.7L engine

22. Remove the secondary chain guide.

⁂ WARNING

The nut on the camshaft sprocket should not be removed for any reason, as the sprocket and camshaft sensor target wheel is serviced as an assembly. If the nut was removed, retorque nut to 44 inch lbs. (5 Nm).

23. Remove the retaining bolt and the camshaft drive gear.

⁂ WARNING

Do not allow the engine to rotate. Severe damage to the valve train can occur.

➡Do not overlook the four smaller bolts at the front of the cylinder head. Do not attempt to remove the cylinder head without removing these four bolts.

⁂ WARNING

Do not hold or pry on the camshaft target wheel for any reason. A damaged target wheel can result in a vehicle no start condition.

24. Remove the 12 cylinder head retaining bolts.
25. Remove the cylinder head and gasket. Discard the gasket.

⁂ WARNING

Do not lay the cylinder head on its gasket sealing surface, due to the design of the cylinder head gasket any distortion to the cylinder head sealing surface may prevent the gasket from properly sealing resulting in leaks.

To install:

⁂ WARNING

The cylinder head bolts are tightened using a torque plus angle procedure. The bolts must be examined BEFORE reuse. If the threads are necked down the bolts should be replaced. Necking can be checked by holding a straight edge against the threads. If all the threads do not contact the scale, the bolt should be replaced.

⁂ WARNING

When cleaning cylinder head and cylinder block surfaces, DO NOT use a metal scraper because the surfaces could be cut or ground. Use only a wooden or plastic scraper.

26. Clean the cylinder head and cylinder block mating surfaces.
27. Position the new cylinder head gasket on the locating dowels.

⁂ WARNING

When installing cylinder head, use care not damage the tensioner arm or the guide arm.

28. Position the cylinder head onto the cylinder block. Make sure the cylinder head seats fully over the locating dowels.

➡The four M8 cylinder head mounting bolts require sealant to be added to them before installing. Failure to do so may cause leaks.

29. Lubricate the cylinder head bolt threads with clean engine oil and install the eight M10 bolts.
30. Coat the four M8 cylinder head bolts with Loctite 242 Lock and Seal Adhesive (or equivalent) then install the bolts.

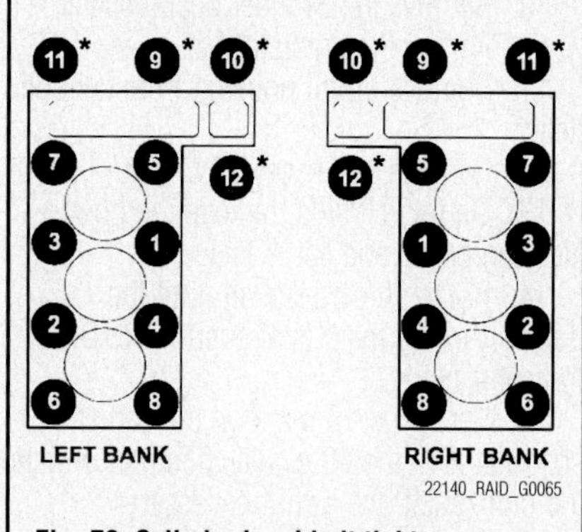

Fig. 78 Cylinder head bolt tighten sequence—3.7L engine

➡The cylinder head bolts are tightened using an angle torque procedure, however, the bolts are not a torque-to-yield design.

31. Tighten the cylinder head bolts, in sequence, as follows:
    - Step 1: Bolts 1–8 to 20 ft. lbs. (27 Nm)
    - Step 2: Bolts 1–8 verify tighten without loosening
    - Step 3: Bolts 9–12 to 10 ft. lbs. (14 Nm)
    - Step 4: Bolts 1–8 plus 90°rotation
    - Step 5: Bolts 1–8 plus 90°rotation
    - Step 6: Bolts 9–12 to 19 ft. lbs. (26 Nm)

⁂ WARNING

The nut on the camshaft sprocket should not be removed for any reason, as the sprocket and camshaft sensor target wheel is serviced as an assembly. If the nut was removed retorque nut to 60 inch lbs. (5 Nm).

32. Position the secondary chain onto the camshaft drive gear, making sure one marked chain link is on either side of the V6 mark on the gear then using Tool 8428 Camshaft wrench, position the gear onto the camshaft.

⁂ WARNING

Remove excess oil from camshaft sprocket retaining bolt before reinstalling bolt. Failure to do so may cause over-torquing of bolt resulting in bolt failure.

33. Install the camshaft drive gear retaining bolt.
34. Install the secondary chain guide.
35. Install the cylinder head access plug.
36. Re-set and install the secondary chain tensioner.
37. Remove Special Tool 8429.
38. Install the timing chain cover.
39. Install the crankshaft damper. Tighten damper bolt 130 ft. lbs. (175 Nm).
40. On the left side, install the power steering pump.
41. Install accessory drive belt.
42. Install the radiator fan and shroud.
43. Install the cylinder head cover.
44. On the left side, install the master cylinder and booster assembly.
45. Install the intake manifold.
46. Install oil fill housing onto cylinder head.
47. Refill the cooling system.
48. Raise the vehicle.
49. Install the exhaust pipe onto the right exhaust manifold.
50. Lower the vehicle.
51. Reconnect battery negative cable.
52. Start the engine and check for leaks.

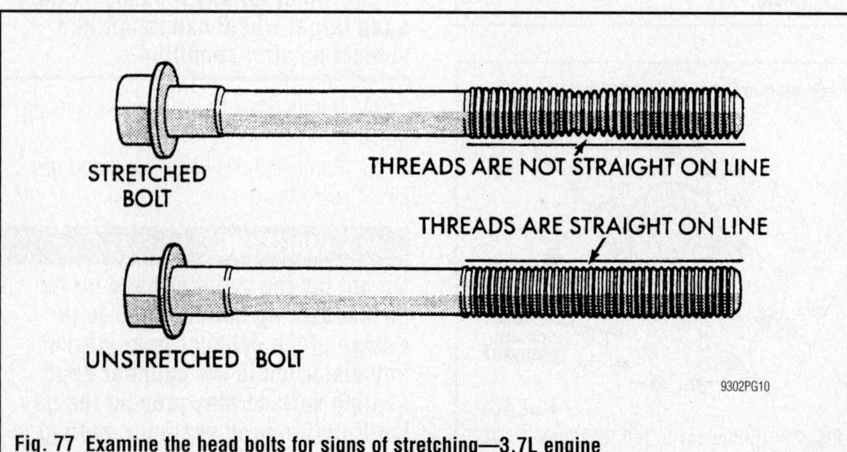

Fig. 77 Examine the head bolts for signs of stretching—3.7L engine

4.7L Engine

See Figures 79 and 80.

1. Before servicing the vehicle, refer to the precautions.
2. Disconnect the negative battery cable.
3. Raise and safely support the vehicle securely on jackstands.
4. Drain the engine coolant.
5. Lower the vehicle.
6. Remove the intake manifold.
7. Remove the cylinder head cover.
8. Remove the fan shroud.
9. Remove oil fill housing from cylinder head.
10. Remove accessory drive belt.
11. Rotate the crankshaft until the damper timing mark is aligned with TDC indicator mark.
12. Verify the V8 mark on the camshaft sprocket is at the 12 o'clock position. Rotate the crankshaft one turn if necessary.
13. Remove the crankshaft damper.

14. Remove the timing chain cover.

15. Lock the secondary timing chains to the idler sprocket using Special Tool 8429.

➡**Mark the secondary timing chain prior to removal to aid in installation.**

16. Mark the secondary timing chain, one link on each side of the V8 mark on the camshaft drive gear .

17. Remove the secondary chain tensioner.

18. Remove the cylinder head access plug.

19. Remove the secondary chain guide.

20. Remove the retaining bolt and the camshaft drive gear.

✱✱ WARNING

Do not allow the engine to rotate. severe damage to the valve train can occur.

➡**Do not overlook the four smaller bolts at the front of the cylinder head. Do not attempt to remove the cylinder head without removing these four bolts.**

✱✱ WARNING

Do not hold or pry on the camshaft target wheel for any reason. A damaged target wheel can result in a vehicle no start condition.

21. Remove the 14 cylinder head retaining bolts using the sequence provided.

22. Remove the cylinder head and gasket. Discard the gasket.

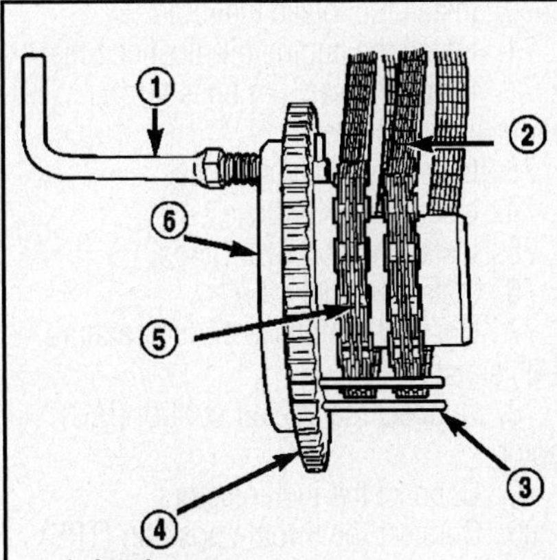

Fig. 79 Lock the secondary timing chains to the idler sprocket using Special Tool 8429

✱✱ WARNING

Do not lay the cylinder head on its gasket sealing surface, due to the design of the cylinder head gasket any distortion to the cylinder head sealing surface may prevent the gasket from properly sealing resulting in leaks.

To install:

✱✱ WARNING

The cylinder head bolts are tightened using a torque plus angle procedure. The bolts must be examined BEFORE reuse. If the threads are necked down the bolts should be replaced. Necking can be checked by holding a straight edge against the threads. If all the threads do not contact the scale, the bolt should be replaced.

✱✱ WARNING

When cleaning cylinder head and cylinder block surfaces, DO NOT use a metal scraper because the surfaces could be cut or ground. Use only a wooden or plastic scraper.

23. Clean the cylinder head and cylinder block mating surfaces.

24. Position the new cylinder head gasket on the locating dowels.

25. Lubricate the cylinder head bolt threads with clean engine oil and install the ten M10 bolts.

26. Coat the four M8 cylinder head bolts with Loctite 242 Lock and Seal Adhesive or equivalent then install the bolts.

27. Tighten the bolts in sequence using the following steps.

- Step 1: Bolts 1-10 to 15 ft. lbs. (20 Nm)
- Step 2: Bolts 1-10 to 35 ft. lbs. (47 Nm)
- Step 3: Bolts 11-14 to 18 ft. lbs. (25 Nm)
- Step 4: Bolts 1-10 plus 90 degrees
- Step 5: Bolts 11-14 to 22 ft. lbs. (30 Nm)

28. Position the secondary chain onto the camshaft drive gear, making sure one marked chain link is on either side of the V8 mark on the gear and position the gear onto the camshaft.

29. Install the camshaft drive gear retaining bolt.

30. Install the secondary chain guide.

31. Install the cylinder head access plug.

32. Re-set and install the secondary chain tensioner.

33. Remove Special Tool 8429.

34. Install the timing chain cover.

35. Install the crankshaft damper.

36. Install accessory drive belt.

37. Install the fan shroud.

38. Install the cylinder head cover.

39. Install the intake manifold.

40. Install oil fill housing onto cylinder head.

41. Refill the cooling system.

42. Raise the vehicle.

43. Install the exhaust pipe onto the right exhaust manifold.

44. Lower the vehicle.

45. Connect the negative battery cable.

46. Start the engine and check for leaks.

ENGINE ASSEMBLY

REMOVAL & INSTALLATION

3.7L Engine

See Figures 81 and 82.

1. Before servicing the vehicle, refer to the precautions.

2. Discharge the A/C system.

3. Drain the cooling system.

4. Release the fuel rail pressure.

5. Disconnect the negative battery cable.

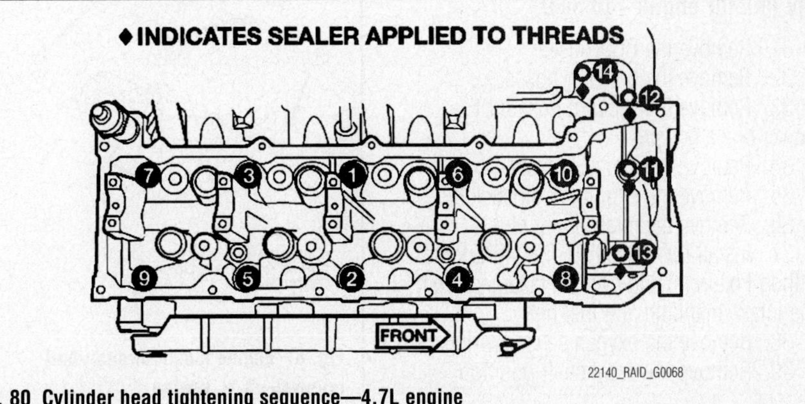

Fig. 80 Cylinder head tightening sequence—4.7L engine

6. Remove the air intake assembly.
7. Remove the upper fan shroud.
8. Remove the accessory drive belt.
9. Remove the viscous fan assembly.
10. Remove the A/C compressor and position out of the way.
11. Remove the alternator and secure away from engine.
12. Remove the power steering pump with lines attached and secure away from engine.

➡Do not remove the phenolic pulley from the power steering pump. It is not required for pump removal.

13. Remove the heater hoses.
14. Remove the upper radiator hose.
15. Remove the lower radiator hose.
16. Remove the transmission oil cooler lines at the radiator.
17. Remove the radiator/cooling module assembly.
18. Remove the engine to body ground straps at the left side of cowl.
19. Disconnect the intake air temperature (IAT) sensor.
20. Disconnect the fuel injectors.
21. Disconnect the throttle position (TPS) switch.
22. Disconnect the idle air control (IAC) motor.
23. Disconnect the engine oil pressure switch.
24. Disconnect the engine coolant temperature (ECT) sensor.
25. Disconnect the manifold absolute pressure map) sensor.
26. Disconnect the camshaft position (CMP) sensor.
27. Disconnect the coil over plugs.
28. Disconnect the crankshaft position sensor.
29. Disconnect the coil over plugs.
30. Remove fuel rail and secure away from engine.

➡It is not necessary to release the quick connect fitting from the fuel supply line for engine removal.

31. Remove the PCV hose.
32. Remove the breather hoses.
33. Remove the vacuum hose for the power brake booster.
34. Remove the knock sensors.
35. Remove the engine oil dipstick tube.
36. Remove the intake manifold.
37. Install the Special Tool 8247 Engine Lifting Fixture, using original fasteners from the intake manifold and fuel rail.
38. Remove the oxygen sensor wiring.
39. Remove the crankshaft position sensor.
40. Remove the engine block heater power cable, if equipped.
41. Disconnect the front driveshaft at the front differential and secure out of the way.
42. Remove the pinion bracket.
43. Remove the starter.
44. Remove the two ground straps from the lower left hand side and one ground strap from the lower right hand side of the engine
45. Remove the structural cover between engine and transmission.
46. Remove the exhaust crossover pipe from exhaust manifolds.
47. Remove the tighten converter bolts, if equipped.

➡Matchmark the bolts for reassembly in the original position.

48. Support the transmission with a suitable jack.
49. Connect a suitable engine hoist to the engine lift plate.
50. Remove the bellhousing-to-engine mounting bolts.
51. Remove the left and right engine mount bolts.
52. Remove the engine from the vehicle.

To install:

53. Position the engine into the vehicle.
54. Install the bellhousing-to-engine mounting bolts. Tighten to 30 ft. lbs. (41 Nm).
55. Install the engine mount bolts and tighten to 70 ft. lbs. (95 Nm).
56. Remove the jack from under the transmission.

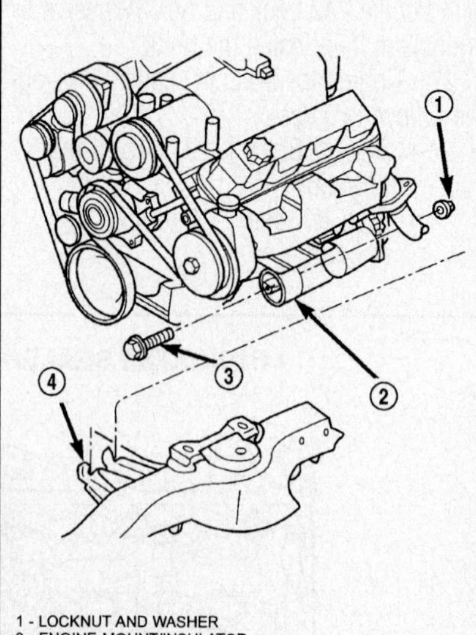

Fig. 81 Engine mount through-bolt removal—3.7L engine

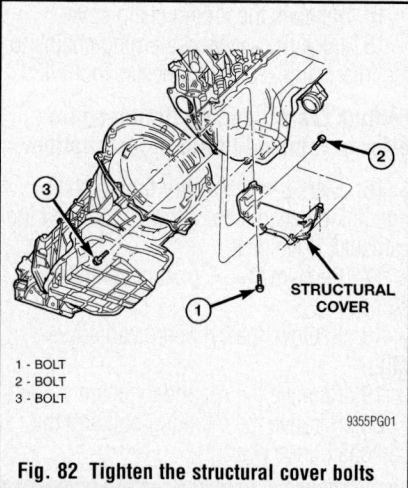

STRUCTURAL COVER

1 - BOLT
2 - BOLT
3 - BOLT

9355PG01

Fig. 82 Tighten the structural cover bolts in this order

57. Remove the Engine Lifting Fixture Tool.
58. Install the tighten converter bolts in their original positions.
59. Install the starter.
60. Install the crankshaft position sensor.
61. Install the engine block heater power cable, if equipped.
62. Install the structural cover.
63. Install the two ground straps from the lower left hand side and one ground strap from the lower right hand side of the engine.
64. Install the pinion bracket.
65. Install the exhaust pipe to the crossover.
66. Install the oxygen sensor wiring.
67. Install the Engine Lifting Fixture.
68. Install the knock sensors.
69. Install the engine to body ground straps at the left side of cowl.
70. Install the intake manifold
71. Install the engine oil dipstick tube.
72. Install the vacuum hose for the power brake booster.
73. Install the breather hoses.
74. Install the PCV hose.
75. Install the coil over plugs.
76. Install the fuel rail.
77. Connect the intake air temperature (IAT) sensor.
78. Connect the idle air control (IAC) motor.
79. Connect the fuel injectors.
80. Connect the throttle position (TPS) switch.
81. Connect the engine oil pressure switch.
82. Connect the engine coolant temperature (ECT) sensor.
83. Connect the manifold absolute pressure (MAP) sensor.
84. Connect the camshaft position (CMP) sensor.

85. Connect the coil over plugs.
86. Connect the crankshaft position sensor.
87. Install the radiator/cooling module assembly
88. Install the lower radiator hose.
89. Install the upper radiator hose.
90. Install the throttle cables.
91. Install the heater hoses.
92. Install the power steering pump
93. Install the alternator.
94. Install the A/C compressor.
95. Install the accessory drive belt.
96. Install the viscous fan assembly.
97. Install the upper fan shroud.
98. Install the radiator core support bracket.
99. Install the air intake assembly
100. Connect the negative battery cable.
101. Fill the engine with oil to the correct level.
102. Fill the cooling system to the correct level.
103. Start the engine and check for leaks.

4.7L Engine

See Figure 83.

1. Before servicing the vehicle, refer to the precautions.
2. Discharge the A/C system.
3. Drain the cooling system.
4. Release the fuel rail pressure.
5. Disconnect the negative battery cable.
6. Remove the ground straps from the lower left and right sides of the engine.
7. Remove the through bolt retaining nut and bolt from left and right side engine mounts.
8. Remove the crankshaft position sensor connector.
9. Remove the exhaust crossover pipe from exhaust manifold.
10. Remove the pinion bracket.
11. Remove structural cover.
12. Remove the starter.
13. Remove the transmission to engine mounting bolts.
14. Remove the engine block heater power cable, if equipped.
15. Remove the throttle body resonator assembly and air inlet hose.
16. Remove the disconnect throttle and cruise control cables.
17. Remove the crankcase breather tubes and breathers.
18. Remove the A/C compressor.
19. Remove the fan assembly, shroud and drive belts.
20. Disconnect transmission oil cooler lines at the radiator.
21. Disconnect upper and lower radiator hoses.
22. Remove radiator, A/C condenser and transmission oil cooler.
23. Remove the alternator.
24. Disconnect heater hoses from heater core and timing chain cover.
25. Remove hoses and tubes from the intake manifold.
26. Disconnect the intake air temperature (IAT) sensor.
27. Disconnect the fuel injectors.
28. Disconnect the throttle position (TPS) switch.
29. Disconnect the idle air control (IAC) motor.
30. Disconnect the engine oil pressure switch.
31. Disconnect the engine coolant temperature (ECT) sensor.
32. Disconnect the manifold absolute pressure (MAP) sensor.
33. Disconnect the camshaft position (CMP) sensor.
34. Disconnect the coil over plugs.
35. Disconnect vacuum lines at the throttle body and intake manifold.
36. Remove power steering pump and move out of the way.
37. Install Special Tools 8400 Lifting Studs, into the cylinder heads.
38. Install Engine Lifting Fixture Special Tool 8347 (3) following these steps.
    a. Holding the lifting fixture at a slight angle, slide the large bore in the front plate over the hex portion of the lifting stud.
    b. Position the two remaining fixture arms onto the two Special Tools 8400 Lifting Studs, in the cylinder heads.
    c. Pull forward and upward on the lifting fixture so that the lifting stud rest in the slotted area below the large bore.
    d. Secure the lifting fixture to the three studs using three 7/16 - 14 N/C locknuts.
    e. Make sure the lifting loop in the lifting fixture is in the last hole (closest to the throttle body) to minimize the angle of engine during removal.
39. Disconnect body ground straps at left and right side cowl.
40. Support transmission with suitable jack.
41. Remove the engine from the vehicle.

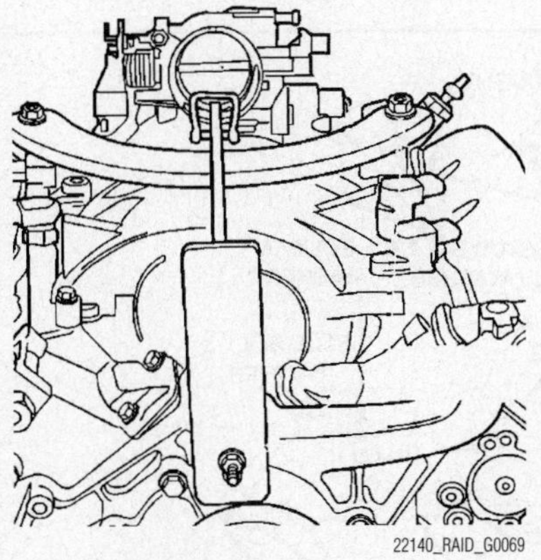

22140_RAID_G0069

Fig. 83 Install Engine Lifting Fixture Special Tool 8347

To install:

42. Position the engine into the vehicle.
43. Install the bellhousing-to-engine mounting bolts. Tighten to 30 ft. lbs. (41 Nm).
44. Install the engine mount bolts and tighten to 70 ft. lbs. (95 Nm).
45. Install the jack from under the transmission.
46. Install the Engine Lifting Fixture Tool.
47. Connect right and left side body ground straps.
48. Position alternator wiring behind the oil dipstick tube and install oil dipstick tube upper mounting bolt.
49. Connect fuel supply line quick connect fitting.
50. Connect vacuum lines at the throttle body and intake manifold.
51. Connect the intake air temperature (IAT) sensor.
52. Connect the idle air control (IAC) motor.
53. Connect the fuel injectors.
54. Connect the throttle position (TPS) switch.
55. Connect the engine oil pressure switch.
56. Connect the engine coolant temperature (ECT) sensor.
57. Connect the manifold absolute pressure (MAP) sensor.
58. Connect the camshaft position (CMP) sensor.
59. Connect the coil over plugs.
60. Install heater hoses to heater core and engine front cover.
61. Install alternator.
62. Install the A/C condenser.
63. Install the radiator.
64. Install the transmission oil cooler.
65. Connect radiator upper and lower hoses.
66. Connect transmission oil cooler lines to the radiator.
67. Install accessory drive belts, fan assembly and fan shroud.
68. Install A/C compressor

69. Install both breathers and connect tube to both crankcase breathers.
70. Connect throttle and cruise control cables.
71. Install throttle body resonator assembly and air inlet hose
72. Raise vehicle and connect crankshaft position sensor.
73. Install the starter.
74. Install the engine block heater power cable, if equipped.
75. Install the structural cover.
76. Install the pinion bracket.
77. Install the exhaust pipe to the crossover.
78. Connect the negative battery cable.
79. Fill the engine with oil to the correct level.
80. Fill the cooling system to the correct level.
81. Start the engine and check for leaks.

EXHAUST MANIFOLD

REMOVAL & INSTALLATION

3.7L Engine

See Figure 84.

1. Before servicing the vehicle, refer to the precautions.
2. Disconnect the negative battery cable.
3. Remove the exhaust manifold heat shields.
4. Remove the exhaust gas recirculation (EGR) tube.
5. Remove the exhaust y-pipe.
6. Remove the exhaust manifolds.

To install:

➡If the exhaust manifold studs came out with the nuts when removing the exhaust manifolds, replace them with new studs.

7. Install the exhaust manifolds. Tighten the fasteners to 20 ft. lbs. (27 Nm), starting with the center nuts and work out to the ends.
8. Install the exhaust y-pipe.
9. Install the EGR tube.
10. Install the exhaust manifold heat shields.
11. Connect the negative battery cable.
12. Start the engine, check for leaks and repair if necessary.

4.7L Engine

1. Before servicing the vehicle, refer to the precautions.
2. Drain the cooling system.
3. Disconnect the negative battery cable.
4. Remove the air intake assembly.
5. Remove the accessory drive belt.
6. Remove the A/C compressor.
7. Remove the A/C accumulator support bracket.
8. Remove the heater hoses.
9. Remove the exhaust manifold heat shields.
10. Remove the exhaust down pipe from the manifold.
11. Remove the starter motor.
12. Remove the exhaust manifolds

To install:

13. Install the exhaust manifolds, using new gaskets. Tighten the bolts to 18 ft. lbs. (25 Nm), starting with the inner bolts and work out to the ends.
14. Install the manifold heat shield. Tighten to 72 inch lbs. (8 Nm) and then back off 45 degrees..
15. Install the starter motor.
16. Install the exhaust down pipe to the manifold.
17. Install the heater hoses.
18. Install the a/c accumulator bracket.
19. Install the a/c compressor.
20. Install the accessory drive belt.
21. Install the air intake assembly.
22. Connect the negative battery cable.
23. Fill the cooling system.
24. Start the engine and check for leaks.

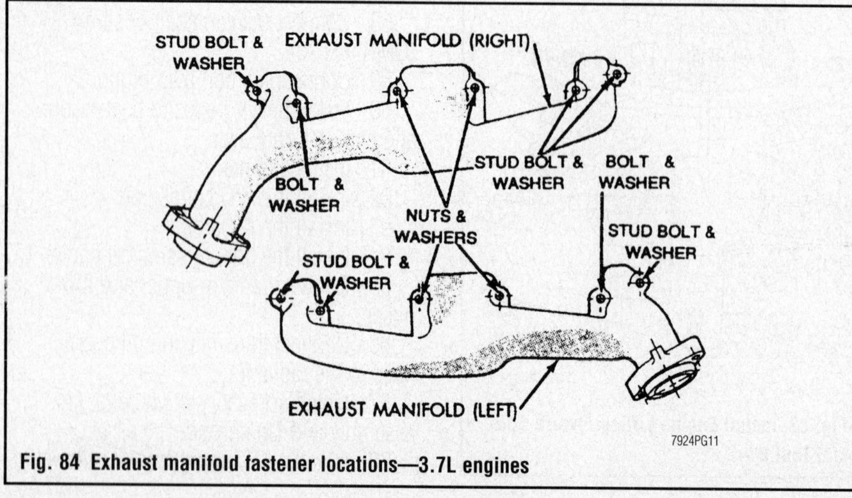

Fig. 84 Exhaust manifold fastener locations—3.7L engines

FLEXPLATE

REMOVAL & INSTALLATION

3.7L Engine

See Figure 85.

1. Remove transmission.
2. Remove attaching bolts
3. Remove flexplate

To install:

4. Position the flexplate onto the crankshaft and install the bolts hand tight.
5. Tighten the flexplate retaining bolts to 70 ft lbs (95 Nm) in a criss-cross sequence.
6. Install transmission.

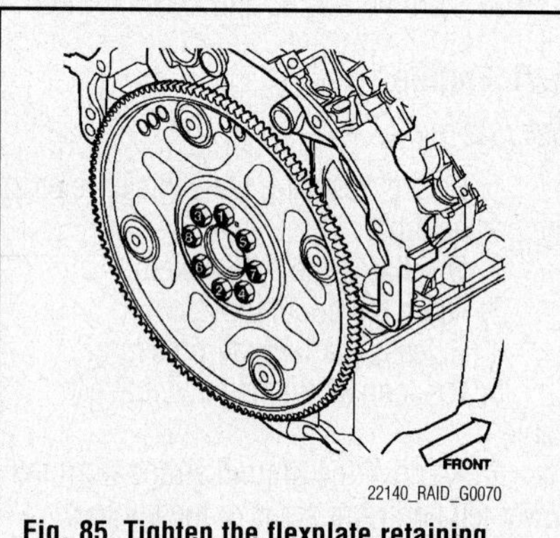

Fig. 85 Tighten the flexplate retaining bolts in a criss-cross sequence

4.7L Engine

See Figure 85.

1. Remove transmission.
2. Remove attaching bolts
3. Remove flexplate

To install:

4. Position the flexplate onto the crankshaft and install the bolts hand tight.
5. Tighten the flexplate retaining bolts to 45 ft lbs (60 Nm) in a criss-cross sequence.
6. Install transmission.

INTAKE MANIFOLD

REMOVAL & INSTALLATION

3.7L Engine

See Figure 86.

1. Before servicing the vehicle, refer to the precautions.
2. Drain the cooling system.

3. Relieve the fuel system pressure.
4. Disconnect the negative battery cable.
5. Remove the air cleaner assembly.
6. Remove the accelerator cable.
7. Remove the cruise control cable.
8. Disconnect the following electrical connectors:
   a. Manifold Absolute Pressure (MAP) sensor
   b. Intake Air Temperature (IAT) sensor
   c. Throttle Position (TP) sensor
   d. Idle Air Control (IAC) valve
9. Remove the Positive Crankcase Ventilation (PCV) valve and hose.
10. Remove the canister purge vacuum line.
11. Remove the brake booster vacuum line.
12. Remove the cruise control servo hose.
13. Remove the accessory drive belt.
14. Disconnect the alternator electrical connections.
15. Remove the engine ground straps.
16. Remove the ignition coil towers.
17. Remove the oil dipstick tube.
18. Remove the fuel line.
19. Remove the fuel rail.
20. Remove the throttle body assembly.
21. Remove the heater hoses.
22. Disconnect the coolant temperature sensor connector.
23. Remove the intake manifold.

➡Remove the fasteners in reverse of the tightening sequence.

To install:

24. Install the intake manifold using new gaskets. Tighten the bolts, in sequence, to 105 inch lbs. (12 Nm).
25. Install the coolant temperature sensor connector.
26. Install the heater hoses.
27. Install the throttle body assembly.

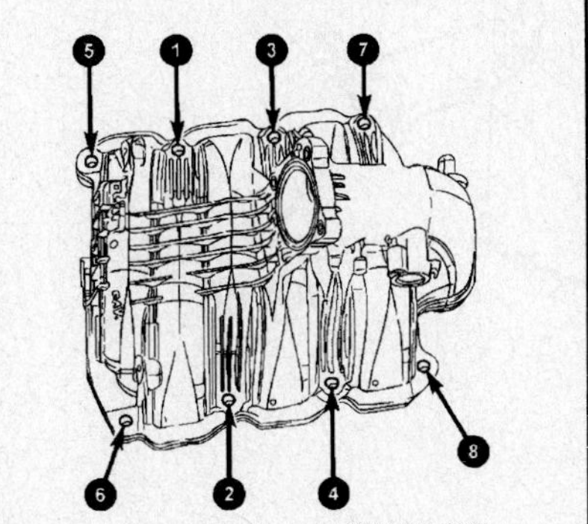

Fig. 86 Intake manifold tighten sequence—3.7L engine

28. Install the fuel rail.
29. Install the fuel line.
30. Install the oil dipstick tube.
31. Install the ignition coil towers.
32. Install the engine ground straps.
33. Connect the alternator electrical connections.
34. Install the accessory drive belt.
35. Install the cruise control servo hose.
36. Install the brake booster vacuum line.
37. Install the canister purge vacuum line.
38. Install the positive crankcase ventilation (PCV) valve and hose.
39. Connect the following electrical connectors:
   a. Idle Air Control (IAC) valve
   b. Throttle Position (TP) sensor
   c. Intake Air Temperature (IAT) sensor
   d. Manifold Absolute Pressure (MAP) sensor
40. Install the cruise control cable.
41. Install the accelerator cable.
42. Install the air cleaner assembly.
43. Connect the negative battery cable.
44. Fill the cooling system.
45. Start the engine and check for leaks.

4.7L Engine

See Figure 87.

1. Before servicing the vehicle, refer to the precautions.
2. Drain the cooling system.
3. Relieve the fuel system pressure.
4. Disconnect the negative battery cable.
5. Remove the wiper module.
6. Remove the air intake assembly.
7. Remove the accelerator cable.
8. Remove the cruise control cable.
9. Remove the Manifold Absolute Pressure (MAP) sensor connector.
10. Remove the Intake Air Temperature (IAT) sensor connector.
11. Remove the Throttle Position Sensor (TPS) connector.
12. Remove the Idle Air Control (IAC) valve connector.
13. Remove the Engine Coolant Temperature (ECT) sensor.
14. Remove the Positive Crankcase Ventilation (PCV) valve and hose.
15. Disconnect the alternator electrical connector.
16. Disconnect the A/C compressor connectors.
17. Remove the canister purge vacuum line.
18. Remove the brake booster vacuum line.
19. Remove the cruise control servo hose.
20. Remove the accessory drive belt.
21. Remove the engine ground straps.
22. Remove the ignition coil towers.
23. Remove the oil dipstick tube.
24. Remove the fuel supply rail.
25. Remove the throttle body assembly and mounting bracket.
26. Remove the heater hoses.
27. Remove the coolant temperature sensor.
28. Remove the intake manifold.

➡Remove the fasteners in reverse of the tightening sequence.

To install:

29. Install the intake manifold using new gaskets. Tighten the bolts in sequence to 105 inch lbs. (12 Nm).
30. Install the coolant temperature sensor.
31. Install the heater hoses.
32. Install the throttle body assembly and mounting bracket.
33. Install the fuel supply rail.
34. Install the oil dipstick tube.
35. Install the ignition coil towers.
36. Install the engine ground straps.
37. Install the accessory drive belt.
38. Install the cruise control servo hose.
39. Install the brake booster vacuum line.
40. Install the canister purge vacuum line.
41. Connect the A/C compressor connectors.
42. Install the alternator electrical connector.
43. Install the PCV valve and hose.
44. Install the ECT sensor.
45. Connect the IAC valve connector.
46. Connect the TPS connector.
47. Connect the IAT sensor connector.
48. Connect the MAP sensor connector.
49. Install the cruise control cable.

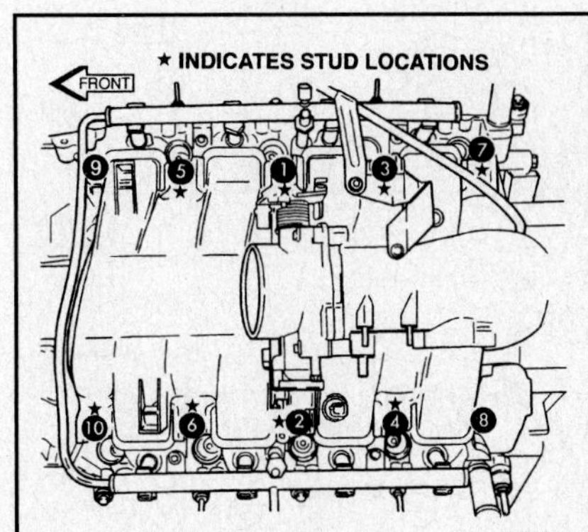

Fig. 87 Intake manifold tighten sequence—4.7L engine

50. Install the accelerator cable.
51. Install the air intake assembly.
52. Install the wiper module.
53. Disconnect the negative battery cable.
54. Fill the cooling system.
55. Start the engine and check for leaks.

OIL PAN

REMOVAL & INSTALLATION

See Figures 88 through 90.

1. Before servicing the vehicle, refer to the precautions.
2. Drain the engine oil.
3. Disconnect the negative battery cable.
4. Install engine support fixture 8354. Do not raise engine at this time.
5. Loosen both left and right side engine mount through bolts. Do not remove bolts.
6. Remove the structural dust cover, if equipped.
7. Remove the front crossmember.
8. Raise engine to provide clearance to remove oil pan.

➡Raise the engine just enough to provide clearance for oil pan removal. Check for proper clearance at fan shroud to fan and cowl to intake manifold.

9. If equipped with 4WD, remove or disconnect the following:
   a. Pinion bracket
   b. Front driveshaft at the front axle
   c. Front axle mounting bolts
   d. Lower the front axle using a suitable jack.
10. Remove the oil pan mounting bolts and oil pan.

➡Do not pry on oil pan or oil pan gasket. Gasket is integral to engine windage tray and does not come out with oil pan.

11. Unbolt oil pump pickup tube and remove tube.

To install:

12. Inspect the integral windage tray and gasket and replace as needed.
13. Clean the oil pan gasket mating surface of the bedplate and oil pan.
14. Position the integrated oil pan gasket/windage tray assembly.
15. Install the oil pickup tube
16. If removed, install stud at position No. 9.
17. Install the mounting bolt and nuts. Tighten nuts to 20 ft. lbs. (28 Nm).
18. Position the oil pan and install the mounting bolts. Tighten the mounting bolts to 11 ft. lbs. (15 Nm) in the sequence shown.
19. Lower the engine into mounts.
20. Install both the left and right side engine mount through bolts. Tighten the nuts to 50 ft. lbs. (68 Nm).
21. Remove the lifting device.
22. If equipped with 4WD, install or connect the following:
   a. Front axle mounting bolts
   b. Pinion bracket
   c. Front driveshaft to front axle
23. Install structural dust cover, if equipped.
24. Install the front crossmember.

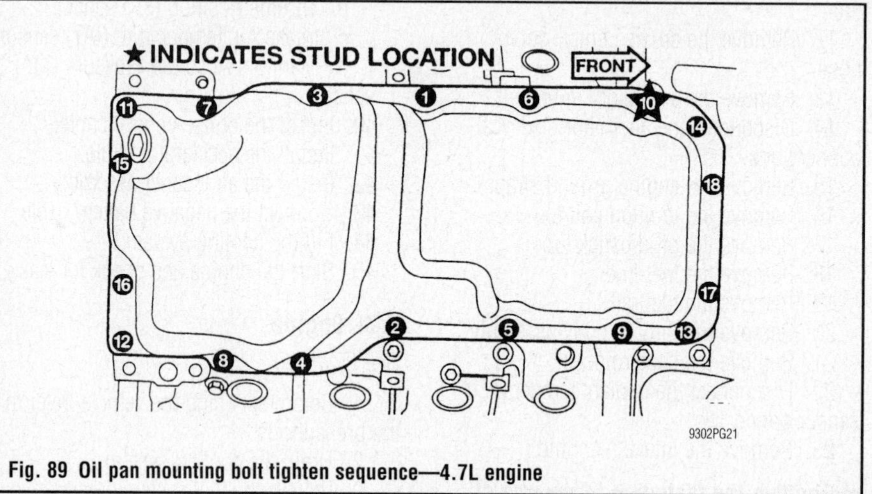

Fig. 89 Oil pan mounting bolt tighten sequence—4.7L engine

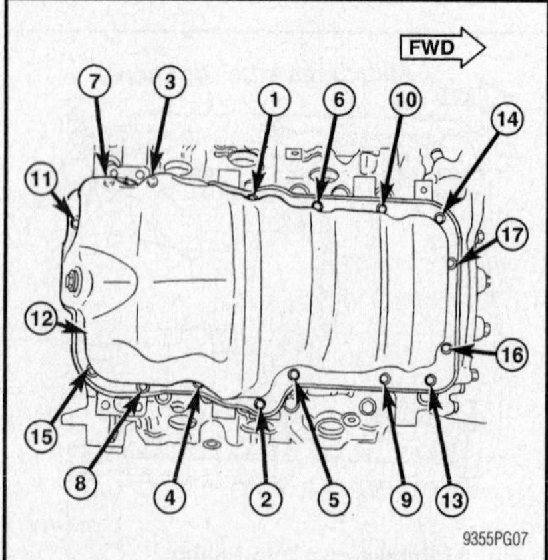

Fig. 88 Oil pan mounting bolt tighten sequence—3.7L Engine

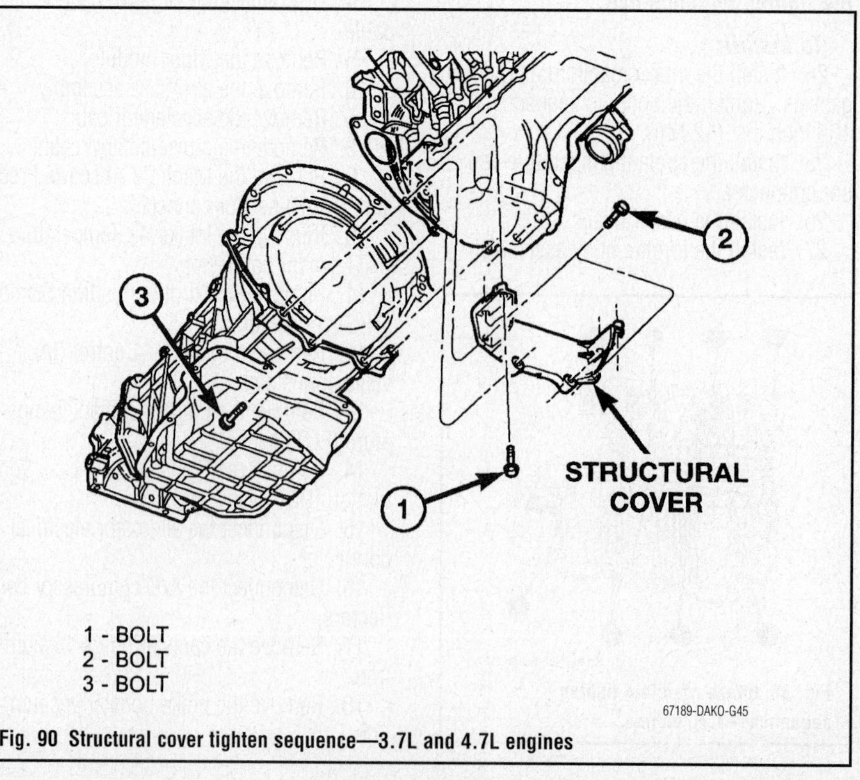

Fig. 90 Structural cover tighten sequence—3.7L and 4.7L engines

25. Refill the engine oil to the correct level.
26. Reconnect the negative battery cable.
27. Start engine and check for leaks.

OIL PUMP

REMOVAL & INSTALLATION

See Figure 91.

1. Before servicing the vehicle, refer to the precautions.
2. Drain the engine oil.
3. Remove the oil pan.
4. Remove the timing chain cover.
5. Remove the timing chains and tensioners.
6. Remove the oil pump.

To install:

7. Remove the oil pump. Tighten the pump bolts, in sequence, to 21 ft. lbs. (28 Nm).
8. Remove the timing chains and tensioners.
9. Remove the timing chain cover.
10. Remove the oil pan.
11. Refill the engine oil to the correct level.

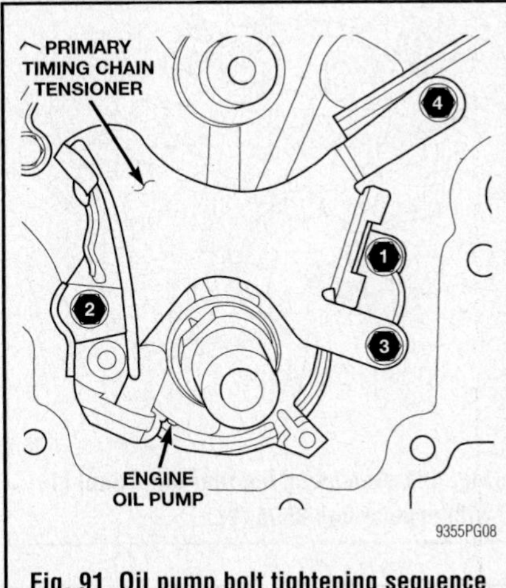

Fig. 91 Oil pump bolt tightening sequence

INSPECTION

See Figures 92 through 97.

⁂ WARNING

The oil pump pressure relief valve and spring should not be removed from the oil pump. If these components are disassembled and or removed from the pump the entire oil pump assembly must be replaced.

1. Clean all parts thoroughly. Mating surface of the oil pump housing should be smooth. If the pump cover is scratched or grooved, the oil pump assembly should be replaced.

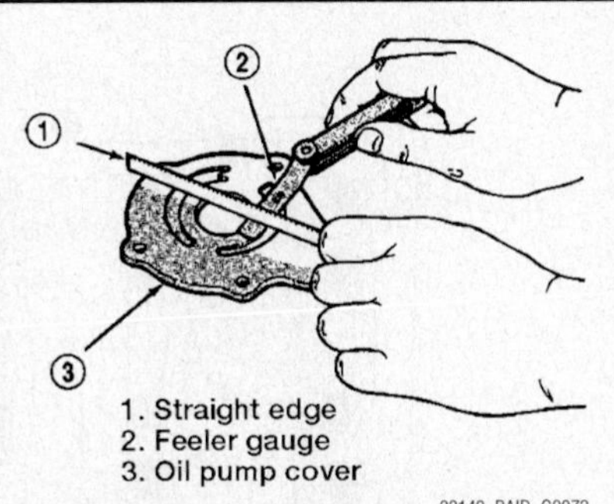

Fig. 92 Lay a straight edge across the pump cover surface. If a 0.001 in. (0.025 mm) feeler gauge can be inserted between the cover and the straight edge, the oil pump assembly should be replaced

2. Lay a straight edge across the pump cover surface. If a 0.001 in. (0.025 mm) feeler gauge can be inserted between the cover and the straight edge, the oil pump assembly should be replaced.
3. Measure the thickness of the outer rotor. If the outer rotor thickness measures at 0.472 in. (12.005 mm) or less the oil pump assembly must be replaced.

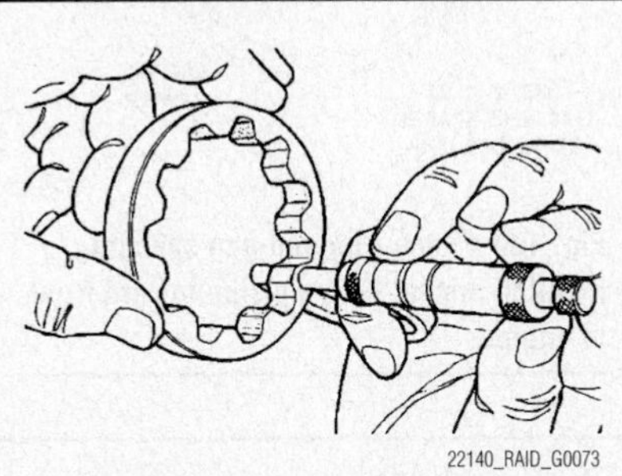

Fig. 93 Measure the thickness of the outer rotor. If the outer rotor thickness measures at 0.472 in. (12.005 mm) or less the oil pump assembly must be replaced

Fig. 94 Measure the thickness of the inner rotor. If the inner rotor thickness measures at 0.472 in. (12.005 mm) or less then the oil pump assembly must be replaced

4. Measure the diameter of the outer rotor. If the outer rotor diameter measures at 3.382 in. (85.925 mm) or less the oil pump assembly must be replaced.
5. Measure the thickness of the inner rotor . If the inner rotor thickness measures at 0.472 in. (12.005 mm) or less then the oil pump assembly must be replaced.
6. Slide outer rotor into the body of the oil pump. Press the outer rotor to one side of the oil pump body and measure clearance between the outer rotor and the body. If the measurement is 0.009 in. (0.235 mm) or more the oil pump assembly must be replaced.
7. Install the inner rotor in the into the oil pump body. Measure the clearance between the inner and outer rotors. If the clearance between the rotors is 0.006 in.

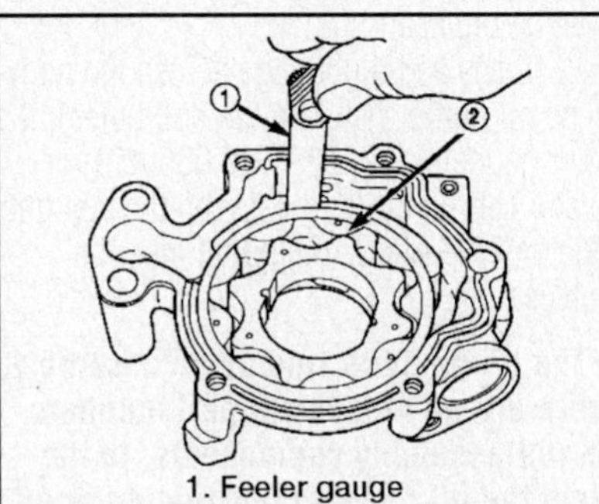

Fig. 95 Slide outer rotor into the body of the oil pump. Press the outer rotor to one side of the oil pump body and measure clearance between the outer rotor and the body. If the measurement is 0.009 in. (0.235 mm) or more the oil pump assembly must be replaced

1. Outer rotor
2. Feeler gauge
3. Inner rotor
22140_RAID_G0076

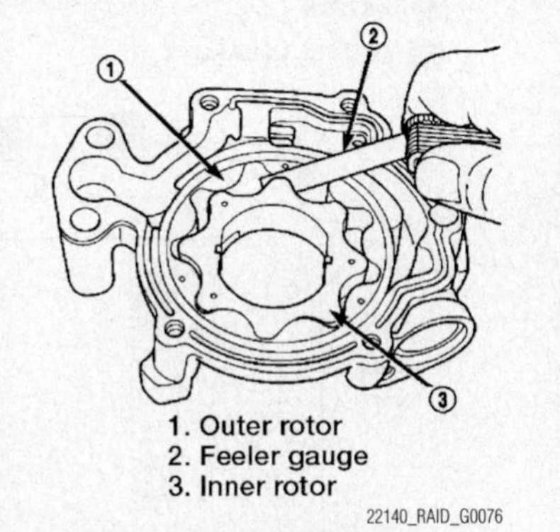

Fig. 96 Install the inner rotor in the into the oil pump body. Measure the clearance between the inner and outer rotors. If the clearance between the rotors is 0.006 in. (0.150 mm) or more the oil pump assembly must be replaced

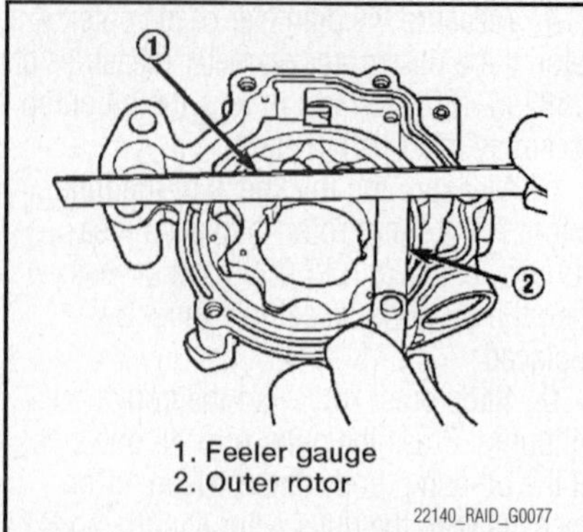

Fig. 97 Place a straight edge across the body of the oil pump (between the bolt holes), if a feeler gauge of 0.0038 in. (0.095 mm) or greater can be inserted between the straightedge and the rotors, the pump must be replaced

(0.150 mm) or more the oil pump assembly must be replaced.

8. Place a straight edge across the body of the oil pump (between the bolt holes), if a feeler gauge of 0.0038 in. (0.095 mm) or greater can be inserted between the straightedge and the rotors, the pump must be replaced.

➡The oil pump is sold as an assembly. There are no Mitsubishi part numbers for sub-assembly components. In the event the oil pump is functioning or out of specification it must be replaced as an assembly.

MAIN BEARING TIGHTEN SEQUENCE

See Figures 98 and 99.

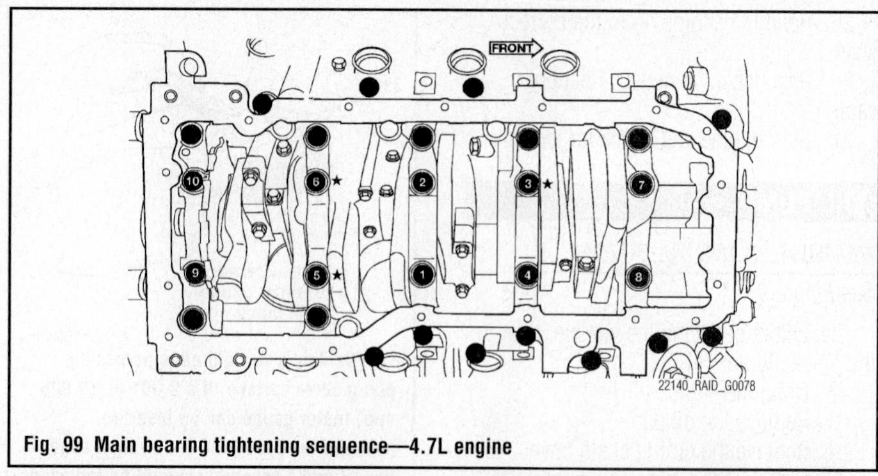

Fig. 99 Main bearing tightening sequence—4.7L engine

PISTON AND RING

POSITIONING

See Figure 100.

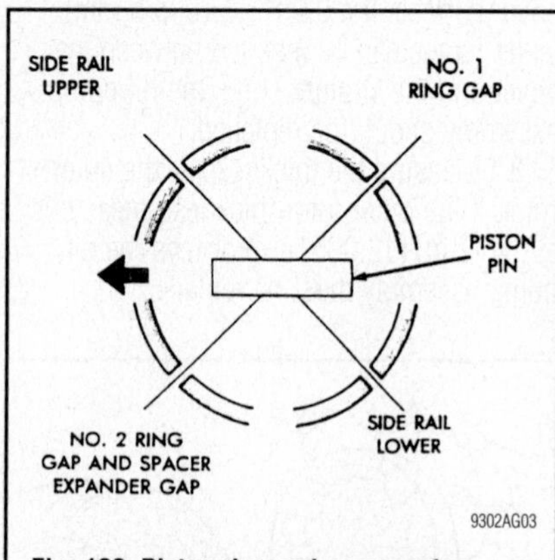

Fig. 100 Piston ring end-gap spacing. Position raised "F" on piston toward front of engine

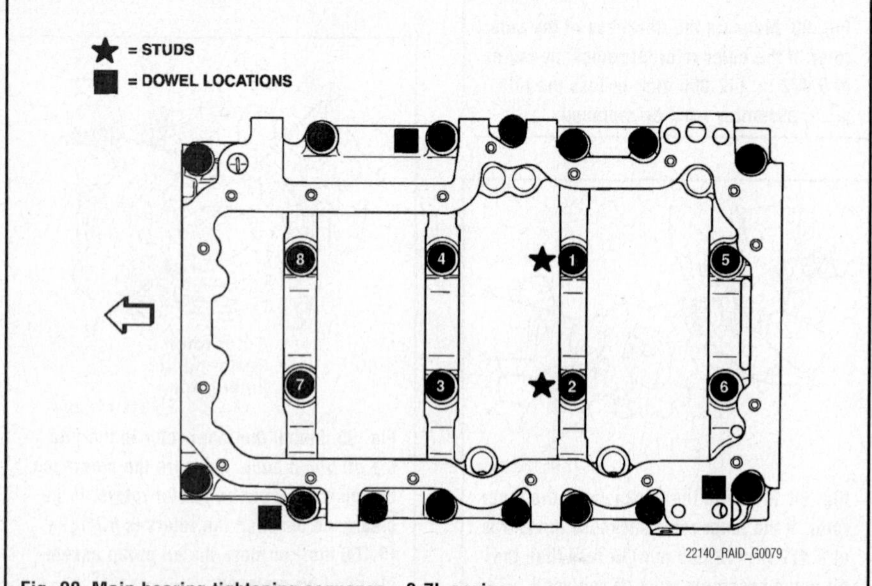

Fig. 98 Main bearing tightening sequence—3.7L engine

REAR MAIN SEAL

REMOVAL & INSTALLATION

See Figures 101 and 102.

1. Before servicing the vehicle, refer to the precautions.
2. Remove the transmission.
3. Remove the flexplate.
4. Thread Oil Seal Remover 8506 into the rear main seal as far as possible and remove the rear main seal.

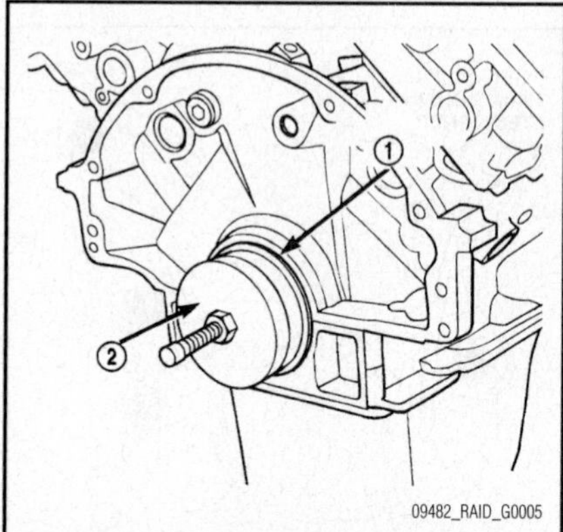

Fig. 101 Removing the rear main seal (1) with special tool 8506 (2)

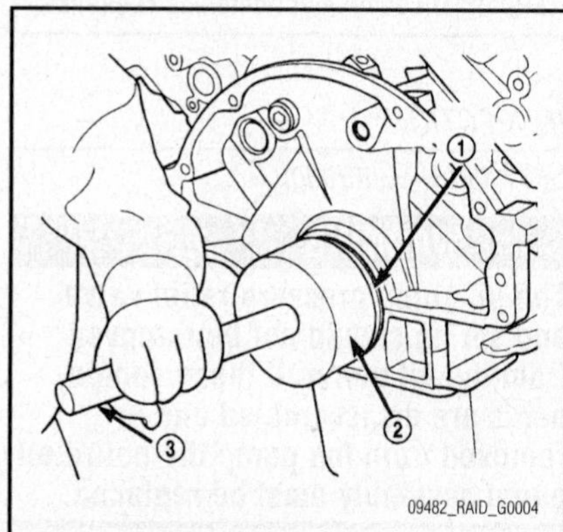

Fig. 102 Installing rear main seal using Special Tool 8349 (2) and C-4171 (3)

To install:

5. Install the Seal Guide 8349-2 onto the crankshaft
6. Install the rear main seal on the seal guide
7. Install the rear main seal, using the Crankshaft Rear Oil Seal Installer 8349 and Driver Handle C-4171; tap it into place until the installer is flush with the cylinder block
8. Install the flexplate. Tighten the bolts to 45 ft. lbs. (60 Nm).
9. Install the transmission.
10. Start the engine and check for leaks.

ROCKER ARMS/SHAFTS

REMOVAL & INSTALLATION

See Figure 103.

1. Before servicing the vehicle, refer to the precautions.
2. Disconnect the negative battery cable.
3. Remove the cylinder head cover.
4. Rotate the crankshaft so that the piston of the cylinder to be serviced is at Bottom Dead Center (BDC) and both valves are closed.
5. Use special tool 8516 to depress the valve and remove the rocker arm.
6. Repeat for each rocker arm to be serviced.

➡Keep valvetrain components in order for reassembly.

To install:

7. Rotate the crankshaft so that the piston of the cylinder to be serviced is at BDC.
8. Compress the valve spring and install each rocker arm in its original position.
9. Repeat for each rocker arm to be installed.

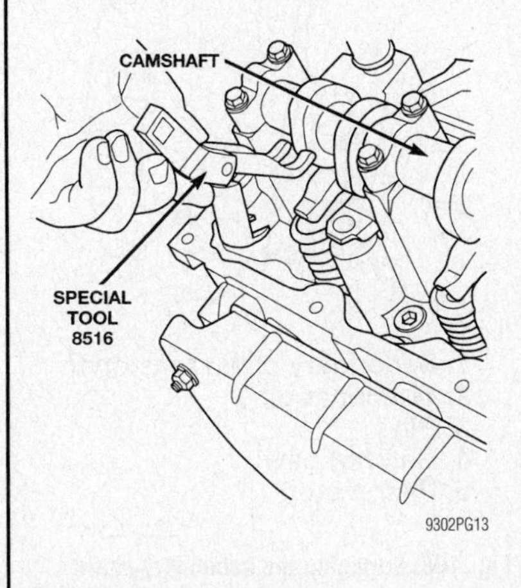

Fig. 103 Rocker arm service

10. Install the cylinder head cover.
11. Connect the negative battery cable.

TIMING CHAIN COVER AND SEAL

REMOVAL & INSTALLATION

3.7L Engine

See Figures 104 through 106, 58, 114.

1. Disconnect the negative battery cable.
2. Drain cooling system.
3. Remove right and left cylinder head covers.
4. Remove radiator fan.
5. Rotate engine until timing mark on crankshaft damper aligns with TDC mark on timing chain cover.
6. Make sure the camshaft sprocket "V6" marks (1) are at the 12 o'clock position (No. 1 TDC exhaust stroke).
7. Remove power steering pump.
8. Remove access plugs (1 and 2) from left and right cylinder heads for access to chain guide fasteners.
9. Remove the oil fill housing to gain access to the right side tensioner arm fastener.
10. Remove crankshaft damper and timing chain cover.

To install:

11. Install timing chain cover and crankshaft damper.
12. Install cylinder head covers.

➡Before installing threaded plug in right cylinder head, the plug must be coated with sealant to prevent leaks.

13. Coat the large threaded access plug with Mitsubishi Thread Sealant with Teflon, then install into the right cylinder head and tighten to 60 ft. lbs. (81 Nm).
14. Install the oil fill housing.
15. Install access plug in left cylinder head.
16. Install power steering pump.
17. Fill cooling system.
18. Connect the negative battery cable.

4.7L Engine

See Figures 115 and 116, 00, 118.

1. Before servicing the vehicle, refer to the precautions.
2. Disconnect the negative battery cable.
3. Drain cooling system.
4. Remove right and left cylinder head covers.
5. Remove radiator fan shroud.
6. Rotate engine until timing mark on crankshaft damper aligns with TDC mark on timing chain cover (No. 1 cylinder exhaust stroke) and the camshaft sprocket "V8" marks are at the 12 o'clock position.
7. Remove power steering pump.
8. Remove access plugs from left and right cylinder heads for access to chain guide fasteners.
9. Remove the oil fill housing to gain access to the right side tensioner arm fastener.
10. Remove crankshaft damper and timing chain cover

To install:

11. Install timing chain cover and crankshaft damper.
12. Install cylinder head covers.

➡Before installing threaded plug in right cylinder head, the plug must be coated with sealant to prevent leaks.

13. Coat the large threaded access plug with Mitsubishi Thread Sealant with Teflon, then install into the right cylinder head and tighten to 60 ft. lbs. (81 Nm).
14. Install the oil fill housing.
15. Install access plug in left cylinder head.
16. Install the radiator fan shroud.
17. Install power steering pump.
18. Fill cooling system.
19. Connect the negative battery cable.

TIMING CHAIN AND SPROCKETS

REMOVAL & INSTALLATION

3.7L Engine

See Figures 104 through 114.

1. Disconnect the negative battery cable.
2. Drain cooling system.
3. Remove right and left cylinder head covers.
4. Remove radiator fan.
5. Rotate engine until timing mark on crankshaft damper aligns with TDC mark on timing chain cover.
6. Make sure the camshaft sprocket "V6" marks (1) are at the 12 o'clock position (No. 1 TDC exhaust stroke).
7. Remove power steering pump.
8. Remove access plugs (1 and 2) from left and right cylinder heads for access to chain guide fasteners.
9. Remove the oil fill housing to gain access to the right side tensioner arm fastener.

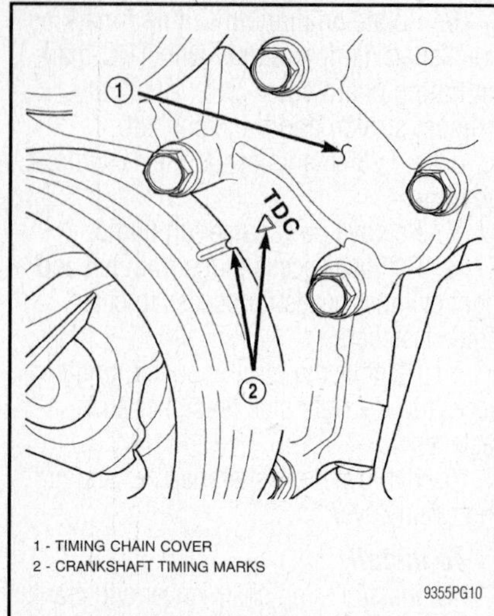

Fig. 104 Crankshaft timing marks—3.7L engine

10. Remove crankshaft damper and timing chain cover.

11. Collapse and pin primary chain tensioner.

✱✱ WARNING

Plate behind left secondary chain tensioner could fall into oil pan. Therefore, cover pan opening.

12. Remove secondary chain tensioners.

13. Remove camshaft position sensor.

✱✱ WARNING

Care should be taken not to damage camshaft target wheel. Do not hold target wheel while loosening or tightening camshaft sprocket. Do not place the target wheel near a magnetic source of any kind. A damaged or magnetized target wheel could cause a vehicle no start condition.

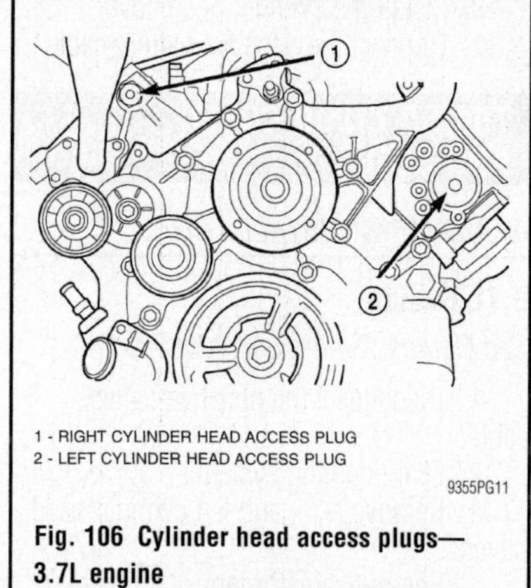

Fig. 106 Cylinder head access plugs—3.7L engine

✱✱ WARNING

Do not forcefully rotate the camshafts or crankshaft independently of each other. Damaging intake valve to piston contact will occur. Ensure negative battery cable is disconnected to guard against accidental starter engagement.

14. Remove left and right camshaft sprocket bolts.

15. While holding the left camshaft steel tube with Camshaft Holder 8428, remove the left camshaft sprocket. Slowly rotate the camshaft approximately 5 degrees clockwise to a neutral position.

16. While holding the right camshaft steel tube with Camshaft Holder 8428, remove the right camshaft sprocket.

17. Remove idler sprocket assembly bolt.

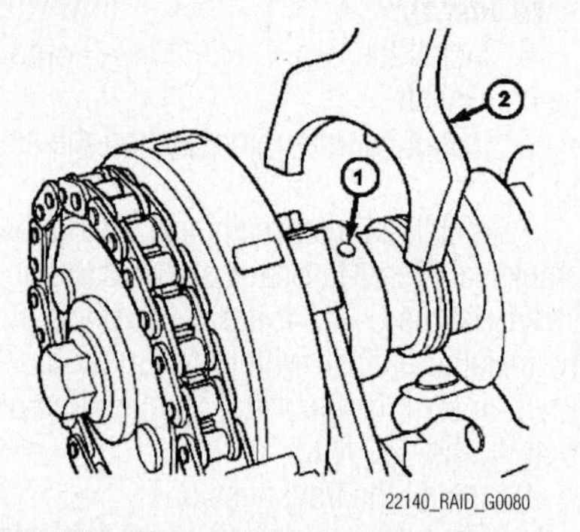

Fig. 107 While holding the camshaft steel tube (1) with Camshaft Holder 8428 (2), remove the camshaft sprocket—3.7L engine

18. Slide the idler sprocket assembly and crank sprocket forward simultaneously to remove the primary and secondary chains.

19. Remove both pivoting tensioner arms and chain guides.

20. Remove primary chain tensioner.

To install:

21. Using a vise, lightly compress the secondary chain tensioner piston until the piston step (5) is flush with the tensioner body. Using a pin or suitable tool, release ratchet pawl (4) by pulling pawl back against spring force through access hole on side of tensioner. While continuing to hold pawl back, Push ratchet device to approximately 2 mm from the tensioner body. Install Tensioner Pins 8514 (2) into hole on front of tensioner. Slowly open vise to transfer piston spring force to lock pin.

22. Position primary chain tensioner over oil pump and insert bolts into lower two holes on tensioner bracket. Tighten bolts to 21 ft. lbs. (28 Nm).

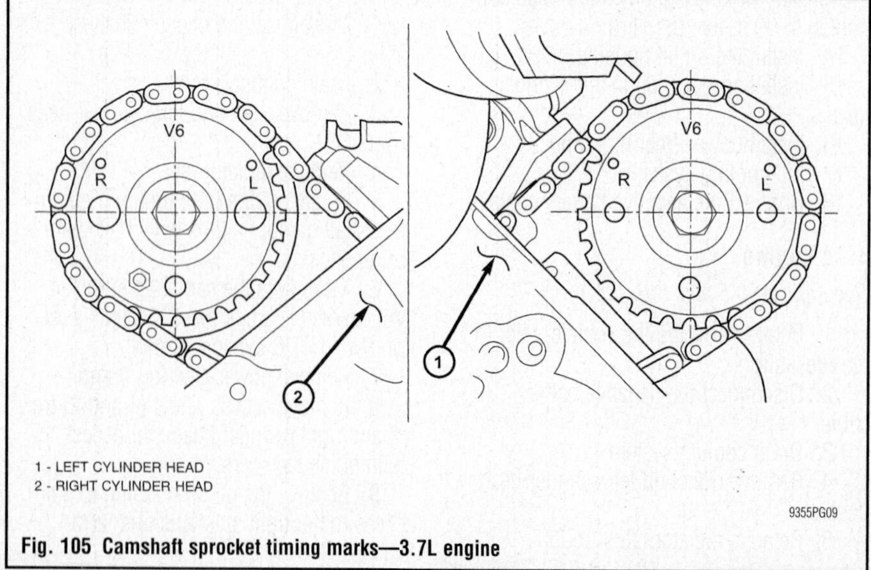

Fig. 105 Camshaft sprocket timing marks—3.7L engine

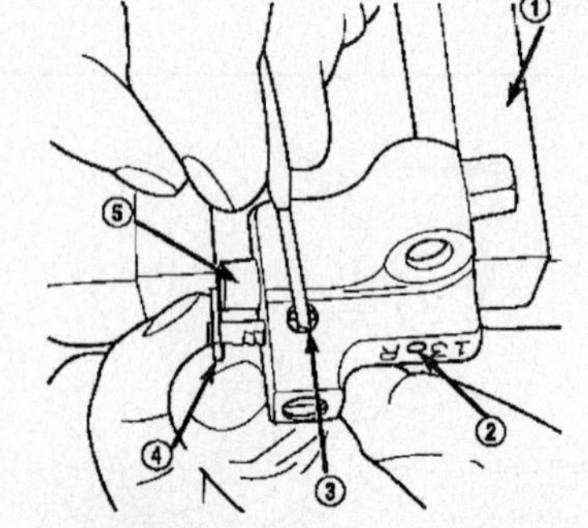

Fig. 108 Servicing the secondary chain tensioner—3.7L engine

23. Install right side chain tensioner arm. Install Torx® bolt. Tighten Torx® bolt to 21 ft. lbs. (28 Nm).

**** CAUTION**

The silver bolts retain the guides to the cylinder heads and the black bolts retain the guides to the engine block.

24. Install the left side chain guide. Tighten the bolts to 21 ft. lbs. (28 Nm).

25. Install left side chain tensioner arm, and Torx® bolt. Tighten Torx® bolt to 21 ft. lbs. (28 Nm).

26. Install the right side chain guide. Tighten the bolts to 21 ft. lbs. (28 Nm).

27. Install both secondary chains onto the idler sprocket. Align two plated links on the secondary chains to be visible through the two lower openings on the idler sprocket (4 o'clock and 8 o'clock). Once the secondary timing chains are installed, install the Secondary Camshaft Chain Holder 8429 to hold chains in place for installation.

28. Align primary chain double plated links with the timing mark at 12 o'clock on the idler sprocket. Align the primary chain single plated link with the timing mark at 6 o'clock on the crankshaft sprocket.

29. Lubricate idler shaft and bushings with clean engine oil.

**** WARNING**

The idler sprocket must be timed to the counterbalance shaft drive gear before the idler sprocket is fully seated.

30. Install all chains, crankshaft sprocket, and idler sprocket as an assembly. After guiding both secondary chains through the block and cylinder head openings, affix

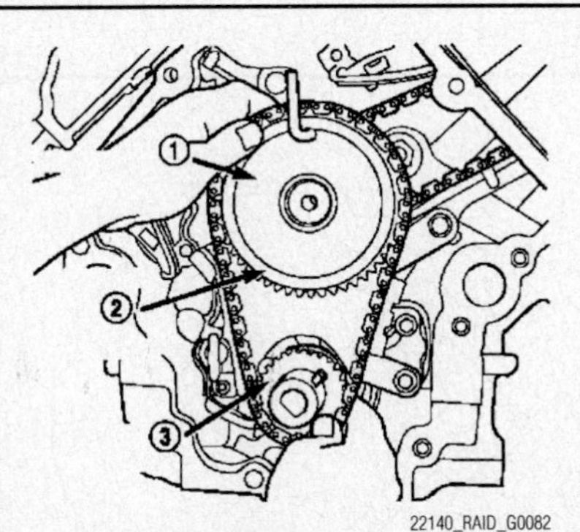

Fig. 109 Install both secondary chains onto the idler sprocket (2) and crankshaft sprocket (3). Then install the Secondary Camshaft Chain Holder 8429 (1) to hold chains in place for installation—3.7L engine

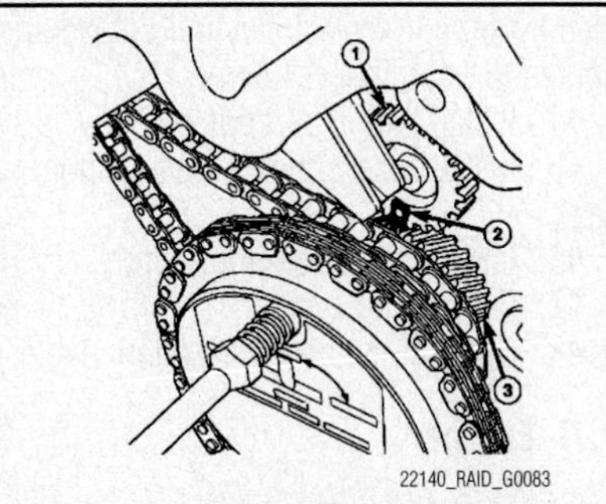

Fig. 110 Align the timing mark (2) on the idler sprocket gear (3) to the timing mark on the counterbalance shaft drive gear (1)—3.7L engine

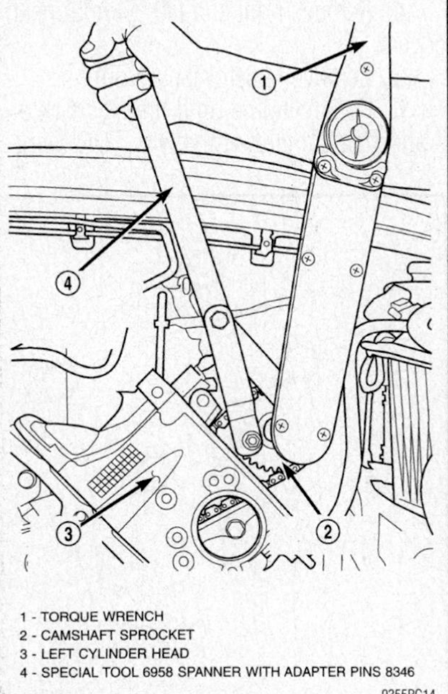

Fig. 111 Tightening the left side camshaft sprocket—3.7L engine

chains with a elastic strap or equivalent. This will maintain tension on chains to aid in installation. Align the timing mark (2) on the idler sprocket gear (3) to the timing mark on the counterbalance shaft drive gear (1), then seat idler sprocket fully. Before installing idler sprocket bolt, lubricate washer with oil, and tighten idler sprocket assembly retaining bolt to 25 ft. lbs. (34 Nm).

➡It will be necessary to slightly rotate camshafts for sprocket installation.

31. Align left camshaft sprocket **L**dot to plated link on chain.

32. Align right camshaft sprocket **R**dot to plated link on chain.

**** WARNING**

Remove excess oil from the camshaft sprocket bolt. Failure to do so can result in over-torque of bolt resulting in bolt failure.

33. Remove Secondary Camshaft Chain Holder 8429, then attach both sprockets to camshafts. Remove excess oil from bolts, then Install sprocket bolts, but do not tighten at this time.

34. Verify that all plated links are aligned with the marks on all sprockets and the "V6" marks on camshaft sprockets are at the 12 o'clock position.

**** CAUTION**

Ensure the plate between the left secondary chain tensioner and block is correctly installed.

35. Install both secondary chain tensioners. Tighten bolts to 250 inch lbs. (28 Nm).

➡Left and right secondary chain tensioners are not common.

36. Remove all locking pins from tensioners.

**** WARNING**

After pulling locking pins out of each tensioner, DO NOT manually extend the tensioner(s) ratchet. Doing so will over tension the chains, resulting in noise and/or high timing chain loads.

37. Using Spanner Wrench 6958, with Adaptor Pins 8346, tighten the left camshaft sprocket bolts to 90 ft. lbs. (122 Nm).

38. Rotate engine two full revolutions. Verify timing marks are at the follow locations:

- Primary chain idler sprocket dot is at 12 o'clock

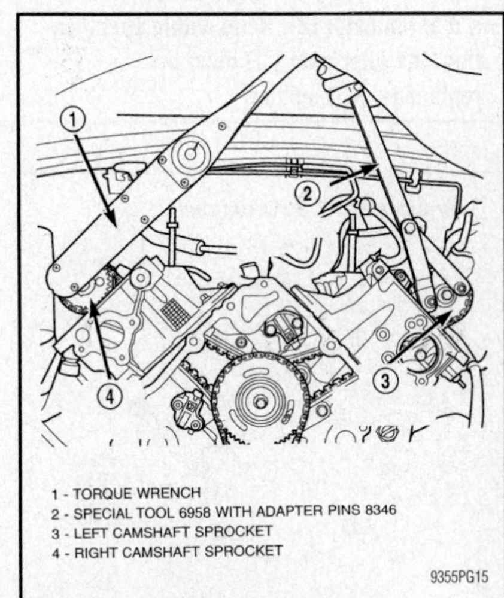

Fig. 112 Tightening the right side camshaft sprocket—3.7L engine

- Primary chain crankshaft sprocket dot is at 6 o'clock
- Secondary chain camshaft sprockets "V6" marks are at 12 o'clock
- Balance shaft drive gear dot is aligned to the idler sprocket gear dot

39. Lubricate all three chains with engine oil.

40. After installing all chains, it is recommended that the idler gear end play be checked. The end play must be within 0.004– 0.010 in. (0.10–0.25 mm). If not within specification, the idler gear must be replaced.

41. Install timing chain cover and crankshaft damper.

42. Install cylinder head covers.

➡Before installing threaded plug in right cylinder head, the plug must be coated with sealant to prevent leaks.

43. Coat the large threaded access plug with Mitsubishi Thread Sealant with Teflon, then install into the right cylinder head and tighten to 60 ft. lbs. (81 Nm).

44. Install the oil fill housing.

45. Install access plug in left cylinder head.

46. Install power steering pump.

47. Fill cooling system.

48. Connect the negative battery cable.

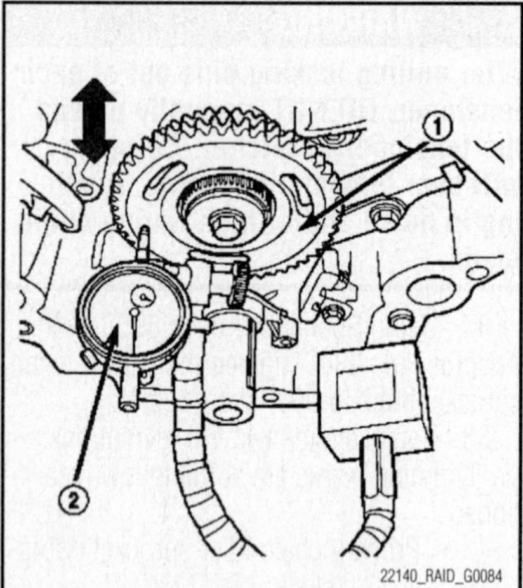

Fig. 113 Check idler gear end play using a dial indicator (2). If not within specification, the idler gear (1) must be replaced—3.7L engine

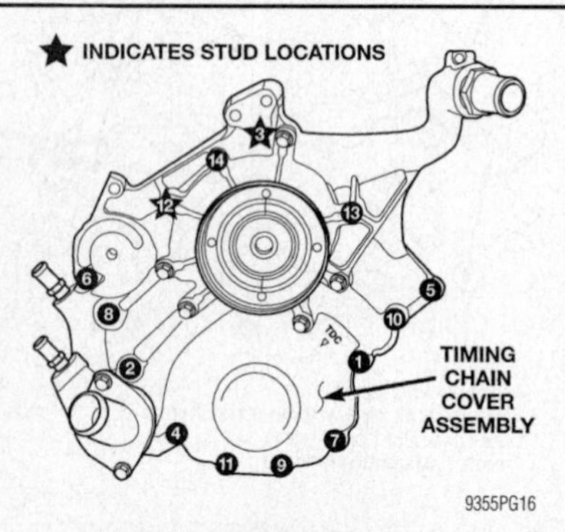

Fig. 114 Timing cover bolt tighten sequence—3.7L engine

4.7L Engine

See Figures 115 through 118.

1. Before servicing the vehicle, refer to the precautions.

2. Disconnect the negative battery cable.

3. Drain cooling system.

4. Remove right and left cylinder head covers.

5. Remove radiator fan shroud.

6. Rotate engine until timing mark on crankshaft damper aligns with TDC mark on timing chain cover (No. 1 cylinder exhaust stroke) and the camshaft sprocket "V8" marks are at the 12 o'clock position.

7. Remove power steering pump.

8. Remove access plugs from left and right cylinder heads for access to chain guide fasteners.

9. Remove the oil fill housing to gain access to the right side tensioner arm fastener.

10. Remove crankshaft damper and timing chain cover

11. Compress the primary timing chain tensioner and install a lock pin.

12. Remove the secondary timing chain tensioners.

13. Remove the camshaft position sensor from the right cylinder head.

Care should be taken not to damage camshaft target wheel. Do not hold target wheel while loosening or tightening camshaft sprocket. Do not place the target wheel near a magnetic source of any kind. A damaged or magnetized target wheel could cause a vehicle no start condition.

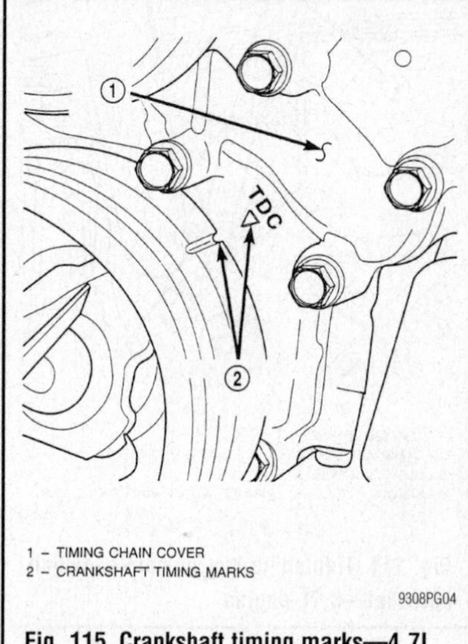

Fig. 115 Crankshaft timing marks—4.7L engine

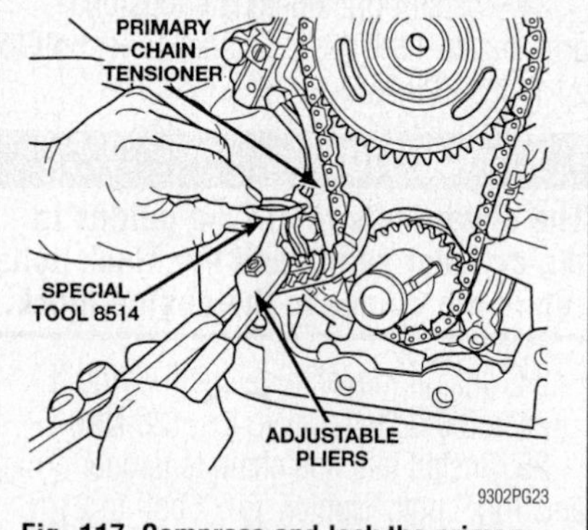

Fig. 117 Compress and lock the primary chain tensioner—4.7L engine

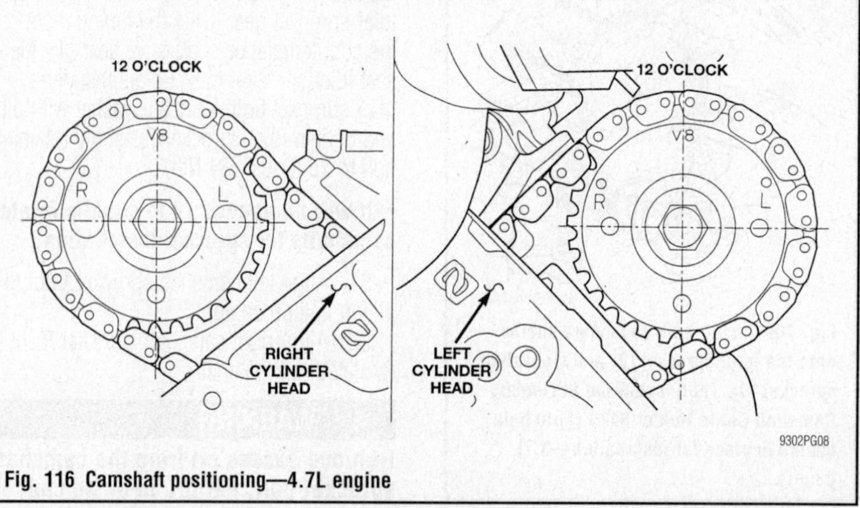

Fig. 116 Camshaft positioning—4.7L engine

WARNING

Do not forcefully rotate the camshafts or crankshaft independently of each other. Damaging intake valve to piston contact will occur. Ensure negative battery cable is disconnected to guard against accidental starter engagement.

14. Remove left and right camshaft sprocket bolts.
15. While holding the left camshaft steel tube with adjustable pliers, remove the left camshaft sprocket. Slowly rotate the camshaft approximately 15 degrees clockwise to a neutral position.
16. While holding the right camshaft steel tube with adjustable pliers, remove the right camshaft sprocket. Slowly rotate the camshaft approximately 45 degrees counterclockwise to a neutral position.
17. Remove idler sprocket assembly bolt.
18. Slide the idler sprocket assembly and crank sprocket forward simultaneously to remove the primary and secondary chains.
19. Remove both pivoting tensioner arms and chain guides.
20. Remove chain tensioner.

To install:

21. Using a vise, lightly compress the secondary chain tensioner piston until the piston step (5) is flush with the tensioner body. Using a pin or suitable tool, release ratchet pawl (4) by pulling pawl back against spring force through access hole on side of tensioner. While continuing to hold pawl back, Push ratchet device to approximately 2 mm from the tensioner body. Install Tensioner Pins 8514 (2) into hole on front of tensioner. Slowly open vise to transfer piston spring force to lock pin.
22. Position primary chain tensioner over oil pump and insert bolts into lower two holes on tensioner bracket. Tighten bolts to 21 ft. lbs. (28 Nm).
23. Install right side chain tensioner arm. Install Torx® bolt. Tighten Torx® bolt to 21 ft. lbs. (28 Nm).

WARNING

The silver bolts retain the guides to the cylinder heads and the black bolts retain the guides to the engine block.

24. Install the left side chain guide. Tighten the bolts to 21 ft. lbs. (28 Nm).
25. Install left side chain tensioner arm, and Torx® bolt. Tighten Torx® bolt to 21 ft. lbs. (28 Nm).
26. Install the right side chain guide. Tighten the bolts to 21 ft. lbs. (28 Nm).
27. Install both secondary chains onto the idler sprocket. Align two plated links on the secondary chains to be visible through the two lower openings on the idler sprocket (4 o'clock and 8 o'clock). Once the secondary timing chains are installed, install the Secondary Camshaft Chain Holder 8429 to hold chains in place for installation.
28. Align primary chain double plated links with the timing mark at 12 o'clock on the idler sprocket. Align the primary chain single plated link with the timing mark at 6 o'clock on the crankshaft sprocket.
29. Lubricate idler shaft and bushings with clean engine oil.

WARNING

The idler sprocket must be timed to the counterbalance shaft drive gear before the idler sprocket is fully seated.

30. Install all chains, crankshaft sprocket, and idler sprocket as an assembly. After guiding both secondary chains through the block and cylinder head openings, affix chains with a elastic strap or equivalent. This will maintain tension on chains to aid in installation. Align the timing mark (2) on the idler sprocket gear (3) to the timing mark on the counterbalance shaft drive gear (1), then seat idler sprocket fully. Before installing idler sprocket bolt, lubricate washer with oil, and tighten idler sprocket assembly retaining bolt to 25 ft. lbs. (34 Nm).

➡**It will be necessary to slightly rotate camshafts for sprocket installation.**

31. Align left camshaft sprocket **L**dot to plated link on chain.
32. Align right camshaft sprocket **R**dot to plated link on chain.

WARNING

Remove excess oil from the camshaft sprocket bolt. Failure to do so can result in over-torque of bolt resulting in bolt failure.

33. Remove Secondary Camshaft Chain Holder 8429, then attach both sprockets to camshafts. Remove excess oil from bolts, then Install sprocket bolts, but do not tighten at this time.
34. Verify that all plated links are aligned with the marks on all sprockets and the "V6" marks on camshaft sprockets are at the 12 o'clock position.

CAUTION

Ensure the plate between the left secondary chain tensioner and block is correctly installed.

35. Install both secondary chain tensioners. Tighten bolts to 250 inch lbs. (28 Nm).

➡**Left and right secondary chain tensioners are not common.**

36. Remove all locking pins from tensioners.

WARNING

After pulling locking pins out of each tensioner, DO NOT manually extend the tensioner(s) ratchet. Doing so will over tension the chains, resulting in noise and/or high timing chain loads.

37. Using Spanner Wrench 6958, with Adaptor Pins 8346, tighten the left camshaft sprocket bolts to 90 ft. lbs. (122 Nm).
38. Rotate engine two full revolutions. Verify timing marks are at the follow locations:
    - Primary chain idler sprocket dot is at 12 o'clock
    - Primary chain crankshaft sprocket dot is at 6 o'clock
    - Secondary chain camshaft sprockets "V6" marks are at 12 o'clock
39. Lubricate all three chains with engine oil.
40. After installing all chains, it is recommended that the idler gear end play be checked. The end play must be within 0.004–0.010 in. (0.10–0.25 mm). If not within specification, the idler gear must be replaced.
41. Install timing chain cover and crankshaft damper.
42. Install cylinder head covers.

➡**Before installing threaded plug in right cylinder head, the plug must be coated with sealant to prevent leaks.**

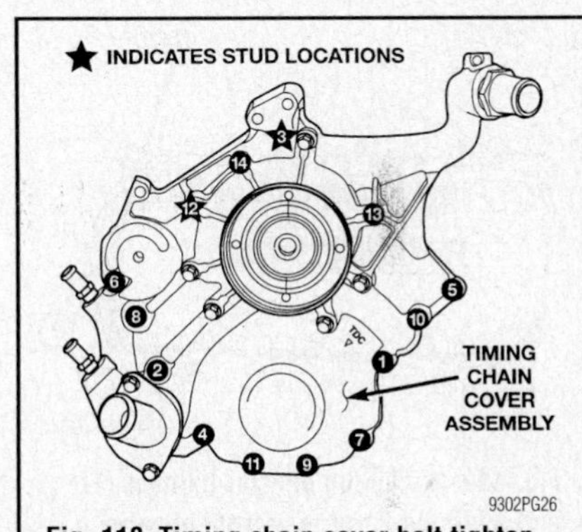

Fig. 118 Timing chain cover bolt tighten sequence—4.7L engine

43. Coat the large threaded access plug with Mitsubishi Thread Sealant with Teflon, then install into the right cylinder head and tighten to 60 ft. lbs. (81 Nm).
44. Install the oil fill housing.
45. Install access plug in left cylinder head.
46. Install the radiator fan shroud.
47. Install power steering pump.
48. Fill cooling system.
49. Connect the negative battery cable.

VALVE COVERS

REMOVAL & INSTALLATION

3.7L Engine

See Figure 119.

1. Before servicing the vehicle, refer to the precautions.
2. Disconnect the negative battery cable.
3. Remove the resonator assembly and air inlet hose.
4. Disconnect injector connectors.
5. Unclip the injector harness and route harness in front of valve cover.
6. Disconnect the left side breather tube and remove breather tube.
7. Remove valve cover mounting bolts
8. Remove valve cover.

To install:

**** WARNING**

Do not use harsh cleaners to clean the cylinder head covers. Severe damage to covers may occur.

➡The gasket may be used again, provided no cuts, tears, or deformation has occurred.

9. Clean cylinder head cover and both sealing surfaces. Inspect and replace gasket as necessary.

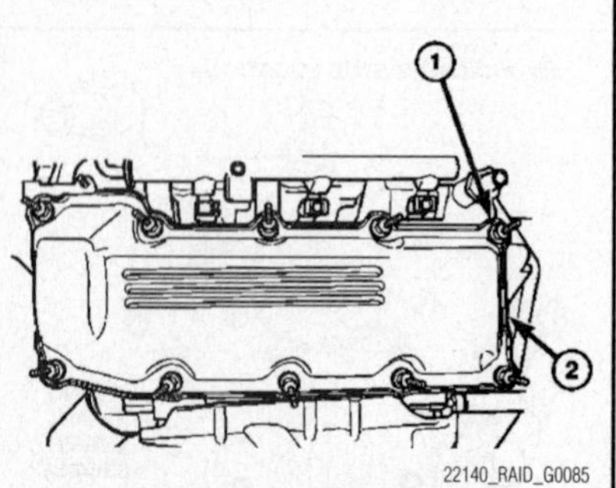

Fig. 119 Unclip the injector harness (1) and route the harness in front of valve cover (2)

10. Tighten cylinder head cover bolts and double ended studs to 105 inch lbs. (12 Nm).
11. Install left side breather and connect breather tube.
12. Connect injector electrical connectors and injector harness retaining clips.
13. Install the resonator and air inlet hose.
14. Connect the negative battery cable.

4.7L Engine

Right Side

See Figure 120.

1. Disconnect the negative battery cable.
2. Remove the air cleaner assembly, resonator assembly and air inlet hose.
3. Drain cooling system.
4. Remove accessory drive belt.
5. Remove air conditioning compressor retaining bolts and move the compressor to the left.
6. Remove heater hoses.
7. Disconnect and remove PCV hose.
8. Remove oil fill tube.
9. Remove the spark plug wires from the secondary spark plugs.
10. Disconnect the right side breather tube and filter assembly.
11. Remove valve cover mounting bolts
12. Remove valve cover.

To install:

**** WARNING**

Do not use harsh cleaners to clean the cylinder head covers. Severe damage to covers may occur.

➡The gasket may be used again, provided no cuts, tears, or deformation has occurred.

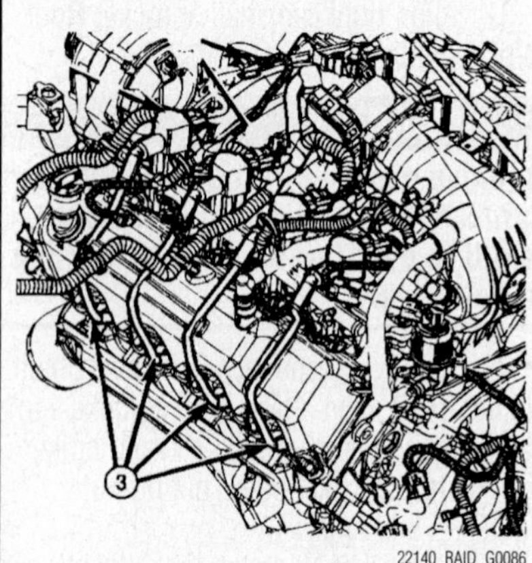

Fig. 120 Remove the spark plug wires from the secondary spark plugs

13. Clean cylinder head cover and both sealing surfaces. Inspect and replace gasket as necessary.
14. Install cylinder head cover and hand start all fasteners.
15. Tighten cylinder head cover bolts and double ended studs to 105 inch lbs. (12 Nm).
16. Install right rear breather tube and filter assembly.
17. Install spark plug wires.
18. Install the oil fill tube.
19. Install PCV hose.
20. Install heater hoses.
21. Install air conditioning compressor retaining bolts.
22. Install accessory drive belt.
23. Fill Cooling system.
24. Install air cleaner assembly, resonator assembly and air inlet hose.
25. Connect the negative battery cable.

Left Side

See Figure 120.

1. Disconnect the negative battery cable.
2. Remove the resonator assembly.
3. Remove the spark plug wires from the secondary spark plugs.
4. Disconnect the left side breather tube and filter assembly.
5. Remove valve cover mounting bolts
6. Remove valve cover.

To install:

**** WARNING**

Do not use harsh cleaners to clean the cylinder head covers. Severe damage to covers may occur.

➡The gasket may be used again, provided no cuts, tears, or deformation has occurred.

7. Clean cylinder head cover and both sealing surfaces. Inspect and replace gasket as necessary.
8. Install cylinder head cover and hand start all fasteners.
9. Tighten cylinder head cover bolts and double ended studs to 105 inch lbs. (12 Nm).
10. Install left breather tube and filter assembly.
11. Install spark plug wires.
12. Install resonator assembly.
13. Connect the negative battery cable.

VALVE LASH

ADJUSTMENT

All gasoline engines use hydraulic lifters. No maintenance or periodic adjustment is required.

ENGINE PERFORMANCE & EMISSION CONTROLS

ACCELERATOR PEDAL POSITION (APP) SENSOR

LOCATION

See Figure 121.

Refer to the accompanying illustration for Accelerator Pedal Position (APP) sensor location.

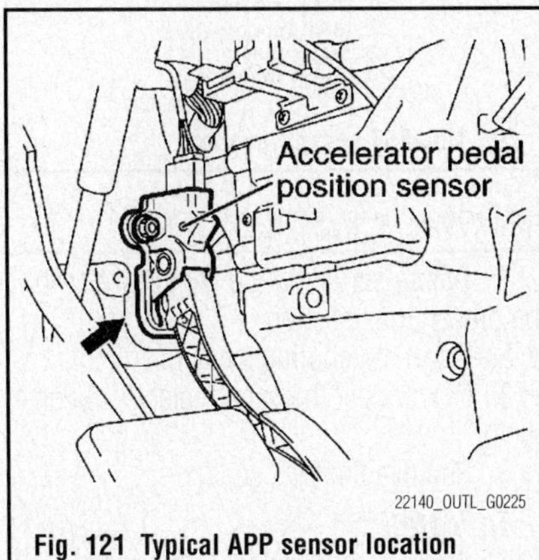

Fig. 121 Typical APP sensor location

REMOVAL & INSTALLATION

1. Before servicing the vehicle, refer to the precautions section.

⁂ CAUTION

The accelerator pedal position sensor should not be moved unnecessarily; it has been precisely adjusted by the manufacturer.

2. Before servicing the vehicle, refer to the precautions.
3. Remove the bottom cover assembly from under the steering wheel.
4. Detach the electrical connector at the sensor.
5. Remove the accelerator pedal assembly from the vehicle.

To install:

6. Install the accelerator pedal and tighten nuts to 111 inch lbs. (13 Nm)
7. Install the accelerator pedal assembly from the vehicle.
8. Connect the electrical connector at the sensor.
9. Install the bottom cover assembly from under the steering wheel.

CAMSHAFT POSITION (CMP) SENSOR

LOCATION

See Figures 122 and 123.

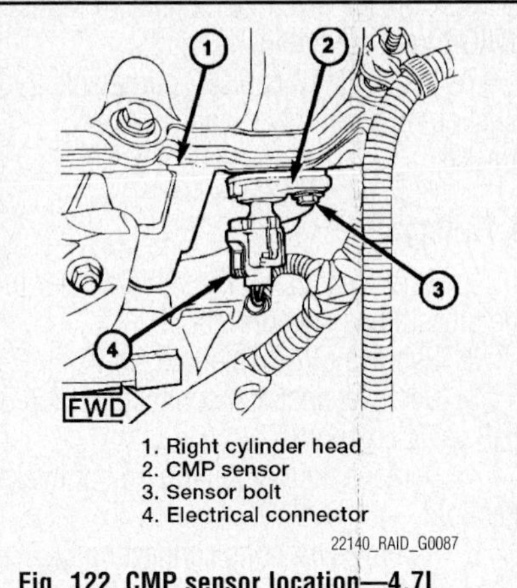

Fig. 122 CMP sensor location—4.7L engine

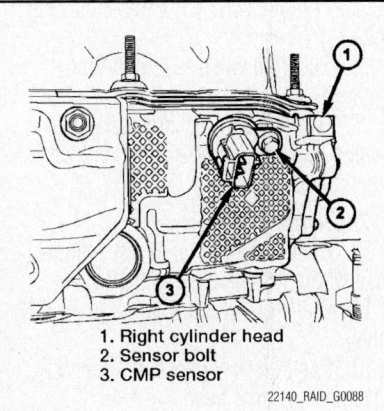

1. Right cylinder head
2. Sensor bolt
3. CMP sensor

22140_RAID_G0088

Fig. 123 CMP sensor location—3.7L engine

Refer to the accompanying illustration for Camshaft Position (CMP) sensor location.

REMOVAL & INSTALLATION

1. Before servicing the vehicle, refer to the precautions section.
2. Disconnect the CMP electrical connector.
3. Remove the CMP sensor mounting bolts.
4. Carefully twist the sensor from the cylinder.

To install:

5. Check the condition of the sensor O-ring.
6. Clean out the machined hole in the cylinder head.
7. Apply a small amount of clean engine oil to the sensor O-ring.
8. Install the CMP sensor into the cylinder head with a slight rocking and twisting action.
9. Install the mounting bolt and tighten to 106 inch lbs. (12 Nm).
10. Connect the electrical connector.

CRANKSHAFT POSITION (CKP) SENSOR

LOCATION

See Figure 124.

Refer to the accompanying illustration for Crankshaft Position (CKP) sensor location.

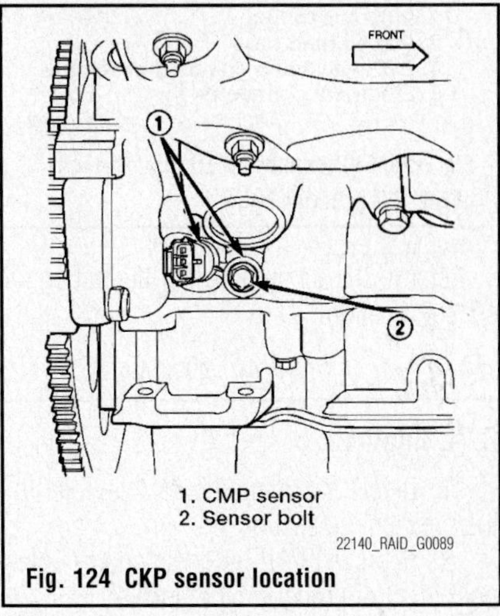

Fig. 124 CKP sensor location

REMOVAL & INSTALLATION

1. Before servicing the vehicle, refer to the precautions section.
2. Raise and safely support the vehicle.
3. Disconnect the sensor electrical connector.
4. Remove the CKP sensor mounting bolt.
5. Carefully twist the CKP sensor from the cylinder block.

To install:

6. Check the condition of the O-ring.
7. Clean out the machined hole in the engine block.
8. Apply a small amount of clean engine oil to the sensor O-ring.
9. Install the CKP sensor into the engine block with a slight rocking and twisting action.
10. Install the mounting bolt and tighten to 21 ft. lbs. (28 Nm).
11. Connect the electrical connector.
12. Lower the vehicle.

EGR VALVE POSITION (EVP) SENSOR

LOCATION

See Figure 125.

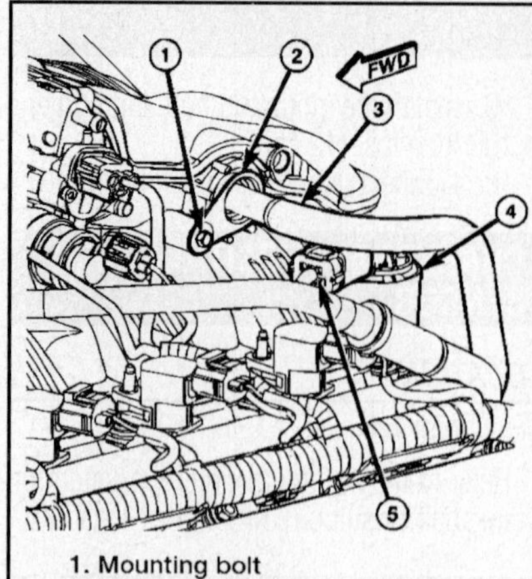

Fig. 125 EVP sensor location at the rear of the left cylinder head

Refer to the accompanying illustration for sensor location.

REMOVAL & INSTALLATION

3.7L Engine

1. Before servicing the vehicle, refer to the precautions section.
2. Remove EVAP purge solenoid. Solenoid lifts from tongue-type bracket.
3. Remove two tube mounting bolts.
4. Remove tube from solenoid. Slip opposite end of tube from intake manifold.
5. Remove gasket located between EGR valve solenoid and tube flange.
6. Disconnect electrical connector at solenoid.
7. Remove two EGR valve solenoid mounting bolts.
8. Remove solenoid from engine.
9. Remove and discard gasket located under EGR solenoid.

To install:

10. Clean gasket area at rear of left cylinder head where it joins base of EGR valve.
11. Clean EGR tube where it joins EGR valve.
12. Position new gasket between EGR valve and cylinder head.
13. Position EGR valve to cylinder head. Install and tighten two bolts. Torque to 80 inch lbs. (9 Nm).
14. Position new gasket between EGR tube flange and EGR valve assembly.
15. Position EGR tube to side of EGR valve. Position end of tube into intake manifold. Install two bolts. Torque to 8.5 ft. lbs. (11 Nm).
16. Connect electrical connector to top of EGR valve solenoid.
17. Install EVAP canister purge solenoid by slipping assembly to tongue-type bracket.

4.7L Engine

1. Before servicing the vehicle, refer to the precautions section.
2. Remove EVAP solenoid.
3. Remove electrical connector at top of EGR valve solenoid.
4. Remove tube mounting bolt at intake manifold.
5. Remove two bolts connecting EGR tube to valve assembly.
6. Remove gasket located between EGR tube flange and EGR valve assembly.
7. Remove two EGR valve mounting bolts.
8. Separate valve assembly from engine.
9. Remove and discard metal gasket located between cylinder head and valve assembly.

To install:

10. Clean area at rear of left cylinder head where it joins base of EGR valve.
11. Clean EGR tube where it joins EGR valve.
12. Position new gasket between EGR valve and cylinder head.
13. Position EGR valve to cylinder head. Install and tighten two bolts. Torque to 80 inch lbs. (9 Nm).
14. Position new gasket between EGR tube flange and EGR valve assembly.
15. Position EGR tube to side of EGR valve and into intake manifold. Install two bolts finger tight (temporarily).
16. Install EGR tube flange bolt at intake manifold. Torque to 8.5 ft. lbs. (11 Nm).
17. Connect electrical connector to top of EGR valve solenoid.
18. Do a final tightening of two EGR tube bolts. Torque to 8.5 ft. lbs. (11 Nm).
19. Install EVAP solenoid.

ENGINE COOLANT TEMPERATURE (ECT) SENSOR

LOCATION

See Figure 126.

Refer to the accompanying illustration for Engine Coolant Temperature (ECT) sensor location.

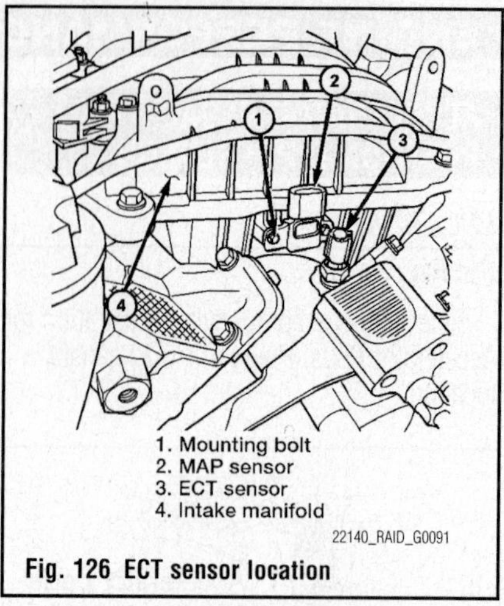

Fig. 126 ECT sensor location

REMOVAL & INSTALLATION

1. Before servicing the vehicle, refer to the precautions section.
2. Drain the cooling system.
3. Disconnect the sensor electrical connector.
4. Remove the ECT sensor.

To install:

5. Apply thread sealant to the sensor threads.
6. Install the ECT sensor into the engine block and tighten the mounting bolt to 8 ft. lbs. (11 Nm).
7. Connect the electrical connector.
8. Refill the cooling system to the correct level.

FUEL LEVEL SENDING UNIT

LOCATION

See Figure 127.

Refer to the accompanying illustration for fuel level sending unit location.

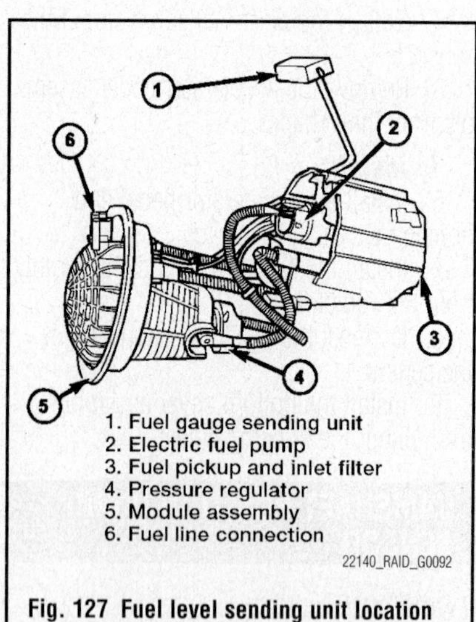

Fig. 127 Fuel level sending unit location

REMOVAL & INSTALLATION

See Figure 128.

1. Before servicing the vehicle, refer to the precautions section.
2. Release fuel system pressure.
3. Drain and remove fuel tank.

➡Note rotational position of module before attempting removal. An indexing arrow is located on top of module for this purpose.

4. Position Lock Ring Remover/Installer 9340 into notches on outside edge of lock ring.
5. Install ½ inch drive breaker bar to Lock Ring Remover/Installer 9340.
6. Rotate breaker bar counter-clockwise to remove lock ring.
7. Remove lock ring. The module will spring up slightly when lock ring is removed.
8. Remove module from fuel tank. Be careful not to bend float arm while removing.

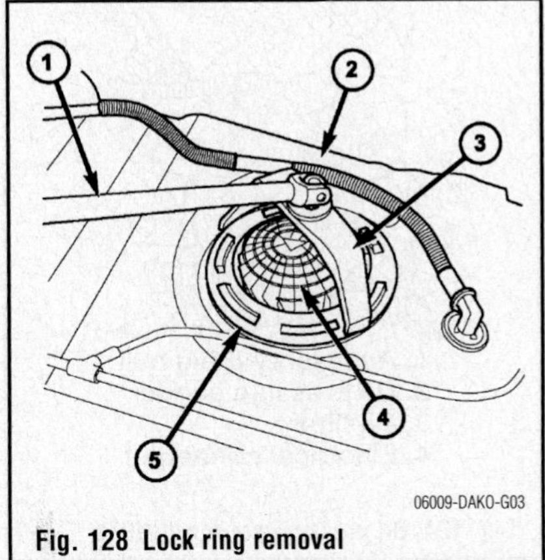

Fig. 128 Lock ring removal

To install:

➡Whenever the fuel pump module is serviced, the rubber seal (gasket) must be replaced.

9. Using a new seal (gasket), position fuel pump module into opening in fuel tank.
10. Position lock ring over top of fuel pump module.
11. Rotate module until embossed alignment arrow points to center alignment mark. This step must be performed to prevent float from contacting side of fuel tank. Also be sure fuel fitting on top of pump module is pointed to drivers side of vehicle.
12. Install Lock Ring Remover/Installer 9340 to lock ring.
13. Install ½ inch drive breaker into Lock Ring Remover/Installer 9340.
14. Tighten lock ring (clockwise) until all seven notches have engaged.
15. Install fuel tank.
16. Fill fuel tank with fuel.
17. Start engine and check for fuel leaks near top of module.

HEATED OXYGEN (HO2S) SENSOR

LOCATION

See Figure 129.

If equipped with a Federal Emission Package, two sensors are used: upstream (referred to as 1/1) and downstream (referred to as 1/2). With this emission package, the upstream sensor (1/1) is located just before the main catalytic converter. The downstream sensor (1/2) is located just after the main catalytic converter.

If equipped with a California Emission Package, 4 sensors are used: 2 upstream (referred to as 1/1 and 2/1) and 2 downstream (referred to as 1/2 and 2/2). With this emission package, the right upstream sensor (2/1) is located in the right exhaust downpipe just before the mini-catalytic converter. The left upstream sensor (1/1) is located in the left exhaust downpipe just before the mini-catalytic converter. The right downstream sensor (2/2) is located in the right exhaust downpipe just after the mini-catalytic converter, and before the main catalytic converter. The left downstream sensor (1/2) is located in the left exhaust downpipe just after the mini-catalytic converter, and before the main catalytic converter.

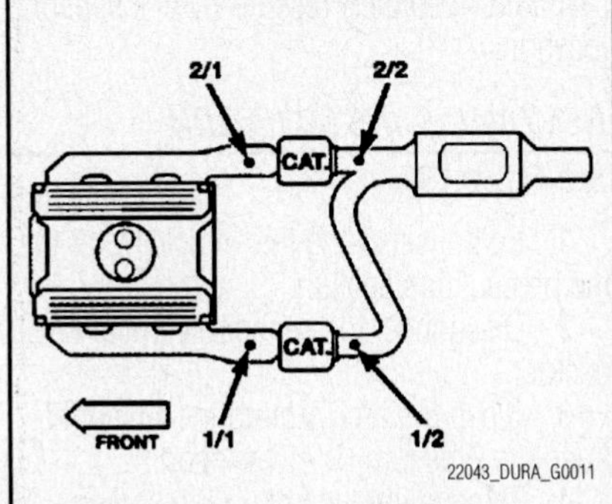

Fig. 129 Common oxygen sensor mounting points

REMOVAL & INSTALLATION

1. Before servicing the vehicle, refer to the precautions section.
2. Raise and safely support the vehicle.
3. Disconnect the wire connector from oxygen sensor.

⁂ WARNING

When disconnecting sensor electrical connector, do not pull directly on wire going into sensor.

4. Remove the sensor with an oxygen sensor removal and installation tool.
5. Clean threads in exhaust pipe using appropriate tap.

To install:

➡Threads of new oxygen sensors are factory coated with anti-seize compound.

⁂ WARNING

Do not add any additional anti-seize compound to the threads of a new oxygen sensor.

6. Install the oxygen sensor and tighten to 22 ft. lbs. (30 Nm).
7. Connect the electrical connector.
8. Lower the vehicle.

INTAKE AIR TEMPERATURE (IAT) SENSOR

LOCATION

See Figure 130.

Refer to the accompanying illustration for Intake Air Temperature (IAT) sensor location.

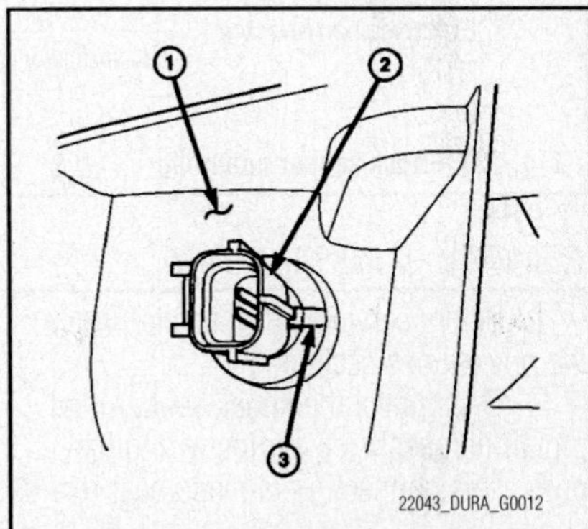

Fig. 130 The IAT sensor (2) is located in the air inlet (1). Lift the tab (3) to remove it

REMOVAL & INSTALLATION

1. Before servicing the vehicle, refer to the precautions section.
2. Disconnect the electrical connector form the Intake Air Temperature (IAT) sensor.
3. Clean any dirt from the air inlet tube at the sensor base.
4. Gently lift the small plastic release tab and rotate the sensor about ¼turn counterclockwise to remove.

To install:

5. Check the condition of the sensor O-ring.
6. Clean the sensor mounting hole.
7. Position the sensor into the intake air tube and rotate clockwise until the release tab clicks into place.
8. Install the electrical connector.

KNOCK SENSOR (KS)

LOCATION

See Figure 131.

Refer to the accompanying illustration for Knock Sensor (KS) location.

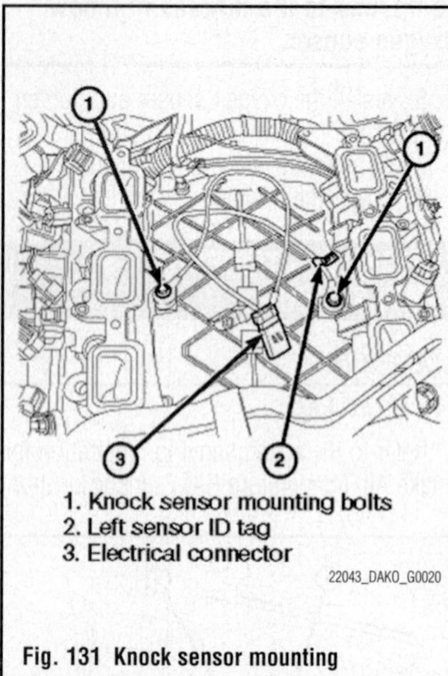

Fig. 131 Knock sensor mounting

REMOVAL & INSTALLATION

1. Before servicing the vehicle, refer to the precautions section.
2. Disconnect the knock sensor dual pigtail harness from engine wiring harness. This connection is made near rear of engine.
3. Remove the intake manifold
4. Remove the Knock Sensor (KS) mounting bolts.
5. Remove the sensors from engine.

To install:

6. Thoroughly clean the KS mounting holes.
7. Install the sensors into the engine block. Tighten the mounting bolts to 15 ft. lbs. (20 Nm).
8. Install the intake manifold.
9. Connect the KS wiring harness to the engine wiring harness at the rear of the engine.

MALFUNCTION INDICATOR LIGHT (MIL)

RESET PROCEDURES

1. Connect an OBD II scan tool to the diagnostic connector.
2. Clear DTCs.
3. The MIL should turn **OFF**.

MANIFOLD ABSOLUTE PRESSURE (MAP) SENSOR

LOCATION

See Figure 132.

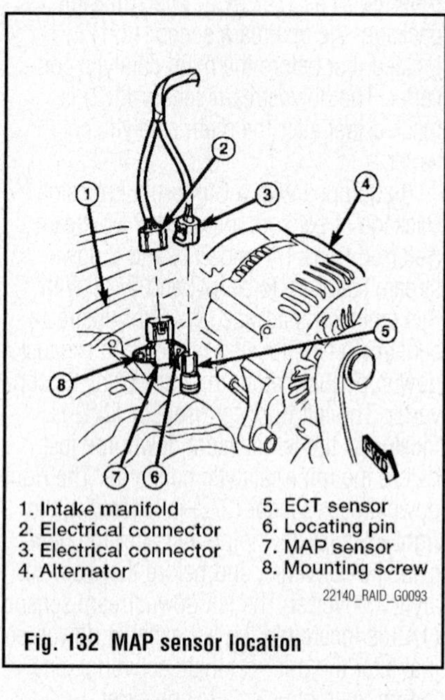

Fig. 132 MAP sensor location

Refer to the accompanying illustration for Manifold Absolute Pressure (MAP) sensor location.

REMOVAL & INSTALLATION

See Figure 133.

1. Before servicing the vehicle, refer to the precautions section.
2. Disconnect the sensor electrical connector.
3. Clean the area around the Manifold Absolute Pressure (MAP) sensor.
4. Remove the two mounting screws.
5. Remove the MAP sensor from the intake manifold.

To install:

6. Inspect the condition of the sensor O-ring and replace if necessary.
7. Position the MAP sensor into the manifold and install the two mounting screws.
8. Connect the electrical connector.

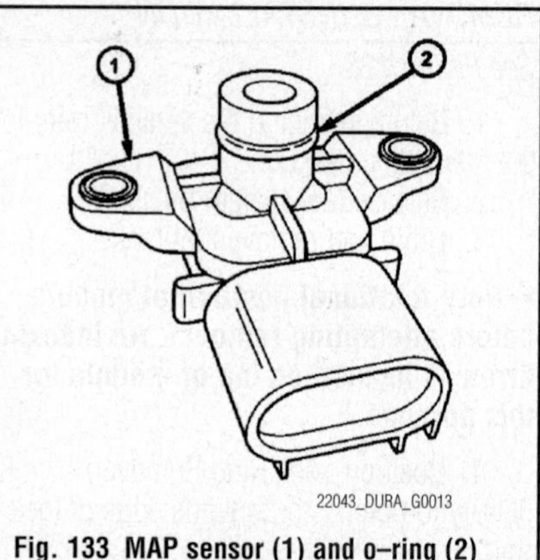

Fig. 133 MAP sensor (1) and o-ring (2)

OIL PRESSURE SENSOR

LOCATION

See Figure 134.

Refer to the accompanying illustration for oil pressure sensor location.

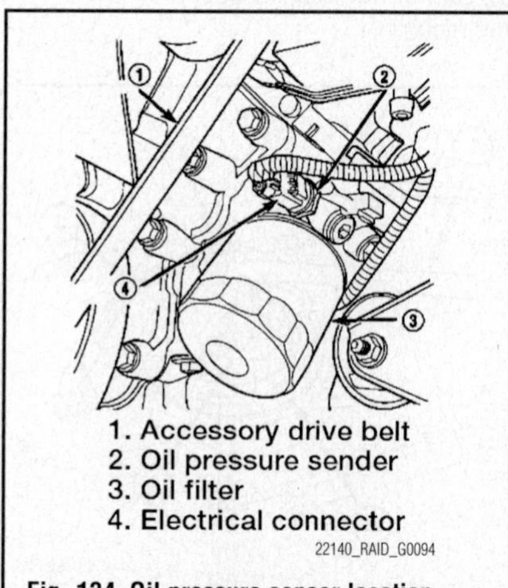

Fig. 134 Oil pressure sensor location

REMOVAL & INSTALLATION

1. Before servicing the vehicle, refer to the precautions.
2. Disconnect the negative battery cable.
3. Raise and safely support the vehicle securely on jackstands.
4. Remove the front splash shield.
5. Disconnect the oil pressure sensor connector.
6. Remove the oil pressure sensor.

To install:

7. Install the oil pressure sensor.
8. Connect the oil pressure sensor connector.
9. Install the front splash shield.
10. Lower the vehicle.
11. Connect the negative battery cable.

POWERTRAIN CONTROL MODULE (PCM)

LOCATION

See Figure 135.

Refer to the accompanying illustration for Powertrain Control Module (PCM) location.

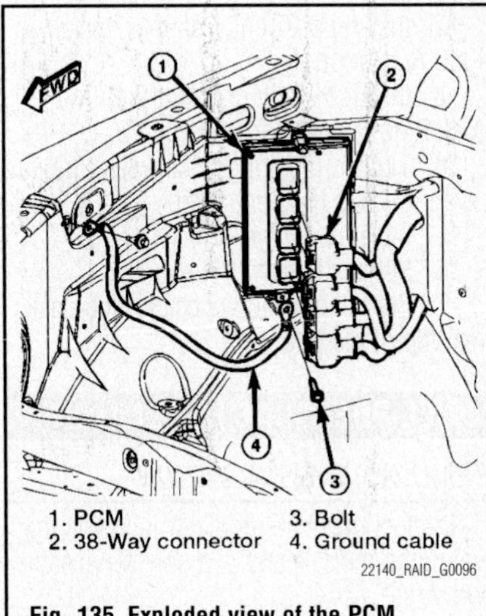

Fig. 135 Exploded view of the PCM mounting

REMOVAL & INSTALLATION

1. Before servicing the vehicle, refer to the precautions section.
2. Disconnect the negative battery cable.
3. Unplug the 38-way connectors from the PCM.

➡A locating pin is used in place of one of the mounting bolts.

4. Pry the clip from the locating pin.
5. Remove the two remaining mounting bolts.
6. Remove the PCM from the vehicle.

To install:

7. Position the PCM to the body and install the two mounting bolts.

➡Position the ground strap in place before tightening the mounting bolts.

8. Install the clip to the locating pin.
9. Tighten the mounting bolts to 35 inch lbs. (4 Nm).
10. Carefully plug in the 38-way connectors to the PCM.
11. Connect the negative battery cable.

THROTTLE POSITION SENSOR (TPS)

LOCATION

See Figure 136.

Refer to the accompanying illustration for Throttle Position Sensor (TPS) location.

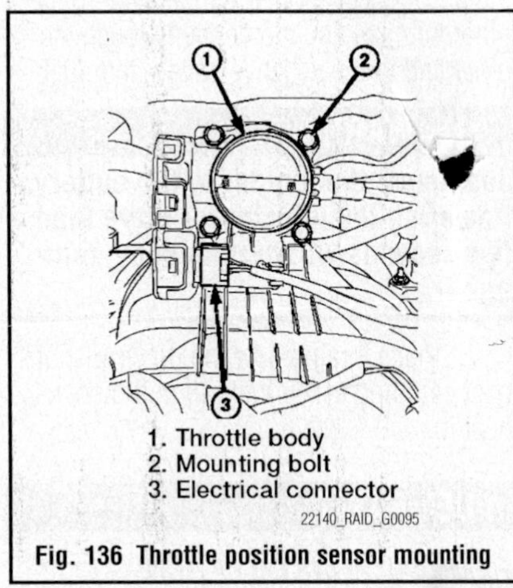

Fig. 136 Throttle position sensor mounting

REMOVAL & INSTALLATION

⁂ WARNING

Never have the ignition key in the ON position when/if checking the throttle body shaft for a binding condition. This may set DTC's.

➡A (factory adjusted) set screw is used to mechanically limit the position of the throttle body throttle plate. Never attempt to adjust the engine idle speed using this screw. All idle speed functions are controlled by the Powertrain Control Module (PCM).

1. Before servicing the vehicle, refer to the precautions.
2. Disconnect the negative battery cable.
3. Remove air intake tube at throttle body flange.
4. Disconnect throttle body electrical connector.
5. Disconnect necessary vacuum lines at throttle body.
6. Remove four throttle body mounting bolts.
7. Remove throttle body from intake manifold.
8. Check condition of old throttle body-to-intake manifold o-ring.

To install:

9. Check condition of throttle body-to-intake manifold O-ring. Replace as necessary.
10. Clean mating surfaces of throttle body and intake manifold.
11. Install O-ring between throttle body and intake manifold.
12. Position throttle body (1) to intake manifold.
13. Install all throttle body mounting bolts (2) finger tight.

⁂ WARNING

The throttle body mounting bolts MUST be torqued to specifications. DO NOT OVER TIGHTEN MOUNTING BOLTS. Over tightening can cause damage to the throttle body, throttle plate, gaskets, bolts and/or the intake manifold. Proper torque of the mounting bolts is critical to normal operation.

14. Tighten mounting bolts in a criss-cross pattern sequence to 65 inch lbs. (7.5 Nm).
15. Install electrical connector.
16. Install necessary vacuum lines.
17. Install air cleaner duct at throttle body.
18. Connect the negative battery cable.
19. Using the diagnostic scan tool, erase all previous DTC's and perform the ETC Relearn function.

VEHICLE SPEED SENSOR (VSS)

LOCATION

The Vehicle Speed Sensor (VSS) is located on the left side of the transmission case.

REMOVAL & INSTALLATION

1. Before servicing the vehicle, refer to the precautions section.
2. Raise and safely support the vehicle.
3. Place a suitable catch pan under the transmission for any fluid.
4. Remove the wiring connector from the output speed sensor.
5. Remove the mounting bolt and remove the speed sensor from the transmission case.

To install:

6. Install the speed sensor into the transmission case and tighten the bolt to 105 inch lbs. (12 Nm).
7. Install the wiring connector to the speed sensor.
8. Verify the proper transmission fluid level and refill as necessary.
9. Lower the vehicle.

FUEL SYSTEM SERVICE PRECAUTIONS

Safety is the most important factor when performing not only fuel system maintenance but any type of maintenance. Failure to conduct maintenance and repairs in a safe manner may result in serious personal injury or death. Maintenance and testing of the vehicle's fuel system components can be accomplished safely and effectively by adhering to the following rules and guidelines.

- To avoid the possibility of fire and personal injury, always disconnect the negative battery cable unless the repair or test procedure requires that battery voltage be applied.
- Always relieve the fuel system pressure prior to disconnecting any fuel system component (injector, fuel rail, pressure regulator, etc.), fitting or fuel line connection. Exercise extreme caution whenever relieving fuel system pressure to avoid exposing skin, face and eyes to fuel spray. Please be advised that fuel under pressure may penetrate the skin or any part of the body that it contacts.
- Always place a shop towel or cloth around the fitting or connection prior to loosening to absorb any excess fuel due to spillage. Ensure that all fuel spillage (should it occur) is quickly removed from engine surfaces. Ensure that all fuel soaked cloths or towels are deposited into a suitable waste container.
- Always keep a dry chemical (Class B) fire extinguisher near the work area.
- Do not allow fuel spray or fuel vapors to come into contact with a spark or open flame.
- Always use a back-up wrench when loosening and tightening fuel line connection fittings. This will prevent unnecessary stress and torsion to fuel line piping.
- Always replace worn fuel fitting O-rings with new. Do not substitute fuel hose or equivalent where fuel pipe is installed.

Before servicing the vehicle, make sure to also refer to the precautions in the beginning of this section as well.

RELIEVING FUEL SYSTEM PRESSURE

1. Before servicing the vehicle, refer to the precautions.
2. Disconnect the negative battery cable.
3. Remove the fuel tank filler cap to release any fuel tank pressure.
4. Remove the fuel pump relay from the PDC.
5. Start and run the engine until it stops.
6. Unplug the connector from any injector and connect a jumper wire from either injector terminal to the positive battery terminal. Connect another jumper wire to the other terminal and momentarily touch the other end to the negative battery terminal.

WARNING

Just touch the jumper to the battery. Powering the injector for more than a few seconds will permanently damage it.

7. Place a rag below the quick-disconnect coupling at the fuel rail and disconnect it.

FUEL FILTER

REMOVAL & INSTALLATION

The fuel filter mounts inside the fuel pump module and is a non-serviceable part.

FUEL PUMP

REMOVAL & INSTALLATION

1. Before servicing the vehicle, refer to the precautions section.
2. Release fuel system pressure.
3. Drain and remove fuel tank.

➡Note rotational position of module before attempting removal. An indexing arrow is located on top of module for this purpose.

4. Position Lock Ring Remover/Installer 9340 into notches on outside edge of lock ring.
5. Install ½ inch drive breaker bar to Lock Ring Remover/Installer 9340.
6. Rotate breaker bar counter-clockwise to remove lock ring.
7. Remove lock ring. The module will spring up slightly when lock ring is removed.
8. Remove module from fuel tank. Be careful not to bend float arm while removing.

To install:

➡Whenever the fuel pump module is serviced, the rubber seal (gasket) must be replaced.

9. Using a new seal (gasket), position fuel pump module into opening in fuel tank.
10. Position lock ring over top of fuel pump module.
11. Rotate module until embossed alignment arrow points to center alignment mark. This step must be performed to prevent float from contacting side of fuel tank. Also be sure fuel fitting on top of pump module is pointed to drivers side of vehicle.
12. Install Lock Ring Remover/Installer 9340 to lock ring.
13. Install ½ inch drive breaker into Lock Ring Remover/Installer 9340.
14. Tighten lock ring (clockwise) until all seven notches have engaged.
15. Install fuel tank.
16. Fill fuel tank with fuel.
17. Start engine and check for fuel leaks near top of module.

FUEL PRESSURE REGULATOR

REMOVAL & INSTALLATION

The fuel pressure regulator is inside the fuel pump module and is a non-serviceable part.

FUEL TANK

REMOVAL & INSTALLATION

See Figure 137.

1. Before servicing the vehicle, refer to the precautions section.
2. Release fuel system pressure.
3. Drain fuel tank.
4. Disconnect vent line from tank.
5. Remove clamp and disconnect fill hose at fuel fill tube.
6. If equipped, remove fuel tank skid plate.
7. Disconnect electrical connector at ESIM switch.
8. Disconnect ESIM, ORVR and EVAP lines at front of tank.
9. Support tank with a hydraulic jack.
10. Remove two fuel tank strap nuts and remove both tank support straps.
11. Carefully lower tank a few inches and disconnect fuel pump module electrical connector at top of tank. To disconnect electrical connector: Push upward on red colored tab to unlock. Push on black colored tab while removing connector.
12. Disconnect fuel line at fuel pump module fitting by pressing on tabs at side of quick-connect fitting.
13. Continue to lower tank for removal.
14. If fuel tank is to be replaced, remove fuel pump module from tank.

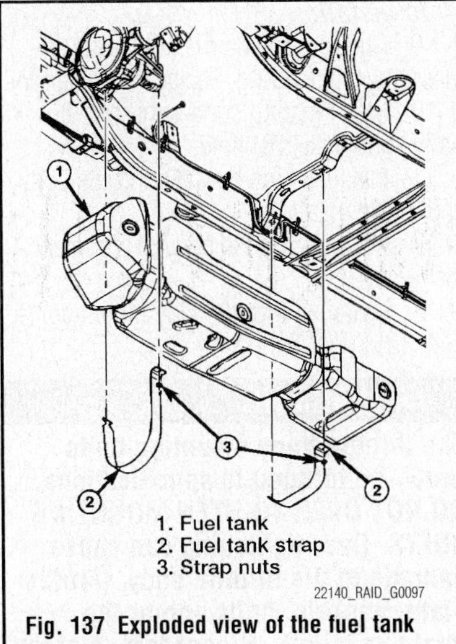

Fig. 137 Exploded view of the fuel tank assembly

To install:

15. If fuel tank is to be replaced, install fuel pump module into tank.
16. Position fuel tank to hydraulic jack.
17. Raise tank until positioned near body.
18. Connect fuel pump module electrical connector at top of tank.
19. Connect fuel line quick-connect fitting to pump module.
20. Continue raising tank until positioned snug to body.
21. Install and position both tank support straps. Install two fuel tank strap nuts and tighten.
22. Connect EVAP, ORVR and ESIM lines at front of tank.
23. Connect electrical connector to ESIM switch.
24. Connect vent line.
25. Connect rubber fill hose to fuel tank fitting and tighten hose clamps.
26. The vapor/vacuum lines and hoses must be firmly connected. Also check the vapor/vacuum lines at the ESIM switch, filter and EVAP canister purge solenoid for damage or leaks. If a leak is present, a Diagnostic Trouble Code (DTC) may be set.
27. If equipped, install fuel tank skid plate.
28. Install plastic liner in front of left-rear tire/wheel.
29. Install left-rear tire/wheel.
30. Lower vehicle.
31. Fill fuel tank with fuel.
32. Start engine and check for fuel leaks near top of module.

FUEL RAIL & INJECTORS

REMOVAL & INSTALLATION

See Figures 138 through 140.

1. Before servicing the vehicle, refer to the precautions.
2. Relieve the fuel system pressure.

**** CAUTION**

The fuel system is under constant pressure even with engine off. Before servicing fuel rail, fuel system pressure must be released.

**** CAUTION**

The left and right fuel rails are replaced as an assembly. Do not attempt to separate rail halves at connector tube. Due to design of tube, it does not use any clamps. Never attempt to install a clamping device of any kind to tube. When removing fuel rail assembly for any reason, be careful not to bend or kink tube.

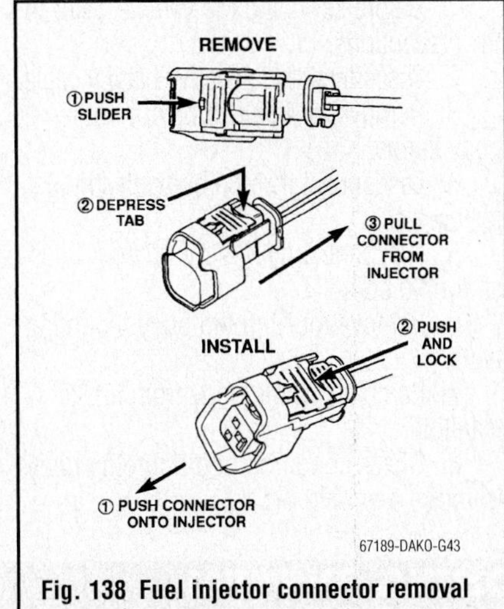

Fig. 138 Fuel injector connector removal

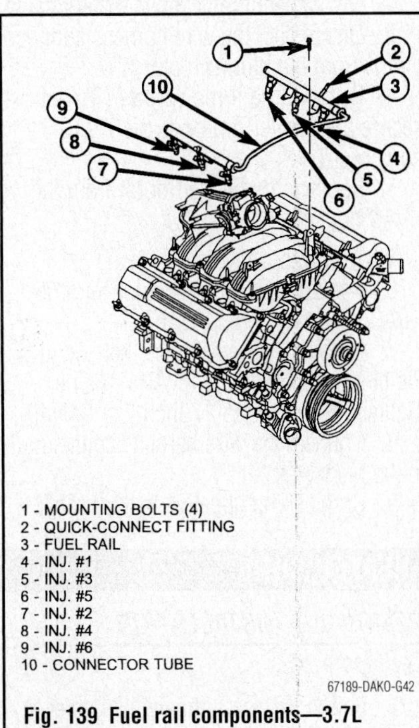

Fig. 139 Fuel rail components—3.7L engine

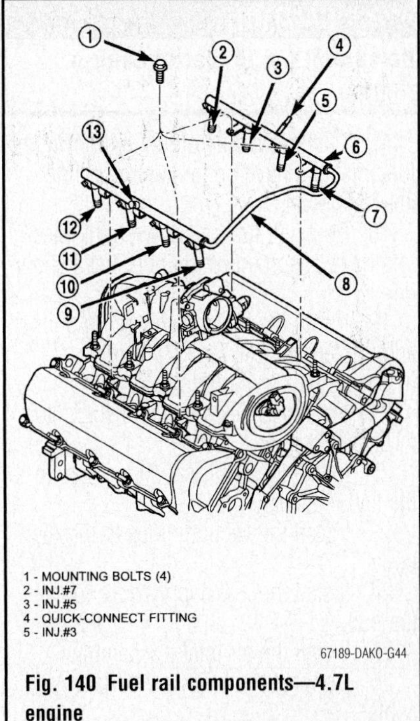

Fig. 140 Fuel rail components—4.7L engine

3. Remove the fuel tank filler tube cap.
4. Remove the negative battery cable.
5. Remove the air intake assembly.
6. Remove the fuel line latch clip and fuel line at fuel rail.
7. Remove the vacuum lines at throttle body.
8. Remove the fuel injector electrical connectors.
9. Remove the electrical connectors at throttle body sensors.
10. Remove the ignition coils.
11. Remove the fuel rail mounting bolts.
12. Remove the fuel rail with injectors attached.
13. Disconnect the clip that retains the fuel injector to fuel rail to remove the injector.

To install:

➡Apply a small amount of clean engine oil to each fuel injector o-ring. This will help in fuel rail installation.

14. Install the injectors and injector clips to fuel rail.
15. Position fuel rail/fuel injector assembly to machined injector openings in cylinder head.
16. Guide each injector into cylinder head.

⁂ CAUTION

Be careful not to tear injector o-rings.

17. Push right side of fuel rail down until fuel injectors have bottomed on cylinder head shoulder.
18. Push left fuel rail down until injectors have bottomed on cylinder head shoulder.
19. Install the fuel rail mounting bolts and tighten to 20 ft. lbs. (27 Nm).
20. Install the ignition coils.
21. Install the electrical connectors to throttle body.
22. Install the electrical connectors to the fuel injectors.
23. Install the vacuum lines to throttle body.
24. Install the fuel supply hose to the fuel rail.
25. Install the air intake assembly.
26. Connect the negative battery cable.
27. Start engine and check for leaks.

IDLE SPEED

ADJUSTMENT

Idle speed is maintained by the Powertrain Control Module (PCM). No adjustment is necessary or possible.

THROTTLE BODY

REMOVAL & INSTALLATION

⁂ WARNING

Never have the ignition key in the ON position when/if checking the throttle body shaft for a binding condition. This may set DTC's.

➡A (factory adjusted) set screw is used to mechanically limit the position of the throttle body throttle plate. Never attempt to adjust the engine idle speed using this screw. All idle speed functions are controlled by the Powertrain Control Module (PCM).

1. Before servicing the vehicle, refer to the precautions.
2. Disconnect the negative battery cable.
3. Remove air intake tube at throttle body flange.
4. Disconnect throttle body electrical connector.
5. Disconnect necessary vacuum lines at throttle body.
6. Remove four throttle body mounting bolts.
7. Remove throttle body from intake manifold.
8. Check condition of old throttle body-to-intake manifold o-ring.

To install:

9. Check condition of throttle body-to-intake manifold O-ring. Replace as necessary.
10. Clean mating surfaces of throttle body and intake manifold.
11. Install O-ring between throttle body and intake manifold.
12. Position throttle body (1) to intake manifold.
13. Install all throttle body mounting bolts (2) finger tight.

⁂ WARNING

The throttle body mounting bolts MUST be torqued to specifications. DO NOT OVER TIGHTEN MOUNTING BOLTS. Over tightening can cause damage to the throttle body, throttle plate, gaskets, bolts and/or the intake manifold. Proper torque of the mounting bolts is critical to normal operation.

14. Tighten mounting bolts in a criss-cross pattern sequence to 65 inch lbs. (7.5 Nm).
15. Install electrical connector.
16. Install necessary vacuum lines.
17. Install air cleaner duct at throttle body.
18. Connect the negative battery cable.
19. Using the diagnostic scan tool, erase all previous DTC's and perform the ETC Relearn function.

HEATING & AIR CONDITIONING SYSTEM

BLOWER MOTOR

REMOVAL & INSTALLATION

See Figure 141.

1. Before servicing the vehicle, refer to the precautions.
2. Disconnect the negative battery cable.
3. Disconnect the wire harness connector (1) from the blower motor (2).
4. Remove the three screws (3) that secure the blower motor to the HVAC housing (4).
5. Remove the blower motor from the HVAC housing.

To install:

6. Position the blower motor into the HVAC housing.
7. Install the three screws that secure the blower motor to the HVAC housing. Tighten the screws to 17 inch lbs. (2 Nm).
8. Connect the wire harness connector to the blower motor.
9. Connect the negative battery cable.

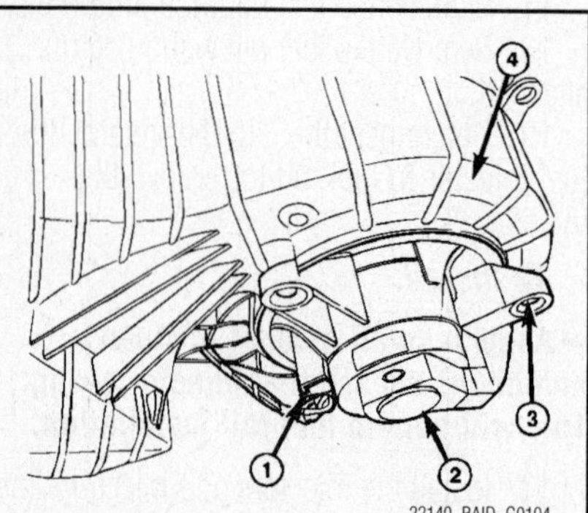

Fig. 141 Disconnect the wire harness connector (1) from the blower motor (2). Remove the three screws (3) that secure the blower motor to the HVAC housing (4)

HEATER CORE

REMOVAL & INSTALLATION

See Figure 142.

1. Before servicing the vehicle, refer to the precautions.
2. Remove the HVAC housing and place it on a workbench.

➡If the foam seal around the heater core is deformed or damaged, the seal must be replaced.

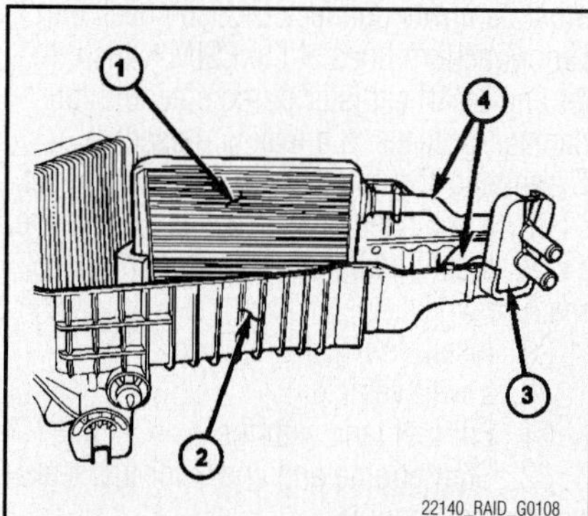

Fig. 142 Heater core assembly (1) inside the HVAC housing (2). If required, remove the rubber flange (3) from around the heater core tubes (4)

3. Carefully lift the heater core out of the lower half of the HVAC housing.

➡If the rubber flange around the heater core tubes is deformed or damaged, the flange must be replaced.

4. If required, remove the rubber flange from around the heater core tubes.

To install:

5. If removed, install the rubber flange over the heater core tubes.

➡Make sure that the foam seal and the rubber flange are properly positioned in the HVAC housing.

6. Carefully install the heater core into the lower half of the HVAC housing.
7. Assemble the HVAC housing.

➡If the heater core is being replaced, flush the cooling system.

8. Install the HVAC housing.

STEERING

POWER RACK & PINION STEERING GEAR

REMOVAL & INSTALLATION

See Figure 143.

1. Before servicing the vehicle, refer to the precautions.
2. Siphon out as much power steering fluid as possible from the pump.
3. Lock the steering wheel.
4. Remove the front wheels.
5. Remove the nuts from the tie rod ends.
6. Disconnect the tie rod ends from the knuckles.
7. Remove the steering gear pinch bolt.
8. Remove the lower steering coupling from the steering gear.
9. Turn the steering gear to the full right position.

➡Protect the end of hoses to prevent contamination to the system and damage to the O-rings.

10. Remove the power steering lines from the gear.
11. Remove the steering gear mounting bolts and nuts.
12. Tip the steering gear assembly forward to allow clearance and move to the right then tip the gear downward on the left side to remove from the vehicle.

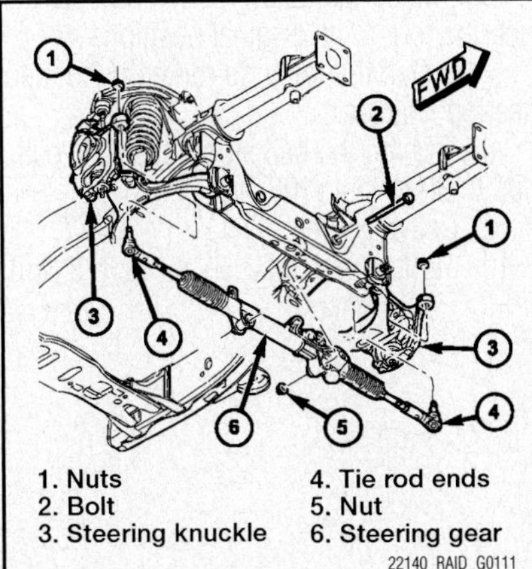

Fig. 143 Power steering gear assembly

To install:

➡Before installing gear inspect bushings and replace if worn or damaged.

13. Install the steering gear assembly to the vehicle and tighten mounting nuts and bolts to 190 ft. lbs. (258 Nm).
14. Install the power steering lines to the steering gear. Tighten the pressure hose to 23 ft. lbs. (31 Nm) and tighten the return hose to 27 ft. lbs. (37 Nm).
15. Slide the shaft coupler onto the steering gear. Install a new bolt and tighten to 42 ft. lbs. (57 Nm).
16. Clean the tie rod end studs and knuckle tapers.
17. Install the tie rod ends into the steering knuckles and tighten the nuts to 60 ft. lbs. (81 Nm).
18. Install the front wheels.
19. Refill the power steering system with fluid to the correct level.
20. Check the wheel alignment and adjust, as necessary.

POWER STEERING PUMP

REMOVAL & INSTALLATION

See Figure 144.

1. Before servicing the vehicle, refer to the precautions.

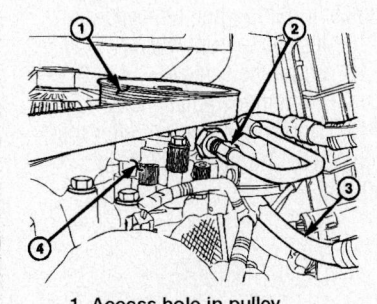

1. Access hole in pulley
2. Pressure hose connection
3. Suction hose connection
4. Power steering pump

22140_RAID_G0112

Fig. 144 Power steering pump connections

2. Drain and siphon the power steering fluid from the pump.
3. Remove the serpentine drive belt.
4. Remove the reservoir return hose at the reservoir.
5. Remove the pressure hose from the pump.
6. Remove 3 pump mounting bolts through pulley access holes.
7. Remove the pump from the engine.

To install:

8. Align the pump with the mounting holes on the engine.
9. Install 3 pump mounting bolts through the pulley access holes. Tighten the bolts to 21 ft. lbs. (28 Nm).
10. Install the pressure hose to the pump. Tighten the tube nut to 23 ft. lbs. (31 Nm).
11. Install reservoir return hose to the reservoir.
12. Install the serpentine drive belt.
13. Fill the power steering pump.

BLEEDING

⁂ WARNING

Mitsubishi Power Steering Fluid + 4 or Mitsubishi ATF+4 Automatic Transmission Fluid is to be used in the power steering system. Both Fluids have the same material standard specifications (MS-9602). No other power steering or automatic transmission fluid is to be used in the system. Damage may result to the power steering pump and system if another fluid is used. Do not overfill the system.

⁂ WARNING

If the air is not purged from the power steering system correctly, pump failure could result.

➡Be sure the vacuum tool used in the following procedure is clean and free of any fluids.

1. Check the fluid level. As measured on the side of the reservoir, the level should

indicate between MAX and MIN when the fluid is at normal ambient temperature. Adjust the fluid level as necessary.

2. Tightly insert Power Steering Cap Adapter 9688, into the mouth of the reservoir.

WARNING

Failure to use a the vacuum pump reservoir (1) may allow power steering fluid to be sucked into the hand vacuum pump.

3. Attach Hand Vacuum Pump C-4207 or equivalent, with reservoir attached, to the Power Steering Cap Adapter.

WARNING

Do not run the engine while vacuum is applied to the power steering system. Damage to the power steering pump can occur.

When performing the following step make sure the vacuum level is maintained during the entire time period.

4. Using Hand Vacuum Pump, apply 20-25 in. Hg (68-85 kPa) of vacuum to the system for a minimum of three minutes.

5. Slowly release the vacuum and remove the special tools.

6. Adjust the fluid level as necessary.

7. Repeat steps until the fluid no longer drops when vacuum is applied.

8. Start the engine and cycle the steering wheel lock-to-lock three times.

WARNING

Do not hold the steering wheel at the stops.

9. Stop the engine and check for leaks at all connections.

10. Check for any signs of air in the reservoir and check the fluid level. If air is present, repeat the procedure as necessary.

STEERING LINKAGE

REMOVAL & INSTALLATION

Inner Tie Rod

See Figure 145.

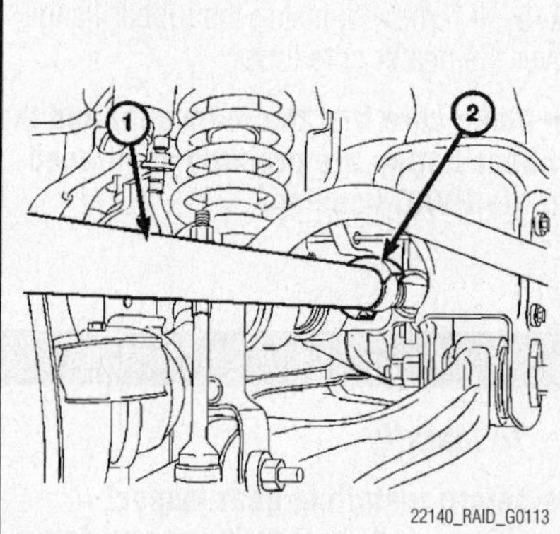

Fig. 145 Remove/install the inner tie rod end using special tool No. 10087-3 socket (1) and No. 10087-2 tube (2)

1. Before servicing the vehicle, refer to the precautions.

2. Remove the outer tie rod end with special tool C-3894-A or equivalent. Count the number of turns when removing, this will aid in setting the toe after installation.

3. Remove the jam nut.

4. Remove the clamps for the inner tie rod boot.

5. Remove the boot.

6. Install the special tool No. 10087-3 socket and No. 10087-2 Tube and remove the inner tie rod end.

To install:

7. Thread the inner tie rod to the rack & pinion.

8. Tighten the inner tie rod end using special tool No. 10087-3 socket and No. 10087-2 tube to 125 ft. lbs. (169 Nm).

9. Install the inner tie rod end boot and clamp to the tie rod and rack.

10. Install the jam nut to the inner tie rod end.

11. Install the outer tie rod the same number of turns when removed.

12. Install the front wheels.

13. Lower the vehicle.

14. Perform toe adjustment.

15. Tighten the jam nut after toe adjustment. Tighten to 37 ft. lbs. (50 Nm).

Outer Tie Rod

See Figure 146.

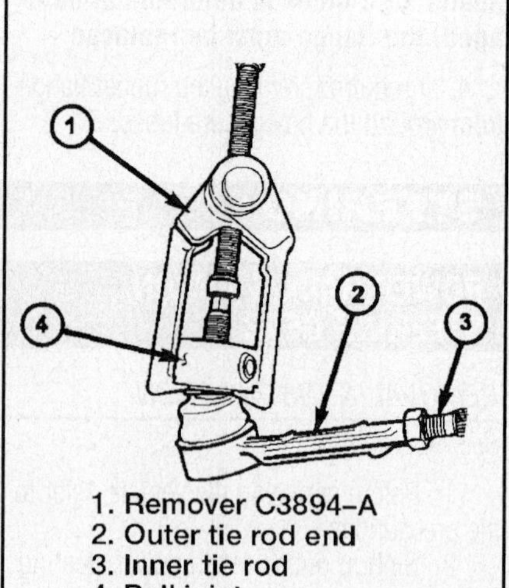

Fig. 146 Separate the outer tie rod end from the knuckle with Remover C3894-A

WARNING

Do not twist the boot anytime during removal or installation.

1. Before servicing the vehicle, refer to the precautions.

2. Loosen the jam nut.

3. Remove the outer tie rod end nut from the ball stud.

4. Separate the outer tie rod end from the knuckle with Remover C3894-A.

5. Unthread the outer tie rod end from the inner tie rod. Count the number of turns when removing the tie rod end, This will give a good starting point when adjusting the toe.

To install:

6. Thread the outer tie rod end onto the inner tie rod, to its original position.

7. Install the outer tie rod end into the steering knuckle.

8. Tighten the ball stud nut on the ball stud to 80 ft. lbs. (108 Nm).

9. Set wheel toe.

10. Tighten jam nut to 55 ft. lbs. (75 Nm).

COIL SPRING

REMOVAL & INSTALLATION

See Figures 147 and 148.

1. Before servicing the vehicle, refer to the precautions.
2. Remove the shock assembly.
3. Install the shock assembly in the Branick 7200T spring removal/installation tool or equivalent press.
4. Compress the spring.
5. Position Special Tool 9362 Wrench on the shock shaft retaining nut. Insert an 8 mm socket though Wrench onto hex located on end of shock shaft. While holding shock shaft from turning, remove nut from shock shaft using Wrench.
6. Remove the upper shock nut.
7. Remove the shock.
8. Remove the shock upper mounting plate.

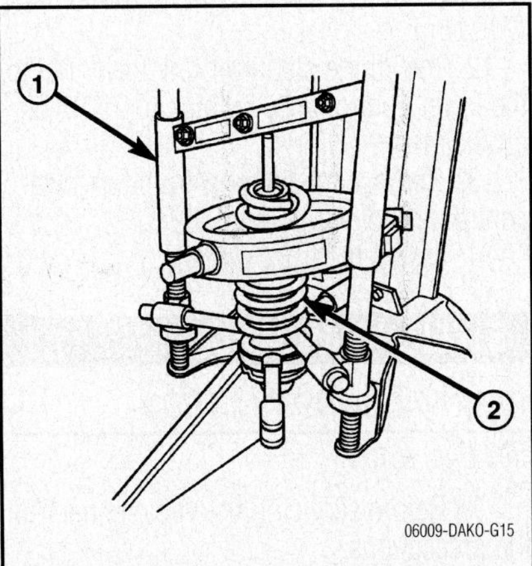

Fig. 147 Strut assembly (2) mounted in the compressor (1)

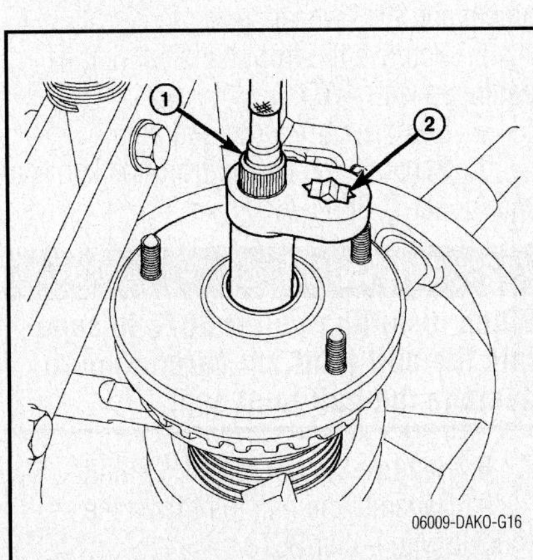

Fig. 148 Position Special Tool 9362 (2), or equivalent, on shock shaft retaining nut (1)

9. Remove and inspect the upper and lower spring isolators.

To install:

10. Install the lower isolator.
11. Install the upper isolator.
12. Position the shock into the coil spring.
13. Install the upper shock mounting plate.
14. Install Wrench (on end of a tighten wrench), Special Tool 9362, on shock shaft retaining nut. Next, insert 8 mm socket though Wrench onto hex located on end of shock shaft. While holding shock shaft from turning, tighten nut using Wrench to 33 ft. lbs. (45 Nm).
15. Install the shock upper mounting nut.
16. Decompress the spring.
17. Remove the shock assembly from the spring compressor tool.
18. Install the shock assembly.

LOWER BALL JOINT

REMOVAL & INSTALLATION

See Figure 149.

1. Before servicing the vehicle, refer to the precautions.
2. Remove the front wheel.
3. Remove the brake caliper and rotor.
4. Remove the outer tie rod retaining nut from the knuckle.
5. Separate the tie rod from the steering knuckle using Special Tool C-3894-A.
6. Remove the upper ball joint nut, then separate the upper ball joint from the knuckle using special tool 8677, or equivalent.
7. Remove the lower ball joint nut, then separate the lower ball joint from the steering knuckle using special tool 8677.
8. Remove the steering knuckle.
9. If equipped with 4WD, move the halfshaft to the side and support the halfshaft out of the way.
10. Remove the snapring from the ball joint flange.

➡Extreme pressure lubrication must be used on the threaded portions of the tool. This will increase the longevity of the tool and insure proper operation during the removal and installation process.

11. Press the ball joint (3) from the lower control arm (4) using special tools C-4212-F (Press) (1), 8445-3 (Driver) (2) and 9604 (Receiver) (5).

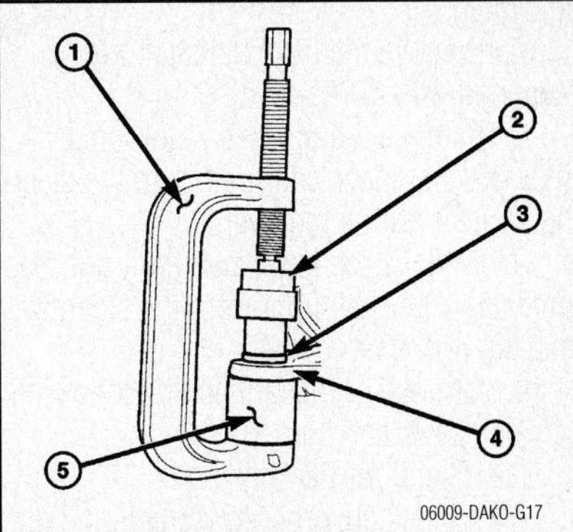

Fig. 149 Install the ball joint (3) into the control arm and press in using special tools C-4212-F (1), 8441-4 (2) and 9654-1 (4).

To install:

➡Extreme pressure lubrication must be used on the threaded portions of the tool. This will increase the longevity of the tool and insure proper operation during the removal and installation process.

12. Install the ball joint (3) into the control arm and press in using special tools C-4212-F (press) (1), 8441-4 (Receiver) (2) and 9654-1 (Driver) (4).
13. Install the snapring around the ball joint flange.
14. Remove the support for the halfshaft and install into position, 4WD models only.
15. Install the steering knuckle.
16. Install the tie rod end into the steering knuckle, then install the retaining nut and tighten to 55 ft. lbs. (75 Nm).
17. Install and tighten the halfshaft nut (if equipped) to 185 ft. lbs. (251 Nm).
18. Install the brake caliper and rotor.
19. Install the front wheel.
20. Check the vehicle ride height.
21. Check the wheel alignment and adjust, as necessary.

LOWER CONTROL ARM

REMOVAL & INSTALLATION

See Figure 150.

1. Before servicing the vehicle, refer to the precautions.
2. Remove the front wheel.
3. Remove the brake caliper assembly and rotor.
4. Disconnect the wheel speed sensor at the wheel well.
5. Remove tie rod end jam nut.

6. Disconnect the tie rod from the knuckle using special tool C-3894-A, or equivalent.

7. Remove the front halfshaft nut, if equipped with 4WD.

8. Remove the upper ball joint nut. Separate the upper ball joint from the steering knuckle with a remover.

9. Remove the lower ball joint nut. Separate the lower ball joint from the steering knuckle with a remover.

10. Remove or disconnect the following:

- Steering knuckle
- Stabilizer bar link
- Shock absorber lower bolt and nut
- Lower control arm bolts, nuts and washers
- Lower control arm

To install:

➡All suspension components should be tightened with the weight of the vehicle on them (curb height).

11. Position the lower control arm at the frame rail brackets. Install the pivot bolts washers and nuts. Tighten the nuts finger-tight.

CAUTION

The ball joint stud taper must be CLEAN and DRY before installing the knuckle. Clean the stud taper with mineral spirits to remove dirt and grease.

12. Install the steering knuckle.

13. Insert the lower ball joint into the steering knuckle. Install and tighten the retaining nut to 95 ft. lbs. (129 Nm).

14. Install shock absorber lower bolt and nut. Tighten to 155 ft. lbs. (210 Nm).

15. Install the front halfshaft nut, if equipped with 4WD.

16. Insert the upper ball joint into the steering knuckle. Install and tighten the retaining nut to 70 ft. lbs. (95 Nm).

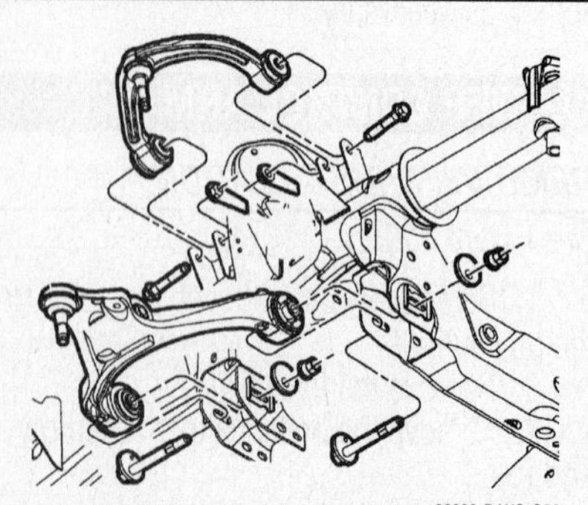

Fig. 150 Lower and upper control arm installation

17. Install the stabilizer bar link and tighten to 75 ft. lbs. (102 Nm).

18. Tighten the lower control arm pivot nut and bolts to 105 ft. lbs. (142 Nm).

19. Insert the outer tie rod end into the steering knuckle. Install and tighten the retaining nut to 55 ft. lbs. (75 Nm).

20. Install the brake caliper and rotor.

21. Install the front wheel.

22. Check the wheel alignment and adjust, as necessary.

TESTING

1. Check the rubber parts for cracks and wear.

2. Check the shock absorber for malfunctions, oil leakage, or abnormal noise.

3. If shock absorber does not perform properly, replace the shock absorber.

SHOCK ABSORBERS

See Figures 151 and 152.

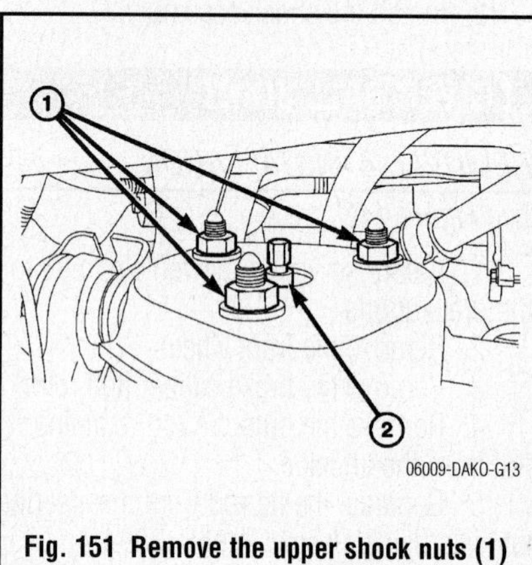

Fig. 151 Remove the upper shock nuts (1)

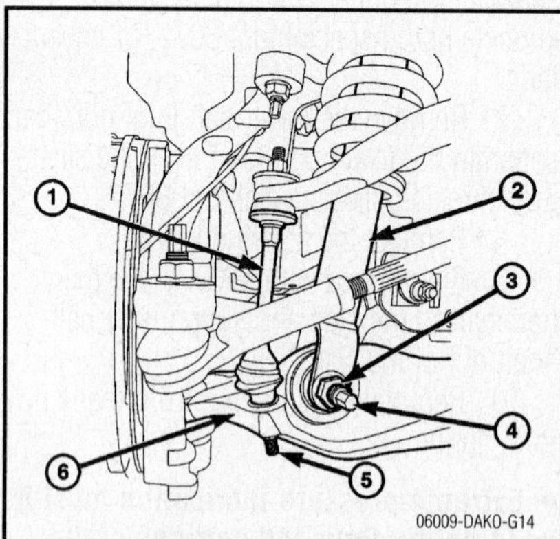

Fig. 152 Stabilizer link (1), strut (2), nut (3), lower strut bolt (4), link lower nut (5), lower control arm (6)

REMOVAL & INSTALLATION

1. Before servicing the vehicle, refer to the precautions.

2. Remove the front wheel.

3. Support the lower control arm outboard end.

4. Remove the upper shock nuts.

5. Remove the stabilizer link lower nut and then separate the stabilizer link from the lower control arm to gain access to the lower shock nut.

6. Remove the lower shock bolt and nut.

7. Remove the shock.

To install:

➡All suspension components should be tightened with the weight of the vehicle on them (curb height).

8. Install the upper part of the shock into the frame bracket.

9. Install the upper nuts. Tighten to 45 ft. lbs. (61 Nm).

10. Install the lower part of the shock into the lower control arm shock bushing.

11. Position shock module clevis to lower control arm. Install bolt so head of bolt is facing rear of vehicle and hand start nut. Tighten the bolt and nut to 155 ft. lbs. (210 Nm).

12. Install the stabilizer link lower nut to the lower control arm. Tighten to 75 ft. lbs. (102 Nm).

13. Remove the support from the lower control arm outboard end.

14. Install the front wheel.

STEERING KNUCKLE

REMOVAL & INSTALLATION

See Figure 150.

1. Before servicing the vehicle, refer to the precautions.

2. Raise and support the vehicle.

3. Remove the wheel and tire assembly.

4. Remove the brake caliper, rotor, shield and ABS wheel speed sensor.

5. Remove the front halfshaft nut, if equipped with 4WD.

6. Remove the tie rod end nut.

7. Separate the tie rod from the knuckle with puller C-3894-A.

WARNING

When installing puller 8677 to separate the ball joint, be careful not to damage the ball joint seal.

8. Remove the upper ball joint nut.

9. Separate the ball joint from the knuckle with puller 867.

10. Install an hydraulic jack to support the lower control arm.

11. Remove the lower ball joint nut.

12. Separate the ball joint from the knuckle with puller 8677 and remove the knuckle.
13. Remove the hub/bearing bolts from the knuckle.
14. Remove the hub/bearing from the steering knuckle.
15. Remove the steering knuckle.

To install:

➡The ball joint stud tapers must be clean and dry before installing the knuckle. Clean the stud tapers with mineral spirits to remove dirt and grease.

16. Install the hub/bearing to the steering knuckle and tighten the bolts to 120 ft. lbs. (163 Nm).
17. Install the knuckle onto the upper and lower ball joints.
18. Install the upper ball joint nut. Tighten the nut to 70 ft. lbs. (95 Nm).
19. Install the lower ball joint nut. Tighten the nut to 95 ft. lbs. (129 Nm)
20. Remove the hydraulic jack from the lower control arm.
21. Install the tie rod end and tighten the nut to 80 ft. lbs. (108 Nm).
22. Install the front halfshaft into the hub/bearing, if equipped with 4WD.
23. Install the halfshaft nut and tighten to 185 ft. lbs. (251 Nm), if equipped with 4WD.
24. Install the ABS wheel speed sensor if equipped and brake shield, rotor and caliper.
25. Install the wheel and tire assembly.
26. Remove the support and lower the vehicle.
27. Perform a wheel alignment.

STABILIZER BAR

REMOVAL & INSTALLATION

See Figure 153.

1. Before servicing the vehicle, refer to the precautions.
2. Raise and support the vehicle.
3. Remove the wheel and tire assembly.
4. Remove the stabilizer link upper nut and remove the retainers and grommets.
5. Remove the stabilizer bar retainer bolts also remove the retainers from the frame sway bar and remove the bar.
6. If necessary, remove the bushings from the stabilizer bar.

To install:

7. If removed, install the bushings on the stabilizer bar.

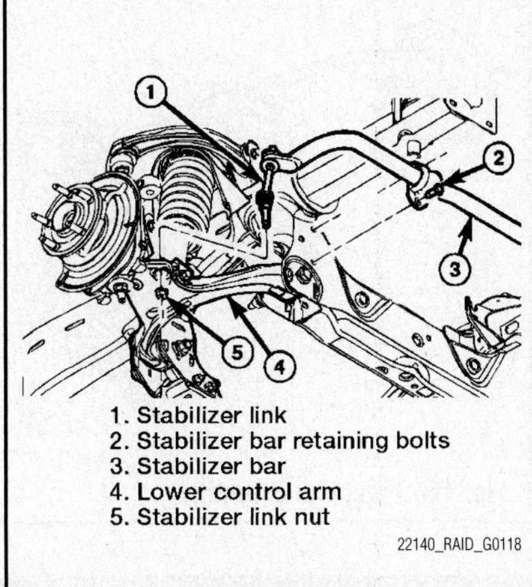

Fig. 153 Front stabilizer bar mounting

8. Position the stabilizer bar on the frame crossmember brackets and install the bracket and bolts finger-tight.

➡Check the alignment of the bar to ensure there is no interference with the either frame rail or chassis component. Spacing should be equal on both sides.

9. Install the stabilizer bar to the stabilizer link and install the grommets and retainers.
10. Install the nuts to the stabilizer link and tighten to 17 ft. lbs. (23 Nm).
11. Tighten the bracket bolts to the frame to 45 ft. lbs. (61 Nm).

UPPER BALL JOINT

REMOVAL & INSTALLATION

These vehicles utilize an upper control arm with an integral ball joint. If the ball joint is damaged or worn, the upper control arm must be replaced.

UPPER CONTROL ARM

REMOVAL & INSTALLATION

See Figure 154.

1. Before servicing the vehicle, refer to the precautions.
2. Raise and support vehicle.
3. Remove front wheel.
4. Remove the nut from upper ball joint.
5. Separate the upper ball joint from the steering knuckle with a ball joint puller.

⁂ WARNING

When installing the tool to separate the ball joint, be careful not to damage the ball joint seal.

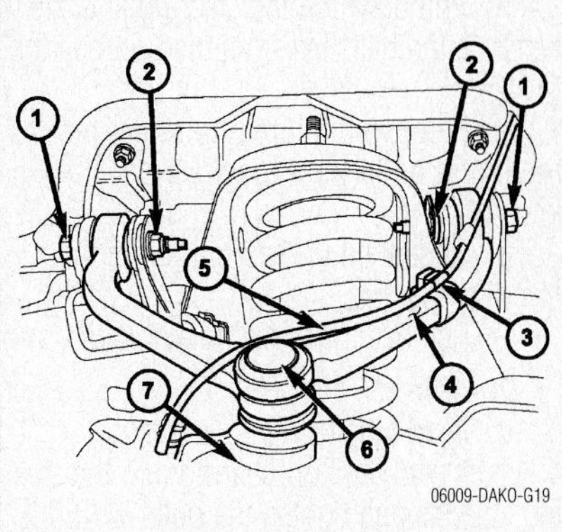

Fig. 154 Upper control arm mounting, 2005 models. (1) bolts, (2) nuts, (3) brackets, (4) control arm, (5) ABS wheel speed wire, (6) ball joint, (7) knuckle

6. Remove the wheel speed sensor wire from the retaining brackets to the upper control arm.
7. Remove the control arm pivot bolts and nuts and remove control arm.

To install:

➡All suspension components should be tightened with the weight of the vehicle on them (curb height).

8. Position the control arm into the frame brackets. Install bolts and nuts. Tighten to 75 ft. lbs. (102 Nm).
9. Reposition the wheel speed wire into the retaining brackets.
10. Insert the ball joint in steering knuckle and tighten ball joint nut to 70 ft. lbs. (95 Nm).
11. Install the front wheel.
12. Check the wheel alignment and adjust, as necessary.

WHEEL HUB AND BEARING

REMOVAL & INSTALLATION

See Figure 155.

1. Before servicing the vehicle, refer to the precautions.
2. Remove the wheel.
3. Remove the brake caliper and rotor.
4. Remove the ABS wheel speed sensor if equipped.
5. Remove the halfshaft nut, if equipped with 4WD.

⁂ WARNING

Do not strike the knuckle with a hammer to remove the tie rod end or the ball joint. Damage to the steering knuckle will occur.

6. Pull down on the steering knuckle to separate the halfshaft from the hub/bearing (4WD).

7. Remove the three hub/bearing mounting bolts from the steering knuckle.

8. Slide the hub/bearing out of the steering knuckle.

9. Remove the brake dust shield.

To install:

10. Install the brake dust shield.

11. Install the hub/bearing into the steering knuckle and tighten the bolts to 120 ft. lbs. (163 Nm).

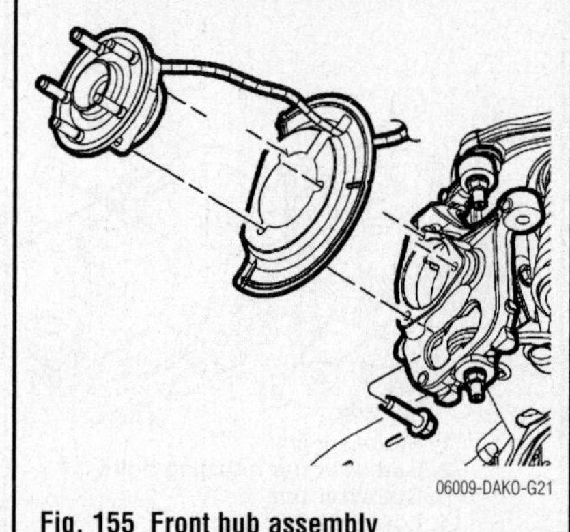

Fig. 155 Front hub assembly

12. Install the brake rotor and caliper.

13. Install the ABS wheel speed sensor if equipped.

14. Install the halfshaft nut, if equipped with 4WD. Tighten to 185 ft. lbs. (251 Nm).

15. Install the wheel.

ADJUSTMENT

The hub/bearing assembly is not adjustable.

SUSPENSION

REAR SUSPENSION

LEAF SPRING

REMOVAL & INSTALLATION

See Figure 156.

1. Before servicing the vehicle, refer to the precautions.

CAUTION

The rear of the vehicle must be lifted only with a jack or hoist. The lift must be placed under the frame rail crossmember located aft of the rear axle. Use care to avoid bending the side rail flange.

2. Raise the vehicle at the frame.

3. Use a hydraulic jack to relieve the axle weight.

4. Remove the rear wheel.

5. Remove the nuts, the U-bolts and spring plate from the axle.

6. Loosen and remove the bolt and then remove the flag nut through the access hole in the bracket from the spring front eye.

1. Shackle nut
2. Shackle
3. Shackle through bolt
4. Front spring mount through bolt
5. Captured nut
6. Spring
7. U-bolt
8. U-bolt retaining nut
9. Bolt

22140_RAID_G0119

Fig. 156 Exploded view of the rear suspension

7. Remove the nut and bolt that attaches the spring shackle to the rear frame bracket.

8. Remove the spring from the vehicle.

9. Remove the shackle from the spring.

To install:

10. Install the spring shackle on the spring finger tight.

11. Position the spring on the rear axle pad. Make sure the spring center bolt is inserted in the pad locating hole.

12. Align front spring eye with the bolt hole in the front frame bracket. Install the spring eye bolt and flag nut through the access hole in the frame and tighten the bolt finger-tight.

13. Align spring shackle eye with the bolt hole in the rear frame bracket. Install the bolt and nut and tighten the spring shackle eye nut finger-tight.

14. Install the U-bolts, spring plate and nuts.

15. Tighten the U-bolt nuts to 110 ft. lbs. (149 Nm).

16. Install the rear wheel.

17. Remove the support stands from under the frame rails. Lower the vehicle until the springs are supporting the weight of the vehicle.

18. Tighten the spring eye pivot bolt and flag nut to 125 ft. lbs. (170 Nm).

19. Tighten the upper shackle bolt and nut and the lower shackle bolt and nut to 125 ft. lbs. (170 Nm).

SHOCK ABSORBER

REMOVAL & INSTALLATION

See Figure 157.

1. Before servicing the vehicle, refer to the precautions.

2. Support rear axle with a suitable jack.

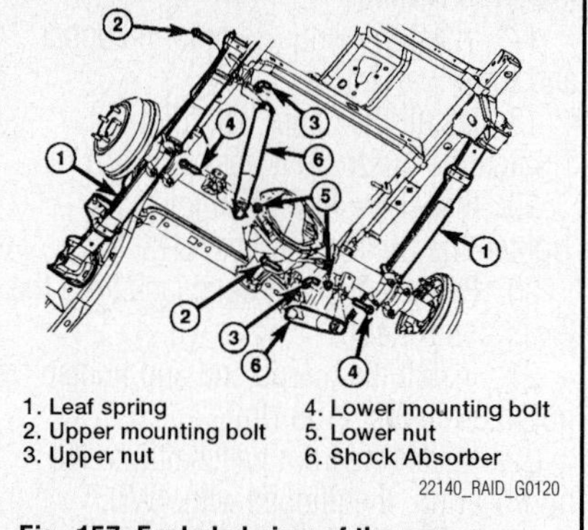

Fig. 157 Exploded view of the rear suspension

3. Remove the shock absorber lower nut and bolt from the axle bracket.

4. Remove the shock absorber upper nut and bolt from the frame bracket and remove the shock absorber.

To install:

5. Install the shock absorber and upper mounting bolt and nut. Tighten the nut to 70 ft. lbs. (95 Nm).

6. Install the shock absorber into the axle bracket. Install the bolt and nut and tighten the nut to 70 ft. lbs. (95 Nm).

TESTING

1. Check the rubber parts for cracks and wear.

2. Check the shock absorber for malfunctions, oil leakage, or abnormal noise.

3. If shock absorber does not perform properly, replace the shock absorber.

WHEEL BEARINGS

REMOVAL & INSTALLATION

Wheel bearing service is covered under Rear Drive Axle, Axle Shaft, Bearing & Seal, Removal & Installation.

MITSUBISHI

Diagnostic Trouble Codes

12

DIAGNOSTIC TROUBLE CODES

OBD II VEHICLE APPLICATIONS

MITSUBISHI

Eclipse
2006-08
- 2.4L I4 .... VIN F
- 3.8L V6 .... VIN T

Eclipse Spyder
2006-08
- 2.4L I4 .... VIN F
- 3.8L V6 .... VIN T

Endeavor
2006-08
- 3.8L V6 .... VIN S

Evolutionr
2006
- 2.0L I4 .... VIN C

2008
- 2.0L I4 .... VIN V

Galant
2006
- 2.4L I4 .... VIN F
- 3.8L V6 .... VIN S

2007
- 2.4L I4 .... VIN F
- 3.8L V6 .... VIN S
- 3.8L V6 .... VIN T

2008
- 2.4L I4 .... VIN F
- 3.8L V6 .... VIN T

Lancer
2006-07
- 2.0L I4 .... VIN E
- 2.4L I4 .... VIN F

2008
- 2.4L I4 .... VIN U

Lancer Sportback
2006
- 2.0L I4 .... VIN F

Outlander
2006
- 2.0L I4 .... VIN F

2007
- 3.0L V6 .... VIN X

2008
- 2.0L I4 .... VIN W
- 3.0L V6 .... VIN X

Raider
2006-08
- 3.7L V6 .... VIN K
- 4.7L V8 .... VIN M

Gas Engine OBD II Trouble Code List (P0xxx Codes)

DTC	Trouble Code Title, Conditions & Possible Causes
DTC: P0011 **2T PCM MIL: YES** **Years:** 2006, 2007, 2008 **Models:** All **Engines:** 2.0L, 2.4L W **Transmissions:** All	**Variable Valve Timing System** Engine started and running for greater than 20 seconds, engine speed is 1200 RPM or greater, engine coolant temperature is greater than 169°F, the phase angle of the intake camshaft is 5 degrees or greater for 5 seconds. **Possible Causes:** • Oil feeder control valve failed • Oil passage of variable valve timing control system clogged • Variable valve timing sprocket operation mechanism stuck • ECM failed
DTC: P0014 **2T PCM MIL: YES** **Years:** 2006, 2007, 2008 **Models:** All **Engines:** 2.0L, 2.4L W **Transmissions:** All	**Exhaust Variable Valve Timing System Target Error** The phase angle of the exhaust camshaft is 5 degrees or more for 5 seconds. **Possible Causes:** • Exhaust engine oil control valve failed. • Oil passage of variable valve timing control system clogged. • Exhaust variable valve timing sprocket operation mechanism stuck. • ECM failed.
DTC: P0016 **2T PCM MIL: YES** **Years:** 2006, 2007, 2008 **Models:** Evolution, Lancer, Outlander **Engines:** 2.0L, 2.4L W **Transmissions:** All	**Crankshaft/Camshaft (Intake) Position Sensor Phasing** The open timing of the intake valve is faster than -15 degrees (ATDC) <Except for California> or -18 degrees (ATDC) <California> for 10 seconds or the open timing of the intake valve is slower than 15 degrees (ATDC) <Except for California> or 12 degrees (ATDC) <California> for 10 seconds. **Possible Causes:** • Timing chain in out of place. • Loose timing chain. • Intake variable valve timing sprocket tooth coming off. • ECM failed.
DTC: P0017 **2T PCM MIL: YES** **Years:** 2006, 2007, 2008 **Models:** Evolution, Lancer, Outlander **Engines:** 2.0L, 2.4L W **Transmissions:** All	**Crankshaft/Camshaft (Exhaust) Position Sensor Phasing** Close timing of the exhaust valve is faster than -15 degrees (ATDC) <Except for California> or -12 degrees (ATDC) <California> for 10 seconds or the close timing of the exhaust valve is slower than 15 degrees (ATDC) <Except for California> or 18 degrees (ATDC) <California> for 10 seconds. **Possible Causes:** • Timing chain in out of place. • Loose timing chain. • Exhaust variable valve timing sprocket tooth coming off. • ECM failed.
DTC: P0031 **2T PCM MIL: YES** **Years:** 2006, 2007, 2008 **Models:** All **Engines:** All **Transmissions:** All	**Heated Oxygen Sensor Heater Control Circuit Low (Sensor 1)** Engine started and running for greater than 2 seconds, battery voltage is between 11 and 16.5 volts, with the front heated oxygen sensor on, current continues to be less than 0.17 amps for 2 seconds, and/or with the front heated oxygen sensor is off, the voltage is less than 2 volts for 2 seconds. **Possible Causes:** • Front heated HO2S circuit open or shorted • PCM has failed
DTC: P0032 **2T PCM MIL: YES** **Years:** 2006, 2007, 2008 **Models:** All **Engines:** All **Transmissions:** All	**Heated Oxygen Sensor Heater Control Circuit High (Sensor 1)** Engine started and running for greater than 2 seconds, battery voltage is between 11 and 16.5 volts, with the front heated oxygen sensor on, current continues to be greater than 10.5 amps for 2 seconds. **Possible Causes:** • Front heated HO2S circuit open or shorted • PCM has failed
DTC: P0037 **2T PCM MIL: YES** **Years:** 2006, 2007, 2008 **Models:** All **Engines:** All **Transmissions:** All	**Heated Oxygen Sensor Heater Control Circuit Low (Sensor 2)** Engine started and running for greater than 2 seconds, battery voltage is between 11 and 16.5 volts, with the rear heated oxygen sensor on, current continues to be less than 0.17 amps for 2 seconds, and/or with the front heated oxygen sensor is off, the voltage is less than 2 volts for 2 seconds. **Possible Causes:** • Rear heated HO2S circuit open or shorted • PCM has failed

DTC	Trouble Code Title, Conditions & Possible Causes
DTC: P0038 **2T PCM MIL: YES** **Years:** 2006, 2007, 2008 **Models:** All **Engines:** All **Transmissions:** All	**Heated Oxygen Sensor Heater Control Circuit High (Sensor 2)** Engine started and running for greater than 2 seconds, battery voltage is between 11 and 16.5 volts, with the rear heated oxygen sensor on, current continues to be greater than 10.5 amps for 2 seconds. **Possible Causes:** • Rear heated HO2S circuit open or shorted • PCM has failed
DTC: P0043 **2T PCM MIL: YES** **Years: 2007, 2008** **Models:** Outlander **Engines:** 3.0L X **Transmissions:** All	**Heated Oxygen Sensor Heater Control Circuit Low** The center exhaust pipe heated oxygen sensor heater voltage has continued to be lower than 2.0 volts for 2 seconds. **Possible Causes:** • Connector damage. • Center exhaust pipe heated oxygen sensor heater failed. • Harness damage • ECM failed.
DTC: P0044 **2T PCM MIL: YES** **Years: 2007, 2008** **Models:** Outlander **Engines:** 3.0L X **Transmissions:** All	**Heated Oxygen Sensor Heater Control Circuit High** The center exhaust pipe heated oxygen sensor heater voltage has continued to behigher than 10.5 amps for 2 seconds. **Possible Causes:** • Center exhaust pipe heated oxygen sensor heater failed. • ECM failed.
DTC: P0051 **2T PCM MIL: YES** **Years:** 2006, 2007, 2008 **Models:** All **Engines:** All **Transmissions:** All	**Heated Oxygen Sensor Heater Control Circuit Low (Bank 2, Sensor 1)** Engine started and running for greater than 2 seconds, battery voltage is between 11 and 16.5 volts, with the front heated oxygen sensor on, current continues to be less than 0.17 amps for 2 seconds, and/or with the front heated oxygen sensor is off, the voltage is less than 2 volts for 2 seconds. **Possible Causes:** • Front heated HO2S circuit open or shorted • PCM has failed
DTC: P0052 **2T PCM MIL: YES** **Years:** 2006, 2007, 2008 **Models:** All **Engines:** All **Transmissions:** All	**Heated Oxygen Sensor Heater Control Circuit High (Bank 2, Sensor 1)** Engine started and running for greater than 2 seconds, battery voltage is between 11 and 16.5 volts, with the front heated oxygen sensor on, current continues to be greater than 10.5 amps for 2 seconds. **Possible Causes:** • Front heated HO2S circuit open or shorted • PCM has failed
DTC: P0057 **2T PCM MIL: YES** **Years:** 2006, 2007, 2008 **Models:** All **Engines:** All **Transmissions:** All	**Heated Oxygen Sensor Heater Control Circuit Low (Bank 2, Sensor 2)** Engine started and running for greater than 2 seconds, battery voltage is between 11 and 16.5 volts, with the rear heated oxygen sensor on, current continues to be less than 0.17 amps for 2 seconds, and/or with the front heated oxygen sensor is off, the voltage is less than 2 volts for 2 seconds. **Possible Causes:** • Rear heated HO2S circuit open or shorted • PCM has failed
DTC: P0058 **2T PCM MIL: YES** **Years:** 2006, 2007, 2008 **Models:** All **Engines:** All **Transmissions:** All	**Heated Oxygen Sensor Heater Control Circuit High (Bank 2, Sensor 2)** Engine started and running for greater than 2 seconds, battery voltage is between 11 and 16.5 volts, with the rear heated oxygen sensor on, current continues to be greater than 10.5 amps for 2 seconds. **Possible Causes:** • Rear heated HO2S circuit open or shorted • PCM has failed
DTC: P0069 **2T PCM MIL: YES** **Years:**: 2006, 2007, 2008 **Models:** All **Engines:** All **Transmissions:** All	**Abnormal Correlation Between Manifold Absolute Pressure Sensor And Barometric Pressure Sensor** Engine off, ignition switch is in OFF position, after two seconds pass from the time when the engine is stopped and engine coolant temperature is higher than 0°C (32°F) and the difference between manifold absolute pressure sensor output and barometric pressure sensor output is more than 9 kPa (2.7 in.Hg) for 2 seconds. **Possible Causes:** • MAP sensor failed • Barometric pressure sensor failed • PCM has failed

DTC	Trouble Code Title, Conditions & Possible Causes
DTC: P0071 **2T PCM MIL: YES** **Years:** 2006, 2007, 2008 **Models:** Raider **Engines:** 3.7L K, 4.7L M **Transmissions:** All	**Ambient Air Temperature Sensor Performance** The PCM compares the ambient, engine coolant, and intake air temperature sensor values. If engine coolant and intake air temperature sensors agree with each other but ambient air temperature does not agree with them, the ambient air temperature sensor is declared as irrational. Three good trips to turn off the MIL. **Possible Causes:** • Signal Circuit High Resistance • Signal Circuit Shorted To Ground • Signal Circuit Shorted To The Sensor Ground Circuit
DTC: P0072 **1T PCM MIL: YES** **Years:**: 2006, 2007, 2008 **Models:** Raider **Engines:** 3.7L K, 4.7L M **Transmissions:** All	**Ambient Air Temperature Sensor Circuit Low** Ambient Temperature Sensor is less than the minimum acceptable voltage. Three good trips to turn off the MIL. **Possible Causes:** • Signal Circuit High Resistance • Signal Circuit Shorted To Ground • Signal Circuit Shorted To The Sensor Ground Circuit
DTC: P0073 **1T PCM MIL: YES** **Years:**: 2006, 2007, 2008 **Models:** Raider **Engines:** 3.7L K, 4.7L M **Transmissions:** All	**Ambient Air Temperature Sensor Circuit High** The Ambient Temperature Sensor voltage is greater than maximum allowable voltage. Three good trips to turn off the MIL. **Possible Causes:** • Signal Circuit OPen • Signal Circuit Shorted To Voltage • Sensor Ground Circuit Open • Ambient Air Temp Sensor • Front Control Module
DTC: P0090 **2T PCM MIL: YES** **Years:** 2006, 2007, 2008 **Models:** Evolution **Engines:** 2.0L C **Transmissions:** All	**Fuel Pressure Solenoid Circuit** Engine being cranked, battery positive voltage is between 10 and 16 volts, the fuel pressure solenoid coil surge voltage (battery positive voltage + 2 volts) is not detected for 0.2 seconds. Fuel pressure solenoid is ON, turbocharger wastegate solenoid is OFF for more than 1 second has elapsed after these conditions have been met the fuel pressure solenoid coil surge voltage is not detected for 1 second. **Possible Causes:** • Fuel pressure solenoid failed • Open or shorted fuel pressure circuit, harness damage and/or connector damage • ECM failed
DTC: P0096 **2T PCM MIL: YES** **Years:** 2006, 2007, 2008 **Models:** Evolution **Engines:** 2.0L V **Transmissions:** All	**Intake Air Temperature Circuit Range/Performance** Changes in the intake air temperature is lower than 1°C (1.8°F). **Possible Causes:** • Intake air temperature sensor 2 failed • Harness damage or connector damage • ECM failed
DTC: P0097 **2T PCM MIL: YES** **Years:**: 2006, 2007, 2008 **Models:** Evolution **Engines:** 2.0L V **Transmissions:** All	**Intake Air Temperature Circuit Low Input** Intake air temperature sensor 2 output voltage has continued to be 0.2 volt or lower [corresponding to an intake air temperature of 115°C (239°F) or higher] for 2 seconds. **Possible Causes:** • Intake air temperature sensor 2 failed • Shorted intake air temperature sensor 2 circuit, or connector damage • ECM failed
DTC: P0098 **2T PCM MIL: YES** **Years:**: 2006, 2007, 2008 **Models:** Evolution **Engines:** 2.0L V **Transmissions:** All	**Intake Air Temperature Circuit High Input** Intake air temperature sensor 2 output voltage has continued to be 4.6 volts or higher [corresponding to an intake air temperature of -40°C (-40°F) or lower] for 2 seconds. **Possible Causes:** • Intake air temperature sensor 2 failed • Open intake air temperature sensor 2 circuit, or connector damage • ECM failed
DTC: P0101 **2T PCM MIL: YES** **Years:**: 2006, 2007, 2008 **Models:** All **Engines:** All **Transmissions:** All	**Mass Airflow Sensor Range/Performance** Engine started, engine speed over 500 RPM, and PCM detected the MAF sensor signal remained at 3.3 Hz or less for 3 seconds. **Possible Causes:** • Intake air leak between the MAF sensor and the throttle body • MAF sensor signal circuit is open or shorted to ground • MAF sensor is damaged or has failed • PCM has failed

DTC	Trouble Code Title, Conditions & Possible Causes
DTC: P0102 **2T PCM MIL: YES** **Years:** 2006, 2007, 2008 **Models:** All **Engines:** All **Transmissions:** All	**Mass Airflow Sensor Circuit Low Input** Engine started, engine speed more than RPM with the TP sensor signal at 1.5v or higher, and the PCM detected the MAF sensor indicated less than 60 Hz for 3 seconds during the PCM test. **Possible Causes:** • MAF sensor signal circuit is open or shorted to ground • MAF sensor is damaged or has failed • PCM has failed
DTC: P0103 **2T PCM MIL: YES** **Years:** 2006, 2007, 2008 **Models:** All **Engines:** All **Transmissions:** All	**Mass Airflow Sensor Circuit High Input** Engine started, engine speed more than RPM with the TP sensor signal at 1.5v or higher, and the PCM detected the MAF sensor indicate more than 800 Hz for 3 seconds in the PCM test. **Possible Causes:** • MAF sensor signal circuit is shorted to system power • MAF sensor is damaged or has failed • PCM has failed
DTC: P0106 **2T PCM MIL: YES** **Years:** 2006, 2007, 2008 **Models:** All **Engines:** All **Transmissions:** All	**Manifold Absolute Pressure Sensor Range/Performance** Key on for less than 350 ms with the engine speed below 250 RPM, and the PCM detected the MAP sensor signal was less 2.196v and more than 0.019v for 300 ms during the PCM test. **Possible Causes:** • MAP sensor signal circuit to the PCM interrupted (intermittent) • MAP sensor is damaged or has failed • PCM has failed
DTC: P0107 **2T PCM MIL: YES** **Years:** 2006, 2007, 2008 **Models:** All **Engines:** All **Transmissions:** All	**Manifold Absolute Pressure Sensor Circuit Low Signal** Engine started, engine speed from 400-1500 RPM, TP sensor less than 1.30v, and the PCM detected the MAP signal was less than 0.02v for 2 seconds during the PCM test. **Possible Causes:** • MAP sensor signal circuit is shorted to ground • MAP sensor power circuit is open (VREF signal from the PCM) • MAP sensor is damaged or has failed • PCM has failed
DTC: P0108 **2T PCM MIL: YES** **Years:** 2006, 2007, 2008 **Models:** All **Engines:** All **Transmissions:** All	**Manifold Air Pressure Sensor Circuit High Input** Engine started and 8 minutes have passed, manifold absolute pressure is 34.6 in. Hg (117 kPa) or higher for 2 seconds. **Possible Causes:** • MAP sensor signal circuit is open between the sensor and PCM • MAP sensor signal circuit is shorted to VREF or system power • MAP sensor ground circuit is open between sensor and PCM • MAP sensor is damaged or has failed • PCM has failed
DTC: P0111 **2T PCM MIL: YES** **Years:** 2006, 2007, 2008 **Models:** All **Engines:** All **Transmissions:** All	**Intake Air Temperature Sensor Range/Performance** Engine coolant temperature greater than 169°F, (76°C), start and driven with vehicle speed greater than 31MPH (50km/h) for more than 60 seconds, then stop vehicle for greater than 30 seconds with a change in the intake air temperature is less than 1.8°F (1°C). **Possible Causes:** • IAT sensor signal circuit is open or shorted to ground • IAT sensor has failed • PCM has failed
DTC: P0112 **2T PCM MIL: YES** **Years:** 2006, 2007, 2008 **Models:** All **Engines:** All **Transmissions:** All	**Intake Air Temperature Sensor Circuit Low Signal** Key on or engine running, and the PCM detected the Intake Air Temperature (IAT) sensor signal was less than 0.2v for 3 seconds. **Possible Causes:** • IAT sensor signal circuit is shorted to ground • IAT sensor is damaged or has failed (it may be shorted) • PCM has failed
DTC: P0113 **2T PCM MIL: YES** **Years:** 2006, 2007, 2008 **Models:** All **Engines:** All **Transmissions:** All	**Intake Air Temperature Sensor Circuit High Signal** Key on or engine running, and the PCM detected the Intake Air Temperature (IAT) signal was more than 4.96v for 3 seconds. **Possible Causes:** • IAT sensor signal circuit is open between sensor and the PCM • IAT sensor signal circuit is shorted to VREF or system power • IAT sensor ground circuit is open between sensor and the PCM • IAT sensor is damaged or has failed (an internal short) • PCM has failed

DTC	Trouble Code Title, Conditions & Possible Causes
DTC: P0116 **2T PCM MIL: YES** **Years:** 2006, 2007, 2008 **Models:** All **Engines:** All **Transmissions:** All	**Engine Coolant Temperature Circuit Range/Performance Problem** Engine started and engine coolant temperature is or was more than 45°F (7°C), the engine coolant temperature fluctuates within 1.8°F (1°C) after 5 minutes have passed since starting the engine. The time is not counted if the intake air temperature is 140°F (60°C) or greater, MAF sensor output frequency is 70 Hz or less and/or during fuel shut-off operation. **Possible Causes:** • ECT sensor signal circuit is open or shorted to ground • ECT sensor has failed • PCM has failed
DTC: P0117 **2T PCM MIL: YES** **Years:** 2006, 2007, 2008 **Models:** All **Engines:** All **Transmissions:** All	**Engine Coolant Temperature Sensor Circuit Low Signal** Key on or engine running, and the PCM detected the Engine Coolant Temperature (ECT) signal indicated less than 0.1v or lower for 2 seconds. **Possible Causes:** • ECT sensor signal circuit is shorted to ground • ECT sensor is damaged or has failed (it may be shorted) • PCM has failed
DTC: P0118 **2T PCM MIL: YES** **Years:** 2006, 2007, 2008 **Models:** All **Engines:** All **Transmissions:** All	**Engine Coolant Temperature Sensor Circuit High Signal** Key on or engine running, and the PCM detected the Engine Coolant Temperature (ECT) signal indicated more than 4.6v or higher for 2 seconds. **Possible Causes:** • ECT sensor signal circuit is open between sensor and the PCM • ECT sensor signal circuit is shorted to VREF or system power • ECT sensor ground circuit is open between sensor and PCM • ECT sensor is damaged or has failed (an internal short) • PCM has failed
DTC: P0121 **2T PCM MIL: YES** **Years:** 2006, 2007, 2008 **Models:** Lancer, Raider **Engines:** 2.0L E, 3.7L K, 4.7L M **Transmissions:** All	**TP Sensor Does Not Agree with MAP Sensor** Engine started, vehicle driven to a speed over 25 MPH at an engine speed over 1500 RPM, TP sensor signal from 3.75-4.71v, and the PCM detected the TP sensor signal indicated from 0.16-0.70v, and that it did not correlate to the MAF sensor signal in the PCM test. **Possible Causes:** • TP sensor signal circuit is open to the PCM (intermittent fault) • TP sensor ground circuit is open (an intermittent fault) • MAP sensor is out of calibration • Throttle body is damaged or throttle linkage is bent or binding • TP sensor is damaged or has failed
DTC: P0122 **2T PCM MIL: YES** **Years:** 2006, 2007, 2008 **Models:** All **Engines:** All **Transmissions:** All	**Throttle Position Sensor Circuit Low Input** Key on or engine running for more than 2 seconds and the PCM detected the sensor signal was less than 0.20v for 2 second. **Possible Causes:** • TP sensor signal circuit is shorted to ground • TP sensor power circuit is open (check VREF from the PCM) • TP sensor has failed or is misadjusted • PCM has failed
DTC: P0123 **2T PCM MIL: YES** Years: 2006, 2007, 2008 **Models:** All **Engines:** All **Transmissions:** All	**Throttle Position Sensor Circuit High Input** Key on or engine running for more than 2 seconds, engine speed is less than 1000 RPM, volumetric efficiency is less than 60 percentand the PCM detected the sensor signal was greater than 2.0v for 2 second. **Possible Causes:** • TP sensor signal circuit is shorted to ground • TP sensor power circuit is open (check VREF from the PCM) • TP sensor has failed or is misadjusted • PCM has failed
DTC: P0125 **1T ECT MIL: YES** **Years:** 2006, 2007, 2008 **Models:** All **Engines:** All **Transmissions:** All	**Insufficient Coolant Temperature for Closed Loop Fuel Control** Engine running for approximately 60 to 300 seconds and engine coolant temperature rises to 45°F (7°C) after start and/or engine coolant temperature decreases from greater than 104°F (40°C) to less than 104°F (40°C) for 5 minutes. **Possible Causes:** • Check the operation of the thermostat (it may be stuck open) • ECT sensor circuit shorted or open • ECT sensor has failed • PCM failed

DTC	Trouble Code Title, Conditions & Possible Causes
DTC: P0128 **2T ECT MIL: YES** **Years:** 2006, 2007, 2008 **Models:** All **Engines:** All **Transmissions:** All	**Coolant Thermostat Malfunction** Cold engine startup (ECT sensor from 14-171°F and the difference in the ECT and IAT sensor signal less than 9°F), engine running with the MAF sensor signal from 50-100 Hz for 300 seconds or less, and the PCM detected the ECT sensor signal did not reach 171°F after a 15-20 minute period had elapsed. **Possible Causes:** • Check the operation of the thermostat (it may be stuck open) • ECT sensor is damaged • PCM has failed
DTC: P0129 **1T ECT MIL: YES** **Years:** 2006, 2007, 2008 **Models:** Raider **Engines:** 3.7L K, 4.7L M **Transmissions:** All	**Barometric Pressure Out-of-Range Low** The PCM senses the voltage from the MAP sensor to be less than 2.2 volts but above 0.04 of a volt for 300 milliseconds. Three good trips to turn off the MIL. MIL will illuminate and the ETC light will flash, if equipped. **Possible Causes:** • 5-Volt supply circuit shorted to battery voltage • 5-Volt supply circuit open • 5-Volt supply circuit shorted to ground • MAP signal circuit open • MAP signal circuit shorted to ground • MAP signal circuit shorted to the sensor ground circuit • MAP sensor • PCM
DTC: P0131 **2T PCM MIL: YES** **Years:** 2006, 2007, 2008 **Models:** All **Engines:** All Engines **Transmissions:** All	**Heated Oxygen Sensor Circuit Low Voltage (Sensor 1)** Engine started and running for greater than 3 minutes, the front heated oxygen sensor signal voltage has continued to be less than 0.2 volts, engine coolant temperature is greater than 169°F, engine speed is greater than 1200 RPM, volumetric efficiency is greater than 25%, MAF sensor output frequency is greater than 75 Hz, at least 20 seconds have passed since fuel shut-off control was canceled, and the changes in the front heated oxygen sensor is less than 0.078 voltage the front heated oxygen sensor output voltage increase beyond 0.2 volts after making the air/fuel ratio 15 percentricher for 10 seconds. **Possible Causes:** • HO2S signal circuit is open between the sensor and the PCM • HO2S signal circuit is shorted to sensor or chassis ground • HO2S signal circuit is shorted to VREF or system power (B+) • HO2S is damaged, contaminated or it has failed • PCM has failed
DTC: P0132 **2T PCM MIL: YES** **Years:** 2006, 2007, 2008 **Models:** All **Engines:** All **Transmissions:** All	**HO2S-11 (Bank 1 Sensor 1) Signal High Signal** Engine started, engine runtime over 2 seconds, and the PCM detected the HO2S-11 signal was more than 1.2v for 2 seconds. **Possible Causes:** • HO2S signal circuit shorted to power (check the heater circuit) • HO2S is damaged or it has failed • PCM has failed
DTC: P0133 **2T O2S2 MIL: YES** **Years:** 2006, 2007, 2008 **Models:** All **Engines:** All **Transmissions:** All	**HO2S-11 (Bank 1 Sensor 1) Slow Response** Engine started, engine runtime over 3 minutes, ECT sensor more than 170°F, vehicle driven to a speed over 24 MPH for 75 seconds, then at idle speed from 512-864 RPM, PSPS signal "off", A/C clutch not cycling, and the PCM detected the HO2S-11 signal that did not reach 670 mv, or it did not switch between 0.39-6.0v engine times in a 6 second period during the Oxygen Sensor Monitor test. **Possible Causes:** • Exhaust leak present in the exhaust manifold or exhaust pipes • HO2S element fuel contamination • HO2S element has deteriorated • PCM has failed
DTC: P0134 **2T O2S1 MIL: YES** **Years:** 2006, 2007, 2008 **Models:** All **Engines:** All **Transmissions:** All	**HO2S-11 (Bank 1 Sensor 1) No Activity Detected** Engine started, engine runtime over 30 seconds, ECT sensor more than 169°F, engine speed is greater than 1200 RPM, volumetric efficiency is between 30 percentand 95%, TPS output voltage is less than 4 volts and the PCM detected the HO2S-11 signal remained fixed near 0.50v for 30 seconds in the Oxygen Sensor Monitor test. **Possible Causes:** • Exhaust leak present in exhaust manifold or exhaust pipes • HO2S element fuel contamination or has deteriorated • HO2S signal circuit or the ground circuit has high resistance • HO2S heater element has failed, or the heater circuit is open • PCM has failed

DTC	Trouble Code Title, Conditions & Possible Causes
DTC: P0137 **2T PCM MIL: YES** **Years:** 2006, 2007, 2008 **Models:** All **Engines:** All **Transmissions:** All	**HO2S-12 (Bank 1 Sensor 2) Signal Low Input** Engine started, ECT sensor signal less than 120°F at engine startup, engine run time over 3 seconds, and the PCM detected the HO2S-12 signal was less than 0.16v during the PCM test. **Possible Causes:** • HO2S signal circuit is open • HO2S signal circuit is shorted to ground • HO2S is damaged or it has failed • PCM has failed
DTC: P0138 **2T PCM MIL: YES** **Years:** 2006, 2007, 2008 **Models:** All **Engines:** All **Transmissions:** All	**HO2S-12 (Bank 1 Sensor 2) Signal High Input** Engine started, engine runtime over 2 seconds and the PCM detected the HO2S-12 signal was more than 1.2v for 2 seconds. **Possible Causes:** • HO2S signal circuit shorted to power (check the heater circuit) • HO2S is damaged or it has failed • PCM has failed
DTC: P0139 **2T O2S1 MIL: YES** **Years:** 2006, 2007, 2008 **Models:** All **Engines:** All **Transmissions:** All	**HO2S-12 Circuit Slow Response (Bank 1 Sensor 2)** Engine started, engine runtime over 10 seconds, ECT sensor more than 169°F, front heated oxygen sensor is active, MAF sensor output frequency is greater than 4000 Hz, engine speed is greater than 1500 RPM, volumetric efficiency is greater than 40 percent, vehicle speed is greater than 19 MPH and all is repeated 3 or more times, the change in the output voltage of the HO2S-12 is less than 0.312 volts. **Possible Causes:** • HO2S signal circuit is open or it has high resistance • HO2S element fuel contamination or has deteriorated • HO2S is damaged or has failed • PCM has failed
DTC: P0140 **2T PCM MIL: YES** **Years:** 2006, 2007, 2008 **Models:** All **Engines:** All **Transmissions:** All	**HO2S-12 (Bank 1 Sensor 2) Circuit No Activity Detected** Engine started, engine runtime over 10 seconds, engine coolant temperature greater than 169°F, front HO2S active, MAF output is greater than 1638 g., engine speed greater than 1500 RPM, volumetric efficiency is greater than 40 percent, vehicle speed greater than 19 MPH, then stop vehicle, repeated 3 or more times, the change in the output voltage in the rear HO2S is less than 0.313 volts. **Possible Causes:** • HO2S-12 failed • HO2S-12 circuit connector damaged • PCM has failed
DTC: P0141 **1T O2S HTR1 MIL: YES** **Years:** 2006, 2007, 2008 **Models:** Raider **Engines:** 3.7L K, 4.7L M **Transmissions:** All	**HO2S-12 (Bank 1 Sensor 2) Heater Circuit Malfunction** Engine started, ECT sensor more than 68°F, system voltage from 11-16v, and the PCM detected the HO2S-12 heater current was less than 0.20v, or it was more than 3.5 amps for 6 seconds in the test. **Possible Causes:** • HO2S heater control circuit is open or shorted to ground • HO2S heater control circuit is shorted to power • HO2S heater power circuit is open (check power from the relay) • HO2S heater is damaged or has failed • PCM has failed
DTC: P0143 **1T O2S HTR1 MIL: YES** **Years:** 2006, 2007, 2008 **Models:** Lancer, Outlander **Engines:** 2.0L U, 3.0L X **Transmissions:** All	**Center Exhaust Pipe Heated Oxygen Sensor Circuit Low** Center exhaust pipe heated oxygen sensor output voltage is lower than 0.2 volt for 2 seconds. **Possible Causes:** • Center exhaust pipe heated oxygen sensor failed. • Connector damage. • Harness damage • ECM failed.
DTC: P0144 **1T O2S HTR1 MIL: YES** **Years:**: 2006, 2007, 2008 **Models:** Lancer, Outlander **Engines:** 2.0L U, 3.0L X **Transmissions:** All	**Center Exhaust Pipe Heated Oxygen Sensor Circuit High** Center exhaust pipe heated oxygen sensor output voltage has continued to be 1.8 volts or higher for 2 seconds. **Possible Causes:** • Connector damage. • Harness damage • ECM failed.

DTC	Trouble Code Title, Conditions & Possible Causes
DTC: P0145 **1T O2S HTR1 MIL: YES** **Years:** 2006, 2007, 2008 **Models:** Lancer, Outlander **Engines:** 2.0L U, 3.0L X **Transmissions:** All	**Center Exhaust Pipe Heated Oxygen Sensor Circuit Slow Response** The center exhaust pipe heated oxygen sensor output voltage does not reach 0.2 volt for 7 seconds from fuel cut start or the center exhaust pipe heated oxygen sensor output voltage does not reach 0.2 volt for 1 second from 0.4 volt while fuel is being shut off. **Possible Causes:** • Center exhaust pipe heated oxygen sensor failed.
DTC: P0146 **1T O2S HTR1 MIL: YES** **Years:** 2006, 2007, 2008 **Models:** Lancer, Outlander **Engines:** 2.0L U, 3.0L X **Transmissions:** All	**Center Exhaust Pipe Heated Oxygen Sensor Circuit No Activity Detected** Change in the output voltage of the center exhaust pipe heated oxygen sensor is lower than 0.313 volt. **Possible Causes:** • Center exhaust pipe heated oxygen sensor failed.
DTC: P0151 **2T PCM MIL: YES** **Years:** 2006, 2007, 2008 **Models:** All **Engines:** All **Transmissions:** All	**HO2S-21 (Bank 2 Sensor 1) Circuit Low Voltage** Engine started and running for greater than 3 minutes, the left front heated oxygen sensor signal voltage has continued to be less than 0.2 volts, engine coolant temperature is greater than 169°F, engine speed is greater than 1200 RPM, volumetric efficiency is greater than 25%, MAF sensor output frequency is greater than 75 Hz, at least 20 seconds have passed since fuel shut-off control was canceled, and the changes in the front heated oxygen sensor is less than 0.078 voltage the front heated oxygen sensor output voltage increase beyond 0.2 volts after making the air/fuel ratio 15 percentricher for 10 seconds. **Possible Causes:** • HO2S signal circuit is open between the sensor and the PCM • HO2S signal circuit is shorted to sensor or chassis ground • HO2S signal circuit is shorted to VREF or system power (B+) • HO2S is damaged, contaminated or it has failed • PCM has failed
DTC: P0152 **2T PCM MIL: YES** **Years:** 2006, 2007, 2008 **Models:** All **Engines:** All **Transmissions:** All	**HO2S-21 (Bank 2 Sensor 1) Signal High Signal** Engine started, engine runtime over 2 minutes at cruise speed, ECT sensor more than 176°F, and the PCM detected the HO2S-21 signal was not less than 4.5v when 5.0v was applied to the HO2S-21 circuit for 3 seconds during the PCM test. **Possible Causes:** • HO2S signal circuit is open • HO2S signal circuit is shorted to the heater power circuit • HO2S is contaminated, damaged or it has failed • PCM has failed
DTC: P0153 **2T PCM MIL: YES** **Years:** 2006, 2007, 2008 **Models:** All **Engines:** All **Transmissions:** All	**HO2S-21 (Bank 2 Sensor 1) Slow Response** Engine started, engine runtime over 3 minutes, ECT sensor more than 170°F, vehicle driven to a speed over 24 MPH for 75 seconds, then at idle speed from 512-864 RPM, PSPS signal "off", A/C clutch not cycling, and the PCM detected the HO2S-11 signal that did not reach 670 mv, or it did not switch between 0.39-6.0v engine times in a 6 second period during the Oxygen Sensor Monitor test. **Possible Causes:** • Exhaust leak present in the exhaust manifold or exhaust pipes • HO2S element fuel contamination • HO2S element has deteriorated • PCM has failed
DTC: P0154 **2T PCM MIL: YES** **Years:** 2006, 2007, 2008 **Models:** All **Engines:** All **Transmissions:** All	**HO2S-11 (Bank 1 Sensor 1) No Activity Detected** Engine started, engine runtime over 2 minutes, ECT sensor more than 176°F, and the PCM detected the HO2S-11 signal remained fixed near 0.50v for 1.5 minutes in the Oxygen Sensor Monitor test. **Possible Causes:** • Exhaust leak present in exhaust manifold or exhaust pipes • HO2S element fuel contamination or has deteriorated • HO2S signal circuit or the ground circuit has high resistance • HO2S heater element has failed, or the heater circuit is open • PCM has failed
DTC: P0157 **2T PCM MIL: YES** **Years:** 2006, 2007, 2008 **Models:** All **Engines:** All **Transmissions:** All	**HO2S-22 (Bank 2 Sensor 2) Signal Low Input** Engine started, ECT sensor signal less than 120°F at engine startup, engine run time over 3 seconds, and the PCM detected the HO2S-22 signal was less than 0.16v during the PCM test. **Possible Causes:** • HO2S signal circuit is open • HO2S signal circuit is shorted to ground • HO2S is damaged or it has failed • PCM has failed

DTC	Trouble Code Title, Conditions & Possible Causes
DTC: P0158 **2T PCM MIL: YES** **Years:** 2006, 2007, 2008 **Models:** All **Engines:** All **Transmissions:** All	**HO2S-22 (Bank 2 Sensor 2) Signal High Input** Engine started, engine runtime over 2 minutes, ECT sensor more than 176°F, and the PCM detected the HO2S-22 signal was more than 1.2v for 3 seconds during the PCM test. **Possible Causes:** • HO2S signal circuit shorted to power (check the heater circuit) • HO2S is damaged or it has failed • PCM has failed
DTC: P0159 **2T PCM MIL: YES** **Years:** 2006, 2007, 2008 **Models:** All **Engines:** All **Transmissions:** All	**HO2S-22 (Bank 2 Sensor 2) Slow Response** Engine started, engine runtime over 3 minutes, ECT sensor more than 170°F, vehicle driven to a speed over 24 MPH for 75 seconds, then at idle speed from 512-864 RPM, PSPS signal "off", A/C clutch not cycling, and the PCM detected the HO2S-21 signal that did not reach 670 mv, or it did not switch between 0.39-6.0v engine times in a 6 second period during the Oxygen Sensor Monitor test. **Possible Causes:** • Exhaust leak present in the exhaust manifold or exhaust pipes • HO2S element fuel contamination • HO2S element has deteriorated • PCM has failed
DTC: P0160 **2T PCM MIL: YES** **Years:** 2006, 2007, 2008 **Models:** All **Engines:** All **Transmissions:** All	**HO2S-22 (Bank 2 Sensor 2) No Activity Detected** Engine started, engine runtime over 2 minutes, ECT sensor more than 176°F, and the PCM detected the HO2S-22 signal remained fixed near 0.50v for 1.5 minutes in the Oxygen Sensor Monitor test. **Possible Causes:** • Exhaust leak present in exhaust manifold or exhaust pipes • HO2S element fuel contamination or has deteriorated • HO2S signal circuit or the ground circuit has high resistance • HO2S heater element has failed, or the heater circuit is open • PCM has failed
DTC: P0161 **1T O2S HTR1 MIL: YES** **Years:** 2006, 2007, 2008 **Models:** Raider **Engines:** 3.7L K, 4.7L M **Transmissions:** All	**HO2S-12 (Bank 2 Sensor 2) Heater Circuit Malfunction** Engine started, ECT sensor more than 68°F, system voltage from 11-16v, and the PCM detected the HO2S-22 heater current was less than 0.20v, or it was more than 3.5 amps for 6 seconds in the test. **Possible Causes:** • HO2S heater control circuit is open or shorted to ground • HO2S heater control circuit is shorted to power • HO2S heater power circuit is open (check power from the relay) • HO2S heater is damaged or has failed • PCM has failed
DTC: P0171 **2T FUEL MIL: YES** **Years:** 2006, 2007, 2008 **Models:** All **Engines:** All **Transmissions:** All	**Fuel System Lean (Bank 1)** DTC P0101, P0102, P0103, P0106, P0107, P0108, P0111, P0112, P0113, P0117, P0118, P0121, P0122, P0123, P0131, P0132, P0133, P0134, P0135, P0138, P0141, P0151, P0152, P0153, P0154, P0155, P0157, P0158, P0159, P0161, P0201-P0206, P0300, P0301-306, P0441, P0500 and P0505 not set, ECT signal over 170°F, vehicle driven at a steady speed of over 62 MPH at a speed over 1500 RPM, and the PCM detected the Long Term fuel trim was +25 percent and the Short Term fuel trim was +12 percent indicating a lean condition during the test. **Possible Causes:** • Air leaks after the MAF sensor, or in the EGR or PCV system • Base engine "mechanical" fault affecting one or more cylinders • Fuel control sensor is out of calibration (i.e., ECT, IAT or MAP) • Fuel delivery system supplying too little fuel during cruise or idle periods (e.g., faulty fuel pump or dirty, restricted fuel filter) • Fuel injector (one or more) dirty or pressure regulator has failed • HO2S is contaminated, deteriorated or it has failed • Vehicle driven low on fuel or until it ran out of fuel

DTC	Trouble Code Title, Conditions & Possible Causes
DTC: P0172 **2T FUEL MIL: YES** **Years:** 2006, 2007, 2008 **Models:** All **Engines:** All **Transmissions:** All	**Fuel System Rich (Bank 1)** DTC P0101, P0102, P0103, P0106, P0107, P0108, P0111, P0112, P0113, P0117, P0118, P0121, P0122, P0123, P0131, P0132, P0133, P0134, P0135, P0138, P0141, P0151, P0152, P0153, P0154, P0155, P0157, P0158, P0159, P0161, P0201-P0206, P0300, P0301-306, P0441, P0500 and P0505 not set, ECT signal over 170°F, vehicle driven at a steady speed of over 62 MPH at a speed over 1500 RPM, and the PCM detected the Long Term fuel trim was -25 percent and the Short Term fuel trim was -12 percent indicating a lean condition during the test. **Possible Causes:** • Base engine "mechanical" fault affecting one or more cylinders • EVAP system component has failed or canister fuel saturated • Exhaust leaks located in front of the A/FS or HO2S location • Fuel control sensor is out of calibration (i.e., ECT, IAT or MAF) • Fuel delivery system supplying too much fuel during cruise or idle periods (e.g., faulty fuel pump, or faulty pressure regulator) • Fuel injector(s) is leaking or stuck partially open (one or more) • HO2S is contaminated, deteriorated or it has failed
DTC: P0174 **2T FUEL MIL: YES** **Years:** 2006, 2007, 2008 **Models:** All **Engines:** All **Transmissions:** All	**Fuel System Lean (Bank 2)** DTC P0101, P0102, P0103, P0106, P0107, P0108, P0111, P0112, P0113, P0117, P0118, P0121, P0122, P0123, P0131, P0132, P0133, P0134, P0135, P0138, P0141, P0151, P0152, P0153, P0154, P0155, P0157, P0158, P0159, P0161, P0201-P0206, P0300, P0301-306, P0441, P0500 and P0505 not set, ECT signal over 170°F, vehicle driven at a steady speed of over 62 MPH at a speed over 1500 RPM, and the PCM detected the Long Term fuel trim was +25 percent and the Short Term fuel trim was +12 percent indicating a lean condition during the test. **Possible Causes:** • Air leaks after the MAF sensor, or in the EGR or PCV system • Base engine "mechanical" fault affecting one or more cylinders • Fuel control sensor is out of calibration (i.e., ECT, IAT or MAP) • Fuel delivery system supplying too little fuel during cruise or idle periods (e.g., faulty fuel pump or dirty, restricted fuel filter) • Fuel injector (one or more) dirty or pressure regulator has failed • HO2S is contaminated, deteriorated or it has failed • Vehicle driven low on fuel or until it ran out of fuel
DTC: P0175 **2T FUEL MIL: YES** **Years:** 2006, 2007, 2008 **Models:** All **Engines:** All **Transmissions:** All	**Fuel System Rich (Bank 2)** DTC P0101, P0102, P0103, P0106, P0107, P0108, P0111, P0112, P0113, P0117, P0118, P0121, P0122, P0123, P0131, P0132, P0133, P0134, P0135, P0138, P0141, P0151, P0152, P0153, P0154, P0155, P0157, P0158, P0159, P0161, P0201-P0206, P0300, P0301-306, P0441, P0500 and P0505 not set, ECT signal over 170°F, vehicle driven at a steady speed of over 62 MPH at a speed over 1500 RPM, and the PCM detected the Long Term fuel trim was -25 percent and the Short Term fuel trim was -12 percent indicating a lean condition during the test. **Possible Causes:** • Base engine "mechanical" fault affecting one or more cylinders • EVAP system component has failed or canister fuel saturated • Exhaust leaks located in front of the A/FS or HO2S location • Fuel control sensor is out of calibration (i.e., ECT, IAT or MAF) • Fuel delivery system supplying too much fuel during cruise or idle periods (e.g., faulty fuel pump, or faulty pressure regulator) • Fuel injector(s) is leaking or stuck partially open (one or more) • HO2S is contaminated, deteriorated or it has failed
DTC: P0181 **2T PCM MIL: YES** **Years:** 2006, 2007, 2008 **Models:** All **Engines:** All **Transmissions:** All	**Fuel Temperature Sensor Range/Performance** Engine started, ECT sensor from 14-97°F at startup, engine running, IAT sensor more than 9°F, vehicle driven to a speed over 17 MPH, ECT sensor more than 140°F, and the PCM detected the Fuel Temperature sensor signal was different than the ECT sensor at startup by more than 27°F during the PCM test. **Possible Causes:** • Fuel temperature sensor signal circuit is open (intermittent) • Fuel temperature sensor signal circuit is shorted to ground • Fuel temperature sensor is damaged or has failed • PCM has failed
DTC: P0182 **2T PCM MIL: YES** **Years:** 2006, 2007, 2008 **Models:** All **Engines:** All **Transmissions:** All	**Fuel Temperature Sensor Circuit Low Input** Key on or engine running, and the PCM detected the Fuel Temperature sensor indicated less than 0.10v for 2 seconds during the PCM test. **Possible Causes:** • Fuel temperature sensor signal circuit is shorted to ground • Fuel temperature sensor is damaged or has failed • PCM has failed

DTC	Trouble Code Title, Conditions & Possible Causes
DTC: P0183 **2T PCM MIL: YES** **Years:** 2006, 2007, 2008 **Models:** All **Engines:** All **Transmissions:** All	**Fuel Temperature Sensor Circuit High Input** Key on or engine running, and the PCM detected the Fuel Temperature sensor indicated more than 4.60v for 2 seconds during the PCM test. **Possible Causes:** • Fuel temperature sensor signal circuit is open • Fuel temperature sensor signal circuit shorted to system power • Fuel temperature sensor ground circuit is open • Fuel temperature sensor is damaged or has failed • PCM has failed
DTC: P0201 **2T PCM MIL: YES** **Years:** 2006, 2007, 2008 **Models:** All **Engines:** All **Transmissions:** All	**Fuel Injector 1 Circuit Malfunction** Engine started, engine speed less than 1000 RPM, TP sensor less than 0.7v, Actuator Tests all "off", and the PCM detected the injector coil surge voltage was too low (i.e., it did not reach system voltage +2v) within 2 seconds after the injector turned "off" during the test. **Possible Causes:** • Injector control circuit is open or shorted to ground • Injector power circuit is open (check power from the MFI relay) • Fuel injector is damaged or has failed
DTC: P0202 **2T PCM MIL: YES** **Years:** 2006, 2007, 2008 **Models:** All **Engines:** All **Transmissions:** All	**Fuel Injector 2 Circuit Malfunction** Engine started, engine speed less than 1000 RPM, TP sensor less than 0.7v, Actuator Tests all "off", and the PCM detected the injector coil surge voltage was too low (i.e., it did not reach system voltage +2v) within 2 seconds after the injector turned "off" during the test. **Possible Causes:** • Injector control circuit is open or shorted to ground • Injector power circuit is open (check power from the MFI relay) • Fuel injector is damaged or has failed
DTC: P0203 **2T PCM MIL: YES** **Years:** 2006, 2007, 2008 **Models:** All **Engines:** All **Transmissions:** All	**Fuel Injector 3 Circuit Malfunction** Engine started, engine speed less than 1000 RPM, TP sensor less than 0.7v, Actuator Tests all "off", and the PCM detected the injector coil surge voltage was too low (i.e., it did not reach system voltage +2v) within 2 seconds after the injector turned "off" during the test. **Possible Causes:** • Injector control circuit is open or shorted to ground • Injector power circuit is open (check power from the MFI relay) • Fuel injector is damaged or has failed
DTC: P0204 **2T PCM MIL: YES** **Years:** 2006, 2007, 2008 **Models:** All **Engines:** All **Transmissions:** All	**Fuel Injector 4 Circuit Malfunction** Engine started, engine speed less than 1000 RPM, TP sensor less than 0.7v, Actuator Tests all "off", and the PCM detected the injector coil surge voltage was too low (i.e., it did not reach system voltage +2v) within 2 seconds after the injector turned "off" during the test. **Possible Causes:** • Injector control circuit is open or shorted to ground • Injector power circuit is open (check power from the MFI relay) • Fuel injector is damaged or has failed
DTC: P0205 **2T PCM MIL: YES** **Years:** 2006, 2007, 2008 **Models:** Eclipse, Spyder, Endeavor, Galant, Outlander **Engines:** 3.8L, 3.0L X **Transmissions:** All	**Fuel Injector 5 Circuit Malfunction** Engine started, engine speed less than 1000 RPM, TP sensor less than 1.1v, Actuator Tests all "off", and the PCM detected the injector coil surge voltage was too low (i.e., it did not reach system voltage +2v) within 2 seconds after the injector turned "off" during the test. **Possible Causes:** • Injector control circuit is open or shorted to ground • Injector power circuit is open (check power from the MFI relay) • Fuel injector is damaged or has failed
DTC: P0206 **2T PCM MIL: YES** **Years:** 2006, 2007, 2008 **Models:** Eclipse, Spyder, Endeavor, Galant, Outlander **Engines:** 3.8L, 3.0L X **Transmissions:** All	**Fuel Injector 6 Circuit Malfunction** Engine started, engine speed less than 1000 RPM, TP sensor less than 1.1v, Actuator Tests all "off", and the PCM detected the injector coil surge voltage was too low (i.e., it did not reach system voltage +2v) within 2 seconds after the injector turned "off" during the test. **Possible Causes:** • Injector control circuit is open or shorted to ground • Injector power circuit is open (check power from the MFI relay) • Fuel injector is damaged or has failed

DTC	Trouble Code Title, Conditions & Possible Causes
DTC: P0207 **2T PCM MIL: YES** **Years:** 2006, 2007, 2008 **Models:** Raider **Engines:** 4.7L M **Transmissions:** All	**Fuel Injector 7 Circuit** Engine started, engine speed less than 1000 RPM, TP sensor less than 1.1v, Actuator Tests all "off", and the PCM detected the injector coil surge voltage was too low (i.e., it did not reach system voltage +2v) within 2 seconds after the injector turned "off" during the test. **Possible Causes:** • Injector control circuit is open or shorted to ground • Injector power circuit is open (check power from the MFI relay) • Fuel injector is damaged or has failed
DTC: P0208 **2T PCM MIL: YES** **Years:** 2006, 2007, 2008 **Models:** Raider **Engines:** 4.7L M **Transmissions:** All	**Fuel Injector 8 Circuit** Engine started, engine speed less than 1000 RPM, TP sensor less than 1.1v, Actuator Tests all "off", and the PCM detected the injector coil surge voltage was too low (i.e., it did not reach system voltage +2v) within 2 seconds after the injector turned "off" during the test. **Possible Causes:** • Injector control circuit is open or shorted to ground • Injector power circuit is open (check power from the MFI relay) • Fuel injector is damaged or has failed
DTC: P0222 **1T PCM MIL: YES** **Years:** 2006, 2007, 2008 **Models:** All **Engines:** All **Transmissions:** All	Throttle Position Sensor (Sub) Circuit Low Input Key on, TPS (sub) output voltage is 2.5 volts or less for 0.5 seconds. **Possible Causes:** • TPS failed • TPS open or shorted • PCM has failed
DTC: P0223 **1T PCM MIL: YES** **Years:** 2006, 2007, 2008 **Models:** All **Engines:** All **Transmissions:** All	**Throttle Position Sensor (Sub) Circuit Low Input** Key on, TPS (sub) output voltage is 4.5 volts or more for 0.5 seconds. **Possible Causes:** • TPS failed • TPS open or shorted • PCM has failed
DTC: P0234 **2T PCM MIL: YES** **Years:** 2006, 2007, 2008 **Models:** Evolution **Engines:** 2.0L **Transmissions:** All	**Turbocharger Wastegate System Malfunction** Engine running, volumetric efficiency is greater than 210–230 percent. **Possible Causes:** • Turbocharger wastegate actuator failed • Charging pressure control system failed • PCM failed
DTC: P0243 **2T PCM MIL: YES** **Years:** 2006, 2007, 2008 **Models:** Evolution **Engines:** 2.0L **Transmissions:** All	**Turbocharger Wastegate Solenoid Circuit Malfunction** Engine started, engine running with the system voltage between 10–16 volts, and the PCM detected the Wastegate solenoid coil surge voltage was not detected with wastegate solenoid is duty cycled between 10 percent and 90 percent too low (system voltage +2v) for 1 second. **Possible Causes:** • Turbocharger wastegate solenoid circuit is open or shorted to ground • Turbocharger wastegate solenoid failed • PCM has failed
DTC: P0247 **2T PCM MIL: YES** **Years:** 2006, 2007, 2008 **Models:** Evolution **Engines:** 2.0L **Transmissions:** All	**Turbocharger Wastegate Solenoid 2 Circuit Malfunction** Engine started, engine running with the system voltage between 10–16 volts, and the PCM detected the Wastegate solenoid coil surge voltage was not detected with wastegate solenoid is duty cycled between 10 percent and 90 percent too low (system voltage +2v) for 1 second. **Possible Causes:** • Turbocharger wastegate solenoid circuit is open or shorted to ground • Turbocharger wastegate solenoid failed • PCM has failed

DTC	Trouble Code Title, Conditions & Possible Causes
DTC: P0300 **2T MISFIRE MIL: YES** **Years:** 2006, 2007, 2008 **Models:** All **Engines:** All **Transmissions:** All	**Random Misfire Detected** DTC P0100, P0105, P0110, P0115, P0120, P0130, P0135, P0136, P0141, P0151, P0155, P0156, P0161, P0440, P0500 and P0505 not set, vehicle driven to a speed of over 2 MPH at a steady throttle with the engine speed from 440-6500 RPM, ECT and IAT sensors more than 14°F, CKP sensor "learn" finished, and the PCM detected multiple cylinders misfiring in over 1.8 percentof engine cycles in 200 revolutions (Catalyst Damaging Misfire); or it detected multiple cylinders misfiring in more than 1.8 percent of engine cycles within 1000 revolutions (High Emissions Misfire) during the Misfire Monitor test. **Note: If the misfire is severe, the MIL will flash on/off on the 1st trip!** **Possible Causes:** • Air leak in the intake manifold, or in the EGR or PCM system • Base engine mechanical fault that affects one or more cylinders • Erratic or interrupted CKP or CMP sensor signals • Fuel delivery component fault that affects one or more cylinders (i.e., a contaminated, dirty or sticking fuel injector) • Ignition system problem (coil or plug) in one or more cylinders
DTC: P0301 **2T MISFIRE MIL: YES** **Years:** 2006, 2007, 2008 **Models:** All **Engines:** All **Transmissions:** All	**Cylinder 1 Misfire Detected** DTC P0100, P0105, P0110, P0115, P0120, P0130, P0135, P0136, P0141, P0151, P0155, P0156, P0161, P0440, P0500 and P0505 not set, vehicle driven to a speed of over 2 MPH at a steady throttle with the engine speed from 440-6500 RPM, ECT and IAT sensors more than 14°F, CKP sensor "learn" finished, and the PCM detected a misfire in one cylinder in over 1.8 percent of engine cycles within 200 revolutions (Catalyst Damaging Misfire); or it detected a misfire in one cylinder in more than 1.8 percent of engine cycles within 1000 revolutions (High Emissions Misfire) in the Misfire Monitor test. **Note: If the misfire is severe, the MIL will flash on/off on the 1st trip!** **Possible Causes:** • Air leak in the intake manifold, or in the EGR or PCM system • Base engine mechanical fault that affects only one cylinder • Fuel delivery component fault that affects only one cylinder (i.e., a contaminated, dirty or sticking fuel injector) • Ignition system problem (coil or plug) that affects one cylinder • TSB 00-11-006 (7/00) has information about this code (V6 only)
DTC: P0302 **2T MISFIRE MIL: YES** **Years:** 2006, 2007, 2008 **Models:** All **Engines:** All **Transmissions:** All	**Cylinder 2 Misfire Detected** DTC P0100, P0105, P0110, P0115, P0120, P0130, P0135, P0136, P0141, P0151, P0155, P0156, P0161, P0440, P0500 and P0505 not set, vehicle driven to a speed of over 2 MPH at a steady throttle with the engine speed from 440-6500 RPM, ECT and IAT sensors more than 14°F, CKP sensor "learn" finished, and the PCM detected a misfire in one cylinder in over 1.8 percent of engine cycles within 200 revolutions (Catalyst Damaging Misfire); or it detected a misfire in one cylinder in more than 1.8 percent of engine cycles within 1000 revolutions (High Emissions Misfire) in the Misfire Monitor test. **Note: If the misfire is severe, the MIL will flash on/off on the 1st trip!** **Possible Causes:** • Air leak in the intake manifold, or in the EGR or PCM system • Base engine mechanical fault that affects only one cylinder • Fuel delivery component fault that affects only one cylinder (i.e., a contaminated, dirty or sticking fuel injector) • Ignition system problem (coil or plug) that affects one cylinder
DTC: P0303 **2T MISFIRE MIL: YES** **Years:** 2006, 2007, 2008 **Models:** All **Engines:** All **Transmissions:** All	**Cylinder 3 Misfire Detected** Engine started, vehicle speed over 1.6 MPH at a steady throttle with the engine speed from 440-6500 RPM for 5 seconds, ECT and IAT sensors more than 14°F, CKP sensor "learn" finished, and the PCM detected a misfire in one cylinder in over 1.8 percent of engine cycles within 200 revolutions (Catalyst Damaging Misfire); or it detected a misfire in one cylinder in more than 1.8 percent of engine cycles within 1000 revolutions (High Emissions Misfire) in the Misfire Monitor test. **Note: If the misfire is severe, the MIL will flash on/off on the 1st trip!** **Possible Causes:** • Air leak in the intake manifold, or in the EGR or PCM system • Base engine mechanical fault that affects only one cylinder • Fuel delivery component fault that affects only one cylinder (i.e., a contaminated, dirty or sticking fuel injector) • Ignition system problem (coil or plug) that affects one cylinder • TSB 00-11-006 (7/00) has information about this code (V6 only)

DTC	Trouble Code Title, Conditions & Possible Causes
DTC: P0304 **2T MISFIRE MIL: YES** **Years:** 2006, 2007, 2008 **Models:** All **Engines:** All **Transmissions:** All **MIL: YES**	**Cylinder 4 Misfire Detected** DTC P0100, P0105, P0110, P0115, P0120, P0130, P0135, P0136, P0141, P0151, P0155, P0156, P0161, P0440, P0500 and P0505 not set, vehicle driven to a speed of over 2 MPH at a steady throttle with the engine speed from 440-6500 RPM, ECT and IAT sensors more than 14°F, CKP sensor "learn" finished, and the PCM detected a misfire in one cylinder in over 1.8 percent of engine cycles within 200 revolutions (Catalyst Damaging Misfire); or it detected a misfire in one cylinder in more than 1.8 percent of engine cycles within 1000 revolutions (High Emissions Misfire) in the Misfire Monitor test. **Note: If the misfire is severe, the MIL will flash on/off on the 1st trip!** **Possible Causes:** • Air leak in the intake manifold, or in the EGR or PCM system • Base engine mechanical fault that affects only one cylinder • Fuel delivery component fault that affects only one cylinder (i.e., a contaminated, dirty or sticking fuel injector) • Ignition system problem (coil or plug) that affects one cylinder
DTC: P0305 **2T MISFIRE MIL: YES** **Years:** 2006, 2007, 2008 **Models:** Eclipse, Spyder, Endeavor, Galant, Outlander **Engines:** 3.8L, 3.0L X **Transmissions:** All	**Cylinder 5 Misfire Detected** DTC P0100, P0105, P0110, P0115, P0120, P0130, P0135, P0136, P0141, P0151, P0155, P0156, P0161, P0440, P0500 and P0505 not set, vehicle driven to a speed of over 2 MPH at a steady throttle with the engine speed from 440-6500 RPM, ECT and IAT sensors more than 14°F, CKP sensor "learn" finished, and the PCM detected a misfire in one cylinder in over 1.8 percent of engine cycles within 200 revolutions (Catalyst Damaging Misfire); or it detected a misfire in one cylinder in more than 1.8 percentof engine cycles within 1000 revolutions (High Emissions Misfire) in the Misfire Monitor test. **Note: If the misfire is severe, the MIL will flash on/off on the 1st trip!** **Possible Causes:** • Air leak in the intake manifold, or in the EGR or PCM system • Base engine mechanical fault that affects only one cylinder • Fuel delivery component fault that affects only one cylinder (i.e., a contaminated, dirty or sticking fuel injector) • Ignition system problem (coil or plug) that affects one cylinder • TSB 00-11-006 (7/00) has information about this code (V6 only)
DTC: P0306 **2T MISFIRE MIL: YES** **Years:** 2006, 2007, 2008 **Models:** Eclipse, Spyder, Endeavor, Galant, Outlander **Engines:** 3.8L, 3.0L X **Transmissions:** All	**Cylinder 6 Misfire Detected** DTC P0100, P0105, P0110, P0115, P0120, P0130, P0135, P0136, P0141, P0151, P0155, P0156, P0161, P0440, P0500 and P0505 not set, vehicle driven to a speed of over 2 MPH at a steady throttle with the engine speed from 440-6500 RPM, ECT and IAT sensors more than 14°F, CKP sensor "learn" finished, and the PCM detected a misfire in one cylinder in over 1.8 percent of engine cycles within 200 revolutions (Catalyst Damaging Misfire); or it detected a misfire in one cylinder in more than 1.8 percent of engine cycles within 1000 revolutions (High Emissions Misfire) in the Misfire Monitor test. **Note: If the misfire is severe, the MIL will flash on/off on the 1st trip!** **Possible Causes:** • Air leak in the intake manifold, or in the EGR or PCM system • Base engine mechanical fault that affects only one cylinder • Fuel delivery component fault that affects only one cylinder (i.e., a contaminated, dirty or sticking fuel injector) • Ignition system problem (coil or plug) that affects one cylinder
DTC: P0307 **2T MISFIRE MIL: YES** **Years:** 2006, 2007, 2008 **Models:** Raider **Engines:** 4.7L M **Transmissions:** All	**Cylinder 7 Misfire Detected** DTC P0100, P0105, P0110, P0115, P0120, P0130, P0135, P0136, P0141, P0151, P0155, P0156, P0161, P0440, P0500 and P0505 not set, vehicle driven to a speed of over 2 MPH at a steady throttle with the engine speed from 440-6500 RPM, ECT and IAT sensors more than 14°F, CKP sensor "learn" finished, and the PCM detected a misfire in one cylinder in over 1.8 percent of engine cycles within 200 revolutions (Catalyst Damaging Misfire); or it detected a misfire in one cylinder in more than 1.8 percent of engine cycles within 1000 revolutions (High Emissions Misfire) in the Misfire Monitor test. **Note: If the misfire is severe, the MIL will flash on/off on the 1st trip!** **Possible Causes:** • Air leak in the intake manifold, or in the EGR or PCM system • Base engine mechanical fault that affects only one cylinder • Fuel delivery component fault that affects only one cylinder (i.e., a contaminated, dirty or sticking fuel injector) • Ignition system problem (coil or plug) that affects one cylinder
DTC: P0308 **2T MISFIRE MIL: YES** **Years:** 2006, 2007, 2008 **Models:** Raider **Engines:** 4.7L M **Transmissions:** All	**Cylinder 8 Misfire Detected** DTC P0100, P0105, P0110, P0115, P0120, P0130, P0135, P0136, P0141, P0151, P0155, P0156, P0161, P0440, P0500 and P0505 not set, vehicle driven to a speed of over 2 MPH at a steady throttle with the engine speed from 440-6500 RPM, ECT and IAT sensors more than 14°F, CKP sensor "learn" finished, and the PCM detected a misfire in one cylinder in over 1.8 percent of engine cycles within 200 revolutions (Catalyst Damaging Misfire); or it detected a misfire in one cylinder in more than 1.8 percent of engine cycles within 1000 revolutions (High Emissions Misfire) in the Misfire Monitor test. **Note: If the misfire is severe, the MIL will flash on/off on the 1st trip!** **Possible Causes:** • Air leak in the intake manifold, or in the EGR or PCM system • Base engine mechanical fault that affects only one cylinder • Fuel delivery component fault that affects only one cylinder (i.e., a contaminated, dirty or sticking fuel injector) • Ignition system problem (coil or plug) that affects one cylinder

DTC	Trouble Code Title, Conditions & Possible Causes
DTC: P0315 **1T PCM MIL: YES** **Years:** 2006, 2007, 2008 **Models:** Raider **Engines:** 3.7L K, 4.7L M **Transmissions:** All	**No Crank Sensor Learned** One of the CKP sensor target windows has more than 2.86 percentvariance from the reference. Three good trips to turn off the MIL. **Possible Causes:** • Tone wheel/pulse ring • Wire harness • Crankshaft position sensor
DTC: P0325 **2T PCM MIL: YES** **Years:** 2006, 2007, 2008 **Models:** All **Engines:** All **Transmissions:** All	**Knock Sensor 1 Circuit Malfunction** Engine started, engine speed from -3000 RPM, and the PCM detected the change in the Knock sensor 1 signal was less than 0.06v, or the signal was over 5.0v during the last 200 revolutions. **Possible Causes:** • Knock sensor signal circuit is open or shorted to ground • Knock sensor signal circuit is shorted to VREF or system power • Knock sensor is damaged or has failed • Verify the Knock Sensor (KS) is tightened its specification • PCM has failed
DTC: P0326 **2T PCM MIL: YES** **Years:** 2006, 2007, 2008 **Models:** Eclipse, Lancer Evolution, Spyder, Endeavor, Galant, Outlander **Engines:** 2.0L, 2.4L W, 3.0L X, 3.8L **Transmissions:** All	**Knock Sensor Circuit Performance** Knock sensor 1 output voltage (knock sensor peak voltage in each 1/3 turn of the crankshaft) has not changed more than 0.06 volt in the last consecutive 200 periods. **Possible Causes:** • Knock sensor 1 failed. • Harness damage. • Connector damage. • ECM failed.
DTC: P0327 **2T PCM MIL: YES** **Years:** 2006, 2007, 2008 **Models:** Eclipse, Lancer Evolution, Spyder, Endeavor, Galant, Outlander **Engines:** 2.0L, 2.4L W, 3.0L X, 3.8L **Transmissions:** All	**Knock Sensor Circuit Low** Knock sensor 1 output voltage has continued to be lower than 0.5 volt for 2 seconds. **Possible Causes:** • Knock sensor 1 failed. • Harness damage. • Connector damage. • ECM failed.
DTC: P0328 **2T PCM MIL: YES** **Years:** 2006, 2007, 2008 **Models:** Eclipse, Lancer Evolution, Spyder, Endeavor, Galant, Outlander **Engines:** 2.0L, 2.4L W, 3.0L X, 3.8L **Transmissions:** All	**Knock Sensor Circuit High** Knock sensor 1 output voltage has continued to be higher than 2.25 volts for 2 seconds. **Possible Causes:** • Knock sensor 1 failed. • Harness damage. • Connector damage. • ECM failed.
DTC: P0330 **2T PCM MIL: YES** **Years:** 2006, 2007, 2008 **Models:** All **Engines:** All **Transmissions:** All	**Knock Sensor 2 Circuit Malfunction** Engine started, engine speed from 3000 RPM, and the PCM detected the change in the Knock sensor 1 signal was less than 0.06v, or the signal was over 5.0v during the last 200 revolutions. **Possible Causes:** • Knock sensor signal circuit is open or shorted to ground • Knock sensor signal circuit is shorted to VREF or system power • Knock sensor is damaged or has failed • Verify the Knock Sensor (KS) is tightened its specification • PCM has failed
DTC: P0335 **1T PCM MIL: YES** **Years:** 2006, 2007, 2008 **Models:** All **Engines:** All **Transmissions:** All	**Crankshaft Position Sensor Circuit Malfunction** Engine cranking, and the PCM did not detect any Crankshaft Position (CKP) sensor signals; or with the engine running, it detected an abnormal pattern of CKP and CMP signal pulses for 2 seconds during the PCM test. **Possible Causes:** • CKP sensor signal circuit is open or shorted to ground • CKP sensor ground circuit is open • CKP sensor power circuit is open (check power to MFI relay) • CKP sensor is damaged or has failed • PCM has failed (the PCM provides the 5v VREF to the sensor)

DTC	Trouble Code Title, Conditions & Possible Causes
DTC: P0340 **2T PCM MIL: YES** **Years:** 2006, 2007, 2008 **Models:** All **Engines:** All **Transmissions:** All	**Camshaft Position Sensor Circuit Malfunction** Engine cranking, and the PCM did not detect any Crankshaft Position (CMP) sensor signals, or with the engine running, the normal pattern of CKP and CMP signals was not detected for 2 seconds. **Possible Causes:** • CMP sensor signal circuit is open or shorted to ground • CMP sensor ground circuit is open • CMP sensor power circuit is open (check power to MFI relay) • CMP sensor is damaged or has failed • PCM has failed (the PCM provides the 5v VREF to the sensor)
DTC: P0365 **2T PCM MIL: YES** **Years:** 2006, 2007, 2008 **Models:** Evolution, Outlander **Engines:** 2.0L, 2.4L W **Transmissions:** All	**Intake Camshaft Position Sensor Circuit** Engine running with engine speed greater than 50 RPM, the normal signal patter has not been input for the cylinder ID from the crankshaft position sensor signal and intake camshaft position sensor signal for 2 seconds. **Possible Causes:** • Intake camshaft position sensor failed • Intake camshaft position sensor circuit shorted or open • PCM has failed
DTC: P0401 **2T EGR MIL: YES** **Years:** 2006, 2007, 2008 **Models:** Lancer **Engines:** 2.0L U **Transmissions:** All	**Insufficient EGR Flow Detected** Engine started, engine runtime over 3 minutes in closed loop, ECT signal over 170°F, vehicle driven to over 3 MPH at a speed of 1952-2400 RPM, MAP signal from 1.80-2.70v, TP sensor from 0.6-1.8v, Short Term fuel trim less than +4.4 percent, and the PCM detected the measured change in the Short Term fuel trim compensation shift was less than 7.4 percent, or the measured change in Short Term fuel trim compensation was more than 20.5 percent during the EGR System test. **Possible Causes:** • EGR solenoid control circuit is open or shorted to ground • EGR vacuum hose to source vacuum is loose or disconnected • EGR exhaust tube is clogged or restricted • EGR valve assembly is damaged or has failed • PCM has failed
DTC: P0403 **2T PCM MIL: YES** **Years:** 2006, 2007, 2008 **Models:** Lancer **Engines:** 2.0L E **Transmissions:** All	**EGR Control Circuit Malfunction** Engine started, battery positive voltage between 10 and 16 volts, and the PCM detected the EGR Control solenoid coil surge voltage (2 volts) is not detected for 0.2 seconds, EGR solenoid duty cycled ON cycle between 10 percent and 90 percent, evaporative emission purge solenoid ON cycle is 0 percent, evaporative emission ventilation is OFF for more than 1 second, the EGR solenoid coil surge voltage is not detected for 1 second when the EGR solenoid is turned OFF. **Possible Causes:** • EGR solenoid control circuit is open or shorted to ground • EGR solenoid power circuit is open (check power from relay) • EGR solenoid is damaged or has failed • PCM has failed
DTC: P0404 **2T PCM MIL: YES** **Years:**: 2006, 2007, 2008 **Models:** Raider **Engines:** 3.7L K, 4.7L M **Transmissions:** All	**EGR Position Sensor Performance** The EGR flow or valve movement is not what is expected. A rationality error has been detected from the EGR Close Position Performance. Two trip fault. **Possible Causes:** • Excessive resistance in the EGR sensor signal circuit • Excessive resistance in the 5-volt supply circuit • Excessive resistance in the EGR solenoid control circuit • Excessive resistance in the sensor ground circuit • EGR solenoid ground circuit open • EGR solenoid assembly • PCM
DTC: P0405 **1T PCM MIL: YES** **Years:** 2006, 2007, 2008 **Models:** Raider **Engines:** 3.7L K, 4.7L M **Transmissions:** All	**EGR Position Sensor Circuit Low** EGR Position Sensor Signal is less than the minimum acceptable value. One trip Fault. **Possible Causes:** • 5-volt supply circuit open • 5-volt supply circuit shorted to ground • EGR position sensor signal circuit shorted to ground • EGR position sensor signal circuit shorted to the sensor ground circuit • Excessive resistance in the EGR signal circuit • EGR solenoid assembly • PCM

DTC	Trouble Code Title, Conditions & Possible Causes
DTC: P0406 **1T PCM MIL: YES** **Years:** 2006, 2007, 2008 **Models:** Raider **Engines:** 3.7L K, 4.7L M **Transmissions:** All	**EGR Position Sensor Circuit High** EGR position sensor signal is greater than the maximum acceptable value. One trip fault. **Possible Causes:** • EGR position sensor signal circuit shorted to the 5-volt supply circuit • EGR position sensor signal circuit shorted to voltage • EGR position sensor signal circuit open • Sensor ground circuit open • EGR solenoid assembly • PCM
DTC: P0420 **2T CAT1 MIL: YES** **Years:** 2006, 2007, 2008 **Models:** Lancer **Engines:** 2.0L **Transmissions:** All	**Catalyst System Efficiency Low (Bank 1)** DTC P0100, P0105, P0110, P0115, P0120, P0130, P0135, P0136, P0141, P0440, P0500 and P0505 not set, engine started, BARO sensor more than 76 kPa, IAT signal more than 14°F, vehicle driven to 45-60 MPH with the throttle open at a speed of less than 3000 RPM for 3-5 minutes, MAF signal from 63-169 Hz, and the PCM detected the switch rate of the rear HO2S-12 and the front HO2S-11 signals were too similar for 90 seconds during the Catalyst Monitor Test. **Possible Causes:** • Air leaks at the exhaust manifold or in the exhaust pipes • Catalytic converter is damaged, contaminated or has failed • Front HO2S or rear HO2S is contaminated with fuel or moisture • Front HO2S and/or the rear HO2S is loose in the mounting hole • Front HO2S older (aged) than the rear HO2S (HO2S is lazy)
DTC: P0421 **2T CAT2 MIL: YES** **Years:** 2006, 2007, 2008 **Models:** Lancer **Engines:** 2.0L E **Transmissions:** All	**Warmup Catalyst Efficiency Low (Bank 1)** DTC P0101, P0102, P0103, P0106, P0107, P0108, P0111, P0112, P0113, P0116, P0117, P0118, P0121, P0122, P0123, P0132, P0133, P0134, P0135, P0137, P0139, P0141 and P0500 not set, IAT sensor more than 14°F, BARO signal over 76 kPa, vehicle driven at 45-60 MPH in closed loop at a steady throttle for 5 minutes, MAF sensor at 69-169 Hz, and the PCM detected switch rate of the rear HO2S and front HO2S signals were too similar for 140 seconds. **Possible Causes:** • Air leaks at the exhaust manifold or in the exhaust pipes • Catalytic converter is damaged, contaminated or has failed • Front HO2S or rear HO2S is contaminated with fuel or moisture • Front HO2S and/or the rear HO2S is loose in the mounting hole • Front HO2S older (aged) than the rear HO2S (HO2S is lazy)
DTC: P0430 **2T CAT1 MIL: YES** **Years:** 2006, 2007, 2008 **Models:** Raider **Engines:** 3.7L K, 4.7L M **Transmissions:** All	**Warmup Catalyst Efficiency Low (Bank 2)** DTC P0106, P0107, P0108, P0111, P0112, P0113, P0117, P0118, P0121, P0122, P0123, P0131, P0132, P0133, P0134, P0135, P0138, P0141, P0201-P0204, P0300, P0301-P0304, P0441, P0500 and P0505 not set, ECT signal more than 170°F, vehicle driven to a speed of 45-60 MPH at an engine speed of 1248-2400 RPM, MAP at 1.5-2.6v, and the PCM detected the switch rate of the rear HO2S-12 reached 70 percent of the switch rate of front HO2S-11 during the test. **Possible Causes:** • Air leaks at the exhaust manifold or in the exhaust pipes • Catalytic converter is damaged, contaminated or has failed • Front HO2S or rear HO2S is contaminated with fuel or moisture • Front HO2S and/or the rear HO2S is loose in the mounting hole • Front HO2S older (aged) than the rear HO2S (HO2S is lazy)
DTC: P0431 **2T CAT2 MIL: YES** **Years:** 2006, 2007, 2008 **Models:** All **Engines:** All **Transmissions:** All	**Warmup Catalyst Efficiency Low (Bank 2)** DTC P0100, P0105, P0110, P0115, P0120, P0130, P0135, P0136, P0141, P0151, P0155, P0156, P0161, P0440, P0500 and P0505 not set, engine started, vehicle driven at 45-60 MPH in closed loop at a steady throttle for 5-10 minutes, BARO signal more than 76 kPa, MAF sensor from 69-169 Hz, IAT sensor more than14°F, and the PCM detected the rear HO2S-22 switch rate was similar to the HO2S-21 switch rate for 140 seconds in the Catalyst Monitor test. **Possible Causes:** • Air leaks at the exhaust manifold or in the exhaust pipes • Catalytic converter is damaged, contaminated or has failed • Front HO2S or rear HO2S is contaminated with fuel or moisture • Front HO2S and/or the rear HO2S is loose in the mounting hole • Front HO2S older (aged) than the rear HO2S (HO2S is lazy)

DTC	Trouble Code Title, Conditions & Possible Causes
DTC: P0440 **2T EVAP MIL: YES** **Years:** 2006, 2007, 2008 **Models:** Raider **Engines:** 3.7L K, 4.7L M **Transmissions:** All	**EVAP Control System Malfunction** Cold engine startup, engine running in closed loop for 3 minutes at idle speed, ECT sensor more than 176°F, then after the EVAP solenoid was commanded "on" for 3 seconds, the PCM detected less than a 3 percentvariation in the Short Term fuel trim during the test. **Possible Causes:** • Charcoal canister is loaded with fuel or moisture • ECT, IAT, MAF, VSS or TP sensor signal is out-of-calibration • Fuel filler cap loose, cross-threaded, incorrect part or damaged • Fuel tank pressure sensor is damaged or has failed • Fuel tank vapor line(s) blocked, damaged or disconnected • Purge Control solenoid valve circuit open or shorted to ground • PCM has failed
DTC: P0441 **2T EVAP MIL: YES** **Years:** 2006, 2007, 2008 **Models:** All **Engines:** All **Transmissions:** All	**EVAP Emission Control System Incorrect Purge Flow** During evaporative emission control system monitoring, 20 seconds after the Evap purge solenoid ON duty cycle is 0 percent, the pressure in the fuel tank is 0.29 psi or less for 0.1 second. **Note: This is a "functionality" test of the EVAP system (flow test).** **Possible Causes:** • Charcoal canister is damaged, clogged or restricted • Purge solenoid circuit is open (fault may be intermittent) • Purge solenoid power circuit is open (check the relay or fuse) • Purge valve vacuum line is clogged, restricted or disconnected • PCM has failed
DTC: P0442 **2T EVAP MIL: YES** **Years:** 2006, 2007, 2008 **Models:** All **Engines:** All **Transmissions:** All	**EVAP Control System Small Leak Detected** Cold engine startup (ECT and IAT sensor signals less than 86°F), BARO sensor more than 75 kPa, engine runtime over 16 minutes, volumetric efficiency from 20-80 percent, ECT sensor more than 140°F, TP sensor from 1-4v, PSP switch "off", vehicle driven to a speed of over 20 MPH at an engine speed over 1600 RPM, then with the EVAP purge and vent solenoids both closed and the pressure rise less than 0.065 psi, the PCM detected the change in internal pressure in the fuel tank was more than 0.122 psi (843 kPa) during the leak test. **Possible Causes:** • Canister Purge valve is damaged, leaking or it has failed • Charcoal canister is loaded with fuel or moisture • Fuel filler cap loose, cross-threaded, incorrect part or damaged • Fuel tank is cracked (leaking), or a leak exists in the 'O' ring • Fuel tank pressure sensor is damaged or has failed • Fuel vapor line(s), fuel pipes or hoses damaged or leaking • PCM has failed
DTC: P0443 **2T PCM MIL: YES** **Years:** 2006, 2007, 2008 **Models:** All **Engines:** All **Transmissions:** All	**EVAP Purge Control Solenoid Circuit Malfunction** Engine started, engine running at cruise speed for 3-5 minutes, system voltage over 10v, and the PCM detected the EVAP Purge Control solenoid coil surge voltage was too low (i.e., it did not reach system voltage +2v) after the EVAP Purge Control solenoid was turned "on" and then "off" during the PCM test. **Possible Causes:** • Purge solenoid circuit control is open or shorted to ground • Purge solenoid power circuit is open (test from the relay) • Purge control solenoid is damaged or has failed • PCM has failed
DTC: P0446 **2T PCM MIL: YES** **Years:** 2006, 2007, 2008 **Models:** All **Engines:** All **Transmissions:** All	**EVAP Vent Control Solenoid Circuit Malfunction** Engine started, system voltage over 10v, and the PCM detected the solenoid surge voltage did not reach system voltage (+2v) within 30 ms after the vent solenoid was commanded OFF during the test. **Possible Causes:** • Vent Control solenoid circuit is open or shorted to ground • Vent Control solenoid power circuit is open (test from the relay) • Vent Control solenoid is damaged or has failed • PCM has failed

DTC	Trouble Code Title, Conditions & Possible Causes
DTC: P0450 **2T PCM MIL: YES** **Years:** 2006, 2007, 2008 **Models:** All **Engines:** All **Transmissions:** All	**Fuel Tank Pressure Sensor Circuit Malfunction** Idle Test Engine running at cruise, followed by a deceleration period to under 1 MPH from over 10 MPH from an engine speed of 2500 RPM with the volumetric efficiency over 55 percent, and the PCM detected a sudden pressure change of over 0.20v occurred 20 times in a 5 ms period. Cruise Test Engine started, IAT sensor from 41-113°F, vehicle driven to a speed of over 19 MPH at an engine speed over 1600 RPM, volumetric efficiency from 20-80 percent, and the PCM detected the FTP sensor signal was more than 4.0v with Purge solenoid driven at a 100 percent duty cycle, or it detected the FTP sensor was less than 1v with the purge solenoid "off". **Possible Causes:** • FTP sensor signal circuit open or shorted to ground • FTP sensor ground circuit is open • FTP sensor power (VREF) circuit is open • FTP sensor is damaged or has failed • PCM has failed
DTC: P0451 **2T PCM MIL: YES** **Years:** 2006, 2007, 2008 **Models:** All **Engines:** All **Transmissions:** All	**Fuel Tank Pressure Sensor Range/Performance** Engine started, IAT sensor from 41-113°F, vehicle driven to a speed over 19 MPH at a speed of over 1600 RPM, volumetric efficiency from 20-80 percent, EVAP purge command at 100 percent, and the PCM detected the FTP sensor signal was over 4.0v; or less than 1.0v for 10 seconds. **Possible Causes:** • Fuel tank pressure sensor vacuum hoses loose or damaged • Fuel tank pressure sensor is damaged or out-of-calibration • PCM has failed
DTC: P0452 **2T PCM MIL: YES** **Years:** 2006, 2007, 2008 **Models:** All **Engines:** All **Transmissions:** All	**Emission Control System Pressure Sensor Circuit Low Input** Engine started, IAT sensor from 41-113°F, vehicle driven to an engine speed of over 1600 RPM, volumetric efficiency from 20-80 percent, EVAP purge commanded "off", and the PCM detected the FTP sensor signal was less than 1.0v for 10 seconds during the test. **Possible Causes:** • Fuel tank pressure sensor signal circuit is shorted to ground • Fuel tank pressure sensor power circuit is open • Fuel tank pressure sensor is damaged or out-of-calibration • PCM has failed
DTC: P0453 **2T PCM MIL: YES** **Years:** 2006, 2007, 2008 **Models:** All **Transmissions:** All **Engines:** All	**Emission Control System Pressure Sensor Circuit High Input** Engine started, IAT sensor from 41-113°F, vehicle driven to an engine speed of over 1600 RPM, volumetric efficiency from 20-80 percent, EVAP purge commanded "off", and the PCM detected the FTP sensor signal was less than 1.0v for 10 seconds during the test. **Possible Causes:** • Fuel tank pressure sensor signal circuit is shorted to ground • Fuel tank pressure sensor power circuit is open • Fuel tank pressure sensor is damaged or out-of-calibration • PCM has failed
DTC: P0455 **2T PCM MIL: YES** **Years:** 2006, 2007, 2008 **Models:** All **Engines:** All **Transmissions:** All	**EVAP Leak Monitor Gross Leak (0.080') Detected** ECT and IAT signals under 86°F at startup, BARO over 75 kPa, ECT signal over 140°F during testing, engine runtime over 16 minutes, TP sensor signal from 1-4v, volumetric efficiency from 20-80 percent, PSP switch "off", vehicle driven to a speed of over 20 MPH at an engine speed over 1600 RPM, then with both the purge and vent solenoids closed and the EVAP pressure rise less than 0.065 psi, the PCM detected the fluctuation of the pressure in fuel tank was less than 324 kPa (0.047 psi) for 20 seconds during the EVAP Leak test. **Possible Causes:** • Canister vent (CV) solenoid may be stuck in open position • EVAP canister tube, EVAP canister purge outlet tube or EVAP return tube disconnected or cracked, or canister is damaged • EVAP canister purge valve stuck closed, or canister damaged • Fuel filler cap missing, loose (not tightened) or the wrong part • Fuel vapor hoses/tubes blocked or restricted, or fuel vapor control valve tube or fuel vapor vent valve assembly blocked • Fuel tank pressure (FTP) sensor has failed (mechanical fault) • Fuel tank control valve is contaminated, damaged or has failed

DTC	Trouble Code Title, Conditions & Possible Causes
DTC: P0455 **2T EVAP MIL: YES** **Years:** 2006, 2007, 2008 **Models:** All **Engines:** All **Transmissions:** All	**EVAP Leak Monitor Gross Leak (0.080') Detected** Cold engine startup (ECT and IAT sensor signals less than 86°F), BARO sensor more than 75 kPa, engine runtime over 16 minutes, ECT sensor more than 140°F, TP sensor from 1-4v, vehicle driven to a speed over 20 MPH with the engine speed over 1600 RPM, then with the EVAP canister vent solenoid closed, and the Leak Detection pump commanded to a specific duty cycle to raise the pressure in the system, the PCM detected the LDP pump continued to operate on and off due to a gross leak in the EVAP system during the test. **Possible Causes:** • Charcoal canister is loaded with fuel or moisture • Canister Vent valve is damaged, leaking or it has failed • Fuel filler cap loose, cross-threaded, incorrect part or missing • Fuel tank is cracked (leaking), or a leak exists in the 'O' ring • Fuel tank filler tube assembly is damaged or leaking • Fuel vapor line(s), fuel pipes or hoses damaged or leaking • PCM has failed
DTC: P0456 **2T PCM MIL: YES** **Years:** 2006, 2007, 2008 **Models:** All **Engines:** All **Transmissions:** All	**EVAP Leak Monitor Very Small Leak (0.020') Detected** Cold engine startup (ECT and IAT sensor signals less than 97°F), BARO sensor more than 75 kPa, fuel temperature less than 97°F, volumetric efficiency from 20-70 percent, vehicle driven and after the fuel tank pressure sensor indicates 1-4v, and the PCM detected the internal fuel tank pressure changed more than 2 kPa in 128 seconds after the fuel tank and vapor line were closed during the test. This small change in pressure indicated a very small leak in the system. **Possible Causes:** • Canister Vent valve is damaged, leaking or it has failed • Fuel filler cap loose, cross-threaded or the incorrect part • Fuel tank is cracked (leaking), or a leak exists in the 'O' ring • Fuel tank filler tube assembly is damaged or leaking • Fuel vapor line(s), fuel pipes or hoses damaged or leaking • PCM has failed
DTC: P0457 **1T PCM MIL: YES** **Years:** 2006, 2007, 2008 **Models:** Raider **Engines:** 3.7L K, 4.7L M **Transmissions:** All	**Loose Fuel Cap** The PCM activates the Evap Purge Solenoid to pull the Evap system into a vacuum to close the ESIM switch. Once the ESIM switch is closed, the PCM turns the Evap Purge Solenoid off to seal the Evap system. If the ESIM switch reopens before the calibrated amount of time after a fuel tank fill, an error is detected. Three good trips to turn off the MIL. **Possible Causes:** • Loose or missing fuel cap
DTC: P0461 **2T PCM MIL: YES** **Years:** 2006, 2007, 2008 **Models:** All **Engines:** All **Transmissions:** All	**Fuel Level Sensor (Main) Circuit Range/Performance** When the fuel consumption calculated from the operation time of the injector amounts to approximately 8 gallons, the diversity of the amount of fuel in the tank calculated from the fuel level sensor is 0.5 gallons or less. **Possible Causes:** • Fuel pump module failed • Fuel level sensor failed • Fuel level sensor circuit shorted or open • PCM has failed
DTC: P0462 **2T PCM MIL: YES** **Years:** 2006, 2007, 2008 **Models:** All **Engines:** All **Transmissions:** All	**Fuel Level Sensor Circuit Low Input** Engine started and running for 2 seconds or more, battery voltage is between 11 and 16.5 volts, fuel level sensor output voltage has continued to be less than 0.3 volts for 2 seconds. **Possible Causes:** • Fuel level sensor failed • Fuel level sensor circuit shorted • PCM has failed
DTC: P0463 **2T PCM MIL: YES** **Years:** 2006, 2007, 2008 **Models:** All **Engines:** All **Transmissions:** All	**Fuel Level Sensor Circuit High Input** Engine started and running for 2 seconds or more, battery voltage is between 11 and 16.5 volts, fuel level sensor output voltage has continued to be greater than 4.6 volts for 2 seconds. **Possible Causes:** • Fuel level sensor failed • Fuel level sensor circuit shorted • PCM has failed

DTC	Trouble Code Title, Conditions & Possible Causes
DTC: P0489 **2T PCM MIL: YES** **Years:** 2006, 2007, 2008 **Models:** Outlander **Engines:** 2.4L W, 3.0L X **Transmissions:** All	**EGR Valve (Stepper Motor) Circuit Malfunction** When the EGR valve is de-energized, the ECM voltage should be 1.5 volts or less for 1.4 seconds. **Possible Causes:** • EGR valve (stepper motor) failed. • Harness damage. • Connector damage. • ECM failed.
DTC: P0490 **2T PCM MIL: YES** **Years:** 2006, 2007, 2008 **Models:** Outlander **Engines:** 2.4L W, 3.0L X **Transmissions:** All	**EGR Valve (Stepper Motor) Circuit Malfunction** When the EGR valve is energized, the ECM voltage should be 6.7 volts or less for 1.4 seconds. **Possible Causes:** • EGR valve (stepper motor) failed. • Harness damage. • Connector damage. • ECM failed.
DTC: P0500 **2T PCM MIL: YES** **Years:** 2006, 2007, 2008 **Models:** All **Engines:** All **Transmissions:** All	**Vehicle Speed Sensor Circuit Malfunction** Engine started, the difference between atmospheric and intake manifold pressure more than 34 kPa, vehicle driven at an engine speed under 3000 RPM for 30 seconds with the throttle valve open and the brakes "off", ECT sensor more than 176°F, gear selector not in P/N, and the PCM detected the vehicle speed was below 1 MPH for 4 seconds. **Possible Causes:** • VSS signal circuit is open or shorted to ground • VSS power circuit is open (check for power from MFI relay) • VSS ground circuit is open between sensor and chassis ground • VSS is damaged or has failed • PCM has failed (the PCM provides 5v on the VSS signal circuit)
DTC: P0501 **2T PCM MIL: YES** **Years:** 2006, 2007, 2008 **Models:** Raider **Engines:** 3.7L K, 4.7L M **Transmissions:** All	**Vehicle Speed Sensor 1 Performance** This code will set if no vehicle speed signal is received from the ABS Module up to 120 seconds for 2 consecutive trips. Three good trips to turn off the MIL. **Possible Causes:** • Active bus or communication DTCS • Tire circumference • PCM
DTC: P0503 **2T PCM MIL: YES** **Years:** 2006, 2007, 2008 **Models:** Raider **Engines:** 3.7L K, 4.7L M **Transmissions:** All	**Vehicle Speed Sensor 1 Performance Erratic** This code will set if no vehicle speed signal is received from the ABS Module up to 120 seconds for 2 consecutive trips. Three good trips to turn off the MIL. **Possible Causes:** • Active bus or communication DTCS • Tire circumference • PCM
DTC: P0506 **2T PCM MIL: YES** **Years:** 2006, 2007, 2008 **Models:** All **Engines:** All **Transmissions:** All	**Idle Speed Control System Lower Than Expected** ECT signal over 180°F, system voltage over 10v, BARO signal over 76 kPa, volumetric efficiency less than 40 percent, IAT signal more than 14°F, and the PCM detected the Actual idle speed was more than 100 RPM lower than the target idle speed for over 12 seconds. **Possible Causes:** • IAC motor control circuit A1, A2, B1 or B2 is open • IAC motor control circuit A1, A2, B1 or B2 is shorted to ground • IAC motor is damaged or has failed (it may be dirty or sticking) • Throttle plate is carbon fouled (it may need to be cleaned) • PCM has failed
DTC: P0507 **2T PCM MIL: YES** **Years:** 2006, 2007, 2008 **Models:** All **Engines:** All **Transmissions:** All	**Idle Speed Control System Higher Than Expected** ECT signal over 171°F, system voltage over 10v, BARO signal over 76 kPa, volumetric efficiency less than 40 percent, IAT signal more than 14°F, and the PCM detected the Actual idle speed was more than 100 RPM higher than the target idle speed for over 12 seconds. **Possible Causes:** • IAC motor control circuit A1, A2, B1 or B2 is open • IAC motor control circuit A1, A2, B1 or B2 is shorted to ground • IAC motor is damaged or has failed (it may be dirty or sticking) • Throttle plate is carbon fouled (it may need to be cleaned) • PCM has failed

DTC	Trouble Code Title, Conditions & Possible Causes
DTC: P0510 **2T PCM MIL: YES** **Years:** 2006, 2007, 2008 **Models:** Lancer **Engines:** 2.0L **Transmissions:** All	**Closed Throttle Position Switch Circuit Malfunction** Engine started, vehicle driven at over 30 MPH and then back to a stop at least 15 times, TP sensor signal over 2.0v at least once, and the PCM detected the CTP switch remained "off" for over 2 seconds. **Possible Causes:** • Closed throttle position switch signal circuit is open or grounded • Closed throttle position switch signal circuit is shorted to power • Closed throttle position switch or TP sensor damaged or failed • PCM has failed
DTC: P0513 **2T PCM MIL: YES** **Years:** 2006, 2007, 2008 **Models:** All **Engines:** All **Transmissions:** All	**Immobilizer Malfunction** Ignition switch ON when an error occurs between the PCM and the Immobilizer ECU for 2 seconds or more. **Possible Causes:** • Harness or connector between the PCM and immobilizer ECU • Immobilizer ECU failed • PCM has failed
DTC: P0522 **1T PCM MIL: YES** **Years:** 2006, 2007, 2008 **Models:** Raider **Engines:** 3.7L K, 4.7L M **Transmissions:** All	**Oil Pressure Too Low** The oil pressure sensor voltage at PCM is not within the calibrated specification. Three good trips to turn off the MIL. **Possible Causes:** • Oil pressure signal circuit open • Oil pressure signal circuit shorted to voltage • Oil pressure signal circuit shorted to ground • Oil pressure switch • PCM
DTC: P0532 **1T PCM MIL: YES** **Years:** 2006, 2007, 2008 **Models:** Raider **Engines:** 3.7L K, 4.7L M **Transmissions:** All	**A/C Pressure Sensor Circuit Low** The PCM detects that the A/C Pressure Transducer input voltage is below the minimum acceptable value. Three good trips to turn off the MIL. **Possible Causes:** • 5 Volt supply circuit open • 5 Volt supply circuit shorted to ground • A/C pressure transducer • A/C pressure signal circuit shorted to ground • A/C pressure signal circuit shorted to the sensor return circuit • Excessive resistance in the A/C pressure signal circuit • FCM
DTC: P0533 **2T PCM MIL: YES** **Years:** 2006, 2007, 2008 **Models:** Raider **Engines:** 3.7L K, 4.7L M **Transmissions:** All	**A/C Pressure Sensor Circuit High** The A/C pressure transducer signal at the PCM goes above 4.92 volts. Three good trips to turn off the MIL. **Possible Causes:** • A/C pressure signal circuit open • Sensor return circuit open • A/C pressure signal circuit shorted to voltage • A/C pressure signal circuit shorted to the 5 volt supply circuit • A/C pressure transducer • FCM
DTC: P0551 **2T PCM MIL: YES** **Years:** 2006, 2007, 2008 **Models:** All **Engines:** All **Transmissions:** All	**Power Steering Pressure Switch Circuit Range/Performance** Engine started, ECT sensor more than 50°F, vehicle driven to an engine speed of 2500 RPM, volumetric efficiency over 55 percent for 2 seconds, then after the vehicle returns to idle speed (800 RPM or less), the PCM detected the Power Steering Pressure switch signal indicated "off" at least 10 times during the PCM Rationality test. **Possible Causes:** • Power steering pressure switch signal circuit is open • Power steering pressure switch signal circuit shorted to ground • Power steering pressure switch is damaged or has failed • PCM has failed
DTC: P0554 **2T PCM MIL: YES** **Years:** 2006, 2007, 2008 **Models:** All **Engines:** All **Transmissions:** All	**Power Steering Pressure Switch Circuit Intermittent** Engine started, BARO signal over 75 kPa, IAT signal more than 14°F, ECT sensor more than 86°F, vehicle driven to a speed over 55 MPH, then back to idle speed at least 10 times, and the PCM detected the PSPS signal remained "on" during the PCM test. **Possible Causes:** • Power steering pressure switch signal circuit is open • Power steering pressure switch signal circuit shorted to ground • Power steering pressure switch is damaged or has failed • PCM has failed

DTC	Trouble Code Title, Conditions & Possible Causes
DTC: P0562 **1T PCM MIL: YES** **Years:** 2006, 2007, 2008 **Models:** Raider **Engines:** 3.7L K, 4.7L M **Transmissions:** All	**Battery Voltage Low** Battery voltage is less than 6 volts. One Trip Fault. **Possible Causes:** • Resistance in the battery positive circuit • Generator field control circuit open • Generator field control circuit shorted to ground • Battery • Generator • PCM
DTC: P0563 **1T PCM MIL: YES** **Years:** 2006, 2007, 2008 **Models:** Raider **Engines:** 3.7L K, 4.7L M **Transmissions:** All	**Battery Voltage High** Battery voltage is 1 volt greater than desired voltage. Battery voltage greater than 15.75 volts. Three good trips to turn off the MIL. **Possible Causes:** • Generator field control circuit shorted to voltage • Generator • PCM
DTC: P0571 **2T PCM MIL: YES** **Years:** 2006, 2007, 2008 **Models:** Raider **Engines:** 3.7L K, 4.7L M **Transmissions:** All	**Brake Switch 1 Performance** If the output of Brake Switch No.1 to the PCM looks like it is not applied, while Brake Lamp Switch Output circuit is applied the fault will mature in 60ms. One Trip Fault. **Possible Causes:** • Brake switch No. 1 signal shorted to ground • Brake switch No. 2 signal circuit open • Brake switch No. 2 signal circuit shorted to ground • Ground circuit open • Stop lamp switch • PCM
DTC: P0572 **2T PCM MIL: YES** **Years:** 2006, 2007, 2008 **Models:** Raider **Engines:** 3.7L K, 4.7L M **Transmissions:** All	**Brake Switch 1 Stuck On** When the PCM recognizes Brake Switch No.1 is mechanically stuck in the low/on position. Three Global Good Trips to Clear. **Possible Causes:** • Brake switch No. 1 signal circuit shorted to ground • Brake switch No. 2 signal circuit open • Stop lamp switch • PCM
DTC: P0573 **2T PCM MIL: YES** **Years:** 2006, 2007, 2008 **Models:** Raider **Engines:** 3.7L K, 4.7L M **Transmissions:** All	**Brake Switch 1 Stuck Off** When the PCM recognizes Brake Switch No.1 is stuck in the high/off position. Three good trips to turn off the MIL. **Possible Causes:** • Brake switch No. 1 signal circuit open • Brake switch No. 2 signal circuit shorted to ground • Ground circuit open • Stop lamp switch • PCM
DTC: P0579 **2T PCM MIL: YES** **Years:** 2006, 2007, 2008 **Models:** Raider **Engines:** 3.7L K, 4.7L M **Transmissions:** All	**Speed Control Switch 1 Performance** Cruise switch voltage output is not out of range but it does not equal any of the values for any of the button positions. **Possible Causes:** • S/C switch No. 1 signal circuit shorted to voltage • S/C switch No. 1 signal circuit open • Switch return circuit open • S/C switch No. 1 signal circuit shorted to ground • S/C switch No. 1 signal circuit shorted to the switch return circuit • Clock spring • Speed control switch • PCM
DTC: P0580 **1T PCM MIL: YES** **Years:** 2006, 2007, 2008 **Models:** Raider **Engines:** 3.7L K, 4.7L M **Transmissions:** All	**Speed Control Switch 1 Circuit Low** Speed control switch input No.1 is below the minimum acceptable voltage at the PCM. One trip fault. **Possible Causes:** • S/C switch No. 1 signal circuit shorted to the S/C switch return • S/C switch No. 1 signal circuit shorted to ground • Clock spring • Speed control switch • PCM

DTC	Trouble Code Title, Conditions & Possible Causes
DTC: P0581 **1T PCM MIL: YES** **Years:** 2006, 2007, 2008 **Models:** Raider **Engines:** 3.7L K, 4.7L M **Transmissions:** All	**Speed Control Switch 1 Circuit High** Speed control switch input above the maximum acceptable voltage at the PCM. One trip fault. **Possible Causes:** • S/C switch No. 1 signal circuit shorted to voltage • S/C switch No. 1 signal circuit open • S/C switch retrun circuit open • Clock spring • Speed control switch • PCM
DTC: P0585 **1T PCM MIL: YES** **Years:** 2006, 2007, 2008 **Models:** Raider **Engines:** 3.7L K, 4.7L M **Transmissions:** All	**Speed Control Switch 1/2 Correlation** Cruise Switch inputs are not coherent with each other. Example: PCM is reading Switch No.1 as Accel and Switch No.2 as Coast at the same time. One trip fault. **Possible Causes:** • S/C switch signal circuit shorted to voltage • Resistance in the S/C switch signal circuit • Resistance in the S/C switch signal return switch circuit • S/C switch No. 1 signal circuit shorted to the S/C switch No. 2 signal circuit • S/C switch signal circuit shorted to ground • Clock spring • Speed control switch • PCM
DTC: P0591 **1T PCM MIL: YES** **Years:** 2006, 2007, 2008 **Models:** Raider **Engines:** 3.7L K, 4.7L M **Transmissions:** All	**Speed Control Switch 2 Performance** Cruise switch voltage output is not out of range but it does not equal any of the values for any of the button positions. One trip fault. **Possible Causes:** • S/C switch No. 2 signal circuit shorted to voltage • S/C switch No. 2 signal circuit open • Switch return circuit open • S/C switch No. 2 signal circuit shorted to ground • S/C switch No. 2 signal circuit shorted to the switch return circuit • Clock spring • Speed control switch • PCM
DTC: P0592 **1T PCM MIL: YES** **Years:** 2006, 2007, 2008 **Models:** Raider **Engines:** 3.7L K, 4.7L M **Transmissions:** All	**Speed Control Switch 1 Circuit Low** Speed control switch input No.1 is below the minimum acceptable voltage at the PCM. One trip fault. **Possible Causes:** • S/C switch No. 1 signal circuit shorted to the S/C switch return • S/C switch No. 1 signal circuit shorted to ground • Clock spring • Speed control switch • PCM
DTC: P0593 **1T PCM MIL: YES** **Years:** 2006, 2007, 2008 **Models:** Raider **Engines:** 3.7L K, 4.7L M **Transmissions:** All	**Speed Control Switch 1 Circuit High** Speed control switch input above the maximum acceptable voltage at the PCM. One trip fault. **Possible Causes:** • S/C switch No. 1 signal circuit shorted to voltage • S/C switch No. 1 signal circuit open • S/C switch retrun circuit open • Clock spring • Speed control switch • PCM
DTC: P0600 **1T PCM MIL: YES** **Years:** 2006, 2007, 2008 **Models:** Raider **Engines:** 3.7L K, 4.7L M **Transmissions:** All	**Serial Communication Link** Internal Bus communication failure between processors. Three Global Good Trips to Clear. **Possible Causes:** • PCM

DTC	Trouble Code Title, Conditions & Possible Causes
DTC: P0601 **1T PCM MIL: YES** **Years:** 2006, 2007, 2008 **Models:** Raider **Engines:** 3.7L K, 4.7L M **Transmissions:** All	**Internal Memory Check Sum Invalid** Internal checksum for software failed, does not match calculated value. One Trip Fault, Three Good Trips to clear. **Possible Causes:** • PCM
DTC: P0603 **1T PCM MIL: YES** **Years:** 2006, 2007, 2008 **Models:** All **Engines:** All **Transmissions:** All	**EEPROM Malfunction** The latest data that was flashed while the ignition switch was in "LOCK" (OFF) position are not stored correctly. **Possible Causes:** • ECM has failed
DTC: P0606 **1T PCM MIL: YES** **Years:** 2006, 2007, 2008 **Models:** All **Engines:** All **Transmissions:** All	**Powertrain Control Module Main Processor Malfunction** Throttle opening degree position is in default position. **Possible Causes:** • ECM has failed
DTC: P0622 **1T PCM MIL: YES** **Years:** 2006, 2007, 2008 **Models:** All **Engines:** All **Transmissions:** All	**Generator FR Terminal Circuit Malfunction** Prohibits generator output suppression control against current consumers. (Operates as a normal generator.). **Possible Causes:** • Generator failed • Open circuit in generator FR terminal circuit, harness damage or connector damage • ECM has failed
DTC: P0627 **1T PCM MIL: YES** **Years:** 2006, 2007, 2008 **Models:** Raider **Engines:** 3.7L K, 4.7L M **Transmissions:** All	**Fuel Pump Relay Circuit** An open or shorted condition is detected in the fuel pump relay control circuit. Three good trips to turn off the MIL. **Possible Causes:** • Internal fused B+ circuit • Fused ignition switch circuit • Fuel pump relay control circuit open • Fuel pump relay control circuit shorted to ground • Fuel pump relay • PCM has failed
DTC: P0630 **1T PCM MIL: YES** **Years:** 2006, 2007, 2008 **Models:** All **Engines:** All **Transmissions:** All	**VIN Malfunction** VIN (current) has not been written. **Possible Causes:** • ECM has failed
DTC: P0632 **2T PCM MIL: YES** **Years:** 2006, 2007, 2008 **Models:** Raider **Engines:** 3.7L K, 4.7L M **Transmissions:** All	**Odometer Not Programmed In PCM** Odometer is not programed into the PCM. Three good trips to turn off the MIL. **Possible Causes:** • Programming mileage into the PCM • PCM has failed
DTC: P0633 **2T PCM MIL: YES** **Years:** 2006, 2007, 2008 **Models:** Raider **Engines:** 3.7L K, 4.7L M **Transmissions:** All	**Skim Secret Key Not Stored In PCM** The SKIM Key information has not been programmed into the PCM. Three good trips to turn off the MIL. **Possible Causes:** • Programming skim key into the PCM • PCM has failed

DTC	Trouble Code Title, Conditions & Possible Causes
DTC: P0638 **1T PCM MIL: YES** **Years:** 2006, 2007, 2008 **Models:** All **Engines:** All **Transmissions:** All	**Throttle Actuator Control Motor Circuit Range/Performance Problem** Key on, battery positive voltage is greater than 8.3 volts, TPS output voltage is between 0.35 and 4.8 volts, drop of TPS output voltage per 100 ms is greater than 0.04 volts and the TPS output voltage has continued to be greater than the target TPS voltage by 0.5 volt or more for 0.5 seconds and/or difference between the TPS output voltage and the target TPS voltage is 1 volt or greater for 4 seconds. **Possible Causes:** • Throttle valve return spring failed • Throttle valve operation failed • Throttle actuator control motor failed • Throttle actuator control motor circuit shorted or open • PCM has failed
DTC: P0642 **1T PCM MIL: YES** **Years:** 2006, 2007, 2008 **Models:** All **Engines:** All **Transmissions:** All	**Throttle Position Sensor Power Supply** Key on, battery positive voltage is greater than 6.3 volts, TPS power voltage should be 4.1 volts or less for 0.5 seconds. **Possible Causes:** • PCM has failed
DTC: P0643 **1T PCM MIL: YES** Years: 2006, 2007, 2008 **Models:** Raider **Engines:** 3.7L K, 4.7L M **Transmissions:** All	**Sensor Reference Voltage 1 Circuit High** When the PCM recognizes the Primary 5-volt Supply circuit voltage is too high. ETC light is flashing. **Possible Causes:** • Primary 5 volt supply shorted to voltage • PCM has failed
DTC: P0645 **1T PCM MIL: YES** **Years:** 2006, 2007, 2008 **Models:** Raider **Engines:** 3.7L K, 4.7L M **Transmissions:** All	**A/C Clutch Relay Circuit** An open or shorted condition is detected in the A/C clutch relay control circuit. Three good trips to turn off the MIL. **Possible Causes:** • Internal fused B+ circuits • A/C clutch relay control circuit open • A/C clutch relay control circuit shorted to ground • A/C clutch relay • PCM has failed
DTC: P0652 **1T PCM MIL: YES** **Years:** 2006, 2007, 2008 **Models:** Raider **Engines:** 3.7L K, 4.7L M **Transmissions:** All	**Sensor Reference Voltage 2 Circuit Low** When the PCM recognizes the Auxiliary 5-volt Supply circuit voltage is too low. ETC light is flashing. **Possible Causes:** • Auxiliary 5 volt supply shorted to ground • Sensor shorted to ground • Cam position sensor • PCM has failed
DTC: P0653 **1T PCM MIL: YES** **Years:** 2006, 2007, 2008 **Models:** Raider **Engines:** 3.7L K, 4.7L M **Transmissions:** All	**Sensor Reference Voltage 2 Circuit High** When the PCM recognizes the Auxiliary 5-volt Supply circuit voltage is too high. ETC light is flashing. **Possible Causes:** • Auxiliary 5 volt supply shorted to voltage • PCM has failed
DTC: P0657 **1T PCM MIL: YES** **Years:** 2006, 2007, 2008 **Models:** All **Engines:** All **Transmissions:** All	**Throttle Actuator Control Motor Relay Circuit Malfunction** Key on, and the electronic controlled throttle valve system should be 4 volts or less for 1 second. **Possible Causes:** • Throttle actuator control motor failed • Throttle actuator control motor circuit shorted or open • PCM has failed
DTC: P0660 **2T PCM MIL: YES** **Years:** 2006, 2007, 2008 **Models:** Eclipse, Spyder, Endeavor, Galant, Outlander **Engines:** 3.8L, 3.0L X **Transmissions:** All	**Intake Manifold Tuning Circuit Malfunction** Engine cranking, battery voltage between 10 and 16 volts, intake manifold tuning solenoid is ON and greater than 1 second has elapsed after conditions have been met, the intake manifold tuning solenoid coil surge voltage (battery voltage 2 volts) is not detected for 0.2 seconds, the PCM monitors for this condition once during the drive cycle and/or intake manifold tuning solenoid coil surge voltage is not detected for 1 second when the intake manifold tuning solenoid is turned off. **Possible Causes:** • Intake manifold tuning solenoid failed • Intake manifold tuning solenoid circuit shorted or open • PCM has failed

DTC	Trouble Code Title, Conditions & Possible Causes
DTC: P0685 **1T PCM MIL: YES** **Years:** 2006, 2007, 2008 **Models:** Raider **Engines:** 3.7L K, 4.7L M **Transmissions: All**	**Auto Shutdown Relay Control Circuit** An open or shorted condition is detected in the ASD relay control circuit. Three good trips to turn off the MIL. **Possible Causes:** • Internal fused B+ circuits • ASD relay control circuit open • ASD relay control circuit shorted to ground • ASD relay • PCM has failed
DTC: P0688 **1T PCM MIL: YES** **Years:** 2006, 2007, 2008 **Models:** Raider **Engines:** 3.7L K, 4.7L M **Transmissions: All**	**Auto Shutdown Relay Sense Circuit Low** No voltage sensed at the PCM when the ASD relay is energized. Three good trips to turn off the MIL. **Possible Causes:** • Internal fused B+ circuits • ASD relay output circuit open • ASD relay • PCM has failed
DTC: P0689 **2T PCM MIL: YES** **Years:** 2006, 2007, 2008 **Models:** Raider **Engines:** 3.7L K, 4.7L M **Transmissions: All**	**Auto Shutdown Relay Sense Circuit Low** The ASD Output circuit voltage drops below an acceptable value at the FCM. The circuit is continuously monitored. **Possible Causes:** • Internal fused B+ circuits • ASD relay output circuit open • ASD relay output circuit shorted to ground • ASD relay • PCM has failed
DTC: P0690 **2T PCM MIL: YES** **Years:** 2006, 2007, 2008 **Models:** Raider **Engines:** 3.7L K, 4.7L M **Transmissions: All**	**Auto Shutdown Relay Sense Circuit High** If the FCM detects high voltage on the ASD Relay Sense circuit for more than 3.5 seconds. **Possible Causes:** • ASD relay output circuit shorted to voltage • ASD relay • FCM has failed • PCM has failed
DTC: P0700 **2T PCM MIL: YES** **Years:** 2006, 2007, 2008 **Models:** Raider **Engines:** 3.7L K, 4.7L M **Transmissions: All**	**Transmission Control Module Signal** Key on or engine running, and the PCM received a signal from the TCM that indicating an internal problem with the TCM had occurred. **Possible Causes:** • Clear the trouble codes and retest for this trouble code. If the same trouble code resets, the TCM has failed and must be replaced to repair this problem.
DTC: P0703 **2T PCM MIL: YES** **Years:** 2006, 2007, 2008 **Models:** Raider **Engines:** 3.7L K, 4.7L M **Transmissions: All**	**Transmission Brake Switch Signal** Key on or engine running, and the PCM detected the Brake Switch signal did not cycle from "high" to "low" as the brake pedal was pressed and then released during the PCM test. **Possible Causes:** • Brake switch signal circuit is open or shorted to ground • Brake switch power circuit is open (check power from the relay) • Brake switch is damaged or has failed • TCM has failed
DTC: P0830 **2T PCM MIL: YES** **Years:** 2006, 2007, 2008 **Models:** All **Engines:** All **Transmissions:** All	**Clutch Pedal Position Switch Circuit Range/Performance** Engine started and vehicle is driven a minimum speed of 19 MPH, the toggling of the signal (high/low) of the clutch pedal position switch is not detected even once. **Possible Causes:** • Clutch pedal position switch failed • Clutch pedal position switch circuit open or shorted • PCM has failed

Gas Engine OBD II Trouble Code List (P1xxx Codes)

DTC	Trouble Code Title, Conditions & Possible Causes
DTC: P1020 **2T PCM MIL: YES** **Years:** 2006, 2007, 2008 **Models:** All **Engines:** All **Transmissions:** All	**Mitsubishi Innovative Valve Timing and Lift Electronic Control System (MIVEC) Performance Problem** Engine started, engine runtime 30 seconds or more, engine coolant temperature 171°F or greater, battery voltage over 10v, engine speed is 3000 RPM or lower and engine oil pressure switch is OFF for 2 seconds, or, engine speed is 4000 RPM or higher and engine oil pressure switch is ON for 2 seconds. **Possible Causes:** • Engine oil pressure switch failed • Engine oil control valve failed • Engine oil pressure switch circuit open or shorted induction control motor circuit is open • PCM has failed
DTC: P01021 **2T PCM MIL: YES** **Years:** 2006, 2007, 2008 **Models:** All **Engines:** All **Transmissions:** All	**Engine Oil Control Valve Circuit** Engine started, battery voltage over 10v, MIVEC operating in the low speed mode the engine control valve circuit to the PCM voltage is less than 1.5 volts for 2 seconds. **Possible Causes:** • Engine oil control valve failed • Engine oil control valve circuit open or shorted • PCM has failed
DTC: P1115 **2T PCM MIL: YES** **Years:** 2006, 2007, 2008 **Models:** Raider **Engines:** 3.7L K, 4.7L M **Transmissions:** All	**General Temperature Rationality** Once the vehicle is soaked for a calibrated engine off time and then driven over calibrated speed and load conditions for some calibrated time, the PCM compares the ambient air, engine coolant, and intake air temperature sensor values. If the values of all the three sensors disagree with one another, a general temperature sensor irrationality is declared. Three good trips to turn off the MIL. **Possible Causes:** • Excessive resistance in the sensor signal circuit • Excessive resistance in the sensor ground circuit • Sensor signal circuit shorted to ground • Sensor signal circuit shorted to voltage • Sensor signal circuit shorted to the sensor ground circuit • Temperature sensor • PCM has failed
DTC: P1128 **2T PCM MIL: YES** **Years:** 2006, 2007, 2008 **Models:** Raider **Engines:** 3.7L K, 4.7L M **Transmissions:** All	**Closed Loop Fueling Not Achieved – Bank 1** Enable conditions are met and the O2 sensor has not been in closed loop control at least once on each of the two consecutive trips, the MIL illuminates and the DTC is set. Three good trips to turn off the MIL. **Possible Causes:** • Restricted fuel supply line • Fuel pump inlet strainer plugged • Fuel pump module • O2 signal circuit • O2 return circuit • 1/1 O2 sensor • MAP sensor • ECT sensor • Engine mechanical problem • Fuel filter/pressure regulator • PCM has failed
DTC: P1129 **2T PCM MIL: YES** **Years:** 2006, 2007, 2008 **Models:** Raider **Engines:** 3.7L K, 4.7L M **Transmissions:** All	**Closed Loop Fueling Not Achieved – Bank 2** Enable conditions are met and the O2 sensor has not been in closed loop control at least once on each of the two consecutive trips, the MIL illuminates and the DTC is set. Three good trips to turn off the MIL. **Possible Causes:** • Restricted fuel supply line • Fuel pump inlet strainer plugged • Fuel pump module • O2 signal circuit • O2 return circuit • 1/1 O2 sensor • MAP sensor • ECT sensor • Engine mechanical problem • Fuel filter/pressure regulator • PCM has failed

DTC	Trouble Code Title, Conditions & Possible Causes
DTC: P1231 **2T PCM MIL: YES** **Years:** 2006, 2007, 2008 **Models:** Evolution, Lancer, Outlander **Engines:** 2.0L, 2.4L W, 3.0L X **Transmissions:** All	**Active Stability Control Plausibility** A torque demand signal from the active stability control is not normal. **Possible Causes:** • Active stability control system failed. • ECM failed.
DTC: P1232 **2T PCM MIL: YES** **Years:** 2006, 2007, 2008 **Models:** Evolution, Lancer, Outlander **Engines:** 2.0L, 2.4L W, 3.0L X **Transmissions:** All	**Fail Safe System** Power supply to the throttle actuator control motor cannot be shut down (though power supply is stopped). **Possible Causes:** • Throttle actuator control motor relay circuit failed. • ECM failed.
DTC: P1233 **2T PCM MIL: YES** **Years:** 2006, 2007, 2008 **Models:** Evolution, Lancer, Outlander **Engines:** 2.0L, 2.4L W, 3.0L X **Transmissions:** All	**Throttle Position Sensor (Main) Plausibility** For 0.4 second, the difference between the actual volumetric efficiency and the volumetric efficiency estimated by the throttle position sensor (main) is 33 percent or more. **Possible Causes:** • Throttle position sensor (main) system failed. • Intake system vacuum leak. • ECM failed.
DTC: P1234 **2T PCM MIL: YES** **Years:** 2006, 2007, 2008 **Models:** Evolution, Lancer, Outlander **Engines:** 2.0L, 2.4L W, 3.0L X **Transmissions:** All	**Throttle Position Sensor (Sub) Plausibility** For 0.4 second, the difference between the actual volumetric efficiency and the volumetric efficiency estimated by the throttle position sensor (sub) is 33 percent or more. **Possible Causes:** • Throttle position sensor (sub) system failed. • Intake system vacuum leak. • ECM failed.
DTC: P1235 **2T PCM MIL: YES** **Years:**: 2006, 2007, 2008 **Models:** Evolution, Lancer, Outlander **Engines:** 2.0L, 2.4L W, 3.0L X **Transmissions:** All	**Mass Airflow Sensor Plausibility** For 0.36 second, the difference between the volumetric efficiency estimated by the (main) throttle position sensor and the volumetric efficiency estimated by the (sub) throttle position sensor is 8.8 percent or less. **Possible Causes:** • Mass airflow sensor system failed. • ECM failed.
DTC: P1236 **2T PCM MIL: YES** **Years:** 2006, 2007, 2008 **Models:** Evolution, Lancer, Outlander **Engines:** 2.0L, 2.4L W, 3.0L X **Transmissions:** All	**A/D Converter** When the input voltage from the accelerator pedal position sensor (sub) is periodically 0 V for 0.45 second, the digital value of the input voltage indicates 0.2 V or more. **Possible Causes:** • ECM failed.
DTC: P1237 **2T PCM MIL: YES** **Years:** 2006, 2007, 2008 **Models:** Evolution, Lancer, Outlander **Engines:** 2.0L, 2.4L W, 3.0L X **Transmissions:** All	**Accelerator Pedal Position Sensor Plausibility** Voltage obtained with the formula given below is 0.4 volt or higher for 0.5 second: accelerator pedal position sensor (main) output voltage - accelerator pedal position sensor (sub) output voltage. **Possible Causes:** • Accelerator pedal position sensor failed. • ECM failed.
DTC: P1238 **2T PCM MIL: YES** **Years:** 2006, 2007, 2008 **Models:** Evolution, Lancer, Outlander **Engines:** 2.0L, 2.4L W, 3.0L X **Transmissions:** All	**Mass Airflow Sensor Plausibility (Torque Monitor)** For 0.5 second, the difference between the actual volumetric efficiency and the volumetric efficiency estimated by the throttle position sensor (sub) is 35 percent or more. **Possible Causes:** • Mass airflow sensor system failed. • ECM failed.

DTC	Trouble Code Title, Conditions & Possible Causes
DTC: P1239 **2T PCM MIL: YES** **Years:** 2006, 2007, 2008 **Models:** Evolution, Lancer, Outlander **Engines:** 2.0L, 2.4L W, 3.0L X **Transmissions:** All	**Engine RPM Plausibility** The engine speed monitored with a 10 degree- cycle pulse is 1,000 PRM or more. **Possible Causes:** • Crankshaft position sensor system failed. • ECM failed.
DTC: P1240 **2T PCM MIL: YES** **Years:** 2006, 2007, 2008 **Models:** Evolution, Lancer, Outlander **Engines:** 2.0L, 2.4L W, 3.0L X **Transmissions:** All	**Ignition Angle** The ignition timing retard angle demand signal is not normal. **Possible Causes:** • ECM failed.
DTC: P1241 **2T PCM MIL: YES** **Years:** 2006, 2007, 2008 **Models:** Evolution, Lancer, Outlander **Engines:** 2.0L, 2.4L W, 3.0L X **Transmissions:** All	**Torque Monitoring** For 1 second, the difference between the actual torque signal and the requested torque signal is 50 N·m (37 ft-lb) or more. **Possible Causes:** • Throttle actuator control motor failed. • Connector damage. • Harness damage. • Vacuum leak. • ECM failed.
DTC: P1242 **2T PCM MIL: YES** **Years:** 2006, 2007, 2008 **Models:** Evolution, Lancer, Outlander **Engines:** 2.0L, 2.4L W, 3.0L X **Transmissions:** All	**Fail Safe Control Monitor** The engine speed is higher than assumed. **Possible Causes:** • ECM failed.
DTC: P1243 **2T PCM MIL: YES** **Years:** 2006, 2007, 2008 **Models:** Evolution, Lancer, Outlander **Engines:** 2.0L, 2.4L W, 3.0L X **Transmissions:** All	**Inquiry/Response Error** ECM can not calculate input data. **Possible Causes:** • ECM failed.
DTC: P1244 **2T PCM MIL: YES** **Years:** 2006, 2007, 2008 **Models:** Evolution, Lancer, Outlander **Engines:** 2.0L, 2.4L W, 3.0L X **Transmissions:** All	**RAM Test for All Area** All RAM data of ECM are defect. **Possible Causes:** • ECM failed.
DTC: P1245 **2T PCM MIL: YES** **Years:** 2006, 2007, 2008 **Models:** Evolution, Lancer, Outlander **Engines:** 2.0L, 2.4L W, 3.0L X **Transmissions:** All	**Cycle RAM Test (Engine)** RAM data (engine) is defect. **Possible Causes:** • ECM failed.
DTC: P1247 **2T PCM MIL: YES** **Years:** 2006, 2007, 2008 **Models:** Evolution, Lancer, Outlander **Engines:** 2.0L, 2.4L W, 3.0L X **Transmissions:** All	**CVT Plausibility** A torque demand signal from TCM is not normal. **Possible Causes:** • Automatic transaxle system failed • ECM failed

DTC	Trouble Code Title, Conditions & Possible Causes
DTC: P1320 **2T PCM MIL: YES** **Years:** 2006, 2007, 2008 **Models:** All **Engines:** All **Transmissions:** All	**Ignition Timing Retard Insufficient** For 10 seconds, the difference between the basic ignition timing and the target/specified ignition timing is 3°CA or less on average during the retard control. **Possible Causes:** • The thermostat is faulty • ECM failed
DTC: P1404 **2T PCM MIL: YES** **Years:** 2006, 2007, 2008 **Models:** Raider **Engines:** 3.7L K, 4.7L M **Transmissions:** All	**EGR Position Sensor Rationality Closed** The EGR flow or valve movement is not what is expected. A rationality error has been detected for the EGR Open Position Performance. Two trip fault. **Possible Causes:** • Excessive resistance in the 5 volt supply circuit • EGR solenoid control circuit open • EGR sensor signal circuit shorted to ground • Ground circuit open • EGR solenoid assembly • PCM failed
DTC: P1501 **1T PCM MIL: YES** **Years:** 2006, 2007, 2008 **Models:** Raider **Engines:** 3.7L K, 4.7L M **Transmissions:** All	**Vehicle Speed Sensor 1/2 Correlation - Drive Wheels** The PCM recognizes rear wheel speed is greater than front wheel speed. One trip fault. **Possible Causes:** • Active bus or communications • Tire circumfrence • PCM failed
DTC: P1502 **2T PCM MIL: YES** Years: 2006, 2007, 2008 **Models:** Raider **Engines:** 3.7L K, 4.7L M **Transmissions:** All	**Vehicle Speed Sensor 1/2 Correlation – Non Drive Wheels** The PCM recognizes front axle speed is greater than rear axle speed. One trip fault. **Possible Causes:** • Active bus or communications • Tire circumfrence • PCM failed
DTC: P1506 **2T PCM MIL: YES** **Years:** 2006, 2007, 2008 **Models:** All **Engines:** All **Transmissions:** All	**Idle Control System RPM Lower Than Expected At Low Temperature** Engine started, engine running under closed loop idle speed control, ECT between 45°F and 106°F, battery voltage greater than 10 volts, power steering pressure switch off, volumetric efficiency is less than 40 percent, barometric pressure is greater than 22.4 in. Hg, IAT is greater than 14°, throttle actuator control motor position is greater than 255 steps, the actual idle speed is greater than 100 RPM lower than target idle speed for 12 seconds. **Possible Causes:** • Throttle valve area dirty • Throttle body assembly failed is damaged or has failed • PCM has failed
DTC: P1507 **2T PCM MIL: YES** **Years:** 2006, 2007, 2008 **Models:** All **Engines:** All **Transmissions:** All	**Idle Control System RPM Higher Than Expected At Low Temperature** Engine started, engine running under closed loop idle speed control, ECT between 45°F and 106°F, battery voltage greater than 10 volts, power steering pressure switch off, volumetric efficiency is less than 40 percent, barometric pressure is greater than 22.4 in. Hg, IAT is greater than 14°, throttle actuator control motor position is 0 steps, the actual idle speed is greater than 200 RPM lower than target idle speed for 12 seconds. **Possible Causes:** • Intake system vacuum leak • Throttle body assembly failed is damaged or has failed • PCM has failed
DTC: P1530 **2T PCM MIL: YES** **Years:** 2006, 2007, 2008 **Models:** Lancer, Evolution **Engines:** 2.0L C, 2.0L E **Transmissions:** All	**A/C1 Switch Circuit Intermittent** Engine started, the repeating ON–OFF switch of the A/C switch 255 times per second. **Possible Causes:** • A/C ECU has failed

DTC	Trouble Code Title, Conditions & Possible Causes
DTC: P1540-41 **2T PCM MIL: YES** **Years:** 2006, 2007, 2008 **Models:** Galant, Outlander **Engines:** 2.4L F, 3.0L X, 3.8L S, 3.8L T **Transmissions:** All	**LIN Communication Check by PCM** The radiator sensor cannot receive the data from the ECM <When P1540 is set>. The ECM cannot receive the data from the radiator sensor <When P1541 is set>. **Possible Causes:** • Tampering of the DOR radiator • Battery failed • Connector damage • Harness damage • ECM failed
DTC: P1543 **2T PCM MIL: YES** **Years:** 2006, 2007, 2008 **Models:** Galant, Outlander **Engines:** 2.4L F, 3.0L X, 3.8L S, 3.8L T **Transmissions:** All	**DOR Radiator Tampering Monitor (Temperature Check)** When the output temperature of the radiator sensor after the thermostat valve is opened is low. **Possible Causes:** • Tampering of the DOR radiator • Engine coolant is insufficient • Thermostat failed • Water pump failed • ECM failed
DTC: P1544 **2T PCM MIL: YES** **Years:** 2006, 2007, 2008 **Models:** Galant, Outlander **Engines:** 2.4L F, 3.0L X, 3.8L S, 3.8L T **Transmissions:** All	**DOR Radiator Tampering Monitor (Temperature Check)** When the output temperature of the radiator sensor after the thermostat valve is opened is high, the sensor is defective. **Possible Causes:** • Tampering of the DOR radiator • Thermostat failed • ECM failed
DTC: P1545 **2T PCM MIL: YES** **Years:** 2006, 2007, 2008 **Models:** Galant, Outlander **Engines:** 2.4L F, 3.0L X, 3.8L S, 3.8L T **Transmissions:** All	**DOR Radiator Tampering Monitor (Temperature Check)** When the output temperature of the radiator sensor after the thermostat valve is opened is rising by 6°C (43°F) or more in 1 second. **Possible Causes:** • Tampering of the DOR radiator • ECM failed
DTC: P1546 **2T PCM MIL: YES** **Years:** 2006, 2007, 2008 **Models:** Galant, Outlander **Engines:** 2.4L F, 3.0L X, 3.8L S, 3.8L T **Transmissions:** All	**DOR Radiator Tampering Monitor (Temperature Check)** When the difference between the radiator sensor output temperature after the engine is started and the coolant temperature is 15°C (59°F) or more. **Possible Causes:** • Tampering of the DOR radiator • ECM failed
DTC: P1547 **2T PCM MIL: YES** **Years:** 2006, 2007, 2008 **Models:** Galant, Outlander **Engines:** 2.4L F, 3.0L X, 3.8L S, 3.8L T **Transmissions:** All	**DOR Radiator Tampering Monitor (Rationality Check)** The temperature difference between thermistor 1 and thermistor 2 is 9°C (48°F) or more. **Possible Causes:** • Tampering of the DOR radiator • ECM failed
DTC: P1572 **1T PCM MIL: YES** **Years:** 2006, 2007, 2008 **Models:** Raider **Engines:** 3.7L K, 4.7L M **Transmissions:** All	**Brake Pedal Stuck On** PCM recognizes the Brake Pedal could not electrically indicate the applied (On) position with both switch inputs. One trip fault. **Possible Causes:** • Brake switch No. 1 signal circuit open • Brake switch No. 2 signal circuit open • Stop lamp switch • PCM failed
DTC: P1573 **1T PCM MIL: YES** **Years:** 2006, 2007, 2008 **Models:** Raider **Engines:** 3.7L K, 4.7L M **Transmissions:** All	**Brake Pedal Stuck Off** PCM recognizes the Brake Pedal could not electronically indicate the released (Off) position with both switches. If P1572 sets, P1573 will also set. One trip fault. **Possible Causes:** • Brake switch No. 1 signal circuit shorted to voltage • Brake switch No. 2 signal circuit shorted to ground • Stop lamp switch • PCM failed

DTC	Trouble Code Title, Conditions & Possible Causes
DTC: P1580 **2T PCM MIL: YES** **Years:** 2006, 2007, 2008 **Models:** Galant **Engines:** 2.4L F **Transmissions:** All	**DOR Radiator Tampering Monitor (Encrypted Massage)** Communication error. **Possible Causes:** • Malfunction of the radiator sensor • Malfunction of the PCM
DTC: P1590 **2T PCM MIL: YES** **Years:** 2006, 2007, 2008 **Models:** Evolution, Lancer, Outlander **Engines:** 2.0L, 2.4L W, 3.0L X **Transmissions:** All	**TCM to ECM Communication Error in Torque Reduction** ECM detects an error in communication between ECM and TCM. **Possible Causes:** • CAN line harness damage or connector damage • ECM has failed
DTC: P1593 **1T PCM MIL: YES** **Years:** 2006, 2007, 2008 **Models:** Raider **Engines:** 3.7L K, 4.7L M **Transmissions:** All	**Speed Control Switch 1 Stuck** One of the S/C Switches is mechanically stuck in the On/Off, Resume/Accel, or Set position for too long. One trip fault. **Possible Causes:** • S/C switch no.1 signal circuit open • S/C switch no.2 signal circuit open • Switch return circuit open • S/C switch no.1 signal circuit shorted to voltage • S/C switch no.2 signal circuit shorted to voltage • S/C switch no.1 signal circuit shorted to ground • S/C switch no.2 signal circuit shorted to ground • S/C switch no.1 signal circuit shorted to the S/C switch return circuit • S/C switch no.2 signal circuit shorted to the S/C switch return circuit • Speed control switch • PCM has failed
DTC: P1602 **2T PCM MIL: YES** **Years:** 2006, 2007, 2008 **Models:** All **Engines:** All **Transmissions:** All	**Communication Malfunction (between PCM Main Processor and System LSI)** Ignition ON, the PCM detects an error in communication with the system LSI 0.07. **Possible Causes:** • PCM has failed
DTC: P1603 **1T PCM MIL: NO** Years: 2006, 2007, 2008 **Models:** All **Engines:** All **Transmissions:** All	**Battery Backup Circuit Malfunction** Engine started, engine running, and the PCM detected the voltage on the Battery Backup circuit was less than 6.0v for 2 seconds. **Possible Causes:** • Battery backup circuit is open (check the power through the fusible link and Dedicated Fuse No. 7 for an open circuit) • Battery terminals corroded, or loose connections • PCM has failed
DTC: P1607 **2T PCM MIL: YES** **Years:** 2006, 2007, 2008 **Models:** Raider **Engines:** 3.7L K, 4.7L M **Transmissions:** All	**PCM Internal Shutdown timer Rationality** This DTC sets if the engine coolant temp does not drop enough or drops too much during engine off time. This DTC may also set if the controller timer is inaccurate. Three good trips to turn off the MIL. **Possible Causes:** • Engine cooling conditions that may set this DTC • Engine coolant level • PCM internal error
DTC: P1618 **1T PCM MIL: YES** **Years:** 2006, 2007, 2008 **Models:** Raider **Engines:** 3.7L K, 4.7L M **Transmissions:** All	**Sensor Reference Voltage 1 Circuit Erratic** When the PCM recognizes the Primary 5-volt Supply circuit voltage is varying too much too quickly. ETC light is flashing. **Possible Causes:** • Primary 5-volt supply shorted to ground • Primary 5-volt supply shorted to voltage • Primary 5-volt supply circuit open • 5-volt sensor • PCM internal error

DTC	Trouble Code Title, Conditions & Possible Causes
DTC: P1628 **1T PCM MIL: YES** **Years:** 2006, 2007, 2008 **Models:** Raider **Engines:** 3.7L K, 4.7L M **Transmissions:** All	**Sensor Reference Voltage 2 Circuit Erratic** When the PCM recognizes the Auxiliary 5-volt Supply circuit voltage is varying too much too quickly. ETC light is flashing. **Possible Causes:** • Primary 5-volt supply shorted to ground • Primary 5-volt supply shorted to voltage • Primary 5-volt supply circuit open • 5-volt sensor • PCM internal error
DTC: P1676 **2T PCM MIL: YES** **Years:** 2006, 2007, 2008 **Models:** Evolution, Lancer, Outlander **Engines:** 2.0L, 2.4W, 3.0LX **Transmissions:** All	**Variant Coding** Vehicle information is not entered into the ECM. **Possible Causes:** • ECM has failed
DTC: P1696 **2T PCM MIL: NO** **Years:** 2006, 2007, 2008 **Models:** Raider **Engines:** 3.7L K, 4.7L M **Transmissions:** All	**PCM EEPROM Write Denied** Key on, and the PCM detected that it was unable to write to one or more EEPROM locations during the initial self-test procedure. **Possible Causes:** • Clear the trouble codes and retest for this trouble code. If the same trouble code resets, the PCM has failed and must be replaced to repair this problem.
DTC: P1697 **2T PCM MIL: NO** **Years:** 2006, 2007, 2008 **Models:** Raider **Engines:** 3.7L K, 4.7L M **Transmissions:** All	**PCM Malfunction - SRI Miles Not Stored** Key on, and the PCM detected that it was unable to write to the Service Reminder Indicator (SRI or EMR) during the initial self-test. **Possible Causes:** • Clear the trouble codes and retest for this trouble code. If the same trouble code resets, the PCM has failed and must be replaced to repair this problem.
DTC: P1897 **1T PCM MIL: NO** **Years:** 2006, 2007, 2008 **Models:** Raider **Engines:** 3.7L K, 4.7L M **Transmissions:** All	**Level 1 RPM Bus Unlock** When the PCM recognizes an internal failure to communicate with the FCM or the CMP and CKP Sensor count periods are too short. ETC light is flashing. **Possible Causes:** • PCM has failed

Gas Engine OBD II Trouble Code List (P2xxx Codes)

DTC	Trouble Code Title, Conditions & Possible Causes
DTC: P2066 **2T PCM MIL: YES** **Years:** 2006, 2007, 2008 **Models:** All **Engines:** All **Transmissions:** All	**Fuel Level Sensor (Sub) Circuit Range/ Performance** When fuel consumption calculated from the operating time of the injector amounts to 8 gallons, the diversity of the amount of the fuel in the tank calculated from the fuel level sensor is 0.5 gallons or less. **Possible Causes:** • Fuel pump module or fuel level sensor (sub) failed • PCM has failed
DTC: P2072 **2T PCM MIL: NO** **Years:** 2006, 2007, 2008 **Models:** Raider **Engines:** 3.7L K, 4.7L M **Transmissions:** All	**Electronic Throttle Contro System – Ice Blockage** The PCM recognizes the Throttle plate is stuck during extremely cold Ambient Temperature operation. The throttle plate goes through a de-icing procedure. If the throttle blade still doesn't move this fault sets. The MIL will not illuminate. ETC light will illuminate. The vehicle will be in Limp home condition, limiting RPM and vehicle speed. **Possible Causes:** • Throttle plate frozen

DTC	Trouble Code Title, Conditions & Possible Causes
DTC: P2096 **2T PCM MIL: NO** **Years:** 2006, 2007, 2008 **Models:** Raider **Engines:** 3.7L K, 4.7L M **Transmissions:** All	**Downstream Fuel Trim System 1 Lean** The conditions that cause this diagnostic to fail is when the upstream O2 sensor becomes biased from an exhaust leak, O2 sensor contamination, or some other extreme operating condition. The downstream O2 sensor is considered to be protected from extreme environments by the catalyst. The PCM monitors the downstream O2 sensor feedback control, called downstream fuel trim, to detect any shift in the upstream O2 sensor target voltage from nominal target voltage. The value of the downstream fuel trim is compared with the lean thresholds. Every time the value exceeds the calibrated threshold, a fail timer is incremented and mass flow through the exhaust is accumulated. If the fail timer and accumulated mass flow exceed the fail thresholds, the test fails and the diagnostic stops running for that trip. If the test fails on consecutive trips, a DTC is set. **Possible Causes:** • Exhaust leak • Fuel delivery system • O2 sensor wiring or connections • Engine mechanical system • PCM has failed
DTC: P2097 **2T PCM MIL: NO** **Years:** 2006, 2007, 2008 **Models:** Raider **Engines:** 3.7L K, 4.7L M **Transmissions:** All	**Downstream Fuel Trim System 1 Rich** The conditions that cause this diagnostic to fail is when the upstream O2 sensor becomes biased from an exhaust leak, O2 sensor contamination, or some other extreme operating condition. The downstream O2 sensor is considered to be protected from extreme environments by the catalyst. The PCM monitors the downstream O2 sensor feedback control, called downstream fuel trim, to detect any shift in the upstream O2 sensor target voltage from nominal target voltage. The value of the downstream fuel trim is compared with the lean thresholds. Every time the value exceeds the calibrated threshold, a fail timer is incremented and mass flow through the exhaust is accumulated. If the fail timer and accumulated mass flow exceed the fail thresholds, the test fails and the diagnostic stops running for that trip. If the test fails on consecutive trips, a DTC is set. **Possible Causes:** • Exhaust leak • Fuel delivery system • O2 sensor wiring or connections • Engine mechanical system • PCM has failed
DTC: P2098 **2T PCM MIL: NO** **Years:** 2006, 2007, 2008 **Models:** Raider **Engines:** 3.7L K, 4.7L M **Transmissions:** All	**Downstream Fuel Trim System 2 Lean** The conditions that cause this diagnostic to fail is when the upstream O2 sensor becomes biased from an exhaust leak, O2 sensor contamination, or some other extreme operating condition. The downstream O2 sensor is considered to be protected from extreme environments by the catalyst. The PCM monitors the downstream O2 sensor feedback control, called downstream fuel trim, to detect any shift in the upstream O2 sensor target voltage from nominal target voltage. The value of the downstream fuel trim is compared with the lean thresholds. Every time the value exceeds the calibrated threshold, a fail timer is incremented and mass flow through the exhaust is accumulated. If the fail timer and accumulated mass flow exceed the fail thresholds, the test fails and the diagnostic stops running for that trip. If the test fails on consecutive trips, a DTC is set. **Possible Causes:** • Exhaust leak • Fuel delivery system • O2 sensor wiring or connections • Engine mechanical system • PCM has failed
DTC: P2099 **2T PCM MIL: NO** **Years:** 2006, 2007, 2008 **Models:** Raider **Engines:** 3.7L K, 4.7L M **Transmissions:** All	**Downstream Fuel Trim System 2 Rich** The conditions that cause this diagnostic to fail is when the upstream O2 sensor becomes biased from an exhaust leak, O2 sensor contamination, or some other extreme operating condition. The downstream O2 sensor is considered to be protected from extreme environments by the catalyst. The PCM monitors the downstream O2 sensor feedback control, called downstream fuel trim, to detect any shift in the upstream O2 sensor target voltage from nominal target voltage. The value of the downstream fuel trim is compared with the lean thresholds. Every time the value exceeds the calibrated threshold, a fail timer is incremented and mass flow through the exhaust is accumulated. If the fail timer and accumulated mass flow exceed the fail thresholds, the test fails and the diagnostic stops running for that trip. If the test fails on consecutive trips, a DTC is set. **Possible Causes:** • Exhaust leak • Fuel delivery system • O2 sensor wiring or connections • Engine mechanical system • PCM has failed

DTC	Trouble Code Title, Conditions & Possible Causes
DTC: P2100 **2T PCM MIL: YES** **Years:** 2006, 2007, 2008 **Models:** All **Engines:** All **Transmissions:** All	**Throttle Actuator Control Motor Circuit Open** Battery voltage is greater than 8.3 volts, the throttle actuator control motor circuit current should be 0.1 amps or less for 0.72 seconds. **Possible Causes:** • Throttle actuator control motor failed • Throttle actuator control motor circuit open • PCM has failed
DTC: P2102 **2T PCM MIL: YES** **Years:** 2006, 2007, 2008 **Models:** Lancer **Engines:** All **Transmissions:** All	**Throttle Actuator Control Motor Circuit Shorted Low** Battery voltage is greater than 8.3 volts, the throttle actuator control motor circuit current should is 12 amps or greater for 0.8 seconds. **Possible Causes:** • Throttle actuator control motor failed • Throttle actuator control motor circuit shorted • PCM has failed
DTC: P2103 **2T PCM MIL: YES** **Years:** 2006, 2007, 2008 **Models:** Lancer **Engines:** All **Transmissions:** All	**Throttle Actuator Control Motor Circuit Shorted High** Battery voltage is greater than 8.3 volts, the throttle actuator control motor circuit current is 8 amps or greater for 0.8 seconds. **Possible Causes:** • Throttle actuator control motor failed • Throttle actuator control motor circuit shorted • PCM has failed
DTC: P2106 **2T PCM MIL: YES** **Years:** 2006, 2007, 2008 **Models:** Raider **Engines:** 3.7L K, 4.7L M **Transmissions:** All	**Electronic Throttle Control System – Forced Limited** This DTC sets for OBDII MIL illumination purposes. This DTC will always have associated DTCs indicating a system failure. Engine speed is being limited and/or throttle motor is power free. ETC light is flashing. **Possible Causes:** • Other DTCs related to ETC have set
DTC: P2107 **1T PCM MIL: YES** **Years:** 2006, 2007, 2008 **Models:** Raider **Engines:** 3.7L K, 4.7L M **Transmissions:** All	**Electronic Throttle Control Module Processor** Internal PCM failure. Module will attempt to reset, so you will be able to hear the throttle relearning. If the condition is continuous, the vehicle may not be driveable. ETC light is flashing. **Possible Causes:** • PCM need to be reprogrammed
DTC: P2108 **1T PCM MIL: YES** **Years:** 2006, 2007, 2008 **Models:** Raider **Engines:** 3.7L K, 4.7L M **Transmissions:** All	**Electronic Throttle Control Module Performance** Internal PCM failure. Customer may experience an extended cranking condition with limited driving and a rough idle. One trip fault and the code will set within 5 seconds. ETC light is flashing. **Possible Causes:** • PCM need to be reprogrammed
DTC: P2110 **1T PCM MIL: YES** **Years:** 2006, 2007, 2008 **Models:** Raider **Engines:** 3.7L K, 4.7L M **Transmissions:** All	**Electronic Throttle Control System – Forced Limited RPM** When the PCM requests to limit engine speed if RPM is too high for 20.5 seconds and before P2118 sets. One trip fault and the code will set within 5 seconds. ETC light is illuminated. **Possible Causes:** • Throttle plate stuck • ECT positive circuit open • ECT negative circuit open • ECT positive circuit shorted to ground • ECT negative circuit shorted to ground • ECT motor/throttle body • PCM
DTC: P2111 **1T PCM MIL: YES** **Years:** 2006, 2007, 2008 **Models:** Raider **Engines:** 3.7L K, 4.7L M **Transmissions:** All	**Electronic Throttle Control – Unable to Close** Just after key on, the throttle is opened and closed to test the system. If the TP Sensor does not return to Limp Home Position at the end of this test, this DTC will set. One trip fault and the code will set within 5 seconds. ETC light is flashing. **Possible Causes:** • Throttle plate stuck above limp home positon • TP sensors both read 2.5 volts • ECT positive circuit shorted to voltage • ECT positive circuit open • ECT negative circuit open • ECT positive circuit shorted to ground • ECT negative circuit shorted to ground • PCM

DTC	Trouble Code Title, Conditions & Possible Causes
DTC: P2112 **1T PCM MIL: YES** **Years:** 2006, 2007, 2008 **Models:** Raider **Engines:** 3.7L K, 4.7L M **Transmissions:** All	**Electronic Throttle Control – Unable to Open** Just after key on, the throttle is opened and closed to test the system. If the TP Sensor does not return to Limp Home Position at the end of this test, this DTC will set. One trip fault and the code will set within 5 seconds. ETC light is flashing. **Possible Causes:** • Throttle plate stuck at or below limp home positon • ECT negative circuit shorted to voltage • ECT positive circuit open • ECT negative circuit open • ECT positive circuit shorted to ground • ECT negative circuit shorted to ground • PCM
DTC: P2115 **1T PCM MIL: YES** **Years:** 2006, 2007, 2008 **Models:** Raider **Engines:** 3.7L K, 4.7L M **Transmissions:** All	**Accelerator Pedal Position Sensor 1 – Minimum Stop Performance** APPS No.1 has failed to achieve the required minimum value during In Plant testing. One trip fault and the code will set within 5 seconds. Engine will only idle. ETC light is illuminated. **Possible Causes:** • In plat test failure • APPS relearn
DTC: P2116 **2T PCM MIL: YES** **Years:** 2006, 2007, 2008 **Models:** Raider **Engines:** 3.7L K, 4.7L M **Transmissions:** All	**Accelerator Pedal Position Sensor 2 – Minimum Stop Performance** APPS No.1 has failed to achieve the required minimum value during In Plant testing. One trip fault and the code will set within 5 seconds. Engine will only idle. ETC light is illuminated. **Possible Causes:** • In plat test failure • APPS relearn
DTC: P2118 **2T PCM MIL: YES** **Years:** 2006, 2007, 2008 **Models:** Raider **Engines:** 3.7L K, 4.7L M **Transmissions:** All	**Electronic Throttle Control Motor Current** When the PCM detects an internal error or a short between the ETC Motor- and ETC Motor + circuits in the ETC Motor Driver. ETC light is flashing. **Possible Causes:** • Throttle plate/bore inspection • ETC positive circuit shorted to voltage • ETC negative circuit shorted to voltage • ETC positive circuit shorted to the negative circuit • ETC positive circuit open • ETC negative circuit open • ETC positive circuit shorted to ground • ETC negative circuit shorted to ground • ETC motor/throttle body • PCM
DTC: P2121 **2T PCM MIL: YES** **Years:** 2006, 2007, 2008 **Models:** Lancer **Engines:** All **Transmissions:** All	**Accelerator Pedal Position Sensor (Main) Circuit Range/Performance Problem** Key on, accelerator pedal position switch ON, accelerator pedal position sensor (sub) output voltage is 1.88 volts or less, the accelerator pedal position sensor (main) output voltage is 1.88 volts or greater for 1 second. **Possible Causes:** • Accelerator pedal position sensor misadjusted or failed • Accelerator pedal position sensor (main) open or shorted • Accelerator pedal position switch misadjusted or failed • Accelerator pedal position switch open or shorted • PCM has failed
DTC: P2122 **2T PCM MIL: YES** **Years:** 2006, 2007, 2008 **Models:** All **Engines:** All **Transmissions:** All	**Accelerator Pedal Position Sensor (Main) Circuit Low Input** Key on, accelerator pedal position sensor (main) output voltage is 0.2 volts or less for 1 second. **Possible Causes:** • Accelerator pedal position sensor misadjusted or failed • Accelerator pedal position sensor (main) open or shorted • PCM has failed
DTC: P2123 **2T PCM MIL: YES** **Years:** 2006, 2007, 2008 **Models:** All **Engines:** All **Transmissions:** All	**Accelerator Pedal Position Sensor (Main) Circuit High Input** Key on, accelerator pedal position sensor (sub) output voltage between 0.2 volts and 2.5, the accelerator pedal position sensor (main) output voltage should be 4.5 volts or greater for 1 second. **Possible Causes:** • Accelerator pedal position sensor misadjusted or failed • Accelerator pedal position sensor (main) open or shorted • PCM has failed

DTC	Trouble Code Title, Conditions & Possible Causes
DTC: P2127 **2T PCM MIL: YES** **Years:** 2006, 2007, 2008 **Models:** All **Engines:** All **Transmissions:** All	**Accelerator Pedal Position Sensor (Main) Circuit Low Input** Key on, accelerator pedal position sensor (sub) output voltage is 0.2 volts or less for 1 second. **Possible Causes:** • Accelerator pedal position sensor misadjusted or failed • Accelerator pedal position sensor (sub) open or shorted • PCM has failed
DTC: P2128 **2T PCM MIL: YES** **Years:** 2006, 2007, 2008 **Models:** All **Engines:** All **Transmissions:** All	**Accelerator Pedal Position Sensor (Main) Circuit High Input** Key on, accelerator pedal position sensor (main) output voltage between 0.2 volts and 2.5, the accelerator pedal position sensor (sub) output voltage should be 4.5 volts or greater for 1 second. **Possible Causes:** • Accelerator pedal position sensor misadjusted or failed • Accelerator pedal position sensor (main) open or shorted • PCM has failed
DTC: P2135 **2T PCM MIL: YES** **Years:** 2006, 2007, 2008 **Models:** All **Engines:** All **Transmissions:** All	**Throttle Position Sensor (Main and Sub) Range/Performance Problem** Key on, TPS (main) output voltage is between 2.5 and 4.8 volts, TPS (sub) output voltage is greater than 2.25 volts, the TPS (sub) output voltage is 4.2 volts or lower. **Possible Causes:** • TPS failed • TPS shorted • PCM has failed
DTC: P2138 **2T PCM MIL: YES** **Years:** 2006, 2007, 2008 **Models:** All **Engines:** All **Transmissions:** All	**Accelerator Pedal Position Sensor (Main and Sub) Range/Performance Problem** Key on, accelerator pedal position sensor (main) output voltage between 0.2 volts and 4.5, the accelerator pedal position sensor (sub) output voltage between 0.2 and 4.5 volts, change of accelerator pedal position sensor (sub) output voltage per 25 ms is less than 0.1 volt, the accelerator pedal position sensor (sub) output voltage minus the accelerator pedal position sensor (main) output voltage should be 1 volt or greater for 1 second and/or accelerator pedal position sensor (main) output voltage minus accelerator pedal position sensor (sub) output voltage should be 1 volt or greater for 0.2 seconds. **Possible Causes:** • Accelerator pedal position sensor misadjusted or failed • Accelerator pedal position sensor (main) open or shorted • PCM has failed
DTC: P2161 **2T PCM MIL: YES** **Years:** 2006, 2007, 2008 **Models:** Raider **Engines:** 3.7L K, 4.7L M **Transmissions:** All	**Vehicle Speed Sensor 2 Erratic** PCM recognizes Vehicle speed input No.2 erratic or high. VSS No.2 is based on the average of the Front Wheel Speeds. One trip fault and the code will set within 5 seconds. No MIL and No ETC light. Cruise is disabled. **Possible Causes:** • Active bus or communication DTCs • Tire Circumfrence • PCM has failed
DTC: P2166 **2T PCM MIL: YES** **Years:** 2006, 2007, 2008 **Models:** Raider **Engines:** 3.7L K, 4.7L M **Transmissions:** All	**Accelerator Pedal Position Sensor 1 – Minimum Stop Performance** APPS No.1 has failed to achieve the required minimum value during In Plant testing. One trip fault and the code will set within 5 seconds. Engine will only idle. ETC light is illuminated. **Possible Causes:** • In plat test failure • APPS relearn
DTC: P2167 **2T PCM MIL: YES** **Years:** 2006, 2007, 2008 **Models:** Raider **Engines:** 3.7L K, 4.7L M **Transmissions:** All	**Accelerator Pedal Position Sensor 2 – Minimum Stop Performance** APPS No.1 has failed to achieve the required minimum value during In Plant testing. One trip fault and the code will set within 5 seconds. Engine will only idle. ETC light is illuminated. **Possible Causes:** • In plat test failure • APPS relearn
DTC: P2172 **2T PCM MIL: YES** **Years:** 2006, 2007, 2008 **Models:** Raider **Engines:** 3.7L K, 4.7L M **Transmissions:** All	**High Airflow / Vacuum Leak Detected** Engine running, the actual intake air amount is greater than the allowable intake air amount for 1.5 seconds. **Possible Causes:** • MAF sensor failed • Throttle valve faulty operation • TPS failed • PCM has failed

DTC	Trouble Code Title, Conditions & Possible Causes
DTC: P2173 **2T PCM MIL: YES** **Years:** 2006, 2007, 2008 **Models:** Raider **Engines:** 3.7L K, 4.7L M **Transmissions:** All	**High Airflow / Vacuum Leak Detected** Engine running, the actual intake air amount is greater than the allowable intake air amount for 1.5 seconds. **Possible Causes:** • MAF sensor failed • Throttle valve faulty operation • TPS failed • PCM has failed
DTC: P2174 **1T PCM MIL: YES** **Years:** 2006, 2007, 2008 **Models:** Raider **Engines:** 3.7L K, 4.7L M **Transmissions:** All	**Low Airflow / Restriction Detected** PCM calculated MAP value is greater than actual MAP value plus an offset value. Three good trips to turn of the mil. ETC light will flash. **Possible Causes:** • Restricted air inlet system • Resistance in the 5-volt supply circuit • 5-volt supply circuit shorted to ground • Resistance in the MAP signal circuit • MAP signal circuit shorted to ground • Resistance in the sensor ground circuit • Resistance in the 5-volt supply circuit • 5-volt supply circuit shorted to ground • Resistance in the TP sensor signal circuit • TP sensor signal circuit shorted to ground • Resistance in the TP sensor return circuit • Map sensor • TP sensor • Powertrain control module (PCM)
DTC: P2175 **2T PCM MIL: YES** **Years:** 2006, 2007, 2008 **Models:** Raider **Engines:** 3.7L K, 4.7L M **Transmissions:** All	**Low Airflow / Restriction Detected** PCM calculated MAP value is greater than actual MAP value plus an offset value. Three good trips to turn of the mil. ETC light will flash. **Possible Causes:** • Restricted air inlet system • Resistance in the 5-volt supply circuit • 5-volt supply circuit shorted to ground • Resistance in the MAP signal circuit • MAP signal circuit shorted to ground • Resistance in the sensor ground circuit • Resistance in the 5-volt supply circuit • 5-volt supply circuit shorted to ground • Resistance in the TP sensor signal circuit • TP sensor signal circuit shorted to ground • Resistance in the TP sensor return circuit • Map sensor • TP sensor • Powertrain control module (PCM)
DTC: P2181 **1T PCM MIL: YES** **Years:** 2006, 2007, 2008 **Models:** Raider **Engines:** 3.7L K, 4.7L M **Transmissions:** All	**Cooling System Performance** PCM recognizes that the ECT has failed its self coherence test. The coolant temp should only change at a certain rate, if this rate is too slow or too fast this fault will set. Three good trips to clear MIL. ETC light will illuminate on first trip failure. **Possible Causes:** • Low coolant level • ECT signal circuit shorted to voltage • ECT signal circuit open • Sensor ground circuit open • ECT signal circuit shorted to ground • ECT signal circuit shorted to the sensor ground • Thermostat • ECT sensor • Powertrain control module (PCM)
DTC: P2195 **2T PCM MIL: YES** **Years:** 2006, 2007, 2008 **Models:** All **Engines:** All **Transmissions:** All	**HO2S Inactive (Bank 1, Sensor 1)** Engine running for more than 20 seconds, engine coolant temperature greater than 44°F, intake air temperature is greater than 13°F, the front HO2S output voltage is less than 0.5 volts for 128 seconds. **Possible Causes:** • Front HO2S deteriorated or failed • PCM has failed

DTC	Trouble Code Title, Conditions & Possible Causes
DTC: P2197 **2T PCM MIL: YES** **Years:** 2006, 2007, 2008 **Models:** All **Engines:** All **Transmissions:** All	**HO2S Inactive (Bank 2, Sensor 1)** Engine running for more than 20 seconds, engine coolant temperature greater than 44°F, intake air temperature is greater than 13°F, the front HO2S output voltage is less than 0.5 volts for 128 seconds. **Possible Causes:** • Front HO2S deteriorated or failed • PCM has failed
DTC: P2228 **2T PCM MIL: YES** **Years:** 2006, 2007, 2008 **Models:** All **Engines:** All **Transmissions:** All	**Barometric Pressure Circuit Low Input** Engine running for 2 seconds or more and battery voltage is greater than 8 volts, barometric pressure sensor output has continued to be 14.6 in. Hg or less (about 15,000 ft above sea level) for 10 seconds. **Possible Causes:** • PCM has failed
DTC: P2229 **2T PCM MIL: YES** **Years:** 2006, 2007, 2008 **Models:** All **Engines:** All **Transmissions:** All	**Barometric Pressure Circuit High Input** Engine running for 2 seconds or more and battery voltage is greater than 8 volts, barometric pressure sensor output has continued to be 33.3 in. Hg or less (about 4,000 ft above sea level) for 10 seconds. **Possible Causes:** • PCM has failed
DTC: P2252 **2T PCM MIL: YES** **Years:** 2006, 2007, 2008 **Models:** All **Engines:** All **Transmissions:** All	**HO2S Offset Circuit Low Voltage** Engine running for 2 seconds or more, the HO2S offset voltage is less than 0.4 volts for 2 seconds. **Possible Causes:** • PCM failed
DTC: P2253 **2T PCM MIL: YES** **Years:** 2006, 2007, 2008 **Models:** All **Engines:** All **Transmissions:** All	**HO2S Offset Circuit High Voltage** Engine running for 2 seconds or more, the HO2S offset voltage is greater than 0.6 volts for 2 seconds. **Possible Causes:** • PCM failed
DTC: P2271 **2T PCM MIL: YES** **Years:** 2006, 2007, 2008 **Models:** Raider **Engines:** 3.7L K, 4.7L M **Transmissions:** All	**HO2S Sensor 1/2 Signal Stuck Rich** The PCM monitors the downstream O2 sensor. If the PCM does not detect a rich to lean switch within a specific time during a decel fuel shutoff event, the monitor will fail. Three good trips to turn off the MIL. **Possible Causes:** • Oxygen sensor wiring or connectors • O2 sensor signal circuit • O2 sensor return circuit • O2 sensor • Powertrain control module (PCM)
DTC: P2273 **2T PCM MIL: YES** **Years:** 2006, 2007, 2008 **Models:** Raider **Engines:** 3.7L K, 4.7L M **Transmissions:** All	**HO2S Sensor 2/2 Signal Stuck Rich** The PCM monitors the downstream O2 sensor. If the PCM does not detect a rich to lean switch within a specific time during a decel fuel shutoff event, the monitor will fail. Three good trips to turn off the MIL. **Possible Causes:** • Oxygen sensor wiring or connectors • O2 sensor signal circuit • O2 sensor return circuit • O2 sensor • Powertrain control module (PCM)
DTC: P2299 **2T PCM MIL: YES** **Years:** 2006, 2007, 2008 **Models:** Raider **Engines:** 3.7L K, 4.7L M **Transmissions:** All	**Brake Pedal Position / Accelerator Pedal Position Incorrect** The PCM recognizes a brake application following the APPS showing a fixed pedal opening. Temporary or permanent. Internally the PCM will reduce throttle opening below driver demand. One trip fault and the code will be set within 5 seconds. ETC light will illuminate, the light will only stay illuminated while DTC is active. **Possible Causes:** • Customer pressing accelerator pedal, then pressing brake pedal, and continues to hold them down simultaneously • Stop lamp switch • APP sensor

DTC	Trouble Code Title, Conditions & Possible Causes
DTC: P2302 **1T PCM MIL: YES** **Years:** 2006, 2007, 2008 **Models:** Raider **Engines:** 3.7L K, 4.7L M **Transmissions:** All	**Ignition Coil 1 Secondary Circuit – Insufficient Ionization** If PCM detects that the secondary ignition burn time is incorrect, too short, or not present, an error is detected. Three good trips to turn off the MIL. **Possible Causes:** • ASD relay output circuit • Coil control circuit open • Coil control circuit shorted to ground • Coil on plug • Powertrain control module (PCM)
DTC: P2305 **2T PCM MIL: YES** **Years:** 2006, 2007, 2008 **Models:** Raider **Engines:** 3.7L K, 4.7L M **Transmissions:** All	**Ignition Coil 2 Secondary Circuit – Insufficient Ionization** If PCM detects that the secondary ignition burn time is incorrect, too short, or not present, an error is detected. Three good trips to turn off the MIL. **Possible Causes:** • ASD relay output circuit • Coil control circuit open • Coil control circuit shorted to ground • Coil on plug • Powertrain control module (PCM)
DTC: P2308 **2T PCM MIL: YES** **Years:** 2006, 2007, 2008 **Models:** Raider **Engines:** 3.7L K, 4.7L M **Transmissions:** All	**Ignition Coil 3 Secondary Circuit – Insufficient Ionization** If PCM detects that the secondary ignition burn time is incorrect, too short, or not present, an error is detected. Three good trips to turn off the MIL. **Possible Causes:** • ASD relay output circuit • Coil control circuit open • Coil control circuit shorted to ground • Coil on plug • Powertrain control module (PCM)
DTC: P2311 **2T PCM MIL: YES** **Years:** 2006, 2007, 2008 **Models:** Raider **Engines:** 3.7L K, 4.7L M **Transmissions:** All	**Ignition Coil 4 Secondary Circuit – Insufficient Ionization** If PCM detects that the secondary ignition burn time is incorrect, too short, or not present, an error is detected. Three good trips to turn off the MIL. **Possible Causes:** • ASD relay output circuit • Coil control circuit open • Coil control circuit shorted to ground • Coil on plug • Powertrain control module (PCM)
DTC: P2314 **2T PCM MIL: YES** **Years:** 2006, 2007, 2008 **Models:** Raider **Engines:** 3.7L K, 4.7L M **Transmissions:** All	**Ignition Coil 5 Secondary Circuit – Insufficient Ionization** If PCM detects that the secondary ignition burn time is incorrect, too short, or not present, an error is detected. Three good trips to turn off the MIL. **Possible Causes:** • ASD relay output circuit • Coil control circuit open • Coil control circuit shorted to ground • Coil on plug • Powertrain control module (PCM)
DTC: P2317 **2T PCM MIL: YES** **Years:** 2006, 2007, 2008 **Models:** Raider **Engines:** 3.7L K, 4.7L M **Transmissions:** All	**Ignition Coil 6 Secondary Circuit – Insufficient Ionization** If PCM detects that the secondary ignition burn time is incorrect, too short, or not present, an error is detected. Three good trips to turn off the MIL. **Possible Causes:** • ASD relay output circuit • Coil control circuit open • Coil control circuit shorted to ground • Coil on plug • Powertrain control module (PCM)

DTC	Trouble Code Title, Conditions & Possible Causes
DTC: P2320 **2T PCM MIL: YES** **Years:** 2006, 2007, 2008 **Models:** Raider **Engines:** 3.7L K, 4.7L M **Transmissions:** All	**Ignition Coil 7 Secondary Circuit – Insufficient Ionization** If PCM detects that the secondary ignition burn time is incorrect, too short, or not present, an error is detected. Three good trips to turn off the MIL. **Possible Causes:** • ASD relay output circuit • Coil control circuit open • Coil control circuit shorted to ground • Coil on plug • Powertrain control module (PCM)
DTC: P2323 **2T PCM MIL: YES** **Years:** 2006, 2007, 2008 **Models:** Raider **Engines:** 3.7L K, 4.7L M **Transmissions:** All	**Ignition Coil 8 Secondary Circuit – Insufficient Ionization** If PCM detects that the secondary ignition burn time is incorrect, too short, or not present, an error is detected. Three good trips to turn off the MIL. **Possible Causes:** • ASD relay output circuit • Coil control circuit open • Coil control circuit shorted to ground • Coil on plug • Powertrain control module (PCM)
DTC: P2423 **2T PCM MIL: YES** **Years:** 2006, 2007, 2008 **Models:** Outlander **Engines:** 3.0L X **Transmissions:** All	**HC Absorber (HC Trap Catalyst) Efficiency Low** After the output voltage of the heated oxygen sensor (rear) reaches 0.5 V, the output voltage of the center exhaust pipe heated oxygen sensor reaches 0.5 V within 1.5 seconds. **Possible Causes:** • HC trap catalyst deterioration within center exhaust pipe. • ECM failed
DTC: P2503 **1T PCM MIL: YES** **Years:** 2006, 2007, 2008 **Models:** Raider **Engines:** 3.7L K, 4.7L M **Transmissions:** All	**Charging System Output Low** The battery sensed voltage is less than the target charging voltage, during engine operation, for a calibrated amount of time. Generator light will illuminate. **Possible Causes:** • Loose accessory drive belt • Excessive resistance in the battery positive circuit • Generator • Powertrain control module (PCM)
DTC: P2504 **2T PCM MIL: YES** **Years:** 2006, 2007, 2008 **Models:** Raider **Engines:** 3.7L K, 4.7L M **Transmissions:** All	**Charging System Output High** The alternator B+ voltage sense circuit voltage reading exceeds the direct Battery B+ sense circuit. The Generator Output terminal in not connected to the Battery B+ post. One trip fault. **Possible Causes:** • Excessive resistance in the battery positive circuit • Powertrain control module (PCM)
DTC: P2567 **2T PCM MIL: YES** **Years:** 2006, 2007, 2008 **Models:** Galant, Outlander **Engines:** 2.4L F, 3.0L X, 3.8L S, 3.8L T **Transmissions:** All	**LIN Communication Check by Radiator Sensor** When the radiator sensor is defective or the PCM is defective **Possible Causes:** • Radiator sensor failed • ECM failed
DTC: P2610 **2T PCM MIL: YES** **Years:** 2006, 2007, 2008 **Models:** Raider **Engines:** 3.7L K, 4.7L M **Transmissions:** All	**PCM Internal Shutdown Timer Rationality Too Fast** The PCM detects that the engine coolant temperature drops a specified amount during the measured engine off time. Three good trips to turn off the MIL. **Possible Causes:** • Excessive resistance in the PCM poer or ground circuits • PCM software update • PCM failed

Gas Engine OBD II Trouble Code List (PU1xxx Codes)

DTC	Trouble Code Title, Conditions & Possible Causes
DTC: U0001 **2T PCM MIL: YES** **Years:** 2006, 2007, 2008 **Models:** Outlander, Raider **Engines:** 3.0X, 3.7L K, 4.7L M **Transmissions:** All	**Buss Off** The PCM loses communication over the CAN C Bus circuit. The circuit is continuously monitored. **Possible Causes:** • CAN C bus failure open or shorted condition • PCM failed
DTC: U0101 **2T PCM MIL: YES** **Years:** 2006, 2007, 2008 **Models:** Outlander, Raider **Engines:** 2.4L W, 3.7L K, 4.7L M **Transmissions:** All	**TCM Time Out** The PCM doesn't receive a Bus Message from the Transmission Control Module for 7 consecutive seconds. The circuit is continuously monitored. Two Trip fault. **Possible Causes:** • CAN C bus open or shorted condition • PCM failed
DTC: U0114 **2T PCM MIL: YES** **Years:** 2006, 2007, 2008 **Models:** Outlander **Engines:** 3.0L X **Transmissions:** All	**AWD-ECU Time Out** Proceed to troubleshoot based on a harness or connector damage on the CAN bus line between the ECM and AWD-ECU, and a failure in the AWD-ECU power supply system. **Possible Causes:** • CAN line harness damage or connector damage • ECU failed
DTC: U0121 **2T PCM MIL: YES** **Years:** 2006, 2007, 2008 **Models:** Outlander, Raider **Engines:** 2.4L W, 3.7L K, 4.7L M **Transmissions:** All	**ABS-ECU Time Out** The PCM doesn't receive an ABS message over the CAN C circuit for 7 consecutive seconds. The circuit is continuously monitored. **Possible Causes:** • CAN C bus open or shorted condition • ABS brake module • PCM failed
DTC: U0141 **2T PCM MIL: YES** **Years:** 2006, 2007, 2008 **Models:** Outlander, Raider **Engines:** 2.4L W, 3.7L K, 4.7L M **Transmissions:** All	**ETACS-ECU Time Out** The PCM doesn't receive a FCM message over the CAN C circuit for 7 consecutive seconds. The circuit is continuously monitored. **Possible Causes:** • CAN C bus open or shorted condition • FCM failed • PCM failed
DTC: U0167 **2T PCM MIL: YES** **Years:** 2006, 2007, 2008 **Models:** Outlander **Engines:** 2.4L W **Transmissions:** All	**Immobilizer Communication Error** Proceed to troubleshoot based on a harness or connector damage on the CAN bus line between the ECM and AWD-ECU, and a failure in the AWD-ECU power supply system. **Possible Causes:** • CAN line harness damage or connector damage. • KOS-ECU/WCM (immobilizer-ECU) failed. • ECM failed.
DTC: U0168 **2T PCM MIL: YES** **Years:** 2006, 2007, 2008 **Models:** Raider **Engines:** 3.7L K, 4.7L M **Transmissions:** All	**Lost Communication With Vehicle Security Control Module** The PCM doesn't receive a SKREEM message over the CAN C circuit for 7 consecutive seconds. The circuit is continuously monitored. **Possible Causes:** • CAN C bus open or shorted condition • No power to vehicle security control module, module unplugged or open fuse • PCM failed
DTC: U1073 **2T PCM MIL: YES** **Years:** 2006, 2007, 2008 **Models:** All **Engines:** All **Transmissions:** All	**Bus Off** Bus off error detected **Possible Causes:** • CAN line harness or connectors damage • PCM failed
DTC: U1102 **2T PCM MIL: YES** **Years:** 2006, 2007, 2008 **Models:** All **Engines:** All **Transmissions:** All	**ABS-ECU Timeout** Engine not cranking or 3 seconds have passed since the engine was cranked and battery voltage is 10 volts or greater, the ABS-ECU was unable to receive a signal through the CAN bus line. **Possible Causes:** • CAN line harness or connectors damage • ABS-ECU power supply system failed • ABS-ECU failed • PCM failed

DTC	Trouble Code Title, Conditions & Possible Causes
DTC: U1108 **2T PCM MIL: YES** **Years:** 2006, 2007, 2008 **Models:** All **Engines:** All **Transmissions:** All	**Combination Meter Timeout** Engine not cranking or 3 seconds have passed since the engine was cranked and battery voltage is 10 volts or greater, the combination meter was unable to receive a signal through the CAN bus line. **Possible Causes:** • CAN line harness or connectors damage • Combination meter power supply system failed • Combination meter failed • PCM failed
DTC: U1109 **2T PCM MIL: YES** **Years:** 2006, 2007, 2008 **Models:** All **Engines:** All **Transmissions:** All	**ETACS–ECU Timeout** Engine not cranking or 3 seconds have passed since the engine was cranked and battery voltage is 10 volts or greater, the ETACS-ECU was unable to receive a signal through the CAN bus line. **Possible Causes:** • CAN line harness or connectors damage • ETACS–ECU power supply system failed • ETACS–ECU failed • PCM failed
DTC: U1110 **2T PCM MIL: YES** **Years:** 2006, 2007, 2008 **Models:** Except Raider **Engines:** All **Transmissions:** All	**A/C–ECU Timeout** Engine not cranking or 3 seconds have passed since the engine was cranked and battery voltage is 10 volts or greater, the A/C-ECU was unable to receive a signal through the CAN bus line. **Possible Causes:** • CAN line harness or connectors damage • A/C–ECU power supply system failed • A/C–ECU failed • PCM failed
DTC: U1110 **2T PCM MIL: YES** **Years:** 2006, 2007, 2008 **Models:** Raider **Engines:** All **Transmissions:** All	**Lost Vehicle Speed Message** The PCM doesn't receive a vehicle speed signal from the Anti-lock brake Module over the CAN C bus. **Possible Causes:** • CAN C bus open or shorted condition • ABS brake module • PCM failed
DTC: U1113 **2T PCM MIL: YES** **Years:** 2006, 2007, 2008 **Models:** Raider **Engines:** All **Transmissions:** All	**Lost A/C Pressure Message** The PCM doesn't receive the a/c pressure signal over the CAN C bus from the FCM. The circuit is continuously monitored. **Possible Causes:** • CAN C bus open or shorted condition • FCM failed • PCM failed
DTC: U1117 **2T PCM MIL: YES** **Years:** 2006, 2007, 2008 **Models:** All **Engines:** All **Transmissions:** All	**Immobilizer–ECU Timeout** Ignition switch is ON, the ETACS-ECU (immobilizer–ECU) was unable to receive a signal through the CAN bus line. **Possible Causes:** • CAN line harness or connectors damage • ETACS–ECU (immobilizer–ECU) power supply system failed • ETACS–ECU (immobilizer–ECU) failed • PCM failed
DTC: U1120 **2T PCM MIL: YES** **Years:** 2006, 2007, 2008 **Models:** Raider **Engines:** All **Transmissions:** All	**Lost Wheel Distance Message** The PCM doesn't receive a wheel distance message from the Anti-lock brake Module or FCM (NON-ABS) over the CAN C bus. **Possible Causes:** • CAN C bus open or shorted condition • ABS brake module • PCM failed
DTC: U1180 **2T PCM MIL: YES** **Years:** 2006, 2007, 2008 **Models:** Outlander **Engines:** 2.4L W **Transmissions:** All	**Combination Meter Time-out** Unable to receive combination meter signals through the CAN bus line for 4 seconds. • CAN line harness damage or connector damage • Combination meter failed • ECM failed

DTC	Trouble Code Title, Conditions & Possible Causes
DTC: U1120 **2T PCM MIL: YES** **Years:** 2006, 2007, 2008 **Models:** Raider **Engines:** All **Transmissions:** All	**Lost Wheel Distance Message** The PCM doesn't receive a wheel distance message from the Anti-lock brake Module or FCM (NON-ABS) over the CAN C bus. **Possible Causes:** • CAN C bus open or shorted condition • ABS brake module • PCM failed
DTC: U1403 **2T PCM MIL: YES** **Years:** 2006, 2007, 2008 **Models:** Raider **Engines:** All **Transmissions:** All	**Implausible Fuel Volume Level Received** The fuel volume message the PCM is receiving is implausible. The circuit is continuously monitored. **Possible Causes:** • CAN B bus open or shorted condition • Cluster module • FCM failed • PCM failed
DTC: U1411 **2T PCM MIL: YES** **Years:** 2006, 2007, 2008 **Models:** Raider **Engines:** All **Transmissions:** All	**Implausible Fuel Volume Signal Received** The fuel volume message the PCM is receiving is implausible. The circuit is continuously monitored. **Possible Causes:** • CAN B bus open or shorted condition • Cluster module • FCM failed • PCM failed
DTC: U1412 **2T PCM MIL: YES** **Years:** 2006, 2007, 2008 **Models:** Raider **Engines:** All **Transmissions:** All	**Implausible Vehicle Speed Signal Received** The PCM gets an implausible signal over the CAN C circuit from the ABS Module. The circuit is continuously monitored. **Possible Causes:** • CAN C bus open or shorted condition • ABS module • PCM failed
DTC: U1413 **2T PCM MIL: YES** **Years:** 2006, 2007, 2008 **Models:** Raider **Engines:** All **Transmissions:** All	**Implausible Odometer Signal Received** The odometer message the PCM is receiving is implausible. The circuit is continuously monitored. **Possible Causes:** • CAN B bus open or shorted condition • Cluster module • FCM failed • PCM failed
DTC: U1417 **2T PCM MIL: YES** **Years:** 2006, 2007, 2008 **Models:** Raider **Engines:** All **Transmissions:** All	**Implausible Left Wheel Distance Signal Received** The PCM gets an implausible signal over the CAN C circuit from the ABS Module. The circuit is continuously monitored. **Possible Causes:** • Vehicle speed sensor fault active in ABS module • CAN C bus open or shorted condition • ABS module • PCM failed
DTC: U1418 **2T PCM MIL: YES** **Years:** 2006, 2007, 2008 **Models:** Raider **Engines:** All **Transmissions:** All	**Implausible Right Wheel Distance Signal Received** The PCM gets an implausible signal over the CAN C circuit from the ABS Module. The circuit is continuously monitored. **Possible Causes:** • Vehicle speed sensor fault active in ABS module • CAN C bus open or shorted condition • ABS module • PCM failed

SUBARU

B9 Tribeca • Tribeca

13

SPECIFICATIONS AND MAINTENANCE CHARTS

ENGINE AND VEHICLE IDENTIFICATION CHART

Engine Code							Model Year	
Code ①	Liters (cc)	Cu. In.	Cyl.	Fuel Sys.	Type	Eng. Mfg.	Code ②	Year
8	3.0 (3000)	183	6	MFI	DOHC	Subaru	6	2006
8	3.0 (3000)	183	6	MFI	DOHC	Subaru	7	2007
9	3.6 (3630)	221	6	MFI	DOHC	Subaru	8	2008

MFI: Multiport Fuel Injection

DOHC: Double Overhead Camshaft

① 6th digit of the VIN

② 10th digit of the VIN

22140_STRI_C0001

GENERAL ENGINE SPECIFICATIONS

Year	Model	Engine Displacement Liters (VIN)	Net Horsepower @ rpm	Net Torque @ rpm (ft. lbs.)	Bore x Stroke (in.)	Com-pression Ratio	Oil Pressure @ rpm
2006	B9 Tribeca	3.0 (8)	250@6600	219@4200	3.51x3.15	10.7:1	20@600
2007	B9 Tribeca	3.0 (8)	250@6600	219@4200	3.51x3.15	10.7:1	20@600
2008	Tribeca	3.6 (9)	256@6000	247@4400	3.62x3.58	10.5:1	15@700

22140_STRI_C0002

ENGINE TUNE-UP SPECIFICATIONS

Year	Model	Engine Displacement Liters (VIN)	Spark Plug Gap (in.)	Ignition Timing (deg.) ①		Fuel Pump (psi)	Idle Speed (rpm)		Valve Clearance ②	
				MT	AT		MT	AT	In.	Ex.
2006	B9 Tribeca	3.0 (8)	0.028-0.031	—	15	③	—	④	⑤	⑥
2007	B9 Tribeca	3.0 (8)	0.028-0.031	—	15	③	—	④	⑤	⑥
2008	Tribeca	3.6 (9)	0.028-0.031	—	15	③	—	⑦	⑤	⑥

NOTE: The Vehicle Emission Control Information label often reflects specification changes made during production. The lable figures must be used if they differ from those in this chart.

① Before Top Dead Center. At idle. +/- 8 degrees.

② Measured with engine cold

③ 49-50.5 at operating temperature

④ 600-700 in N. 720-820 in N with AC on.

⑤ 0.0079 +0.0016/-0.0024

⑥ 0.0138 +/-0.0020

⑦ 600-800 in N. 705-905 in N with AC on.

22140_STRI_C0003

CAPACITIES

Year	Model	Engine Displacement Liters (VIN)	Engine Oil with Filter (qts.)	Transmission (pts.) 5-Spd	Transmission (pts.) Auto.	Transfer Case (pts.)	Drive Axle Front ① (pts.)	Drive Axle Rear (pts.)	Fuel Tank (gal.)	Cooling System (qts.)
2006	B9 Tribeca	3.0 (8)	6.0	–	20.8	–	2.4	1.4	16.9	7.8
2007	B9 Tribeca	3.0 (8)	6.0	–	20.8	–	2.4	1.4	16.9	7.8
2008	Tribeca	3.6 (9)	6.9	–	20.8	–	2.4	1.4	16.9	8.0

Note: All capacities are approximate. Add fluid gradually and check to be sure a proper fluid level is obtained.

① A/T differential

22140_STRI_C0005

FLUID SPECIFICATIONS

Year	Model	Engine Displacement Liters	Engine ID/VIN	Engine Oil	Auto. Trans.	Drive Axle	Power Steering Fluid	Brake Master Cylinder	Engine Coolant
2006	B9 Tribeca	3.0	8	5W-30	Subaru ATF	75W-90	Dexron III	DOT 3	Subaru Coolant
2007	B9 Tribeca	3.0	8	5W-30	Subaru ATF	75W-90	Dexron III	DOT 3	Subaru Coolant
2008	Tribeca	3.0	9	5W-30	Subaru ATF	75W-90	Dexron III	DOT 3	Subaru Coolant

DOT: Department Of Transpotation

22140_STRI_C0004

VALVE SPECIFICATIONS

All measurements are given in inches.

Year	Engine Displacement Liters (VIN)	Seat Angle (deg.)	Face Angle (deg.)	Spring Test Pressure (lbs. @ in.)	Spring Installed Height (in.)	Stem-to-Guide Clearance (in.) Intake	Stem-to-Guide Clearance (in.) Exhaust	Stem Diameter (in.) Intake	Stem Diameter (in.) Exhaust
2006	3.0 (8)	90	NA	NA	①	0.0012-0.0022	0.0016-0.0026	002148-0.2154	0.2144-0.2150
2007	3.0 (8)	90	NA	NA	①	0.0012-0.0022	0.0016-0.0026	002148-0.2154	0.2144-0.2150
2008	3.6 (9)	90	NA	NA	②	0.0012-0.0022	0.0016-0.0026	002148-0.2154	0.2144-0.2150

NA: Not Available

① Free length: Intake inner 1.557 in., outer 1.621 in. exhaust 1.824 in.

② Free length: Intake 1.6342 in., exhaust 1.6342 in.

22140_STRI_C0007

CAMSHAFT SPECIFICATIONS

All measurements are given in inches.

Year	Engine Displacement Liters (VIN)	Journal Dia.	Brg. Oil Clearance	Shaft End-play	Runout	Lobe Height Intake	Lobe Height Exhaust
2006	3.0 (8)	①	0.0015-0.0028	②	NA	③	1.6398-1.6437
2007	3.0 (8)	①	0.0015-0.0028	②	NA	③	1.6398-1.6437
2008	3.6 (9)	①	0.0015-0.0028	②	NA	1.8071-1.8110	1.7579-1.7618

NA: Not Available

① Front: 1.4939-1.4946
Except front: 1.0215-1.0222

② Intake: 0.0030-0.0053
Exhaust: 0.0012-0.0035

③ High: 1.6571-1.6610
Low 1: 1.5016-1.5055
Low 2: 1.3756-1.3795

22140_STRI_C0006

CRANKSHAFT AND CONNECTING ROD SPECIFICATIONS

All measurements are given in inches.

Year	Engine Displacement Liters (VIN)	Crankshaft Main Brg. Journal Dia.	Crankshaft Main Brg. Oil Clearance	Crankshaft Shaft End-play	Crankshaft Thrust on No.	Connecting Rod Journal Diameter	Connecting Rod Oil Clearance	Connecting Rod Side Clearance
2006	3.0 (8)	2.5194-2.5200	0.0004-0.0012	0.0012-0.0045	NA	NA	0.0006-0.0017	0.0028-0.0130
2007	3.0 (8)	2.5194-2.5200	0.0004-0.0012	0.0012-0.0045	NA	NA	0.0006-0.0017	0.0028-0.0130
2008	3.6 (9)	2.5194-2.5200	0.0004-0.0012	0.0012-0.0045	NA	NA	0.0006-0.0017	0.0028-0.0130

NA: Not Available

22140_STRI_C0008

PISTON AND RING SPECIFICATIONS

All measurements are given in inches.

Year	Engine Displacement Liters (VIN)	Piston Clearance	Ring Gap Top Compression	Ring Gap Bottom Compression	Ring Gap Oil Control	Ring Side Clearance Top Compression	Ring Side Clearance Bottom Compression	Ring Side Clearance Oil Control
2006	3.0 (8)	0.0004	0.0079-0.0138	0.0138-0.0197	0.0079-0.0236	0.0016-0.0031	0.0012-0.0028	0.0018-0.0049
2007	3.0 (8)	0.0004	0.0079-0.0138	0.0138-0.0197	0.0079-0.0236	0.0016-0.0031	0.0012-0.0028	0.0018-0.0049
2008	3.6 (9)	0.0004	0.0079-0.0138	0.0138-0.0197	0.0079-0.0236	0.0016-0.0031	0.0012-0.0028	0.0018-0.0049

22140_STRI_C0009

TORQUE SPECIFICATIONS

All readings in ft. lbs.

Year	Engine Displacement Liters (VIN)	Cylinder Head Bolts	Main Bearing Bolts	Rod Bearing Bolts	Crankshaft Damper Bolts	Flywheel Bolts	Manifold Intake	Manifold Exhaust	Spark Plugs	Oil Pan Drain Plug
2006	3.0 (8)	①	②	39	131	60	18	52	15	33
2007	3.0 (8)	①	②	39	131	60	18	52	15	33
2008	3.6 (9)	①	②	39	144	60	18	52	15	33

① Step 1: Tighten all bolts to 14 ft. lbs.
Step 2: Tighten all bolts to 37 ft. lbs.
Step 3: Loosen all bolts 180 degrees, and again 180 degrees
Step 4: Tighten all bolts to 14 ft. lbs.
Step 5: Bolts (1-4) 35.4 ft. lbs.
Step 6: Bolts (5-8) 33 ft. lbs.
Step 7: + 90 degrees
Step 8: Bolts (1-4) 45 degrees

② Step 1: Bolts (1-11 and 13) 18 ft. lbs.
Bolts (12 and 14) 14. ft. lbs.
Step 2: Retighten bolts (1-11 and 13) 18. ft. lbs.
Retighten bolts (12 and 14) 14. ft. lbs.
Step 3: + 90 degrees
Step 4: Upper bolt to cylinder block 18 ft. lbs.
Step 7: + 90 degrees
Step 8: Bolts (1-4) 45 degrees

22140_STRI_C0010

WHEEL ALIGNMENT

Year	Model		Caster Range (+/-Deg.)	Caster Preferred Setting (Deg.)	Camber Range (+/-Deg.)	Camber Preferred Setting (Deg.)	Toe-in (in.)
2006	B9 Tribeca	F	—	①	—	②	③
		R	—	—	—	④	⑤
2007	B9 Tribeca	F	—	①	—	②	③
		R	—	—	—	④	⑤
2008	Tribeca	F	—	①	—	②	③
		R	—	—	—	④	⑤

① Referential value: 4 degrees 04'
② Difference between right and left is 45' or less: 0 degrees 00'
③ 0 +/-0.12: toe angle (sum of both wheels) 0 degrees +/-0 degrees 14'
④ Difference between right and left is 45' or less: -0 degrees 31'
⑤ 0.08 +/-0.08: toe angle (sum of both wheels) 0 degrees +/-0 degrees 14'

22140_STRI_C0011

TIRE, WHEEL AND BALL JOINT SPECIFICATIONS

		OEM Tires		Tire Pressures (psi)		Wheel	Ball Joint	Lug
Year	Model	Standard	Optional	Front	Rear	Size	Inspection	Nut
2006	B9 Tribeca	P255/55 R18	None	33	32	18 x 8JJ	0.012	81
2007	B9 Tribeca	P255/55 R18	None	33	32	18 x 8JJ	0.012	81
2008	Tribeca	P255/55 R18	T165/80 R17	33	32	18 x 8JJ ①	0.012	②

OEM: Original Equipment Manufacturer

PSI: Pounds Per Square Inch

① 17 inch wheel size 17 x 4T and tire (psi) is 60

② Chromed wheel 111 ft. lbs. Other than chrome 88 ft. lbs.

22140_STRI_C0012

BRAKE SPECIFICATIONS

All measurements in inches unless noted

			Brake Disc			Brake Drum Diameter			Minimum Lining Thickness		Brake Caliper	
Year	Model		Original Thickness	Minimum Thickness	Maximum Runout	Original Inside Diameter	Max. Wear Limit	Maximum Machine Diameter	Front	Rear	Bracket Bolts (ft. lbs.)	Mounting Bolts (ft. lbs.)
2006	B9 Tribeca	F	1.180	1.100	0.0020	NA	NA	NA	0.059	—	89	20
		R	0.710	0.630	0.0020	NA	NA	NA	—	0.059	49	20
2007	B9 Tribeca	F	1.180	1.100	0.0020	NA	NA	NA	0.059	—	89	20
		R	0.710	0.630	0.0020	NA	NA	NA	—	0.059	49	20
2008	Tribeca	F	1.180	1.100	0.0020	NA	NA	NA	0.059	—	89	20
		R	0.710	0.630	0.0020	NA	NA	NA	—	0.059	49	20

NA: Not Available

22140_STRI_C0013

SCHEDULED MAINTENANCE INTERVALS

2006-07 B9 TRIBECA, 2008 TRIBECA

TO BE SERVICED	TYPE OF SERVICE	VEHICLE MILEAGE INTERVAL (x1000)												
		7.5	15	22.5	30	37.5	45	52.5	60	67.5	75	82.5	90	97.5
Engine oil & filter ①	R	✓	✓	✓	✓	✓	✓	✓	✓	✓	✓	✓	✓	✓
Brake lines	S/I		✓		✓		✓		✓		✓		✓	
Disc brake pads & discs, front & rear axle boots & axle shaft joint portions	S/I		✓		✓		✓		✓		✓		✓	
Parking brake	S/I		✓		✓		✓		✓		✓		✓	
Steering & suspension	S/I		✓		✓		✓		✓		✓		✓	
A/C cabin filter	R		✓		✓		✓		✓		✓		✓	
Air filter element	R				✓				✓				✓	
Engine coolant	R				✓				✓				✓	
Spark plugs	R				✓				✓				✓	
Automatic transmission fluid & filter	S/I				✓				✓				✓	
Brake fluid	R				✓				✓				✓	
Brake linings & drums	S/I				✓				✓				✓	
Camshaft drive belt	S/I				✓				✓				✓	
Coolant level, hoses & clamps	S/I				✓				✓				✓	
Drive belts	S/I				✓				✓				✓	
Fuel system, hoses & connections	S/I				✓				✓				✓	
Transmission and/or differential gear fluid	S/I				✓								✓	
Tires (rotate)	S/I	✓	✓	✓	✓	✓	✓	✓	✓	✓	✓	✓	✓	✓
Front & rear wheel bearing	I								✓					

R: Replace S/I: Service or Inspect

① First oil change 3,000

FREQUENT OPERATION MAINTENANCE (SEVERE SERVICE)

If a vehicle is operated under any of the following conditions it is considered severe service:

- Extremely dusty areas.

- 50% or more of the vehicle operation is in 32°C (90°F) or higher temperatures, or constant operation in temperatures below 0°C (32°F).

- Prolonged idling (vehicle operation in stop and go traffic).

- Frequent short running periods (engine does not warm to normal operating temperatures).

- Police, taxi, delivery usage or trailer towing usage.

Oil & oil filter change: change every 3750 miles.

Air filter element: service or inspect every 15,000 miles.

Automatic transmission fluid: service or inspect every 15,000 miles.

Brake linings & drums: service or inspect every 15,000 miles.

Coolant level, hoses & clamps: service or inspect every 15,000 miles.

Drive belts: service or inspect every 15,000 miles.

Transmission/differential gear oil: service or inspect every 15,000 miles.

Front & rear wheel bearing: service or inspect every 30,000 miles.

Suspension & Steering: inspect every 7,500 miles.

Axle boots & Joints: inspect every 7,500 miles.

Note: Inspect SRS system every 10 years.

22140_STRI_C0014

PRECAUTIONS

Before servicing any vehicle, please be sure to read all of the following precautions, which deal with personal safety, prevention of component damage, and important points to take into consideration when servicing a motor vehicle:

- Never open, service or drain the radiator or cooling system when the engine is hot; serious burns can occur from the steam and hot coolant.
- Observe all applicable safety precautions when working around fuel. Whenever servicing the fuel system, always work in a well-ventilated area. Do not allow fuel spray or vapors to come in contact with a spark, open flame, or excessive heat (a hot drop light, for example). Keep a dry chemical fire extinguisher near the work area. Always keep fuel in a container specifically designed for fuel storage; also, always properly seal fuel containers to avoid the possibility of fire or explosion. Refer to the additional fuel system precautions later in this section.
- Fuel injection systems often remain pressurized, even after the engine has been turned **OFF**. The fuel system pressure must be relieved before disconnecting any fuel lines. Failure to do so may result in fire and/or personal injury.
- Brake fluid often contains polyglycol ethers and polyglycols. Avoid contact with the eyes and wash your hands thoroughly after handling brake fluid. If you do get brake fluid in your eyes, flush your eyes with clean, running water for 15 minutes. If eye irritation persists, or if you have taken brake fluid internally, IMMEDIATELY seek medical assistance.
- The EPA warns that prolonged contact with used engine oil may cause a number of skin disorders, including cancer. You should make every effort to minimize your exposure to used engine oil. Protective gloves should be worn when changing oil. Wash your hands and any other exposed skin areas as soon as possible after exposure to used engine oil. Soap and water, or waterless hand cleaner should be used.
- All new vehicles are now equipped with an air bag system, often referred to as a Supplemental Restraint System (SRS) or Supplemental Inflatable Restraint (SIR) system. The system must be disabled before performing service on or around system components, steering column, instrument panel components, wiring and sensors. Failure to follow safety and disabling procedures could result in accidental air bag deployment, possible personal injury and unnecessary system repairs.
- Always wear safety goggles when working with, or around, the air bag system. When carrying a non-deployed air bag, be sure the bag and trim cover are pointed away from your body. When placing a non-deployed air bag on a work surface, always face the bag and trim cover upward, away from the surface. This will reduce the motion of the module if it is accidentally deployed. Refer to the additional air bag system precautions later in this section.
- Clean, high quality brake fluid from a sealed container is essential to the safe and proper operation of the brake system. You should always buy the correct type of brake fluid for your vehicle. If the brake fluid becomes contaminated, completely flush the system with new fluid. Never reuse any brake fluid. Any brake fluid that is removed from the system should be discarded. Also, do not allow any brake fluid to come in contact with a painted surface; it will damage the paint.
- Never operate the engine without the proper amount and type of engine oil; doing so WILL result in severe engine damage.
- Timing belt maintenance is extremely important. Many models utilize an interference-type, non-freewheeling engine. If the timing belt breaks, the valves in the cylinder head may strike the pistons, causing potentially serious (also time-consuming and expensive) engine damage. Refer to the maintenance interval charts for the recommended replacement interval for the timing belt, and to the timing belt section for belt replacement and inspection.
- Disconnecting the negative battery cable on some vehicles may interfere with the functions of the on-board computer system(s) and may require the computer to undergo a relearning process once the negative battery cable is reconnected.
- When servicing drum brakes, only disassemble and assemble one side at a time, leaving the remaining side intact for reference.
- Only an MVAC-trained, EPA-certified automotive technician should service the air conditioning system or its components.

BRAKES — ANTI-LOCK BRAKE SYSTEM (ABS)

GENERAL INFORMATION

When wheel slip is detected during a brake application, an ABS event occurs. During antilock braking, hydraulic pressure in the individual wheel circuits is controlled to prevent any wheel from slipping. A separate hydraulic line and specific solenoid valves are provided for each wheel. The ABS can decrease, hold, or increase hydraulic pressure to each wheel. During antilock braking, a series of rapid pulsations is felt in the brake pedal. These pulsations are caused by the rapid changes in position of the individual solenoid valves as the Vehicle Dynamics Control Module (VDCCM) & Hydraulic Control Unit (H/U) responds to wheel speed sensor inputs and attempts to prevent wheel slip. These pedal pulsations are present only during antilock braking and stop when normal braking is resumed or when the vehicle comes to a stop. A ticking or popping noise may also be heard as the solenoid valves cycle rapidly. During antilock braking on dry pavement, intermittent chirping noises may be heard as the tires approach slipping. These noises and pedal pulsations are considered normal during antilock operation.

Vehicles equipped with ABS may be stopped by applying normal force to the brake pedal. Brake pedal operation during normal braking is no different than that of previous non-ABS systems. Maintaining a constant force on the brake pedal provides the shortest stopping distance while maintaining vehicle

SPEED SENSORS

REMOVAL & INSTALLATION

Front

See Figure 1.

1. Disconnect the ground cable from the battery.
2. Disconnect the ABS wheel speed sensor connector in the engine compartment.
3. Remove the (1) sensor harness from the (2) clip.
4. Remove the clip which secures the sensor harness to the front strut.

⁂ CAUTION

Be careful not to damage the sensor.

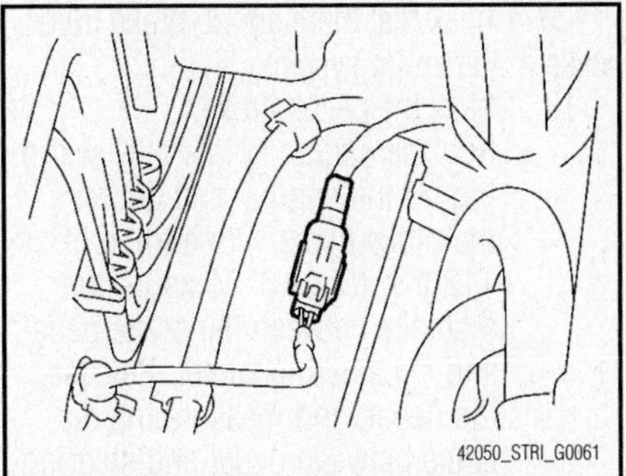

Fig. 1 Disconnect the ABS front wheel speed sensor connector in the engine compartment

⁂ CAUTION

Do not apply excessive force to the sensor harness.

To install:

5. Install in the reverse order of removal. Tighten to 60 inch lbs. (7.5 Nm).

⁂ CAUTION

Be careful not to damage the sensor.

➡ Check the identification (mark) on the harness to make sure there is no warpage. (W1 white)

➡ Check if the harness is not pulled and does not come in contact with the suspension or body during steering wheel effort.

Rear

See Figure 2.

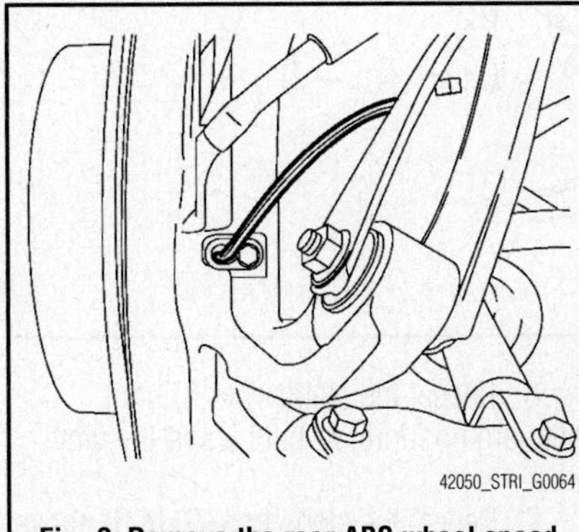

Fig. 2 Remove the rear ABS wheel speed sensor from the rear housing

1. Disconnect the ground cable from the battery.
2. Disconnect the connector from the rear ABS wheel speed sensor.
3. Remove the sensor harness from the rear arm clip.
4. Remove the rear ABS wheel speed sensor from the rear housing.

⁂ CAUTION

Be careful not to damage the sensor.

⁂ CAUTION

Do not apply excessive force to the sensor harness.

To install:

5. Install in the reverse order of removal. Tighten to 60 inch lbs. (7.5 Nm).

⁂ CAUTION

Be careful not to damage the sensor.

➡ Check the identification (mark) on the harness to make sure there is no warpage. [W3 (white)]

BRAKES — BLEEDING THE BRAKE SYSTEM

BLEEDING PROCEDURE

⁂ CAUTION

Do not let brake fluid come into contact with the painted surface of the vehicle body. Wash away with water immediately and wipe off if it is spilled by accident.

⁂ CAUTION

Avoid mixing brake fluids of different brands to prevent fluid performance from degrading.

⁂ CAUTION

Be careful not to allow dirt or dust to enter the reservoir tank.

Master Cylinder

See Figure 3.

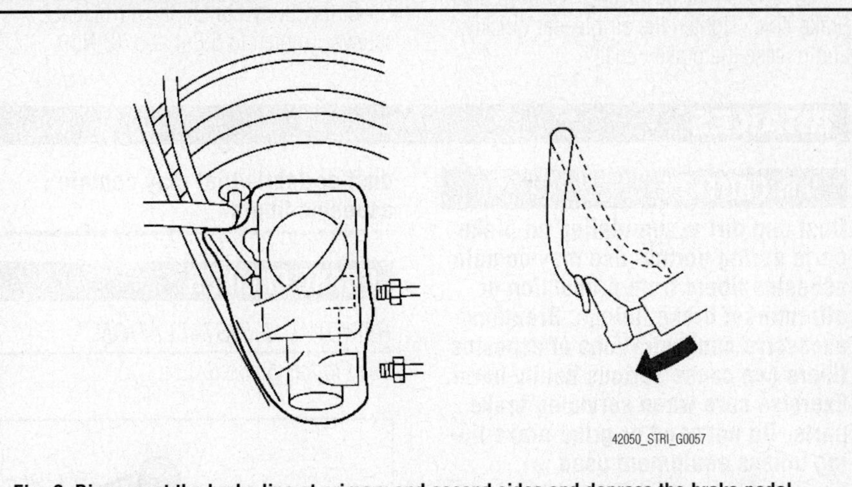

Fig. 3 Disconnect the brake line at primary and second sides and depress the brake pedal

➡ When the master cylinder is disassembled or the reservoir tank is empty, bleed the master cylinder. If bleeding the master cylinder is not necessary, omit the following procedures, and perform bleeding of the brake line.

1. Fill the reservoir tank of the master cylinder with brake fluid.

➡ While bleeding air, keep the reservoir tank filled with brake fluid to prevent entry of air.

2. Disconnect the brake line at primary and secondary sides.
3. Wrap the master cylinder with a plastic bag.
4. Depress the brake pedal slowly and hold it.
5. Plug the outlet plug with your finger, and then release the brake pedal.
6. Repeat the step 4) and 5) several times.
7. Remove the plastic bag.
8. Install the brake pipe to the master cylinder. Tighten to 14.0 ft. lbs. (19 Nm).
9. Bleed air from the brake line.

Brake Line

See Figure 4.

1. When the master cylinder is disassembled or the reservoir tank is empty, bleed the master cylinder before bleeding the brake line.
2. Fill the reservoir tank of the master cylinder with brake fluid.

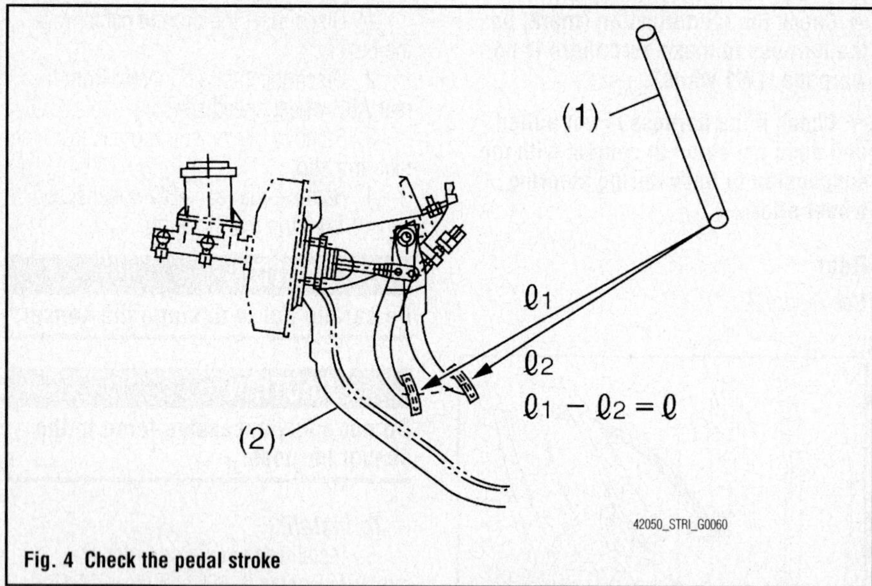

Fig. 4 Check the pedal stroke

➡ While bleeding air, keep the reservoir tank filled with brake fluid to prevent entry of air.

3. Attach one end of the vinyl tube to the air bleeder and the other end to the brake fluid container.
4. Depress the brake pedal several times, and keep it pressed.
5. Loosen the air bleeder screw to drain brake fluid. Tighten the air bleeder quickly, and release the brake pedal.
6. Repeat the steps 4) to 5) until there are no more air bubbles in the vinyl tube.
7. Repeat the steps from 2) to 6) above to bleed air from each wheel.

➡ Perform the operation in the order from closest wheel cylinder to the master cylinder.

8. Securely tighten the air bleeder screws. Tighten to 5.8 ft. lbs. (8 Nm)
9. Check that there are no brake fluid leaks in the entire system.
10. Check the pedal stroke.
    - Run the engine at idle after warming up the engine, and depress the brake pedal with a force of 112 lbs. (500 N). Measure the distance between the brake pedal and (1) steering wheel. Release the pedal, and measure the distance between pedal and steering wheel again.
    - When depressing the pedal with a force of 112 lbs. (500 N). Less than 4.53 inch (115 mm)
11. If the distance is more than specification, there is a possibility of air being caught in the brake line. Bleed the brake line of all air until the pedal stroke meets the specification.
12. Operate the hydraulic control unit in the sequence control mode.
13. Check the pedal stroke again.
14. If the distance is more than specification, there is a possibility of air being caught in the hydraulic unit. Repeat above steps 2) to 9) until the pedal stroke meets the specification.
15. Fill brake fluid up to the "MAX" level of the reservoir tank.
16. Test run the vehicle and ensure that the brakes operate normally.

BRAKES

FRONT DISC BRAKES

⁂ CAUTION

Dust and dirt accumulating on brake parts during normal use may contain asbestos fibers from production or aftermarket brake linings. Breathing excessive concentrations of asbestos fibers can cause serious bodily harm. Exercise care when servicing brake parts. Do not sand or grind brake lining unless equipment used is designed to contain the dust residue. Do not clean brake parts with compressed air or by dry brushing. Cleaning should be done by dampening the brake components with a fine mist of water, then wiping the brake components clean with a dampened cloth. Dispose of cloth and all residue containing asbestos fibers in an impermeable container with the appropriate label. Follow practices prescribed by the Occupational Safety and Health Administration (OSHA) and the Environmental Protection Agency (EPA) for the handling, processing, and disposing of dust or debris that may contain asbestos fibers.

BRAKE CALIPER

REMOVAL & INSTALLATION

See Figures 5 and 6.

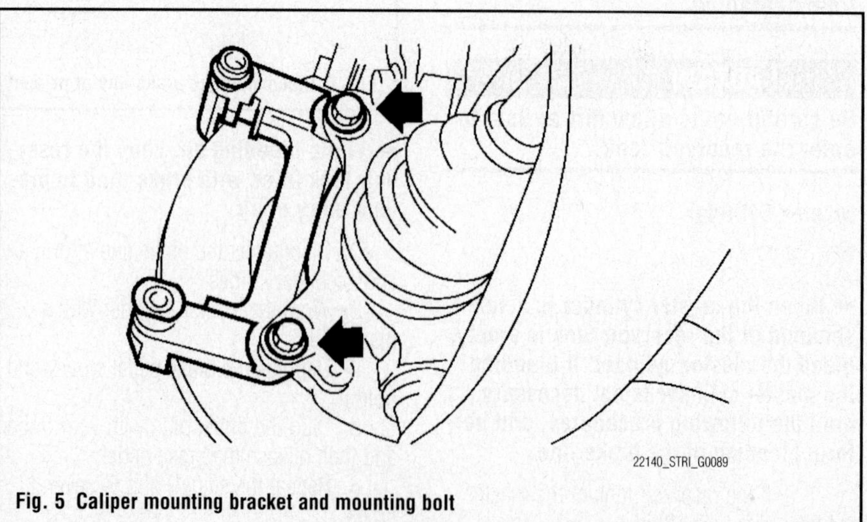

Fig. 5 Caliper mounting bracket and mounting bolt

1. Before servicing the vehicle, refer to the Precautions Section.
2. Loosen the wheel nuts.
3. Raise and support the vehicle safely.
4. Remove the tire and wheel.
5. Remove the union bolt. Disconnect the brake line from the brake caliper. Be sure

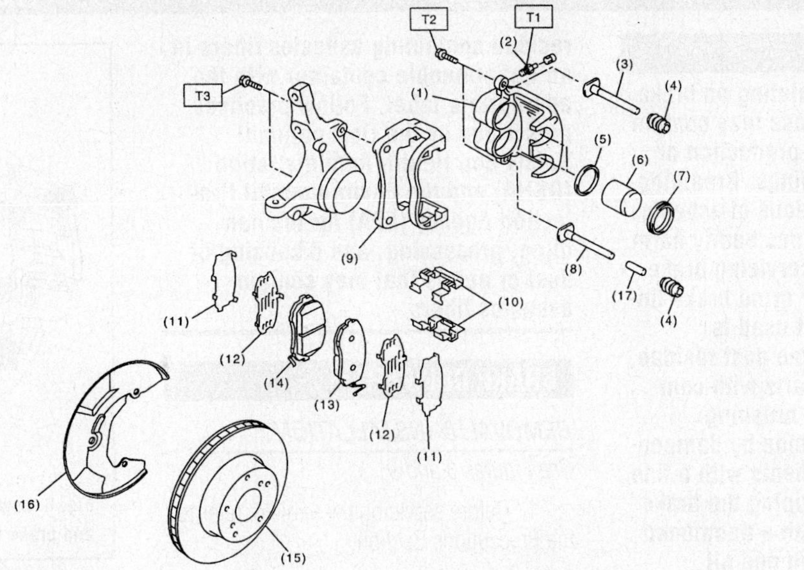

T1
T2
T3

(1)	Caliper body	(9)	Support	(16)	Disc cover
(2)	Air bleeder screw	(10)	Pad clip	(17)	Bushing
(3)	Guide pin (Green)	(11)	Outer shim		
(4)	Pin boot	(12)	Inner shim		
(5)	Piston seal	(13)	Pad (Outside)		
(6)	Piston	(14)	Pad (Inside)		
(7)	Piston boot	(15)	Disc rotor		
(8)	Lock pin (Yellow)				

Tightening torque: N·m (kgf-m, ft-lb)
T1: 8 (0.8, 5.8)
T2: 27 (2.8, 19.9)
T3: 120 (12.2, 88.5)

09490_TRIB_G0018

Fig. 6 Front disc brake and related components

to properly catch the fluid to avoid damage to painted surfaces and improper disposal.

6. Remove the bolt retaining the lock pin to the caliper body.
7. Remove the caliper from its mounting.

To install:

8. Installation is the reverse of the removal procedure.
9. Tighten the support bolts to 88 ft. lbs. (120 Nm).
10. Check the brake fluid level, correct as required.
11. Bleed the hydraulic system, as required.

DISC BRAKE PADS

REMOVAL & INSTALLATION

See Figure 7.

1. Before servicing the vehicle, refer to the Precautions Section.
2. Loosen the wheel nuts.
3. Raise and support the vehicle safely.
4. Remove the tire and wheel.
5. Remove the caliper bolt.
6. Raise the caliper body and properly support it.

➡ **Do not disconnect the brake fluid line from the caliper.**

7. Remove the pads.

To install:

8. Installation is the reverse of the removal procedure.
9. Apply a thin coat of Molykote M7439, or equivalent to the pad clip.
10. Apply a thin coat of Molykote AS-880N (part number K0779YA010) or equivalent to the contact surface between the pad and pad inner shim.
11. Apply a thin coat of Molykote AS-880N (part number K0779YA010) or equivalent to the three contact surfaces between the inner shim and outer shim of the outer pads.

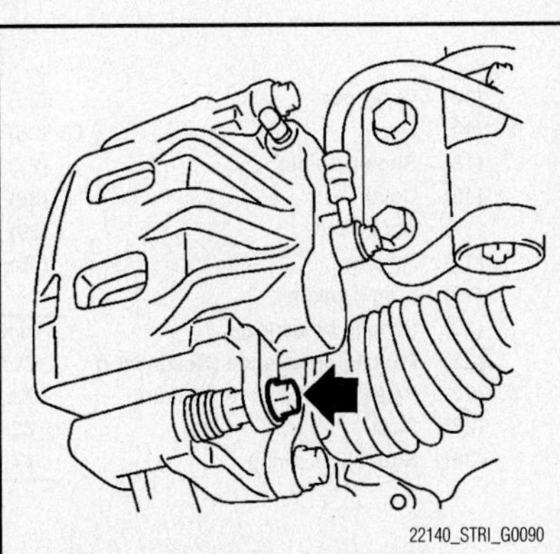

Fig. 7 Caliper mounting bolt view

12. Check the brake fluid level, correct as required.
13. Bleed the hydraulic system, as required.

ROTOR

REMOVAL & INSTALLATION

1. Lift up the vehicle, and remove the front wheels.
2. Remove the caliper body and the support from housing, and suspend it from the strut using a wire.
3. Remove the disc rotor.

➡ **If it is difficult to remove the disc rotor from the hub, drive an 8 mm bolt into the threaded section of the rotor, then remove the rotor.**

4. Remove mud and foreign matter from the caliper body assembly and support.

To install:

5. Install the disc rotor.
6. Install the caliper body and the support to housing.
7. Tighten the support bolts to 88 ft. lbs. (120 Nm).
8. Install the front wheels and lower vehicle.

❋❋ CAUTION

Dust and dirt accumulating on brake parts during normal use may contain asbestos fibers from production or aftermarket brake linings. Breathing excessive concentrations of asbestos fibers can cause serious bodily harm. Exercise care when servicing brake parts. Do not sand or grind brake lining unless equipment used is designed to contain the dust residue. Do not clean brake parts with compressed air or by dry brushing. Cleaning should be done by dampening the brake components with a fine mist of water, then wiping the brake components clean with a dampened cloth. Dispose of cloth and all residue containing asbestos fibers in an impermeable container with the appropriate label. Follow practices prescribed by the Occupational Safety and Health Administration (OSHA) and the Environmental Protection Agency (EPA) for the handling, processing, and disposing of dust or debris that may contain asbestos fibers.

BRAKE CALIPER

REMOVAL & INSTALLATION

See Figures 8 and 9.

1. Before servicing the vehicle, refer to the Precautions Section.

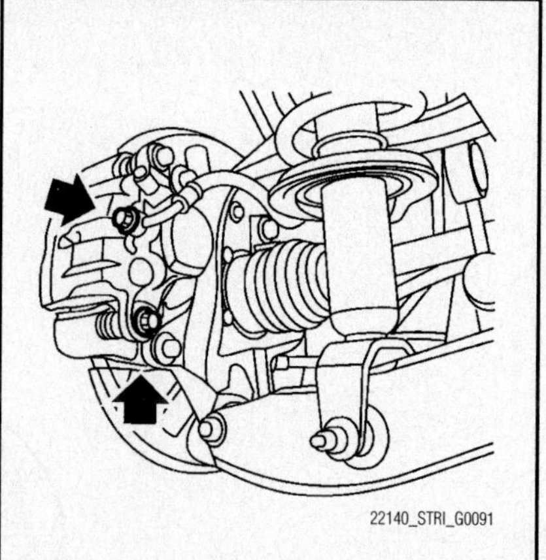

Fig. 8 Lower caliper slide mounting bolt and brake line banjo bolt

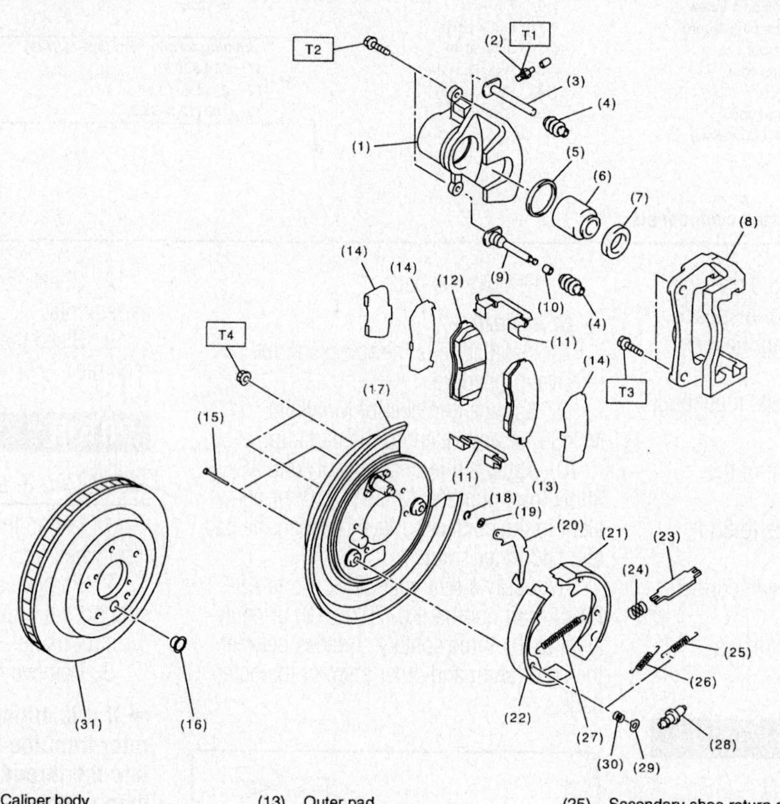

Fig. 9 Rear disc brake and related components

2. Loosen the wheel nuts.
3. Raise and support the vehicle safely.
4. Remove the tire and wheel.
5. Disconnect the brake line from the brake caliper. Be sure to properly catch the fluid to avoid damage to painted surfaces and improper disposal.
6. Remove the caliper retaining bolts. Remove the caliper from its mounting.

To install:

7. Installation is the reverse of the removal procedure.
8. Tighten the caliper retaining bolts to 20 ft. lbs. (36 Nm).
9. Check the brake fluid level, correct as required.
10. Bleed the hydraulic system, as required.

DISC BRAKE PADS

REMOVAL & INSTALLATION

See Figures 10.

1. Before servicing the vehicle, refer to the Precautions Section.
2. Loosen the wheel nuts.
3. Raise and support the vehicle safely.
4. Remove the tire and wheel.
5. Remove the brake hose bracket.
6. Remove the caliper bolt.
7. Raise the caliper body and properly support it.

➡ Do not disconnect the brake fluid line from the caliper.

8. Remove the pads.

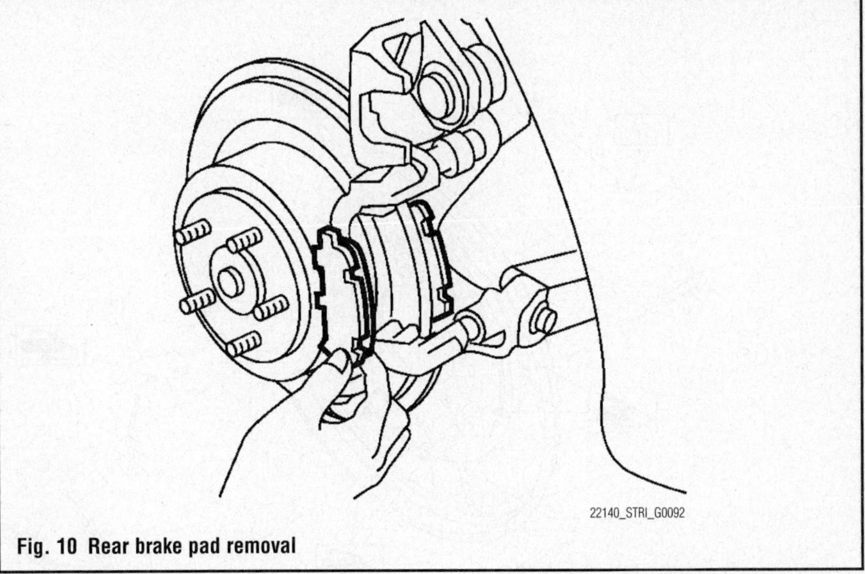

Fig. 10 Rear brake pad removal

To install:

9. Installation is the reverse of the removal procedure.
10. Apply a thin coat of Molykote M7439, or equivalent to the pad clip.
11. Apply a thin coat of Molykote AS-880N (part number K0779YA010) or equivalent to the contact surface between the pad and pad inner shim.
12. Check the brake fluid level, correct as required.
13. Bleed the hydraulic system, as required.

ROTOR

REMOVAL & INSTALLATION

1. Lift up the vehicle, and then remove the rear wheels.
2. Release the parking brake.
3. Remove the two mounting bolts, and remove the rear disc brake assembly.
4. Suspend the rear disc brake assembly so that the hose is not stretched.
5. Remove the rear disc brake rotor.

➡ If the disc rotor is difficult to remove loosen the parking brake shoes. If it is difficult to remove the disc rotor from the hub, drive an 8 mm bolt into the threads of the rotor, then remove the rotor.

To install:

6. Install in the reverse order of removal.
7. Tighten the two mounting bolts to 49 ft. lbs. (66 Nm).
8. Adjust the parking brake.

BRAKES — PARKING BRAKE

PARKING BRAKE CABLES

ADJUSTMENT

Adjust the pedal stroke as follows:

a. Adjust the shoe clearance before adjusting pedal stroke.
b. Operate the parking brake pedal 3 to 4 times.
c. Remove the adjuster cap from the parking brake pedal.
d. Turn the adjusting nut until the pedal stroke is at the specified value.

➡ The pedal stroke when stepped on with a force of 67 lbs. (300 N) should be about 5–6 notches.

1. Check there is no brake drag.
2. Install the adjuster cap.

PARKING BRAKE SHOES

REMOVAL & INSTALLATION

See Figure 11.

1. Release the parking brake.
2. Lift up the vehicle, and then remove the rear wheels.
3. Remove the two mounting bolts and remove the disc brake caliper assembly.
4. Suspend the rear disc brake caliper assembly so that the brake hose is not stretched.
5. Remove the rear disc brake rotor.

➡ If the disc rotor is difficult to remove loosen the parking brake shoes. If it is difficult to remove the disc rotor from the hub, drive an 8 mm bolt into the threads of the rotor, then remove the rotor.

6. Remove the shoe return spring.
7. Remove the brake shoe cup and brake shoe spring, then remove the primary brake shoe.
8. Remove the strut and strut spring.
9. Remove the adjuster assembly.
10. Remove the brake shoe cup and brake shoe spring, then remove the secondary brake shoe.

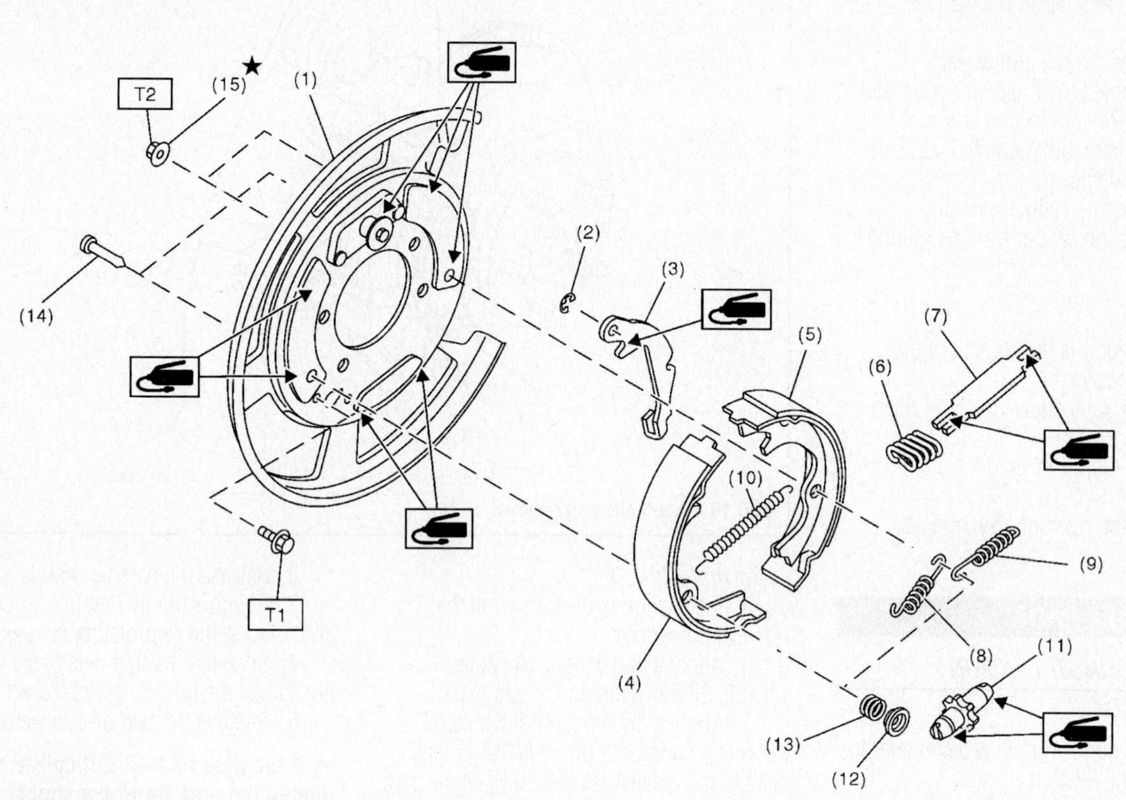

(1)	Back plate	(8)	Primary return spring	(14)	Shoe hold pin
(2)	Retainer	(9)	Secondary return spring	(15)	Self locking nut (with WAX)
(3)	Lever	(10)	Adjusting spring		
(4)	Parking brake shoe (Primary)	(11)	Adjuster		
(5)	Parking brake shoe (Secondary)	(12)	Brake shoe cup		
(6)	Strut spring	(13)	Brake shoe spring		
(7)	Strut				

Tightening torque:N·m (kgf-m, ft-lb)

T1: 8 (0.8, 5.9)

T2: 75 (7.6, 55.33)

22140_STRI_G0095

Fig. 11 Exploded view of parking brake shoe components

11. Remove the parking brake cable from lever.

12. Remove the retainer from the secondary side brake shoe. Remove the parking lever and washer from brake shoe.

To install:

CAUTION

Be sure the lining surface is free from oil and grease.

13. Apply Brake Grease (Part No. 003602002) to the following locations:
- Six contact surfaces of brake shoe rim and back plate gasket
- Contact surface of the brake shoe and the anchor pin
- Contact surface of the lever and strut
- Contact surface of the brake shoe and the adjuster assembly
- Contact surface of the brake shoe and the strut
- Contact surface of the lever and brake shoe

14. Install the parking lever to the secondary side brake shoe, then clamp and lock the retainer securely.

15. Attach the parking brake cable to the parking brake lever.

16. Attach the adjuster assembly and adjusting spring to the brake shoe.

➡ Install the adjuster assembly so that the screw section will be towards the rear of the vehicle.

17. Check the parking brake cable is not fallen from the cable guide.

18. Install the brake shoe to the back plate using shoe hold pins, brake shoe spring and brake shoe cup.

19. Install the strut and strut spring to the brake shoe.

➡ Install the strut spring so that it is placed on the front of vehicle.

20. Install the primary side return spring, then the secondary side return spring.

21. Adjust the parking brake.

When replacing with a new brake shoe, drive the vehicle to break-in the parking brake lining as follows:

a. Drive the vehicle at about 35 km/h (22 MPH).

b. lightly step on the parking brake pedal.

c. Drive the vehicle for about 200 m (0.12 mile) in this condition.

d. Wait 5 to 10 minutes for the parking brake to cool down. Repeat again from step (1).

e. After breaking-in, re-adjust the parking brakes.

ADJUSTMENT

See Figure 12.

1. Return the parking brake lever fully.
2. Remove the adjusting hole cover from the disc rotor.
3. Turn the adjusting screw using a flat tip screwdriver until the brake shoe is in close contact with the disc rotor.
4. Turn back (downward) the adjusting screw 3 to 4 notches.

⁂ CAUTION

Check that there is no brake drag.

5. Install the adjusting hole cover to the disc rotor.
6. Adjust the parking brake pedal stroke.

Adjust the pedal stroke as follows:

a. Adjust the shoe clearance before adjusting pedal stroke.

b. Operate the parking brake pedal 3 to 4 times.

c. Remove the adjuster cap from the parking brake pedal.

d. Turn the adjusting nut until the pedal stroke is at the specified value.

➡ The pedal stroke when stepped on with a force of 67 lbs. (300 N) should be about 5–6 notches.

7. Check there is no brake drag.
8. Install the adjuster cap.

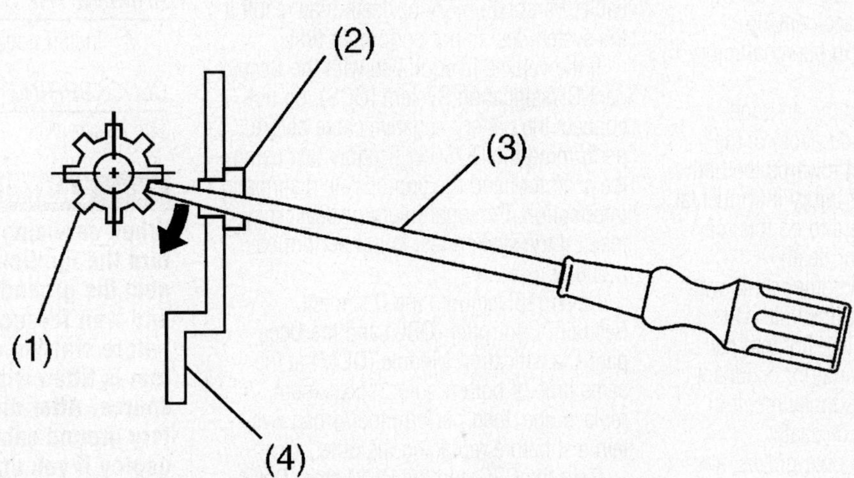

Fig. 12 Parking brake adjusting wheel view

CHASSIS ELECTRICAL **AIR BAG (SUPPLEMENTAL RESTRAINT SYSTEM)**

GENERAL INFORMATION

SERVICE PRECAUTIONS

Disconnect and isolate the battery negative cable before beginning any airbag system component diagnosis, testing, removal, or installation procedures. Allow system capacitor to discharge for two minutes before beginning any component service. This will disable the airbag system. Failure to disable the airbag system may result in accidental airbag deployment, personal injury, or death.

Do not place an intact undeployed airbag face down on a solid surface. The airbag will propel into the air if accidentally deployed and may result in personal injury or death.

When carrying or handling an undeployed airbag, the trim side (face) of the airbag should be pointing towards the body to minimize possibility of injury if accidental deployment occurs. Failure to do this may result in personal injury or death.

Replace airbag system components with OEM replacement parts. Substitute parts may appear interchangeable, but internal differences may result in inferior occupant protection. Failure to do so may result in occupant personal injury or death.

Wear safety glasses, rubber gloves, and long sleeved clothing when cleaning powder residue from vehicle after an airbag deployment. Powder residue emitted from a deployed airbag can cause skin irritation. Flush affected area with cool water if irritation is experienced. If nasal or throat irritation is experienced, exit the vehicle for fresh air until the irritation ceases. If irritation continues, see a physician.

Do not use a replacement airbag that is not in the original packaging. This may result in improper deployment, personal injury, or death.

The factory installed fasteners, screws and bolts used to fasten airbag components have a special coating and are specifically designed for the airbag system. Do not use substitute fasteners. Use only original equipment fasteners listed in the parts catalog when fastener replacement is required.

During, and following, any child restraint anchor service, due to impact event or vehicle repair, carefully inspect all mounting hardware, tether straps, and anchors for proper installation, operation, or damage. If a child restraint anchor is found damaged in any way, the anchor must be replaced. Failure to do this may result in personal injury or death.

Deployed and non-deployed airbags may or may not have live pyrotechnic material within the airbag inflator.

Do not dispose of driver/passenger/curtain airbags or seat belt tensioners unless you are sure of complete deployment. Refer to the Hazardous Substance Control System for proper disposal.

Dispose of deployed airbags and tensioners consistent with state, provincial, local, and federal regulations.

After any airbag component testing or service, do not connect the battery negative cable. Personal injury or death may result if the system test is not performed first.

If the vehicle is equipped with the Occupant Classification System (OCS), do not connect the battery negative cable before performing the OCS Verification Test using the scan tool and the appropriate diagnostic information. Personal injury or death may result if the system test is not performed properly.

Never replace both the Occupant Restraint Controller (ORC) and the Occupant Classification Module (OCM) at the same time. If both require replacement, replace one, then perform the Airbag System test before replacing the other.

Both the ORC and the OCM store Occupant Classification System (OCS) calibration data, which they transfer to one another when one of them is replaced. If both are replaced at the same time, an irreversible fault will be set in both modules and the OCS may malfunction and cause personal injury or death.

If equipped with OCS, the Seat Weight Sensor is a sensitive, calibrated unit and must be handled carefully. Do not drop or handle roughly. If dropped or damaged, replace with another sensor. Failure to do so may result in occupant injury or death.

If equipped with OCS, the front passenger seat must be handled carefully as well. When removing the seat, be careful when setting on floor not to drop. If dropped, the sensor may be inoperative, could result in occupant injury, or possibly death.

If equipped with OCS, when the passenger front seat is on the floor, no one should sit in the front passenger seat. This uneven force may damage the sensing ability of the seat weight sensors. If sat on and damaged, the sensor may be inoperative, could result in occupant injury, or possibly death.

DISARMING THE SYSTEM

1. Turn the ignition switch **OFF**.
2. Disconnect the ground cable from the battery and wait for at least 20 seconds before starting work.

WARNING

The airbag system is fitted with a backup power source. After disconnecting the battery ground cable, the airbag may deploy if you do not wait for more than 20 seconds before starting the service of airbag system.

ARMING THE SYSTEM

1. Install negative battery cable.

CLOCKSPRING CENTERING

See Figure 13.

CAUTION

When servicing a vehicle, be sure to turn the ignition switch OFF , disconnect the ground cable from battery, and wait for more than 20 seconds before starting work. The airbag system is fitted with a backup power source. After disconnecting the battery ground cable, the airbag may deploy if you do not wait for more than 20 seconds before starting the service of airbag system.

1. Turn the ignition switch **OFF**.
2. Disconnect the ground cable from battery and wait for at least 20 seconds before starting work.
3. Check that front wheels are positioned in straight ahead direction.
4. Turn the clock spring pin (A) clockwise until it stops.
5. Turn the clock spring connector pins (A) approx. 3.25 turns until the marks are aligned.

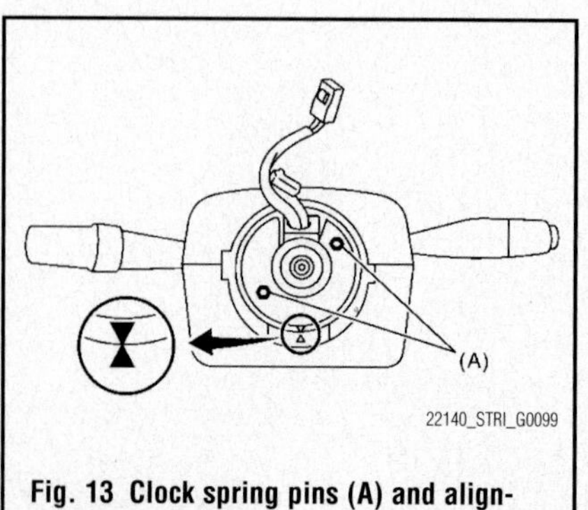

Fig. 13 Clock spring pins (A) and alignment marks shown

AUTOMATIC TRANSMISSION ASSEMBLY

REMOVAL & INSTALLATION

See Figure 14.

1. Before servicing the vehicle, refer to the Precautions Section.
2. Open the hood to the full open position.
3. Disconnect the negative battery cable.
4. Remove the collector cover.
5. Remove the air intake chamber. Remove the air breather hose.
6. Remove the starter.
7. Disconnect the front oxygen sensor and the transmission harness connector.
8. Disconnect the engine harness connectors and remove the rear engine hanger.
9. Remove the vacuum pipe and hose assembly. Remove the brake vacuum pump.
10. Remove the throttle body retaining bolts and slide the throttle body over.
11. To separate the torque converter clutch from the drive plate, remove the service hole plug. Remove the bolts that retain the flex plate to the torque converter. When rotating the engine do so in the proper direction of rotation. Install tool ST49827720, or equivalent, to the torque converter case to hold the assembly in place.
12. Remove the pitching stopper. Remove the pitching stopper bracket.
13. Position engine support tool ST-41099AC010 and ST-41099AC020 or equivalent to hold the engine assembly in place.
14. Remove the upper side transmission to engine retaining bolts.
15. Raise and support the vehicle safely. Remove the undercover.
16. Remove the front, and rear exhaust pipe and muffler. Remove the heat shield cover.
17. Drain the transmission. Remove the oil charge pipe.
18. Disconnect the connector from the number 1 turbine speed sensor. Remove the number one turbine speed sensor from the transmission body. Be sure to cover the hole to prevent dirt from entering.

➡ **Failure to follow this procedure may cause interference between the vehicle body and the sensor when removing the transmission.**

19. Remove the driveshaft.
20. Remove the shift select cable.
21. Disconnect the hose from the transmission inlet and outlet fluid lines.
22. Remove the front crossmember support plate.
23. Remove the two clutch housing cover retaining bolts. Remove the front stabilizer clamp.
24. Remove the bolts that retain the front ball joint to the housing.
25. Pull out the front halfshaft from the transmission. Place a cloth between the tool and the transmission to avoid damaging the side retainer of the transmission.
26. Position a transmission jack under the transmission.

➡ **Make sure that the support plates of the transmission jack do not contact the oil pan.**

27. Remove the transmission rear crossmember retaining bolts. Remove the crossmember from the vehicle.
28. Remove the transmission to engine lower retaining bolts.
29. Remove the transmission from the vehicle. Turn the engine support fixture assembly to the left and lower the rear of the engine for easy removal.

➡ **Move the transmission and torque converter away from the engine as an assembly.**

To install:

30. Replace the front differential side retainer oil seal.
31. Attach the ST-498277200 to converter case to hold converter in place.
32. Install the transmission onto the transmission jack and secure.
33. Install the transmission to the engine.
34. Install the engine mounting bolts and nuts (lower side). Tighten the bolts and nuts to 37 ft. lbs. (50 Nm).
35. Install the transmission rear crossmember. Tighten the bolts to 55 ft. lbs. (75 Nm).

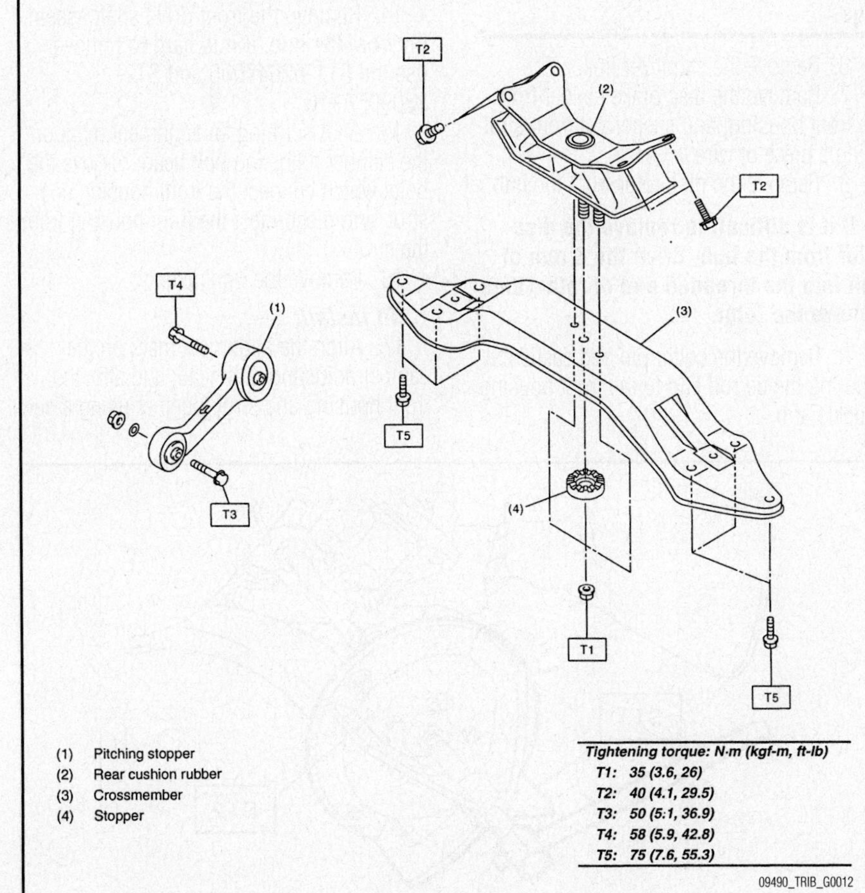

Fig. 14 Automatic transmission crossmember and related components

36. Remove the transmission jack.
37. Lower the vehicle.
38. Install the engine mounting bolts (upper side). Tighten the bolts to 37 ft. lbs. (50 Nm).
39. Remove the ST-498277200 from converter case.
40. Install the torque converter to drive plate.
41. Install the bolts which hold torque converter to drive plate.
42. Install all four bolts by rotating the crank pulley a little at a time. Tighten the bolts to 18 ft. lbs. (25 Nm).
43. Install the starter motor.
44. Install the pitching stopper bracket and tighten bolts to 30 ft. lbs. (41 Nm).
45. Install the pitching stopper and tighten to 37 ft. lbs. (50 Nm).
46. Lift up the vehicle.
47. Install oil seal protector ST-28399SA000 then install front drive shaft into the transmission.

CAUTION

Replace the circlip of drive shaft with a new part.

48. Install the front drive shaft into transmission, remove the ST and insert the drive shaft securely.
49. Install the inlet and outlet hoses to the ATF inlet and outlet pipes.
50. Insert the ball joint into housing.
51. Install the front stabilizer clamp.
52. Install the clutch housing cover securing bolts.
53. Install the front crossmember support plate.
54. Install the propeller shaft.
55. Install the shift select cable.
56. Install the turbine speed sensor 1 and harness, and then connect the connector. Tighten mounting bolt to 5 ft. lbs. (7 Nm).
57. Install the oil charge pipe. Use new bolts and tighten to 28 ft. lbs. (38 Nm).
58. Install the heat shield cover.
59. Install the front exhaust pipe, rear exhaust pipe and muffler.
60. Install the undercover.
61. Lower the vehicle.
62. Connect the front oxygen A/F sensor, rear oxygen sensor and transmission harness connector.
63. Attach the vacuum pump and vacuum pipe & hose.
64. Install the engine hanger rear, and then connect the engine harness connector.
65. Refill transmission with ATF.
66. Install the air breather hose.
67. Install the throttle body to intake manifold and tighten to 6 ft. lbs. (8 Nm).
68. Install the air intake chamber.
69. Install the collector cover.
70. Connect the battery ground cable to the battery.

➡ **Use Subaru Select Monitor and Perform Clear Memory 2 operation.**

71. Road test vehicle, check for leaks and shift quality.

FRONT HALFSHAFT

REMOVAL & INSTALLATION

See Figures 15 and 16.

1. Before servicing the vehicle, refer to the Precautions Section.
2. Disconnect the ground cable from the battery.
3. Lift-up the vehicle, and remove the front wheels.
4. Lift the crimped section of axle nut.
5. Remove the axle nut using a socket wrench while depressing the brake pedal.

CAUTION

Remove the wheel before loosening the axle nut. Failure to follow this rule may damage the wheel bearings.

6. Remove the stabilizer link.
7. Remove the disc brake caliper from the front housing, and suspend it from strut using a piece of wire.
8. Remove the disc rotor from the hub.

➡ **If it is difficult to remove the disc rotor from the hub, drive the 8 mm of bolt into the threaded end of rotor, then remove the rotor.**

9. Remove the cotter pin and castle nut securing the tie rod end to the front housing knuckle arm.

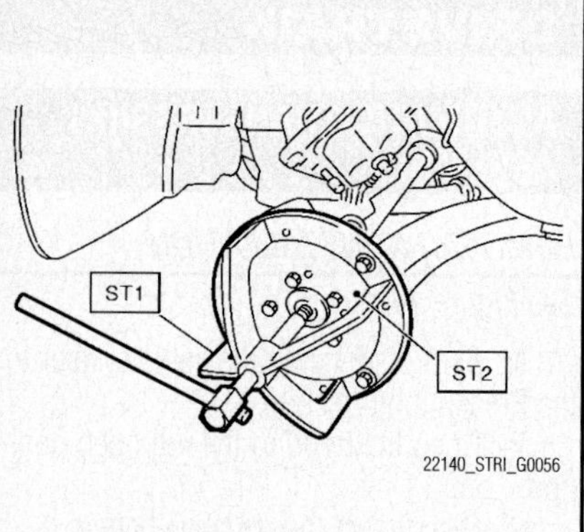

Fig. 16 Special tools shown for removal of axle shaft

10. Using a puller, remove the tie-rod ball joint from knuckle arm.

CAUTION

When removing tie-rod, do not hit the tie-rod end with hammer.

11. Remove the ABS wheel speed sensor assembly and harness.
12. Remove the front arm ball joint from the front housing.
13. Remove the PTJ from transmission.
14. Remove the front drive shaft assembly from the hub. If it is hard to remove, use the ST1-92647000 and ST2-28099PA110.
15. After scribing an alignment mark on the camber adjusting bolt head, remove the bolts which connect the front housing and strut, and disconnect the front housing from the strut.
16. Remove the front axle.

To install:

17. Align the alignment mark on the camber adjusting bolt head, and affix the front housing and strut together using a new

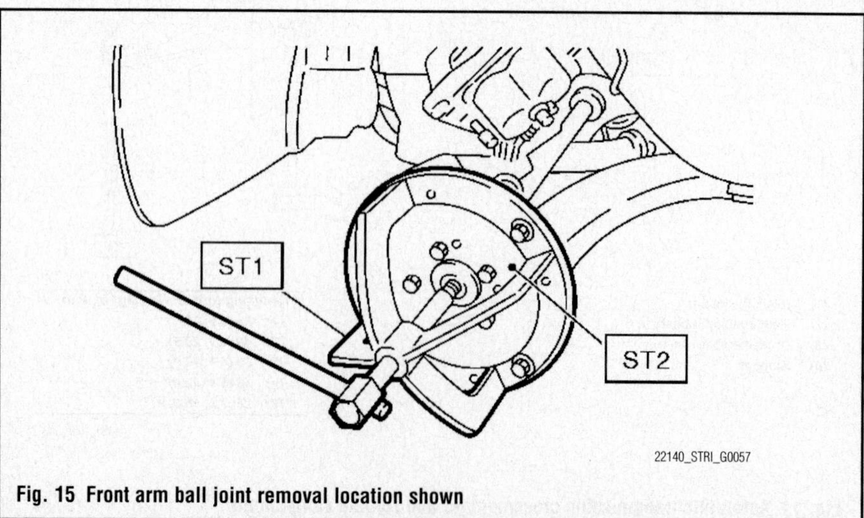

Fig. 15 Front arm ball joint removal location shown

self locking nut. Tighten to 129 ft. lbs. (175 Nm).

18. Install oil seal protector ST-28399SA000 then install front drive shaft into the transmission.

19. Install the front arm ball joint to the front housing. Tighten to 37 ft. lbs. (50 Nm).

20. Install the ABS wheel speed sensor on the front housing. Tighten to 6 ft. lbs. (7.5 Nm).

21. Install the disc rotor to hub.

22. Install the disc brake caliper to the front housing. Tighten mounting bolts to 88 ft. lbs. (120 Nm).

23. Install the stabilizer link.

24. Connect the tie-rod end ball joint to the knuckle arm with a castle nut. Tighten nut to 20 ft. lbs. (27 Nm).

25. Tighten the castle nut to specified torque and tighten further within 60° until the pin hole is aligned with the slot in nut. Bend the cotter pin to lock.

26. While depressing the brake pedal, tighten a new axle nut (olive color) to 177 ft. lbs. (240 Nm) and lock it securely.

⁂ CAUTION

Be sure to tighten the axle nut to specified torque. Do not over tighten as this may damage the wheel bearing.

27. After tightening the axle nut, lock it securely.

28. Install the wheel and tighten the wheel nuts to 81 ft. lbs. (110 Nm).

CV-BOOTS INSPECTION

See Figures 17 through 21.

1. Before servicing the vehicle, refer to the Precautions Section.
2. Place alignment marks on the shaft and outer race.
3. Remove the inner boot band and boot.
4. Remove the circlip from the inner joint outer race using a suitable prytool.
5. Outer race from the shaft assembly and wipe off the grease.
6. Place alignment marks on the free ring and trunnion as shown in the illustration.
7. Remove the free ring from the trunnion.
8. Place alignment marks on the trunnion and shaft as shown in the illustration.
9. Remove the snap ring and trunnion.
10. Place the shaft in a vise between wooden blocks.

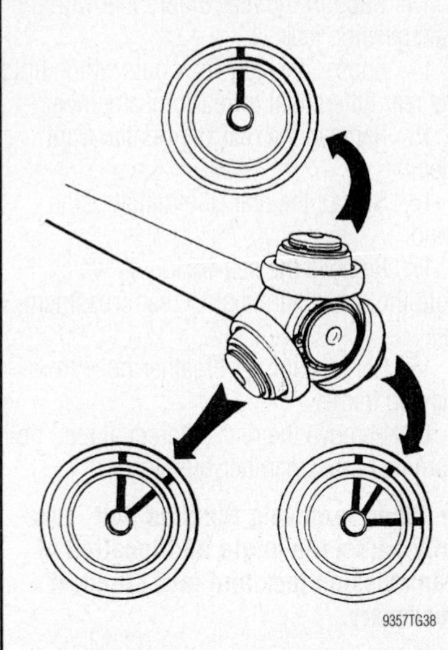

Fig. 17 Place alignment marks on the free ring and trunnion as shown—Front halfshaft

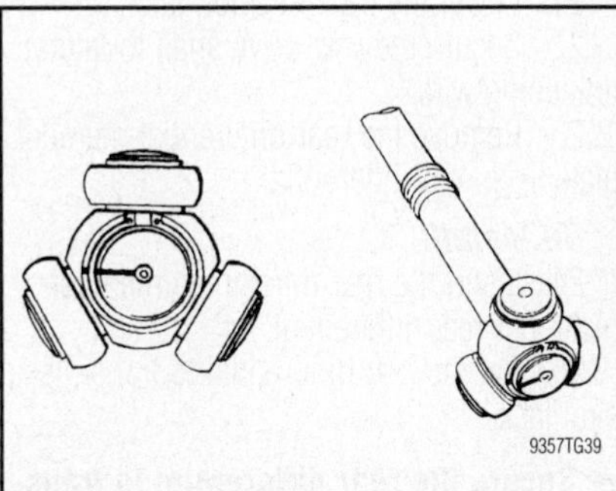

Fig. 18 Place alignment marks on the trunnion and shaft as shown—Front halfshaft

11. Using a suitable pry tool, raise the outer boot band claws.
12. Cut and remove the boot.

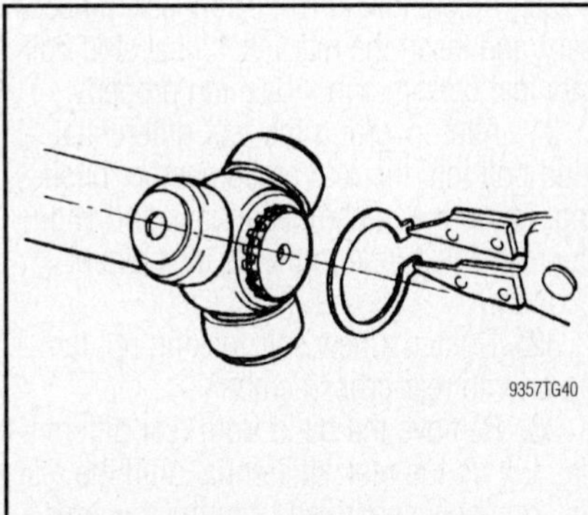

Fig. 19 Remove the snap ring and trunnion—Front halfshaft

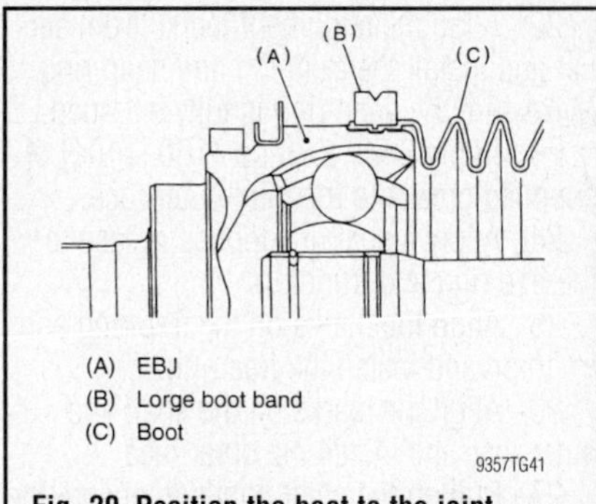

Fig. 20 Position the boot to the joint groove, and attach the large boot band as shown—Front halfshaft

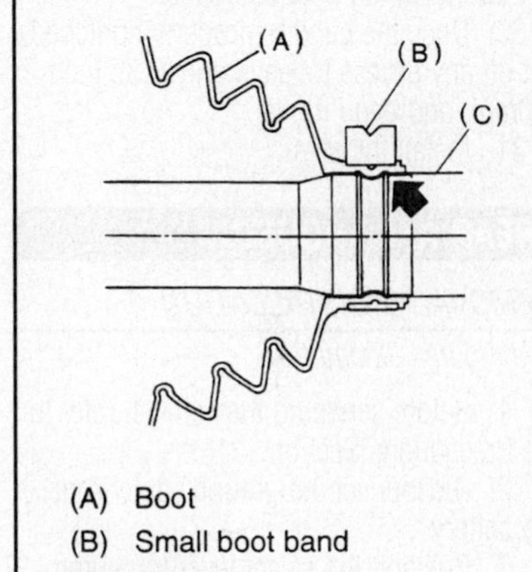

Fig. 21 Position the boot to the shaft groove, and attach the small boot band as shown—Front halfshaft

13. Only the boot can be replaced, the joint is not serviceable and must be replaced if damaged.

To install:

14. Place the half shaft in a vise.
15. Place the outer boot and small band on the shaft.
16. Apply 3.53–3.88 oz. (100–110g) of supplied grease to the joint.
17. Apply 1.06–1.41 oz. (30–40g) of supplied grease to the whole inner surface of the boot, and apply some grease to the shaft.
18. Install the boot to the joint groove, and attach the large boot band as shown.
19. Install the boot to the shaft groove, and attach the small boot band as shown.
20. Use boot band pliers tool ST-925091000 to tighten the large and small bands
21. Place the inner boot on the center of the shaft.

22. Align the alignment marks from earlier and install the trunnion and snap ring. Make sure the snap ring is fully engaged.
23. Apply 3.53–3.88 oz. (100–110g) of supplied grease to the joint outer race.
24. Apply a coat of supplied grease to the free ring and trunnion.
25. Align the marks on the free ring and trunnion and install the free ring.
26. Align the marks on the shaft and outer race and install the outer race.
27. Pull on the shaft lightly to ensure the circlip is completely engaged.
28. Apply an even coat 1.06–1.41 oz. (30–40g) of the supplied grease to the entire inner surface of the boot.
29. Install the boot and band.
30. Once the band is properly tightened, cut off any excess to leave only 0.39 inch (10mm) and bend it over.
31. Install the shaft.

REAR AXLE HOUSING

REMOVAL & INSTALLATION

See Figures 22 and 23.

1. Before servicing the vehicle, refer to the Precautions Section.
2. Disconnect the ground cable from the battery.
3. Position the select lever to neutral.
4. Release the parking brake.
5. Lift up the vehicle.
6. Remove the rear exhaust pipe and muffler.
7. Remove the heat shield cover.
8. Remove the propeller shaft.
9. Remove the connector from the rear differential oil temperature switch.
10. Prepare the transmission jack and band.
11. Remove the DOJ of rear drive shaft from rear differential.
12. Remove the nuts which hold the rear differential front member.

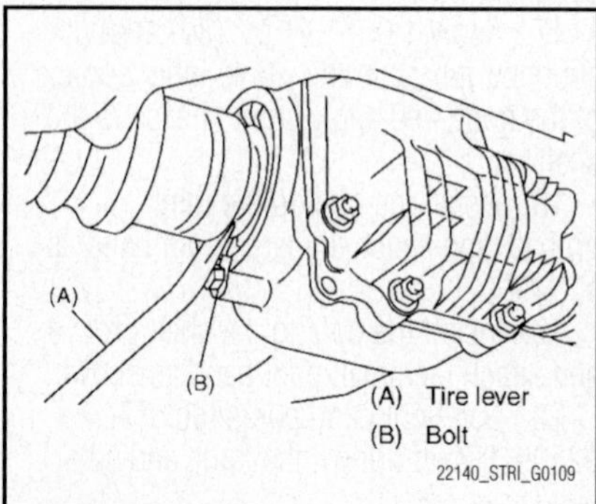

Fig. 22 Drive axle removal, tire lever (A) bolt (B)

13. Support the rear differential with the transmission jack.
14. Loosen the self-lock nuts which hold the rear differential to rear crossmember.
15. Remove the rear differential front member.
16. Secure the rear differential using band.
17. Remove the self-lock nuts which hold the rear differential to rear crossmember.
18. Remove the air breather hose from the sub frame.
19. Remove the rear differential stud bolt from rear crossmember bushing.

➡ When removing the stud bolt, carefully adjust the angle and location of transmission jack and jack stand, if necessary.

20. Pull out the axle shaft from rear differential.

➡ If it is difficult to remove the axle shaft from rear differential, remove it using tire lever.

21. Lower the transmission jack.
22. Secure the rear drive shaft to lateral link using wire.
23. Remove the rear differential member plate from rear differential.

To install:

24. Insert the rear differential member plate into rear differential.
25. Set the rear differential to transmission jack.

➡ Secure the rear differential to transmission jack using band.

26. Attach the ST-28099PA090 seal protector to rear differential.
27. Insert the shaft until the spline portion comes inside the side oil seal.
28. Remove ST from rear differential.
29. Push the rear differential to insert the axle shaft into rear differential.
30. Adjust the transmission jack, if necessary, and insert the rear differential stud bolt into rear crossmember bushing properly.
31. After inserting the rear differential stud bolt into the rear crossmember bushing, lift up the transmission jack and align the rear differential to the its attachment position.
32. Tighten a new self-locking nut temporarily to rear crossmember.
33. Remove the band from rear differential. Lift up the rear differential until the rear differential is separated from the transmission jack.
34. Install the rear differential front member with new self-locking nuts. Tighten

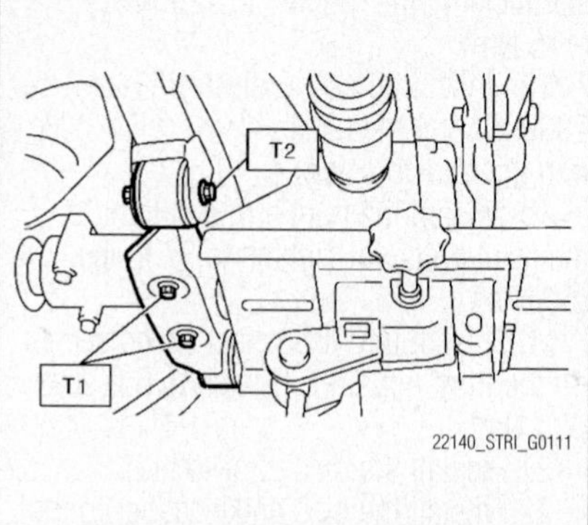

Fig. 23 Front member and (T1) and (T2) locations shown

(T1) to 37 ft. lbs. (50 Nm). Tighten (T2) to 81 ft. lbs. (110 Nm).
35. Tighten the self-locking nut to 51 ft. lbs. (70 Nm).
36. Lower the transmission jack.
37. Install the air breather hose to the sub frame.
38. Install the propeller shaft.
39. Install the heat shield cover.
40. Install the rear exhaust pipe and muffler.
41. Fill rear differential with approved fluid.

REAR AXLE SHAFT SEAL

REMOVAL & INSTALLATION

See Figure 24.

1. Before servicing the vehicle, refer to the Precautions Section.
2. Remove the rear differential.
3. Remove the rear differential side oil seal using a screwdriver wrapped with vinyl tape to prevent the side retainer from scratches.

To install:

4. Using the ST-398437700, install the oil seal to the side retainer.
5. Install the rear differential.

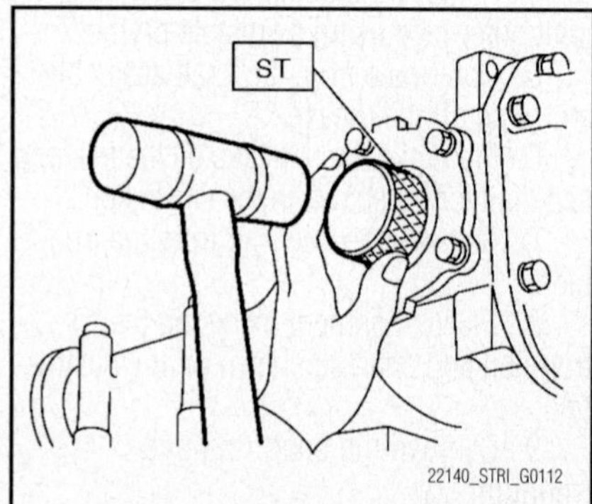

Fig. 24 Rear differential axle side oil seal installation

REAR HALFSHAFT

REMOVAL & INSTALLATION

See Figure 25.

1. Before servicing the vehicle, refer to the Precautions Section.
2. Disconnect the negative battery cable.
3. Raise and support the vehicle safely.
4. Remove the wheels and tires.
5. Unlock the axle nut. Depress the parking brake. Remove the axle nut.

➡ **Be sure to loosen the axle nut after removing the tire and wheel from the vehicle. Failure to do this may damage the wheel bearings.**

6. Remove the rear differential assembly, VA-type.
7. Remove the axle nut and rear halfshaft.

➡ **Use axle shaft puller tools ST-92647000 and ST-28099PA110, or equivalent, if it is difficult to remove the halfshaft. Do not hammer the halfshaft to remove it. Be careful not to damage the oil seal or magnetic encoder.**

To install:

8. Insert the EBJ into hub splines.

⁂ CAUTION

Be careful not to damage the magnetic encoder. Do not get closer the tool which charged magnetism to magnetic encoder.

9. Draw the rear drive shaft into specified position.
10. Tighten the axle nut temporarily.
11. Install the rear differential assembly, VA-type.
12. While applying the parking brake and depressing the brake pedal, tighten a new axle nut to (240 Nm) and lock it securely.

⁂ CAUTION

Be sure to tighten the axle nut to specified torque. Do not over tighten as this may damage the wheel bearing.

13. Lock the axle nut securely.
14. Install the rear wheel.
15. For chrome wheels tighten to 111 ft. lbs. (150 Nm).
16. All other wheels tighten to 89 ft. lbs. (120 Nm).

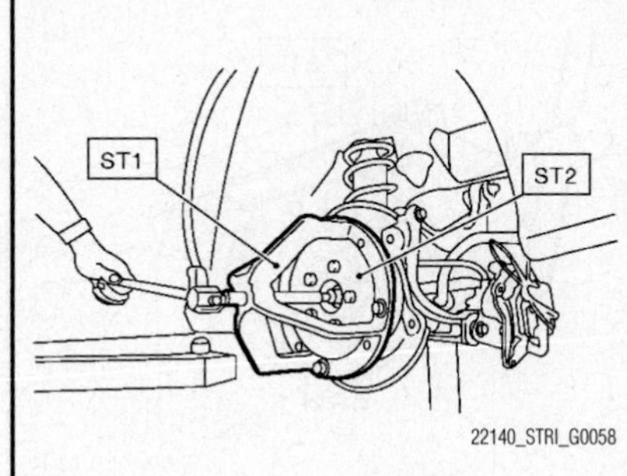

Fig. 25 Axle shaft puller tools ST-92647000 and ST-28099PA110

CV-JOINTS OVERHAUL

See Figure 26.

The Double Offset Joint (DOJ), is the only part of the assembly that can be replaced, if any of the other components are defective then the shaft should be replaced.

1. Before servicing the vehicle, refer to the Precautions Section.
2. Straighten the bent claw of the large clamp at the Double Offset Joint (DOJ) end of the boot.
3. Loosen the band using pliers being careful not to damage the bolt.
4. Remove the small boot band from the DOJ using the same technique.
5. Remove the boot from the large end of the DOJ outer race.
6. Remove the round circlip using a suitable prytool from the neck of the joint outer race.
7. Remove the joint outer race.

⁂ CAUTION

The grease used is for CV–Joints, do not replace with another type of grease

8. Clean the grease and remove the balls. Be careful not to lose any of the 6 balls.
9. Turn the cage by a half pitch to the track groove of the inner race and shift the cage.
10. Remove the snap ring that secures the inner race to the shaft and remove the inner race.
11. Take the cage from the shaft and remove the boot.
12. The other boots may be removed in the same manner as the DOJ boot.
13. Wrap the shaft splines with tape to prevent damage.

To install:

14. The following grease must be used during assembly:

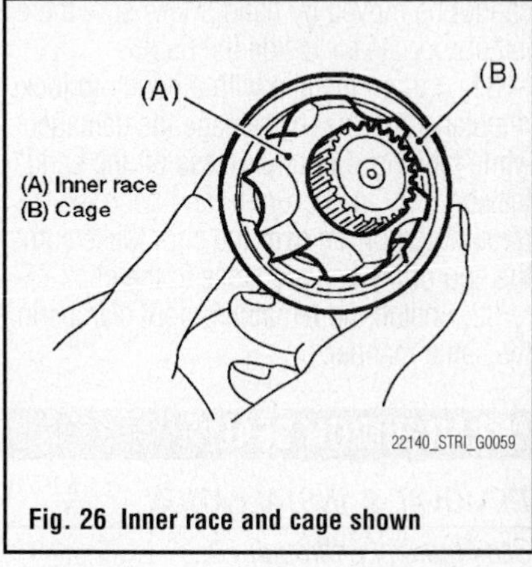

Fig. 26 Inner race and cage shown

15. DOJ side: NKG205 (part number 28495AG00A).
16. Install the BJ or EBJ boots and fill it with 2.12–2.47 oz. (60–70g) of grease.
17. Place the DOJ boot on the center of the shaft.
18. Insert the DOJ cage onto the shaft. Make sure to insert the cage with the cut–out portion facing shaft end.
19. Install the inner race on the shaft and fasten with the snap ring. Make sure the snap ring is firmly engaged.
20. Install the cage (previously fitted) to the inner race on the shaft. Fit the cage with the protruded part aligned with the track on the inner race and then turn a half pitch.
21. Fill the DOJ inner race with 2.82–3.17 oz. (80–90g) of grease.
22. Apply a coat of grease to the cage pocket and 6 balls.
23. Insert the 6 balls.
24. Align the outer race track and ball positions and place where the shaft, inner race, cage and balls were located prior to removal and then install the outer race.
25. Install the circlip into the groove on the outer race.

➡ **Make sure the balls, cage and inner race are fully seated. make sure not to place the matched position of the circlip in the ball groove of the outer race. Pull the shaft lightly to make sure the circlip is fully engaged.**

26. Apply an even coat 0.71–1.06 oz. (20–30g) of grease to the entire inner surface of the boot and to the shaft.
27. Make sure the boot is free from any dirt or foreign materials prior to installation.
28. Place the outer race of the boot at the center of its travel.
29. Put a band through the boot clip and wind twice in alignment with the groove on the boot.
30. Pinch the end of the band using tool ST 9250910000 and tighten securely until it

cannot be moved by hand. Make sure there is appropriate air inside the boot.

31. Tap on the clip with a punch to lock it making sure not to damage the damaged while tapping. Cut any excess off the band leaving 0.39 inch (10mm) and bend the remaining portion over the clip. Make sure the end of the band is close to the clip.

32. Install the remaining boot clamps in the same manner.

REAR PINION SEAL

REMOVAL & INSTALLATION

See Figures 27 through 29.

1. Before servicing the vehicle, refer to the Precautions Section.

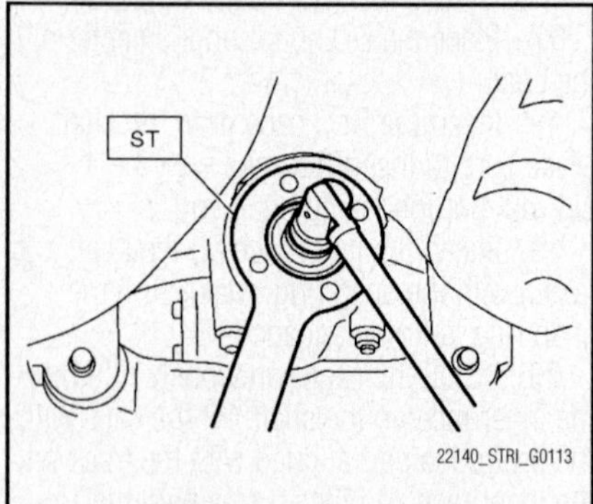

Fig. 27 Removing the self locking nut using ST-498427200 Flange wrench.

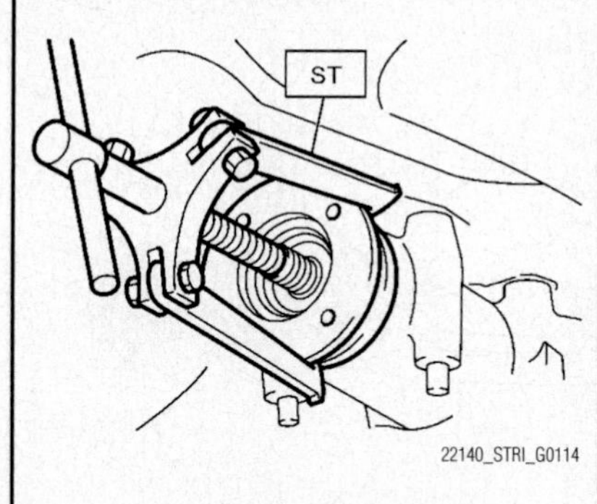

Fig. 28 Removal of companion flange with ST-399703600 Puller Assembly

2. Release the parking brake.
3. Remove the oil drain plug, and drain gear oil.
4. Install the oil drain plug. Use a new metal gasket and tighten to 22 ft. lbs. (29 Nm).
5. Jack-up the rear wheel and support the body with rigid racks.
6. Remove the rear exhaust pipe and muffler.
7. Remove the heat shield cover.
8. Remove the propeller shaft.
9. Remove the self-locking nut while holding the companion flange with ST-498427200 Flange wrench.
10. Extract the companion flange using ST-399703600 Puller Assembly.
11. Remove the oil seal using ST-398527700 puller or screwdriver.

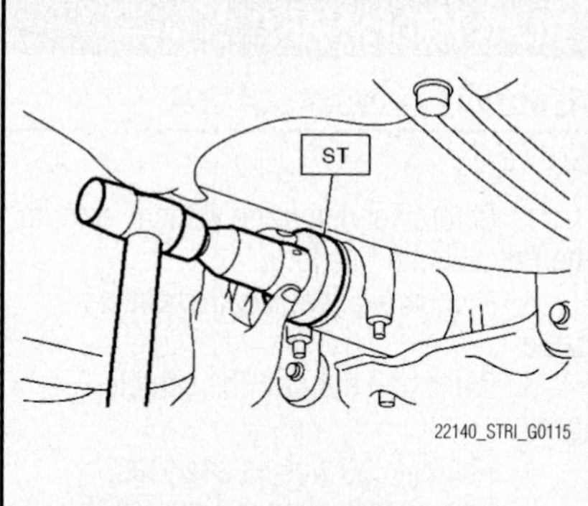

Fig. 29 Pinion oil seal installation with ST-498447120 oil seal installer

To install:

12. Install a new pinion oil seal with ST-498447120 oil seal installer.
13. Install the companion flange.

➡ **Use a plastic hammer to install companion flange.**

14. Apply Lock Tite THREE BOND 1324 (Part No. 004403042) to the drive pinion shaft screw threads and the nut seat surface.

➡ **Use a new self-locking nut.**

15. Tighten the self-locking nut to 141 ft. lbs. (191 Nm) so that the rotating resistance of companion flange becomes the same as that of before oil seal replacement.
16. Hereafter, reassemble in the reverse order of disassembly.

ENGINE COOLING

ENGINE FAN

REMOVAL & INSTALLATION

See Figures 30 through 33.

1. Before servicing the vehicle, refer to the Precautions Section.
2. Set the vehicle on a lift.
3. Remove the collector cover.

Fig. 30 Remove the bolts from the underside of the radiator stay

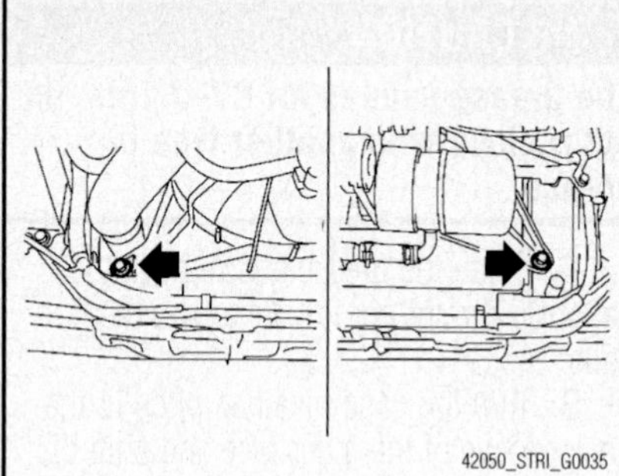

Fig. 31 Remove the bolts from the underside of the radiator fan shroud

4. Disconnect the ground cable from battery.
5. Lift up the vehicle.
6. Remove the undercover.
7. Remove the bolts on the underside of the radiator stay.
8. Remove the bolts on the underside of the radiator fan shroud.
9. Lower the vehicle.

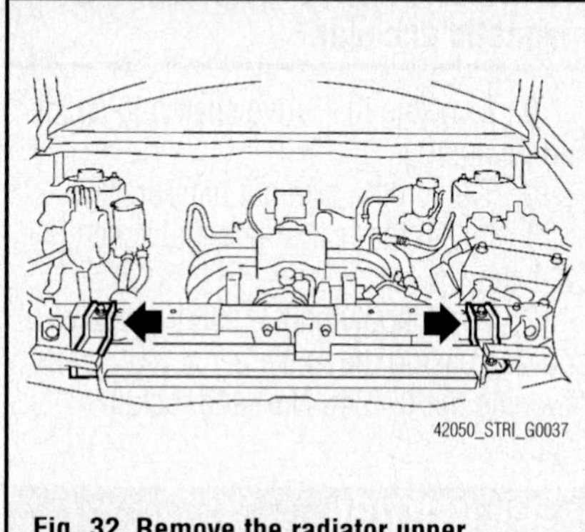

Fig. 32 Remove the radiator upper brackets

10. Remove the air intake duct.
11. Remove the front upper cover.
12. Remove the radiator upper brackets.
13. Remove the radiator stay by performing the following:
    - Remove the latch
    - Remove the radiator hose bracket
    - Remove the clip holding the harness

Fig. 33 Remove the reservoir tank

- Remove the bolts on the left side of the radiator stay
- Remove the bolts on the right side of the radiator stay

14. Disconnect the connector from radiator fan control unit.
15. Remove the reservoir tank.
16. Remove the bolts on the upper side of the radiator fan shroud.
17. Remove the radiator fan shroud.

➡ **When pulling the radiator fan shroud up, be careful not to hit and damage the radiator and ATF hoses.**

To install:

18. Install the radiator fan shroud. Tighten the bolts on the top side of the radiator fan shroud to 60 inch lbs. (7 Nm).
19. Install the reservoir tank.

RADIATOR

REMOVAL & INSTALLATION

See Figures 34 through 39.

✱✱ WARNING

The radiator is pressurized. Wait until engine cools down before working on the radiator.

Fig. 34 Remove the bolts on the underside of the radiator stay

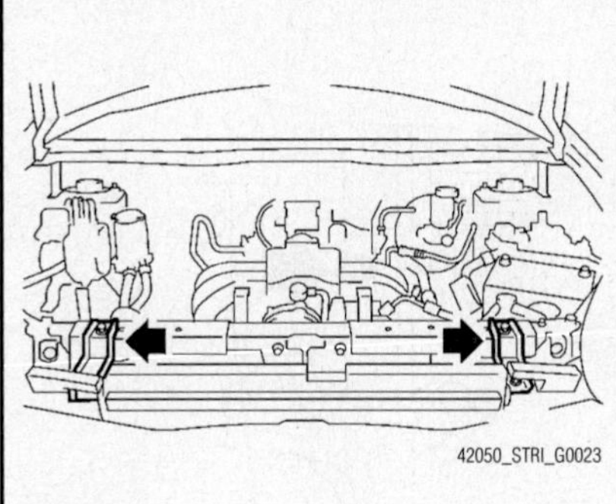

Fig. 35 Remove the radiator upper brackets

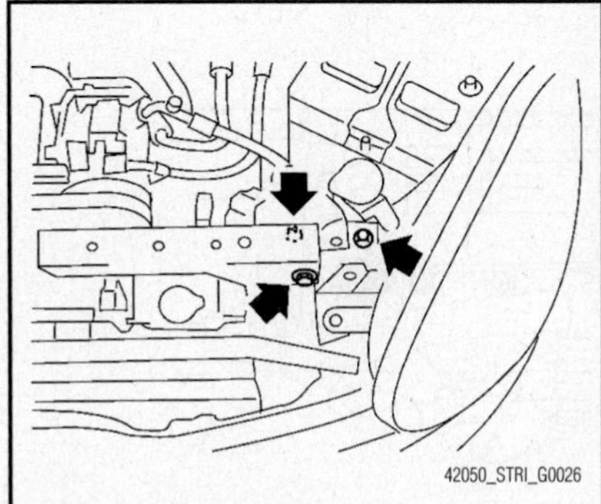

Fig. 36 Remove the bolts from the left side of the radiator stay

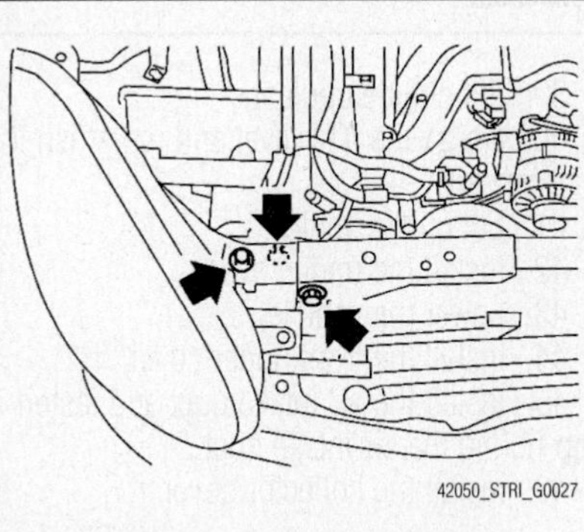

Fig. 37 Remove the bolts from the right side of the radiator stay

1. Before servicing the vehicle, refer to the Precautions Section.
2. Set the vehicle on a lift.
3. Remove the collector cover.
4. Disconnect the ground cable from battery.
5. Lift-up the vehicle.
6. Remove the undercover.
7. Drain engine coolant completely.
8. Disconnect the radiator hose from radiator.
9. Remove the bolts on the underside of the radiator stay.
10. Lower the vehicle.

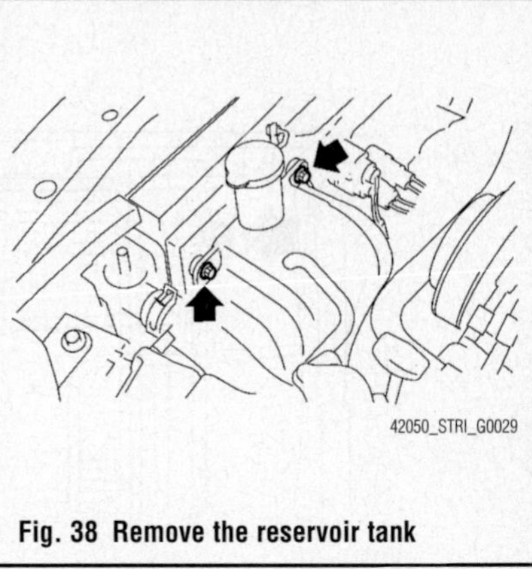

Fig. 38 Remove the reservoir tank

11. Remove the air intake duct.
12. Remove the front upper cover.
13. Remove the radiator upper brackets.
14. Remove the radiator stay by performing the following:
    a. Remove the latch
    b. Remove the radiator hose bracket
    c. Remove the clip holding the harness
    d. Remove the bolts on the left side of the radiator stay
    e. Remove the bolts on the right side of the radiator stay
15. Disconnect the connector from radiator fan control unit.
16. Remove the reservoir tank.
17. Disconnect the radiator hose from radiator.
18. Disconnect the ATF cooler hose from the radiator.

➡ **Plug the ATF pipe to prevent ATF leaks.**

19. Remove the front bumper.
20. Remove the bolts which hold the radiator and condenser.
21. Remove the clip holding the ambient temperature sensor harness.
22. Lift the radiator up and away from vehicle.

To install:

23. Install the radiator lower cushion to the vehicle.
24. Install the radiator to vehicle.

➡ **Insert the pins on the lower side of radiator into the radiator lower cushions on the vehicle side.**

25. Hold the harness of the ambient temperature sensor with a clip.
26. Hold the radiator and condenser with bolts. Tighten to 60 inch lbs. (7 Nm).
27. Install the front bumper. Tighten the front bumper beam to 24 ft. lbs. (33 Nm).
28. Connect the ATF cooler hoses.

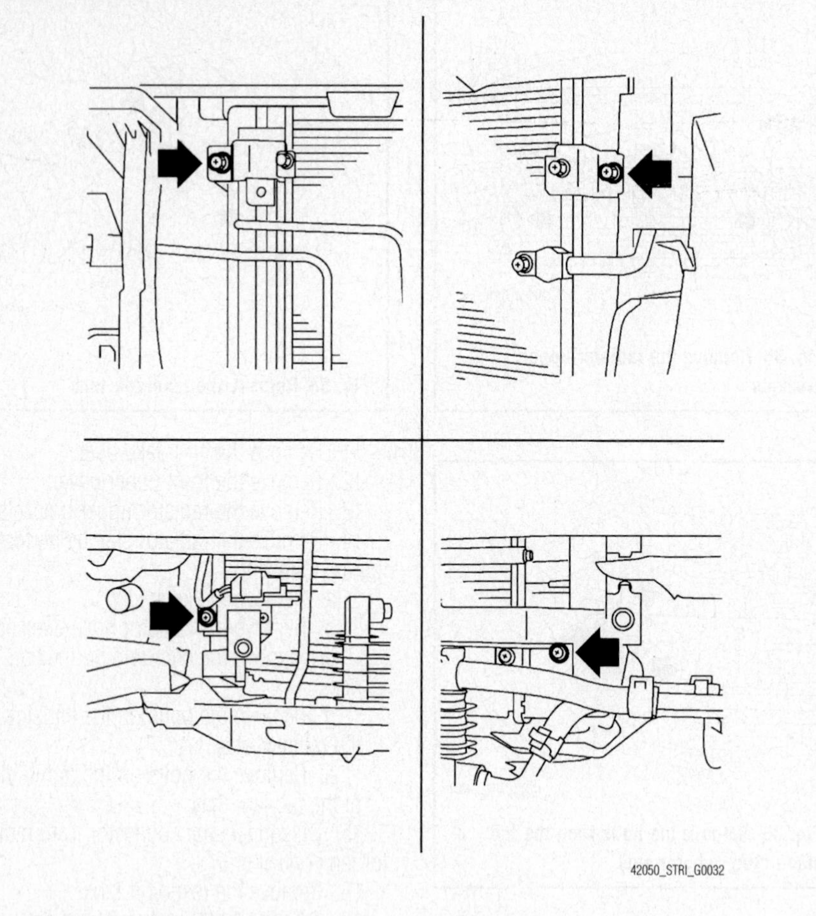

Fig. 39 Remove the bolts holding the radiator and condenser

29. Connect the radiator hose.
30. Install the reservoir tank.
31. Connect the connector of the radiator fan control unit.
32. Install the radiator stay by performing the following:
   - Install the latch. Torque to 25 ft. lbs. (33 Nm)
   - Install the radiator hose bracket
   - Hold the harness to the radiator stay with a clip
   - Tighten the bolts on the left side of the radiator stay. Tighten to 13 ft. lbs. (18 Nm)
   - Tighten the bolts on the right side of the radiator stay. Tighten to 13 ft. lbs. (18 Nm)
33. Install the radiator upper brackets. Tighten to 9 ft. lbs. (12 Nm)
34. Lift up the vehicle.
35. Connect the radiator hose.
36. Tighten the bolts on the lower side of the radiator stay. Tighten to 13.3 ft. lbs. (18 Nm).
37. Lower the vehicle.
38. Connect the ground cable to battery.
39. Fill engine coolant.
40. Check the ATF level and replenish it if necessary.
41. Lift up the vehicle.
42. Install the undercover.
43. Lower the vehicle.
44. Install the front upper cover.
45. Install the air intake duct and fasten clip holing the air intake duct.
46. Install the collector cover.

THERMOSTAT

REMOVAL & INSTALLATION

See Figure 40.

1. Before servicing the vehicle, refer to the Precautions Section.
2. Set the vehicle on a lift.
3. Lift up the vehicle.
4. Remove the undercover.
5. Drain engine coolant completely.
6. Disconnect the radiator hose from the thermostat cover.
7. Remove the thermostat cover, and then remove the thermostat.
8. Install a gasket to thermostat.

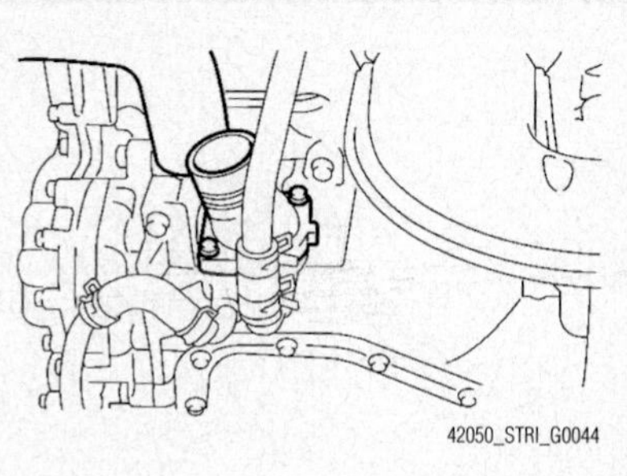

Fig. 40 Remove the thermostat cover and the thermostat

➡ **Use a new gasket.**

9. Install the thermostat and thermostat cover.

➡ **The thermostat must be installed with the jiggle pin facing to the up side. Tighten to 60 inch lbs. (6.4 Nm).**

10. Connect the radiator hose to thermostat cover.
11. Install the undercover.
12. Lower the vehicle.
13. Fill the engine with coolant.

WATER PUMP

REMOVAL & INSTALLATION

See Figures 41 and 42.

1. Before servicing the vehicle, refer to the Precautions Section.
2. Disconnect the negative battery cable.
3. Drain the cooling system. Remove the radiator.
4. Position a suitable tool on the belt tension assembly mounting bolt. Rotate the tool clockwise and loosen the drive belt. Remove the drive belt. Remove the drive belt cover.

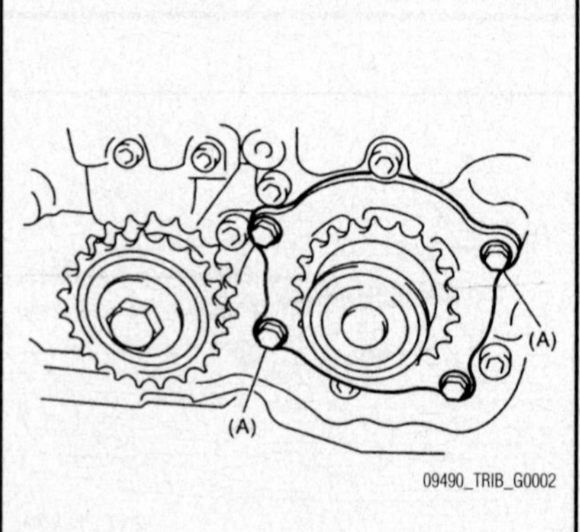

Fig. 41 Water pump location—3.0L engine

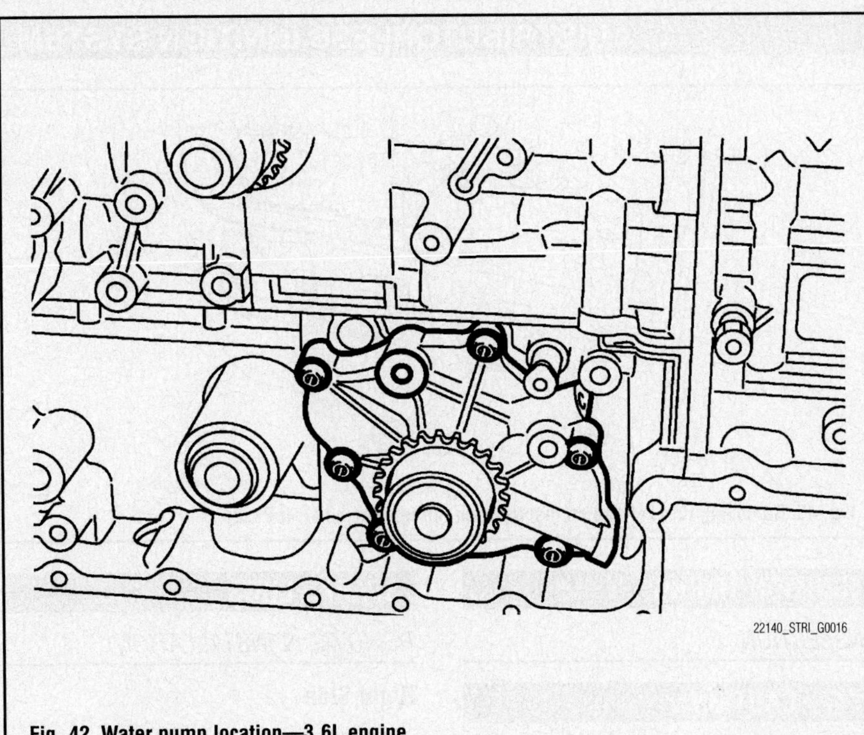

Fig. 42 Water pump location—3.6L engine

➡ Upon installation make sure that the automatic belt tensioner indicator, indicates proper belt alignment.

5. Remove the front timing chain cover. Remove the timing chain assembly.
6. Remove the water pump retaining bolts. Remove the water pump from the engine.

➡ If the pump cannot be removed easily; screw in a bolt (A) to the threaded end and remove the water pump.

To install:

7. Installation is the reverse of the removal procedure.
8. Be sure to use a new O ring. Apply engine coolant to the O ring, prior to installation.
9. Torque the water pump retaining bolts to 60 inch lbs. (6.4 Nm).
10. Fill the radiator with the proper grade and type engine coolant.
11. Start the engine and check for leaks. Correct as required.

ENGINE ELECTRICAL

CHARGING SYSTEM

ALTERNATOR

REMOVAL & INSTALLATION

See Figure 43.

1. Before servicing the vehicle, refer to the Precautions Section.
2. Disconnect the negative battery cable.
3. Remove the collector cover.
4. Disconnect the electrical connector and the terminal from the alternator.
5. Position a suitable tool on the belt tension assembly mounting bolt. Rotate the tool clockwise and loosen the drive belt. Remove the drive belt. Remove the drive belt cover.

Fig. 43 Alternator and retaining bolt locations shown

➡ Upon installation make sure that the automatic belt tensioner indicator indicates proper belt alignment.

6. Remove the alternator retaining bolts. Remove the alternator from the vehicle.

To install:

7. Installation is the reverse of the removal procedure.
8. Tighten the retaining bolts to 18 ft. lbs. (25 Nm).

FIRING ORDER

The firing order is: 1–6–3–2–5–4.

IGNITION COIL

REMOVAL & INSTALLATION

Right Side

See Figure 44.

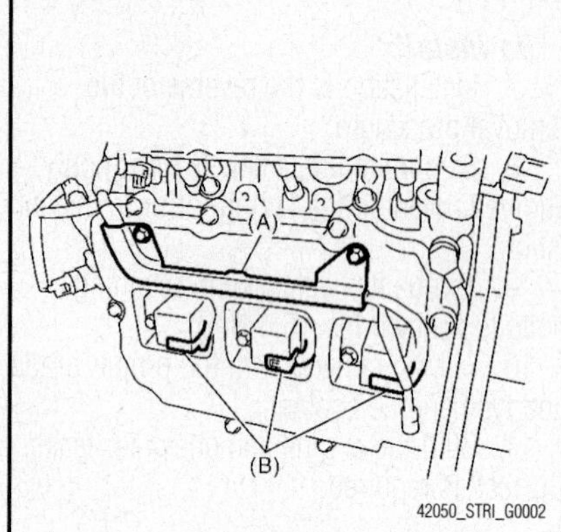

Fig. 44 Removing bracket and connector from ignition coil—right side

1. Remove the collector cover.
2. Disconnect the ground cable from battery.
3. Remove the air cleaner case.
4. Remove the (A) bracket.
5. Disconnect the (B) connector from ignition coil.
6. Remove the ignition coil.

➡ **Turn the #5 ignition coil to remove it.**

7. Installation is the reverse of the removal procedure. Tighten the ignition coil to 12 ft. lbs. (16 Nm)

Left Side

See Figure 45.

1. Remove the collector cover.
2. Remove the battery and battery carrier.
3. Remove the (A) bracket.
4. Disconnect the (B) connector from ignition coil.
5. Remove the ignition coil.

➡ **Turn the no. 6 ignition coil to remove it.**

6. Installation is the reverse of the removal procedure. Tighten the ignition coil to 12 ft. lbs. (16 Nm)

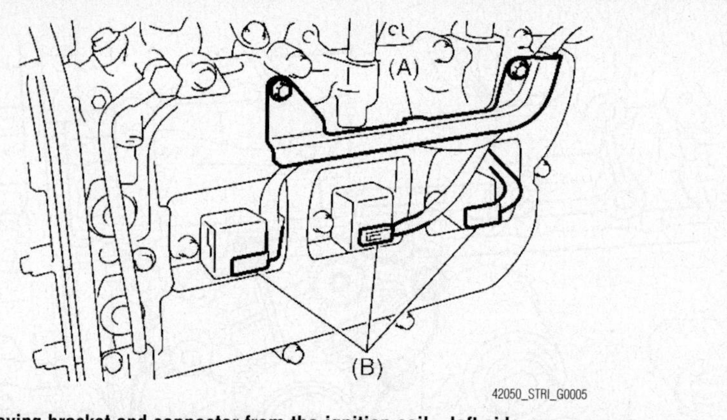

Fig. 45 Removing bracket and connector from the ignition coil—left side

IGNITION TIMING

INSPECTION

✻✻ CAUTION

After warming—up, engine becomes very hot. Be careful not to burn yourself at measurement.

1. Before checking the ignition timing, check the following item:
   - Check the air cleaner element is free from clogging, spark plugs are in good condition, and hoses are connected properly
   - Check the malfunction indicator light does not illuminate
2. Idle the engine.
3. Stop the engine, and turn the ignition switch **OFF**.
4. Insert the cartridge to Subaru Select Monitor.
5. Connect the Subaru Select Monitor to data link connector.
6. Turn the ignition switch **ON** and Subaru Select Monitor switch **ON**.
7. Select **Each System Check** in Main Menu.
8. Select **Engine** in Selection Menu.
9. Select **Current Data Display & Save** in Engine Control System Diagnosis.
10. Select **Data Display** in Data Display Menu.
11. Start the engine and check the ignition timing at idle speed. Ignition timing (BTDC/RPM): 15° +/- 8° /650 to 700
12. If the timing is not correct, check the ignition control system.

ADJUSTMENT

The ignition timing is controlled by the Engine Control Module (ECM). No adjustment is necessary or possible.

SPARK PLUGS

REMOVAL & INSTALLATION

Right Side

See Figure 44.

✻✻ CAUTION

All spark plugs installed on an engine must be of the same heat range.

1. Remove the collector cover.
2. Disconnect the ground cable from battery.
3. Remove the air cleaner case.
4. Remove the (A) bracket.
5. Disconnect the (B) connector from ignition coil.
6. Remove the ignition coil.

➡ **Turn the #5 ignition coil to remove it.**

7. Remove the spark plug with a spark plug socket.

To install:

To install reverse the removal procedures

- Tighten the spark plug to 15 ft. lbs. (21 Nm)
- Tighten the ignition coil to 12 ft. lbs. (16 Nm)

➡ **The tightening torque described above should be applied to only new spark plugs without oil on their threads. In case their threads are lubricated, the torque should be reduced by approximately ⅓ of the specified torque in order to avoid over—stressing.**

Left Side

See Figure 45.

1. Remove the collector cover.
2. Remove the battery and battery carrier.
3. Remove the (A) bracket.

4. Disconnect the (B) connector from ignition coil.

5. Remove the ignition coil.

➡ **Turn the no. 6 ignition coil to remove it.**

6. Remove the spark plug with a spark plug socket.

To install:

7. To install reverse the removal procedures

- Tighten the spark plug to 15 ft. lbs. (21 Nm)
- Tighten the ignition coil to 12 ft. lbs. (16 Nm)

➡ **The tightening torque described above should be applied to only new spark plugs without oil on their threads. In case their threads are lubricated, the torque should be reduced by approximately ⅓ of the specified torque in order to avoid over—stressing.**

ENGINE ELECTRICAL

STARTING SYSTEM

STARTER

REMOVAL & INSTALLATION

See Figure 46.

1. Before servicing the vehicle, refer to the Precautions Section.

2. Disconnect the negative battery cable.

3. Remove the collector cover.

4. Remove the air intake chamber.

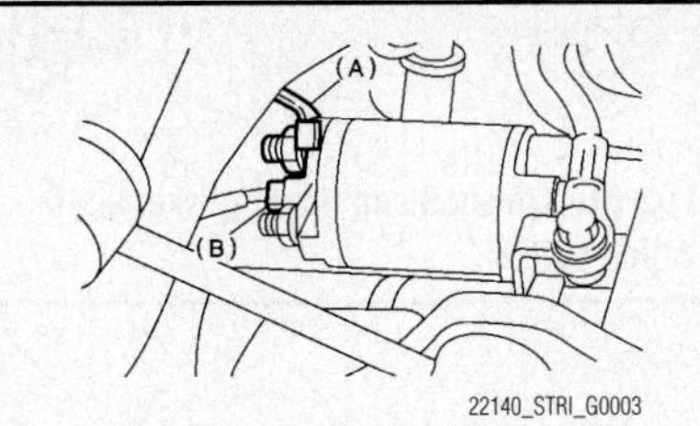

Fig. 46 Starter terminal A and connector B shown

5. Disconnect the electrical connectors from the starter.

6. Remove the starter retaining bolts. Remove the starter from the vehicle.

To install:

7. Installation is the reverse of the removal procedure.

8. Torque the starter retaining bolts to 37 ft. lbs.

ENGINE MECHANICAL

➡ **Disconnecting the negative battery cable may interfere with the functions of the on board computer systems and may require the computer to undergo a relearning process, once the negative battery cable is reconnected.**

ACCESSORY DRIVE BELTS

ACCESSORY BELT ROUTING

See Figure 47.

Fig. 47 Accessory belt routing 3.0L and 3.6L engines

INSPECTION

Inspect the drive belt for signs of glazing or cracking. A glazed belt will be perfectly smooth from slippage, while a good belt will have a slight texture of fabric visible. Cracks will usually start at the inner edge of the belt and run outward. All worn or damaged drive belts should be replaced immediately.

ADJUSTMENT

The accessory drive belt is self adjusting.

REMOVAL & INSTALLATION

See Figure 48.

1. Install the tool to belt tension adjuster assembly installation bolt.

Fig. 48 Removing accessory drive belt

2. Rotate the tool clockwise and loosen the belt to remove.

3. Remove the belt cover.

4. To install reverse the removal procedure.

CAMSHAFT AND VALVE LIFTERS

INSPECTION

1. Before servicing the vehicle, refer to the Precautions Section.

2. Remove the camshaft from the engine.

3. Check the camshaft bearing journals for damage and binding.

4. If the journals are binding, check the cylinder head for damage.

5. Check the cylinder head for clogged oil holes.

6. Check the camshaft surface for abnormal wear and damage. Replace the camshaft, as required.

7. Measure the camshaft lobe surface and replace the camshaft if not within specification.

8. Measure the camshaft journal diameter and replace the camshaft if not within specification.

9. Measure the camshaft run out and replace the camshaft if not within specification.

REMOVAL & INSTALLATION

3.0L Engine

See Figures 49 through 58.

1. Before servicing the vehicle, refer to the Precautions Section.

2. Disconnect the negative battery cable.

3. Remove the crankshaft pulley cover.
4. Remove the crankshaft pulley bolt.
5. Lock the crankshaft in place using tool ST-499977100, or equivalent.
6. Remove the crankshaft pulley.
7. Remove the front timing chain cover.

➡ Bolts are three different sizes. Be careful to install the correct bolt in the correct hole.

8. Remove the timing chain.
9. Remove the camshaft sprocket.

➡ Be sure to lock the camshaft in place using tool ST-499977500, or equivalent.

10. Remove the crankshaft sprocket.
11. Remove the oil pump.
12. Remove the water pump.
13. Remove the rear timing chain cover retaining bolts. Remove the rear timing chain cover.

➡ There are seven different size bolts. Be sure not to confuse them on installation.

14. Disconnect the oil pipe.
15. Remove the rocker cover retaining bolts. Remove the rocker cover. Discard the gasket.
16. Remove the plugs, see illustration for location.
17. Loosen the camshaft cap bolts in the proper sequence. Remove the camshaft caps and remove the camshaft.

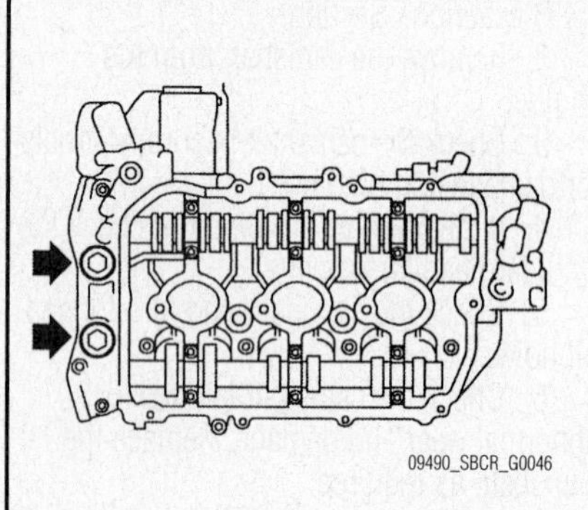

Fig. 49 Camshaft plug bolt location

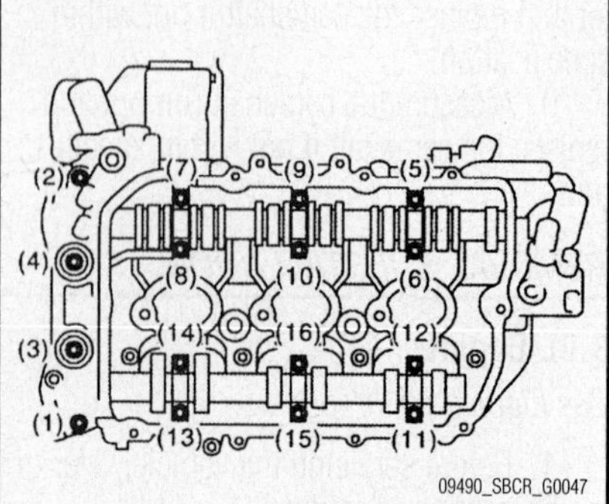

Fig. 50 Camshaft bolt loosening sequence

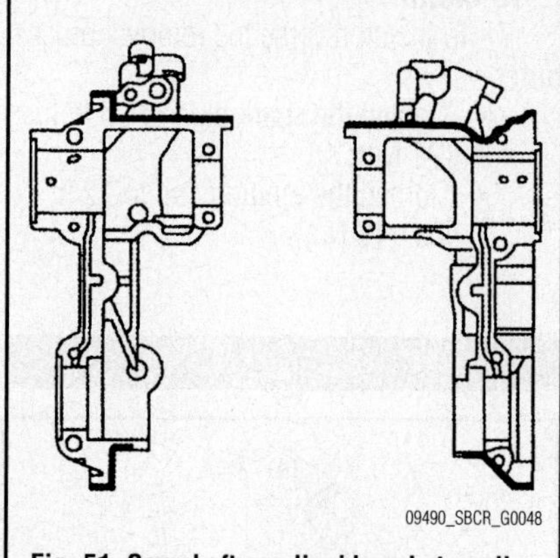

Fig. 51 Camshaft cap liquid gasket application points

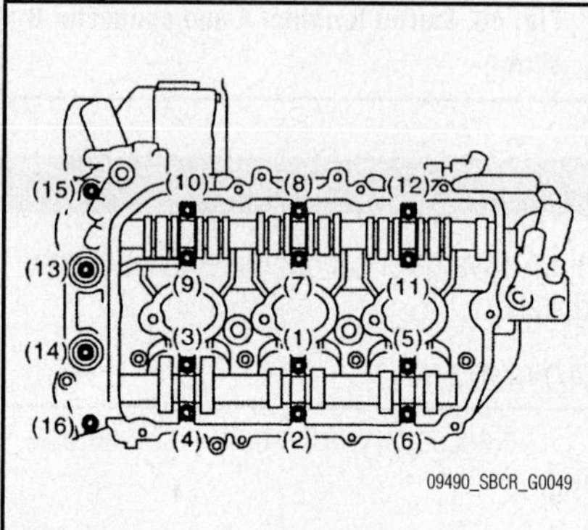

Fig. 52 Camshaft bolt tightening sequence

To install:

18. Apply a coat of engine oil to the journals on the camshafts and place the camshafts into position.
19. To install the camshaft cap, apply a small amount of liquid gasket to the mating surface of the cap. Do not apply an excessive amount of sealant, as it will squish out and flow toward the cam journal resulting in engine seizure.
20. Apply a thin coat of engine oil to the cap bearing surface and install the cap. Tighten cap bolts 1 through 12 to 12 ft. lbs. (16 Nm) and bolts 13 through 16 to 84 inch lbs. (9.5 Nm) in the proper sequence. Install the plugs and torque to 44 ft. lbs. (60 Nm).
21. Apply fluid gasket maker of the of the cylinder heads and valve covers.

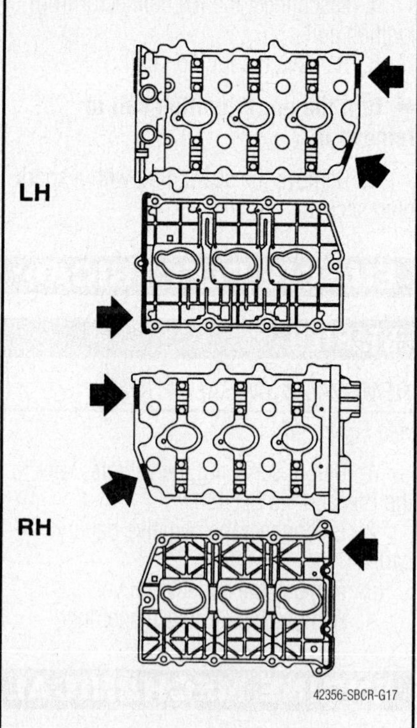

Fig. 53 Apply fluid gasket maker of the of the cylinder heads and valve covers as shown

⁂ CAUTION

Do not apply too much gasket maker. This may cause excess fluid gasket maker to come out and flow towards the camshaft journal resulting in engine damage.

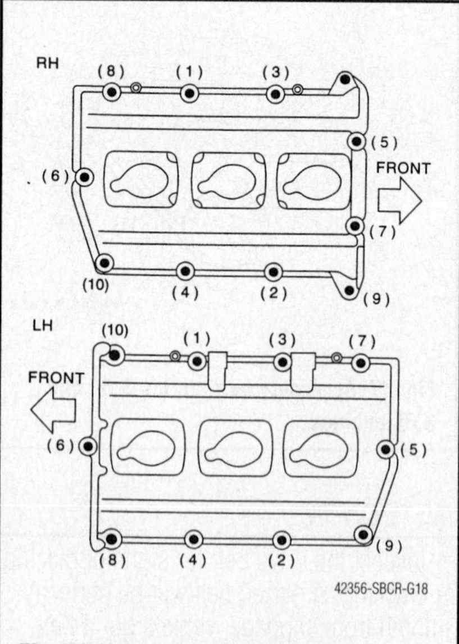

Fig. 54 Valve cover bolt tightening sequence

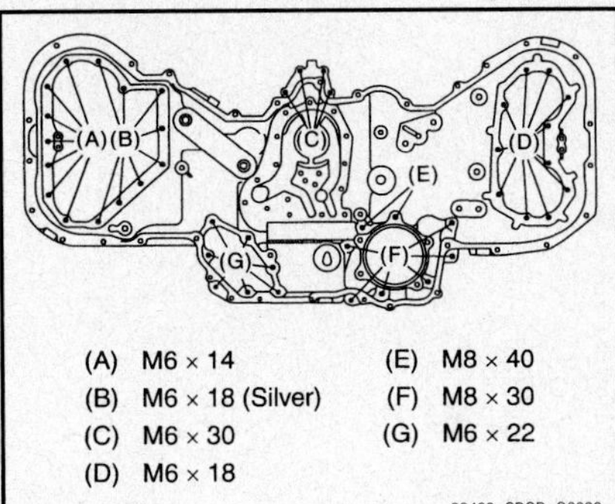

Fig. 55 Rear timing chain bolt sizes and locations

22. Tighten the valve cover bolts to 60 inch lbs. (8 Nm) and in the proper sequence.

23. Install the rear chain cover as follows:

➡ **There are several size bolts used, refer to the illustration for size and locations.**

a. Rear chain cover gasket and clean the mating surfaces.

b. Apply liquid gasket maker to the mating surfaces of the cover. Refer to the illustration for gasket maker application and diameter.

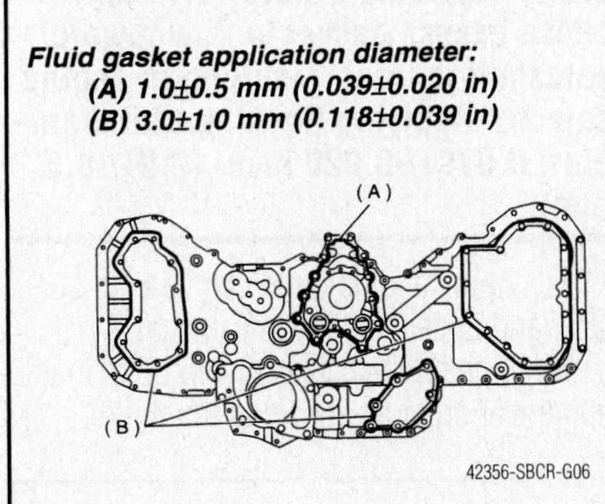

Fluid gasket application diameter:
(A) 1.0±0.5 mm (0.039±0.020 in)
(B) 3.0±1.0 mm (0.118±0.039 in)
(A)
(B)
42356-SBCR-G06

Fig. 56 Rear timing chain cover sealant application

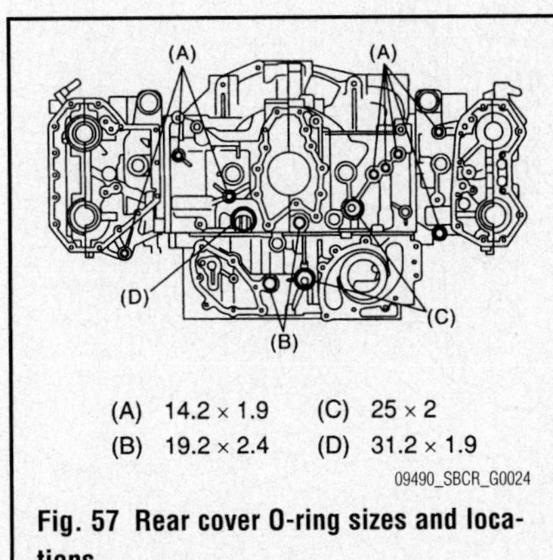

Fig. 57 Rear cover O-ring sizes and locations

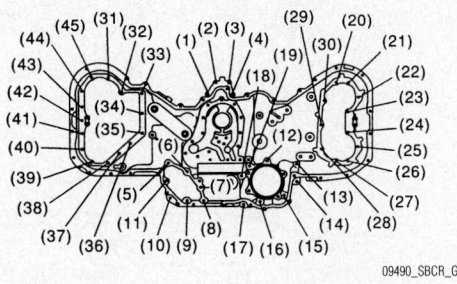

(1) — (11)	9 N·m (0.9 kgf-m, 6.5 ft-lb)
(12) — (19)	20 N·m (2.0 kgf-m, 14 ft-lb)
(20) — (30)	9 N·m (0.9 kgf-m, 6.5 ft-lb)
(31) — (38)	12 N·m (1.2 kgf-m, 8.7 ft-lb)
(39) — (45)	9 N·m (0.9 kgf-m, 6.5 ft-lb)

09490_SBCR_G0025

Fig. 58 Rear cover bolt tightening sequence and torque specifications

c. Install new O-rings. Refer to the illustration for O-ring location and size

d. Install the rear chain cover and temporarily tighten the bolts, refer to the illustration for size and locations.

e. Tighten the cover bolts in the sequence illustrated to the specifications shown in the illustration.

24. Continue the installation in the reverse order of the removal procedure.

3.6L Engine

See Figures 59 through 66.

1. Before servicing the vehicle, refer to the Precautions Section.
2. Remove the engine unit from vehicle.
3. Remove the crank pulley.
4. Remove the chain cover.
5. Remove the timing chain assembly.
6. Remove the cam sprocket.
7. Remove the crank sprocket.
8. Remove the exhaust oil flow control solenoid valve.
9. Remove the camshaft position sensor.
10. Loosen the rocker cover bolt in the numerical order as shown in the figure, and then remove the rocker cover.
11. Loosen the camshaft cap bolts equally, a little at a time in numerical sequence shown in the figure.
12. Remove the camshaft caps and camshaft RH.
13. Similarly, remove the camshafts LH and related parts.

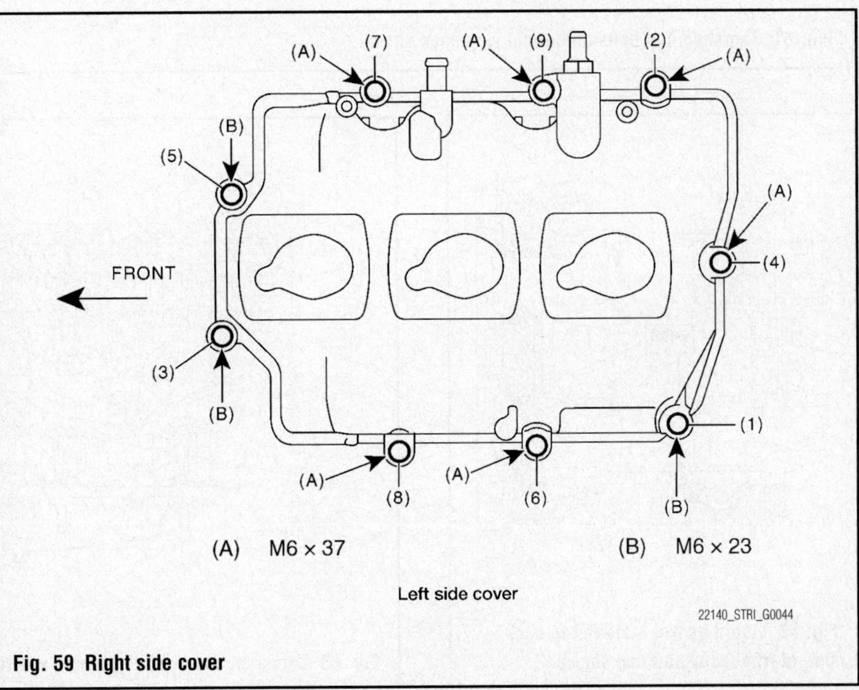

Fig. 59 Right side cover

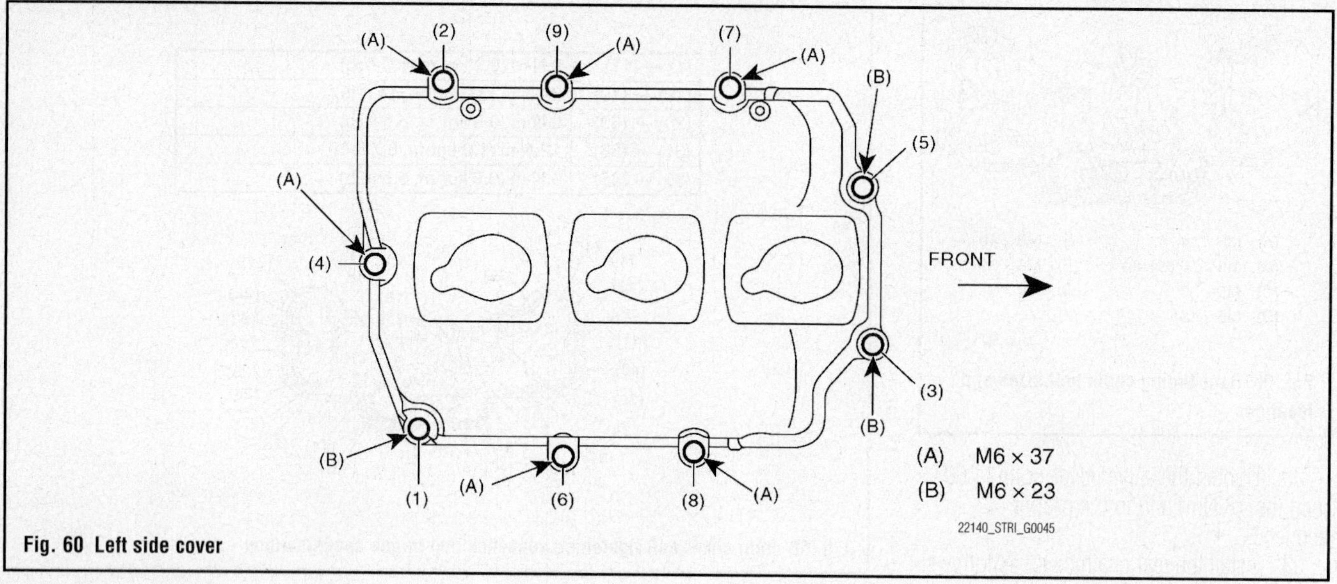

Fig. 60 Left side cover

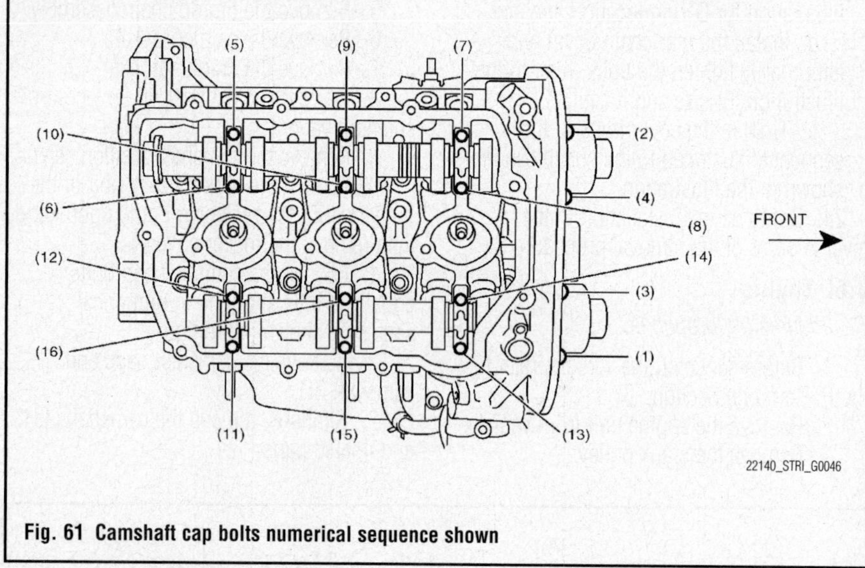

Fig. 61 Camshaft cap bolts numerical sequence shown

To install:

14. Apply engine oil to camshaft journals, and install the camshaft.
15. Install the camshaft cap.
16. Apply liquid gasket sparingly to back side of front camshaft cap as shown in the figure.

CAUTION

Do not apply liquid gasket excessively. Applying excessively may cause excess gasket to flow toward camshaft journal, resulting in engine seizure. Applying liquid gasket diameter: 0.079+/-0.020 inch. (2.0+/-0.5 mm).

17. Apply engine oil to cap bearing surface, and install the cap to camshaft.
18. Tighten the camshaft cap bolts in the numerical order as shown in the figure.

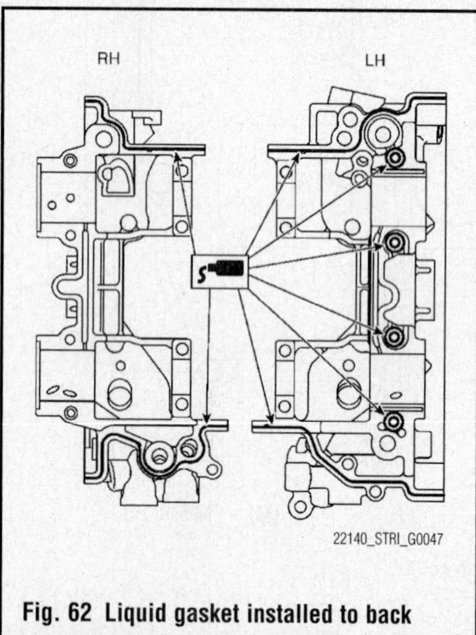

Fig. 62 Liquid gasket installed to back side of front camshaft cap shown

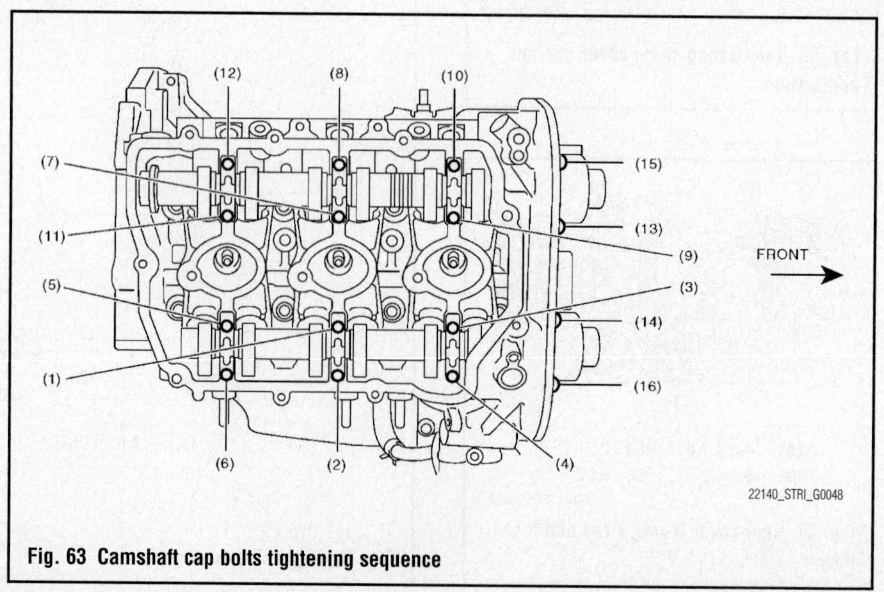

Fig. 63 Camshaft cap bolts tightening sequence

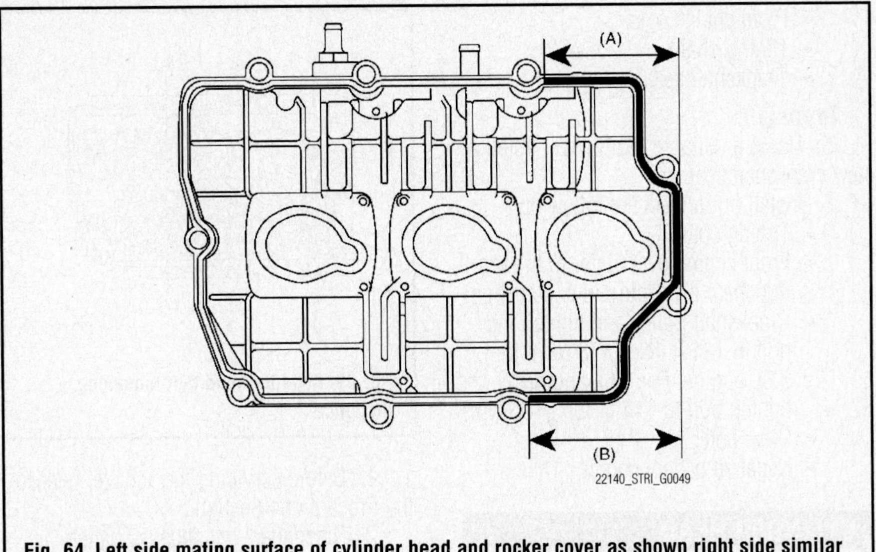

Fig. 64 Left side mating surface of cylinder head and rocker cover as shown right side similar

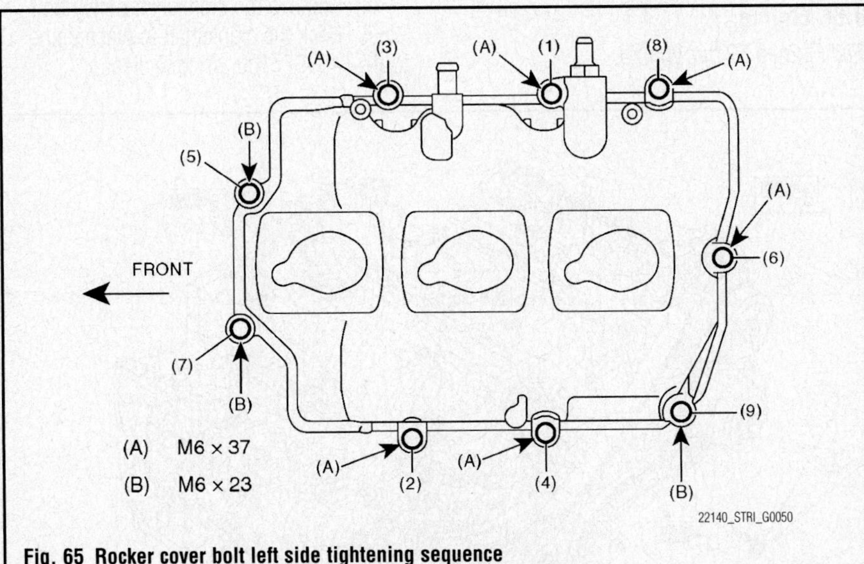

Fig. 65 Rocker cover bolt left side tightening sequence

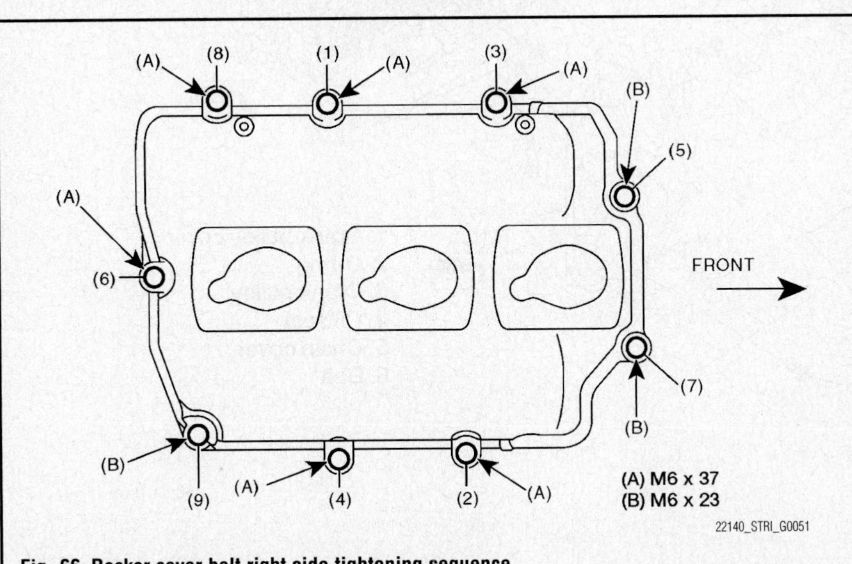

Fig. 66 Rocker cover bolt right side tightening sequence

19. Install the rocker cover gasket to the rocker cover.
20. Apply liquid gasket sparingly to the mating surface of cylinder head and rocker cover as shown in the figure.
21. Apply liquid gasket THREE BOND 1217G (Part No. K0877Y0100) or equivalent. Applying liquid gasket diameter: 0.138+/-0.020 inch. (3.5+/-0.5 mm).
22. Tighten the rocker cover bolts in the numerical order as shown in the figure.
23. Install the crank sprocket.
24. Install the cam sprocket.
25. Install the timing chain assembly.
26. Install the chain cover.
27. Install the crank pulley.
28. Install the engine unit to vehicle.
29. Install engine oil.
30. Refill coolant and bleed cooling system of air.
31. Confirm that there are no coolant leaks.

CRANKSHAFT DAMPER

REMOVAL & INSTALLATION

See Figure 67.

1. Remove the drive belt.
2. Before servicing the vehicle, refer to the Precautions Section.
3. Remove the crank pulley cover.
4. Remove the crank pulley bolt. To lock the crankshaft, use ST1-18355AA000 pulley wrench and ST2-18334AA000 pulley wrench pin set.
5. Remove the crank pulley damper.

To install:

6. Install the crank pulley.
7. Install the crank pulley bolt. To lock the crankshaft, use ST1-18355AA000 pulley wrench and ST2-18334AA000 pulley wrench pin set.
8. Clean the crankshaft thread using compressed air.
9. Apply engine oil to the crank pulley bolt seat and thread.

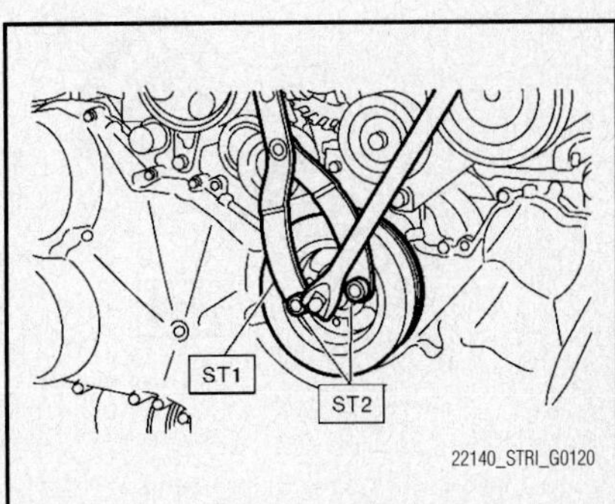

Fig. 67 Crank pulley damper removal with special tools

10. Tighten the crank pulley bolts to 144ft. lbs. (195 Nm).
11. Install the crank pulley cover. Assemble the O-ring to crank pulley cover and tighten to 5 ft lbs. (6.4 Nm).
12. Install the drive belt.

CRANKSHAFT FRONT SEAL

REMOVAL & INSTALLATION

See Figure 68.

1. Before servicing the vehicle, refer to the Precautions Section.
2. Remove or disconnect the following:
   - Negative battery cable
   - Drive belt
3. Secure the crankshaft pulley with tool No. 499977100.
4. Remove or disconnect the following:
   - Crankshaft pulley bolt and pulley
   - Front chain cover
   - Timing chain
   - Crankshaft seal

To install:

5. Using a suitable seal driver, install a new crankshaft seal.
6. Install or connect the following:
   - Timing chain
   - Front chain cover. Refer to the timing chain procedure in this section.
   - Crankshaft pulley and tighten the bolt to 131 ft. lbs. (178 Nm). for 3.0L engine. For 3.6L engines tighten bolt to 144 ft. lbs. (195 Nm)
   - Drive belt
   - Negative battery cable

CYLINDER HEAD

REMOVAL & INSTALLATION

3.0L Engine

See Figures 69 through 73.

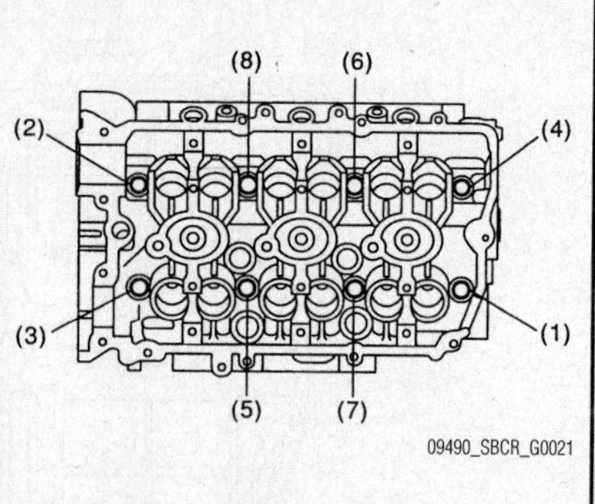

Fig. 69 Cylinder head bolt loosening sequence

1. Before servicing the vehicle, refer to the Precautions Section.
2. Disconnect the negative battery cable.
3. Remove the crankshaft pulley cover.
4. Remove the crankshaft pulley bolt.
5. Lock the crankshaft in place using tool ST499977100, or equivalent.

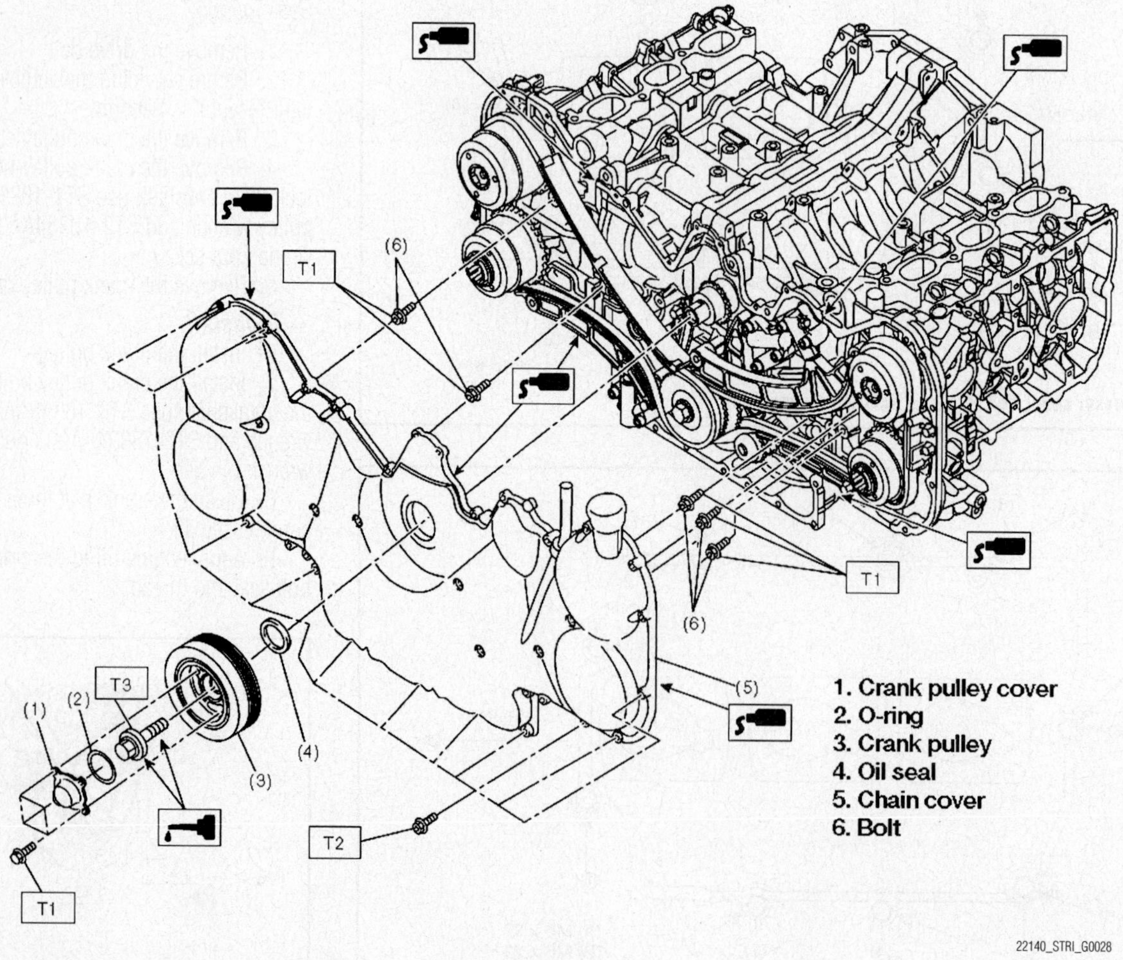

1. Crank pulley cover
2. O-ring
3. Crank pulley
4. Oil seal
5. Chain cover
6. Bolt

22140_STRI_G0028

Fig. 68 Front engine view 3.6L engine shown

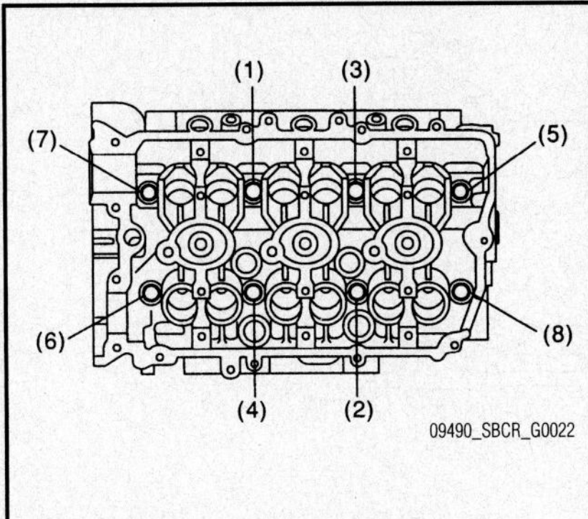

Fig. 70 Cylinder head bolt tightening sequence

6. Remove the crankshaft pulley.
7. Remove the front timing chain cover.

➡ **Bolts are three different sizes. Be careful to install the correct bolt in the correct hole.**

8. Remove the timing chain.
9. Remove the camshaft sprocket.

➡ **Be sure to lock the camshaft in place using tool ST499977500, or equivalent.**

10. Remove the crankshaft sprocket.
11. Remove the oil pump.
12. Remove the water pump.
13. Remove the rear timing chain cover retaining bolts. Remove the rear timing chain cover.

➡ **There are seven different size bolts. Be sure not to confuse them on installation.**

14. Remove the camshafts.
15. Remove the cylinder head bolts in the proper sequence. Leave bolts 2 and 4 connected by a few threads to prevent the head from falling. Tap the head with a plastic mallet to separate it from the block.
16. Remove bolts 2 and 4 from the cylinder head. Remove the cylinder head from the engine. Discard the gasket.
17. Clean all gasket material from both mating surfaces.

To install:

18. Installation is the reverse of the removal procedure.
19. Apply a thin coat of clean engine oil to the washers and cylinder head bolts.
20. Tighten the cylinder head retaining bolts to specification and in the proper sequence.
21. Install the rear chain cover as follows:

➡ **There are several size bolts used, refer to the illustration for size and locations.**

Fluid gasket application diameter:
(A) 1.0±0.5 mm (0.039±0.020 in)
(B) 3.0±1.0 mm (0.118±0.039 in)

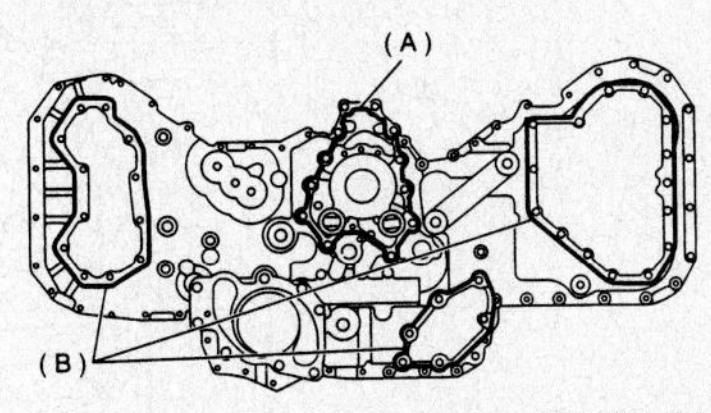

Fig. 71 Rear timing chain cover sealant application

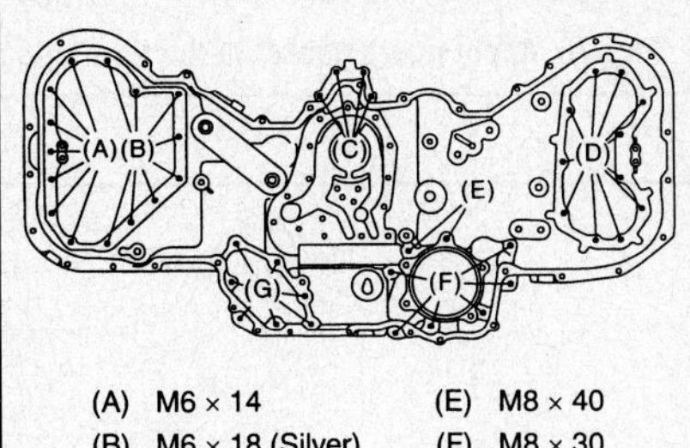

(A) M6 × 14
(B) M6 × 18 (Silver)
(C) M6 × 30
(D) M6 × 18
(E) M8 × 40
(F) M8 × 30
(G) M6 × 22

09490_SBCR_G0023

Fig. 72 Rear timing chain bolt sizes and locations

a. Rear chain cover gasket and clean the mating surfaces.
b. Apply liquid gasket maker to the mating surfaces of the cover. Refer to the illustration for gasket maker application and diameter.
c. Install new O-rings. Refer to the illustration for O-ring location and size
d. Install the rear chain cover and temporarily tighten the bolts, refer to the illustration for size and locations.
e. Tighten the cover bolts in the sequence illustrated to the specifications shown in the illustration.

22. Start the engine and allow it to reach operating temperature.
23. Check for leaks, correct as required.

3.6L Engine

See Figures 74 through 81.

1. Before servicing the vehicle, refer to the Precautions Section.
2. Remove the crank pulley.
3. Remove the chain cover.
4. Remove the timing chain assembly.
5. Remove the cam sprocket. To lock the camshaft, use the ST499977100, or equivalent.
6. Remove the crank sprocket.
7. Remove the seal bolts as shown in the figure.

(1) — (11)	9 N·m (0.9 kgf-m, 6.5 ft-lb)
(12) — (19)	20 N·m (2.0 kgf-m, 14 ft-lb)
(20) — (30)	9 N·m (0.9 kgf-m, 6.5 ft-lb)
(31) — (38)	12 N·m (1.2 kgf-m, 8.7 ft-lb)
(39) — (45)	9 N·m (0.9 kgf-m, 6.5 ft-lb)

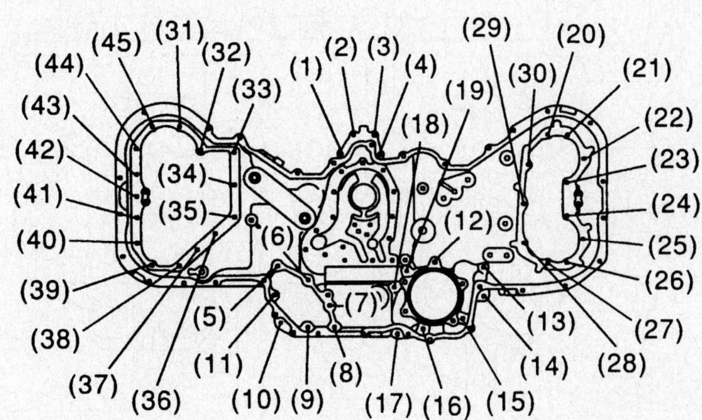

Fig. 73 Rear cover bolt tightening sequence and torque specifications

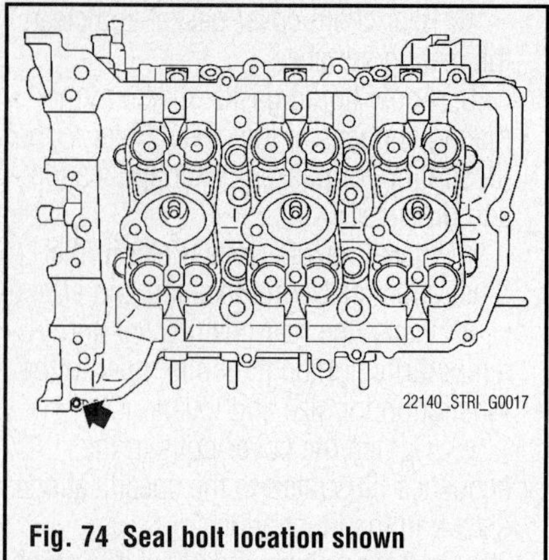

Fig. 74 Seal bolt location shown

8. Remove the cylinder head bolts in the numerical order as shown in the figure. Leave bolts (2) and (4) engaged by three or four threads to prevent the cylinder head from falling.

9. While tapping the cylinder head with a plastic hammer, separate it from cylinder block.

10. Remove the bolts (2) and (4) to remove cylinder head.

11. Remove the cylinder head gasket.

CAUTION

Be careful not to scratch the mating surface of cylinder head and cylinder block.

12. Similarly, remove the cylinder head RH.

To install:

13. Apply liquid gasket THREE BOND 1217G (K0877Y0100) or equivalent to the mating surface of cylinder block as shown in the figure.

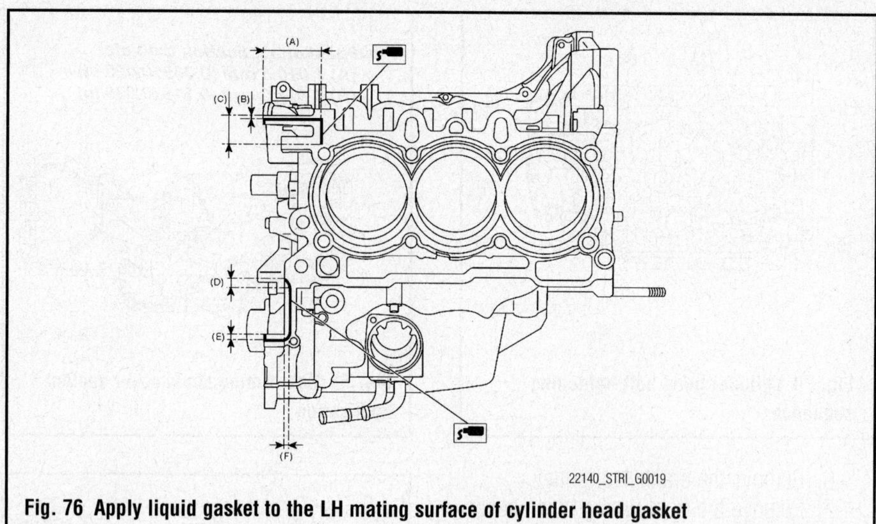

Fig. 76 Apply liquid gasket to the LH mating surface of cylinder head gasket

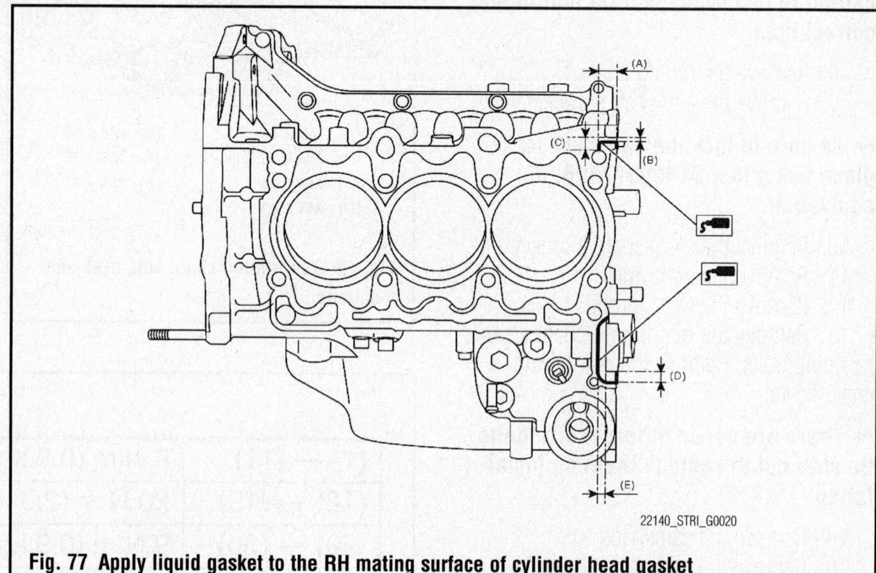

Fig. 77 Apply liquid gasket to the RH mating surface of cylinder head gasket

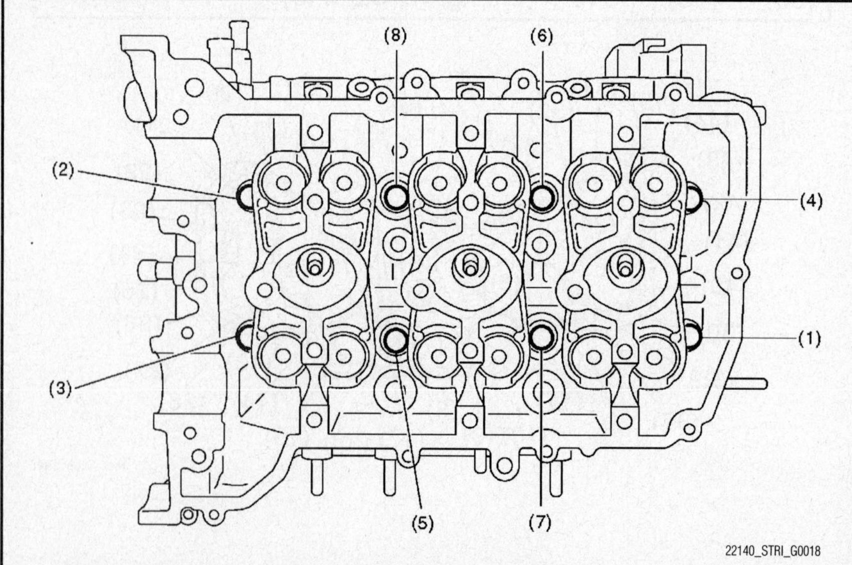

Fig. 75 Cylinder head bolt loosening sequence

CAUTION

Do not apply liquid gasket excessively. If applying excessively, remove the excess liquid gasket that flowed out.

➡ Install within 5 min. after applying liquid gasket.

14. Install the cylinder head gaskets LH and RH on cylinder block.

15. Apply liquid gasket to the mating surface of cylinder head gasket as shown in the figure.

16. Install the cylinder head to cylinder block.

17. Tighten the cylinder head bolts, as follows:

- Apply a thin coat of engine oil to washers and cylinder head bolt threads.

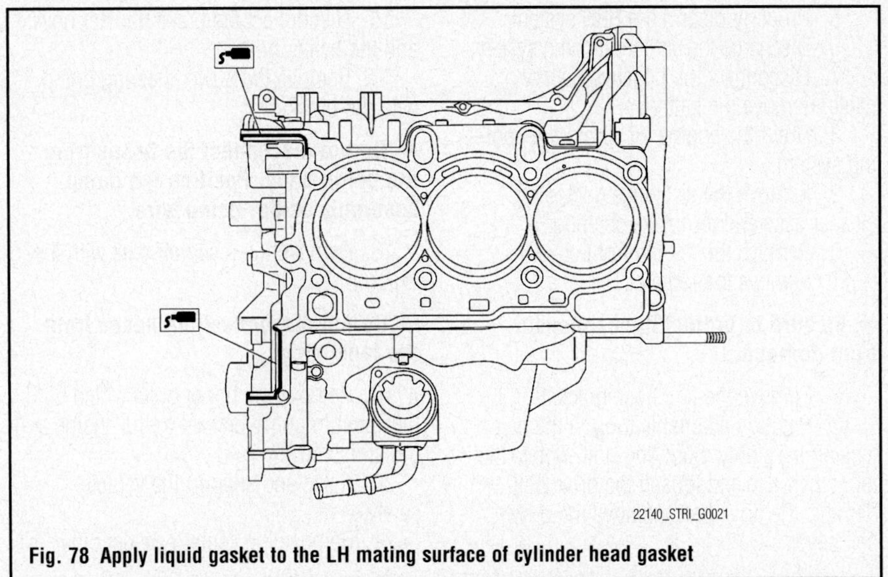

Fig. 78 Apply liquid gasket to the LH mating surface of cylinder head gasket

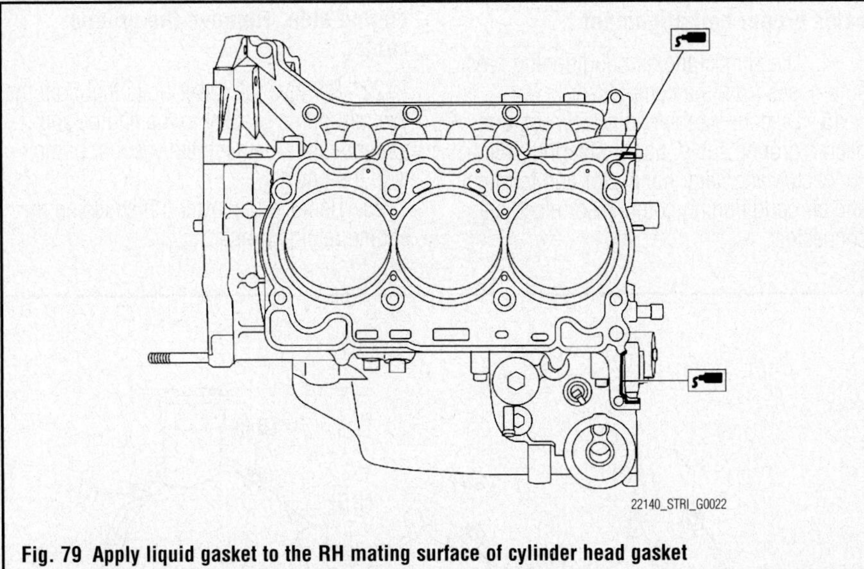

Fig. 79 Apply liquid gasket to the RH mating surface of cylinder head gasket

- Install the cylinder head to cylinder block, and then tighten the bolts with torque of 15 ft. lbs. (20 Nm) in numerical sequence as shown in the figure.
- Tighten the bolts with torque of 37 ft. lbs. (50 Nm) in numerical sequence as shown in the figure.
- Loosen all the bolts by 180° in reverse order of installation, and loosen again by 180° in the same order.
- Tighten the bolts with torque of 15 ft. lbs. (20 Nm) in numerical sequence as shown in the figure.
- Tighten the bolts (1)–(4) with torque of 35 ft. lbs. (48 Nm) in numerical sequence.
- Tighten the bolts (5)–(8) with torque of 33 ft. lbs. (44 Nm) in numerical sequence.
- Tighten all bolts 90° in the numerical order as shown in the figure.
- Tighten the bolt (1)–(4) by 45° in the numerical order.

➡ Apply seal material on the bolt threads before installing the seal bolts.

18. Install the seal bolts and tighten to 5 ft. lbs. (6.4 Nm).
19. Install the crank sprocket.
20. Install the camshaft.
21. Install the timing chain assembly.
22. Apply liquid gasket to the mating surface of chain cover.
23. Install the chain cover and temporarily tighten the bolts.
24. Tighten the bolts in the numerical order as shown in the figure to 7 ft. lbs. (10 Nm).

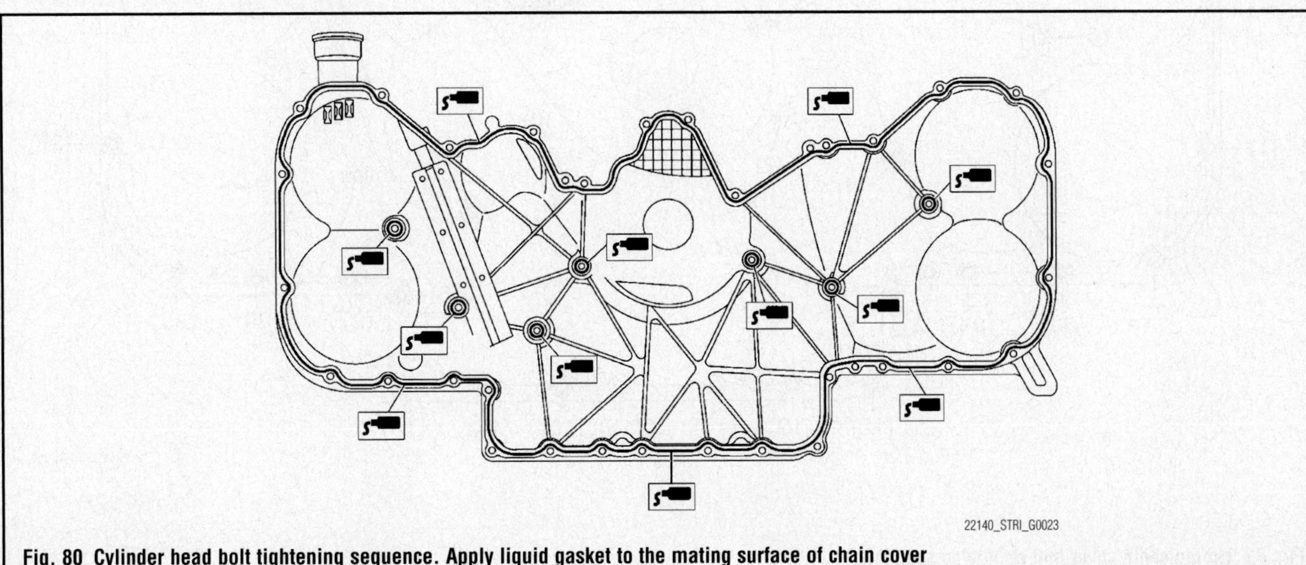

Fig. 80 Cylinder head bolt tightening sequence. Apply liquid gasket to the mating surface of chain cover

25. Install the crank pulley and tighten bolts to 144 ft. lbs. (195 Nm).
26. Install the crank pulley cover.
27. Install the engine unit to vehicle.

ENGINE ASSEMBLY

REMOVAL & INSTALLATION

3.0L Engine

CAUTION

Whenever working near any of the Supplemental Restraint System (SRS) components, such as the impact sensors, the air bag module, steering column and instrument panel, properly disable the SRS.

1. Before servicing the vehicle, refer to the Precautions Section.
2. Open the hood to the full open position.

➡ **Change the bolt mounting position of the hood hinge from the top mounting to the bottom.**

3. Remove the collector cover.
4. Properly relieve the fuel system pressure. Remove the fuel cap.
5. Properly disarm the SRS system.
6. Discharge the air conditioning system.
7. Disconnect the negative battery cable. Remove the battery.
8. Drain the engine oil. Drain the cooling system.
9. Remove the air intake duct, air cleaner case and air intake chamber.
10. Remove the engine front cover.
11. Remove the radiator.

➡ **Be sure to protect the condenser from damage.**

12. Remove the fuel hose bracket.
13. Position a suitable tool on the belt tension assembly mounting bolt. Rotate the tool clockwise and loosen the drive belt. Remove the drive belt. Remove the drive belt cover.

➡ **Upon installation make sure that the automatic belt tensioner indicator, indicates proper belt alignment.**

14. Disconnect the air conditioning pressure hoses from the compressor.
15. Disconnect the engine harness connector, ground cable, power steering switch connector, alternator connector and terminal and air conditioning compressor electrical connector.
16. Disconnect the brake booster hose and the heater hoses.
17. Remove the power steering pump with the bracket.

➡ **Do not disconnect the hoses from the pump body. Position the pump assembly aside, using wire.**

18. Remove the reservoir tank with the bracket.

➡ **Do not disconnect the hoses from the tank body.**

19. Remove the upper bolts which hold the vacuum pump bracket to the engine and transmission.
20. Raise and support the vehicle safely.
21. Remove the center exhaust pipe.

➡ **Do not let the front exhaust pipe interfere with the coolant pipes on the engine side. Remove the ground cable.**

22. Remove the lower bolts that hold the vacuum pump bracket to the engine and transmission. Remove the vacuum pump with the bracket.
23. Remove the lower transmission to engine retaining nuts.

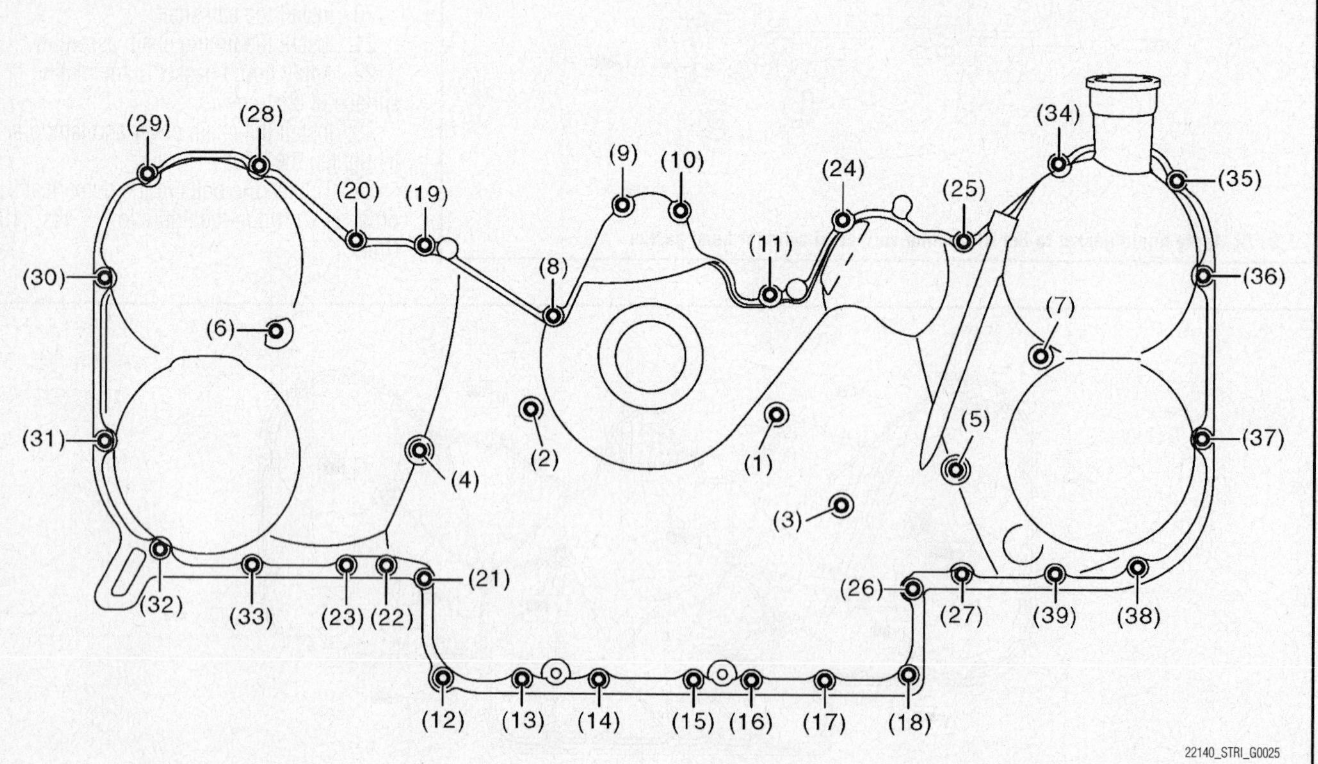

Fig. 81 Timing chain cover bolt tightening sequence

24. Remove the front cushion rubber onto front crossmember retaining nuts.

25. Lower the vehicle.

26. Remove the service plug hole cap. Remove the torque converter to drive plate retaining bolts.

27. Remove the pitching stopper. Disconnect the fuel hoses from the fuel pipes.

28. Position the engine lifting fixture in place. Support the transmission using a suitable jack.

➡ **Before separating the engine from the transmission, check to be sure nothing has been overlooked that will stop the engine from being removed. This is very important in order to facilitate reinstallation and because the transmission lowers under its own weight.**

29. Separate the engine from the transmission.

30. Remove the starter.

31. Install tool ST498277200, with will hold the torque converter in place.

32. Remove the upper side bolts that retain the transmission to the engine.

33. Slightly raise the engine. Raise the transmission, using the suitable jack. Move the engine horizontally until the main shaft is withdrawn from the clutch cover. Carefully remove the engine from the vehicle.

➡ **Be careful not to damage adjacent body parts or panels with the crank pulley, oil level gauge etc.**

34. Remove the front cushion rubber mounts.

To install:

35. Installation is the reverse of the removal procedure.

36. Torque the upper transmission to engine retaining bolts to 37 ft. lbs. (50 Nm).

37. Torque the lower transmission to engine retaining bolts to 37 ft. lbs. (50 Nm).

38. Tighten the torque converter to drive plate bolts to 18 ft. lbs. (25 Nm).

39. Fill the engine with the proper grade and type engine oil.

40. Fill and bleed the cooling system.

41. Charge the air conditioning system.

42. Check the automatic transmission fluid level, correct as required using the proper grade and type transmission fluid.

43. Start the engine and allow it to reach normal operating temperature. Check for leaks and correct as required.

3.6L Engine

See Figure 82.

1. Before servicing the vehicle, refer to the Precautions Section.

2. Set the vehicle on a lift.

3. Change the bolt mounting position from (A) to (B), then completely open the front hood.

4. Remove the collector cover.

5. Collect the refrigerant from A/C system.

6. Release the fuel pressure.

7. Remove the battery from vehicle.

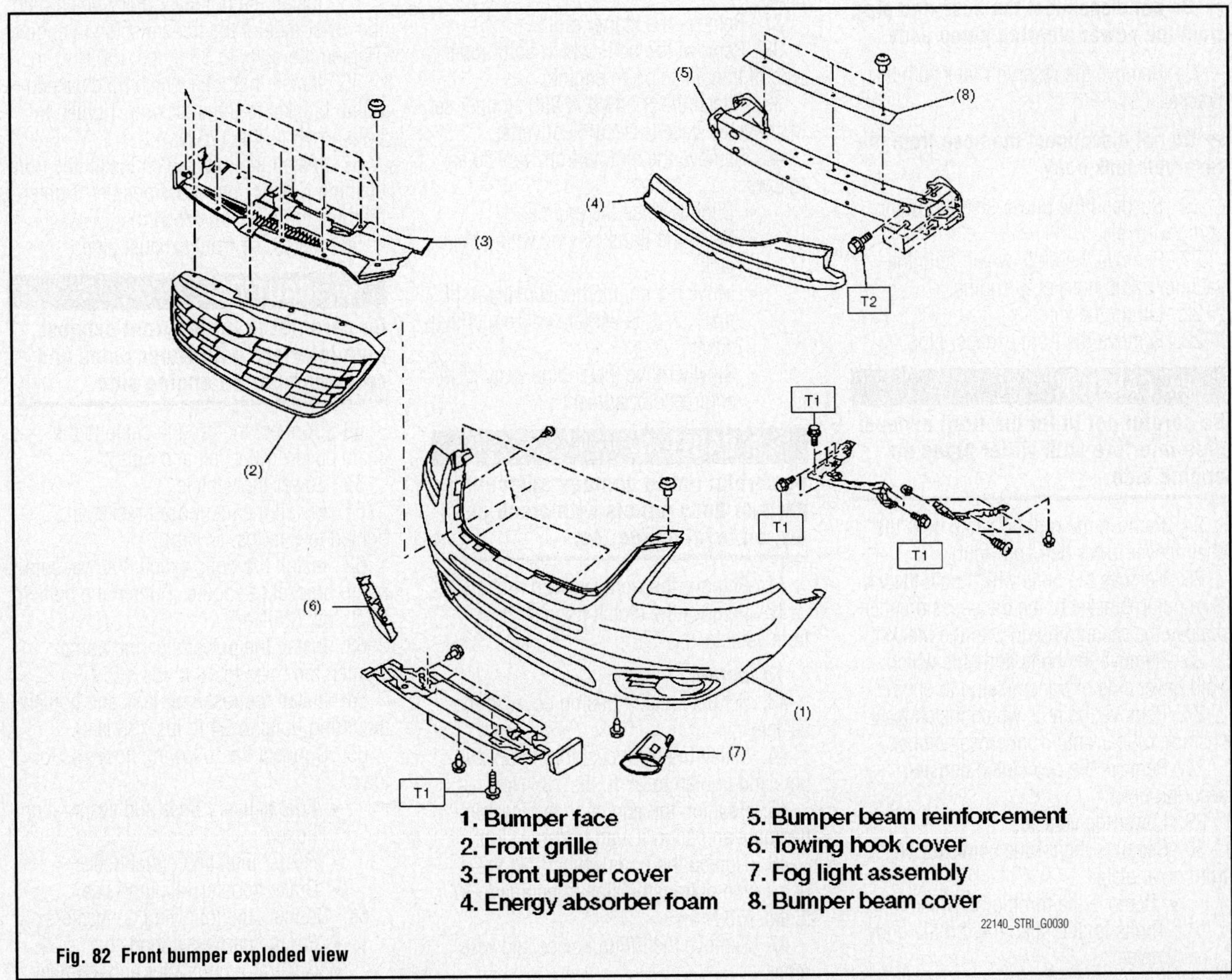

Fig. 82 Front bumper exploded view

8. Remove the air intake duct, air cleaner case and air intake chamber.
9. Remove the front upper cover.
10. Remove the front bumper.
11. Remove the radiator from the vehicle.

➡ **Protect the condenser so that it will not be damaged.**

12. Remove the fuel hose bracket.
13. Remove the V-belt.
14. Disconnect the A/C pressure hoses from the A/C compressor.
15. Remove the engine ground terminal.
16. Disconnect the following connectors:
   - Engine harness connector
   - Generator connector, terminal and A/C compressor connector
   - Power steering switch connector
17. Disconnect the following hoses:
   - Brake booster vacuum hose
   - Heater inlet and outlet hoses
18. Remove the power steering pump with the bracket.

➡ **Do not disconnect the hose and pipe from the power steering pump body.**

19. Remove the reservoir tank with the bracket.

➡ **Do not disconnect the hose from the reservoir tank body.**

20. Suspend the power steering pump using wire, etc.
21. Remove the bolts which hold the vacuum pump bracket to engine.
22. Lift up the vehicle.
23. Remove the front exhaust pipe.

⁂ CAUTION

Be careful not to let the front exhaust pipe interfere with water pipes on engine side.

24. Remove the ground cable from the chain cover lower (left and right).
25. Remove the bolts which hold the vacuum pump bracket to the transmission, then remove the vacuum pump with the bracket.
26. Remove the bolts and nuts which hold lower side of transmission to engine.
27. Remove the nuts which install front cushion rubber onto front crossmember.
28. Remove the two clutch housing securing bolts.
29. Lower the vehicle.
30. Separate the torque converter clutch from drive plate:
   - Remove the throttle body securing bolts to and slide the throttle body over.
   - Remove the service hole plug.
   - Remove the bolts which hold torque converter clutch to drive plate.
   - Remove other bolts while rotating the crankshaft using socket wrench.
31. Remove the pitching stopper mount.
32. Disconnect the fuel delivery hose and evaporation hose. Use quick connector release tool ST 42099AE000 to remove fuel hose.

⁂ CAUTION

Be careful not to spill fuel. Catch the fuel from hoses using a container or cloth.

33. Remove the radiator center bracket.
34. Support the engine with a lifting device and wire ropes.
35. Support the transmission with a garage jack. Doing this is very important to prevent the transmission from lowering due to its own weight.
36. Separation of engine and transmission.
37. Remove the starter motor.
38. Remove the bolts which hold upper side of transmission to engine.
39. Attach the ST 498277200 stopper set to converter case to secure converter.
40. Remove the engine from vehicle as follows:
   - Slightly raise the engine.
   - Raise the transmission with garage jack.
   - Move the engine horizontally until main shaft is withdrawn from clutch cover.
   - Slowly move the engine away from engine compartment.

⁂ CAUTION

Be careful not to damage adjacent parts or body panels with crank pulley, oil level gauge, etc.

41. Remove the front cushion rubbers.
42. Remove the clutch housing cover from vehicle.

To install:

43. Set the clutch housing cover to the vehicle.
44. Install the front cushion mount rubbers and tighten to 26 ft. lbs. (35 Nm).
45. Position the engine in engine compartment and align it with transmission.
46. Tighten the bolts which hold the upper side of transmission to engine to 37 ft. lbs. (50 Nm).
47. Remove the lifting device and wire ropes.
48. Remove the garage jack.
49. Install the radiator center bracket and tighten to 13 ft. lbs. (18 Nm).
50. Install the pitching stopper mount and tighten engine side bolt to 43 ft. lbs. (58 Nm). Tighten opposite end to 37 ft. lbs. (50 Nm).
51. Remove the ST 498277200 stopper set from converter case.
52. Install the starter.
53. Install the torque converter clutch to drive plate as follows:
   - Tighten the bolts which hold torque converter clutch to drive plate to 18 ft. lbs. (25 Nm).
   - Tighten other bolts to 18 ft. lbs. (25 Nm) while rotating the crankshaft using socket wrench.
   - Install the service hole plug to prevent foreign matter from being mixed.
   - Install the throttle body to intake manifold and tighten to 6 ft. lbs. (8 Nm).
54. Lift up the vehicle.
55. Install the bolts and nuts which hold the lower side of the transmission to engine. Tighten the bolts to 37 ft. lbs. (50 Nm).
56. Install the bolts which hold vacuum pump bracket to transmission. Tighten the bolts to 18 ft. lbs. (25 Nm).
57. Install the nuts which install the front cushion rubber onto crossmember. Tighten the nuts to 55 ft. lbs. (75 Nm).
58. Install the front exhaust pipe.

⁂ CAUTION

Be care not to let the front exhaust pipe interfere with water pipes and crossmember on engine side.

59. Connect the ground cable to the chain cover lower (left and right).
60. Lower the vehicle.
61. Install the fuel hose bracket and tighten to 5 ft. lbs. (7 Nm).
62. Install the bolts which hold vacuum pump bracket to engine. Tighten the bolts to 18 ft. lbs. (25 Nm)
63. Install the power steering pump. Tighten the bolts to 18 ft. lbs. (25 Nm).
64. Install the reservoir tank and tighten mounting bolts to 24 ft. lbs. (33 Nm).
65. Connect the following hoses as follows:
   - Fuel delivery hose and evaporation hose
   - Heater inlet and outlet hoses
   - Brake booster vacuum hose
66. Connect the following connectors:
   - Engine harness connector
   - Generator connector and terminal

- A/C compressor connector
- Power steering switch connector

67. Connect the engine ground terminal and tighten to 10 ft. lbs. (14 Nm).
68. Install the A/C pressure hoses.
69. Install the V-belt.
70. Install the radiator to vehicle.
71. Install the front bumper.
72. Install the fuse of fuel pump to the main fuse box.
73. Install the battery to vehicle.
74. Refill engine coolant and bleed system.
75. Check the ATF level and replenish it if necessary.
76. Vacuum and Charge the A/C system with refrigerant.
77. Install the front upper cover.
78. Install the air intake duct, air cleaner case and air intake chamber.
79. Install the collector cover.
80. Change the bolt mounting position from (B) to (A), then close the front hood.
81. Lower the vehicle from the lift.
82. Test drive vehicle and check for any leaks repair as needed.

EXHAUST MANIFOLD

REMOVAL & INSTALLATION

Due to the unique design of the Subaru engine, an exhaust manifold is not used. The exhaust enters directly into the front Y-pipe from the cylinder heads.

Front Y-pipe

See Figure 83.

1. Before servicing the vehicle, refer to the Precautions Section.
2. Remove or disconnect the following:
   - Negative battery cable
   - Air cleaner case, if necessary
   - Front Oxygen Sensor (O_2S)
   - Front undercover
   - Rear O_2S electrical connector
   - Y-pipe-to-rear pipe mounting nuts and separate the Y-pipe from the rear pipe
   - Bolts that secure the front Y-pipe to the cylinder head
   - Y-pipe from the hanger bracket
   - Front exhaust pipe from the catalytic converter and discard the gaskets

To install:

3. Clean all gasket surfaces completely.
4. Install or connect the following:
   - New gaskets
   - Front catalytic converter to front exhaust pipe

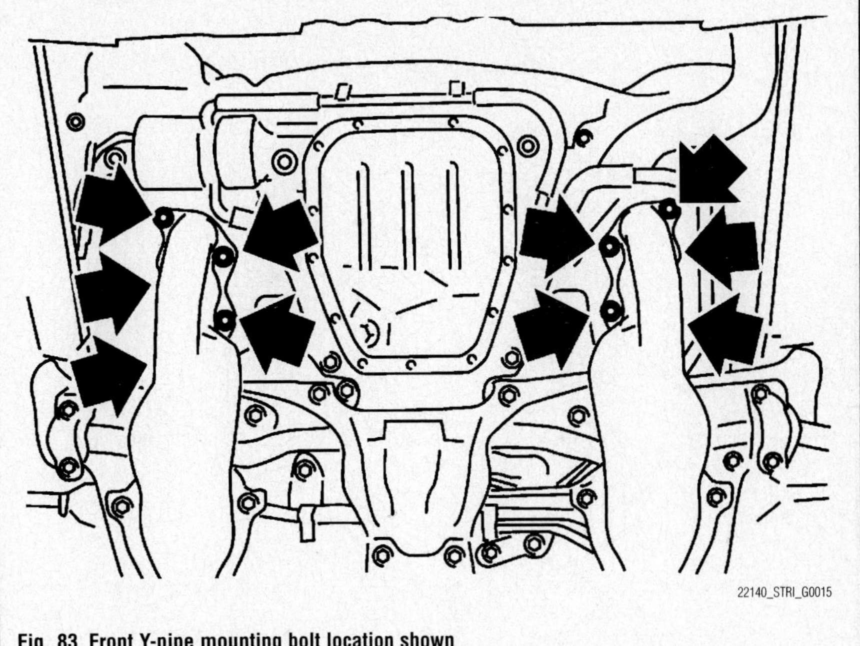

Fig. 83 Front Y-pipe mounting bolt location shown

   - Y—pipe. Temporarily tighten the bolt that holds the center exhaust pipe to the hanger bracket.
   - Y-pipe, to the cylinder head. Tighten the mounting bolts to 52 ft. lbs. (70 Nm).
   - Y-pipe to the rear pipe. Tighten the retainers.
5. Tighten the center exhaust pipe to hanger bracket.
   - Rear O_2S electrical connector
   - Front O_2S electrical connector
   - Front undercover, if equipped
   - Air cleaner case, if removed
   - Negative battery cable
6. Start the engine and check for exhaust leaks.

FLEXPLATE

REMOVAL & INSTALLATION

1. Before servicing the vehicle, refer to the Precautions Section.
2. Separate the torque converter from drive plate.
3. Remove the transmission assembly.
4. Remove flexplate bolts in a star sequence.

To install:

5. Clean flexplate bolts install thread locker.
6. Install flexplate hand tighten bolts.
7. Tighten flexplate mounting bolts in a star pattern to 60 ft. lbs. (81 Nm).
8. Install transmission assembly.
9. Carefully install the torque converter to drive plate.
10. Install the bolts which hold the torque converter to drive plate and tighten to 18 ft. lbs. (25 Nm).

INTAKE MANIFOLD

REMOVAL & INSTALLATION

3.0L Engine

See Figure 84.

1. Before servicing the vehicle, refer to the Precautions Section.
2. Properly relieve the fuel system pressure. Remove the fuel cap.
3. Disconnect the negative battery cable. Drain the engine coolant.
4. Remove the air cleaner case and air intake chamber.
5. Remove the alternator.
6. Disconnect the electrical connector from the throttle body. Disconnect the engine coolant hoses from the throttle body.
7. Disconnect the engine harness connector. Disconnect the PCV hose. Disconnect the brake booster hose. Disconnect the fuel hoses from the fuel pipe.
8. Remove the left side fuel line protector. Remove the engine harness from the left side fuel injector pipe. Remove the bolts which hold the fuel injector pipe to the left cylinder head.
9. Remove the right side fuel line protector. Remove the engine harness from the right side fuel injector pipe. Remove the bolts which hold the fuel injector pipe to the right cylinder head.

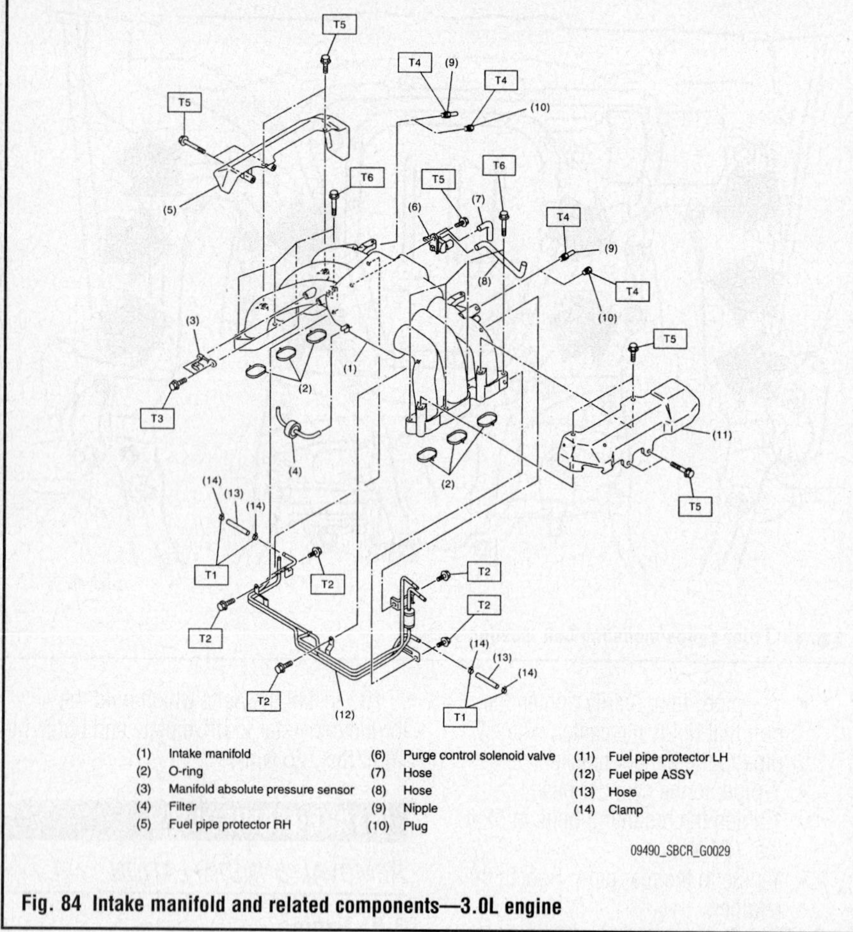

Fig. 84 Intake manifold and related components—3.0L engine

10. Remove the right and left intake manifold retaining bolts. Remove the intake manifold from the engine.

To install:

11. Installation is the reverse of the removal procedure.
12. Be sure to use new intake manifold O-rings. Torque the manifold retaining bolts to specification and in alternating sequence.
13. Be sure to fill the cooling system with the proper grade and type engine coolant.
14. Start the engine and check for leaks, correct as required.

3.6L Engine

See Figures 85 and 86.

1. Before servicing the vehicle, refer to the Precautions Section.
2. Remove the collector cover.
3. Release the fuel pressure.
4. Disconnect the ground cable from the battery.
5. Open the fuel filler flap lid, and remove the fuel filler cap.
6. Remove the air cleaner case and air intake chamber.
7. Remove the generator.
8. Disconnect the connector from the throttle body.
9. Disconnect the engine harness connector.
10. Disconnect the engine coolant hoses from the throttle body.
11. Disconnect the PCV hose.
12. Disconnect the brake booster hose.
13. Disconnect the fuel hoses from the fuel pipe.
    a. Disconnect the connector of fuel pipe by pushing the ST in the direction of arrow.
    b. Remove the clip, and disconnect the evaporation hose from the pipe.

CAUTION

Be careful not to spill fuel and catch the fuel from hoses using a container or cloth.

14. Remove the fuel pipe protector LH.
15. Remove the engine harness from the fuel injector pipe LH.
16. Disconnect the connectors of the oil temperature sensor, engine coolant temperature sensor and intake oil flow control solenoid valve LH.
17. Remove the bolts which hold fuel injector pipe LH to cylinder head.
18. Disconnect the connector of intake oil flow control solenoid valve RH.
19. Remove the fuel pipe protector RH.
20. Remove the engine harness from the fuel injector pipe RH.
21. Remove the bolts which hold the fuel injector pipe RH to the cylinder head.
22. Remove the bolts which hold the intake manifold to the cylinder head.
23. Remove the EGR pipe fixing bolt.
24. Remove the intake manifold.

To install:

25. Use new O-rings and install the intake manifold onto cylinder heads.
26. Install and tighten the intake mounting bolts to 18.4 ft. lbs. (25 Nm).

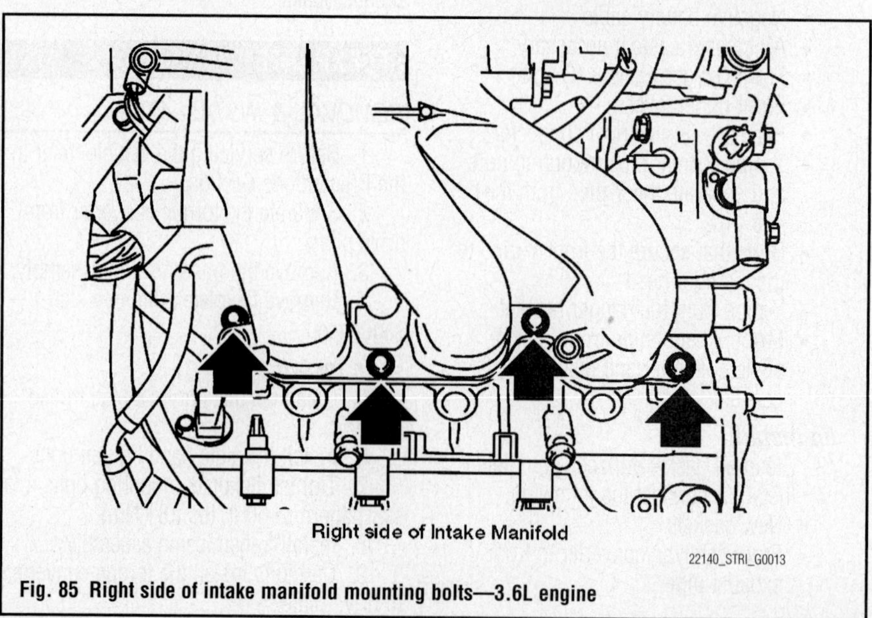

Fig. 85 Right side of intake manifold mounting bolts—3.6L engine

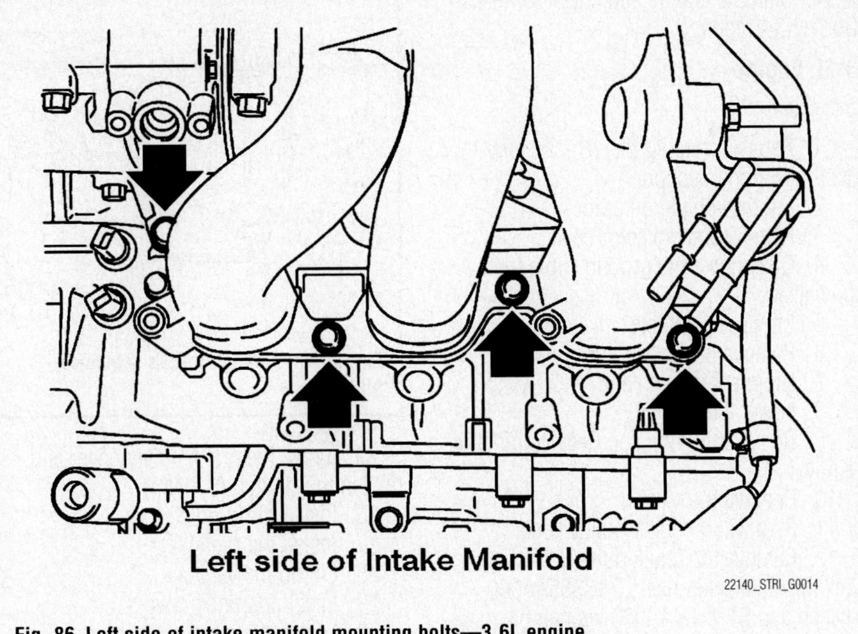

Fig. 86 Left side of intake manifold mounting bolts—3.6L engine

27. Use new gasket and install the EGR pipe fixing bolt. Tighten the bolt to 4.7 ft. lbs. (6.4 Nm).
28. Use new O-rings and install the bolts which hold fuel injector pipe RH to cylinder head. Tighten the bolts to 14 ft. lbs. (19 Nm).
29. Install the engine harness to fuel injector pipe RH.
30. Install the fuel pipe protector RH. Tighten the bolts to 14 ft. lbs. (19 Nm).
31. Connect the connector of intake oil flow control solenoid valve RH.
32. Install the bolts which hold fuel injector pipe LH to cylinder head.
33. Use new O-rings and install the bolts which hold fuel injector pipe LH to cylinder head. Tighten the bolts to 14 ft. lbs. (19 Nm).
34. Connect the connectors of the oil temperature sensor, engine coolant temperature sensor and intake oil flow control solenoid valve.
35. Install the engine harness to fuel injector pipe LH.
36. Install the fuel pipe protector LH. Tighten the bolts to 14 ft. lbs. (19 Nm).
37. Connect the fuel hoses to fuel pipe.
38. Connect the brake booster hose.
39. Connect the PCV hose.
40. Connect the engine coolant hoses to throttle body.
41. Connect the connector to throttle body.
42. Connect the engine harness connector.
43. Install the generator.
44. Install the air cleaner case and air intake chamber.
45. Install the fuse of fuel pump to the main fuse box.
46. Connect the battery ground cable to the battery.
47. Install the collector cover.

OIL PAN

REMOVAL & INSTALLATION

See Figures 87 and 88.

➡ **If removing the upper oil pan, the engine must first be removed from the vehicle.**

1. Before servicing the vehicle, refer to the Precautions Section.
2. Disconnect the negative battery cable.
3. Raise and support the vehicle safely.
4. Remove the undercover. Drain the engine oil.
5. Remove the lower oil pan retaining bolts.
6. Insert an oil pan gasket cutter tool into the gap between the upper oil pan and the lower oil pan. Remove the lower oil pan from the engine.

➡ **Do not use a screwdriver or similar tool in place of the cutter tool.**

7. Remove the oil strainer, if required.

To install:

8. Be sure to clean the old gasket material from the mating surfaces.
9. Replace the O-ring and install the oil strainer. Tighten the bolt to 4.7 ft. lbs. (6.4 Nm).

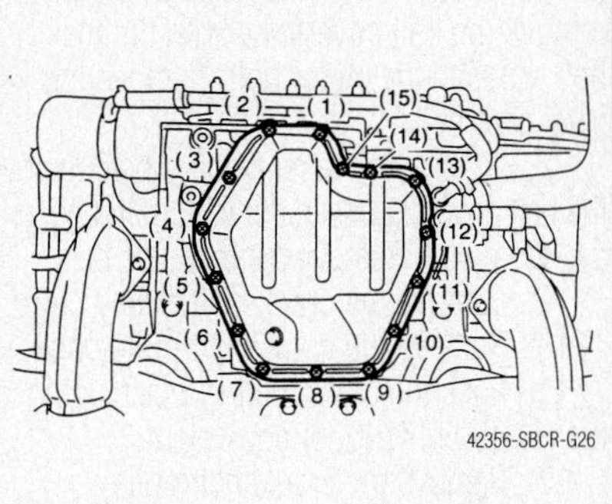

Fig. 87 Oil pan bolt torque sequence—3.0L engine

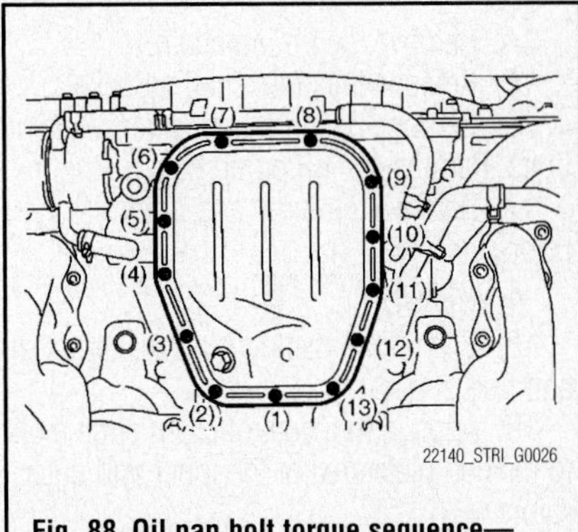

Fig. 88 Oil pan bolt torque sequence—3.6L engine

10. Apply a continuous bead (0.039 inch thick) of liquid gasket, part number K0877YA018 or equivalent, to the mating surfaces. Install the oil pan.
11. Tighten the oil pan lower installing bolts in the numerical order as shown in the figure to 5 ft. lbs. (6.4 Nm).
12. Continue the installation in the reverse order of the removal procedure.
13. Be sure to fill the engine with the correct grade and type engine oil.
14. Start the engine and check for leaks. Correct as required.

OIL PUMP

REMOVAL & INSTALLATION

3.0L Engine

See Figure 89.

1. Before servicing the vehicle, refer to the Precautions Section.
2. Disconnect the negative battery cable. Drain the cooling system.
3. Remove the collector cover.
4. Raise and support the vehicle safely.
5. Remove the undercover.
6. Lower the vehicle. Remove the radiator.

7. To remove the front side V belt, remove the belt covers. Loosen the lock bolt. Loosen the slider bolt. Remove the front side belt.

8. To remove the rear side V belt, remove the belt covers. Loosen the lock bolt. Loosen the slider bolt. Remove the rear side belt.

9. Remove the crankshaft pulley cover.

10. Remove the crankshaft pulley bolt.

11. Lock the crankshaft in place using tool ST499977100, or equivalent.

12. Remove the crankshaft pulley.

13. Remove the front timing chain cover.

➡ Bolts are three different sizes. Be careful to install the correct bolt in the correct hole.

14. Remove the timing chain.

15. Remove the crankshaft sprocket.

16. Remove the oil pump cover retaining bolts. Remove the oil pump cover.

17. Remove the inner and outer rotors.

To install:

18. Be sure all mating surfaces are clean and free of dirt.

19. Apply a thin coat of clean engine oil to the complete area of the inner and outer rotors.

20. Position the inner rotor in place. Position the outer rotor in place.

21. Install the pump cover. Tighten the retaining bolts to 60 inch lbs. (6 Nm) and in the proper sequence.

➡ Make sure that the bolts are installed in the correct positions.

22. Continue the installation in the reverse order of the removal procedure.

23. Be sure to fill the cooling system and engine oil with the proper grade oil and type coolant.

24. Start the engine and check for leaks. Correct, as required.

3.6L Engine

See Figure 90.

1. Before servicing the vehicle, refer to the Precautions Section.

2. Set the vehicle on a lift.

3. Remove the collector cover.

4. Disconnect the ground cable from the battery.

5. Lift up the vehicle.

6. Remove the undercover.

7. Drain engine coolant.

8. Lower the vehicle.

9. Remove the radiator. Refer to radiator removal.

10. Remove the V-belt.

11. Remove the crank pulley cover.

12. Remove the crank pulley bolt. To lock the crankshaft, use ST 18355AA000 wrench and ST 1834AA000 wrench pin set.

13. Remove the chain cover.

14. Remove the timing chain. Refer to timing chain removal.

15. Remove the oil pan.

16. Remove the oil pump.

17. Remove the O-ring from the oil pan upper.

To install:

18. Apply the engine oil to the O-ring and install it to the oil pan upper.

19. Install the oil pump and tighten mounting bolts to 10 ft. lbs. (13 Nm).

20. Install the oil pan lower.

21. Tighten the oil pan lower installing bolts in the numerical order as shown in the figure to 5 ft. lbs. (6.4 Nm).

22. Install the timing chain.

23. Install the chain cover.

24. Install the crank pulley and tighten to 144 ft. lbs. (195 Nm).

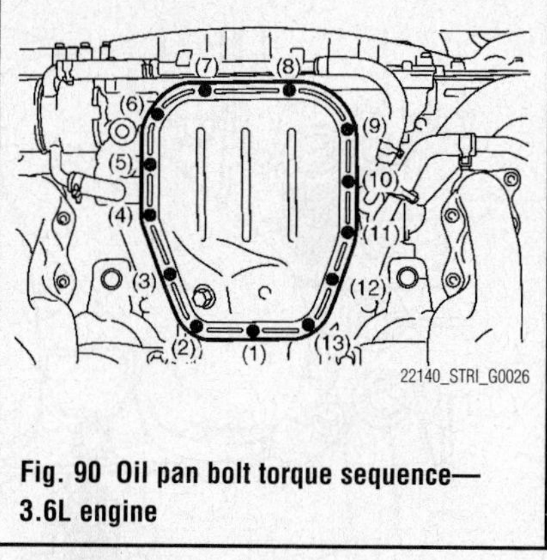

Fig. 90 Oil pan bolt torque sequence—3.6L engine

25. Install the crank pulley cover and tighten to 5 ft. lbs. (6.4 Nm).

26. Install the V-belt.

27. Install the radiator.

28. Lift up the vehicle.

29. Install the undercover.

30. Lower the vehicle.

31. Be sure to fill the cooling system and engine oil with the proper grade oil and type coolant.

32. Start the engine and check for leaks. Correct, as required.

INSPECTION

Inspect the oil pump exterior for the following:

- Cracks, damage or oil leakage
- Play of pulley shaft

Inspect the oil pump interior for the following:

- Faulty or seized pump vane
- Bends in the shaft or damage to bearing
- Visually check the strainer part for clogging

PISTON AND RING

POSITIONING

See Figures 91 through 93.

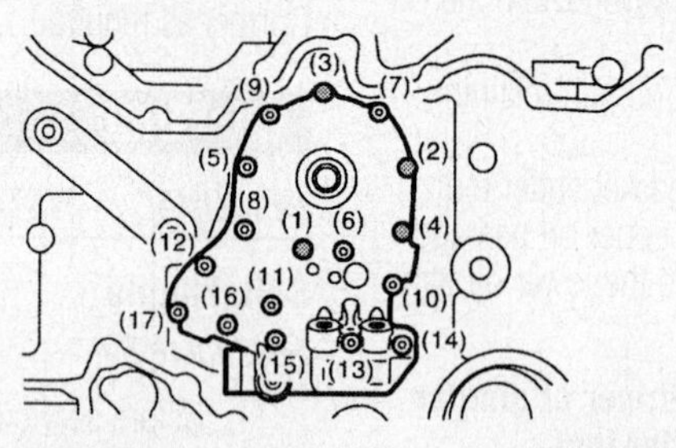

Bolt installing position	Bolt dimension
(1) and (3)	6 x 14 x 14
(2) and (4)	6 x 35 x 18
(5), (6), (7), (8), (9), (10) and (11)	6 x 35 x 15
(12), (15), (16) and (17)	6 x 16 x 16
(13) and (14)	6 x 26 x 15

09490_SBCR_G0059

Fig. 89 Oil pump tightening sequence and bolt location—3.0L engine

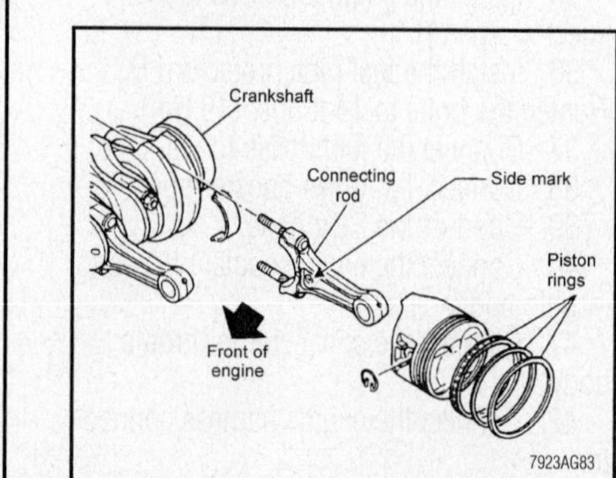

Fig. 91 Piston and connecting rod assembly positioning

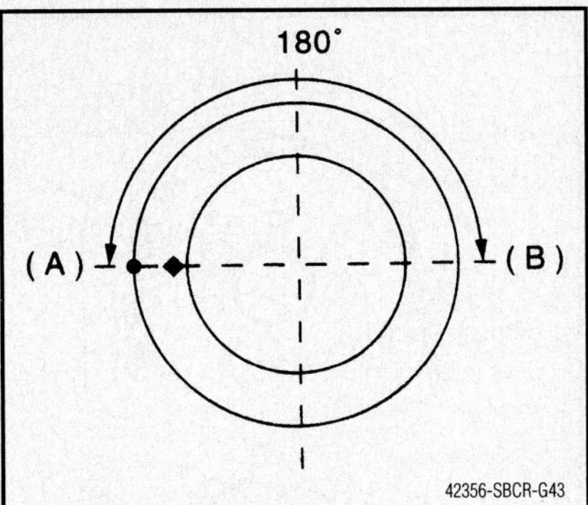

Fig. 92 Top ring end-gap (A), second ring gap (B)

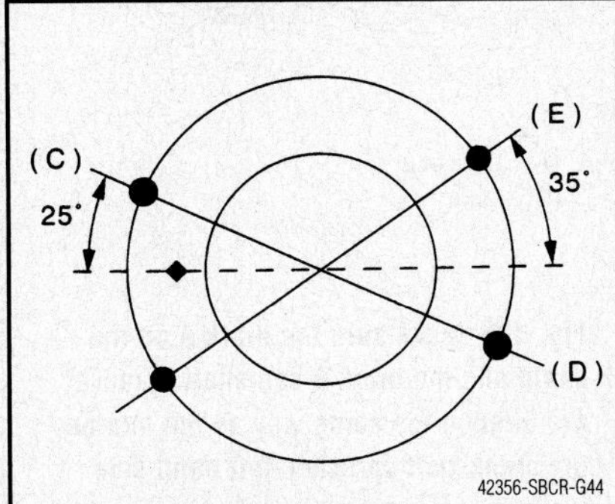

Fig. 93 Upper rail end-gap (C), expander gap (D) and lower rail gap (E)

REAR MAIN SEAL

REMOVAL & INSTALLATION

See Figure 94.

1. Before servicing the vehicle, refer to the Precautions Section.
2. Remove or disconnect the following:
   - Engine from the vehicle
   - Clutch assembly/flywheel using the Clutch Disc Guide tool 499747000, if equipped with a manual transmission

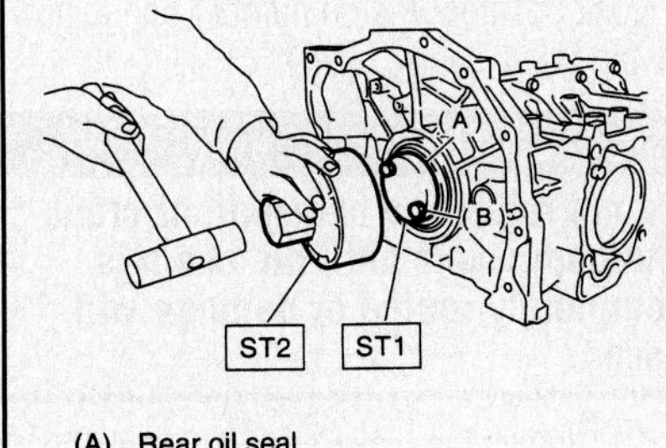

Fig. 94 Installing the rear main seal using oil seal guide ST1 499597100 and ST2 499598200

   - Torque converter flexplate from the crankshaft, if equipped with an automatic transmission
   - Oil seal from the cylinder block using a small prybar

To install:

3. Install or connect the following:
   - New oil seal by pressing it into the cylinder block using the appropriate driver and hammer
   - Flywheel housing using new gaskets and sealant where necessary.
   - Flywheel and tighten the bolts to specification.
   - Engine

TIMING CHAIN, SPROCKETS, FRONT COVER AND SEAL

REMOVAL & INSTALLATION

3.0L Engine

See Figures 95 through 108.

1. Before servicing the vehicle, refer to the Precautions Section.
2. Disconnect the negative battery cable.
3. Remove the crankshaft pulley cover.
4. Remove the crankshaft pulley bolt.
5. Lock the crankshaft in place using tool ST499977100, or equivalent.

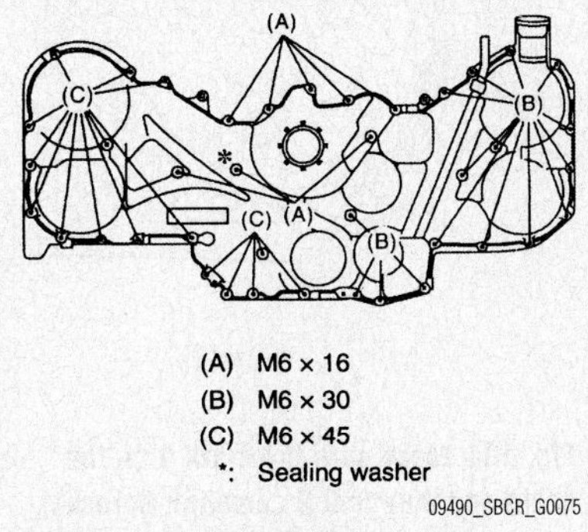

Fig. 95 Front cover bolt sizes and locations—3.0L engine

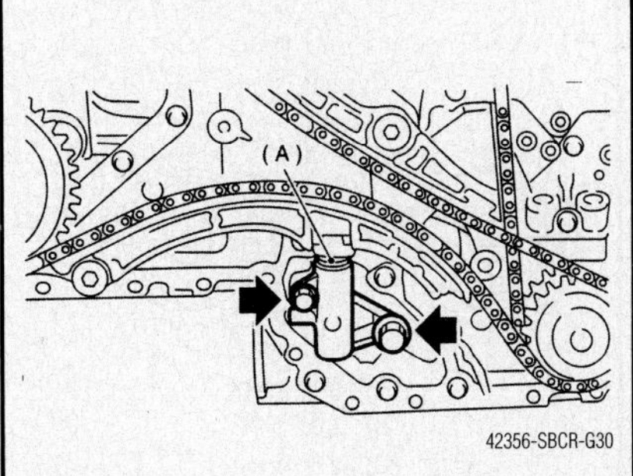

Fig. 96 The chain tensioner plunger A does not come out—right hand side

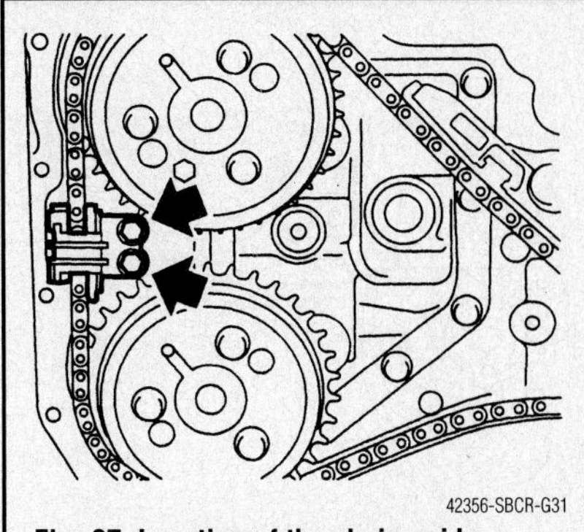

Fig. 97 Location of the chain guide between the cams—right hand side

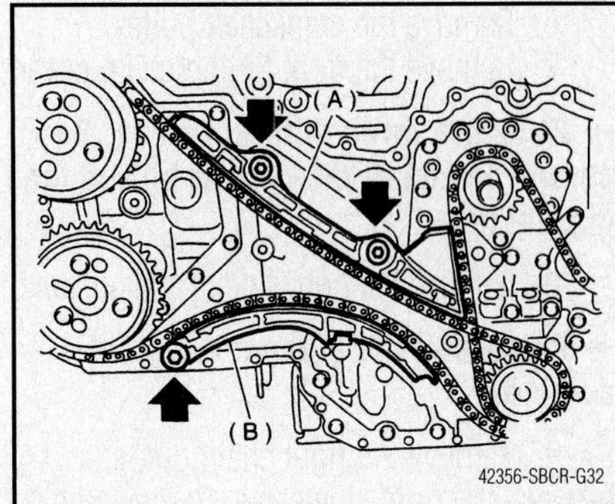

Fig. 98 Location of the chain guide—right hand side

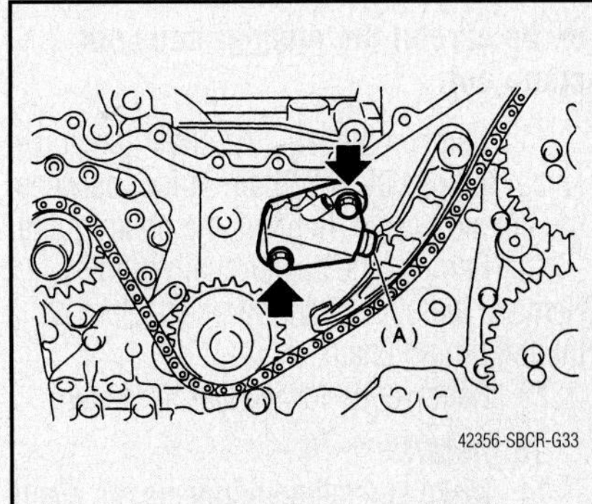

Fig. 99 The chain tensioner plunger A does not come out—left hand side

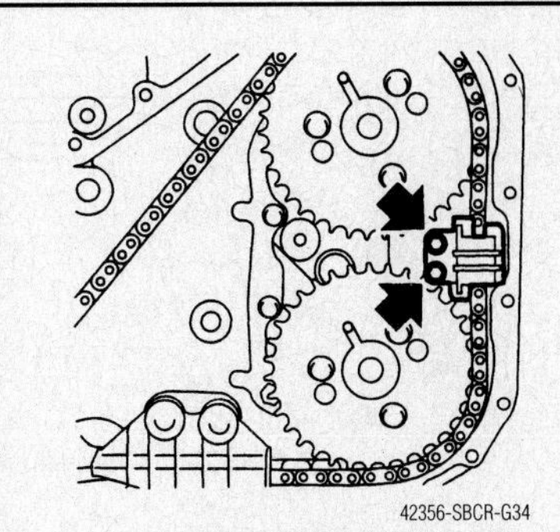

Fig. 100 Location of the chain guide between the cams—left hand side

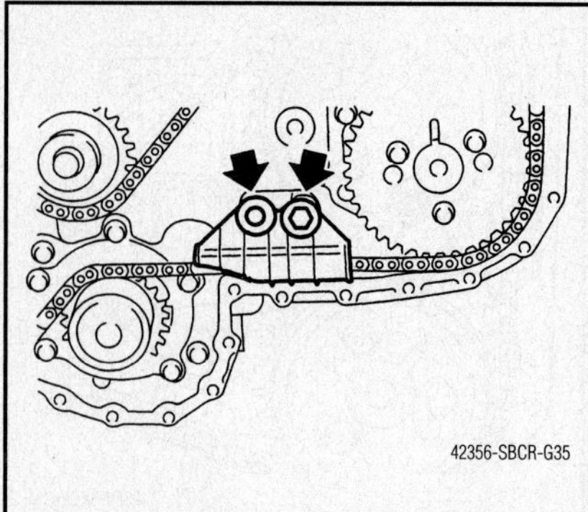

Fig. 101 Location of the chain guide—left hand side

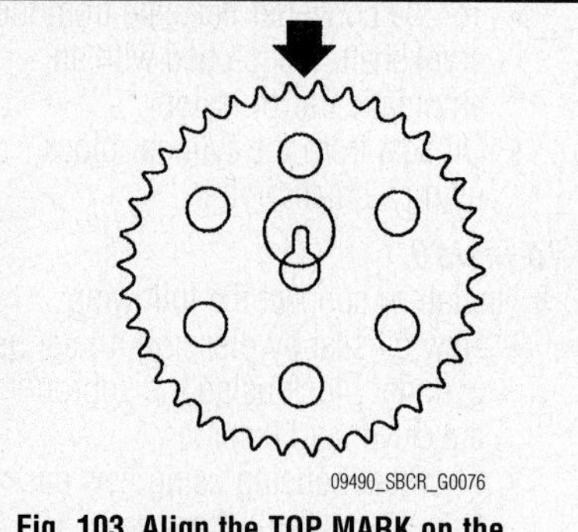

Fig. 103 Align the TOP MARK on the camshaft sprocket to the 12 o'clock position

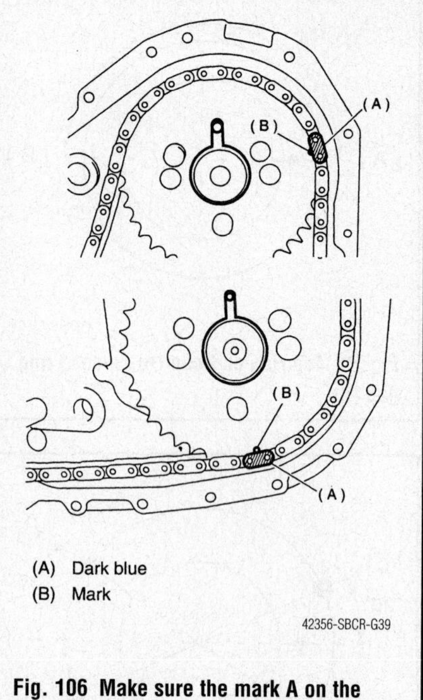

Fig. 106 Make sure the mark A on the chain and the mark B camshaft sprocket are aligned the same way as the one on the crankshaft sprocket–left hand side

6. Remove the crankshaft pulley.
7. Remove the front timing chain cover.

➡ Bolts are three different sizes. Be careful to install the correct bolt in the correct hole.

8. Remove the right side chain tensioner.

➡ Be careful the plunger does not come out.

9. Remove the right chain guide, between the right side cams. Remove the right side chain guide. Remove the right side chain tensioner lever. Remove the right timing chain.
10. Remove the left side chain tensioner.

➡ Be careful the plunger does not come out.

11. Remove the left side chain tensioner lever. Remove the left chain guide, between the left side cams. Remove the chain guide.
12. Remove the center chain guide. Remove the upper idler sprocket. Remove the left timing chain.
13. Remove the lower idler sprocket.

To install:

14. Make sure all components are clean. Apply oil to the chain guide, tensioner lever and idler sprockets.

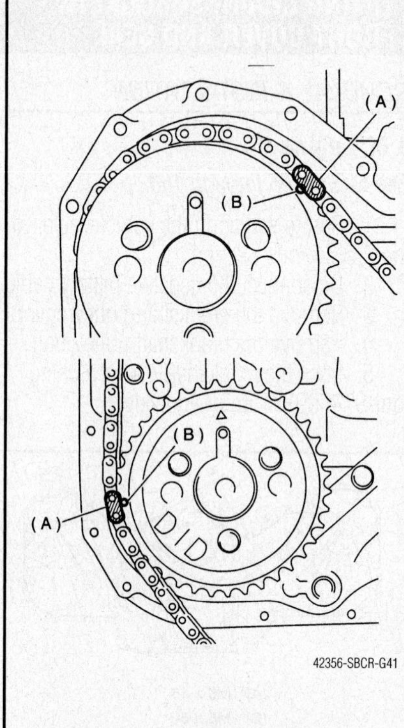

Fig. 104 Make sure the mark A on the chain and the mark B camshaft sprocket are aligned the same way as the one on the crankshaft sprocket–right hand side

15. Place the screw, spring, pin and tension rod into the tensioner body.
16. While pressing the tensioner onto a rubber mat, twist it to the left and right to shorten the rod. Place a thin pin into the holes between the rod and body to hold it in place. Always perform this task on a rubber mat.
17. Using the crankshaft socket tool, align the **TOP MARK** on the crankshaft sprocket to the 9 o'clock position, as shown in the illustration.
18. Align the key groove on the exhaust camshaft sprocket to the 12 o'clock position, as shown in the illustration.
19. Align the intake camshaft sprocket, as shown in the illustration.
20. Turn the crankshaft sprocket clockwise; align the **TOP MARK** to the 12 o'clock position. Piston number one is now at Top Dead Center (TDC).

⁂ CAUTION

Do not rotate the camshaft or crankshaft sprockets until the chain is completely routed or damage will occur.

21. Install the lower idler sprocket and tighten the bolt to 51 ft. lbs. (69 Nm).
22. Install the left side timing chain, align the mark "B" on the crankshaft sprocket with the matching mark "A" on the timing chain.

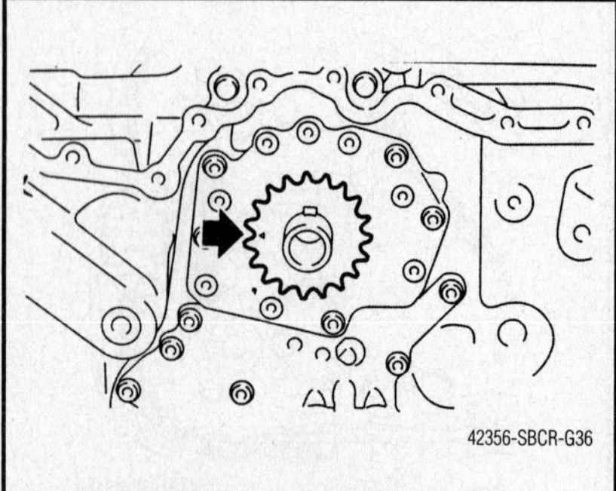

Fig. 102 Align the TOP MARK on the crankshaft sprocket to the 9 o'clock position

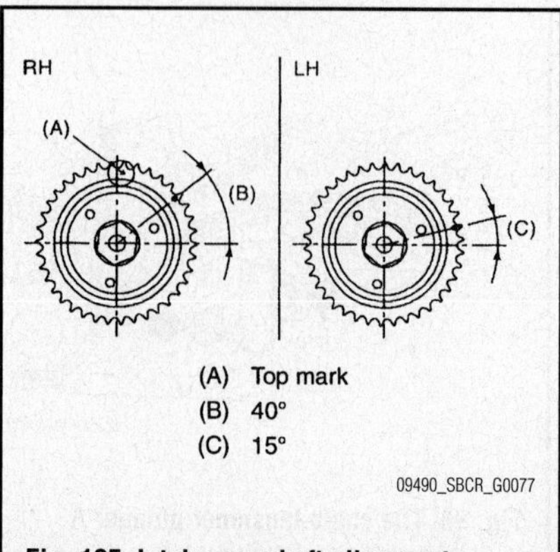

Fig. 105 Intake camshaft alignment

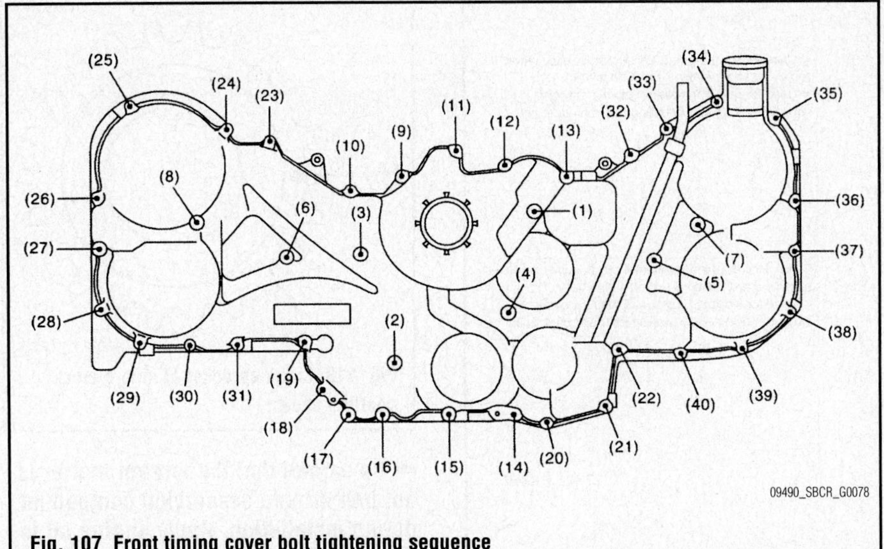

Fig. 107 Front timing cover bolt tightening sequence

23. Route the left side timing chain onto the lower idler sprocket, water pump, exhaust cam sprocket and the intake cam sprocket in that order.

24. Make sure the mark "A" on the chain and the mark "B" camshaft sprocket are aligned the same way as the one on the crankshaft sprocket or damage will occur.

25. Install the upper chain idler and tighten the bolt to 51 ft. lbs. (69 Nm).

26. Install the left side chain guide, between the camshafts. Tighten the bolt to 60 inch lbs. (6 Nm) using a NEW bolt.

27. Install the left side chain guide. Tighten the bolts to 12 ft. lbs. (16 Nm). Install the tensioner lever on the left side and tighten the bolt to 12 ft. lbs. (16 Nm). Install the chain tensioner on the left side and tighten the bolts to 12 ft. lbs. (16 Nm).

28. Install the right side timing chain. Align the marks of the left and right timing chains on the lower idler sprocket. Install the right side timing chain to the right side intake camshaft sprocket and the right side exhaust camshaft sprocket in this order.

29. Make sure the mark "A" on the chain and the mark "B" camshaft sprocket are aligned the same way as the one on the crankshaft sprocket or damage will occur.

30. Install the right side chain guide. Install the right side chain tensioner lever and tighten the bolts to 12 ft. lbs. (16 Nm). Install the right side chain guide and tighten the NEW bolt to 60 inch lbs. (6 Nm). Install the right side chain tensioner and tighten the bolts to 12 ft. lbs. (16 Nm).

31. Adjust the clearance between the chain guide on the right side and the center chain guide so that there is range between 0.331–0.339 inch (8.4–8.6mm).

32. Install the center chain guide and tighten the NEW bolt to 72 inch lbs. (8 Nm).

33. Check the match marks on each sprocket and corresponding timing chain are correct, remove the stopper from the tensioner.

34. Clean the mating surfaces on the front timing cover. Apply a bead of liquid gasket 0.020 inch in diameter to the mating surface of the front timing chain cover. Install the timing chain cover. Torque the retaining bolts to 60 inch lbs. and in the proper sequence.

35. Continue the installation in the reverse order of the removal procedure.

3.6L Engine

See Figures 109 through 122.

1. Before servicing the vehicle, refer to the Precautions Section.
2. Drain the engine oil.
3. Remove the radiator.
4. Remove the V-belt.
5. Remove the crank pulley.
6. Remove the bolts which hold oil cooler pipe to chain cover.
7. Remove the chain cover.

➡ **Chain cover installation bolt has three different sizes. To prevent the confusion in installation, keep these bolts on container individually.**

8. Remove the chain tensioner RH.

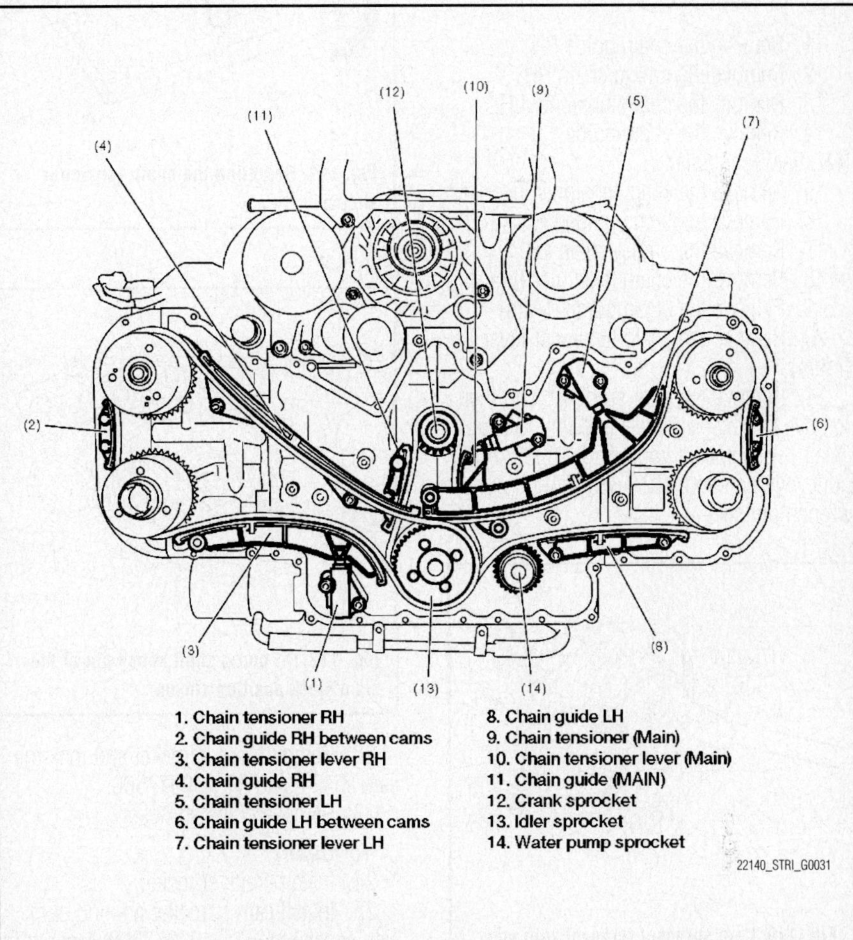

Fig. 108 Timing chain assembly parts location—3.6L engine

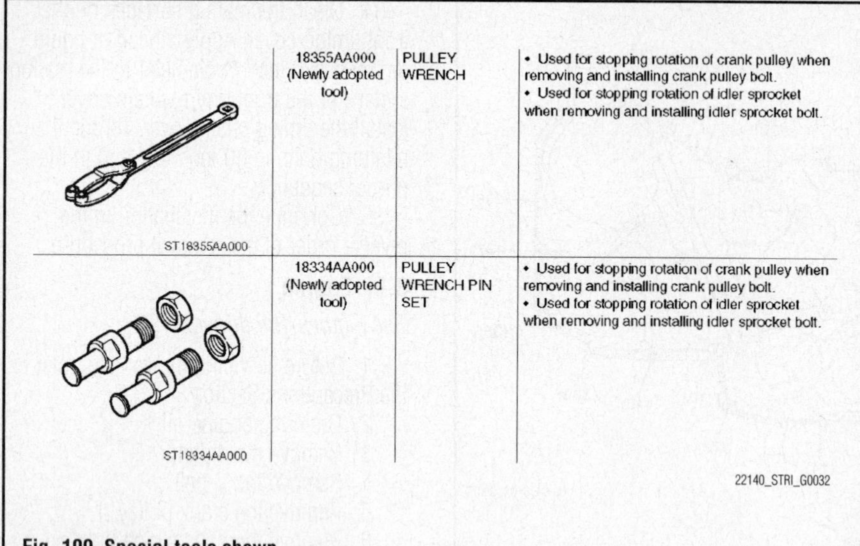

Tool	Tool number	Tool name	Remarks
ST18355AA000	18355AA000 (Newly adopted tool)	PULLEY WRENCH	• Used for stopping rotation of crank pulley when removing and installing crank pulley bolt. • Used for stopping rotation of idler sprocket when removing and installing idler sprocket bolt.
ST18334AA000	18334AA000 (Newly adopted tool)	PULLEY WRENCH PIN SET	• Used for stopping rotation of crank pulley when removing and installing crank pulley bolt. • Used for stopping rotation of idler sprocket when removing and installing idler sprocket bolt.

Fig. 109 Special tools shown

➡ During removal of chain tensioner RH from chain tensioner lever RH, press the plunger by hand to prevent it from popping out.

9. Remove the chain guide (RH: between cams).
10. Remove the chain tensioner lever RH.
11. Remove the chain guide RH.
12. Remove the timing chain RH.
13. Remove the chain tensioner LH.
14. Remove the chain guide (LH: between cams).
15. Remove the chain tensioner lever LH.
16. Remove the chain guide LH.
17. Remove the timing chain LH.
18. Remove the chain tensioner (Main).
19. Remove the chain guide (Main).
20. Remove the chain tensioner lever (Main).
21. Remove the idler sprocket, and then remove the idler sprocket and timing chain (Main). Use special tools ST1-18355AA000 pulley wrench and ST2-18354AA000 pulley wrench pin set.

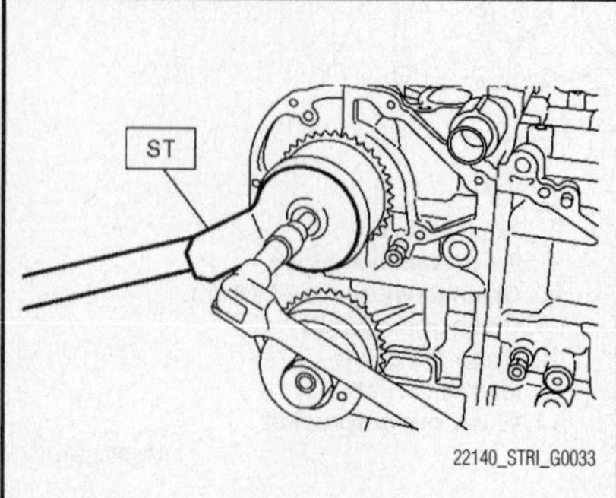

Fig. 110 Cam sprocket removal with special tool ST-499977500 shown

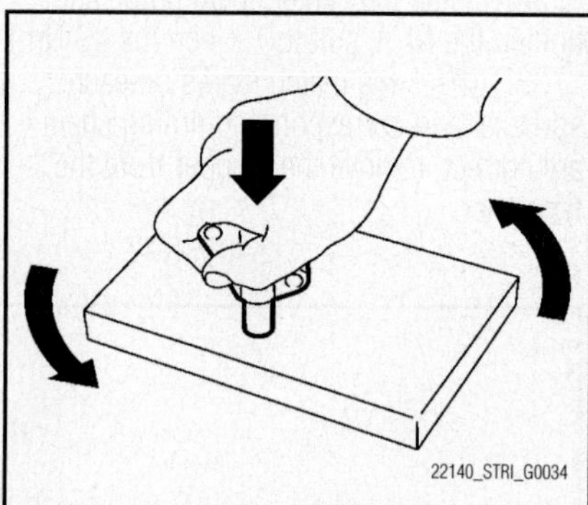

Fig. 111 Resetting the chain tensioner shown

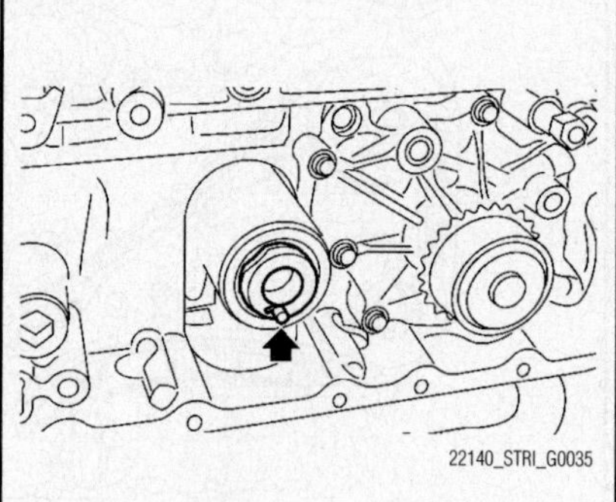

Fig. 112 Oil pump shaft knock pin at the six o'clock position shown

22. Remove cam sprocket and lock the cam shaft using ST-499977500
23. Remove crank sprocket.

To install:

24. Install crank sprocket.
25. Install cam sprocket holding cam with special tool ST-499977500. Tighten the bolt to 22 ft. lbs. (29 Nm).

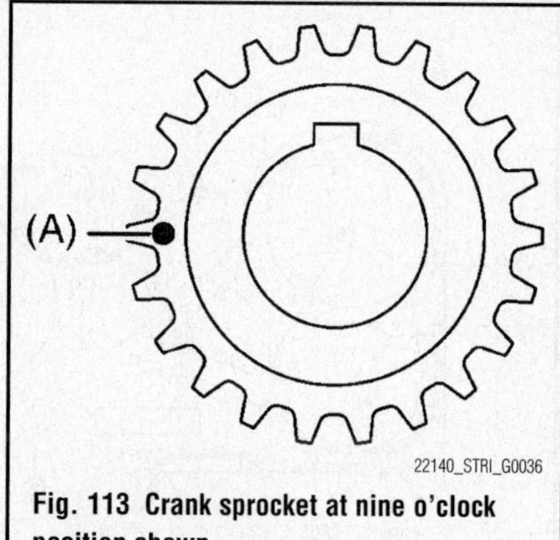

Fig. 113 Crank sprocket at nine o'clock position shown

➡ Be careful that the foreign matter is not into or onto assembled component during installation. Apply engine oil to the all timing chain components.

26. The preparation for chain tensioner installation is as follows:
    - Insert the screw, spring pin and plunger into tensioner body.
    - Rotate the rubber mat counterclockwise while pressing the chain tensioner from upper side by hand as shown in the figure.
    - Insert the stopper pin into the chain tensioner body hole.
27. Set the oil pump shaft knock pin at the six o'clock position as shown in the figure.
28. Using the ST, align the "Top mark" on crank sprocket to nine o'clock position as shown in the figure.
29. Align the intake cam sprocket to 12 o'clock position as shown in the figure.
30. Align the exhaust cam sprocket to 12 o'clock position as shown in the figure.
31. Using the ST, align the "Top mark" on crank sprocket to 12 o'clock position as shown in the figure.

➡ Crank key is at 3 o'clock position and (Number 1 piston is at TDC).

32. Install the chain guide (Main) and tighten bolts to 12 ft. lbs. (16 Nm).
33. Install the idler sprocket and timing chain (Main) as follows:
    - Align the timing chain mark (gold) to the idler sprocket timing mark position.
    - Align the idler sprocket timing mark to the six o'clock position, and then install the idler sprocket and timing chain.
    - Confirm the timing chain mark (gold) is set to the crank sprocket 12 o'clock position.
    - Install the idler sprocket bolt to 88 ft. lbs. (120 Nm).

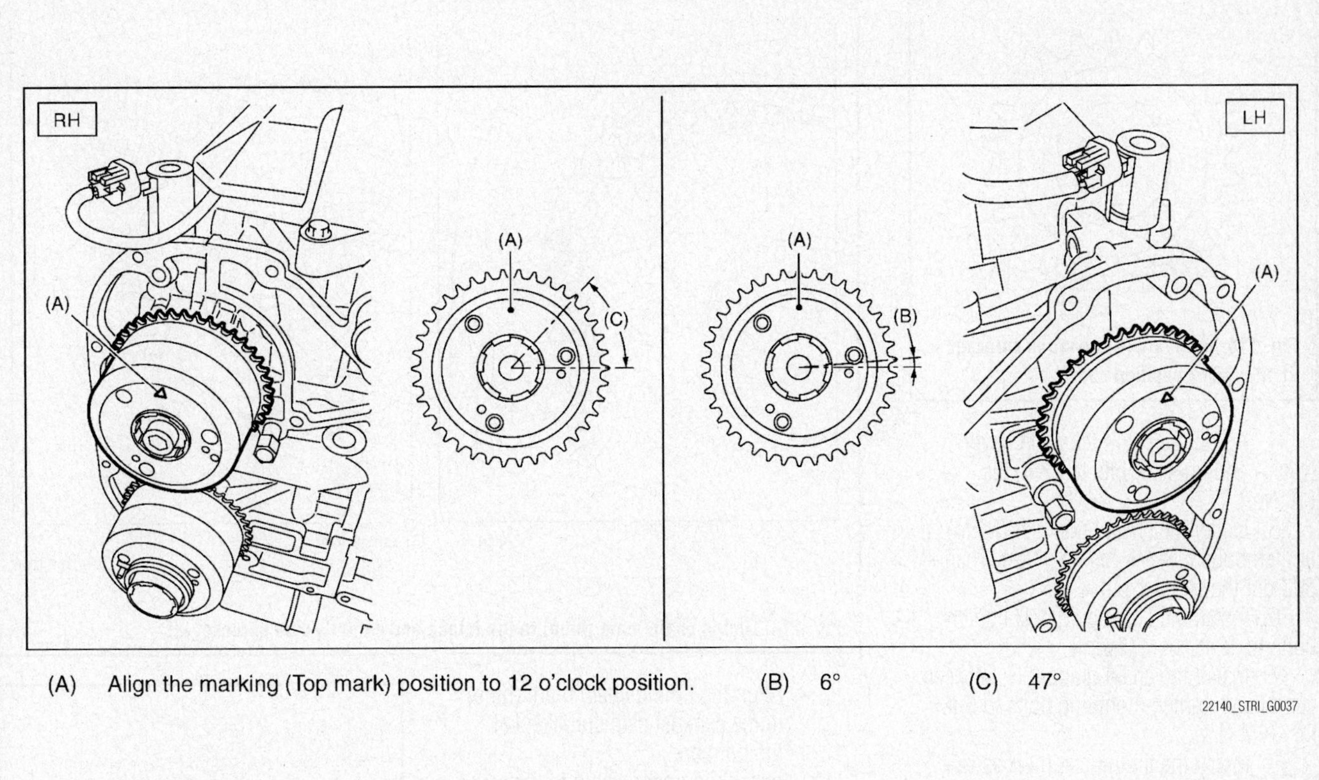

(A) Align the marking (Top mark) position to 12 o'clock position. (B) 6° (C) 47°

22140_STRI_G0037

Fig. 114 Intake cam sprocket at 12 o'clock position as shown

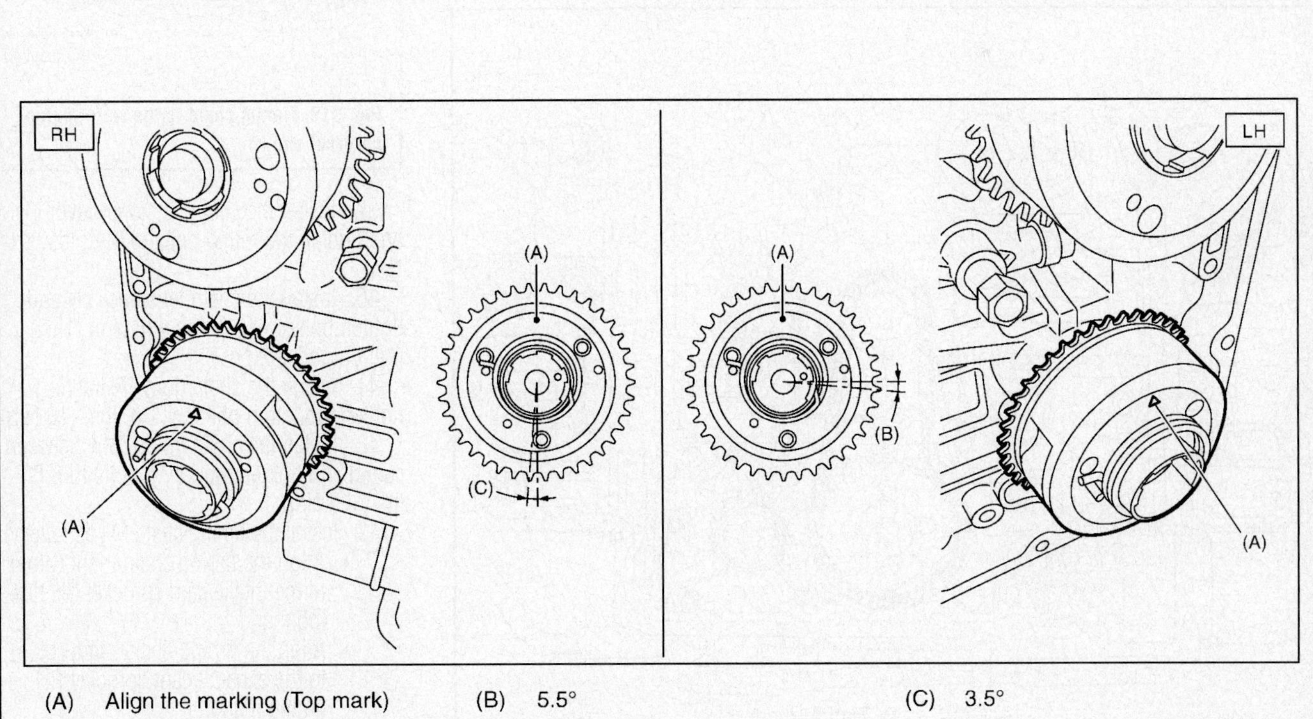

(A) Align the marking (Top mark) (B) 5.5° (C) 3.5°

22140_STRI_G0038

Fig. 115 Exhaust cam sprocket at 12 o'clock position as shown

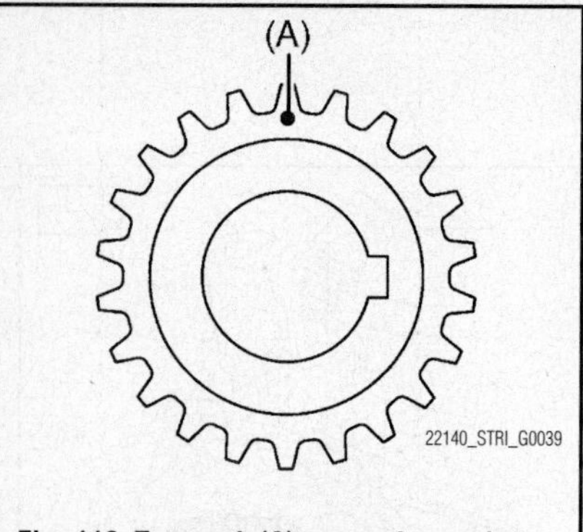

Fig. 116 Top mark (A) on crank sprocket at 12 o'clock position shown

34. Install the chain tensioner lever (Main) and tighten bolts to 12 ft. lbs. (16 Nm).

35. Install the chain tensioner (Main) tighten bolts to 12 ft. lbs. (16 Nm). Then pull out the stopper pin.

36. Install the chain guide LH tighten bolts to 12 ft. lbs. (16 Nm).

37. Install the chain guide (LH: between cams) and tighten mounting bolts to 5 ft. lbs. (6.4 Nm).

38. Install the timing chain LH as follows:

- Align the timing chain mark (blue) to the intake cam sprocket LH timing mark.
- Align the timing chain mark (blue) to the exhaust cam sprocket LH timing mark.
- Install the timing chain to the water pump sprocket.
- Align the idler sprocket timing mark to the timing chain mark (gold).

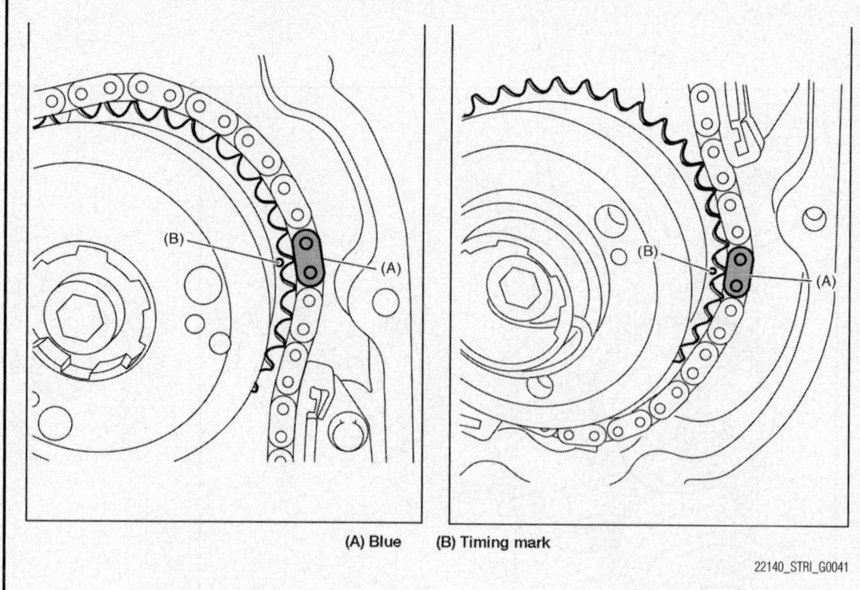

Fig. 118 Timing chain mark (blue) to the intake and exhaust cam sprocket LH

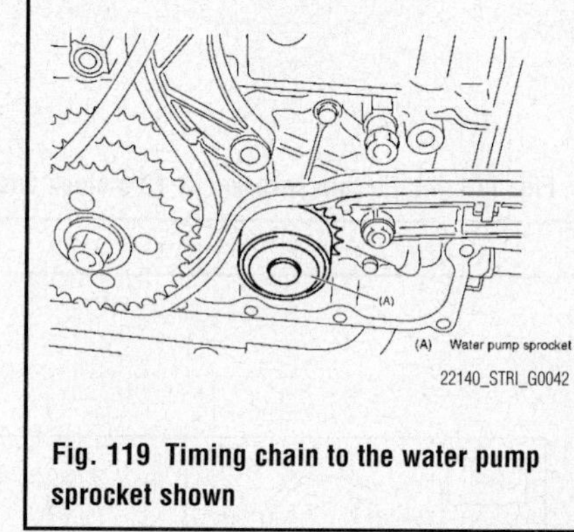

Fig. 119 Timing chain to the water pump sprocket shown

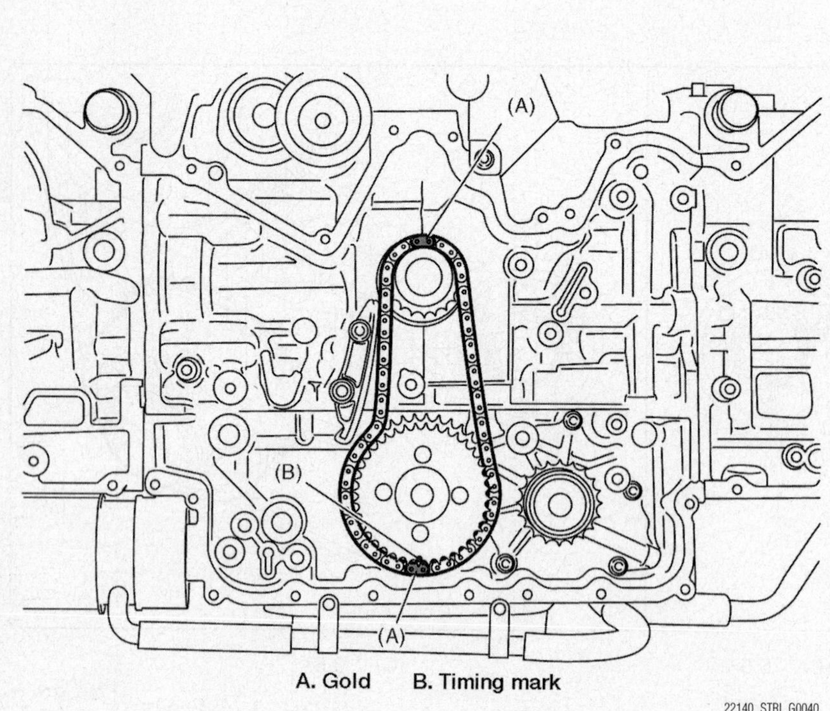

Fig. 117 Main timing chain marks shown

39. Install the chain tensioner lever LH and tighten mounting bolts to12 ft. lbs. (16 Nm).

40. Install the chain tensioner LH and tighten bolts to 12 ft. lbs. (16 Nm). Then pull out the stopper pin.

41. Install the chain guide RH and tighten mounting bolts to12 ft. lbs. (16 Nm).

42. Install the chain guide (RH: between cams) and tighten mounting bolts to 5 ft. lbs. (6.4 Nm).

43. Install the timing chain RH as follows:

- Align the timing chain mark (blue) to the intake cam sprocket RH timing mark.
- Align the timing chain mark (blue) to the exhaust cam sprocket RH timing mark.
- Install the timing chain to the water pump sprocket.
- Align the idler sprocket timing mark to the timing chain mark (gold).

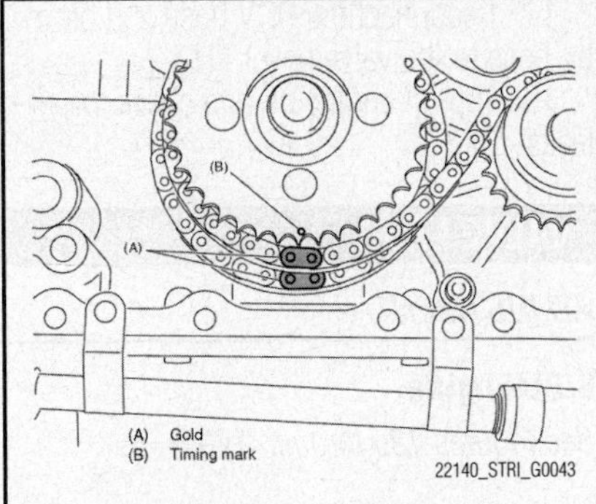

Fig. 120 Sprocket timing mark to the timing chain mark shown

44. Install the chain tensioner lever RH and tighten mounting bolts to12 ft. lbs. (16 Nm).

45. Install the chain tensioner RH and tighten bolts to 12 ft. lbs. (16 Nm). Then pull out the stopper pin.

46. Confirm the following after installation:

- Confirm that the idler sprocket timing mark and three timing chain marks (gold) are aligned.
- Confirm that the crank sprocket 12 o'clock position and timing chain (Main) mark (gold) are aligned.
- Confirm that the LH cam sprocket timing mark and timing chain mark (blue) are aligned.
- Confirm that the RH cam sprocket timing mark and timing chain mark (blue) are aligned.
- Confirm that the bolts are tightened to the specified tightening torque.
- Confirm that there are no abnormalities, while turning the crankshaft in the engine rotating direction using ST.

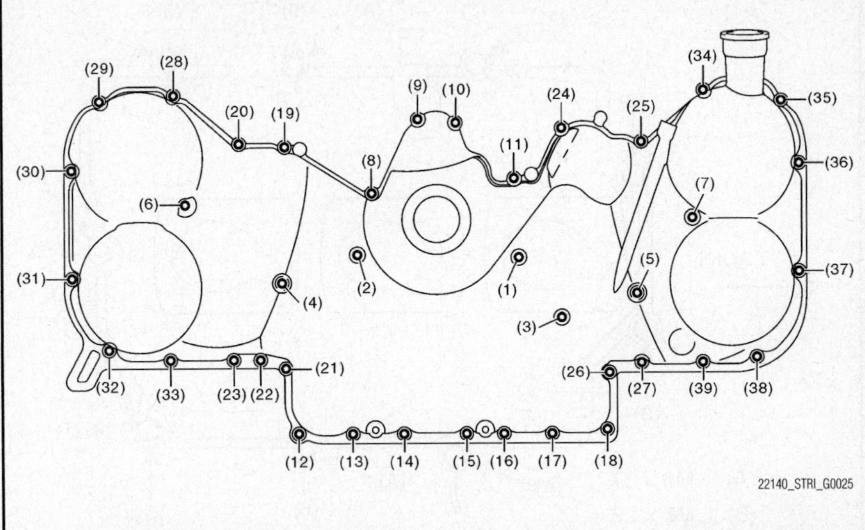

Fig. 122 Timing chain cover bolt tightening sequence

47. Apply liquid gasket to the mating surface of chain cover.

48. Install the chain cover and temporarily tighten the bolts.

49. Tighten the bolts in the numerical order as shown in the figure to 7 ft. lbs. (10 Nm).

50. Install the crank pulley.

51. Install the V-belt.

52. Fill engine oil.

53. Install the radiator.

54. Refill coolant and bleed cooling system of air.

55. Confirm that there are no oil leakage at the chain cover mating surface.

56. Confirm that there are no coolant leaks.

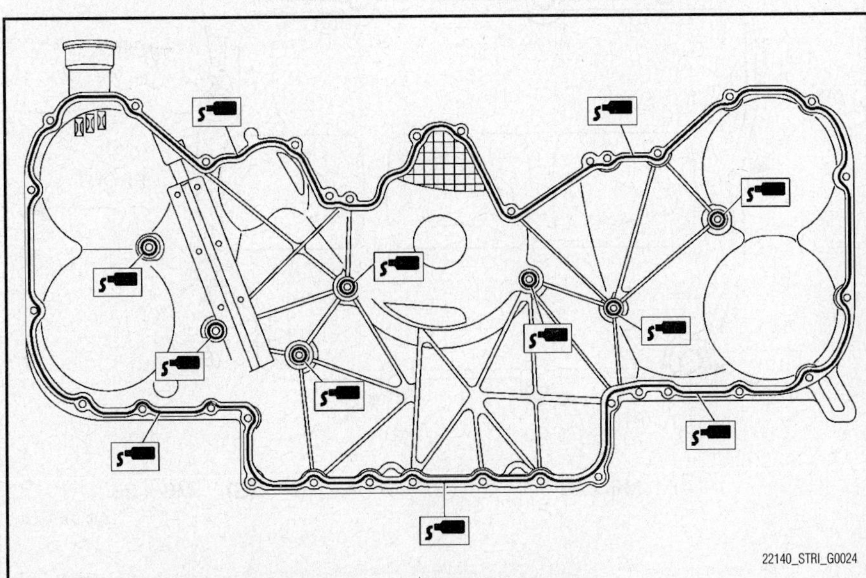

Fig. 121 Apply liquid gasket to the mating surface of chain cover

VALVE COVERS

REMOVAL & INSTALLATION

Right Side

See Figure 123.

1. Before servicing the vehicle, refer to the Precautions Section.
2. Set the vehicle on a lift.
3. Remove the collector cover.
4. Disconnect the ground cable from the battery.
5. Lift up the vehicle.
6. Remove the undercover.
7. Lower the vehicle.
8. When removing RH side valve cover remove the air intake duct and air cleaner case. And remove the fuel pipe protector.
9. Disconnect the connector of oil pressure switch.
10. Remove the ignition coils.
11. Loosen the bolts cover bolt in the numerical order as shown in the figure, and then remove the valve cover.

To install:

12. Clean valve cover and head contact surface.
13. Install the valve cover gasket to the rocker cover.
14. Apply liquid gasket sparingly to the mating surface of cylinder head and rocker cover as shown in the figure.

CAUTION

Do not apply liquid gasket excessively. Applying excessively may cause excess gasket to flow toward camshaft journal, resulting in engine seizure.

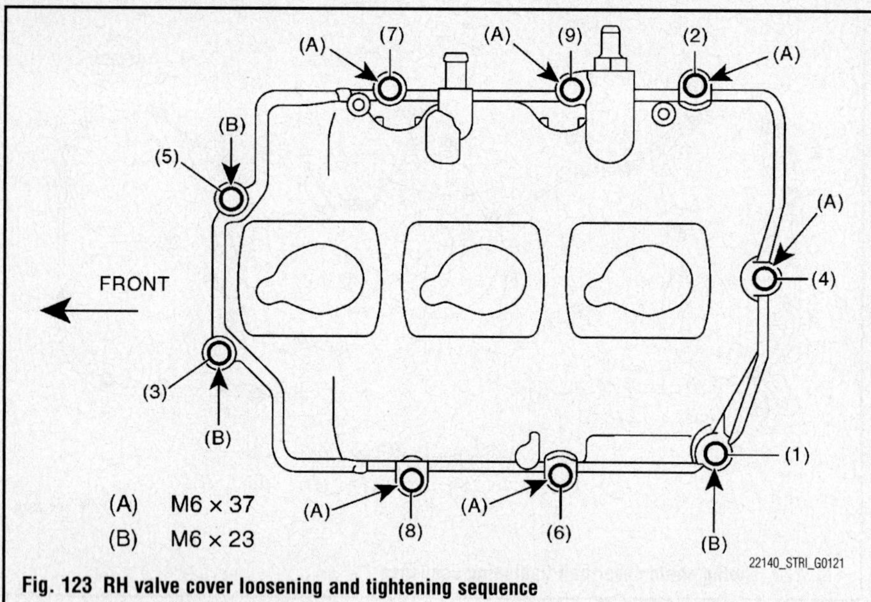

Fig. 123 RH valve cover loosening and tightening sequence

15. Tighten the valve cover bolts to 5 ft. lbs. (6.4 Nm) in the numerical sequence as shown in the figure.

16. Install ignition coils and tighten mounting bolts to 12 ft. lbs. (16 Nm).

17. Reconnect the connector of oil pressure switch.

18. Install air intake duct and air cleaner case. Install the fuel pipe protector.

19. Install the collector cover.

20. Reconnect the ground cable from the battery

Left Side

See Figure 124.

1. Before servicing the vehicle, refer to the Precautions Section.

2. When removing the LH side valve cover. Disconnect the battery cable, and then remove the battery and battery carrier. Disconnect the PCV hose and blow-by hose from the valve cover LH.

3. Set the vehicle on a lift.

4. Remove the fuel pipe protector LH.

5. Remove the ignition coils.

6. Loosen the valve cover bolts in the numerical order as shown in the figure, and then remove the valve cover.

To install:

7. Clean valve cover and head contact surface.

8. Install the valve cover gasket to the rocker cover.

9. Apply liquid gasket sparingly to the mating surface of cylinder head and rocker cover.

CAUTION

Do not apply liquid gasket excessively. Applying excessively may cause excess gasket to flow toward camshaft journal, resulting in engine seizure.

10. Tighten the valve cover bolts to 5 ft. lbs. (6.4 Nm) in the numerical sequence as shown in the figure.

11. Install ignition coils and tighten mounting bolts to 12 ft. lbs. (16 Nm).

12. Install the fuel pipe protector LH.

13. Reconnect the battery cable, and then install the battery and battery carrier.

14. Reconnect the PCV hose and blow-by hose to the valve cover LH.

15. Top off engine oil and check for leaks.

VALVE LASH

ADJUSTMENT

3.0L Engine

See Figures 125 through 129.

➡ The valve adjustment should be performed while the engine is cold.

1. Before servicing the vehicle, refer to the precautions in the beginning of this section.

2. Raise and support the vehicle safely.

3. Remove the undercover.

4. Lower the vehicle.

5. Remove the collector cover.

6. Disconnect the negative battery cable.

7. On the right side, remove the air intake duct and air cleaner case. Remove the fuel tank protector. Disconnect the oil pressure switch electrical connector. Remove the ignition coil.

8. On the left side, remove the battery and battery carrier. Disconnect the PCV hose from the rocker cover. Remove the ignition coil.

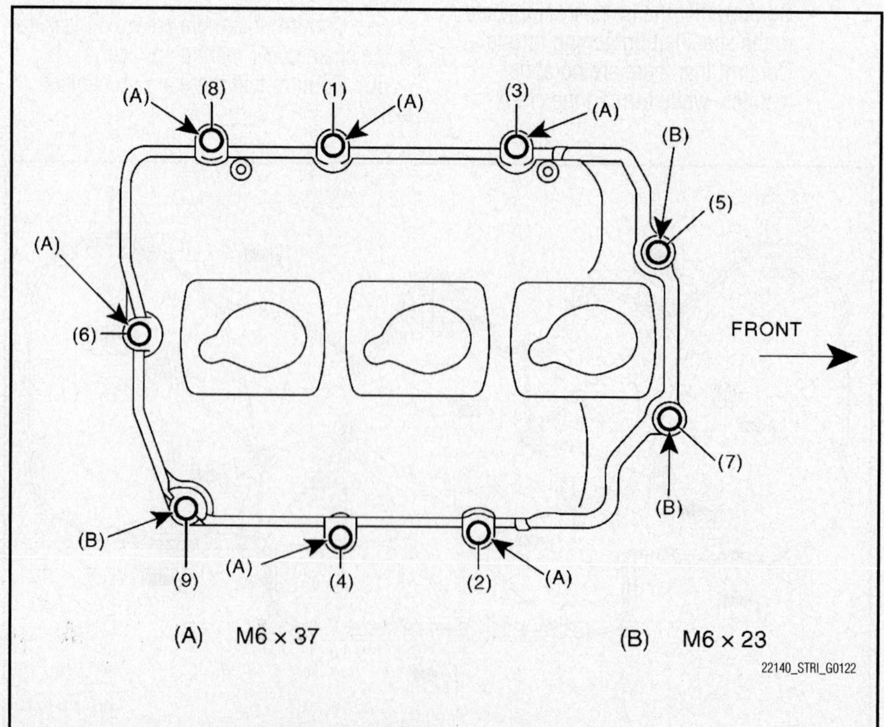

Fig. 124 LH valve cover loosening and tightening sequence

(1) Valve clearance (Intake side)
(2) Valve clearance (Exhaust side)
(3) High lift cam
(4) Low lift cam

09490_SBCR_G0053

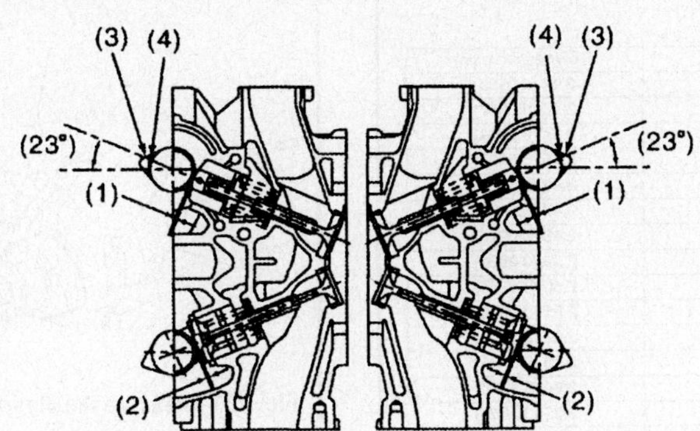

Fig. 125 Valve adjustment crankshaft positioning—3.0L engine

9. Remove the rocker cover retaining bolts. Remove the rocker covers from the engine.

10. Rotate the crankshaft clockwise until the cam is set in position, see illustration.

11. Using a feeler gauge measure and record the clearance of the intake and exhaust valve.

➡ **Measure it within the range of +/- 30 degrees from the specified position, shown in the illustration. Measure it in the low cam for the intake side. Insert the feeler gauge in a horizontally as possible with respect to the valve lifter.**

12. Further turn the crankshaft pulley clockwise and then measure and record the valve clearance again.

13. If adjustment is required, remove the camshafts.

14. Remove and measure the thickness of the valve lifter. Select a suitable shim, using the shim selection chart.

15. Install the replacement shim to the lifter.

16. After all shims have been adjusted, inspect the valve clearances again.

17. After completion, install all removed components.

Unit: (mm)
S = (V + T) – 0.35
S: Valve lifter thickness required V: Measured valve clearance T: Valve lifter thickness to be used

09490_SBCR_G0054

Fig. 126 Use this table to help you select a suitable exhaust valve shim

Unit: (mm)
S = (V + T) – 0.20
S: Required shim thickness V: Measured valve clearance T: Shim thickness to be used

09490_SBCR_G0055

Fig. 127 Use this table to help you select a suitable intake valve shim

Part No.	Thickness mm (in)
13228AD180	4.32 (0.1701)
13228AD190	4.34 (0.1709)
13228AD200	4.36 (0.1717)
13228AD210	4.38 (0.1724)
13228AD220	4.40 (0.1732)
13228AD230	4.42 (0.1740)
13228AD240	4.44 (0.1748)
13228AD250	4.46 (0.1756)
13228AD260	4.48 (0.1764)
13228AD270	4.50 (0.1772)
13228AD280	4.52 (0.1780)
13228AD290	4.54 (0.1787)
13228AD300	4.56 (0.1795)
13228AD310	4.58 (0.1803)
13228AD320	4.60 (0.1811)
13228AC580	4.62 (0.1819)
13228AC590	4.63 (0.1823)
13228AC600	4.64 (0.1827)
13228AC610	4.65 (0.1831)
13228AC620	4.66 (0.1835)
13228AC630	4.67 (0.1839)
13228AC640	4.68 (0.1843)
13228AC650	4.69 (0.1846)
13228AC660	4.70 (0.1850)
13228AC670	4.71 (0.1854)
13228AC680	4.72 (0.1858)
13228AC690	4.73 (0.1862)
13228AC700	4.74 (0.1866)
13228AC710	4.75 (0.1870)
13228AC720	4.76 (0.1874)
13228AC730	4.77 (0.1878)
13228AC740	4.78 (0.1882)
13228AC750	4.79 (0.1886)
13228AC760	4.80 (0.1890)
13228AC770	4.81 (0.1894)
13228AC780	4.82 (0.1898)
13228AC790	4.83 (0.1902)
13228AC800	4.84 (0.1906)
13228AC810	4.85 (0.1909)
13228AC820	4.86 (0.1913)
13228AC830	4.87 (0.1917)
13228AC840	4.88 (0.1921)
13228AC850	4.89 (0.1925)

Part No.	Thickness mm (in)
13228AC860	4.90 (0.1929)
13228AC870	4.91 (0.1933)
13228AC880	4.92 (0.1937)
13228AC890	4.93 (0.1941)
13228AC900	4.94 (0.1945)
13228AC910	4.95 (0.1949)
13228AC920	4.96 (0.1953)
13228AC930	4.97 (0.1957)
13228AC940	4.98 (0.1961)
13228AC950	4.99 (0.1965)
13228AC960	5.00 (0.1969)
13228AC970	5.01 (0.1972)
13228AC980	5.02 (0.1976)
13228AC990	5.03 (0.1980)
13228AD000	5.04 (0.1984)
13228AD010	5.05 (0.1988)
13228AD020	5.06 (0.1992)
13228AD030	5.07 (0.1996)
13228AD040	5.08 (0.2000)
13228AD050	5.09 (0.2004)
13228AD060	5.10 (0.2008)
13228AD070	5.11 (0.2012)
13228AD080	5.12 (0.2016)
13228AD090	5.13 (0.2020)
13228AD100	5.14 (0.2024)
13228AD110	5.15 (0.2028)
13228AD120	5.16 (0.2032)
13228AD130	5.17 (0.2035)
13228AD140	5.18 (0.2039)
13228AD150	5.19 (0.2043)
13228AD160	5.20 (0.2047)
13228AD170	5.21 (0.2051)
13228AD330	5.23 (0.2059)
13228AD340	5.25 (0.2067)
13228AD350	5.27 (0.2075)
13228AD360	5.29 (0.2083)
13228AD370	5.31 (0.2091)
13228AD380	5.33 (0.2098)
13228AD390	5.35 (0.2106)
13228AD400	5.37 (0.2114)
13228AD410	5.39 (0.2122)
13228AD420	5.41 (0.2130)
13228AD430	5.43 (0.2138)
13228AD440	5.45 (0.2146)
13228AD450	5.47 (0.2154)
13228AD460	5.49 (0.2161)
13228AD470	5.51 (0.2169)
13228AD480	5.53 (0.2177)
13228AD490	5.55 (0.2185)
13228AD500	5.57 (0.2193)
13228AD510	5.59 (0.2201)

09490_SBCR_G0056

Fig. 128 Exhaust valve adjusting shim chart

Part No.	Thickness mm (in)
13218AK890	1.92 (0.0756)
13218AK900	1.94 (0.0764)
13218AK910	1.96 (0.0772)
13218AK920	1.98 (0.0780)
13218AK930	2.00 (0.0787)
13218AK940	2.02 (0.0795)
13218AK950	2.04 (0.0803)
13218AK960	2.06 (0.0811)
13218AK970	2.07 (0.0815)
13218AK980	2.08 (0.0819)
13218AK990	2.09 (0.0823)
13218AL000	2.10 (0.0827)
13218AL010	2.11 (0.0831)
13218AL020	2.12 (0.0835)
13218AL030	2.13 (0.0839)
13218AL040	2.14 (0.0843)
13218AL050	2.15 (0.0846)
13218AL060	2.16 (0.0850)
13218AL070	2.17 (0.0854)
13218AL080	2.18 (0.0858)
13218AL090	2.19 (0.0862)
13218AL100	2.20 (0.0866)
13218AL110	2.21 (0.0870)
13218AL120	2.22 (0.0874)
13218AL130	2.23 (0.0878)
13218AL140	2.24 (0.0882)
13218AL150	2.25 (0.0886)
13218AL160	2.26 (0.0890)
13218AL170	2.27 (0.0894)
13218AL180	2.28 (0.0898)
13218AL190	2.29 (0.0902)
13218AL200	2.30 (0.0906)
13218AL210	2.31 (0.0909)
13218AL220	2.32 (0.0913)
13218AL230	2.33 (0.0917)
13218AL240	2.34 (0.0921)
13218AL250	2.35 (0.0925)
13218AL260	2.36 (0.0929)
13218AL270	2.37 (0.0933)
13218AL280	2.38 (0.0937)
13218AL290	2.39 (0.0941)
13218AL300	2.40 (0.0945)
13218AL310	2.41 (0.0949)

Part No.	Thickness mm (in)
13218AL320	2.42 (0.0953)
13218AL330	2.43 (0.0957)
13218AL340	2.44 (0.0961)
13218AL350	2.45 (0.0965)
13218AL360	2.46 (0.0969)
13218AL370	2.47 (0.0972)
13218AL380	2.48 (0.0976)
13218AL390	2.49 (0.0980)
13218AL400	2.50 (0.0984)
13218AL410	2.51 (0.0988)
13218AL420	2.52 (0.0992)
13218AL430	2.53 (0.0996)
13218AL440	2.54 (0.1000)
13218AL450	2.55 (0.1004)
13218AL460	2.56 (0.1008)
13218AL470	2.57 (0.1012)
13218AL480	2.58 (0.1016)
13218AL490	2.59 (0.1020)
13218AL500	2.60 (0.1024)
13218AL510	2.61 (0.1028)
13218AL520	2.62 (0.1032)
13218AL530	2.64 (0.1039)
13218AL540	2.66 (0.1047)
13218AL550	2.68 (0.1055)
13218AL560	2.70 (0.1063)
13218AL570	2.72 (0.1071)
13218AL580	2.74 (0.1079)
13218AL590	2.76 (0.1087)

09490_SBCR_G0057

Fig. 129 Intake valve adjusting shim chart

3.6L Engine

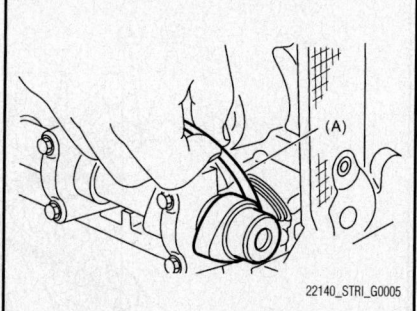

See Figures 130 through 139.

➡ Adjustment of valve clearance should be performed while engine is cold. Do not wear gloves during removal and installation of valve lifter. Do not use valve lifters that were dropped or otherwise exposed to strong impacts.

1. Before servicing the vehicle, refer to the precautions in the beginning of this section.
2. Set the vehicle on a lift.
3. Remove the collector cover.
4. Disconnect the ground cable from the battery.
5. Lift up the vehicle.
6. Remove the undercover.
7. Lower the vehicle.
8. When inspecting RH side cylinders:
   a. Remove the air intake duct and air cleaner case.

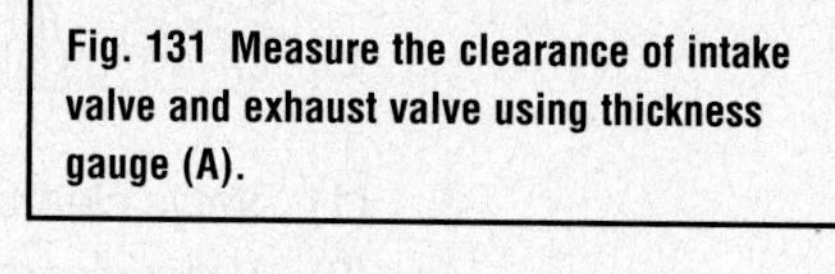

Fig. 131 Measure the clearance of intake valve and exhaust valve using thickness gauge (A).

   b. Remove the fuel pipe protector RH.
   c. Disconnect the connector of oil pressure switch.
   d. Remove the ignition coil.
   e. Remove the rocker cover RH.
9. When inspecting LH side cylinders:
   a. Disconnect the battery cable, and then remove the battery and battery carrier.
   b. Disconnect the PCV hose and blow-by hose from the rocker cover LH.
   c. Remove the fuel pipe protector LH.
   d. Remove the ignition coil.
   e. Remove the rocker cover LH.
10. Turn the crankshaft clockwise until the cam is set to position shown in the figure.
11. Measure the clearance of intake valve and exhaust valve using thickness gauge (A).

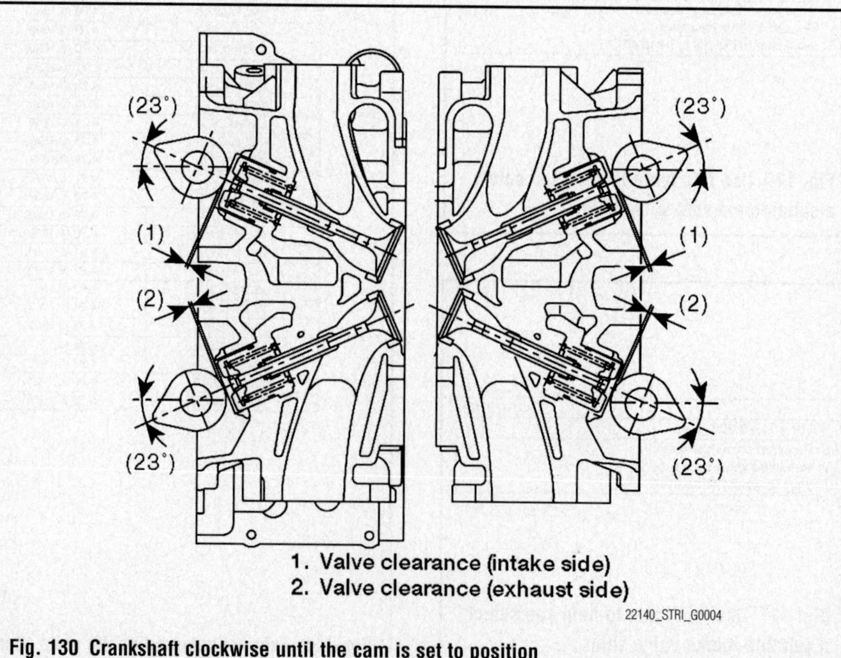

Fig. 130 Crankshaft clockwise until the cam is set to position

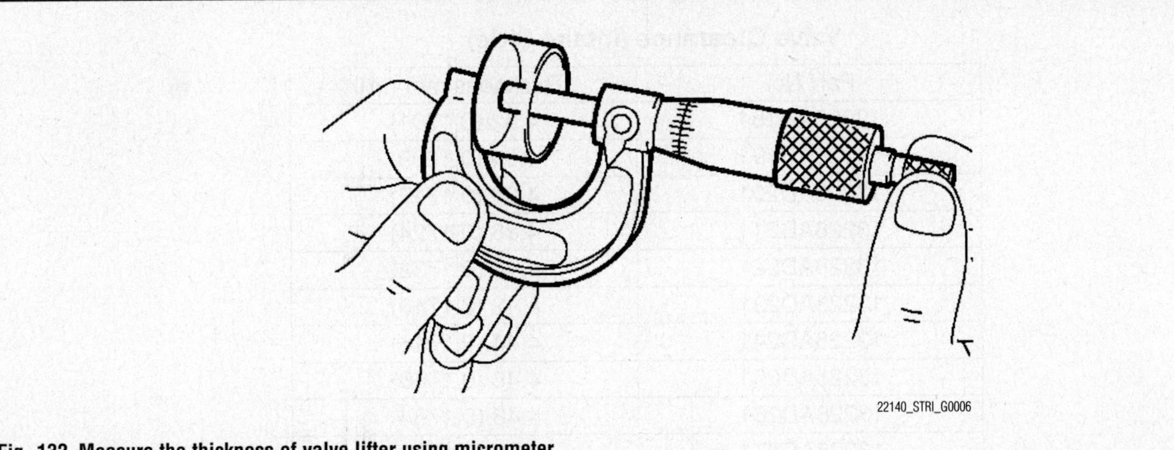

Fig. 132 Measure the thickness of valve lifter using micrometer

Use this table to help you select a suitable Intake valve shim

Unit: mm (in)
S = (V + T) – 0.20 (0.0079)
S: Valve lifter thickness required V: Measured valve clearance T: Valve lifter thickness to be used

22140_STRI_G0007

Fig. 133 Use this table to help you select a suitable intake valve shim

Use this table to help you select a suitable Exhaust valve shim

Unit: mm (in)
S = (V + T) – 0.35 (0.0138)
S: Valve lifter thickness required V: Measured valve clearance T: Valve lifter thickness to be used

22140_STRI_G0010

Fig. 134 Use this table to help you select a suitable exhaust valve shim

Valve Clearance (Intake Side)

Part No.	Thickness mm (in)
13228AD181	4.32 (0.1701)
13228AD191	4.34 (0.1709)
13228AD201	4.36 (0.1717)
13228AD211	4.38 (0.1724)
13228AD221	4.40 (0.1732)
13228AD231	4.42 (0.1740)
13228AD241	4.44 (0.1748)
13228AD251	4.46 (0.1756)
13228AD261	4.48 (0.1764)
13228AD271	4.50 (0.1772)
13228AD281	4.52 (0.1780)
13228AD291	4.54 (0.1787)
13228AD301	4.56 (0.1795)
13228AD311	4.58 (0.1803)
13228AD321	4.60 (0.1811)
13228AC581	4.62 (0.1819)
13228AC591	4.63 (0.1823)
13228AC601	4.64 (0.1827)
13228AC611	4.65 (0.1831)
13228AC621	4.66 (0.1835)
13228AC631	4.67 (0.1839)
13228AC641	4.68 (0.1843)
13228AC651	4.69 (0.1846)
13228AC661	4.70 (0.1850)
13228AC671	4.71 (0.1854)
13228AC681	4.72 (0.1858)
13228AC691	4.73 (0.1862)
13228AC701	4.74 (0.1866)
13228AC711	4.75 (0.1870)
13228AC721	4.76 (0.1874)
13228AC731	4.77 (0.1878)
13228AC741	4.78 (0.1882)
13228AC751	4.79 (0.1886)
13228AC761	4.80 (0.1890)
13228AC771	4.81 (0.1894)
13228AC781	4.82 (0.1898)
13228AC791	4.83 (0.1902)
13228AC801	4.84 (0.1906)
13228AC811	4.85 (0.1909)
13228AC821	4.86 (0.1913)
13228AC831	4.87 (0.1917)
13228AC841	4.88 (0.1921)
13228AC851	4.89 (0.1925)

22140_STRI_G0008

Fig. 135 Intake valve adjusting shim chart

Valve Clearance (Intake side)

Part No.	Thickness mm (in)
13228AC861	4.90 (0.1929)
13228AC871	4.91 (0.1933)
13228AC881	4.92 (0.1937)
13228AC891	4.93 (0.1941)
13228AC901	4.94 (0.1945)
13228AC911	4.95 (0.1949)
13228AC921	4.96 (0.1953)
13228AC931	4.97 (0.1957)
13228AC941	4.98 (0.1961)
13228AC951	4.99 (0.1965)
13228AC961	5.00 (0.1969)
13228AC971	5.01 (0.1972)
13228AC981	5.02 (0.1976)
13228AC991	5.03 (0.1980)
13228AD001	5.04 (0.1984)
13228AD011	5.05 (0.1988)
13228AD021	5.06 (0.1992)
13228AD031	5.07 (0.1996)
13228AD041	5.08 (0.2000)
13228AD051	5.09 (0.2004)
13228AD061	5.10 (0.2008)
13228AD071	5.11 (0.2012)
13228AD081	5.12 (0.2016)
13228AD091	5.13 (0.2020)
13228AD101	5.14 (0.2024)
13228AD111	5.15 (0.2028)
13228AD121	5.16 (0.2032)
13228AD131	5.17 (0.2035)
13228AD141	5.18 (0.2039)
13228AD151	5.19 (0.2043)
13228AD161	5.20 (0.2047)
13228AD171	5.21 (0.2051)
13228AD331	5.23 (0.2059)
13228AD341	5.25 (0.2067)
13228AD351	5.27 (0.2075)
13228AD361	5.29 (0.2083)
13228AD371	5.31 (0.2091)
13228AD381	5.33 (0.2098)
13228AD391	5.35 (0.2106)
13228AD401	5.37 (0.2114)
13228AD411	5.39 (0.2122)
13228AD421	5.41 (0.2130)
13228AD431	5.43 (0.2138)
13228AD441	5.45 (0.2146)
13228AD451	5.47 (0.2154)
13228AD461	5.49 (0.2161)
13228AD471	5.51 (0.2169)
13228AD481	5.53 (0.2177)
13228AD491	5.55 (0.2185)
13228AD501	5.57 (0.2193)
13228AD511	5.59 (0.2201)

22140_STRI_G0009

Fig. 136 Intake valve adjusting shim chart

Valve Clearance (Exhaust Side)

Part No.	Thickness mm (in)
13228AD081	5.12 (0.2016)
13228AD091	5.13 (0.2020)
13228AD101	5.14 (0.2024)
13228AD111	5.15 (0.2028)
13228AD121	5.16 (0.2032)
13228AD131	5.17 (0.2035)
13228AD141	5.18 (0.2039)
13228AD151	5.19 (0.2043)
13228AD161	5.20 (0.2047)
13228AD171	5.21 (0.2051)
13228AD331	5.23 (0.2059)
13228AD341	5.25 (0.2067)
13228AD351	5.27 (0.2075)
13228AD361	5.29 (0.2083)
13228AD371	5.31 (0.2091)
13228AD381	5.33 (0.2098)
13228AD391	5.35 (0.2106)
13228AD401	5.37 (0.2114)
13228AD411	5.39 (0.2122)
13228AD421	5.41 (0.2130)
13228AD431	5.43 (0.2138)
13228AD441	5.45 (0.2146)
13228AD451	5.47 (0.2154)
13228AD461	5.49 (0.2161)
13228AD471	5.51 (0.2169)
13228AD481	5.53 (0.2177)
13228AD491	5.55 (0.2185)
13228AD501	5.57 (0.2193)
13228AD511	5.59 (0.2201)

22140_STRI_G0011

Fig. 137 Exhaust valve adjusting shim chart

Valve Clearance (Exhaust Side)

Part No.	Thickness mm (in)
13228AD181	4.32 (0.1701)
13228AD191	4.34 (0.1709)
13228AD201	4.36 (0.1717)
13228AD211	4.38 (0.1724)
13228AD221	4.40 (0.1732)
13228AD231	4.42 (0.1740)
13228AD241	4.44 (0.1748)
13228AD251	4.46 (0.1756)
13228AD261	4.48 (0.1764)
13228AD271	4.50 (0.1772)
13228AD281	4.52 (0.1780)
13228AD291	4.54 (0.1787)
13228AD301	4.56 (0.1795)
13228AD311	4.58 (0.1803)
13228AD321	4.60 (0.1811)

22140_STRI_G0012

Fig. 138 Exhaust valve adjusting shim chart

Right side of Intake Manifold
22140_STRI_G0013

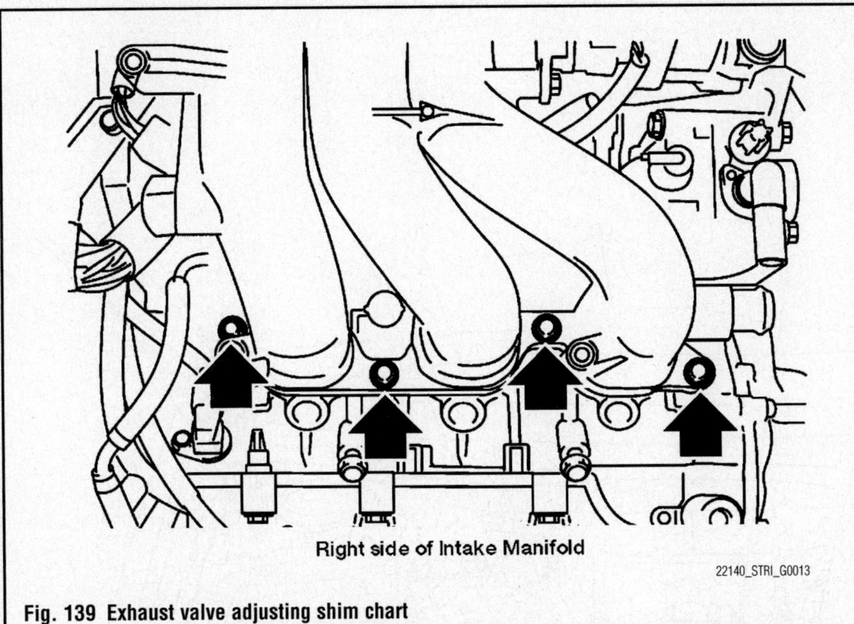

Fig. 139 Exhaust valve adjusting shim chart

➡ Measure it within the range of +/-30° from specified position shown in the figure. Insert a thickness gauge in a direction as horizontal as possible with respect to the valve lifter. If the measured value is not within specification, take notes of the value in order to adjust the valve clearance later on.

12. If adjustment is required, remove the camshafts.
13. Remove and measure the thickness of the valve lifter. Select a suitable shim, using the shim selection chart.

To install:

14. Install the replacement shim to the lifter.
15. After all shims have been adjusted, inspect the valve clearances again.
16. After completion, install all removed components.

ENGINE PERFORMANCE & EMISSION CONTROLS

COMPONENT LOCATIONS

See Figures 140 through 142.

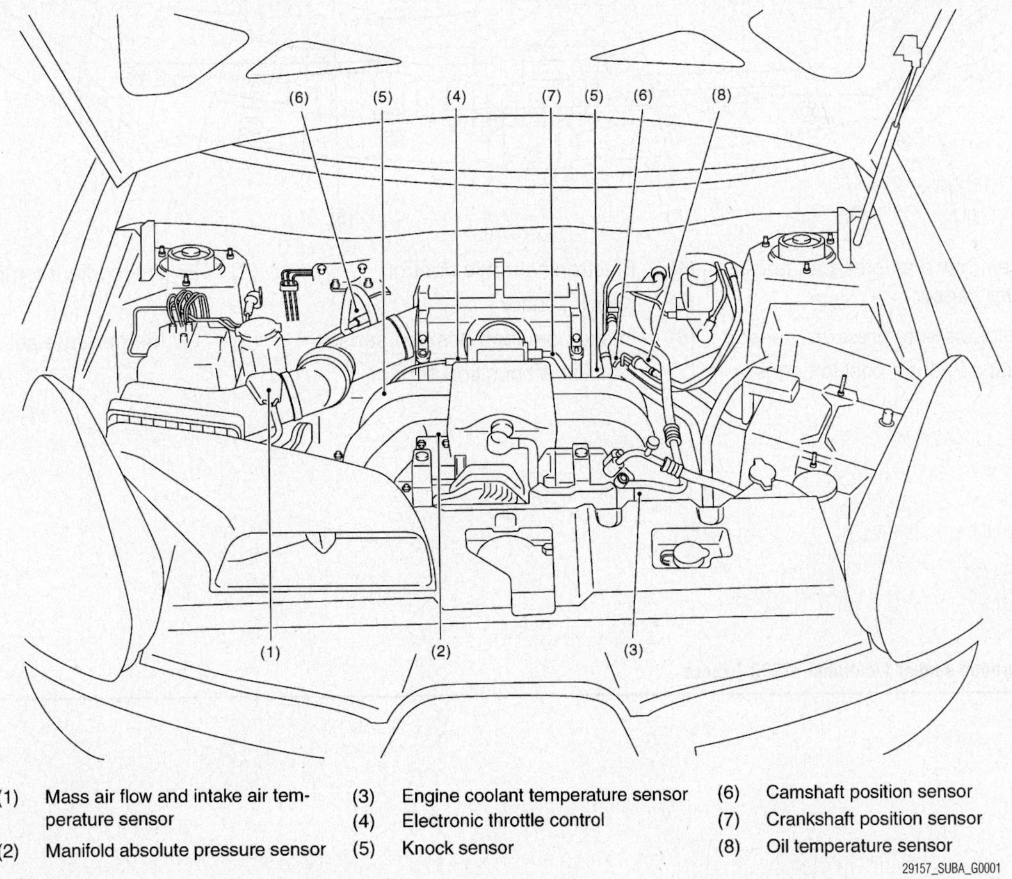

(6)
(5)
(4)
(7)
(5)
(6)
(8)
(1)
(2)
(3)
(1) Mass air flow and intake air temperature sensor
(2) Manifold absolute pressure sensor
(3) Engine coolant temperature sensor
(4) Electronic throttle control
(5) Knock sensor
(6) Camshaft position sensor
(7) Crankshaft position sensor
(8) Oil temperature sensor
29157_SUBA_G0001

Fig. 140 Underhood sensor locations—2006–07 B9 Tribeca

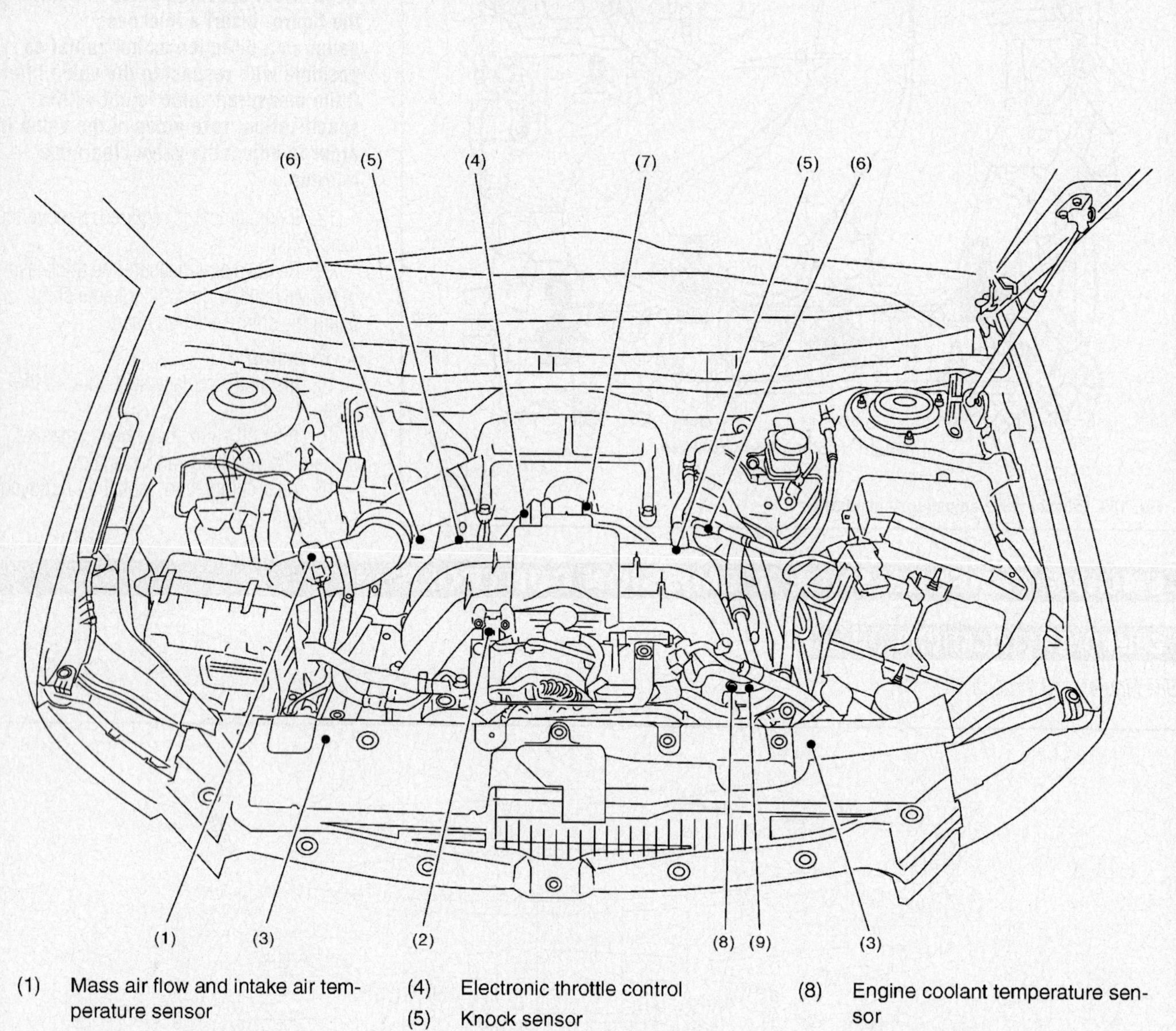

(1)	Mass air flow and intake air temperature sensor	(4)	Electronic throttle control	(8)	Engine coolant temperature sensor
		(5)	Knock sensor		
(2)	Manifold absolute pressure sensor	(6)	Intake camshaft position sensor	(9)	Oil temperature sensor
(3)	Exhaust camshaft position sensor	(7)	Crankshaft position sensor		

22140_STRI_G0123

Fig. 141 Underhood sensor locations—2008 Tribeca

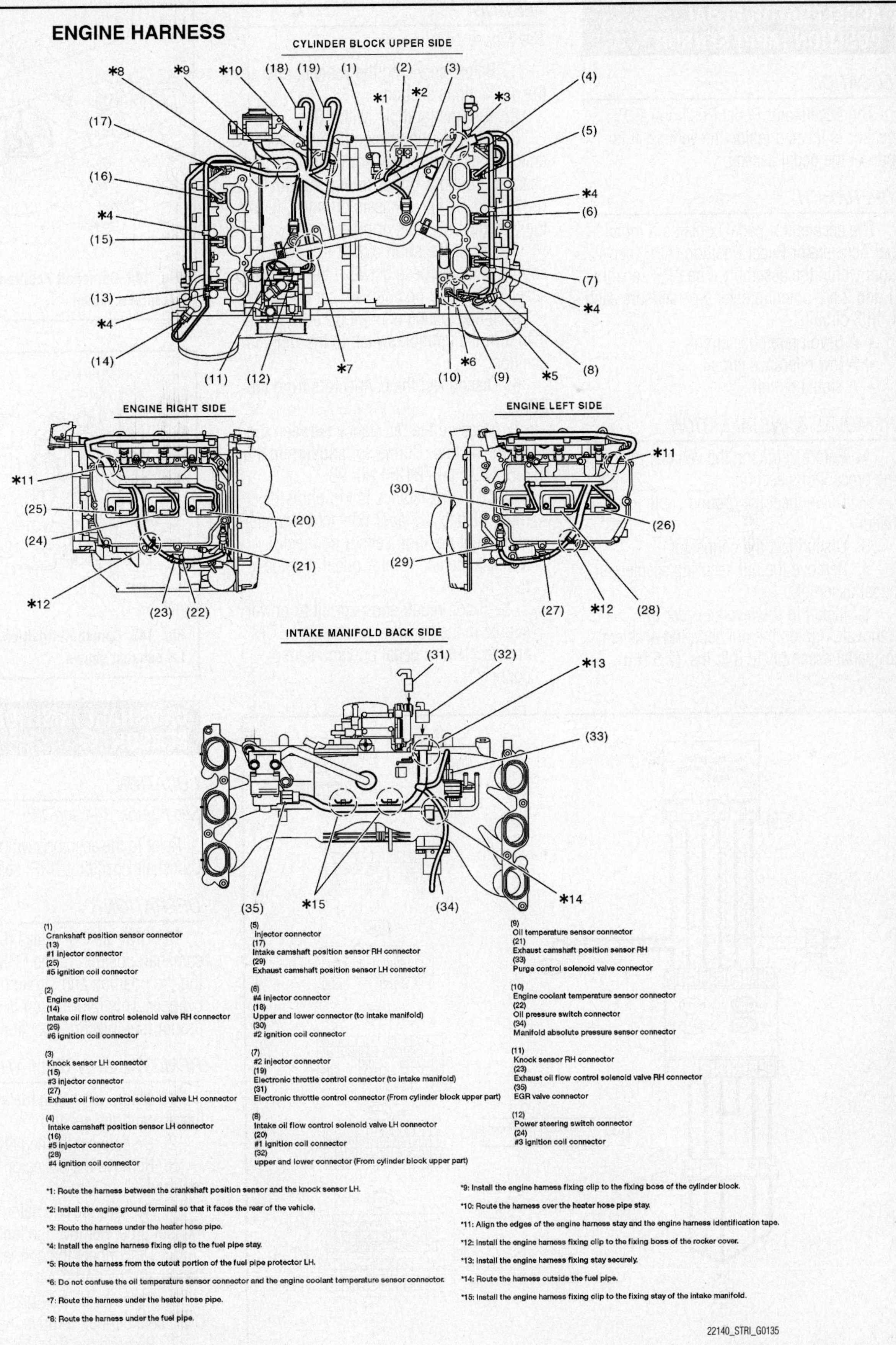

Fig. 142 Engine harness connector view—2008 Tribeca

ACCELERATOR PEDAL POSITION (APP) SENSOR

LOCATION

The Accelerator Pedal Position (APP) sensor is located inside the vehicle. It is part of the pedal assembly.

OPERATION

The accelerator pedal contains 2 individual Accelerator Pedal Position (APP) sensors within the assembly. The APP sensors 1 and 2 are potentiometer type sensors each with 3 circuits:

- A 5-volt reference circuit
- A low reference circuit
- A signal circuit

REMOVAL & INSTALLATION

1. Before servicing the vehicle, refer to the precautions section.
2. Disconnect the ground cable from battery.
3. Disconnect the connector.
4. Remove the nut securing accelerator pedal assembly.
5. Install in the reverse order of removal. Tighten the nut securing accelerator pedal assembly to 6 ft. lbs. (7.5 Nm).

TESTING

See Figure 143.

1. Before servicing the vehicle, refer to the precautions section.
2. Turn the ignition switch **ON**.
3. Measure the voltage between accelerator pedal position sensor connector and chassis ground. Connector and terminal (B315) No. 6 (+)—Chassis ground (–): Is the voltage 4.85 volts or more.
4. Repair the short circuit to power source in the harness between the ECM and accelerator pedal position sensor connector.
5. If the voltage is not 4.85 or more than turn the ignition switch to the OFF position.
6. Disconnect the connectors from the ECM.
7. Measure the resistance between the ECM connectors. Connector and terminal (B135) No. 21—(B135) No. 23:
   a. If the resistance is 1m ohms or more. Repair the poor contact of accelerator pedal position sensor connector. Replace the accelerator pedal if defective.
   b. If not, repair short circuit to power source in the harness between the ECM and accelerator pedal position sensor connector.

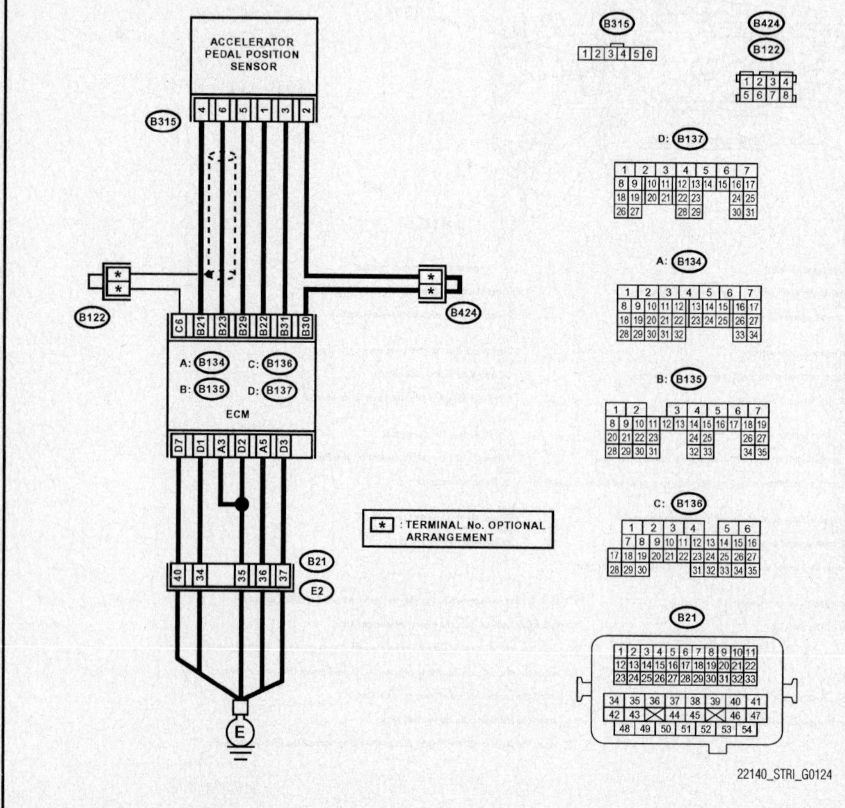

Fig. 143 Accelerator Pedal Position (APP) Sensor circuit view

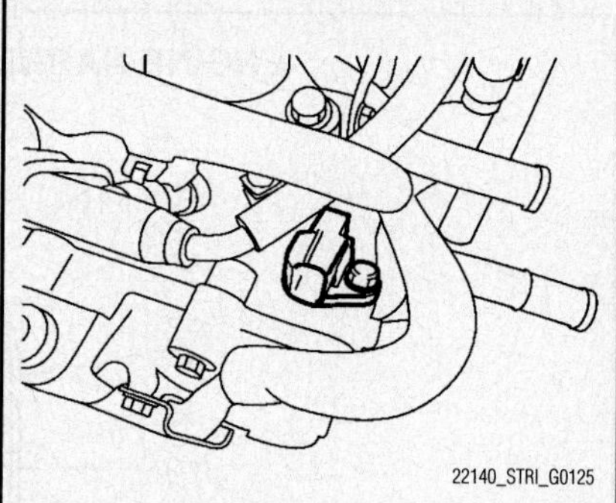

Fig. 144 Camshaft Position (CMP) Sensor LH intake shown

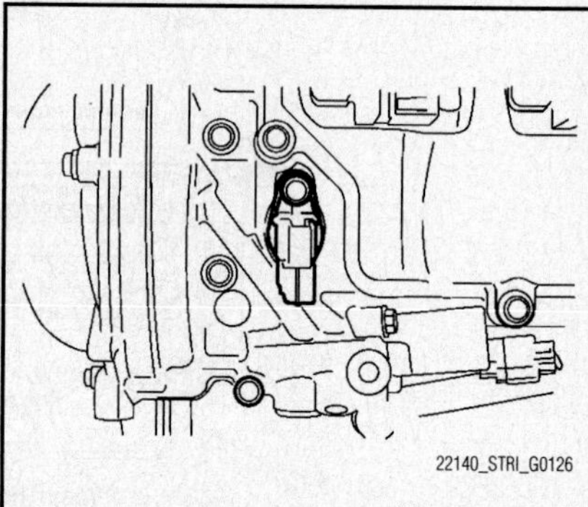

Fig. 145 Camshaft Position (CMP) Sensor LH exhaust shown

CAMSHAFT POSITION (CMP) SENSOR

LOCATION

See Figures 144 and 145.

Refer to the accompanying illustrations for Camshaft Position (CMP) sensor location.

OPERATION

The CMP sensor relays the relative camshaft position to the ECM for determining the position and stroke of the No. 1 cylinder. This information is required for proper fuel injection functioning.

REMOVAL & INSTALLATION

1. Before servicing the vehicle, refer to the precautions section.
2. Disconnect the negative battery cable.
3. Remove the collector cover, as required.
4. Remove the alternator harness from the fuel pipe. Remove the fuel pipe protector.
5. Disconnect the connector from the Camshaft Position (CMP) sensor.
6. Remove the bolt that retains the CMP sensor, then remove the sensor.
7. Remove the RH CMP sensor using the same procedure as the LH sensor.

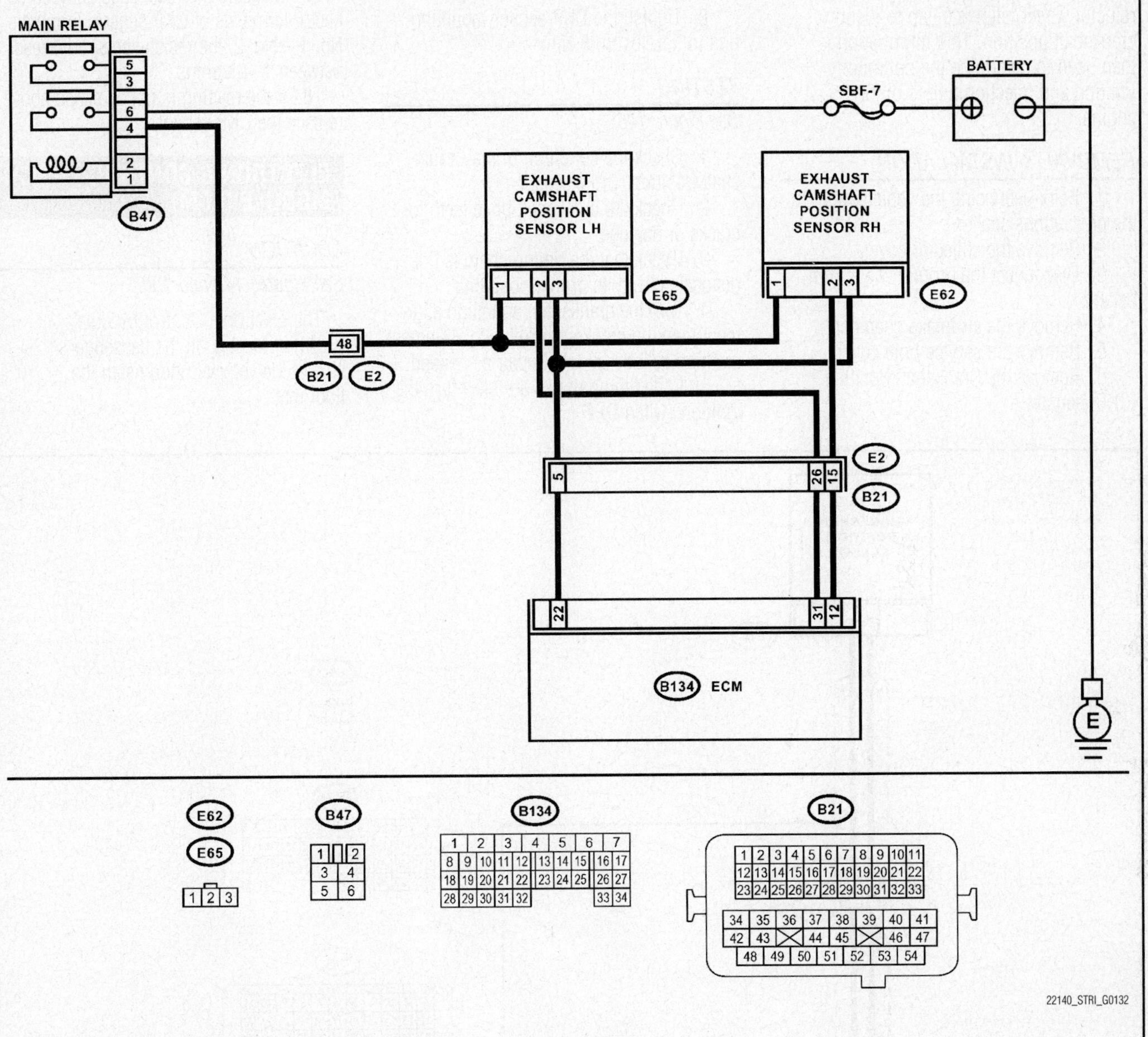

Fig. 146 Camshaft Position (CMP) sensor circuit view

8. Installation is the reverse of the removal procedure. Tighten the camshaft sensor mounting bolt to 5 ft. lbs. (6.8 Nm).

TESTING

See Figure 146.

1. Using the Subaru scan tool, check the sensor output waveform.
2. Measurement is taken using connector B134 terminal 11 for the left side and connector B134 terminal 21 for the right side.
3. If abnormality is found, replace the sensor.

Fig. 147 Crankshaft Position (CKP) sensor location view

CRANKSHAFT POSITION (CKP) SENSOR

LOCATION

See Figure 147.

The Crankshaft Position (CKP) sensor is located in the front of the engine near the crankshaft pulley.

OPERATION

The Crankshaft Position (CKP) sensor sends a digital signal, which represents an image of the crankshaft reluctor wheel, to the ECM as each tooth on the wheel rotates past the CKP sensor. The ECM uses each

CKP signal pulse to determine crankshaft speed and decodes the crankshaft reluctor wheel reference gap to identify crankshaft position. This information is then used to determine the optimum ignition and injection points of the engine.

REMOVAL & INSTALLATION

1. Before servicing the vehicle, refer to the precautions section.
2. Remove the collector cover.
3. Disconnect the ground cable from battery.
4. Remove the air intake chamber.
5. Remove the service hole cover.
6. Remove the Crankshaft Position (CKP) sensor.
7. Install in the reverse order of removal.
8. Tighten the CKP sensor mounting bolt to 5 ft. lbs. (6.8 Nm).

TESTING

See Figure 148.

1. Check the condition of the Crankshaft Position (CKP) sensor.
2. Check the crankshaft plate teeth for cracks or damage.
3. Check that the timing chain is not dislocated from its proper position.
4. Turn the crankshaft, and align alignment mark on crank sprocket with alignment mark on cylinder block. Repair as needed.
5. If the timing marks are okay turn the ignition switch **OFF**.
6. Remove the CKP sensor.
7. Measure the resistance between connector terminals of CKP sensor. Terminals No. 1—No. 2: the resistance should read between 1–4 k ohms.
8. If the reading is not as stated above replace the CKP sensor.

ELECTRONIC CONTROL MODULE (ECM)

LOCATION

See Figures 149 and 150.

The Electronic Control Module (ECM) is located on the passenger's side of the vehicle, underneath the floor mat.

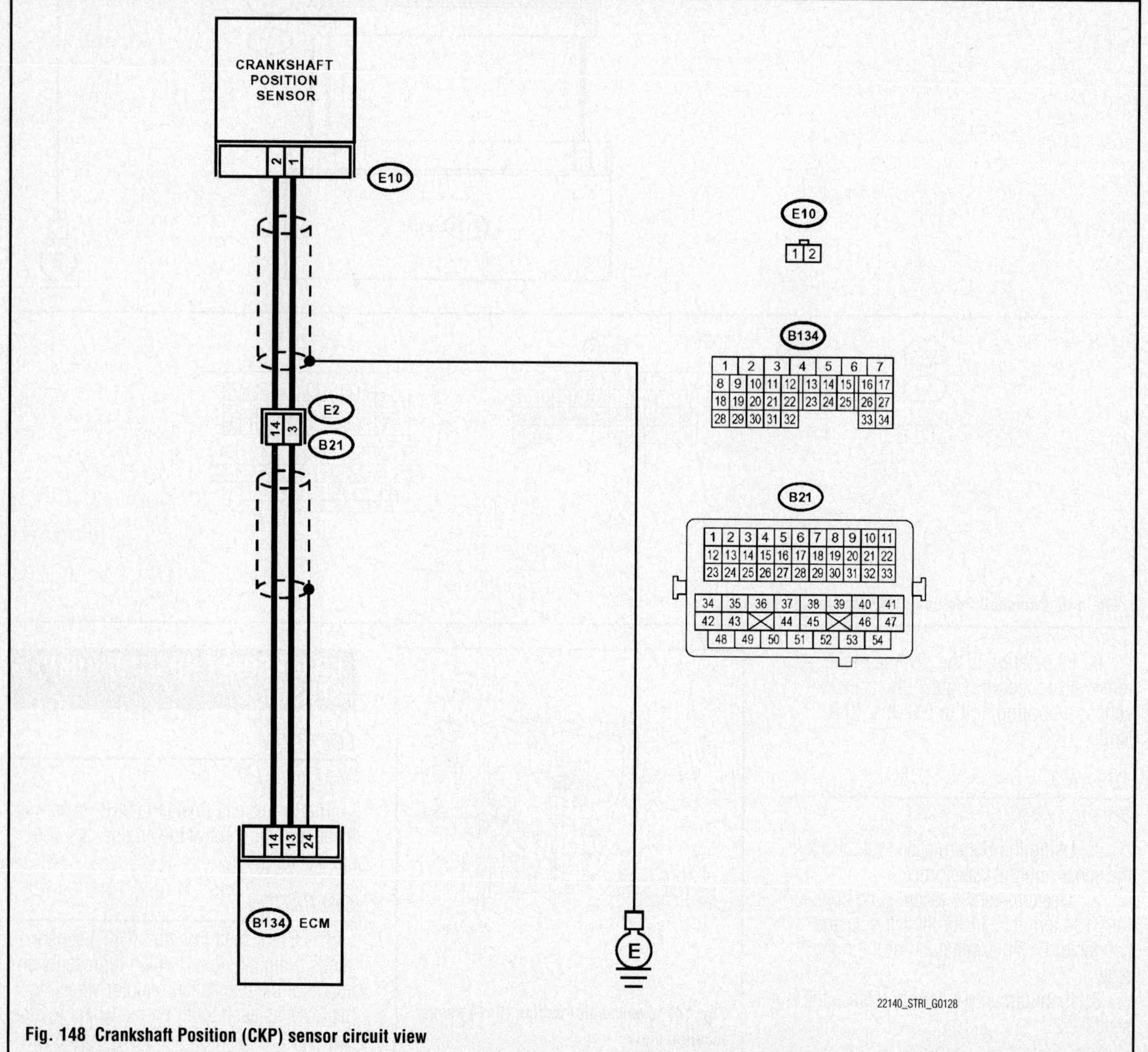

Fig. 148 Crankshaft Position (CKP) sensor circuit view

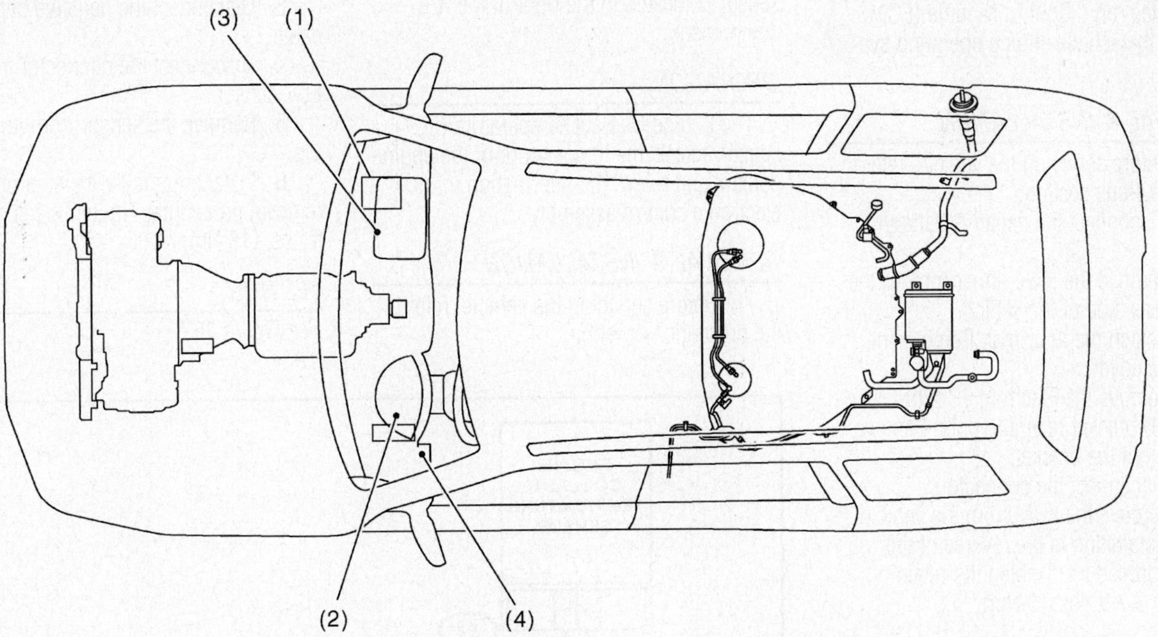

(1) Engine control module (ECM)
(2) Malfunction indicator light
(3) Test mode connector
(4) Data link connector

Fig. 149 ECM and related components—2006–07 B9 Tribeca

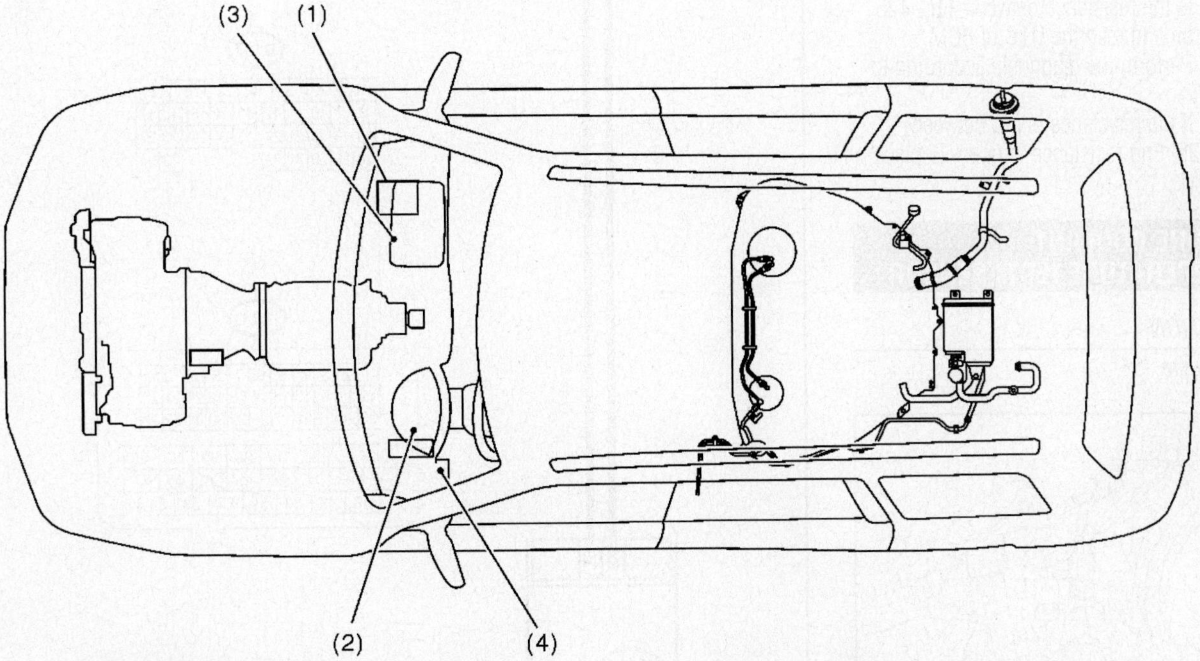

(1) Engine control module (ECM)
(2) Malfunction indicator light
(3) Delivery (test) mode connector
(4) Data link connector

Fig. 150 ECM and related components—2008 Tribeca

OPERATION

The Electronic Control Module (ECM) controls the vehicle engine operating system.

REMOVAL & INSTALLATION

1. Before servicing the vehicle, refer to the precautions section.
2. Disconnect the negative battery cable.
3. Remove the lower inner trim on the passenger's side of the vehicle.
4. Detach the floor mat. Remove the protective cover.
5. Remove the Electronic Control Module (ECM) bracket retaining nuts. Remove the clip from the bracket.
6. Disconnect the connectors.
7. Remove the ECM from the vehicle.
8. Installation is the reverse of the removal procedure. Tighten the retaining screws to 3.7 ft. lbs. (5 Nm).

➡ When replacing the ECM, be careful not to use the wrong part number, as damage to the injection system could occur.

TESTING

1. Disconnect the Electronic Control Module (ECM) connector (B136).
2. Measure the resistance between ECM terminals No. 27 and No. 35.
3. Is the resistance between 115–125 than scan and read the DTC of ECM.
4. Perform the diagnosis according to the DTC.
5. If the resistance is not between 115–125. End resistance is open. Replace the ECM.

ENGINE COOLANT TEMPERATURE (ECT) SENSOR

LOCATION

See Figure 151.

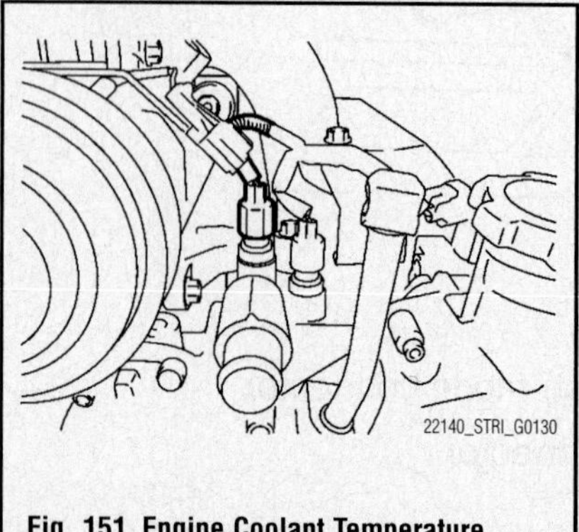

Fig. 151 Engine Coolant Temperature (ECT) sensor location view

The Engine Coolant Temperature (ECT) sensor is located to the right of the A/C compressor.

OPERATION

The Engine Coolant Temperature (ECT) sensor detects the temperature of the engine coolant and relays the information to the electronic control assembly.

REMOVAL & INSTALLATION

1. Before servicing the vehicle, refer to the precautions section.
2. Remove the collector cover.
3. Disconnect the negative battery cable.
4. Disconnect the connector from the ECT sensor.
5. Remove the sensor from its mounting.
6. Installation is the reverse of the removal procedure. Tighten the sensor to 13 ft. lbs. (18 Nm).

TESTING

See Figure 152.

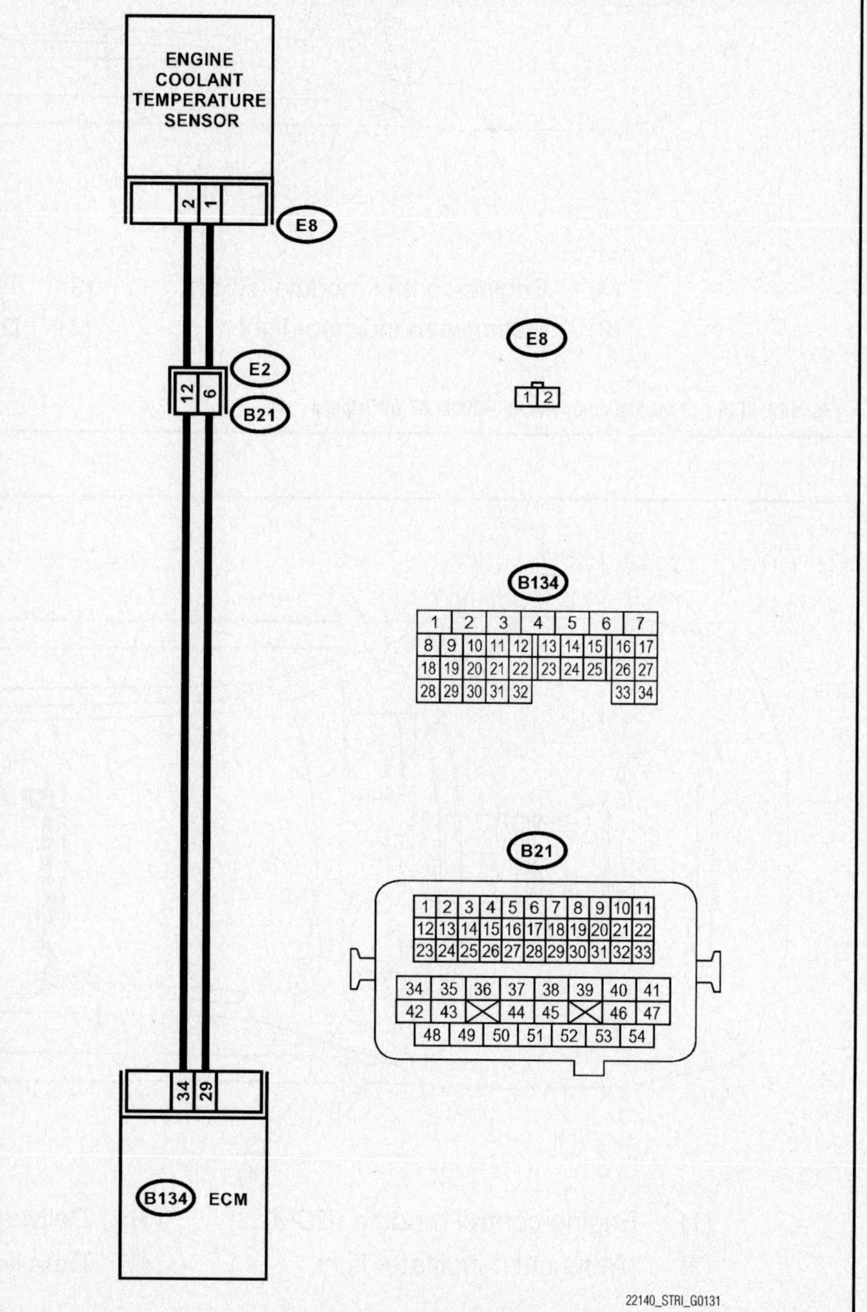

ENGINE COOLANT TEMPERATURE SENSOR
E8 2 1
E2 12 6
B21
E8
B134
B21
34 29
B134 ECM
22140_STRI_G0131

Fig. 152 Engine Coolant Temperature (ECT) Sensor Circuit view

1. Measure the resistance between the ECT sensor terminals No.1 and No.2 when the engine coolant is cold and after warmed-up.
2. Is the resistance of the ECT sensor different between when engine coolant is cold and after warmed-up, check for poor contact in ECT and ECM connector.
3. If the resistance does not change between cold and warmed-up engine. Replace the ECT sensor.

ENGINE OIL TEMPERATURE (EOT) SENSOR

LOCATION

See Figure 153.

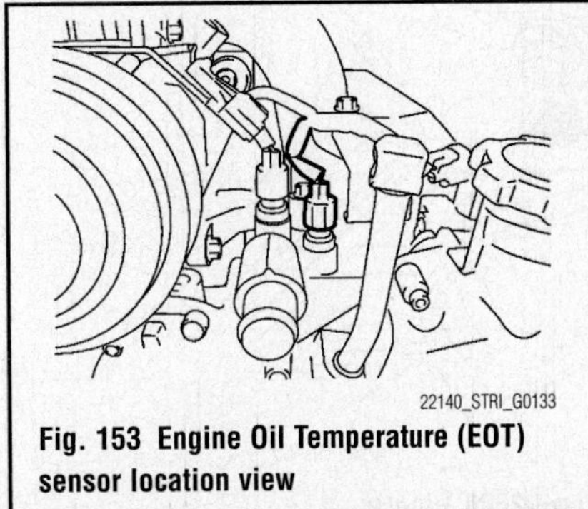

Fig. 153 Engine Oil Temperature (EOT) sensor location view

The Engine Oil Temperature (EOT) sensor is located to the right of the A/C compressor next to the Engine Coolant Temperature (ECT) sensor.

OPERATION

Engine Oil Temperature (EOT) sensor detects the temperature of the engine oil and relays the information to the electronic control assembly.

REMOVAL & INSTALLATION

1. Before servicing the vehicle, refer to the precautions section.
2. Remove the collector cover.
3. Disconnect the ground cable from battery.
4. Disconnect the connector from Engine Oil Temperature (EOT) sensor.
5. Remove the EOT sensor.
6. Install in the reverse order of removal. Tighten the EOT sensor to 16 ft. lbs. (22 Nm).

TESTING

See Figure 154.

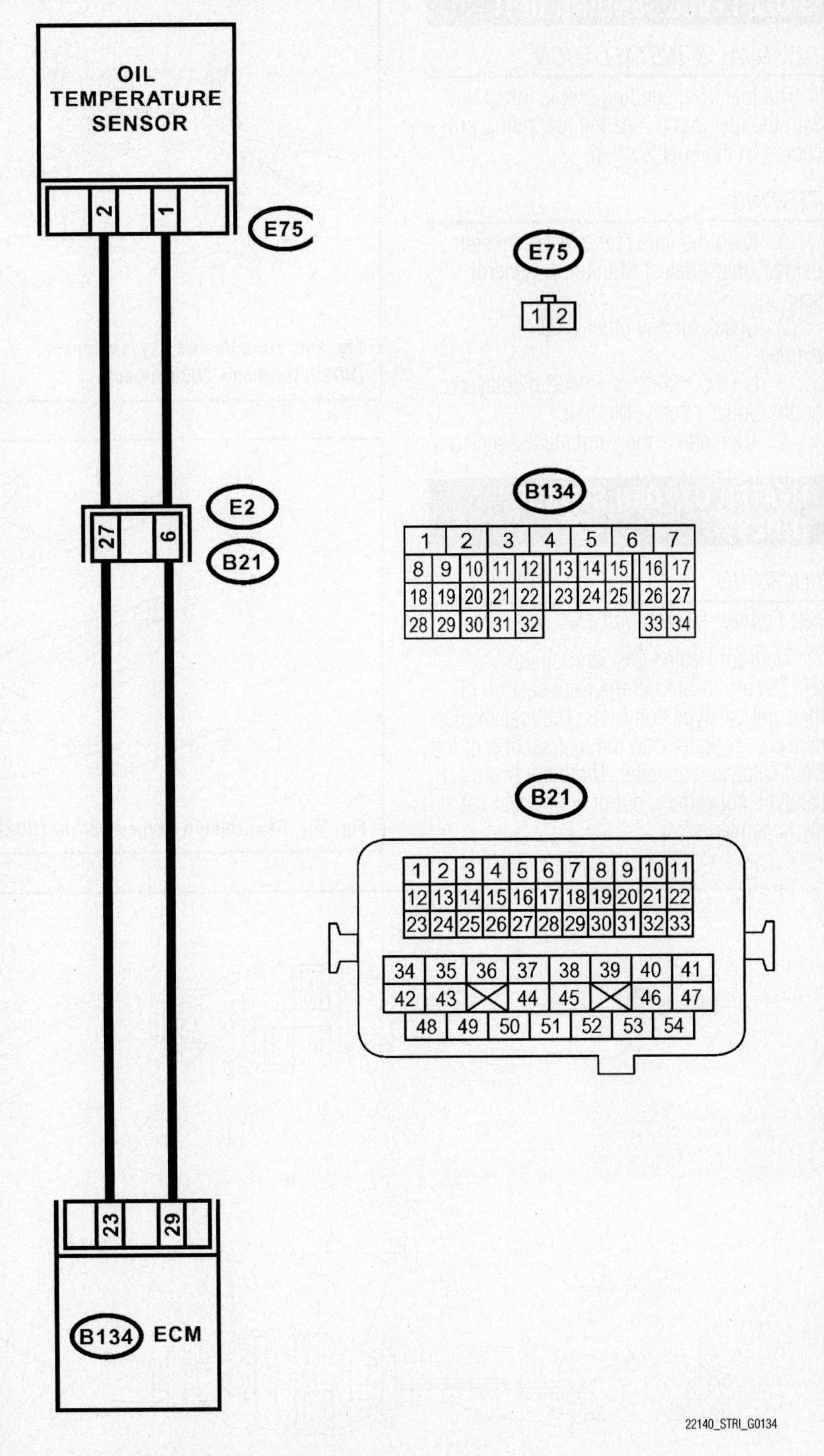

Fig. 154 Engine Oil Temperature (EOT) sensor circuit view

1. Start the engine.
2. Read the data of the Engine Oil Temperature (EOT) sensor signal using the Subaru Select Monitor or general scan tool.
3. Is the oil temperature 419° F (215° C) or more. If not the problem may be intermittent. In this case, there may be a temporary connector contact failure.
4. If the oil temperature is 419° F (215°C) or more. Turn the ignition switch **OFF**.
5. Disconnect the connector from the ECM and EOT sensor.
6. Measure the resistance between ECM and chassis ground. Connector (B134) terminal No. 23
7. If the resistance is 1m ohms or more replace the EOT sensor.
8. If not repair the ground short circuit of harness between ECM and the EOT sensor.

FUEL LEVEL SENDING UNIT

REMOVAL & INSTALLATION

The fuel level sending unit is integrated with the fuel pump. See the fuel pump procedure in the Fuel Section.

TESTING

1. Read the data of the fuel level sensor using Subaru Select Monitor or general scan tool.
2. Check for any other DTC on display.
3. Is DTC P0462 or P0463 displayed on the Subaru Select Monitor.
4 Then check the combination meter.

HEATED OXYGEN SENSOR (HO2S)

LOCATION

See Figures 155 through 157.

The front Heated Oxygen Sensors (HO2S) are located in the front section of the front catalytic converter. The rear oxygen sensors are located in the rear section of the front catalytic converter. There are two front catalytic converters, but only one rear catalytic converter.

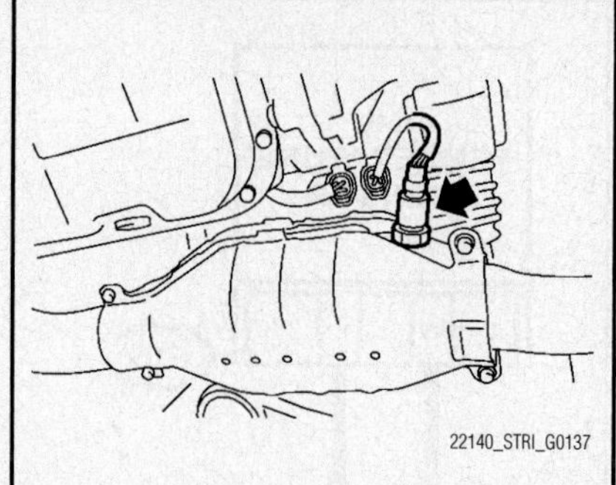

Fig. 156 Front Heated Oxygen Sensor (HO2S) location—2008 Tribeca

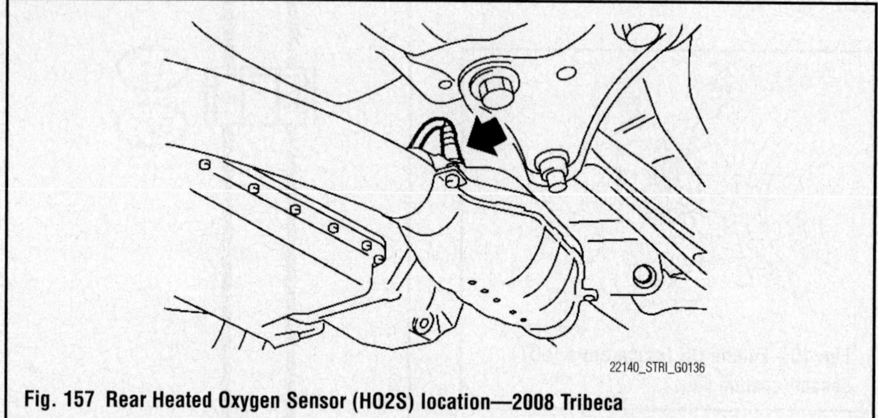

Fig. 157 Rear Heated Oxygen Sensor (HO2S) location—2008 Tribeca

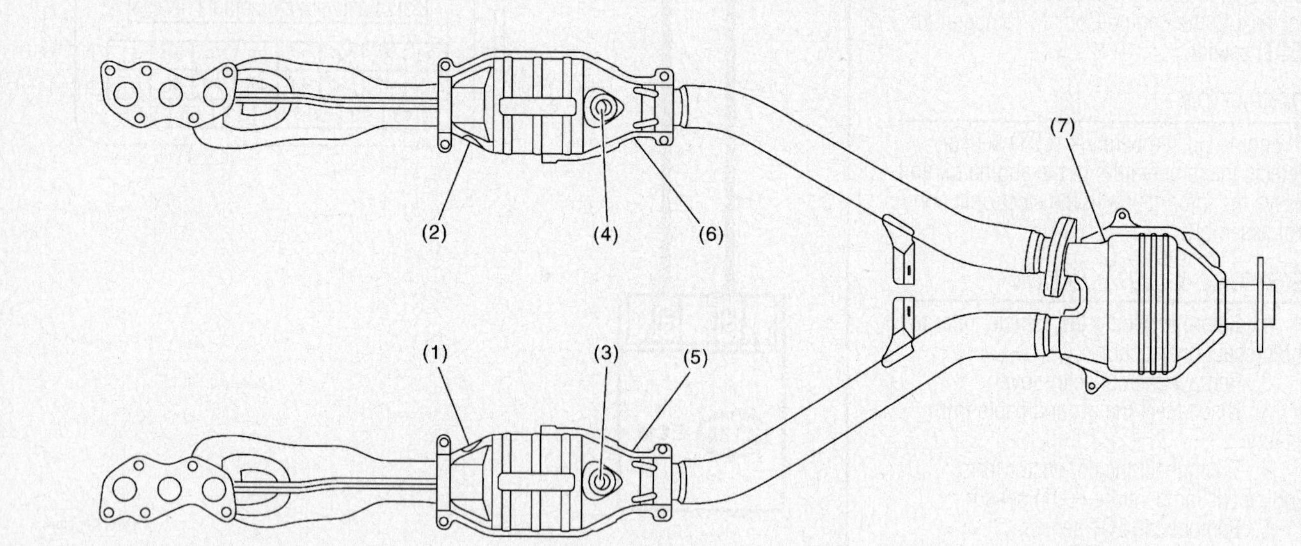

Fig. 155 Heated Oxygen Sensor (HO2S) location—2006-07 B9 Tribeca

OPERATION

The Heated Oxygen Sensors (HO2S) supplies the electronic control assembly with a signal which indicates either a rich or lean mixture condition, during the engine operation.

REMOVAL & INSTALLATION

Front Sensor

1. Before servicing the vehicle, refer to the precautions section.
2. Disconnect the negative battery cable.

3. Raise and support the vehicle safely.

4. Disconnect the connector.

5. Remove the Heated Oxygen Sensor (HO2S).

To install:

6. Installation is the reverse of the removal procedure.

➡ Apply anti-seize compound to the threaded portion of the sensor, prior to installation. Never apply anti-seize compound to the protector of the sensor.

7. Tighten the sensor to 16 ft. lbs. (21 Nm).

Rear Sensor

1. Before servicing the vehicle, refer to the precautions section.

2. Disconnect the negative battery cable.

3. Raise and support the vehicle safely.

4. Disconnect the connector. Remove the clip by pulling it out from the upper side of the crossmember.

5. Remove the Heated Oxygen Sensor (HO2S).

To install:

6 Installation is the reverse of the removal procedure.

➡ Apply anti-seize compound to the threaded portion of the sensor, prior to installation. Never apply anti-seize compound to the protector of the sensor.

7 Tighten the sensor to 16 ft. lbs. (21 Nm).

TESTING

Front Sensor

See Figure 158.

1. Measure the resistance between the Heated Oxygen Sensor (HO2S) connector terminals (terminals 1 and 2).

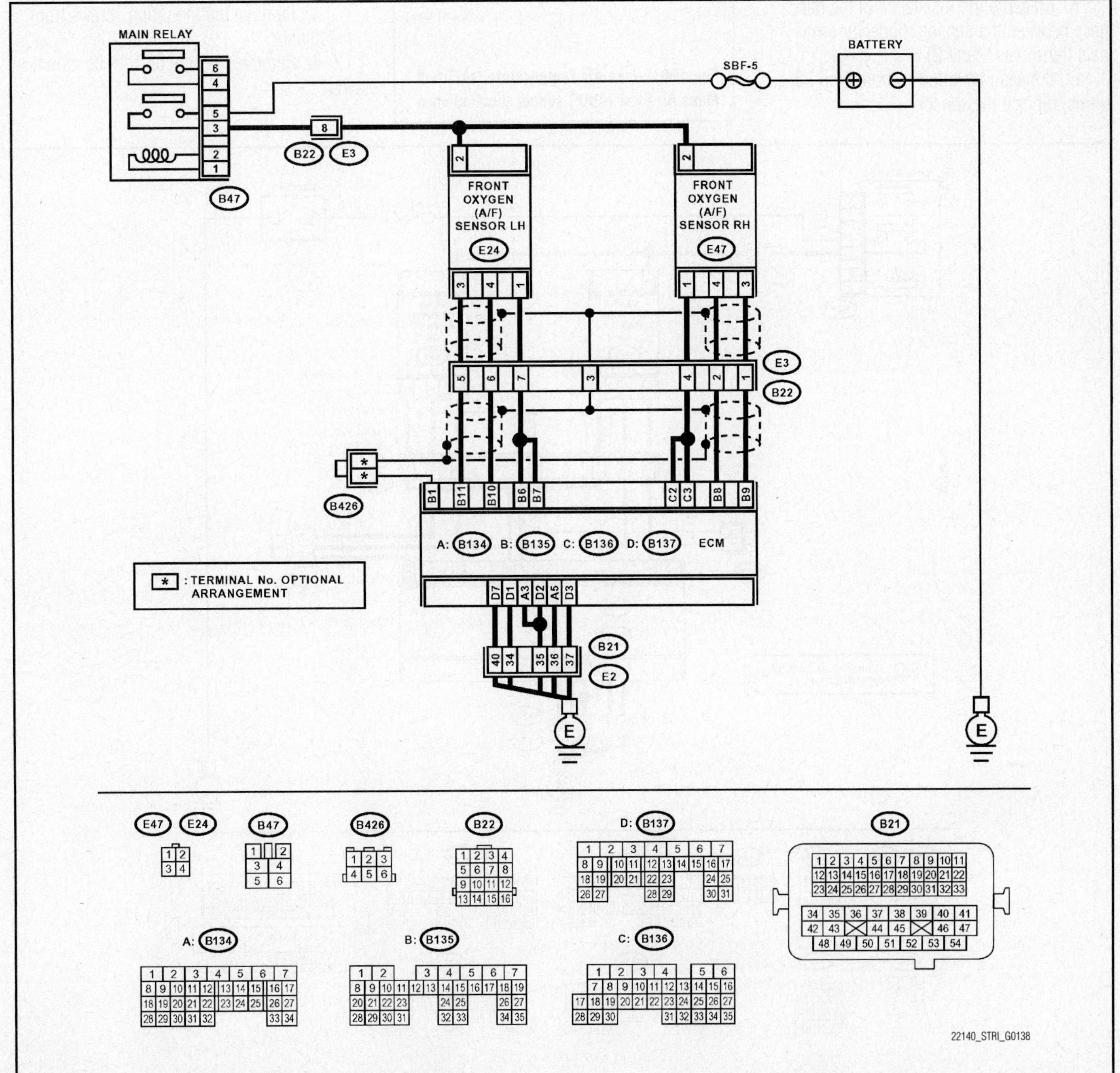

Fig. 158 Front Heated Oxygen Sensor (HO2S) circuit view

2. If resistance is more than 5 ohms, replace the sensor.

Rear Sensor

See Figure 159.

Perform a visual inspection of the Heated Oxygen Sensor (HO2S) as follows:

1. If the sensor tip has a black/sooty deposit, this may indicate a rich fuel mixture.
2. If the sensor tip has a white, gritty deposit, this may indicate an internal coolant leak.
3. If the sensor tip has a brown deposit, this could indicate oil consumption.
4. Turn the ignition switch OFF.
5. Measure the resistance of the harness between the sensor connector terminals (terminals 1 and 2).
6. If measured value is more than 30 ohms, replace the sensor.

INTAKE AIR TEMPERATURE (IAT) SENSOR

LOCATION

See Figure 160.

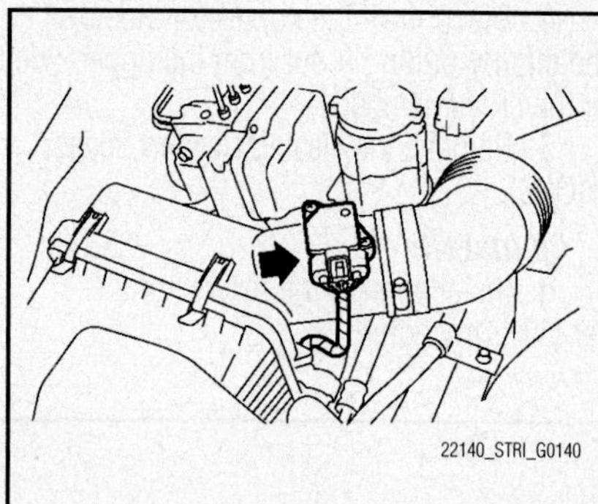

Fig. 160 Intake Air Temperature (IAT) and Mass Air Flow (MAF) sensor location view

The Intake Air Temperature (IAT) sensor is mounted in the intake air hose of the air cleaner assembly. This sensor is combined with the Mass Air Flow (MAF) sensor.

OPERATION

The sensor provides a signal to the ECM for incoming air temperature. This information is used to adjust the injector pulse width and in turn the air/fuel ratio.

REMOVAL & INSTALLATION

1. Before servicing the vehicle, refer to the precautions section.
2. Disconnect the negative battery cable.
3. Remove the mounting screws from the sensor.
4. Remove the sensor from its mounting.

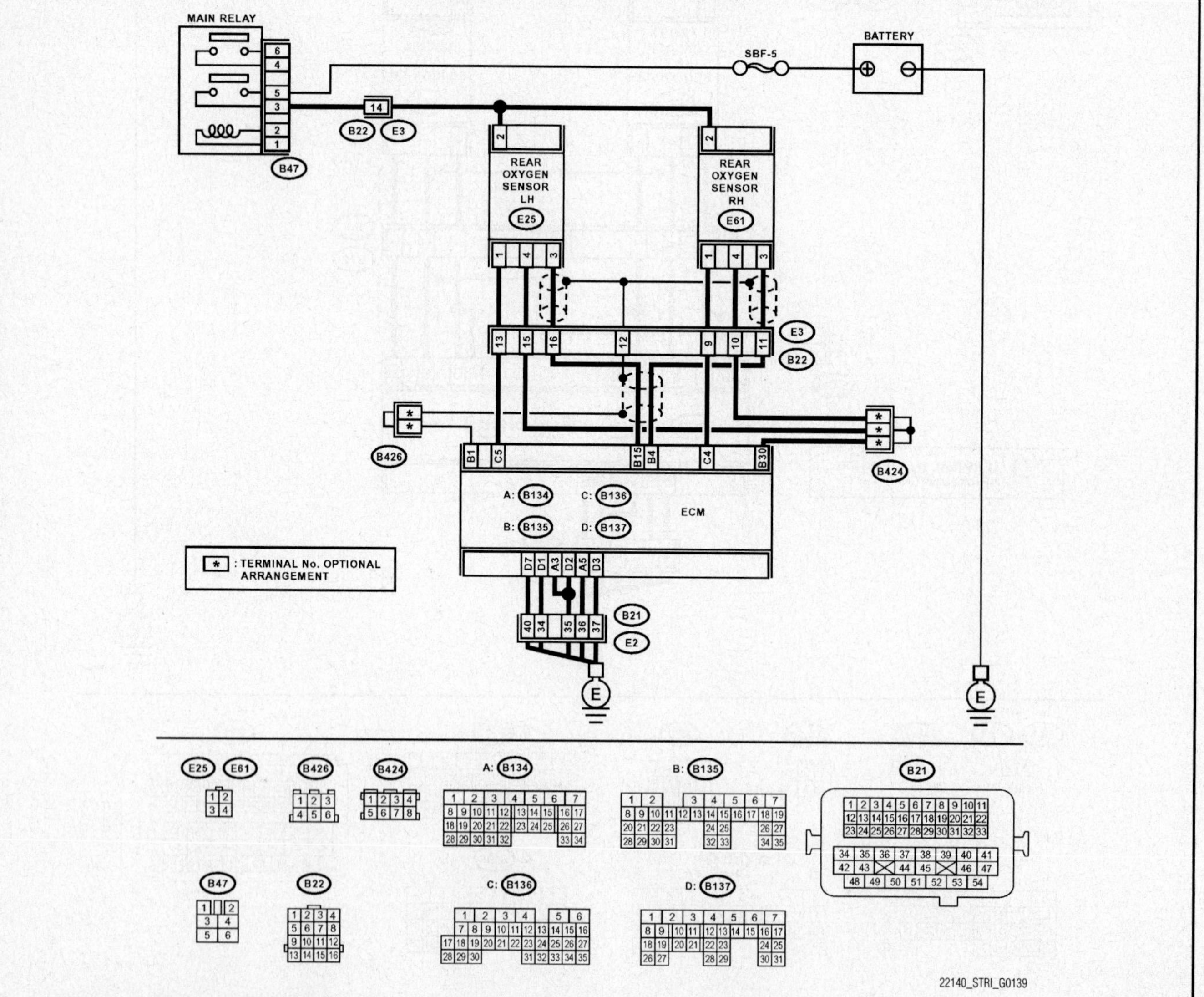

22140_STRI_G0139

Fig. 159 Front oxygen sensor circuit view

5. Installation is the reverse of the removal procedure.

6. Tighten the mounting screws to 0.7 ft. lbs. (1 Nm).

TESTING

1. Start the engine and allow it to reach operating temperature.

2. Using the Subaru scan tool, read the sensor signal data.

3. If the measured value is less than -40 degrees F, replace the sensor.

KNOCK SENSOR (KS)

LOCATION

See Figure 161.

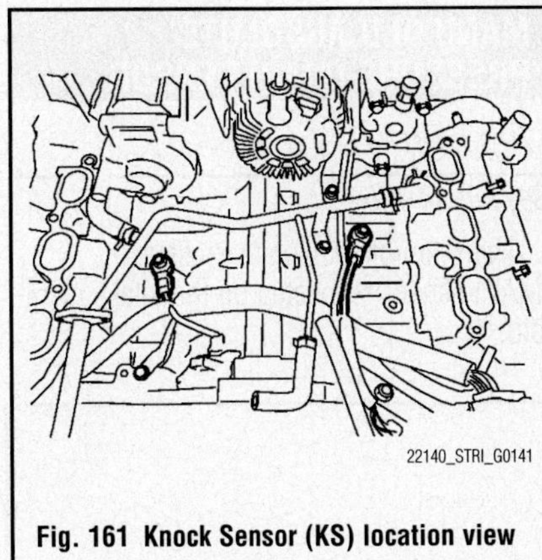

Fig. 161 Knock Sensor (KS) location view

The Knock Sensor (KS) is located under the intake manifold.

OPERATION

The KS is used to detect engine vibrations caused by preignition or detonation and provides information to the ECM, which then retards the timing to eliminate detonation.

REMOVAL & INSTALLATION

See Figure 162.

1. Before servicing the vehicle, refer to the precautions section.

2. Disconnect the negative battery cable.

3. Remove the collector cover, as required.

4. Remove the intake manifold.

5. Disconnect the sensor connector.

6. Remove the sensor from its mounting.

To install:

7. Installation is the reverse of the removal procedure.

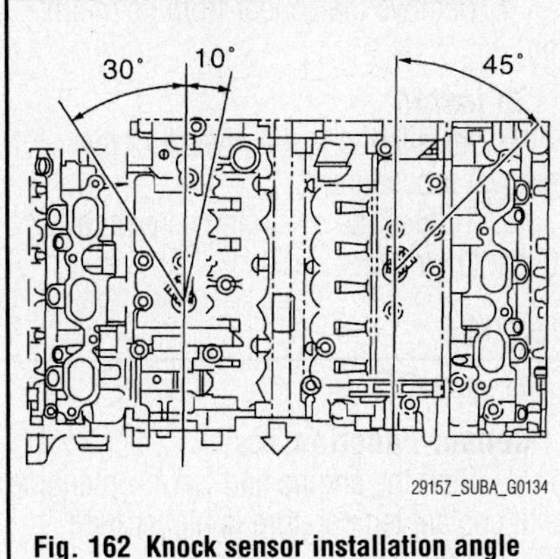

Fig. 162 Knock sensor installation angle

8. Refer to the illustration for proper sensor installation angle.

9. Tighten the sensor to 18 ft. lbs. (25 Nm).

TESTING

See Figure 163.

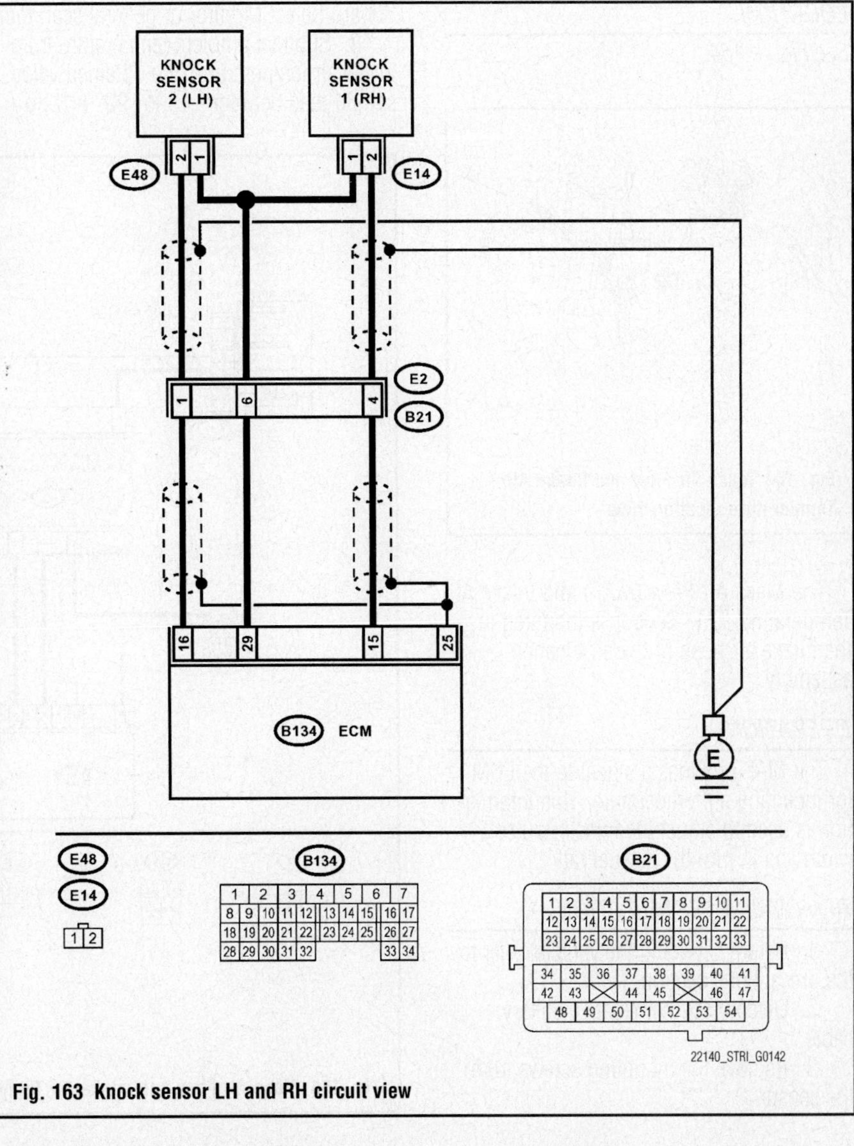

Fig. 163 Knock sensor LH and RH circuit view

1. Remove the sensor from the vehicle.

2. Measure the resistance between the connector terminals of the sensor, (terminal 1 and 2).

3. If measured value is more than 600k ohms, replace the sensor.

MALFUNCTION INDICATOR LIGHT (MIL)

RESET PROCEDURES

Subaru Select Monitor (OBD MODE)

1. On the (Main Menu) display screen, select the (Each System Check).

2. On the (System Selection Menu) display screen, select the (Engine Control System).

3. Select (OK) after the information of engine type has been displayed.

4. On the (Engine Diagnosis) display screen, select the (OBD System).

5. On the «OBD Menu» display screen, select the (Clear Diagnostic Code).

6. When the (Clear Diagnostic Code) is shown on the screen, click the (Yes) button.

7. When (Done) and (Turn ignition switch **OFF**) are shown on the display screen, turn the ignition switch **OFF**.

General Scan Tool

1. For procedures clearing memory using the general scan tool, refer to the general scan tool operation manual.

➡ Initial diagnosis of electronic throttle control is performed after memory clearance. For this reason, start the engine after 10 seconds or more have elapsed since the ignition switch was turned to ON.

➡ The Malfunction Indicator Light (MIL) must be reset with a scan tool.

MASS AIR FLOW (MAF) SENSOR

LOCATION

See Figure 164.

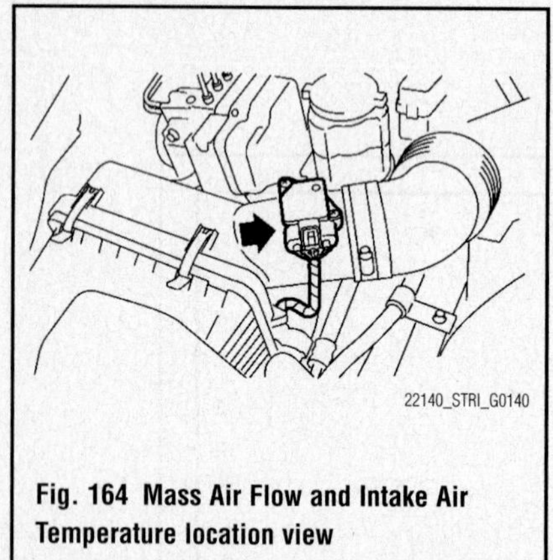

Fig. 164 Mass Air Flow and Intake Air Temperature location view

The Mass Air Flow (MAF) and Intake Air Temperature (IAT) Sensor is mounted in the intake air hose of the air cleaner assembly.

OPERATION

The MAF provides a signal to the ECM for incoming air temperature. This information is used to adjust the injector pulse width and in turn the air/fuel ratio.

REMOVAL & INSTALLATION

1. Before servicing the vehicle, refer to the precautions section.

2. Disconnect the negative battery cable.

3. Remove the mounting screws from the sensor.

4. Remove the sensor from its mounting.

To install:

5. Installation is the reverse of the removal procedure.

6. Tighten the mounting screws to 0.7 ft. lbs. (1 Nm).

TESTING

See Figure 165.

Sensor Function Test

1. Start the engine and warm-up engine until coolant temperature is higher than 167° F (75° C).

2. Place the select lever in "P" range or "N" range.

3. Turn the A/C switch **OFF**.

4. Turn all the accessory switches to OFF.

5. Open the front hood.

6. Measure the ambient temperature.

7. Read the data of mass air flow and intake air temperature sensor signal using Subaru Select Monitor or general scan tool.

8. Subtract ambient temperature from intake air temperature. The obtained value should read between -18 to -90° (-10 to -50° C). If the reading was within specifications, check for faulty contact of the MAF and ECM connectors.

9. If the reading was not as stated above, suspect faulty mass air flow and intake air temperature sensor.

Power Supply Test

10. Turn the ignition switch **OFF**.

11. Disconnect the connector from the mass air flow and intake air temperature sensor.

12. Turn the ignition switch **ON**.

13. Measure the voltage between mass air flow and intake air temperature sensor connector (B3) terminal No.1 and engine ground

14. If the voltage does not read 10 volts or more, check for an open circuit.

MANIFOLD ABSOLUTE PRESSURE (MAP) SENSOR

LOCATION

See Figure 166.

The Manifold Absolute Pressure (MAP) sensor is located on the intake manifold.

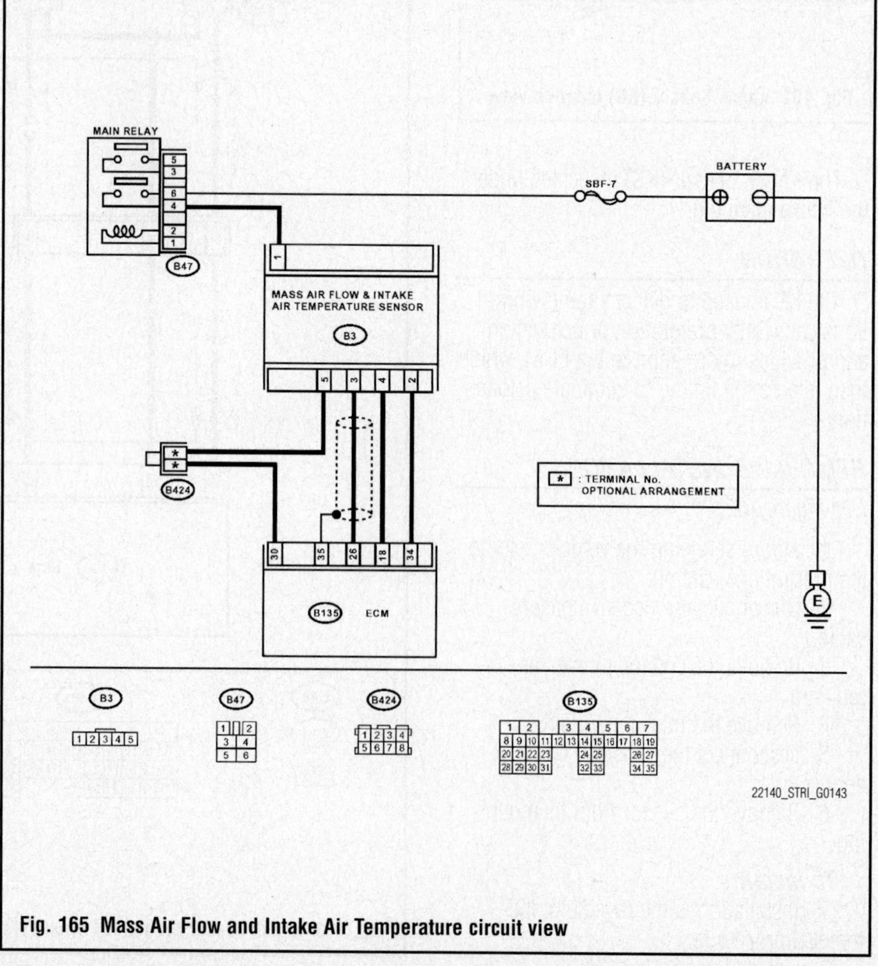

Fig. 165 Mass Air Flow and Intake Air Temperature circuit view

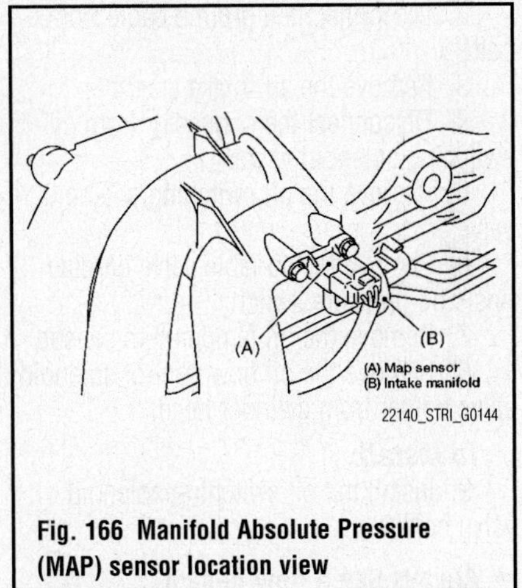

Fig. 166 Manifold Absolute Pressure (MAP) sensor location view

OPERATION

The Manifold Absolute Pressure (MAP) sensor signals the ECM of changes in intake manifold pressure which result from engine load, speed and atmospheric pressure changes.

REMOVAL & INSTALLATION

1. Before servicing the vehicle, refer to the precautions section.
2. Disconnect the negative battery cable.
3. Remove the collector cover, if equipped.
4. Disconnect the connector from the sensor.
5. Remove the filter assembly from the intake manifold, if equipped.
6. Remove the MAP sensor from its mounting.

To install:

7. Installation is the reverse of the removal procedure.
8. Be sure to use new O-rings.
9. Torque the retaining bolts to 5 ft. lbs. (6.4 Nm).

TESTING

See Figure 167.

1. Turn the ignition switch **OFF**.
2. Disconnect the connectors from the ECM.
3. Measure the resistance of the harness between the ECM and the sensor (connector B134 terminal 29 and connector E28 terminal1)
4. If the resistance is less than 1 ohm, replace the sensor.
5. Check poor contact of manifold absolute pressure sensor connector.
6. If connector contact are okay, replace MAP sensor.

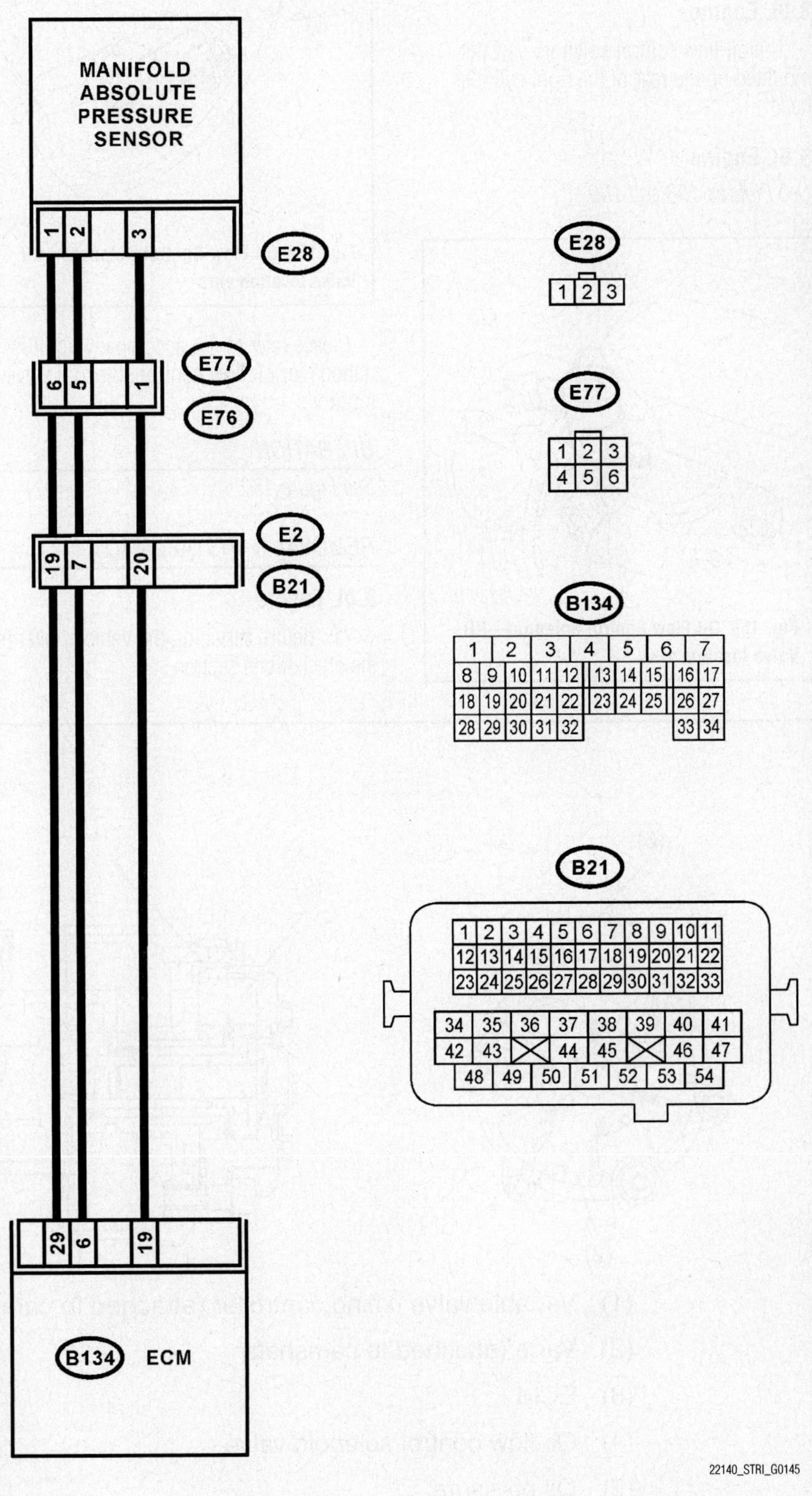

22140_STRI_G0145

Fig. 167 Manifold Absolute Pressure (MAP) sensor circuit view

OIL FLOW CONTROL SOLENOID

LOCATION

3.0L Engine

The oil flow control solenoid valve is mounted on the rear of the right cylinder head.

3.6L Engine

See Figures 168 and 169.

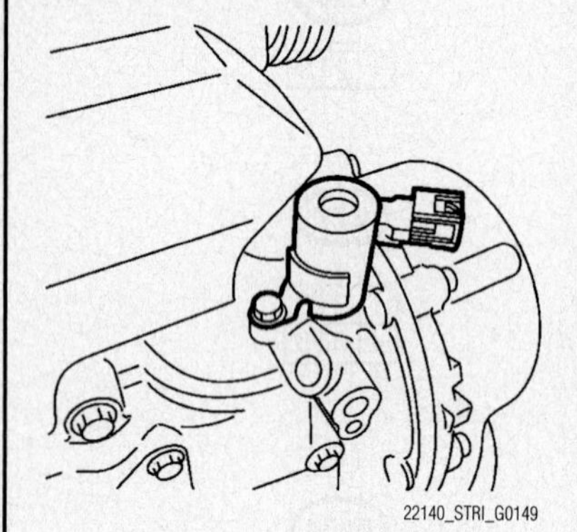

Fig. 168 Oil Flow Control Solenoid—RH Valve location view

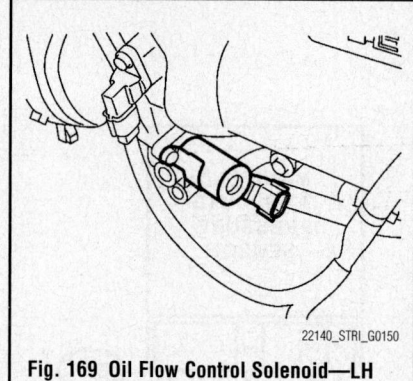

22140_STRI_G0150

Fig. 169 Oil Flow Control Solenoid—LH Valve location view

Please refer to the accompanying illustrations for oil flow control solenoid valve locations.

OPERATION

See Figure 170.

REMOVAL & INSTALLATION

3.0L Engine

1. Before servicing the vehicle, refer to the precautions section.
2. Disconnect the ground cable from battery.
3. Remove the air intake chamber.
4. Disconnect the connector from oil switching solenoid valve.
5. Remove the oil switching solenoid valve.
6. Remove the variable valve lift diagnosis oil pressure switch.
7. Remove the oil temperature sensor.
8. Remove the oil flow control solenoid valve holder from cylinder head.

To install:

9. Install the oil switching solenoid valve holder.

➡ **Always use a new gasket.**

10. Tighten the mounting bolts to 5 ft. lbs. (6.4 Nm).
11. Install the oil temperature sensor.
12. Install the variable valve lift diagnosis oil pressure switch.
13. Install the oil switching solenoid valve.
14. Connect the connector to oil switching solenoid valve.
15. Install the air intake chamber.

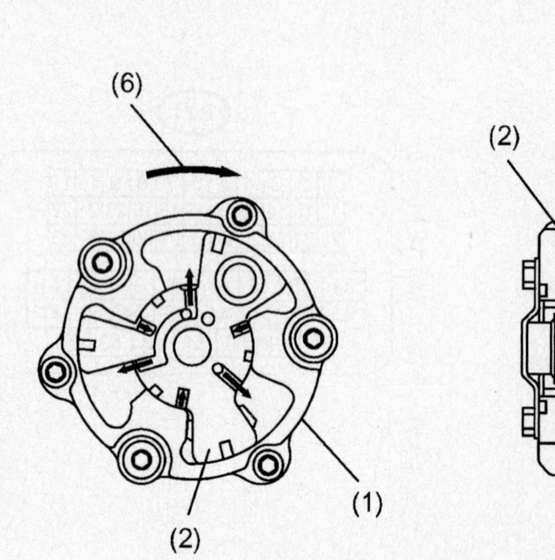

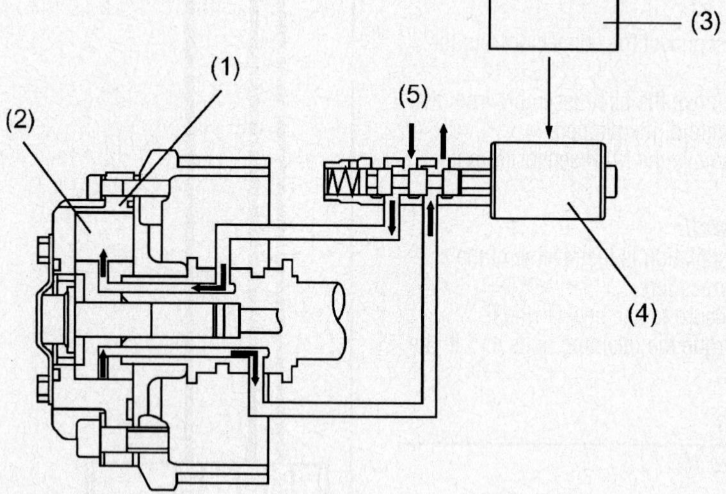

(1) Variable valve timing controller (attached to camshaft sprocket)
(2) Vane (attached to camshaft)
(3) ECM
(4) Oil flow control solenoid valve
(5) Oil pressure
(6) Turns in advance direction

22140_STRI_G0148

Fig. 170 Oil Flow Control Solenoid Valve and related parts

3.6L Engine

Intake Side

1. Before servicing the vehicle, refer to the precautions section.
2. Remove the collector cover.
3. Disconnect the ground cable from battery.
4. Remove the generator harness from the fuel protector RH.
5. Disconnect the connector from intake oil flow control solenoid valve RH.
6. Remove the intake oil flow control solenoid valve RH.
7. Remove the intake oil flow control solenoid valve LH in the same procedure as RH.

To install:

8. Install in the reverse order of removal.
9. Tighten the oil flow control solenoid valve mounting bolts to 5 ft. lbs. (6.4 Nm).

Exhaust Side

1. Disconnect the ground cable from battery.
2. Lift up the vehicle.
3. Remove the under cover
4. Disconnect the connector from exhaust oil flow control solenoid valve LH.
5. Remove the exhaust oil flow control solenoid valve LH.
6. Remove the exhaust oil flow control solenoid valve RH in the same procedure as LH.

To install:

7. Install in the reverse order of removal.
8. Tighten the oil flow control solenoid valve mounting bolts to 5 ft. lbs. (6.4 Nm).

TESTING

See Figure 171.

1. Measure the resistance between oil flow control solenoid valve terminals No. 1 and No. 2.
2. If the reading is not between 6–12 ohms, replace the oil flow control solenoid valve
3. If the reading is between 6–12 ohms. Check for poor contact of the ECM and oil flow control solenoid valve connectors.

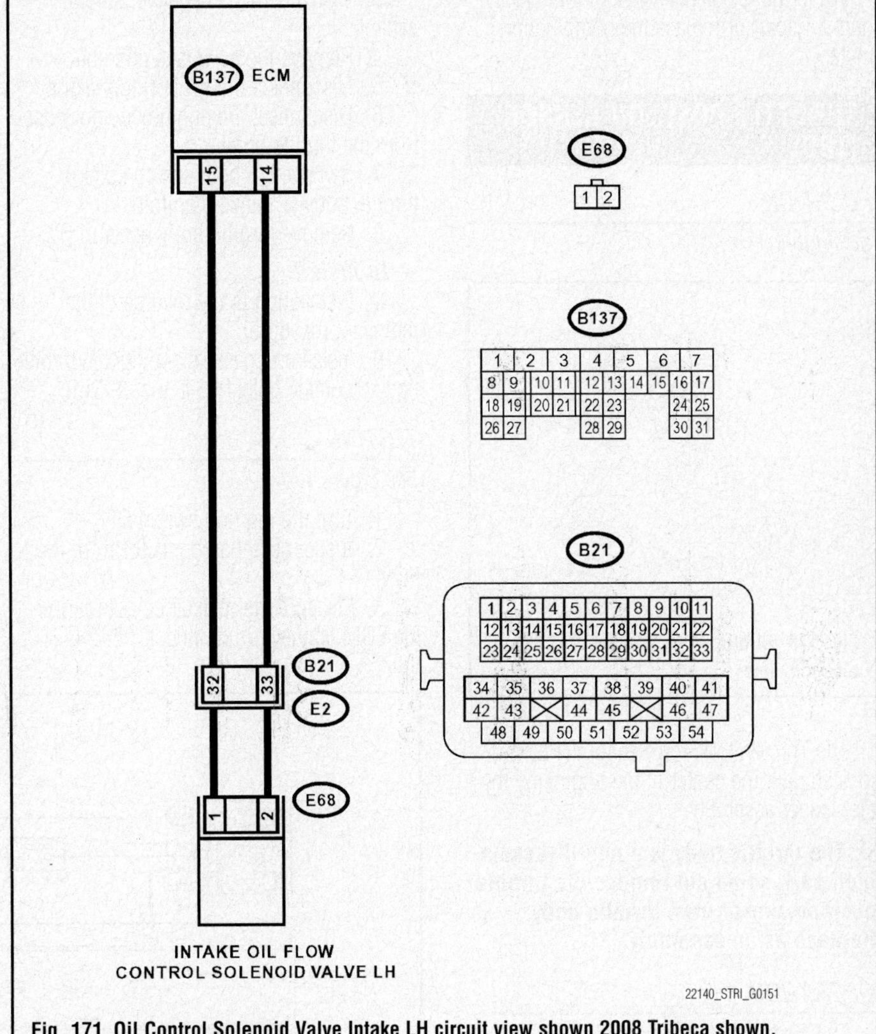

Fig. 171 Oil Control Solenoid Valve Intake LH circuit view shown 2008 Tribeca shown, 2006–07 B9 Tribeca similar

OIL PRESSURE SENSOR

REMOVAL & INSTALLATION

See Figure 172.

1. Set the vehicle on a lift.
2. Disconnect the ground cable from battery.
3. Lift-up the vehicle.
4. Remove the undercover.
5. Disconnect the terminal from oil pressure switch.
6. Remove the oil pressure switch.

To install:

7. Apply liquid gasket THREE BOND 1324 (Part No. 004403042) or equivalent to the oil pressure switch threads.
8. Install the oil pressure switch. Tighten to 18 ft. lbs. (25 Nm)
9. Connect the terminal of the oil pressure switch.
10. Install the undercover.

TESTING

1. Turn the ignition switch **OFF**.
2. Disconnect the connector from oil pressure sensor.
3. Turn the ignition switch **ON**.
4. Measure the voltage of harness between oil pressure sensor connector and chassis ground.
5. Is the voltage 10 volts or more, replace the oil pressure switch.
6. If not, turn the ignition switch **OFF**.
7. Remove the combination meter.
8. Measure the resistance of combination meter.
9. Is the resistance less than 10 ohms, repair poor connection or open circuit

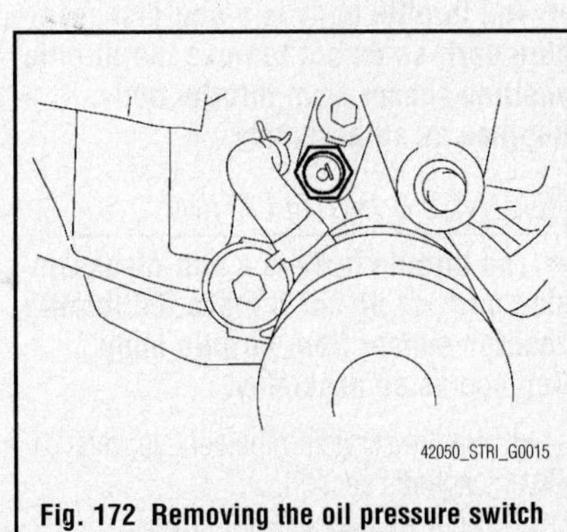

Fig. 172 Removing the oil pressure switch

10. If the resistance is more than 10 ohms. Repair or replace the combination meter.

THROTTLE POSITION SENSOR (TPS)

LOCATION

See Figure 173.

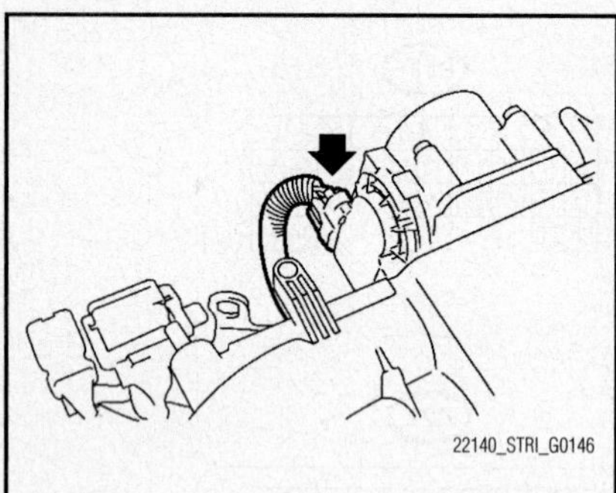

Fig. 173 Throttle Position Sensor (TPS) location view

The Throttle Position Sensor (TPS) is located near the center of the engine by the air cleaner assembly.

➡ The throttle body is a non-disassembled part, so do not remove the throttle position sensor from throttle body. Replace as an assembly.

OPERATION

The Throttle Position Sensor (TPS) provides a signal to the ECM that is related to the relative throttle plate position. As the throttle plate moves in relation to driving conditions, a signal is sent to the control unit which adjusts the injector pulse width and air/fuel ratio. As the throttle plate is opened further, more air is taken into the combustion chambers, and as a result the relative fuel demand of the engine changes. The throttle position sensor relays this information to the ECM which in alters the fuel amount.

➡ The throttle body is a non-disassembled part, so do not remove the throttle position sensor from throttle body. Replace as an assembly.

REMOVAL & INSTALLATION

➡ The throttle body is a non-disassembled part, so do not remove the throttle position sensor from throttle body. Replace as an assembly.

1. Before servicing the vehicle, refer to the precautions section.
2. Remove the collector cover.
3. Disconnect the negative battery cable.
4. Remove the air intake chamber.
5. Disconnect the sensor connector.
6. Disconnect the engine coolant hoses from the throttle body.
7. Remove the bolts which secure throttle body to intake manifold.
8. Remove throttle body assembly.

To install:

9. Installation is the reverse of the removal procedure.
10. Install new gaskets and tighten throttle body mounting bolts to 6 ft. lbs. (8 Nm).

TESTING

See Figure 174.

1. Turn the ignition switch OFF.
2. Disconnect the connector from the ECM.
3. Measure the resistance between the ECM connectors (connector B134 terminal 18 and connector B134 terminal 19).
4. If the measured value is more than 1m ohms, repair the electronic throttle control.

VEHICLE SPEED SENSOR (VSS)

LOCATION

See Figure 175 and 176.

This vehicle uses a front and rear Vehicle Speed Sensor (VSS). Both sensors are mounted on the transaxle.

OPERATION

The Vehicle Speed Sensor (VSS) sends a signal to the ECM to control. The ECM uses this information to control transmission shift patterns. The vehicle speed sensor is also used to send a speed signal to the speed control servo of the cruise control system.

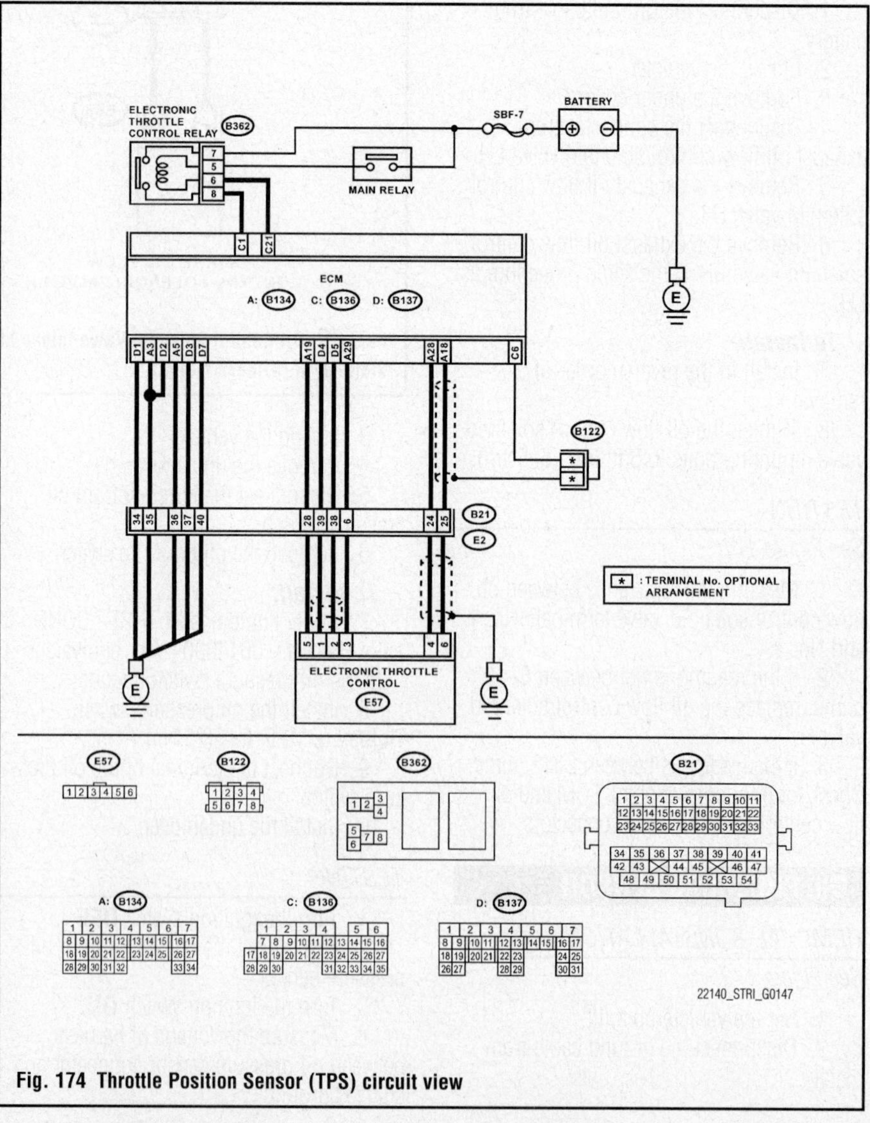

Fig. 174 Throttle Position Sensor (TPS) circuit view

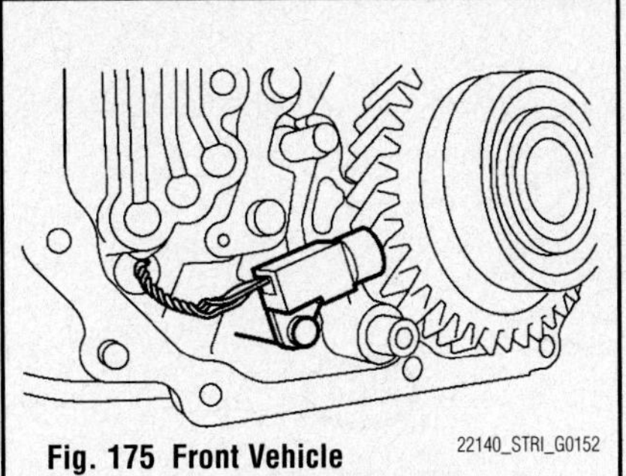

Fig. 175 Front Vehicle Speed Sensor (VSS) location view

REMOVAL & INSTALLATION

1. Raise and support the vehicle safely.
2. Place a drip pan below the speed sensor to catch any spilled fluid.
3. Disconnect the connector from the Vehicle Speed Sensor (VSS).
4. Remove the VSS from its mounting.

To install:

5. Installation is the reverse of the removal procedure.
6. Replace any lost fluid.

TESTING

See Figures 177 and 178.

1. Turn the ignition switch **OFF**.
2. Disconnect the connectors from TCM and transmission.
3. Measure the resistance of harness between TCM connector and transmission VSS connectors.
4. If the resistance is not less than 1 ohm, repair the open circuit of harness between TCM and transmission VSS connector.

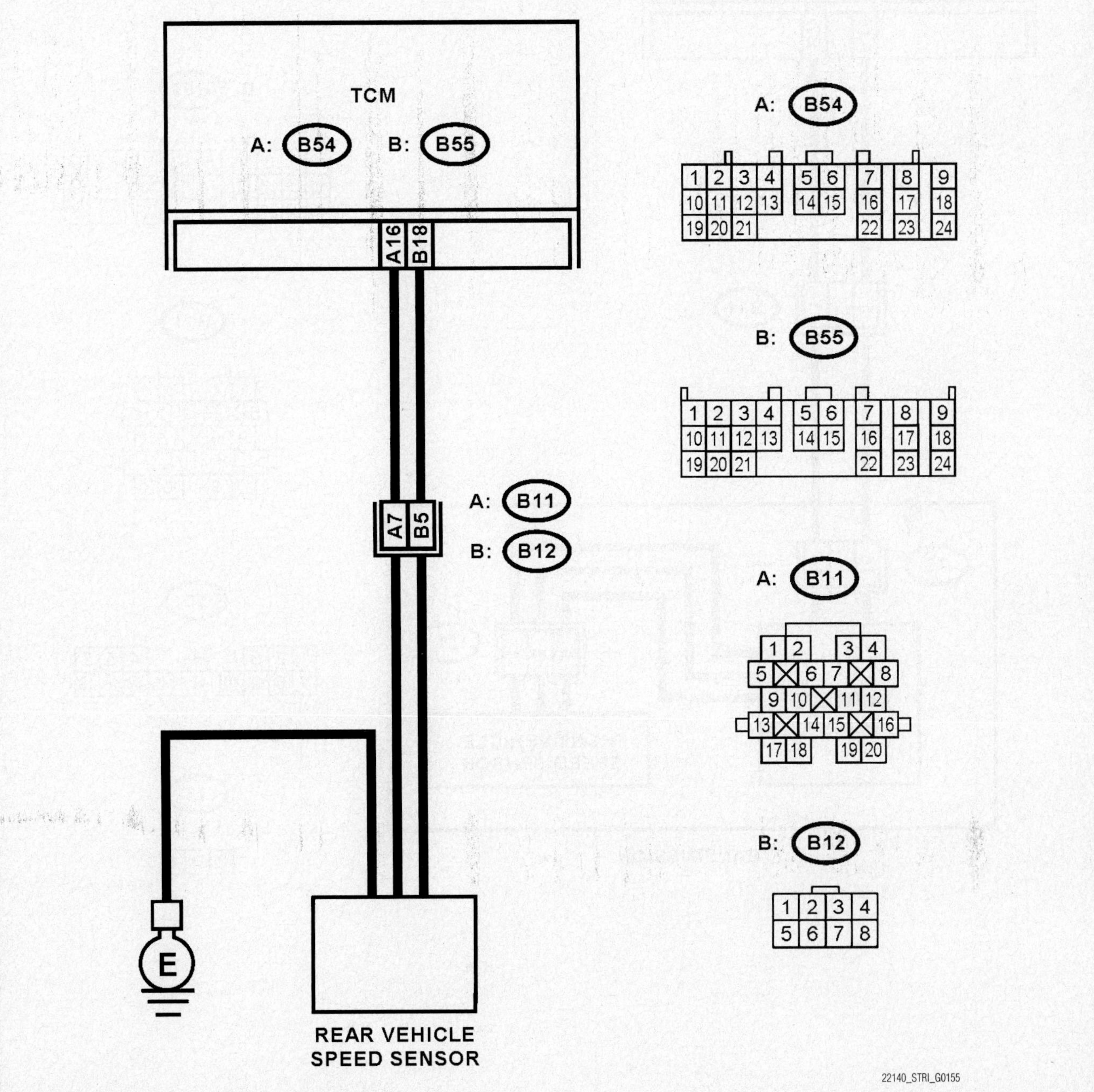

Fig. 176 Rear Vehicle Speed Sensor (VSS) location view

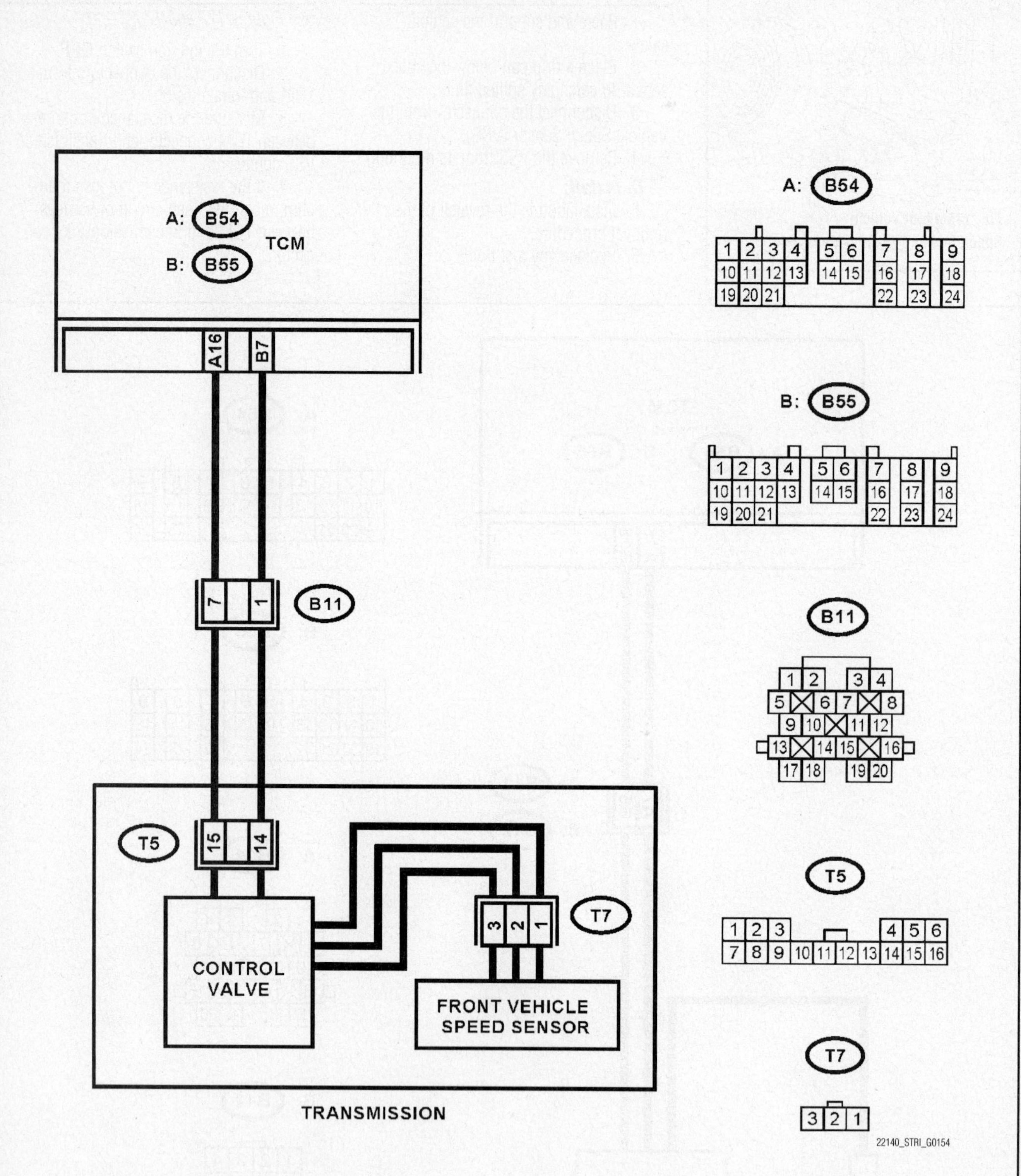

Fig. 177 Front Vehicle Speed Sensor (VSS) circuit view

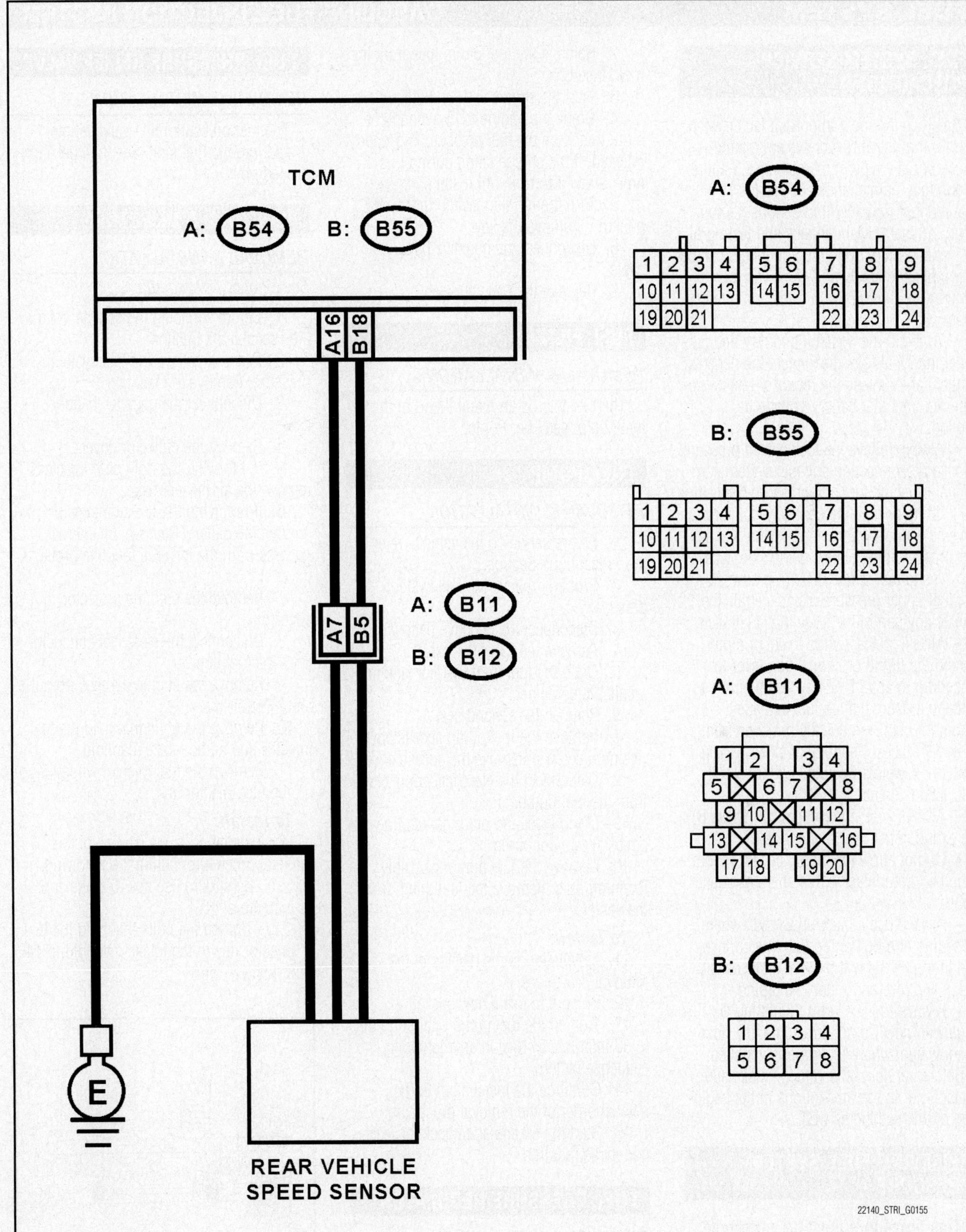

Fig. 178 Rear Vehicle Speed Sensor (VSS) circuit view

FUEL SYSTEM SERVICE PRECAUTIONS

Safety is the most important factor when performing not only fuel system maintenance but any type of maintenance. Failure to conduct maintenance and repairs in a safe manner may result in serious personal injury or death. Maintenance and testing of the vehicle's fuel system components can be accomplished safely and effectively by adhering to the following rules and guidelines.

- To avoid the possibility of fire and personal injury, always disconnect the negative battery cable unless the repair or test procedure requires that battery voltage be applied.
- Always relieve the fuel system pressure prior to disconnecting any fuel system component (injector, fuel rail, pressure regulator, etc.), fitting or fuel line connection. Exercise extreme caution whenever relieving fuel system pressure to avoid exposing skin, face and eyes to fuel spray. Please be advised that fuel under pressure may penetrate the skin or any part of the body that it contacts.
- Always place a shop towel or cloth around the fitting or connection prior to loosening to absorb any excess fuel due to spillage. Ensure that all fuel spillage (should it occur) is quickly removed from engine surfaces. Ensure that all fuel soaked cloths or towels are deposited into a suitable waste container.
- Always keep a dry chemical (Class B) fire extinguisher near the work area.
- Do not allow fuel spray or fuel vapors to come into contact with a spark or open flame.
- Always use a back-up wrench when loosening and tightening fuel line connection fittings. This will prevent unnecessary stress and torsion to fuel line piping.
- Always replace worn fuel fitting O-rings with new Do not substitute fuel hose or equivalent where fuel pipe is installed.

Before servicing the vehicle, make sure to also refer to the precautions in the beginning of this section as well.

RELIEVING FUEL SYSTEM PRESSURE

➡ This procedure must be performed prior to servicing any component of the fuel injection system.

1. Before servicing the vehicle, refer to the Precautions Section.
2. Remove the fuel pump fuse from the main fuse box.
3. Start the engine and run until it stalls.
4. Crank the engine for 5 seconds or more to ensure the fuel pressure is properly relieved. If the engine starts during this time, allow it to run until it stalls.
5. Turn the ignition switch to the OFF position. Remove the key.
6. Disconnect the negative battery cable.
7. Remove the fuel cap.

FUEL FILTER

REMOVAL & INSTALLATION

The fuel filter is an integral part of the fuel pump, see Fuel Pump.

FUEL PUMP

REMOVAL & INSTALLATION

1. Before servicing the vehicle, refer to the Precautions Section.
2. Properly relieve the fuel system pressure.
3. Disconnect the negative battery cable.
4. Remove the fuel filler cap.
5. Drain the fuel from the fuel tank into a suitable container.
6. Remove the second seat.
7. Remove the fuel pump access cover retaining screws. Remove the access cover.
8. Disconnect the electrical connector from the fuel pump.
9. Disconnect and plug the fuel line hoses at the fuel pump.
10. Remove the fuel pump assembly retaining nuts. Remove the fuel pump from the vehicle.

To install:

11. Installation is the reverse of the removal procedure.
12. Be sure to use a new gasket.
13. Tighten the fuel pump retaining nuts to 36 inch lbs, (4 Nm) in an alternating sequence pattern.
14. Continue the installation in the reverse order of the removal procedure.
15. Start the engine and check for leaks, correct as required.

FUEL PRESSURE REGULATOR

REMOVAL & INSTALLATION

The fuel pressure regulator is integrated with the fuel pump. See the Fuel Pump procedure

FUEL PRESSURE RELIEF VALVE

REMOVAL & INSTALLATION

The fuel pressure relief valve is integrated into the fuel tank. See the Fuel Tank procedure

FUEL RAIL & INJECTORS

REMOVAL & INSTALLATION

See Figures 179 and 180.

1. Before servicing the vehicle, refer to the Precautions Section.
2. Properly relieve the fuel system pressure. Remove the fuel cap.
3. Disconnect the negative battery cable.
4. Remove the collector cover.
5. If removing the right side injectors, remove the air cleaner case.
6. If removing the left side injectors, remove the battery. Remove the alternator harness from the left side fuel line protector cover.
7. Remove the fuel line protector cover(s).
8. Disconnect the electrical connector from the fuel injector.
9. Remove the engine harness from the fuel injector line.
10. Remove the bolts that hold the fuel injector line to the intake manifold.
11. Remove the fuel injector while lifting up the fuel injector line.

To install:

12. Installation is the reverse of the removal procedure, noting the following:
    a. Be sure to use new O-rings and insulators.
    b. Tighten the bolts that hold the fuel injector line to the intake manifold to 14 ft. lbs. (19 Nm).

Fig. 179 Fuel injector line mounting bolt shown

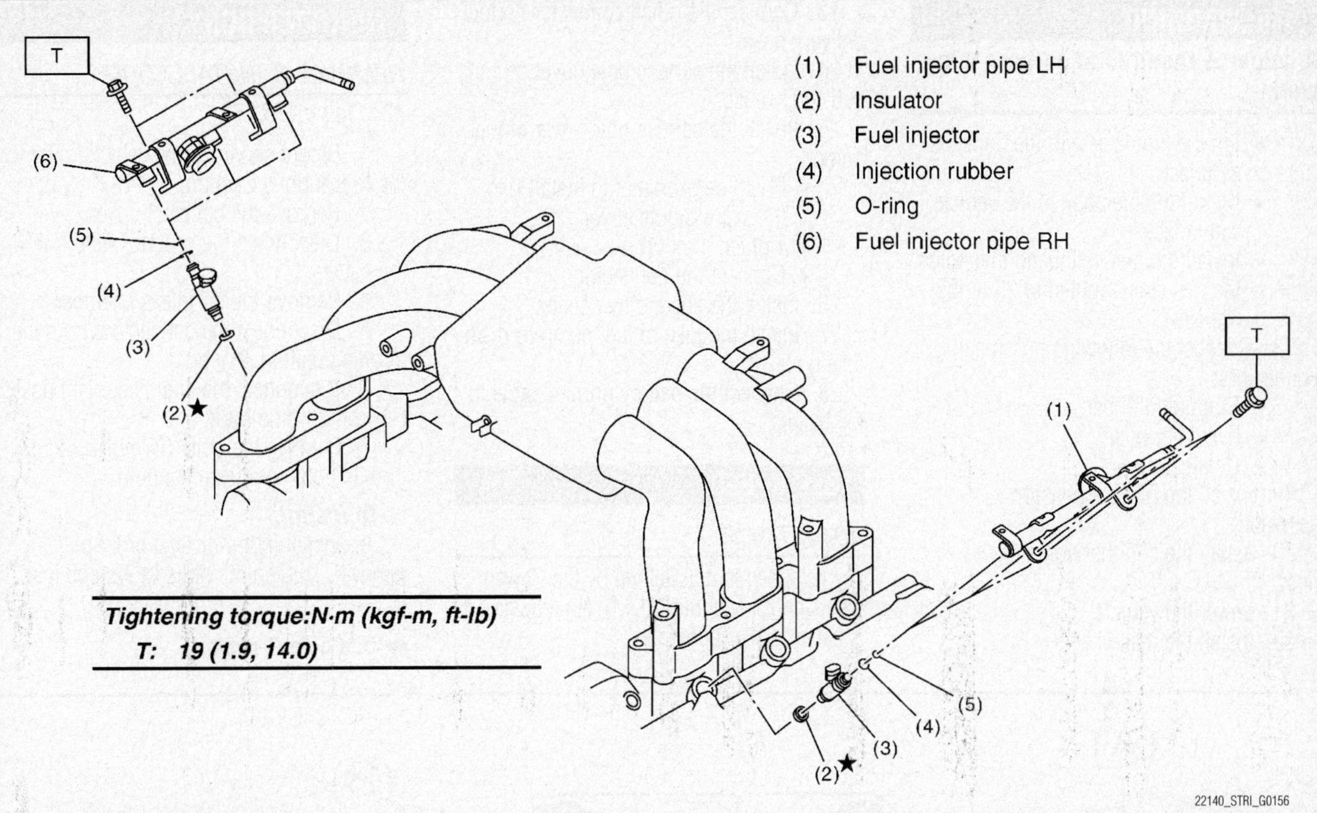

(1) Fuel injector pipe LH
(2) Insulator
(3) Fuel injector
(4) Injection rubber
(5) O-ring
(6) Fuel injector pipe RH

Tightening torque:N·m (kgf-m, ft-lb)
T: 19 (1.9, 14.0)

22140_STRI_G0156

Fig. 180 Fuel rail and injector view

c. Tighten the fuel line protector cover bolts to 14 ft. lbs. (19 Nm).

d. Start the engine and check for leaks, correct as required.

FUEL TANK

REMOVAL & INSTALLATION

See Figures 181 and 182.

1. Before servicing the vehicle, refer to the Precautions Section.
2. Set the vehicle on a lift.
3. Release the fuel pressure by performing the following:
   - Remove the fuse of fuel pump from main fuse box
   - Start the engine and run until it stalls
   - After the engine stalls, crank it for five more seconds
   - Turn the ignition switch **OFF**
4. Disconnect the ground cable from battery.
5. Disconnect the ground cable from battery.
6. Drain fuel from fuel tank.
7. Remove the second row seats.
8. Remove the service hole cover of fuel pump.
   - Disconnect the connector
   - Remove the bolt
   - Push the grommet down and remove service hole cover
9. Remove the service hole cover of fuel sub level sensor.
10. Disconnect the quick connector on the fuel delivery hose.
11. Remove the rear wheels.
12. Lift-up the vehicle.
13. Remove the clip holding the harness.
14. Remove the rear suspension assembly.
15. Disconnect the following hoses and connectors:
    - Evaporation hose (A)
    - Fuel filler hose (C)
    - Connector (B)

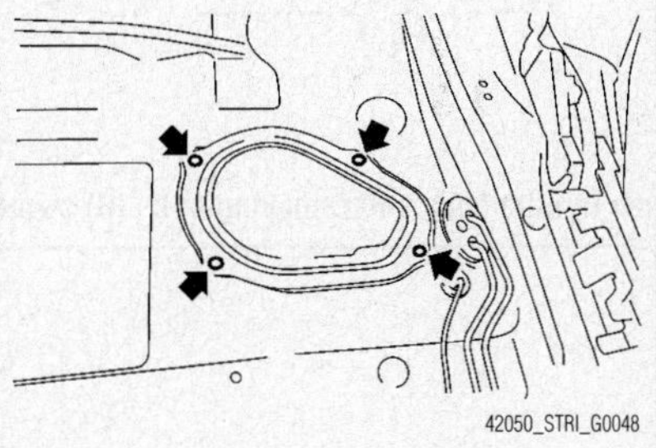

42050_STRI_G0048

Fig. 181 Remove the service hole cover of the fuel sub level sensor

⁂ WARNING

A helper is required to perform this work.

16. Remove the fuel tank from the vehicle, as follows:
    - Support the fuel tank with the transmission jack
    - Remove the fuel tank band that holds the fuel tank
    - Remove the fuel tank from the vehicle

To install:

17. Install the fuel tank to the vehicle.

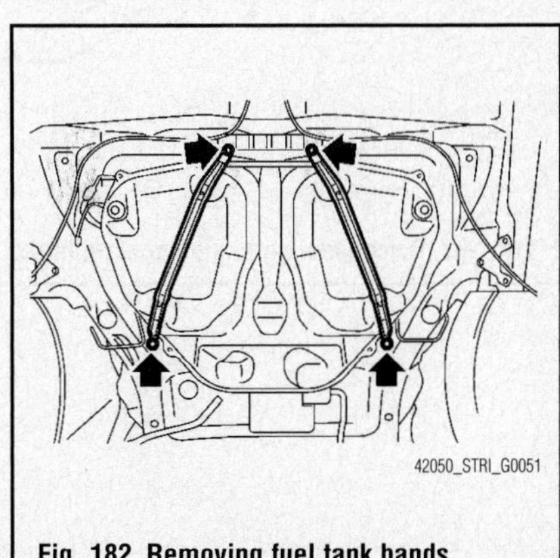

Fig. 182 Removing fuel tank bands

WARNING

A helper is required to perform this work.

- Lift the fuel tank with the transmission jack
- Mount the fuel tank to the vehicle with a jack
- Install the fuel tank band that holds the fuel tank. Tighten to 25 ft. lbs. (33 Nm)

18. Connect the following hoses and connectors:
    - Evaporation hose
    - Fuel filler hose
    - Connector
19. Install the rear suspension assembly.
20. Install the clip that holds the harness.
21. Lower the vehicle.
22. Install the rear wheels.
23. Connect the quick connector on fuel delivery hose.
24. Install the service hole cover of fuel sub level sensor.
25. Install the service hole cover of fuel pump:
    - Push the grommet to install it to the second hole cover
    - Tighten the bolt
    - Connect the connector
26. Install the second row seats.
27. Install the fuse of fuel pump to main fuse box.
28. Connect the battery ground cable to the battery.

IDLE SPEED

ADJUSTMENT

Idle speed is maintained by the Powertrain Control Module (PCM). No adjustment is necessary or possible.

THROTTLE BODY

REMOVAL & INSTALLATION

See Figure 183.

1. Before servicing the vehicle, refer to the Precautions Section.
2. Remove the collector cover.
3. Disconnect the ground cable from the battery.
4. Remove the air intake chamber.
5. Disconnect the connectors from the throttle position sensor.
6. Disconnect the engine coolant hoses (A) from throttle body.
7. Remove the bolts (B) which secure throttle body to intake manifold.

To install:

8. Install in the reverse order of removal. Tighten the bolts to 72 inch lbs. (8 Nm).

➡ **Use new O-rings.**

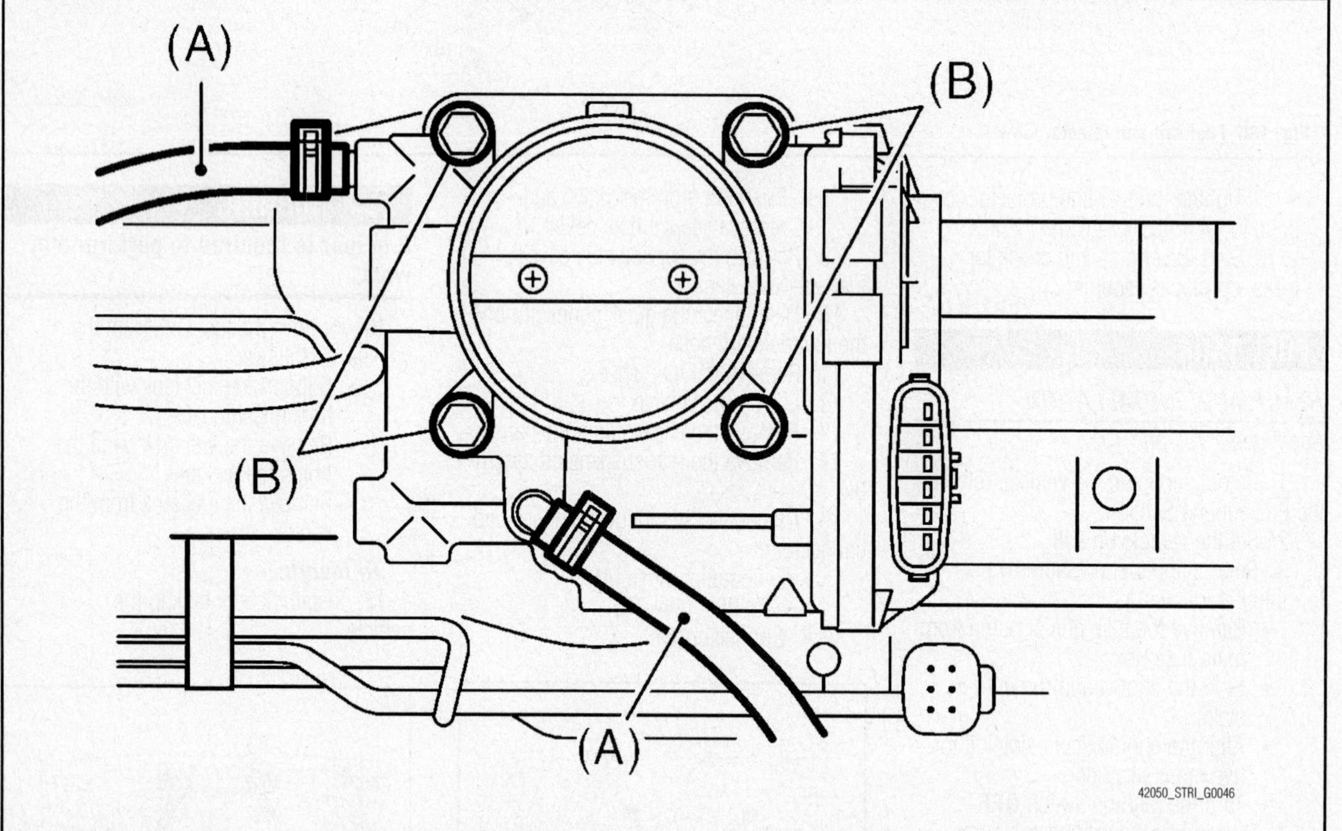

Fig. 183 Disconnecting engine coolant hoses (A) from throttle body and removing bolts (B) securing throttle body to the intake manifold

HEATING & AIR CONDITIONING SYSTEM

BLOWER MOTOR

REMOVAL & INSTALLATION

See Figure 184.

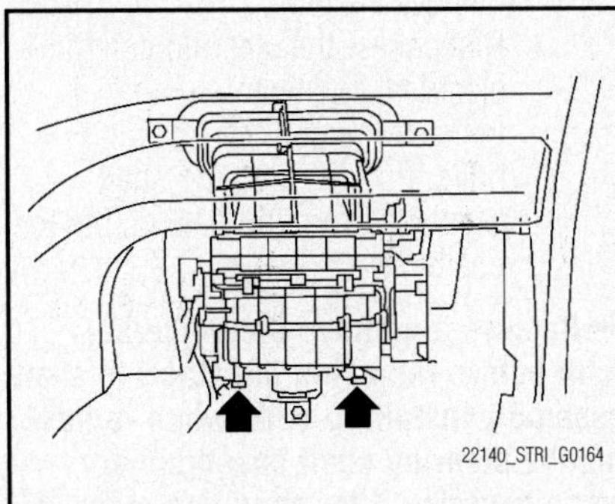

Fig. 184 Blower motor mounting screws shown

1. Before servicing the vehicle, refer to the Precautions Section.
2. Disconnect the ground cable from battery.
3. Remove the glove box lower cover.
4. Disconnect the blower motor electrical connector.
5. Loosen the screws to remove the blower motor.

To install:

6. Install the blower motor and mounting screws.
7. Reconnect the blower motor electrical connector
8. Install the glove box lower cover.
9. Connect the negative battery cable.

HEATER CORE

REMOVAL & INSTALLATION

See Figures 185 through 192.

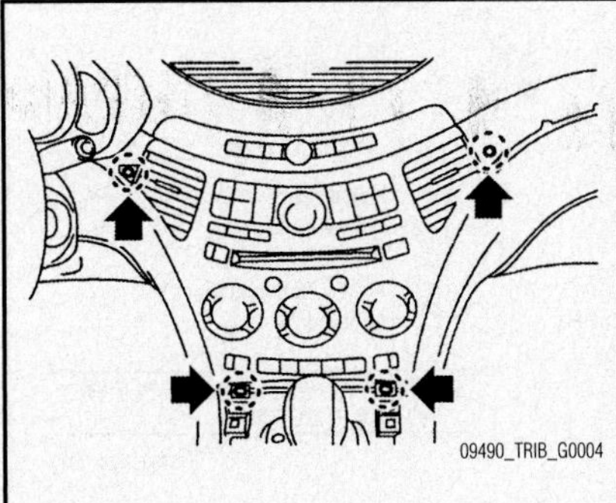

Fig. 185 Front control panel retaining clip locations

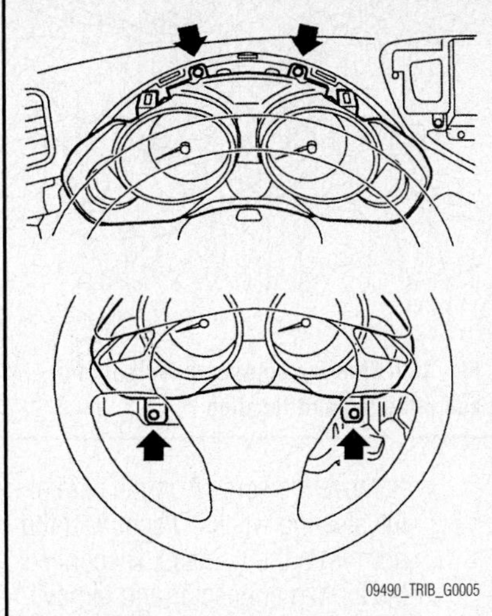

Fig. 186 Instrument cluster retaining screw locations

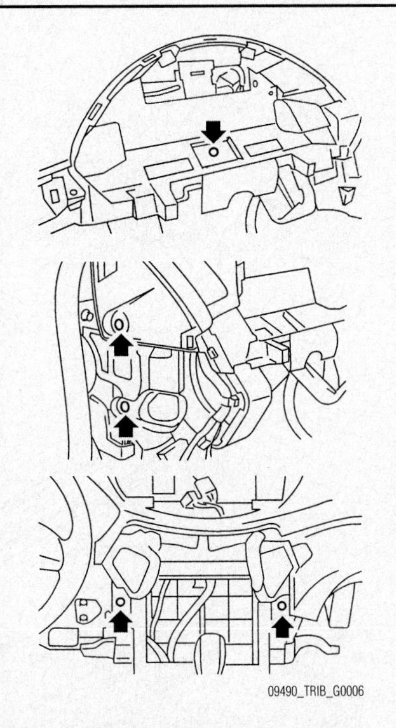

Fig. 187 Upper instrument panel retaining screw locations

➡ Position the front wheels in the straight ahead position.

1. Before servicing the vehicle, refer to the Precautions Section.
2. Disarm the SRS system. Wait at least 20 seconds before starting any repair work.

➡ Air bag connectors are colored yellow. Do not use electrical equipment on these circuits. Be careful not to damage the airbag system harness when servicing.

3. Disconnect the negative battery cable.
4. Drain the engine coolant. Discharge the air conditioning system.
5. Remove the bolts securing the expansion valve and pipe.
6. Disconnect and plug the heater hoses.
7. To remove the instrument panel upper:
   - Remove the front door scuff plate and pillar lower trim
   - Remove the left side instrument panel side cover
   - Remove the clips, and remove the left side console lower panel
   - Remove the clips, and remove the instrument panel lower under cover
   - Remove the clips and hooks, disconnect the connectors, and then remove the instrument panel lower cover
   - Remove the glove box
   - Remove the front console panel
   - Remove the instrument panel ornament and the driver's and passenger's side inner panels
   - Remove the screws, and pull out the control panel. Disconnect the electrical harness and remove the control panel.
   - Remove the instrument panel lower cover
   - Remove the console ring indicator and front console panel cover. The ring indicator can be removed by inserting the clip remover tool or equivalent into the positions indicated by the arrows in the illustration. Remove the screws. Remove the bolts inside the upper pocket. Pull out the console lower pocket and remove the console box.
   - Remove the screws and clips and remove the console side upper panel
   - Remove the screws and pull out the front control panel
   - Disconnect the harness connectors and remove the front control panel
   - Remove the audio system retaining screws and partially pull the assembly forward. Disconnect the electrical connectors and antenna feed cable. Remove the radio assembly.
   - Position the tilt steering in the lowest detent

- Remove the instrument cluster hood. Remove the instrument cluster retaining screws and pull the cluster forward. Disconnect the electrical connector and remove the assembly
- Lift the upper grille air vent and remove the catch on the hazard side switch
- Hold the side of the monitor and remove the catch on the upper right and left sides
- Disconnect the connectors and remove the upper air vent grille
- Remove the multifunction display or navigation monitor and warning box
- Remove the front upper pillar trim
- Remove the passenger's side air bag module retaining bolts
- Remove the screw in the center of the instrument cluster assembly housing. Remove the retaining screws on the right and left side of the instrument panel. Remove the screws in the instrument panel center
- Be sure that the upper instrument panel is removed from the steering support beam
- Disconnect the necessary electrical connectors and remove the assembly from the vehicle body
- Remove the air vent grille
- Remove the passenger's side air bag module retaining screws. Remove the air bag module.

8. To remove the steering support beam:

- Remove the knee guard plate retaining bolts, remove the knee guard plate

➡ **The steering column may have to be removed from the vehicle for this procedure. Be sure the front wheels are in the straight ahead position.**

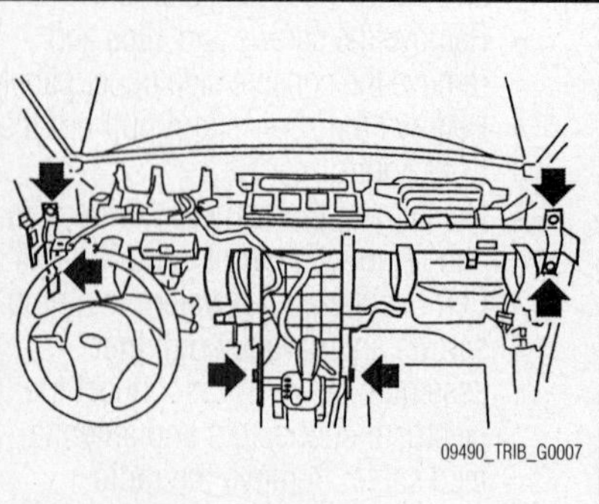

Fig. 188 Steering support beam retaining bolt locations

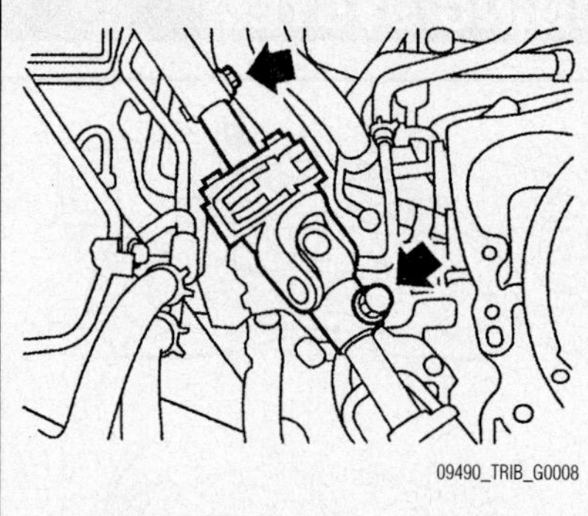

Fig. 189 Steering column universal joint and retaining bolt location

- Remove the screws on the side of the steering wheel. Disconnect the horn from the harness. Disconnect the air bag connector and remove the air bag module.
- Place alignment marks on the steering wheel and the shaft. Remove the retaining nut. Using the proper tool, remove the steering wheel.
- Place alignment marks on the universal joint. Remove the joint bolt. Remove the joint.
- Disconnect the steering column electrical connectors. Remove the steering column retaining bolts. Pull the steering shaft assembly from the hole on the toe board.

➡ **Be sure to remove the universal joint before removing the steering shaft assembly installing bolts when removing the steering shaft assembly or when lowering it for servicing other components.**

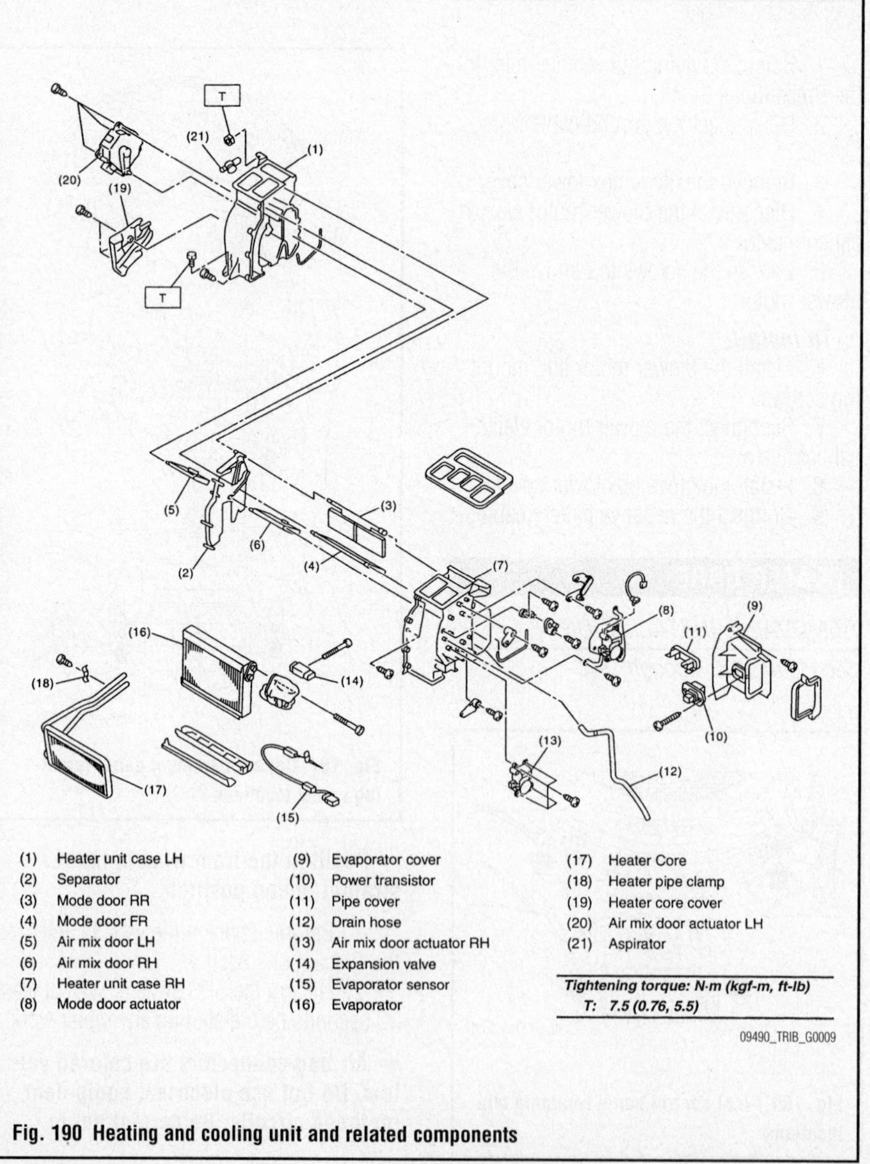

Fig. 190 Heating and cooling unit and related components

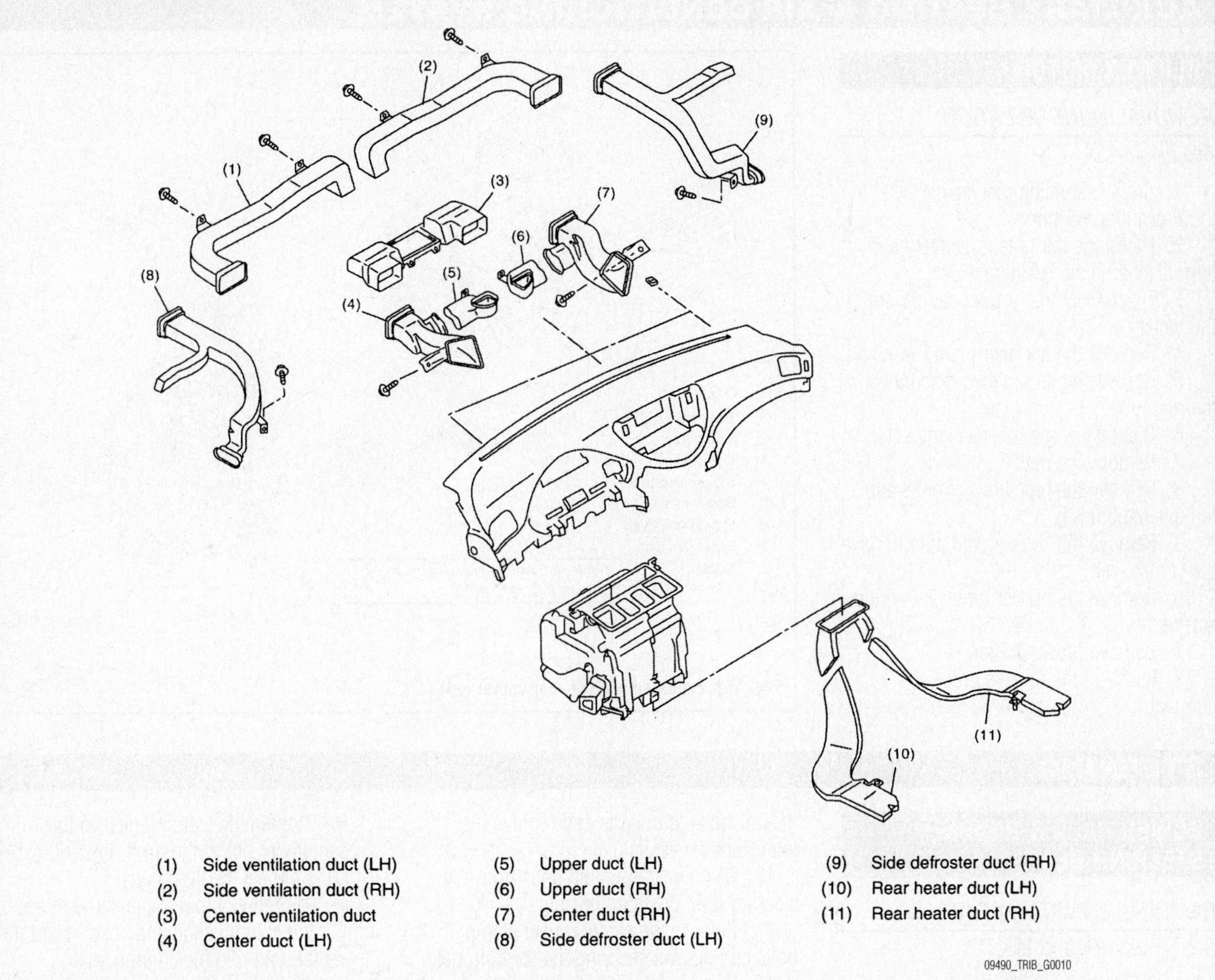

(1) Side ventilation duct (LH)
(2) Side ventilation duct (RH)
(3) Center ventilation duct
(4) Center duct (LH)
(5) Upper duct (LH)
(6) Upper duct (RH)
(7) Center duct (RH)
(8) Side defroster duct (LH)
(9) Side defroster duct (RH)
(10) Rear heater duct (LH)
(11) Rear heater duct (RH)

09490_TRIB_G0010

Fig. 191 Heater duct routing

- Remove the bolts and remove the steering support beam

9. Disconnect the electrical connectors from the air conditioning control module, intake door actuator, blower motor, power transistor and blower resistor.

10. Lift up the floor mat. Loosen the bolt and nut and remove the blower motor assembly.

11. Disconnect the actuator connector. Remove the bolts and nuts and remove the heater/cooling unit.

12. Remove the screws and remove the heater core cover and pipe clamp. Remove the heater core.

To install:

13. Installation is the reverse of the removal procedure.

14. Tighten the steering support beam retaining bolts to 18 ft. lbs. (25 Nm).

15. Tighten the upper instrument panel retaining bolts to 18 ft. lbs. (25 Nm).

16. When installing the steering column be sure to align the cutout portion at the serrated section on the column shaft and yoke, and then install the universal joint into the column shaft. Torque the bolt to 17.4 ft. lbs. (24 Nm).

17. Tighten the steering shaft to instrument panel retaining bolts to 18 ft. lbs. (25 Nm).

18. After installing the roll connector, be sure to check for proper adjustment:

- Check that the front wheels are in the straight ahead position
- Turn the roll connector pin (A) clockwise until it stops
- Turn the roll connector pins (A) approximately 3.25 turns until the triangle marks are aligned

19. Tighten the steering wheel retaining nut to 33 ft. lbs. (45 Nm).

20. Refill the cooling system with the proper grade and type coolant.

21. Vacuum, leak test and recharge A/C system. For single A/C models refrigerant amount is (1.26 lbs–1.39 lbs.) For dual A/C models (1.85 lbs–1.98 lbs.)

22. Start the engine and check for leaks, correct as required.

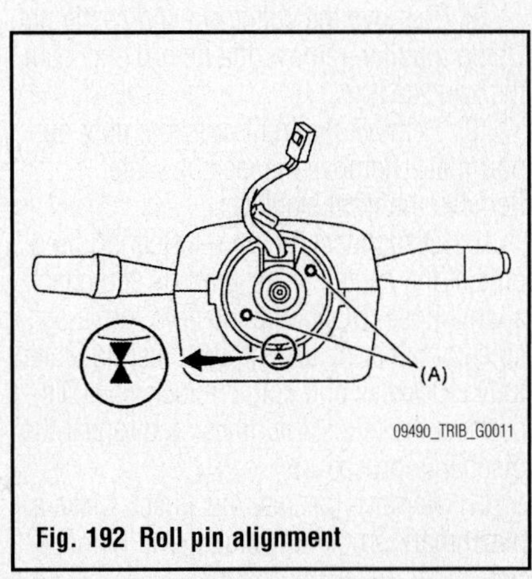

Fig. 192 Roll pin alignment

AUXILIARY HEATING & AIR CONDITIONING SYSTEM

BLOWER MOTOR

REMOVAL & INSTALLATION

See Figure 193.

1. Before servicing the vehicle, refer to the Precautions Section.
2. Using the refrigerant recovery system, discharge the refrigerant.
3. Disconnect the ground cable from the battery.
4. Remove the left rear quarter trim.
5. Remove the bolts, then disconnect the rear tube.
6. Disconnect the harness connector.
7. Remove the nuts.
8. Remove the pipe bracket bolts and remove the bracket.
9. Remove the screws, and then remove the cooler unit.
10. Remove the blower motor mounting screws.
11. Remove blower motor.
12. Install in the reverse order of removal.

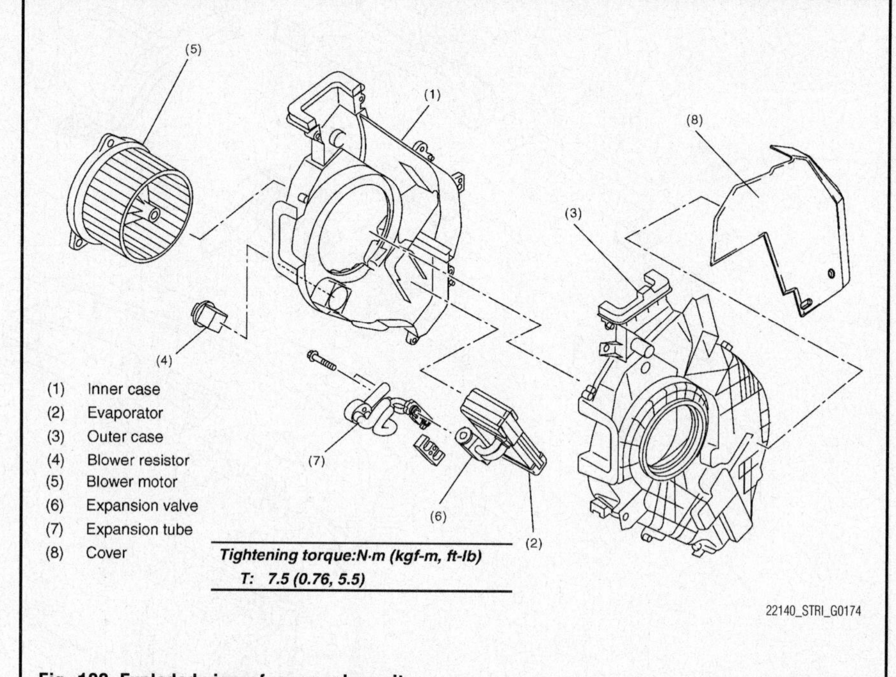

Fig. 193 Exploded view of rear cooler unit

STEERING

POWER RACK & PINION STEERING GEAR

REMOVAL & INSTALLATION

See Figures 194 and 195.

1. Before servicing the vehicle, refer to the Precautions Section.
2. Disconnect the negative battery cable.
3. Loosen the front wheel nut.
4. Raise and support the vehicle safely.
5. Remove the tires and wheels.
6. Remove the undercover.
7. Remove the front exhaust pipe assembly.
8. Remove the cotter pin and castle nut. Using a puller, remove the tie rod end from the knuckle arm.
9. Remove the front crossmember support plate. Remove the jack up plate. Remove the front stabilizer.
10. Disconnect the power steering fluid pipe at the center of the gearbox and attach a vinyl hose. Discharge the fluid into a suitable container by turning the steering wheel fully clockwise and counterclockwise. Disconnect the other fluid lines, and repeat the discharge procedure.
11. Remove the steering wheel. Make a matchmark on the universal joint. Remove the universal joint bolts and remove the joint from the vehicle.
12. Disconnect the fluid lines from the steering gear, pressure hose first.
13. Remove the steering gear clamp bolts and bracket securing the steering gear.
14. Remove the steering gear from the vehicle.

To install:

15. Insert the steering gear into the crossmember. Be careful not to damage the gearbox boot.
16. Tighten the steering gear to the crossmember bracket to 44 ft. lbs. (60 Nm).
17. Connect the fluid lines.
18. Align the cutout at the serrated section of the column shaft and yoke. Insert the universal joint into the column shaft.
19. Align the mating marks and insert the universal joint to serrated section of the steering gear assembly. Tighten the bolt to 18 ft. lbs. (25 Nm).
20. Continue the installation in the reverse order of the removal procedure.

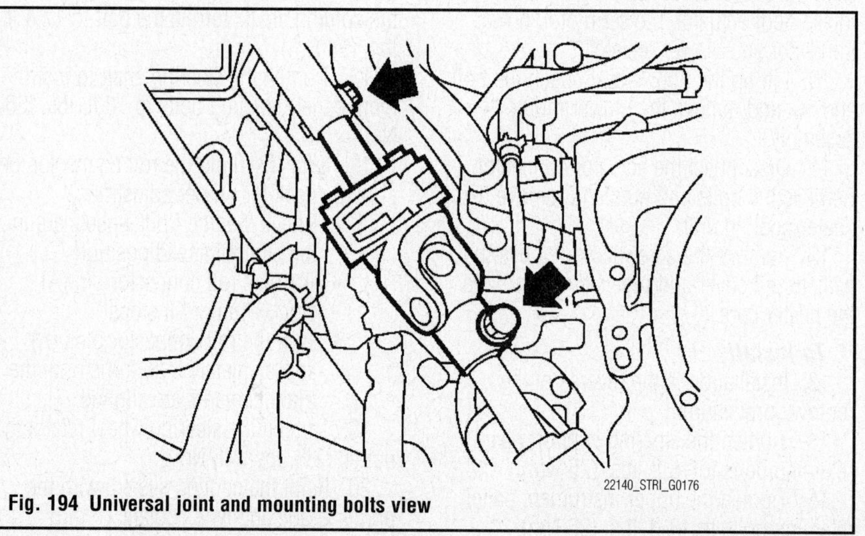

Fig. 194 Universal joint and mounting bolts view

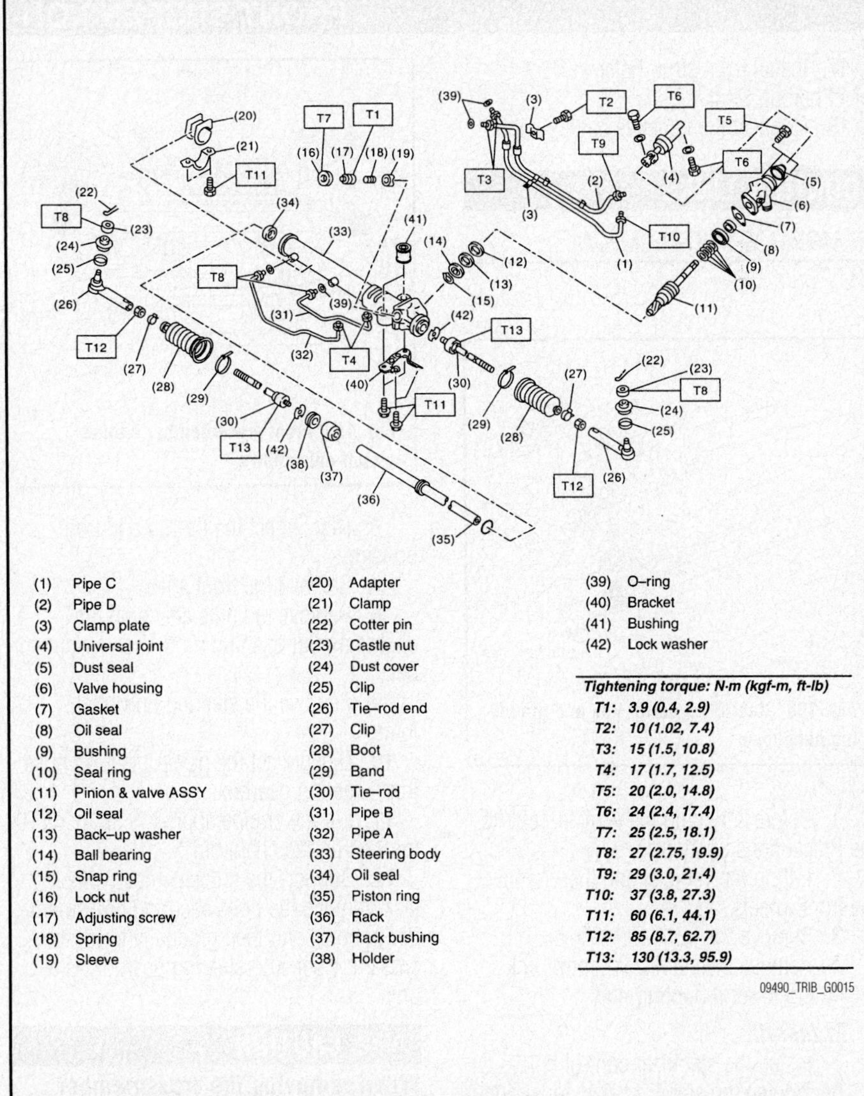

Fig. 195 Power steering gear and related components

21. Fill the power steering system with the proper grade and type fluid.
22. Start the engine and check for leaks. Correct as required.

POWER STEERING PUMP

REMOVAL & INSTALLATION

See Figure 196.

1. Before servicing the vehicle, refer to the Precautions Section.
2. Disconnect the ground cable from the battery.
3. Remove the air intake duct.
4. Remove the pulley belt cover.
5. Remove the drive belt.
6. Disconnect the connector from power steering pump switch.
7. Disconnect the (2) pressure hose and (1) suction hose from power steering pump.

CAUTION

Do not spill power steering fluid.

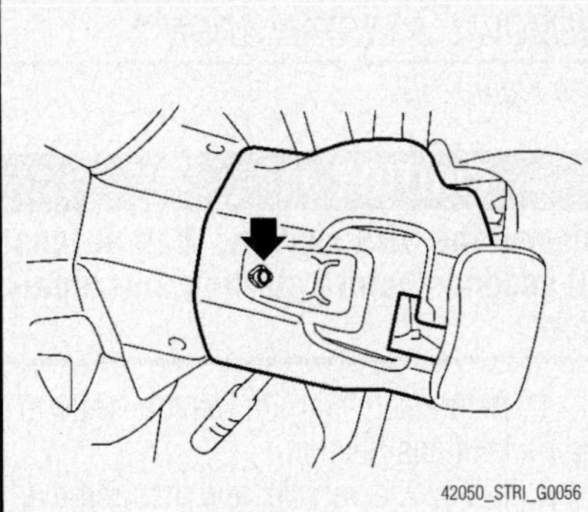

Fig. 196 Disconnecting the pressure and suction hoses from the power steering pump

CAUTION

To prevent foreign matter from entering the hose and pipe, cover the open ends with clean cloth.

8. Remove the installation bolt of the power steering pump bracket.
9. Place the oil pump bracket in a vise, and remove the two bolts from the front side of the oil pump.

CAUTION

When securing the oil pump bracket in a vice, hold the oil pump bracket with the least possible force between two pieces of wood.

10. Remove the bolt from the rear side of oil pump.
11. Disassemble the oil pump and bracket by inserting a flat tip screwdriver

To install

12. Install in the reverse order of removal.
13. Tighten oil pump to bracket bolts to 12 ft. lbs. (15.7 Nm).
14. Tighten the bolt from the rear side of oil pump to 27 ft. lbs. (36.8 Nm).
15. Tighten steering hose eye bolts to 30 ft. lbs. (40 Nm).
16. Replace the fluid and perform air purge.

CAUTION

Never start the engine before filling with fluid, as doing so may cause the vane pump to seize.

BLEEDING

1. Before servicing the vehicle, refer to the Precautions Section.
2. Lift up the vehicle.
3. Add the specified fluid to reservoir tank at "MAX" level.
4. Turn the steering wheel slowly from lock to lock until the bubbles stop appearing on oil surface while keeping the fluid at the "MAX" level

➡ **Normally bubbles will stop appearing after turning the steering wheel from lock to lock three times.**

5. In case bubbles do not stop appearing in the tank, or a grinding noise is generated from the oil pump leave it about half an hour and then do the step all over again.
6. Lower the vehicle, and then idle the engine.
7. Continue to turn the steering wheel from lock to lock until the bubbles stop appearing and change of the fluid level is within 0.12 inch (3 mm).

COIL SPRING

REMOVAL & INSTALLATION

See Figure 197.

1. Before servicing the vehicle, refer to the Precautions Section.
2. Disconnect the negative battery cable.
3. Raise and support the vehicle safely.
4. Remove front strut, Refer to front MacPherson Strut.
5. Using a coil spring compressor, compress the coil spring.
6. Using a strut mount socket remove the self-locking nut.
7. Remove the strut mount, spacer and upper spring seat from strut.
8. Gradually decrease the compression force of compressor, and remove the coil spring.
9. Remove the dust cover and helper spring.

To install:

10. Using a coil spring compressor, compress the coil spring.
11. Set the coil spring correctly so that its end face seats well in the spring seat as shown in the figure.
12. Install the helper and dust cover to the piston rod.
13. Pull the piston rod fully upward, and install the spring seat.
14. Install spacer and the strut mount to piston rod, and tighten a new self-locking nut temporarily.
15. Using a hexagon wrench to prevent strut rod from turning, tighten the new self-locking nut to 41 ft. lbs. (55 Nm), with a strut mount socket.
16. Loosen the coil spring compressor carefully.

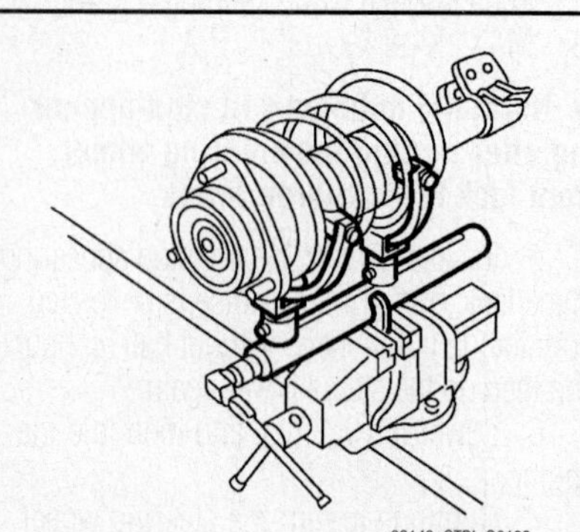

Fig. 197 Coil spring being compressed by spring compressor tool

17. Install front strut, Refer to front MacPherson Strut.
18. Connect the negative battery cable.

CONTROL LINKS

REMOVAL & INSTALLATION

See Figure 198.

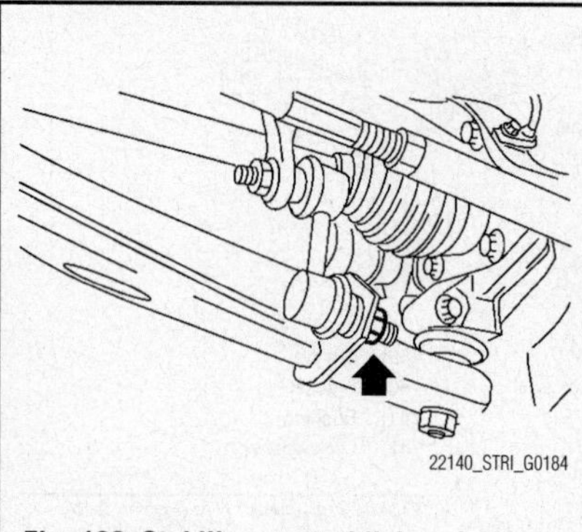

Fig. 198 Stabilizer control link and mounting nut shown

1. Before servicing the vehicle, refer to the Precautions Section.
2. Lift up the vehicle, and then remove the front wheels.
3. Remove the front under cover.
4. Remove the stabilizer control link upper and lower mounting nuts.

To install:

5. Install the stabilizer control link.
6. Tighten the stabilizer link upper and lower mounting nuts to 44 ft. lbs. (60 Nm).
7. Install front wheels, and lower the vehicle.

CROSSMEMBER

REMOVAL & INSTALLATION

See Figure 199.

⁂ CAUTION

Disarm the SRS system. Wait at least 20 seconds before starting any repair work.

1. Before servicing the vehicle, refer to the Precautions Section.
2. Lift up the vehicle, and then remove the front wheels.
3. Remove the front under cover.
4. Remove the front crossmember support plate.
5. Remove the front stabilizer.

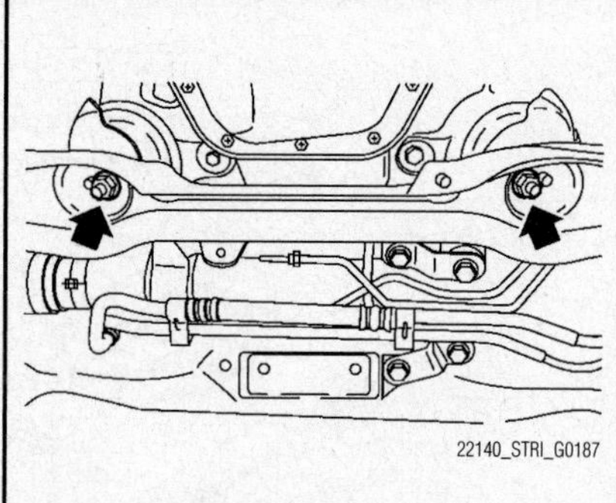

Fig. 199 Front crossmember engine mount nuts shown

6. Disconnect the tie-rod end from housing.
7. Remove the front arm.
8. Remove the nuts attaching the engine mount cushion rubber to crossmember.
9. Remove the steering universal joint.
10. Disconnect the power steering hose from steering gearbox.
11. Lift the engine approx. 0.39 inch (10 mm) using the chain block.
12. Support the crossmember with a jack, remove the bolts securing crossmember to body, and then gradually lower the crossmember with steering gearbox as a unit.

⁂ CAUTION

When removing the crossmember downward, be careful that the tie-rod end does not interfere with drive shaft boot.

To install:

13. Install in the reverse order of removal, noting the following tightening specifications:
    - Crossmember to body bolts to 70 ft. lbs. (95 Nm).
    - Engine mounting to crossmember nuts to 62.7 ft. lbs. (85 Nm).
    - Front arm to crossmember bolts to 70 ft. lbs. (95 Nm).
    - Front arm to support plate bolts to 65 ft. lbs. (88 Nm).
    - Support plate to body bolts to 111 ft. lbs. (150 Nm).
    - Tie-rod end to housing to 20 ft. lbs. (27 Nm).
    - Crossmember to body bolts to 70 ft. lbs. (95 Nm). After tightening to the specified torque, tighten the castle nut further but within 60°

until the hole in the ball stud is aligned with a slot in castle nut.
- Universal joint to 17 ft. lbs. (24 Nm).
- Stabilizer link to 44 ft. lbs. (60 Nm).
- Stabilizer clamp to 18 ft. lbs. (25 Nm).
- Power steering hose to Steering gearbox 11 ft. lbs. (15 Nm).

14. Purge air from the power steering system.
15. Inspect the wheel alignment and adjust if necessary.

LOWER BALL JOINT

REMOVAL & INSTALLATION

See Figure 200.

1. Before servicing the vehicle, refer to the Precautions Section.
2. Raise and support the vehicle safely.
3. Remove the tire and wheel.
4. Remove the stabilizer bar brackets and bushings (both sides).
5. Remove the cotter pin from the ball stud. Remove the castle nut. Extract the ball stud from the front arm.
6. Remove the bolt securing the ball joint to the housing. Extract the ball joint from the housing.

To install:

7. Installation is the reverse of the removal procedure.
8. Install the ball joint to the front arm and tighten the castle nut to 33.2 ft. lbs. (45 Nm). Tighten the castle nut an additional 60 degrees until the slot in the castle nut is aligned with the cotter pin hole in the ball joint.
9. Install the stabilizer bracket. Tighten to 18 ft. lbs. (25 Nm).
10. Always fully tighten the rubber bushings when the wheels are in full contact with the ground and the vehicle is at curb height.

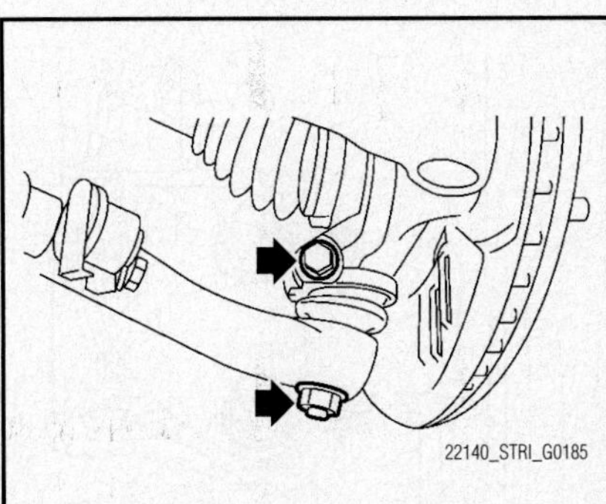

Fig. 200 Ball joint mounting nut and bolt shown

11. Check and adjust alignment, as required.

LOWER CONTROL ARM

REMOVAL & INSTALLATION

1. Before servicing the vehicle, refer to the Precautions Section.
2. Raise and support the vehicle safely.
3. Remove the tire and wheel.
4. Remove the front crossmember support plate.
5. Remove the stabilizer bar.
6. Remove the ball joint from the front arm.
7. Remove the nut securing the front arm to the crossmember.

➡ Do not remove the bolt.

8. Remove the front arm support plate.
9. Remove the bolt securing the front arm to the crossmember and pull the front arm out of the crossmember.
10. To remove the stud bolt, use tool ST-20299AG020.

➡ Do not remove the stud bolt unless it is necessary. Always replace the removed parts with new ones.

To install:

11. Installation is the reverse of the removal procedure.
12. Tighten the stud bolt to 81 ft. lbs. (110 Nm), if removed.
13. Tighten the support plate to front arm bolts to 107 ft. lbs. (145 Nm).
14. Tighten the support plate to body bolts to 111 ft. lbs. (150 Nm).
15. Always fully tighten the rubber bushings when the wheels are in full contact with the ground and the vehicle is at curb height.
16. Check and adjust alignment, as required.

CONTROL ARM BUSHING REPLACEMENT

See Figure 201.

1. Remove the control arm from the vehicle.
2. Mount the control arm in a soft jawed vise.
3. Use either a press or a control arm bushing fixture (C-clamp like tool) along with a slotted washer and a piece of pipe (slightly larger than the bushing) and press out the old bushing.
4. Clean the inside bushing contact surfaces of rust and old rubber.

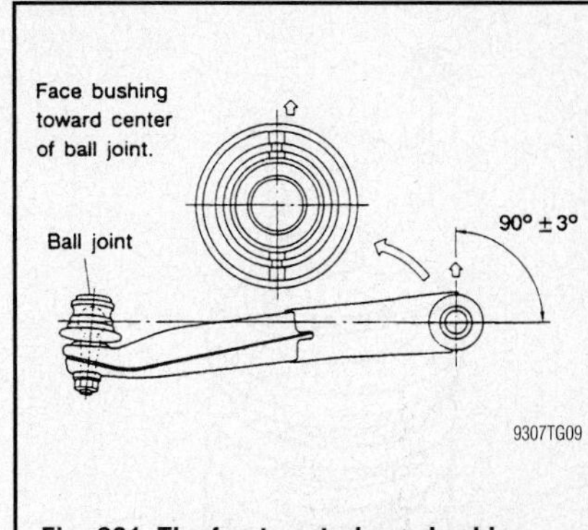

Fig. 201 The front control arm bushing must be installed in the proper direction

To install:

5. Apply a light coating of grease to both the replacement busing and bushing contact surfaces on the control arm.
6. Align the bushing.
7. Install the bushing using the press tool. A bushing install clamp can also be used to compress the bushing into the control arm.
8. Install the control arm on the vehicle.

MACPHERSON STRUT

REMOVAL & INSTALLATION

See Figure 202.

1. Before servicing the vehicle, refer to the Precautions Section.
2. Disconnect the negative battery cable.
3. Raise and support the vehicle safely.
4. Remove the tire and wheel.
5. Remove the bolt retaining the brake hose to the strut.
6. Make and alignment mark on the camber adjusting bolt which secures the strut to the housing.
7. Remove the clip retaining the ABS wheel speed sensor harness.
8. Remove the two bolts retaining the strut to the housing.

➡ While holding the head of the adjusting bolt, loosen the self locking nut.

9. Remove the three upper strut retaining nuts.
10. Remove the strut from the vehicle.

To install:

11. Installation is the reverse of the removal procedure.
12. Tighten the upper retaining nuts to 15 ft. lbs. (20 Nm). Tighten the lower retaining bolts to 129 ft. lbs. (174.9 Nm)
13. Position the alignment mark on the camber adjusting bolt with the alignment

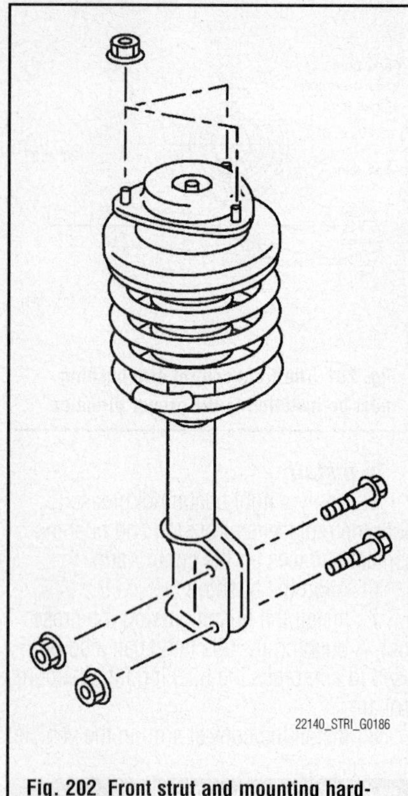

Fig. 202 Front strut and mounting hardware shown

mark on the lower side of the strut. Install using a new self locking nut.

➡ **While holding the head of the adjusting bolt, tighten the self locking nut.**

14. Check and adjust wheel alignment, as necessary.

OVERHAUL

See Figure 203.

1. Before servicing the vehicle, refer to the Precautions Section.
2. Remove the strut from the vehicle.
3. Using a coil spring compressor tool, carefully compress the spring. Remove the self locking nut.
4. Remove the strut mount, upper spring and rubber seat from the strut.
5. Gradually decrease the compression force of the spring compressor tool. Remove the coil spring.
6. Remove the dust cover and helper spring.
7. Check for the presence of air in the damping force generating mechanism.
8. Using the spring compression tool, compress the coil spring.

➡ **Be sure to properly install the coil spring.**

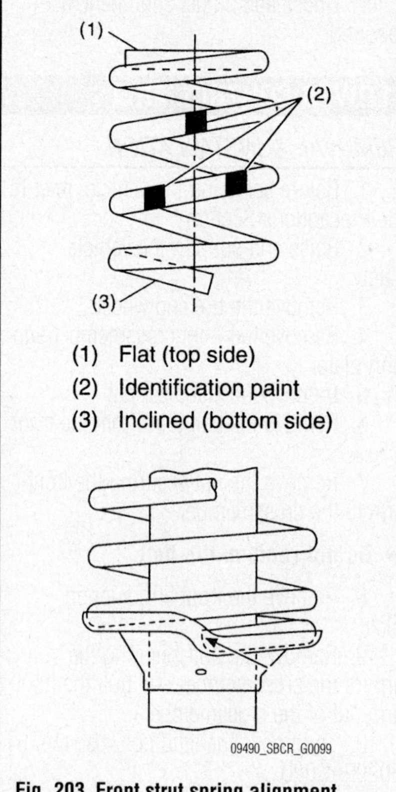

Fig. 203 Front strut spring alignment

9. Position the coil spring so that its end face fits good into the spring seat.
10. Install the helper spring and dust cover to the piston rod.
11. Pull the piston rod fully upward, and install the rubber seat and spring seat.
12. Install the strut mount to the piston rod, and then tighten the self locking nut, temporarily. Be sure to use a new self locking nut.
13. Use a hexagon wrench to prevent the strut rod from turning. Tighten the self locking nut to 41 ft. lbs. (55 Nm).
14. Carefully loosen the coil spring.

STEERING KNUCKLE

REMOVAL & INSTALLATION

See Figures 204 through 206.

1. Before servicing the vehicle, refer to the Precautions Section.
2. Raise and support the vehicle safely.
3. Remove the tire and wheels.
4. Remove the front under cover.
5. Remove the stabilizer link.
6. Remove the disc brake caliper from the front housing, and suspend it from strut using a piece of wire.
7. Remove the disc rotor from the hub.
8. Remove the cotter pin and castle nut securing the tie—rod end to the front housing knuckle arm.

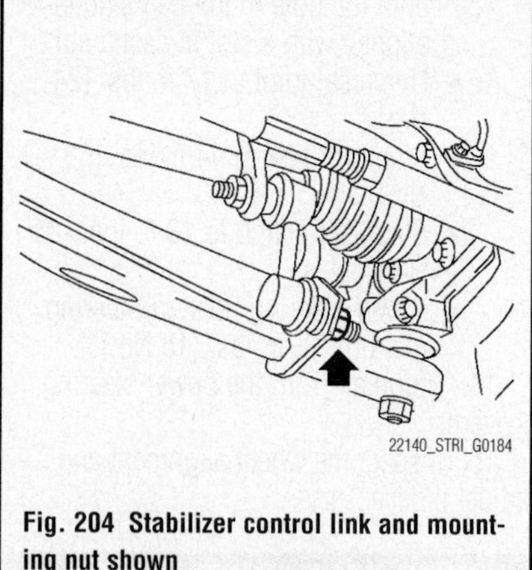

Fig. 204 Stabilizer control link and mounting nut shown

9. Using a puller, remove the tie rod ball joint from knuckle arm.
10. Remove the ABS wheel speed sensor assembly and harness.
11. Remove the lower strut mounting bolts.
12. Remove the front arm ball joint from the front housing.
13. Remove the front drive shaft assembly from the steering knuckle hub.
14. After scribing an alignment mark on the camber adjusting bolt head, remove the bolts which connect the front housing and

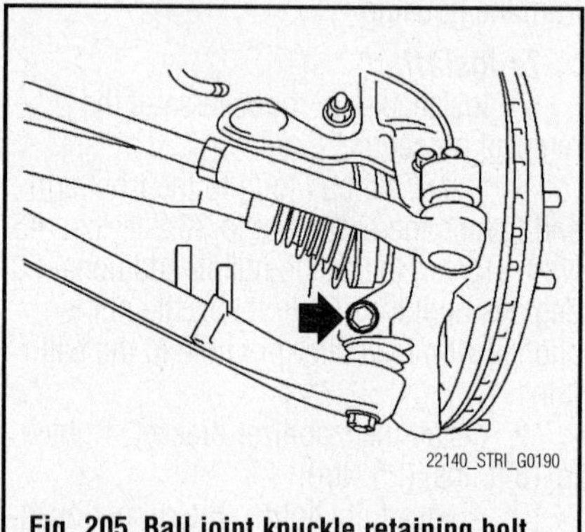

Fig. 205 Ball joint knuckle retaining bolt shown

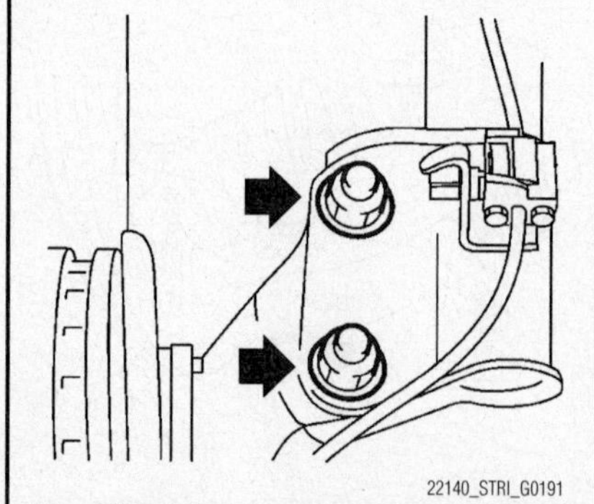

Fig. 206 Strut to knuckle retaining bolts shown

strut, and disconnect the steering knuckle from the strut.

15. Remove the front steering knuckle hub.

To install:

16. Install the front steering knuckle hub.
17. Align the alignment mark on the camber adjusting bolt head, and affix the front housing and strut together using a new self—locking nut. Tighten to 129 ft. lbs. (175 Nm).
18. Install the front drive shaft.
19. Install the front arm ball joint to the front housing. Tighten mounting bolt to 37 ft. lbs. (50 Nm).
20. Install the ABS wheel speed sensor on the front housing. Tighten to 5.5 ft. lbs. (7.5 Nm).
21. Install the disc rotor to hub.
22. Install the disc brake caliper to the front housing. Tighten to 88.5 ft. lbs. (120 Nm).
23. Install the stabilizer link.
24. Connect the tie-rod end ball joint to the knuckle arm with a castle nut. Tighten to 20 ft. lbs. (27 Nm).
25. Tighten the castle nut to specified torque and tighten further within 60° until the pin hole is aligned with the slot in nut. Bend the cotter pin to lock.
26. While depressing the brake pedal, tighten a new axle nut to 177 ft. lbs. (240 Nm). and lock it securely.
27. Install the tire and wheels.
28. Lower the vehicle.
29. Inspect the wheel alignment and adjust if necessary.

STABILIZER BAR

REMOVAL & INSTALLATION

See Figure 207.

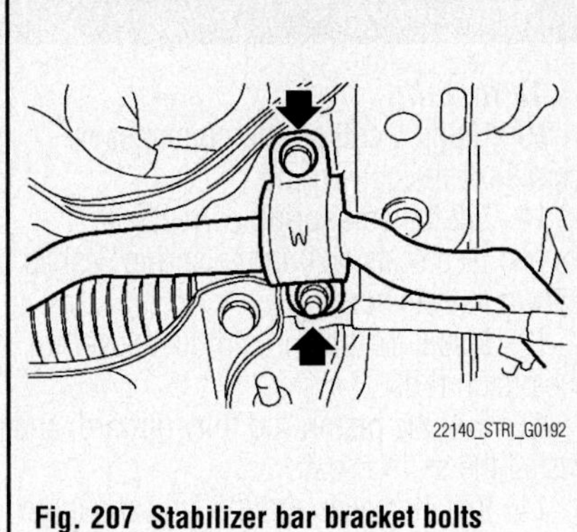

Fig. 207 Stabilizer bar bracket bolts shown

1. Before servicing the vehicle, refer to the Precautions Section.
2. Raise and support the vehicle safely.
3. Remove the tire and wheels.
4. Remove the front under cover.
5. Remove the front crossmember support plate.
6. Remove the stabilizer bar link.
7. Remove the stabilizer bar bracket bolts and bushings.
8. Remove the stabilizer bar from the vehicle.

To install:

9. Installation is the reverse of the removal procedure.
10. Be sure to use new self locking nuts, as required.
11. Install the rubber bushing, so that the paint mark on the stabilizer bar is on the left side of the vehicle.
12. Install the stabilizer bushing (front crossmember side), while aligning it with the paint mark on the stabilizer bar.
13. Tighten the stabilizer link bolts to 44 ft. lbs. (60 Nm). Tighten the stabilizer bar clamp to 18 ft. lbs. (25 Nm).
14. Always fully tighten the rubber bushings when the wheels are in full contact with the ground and the vehicle is at curb height.

WHEEL HUB AND BEARING

REMOVAL & INSTALLATION

See Figure 208.

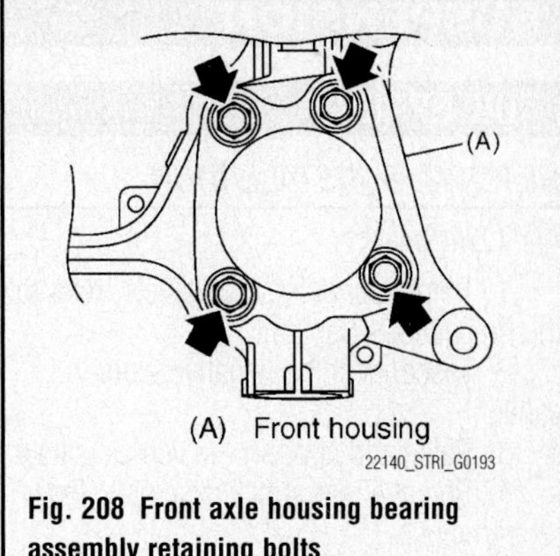

Fig. 208 Front axle housing bearing assembly retaining bolts

1. Before servicing the vehicle, refer to the Precautions Section.
2. Disconnect the ground cable from the battery.
3. Lift up the vehicle, and remove the front wheels.
4. Lift the crimped section of axle nut.
5. Remove the axle nut using a socket wrench while depressing the brake pedal.
6. Remove the disc brake caliper from the front housing, and suspend it from strut using a piece of wire.
7. Remove the disc rotor from the hub.
8. Remove four bolts from the front housing.
9. Remove the front hub unit bearing. If it is hard to remove, use a puller.

To install:

10. Place the disc cover between front housing and front hub unit, and tighten the four bolts to 48 ft. lbs. (65 Nm).
11. Tighten the axle nut temporarily.
12. Install the disc rotor to hub.
13. Install the disc brake caliper to the front housing and tighten mounting bolts to 88.5 ft. lbs. (120 Nm).
14. While depressing the brake pedal, tighten a new axle nut to 177 ft. lbs. (240 Nm). and lock it securely.
15. Install the tire and wheels.
16. Lower the vehicle.
17. Inspect the wheel alignment and adjust if necessary.

ADJUSTMENT

The wheel bearing is of a sealed design, and no adjustment is possible.

COIL SPRING

REMOVAL & INSTALLATION

See Figure 209.

1. Before servicing the vehicle, refer to the Precautions Section.
2. Disconnect the negative battery cable.
3. Raise and support the vehicle safely.
4. Remove rear shock assembly, Refer to rear shock.
5. Using a coil spring compressor, compress the coil spring.
6. Using a strut mount socket remove the self-locking nut.
7. Remove the strut mount, spacer and upper spring seat from strut.
8. Gradually decrease the compression force of compressor, and remove the coil spring.
9. Remove the dust cover and helper spring.

To install:

10. Using a coil spring compressor, compress the coil spring.
11. Set the coil spring correctly so that its end face seats well in the spring seat as shown in the figure.
12. Install the helper and dust cover to the piston rod.
13. Pull the piston rod fully upward, and install the spring seat.
14. Install spacer and the strut mount to piston rod, and tighten a new self—locking nut temporarily.
15. Using a hexagon wrench to prevent strut rod from turning, tighten the new self-locking nut to 26 ft. lbs. (35 Nm), with a strut mount socket.
16. Loosen the coil spring compressor carefully.
17. Install rear shock assembly, Refer to rear shock.
18. Connect the negative battery cable.

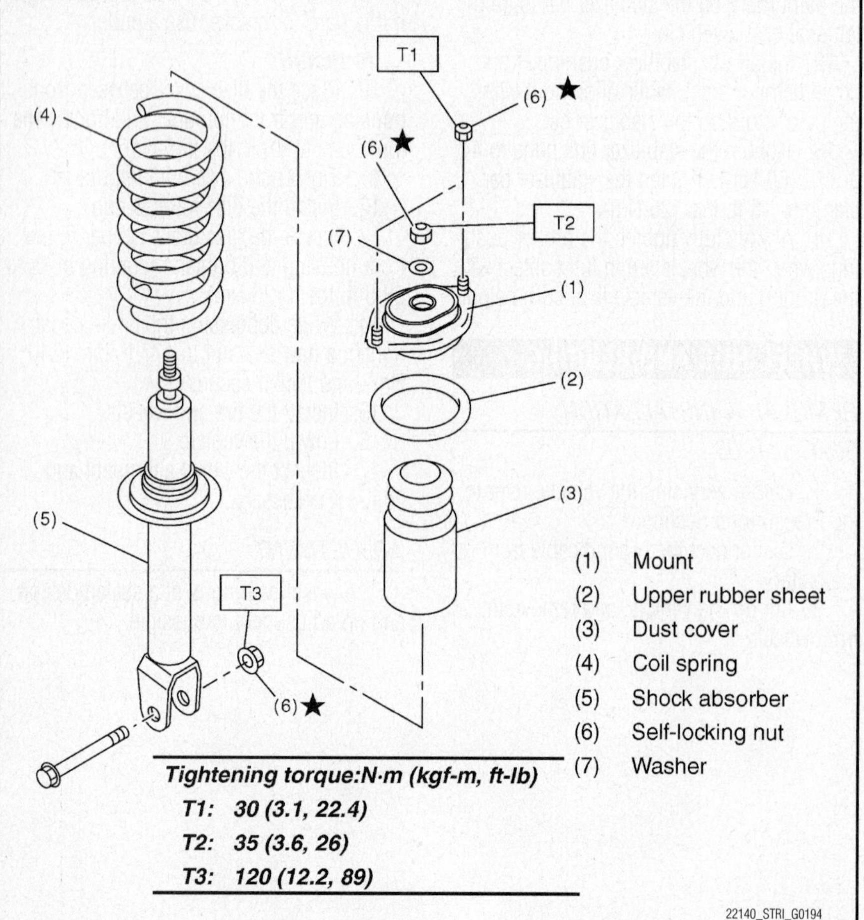

Fig. 209 Exploded view of rear shock assembly

LATERAL LINK

REMOVAL & INSTALLATION

Front

See Figure 210.

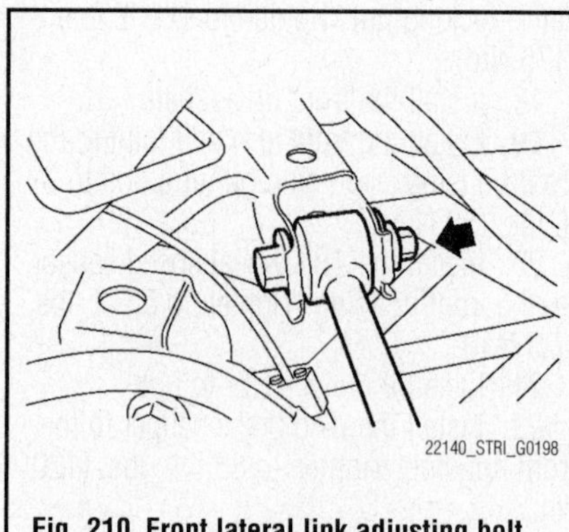

Fig. 210 Front lateral link adjusting bolt and nut

1. Before servicing the vehicle, refer to the Precautions Section.
2. Lift-up the vehicle, and then remove the rear wheels.
3. Remove the snap pin and nut.
4. Use a puller to detach the ball joint.
5. Scribe an alignment mark on the front lateral link adjustment bolt and the rear sub frame.

⁂ CAUTION

When removing the adjusting bolt, loosen the nut with the bolt head secured.

6. Remove the adjusting bolt, and then remove the front lateral link.

To install:

7. Installation is the reverse of the removal procedure.
8. Be sure to use new bolts and nuts, as required.
9. Always fully tighten the rubber bushings when the wheels are in full contact with the ground and the vehicle is at curb height.
10. Tighten the front lateral link to sub frame to 89 ft. lbs. (120 Nm).
11. Tighten the front lateral link to rear axle housing to 20 ft. lbs. (27 Nm).
12. Check and adjust the wheel alignment, as necessary.

Rear

See Figure 211.

1. Before servicing the vehicle, refer to the Precautions Section.

Fig. 211 Rear lateral link removal points shown

2. Lift-up the vehicle, and then remove the rear wheels.
3. Remove the nut and detach the stabilizer link.
4. Remove the bolts on the bottom side of the shock absorber.
5. Remove the bolts and remove the rear lateral link.

To install:

6. Installation is the reverse of the removal procedure.
7. Be sure to use new bolts and nuts, as required.
8. Always fully tighten the rubber bushings when the wheels are in full contact with the ground and the vehicle is at curb height.
9. Tighten the rear lateral link to 89 ft. lbs. (120 Nm).
10. Tighten the shock absorber to 89 ft. lbs. (120 Nm).
11. Tighten the stabilizer link to 44 ft. lbs. (60 Nm).
12. Check and adjust the wheel alignment, as necessary.

REAR TRAILING LINK

REMOVAL & INSTALLATION

See Figure 212.

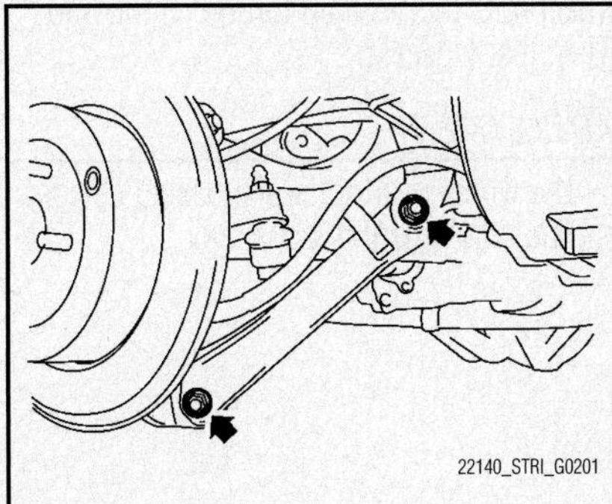
Fig. 212 Rear trailing link removal points shown

1. Before servicing the vehicle, refer to the Precautions Section.
2. Lift-up the vehicle, and then remove the rear wheels.
3. Remove the bracket, and remove the parking brake cable from the guide.
4. Remove the ABS wheel speed sensor harness from the trailing link
5. Remove the trailing link

To install:

6. Installation is the reverse of the removal procedure, noting the following:
   a. Be sure to use new bolts and nuts, as required.
   b. Always fully tighten the rubber bushings when the wheels are in full contact with the ground and the vehicle is at curb height.
   c. Tighten the trailing link mounting bolts to 59 ft. lbs. (80 Nm).
   d. Tighten the parking brake cable bracket bolt to 24 ft. lbs. (33 Nm).
   e. Check and adjust the wheel alignment, as necessary.

SHOCK ABSORBER

REMOVAL & INSTALLATION

See Figure 213.

1. Before servicing the vehicle, refer to the Precautions Section.
2. Remove the strut cap (of the quarter trim).

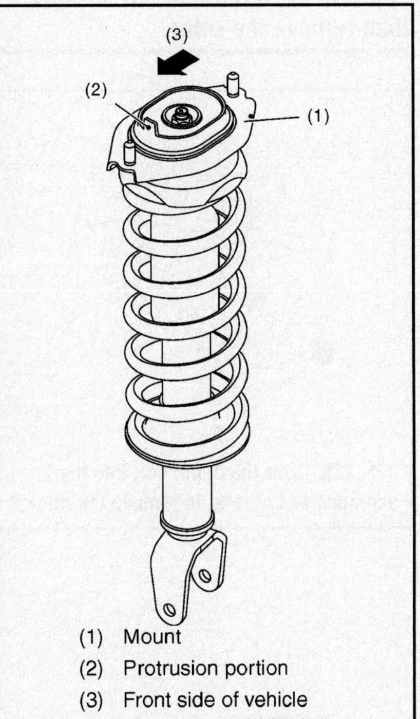

Fig. 213 Rear shock assembly view

3. Loosen the rear wheel lug nuts.
4. Raise and support the vehicle safely.
5. Remove the tire and wheel.
6. Remove the nut and detach the rear stabilizer link.
7. Using a jack, support the shock absorber.
8. Remove the bolts on the bottom of the shock absorber.
9. Detach the rear lateral link.
10. Remove the nuts that retain the shock absorber mount to the vehicle.
11. Remove the shock absorber from the vehicle.

To install:

12. Installation is the reverse of the removal procedure.
13. Be sure to use new bolts and nuts, as required.
14. Always fully tighten the rubber bushings when the wheels are in full contact with the ground and the vehicle is at curb height.
15. Check and adjust the wheel alignment, as necessary.

STABILIZER BAR

REMOVAL & INSTALLATION

See Figure 214.

1. Before servicing the vehicle, refer to the Precautions Section.
2. Loosen the rear wheel lug nuts.
3. Raise and support the vehicle safely.
4. Remove the tire and wheel.
5. Remove the stabilizer link.
6. Remove the clamp and bushing bolts which secure the stabilizer bar. Remove the clamps and bushings.
7. Remove the stabilizer bar from the vehicle.

To install:

8. Installation is the reverse of the removal procedure.

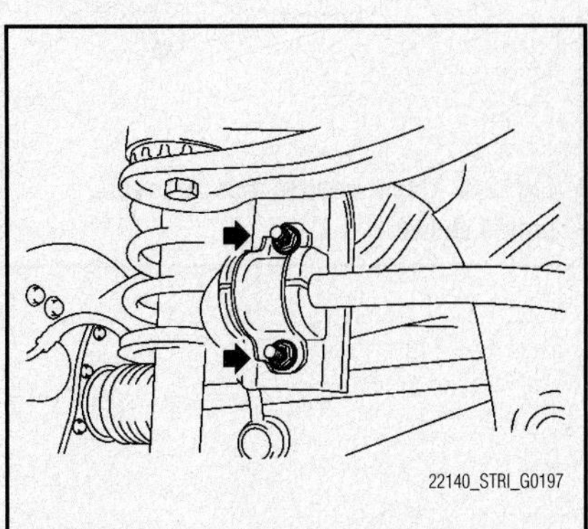
Fig. 214 Stabilizer bar retaining nuts shown

9. Be sure that the stabilizer bar and the bushings have the same markings and color.

10. Be sure to use new bolts and nuts, as required.

11. Tighten the stabilizer link retaining bolts to 44 ft. lbs (60 Nm).

12. Tighten the stabilizer clamp retaining bolts to 28 ft. lbs (38 Nm).

13. Always fully tighten the rubber bushings when the wheels are in full contact with the ground and the vehicle is at curb height.

14. Check and adjust the wheel alignment, as necessary.

UPPER CONTROL ARM

REMOVAL & INSTALLATION

See Figure 215.

1. Before servicing the vehicle, refer to the Precautions Section.
2. Loosen the wheel nuts.
3. Raise and support the vehicle safely.
4. Remove the tire and wheel.
5. As required, support the assembly before removing the upper control arm.
6. Remove the bolts. Remove the upper control arm from the vehicle.

To install:

7. Installation is the reverse of the removal procedure.
8. Be sure to use new bolts and nuts, as required.
9. Tighten the upper arm to rear sub frame retaining bolts to 111 ft. lbs (150 Nm).

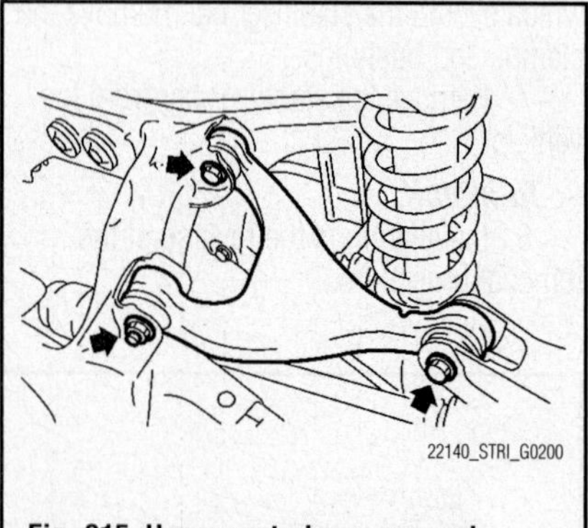

Fig. 215 Upper control arm removal points shown

10. Tighten the upper arm to rear housing retaining bolts to 89 ft. lbs (120 Nm).

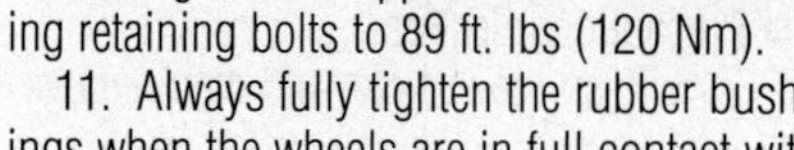

11. Always fully tighten the rubber bushings when the wheels are in full contact with the ground and the vehicle is at curb height.

12. Check and adjust the wheel alignment, as necessary.

WHEEL HUB AND BEARING

REMOVAL & INSTALLATION

See Figure 216.

1. Disconnect the ground cable from the battery.
2. Lift-up the vehicle, and remove the front wheels.
3. Lift the crimped section of axle nut.
4. Remove the axle nut using a socket wrench while depressing the brake pedal.

✻✻ CAUTION

Remove the wheel before loosening the axle nut. Failure to follow this rule may damage the wheel bearings.

5. Remove the disc brake caliper from the front housing, and suspend it from strut using a piece of wire.
6. Remove the disc rotor from the hub.

✻✻ CAUTION

If it is difficult to remove the disc rotor from the hub, drive the 8 mm of bolt into the threaded end of rotor, then remove the rotor.

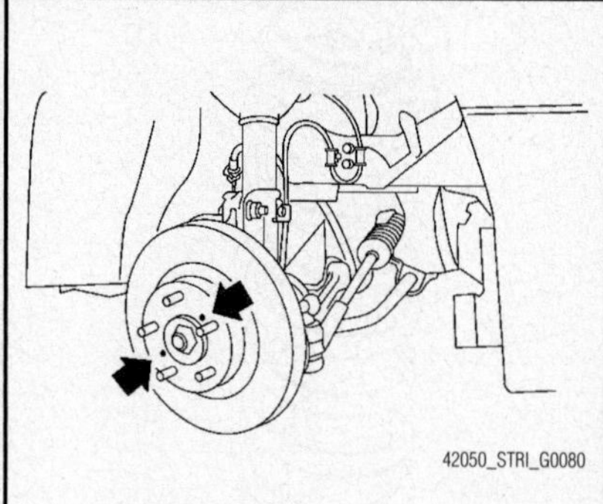

Fig. 216 Drive the 8 mm bolt into the threaded end of rotor to remove the rotor

7. Remove four bolts from the front housing.

✻✻ CAUTION

Do not get close to the tool which charged magnetism to magnetic encoder.

✻✻ CAUTION

Be careful not to damage the magnetic encoder.

8. Remove the front hub unit bearing. If it is hard to remove, use the ST.

To install:

9. Place the disc cover between front housing and front hub unit, and tighten the four bolts. Tighten to 47.9 ft. lbs. (65 Nm).
10. Install the front drive shaft.
11. Tighten the axle nut temporarily.
12. Install the disc rotor to hub.
13. Install the disc brake caliper to the front housing. Tighten to 89 ft. lbs. (120 Nm).
14. While depressing the brake pedal, tighten a new axle nut (olive color) to the specified torque and lock it securely. Tighten to 177 ft. lbs. (240 Nm).

✻✻ CAUTION

Install the wheel after installation of axle nut. Failure to follow this rule may damage the wheel bearing.

✻✻ CAUTION

Be sure to tighten the axle nut to specified torque. Do not over tighten it as this may damage the wheel bearing.

15. After tightening the axle nut, lock it securely.
16. Install the wheel and tighten the wheel nuts to specified torque. Tighten to 81 ft. lbs. (110 Nm).

ADJUSTMENT

The wheel bearing is of a sealed design, and no adjustment is possible.

SUBARU

Baja • Impreza • Legacy • Outback

14

SPECIFICATIONS AND MAINTENANCE CHARTS

ENGINE AND VEHICLE IDENTIFICATION CHART

Engine Code							Model Year	
Code ①	Liters (cc)	Cu. In.	Cyl.	Fuel Sys.	Type	Eng. Mfg.	Code ②	Year
6	2.5 (2457)	150	4	MFI	DOHC	Subaru	6	2006
6	2.5 (2457)	150	4	MFI	SOHC	Subaru	7	2007
8	3.0 (3000)	183	6	MFI	DOHC	Subaru	8	2008

MFI: Multiport Fuel Injection

SOHC: Single Overhead Camshaft

DOHC: Double Overhead Camshaft

① 6th digit of the VIN

② 10th digit of the VIN

22140_SUBA_C0001

GENERAL ENGINE SPECIFICATIONS

Year	Model	Engine Displacement Liters (cc)	Engine ID/VIN	Net Horsepower @ rpm	Net Torque @ rpm (ft. lbs.)	Bore x Stroke (in.)	Com-pression Ratio	Oil Pressure @ rpm
2006	Impreza	2.5 (2457)	6	165@5600	166@4000	3.92x3.11	10.0:1	14@600
	Outback	2.5 (2457)	6	165@5600	166@4000	3.92x3.11	10.0:1	14@600
	Outback	3.0 (3000)	8	212@6000	210@4400	3.51x3.15	10.7:1	14@600
	Baja	2.5 (2457)	6	165@5600	166@4000	3.92x3.11	10.0:1	14@600
	Legacy	2.5 (2457)	6	165@5600	166@4000	3.92x3.11	10.0:1	14@600
2007	Impreza	2.5 (2457)	6	165@5600	166@4000	3.92x3.11	10.0:1	14@600
	Outback	2.5 (2457)	6	165@5600	166@4000	3.92x3.11	10.0:1	14@600
	Outback	3.0 (3000)	8	212@6000	210@4400	3.51x3.15	10.7:1	14@600
	Legacy	2.5 (2457)	6	165@5600	166@4000	3.92x3.11	10.0:1	14@600
2008	Impreza	2.5 (2457)	6	165@5600	166@4000	3.92x3.11	10.0:1	14@600
	Outback	2.5 (2457)	6	165@5600	166@4000	3.92x3.11	10.0:1	14@600
	Outback	3.0 (3000)	8	212@6000	210@4400	3.51x3.15	10.7:1	14@600
	Legacy	2.5 (2457)	6	165@5600	166@4000	3.92x3.11	10.0:1	14@600
	Legacy	3.0 (3000)	8	212@6000	210@4400	3.51x3.15	10.7:1	14@600

22140_SUBA_C0002

ENGINE TUNE-UP SPECIFICATIONS

Year	Model	Engine Displacement Liters	Spark Plug Gap (in.)	Ignition Timing (deg.) ①		Fuel Pump (psi)	Idle Speed (rpm)		Valve Clearance ②	
				MT	AT		MT	AT	In.	Ex.
2006	Impreza	2.5	0.028-0.031	12 ③	17 ③	33-38	650-750	650-750	0.0079-0.0008	0.0138-0.0008
	Outback	2.5	0.028-0.031	12 ③	17 ③	34-38	650-750	650-750	0.0079-0.0008	0.0138-0.0008
	Outback	3.0	0.028-0.031	15 ④	15 ④	40-45	650-750	650-750	0.0079-0.0016	0.0138-0.0020
	Baja	2.5	0.028-0.031	12 ③	17 ③	34-38	650-750	650-750	0.0079-0.0008	0.0138-0.0008
	Legacy	2.5	0.028-0.031	12 ③	17 ③	34-38	650-750	650-750	0.0079-0.0008	0.0138-0.0008
2007	Impreza	2.5	0.028-0.031	12 ③	17 ③	34-38	650-750	650-750	0.0071-0.0087	0.0090-0.0106
	Outback	2.5	0.028-0.031	12 ③	17 ③	34-38	650-750	650-750	0.0071-0.0087	0.0090-0.0106
	Outback	3.0	0.028-0.031	15 ④	15 ④	40-45	650-750	650-750	0.0079-0.0016	0.0138-0.0020
	Legacy	2.5	0.028-0.031	12 ③	17 ③	34-38	650-750	600-800	0.0071-0.0087	0.0090-0.0106
2008	Impreza	2.5	0.028-0.031	12 ③	17 ③	34-38	650-750	650-750	0.0071-0.0087	0.0090-0.0106
	Outback	2.5	0.028-0.031	12 ③	17 ③	34-38	650-750	650-750	0.0071-0.0087	0.0090-0.0106
	Outback	3.0	0.028-0.031	15 ④	15 ④	49-51	650-750	650-750	0.0079-0.0016	0.0138-0.0020
	Legacy	2.5	0.028-0.031	12 ③	17 ③	34-38	600-800	650-750	0.0071-0.0087	0.0090-0.0106
	Legacy	3.0	0.028-0.031	15 ④	15 ④	49-51	650-750	650-750	0.0079-0.0016	0.0138-0.0020

Note: The Vehicle Emission Control Information label often reflects specification changes made during production. The label figures must be used if they differ from those in this chart.

① Before Top Dead Center

② Measured with engine cold

③ Plus or minus (10 deg.) at 750 rpm

④ Plus or minus (8 deg.) at 650 rpm

22140_SUBA_C0003

CAPACITIES

Year	Model	Engine Displacement Liters	Engine Oil with Filter (qts.)	Transmission (pts.) 5-Spd	Transmission (pts.) Auto.	Transfer Case (pts.)	Drive Axle Front ① (pts.)	Drive Axle Rear (pts.)	Fuel Tank (gal.)	Cooling System (qts.)
2006	Impreza	2.5	4.2	7.4	20.0	—	2.6	1.6	16.9	7.4
	Baja	2.5	4.2	7.4	20.0	—	2.6	1.6	16.9	7.4
	Outback	3.0	5.7	7.4	20.2	—	2.6	1.6	16.9	7.4
	Legacy	2.5	4.2	7.4	20.0	—	2.6	1.6	16.9	6.9
2007	Impreza	2.5	4.2	7.4	20.0	—	2.6	1.6	16.9	7.4
	Outback	2.5	4.2	7.4	20.0	—	2.6	1.6	16.9	7.4
	Outback	3.0	5.7	7.4	20.2	—	2.6	1.6	16.9	7.4
	Legacy	2.5	4.2	7.4	20.0	—	2.6	1.6	16.9	6.9
2008	Impreza	2.5	4.2	7.4	20.0	—	2.6	1.6	16.9	7.4
	Outback	2.5	4.2	7.4	20.0	—	2.6	1.6	16.9	7.4
	Outback	3.0	5.7	7.4	20.2	—	2.6	1.6	16.9	7.4
	Legacy	2.5	4.2	7.4	20.0	—	2.6	1.6	16.9	6.9

Note: All capacities are approximate. Add fluid gradually and check to be sure a proper fluid level is obtained.

22140_SUBA_C0004

FLUID SPECIFICATIONS

Year	Model	Engine Displacement Liters	Engine ID/VIN	Engine Oil	Auto. Trans.	Drive Axle	Power Steering Fluid	Brake Master Cylinder	Engine Coolant
2006	Impreza	2.5	6	5W-30	Subaru ATF HP	75W-90	Dexron III	DOT 3	Subaru Coolant
		3.0	8	5W-30	Subaru ATF HP	75W-90	Dexron III	DOT 3	Subaru Coolant
	Baja	2.5	6	5W-30	Dexron III	75W-90	Dexron III	DOT 3	Subaru Coolant
	Outback	2.5	6	5W-30	Subaru ATF HP	75W-90	Dexron III	DOT 3	Subaru Coolant
		3.0	8	5W-30	Subaru ATF HP	75W-90	Dexron III	DOT 3	Subaru Coolant
	Legacy	2.5	6	5W-30	Subaru ATF HP	75W-90	Dexron III	DOT 3	Subaru Coolant
		3.0	8	5W-30	Subaru ATF HP	75W-90	Dexron III	DOT 3	Subaru Coolant
2007	Impreza	2.5	6	5W-30	Subaru ATF HP	75W-90	Dexron III	DOT 3	Subaru Coolant
		3.0	8	5W-30	Subaru ATF HP	75W-90	Dexron III	DOT 3	Subaru Coolant
	Outback	2.5	6	5W-30	Subaru ATF HP	75W-90	Dexron III	DOT 3	Subaru Coolant
		3.0	8	5W-30	Subaru ATF HP	75W-90	Dexron III	DOT 3	Subaru Coolant
	Legacy	2.5	6	5W-30	Subaru ATF HP	75W-90	Dexron III	DOT 3	Subaru Coolant
		3.0	8	5W-30	Subaru ATF HP	75W-90	Dexron III	DOT 3	Subaru Coolant
2008	Impreza	2.5	6	5W-30	Subaru ATF HP	75W-90	Dexron III	DOT 3	Subaru Coolant
		3.0	8	5W-30	Subaru ATF HP	75W-90	Dexron III	DOT 3	Subaru Coolant
	Outback	2.5	6	5W-30	Subaru ATF HP	75W-90	Dexron III	DOT 3	Subaru Coolant
		3.0	8	5W-30	Subaru ATF HP	75W-90	Dexron III	DOT 3	Subaru Coolant
	Legacy	2.5	6	5W-30	Subaru ATF HP	75W-90	Dexron III	DOT 3	Subaru Coolant
		3.0	8	5W-30	Subaru ATF HP	75W-90	Dexron III	DOT 3	Subaru Coolant

DOT: Department Of Transpotation

22140_SUBA_C0005

VALVE SPECIFICATIONS

Year	Engine Displacement Liters	Seat Angle (deg.)	Face Angle (deg.)	Spring Test Pressure (lbs. @ in.)	Spring Installed Height (in.)	Stem-to-Guide Clearance (in.) Intake	Stem-to-Guide Clearance (in.) Exhaust	Stem Diameter (in.) Intake	Stem Diameter (in.) Exhaust
2006	2.5	45	45	102 - 118@ 1.315	①	0.0014- 0.0024	0.0016- 0.0026	0.2343- 0.2348	0.2341- 0.2346
	3.0	45	45	102 - 118@ 1.315	②	0.0012- 0.0022	0.0016- 0.0026	002148- 0.2154	0.2148- 0.2150
2007	2.5	45	45	102 - 118@ 1.315	①	0.0014- 0.0024	0.0016- 0.0026	0.2343- 0.2348	0.2341- 0.2346
	3.0	45	45	102 - 118@ 1.315	②	0.0012- 0.0022	0.0016- 0.0026	002148- 0.2154	0.2148- 0.2150
2008	2.5	45	45	102 - 118@ 1.315	①	0.0014- 0.0024	0.0016- 0.0026	0.2343- 0.2348	0.2341- 0.2346
	3.0	45	45	102 - 118@ 1.315	②	0.0012- 0.0022	0.0016- 0.0026	002148- 0.2154	0.2148- 0.2150

① Free length: 1.8913 in.

② Free length: 1.8421 in.

22140_SUBA_C0007

CAMSHAFT AND BEARING SPECIFICATIONS CHART

All measurements are given in inches.

Year	Engine Displ. Liters	Engine ID/VIN	Journal Dia.	Brg. Oil Clearance	Shaft End-play	Runout	Journal Bore	Lobe Height Intake	Lobe Height Exhaust
2006	2.5 SOHC	6	1.2570- 1.2577	0.0022- 0.0035	0.0012- 0.0035	0.0024	NA	1.8330- 1.8370	1.8410- 1.8440
	2.5 DOHC	6	1.0614- 1.0620	0.0022- 0.0035	0.0027- 0.0047	0.0024	NA	1.5778- 1.5817	1.5468- 1.5507
	3.0	8	1.0215- 1.0222	0.0015- 0.0028-	NA	0.0024	NA	1.6571- 1.6610	1.5925- 1.5965
2007	2.5 SOHC	6	1.2570- 1.2577	0.0022- 0.0035	0.0012- 0.0035	0.0024	NA	1.8330- 1.8370	1.8410- 1.8440
	2.5 DOHC	6	1.0614- 1.0620	0.0022- 0.0035	0.0027- 0.0047	0.0024	NA	1.5778- 1.5817	1.5468- 1.5507
	3.0	8	1.0215- 1.0222	0.0015- 0.0028-	NA	0.0024	NA	1.6571- 1.6610	1.5925- 1.5965
2008	2.5 SOHC	6	1.2570- 1.2577	0.0022- 0.0035	0.0012- 0.0035	0.0024	NA	1.8330- 1.8370	1.8410- 1.8440
	2.5 DOHC	6	1.0614- 1.0620	0.0022- 0.0035	0.0027- 0.0047	0.0024	NA	1.5778- 1.5817	1.5468- 1.5507
	3.0	8	1.0215- 1.0222	0.0015- 0.0028-	NA	0.0024	NA	1.6571- 1.6610	1.5925- 1.5965

NA: Not Available

① Intake Journals 4 and 5: 0.0010 - 0.0022 in.

All Others: 0.0010 - 0.0024 in.

22140_SUBA_C0006

CRANKSHAFT AND CONNECTING ROD SPECIFICATIONS

All measurements are given in inches.

Year	Engine Displacement Liters	Crankshaft Main Brg. Journal Dia.	Crankshaft Main Brg. Oil Clearance	Crankshaft Shaft End-play	Crankshaft Thrust on No.	Connecting Rod Journal Diameter	Connecting Rod Oil Clearance	Connecting Rod Side Clearance
2006	2.5	2.3619-2.3625	①	0.0012-0.0098	3	1.8891-1.8898	0.0005-0.0015	0.0028-0.0130
	3.0	2.3619-2.3625	②	0.0012-0.0098	3	1.8891-1.8898	0.0009-0.0020	0.0028-0.0130
2007	2.5	2.3619-2.3625	①	0.0012-0.0098	3	1.8891-1.8898	0.0005-0.0015	0.0028-0.0130
	3.0	2.3619-2.3625	②	0.0012-0.0098	3	1.8891-1.8898	0.0009-0.0020	0.0028-0.0130
2008	2.5	2.3619-2.3625	①	0.0012-0.0098	3	1.8891-1.8898	0.0005-0.0015	0.0028-0.0130
	3.0	2.3619-2.3625	②	0.0012-0.0098	3	1.8891-1.8898	0.0009-0.0020	0.0028-0.0130

① Journals 1 and 5: 0.0001 - 0.0016 in.
Journals 2 and 4: 0.0004 - 0.0014 in.
Journal 3: 0.0004 - 0.0014 in.

② Journals 1 and 5: 0.0001 - 0.0016 in.
Journals 2 and 4: 0.0004 - 0.0018 in.
Journal 3: 0.0004 - 0.0016 in.

22140_SUBA_C0008

PISTON AND RING SPECIFICATIONS

All measurements are given in inches.

Year	Engine Displacement Liters	Piston Clearance	Ring Gap Top Compression	Ring Gap Bottom Compression	Ring Gap Oil Control	Ring Side Clearance Top Compression	Ring Side Clearance Bottom Compression	Ring Side Clearance Oil Control
2006	2.5	0.0004-0.0012	0.0079-0.0138	0.0146-0.0250	0.0079-0.0197	0.0016-0.0031	0.0012-0.0028	NA
	3.0	0.0004-0.0012	0.0079-0.0138	0.0138-0.0197	0.0079-0.0236	0.0016-0.0031	0.0012-0.0028	NA
2007	2.5	0.0004-0.0012	0.0079-0.0138	0.0146-0.0250	0.0079-0.0197	0.0016-0.0031	0.0012-0.0028	NA
	3.0	0.0004-0.0012	0.0079-0.0138	0.0138-0.0197	0.0079-0.0236	0.0016-0.0031	0.0012-0.0028	NA
2008	2.5	0.0004-0.0012	0.0079-0.0138	0.0146-0.0250	0.0079-0.0197	0.0016-0.0031	0.0012-0.0028	NA
	3.0	0.0004-0.0012	0.0079-0.0138	0.0138-0.0197	0.0079-0.0236	0.0016-0.0031	0.0012-0.0028	NA

NA: Not Available

22140_SUBA_C0009

TORQUE SPECIFICATIONS

All readings in ft. lbs.

Year	Engine Displacement Liters	Cylinder Head Bolts	Main ① Bearing Bolts	Rod Bearing Bolts	Crankshaft Damper Bolts	Flywheel Bolts	Manifold Intake	Manifold Exhaust	Spark Plugs	Drain Plug
2006	2.5	②	③	38	133	51 - 55	18	26	15	33
	3.0	④	14	39	131	60	18	26	15	33
2007	2.5	②	③	38	133	51 - 55	18	26	15	33
	3.0	④	14	39	131	60	18	26	15	33
2008	2.5	②	③	38	133	51 - 55	18	26	15	33
	3.0	④	14	39	131	60	18	26	15	33

① Engine block connecting bolts

② Step 1: Tighten all bolts to 22 ft. lbs.
Step 2: Tighten all bolts to 51 ft. lbs.
Step 3: Loosen all botls 180 degrees.
Step 4: Repeat Step 3.
Step 5: Tighten bolts 1 and 2 to 25 ft. lbs.
Step 6: Tighten bolts 3, 4, 5 and 6 to 11 ft. lbs.
Step 7: Tighten all bolts 80 to 90 degrees.
Step 8: Repeat Step 7. Do not exceed 180 degrees total tightening.

③ Split engine case connecting bolts:
Short bolts: 17-20 ft. lbs.
Long bolts: 33-37 ft. lbs.
Smaller short bolts (if used) 5 ft. lbs.

④ Step 1: Tighten all bolts to 14 ft. lbs.
Step 2: Tighten all bolts to 37 ft. lbs.
Step 3: Loosen all bolts 180 degrees the an additional 180 degrees in two steps in the reverse order of tightening sequence.
Step 4: Tighten all bolts to 18 ft. lbs.
Step 5: Tighten all bolts to 18 ft. lbs.
Step 6: Tighten bolts 90 degrees
Step 7: Tighten bolts 1, 2, 3 and 4 90 degrees.
Step 8: Tighten bolts 5, 6, 7 and 8 45 degrees.

22140_SUBA_C0010

WHEEL ALIGNMENT

Year	Model		Caster		Camber		
			Range (+/-Deg.)	Preferred Setting (Deg.)	Range (+/-Deg.)	Preferred Setting (Deg.)	Toe-in (in.)
2006	Impreza Sedan	F	1.00	+3.25	0.45	-0.15	0+/-0.12
		R	—	—	0.45	-1.25	0+/-0.12
	Impreza Sedan	F	1.00	+3.30	0.45	-0.25	0+/-0.12
	Turbo	R	—	—	0.45	-1.25	0+/-0.12
	Impreza Sedan	F	1.00	+4.50	0.45	-0.30	0+/-0.12
	STI	R	—	—	0.45	-1.15	0+/-0.12
	Impreza Wagon	F	1.00	+3.25	0.45	-0.10	0+/-0.12
		R	—	—	0.45	-1.15	0+/-0.12
	Impreza Wagon	F	1.00	+3.30	0.45	-0.20	0+/-0.12
	Turbo	R	—	—	0.45	-1.20	0+/-0.12
	Legacy Sedan	F	0.75	+3.05	0.50	-0.05	0+/-0.12
		R	—		0.75	-0.50	0+/-0.12
	Legacy Wagon	F	0.75	+2.05	0.50	-0.05	0+/-0.12
		R	—		0.75	-0.20	0+/-0.12
	Baja	F	1.00	+3.05	0.50	+0.40	0+/-0.12
		R	—	—	0.75	+0.05	0+/-0.12
	Outback	F	0.75	+2.45	0.50	0.05	0+/-0.12
		R	—		0.75	-0.17	0+/-0.12
2007	Impreza Sedan	F	1.00	+3.25	0.45	-0.15	0+/-0.12
		R	—	—	0.45	-1.25	0+/-0.12
	Impreza Sedan	F	1.00	+3.30	0.45	-0.25	0+/-0.12
	Turbo	R	—	—	0.45	-1.25	0+/-0.12
	Impreza Sedan	F	1.00	+4.50	0.45	-0.30	0+/-0.12
	STI	R	—	—	0.45	-1.15	0+/-0.12
	Impreza Wagon	F	1.00	+3.25	0.45	-0.10	0+/-0.12
		R	—	—	0.45	-1.15	0+/-0.12
	Impreza Wagon	F	1.00	+3.30	0.45	-0.20	0+/-0.12
	Turbo	R	—	—	0.45	-1.20	0+/-0.12
	Legacy Sedan	F	0.75	+3.05	0.50	-0.05	0+/-0.12
		R	—		0.75	-0.50	0+/-0.12
	Legacy Wagon	F	0.75	+2.05	0.50	-0.05	0+/-0.12
		R	—		0.75	-0.20	0+/-0.12
	Outback	F	0.75	+2.45	0.50	0.05	0+/-0.12
		R	—		0.75	-0.17	0+/-0.12
2008	Impreza Sedan	F	1.00	+3.25	0.45	-0.15	0+/-0.12
		R	—	—	0.45	-1.25	0+/-0.12
	Impreza Sedan	F	1.00	+3.30	0.45	-0.25	0+/-0.12
	Turbo	R	—	—	0.45	-1.25	0+/-0.12
	Impreza Sedan	F	1.00	+4.50	0.45	-0.30	0+/-0.12
	STI	R	—	—	0.45	-1.15	0+/-0.12
	Impreza Wagon	F	1.00	+3.25	0.45	-0.10	0+/-0.12
		R	—	—	0.45	-1.15	0+/-0.12
	Impreza Wagon	F	1.00	+3.30	0.45	-0.20	0+/-0.12
	Turbo	R	—	—	0.45	-1.20	0+/-0.12
	Legacy Sedan	F	0.75	+3.05	0.50	-0.05	0+/-0.12
		R	—		0.75	-0.50	0+/-0.12
	Legacy Wagon	F	0.75	+2.05	0.50	-0.05	0+/-0.12
		R	—		0.75	-0.20	0+/-0.12
	Outback	F	0.75	+2.45	0.50	0.05	0+/-0.12
		R	—		0.75	-0.17	0+/-0.12

22140_SUBA_C0011

TIRE, WHEEL AND BALL JOINT SPECIFICATIONS

Year	Model	OEM Tires		Tire Pressures (psi)		Wheel	Ball Joint	Lug Nut
		Standard	Optional	Front	Rear	Size	Inspection	(ft. lbs.)
2006	Impreza	P205/55R16	P205/55R17	35	35	6.5-JJ	0.012 in.	66
	Outback	P205/55R16	P205/55R17 P225/55R17 T155/70 D17	35	35	6.5-JJ 7-JJ 17x4T	0.012 in.	66
	Impreza WRX STI	P215/45ZR17	P225/45R17 P215/45R17	35	35	6-JJ 7-JJ	0.012 in.	66
	Baja	P225/60R16	P195/60R15	33	33	6.5-JJ	0.012 in.	81
	Legacy	P205/55R17 P205/50R17	P215/45ZR17 P215/45R18	35	35	7-JJ	0.012 in.	81
2007	Impreza	P205/55R16 P205/50R17 215/45ZR18 P225/55R17	T135/70 D17 T135/80R16	35	35	6.5-JJ 7-JJ	0.012 in.	74
	Outback	P205/55R16	T145/80R16	35	35	6.5-JJ	0.012 in.	74
	Impreza WRX	P215/45R17	None	35	35	7-JJ		74
	Impreza STI	P225/45R17	P205/55R16 T135/70 D17	35	35	8-JJ 17x4T	0.012 in.	74
	Legacy	P205/55R16 P205/50R17 215/45ZR18 P225/55R17	T135/70 D17 T135/80R16	35	35	6.5-JJ 7-JJ	0.012 in.	86
2008	Impreza	P205/55R16 P205/50R17 215/45ZR17 215/45ZR18	T135/70 D17 T135/80R16	35	35	6.5-JJ 7-JJ	0.012 in.	74
	Outback	P225/60R16 P225/55R17	T145/80R16	35	35	6.5-JJ	0.012 in.	74
	Impreza WRX	P215/45R17	None	35	35	7-JJ		74
	Impreza STI	P225/45R17	P205/55R16 T135/70 D17	35	35	8-JJ 17x4T	0.012 in.	74
	Legacy	P205/55R16 P205/50R17 215/45ZR17 215/45ZR18	T135/70 D17 T135/80R16	35	35	6.5-JJ 7-JJ	0.012 in.	86

OEM: Original Equipment Manufacturer

PSI: Pounds Per Square Inch

22140_SUBA_C0012

BRAKE SPECIFICATIONS

All measurements in inches unless noted

			Brake Disc			Brake Drum Diameter			Minimum Lining Thickness		Brake Caliper	
Year	Model		Original Thickness	Minimum Thickness	Maximum Runout	Original Inside Diameter	Max. Wear Limit	Maximum Machine Diameter	Front	Rear	Bracket Bolts (ft. lbs.)	Mounting Bolts (ft. lbs.)
2006	Impreza	F	0.940 ①	0.870 ②	0.003	NA	NA	NA	0.059	—	59	20
		R	0.390 ③	0.335 ④	0.003	NA	NA	NA	—	0.059	39	28
	WRX	F	0.940 ①	0.870 ②	0.003	NA	NA	NA	0.059	—	59	20
		R	0.390 ③	0.335 ④	0.003	NA	NA	NA	—	0.059	39	28
	STI	F	0.940 ①	0.870 ②	0.003	NA	NA	NA	0.059	—	59	20
		R	0.390 ③	0.335 ④	0.003	NA	NA	NA	—	0.059	39	28
	Outback	F	0.940 ①	0.870 ②	0.003	NA	NA	NA	0.059	—	59	20
		R	0.390 ③	0.335 ④	0.003	NA	NA	NA	—	0.059	39	28
	Baja	F	0.940	0.870	0.003	NA	NA	NA	0.059	—	58	29
		R	0.390	0.335	0.003	NA	NA	NA	—	0.059	58	29
	Legacy	F	0.940 ①	0.870 ②	0.003	NA	NA	NA	0.059	—	59	20
		R	0.390 ③	0.335 ④	0.003	NA	NA	NA	—	0.059	39	28
2007	Impreza	F	0.940 ①	0.870 ②	0.003	NA	NA	NA	0.059	—	59	20
		R	0.390 ③	0.335 ④	0.003	NA	NA	NA	—	0.059	39	28
	WRX	F	0.940 ①	0.870 ②	0.003	NA	NA	NA	0.059	—	59	20
		R	0.390 ③	0.335 ④	0.003	NA	NA	NA	—	0.059	39	28
	STI	F	0.940 ①	0.870 ②	0.003	NA	NA	NA	0.059	—	59	20
		R	0.390 ③	0.335 ④	0.003	NA	NA	NA	—	0.059	39	28
	Outback	F	0.940 ①	0.870 ②	0.003	NA	NA	NA	0.059	—	89	20
		R	0.390 ③	0.335 ④	0.003	NA	NA	NA	—	0.059	47	27 ⑤
	Legacy	F	0.940 ①	0.870 ②	0.003	NA	NA	NA	0.059	—	89	20
		R	0.390 ③	0.335 ④	0.003	NA	NA	NA	—	0.059	47	27 ⑤
2008	Impreza	F	0.940 ①	0.870 ②	0.003	NA	NA	NA	0.059	—	59	20
		R	0.390 ③	0.335 ④	0.003	NA	NA	NA	—	0.059	39	28
	WRX	F	0.940 ①	0.870 ②	0.003	—	NA	—	—	—	59	20
		R	0.390 ③	0.335 ④	0.003	10.000	10.080	10.080	0.177	0.177	39	28
	STI	F	0.940 ①	0.870 ②	0.003	NA	NA	NA	0.059	—	59	20
		R	0.390 ③	0.335 ④	0.003	NA	NA	NA	—	0.059	39	28
	Outback	F	0.940 ①	0.870 ②	0.003	NA	NA	NA	0.059	—	89	20
		R	0.390 ③	0.335 ④	0.003	NA	NA	NA	—	0.059	47	27 ⑤
	Legacy	F	0.940 ①	0.870 ②	0.003	NA	NA	NA	0.059	—	89	20
		R	0.390 ③	0.335 ④	0.003	NA	NA	NA	—	0.059	47	27 ⑤

NA: Not Available

① 17 inch type rotors original thickness: 1.18

② 17 inch type rotors minimum thickness: 1.10

③ Vented type rotors original thickness: 0.71

④ Vented type rotors minimum thickness: 0.63

④ Solid type rotors tighten to 20 ft. lbs.

22140_SUBA_C0013

SCHEDULED MAINTENANCE INTERVALS
SUBARU—IMPREZA, OUTBACK, LEGACY & BAJA

TO BE SERVICED	TYPE OF SERVICE	VEHICLE MILEAGE INTERVAL (x1000)												
		7.5	15	22.5	30	37.5	45	52.5	60	67.5	75	82.5	90	97.5
Engine oil & filter ① ②	R	✓	✓	✓	✓	✓	✓	✓	✓	✓	✓	✓	✓	✓
Brake lines	S/I		✓		✓		✓		✓		✓		✓	
Clutch & hill holder system	S/I		✓		✓		✓		✓		✓		✓	
Disc brake pads & discs, front & rear axle boots & axle shaft joint portions	S/I		✓		✓		✓		✓		✓		✓	
Parking brake	S/I		✓		✓		✓		✓		✓		✓	
Steering & suspension	S/I		✓		✓		✓		✓		✓		✓	
Air filter element	R				✓				✓				✓	
Engine coolant	R				✓				✓				✓	
Fuel filter	R								✓				✓	
Spark plugs	R				✓				✓					
Automatic transmission fluid & filter	S/I				✓				✓				✓	
Brake fluid	S/I				✓				✓				✓	
Camshaft drive belt ③	S/I				✓				✓				✓	
Coolant level, hoses & clamps	S/I				✓				✓				✓	
Drive belts	S/I				✓				✓				✓	
Fuel system, hoses & connections	S/I				✓				✓				✓	
Transmission and/or differential gear fluid	S/I				✓								✓	
Front & rear wheel bearing	S/I								✓					

R: Replace S/I: Service or Inspect

① 3.0L engine change every 3000 miles.

② 2.5L turbo models change every 3875 miles or 3.75 months.

③ Replace camshaft drive belt at 105,000 miles or 105 months

FREQUENT OPERATION MAINTENANCE (SEVERE SERVICE)

If a vehicle is operated under any of the following conditions it is considered severe service:

- Extremely dusty areas.

- 50% or more of the vehicle operation is in 32°C (90°F) or higher temperatures, or constant operation in temperatures below 0°C (32°F).

- Prolonged idling (vehicle operation in stop and go traffic).

- Frequent short running periods (engine does not warm to normal operating temperatures).

- Police, taxi, delivery usage or trailer towing usage.

Oil & oil filter change: change every 3750 miles.

Clutch & hill holder system: service or inspect every 7500 miles.

Disc brake pads & discs, front & rear axle boots & axle shaft joint portions: service or inspect every 7500 miles.

Steering & suspension: service or inspect every 7500 miles.

Air filter element: service or inspect every 15,000 miles.

Automatic transmission fluid: service or inspect every 15,000 miles.

Brake linings & drums: service or inspect every 15,000 miles.

Coolant level, hoses & clamps: service or inspect every 15,000 miles.

Drive belts: service or inspect every 15,000 miles.

Transmission/differential gear oil service or inspect every 15,000 miles.

Front & rear wheel bearing repack: service or inspect every 30,000 miles.

22140_SUBA_C0014

PRECAUTIONS

Before servicing any vehicle, please be sure to read all of the following precautions, which deal with personal safety, prevention of component damage, and important points to take into consideration when servicing a motor vehicle:

- Never open, service or drain the radiator or cooling system when the engine is hot; serious burns can occur from the steam and hot coolant.
- Observe all applicable safety precautions when working around fuel. Whenever servicing the fuel system, always work in a well-ventilated area. Do not allow fuel spray or vapors to come in contact with a spark, open flame, or excessive heat (a hot drop light, for example). Keep a dry chemical fire extinguisher near the work area. Always keep fuel in a container specifically designed for fuel storage; also, always properly seal fuel containers to avoid the possibility of fire or explosion. Refer to the additional fuel system precautions later in this section.
- Fuel injection systems often remain pressurized, even after the engine has been turned **OFF**. The fuel system pressure must be relieved before disconnecting any fuel lines. Failure to do so may result in fire and/or personal injury.
- Brake fluid often contains polyglycol ethers and polyglycols. Avoid contact with the eyes and wash your hands thoroughly after handling brake fluid. If you do get brake fluid in your eyes, flush your eyes with clean, running water for 15 minutes. If eye irritation persists, or if you have taken brake fluid internally, IMMEDIATELY seek medical assistance.
- The EPA warns that prolonged contact with used engine oil may cause a number of skin disorders, including cancer. You should make every effort to minimize your exposure to used engine oil. Protective gloves should be worn when changing oil. Wash your hands and any other exposed skin areas as soon as possible after exposure to used engine oil. Soap and water, or waterless hand cleaner should be used.
- All new vehicles are now equipped with an air bag system, often referred to as a Supplemental Restraint System (SRS) or Supplemental Inflatable Restraint (SIR) system. The system must be disabled before performing service on or around system components, steering column, instrument panel components, wiring and sensors. Failure to follow safety and disabling procedures could result in accidental air bag deployment, possible personal injury and unnecessary system repairs.
- Always wear safety goggles when working with, or around, the air bag system. When carrying a non-deployed air bag, be sure the bag and trim cover are pointed away from your body. When placing a non-deployed air bag on a work surface, always face the bag and trim cover upward, away from the surface. This will reduce the motion of the module if it is accidentally deployed. Refer to the additional air bag system precautions later in this section.
- Clean, high quality brake fluid from a sealed container is essential to the safe and proper operation of the brake system. You should always buy the correct type of brake fluid for your vehicle. If the brake fluid becomes contaminated, completely flush the system with new fluid. Never reuse any brake fluid. Any brake fluid that is removed from the system should be discarded. Also, do not allow any brake fluid to come in contact with a painted surface; it will damage the paint.
- Never operate the engine without the proper amount and type of engine oil; doing so WILL result in severe engine damage.
- Timing belt maintenance is extremely important. Many models utilize an interference-type, non-freewheeling engine. If the timing belt breaks, the valves in the cylinder head may strike the pistons, causing potentially serious (also time-consuming and expensive) engine damage. Refer to the maintenance interval charts for the recommended replacement interval for the timing belt, and to the timing belt section for belt replacement and inspection.
- Disconnecting the negative battery cable on some vehicles may interfere with the functions of the on-board computer system(s) and may require the computer to undergo a relearning process once the negative battery cable is reconnected.
- When servicing drum brakes, only disassemble and assemble one side at a time, leaving the remaining side intact for reference.
- Only an MVAC-trained, EPA-certified automotive technician should service the air conditioning system or its components.

BRAKES

ANTI-LOCK BRAKE SYSTEM (ABS)

GENERAL INFORMATION

See Figure 1.

When wheel slip is detected during a brake application, an ABS event occurs. During antilock braking, hydraulic pressure in the individual wheel circuits is controlled to prevent any wheel from slipping. A separate hydraulic line and specific solenoid valves are provided for each wheel. The ABS can decrease, hold, or increase hydraulic pressure to each wheel. During antilock braking, a series of rapid pulsations is felt in the brake pedal. These pulsations are caused by the rapid changes in position of the individual solenoid valves as the Vehicle Dynamics Control Module (VDCCM) & Hydraulic Control Unit (H/U) responds to wheel speed sensor inputs and attempts to prevent wheel slip. These pedal pulsations are present only during antilock braking and stop when normal braking is resumed or when the vehicle comes to a stop. A ticking or popping noise may also be heard as the solenoid valves cycle rapidly. During antilock braking on dry pavement, intermittent chirping noises may be heard as the tires approach slipping. These noises and pedal pulsations are considered normal during antilock operation.

Vehicles equipped with ABS may be stopped by applying normal force to the brake pedal. Brake pedal operation during normal braking is no different than that of previous non-ABS systems. Maintaining a constant force on the brake pedal provides the shortest stopping distance while maintaining vehicle.

WHEEL SPEED SENSORS

REMOVAL & INSTALLATION

Front

Baja Models

1. Before servicing the vehicle, refer to the Precautions Section.
2. Disconnect front ABS sensor connector located in engine compartment.
3. Remove bolts which secure sensor harness to strut.
4. Remove bolts which secure sensor harness to body.
5. Remove bolts which secure front ABS sensor to housing, and remove front ABS sensor.

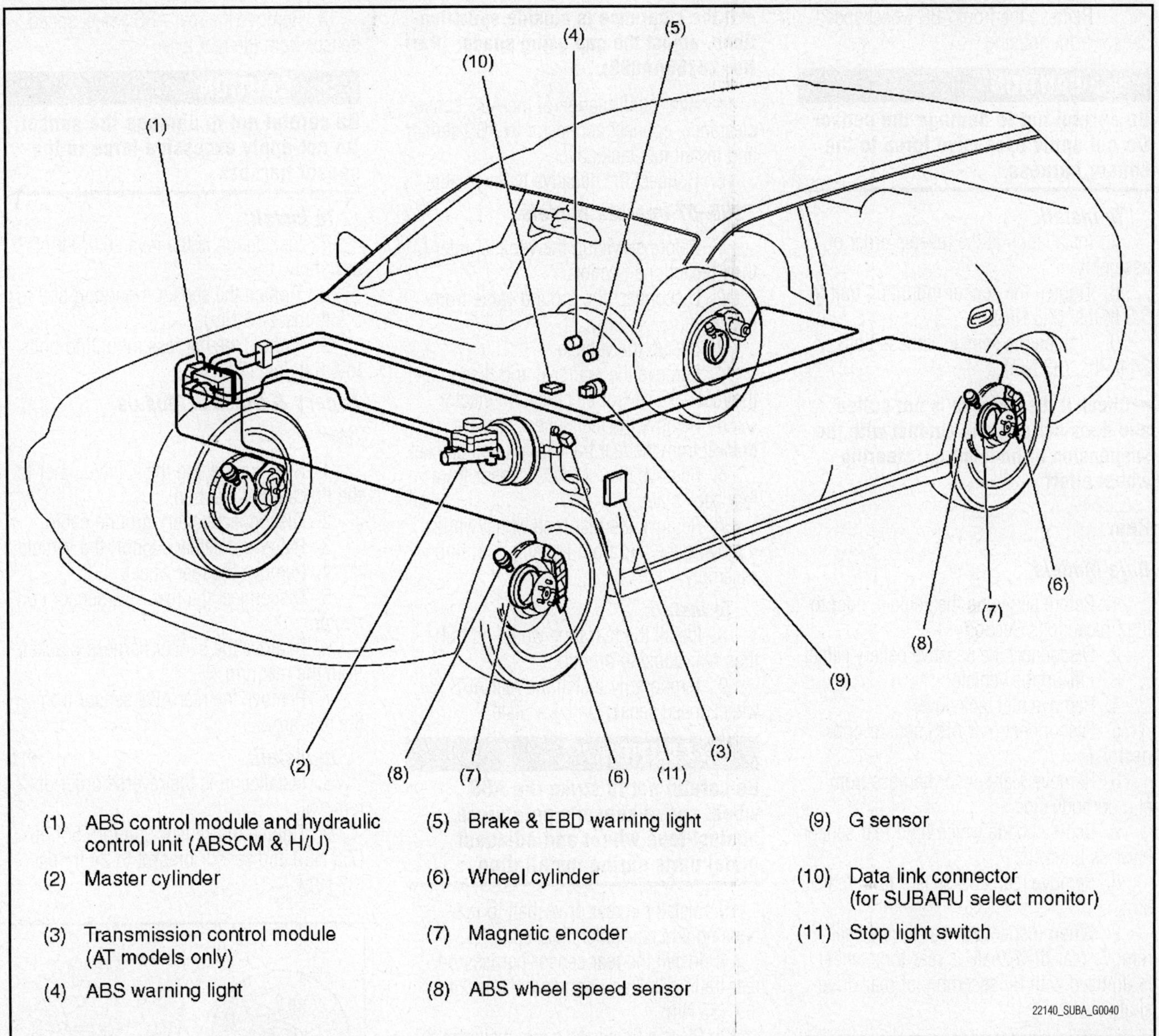

(1) ABS control module and hydraulic control unit (ABSCM & H/U)
(2) Master cylinder
(3) Transmission control module (AT models only)
(4) ABS warning light
(5) Brake & EBD warning light
(6) Wheel cylinder
(7) Magnetic encoder
(8) ABS wheel speed sensor
(9) G sensor
(10) Data link connector (for SUBARU select monitor)
(11) Stop light switch

22140_SUBA_G0040

Fig. 1 Exploded view of the ABS components

⁂ CAUTION

Be careful not to damage pole piece located at tip of the sensor and teeth faces during removal.

⁂ CAUTION

Do not pull sensor harness during removal.

To install:

6. Installation is the reverse order of assembly.
7. Place a thickness gauge between the ABS sensor's and tone wheel's tooth face. Standard clearance should be 0.012–0.031 inch. (0.3–0.8 mm). Once clearance is obtained over the entire perimeter, tighten the ABS sensor on housing to 24 ft. lbs. (32 Nm).
8. After confirmation of the ABS sensor clearance, connect connector to ABS sensor.
9. Connect the negative battery cable.

Impreza, Legacy & Outback Models

See Figure 2.

1. Before servicing the vehicle, refer to the Precautions Section.
2. Disconnect the battery ground cable from the battery.
3. Disconnect the ABS wheel speed sensor connector located next to the front strut mounting house in the engine compartment.
4. Remove the sensor harness bracket.
5. Remove the bolts which secure the sensor harness to the front strut.

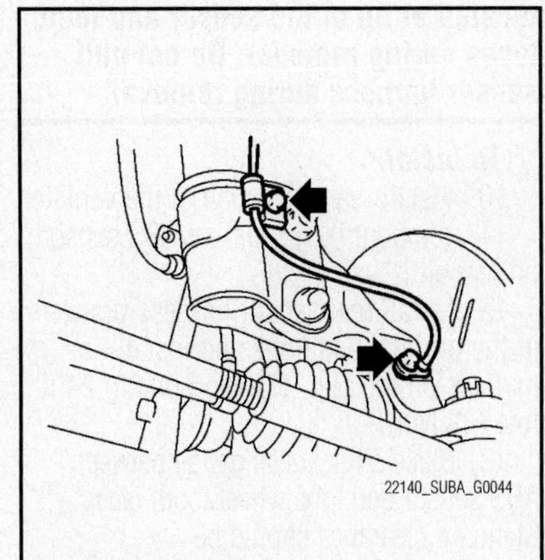

Fig. 2 Front wheel speed sensor and mounting bolt location

6. Remove the front ABS wheel speed sensor from housing.

CAUTION

Be careful not to damage the sensor. Do not apply excessive force to the sensor harness.

To install:

7. Installation is the reverse order of assembly.
8. Tighten the sensor mounting bolt to 5.5 ft. lbs. (7.5 Nm).
9. Tighten the sensor harness bolts to 24 ft. lbs. (33 Nm).

➡Check if the harness is not pulled and does not come in contact with the suspension or body during steering wheel effort.

Rear

Baja Models

1. Before servicing the vehicle, refer to the Precautions Section.
2. Disconnect the negative battery cable.
3. Lift-up the vehicle.
4. Remove fuel tank cover.
5. Disconnect rear ABS sensor connector.
6. Remove rear sensor harness from clip on body side.
7. Remove bolts which hold rear sensor harness brackets.
8. Remove rear ABS sensor from rear arm.
9. When inspecting rear tone wheel, remove rear driveshaft as rear tone wheel is unitized with BJ assembly of rear driveshaft.

CAUTION

Be careful not to damage pole piece located at tip of the sensor and teeth faces during removal. Do not pull sensor harness during removal.

To install:

10. Install rear driveshaft to the vehicle.
11. Temporarily install rear ABS sensor on rear arm.
12. Install rear sensor harness brackets in the original positions and install harness on the clip. Tighten bolts to 24 ft. lbs. (33 Nm).
13. Place a thickness gauge between ABS sensor and tone wheel tooth face. Standard clearance should be 0.0173–0.0370 inch. (0.44–0.94 mm). After standard clearance is obtained (over the entire perimeter, tighten ABS sensor on rear arm to 24 ft. lbs. (33 Nm).

➡If the clearance is outside specifications, adjust the gap using spacer (Part No. 26755AA000).

14. After confirmation of the ABS sensor clearance, connect connector to ABS sensor and install fuel tank cover
15. Connect the negative battery cable.

2006–07 Impreza Models

1. Before servicing the vehicle, refer to the Precautions Section.
2. Disconnect the ground cable from battery.
3. Lift-up the vehicle.
4. Remove the rear seat and disconnect the rear ABS sensor electrical connector.
5. Remove the rear sensor harness bracket from the rear trailing link and bracket.
6. Remove the rear ABS sensor from rear back plate.
7. Remove the rear tone wheel while removing the hub from housing and hub assembly.

To install:

8. Install the rear tone wheel on hub, then rear housing on hub.
9. Temporarily install the rear ABS wheel speed sensor on back plate.

CAUTION

Be careful not to strike the ABS wheel speed sensor's pole piece against tone wheel and adjacent metal parts during installation.

10. Install the rear driveshaft to rear housing and rear differential spindle.
11. Install the rear sensor harness on rear trailing link and tighten bolts to 24 ft. lbs. (33 Nm).
12. Place a thickness gauge between the ABS sensor's and tone wheel's tooth face. Standard clearance should be 0.028–0.047 in. (0.7–1.2 mm).
13. Once clearance is obtained over the entire perimeter, tighten the ABS sensor on housing to 24 ft. lbs. (32 Nm).
14. After confirmation of the ABS wheel speed sensor clearance, connect the connector to ABS wheel speed sensor.
15. Connect the negative battery cable.

2008 Impreza Models

1. Before servicing the vehicle, refer to the Precautions Section.
2. Disconnect the negative battery cable.
3. Disconnect the connector from the rear ABS wheel speed sensor.
4. Remove the sensor harness clamp of the rear sub frame.
5. Remove the sensor harness bracket from the upper arm.
6. Remove the rear ABS wheel speed sensor from the rear axle.

CAUTION

Be careful not to damage the sensor. Do not apply excessive force to the sensor harness.

To install:

7. Installation is the reverse order of assembly.
8. Tighten the sensor mounting bolt to 5.5 ft. lbs. (7.5 Nm).
9. Tighten the harness mounting bolts to 5.5 ft. lbs. (7.5 Nm).

Legacy & Outback Models

See Figure 3.

1. Before servicing the vehicle, refer to the Precautions Section.
2. Disconnect battery ground cable.
3. Raise and safely support the vehicle.
4. Remove the rear wheel.
5. Disconnect the rear ABS sensor connector.
6. Remove the sensor harness bracket from the rear arm.
7. Remove the rear ABS sensor from the rear arm.

To install:

8. Installation is the reverse order of removal.
9. Tighten the sensor bolt to 5.5 ft. lbs. (7.5 Nm) and sensor bracket to 24 ft. lbs. (33 Nm).

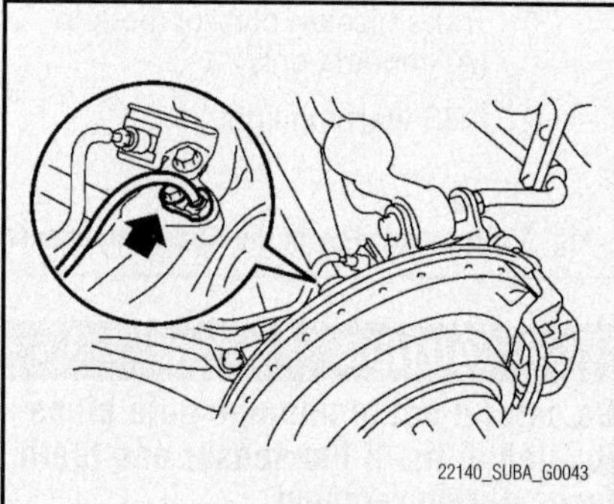

Fig. 3 Rear wheel speed sensor location 2006–08 Legacy and Outback

WHEEL SPEED SENSOR RINGS (TOOTHED RINGS)

REMOVAL & INSTALLATION

Front

Baja and 2006–07 Impreza Models

Refer to Front Driveshaft, because front tone wheel is integrated with front driveshaft.

2008 Impreza and 2006–08 Legacy & Outback Models

Refer to Front Hub Bearing for removal, because the front magnetic encoder is integrated with front hub bearing.

Rear

Baja and 2006–07 Impreza Models

Refer to Rear Driveshaft, because rear tone wheel is integrated with rear driveshaft.

2008 Impreza and 2006–08 Legacy & Outback Models

Refer to Front Hub Bearing for removal, because the front magnetic encoder is integrated with front hub bearing.

BRAKES

BLEEDING THE BRAKE SYSTEM

BLEEDING PROCEDURE

See Figures 4 and 5.

1. Raise the vehicle and safely support it.
2. Remove both front and rear wheels.
3. Draw out the brake fluid from master cylinder with syringe.
4. Refill reservoir tank with recommended brake fluid.
5. Install one end of a vinyl tube onto the air bleeder and insert the other end of the tube into a container to collect the brake fluid.
6. Instruct your co-worker to depress the brake pedal slowly two or three times and then hold it depressed.
7. Loosen bleeder screw approximately 1/4 turn until a small amount of brake fluid drains into container, and then quickly tighten screw.
8. Repeat again from the two former procedures above until there are no air bubbles in drained brake fluid and new fluid flows through vinyl tube.

➡Add brake fluid as necessary while performing the air bleed operation, in order to prevent the tank from running short of brake fluid.

9. After completing the bleeding operation, hold brake pedal depressed and tighten screw and install bleeder cap.

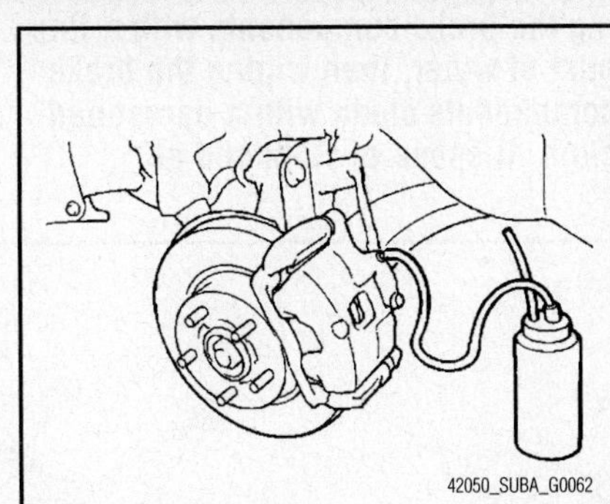

Fig. 4 Connect one end of the tube onto the air bleeder.

10. Bleed air from each wheel cylinder using the same procedures as described above.
11. Depress brake pedal and hold it there for approximately 20 seconds. At this time check pedal to see if it shows any unusual movement. Visually inspect bleeder screws and brake pipe joints to make sure that there is no fluid leakage.
12. Install the wheels, and drive car for a short distance (between 1–2 miles to make sure that brakes are operating properly.

BLEEDING THE ABS SYSTEM

Refer to bleeding the brake system.

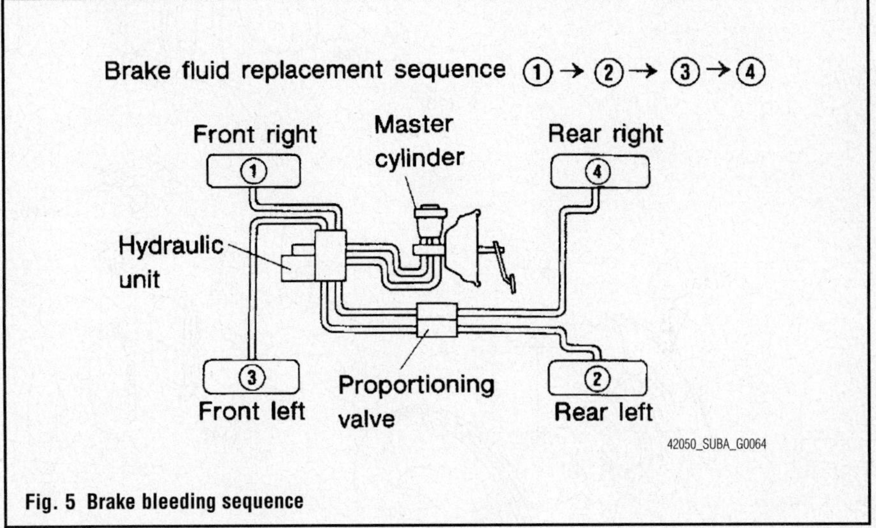

Fig. 5 Brake bleeding sequence

See Figures 6 through 10.

⁂ CAUTION

Dust and dirt accumulating on brake parts during normal use may contain asbestos fibers from production or aftermarket brake linings. Breathing excessive concentrations of asbestos fibers can cause serious bodily harm. Exercise care when servicing brake parts. Do not sand or grind brake lining unless equipment used is designed to contain the dust residue. Do not clean brake parts with compressed air or by dry brushing. Cleaning should be done by dampening the brake components with a fine mist of water, then wiping the brake components clean with a dampened cloth. Dispose of cloth and all residue containing asbestos fibers in an impermeable container with the appropriate label. Follow practices prescribed by the Occupational Safety and Health Administration (OSHA) and the Environmental Protection Agency (EPA) for the handling, processing, and disposing of dust or debris that may contain asbestos fibers.

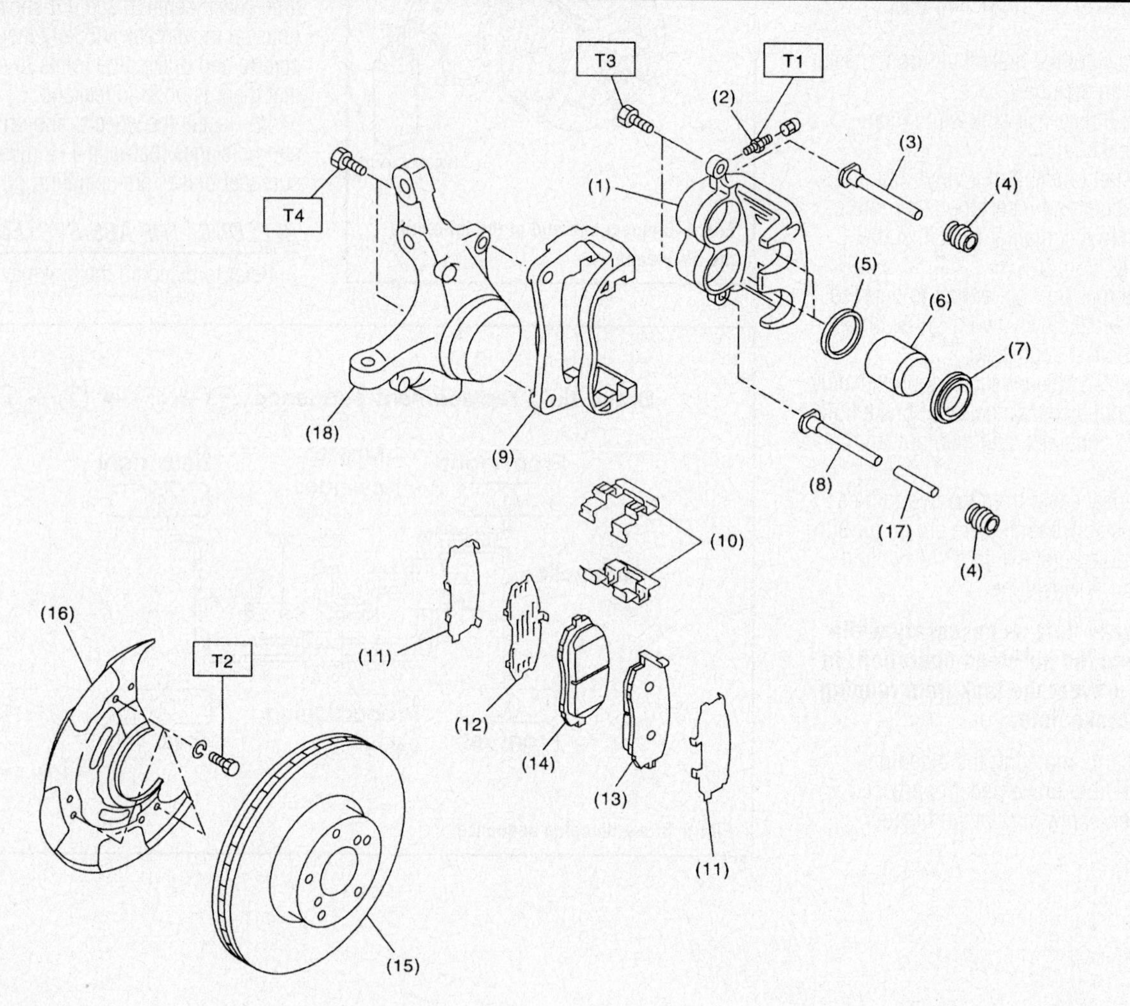

(1)	Caliper body	(9)	Support	(17)	Bushing
(2)	Air bleeder screw	(10)	Pad clip	(18)	Housing
(3)	Guide pin (Green)	(11)	Outer shim		
(4)	Pin boot	(12)	Inner shim		
(5)	Piston seal	(13)	Pad (Outside)		
(6)	Piston	(14)	Pad (Inside)		
(7)	Piston boot	(15)	Disc rotor		
(8)	Lock pin (Yellow)	(16)	Disc cover		

Tightening torque: N·m (kgf-m, ft-lb)

T1: 8 (0.8, 5.8)
T2: 18 (1.8, 13.0)
T3: 26.5 (2.7, 19.5)
T4: 80 (8.2, 59)

09490_SBCR_G0104

Fig. 6 Front disc brakes and related components 15 inch type—Impreza and WRX

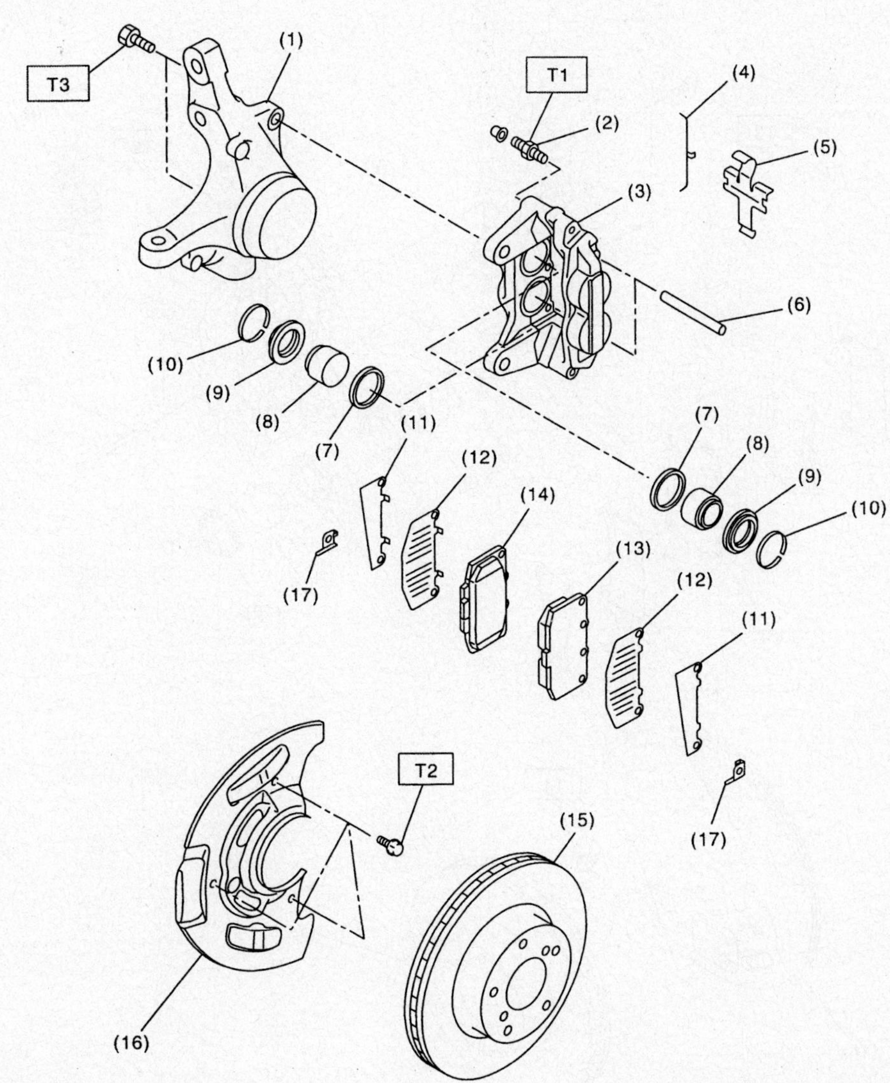

(1) Housing
(2) Air bleeder screw
(3) Caliper body
(4) M clip
(5) Cross spring
(6) Pad pin
(7) Piston seal
(8) Piston
(9) Piston boot
(10) Boot ring
(11) Outer shim
(12) Inner shim
(13) Pad (Outside)
(14) Pad (Inside)
(15) Disc rotor
(16) Disc cover
(17) Spacer

Tightening torque: N·m (kgf-m, ft-lb)

T1: 8 (0.8, 5.8)
T2: 18 (1.8, 13.0)
T3: 80 (8.2, 59)

09490_SBCR_G0105

Fig. 7 Front disc brakes and related components 16 inch type—Impreza and WRX

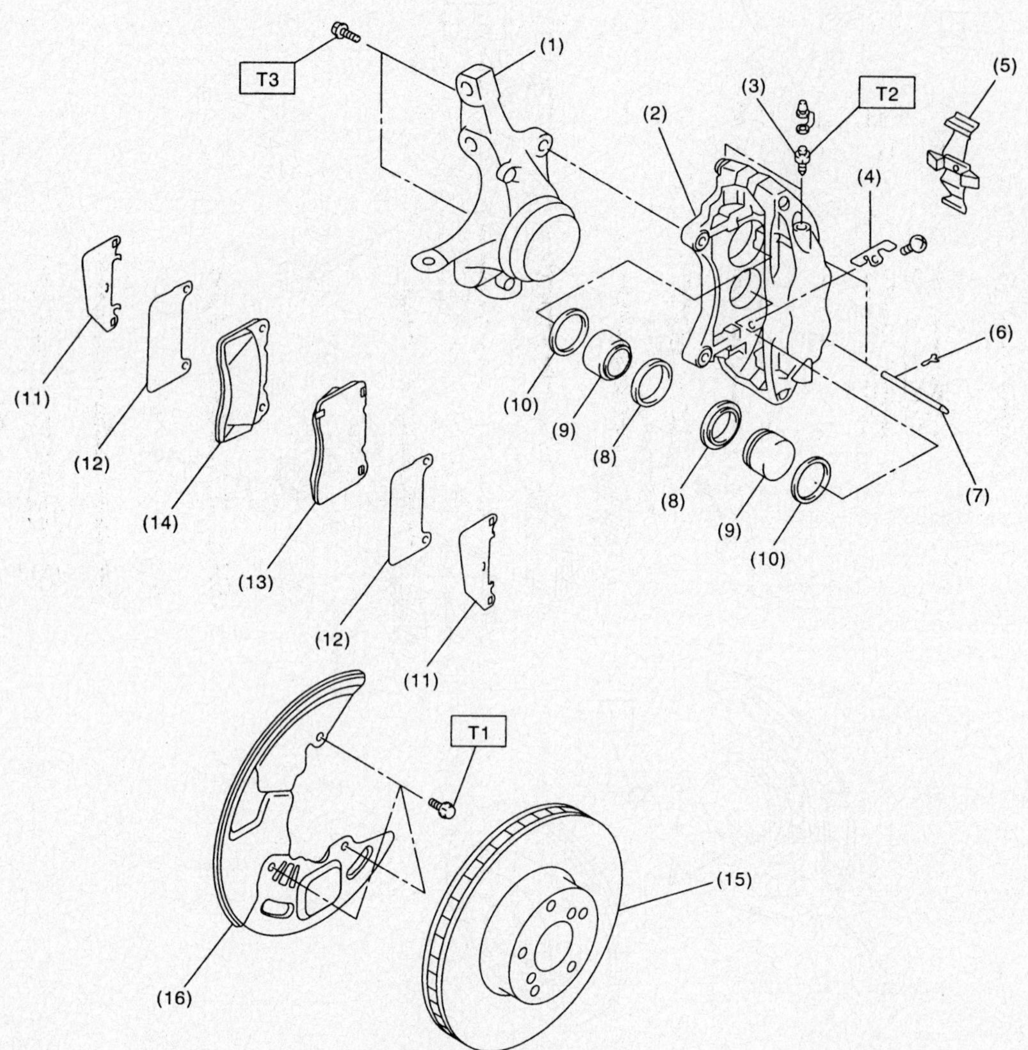

(1)	Housing	(8)	Piston boot	(15)	Disc rotor
(2)	Caliper body	(9)	Piston	(16)	Disc cover
(3)	Air bleeder screw	(10)	Piston seal		
(4)	Guide plate	(11)	Pad shim (Outside)		
(5)	Cross spring	(12)	Pad shim (Inside)		
(6)	Clip	(13)	Pad (Outside)		
(7)	Pad pin	(14)	Pad (Inside)		

Tightening torque: N·m (kgf-m, ft-lb)
T1: 18 (1.8, 13.0)
T2: 20 (2.0, 14.5)
T3: 155 (15.8, 114.3)

09490_SBCR_G0106

Fig. 8 Front disc brakes and related components 17 inch type—Impreza and WRX

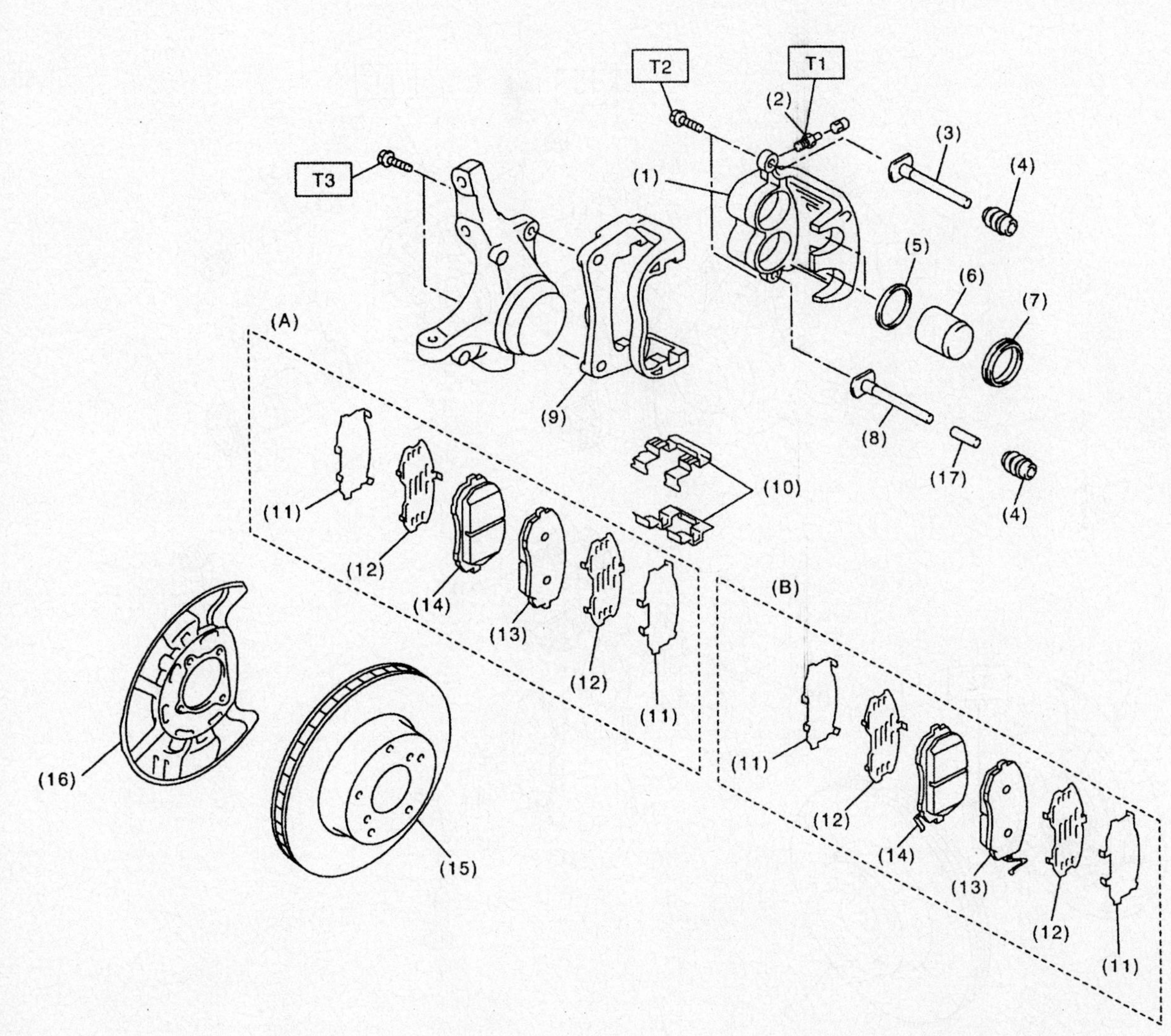

(A) 16-inch type
(B) 17-inch type

(1) Caliper body
(2) Air bleeder screw
(3) Guide pin (Green)
(4) Pin boot
(5) Piston seal
(6) Piston
(7) Piston boot
(8) Lock pin (Yellow)
(9) Support
(10) Pad clip
(11) Outer shim
(12) Inner shim
(13) Pad (Outside)
(14) Pad (Inside)
(15) Disc rotor
(16) Disc cover
(17) Bushing

Tightening torque: N·m (kgf-m, ft-lb)

T1: 8 (0.8, 5.8)
T2: 27 (2.8, 19.9)
T3: 80 (8.2, 59)

09490_SBCR_G0107

Fig. 9 Front disc brakes and related components—Outback and Legacy

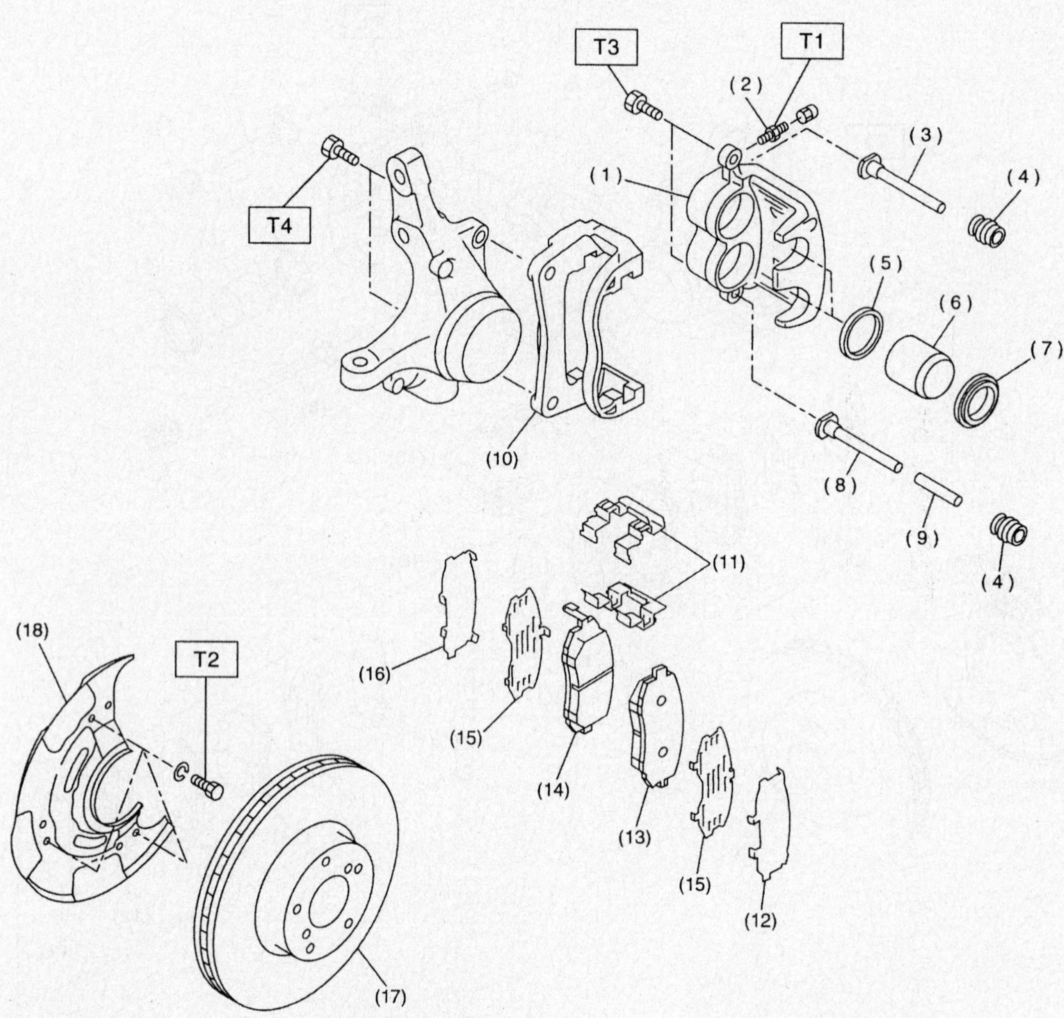

(1)	Caliper body	(9)	Bushing	(17)	Disc rotor
(2)	Air bleeder screw	(10)	Support	(18)	Disc cover
(3)	Guide pin (Green)	(11)	Pad clip		
(4)	Pin boot	(12)	Outer shim		
(5)	Piston seal	(13)	Pad (Outside)		
(6)	Piston	(14)	Pad (Inside)		
(7)	Piston boot	(15)	Rubber coated shim		
(8)	Lock pin (Yellow)	(16)	Inner shim		

Tightening torque: N·m (kgf-m, ft-lb)

T1: 8 (0.8, 5.8)
T2: 18 (1.8, 13.0)
T3: 37 (3.8, 27.5)
T4: 80 (8.2, 59)

09490_SBCR_G0108

Fig. 10 Front disc brakes and related components—2006 Baja

BRAKE CALIPER

REMOVAL & INSTALLATION

See Figure 11.

1. Before servicing the vehicle, refer to the Precautions Section.
2. Set the vehicle on a lift.
3. Loosen the wheel nuts.
4. Lift up the vehicle, and remove the front wheels.
5. Remove the union bolt, and disconnect the brake hose from the caliper body assembly.
6. Remove the bolt securing the lock pin to caliper body.
7. Raise the caliper body, and then move it toward vehicle center to separate it from the support.
8. Remove the support from housing.

➡Remove the support only when replacing the rotor or support. It need not be removed when servicing the caliper body assembly.

9. Remove mud and foreign matter from the caliper body assembly and the support.

To install:

10. Apply a thin coat of Molykote M7439 (Part No. 003602001) to the support.
11. Apply a thin coat of Molykote M7439 (Part No. 003602001) to the contact surface between the pad and pad clip.
12. Apply a thin coat of Molykote AS-880N (Part No. K0777YA010) to both surfaces of the inner shim.
13. Install the pad to support.
14. Tightening specifications are as follows:
    - 15-inch type tighten the support housing to 59 ft. lbs. (80 Nm)
    - 15-inch type caliper body support housing to 20 ft. lbs. (26 Nm)
    - 16-inch type caliper body support housing to 59 ft. lbs. (80 Nm)
    - 17-inch type install the caliper body assembly to the housing and tighten to 114 ft. lbs. (155 Nm).
15. Connect the brake hose using a new brake hose gasket and tighten to 13 ft. lbs. (18 Nm).
16. Bleed the hydraulic system.

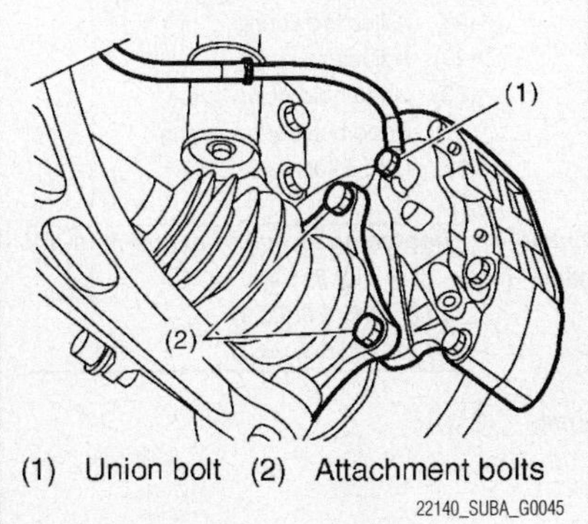

Fig. 11 Rear brake caliper view

DISC BRAKE PADS

REMOVAL & INSTALLATION

See Figures 12, 13 and 14.

1. Before servicing the vehicle, refer to the Precautions Section.
2. Raise and safely support the vehicle.
3. Loosen the wheel nuts.
4. Remove the wheel.
5. Raise and support the vehicle safely.
6. Remove the tire and wheel.
7. On 15-inch type, remove the lower caliper bolt. Raise the caliper body upward and support it. Do not disconnect the fluid line.
8. On 16-inch type, remove the "M" clip. Remove the pad pins and cross spring. Expand the pads and then push the piston back.
9. On 17-inch type, remove the clip. Remove the pad pins and cross spring. Expand the pads and then push the piston back.
10. Remove the disc brake pads.

To install:

11. Installation is the reverse of the removal procedure.
12. On 15-inch type, apply a thin coat of Molykote AS-880N (part number K0779YA010) or equivalent to the frictional portion between the pad and pad clip, and the pad and pad inner shim.

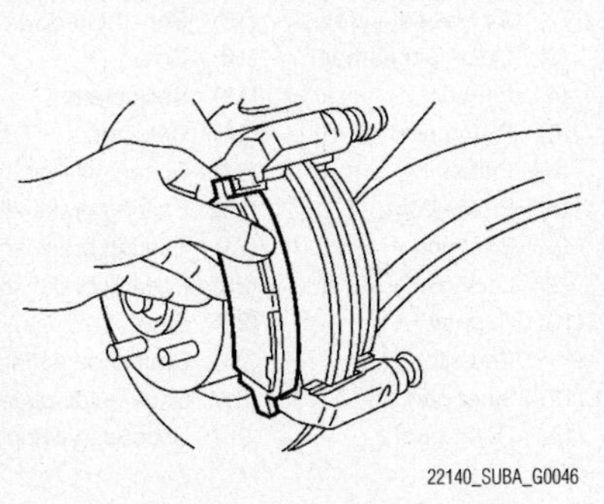

Fig. 12 15-inch type pads shown

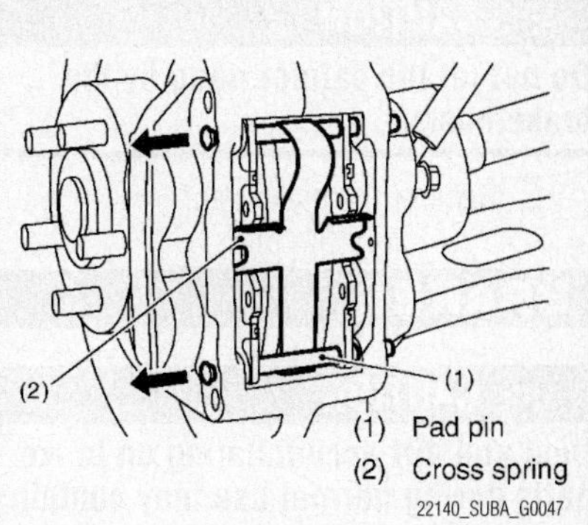

Fig. 13 16-inch type pads shown

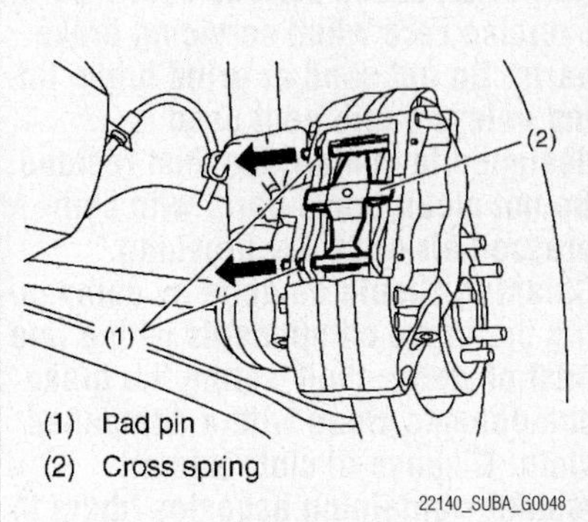

Fig. 14 17-inch type pads shown

13. On 15-inch type tighten the caliper mounting bolts to 20 ft. lbs. (26 Nm)
14. On 16-inch type, apply a thin coat of Molykote AS-880N (part number K0779YA010) or equivalent to the frictional portion between the pad and pad inner shim.
15. On 17-inch type, apply a thin coat of Molykote AS-880N (part number K0779YA010) or equivalent to the frictional portion between the pad and pad shim.
16. Check the brake fluid level, correct as required.
17. Bleed the hydraulic system, as required.

ROTOR

REMOVAL & INSTALLATION

1. Before servicing the vehicle, refer to the Precautions Section.
2. Raise and safely support the vehicle.
3. Remove the wheel.
4. Remove the caliper assembly, and do not disconnect the brake lines. Hang the caliper securely from the strut using mechanics wire. Refer to "Brake Caliper, Removal & Installation" to remove the caliper assembly.

✻✻ CAUTION

Do not let the caliper hang by the brake hoses.

5. Remove the brake rotor.

➡If the rotor is seized and is difficult to remove, install an 8 mm bolt into the threaded service holes (B) on the hub and tighten.

To install:

6. Installation is the reverse order of removal.

BRAKES — REAR DISC BRAKES

✻✻ CAUTION

Dust and dirt accumulating on brake parts during normal use may contain asbestos fibers from production or aftermarket brake linings. Breathing excessive concentrations of asbestos fibers can cause serious bodily harm. Exercise care when servicing brake parts. Do not sand or grind brake lining unless equipment used is designed to contain the dust residue. Do not clean brake parts with compressed air or by dry brushing. Cleaning should be done by dampening the brake components with a fine mist of water, then wiping the brake components clean with a dampened cloth. Dispose of cloth and all residue containing asbestos fibers in an impermeable container with the appropriate label. Follow practices prescribed by the Occupational Safety and Health Administration (OSHA) and the Environmental Protection Agency (EPA) for the handling, processing, and disposing of dust or debris that may contain asbestos fibers.

BRAKE CALIPER

REMOVAL & INSTALLATION

Baja Models

See Figure 15.

1. Before servicing the vehicle, refer to the Precautions Section.
2. Lift-up vehicle and remove wheels.
3. Disconnect brake hose from caliper body assembly.
4. Remove bolt securing lock pin to caliper body.
5. Raise caliper body and move it toward vehicle center to separate it from support.
6. Remove support from back plate.

➡Remove support only when replacing it or the rotor. It need not be removed when servicing caliper body assembly.

7. Clean mud and foreign particles from caliper body assembly and support.

✻✻ CAUTION

Be careful not to allow foreign particles to enter inlet (at brake hose connector).

To install:

8. Install disc rotor on hub.
9. Install support on back plate and tighten to 58 ft. lbs. (78 Nm).

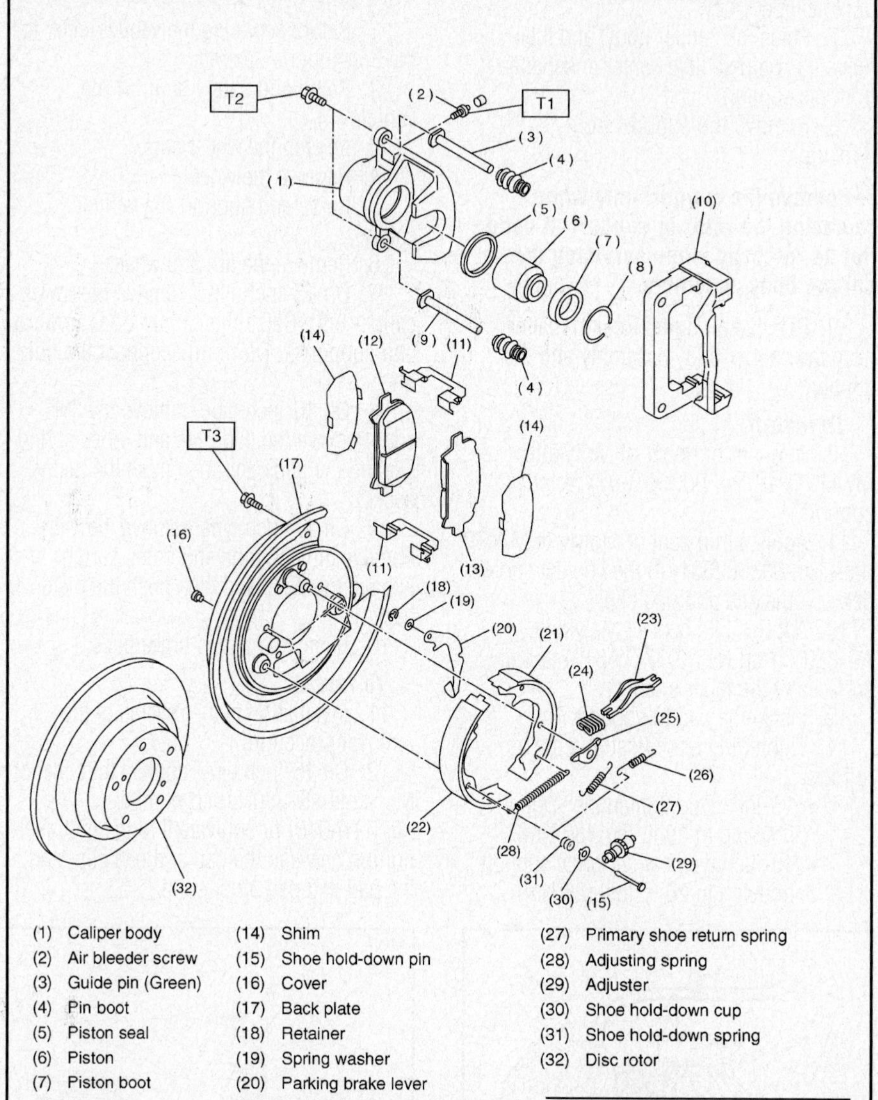

Fig. 15 Exploded view of rear disc brake system

10. Apply thin coat of Molykote AS880N (Part No. 26298AC000) to the frictional portion between pad and pad clip.
11. Install pads on support.
12. Install caliper body on support and tighten to 29 ft. lbs. (39 Nm).
13. Connect the brake hose using a new brake hose gasket and tighten to 13 ft. lbs. (18 Nm).
14. Bleed air from brake system.

2006 Impreza Models

See Figure 16.

14–type:

1. Before servicing the vehicle, refer to the Precautions Section.
2. Set the vehicle on a lift.
3. Loosen the wheel nuts.
4. Lift-up the vehicle, and then remove the wheels.
5. Disconnect the brake hose from caliper body assembly.
6. Remove the bolt securing lock pin to caliper body.
7. Raise the caliper body and move it toward vehicle center to separate it from support.
8. Remove the support from housing.

➡Remove the support only when replacing it or the disc rotor. It need not be removed when servicing the caliper body assembly

9. Clean mud and foreign particles from the caliper body assembly and support.

⁂ CAUTION

Be careful not to allow foreign particles to enter brake hose connector.

15–type:

1. Set the vehicle on a lift.
2. Loosen the wheel nuts.
3. Lift-up the vehicle, and then remove the wheels.
4. Remove the pads from caliper body.
5. Disconnect the brake hose from caliper body assembly.
6. Remove the caliper body from housing.
7. Clean mud and foreign particles from the caliper body assembly.

17–type:

1. Set the vehicle on a lift.
2. Loosen the wheel nuts.
3. Lift-up the vehicle, and then remove the wheels.
4. Remove the brake pads from caliper body.
5. Disconnect the brake hose from caliper body.
6. Remove the caliper body from housing.
7. Clean mud and foreign particles from the caliper body.

To install:

14–type:

1. Install the support on housing and tighten to 39 ft. lbs. (53 Nm).
2. Apply thin coat of Molykote AS-880N (Part No. K0777YA010) to the frictional portion between pad and pad clip.
3. Install the pads on support.
4. Install the caliper body on support 27.5ft. lbs. (37 Nm).
5. Replace the brake hose gaskets with new ones, and then connect the brake hose and tighten to 13 ft. lbs. (18 Nm).
6. Bleed air from the brake system.

15–type:

1. Install the caliper body on housing and tighten to 39 ft. lbs. (53 Nm).
2. Apply a thin coat of Molykote AS-880N (Part No. K0777YA010) to the frictional portion between pad and pad clip.
3. Install the pads on caliper body.

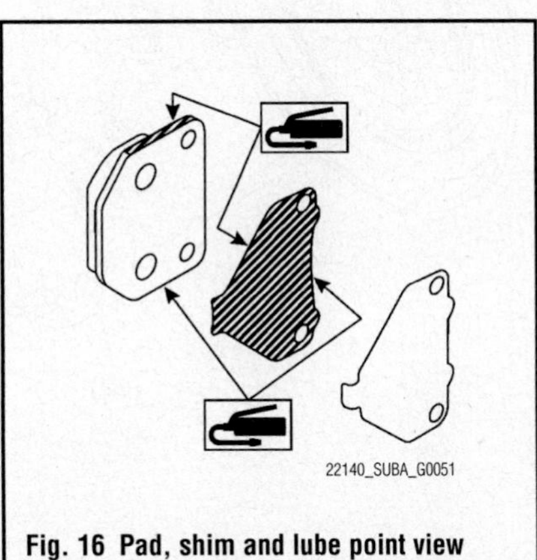

Fig. 16 Pad, shim and lube point view

4. Replace the brake hose gaskets with new ones, and then connect the brake hose and tighten to 13 ft. lbs. (18 Nm).
5. Bleed air from the brake system.

17–type:

1. Install the caliper body on housing and tighten to 48 ft. lbs. (65 Nm).
2. Install the pads on caliper body.
3. Replace the brake hose gaskets with new ones, and then connect the brake hose and tighten to 13 ft. lbs. (18 Nm).
4. Bleed air from the brake system.

2007–08 Impreza Models

See Figure 17.

1. Before servicing the vehicle, refer to the Precautions Section.
2. Lift up the vehicle, then remove the rear wheels.
3. Remove the caliper body.
4. Remove mud and foreign matter from the caliper body assembly.

⁂ CAUTION

Be careful not to allow foreign matter to enter the brake hose connector.

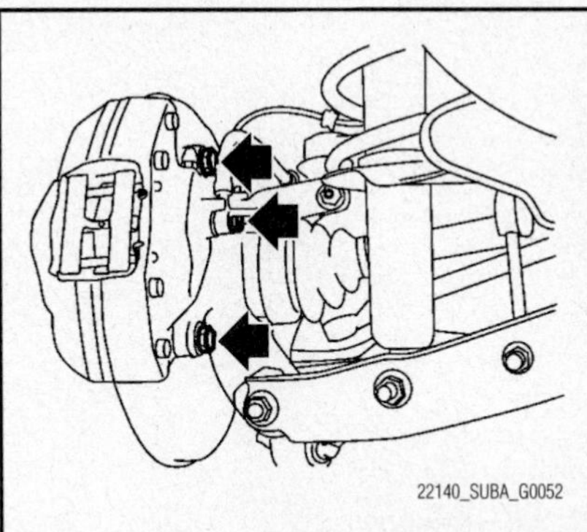

Fig. 17 Rear brake caliper mounting bolts and brake hose banjo bolt

To install:

5. Install the caliper body to the housing and tighten to 48 ft. lbs. (65 Nm).
6. Apply a thin coat of Molykote AS880N (Part No. K0777YA010) or the pad kit grease to both surfaces of the pad side and pad inner shim.
7. Install the pads to the caliper body.
8. Install the cross spring and pad pins.
9. Install the clips.
10. Replace the brake hose gaskets with new ones, and then connect the brake hose and tighten to 13 ft. lbs. (18 Nm).
11. Bleed air from the brake system.

Legacy & Outback Models

See Figure 18.

1. Before servicing the vehicle, refer to the Precautions Section.
2. Raise and safely support the vehicle.
3. Loosen the lug nuts and remove the rear wheels.
4. Disconnect the brake hose from caliper body assembly.
5. Remove the caliper lower bolt.
6. Raise the caliper body, and then move it toward vehicle center to separate it from the support.

➡Remove the support only when replacing the rotor or support. It need not be removed when servicing the caliper body assembly.

7. Remove the support from housing.
8. Remove mud and foreign matter from the caliper body assembly and the support.

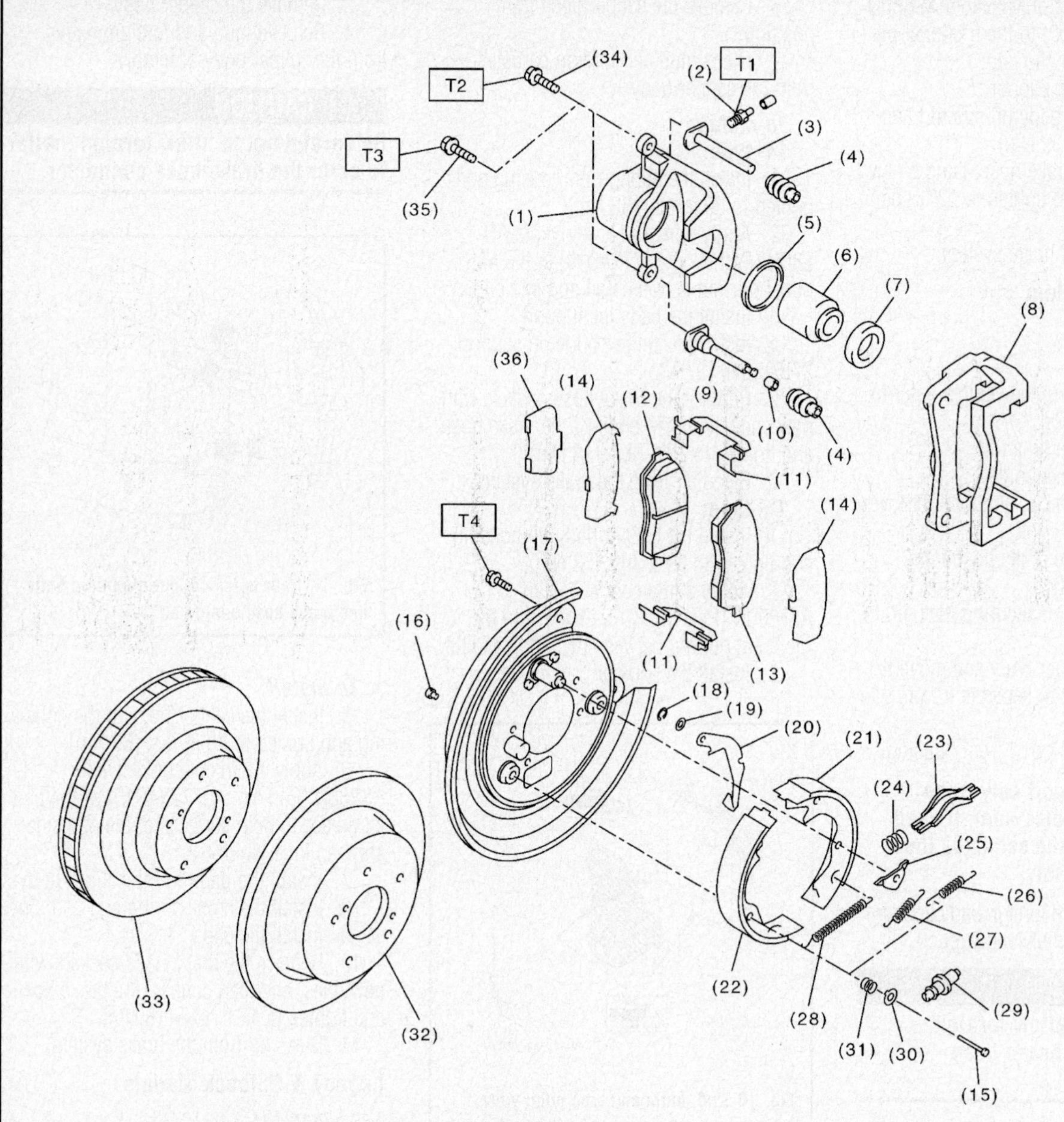

(1)	Caliper body	(15)	Shoe hold-down pin	(29)	Adjuster
(2)	Air bleeder screw	(16)	Cover	(30)	Shoe hold-down cup
(3)	Guide pin (black)	(17)	Back plate	(31)	Shoe hold-down spring
(4)	Pin boot	(18)	Retainer	(32)	Disc rotor (solid type)
(5)	Piston seal	(19)	Spring washer	(33)	Disc rotor (ventilated type)
(6)	Piston	(20)	Parking brake lever	(34)	Bolt (for solid disc brake)
(7)	Piston boot	(21)	Parking brake shoe (secondary)	(35)	Bolt (for ventilated disc brake)
(8)	Support	(22)	Parking brake shoe (primary)	(36)	Shim (for solid disc brake)
(9)	Lock pin (silver)	(23)	Strut		
(10)	Bushing	(24)	Strut shoe spring		
(11)	Pad clip	(25)	Shoe guide plate		
(12)	Inner pad	(26)	Secondary shoe return spring		
(13)	Outer pad	(27)	Primary shoe return spring		
(14)	Shim	(28)	Adjusting spring		

Tightening torque:N·m (kgf-m, ft-lb)

T1: 8 (0.8, 5.8)
T2: 27 (2.8, 19.9)
T3: 37 (3.7, 27.2)
T4: 66 (6.7, 48.7)

22140_SUBA_G0053

Fig. 18 Legacy and Outback rear brake exploded view

CAUTION

Be careful not to allow foreign matter to enter the brake hose connector.

To install:

9. Install the support to the housing. Tighten to 39 ft. lbs. (53 Nm) for 2006 models. For 2007–08 models tighten to 49 ft. lbs. (66 Nm).
10. Apply a thin coat of Molykote M7439 (Part No. 003602001) to the pad clip.
11. Apply a thin coat of Molykote AS880N (Part No. K0777YA010) to the contact surface between the pad and shim.
12. Install the pad to support.
13. Install the caliper body to the support.
14. Tighten the caliper body to support mounting bolts as follows:
    - Solid disc brake models 20 ft. lbs. (27 Nm).
    - Ventilated disc brake models 27 ft. lbs. (37 Nm).
15. Connect the brake hose using a new brake hose gasket and tighten to 13 ft. lbs. (18 Nm).
16. Bleed air from the brake system.

DISC BRAKE PADS

REMOVAL & INSTALLATION

Baja Models

See Figure 19.

1. Before servicing the vehicle, refer to the Precautions Section.
2. Loosen wheel nuts, jack-up vehicle, support it with safety stands, and remove wheel.
3. Remove bottom bolt.
4. Raise caliper body and suspend it securely.
5. Remove pad from support.

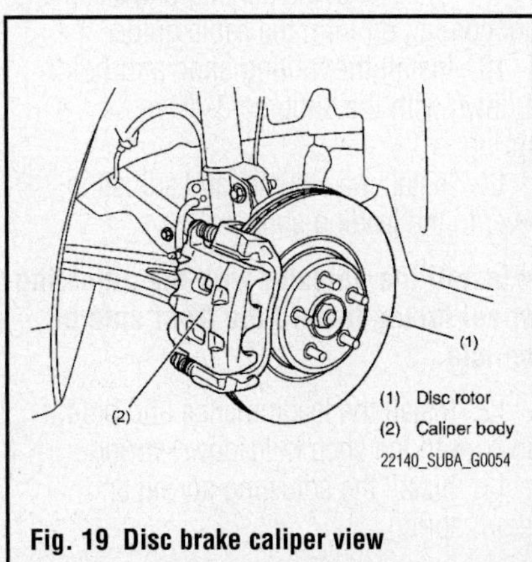

Fig. 19 Disc brake caliper view

To install:

6. Apply thin coat of Molykote M7439 (Part No. 003608001) to the frictional portion between pad and pad clip.
7. Check disc rotor thickness and runout.
8. Install pad on support.
9. Install caliper body on support and tighten to 29 ft. lbs. (39 Nm).
10. Depress brake pedal several times.
11. Check that brake fluid level is at max. line.

2006–07 Impreza Models

See Figures 20 and 21.

1. Before servicing the vehicle, refer to the Precautions Section.
2. Loosen the wheel nuts.
3. Disconnect the negative battery cable.
4. Raise and support the vehicle safely.
5. Remove the tire and wheel.
6. On 14 inch type, remove the lower caliper bolt. Raise the caliper body upward and support it. Do not disconnect the fluid line.
7. On 15 inch type, remove the "M" clip. Remove the pad pins and cross spring. Expand the pads and then push the piston back.
8. On 17 inch type, remove the clip. Remove the pad pins and cross spring. Expand the pads and then push the piston back.
9. Remove the disc brake pads. On 15 inch type also remove the shims.

To install:

10. Installation is the reverse of the removal procedure.
11. On 14 inch wheel, apply a thin coat of Molykote AS-880N (part number K0779YA010) or equivalent to the frictional portion between the pad and pad clip. Tighten the caliper body to support bolts to 27.5 ft. lbs. (37 Nm).
12. On 15 and 17 inch types, apply a thin coat of Molykote AS-880N (part number K0779YA010) or equivalent to the frictional portion between the pad and pad inner shim.
13. Check the brake fluid level, correct as required.
14. Bleed the hydraulic system, as required.

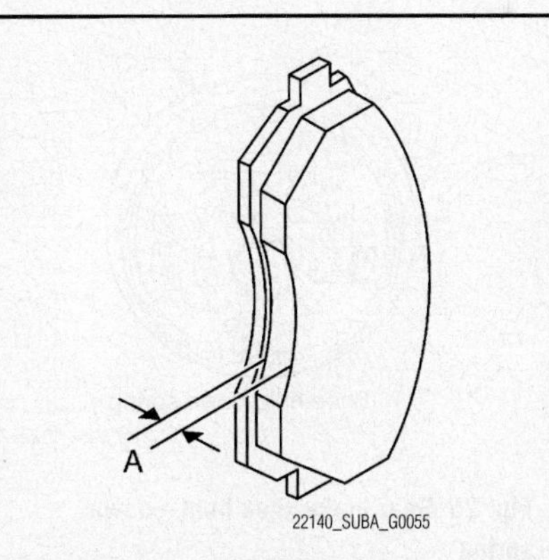

Fig. 20 14 inch type brake pad

A
22140_SUBA_G0056

Fig. 21 15–17 inch type brake pad

2008 Impreza Models

1. Before servicing the vehicle, refer to the Precautions Section.
2. Lift up the vehicle, then remove the rear wheels.
3. Disconnect the negative battery cable.
4. Remove the pad pins and cross spring.
5. Spread apart the pads and push back the piston.
6. Remove the pads.

To install:

7. Apply a thin coat of Molykote AS880N (Part No. K0777YA010) or the pad kit grease to both surfaces of the pad side and pad shim.
8. Install the pads to the caliper body.
9. Install the cross spring and pad pins.
10. Install the clips.

Legacy & Outback Models

See Figure 22.

1. Before servicing the vehicle, refer to the Precautions Section.
2. Lift up the vehicle, then remove the rear wheels.
3. Remove the caliper bolt.
4. Raise the caliper body and support it.
5. Remove the pad.

To install:

6. Apply a thin coat of Molykote M7439 (Part No. K0777YA000) to the pad clip.
7. Apply a thin coat of Molykote AS880N (Part No. K0779YA010) to the contact surface between the pad and shim.

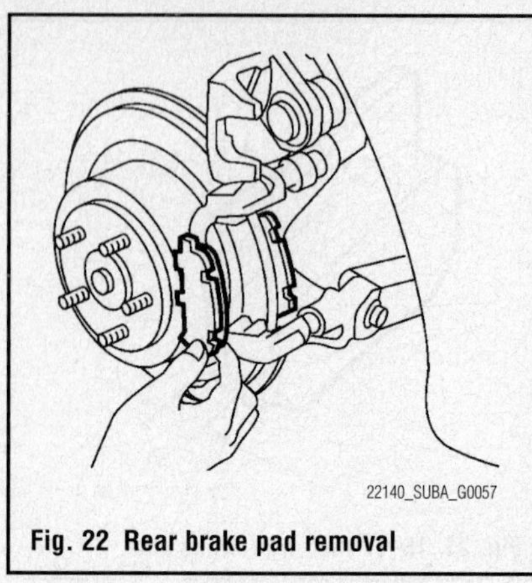

Fig. 22 Rear brake pad removal

8. Install the pad to support.
9. Install the caliper body to the support.
10. Tighten the caliper body to support bolts as follows:
    - Solid disc brake models 20 ft. lbs. (27 Nm).
    - Ventilated disc brake models 27 ft. lbs. (37 Nm).

ROTOR

REMOVAL & INSTALLATION

See Figures 23 and 24.

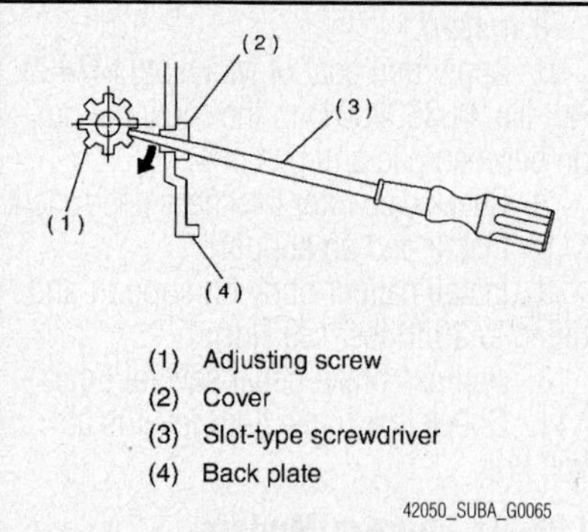

Fig. 23 Turn the adjusting screw to back off the brake shoe to remove the rotor.

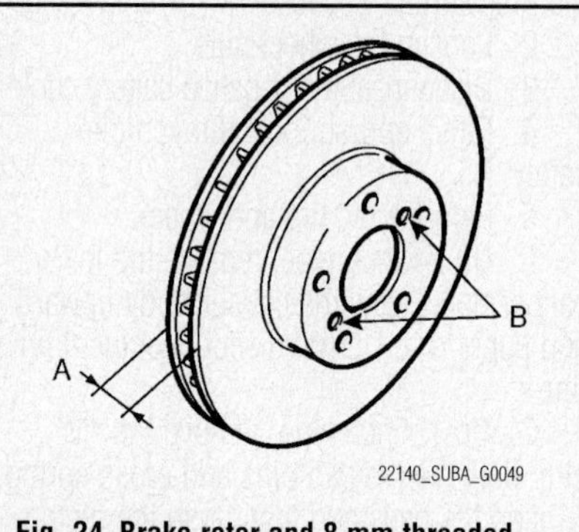

Fig. 24 Brake rotor and 8 mm threaded holes shown

1. Raise and safely support the vehicle.
2. Remove the wheel.
3. Remove the caliper assembly, and do not disconnect the brake lines. Hang the caliper securely from the strut using mechanics wire. Refer to "Brake Caliper, Removal & Installation" to remove the caliper assembly.

CAUTION

Do not let the caliper hang by the brake hoses.

4. Pull down and release the parking brake.
5. Remove the rotor.
6. If the rotor is difficult to remove, try the following methods to remove:
    a. Turn the adjusting screw using a slot-type screwdriver until brake shoe is away enough from the disc rotor.
    b. Install an 8 mm bolt into the services holes on the hub and tighten.
7. Remove the brake rotor.

To install:

8. Installation is the reverse order of removal.
9. Adjust the parking brake.

BRAKES

REAR DRUM BRAKES

BRAKE DRUM

REMOVAL & INSTALLATION

2008 Impreza Models

1. Before servicing the vehicle, refer to the Precautions Section.
2. Lift up the vehicle, then remove the rear wheels.
3. Release the parking brake.
4. Remove the brake drum from brake assembly.

➡If the brake drum is difficult to remove, drive it out by installing an 8-mm bolt into bolt hole in brake drum.

To install:

5. Installation is the reverse order of removal.
6. Adjust the brake shoes and parking brake.

BRAKE SHOES

REMOVAL & INSTALLATION

2008 Impreza Models

See Figure 25.

1. Before servicing the vehicle, refer to the Precautions Section.
2. Remove the rear drum brake drum.
3. Remove the shoe hold-down spring.
4. Remove upper shoe return spring, lower shoe return spring and adjuster, then remove the leading shoe
5. Remove the shoe hold-down spring and trailing shoe.
6. Disconnect the parking brake cable from the parking brake lever.

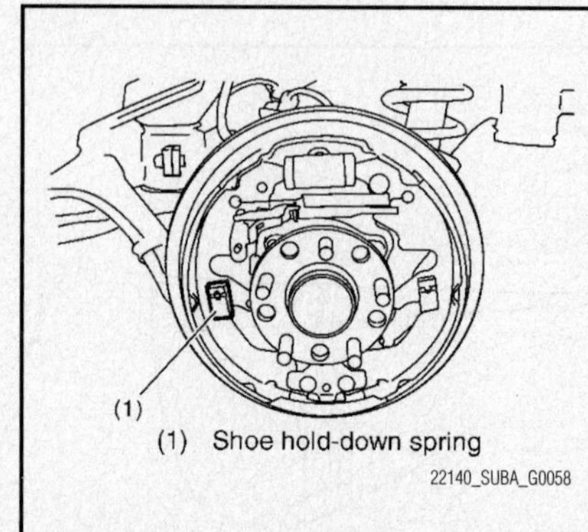

Fig. 25 Rear brake shoe hold—down spring

To install:

7. Apply brake grease to the following locations:
    - Six contact surfaces of the brake shoe rim and back plate
    - Contact surface of the brake shoe and the wheel cylinder
    - Contact surface of the brake shoe and the adjuster
    - Contact surface of lever and brake shoe
8. Connect the parking brake cable to the parking brake lever.
9. Check that the parking brake cable is not coming off from the cable guide.
10. Install the trailing shoe and hold it down with the shoe hold-down spring.
11. Install the adjuster and adjusting lever to the leading shoe.

➡Install the adjuster with the adjusting wheel facing toward the front side of vehicle.

12. Install the leading shoe and hold it down with the shoe hold-down spring.
13. Install the adjusting spring and return spring.

14. Install the rear brake drum.
15. Adjust the brake shoes and parking brake.
16. Road test and burnish brake shoes.
17. Re—adjust brake shoes if needed.

ADJUSTMENT

2008 Impreza Models

1. Return the parking brake lever fully.
2. Remove the adjusting hole cover.
3. Turn the adjusting screw using a flat tip screwdriver until the brake shoe is in close contact with the brake drum.
4. Turn back the adjusting screw 3 to 4 notches.
5. Install the adjusting hole cover.

BRAKES — PARKING BRAKE

PARKING BRAKE CABLES

ADJUSTMENT

See Figure 26.

1. Remove the console cover.
2. Forcefully pull the parking brake lever 3 to 5 times.
3. Adjust parking brake cable by turning the adjusting nut until the parking brake lever stroke is set at 7 to 8 notches with operating force of 44 lbs. (196 N).
4. Tighten lock nut.
5. Install the console cover.

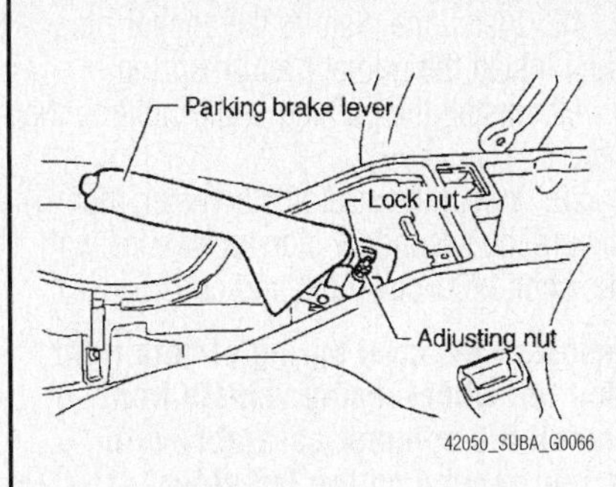

Fig. 26 Turn the adjusting nut to adjust the parking brake cable.

PARKING BRAKE SHOES

REMOVAL & INSTALLATION

See Figure 27.

1. Before servicing the vehicle, refer to the Precautions Section.
2. Release the parking brake.
3. Remove the two mounting bolts and remove the brake caliper assembly.
4. Suspend the brake caliper assembly so that the hose is not stretched.
5. Remove the disc rotor.

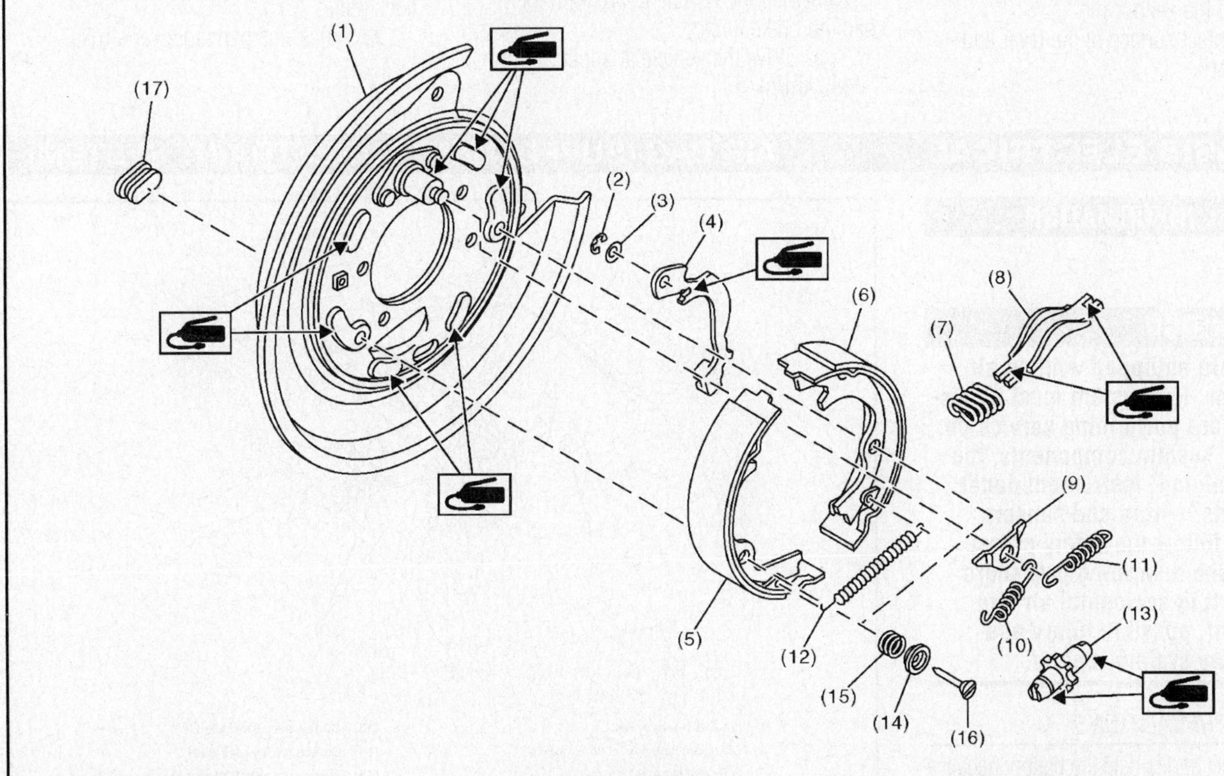

Fig. 27 Exploded view of parking brake components

6. If disc rotor is seized on the hub, drive the disc rotor out by pushing two 8 mm bolts in holes B on the rotor.
7. Remove the shoe return spring from the parking brake assembly.
8. Remove the front shoe hold down spring and pin.
9. Remove the strut and strut spring.
10. Remove the adjuster assembly from the parking brake assembly.
11. Remove the brake shoe.
12. Remove the rear shoe hold down spring and pin with pliers.
13. Remove the parking brake cable from lever.
14. Using a flat tip screwdriver, raise the retainer. Remove the parking lever and washer from brake shoe.

To install:

15. Apply brake grease to the following locations:
   - Six contact surfaces of the shoe rim and back plate gasket
   - Contact surface of the shoe wave and the anchor pin
   - Contact surface of the lever and strut
   - Contact surface of the shoe wave and the adjuster assembly
   - Contact surface of the shoe wave and the strut
   - Contact surface of the lever and the shoe wave
16. Insert the primary side brake shoe into the anchor pin groove.
17. Secure the brake shoe with the shoe hold-down pin and cup.
18. Install the plate to the anchor pin, then install the primary return spring.
19. Install the parking brake cable to the lever.
20. Install the strut and adjuster, then secure the secondary side brake shoe with the shoe hold-down pin and cup.

➡Install the strut spring of both right and left wheel facing vehicle front. Install the adjuster assembly with screw section on the left side.

21. Install the secondary return spring and the adjusting spring.
22. Adjust the parking brake.
23. Drive the vehicle to break-in the parking brake lining.
   a. Drive the vehicle at about 22 mph (35 km/h).
   b. With the parking brake release button pushed in, pull the parking brake lever gently.
   c. Drive the vehicle for about 0.12 miles (200 m) in this condition.
   d. Wait 5 to 10 minutes for the parking brake to cool down. Repeat again from step (A).
   e. After breaking-in, re—adjust the parking brakes.

ADJUSTMENT

1. Before servicing the vehicle, refer to the Precautions Section.
2. Return the parking brake lever fully.
3. Remove the adjusting hole cover from the back plate.
4. Turn the adjusting screw using a flat tip screwdriver until the brake shoe is in close contact with the disc rotor.
5. Turn back (downward) the adjusting screw 3 to 4 notches. Check there is no brake drag.
6. Install the adjusting hole cover to the back plate.
7. Adjust the parking lever stroke.

CHASSIS ELECTRICAL

AIR BAG (SUPPLEMENTAL RESTRAINT SYSTEM)

GENERAL INFORMATION

See Figure 28.

⁂ CAUTION

Vehicles are equipped with an air bag system. The system must be disarmed before performing service on, or around, system components, the steering column, instrument panel components, wiring and sensors. Failure to follow the safety precautions and the disarming procedure could result in accidental air bag deployment, possible injury and unnecessary system repairs.

SERVICE PRECAUTIONS

Disconnect and isolate the battery negative cable before beginning any airbag system component diagnosis, testing, removal, or installation procedures. Allow system capacitor to discharge for two minutes before beginning any component service. This will disable the airbag system. Failure to disable the airbag system may result in accidental airbag deployment, personal injury, or death.

Do not place an intact undeployed airbag face down on a solid surface. The airbag will

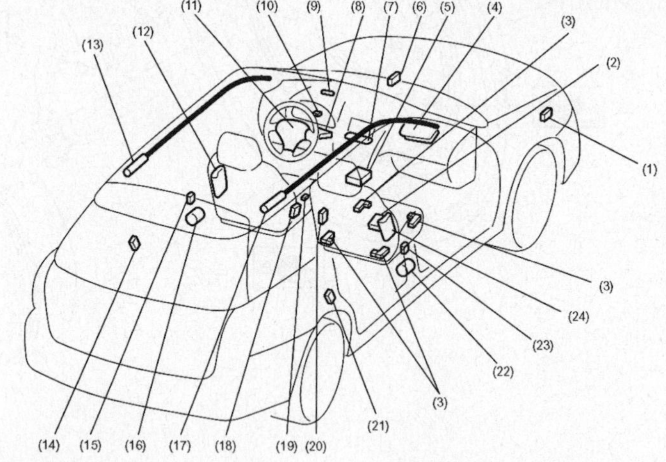

(1) Front sub sensor RH
(2) Occupant detection control module
(3) Load cell sensors
(4) Passenger's seat airbag module
(5) Airbag control module
(6) Front sub sensor LH
(7) Passenger's airbag ON/OFF indicator light
(8) Body integrated control module
(9) Passenger's seat belt warning light
(10) Airbag warning light
(11) Driver's seat airbag module
(12) Side airbag module LH
(13) Curtain airbag module LH
(14) Curtain airbag sensor LH
(15) Side inside airbag sensor LH
(16) Pretensioner LH
(17) Curtain airbag module RH
(18) Buckle switch LH
(19) Seat position sensor LH
(20) Buckle switch RH
(21) Curtain airbag sensor RH
(22) Pretensioner RH
(23) Side airbag sensor RH
(24) Side airbag module RH

22140_SUBA_G0067

Fig. 28 SRS system components—2008 Outback shown

propel into the air if accidentally deployed and may result in personal injury or death.

When carrying or handling an undeployed airbag, the trim side (face) of the airbag should be pointing towards the body to minimize possibility of injury if accidental deployment occurs. Failure to do this may result in personal injury or death.

Replace airbag system components with OEM replacement parts. Substitute parts may appear interchangeable, but internal differences may result in inferior occupant protection. Failure to do so may result in occupant personal injury or death.

Wear safety glasses, rubber gloves, and long sleeved clothing when cleaning powder residue from vehicle after an airbag deployment. Powder residue emitted from a deployed airbag can cause skin irritation. Flush affected area with cool water if irritation is experienced. If nasal or throat irritation is experienced, exit the vehicle for fresh air until the irritation ceases. If irritation continues, see a physician.

Do not use a replacement airbag that is not in the original packaging. This may result in improper deployment, personal injury, or death.

The factory installed fasteners, screws and bolts used to fasten airbag components have a special coating and are specifically designed for the airbag system. Do not use substitute fasteners. Use only original equipment fasteners listed in the parts catalog when fastener replacement is required.

During, and following, any child restraint anchor service, due to impact event or vehicle repair, carefully inspect all mounting hardware, tether straps, and anchors for proper installation, operation, or damage. If a child restraint anchor is found damaged in any way, the anchor must be replaced. Failure to do this may result in personal injury or death.

Deployed and non-deployed airbags may or may not have live pyrotechnic material within the airbag inflator.

Do not dispose of driver/passenger/curtain airbags or seat belt tensioners unless you are sure of complete deployment. Refer to the Hazardous Substance Control System for proper disposal.

Dispose of deployed airbags and tensioners consistent with state, provincial, local, and federal regulations.

After any airbag component testing or service, do not connect the battery negative cable. Personal injury or death may result if the system test is not performed first.

If the vehicle is equipped with the Occupant Classification System (OCS), do not connect the battery negative cable before performing the OCS Verification Test using the scan tool and the appropriate diagnostic information. Personal injury or death may result if the system test is not performed properly.

Never replace both the Occupant Restraint Controller (ORC) and the Occupant Classification Module (OCM) at the same time. If both require replacement, replace one, then perform the Airbag System test before replacing the other.

Both the ORC and the OCM store Occupant Classification System (OCS) calibration data, which they transfer to one another when one of them is replaced. If both are replaced at the same time, an irreversible fault will be set in both modules and the OCS may malfunction and cause personal injury or death.

If equipped with OCS, the Seat Weight Sensor is a sensitive, calibrated unit and must be handled carefully. Do not drop or handle roughly. If dropped or damaged, replace with another sensor. Failure to do so may result in occupant injury or death.

If equipped with OCS, the front passenger seat must be handled carefully as well. When removing the seat, be careful when setting on floor not to drop. If dropped, the sensor may be inoperative, could result in occupant injury, or possibly death.

If equipped with OCS, when the passenger front seat is on the floor, no one should sit in the front passenger seat. This uneven force may damage the sensing ability of the seat weight sensors. If sat on and damaged, the sensor may be inoperative, could result in occupant injury, or possibly death.

DISARMING THE SYSTEM

1. Be sure to position the front wheels in the straight ahead position.
2. Disconnect the negative battery cable. Tape the battery cable for added protection.
3. Wait more than 20 seconds before starting work.

ARMING THE SYSTEM

1. Connect the negative battery cable.

CLOCKSPRING CENTERING

See Figure 29.

✱✱ CAUTION

When servicing a vehicle, be sure to turn the ignition switch to OFF, disconnect the ground cable from battery, and wait for more than 20 seconds before starting work. The airbag system is fitted with a backup power source. After disconnecting the battery ground cable, the airbag may deploy if you do not wait for more than 20 seconds before starting the service of airbag system.

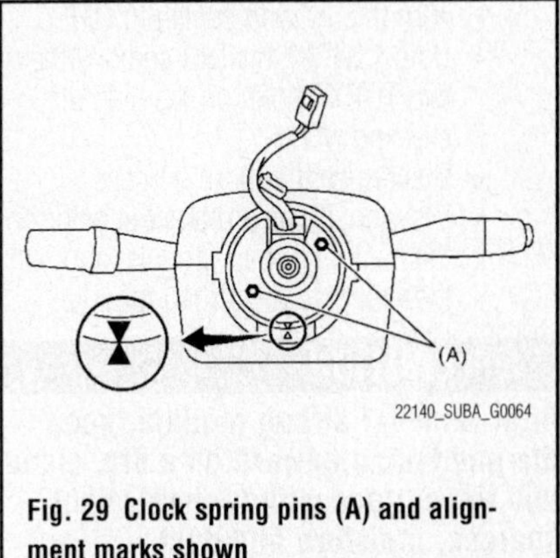

Fig. 29 Clock spring pins (A) and alignment marks shown

1. Turn the ignition switch to OFF.
2. Disconnect the ground cable from battery and wait for at least 20 seconds before starting work.
3. Check that front wheels are positioned in straight ahead direction.
4. Turn the clock spring pin (A) clockwise until it stops.
5. Turn the clock spring connector pins (A) approx. 3.25 turns until the marks are aligned.

CLOCKSPRING REMOVAL & INSTALLATION

See Figure 30.

✱✱ CAUTION

When servicing a vehicle, be sure to turn the ignition switch to OFF, disconnect the ground cable from battery, and wait for more than 20 seconds before starting work. The airbag system is fitted with a backup power source. After disconnecting the battery ground cable, the airbag may deploy if you do not wait for more than 20 seconds before starting the service of airbag system.

1. Turn the ignition switch to OFF.
2. Disconnect the ground cable from battery and wait for at least 20 seconds before starting work.
3. Remove the driver's airbag module as follows:
   - Position the front wheels straight ahead. (After moving a vehicle 5 m (16 ft) or more with front wheels positioned straight ahead, make sure that the vehicle moves straight ahead.)

- Turn the ignition switch to OFF.
- Using TORX® bit T30, remove the two TORX® bolts on the side of steering wheel.
- Disconnect the horn harness.
- Disconnect the airbag connector on the back of airbag module, and then remove the airbag module.

⁂ CAUTION

After removal airbag module, keep the pad facing upward on a dry, clean and flat surface without heat, light sources, moisture and dust.

4. Place alignment marks on the steering wheel and steering shaft.
5. Remove the steering wheel nut, and then draw out the steering wheel from shaft using steering puller.

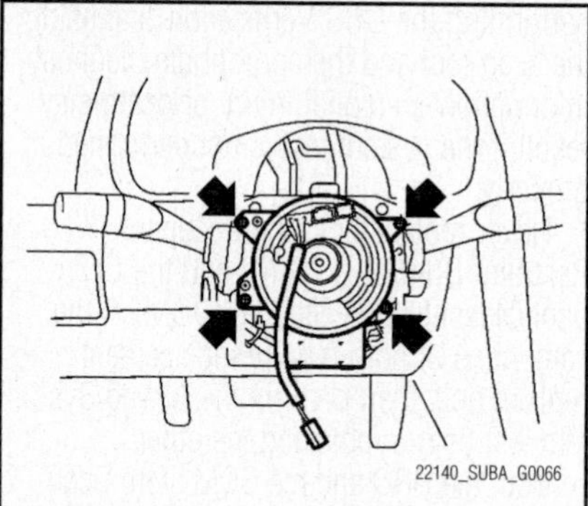

Fig. 30 Clock spring and retaining screws shown

6. Remove the steering column cover.
7. Remove the screws, and then remove the clock spring assembly.

To install:

8. Align the center position of the clock spring.
9. Install in the reverse order of removal.
10. Align the alignment marks on the steering wheel and steering shaft. Install and tighten the shaft nut to 33 ft. lbs. (45 Nm).
11. Tighten air bag module retaining screws to 7 ft. lbs. (10 Nm).

➡Column cover-to-steering wheel clearance: 0.08–0.16 inch. (2–4 mm).

⁂ CAUTION

Insert the roll connector guide pin into the guide hole on the lower end of the steering wheel surface. Avoid damaging the pin.

DRIVE TRAIN

AUTOMATIC TRANSMISSION ASSEMBLY

REMOVAL & INSTALLATION

4AT Automatic Transmission

See Figures 31 and 32.

1. Before servicing the vehicle, refer to the Precautions Section.
2. Set the vehicle on a lift.
3. Open the front hood and support with the hood stay.
4. Disconnect the ground cable from the battery.
5. Remove the air intake duct.
6. Remove the air intake chamber.
7. Remove the air intake chamber stay.
8. Disconnect the following connectors:
   - Transmission harness connectors
   - Transmission ground terminal
9. Remove the starter.
10. Remove the pitching stopper.
11. Separate the torque converter assembly from the drive plate.

⁂ CAUTION

Be careful not to damage the mounting bolts. Be careful not to drop bolts into the converter case.

12. Attach the ST-498277200 converter holder to the converter case.
13. Remove the ATF level gauge. Plug the opening to prevent entry of foreign particles into the transmission fluid.
14. Remove the throttle body.
15. Disconnect the engine harness, then remove the harness connector from the engine harness bracket.
16. Remove the engine harness bracket.
17. Remove the pitching stopper bracket.
18. Install the ST-41099AC000 engine support assembly.
19. Remove the bolts which hold upper side of transmission to engine.
20. Lift up the vehicle.
21. Remove the undercover.
22. Remove the front, center and rear exhaust pipes and the muffler.
23. Remove the heat shield cover.
24. Remove the drain plug to drain the transmission fluid..
25. Disconnect the ATF cooler hoses from the pipes of the transmission side, and remove the oil charge pipe.
26. Remove the propeller shaft.
27. Remove the shift select cable.
28. Remove the brackets (two) which hold front stabilizer.

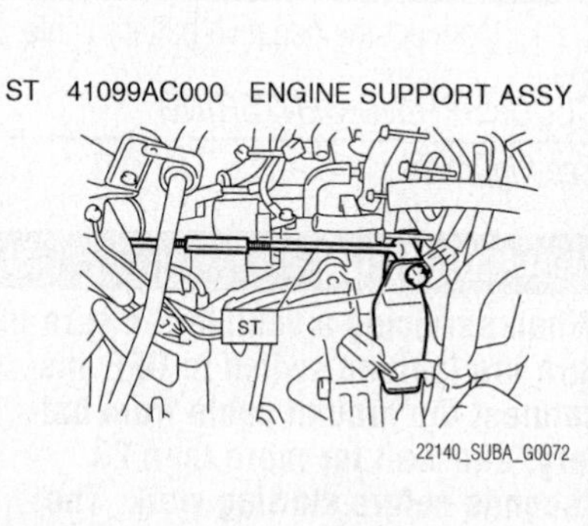

Fig. 31 Engine support assembly

29. Remove the bolt securing the ball joint of the front arm to housing.
30. Pull out the front driveshaft from the transmission.
    a. Using a tire lever or a crow bar, etc., pull out until the front driveshaft transmission side joint slides move smoothly.
    b. Hold the transmission side joint of the front driveshaft by hand and extract the housing from the transmission while pressing the housing outward, so as not to stretch the boot.
31. Remove the bolts which hold the clutch housing cover.
32. Remove the bolts and nuts which hold lower side of transmission to engine.
33. Place the transmission jack under the transmission.

➡Make sure that the support plates of transmission jack do not touch the oil pan.

34. Remove the transmission rear crossmember from the vehicle.
35. While lowering the transmission jack gradually, fully retract the engine support, and then tilt the engine rearward.
36. Retract the support until the clearance between front crossmember and converter case becomes approx. 0.39 inch (10 mm).
37. Remove the transmission.
38. Remove the rear cushion rubber from the transmission assembly.

To install:

39. Replace the differential side oil seal with a new part.

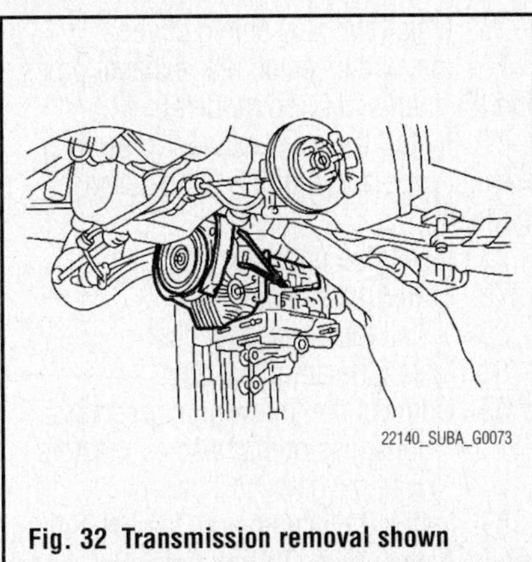

Fig. 32 Transmission removal shown

40. Install the rear cushion rubber to the transmission assembly. Tighten to 29.5 ft. lbs. (40 Nm).
41. Attach the ST-498277200 to the converter case.
42. Install the transmission onto the engine as follows:
    - Lift up the transmission gradually using transmission jack.
    - Insert the engine side stud bolt into the transmission bolt hole.
    - While raising the transmission jack gradually, turn the screw of engine support, then tilt the engine forward and connect.
43. Install the transmission rear crossmember. Tighten the center insulator mount to 26 ft. lbs. (35 Nm).
44. Tighten outer crossmember bolts to 55 ft. lbs. (75 Nm).
45. Remove the transmission jack.
46. Tighten the bolts and nuts which hold the lower side of transmission to the engine to 37 ft. lbs. (50 Nm).
47. Install the clutch housing cover bolts.
48. Lower the lift.
49. Connect the engine and transmission.
50. Remove the ST from converter case.
51. Install the starter.
52. Tighten the bolts which hold the upper side of the transmission to the engine and tighten to 37 ft. lbs. (50 Nm).
53. Install the torque converter assembly to the drive plate.
54. Tighten the bolts which hold the torque converter assembly to the drive plate.
55. Place the wrench on the crank pulley bolt, and remove all the bolts while rotating the crank pulley a little bit at a time. Tighten to 18 ft. lbs. (25 Nm).
56. Fit the plug to service hole.
57. Install the V-belt cover.
58. Remove the ST-41099AC000 engine support assembly.
59. Install the engine harness bracket. Tighten to 12 ft. lbs. (16 Nm).
60. Install the harness connector to engine harness bracket, then connect the harness.
61. Install the pitching stopper bracket and tighten to 30 ft. lbs. (41 Nm).
62. Install the throttle body.
63. Install the pitching stopper and tighten to 37 ft. lbs. (50 Nm).
64. Lift up the vehicle.
65. Replace the circlip of the front driveshaft with a new part.
66. Apply grease to the oil seal lip.
67. Attach the ST-28399SA010 seal protector to side retainer.
68. Align and insert the spline of the front driveshaft to the splines of the differential bevel gear, and remove the ST.
69. Insert the front driveshaft into the transmission securely by pressing the front housing from the outside.
70. Install the ball joint into the housing. Tighten the attachment bolts to 37 ft. lbs. (50 Nm).
71. Install the stabilizer to the crossmember.
72. Always tighten the rubber bushing locations with wheels in full contact with the ground and the vehicle at curb weight. Tighten to 18 ft. lbs. (25 Nm).
73. Install the shift select cable onto select lever.
74. Install the oil charge pipe, and connect the ATF cooler hoses to the pipe.
75. Install the propeller shaft.
76. Install the heat shield cover.
77. Install the rear exhaust pipe and muffler.
78. Install the front and center exhaust pipe.
79. Install the undercover.
80. Lower the lift.
81. Install the ATF level gauge.
82. Connect the following connectors:
    - Transmission harness connectors
    - Transmission ground terminal
83. Install the air intake chamber stay and tighten to 12 ft. lbs. (16 Nm).
84. Install the air intake chamber.
85. Install the air intake duct.
86. From the oil charge pipe, add ATF until the fluid level is between the upper and lower level on the "COLD" side of the level gauge.
87. Lower the vehicle from the lift.
88. Check the differential gear oil level.
89. Check the select lever operation.
90. Bleed the air of control valve.
91. Check the ATF level.
92. Execute the learning control promotion.
93. Perform the road test and check for leaks.

5AT Automatic Transmission

See Figure 33.

1. Before servicing the vehicle, refer to the Precautions Section.
2. Set the vehicle on a lift.
3. Open the front hood and support with the hood stay.
4. Disconnect the ground cable from the battery.
5. Remove the collector cover.
6. Remove the intercooler. (Turbo models)
7. Remove the air intake chamber. (Non-turbo models)
8. Remove the air cleaner case. (Non-turbo models)
9. Remove the air breather hose.
10. Remove the starter.
11. Front oxygen (A/F) sensor.
12. Transmission harness connectors.
13. Remove the intercooler stay and engine hanger rear. (Turbo models)
14. Disconnect the engine harness connectors, and then remove the engine hanger rear. (Non-turbo model)
15. Remove the water by-pass pipe. (Turbo models)
16. Separate the torque converter from drive plate as follows:
    - Remove the service hole plug.
    - Remove the bolts which hold torque converter to drive plate.
    - Remove the four bolts by rotating the clamp pulley a little at a time.
    - Make sure the torque converter moves freely by rotating with finger through the starter installation hole.
17. Attach the ST-498277200 converter holder to the converter case.
18. Remove the pitching stopper.
19. Remove the pitching stopper bracket.
20. Install the ST-41099AC000 engine support assembly.
21. Remove the air intake duct. (Turbo models)
22. Remove the air cleaner case. (Turbo models)
23. Remove the transmission mounting bolt (upper side).
24. Lift-up the vehicle. (Turbo models)
25. Remove the undercover. (Turbo models)
26. Remove the center and rear exhaust pipes and the muffler. (Turbo models)
27. Remove the front exhaust pipe, rear exhaust pipe and muffler. (Non-turbo model)

28. Remove the heat shield cover.
29. Remove the drain plug to drain transmission fluid.
30. Remove the oil charge pipe.
31. Disconnect the connector from turbine speed sensor 1.
32. Remove the turbine speed sensor 1 connector mounting bolt and rotate the sensor by 180°.

CAUTION

Failure to follow this procedure may cause the interference between vehicle body and sensor while removing/installing transmission, and resulting in damage.

33. Remove the propeller shaft.
34. Remove the shift select cable.
35. Disconnect the hose from the ATF inlet and outlet pipes.
36. Remove the front crossmember support plate.
37. Remove the two clutch housing cover securing bolts.
38. Remove the front stabilizer bracket.
39. Remove the bolts which secure front ball joint to the housing.
40. Pull out the driveshaft from transmission.
41. Set the transmission jack under the transmission.
42. Remove the rear crossmember.
43. Remove the transmission mounting bolt (lower side).
44. Remove the transmission.

➡Turn the engine support assembly from the vehicle under body to the left (to shorten the engine support length), and lower the rear of the engine for easy disassembly. Be careful not to allow breather pipe and etc. to touch the vehicle body when detaching the automatic transmission assembly by pulling it backward.

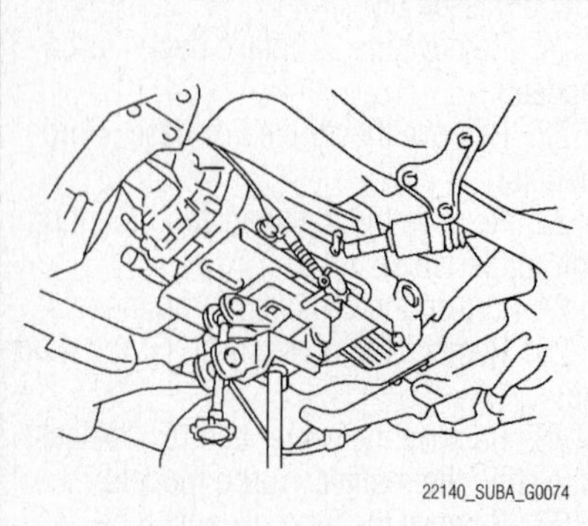

Fig. 33 Transmission jack set into place

To install:

45. Attach the ST-498277200 to the converter case.
46. Install the transmission onto the engine.
47. Lift-up the transmission gradually using transmission jack. Engage at the spline.
48. Install the engine mounting bolt (lower side) and tighten to 37 ft. lbs. (50 Nm).
49. Install the transmission rear crossmember and tighten the mounting bolts to 55 ft. lbs. (75 Nm).
50. Remove the transmission jack.
51. Lower the vehicle.
52. Install the engine mounting bolt (upper side) and tighten to 37 ft. lbs. (50 Nm).
53. Remove the ST from converter case.
54. Install the starter.
55. Install the torque converter to drive plate.
   a. Install the bolts which hold torque converter to drive plate.
   b. Install all four bolts by rotating the crank pulley a little at a time. Tighten to 18 ft. lbs. (25 Nm).
   c. Install the service hole.
56. Remove the ST-41099AC000 engine support assembly.
57. Install the pitching stopper.
58. Install the pitching stopper bracket and tighten to 30 ft. lbs. (41 Nm).
59. Install the pitching stopper and tighten to 37 ft. lbs. (50 Nm).
60. Lift-up the vehicle.
61. Replace the front differential side retainer oil seal.
62. Replace the circlip of the front driveshaft with a new part.
63. Apply grease to the oil seal lip.
64. Attach the ST-28399SA010 seal protector to side retainer.
65. Align and insert the spline of the front driveshaft to the splines of the differential bevel gear, and remove the ST-28399SA010.
66. Install the inlet and outlet hoses to the ATF inlet and outlet pipes.
67. Insert the ball joint into housing.
68. Install the front stabilizer bracket.
69. Install the clutch housing cover securing bolts.
70. Install the front cross support plate.
71. Install the propeller shaft.
72. Install the shift select cable.
73. Install the turbine speed sensor 1 and harness, and then connect the connector. Tighten the mounting bolt to 5 ft. lbs. (7 Nm).
74. Install the oil charge pipe and tighten to 30 ft. lbs. (41 Nm).
75. Install the heat shield cover.
76. Install the center, rear exhaust pipes and the muffler. (Turbo models)
77. Install the front exhaust pipe, rear exhaust pipe and muffler. (Non-turbo models)
78. Install the undercover.
79. Lower the vehicle.
80. Install the air cleaner case.
81. Install the air intake duct.
82. Connect the following connectors:
   - Transmission harness connectors
   - Front oxygen (A/F) sensor
83. Install the intercooler stay RH and engine hanger rear. (Turbo models)
84. Install the engine hanger rear, and then connect the engine harness connector. (Non-turbo models)
85. Install the water by-pass pipe. (Turbo models)
86. Pour ATF from the oil charge pipe.
87. Install the air breather hose.
88. Install the intercooler. (Turbo models)
89. Install the air intake chamber. (Non-turbo models)
90. Install the air cleaner case. (Non-turbo models)
91. Install the collector cover.
92. Connect the ground cable to battery.
93. Perform Clear Memory 2 operation.
   a. Connect the Subaru Select Monitor to the data link connector.
   b. Turn the ignition switch to ON (engine OFF) and turn Subaru Select Monitor switch to ON.
   c. Ensure that the select lever is in the "P" range.
   d. On the «Main Menu» display screen, select the {Each System Check} and press the [YES] key.
   e. On the «System Selection Menu» display screen, select the {Transmission} and press the [YES] key.
   f. Press the [YES] key after the information of transmission type is displayed.
   g. On the «Transmission Diagnosis» display screen, select the {Clear Memory 2} and press the [YES] key.
94. Perform the inspection with driving the vehicle at the end of repair work, and make sure there is no faulty as below:
   - Excessive shift shock
   - Oil leakage from the transmission body, etc.
   - Occurrence of noise caused by interference etc.

➡If excessive shift shock is felt, execute advance operation of learning control.

MANUAL TRANSMISSION ASSEMBLY

REMOVAL & INSTALLATION

5MT Manual Transmission

See Figures 34, 35 and 36.

1. Before servicing the vehicle, refer to the Precautions Section.
2. Open the hood to the full open position.
3. Disconnect the negative battery cable.
4. Drain the transmission fluid.
5. On non turbocharged engine, remove the air intake duct and cleaner case. Remove the air cleaner case stay.
6. On turbocharged engine, remove the intercooler assembly.
7. Disconnect the neutral position switch connector and the backup light switch connector.
8. Disconnect the VSS sensor electrical connector.
9. Remove the starter.
10. Remove the clutch operating cylinder from the transmission and suspend on a wire.
11. Remove the pitching stopper. Position engine support tool ST-41099AC000, or equivalent to hold the engine assembly in place.
12. Remove the bolts which hold upper side of transmission to engine.
13. Lift the vehicle.
14. Remove the front and center exhaust pipes. (Non-turbo models)
15. Remove the center exhaust pipe. (Turbo models)
16. Remove the rear exhaust pipe and muffler.
17. Remove the heat shield cover. (If equipped)
18. Remove the propeller shaft.
19. Remove the gear shift rod and the stay from the transmission.
    a. Disconnect the stay from the transmission.
    b. Remove the gear shift rod from the transmission.
20. Remove the stabilizer link from the front arm.
21. Remove the bolt securing the ball joint of the front arm to the housing, then separate the front arms and the housing.
22. Using a crowbar, remove the left and right front driveshaft from the transmission.
23. Remove the bolts and nuts which hold lower side of transmission to engine.
24. Place the transmission jack under the transmission.

✱✱ CAUTION

Always support the transmission case with a transmission jack.

25. Remove the transmission rear crossmember from the vehicle.
26. Tighten the turnbuckle of the ST-41099AC000 while lowering the transmission jack to tilt the engine assembly towards the back.
27. Remove the transmission.

➡Move the transmission jack towards the rear until the main shaft is withdrawn from the clutch disc.

28. Separate the transmission assembly from the rear cushion rubber.

To install:

29. Replace the differential side retainer oil seal.
30. Install the rear cushion rubber to the transmission assembly and tighten to 26 ft. lbs. (35 Nm).
31. Install the transmission onto the engine.
    a. Lift-up the transmission gradually using a transmission jack.
    b. Engage at the spline section.
32. Loosen the turnbuckle of the ST-41099AC000 while raising the transmission jack to return the engine to its original position.
33. Install the transmission rear crossmember.
34. Take out the transmission jack.
35. Tighten the bolts and nuts which hold the lower side of transmission to the engine to 37 ft. lbs. (50 Nm).
36. Connect the transmission to the engine.
37. Install the starter
38. Tighten the bolts which hold the upper side of the transmission to the engine to 37 ft. lbs. (50 Nm).
39. Remove the ST-41099AC000 engine support tool, or equivalent.
40. Install the pitching stopper. Tighten to 37 ft. lbs. (50 Nm).
41. Lift the vehicle.
42. Install the front driveshaft into the transmission. Use seal protector ST-28399SA010 or equivalent.
43. Insert the ball joints of the front arm into the housing, then tighten the installing bolts to 37 ft. lbs. (50 Nm).
44. Attach the stabilizer link to the front arm and tighten to 33 ft. lbs. (45 Nm).
45. Attach the gear shift rod and stay.
    a. Attach the gear shift rod to the transmission.
    b. Attach the stay to the transmission. Tighten to 13 ft. lbs. (18 Nm).
46. Install the propeller shaft.
47. Install the heat shield cover. (If equipped)
48. Install the rear exhaust pipe and muffler.
49. Install the front exhaust pipe and the center exhaust pipe. (Non-turbo models)
50. Install the center exhaust pipe. (Turbo models)
51. Install the operating cylinder and tighten to 27 ft. lbs. (37 Nm).

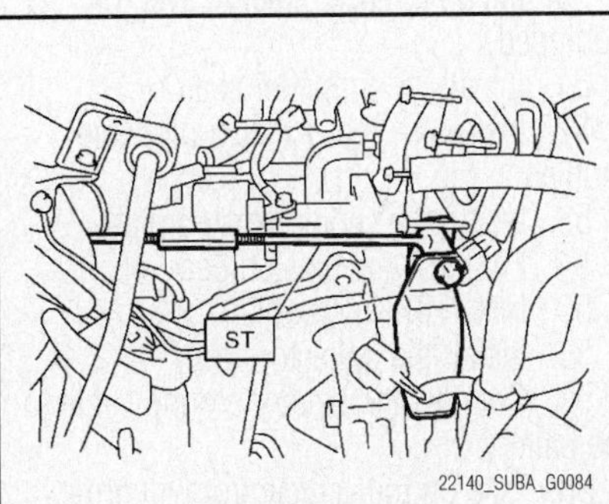

Fig. 34 ST-41099AC000 engine support

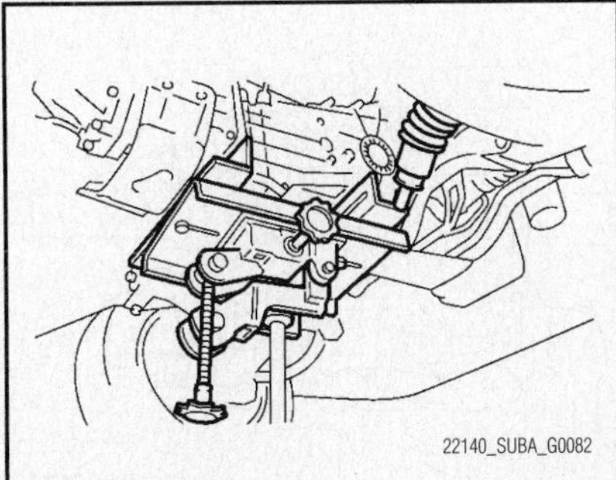

Fig. 35 Transmission jack supporting the transmission

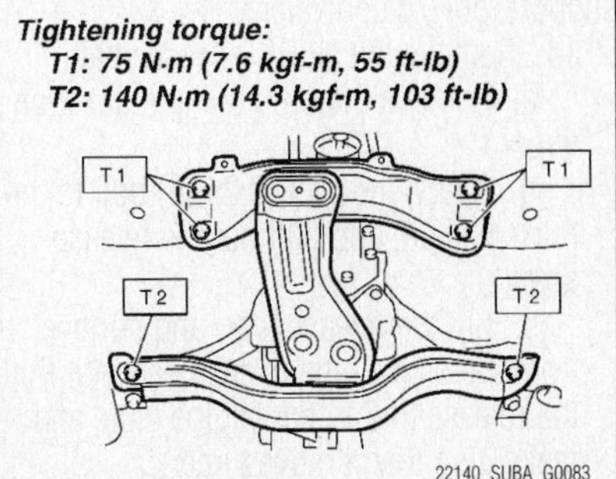

Fig. 36 Transmission rear crossmember and mounting bolts

52. Connect the following connectors.
- Transmission ground cable, tighten to 9 ft. lbs. (13 Nm).
- Neutral position switch connector
- Back-up light switch connector

53. Fill transmission gear oil through the transmission level gauge hole.
54. Install the air intake chamber stay and tighten to 12 ft. lbs. (16 Nm). (Non-turbo models)
55. Install the air intake chamber and air cleaner case. (Non-turbo model)
56. Install the intercooler. (Turbo models)
57. Connect the battery ground cable to the battery.
58. Take off the vehicle from lift arms.
59. Start the engine and check for leaks, correct as required.
60. Road test the vehicle.

6MT Manual Transmission

See Figure 37.

1. Before servicing the vehicle, refer to the Precautions Section.
2. Set the vehicle on a lift. Open the front hood, and support it with the stay.
3. Disconnect the ground cable from battery.
4. Remove the collector cover.
5. Remove the intercooler.
6. Remove the front wheels.
7. Disconnect the following harness connectors:
- Neutral position switch backup light switch connector
- Rear oxygen sensor connector

8. Remove the engine hanger rear.
9. Disconnect the ground cable on the upper side of the transmission case and body.
10. Remove the starter assembly.
11. Remove the operating cylinder from the transmission.

➡Hang the removed operating cylinder with a piece of wire.

12. Remove the pitching stopper and pitching stopper bracket.
13. Install the ST-41099AC000 engine support tool, or equivalent.
14. Remove the clutch release shaft.
 a. Remove the plug using a hexagon wrench.
 b. Attach a 6 mm (0.24 in) bolt to the release shaft, and pull out the release shaft.
 c. Lift the release fork, and remove from the claw of the release bearing. Pull the release fork to the engine side, and make it so that it moves freely.
15. Remove the bolts which hold upper side of transmission to engine.
16. Lift the vehicle.
17. Remove the center exhaust pipe.
18. Remove the rear exhaust pipe and muffler.
19. Remove the heat shield cover. (If equipped)
20. Remove the propeller shaft.
21. Remove the front stabilizer link.
22. Remove the ball joint of front arm from the housing.
23. Remove the front driveshaft.
24. Set the transmission jack under the transmission, and remove the front crossmember and rear crossmember.
25. Move the transmission to the right side of the vehicle, and remove the joint COMPL, stay bolts and reverse check cable.

➡If the transmission is not moved aside, the joint COMPL and stay bolts may contact the body and cause damage.

26. Tighten the turnbuckle of the ST-41099AC000 to tilt the engine assembly towards the back.
27. Remove the bolts holding the bottom of transmission to the engine, and remove the transmission from the vehicle.

To install:

28. Set the release fork, release bearing and release shaft to the transmission.
29. Replace the front differential side retainer oil seal.
30. Remove the oil seal by using flat tip screwdriver etc.
31. Apply gear oil to the lip of new oil seals.
32. Install a new oil seal using ST-18675AA000.

➡Be sure to replace the differential side oil seal after the procedure of removing front driveshaft from transmission.

33. Loosen the turnbuckle of ST-41099AC000 to return the engine to its original position.

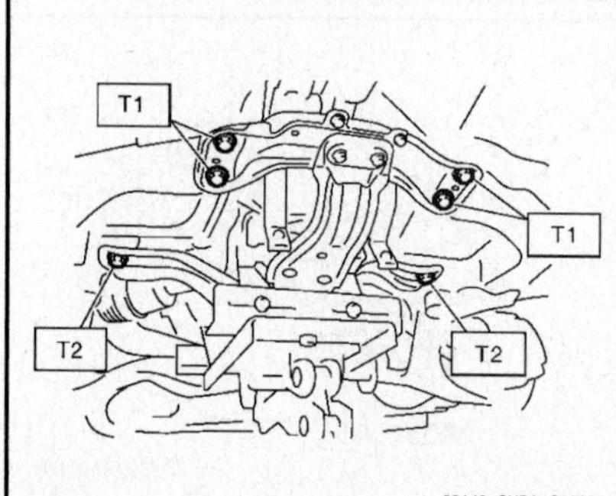

Fig. 37 Transmission front and rear crossmember and mounting bolts

34. Install the transmission.
35. Tighten the bolts and nuts which hold the lower side of transmission to the engine to 37 ft. lbs. (50 Nm).
36. Move the transmission to the right side of the vehicle, and attach the joint COMPL, stay bolts and reverse check cable.
37. Install the front crossmember and rear crossmember.
38. Tighten the bolts which hold the upper side of the transmission to the engine to 37 ft. lbs. (50 Nm).
39. Make sure that the release bearing is completely inserted.

➡Push the release fork towards the operating cylinder side until a clicking sound is heard. Pull the release fork towards the engine side. If the release fork is not in contact with the case, the setting is complete. Confirm that the boot cover is set securely.

40. Install the pitching stopper bracket and tighten to 30 ft. lbs. (41 Nm).
41. Install the pitching stopper and tighten to 37 ft. lbs. (50 Nm).
42. Install the clutch slave cylinder. Tighten mounting bolts to 30 ft. lbs. (41 Nm).
43. Install the starter assembly.
44. Attach the ground cable to the transmission and body.
45. Connect the following harness connectors:
- Neutral position switch backup light switch connector
- Rear oxygen sensor connector

46. Attach the engine hanger rear.
47. Install the front driveshafts into the transmission. Use a seal protector ST-28399SA010 and replace the circlip of driveshaft with a new part.
48. Install the ball joint of the front arm. Tighten bolt to 37 ft. lbs. (50 Nm).
49. Install the front stabilizer links and tighten nuts to 33 ft. lbs. (45 Nm).
50. Install the heat shield cover. (If equipped)
51. Install the propeller shaft.
52. Install the rear exhaust pipe and muffler.
53. Install the center exhaust pipe.
54. Fill the transmission gear oil.
55. Install the intercooler.
56. Install the collector cover.
57. Connect the battery ground cable to the battery.
58. Take off the vehicle from lift arms.
59. Start the engine and check for leaks, correct as required.
60. Road test the vehicle.

CLUTCH DRIVEN DISC & PRESSURE PLATE

REMOVAL & INSTALLATION

See Figures 38, 39 and 40.

1. Before servicing the vehicle, refer to the Precautions Section.
2. Disconnect the negative battery cable.
3. Remove the transmission.
4. Install tool ST499747100, or equivalent on the flywheel.
5. Remove the clutch cover and clutch disc. Do not disassemble

➡**Be sure to put alignment marks on the flywheel and clutch cover before removing the clutch cover.**

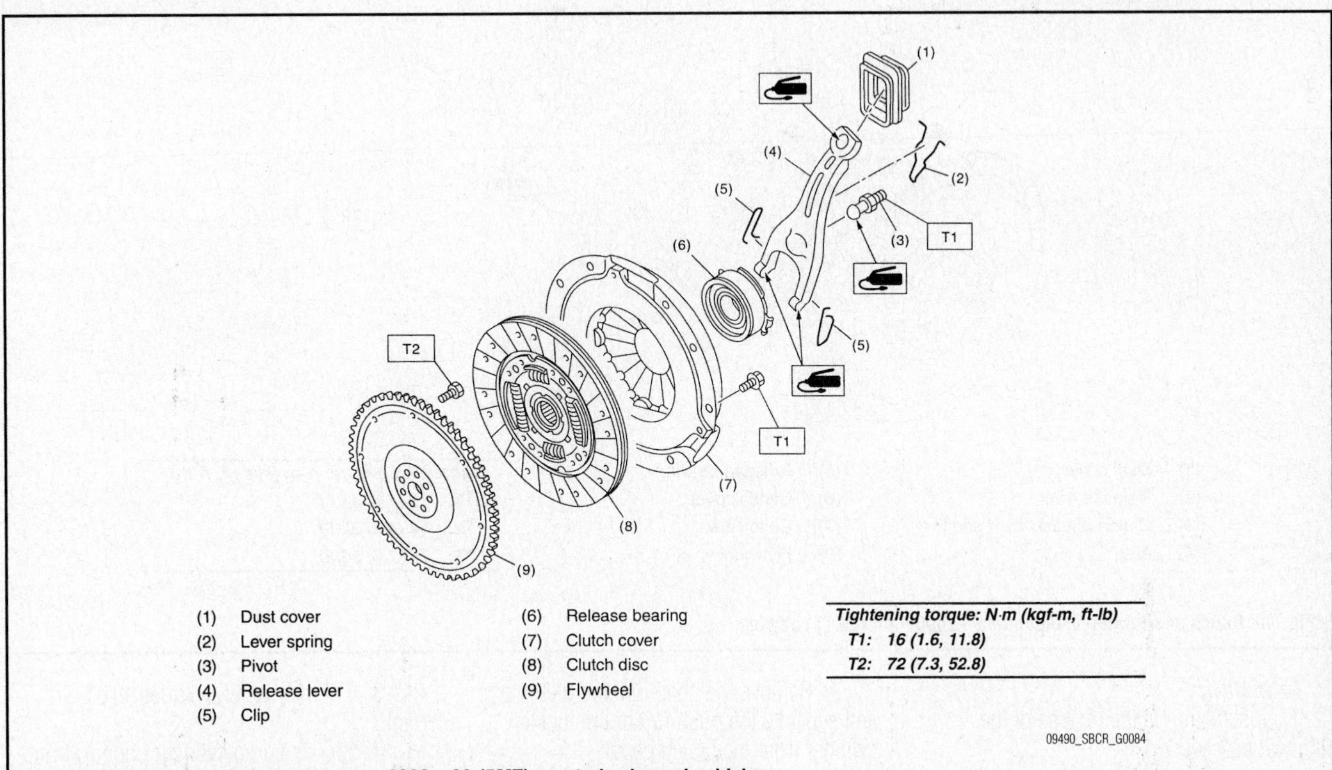

Fig. 38 Clutch and related components—2006—08 (5MT) non-turbocharged vehicles

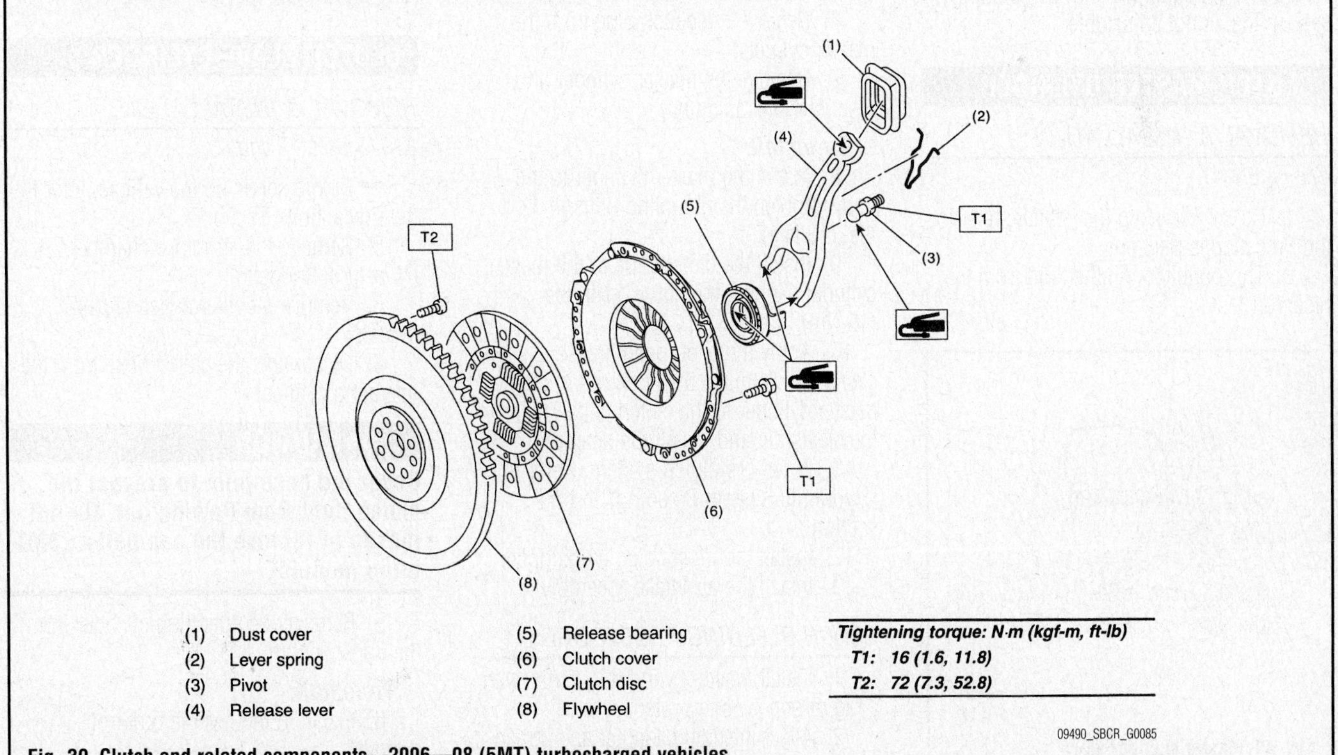

Fig. 39 Clutch and related components—2006—08 (5MT) turbocharged vehicles

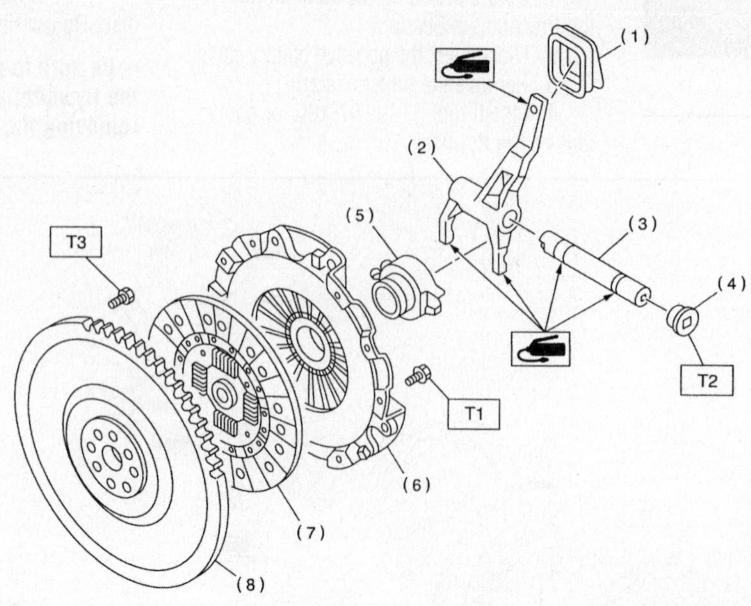

Fig. 40 Clutch and related components—2006—08 (6MT) vehicles

To install:

6. Installation is the reverse of the removal procedure.

ADJUSTMENTS

Models are equipped with a hydraulic system that is not adjustable.

CLUTCH MASTER CYLINDER

REMOVAL & INSTALLATION

See Figure 41.

1. Before servicing the vehicle, refer to the Precautions Section.
2. Thoroughly drain the fluid from the reservoir tank.

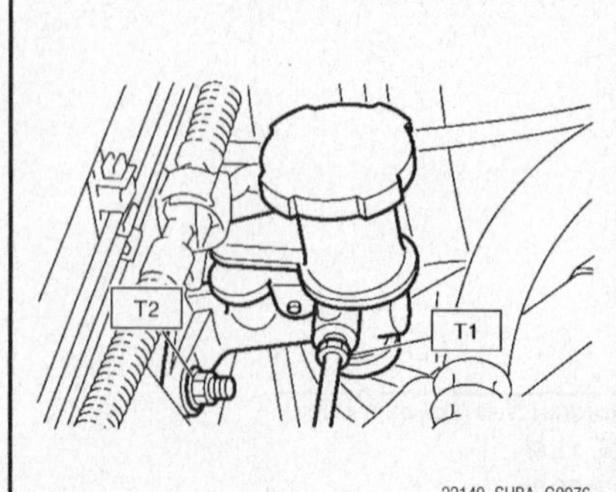

Fig. 41 Master cylinder view

3. Remove the snap pin and clevis pin, and separate the push rod of the master cylinder from the clutch pedal.
4. Remove the air intake assembly.
5. Remove the intercooler, if equipped. (Turbo models)
6. Remove the clutch pipe from the master cylinder.
7. Remove the master cylinder and reservoir tank as a unit.

To install:

8. Install the master cylinder to the body. Tighten the mounting bolts to 13 ft. lbs. (18 Nm).
9. Install the clutch pipe to the master cylinder. Tighten the nut to 11 ft. lbs. (15 Nm).
10. Apply multi-purpose grease to the clevis pin. Connect the push rod of the master cylinder to the clutch pedal. Insert the clevis pin and install the snap pin.
11. Bleed the clutch system. Refer to "Hydraulic System Bleeding" in this section.
12. Install the intercooler, if equipped.
13. Install the air intake assembly.

BENCH BLEEDING PROCEDURE

1. Install master cylinder in a vise with a rag or soft jaws to protect cylinder.
2. Attach bleeding adapter and thread into master cylinder.
3. Fill the master cylinder reservoir tank.
4. Slowly pump cylinder push rod until no signs of air bubbles are present.
5. Remove bleeding equipment and plug brake pipe line.

CLUTCH SLAVE CYLINDER

REMOVAL & INSTALLATION

See Figures 42 and 43.

1. Before servicing the vehicle, refer to the Precautions Section.
2. Remove the air intake chamber. (Non-turbo models)
3. Remove the intercooler. (Turbo models)
4. Disconnect the clutch hose from the operating cylinder.

**** CAUTION**

Cover the hose joint to prevent the clutch fluid from flowing out. Do not loosen or remove the cap bolts. (5MT turbo models)

5. Remove the operating cylinder from the transmission.

To install:

6. Install in the reverse order of removal.

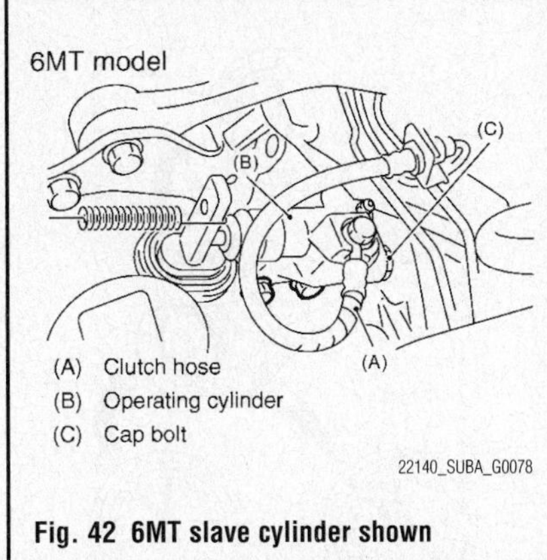

Fig. 42 6MT slave cylinder shown

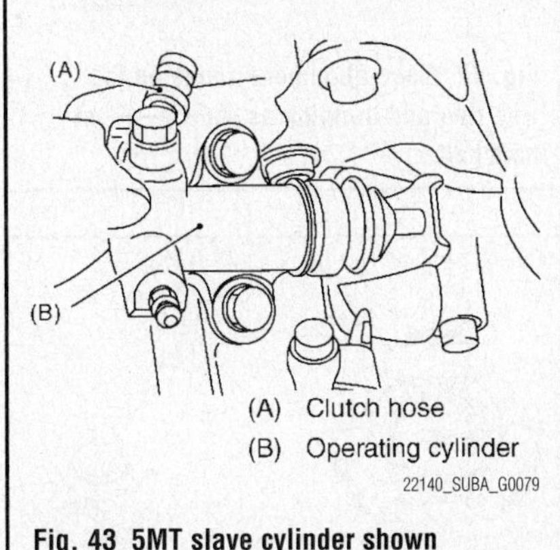

Fig. 43 5MT slave cylinder shown

➡Before installing the operating cylinder, apply grease (KOPR-KOTE: Part No. 003603001) to the contact point of the release lever and operating cylinder.

7. Tighten the slave cylinder mounting bolts as follows:
   - 5MT models 27 ft. lbs. (37 Nm).
   - 6MT models 30 ft. lbs. (41 Nm).
8. Tighten the feed hose bolt to 13 ft. lbs. (18 Nm).

CLUTCH HYDRAULIC SYSTEM BLEEDING

BLEEDING PROCEDURE

5MT Manual Transmissions

See Figure 44.

1. On non turbocharged engine, remove the air intake chamber.
2. On turbocharged engine, remove the intercooler.
3. Connect a vinyl tube to the air bleeder on the master cylinder. Put the other end in a jar with clean clutch fluid.
4. Slowly depress the clutch pedal and keep it depressed. Open the air bleeder to discharge air and fluid.

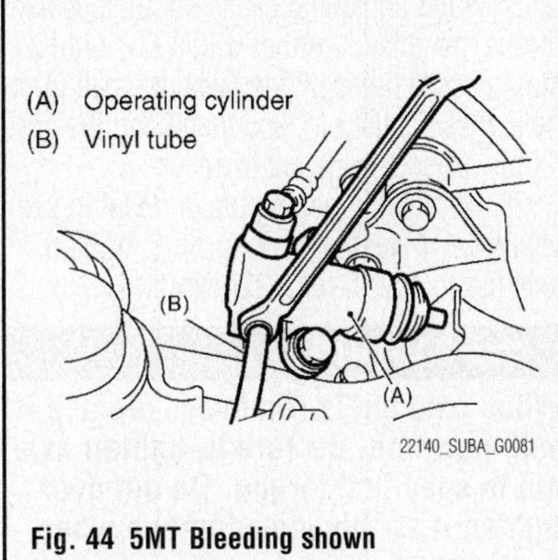

Fig. 44 5MT Bleeding shown

5. Release the air bleeder for one or two seconds. With the bleeder closed, slowly release the clutch pedal.
6. Repeat the procedure until there are no more air bubbles in the vinyl tube.
7. Tighten the air bleeder.
8. Connect a vinyl tube to the air bleeder on the clutch operating (slave) cylinder. Put the other end in a jar with clean clutch fluid.
9. Slowly depress the clutch pedal and keep it depressed. Open the air bleeder to discharge air and fluid.
10. Release the air bleeder for one or two seconds. With the bleeder closed, slowly release the clutch pedal.
11. Repeat the procedure until there are no more air bubbles in the vinyl tube.
12. Tighten the air bleeder.
13. After depressing the clutch pedal, make sure that there are no leaks in the entire system
14. Recheck to ensure that the clutch is operating correctly.

6MT Manual Transmissions

See Figure 45.

1. Remove the intercooler.
2. Remove the clutch operating cylinder. Do not remove the clutch hose.

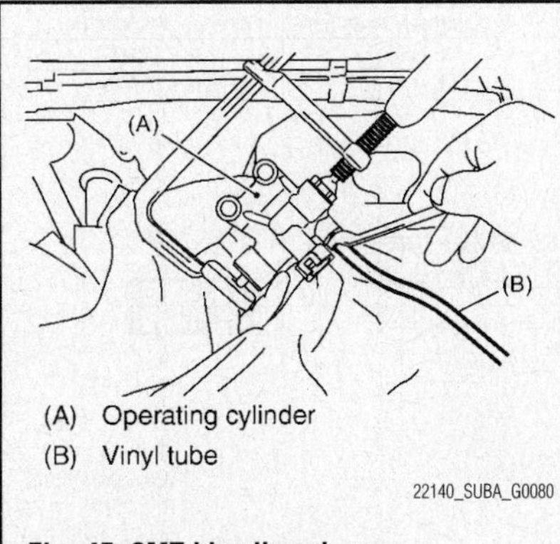

Fig. 45 6MT bleeding shown

3. Using a service clamp, fix the piston to avoid the piston from jumping out.
4. Connect a vinyl tube to the air bleeder on the clutch operating (slave) cylinder. Put the other end in a jar with clean clutch fluid.
5. Slowly depress the clutch pedal and keep it depressed. Open the air bleeder to discharge air and fluid.
6. Release the air bleeder for one or two seconds. With the bleeder closed, slowly release the clutch pedal.

➡Set the air breather screw position higher than the tip of the operating cylinder when performing this procedure.

7. Repeat the procedure until there are no more air bubbles in the vinyl tube.
8. Tighten the air bleeder.
9. After depressing the clutch pedal, make sure that there are no leaks in the entire system
10. Recheck to ensure that the clutch is operating correctly.

TRANSFER CASE ASSEMBLY

REMOVAL & INSTALLATION

The transfer case is removed as an assembly with the transmission. Refer to "Transmission Assembly, Removal & Installation" procedures in this section.

FRONT HALFSHAFT

REMOVAL & INSTALLATION

See Figure 46.

1. Before servicing the vehicle, refer to the Precautions Section.
2. Disconnect the negative battery cable.
3. Jack-up vehicle, support it with safety stands, and remove front wheels.
4. Unlock axle nut.
5. Remove axle nut while depressing brake pedal to prevent front driveshaft from turning.

✲✲ CAUTION

Be sure to loose and retighten axle nut after removing wheel from vehicle. Failure to follow this rule may damage wheel bearings.

6. Remove stabilizer link.
7. Remove disc brake caliper from housing, and suspend it from strut using a wire.
8. Remove disc rotor from hub.
9. If disc rotor seizes up within hub, drive disc rotor out by installing an 8-mm bolt in screw hole on the rotor.

10. Remove cotter pin and castle nut which secure tie-rod end to housing knuckle arm.
11. Using a puller, remove tie-rod ball joint from knuckle arm.
12. Remove ABS sensor assembly and harness.
13. Remove bolt which secures sensor harness to strut.
14. Remove transverse link ball joint from housing.
15. Remove inner joint from transmission spindle.
16. Remove front driveshaft assembly from hub. If it is hard to remove, use a ST-926470000 puller.

CAUTION

Be careful not to damage oil seal lip and tone wheel when removing front driveshaft. If front driveshaft is removed, replace inner oil seal with new one.

17. After scribing an alignment mark on camber adjusting bolt head, remove bolts which connect housing and strut, and disconnect housing from strut.

To install:

18. While aligning alignment mark on camber adjusting bolt head, connect housing and strut. Tighten to 130 ft. lbs. (177 Nm).
19. Install front driveshaft.
20. Install transverse link ball joint to housing and tighten to 36 ft. lbs. (49 Nm).
21. Install ABS sensor harness on strut.
22. Install ABS sensor on housing. Tighten the mounting bolt to 24 ft. lbs. (32 Nm).
23. Install disc rotor on hub.
24. Install disc brake caliper on housing and tighten to 58 ft. lbs. (78 Nm).
25. Connect stabilizer link.
26. Install tie-rod end and tighten castle nut to 20 ft. lbs. (27 Nm).
27. After tightening castle nut to specified torque, retighten it further within 60° until a slot in castle nut is aligned with the ball joint hole. Insert cotter pin, and then bend the cotter pin around castle nut to secure it.
28. While depressing brake pedal to prevent front driveshaft from turning, tighten axle nut to 162 ft. lbs. (220 Nm).

CAUTION

When axle nut is removed, replace it with new one. Be sure to tighten axle nut to specified torque. Do not over tighten it as this may damage wheel bearing.

29. After tightening axle nut, lock it securely.
30. Install wheel and tighten wheel nuts.

CV-JOINTS OVERHAUL

See Figures 47 through 50.

1. Place alignment marks on the shaft and outer race.
2. Remove the inner boot band and boot.
3. Remove the circlip from the inner joint outer race using a suitable pry tool.
4. Outer race from the shaft assembly and wipe off the grease.
5. Place alignment marks on the free ring and trunnion as shown in the illustration.
6. Remove the free ring from the trunnion.
7. Place alignment marks on the trunnion and shaft as shown in the illustration.
8. Remove the snapring and trunnion.
9. Place the shaft in a vise between wooden blocks.
10. Using a suitable pry tool, raise the outer boot band claws.
11. Cut and remove the boot.
12. Only the boot can be replaced, the joint is not serviceable and must be replaced if damaged.

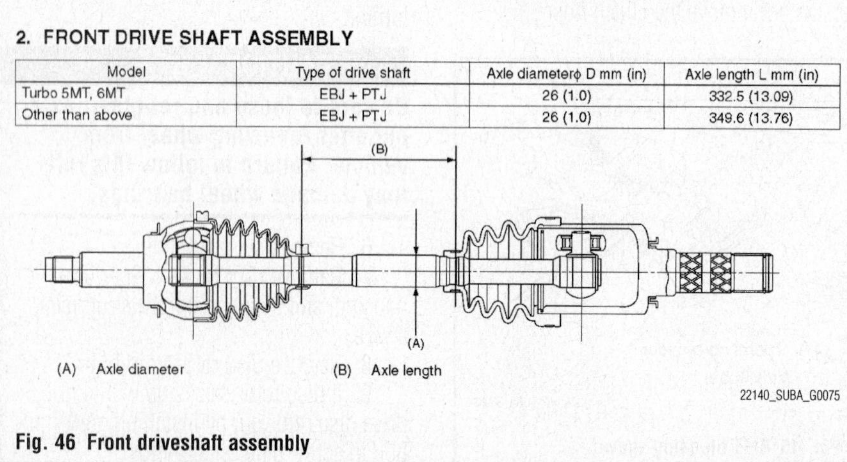

2. FRONT DRIVE SHAFT ASSEMBLY

Model	Type of drive shaft	Axle diameterφ D mm (in)	Axle length L mm (in)
Turbo 5MT, 6MT	EBJ + PTJ	26 (1.0)	332.5 (13.09)
Other than above	EBJ + PTJ	26 (1.0)	349.6 (13.76)

Fig. 46 Front driveshaft assembly

Fig. 47 Place alignment marks on the free ring and trunnion as shown—Front halfshaft

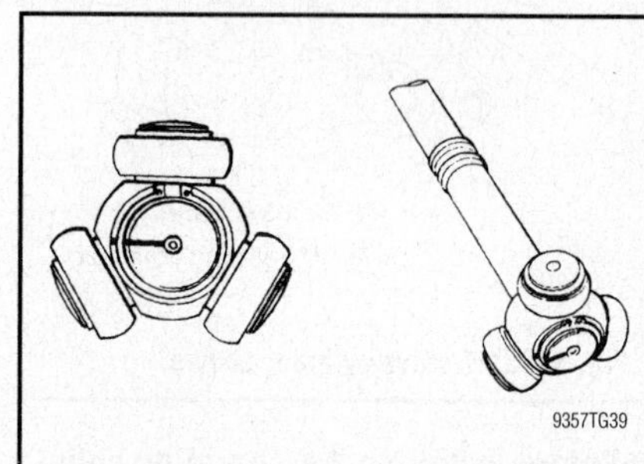

Fig. 48 Place alignment marks on the trunnion and shaft as shown—Front halfshaft

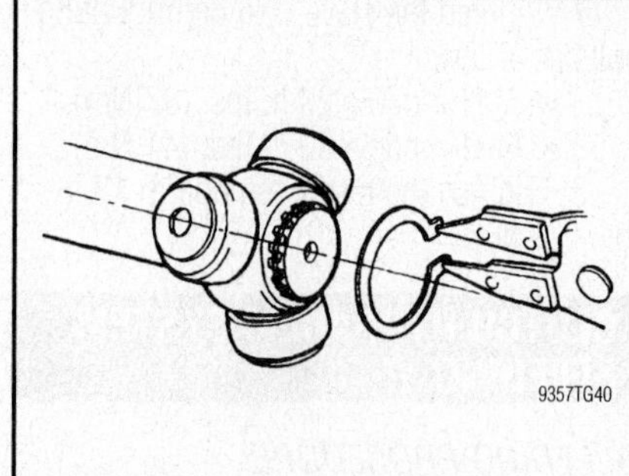

Fig. 49 Remove the snapring and trunnion—Front halfshaft

To install:

13. Place the half shaft in a vise.
14. Place the outer boot and small band on the shaft.
15. Apply 2.12–2.47 oz. (60–70g) of supplied grease to the joint.
16. Apply 0.71–1.06 oz. (20–30g) of supplied grease to the whole inner surface of the boot, and apply some grease to the shaft.

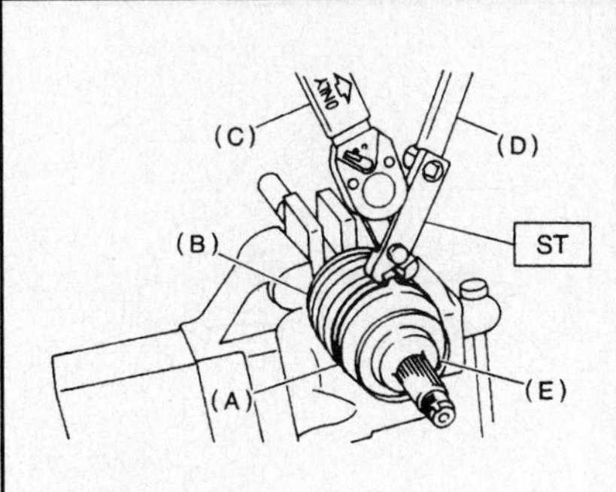

Fig. 50 Using boot band pliers tool 28099A000 tighten the boot bands—Front halfshaft

17. Install the boot to the joint groove, and attach the large boot band as shown.
18. Install the boot to the shaft groove, and attach the small boot band as shown.
19. Using boot band pliers tool 28099A000 to tighten the large band to 116 ft. lbs. (157 Nm) and the small band to 98 ft. lbs. (133 Nm).
20. Place the inner boot on the center of the shaft.
21. Align the alignment marks from earlier and install the trunnion and snapring. Make sure the snapring is fully engaged.
22. Apply 3.53–3.88 oz. (100–110g) of supplied grease to the joint outer race.
23. Apply a coat of supplied grease to the free ring and trunnion.
24. Align the marks on the free ring and trunnion and install the free ring.
25. Align the marks on the shaft and outer race and install the outer race.
26. Pull on the shaft lightly to ensure the circlip is completely engaged.
27. Apply an even coat 1.06–1.41 oz. (30–40g) of the supplied grease to the entire inner surface of the boot.
28. Install the boot and band.
29. Once the band is properly tightened, cut off any excess to leave only 0.39 inch (10mm) and bend it over.
30. Install the shaft.

REAR AXLE HOUSING

REMOVAL & INSTALLATION

T-Type Rear Differential

See Figure 51 and 52.

1. Before servicing the vehicle, refer to the Precautions Section.
2. Disconnect the ground cable from the battery.
3. Shift the select lever or gear shift lever to neutral.
4. Release the parking brake.
5. Lift up the vehicle.
6. Remove the rear exhaust pipe and muffler.
7. Remove the heat shield cover.
8. Remove the propeller shaft.
9. Prepare the transmission jack and band.
10. Loosen the self-lock nuts which hold the rear differential to rear crossmember.
11. Remove the DOJ of rear driveshaft from rear differential using ST-28099PA100.
12. Remove the rear differential front member.
13. Support the rear differential with the transmission jack.
14. Secure the rear differential using band.
15. Remove the self-lock nuts which hold the rear differential to the crossmember.
16. Remove the rear differential stud bolt from rear crossmember bushing.

➡When removing the stud bolt, carefully adjust the angle and location of transmission jack and jack stand, if necessary.

17. Lower the transmission jack stand after removing the rear differential stud bolt from the rear crossmember. Rear driveshaft should not come into contact with the lateral link bolt.
18. Pull out the axle shaft from rear differential.
19. Lower the transmission jack.
20. Secure the rear driveshaft to lateral link using wire.
21. Remove the rear differential member plate from rear differential.

To install:

22. Install the rear differential member plate to the rear differential.
23. Set the rear differential to transmission jack.

➡Secure the rear differential to transmission jack using band.

24. Insert the spline shaft until the spline portion comes inside the side oil seal. Use seal protector ST-28099PA090.
25. Remove ST from rear differential.
26. Push the rear differential to insert the axle shaft into rear differential.
27. Adjust the transmission jack, if necessary, and insert the rear differential stud bolt into rear crossmember bushing properly.
28. After inserting the rear differential stud bolt into the rear crossmember bushing, lift up the transmission jack and align the rear differential to its attachment position.
29. Tighten a new self-locking nut temporarily to rear crossmember.
30. Remove the band from rear differential. Lift up the rear differential until the rear differential is separated from the transmission jack.
31. Install the rear differential front member with a new self-locking nut. Tighten to specifications.
32. Tighten the self-locking nut. Tighten to 51 ft. lbs. (70 Nm).
33. Lower the transmission jack.
34. Install the propeller shaft.
35. Install the heat shield cover.
36. Install the rear exhaust pipe and muffler.
37. After installing the rear differential carrier to the vehicle, remove the filler plug, and refill with gear oil up to the lower lip of the plug hole.
38. Oil capacity specifications:
    a. Except for GT specifications B 6MT model: 0.8 qt. (0.8 L)
    b. GT specifications B 6MT model: 1.1 qt. (1.0 L)
39. Tighten the filler plug.

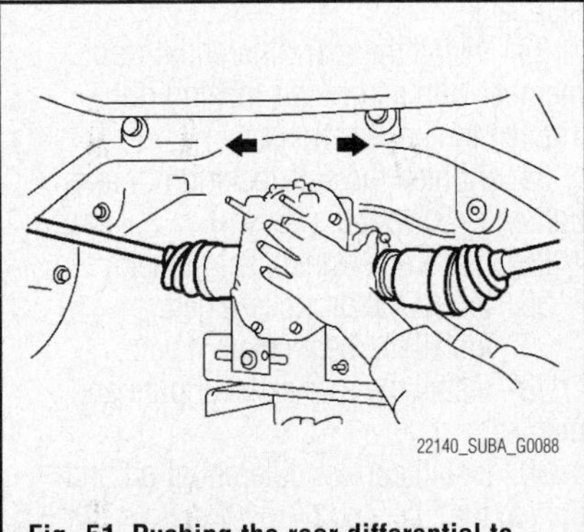

Fig. 51 Pushing the rear differential to insert the axle shaft

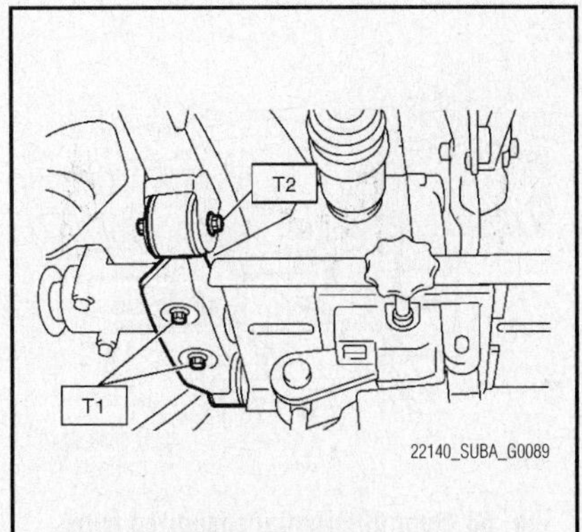

Fig. 52 Rear differential front member

40. Road test and check for noise and leaks.

VA-Type Rear Differential

See Figures 53, 54 and 55

1. Before servicing the vehicle, refer to the Precautions Section.
2. Disconnect the ground cable from the battery.
3. Shift the select lever or gear shift lever to neutral.
4. Release the parking brake.
5. Lift up the vehicle.
6. Remove the rear exhaust pipe and muffler.
7. Remove the heat shield cover.
8. Remove the propeller shaft.
9. Prepare the transmission jack and band.
10. Loosen the self-lock nuts which hold the rear differential to rear crossmember.
11. Remove the DOJ of rear driveshaft from rear differential.
12. Remove the nuts which hold the rear differential front member.
13. Support the rear differential with the transmission jack.
14. Remove the rear differential front member.
15. Secure the rear differential using band.
16. Remove the self-lock nuts which hold the rear differential to rear crossmember.
17. Remove the rear differential stud bolt from rear crossmember bushing.

➡When removing the stud bolt, carefully adjust the angle and location of transmission jack and jack stand, if necessary

18. Lower the transmission jack stand after removing the rear differential stud bolt from the rear crossmember. Rear driveshaft should not come into contact with the lateral link bolt.

Fig. 53 Rear differential supported with transmission jack

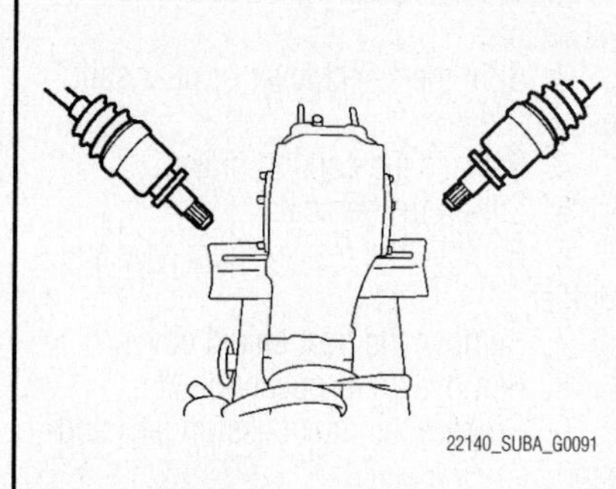

Fig. 54 Axle shaft removal from rear differential shown

19. Pull out the axle shafts from the rear differential.
20. Lower the transmission jack.
21. Secure the rear driveshaft to lateral link using wire.
22. Remove the rear differential member plate from rear differential.

To install:

23. Insert the rear differential member plate into rear differential.
24. Set the rear differential to transmission jack.
25. Attach the ST to rear differential.
26. Insert the spline shaft until the spline portion comes inside the side oil seal.
27. Remove ST from rear differential.
28. Push the rear differential to insert the axle shaft into rear differential.
29. Adjust the transmission jack, if necessary, and insert the rear differential stud bolt into rear crossmember bushing properly.
30. After inserting the rear differential stud bolt into the rear crossmember bushing, lift up the transmission jack and align the rear differential to its attachment position.
31. Tighten a new self-locking nut temporarily to rear crossmember.
32. Remove the band from rear differential. Lift up the rear differential until the rear differential is separated from the transmission jack.
33. Install the rear differential front member with a new self-locking nut. Tighten to specifications.
34. Tighten the self-locking nut and tighten to 51 ft. lbs. (70 Nm).
35. Lower the transmission jack.
36. Install the propeller shaft.
37. Install the heat shield cover.
38. Install the rear exhaust pipe and muffler.
39. Install correct differential oil and tighten the filler plug.
40. Road test and check for noise and leaks.

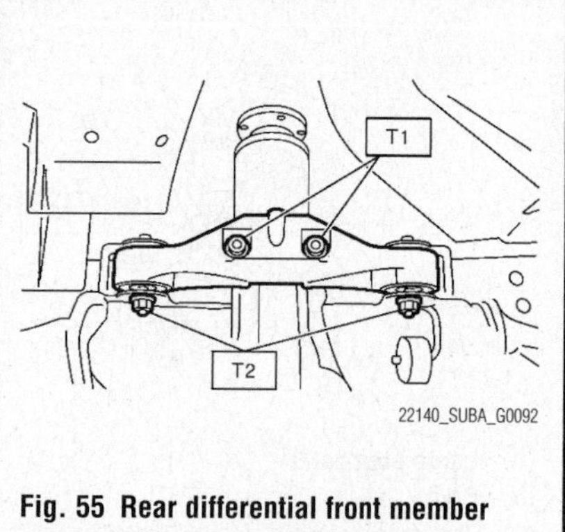

Fig. 55 Rear differential front member and locking nuts shown

REAR AXLE SEAL

REMOVAL & INSTALLATION

See Figure 56.

1. Before servicing the vehicle, refer to the Precautions Section.
2. Remove the rear differential.
3. Remove the rear differential side oil seal using a screwdriver wrapped with vinyl tape to prevent the side retainer from scratches.

To install:

4. Using the ST, install the oil seal to the side retainer.
5. Install the rear differential.

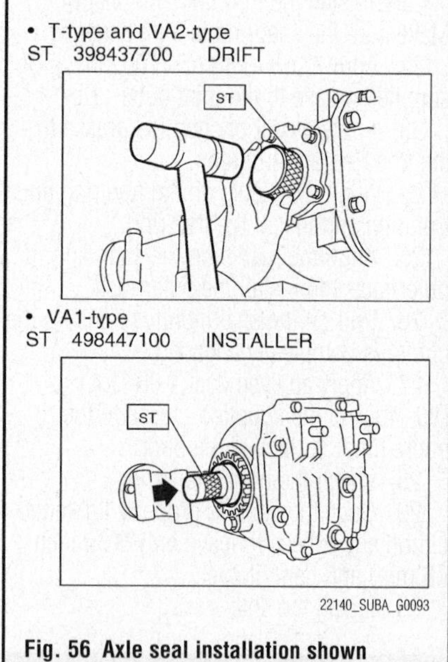

Fig. 56 Axle seal installation shown

REAR HALFSHAFT

REMOVAL & INSTALLATION

Baja Models

See Figure 57.

1. Before servicing the vehicle, refer to the Precautions Section.
2. Disconnect the negative battery cable.
3. Raise and support the vehicle safely.
4. Remove the wheels and tires.
5. Unlock the axle nut. Depress the parking brake. Remove the axle nut.

➡Be sure to loosen and retighten the axle nut after removing the tire and wheel assembly from the vehicle. Failure to follow this rule may damage the wheel bearings.

6. Remove the rear differential assembly.
7. Remove the axle nut and rear halfshaft.

➡Use axle shaft puller tools ST92647000 and ST28099PA110, or equivalent, if it is difficult to remove the halfshaft. Do not hammer the halfshaft to remove it. Be careful not to damage the oil seal or magnetic encoder.

To install:

8. Installation is the reverse of the removal procedure.
9. Tighten the axle nut to 177 ft. lbs. with the parking brake applied.

➡Install the tire and wheel assembly after installation of the axle nut. Failure to follow this order, may cause damage to the wheel bearings.

Impreza Models

See Figure 58.

1. Before servicing the vehicle, refer to the Precautions Section.
2. Disconnect the negative battery cable.
3. Raise and support the vehicle safely.
4. Remove the wheels and tires.
5. Unlock the axle nut. Depress the parking brake. Remove the axle nut.

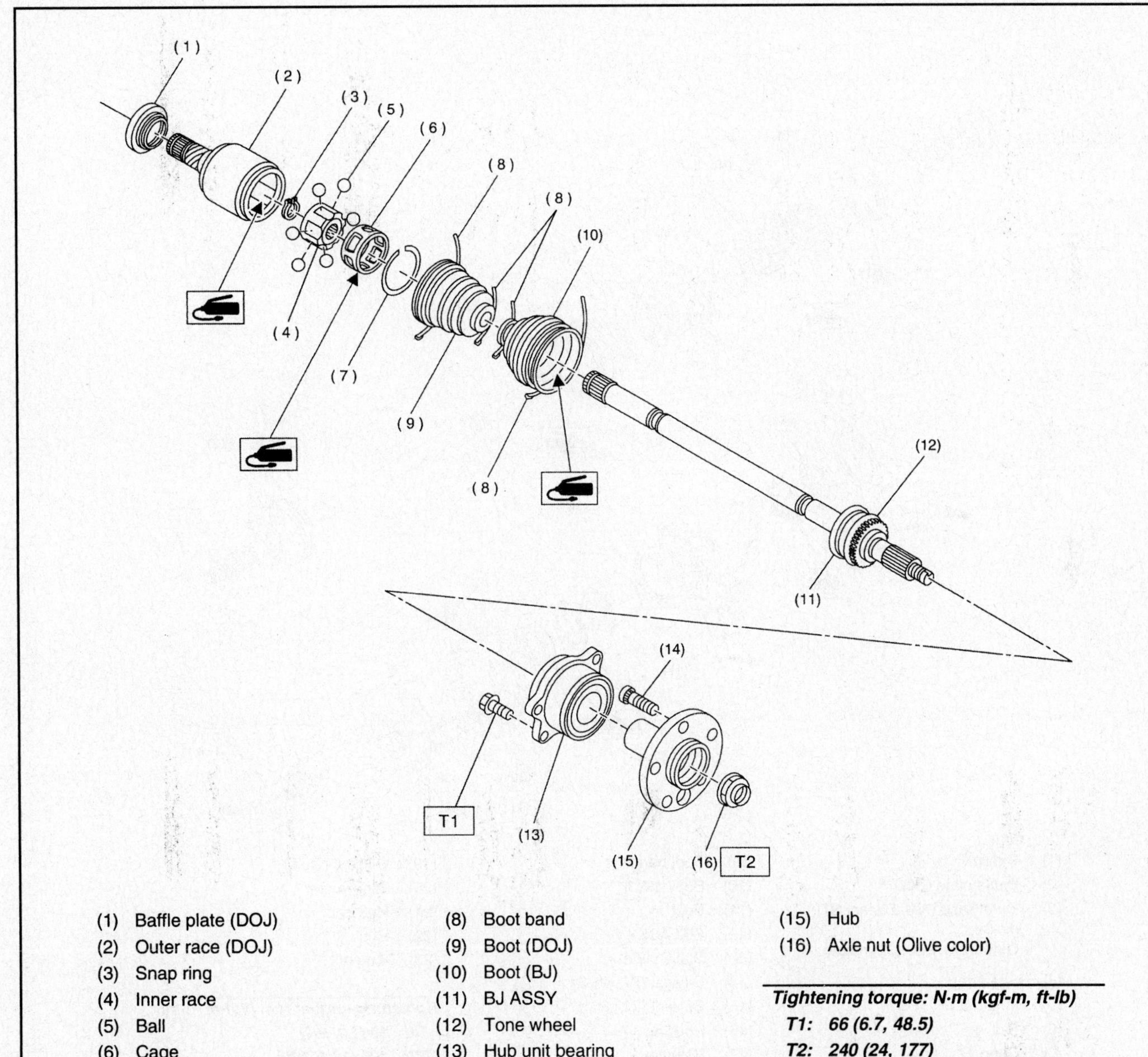

Fig. 57 Rear halfshaft and related components—2006 Baja

➡**Remove the axle nut with vehicle weight NOT applied to the axle.**

6. Disconnect the stabilizer link. Remove the bolt that retains the trailing link to the housing.
7. Remove the bolts which secure the front lateral link and rear lateral link to the housing.
8. Remove the ABS wheel speed sensor from the back plate.
9. Remove the halfshaft from the rear axle. If it is hard to remove, remove the brake disk rotor.

➡**Use axle shaft puller tools ST92647000 and ST28099PA110, or equivalent, if it is difficult to remove the halfshaft. Do not hammer the halfshaft to remove it. Be careful not to damage the oil seal or wheel tone.**

10. Remove the halfshaft from the differential using tool ST28099PA100.

➡**Using the tool will prevent damage to the side bearing retainer.**

To install:

11. Installation is the reverse of the removal procedure.
12. Tighten the axle nut to 140 ft. lbs. with the parking brake applied.

Legacy & Outback Models

See Figure 59.

1. Disconnect the negative battery cable.
2. Raise and support the vehicle safely.
3. Remove the wheels and tires.
4. Unlock the axle nut. Depress the parking brake. Remove the axle nut.

➡**Be sure to loosen the axle nut after removing the tire and wheel from the vehicle. Failure to do this may damage the wheel bearings.**

5. Remove the rear differential assembly.

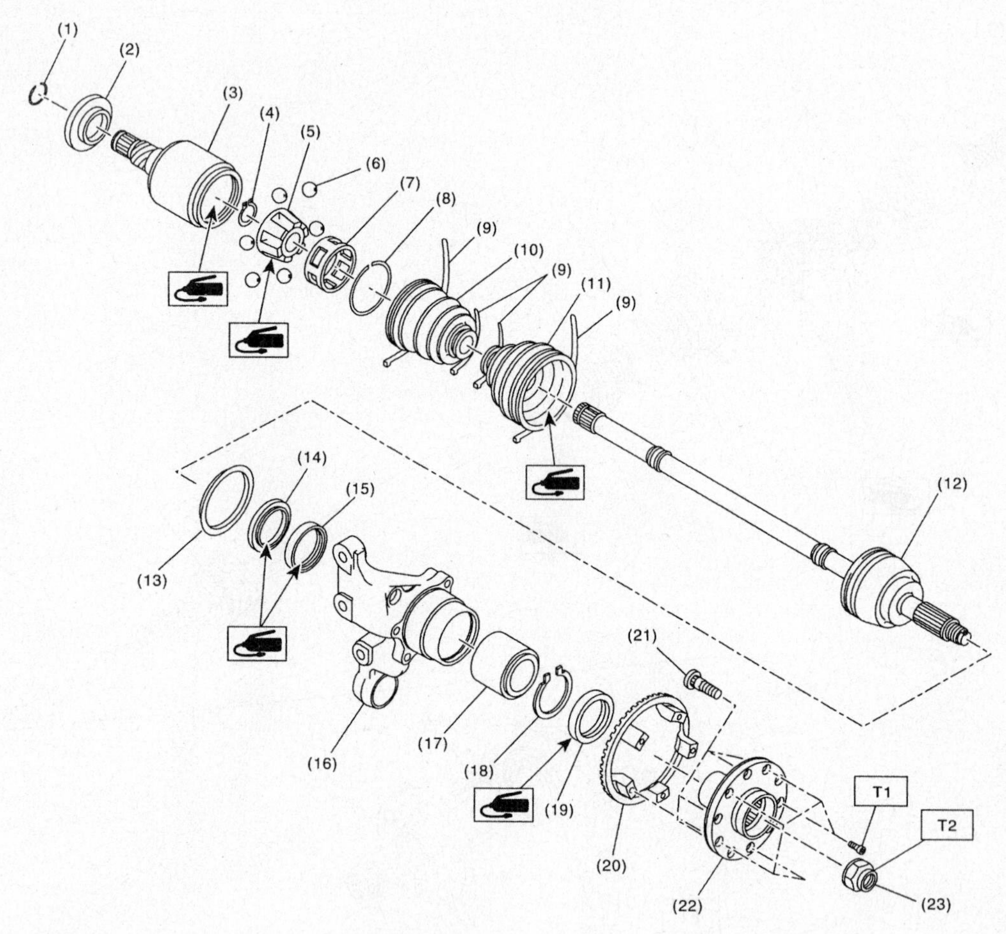

(1) Circlip
(2) Baffle plate (DOJ)
(3) Outer race DOJ: Except STI model
Outer race EDJ: STI model
(4) Snap ring
(5) Inner race
(6) Ball
(7) Cage
(8) Snap ring
(9) Boot band
(10) Boot (DOJ)
(11) Boot
(12) EBJ ASSY
(13) Baffle plate
(14) Oil seal (IN. No. 2)
(15) Oil seal (IN.)
(16) Housing
(17) Bearing
(18) Snap ring
(19) Oil seal (OUT)
(20) Tone wheel
(21) Hub bolt
(22) Hub
(23) Axle nut

Tightening torque: N·m (kgf-m, ft-lb)
T1: 13 (1.3, 9.4)
T2: 190 (19.4, 140)

09490_SBCR_G0092

Fig. 58 Rear halfshaft and related components

6. Remove the axle nut and rear half-shaft.

➡**Use axle shaft puller tools ST92647000 and ST28099PA110, or equivalent, if it is difficult to remove the halfshaft. Do not hammer the half-shaft to remove it. Be careful not to damage the oil seal or magnetic encoder.**

To install:

7. Installation is the reverse of the removal procedure.

8. Tighten the axle nut to 177 ft. lbs. with the parking brake applied.

➡**Install the tire and wheel assembly after installation of the axle nut. Failure to follow this order, may cause damage to the wheel bearings.**

CV-JOINTS OVERHAUL

See Figure 57.

The Double Offset Joint (DOJ), is the only part of the assembly that can be replaced, if any of the other components are defective then the shaft should be replaced.

1. Straighten the bent claw of the large clamp at the Double Offset Joint (DOJ) end of the boot.
2. Loosen the band using pliers being careful not to damage the bolt.
3. Remove the small boot band from the DOJ using the same technique.
4. Remove the boot from the large end of the DOJ outer race.
5. Remove the round cir-clip using a suitable pry-tool from the neck of the joint outer race.
6. Remove the joint outer race.

CAUTION

The grease used is for CV–Joints, do not replace with another type of grease

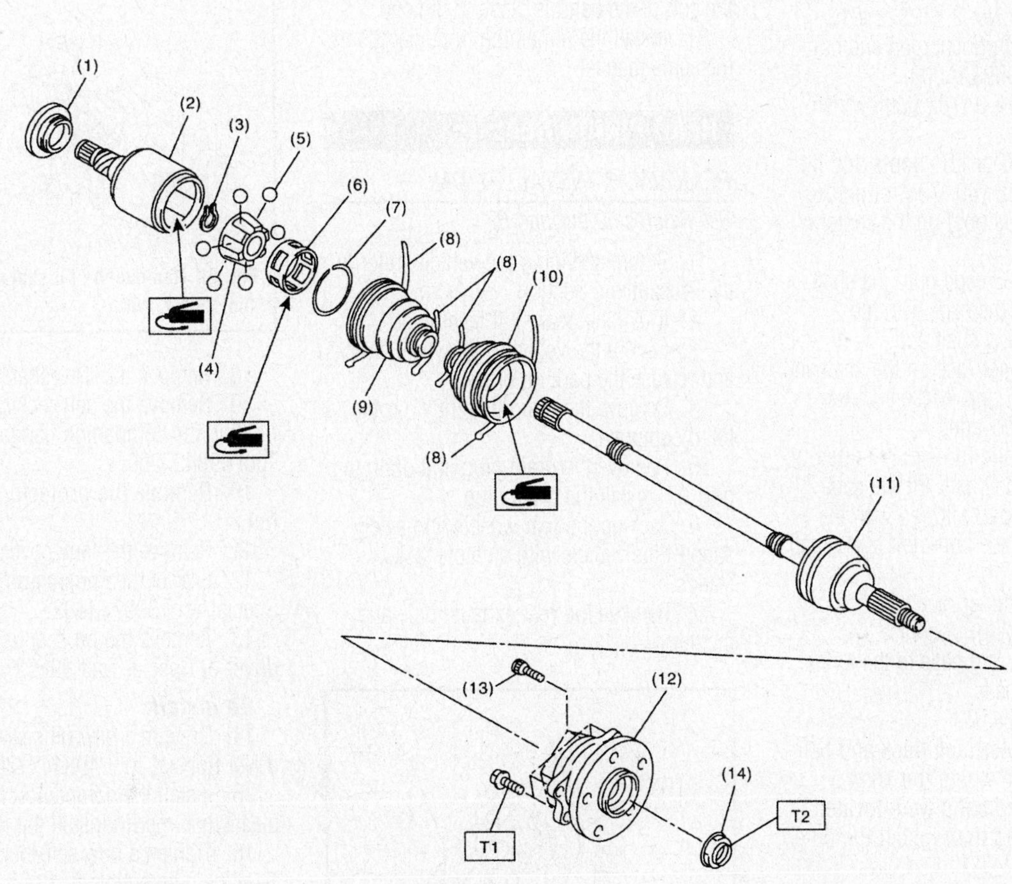

Fig. 59 Rear halfshaft and related components 2006–08 Legacy and Outback models

7. Clean the grease and remove the balls. Be careful not to lose any of the 6 balls.
8. Turn the cage by a half pitch to the track groove of the inner race and shift the cage.
9. Remove the snap ring that secures the inner race to the shaft and remove the inner race.
10. Take the cage from the shaft and remove the boot.
11. The other boots may be removed in the same manner as the DOJ boot.
12. Wrap the shaft splines with tape to prevent damage.

To install:

13. The following grease must be used during assembly:
   a. BJ side (Non-turbocharged engine): Molylex No. 2 #723223010.
   b. EBJ side (Turbocharged engine): NTG2218 No. 28093AA000.
14. DOJ side: VU–3A702 (Yellow) No. 23223GA050.
15. Install the BJ or EBJ boots and fill it with 2.12–2.47 oz. (60–70g) of grease.
16. Place the DOJ boot on the center of the shaft.
17. Insert the DOJ cage onto the shaft. Make sure to insert the cage with the cut–out portion facing shaft end.
18. Install the inner race on the shaft and fasten with the snap ring. Make sure the snap ring is firmly engaged.
19. Install the cage (previously fitted) to the inner race on the shaft. Fit the cage with the protruded part aligned with the track on the inner race and then turn a half pitch.
20. Fill the DOJ inner race with 2.82–3.17 oz. (80–90g) of grease.
21. Apply a coat of grease to the to the cage pocket and 6 balls.
22. Insert the 6 balls.
23. Align the outer race track and ball positions and place where the shaft, inner race, cage and balls were located prior to removal and then install the outer race.
24. Install the circlip into the groove on the outer race.

➡Make sure the balls, cage and inner race are fully seated. make sure not to place the matched position of the circlip in the ball groove of the outer race. Pull the shaft lightly to make sure the circlip is fully engaged.

25. Apply an even coat 0.71–1.06 oz. (20–30g) of grease to the entire inner surface of the boot and to the shaft.
26. Make sure the boot is free from any dirt or foreign materials prior to installation.
27. Place the outer race of the boot at the center of its travel.
28. Put a band through the boot clip and wind twice in alignment with the groove on the boot.
29. Pinch the end of the band using tool ST 9250910000 and tighten securely until it cannot be moved by hand. Make sure there is appropriate air inside the boot.
30. Tap on the clip with a punch to lock it making sure not to damage the damaged while tapping. Cut any excess off the band leaving 0.39 inch (10mm) and bend the remaining portion over the clip. Make sure the end of the band is close to the clip.
31. Install the remaining boot clamps in the same manner.

REAR PINION SEAL

REMOVAL & INSTALLATION

See Figures 60 through 62

1. Before servicing the vehicle, refer to the Precautions Section.
2. Move the gear shift level to "N".
3. Secure the vehicle with wheel chocks and release the parking brake.
4. Remove the rear differential protector, if equipped.
5. Remove the drain plug and drain the gear oil. Install the drain plug.
6. Jack up the rear wheels and safely support the vehicle with suitable jack stands.
7. Remove the rear exhaust pipe and muffler.

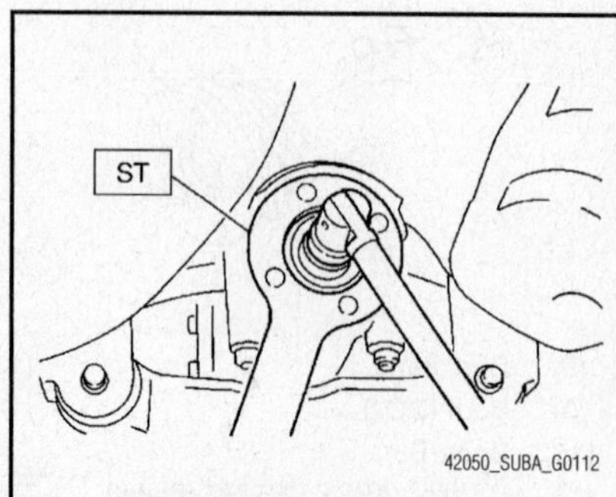

Fig. 60 Hold the companion flange with Special Tool 498427200 and remove the pinion nut.

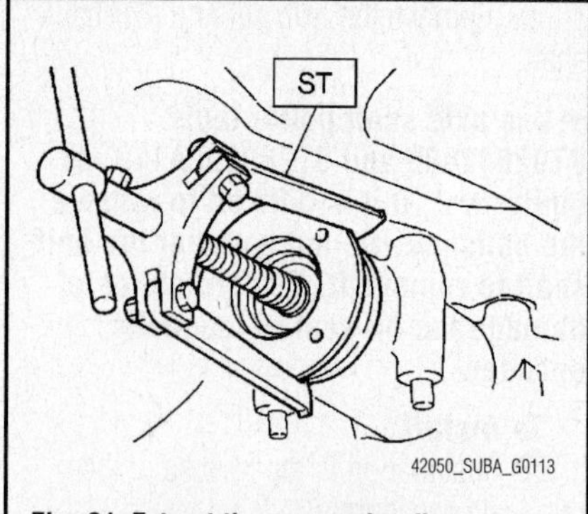

Fig. 61 Extract the companion flange using Special Tool 399703600

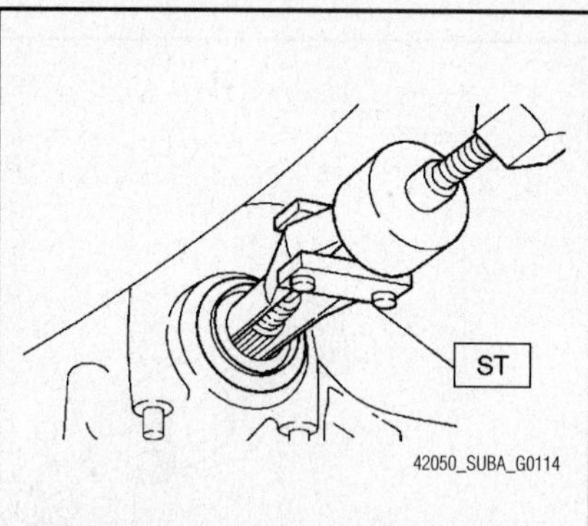

Fig. 62 Remove the oil seal using Special Tool 398527700

8. Remove the driveshaft.
9. Remove the self-locking nut while holding the companion flange with Special Tool 498427200.
10. Remove the protector mounting nut.
11. Remove the tank cover.
12. Extract the companion flange using Special Tool 399703602.
13. Remove the oil seal using a screw driver or Special Tool 398527700.

To install:

14. Drive in a new oil seal using a seal driver Special Tool 498447120.
15. Install the companion flange. Use a plastic hammer to install the flange.
16. Tighten a new self-locking nut for T—type differentials to 134 ft. lbs. (181 Nm). For VA—type tighten to 141 ft. lbs. (191 Nm). Use Special Tool 498427200 Flange Wrench to hold the companion flange.
17. The remainder of the installation is the reverse order of removal.
18. Refill the differential with fluid to the correct level.

ENGINE COOLING

ENGINE FAN

REMOVAL & INSTALLATION

2.5L Engine

1. Before servicing the vehicle, refer to the Precautions Section.
2. Disconnect the negative battery cable.
3. Raise and safely support the vehicle.
4. Disconnect the fan electrical connector.
5. Remove the transmission cooler hose from the clip of the radiator fan motor assembly, if equipped.
6. Lower the vehicle.
7. Disconnect the overflow hose and reservoir tank.
8. Remove the fan shroud mounting bolts.
9. Remove the main fan assembly.

To install:

10. Installation is the reverse order of assembly.
11. Tighten the fan shroud mounting bolts to 6 ft. lbs. (7.5 Nm).

3.0L Engine

See Figure 63.

1. Before servicing the vehicle, refer to the Precautions Section.
2. Disconnect the negative battery cable.
3. Remove the hood stay holder.
4. Remove the air intake assembly.
5. Disconnect the connector from the radiator fan control unit.
6. Raise and safely support the vehicle.
7. Remove the engine undercover.
8. Drain the engine coolant.
9. Remove the transmission cooler hose from the clip of the radiator fan motor assembly, if equipped.
10. Remove the radiator fan motor harness from the clip.
11. Lower the vehicle.
12. Remove the reservoir tank.
13. Remove the radiator sub-fan assembly.
14. Remove the radiator fan assembly.

➡When removing the main fan assembly, lifting it straight up will cause the main fan shroud contacts to inlet part of engine coolant. To avoid contacting it, move the main fan assembly to sub fan assembly side before removal.

To install:

15. Installation is the reverse order of assembly.

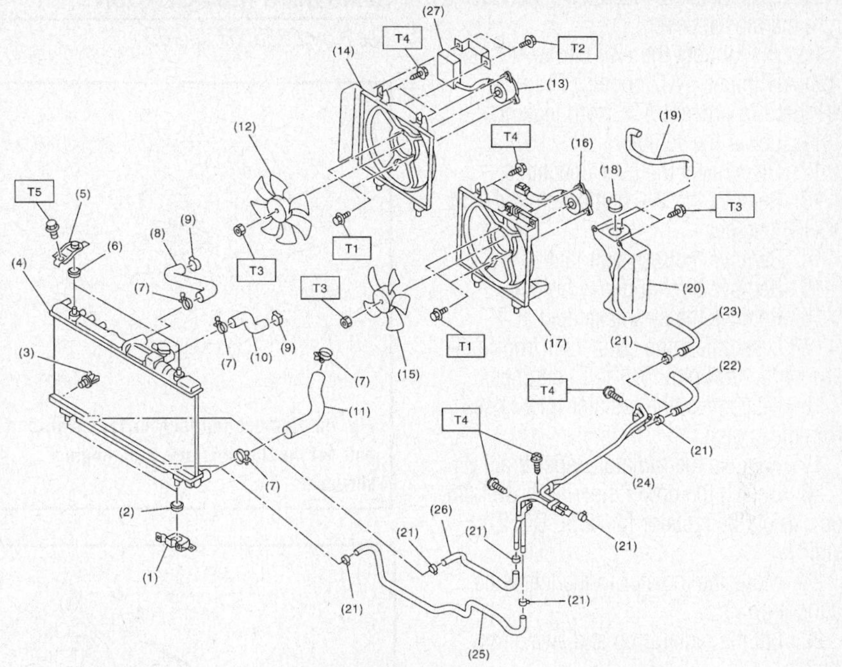

(1)	Radiator lower bracket	(13)	Radiator sub fan motor	(24)	ATF pipe
(2)	Radiator lower cushion	(14)	Radiator sub fan shroud	(25)	ATF hose C
(3)	Engine coolant drain plug	(15)	Radiator main fan	(26)	ATF hose D
(4)	Radiator	(16)	Radiator main fan motor	(27)	Radiator fan control unit
(5)	Radiator upper bracket	(17)	Radiator main fan shroud		
(6)	Radiator upper cushion	(18)	Engine coolant reservoir tank cap		
(7)	Clamp	(19)	Over flow hose		
(8)	Radiator hose A	(20)	Engine coolant reservoir tank		
(9)	Clamp	(21)	ATF hose clamp		
(10)	Radiator hose B	(22)	ATF hose A		
(11)	Radiator hose C	(23)	ATF hose B		
(12)	Radiator sub fan				

Tightening torque:N·m (kgf-m, ft-lb)

T1:	***3.8 (0.39, 2.8)***
T2:	***5.4 (0.55, 4.0)***
T3:	***6.2 (0.63, 4.6)***
T4:	***7.5 (0.76, 5.5)***
T5:	***12 (1.2, 8.9)***

22140_SUBA_G0100

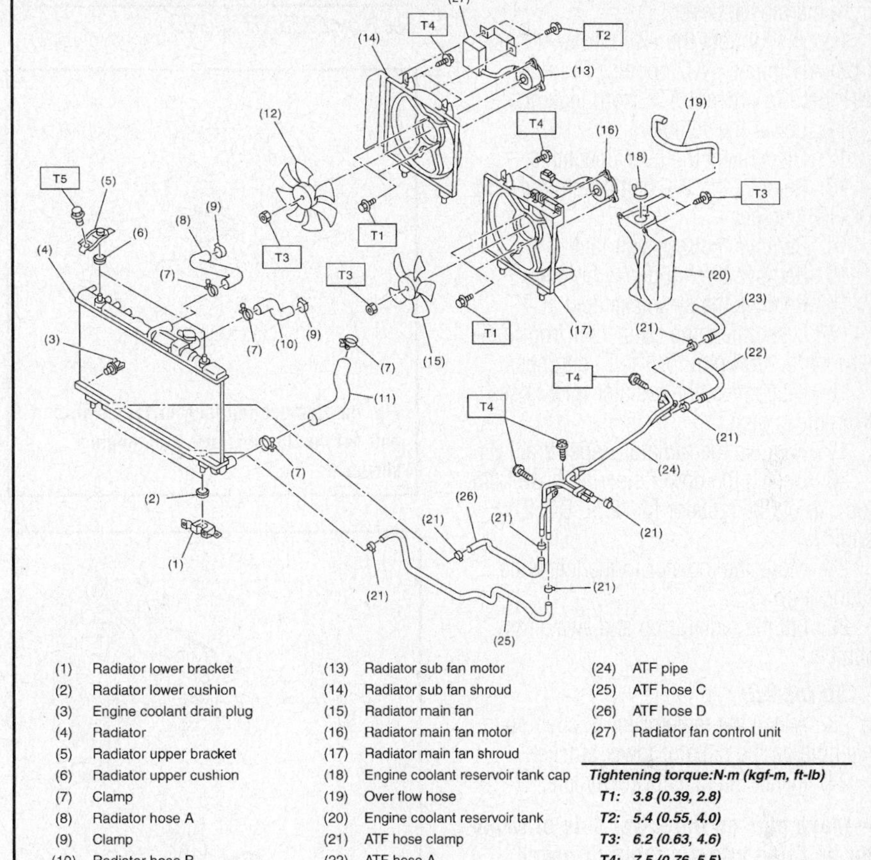

Fig. 63 Exploded view of radiator and fan system

16. Tighten the fan shroud mounting bolts to 6 ft. lbs. (7.5 Nm).
17. Refill the engine cooling system to the correct level.

RADIATOR

REMOVAL & INSTALLATION

See Figures 64 and 65

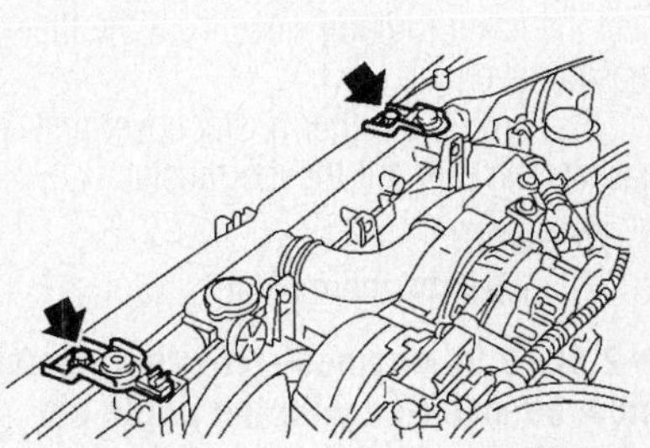

22140_SUBA_G0101

Fig. 64 Radiator upper bracket and mounting bolts shown

1. Before servicing the vehicle, refer to the Precautions Section.
2. Set the vehicle on a lift.
3. Disconnect the ground cable from battery.
4. Lift-up the vehicle.
5. Remove the undercover.
6. Drain engine coolant completely.
7. Remove the radiator undercover.

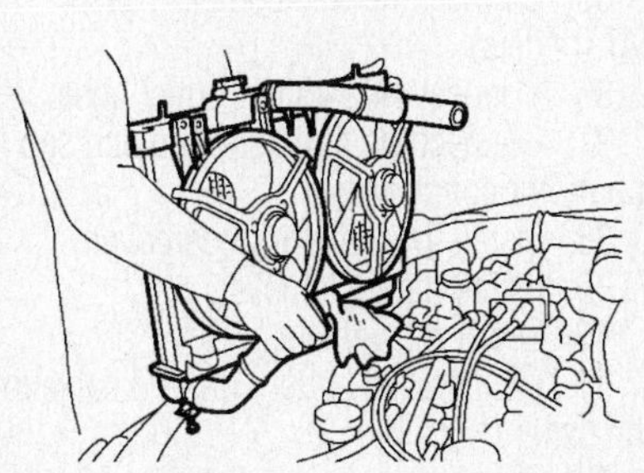

22140_SUBA_G0102

Fig. 65 Radiator removal shown

8. Disconnect the radiator main and sub fan motor connectors.
9. Disconnect the radiator outlet hose from thermostat cover.
10. Disconnect the ATF cooler hoses from ATF pipes. (AT models) Plug the ATF pipe to prevent ATF from leaking.
11. Lower the vehicle.
12. Disconnect the over flow hose.
13. Remove the V—belt covers for (2.5L DOHC) engines.
14. Remove the reservoir tank.
15. Remove the hood stay holder.
16. Remove the air intake duct.
17. Disconnect the connector from radiator fan control unit for (3.0L) engines.
18. Disconnect the radiator inlet hoses from the engine.
19. Remove the radiator upper brackets.
20. Detach the power steering hose from the clip on the radiator for (2.5L DOHC) engines.
21. Move the radiator to the left while lifting it upward.
22. Lift the radiator up and away from vehicle.

To install:

23. Attach the radiator lower cushion to the hole on the radiator lower bracket.
24. Install the radiator to vehicle.

➡Make pins on the lower side of radiator be fitted into the radiator lower cushions on body side.

25. Install the radiator upper brackets and tighten the bolts 9 ft. lbs. (12 Nm).
26. Attach the power steering hose to the radiator for (2.5L DOHC) engines.
27. Connect the radiator inlet hose.
28. Install the air intake duct.
29. Connect the connector from radiator fan control unit for (3.0L) engines.
30. Install the hood stay holder.
31. Install the reservoir tank.
32. Connect the over flow hose.
33. Install the V—belt covers for (2.5L DOHC) engines.
34. Lift-up the vehicle.
35. Connect the ATF cooler hoses. (AT models)
36. Connect the radiator outlet hose.
37. Connect the radiator main and sub fan motor connectors.
38. Install the radiator undercover.
39. Install the undercover.
40. Lower the vehicle.
41. Connect the battery ground cable to the battery.
42. Fill engine coolant and bleed the cooling system.
43. Check the ATF level.

THERMOSTAT

REMOVAL & INSTALLATION

See Figures 66, 67 and 68.

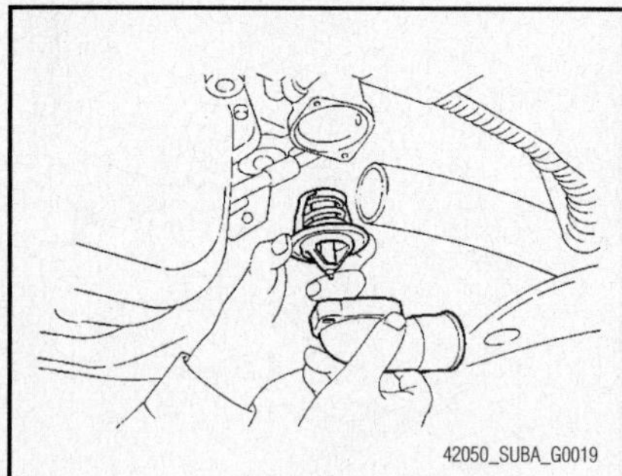

Fig. 66 Remove the thermostat cover, and pull out the thermostat—2.5L engine shown

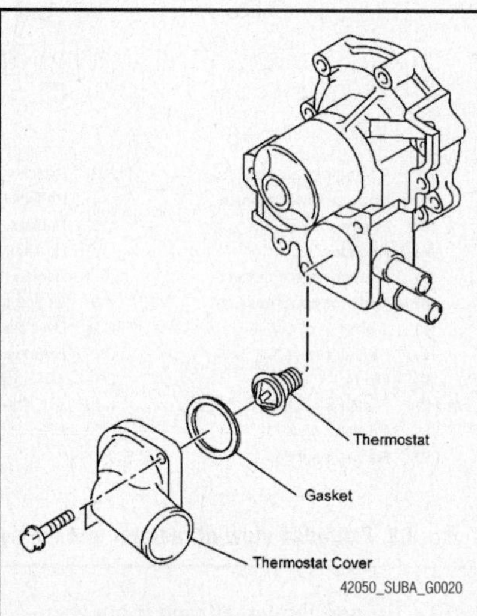

Fig. 67 Exploded view of the thermostat and related components—2.5L Turbo engine shown

1. Before servicing the vehicle, refer to the Precautions Section.
2. Raise and safely support the vehicle.
3. Remove the engine undercover.
4. Drain the engine coolant.
5. Loosen the hose clamp and disconnect the lower radiator hose from the thermostat cover.
6. Remove the thermostat cover and gasket, and pull out the thermostat.

To install:

7. Install the thermostat.

➡2.5L, 3.0L engines—The thermostat must be installed with the jiggle pin upward.

➡2.5L Turbo engines—The thermostat must be install with the jiggle pin facing front.

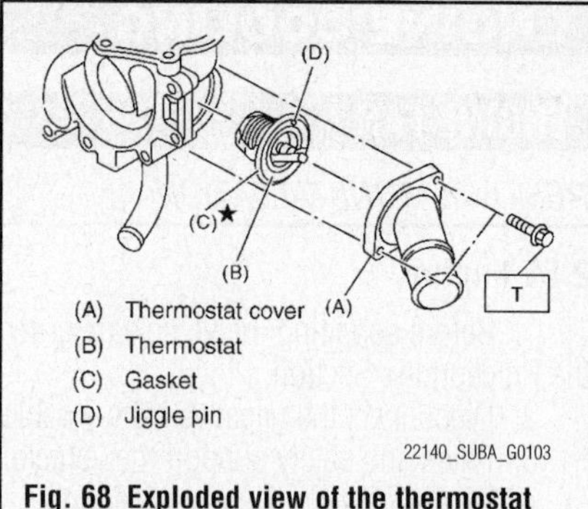

Fig. 68 Exploded view of the thermostat and related components—3.0L engine shown

8. Install the thermostat cover together with a new gasket. For 2.5L engines tighten to 9 ft. lbs. (12 Nm). For 3.0L engines tighten to 5 ft. lbs. (6.4 Nm).
9. Install the lower radiator hose to the thermostat cover and tighten the hose clamp.
10. Install the engine undercover.
11. Lower the vehicle.
12. Refill the engine cooling system to the correct level.
13. Start the engine and check for leaks.

WATER PUMP

REMOVAL & INSTALLATION

2.5L Engine

See Figures 69 and 70.

1. Before servicing the vehicle, refer to the Precautions Section.
2. Remove or disconnect the following:
   - Negative battery cable
   - Engine undercover, if equipped
3. Drain the coolant into a suitable container.
4. Remove or disconnect the following:
   - Radiator fan connectors
   - Radiator outlet and heater hoses
   - Heater bypass hose or overflow hose, if equipped
   - Reservoir tank, on Legacy
   - Radiator fan motor assembly
   - Accessory drive belts
   - Timing belt
   - Belt tension adjuster
   - Belt idler No. 2
   - Camshaft Position (CMP) sensor
   - Left side camshaft pulley
   - Left side rear timing belt cover
   - Tensioner bracket
   - Radiator and heater hoses from water pump
   - Water pump retainer bolts
   - Water pump

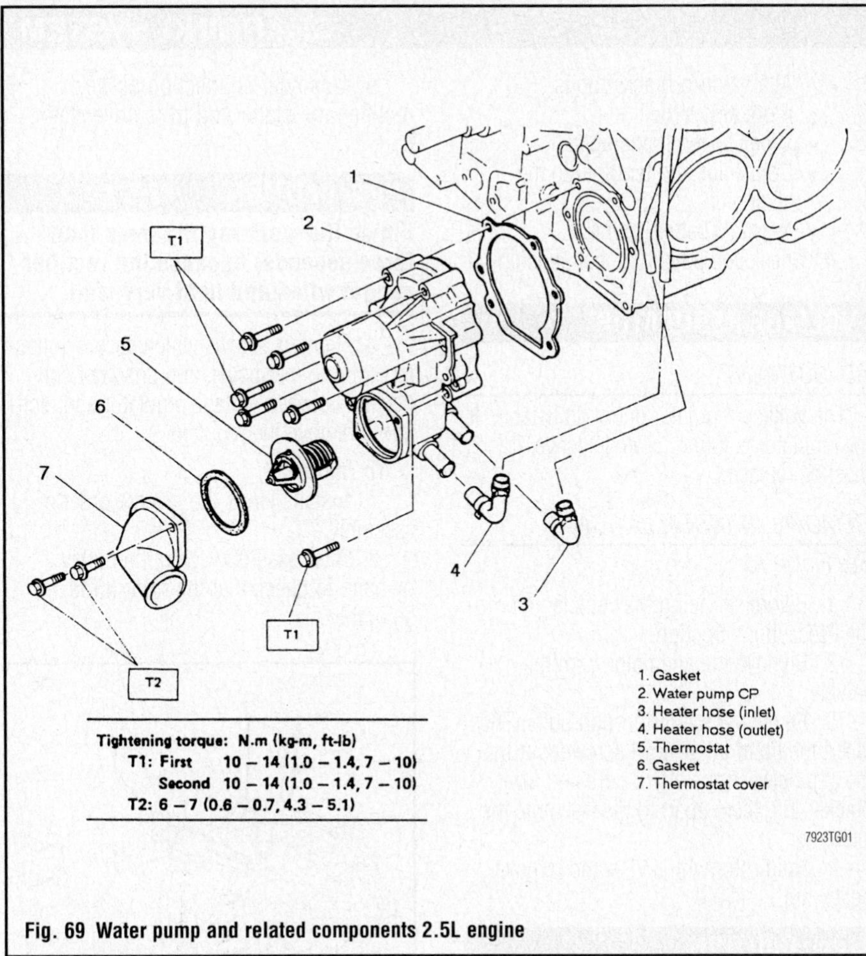

Fig. 69 Water pump and related components 2.5L engine

5. Inspect the radiator hoses for deterioration and replace as necessary.

To install:

6. Clean the gasket mating surfaces thoroughly. Always use new gaskets during installation.
7. Install or connect the following:
8. Tighten the pump bolts in sequence to 8.9 ft. lbs. (12 Nm). After tightening the bolts once, retighten to the same specification again.
   - Radiator heater hoses to water pump
   - Tensioner bracket and tighten to 18 ft. lbs. (25 Nm)
   - Left side rear timing belt cover
   - Left side camshaft pulley(s). Tighten to 58 ft. lbs. (78 Nm) on non turbocharged engine and 72 ft. lbs. (98 Nm) on turbocharged engine.
   - CMP sensor
   - Belt idler No. 2 and tighten to 29 ft. lbs. (39 Nm)
   - Belt tension adjuster
   - Timing belt
   - Accessory drive belts
   - Radiator fan assembly
   - Reservoir tank, if removed
   - Heater bypass hose or overflow hose, if equipped
   - Air intake duct
   - Radiator outlet and heater hoses
   - Radiator fan connectors
   - Engine undercover, if removed
9. Fill the system with coolant and connect the negative battery cable.

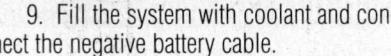

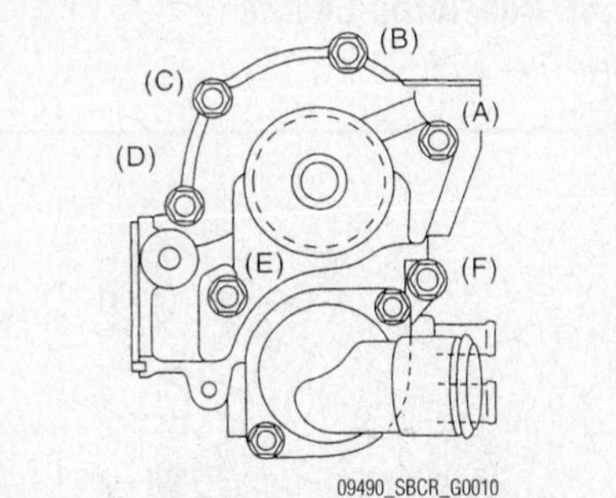

Fig. 70 Tighten the water pump bolts in two steps using the following sequence—2.5L engine

10. Start the engine and allow it to reach operating temperature.
11. Check for leaks.

3.0L Engine

See Figure 71.

1. Before servicing the vehicle, refer to the Precautions Section.
2. Remove or disconnect the following:
   - Negative battery cable
   - Engine undercover, if equipped
3. Drain the coolant into a suitable container.
4. Remove or disconnect the following:
   - Radiator
   - Accessory drive belts
   - Timing chain
   - Water pump retainer bolts
   - Water pump
5. Inspect the radiator hoses for deterioration and replace as necessary.

To install:

6. Clean the gasket mating surfaces thoroughly. Always use new gaskets during installation.
7. Apply coolant to the new O-ring before installation
8. Install or connect the following:
   - Water pump with a new O-ring, tighten the bolts to 5 ft. lbs. (7 Nm).
   - Timing chain
   - Front chain cover
   - Accessory drive belts
   - Radiator
   - Engine undercover, if removed
9. Fill the system with coolant and connect the negative battery cable.
10. Start the engine and allow it to reach operating temperature.
11. Check for leaks.

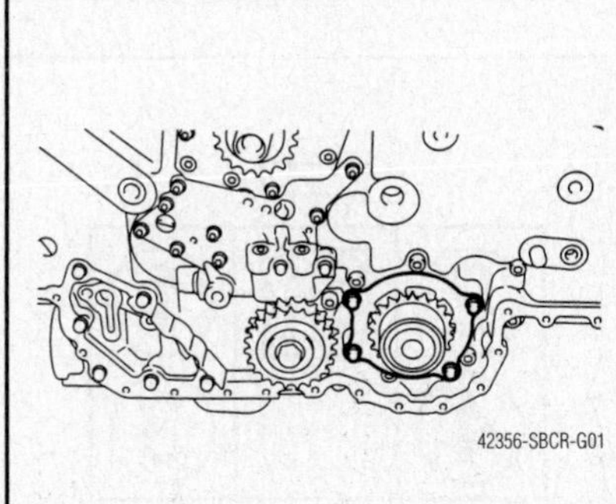

Fig. 71 Water pump location—3.0L engine

ENGINE ELECTRICAL — CHARGING SYSTEM

ALTERNATOR

REMOVAL & INSTALLATION

See Figure 72.

1. Before servicing the vehicle, refer to the Precautions Section.
2. Remove or disconnect the following:
   - Negative battery cable
   - Connector and terminal from the alternator
   - V-belt cover, if equipped
   - Front side V-belt
   - Alternator to bracket bolts
   - Alternator from the vehicle

To install:

3. Install or connect the following:
   - Alternator into the vehicle
   - Alternator to bracket bolts
   - Front side V-belt
   - V-belt cover, if equipped
   - Connector and terminal to the alternator
   - Negative battery cable
4. Check and adjust the belt tension.

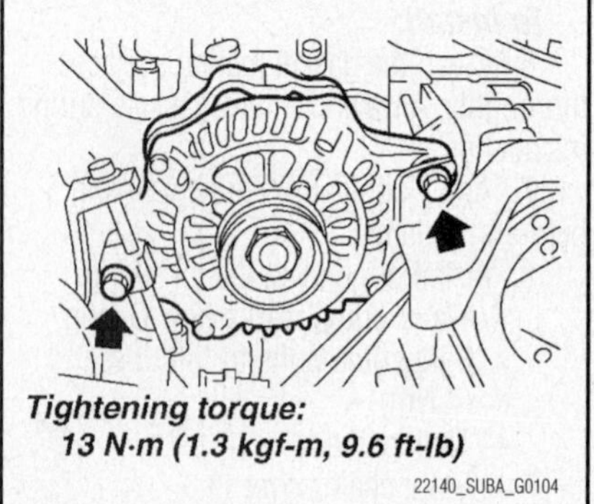

Fig. 72 Alternator and mounting bolts

VOLTAGE REGULATOR

ADJUSTMENT

The voltage regulator is not adjustable. If the regulator is found to be defective, it must be replaced.

REMOVAL & INSTALLATION

See Figure 73.

1. Before servicing the vehicle, refer to the Precautions Section.
2. Remove the alternator from the vehicle.
3. Remove the four through bolts. Then insert the tip of a flat-head screwdriver into the gap between the stator core and front bracket. Pry them apart to disassemble the alternator.
4. Hold rotor with a vise and remove pulley nut.

⁂ CAUTION

When holding rotor with vise, insert aluminum plates or wood pieces on the contact surfaces of the vise to prevent rotor from damage.

5. Unsolder connection between rectifier and stator coil to remove stator coil.

⁂ CAUTION

Finish the work rapidly (less than three seconds) because the rectifier cannot withstand heat very well.

6. Remove screws which secure voltage regulator to rear cover, and unsolder connection between voltage regulator and rectifier to remove the regulator.

To install:

7. Installation is the reverse order of assembly.
8. After assembly, turn the pulley by hand to check that the rotor turns smoothly.

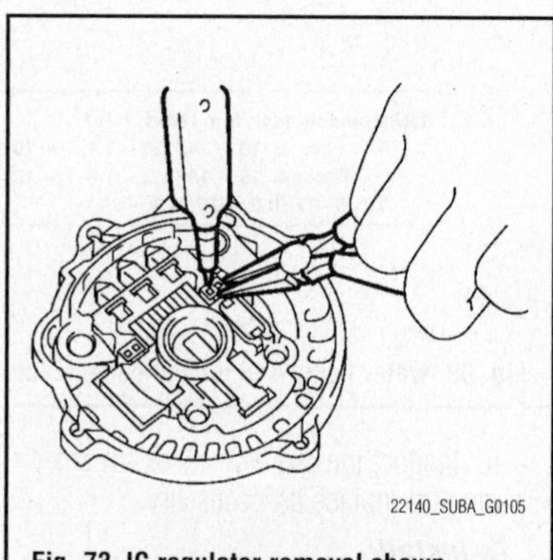

Fig. 73 IC regulator removal shown

ENGINE ELECTRICAL — DISTRIBUTORLESS IGNITION SYSTEM

FIRING ORDER

See Figure 74.

3.0L Engine Firing Order: 1–6–3–2–5–4

Fig. 74 2.5L Engine Firing order: 1–3–2–4 Distributorless ignition system

IGNITION COIL PACK

REMOVAL & INSTALLATION

2.5L Non-Turbo Engine

See Figure 75.

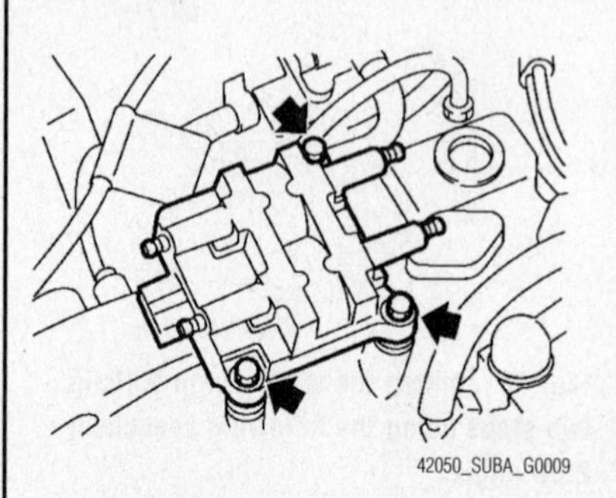

Fig. 75 Location of the ignition coil pack mounting bolts—2.5L Non-Turbo engine

1. Before servicing the vehicle, refer to the Precautions Section.
2. Disconnect the negative battery cable.
3. Disconnect the spark plug cables from the ignition coil pack.
4. Disconnect the electrical connector from the ignition coil pack.
5. Remove the ignition coil pack assembly.

To install:

6. Installation is the reverse order of removal.
7. Tighten the ignition coil pack mounting bolts to 5 ft. lbs. (6 Nm).

IGNITION COIL

REMOVAL & INSTALLATION

Right Side

See Figure 76.

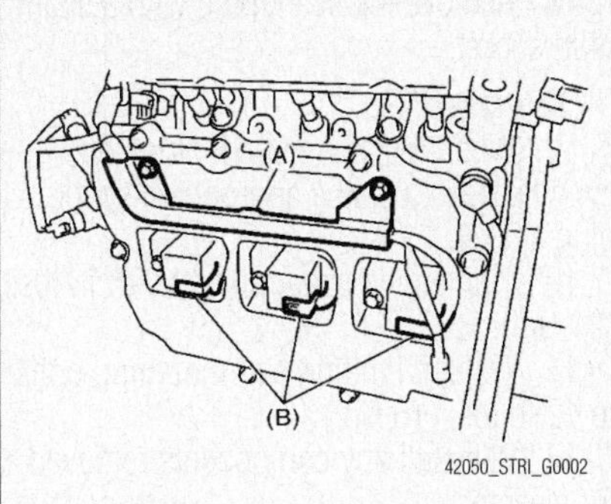

Fig. 76 Removing bracket and connector from ignition coil—right side

1. Before servicing the vehicle, refer to the Precautions Section.
2. Remove the collector cover.
3. Disconnect the ground cable from battery.
4. Remove the air cleaner case.
5. Remove the (A) bracket.
6. Disconnect the (B) connector from ignition coil.
7. Remove the ignition coil.

➡Turn the #5 ignition coil to remove it.

To install:

To install reverse the removal procedures

- Tighten the ignition coil to 12 ft. lbs. (16 Nm)

Left Side

See Figure 77.

1. Remove the collector cover.
2. Remove the battery and battery carrier.
3. Remove the (A) bracket.
4. Disconnect the (B) connector from ignition coil.
5. Remove the ignition coil.

➡Turn the no. 6 ignition coil to remove it.

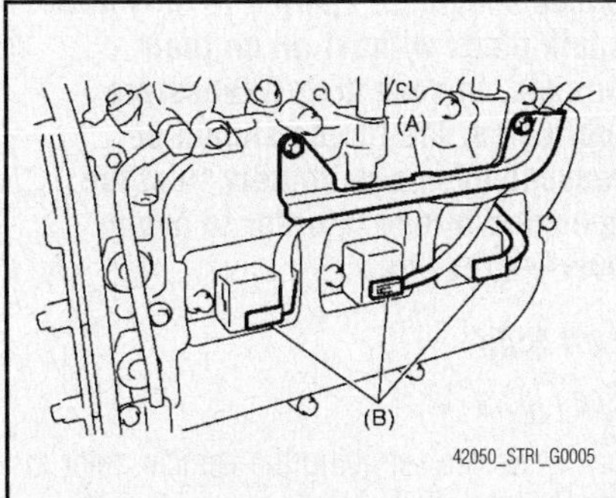

Fig. 77 Removing bracket and connector from the ignition coil—left side

To install:

6. To install reverse the removal procedures
   - Tighten the ignition coil to 12 ft. lbs. (16 Nm)

IGNITION TIMING

INSPECTION

2.5L and 3.0L Engines

See Figure 78.

METHOD WITH SUBARU SELECT MONITOR

Before checking the ignition timing, check the following:

- Ensure the air cleaner element is free from clogging, spark plugs are in good condition, and that hoses are connected properly.
- Ensure the malfunction indicator light does not illuminate.

1. Idle the engine.
2. Stop the engine, and turn the ignition switch to OFF.
3. Insert the cartridge to the Subaru Select Monitor.
4. Connect the Subaru Select Monitor to data link connector.
5. Turn the ignition switch to ON, and Subaru Select Monitor power switch to ON.
6. Select Each System Check in the Main Menu.
7. Select Engine in the Selection Menu.
8. Select Current Data Display & Save in the Engine Control System Diagnosis.
9. Select Data Display in the Data Display Menu.
10. Start the engine, and read the ignition timing by idle speed.
11. If the timing is not correct, check the ignition control system.
12. Ignition timing is as follows:
    - 10° +/- 8°650 rpm (MT Models)
    - 15° +/- 8°700 rpm (AT Models)
    - 17° +/- 10°700 rpm (STI Model)

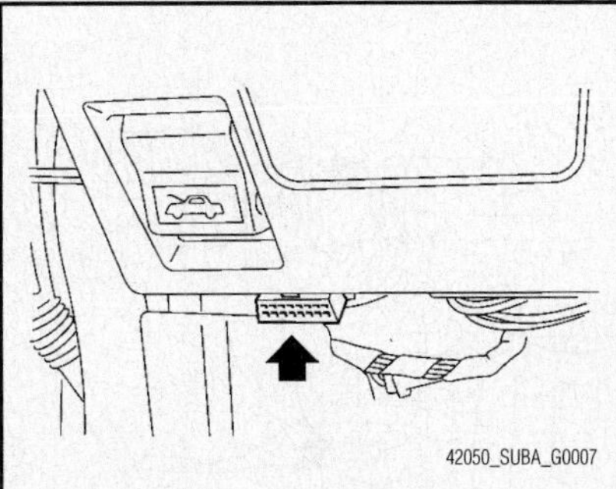

Fig. 78 Location of the data link connector under the driver's side instrument panel.

METHOD WITH TIMING LIGHT

1. Before checking the ignition timing, check the following item:
2. Check the air cleaner element is free from clogging, spark plugs are in good condition, and hoses are connected properly.
3. Check the malfunction indicator light does not illuminate.
4. Warm-up the engine.
5. Stop the engine, and turn the ignition switch to OFF.
6. Remove the air intake duct.
7. Disconnect the connectors of the mass air flow and intake air temperature sensor.
8. Remove the air cleaner case and element.
9. Connect the timing light to the power wire of #1 ignition coil.
10. Attach the air cleaner case, element and connector of mass air flow and intake air temperature sensor.
11. Start the engine, turn the timing light to the crank pulley, and check the ignition timing by means of crank pulley indicator.
12. If the timing is not correct, check the ignition control system.
13. Ignition timing is as follows:
    - 10° +/- 8°650 rpm (MT Models)
    - 15° +/- 8°700 rpm (AT Models)
    - 17° +/- 10°700 rpm (STI Model)

ADJUSTMENT

The ignition timing is controlled by the Powertrain Control Module (PCM). No adjustment is necessary or possible.

SPARK PLUGS

REMOVAL & INSTALLATION

2.5L SOHC Engines

See Figures 79 and 80

1. Before servicing the vehicle, refer to the Precautions Section.
2. Disconnect the negative battery cable.
3. Remove the following components if necessary for access to remove the spark plugs:
   - Battery
   - Air intake assembly
   - Windshield washer tank (Baja Only)

⁂ CAUTION

Do not disconnect the washer tank hoses.

4. Gripping the spark plug wire by the boot, remove the spark plug wires.
5. Remove the spark plugs.

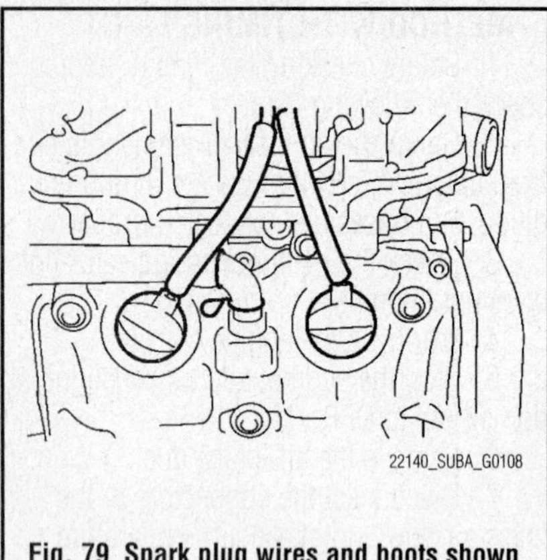

Fig. 79 Spark plug wires and boots shown

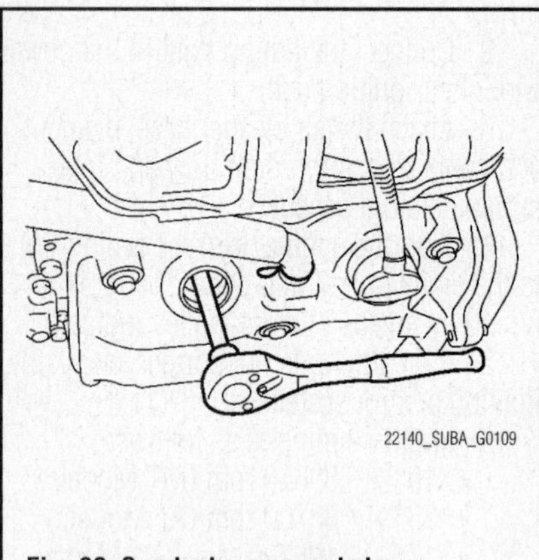

Fig. 80 Spark plug removal shown

To install:

6. When installing spark plugs on cylinder head, use the appropriate spark plug socket. Tighten the spark plugs to 15 ft. lbs. (21 Nm).
7. Connect the spark plug wires.
8. Reinstall any components removed for clearance.
9. Connect the negative battery cable.

2.5L DOHC Turbo Engines

Right Side

See Figures 81 and 80.

1. Before servicing the vehicle, refer to the Precautions Section.
2. Remove the collector cover.
3. Disconnect the ground cable from the battery.
4. Remove the air cleaner case.
5. Disconnect the connector from ignition coil.
6. Remove the ignition coil.

➡Turn number 3 ignition coil by 180° to remove it.

7. Remove the spark plug with a spark plug socket.

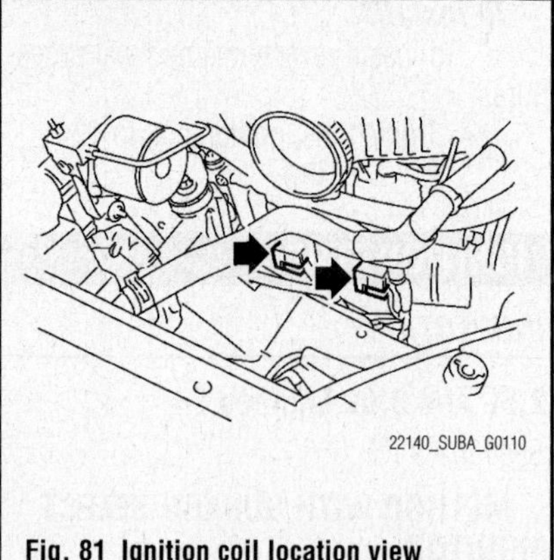

Fig. 81 Ignition coil location view

To install:

8. When installing spark plugs on cylinder head, use the appropriate spark plug socket.
9. Tighten the spark plugs to 15 ft. lbs. (21 Nm).
10. Tighten ignition coil mounting bolts to 12 ft. lbs. (16 Nm).
11. Reinstall any components removed for clearance.
12. Connect the negative battery cable.

Left Side

See Figures 82 and 80

1. Before servicing the vehicle, refer to the Precautions Section.
2. Remove the collector cover.
3. Remove the battery and battery carrier.
4. Disconnect the secondary air pump duct from the secondary air pump.
5. Remove the bolts that attach the secondary air pump duct to the rocker cover (LH), and raise the secondary air pump duct.
6. Disconnect the connector from ignition coil.
7. Remove the ignition coil.

➡Turn number 4 ignition coil by 180° to remove it.

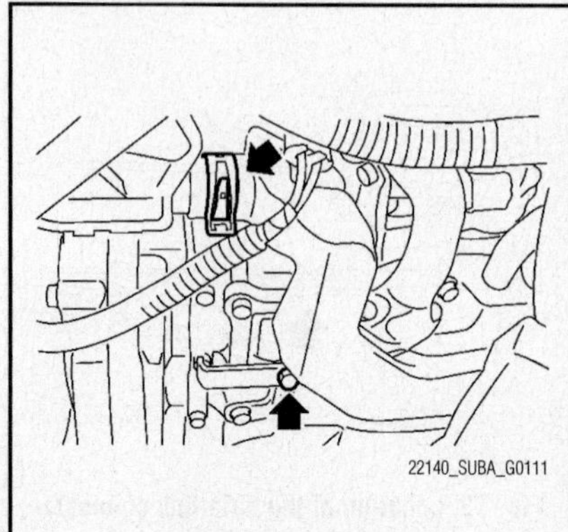

Fig. 82 Secondary air pump duct view

8. Remove the spark plug with a spark plug socket.

To install:

9. When installing spark plugs on cylinder head, use the appropriate spark plug socket.
10. Tighten the spark plugs to 15 ft. lbs. (21 Nm).
11. Tighten ignition coil mounting bolts to 12 ft. lbs. (16 Nm).
12. Reinstall any components removed for clearance.
13. Connect the negative battery cable.

3.0L Engine

Right Side

See Figure 76.

⁂ CAUTION

All spark plugs installed on an engine must be of the same heat range.

1. Before servicing the vehicle, refer to the Precautions Section.
2. Remove the collector cover.
3. Disconnect the ground cable from battery.
4. Remove the air cleaner case.
5. Remove the (A) bracket.
6. Disconnect the (B) connector from ignition coil.
7. Remove the ignition coil.

➡Turn the #5 ignition coil to remove it.

8. Remove the spark plug with a spark plug socket.

To install:

9. To install reverse the removal procedures, and note the following:
   - Tighten the spark plug to 15 ft. lbs. (21 Nm)
   - Tighten the ignition coil to 12 ft. lbs. (16 Nm)

➡The tightening torque described above should be applied to only new spark plugs without oil on their threads. In case their threads are lubricated, the torque should be reduced by approximately ⅓ of the specified torque in order to avoid over—stressing.

Left Side

See Figure 77.

1. Before servicing the vehicle, refer to the Precautions Section.
2. Remove the collector cover.
3. Remove the battery and battery carrier.
4. Remove the (A) bracket.

5. Disconnect the (B) connector from ignition coil.
6. Remove the ignition coil.

➡**Turn the no. 6 ignition coil to remove it.**

7. Remove the spark plug with a spark plug socket.

To install:

8. To install reverse the removal procedures, and note the following:
   - Tighten the spark plug to 15 ft. lbs. (21 Nm)
   - Tighten the ignition coil to 12 ft. lbs. (16 Nm)

➡**The tightening torque described above should be applied to only new spark plugs without oil on their threads. In case their threads are lubricated, the torque should be reduced by approximately ⅓ of the specified torque in order to avoid over—stressing.**

ENGINE ELECTRICAL

STARTING SYSTEM

STARTER

REMOVAL & INSTALLATION

1. Before servicing the vehicle, refer to the Precautions Section.
2. Disconnect the negative battery cable.
3. Remove the air intake chamber, on non turbocharged engine.
4. Remove the intercooler, on turbocharged engine.
5. Remove the air intake chamber stay, on non turbocharged engine.
6. Disconnect the electrical connectors from the starter.
7. Remove the starter retaining bolts. Remove the starter from the vehicle.

To install:

8. Installation is the reverse of the removal procedure.
9. Torque the starter retaining bolts to 37 ft. lbs. (50 Nm).

SOLENOID OR RELAY REPLACEMENT

1. Before servicing the vehicle, refer to the Precautions Section.
2. Remove the starter from the vehicle.
3. Disconnect the lead wire from the solenoid switch.
4 Remove the through-bolts from the end frame.
5. Remove the yoke from the solenoid switch.
6. Remove the mounting screws that hold the housing to the solenoid switch.
7. Separate the solenoid switch from the housing.
8. Installation is the reverse order of removal.

ENGINE MECHANICAL

ACCESSORY DRIVE BELTS

ACCESSORY BELT ROUTING

See Figures 83 and 84.

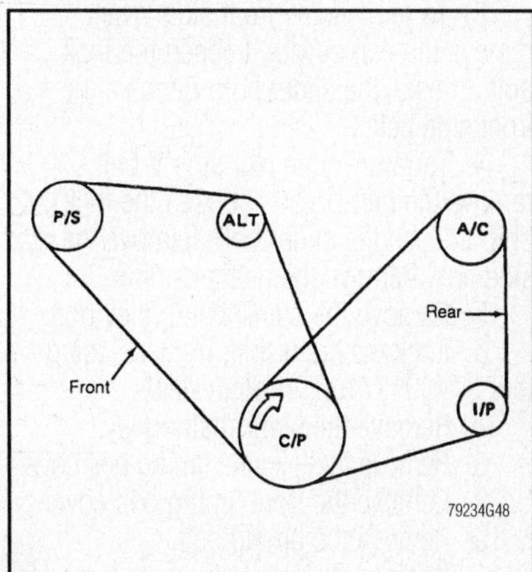

Fig. 83 Accessory drive belt routing—2.5L engines

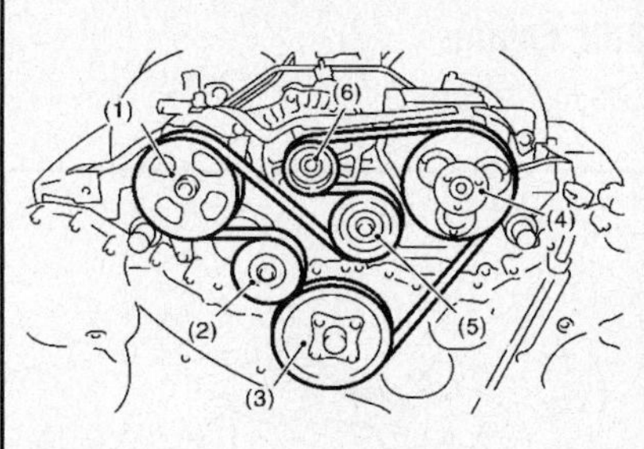

Fig. 84 Accessory drive belt routing—3.0L engine

INSPECTION

Inspect the drive belt for signs of glazing or cracking. A glazed belt will be perfectly smooth from slippage, while a good belt will have a slight texture of fabric visible. Cracks will usually start at the inner edge of the belt and run outward. All worn or damaged drive belts should be replaced immediately.

ADJUSTMENT

2.5L Engines

See Figures 85 and 86

➡**The belt tension may be checked with or without a belt tension gauge.**

1. If using a belt tension gauge, the tension should be as follows:
   a. Used front belt: 110.2–143.9 lbs. (490–640 N)

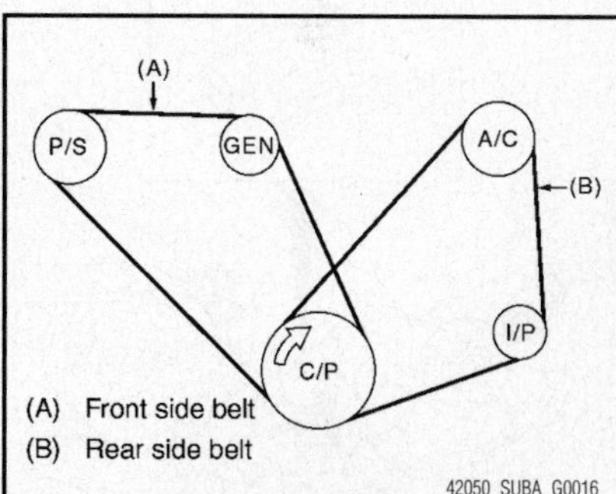

Fig. 85 Location to check the belt tension if you are using a belt tension gauge—2.5L engine

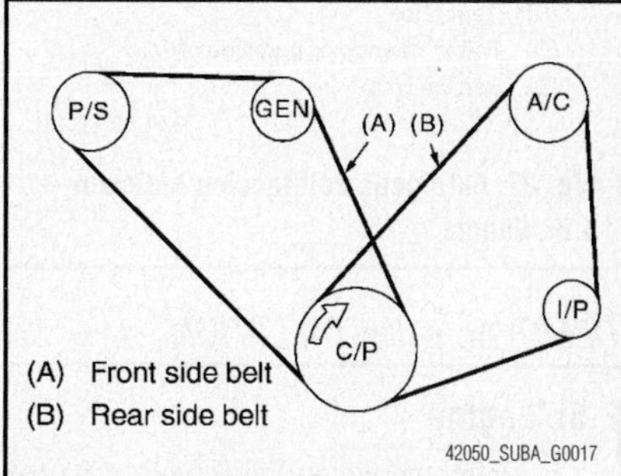

Fig. 86 Location to check to the belt tension if you are not using a belt tension gauge—2.5L engine

b. New front belt: 143–175 lbs. (640–780 N)

c. Used rear belt: 78.7–101.2 lbs. (350–450 N)

d. New rear belt: 166–198 lbs. (740–880 N)

2. If you are not using a belt tension gauge, using 22 lbs. (98 Nm) of force, push on the belt as shown in the illustration. The total distance the belt travels up and down should be:

a. Used front belt: 0.354–0.433 in. (9–11 mm)

b. New front belt: 0.276–0.354 in. (7–9 mm)

c. Used rear belt: 0.354–0.394 in. (9.0–10.0 mm)

d. New rear belt: 0.295–0.335 in. (7.5–8.5 mm)

3.0L Engine

See Figure 87.

1. With the belt installed, check that the automatic belt tension indicator is within the range.
2. Replace the belt if the indicator is outside of the service limit.

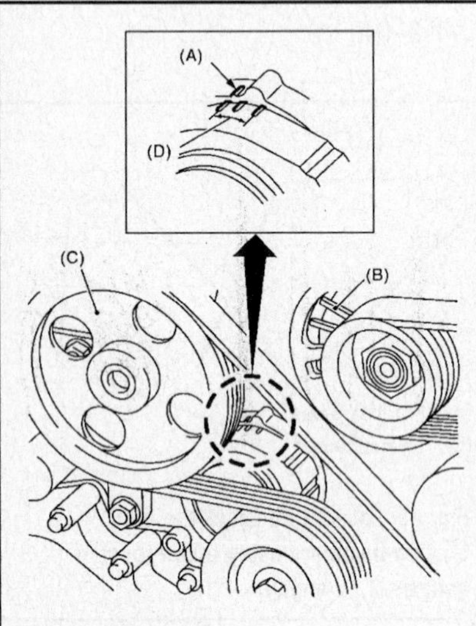

Fig. 87 Automatic belt tension indicator—3.0L Engine

REMOVAL & INSTALLATION

2.5L Engine

1. Before servicing the vehicle, refer to the Precautions Section.
2. Remove the accessory drive belt cover, if equipped.
3. Loosen the lock bolt (bottom) of the front belt tensioner.
4. Loosen the slider bolt (top) of the front belt tensioner.
5. Remove the front side belt.
6. Loosen the lock bolt (bottom) of the rear belt tensioner.
7. Loosen the slider bolt (top) of the rear belt tensioner.
8. Remove the rear side belt.

To install:

➡Wipe off any oil or water on the belt and pulley.

9. Install the rear side belt, then tighten the slider bolt until the belt reaches the specified tension. Please refer to the following topic: “Accessory Drive Belts”, “Adjustment.”
10. Tighten the lock nut of the rear belt tensioner to 18 ft. lbs. (25 Nm).
11. Install the front side belt, then tighten the slider bolt until the belt reaches the specified tension. Please refer to the following topic: “Accessory Drive Belts”, “Adjustment.”
12. Tighten the lock nut of the front belt tensioner to 17 ft. lbs. (23 Nm).

3.0L Engine

See Figure 88.

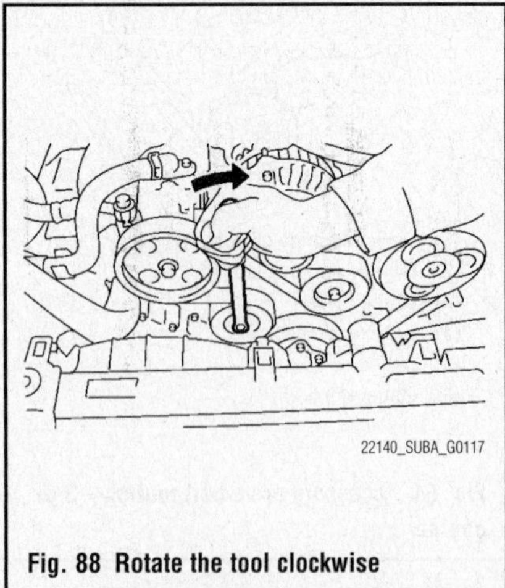

Fig. 88 Rotate the tool clockwise

1. Before servicing the vehicle, refer to the Precautions Section.
2. Remove the collector cover.
3. Install the tool to belt tension adjuster assembly installation bolt.
4. Rotate the tool clockwise and loosen the V—belt to remove.
5. Remove the accessory drive belt.

To install:

6. Installation is the reverse order of removal.

CAMSHAFT AND VALVE LIFTERS

INSPECTION

1. Remove the camshaft from the engine.
2. Check the camshaft bearing journals for damage and binding.
3. If the journals are binding, check the cylinder head for damage.
4. Check the cylinder head for clogged oil holes.
5. Check the camshaft surface for abnormal wear and damage. Replace the camshaft, as required.
6. Measure the camshaft lobe surface and replace the camshaft if not within specification.
7. Measure the camshaft journal diameter and replace the camshaft if not within specification.
8. Measure the camshaft run out and replace the camshaft if not within specification.

REMOVAL & INSTALLATION

2.5L SOHC Engine

See Figures 89 through 96.

1. Before servicing the vehicle, refer to the Precautions Section.
2. Disconnect the negative battery cable.
3. To remove the front side V belt, remove the belt covers. Loosen the lock bolt. Loosen the slider bolt. Remove the front side belt.
4. To remove the rear side V belt, remove the belt covers. Loosen the lock bolt. Loosen the slider bolt. Remove the rear side belt. Remove the belt tensioner.
5. Remove the crankshaft pulley bolt.
6. Lock the crankshaft in place using tool ST499977100, or equivalent.
7. Remove the crankshaft pulley.
8. Remove the left side timing belt cover.
9. Remove the front timing belt cover.
10. Remove the timing belt.
11. Remove the camshaft position sensor.
12. Remove the camshaft sprockets.

➡Be sure to lock the camshaft in place using tool ST18231AA010, or equivalent.

13. Remove the left and right timing belt N02 covers.

➡Do not damage or lose the rubber seal when removing the covers.

14. Remove the tensioner bracket. Remove the camshaft position sensor support, on the left side.

15. Remove the oil level gauge guide, on the left side.

16. Remove the valve rocker arm assembly.

17. Remove camshaft cap retaining bolts "A" and "B" in the proper sequence.

18. Loosen camshaft cap bolts "C" through "J" all the way in the proper sequence.

19. Remove camshaft cap bolts "K" through "P" in the proper sequence using a Torx® head bit.

20. Remove the camshaft caps.

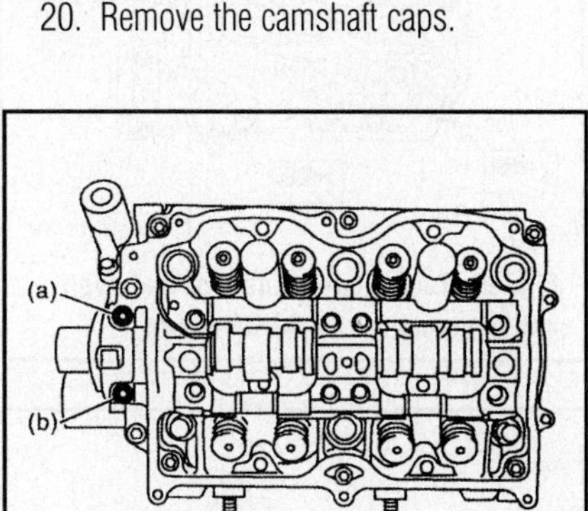

Fig. 89 Camshaft bolt loosening sequence—2.5L SOHC engine

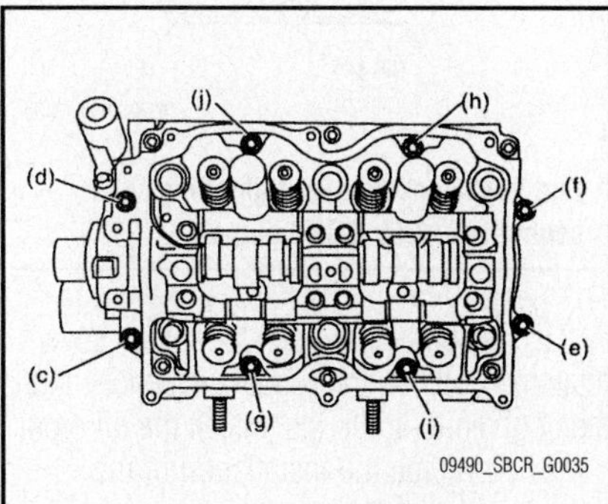

Fig. 90 Camshaft bolt loosening sequence—2.5L SOHC engine

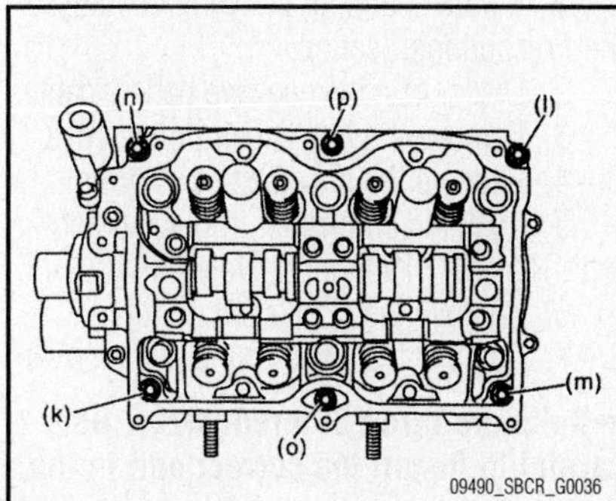

Fig. 91 Camshaft bolt loosening sequence—2.5L SOHC engine

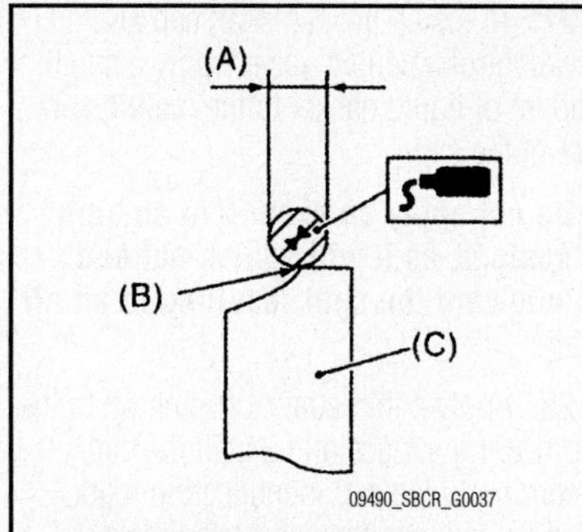

Fig. 92 Camshaft cap sealant application—2005 2.5L SOHC engine

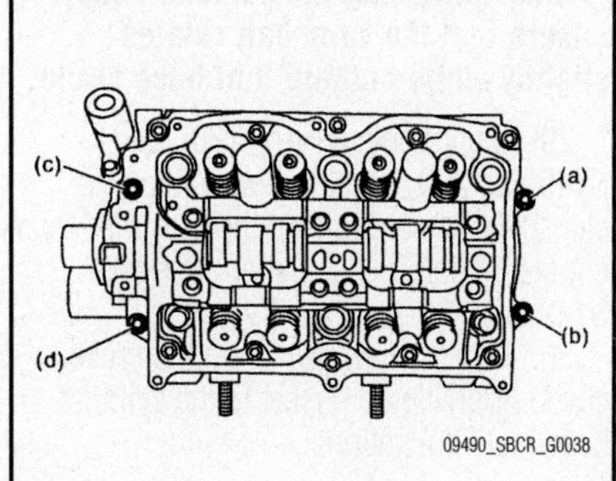

Fig. 93 Camshaft bolt tightening sequence—2005 2.5L SOHC engine

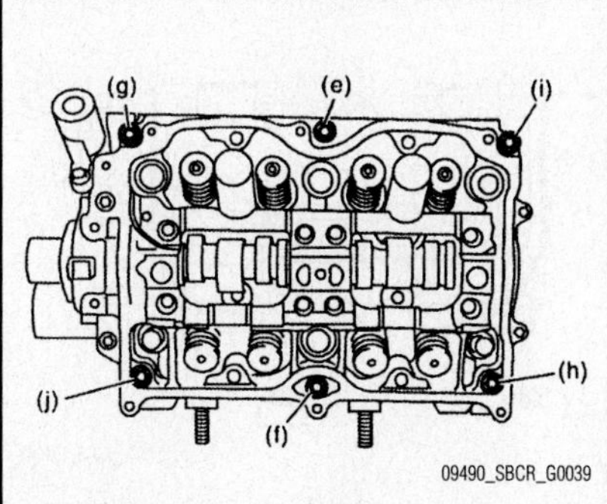

Fig. 94 Camshaft bolt tightening sequence—2005 2.5L SOHC engine

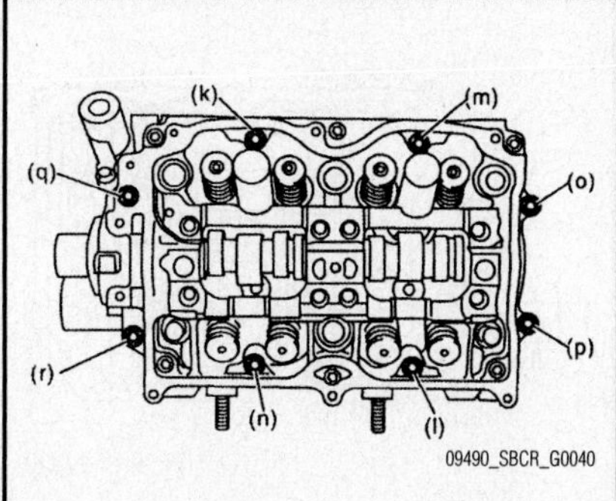

Fig. 95 Camshaft bolt tightening sequence—2005 2.5L SOHC engine

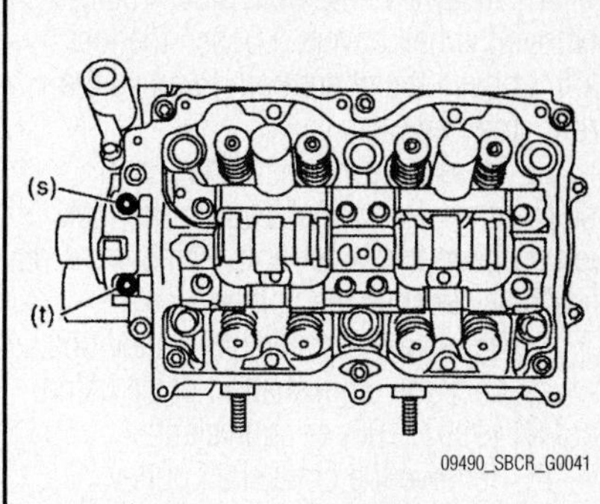

Fig. 96 Camshaft bolt tightening sequence—2005 2.5L SOHC engine

21. Remove the camshaft. Remove the oil seal. Remove the plug from the rear side of the camshaft.

➡Do not remove the oil seal unless necessary.

To install:

➡Lubricate the camshaft journals with clean engine oil prior to installation.

22. Install the camshaft into the cylinder head.

23. Apply liquid gasket to the mating surfaces of the camshaft cap.

24. Apply a bead of sealant (0.12 inch in diameter) along the edge of the camshaft cap mating surface. Install with 20 minutes after applying the sealant.

25. Temporarily tighten the bolts "A" through "D" in the proper sequence.

26. Install the valve rocker arm assembly. Tighten Torx® head bolts "E" through "J" in the proper sequence to 13 ft. lbs.

27. Tighten bolts "K" through "R" in the proper sequence to 7.2 ft. lbs.

28. Tighten bolts "S" through "T" in the proper sequence to 7.2 ft. lbs.

➡Be sure to use a new seal washer.

29. Using tools ST499597000 and ST499587500, install a new seal on the camshaft. Be sure to coat the seal with clean engine oil before installation. Use tool ST499587700 and install the plug.

30. Adjust the valve clearance.

31. Continue the installation in the reverse order of the removal procedure.

2.5L DOHC Engine

See Figures 97 through 100.

1. Before servicing the vehicle, refer to the Precautions Section.

2. Disconnect the negative battery cable.

3. Remove the collector cover.

4. To remove the front side V belt, remove the belt covers. Loosen the lock bolt. Loosen the slider bolt. Remove the front side belt.
5. To remove the rear side V belt, remove the belt covers. Loosen the lock bolt. Loosen the slider bolt. Remove the rear side belt. Remove the belt tensioner.
6. Remove the crankshaft pulley bolt.
7. Lock the crankshaft in place using tool ST499977100, or equivalent.
8. Remove the crankshaft pulley.
9. Remove the left side timing belt cover.
10. Remove the right side timing belt cover.
11. Remove the front timing belt cover.
12. Remove the timing belt.
13. Remove the camshaft position sensor.
14. Remove the camshaft sprockets.

➡Be sure to lock the camshaft in place using tool ST499207400, or equivalent.

15. Lock the crankshaft in place using tool ST499977100, or equivalent.
16. Remove the crankshaft pulley.
17. Remove the tensioner bracket. Remove the right and left timing belt No.2 covers.
18. Remove the spark plug wires. Remove the oil level gauge, on the left side.
19. Remove the rocker cover and gasket. Remove the oil pipe.
20. Loosen the oil flow control solenoid valve assembly and the intake camshaft cap bolts equally and in the proper sequence.
21. Loosen the exhaust camshaft cap bolts equally and in the proper sequence.
22. Remove the oil flow control solenoid valve assembly, intake camshaft cap and camshaft.
23. Remove the exhaust camshaft caps and camshaft.

➡Arrange the camshafts caps so that they can be installed in their original positions.

To install:

➡Lubricate the camshaft journals with clean engine oil prior to installation.

24. Install the camshaft so that the valves are closed or in contact with the "base circle" of the cam lobe.
25. If the camshafts are positioned as shown in the, the camshafts need to be rotated at a minimum to align the timing belt during installation.
26. The right hand camshaft need not be rotated when set at the position illustrated, the left hand intake camshaft should be rotated 80 degrees clockwise. The left hand exhaust camshaft should be rotated 45 degrees counterclockwise.
27. To install the camshaft cap and oil flow control solenoid valve, apply a small amount of liquid gasket to the mating surface of the cap.

➡Do not apply an excessive amount of sealant, as it will squish out and flow toward the seal resulting in an oil leak.

28. Apply a thin coat of engine oil to the cap bearing surface and install the cap according to the cap identification mark. Gradually tighten the cap bolts in two stages, first to 7.2 ft. lbs. and then to 14.5 ft. lbs. in the proper sequence.

➡After tightening the camshaft cap, ensure that the camshaft rotates slightly while holding it at base circle.

29. Using tools ST49587600 and ST499597200, install a new seal on the camshaft. Be sure to coat the seal with clean engine oil before installation. Use tool ST499587700 and install the plug.
30. Install a new gasket on the rocker cover. Apply liquid gasket to the cylinder head (see illustration).

➡Apply an extra amount of liquid gasket around the semicircular plugs, 5mm or more.

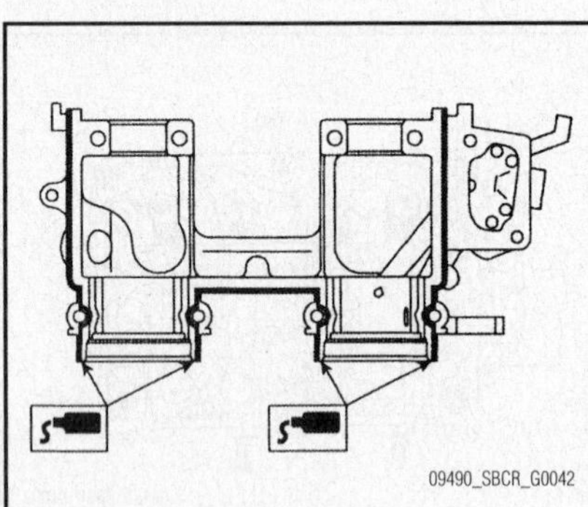

Fig. 97 Camshaft cap liquid gasket application points—2.5L DOHC engine

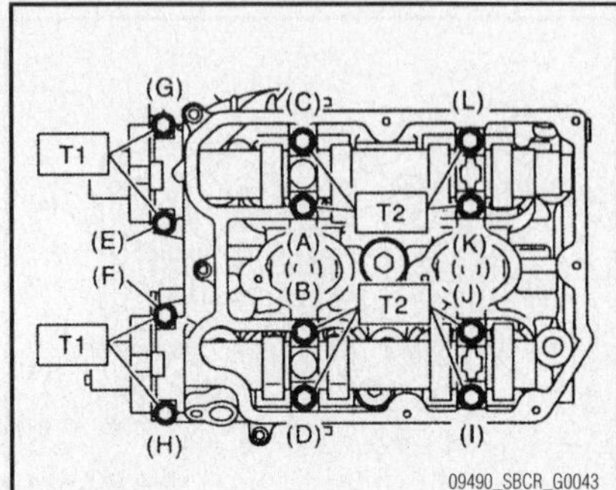

Fig. 98 Camshaft bolt tightening sequence—2.5L DOHC engine

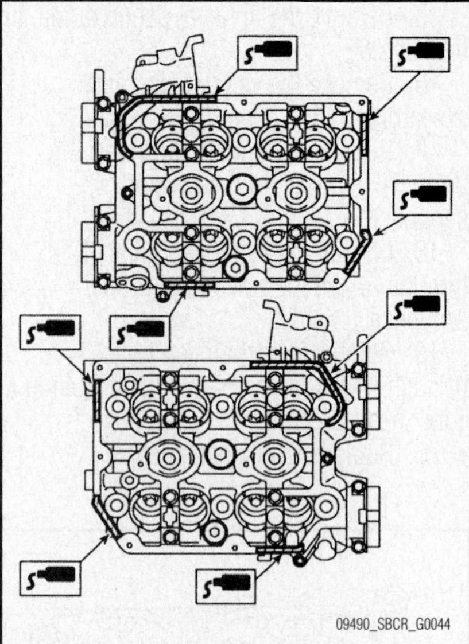

Fig. 99 Rocker cover liquid gasket application points—2.5L DOHC engine

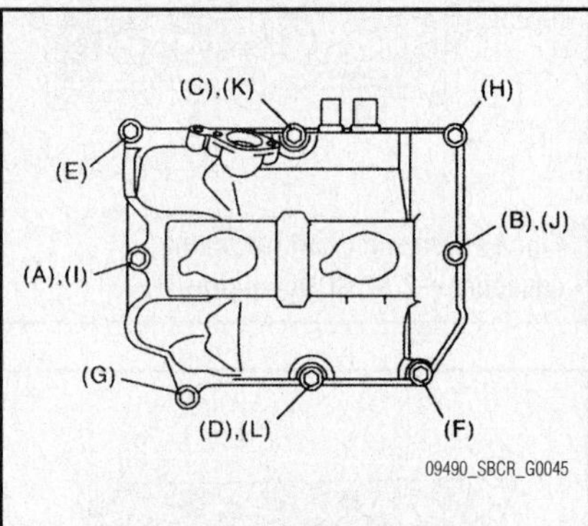

Fig. 100 Rocker cover tightening sequence—2.5L DOHC engine

31. Temporarily tighten the rocker cover retaining bolts, in the proper sequence, and then tighten to 4.7 ft. lbs. Install the oil pipe.
32. Continue the installation in the reverse order of the removal procedure.

3.0L Engine

See Figures 101 through 110.

1. Before servicing the vehicle, refer to the Precautions Section.
2. Disconnect the negative battery cable.
3. Remove the crankshaft pulley cover.
4. Remove the crankshaft pulley bolt.
5. Lock the crankshaft in place using tool ST499977100, or equivalent.
6. Remove the crankshaft pulley.
7. Remove the front timing chain cover.

➡Bolts are three different sizes. Be careful to install the correct bolt in the correct hole.

8. Remove the timing chain.
9. Remove the camshaft sprocket.

➡Be sure to lock the camshaft in place using tool ST499977500, or equivalent.

10. Remove the crankshaft sprocket.
11. Remove the oil pump.
12. Remove the water pump.
13. Remove the rear timing chain cover retaining bolts. Remove the rear timing chain cover.

➡There are seven different size bolts. Be sure not to confuse them on installation.

14. Disconnect the oil pipe.
15. Remove the rocker cover retaining bolts. Remove the rocker cover. Discard the gasket.
16. Remove the plugs, see illustration for location.
17. Loosen the camshaft cap bolts in the proper sequence. Remove the camshaft caps and remove the camshaft.

To install:

18. Apply a coat of engine oil to the journals on the camshafts and place the camshafts into position.
19. To install the camshaft cap, apply a small amount of liquid gasket to the mating surface of the cap. Do not apply an excessive amount of sealant, as it will squish out and flow toward the cam journal resulting in engine seizure.
20. Apply a thin coat of engine oil to the cap bearing surface and install the cap. Tighten cap bolts 1 through 12 to 12 ft. lbs. and bolts 13 through 16 to 7.2 ft. lbs. in the

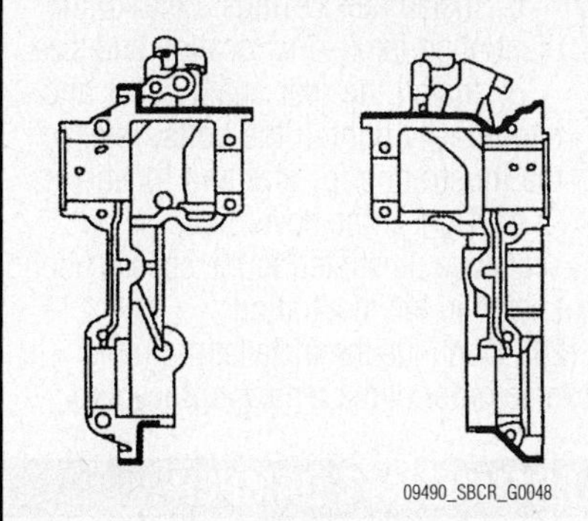

Fig. 103 Camshaft cap liquid gasket application points—3.0L engine

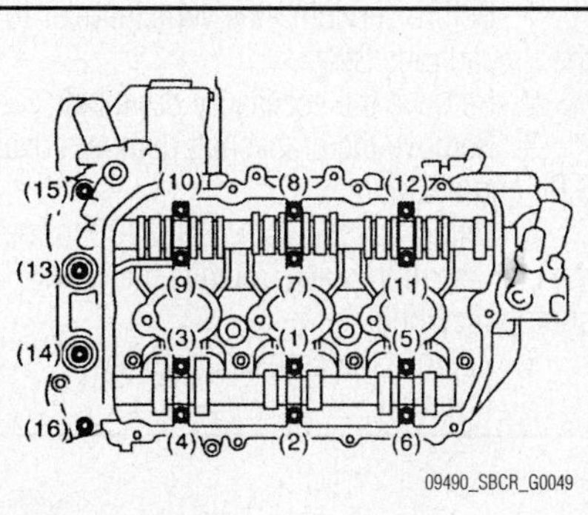

Fig. 104 Camshaft bolt tightening sequence—3.0L engine

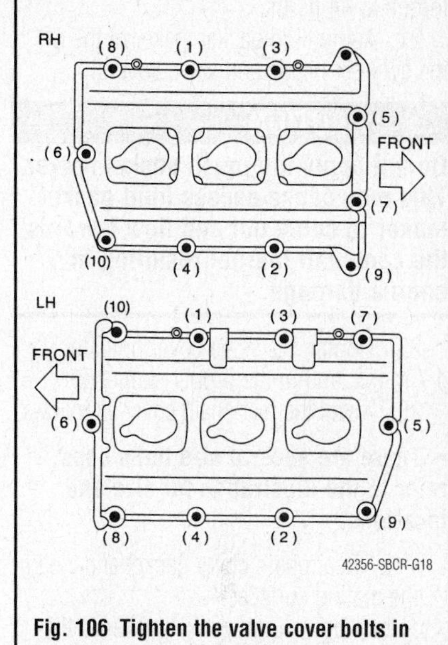

Fig. 106 Tighten the valve cover bolts in this sequence—3.0L engine

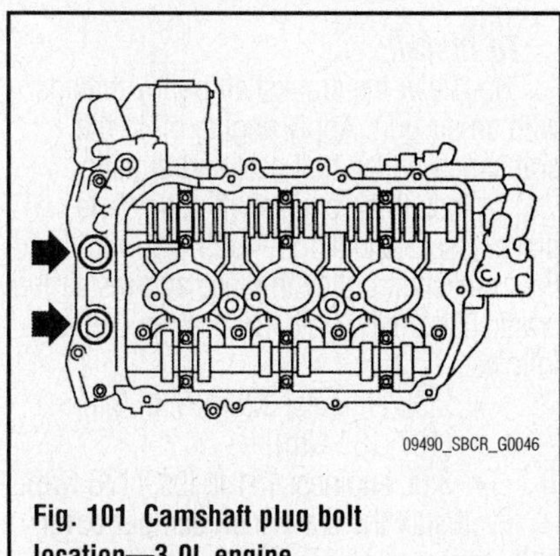

Fig. 101 Camshaft plug bolt location—3.0L engine

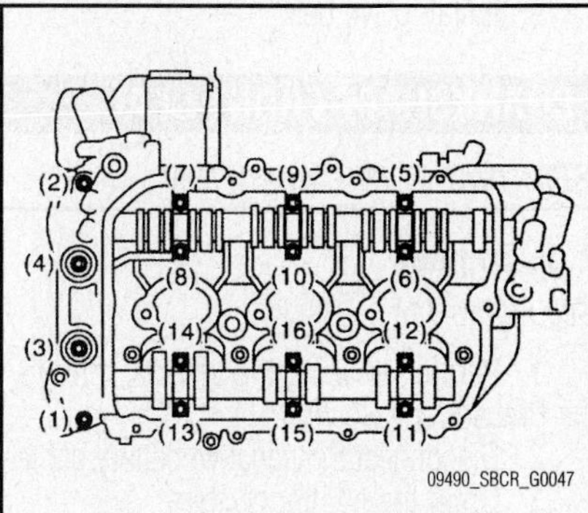

Fig. 102 Camshaft bolt loosening sequence—3.0L engine

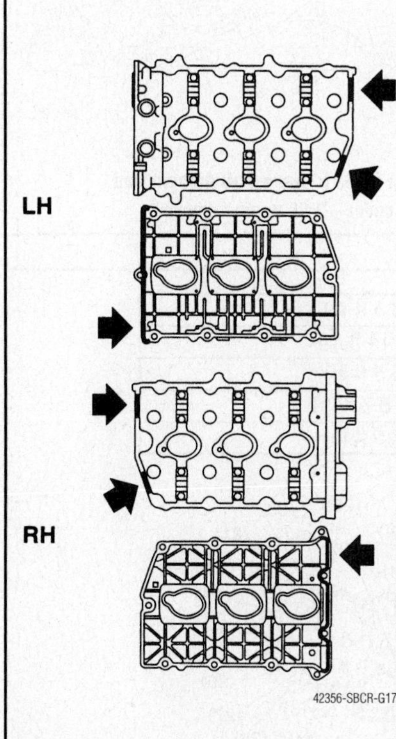

Fig. 105 Apply fluid gasket maker on the cylinder heads and valve covers as shown—3.0L engine

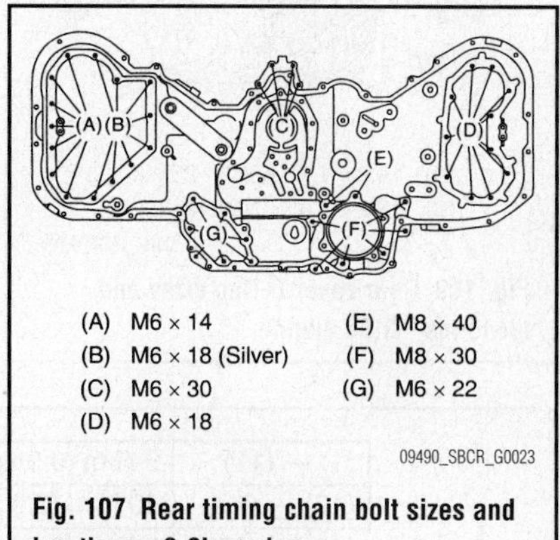

Fig. 107 Rear timing chain bolt sizes and locations—3.0L engine

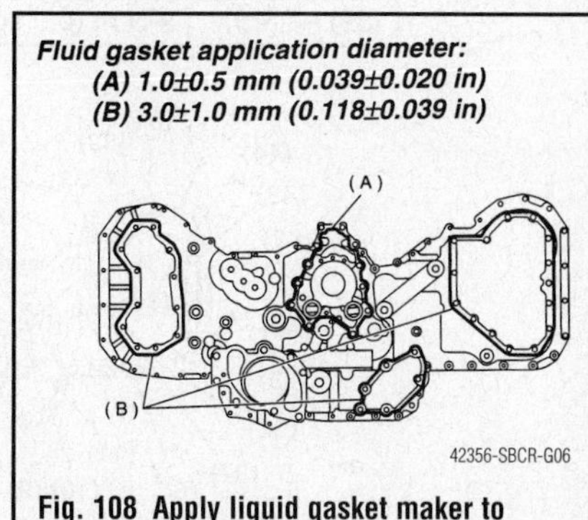

Fig. 108 Apply liquid gasket maker to the mating surfaces of the cover as shown—3.0L engine

proper sequence. Install the plugs and torque to 44 ft. lbs.

21. Apply fluid gasket maker of the of the cylinder heads and valve covers.

✻✻ CAUTION

Do not apply too much gasket maker. This may cause excess fluid gasket maker to come out and flow towards the camshaft journal resulting in engine damage.

22. Tighten the valve cover bolts to 4.7 ft. lbs. and in the proper sequence.

23. Install the rear chain cover as follows:

➡There are several size bolts used, refer to the illustration for size and locations.

a. Rear chain cover gasket and clean the mating surfaces.

b. Apply liquid gasket maker to the mating surfaces of the cover. Refer to the illustration for gasket maker application and diameter.

c. Install new O-rings. Refer to the illustration for O-ring location and size

d. Install the rear chain cover and temporarily tighten the bolts, refer to the illustration for size and locations.

e. Tighten the cover bolts in the sequence illustrated to the specifications shown in the illustration.

24. Continue the installation in the reverse order of the removal procedure.

CRANKSHAFT DAMPER

REMOVAL & INSTALLATION

See Figures 111, 112 and 113.

1. Before servicing the vehicle, refer to the Precautions Section.
2. Remove the accessory drive belt.
3. Remove the crankshaft damper cover, 3.0L engines only.
4. Using Special Tool 49997700 to lock the crankshaft in place, remove the crankshaft pulley bolt.
5. Remove the crankshaft pulley.

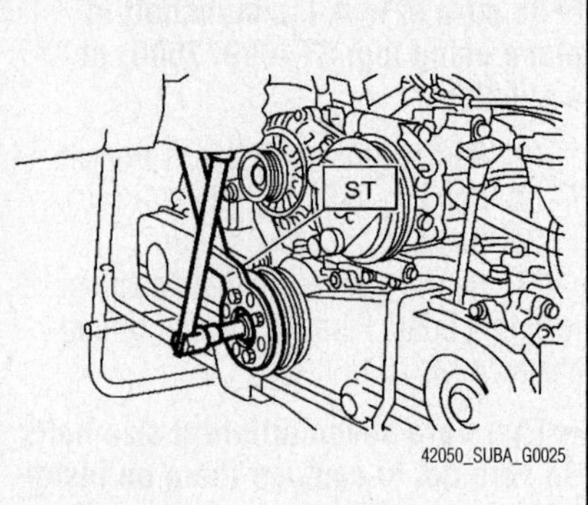

Fig. 112 Special Tool 499977100–2.5L engine shown

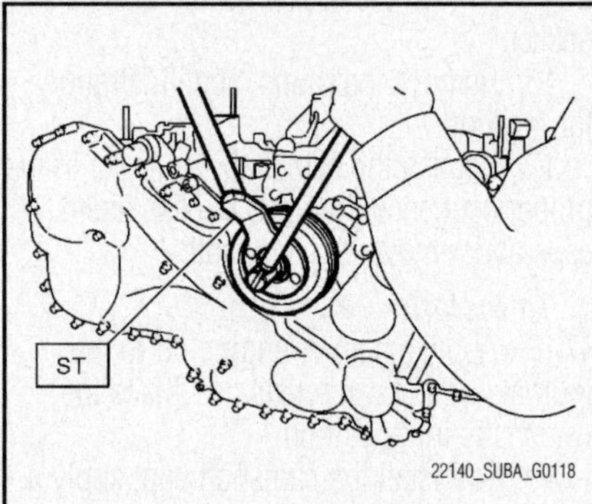

Fig. 113 Special Tool 499977100–3.0L engine shown

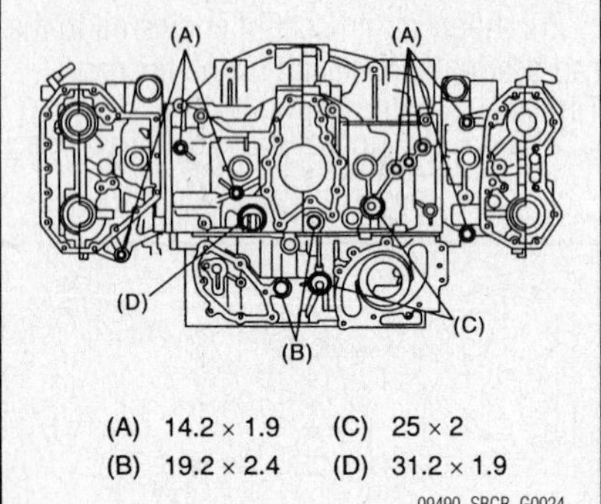

Fig. 109 Rear cover O-ring sizes and locations—3.0L engine

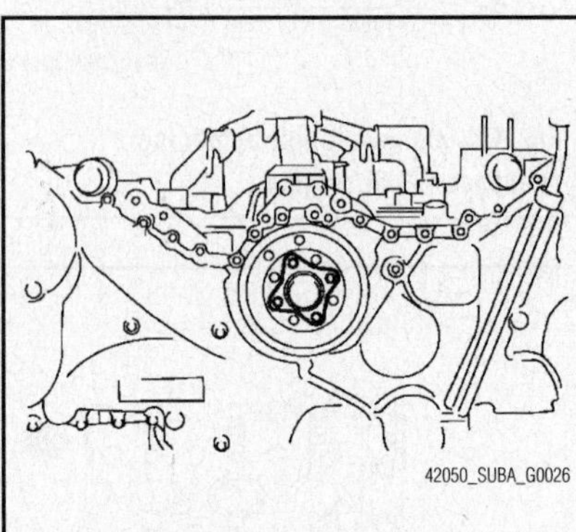

Fig. 111 Crankshaft damper and cover—3.0L engine shown

(1) — (11)	9 N·m (0.9 kgf-m, 6.5 ft-lb)
(12) — (19)	20 N·m (2.0 kgf-m, 14 ft-lb)
(20) — (30)	9 N·m (0.9 kgf-m, 6.5 ft-lb)
(31) — (38)	12 N·m (1.2 kgf-m, 8.7 ft-lb)
(39) — (45)	9 N·m (0.9 kgf-m, 6.5 ft-lb)

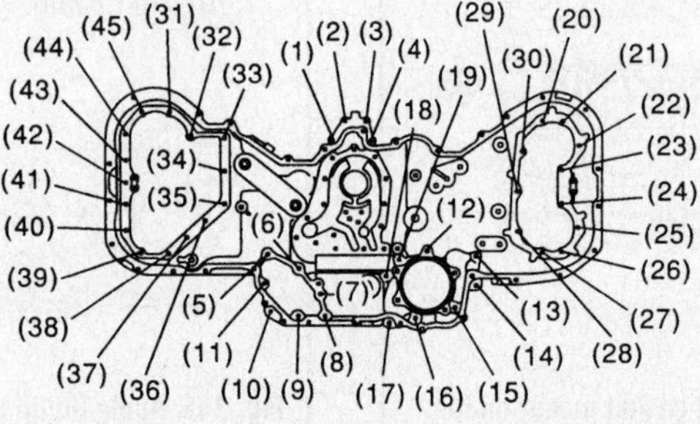

Fig. 110 Rear cover bolt tightening sequence and torque specifications—3.0L engine

To install:

6. Clean the crankshaft pulley threads with an air gun. Apply engine oil to the crankshaft pulley bolt seat and threads.

7. Install the crankshaft pulley into place. Use Special Tool 499977100 to lock the crankshaft pulley in place and install the crankshaft pulley bolt. Tighten the bolt as follows

- 2.5L engines: 123-137 ft. lbs. (167-187 Nm).
- 3.0L engines: 131 ft. lbs. (178 Nm).

8. Install the crankshaft damper cover, 3.0L engines only. Tighten to 4.7 ft. lbs. (6.4 Nm).

9. Install drive belt.

CRANKSHAFT FRONT SEAL

REMOVAL & INSTALLATION

2.5L Engines

See Figure 114.

1. Before servicing the vehicle, refer to the Precautions Section.
2. Disconnect the negative battery cable.
3. Drain the cooling system.
4. On the DOHC engine, remove the collector cover.

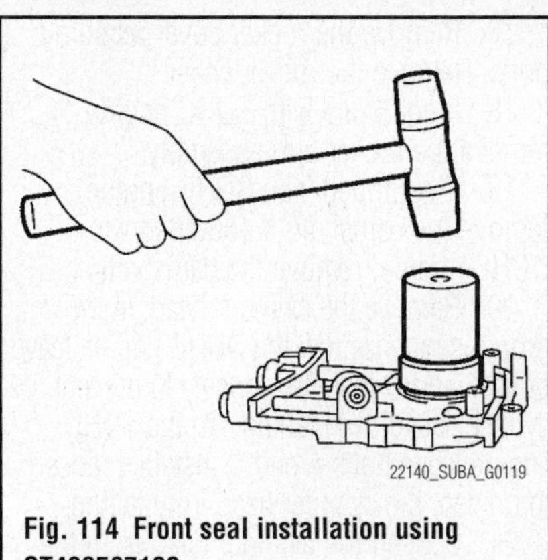

Fig. 114 Front seal installation using ST499587100

5. Raise and support the vehicle safely.
6. Remove the undercover.
7. On the DOHC engine, remove the bolts which retain the water pipe of the oil cooler to the oil pump. Remove the water pipe and hoses between the oil cooler and the water pump.
8. Lower the vehicle. Remove the radiator.
9. To remove the front side V belt, remove the belt covers. Loosen the lock bolt. Loosen the slider bolt. Remove the front side belt.
10. To remove the rear side V belt, remove the belt covers. Loosen the lock bolt. Loosen the slider bolt. Remove the rear side belt.
11. On the SOHC engine remove the belt tensioner.
12. On the DOHC engine remove the rear side V belt tensioner
13. Remove the crankshaft position sensor.
14. Remove the crankshaft pulley bolt.
15. Lock the crankshaft in place using tool ST499977100 for the SOHC engine and tool ST499207400 for the DOHC engine, or equivalent.
16. Remove the crankshaft pulley.
17. Remove the water pump.
18. If equipped, remove the timing belt guide. Remove the crankshaft sprocket.
19. Remove the oil pump retaining bolts.

➡When disassembling and checking the oil pump, loosen the relief valve plug before removing the oil pump from its mounting.

20. Using a flat tip tool remove the oil pump from the engine.
21. Carefully remove the front oil seal.

To install:

22. Replace the front oil seal with a new part using ST499587100.
23. Be sure all mating surfaces are clean and free of dirt.
24. Apply liquid gasket part number 004403007, or equivalent to the mating surfaces of the oil pump.
25. Be sure to replace the O-ring with a new one.
26. Apply a thin coat of clean engine oil to the inside of the oil seal.
27. Position the oil pump to its mounting, aligning the notched area with the crankshaft and push the pump straight.

➡Be sure that the oil seal lip is not folded.

28. Install the oil pump. Apply liquid gasket part number 004403042, or equivalent to the three retaining bolt threads. Install the bolts and tighten to 4.7 ft. lbs.
29. Continue the installation in the reverse order of the removal procedure.
30. Be sure to fill the cooling system with the proper grade and type coolant.
31. Start the engine and check for leaks. Correct, as required.

3.0L Engine

Refer to the Oil Pump Removal & Installation procedure in this section.

CYLINDER HEAD

REMOVAL & INSTALLATION

2.5L Engines

See Figures 115 through 120.

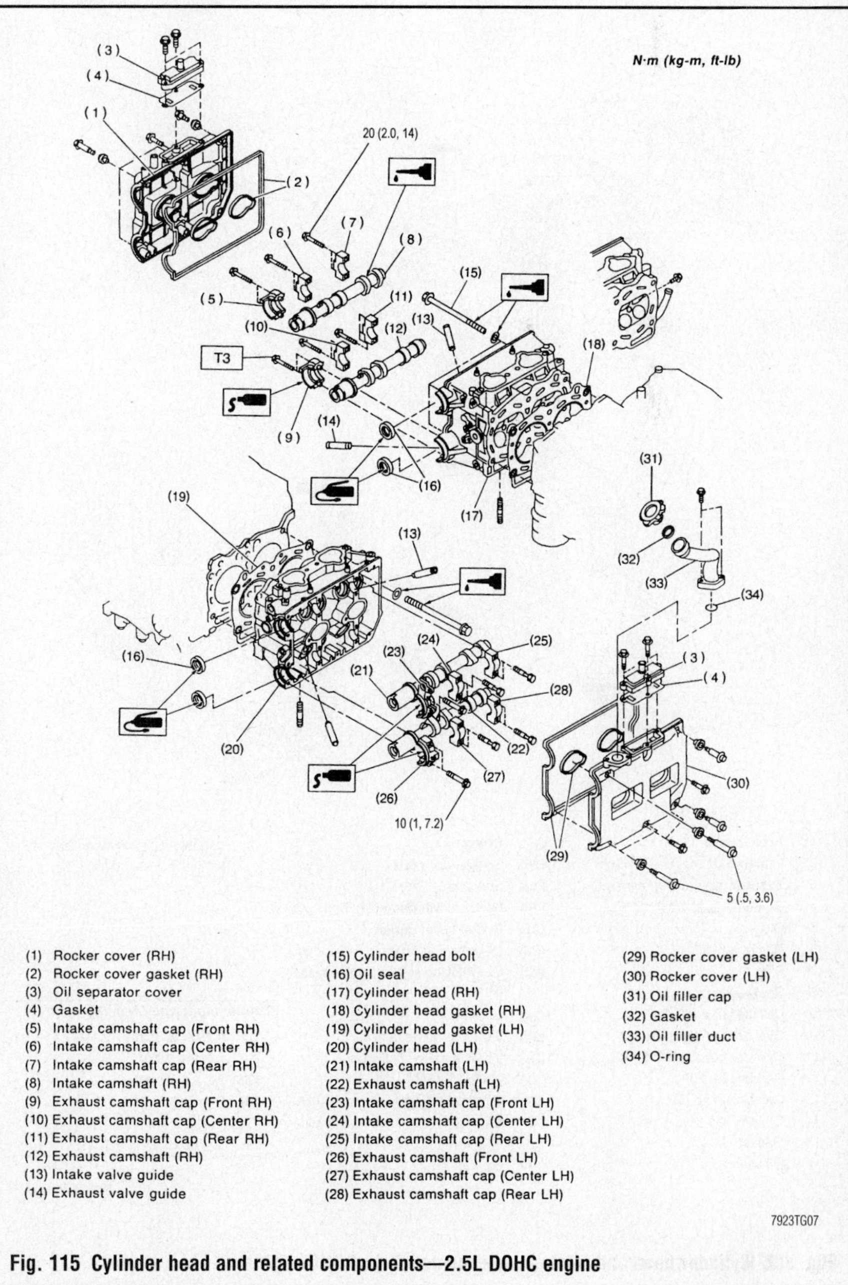

Fig. 115 Cylinder head and related components—2.5L DOHC engine

1. Before servicing the vehicle, refer to the Precautions Section.
2. Disconnect the negative battery cable.
3. If equipped with DOHC engine, remove the collector cover.
4. To remove the front side V belt, remove the belt covers. Loosen the lock bolt. Loosen the slider bolt. Remove the front side belt.
5. To remove the rear side V belt, remove the belt covers. Loosen the lock bolt. Loosen the slider bolt. Remove the rear side belt. Remove the belt tensioner.
6. Remove the crankshaft pulley bolt.
7. Lock the crankshaft in place using tool ST499977100, or equivalent.
8. Remove the crankshaft pulley.
9. Remove the left side timing belt cover.
10. If equipped with DOHC engine, remove the right side timing belt cover.
11. Remove the front timing belt cover.
12. Remove the timing belt.
13. Remove the camshaft position sensor.
14. Remove the camshaft sprockets.

➡Be sure to lock the camshaft in place using tool ST18231AA010 (SOHC engine) and ST499207400 (DOHC engine), or equivalent.

15. Remove the intake manifold.
16. If equipped with SOHC engine, remove the bolt that retains the air conditioning compressor bracket to the cylinder head.
17. Remove the rocker cover retaining bolts. Remove the rocker cover.
18. If equipped with SOHC engine, remove the rocker arm assembly.
19. If equipped with SOHC engine remove the camshaft. If equipped with DOHC engine, remove the camshafts.
20. Remove the cylinder head, in the proper sequence. On the DOHC engine leave bolts A and D installed loosely to prevent the cylinder head from falling. On the SOHC engine leave bolts A and C installed loosely to prevent the cylinder head from falling.
21. Loosen the cylinder head from the block using a plastic-faced hammer, if needed.
22. Remove bolts A and D on the DOHC engine and bolts A and C on the SOHC engine. Remove the cylinder head from the engine. Discard the gasket
23. Clean all gasket material from both mating surfaces.

To install:

24. Installation is the reverse of the removal procedure.
25. Apply a thin coat of clean engine oil to the washers and cylinder head bolts.
26. Tighten the cylinder head retaining bolts to specification and in the proper sequence.

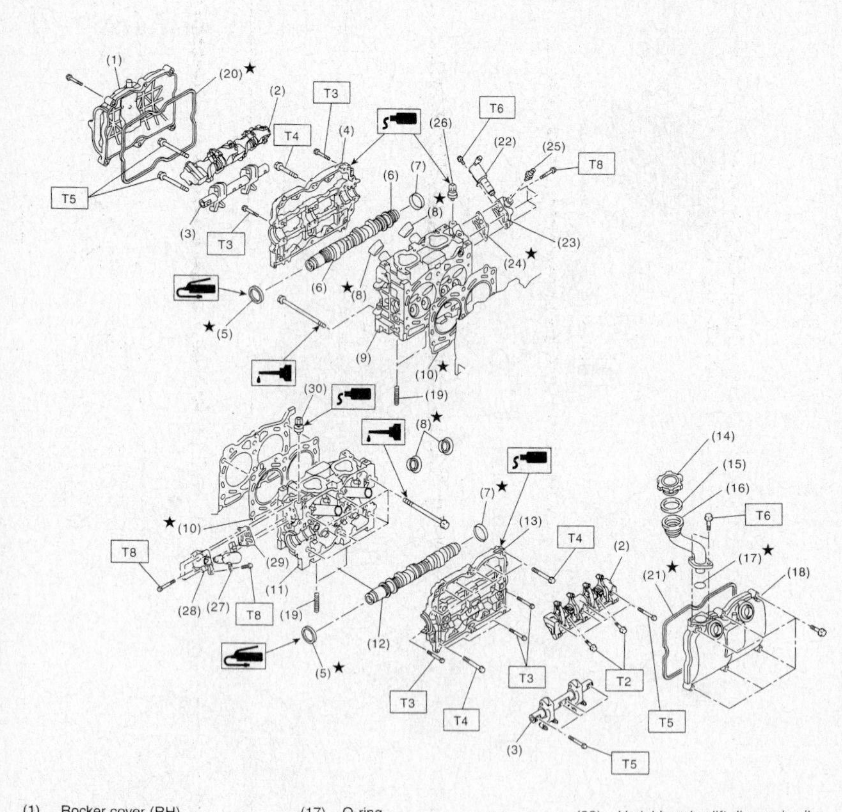

(1)	Rocker cover (RH)	(17)	O-ring	(30)	Variable valve lift diagnosis oil pressure switch (LH)
(2)	Intake valve rocker assembly	(18)	Rocker cover (LH)		
(3)	Exhaust valve rocker assembly	(19)	Stud bolt		
(4)	Camshaft cap (RH)	(20)	Rocker cover gasket (RH)		
(5)	Oil seal	(21)	Rocker cover gasket (LH)		
(6)	Camshaft (RH)	(22)	Oil switching solenoid valve (RH)		
(7)	Plug	(23)	Oil switching solenoid valve holder (RH)		
(8)	Spark plug pipe gasket				
(9)	Cylinder head (RH)	(24)	Gasket		
(10)	Cylinder head gasket	(25)	Oil temperature sensor		
(11)	Cylinder head (LH)	(26)	Variable valve lift diagnosis oil pressure switch (RH)		
(12)	Camshaft (LH)				
(13)	Camshaft cap (LH)	(27)	Oil switching solenoid valve (LH)		
(14)	Oil filler cap	(28)	Oil switching solenoid valve holder (LH)		
(15)	Gasket				
(16)	Oil filler duct	(29)	Gasket		

Tightening torque: N·m (kgf-m, ft-lb)	
T3:	*9.75 (1.0, 7.2)*
T4:	*18 (1.8, 13.0)*
T5:	*25 (2.5, 18.1)*
T6:	*6.4 (0.65, 4.7)*
T7:	*8 (0.8, 5.9)*
T8:	*10 (1.0, 7.4)*

09490_SBCR_G0016

Fig. 116 Cylinder head and related components—2.5L SOHC engine

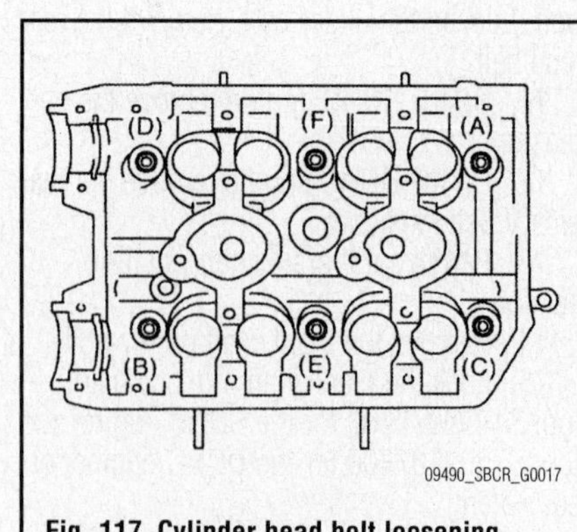

Fig. 117 Cylinder head bolt loosening sequence—2.5L DOHC engine

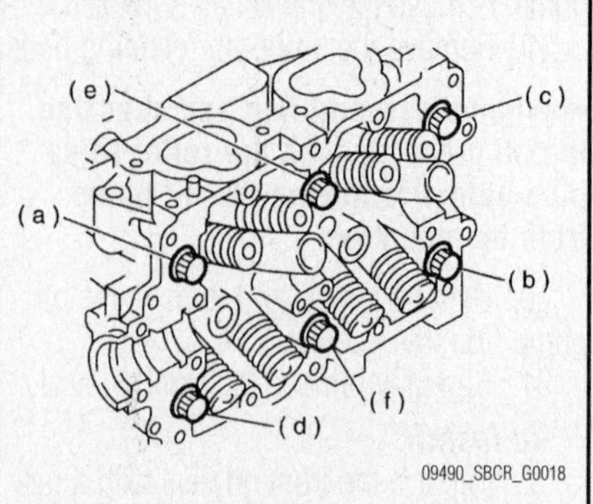

Fig. 118 Cylinder head bolt loosening sequence—2.5L SOHC engine

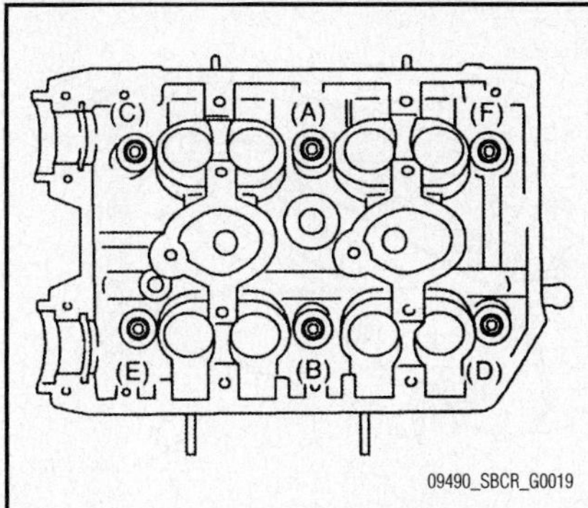

Fig. 119 Cylinder head bolt tightening sequence—2.5L DOHC engine

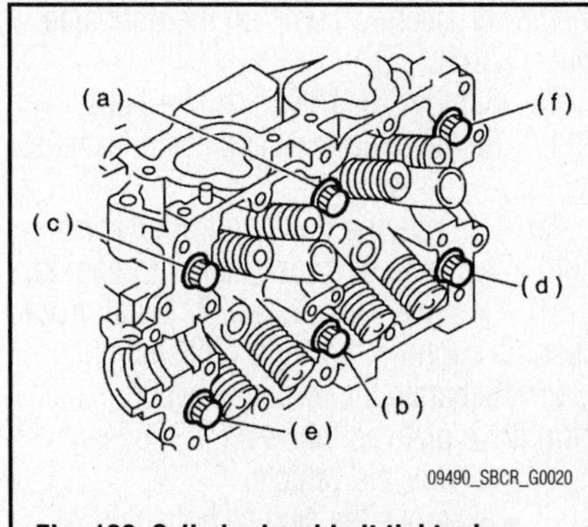

Fig. 120 Cylinder head bolt tightening sequence—2.5L SOHC engine

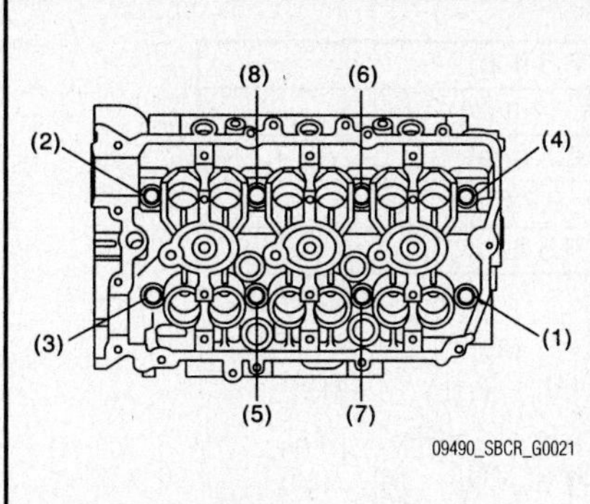

Fig. 121 Cylinder head bolt loosening sequence—3.0L engine

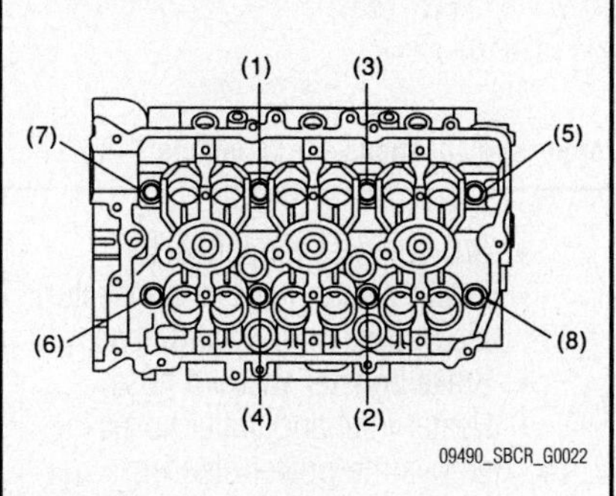

Fig. 122 Cylinder head bolt tightening sequence—3.0L engine

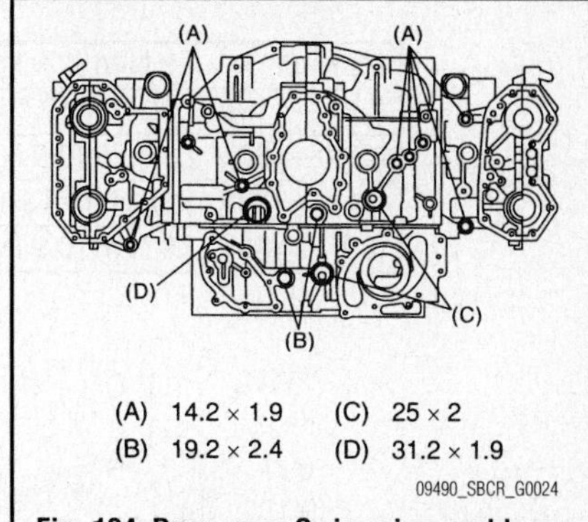

Fig. 124 Rear cover O-ring sizes and locations—3.0L engine

27. Start the engine and allow it to reach operating temperature.

28. Check for leaks, correct as required.

3.0L Engine

See Figures 121 through 125.

1. Before servicing the vehicle, refer to the Precautions Section.
2. Remove the crankshaft pulley cover.
3. Remove the crankshaft pulley bolt.
4. Lock the crankshaft in place using tool ST499977100, or equivalent.
5. Remove the crankshaft pulley.
6. Remove the front timing chain cover.

➡Bolts are three different sizes. Be careful to install the correct bolt in the correct hole.

7. Remove the timing chain.
8. Remove the camshaft sprocket.

➡Be sure to lock the camshaft in place using tool ST499977500, or equivalent.

9. Remove the crankshaft sprocket.
10. Remove the oil pump.
11. Remove the water pump.
12. Remove the rear timing chain cover retaining bolts. Remove the rear timing chain cover.

➡There are seven different size bolts. Be sure not to confuse them on installation.

13. Remove the camshafts.
14. Remove the cylinder head bolts in the proper sequence. Leave bolts 2 and 4 connected by a few threads to prevent the head from falling. Tap the head with a plastic mallet to separate it from the block.
15. Remove bolts 2 and 4 from the cylinder head. Remove the cylinder head from the engine. Discard the gasket.
16. Clean all gasket material from both mating surfaces.

To install:

17. Installation is the reverse of the removal procedure.
18. Apply a thin coat of clean engine oil to the washers and cylinder head bolts.
19. Tighten the cylinder head retaining bolts to specification and in the proper sequence.
20. Install the rear chain cover as follows:

➡There are several size bolts used, refer to the illustration for size and locations.

a. Rear chain cover gasket and clean the mating surfaces.

b. Apply liquid gasket maker to the mating surfaces of the cover. Refer to the illustration for gasket maker application and diameter.

c. Install new O-rings. Refer to the illustration for O-ring location and size

d. Install the rear chain cover and temporarily tighten the bolts, refer to the illustration for size and locations.

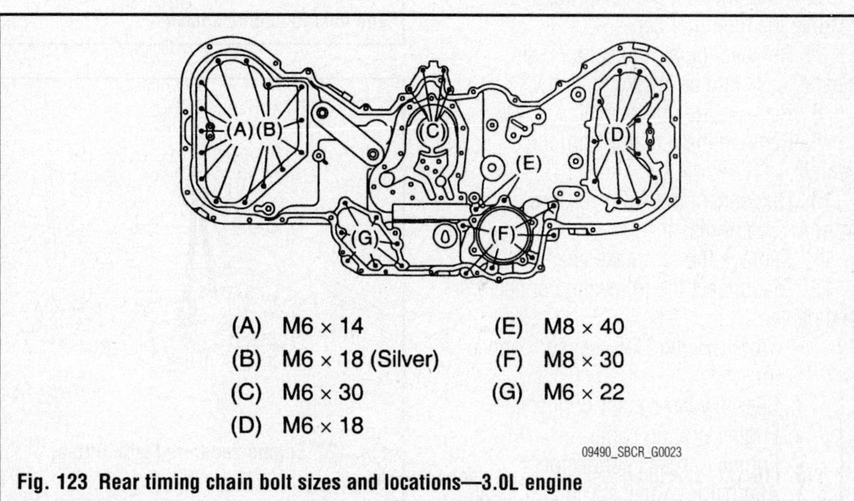

Fig. 123 Rear timing chain bolt sizes and locations—3.0L engine

(1) — (11)	9 N·m (0.9 kgf-m, 6.5 ft-lb)
(12) — (19)	20 N·m (2.0 kgf-m, 14 ft-lb)
(20) — (30)	9 N·m (0.9 kgf-m, 6.5 ft-lb)
(31) — (38)	12 N·m (1.2 kgf-m, 8.7 ft-lb)
(39) — (45)	9 N·m (0.9 kgf-m, 6.5 ft-lb)

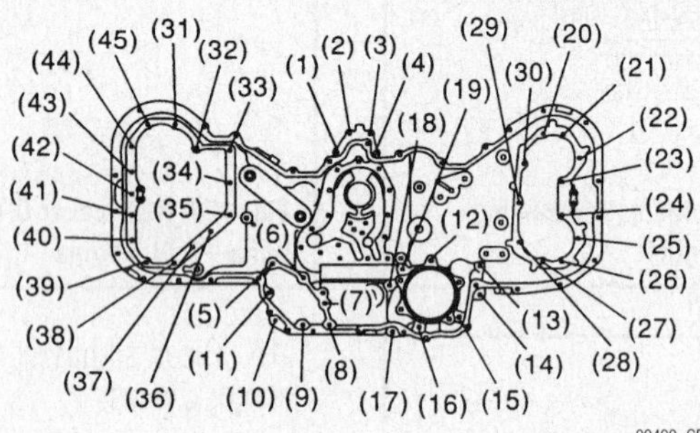

Fig. 125 Rear cover bolt tightening sequence and torque specifications—3.0L engine

e. Tighten the cover bolts in the sequence illustrated to the specifications shown in the illustration.

21. Start the engine and allow it to reach operating temperature.
22. Check for leaks, correct as required.

ENGINE ASSEMBLY

REMOVAL & INSTALLATION

2.5L Engines

See Figures 126 through 129.

1. Before servicing the vehicle, refer to the Precautions Section.
2. Open the front hood fully and support with the front food stay.
3. Collect the refrigerant from the A/C system.
4. Release the fuel pressure.
5. Disconnect the ground cable from the battery.
6. Open the fuel filler flap lid, and remove the fuel filler cap.
7. Remove the air intake duct, air cleaner case and air intake chamber.
8. Remove the undercover.
9. Remove the radiator from the vehicle.
10. Disconnect the A/C pressure hoses from A/C compressor.
11. Remove the air intake chamber stay.
12. Disconnect the following connectors and cables.
    - Front oxygen (A/F) sensor connector
    - Rear oxygen sensor connector
    - Engine ground cable
    - Engine harness connectors
    - Alternator connector and terminal
    - A/C compressor connector
    - Power steering switch connector
13. Disconnect the following hoses.
    - Brake booster vacuum hose
    - Heater inlet and outlet hoses
14. Remove the power steering pump.
15. Remove the bolts which secure the power steering pump to the bracket. Place the power steering pump on the right side wheel apron.

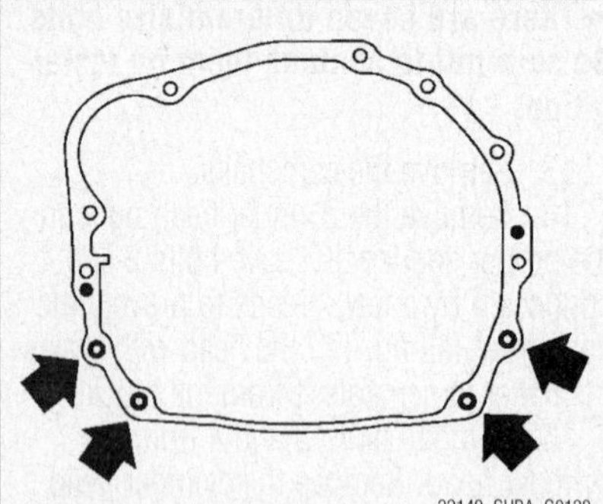

Fig. 126 Lower side of transmission bolts and nuts to engine shown

Fig. 127 Engine supported with lifting device

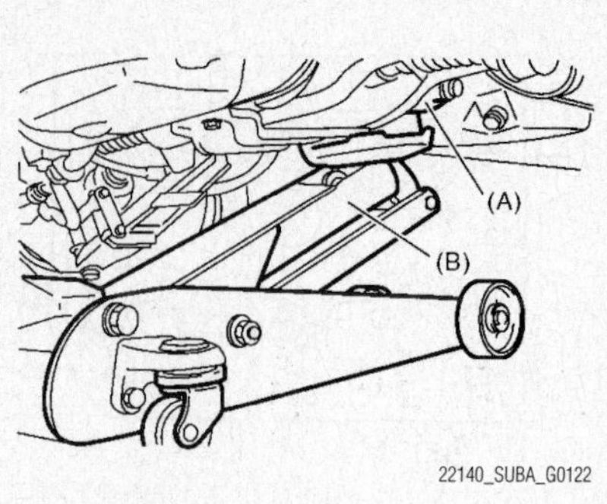

Fig. 128 Transmission supported with garage jack

16. Remove the front side V—belts.
17. Remove the front and center exhaust pipes.
18. Remove the bolts and nuts which hold lower side of transmission to engine.
19. Remove the nuts which install front cushion rubber onto front crossmember.
20. Separate the torque converter clutch from drive plate as follows: (AT models)
    - Lower the vehicle.
    - Remove the service hole plug.
    - Remove the bolts which hold torque converter clutch to drive plate.
    - Remove all bolts while turning the crankshaft with a socket wrench.
21. Remove the pitching stopper.
22. Disconnect the fuel hoses from fuel pipe.
23. Remove the clip and disconnect the evaporation hose from the pipe.
24. Support the engine with a lifting device.
25. Support the transmission with a garage jack.

✲✲ CAUTION

Before removing the engine away from transmission, check to be sure no work has been overlooked.

26. Remove the starter.
27. Remove the bolts which hold upper side of transmission to engine.
28. Attach the ST498277200 to the converter case to keep converter from slipping out . (AT models)
29. Carefully remove the engine from vehicle as follows:
    - Slightly raise the engine.
    - Raise the transmission with garage jack.
    - Move the engine horizontally until main shaft is withdrawn from clutch cover.

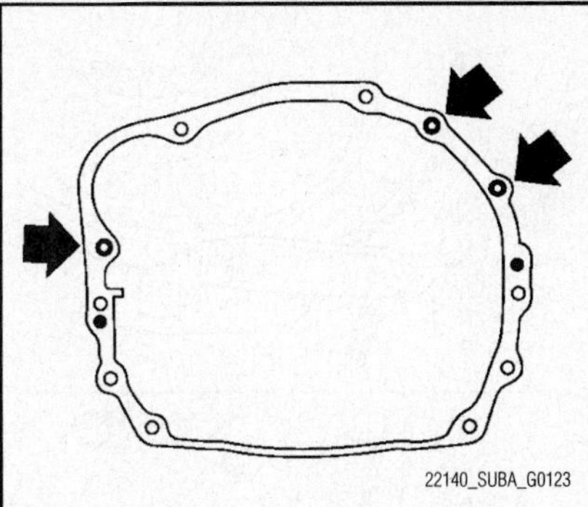

Fig. 129 Upper side of transmission bolts and nuts to engine shown

- Slowly move the engine away from engine compartment.

➡Be careful not to damage adjacent parts or body panels with crank pulley, oil level gauge, etc.

30. Remove the front cushion rubber engine mounts.

To install:

31. Install the front cushion rubber mounts to engine, tighten to 26 ft. lbs. (35 Nm).
32. Position the engine in engine compartment and align it with transmission.
33. Apply a small amount of grease to splines of main shaft. (MT models)
34. Tighten the bolts which hold upper side of transmission to engine, to 37 ft. lbs. (50 Nm).
35. Remove the lifting.
36. Remove the garage jack.
37. Install the pitching stopper.
38. Remove the ST498277200 from converter case. (AT models)
39. Install the starter.
40. Tighten the bolts which hold torque converter clutch to drive plate. (AT models)
41. Tighten other bolts while rotating the crankshaft using a socket wrench.
42. Tighten the bolts to 18 ft. lbs. (25 Nm).
43. Attach the service hole plug to prevent entry of foreign objects.
44. Install the power steering pump.
45. Install the power steering pump to the bracket, and tighten the bolts to 16 ft. lbs. (22 Nm).
46. Connect the power steering switch connector.
47. Install and adjust the front side belt.
48. Lift up the vehicle.
49. Tighten the bolts and nuts which hold the lower side of the transmission to engine. Tighten to 37 ft. lbs. (50 Nm).
50. Tighten the nuts which install the front cushion rubber onto crossmember to 55 ft. lbs. (75 Nm).

➡Make sure the front cushion rubber mounting bolts and locator are securely installed.

51. Install the front and center exhaust pipe.
52. Lower the vehicle.
53. Connect the following hoses:
    - Fuel delivery hose and evaporation hose
    - Heater inlet and outlet hoses
    - Brake booster vacuum hose
54. Connect the following connectors:
    - Front oxygen (A/F) sensor connector
    - Rear oxygen sensor connector
    - Engine harness connectors
    - Alternator connector and terminal
    - A/C compressor connector
55. Install the air intake chamber stay and tighten to 12 ft. lbs. (16 Nm).
56. Tighten the engine ground cable to 10 ft. lbs. (14 Nm).
57. Install the A/C pressure hoses.
58. Install the radiator to vehicle.
59. Install the air intake duct, air cleaner case and air intake chamber.
60. Install the undercover.
61. Install the battery in the vehicle, and connect cables.
62. Fill engine coolant and bleed the cooling system.
63. Check the ATF level and replenish it if necessary.
64. Vacuum and charge the A/C system with refrigerant. Test for leaks at connections.
65. Remove the front hood stay, and close the front hood.
66. Lower the vehicle from the lift.
67. Road test and check for leaks.

3.0L Engine

See Figures 127, 128 and 130.

1. Before servicing the vehicle, refer to the Precautions Section.
2. Set the vehicle on a lift.
3. Open the front hood fully and support with the front hood stay.
4. Remove the collector cover.
5. Collect the refrigerant from the A/C system.
6. Release the fuel pressure.
7. Remove the fuel filler cap.
8. Remove the battery from vehicle.
9. Remove the air intake duct, air cleaner case and air intake chamber.
10. Remove the radiator from vehicle.
11. Remove the V-belts.
12. Disconnect the A/C pressure hoses from the A/C compressor.
13. Remove the engine ground terminal.
14. Disconnect the following connectors:
    - Engine harness connectors
    - Generator connector and terminal
    - A/C compressor connector
    - Power steering switch connector
15. Disconnect the following hoses.
    - Brake booster vacuum hose
    - Heater inlet and outlet hoses
    - Pressure regulator vacuum hose
16. Remove the power steering pump together with the bracket.

➡Do not disconnect the hose and pipe from the pump body.

17. Place the power steering pump on the right side wheel apron.
18. Lift up the vehicle.
19. Remove the undercover.
20. Remove the front exhaust pipe.
21. Remove the ground cable.
22. Remove the bolts and nuts which hold lower side of transmission to engine.
23. Remove the nuts which install front cushion rubber onto front crossmember.
24. Separate the torque converter clutch from drive plate as follows:
    - Lower the vehicle.
    - Remove the service hole plug.
    - Remove the bolts which hold torque converter clutch to drive plate.
    - Remove other bolts while rotating the crankshaft using socket wrench.
25. Remove the pitching stopper.
26. Disconnect the fuel delivery hose and evaporation hose.
27. Disconnect the connector of fuel pipe by pushing the ST in the direction of arrow.
28. Remove the clip, and disconnect the evaporation hose from the pipe.

⁂ CAUTION

Be careful not to spill fuel. Catch the fuel from hoses using a container or cloth.

29. Support the engine with a lifting device.
30. Support the transmission with a garage jack.

➡Before removing the engine away from transmission, check to be sure no work has been overlooked.

31. Remove the starter.
32. Remove the bolts which hold upper side of transmission to engine.
33. Separation of engine and transmission.
34. Attach the ST498277200 to the converter case to keep converter from slipping out .

35. Slightly raise the engine.
36. Raise the transmission with garage jack.
37. Slowly move the engine away from engine compartment.
38. Remove the engine from vehicle.
39. Remove the front cushion rubbers.

To install:

40. Install the front cushion rubbers and tighten to 26 ft. lbs. (35 Nm).
41. Position the engine in engine compartment and align it with transmission.

⁂ CAUTION

Be careful not to damage adjacent parts or body panels with crank pulley, oil level gauge, etc.

42. Tighten the bolts which hold upper side of transmission to engine, to 37 ft. lbs. (50 Nm).
43. Remove the lifting device and wire ropes.
44. Remove the garage jack.
45. Install the pitching stopper.
46. Remove the ST498277200 from converter case.
47. Install the starter.
48. Install the torque converter clutch to drive plate.
49. Tighten the bolts which hold torque converter clutch to drive plate.
50. Tighten other bolts while rotating the crankshaft using socket wrench.
51. Tighten the bolts to 18 ft. lbs. (25 Nm).
52. Install the service hole plug to prevent getting foreign matter inside.
53. Install the power steering pump.
54. Install the power steering pump, tighten as follows:
    - Tighten bolt (A) to 18 ft. lbs. (25 Nm).
    - Tighten bolt (B) to 24 ft. lbs. (33 Nm).
55. Lift up the vehicle.

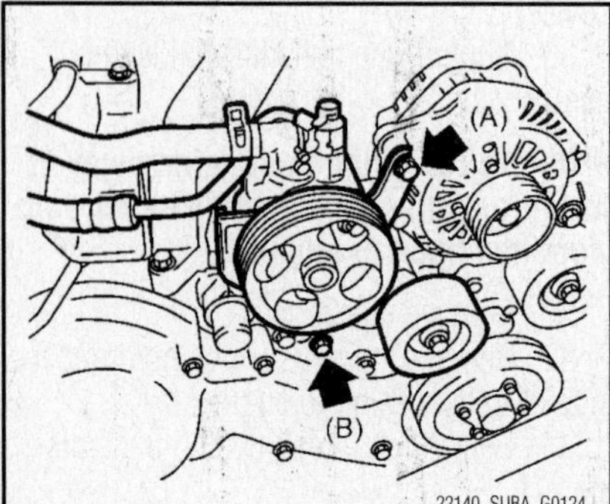

Fig. 130 Power steering pump and mounting bolts

56. Tighten the bolts and nuts which hold the lower side of the transmission to engine. Tighten to 37 ft. lbs. (50 Nm).
57. Tighten the nuts which install the front cushion rubber onto crossmember to 63 ft. lbs. (85 Nm).

➡Make sure the front cushion rubber mounting bolts and locator are securely installed.

58. Install the front exhaust pipe.
59. Connect the ground cable.
60. Install the undercover.
61. Lower the vehicle.
62. Connect the following hoses:
    - Fuel delivery hose and evaporation hose
    - Heater inlet and outlet hoses
    - Brake booster vacuum hose
    - Pressure regulator vacuum hose
63. Connect the following connectors:
    - Engine harness connectors
    - Alternator connector and terminal
    - A/C compressor connector
    - Power steering switch connector
64. Tighten the engine ground cable to 10 ft. lbs. (14 Nm).
65. Install the A/C pressure hoses.
66. Install drive belt.
67. Install the radiator to vehicle.
68. Install the air intake duct, air cleaner case and air intake chamber.
69. Install the undercover.
70. Install the battery in the vehicle, and connect cables.
71. Fill engine coolant and bleed the cooling system.
72. Check the ATF level and replenish it if necessary.
73. Vacuum and charge the A/C system with refrigerant. Test for leaks at connections.
74. Remove the front hood stay, and close the front hood.
75. Lower the vehicle from the lift.
76. Road test and check for leaks.

EXHAUST MANIFOLD

REMOVAL & INSTALLATION

Due to the unique design of the Subaru engine, an exhaust manifold is not used. The exhaust enters directly into the front Y—pipe.

FLYWHEEL

REMOVAL & INSTALLATION

See Figures 131 and 132.

1. Before servicing the vehicle, refer to the Precautions Section.

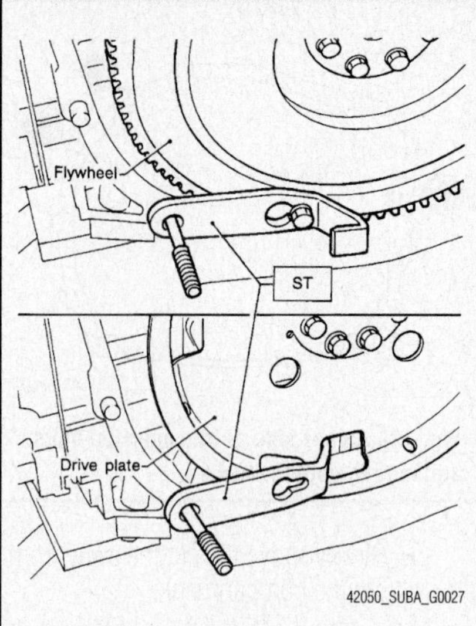

Fig. 131 Use Special Tool 498497100 to lock the crankshaft into place and remove the flywheel

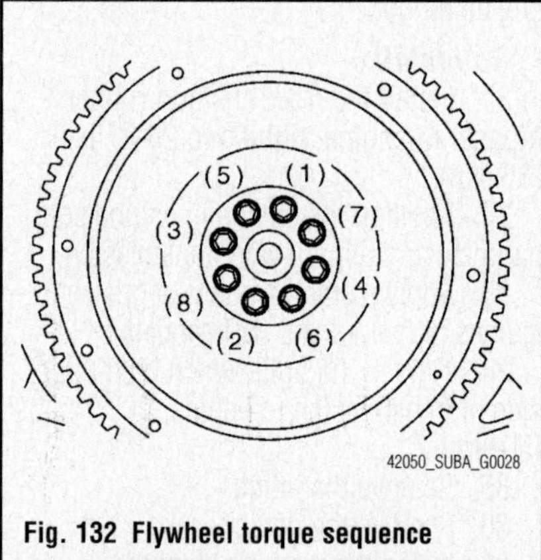

Fig. 132 Flywheel torque sequence

2. Remove the engine and place on a suitable engine stand. Refer to "Engine Assembly, Removal & Installations" procedure in this section.
3. If equipped with a manual transmission, remove the clutch housing cover.
4. Using Special Tool 498497100 to lock the crankshaft into place, remove the flywheel (manual transmission) or driveplate (automatic transmission).

To install:

5. Install the flywheel (M/T) or drive plate (A/T) and use Special Tool 498497100 to lock the crankshaft into place. Tighten the bolts in the sequence shown as follows:
    a. Except 3.0L engine: 51–55 ft. lbs. (69–75 Nm).
    b. 3.0L engine: 60 ft. lbs. (81 Nm).
6. Install the clutch housing cover, if equipped with manual transmission.
7. Install the engine assembly into the vehicle.

INTAKE MANIFOLD

REMOVAL & INSTALLATION

2.5L Engines

See Figures 133 and 134.

1. Before servicing the vehicle, refer to the Precautions Section.
2. Properly relieve the fuel system pressure. Remove the fuel cap.
3. Disconnect the negative battery cable. Drain the engine coolant.
4. If equipped, remove the undercover.
5. Remove the air intake duct, air cleaner case and air intake chamber.
6. If equipped, remove the intercooler.
7. Remove the alternator.
8. If equipped with DOHC engine, remove the coolant filler tank.
9. If equipped with SOHC engine, remove the spark plug wires.
10. Disconnect the engine coolant hoses from the throttle body. Disconnect the brake booster hose.
11. Disconnect the PCV hose from the intake manifold. Disconnect the engine harness electrical connectors from the bulkhead harness connectors.
12. Disconnect the engine coolant temperature sensor electrical connector, knock sensor electrical connector and crankshaft position sensor connector.

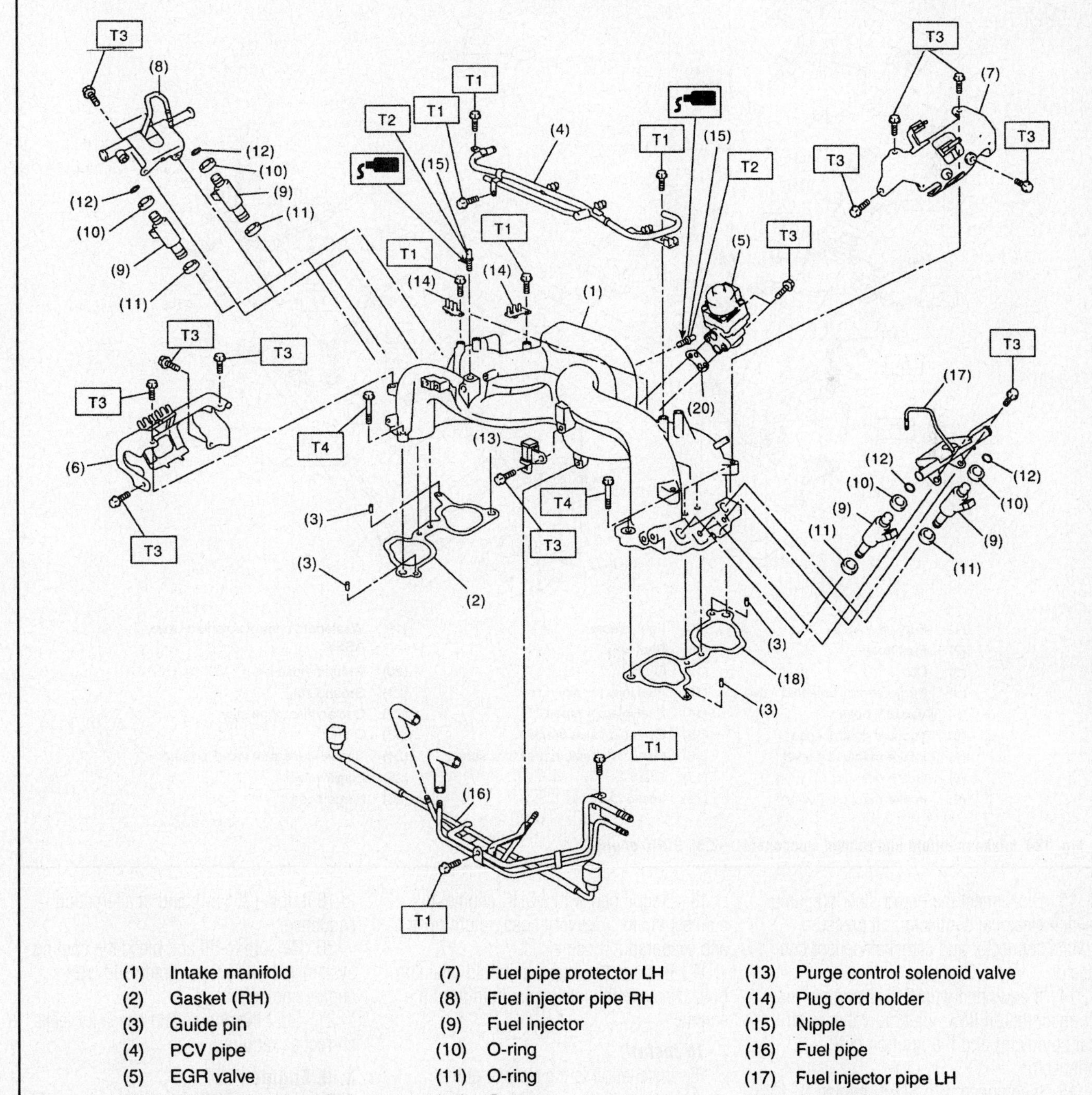

Fig. 133 Intake manifold and related components—2.5L SOHC engine

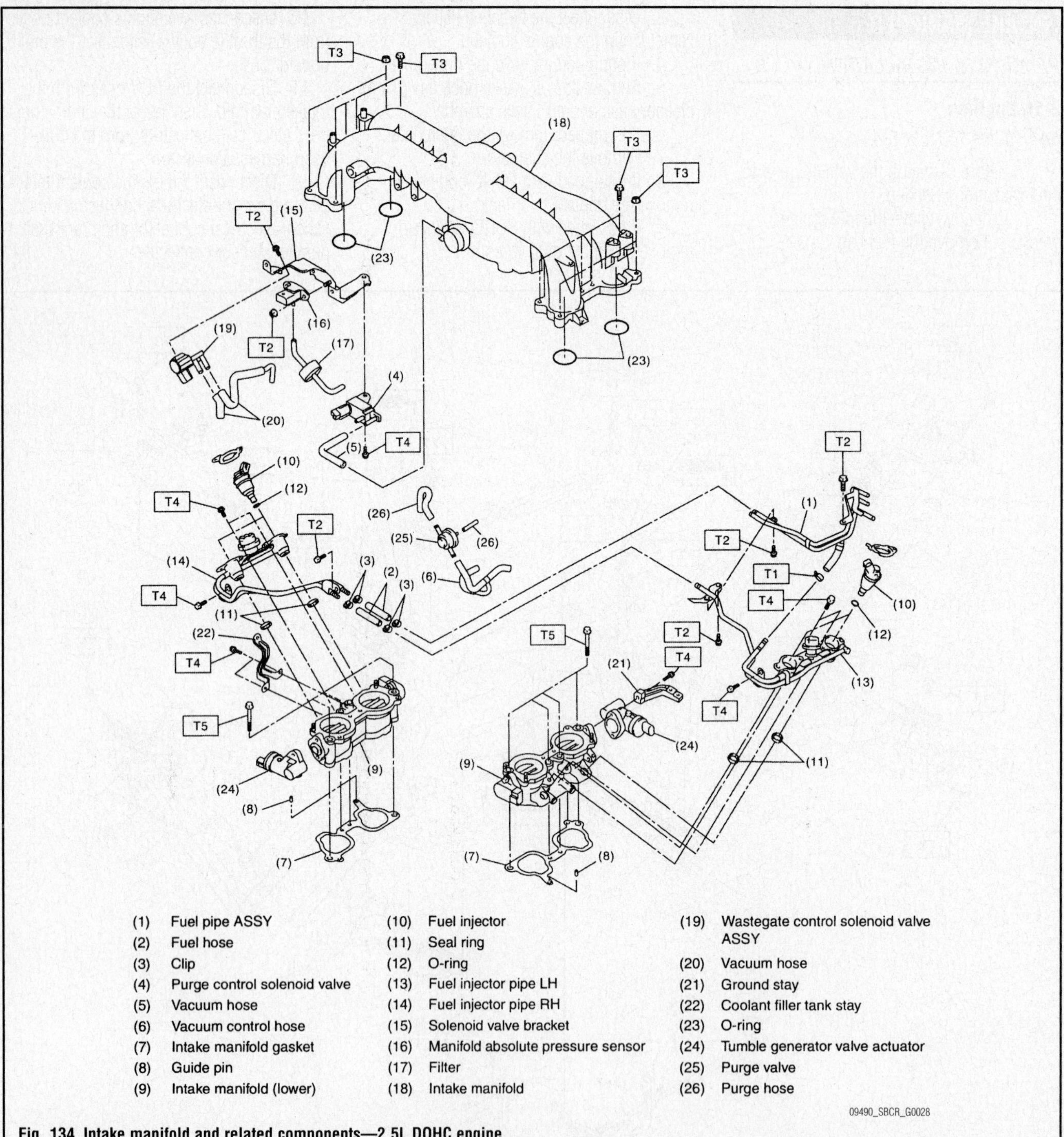

Fig. 134 Intake manifold and related components—2.5L DOHC engine

13. Disconnect the power steering pump switch electrical connector, oil pressure switch connector and camshaft sensor connector.

14. If equipped with DOHC engine, disconnect the oil flow solenoid valve electrical connector and the ignition coil connector.

15. If equipped with SOHC engine, remove the EGR pipe from the intake manifold and disconnect the fuel lines from the fuel pipe.

16. If equipped with DOHC engine, disconnect the fuel delivery hose, return hose and evaporation hose.

17. Remove the intake manifold retaining bolts. Remove the intake manifold from the engine.

To install:

18. Installation is the reverse of the removal procedure.

19. Be sure to use new intake manifold gaskets. Tighten the manifold retaining bolts to 18 ft. lbs. (25 Nm). and in alternating sequence.

20. Be sure to fill and bleed the cooling system with the proper grade and type engine coolant.

21. Start the engine and check for leaks, correct as required.

3.0L Engine

See Figure 135.

1. Before servicing the vehicle, refer to the Precautions Section.

2. Properly relieve the fuel system pressure. Remove the fuel cap.

3. Disconnect the negative battery cable. Drain the engine coolant.

4. Remove the air cleaner case and air intake chamber.

5. Remove the alternator.

6. Disconnect the electrical connector from the throttle body. Disconnect the engine coolant hoses from the throttle body.

7. Disconnect the engine harness connector. Disconnect the PCV hose. Disconnect the brake booster hose. Disconnect the fuel hoses from the fuel pipe.

8. Remove the left side fuel line protector. Remove the engine harness from the left side fuel injector pipe. Remove the bolts which hold the fuel injector pipe to the left cylinder head.

9. Remove the right side fuel line protector. Remove the engine harness from the right side fuel injector pipe. Remove the bolts which hold the fuel injector pipe to the right cylinder head.

10. Remove the right and left intake manifold retaining bolts. Remove the intake manifold from the engine.

To install:

11. Installation is the reverse of the removal procedure.

12. Be sure to use new intake manifold O-rings. Torque the manifold retaining bolts to specification and in alternating sequence.

13. Be sure to fill the cooling system with the proper grade and type engine coolant.

14. Start the engine and check for leaks, correct as required.

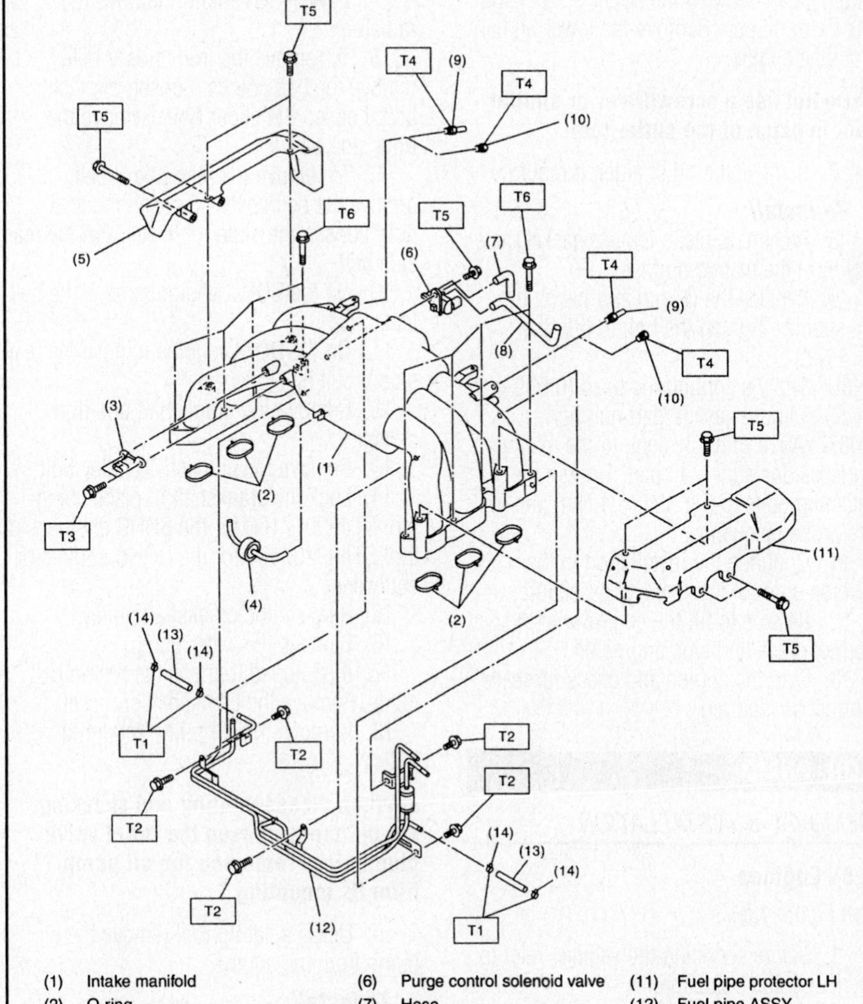

Fig. 135 Intake manifold and related components—3.0L engine

OIL PAN

REMOVAL & INSTALLATION

2.5L Engines

See Figures 136 and 137.

1. Before servicing the vehicle, refer to the Precautions Section.

2. Disconnect the negative battery cable.

3. Raise and support the vehicle safely.

4. Remove the front tires and wheels.

5. Lower the vehicle.

6. Remove the air intake duct and the air cleaner case. Remove the air intake chamber.

7. Remove the pitching stopper.

8. Remove the hood stay holder and the radiator upper brackets.

9. Properly support the engine with a lifting device and wire ropes.

10. Lift the vehicle and support it safely.

➡When lifting the vehicle, raise the wire ropes at the same time.

11. Remove the undercover.

12. Drain the engine oil.

13. Remove the front and center exhaust pipes.

14. Remove the nuts which retain the front cushion rubber onto the front crossmember.

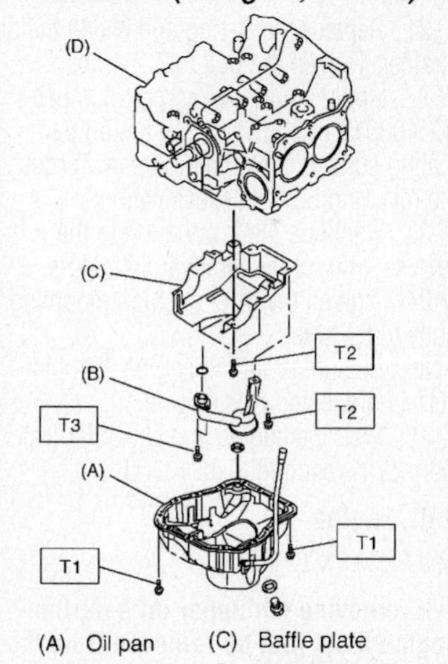

Fig. 136 Exploded view of oil pan, strainer, baffle plate and cylinder block

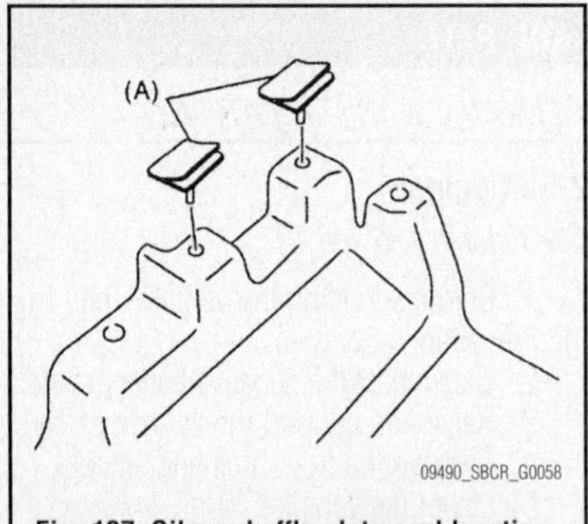

Fig. 137 Oil pan baffle plate seal location and positioning—2.5L engine

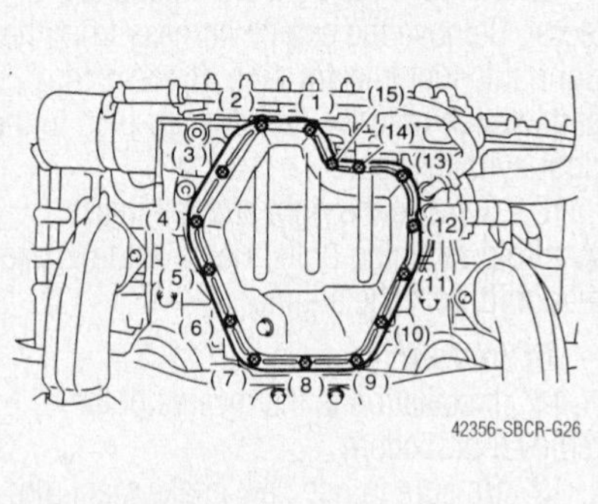

Fig. 138 Oil pan bolt torque sequence—3.0L engine

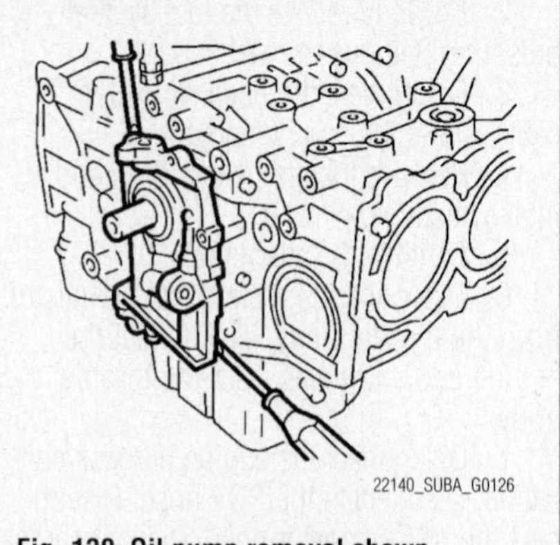

Fig. 139 Oil pump removal shown

15. Remove the bolts that retain the oil pan to the cylinder block, with the engine in the raised position.
16. Insert an oil pan gasket cutter tool into the gap between the cylinder block and the oil pan. Remove the oil pan from the engine.

➡Do not use a screwdriver or similar tool in place of the cutter tool.

17. Remove the oil strainer, if required. Remove the baffle plate, if required.

To install:

18. Be sure to clean the old gasket material from the mating surfaces.
19. Apply a continuous bead of sealer to a new oil pan gasket.
20. Make sure that the seals (A) are installed securely on the baffle plate and in the direction shown in the illustration. Install the baffle plate; tighten the retaining bolts to 4.7 ft. lbs.
21. Replace the O-ring and install the oil strainer. Tighten the bolt to 7.2 ft. lbs.
22. Apply liquid gasket, part number 004403012 or equivalent, to the oil pan mating surface. Install the oil pan. Torque the retaining bolts to specification.
23. Continue the installation in the reverse order of the removal procedure.
24. Tighten the front cushion mounting bolts to 63 ft. lbs.
25. Be sure to fill the engine with the correct grade and type engine oil.
26. Start the engine and check for leaks. Correct as required.

3.0L Engine

See Figure 138.

➡If removing the upper oil pan, the engine must first be removed from the vehicle.

1. Before servicing the vehicle, refer to the Precautions Section.
2. Disconnect the negative battery cable.
3. Raise and support the vehicle safely.
4. Remove the undercover. Drain the engine oil.
5. Remove the lower oil pan retaining bolts.
6. Insert an oil pan gasket cutter tool into the gap between the upper oil pan and the lower oil pan. Remove the lower oil pan from the engine.

➡Do not use a screwdriver or similar tool in place of the cutter tool.

7. Remove the oil strainer, if required.

To install:

8. Be sure to clean the old gasket material from the mating surfaces.
9. Replace the O-ring and install the oil strainer. Tighten the bolt to 5 ft. lbs (6.4 Nm)
10. Apply a continuous bead (0.039 inch thick) of liquid gasket, part number K0877YA018 or equivalent, to the mating surfaces. Install the oil pan. Tighten the retaining bolts to 5 ft. lbs (6.4 Nm) and in the proper sequence.
11. Continue the installation in the reverse order of the removal procedure.
12. Be sure to fill the engine with the correct grade and type engine oil.
13. Start the engine and check for leaks. Correct as required.

OIL PUMP

REMOVAL & INSTALLATION

2.5L Engines

See Figure 139.

1. Before servicing the vehicle, refer to the Precautions Section.
2. Disconnect the negative battery cable. Drain the cooling system.
3. On the DOHC engine, remove the collector cover.
4. Raise and support the vehicle safely.
5. Remove the undercover.
6. On the DOHC engine, remove the bolts which retain the water pipe of the oil cooler to the oil pump. Remove the water pipe and hoses between the oil cooler and the water pump.
7. Lower the vehicle. Remove the radiator.
8. To remove the front side V belt, remove the belt covers. Loosen the lock bolt. Loosen the slider bolt. Remove the front side belt.
9. To remove the rear side V belt, remove the belt covers. Loosen the lock bolt. Loosen the slider bolt. Remove the rear side belt.
10. On the SOHC engine remove the belt tensioner.
11. On the DOHC engine remove the rear side V belt tensioner
12. Remove the crankshaft position sensor.
13. Remove the crankshaft pulley bolt.
14. Lock the crankshaft in place using tool ST499977100 for the SOHC engine and tool ST499207400 for the DOHC engine, or equivalent.
15. Remove the crankshaft pulley.
16. Remove the water pump.
17. If equipped, remove the timing belt guide. Remove the crankshaft sprocket.
18. Remove the oil pump retaining bolts.

➡When disassembling and checking the oil pump, loosen the relief valve plug before removing the oil pump from its mounting.

19. Using a flat tip tool remove the oil pump from the engine.

To install:

20. Be sure all mating surfaces are clean and free of dirt.
21. Apply liquid gasket part number 004403007, or equivalent to the mating surfaces of the oil pump.

22. Be sure to replace the O-ring with a new one.

23. Apply a thin coat of clean engine oil to the inside of the oil seal.

24. Position the oil pump to its mounting, aligning the notched area with the crankshaft and push the pump straight.

➡Be sure that the oil seal lip is not folded.

25. Install the oil pump. Apply liquid gasket part number 004403042, or equivalent to the three retaining bolt threads. Install the bolts and tighten to 5 ft. lbs (6.4 Nm)

26. Continue the installation in the reverse order of the removal procedure.

27. Be sure to fill the cooling system with the proper grade and type coolant.

28. Start the engine and check for leaks. Correct, as required.

3.0L Engine

See Figure 140.

1. Before servicing the vehicle, refer to the Precautions Section.

2. Disconnect the negative battery cable. Drain the cooling system.

3. Remove the collector cover.

4. Raise and support the vehicle safely.

5. Remove the undercover.

6. Lower the vehicle. Remove the radiator.

7. To remove the front side V belt, remove the belt covers. Loosen the lock bolt. Loosen the slider bolt. Remove the front side belt.

8. To remove the rear side V belt, remove the belt covers. Loosen the lock bolt. Loosen the slider bolt. Remove the rear side belt.

9. Remove the crankshaft pulley cover.

10. Remove the crankshaft pulley bolt.

11. Lock the crankshaft in place using tool ST499977100, or equivalent.

12. Remove the crankshaft pulley.

13. Remove the front timing chain cover.

➡Bolts are three different sizes. Be careful to install the correct bolt in the correct hole.

14. Remove the timing chain.

15. Remove the crankshaft sprocket.

16. Remove the oil pump cover retaining bolts. Remove the oil pump cover.

17. Remove the inner and outer rotors.

To install:

18. Be sure all mating surfaces are clean and free of dirt.

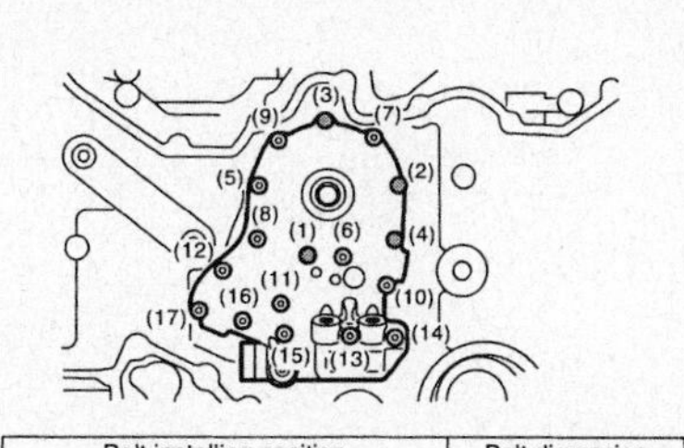

Bolt installing position	Bolt dimension
(1) and (3)	6 x 14 x 14
(2) and (4)	6 x 35 x 18
(5), (6), (7), (8), (9), (10) and (11)	6 x 35 x 15
(12), (15), (16) and (17)	6 x 16 x 16
(13) and (14)	6 x 26 x 15

09490_SBCR_G0059

Fig. 140 Oil pump tightening sequence and bolt location—3.0L engine

19. Apply a thin coat of clean engine oil to the complete area of the inner and outer rotors.

20. Position the inner rotor in place. Position the outer rotor in place.

21. Install the pump cover. Tighten the retaining bolts to 5 ft. lbs (6.4 Nm). and in the proper sequence.

➡Make sure that the bolts are installed in the correct positions.

22. Continue the installation in the reverse order of the removal procedure.

23. Be sure to fill the cooling system with the proper grade and type coolant.

24. Start the engine and check for leaks. Correct, as required.

INSPECTION

See Figures 141, 142 and 143.

1. Check the oil pump case for worn shaft hole, clogged oil passage, worn rotor chamber, cracks, and other faults.

2. Check the oil seal lips for deformation, hardening, wear, etc. and replace if defective.

3. Measure the tip clearance of rotors. If clearance exceeds the standard, replace the rotors as a matched set.

a. Except 3.0L Engine: 0.0016–0.0055 in. (0.04–0.14 mm)

b. 3.0L Engine: 0.0016–0.0055 in. (0.04–0.14 mm)

4. Measure the clearance between the outer rotor and oil pump rotor housing. If clearance exceeds the standard, replace the rotor.

a. Except 3.0L Engine: 0.0039–0.0069 in. (0.10–0.18 mm)

b. 3.0L Engine: 0.0043–0.0069 in. (0.11–0.18 mm)

5. Measure the clearance between the oil pump inner rotor and pump cover. If clearance exceeds the standard, replace the rotor or pump body.

a. Except 3.0L Engine: 0.0008–0.0028 in. (0.02–0.07 mm)

b. 3.0L Engine: 0.0008–0.0018 in. (0.02–0.05 mm)

6. Check the valve for fitting condition and damage, and the relief valve spring for damage and deterioration. Replace the parts if defective.

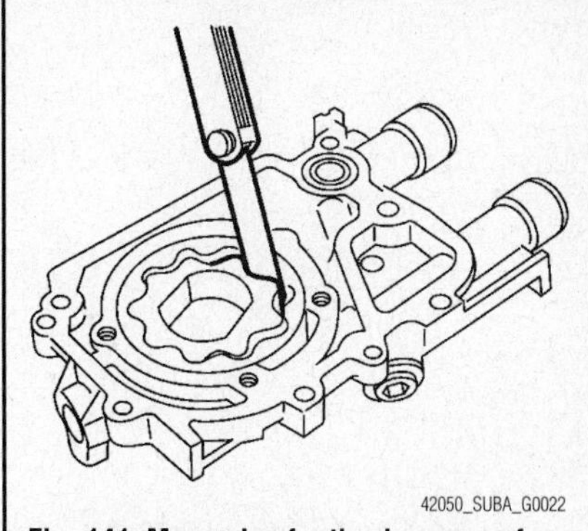

42050_SUBA_G0022

Fig. 141 Measuring for tip clearance of the rotors of the oil pump.

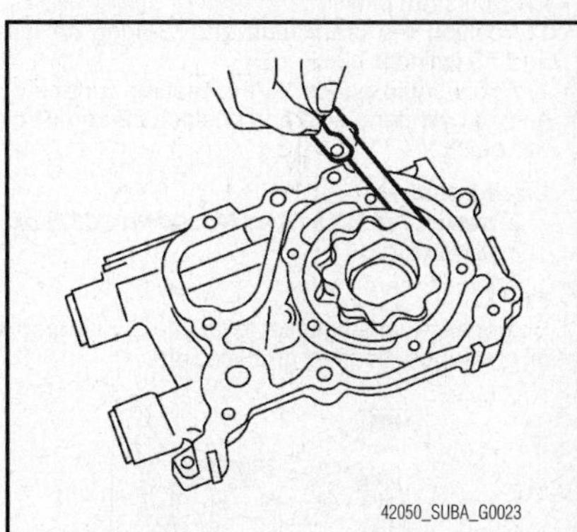

42050_SUBA_G0023

Fig. 142 Measuring the case clearance of the oil pump.

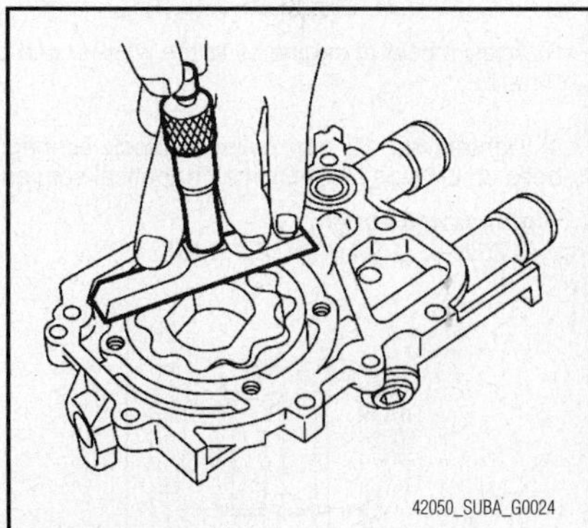

42050_SUBA_G0024

Fig. 143 Measuring the side clearance of the oil pump.

MAIN BEARING TORQUE SEQUENCE

See Figures 144, 145 and 146.

NOTE:
Remove oil on the mating surface of bearing and cylinder block before installation. Apply engine oil to crankshaft pins.

1) Position the crankshaft and O-rings on the #1 and #3 cylinder block.

2) Apply liquid gasket to the mating surface of #1 and #3 cylinder block, and position #2 and #4 cylinder block.

Liquid gasket:
Three bond 1215 (Part No. 004403007) or equivalent

NOTE:
Do not allow liquid gasket to jut into O-ring grooves, oil passages, bearing grooves, etc.

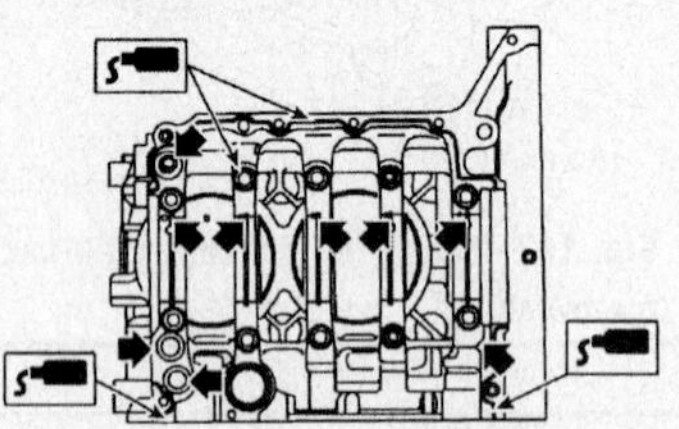

3) Apply a coat of engine oil to the washer and bolt thread.

4) Tighten the 10 mm cylinder block connecting bolts on LH side (A — D) in alphabetical sequence.

Tightening torque:
9.75 N·m (1.0 kgf-m, 7.2 ft-lb)

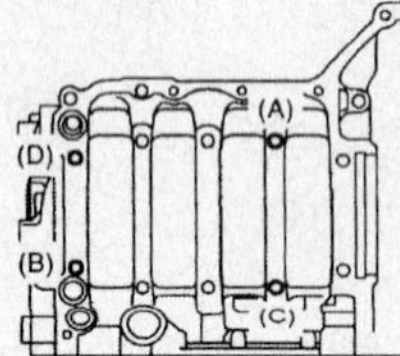

5) Tighten the 10 mm cylinder block connecting bolts on RH side (E — J) in alphabetical sequence.

Tightening torque:
9.75 N·m (1.0 kgf-m, 7.2 ft-lb)

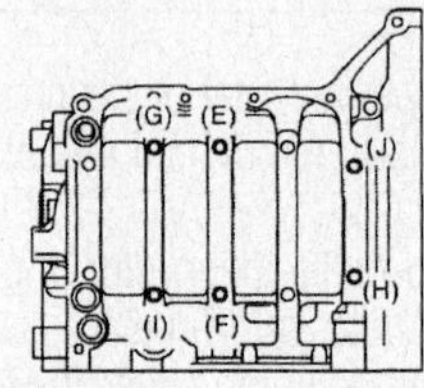

6) Further tighten the LH side bolts (A — D) in alphabetical sequence.

Tightening torque:
18 N·m (1.8 kgf-m, 13 ft-lb)

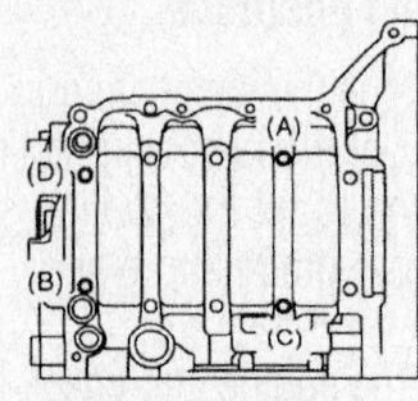

7) Further tighten the RH side bolts (E — J) in alphabetical sequence.

Tightening torque:
18 N·m (1.8 kgf-m, 13 ft-lb)

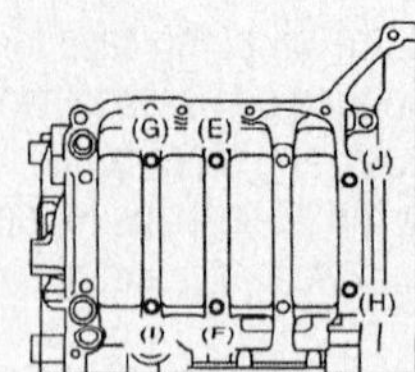

8) Further tighten the LH side bolts (A — D) in alphabetical sequence.

- (A), (C): Angle tightening

Tightening angle:
90°

- (B), (D): Torque tightening

Tightening torque:
40 N·m (4.1 kgf-m, 29.6 ft-lb)

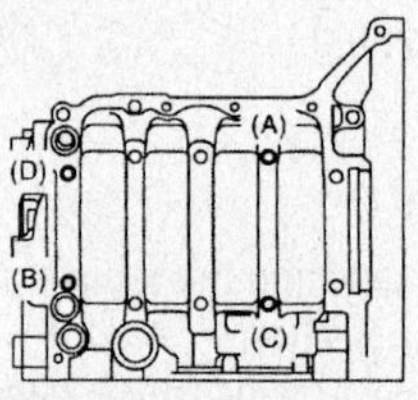

9) Tighten the RH side bolts (E — J) 90° further in alphabetical sequence.

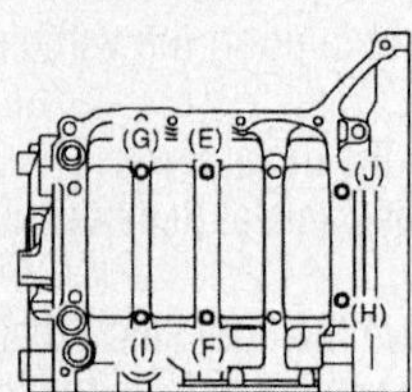

10) Tighten the 8 mm and 6 mm cylinder block connecting bolts on LH side (A — H) in alphabetical sequence.

Tightening torque:
(A) — (G): 25 N·m (2.5 kgf-m, 18.1 ft-lb)
(H): 6.4 N·m (0.65 kgf-m, 4.7 ft-lb)

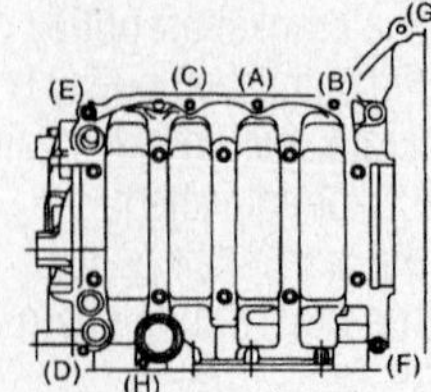

09490_SBCR_G0004

Fig. 144 Main bearing torque sequence—2.5L SOHC engine.

1) Remove oil on the mating surface of bearing and cylinder block before installation. Apply engine oil to crankshaft pins.
2) Position the crankshaft and O-rings on the #1 and #3 cylinder block.
3) Apply liquid gasket to the mating surface of #1 and #3 cylinder blocks, and position cylinder block #2 and #4.

NOTE:
Do not allow liquid gasket to run over to O-ring grooves, oil passages, bearing grooves, etc.

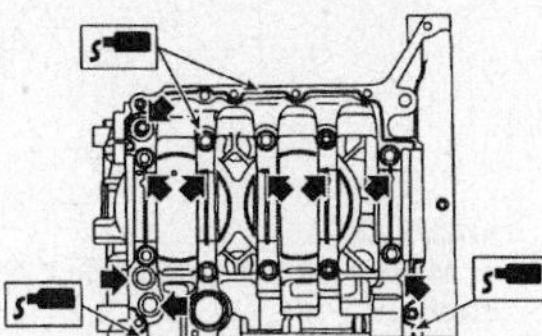

4) Apply a coat of engine oil to the washer and bolt thread.
5) Tighten the 10 mm cylinder block connecting bolts on LH side (A — D) in alphabetical sequence.

Tightening torque:
10 N·m (1.0 kgf-m, 7.2 ft-lb)

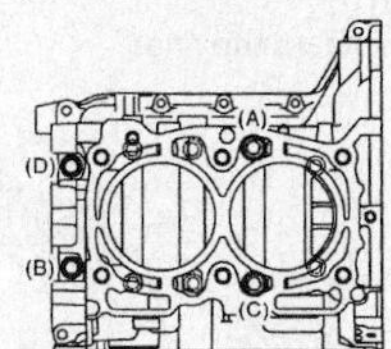

6) Tighten the 10 mm cylinder block connecting bolts on RH side (E — J) in alphabetical sequence.

Tightening torque:
10 N·m (1.0 kgf-m, 7.2 ft-lb)

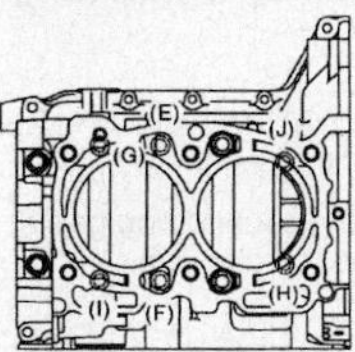

7) Further tighten the LH side bolts (A — D) in alphabetical sequence.

Tightening torque:
18 N·m (1.8 kgf-m, 13.0 ft-lb)

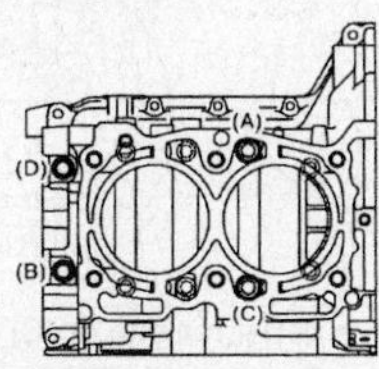

8) Further tighten the RH side bolts (E — J) in alphabetical sequence.

Tightening torque:
18 N·m (1.8 kgf-m, 13.0 ft-lb)

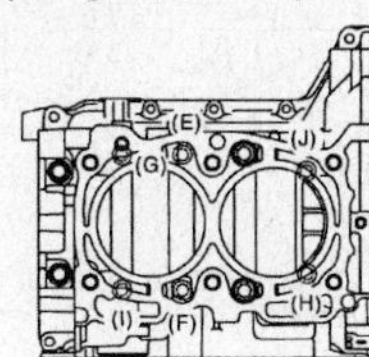

9) Further tighten the LH side bolts (A — D) in alphabetical sequence.
(A), (C): 90°
(B), (D): 40 N·m (4.1 kgf-m, 29.5 ft-lb)

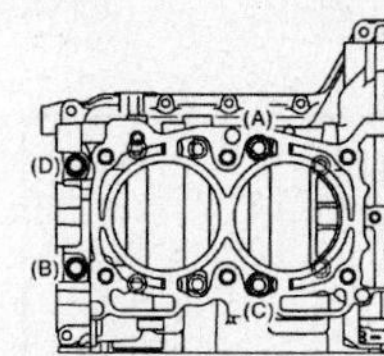

10) Tighten the RH side bolts (E — J) 90° further in alphabetical sequence.

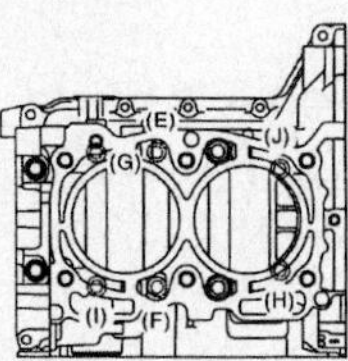

11) Tighten the 8 mm and 6 mm cylinder block connecting bolts on LH side (A — H) in alphabetical sequence.

Tightening torque:
(A) — (G): 25 N·m (2.5 kgf-m, 18.1 ft-lb)
(H): 6.4 N·m (0.65 kgf-m, 4.7 ft-lb)

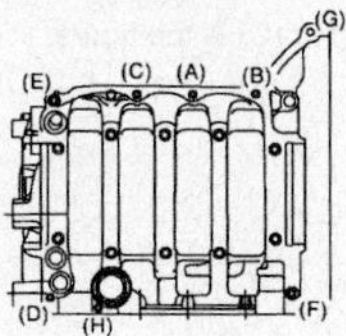

09490_SBCR_G0007

Fig. 145 Main bearing torque sequence—2.5L DOHC engine

1) After setting the cylinder block to ST, install the crankshaft bearing.

NOTE:
Remove oil on the mating surface of bearing and cylinder block before installation. Apply a coat of engine oil to crankshaft pins.
2) Position the crankshaft and connecting rod on #2, #4 and #6 cylinder block.
3) Apply liquid gasket to the mating surface of the #1, #3 and #5 cylinder blocks, and position it on #2, #4 and #6 cylinder blocks.

NOTE:
Do not allow liquid gasket to run over to oil passages, bearing grooves, etc.

Applying liquid gasket diameter:
1.0±0.2 mm (0.039±0.008 in)

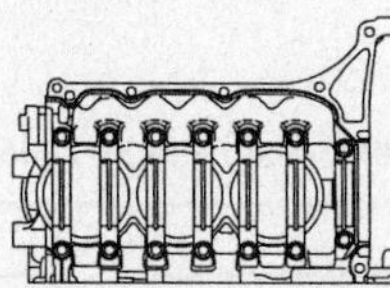

4) Apply a coat of engine oil to the washer and bolt thread.
5) Tighten all bolts in the numerical order as shown in the figure.

Tightening torque:
(1) — (11), (13): 25 N·m (2.5 kgf-m, 18 ft-lb)
(12), (14): 20 N·m (2.0 kgf-m, 14 ft-lb)

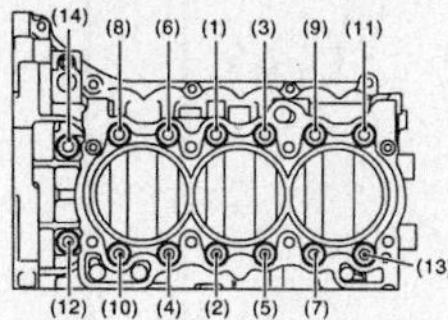

6) Retighten all bolts in the numerical order as shown in the figure.

Tightening torque:
(1) — (11), (13): 25 N·m (2.5 kgf-m, 18.4 ft-lb)
(12), (14): 20 N·m (2.0 kgf-m, 14 ft-lb)

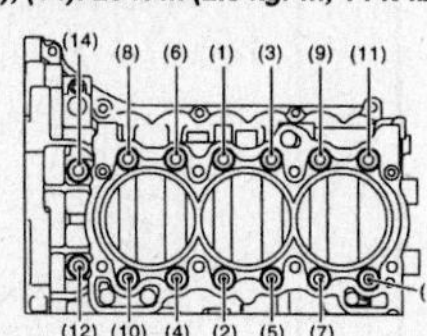

7) Tighten all bolts 90° — 110° in the numerical order as shown in the figure.

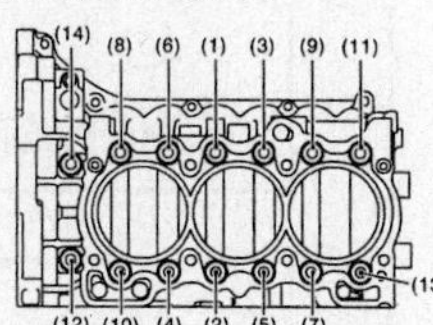

8) Install the upper bolt to cylinder block.

Tightening torque:
25 N·m (2.5 kgf-m, 18 ft-lb)

NOTE:
Remove the liquid gasket which is running over to sealing surface between cylinder block and rear chain cover, cylinder block and oil pan upper, after tightening the bolts which combine the cylinder block.

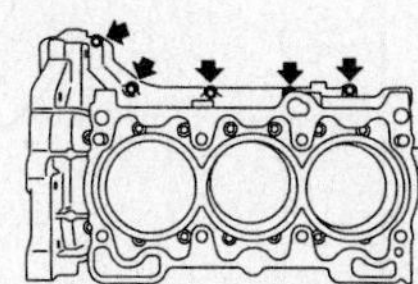

09490_SBCR_G0009

Fig. 146 Main bearing torque sequence—3.0L engine

PISTON AND RING

POSITIONING

See Figures 147 through 150.

Position the top ring gap at (A) or (B) in the figure.

Position the second ring gap at 180° on the reverse side the top ring gap.

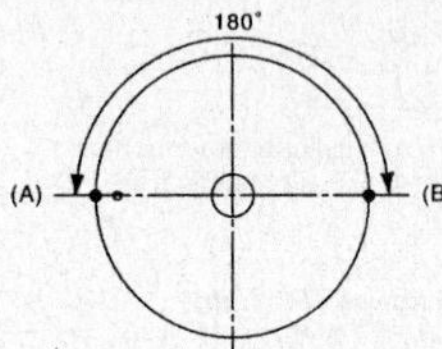

Position the upper rail gap at (C) in the figure.

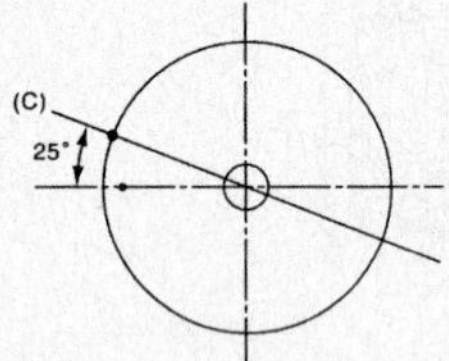

Align the upper rail spin stopper (D) to the side hole (E) on the piston.

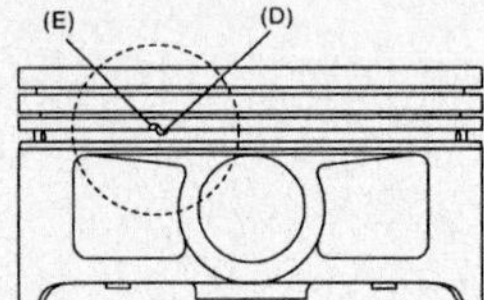

Position the expander gap at (F) in the figure on the 180° opposite direction of (C).

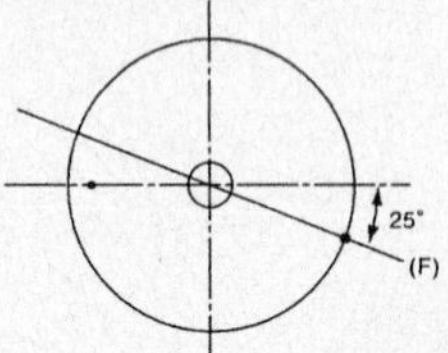

Position the lower rail gap at (G) in the figure.

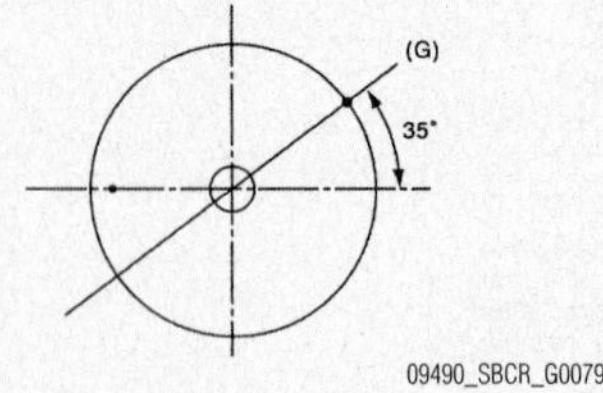

Fig. 147 Piston ring alignment and positioning—2005 2.5L SOHC engine

Position the top ring gap at (A) or (B) in the figure.

NOTE:
Assemble so that the piston ring mark "R" faces the upper side of the piston.

Position the second ring gap at 180° on the reverse side the top ring gap.

NOTE:
Assemble so that the piston ring mark "R" faces the upper side of the piston.

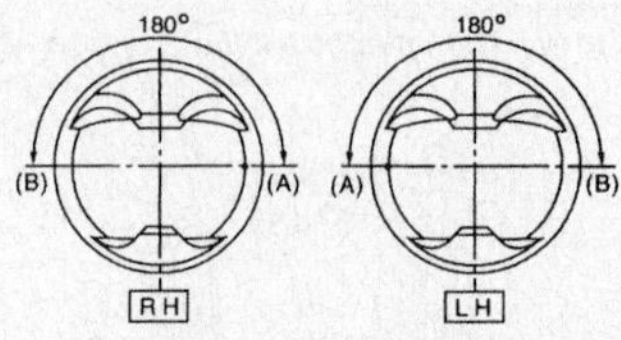

Position the upper rail gap at (C) in the figure.

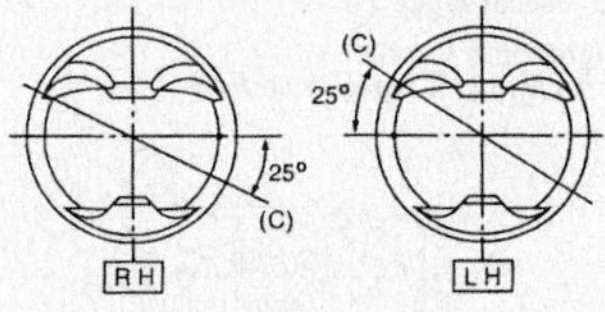

Align the upper rail spin stopper (E) to the side hole (D) on the piston.

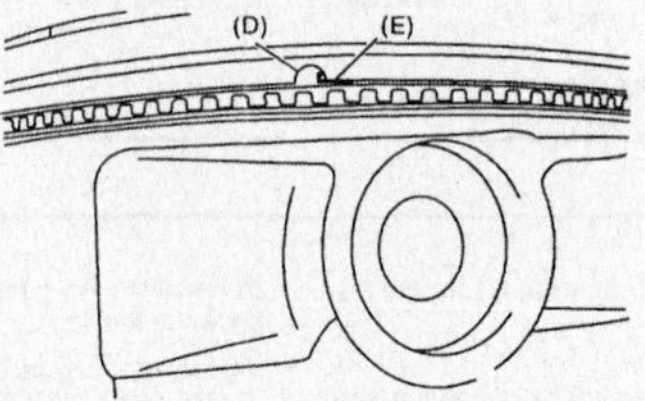

Position the expander gap at (F) in the figure.

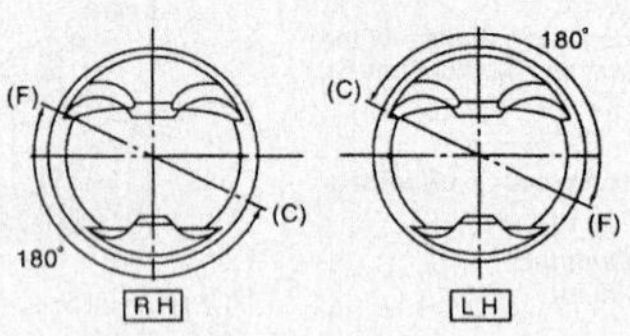

Position the lower rail gap at (G) in the figure.

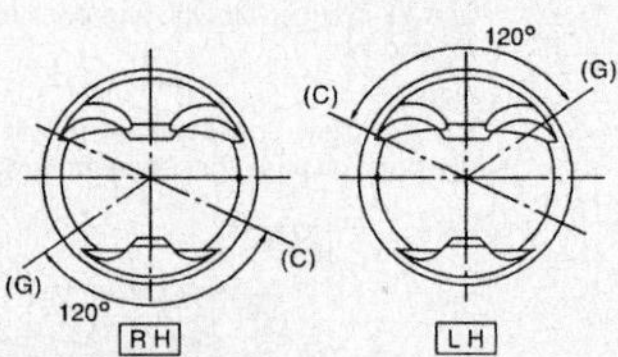

CAUTION:
- **Make sure ring gaps do not face the same direction.**
- **Make sure ring gaps are not within the piston skirt area.**

Install the snap ring.

Install the snap rings in the piston holes located opposite to the service holes in cylinder block when positioning all pistons in corresponding cylinders.

NOTE:
Use new snap rings.

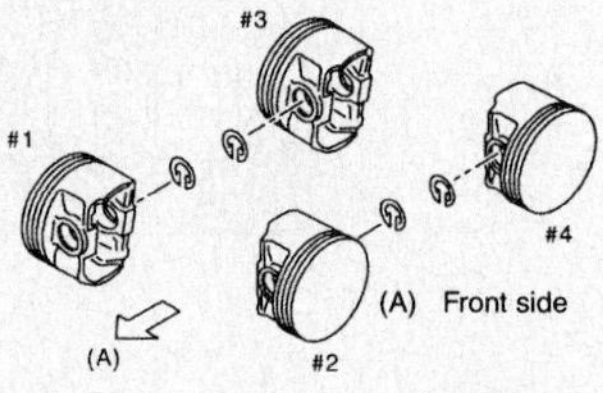

CAUTION:
Piston front mark faces towards the front of engine.

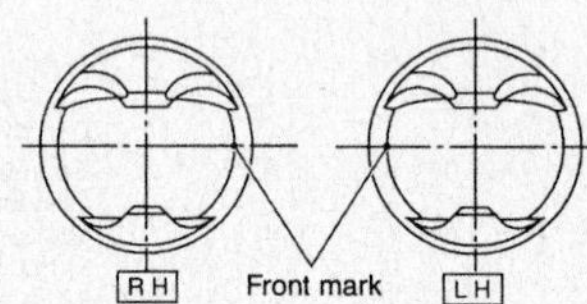

Fig. 148 Piston ring alignment and positioning—2005 2.5L DOHC engine

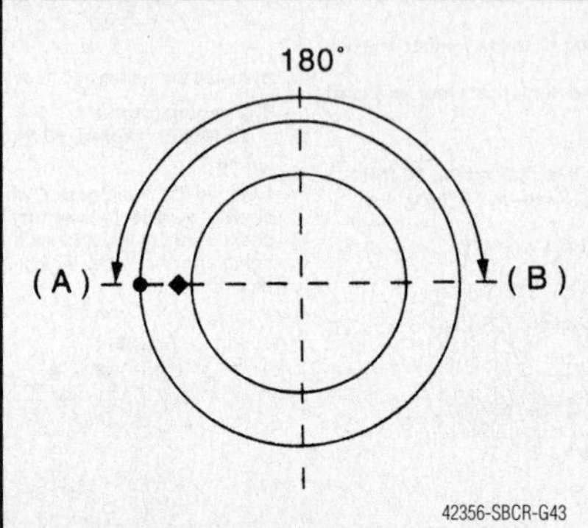

Fig. 149 Top ring end-gap (A), second ring gap (B)—3.0L engine

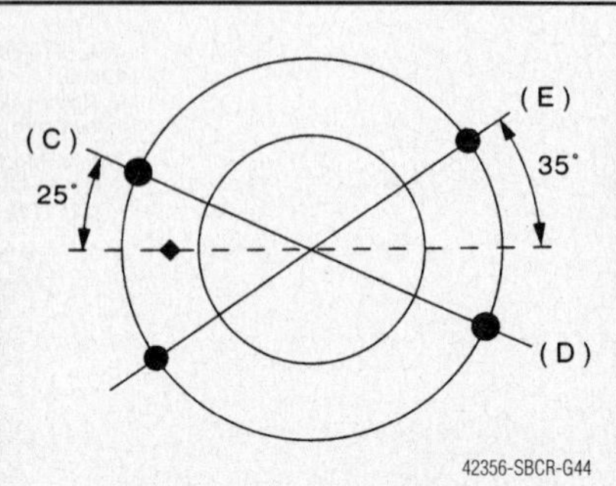

Fig. 150 Upper rail end-gap (C), expander gap (D) and lower rail gap (E)—3.0L engine

REAR MAIN SEAL

REMOVAL & INSTALLATION

See Figure 151.

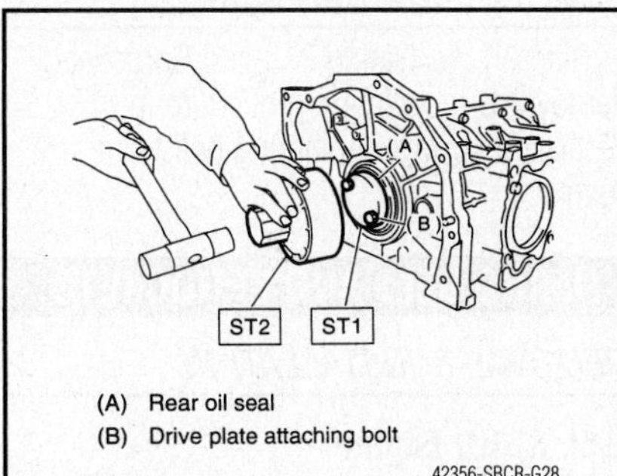

Fig. 151 Installing the rear main seal using oil seal guide ST1 499597100 and ST2 499598200

1. Before servicing the vehicle, refer to the Precautions Section.
2. Remove or disconnect the following:
   - Engine from the vehicle
   - Clutch assembly/flywheel using the Clutch Disc Guide tool 499747000, if equipped with a manual transmission
   - Torque converter flexplate from the crankshaft, if equipped with an automatic transmission
   - Oil seal from the cylinder block using a small pry bar

To install:

3. Install or connect the following:
   - New oil seal by pressing it into the cylinder block using the appropriate driver and hammer
   - Flywheel housing using new gaskets and sealant where necessary.
   - Flywheel and tighten the bolts to specification.
   - Engine

ROCKER ARMS/SHAFTS

REMOVAL & INSTALLATION

2.5L SOHC Engine

See Figures 152 through 155.

1. Before servicing the vehicle, refer to the Precautions Section.
2. Disconnect the negative battery cable.
3. To remove the front side V belt, remove the belt covers. Loosen the lock bolt. Loosen the slider bolt. Remove the front side belt.
4. To remove the rear side V belt, remove the belt covers. Loosen the lock bolt. Loosen the slider bolt. Remove the rear side belt. Remove the belt tensioner.
5. Remove the crankshaft pulley bolt.
6. Lock the crankshaft in place using tool ST499977100, or equivalent.
7. Remove the crankshaft pulley.
8. Remove the left side timing belt cover.
9. Remove the front timing belt cover.
10. Remove the timing belt.
11. Remove the camshaft position sensor.
12. Remove the camshaft sprockets, No.1 and No.2.

➡Be sure to lock the camshaft in place using tool ST18231AA010, or equivalent.

13. Disconnect the PCV valve hose. Remove the rocker cover retaining bolts. Remove the rocker cover from the engine.
14. Use tool St18258AA000, or equivalent, to rotate the spring stopper in direction of the arrow, see illustration, and remove the spring stopper from the adjuster spring.
15. Remove bolts "A" through "J" in sequence. Remove the assembly from the engine.

➡Leave two or three threads on bolts "I" and "J" engaged in order to retain the rocker arm assembly.

16. Position tool ST18354AA000 on the rocker arm assembly. Remove the assembly from the engine.

To install:

17. Install the rocker arm assembly on its mounting.
18. Temporarily tighten the bolts equally in the proper sequence.

➡Do not temporarily tighten bolts "I" and "J". Position tool ST18354AA000 to the rocker arm assembly.

19. Tighten bolts "A" through "H" to specification 18 ft. lbs. (25 Nm).
20. Tighten bolts "I" through "J" to specification 4.3 ft. lbs. (5 Nm).
21. Use tool ST18258AA000 to rotate the spring stopper in the opposite direction of the arrow, see illustration, to fasten the adjuster pin.
22. Adjust the valve clearance.
23. Use a new gasket and install the rocker arm cover. Tighten the retaining bolts to 5 ft. lbs (6.4 Nm). and in the proper sequence. Recheck and tighten again to 5 ft. lbs (6.4 Nm) in the proper sequence.

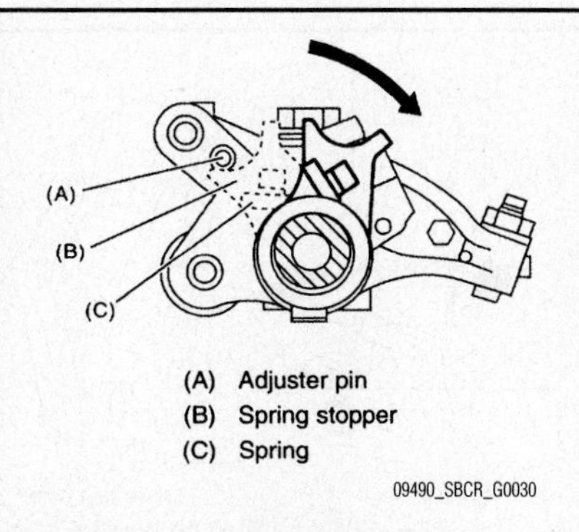

Fig. 152 Rocker arm spring stopper removal—2.5L SOHC engine

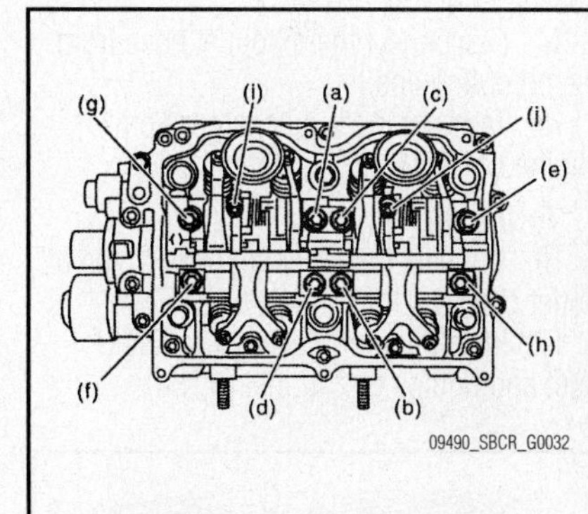

Fig. 154 Rocker arm assembly bolt tightening sequence—2.5L SOHC engine

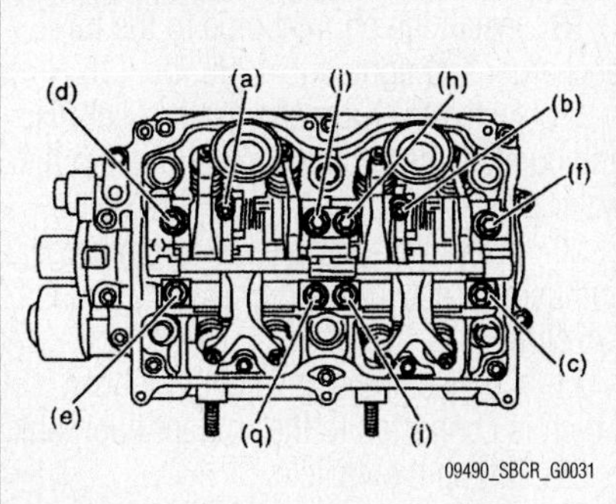

Fig. 153 Rocker arm assembly bolt loosening sequence—2.5L SOHC engine

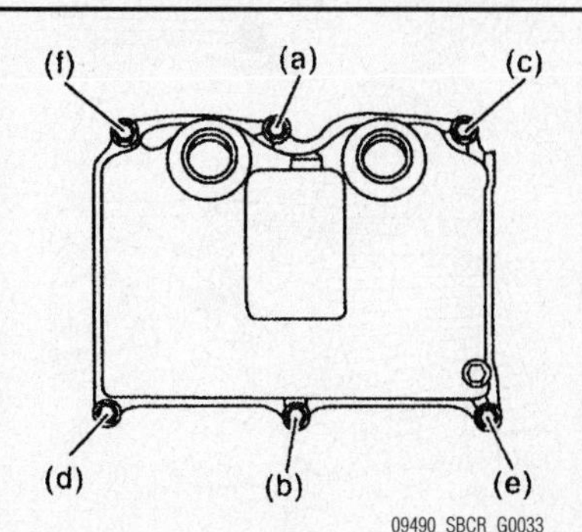

Fig. 155 Rocker arm cover bolt tightening sequence—2.5L SOHC engine

24. Continue the installation in the reverse order of the removal procedure.

TURBOCHARGER

REMOVAL & INSTALLATION

2.5L DOHC Engine

See Figures 156 through 159.

1. Before servicing the vehicle, refer to the Precautions Section.
2. Set the vehicle on a lift.
3. Remove the collector cover.
4. Disconnect the ground cable from the battery.
5. Remove the intercooler.
6. Remove the center exhaust pipe.
7. Lower the vehicle.
8. Separate the turbocharger joint pipe from turbocharger.
9. Disconnect the engine coolant hose which is connected to coolant filler tank.
10. Loosen the clamp which secures turbocharger to intake duct.
11. Remove the bolts which secure oil inlet pipe bracket to turbocharger.
12. Remove the oil inlet pipe from the turbocharger.
13. Disconnect the engine coolant hose from the water pipe, and remove the turbocharger bracket RH.
14. Disconnect the oil outlet hose from the oil outlet pipe.
15. Take out the turbocharger from engine compartment.

To install:

16. Connect the oil outlet hose to the oil outlet pipe.
17. Install the turbocharger to intake duct and tighten to 2 ft. lbs. (3 Nm).

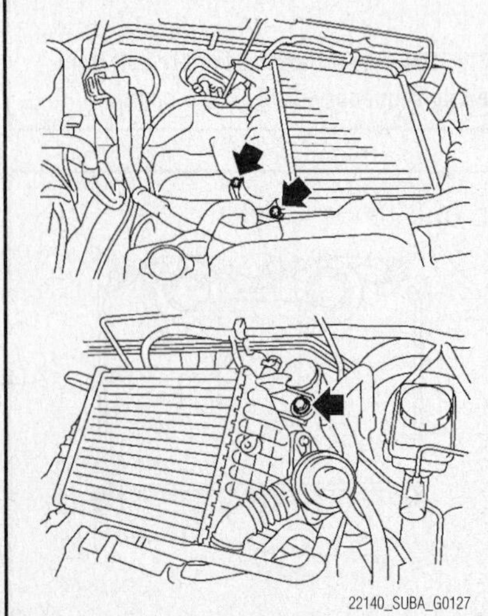

Fig. 156 Intercooler mounting bolts location shown

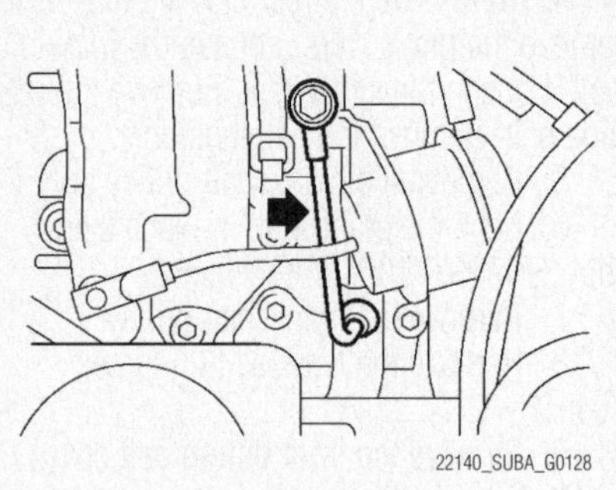

Fig. 157 Remove the oil inlet pipe from the turbocharger

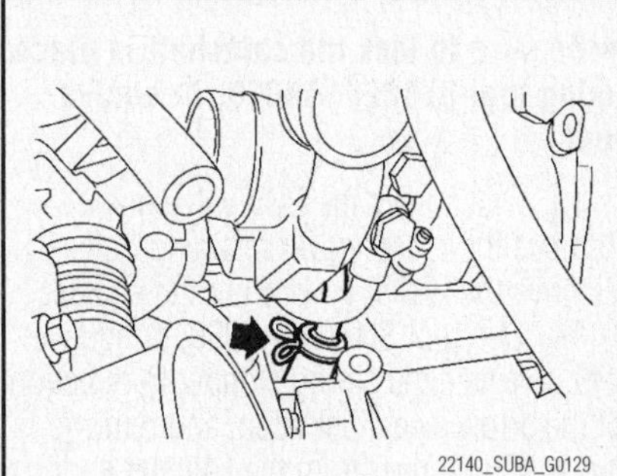

Fig. 158 Disconnect the oil outlet hose from the oil outlet pipe

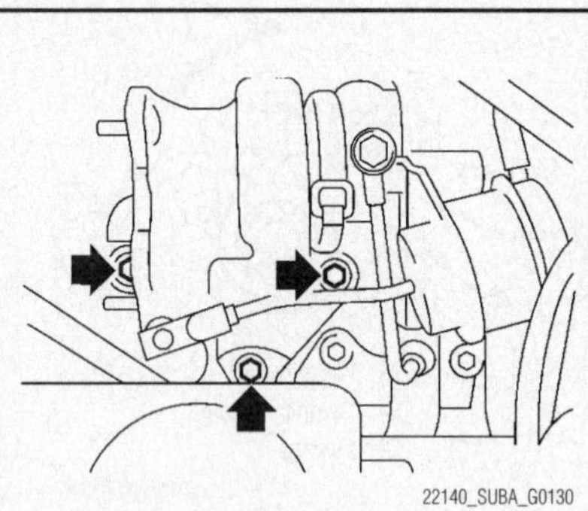

Fig. 159 Install the joint pipe to turbocharger

18. Install the oil inlet pipe to the turbocharger and tighten to 12 ft. lbs. (16 Nm).
19. Install the turbocharger bracket RH, and connect the engine coolant hose to the water pipe. Tighten to 24 ft. lbs. (33 Nm).
20. Install the joint pipe to turbocharger and using a new gasket tighten to 25 ft. lbs. (35 Nm).
21. Connect the engine coolant hose which is connected to the coolant filler tank.
22. Lift up the vehicle.
23. Install the center exhaust pipe.
24. Lower the vehicle.
25. Install the intercooler and tighten mounting bolts to 12 ft. lbs. (16 Nm).
26. Install the collector cover.
27. Connect the ground cable to battery.

TIMING BELT FRONT COVER

REMOVAL & INSTALLATION

Refer to the Timing Belt and Sprockets, Removal & Installation procedure in this section to remove the timing belt front cover.

TIMING BELT AND SPROCKETS

REMOVAL & INSTALLATION

2.5L SOHC Engine

See Figures 160 through 164.

1. Before servicing the vehicle, refer to the Precautions Section.
2. Disconnect the negative battery cable.
3. To remove the front side V belt, remove the belt covers. Loosen the lock bolt. Loosen the slider bolt. Remove the front side belt.
4. To remove the rear side V belt, remove the belt covers. Loosen the lock bolt. Loosen the slider bolt. Remove the rear side belt. Remove the belt tensioner.
5. Remove the crankshaft pulley bolt.
6. Lock the crankshaft in place using tool ST499977100, or equivalent.
7. Remove the crankshaft pulley.
8. Remove the left side timing belt cover.
9. Remove the front timing belt cover.
10. If equipped with manual transmission, remove the timing belt guide.

➡If the belt is going to be reused and the alignment mark on the belt is not readable, put a new mark on the belt to indicate the direction of rotation. Using tool ST499987500, turn the crankshaft to align the mark of the sprocket "A" to the cylinder mark notch "B". Ensure that the right side cam sprocket mark "C", cam cap and cylinder head matching surface "D" or left side cam sprocket mark "E", timing belt cover notch "F" are properly aligned. Paint an alignment mark on the belt in relation to the crankshaft sprocket and camshaft sprockets. Z1 measurement is 46.8 teeth. Z2 measurement is 43.7 teeth.

11. Remove both the number two belt idlers. Remove the timing belt from the engine.
12. Remove the number one belt idler. Remove the automatic belt tension adjuster assembly.

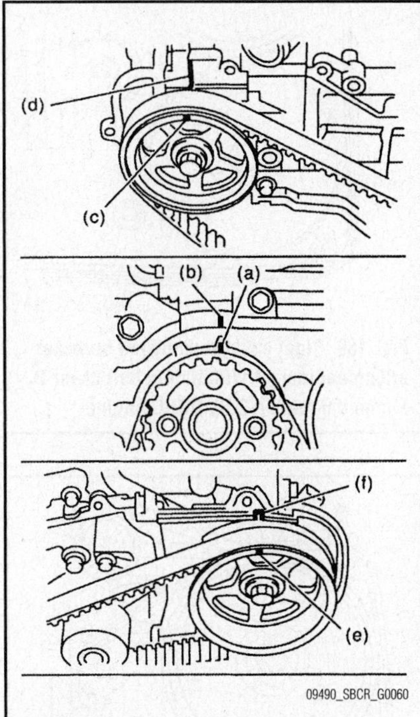

Fig. 160 Timing belt alignment—2.5L SOHC engine

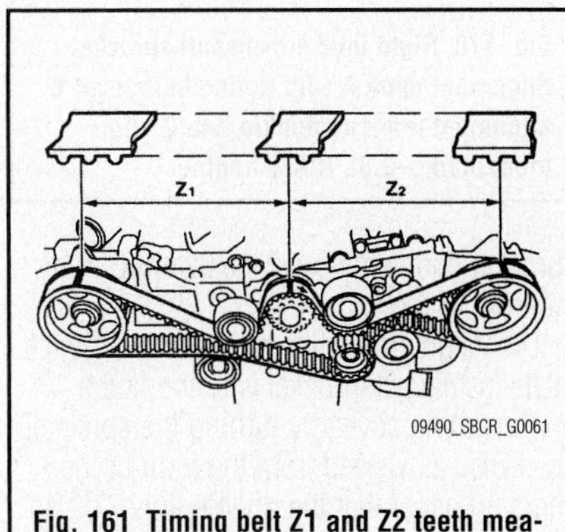

Fig. 161 Timing belt Z1 and Z2 teeth measurement—2.5L SOHC engine

To install:

13. Attach the automatic belt tension adjuster assembly to a vertical pressing tool.

➡Always use a vertical type pressing tool to move the adjuster rod downward. Do not use a lateral type vise. Push the adjuster rod vertically. Press in the push adjuster rod gradually, which should take three minutes or more. Do not allow pressure to exceed 2,205 lb. force.

14. Slowly move the adjuster rod down until the adjuster rod is aligned with the stopper pin hole in the cylinder.

➡Press the adjuster rod as far as the end surface of the cylinder. Do not press the adjuster rod into the cylinder. Doing so may damage the cylinder.

15. Using a 0.08 inch stopper pin, insert it into the stopper pin hole in the cylinder. Secure the adjuster rod.

➡Do not release the press pressure until the stopper pin is completely inserted in the hole.

16. Install the automatic belt tensioner assembly. Tighten the retaining bolt to 28.9 ft. lbs.

17. Install the belt idler number one. Tighten the retaining bolt to 28.9 ft. lbs.

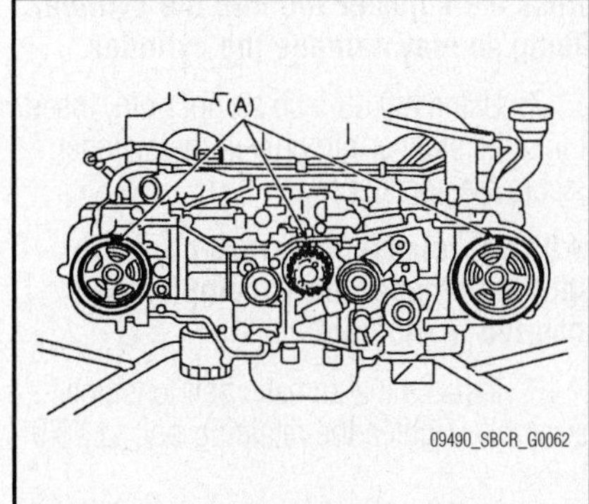

Fig. 162 Timing mark alignment position A—2.5L SOHC engine

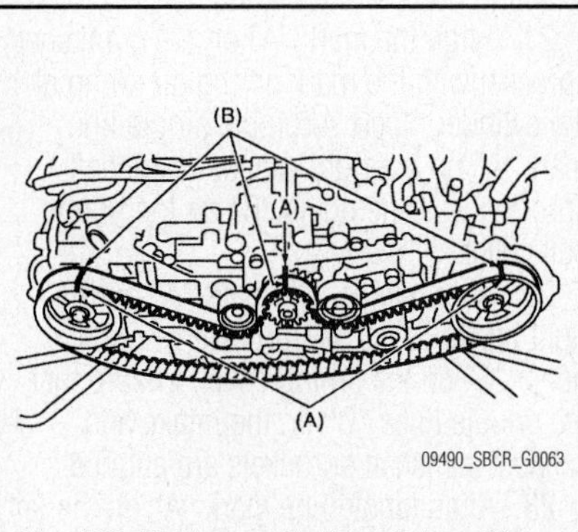

Fig. 163 Timing mark alignment position B—2.5L SOHC engine

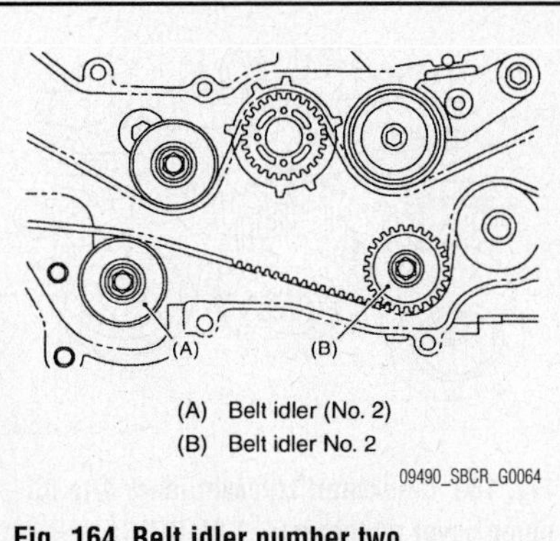

Fig. 164 Belt idler number two locations—2.5L SOHC engine

18. Turn the number one and number two camshaft sprockets, using tool ST499207100 or tool ST18231AA010 and position the alignment marks "A" on each at the highest position.

19. While aligning the alignment mark "B" on the timing belt with mark "A" on the sprockets, position the timing belt properly.

20. Install both belt idler number two's. Tighten the retaining bolt to 28.9 ft. lbs.

21. After checking to be sure the marks on the timing belt and the camshaft sprockets are aligned remove the stopper pin from the belt tension adjuster.

22. Install the timing belt guide, if equipped with manual transmission. Temporarily tighten the bolts. Check and adjust the clearance between the belt and the guide. It should be 0.039 +/- 0.020 inch. Tighten the bolts to 7.2 ft. lbs.

23. Continue the installation in the reverse order of the removal procedure.

2.5L DOHC Engine

See Figures 165 through 176.

1. Disconnect the negative battery cable.
2. Remove the collector cover.
3. To remove the front side V belt, remove the belt covers. Loosen the lock bolt. Loosen the slider bolt. Remove the front side belt.
4. To remove the rear side V belt, remove the belt covers. Loosen the lock bolt. Loosen the slider bolt. Remove the rear side belt. Remove the belt tensioner.
5. Remove the crankshaft pulley bolt.
6. Lock the crankshaft in place using tool ST499977100, or equivalent.
7. Remove the crankshaft pulley.
8. Remove the left side timing belt cover.
9. Remove the right side timing belt cover.
10. Remove the front timing belt cover.
11. Remove the timing belt guide, if equipped.

➡If the belt is going to be reused and the alignment mark on the belt is not readable, put a new mark on the belt to indicate the direction of rotation. Using tool ST499987500, turn the crankshaft to align the mark on the crankshaft sprocket, intake camshaft sprocket (left), exhaust camshaft sprocket (left), intake camshaft sprocket (right), exhaust camshaft sprocket (right) with the notches of the timing belt cover and cylinder block. Paint an alignment mark on the belts in relation to the camshaft sprockets. Z1 measurement is 54.4 teeth. Z2 measurement is 51.0 teeth and Z3 measurement is 28.0 teeth.

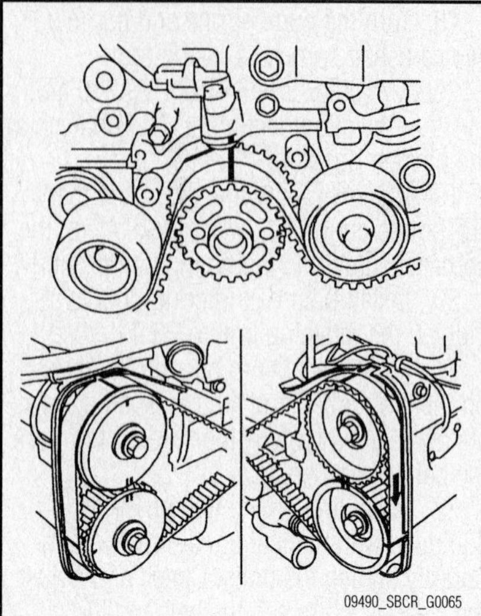

Fig. 165 Timing belt alignment—2.5L DOHC engine

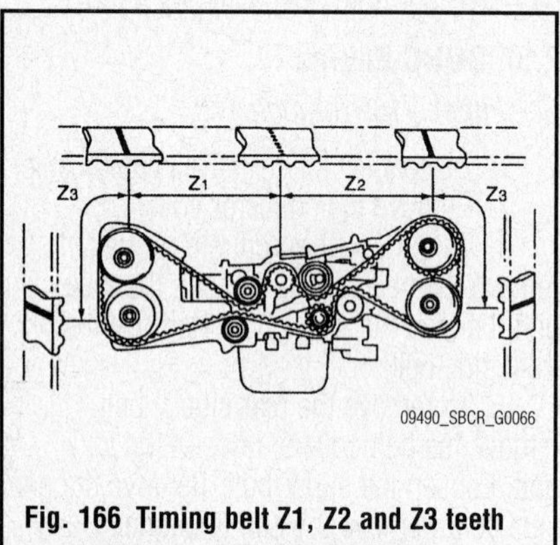

Fig. 166 Timing belt Z1, Z2 and Z3 teeth measurement—2.5L DOHC engine

12. Remove the belt idler belt "A". Remove the timing belt.
13. Remove the belt idlers belt "B" and "C".
14. Remove the belt idler number two. Remove the automatic belt tension adjuster assembly.

To install:

15. Attach the automatic belt tension adjuster assembly to a vertical pressing tool.

➡Always use a vertical type pressing tool to move the adjuster rod downward. Do not use a lateral type vise. Push the adjuster rod vertically. Press in the push adjuster rod gradually, which should take three minutes or more. Do not allow pressure to exceed 2,205 lb. force.

16. Slowly move the adjuster rod down until the adjuster rod is aligned with the stopper pin hole in the cylinder.

➡Press the adjuster rod as far as the end surface of the cylinder. Do not

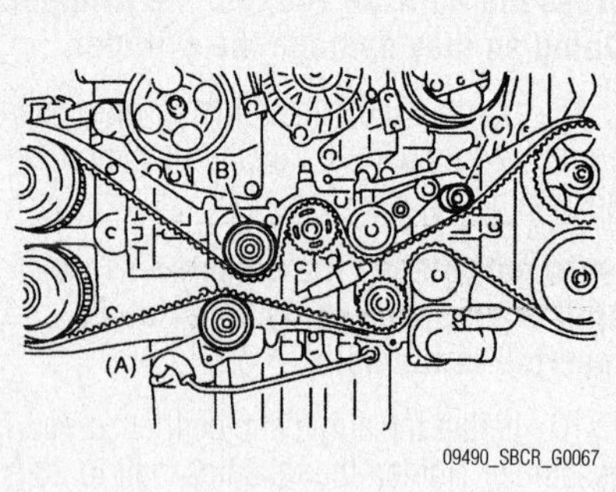

Fig. 167 Belt idler identification and location—2.5L DOHC engine

press the adjuster rod into the cylinder. Doing so may damage the cylinder.

17. Using a 0.08 inch stopper pin, insert it into the stopper pin hole in the cylinder. Secure the adjuster rod.

➡Do not release the press pressure until the stopper pin is completely inserted in the hole.

18. Install the automatic belt tensioner assembly. Tighten the retaining bolt to 28.9 ft. lbs.
19. Install the belt idler number two. Tighten the retaining bolt to 28.9 ft. lbs.
20. Install the belt idlers. Tighten the retaining bolts to 28.9 ft. lbs.
21. Align the mark "A" on the crankshaft sprocket with the mark on the oil pump at the cylinder block. Align the single line mark "A" on the right exhaust camshaft sprocket with the notch "B" on the timing belt cover.
22. Align single line mark "A" on the right intake camshaft sprocket with the notch "B" on the timing cover. Ensure that the double lines "C" on the intake and exhaust camshaft sprockets are aligned.
23. Align single line mark "A" on the left exhaust camshaft sprocket with the notch "B" on the timing cover by turning the

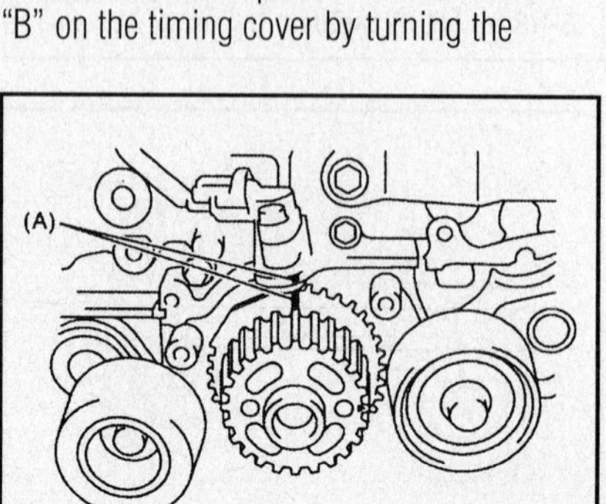

Fig. 168 Crankshaft sprocket mark A to oil pump cover alignment—2.5L DOHC engine

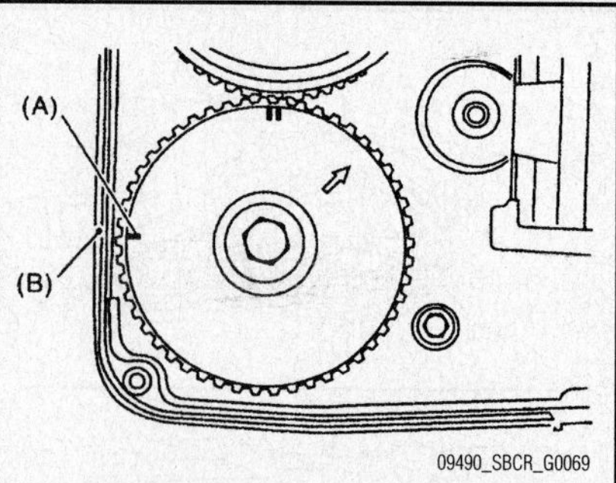

Fig. 169 Right exhaust camshaft sprocket alignment mark A with timing belt cover B alignment mark—2.5L DOHC engine

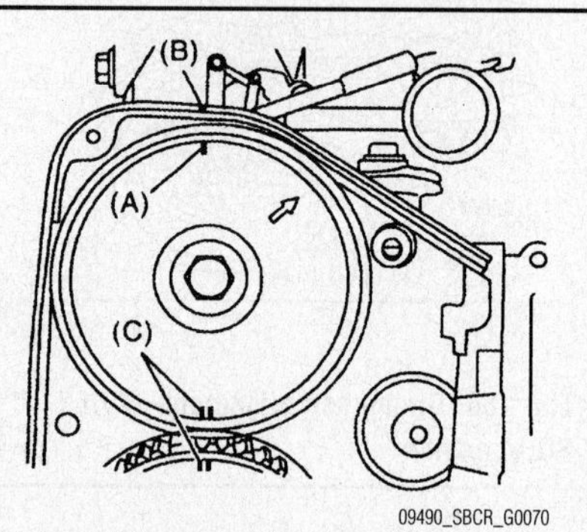

Fig. 170 Right intake camshaft sprocket alignment mark A with timing belt cover B alignment mark an double line C alignment mark—2.5L DOHC engine

sprocket counterclockwise as viewed from the front of the engine.

24. Align single line mark "A" on the left intake camshaft sprocket with the notch "B" on the timing cover, by turning the sprocket clockwise as viewed from the front of the engine. Ensure that the double lines "C" on the intake and exhaust camshaft sprockets are aligned.
25. Make sure that the camshaft and crankshaft sprockets are positioned properly.

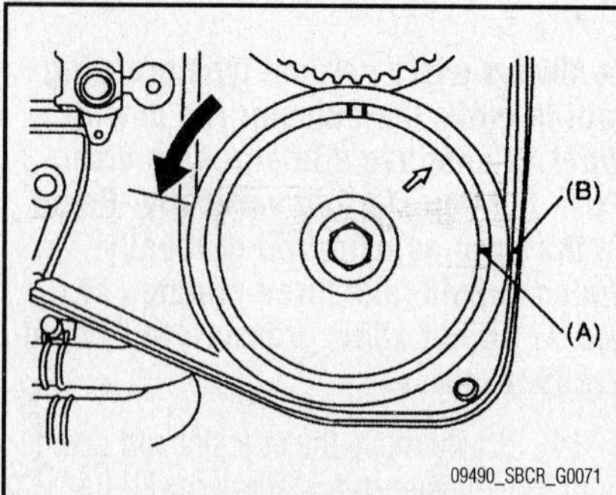

Fig. 171 Left exhaust camshaft sprocket alignment mark A with timing belt cover B alignment mark—2.5L DOHC engine

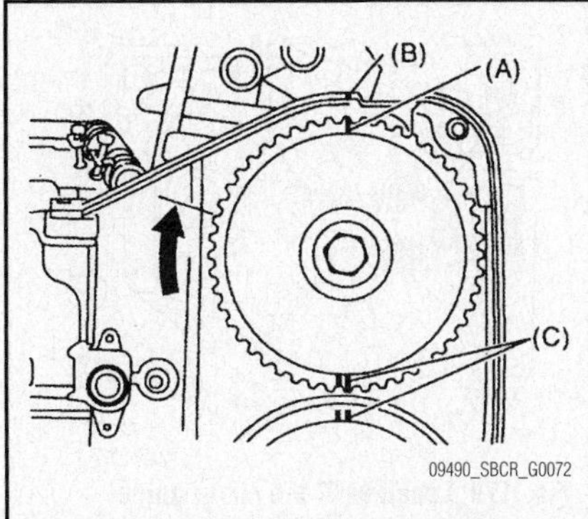

Fig. 172 Left intake camshaft sprocket alignment mark A with timing belt cover B alignment mark an double line C alignment mark—2.5L DOHC engine

➡The intake and exhaust camshafts on this engine can be rotated independently with the timing belt removed. By looking at the illustration it will show you that if the intake and exhaust valve are lift together the heads will hit each other and bend.

➡When the timing belts are not installed, 4 camshafts are held at "zero lift" position, where all cams on the camshafts do not push the intake and exhaust valves down (under this condition all valves remain unlifted). When the camshafts are rotated to install the timing belts, No. 2 intake and No. 4 exhaust cam of the left hand camshafts are held to push their corresponding valves down. Under this condition these valves are held lifted. The right side camshafts are held in so that their cams do not push the valves down. The left hand camshafts must be rotated from the "zero lift" position to the position where the timing belt is to be installed at as small an angle as possible, in order to prevent mutual interference of intake and exhaust valve heads. Do not allow the camshafts to rotate in the direction illustrated as this causes both the intake and exhaust valves to lift off at the same time with will cause valve damage.

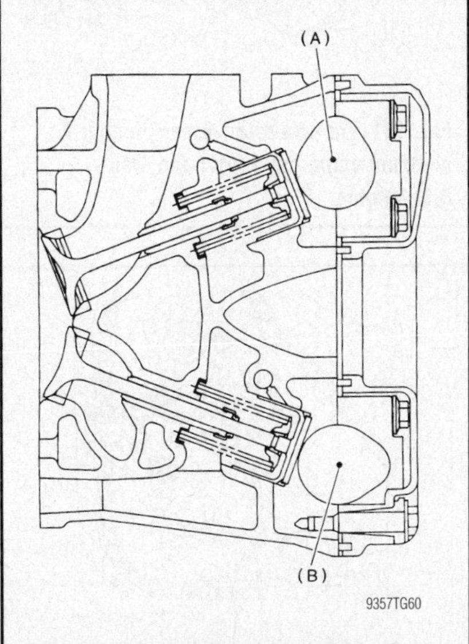

Fig. 173 If the intake and exhaust valve are lift together the heads will hit each other and bend—2.5L DOHC engine

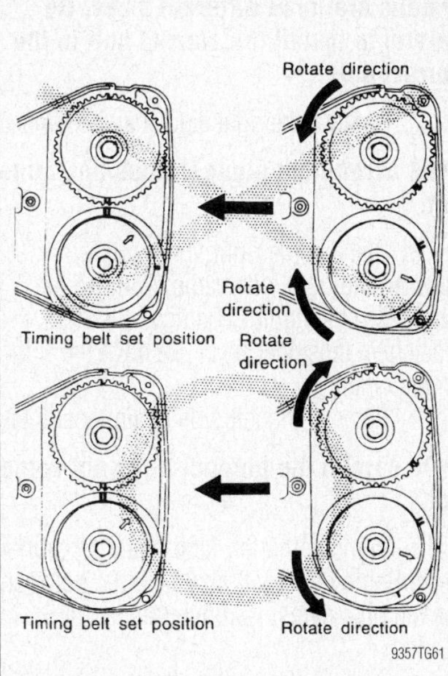

Fig. 174 Do not allow the camshafts to rotate in the direction shown as this causes both the intake and exhaust valves to lift off at the same time with will cause valve damage on DOHC engines

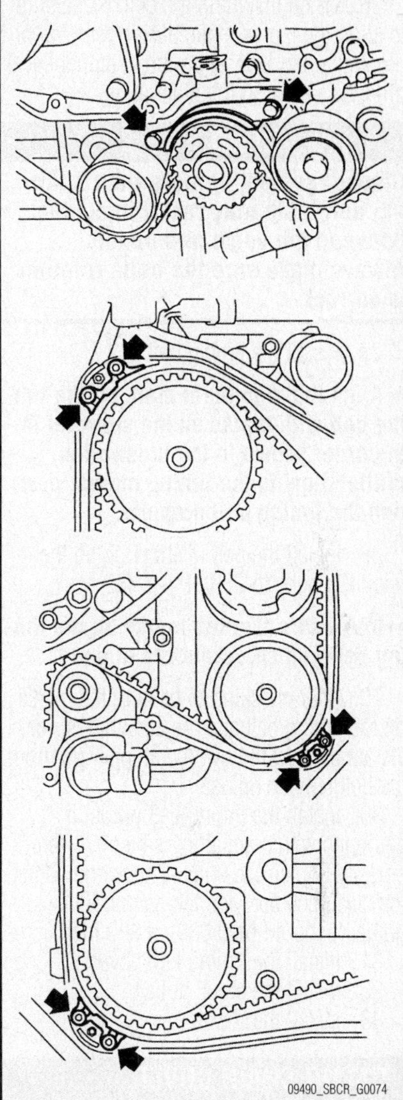

Fig. 176 Timing belt guide bolt location—2.5L DOHC engine

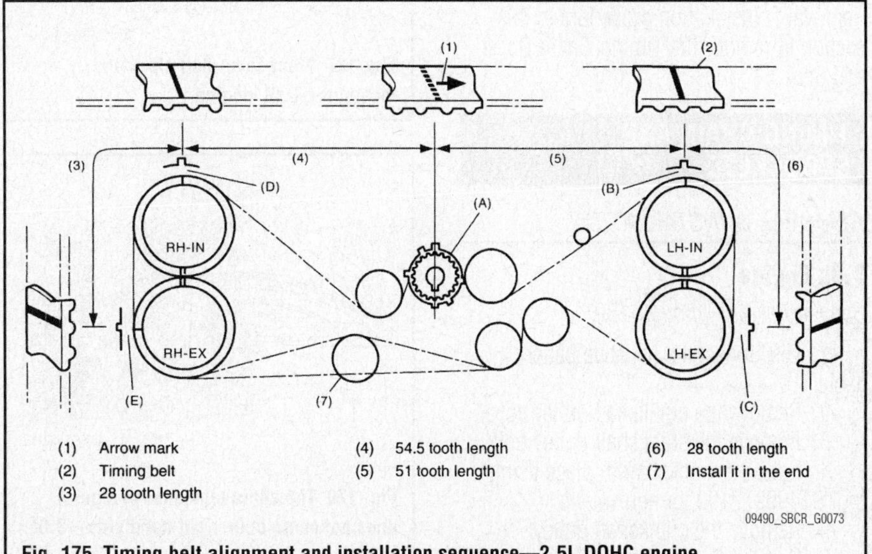

Fig. 175 Timing belt alignment and installation sequence—2.5L DOHC engine

26. When installing the belt, make sure to align the marks made during removal or if using a new belt, align the in alphabetical order as shown in the illustration.

WARNING

Disengagement of more than 3 timing belt teeth may result in contact between the valve and piston. Always make sure the belts rotation is correct.

27. Install the timing belt.

➡Align the alignment mark on the timing belt with marks on the sprocket in the order shown in the illustration. While aligning the timing marks, position the timing belt properly.

28. Install the belt idlers. Tighten the retaining bolts to 28.9 ft. lbs. (39 Nm).

➡Make sure that the marks on the timing belt and sprockets are aligned.

29. After checking to be sure the marks on the timing belt and the camshaft sprockets are aligned remove the stopper pin from the belt tension adjuster.
30. Install the timing belt guide, if equipped with manual transmission. Temporarily tighten the bolts. Check and adjust the clearance between the belt and the guide. It should be 0.039/0.020 inch.
31. Install the timing belt cover.
32. Install the crank pulley.
33. Install the V-belts.

TIMING CHAIN COVER AND SEAL

REMOVAL & INSTALLATION

Refer to the Timing Chain and Sprockets, Removal & Installation procedure in this section to remove the Timing Chain Cover and Seal.

TIMING CHAIN AND SPROCKETS

REMOVAL & INSTALLATION

3.0L Engine

See Figures 177 through 189.

1. Disconnect the negative battery cable.
2. Remove the crankshaft pulley cover.
3. Remove the crankshaft pulley bolt.
4. Lock the crankshaft in place using tool ST499977100, or equivalent.
5. Remove the crankshaft pulley.
6. Remove the front timing chain cover.

➡Bolts are three different sizes. Be careful to install the correct bolt in the correct hole.

7. Remove the right side chain tensioner.

➡Be careful the plunger does not come out.

8. Remove the right chain guide, between the right side cams. Remove the right side chain guide. Remove the right side chain tensioner lever. Remove the right timing chain.
9. Remove the left side chain tensioner.

➡Be careful the plunger does not come out.

10. Remove the left side chain tensioner lever. Remove the left chain guide, between the left side cams. Remove the chain guide.
11. Remove the center chain guide. Remove the upper idler sprocket. Remove the left timing chain.
12. Remove the lower idler sprocket.

To install:

13. Make sure all components are clean. Apply oil to the chain guide, tensioner lever and idler sprockets.
14. Place the screw, spring, pin and tension rod into the tensioner body.

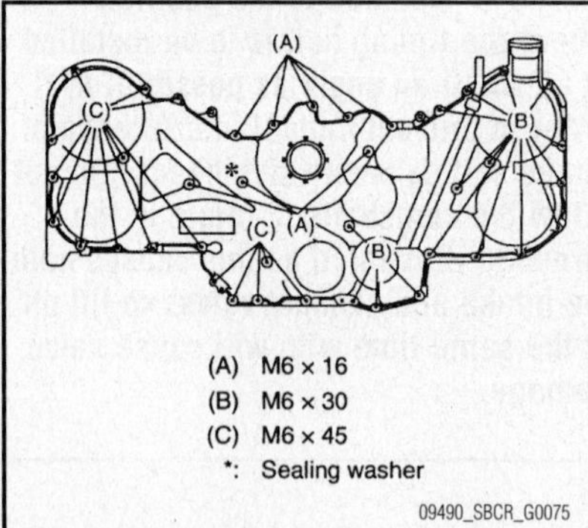

Fig. 177 Front cover bolt sizes and locations—3.0L engine

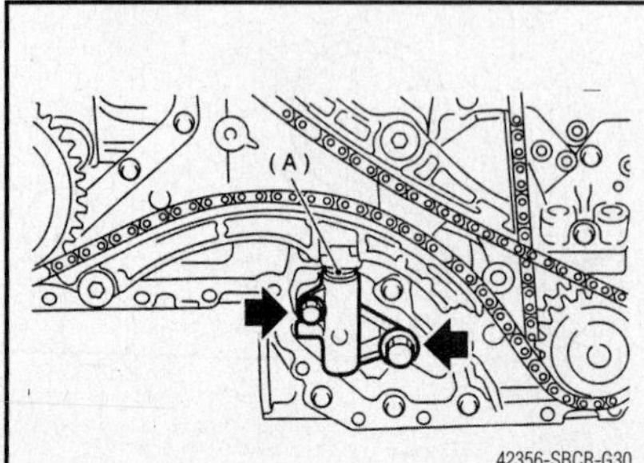

Fig. 178 The chain tensioner plunger A does not come out—right hand side—3.0L engine

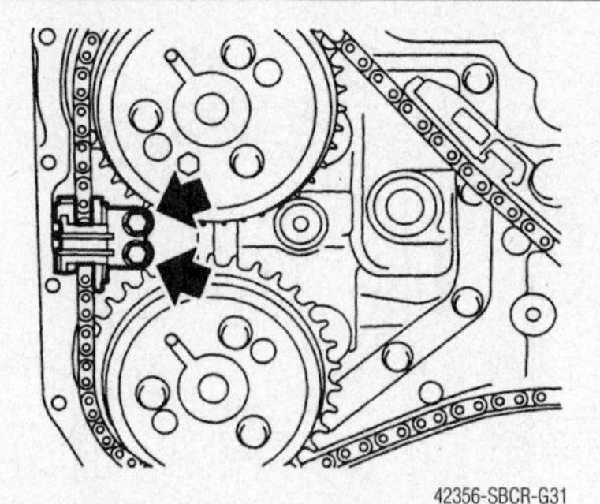

Fig. 179 Location of the chain guide between the cams—right hand side—3.0L engine

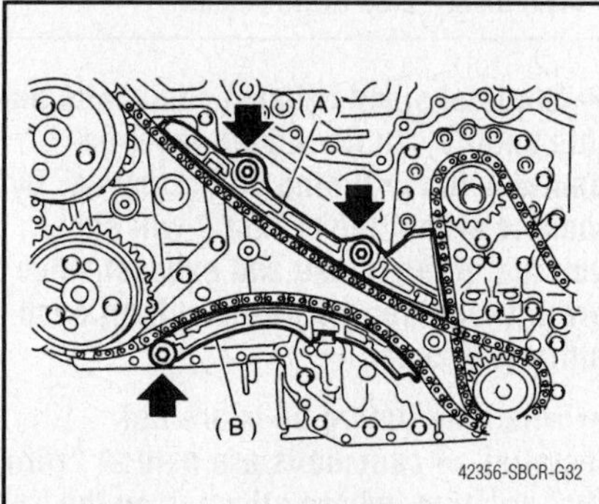

Fig. 180 Location of the chain guide—right hand side—3.0L engine

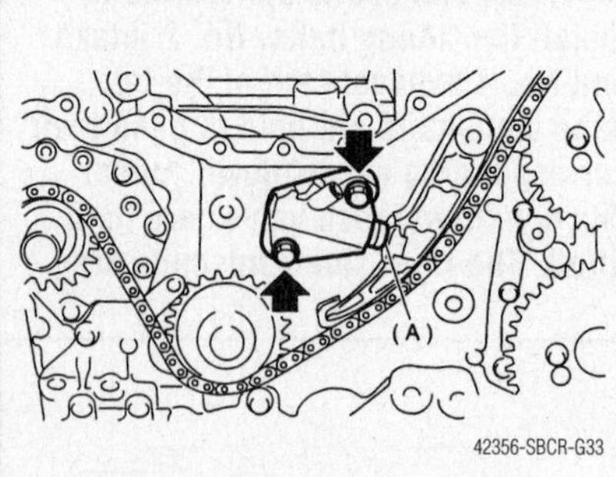

Fig. 181 The chain tensioner plunger A does not come out—left hand side—3.0L engine

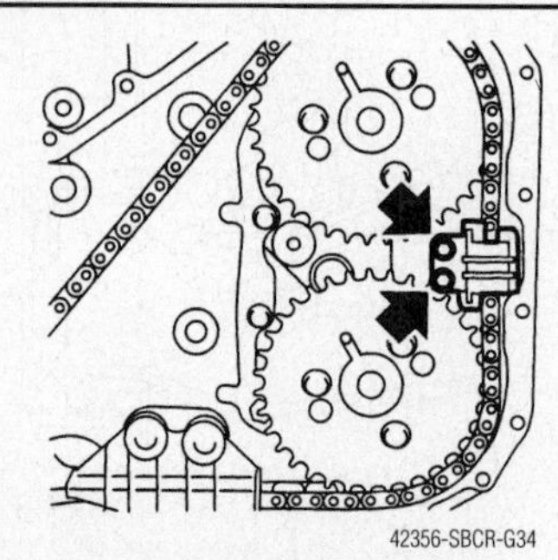

Fig. 182 Location of the chain guide between the cams—left hand side—3.0L engine

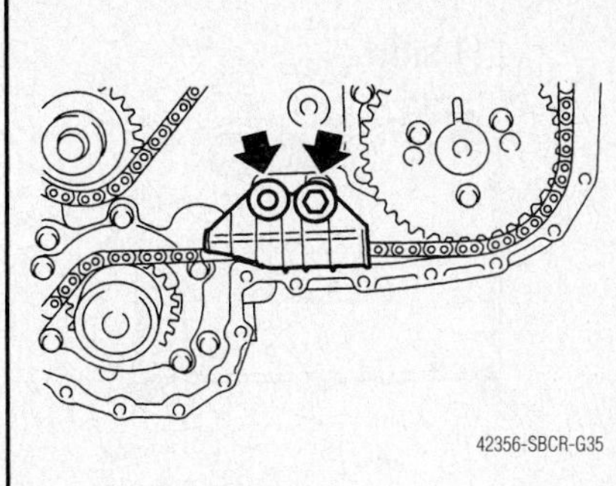

Fig. 183 Location of the chain guide—left hand side—3.0L engine

15. While pressing the tensioner onto a rubber mat, twist it to the left and right to shorten the rod. Place a thin pin into the holes between the rod and body to hold it in place. Always perform this task on a rubber mat.

16. Using the crankshaft socket tool, align the **TOP MARK** on the crankshaft sprocket to the 9 O'clock position, as shown in the illustration.

17. Align the key groove on the exhaust camshaft sprocket to the 12 O'clock position, as shown in the illustration.

18. Align the intake camshaft sprocket, as shown in the illustration.

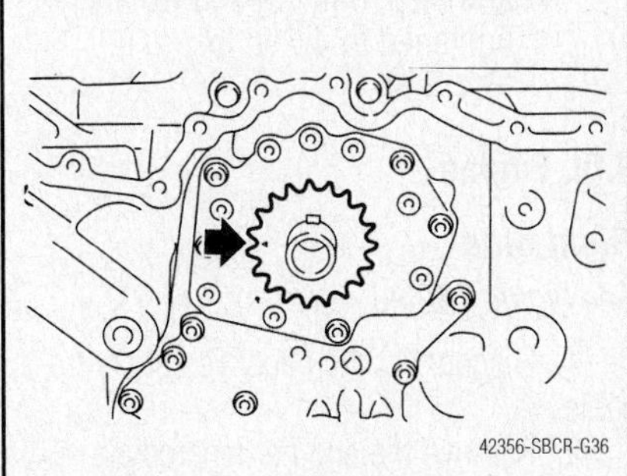

Fig. 184 Align the TOP MARK on the crankshaft sprocket to the 9 O'clock position—3.0L engine

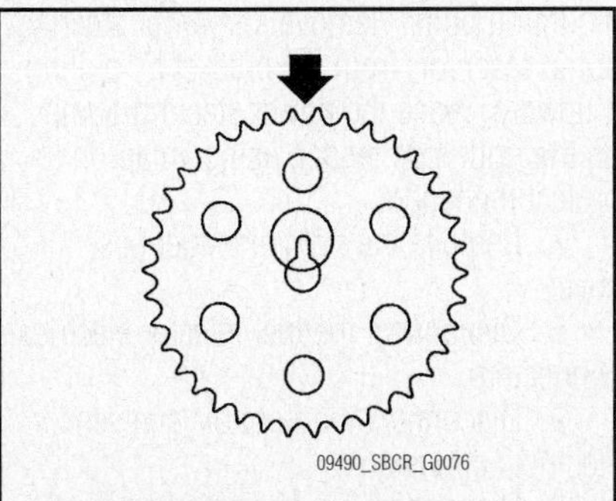

Fig. 185 Align the TOP MARK on the camshaft sprocket to the 12 O'clock position—2005 3.0L engine

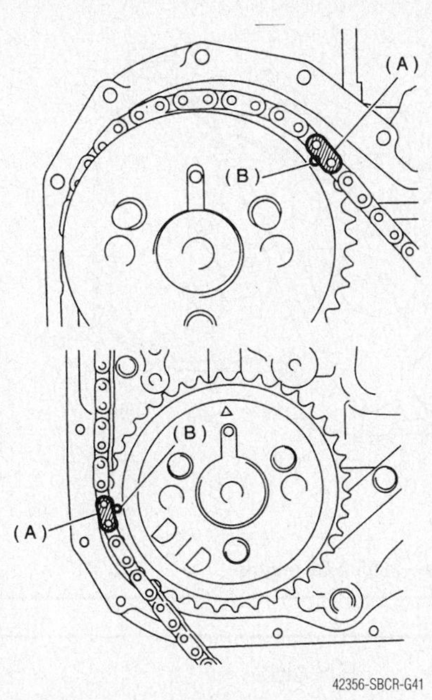

Fig. 186 Make sure the mark A on the chain and the mark B camshaft sprocket are aligned the same way as the one on the crankshaft sprocket–right hand side—3.0L engine

19. Turn the crankshaft sprocket clockwise; align the **TOP MARK** to the 12 O'clock position. Piston number one is now at Top Dead Center (TDC).

CAUTION

Do not rotate the camshaft or crankshaft sprockets until the chain is completely routed or damage will occur.

20. Install the lower idler sprocket and tighten the bolt to 50.6 ft. lbs. (69 Nm).

21. Install the left side timing chain, align the mark "B" on the crankshaft sprocket with the matching mark "A" on the timing chain.

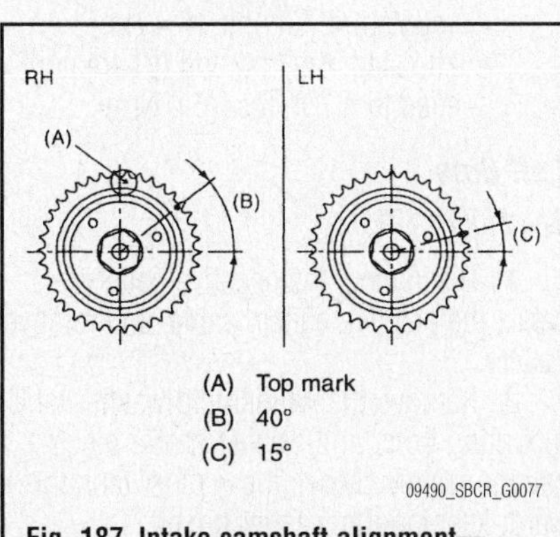

Fig. 187 Intake camshaft alignment—2005 3.0L engine

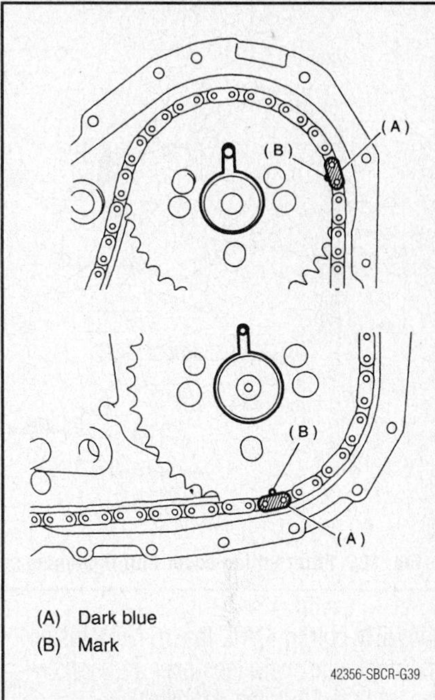

Fig. 188 Make sure the mark A on the chain and the mark B camshaft sprocket are aligned the same way as the one on the crankshaft sprocket–left hand side—3.0L engine

22. Route the left side timing chain onto the lower idler sprocket, water pump, exhaust cam sprocket and the intake cam sprocket in that order.

23. Make sure the mark "A" on the chain and the mark "B" camshaft sprocket are aligned the same way as the one on the crankshaft sprocket or damage will occur.

24. Install the upper chain idler and tighten the bolt to 50.6 ft. lbs. (69 Nm).

25. Install the left side chain guide, between the camshafts. Tighten the bolt to 4.7 ft. lbs. (6 Nm) using a NEW bolt.

26. Install the left side chain guide. Tighten the bolts to 12 ft. lbs. (16 Nm). Install the tensioner lever on the left side and tighten the bolt to 12 ft. lbs. (16 Nm). Install the chain tensioner on the left side and tighten the bolts to 12 ft. lbs. (16 Nm).

27. Install the right side timing chain. Align the marks of the left and right timing chains on the lower idler sprocket. Install the right side timing chain to the right side intake camshaft sprocket and the right side exhaust camshaft sprocket in this order.

28. Make sure the mark "A" on the chain and the mark "B" camshaft sprocket are aligned the same way as the one on the crankshaft sprocket or damage will occur.

29. Install the right side chain guide. Install the right side chain tensioner lever and tighten the bolts to 12 ft. lbs. (16 Nm). Install the right side chain guide and tighten

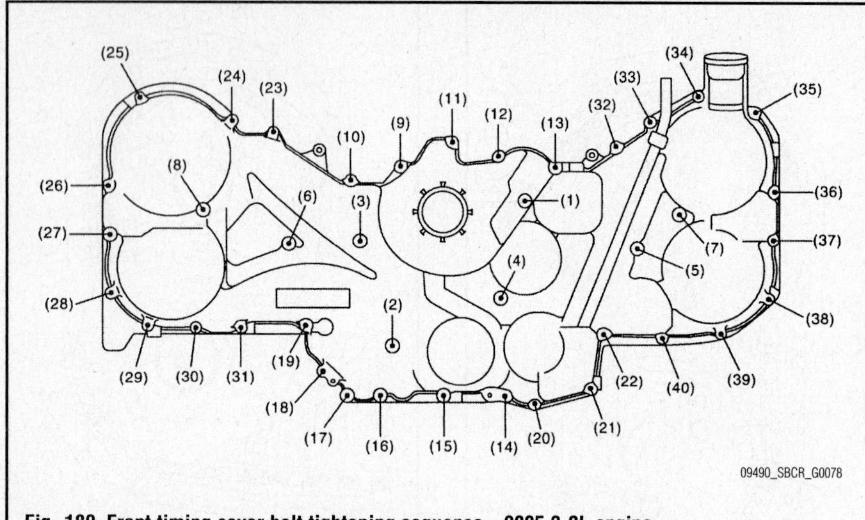

Fig. 189 Front timing cover bolt tightening sequence—2005 3.0L engine

the NEW bolt to 4.7 ft. lbs. (6 Nm). Install the right side chain tensioner and tighten the bolts to 12 ft. lbs. (16 Nm).

30. Adjust the clearance between the chain guide on the right side and the center chain guide so that there is range between 0.331–0.339 inch (8.4–8.6mm).

31. Install the center chain guide and tighten the NEW bolt to 5.8 ft. lbs. (8 Nm).

32. Check the match marks on each sprocket and corresponding timing chain are correct, remove the stopper from the tensioner.

33. Clean the mating surfaces on the front timing cover. Apply a bead of liquid gasket 0.020 inch in diameter to the mating surface of the front timing chain cover. Install the timing chain cover. Torque the retaining bolts to 4.8 ft. lbs. and in the proper sequence.

34. Continue the installation in the reverse order of the removal procedure.

VALVE COVERS

REMOVAL & INSTALLATION

2.5L Engines

Right Side

See Figure 190.

1. Remove the engine undercover.
2. Disconnect the negative battery cable.
3. Remove the left side timing belt front cover. Refer to "Timing Belt and Sprockets, Removal & Installation" procedure in this section to remove the timing belt front cover.
4. Remove the air intake assembly.
5. Disconnect the spark plug cables from the Nos. 1 and 3 cylinders. (SOHC Engines)

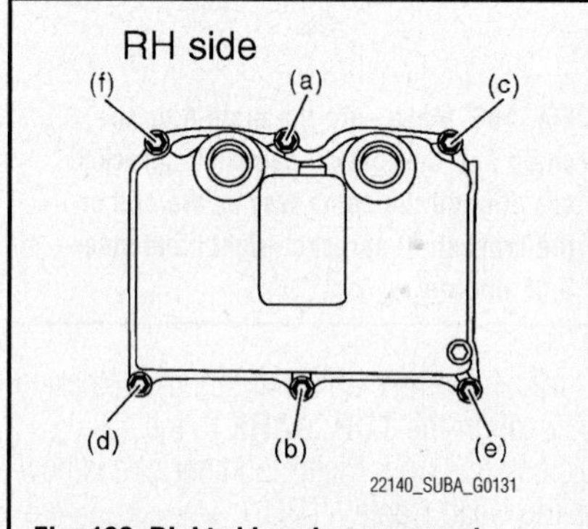

Fig. 190 Right side valve cover and sequence shown 2.5L engine

6. Remove ignition coils from the Nos. 1 and 3 cylinders. (DOHC Engines)
7. Disconnect the PCV hose from the valve cover.
8. Remove the right side valve cover.

To install:

9. Installation is the reverse order of removal.
10. Temporarily tighten the bolts in alphabetical sequence as shown in figure, then tighten the bolt in 2 steps.
    - 1st step: 4.7 ft. lbs. (6.4 Nm)
    - 2nd step: only (a) and (b) are tightened to 4.7 ft. lbs. (6.4 Nm)

Left Side

See Figure 191.

1. Disconnect the negative battery cable, the positive battery cable, and remove battery.
2. Remove the windshield washer tank mounting bolts and washer motor electrical connectors. Move the washer tank forward, keeping the washer hoses connected, and secure the tank with mechanics wire.

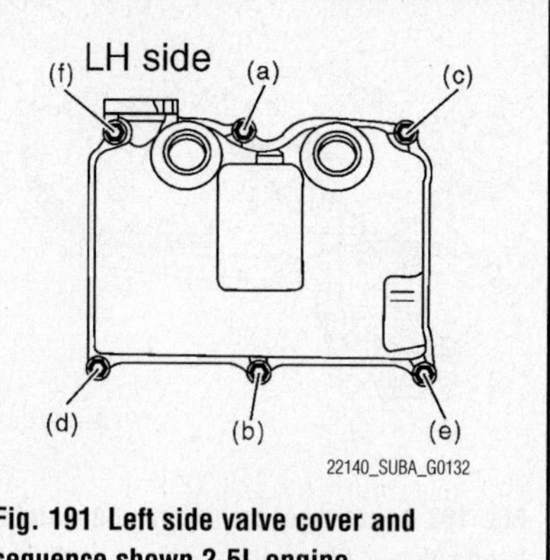

Fig. 191 Left side valve cover and sequence shown 2.5L engine

3. Disconnect the spark plug cables from the Nos. 2 and 4 cylinders. (SOHC Engines)
4. Remove ignition coils from the Nos. 2 and 4 cylinders. (DOHC Engines)
5. Disconnect the PCV hose from the valve cover.
6. Remove the left side valve cover.

To install:

7. Installation is the reverse order of removal.
8. Temporarily tighten the bolts in alphabetical sequence as shown in figure, then tighten the bolt in 2 steps.
    - 1st step: 4.7 ft. lbs. (6.4 Nm)
    - 2nd step: only (a) and (b) are tightened to 4.7 ft. lbs. (6.4 Nm)

3.0L Engine

Right Side

See Figure 192.

1. Disconnect the negative battery cable.
2. Remove the engine undercover.
3. Remove the air intake assembly.
4. Remove the accessory drive belts.
5. Remove the power steering hose from the mounting bracket.
6. Remove the power steering pump mounting bolts. Remove the power steering pump reservoir from the bracket by pulling it upward. Place the power steering pump on the right side fender, using an apron to protect the fender.
7. Remove the right side fuel hose cover.
8. Disconnect the fuel injector electrical connectors.
9. Disconnect the front oxygen sensor electrical connector.
10. Disconnect the oil pressure sensor electrical connector.
11. Remove the ignition coils.
12. Remove the right side valve cover.

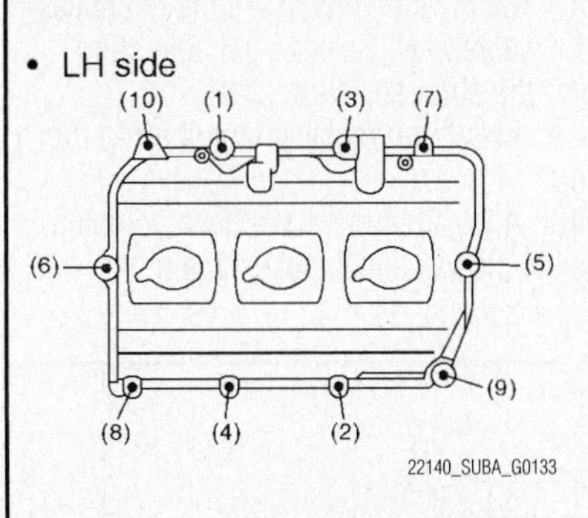

Fig. 192 Right side valve cover and sequence shown—3.0L engine

To install:

13. Installation is the reverse order of removal.
14. Install a new gasket and tighten in sequence to 5 ft. lbs. (6.4 Nm).

Left Side

See Figure 193.

1. Disconnect the negative battery cable, the positive battery cable, and remove battery.
2. Remove the windshield washer tank mounting bolts and washer motor electrical connectors. Move the washer tank upward, keeping the washer hoses connected, and secure the tank with mechanics wire.
3. Disconnect the PCV and blow-by hose from the left side valve cover.
4. Remove the left side fuel hose cover.
5. Disconnect the fuel injector electrical connectors.
6. Disconnect the front oxygen sensor connector.
7. Remove the ignition coils.
8. Remove the left side valve cover.

To install:

9. Installation is the reverse order of removal.
10. Install a new gasket and tighten in sequence to 5 ft. lbs. (6.4 Nm).

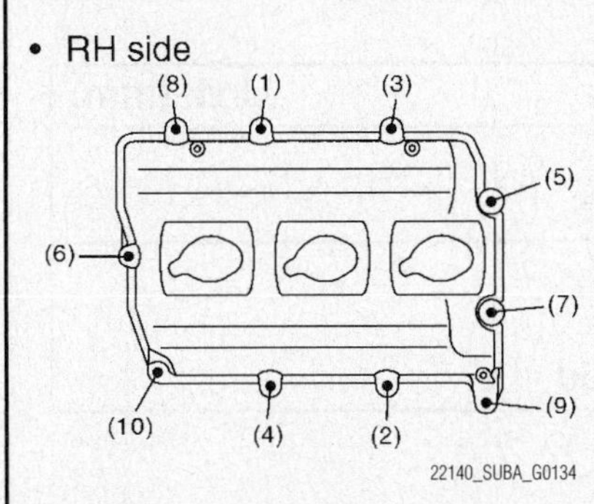

Fig. 193 Left side valve cover and sequence shown—3.0L engine

VALVE LASH

ADJUSTMENT

2.5L SOHC Engine

See Figure 194.

➡The valve adjustment should be performed while the engine is cold.

1. Raise and support the vehicle safely.
2. Remove the undercover.
3. Lower the vehicle.
4. Disconnect the negative battery cable.
5. To remove the front side V belt, remove the belt covers. Loosen the lock bolt. Loosen the slider bolt. Remove the front side belt.
6. To remove the rear side V belt, remove the belt covers. Loosen the lock bolt. Loosen the slider bolt. Remove the rear side belt. Remove the belt tensioner.
7. Remove the crankshaft pulley bolt.
8. Lock the crankshaft in place using tool ST499977100, or equivalent.
9. Remove the crankshaft pulley.
10. Remove the left side timing belt cover.
11. Remove the fuel injector.
12. Remove the rocker cover.
13. Position the number one piston at TDC of the compression stroke.

➡When the arrow (see illustration) on the camshaft sprocket (left side) comes exactly to the top, number one cylinder piston is at TDC of the compression stroke.

14. Measure the valve clearance, using a feeler gauge.
15. If adjustment is needed, loosen the valve rocker nut and screw. Position the feeler gauge.

➡Insert the feeler gauge in a horizontally as possible with respect to the valve stem end face. Adjust the exhaust valve clearance while lifting up the vehicle.

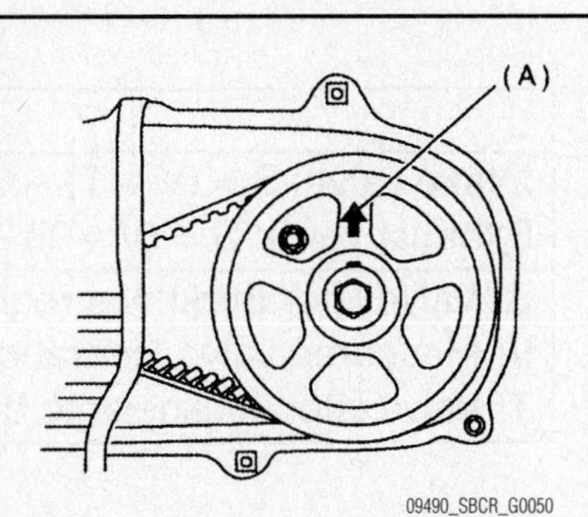

Fig. 194 TDC alignment—2.5L SOHC engine

16. While noting the valve clearance, tighten the rocker adjusting screw.
17. When the proper valve clearance is obtained, tighten the valve rocker nut to 7.2 ft. lbs.
18. Adjust the valve clearance on the remaining cylinders, following the above procedure.

➡Be sure to position the pistons to their respective TDC positions on the compression stroke, before checking and adjusting the valves. By rotating the crankshaft pulley clockwise every 180 degrees from the state that number one piston is on TDC of the compression stroke, the remaining pistons come to TDC of the compression stroke in the following order, #3, #2, and #4.

19. After adjustment, replace any removed components.
20. Be sure to use new gaskets and seals, as required.

2.5L DOHC Engine

See Figures 195 through 201.

➡The valve adjustment should be performed while the engine is cold.

1. Raise and support the vehicle safely.
2. Remove the undercover.
3. Lower the vehicle.
4. Remove the collector cover.
5. Disconnect the negative battery cable.
6. Remove the air intake duct.
7. Remove the bolt that retains the right side timing belt cover. Remove the remaining bolts and remove the right side timing belt cover.
8. Disconnect the ignition coil electrical connector. Remove the ignition coil.
9. Position a suitable container under the vehicle.
10. Disconnect the PCV hose from the rocker cover. Remove the rocker cover retaining bolts. Remove the rocker cover from the vehicle.
11. Position the number one piston at TDC of the compression stroke.
12. Using a feeler gauge, measure and record the clearance of the number one cylinder intake and the number three cylinder exhaust valves.

➡Insert the feeler gauge in a horizontally as possible with respect to the valve lifter. Measure and record the exhaust valve clearance while lifting up the vehicle.

13. Rotate the crankshaft pulley clockwise until the arrow mark on the camshaft is positioned as shown to measure and record the clearance on the number two exhaust and number three intake valves.

14. Rotate the crankshaft pulley clockwise until the arrow mark on the camshaft is positioned as shown to measure and record the number two intake and number four exhaust valves.

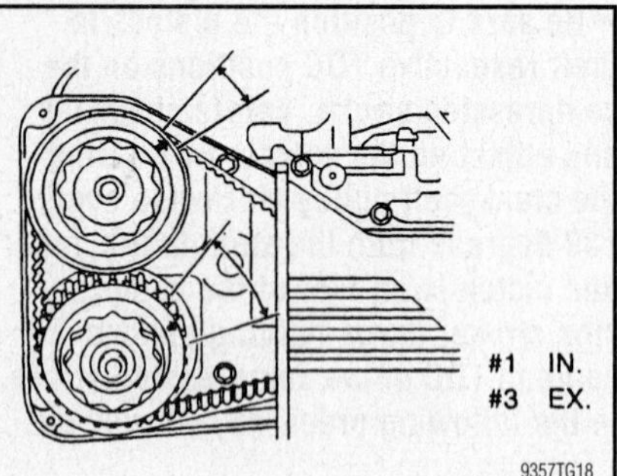

Fig. 195 Turn the crankshaft pulley clockwise until the arrow mark on the camshaft is positioned as shown to measure the No. 1 intake and No. 3 exhaust valves—2.5L DOHC engine

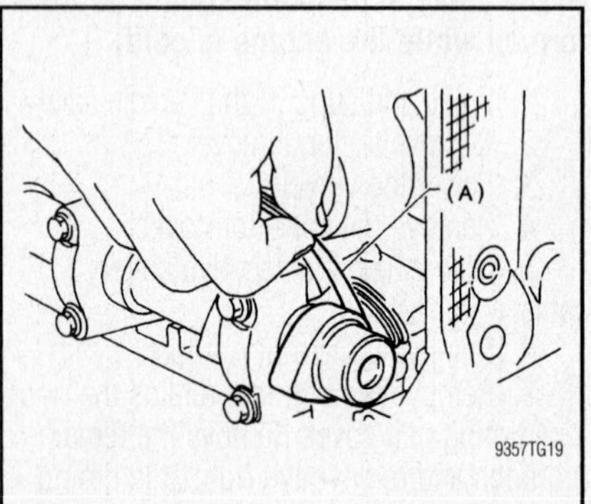

Fig. 196 Use a feeler gauge to inspect the valve clearance—2.5L DOHC engine

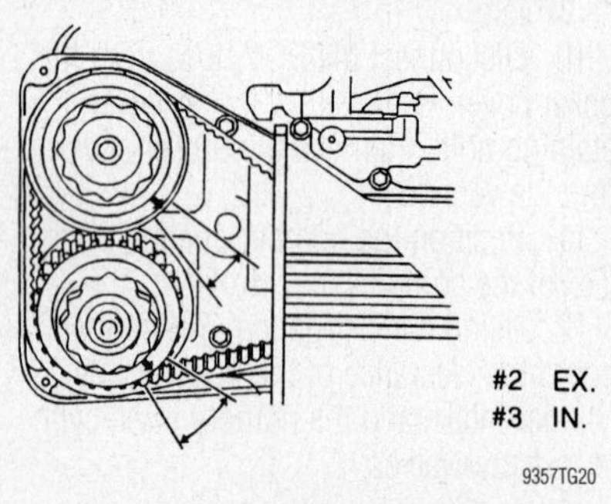

Fig. 197 Turn the crankshaft pulley clockwise until the arrow mark on the camshaft is positioned as shown to measure the No. 2 exhaust and No. 3 intake valves—2.5L DOHC engine

15. Rotate the crankshaft pulley clockwise until the arrow mark on the camshaft is positioned as shown to measure and record the number one exhaust and number four intake valves.

16. If adjustment is required, remove the camshafts.

17. Remove and measure the thickness of the valve lifter. Select a suitable shim, using the shim selection chart.

18. Install the replacement shim to the lifter.

19. After all shims have been adjusted, inspect the valve clearances again.

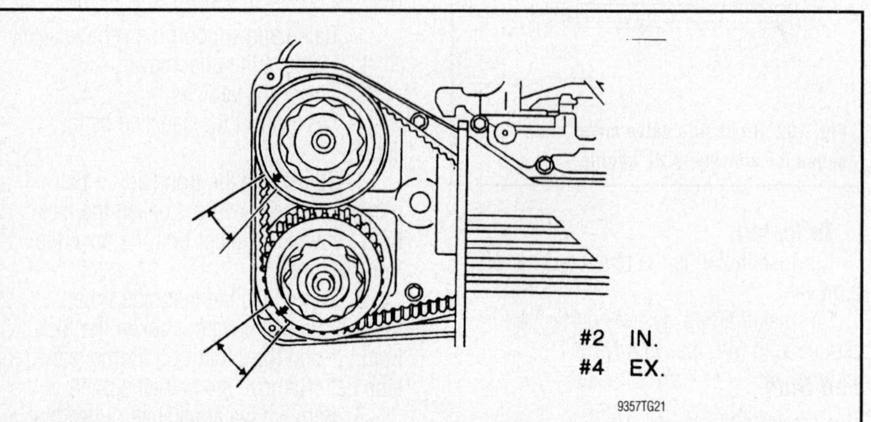

Fig. 198 Turn the crankshaft pulley clockwise until the arrow mark on the camshaft is positioned as shown to measure the No. 2 intake and No. 4 exhaust valves—2.5L DOHC engine

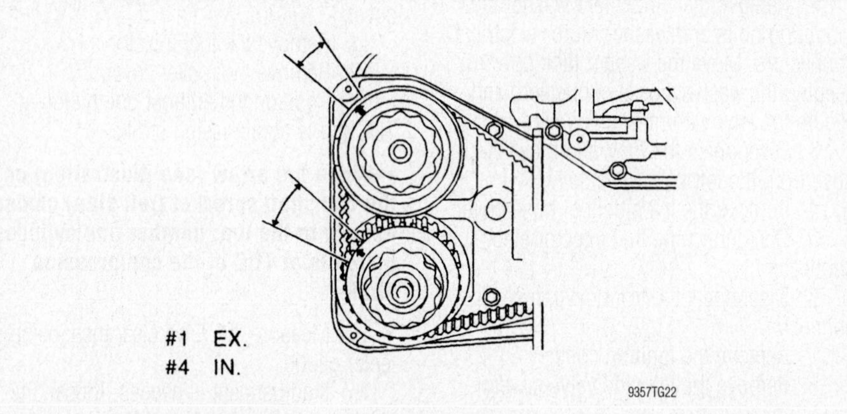

Fig. 199 Turn the crankshaft pulley clockwise until the arrow mark on the camshaft is positioned as shown to measure the No. 1 exhaust and No. 4 intake valves—2.5L DOHC engine

Unit: (mm)
Intake valve: S = (V + T) – 0.20 Exhaust valve: S = (V + T) – 0.35
S: Valve lifter thickness required V: Measured valve clearance T: Valve lifter thickness to be used

09490_SBCR_G0051

Fig. 200 Use this table to help you select a suitable shim—2.5L DOHC engine

Part No.	Thickness mm (in)
13228 AB102	4.68 (0.1843)
13228 AB112	4.69 (0.1846)
13228 AB122	4.70 (0.1850)
13228 AB132	4.71 (0.1854)
13228 AB142	4.72 (0.1858)
13228 AB152	4.73 (0.1862)
13228 AB162	4.74 (0.1866)
13228 AB172	4.75 (0.1870)
13228 AB182	4.76 (0.1874)
13228 AB192	4.77 (0.1878)
13228 AB202	4.78 (0.1882)
13228 AB212	4.79 (0.1886)
13228 AB222	4.80 (0.1890)
13228 AB232	4.81 (0.1894)
13228 AB242	4.82 (0.1898)
13228 AB252	4.83 (0.1902)
13228 AB262	4.84 (0.1906)
13228 AB272	4.85 (0.1909)
13228 AB282	4.86 (0.1913)
13228 AB292	4.87 (0.1917)
13228 AB302	4.88 (0.1921)
13228 AB312	4.89 (0.1925)
13228 AB322	4.90 (0.1929)
13228 AB332	4.91 (0.1933)
13228 AB342	4.92 (0.1937)
13228 AB352	4.93 (0.1941)
13228 AB362	4.94 (0.1945)
13228 AB372	4.95 (0.1949)
13228 AB382	4.96 (0.1953)
13228 AB392	4.97 (0.1957)
13228 AB402	4.98 (0.1961)
13228 AB412	4.99 (0.1965)
13228 AB422	5.00 (0.1969)
13228 AB432	5.01 (0.1972)
13228 AB442	5.02 (0.1976)
13228 AB452	5.03 (0.1980)
13228 AB462	5.04 (0.1984)
13228 AB472	5.05 (0.1988)
13228 AB482	5.06 (0.1992)
13228 AB492	5.07 (0.1996)
13228 AB502	5.08 (0.2000)
13228 AB512	5.09 (0.2004)
13228 AB522	5.10 (0.2008)
13228 AB532	5.11 (0.2012)
13228 AB542	5.12 (0.2016)
13228 AB552	5.13 (0.2020)
13228 AB562	5.14 (0.2024)
13228 AB572	5.15 (0.2028)
13228 AB582	5.16 (0.2031)
13228 AB592	5.17 (0.2035)
13228 AB602	5.18 (0.2039)
13228 AB612	5.19 (0.2043)

Part No.	Thickness mm (in)
13228 AB622	5.20 (0.2047)
13228 AB632	5.21 (0.2051)
13228 AB642	5.22 (0.2055)
13228 AB652	5.23 (0.2059)
13228 AB662	5.24 (0.2063)
13228 AB672	5.25 (0.2067)
13228 AB682	5.26 (0.2071)
13228 AB692	5.27 (0.2075)
13228 AB702	4.38 (0.1724)
13228 AB712	4.40 (0.1732)
13228 AB722	4.42 (0.1740)
13228 AB732	4.44 (0.1748)
13228 AB742	4.46 (0.1756)
13228 AB752	4.48 (0.1764)
13228 AB762	4.50 (0.1771)
13228 AB772	4.52 (0.1780)
13228 AB782	4.54 (0.1787)
13228 AB792	4.56 (0.1795)
13228 AB802	4.58 (0.1803)
13228 AB812	4.60 (0.1811)
13228 AB822	4.62 (0.1819)
13228 AB832	4.64 (0.1827)
13228 AB842	4.66 (0.1835)
13228 AB852	5.29 (0.2083)
13228 AB862	5.31 (0.2091)
13228 AB872	5.33 (0.2098)
13228 AB882	5.35 (0.2106)
13228 AB892	5.37 (0.2114)
13228 AB902	5.39 (0.2122)
13228 AB912	5.41 (0.2123)
13228 AB922	5.43 (0.2138)
13228 AB932	5.45 (0.2146)
13228 AB942	5.47 (0.2154)
13228 AB952	5.49 (0.2161)
13228 AB962	5.51 (0.2169)
13228 AB972	5.53 (0.2177)
13228 AB982	5.55 (0.2185)
13228 AB992	5.57 (0.2193)
13228 AC002	5.59 (0.2201)
13228 AC012	5.61 (0.2209)
13228 AC022	5.63 (0.2217)
13228 AC032	5.65 (0.2224)

09490_SBCR_G0052

Fig. 201 Valve adjusting shim chart—2.5L DOHC engine

20. After completion, install all removed components.

3.0L Engine

See Figures 202 through 206.

➡The valve adjustment should be performed while the engine is cold.

1. Raise and support the vehicle safely.
2. Remove the undercover.
3. Lower the vehicle.
4. Remove the collector cover.
5. Disconnect the negative battery cable.
6. On the right side, remove the air intake duct and air cleaner case. Remove the fuel tank protector. Disconnect the oil pressure switch electrical connector. Remove the ignition coil.
7. On the left side, remove the battery and battery carrier. Disconnect the PCV hose from the rocker cover. Remove the ignition coil.
8. Remove the rocker cover retaining bolts. Remove the rocker covers from the engine.
9. Rotate the crankshaft clockwise until the cam is set in position, see illustration.
10. Using a feeler gauge measure and record the clearance of the intake and exhaust valve.

➡Measure it within the range of +/- 30 degrees from the specified position, shown in the illustration. Measure it in the low cam for the intake side. Insert the feeler gauge in a horizontally as possible with respect to the valve lifter.

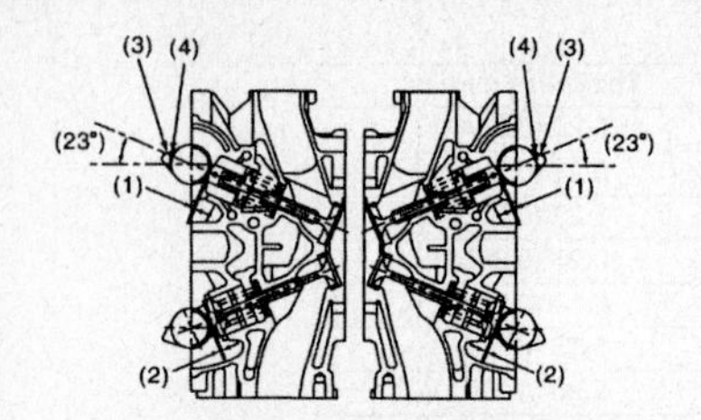

(1) Valve clearance (Intake side)
(2) Valve clearance (Exhaust side)
(3) High lift cam
(4) Low lift cam

09490_SBCR_G0053

Fig. 202 Valve adjustment crankshaft positioning—3.0L engine

Unit: (mm)
S = (V + T) − 0.35
S: Valve lifter thickness required V: Measured valve clearance T: Valve lifter thickness to be used

09490_SBCR_G0054

Fig. 203 Use this table to help you select a suitable exhaust valve shim—3.0L engine

Unit: (mm)
S = (V + T) − 0.20
S: Required shim thickness V: Measured valve clearance T: Shim thickness to be used

09490_SBCR_G0055

Fig. 204 Use this table to help you select a suitable intake valve shim—3.0L engine

11. Further turn the crankshaft pulley clockwise and then measure and record the valve clearance again.
12. If adjustment is required, remove the camshafts.
13. Remove and measure the thickness of the valve lifter. Select a suitable shim, using the shim selection chart.
14. Install the replacement shim to the lifter.
15. After all shims have been adjusted, inspect the valve clearances again.
16. After completion, install all removed components.

Part No.	Thickness mm (in)
13228AD180	4.32 (0.1701)
13228AD190	4.34 (0.1709)
13228AD200	4.36 (0.1717)
13228AD210	4.38 (0.1724)
13228AD220	4.40 (0.1732)
13228AD230	4.42 (0.1740)
13228AD240	4.44 (0.1748)
13228AD250	4.46 (0.1756)
13228AD260	4.48 (0.1764)
13228AD270	4.50 (0.1772)
13228AD280	4.52 (0.1780)
13228AD290	4.54 (0.1787)
13228AD300	4.56 (0.1795)
13228AD310	4.58 (0.1803)
13228AD320	4.60 (0.1811)
13228AC580	4.62 (0.1819)
13228AC590	4.63 (0.1823)
13228AC600	4.64 (0.1827)
13228AC610	4.65 (0.1831)
13228AC620	4.66 (0.1835)
13228AC630	4.67 (0.1839)
13228AC640	4.68 (0.1843)
13228AC650	4.69 (0.1846)
13228AC660	4.70 (0.1850)
13228AC670	4.71 (0.1854)
13228AC680	4.72 (0.1858)
13228AC690	4.73 (0.1862)
13228AC700	4.74 (0.1866)
13228AC710	4.75 (0.1870)
13228AC720	4.76 (0.1874)
13228AC730	4.77 (0.1878)
13228AC740	4.78 (0.1882)
13228AC750	4.79 (0.1886)
13228AC760	4.80 (0.1890)
13228AC770	4.81 (0.1894)
13228AC780	4.82 (0.1898)
13228AC790	4.83 (0.1902)
13228AC800	4.84 (0.1906)
13228AC810	4.85 (0.1909)
13228AC820	4.86 (0.1913)
13228AC830	4.87 (0.1917)
13228AC840	4.88 (0.1921)
13228AC850	4.89 (0.1925)

Part No.	Thickness mm (in)
13228AC860	4.90 (0.1929)
13228AC870	4.91 (0.1933)
13228AC880	4.92 (0.1937)
13228AC890	4.93 (0.1941)
13228AC900	4.94 (0.1945)
13228AC910	4.95 (0.1949)
13228AC920	4.96 (0.1953)
13228AC930	4.97 (0.1957)
13228AC940	4.98 (0.1961)
13228AC950	4.99 (0.1965)
13228AC960	5.00 (0.1969)
13228AC970	5.01 (0.1972)
13228AC980	5.02 (0.1976)
13228AC990	5.03 (0.1980)
13228AD000	5.04 (0.1984)
13228AD010	5.05 (0.1988)
13228AD020	5.06 (0.1992)
13228AD030	5.07 (0.1996)
13228AD040	5.08 (0.2000)
13228AD050	5.09 (0.2004)
13228AD060	5.10 (0.2008)
13228AD070	5.11 (0.2012)
13228AD080	5.12 (0.2016)
13228AD090	5.13 (0.2020)
13228AD100	5.14 (0.2024)
13228AD110	5.15 (0.2028)
13228AD120	5.16 (0.2032)
13228AD130	5.17 (0.2035)
13228AD140	5.18 (0.2039)
13228AD150	5.19 (0.2043)
13228AD160	5.20 (0.2047)
13228AD170	5.21 (0.2051)
13228AD330	5.23 (0.2059)
13228AD340	5.25 (0.2067)
13228AD350	5.27 (0.2075)
13228AD360	5.29 (0.2083)
13228AD370	5.31 (0.2091)
13228AD380	5.33 (0.2098)
13228AD390	5.35 (0.2106)
13228AD400	5.37 (0.2114)
13228AD410	5.39 (0.2122)
13228AD420	5.41 (0.2130)
13228AD430	5.43 (0.2138)
13228AD440	5.45 (0.2146)
13228AD450	5.47 (0.2154)
13228AD460	5.49 (0.2161)
13228AD470	5.51 (0.2169)
13228AD480	5.53 (0.2177)
13228AD490	5.55 (0.2185)
13228AD500	5.57 (0.2193)
13228AD510	5.59 (0.2201)

09490_SBCR_G0056

Fig. 205 Exhaust valve adjusting shim chart—3.0L engine

Part No.	Thickness mm (in)
13218AK890	1.92 (0.0756)
13218AK900	1.94 (0.0764)
13218AK910	1.96 (0.0772)
13218AK920	1.98 (0.0780)
13218AK930	2.00 (0.0787)
13218AK940	2.02 (0.0795)
13218AK950	2.04 (0.0803)
13218AK960	2.06 (0.0811)
13218AK970	2.07 (0.0815)
13218AK980	2.08 (0.0819)
13218AK990	2.09 (0.0823)
13218AL000	2.10 (0.0827)
13218AL010	2.11 (0.0831)
13218AL020	2.12 (0.0835)
13218AL030	2.13 (0.0839)
13218AL040	2.14 (0.0843)
13218AL050	2.15 (0.0846)
13218AL060	2.16 (0.0850)
13218AL070	2.17 (0.0854)
13218AL080	2.18 (0.0858)
13218AL090	2.19 (0.0862)
13218AL100	2.20 (0.0866)
13218AL110	2.21 (0.0870)
13218AL120	2.22 (0.0874)
13218AL130	2.23 (0.0878)
13218AL140	2.24 (0.0882)
13218AL150	2.25 (0.0886)
13218AL160	2.26 (0.0890)
13218AL170	2.27 (0.0894)
13218AL180	2.28 (0.0898)
13218AL190	2.29 (0.0902)
13218AL200	2.30 (0.0906)
13218AL210	2.31 (0.0909)
13218AL220	2.32 (0.0913)
13218AL230	2.33 (0.0917)
13218AL240	2.34 (0.0921)
13218AL250	2.35 (0.0925)
13218AL260	2.36 (0.0929)
13218AL270	2.37 (0.0933)
13218AL280	2.38 (0.0937)
13218AL290	2.39 (0.0941)
13218AL300	2.40 (0.0945)
13218AL310	2.41 (0.0949)

Part No.	Thickness mm (in)
13218AL320	2.42 (0.0953)
13218AL330	2.43 (0.0957)
13218AL340	2.44 (0.0961)
13218AL350	2.45 (0.0965)
13218AL360	2.46 (0.0969)
13218AL370	2.47 (0.0972)
13218AL380	2.48 (0.0976)
13218AL390	2.49 (0.0980)
13218AL400	2.50 (0.0984)
13218AL410	2.51 (0.0988)
13218AL420	2.52 (0.0992)
13218AL430	2.53 (0.0996)
13218AL440	2.54 (0.1000)
13218AL450	2.55 (0.1004)
13218AL460	2.56 (0.1008)
13218AL470	2.57 (0.1012)
13218AL480	2.58 (0.1016)
13218AL490	2.59 (0.1020)
13218AL500	2.60 (0.1024)
13218AL510	2.61 (0.1028)
13218AL520	2.62 (0.1032)
13218AL530	2.64 (0.1039)
13218AL540	2.66 (0.1047)
13218AL550	2.68 (0.1055)
13218AL560	2.70 (0.1063)
13218AL570	2.72 (0.1071)
13218AL580	2.74 (0.1079)
13218AL590	2.76 (0.1087)

09490_SBCR_G0057

Fig. 206 Intake valve adjusting shim chart—3.0L engine

ENGINE PERFORMANCE & EMISSION CONTROLS COMPONENT LOCATIONS

See Figures 207 through 213.

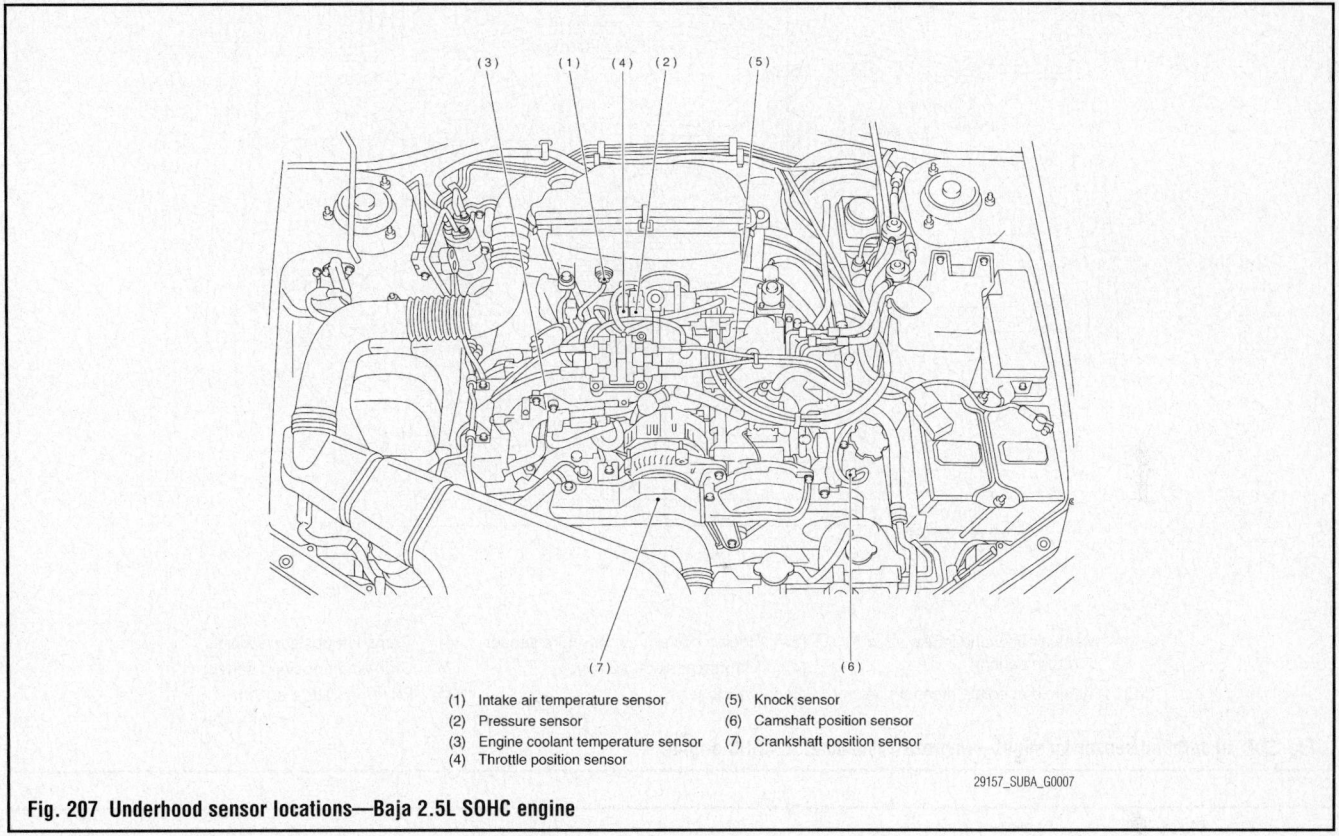

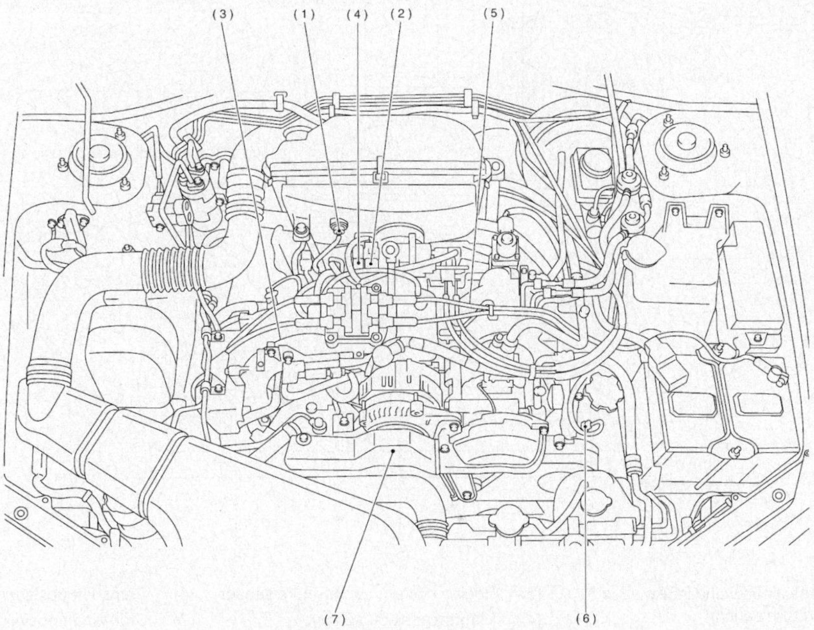

(1) Intake air temperature sensor
(2) Pressure sensor
(3) Engine coolant temperature sensor
(4) Throttle position sensor
(5) Knock sensor
(6) Camshaft position sensor
(7) Crankshaft position sensor

29157_SUBA_G0007

Fig. 207 Underhood sensor locations—Baja 2.5L SOHC engine

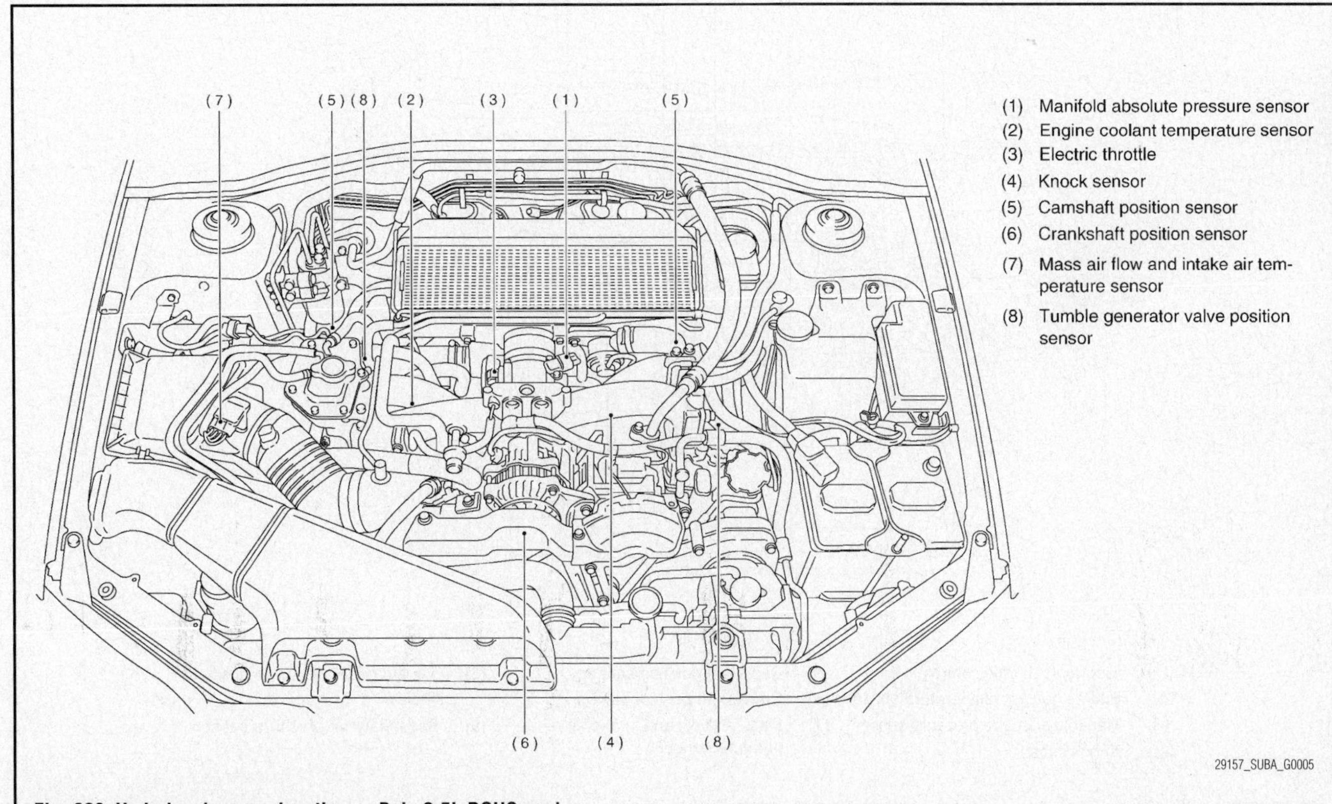

Fig. 208 Underhood sensor locations—Baja 2.5L DOHC engine

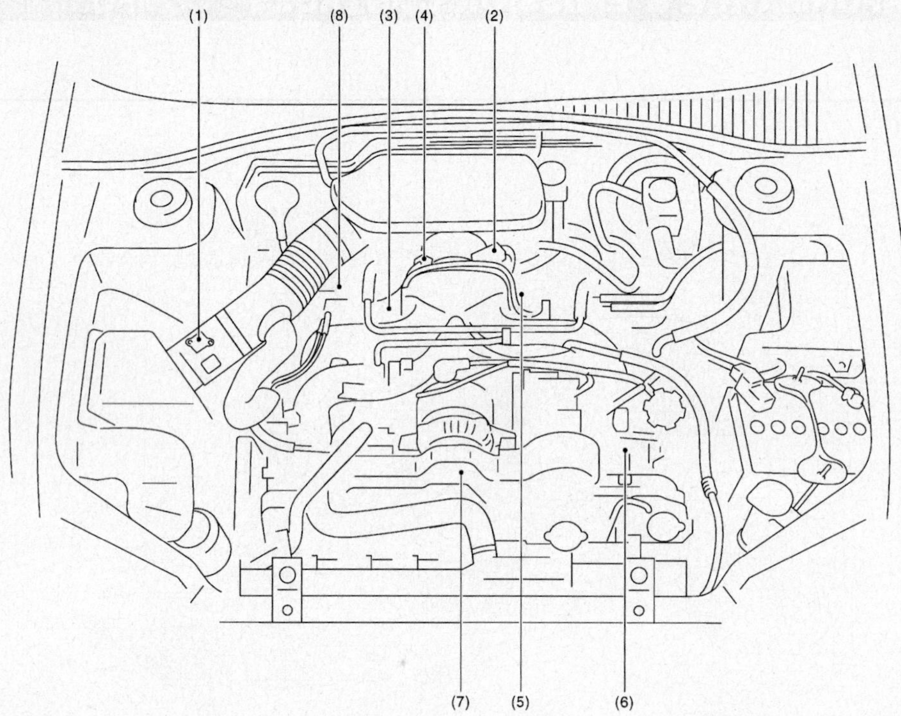

(1) Mass air flow and intake air temperature sensor
(2) Manifold absolute pressure sensor
(3) Engine coolant temperature sensor
(4) Throttle position sensor
(5) Knock sensor
(6) Camshaft position sensor
(7) Crankshaft position sensor
(8) Oil temperature sensor

29157_SUBA_G0066

Fig. 209 Underhood sensor locations—Impreza 2006–08 2.5L SOHC engine

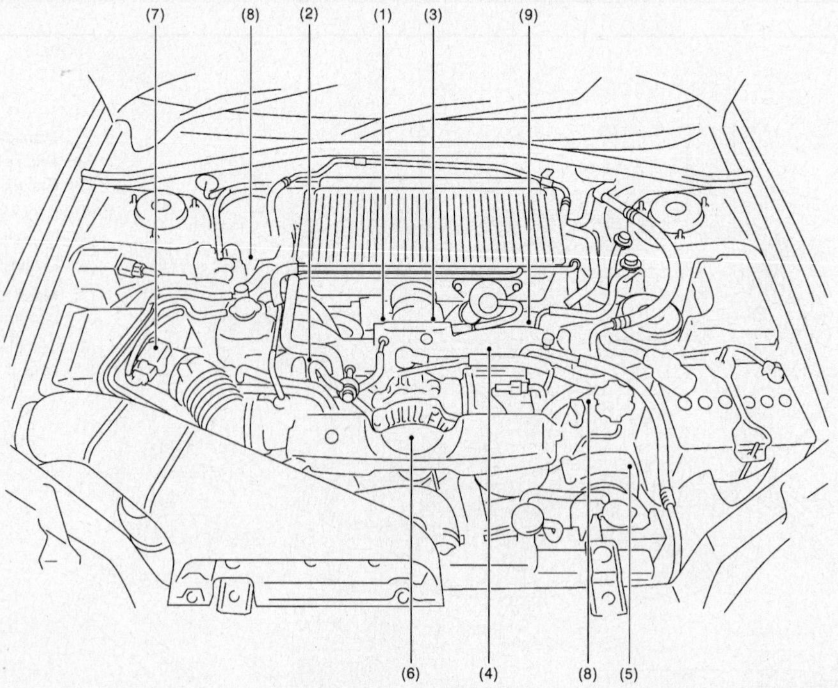

(1) Electronic throttle control
(2) Engine coolant temperature sensor
(3) Manifold absolute pressure sensor
(4) Knock sensor
(5) Camshaft position sensor
(6) Crankshaft position sensor
(7) Mass air flow and intake air temperature sensor
(8) Tumble generator valve position sensor
(9) Secondary air pressure sensor

29157_SUBA_G0058

Fig. 210 Underhood sensor locations—Impreza 2006–08 2.5L DOHC engine

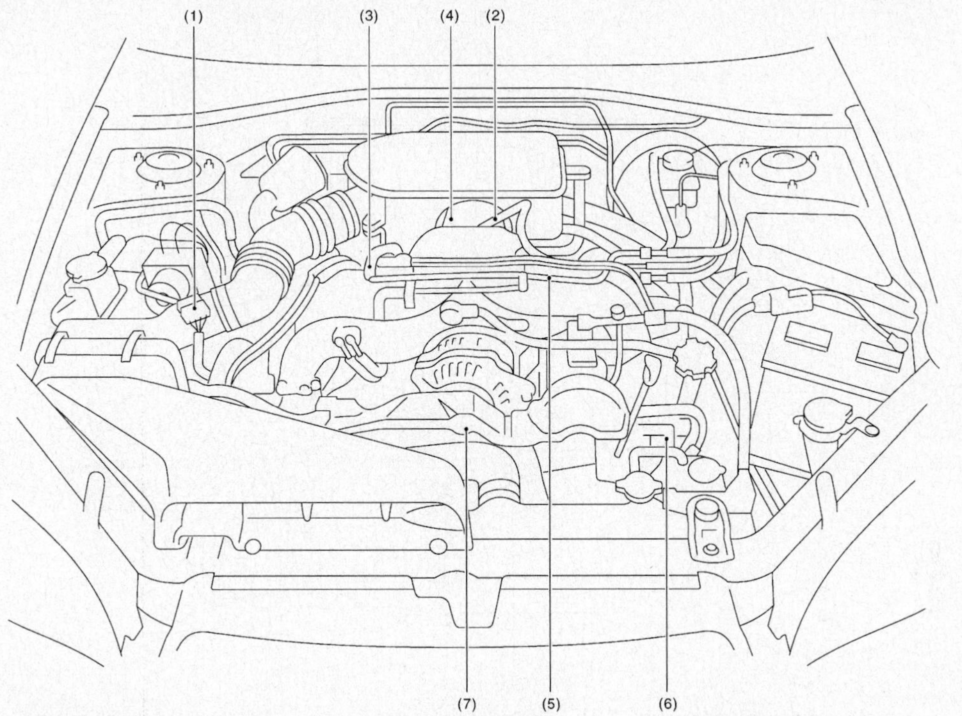

(1) Mass air flow and intake air temperature sensor
(2) Manifold absolute pressure sensor
(3) Engine coolant temperature sensor
(4) Electronic throttle control
(5) Knock sensor
(6) Camshaft position sensor
(7) Crankshaft position sensor

29157_SUBA_G0109

Fig. 211 Underhood sensor locations—Legacy and Outback 2006–08 2.5L SOHC engine

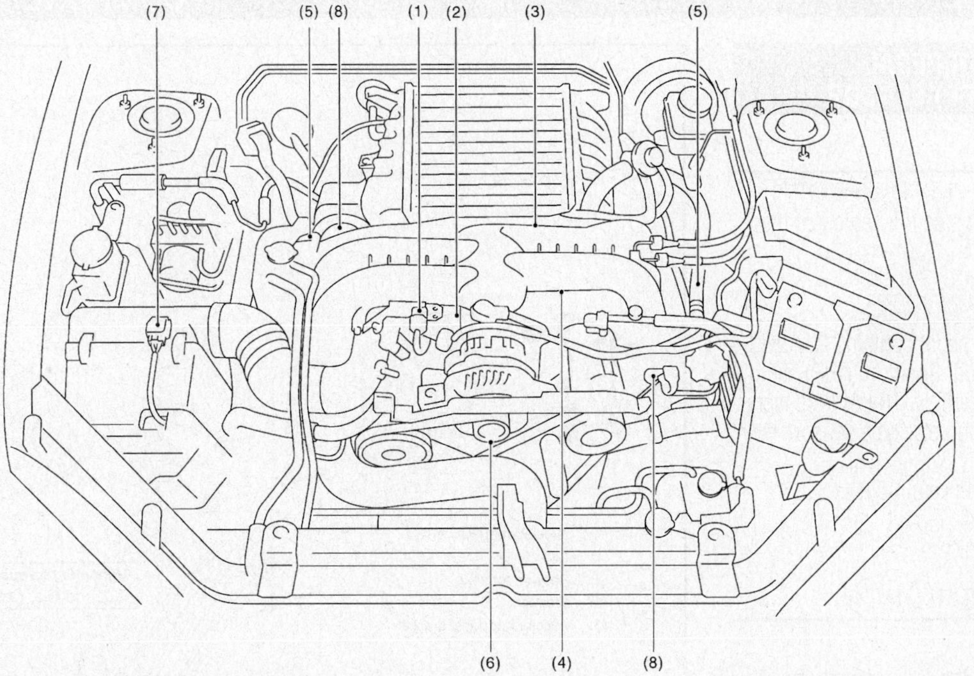

(1) Manifold absolute pressure sensor
(2) Engine coolant temperature sensor
(3) Electronic throttle control
(4) Knock sensor
(5) Intake camshaft position sensor
(6) Crankshaft position sensor
(7) Mass air flow and intake air temperature sensor
(8) Tumble generator valve position sensor

29157_SUBA_G0092

Fig. 212 Underhood sensor locations—Legacy and Outback 2006–08 2.5L DOHC engine

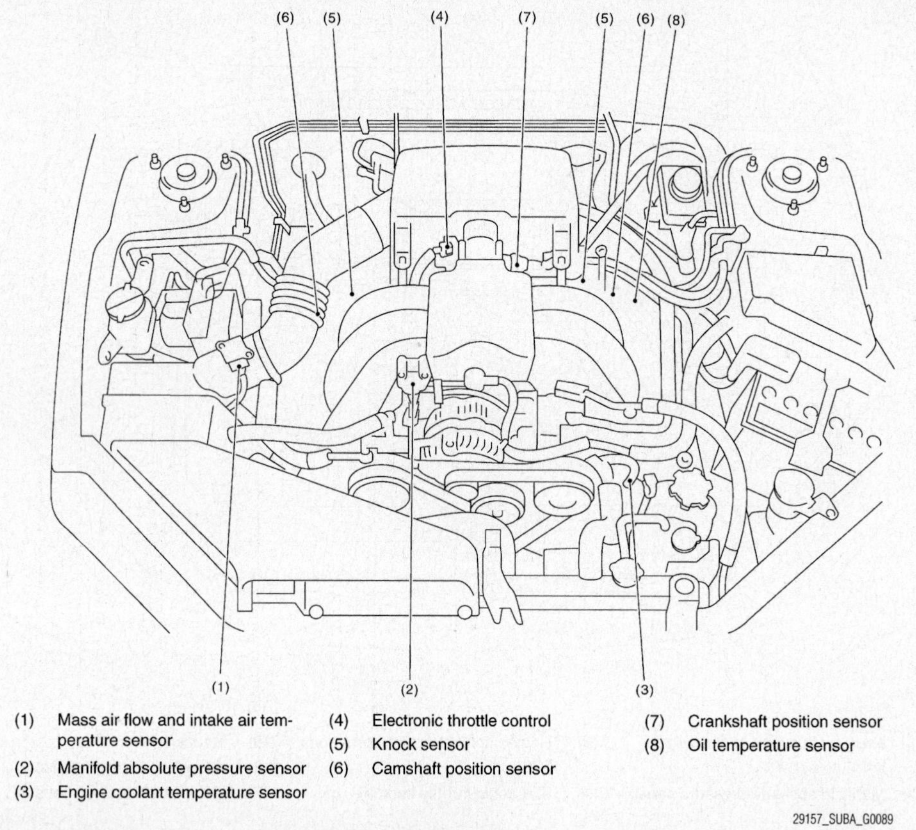

(1) Mass air flow and intake air temperature sensor
(2) Manifold absolute pressure sensor
(3) Engine coolant temperature sensor
(4) Electronic throttle control
(5) Knock sensor
(6) Camshaft position sensor
(7) Crankshaft position sensor
(8) Oil temperature sensor

29157_SUBA_G0089

Fig. 213 Underhood sensor locations—Legacy and Outback 2006–08 3.0L engine

COMPONENT TESTING

ACCELERATOR PEDAL POSITION (APP) SENSOR

LOCATION

The Accelerator Pedal Position (APP) sensor is located inside the vehicle at the accelerator pedal.

OPERATION

The accelerator pedal contains 2 individual Accelerator Pedal Position (APP) sensors within the assembly. The APP sensors 1 and 2 are potentiometer type sensors each with 3 circuits:

- A 5-volt reference circuit
- A low reference circuit
- A signal circuit

REMOVAL & INSTALLATION

See Figure 214.

1. Disconnect the ground cable from the battery.
2. Disconnect the connector.
3. Remove the nut securing accelerator pedal assembly.

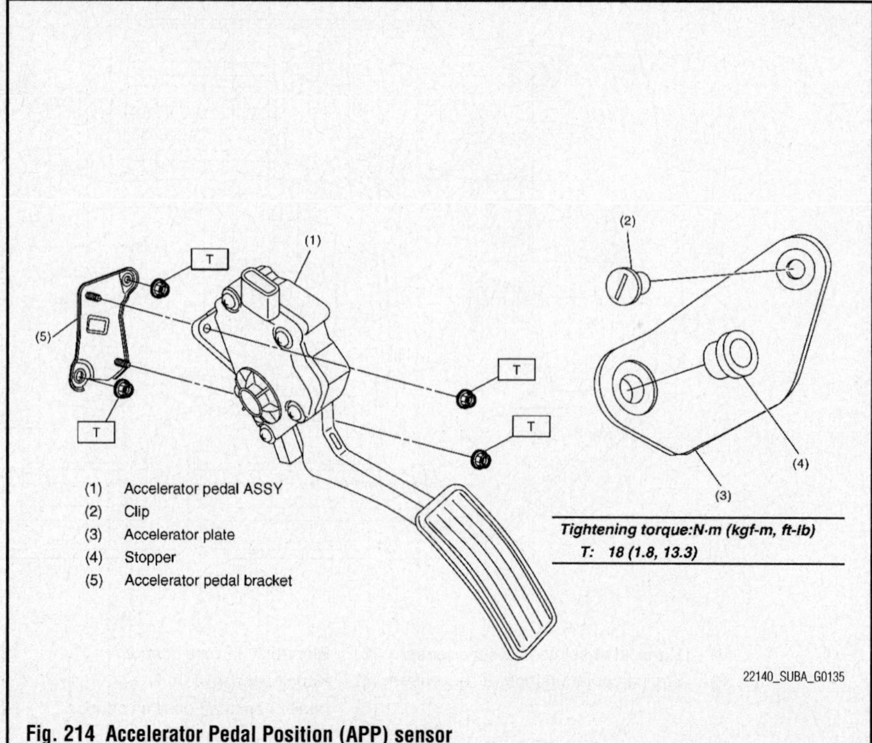

Fig. 214 Accelerator Pedal Position (APP) sensor

To install:

4. Install in the reverse order of removal.
5. Tighten to 13 ft. lbs. (18 Nm).

TESTING

1. Measure the voltage between accelerator pedal position sensor connector and chassis ground as follows:
   - (B315) No. 6 (+)—Chassis ground (-)
   - (B315) No. 3 (+)—Chassis ground (-)
2. Is the difference in measured values for the main accelerator pedal position sensor signal and the sub accelerator pedal position sensor signal 0 volts. If so replace the APP sensor.
3. If not check and repair the following items:
   - Open circuit of harness between the ECM and accelerator pedal position sensor connector.
   - Ground short circuit of harness between the ECM and accelerator pedal position sensor connectors.

CAMSHAFT POSITION (CMP) SENSOR

LOCATION

See Figures 215, 216 and 217.

T1
(4)
T2
T1
(1)
T1
(2)
(3)
(1) Crankshaft position sensor
(2) Knock sensor
(3) Camshaft position sensor LH
(4) Camshaft position sensor RH
Tightening torque:N·m (kgf-m, ft-lb)
T1: 6.4 (0.65, 4.7)
T2: 24 (2.4, 17.7)
22140_SUBA_G0138

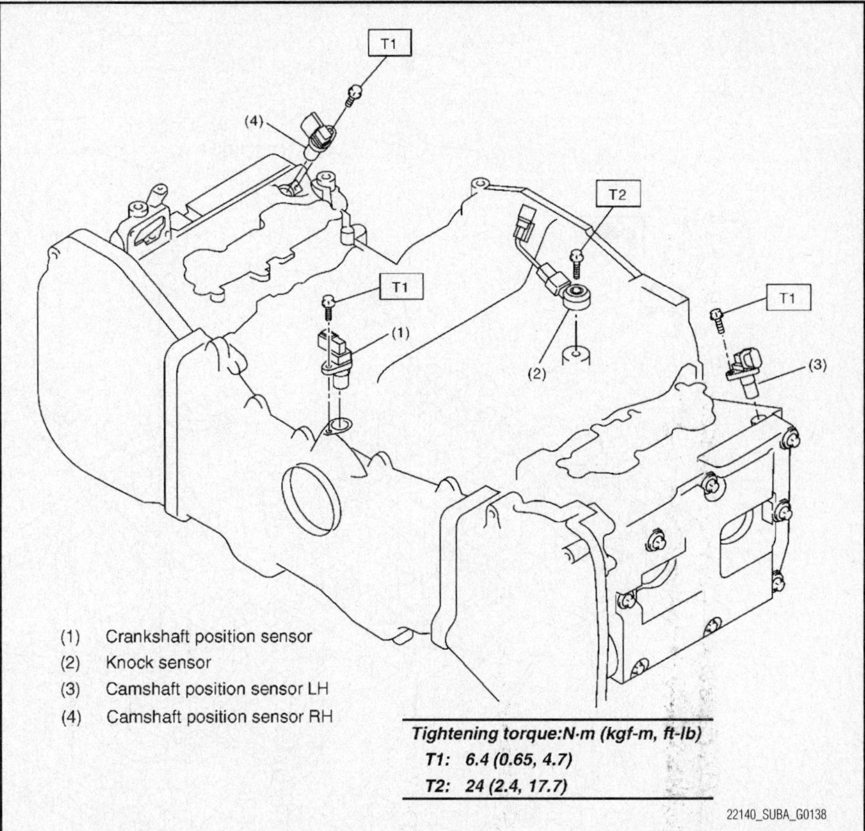

Fig. 216 Camshaft Position (CMP) sensor & Crankshaft Position (CKP) sensor locations—2.5L DOHC engine

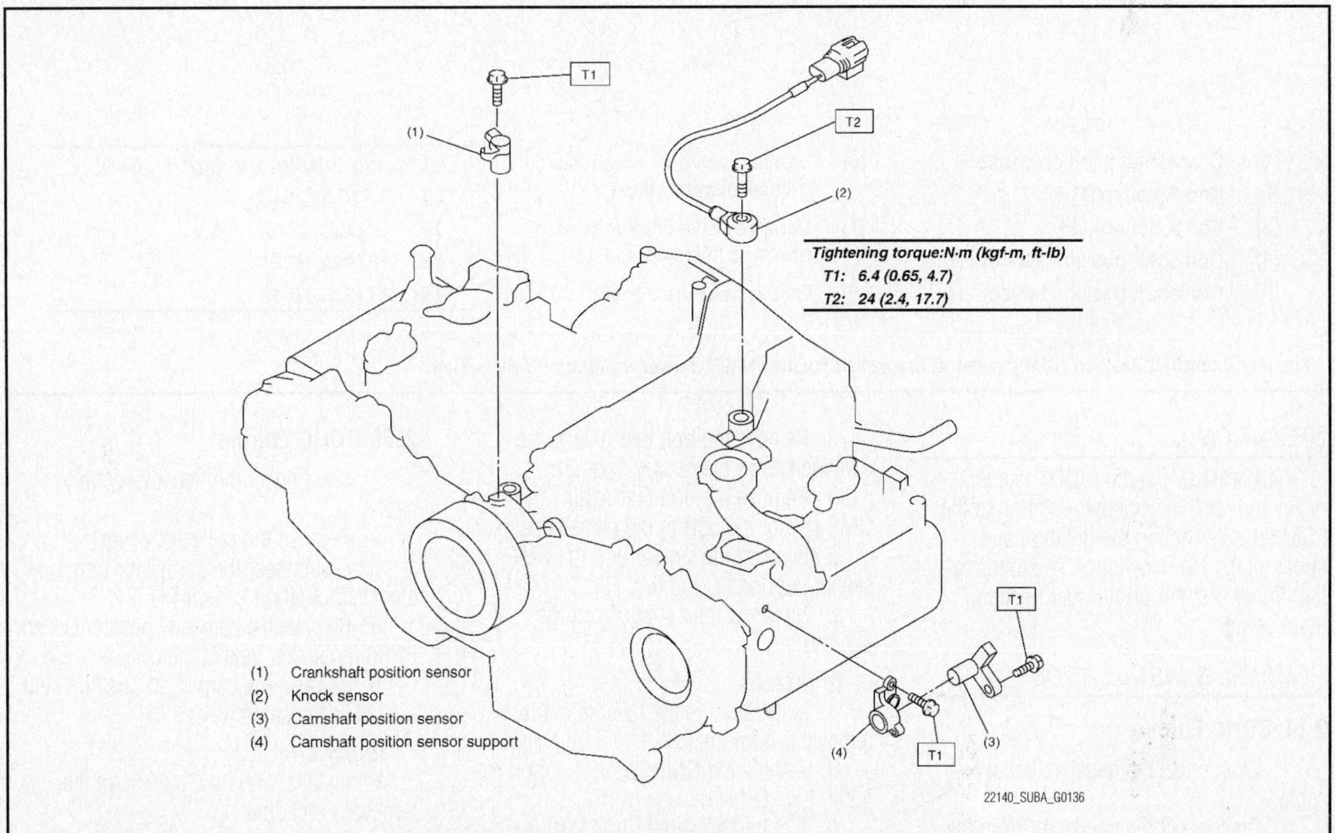

T1
(1)
T2
(2)
Tightening torque:N·m (kgf-m, ft-lb)
T1: 6.4 (0.65, 4.7)
T2: 24 (2.4, 17.7)
T1
(1) Crankshaft position sensor
(2) Knock sensor
(3) Camshaft position sensor
(4) Camshaft position sensor support
(4)
T1
(3)
22140_SUBA_G0136

Fig. 215 Camshaft Position (CMP) sensor & Crankshaft Position (CKP) sensor locations—2.5L SOHC engine

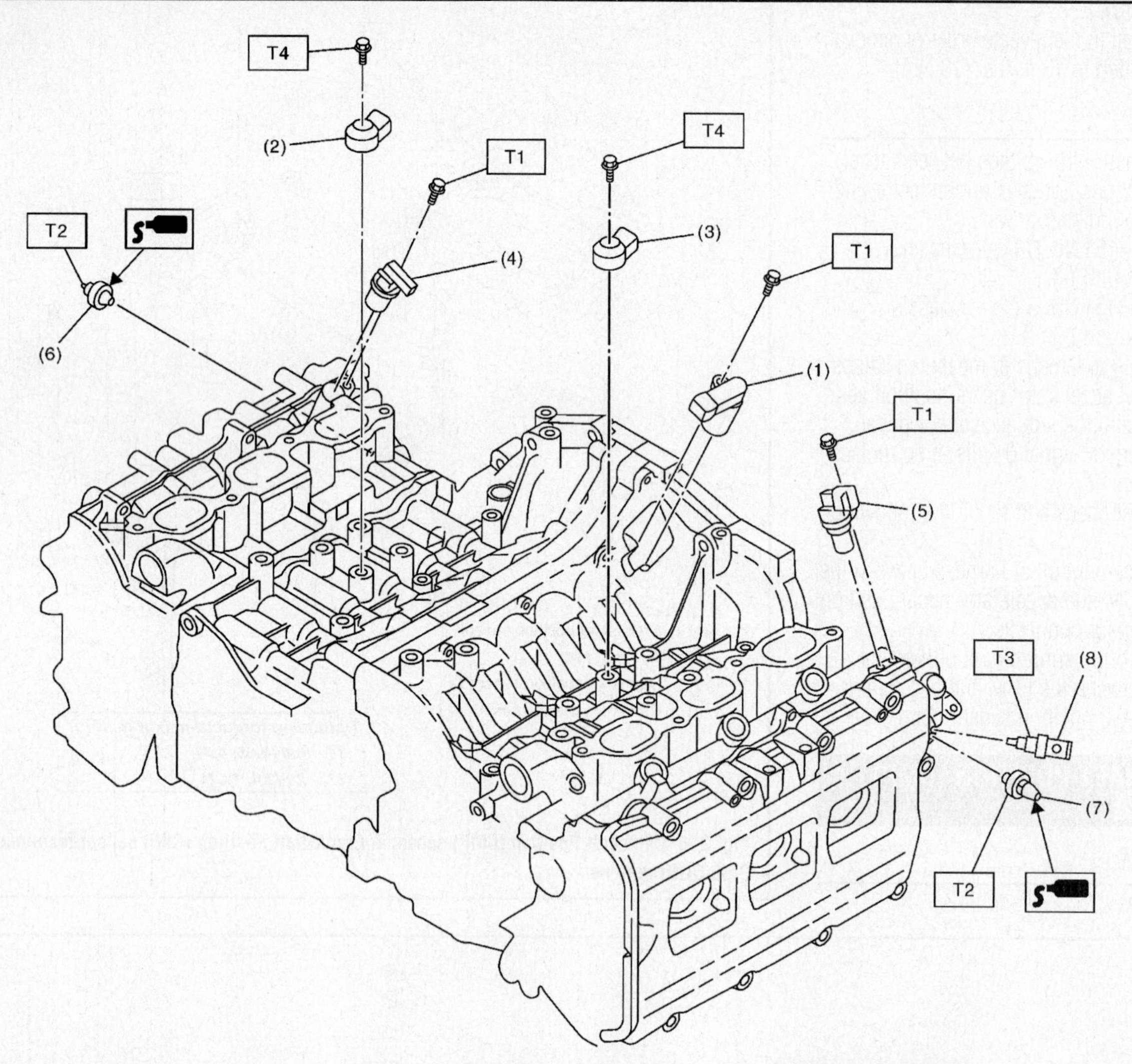

Fig. 217 Camshaft Position (CMP) sensor & Crankshaft Position (CKP) sensor locations—3.0L engine

OPERATION

The Camshaft Position (CMP) sensor relays the relative camshaft position to the ECM for determining the position and stroke of the No. 1 cylinder. This information is required for proper fuel injection functioning.

REMOVAL & INSTALLATION

2.5L SOHC Engine

1. Disconnect the negative battery cable.
2. Disconnect the connector from the Camshaft Position (CMP) sensor.
3. Remove the bolt that retains the CMP sensor to the sensor support.
4. Remove the bolt that retains the CMP sensor support to the camshaft cap.
5. Remove the sensor and the CMP sensor support as a unit.
6. Separate the sensor from the support.

To install:

7. Installation is the reverse of the removal procedure.
8. Tighten the CMP sensor support to 4.7 ft. lbs. (6.4 Nm).
9. Tighten the sensor to 4.7 ft. lbs. (6.4 Nm).

2.5L DOHC Engine

1. Disconnect the negative battery cable.
2. Remove the collector cover.
3. Disconnect the connector from camshaft position sensor RH.
4. Remove the camshaft position sensor RH from the rear side of the cylinder head.
5. Remove the cam shaft position sensor LH in the same way as RH.

To install:

6. Installation is the reverse of the removal procedure.
7. Tighten the sensor to 4.7 ft. lbs. (6.4 Nm).

3.0L Engine

1. Disconnect the negative battery cable.
2. Remove the collector cover, as required.
3. Remove the alternator harness from the fuel pipe. Remove the fuel pipe protector.
4. Disconnect the connector from the sensor.
5. Remove the bolt that retains the sensor.
6. Remove the sensor.

To install:

7. Installation is the reverse of the removal procedure.
8. Tighten the sensor to 4.7 ft. lbs. (6.4 Nm).

TESTING

2.5L SOHC Engine

1. Remove the sensor from the vehicle.
2. Measure the resistance between the two connector terminals of the sensor, (terminals 1 and 2).
3. If measured value is not within 1–4 k ohms, replace the sensor.

2.5L DOHC Engine

1. Using the Subaru scan tool, check the sensor output waveform.
2. With the engine OFF and the ignition switch ON the signal (V) should be 0–0.9.
3. Measurement is taken using connector B135 terminal 8 for the left side and connector B135 terminal 9 for the right side.
4. If abnormality is found, replace the sensor.

3.0L Engine

1. Using the Subaru scan tool, check the sensor output waveform.
2. The power supply measurement is taken using connector B134 terminal 21 and for ground using connector B134 terminal 21, for the left side and power supply measurement is taken using connector B134 terminal 11 and for ground using connector B134 terminal 22, for the right side.
3. If abnormality is found, replace the sensor.

CRANKSHAFT POSITION (CKP) SENSOR

LOCATION

See Figures 215, 216 and 217.

Refer to the illustrations under Camshaft Position (CMP) Sensor for Crankshaft Position (CKP) sensor locations.

OPERATION

The Crankshaft Position (CKP) sensor is a variable reluctance sensor which is used to inform the ECM when the No.1 piston is at top dead center. This information is used for controlling and adjusting ignition and fuel injector timing.

REMOVAL & INSTALLATION

Baja

1. Disconnect the negative battery cable.
2. Remove the bolt that retains the sensor in place at the cylinder block.
3. Remove the sensor from its mounting.
4. Disconnect the connector from the sensor.

To install:

5. Installation is the reverse of the removal procedure.
6. Tighten the sensor to 4.7 ft. lbs. (6.4 Nm).

Impreza, Legacy, and Outback

2.5L Engines

1. Remove the collector cover.
2. Disconnect the ground cable from battery.
3. Remove the alternator.
4. Remove the bolt which installs crankshaft position sensor to cylinder block.
5. Remove the crankshaft position sensor, and then disconnect the connector from it.

To install:

6. Installation is the reverse of the removal procedure.
7. Tighten the sensor to 4.7 ft. lbs. (6.4 Nm).

3.0L Engine

1. Before servicing the vehicle, refer to the precautions section.
2. Remove the collector cover.
3. Disconnect the ground cable from battery.
4. Remove the air intake chamber.
5. Remove the service hole cover.
6. Remove the crankshaft position sensor.
7. Disconnect the connector from crankshaft position sensor.

To install:

8. Install in the reverse order of removal.
9. Tighten the crankshaft sensor mounting bolt to 5 ft. lbs. (6.8 Nm).

TESTING

1. Remove the sensor from the vehicle.
2. Measure the resistance between the two connector terminals of the sensor, (terminals 1 and 2).
3. If the measured value is not within 1–4 k ohms, replace the sensor.

EGR VALVE

LOCATION

On the 2.5L SOHC engine and 3.0L engine, the Exhaust Gas Recirculation Valve (EGR) valve is mounted on the intake manifold. The DOHC engine does not use a conventional EGR valve.

OPERATION

The EGR valve controls the flow of exhaust gases. The ECM monitors the flow and regulates the valve accordingly.

REMOVAL & INSTALLATION

1. Disconnect the negative battery cable.
2. Disconnect the connector.
3. Remove the valve from the intake manifold.

To install:

4. Installation is the reverse of the removal procedure.
5. Tighten the retaining bolts to 14 ft. lbs. (19 Nm).

TESTING

See Figure 218.

1. Disconnect the connector from the ECM.
2. Measure the resistance between the ECM connector and chassis ground.

➡If DTC code P1492 was set, use connector B134 and terminal 18/chassis ground. If DTC code P1494 was set, use connector B134 and terminal 17/chassis ground. If DTC code P1496 was set, use connector B134 and terminal 16/chassis ground. If DTC code P1498 was set, use connector B134 and terminal 15/chassis ground.

3. If measured value is more than 1 M ohm, check the ECM connector and the EGR solenoid for poor contact.
4. If contact is good, replace the EGR solenoid valve.

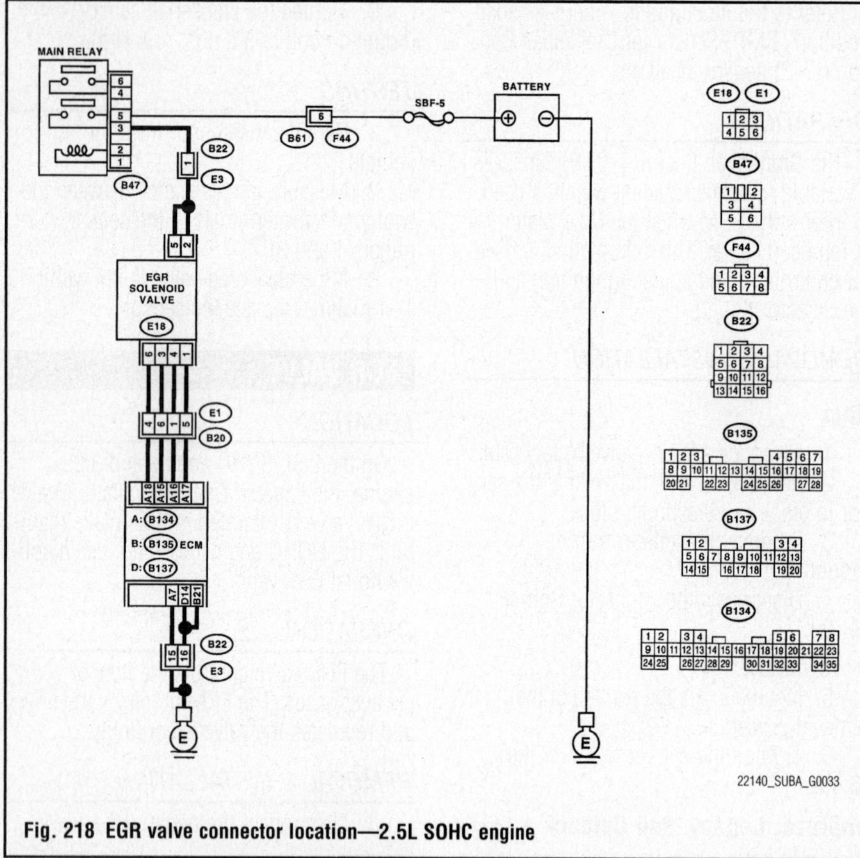

Fig. 218 EGR valve connector location—2.5L SOHC engine

ELECTRONIC CONTROL MODULE (ECM)

LOCATION

See Figure 219.

The ECM is located on the passenger's side of the vehicle, underneath the floor mat.

OPERATION

The Electronic Control Module (ECM) controls the vehicle engine operating system.

REMOVAL & INSTALLATION

See Figure 220.

1. Disconnect the negative battery cable.

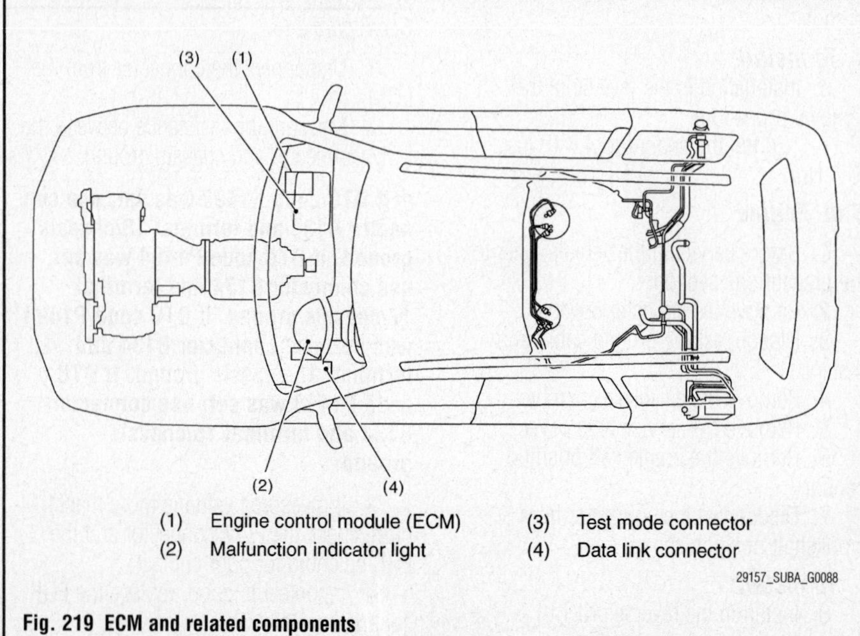

Fig. 219 ECM and related components

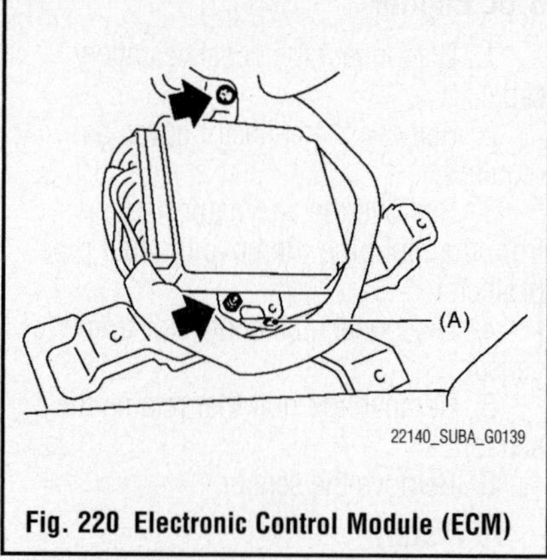

Fig. 220 Electronic Control Module (ECM)

2. Remove the lower inner trim on the passenger's side of the vehicle.
3. Detach the floor mat. Remove the protective cover.
4. Remove the Electronic Control Module (ECM) bracket retaining nuts. Remove the clip (A) from the bracket.
5. Disconnect the connectors.
6. Remove the ECM from the vehicle.

To install:

7. Installation is the reverse of the removal procedure.
8. Tighten the retaining screws to 5.5 ft. lbs. (7.5 Nm).

➡When replacing the ECM, be careful not to use the wrong part number, as damage to the injection system could occur.

TESTING

Scan to verify communication with ECM. If no communication, check power and grounds to the ECM.

ENGINE COOLANT TEMPERATURE (ECT) SENSOR

LOCATION

See Figures 221 and 222.

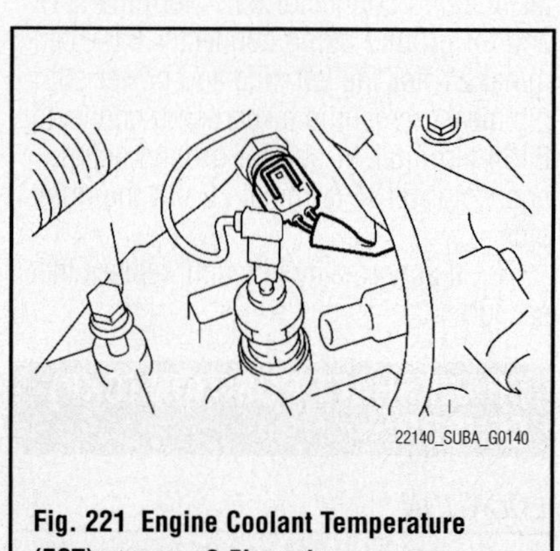

Fig. 221 Engine Coolant Temperature (ECT) sensor—2.5L engine

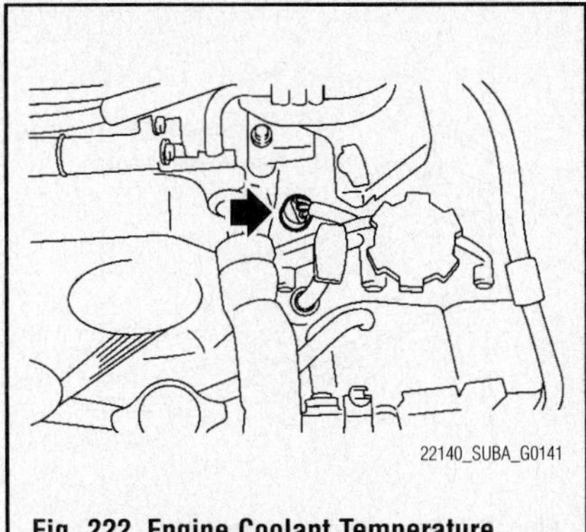

Fig. 222 Engine Coolant Temperature (ECT) sensor—3.0L engine

The Engine Coolant Temperature (ECT) sensor is located by the heater outlet fitting or in a cooling passage on the engine, depending upon the particular vehicle. Refer below for engine location view.

OPERATION

The Engine Coolant Temperature (ECT) sensor detects the temperature of the engine coolant and relays the information to the electronic control assembly.

REMOVAL & INSTALLATION

2.5L Engines

1. Disconnect the negative battery cable.
2. Remove the alternator, as required.
3. Remove the air intake duct and air cleaner case, as required.
4. Disconnect the connector from the sensor.
5. Drain the cooling system, as required.
6. Remove the sensor from its mounting.

To install:

7. Installation is the reverse of the removal procedure.
8. Tighten the sensor to 13.3 ft. lbs. (18 Nm).

3.0L Engine

1. Disconnect the negative battery cable.
2. Disconnect the ground cable from battery.
3. Disconnect the connectors from the engine coolant temperature sensor.
4. Remove the engine coolant temperature sensor.

To install:

5. Installation is the reverse of the removal procedure.
6. Tighten the sensor to 13.3 ft. lbs. (18 Nm).

TESTING

See Figure 223.

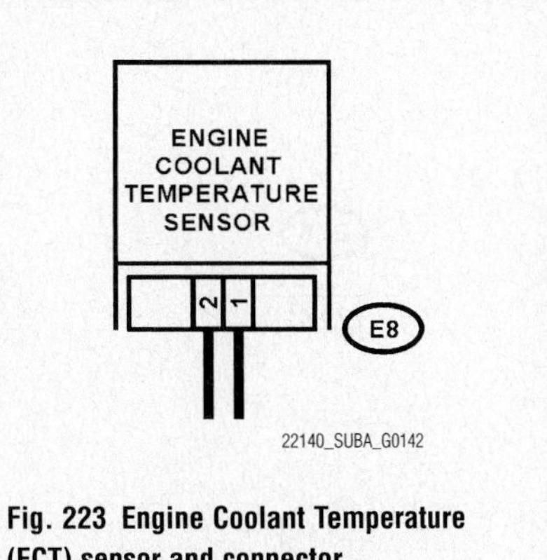

Fig. 223 Engine Coolant Temperature (ECT) sensor and connector

1. Measure the resistance between engine coolant temperature sensor terminals No.1 and No.2 when the engine coolant is cold and after warmed-up.
2. Is the resistance of engine coolant temperature sensor different between when engine coolant is cold and after warmed-up, check for poor contact in ECT and ECM connector.
3. If the resistance does not change between cold and warmed-up engine. Replace the engine coolant temperature sensor.

ENGINE OIL TEMPERATURE (EOT) SENSOR

LOCATION

See Figures 224 and 225.

Refer below for location views of the Engine Oil Temperature (EOT) Sensor.

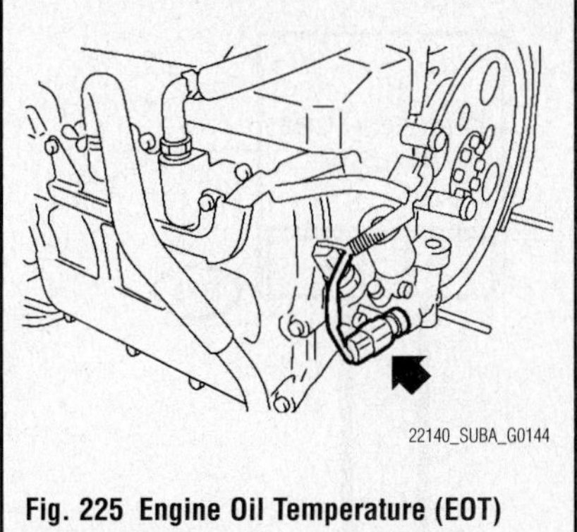

Fig. 225 Engine Oil Temperature (EOT) sensor location view—3.0L engine

OPERATION

This component detects the temperature of the engine oil and relays the information to the electronic control assembly.

REMOVAL & INSTALLATION

1. Before servicing the vehicle, refer to the precautions section.
2. Remove the collector cover.
3. Disconnect the ground cable from battery.
4. Disconnect the connector from oil temperature sensor.
5. Remove the oil temperature sensor.

To install:

6. Install in the reverse order of removal.
7. Tighten the EOT sensor to 16 ft. lbs. (22 Nm).

TESTING

See Figure 226.

1. Start the engine.

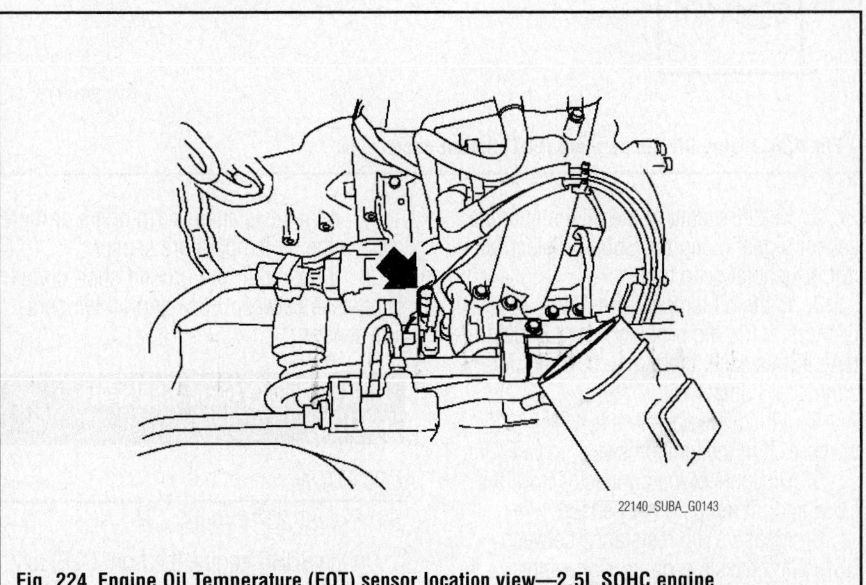

Fig. 224 Engine Oil Temperature (EOT) sensor location view—2.5L SOHC engine

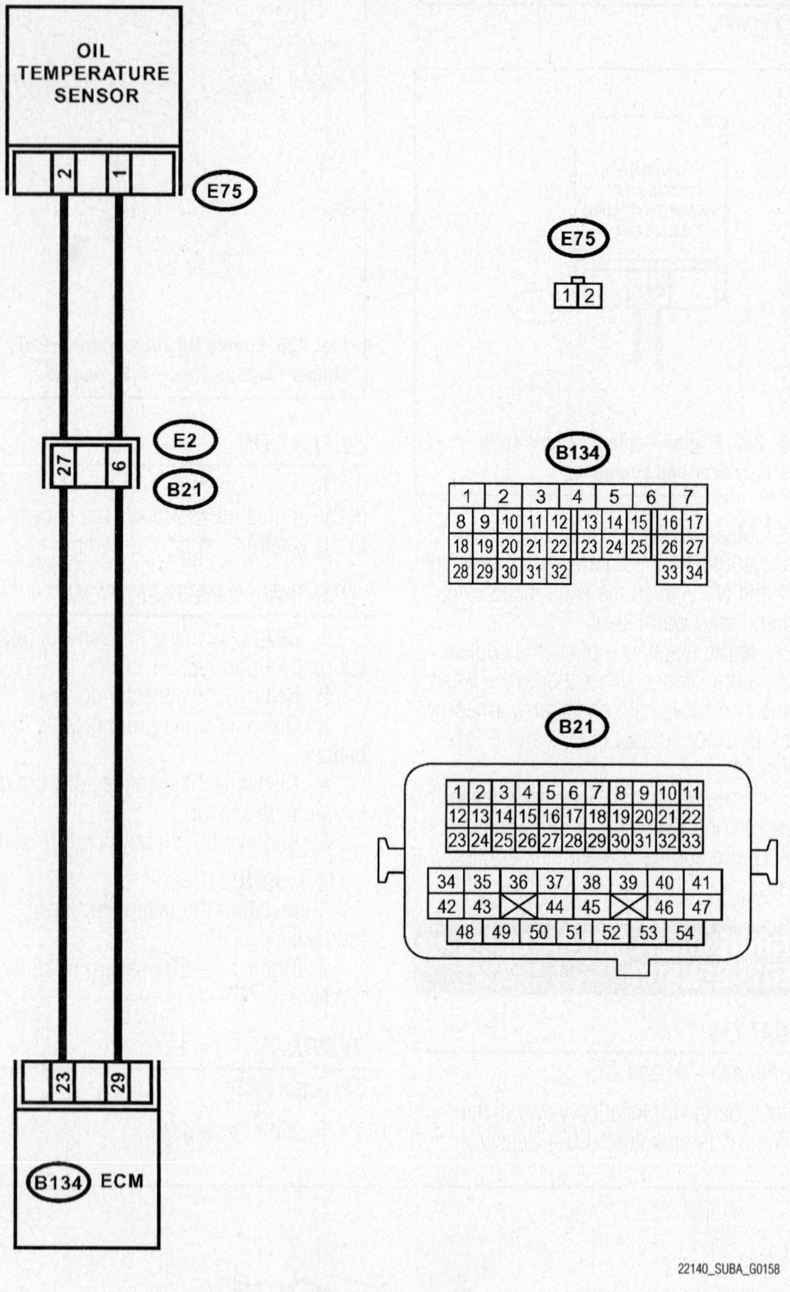

OIL TEMPERATURE SENSOR

E75

E2

B21

B134

ECM

22140_SUBA_G0158

Fig. 226 Engine Oil Temperature (EOT) Sensor circuit view

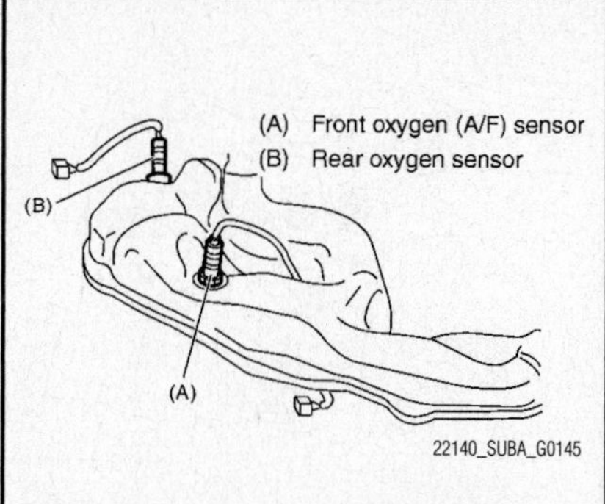

Fig. 227 Front and rear Heated Oxygen Sensor (HO2S) view —2.5L SOHC engine

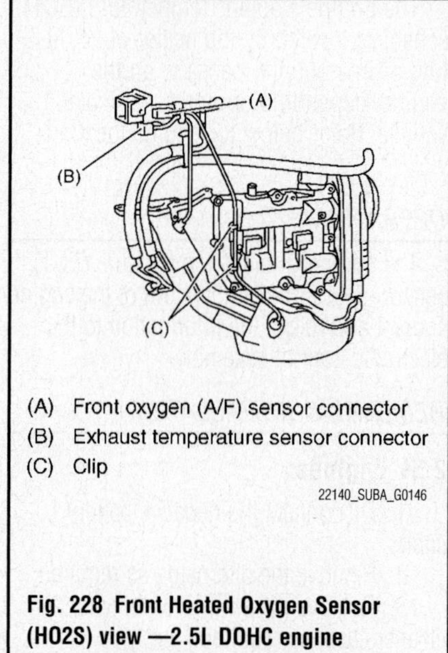

Fig. 228 Front Heated Oxygen Sensor (HO2S) view —2.5L DOHC engine

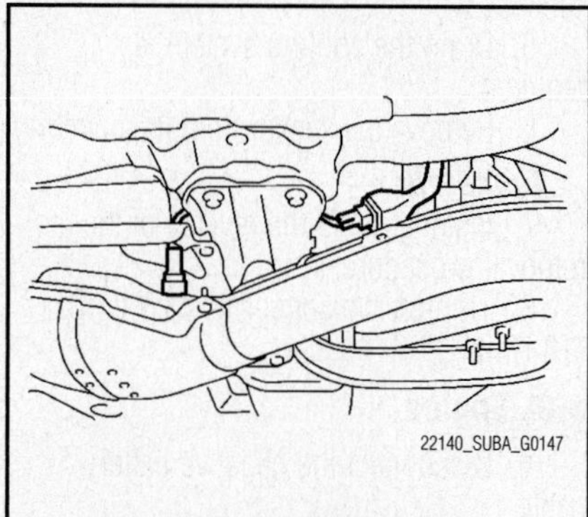

Fig. 229 Rear Heated Oxygen Sensor (HO2S) view—2.5L DOHC engine

2. Read the data of the oil temperature sensor signal using the Subaru Select Monitor or general scan tool.

3. Is the oil temperature 419°F (215°C) or more. If not the problem may be intermittent. In this case, there may be a temporary connector contact failure.

4. If the oil temperature is 419°F (215°C) or more. Turn the ignition switch to OFF.

5. Disconnect the connector from the ECM and oil temperature sensor.

6. Measure the resistance between ECM and chassis ground. Connector (B134) terminal No. 23

7. If the resistance is 1m ohms or more replace the oil temperature sensor.

8. If not repair the ground short circuit of harness between ECM and oil temperature sensor.

HEATED OXYGEN SENSOR (HO2S)

LOCATION

See Figures 227 through 231.

On the SOHC engine the front (A/F) oxygen sensor is located in the front section of the front catalytic converter. The rear oxygen sensor is located in the rear section of the front catalytic converter. On the DOHC engine the front oxygen sensor is located in the front section of the exhaust system, just past the crossover pipe. The rear oxygen sensor is located in the front section of the catalytic converter.

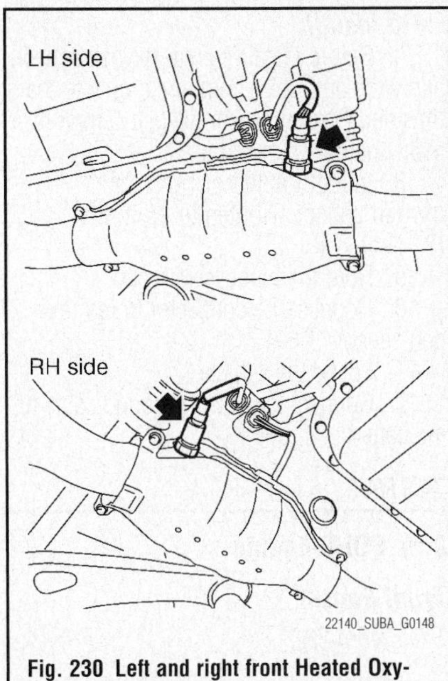

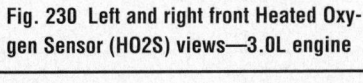

Fig. 230 Left and right front Heated Oxygen Sensor (HO2S) views—3.0L engine

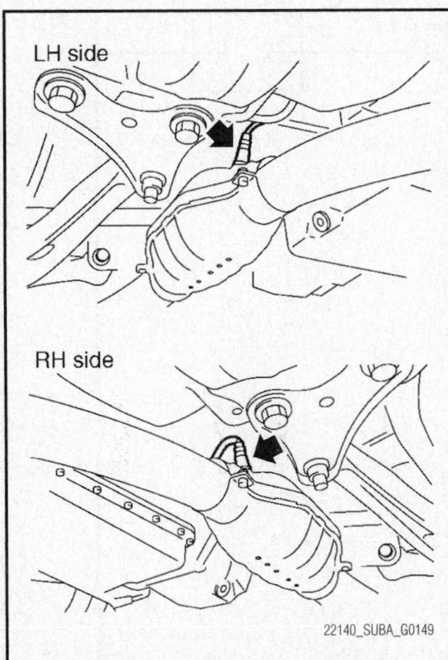

Fig. 231 Left and right rear Heated Oxygen Sensor (HO2S) views—3.0L engine

OPERATION

The Heated Oxygen Sensor (HO2S) supplies the electronic control assembly with a signal which indicates either a rich or lean mixture condition, during the engine operation.

REMOVAL & INSTALLATION

2.5L SOHC Engine

Front Sensor

1. Disconnect the negative battery cable.
2. Remove the clip fastening the harness and disconnect the front oxygen (A/F) sensor connector.
3. Lift-up the vehicle.
4. Remove the undercover.
5. Apply spray-type lubricant to the threaded portion of front oxygen (A/F) sensor, and leave it for one minute or more.
6. Remove the front oxygen (A/F) sensor.

**** CAUTION**

When removing the front oxygen (A/F) sensor, wait until exhaust pipe cools, because it can damage the exhaust pipe.

To install:

7. Before installing front oxygen (A/F) sensor, apply anti-seize compound only to the threaded portion of front oxygen (A/F) sensor to make the next removal easier.
8. Install the front oxygen (A/F) sensor and tighten to 15 ft. lbs. (21 Nm).
9. Install the undercover.
10. Lower the vehicle.
11. Connect the connector of front oxygen (A/F) sensor connector and fasten the harness with clips.
12. Connect the battery ground cable to the battery.

Rear Sensor

1. Disconnect the negative battery cable.
2. Remove the clip fastening the harness and disconnect the rear oxygen sensor connector.
3. Lift-up the vehicle.
4. Remove the undercover.
5. Apply spray-type lubricant to the threaded portion of rear oxygen sensor, and leave it for one minute or more.
6. Remove the rear oxygen sensor.

To install:

7. Before installing rear oxygen sensor, apply the anti-seize compound only to the threaded portion of rear oxygen sensor to make the next removal easier.
8. Install the rear oxygen sensor and tighten to 15 ft. lbs. (21 Nm).
9. Install the undercover.
10. Lower the vehicle.
11. Connect the connector to rear oxygen sensor and mount the harness clips to the bracket for fastening.
12. Connect the battery ground cable to the battery.

2.5L DOHC Engine

Front Sensor

1. Disconnect the negative battery cable.
2. Disconnect the connector from front oxygen (A/F) sensor.
3. Disconnect the engine harness fixed by a clip, from the bracket.
4. Remove the front right side wheel.
5. Lift-up the vehicle.
6. Remove the service hole cover.
7. Apply spray-type lubricant to the threaded portion of front oxygen (A/F) sensor, and leave it for one minute or more.
8. Remove the front oxygen (A/F) sensor.

**** CAUTION**

When removing the oxygen (A/F) sensor, wait until exhaust pipe cools, otherwise it will damage the exhaust pipe.

To install:

9. Before installing front oxygen (A/F) sensor, apply the anti-seize compound only to the threaded portion of front oxygen (A/F) sensor. This facilitates the next removal.
10. Install the front oxygen (A/F) sensor and tighten to 22 ft. lbs. (30 Nm).
11. Install the service hole cover.
12. Lower the vehicle.
13. Install the front right side wheel.
14. Connect the engine harness to the bracket using a clip.
15. Connect the connector of front oxygen (A/F) sensor.
16. Connect the battery ground cable to the battery.

Rear Sensor

1. Disconnect the negative battery cable.
2. Lift-up the vehicle.
3. Disconnect the connector from the rear oxygen sensor.
4. Apply spray-type lubricant to the threaded portion of rear oxygen sensor, and leave it for one minute or more.
5. Remove the rear oxygen sensor.

**** CAUTION**

When removing the rear oxygen sensor, wait until exhaust pipe cools, otherwise it may damage the exhaust pipe.

To install:

6. Before installing rear oxygen sensor, apply the anti-seize compound only to the threaded portion of rear oxygen sensor to make the next removal easier.
7. Install the rear oxygen sensor and tighten to 15 ft. lbs. (21 Nm).
8. Connect the connector to rear oxygen sensor.

9. Lower the vehicle.
10. Connect the battery ground cable to the battery.

3.0L Engine

Front Sensor

1. Disconnect the negative battery cable.
2. Lift-up the vehicle.
3. Disconnect the connector of the left or right front oxygen (A/F) sensor.
4. Remove the front oxygen (A/F) sensor.

To install:

5. Before installing front oxygen (A/F) sensor, apply anti-seize compound only to the threaded portion of front oxygen (A/F) sensor to make the next removal easier.
6. Install the left or right front oxygen (A/F) sensor. Tighten to 15 ft. lbs. (21 Nm).
7. Connect the connector of front oxygen (A/F) sensor.
8. Lower the vehicle.
9. Connect the battery ground cable to the battery.

Rear Sensor

1. Set the vehicle on a lift.
2. Disconnect the ground cable from battery.
3. Lift-up the vehicle.
4. Disconnect the connector of the left or right rear oxygen sensor.
5. Remove the clip holding the harness.
6. Remove the rear oxygen sensor.

CAUTION

When removing the rear oxygen sensor, wait until exhaust pipe cools, otherwise it will damage the exhaust pipe.

To install:

7. Before installing rear oxygen sensor, apply the anti-seize compound only to the threaded portion of rear oxygen sensor to make the next removal easier.
8. Install the left or right rear oxygen sensor. Tighten to 15 ft. lbs. (21 Nm).
9. Hold the harness with clip.
10. Connect the connector to rear oxygen sensor.
11. Lower the vehicle.
12. Connect the battery ground cable to the battery.

TESTING

2.5L SOHC Engine

Front Sensor

See Figure 232.

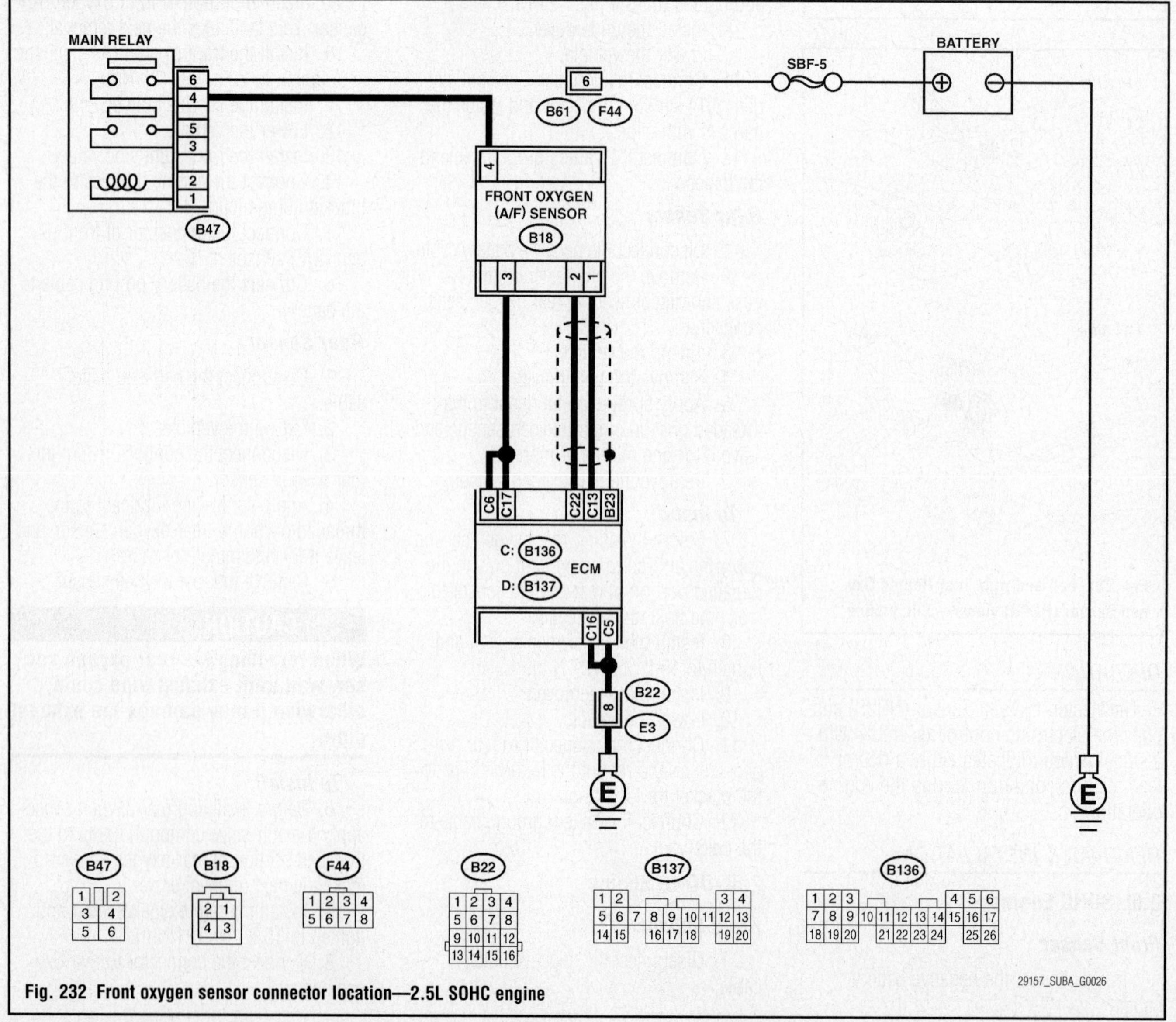

Fig. 232 Front oxygen sensor connector location—2.5L SOHC engine

1. Perform a visual inspection of the sensor as follows:
   - If the sensor tip has a black/sooty deposit, this may indicate a rich fuel mixture.
   - If the sensor tip has a white, gritty deposit, this may indicate an internal coolant leak.
   - If the sensor tip has a brown deposit, this could indicate oil consumption
2. Turn the ignition switch to OFF.
3. Disconnect the connectors from the ECM and the sensor.
4. Measure the resistance of the harness between the ECM and the sensor (connector B136 terminal 13/connector E18 terminal 1 and connector B136 terminal 22/connector B18 terminal 2).
5. If measured value is less than 1 ohm, replace the sensor.

Rear Sensor

See Figure 233.

1. Perform a visual inspection of the sensor as follows:
   - If the sensor tip has a black/sooty deposit, this may indicate a rich fuel mixture.
   - If the sensor tip has a white, gritty deposit, this may indicate an internal coolant leak.
   - If the sensor tip has a brown deposit, this could indicate oil consumption.
2. Turn the ignition switch to OFF.
3. Disconnect the connectors from the sensor.
4. Turn the ignition switch ON.
5. Measure the voltage between the sensor harness connector and engine ground (connector T6 terminal 4+/engine ground -).
6. If measured value is within 0.2–0.5 volt, replace the sensor.

2.5L DOHC Engine

Front Sensor

See Figure 234.

1. Perform a visual inspection of the sensor as follows:
   - If the sensor tip has a black/sooty deposit, this may indicate a rich fuel mixture.
   - If the sensor tip has a white, gritty deposit, this may indicate an internal coolant leak.

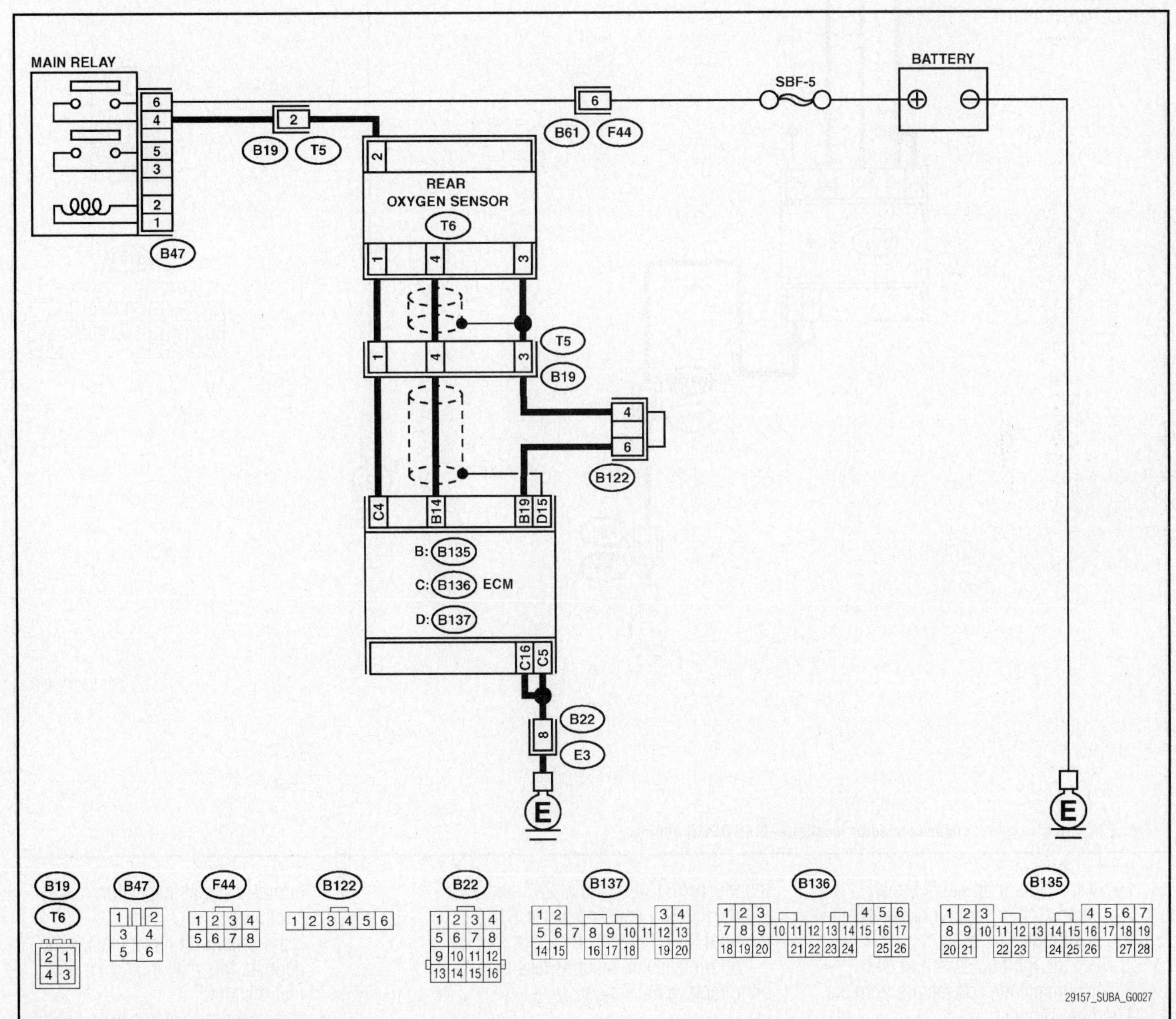

Fig. 233 Rear oxygen sensor connector location—2.5L SOHC engine

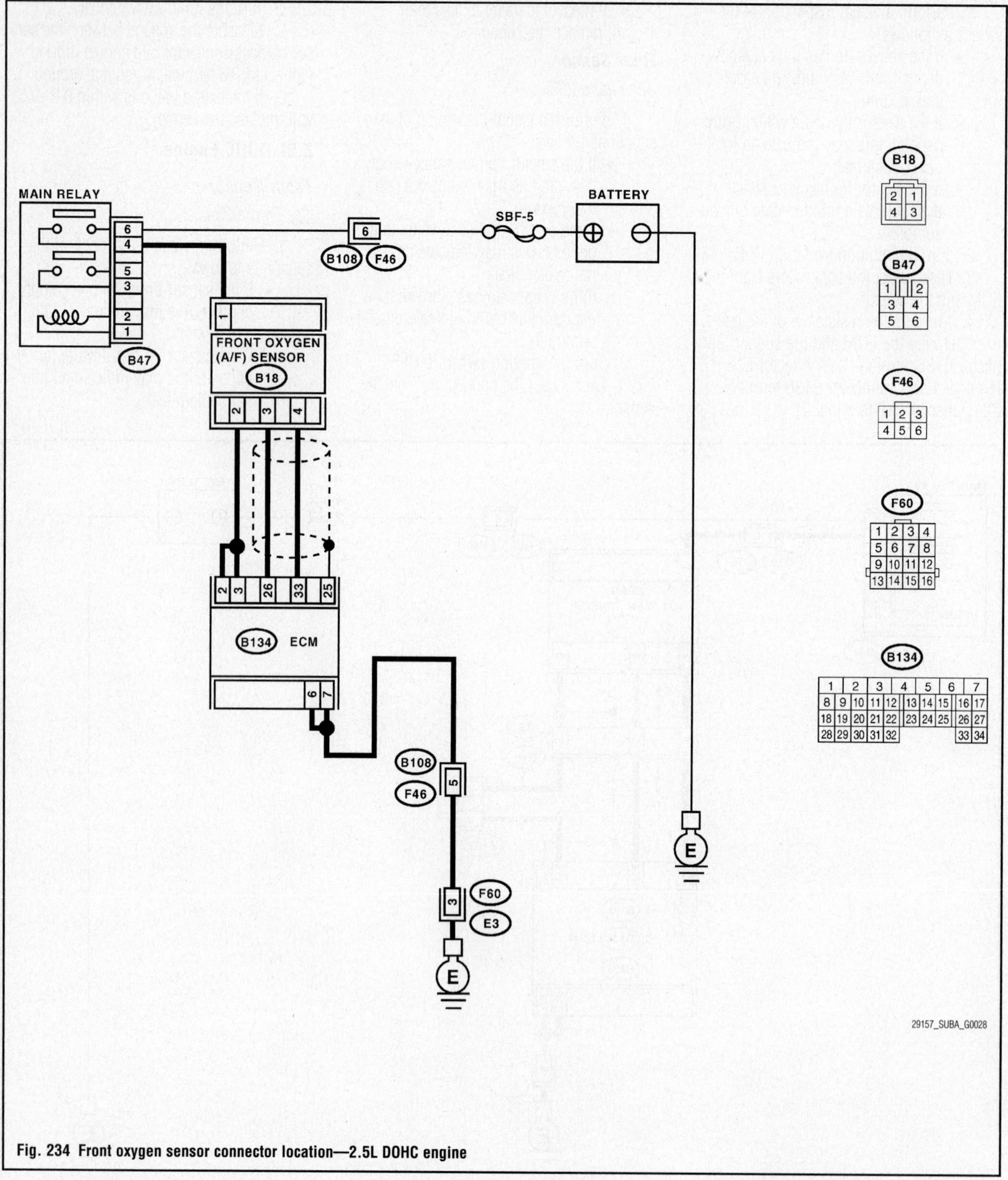

Fig. 234 Front oxygen sensor connector location—2.5L DOHC engine

- If the sensor tip has a brown deposit, this could indicate oil consumption

2. Turn the ignition switch to OFF.
3. Disconnect the connectors from the ECM and the sensor.
4. Measure the resistance of the harness between the ECM and the sensor (connector B134 terminal 26/connector B18 terminal 3 and connector B134 terminal 33/connector B18 terminal 4).
5. If measured value is less than 1 ohm, replace the sensor.

Rear Sensor

See Figure 235.

1. Perform a visual inspection of the sensor as follows:
   - If the sensor tip has a black/sooty deposit, this may indicate a rich fuel mixture.
   - If the sensor tip has a white, gritty deposit, this may indicate an internal coolant leak.

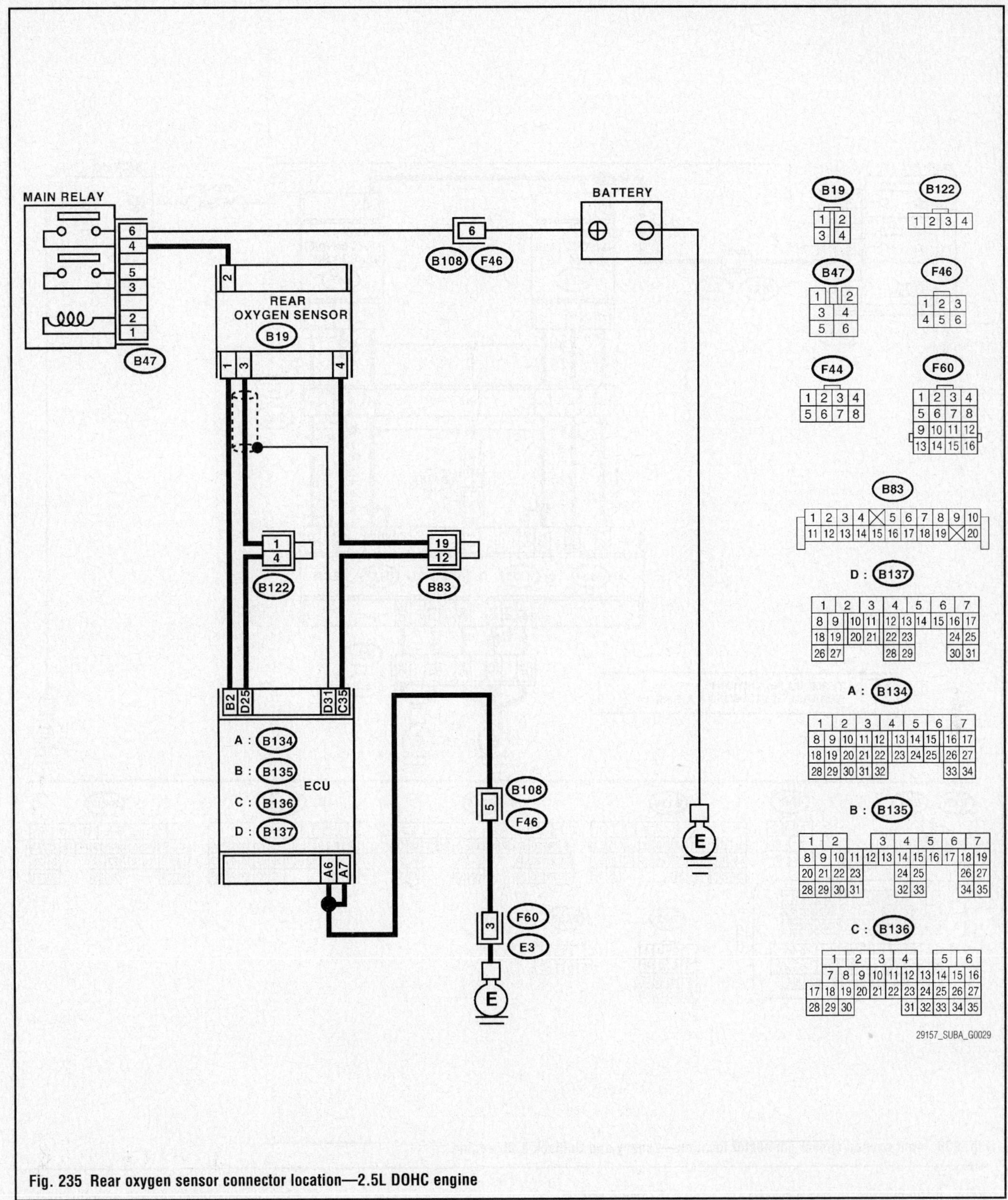

Fig. 235 Rear oxygen sensor connector location—2.5L DOHC engine

- If the sensor tip has a brown deposit, this could indicate oil consumption

2. Using the Subaru scan tool, read the sensor signal data. If the measured value is not within 0.2–0.4 volt, replace the sensor.

3.0L Engine

Front Sensor

See Figure 236.

1. Perform a visual inspection of the sensor as follows:

- If the sensor tip has a black/sooty deposit, this may indicate a rich fuel mixture.
- If the sensor tip has a white, gritty deposit, this may indicate an internal coolant leak.

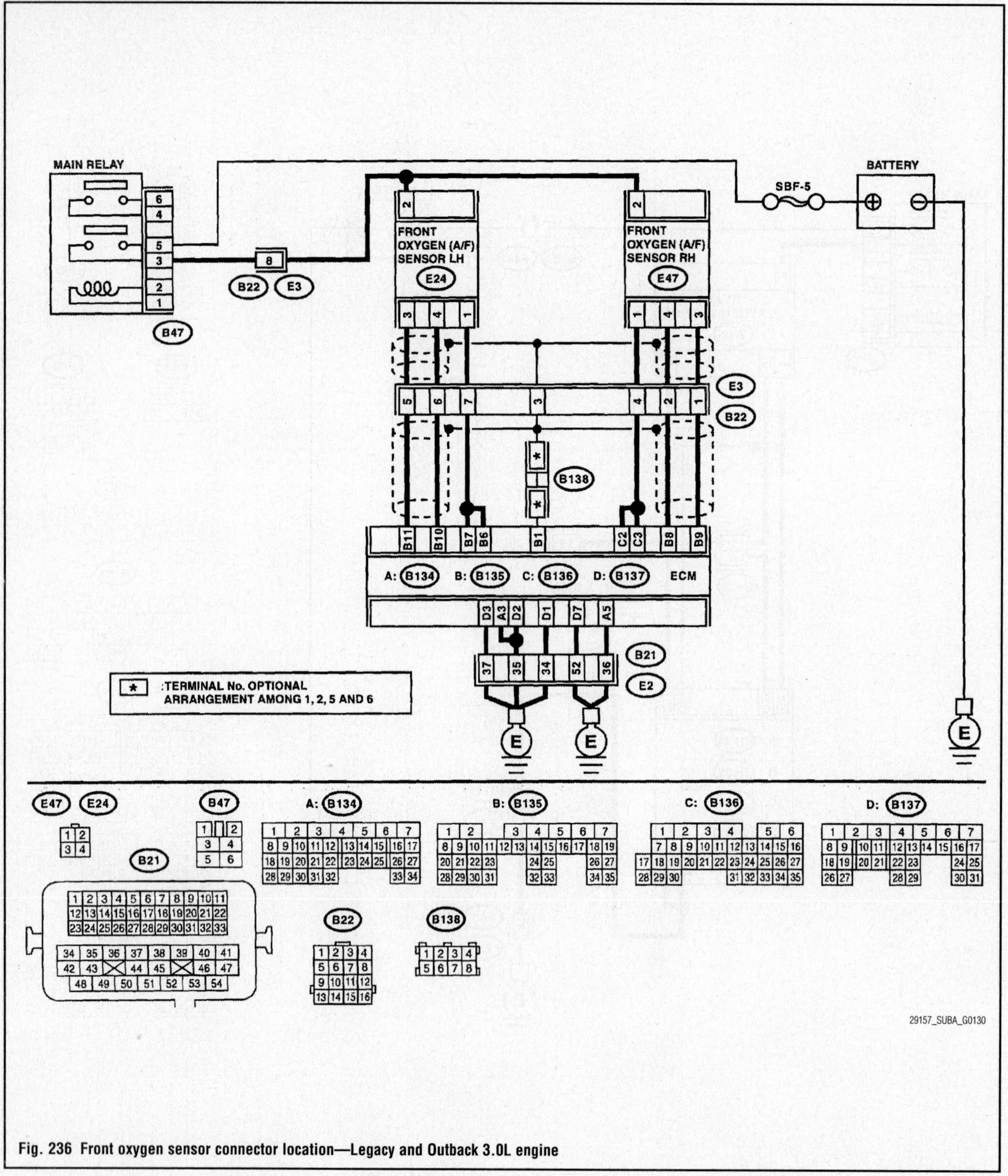

Fig. 236 Front oxygen sensor connector location—Legacy and Outback 3.0L engine

- If the sensor tip has a brown deposit, this could indicate oil consumption

2. Measure the resistance between the sensor connector terminals (terminals 1 and 2).

3. If resistance is more than 5 ohm, replace the sensor.

Rear Sensor

See Figure 237.

1. Perform a visual inspection of the sensor as follows:
   - If the sensor tip has a black/sooty deposit, this may indicate a rich fuel mixture.
   - If the sensor tip has a white, gritty deposit, this may indicate an internal coolant leak.
   - If the sensor tip has a brown deposit, this could indicate oil consumption
2. Turn the ignition switch OFF.

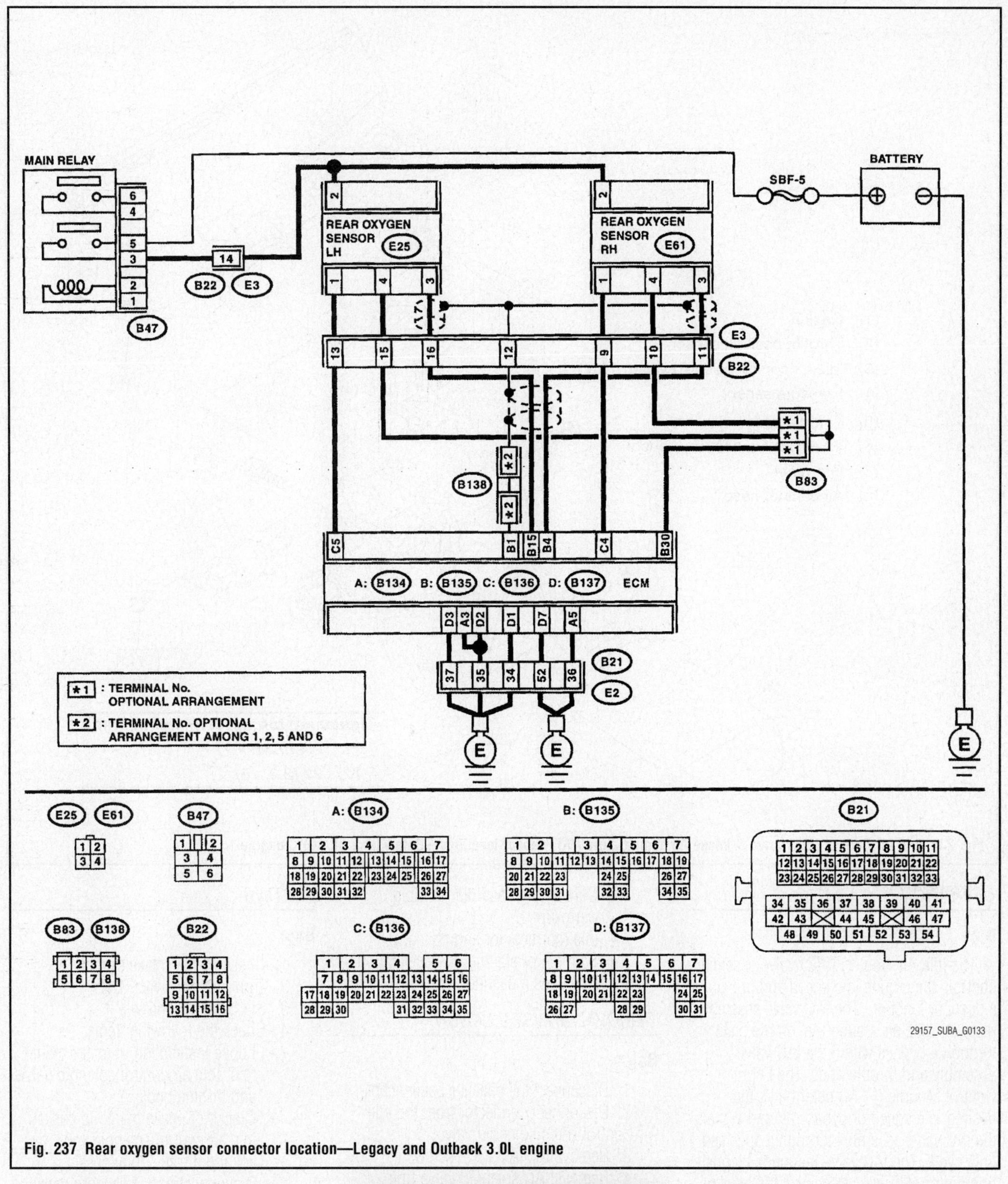

Fig. 237 Rear oxygen sensor connector location—Legacy and Outback 3.0L engine

3. Measure the resistance of the harness between the sensor connector terminals (terminals 1 and 2).

4. If measured value is more than 30ohm, replace the sensor.

IDLE AIR CONTROL (IAC) VALVE

LOCATION

Baja

See Figure 238.

The Idle Air Control (IAC) valve location and related parts is located in the accompanying illustration.

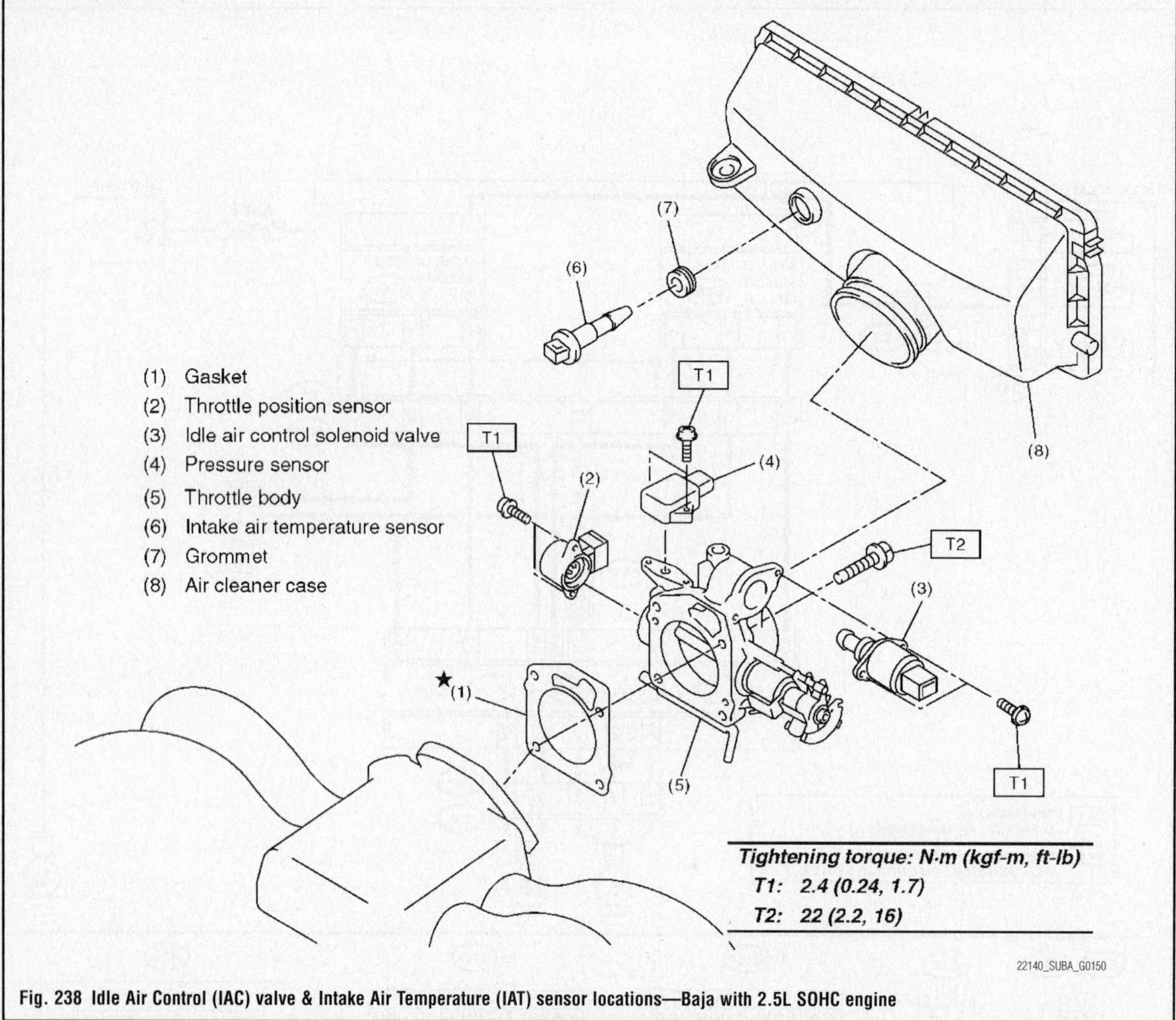

Fig. 238 Idle Air Control (IAC) valve & Intake Air Temperature (IAT) sensor locations—Baja with 2.5L SOHC engine

OPERATION

Baja

The Idle Air Control (IAC) valve assembly controls the engine idle speed and provides a dashpot function. The IAC valve assembly meters intake air around the throttle plate through a bypass within the IAC valve assembly and throttle body. The Engine Control Module (ECM) determines the desired idle speed or bypass air and signals the IAC valve assembly through a specified duty cycle. The IAC valve responds by positioning the IAC valve to control the amount of bypassed air. The ECM monitors engine RPM and increases or decreases the IAC duty cycle in order to achieve the desired RPM.

1. The ECM uses the IAC valve assembly to control:
   - No touch start
   - Cold engine fast idle for rapid warm-up
   - Idle (corrects for engine load)
   - Stumble or stalling on deceleration (provides a dashpot function)

REMOVAL & INSTALLATION

Baja

1. Disconnect the negative battery cable.
2. Disconnect connector from the Idle Air Control (IAC) solenoid valve.
3. Remove the IAC valve screws
4. Remove IAC solenoid valve from throttle body.

To install:

5. Install in the reverse order of removal, noting the following:
   a. Always use a new gasket.
   b. Tighten the IAC mounting screws to 1.7 ft. lbs. (2.4 Nm).

TESTING

Baja

1. Check the air intake system.
2. Turn ignition switch **ON**.
3. Start engine, and idle it.
4. Check the following items:
   - Loose installation of intake manifold, idle air control solenoid valve and throttle body
   - Cracks of intake manifold gasket, idle air control solenoid valve gasket and throttle body gasket
   - Disconnections of vacuum hoses
5. Check throttle cable for binding.
6. Turn ignition switch to OFF.
7. Remove idle air control solenoid valve from throttle body.
8. Confirm that there are no foreign particles in by-pass air line.
9. Check current and ground to IAC.

10. If no problems could be found suspect faulty IAC valve.

INTAKE AIR TEMPERATURE (IAT) SENSOR

LOCATION

See Figure 238.

The sensor is mounted in the intake air hose of the air cleaner assembly on Baja Models. On the DOHC and 3.0L engines this sensor is combined with the mass air flow sensor.

OPERATION

The sensor provides a signal to the ECM for incoming air temperature. This information is used to adjust the injector pulse width and in turn the air/fuel ratio.

REMOVAL & INSTALLATION

1. Disconnect the negative battery cable.
2. Disconnect the connector from the sensor.
3. Remove the sensor from its mounting.
4. Installation is the reverse of the removal procedure.

TESTING

1. Start the engine and allow it to reach operating temperature.
2. Using the Subaru scan tool, read the sensor signal data. If the measured value is within 167–203 degrees F, replace the sensor.

KNOCK SENSOR (KS)

LOCATION

2.5L Engine

See Figure 239.

The Knock Sensor (KS) is located at the top right (driver's) side of the engine and is positioned on the cylinder block.

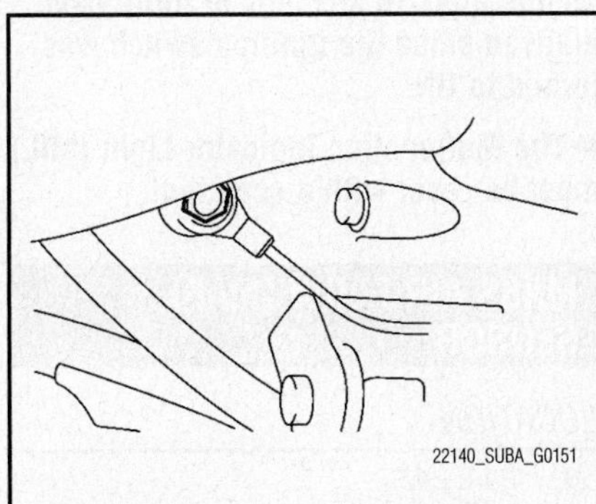

Fig. 239 Knock Sensor (KS) location view—2.5L engine

3.0L Engine

See Figure 240.

Fig. 240 Knock Sensor (KS) location view—3.0L engine

On the 3.0L engine the sensor is located under the intake manifold.

OPERATION

The Knock Sensor (KS) is used to detect engine vibrations caused by pre-ignition or detonation and provides information to the ECM, which then retards the timing to eliminate detonation.

REMOVAL & INSTALLATION

2.5L Engines

1. Disconnect the negative battery cable.
2. Remove the air cleaner case.
3. On DOHC engine, remove the intercooler.
4. Disconnect the sensor connector.
5. Remove the sensor from its mounting.

To install:

6. Installation is the reverse of the removal procedure.
7. Tighten the sensor to 18 ft. lbs. (24 Nm).

➡The extraction area of the knock sensor wire must be positioned at a 60 degree angle relative to the engine rear.

3.0L Engine

See Figure 241.

1. Disconnect the negative battery cable.
2. Remove the collector cover, as required.
3. Remove the intake manifold.
4. Disconnect the sensor connector.
5. Remove the sensor from its mounting.

To install:

6. Installation is the reverse of the removal procedure.
7. Refer to the illustration for proper sensor installation angle.
8. Tighten the sensor to 18 ft. lbs. (24 Nm).

TESTING

See Figure 242.

1. Remove the sensor from the vehicle.
2. Measure the resistance between the connector terminals of the sensor, (terminal 1 and 2).
3. If measured value is more than 600k ohms, replace the sensor.

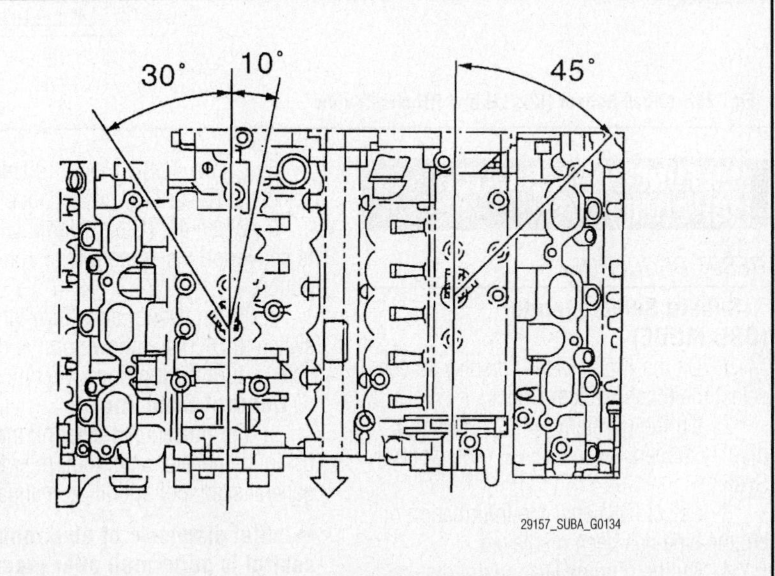

Fig. 241 Knock sensor installation angle—Legacy and Outback 3.0L engine

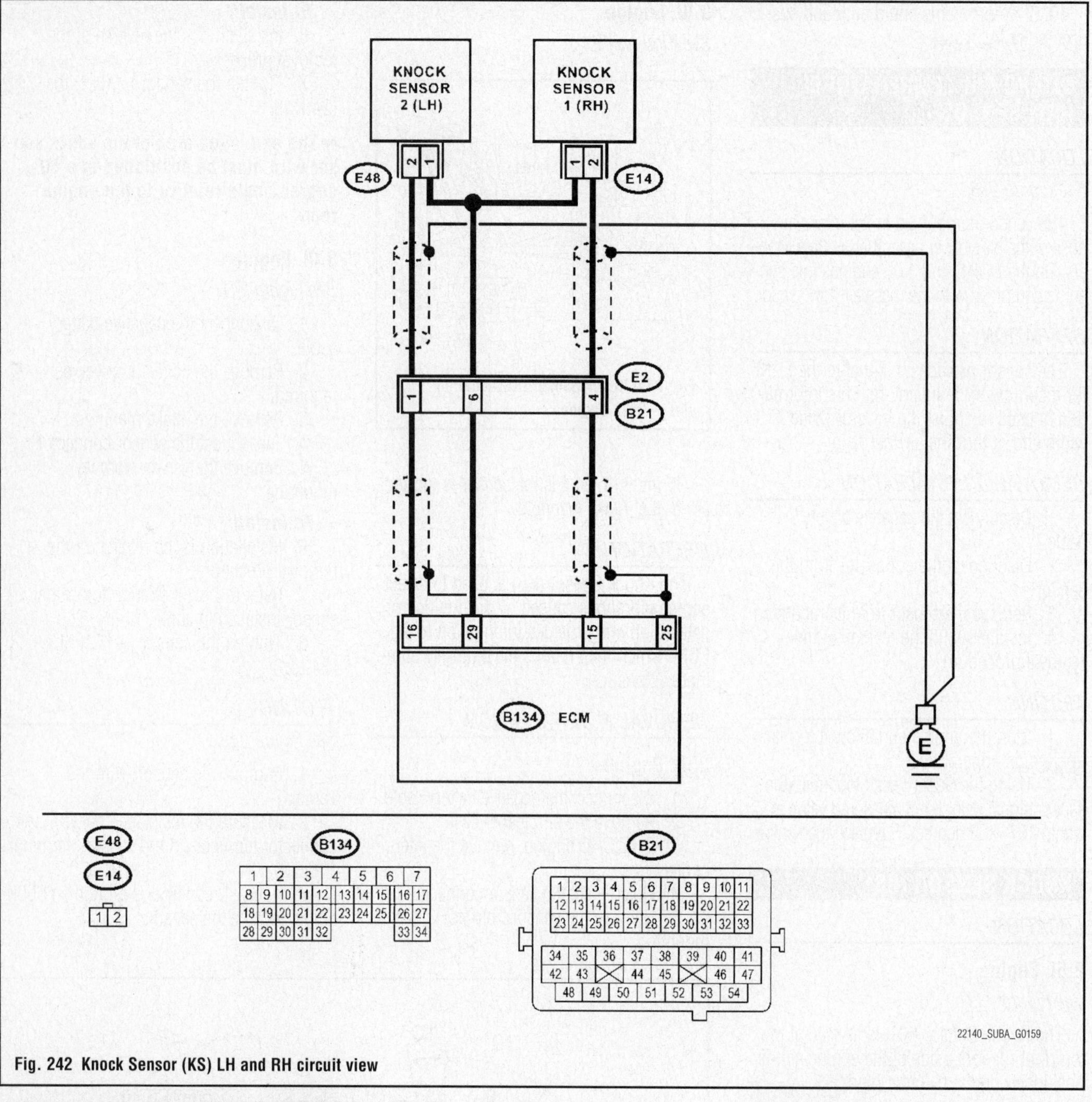

Fig. 242 Knock Sensor (KS) LH and RH circuit view

MALFUNCTION INDICATOR LIGHT (MIL)

RESET PROCEDURES

Subaru Select Monitor (OBD MODE)

1. On the (Main Menu) display screen, select the (Each System Check).
2. On the (System Selection Menu) display screen, select the (Engine Control System).
3. Select (OK) after the information of engine type has been displayed.
4. On the (Engine Diagnosis) display screen, select the (OBD System).
5. On the «OBD Menu» display screen, select the (Clear Diagnostic Code).
6. When the (Clear Diagnostic Code) is shown on the screen, click the (Yes) button.
7. When (Done) and (Turn ignition switch to OFF) are shown on the display screen, turn the ignition switch to OFF.

General Scan Tool

1. For procedures clearing memory using the general scan tool, refer to the general scan tool operation manual.

➡Initial diagnosis of electronic throttle control is performed after memory clearance. For this reason, start the engine after 10 seconds or more have elapsed since the ignition switch was turned to ON.

➡The Malfunction Indicator Light (MIL) must be reset with a scan tool.

MASS AIR FLOW (MAF) SENSOR

LOCATION

See Figure 243.

The MAF sensor is mounted in the intake air hose of the air cleaner assembly. This sensor is combined with the Intake Air

Fig. 243 Mass Air Flow (MAF) sensor location view—3.0L engine shown, 2.5L engine similar

Temperature (IAT) sensor on all models except the Baja.

OPERATION

The Mass Air Flow (MAF) sensor provides a signal to the ECM for incoming air temperature. This information is used to adjust the injector pulse width and in turn the air/fuel ratio.

REMOVAL & INSTALLATION

1. Disconnect the negative battery cable.
2. Remove the collector cover, if equipped.
3. Disconnect the connector from the sensor.
4. Remove the filter assembly from the intake manifold, if equipped.
5. Remove the sensor from its mounting.

To install:

6. Installation is the reverse of the removal procedure.
7. Tighten the retaining bolts to 0.7 ft. lbs. (1 Nm).

TESTING

Sensor Function Test

1. Start the engine and warm-up engine until coolant temperature is higher than 167°F (75°C).
2. Place the select lever in "P" range or "N" range.
3. Turn the A/C switch to OFF.
4. Turn all the accessory switches to OFF.
5. Open the front hood.
6. Measure the ambient temperature.
7. Read the data of mass air flow and intake air temperature sensor signal using Subaru Select Monitor or general scan tool.
8. Subtract ambient temperature from intake air temperature. The obtained value should read between -18 to -90° (-10 to -50°C). If the reading was within specifications, check for faulty contact of the MAF and ECM connectors.
9. If the reading was not as stated above, suspect faulty mass air flow and intake air temperature sensor.

MANIFOLD ABSOLUTE PRESSURE (MAP) SENSOR

LOCATION

2.5L SOHC engine

See Figure 244.

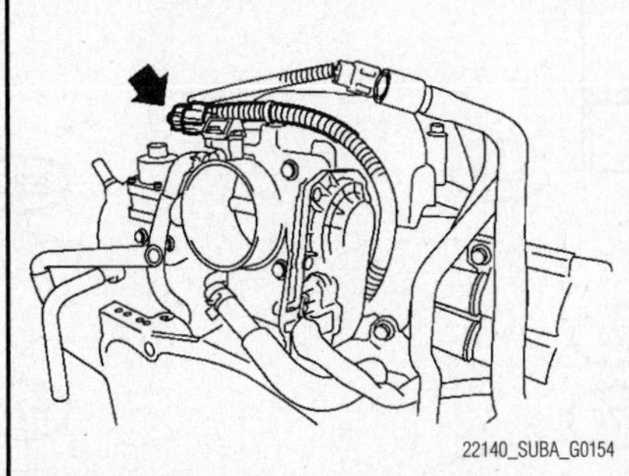

Fig. 244 Manifold Absolute Pressure (MAP) Sensor location—2.5L SOHC engine

The Manifold Absolute Pressure (MAP) sensor for the 2.5L SOHC engine is located on the throttle body unit.

2.5L DOHC engine

See Figure 245.

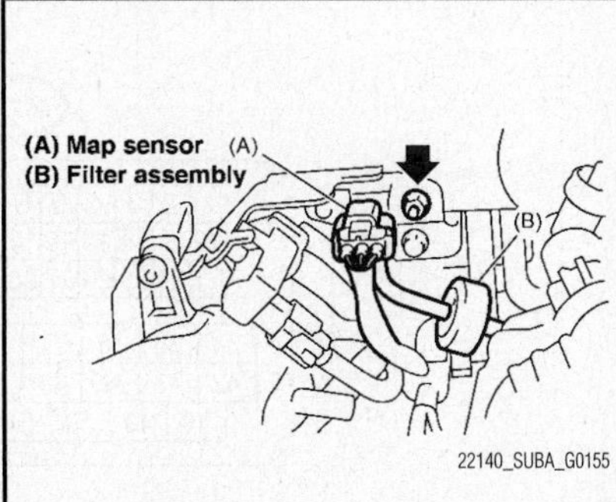

Fig. 245 Manifold Absolute Pressure (MAP) sensor location—2.5L DOHC engine

The Manifold Absolute Pressure (MAP) sensor for the 2.5L DOHC engine is located on the solenoid valve bracket.

3.0L Engine

See Figure 246.

The Manifold Absolute Pressure (MAP) sensor for the 3.0L engine is located on the intake manifold.

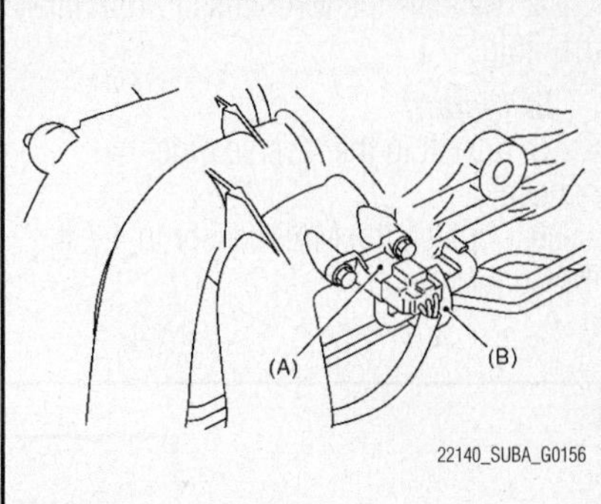

Fig. 246 Manifold Absolute Pressure (MAP) Sensor location—3.0L engine

OPERATION

The Manifold Absolute Pressure (MAP) sensor monitors and signals the ECM of changes in intake manifold pressure which result from engine load, speed and atmospheric pressure changes.

REMOVAL & INSTALLATION

2.5L SOHC engine

1. Disconnect the ground cable from the battery.
2. Disconnect the connector from Manifold Absolute Pressure (MAP) sensor.
3. Remove the MAP sensor from throttle body.

To install:

4. Install in the reverse order of removal, noting the following:
   a. Use new O-rings.
   b. Tighten the MAP sensor to 1.5 ft. lbs. (2 Nm).

2.5L DOHC engine

1. Remove the collector cover.
2. Disconnect the ground cable from battery.
3. Disconnect the connector from Manifold Absolute Pressure (MAP) sensor, and remove the filter assembly from intake manifold.
4. Remove the MAP sensor from the solenoid valve bracket.

To install:

5. Install in the reverse order of removal.
6. Tighten the MAP sensor to 4.7 ft. lbs. (6.4 Nm).

3.0L Engine

1. Remove the collector cover.
2. Disconnect the ground cable from the battery.
3. Disconnect the connector from the Manifold Absolute Pressure (MAP) sensor, and remove the filter assembly from the intake manifold.

4. Remove the MAP sensor from intake manifold.

To install:

5. Install in the reverse order of removal.
6. Tighten the MAP sensor to 4.7 ft. lbs. (6.4 Nm).

TESTING

See Figure 247.

1. Turn the ignition switch **OFF**.
2. Disconnect the connectors from the ECM.
3. Measure the resistance of the harness between the ECM and the sensor (connector B134 terminal 29 and connector E28 terminal1)
4. If the resistance is less than 1 ohm, replace the sensor.
5. Check poor contact of manifold absolute pressure sensor connector.
6. If connector contact are okay, replace MAP sensor.

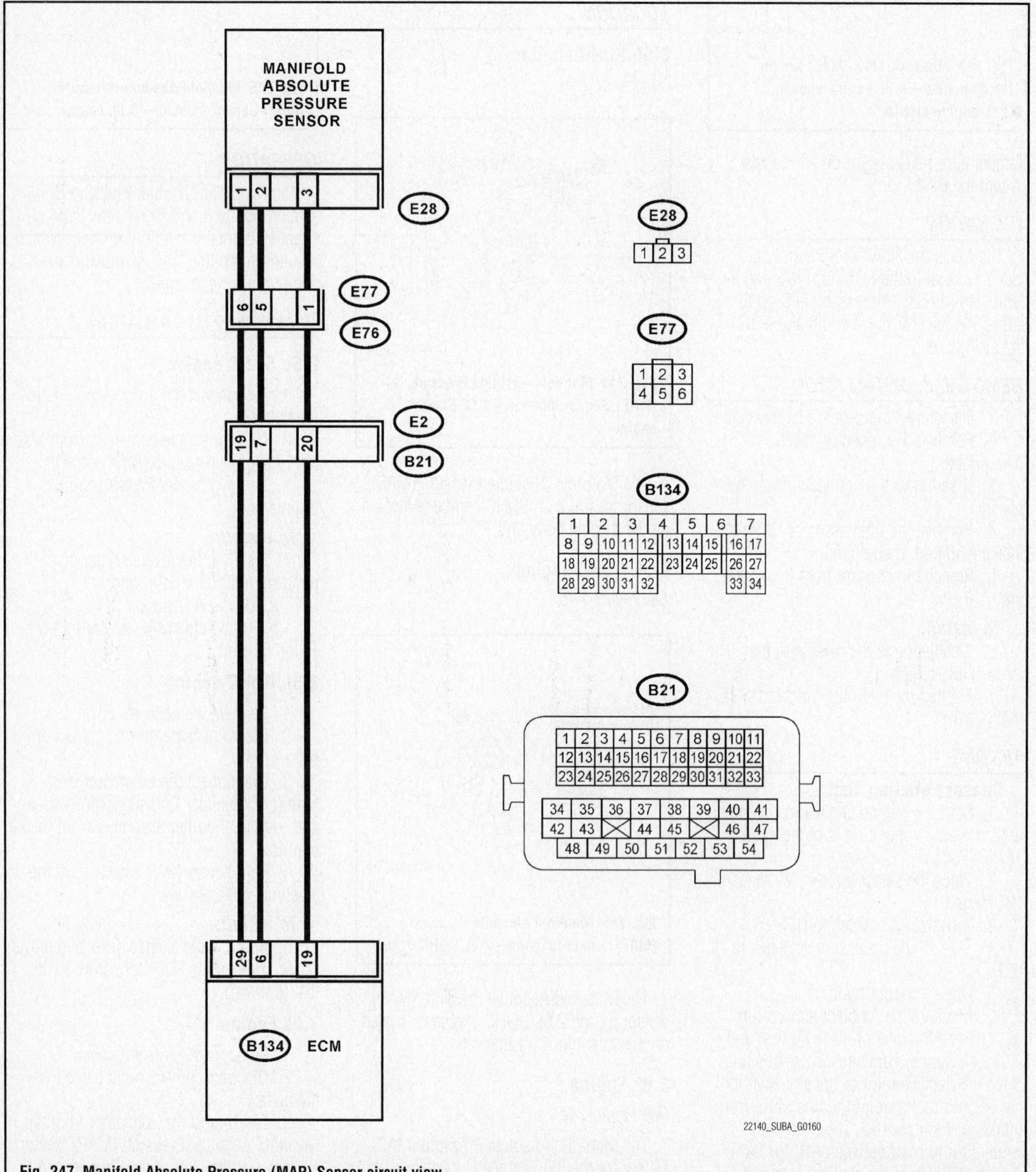

Fig. 247 Manifold Absolute Pressure (MAP) Sensor circuit view

THROTTLE POSITION SENSOR (TPS)

LOCATION

Baja

See Figure 248.

On the Baja, the Throttle Position Sensor (TPS) is mounted on the throttle body.

Impreza, Outback and Legacy

On these models, the throttle body is a non-disassembled part, so do not remove the Throttle Position Sensor (TPS) from throttle body. Refer to throttle body for removal and installation procedure.

OPERATION

The Throttle Position Sensor (TPS) provides a signal to the ECM that is related to the relative throttle plate position. As the throttle plate moves in relation to driving conditions, a signal is sent to the control unit which adjusts the injector pulse width and air/fuel ratio. As the throttle plate is opened further, more air is taken into the combustion chambers, and as a result the relative fuel demand of the engine changes. The TPS sensor relays this information to the ECM which in alters the fuel amount.

REMOVAL & INSTALLATION

Baja

See Figure 249.

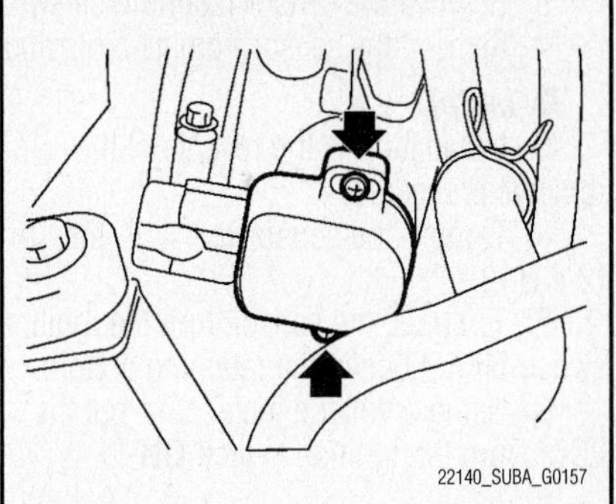

Fig. 249 Throttle Position Sensor (TPS) and retaining screws

1. Disconnect the negative battery cable.
2. Disconnect the sensor connector.

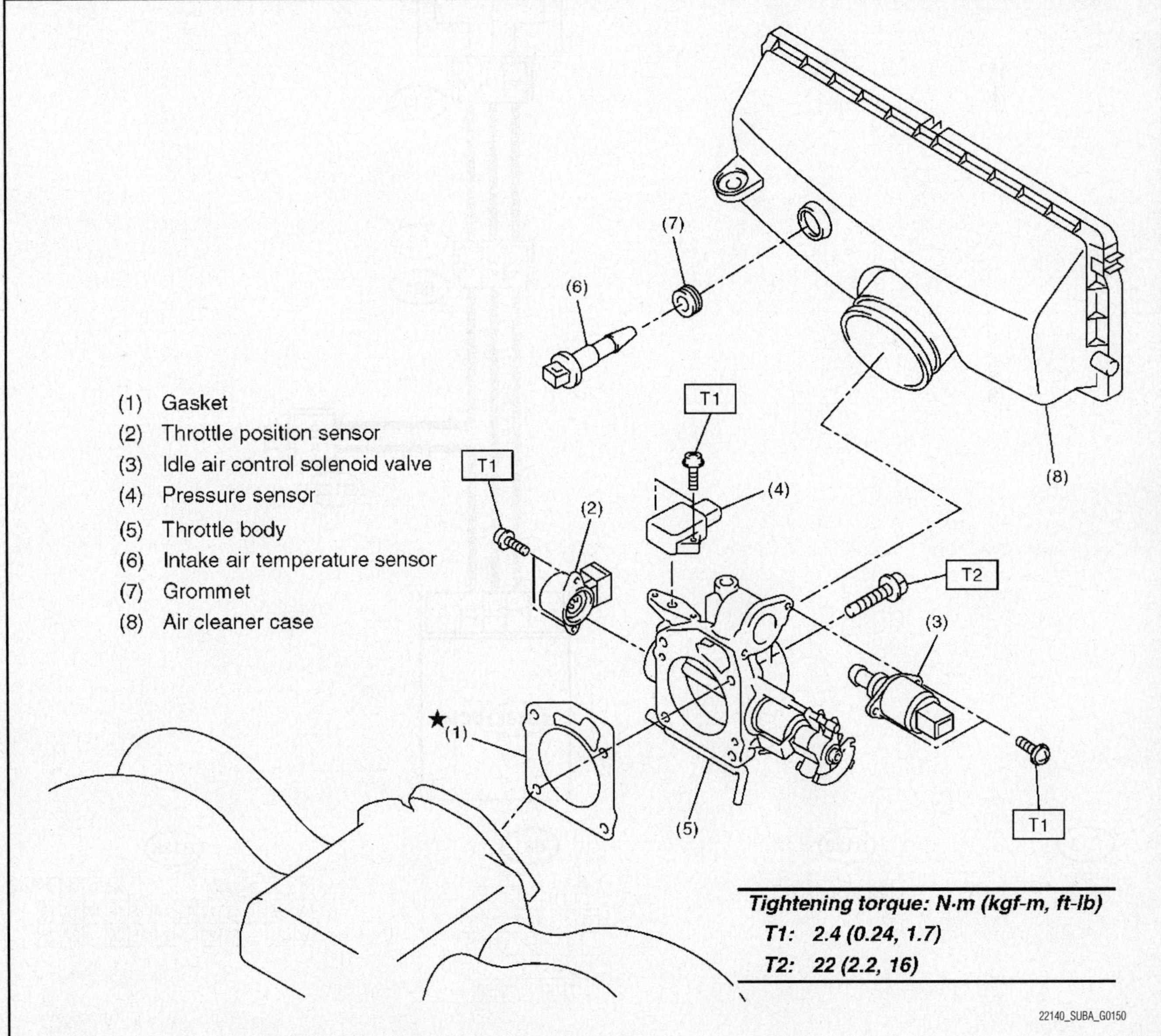

(1) Gasket
(2) Throttle position sensor
(3) Idle air control solenoid valve
(4) Pressure sensor
(5) Throttle body
(6) Intake air temperature sensor
(7) Grommet
(8) Air cleaner case

Tightening torque: N·m (kgf-m, ft-lb)
T1: 2.4 (0.24, 1.7)
T2: 22 (2.2, 16)

22140_SUBA_G0150

Fig. 248 Throttle Position Sensor (TPS) location view—2.5L SOHC engine–2006 Baja

3. Remove the sensor retaining screws.
4. Remove the sensor from its mounting.

To install:

5. Installation is the reverse of the removal procedure.
6. Tighten the sensor to 1.7 ft. lbs. (2.4 Nm).
7. To adjust the sensor, turn the ignition switch **OFF**. Loosen the retaining screws.
8. Using a voltage meter, take out the ECM. Turn the ignition switch **ON**.
9. Adjust the TPS to the proper position to allow the voltage signal to the ECM to be within specification.
10. Specification is 0.45–0.55 volt. Specification is measured at connector B135 and terminal 13.
11. Tighten the retaining screws.

TESTING

2.5L SOHC Engine

See Figure 250.

1. Visually inspect the throttle linkage and throttle for binding and sticking.
2. Measure the resistance of the harness between the TPS connector and engine ground, (connector E13 terminal 3 and ground).
3. If the measured value is more than 1 M ohm, replace the sensor.

2.5L DOHC Engines

See Figure 251.

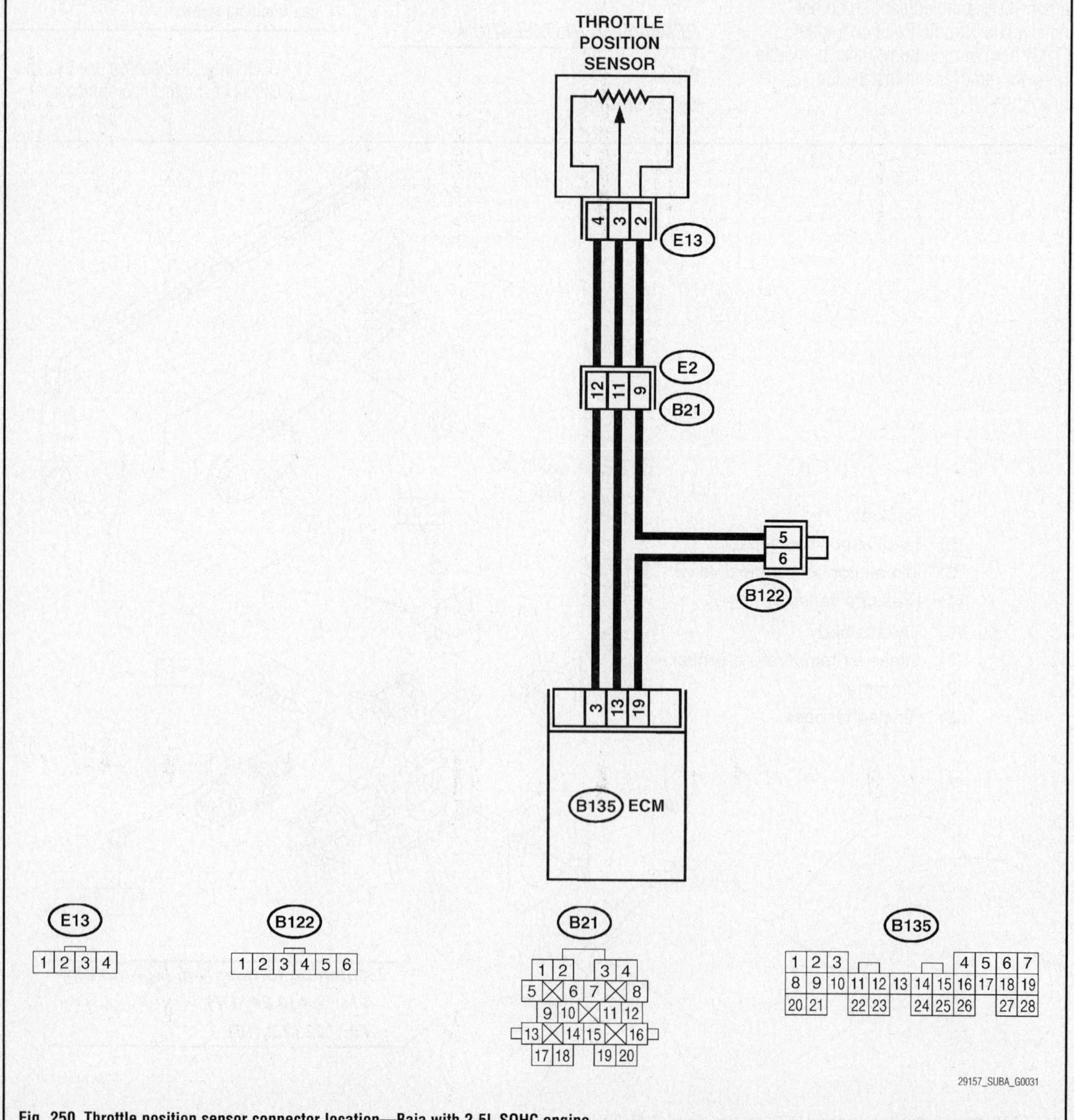

Fig. 250 Throttle position sensor connector location—Baja with 2.5L SOHC engine

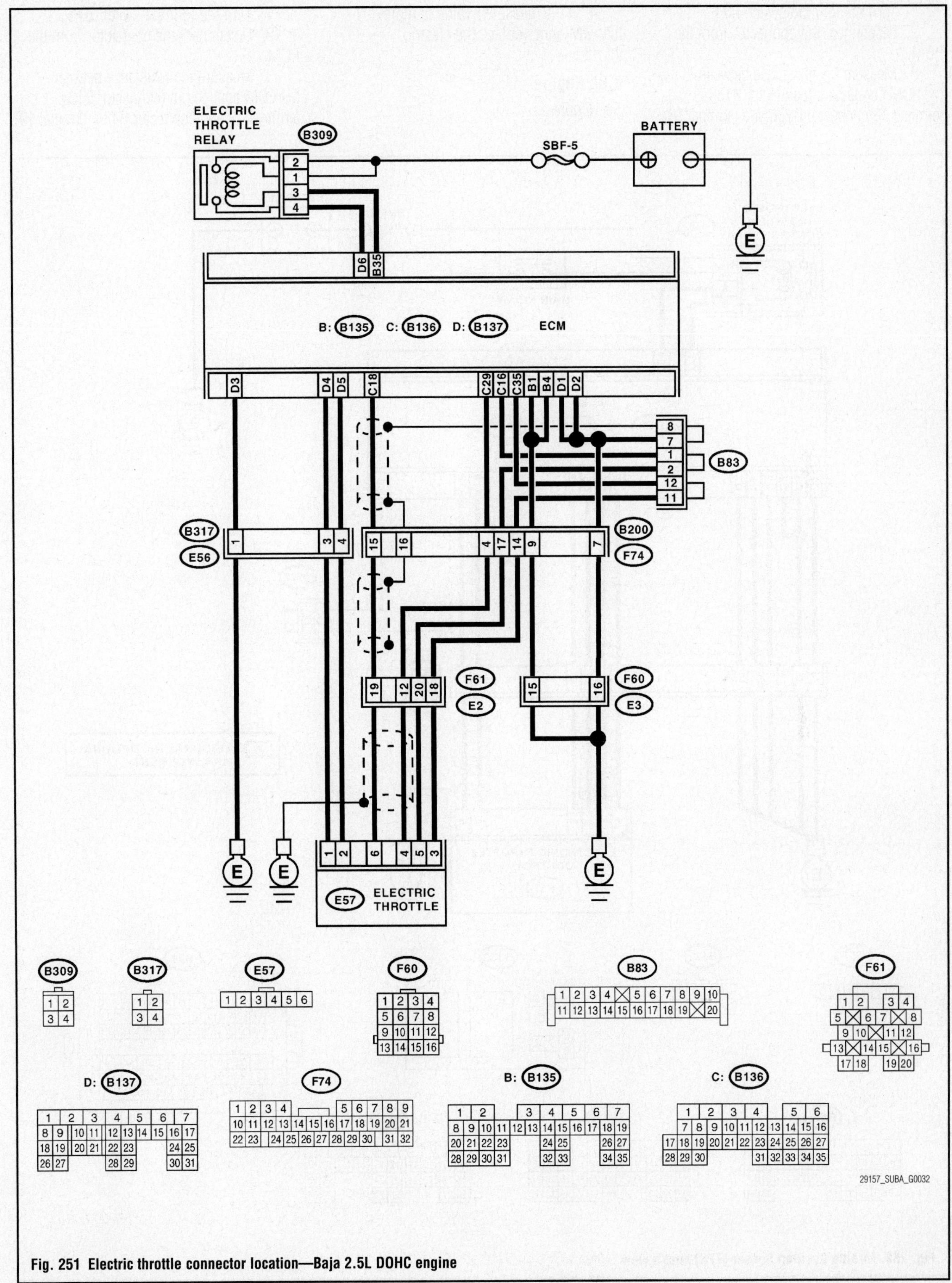

Fig. 251 Electric throttle connector location—Baja 2.5L DOHC engine

1. Turn the ignition switch **OFF**.
2. Disconnect the connector from the ECM.
3. Measure the resistance between the ECM connectors (connector B136 terminal 18/connector B136 and terminal 16).
4. If the measured value is more than 1M ohm, replace the electric throttle.

3.0L Engine

See Figure 252.

1. Turn the ignition switch **OFF**.
2. Disconnect the connector from the ECM.
3. Measure the resistance between the ECM connectors (connector B134 terminal 18 and connector B134 terminal 19).

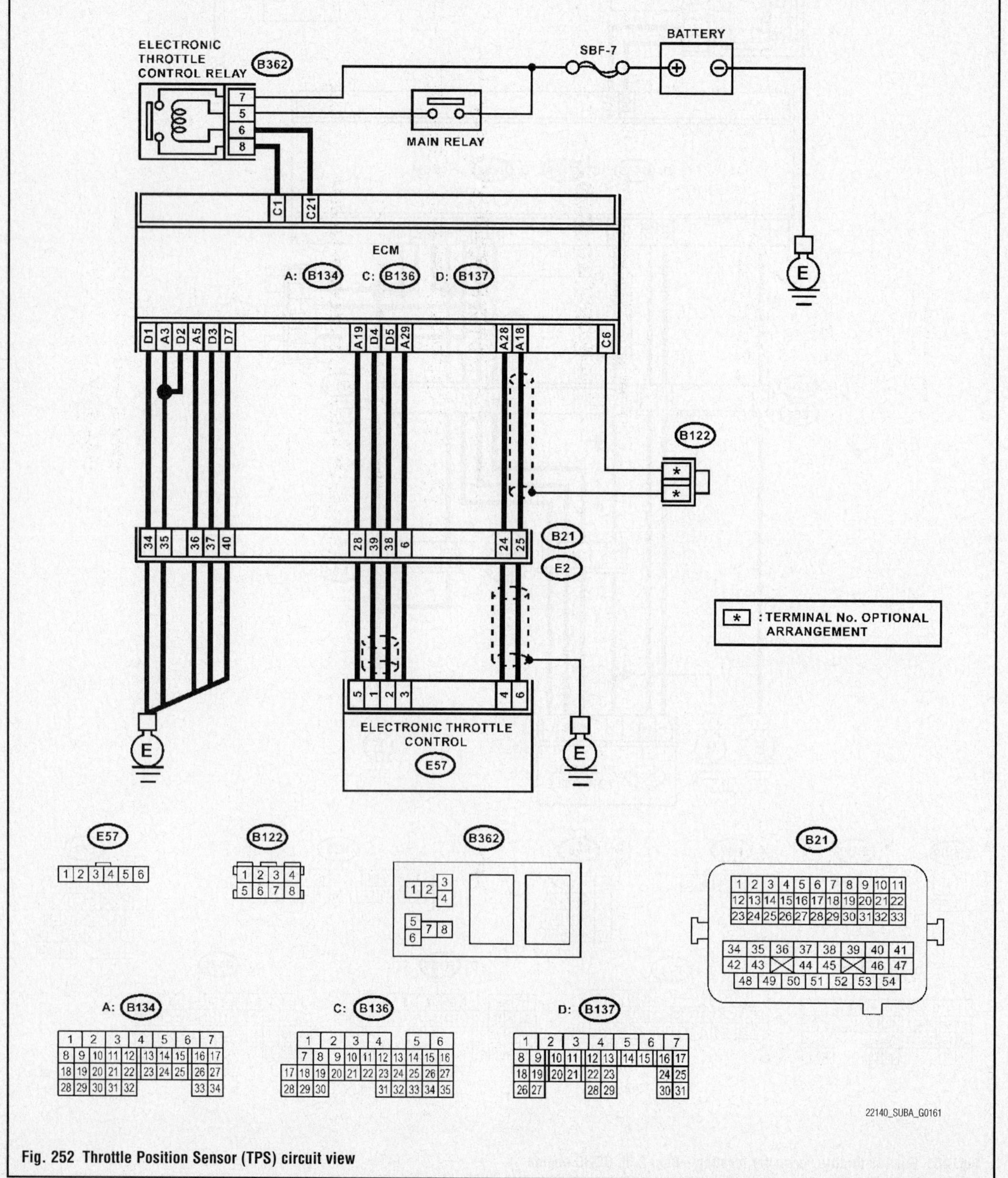

Fig. 252 Throttle Position Sensor (TPS) circuit view

4. If the measured value is more than 1m ohms, repair the electronic throttle control.

VEHICLE SPEED SENSOR (VSS)

LOCATION

See Figures 253 and 254.

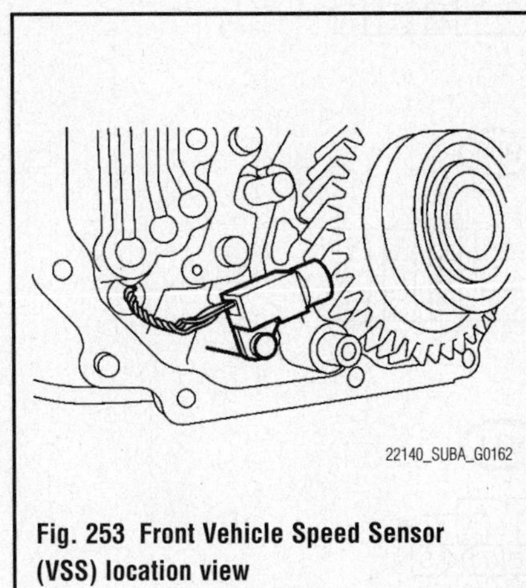

Fig. 253 Front Vehicle Speed Sensor (VSS) location view

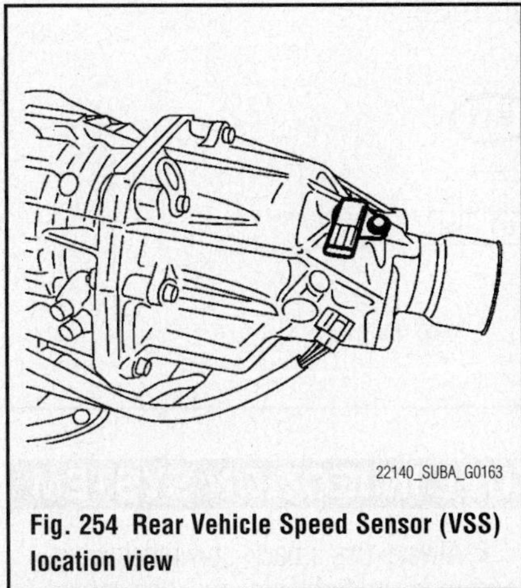

Fig. 254 Rear Vehicle Speed Sensor (VSS) location view

This vehicle uses a front and rear Vehicle Speed Sensor (VSS). Both sensors are mounted on the transaxle.

OPERATION

The Vehicle Speed Sensor (VSS) sends a signal to the ECM to control. The ECM uses this information to control transmission shift patterns. The vehicle speed sensor is also used to send a speed signal to the speed control servo of the cruise control system.

REMOVAL & INSTALLATION

1. Raise and support the vehicle safely.
2. Place a drip pan below the speed sensor to catch any spilled fluid.
3. Disconnect the connector.
4. Remove the sensor from its mounting.

To install:

5. Installation is the reverse of the removal procedure.
6. Replace any lost fluid.

TESTING

See Figures 255 and 256.

1. Turn the ignition switch to OFF.
2. Disconnect the connectors from TCM and transmission.
3. Measure the resistance of harness between TCM connector and transmission Vehicle Speed Sensor (VSS) connectors.
4. If the resistance is not less than 1 ohm, repair the open circuit of harness between TCM and transmission VSS connector.

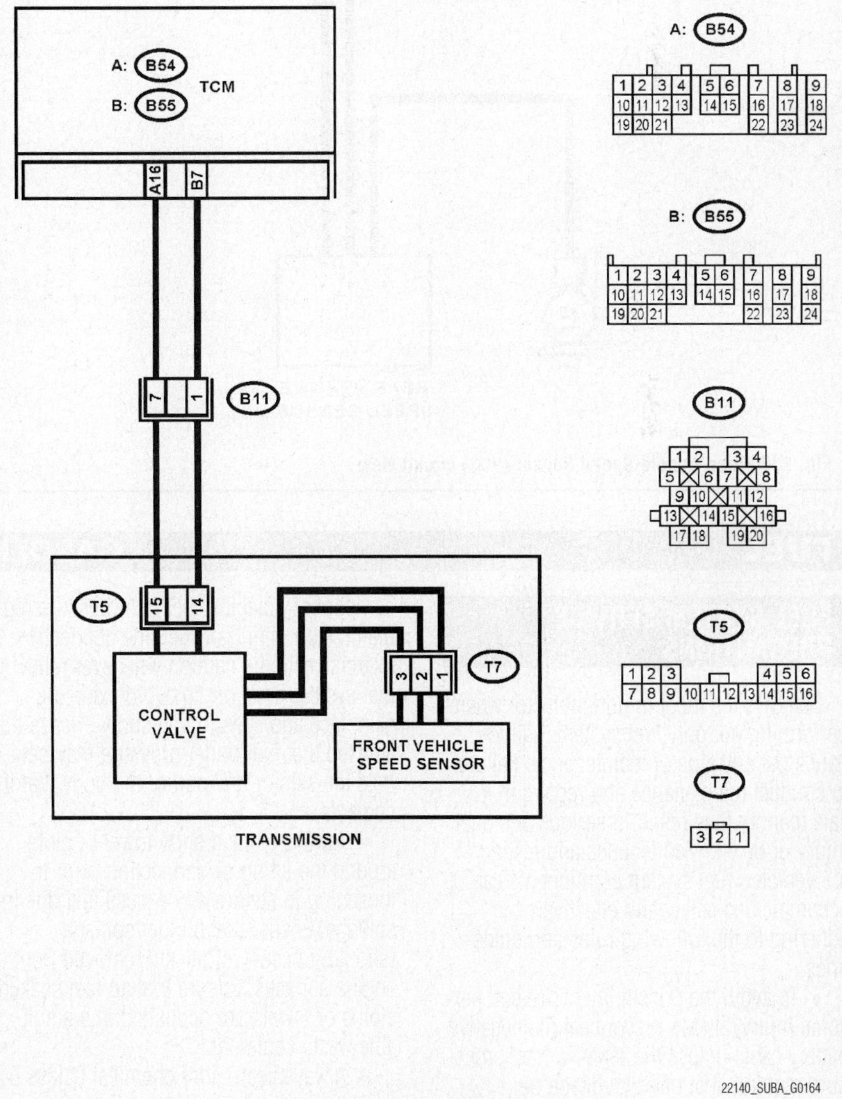

A: B54
B: B55
TCM
A16
B7
B11
T5
15
14
CONTROL VALVE
T7
3 2 1
FRONT VEHICLE SPEED SENSOR
TRANSMISSION
22140_SUBA_G0164

Fig. 255 Front Vehicle Speed Sensor (VSS) circuit view

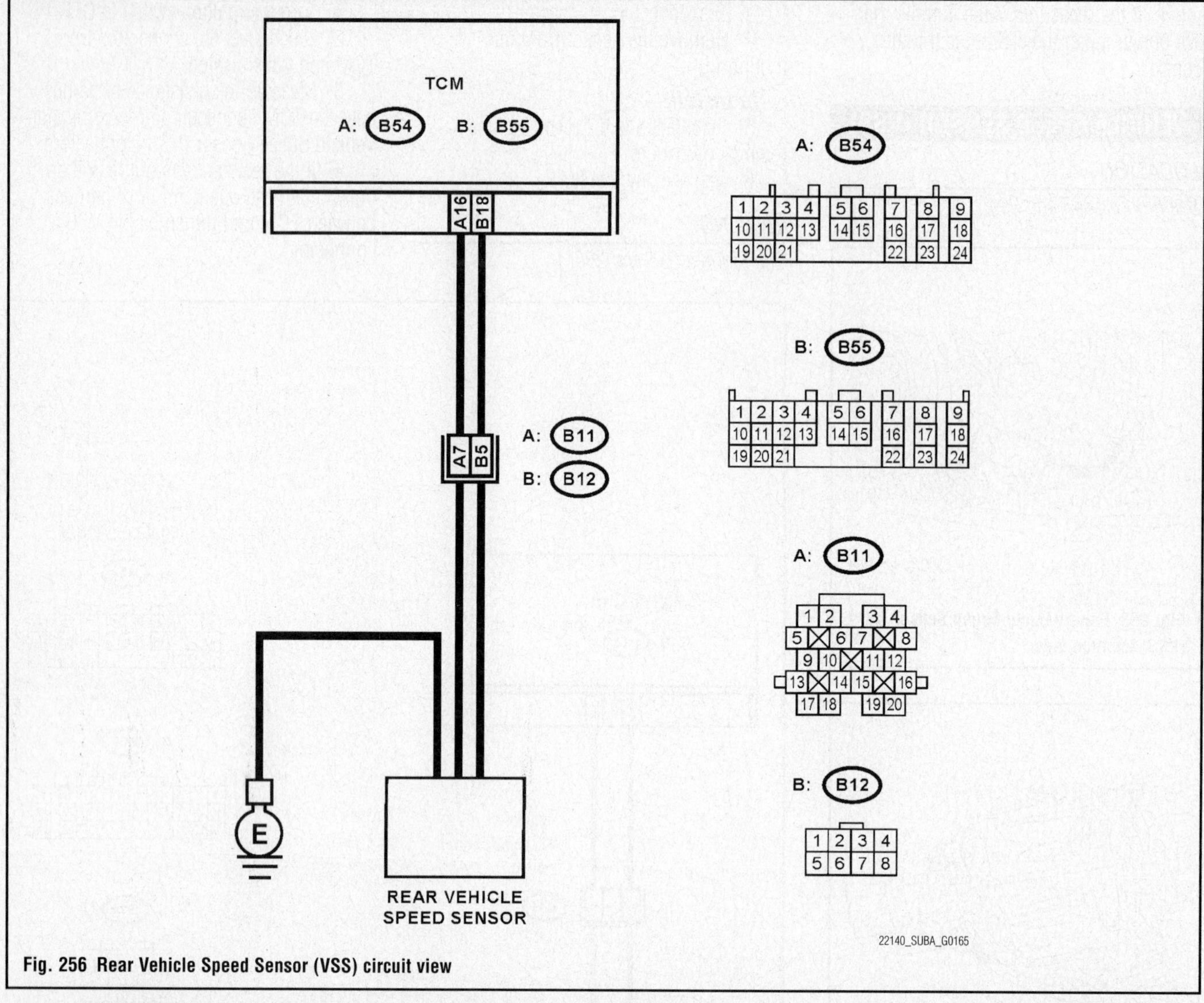

Fig. 256 Rear Vehicle Speed Sensor (VSS) circuit view

FUEL

GASOLINE FUEL INJECTION SYSTEM

FUEL SYSTEM SERVICE PRECAUTIONS

Safety is the most important factor when performing not only fuel system maintenance but any type of maintenance. Failure to conduct maintenance and repairs in a safe manner may result in serious personal injury or death. Maintenance and testing of the vehicle's fuel system components can be accomplished safely and effectively by adhering to the following rules and guidelines.

- To avoid the possibility of fire and personal injury, always disconnect the negative battery cable unless the repair or test procedure requires that battery voltage be applied.
- Always relieve the fuel system pressure prior to disconnecting any fuel system component (injector, fuel rail, pressure regulator, etc.), fitting or fuel line connection. Exercise extreme caution whenever relieving fuel system pressure to avoid exposing skin, face and eyes to fuel spray. Please be advised that fuel under pressure may penetrate the skin or any part of the body that it contacts.
- Always place a shop towel or cloth around the fitting or connection prior to loosening to absorb any excess fuel due to spillage. Ensure that all fuel spillage (should it occur) is quickly removed from engine surfaces. Ensure that all fuel soaked cloths or towels are deposited into a suitable waste container.
- Always keep a dry chemical (Class B) fire extinguisher near the work area.
- Do not allow fuel spray or fuel vapors to come into contact with a spark or open flame.
- Always use a back-up wrench when loosening and tightening fuel line connection fittings. This will prevent unnecessary stress and torsion to fuel line piping.
- Always replace worn fuel fitting O-rings with new. Do not substitute fuel hose or equivalent where fuel pipe is installed.

Before servicing the vehicle, make sure to also refer to the precautions in the beginning of this section as well.

RELIEVING FUEL SYSTEM PRESSURE

➡This procedure must be performed prior to servicing any component of the fuel injection system.

1. On Legacy, Outback and Baja models, remove the fuel pump fuse from the main fuse box.

2. On Impreza and WRX models, disconnect the fuel pump connector from the fuel pump relay.

3. Start the engine and run until it stalls.

4. Crank the engine for 5 seconds or more to ensure the fuel pressure is properly relieved. If the engine starts during this time, allow it to run until it stalls.

5. Turn the ignition switch to the **OFF** position. Remove the key.

6. Disconnect the negative battery cable.

7. Remove the fuel cap.

FUEL FILTER

REMOVAL & INSTALLATION

The fuel filter is an integral part of the fuel pump assembly.

FUEL INJECTORS

REMOVAL & INSTALLATION

Baja

2.5L SOHC Engine

See Figures 257 and 258.

⁂ CAUTION

Place "NO OPEN FLAMES" signs near the working area.

1. Before servicing the vehicle, refer to the Precautions Section.

2. Properly relieve the fuel system pressure. Remove the fuel cap.

3. Disconnect the negative battery cable.

4. If removing the right side injectors, remove the resonator chamber. Remove the number one and three spark plug wires.

5. If removing the right side injectors, remove the front side V belt, remove the belt covers. Loosen the lock bolt. Loosen the slider bolt. Remove the front side belt. Remove the bolts that retain the power steering hoses to the intake manifold protector. Do not disconnect the power steering hoses. Remove the bolts which retain the power steering pump to the bracket. Disconnect the power steering pump switch connector. Remove the reservoir tank from the bracket, by pulling upward. Place the power steering pump and tank assembly to the side.

6. If removing the left side injectors, remove the two bolts that retain the washer fluid tank to its mounting. Disconnect the electrical connector from the front window washer motor. Disconnect the electrical connector and the washer hose from the rear gate glass washer motor. Position the tank to the side. Remove the number two and four spark plug wires.

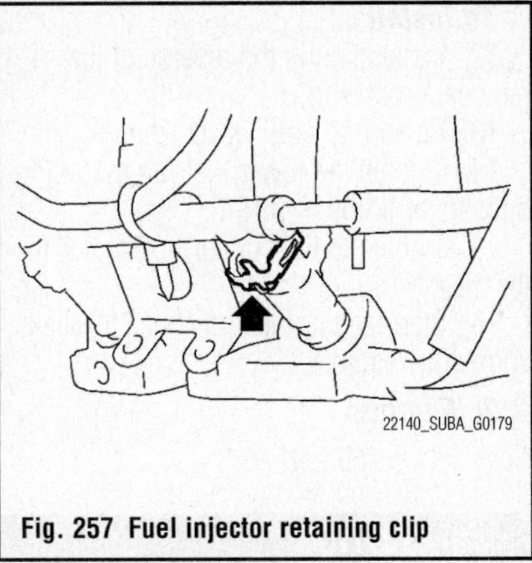

Fig. 257 Fuel injector retaining clip

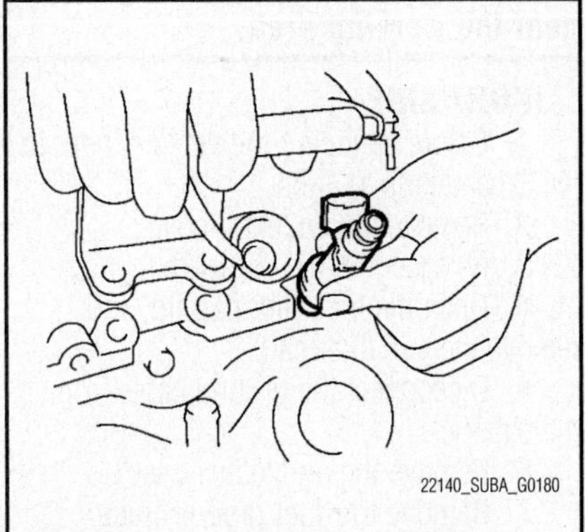

Fig. 258 Fuel injector removal shown

7. Remove the fuel line protector covers.

8. Disconnect the electrical connector from the fuel injector.

9. Remove the bolts that hold the fuel injector line to the intake manifold.

10. Remove the fuel injector retaining clip.

11. Remove the fuel injector while lifting up the fuel injector line.

To install:

12. Installation is the reverse of the removal procedure.

13. Be sure to use new O-rings.

14. Tighten fuel rail mounting bolts to 14 ft. lbs. (19 Nm).

15. Start the engine and check for leaks, correct as required.

2.5L DOHC Engine

See Figure 259.

⁂ CAUTION

Place "NO OPEN FLAMES" signs near the working area.

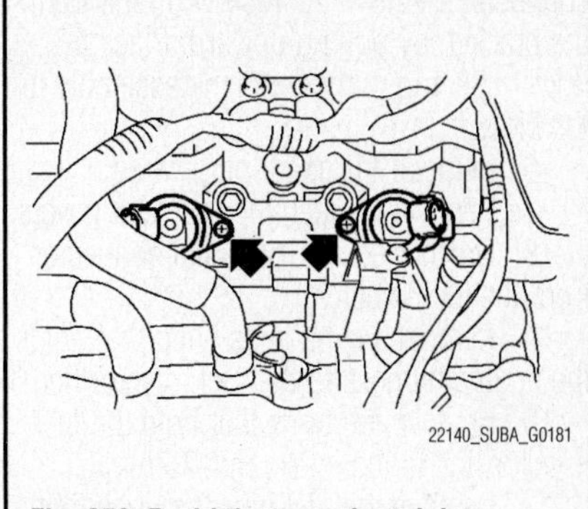

Fig. 259 Fuel injector and retaining screws shown

1. Before servicing the vehicle, refer to the Precautions Section.

2. Properly relieve the fuel system pressure. Remove the fuel cap.

3. Disconnect the negative battery cable.

4. If removing the left side injectors, remove the intake manifold.

5. Disconnect the electrical connector from the fuel injector.

6. Remove the screw and then remove the fuel injector.

To install:

7. Installation is the reverse of the removal procedure.

8. Be sure to use new O-rings.

9. Start the engine and check for leaks, correct as required.

Impreza, Legacy & Outback

2.5L Engines

See Figures 260, 261 and 262.

⁂ CAUTION

Place "NO OPEN FLAMES" signs near the working area.

1. Before servicing the vehicle, refer to the Precautions Section.

2. Properly relieve the fuel system pressure. Remove the fuel cap.

3. Disconnect the negative battery cable.

4. If removing the right side injectors, remove the air intake chamber and air cleaner case. To remove the front side V belt, remove the belt covers. Loosen the lock bolt. Loosen the slider bolt. Remove the front side belt.

5. If removing the right side injectors, remove the bolts that retain the power steering hoses to the intake manifold protector. Do not disconnect the power steering hoses. Remove the bolts which retain the power steering pump to the bracket. Disconnect the power steering pump switch

connector. Remove the reservoir tank from the bracket, by pulling upward. Place the power steering pump and tank assembly to the side.

6. Remove the spark plug wires.
7. Remove the fuel line protector covers.
8. Disconnect the electrical connector from the fuel injector.
9. Remove the harness band that holds the engine harness to the fuel injector line.
10. Remove the bolts that hold the fuel injector line to the intake manifold.
11. Remove the fuel injector while lifting up the fuel injector line.

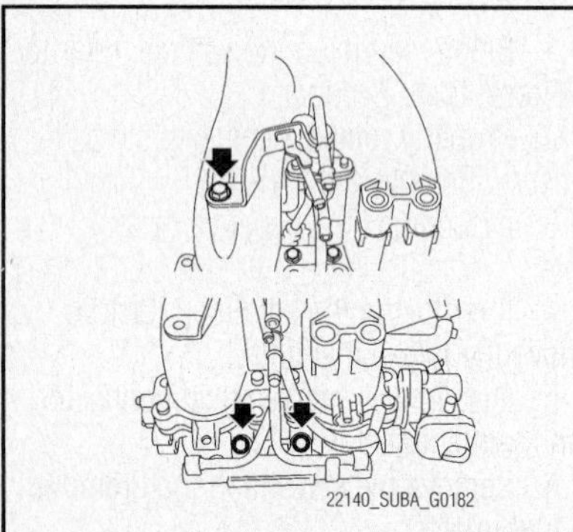

Fig. 260 Fuel injector line mounting bolts LH shown

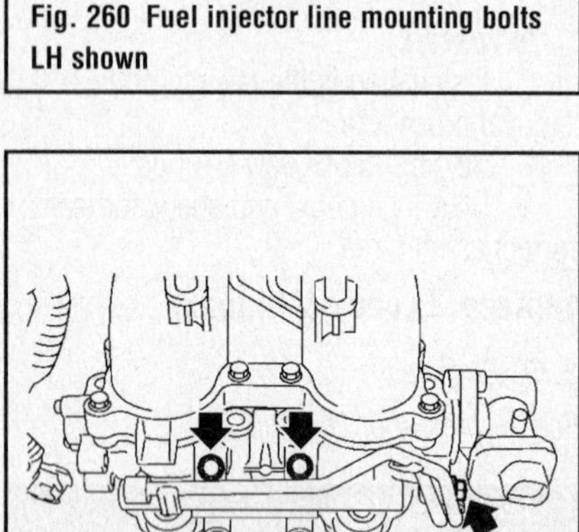

Fig. 261 Fuel injector line mounting bolts RH shown

Fig. 262 Fuel injector removal while lifting up the fuel line

To install:

12. Installation is the reverse of the removal procedure.
13. Be sure to use new O-rings.
14. Tighten the lower fuel rail mounting bolts to 14 ft. lbs. (19 Nm).
15. Tighten the LH upper bolt to 4.7 ft. lbs. (6.4 Nm).
16. Start the engine and check for leaks, correct as required.

3.0L Engines

See Figures 263 and 264.

⁂ CAUTION

Place "NO OPEN FLAMES" signs near the working area.

RIGHT SIDE

1. Before servicing the vehicle, refer to the Precautions Section.
2. Remove the collector cover.
3. Release the fuel pressure.
4. Open the fuel filler flap lid, and remove the fuel filler cap.
5. Disconnect the ground cable from the battery.
6. Remove the air cleaner case.
7. Remove the fuel pipe protector RH.

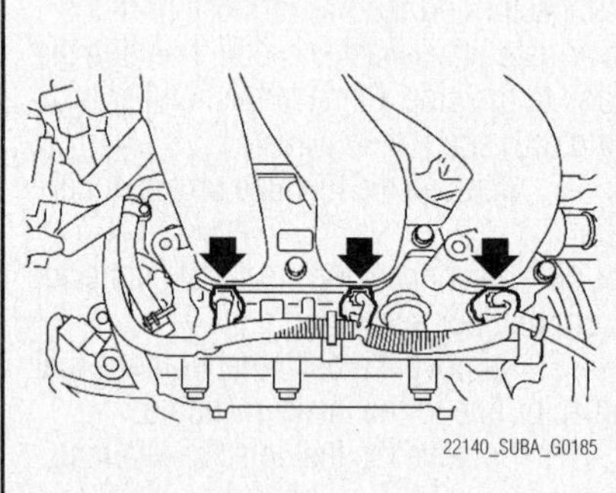

Fig. 263 Fuel injector connector RH shown

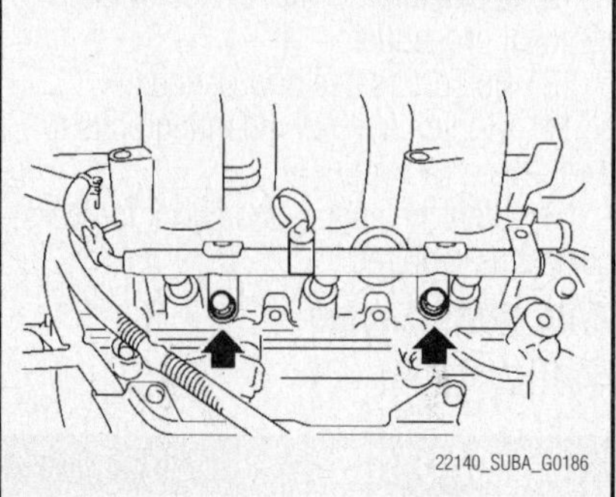

Fig. 264 Fuel injector rail mounting bolts RH shown

8. Disconnect the connector from fuel injector.
9. Remove the engine harness from the fuel injector pipe RH.
10. Remove the bolts which hold fuel injector pipe to the cylinder head.
11. Remove the fuel injector while lifting up the fuel injector pipe.

LEFT SIDE

12. Remove the generator harness from the fuel pipe protector LH.
13. Remove the fuel pipe protector LH.
14. Disconnect the connector from fuel injector.
15. Remove the engine harness from the fuel injector pipe LH.
16. Remove the bolts which hold fuel injector pipe to the cylinder head.
17. Remove the fuel injector while lifting up the fuel injector pipe.

To install:

18. Install in the reverse order of removal.
19. Use new O-rings.
20. Tighten for rail mounting bolts to 14 ft. lbs. (19 Nm).
21. Tighten the fuel pipe protectors to 14 ft. lbs. (19 Nm).

FUEL PUMP

REMOVAL & INSTALLATION

Baja Models

⁂ CAUTION

When the fuel gauge pointer is at two third or more, the fuel may spill out. Be sure to drain the fuel before the operation. Be careful not to spill fuel.

1. Before servicing the vehicle, refer to the Precautions Section.
2. Properly relieve the fuel system pressure.
3. Disconnect the negative battery cable.
4. Remove the fuel filler cap.
5. Raise and support the vehicle safely.
6. On the DOHC engine, remove the front side fuel tank cover.
7. Drain the fuel from the fuel tank into a suitable container.

➡Vehicles equipped with the DOHC engine are equipped with a fuel tank drain plug. After draining the fuel tank, use a new gasket and replace the plug. Tighten the plug bolt to 19.2 ft. lbs.

8. Lower the vehicle.
9. Remove the rear seat. Turn the floor mat up.

10. Remove the fuel pump access cover retaining screws. Remove the access cover.
11. Disconnect the electrical connector from the fuel pump.
12. Disconnect and plug the fuel line hoses at the fuel pump.
13. Remove the fuel pump assembly retaining nuts. Remove the fuel pump from the vehicle.

To install:

14. Installation is the reverse of the removal procedure.
15. Be sure to use a new gasket and retainer.
16. Tighten the fuel pump retaining nuts to 4.3 ft. lbs. (5.9 Nm) in an alternating sequence pattern.
17. Continue the installation in the reverse order of the removal procedure.
18. Start the engine and check for leaks, correct as required.

Impreza, Legacy & Outback Models

See Figure 265.

⁂ CAUTION

When the fuel gauge pointer is at two third or more, the fuel may spill out. Be sure to drain the fuel before the operation. Be careful not to spill fuel.

1. Before servicing the vehicle, refer to the Precautions Section.
2. Release the fuel pressure.
3. Drain fuel.
4. Disconnect the ground cable from the battery.
5. Remove the rear seat.
6. Remove the service hole cover.
7. Disconnect the connector from fuel pump.
8. Disconnect the quick connector, then disconnect the fuel delivery tube and jet pump tube.
9. Remove the nuts which install fuel pump assembly onto fuel tank.
10. Remove the fuel pump assembly from the fuel tank.

To install:

11. Install in the reverse order of removal while being careful of the following:
    - Make sure the sealing portion is free from fuel or foreign matter before installation.
    - When assembling, point the protrusion (A) of the gasket towards the front of the vehicle.
    - Insert the protrusion (B) of the gasket into the upper plate.

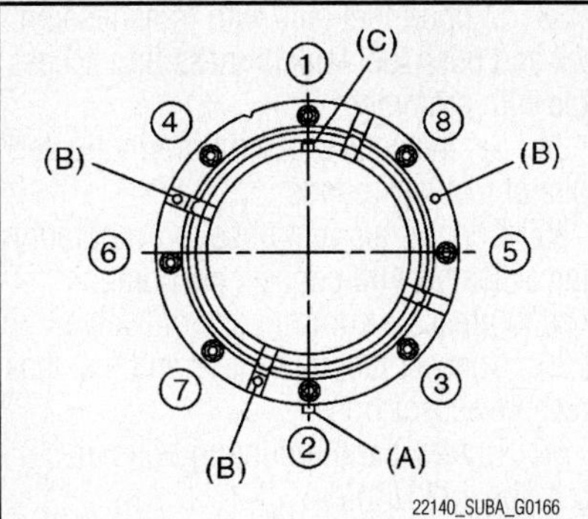

Fig. 265 Fuel pump assembly tightening procedure

- Align the protrusion (C) of the fuel pump assembly to the cut out in the upper plate.

12. Tighten the nuts to 3.2 ft. lbs. (4.4 Nm) in the order as shown in the figure below.

FUEL PRESSURE REGULATOR

REMOVAL & INSTALLATION

Baja Models

See Figure 266.

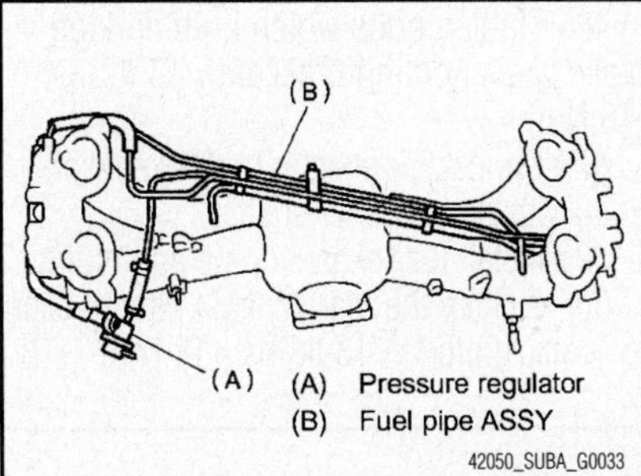

Fig. 266 Location of the pressure regulator on the intake manifold—Baja with 2.5L engine

⁂ CAUTION

Be careful not to spill fuel.

1. Before servicing the vehicle, refer to the Precautions Section.
2. Properly relieve the fuel pressure.
3. Remove the intake manifold. Refer to the Intake Manifold, Removal & Installation procedure in the Engine Mechanical Section.
4. Remove the fuel pressure regulator and fuel lines from the intake manifold.
5. Installation is the reverse order of removal.

FUEL TANK

REMOVAL & INSTALLATION

Baja Models

See Figures 267, 268 and 269.

⁂ CAUTION

Place "NO OPEN FLAMES" signs near the working area.

1. Before servicing the vehicle, refer to the Precautions Section.
2. Set vehicle on the lift.
3. Release fuel pressure.
4. Drain fuel from fuel tank.
5. Remove holder clip which secures fuel tank cord on bracket.
6. Disconnect connector of fuel tank cord to rear harness.
7. Push grommet which holds fuel tank cord on service hole cover into body side.
8. Separate the quick connectors from fuel delivery and return hoses.
9. Remove console box.
10. Remove parking brake bracket and disconnect parking brake cable from equalizer.
11. Remove parking brake cable.
12. Separate the quick connector of evaporation pipe
13. Remove wheel nuts from rear wheels.
14. Lift-up the vehicle.
15. Remove rear wheel.
16. Remove front side fuel tank cover.
17. Remove rear exhaust pipe and muffler.

➡To facilitate removal, apply a coat of SUBARU CRC to matching area of rubber cushions in advance.

18. Separate rear exhaust pipe and center exhaust pipe.
19. Remove left and right rubber cushions.
20. Remove front rubber cushion and detach muffler assembly.
21. Remove propeller shaft.
22. Disconnect connector from ABS sensor.
23. Remove bolts which hold parking brake cable holding bracket.
24. Remove parking brake cable from cabin by forcibly pulling it backward.
25. Remove bolts which hold parking brake cable holding bracket.
26. Remove bolts which hold rear brake hoses holding bracket.
27. Remove rear brake caliper, then tie it up to the body side of the vehicle.

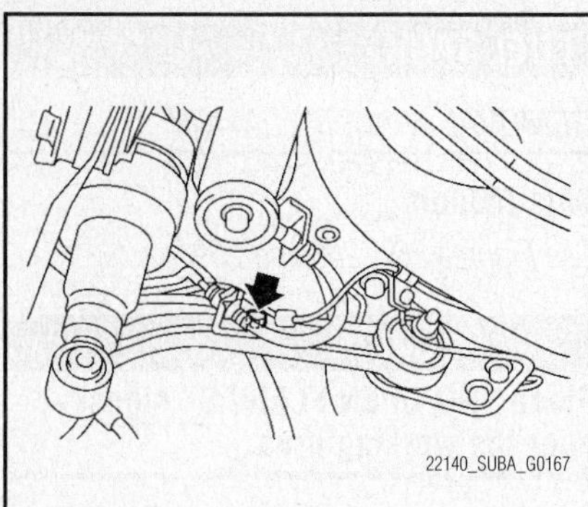

Fig. 267 Rear brake hoses holding bracket and bolt

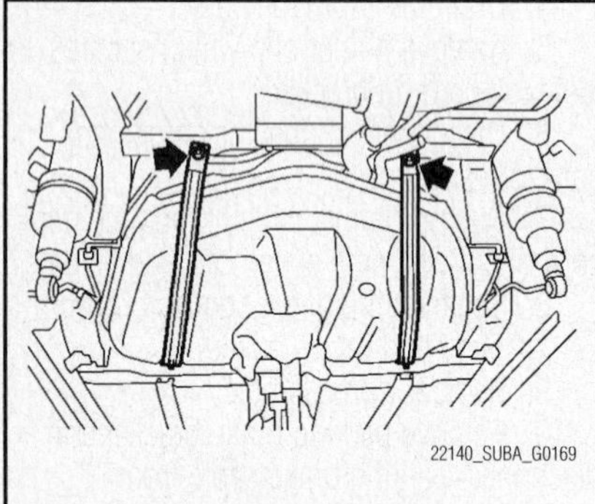

Fig. 268 Support fuel tank with transmission jack, remove bolts as shown

✽✽ CAUTION

A helper is required to perform this work.

28. Remove rear suspension assembly as follows:
    - Support rear differential with transmission jack.
    - Remove bolt which holds rear shock absorber to rear suspension arm.
    - Remove bolts which secure rear suspension assembly to body.
    - Remove rear suspension assembly.
29. Remove rear side fuel tank cover.
30. Disconnect fuel filler hose and fuel tank pressure sensor hose.
31. Disconnect air vent hose from evaporation pipe assembly and disconnect evaporation hose from pressure control solenoid valve.
32. Support fuel tank with transmission jack, remove bolts from bands and dismount fuel tank from the vehicle.

To install:

✽✽ CAUTION

A helper is required to perform this work.

33. Support fuel tank with transmission jack and push fuel tank harness into access hole with grommet.
34. Set fuel tank and temporarily tighten bolts of fuel tank bands.
35. Connect air vent hose to evaporation pipe assembly and connect evaporation hose to pressure control solenoid valve.
36. Connect fuel filler hose and fuel tank pressure sensor hose .
37. Tighten band mounting bolts to 25 ft. lbs. (33 Nm).
38. Install rear side fuel tank cover and tighten to 13 ft. lbs. (18 Nm).
39. Install rear suspension assembly.
40. Support rear suspension assembly and then tighten bolts which secure rear suspension assembly.
41. Tighten assembly bots as follows:
    - T1: 127 ft. lbs. (172 Nm)
    - T2: 80 ft. lbs. (108 Nm)
    - T3: 48 ft. lbs. (66 Nm)
42. Tighten bolt which holds rear shock absorber to rear suspension arm, to 116 ft. lbs. (157 Nm)
43. Install rear brake caliper.
44. Tighten bolts which hold rear brake caliper to 25 ft. lbs. (33 Nm).
45. Install parking brake cable to cabin by forcibly pushing it forward.
46. Tighten bolts which hold parking brake cable holding bracket to 13 ft. lbs. (18 Nm).
47. Connect connector to ABS sensor.
48. Install propeller shaft.
49. Install rear exhaust pipe and muffler.
50. Tighten the rear exhaust pipe to center exhaust pipe to 13 ft. lbs. (18 Nm).

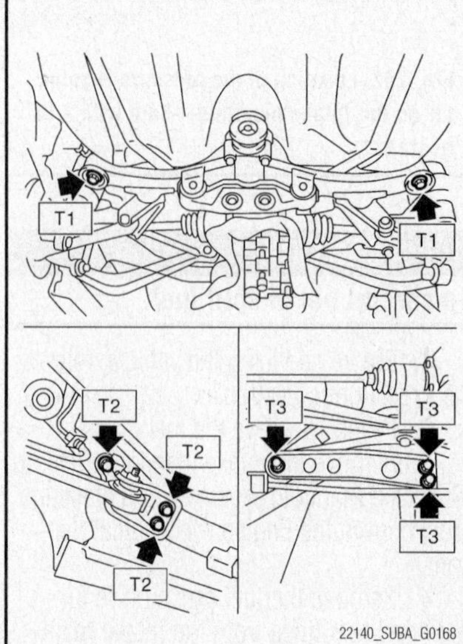

Fig. 269 Rear suspension assembly bolt locations

51. Install front side fuel tank cover and tighten bolts to 13 ft. lbs. (18 Nm).
52. Install rear wheel.
53. Lower the vehicle.
54. Tighten wheel nuts to rear wheel.
55. Install parking brake cable.
56. Install console box.
57. Connect fuel hoses and hold them with quick connector.
58. Connect evaporation pipe and hold it with quick connector.
59. Install pipe protector.
60. Connect fuel jet pump hose.
61. Connect connector to fuel sub level sensor.
62. Install sub service hole cover.
63. Connect connectors to fuel tank cord and plug service hole with grommet.
64. Install holder clip which secures fuel tank cord on bracket.
65. Set rear seat and floor mat.
66. Connect connector to fuel pump relay.
67. Adjust parking brake lever stroke.
68. Check wheel alignment and adjust if necessary.

Impreza Models

See Figures 270 through 273.

✽✽ CAUTION

Place "NO OPEN FLAMES" signs near the working area.

1. Before servicing the vehicle, refer to the Precautions Section.
2. Set vehicle on the lift.
3. Release fuel pressure.
4. Drain fuel from fuel tank.
5. Disconnect the ground cable from the battery.
6. Remove the rear seat.
7. Disconnect fuel pump connector, and remove clip.
8. Remove the bolts.
9. Push the grommet down and remove the service hole cover.
10. Remove the service hole cover of fuel sub level sensor.
11. Disconnect the quick connector on the fuel delivery tube.
12. Remove the rear wheels.
13. Lift up the vehicle.
14. Remove the rear ABS wheel speed sensor from the rear housing.
15. Remove the bolt holding the rear brake hose bracket.
16. Remove the rear brake caliper and tie it to the body side of the vehicle.
17. Remove the parking brake cable from the parking brake assembly.
18. Remove the rear exhaust pipe.

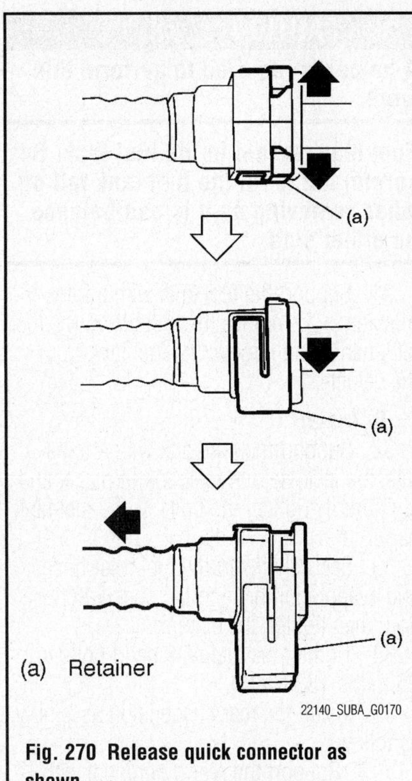

Fig. 270 Release quick connector as shown

19. Remove the propeller shaft.
20. Remove the heat shield cover and fuel tank protector.
21. Disconnect the connector from the rear ABS wheel speed sensor.
22. Remove the bolts securing the parking brake cable clamp.
23. Disconnect drain hose from the canister drain connector.

CAUTION

A helper is required to perform this work.

24. Remove the rear suspension assembly as follows:
    - Support the rear differential with the transmission jack.
    - Remove the bolts which hold the rear shock absorber to the rear suspension arm.
    - Remove the bolts which secure the rear suspension assembly to the body.
    - Remove the rear suspension assembly.
25. Disconnect evaporation hose from connector.
26. Disconnect the quick connector of the evaporation hose from the evaporation pipe as shown in the figure.
27. Disconnect the fuel filler hose and evaporation hose from the fuel filler pipe assembly.
28. Support the fuel tank with a transmission jack, remove the bolts from the fuel tank band, and remove the fuel tank from the vehicle.

To install:

29. Support the fuel tank with a transmission jack, set the fuel tank in place, and temporarily tighten the bolts of the fuel tank band.
30. Securely insert the fuel filler hose and evaporation hose into the specified position, then attach the clamp and clip as shown in the figure.
31. Connect the quick connector of the evaporation hose to the evaporation pipe.

CAUTION

Check that there is no damage or dust on the quick connector. If necessary, clean seal surface of pipe. When connecting the quick connector, insert the pipe all the way in securely, then operate the push lock. If it is not possible to perform the push lock operation of the retainer, recheck whether the pipe is securely inserted. Confirm that the quick connector is securely connected.

32. Connect evaporation hose to connector.

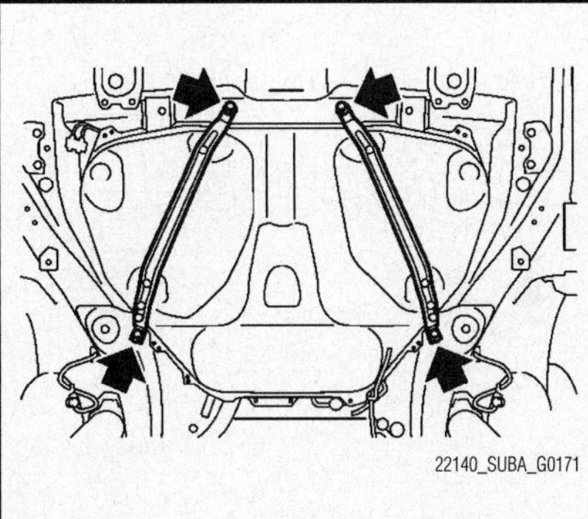

Fig. 271 Support with a transmission jack, remove the bolts as shown

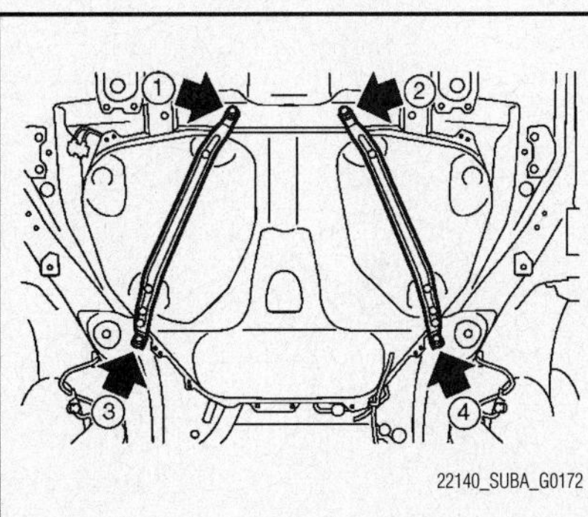

Fig. 272 Fuel tank bolt tightening sequence

33. Tighten the fuel tank band bolts in the order shown in the figure to 25 ft. lbs. (33 Nm).
34. Support the rear differential with the transmission jack.
35. Install the rear suspension assembly.
36. Support the rear suspension assembly, and tighten the bolts which secure the rear suspension assembly to the body.
37. Tighten the rear suspension bolts as follows:
    - T1: 52 ft. lbs. (70 Nm)
    - T2: 148 ft. lbs. (200 Nm)
38. Install the bolts which hold the rear shock absorber to the rear suspension arm. Tighten to 86 ft. lbs. (117 Nm).(120 Nm).
39. Connect drain hose to canister drain connector.
40. Tighten the bolts holding the parking brake cable clamp to 13 ft. lbs. (18 Nm)
41. Connect the connector to the rear ABS wheel speed sensor.
42. Install the heat shield cover.
43. Install the fuel tank protector. Tighten the nuts to 7 ft. lbs. (9 Nm). Tighten the bolts to 13 ft. lbs. (18 Nm).
44. Install the propeller shaft.
45. Install the rear exhaust pipe.
46. Lower the vehicle.
47. Connect the parking brake cable to the parking brake assembly.
48. Install the rear brake caliper.
49. Tighten the bolts which hold the rear brake hose bracket to 25 ft. lbs. (33 Nm).
50. Attach the rear ABS vehicle speed sensor to the rear housing. Tighten the mounting bolt to 6 ft. lbs. (8 Nm).
51. Install the rear wheels.
52. Connect quick connector of the fuel delivery tube.
53. Install the service hole cover of fuel sub level sensor.
54. Attach the service hole cover of the fuel pump, and secure the connector and clip.
55. Install the rear seat.

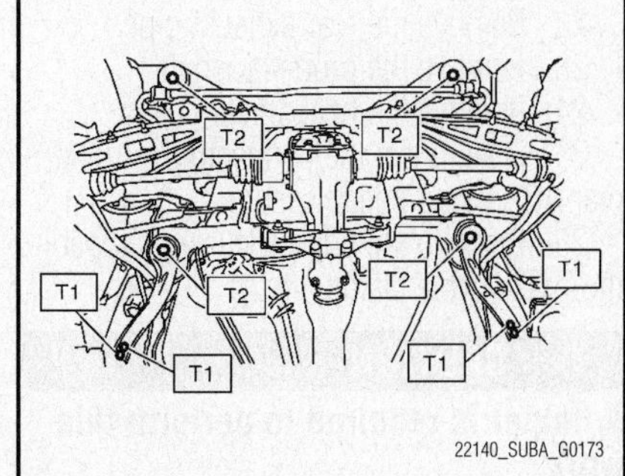

Fig. 273 Tighten the rear suspension bolts as shown

56. Install the fuse of fuel pump to main fuse box.

57. Connect the ground cable to the battery.

58. Inspect the wheel alignment and adjust if necessary.

Legacy & Outback Models

See Figures 274 through 277.

**** CAUTION**

Place "NO OPEN FLAMES" signs near the working area.

1. Before servicing the vehicle, refer to the Precautions Section.
2. Set vehicle on the lift.
3. Release fuel pressure.
4. Drain fuel from fuel tank.
5. Remove the rear seat.
6. Remove the service hole cover of fuel pump.
7. Disconnect the connector from fuel pump.
8. Remove the connector and clip.
9. Remove the bolts.
10. Push the grommet down and remove the service hole cover.
11. Disconnect connector from fuel sub level sensor.
12. Disconnect the quick connector on the fuel delivery hose.
13. Remove the trunk room trim. (Sedan models)
14. Remove the rear quarter trim. (Wagon models)
15. Remove the pipe protector.
16. Remove the grommet and disconnect the quick connector of the evaporation pipe.
17. Remove the rear wheels.
18. Lift up the vehicle.
19. Remove the bolts which secure the rear brake hose mounting bracket.
20. Remove the rear brake caliper and tie it to the body side of the vehicle.
21. Remove the parking brake cable from the parking brake assembly.
22. Remove the rear exhaust pipe.
23. Remove the propeller shaft.
24. Remove the heat shield cover.
25. Disconnect the connector from the rear ABS wheel speed sensor.
26. Remove the bolts securing the parking brake cable clamp.

**** CAUTION**

A helper is required to perform this work.

27. Remove the rear suspension assembly as follows:
    - Support the rear differential with the transmission jack.
    - Remove the bolts which hold the rear shock absorber to the rear suspension arm.
    - Remove the bolts which secure the rear suspension assembly to the body.
    - Remove the rear suspension assembly.
28. Disconnect the connector.
29. Disconnect the evaporation hose.
30. Disconnect the fuel filler hose and evaporation hose.

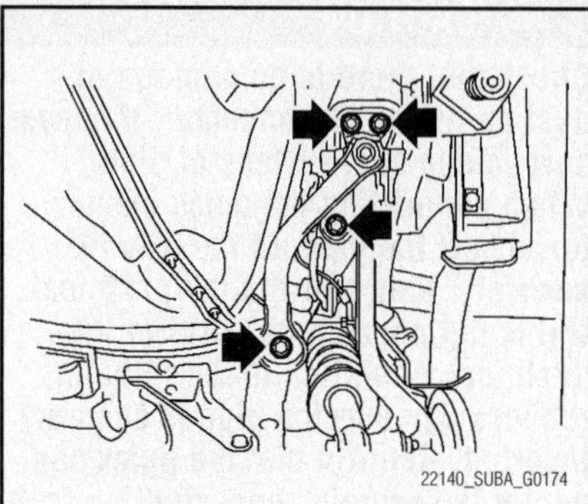

Fig. 274 Bolts which secure the rear suspension view (1)

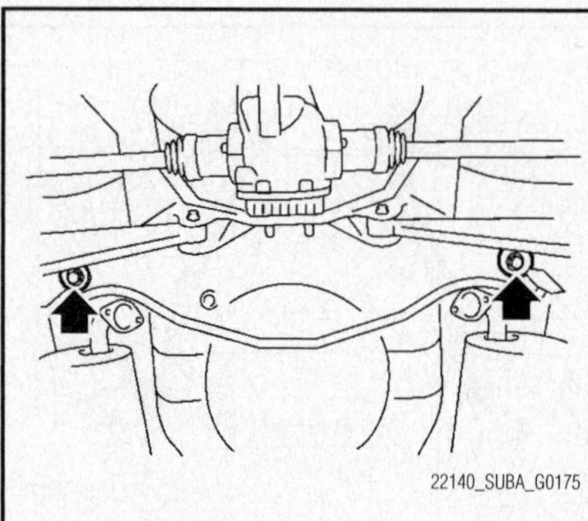

Fig. 275 Bolts which secure the rear suspension view (2)

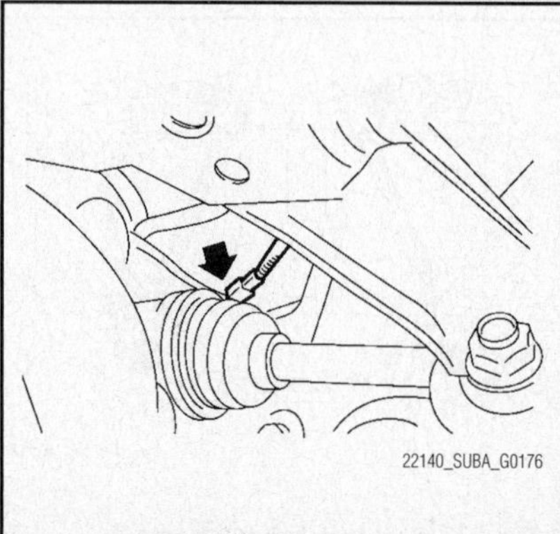

Fig. 276 Connector view

**** CAUTION**

A helper is required to perform this work.

Fuel may remain in the fuel tank. Be careful not to let the fuel tank fall off when removing as it is bad balance on either side.

31. Support the fuel tank with a transmission jack, remove the bolts from the fuel tank band, and remove the fuel tank from the vehicle.

To install:

32. Support the fuel tank with a transmission jack, set the fuel tank in place, and temporarily tighten the bolts of the fuel tank band.
33. Securely insert the fuel filler hose and evaporation hose to the specified position, then tighten the clamp.
34. Tighten the fuel tank band bolts to 25 ft. lbs. (33 Nm).
35. Install the rear suspension assembly as follows:
    - Support the rear differential with the transmission jack.
    - Support the rear suspension assembly, and tighten the bolts which secure the rear suspension assembly to the body.
    - Tighten the bolts which hold the rear shock absorber to the rear suspension arm.

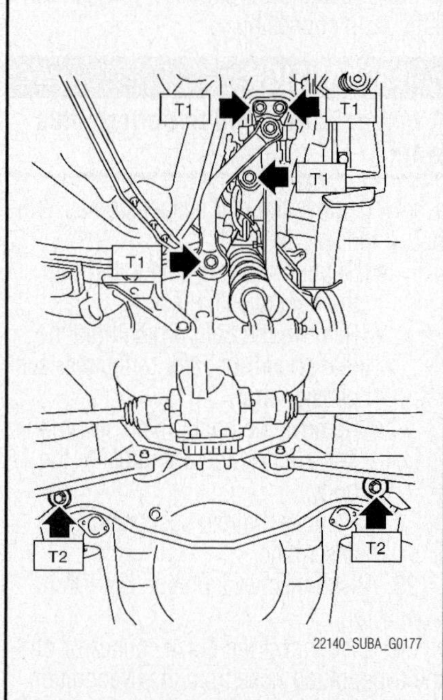

Fig. 277 Tighten the suspension bolts T1: 92 ft. lbs. (125 Nm) and T1: 129 ft. lbs. (175 Nm) as shown

36. Tighten the suspension bolts T1: 92 ft. lbs. (125 Nm) and T1: 129 ft. lbs. (175 Nm).
37. Install the bolts which hold the rear shock absorber to the rear suspension arm. Tighten to 46 ft. lbs. (62 Nm).
38. Tighten the bolts holding the parking brake cable clamp to 13 ft. lbs. (18 Nm).
39. Connect the connector to the rear ABS wheel speed sensor.
40. Install the heat shield cover.
41. Install the propeller shaft.
42. Install the rear exhaust pipe.
43. Lower the vehicle.
44. Connect the parking brake cable to the parking brake assembly.
45. Install the rear brake caliper.
46. Tighten the bolts which secure the rear brake hose mounting bracket to 25 ft. lbs. (33 Nm).
47. Install the rear wheels.
48. Connect the quick connector of the evaporation pipe.
49. Install the pipe protector.
50. Install the trunk room trim. (Sedan model)
51. Install the rear quarter trim. (Wagon model)
52. Connect connector to the fuel sub level sensor.
53. Connect the quick connector of the fuel delivery hose.
54. Install the service hole cover of fuel sub level sensor.
55. Connect connector, and install clip.
56. Connect the connector to the fuel pump.
57. Install the service hole cover of fuel pump.
58. Install the rear seat.
59. Install the fuse of fuel pump to the main fuse box.
60. Connect the battery ground cable to battery.
61. Inspect the wheel alignment and adjust if necessary.

IDLE SPEED

ADJUSTMENT

The idle speed cannot be adjusted manually, because the idle speed is automatically adjusted.

THROTTLE BODY

REMOVAL & INSTALLATION

Baja SOHC 2.5L engine

See Figure 278.

1. Before servicing the vehicle, refer to the Precautions Section.

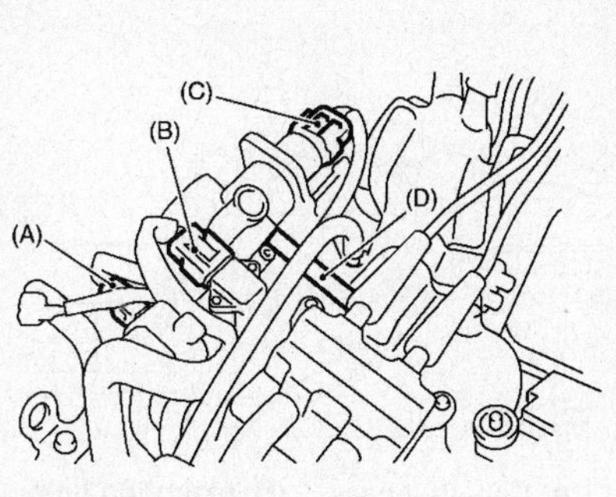

Fig. 278 Throttle body sensor locations—Baja 2.5L SOHC engine

2. Disconnect the negative battery cable.
3. Remove the air intake assembly.
4. Disconnect the accelerator cable.
5. Disconnect the cruise control cable, if equipped.
6. Disconnect the throttle position sensor electrical connector.
7. Disconnect the idle air control solenoid valve and air bypass hose.
8. Disconnect the Manifold Absolute Pressure (MAP) sensor.
9. Disconnect the coolant hoses from the throttle body.
10. Remove the throttle body mounting bolts.
11. Remove the throttle body and gasket.

To install:

12. Install the throttle body with a new gasket. Tighten the mounting bolts to 16 ft. lbs. (22 Nm).
13. The remainder of the installation is the reverse order of removal.

Legacy & Outback

2.5L SOHC Engines

See Figure 279.

1. Before servicing the vehicle, refer to the Precautions Section.
2. Disconnect the negative battery cable.
3. Disconnect the throttle position sensor and manifold absolute pressure (MAP) sensor.
4. Disconnect the coolant hoses from the throttle body.
5. Remove the throttle body mounting bolts.
6. Remove the throttle body and gasket.

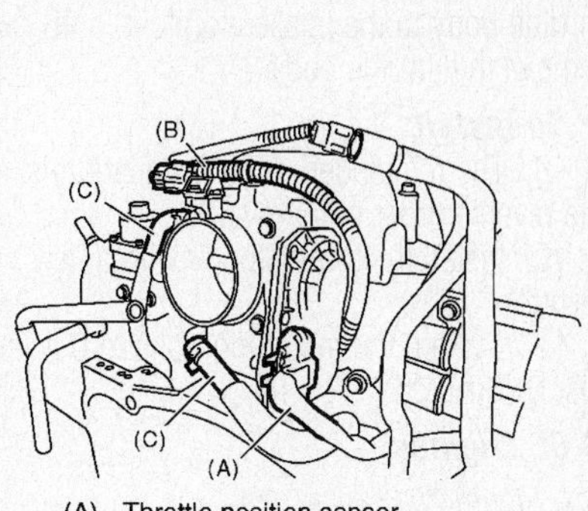

Fig. 279 Throttle body view SOHC—2.5L engine

To install:

7. Install the throttle body with a new gasket.
8. Tighten the mounting bolts to 6 ft. lbs. (8 Nm).
9. The remainder of the installation is the reverse order of removal.

2.5L DOHC Engines

See Figure 280.

1. Before servicing the vehicle, refer to the Precautions Section.
2. Disconnect the negative battery cable.
3. Remove the intercooler.
4. Disconnect the connectors from the throttle position sensor.
5. Remove the bolts which secure the air by-pass pipe and PCV pipe to the intake manifold, and loosen the clamp which connects the throttle body and duct.
6. Remove the duct from the throttle body.
7. Disconnect the coolant hoses from the throttle body.

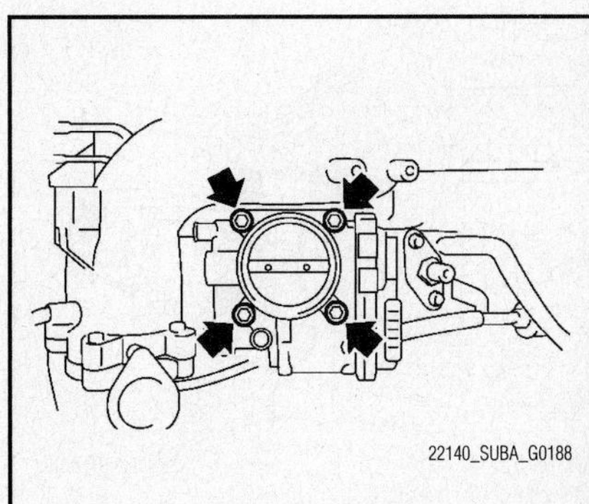

Fig. 280 Remove the bolts which secure the throttle body

8. Remove the bolts which secure the throttle body to the intake manifold, and remove the throttle body.

To install:

9. The remainder of the installation is the reverse order of removal.
10. Install the throttle body with a new gasket.
11. Tighten the mounting bolts to 6 ft. lbs. (8 Nm).

3.0L Engines

See Figure 281.

1. Before servicing the vehicle, refer to the Precautions Section.
2. Disconnect the negative battery cable.

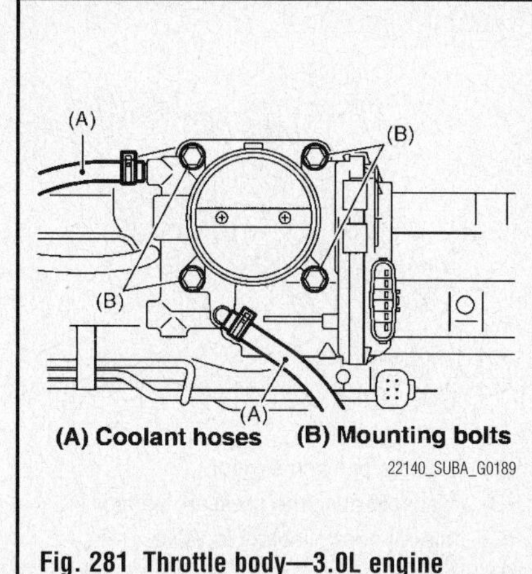

Fig. 281 Throttle body—3.0L engine

3. Remove the collector cover.
4. Remove the air intake chamber.
5. Disconnect the connector from throttle body.
6. Disconnect the engine coolant hoses (A) from the throttle body.
7. Remove the bolts (B) which secure throttle body to intake manifold.

To install:

8. The remainder of the installation is the reverse order of removal.
9. Install the throttle body with a new gasket.
10. Tighten the mounting bolts to 6 ft. lbs. (8 Nm).

HEATING & AIR CONDITIONING SYSTEM

BLOWER MOTOR

REMOVAL & INSTALLATION

Baja Models

See Figure 282.

1. Before servicing the vehicle, refer to the Precautions Section.
2. Disconnect the negative battery cable.
3. Remove the glove box.
4. Remove the wiring harness bracket mounting bolts.
5. Remove the nuts of the keyless unit stay and cruise unit stay.
6. Disconnect the sunroof electrical connector.
7. Disconnect the servo connector.
8. Disconnect the blower motor electrical harness.
9. Remove the blower motor mounting screws.
10. Disconnect the aspirator pipe (A).
11. Remove the blower motor unit.
12. Installation is the reverse order of removal.

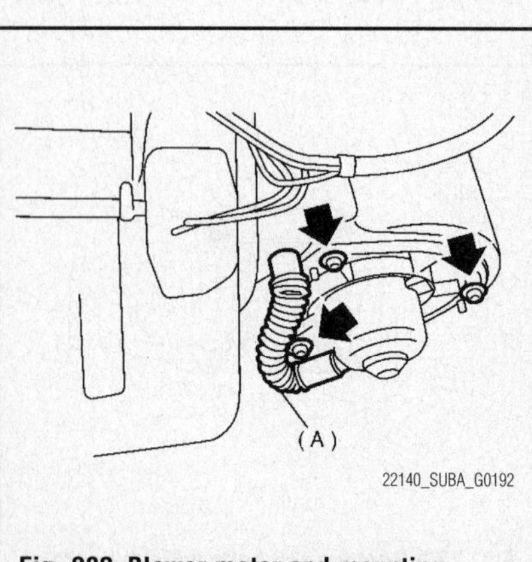

Fig. 282 Blower motor and mounting screws

2006–07 Impreza models

See Figure 283.

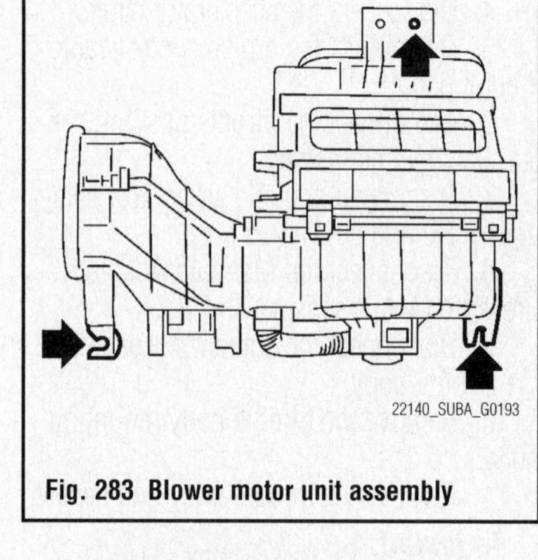

Fig. 283 Blower motor unit assembly

1. Before servicing the vehicle, refer to the Precautions Section.
2. Disconnect the negative battery cable.
3. Remove the glove box.
4. Loosen the nut to remove the support beam stay.
5. Disconnect the blower motor wiring harness.
6. Disconnect the blower resistor connector.
7. Loosen the bolt and nut to remove the blower motor unit assembly.
8. Installation is the reverse order of removal.

2008 Impreza models

See Figure 282.

1. Before servicing the vehicle, refer to the Precautions Section.
2. Disconnect the battery ground cable from the battery.
3. Turn up the floor mat near the blower motor.
4. Loosen the screw to remove the blower motor.
5. Installation is the reverse order of removal.

Legacy & Outback Models

See Figure 282.

1. Before servicing the vehicle, refer to the Precautions Section.
2. Disconnect the negative battery cable.
3. Remove the glove box lower cover.
4. Disconnect the blower motor wiring harness.
5. Remove the mounting screws and remove the blower motor.
6. Installation is the reverse order of removal.

HEATER CORE

REMOVAL & INSTALLATION

Baja Models

See Figure 284.

1. Before servicing the vehicle, refer to the Precautions Section.
2. Disconnect the negative battery cable.
3. Drain the cooling system.
4. Remove the air intake system.
5. Release the heater hose clamps in the engine compartment to remove the heater hoses.
6. Remove the A/C unit.
7. Using a Torx® wrench, remove the airbag control unit.
8. Disconnect the connector of the air bag main wiring harness near the steering support beam.

Fig. 284 Removing the heater core from the heater unit—Baja

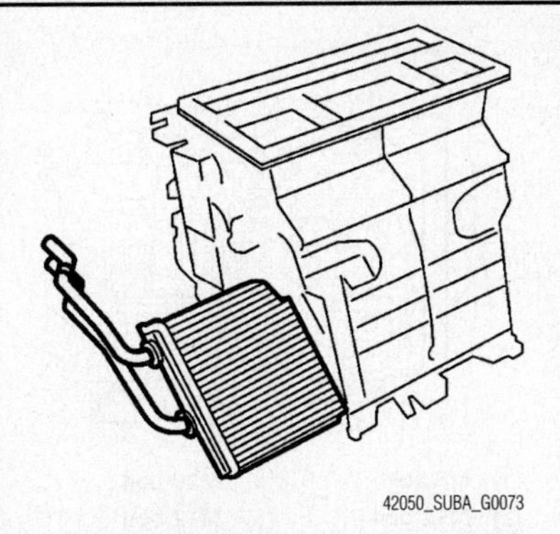

Fig. 285 Removing the heater core from the heater and cooling unit—2006–08 Impreza

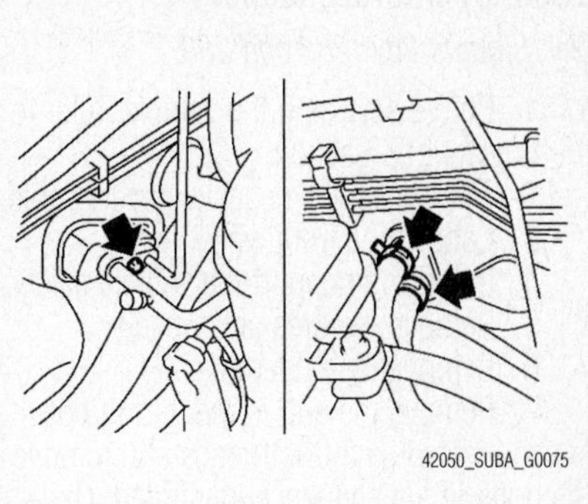

Fig. 286 Remove the bolt securing the expansion valve (left) and disconnect the heater hoses (right).

9. Loosen the bolts and nuts of the support beam to remove the support beam.
10. Disconnect the servo connector.
11. Loosen the bolts and nuts of the heater unit to remove.
12. Separate the heater unit case to remove the heater core.
13. Installation is the reverse of removal.
14. Fill the cooling system to the correct level.

Impreza Models

See Figure 285.

1. Before servicing the vehicle, refer to the Precautions Section.
2. Disconnect the negative battery cable.
3. Discharge the air conditioning system.
4. Drain the cooling system.
5. Remove the bolts securing the expansion valve and pipe in the engine compartment. Release the heater hose clamps in the engine compartment to remove the heater hoses.
6. Remove the instrument panel.
7. Remove the support beam.
8. Remove the blower motor assembly.
9. Loosen the bolts and nuts to remove the heater and cooling unit.
10. Loosen the screws to remove the heater core cover.
11. Remove the heater core.
12. Installation is the reverse order of removal.

Legacy & Outback Models

See Figure 286.

1. Before servicing the vehicle, refer to the Precautions Section.
2. Disconnect the negative battery cable.
3. Discharge the air conditioning system.
4. Drain the cooling system.
5. Remove the bolts securing the expansion valve and pipe in the engine compartment. Release the heater hose clamps in the engine compartment to remove the heater hoses.
6. Remove the instrument panel.
7. Remove the support beam.
8. Remove the blower motor assembly.
9. Disconnect the actuator connector.
10. Loosen the bolts and nuts of the heater unit to remove.
11. Separate the heater unit case to remove the heater core.
12. Installation is the reverse of removal.
13. Fill the cooling system to the correct level.
14. Charge the air conditioning system.

STEERING

POWER RACK & PINION STEERING GEAR

REMOVAL & INSTALLATION

Baja Models

1. Before servicing the vehicle, refer to the Precautions Section.
2. Disconnect the negative battery cable.
3. Loosen the front wheel nut.
4. Raise and support the vehicle safely.
5. Remove the tires and wheels.
6. Remove the front exhaust pipe assembly.
7. Remove the cotter pin and castle nut. Using a puller, remove the tie rod end from the knuckle arm.
8. Remove the jack up plate. Remove the front stabilizer.
9. Disconnect the power steering fluid pipe at the center of the gearbox and attach a vinyl hose. Discharge the fluid into a suitable container by turning the steering wheel fully clockwise and counterclockwise. Disconnect the other fluid line, and repeat the discharge procedure.
10. Remove the steering wheel. Make a match mark on the universal joint. Remove the universal joint bolts and remove the joint from the vehicle.
11. Disconnect the fluid lines from the steering gear, return hose first.
12. Remove the steering gear clamp bolts securing the steering gear to the crossmember. Remove the steering gear from the vehicle.

To install:

13. Insert the steering gear into the crossmember. Be careful not to damage the gearbox boot.
14. Tighten the steering gear to the crossmember bracket to 43 ft. lbs. (58 Nm).
15. Connect the fluid lines.
16. Align the cutout at the serrated section of the column shaft and yoke. Insert the universal joint into the column shaft.
17. Align the mating marks and insert the universal joint to serrated section of the steering gear assembly. Tighten the bolt to 17 ft. lbs. (23 Nm).
18. Continue the installation in the reverse order of the removal procedure.
19. Fill the power steering system with the proper grade and type fluid.
20. After adjusting toe-in and steering angle, tighten the lock nut on tie-rod end.
21. Start the engine and check for leaks. Correct as required.

2006–07 Impreza Models

See Figures 287, 288 and 289.

1. Before servicing the vehicle, refer to the Precautions Section.
2. Disconnect the negative battery cable.
3. Loosen the front wheel nut.
4. Raise and support the vehicle safely.
5. Remove the tires and wheels.
6. Remove the undercover.
7. Remove the sub frame. Leave bolt (1) connected by a few threads and remove the bolts in the sequence illustrated. Once the other bolts are removed, remove bolt (1) and the sub frame.
8. Remove the front exhaust pipe.
9. Remove the cotter pin and castle nut. Using a puller, remove the tie rod end from the knuckle arm.
10. Remove the jack up plate. Remove the front stabilizer.
11. Disconnect the power steering fluid pipe at the center of the gearbox and attach a vinyl hose. Discharge the fluid into a suitable container by turning the steering wheel fully clockwise and counterclockwise. Disconnect the other fluid line, and repeat the discharge procedure.
12. Remove the steering wheel.
13. Make a match mark on the universal joint. Remove the universal joint bolts and remove the joint from the vehicle.
14. Disconnect the fluid lines from the steering gear, pressure hose first.
15. Remove the steering gear retaining bolts and clamps securing the steering gear to the crossmember.
16. Remove the steering gear from the vehicle.

To install:

17. Insert the steering gear into the cross-member. Be careful not to damage the gearbox boot.
18. Tighten the steering gear to the cross-member bracket to 44.3 ft. lbs. (60 Nm).
19. Connect the fluid lines.

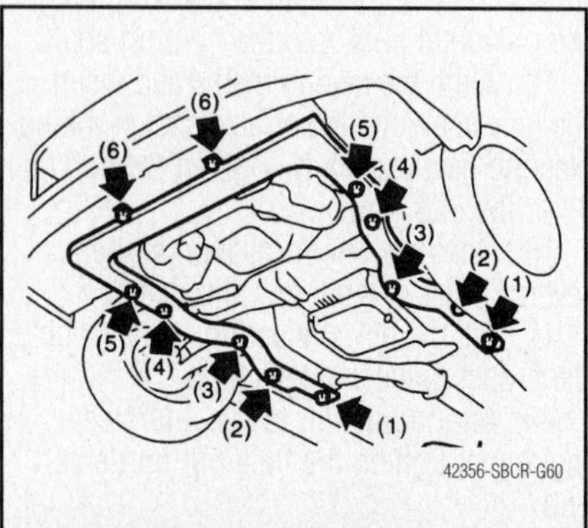

Fig. 287 Sub-frame bolt removal sequence—2006–07 Impreza models

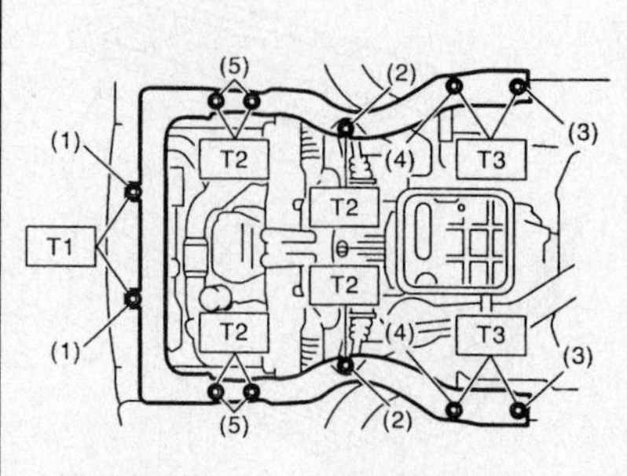

Fig. 288 Sub frame bolt tightening sequence—2006–07 Impreza models

20. Align the cutout at the serrated section of the column shaft and yoke. Insert the universal joint into the column shaft.
21. Align the mating marks and insert the universal joint to serrated section of the steering gear assembly. Tighten the bolt to 17.4 ft. lbs. (24 Nm).
22. After adjusting toe-in and steering angle, tighten the lock nut on tie-rod end.
23. Continue the installation in the reverse order of the removal procedure.
24. When installing the sub frame be sure to torque the bolts to specification and in the proper sequence.
25. Fill the power steering system with the proper grade and type fluid.
26. Start the engine and check for leaks. Correct as required.

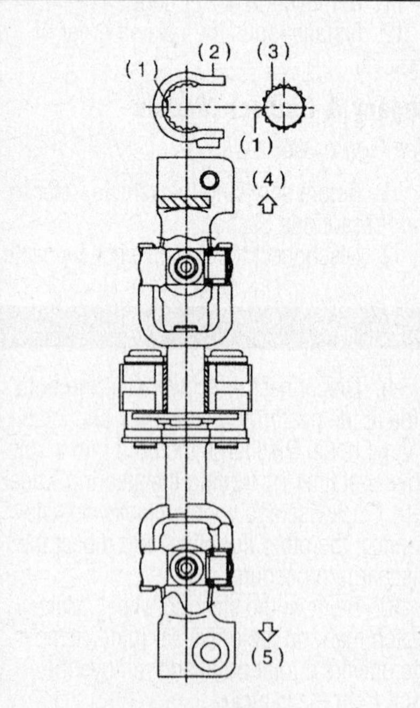

Fig. 289 Steering column joint alignment—2006–07 Impreza models

2008 Impreza Models

See Figures 290 and 291.

1. Before servicing the vehicle, refer to the Precautions Section.
2. Set the vehicle on a lift.
3. Disconnect the ground cable from the battery.
4. Loosen the front wheel nuts.
5. Lift up the vehicle, and remove the front wheels.
6. Remove the undercover.
7. After pulling off the cotter pin and removing the castle nut, use a puller to remove the tie-rod end from the knuckle arm.
8. Remove the front crossmember support plate and front stabilizer.
9. Remove the one pipe joint at the center of the gearbox, and connect the vinyl hose to the pipe and the joint. Discharge the fluid by turning the steering wheel fully clockwise and counterclockwise. Discharge the fluid similarly from other pipes.
10. Remove the steering wheel

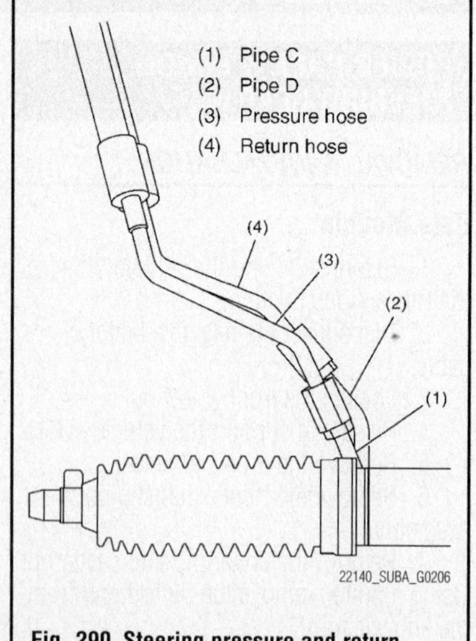

Fig. 290 Steering pressure and return hoses

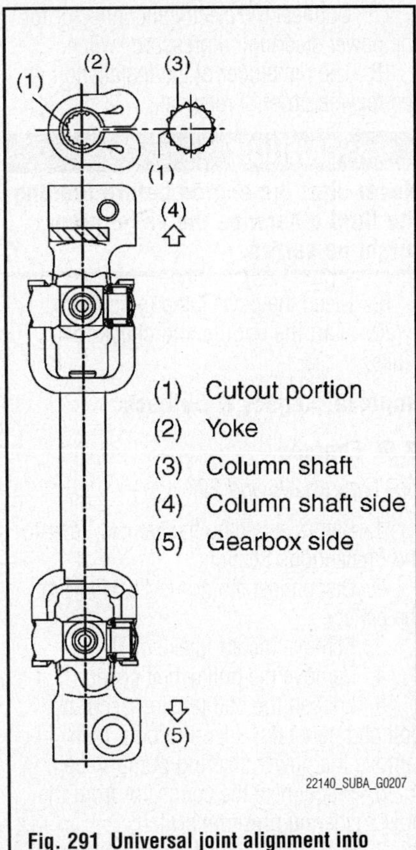

Fig. 291 Universal joint alignment into column shaft

11. Place alignment marks on universal joint.
12. Remove the universal joint bolt, and then remove the universal joint.
13. Disconnect the pipe (C) from pressure hose first, then disconnect pipe (D) from the return hose.
14. Remove the clamp bolts securing the gearbox to the crossmember, and remove the clamp.
15. Remove the bolts which secure the gearbox bracket, and remove the bracket and gearbox.

To install:

16. Insert the gearbox into crossmember, being careful not to damage gearbox boot.
17. Install the gearbox and bracket. Temporarily tighten the bolts.
18. Insert bolts through the clamp to temporarily tighten the gearbox to the crossmember bracket.
19. Tighten the steering gear bracket bolts to 44 ft. lbs. (60 Nm).
20. Connect pipe to the return hose, then connect pipe to the pressure hose. Tighten hose connections to 11 ft. lbs. (15 Nm).
21. Install the universal joint.
22. Align the cutout portion at serrated section of the column shaft and yoke, then install the universal joint into column shaft.
23. Connect the tie-rod end and knuckle arm. Tighten the castle nut to 20 ft. lbs. (27 Nm).
24. After tightening the castle nut to the specified tightening torque, tighten it further within 60° until the cotter pin hole is aligned with slot in the nut. Fit the cotter pin into the nut, and then bend the pin to lock.
25. Install the front stabilizer.
26. Install the front crossmember support plate.
27. Install the undercover.
28. Install the front wheels.
29. Lower the vehicle.
30. Remove the steering wheel.
31. Align the center position of the roll connector.
32. Install the steering wheel.
33. Connect the ground cable to the battery.
34. Pour fluid into the oil tank, and perform air bleeding.
35. Check for fluid leaks.
36. Check the fluid level in oil tank.
37. After adjusting toe-in and steering angle, tighten the lock nut on tie-rod end to 63 ft. lbs. (85 Nm).

➡When adjusting toe-in, hold the boot to prevent it from being rotated or twisted. If it becomes twisted, straighten it.

Legacy & Outback Models

See Figures 292 and 293.

1. Before servicing the vehicle, refer to the Precautions Section.
2. Set the vehicle on a lift.
3. Disconnect the ground cable from battery.
4. Loosen the front wheel nuts.
5. Lift-up the vehicle, and remove the front wheels.
6. Remove the undercover.
7. Remove the front exhaust pipe assembly. (Non-turbo model)
8. After pulling off the cotter pin and removing the castle nut, use a puller to remove the tie-rod end from the knuckle arm.
9. Remove the front crossmember support plate, jack-up plate and front stabilizer.
10. Remove the one pipe joint at the center of the gearbox, and connect the vinyl hose to the pipe and the joint. Discharge the fluid by turning the steering wheel fully clockwise and counterclockwise. Discharge the fluid similarly from other pipes.
11. Remove the steering wheel
12. Place alignment marks on universal joint.
13. Remove the universal joint bolt, and then remove the universal joint.

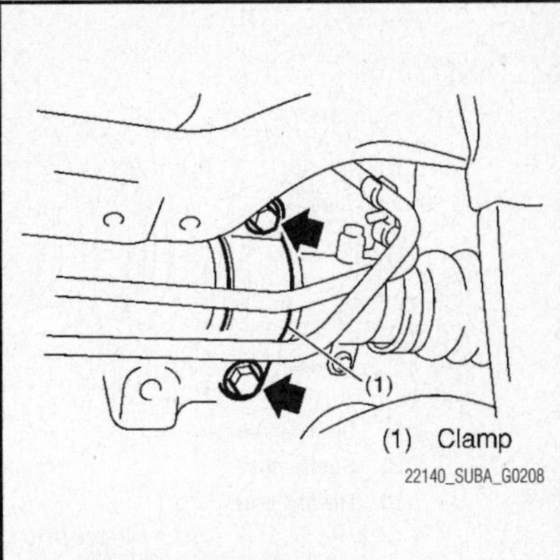

Fig. 292 Steering gear clamp and mounting bolts—Legacy & Outback

14. Disconnect the pipe from pressure hose first, then disconnect pipe from the return hose.
15. Remove the clamp bolts securing the gearbox to the crossmember, and remove the clamp.
16. Remove the bolts which secure the gearbox bracket, and remove the bracket and gearbox.

To install:

17. Insert the gearbox into crossmember, being careful not to damage gearbox boot.
18. Install the gearbox and bracket. Temporarily tighten the bolts.
19. Insert bolts through the clamp to temporarily tighten the gearbox to the crossmember bracket.
20. Tighten the steering gear bracket bolts to 44 ft. lbs. (60 Nm).
21. Connect pipe to the return hose, then connect pipe to the pressure hose. Tighten hose connections to 11 ft. lbs. (15 Nm).
22. Install universal joint as follows:
    - Align the cutout portion at serrated section of the column shaft and yoke, then install the universal joint into column shaft.
    - Install the universal joint to the serrations of gearbox assembly by matching alignment marks.
    - Tighten the bolt to 17 ft. lbs (24 Nm).

⁂ CAUTION

Excessively large tightening torque of universal joint bolts may lead to heavy steering wheel operation.

23. Connect the tie-rod end and knuckle arm. Tighten the castle nut to 20 ft. lbs. (27 Nm).
24. After tightening the castle nut to the specified tightening torque, tighten it further within 60° until the cotter pin hole is

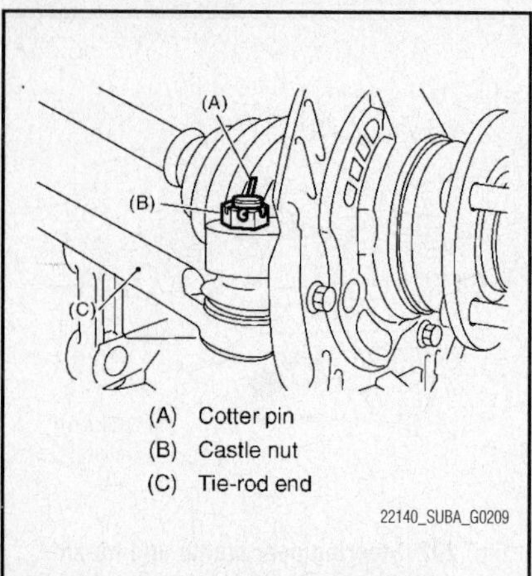

Fig. 293 Connect the tie-rod end and knuckle arm—Legacy & Outback

aligned with slot in the nut. Fit the cotter pin into the nut, and then bend the pin to lock.

25. Install the front stabilizer.
26. Install the front crossmember support plate and jack-up plate.
27. Install the front exhaust pipe assembly. (Non-turbo model)
28. Install the undercover.
29. Install the front wheels.
30. Lower the vehicle.
31. Remove the steering wheel.
32. Align the center position of the roll connector.
33. Install the steering wheel.
34. Connect the ground cable to the battery.
35. Pour fluid into the oil tank, and perform air bleeding.
36. Check for fluid leaks.
37. Check the fluid level in oil tank.
38. After adjusting toe-in and steering angle, tighten the lock nut on tie-rod end to 63 ft. lbs. (85 Nm).

➡When adjusting toe-in, hold the boot to prevent it from being rotated or twisted. If it becomes twisted, straighten it.

POWER STEERING PUMP

REMOVAL & INSTALLATION

Baja Models

See Figures 294 and 295.

1. Before servicing the vehicle, refer to the Precautions Section.
2. Disconnect the negative battery cable.
3. Drain about ⅓ of the power steering fluid from the power steering reservoir.
4. Remove the accessory drive belt cover.
5. Loosen the lock bolt and slider bolt, and remove the power steering pump belt.
6. Disconnect the electrical connector form the power steering pump switch.
7. Disconnect pressure and return hoses from the power steering pump.
8. Remove power steering pump bracket mounting bolts.
9. Place the power steering pump in a vise and remove the bolts from the front side of the pump.
10. Remove the socket from the pump.
11. Remove the bolt from the rear side of the power steering pump to remove it completely from the mounting bracket.

To install:

12. Install the mounting bracket to the power steering pump. Tighten the rear bolt to 28 ft. lbs. (37 Nm).
13. Install the socket to the pump and tighten to 5 ft. lbs. (7 Nm).
14. Install the bolts on the front side of the pump. Tighten to 12 ft. lbs. (16 Nm).
15. Install the power steering pump bracket assembly into the vehicle. Tighten the mounting bolts to 16 ft. lbs. (22 Nm).
16. Connect the pipes that were removed. Tighten the joint nuts to 29 ft. lbs. (39 Nm).

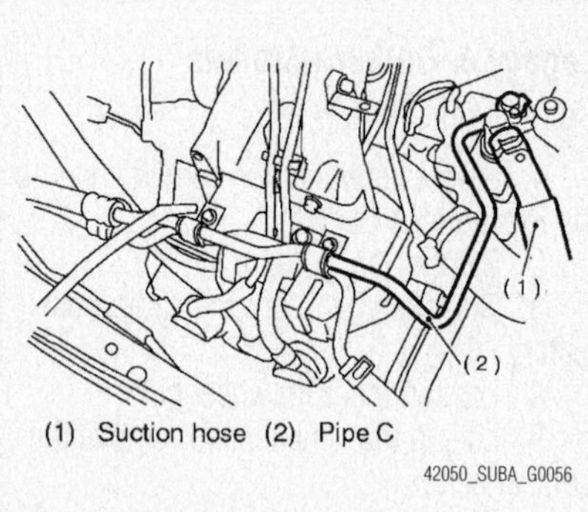

Fig. 294 Pressure (Pipe 'C') and return (suction) hose component locations—Baja

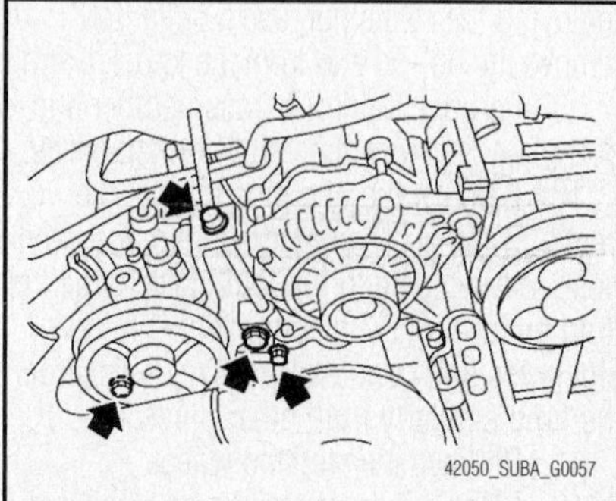

Fig. 295 Power steering pump mounting bolt locations—Baja

17. Connect the electrical connector for the power steering oil pressure switch.
18. The remainder of the installation is the reverse order of removal.

⁂ CAUTION

Never start the engine before feeding the fluid otherwise the vane pump might be seized.

19. Bleed the power steering system.
20. Start the engine and check for leaks.

Impreza, Legacy & Outback

2.5L Engines

See Figures 296 and 297.

1. Before servicing the vehicle, refer to the Precautions Section.
2. Disconnect the ground cable from the battery.
3. Remove the air intake duct.
4. Remove the pulley belt cover.
5. Loosen the belt tension securing bolt and generator securing bolt, then remove the power steering pump V-belt.
6. Disconnect the connector from the power steering pressure switch.
7. Disconnect the pressure hose and suction hose from the oil pump.

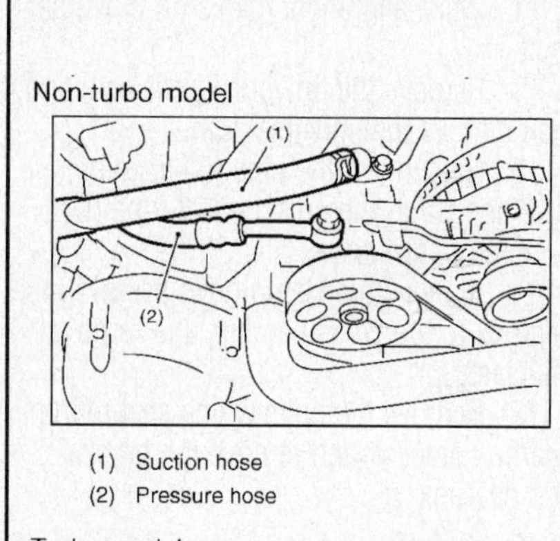

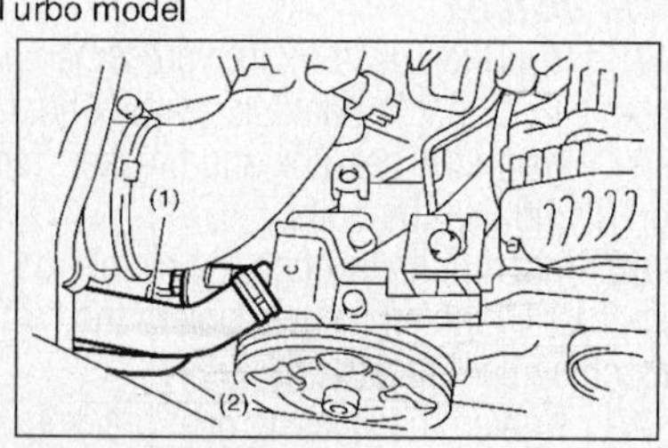

Fig. 296 Pressure hose and suction hose—2.5L turbo and non-turbo engines

CAUTION

Prevent foreign matter from entering the hose and pipe, cover the open ends with clean cloth.

8. Remove the installation bolt of the power steering pump bracket.
9. Place the oil pump bracket in a vise, and remove the two bolts from the front side of the oil pump.

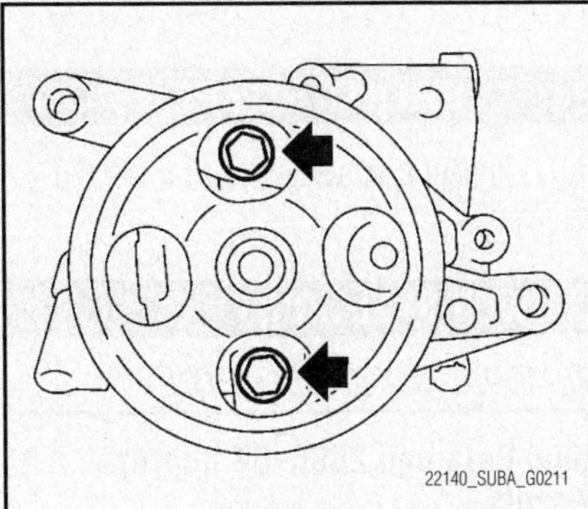

Fig. 297 Front side oil pump to bracket bolts

10. Remove the bolt from the rear side of oil pump.
11. Disassemble the oil pump and bracket by inserting a flat tip screwdriver as shown in the figure.
12. Remove the oil pump.

To install:

13. Install the oil pump to bracket.
14. Place the oil pump bracket in a vise.
15. Tighten the bushing using a 12.7 mm (½) type, 14 and 21 mm box wrench until it is in contact with the oil pump mounting surface.

CAUTION

When securing the oil pump bracket in a vice, hold the oil pump bracket with the least possible force between two pieces of wood.

16. Tighten the two front bolts which hold the oil pump to the bracket to 12 ft. lbs. (16 Nm).
17. Tighten the rear pump to bracket bolt to 27.5 ft. lbs. (37 Nm).
18. Attach the installation bolts of the power steering pump bracket.
19. Connect the pressure hose and suction hose. Tighten the pressure hose eye bolt to 29.5 ft. lbs. (40 Nm).
20. Connect the power steering pressure switch connector.
21. Install the V-belts to the oil pump.
22. Check the tension of the V-belt.
23. Tighten the belt tension bolt to 18.4 ft. lbs. (25 Nm).
24. Install the pulley belt cover.
25. Install the air intake duct.
26. Connect the battery ground cable to the battery.
27. Fill with the specified power steering fluid. (ATF DEXRON III®)

CAUTION

Never start the engine before feeding the fluid otherwise the vane pump might be seized.

28. Bleed the power steering system.
29. Start the engine and check for leaks.

3.0L Engines

See Figures 297 and 298.

1. Disconnect the ground cable from the battery.
2. Remove the cover of the pulley belt.
3. Remove the drive belts.
4. Remove the power steering pressure switch connector.
5. Remove the tensioner adjuster.
6. Disconnect the pressure hose and suction hose from the oil pump.

CAUTION

Prevent foreign matter from entering the hose and pipe, cover the open ends with clean cloth.

7. Remove the bolts which holds the power steering pump bracket.
8. Place the oil pump bracket in a vise, and remove the two bolts from the front side of the oil pump.

CAUTION

When securing the oil pump bracket in a vice, hold the oil pump bracket with the least possible force between two pieces of wood.

9. Remove the bolt from the back side of the oil pump.
10. Remove the oil pump from the bracket.

To install:

11. Install the oil pump to bracket.
12. Tighten the bushing using a 12.7 mm (½) type, 14 and 21 mm box wrench until it is in contact with the oil pump mounting surface.

CAUTION

When securing the oil pump bracket in a vice, hold the oil pump bracket with the least possible force between two pieces of wood.

Tightening torque:
15.7 N·m (1.6 kgf-m, 11.6 ft-lb)

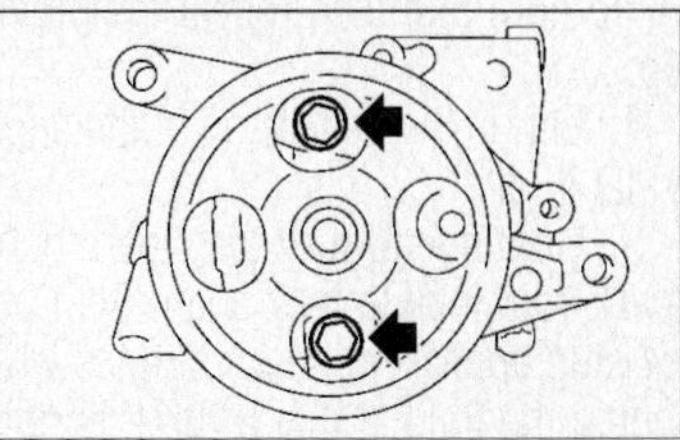

Tightening torque:
37.3 N·m (3.8 kgf-m, 27.5 ft-lb)

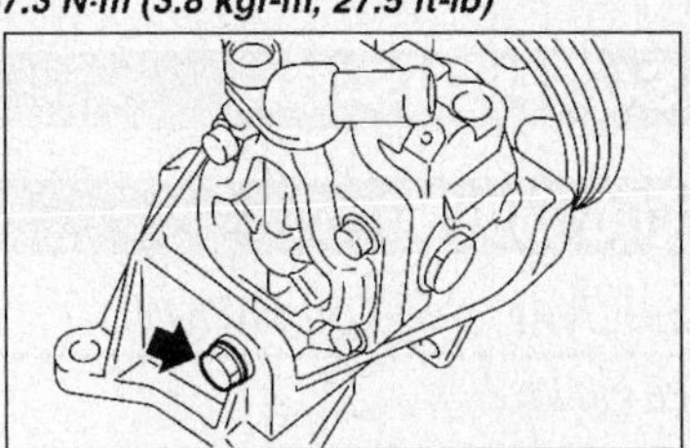

Fig. 298 Pump to bracket bolt location, front and rear

13. Tighten the two front bolts which hold the oil pump to the bracket to 12 ft. lbs. (16 Nm).
14. Tighten the rear pump to bracket bolt to 27.5 ft. lbs. (37 Nm).
15. Attach the installation bolts of the power steering pump bracket.
16. Connect the pressure hose and suction hose. Tighten the pressure hose eye bolt to 29.5 ft. lbs. (40 Nm).
17. Connect the power steering pressure switch connector.
18. Install the tensioner adjuster.
19. Install the drive belts.
20. Install the cover of the pulley belt.
21. Connect the battery ground cable to the battery.
22. Fill with the specified power steering fluid. (ATF DEXRON III®)

CAUTION

Never start the engine before feeding the fluid otherwise the vane pump might be seized.

23. Bleed the power steering system.
24. Start the engine and check for leaks.

BLEEDING

1. Fill the power steering fluid reservoir about half way with the specified fluid.
2. Continue to turn the steering wheel slowly from lock to lock until bubbles stop appearing on oil surface while keeping the fluid at that level.

3. If turning the steering wheel in low fluid level condition, air will be sucked in pipe. In this case, leave it about half an hour and then repeat the previous step.

4. Lift up the vehicle, start the engine and let it idle.

5. Continue to turn the steering wheel slowly from lock to lock again until bubbles stop appearing on oil surface while keeping the fluid at that level. It is normal that bubbles stop appearing after three times turning of steering wheel from lock to lock.

6. In case the bubbles do not stop appearing in the tank, leave it about half an hour and then begin the process again.

7. Lower the vehicle, and then idle the engine.

8. Continue to turn the steering wheel from lock to lock until bubbles stop appearing and change of the fluid level is within 3 mm (0.12 in).

9. In case the following happens, leave it about half an hour and then do step 5–8 again.

a. The fluid level changes over 3 mm (0.12 in).

b. Bubbles remain on the upper surface of the fluid.

c. Grinding noise is generated from oil pump.

10. Check the fluid leakage after turning steering wheel from lock to lock with engine running.

SUSPENSION

FRONT SUSPENSION

COIL SPRING

REMOVAL & INSTALLATION

See Figure 299.

1. Before servicing the vehicle, refer to the Precautions Section.

2. Disconnect the negative battery cable.

3. Raise and support the vehicle safely.

4. Remove front strut, Refer to front MacPherson Strut.

5. Using a coil spring compressor, compress the coil spring.

6. Using a strut mount socket remove the self-locking nut.

7. Remove the strut mount, spacer and upper spring seat from strut.

8. Gradually decrease the compression force of compressor, and remove the coil spring.

9. Remove the dust cover and helper spring.

To install:

10. Using a coil spring compressor, compress the coil spring.

11. Set the coil spring correctly so that its end face seats well in the spring seat as shown in the figure.

12. Install the helper and dust cover to the piston rod.

13. Pull the piston rod fully upward, and install the spring seat.

14. Install spacer and the strut mount to piston rod, and tighten a new self-locking nut temporarily.

15. Using a hexagon wrench to prevent strut rod from turning, tighten the new self-locking nut to 41 ft. lbs. (55 Nm), with a strut mount socket.

16. Loosen the coil spring compressor carefully.

17. Install front strut. "Refer to front MacPherson Strut".

18. Connect the negative battery cable.

CONTROL LINKS

REMOVAL & INSTALLATION

See Figure 300.

1. Before servicing the vehicle, refer to the Precautions Section.

2. Lift up the vehicle, and then remove the front wheels.

3. Remove the front undercover.

4. Remove the stabilizer control link upper and lower mounting nuts.

To install:

5. Install the stabilizer control link.

6. Tighten the stabilizer link upper and lower mounting nuts to 44 ft. lbs. (60 Nm).

7. Install front wheels, and lower the vehicle.

LOWER BALL JOINT

REMOVAL & INSTALLATION

2006 Baja and 2006–07 Impreza Models

See Figure 301.

1. Before servicing the vehicle, refer to the Precautions Section.

2. Disconnect the negative battery cable.

3. Raise and support the vehicle safely.

4. Remove the tire and wheel.

5. Remove the cotter pin from the ball stud.

6. Remove the castle nut.

7. Extract the ball stud from the transverse link.

8. Remove the bolt securing the ball joint to the housing.

9. Extract the ball joint from the housing.

To install:

10. Installation is the reverse of the removal procedure.

11. Install the ball joint to the transverse link arm and tighten to 37 ft. lbs. (50 Nm).

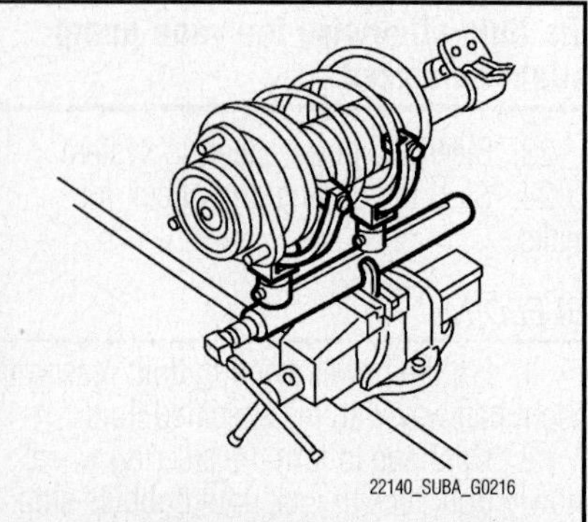

Fig. 299 Coil spring being compressed by spring compressor tool

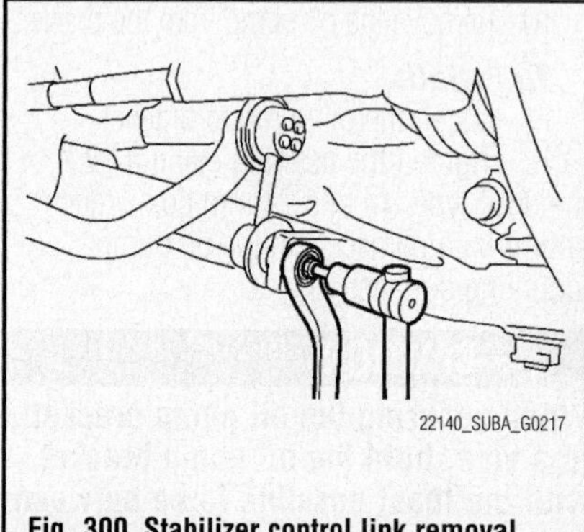

Fig. 300 Stabilizer control link removal shown

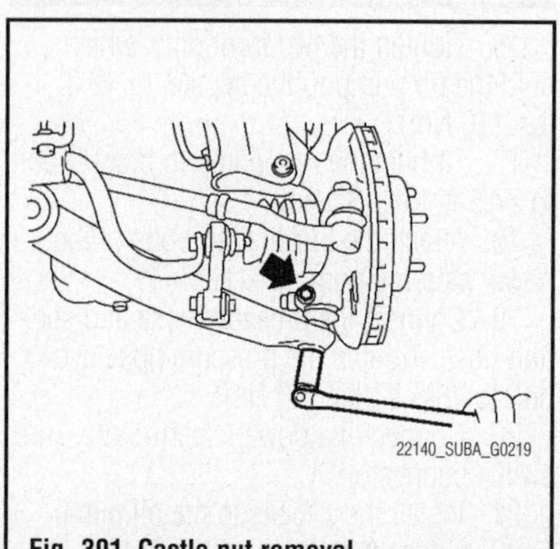

Fig. 301 Castle nut removal

12. Install the castle nut and tighten to 29 ft. lbs. (39 Nm). Tighten the castle nut an additional 60 degrees until the slot in the castle nut is aligned with the cotter pin hole in the ball joint.
13. For Sedan turbo and STI models, tighten the castle nut to 22 ft. lbs. (30 Nm). Tighten the castle nut an additional 60 degrees until the slot in the castle nut is aligned with the cotter pin hole in the ball joint.
14. Check and adjust alignment, as required.
15. Install the front wheel.

2008 Impreza and 2006–08 Legacy & Outback Models

See Figure 302.

1. Before servicing the vehicle, refer to the Precautions Section.
2. Disconnect the negative battery cable.
3. Raise and support the vehicle safely.
4. Remove the tire and wheel.
5. Remove the stabilizer bar brackets and bushings (both sides).
6. Remove the cotter pin from the ball stud. Remove the castle nut.
7. Extract the ball stud from the transverse link.
8. Remove the bolt securing the ball joint to the housing.
9. Extract the ball joint from the housing.

To install:

10. Installation is the reverse of the removal procedure.
11. Install the ball joint to the transverse link arm and tighten to 37 ft. lbs. (50 Nm).
12. Install and tighten the castle nut to 22 ft. lbs. (30 Nm), if the front arm is aluminum and 28.8 ft. lbs. (39 Nm), if the front arm is steel. Tighten the castle nut an additional 60 degrees until the slot in the castle nut is aligned with the cotter pin hole in the ball joint.

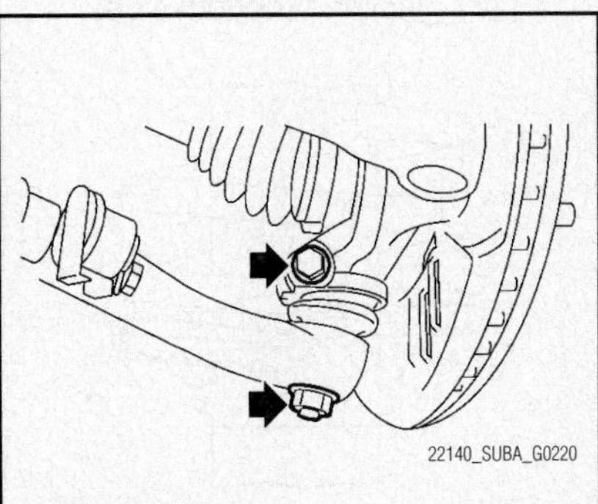

Fig. 302 Bolt securing the ball joint, and castle nut shown

13. The stabilizer bracket has a set orientation. Install it with the arrow mark facing the upper side of the vehicle.
14. Always fully tighten the rubber bushings when the wheels are in full contact with the ground and the vehicle is at curb height.
15. Tighten the stabilizer bracket bolts to 18 ft. lbs. (25 Nm).
16. Check and adjust alignment, as required.

LOWER CONTROL ARM

REMOVAL & INSTALLATION

Baja and 2006–07 Impreza Models

See Figure 303.

1. Before servicing the vehicle, refer to the Precautions Section.
2. Disconnect the negative battery cable.
3. Raise and support the vehicle safely.
4. Remove the tire and wheel.
5. Remove the sub frame, on Impreza and WRX.
6. Disconnect the stabilizer link from the transverse link.
7. Remove the bolt securing the ball joint of the transverse link to housing.
8. Remove the nut (do not remove the bolt) securing the transverse link to the crossmember.
9. Remove the two bolts securing the bushing bracket of the transverse link to the vehicle body at the rear bushing.
10. Remove the ball joint from the housing.
11. Remove the bolt securing the transverse link to the crossmember. Remove the transverse link from the crossmember.

To install:

12. Install the transverse link to its mounting.
13. Temporarily tighten the two bolts used to secure the rear bushing of the transverse link.

➡These bolts should be tightened so that they can still move back and forth in the oblong shaped hole in the bracket, which holds the bushing.

14. Continue the installation in the reverse order of the removal procedure.
15. Always fully tighten the rubber bushings when the wheels are in full contact with the ground and the vehicle is at curb height.
16. Tighten the transverse link rear bushing to body to 184 ft. lbs. (250 Nm) on Impreza and WRX.
17. Tighten the transverse link rear bushing to body to 181 ft. lbs. (250 Nm) on Baja.

➡Move the rear bushing back and forth until the transverse link to rear bushing clearance is established, before tightening.

18. Check and adjust alignment, as required.

2008 Impreza and 2006–08 Legacy & Outback Models

See Figures 304 and 305.

1. Before servicing the vehicle, refer to the Precautions Section.
2. Lift up the vehicle, and then remove the front wheels.
3. Remove the front crossmember support plate.
4. Remove the front stabilizer.

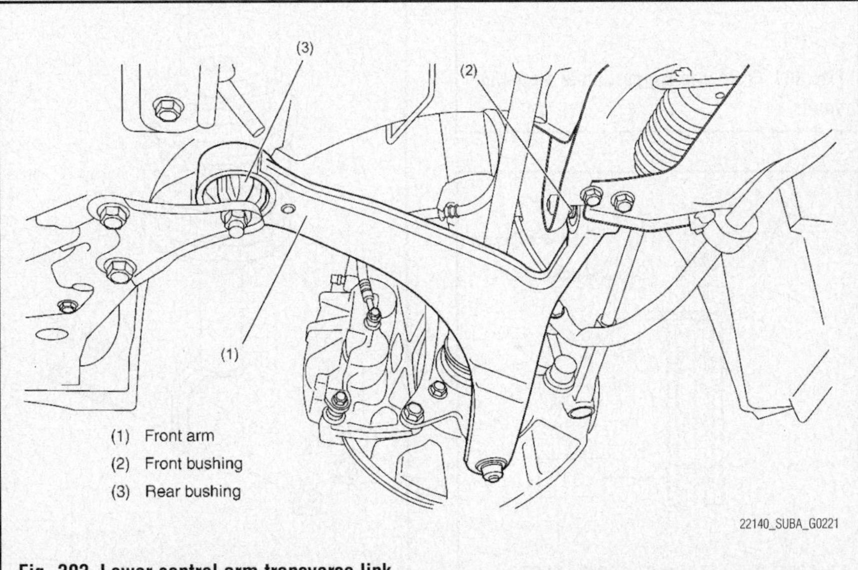

Fig. 303 Lower control arm transverse link

5. Remove the ball joint of front arm.
6. Remove the nut securing the front arm to crossmember. (Do not remove the bolt.)
7. Remove the front arm support plate.
8. Remove the bolt securing front arm to crossmember and pull the front arm out of the crossmember.
9. To remove the stud bolt, use the ST-20299AG020

CAUTION

Do not remove the stud bolt without necessity. Always replace the parts with new parts when removed.

To install:

10. Using the ST-20299AG020, install the stud bolt. Tighten to 81 ft. lbs. (119 Nm).
11. Using new bolts and self-locking nuts, temporarily tighten the front arm to crossmember.
12. Secure the front arm to body, and then install the support plate with new bolts and self-locking nuts.
13. Tighten as follows:
    - Support plate to Front arm: 65 ft. lbs. (88 Nm)
    - Support plate to Body: 111 ft. lbs. (150 Nm)

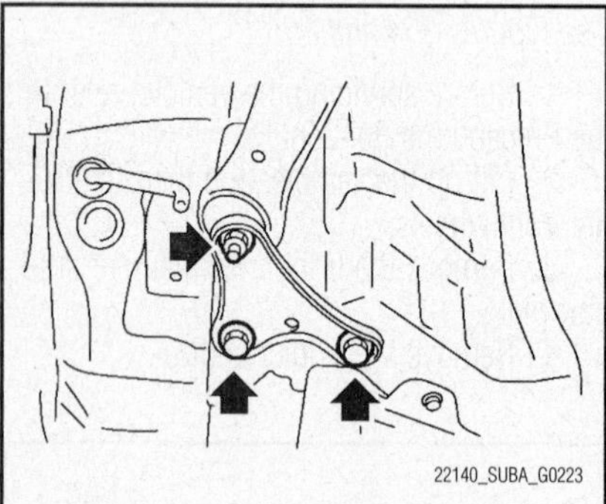

Fig. 304 Front arm support plate mounting points

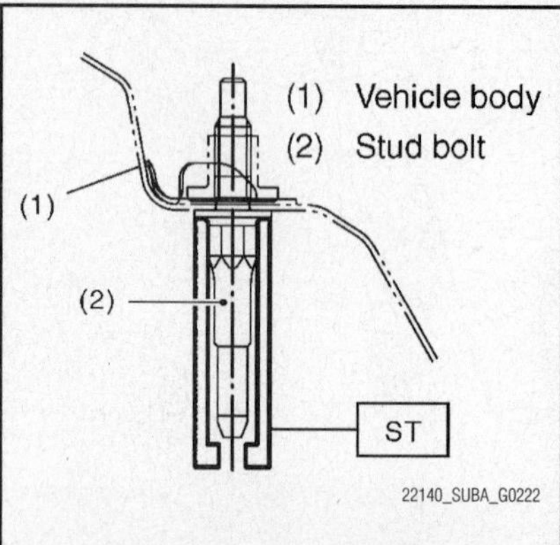

Fig. 305 Special tool ST-20299AG020

14. Install the ball joint into housing, and tighten retaining bolt to 37 ft. lbs. (50 Nm).
15. Install the stabilizer.
16. Lower the vehicle from lift, and tighten the bolt which secures the front arm to crossmember with wheels in full contact with the ground and the vehicle at curb weight.
17. Tighten the stabilizer as follows:
    - Stabilizer link: 33 ft. lbs. (45 Nm).
    - Stabilizer bracket:18 ft. lbs. (25 Nm).
18. Check and adjust alignment, as required.

MACPHERSON STRUT

REMOVAL & INSTALLATION

See Figure 306.

1. Disconnect the negative battery cable.
2. Raise and support the vehicle safely.
3. Remove the tire and wheel.
4. Remove the bolt retaining the brake hose to the strut.
5. Make and alignment mark on the camber adjusting bolt which secures the strut to the housing.
6. Remove the bolt retaining the ABS wheel speed sensor harness.
7. Remove the two bolts retaining the strut to the housing.

➡While holding the head of the adjusting bolt, loosen the self locking nut.

8. Remove the three upper strut retaining nuts.
9. Remove the strut from the vehicle.

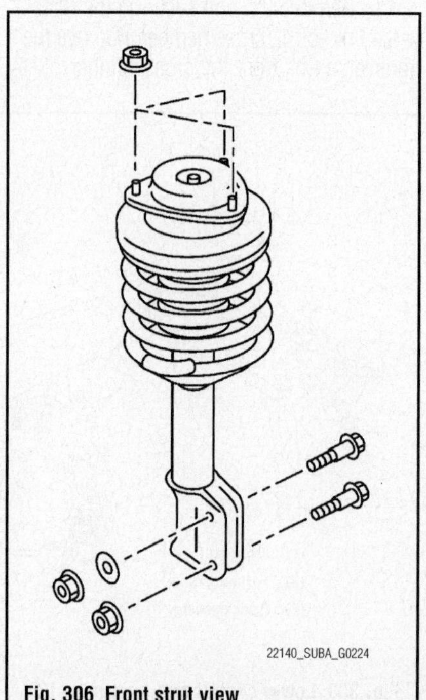

Fig. 306 Front strut view

To install:

10. Installation is the reverse of the removal procedure.
11. Tighten the upper retaining nuts to 14.5 ft. lbs. (20 NM). Tighten the lower retaining bolts to 129 ft. lbs. on the Impreza and WRX. Tighten the lower retaining bolts to 112.1 ft. lbs. (152 Nm) on the Legacy, Outback and Baja.
12. Position the alignment mark on the camber adjusting bolt with the alignment mark on the lower side of the strut. Install using a new self locking nut.
13. Install the ABS wheel speed sensor harness, and brake hose mounting bolts. Tighten both to 24 ft. lbs. (33 Nm).

➡While holding the head of the adjusting bolt, tighten the self locking nut.

14. Check and adjust wheel alignment, as necessary.

OVERHAUL

See Figure 307.

1. Remove the strut from the vehicle.
2. Using a coil spring compressor tool, carefully compress the spring. Remove the self locking nut.
3. Remove the strut mount, upper spring and rubber seat from the strut.

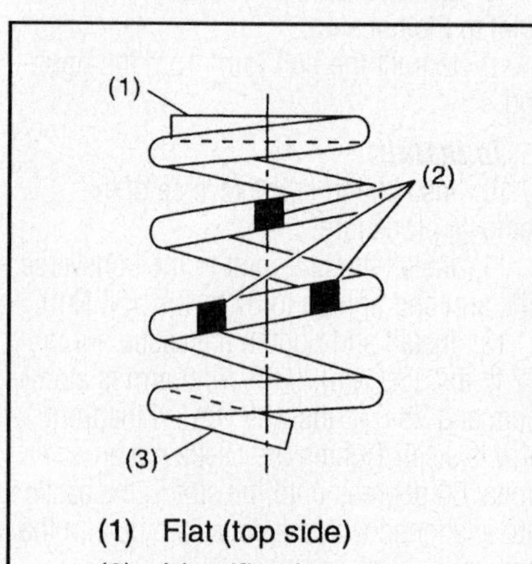

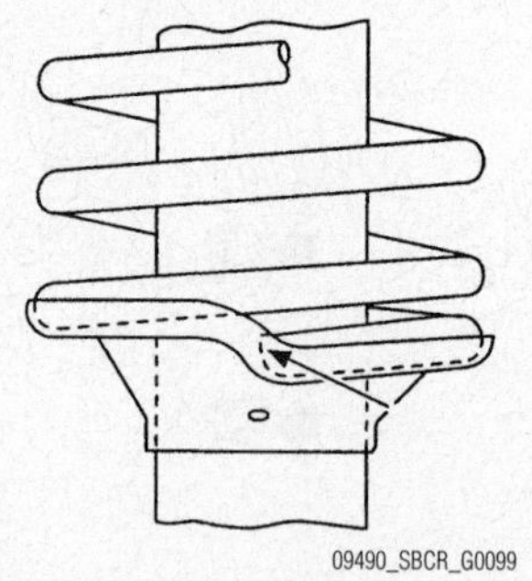

Fig. 307 Front strut spring alignment

4. Gradually decrease the compression force of the spring compressor tool. Remove the coil spring.
5. Remove the dust cover and helper spring.
6. Check for the presence of air in the damping force generating mechanism.
7. Using the spring compression tool, compress the coil spring.

➡Be sure to properly install the coil spring.

8. Position the coil spring so that its end face fits good into the spring seat.
9. Install the helper spring and dust cover to the piston rod.
10. Pull the piston rod fully upward, and install the rubber seat and spring seat.
11. Install the strut mount to the piston rod, and then tighten the self locking nut, temporarily. Be sure to use a new self locking nut.
12. Use a hexagon wrench to prevent the strut rod from turning. Tighten the self locking nut to 41 ft. lbs.
13. Carefully loosen the coil spring.

STEERING KNUCKLE

REMOVAL & INSTALLATION

See Figures 308 through 311.

1. Before servicing the vehicle, refer to the Precautions Section.
2. Raise and support the vehicle safely.
3. Remove the tire and wheels.
4. Remove the front undercover.
5. Remove the stabilizer link.
6. Remove the disc brake caliper from the front housing, and suspend it from strut using a piece of wire.
7. Remove the disc rotor from the hub.
8. Remove the cotter pin and castle nut securing the tie—rod end to the front housing knuckle arm.
9. Using a puller, remove the tie rod ball joint from knuckle arm.

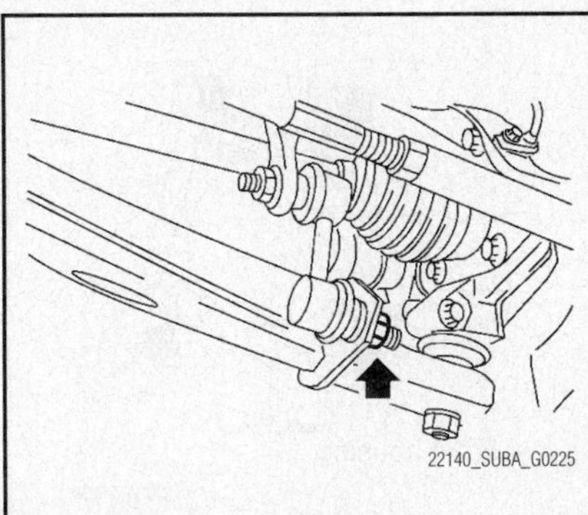

Fig. 308 Stabilizer control link and mounting nut shown

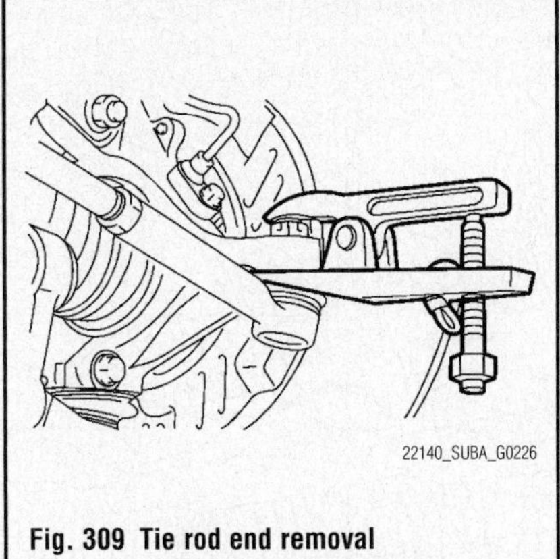

Fig. 309 Tie rod end removal

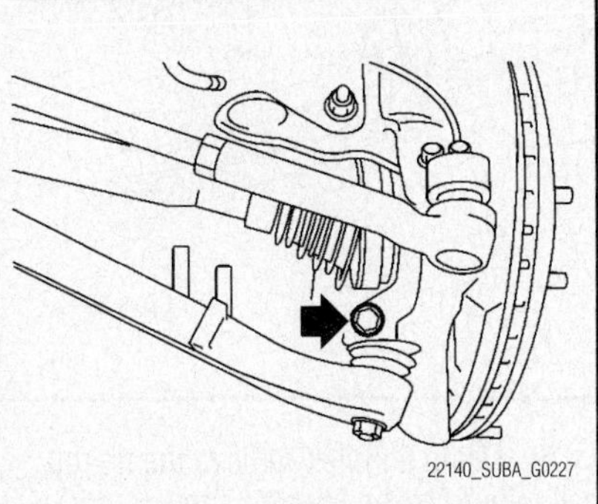

Fig. 310 Ball joint knuckle retaining bolt shown

10. Remove the ABS wheel speed sensor assembly and harness.
11. Remove the lower strut mounting bolts.
12. Remove the front arm ball joint from the front housing.
13. Remove the front driveshaft assembly from the steering knuckle hub.
14. After scribing an alignment mark on the camber adjusting bolt head, remove the bolts which connect the front housing and strut, and disconnect the steering knuckle from the strut.
15. Remove the front steering knuckle hub.

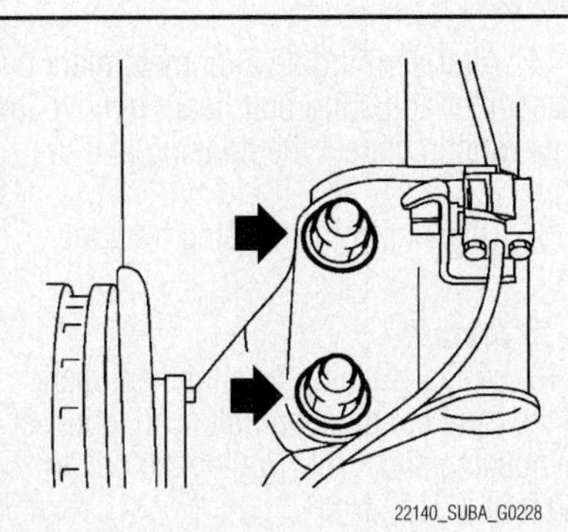

Fig. 311 Strut to knuckle retaining bolts shown

To install:

16. Install the front steering knuckle hub.
17. Align the alignment mark on the camber adjusting bolt head, and affix the front housing and strut together using a new self—locking nut. Tighten to 129 ft. lbs. (175 Nm).
18. Install the front driveshaft.
19. Install the front arm ball joint to the front housing. Tighten mounting bolt to 37 ft. lbs. (50 Nm).
20. Install the ABS wheel speed sensor on the front housing. Tighten to 5.5 ft. lbs. (7.5 Nm).
21. Install the disc rotor to hub.
22. Install the disc brake caliper to the front housing. Tighten to 88.5 ft. lbs. (120 Nm).
23. Install the stabilizer link.
24. Connect the tie-rod end ball joint to the knuckle arm with a castle nut. Tighten to 20 ft. lbs. (27 Nm).
25. Tighten the castle nut to specified torque and tighten further within 60° until the pin hole is aligned with the slot in nut. Bend the cotter pin to lock.
26. While depressing the brake pedal, tighten a new axle nut to 177 ft. lbs. (240 Nm). and lock it securely.
27. Install the tire and wheels.
28. Lower the vehicle.
29. Inspect the wheel alignment and adjust if necessary.

STABILIZER BAR

REMOVAL & INSTALLATION

See Figure 312.

1. Before servicing the vehicle, refer to the Precautions Section.
2. Raise and support the vehicle safely.
3. Remove the tire and wheels.
4. Remove the front undercover.
5. Remove the front crossmember support plate.
6. Remove the stabilizer bar link.
7. Remove the stabilizer bar bracket bolts and bushings.
8. Remove the stabilizer bar from the vehicle.

To install:

9. Installation is the reverse of the removal procedure.
10. Be sure to use new self locking nuts, as required.
11. Install the rubber bushing, so that the paint mark on the stabilizer bar is on the left side of the vehicle.
12. Install the stabilizer bushing (front crossmember side), while aligning it with the paint mark on the stabilizer bar.

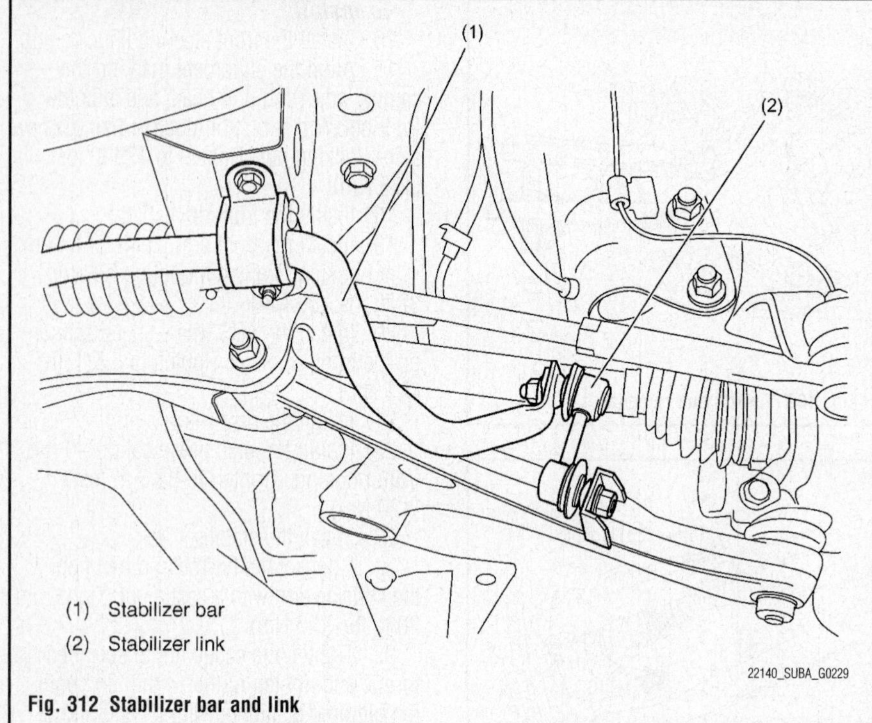

Fig. 312 Stabilizer bar and link

13. Tighten the stabilizer link bolts to 44 ft. lbs. (60 Nm). Tighten the stabilizer bar clamp to 18 ft. lbs. (25 Nm).
14. Always fully tighten the rubber bushings when the wheels are in full contact with the ground and the vehicle is at curb height.

WHEEL HUB AND BEARING (SEALED UNIT)

REMOVAL & INSTALLATION

Baja Models

1. Disconnect the negative battery cable.
2. Raise and support the vehicle safely.
3. Remove the tire and wheel.
4. Remove the crimped section of the axle nut. Depress the brake pedal and remove the axle nut.

➡Be sure to loosen the axle nut after removing the tire and wheel from the vehicle. Failure to do this may damage the wheel bearings.

5. Remove the stabilizer link.
6. Remove the disc brake caliper from its mounting as suspend it to the side, with wire. Do not disconnect the brake line.
7. Remove the rotor.

➡If the rotor is seized within the hub, remove the rotor by installing an 8mm bolt in the screw hole on the rotor.

8. Remove the cotter pin and castle nut which secure the tie rod end to the housing knuckle arm.
9. Using a puller, remove the tie rod ball joint from the knuckle arm.
10. Remove the ABS wheel speed sensor assembly and harness. Remove the bolt that retains the sensor harness to the strut.
11. Remove the transverse link ball joint from the housing.
12. Remove the inner joint from the transmission spindle.

➡Do not pull the inner joint when removing the front halfshaft.

13. Remove the front halfshaft from the hub. Use tools ST92647000 and ST28099PA100 if necessary.

➡Be careful not to damage the oil seal lip and tone wheel when removing the halfshaft. If the front halfshaft is removed, replace the inner seal with a new one.

14. After scribing an alignment mark on the camber adjusting bolt head, remove the bolts which connect the housing and the strut.
15. Disconnect the housing from the strut.

To install:

16. While aligning the alignment mark on the camber adjusting bolt head, connect the housing and strut. Tighten the bolt to 130 ft. lbs. (177 Nm).

➡When the self locking nut is removed, replace it with a new one.

17. Install the halfshaft.
18. Continue the installation in the reverse order of the removal procedure.
19. Be sure to use a new axle nut. While depressing the brake pedal, tighten the axle nut to 162 ft. lbs. (220 Nm). Lock it securely in place.

➡Install the tire and wheel after installation of the axle nut. Failure to do this may result in wheel bearing damage. Do not over tighten, as this too could cause wheel bearing damage.

20. Continue the installation in the reverse order of the removal procedure.

Impreza, Legacy & Outback Models

See Figure 313.

➡It may be necessary to remove the front halfshaft from the vehicle.

1. Disconnect the negative battery cable.
2. Raise and support the vehicle safely.
3. Remove the tire and wheel.
4. Remove the crimped section of the axle nut. Depress the brake pedal and remove the axle nut.

➡Be sure to loosen the axle nut after removing the tire and wheel from the vehicle. Failure to do this may damage the wheel bearings.

5. Remove the disc brake caliper from its mounting as suspend it to the side, with wire. Do not disconnect the brake line.
6. Remove the rotor.

➡If the rotor is seized within the hub, remove the rotor by installing an 8mm bolt in the screw hole on the rotor.

7. Remove the ABS wheel speed sensor assembly and harness.
8. Remove the four bolts from the housing.
9. Remove the front hub unit bearing. Use tools ST92647000 and ST28099PA100 if necessary.

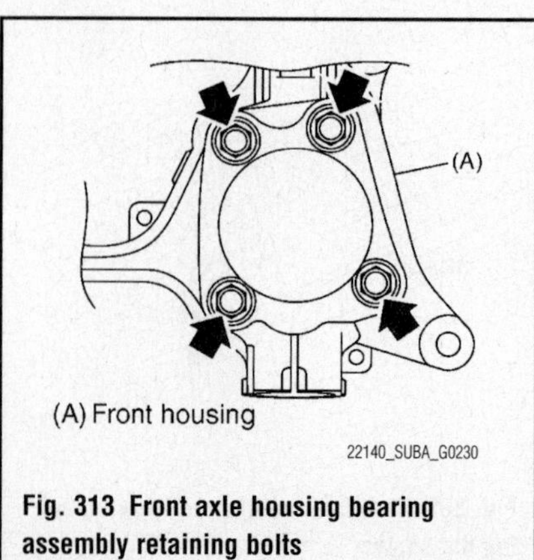

Fig. 313 Front axle housing bearing assembly retaining bolts

To install:

10. Tighten the front hub unit bearing to housing bolts to 47.9 ft. lbs. (65 Nm).

11. Be sure to use a new axle nut. Tighten the axle nut, temporarily.

12. Install the rotor. Install the caliper.

13. While depressing the brake pedal, tighten the axle nut (olive color) to 162 ft. lbs. (220 Nm). Lock it securely in place.

➡Install the tire and wheel after installation of the axle nut. Failure to do this may result in wheel bearing damage. Do not over tighten, as this too could cause wheel bearing damage.

14. Continue the installation in the reverse order of the removal procedure.

ADJUSTMENT

The wheel bearings are of a sealed unit and are not adjustable.

REAR SUSPENSION

COIL SPRING

REMOVAL & INSTALLATION

Baja, Legacy & Outback models

See Figure 314.

1. Before servicing the vehicle, refer to the Precautions Section.
2. Disconnect the negative battery cable.
3. Raise and support the vehicle safely.
4. Remove rear shock assembly, Refer to rear shock.
5. Using a coil spring compressor, compress the coil spring.
6. Using a strut mount socket remove the self-locking nut.
7. Remove the strut mount, spacer and upper spring seat from strut.
8. Gradually decrease the compression force of compressor, and remove the coil spring.
9. Remove the dust cover and helper spring.

To install:

10. Using a coil spring compressor, compress the coil spring.
11. Set the coil spring correctly so that its end face seats well in the spring seat as shown in the figure.
12. Install the helper and dust cover to the piston rod.
13. Pull the piston rod fully upward, and install the spring seat.
14. Install spacer and the strut mount to piston rod, and tighten a new self—locking nut temporarily.
15. Using a hexagon wrench to prevent strut rod from turning, tighten the new self-locking nut to 26 ft. lbs. (35 Nm), with a strut mount socket.
16. Loosen the coil spring compressor carefully.
17. Install rear shock assembly, Refer to rear shock.
18. Connect the negative battery cable.

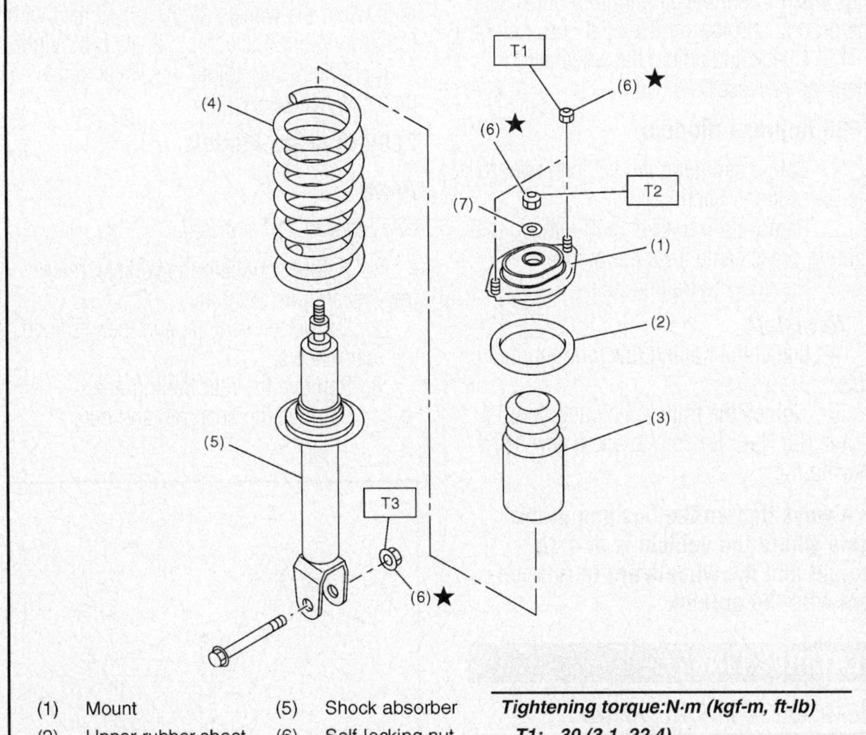

Fig. 314 Exploded view of rear shock assembly

2006–07 Impreza Models

See Figure 299.

1. Before servicing the vehicle, refer to the Precautions Section.
2. Disconnect the negative battery cable.
3. Raise and support the vehicle safely.
4. Remove rear strut, Refer to rear MacPherson Strut.
5. Using a coil spring compressor, compress the coil spring.
6. Using a strut mount socket remove the self-locking nut.
7. Remove the strut mount, spacer and upper spring seat from strut.
8. Gradually decrease the compression force of compressor, and remove the coil spring.
9. Remove the dust cover and helper spring.

To install:

10. Using a coil spring compressor, compress the coil spring.
11. Set the coil spring correctly so that its end face seats well in the spring seat as shown in the figure.
12. Install the helper and dust cover to the piston rod.
13. Pull the piston rod fully upward, and install the spring seat.
14. Install spacer and the strut mount to piston rod, and tighten a new self-locking nut temporarily.
15. Using a hexagon wrench to prevent strut rod from turning, tighten the new self-locking nut to 41 ft. lbs. (55 Nm), with a strut mount socket.

16. Loosen the coil spring compressor carefully.
17. Install rear strut. "Refer to rear MacPherson Strut".
18. Connect the negative battery cable.

FRONT LINK

REMOVAL & INSTALLATION

Baja, Legacy & Outback Models

See Figures 315 and 316.

1. Before servicing the vehicle, refer to the Precautions Section.
2. Loosen wheel nuts. Lift-up vehicle and remove wheel.
3. Use transmission jack to support rear arm horizontally.
4. Remove bolt securing link front to sub frame.
5. Remove bolts which secure link front to rear arm and detach link front.

➡Link front bushing cannot be replaced alone. Always replace link front and bushing as a single unit.

To install:

6. Installation is the reverse of the removal procedure.

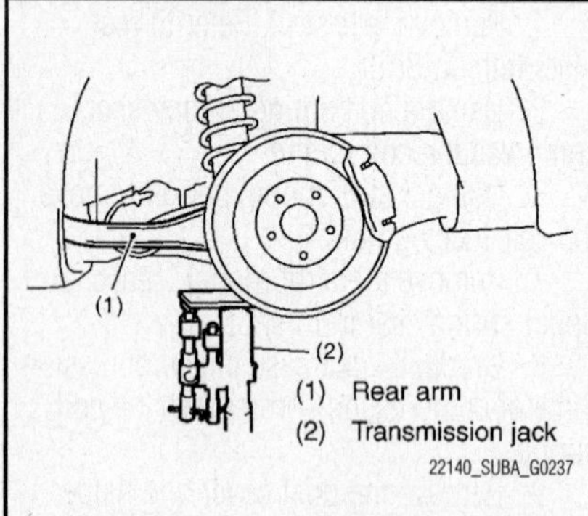

Fig. 315 Use transmission jack to support rear arm

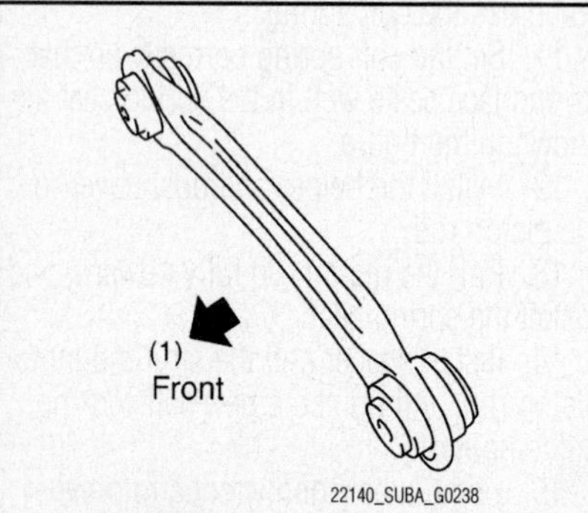

Fig. 316 Install front link with protruding side facing the front side of the vehicle

7. Tighten bolts which secure link to 89 ft. lbs. (120 Nm).
8. Check and adjust the wheel alignment, as necessary.

TRAILING LINK

REMOVAL & INSTALLATION

2006–07 Impreza Models

1. Before servicing the vehicle, refer to the Precautions Section.
2. Loosen the wheel nuts.
3. Raise and support the vehicle safely.
4. Remove the tire and wheel.
5. Remove both the rear parking brake clamp and the ABS wheel speed sensor harness.
6. Remove the bolt which secures the trailing link to the trailing link bracket.
7. Remove the bolt which secures the trailing link to the rear housing.
8. Remove the trailing link from the vehicle.

To install:

9. Installation is the reverse of the removal procedure.
10. Be sure to use new bolts and nuts, as required.
11. Always fully tighten the rubber bushings when the wheels are in full contact with the ground and the vehicle is at curb height.
12. Check and adjust the wheel alignment, as necessary.

2008 Impreza Models

1. Before servicing the vehicle, refer to the Precautions Section.
2. Remove the bracket, and remove the parking brake cable from the guide.
3. Remove the trailing link.

To install:

4. Install the trailing link, and mounting nuts.
5. Tighten the trailing link nuts to 66.4 ft. lbs. (90 Nm). Be sure to use a new self-locking nut.

➡Always tighten the bushing in the state where the vehicle is at curb weight and the wheels are in full contact with the ground.

LATERAL LINK

REMOVAL & INSTALLATION

2006–07 Impreza Models

1. Before servicing the vehicle, refer to the Precautions Section.
2. Loosen the wheel nuts.
3. Raise and support the vehicle safely.
4. Remove the tire and wheel.
5. Remove the stabilizers.
6. Remove the ABS wheel speed sensor harness from the trailing arm.
7. Remove the bolt securing the trailing link to the rear housing.
8. Remove the bolts which secure the lateral link assembly to the rear housing.
9. Remove the DOJ from the rear differential, using tool ST28099PA100 or equivalent.

➡The side spline snapring comes out together with the shaft. Be careful not to damage the side bearing retainer when using the special tool.

10. Scribe an alignment mark on the rear lateral link adjusting bolt and crossmember.
11. Remove the bolts securing the front and rear lateral links to the crossmember, detach the lateral links.

➡To loosen the adjusting bolt, always loosen the nut while holding the head of the adjusting bolt.

To install:

12. Installation is the reverse of the removal procedure.
13. Be sure to use new bolts and nuts, as required.
14. Always fully tighten the rubber bushings when the wheels are in full contact with the ground and the vehicle is at curb height.
15. Check and adjust the wheel alignment, as necessary.

2008 Impreza Models

Front

See Figures 317 and 318.

1. Before servicing the vehicle, refer to the Precautions Section.
2. Lift up the vehicle, and then remove the rear wheels.
3. Remove the rear trailing link.
4. Remove the snap pin and nut.

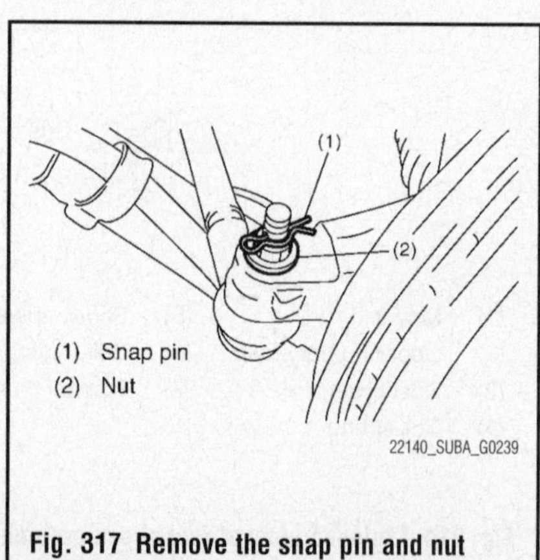

Fig. 317 Remove the snap pin and nut

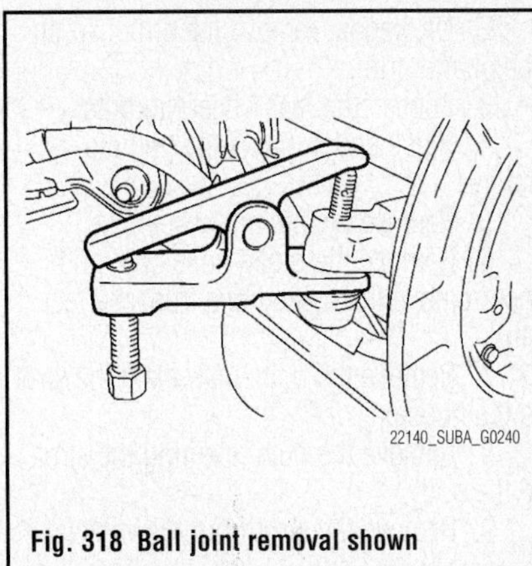
Fig. 318 Ball joint removal shown

5. Using a puller, remove the ball joint.
6. Scribe an alignment mark on the front lateral link adjusting bolt and rear sub frame.
7. Remove the adjusting bolt, and remove the front lateral link.

To install:

8. Install the front lateral link, and adjusting bolt. (Match alignment marks)
9. Install the remove the ball joint.
10. Install the rear trailing link.
11. Observe the following tightening specifications:
   - Front lateral link to sub-frame 74 ft. lbs. (100 Nm).
   - Front lateral link to rear axle housing 44 ft. lbs. (60 Nm).
12. Inspect the wheel alignment and adjust if necessary.

Rear

See Figure 319.

1. Before servicing the vehicle, refer to the Precautions Section.
2. Lift up the vehicle, and then remove the rear wheels.
3. Remove the nut and disconnect the stabilizer link.
4. Remove the shock absorber lower bolt.

Fig. 319 Rear lateral link, and bolt locations

5. Remove the bolt, and remove rear lateral link.

To install:

6. Install rear lateral link, and bolt.
7. Install shock absorber lower bolt.
8. Connect the stabilizer link.
9. Observe the following tightening specifications:
   - Rear lateral link 89 ft. lbs. (120 Nm).
   - Shock absorber lower bolt 89 ft. lbs. (120 Nm).
   - Stabilizer link 33.2 ft. lbs. (45 Nm).
10. Inspect the wheel alignment and adjust if necessary.

REAR LINK

REMOVAL & INSTALLATION

Baja, Legacy & Outback Models

1. Before servicing the vehicle, refer to the Precautions Section.
2. Loosen wheel nuts. Lift-up vehicle and remove wheel.
3. Remove bolt securing stabilizer clamps to sub frame.
4. Remove support sub frame RH. (When removing RH side link rear.)
5. Remove stabilizer link.
6. Use transmission jack to support rear arm horizontally.
7. Remove bolt securing link rear to rear arm.
8. Scribe an alignment mark on link rear adjusting bolt and sub frame.
9. Remove bolts securing link rear to sub frame, detach link rear.

**** CAUTION**

To loosen adjusting bolt, always loosen nut while holding the head of adjusting bolt.

10. Installation is the reverse of the removal procedure.
11. Tighten bolts securing link rear to sub frame 129 ft. lbs. (175 Nm).
12. Check and adjust the wheel alignment, as necessary.

REAR ARM

REMOVAL & INSTALLATION

Baja, Legacy & Outback Models

See Figure 320.

1. Loosen the wheel nuts.
2. Raise and support the vehicle safely.
3. Remove the tire and wheel.
4. Remove the sub frame support arm.
5. Remove the bearing unit.

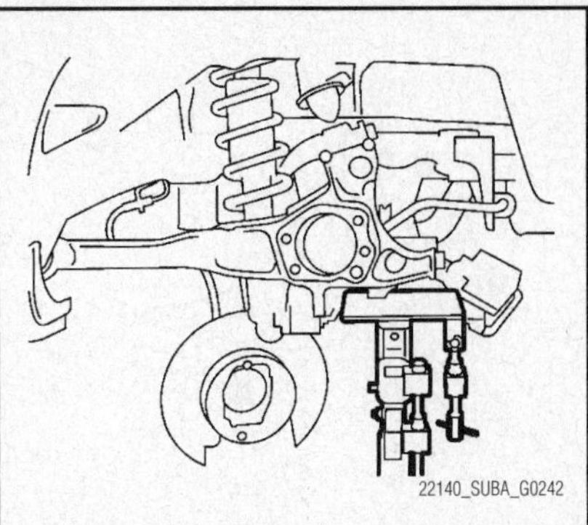
Fig. 320 Supporting the rear arm assembly using a transmission jack

6. Hang the backing plate from the sub frame.
7. Remove the bolt that secures the parking brake cable clamp to the rear arm bracket.
8. Remove the bolt which holds the brake hose bracket and the ABS wheel speed sensor bracket to the rear arm.
9. Remove the bolts which secure the brake hose bracket to the rear arm. Remove the bolts which secure the ABS wheel speed sensor to the rear arm.
10. Remove the stabilizer link from the rear arm. Remove the shock absorber from the rear arm.
11. Support the rear arm assembly, horizontally using a transmission jack.
12. Remove the nuts that retain the rear arm to the bracket. Remove the rear arm bracket.
13. Loosen the nut which holds the front link to the rear arm.
14. Loosen the nut which holds the rear link to the rear arm.
15. Loosen the nut which holds the upper link to the rear arm.
16. Remove the rear arm from the vehicle.

To install:

17. Installation is the reverse of the removal procedure.
18. Be sure to use new bolts and nuts, as required.
19. Always fully tighten the rubber bushings when the wheels are in full contact with the ground and the vehicle is at curb height.
20. Check and adjust the wheel alignment, as necessary.

UPPER ARM

REMOVAL & INSTALLATION

2008 Impreza Models

See Figure 321.

1. Before servicing the vehicle, refer to the Precautions Section.

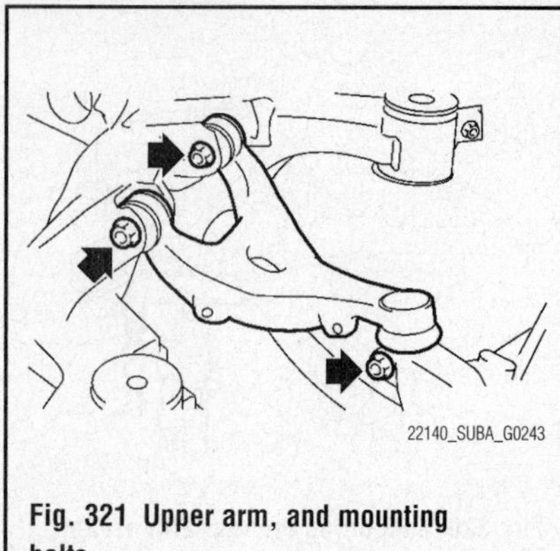
Fig. 321 Upper arm, and mounting bolts

2. Separate the front exhaust pipe and rear exhaust pipe
3. Remove the rear exhaust pipe and muffler.
4. Remove the propeller shaft.
5. Remove the brake hose bracket.
6. Remove the rear disc brake caliper and suspend it from the shock absorber. (disc brake model)
7. Remove the rear parking brake cable from the parking brake assembly
8. Disconnect the ABS wheel speed sensor connector.
9. Disconnect the drain hose from the canister drain connector.
10. Remove the shock absorber lower bolt.
11. Support the sub frame using a jack.
12. Remove the support plate.
13. Remove the rear sub frame.
14. Remove the rear ABS wheel speed sensor bracket.
15. Remove the bolts, then remove the upper arm.

To install:

16. Install in the reverse order of removal.
17. Tighten the upper arm to rear sub frame to 111 ft. lbs. (150 Nm).
18. Tighten the upper arm to rear housing to 66.4 ft. lbs. (90 Nm).
19. Tighten the rear ABS wheel speed sensor bracket to 5.5 ft. lbs. (7.5 Nm).
20. Inspect the wheel alignment and adjust if necessary.

➡Always tighten the bushing when the arm is positioned in the state where the vehicle is at curb weight and the wheels are in full contact with the ground.

UPPER LINK

REMOVAL & INSTALLATION

Baja, Legacy & Outback Models

See Figure 322.

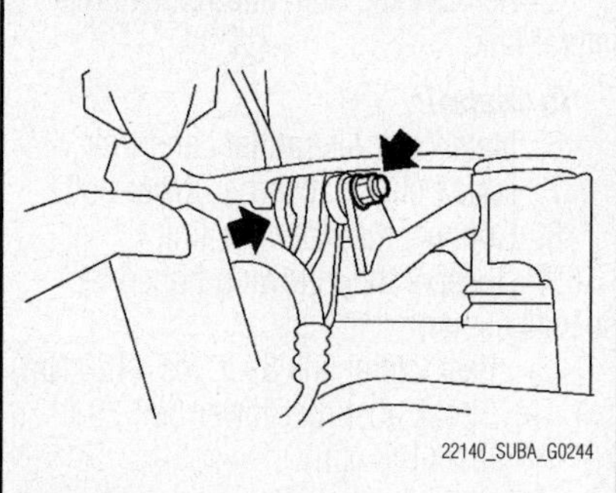
Fig. 322 Bolt securing upper link to sub frame

1. Before servicing the vehicle, refer to the Precautions Section.
2. Loosen wheel nuts. Lift-up vehicle and remove wheel.
3. Use transmission jack to support rear arm horizontally.
4. Remove bolt securing upper link to sub frame.
5. Remove bolts which secure upper link to rear arm and detach link upper.

To install:

6. Installation is the reverse of the removal procedure.

➡Using transmission jack, support rear arm horizontally, install upper link and tighten.

7. Tighten upper link mounting bolts to 88 ft. lbs. (120 Nm).
8. Check and adjust the wheel alignment, as necessary.

MACPHERSON STRUTS

REMOVAL & INSTALLATION

2006–07 Impreza Models

See Figure 323.

1. On sedan, remove the rear seat cushion and backrest.
2. On wagon, remove the strut cap on the quarter trim.
3. Loosen the rear wheel lug nuts.
4. Raise and support the vehicle safely.
5. Remove the tire and wheel.
6. Remove the brake hose clip, and then remove the brake hose from the rear strut.
7. Remove the bolts that retain the strut to the housing.
8. Remove the nuts retaining the strut to the body.
9. Remove the strut from the vehicle.

To install:

10. Installation is the reverse of the removal procedure.
11. Be sure to use new locknuts, as required.
12. Do not subject the ABS wheel speed sensor to excessive tension.
13. Check and adjust the wheel alignment, as necessary.

TESTING

See Figure 324.

1. Check for oil leakage.
2. Move the piston rod up and down to check it operates smoothly without any binding.
3. Measure the play as follows:
   a. Fix outer shell in a vice and fully extend the rod.
   b. Set a dial gauge at 0.39 in (10 mm) from the end of the rod, then apply a force of W 4 lb (20 N) to threaded portion.
   c. With the force of 4 lb (20 N) applied, read dial gauge indication..
   d. Apply a force of 4 lb (20 N) in the opposite direction of "W", then read dial gauge indication.
   e. If the difference of the two is more than 0.031 in. (0.8 mm), replace the strut.

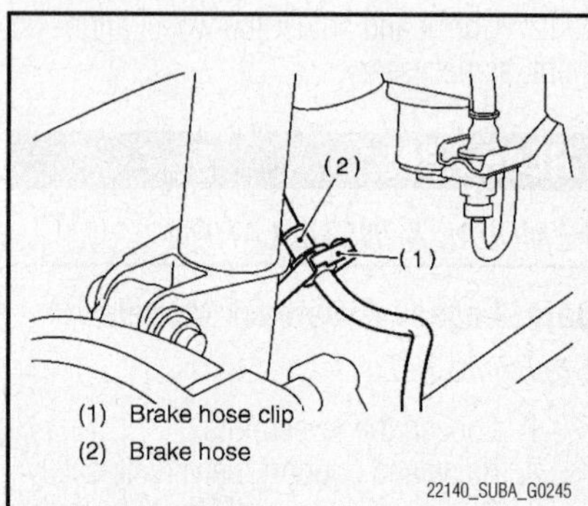

Fig. 323 Brake hose clip, and the brake hose shown

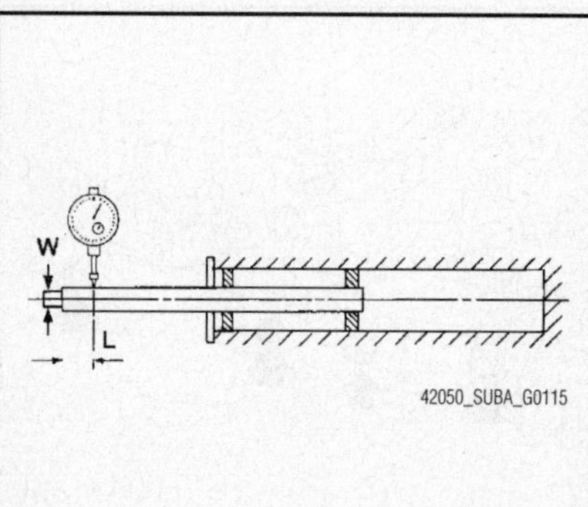

Fig. 324 Use a dial gauge to check the free play of the strut

OVERHAUL

1. Remove the strut from the vehicle.
2. Using a coil spring compressor tool, carefully compress the spring. Remove the self locking nut.
3. Remove the strut mount, upper spring and rubber seat from the strut.
4. Gradually decrease the compression force of the spring compressor tool. Remove the coil spring.
5. Remove the dust cover and helper spring.
6. Check for the presence of air in the damping force generating mechanism.
7. Using the spring compression tool, compress the coil spring.

➡Be sure to properly install the coil spring.

8. Position the coil spring so that its end face fits good into the spring seat.
9. Install the helper spring and dust cover to the piston rod.
10. Pull the piston rod fully upward, and install the rubber seat and spring seat.
11. Install the strut mount to the piston rod, and then tighten the self locking nut, temporarily. Be sure to use a new self locking nut.
12. Use a hexagon wrench to prevent the strut rod from turning. Tighten the self locking nut to 41 ft. lbs.
13. Carefully loosen the coil spring.

STABILIZER BAR

REMOVAL & INSTALLATION

See Figure 325.

1. Loosen the rear wheel lug nuts.
2. Raise and support the vehicle safely.
3. Remove the tire and wheel.
4. Remove the bolts that secure the stabilizer link to the rear arm.
5. Remove the bolts which secure the stabilizer bar to the sub frame.
6. Remove the stabilizer bar from the vehicle.

To install:

7. Installation is the reverse of the removal procedure.
8. Be sure that the stabilizer bar and the bushings have the same identification markings and/or colors.
9. Be sure to use new bolts and nuts, as required.
10. Always fully tighten the rubber bushings when the wheels are in full contact with the ground and the vehicle is at curb height.
11. Check and adjust the wheel alignment, as necessary.

SHOCK ABSORBER

REMOVAL & INSTALLATION

Baja, Legacy & Outback Models

See Figure 326.

1. Before servicing the vehicle, refer to the Precautions Section.
2. Remove the luggage floor mat. (Wagon model)
3. Roll up the trunk side trim. (Sedan model)
4. Lift up the vehicle, and then remove the rear wheels.
5. Remove the bolts which secure the shock absorber to rear arm.
6. Using a jack, support the shock absorber.
7. Remove the bolts on the bottom of the shock absorber.
8. Remove the nuts which secure the shock absorber mount to vehicle.
9. Remove the shock absorber from the vehicle.

To install:

10. Installation is the reverse of the removal procedure.
11. Be sure to use new bolts and nuts, as required.
12. Tighten upper mounting nuts to 22 ft. lbs. (30 Nm).
13. For 2006—models tighten lower mounting bolt to 118 ft. lbs. (160 Nm).
14. For 2007–08 models, tighten lower mounting bolt to 46 ft. lbs. (62 Nm).
15. Always fully tighten the rubber bushings when the wheels are in full contact with the ground and the vehicle is at curb height.
16. Check and adjust the wheel alignment, as necessary.

WHEEL HUB AND BEARING (SEALED UNIT)

REMOVAL & INSTALLATION

Baja, Legacy & Outback Models

See Figures 327 and 328.

1. Before servicing the vehicle, refer to the Precautions Section.
2. Disconnect the ground cable from the battery.
3. Lift-up the vehicle, and remove the front wheels.
4. Lift the crimped section of axle nut.
5. Remove the axle nut using a socket wrench while depressing the brake pedal.

⁂ CAUTION

Remove the wheel before loosening the axle nut. Failure to follow this rule may damage the wheel bearings.

6. Remove the disc brake caliper from the front housing, and suspend it from strut using a piece of wire.
7. Remove the disc rotor from the hub.

⁂ CAUTION

If it is difficult to remove the disc rotor from the hub, drive the 8 mm of bolt into the threaded end of rotor, then remove the rotor.

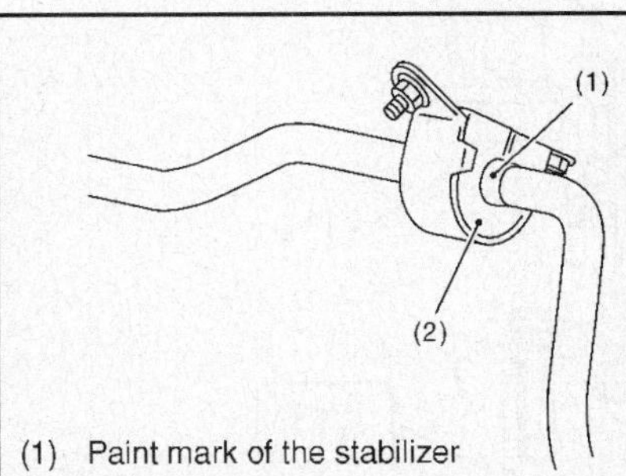

Fig. 325 Stabilizer bar and bushing identification markings

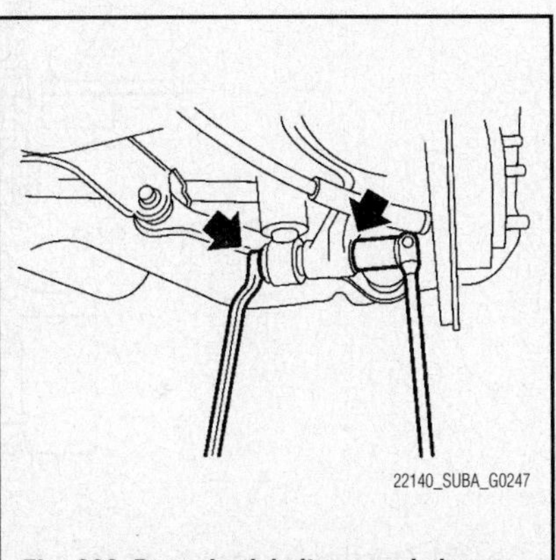

Fig. 326 Rear shock bolt removal shown

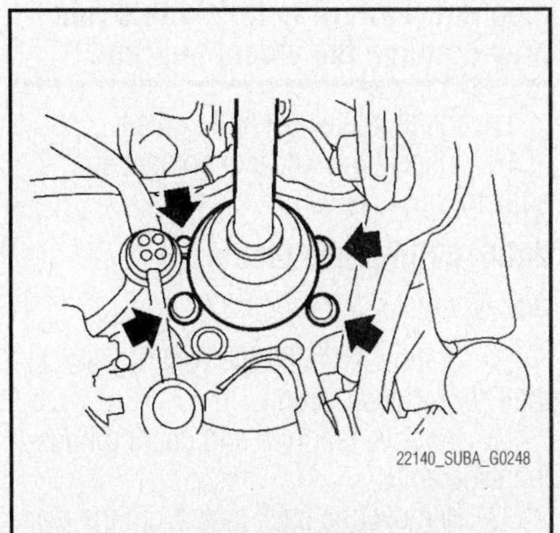

Fig. 327 Four bolts shown at rear housing

8. Remove four bolts from the housing.

CAUTION

Do not get close to the tool which charged magnetism to magnetic encoder.

CAUTION

Be careful not to damage the magnetic encoder.

9. Remove the rear hub unit bearing. If it is hard to remove, use a puller.

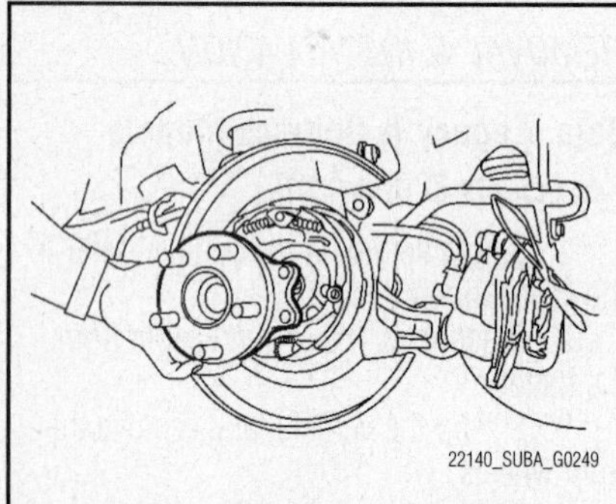

22140_SUBA_G0249

Fig. 328 Rear hub and bearing assembly removal

To install:

10. Place the disc cover between front housing and front hub unit, and tighten the four bolts. Tighten to 47.9 ft. lbs. (65 Nm).
11. Install the front driveshaft.
12. Tighten the axle nut temporarily.
13. Install the four bolts to the housing and tighten to 48 ft. lbs. (65 Nm).
14. Install the disc rotor to hub.
15. Install the disc brake caliper to the rear housing. 48 ft. lbs. (65 Nm).
16. While depressing the brake pedal, tighten a new axle nut to 177 ft. lbs. (240 Nm) and lock it securely.

CAUTION

Install the wheel after installation of axle nut. Failure to follow this rule may damage the wheel bearing.

17. Install the wheel and tighten.
18. Inspect the wheel alignment and adjust if necessary.

2006–07 Impreza Models

See Figures 329 through 331.

1. Before servicing the vehicle, refer to the Precautions Section.
2. Remove rear hub and bearing housing assembly.
3. Remove the back plate from the rear housing

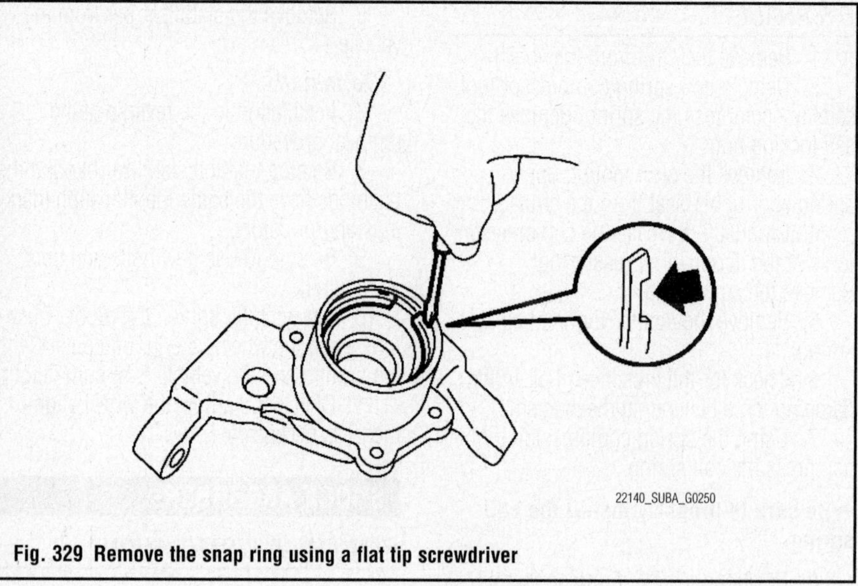

22140_SUBA_G0250

Fig. 329 Remove the snap ring using a flat tip screwdriver

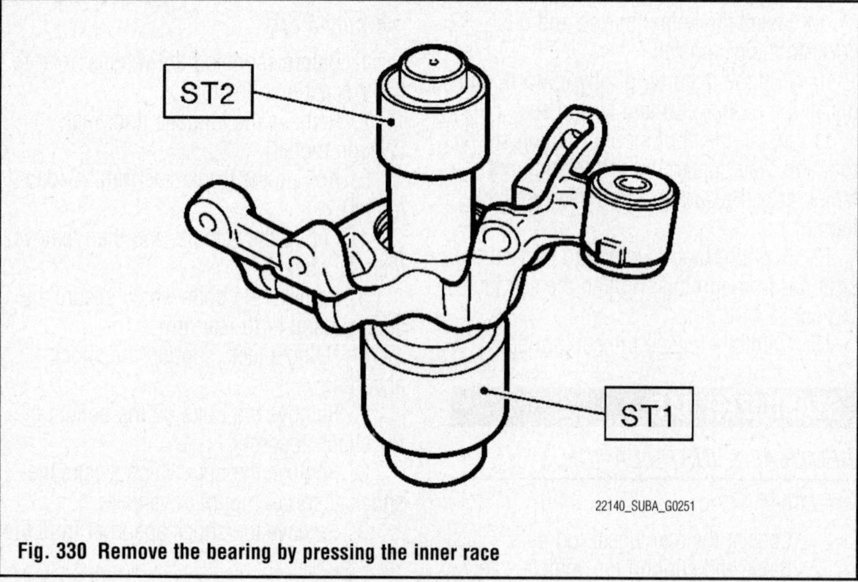

22140_SUBA_G0251

Fig. 330 Remove the bearing by pressing the inner race

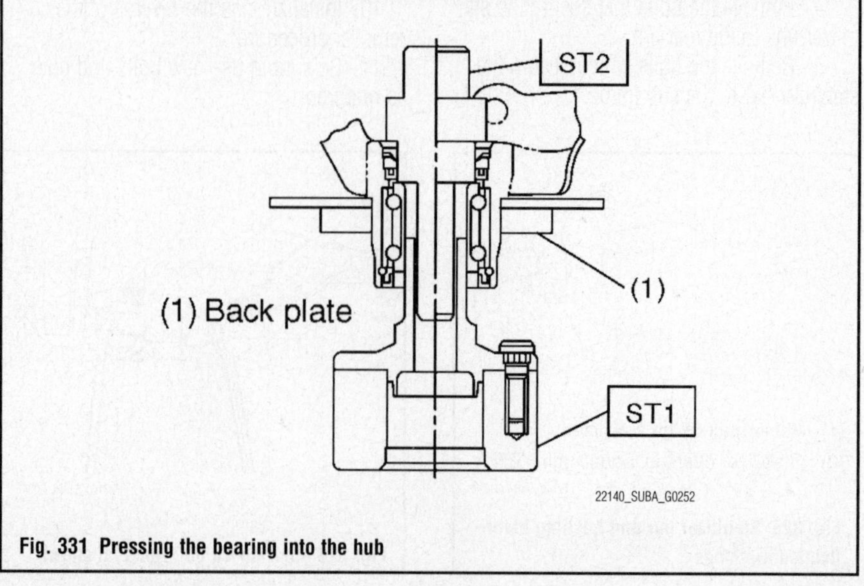

22140_SUBA_G0252

Fig. 331 Pressing the bearing into the hub

4. Remove the outer and inner oil seals using a flat tip screwdriver.
5. Remove the snap ring using a flat tip screwdriver.

➡**Be careful not to damage housing during removal.**

6. Using ST1 and ST2, remove the bearing by pressing the inner race.
7. Remove the tone wheel bolts and remove the tone wheel from the hub.
8. Using ST, press the hub bolt out.

⁂ CAUTION

Be careful not to hammer the hub bolts. Doing so may deform the hub.

To install:

➡**When the hub is removed from the housing, replace the bearing set and oil seal.**

9. Remove the foreign particles (dust, rust, etc.) from the mating surfaces of hub tone wheel, and install the tone wheel to hub.
10. Clean the housing interior completely. Using ST1 and ST2, press the bearing into the housing.
11. Be careful not to remove the plastic lock from the inner race when installing the bearings.
12. Using snap ring pliers, securely install the snap ring.
13. Using the ST1 and ST2, press the outer oil seal unit until it comes in contact with snap ring.
14. Invert both ST1 and housing (upside down).
15. Using ST2, press the inner oil seal into the housing until it touches the bottom.
16. Using ST1 and ST2, press the sub seal into place.
17. Apply sufficient grease to the oil seal lip.
18. Install back plate to the rear housing.
19. Using ST1 and ST2, press the bearing into the hub.
20. Install the rear hub and bearing housing assembly.

2008 Impreza Models

See Figures 332 and 333.

1. Before servicing the vehicle, refer to the Precautions Section.
2. Disconnect the ground cable from the battery.
3. Lift up the vehicle, and then remove the rear wheels.
4. Lift the crimped section of axle nut.
5. Remove the axle nut using a socket wrench while depressing the brake pedal.

⁂ CAUTION

Remove the wheel before loosening the axle nut. Failure to follow this rule may damage the wheel bearings.

6. Remove the disc brake caliper from the rear housing, and suspend it from the vehicle using a string.
7. Remove the rear disc rotor.
8. Remove the four bolts from the rear housing.
9. Remove the rear hub unit bearing.

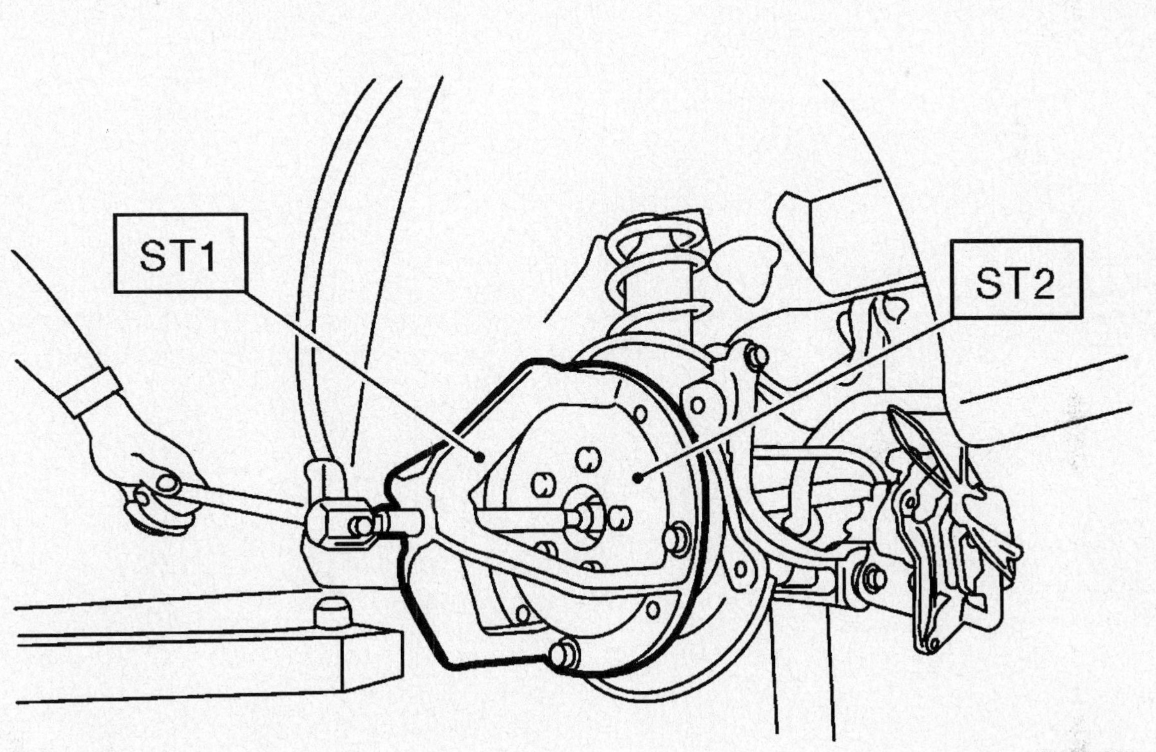

Fig. 332 Rear hub removal with the help of a puller

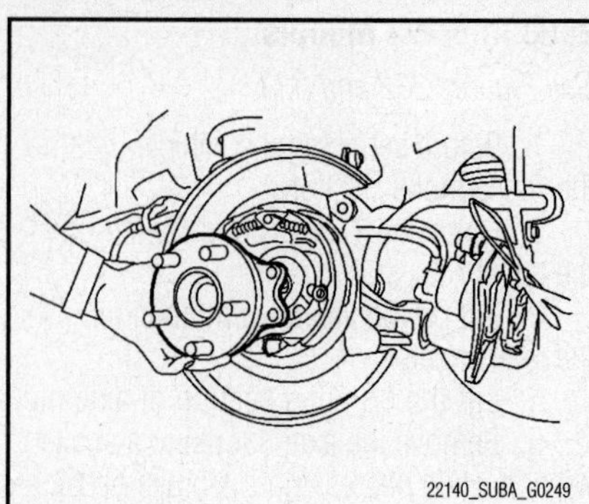

Fig. 333 Rear hub and bearing assembly removal

➡If it is hard to remove, use a puller.

To install:

10. Aligning with the mounting hole of the rear brake back plate, temporarily tighten the rear hub unit bearing to the rear housing.
11. Tighten the four bolts of the rear housing to 48 ft. lbs. (65 Nm).
12. Tighten the new axle nut temporarily.
13. Install the rear disc rotor.
14. Install the disc brake caliper on the rear housing. Tighten to 48 ft. lbs. (65 Nm).
15. While depressing the brake pedal, tighten a new axle nut to 140 ft. lbs. (190 Nm) and lock it securely.

✲✲ CAUTION

Do not install wheel and let it touch the ground before tightening the axle nut. Failure to follow this rule may damage the axle bearing. Do not over tighten the nuts as this may damage the axle bearing.

ADJUSTMENT

The wheel bearings are not adjustable.

SUBARU

Forester

15

BRAKES 15-11

CHASSIS ELECTRICAL 15-20

DRIVETRAIN 15-22

ENGINE COOLING...........15-31

ENGINE ELECTRICAL.......15-36

ENGINE MECHANICAL......15-41

SPECIFICATIONS AND MAINTENANCE CHARTS

ENGINE AND VEHICLE IDENTIFICATION

Engine Code							Model Year	
Code ①	Liters	Cu. In.	Cyl.	Fuel Sys.	Type	Eng. Mfg.	Code ②	Year
6	2.5L	150	4	MFI	DOHC	Subaru	6	2006
6 ③	2.5L	150	4	MFI	DOHC	Subaru	7	2007
6	2.5L	150	4	MFI	SOHC	Subaru	8	2008

MFI: Multiport Fuel Injection

DOHC: Double Overhead Camshafts

SOHC: Single Overhead Camshaft

① 6th digit of the VIN

② 10th digit of the VIN

③ Turbocharged

22140_FORE_C0001

GENERAL ENGINE SPECIFICATIONS

Year	Model	Engine Displacement Liters (VIN)	Net Horsepower @ rpm	Net Torque @ rpm (ft. lbs.)	Bore x Stroke (in.)	Com-pression Ratio	Oil Pressure psi @ rpm
2006	Forester	2.5 (6) ①	173@5600	166@4400	3.92x3.11	10.0:1	14@600
		2.5 (6) ②	230@5600	235@3600	3.92x3.11	8.2:1	14@600
2007	Forester	2.5 (6) ①	173@5600	166@4400	3.92x3.11	10.0:1	14@600
		2.5 (6) ②	230@5600	235@3600	3.92x3.11	8.2:1	14@600
2008	Forester	2.5 (6) ①	173@5600	166@4400	3.92x3.11	10.0:1	14@600
		2.5 (6) ②	230@5600	235@3600	3.92x3.11	8.2:1	14@600

① SOHC

② DOHC

22140_FORE_C0002

ENGINE TUNE-UP SPECIFICATIONS

Year	Engine Displacement Liters (VIN)	Spark Plugs Gap (in.)	Ignition Timing (deg.) ① ② MT	Ignition Timing (deg.) ① ② AT	Fuel Pump (psi)	Idle Speed (rpm) MT	Idle Speed (rpm) AT	Valve Clearance ③ In.	Valve Clearance ③ Ex.
2005	2.5 (6) ④	0.039-0.043	⑤	⑥	⑦	550-750	600-800	0.0063-0.0095	0.0082-0.0114
	2.5 (6) ⑧	0.028-0.031	⑨	⑨	⑩	600-800	600-800	0.0055-0.0095	0.0118-0.0158
2007	2.5 (6) ④	0.039-0.043	⑤	⑥	⑦	550-750	600-800	0.0063-0.0095	0.0082-0.0114
	2.5 (6) ⑧	0.028-0.031	⑨	⑨	⑩	600-800	600-800	0.0055-0.0095	0.0118-0.0158
2008	2.5 (6) ④	0.039-0.043	⑤	⑥	⑦	550-750	600-800	0.0063-0.0095	0.0082-0.0114
	2.5 (6) ⑧	0.028-0.031	⑨	⑨	⑩	600-800	600-800	0.0055-0.0095	0.0118-0.0158

① At idle
② BTDC: before top dead center
③ With engine cold
④ SOHC
⑤ 10 degrees +/-8 at 650rpm
⑥ 15 degrees +/-8 at 700rpm
⑦ 41-46 while disconnecting pressure regulator vacuum hose from intake manifold
33-38 after connecting pressure regulator vacuum hose
⑧ DOHC
⑨ 17 degrees +/-10 at 700rpm
⑩ 48-53 while disconnecting pressure regulator vacuum hose from intake manifold
40-45 after connecting pressure regulator vacuum hose

22140_FORE_C0005

CAPACITIES

Year	Model	Engine Displacement Liters (VIN)	Engine Oil with Filter (qts.)	Transmission (pts.) 4-Spd	Transmission (pts.) 5-Spd	Transmission (pts.) Auto.	Transfer Case (pts.)	Drive Axle Front (pts.)	Drive Axle Rear (pts.)	Fuel Tank (gal.)	Cooling System (qts.)
2006	Forester	2.5 (6)	4.2	—	7.4	19.6	—	2.6 ①	1.6	15.9	②
2007	Forester	2.5 (6)	4.2	—	7.4	19.6	—	2.6 ①	1.6	15.9	③
2008	Forester	2.5 (6)	4.2	—	7.4	19.6	—	2.6 ①	1.6	15.9	③

① Automatic transmission differential
② Manual transmission: 7.3
Automatic transmission: 7.2
③ SOHC: manual transmission: 7.3. Automatic transmission: 7.2
DOHC: manual transmission: 7.8. Automatic transmission: 7.7

22140_FORE_C0004

FLUID SPECIFICATIONS

Year	Model	Engine Displacement Liters	Engine ID/VIN	Engine Oil	Auto. Trans.	Drive Axle	Power Steering Fluid	Brake Master Cylinder	Engine Coolant
2006	Forester	2.5L ①	6	5W-30	Subaru ATF HP	75W-90	Dexron III	DOT 3	Subaru Coolant
		2.5L	6	5W-30	Subaru ATF HP	75W-90	Dexron III	DOT 3	Subaru Coolant
2007	Forester	2.5L ①	6	5W-30	Subaru ATF HP	75W-90	Dexron III	DOT 3	Subaru Coolant
		2.5L	6	5W-30	Subaru ATF HP	75W-90	Dexron III	DOT 3	Subaru Coolant
2008	Forester	2.5L ①	6	5W-30	Subaru ATF HP	75W-90	Dexron III	DOT 3	Subaru Coolant
		2.5L	6	5W-30	Subaru ATF HP	75W-90	Dexron III	DOT 3	Subaru Coolant

DOT: Department Of Transpotation

① DOHC

22140_FORE_C0003

VALVE SPECIFICATIONS

All measurements are given in inches.

Year	Engine Displacement Liters (VIN)	Seat Angle (deg.)	Face Angle (deg.)	Spring Test Pressure (lbs. @ in.)	Spring Installed Height (in.)	Stem-to-Guide Clearance (in.)		Stem Diameter (in.)	
						Intake	Exhaust	Intake	Exhaust
2006	2.5 (6) ①	90	NA	102 - 118@ 1.315	②	0.0014-0.0024	0.0016-0.0026	0.2343-0.2348	0.2341-0.2346
	2.5 (6) ③	90	NA	102 - 118@ 1.315	⑤	0.0012-0.0022	0.0016-0.0026	0.2344-0.2350	0.2341-0.2346
2007	2.5 (6) ①	90	NA	102 - 118@ 1.315	②	0.0014-0.0024	0.0016-0.0026	0.2343-0.2348	0.2341-0.2346
	2.5 (6) ③	90	NA	102 - 118@ 1.315	⑤	0.0012-0.0022	0.0016-0.0026	0.2344-0.2350	0.2341-0.2346
2008	2.5 (6) ①	90	NA	102 - 118@ 1.315	②	0.0014-0.0024	0.0016-0.0026	0.2343-0.2348	0.2341-0.2346
	2.5 (6) ③	90	NA	102 - 118@ 1.315	⑤	0.0012-0.0022	0.0016-0.0026	0.2344-0.2350	0.2341-0.2346

NA: Not Available

① SOHC

② Free length: 2.173 in.

③ DOHC

④ Free length: 1.863 in.
Lift: 95.8-110.0@1.043

22140_FORE_C0006

CAMSHAFT SPECIFICATIONS

All measurements are given in inches.

Year	Engine Displacement Liters (VIN)	Journal Diameter	Brg. Oil Clearance	Shaft End-play ①	Runout	Lobe Height	
						Intake	Exhaust
2006	2.5 (6) ②	1.2570-1.2577	0.0022-0.0035	0.0012-0.0035	NA	1.5545-1.5585	③
	2.5 (6) ④	⑤	0.0015-0.0028	0.0027-0.0047	NA	1.8330-1.8370	1.8410-1.8440
2007	2.5 (6) ②	1.2570-1.2577	0.0022-0.0035	0.0012-0.0035	NA	1.5545-1.5585	③
	2.5 (6) ④	⑤	0.0015-0.0028	0.0027-0.0047	NA	1.8330-1.8370	1.8410-1.8440
2008	2.5 (6) ②	1.2570-1.2577	0.0022-0.0035	0.0012-0.0035	NA	1.5545-1.5585	③
	2.5 (6) ④	⑤	0.0015-0.0028	0.0027-0.0047	NA	1.8330-1.8370	1.8410-1.8440

NA: Not Available

① Side clearance

② SOHC

③ California: 1.5686-1.5726
Except California: 1.5638-1.5677

④ DOHC

⑤ Front: 1.4939-1.4946
Except front: 1.1790-1.1796

22140_FORE_C0007

CRANKSHAFT AND CONNECTING ROD SPECIFICATIONS

All measurements are given in inches.

Year	Engine Displacement Liters (VIN)	Crankshaft				Connecting Rod		
		Main Brg. Journal Dia.	Main Brg. Oil Clearance	Shaft End-play	Thrust on No.	Journal Diameter	Oil Clearance	Side Clearance
2006	2.5 (6) ①	2.3619-2.3625	0.0004-0.0012	0.0012-0.0045	NA	NA	0.0006-0.0017	0.0028-0.0130
	2.5 (6) ③	2.3619-2.3625	0.0004-0.0012	0.0012-0.0045	NA	NA	0.0007-0.0018	0.0028-0.0130
2007	2.5 (6) ①	2.3619-2.3625	0.0004-0.0012	0.0012-0.0045	NA	NA	0.0006-0.0017	0.0028-0.0130
	2.5 (6) ③	2.3619-2.3625	0.0004-0.0012	0.0012-0.0045	NA	NA	0.0007-0.0018	0.0028-0.0130
2008	2.5 (6) ①	2.3619-2.3625	0.0004-0.0012	0.0012-0.0045	NA	NA	0.0006-0.0017	0.0028-0.0130
	2.5 (6) ③	2.3619-2.3625	0.0004-0.0012	0.0012-0.0045	NA	NA	0.0007-0.0018	0.0028-0.0130

NA: Not Available

① SOHC

② DOHC

22140_FORE_C0008

PISTON AND RING SPECIFICATIONS

All measurements are given in inches.

Year	Engine Displacement Liters (VIN)	Piston Clearance	Ring Gap Top Compression	Ring Gap Bottom Compression	Ring Gap Oil Control	Ring Side Clearance Top Compression	Ring Side Clearance Bottom Compression	Ring Side Clearance Oil Control
2006	2.5 (6) ①	0.0004-0.0012	0.0079-0.0138	0.0144-0.0203	0.0079-0.0197	0.0016-0.0031	0.0012-0.0028	NA
	2.5 (6) ②	0.0004-0.0012	0.0079-0.0098	0.0150-0.0203	0.0079-0.0197	0.0016-0.0031	0.0012-0.0028	NA
2007	2.5 (6) ①	0.0004-0.0012	0.0079-0.0138	0.0144-0.0203	0.0079-0.0197	0.0016-0.0031	0.0012-0.0028	NA
	2.5 (6) ②	0.0004-0.0012	0.0079-0.0098	0.0150-0.0203	0.0079-0.0197	0.0016-0.0031	0.0012-0.0028	NA
2008	2.5 (6) ①	0.0004-0.0012	0.0079-0.0138	0.0144-0.0203	0.0079-0.0197	0.0016-0.0031	0.0012-0.0028	NA
	2.5 (6) ②	0.0004-0.0012	0.0079-0.0098	0.0150-0.0203	0.0079-0.0197	0.0016-0.0031	0.0012-0.0028	NA

NA: Not Available

① SOHC

② DOHC

22140_FORE_C0009

TORQUE SPECIFICATIONS

All readings in ft. lbs.

Year	Engine Displacement Liters (VIN)	Cylinder Head Bolts	Main ① Bearing Bolts	Rod Bearing Bolts	Crankshaft Damper Bolts	Flywheel Bolts	Manifold Intake	Manifold Exhaust	Spark Plugs	Oil Pan Drain Plug
2006	2.5 (6) ②	③	④	33.3	132.7	52.8	18	19-26	13-17	33
	2.5 (6) ⑤	⑥	④	38.4	132.7	⑦	18	⑧	13-17	33
2007	2.5 (6) ②	③	④	33.3	132.7	52.8	18	19-26	13-17	33
	2.5 (6) ⑤	⑥	④	38.4	132.7	⑦	18	⑧	13-17	33
2008	2.5 (6) ②	③	④	33.3	132.7	52.8	18	19-26	13-17	33
	2.5 (6) ⑤	⑥	④	38.4	132.7	⑦	18	⑧	13-17	33

① Engine block connecting bolts

② SOHC

③ Step 1: Tighten all bolts to 22 ft. lbs.
Step 2: Tighten all bolts to 51 ft. lbs.
Step 3: Loosen all bolts 180 degrees, and again 180 degrees
Step 4: Tighten all bolts to 36 ft. lbs.
Step 5: + 80-90 degrees
Step 6: + 40-45 degrees

④ Step 1: Left side 10mm bolts (A-D) 7.2 ft. lbs.
Step 2: Right side 10mm bolts (E-J) 7.2 ft. lbs.
Step 3: Left side 10mm bolts (A-D) 13 ft. lbs.
Step 4: Right side 10mm bolts (E-J) 13 ft. lbs.
Step 5: Left side bolts (A and C) +90 degrees, bolts (B and D) 29.5 ft. lbs.
Step 6: Right side 10mm bolts (E-J) + 90 degrees
Step 7: Left side cylinder block connecting bolts (A-G) 18.1 ft. lbs., bolt (H) 4.7 ft. lbs.

⑤ DOHC

⑥ Step 1: Tighten all bolts to 21.4 ft. lbs.
Step 2: Tighten all bolts to 51 ft. lbs.
Step 3: Loosen all bolts 180 degrees, and again 180 degrees
Step 4: Tighten all bolts to 36 ft. lbs.
Step 5: + 80-90 degrees
Step 6: + 40-45 degrees

⑦ Automatic: 53.1 ft. lbs.
Manual: 38.4 ft. lbs.

⑧ Exhaust pipe to manifold 26 ft. lbs.
Right upper cover 5.5 ft. lbs.
Front exhaust pipe 13.7 ft. lbs.
Right manifold to turbocharger joint pipe 26-28 ft. lbs.
Lower covers 13.7 ft. lbs.

22140_FORE_C0010

WHEEL ALIGNMENT

Year	Model		Caster Range (+/-Deg.)	Caster Preferred Setting (Deg.)	Camber Range (+/-Deg.)	Camber Preferred Setting (Deg.)	Toe-in (in.)
2006	Forester	F	—	①	0.50	-0.25	0+/-0.12
		R	—	—	0.75	②	③
2007	Forester	F	—	①	0.50	-0.25	0+/-0.12
		R	—	—	0.75	②	③
2008	Forester	F	—	①	0.50	-0.25	0+/-0.12
		R	—	—	0.75	②	③

① Reference: 3 degrees 03'

② SOHC: -0 degrees 50'. DOHC: -0 degrees 55'

③ 0.079 +/-0.12 inch. Toe angle (sum of both wheels) 0 degrees 10' +/-0 degrees 15'

22140_FORE_C0011

TIRE, WHEEL AND BALL JOINT SPECIFICATIONS

Year	Model	OEM Tires Standard	OEM Tires Optional	Tire Pressures (psi) Front	Tire Pressures (psi) Rear	Wheel Size	Ball Joint Inspection	Lug Nut
2006	Forester	P215/60R16 94H	—	①	①	16X6.5 JJ	0.012 in. ②	66
2007	Forester	P215/60R16 94H	—	①	①	16X6.5 JJ	0.012 in. ②	74
		P215/55R17 93H	—	①	①	17X7 JJ	0.012 in. ②	74
2008	Forester	P215/60R16 94H	—	①	①	16X6.5 JJ	0.012 in. ②	74
		P215/55R17 93H	—	①	①	17X7 JJ	0.012 in. ②	74

OEM: Original Equipment Manufacturer

PSI: Pounds Per Square Inch

STD: Standard

OPT: Optional

① See drivers door placard

② Apply 154 lbs. vertical force

22140_FORE_C0012

BRAKE SPECIFICATIONS

All measurements in inches unless noted

Year	Model		Brake Disc Original Thickness	Brake Disc Minimum Thickness	Brake Disc Maximum Runout	Brake Drum Diameter Original Inside Diameter	Brake Drum Diameter Max. Wear Limit	Brake Drum Diameter Maximum Machine Diameter	Minimum Lining Thickness Front	Minimum Lining Thickness Rear	Brake Caliper Bracket Bolts (ft. lbs.)	Brake Caliper Mounting Bolts (ft. lbs.)
2006	Forester	F	0.940	0.870	0.0030	—	—	—	0.059	0.059	59	19
		R	0.390	0.335	0.0028	9.00 ①	9.08 ②	NA	—	0.059	38	28
2007	Forester	F	0.940	0.870	0.0030	—	—	—	0.059	0.059	59	19
		R	0.390	0.335	0.0028	9.00 ①	9.08 ②	NA	—	0.059	38	28
2008	Forester	F	0.940	0.870	0.0030	—	—	—	0.059	0.059	59	19
		R	0.390	0.335	0.0028	9.00 ①	9.08 ②	NA	—	0.059	38	28

NA: Not Available

① Parking brake drum on vehicles with rear disc brakes: 6.69

② Parking brake drum on vehicles with rear disc brakes: 6.73

22140_FORE_C0013

SCHEDULED MAINTENANCE INTERVALS

2006-08 SUBARU—FORESTER

TO BE SERVICED	SERVICE	VEHICLE MILEAGE INTERVAL (x1000)																
		3	7.5	15	22.5	30	37.5	45	52.5	60	67.5	75	82.5	90	97.5	105	112.5	120
Accessory drive belts	R									✓								✓
Accessory drive belts	S/I					✓				✓						✓		
Air condition filter	S/I		✓	✓	✓	✓	✓	✓	✓	✓	✓	✓	✓	✓	✓	✓	✓	✓
Air cleaner filter	R					✓				✓				✓				✓
Automatic transmission fluid	S/I					✓				✓				✓				✓
Axle shaft joints	S/I			✓		✓		✓		✓		✓		✓		✓		✓
Brake fluid	R					✓				✓				✓				✓
Brake system lines	S/I			✓		✓		✓		✓		✓		✓		✓		✓
Clutch operation	S/I			✓		✓		✓		✓		✓		✓		✓		✓
Disc brake pads & rotors	S/I			✓		✓		✓		✓		✓		✓		✓		✓
Drums brake linings & drums	S/I					✓				✓				✓				✓
Engine coolant	R					✓				✓				✓				✓
Engine cooling system, hoses & connections	S/I					✓				✓				✓				✓
Engine oil & filter	R	✓	✓	✓	✓	✓	✓	✓	✓	✓	✓	✓	✓	✓	✓	✓	✓	✓
Front & rear axle boots	S/I			✓		✓		✓		✓		✓		✓		✓		✓
Fuel Filter	S/I					✓				✓				✓				
Front & rear wheel bearings	S/I			✓		✓		✓		✓		✓		✓		✓		✓
Parking & service brake systems' operation	S/I			✓		✓		✓		✓		✓		✓		✓		✓
Spark plugs (non turbo)	R					✓				✓				✓				✓
Spark plugs (turbo)	R									✓								✓
Steering & suspension	S/I			✓		✓		✓		✓		✓		✓		✓		✓
Timing belt	R															✓		
Timing belt	S/I					✓				✓				✓				
Tire rotation	S/I		✓	✓	✓	✓	✓	✓	✓	✓	✓	✓	✓	✓	✓	✓	✓	✓
Transmission & differential fluid levels	S/I					✓				✓				✓				✓
Valve clearance	S/I															✓		

R: Replace S/I: Inspect and service, if needed L: Lubricate

FREQUENT OPERATION MAINTENANCE (SEVERE SERVICE)

If a vehicle is operated under any of the following conditions it is considered severe service

- **Towing a trailer or using a camper or car-top carrier.**
- **Repeated short trips of less than 5 miles in temperatures below freezing, or trips of less than 10 miles in any temperature.**
- **Extensive idling or low-speed driving for long distances as in heavy commercial use, such as delivery, taxi or police cars.**
- **Operating on rough, muddy or salt-covered roads, or extensive mountain driving.**
- **Operating on unpaved or dusty roads.**
- **Driving in extremely hot (over 90°) conditions.**

Engine oil and filter: replace every 3000 miles or 3 months, whichever occurs first.

Fuel system, hoses & connections: inspect every 7500 miles or 7.5 months, whichever occurs first.

Transmission & differential fluid: replace every 15,000 miles.

Automatic transmission fluid: replace every 15,000 miles.

Brake fluid: replace every 15,000 miles.

Disc brake pads & rotors: inspect every 7500 miles or 7.5 months, whichever occurs first.

Front & rear axle boots: inspect every 7500 miles or 7.5 months, whichever occurs first.

Axle shaft boots: inspect every 7500 miles or 7.5 months, whichever occurs first.

Drum brake linings & drums: inspect every 7500 miles or 7.5 months, whichever occurs first.

Brake lines: inspect every 7500 miles or 7.5 months, whichever occurs first.

Parking & service brake system operation: inspect every 7500 miles or 7.5 months, whichever occurs first.

Clutch operation: inspect every 7500 miles or 7.5 months, whichever occurs first.

22140_FORE_C0014

PRECAUTIONS

Before servicing any vehicle, please be sure to read all of the following precautions, which deal with personal safety, prevention of component damage, and important points to take into consideration when servicing a motor vehicle:

- Never open, service or drain the radiator or cooling system when the engine is hot; serious burns can occur from the steam and hot coolant.
- Observe all applicable safety precautions when working around fuel. Whenever servicing the fuel system, always work in a well-ventilated area. Do not allow fuel spray or vapors to come in contact with a spark, open flame, or excessive heat (a hot drop light, for example). Keep a dry chemical fire extinguisher near the work area. Always keep fuel in a container specifically designed for fuel storage; also, always properly seal fuel containers to avoid the possibility of fire or explosion. Refer to the additional fuel system precautions later in this section.
- Fuel injection systems often remain pressurized, even after the engine has been turned **OFF**. The fuel system pressure must be relieved before disconnecting any fuel lines. Failure to do so may result in fire and/or personal injury.
- Brake fluid often contains polyglycol ethers and polyglycols. Avoid contact with the eyes and wash your hands thoroughly after handling brake fluid. If you do get brake fluid in your eyes, flush your eyes with clean, running water for 15 minutes. If eye irritation persists, or if you have taken brake fluid internally, IMMEDIATELY seek medical assistance.
- The EPA warns that prolonged contact with used engine oil may cause a number of skin disorders, including cancer. You should make every effort to minimize your exposure to used engine oil. Protective gloves should be worn when changing oil. Wash your hands and any other exposed skin areas as soon as possible after exposure to used engine oil. Soap and water, or waterless hand cleaner should be used.
- All new vehicles are now equipped with an air bag system, often referred to as a Supplemental Restraint System (SRS) or Supplemental Inflatable Restraint (SIR) system. The system must be disabled before performing service on or around system components, steering column, instrument panel components, wiring and sensors. Failure to follow safety and disabling procedures could result in accidental air bag deployment, possible personal injury and unnecessary system repairs.
- Always wear safety goggles when working with, or around, the air bag system. When carrying a non-deployed air bag, be sure the bag and trim cover are pointed away from your body. When placing a non-deployed air bag on a work surface, always face the bag and trim cover upward, away from the surface. This will reduce the motion of the module if it is accidentally deployed. Refer to the additional air bag system precautions later in this section.
- Clean, high quality brake fluid from a sealed container is essential to the safe and proper operation of the brake system. You should always buy the correct type of brake fluid for your vehicle. If the brake fluid becomes contaminated, completely flush the system with new fluid. Never reuse any brake fluid. Any brake fluid that is removed from the system should be discarded. Also, do not allow any brake fluid to come in contact with a painted surface; it will damage the paint.
- Never operate the engine without the proper amount and type of engine oil; doing so WILL result in severe engine damage.
- Timing belt maintenance is extremely important. Many models utilize an interference-type, non-freewheeling engine. If the timing belt breaks, the valves in the cylinder head may strike the pistons, causing potentially serious (also time-consuming and expensive) engine damage. Refer to the maintenance interval charts for the recommended replacement interval for the timing belt, and to the timing belt section for belt replacement and inspection.
- Disconnecting the negative battery cable on some vehicles may interfere with the functions of the on-board computer system(s) and may require the computer to undergo a relearning process once the negative battery cable is reconnected.
- When servicing drum brakes, only disassemble and assemble one side at a time, leaving the remaining side intact for reference.
- Only an MVAC-trained, EPA-certified automotive technician should service the air conditioning system or its components.

BRAKES — ANTI-LOCK BRAKE SYSTEM (ABS)

GENERAL INFORMATION

When wheel slip is detected during a brake application, an ABS event occurs. During antilock braking, hydraulic pressure in the individual wheel circuits is controlled to prevent any wheel from slipping. A separate hydraulic line and specific solenoid valves are provided for each wheel. The ABS can decrease, hold, or increase hydraulic pressure to each wheel. During antilock braking, a series of rapid pulsations is felt in the brake pedal. These pulsations are caused by the rapid changes in position of the individual solenoid valves as the ABS Control Module and Hydraulic Control Unit (ABSCM&H/U) responds to wheel speed sensor inputs and attempts to prevent wheel slip. These pedal pulsations are present only during antilock braking and stop when normal braking is resumed or when the vehicle comes to a stop. A ticking or popping noise may also be heard as the solenoid valves cycle rapidly. During antilock braking on dry pavement, intermittent chirping noises may be heard as the tires approach slipping. These noises and pedal pulsations are considered normal during antilock operation. Vehicles equipped with ABS may be stopped by applying normal force to the brake pedal. Brake pedal operation during normal braking is no different than that of previous non-ABS systems. Maintaining a constant force on the brake pedal provides the shortest stopping distance while maintaining vehicle.

PRECAUTIONS

- Certain components within the ABS system are not intended to be serviced or repaired individually.
- Do not use rubber hoses or other parts not specifically specified for and ABS system. When using repair kits, replace all parts included in the kit. Partial or incorrect repair may lead to functional problems and require the replacement of components.
- Lubricate rubber parts with clean, fresh brake fluid to ease assembly. Do not use shop air to clean parts; damage to rubber components may result.

• Use only DOT 3 brake fluid from an unopened container.

• If any hydraulic component or line is removed or replaced, it may be necessary to bleed the entire system.

• A clean repair area is essential. Always clean the reservoir and cap thoroughly before removing the cap. The slightest amount of dirt in the fluid may plug an orifice and impair the system function. Perform repairs after components have been thoroughly cleaned; use only denatured alcohol to clean components. Do not allow ABS components to come into contact with any substance containing mineral oil; this includes used shop rags.

• The Anti-Lock control unit is a microprocessor similar to other computer units in the vehicle. Ensure that the ignition switch is **OFF** before removing or installing controller harnesses. Avoid static electricity discharge at or near the controller.

• If any arc welding is to be done on the vehicle, the control unit should be unplugged before welding operations begin.

BLEEDING THE ABS SYSTEM

Refer to manual brake system. "Brake Bleeding Procedure"

WHEEL SPEED SENSORS

REMOVAL & INSTALLATION

Front

See Figure 1.

1. Before servicing the vehicle, refer to the Precautions Section.
2. Disconnect the ground cable from the battery.
3. Disconnect the ABS wheel speed sensor connector located next to the front strut mounting house in the engine compartment.
4. Remove the bolts which secure the sensor harness to the strut.
5. Remove the bolts which secure the sensor harness to the body.
6. Remove the bolts which secure front ABS wheel speed sensor to the housing, and remove the front ABS wheel speed sensor.

CAUTION

Be careful not to damage the pole piece and the face of the teeth located at tip of the sensor during removal. Do not pull on the sensor harness during removal.

To install:

7. Temporarily install the front ABS wheel speed sensor on the housing.
8. Install the front ABS wheel speed sensor on the strut and the wheel apron bracket. Tighten mounting bolt for bracket to 25 ft. lbs. (33 Nm).
9. Check the clearance of the sensor.
10. ABS wheel speed sensor gap standard value is as follows:
    a. 0.012–0.031 inch (0.3–0.8mm)
11. If clearance is outside of the standard value, readjust by using the spacer (Part No. 26755AA000).
12. After confirmation of the ABS wheel speed sensor clearance, connect the connector to the ABS wheel speed sensor.
13. Connect the ground cable to the battery.

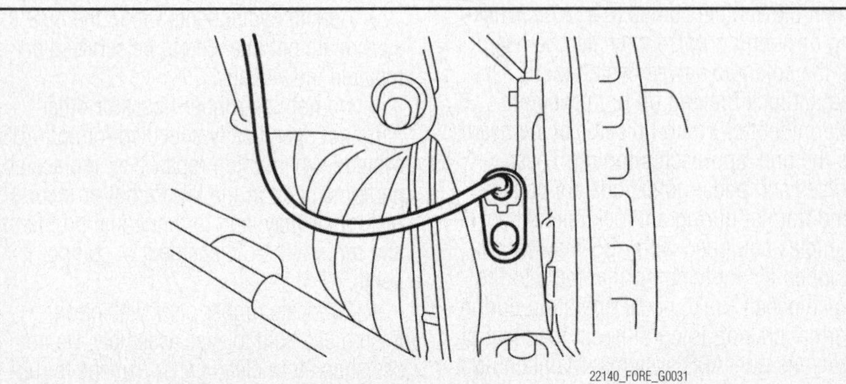

Fig. 1 Front ABS wheel speed sensor view

Rear

See Figure 2.

1. Disconnect the ground cable from the battery.
2. Lift up the vehicle.
3. Remove the rear seat and disconnect the rear ABS wheel speed sensor connector.
4. Remove the rear sensor harness bracket from the rear trailing link and bracket.
5. Remove the rear ABS wheel speed sensor from the back plate.

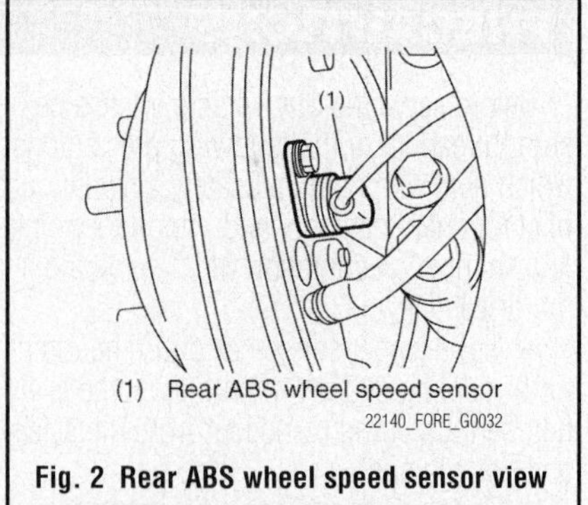

Fig. 2 Rear ABS wheel speed sensor view

6. Remove the rear tone wheel when removing the hub from the housing and hub assembly.

To install:

7. Attach the hub to the rear tone wheel and attach the rear housing.
8. Temporarily attach the rear ABS wheel speed sensor to the back plate.

CAUTION

Be careful not to hit the ABS wheel speed sensor pole piece and tone wheel against adjacent metal parts during installation.

9. Install the rear driveshaft to the rear housing and rear differential spindle.
10. Install the rear sensor harness on the rear trailing link. Tighten the harness bracket mounting bolt to 25 ft. lbs. (33 Nm).
11. Check the clearance of the sensor.
12. ABS wheel speed sensor gap standard value is as follows:
    a. 0.028–0.047 inch (0.7–1.2mm)
13. When the clearance is within standard values, tighten the ABS wheel speed sensor to the back plate to 25 ft. lbs. (33 Nm).
14. If clearance is outside of the standard value, readjust by using the spacer (Part No. 26755AA000).
15. After confirmation of the ABS wheel speed sensor clearance, connect the connector to the ABS wheel speed sensor.
16. Connect the ground cable to the battery.

WHEEL SPEED SENSOR RINGS (TOOTHED RINGS)

REMOVAL & INSTALLATION

Front

The front tone wheel is integrated with the front driveshaft.

Rear

The rear tone wheel is attached to the rear hub. Refer to Rear Axle.

BLEEDING PROCEDURE

See Figures 3 through 5.

1. Before servicing the vehicle, refer to the Precautions Section.
2. Raise the vehicle and safely support it.
3. Remove both front and rear wheels.
4. Draw out the brake fluid from master cylinder with syringe.
5. Refill reservoir tank with recommended brake fluid.
6. Install one end of a vinyl tube onto the air bleeder and insert the other end of the tube into a container to collect the brake fluid.
7. Instruct your co-worker to depress the brake pedal slowly two or three times and then hold it depressed.
8. Loosen bleeder screw approximately ¼ turn until a small amount of brake fluid drains into container, and then quickly tighten screw.
9. Repeat again from the two former procedures above until there are no air bubbles in drained brake fluid and new fluid flows through vinyl tube.

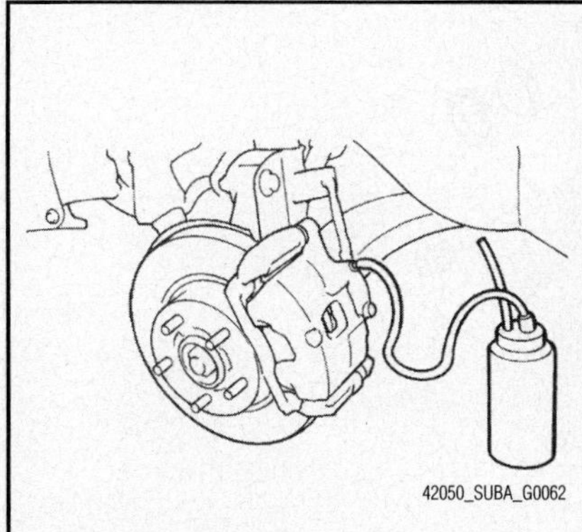

Fig. 3 Connect one end of the tube onto the air bleeder

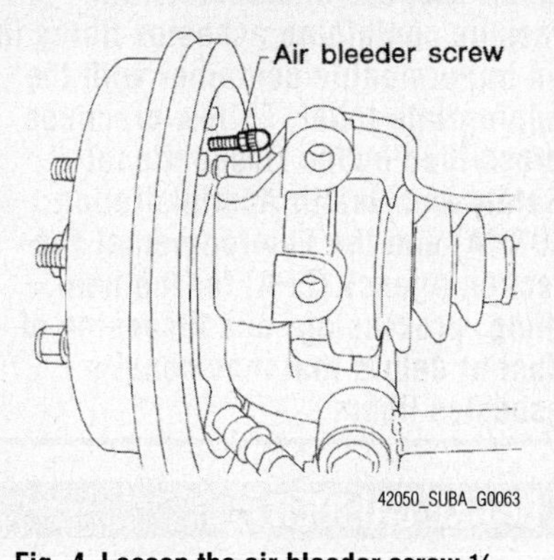

Fig. 4 Loosen the air bleeder screw ¼ turn to bleed the system

➡Add brake fluid as necessary while performing the air bleed operation, in order to prevent the tank from running short of brake fluid.

10. After completing the bleeding operation, hold brake pedal depressed and tighten screw and install bleeder cap.
11. Bleed air from each wheel cylinder using the same procedures as described above.
12. Depress brake pedal and hold it there for approximately 20 seconds. At this time check pedal to see if it shows any unusual movement. Visually inspect bleeder screws and brake pipe joints to make sure that there is no fluid leakage.
13. Install the wheels, and drive car for a short distance (between 1–2 miles to make sure that brakes are operating properly.

BLEEDING THE ABS SYSTEM

Refer to manual brake system. "Brake Bleeding Procedure"

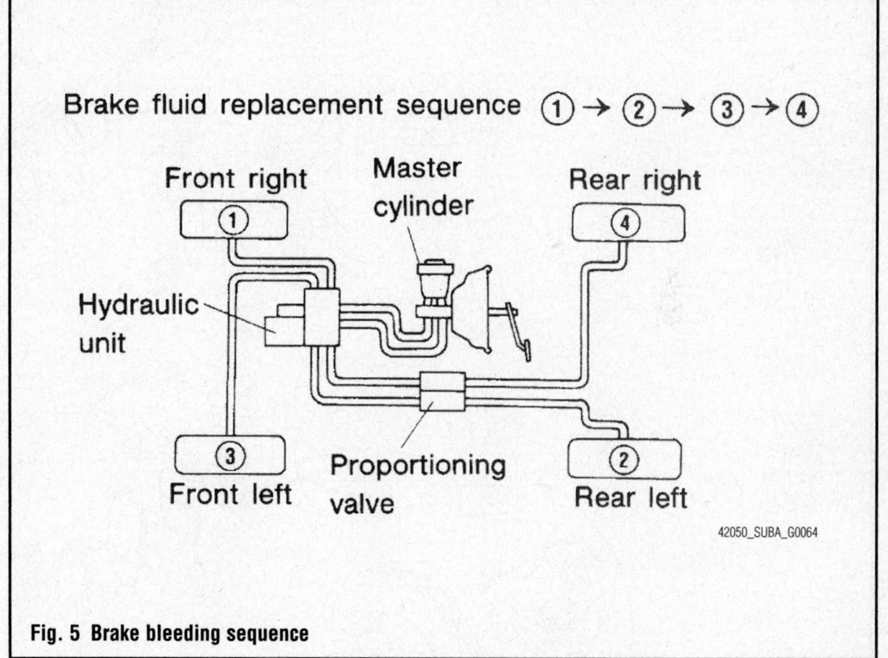

Fig. 5 Brake bleeding sequence

See Figure 6.

⁂ CAUTION

Dust and dirt accumulating on brake parts during normal use may contain asbestos fibers from production or aftermarket brake linings. Breathing excessive concentrations of asbestos fibers can cause serious bodily harm. Exercise care when servicing brake parts. Do not sand or grind brake lining unless equipment used is designed to contain the dust residue. Do not clean brake parts with compressed air or by dry brushing. Cleaning should be done by dampening the brake components with a fine mist of water, then wiping the brake components clean with a dampened cloth. Dispose of cloth and all residue containing asbestos fibers in an impermeable container with the appropriate label. Follow practices prescribed by the Occupational Safety and Health Administration (OSHA) and the Environmental Protection Agency (EPA) for the handling, processing, and disposing of dust or debris that may contain asbestos fibers

BRAKE CALIPER

REMOVAL & INSTALLATION

See Figure 7.

1. Before servicing the vehicle, refer to the Precautions Section.

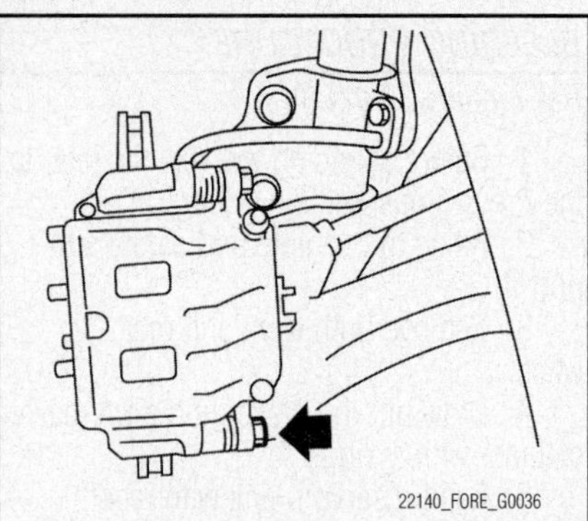

Fig. 7 Brake caliper mounting bolt securing the lock pin (yellow)

2. Loosen the wheel nuts.
3. Raise and support the vehicle safely.
4. Remove the tire and wheel.

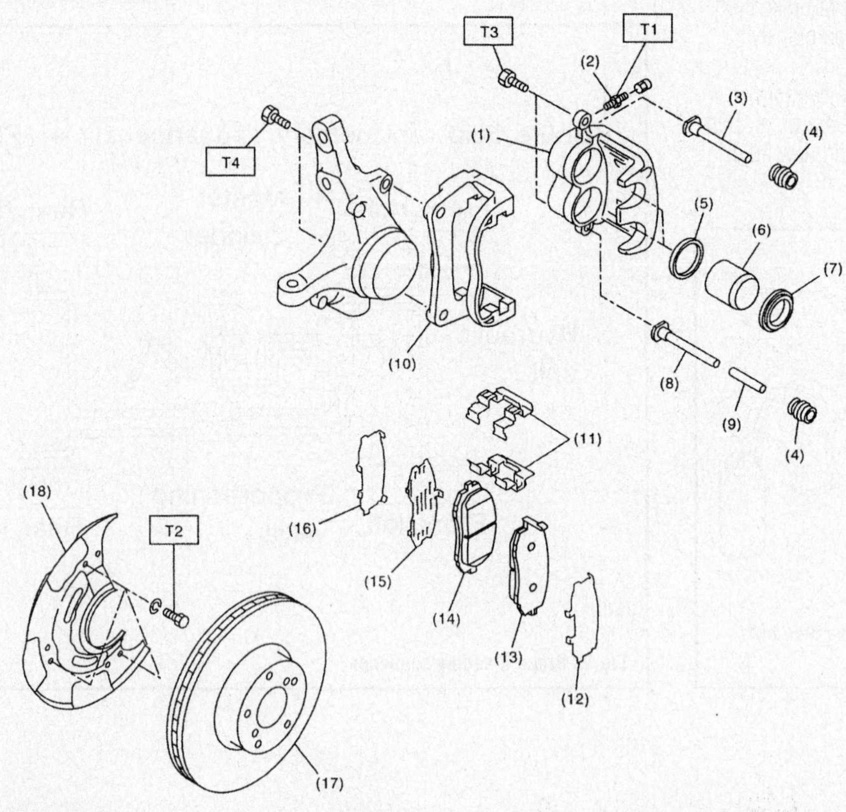

(1)	Caliper body	(9)	Bushing	(17)	Disc rotor
(2)	Air bleeder screw	(10)	Support	(18)	Disc cover
(3)	Guide pin (Green)	(11)	Pad clip		
(4)	Pin boot	(12)	Outer shim		
(5)	Piston seal	(13)	Outer pad		
(6)	Piston	(14)	Inner pad		
(7)	Piston boot	(15)	Rubber coat shim		
(8)	Lock pin (Yellow)	(16)	Inner shim		

Tightening torque: N·m (kgf-m, ft-lb)

T1: 8 (0.8, 5.8)
T2: 18 (1.8, 13.0)
T3: 26 (2.7, 19.2)
T4: 80 (8.2, 59)

09490_FORE_G0017

Fig. 6 Front disc brake and related components

5. Remove the brake hose mounting bolt. Disconnect the brake line from the brake caliper. Be sure to properly catch the fluid to avoid damage to painted surfaces and improper disposal.

6. Remove the bolt securing the lock pin (yellow) to caliper body assembly.

7. Raise the caliper body assembly and move it toward the center of the vehicle to separate it from the support.

8. Remove the caliper from its mounting.

To install:

9. Installation is the reverse of the removal procedure.

10. Apply a thin coat of Molykote M7439 (Part No. 003602001) to the support.

11. Apply a thin coat of Molykote AS880N (Part No. K0777YA010) to the pad clip and both sides of inner shim.

12. Tighten the caliper retaining bolts to 19 ft. lbs. (26 Nm).

CAUTION

When connecting the brake hose, do not twist it.

13. Be sure to use new brake hose gaskets. Tighten the banjo bolt to 13 ft. lbs. (18 Nm).

14. Bleed the hydraulic system, as required.

15. Check the brake fluid level, correct as required.

16. Test drive and burnish brake system.

DISC BRAKE PADS

REMOVAL & INSTALLATION

See Figure 8.

1. Before servicing the vehicle, refer to the Precautions Section.

2. Set the vehicle on a lift.

3. Loosen the wheel nuts.

4. Lift up the vehicle, and remove the front wheels.

5. Remove the lower caliper bolts.

6. Raise the caliper body and support it.

7. Remove the pad.

8. If the brake pad is difficult to remove, proceed as follows:

- Remove the caliper body from support.

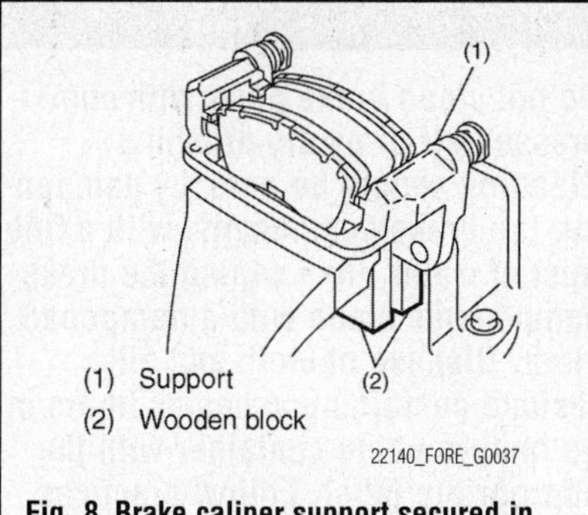

Fig. 8 Brake caliper support secured in vise

- Remove the support.
- Place the support between wooden blocks in the vise.
- Apply a rod with less than 0.47 inch (12mm) diameter to the shaded area of brake pad, and strike the rod with a hammer to remove brake pad.

To install:

9. Compress the caliper pistons.

➡If it is difficult to push the pistons during pad replacement, loosen the air bleeder to facilitate work.

10. Apply a thin coat of Molykote M7439 (Part No. 003602001) to the support.

11. Apply a thin coat of Molykote AS880N (Part No. K0777YA010) to the pad clip and both sides of inner shim.

12. Install the pad to support.

13. Install the caliper body to the support. Tighten the caliper retaining bolts to 19 ft. lbs. (26 Nm).

CAUTION

When connecting the caliper, do not twist brake hose.

14. Bleed the hydraulic system, as required.

15. Check the brake fluid level, correct as required.

16. Test drive and burnish brake system.

ROTOR

REMOVAL & INSTALLATION

See Figure 9.

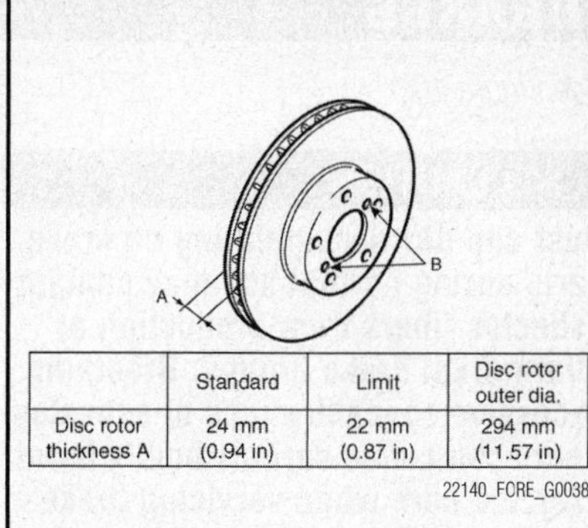

	Standard	Limit	Disc rotor outer dia.
Disc rotor thickness A	24 mm (0.94 in)	22 mm (0.87 in)	294 mm (11.57 in)

Fig. 9 Front brake rotor thickness (A) service holes (B)

1. Before servicing the vehicle, refer to the Precautions Section.

2. Loosen the wheel nuts.

3. Raise and support the vehicle safely.

4. Remove the front wheel.

5. Remove the caliper assembly and suspend it from the strut using mechanic's wire.

➡Do not disconnect the brake line.

CAUTION

Do not let the caliper hang by the brake hoses.

6. Remove the rotor.

7. Remove mud and foreign matter from caliper body assembly, support and rotor contact surface.

➡If the rotor is seized and is difficult to remove, install an 8 mm bolt into the services holes on the hub and tighten.

To install:

8. Install the disc rotor. Disc rotor runout limit is 0.0030 inch (0.075mm).

9. Remove mechanic's wire, and lower the caliper into position.

10. Install the caliper body and the support to housing. Tighten the mounting bolts to 59 ft. lbs. (80 Nm).

11. Install the wheel.

12. Test drive and burnish brake system.

See Figure 10.

CAUTION

Dust and dirt accumulating on brake parts during normal use may contain asbestos fibers from production or aftermarket brake linings. Breathing excessive concentrations of asbestos fibers can cause serious bodily harm. Exercise care when servicing brake parts. Do not sand or grind brake lining unless equipment used is designed to contain the dust residue. Do not clean brake parts with compressed air or by dry brushing. Cleaning should be done by dampening the brake components with a fine mist of water, then wiping the brake components clean with a dampened cloth. Dispose of cloth and all residue containing asbestos fibers in an impermeable container with the appropriate label. Follow practices prescribed by the Occupational Safety and Health Administration (OSHA) and the Environmental Protection Agency (EPA) for the handling, processing, and disposing of dust or debris that may contain asbestos fibers.

BRAKE CALIPER

REMOVAL & INSTALLATION

See Figure 11.

1. Before servicing the vehicle, refer to the Precautions Section.
2. Loosen the wheel nuts.

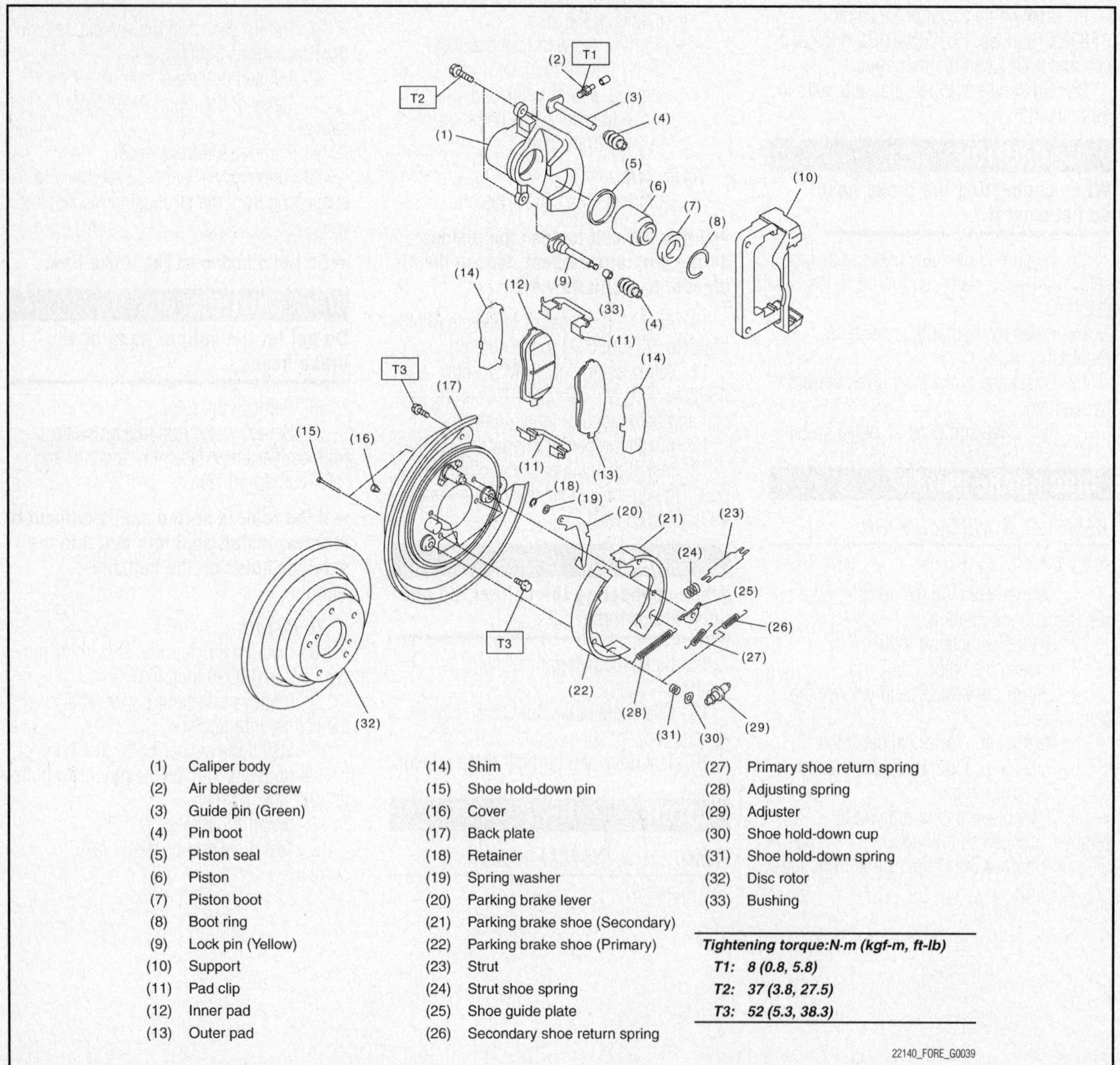

Fig. 10 Exploded view of rear disc brake system

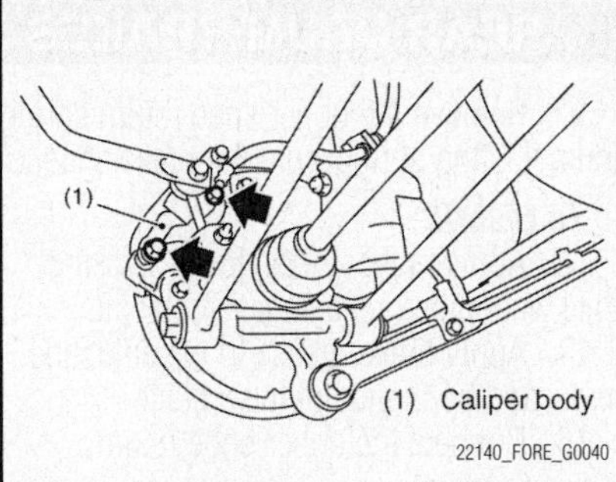

Fig. 11 Caliper body (1), bolt securing the lock pin (yellow) and brake hose mounting bolt

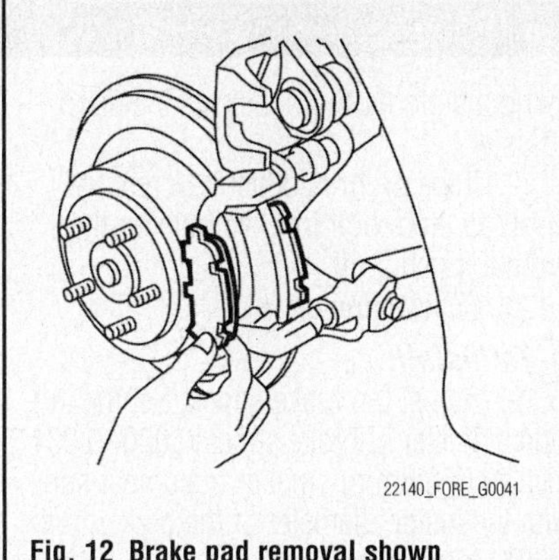

Fig. 12 Brake pad removal shown

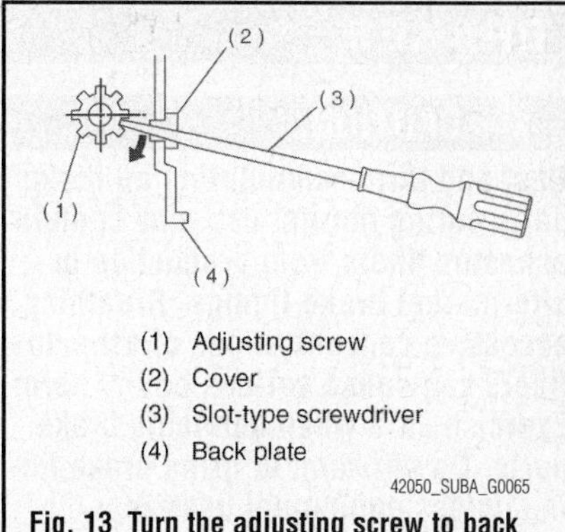

Fig. 13 Turn the adjusting screw to back off the brake shoe to remove the rotor.

3. Raise and support the vehicle safely.
4. Remove the tire and wheel.
5. Disconnect the brake hose from the brake caliper. Be sure to properly catch the fluid to avoid damage to painted surfaces and improper disposal.
6. Remove the bolt securing the lock pin (yellow) to caliper body assembly.
7. Raise the caliper body assembly and move it toward the center of the vehicle to separate it from the support.
8. Remove the caliper from its mounting.

To install:

9. Installation is the reverse of the removal procedure.
10. Tighten the caliper retaining bolts to 28 ft. lbs. (37 Nm).

**** CAUTION**

When connecting the brake hose, do not twist it.

11. Be sure to use new brake hose gaskets. Tighten banjo bolt to 13 ft. lbs. (18 Nm).
12. Bleed the hydraulic system, as required.
13. Check the brake fluid level, correct as required.
14. Test drive and burnish brake system.

DISC BRAKE PADS

REMOVAL & INSTALLATION

See Figure 12.

1. Before servicing the vehicle, refer to the Precautions Section.
2. Set the vehicle on a lift.
3. Loosen the wheel nuts.
4. Lift up the vehicle, and then remove the rear wheels.
5. Remove the lower caliper bolts.
6. Raise the caliper body and support it.
7. Remove brake pads.
8. If the brake pad is difficult to remove, proceed as follows:
   - Remove the caliper body from support.
   - Remove the support.
   - Place the support between wooden blocks in the vise.
   - Apply a rod with less than 0.47 inch (12mm) diameter to the shaded area of brake pad, and strike the rod with a hammer to remove brake pad.

To install:

9. Compress the caliper piston.

➡If it is difficult to push the pistons during pad replacement, loosen the air bleeder to facilitate work.

10. Apply a thin coat of Molykote M7439 (Part No. 003602001) to the support.
11. Apply a thin coat of Molykote AS880N (Part No. K0777YA010) to the pad clip and both sides of inner shim.
12. Install the pad to support.
13. Install the caliper body to the support. Tighten the caliper retaining bolts to 28 ft. lbs. (37 Nm).

**** CAUTION**

When connecting the caliper, do not twist brake hose.

14. Bleed the hydraulic system, as required.
15. Check the brake fluid level, correct as required.
16. Test drive and burnish brake system.

ROTOR

REMOVAL & INSTALLATION

See Figure 13.

1. Before servicing the vehicle, refer to the Precautions Section.
2. Raise and safely support the vehicle.
3. Remove the wheel.
4. Remove the caliper assembly, and do not disconnect the brake lines. Hang the caliper securely from the strut using mechanics wire. Refer to "Brake Caliper, Removal & Installation" to remove the caliper assembly.

**** CAUTION**

Do not let the caliper hang by the brake hoses.

5. Pull down and release the parking brake.
6. Remove the rotor.
7. If the rotor is difficult to remove, try the following methods to remove:
   a. Turn the adjusting screw using a slot-type screwdriver until brake shoe is away enough from the disc rotor.
   b. Install an 8 mm bolt into the services holes on the hub and tighten.
8. Installation is the reverse order of removal.
9. Adjust the parking brake.
10. Bleed the hydraulic system, as required.
11. Check the brake fluid level, correct as required.
12. Test drive and burnish brake system.

CAUTION

Dust and dirt accumulating on brake parts during normal use may contain asbestos fibers from production or aftermarket brake linings. Breathing excessive concentrations of asbestos fibers can cause serious bodily harm. Exercise care when servicing brake parts. Do not sand or grind brake lining unless equipment used is designed to contain the dust residue. Do not clean brake parts with compressed air or by dry brushing. Cleaning should be done by dampening the brake components with a fine mist of water, then wiping the brake components clean with a dampened cloth. Dispose of cloth and all residue containing asbestos fibers in an impermeable container with the appropriate label. Follow practices prescribed by the Occupational Safety and Health Administration (OSHA) and the Environmental Protection Agency (EPA) for the handling, processing, and disposing of dust or debris that may contain asbestos fibers.

BRAKE DRUM

REMOVAL & INSTALLATION

See Figure 14.

1. Before servicing the vehicle, refer to the Precautions Section.
2. Loosen the wheel nuts.
3. Raise and support the vehicle safely.
4. Remove the rear wheels.
5. Release the parking brake.
6. If necessary, remove the adjusting hole cover from the backing plate and using a suitable tool, back off the shoe adjuster.
7. If the drum is difficult to remove, insert an 8mm bolt into the hole on the drum to push it off.
8. Remove the drum.

To install:

9. Adjust the brake shoes. Set the outside diameter of brake shoes 0.020–0.031 inch (0.5–0.8mm) smaller in comparison with the inside diameter of the brake drum.
10. Install the drum.
11. Install the rear wheels.
12. Bleed the hydraulic system, as required.
13. Check the brake fluid level, correct as required.
14. Test drive and burnish brake system.

BRAKE SHOES

REMOVAL & INSTALLATION

See Figure 15.

1. Before servicing the vehicle, refer to the Precautions Section.
2. Loosen the wheel nuts.
3. Raise and safely support the vehicle.
4. Remove the rear wheel.
5. Release the parking brake.
6. Remove the brake drum from the brake assembly.
7. Hold the hold-down pin by securing rear of back plate with your hand.
8. Disconnect the hold-down cup from hold-down pin by rotating hold-down cup.
9. Disconnect the lower shoe return spring from shoes.
10. Remove the shoes one by one from back plate with the adjuster.
11. Disconnect the parking brake cable from parking lever.
12. Remove the upper shoe return spring and adjusting spring from the brake shoe.

To install:

13. Apply brake grease to the backing plate where the brake shoes contact it.
14. Apply brake grease to the adjusting screw and both ends of the adjuster.
15. Connect the upper shoe return spring to the shoes.
16. Connect the parking brake cable to the parking lever.
17. While positioning the shoes one at a time in the groove on the wheel cylinder, secure the shoes.
18. Secure the shoes by connecting the hold-down cup to the hold-down pin.
19. Connect the lower shoe return spring to both shoes.
20. Set the outside diameter of the shoes less than 0.020–0.031 in. (0.5–0.8mm) compared to the inside diameter of the drum.
21. Install the drum and adjust the brake shoes.
22. Install the wheels and lower the vehicle.
23. Bleed the hydraulic system, as required.
24. Check the brake fluid level, correct as required.
25. Test drive and burnish brake system.

ADJUSTMENT

See Figure 16.

1. Remove the adjusting hole cover from back plate.
2. Turn adjusting screw using a slot-type screwdriver until brake shoe is in close contact with the drum.
3. Turn back (downward) the adjusting screw 3 or 4 notches.
4. Install the adjusting hole cover to back plate.

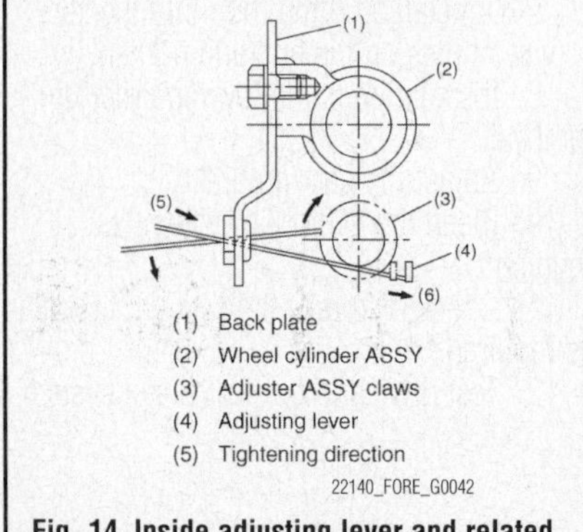

Fig. 14 Inside adjusting lever and related parts

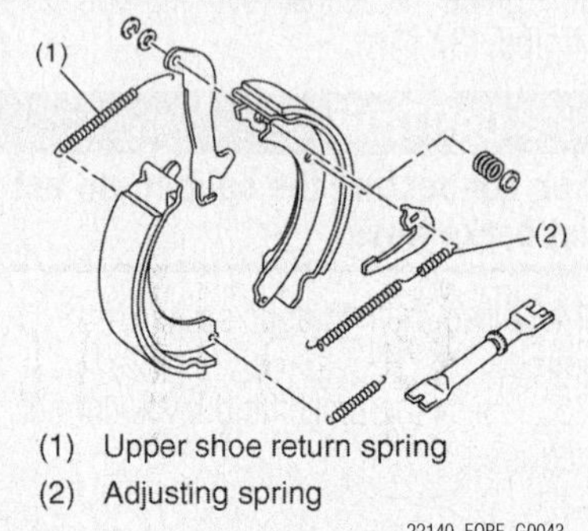

Fig. 15 Rear brake shoes and related parts

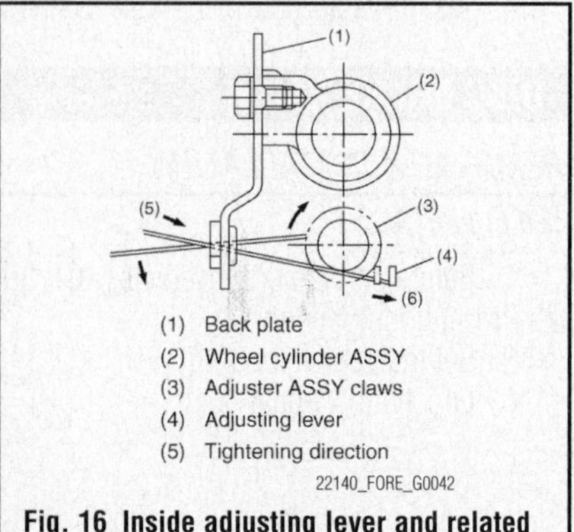

Fig. 16 Inside adjusting lever and related parts

PARKING BRAKE CABLES

ADJUSTMENT

See Figure 17.

1. Before servicing the vehicle, refer to the Precautions Section.
2. Remove the console cover.
3. Forcibly pull the parking brake lever 3 to 5 times.
4. Adjust the parking brake lever by turning adjuster until parking brake lever stroke is set at 7 to 8 notches with operating force of 44 lbs. (196 N).
5. Tighten the lock nut.
6. Install the console cover.

PARKING BRAKE SHOES

REMOVAL & INSTALLATION

See Figures 18 and 19.

1. Before servicing the vehicle, refer to the Precautions Section.
2. Release the parking brake.
3. Raise and safely support the vehicle.
4. Remove the two mounting bolts and remove the brake caliper assembly. Suspend the brake caliper assembly so that the hose is not stretched.
5. Remove the brake rotor.

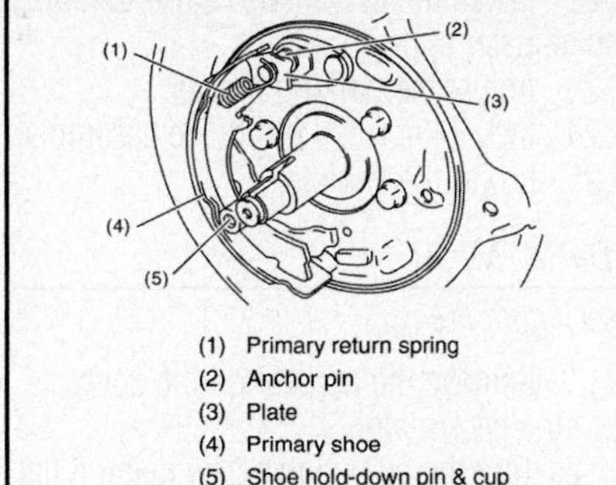

Fig. 18 Assembled primary components of the parking brake

6. If the disc rotor is difficult to remove, try the following two methods in order:
   a. Turn the adjusting screw using a flat tip screwdriver until the brake shoe moves adequately away from the disc rotor.
   b. If disc rotor is seized up on the hub, drive the disc rotor out by pushing two 8 mm bolts in holes B on the rotor.
7. Remove the shoe return spring from the parking brake assembly.
8. Using pliers, remove the front shoe hold-down spring and pin.
9. Remove the strut and strut spring.
10. Remove the adjuster assembly from the parking brake assembly.

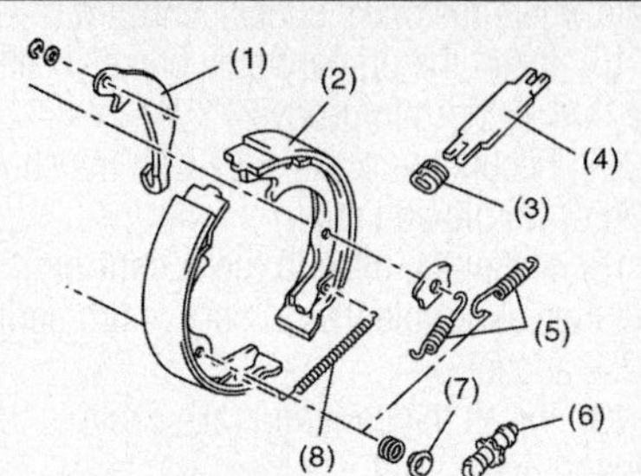

Fig. 19 Exploded view of the parking brake secondary components

11. Remove the brake shoe.
12. Remove the rear shoe hold-down spring and pin with pliers.
13. Remove the parking brake cable from the parking brake lever.
14. Using a standard screwdriver, raise the retainer. Remove the parking brake lever and washer from the brake shoe.

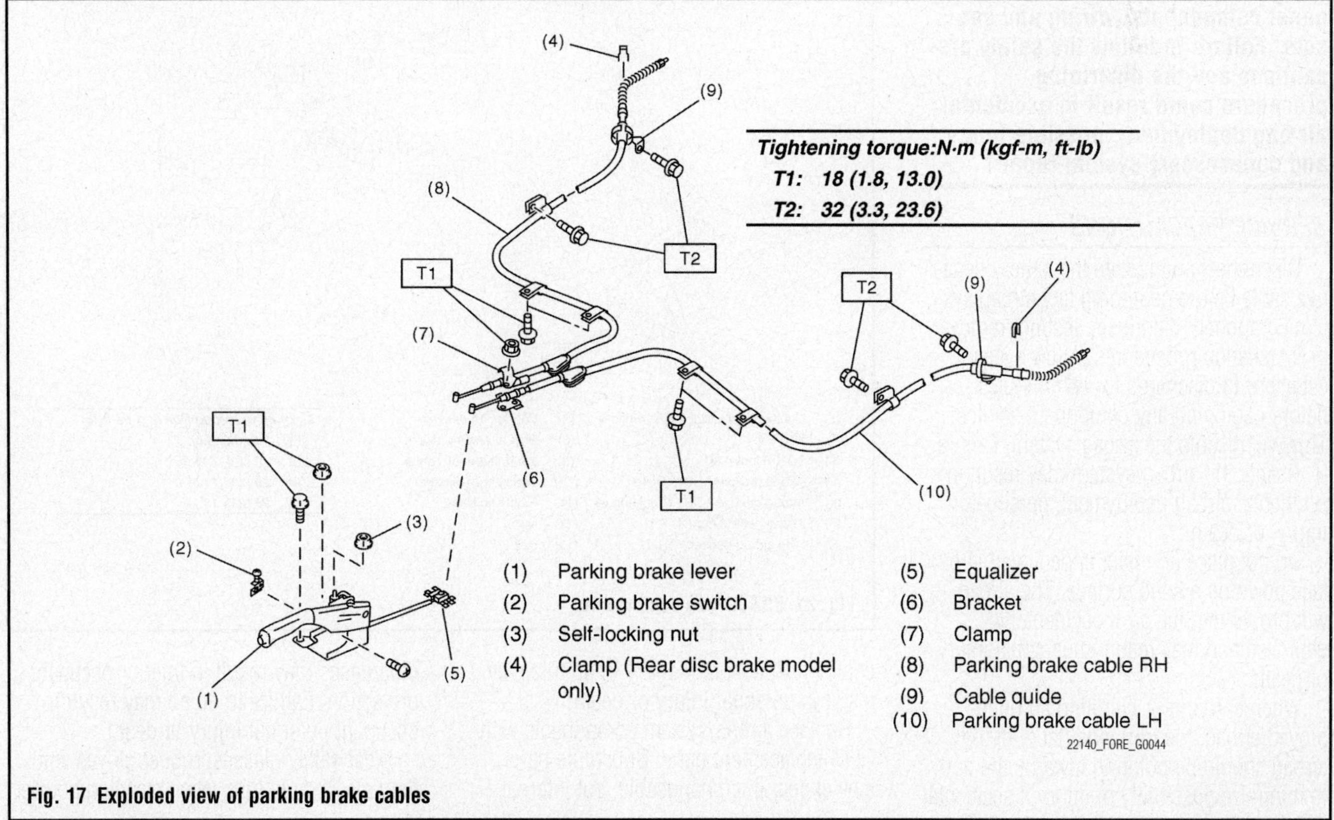

Fig. 17 Exploded view of parking brake cables

To install:

15. Apply brake grease to the backing plate where the brake shoes contact it.
16. Insert the primary side brake shoe into the anchor pin groove.
17. Secure the brake shoe with the shoe-hold down pin and cup.
18. Install the plate to the anchor pin, and then assemble the primary return spring to the anchor pin.
19. Install the parking brake cable to the parking brake lever.
20. Assemble the strut and adjuster, and then secure the secondary side brake shoe with the shoe hold-down pin and cup.

➡Install the strut spring of both right and left wheel facing vehicle front.

21. Install the adjuster assembly with screw on left side.
22. Install the secondary return spring and adjuster spring.
23. Adjust the parking brakes.
24. Install the rotor and brake assembly.
25. Lower the vehicle.

ADJUSTMENT

See Figure 20.

1. Remove the adjusting hole cover from the back plate.
2. Turn the adjusting screw using a flat tip screwdriver until the brake shoe is in close contact with the disc rotor.
3. Turn back (downward) the adjusting screw 3 to 4 notches.
4. Install the adjusting hole cover to the back plate.

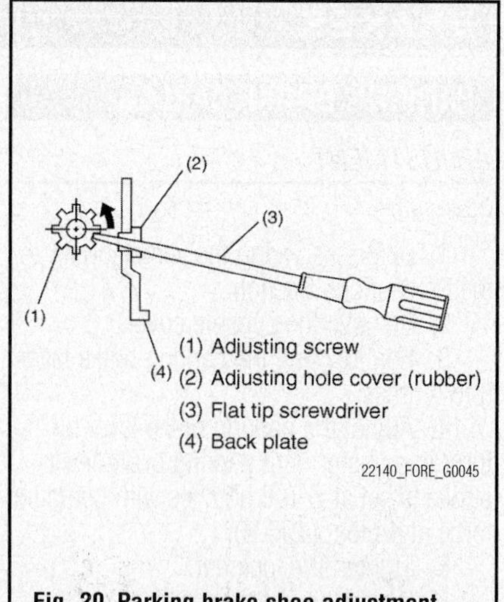

Fig. 20 Parking brake shoe adjustment shown

CHASSIS ELECTRICAL

AIR BAG (SUPPLEMENTAL RESTRAINT SYSTEM)

See Figure 21.

GENERAL INFORMATION

⁂ CAUTION

Some vehicles are equipped with an air bag system. The system must be disarmed before performing service on, or around, system components, the steering column, instrument panel components, wiring and sensors. Failure to follow the safety precautions and the disarming procedure could result in accidental air bag deployment, possible injury and unnecessary system repairs.

SERVICE PRECAUTIONS

Disconnect and isolate the battery negative cable before beginning any airbag system component diagnosis, testing, removal, or installation procedures. Allow system capacitor to discharge for two minutes before beginning any component service. This will disable the airbag system. Failure to disable the airbag system may result in accidental airbag deployment, personal injury, or death.

Do not place an intact undeployed airbag face down on a solid surface. The airbag will propel into the air if accidentally deployed and may result in personal injury or death.

When carrying or handling an undeployed airbag, the trim side (face) of the airbag should be pointing towards the body to minimize possibility of injury if accidental deployment occurs. Failure to do this may result in personal injury or death.

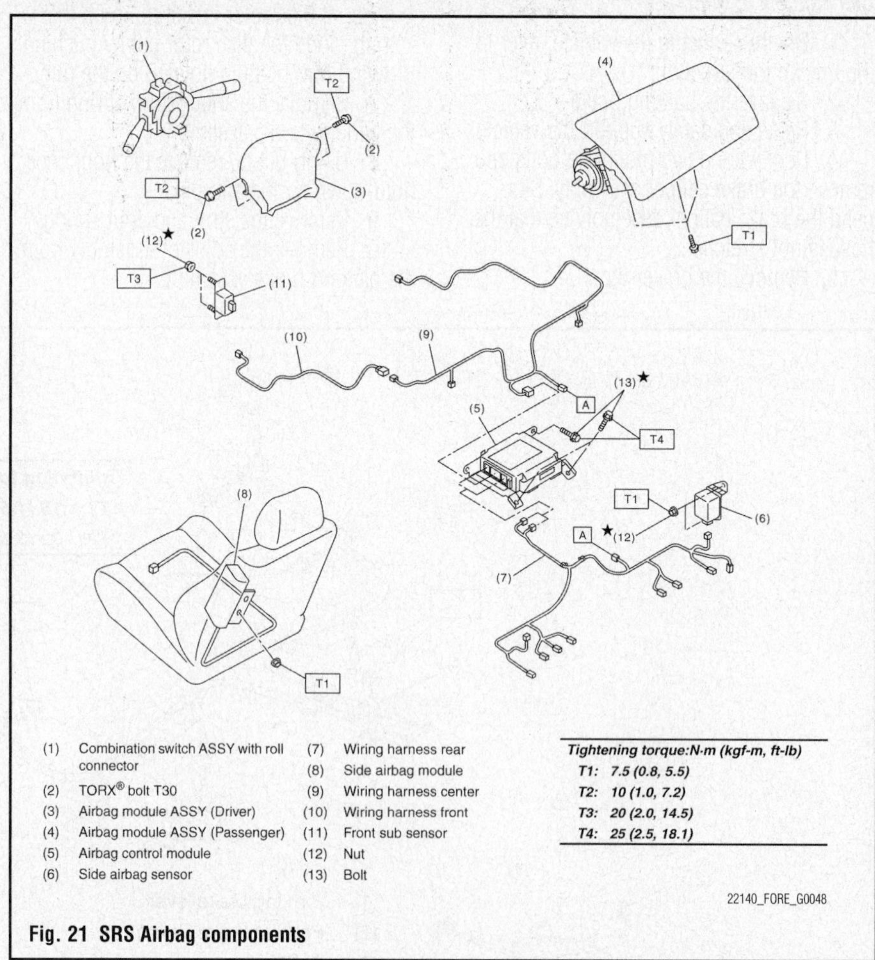

Fig. 21 SRS Airbag components

Replace airbag system components with OEM replacement parts. Substitute parts may appear interchangeable, but internal differences may result in inferior occupant protection. Failure to do so may result in occupant personal injury or death.

Wear safety glasses, rubber gloves, and long sleeved clothing when cleaning powder residue from vehicle after an airbag deployment. Powder residue emitted from a deployed airbag can cause skin irritation. Flush affected area with cool water if irritation is experienced. If nasal or throat irritation is experienced, exit the vehicle for fresh air until the irritation ceases. If irritation continues, see a physician.

Do not use a replacement airbag that is not in the original packaging. This may result in improper deployment, personal injury, or death.

The factory installed fasteners, screws and bolts used to fasten airbag components have a special coating and are specifically designed for the airbag system. Do not use substitute fasteners. Use only original equipment fasteners listed in the parts catalog when fastener replacement is required.

During, and following, any child restraint anchor service, due to impact event or vehicle repair, carefully inspect all mounting hardware, tether straps, and anchors for proper installation, operation, or damage. If a child restraint anchor is found damaged in any way, the anchor must be replaced. Failure to do this may result in personal injury or death.

Deployed and non-deployed airbags may or may not have live pyrotechnic material within the airbag inflator.

Do not dispose of driver/passenger/curtain airbags or seat belt tensioners unless you are sure of complete deployment. Refer to the Hazardous Substance Control System for proper disposal.

Dispose of deployed airbags and tensioners consistent with state, provincial, local, and federal regulations.

After any airbag component testing or service, do not connect the battery negative cable. Personal injury or death may result if the system test is not performed first.

If the vehicle is equipped with the Occupant Classification System (OCS), do not connect the battery negative cable before performing the OCS Verification Test using the scan tool and the appropriate diagnostic information. Personal injury or death may result if the system test is not performed properly.

Never replace both the Occupant Restraint Controller (ORC) and the Occupant Classification Module (OCM) at the same time. If both require replacement, replace one, then perform the Airbag System test before replacing the other.

Both the ORC and the OCM store Occupant Classification System (OCS) calibration data, which they transfer to one another when one of them is replaced. If both are replaced at the same time, an irreversible fault will be set in both modules and the OCS may malfunction and cause personal injury or death.

If equipped with OCS, the Seat Weight Sensor is a sensitive, calibrated unit and must be handled carefully. Do not drop or handle roughly. If dropped or damaged, replace with another sensor. Failure to do so may result in occupant injury or death.

If equipped with OCS, the front passenger seat must be handled carefully as well. When removing the seat, be careful when setting on floor not to drop. If dropped, the sensor may be inoperative, could result in occupant injury, or possibly death.

If equipped with OCS, when the passenger front seat is on the floor, no one should sit in the front passenger seat. This uneven force may damage the sensing ability of the seat weight sensors. If sat on and damaged, the sensor may be inoperative, could result in occupant injury, or possibly death.

DISARMING THE SYSTEM

1. Be sure to position the front wheels in the straight ahead position.
2. Disconnect the negative battery cable. Tape the battery cable for added protection.
3. Wait more than 20 seconds before starting work.

ARMING THE SYSTEM

Connect the negative battery cable.

CLOCKSPRING CENTERING

See Figure 22.

1. Check that front wheels are positioned in straight ahead direction.
2. Turn the clockspring pin (A) clockwise until it stops.
3. Turn the clockspring pins (A) approx. 3.25 turns until marks are aligned.

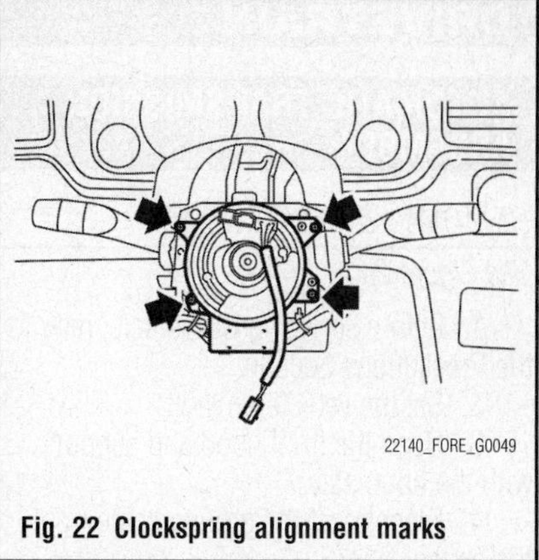

Fig. 22 Clockspring alignment marks

REMOVAL & INSTALLATION

See Figure 23.

1. Before servicing the vehicle, refer to the Precautions Section.
2. Turn the ignition switch **OFF**.
3. Disconnect the ground cable from battery and wait for at least 20 seconds before starting work.
4. Remove the driver's airbag module.
5. Remove the steering wheel nut, and then draw out the steering wheel from the shaft using a steering puller.
6. Remove the steering column cover.
7. Remove the screws, and then remove the clockspring.
8. Install in the reverse order of removal.

CAUTION

Before installing steering wheel, be sure to adjust the direction of roll connector with steering.

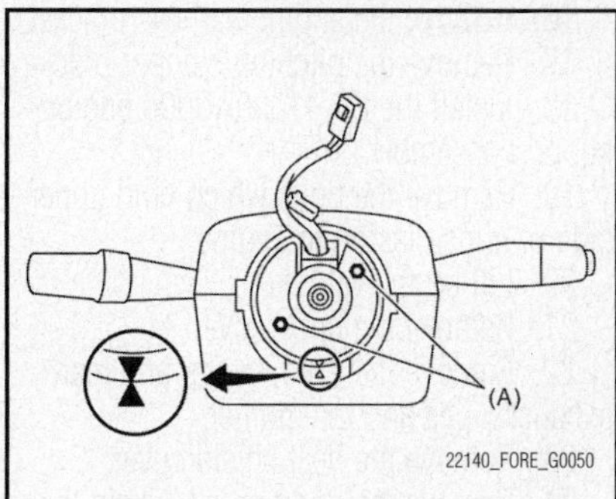

Fig. 23 Clockspring, and mounting screws shown

DRIVETRAIN

AUTOMATIC TRANSMISSION ASSEMBLY

REMOVAL & INSTALLATION

See Figure 24.

1. Before servicing the vehicle, refer to the Precautions Section.
2. Set the vehicle on a lift.
3. Open the front hood and support with the hood stay.
4. Disconnect the ground cable from the battery.
5. Remove the air intake duct.
6. Remove the air intake chamber.
7. Remove the air intake chamber stay.
8. Disconnect the following connectors:
   - Transmission harness connectors
   - Transmission ground terminal
9. Remove the starter.
10. Remove the pitching stopper.
11. Separate the torque converter assembly from the drive plate.

⁂ CAUTION

Be careful not to damage the mounting bolts. Be careful not to drop bolts into the converter case.

12. Attach the ST-498277200 converter holder to the converter case.
13. Remove the ATF level gauge. Plug the opening to prevent entry of foreign particles into the transmission fluid.
14. Remove the throttle body.
15. Disconnect the engine harness, then remove the harness connector from the engine harness bracket.
16. Remove the engine harness bracket.
17. Remove the pitching stopper bracket.
18. Install the ST-41099AC000 engine support assembly.
19. Remove the bolts which hold upper side of transmission to engine.
20. Lift up the vehicle.
21. Remove the undercover.
22. Remove the front, center and rear exhaust pipes and the muffler.
23. Remove the heat shield cover.
24. Remove the drain plug to drain the transmission fluid..
25. Disconnect the ATF cooler hoses from the pipes of the transmission side, and remove the oil charge pipe.
26. Remove the propeller shaft.
27. Remove the shift select cable.
28. Disconnect the stabilizer link from the transverse link.
29. Remove the bolt securing ball joint of the transverse link to housing.
30. Pull out the front driveshaft from the transmission.
    a. Using a tire lever or a crow bar, etc., pull out until the front driveshaft transmission side joint slides move smoothly.
    b. Hold the transmission side joint of the front driveshaft by hand and extract the housing from the transmission while pressing the housing outward, so as not to stretch the boot.
31. Remove the bolts which hold the clutch housing cover.
32. Remove the bolts and nuts which hold lower side of transmission to engine.
33. Place the transmission jack under the transmission.

➡Make sure that the support plates of transmission jack do not touch the oil pan.

34. Remove the transmission rear crossmember from the vehicle.
35. While lowering the transmission jack gradually, fully retract the engine support, and then tilt the engine rearward.
36. Retract the support until the clearance between front crossmember and converter case becomes approx. 0.39 inch (10mm).
37. Remove the transmission.
38. Remove the rear cushion rubber from the transmission assembly.

To install:

39. Replace the differential side oil seal with a new part.
40. Install the rear cushion rubber to the transmission assembly. Tighten to 29.5 ft. lbs. (40 Nm).
41. Attach the ST-498277200 to the converter case.
42. Install the transmission onto the engine as follows:
    - Lift up the transmission gradually using transmission jack.
    - Insert the engine side stud bolt into the transmission bolt hole.
    - While raising the transmission jack gradually, turn the screw of engine support, then tilt the engine forward and connect.
43. Install the transmission rear crossmember. Tighten the center insulator mount to 26 ft. lbs. (35 Nm).
44. Tighten outer crossmember bolts to 55 ft. lbs. (75 Nm).
45. Remove the transmission jack.
46. Tighten the bolts and nuts which hold the lower side of transmission to the engine to 37 ft. lbs. (50 Nm).
47. Install the clutch housing cover bolts.
48. Lower the lift.
49. Connect the engine and transmission.
50. Remove the ST from converter case.
51. Install the starter.
52. Tighten the bolts which hold the upper side of the transmission to the engine and tighten to 37 ft. lbs. (50 Nm).

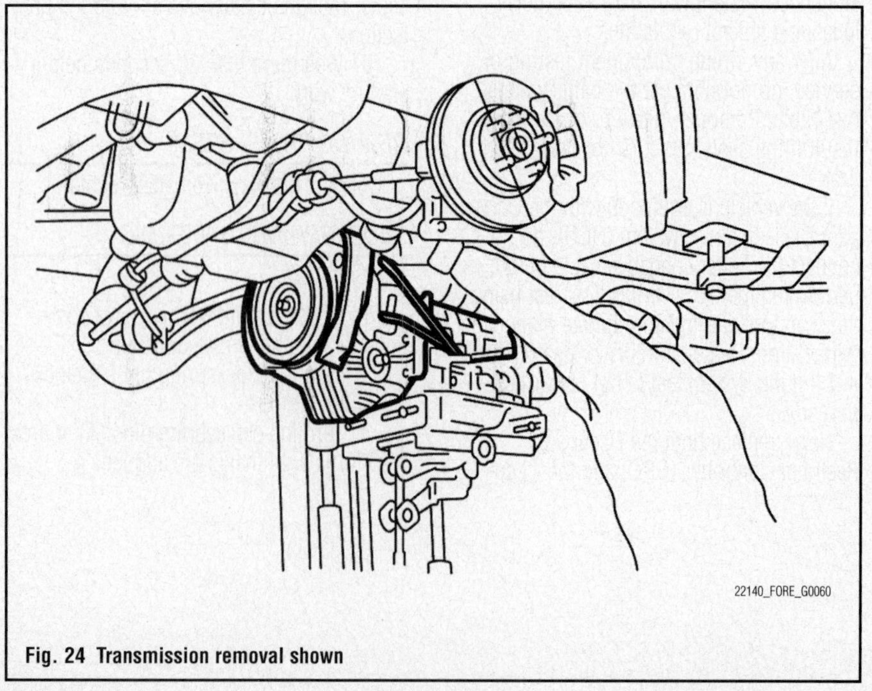

Fig. 24 Transmission removal shown

53. Install the torque converter assembly to the drive plate.
54. Tighten the bolts which hold the torque converter assembly to the drive plate.
55. Place the wrench on the crank pulley bolt, and remove all the bolts while rotating the crank pulley a little bit at a time. Tighten to 18 ft. lbs. (25 Nm).
56. Fit the plug to service hole.
57. Install the V-belt cover.
58. Remove the ST-41099AC000 engine support assembly.
59. Install the engine harness bracket. Tighten to 12 ft. lbs. (16 Nm).
60. Install the harness connector to engine harness bracket, then connect the harness.
61. Install the pitching stopper bracket and tighten to 30 ft. lbs. (41 Nm).
62. Install the throttle body.
63. Install the pitching stopper and tighten to 37 ft. lbs. (50 Nm).
64. Lift up the vehicle.
65. Replace the circlip of the front driveshaft with a new part.
66. Apply grease to the oil seal lip.
67. Attach the ST-28399SA010 seal protector to side retainer.
68. Align and insert the spline of the front driveshaft to the splines of the differential bevel gear, and remove the ST.
69. Insert the front driveshaft into the transmission securely by pressing the front housing from the outside.
70. Install the ball joint into the housing. Tighten the attachment bolts to 37 ft. lbs. (50 Nm).
71. Install the stabilizer to the crossmember.
72. Always tighten the rubber bushing locations with wheels in full contact with the ground and the vehicle at curb weight. Tighten to 18 ft. lbs. (25 Nm).
73. Install the shift select cable onto select lever.
74. Install the oil charge pipe, and connect the ATF cooler hoses to the pipe.
75. Install the propeller shaft.
76. Install the heat shield cover.
77. Install the rear exhaust pipe and muffler.
78. Install the front and center exhaust pipe.
79. Install the undercover.
80. Lower the lift.
81. Install the ATF level gauge.
82. Connect the following connectors:
- Transmission harness connectors
- Transmission ground terminal

83. Install the air intake chamber stay and tighten to 12 ft. lbs. (16 Nm).
84. Install the air intake chamber.
85. Install the air intake duct.
86. From the oil charge pipe, add ATF until the fluid level is between the upper and lower level on the "COLD" side of the level gauge.
87. Lower the vehicle from the lift.
88. Check the differential gear oil level.
89. Check the select lever operation.
90. Bleed the air of control valve.
91. Check the ATF level.
92. Execute the learning control promotion.
93. Perform the road test and check for leaks.

MANUAL TRANSMISSION ASSEMBLY

REMOVAL & INSTALLATION

See Figures 25 and 26.

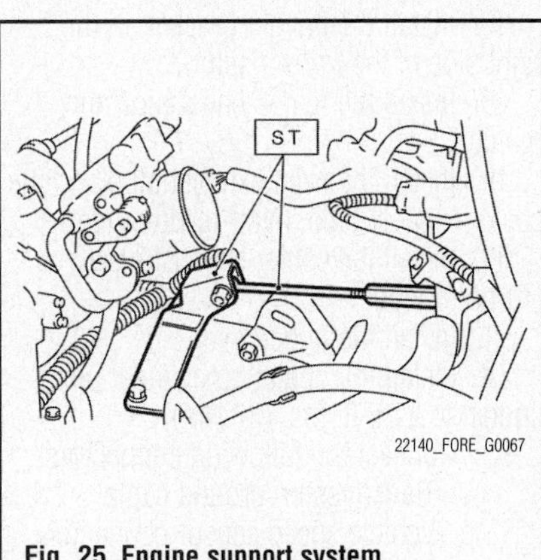

Fig. 25 Engine support system

1. Before servicing the vehicle, refer to the Precautions Section.
2. Set the vehicle on a lift.
3. Open the front hood fully, and support with stay.
4. Disconnect the ground cable from the battery.
5. Drain transmission gear oil completely.
6. Remove the intake boot and air intake chamber. (Non-turbo model)
7. Remove the air intake chamber stay. (Non-turbo model)
8. Remove the intercooler. (Turbo model)
9. Disconnect the following connectors and transmission cables.
    a. Neutral position switch connector
    b. Back-up light switch connector
10. Remove the starter.
11. Remove the operating cylinder from transmission, and suspend on a wire.
12. Remove the pitching stopper.
13. Install engine support ST-41099AA020 and bracket ST-41099AA010.
14. Remove the bolts which hold upper side of transmission to engine.
15. Remove the front and center exhaust pipes. (Non-turbo model)
16. Remove the center exhaust pipe. (Turbo model)
17. Remove the rear exhaust pipe and muffler.
18. Remove the hanger bracket from the right side of transmission.
19. Remove the propeller shaft.

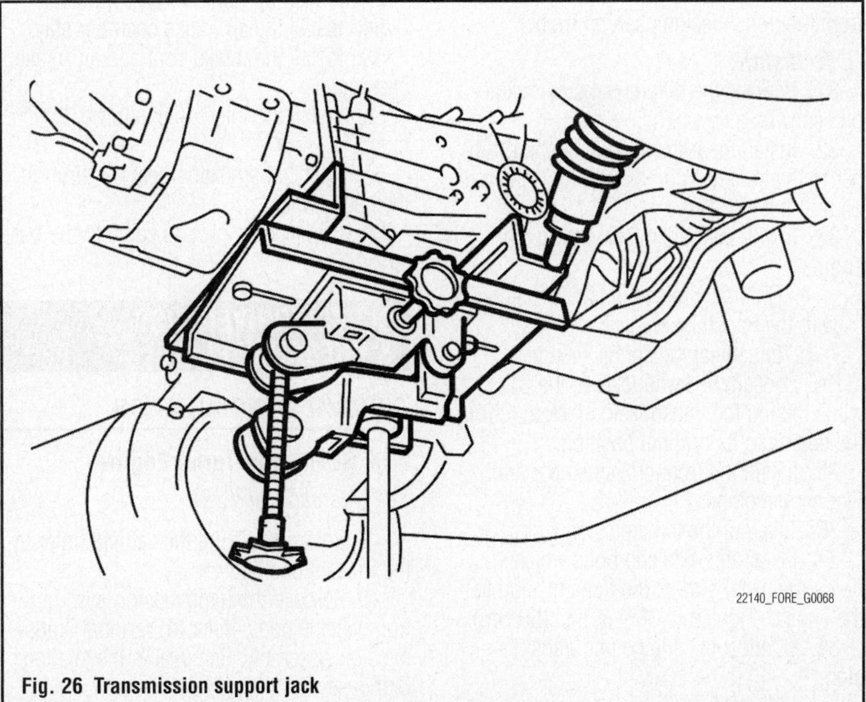

Fig. 26 Transmission support jack

20. Remove the gear shift rod and the stay from the transmission.

a. Disconnect the stay from the transmission.

b. Remove the gear shift rod from the transmission.

21. Separate the stabilizer link from the transverse link.

22. Remove the bolt securing the ball joint of the transverse link to the housing, then separate the transverse link and the housing.

23. Using the ST-28399SA000 pry tool, remove the front driveshaft from the transmission side.

24. Hold the transmission side joint (AARi) of front driveshaft by hand and extract the housing from the transmission by pressing it outside so as not to stretch the boot on the AARi side.

25. Remove the bolts and nuts which hold lower side of transmission to engine.

26. Place the transmission jack under the transmission.

⁑ CAUTION

Always support the transmission case with a transmission jack.

27. Remove the front crossmember and rear crossmember from the vehicle.

28. Tighten the turnbuckle of the ST while lowering the transmission jack to tilt the engine assembly towards the back.

29. Remove the transmission. Move the transmission jack towards the rear until main shaft is withdrawn from the clutch cover.

30. Separate the transmission assembly from the transmission cushion rubber.

To install:

31. Replace the differential side retainer oil seal with a new part.

32. Install the transmission cushion rubber to the transmission assembly. Tighten to 25.8 ft. lbs. (32 Nm).

33. Install the transmission onto the engine.

a. Gradually raise the transmission with the transmission jack.

b. Engage at the spline section.

34. Loosen the turnbuckle of the ST while raising the transmission jack to return the engine to its original position.

35. Install the front crossmember and rear crossmember.

36. Take out the transmission jack.

37. Install the nuts and bolts which holds the lower side of the transmission to the engine. Tighten to 36.9 ft. lbs. (50 Nm).

38. Connect the engine and transmission.

a. Install the starter.

b. Install the bolts which holds the upper side of the transmission to the engine. Tighten to 36.9 ft. lbs. (50 Nm).

39. Remove the engine support ST-41099AA020 and bracket ST-41099AA010.

40. Install the pitching stopper.

41. Lift up the vehicle.

42. Install the front driveshaft into the transmission.

43. Insert the ball joints of the transverse link into the housing, then tighten the installing bolts to 36.9 ft. lbs. (50 Nm).

44. Attach the stabilizer link to the transverse link. Tighten to 33.2 ft. lbs. (45 Nm).

45. Attach the gear shift rod and stay.

a. Attach the gear shift rod to the transmission and tighten to 9 ft. lbs. (12 Nm).

b. Attach the stay to the transmission and tighten to 13.3 ft. lbs. (18 Nm).

46. Install the propeller shaft

47. Install the hanger bracket on the right side of the transmission.

48. Install the rear exhaust pipe and muffler.

49. Install the front exhaust pipe and the center exhaust pipe. (Non-turbo model)

50. Install the center exhaust pipe. (Turbo model)

51. Install the undercover.

52. Install the operating cylinder and tighten to 27.5 ft. lbs. (37 Nm).

53. Connect the following connectors.

a. Transmission ground cable

b. Vehicle speed sensor connector

c. Neutral position switch connector

d. Back-up light switch connector

54. Install the air intake chamber stay.

55. Install the intake boot and air intake chamber

56. Install the intercooler. (Turbo model)

57. Pour transmission gear oil through the gauge hole.

58. Connect the ground cable to the battery.

CLUTCH DRIVEN DISC & PRESSURE PLATE

REMOVAL & INSTALLATION

2.5L SOHC Non-Turbo Engine

See Figures 27 and 28.

1. Before servicing the vehicle, refer to the Precautions Section.

2. Remove the transmission assembly from vehicle body. Refer to "Manual Transmission Assembly, Removal & Installation" in this section.

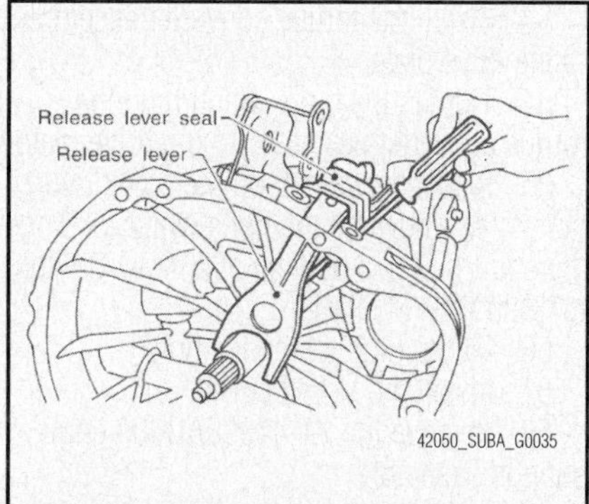

Fig. 27 Remove the release lever retainer spring using a screwdriver

3. Remove the two clips from the clutch release lever and remove the clutch release bearing.

4. Remove the clutch release lever seal (dust cover).

5. Remove the clutch release lever retainer spring from the clutch release lever point with a screwdriver by accessing it through clutch housing clutch release lever hole.

6. Remove clutch release lever.

7. Remove the pressure plate and clutch disc.

To install:

8. Insert the alignment tool into the clutch disc and attach to the flywheel by inserting the end into pilot bearing.

9. Install the clutch cover to the flywheel and tightened the bolts in a criss-cross order. Tighten to 11.6 ft. lbs. (16 Nm).

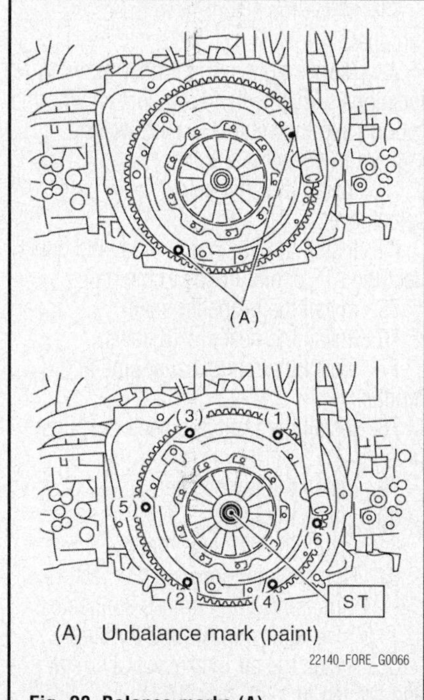

Fig. 28 Balance marks (A)

➡When installing the pressure plate to the flywheel, position in order to make the gap between in balance marks (0 paint mark) of flywheel and clutch cover 120°or more. (The in balance marks show the residual in balance direction.) Note the front and rear of the clutch disc when installing.

10. Before or during assembling, lubricate the following points with a light coat of grease:
   - Inner groove of release bearing
   - Contact surface of lever and pivot
   - Contact surface of lever and bearing
   - Transmission main shaft spline
11. While pushing the fork to pivot and twisting it to both sides, fit retainer spring onto the constricted portion of pivot.

➡Apply grease to the contact point of release lever and slave cylinder.

12. Confirm that retainer spring is securely fitted by observing it through the main case hole.
13. Install the release bearing and fasten it with two clips.
14. Install the release lever seal.
15. Install the manual transmission assembly.

2.5L DOHC Turbo Engine

See Figure 29.

1. Before servicing the vehicle, refer to the Precautions Section.
2. Remove the transmission assembly. Refer to "Manual Transmission Assembly, Removal & Installation" in this section.
3. Remove the clutch release lever from the transmission.
4. Put the clutch release bearing in the engine side.
5. Remove the clutch release bearing from the clutch cover using a flat tip screwdriver.
6. Remove the pressure plate and clutch disc.

To install:

7. Insert the alignment tool into the clutch disc and attach to the flywheel by inserting the end into pilot bearing.
8. Install the clutch cover to the flywheel and tightened the bolts in a criss-cross order. Tighten to 11.6 ft. lbs. (16 Nm).
9. Install the release bearing on the transmission.
10. Insert the release fork into the release bearing tab.

➡Apply grease to the contact point of release lever and slave cylinder.

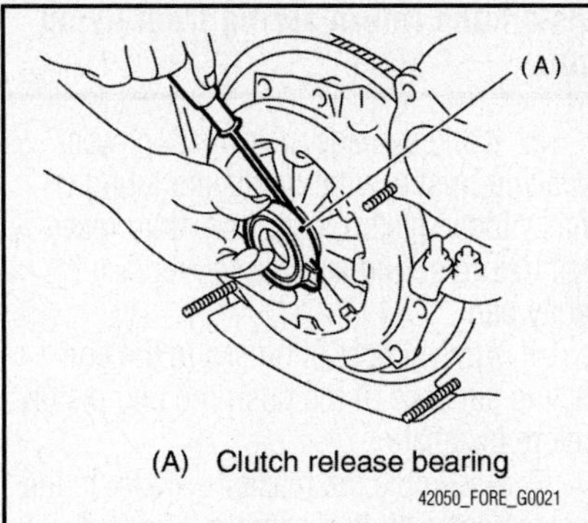

Fig. 29 Remove the clutch release bearing using a screwdriver—Turbo Models

11. Insert the release fork shaft into the release fork.
12. Tighten the plug.
13. Move the release fork to ensure that the release bearing and release fork move slowly.
14. Install the manual transmission assembly.

ADJUSTMENTS

This vehicle is equipped with a hydraulic clutch that is self-adjusting, therefore no adjustment is possible or necessary.

CLUTCH MASTER CYLINDER

REMOVAL & INSTALLATION

See Figures 30 and 31.

1. Before servicing the vehicle, refer to the Precautions Section.
2. Thoroughly drain the fluid from the reservoir tank.
3. Remove the snap pin and clevis pin, and separate the push rod of the master cylinder from the clutch pedal.
4. Remove the air intake assembly.
5. Remove the intercooler, if equipped.
6. Remove the clutch pipe from the master cylinder.

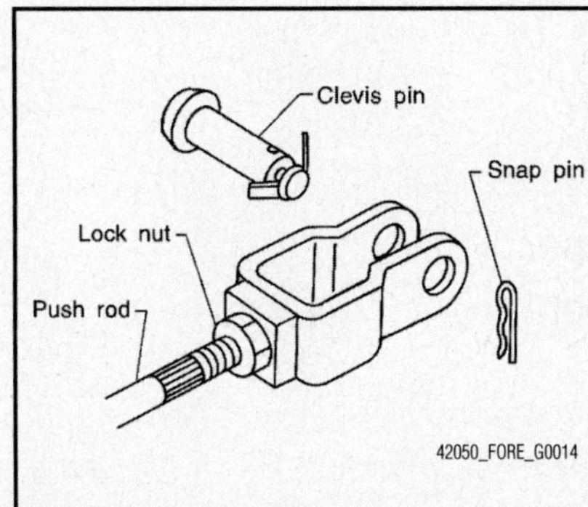

Fig. 30 Separate the push rod from the clutch pedal.

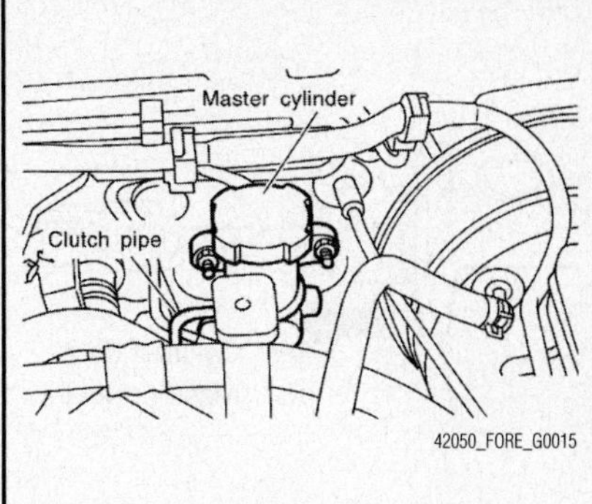

Fig. 31 Location of the master cylinder in the engine compartment

7. Remove the master cylinder assembly.

To install:

8. Install the master cylinder to the body. Tighten the mounting bolts to 13 ft. lbs. (18 Nm).
9. Install the clutch pipe to the master cylinder. Tighten the nut to 11 ft. lbs. (15 Nm).
10. Apply multi-purpose grease to the clevis pin. Connect the push rod of the master cylinder to the clutch pedal. Insert the clevis pin and install the snap pin.
11. Bleed the clutch system. Refer to "Hydraulic System Bleeding" in this section.
12. Install the intercooler.
13. Install the air intake assembly.

BENCH BLEEDING PROCEDURE

1. Install clutch master cylinder in soft jawed vise.
2. Connect a vinyl tube to the clutch pipe fitting and put the other end in the master cylinder reservoir tank.
3. Pump the piston rod until no bubbles are seen, and proper resistance is felt.

OVERHAUL

See Figures 32 through 34.

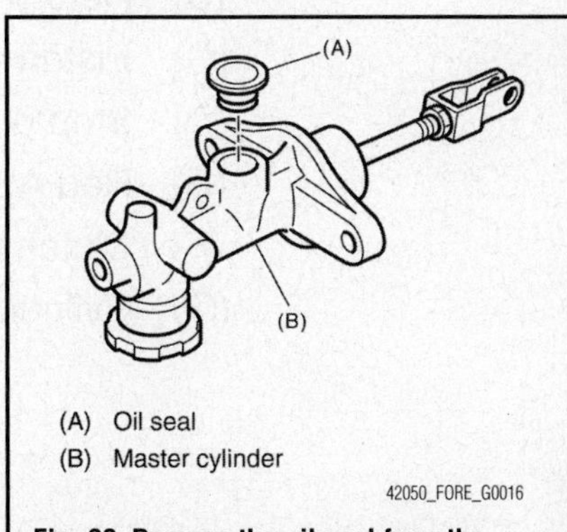

Fig. 32 Remove the oil seal from the master cylinder

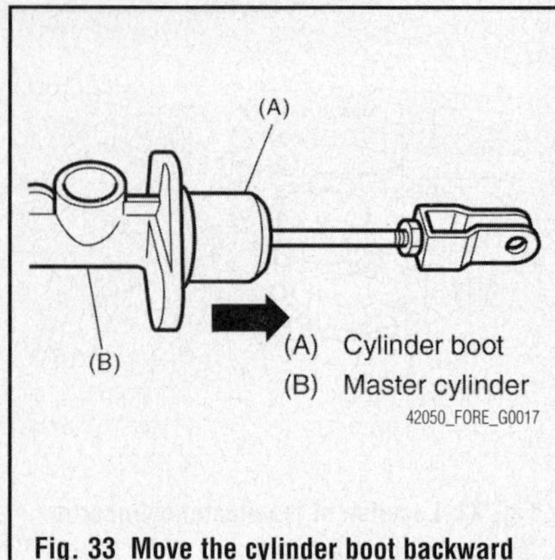

Fig. 33 Move the cylinder boot backward

1. Remove the straight pin and reservoir tank.
2. Remove the oil seal.
3. Move the cylinder boot backward.
4. Remove the snap ring.

CAUTION

Be careful when removing the snap ring to prevent the rod, washer, and piston and return spring from flying out.

5. If any damage, deformation, wear, swelling, rust or other faults are found on the cylinder, piston, push rod, fluid reservoir, return spring and gasket, replace the faulty part.
6. Apply a coat of grease to the contacting surfaces of the push rod and piston before installation.
7. Assemble the master cylinder in the reverse order of disassembly.

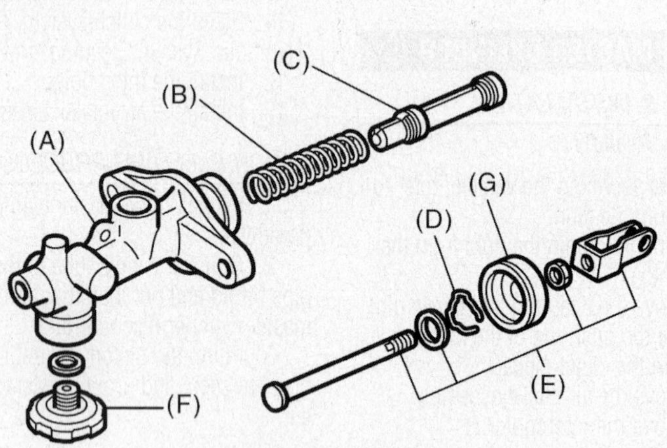

Fig. 34 Exploded view of the master cylinder

CLUTCH SLAVE CYLINDER

REMOVAL & INSTALLATION

See Figures 35 and 36.

1. Before servicing the vehicle, refer to the Precautions Section.
2. Remove the air intake assembly.
3. Remove the intercooler, if equipped.
4. Remove the clutch hose from the slave cylinder.

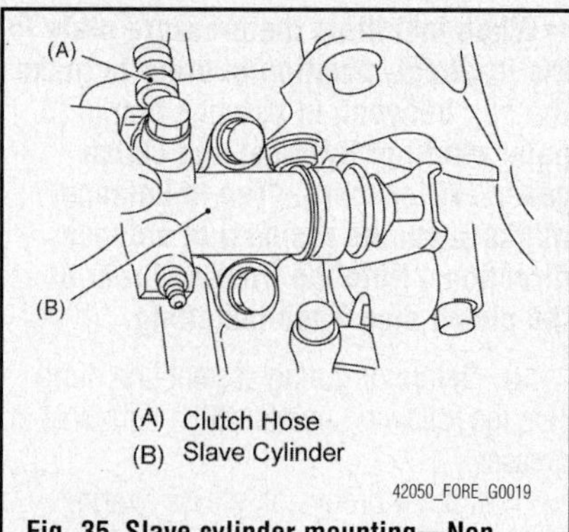

Fig. 35 Slave cylinder mounting—Non-Turbo Models

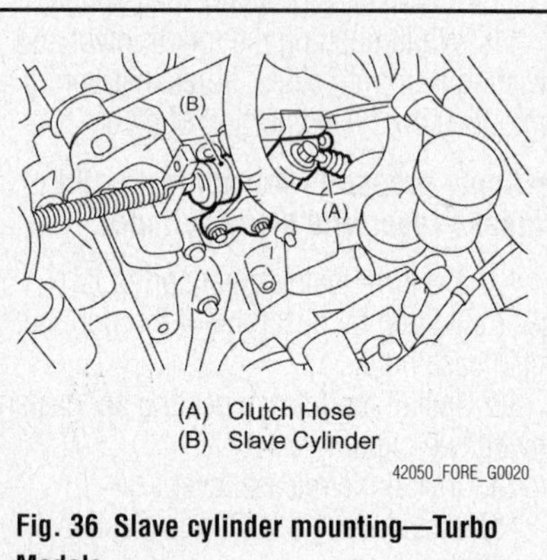

Fig. 36 Slave cylinder mounting—Turbo Models

➡Immediately plug the hose to prevent fluid from spilling.

5. Remove the slave cylinder from the transmission.

To install:

6. Apply multi-purpose grease to the contact point of the release lever and slave cylinder.
7. Install the slave cylinder and tighten the mounting bolts to 28 ft. lbs. (37 Nm).
8. Connect the clutch hose to the slave cylinder. Tighten the nut to 13 ft. lbs. (18 Nm).
9. Bleed the clutch system. Refer to "Hydraulic System Bleeding" in this section.
10. Install the intercooler.
11. Install the air intake assembly.

CLUTCH HYDRAULIC SYSTEM BLEEDING

See Figure 37.

➡To properly bleed the system, it must be bled at the slave cylinder and at the damper. Each of these has an air bleeder on it.

1. Before servicing the vehicle, refer to the precautions in the beginning of this section.

2. Connect a vinyl tube to the air bleeder on the damper and put the other end in a jar with clean clutch fluid.

➡Do not let the fluid level fall too low in the master cylinder. Do not release the pedal with the bleeder open.

3. With the help of an assistant depressing the clutch pedal, slowly open the bleeder valve. Close the bleeder valve and release the pedal. Repeat this process until no air bubbles appear in the jar.

4. Move the tube to the bleeder on the slave cylinder and repeat the process. Check the operation of the clutch after the bleed procedure is complete.

TRANSFER CASE ASSEMBLY

REMOVAL & INSTALLATION

The transfer case is removed as an assembly with the transmission. Refer to Transmission Assembly, Removal & Installation in this section.

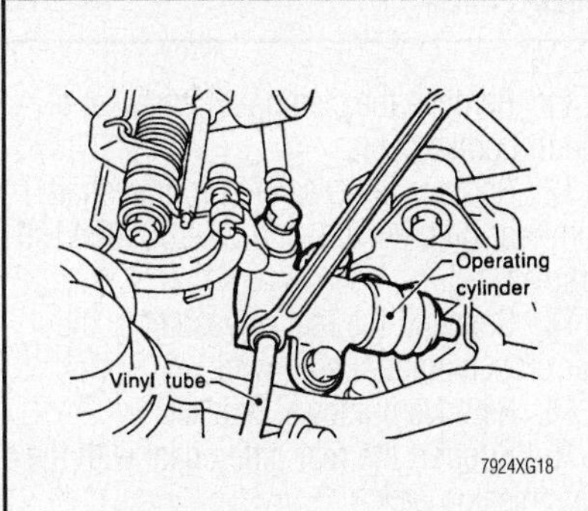

Fig. 37 Bleeding the hydraulic clutch at the slave cylinder

FRONT HALFSHAFTS

REMOVAL & INSTALLATION

See Figure 38.

1. Before servicing the vehicle, refer to the Precautions Section.

2. Disconnect ground cable from battery.

3. Jack-up vehicle, support it with safety stands, and remove front wheels.

4. Unlock axle nut.

5. Remove axle nut while depressing brake pedal to prevent front driveshaft from turning.

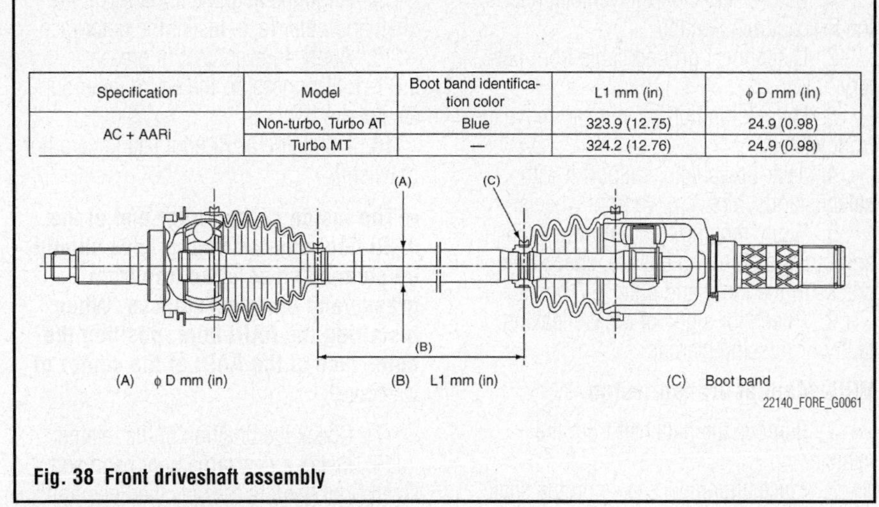

Specification	Model	Boot band identification color	L1 mm (in)	ϕ D mm (in)
AC + AARi	Non-turbo, Turbo AT	Blue	323.9 (12.75)	24.9 (0.98)
	Turbo MT	—	324.2 (12.76)	24.9 (0.98)

Fig. 38 Front driveshaft assembly

⁂ CAUTION

Be sure to loose and retighten axle nut after removing wheel from vehicle. Failure to follow this rule may damage wheel bearings.

6. Remove stabilizer link.

7. Remove disc brake caliper from housing, and suspend it from strut using a wire.

8. Remove disc rotor from hub.

9. If disc rotor seizes up within hub, drive disc rotor out by installing an 8-mm bolt in screw hole on the rotor.

10. Remove cotter pin and castle nut which secure tie-rod end to housing knuckle arm.

11. Using a puller, remove tie-rod ball joint from knuckle arm.

12. Remove ABS sensor assembly and harness.

13. Remove bolt which secures sensor harness to strut.

14. Remove transverse link ball joint from housing.

15. Remove inner joint from transmission spindle.

16. Remove front driveshaft assembly from hub. If it is hard to remove, use a ST-926470000 puller.

⁂ CAUTION

Be careful not to damage oil seal lip and tone wheel when removing front driveshaft. If front driveshaft is removed, replace inner oil seal with new one.

17. After scribing an alignment mark on camber adjusting bolt head, remove bolts which connect housing and strut, and disconnect housing from strut.

To install:

18. While aligning alignment mark on camber adjusting bolt head, connect housing and strut. Tighten to 130 ft. lbs. (177 Nm).

19. Install front driveshaft.

20. Install transverse link ball joint to housing and tighten to 36 ft. lbs. (49 Nm).

21. Install ABS sensor harness on strut.

22. Install ABS sensor on housing. Tighten the mounting bolt to 24 ft. lbs. (32 Nm).

23. Install disc rotor on hub.

24. Install disc brake caliper on housing and tighten to 58 ft. lbs. (78 Nm).

25. Connect stabilizer link.

26. Install tie-rod end and tighten castle nut to 20 ft. lbs. (27 Nm).

27. After tightening castle nut to specified torque, retighten it further within 60° until a slot in castle nut is aligned with the ball joint hole. Insert cotter pin, and then bend the cotter pin around castle nut to secure it.

28. While depressing brake pedal to prevent front driveshaft from turning, tighten axle nut to 162 ft. lbs. (220 Nm).

⁂ CAUTION

When axle nut is removed, replace it with new one. Be sure to tighten axle nut to specified torque. Do not over tighten it as this may damage wheel bearing.

29. After tightening axle nut, lock it securely.

30. Install wheel and tighten wheel nuts.

CV-BOOTS INSPECTION

With Automatic Transmission

See Figure 39.

1. Before servicing the vehicle, refer to the Precautions Section.
2. Disconnect ground cable from battery.
3. Unlock ignition so you can move the front wheels.
4. Jack-up vehicle, support it with safety stands, and remove front wheels.
5. While moving the front wheels in and out and spinning the tire, check for cracks in the inner and outer CV-boots.
6. Check for signs of grease leaking, faulty or missing clamps.

With Manual Transmission

1. Remove the halfshaft from the vehicle.
2. Place alignment marks on the shaft and outer race.
3. Remove the AARi boot band and boot.
4. Remove the retainer from the AARi outer race.
5. Remove the AARi outer race from the shaft assembly. Wipe off the grease.

➡This is special grease. Do not confuse it with other greases.

6. Place an alignment mark on the trunion and shaft.
7. Remove the snapring and trunion.
8. Remove the spider.
9. Remove the AARi boot.

➡Be sure to use the proper grease, One Louver C (part number 28395SA000).

10. Place the AARi boot and retainer to the shaft and position it at the center of the shaft.
11. Align the alignment marks and install the trunion on the shaft.
12. Fill 3.53–3.88 ounces of grease into the interior of the AARi outer race.
13. Apply a coat of grease to the free ring and trunion.

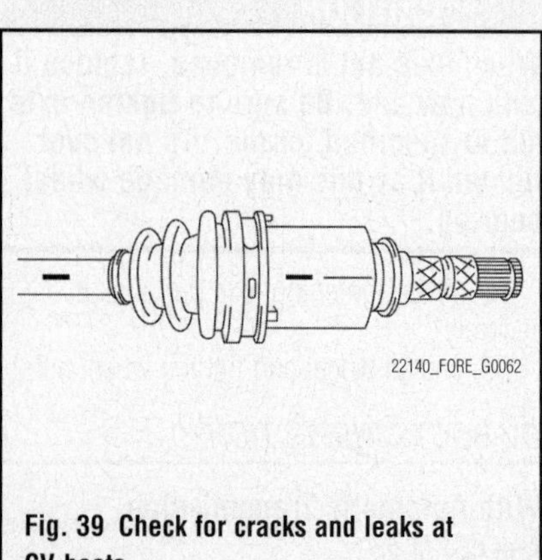

Fig. 39 Check for cracks and leaks at CV-boots

14. Align the alignment marks on the shaft and outer race. Install the outer race.
15. Apply an even coat of grease, 1.06–1.41 ounces, to the entire inner surface of the boot.
16. Install the AARi boot taking care not to twist it.

➡The inside of the larger end of the AARi boot and the boot groove should be cleaned so as to be free from grease and other substances. When installing the AARi boot, position the outer race of the AARi at the center of its travel.

17. Check the position of the retainer.
18. Install a new large boot band and small boot band in the specified position.
19. Tighten the band using tool ST280099AC000. Clearance at the crimped section of the large boot band should be 0.04 inch, or less. Clearance at the crimped section of the small boot band should be 0.04 inch, or less.

➡Extend and retract the AARi to provide equal grease coating.

REAR AXLE HOUSING

REMOVAL & INSTALLATION

See Figures 40 through 42.

1. Before servicing the vehicle, refer to the Precautions Section.
2. Set the vehicle on a lift.
3. Disconnect the ground cable from the battery.
4. Shift the select lever or gear shift lever to neutral.
5. Release the parking brake.
6. Loosen the wheel nuts.
7. Lift up the vehicle.
8. Remove the wheels.
9. Remove the rear exhaust pipe and muffler.
10. Remove the propeller shaft.

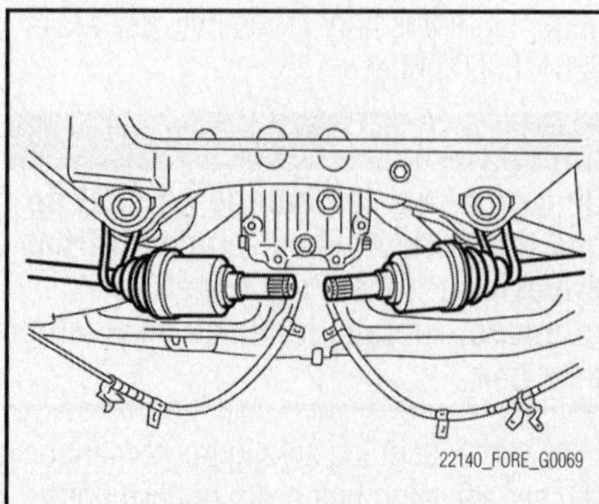

Fig. 40 Rear driveshaft suspended using wire

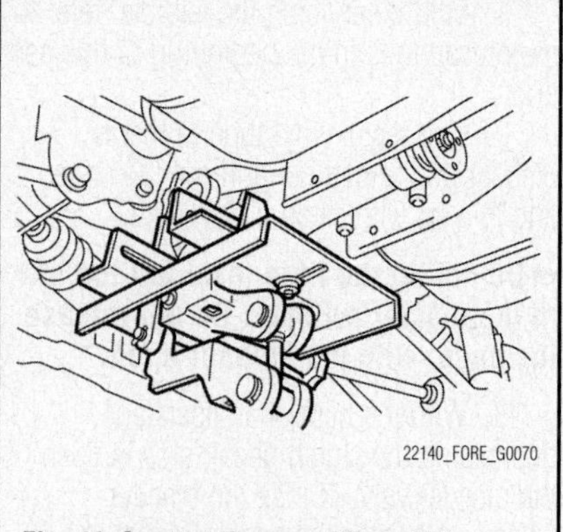

Fig. 41 Support the rear differential

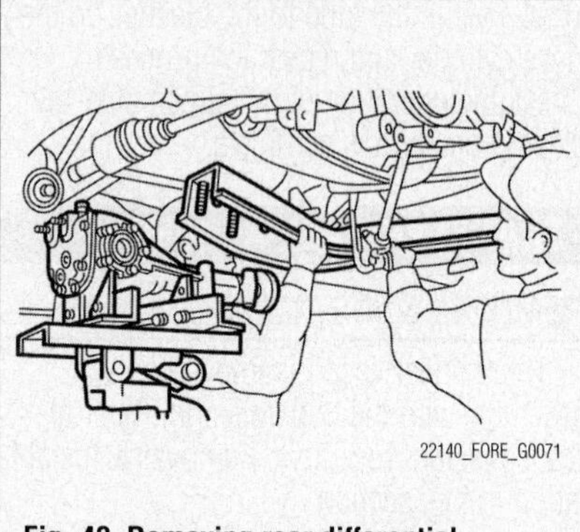

Fig. 42 Removing rear differential from vehicle

11. Remove the clamps and bracket of parking brake cable.
12. Remove the DOJ of rear driveshaft from rear differential using ST-28099PA100 remover.
13. Suspend the rear driveshaft to the rear crossmember using wire.
14. Remove the lower bracket.
15. Support the rear differential with the transmission jack.
16. Remove the self-locking nuts and bolts.
17. Remove the bolts which secure the front differential member to the vehicle. Loosen the bolt A first, then remove the bolt B.

➡Instruct your co-worker to hold the front differential member so that it will not fall.

18. Remove the bolt A.
19. While slowly lowering the transmission jack, move the rear differential forward, and remove the front differential member and the rear differential from vehicle.
20. Remove the rear differential from the front differential member.

To install:

21. Install in the reverse order of removal.

22. Position the front differential member with the vehicle by passing the member under the parking brake cable and securing it to rear differential.

23. Install new rear differential side oil seal.

24. Install the DOJ of the driveshaft into the rear differential using oil seal protector ST-28099PA090.

25. After installing, fill the differential carrier with gear oil. Approximately 0.8 qt. (0.8L). GL-5–75w90 up to the bottom of the filler plug hole.

REAR AXLE SHAFT SEAL

REMOVAL & INSTALLATION

See Figures 43 and 44.

1. Before servicing the vehicle, refer to the Precautions Section.
2. Disconnect the ground cable from the battery.
3. Shift the select lever or gear shift lever to neutral.
4. Release the parking brake.
5. Loosen the wheel nuts on both sides.
6. Lift up the vehicle.
7. Remove the wheels.
8. Remove the rear exhaust pipe and muffler.
9. Remove the DOJ of rear driveshaft from rear differential.
10. Remove the ABS wheel speed sensor cable clamp and the parking brake cable clamp from bracket.
11. Remove the ABS wheel speed sensor cable clamp from the trailing link.
12. Remove the ABS wheel speed sensor cable clamp and the parking brake cable guide from the trailing link.
13. Remove the rear stabilizer link.
14. Remove the bolts which secure the trailing link to the housing.
15. Remove the bolts which secure the front and rear lateral links to the rear housing.

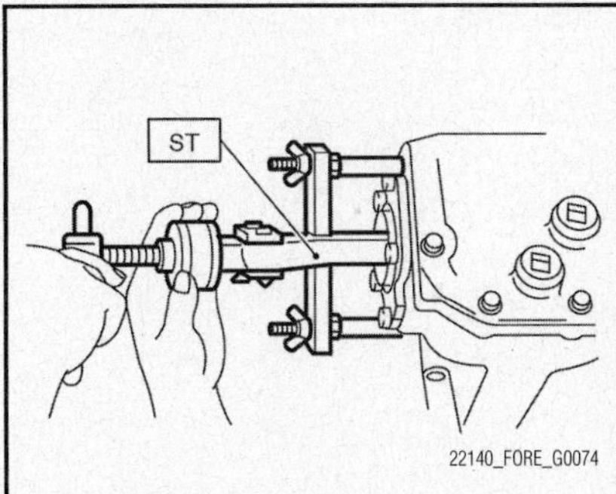

Fig. 43 Oil seal removal shown with ST-398527700

Fig. 44 Oil seal installation shown with ST-398437700

16. Remove the DOJ from the rear differential by using ST.
17. Suspend the rear driveshaft to the rear crossmember using wire.
18. Using the ST-398527700 remove the oil seal.

To install:

19. Install a new side oil seal using the ST-398437700.
20. Insert the DOJ into rear differential.
    a. Attach the ST to rear differential.
    b. Install the spline shaft until the spline portion is inside the side oil seal using ST-28099PA090 oil seal protector.
    c. Remove the ST-28099PA090.
21. Hereafter, reassemble in the reverse order of disassembly.

REAR HALFSHAFT

REMOVAL & INSTALLATION

1. Before servicing the vehicle, refer to the Precautions Section.
2. Disconnect the negative battery cable.
3. Raise and support the vehicle safely.
4. Remove the wheels and tires.
5. Unlock the axle nut. Depress the parking brake. Remove the axle nut.

➡Be sure to loosen the axle nut after removing the tire and wheel from the vehicle. Failure to do this may damage the wheel bearings.

6. Remove the rear stabilizer link. Remove the bolts that retain the trailing link to the housing.
7. Remove the bolts that retain the front and rear lateral link to the housing.
8. Remove the rear ABS wheel speed sensor from the backing plate.
9. Remove the rear halfshaft.

➡Use axle shaft puller tools ST92647000 and ST28099PA110, or equivalent, if it is difficult to remove the halfshaft. Do not hammer the half-shaft to remove it. Be careful not to damage the oil seal or magnetic encoder.

To install:

10. Installation is the reverse of the removal procedure.
11. Tighten the axle nut to 140 ft. lbs. (190 Nm), with the parking brake applied.

➡Install the tire and wheel assembly after installation of the axle nut. Failure to follow this order, may cause damage to the wheel bearings.

CV-JOINTS OVERHAUL

The Double Offset Joint (DOJ), is the only part of the assembly that can be replaced, if any of the other components are defective then the shaft should be replaced.

1. Before servicing the vehicle, refer to the Precautions Section.
2. Remove the halfshaft from the vehicle.
3. Straighten the bent claw of the large clamp at the Double Offset Joint (DOJ) end of the boot.
4. Loosen the band using pliers being careful not to damage the bolt.
5. Remove the small boot band from the DOJ using the same technique.
6. Remove the boot from the large end of the DOJ outer race.
7. Remove the round circlip using a suitable pry tool from the neck of the joint outer race.
8. Remove the joint outer race.

✲✲ CAUTION

The grease used is for CV–Joints, do not replace with another type of grease

9. Clean the grease and remove the balls. Be careful not to lose any of the 6 balls.
10. Turn the cage by a half pitch to the track groove of the inner race and shift the cage.
11. Remove the snapring that secures the inner race to the shaft and remove the inner race.
12. Take the cage from the shaft and remove the boot.
13. The other boots may be removed in the same manner as the DOJ boot.
14. Wrap the shaft splines with tape to prevent damage.

➡Be sure to use the proper grease. Use NTG2218-M (part number 28395AG010) for the EBJ side and NKG205 (part number 28495AG000) for the DOJ side.

15. Install the BJ or EBJ boots and fill it with 2.12–2.47 ounces of grease.
16. Place the DOJ boot on the center of the shaft.
17. Insert the DOJ cage onto the shaft. Make sure to insert the cage with the cut–out portion facing shaft end.
18. Install the inner race on the shaft and fasten with the snap ring. Make sure the snap ring is firmly engaged.
19. Install the cage (previously fitted) to the inner race on the shaft. Fit the cage with the protruded part aligned with the track on the inner race and then turn a half pitch.
20. Fill the DOJ inner race with 2.82–3.17 ounces of grease.
21. Apply a coat of grease to the cage pocket and 6 balls.
22. Insert the 6 balls.
23. Align the outer race track and ball positions and place where the shaft, inner race, cage and balls were located prior to removal and then install the outer race.
24. Install the circlip into the groove on the outer race.

➡Make sure the balls, cage and inner race are fully seated. Make sure not to place the matched position of the circlip in the ball groove of the outer race. Pull the shaft lightly to make sure the circlip is fully engaged.

25. Apply an even coat 0.71–1.06 ounces of grease to the entire inner surface of the boot and to the shaft.
26. Make sure the boot is free from any dirt or foreign materials prior to installation.
27. Place the outer race of the boot at the center of its travel.
28. Put a band through the boot clip and wind twice in alignment with the groove on the boot.
29. Pinch the end of the band using tool ST 9250910000 and tighten securely until it cannot be moved by hand. Make sure there is appropriate air inside the boot.
30. Tap on the clip with a punch to lock it making sure not to damage the damaged while tapping. Cut any excess off the band leaving 0.39 inch and bend the remaining portion over the clip. Make sure the end of the band is close to the clip.
31. Install the remaining boot clamps in the same manner.

REAR PINION SEAL

REMOVAL & INSTALLATION

See Figures 45 through 47.

1. Before servicing the vehicle, refer to the Precautions Section.
2. Move the gear shift level to "N".
3. Secure the vehicle with wheel chocks and release the parking brake.

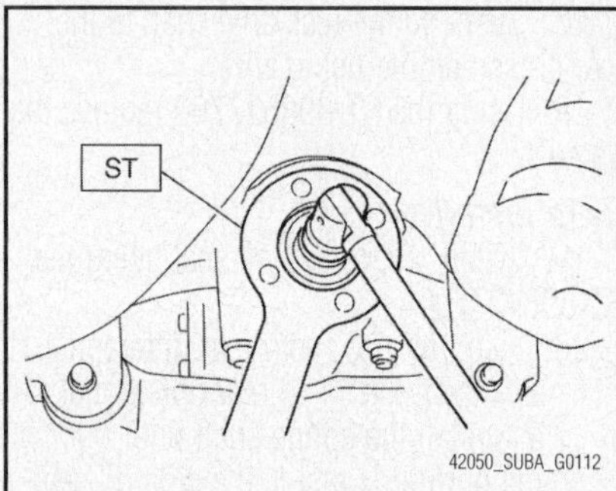

Fig. 45 Hold the companion flange with Special Tool 498427200 and remove the pinion nut.

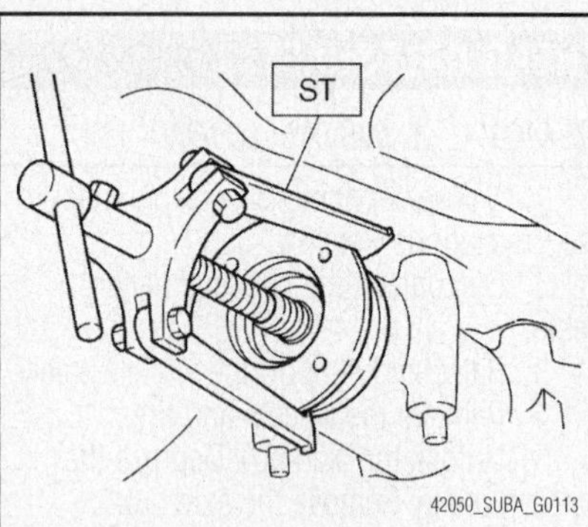

Fig. 46 Extract the companion flange using Special Tool 399703602.

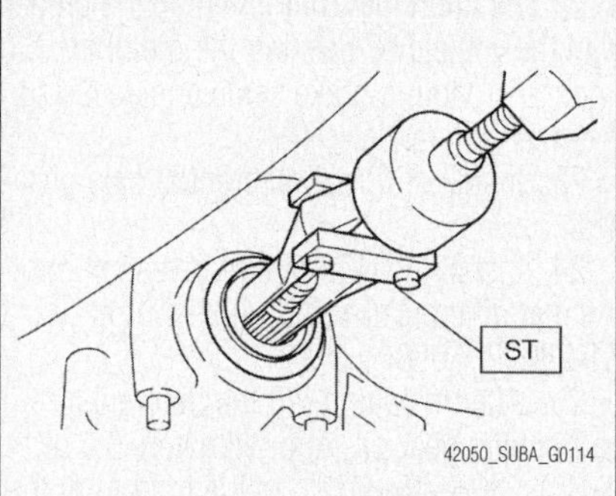

Fig. 47 Remove the oil seal using Special Tool 499705401.

4. Remove the rear differential protector, if equipped.
5. Remove the drain plug and drain the gear oil. Install the drain plug.
6. Jack up the rear wheels and safely support the vehicle with suitable jack stands.
7. Remove the rear exhaust pipe and muffler.
8. Remove the driveshaft.
9. Remove the self-locking nut while holding the companion flange with Special Tool 498427200.
10. Remove the protector mounting nut.
11. Remove the tank cover.
12. Extract the companion flange using Special Tool 399703602.
13. Remove the oil seal using Special Tool 499705401.

To install:

14. Drive in a new oil seal using Special Tool 498447120.
15. Install the companion flange. Use a plastic hammer to install the flange.
16. Tighten a new self-locking nut to 134 ft. lbs. (181 Nm) using Special Tool 498427200 Flange Wrench to hold the companion flange.
17. The remainder of the installation is the reverse order of removal.
18. Refill the differential with fluid to the correct level.

ENGINE COOLING

ENGINE FAN

REMOVAL & INSTALLATION

Main Fan Motor

2.5L SOHC Non-Turbo Engines

See Figures 48 through 51.

1. Before servicing the vehicle, refer to the Precautions Section.
2. Disconnect the negative battery cable.
3. Raise and safely support the vehicle.
4. Remove the engine undercover attaching bolts.
5. Drain approximately 1 quart of engine coolant.
6. Disconnect the fan electrical connectors.
7. Remove the transmission cooler hose from the clip of the radiator fan motor assembly, if equipped.
8. Lower the vehicle.
9. Remove the accessory drive belt cover.
10. Disconnect the overflow hose.
11. Remove the reservoir tank.
12. Remove the three upper fan assembly mounting bolts.
13. Disconnect the upper radiator hose from the radiator.
14. Detach the power steering hose from the clip on the radiator.
15. Lift the engine fan assembly up and out of the engine compartment.
16. Remove the clip which holds motor connector onto the shroud.
17. Remove the nut which secures the fan to the fan motor.
18. Remove the bolts which hold fan motor onto shroud.

To install:

19. Installation is the reverse order of assembly.

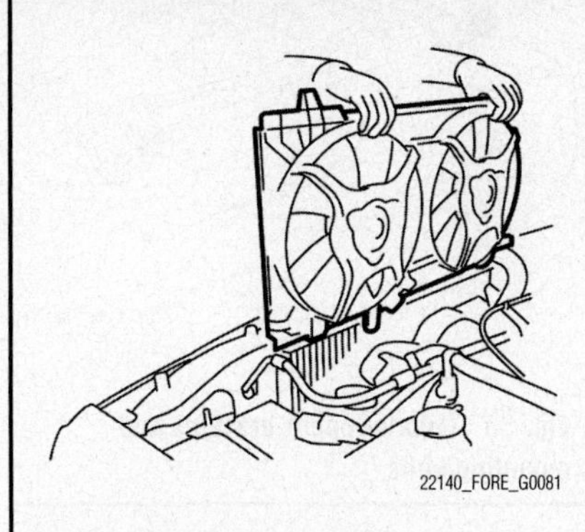

Fig. 49 Engine fan assembly removal

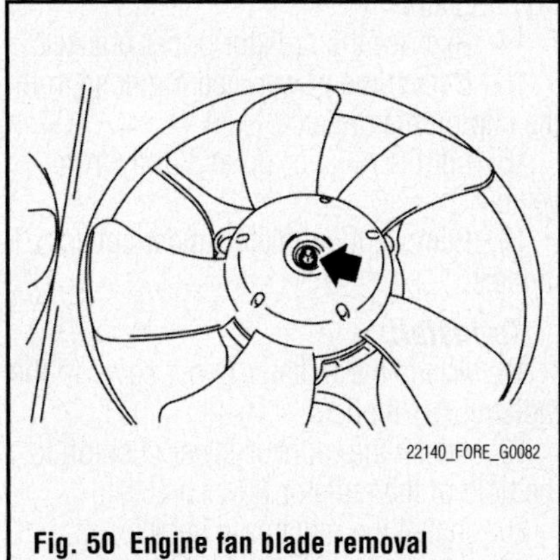

Fig. 50 Engine fan blade removal

20. Tighten fan motor mounting bolts to 3.3 ft. lbs. (4.4 Nm).
21. Tighten fan blade mounting nut to 2.5 ft. lbs. (3.4 Nm).
22. Tighten the fan assembly mounting bolts to 3.6 ft. lbs. (4.9 Nm).

2.5L DOHC Turbo Engines

See Figure 51.

1. Before servicing the vehicle, refer to the Precautions Section.
2. Set the vehicle on a lift.

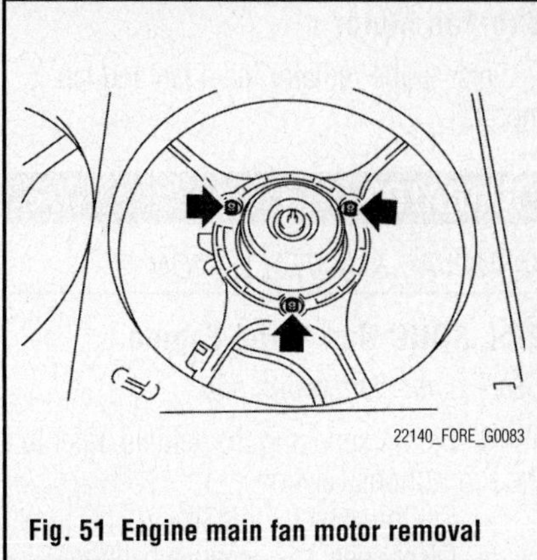

Fig. 51 Engine main fan motor removal

3. Disconnect the negative battery cable.
4. Remove the collector cover.
5. Remove the upper radiator bracket, if equipped with automatic transmission.
6. Raise and safely support the vehicle.
7. Remove the engine undercover.
8. Drain approximately 1 quart of engine coolant.
9. Disconnect the fan electrical connectors.
10. Remove the radiator undercover mounting bolt, if equipped with automatic transmission.
11. Pull the lower radiator hose upward to separate the radiator from the radiator under cover. Remove the automatic transmission hose from the two clips on the fan motor assembly.
12. Lower the vehicle.
13. Remove the reservoir tank.
14. Remove the mounting bolts from radiator main fan motor assembly.
15. Disconnect the radiator inlet hose and the over flow hose.
16. Disconnect the power steering hose from the clip on the radiator.
17. Disconnect the upper radiator hose from the radiator.
18. Lift the engine fan assembly up and out of the engine compartment.
19. Remove the clip which holds motor connector onto the shroud.
20. Remove the nut which secures the fan to the fan motor.
21. Remove the bolts which hold fan motor onto shroud.

To install:

22. Installation is the reverse order of assembly.
23. Tighten fan motor mounting bolts to 3.3 ft. lbs. (4.4 Nm).
24. Tighten fan blade mounting nut to 2.5 ft. lbs. (3.4 Nm).

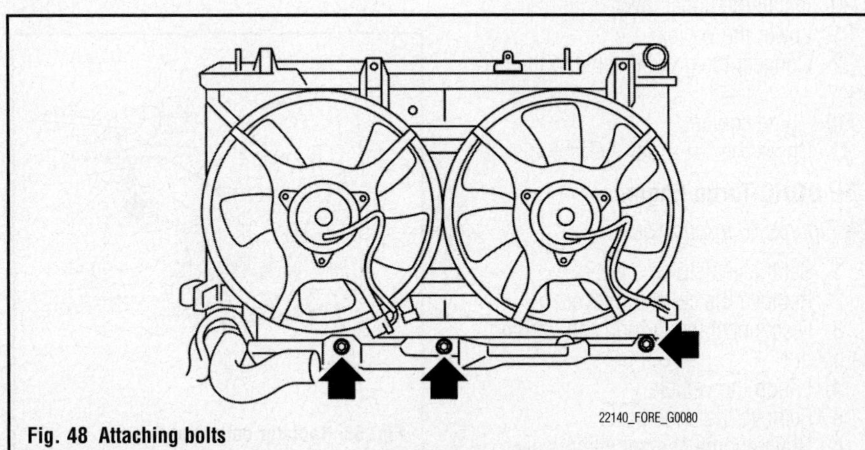

Fig. 48 Attaching bolts

25. Tighten the fan assembly mounting bolts to 3.6 ft. lbs. (4.9 Nm).

Sub-Fan Motor

Refer to the radiator main fan and fan motor.

RADIATOR

REMOVAL & INSTALLATION

2.5L SOHC Non-Turbo Engine

See Figures 53 through 55.

1. Before servicing the vehicle, refer to the Precautions Section.
2. Set the vehicle on a lift.
3. Disconnect the negative battery cable.
4. Lift up the vehicle.
5. Remove the undercover.
6. Drain engine coolant completely.
7. Disconnect the connectors of radiator main fan (A) and sub fan (B) motor.
8. Disconnect the radiator outlet hose from thermostat cover.
9. Disconnect the ATF hoses from ATF pipes, A/T models. Apply the cap to the ATF pipe to prevent ATF leaks.
10. Lower the vehicle.
11. Disconnect the over flow hose.

Fig. 53 Connector locations of radiator main fan (A) and sub fan (B)

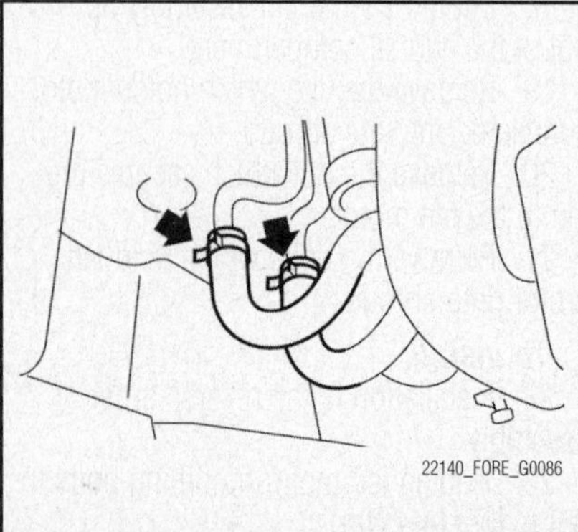

Fig. 54 ATF hose clamps and ATF pipes

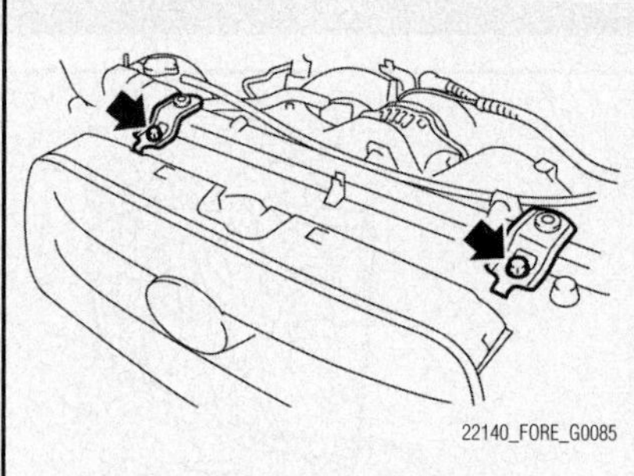

Fig. 55 Radiator upper brackets and mounting bolts

12. Remove the reservoir tank.
13. Disconnect the radiator inlet hose from radiator.
14. Remove the radiator upper brackets.
15. Detach the power steering hose from the clip on the radiator.
16. Lift the radiator up and away from vehicle.
17. Remove the radiator under cover, A/T models.

To install:

18. Attach the radiator under cover to the radiator, A/T models.
19. Attach the radiator lower cushion to the hole of the radiator lower bracket.
20. Install the radiator to vehicle.
21. Install the radiator upper brackets and tighten the bolts to 13.3 ft. lbs. (18 Nm).
22. Attach the power steering hose to the radiator.
23. Connect the radiator inlet hose.
24. Install the reservoir tank.
25. Connect the over flow hose.
26. Lift up the vehicle.
27. Connect the ATF hoses, A/T models.
28. Connect the radiator outlet hose.
29. Connect the connectors to the radiator main fan motor (A) and sub fan motor (B).
30. Install the undercover.
31. Lower the vehicle.
32. Connect the ground cable to the battery.

Fill engine coolant.

33. Check the ATF level, A/T models.

2.5L DOHC Turbo Engine

See Figures 56 through 58.

1. Set the vehicle on a lift.
2. Remove the collector cover.
3. Disconnect the ground cable from the battery.
4. Lift up the vehicle.
5. Remove the undercover.
6. Drain engine coolant completely.
7. Disconnect the connectors of radiator main fan and sub fan motor.
8. Disconnect the radiator outlet hose from thermostat cover.
9. Disconnect the ATF hoses from ATF pipes, A/T models.

Apply the cap to the ATF pipe to prevent ATF leaks.

10. Lower the vehicle.
11. Remove the reservoir tank.
12. Disconnect the radiator inlet hose, air breather hose and the overflow hose from radiator.
13. Remove the radiator upper brackets.
14. Disconnect the power steering hose from the clip on the radiator.
15. Lift the radiator up and away from vehicle.
16. Remove the radiator under cover, A/T models.

To install:

17. Attach the radiator under cover to the radiator, A/T models.
18. Attach the radiator lower cushion to the hole of the radiator lower bracket.
19. Install the radiator to vehicle.

➡Make pins on the lower side of radiator be fitted into the radiator lower cushions on body side.

20. Install the radiator upper brackets and tighten the bolts to 13.3 ft. lbs. (18 Nm).
21. Attach the power steering hose to the radiator.
22. Connect the radiator inlet hose, air breather hose and the overflow hose (B).
23. Install the reservoir tank.
24. Lift up the vehicle.
25. Connect the ATF hoses, A/T models.
26. Connect the radiator outlet hose.
27. Connect the connectors to the radiator main fan motor (A) and sub fan motor (B).
28. Install the undercover.
29. Lower the vehicle.

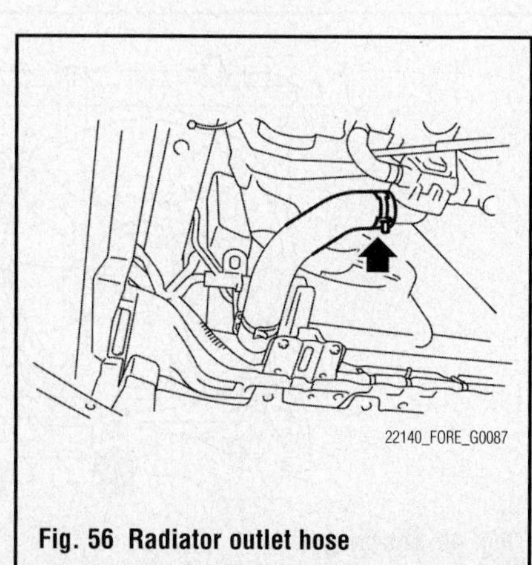

Fig. 56 Radiator outlet hose

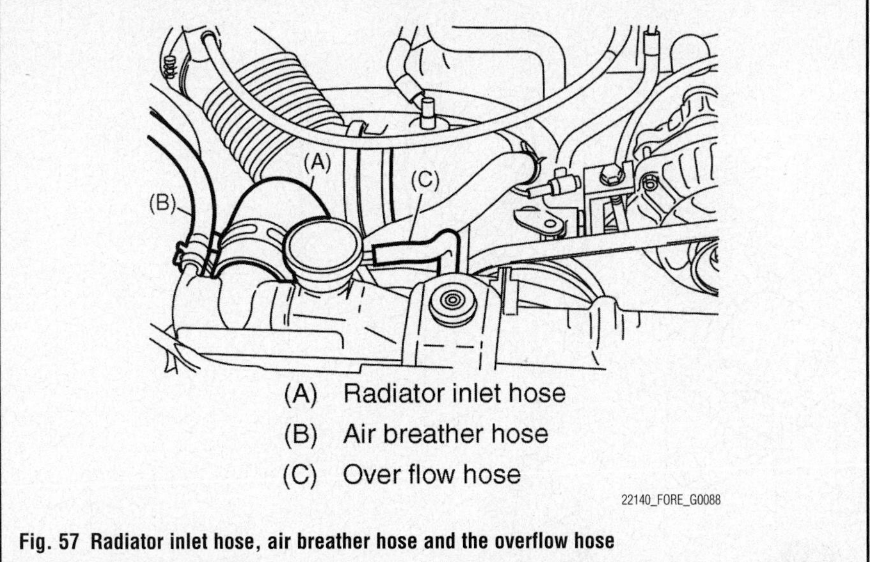

Fig. 57 Radiator inlet hose, air breather hose and the overflow hose

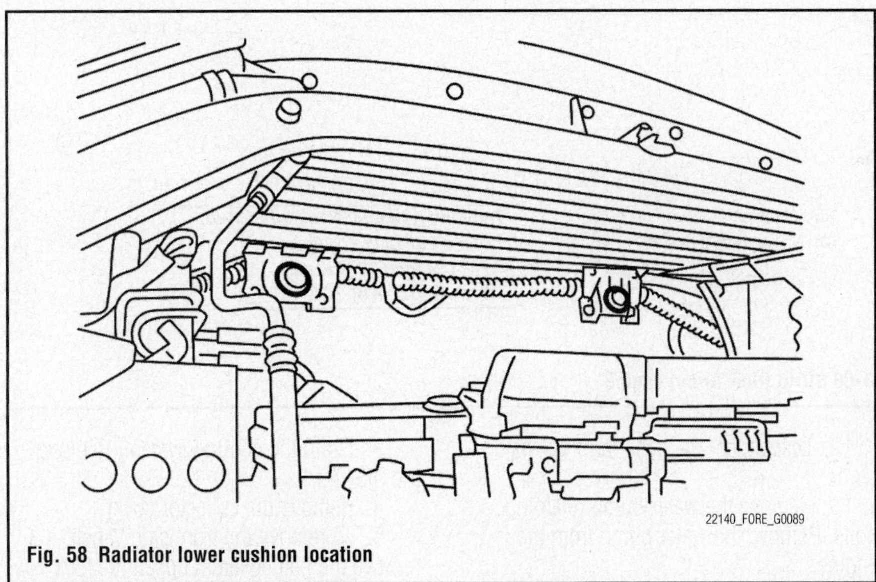

Fig. 58 Radiator lower cushion location

30. Connect the ground cable to the battery.
31. Fill engine coolant.
32. Check the ATF level.
33. Install the collector cover.

THERMOSTAT

REMOVAL & INSTALLATION

See Figures 59 through 61.

1. Raise and safely support the vehicle.
2. Remove the engine undercover.
3. Drain the engine coolant.
4. Loosen the hose clamp and disconnect the lower radiator hose from the thermostat cover.
5. Remove the thermostat cover and gasket, and pull out the thermostat.

To install:

6. Install the thermostat in the intake manifold.

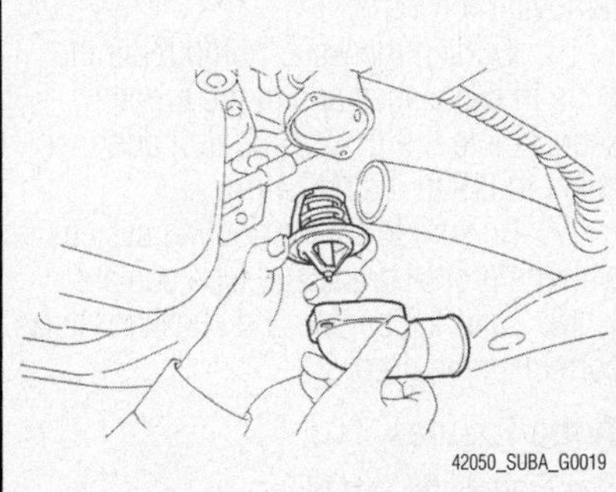

Fig. 59 Remove the thermostat cover, and pull out the thermostat—2.5L Non-Turbo engine shown

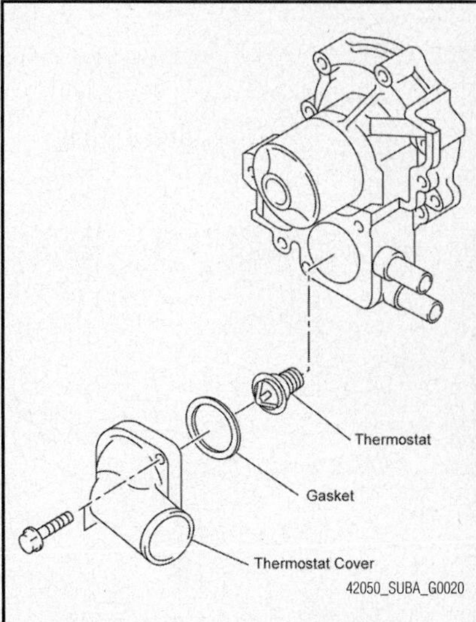

Fig. 60 Exploded view of the thermostat and related components—2.5L Turbo engine shown

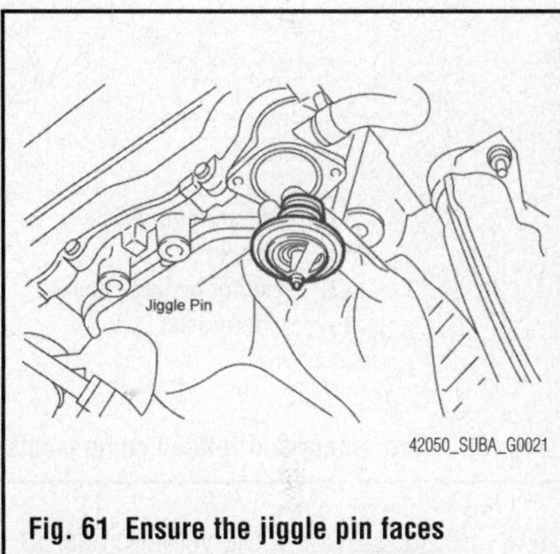

Fig. 61 Ensure the jiggle pin faces upward—2.5L Non-Turbo Engines

➡On non-Turbo engines, the thermostat must be installed with the jiggle pin upward. On turbo engines, the thermostat must be install with the jiggle pin facing front.

7. Install the thermostat cover together with a new gasket.
8. Install the lower radiator hose to the thermostat cover and tighten the hose clamp.
9. Install the engine undercover.
10. Lower the vehicle.
11. Refill the engine cooling system to the correct level.
12. Start the engine and check for leaks.

WATER PUMP

REMOVAL & INSTALLATION

Non-Turbo Engines

See Figures 62 and 63.

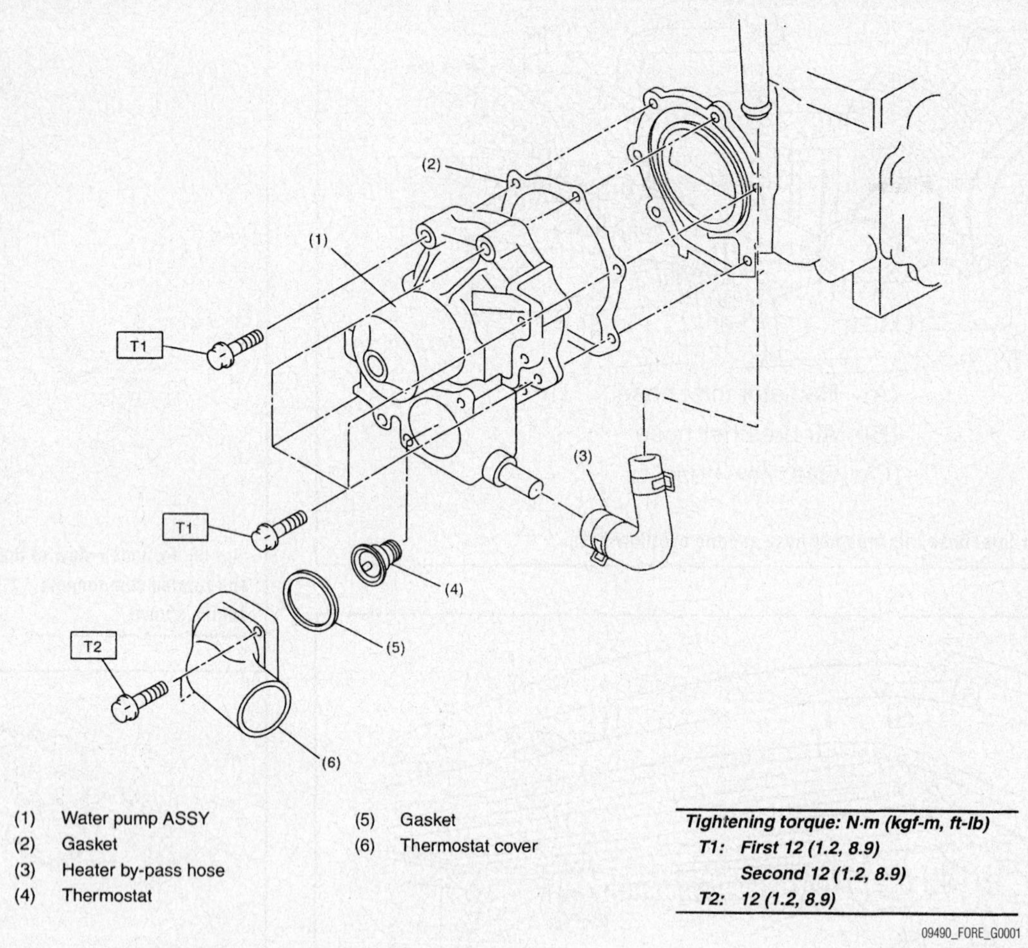

(1) Water pump ASSY
(2) Gasket
(3) Heater by-pass hose
(4) Thermostat
(5) Gasket
(6) Thermostat cover

Tightening torque: N·m (kgf-m, ft-lb)
T1: First 12 (1.2, 8.9)
Second 12 (1.2, 8.9)
T2: 12 (1.2, 8.9)

09490_FORE_G0001

Fig. 62 Water pump and related components—2006–08 SOHC (Non-Turbo) engine

1. Before servicing the vehicle, refer to the Precautions Section.
2. Disconnect the negative battery cable.
3. Drain the cooling system. Remove the radiator.
4. To remove the front side V-belt, remove the belt covers. Loosen the lock bolt. Loosen the slider bolt. Remove the front side belt.
5. To remove the rear side V-belt, remove the belt covers. Loosen the lock bolt. Loosen the slider bolt. Remove the rear side belt. Remove the belt tensioner.
6. Remove the crankshaft pulley bolt.
7. Lock the crankshaft in place using tool ST499977100, or equivalent.
8. Remove the crankshaft pulley.
9. Remove the timing belt.
10. Remove the automatic belt tensioner adjuster. Remove the belt idler number 2.
11. Using tool ST18231AA010, remove the left side cam sprocket. Remove the left side belt cover number 2. Remove the tensioner bracket.
12. Disconnect the hose from the water pump.
13. Remove the water pump retaining bolts. Remove the water pump from the engine.

To install:

14. Clean the gasket mating surfaces thoroughly. Always use new gaskets during installation.
15. Installation is the reverse of the removal procedure.
16. Tighten the water pump retaining bolts in two stages and in the proper sequence to 8.9 ft. lbs. (12 Nm) and then again to 8.9 ft. lbs. (12 Nm).
17. Be sure to fill the cooling system with the proper grade and type coolant.
18. Start the engine and check for leaks, correct as required.

Turbo Engines

See Figures 64 and 65.

1. Before servicing the vehicle, refer to the Precautions Section.
2. Disconnect the negative battery cable.
3. Drain the cooling system. Remove the radiator.
4. Remove the collector cover.
5. To remove the front side V-belt, remove the belt covers. Loosen the lock bolt. Loosen the slider bolt. Remove the front side belt.
6. To remove the rear side V-belt, remove the belt covers. Loosen the lock bolt. Loosen the slider bolt. Remove

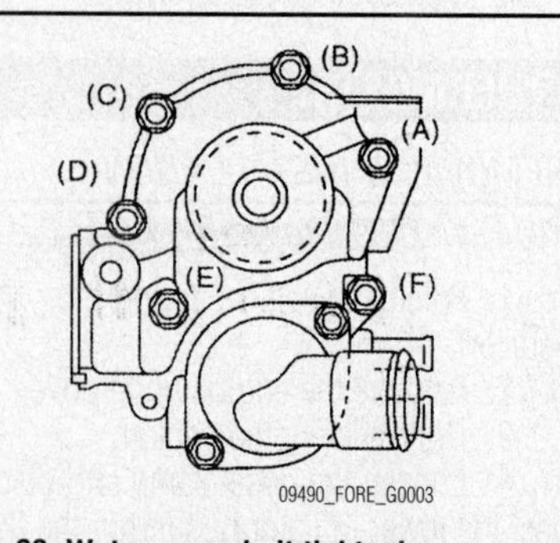

Fig. 63 Water pump bolt tightening sequence—2006–08 SOHC (Non-Turbo) engine

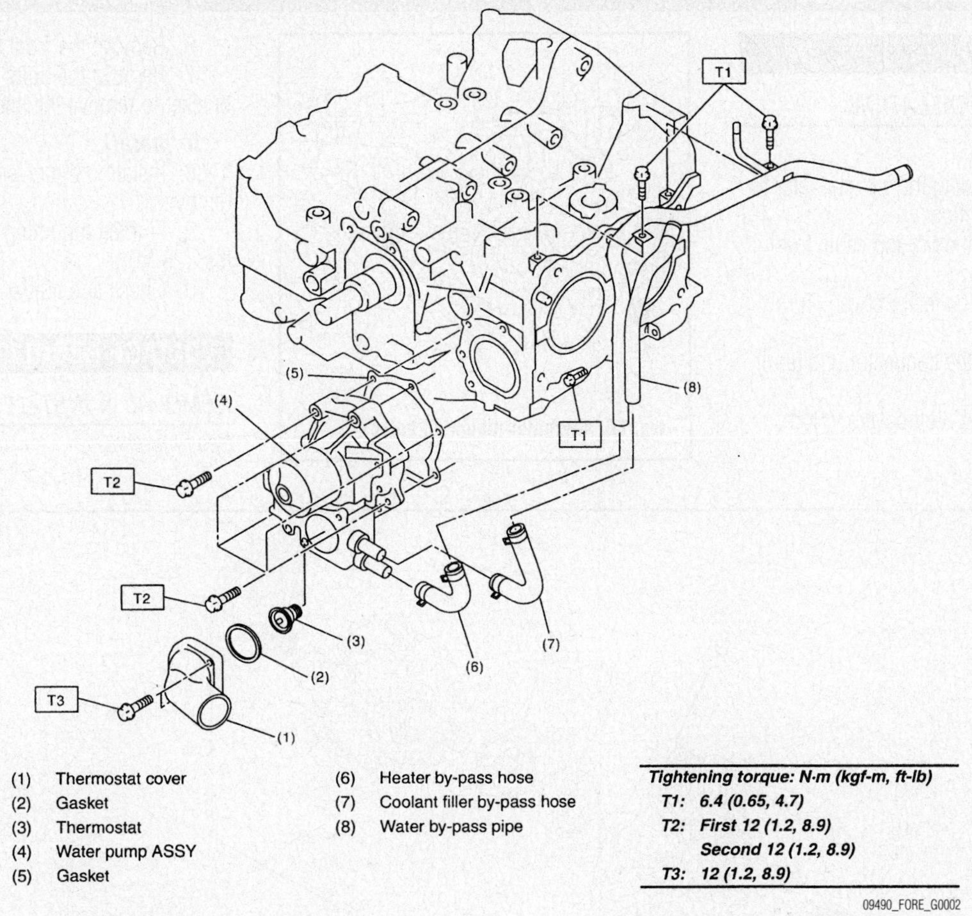

T1
T1
T2
T2
T3
(1)
(2)
(3)
(4)
(5)
(6)
(7)
(8)
(1) Thermostat cover
(2) Gasket
(3) Thermostat
(4) Water pump ASSY
(5) Gasket
(6) Heater by-pass hose
(7) Coolant filler by-pass hose
(8) Water by-pass pipe
Tightening torque: N·m (kgf-m, ft-lb)
T1: 6.4 (0.65, 4.7)
T2: First 12 (1.2, 8.9)
Second 12 (1.2, 8.9)
T3: 12 (1.2, 8.9)
09490_FORE_G0002

Fig. 64 Water pump and related components—2006–08 DOHC (Turbo) engine

the rear side belt. Remove the belt tensioner.

7. Remove the crankshaft pulley bolt.
8. Lock the crankshaft in place using tool ST499977100, or equivalent.
9. Remove the crankshaft pulley.
10. Remove the timing belt.
11. Remove the automatic belt tensioner adjuster. Remove the idler belt. Remove the belt idler number 2.
12. Remove the camshaft position sensor.
13. Using tool ST499977500, remove the left side cam sprocket. Remove the left side belt cover number 2. Remove the tensioner bracket.
14. Disconnect the hose from the water pump.
15. Remove the water pump retaining bolts. Remove the water pump from the engine.

To install:

16. Clean the gasket mating surfaces thoroughly. Always use new gaskets during installation.
17. Installation is the reverse of the removal procedure.
18. Tighten the water pump retaining bolts in two stages and in the proper sequence to 8.9 ft. lbs. (12 Nm) and then again to 8.9 ft. lbs. (12 Nm).
19. Be sure to fill the cooling system with the proper grade and type coolant.
20. Start the engine and check for leaks, correct as required.

(A)
(B)
(C)
(D)
(E)
(F)
09490_FORE_G0003

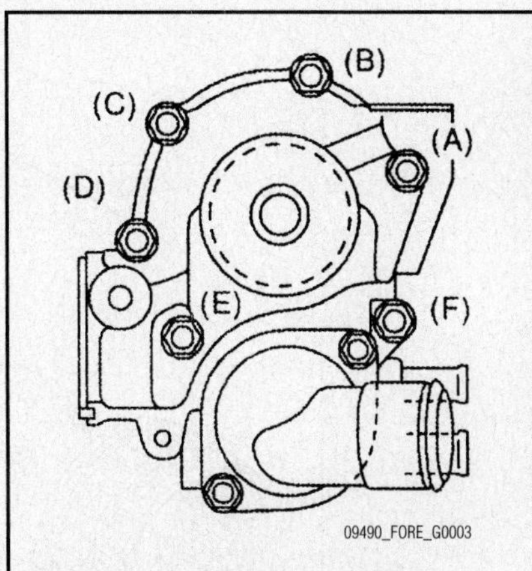

Fig. 65 Water pump bolt tightening sequence—2006–08 DOHC (Turbo) engine

ALTERNATOR

REMOVAL & INSTALLATION

See Figure 66.

1. Before servicing the vehicle, refer to the Precautions Section.
2. Disconnect the ground cable from the battery.
3. Remove the collector cover. (Turbo model)
4. Disconnect the connector and terminal from alternator.
5. Remove the V-belt covers. (Non-turbo model)

Fig. 66 Alternator mounting bolts

6. Remove the front side belts.
7. Remove the bolts which install the bracket to remove the alternator.

To install:

8. Install in the reverse order of removal.
9. Tighten mounting bolts to 18.4 ft. lbs. (25 Nm).
10. Check and adjust the V-belt tension.

VOLTAGE REGULATOR

REMOVAL & INSTALLATION

See Figures 67.

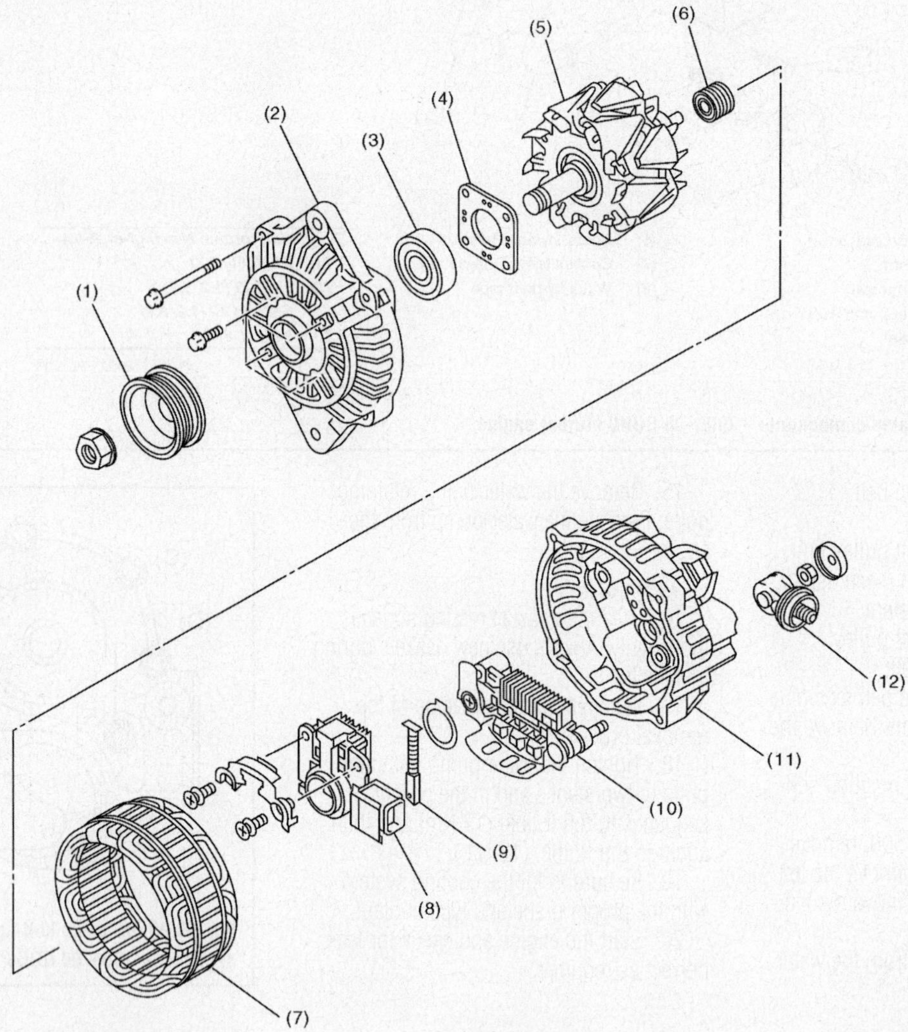

(1)	Pulley	(5)	Rotor	(9)	Brush
(2)	Front cover	(6)	Bearing	(10)	Rectifier
(3)	Ball bearing	(7)	Stator coil	(11)	Rear cover
(4)	Bearing retainer	(8)	IC regulator with brush	(12)	Terminal

22140_FORE_G0091

Fig. 67 Exploded view of alternator

1. Remove the alternator from the vehicle.
2. Remove the four through bolts. Then insert the tip of a flat-head screwdriver into the gap between the stator core and front bracket. Pry them apart to disassemble the alternator.
3. Hold rotor with a vise and remove pulley nut.

CAUTION

When holding rotor with vise, insert aluminum plates or wood pieces on the contact surfaces of the vise to prevent rotor from damage.

4. Unsolder connection between rectifier and stator coil to remove stator coil.

CAUTION

Finish the work rapidly (less than three seconds) because the rectifier cannot withstand heat very well.

5. Remove screws which secure voltage regulator to rear cover, and unsolder connection between voltage regulator and rectifier to remove the regulator.
6. Installation is the reverse order of assembly. After assembly, turn the pulley by hand to check that the rotor turns smoothly

ENGINE ELECTRICAL — IGNITION SYSTEM

FIRING ORDER

See Figure 68.

Fig. 68 2.5L Engine
Firing order: 1–3–2–4
Distributorless ignition system

IGNITION COIL

REMOVAL & INSTALLATION

See Figure 69.

1. Disconnect the ground cable from the battery.
2. Remove the collector cover.
3. Remove the air cleaner case (RH side).
4. Remove the secondary air pump for LH side
5. Disconnect the connector from ignition coil.

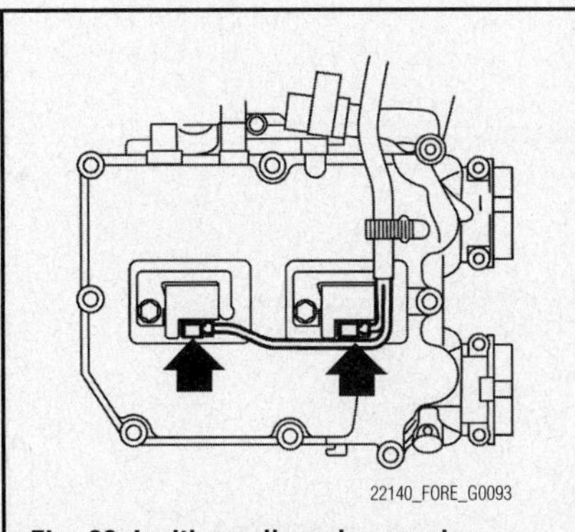

Fig. 69 Ignition coils and connectors shown

6. Remove the ignition coil.

To install:

7. Install in the reverse order of removal.
8. Tighten the ignition coil mounting bolt to 12 ft. lbs. (16 Nm).

TESTING

See Figure 70.

1. Using a spark plug tester check for spark at each cylinder.
2. If spark is not present.
3. Turn the ignition switch **OFF**.
4. Disconnect the connector from ignition coil.
5. Turn the ignition switch **ON**.
6. Measure the power supply voltage between ignition coil connector and engine ground.
7. The voltage reading should be more than 10 volts. If not repair open circuit.

IGNITION COIL PACK

REMOVAL & INSTALLATION

See Figure 71.

1. Disconnect the negative battery cable.
2. Disconnect the spark plug wires from the ignition coil.

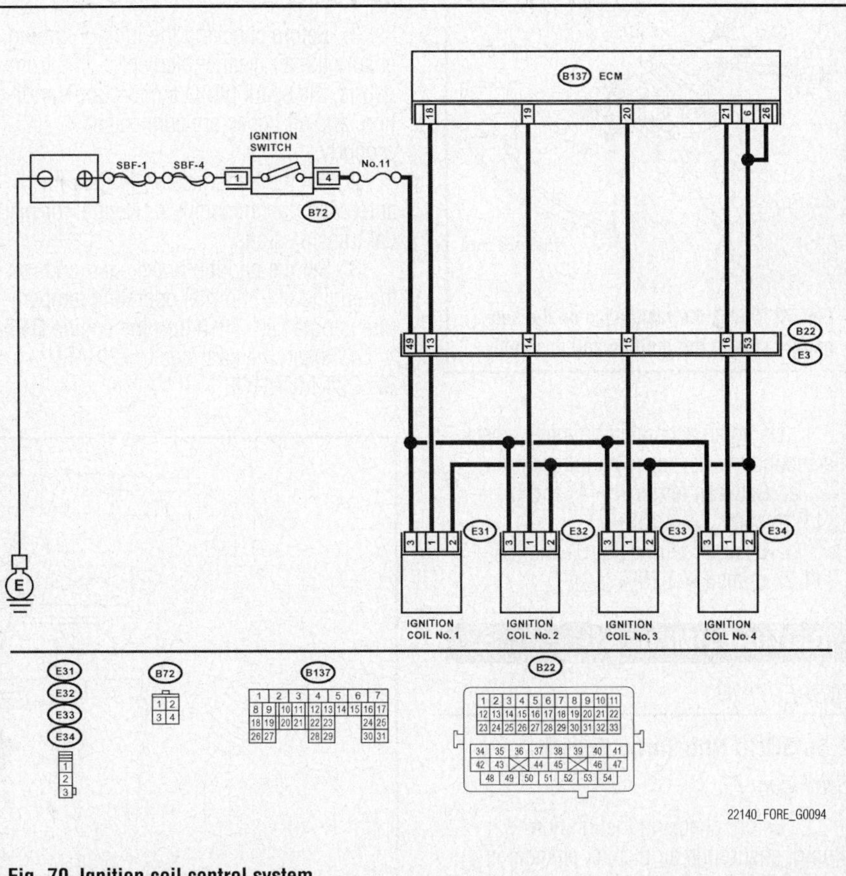

Fig. 70 Ignition coil control system

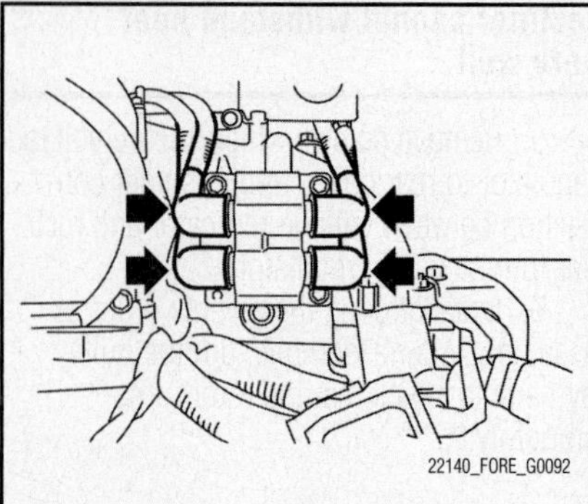

Fig. 71 Spark plug wires at ignition coil assembly

3. Disconnect the electrical connector from the ignition coil.

4. Unbolt and remove the ignition coil and igniter assembly.

To install:

5. Installation is the reverse order of assembly.

6. Tighten the ignition coil mounting bolts to 5 ft. lbs. (7 Nm).

TESTING

See Figure 72.

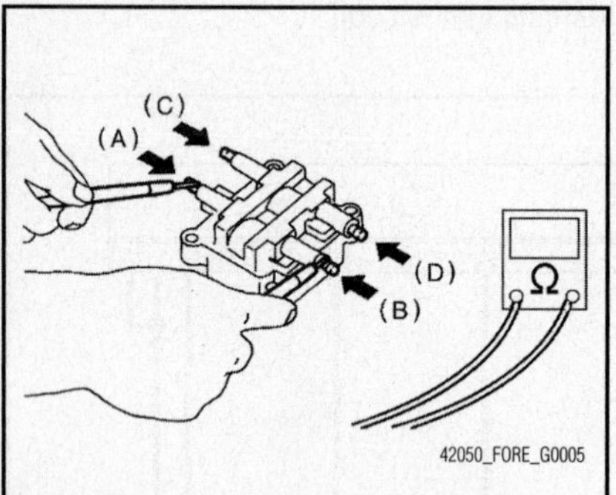

Fig. 72 Check the resistance on the secondary side of the ignition coil assembly

1. Using an accurate ohmmeter, check the secondary resistance as follows:

a. Between terminals (A) and (B)—11.2k ohms (+/-) 15%

b. Between terminals (C) and (D)—11.2k ohms (+/-) 15%

IGNITION TIMING

INSPECTION

2.5L SOHC Non-Turbo Engine

See Figure 73.

1. Before checking the ignition timing, ensure the air cleaner element is free from debris, the spark plugs are in good condition, and all hoses are connected properly.

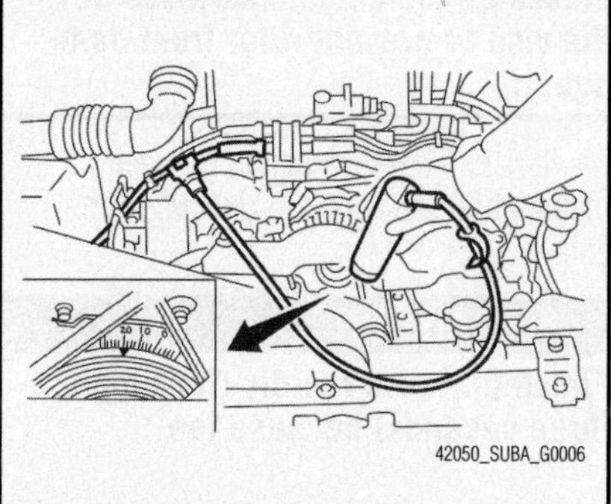

Fig. 73 Using a conventional timing light to check the ignition timing

2. Set the transmission in Park for automatic transmissions or Neutral for manual transmissions.

3. Set the parking brake, start and run the engine until normal operating temperature is obtained. Then turn the engine **OFF**.

4. Connect a conventional power timing light to the No. 1 cylinder spark plug wire. Start the engine and run at idle.

5. Aim the timing light at the timing mark located near the crankshaft pulley.

2.5L DOHC Turbo Engine

See Figure 74.

1. Before checking the ignition timing, ensure the air cleaner element is free from debris, the spark plugs are in good condition, and all hoses are connected properly.

2. Set the transmission in Park for automatic transmissions or Neutral for manual transmissions.

3. Set the parking brake, start and run the engine until normal operating temperature is obtained. Then turn the engine **OFF**.

4. Insert the cartridge to SUBARU SELECT MONITOR.

5. Connect SUBARU SELECT MONITOR to the data link connector.

6. Turn ignition switch to ON, and SUBARU SELECT MONITOR switch to ON.

7. Select "2. Each System Check" in Main Menu.

8. Select "Engine Control System" in Selection Menu.

9. Select "1. Current Data Display & Save" in Engine Control System Diagnosis.

10. Select "1.12 Data Display" in Data Display Menu.

11. Start engine at idle speed and check the ignition timing.

ADJUSTMENT

The ignition timing is controlled by the Electronic Control Module (ECM). No adjustment is necessary or possible.

SPARK PLUGS

REMOVAL & INSTALLATION

2.5L SOHC Non-Turbo Engine

See Figure 75.

1. Disconnect the negative battery cable.

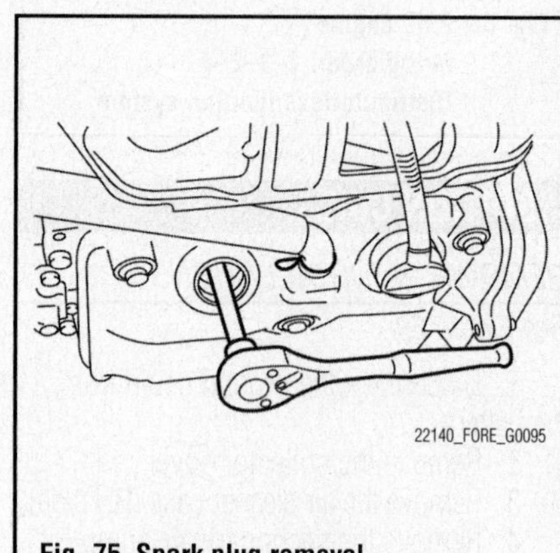

Fig. 75 Spark plug removal

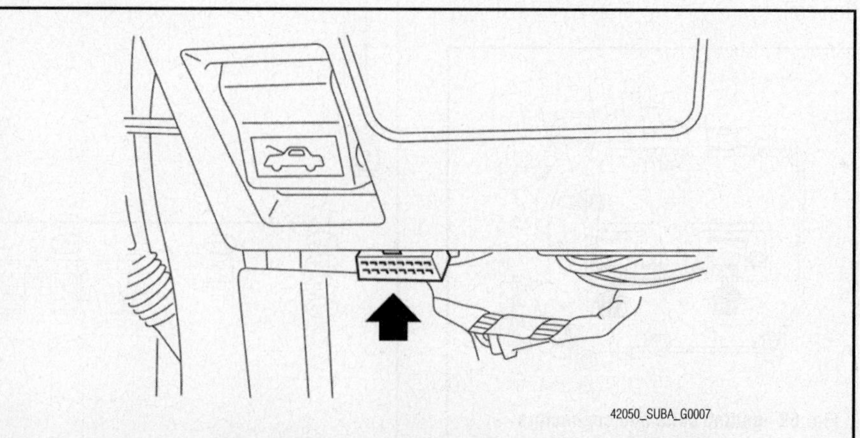

Fig. 74 Location of the data link connector under the driver's side instrument panel.

2. Disconnect the mass air flow sensor connector.

3. Remove the air intake assembly.

4. Remove the spark plug cable by the grasping the boot firmly.

5. Using a spark plug socket with an extension and universal joint, remove the spark plug.

To install:

6. Install the spark plug and tighten to 15 ft. lbs. (20 Nm).

7. The remainder of the installation is the reverse order of removal.

2.5L DOHC Turbo Engine

See Figure 76.

1. Disconnect the negative battery cable.

2. Remove the collector cover.

3. Remove the secondary air pump (LH side).

4. Remove the air cleaner case (RH side).

5. Disconnect the electrical connector form the ignition coil.

6. Unbolt and remove the ignition coil.

7. Remove the spark plug with a spark plug socket.

8. Using a spark plug socket with an extension and universal joint, remove the spark plug.

To install:

9. Install the spark plug and tighten to 15 ft. lbs. (20 Nm).

10. Install the ignition coil and tighten to 12 ft. lbs. (16 Nm).

11. The remainder of the installation is the reverse order of removal.

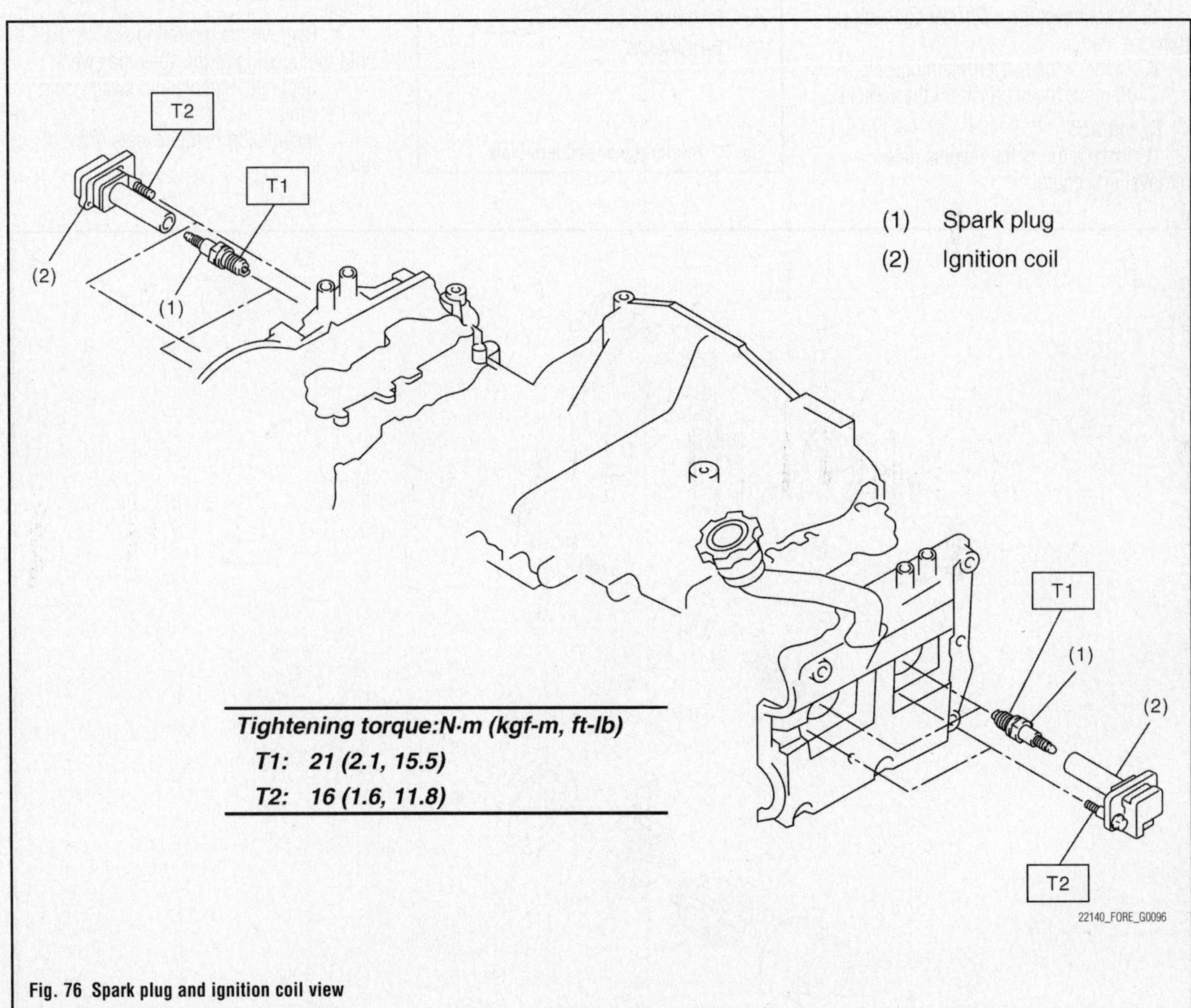

Fig. 76 Spark plug and ignition coil view

STARTER

REMOVAL & INSTALLATION

See Figure 77.

1. Disconnect the negative battery cable.
2. Remove the air intake chamber, if equipped with a non-turbocharged engine.
3. Remove the intercooler, if equipped with a turbocharged engine.
4. Remove the air intake chamber stay, if equipped with a non turbocharged engine.
5. Disconnect the electrical connectors from the starter.
6. Remove the starter retaining bolts.
7. Remove the starter from the vehicle.

To install:

8. Installation is the reverse of the removal procedure.

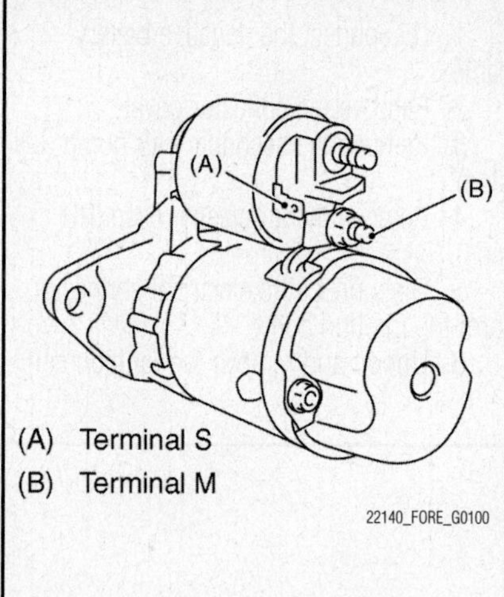

Fig. 77 Starter motor and terminals

9. Torque the starter retaining bolts to 37 ft. lbs. (50 Nm).

SOLENOID OR RELAY REPLACEMENT

See Figure 78.

1. Remove the starter from the vehicle.
2. Disconnect the lead wire from the solenoid switch.
3. Remove the through-bolts from the end frame.
4. Remove the yoke from the solenoid switch.
5. Remove the mounting screws that hold the housing to the solenoid switch.
6. Separate the solenoid switch from the housing.
7. Installation is the reverse order of removal.

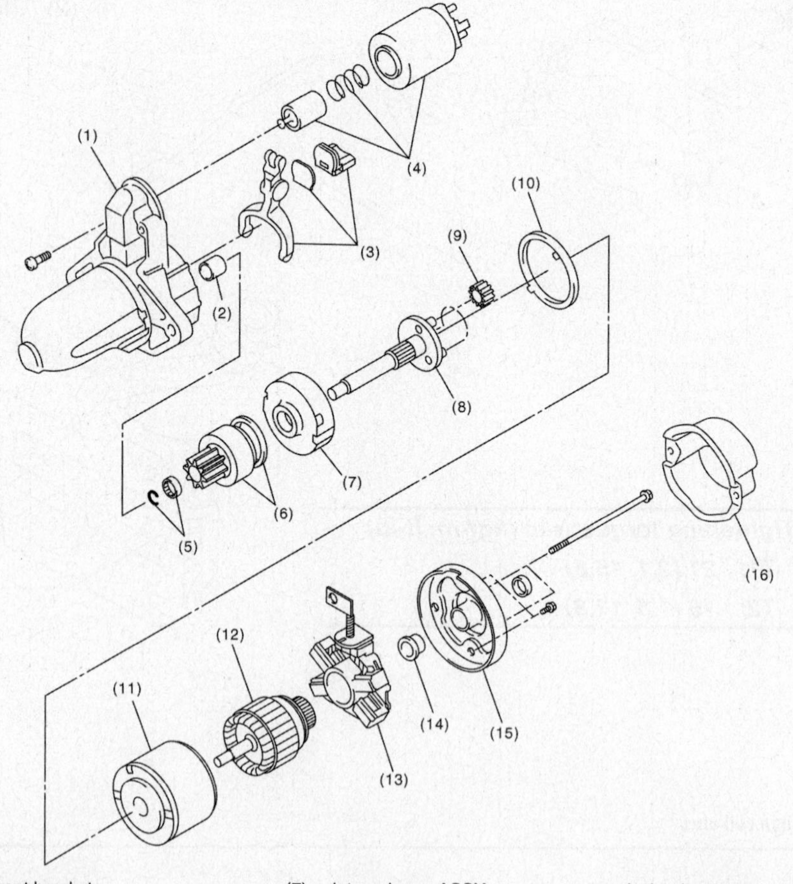

Fig. 78 Starter exploded view

Wear appropriate work clothing, including a cap, protective goggles and protective shoes when performing any work.

Remove contamination including dirt and corrosion before removal, installation or disassembly.

Keep the disassembled parts in order and protect them from dust and dirt.

Before removal, installation or disassembly, be sure to clarify the failure. Avoid unnecessary removal, installation, disassembly and replacement.

Vehicle components are extremely hot after driving. Be wary of receiving burns from heated parts.

Be sure to tighten fasteners including bolts and nuts to the specified torque.

Place shop jacks or rigid racks at the specified points.

Before disconnecting connectors of sensors or units, be sure to disconnect the ground cable from the battery.

All parts should be thoroughly cleaned, paying special attention to engine oil passages, pistons and bearings.

Rotating parts and sliding parts such as piston, bearing and gear should be coated with oil prior to assembly. Be careful not to let oil, grease or coolant contact the timing belt, clutch disc and flywheel.

All removed parts, if to be reused, should be reinstalled in the original positions and directions.

Bolts, nuts and washers should be replaced with new parts as required.

Even if necessary inspections have been made in advance, proceed with assembly work while making rechecks.

Remove or install the engine in an area where chain hoists, lifting devices, etc. are available for ready use.

Be sure not to damage coated surfaces of body panels with tools, or not to stain seats and windows with coolant or oil. Place a cover over fender, as required, for protection.

Prior to starting work, prepare the following:

1. Service tools, clean cloth, containers to catch coolant and oil, wire ropes, chain hoist, transmission jacks, etc.
2. Lift up or lower the vehicle when necessary. Make sure to support the correct positions.

ACCESSORY DRIVE BELTS

ACCESSORY BELT ROUTING

See Figure 79.

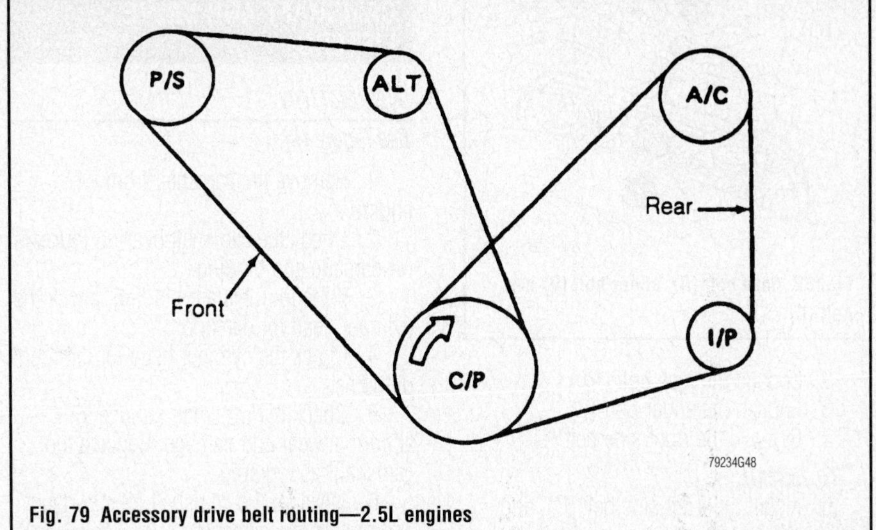

Fig. 79 Accessory drive belt routing—2.5L engines

INSPECTION

Inspect the drive belt for signs of glazing or cracking. A glazed belt will be perfectly smooth from slippage, while a good belt will have a slight texture of fabric visible. Cracks will usually start at the inner edge of the belt and run outward. All worn or damaged drive belts should be replaced immediately.

ADJUSTMENT

See Figures 80 and 81.

1. Loosen the lock bolt of the belt you are adjusting. Turn the slider bolt until the correct tension is achieved. Tighten the lock bolt.

The belt tension may be check with or without a belt tension gauge.

2. If using a belt tension gauge, the tension should be as follows:
   a. Used front belt: 110.2–143.9 lbs. (490–640 N)
   b. New front belt: 143–175 lbs. (640–780 N)

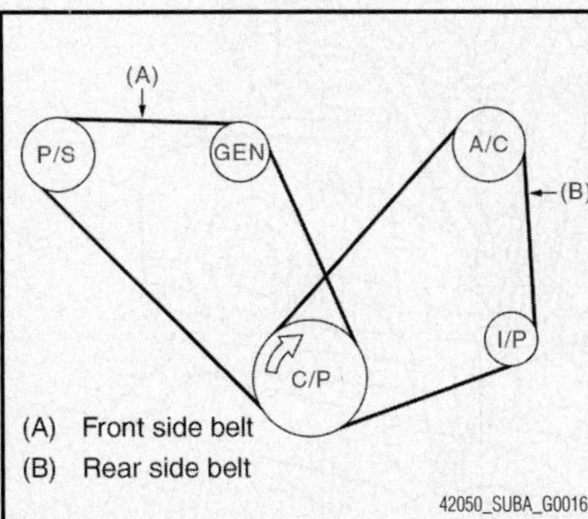

Fig. 80 Location to check the belt tension if you are using a belt tension gauge

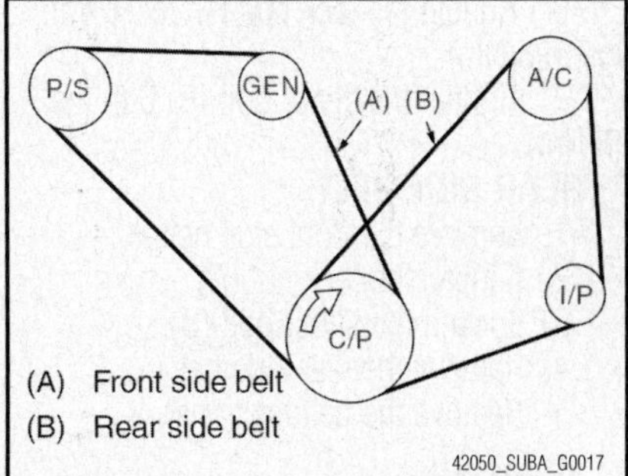

Fig. 81 Location to check to the belt tension if you are not using a belt tension gauge

   c. Used rear belt: 78.7–101.2 lbs. (350–450 N)
   d. New rear belt: 166–198 lbs. (740–880 N)

3. If you are not using a belt tension gauge, using 22 lbs. (98 Nm) of force, push on the belt as shown in the illustration. The total distance the belt travels up and down should be:
   a. Used front belt: 0.354–0.433 in. (9–11mm)
   b. New front belt: 0.276–0.354 in. (7–9mm)
   c. Used rear belt: 0.354–0.394 in. (9.0–10.0mm)
   d. New rear belt: 0.295–0.335 in. (7.5–8.5mm)

REMOVAL & INSTALLATION

See Figures 82 and 83.

FRONT SIDE BELT:

1. Remove the accessory drive belt cover, if equipped.

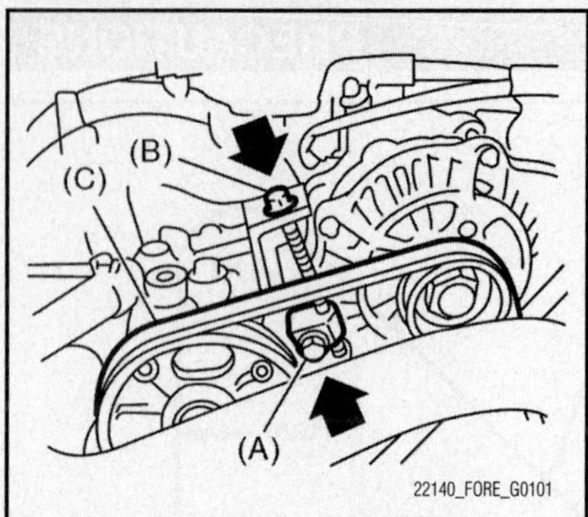

Fig. 82 Lock bolt (A), slider bolt (B) and belt (C)

2. Loosen the lock bolt (A).
3. Loosen the slider bolt (B).
4. Remove the front side belt (C).

To install:

5. Install the front side belt (C), and tighten the slider bolt so as to obtain the specified belt tension.
6. Tighten the lock bolt (A) to 18.4 ft. lbs. (25 Nm).
7. Tighten the slider bolt (B) 6 ft. lbs. (8 Nm).

REAR SIDE BELT:

1. Remove the front side belts.
2. Loosen the lock nut (A).
3. Loosen the slider bolt (B).
4. Remove the rear side belt.
5. Remove the belt tensioner.

To install:

6. Install the belt tensioner.
7. Install a rear side belt, and tighten the slider bolt (B) so as to obtain the specified belt tension.
8. Tighten the lock nut (A) to 17 ft. lbs. (23 Nm).

CAMSHAFT AND VALVE LIFTERS

INSPECTION

See Figure 84.

1. Remove the camshaft from the engine.
2. Check the camshaft bearing journals for damage and binding.
3. If the journals are binding, check the cylinder head for damage.
4. Check the cylinder head for clogged oil holes.
5. Check the camshaft surface for abnormal wear and damage. Replace the camshaft, as required.
6. Measure the camshaft lobe surface and replace the camshaft if not within specification.
7. Measure the camshaft journal diameter and replace the camshaft if not within specification.
8. Measure the camshaft run out and replace the camshaft if not within specification.

Fig. 84 Checking camshaft for wear

REMOVAL & INSTALLATION

2.5L SOHC Non-Turbo Engine

See Figures 85 through 92.

1. Disconnect the negative battery cable.
2. To remove the front side V-belt, remove the belt covers. Loosen the lock bolt. Loosen the slider bolt. Remove the front side belt.
3. To remove the rear side V-belt, remove the belt covers. Loosen the lock bolt. Loosen the slider bolt. Remove the rear side belt. Remove the belt tensioner.
4. Remove the crankshaft pulley bolt.
5. Lock the crankshaft in place using tool ST499977100, or equivalent.
6. Remove the crankshaft pulley.
7. Remove the left side timing belt cover.
8. Remove the front timing belt cover.
9. Remove the timing belt.
10. Remove the camshaft position sensor.
11. Remove the camshaft sprockets.

➡Be sure to lock the camshaft in place using tool ST18231AA010, or equivalent.

12. Remove the left and right timing belt N02 covers.

➡Do not damage or lose the rubber seal when removing the covers.

13. Remove the tensioner bracket. Remove the camshaft position sensor support, on the left side.
14. Remove the oil level gauge guide, on the left side.

Fig. 83 Lock nut (A), slider bolt (B).

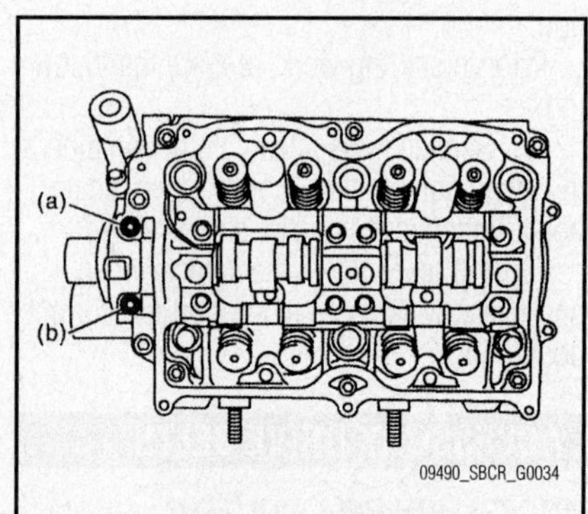

Fig. 85 Camshaft bolt loosening sequence—2006–08 SOHC engine

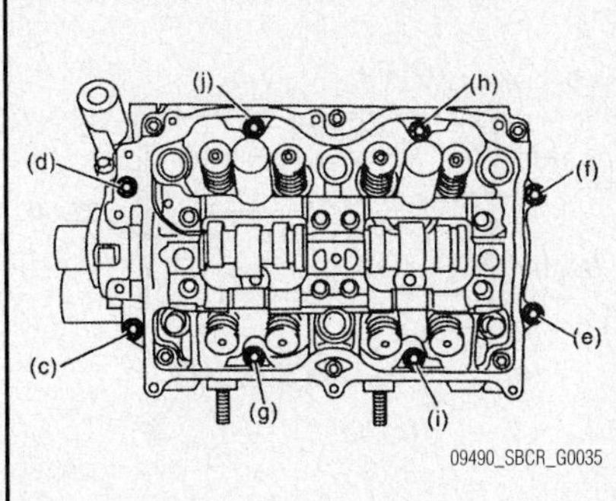

Fig. 86 Camshaft bolt loosening sequence—2006-08 SOHC engine

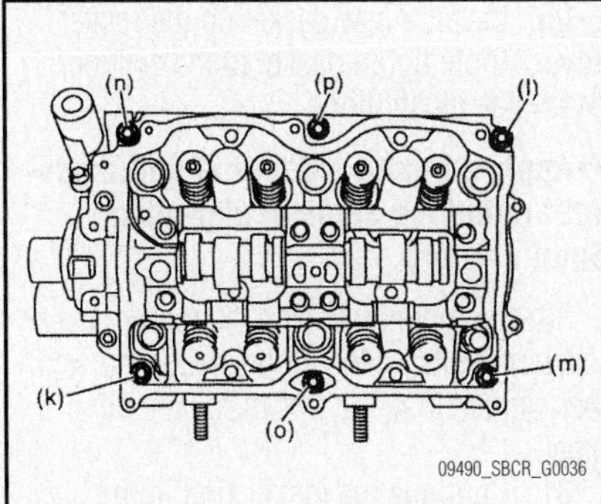

Fig. 87 Camshaft bolt loosening sequence—2006-08 SOHC engine

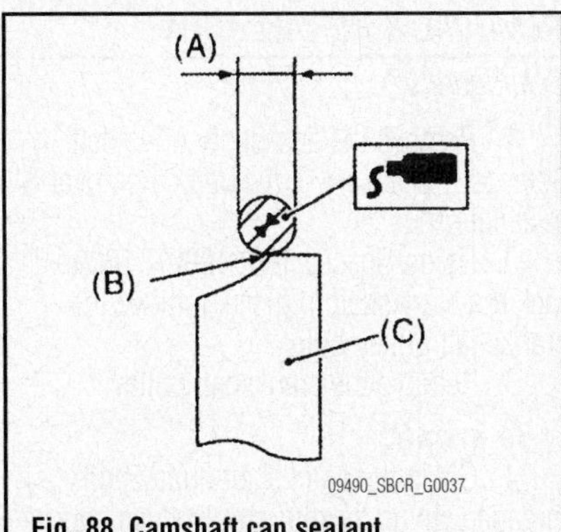

Fig. 88 Camshaft cap sealant application—2006-08 SOHC engine

15. Remove the valve rocker arm assembly.

16. Remove camshaft cap retaining bolts "A" and "B" in the proper sequence.

17. Loosen camshaft cap bolts "C" through "J" all the way in the proper sequence.

18. Remove camshaft cap bolts "K" through "P" in the proper sequence using a Torx® head bit.

19. Remove the camshaft caps.

20. Remove the camshaft. Remove the oil seal. Remove the plug from the rear side of the camshaft.

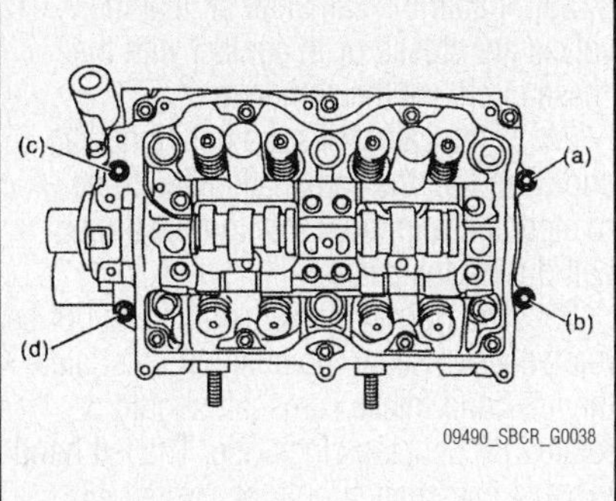

Fig. 89 Camshaft bolt tightening sequence—2006-08 SOHC engine

➡**Do not remove the oil seal unless necessary.**

To install:

➡**Lubricate the camshaft journals with clean engine oil prior to installation.**

21. Install the camshaft into the cylinder head.

22. Apply liquid gasket to the mating surfaces of the camshaft cap.

23. Apply a bead of sealant (0.12 inch in diameter) along the edge of the camshaft cap mating surface. Install with 20 minutes after applying the sealant.

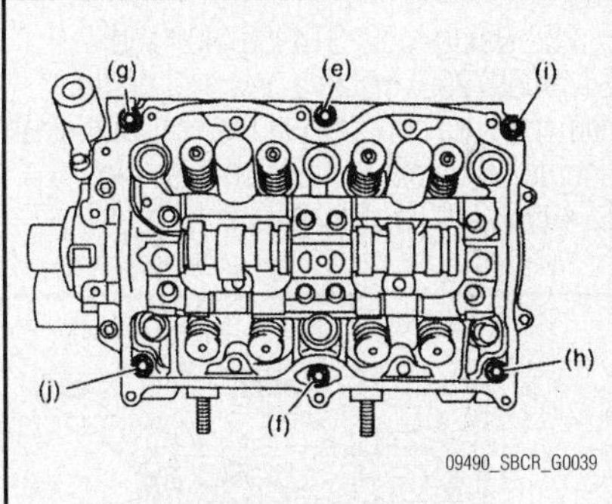

Fig. 90 Camshaft bolt tightening sequence—2006-08SOHC engine

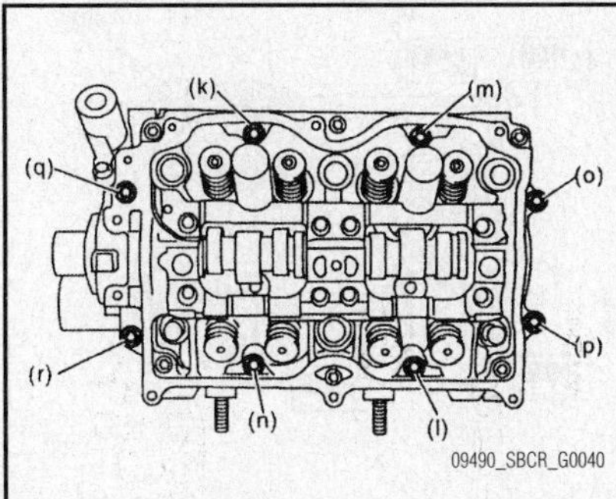

Fig. 91 Camshaft bolt tightening sequence—2006-08 SOHC engine

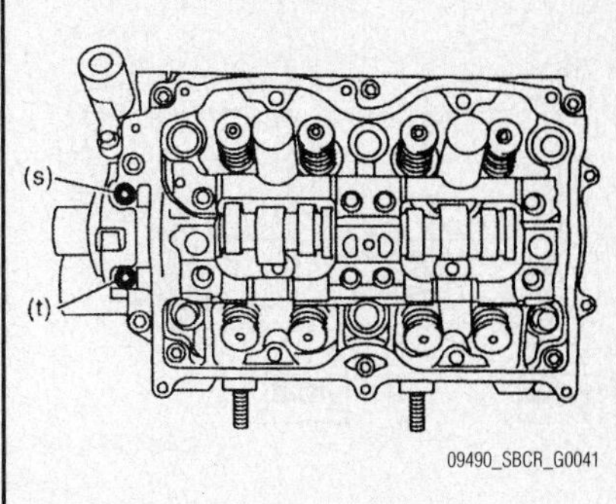

Fig. 92 Camshaft bolt tightening sequence—2006-08 SOHC engine

24. Temporarily tighten the bolts "A" through "D" in the proper sequence.

25. Install the valve rocker arm assembly. Tighten Torx® head bolts "E" through "J" in the proper sequence to 13 ft. lbs. (18 Nm).

26. Tighten bolts "K" through "R" in the proper sequence to 7.2 ft. lbs. (9.75 Nm).

27. Tighten bolts "S" through "T" in the proper sequence to 7.2 ft. lbs. (9.75 Nm).

➡**Be sure to use a new seal washer.**

28. Using tools ST499597000 and ST499587500, install a new seal on the camshaft. Be sure to coat the seal with clean engine oil before installation. Use tool ST499587700 and install the plug.

29. Adjust the valve clearance.

30. Continue the installation in the reverse order of the removal procedure.

2.5L DOHC Turbo Engine

See Figures 93 through 96.

1. Disconnect the negative battery cable.
2. Remove the collector cover.
3. To remove the front side V-belt, remove the belt covers. Loosen the lock bolt. Loosen the slider bolt. Remove the front side belt.
4. To remove the rear side V-belt, remove the belt covers. Loosen the lock bolt. Loosen the slider bolt. Remove the rear side belt. Remove the belt tensioner.
5. Remove the crankshaft pulley bolt.
6. Lock the crankshaft in place using tool ST499977100, or equivalent.
7. Remove the crankshaft pulley.
8. Remove the left side timing belt cover.
9. Remove the right side timing belt cover.
10. Remove the front timing belt cover.
11. Remove the timing belt.
12. Remove the camshaft position sensor.
13. Remove the camshaft sprockets.

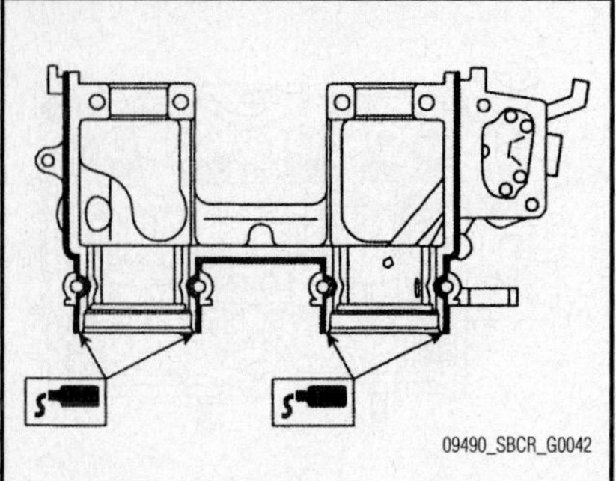

Fig. 93 Camshaft cap liquid gasket application points—2006–08 DOHC engine

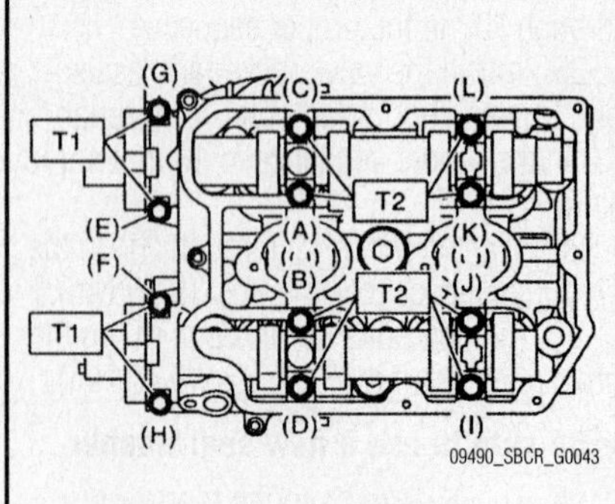

Fig. 94 Camshaft bolt tightening sequence—2006–08 DOHC engine

➡Be sure to lock the camshaft in place using tool ST499207400, or equivalent.

14. Lock the crankshaft in place using tool ST499977100, or equivalent.
15. Remove the crankshaft pulley.
16. Remove the tensioner bracket. Remove the right and left timing belt No.2 covers.
17. Remove the spark plug wires. Remove the oil level gauge, on the left side.
18. Remove the rocker cover and gasket. Remove the oil pipe.
19. Loosen the oil flow control solenoid valve assembly and the intake camshaft cap bolts equally and in the proper sequence.
20. Loosen the exhaust camshaft cap bolts equally and in the proper sequence.
21. Remove the oil flow control solenoid valve assembly, intake camshaft cap and camshaft.
22. Remove the exhaust camshaft caps and camshaft.

➡Arrange the camshafts caps so that they can be installed in their original positions.

To install:

➡Lubricate the camshaft journals with clean engine oil prior to installation.

23. Install the camshaft so that the valves are closed or in contact with the "base circle" of the cam lobe.
24. If the camshafts are positioned as shown in the, the camshafts need to be rotated at a minimum to align the timing belt during installation.
25. The right hand camshaft need not be rotated when set at the position illustrated, the left hand intake camshaft should be rotated 80 degrees clockwise. The left hand exhaust camshaft should be rotated 45 degrees counterclockwise.
26. To install the camshaft cap and oil flow control solenoid valve, apply a small amount of liquid gasket to the mating surface of the cap.

Do not apply an excessive amount of sealant, as it will squish out and flow toward the seal resulting in an oil leak.

27. Apply a thin coat of engine oil to the cap bearing surface and install the cap according to the cap identification mark. Gradually tighten the cap bolts in two stages, first to 7.2 ft. lbs. (9.75 Nm). and then to 14.8 ft. lbs. (20 Nm) in the proper sequence.

➡After tightening the camshaft cap, ensure that the camshaft rotates slightly while holding it at base circle.

28. Using tools ST49587600 and ST499597200, install a new seal on the camshaft. Be sure to coat the seal with clean engine oil before installation. Use tool ST499587700 and install the plug.

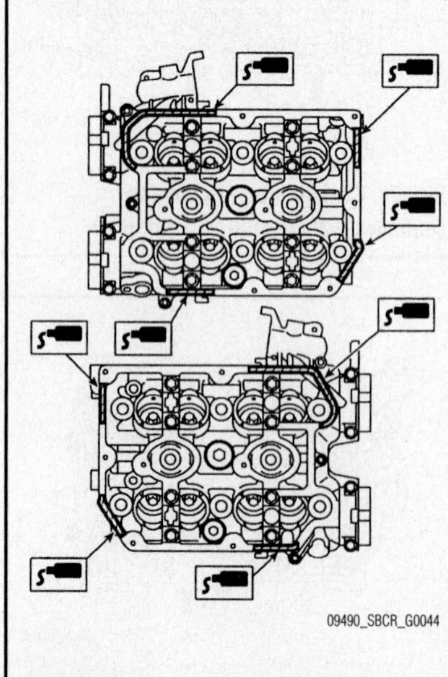

Fig. 95 Rocker cover liquid gasket application points—2006–08 DOHC engine

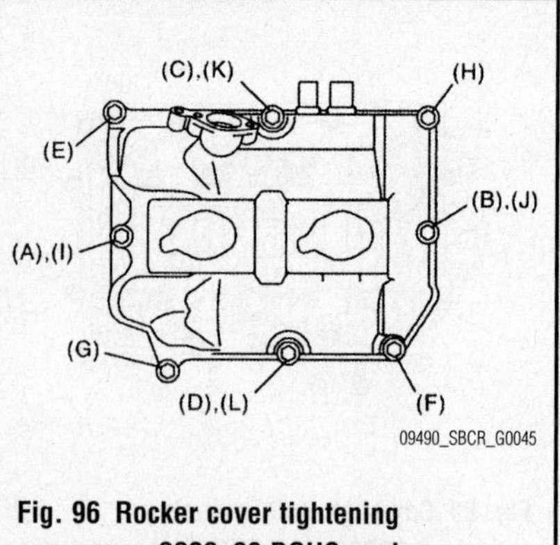

Fig. 96 Rocker cover tightening sequence—2006–08 DOHC engine

29. Install a new gasket on the rocker cover. Apply liquid gasket to the cylinder head (see illustration).

➡Apply an extra amount of liquid gasket around the semicircular plugs, 5mm or more.

30. Temporarily tighten the rocker cover retaining bolts, in the proper sequence, and then tighten to 4.7 ft. lbs. Install the oil pipe.
31. Continue the installation in the reverse order of the removal procedure.

CRANKSHAFT DAMPER

REMOVAL & INSTALLATION

See Figure 97.

1. Remove the accessory drive belt. Refer to "Accessory Drive Belts, Removal & Installation".
2. Using Special Tool 499977100 to lock the crankshaft in place, remove the crankshaft pulley bolt.
3. Remove the crankshaft pulley.

To install:

4. Clean the crankshaft pulley bolt threads with an air gun. Apply clean engine oil to the crankshaft pulley bolt seat and threads.

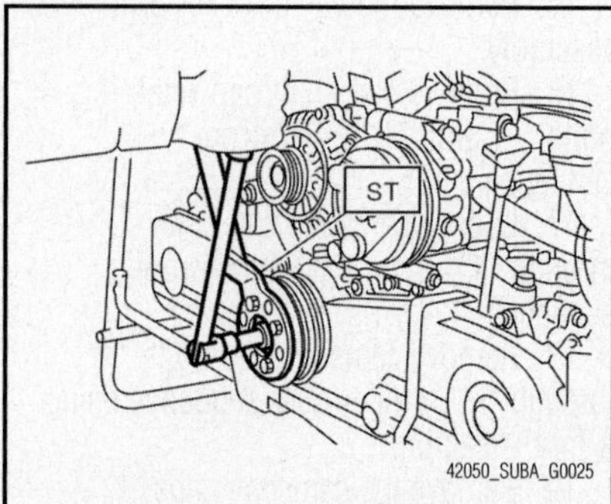

Fig. 97 Use Special Tool 49997700 to lock the crankshaft pulley into place.

5. Install the crankshaft pulley and pulley bolt, and temporarily tighten the pulley bolt.

6. Lock the crankshaft in place with Special Tool 499977100 and tighten the crankshaft pulley bolt to 132 ft. lbs. (180 Nm).

7. The remainder of the installation is the reverse order of removal.

CRANKSHAFT FRONT SEAL

REMOVAL & INSTALLATION

Refer to Oil Pump, "Removal & Installation"

CYLINDER HEAD

REMOVAL & INSTALLATION

2.5L SOHC Non-Turbo Engine

See Figures 98 through 101.

1. Before servicing the vehicle, refer to the precautions in the beginning of this section.
2. Properly relieve the fuel system pressure.
3. Remove or disconnect the following:
   - Negative battery cable
   - Oxygen (O_2S) sensor. If equipped with California emissions, disconnect the rear O_2S sensor.
   - Engine undercover
   - Exhaust Y-pipe and lower it just enough to clear the studs in the heads. Do not allow the Y-pipe to hang without support.
   - Accessory drive belts
   - Engine accessories and brackets from the side of the engine the cylinder head is being removed
   - Connector bracket attaching bolt, if necessary
4. On the left cylinder head, remove the CMP sensor.

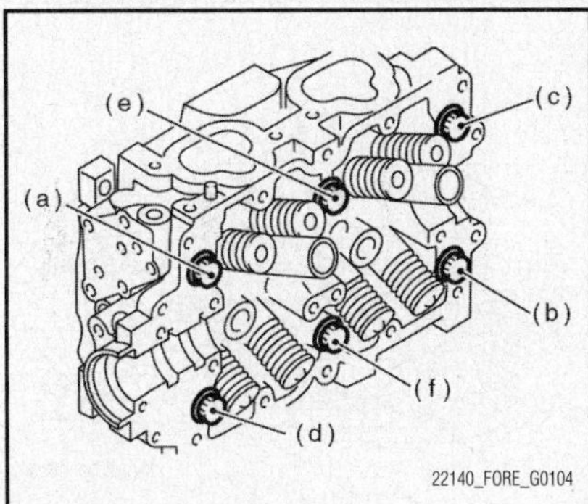

Fig. 98 Cylinder head bolt loosening sequence–Non-Turbo engine

5. Remove or disconnect the following:
   - Fuel pipes
   - Intake manifold and gasket
   - Timing belt, camshaft sprockets, and related components
   - Valve covers, camshafts and related components
   - Oil dipstick tube attaching bolt on the left cylinder head
   - Bolt attaching the A/C compressor to the head
   - Cylinder head bolts in the proper sequence. Leave bolts C and F installed loosely to prevent the cylinder head from falling.
6. Separate the cylinder head from the block. Use a plastic-faced hammer, if needed.
7. Remove bolts C and F. Remove the cylinder head and gasket.
8. Clean all gasket material from both mating surfaces.

To install:

9. Inspect the cylinder head for warpage. Warpage should not exceed 0.0020 in. (0.05mm).
10. Install the cylinder head(s) on the block using new gaskets. Secure in place with the mounting bolts. Coat each bolts with clean engine oil, and hand-tighten.
11. Tighten the cylinder head bolts as follows:
    a. Step 1: Torque all the bolts in sequence to 22 ft. lbs. (29 Nm).
    b. Step 2: Torque all the bolts in sequence to 51 ft. lbs. (69 Nm).
    c. Step 3: Loosen all bolts in sequence by 180 degrees, then loosen an additional 180 degrees.

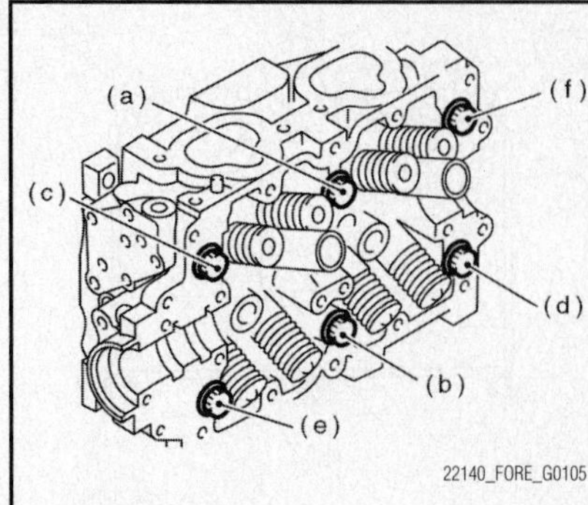

Fig. 100 Cylinder head torque sequence—Non-Turbo engine

    d. Step 4: Tighten all bolts in sequence: 40–45 degrees.

CAUTION

Do not tighten the bolts more that 45 degrees in step 4.

    e. Step 5: Tighten bolts A and B in sequence: 40–45 degrees.

WARNING

Do not exceed 90 degrees total tightening.

12. Install or connect the following:
    - Oil dipstick tube attaching bolt on the left cylinder head
    - Camshafts and related components
    - Camshaft sprocket, timing belt, and related components
    - Intake manifold
    - Fuel delivery pipes
13. On the left cylinder head, install the CMP sensor.

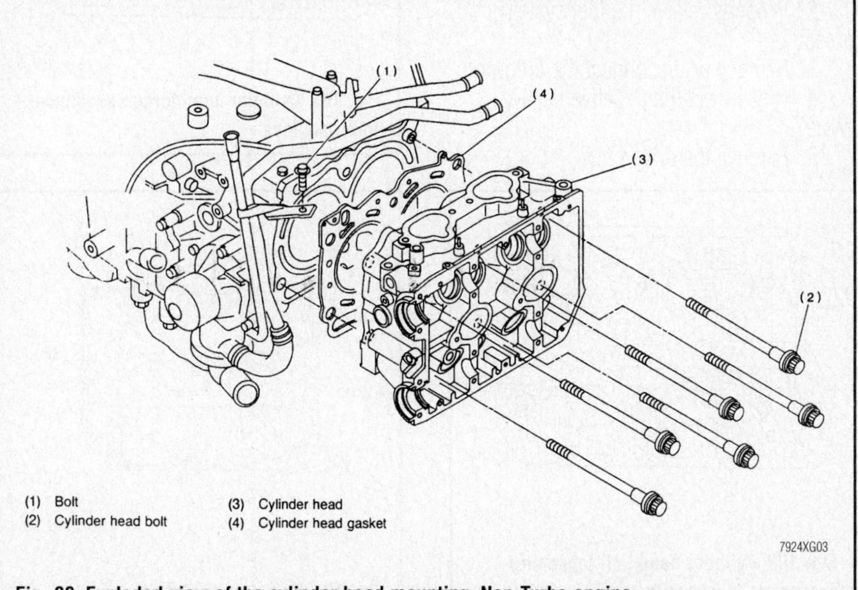

Fig. 99 Exploded view of the cylinder head mounting–Non-Turbo engine

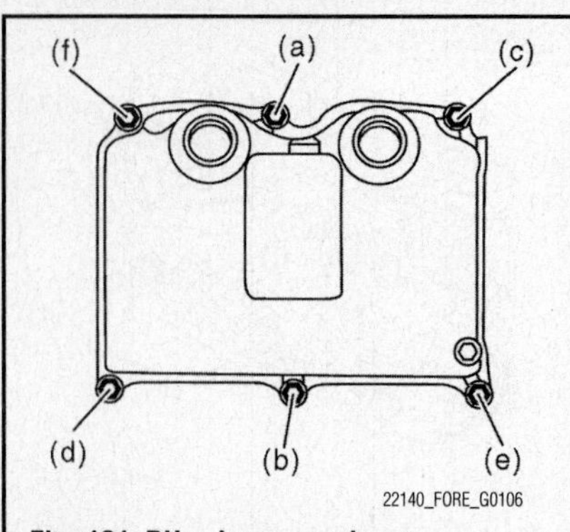

Fig. 101 RH valve cover shown, LH similar

14. Tighten the bolts to the valve covers in two stages in alphabetical sequence as shown in figure:
    - First stage: 4.7 ft. lbs. (6.4 Nm).
    - Second stage: 4.7 ft. lbs. (6.4 Nm).
15. Install or connect the following:
    - Connector bracket attaching bolt
    - Spark plug wires
    - Engine accessories and brackets
    - Accessory drive belts
    - Exhaust Y-pipe. Torque the fasteners to 19–26 ft. lbs. (25–35 Nm).
    - Engine undercover
    - Front O_2S and rear O_2S, if removed.
    - Negative battery cable
16. Start the engine and allow it to reach operating temperature.
17. Check for leaks.

2.5L DOHC Turbo Engine

See Figures 102 through 104.

1. Before servicing the vehicle, refer to the precautions in the beginning of this section.
2. Properly relieve the fuel system pressure.
3. Remove or disconnect the following:
4. Disconnect the negative battery cable.
5. Remove the drive belts.

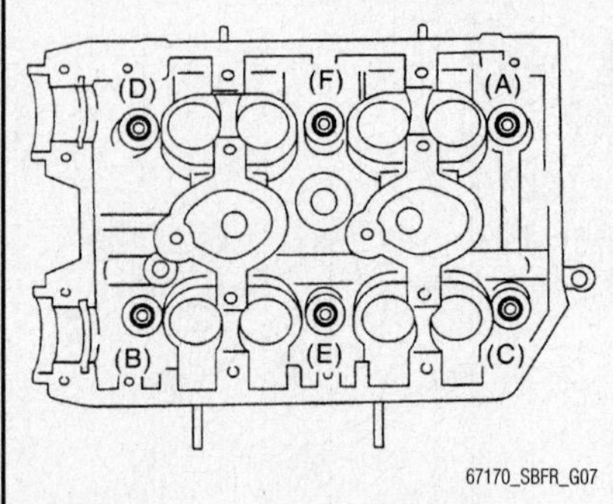

Fig. 102 Cylinder head bolt loosening sequence–Turbo engines

6. Remove the crankshaft pulley.
7. Remove the timing belt cover, timing belt assembly and camshaft sprockets.
8. Remove the intake manifold.
9. Remove the bolt attaching the A/C compressor bracket to the head.
10. Remove the camshaft.
11. Remove the bolts in the sequence illustrated. Leave bolts **A** and **D** installed loosely to prevent the cylinder head from falling.
12. Separate the cylinder head from the block. Use a plastic-faced hammer, if needed.
13. Remove bolts A and D. Remove the cylinder head and gasket.
14. Clean all gasket material from both mating surfaces.

To install:

15. Inspect the cylinder head for warpage. Warpage should not exceed 0.0020 in. (0.05mm).
16. Install the cylinder head(s) on the block using new gaskets. Secure in place with the mounting bolts. Coat each bolts with clean engine oil, and hand-tighten.
17. Tighten the cylinder head bolts as follows:
    a. Step 1: Torque all the bolts in sequence to 22 ft. lbs. (29 Nm).

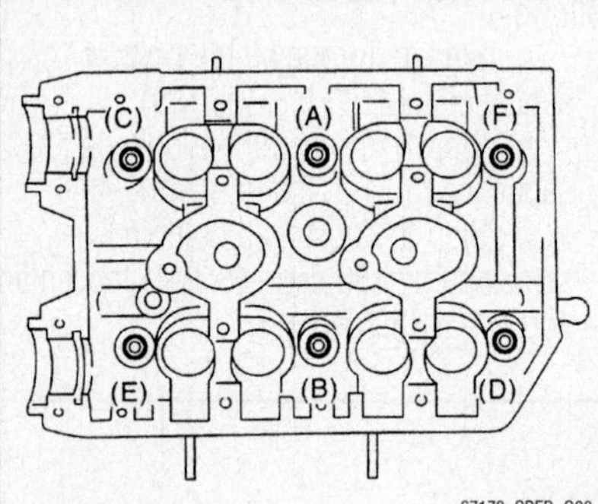

Fig. 103 Cylinder head torque sequence—Turbo engines

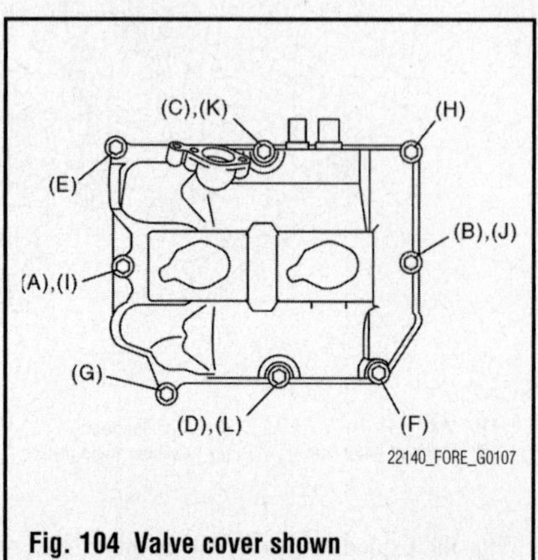

Fig. 104 Valve cover shown

    b. Step 2: Torque all the bolts in sequence to 51 ft. lbs. (69 Nm).
    c. Step 3: Loosen all bolts in sequence by 180 degrees, then loosen an additional 180 degrees.
    d. Step 4: Tighten all bolts in sequence to 36 ft. lbs. (49 Nm).
    e. Step 5: Tighten all bolts in sequence an additional 80–90 degrees
    f. Step 6: All bolts: turn an additional 40–45 degrees.
    g. Step 7: Bolts **A** and **B** an additional 40–45 degrees.

WARNING

Do not exceed 90 degrees total tightening in the previous two steps.

18. Temporarily tighten the valve cover tightening bolts in alphabetical sequence shown in the figure, and then tighten to 4.7 ft. lbs. (6.4 Nm) in alphabetical sequence.
19. Install the remaining components in the reverse order of removal.
20. Start the engine and allow it to reach operating temperature.
21. Check for leaks.

ENGINE ASSEMBLY

REMOVAL & INSTALLATION

2.5L Non-Turbo Engine

See Figures 105 and 106.

1. Before servicing the vehicle, refer to the precautions in the beginning of this section.
2. Open the hood to the full open position and support it with the hood stay.
3. Properly relieve the fuel system pressure. Remove the fuel cap.
4. Properly disarm the SRS system.
5. Discharge the air conditioning system.
6. Disconnect the battery cables and remove battery from vehicle.

Fig. 105 Engine lifting fixture in place

7. Drain the engine oil. Drain the cooling system.

8. Remove the air intake duct, air cleaner case and air intake chamber.

9. Remove the collector cover.

10. Remove the radiator.

11. Disconnect the air conditioning pressure hoses from the compressor.

12. Disconnect the front and rear oxygen sensor electrical connectors, the engine ground cable, the engine harness connector, the alternator connector and terminal, the air conditioning compressor electrical connector and the power steering switch connector.

13. Disconnect the brake booster hose and the heater hoses.

14. Remove the front side V-belt. Remove the power steering pipe and bracket. Remove the reservoir tank. Remove the power steering pump and position it aside.

15. Remove the front and center exhaust pipes.

16. Remove the lower transmission to engine retaining nuts.

17. Remove the front cushion rubber onto front crossmember retaining nuts.

18. If equipped with automatic transmission, remove the service plug hole cap. Remove the torque converter to drive plate retaining bolts.

19. Remove the pitching stopper. Disconnect the fuel hoses from the fuel pipes.

20. Remove the clip and disconnect the evaporator hose from the pipe.

21. Position the engine lifting fixture in place. Support the transmission using a suitable jack.

➡Before separating the engine from the transmission, check to be sure nothing has been overlooked that will stop the engine from being removed. This is very important in order to facilitate reinstallation and because the transmission lowers under its own weight.

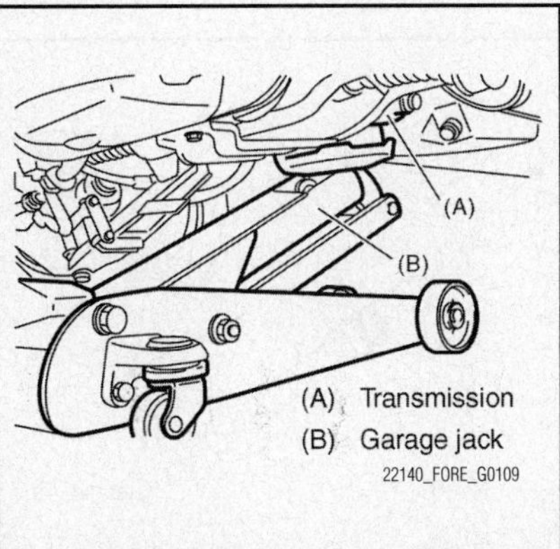

Fig. 106 Supporting the transmission

22. Separate the engine from the transmission.

23. Remove the starter.

24. If equipped with automatic transmission, install tool ST498277200, with will hold the torque converter in place.

25. Remove the upper right side bolts that retain the transmission to the engine.

26. Slightly raise the engine. Raise the transmission, using the suitable jack. Move the engine horizontally until the mainshaft is withdrawn from the clutch cover. Carefully remove the engine from the vehicle.

To install:

27. Installation is the reverse of the removal procedure.

28. Torque the upper transmission to engine retaining bolts to 36.9 ft. lbs. (50 Nm).

29. Torque the lower transmission to engine retaining bolts to 36.9 ft. lbs. (50 Nm).

30. If equipped, tighten the torque converter to drive plate bolts to 18 ft. lbs. (24 Nm).

31. Fill the engine with the proper grade and type engine oil.

32. Fill and bleed the cooling system.

33. Charge the air conditioning system.

34. If equipped. Adjust the clutch cable as required.

35. If equipped, check the automatic transmission fluid level, correct as required using the proper grade and type transmission fluid.

36. Start the engine and allow it to reach normal operating temperature. Check for leaks and correct as required.

2.5L DOHC Turbo Engine

See Figure 107.

1. Before servicing the vehicle, refer to the precautions in the beginning of this section.

2. Open the hood to the full open position and support it with the hood stay.

3. Remove the collector cover.

4. Properly relieve the fuel system pressure. Remove the fuel cap.

5. Properly disarm the SRS system.

6. Discharge the air conditioning system.

7. Disconnect the negative battery cable.

8. Drain the engine oil. Drain the cooling system.

9. Remove the radiator. Remove the coolant filler tank.

10. Remove the secondary air pump.

11. Disconnect the air conditioning pressure hoses from the compressor.

12. Remove the intercooler.

13. Disconnect the right and left engine ground cables, the engine harness connector, the alternator connector and terminal, the air conditioning compressor electrical connector and the power steering switch connector.

14. Disconnect the brake booster hose and the heater hoses.

15. Remove the front side V-belt. Remove the power steering pipe and bracket. Remove the reservoir tank. Remove the power steering pump and position it aside.

16. Raise and support the vehicle safely. Remove the transmission cooler line from the frame, if equipped.

17. Remove the center exhaust pipe.

18. Remove the lower transmission to engine retaining nuts.

19. Remove the front cushion rubber onto front crossmember retaining nuts.

20. If equipped with automatic transmission, remove the service plug hole cap. Remove the torque converter to drive plate retaining bolts.

21. Remove the pitching stopper. Disconnect the fuel hoses from the fuel pipes.

22. Position the engine lifting fixture in place.

23. Support the transmission using a suitable jack.

➡Before separating the engine from the transmission, check to be sure nothing has been overlooked that will stop the engine from being removed. This is very important in order to facilitate reinstallation and because the transmission lowers under its own weight.

24. Separate the engine from the transmission.

25. Remove the starter.

26. If equipped with automatic transmission, install tool ST498277200, with will hold the torque converter in place.

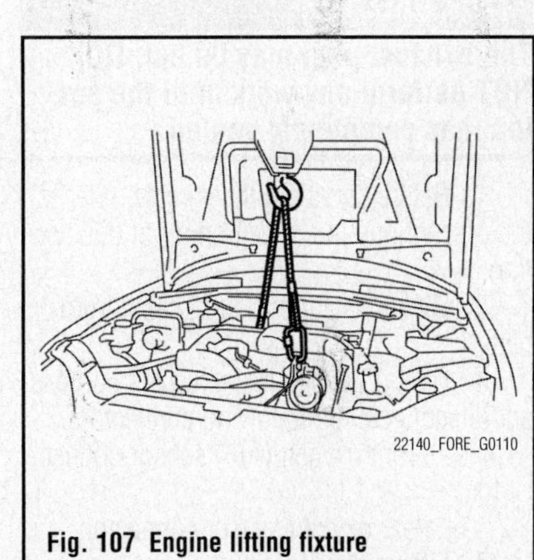

Fig. 107 Engine lifting fixture

27. Remove the upper right side bolts that retain the transmission to the engine.

28. Slightly raise the engine. Raise the transmission, using the suitable jack. Move the engine horizontally until the mainshaft is withdrawn from the clutch cover. Carefully remove the engine from the vehicle.

To install:

29. Installation is the reverse of the removal procedure.

30. Torque the upper transmission to engine retaining bolts to 36.9 ft. lbs. (50 Nm).

31. Torque the lower transmission to engine retaining bolts to 36.9 ft. lbs. (50 Nm).

32. If equipped, tighten the torque converter to drive plate bolts to 18 ft. lbs. (24 Nm).

33. Fill the engine with the proper grade and type engine oil.

34. Fill and bleed the cooling system.

35. Charge the air conditioning system.

36. If equipped. Adjust the clutch cable as required.

37. If equipped, check the automatic transmission fluid level, correct as required using the proper grade and type transmission fluid.

38. Start the engine and allow it to reach normal operating temperature. Check for leaks and correct as required.

EXHAUST MANIFOLD

Due to the unique design of the Subaru engine an exhaust manifold is not used. The exhaust enters directly into the front Y-pipe.

REMOVAL & INSTALLATION

2.5L SOHC Non-Turbo Engine

See Figure 108.

**** CAUTION**

The exhaust pipe may be hot; DO NOT perform any work until the system has completely cooled.

1. Before servicing the vehicle, refer to the precautions in the beginning of this section.

2. Disconnect the ground cable from battery.

3. Unclip the clip fastening the harness and disconnect the following connectors.

 a. Front oxygen (A/F) sensor connector.

 b. Rear oxygen sensor connector.

4. Lift-up the vehicle.

5. Separate the center exhaust pipe from rear exhaust pipe.

6. Remove the undercover.

7. Remove the nuts which hold front exhaust pipe onto cylinder heads.

**** CAUTION**

Be careful not to pull down the front and center exhaust pipe assembly.

8. Remove the bolt which holds center exhaust pipe to hanger bracket.

9. Remove the front and center exhaust pipe assembly from the vehicle.

10. Separate the front exhaust pipe from center exhaust pipe.

11. Remove the front oxygen (A/F) sensor and rear oxygen sensor.

To install:

12. Install the front oxygen (A/F) sensor and rear oxygen sensor to the front exhaust pipe

13. Install the front exhaust pipe to center exhaust pipe. Use a new gasket and tighten to 30 ft. lbs. (40 Nm).

14. Install the front and center exhaust pipe assembly to the vehicle.

15. Temporarily tighten the nuts which hold front exhaust pipe to cylinder heads.

16. Tighten the bolt which holds the center exhaust pipe to exhaust pipe. Use a new gasket and tighten to 13 ft. lbs. (18 Nm).

17. Install the bolt which holds center exhaust pipe to hanger bracket. Tighten to 26 ft. lbs. (35 Nm).

18. Install the nuts which hold front exhaust pipe to cylinder heads.

19. Use a new gasket and tighten to 22.4 ft. lbs. (30 Nm).

20. Install the undercover.

21. Lower the vehicle.

22. Connect the following connectors.

 c. Front oxygen (A/F) sensor connector

 d. Rear oxygen sensor connector

23. Connect the ground cable to battery.

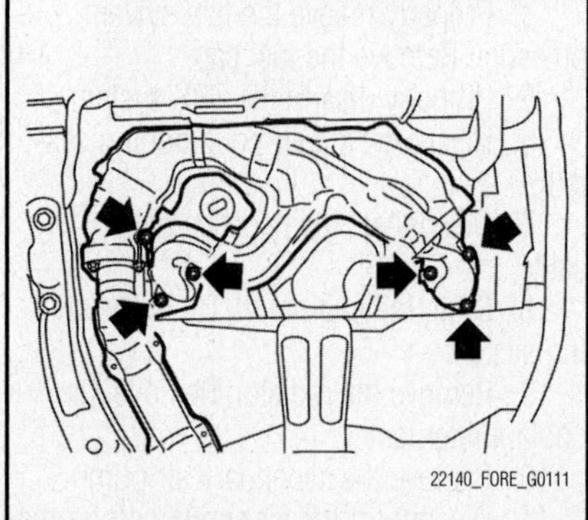

Fig. 108 Exhaust pipe to cylinder heads

2.5L DOHC Turbo Engine

See Figures 109 and 110.

**** CAUTION**

The exhaust pipe may be hot; DO NOT perform any work until the system has completely cooled.

1. Before servicing the vehicle, refer to the precautions in the beginning of this section.

2. Set the vehicle on a lift.

3. Disconnect the ground cable from battery.

4. Remove the collector cover.

5. Remove the front oxygen (A/F) sensor.

6. Remove the undercover.

7. Remove the exhaust manifold lower cover (RH) and the exhaust manifold cover (LH).

8. Remove the bolts and nuts which hold front exhaust pipe assembly onto turbocharger joint pipe.

9. While holding the front exhaust pipe assembly with one hand, remove the nuts which hold the front exhaust pipe assembly to cylinder head exhaust port.

10. Remove the front exhaust pipe assembly.

11. Remove the covers from exhaust manifold and front exhaust pipe.

12. Separate the front exhaust pipe from exhaust manifolds.

To install:

13. Assemble the front exhaust pipe and the exhaust manifold. Use a new gasket and tighten to 31.3 ft. lbs. (42.5 Nm).

14. Install the front exhaust pipe cover. Tighten to 5.5 ft. lbs. (7.5 Nm).

15. Install the exhaust manifold upper cover (RH). Tighten to 14 ft. lbs. (19 Nm).

16. Install the front exhaust pipe assembly. Use a new gasket and tighten to 28.9 ft. lbs. (40 Nm).

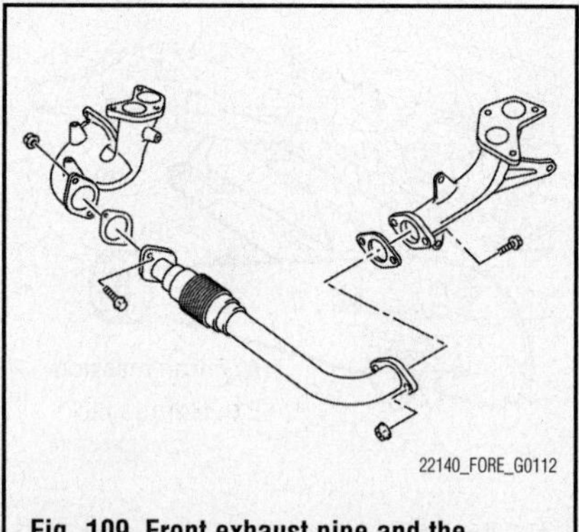

Fig. 109 Front exhaust pipe and the exhaust manifold

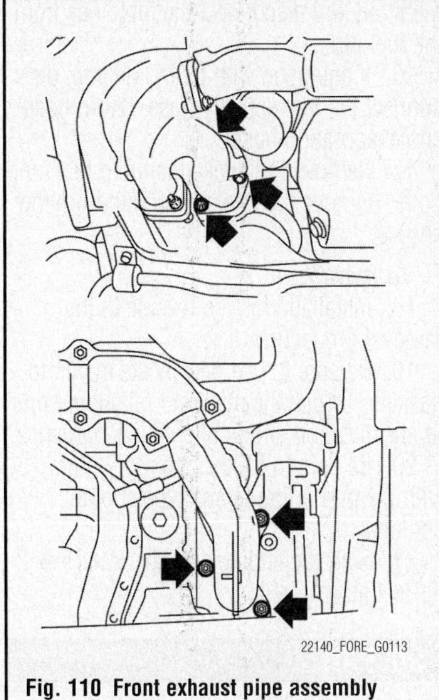

Fig. 110 Front exhaust pipe assembly

17. Connect the exhaust manifold (RH) to turbocharger joint pipe. Use a new gasket and tighten to 31.3 ft. lbs. (42.5 Nm).
18. Install the exhaust manifold lower cover (RH) and the exhaust manifold cover (LH).
19. Install the front oxygen (A/F) sensor.
20. Install the undercover.
21. Lower the vehicle.
22. Connect the ground cable to battery.
23. Install the collector cover.

FLYWHEEL

REMOVAL & INSTALLATION

See Figures 111 and 112.

1. Remove the transmission assembly. Refer to "Transmission Assembly, Removal & Installations" procedure in this section.
2. If equipped with a manual transmission, remove the clutch housing cover.
3. Using Special Tool 498497100 to lock the crankshaft into place, remove the flywheel (manual transmission) or driveplate (automatic transmission).

To install:

4. Install the flywheel (M/T) or drive plate (A/T) and use Special Tool 498497100 to lock the crankshaft into place.
5. Tighten the bolts in the sequence shown to 52.8 ft. lbs. (72 Nm).
6. Install the clutch housing cover, if equipped with manual transmission.
7. Install the transmission assembly.

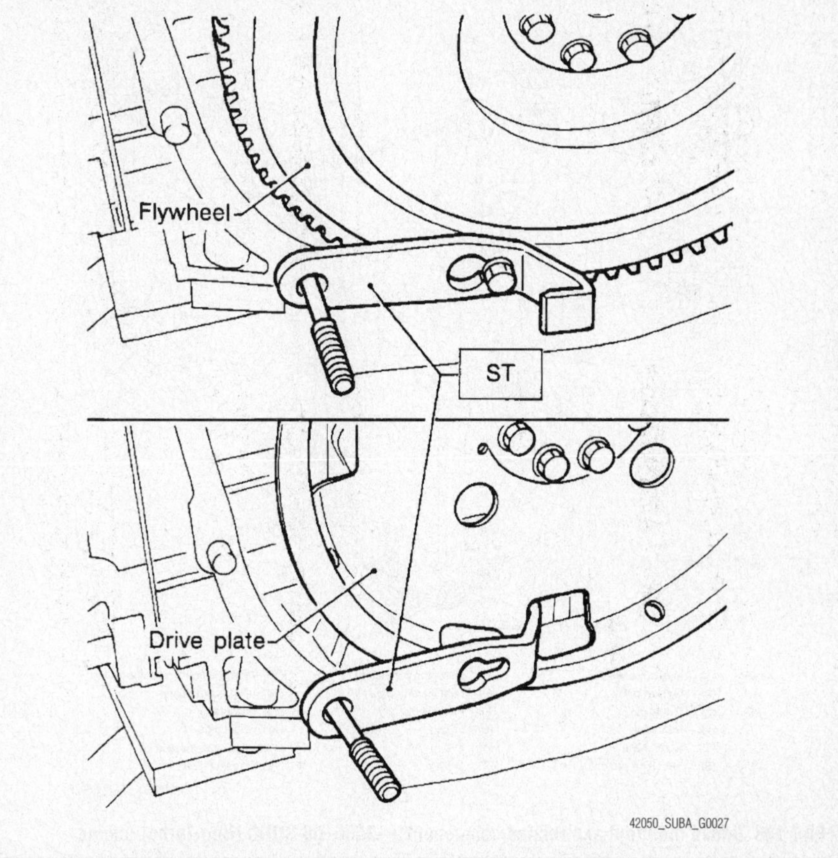

Fig. 111 Use Special Tool 498497100 to lock the crankshaft into place and remove the flywheel

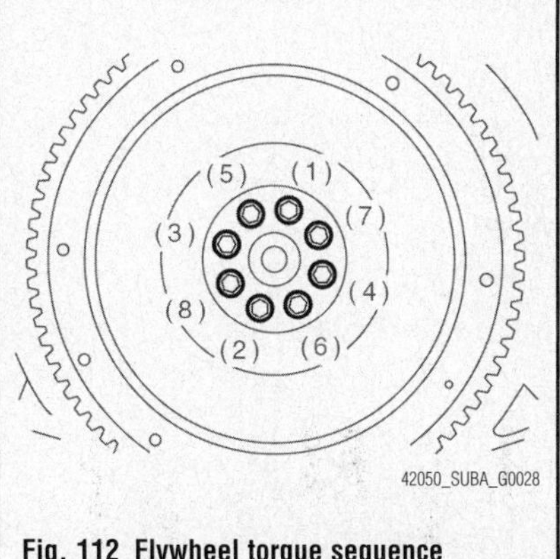

Fig. 112 Flywheel torque sequence

INTAKE MANIFOLD

REMOVAL & INSTALLATION

See Figures 113 and 114.

1. Before servicing the vehicle, refer to the Precautions Section.
2. Properly relieve the fuel system pressure. Remove the fuel cap.
3. Disconnect the negative battery cable. Drain the engine coolant.
4. If equipped, remove the undercover.
5. Remove the air intake duct, air cleaner case and air intake chamber.
6. If equipped, remove the intercooler.
7. Remove the alternator.
8. If equipped with DOHC engine, remove the coolant filler tank.
9. If equipped with SOHC engine, remove the spark plug wires.
10. Disconnect the engine coolant hoses from the throttle body. Disconnect the brake booster hose.
11. Disconnect the PCV hose from the intake manifold. Disconnect the engine harness electrical connectors from the bulkhead harness connectors.
12. Disconnect the engine coolant temperature sensor electrical connector, knock sensor electrical connector and crankshaft position sensor connector.
13. Disconnect the power steering pump switch electrical connector, oil pressure switch connector and camshaft sensor connector.
14. If equipped with DOHC engine, disconnect the oil flow solenoid valve electrical connector and the ignition coil connector.
15. If equipped with SOHC engine, remove the EGR pipe from the intake

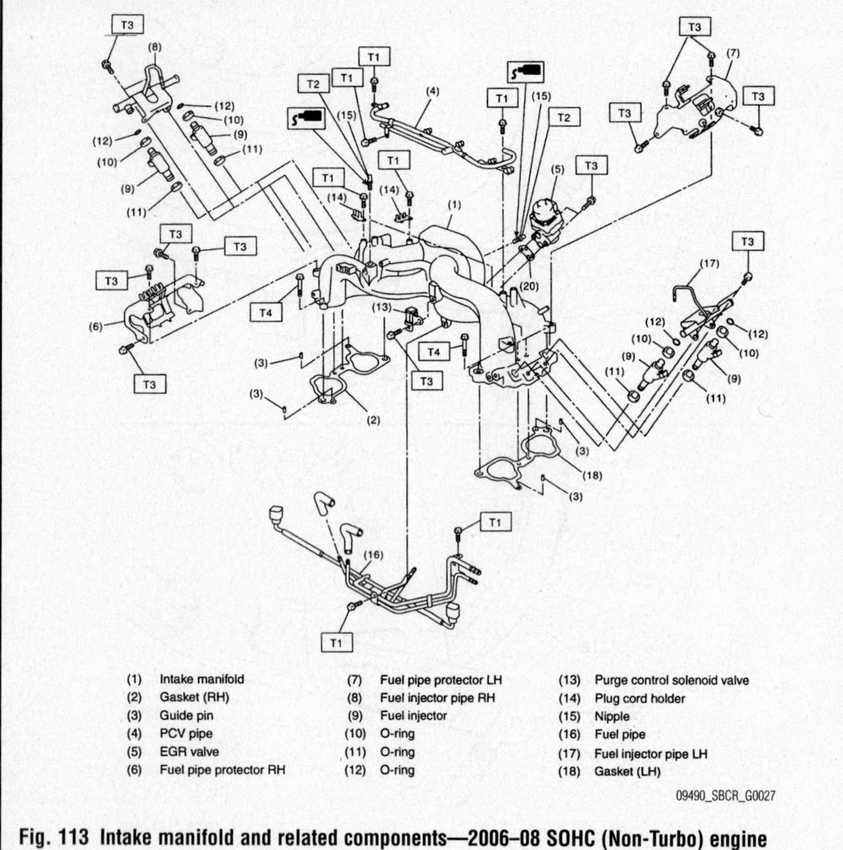

Fig. 113 Intake manifold and related components—2006–08 SOHC (Non-Turbo) engine

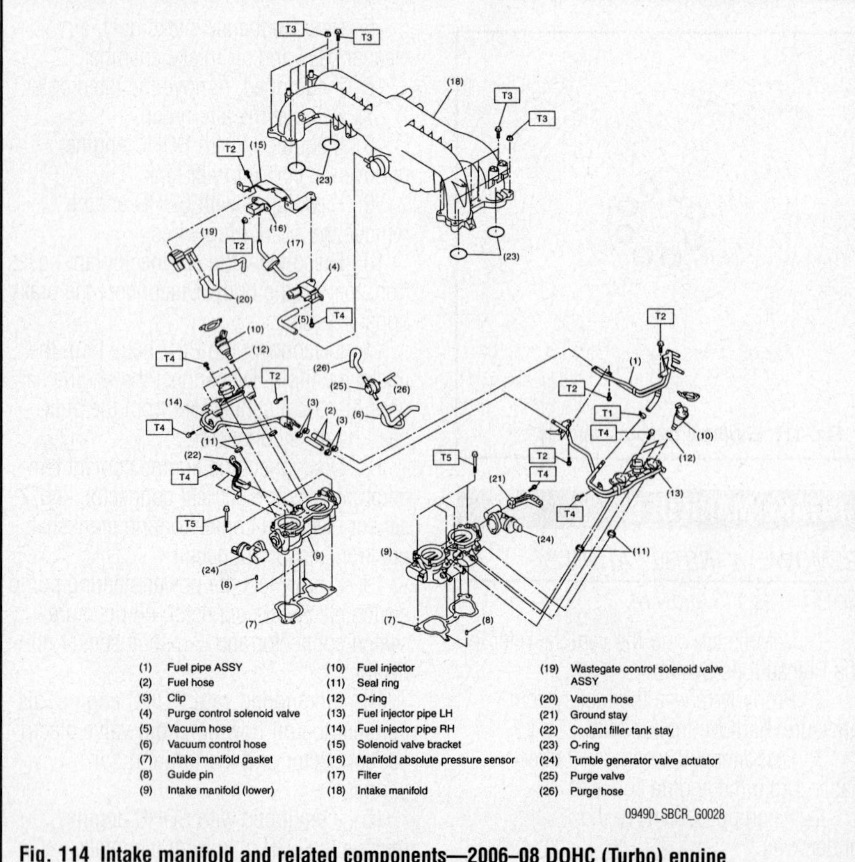

Fig. 114 Intake manifold and related components—2006–08 DOHC (Turbo) engine

manifold and disconnect the fuel lines from the fuel pipe.

16. If equipped with DOHC engine, disconnect the fuel delivery hose, return hose and evaporation hose.

17. Remove the intake manifold retaining bolts. Remove the intake manifold from the engine.

To install:

18. Installation is the reverse of the removal procedure.

19. Be sure to use new intake manifold gaskets. Torque the manifold retaining bolts to specification and in alternating sequence.

20. Be sure to fill the cooling system with the proper grade and type engine coolant.

21. Start the engine and check for leaks, correct as required.

OIL PAN

REMOVAL & INSTALLATION

2.5L SOHC Non-Turbo Engine

See Figure 115.

1. Before servicing the vehicle, refer to the Precautions Section.

2. Disconnect the negative battery cable.

3. Raise and support the vehicle safely.

4. Remove the front tires and wheels.

5. Lower the vehicle.

6. Remove the air intake duct and the air cleaner case. Remove the air intake chamber.

7. Remove the pitching stopper.

8. Remove the hood stay holder and the radiator upper brackets.

9. Properly support the engine with a lifting device and wire ropes.

10. Lift the vehicle and support it safely.

➡When lifting the vehicle, raise the wire ropes at the same time.

11. Remove the undercover.

12. Drain the engine oil.

13. Remove the front and center exhaust pipes.

14. Remove the nuts which retain the front cushion rubber onto the front crossmember.

15. Remove the bolts that retain the oil pan to the cylinder block, with the engine in the raised position.

16. Insert an oil pan gasket cutter tool into the gap between the cylinder block and the oil pan. Remove the oil pan from the engine.

➡Do not use a screwdriver or similar tool in place of the cutter tool.

17. Remove the oil strainer, if required. Remove the baffle plate, if required.

To install:

18. Be sure to clean the old gasket material from the mating surfaces.
19. Apply a continuous bead of sealer to a new oil pan gasket.
20. Make sure that the seals (A) are installed securely on the baffle plate and in the direction shown in the illustration. Install the baffle plate; tighten the retaining bolts to 4.7 ft. lbs. (6.4 Nm).
21. Replace the O-ring and install the oil strainer. Tighten the bolt to 7 ft. lbs. (10 Nm).
22. Apply liquid gasket, part number 004403012 or equivalent, to the oil pan mating surface. Install the oil pan. Torque the retaining bolts to specification.
23. Continue the installation in the reverse order of the removal procedure.
24. Torque the front cushion mounting bolts to 63 ft. lbs.
25. Be sure to fill the engine with the correct grade and type engine oil.
26. Start the engine and check for leaks. Correct as required.

2.5L DOHC Turbo Engine

1. Before servicing the vehicle, refer to the Precautions Section.
2. Disconnect the negative battery cable.
3. Raise and support the vehicle safely.
4. Remove the front tires and wheels.
5. Remove the collector cover.
6. Lower the vehicle.
7. Disconnect the connector from the MAF sensor.
8. Remove the air intake boot and air cleaner upper cover.
9. Remove the intercooler.
10. Remove the pitching stopper.

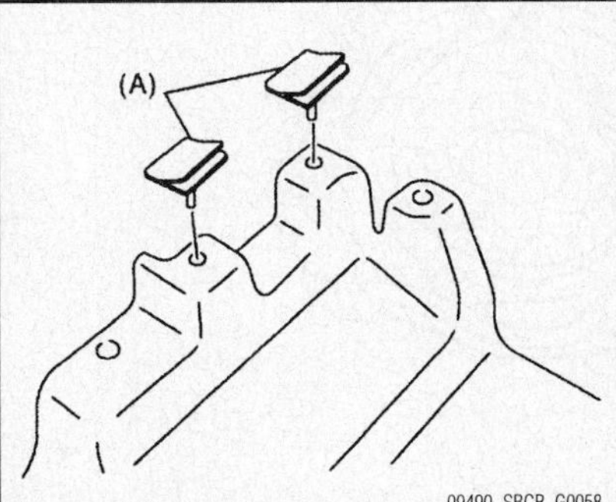

Fig. 115 Oil pan baffle plate seal location and positioning—2006-08

11. Remove the radiator upper brackets.
12. Properly support the engine with a lifting device and wire ropes.
13. Lift the vehicle and support it safely.

➡When lifting the vehicle, raise the wire ropes at the same time.

14. Remove the undercover.
15. Drain the engine oil.
16. Remove the front exhaust pipe.
17. Remove the nuts which retain the front cushion rubber onto the front crossmember.
18. Remove the bolts that retain the oil pan to the cylinder block, with the engine in the raised position.
19. Insert an oil pan gasket cutter tool into the gap between the cylinder block and the oil pan. Remove the oil pan from the engine.

➡Do not use a screwdriver or similar tool in place of the cutter tool.

20. Remove the oil strainer, if required. Remove the baffle plate, if required.

To install:

21. Be sure to clean the old gasket material from the mating surfaces.
22. Apply a continuous bead of sealer to a new oil pan gasket.
23. Make sure that the seals (A) are installed securely on the baffle plate and in the direction shown in the illustration. Install the baffle plate; tighten the retaining bolts to 4.7 ft. lbs. (6.4 Nm).
24. Replace the O-ring and install the oil strainer. Tighten the bolt to 7 ft. lbs. (10 Nm)
25. Apply liquid gasket, part number 004403012 or equivalent, to the oil pan mating surface. Install the oil pan. Torque the retaining bolts to specification.
26. Continue the installation in the reverse order of the removal procedure.
27. Torque the front cushion mounting bolts to 63 ft. lbs. (85 Nm).
28. Be sure to fill the engine with the correct grade and type engine oil.
29. Start the engine and check for leaks. Correct as required.

OIL PUMP

REMOVAL & INSTALLATION

See Figures 116 and 117.

1. Before servicing the vehicle, refer to the Precautions Section.
2. Disconnect the negative battery cable. Drain the cooling system.

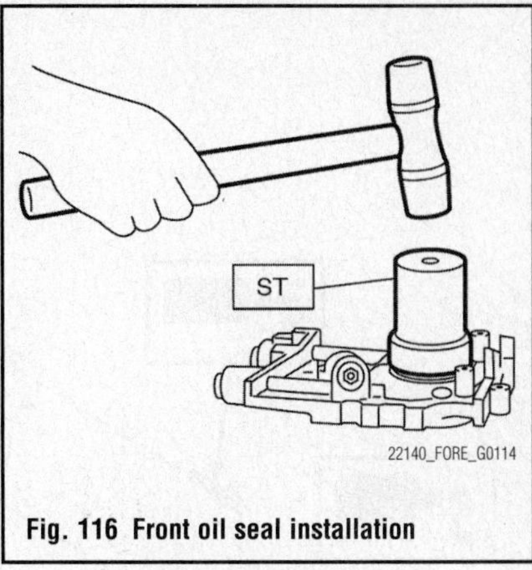

Fig. 116 Front oil seal installation

3. On the DOHC (Turbo) engine, remove the collector cover.
4. Raise and support the vehicle safely.
5. Remove the undercover.
6. On the DOHC engine, remove the bolts which retain the water pipe of the oil cooler to the oil pump. Remove the water pipe and hoses between the oil cooler and the water pump.
7. Lower the vehicle.
8. Remove the radiator.
9. To remove the front side V-belt, remove the belt covers. Loosen the lock bolt. Loosen the slider bolt. Remove the front side belt.
10. To remove the rear side V-belt, remove the belt covers. Loosen the lock bolt. Loosen the slider bolt. Remove the rear side belt.
11. On the SOHC engine remove the belt tensioner.
12. On the DOHC engine remove the rear side V-belt tensioner
13. Remove the crankshaft position sensor.
14. Remove the crankshaft pulley bolt.
15. Lock the crankshaft in place using tool ST499977100 for the SOHC engine and tool ST499207400 for the DOHC engine, or equivalent.
16. Remove the crankshaft pulley.
17. Remove the water pump.
18. If equipped, remove the timing belt guide. Remove the crankshaft sprocket.
19. Remove the oil pump retaining bolts.

➡When disassembling and checking the oil pump, loosen the relief valve plug before removing the oil pump from its mounting.

20. Using a flat tip tool remove the oil pump from the engine.

To install:

21. Be sure all mating surfaces are clean and free of dirt.

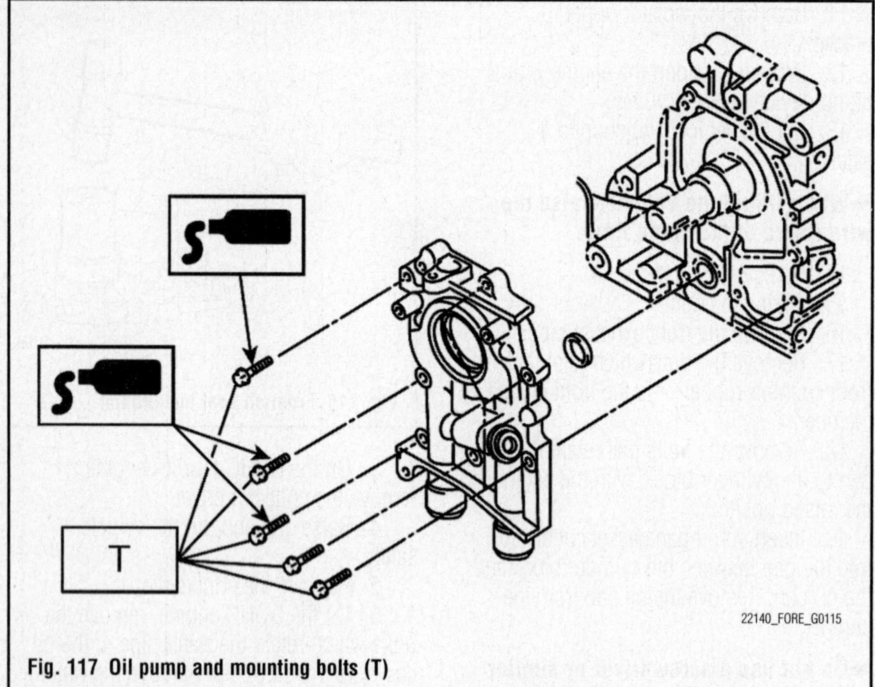

Fig. 117 Oil pump and mounting bolts (T)

22. Apply liquid gasket part number 004403007, or equivalent to the mating surfaces of the oil pump.
23. Be sure to replace the O-ring with a new one.
24. Using the ST-499587100, install the front oil seal.
25. Apply a thin coat of clean engine oil to the inside of the oil seal.
26. Position the oil pump to its mounting, aligning the notched area with the crankshaft and push the pump straight.

➡**Be sure that the oil seal lip is not folded.**

27. Install the oil pump. Apply liquid gasket part number 004403042, or equivalent to the three retaining bolt threads.
28. Install the bolts and tighten to 4.7 ft. lbs. (6.4 Nm).
29. Continue the installation in the reverse order of the removal procedure.
30. Be sure to fill the cooling system with the proper grade and type coolant.
31. Start the engine and check for leaks. Correct, as required.

INSPECTION

See Figures 118 through 120.

1. Check the oil pump case for worn shaft hole, clogged oil passage, worn rotor chamber, cracks, and other faults.
2. Check the oil seal lips for deformation, hardening, wear, etc. and replace if defective.

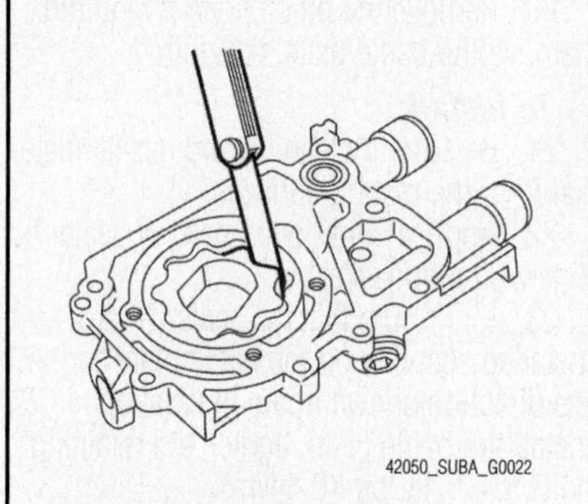

Fig. 118 Measuring for tip clearance of the rotors of the oil pump.

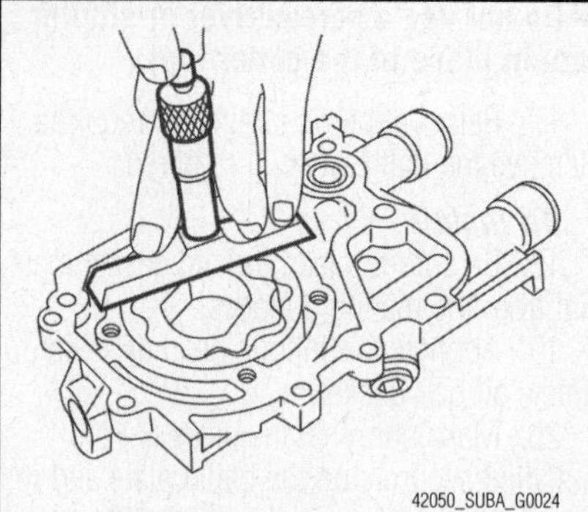

Fig. 120 Measuring the side clearance of the oil pump.

3. Measure the tip clearance of rotors. If clearance exceeds the standard, replace the rotors as a matched set.
   a. 0.0016–0.0055 in. (0.04–0.14mm)
4. Measure the clearance between the outer rotor and oil pump rotor housing. If clearance exceeds the standard, replace the rotor.
   a. 0.0039–0.0069 in. (0.10–0.18mm)
5. Measure the clearance between the oil pump inner rotor and pump cover. If clearance exceeds the standard, replace the rotor or pump body.
   a. 0.0008–0.0028 in. (0.02–0.07mm)
6. Check the valve for fitting condition and damage, and the relief valve spring for damage and deterioration. Replace the parts if defective.

MAIN BEARING TORQUE SEQUENCE

See Figures 121 and 122.

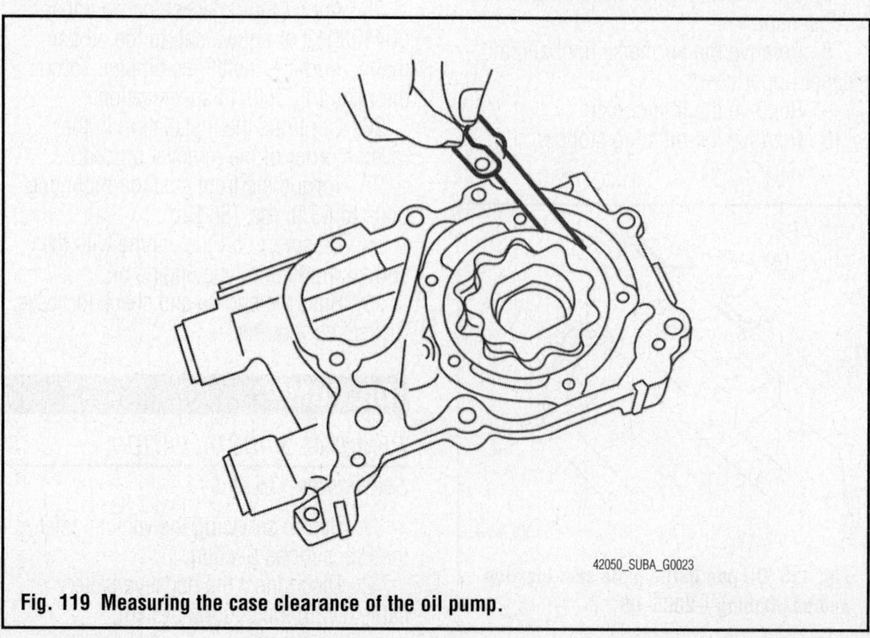

Fig. 119 Measuring the case clearance of the oil pump.

NOTE:
Remove oil on the mating surface of bearing and cylinder block before installation. Apply engine oil to crankshaft pins.
1) Position the crankshaft and O-rings on the #1 and #3 cylinder block.
2) Apply liquid gasket to the mating surface of #1 and #3 cylinder block, and position #2 and #4 cylinder block.

Liquid gasket:
Three bond 1215 (Part No. 004403007) or equivalent

NOTE:
Do not allow liquid gasket to jut into O-ring grooves, oil passages, bearing grooves, etc.

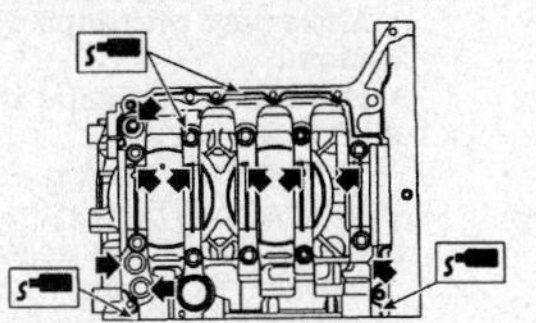

3) Apply a coat of engine oil to the washer and bolt thread.

4) Tighten the 10 mm cylinder block connecting bolts on LH side (A — D) in alphabetical sequence.

Tightening torque:
9.75 N·m (1.0 kgf-m, 7.2 ft-lb)

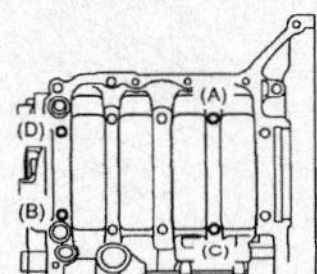

5) Tighten the 10 mm cylinder block connecting bolts on RH side (E — J) in alphabetical sequence.

Tightening torque:
9.75 N·m (1.0 kgf-m, 7.2 ft-lb)

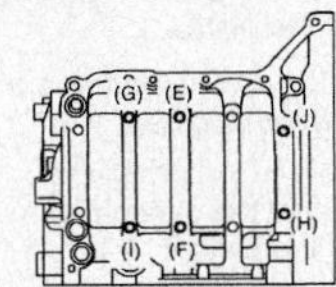

6) Further tighten the LH side bolts (A — D) in alphabetical sequence.

Tightening torque:
18 N·m (1.8 kgf-m, 13 ft-lb)

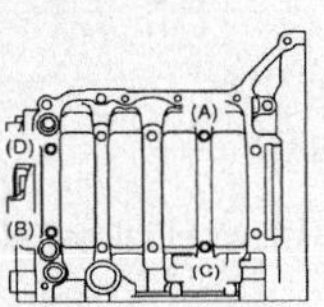

7) Further tighten the RH side bolts (E — J) in alphabetical sequence.

Tightening torque:
18 N·m (1.8 kgf-m, 13 ft-lb)

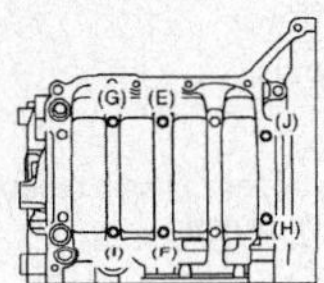

8) Further tighten the LH side bolts (A — D) in alphabetical sequence.
• (A), (C): Angle tightening

Tightening angle:
90°
• (B), (D): Torque tightening

Tightening torque:
40 N·m (4.1 kgf-m, 29.6 ft-lb)

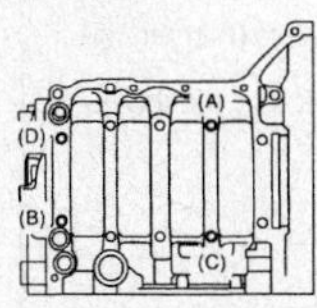

9) Tighten the RH side bolts (E — J) 90° further in alphabetical sequence.

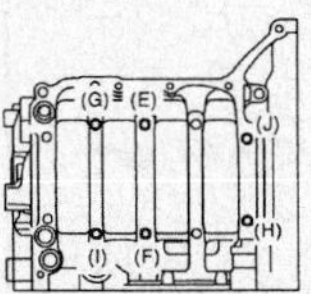

10) Tighten the 8 mm and 6 mm cylinder block connecting bolts on LH side (A — H) in alphabetical sequence.

Tightening torque:
(A) — (G): 25 N·m (2.5 kgf-m, 18.1 ft-lb)
(H): 6.4 N·m (0.65 kgf-m, 4.7 ft-lb)

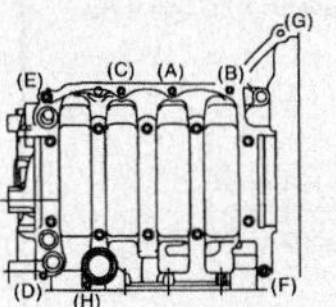

09490_SBCR_G0004

Fig. 121 Main bearing torque sequence—2006–08 SOHC engine

1) Remove oil on the mating surface of bearing and cylinder block before installation. Apply engine oil to crankshaft pins.
2) Position the crankshaft and O-rings on the #1 and #3 cylinder block.
3) Apply liquid gasket to the mating surface of #1 and #3 cylinder blocks, and position cylinder block #2 and #4.

NOTE:
Do not allow liquid gasket to run over to O-ring grooves, oil passages, bearing grooves, etc.

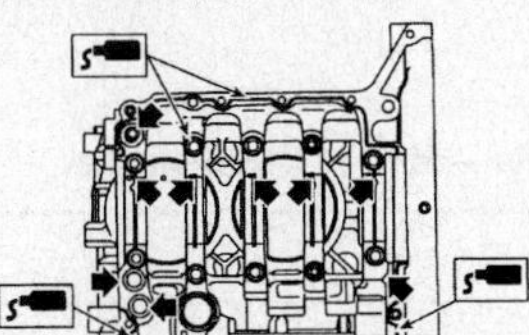

4) Apply a coat of engine oil to the washer and bolt thread.
5) Tighten the 10 mm cylinder block connecting bolts on LH side (A — D) in alphabetical sequence.

Tightening torque:
10 N·m (1.0 kgf-m, 7.2 ft-lb)

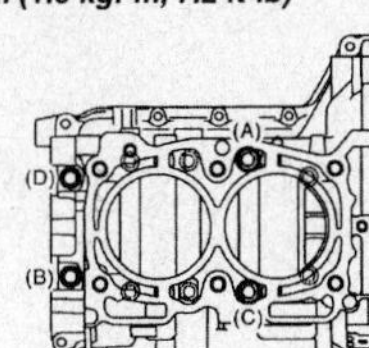

6) Tighten the 10 mm cylinder block connecting bolts on RH side (E — J) in alphabetical sequence.

Tightening torque:
10 N·m (1.0 kgf-m, 7.2 ft-lb)

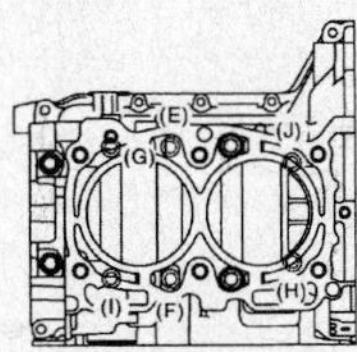

7) Further tighten the LH side bolts (A — D) in alphabetical sequence.

Tightening torque:
18 N·m (1.8 kgf-m, 13.0 ft-lb)

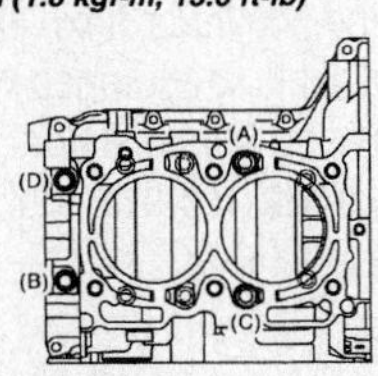

8) Further tighten the RH side bolts (E — J) in alphabetical sequence.

Tightening torque:
18 N·m (1.8 kgf-m, 13.0 ft-lb)

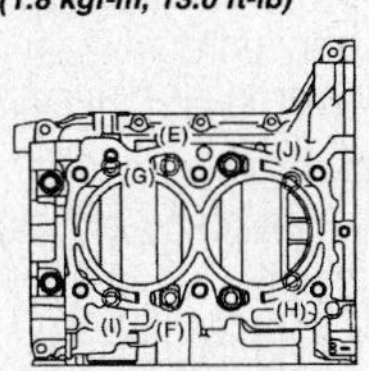

9) Further tighten the LH side bolts (A — D) in alphabetical sequence.
(A), (C): 90°
(B), (D): 40 N·m (4.1 kgf-m, 29.5 ft-lb)

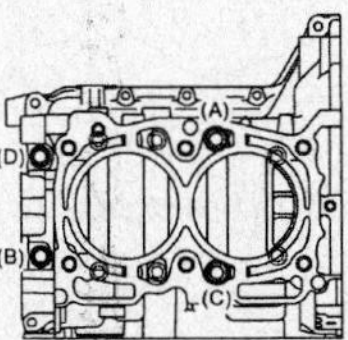

10) Tighten the RH side bolts (E — J) 90° further in alphabetical sequence.

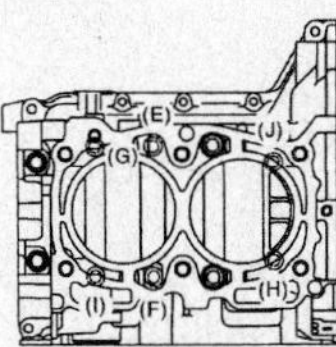

11) Tighten the 8 mm and 6 mm cylinder block connecting bolts on LH side (A — H) in alphabetical sequence.

Tightening torque:
(A) — (G): 25 N·m (2.5 kgf-m, 18.1 ft-lb)
(H): 6.4 N·m (0.65 kgf-m, 4.7 ft-lb)

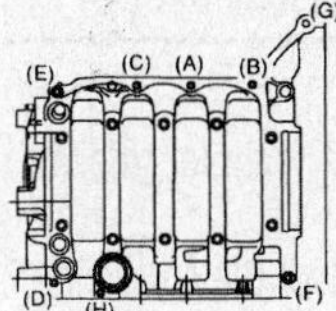

09490_SBCR_G0007

Fig. 122 Main bearing torque sequence—2006–08 DOHC engine

PISTON AND RING

POSITIONING

See Figures 123 and 124.

Position the top ring gap at (A) or (B) in the figure.
Position the second ring gap at 180° on the reverse side the top ring gap.

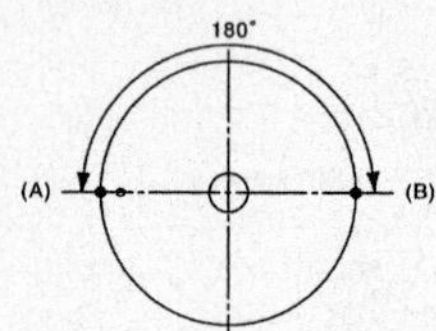

Position the upper rail gap at (C) in the figure.

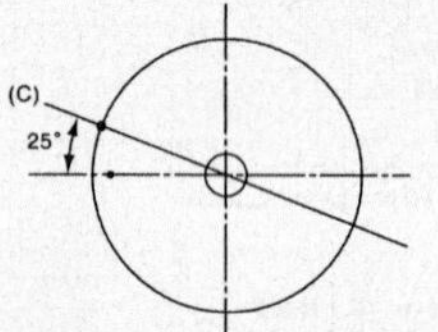

Align the upper rail spin stopper (D) to the side hole (E) on the piston.

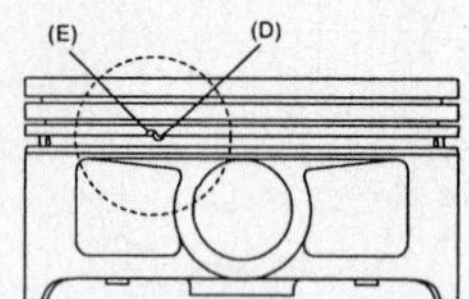

Position the expander gap at (F) in the figure on the 180° opposite direction of (C).

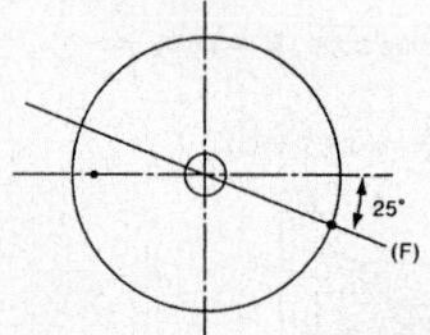

Position the lower rail gap at (G) in the figure.

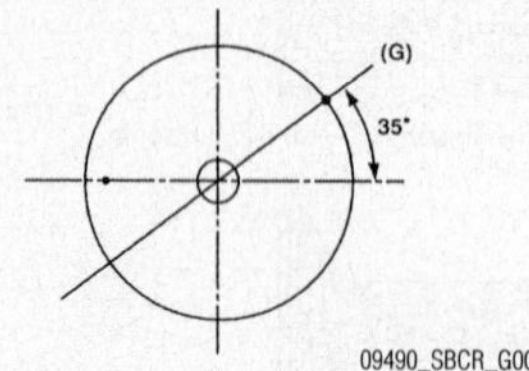

09490_SBCR_G0079

Fig. 123 Piston ring alignment and positioning—2006–08 SOHC engine

Position the top ring gap at (A) or (B) in the figure.

NOTE:
Assemble so that the piston ring mark "R" faces the upper side of the piston.

Position the second ring gap at 180° on the reverse side the top ring gap.

NOTE:
Assemble so that the piston ring mark "R" faces the upper side of the piston.

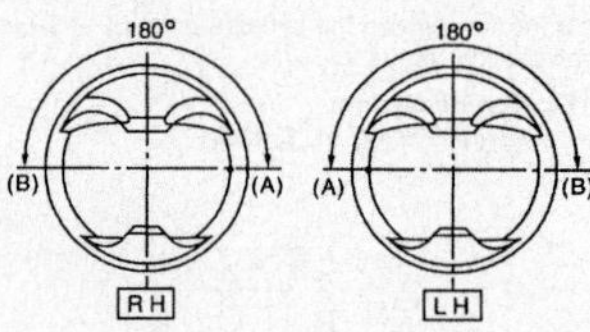

Position the upper rail gap at (C) in the figure.

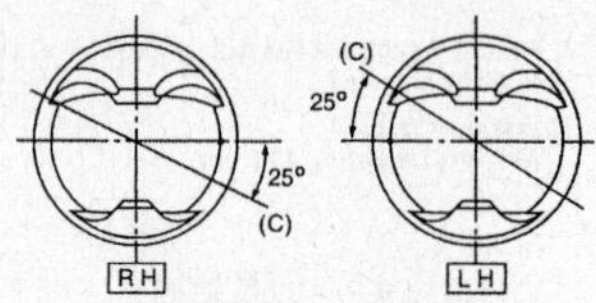

Align the upper rail spin stopper (E) to the side hole (D) on the piston.

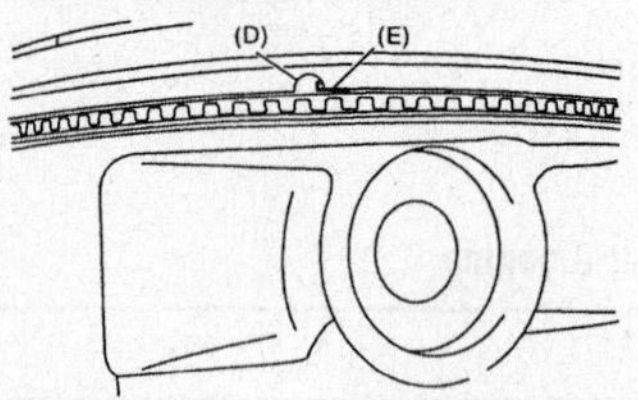

Position the expander gap at (F) in the figure.

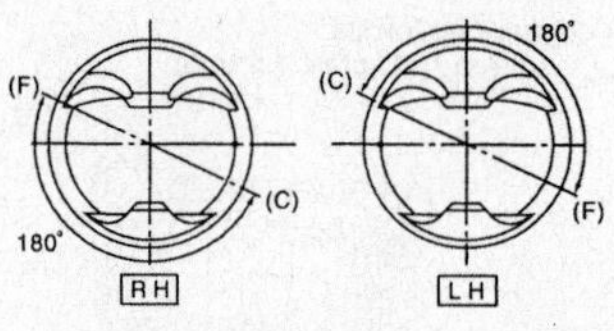

Position the lower rail gap at (G) in the figure.

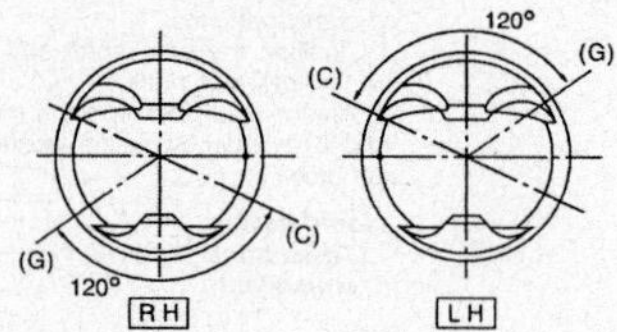

CAUTION:
- **Make sure ring gaps do not face the same direction.**
- **Make sure ring gaps are not within the piston skirt area.**

Install the snap ring.
Install the snap rings in the piston holes located opposite to the service holes in cylinder block when positioning all pistons in corresponding cylinders.

NOTE:
Use new snap rings.

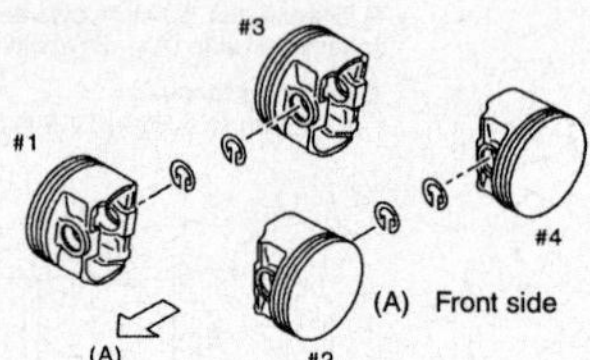

CAUTION:
Piston front mark faces towards the front of engine.

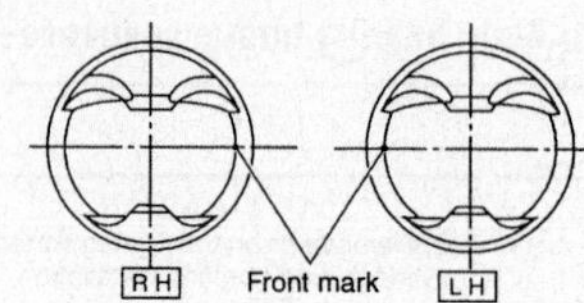

09490_SBCR_G0080

Fig. 124 Piston ring alignment and positioning—2006–08 DOHC engine

REAR MAIN SEAL

REMOVAL & INSTALLATION

See Figure 125.

1. Before servicing the vehicle, refer to the Precautions Section.
2. Remove or disconnect the following:
   - Engine from the vehicle
   - Clutch assembly/flywheel, if equipped with manual transmission
   - Torque converter flexplate from the crankshaft, if equipped with an automatic transmission
3. Using a seal removal tool, pry the oil seal from the housing.

To install:

4. Utilizing the appropriate seal installer, install or connect the following:
   - New oil seal and press it into the housing using oil seal guide ST 499597100 and installer ST 499587200.
   - Clutch assembly/flywheel, if equipped
   - Flywheel/flexplate and tighten the bolts to 53 ft. lbs. (72 Nm), if equipped
   - Engine into the vehicle

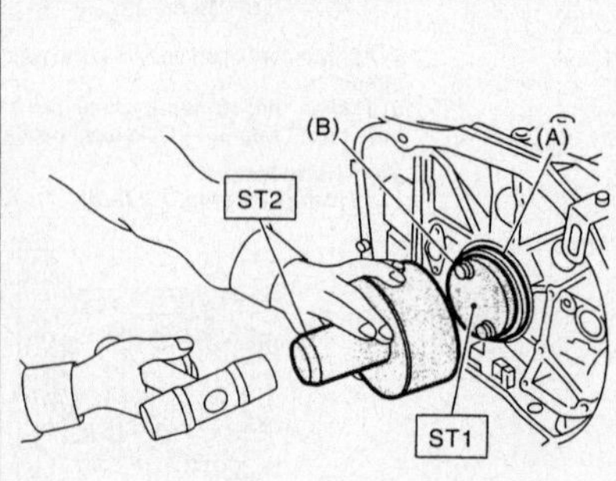

(A) Rear oil seal
(B) Flywheel attaching bolt

67170-SBFR-G27

Fig. 125 Installing the rear main seal

ROCKER ARMS/SHAFTS

REMOVAL & INSTALLATION

2.5L SOHC Non-Turbo Engine

See Figures 126 through 129.

1. Before servicing the vehicle, refer to the Precautions Section.
2. Disconnect the negative battery cable.
3. To remove the front side V-belt, remove the belt covers. Loosen the lock bolt. Loosen the slider bolt. Remove the front side belt.

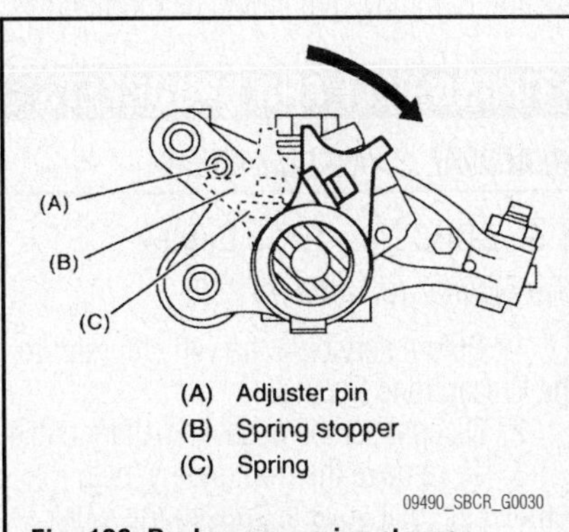

Fig. 126 Rocker arm spring stopper removal—SOHC engine

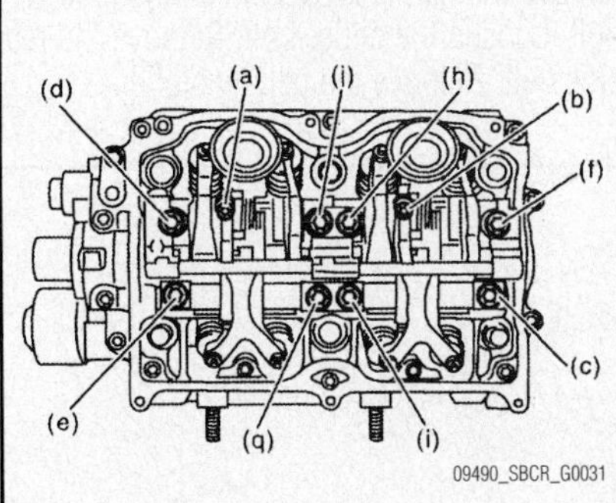

Fig. 127 Rocker arm assembly bolt loosening sequence—SOHC engine

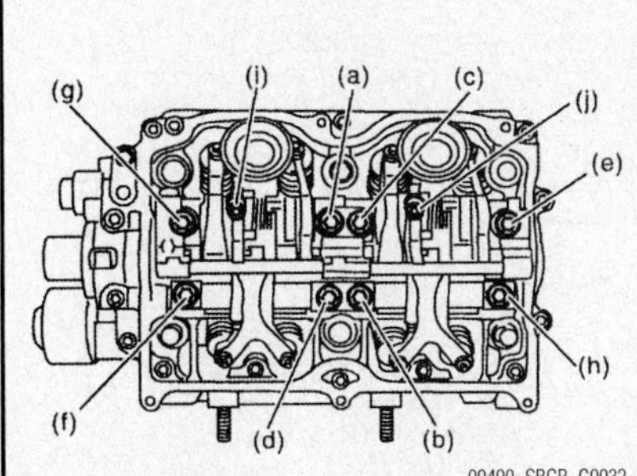

Fig. 128 Rocker arm assembly bolt tightening sequence—SOHC engine

4. To remove the rear side V-belt, remove the belt covers. Loosen the lock bolt. Loosen the slider bolt. Remove the rear side belt. Remove the belt tensioner.
5. Remove the crankshaft pulley bolt.
6. Lock the crankshaft in place using tool ST499977100, or equivalent.
7. Remove the crankshaft pulley.
8. Remove the left side timing belt cover.
9. Remove the front timing belt cover.
10. Remove the timing belt.
11. Remove the camshaft position sensor.
12. Remove the camshaft sprockets, No.1 and No.2.

➡Be sure to lock the camshaft in place using tool ST18231AA010, or equivalent.

13. Disconnect the PCV valve hose. Remove the rocker cover retaining bolts. Remove the rocker cover from the engine.
14. Use tool St18258AA000, or equivalent, to rotate the spring stopper in direction of the arrow, see illustration, and remove the spring stopper from the adjuster spring.
15. Remove bolts "A" through "J" in sequence. Remove the assembly from the engine.

➡Leave two or three threads on bolts "I" and "J" engaged in order to retain the rocker arm assembly.

16. Position tool ST18354AA000 on the rocker arm assembly. Remove the assembly from the engine.

To install:

17. Install the rocker arm assembly on its mounting.
18. Temporarily tighten the bolts equally in the proper sequence.

➡Do not temporarily tighten bolts "I" and "J". Position tool ST18354AA000 to the rocker arm assembly.

19. Tighten bolts "A" through "H" to specification (18.1 ft. lbs.).
20. Tighten bolts "I" through "J" to specification (5.9 ft. lbs.).
21. Use tool ST18258AA000 to rotate the spring stopper in the opposite direction of the arrow, see illustration, to fasten the adjuster pin.
22. Adjust the valve clearance.
23. Use a new gasket and install the rocker arm cover. Torque the retaining bolts to 4.7 ft. lbs and in the proper sequence. Recheck and torque to 4.7 ft. lbs in the proper sequence.
24. Continue the installation in the reverse order of the removal procedure.

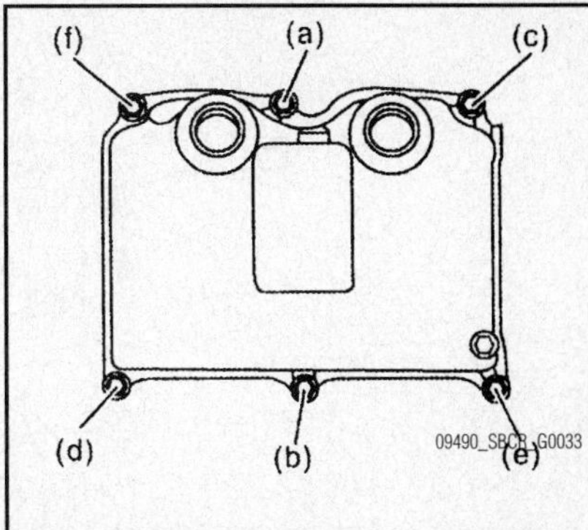

Fig. 129 Rocker arm cover bolt tightening sequence—SOHC engine

TURBOCHARGER

REMOVAL & INSTALLATION

See Figure 130.

1. Before servicing the vehicle, refer to the Precautions Section.
2. Set the vehicle on a lift.
3. Remove the collector cover.
4. Disconnect the ground cable from the battery.
5. Remove the intercooler.
6. Remove the center exhaust pipe.
7. Lower the vehicle.
8. Separate the turbocharger joint pipe from turbocharger.
9. Disconnect the engine coolant hose which is connected to coolant filler tank.
10. Loosen the clamp which secures turbocharger to intake duct.
11. Remove the bolts which secure oil inlet pipe bracket to turbocharger.
12. Remove the oil inlet pipe from the turbocharger.
13. Disconnect the engine coolant hose from the water pipe, and remove the turbocharger bracket RH.
14. Disconnect the oil outlet hose from the oil outlet pipe.
15. Take out the turbocharger from engine compartment.

To install:

16. Connect the oil outlet hose to the oil outlet pipe.
17. Install the turbocharger to intake duct and tighten to 2 ft. lbs. (3 Nm).
18. Install the oil inlet pipe to the turbocharger and tighten to 12 ft. lbs. (16 Nm).
19. Install the turbocharger bracket RH, and connect the engine coolant hose to the water pipe. Tighten to 24 ft. lbs. (33 Nm).
20. Install the joint pipe to turbocharger and using a new gasket tighten to 25 ft. lbs. (35 Nm).
21. Connect the engine coolant hose which is connected to the coolant filler tank.

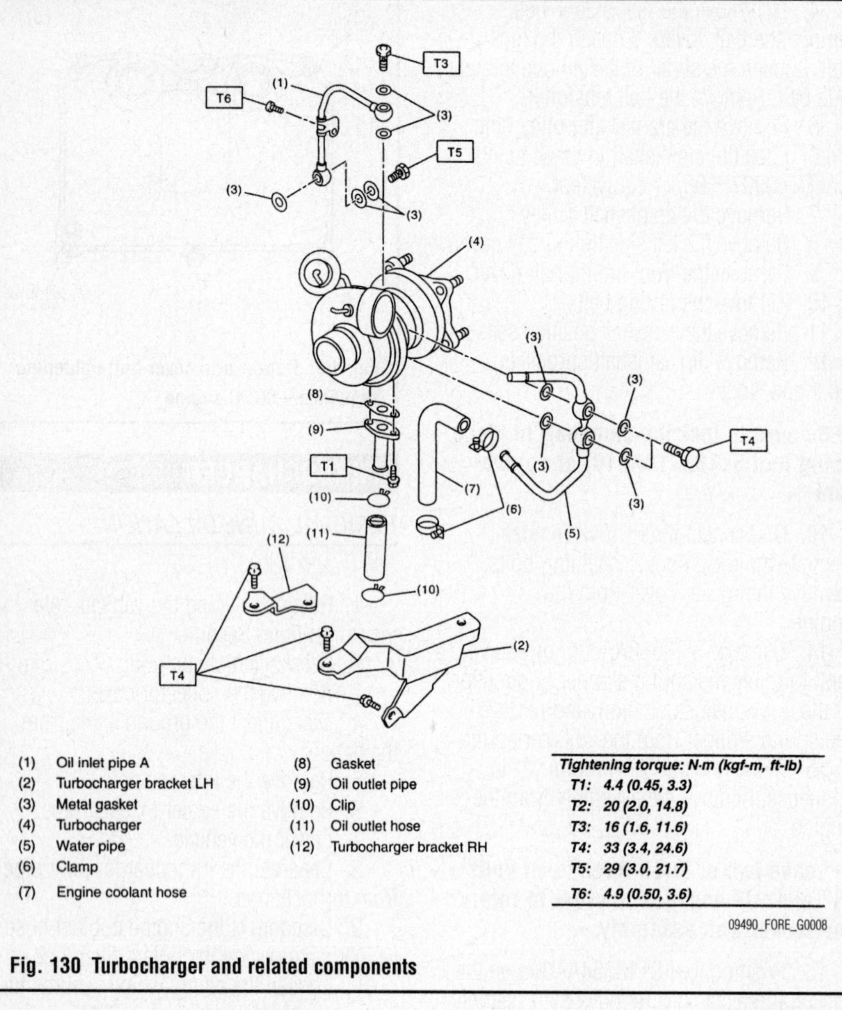

Fig. 130 Turbocharger and related components

22. Lift up the vehicle.
23. Install the center exhaust pipe.
24. Lower the vehicle.
25. Install the intercooler and tighten mounting bolts to 12 ft. lbs. (16 Nm).
26. Install the collector cover.
27. Connect the ground cable to battery.

TURBOCHARGER INTERCOOLER

REMOVAL & INSTALLATION

See Figure 131.

1. Before servicing the vehicle, refer to the Precautions Section.
2. Disconnect the negative battery cable.
3. Disconnect the PVC hose from the PVC pipe.
4. Remove the air bypass valve from the intercooler.
5. Remove the bolts which retain the intercooler to the bracket.
6. Loosen the clamps which connect the turbocharger and intercooler duct.
7. Loosen the clamps which connect the throttle body and the intercooler.
8. Separate the intercooler duct from the turbocharger.
9. Remove the intercooler sensor from the throttle body.
10. Installation is the reverse of the removal procedure.

TIMING BELT FRONT COVER

REMOVAL & INSTALLATION

Refer to "Timing Belt and Sprockets, Removal & Installation" procedure in this section to remove the timing belt front cover.

TIMING BELT AND SPROCKETS

REMOVAL & INSTALLATION

2.5L SOHC Non-Turbo Engine

See Figures 132 through 136.

1. Before servicing the vehicle, refer to the Precautions Section.
2. Disconnect the negative battery cable.
3. To remove the front side V-belt, remove the belt covers. Loosen the lock bolt. Loosen the slider bolt. Remove the front side belt.
4. To remove the rear side V-belt, remove the belt covers. Loosen the lock bolt. Loosen the slider bolt. Remove the rear side belt. Remove the belt tensioner.

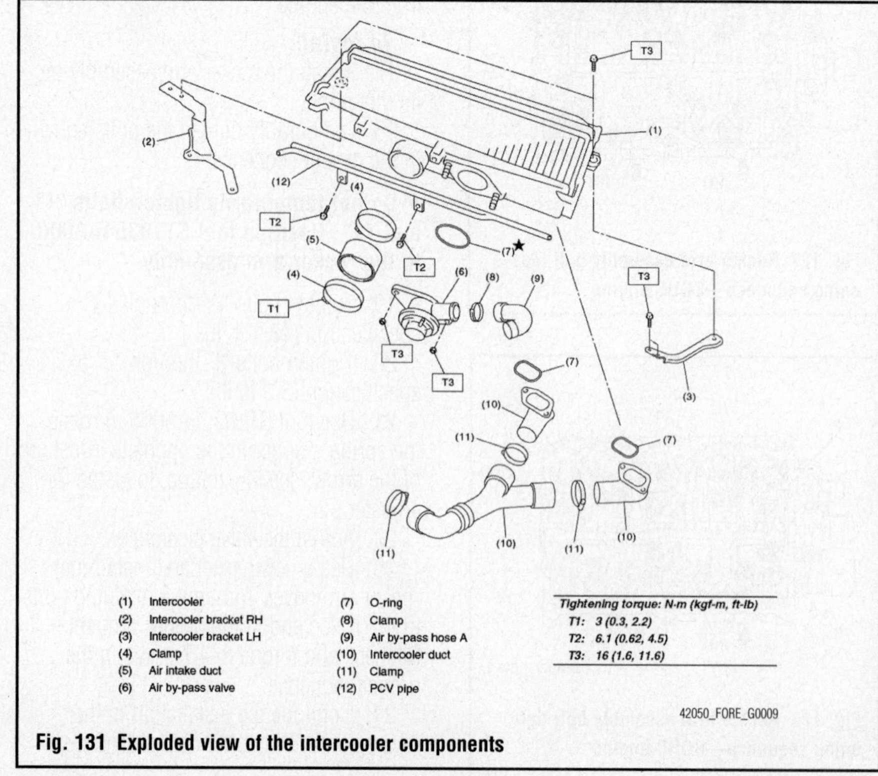

Fig. 131 Exploded view of the intercooler components

5. Remove the crankshaft pulley bolt.
6. Lock the crankshaft in place using tool ST499977100, or equivalent.
7. Remove the crankshaft pulley.
8. Remove the left side timing belt cover.
9. Remove the front timing belt cover.
10. If equipped with manual transmission, remove the timing belt guide.

➡If the belt is going to be reused and the alignment mark on the belt is not readable, put a new mark on the belt to indicate the direction of rotation. Using tool ST499987500, turn the crankshaft to align the mark of the sprocket "A" to the cylinder mark notch "B". Ensure that the right side cam sprocket mark "C", cam cap and cylinder head matching surface "D" or left side cam sprocket mark "E", timing belt cover notch "F" are properly aligned. Paint an alignment mark on the belt in relation to the crankshaft sprocket and camshaft sprockets. Z1 measurement is 46.8 teeth. Z2 measurement is 43.7 teeth.

11. Remove both the number two belt idlers. Remove the timing belt from the engine.
12. Remove the number one belt idler. Remove the automatic belt tension adjuster assembly.

To install:

13. Attach the automatic belt tension adjuster assembly to a vertical pressing tool.

➡Always use a vertical type pressing tool to move the adjuster rod downward. Do not use a lateral type vise. Push the adjuster rod vertically. Press in the push adjuster rod gradually, which should take three minutes or more. Do not allow pressure to exceed 2,205 lb. force.

14. Slowly move the adjuster rod down until the adjuster rod is aligned with the stopper pin hole in the cylinder.

➡Press the adjuster rod as far as the end surface of the cylinder. Do not press the adjuster rod into the cylinder. Doing so may damage the cylinder.

15. Using a 0.08 inch stopper pin, insert it into the stopper pin hole in the cylinder. Secure the adjuster rod.

➡Do not release the press pressure until the stopper pin is completely inserted in the hole.

16. Install the automatic belt tensioner assembly. Tighten the retaining bolt to 28.9 ft. lbs.
17. Install the belt idler number one. Tighten the retaining bolt to 28.9 ft. lbs.
18. Turn the number one and number two camshaft sprockets, using tool ST499207100 or tool ST18231AA010 and position the alignment marks "A" on each at the highest position.
19. While aligning the alignment mark "B" on the timing belt with mark "A" on the sprockets, position the timing belt properly.
20. Install both belt idler number two's. Tighten the retaining bolt to 28.9 ft. lbs.
21. After checking to be sure the marks on the timing belt and the camshaft sprockets are aligned remove the stopper pin from the belt tension adjuster.
22. Install the timing belt guide, if equipped with manual transmission. Temporarily tighten the bolts. Check and adjust the clearance between the belt and the guide. It should be 0.039 +/- 0.020 inch. Tighten the bolts to 7.2 ft. lbs.
23. Continue the installation in the reverse order of the removal procedure.

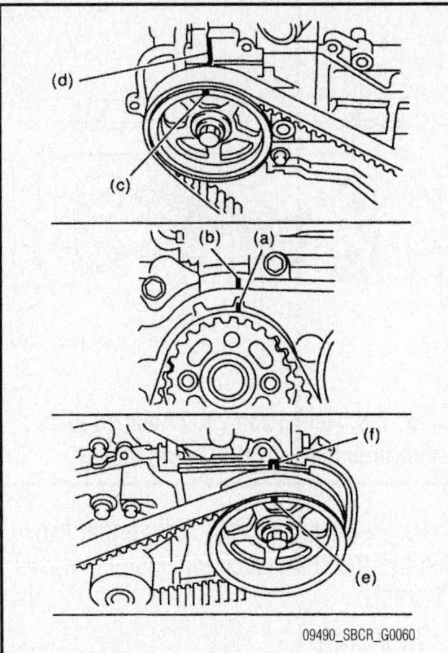

Fig. 132 Timing belt alignment—2006-08 SOHC engine

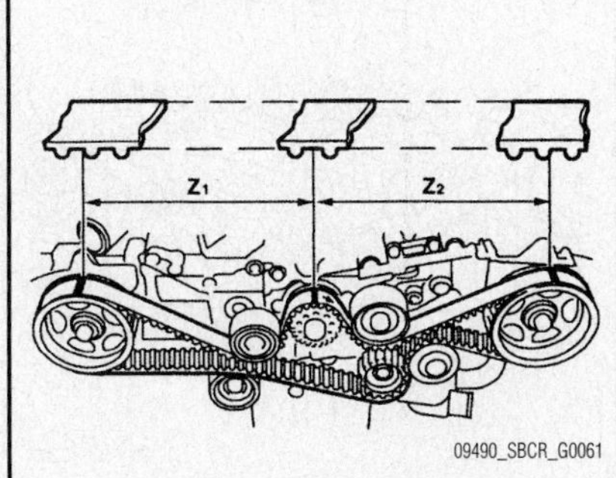

Fig. 133 Timing belt Z1 and Z2 teeth measurement—2006-08 SOHC engine

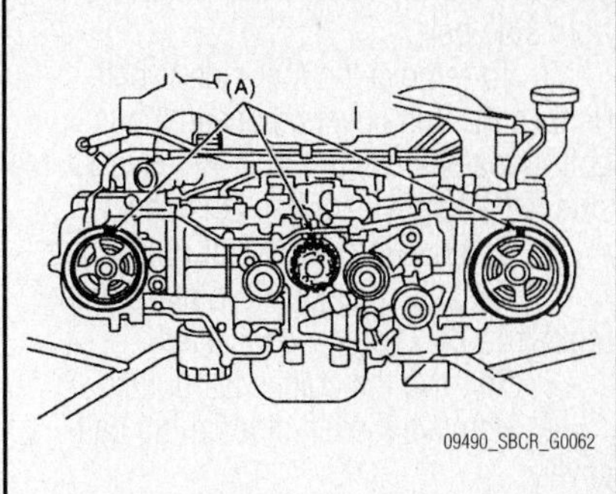

Fig. 134 Timing mark alignment position A—2006-08 SOHC engine

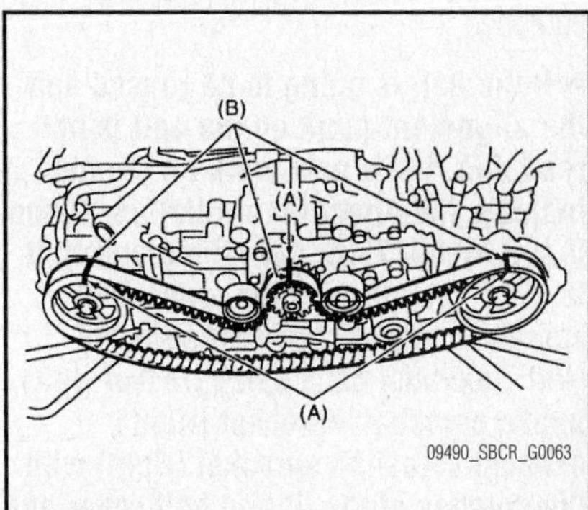

Fig. 135 Timing mark alignment position B—2006-08 SOHC engine

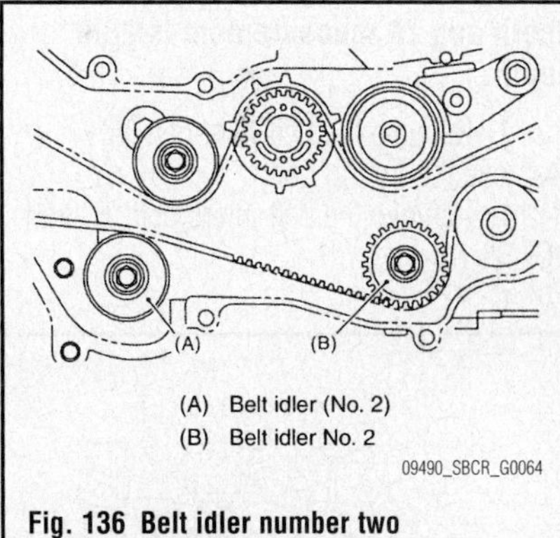

Fig. 136 Belt idler number two locations—2006-08 SOHC engine

2.5L DOHC Turbo Engine

See Figures 137 through 148.

1. Before servicing the vehicle, refer to the Precautions Section.
2. Disconnect the negative battery cable.
3. Remove the collector cover.
4. To remove the front side V-belt, remove the belt covers. Loosen the lock

bolt. Loosen the slider bolt. Remove the front side belt.

5. To remove the rear side V-belt, remove the belt covers. Loosen the lock bolt. Loosen the slider bolt. Remove the rear side belt. Remove the belt tensioner.

6. Remove the crankshaft pulley bolt.

7. Lock the crankshaft in place using tool ST499977100, or equivalent.

8. Remove the crankshaft pulley.

9. Remove the left side timing belt cover.

10. Remove the right side timing belt cover.

11. Remove the front timing belt cover.

12. Remove the timing belt guide, if equipped.

➡If the belt is going to be reused and the alignment mark on the belt is not readable, put a new mark on the belt to indicate the direction of rotation. Using tool ST499987500, turn the crankshaft to align the mark on the crankshaft sprocket, intake camshaft sprocket (left), exhaust camshaft sprocket (left), intake camshaft sprocket (right), exhaust camshaft sprocket (right) with the notches of the timing belt cover and cylinder block. Paint an alignment mark on the belts in relation to the camshaft sprockets. Z1 measurement is 54.4 teeth. Z2 measurement is 51.0 teeth and Z3 measurement is 28.0 teeth.

13. Remove the belt idler belt "A". Remove the timing belt.

14. Remove the belt idlers belt "B" and "C".

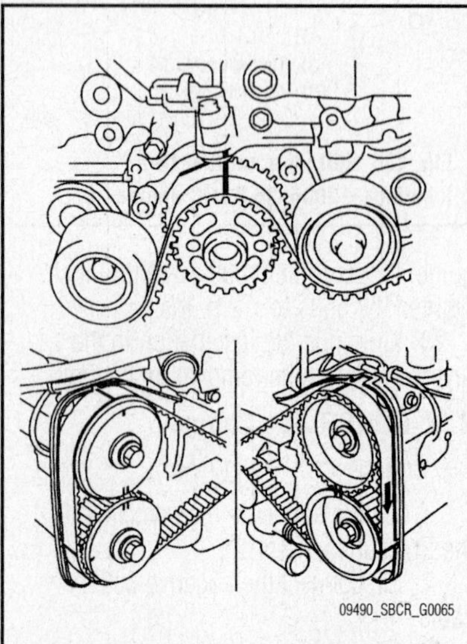

Fig. 137 Timing belt alignment—DOHC (Turbo) engine

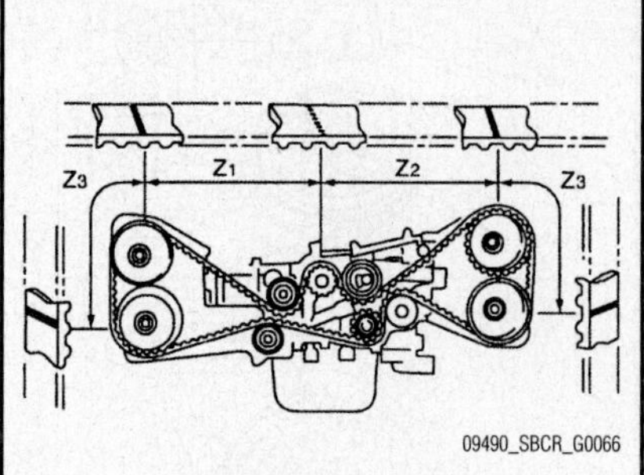

Fig. 138 Timing belt Z1, Z2 and Z3 teeth measurement—DOHC (Turbo) engine

15. Remove the belt idler number two. Remove the automatic belt tension adjuster assembly.

To install:

16. Attach the automatic belt tension adjuster assembly to a vertical pressing tool.

➡Always use a vertical type pressing tool to move the adjuster rod downward. Do not use a lateral type vise. Push the adjuster rod vertically. Press in the push adjuster rod gradually, which should take three minutes or more. Do not allow pressure to exceed 2,205 lb. force.

17. Slowly move the adjuster rod down until the adjuster rod is aligned with the stopper pin hole in the cylinder.

➡Press the adjuster rod as far as the end surface of the cylinder. Do not press the adjuster rod into the cylinder. Doing so may damage the cylinder.

18. Using a 0.08 inch stopper pin, insert it into the stopper pin hole in the cylinder. Secure the adjuster rod.

➡Do not release the press pressure until the stopper pin is completely inserted in the hole.

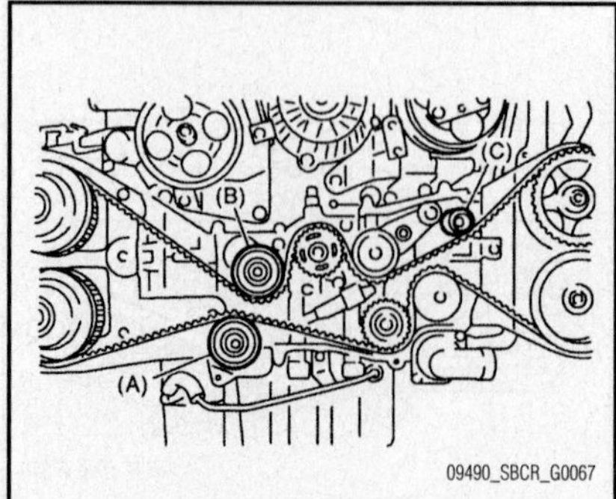

Fig. 139 Belt idler identification and location—DOHC (Turbo) engine

19. Install the automatic belt tensioner assembly. Tighten the retaining bolt to 28.9 ft. lbs.

20. Install the belt idler number two. Tighten the retaining bolt to 28.9 ft. lbs.

21. Install the belt idlers. Tighten the retaining bolts to 28.9 ft. lbs.

22. Align the mark "A" on the crankshaft sprocket with the mark on the oil pump at the cylinder block.

23. Align the single line mark "A" on the right exhaust camshaft sprocket with the notch "B" on the timing belt cover.

24. Align single line mark "A" on the right intake camshaft sprocket with the notch "B" on the timing cover. Ensure that the double lines "C" on the intake and exhaust camshaft sprockets are aligned.

25. Align single line mark "A" on the left exhaust camshaft sprocket with the notch "B" on the timing cover by turning the sprocket counterclockwise as viewed from the front of the engine.

26. Align single line mark "A" on the left intake camshaft sprocket with the notch "B" on the timing cover, by turning the sprocket clockwise as viewed from the front of the engine. Ensure that the double lines "C" on the intake and exhaust camshaft sprockets are aligned.

27. Make sure that the camshaft and crankshaft sprockets are positioned properly.

➡The intake and exhaust camshafts on this engine can be rotated independently with the timing belt removed. By looking at the illustration it will show you that if the intake and exhaust valve are lift together the heads will hit each other and bend.

➡When the timing belts are not installed, 4 camshafts are held at "zero lift" position, where all cams on the camshafts do not push the intake and exhaust valves down (under this condition all valves remain unlifted). When the camshafts are rotated to install the timing belts, No. 2 intake and No. 4 exhaust cam of the left hand camshafts are held to push their corresponding valves down. Under this condition these valves are held lifted. The right side camshafts are held in so that their cams do not push the valves down. The left hand camshafts must be rotated from the "zero lift" position to the position where the timing belt is to be installed at as small an angle as possible, in order to prevent mutual

Fig. 140 Crankshaft sprocket mark A to oil pump cover alignment—DOHC (Turbo) engine

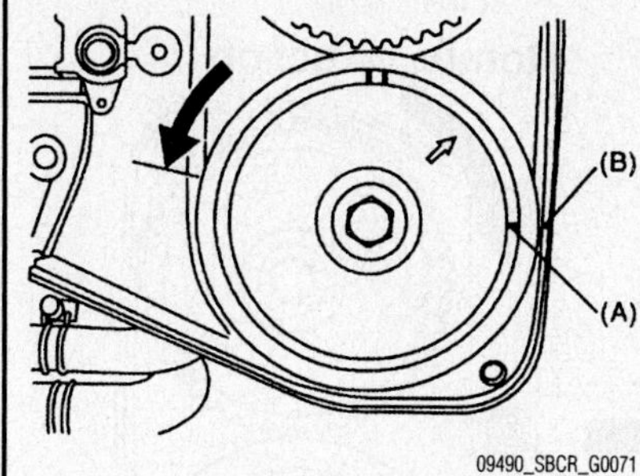

Fig. 143 Left exhaust camshaft sprocket alignment mark A with timing belt cover B alignment mark—DOHC (Turbo) engine

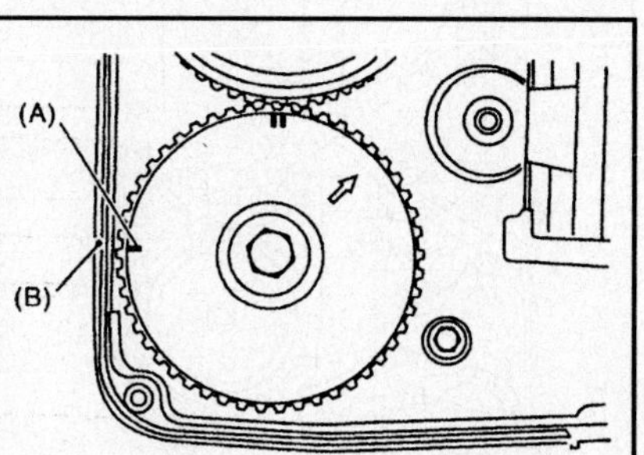

Fig. 141 Right exhaust camshaft sprocket alignment mark A with timing belt cover B alignment mark—DOHC (Turbo) engine

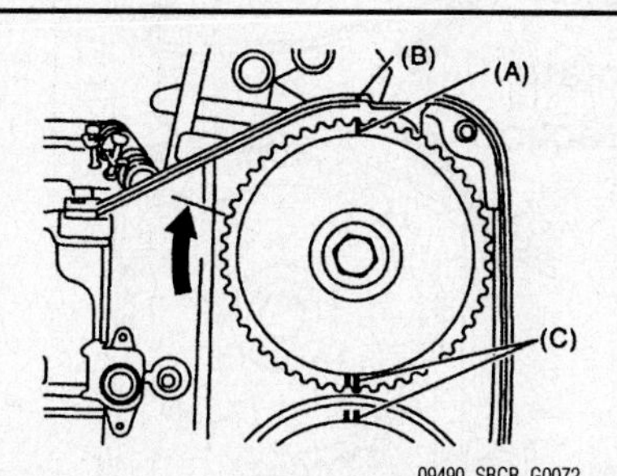

Fig. 144 Left intake camshaft sprocket alignment mark A with timing belt cover B alignment mark an double line C alignment mark—DOHC (Turbo) engine

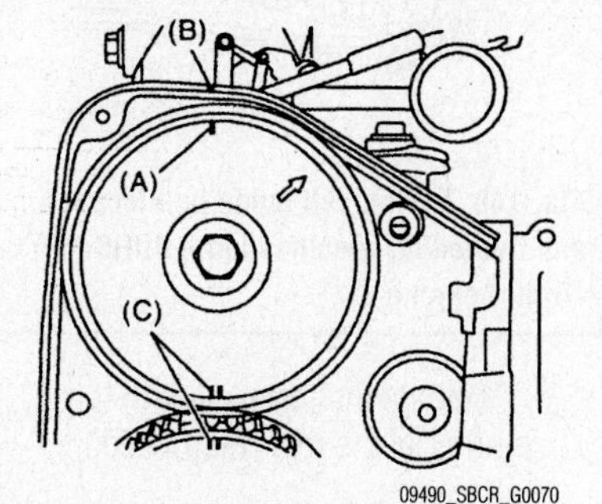

Fig. 142 Right intake camshaft sprocket alignment mark A with timing belt cover B alignment mark an double line C alignment mark—DOHC (Turbo) engine

interference of intake and exhaust valve heads. Do not allow the camshafts to rotate in the direction illustrated as this causes both the intake and exhaust valves to lift off at the same time with will cause valve damage.

28. When installing the belt, make sure to align the marks made during removal or if using a new belt, align the in alphabetical order as shown in the illustration.

✻✻ WARNING

Disengagement of more than 3 timing belt teeth may result in contact between the valve and piston. Always make sure the belts rotation is correct.

29. Install the timing belt.

➡Align the alignment mark on the timing belt with marks on the sprocket in the order shown in the illustration. While aligning the timing marks, position the timing belt properly.

30. Install the belt idlers. Tighten the retaining bolts to 28.9 ft. lbs.

➡Make sure that the marks on the timing belt and sprockets are aligned.

31. After checking to be sure the marks on the timing belt and the camshaft sprockets are aligned remove the stopper pin from the belt tension adjuster.

32. Install the timing belt guide, if equipped with manual transmission. Temporarily tighten the bolts. Check and adjust the clearance between the belt and the guide. It should be 0.039 +/- 0.020 inch. Tighten the guide bolts to specification, see illustration.

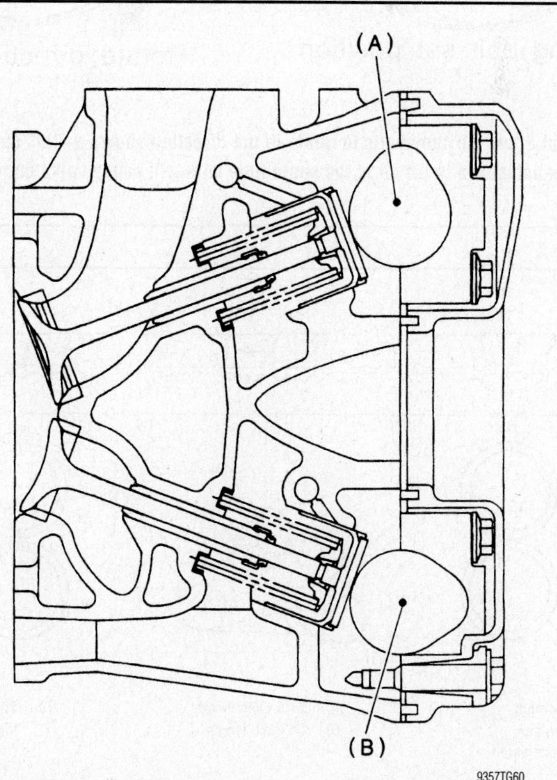

Fig. 145 If the intake and exhaust valve are lifted together the heads will hit each other and bend—DOHC (Turbo) engine

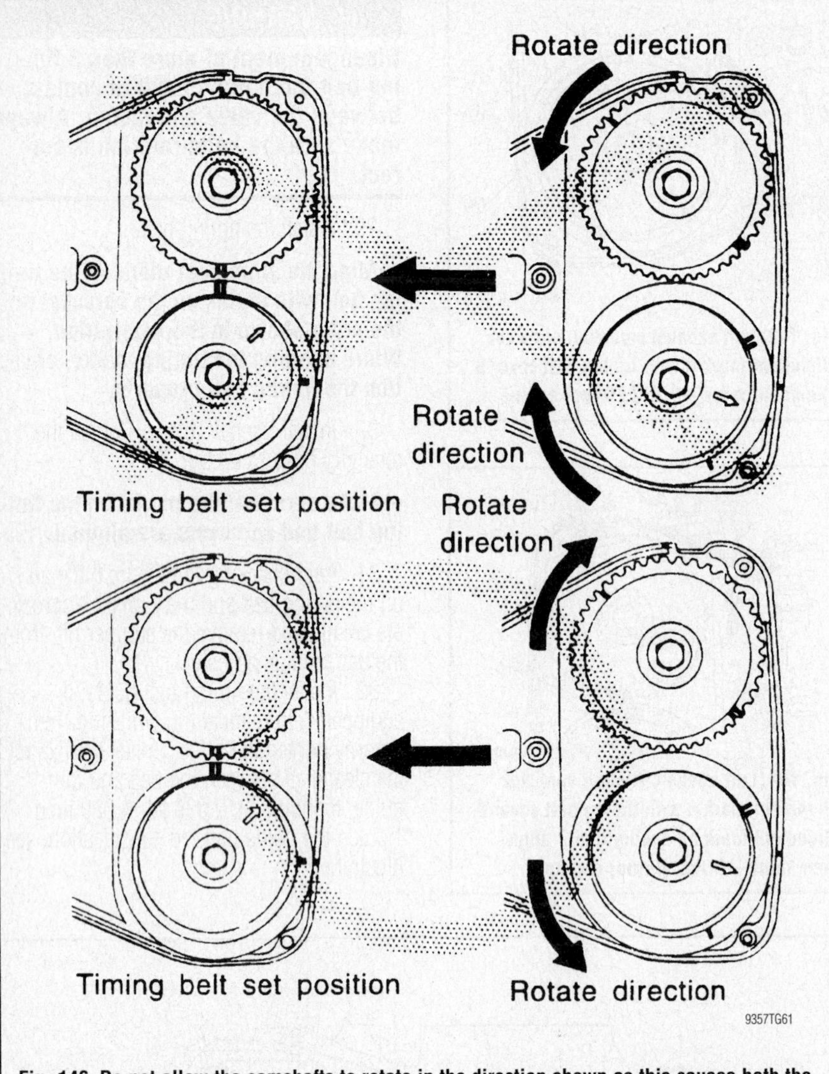

Fig. 146 Do not allow the camshafts to rotate in the direction shown as this causes both the intake and exhaust valves to lift off at the same time with will cause valve damage—DOHC (Turbo) engine

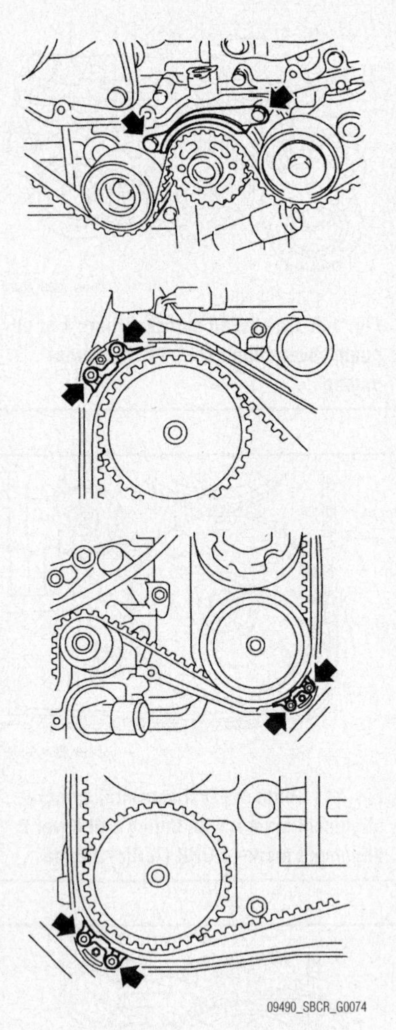

Fig. 148 Timing belt guide bolt location and tightening specification—DOHC (Turbo) engine

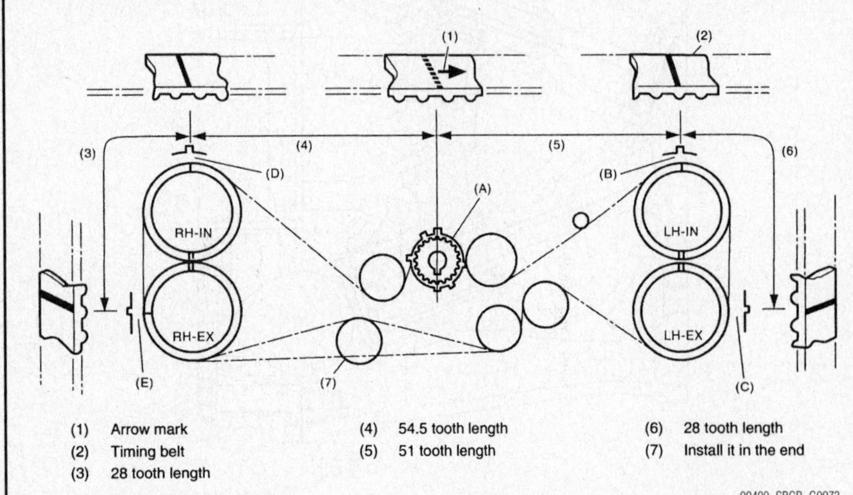

Fig. 147 Timing belt alignment and installation sequence— DOHC (Turbo) engine

33. Continue the installation in the reverse order of the removal procedure.

VALVE COVERS

REMOVAL & INSTALLATION

2.5L SOHC Non-Turbo Engine

Right Side

See Figure 149.

1. Before servicing the vehicle, refer to the Precautions Section.
2. Disconnect the negative battery cable.
3. Remove the bolts which secure the upper and center side of the timing belt cover.
4. Raise and safely support the vehicle.
5. Remove the engine undercover.
6. Remove the bolt which secures the

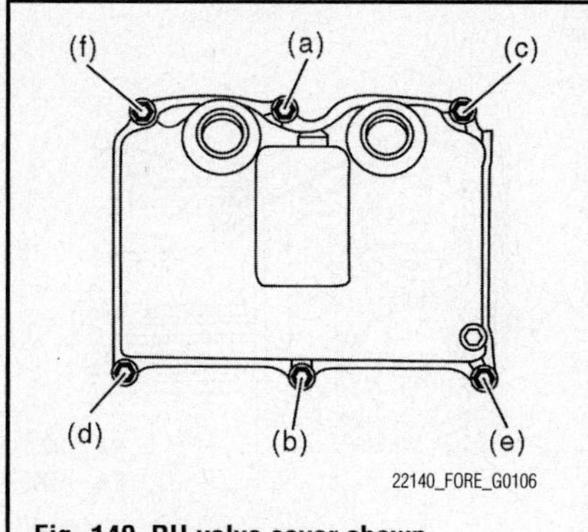

Fig. 149 RH valve cover shown, LH similar

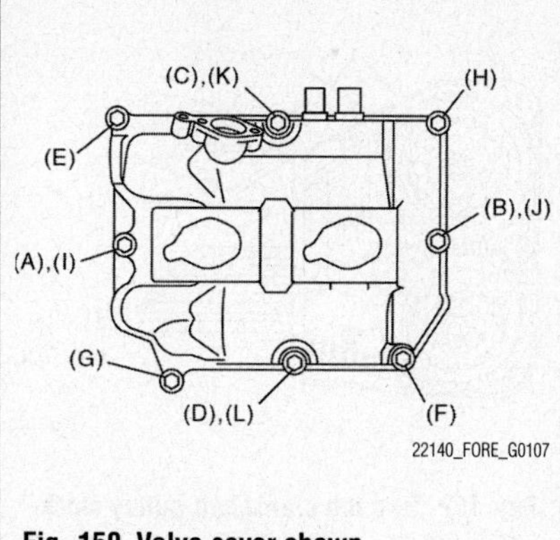

Fig. 150 Valve cover shown

underside of the timing belt cover, then remove the timing belt cover.

7. Disconnect the spark plug wires.
8. Disconnect the PCV hose from the rocker cover.
9. Remove the rocker cover.

To install:

10. Installation is the reverse order of removal.
11. Tighten the bolts to the valve covers in two stages in alphabetical sequence as shown in figure:
    - First stage: 4.7 ft. lbs. (6.4 Nm).
    - Second stage: 4.7 ft. lbs. (6.4 Nm).

Left Side

1. Before servicing the vehicle, refer to the Precautions Section.
2. Disconnect the negative battery cable.
3. Remove the bolts which secure the upper and center side of the timing belt cover.
4. Raise and safely support the vehicle.
5. Remove the engine undercover.
6. Remove the bolt which secures the underside of the timing belt cover, then remove the timing belt cover.
7. Lower the vehicle.
8. Disconnect the spark plug wires.
9. Disconnect the PCV hose from the rocker cover.
10. Remove the rocker cover.
11. Installation is the reverse order of removal.

2.5L DOHC Turbo Engine

Right Side

See Figure 150.

1. Before servicing the vehicle, refer to the Precautions Section.
2. Raise and safely support the vehicle.
3. Disconnect the negative battery cable.
4. Remove the air intake assembly.
5. Remove the bolts which secure the upper and center side of the timing belt cover.
6. Remove the engine undercover.
7. Remove the bolt which secures the underside of the timing belt cover, then remove the timing belt cover.
8. Lower the vehicle.
9. Pull out the engine wiring harness connector with the bracket from the upper air cleaner cover.
10. Remove the air intake assembly.
11. Disconnect the ignition coil connector and remove the ignition coils.
12. Disconnect the PCV hose from the rocker cover.
13. Remove the rocker cover.

To install:

14. Installation is the reverse order of removal.
15. Temporarily tighten the valve cover tightening bolts in alphabetical sequence shown in the figure, and then tighten to 4.7 ft. lbs. (6.4 Nm) in alphabetical sequence.

Left Side

1. Before servicing the vehicle, refer to the Precautions Section.
2. Raise and safely support the vehicle.
3. Disconnect the negative battery cable.
4. Remove the air intake assembly.
5. Remove the bolts which secure the upper and center side of the timing belt cover.
6. Remove the engine undercover.
7. Remove the bolt which secures the underside of the timing belt cover, then remove the timing belt cover.
8. Lower the vehicle.
9. Remove the battery and battery tray.
10. Remove the bolt which secures the engine wiring harness onto the body.
11. Disconnect the washer motor electrical connectors. Remove the washer tank mounting bolts, and move the tank upwards and secure out of the way.
12. Disconnect the ignition coil connector and remove the ignition coils.
13. Disconnect the PCV hose from the rocker cover.
14. Remove the rocker cover.
15. Installation is the reverse order of removal. Tighten the rocker cover bolts to 4.7 ft. lbs. (6.4 Nm).

VALVE LASH

ADJUSTMENT

2.5L SOHC Non-Turbo Engine

See Figure 151.

➡The valve adjustment should be performed while the engine is cold.

1. Before servicing the vehicle, refer to the Precautions Section.
2. Before servicing the vehicle, refer to the precautions in the beginning of this section.
3. Raise and support the vehicle safely.
4. Remove the undercover.
5. Lower the vehicle.
6. Disconnect the negative battery cable.
7. To remove the front side V-belt, remove the belt covers. Loosen the lock bolt. Loosen the slider bolt. Remove the front side belt.
8. To remove the rear side V-belt, remove the belt covers. Loosen the lock bolt. Loosen the slider bolt. Remove the rear side belt. Remove the belt tensioner.
9. Remove the crankshaft pulley bolt.
10. Lock the crankshaft in place using tool ST499977100, or equivalent.
11. Remove the crankshaft pulley.
12. Remove the left side timing belt cover.
13. Remove the fuel injector.

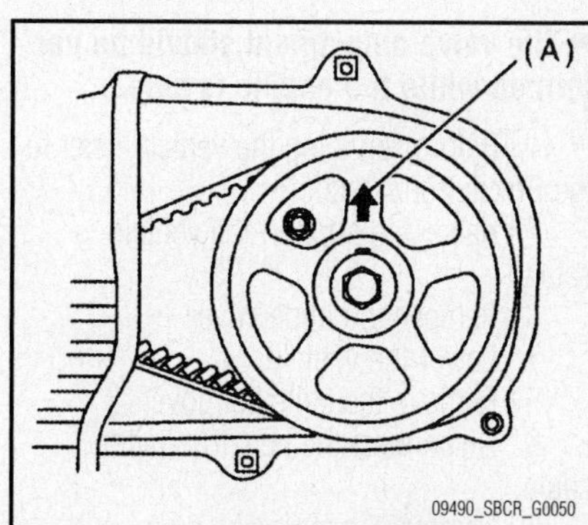

Fig. 151 TDC alignment—2006-08 SOHC engine

14. Remove the rocker cover.
15. Position the number one piston at TDC of the compression stroke.

➡When the arrow (see illustration) on the camshaft sprocket (left side) comes exactly to the top, number one cylinder piston is at TDC of the compression stroke.

16. Measure the valve clearance, using a feeler gauge.
17. If adjustment is needed, loosen the valve rocker nut and screw. Position the feeler gauge.

➡Insert the feeler gauge in a horizontally as possible with respect to the valve stem end face. Adjust the exhaust valve clearance while lifting up the vehicle.

18. While noting the valve clearance, tighten the rocker adjusting screw.
19. When the proper valve clearance is obtained, tighten the valve rocker nut to 7.2 ft. lbs.
20. Adjust the valve clearance on the remaining cylinders, following the above procedure.

➡Be sure to position the pistons to their respective TDC positions on the compression stroke, before checking and adjusting the valves. By rotating the crankshaft pulley clockwise every 180 degrees from the state that number one piston is on TDC of the compression stroke, the remaining pistons come to TDC of the compression stroke in the following order, No. 3, No. 2, and No. 4.

21. After adjustment, replace any removed components.
22. Be sure to use new gaskets and seals, as required.

2.5L DOHC Turbo Engine

See Figures 152 through 158.

➡The valve adjustment should be performed while the engine is cold.

1. Before servicing the vehicle, refer to the Precautions Section.
2. Raise and support the vehicle safely.
3. Remove the undercover.
4. Lower the vehicle.
5. Remove the collector cover.
6. Disconnect the negative battery cable.
7. Remove the air intake duct.
8. Remove the bolt that retains the right side timing belt cover. Remove the remaining bolts and remove the right side timing belt cover.
9. Disconnect the ignition coil electrical connector. Remove the ignition coil.
10. Position a suitable container under the vehicle.
11. Disconnect the PCV hose from the rocker cover. Remove the rocker cover retaining bolts. Remove the rocker cover from the vehicle.

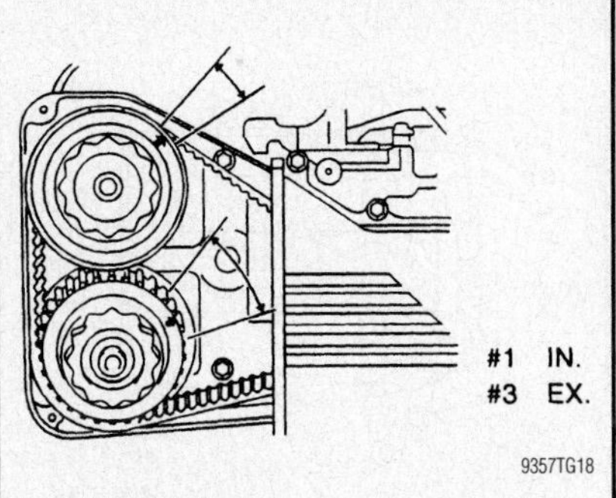

Fig. 152 Turn the crankshaft pulley clockwise until the arrow mark on the camshaft is positioned as shown to measure the No. 1 intake and No. 3 exhaust valves—DOHC engine

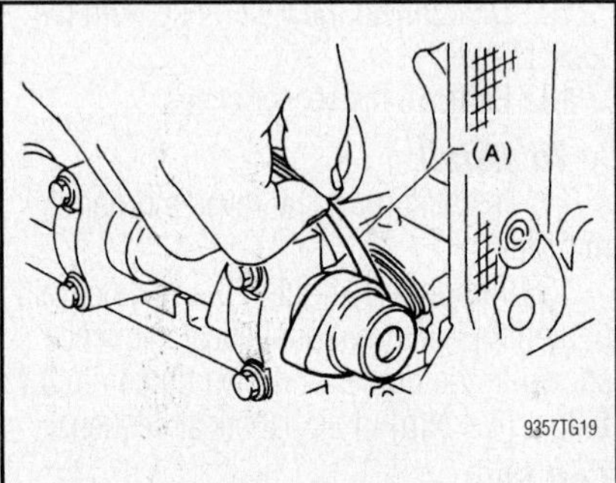

Fig. 153 Use a feeler gauge to inspect the valve clearance—DOHC engine

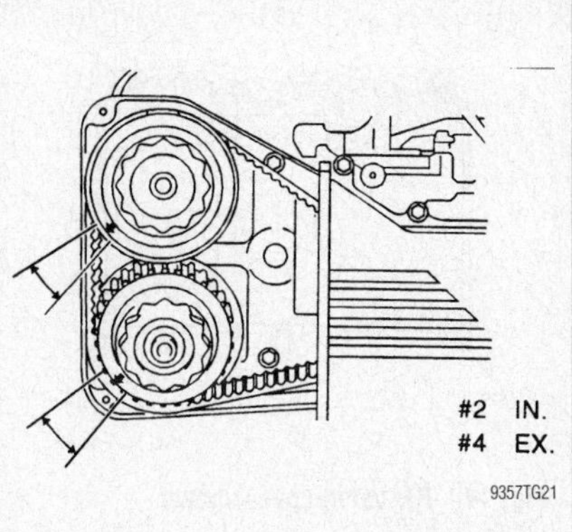

Fig. 155 Turn the crankshaft pulley clockwise until the arrow mark on the camshaft is positioned as shown to measure the No. 2 intake and No. 4 exhaust valves—DOHC engine

12. Position the number one piston at TDC of the compression stroke.
13. Using a feeler gauge, measure and record the clearance of the number one cylinder intake and the number three cylinder exhaust valves.

➡Insert the feeler gauge in a horizontally as possible with respect to the valve lifter. Measure and record the exhaust valve clearance while lifting up the vehicle.

14. Rotate the crankshaft pulley clockwise until the arrow mark on the camshaft is positioned as shown to measure and record the clearance on the number two exhaust and number three intake valves.
15. Rotate the crankshaft pulley clockwise until the arrow mark on the camshaft is positioned as shown to measure and record the number two intake and number four exhaust valves.

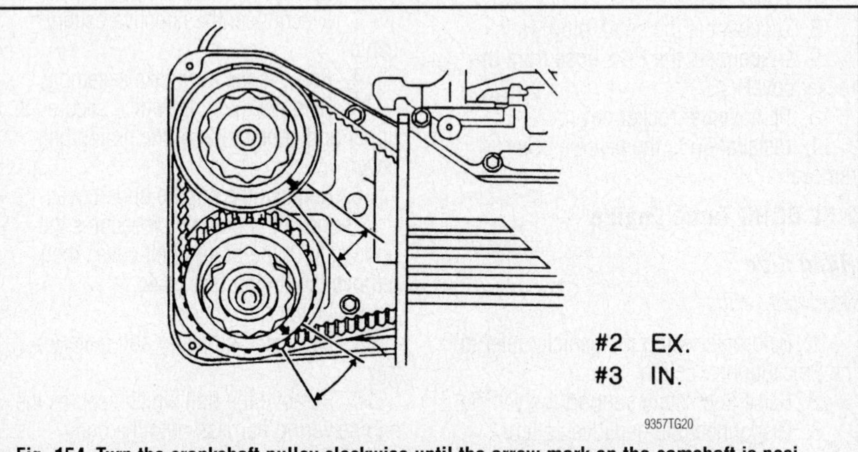

Fig. 154 Turn the crankshaft pulley clockwise until the arrow mark on the camshaft is positioned as shown to measure the No. 2 exhaust and No. 3 intake valves—DOHC engine

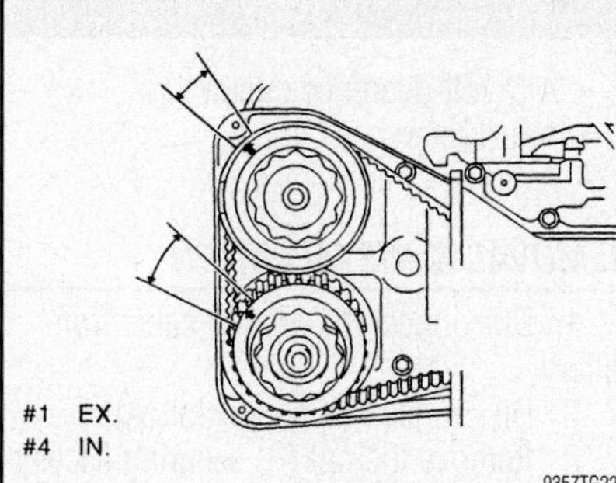

Fig. 156 Turn the crankshaft pulley clockwise until the arrow mark on the camshaft is positioned as shown to measure the No. 1 exhaust and No. 4 intake valves—DOHC engine

Unit: (mm)
Intake valve: S = (V + T) − 0.20 Exhaust valve: S = (V + T) − 0.35
S: Valve lifter thickness required V: Measured valve clearance T: Valve lifter thickness to be used

09490_SBCR_G0051

Fig. 157 Use this table to help you select a suitable shim—DOHC engine

16. Rotate the crankshaft pulley clockwise until the arrow mark on the camshaft is positioned as shown to measure and record the number one exhaust and number four intake valves.

17. If adjustment is required, remove the camshafts.

18. Remove and measure the thickness of the valve lifter. Select a suitable shim, using the shim selection chart.

19. Install the replacement shim to the lifter.

20. After all shims have been adjusted, inspect the valve clearances again.

21. After completion, install all removed components.

Part No.	Thickness mm (in)
13228 AB102	4.68 (0.1843)
13228 AB112	4.69 (0.1846)
13228 AB122	4.70 (0.1850)
13228 AB132	4.71 (0.1854)
13228 AB142	4.72 (0.1858)
13228 AB152	4.73 (0.1862)
13228 AB162	4.74 (0.1866)
13228 AB172	4.75 (0.1870)
13228 AB182	4.76 (0.1874)
13228 AB192	4.77 (0.1878)
13228 AB202	4.78 (0.1882)
13228 AB212	4.79 (0.1886)
13228 AB222	4.80 (0.1890)
13228 AB232	4.81 (0.1894)
13228 AB242	4.82 (0.1898)
13228 AB252	4.83 (0.1902)
13228 AB262	4.84 (0.1906)
13228 AB272	4.85 (0.1909)
13228 AB282	4.86 (0.1913)
13228 AB292	4.87 (0.1917)
13228 AB302	4.88 (0.1921)
13228 AB312	4.89 (0.1925)
13228 AB322	4.90 (0.1929)
13228 AB332	4.91 (0.1933)
13228 AB342	4.92 (0.1937)
13228 AB352	4.93 (0.1941)
13228 AB362	4.94 (0.1945)
13228 AB372	4.95 (0.1949)
13228 AB382	4.96 (0.1953)
13228 AB392	4.97 (0.1957)
13228 AB402	4.98 (0.1961)
13228 AB412	4.99 (0.1965)
13228 AB422	5.00 (0.1969)
13228 AB432	5.01 (0.1972)
13228 AB442	5.02 (0.1976)
13228 AB452	5.03 (0.1980)
13228 AB462	5.04 (0.1984)
13228 AB472	5.05 (0.1988)
13228 AB482	5.06 (0.1992)
13228 AB492	5.07 (0.1996)
13228 AB502	5.08 (0.2000)
13228 AB512	5.09 (0.2004)
13228 AB522	5.10 (0.2008)
13228 AB532	5.11 (0.2012)
13228 AB542	5.12 (0.2016)
13228 AB552	5.13 (0.2020)
13228 AB562	5.14 (0.2024)
13228 AB572	5.15 (0.2028)
13228 AB582	5.16 (0.2031)
13228 AB592	5.17 (0.2035)
13228 AB602	5.18 (0.2039)
13228 AB612	5.19 (0.2043)

Part No.	Thickness mm (in)
13228 AB622	5.20 (0.2047)
13228 AB632	5.21 (0.2051)
13228 AB642	5.22 (0.2055)
13228 AB652	5.23 (0.2059)
13228 AB662	5.24 (0.2063)
13228 AB672	5.25 (0.2067)
13228 AB682	5.26 (0.2071)
13228 AB692	5.27 (0.2075)
13228 AB702	4.38 (0.1724)
13228 AB712	4.40 (0.1732)
13228 AB722	4.42 (0.1740)
13228 AB732	4.44 (0.1748)
13228 AB742	4.46 (0.1756)
13228 AB752	4.48 (0.1764)
13228 AB762	4.50 (0.1771)
13228 AB772	4.52 (0.1780)
13228 AB782	4.54 (0.1787)
13228 AB792	4.56 (0.1795)
13228 AB802	4.58 (0.1803)
13228 AB812	4.60 (0.1811)
13228 AB822	4.62 (0.1819)
13228 AB832	4.64 (0.1827)
13228 AB842	4.66 (0.1835)
13228 AB852	5.29 (0.2083)
13228 AB862	5.31 (0.2091)
13228 AB872	5.33 (0.2098)
13228 AB882	5.35 (0.2106)
13228 AB892	5.37 (0.2114)
13228 AB902	5.39 (0.2122)
13228 AB912	5.41 (0.2123)
13228 AB922	5.43 (0.2138)
13228 AB932	5.45 (0.2146)
13228 AB942	5.47 (0.2154)
13228 AB952	5.49 (0.2161)
13228 AB962	5.51 (0.2169)
13228 AB972	5.53 (0.2177)
13228 AB982	5.55 (0.2185)
13228 AB992	5.57 (0.2193)
13228 AC002	5.59 (0.2201)
13228 AC012	5.61 (0.2209)
13228 AC022	5.63 (0.2217)
13228 AC032	5.65 (0.2224)

09490_SBCR_G0052

Fig. 158 Valve adjusting shim chart—DOHC engine

ENGINE PERFORMANCE & EMISSION CONTROLS

COMPONENT LOCATIONS

See Figures 159 and 160.

ACCELERATOR PEDAL POSITION (APP) SENSOR

LOCATION

See Figure 161.

The Accelerator Pedal Position (APP) Sensor is located inside the vehicle at the accelerator pedal.

OPERATION

The accelerator pedal contains 2 individual Accelerator Pedal Position (APP) sensors within the assembly. The APP sensors 1 and 2 are potentiometer type sensors each with 3 circuits:

- A 5-volt reference circuit
- A low reference circuit
- A signal circuit

REMOVAL & INSTALLATION

1. Disconnect the ground cable from battery.
2. Disconnect the connector (A).
3. Remove the nut (B) securing accelerator pedal assembly.

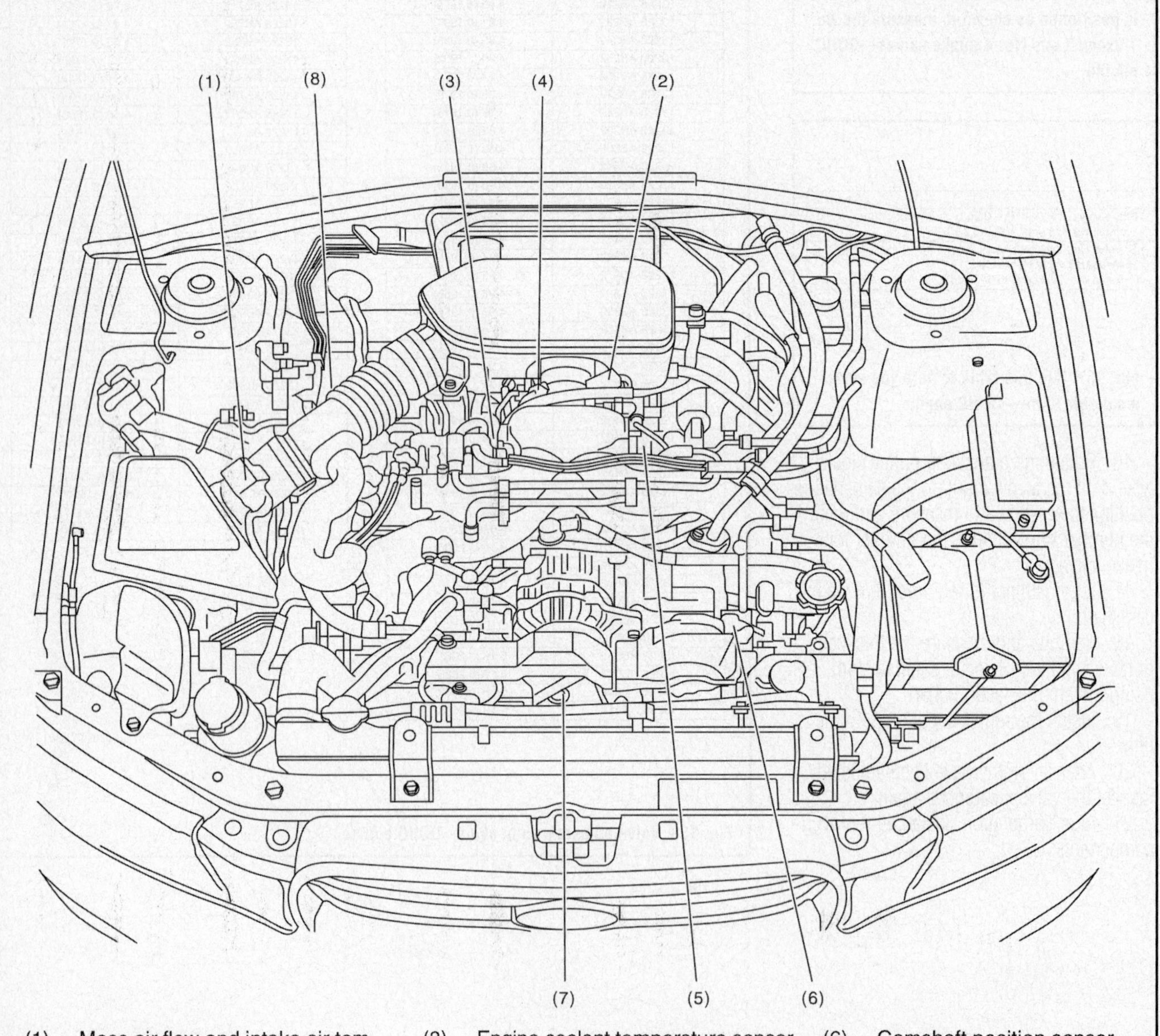

Fig. 159 Underhood sensor locations—2006–08 2.5L SOHC engine

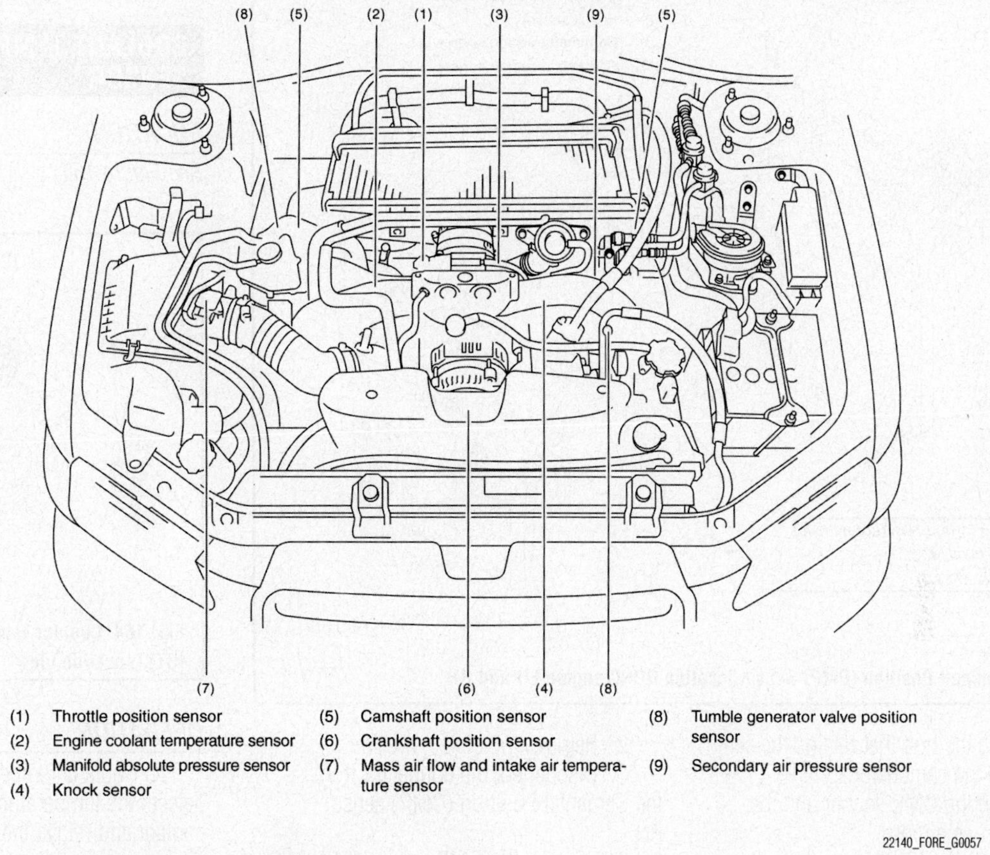

(8) (5) (2) (1) (3) (9) (5)
(7) (6) (4) (8)
(1) Throttle position sensor
(2) Engine coolant temperature sensor
(3) Manifold absolute pressure sensor
(4) Knock sensor
(5) Camshaft position sensor
(6) Crankshaft position sensor
(7) Mass air flow and intake air temperature sensor
(8) Tumble generator valve position sensor
(9) Secondary air pressure sensor
22140_FORE_G0057

Fig. 160 Underhood sensor locations—2006–08 2.5L DOHC engine

(1)
(1) Accelerator pedal position sensor
22140_FORE_G0116

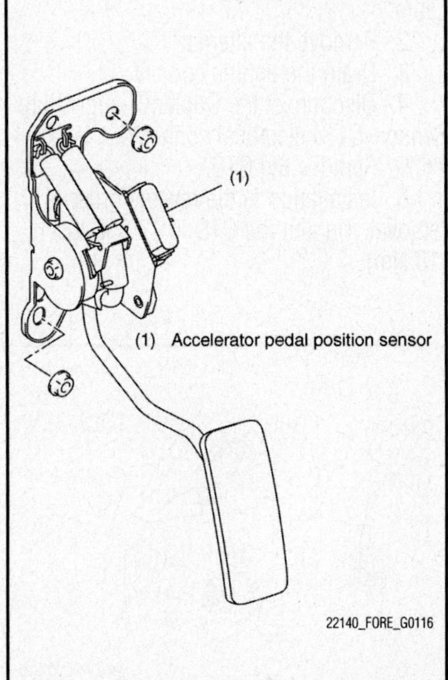

Fig. 161 Accelerator Pedal Position (APP) Sensor view

To install:

4. Install in the reverse order of removal.

5. Tighten the securing nut to 13 ft. lbs. (18 Nm).

TESTING

1. Measure the voltage between accelerator pedal position sensor connector and chassis ground as follows:
   - (B315) No. 6 (+)—Chassis ground (–)
   - (B315) No. 3 (+)—Chassis ground (–)

2. Is the difference in measured values for the main accelerator pedal position sensor signal and the sub accelerator pedal position sensor signal 0 volts. If so replace the APP sensor.

3. If not check and repair the following items:
   - Open circuit of harness between the ECM and accelerator pedal position sensor connector.
   - Ground short circuit of harness between the ECM and accelerator pedal position sensor connectors.

CAMSHAFT POSITION (CMP) SENSOR

LOCATION

See Figures 162 and 163.

Refer to the accompanying figures for Camshaft Position (CMP) sensor location.

OPERATION

The Camshaft Position (CMP) sensor relays the relative camshaft position to the ECM for determining the position and stroke of the No. 1 cylinder. This information is required for proper fuel injection functioning.

REMOVAL & INSTALLATION

2.5L DOHC Non-Turbo Engine

1. Disconnect the negative battery cable.

2. Disconnect the connector from the sensor.

3. Remove the bolt that retains the Camshaft Position (CMP) sensor to the sensor support.

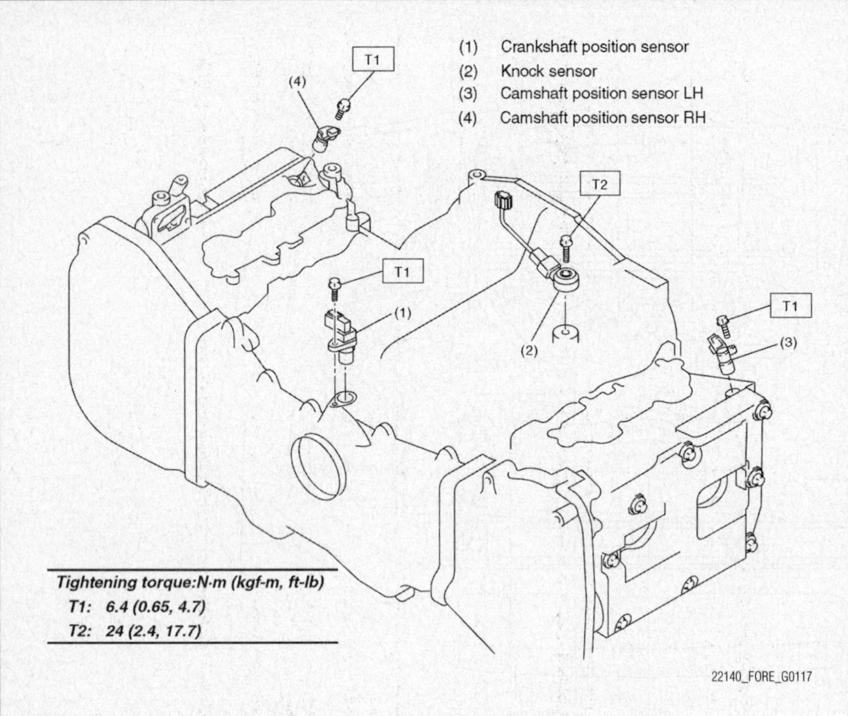

Fig. 162 Camshaft Position (CMP) Sensor location DOHC engine LH and RH

4. Remove the bolt that retains the sensor support to the camshaft cap.
5. Remove the CMP sensor and the sensor support as a unit.
6. Separate the CMP sensor from the support.

To install:

7. Installation is the reverse of the removal procedure.
8. Tighten the CMP sensor support to 4.7 ft. lbs. (6.4 Nm).
9. Tighten the CMP sensor to 4.7 ft. lbs. (6.4 Nm).

2.5L DOHC Turbo Engine

1. Disconnect the negative battery cable.

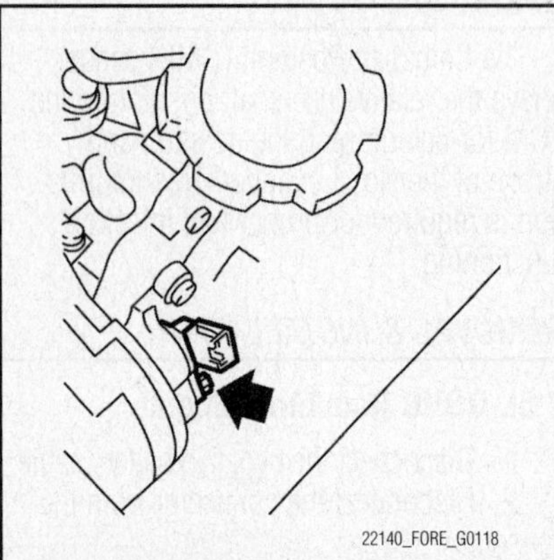

Fig. 163 Camshaft Position (CMP) Sensor location SOHC engine

2. Remove the collector cover.
3. Disconnect the connector from the Camshaft Position (CMP) sensor RH.
4. Remove the CMP sensor RH from the rear side of the cylinder head.
5. Remove the CMP sensor LH in the same way as RH.

To install:

6. Installation is the reverse of the removal procedure.
7. Tighten the sensor to 4.7 ft. lbs. (6.4 Nm).

TESTING

2.5L SOHC Non-Turbo Engine

1. Remove the sensor from the vehicle.
2. Measure the resistance between the two connector terminals of the sensor, (terminals 1 and 2).
3. If measured value is not within 1–4 k ohms, replace the sensor.

2.5L DOHC Turbo Engine

1. Using the Subaru scan tool, check the sensor output waveform.
2. With the engine OFF and the ignition switch ON the signal (V) should be 0–0.9.
3. On 2006 vehicles measurement is taken using connector B134 terminal 21 for the left side and connector B134 terminal 11 for the right side.
4. If abnormality is found, replace the sensor.

COOLANT TEMPERATURE SENSOR

LOCATION

See Figure 164.

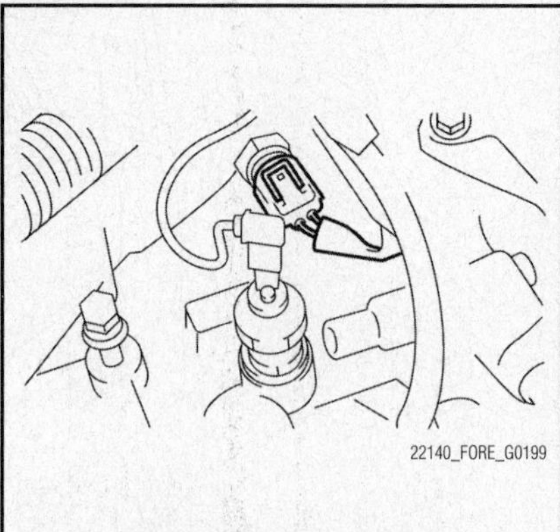

Fig. 164 Coolant Temperature Sensor (CTS) location view

OPERATION

The Coolant Temperature Sensor (CTS) detects the temperature of the engine coolant and relays the information to the electronic control assembly.

REMOVAL & INSTALLATION

See Figure 165.

1. Disconnect the negative battery cable.
2. Remove the alternator.
3. Drain the engine coolant.
4. Disconnect the Coolant Temperature Sensor (CTS) electrical connector.
5. Remove the CTS.
6. Installation is the reverse order of removal. Tighten the CTS to 13.3 ft. lbs. (18 Nm).

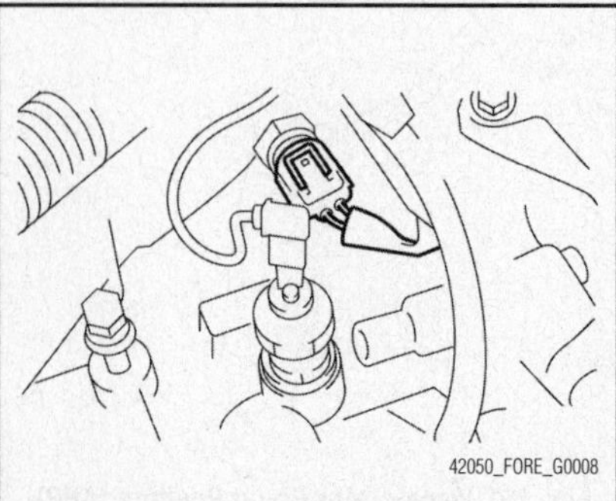

Fig. 165 Coolant temperature sensor location—2.5L Turbo Engine shown

TESTING

Subaru's hand-held tester or an equivalent OBD-II scan tool must be used in order to check the operation of the Coolant Temperature Sensor (CTS). For the detailed operation procedures, refer to the OBD-II General Scan Tool instruction manual.

Start the car and allow it to run to normal operating temperature. If the scan tool data shows the temperature less than 140°F (60°C), the sensor should be replaced.

CRANKSHAFT POSITION (CKP) SENSOR

LOCATION

See Figures 166 and 167.

The Crankshaft Position (CKP) sensor is located in the front of the engine near the crankshaft pulley.

OPERATION

The Crankshaft Position (CKP) sensor is a variable reluctance sensor which is used to inform the ECM when the No.1 piston is at top dead center. This information is used for controlling and adjusting ignition and fuel injector timing.

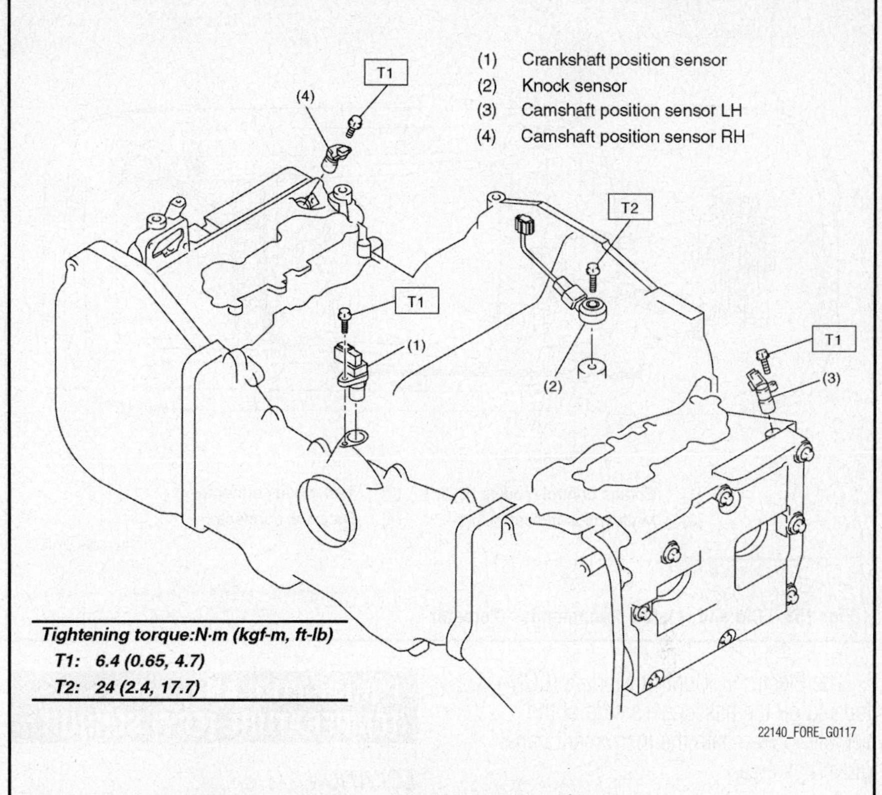

Fig. 167 Crankshaft Position (CKP) Sensor location DOHC engine

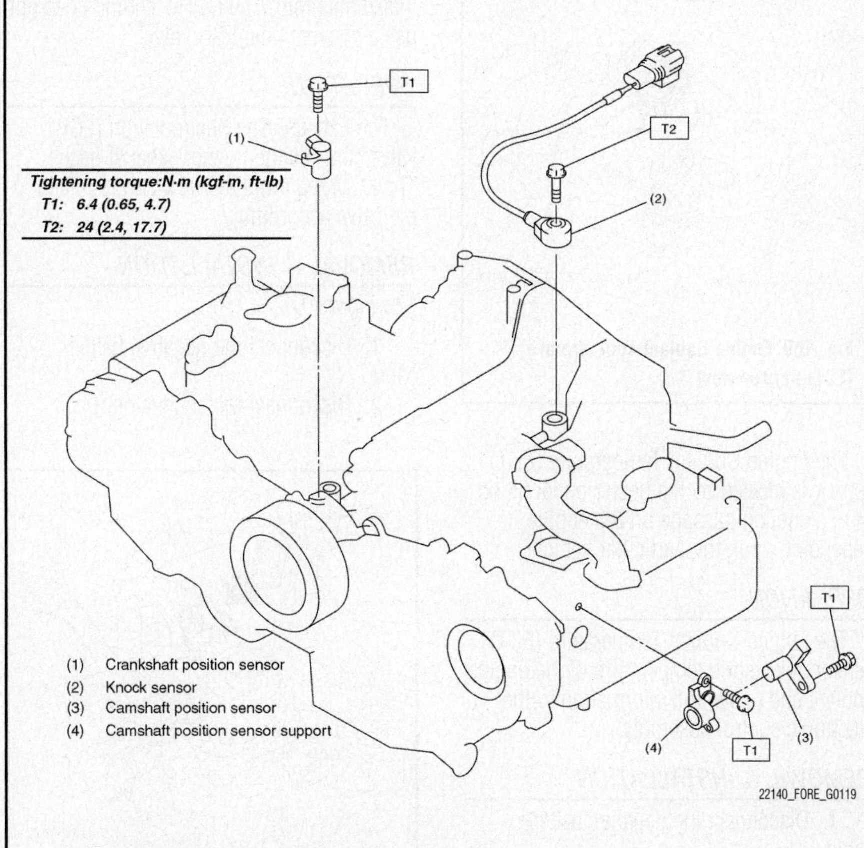

Fig. 166 Crankshaft Position (CKP) Sensor SOHC engine

REMOVAL & INSTALLATION

1. Disconnect the negative battery cable.
2. Remove the collector cover, as required.
3. Remove the bolt that retains the Crankshaft Position (CKP) sensor in place at the cylinder block.
4. Remove the sensor from its mounting.
5. Disconnect the connector from the sensor.

To install:

6. Installation is the reverse of the removal procedure.
7. Tighten the sensor to 4.7 ft. lbs. (6.4 Nm).

TESTING

1. Remove the Crankshaft Position (CKP) sensor from the vehicle.
2. Measure the resistance between the two connector terminals of the sensor, (terminals 1 and 2).
3. If measured value is not within 1–4 k ohms, replace the sensor.

ELECTRONIC CONTROL MODULE (ECM)

LOCATION

See Figure 168.

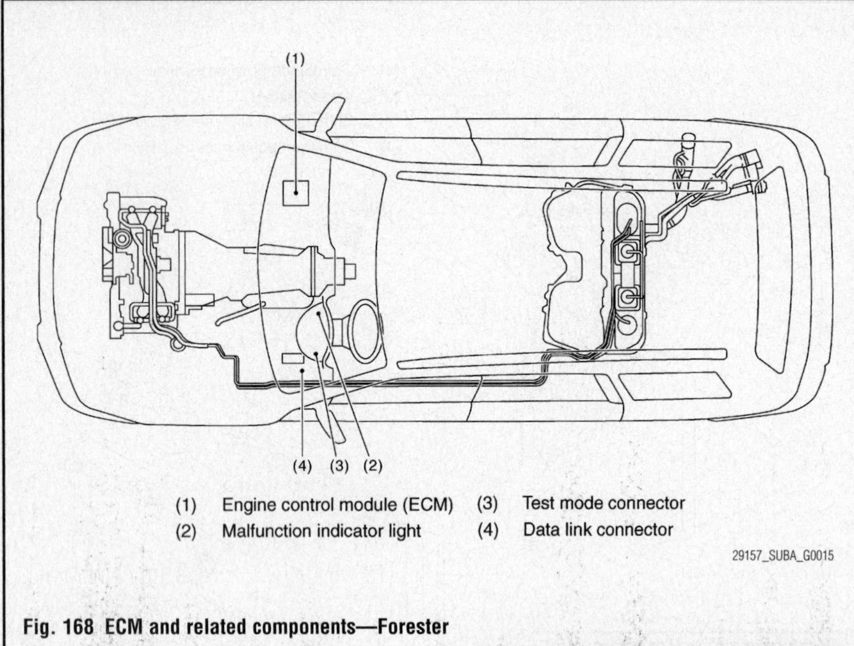

Fig. 168 ECM and related components—Forester

The Electronic Control Module (ECM) is located on the passenger's side of the vehicle, underneath the instrument panel glove box area.

OPERATION

The Electronic Control Module (ECM) controls the vehicle engine operating system.

REMOVAL & INSTALLATION

1. Disconnect the negative battery cable.
2. Remove the lower inner trim on the passenger's side of the vehicle.
3. Detach the floor mat. Remove the protective cover.
4. Remove the Electronic Control Module (ECM) bracket retaining nuts. Remove the clip from the bracket.
5. Disconnect the connectors.
6. Remove the ECM from the vehicle.

To install:

7. Installation is the reverse of the removal procedure.
8. Tighten the retaining screws to 5.5 ft. lbs. (7.5 Nm).

➡When replacing the ECM, be careful not to use the wrong part number, as damage to the injection system could occur.

TESTING

Scan to verify communication with ECM. If no communication, check power and grounds to the ECM.

ENGINE COOLANT TEMPERATURE (ECT) SENSOR

LOCATION

See Figure 169.

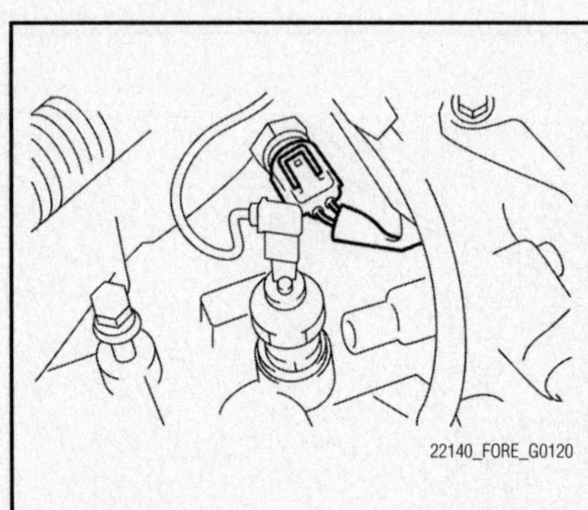

Fig. 169 Engine Coolant Temperature (ECT) Sensor view

The Engine Coolant Temperature (ECT) sensor is located by the heater outlet fitting or in a cooling passage on the engine, depending upon the particular vehicle.

OPERATION

The Engine Coolant Temperature (ECT) sensor detects the temperature of the engine coolant and relays the information to the electronic control assembly.

REMOVAL & INSTALLATION

1. Disconnect the negative battery cable.
2. Remove the alternator, as required.
3. Remove the air intake duct and air cleaner case.
4. Disconnect the connector from the Engine Coolant Temperature (ECT) sensor.
5. Drain the cooling system, as required.
6. Remove the sensor from its mounting.

To install:

7. Installation is the reverse of the removal procedure.
8. Tighten the sensor to 13.3 ft. lbs. (18 Nm).

TESTING

1. Turn the ignition switch **OFF**.
2. Disconnect the connector from the sensor.
3. Turn the ignition switch ON.
4. Using the Subaru scan tool, read the sensor signal data. If the measured value is less than -40 degrees F, replace the sensor.

EXHAUST GAS RECIRCULATION VALVE

LOCATION

On the SOHC engine the Exhaust Gas Recirculation (EGR) valve is mounted on the intake manifold. The DOHC engine does not use a conventional EGR valve.

OPERATION

The Exhaust Gas Recirculation (EGR) valve controls the flow of exhaust gases. The ECM monitors the flow and regulates the valve accordingly.

REMOVAL & INSTALLATION

See Figure 170.

1. Disconnect the negative battery cable.
2. Disconnect the connector.

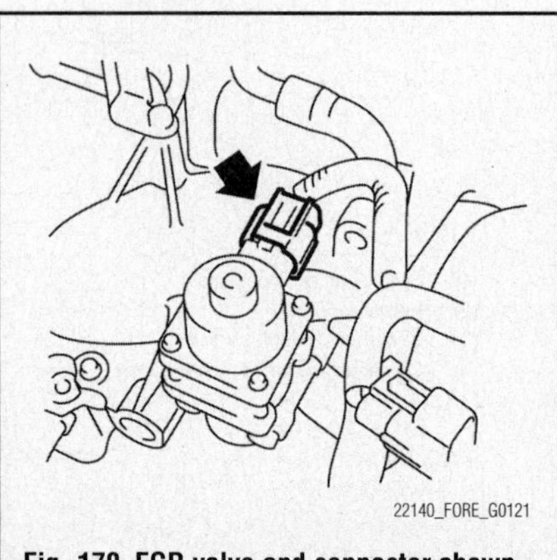

Fig. 170 EGR valve and connector shown

3. Remove the Exhaust Gas Recirculation (EGR) valve from the intake manifold.

To install:

4. Installation is the reverse of the removal procedure.
5. Tighten the retaining bolts to 14.0 ft. lbs. (19 Nm).

TESTING

See Figure 171.

1. Disconnect the connector from the ECM.
2. Measure the resistance between the ECM connector and chassis ground.
3. If DTC code P1492 was set, use connector B134 and terminal 10/chassis ground. If DTC code P1494 was set, use connector B134 and terminal 9/chassis ground. If DTC code P1496 was set, use connector B134 and terminal 8/chassis ground. If DTC code P1498 was set, use connector B134 and terminal 20/chassis ground.
4. If measured value is more than 1 M ohm, check the ECM connector and the EGR solenoid for poor contact.
5. If contact is good, replace the EGR solenoid valve.

ENGINE OIL TEMPERATURE (EOT) SENSOR

LOCATION

See Figure 172.

The Engine Oil Temperature (EOT) sensor is located in the engine block, just below coil assembly.

Fig. 172 Engine Oil Temperature (EOT) sensor location

OPERATION

The Engine Oil Temperature (EOT) detects the temperature of the engine oil and relays the information to the electronic control assembly.

REMOVAL & INSTALLATION

1. Before servicing the vehicle, refer to the precautions section.
2. Remove the collector cover.
3. Disconnect the ground cable from battery.
4. Disconnect the connector from Engine Oil Temperature (EOT) sensor.
5. Remove the EOT sensor.

To install:

6. Install in the reverse order of removal.
7. Tighten the EOT sensor to 16 ft. lbs. (22 Nm).

TESTING

1. Start the engine.
2. Read the data of the oil temperature sensor signal using the Subaru Select Monitor or general scan tool.
3. Is the oil temperature 419°F (215°C) or more. If not the problem may be intermittent. In this case, there may be a temporary connector contact failure.
4. If the oil temperature is 419°F (215°C) or more. Turn the ignition switch **OFF**.
5. Disconnect the connector from the ECM and oil temperature sensor.
6. Measure the resistance between ECM and chassis ground. Connector (B134) terminal No. 23
7. If the resistance is 1m ohms or more replace the oil temperature sensor.
8. If not repair the ground short circuit of harness between ECM and oil temperature sensor.

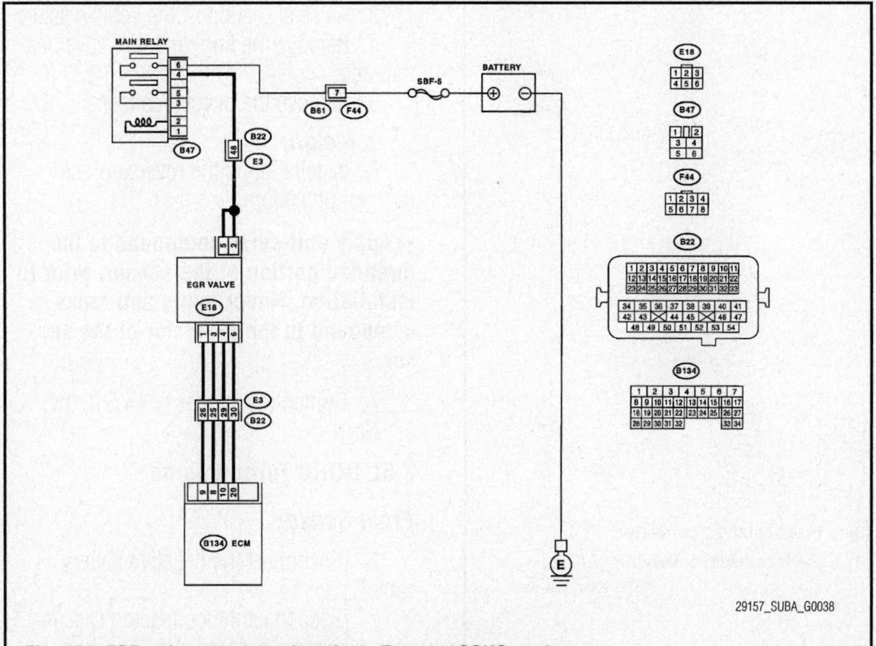

Fig. 171 EGR valve connector location—Forester SOHC engine

FUEL LEVEL SENDING UNIT

LOCATION

The fuel level sending unit is located in the fuel tank.

OPERATION

The fuel level sensor relays a signal to the combination meter for correct fuel indication.

REMOVAL & INSTALLATION

See Figure 173.

➡The fuel level sending unit is built into the fuel pump assembly.

1. Relieve the fuel system pressure.
2. Disconnect the negative battery cable.

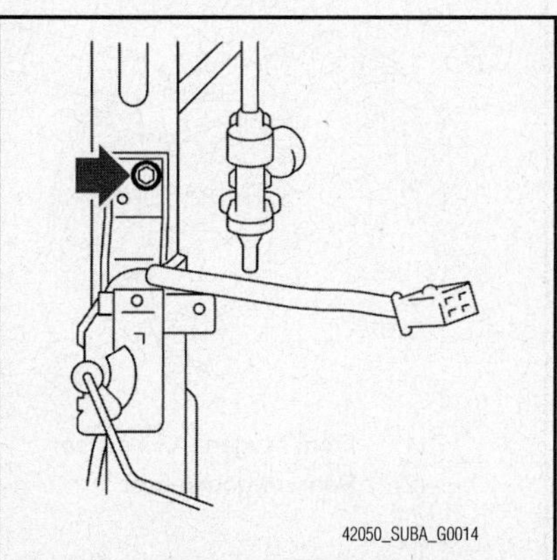

Fig. 173 Remove the bolt that holds the sending unit to the mounting bracket

3. Remove the fuel pump. For additional information, please refer to the Fuel Pump Removal & Installation procedure in the Fuel Systems Section.

4. Disconnect the electrical connector from the sending unit.

5. Disconnect the fuel temperature sensor, non-turbo models only.

6. Remove the bolt that holds the fuel level sending unit on the mounting bracket.

7. Installation is the reverse order of removal.

TESTING

1. Remove the fuel pump assembly.

2. With the float set to the full position, using an ohmmeter, measure the resistance between the sending unit and terminal Nos. 2 and 3.

3. Resistance should be between 0.5–2.5 ohms. Replace the fuel level sending unit if measurement is outside of the specification.

4. Using an ohmmeter, measure the resistance between terminal Nos. 2 and 3 while moving the float up and down. If the resistance does not change smoothly, replace the fuel level sending unit.

HEATED OXYGEN SENSOR (HO2S)

LOCATION

See Figures 174 and 175.

On vehicles with SOHC engine, the front oxygen sensor is located in the front section of the front catalytic converter. The rear oxygen sensor is located in the rear section of the front catalytic converter. The DOHC engine the front oxygen sensor is located in the front section of the exhaust system, just past the crossover pipe. The rear oxygen sensor is located in the front section of the rear catalytic converter.

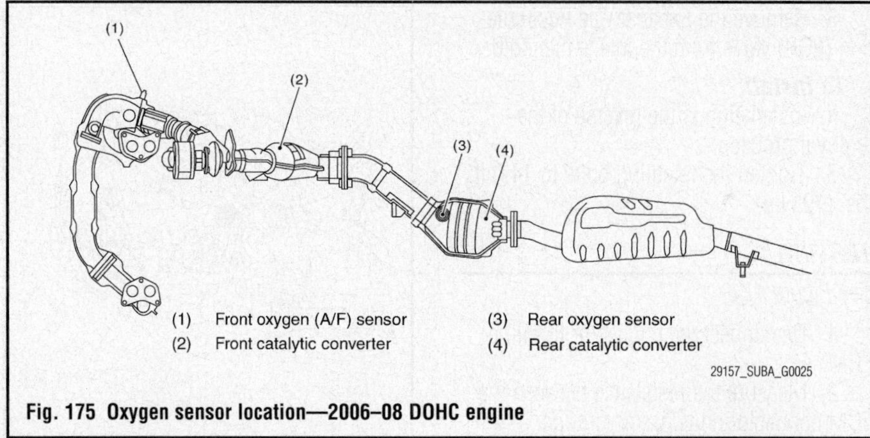

Fig. 175 Oxygen sensor location—2006–08 DOHC engine

OPERATION

The exhaust gas oxygen sensor supplies the electronic control assembly with a signal which indicates either a rich or lean mixture condition, during the engine operation.

REMOVAL & INSTALLATION

2.5L SOHC Non-Turbo Engine

Front Sensor

1. Disconnect the negative battery cable.

2. Disconnect the clip fastening the harness. Disconnect the connector.

3. Raise and support the vehicle safely.

4. Remove the engine under cover, as required.

5. Remove the oxygen sensor.

To install:

6. Installation is the reverse of the removal procedure.

➡Apply anti-seize compound to the threaded portion of the sensor, prior to installation. Never apply anti-seize compound to the protector of the sensor.

7. Tighten the sensor to 15.5 ft. lbs. (21 Nm).

Rear Sensor

1. Disconnect the negative battery cable.

2. Disconnect the clip fastening the harness. Disconnect the connector.

3. Raise and support the vehicle safely.

4. Remove the engine under cover, as required.

5. Remove the oxygen sensor.

To install:

6. Installation is the reverse of the removal procedure.

➡Apply anti-seize compound to the threaded portion of the sensor, prior to installation. Never apply anti-seize compound to the protector of the sensor.

7. Tighten the sensor to 15.5 ft. lbs. (21 Nm).

2.5L DOHC Turbo Engine

Front Sensor

1. Disconnect the negative battery cable.

2. Disconnect the connector. Disconnect the engine harness clip.

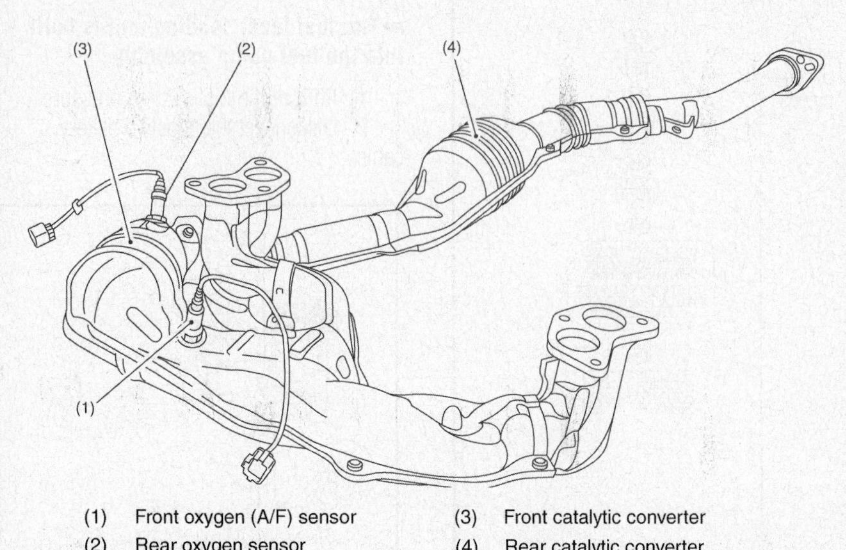

Fig. 174 Oxygen sensor location—2006–08 SOHC engine

3. Raise and support the vehicle safely.
4. Remove the tire and wheel.
5. Remove the service hole cover.
6. Remove the oxygen sensor.

To install:

7. Installation is the reverse of the removal procedure.

➡Apply anti-seize compound to the threaded portion of the sensor, prior to installation. Never apply anti-seize compound to the protector of the sensor.

8. Tighten the sensor to 22.1 ft. lbs. (30 Nm).

Rear Sensor

1. Disconnect the negative battery cable.
2. Raise and support the vehicle safely.
3. Disconnect the connector.
4. Remove the clip by pulling it out from the upper side of the crossmember.
5. Remove the oxygen sensor.

To install:

6. Installation is the reverse of the removal procedure.

➡Apply anti-seize compound to the threaded portion of the sensor, prior to installation. Never apply anti-seize compound to the protector of the sensor.

7. Tighten the sensor to 22.1 ft. lbs. (30 Nm).

TESTING

2.5L SOHC Non-Turbo Engine

Front Sensor

See Figure 176.

Perform a visual inspection of the sensor as follows:

1. If the sensor tip has a black/sooty deposit, this may indicate a rich fuel mixture.
2. If the sensor tip has a white, gritty deposit, this may indicate an internal coolant leak.
3. If the sensor tip has a brown deposit, this could indicate oil consumption.
4. Measure the resistance between the sensor and the connector terminals (connector B262/terminal 2 and connector B262/terminal 1) and (connector B262/terminal 3 and connector B262/terminal 4).
5. Measure the resistance between the sensor and the connector terminals (terminals 1 and 4).
6. If measured value is more than 5 ohm, replace the sensor.

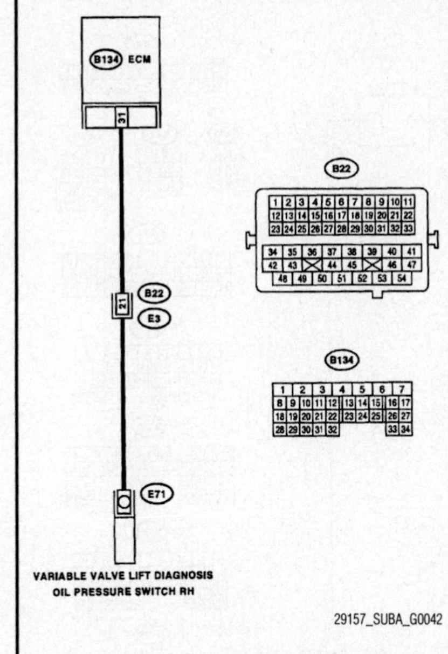

Fig. 176 Front oxygen sensor connector location—2006–08 SOHC engine

Rear Sensor

See Figure 177.

Perform a visual inspection of the sensor as follows:

1. If the sensor tip has a black/sooty deposit, this may indicate a rich fuel mixture.
2. If the sensor tip has a white, gritty deposit, this may indicate an internal coolant leak.
3. If the sensor tip has a brown deposit, this could indicate oil consumption.
4. Turn the ignition switch **OFF**.
5. Disconnect the connectors from the sensor.
6. Turn the ignition switch **ON**.
7. Measure the voltage between the sensor harness connector terminals (terminals 1 and 2). If measured value is more than 30 ohms, replace the sensor.

2.5L DOHC Turbo Engine

Front Sensor

See Figure 178.

Perform a visual inspection of the sensor as follows:

1. If the sensor tip has a black/sooty deposit, this may indicate a rich fuel mixture.
2. If the sensor tip has a white, gritty deposit, this may indicate an internal coolant leak.
3. If the sensor tip has a brown deposit, this could indicate oil consumption.
4. Measure the resistance of the harness between the sensor connector terminals (terminals 1 and 2).

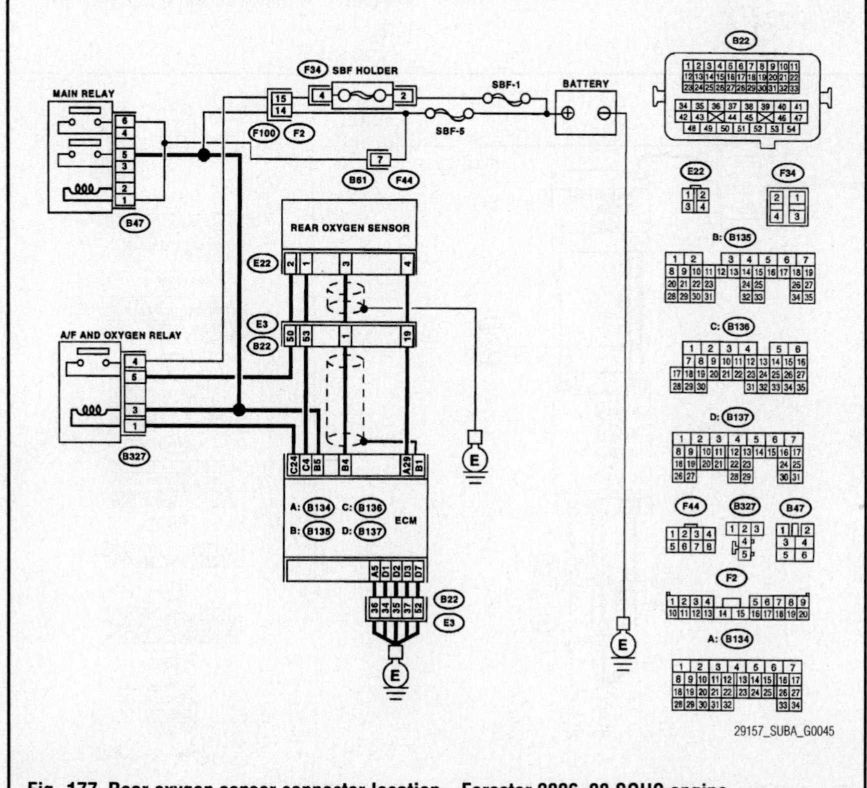

Fig. 177 Rear oxygen sensor connector location—Forester 2006–08 SOHC engine

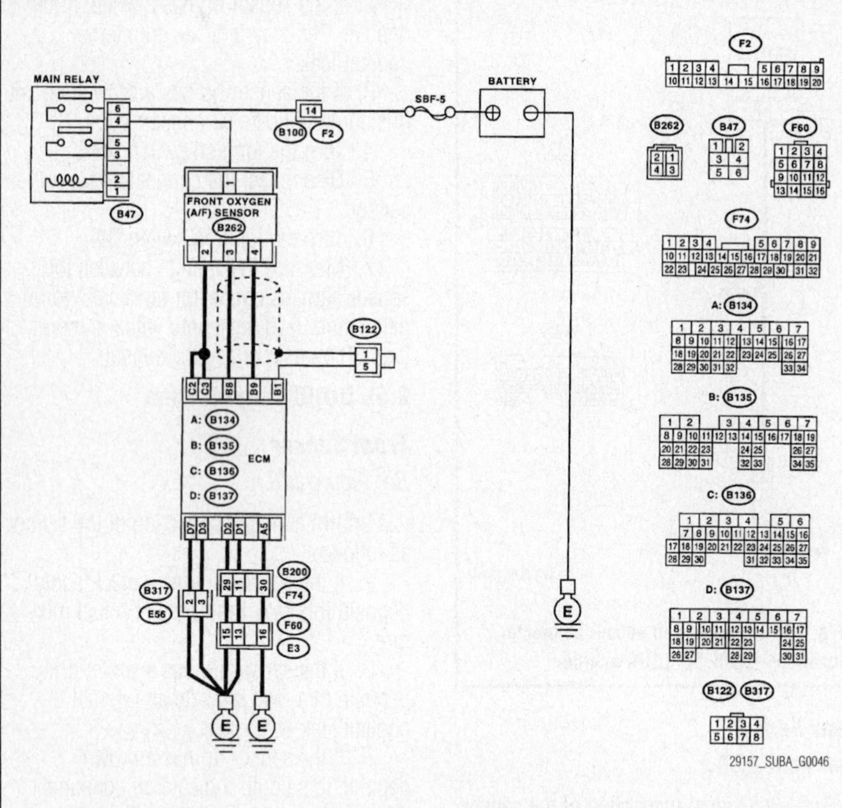

Fig. 178 Front oxygen sensor connector location—2006–08 DOHC engine

5. If measured value is more than 5 ohm, replace the sensor.

Rear Sensor

See Figure 179.

Perform a visual inspection of the sensor as follows:

1. If the sensor tip has a black/sooty deposit, this may indicate a rich fuel mixture.

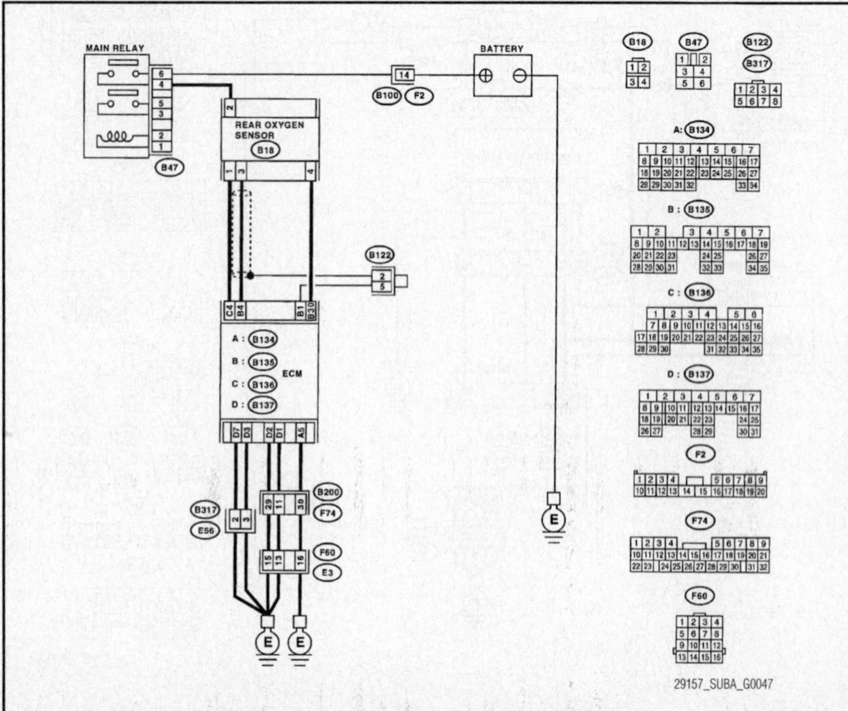

Fig. 179 Rear oxygen sensor connector location—2006–08 DOHC engine

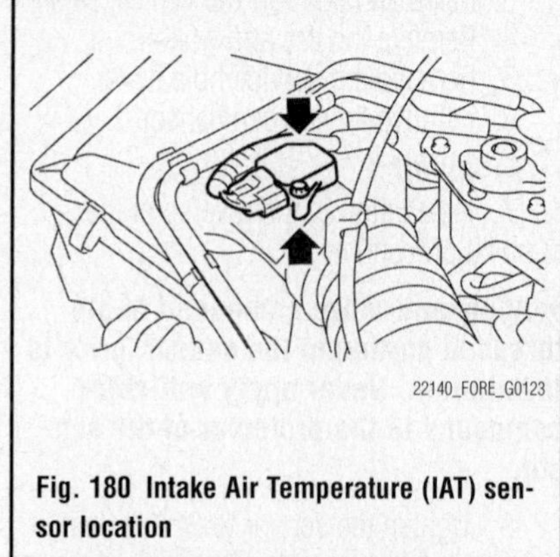

Fig. 180 Intake Air Temperature (IAT) sensor location

2. If the sensor tip has a white, gritty deposit, this may indicate an internal coolant leak.
3. If the sensor tip has a brown deposit, this could indicate oil consumption.
4. Turn the ignition switch OFF.
5. Measure the resistance of the harness between the sensor connector terminals (terminals 1 and 2).
6. If measured value is more than 30ohm, replace the sensor.

INTAKE AIR TEMPERATURE (IAT) SENSOR

See Figure 180.

LOCATION

The Intake Air Temperature (IAT) sensor is mounted in the intake air hose of the air cleaner assembly. This sensor is combined with the Mass Air Flow (MAF) sensor.

OPERATION

The sensor provides a signal to the ECM for incoming air temperature. This information is used to adjust the injector pulse width and in turn the air/fuel ratio.

REMOVAL & INSTALLATION

1. Disconnect the negative battery cable.
2. Disconnect the connector from the Mass Air Flow (MAF) and Intake Air Temperature (IAT) sensor.
3. Remove the MAF and IAT sensor.

To install:

4. Installation is the reverse of the removal procedure.
5. Tighten the sensor mounting screws to 0.74 ft. lbs. (1.0 Nm).

TESTING

See Figure 181.

1. Use a Scan tool to check the sensor signal data as follows:

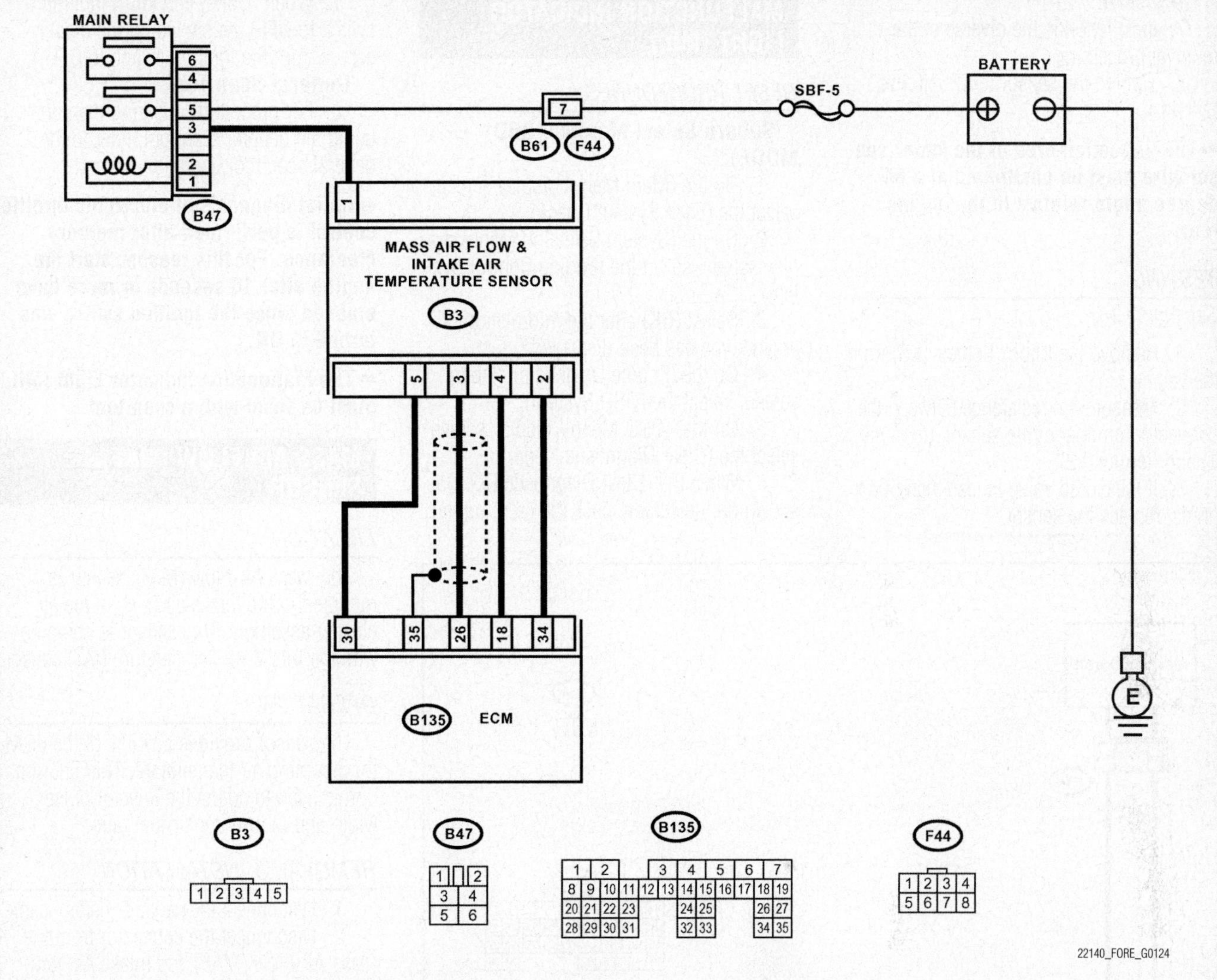

MAIN RELAY
BATTERY
SBF-5
B61 F44
B47
MASS AIR FLOW & INTAKE AIR TEMPERATURE SENSOR
B3
B135 ECM
B3 B47 B135 F44
22140_FORE_G0124

Fig. 181 Mass Air Flow (MAF) and Intake Air Temperature (IAT) sensor

- Start the engine and allow it to reach operating temperature.
- Using the Subaru scan tool, read the sensor signal data.
- If the measured value is less than -40 degrees F, replace the sensor.

2. Check the power supply circuit of the sensor as follows:

- Turn the ignition switch **OFF**.
- Disconnect the connector from the mass air flow and intake air temperature sensor.
- Turn the ignition switch **ON**.
- Measure the voltage between mass air flow and intake air temperature sensor connector and engine ground. Connector & terminal (B3) No. 1 (Pos)—Engine ground (Neg).
- Voltage reading should be 10 volts or more, if not repair the faulty circuit.

KNOCK SENSOR (KS)

LOCATION

The Knock Sensor (KS) is located at the top right (driver's) side of the engine and is positioned on the cylinder block.

OPERATION

The Knock Sensor (KS) sensor is used to detect engine vibrations caused by pre-ignition or detonation and provides information to the ECM, which then retards the timing to eliminate detonation.

REMOVAL & INSTALLATION

See Figure 182.

1. Disconnect the negative battery cable.
2. Remove the air intake chamber.
3. Remove the collector cover, as required.
4. On DOHC engine, remove the intercooler.
5. Disconnect the Knock Sensor (KS) connector.
6. Remove the KS from its mounting.

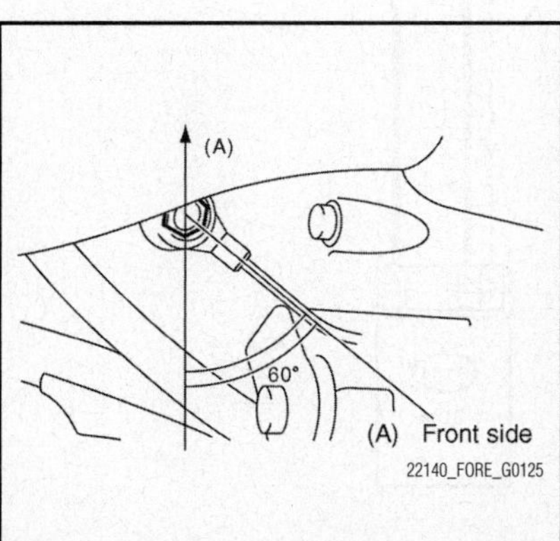

Fig. 182 Knock Sensor (KS) positioned at a 60 degree angle

To install:

7. Installation is the reverse of the removal procedure.

8. Tighten the sensor to 17.7 ft. lbs. (24 Nm).

➡**The extraction area of the knock sensor wire must be positioned at a 60 degree angle relative to the engine rear.**

TESTING

See Figure 183.

1. Remove the Knock Sensor (KS) from the vehicle.

2. Measure the resistance between the connector terminal of the sensor, (terminal 1) and (terminal 2).

3. If measured value is less than 500 k ohms, replace the sensor.

MALFUNCTION INDICATOR LIGHT (MIL)

RESET PROCEDURES

Subaru Select Monitor (OBD MODE):

1. On the (Main Menu) display screen, select the (Each System Check).

2. On the (System Selection Menu) display screen, select the (Engine Control System).

3. Select (OK) after the information of engine type has been displayed.

4. On the (Engine Diagnosis) display screen, select the (OBD System).

5. On the «OBD Menu» display screen, select the (Clear Diagnostic Code).

6. When the (Clear Diagnostic Code) is shown on the screen, click the (Yes) button.

7. When (Done) and (Turn ignition switch to OFF) are shown on the display screen, turn the ignition switch to OFF.

General Scan Tool:

8. For procedures clearing memory using the general scan tool, refer to the general scan tool operation manual.

➡**Initial diagnosis of electronic throttle control is performed after memory clearance. For this reason, start the engine after 10 seconds or more have elapsed since the ignition switch was turned to ON.**

➡**The Malfunction Indicator Light (MIL) must be reset with a scan tool.**

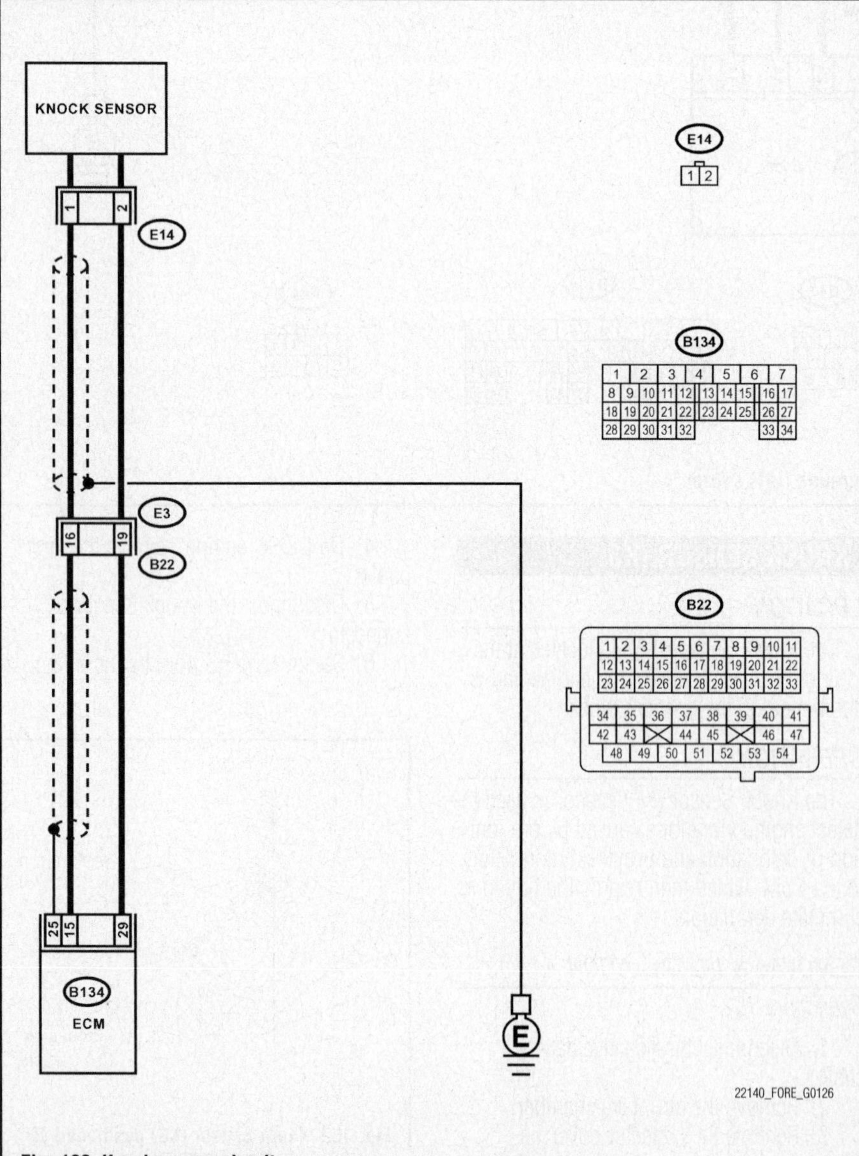

Fig. 183 Knock sensor circuit

MASS AIR FLOW (MAF) SENSOR

LOCATION

The Mass Air Flow (MAF) sensor is mounted in the intake air hose of the air cleaner assembly. This sensor is combined with the Intake Air Temperature (IAT) sensor.

OPERATION

The sensor provides a signal to the ECM for incoming air temperature. This information is used to adjust the injector pulse width and in turn the air/fuel ratio.

REMOVAL & INSTALLATION

1. Disconnect the negative battery cable.

2. Disconnect the connector from the Mass Air Flow (MAF) and Intake Air Temperature (IAT) sensor.

3. Remove the MAF and IAT sensor.

To install:

4. Installation is the reverse of the removal procedure.

5. Tighten the sensor mounting screws to 0.74 ft. lbs. (1.0 Nm).

TESTING

See Figure 184.

1. Use a Scan tool to check the sensor signal data as follows:
    - Start the engine and allow it to reach operating temperature.
    - Using the Subaru scan tool, read the sensor signal data.
    - If the measured value is less than -40 degrees F, replace the sensor.

2. Check the power supply circuit of the sensor as follows:
    - Turn the ignition switch **OFF**.
    - Disconnect the connector from the mass air flow and intake air temperature sensor.

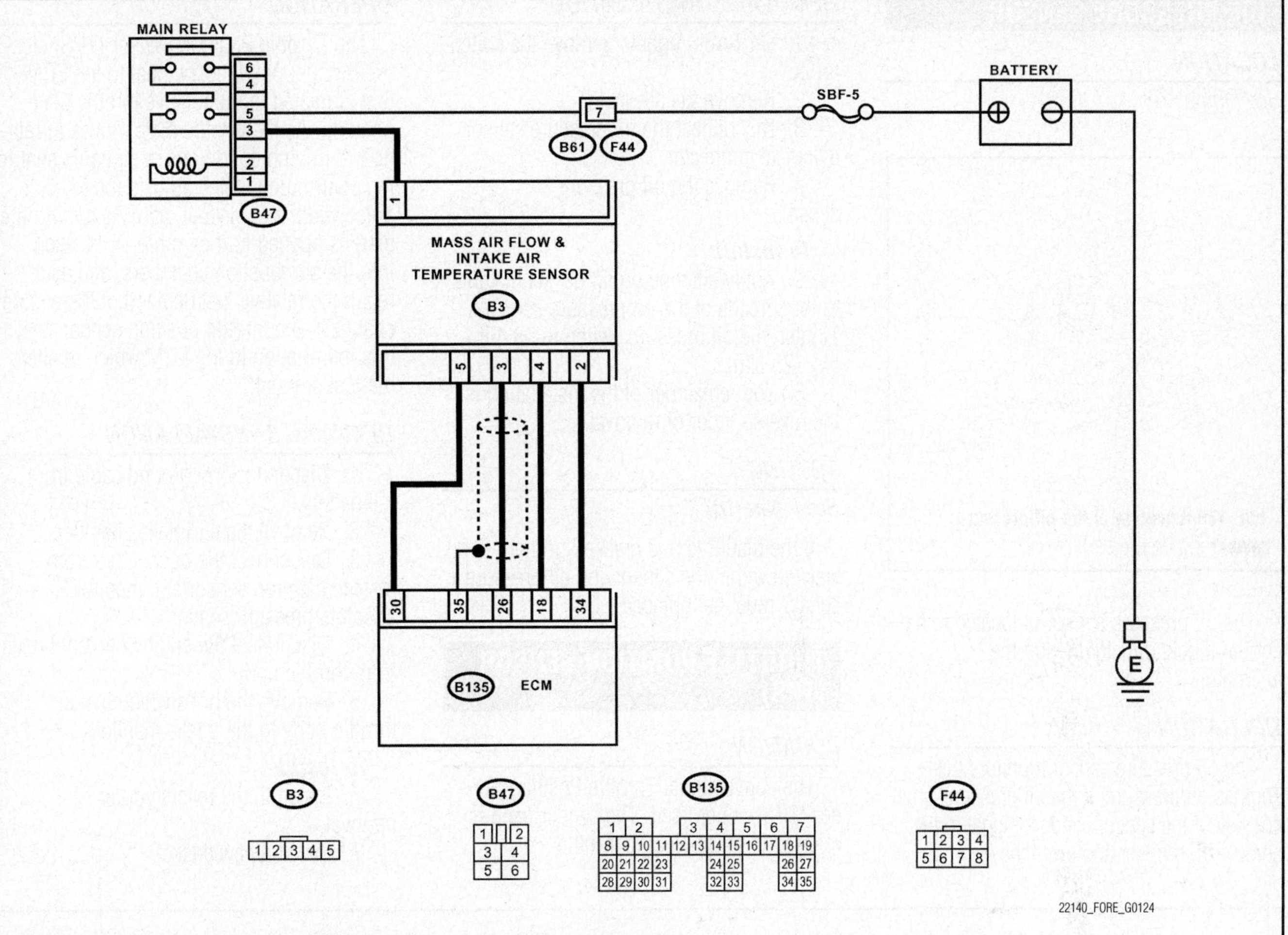

Fig. 184 Mass Air Flow (MAF) and Intake Air Temperature (IAT) sensor

- Turn the ignition switch **ON**.
- Measure the voltage between mass air flow and intake air temperature sensor connector and engine ground. Connector & terminal (B3) No. 1 (Pos)—Engine ground (Neg).
- Voltage reading should be 10 volts or more, if not repair the faulty circuit.

MANIFOLD ABSOLUTE PRESSURE (MAP) SENSOR

LOCATION

See Figure 185.

The sensor is located on the throttle body unit.

OPERATION

The Manifold Absolute Pressure (MAP) sensor monitors and signals the ECM of changes in intake manifold pressure which result from engine load, speed and atmospheric pressure changes.

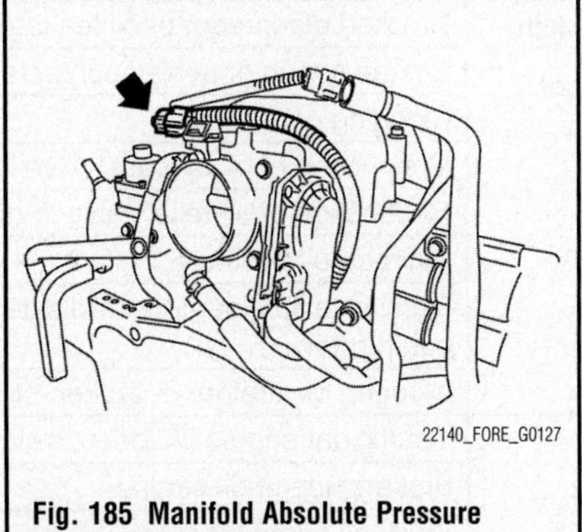

Fig. 185 Manifold Absolute Pressure (MAP) sensor location

REMOVAL & INSTALLATION

1. Disconnect the negative battery cable.
2. Disconnect the connector from the sensor.
3. Remove the Manifold Absolute Pressure (MAP) sensor from its mounting.

To install:

4. Installation is the reverse of the removal procedure, noting the following:
   a. Be sure to use new O-rings.
   b. Tighten the retaining screw to 1.5 ft. lbs. (2.0 Nm).

TESTING

1. Start the engine and warm-up engine until coolant temperature is higher than 167°F (75°C).
2. For AT models, set the select lever to the "P" or "N" range, and for MT models, place the shift lever in the neutral position.
3. Turn the A/C switch to OFF.
4. Turn all the accessory switches to OFF.
5. Read the data of intake manifold pressure sensor signal using Subaru Select Monitor or general scan tool.
6. The measured value should read 73.3–106.6 kPa (550–800 mmHg, 21.65–31.50 in Hg) when the ignition is turned ON, and 20.0–46.7 kPa (150–350 mm Hg, 5.91–13.78 in Hg) during idling.
7. If the readings are not as stated, replace the MAP sensor.

➡Be sure to thoroughly check the air intake system for vacuum leaks.

OIL PRESSURE SENSOR

LOCATION

See Figure 186.

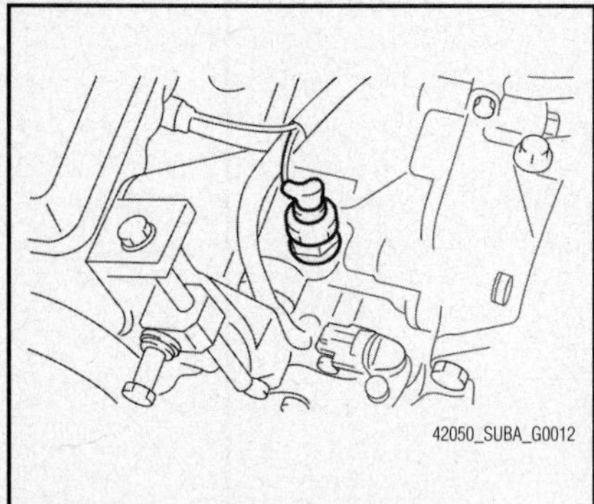

Fig. 186 Location of the oil pressure sensor

The oil pressure sensor is located in the engine block, directly behind the alternator.

OPERATION

The oil pressure sensor monitors the engines oil pressure. If the oil pressure is not consistent the sensor sends a signal to the combination meter and turns the light on.

REMOVAL & INSTALLATION

1. On Turbo models, remove the collector cover.
2. Remove the alternator.
3. Disconnect the oil pressure sensor electrical connector.
4. Remove the oil pressure sensor.

To install:

5. Apply silicone liquid gasket material to the threads of the oil pressure sensor. Tighten the oil pressure switch to 18.4 ft. lbs. (25 Nm).
6. The remainder of the installation is the reverse order of removal.

TESTING

See Figure 187.

If the diaphragm is cracked or there is oil leakage within the sensor, the oil pressure sensor must be replaced.

THROTTLE POSITION SENSOR (TPS)

LOCATION

The conventional Throttle Position Sensor (TPS) is not used. The sensor used is referred to as an electric throttle.

OPERATION

The Throttle Position Sensor (TPS) electric throttle, provides a signal to the ECM that is related to the relative throttle plate position. As the throttle plate moves in relation to driving conditions, a signal is sent to the control unit which adjusts the injector pulse width and air/fuel ratio. As the throttle plate is opened further, more air is taken into the combustion chambers, and as a result the relative fuel demand of the engine changes. The throttle position sensor relays this information to the ECM which in alters the fuel amount.

REMOVAL & INSTALLATION

1. Disconnect the ground cable from the battery.
2. Remove the air intake chamber.
3. Disconnect the connectors from throttle position sensor and manifold absolute pressure sensor.
4. Disconnect the engine coolant hoses from throttle body.
5. Remove the bolts which install throttle body to the intake manifold.

To install:

6. Install in the reverse order of removal.
7. Install a new gasket.

Trouble	Possible cause		Corrective action
1. Warning light remains on.	1) Oil pressure switch failure	Cracked diaphragm or oil leakage within switch	Replace.
		Broken spring or seized contacts	Replace.
	2) Low oil pressure	Clogging of oil filter	Replace.
		Malfunction of oil by-pass valve in oil filter	Clean or replace.
		Malfunction of oil relief valve in oil pump	Clean or replace.
		Clogged oil passage	Clean.
		Excessive tip clearance and side clearance of oil pump rotor	Replace.
		Clogged oil strainer or broken pipe	Clean or replace.
	3) No oil pressure	Insufficient engine oil (degradation, etc.)	Replace.
		Broken pipe of oil strainer	Replace.
		Stuck oil pump rotor	Replace.
2. Warning light does not illuminate.	1) Malfunction of combination meter		Replace.
	2) Poor contact of switch contact points		Replace.
	3) Disconnection of wiring		Repair.
3. Warning light flickers momentarily.	1) Poor contact of terminals		Repair.
	2) Defective wiring harness		Repair.
	3) Low oil pressure		Check for the same possible causes as listed 1. — 2).

22140_FORE_G0128

Fig. 187 Engine lubrication system general inspection chart

8. Tighten throttle body mounting bolts to 6 ft. lbs. (8 Nm).

TESTING

See Figures 188 and 189.

1. Visually inspect the throttle linkage and throttle for binding and sticking.
2. Turn the ignition switch **ON**.
3. Measure the voltage between electronic throttle control connector and engine ground. Connector & terminal (E57) No. 5 (Pos.)—Engine ground (Neg.)
4. The voltage reading should be 4.5–5.5 volts.
5. If the voltage reading was not as stated, repair poor contact in ECM connector.
6. Replace the ECM if defective.

TUMBLE GENERATOR VALVE ACTUATOR

LOCATION

A tumble generator valve is provided on each engine bank, between the intake manifold and intake air ports.

OPERATION

See Figure 190.

A tumble generator valve is provided on each engine bank, between the intake manifold and intake air ports. The right bank tumble generator valve has butterfly valves for the #2 and #4 cylinders. The two butterfly valves in each tumble generator valve are fitted on a single shaft that is driven by an actuator.

The tumble generator valves are controlled by the ECM according to the coolant temperature and the time elapsed after start of the engine. When the engine is started, the butterfly valves are moved to the closing ends. In this state, the intake air flows at very high speeds passing through narrowed passages in the directions determined by the individual intake air ports in the cylinder head. This creates tumbling air motions in the cylinders, which enable lean mixtures to be ignited and thus harmful exhaust emissions to be reduced during engine start. The tumble generator valves are fully open when the engine is operating at an ordinary driving speed, allowing intake air to flow without being changed in direction and velocity.

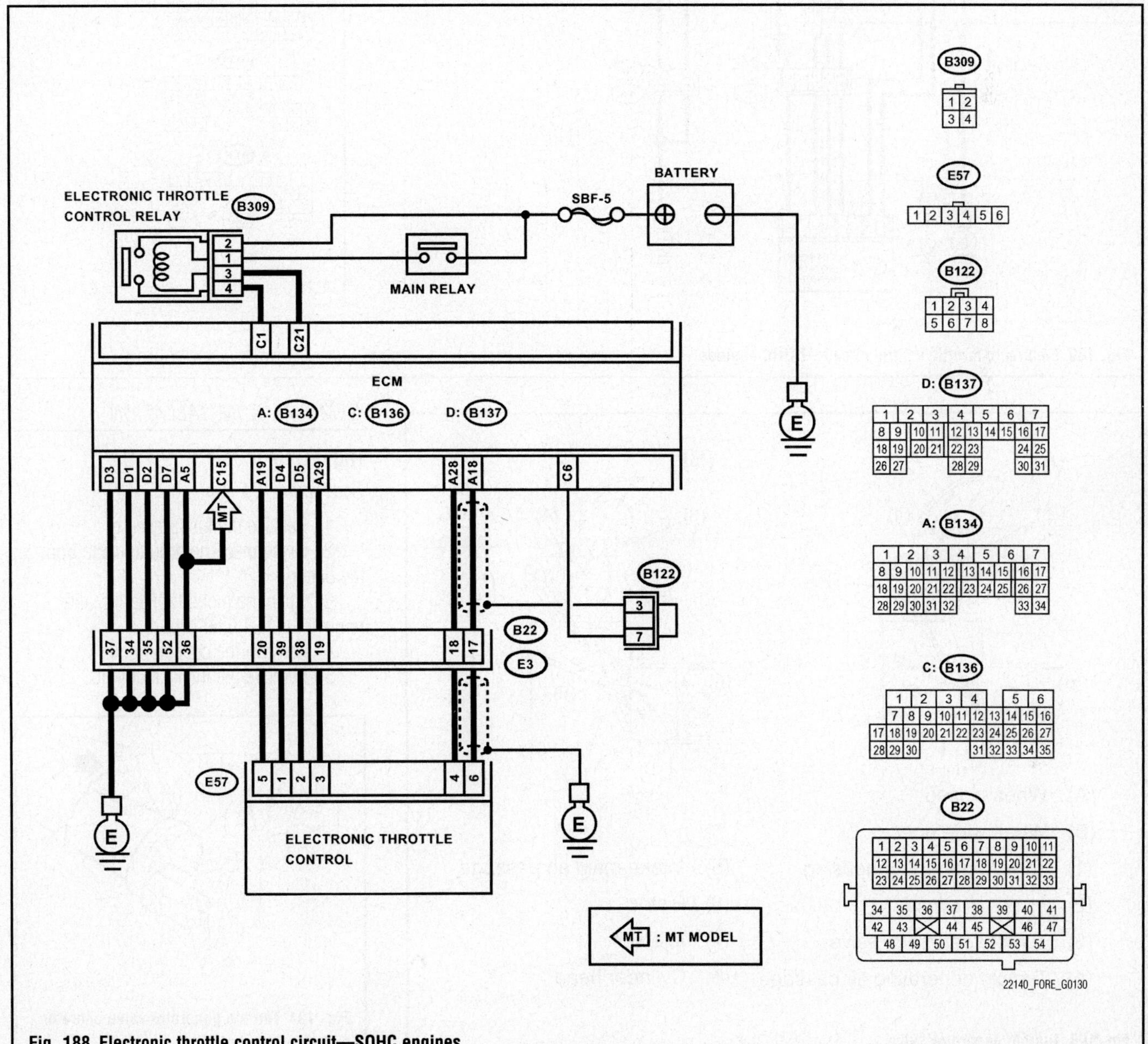

Fig. 188 Electronic throttle control circuit—SOHC engines

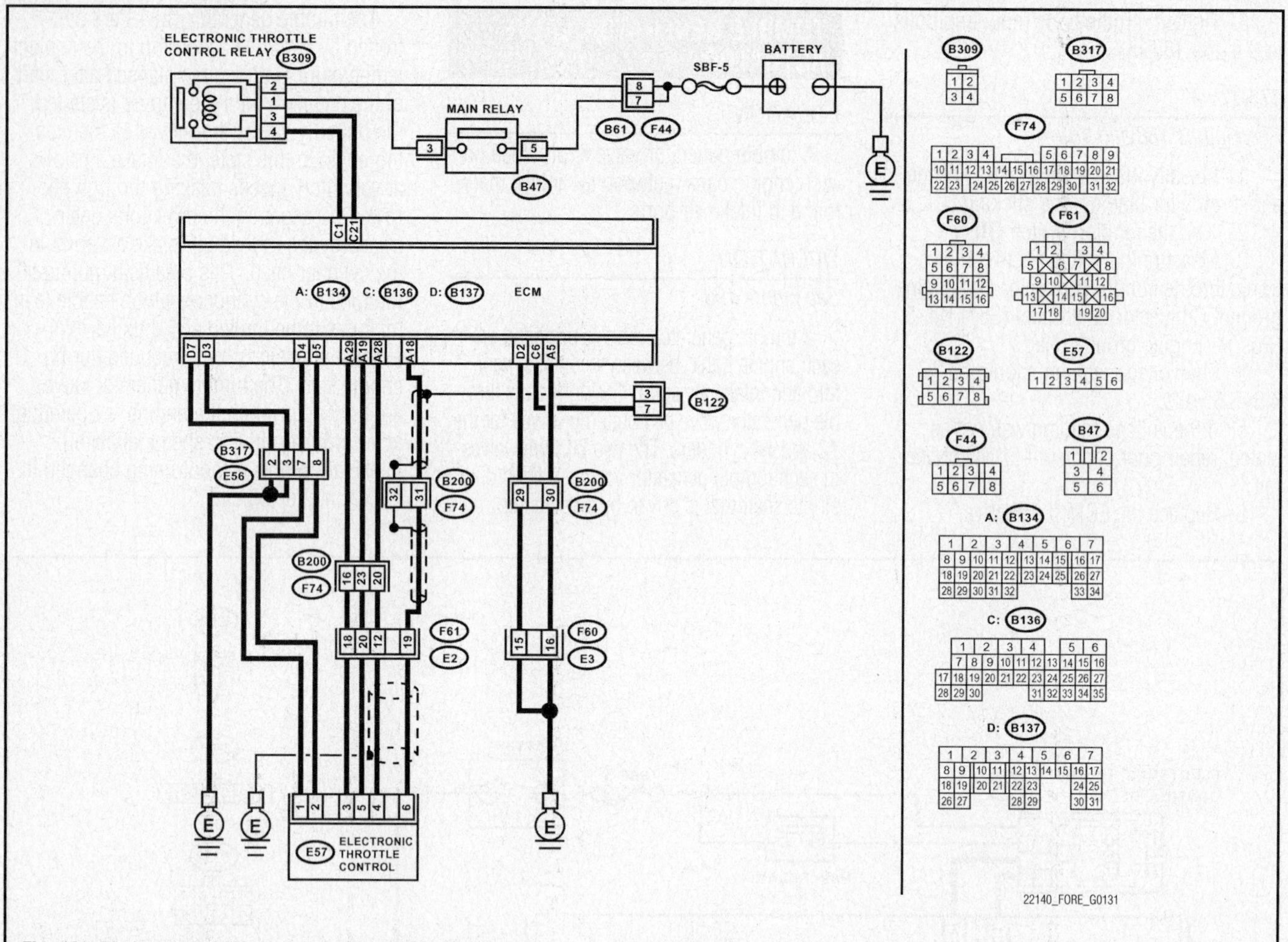

Fig. 189 Electronic throttle control circuit—DOHC engines

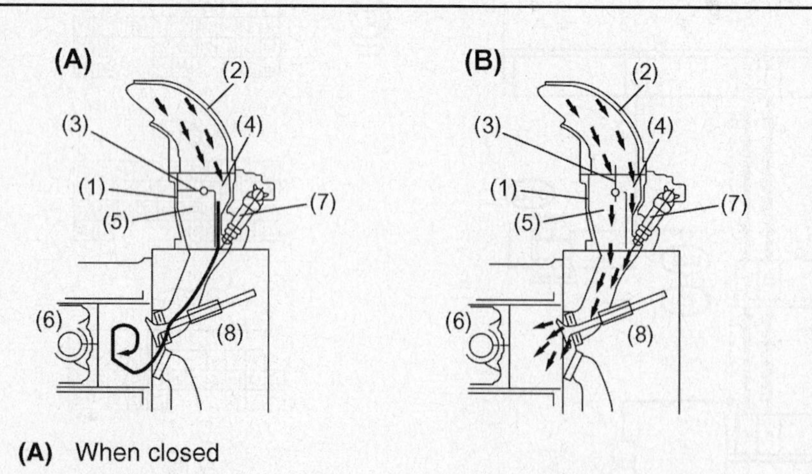

(A) When closed

(B) When wide open

(1) Tumble generator housing
(2) Intake manifold
(3) Tumble generator valve
(4) Tumble generating air passage
(5) Intake main air passage
(6) Piston
(7) Injector
(8) Cylinder head

22140_FORE_G0134

Fig. 190 Tumble generator valve

REMOVAL & INSTALLATION

Right Side

See Figure 191.

1. Release the fuel pressure.
2. Disconnect the ground cable from the battery.
3. Open the fuel filler flap lid, and remove the fuel filler cap.
4. Remove the collector cover.
5. Remove the intake manifold.

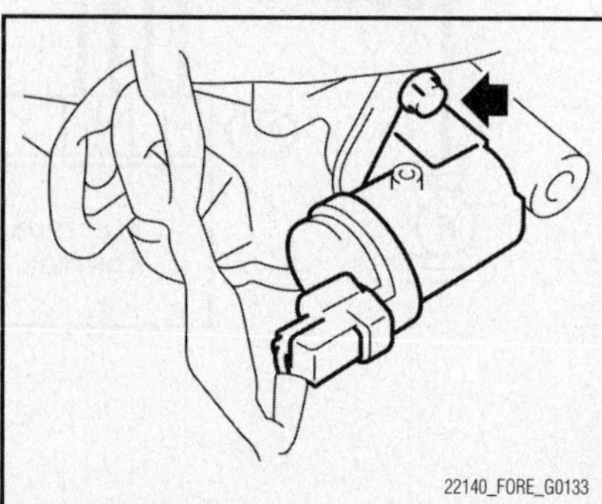

22140_FORE_G0133

Fig. 191 Tumble generator valve actuator and mounting bolt

6. Disconnect the connector from tumble generator valve actuator.
7. Remove the tumble generator valve actuator.

To install:

8. Install in the reverse order of removal.
9. Tighten valve retaining bolt to 4.4 ft. lbs. (6 Nm).

Left Side

1. Disconnect the ground cable from the battery.
2. Remove the collector cover.
3. Remove the secondary air pump.
4. Disconnect the connector from tumble generator valve actuator.
5. Remove the tumble generator valve actuator.

To install:

6. Install in the reverse order of removal.
7. Tighten valve retaining bolt to 4.4 ft. lbs. (6 Nm).

TESTING

See Figure 192.

1. Turn the ignition switch **OFF**.
2. Disconnect the connector from tumble generator valve connector.
3. Turn the ignition switch **ON**.
4. Measure the voltage between tumble generator valve actuator and chassis ground. Connector and terminal (E54) No. 1 (Pos.)—Engine ground (Neg.)
5. The voltage should read more than 4.5 volts, if not check for the following problems:
   - Open circuit of harness between tumble generator valve position sensor and ECM connector
   - Poor contact of tumble generator valve position sensor connector
   - Poor contact in ECM connector
   - Poor contact of coupling connector
   - Poor contact of joint connector

VEHICLE SPEED SENSOR (VSS)

LOCATION

See Figure 193.

The Forester uses a front and rear Vehicle Speed Sensor (VSS). Both sensors are mounted on the transmission.

OPERATION

The Vehicle Speed Sensor (VSS) sends a signal to the ECM to control. The ECM uses this information to control transmission shift patterns. The vehicle speed sensor is also used to send a speed signal to the speed control servo of the cruise control system.

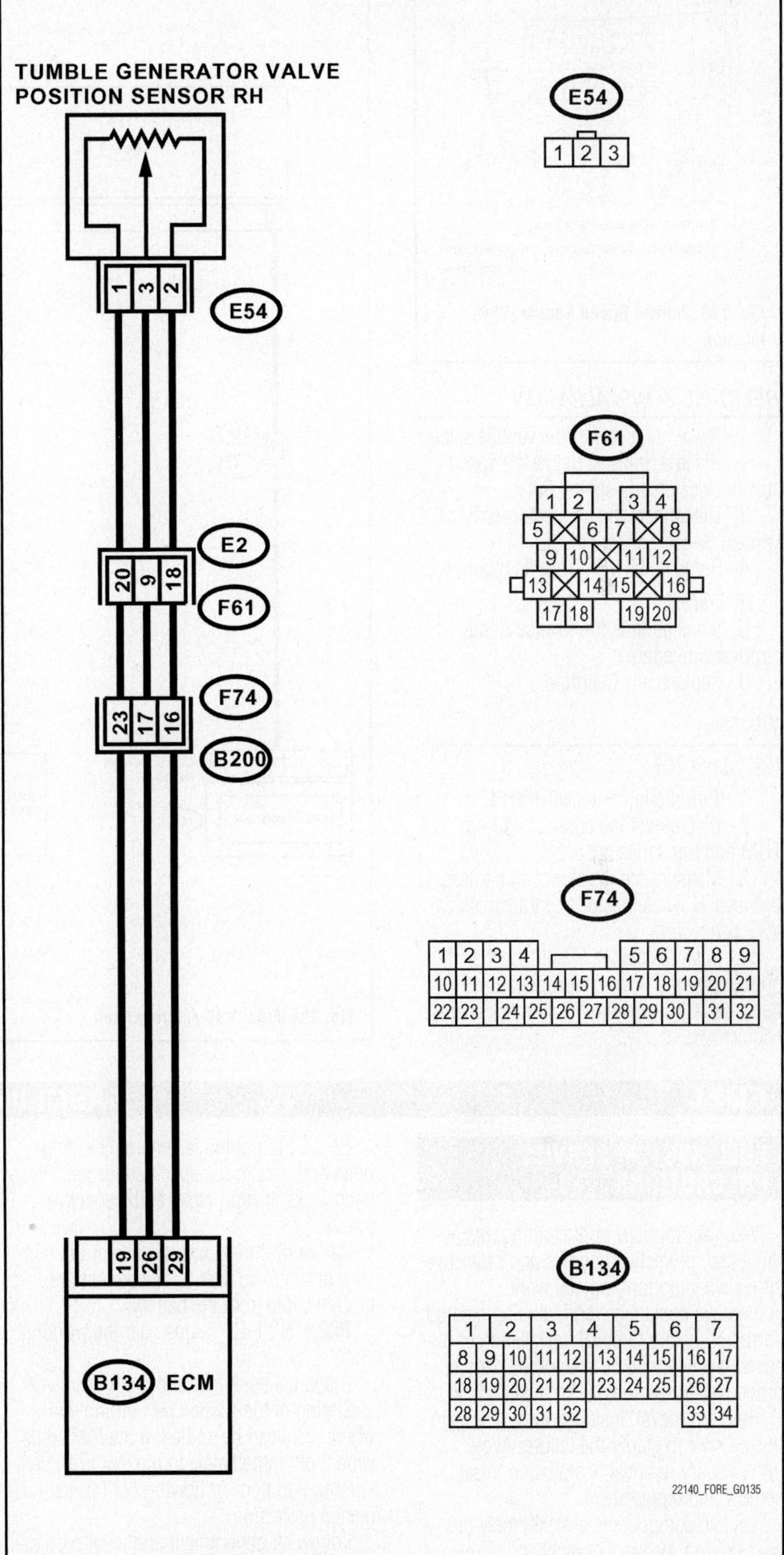

Fig. 192 Tumble generator valve position sensor circuit

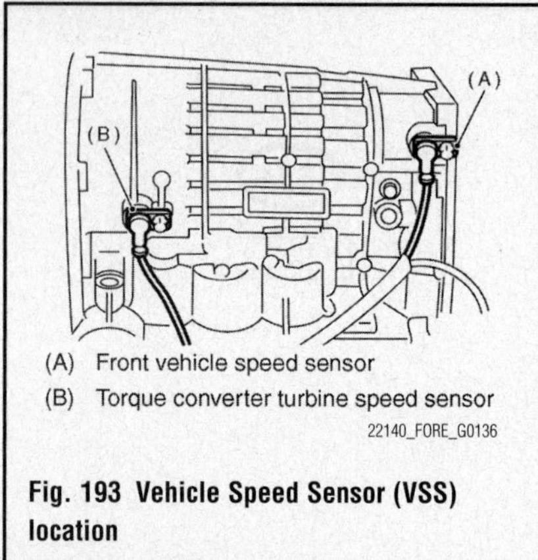

Fig. 193 Vehicle Speed Sensor (VSS) location

REMOVAL & INSTALLATION

1. Raise and support the vehicle safely.
2. Place a drip pan below the speed sensor to catch any spilled fluid.
3. Disconnect the connector from the Vehicle Speed Sensor (VSS).
4. Remove the VSS from its mounting.

To install:

5. Installation is the reverse of the removal procedure.
6. Replace any lost fluid.

TESTING

See Figure 194.

1. Turn the ignition switch **OFF**.
2. Disconnect the connectors from TCM and transmission.
3. Measure the resistance of harness between TCM connector and transmission VSS connectors.
4. If the resistance is not less than 1 ohm, repair the open circuit of harness between TCM and transmission VSS connector.

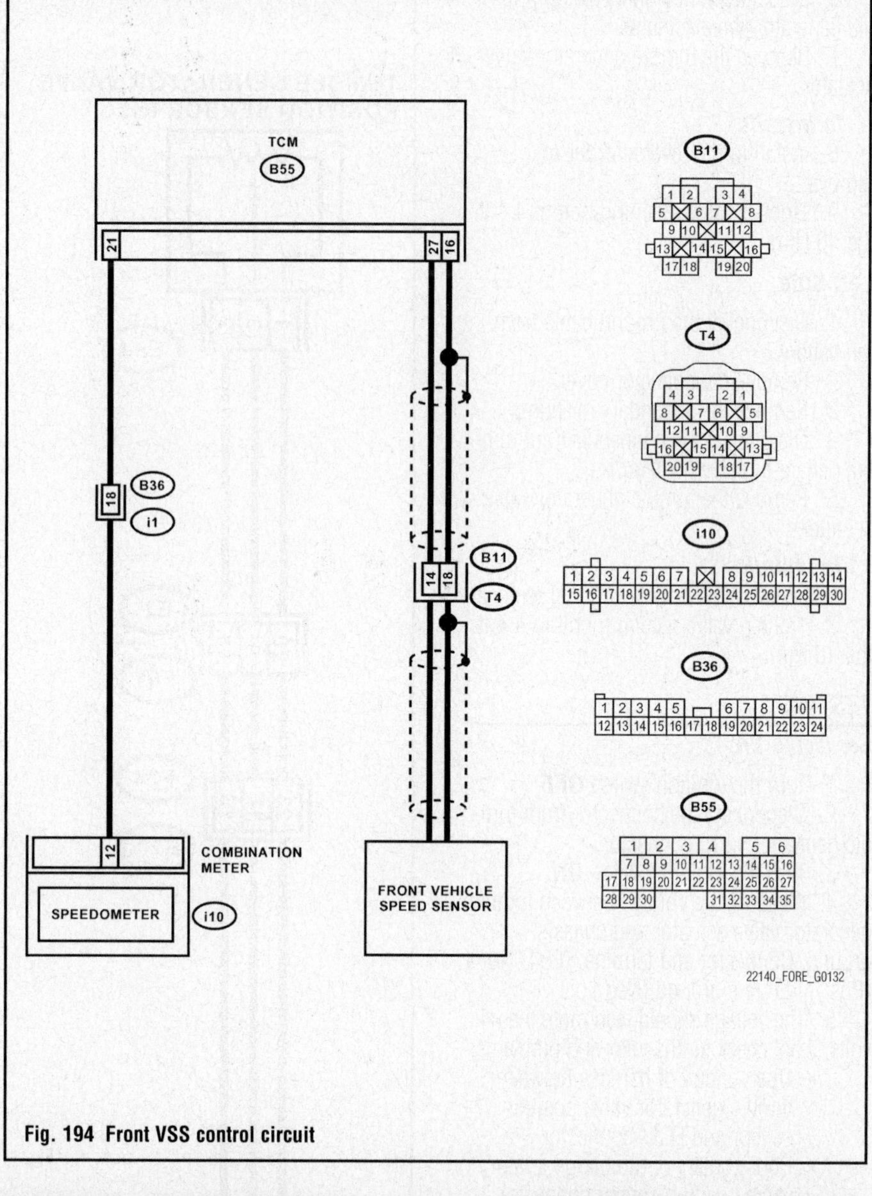

Fig. 194 Front VSS control circuit

FUEL — GASOLINE FUEL INJECTION SYSTEM

FUEL SYSTEM SERVICE PRECAUTIONS

Wear appropriate work clothing, including a cap, protective goggles and protective shoes when performing any work.

Remove contamination including dirt and corrosion before removal, installation or disassembly. Keep the disassembled parts in order and protect them from dust and dirt.

Before removal, installation or disassembly, be sure to clarify the failure. Avoid unnecessary removal, installation, disassembly and replacement.

Vehicle components are extremely hot after driving. Be wary of receiving burns from heated parts.

Be sure to tighten fasteners including bolts and nuts to the specified torque. Place shop jacks or rigid racks at the specified points.

Before disconnecting connectors of sensors or units, be sure to disconnect the ground cable from the battery.

Place "NO FIRE" signs near the working area.

Prepare a container and cloth to prevent scattering of fuels when performing work where fuels can be spilled. If the fuel spills, wipe it off immediately to prevent from penetrating into floor or flowing out for environmental protection.

Follow all government and local regulations concerning disposal of refuse when disposing fuel.

DRAINING FUEL TANK

1. Place "NO FIRE" signs near the working area.
2. Before servicing the vehicle, refer to the Precautions Section.

➡This procedure must be performed prior to servicing any component of the fuel injection system.

3. Disconnect the fuel pump connector from the fuel pump relay.
4. Release the fuel pressure.
5. Disconnect the ground cable from the battery.
6. Remove the fuel pump.
7. Drain fuel from the fuel pump attachment section using pump, etc.

⁂ CAUTION

Be sure to use an anti-gasoline pump.

8 Install the fuel pump.

RELIEVING FUEL SYSTEM PRESSURE

See Figure 195.

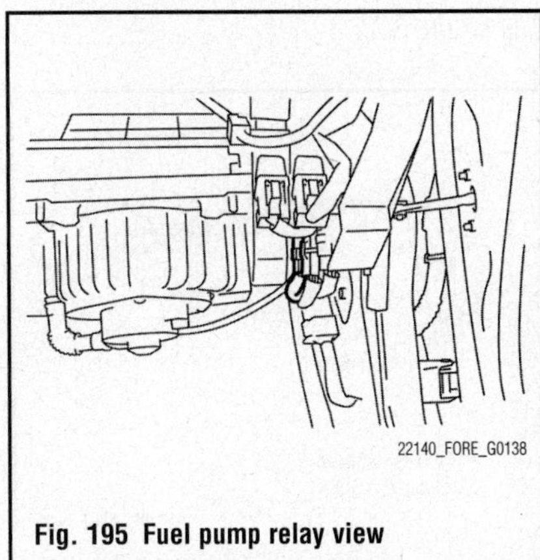

Fig. 195 Fuel pump relay view

1. Before servicing the vehicle, refer to the Precautions Section.

➡This procedure must be performed prior to servicing any component of the fuel injection system.

2. Disconnect the fuel pump connector from the fuel pump relay.
3. Start the engine and let it stall.
4. Crank the engine for 5 seconds or more to ensure the fuel pressure is properly relieved. If the engine starts during this time, allow it to run until it stalls.
5. After performing the required service, connect the fuel pump harness.

FUEL CUT VALVE

REMOVAL & INSTALLATION

See Figure 196.

➡This procedure must be performed prior to servicing any component of the fuel injection system.

1. Before servicing the vehicle, refer to the Precautions Section.
2. Disconnect the fuel pump connector from the fuel pump relay.
3. Properly relieve the fuel system pressure.
4. Remove the fuel tank.
5. Remove the protect cover.
6. Move the clip and disconnect the evaporation hose from the fuel cut valve.

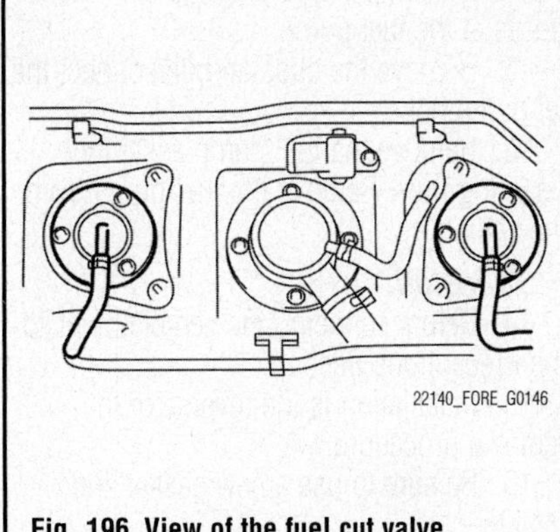

Fig. 196 View of the fuel cut valve

7. Remove the bolts which install the fuel cut valve.

To install:

8. Install in the reverse order of removal while being careful of the following.
9. Make sure the sealing portion is free from fuel or foreign matter before installation.
10. Use a new gasket.
11. Tighten the mounting bolts to 3.3 ft. lbs. (4.4 Nm)

FUEL DAMPER VALVE

REMOVAL & INSTALLATION

See Figure 197.

1. Place "NO FIRE" signs near the working area.
2. Before servicing the vehicle, refer to the Precautions Section.

➡This procedure must be performed prior to servicing any component of the fuel injection system.

3. Disconnect the fuel pump connector from the fuel pump relay.
4. Release the fuel pressure.
5. Remove the fuel damper valve from the fuel delivery line (A) and the fuel return line (B).

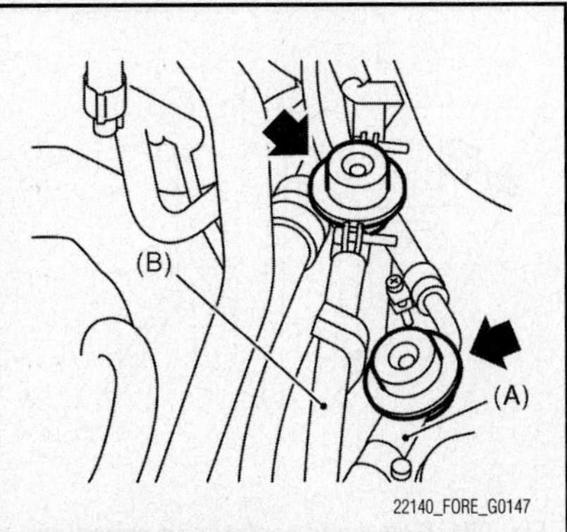

Fig. 197 Fuel damper valve, fuel delivery line (A), and fuel return line (B)

⁂ CAUTION

If fuel hoses or clamps are damaged, replace them with new parts.

To install:

6. Install in the reverse order of removal.
7. Tighten the clamps to 0.94 ft. lbs. (1.25 Nm).

FUEL FILTER

REMOVAL & INSTALLATION

See Figures 198 and 199.

➡The fuel filter is an integral part of the fuel pump assembly.

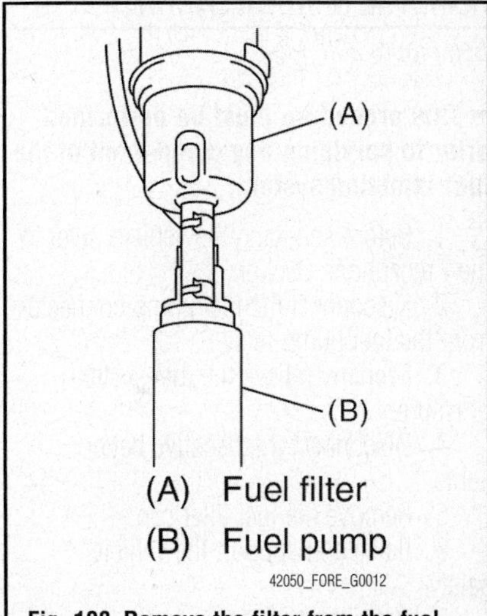

Fig. 198 Remove the filter from the fuel pump

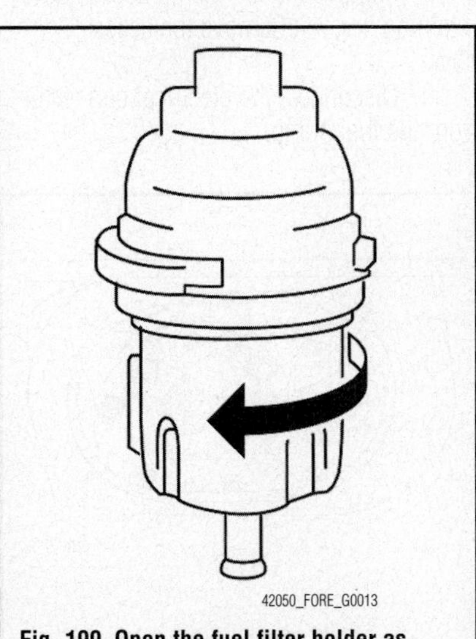

Fig. 199 Open the fuel filter holder as shown

➡This procedure must be performed prior to servicing any component of the fuel injection system.

1. Before servicing the vehicle, refer to the Precautions Section.
2. Disconnect the fuel pump connector from the fuel pump relay.
3. Properly relieve the fuel system pressure.
4. Disconnect the negative battery cable.
5. Remove the fuel pump assembly.
6. Separate the fuel filter from the fuel pump.
7. Turn the filter holder in the direction shown and then remove the filter.
8. Installation is the reverse of the removal procedure.

FUEL PUMP

REMOVAL & INSTALLATION

See Figure 200.

➡This procedure must be performed prior to servicing any component of the fuel injection system.

1. Before servicing the vehicle, refer to the Precautions Section.
2. Disconnect the fuel pump connector from the fuel pump relay.
3. Properly relieve the fuel system pressure.
4. Disconnect the negative battery cable.
5. Remove the fuel filler cap.
6. Raise and support the vehicle safely.
7. Drain the fuel tank.
8. Remove the luggage floor mat.
9. Remove the fuel pump access cover retaining screws. Remove the access cover.
10. Disconnect the electrical connector from the fuel pump.
11. Disconnect and plug the fuel line hoses at the fuel pump.
12. Remove the clips and disconnect the jet pump hose.
13. Remove the fuel pump assembly retaining nuts. Remove the fuel pump from the vehicle.

To install:

14. Before servicing the vehicle, refer to the Precautions Section.
15. Installation is the reverse of the removal procedure.
16. Be sure to use a new gasket and retainer.
17. Tighten the fuel pump retaining nuts to 3.3 ft. lbs. (4.4 Nm). in an alternating sequence pattern.
18. Continue the installation in the reverse order of the removal procedure.
19. Start the engine and check for leaks, correct as required.

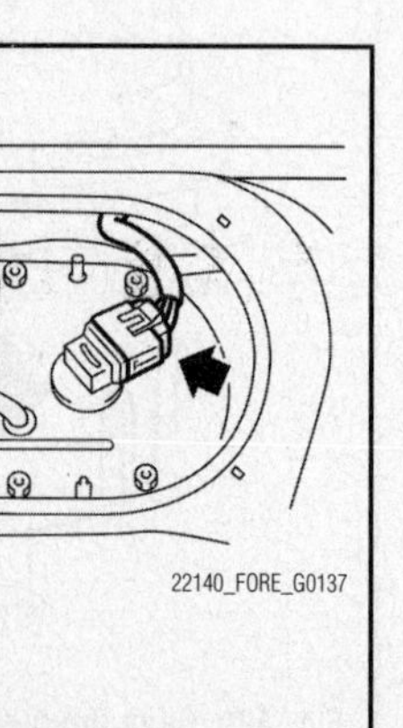

Fig. 200 Fuel pump electrical connector

FUEL PUMP CONTROL UNIT

REMOVAL & INSTALLATION

2.5L DOHC Turbo Engine

See Figure 201.

1. Disconnect the ground cable from the battery.
2. Remove the rear seat.
3. Remove the side sill rear cover.
4. Remove the rear rail trim.
5. Loosen the screws to remove the strut cap.
6. Remove the rear seat belt shoulder anchor.
7. Loosen the screws and clips to remove the rear quarter upper trim.
8. Remove the rear skirt trim.
9. Loosen the screws and clips to remove the rear quarter lower trim.
10. Disconnect the connector from fuel pump control unit.
11. Remove the fuel pump control unit.

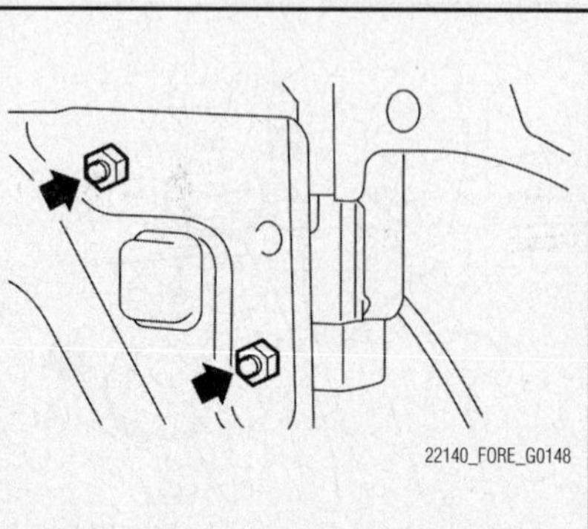

Fig. 201 Fuel pump control unit mounting bolts

To install:

12. Install in the reverse order of removal.
13. Tighten the fuel pump control unit mounting nuts to 3.7 ft. lbs. (5 Nm).

FUEL PRESSURE REGULATOR

REMOVAL & INSTALLATION

2.5L DOHC Turbo Engine

See Figure 202.

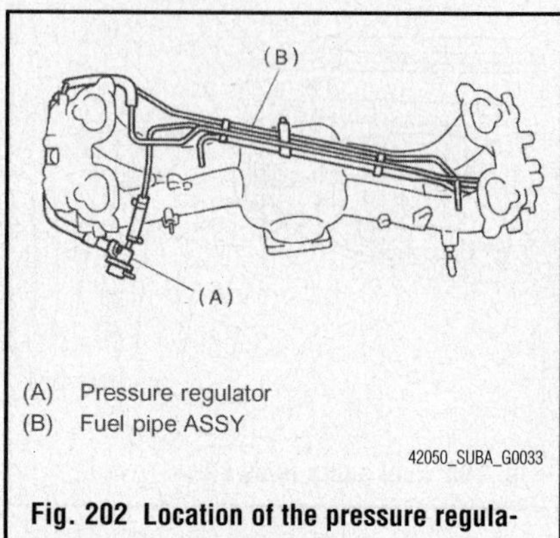

Fig. 202 Location of the pressure regulator on the intake manifold

➡This procedure must be performed prior to servicing any component of the fuel injection system.

1. Before servicing the vehicle, refer to the Precautions Section.
2. Disconnect the fuel pump connector from the fuel pump relay.
3. Properly relieve the fuel system pressure.
4. Remove the intake manifold. Refer to "Intake Manifold, Removal & Installation" to remove the intake manifold.
5. Remove the fuel pressure regulator and fuel lines from the intake manifold.

To install:

6. Installation is the reverse order of removal.
7. Tighten the regulator mounting bolt to 4.7ft. lbs. (6.4 Nm).

FUEL TANK

REMOVAL & INSTALLATION

See Figures 203 through 205.

⁂ CAUTION

Place "NO FIRE" signs near the working area. Be careful not to spill fuel.

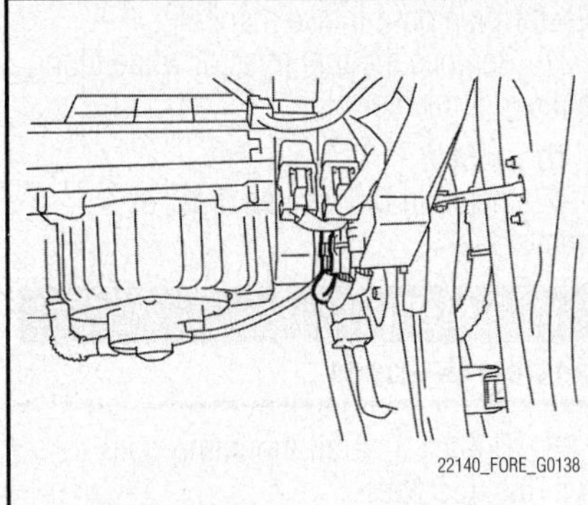

Fig. 203 Fuel pump relay view next to blower motor

➡This procedure must be performed prior to servicing any component of the fuel injection system.

1. Before servicing the vehicle, refer to the Precautions Section.
2. Disconnect the fuel pump connector from the fuel pump relay.
3. Properly relieve the fuel system pressure.
4. Remove the fuel cap.
5. Set the vehicle on a lift.
6. Drain fuel from fuel tank.
7. Remove the rear seat.
8. Disconnect the connector (A) between the fuel tank cord and the rear harness.
9. Push the grommet (B) which holds the fuel tank cord on the floor panel into under the body.
10. Remove the rear crossmember.
11. Remove the canisters.
12. Disconnect the connector from the pressure control solenoid valve.
13. Loosen the clamp and disconnect the fuel filler hose (A) and evaporation hose (B) from the fuel filler pipe.
14. Move the retainer for fuel delivery tube, and disconnect the quick connector.
15. Support the fuel tank with transmission jack, remove the bolts from bands and dismount the fuel tank from the vehicle.

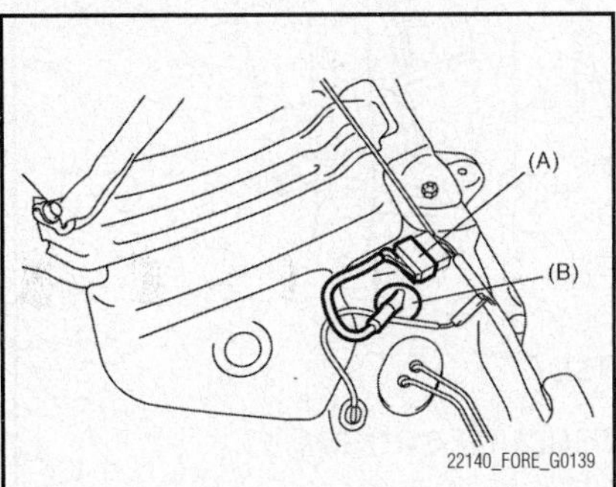

Fig. 204 Connector (A), and grommet (B) shown

CAUTION

A helper is required to perform this work. Fuel may remain in the fuel tank. Be careful not to lose balance and drop the fuel tank when removing it, as this remaining fuel may offset the weight balance of the tank.

To install:

16. Support the fuel tank with transmission jack and push the fuel tank harness into access hole along with the grommet.
17. Set the fuel tank and temporarily tighten the bolts of fuel tank bands.
18. Insert the fuel filler hose (A) approx. 35 to 40 mm (1.38 to 1.57 in) over the lower end of fuel filler pipe and tighten the clamp.
19. Insert the evaporation hose (B) into the lower end of evaporation pipe, and secure the clamps and clips.
20. Connect the fuel hoses, and hold them with clips and quick connector.

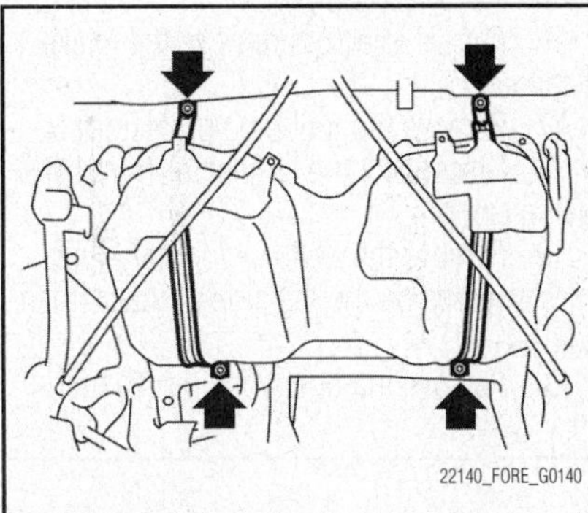

Fig. 205 Mounting bolts, and fuel tank bands

21. Connect the connector to pressure control solenoid valve.
22. Install the canister.
23. Tighten the band mounting bolts to 24.3 ft. lbs. (33 Nm).
24. Install the rear crossmember.
25. Connect the connector (A) to the fuel tank cord and plug the service hole with grommet (B).
26. Set the rear seat and floor mats.
27. Connect the connector to fuel pump relay.

FUEL RAIL & INJECTORS

REMOVAL & INSTALLATION

2.5L SOHC Non-Turbo Engine

See Figure 206.

CAUTION

Place "NO FIRE" signs near the working area. Be careful not to spill fuel.

➡This procedure must be performed prior to servicing any component of the fuel injection system.

1. Before servicing the vehicle, refer to the Precautions Section.
2. Disconnect the fuel pump connector from the fuel pump relay.
3. Properly relieve the fuel system pressure.
4. Remove the fuel cap.
5. If removing the right side injectors, remove the air intake duct and air cleaner case.
6. Remove the number one and three spark plug wires.
7. If removing the right side injectors, remove the front side V-belt, remove the belt covers. Loosen the lock bolt. Loosen the slider bolt.
8. Remove the front side belt.
9. Remove the bolts that retain the power steering hoses to the intake manifold protector. Do not disconnect the power steering hoses.
10. Remove the bolts which retain the power steering pump to the bracket.
11. Disconnect the power steering pump switch connector.
12. Remove the reservoir tank from the

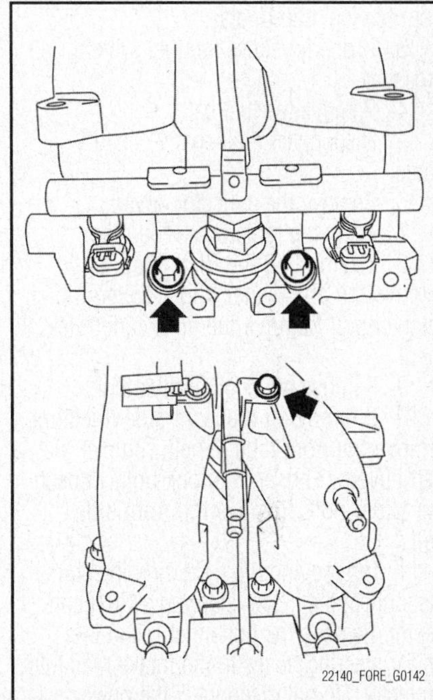

Fig. 206 Bolts that hold the fuel injector line

bracket, by pulling upward. Place the power steering pump and tank assembly to the side.

13. If removing the left side injectors, remove the battery. Remove the number two and four spark plug wires.

14. Remove the fuel line protector cover(s).

15. Disconnect the electrical connector from the fuel injector.

16. Remove the bolts that hold the fuel injector line to the intake manifold.

17. Remove the fuel injector retaining clip. Remove the fuel injector while lifting up the fuel injector line.

To install:

18. Installation is the reverse of the removal procedure.

19. Be sure to use new O-rings.

20. Start the engine and check for leaks, correct as required.

2.5L SOHC Turbo Engine

See Figure 207.

✲✲ CAUTION

Place "NO FIRE" signs near the working area. Be careful not to spill fuel.

➡This procedure must be performed prior to servicing any component of the fuel injection system.

1. Before servicing the vehicle, refer to the Precautions Section.
2. Disconnect the fuel pump connector from the fuel pump relay.
3. Properly relieve the fuel system pressure.
4. Remove the fuel cap.
5. Disconnect the negative battery cable.
6. Remove the collector cover.
7. Drain the radiator, as required.
8. If removing the RH injectors, remove the air cleaner upper cover and air intake boot. Remove the air cleaner element.
9. Remove the coolant filler tank.
10. If removing the right side injectors, remove the front side V-belt, remove the belt covers. Loosen the lock bolt. Loosen the slider bolt. Remove the front side belt.
11. If removing the right side injectors, disconnect the power steering switch connector. Remove the bolts that retain the power steering hoses to the intake manifold protector. Do not disconnect the power steering hoses. Remove the bolts which retain the power steering pump to the

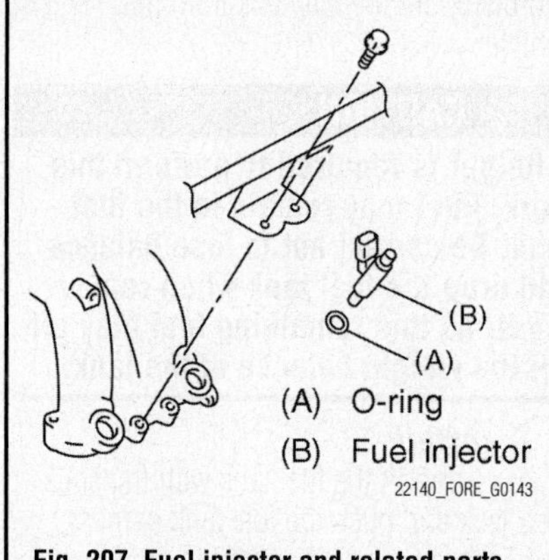

Fig. 207 Fuel injector and related parts

bracket. Remove the reservoir tank from the bracket, by pulling upward. Place the power steering pump and tank assembly to the right side.

12. Remove the fuel pipe protector RH.
13. Disconnect the connector from fuel injector.
14. Remove the bolt which holds fuel injector pipe onto intake manifold.
15. Remove the fuel injector while lifting up the fuel injector pipe.
16. For LH injectors remove the intake manifold.
17. Remove the fuel pipe protector LH.
18. Disconnect the connector from LH fuel injector.
19. Remove the bolts which hold fuel injector pipes on the left side of intake manifold.
20. Remove the bolt which holds fuel injector pipe onto intake manifold.
21. Remove the fuel injector while lifting up the fuel injector pipe.

To install:

22. Install in the reverse order of removal.

✲✲ CAUTION

Use new O—rings.

23. Tighten fuel rail mounting bolts to 14 ft. lbs. (19 Nm).
24. Tighten the fuel pipe protectors to 14 ft. lbs. (19 Nm).
25. Start the engine and check for leaks, correct as required.

IDLE SPEED

ADJUSTMENT

Idle speed is maintained by the Electronic Control Module (ECM). No adjustment is necessary or possible.

THROTTLE BODY

REMOVAL & INSTALLATION

2.5L SOHC Non-Turbo Engine

See Figure 208.

1. Before servicing the vehicle, refer to the Precautions Section.
2. Disconnect the ground cable from battery.
3. Remove the air intake chamber.

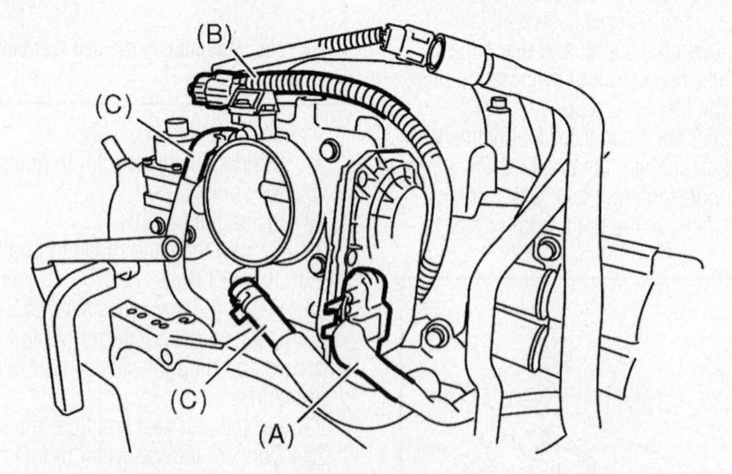

Fig. 208 2.5L SOHC Non-Turbo Engine throttle body

4. Disconnect the connectors from throttle position sensor and manifold absolute pressure sensor.

5. Disconnect the engine coolant hoses from throttle body.

6. Remove the bolt which installs throttle body on intake manifold.

To install:

7. Install the throttle body with a new gasket.

8. Tighten throttle body mounting bolts to ft. lbs. (8 Nm).

9. The remainder of the installation is the reverse order of removal.

2.5L DOHC Turbo Engine

See Figure 209.

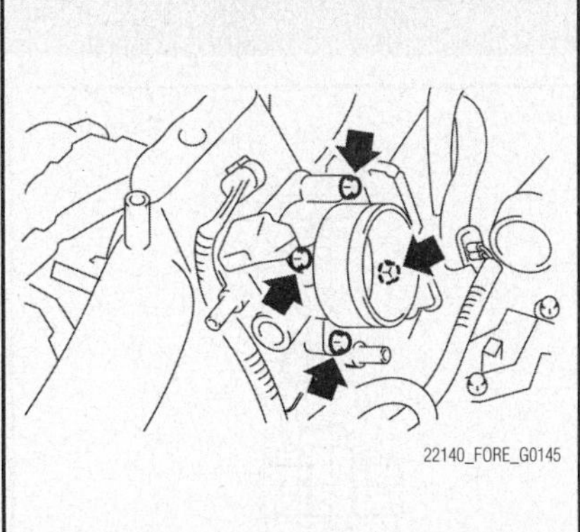

Fig. 209 Throttle body and mounting bolts

1. Disconnect the negative battery cable.

2. Remove the intercooler.

3. Disconnect the throttle position sensor and manifold absolute pressure (MAP) sensor.

4. Disconnect the coolant hoses from the throttle body.

5. Remove the throttle body mounting bolts.

6. Remove the throttle body and gasket.

To install:

7. Install the throttle body with a new gasket.

8. Tighten the mounting bolts to 6 ft. lbs. (8 Nm).

9. The remainder of the installation is the reverse order of removal.

HEATING & AIR CONDITIONING SYSTEM

BLOWER MOTOR

REMOVAL & INSTALLATION

See Figures 210 and 211.

1. Before servicing the vehicle, refer to the Precautions Section.

2. Disconnect the negative battery cable.

3. Remove the glove box.

4. Loosen the nut to remove the support beam stay.

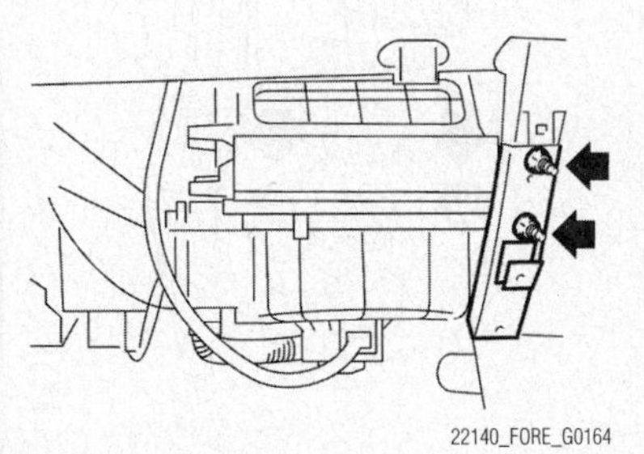

5. Disconnect the blower motor wiring harness.

6. Disconnect the blower resistor connector.

7. Loosen the bolt and nut to remove the blower motor unit assembly.

8. Installation is the reverse order of removal.

Fig. 210 Support beam stay and mounting nuts

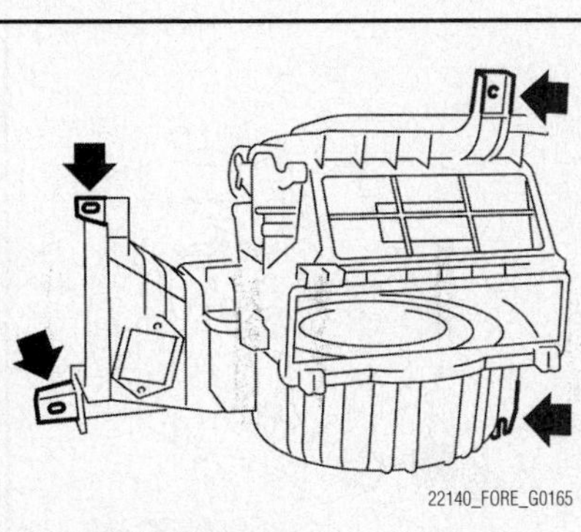

Fig. 211 Blower motor assembly and mounting locations

HEATER CORE

REMOVAL & INSTALLATION

See Figures 212 and 213.

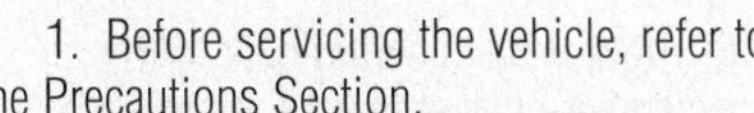

1. Before servicing the vehicle, refer to the Precautions Section.

2. Disconnect the ground cable from the battery.

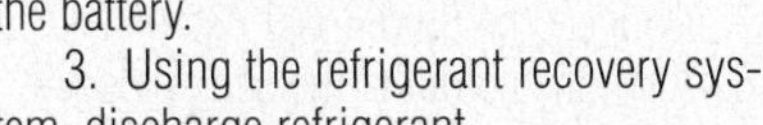

3. Using the refrigerant recovery system, discharge refrigerant.

4. Drain coolant from the radiator.

5. Remove the bolts securing expansion valve and pipe in engine compartment. Release the heater hose clamps in engine compartment to disconnect the hoses.

6. Remove the instrument panel.

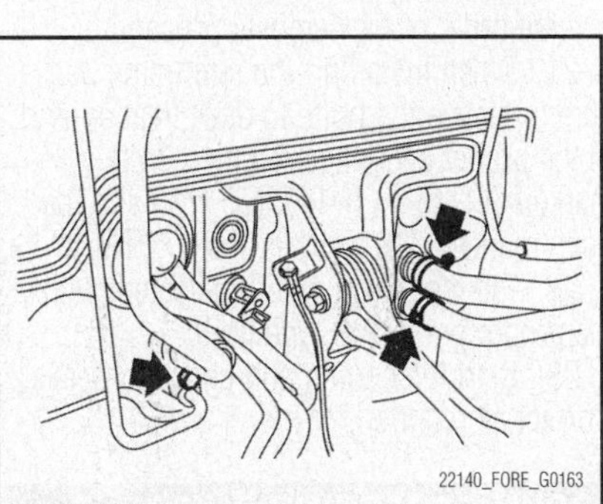

Fig. 212 Remove expansion valve, pipe and disconnect the heater hoses

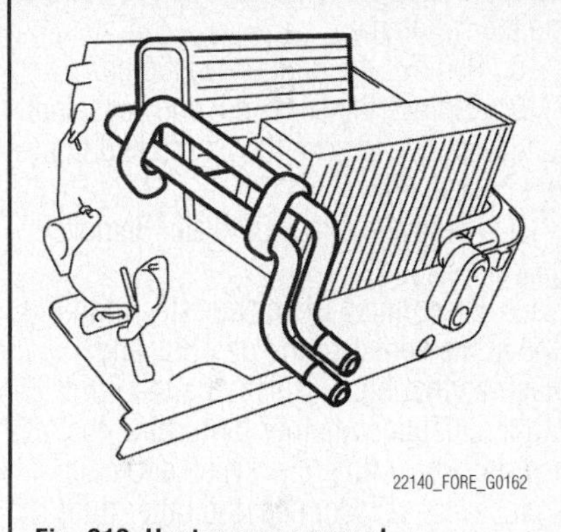

Fig. 213 Heater core removal

7. Remove the support beam.

8. Remove the blower motor unit assembly.

9. Disconnect the servo motor connector.

10. Loosen the bolt and nuts and remove the heater and cooling unit.

11. Open the heater core pipe cover.

12. Loosen the screws to remove the mode actuator.

13. Loosen the screws to remove the foot duct.

14. Loosen the screws to remove evaporator cover.

15. Loosen the screws to remove lower case.

16. Remove the heater core.

To install:

17. Install in the reverse order of removal.

18. Tighten mounting screw to 5.5 ft. lbs. (7.5 Nm).

19. Vacuum and recharge A/C system.

20. Fill and bleed cooling system.

21. Check for leaks

STEERING

POWER RACK & PINION STEERING GEAR

REMOVAL & INSTALLATION

See Figures 214 and 215.

1. Before servicing the vehicle, refer to the Precautions Section.
2. Disconnect the negative battery cable.
3. Loosen the front wheel nut.
4. Raise and support the vehicle safely.
5. Remove the tires and wheels.
6. Remove the undercover.
7. Remove the sub frame. Leave bolt (1) connected by a few threads and remove the bolts in the sequence illustrated. Once the other bolts are removed, remove bolt (1) and the sub frame.
8. Remove the front exhaust pipe.
9. Remove the cotter pin and castle nut. Using a puller, remove the tie rod end from the knuckle arm.
10. Remove the jack up plate. Remove the front stabilizer.
11. Disconnect the power steering fluid pipe at the center of the gearbox and attach a vinyl hose. Discharge the fluid into a suitable container by turning the steering wheel fully clockwise and counterclockwise. Disconnect the other fluid line, and repeat the discharge procedure.
12. Remove the steering wheel. Make a matchmark on the universal joint. Remove the universal joint bolts and remove the joint from the vehicle.
13. Disconnect the fluid lines from the steering gear, pressure hose first.
14. Remove the steering gear retaining bolts and clamps securing the steering gear to the crossmember. Remove the steering gear from the vehicle.

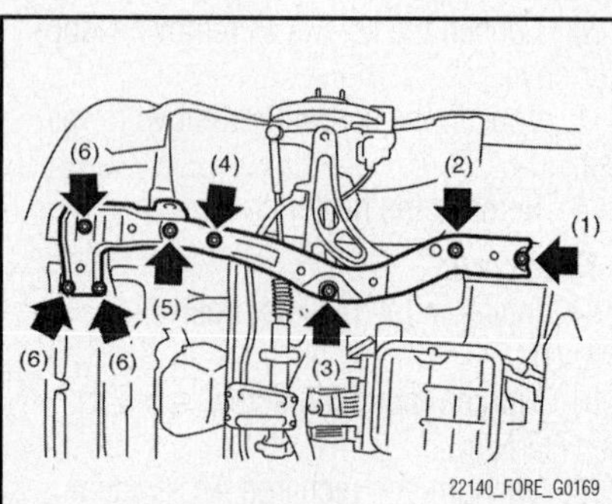

Fig. 214 Sub frame and mounting bolt view

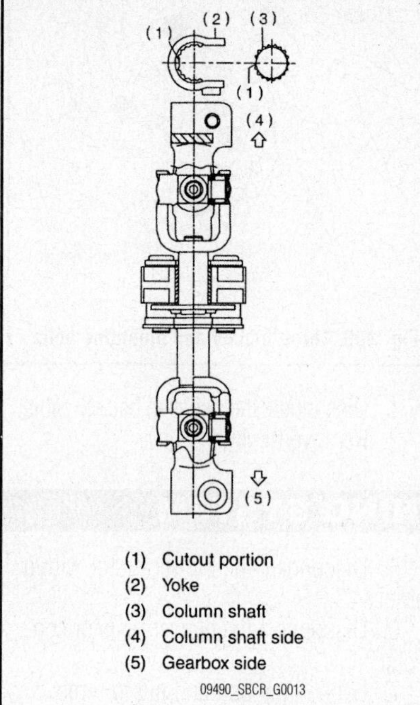

Fig. 215 Steering column joint alignment

To install:

15. Insert the steering gear into the crossmember.

➡**Be careful not to damage the gearbox boot.**

16. Tighten the steering gear to the crossmember bracket to 44.1 ft. lbs. (55 Nm).
17. Connect the fluid lines.
18. Align the cutout at the serrated section of the column shaft and yoke. Insert the universal joint into the column shaft.
19. Align the mating marks and insert the universal joint to serrated section of the steering gear assembly. Tighten the bolt to 17.4 ft. lbs.
20. Continue the installation in the reverse order of the removal procedure.
21. When installing the sub frame be sure to torque the bolts to specification and in the proper sequence. Tighten bolts marked (T1) to 41 ft. lbs. (55 Nm) and the bolts marked (T2) to 52 ft. lbs. (71 Nm).
22. Fill the power steering system with the proper grade and type fluid.
23. Start the engine and check for leaks. Correct as required.

POWER STEERING PUMP

REMOVAL & INSTALLATION

See Figures 216 through 219.

1. Before servicing the vehicle, refer to the Precautions Section.
2. Disconnect the negative battery cable.
3. Drain about ⅓ of the power steering fluid from the power steering reservoir.
4. Remove the accessory drive belt cover.
5. Loosen the lock bolt and slider bolt, and remove the power steering pump belt.
6. Disconnect the electrical connector form the power steering pump switch.

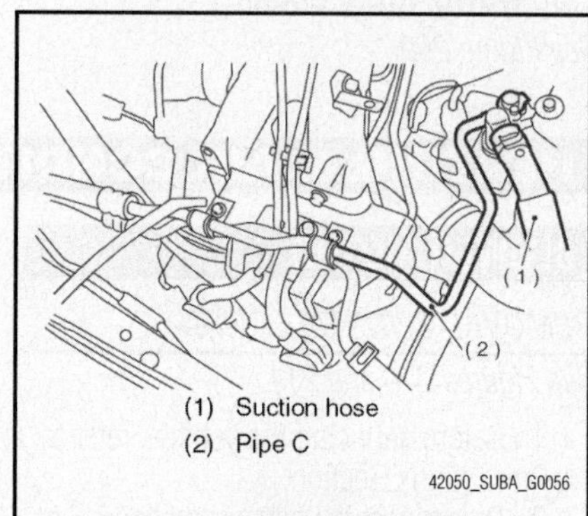

Fig. 216 Pressure (Pipe 'C') and return (suction) hose component locations

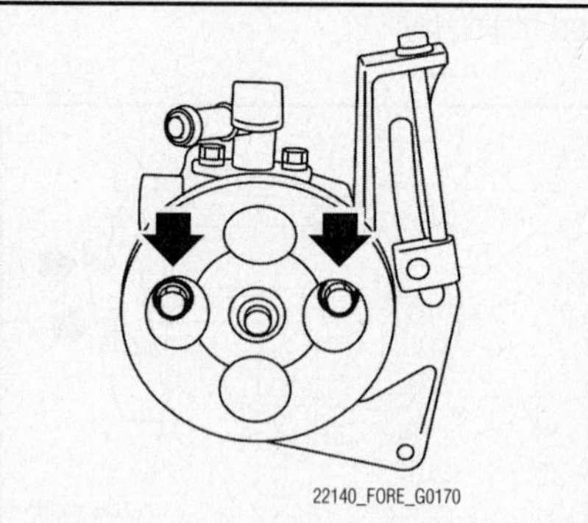

Fig. 217 Front power steering mounting bolts

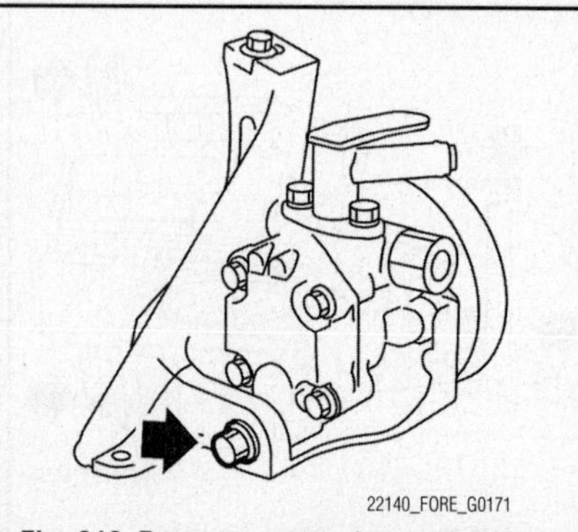

Fig. 218 Rear power steering mounting bolt

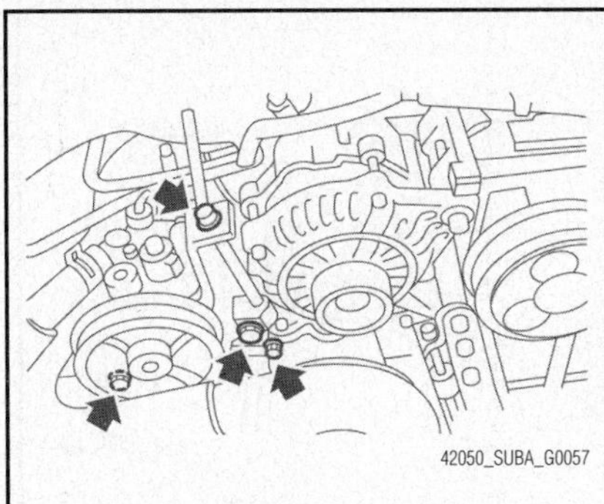

Fig. 219 Power steering pump mounting bolt locations—2.5L engines

7. Disconnect pressure and return hoses from the power steering pump.
8. Remove power steering pump bracket mounting bolts.
9. Place the power steering pump in a vise and remove the bolts from the front side of the pump.
10. Remove the socket from the pump.
11. Remove the bolt from the rear side of the power steering pump to remove it completely from the mounting bracket.

To install:

12. Install the mounting bracket to the power steering pump. Tighten the rear bolt to 28 ft. lbs. (37 Nm).
13. Install the socket to the pump and tighten to 5 ft. lbs. (7 Nm).
14. Install the bolts on the front side of the pump. Tighten to 12 ft. lbs. (16 Nm).
15. Install the power steering pump bracket assembly into the vehicle. Tighten the mounting bolts to 16 ft. lbs. (22 Nm).
16. Connect the pipes that were removed. Tighten the joint nuts to 29 ft. lbs. (39 Nm).
17. Connect the electrical connector for the power steering oil pressure switch.
18. The remainder of the installation is the reverse order of removal.
19. Bleed the power steering system.
20. Start the engine and check for leaks.

BLEEDING

1. Before servicing the vehicle, refer to the Precautions Section.
2. Fill the power steering fluid reservoir about half way with the specified fluid.
3. Continue to turn the steering wheel slowly from lock to lock until bubbles stop appearing on oil surface while keeping the fluid at that level.
4. If turning the steering wheel in low fluid level condition, air will be sucked in pipe. In this case, leave it about half an hour and then repeat the previous step.
5. Lift up the vehicle, start the engine and let it idle.
6. Continue to turn the steering wheel slowly from lock to lock again until bubbles stop appearing on oil surface while keeping the fluid at that level. It is normal that bubbles stop appearing after three times turning of steering wheel from lock to lock.
7. In case the bubbles do not stop appearing in the tank, leave it about half an hour and then begin the process again.
8. Lower the vehicle, and then idle the engine.
9. Continue to turn the steering wheel from lock to lock until bubbles stop appearing and change of the fluid level is within 3 mm (0.12 in).
10. In case the following happens, leave it about half an hour and then do step 5–8 again.
    a. The fluid level changes over 3 mm (0.12 in).
    b. Bubbles remain on the upper surface of the fluid.
    c. Grinding noise is generated from oil pump.
11. Check the fluid leakage after turning steering wheel from lock to lock with engine running.

STEERING LINKAGE

REMOVAL & INSTALLATION

Tie-Rods

See Figures 220 through 229.

1. Before servicing the vehicle, refer to the Precautions Section.
2. Remove the rack and pinion gearbox and secure it in a vice using Special Tool 926200000.
3. Remove tie-rod end and lock nut from gearbox.
4. Remove the tie-rod end plate, if equipped

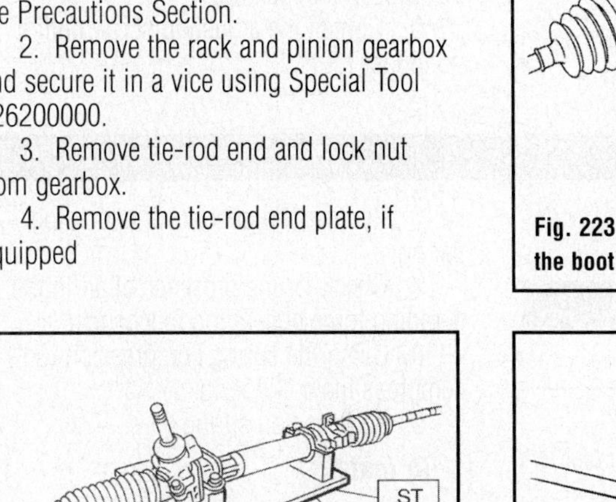
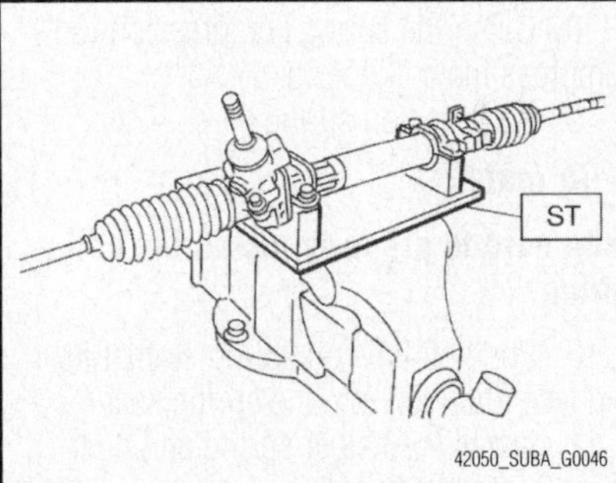

Fig. 220 Mount the steering rack in a stand 92623000 and mount the stand in a vise.

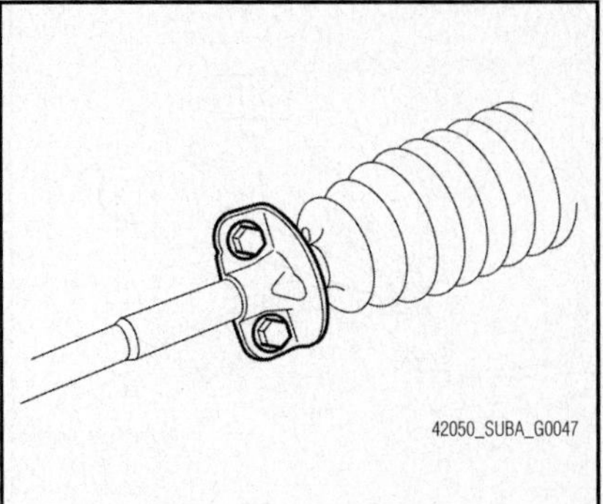

Fig. 221 Remove the tie-rod end plate if equipped

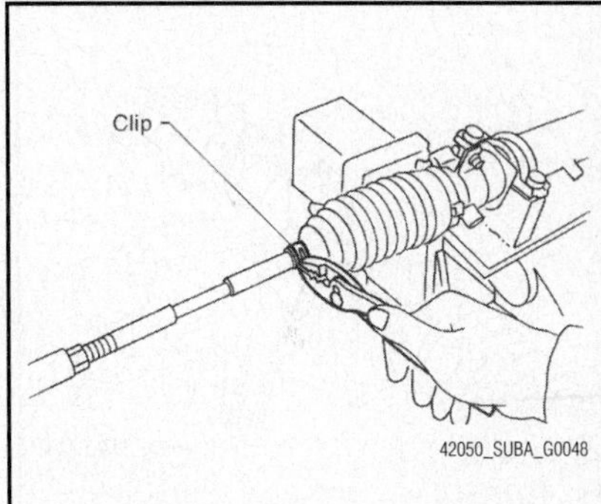

Fig. 222 Remove the small clip from the boot.

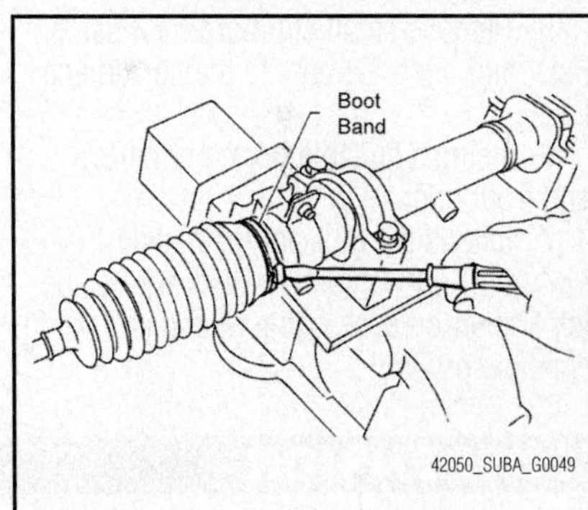

Fig. 223 Remove the boot band to remove the boot.

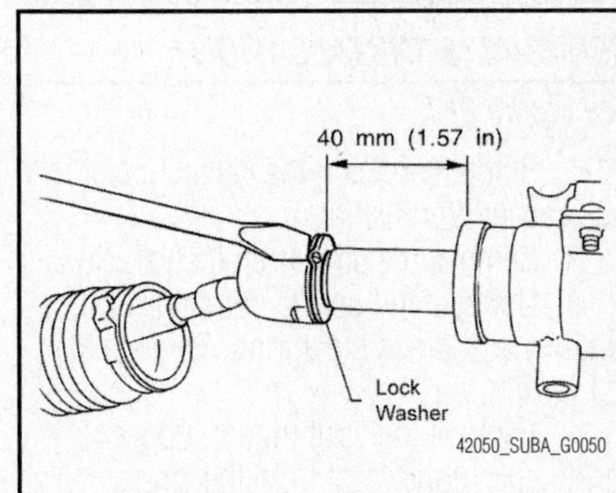

Fig. 224 Use a pry tool to remove the lock washer.

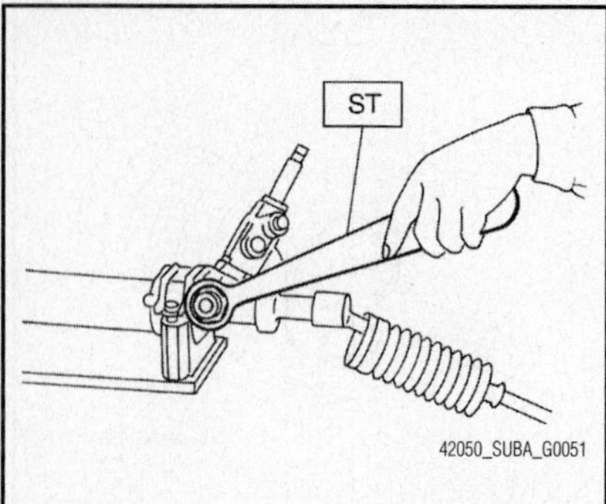

Fig. 225 Use a spanner wrench to loosen the lock nut.

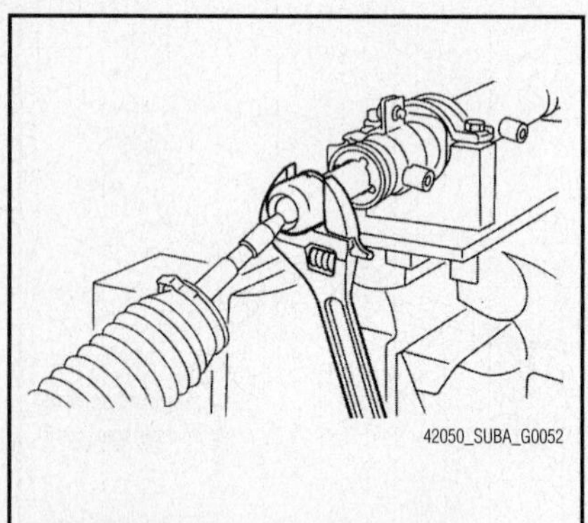

Fig. 226 Use an adjustable wrench to remove the tie-rod.

5. Remove small clip from boot using pliers, and slide the boot to the tie-rod end side.
6. Using a suitable pry tool, remove band from boot.
7. Extend the rack approximately 1.57 in. (40mm) out. Unlock the lock wire at the lock washer on each of the tie rod end using a suitable pry tool.

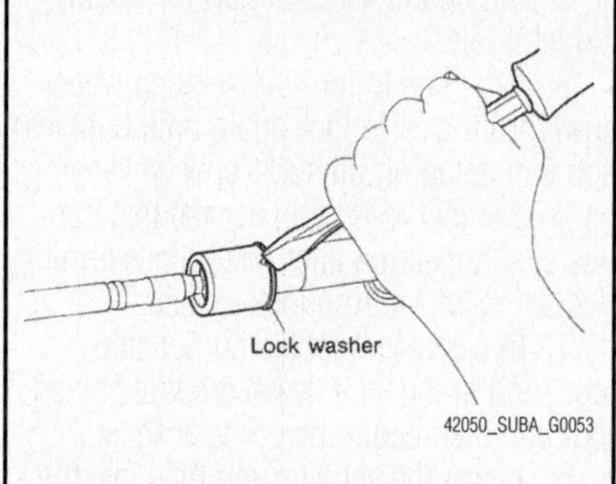

Fig. 227 Stake the lock washer with a hammer and chisel.

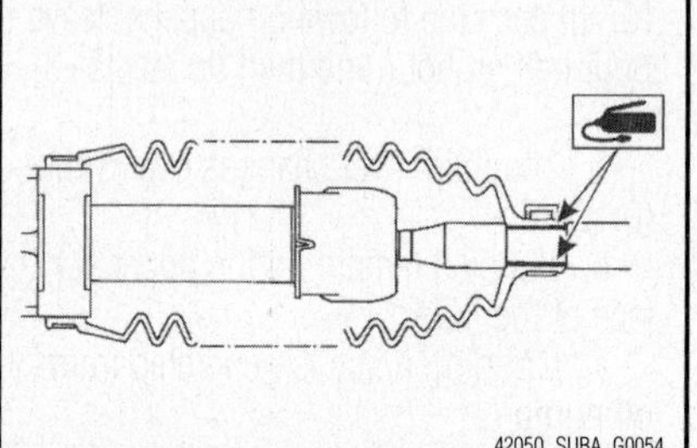

Fig. 228 Apply grease to the tie-rod groove as shown, then install the boot so it sits securely in the grooves.

CAUTION

Be careful not to scratch rack surface as oil leaks may result.

8. Using Special Tool 92623000 Spanner, loosen the lock nut.
9. Tighten the adjusting screw until it no longer turns.

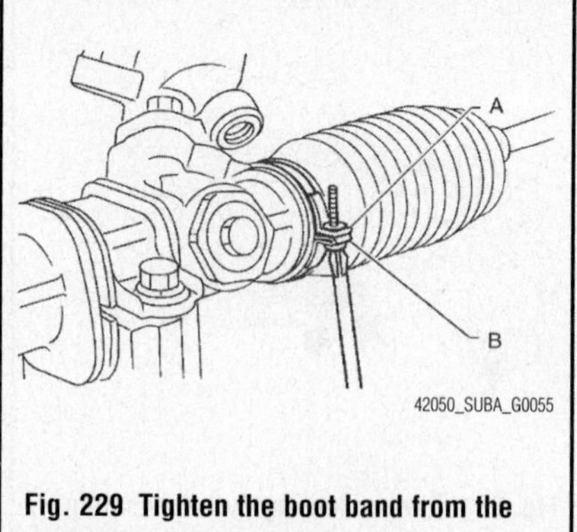

Fig. 229 Tighten the boot band from the underside until 'A' and 'B' meet.

10. Using an adjustable wrench, remove the tie rod.

To install:

11. Install the lock washer and tighten the tie-rods into the rack ends.
12. Stake the lock washer with a chisel.
13. Install the boot onto the housing. Ensure the boot sits securely in the groove on the gearbox and rod.

➡Before installing the boot, apply grease to the groove of the tie rod.

14. Using a screwdriver, tighten the screw of the boot band until the ends 'A' and 'B' of the band come into contact with each other.

➡Always tighten the band from the underside of the steering gear box.

15. Install the small clip on the boot end.
16. Install the lock plate, if equipped.
17. Screw in the lock nut and tie-rod end to the screwed portion of the tie-rod.
18. Install the rack and pinion gearbox into the vehicle.

SUSPENSION

FRONT SUSPENSION

See Figure 230.

COIL SPRING

REMOVAL & INSTALLATION

See Figure 231.

1. Before servicing the vehicle, refer to the Precautions Section.
2. Remove the strut from the vehicle.
3. Using a coil spring compressor tool, carefully compress the spring. Remove the self locking nut.
4. Remove the strut mount, upper spring and rubber seat from the strut.
5. Gradually decrease the compression force of the spring compressor tool. Remove the coil spring.
6. Remove the dust cover and helper spring.
7. Check for the presence of air in the damping force generating mechanism.
8. Using the spring compression tool, compress the coil spring.
9. Remove coil spring.

To install:

➡Be sure to properly install the coil spring.

10. Position the coil spring so that its end face fits good into the spring seat.
11. Install the helper spring and dust cover to the piston rod.
12. Pull the piston rod fully upward, and install the rubber seat and spring seat.
13. Install the strut mount to the piston rod, and then tighten the self locking nut, temporarily. Be sure to use a new self locking nut.
14. Use a hexagon wrench to prevent the strut rod from turning.
15. Tighten the self locking nut to 41 ft. lbs. (55 Nm).
16. Carefully loosen the coil spring.

FRONT CROSSMEMBER

REMOVAL & INSTALLATION

See Figure 232.

1. Before servicing the vehicle, refer to the Precautions Section.
2. Lift up the vehicle, and remove the front wheels.

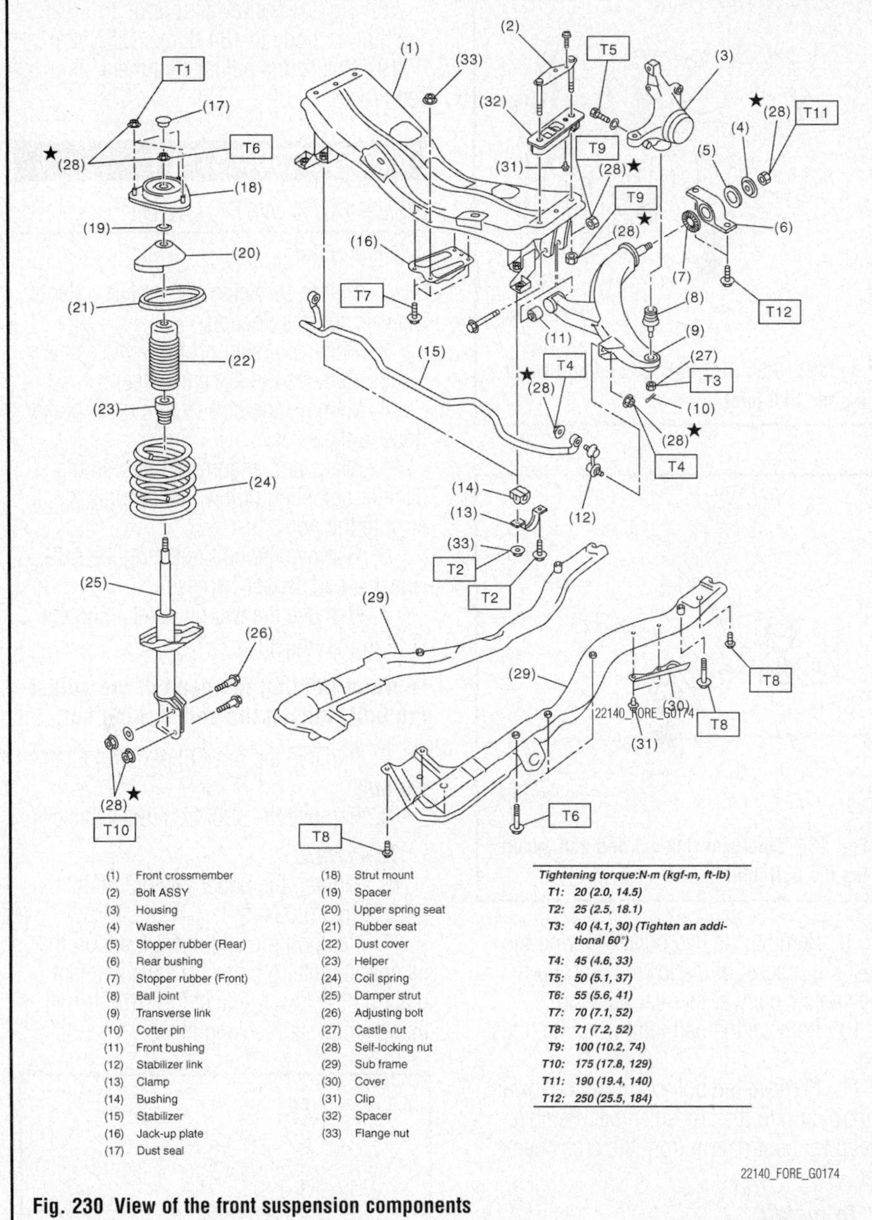

Fig. 230 View of the front suspension components

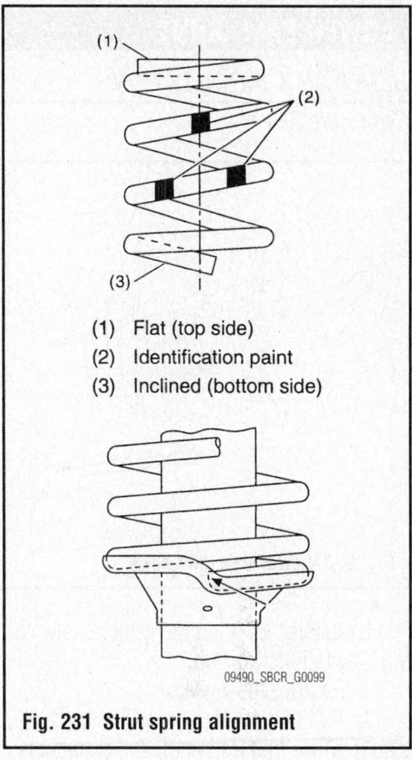

Fig. 231 Strut spring alignment

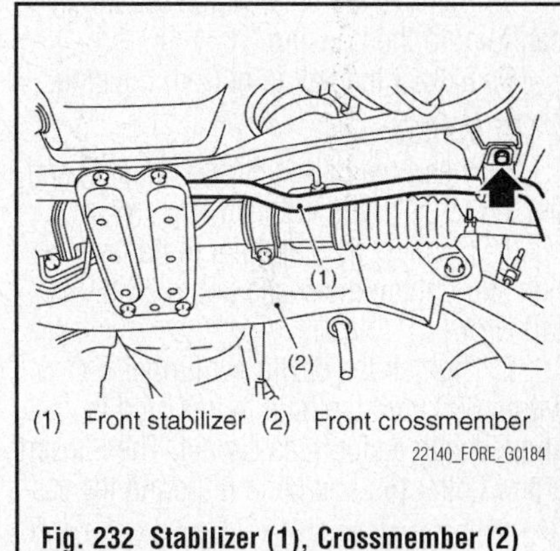

Fig. 232 Stabilizer (1), Crossmember (2) and jack-up plate

3. Remove the sub frame.
4. Remove the stabilizer and jack-up plate.
5. Disconnect the tie-rod end from the housing.
6. Remove the front exhaust pipe.
7. Remove the front transverse link from the front crossmember and body.
8. Remove the bolts attaching the engine mount cushion rubber to crossmember.
9. Remove the steering universal joint.
10. Disconnect the power steering pipe from steering gearbox.
11. Lift the engine approx. 10 mm (0.39 in) using a chain block.
12. Support the crossmember with a jack, remove the nuts securing the crossmember to body and lower the crossmember gradually along with the steering gearbox.

CAUTION

When pulling the crossmember downward to remove, be careful that the tie-rod end does not interfere with SFJ boot.

To install:

13. Install in the reverse order of removal.

➡Always tighten bushings with wheels in full contact with the ground and the vehicle at curb weight.

14. Tightening specification are as follows:

- Transverse link bushing to crossmember: 74 ft. lbs. (100 Nm).
- Stabilizer to bushing: 18 ft. lbs. (25 Nm).
- Tie-rod end to housing: 20 ft. lbs. (27 Nm).
- Front cushion rubber to crossmember: 63 ft. lbs. (85 Nm)
- Universal joint to pinion shaft: 17.4 ft. lbs. (24 Nm).
- Crossmember to body: 74 ft. lbs. (100 Nm).

15. Purge air from the power steering system.
16. Inspect the wheel alignment and adjust if necessary.

LOWER BALL JOINT

REMOVAL & INSTALLATION

See Figure 233.

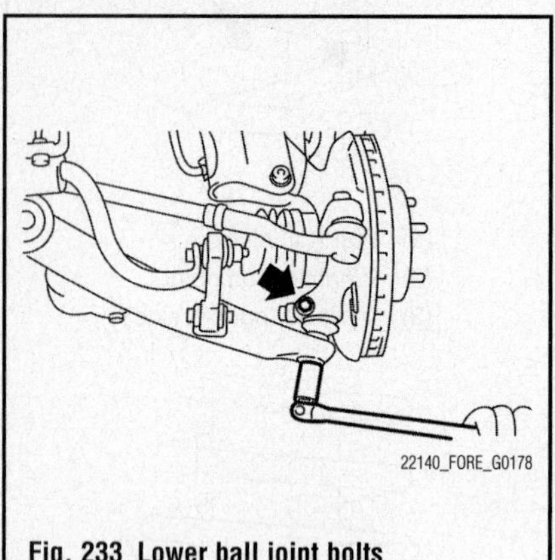

Fig. 233 Lower ball joint bolts

1. Before servicing the vehicle, refer to the Precautions Section.
2. Remove the wheels.
3. Pull out the cotter pin from the ball stud, remove the castle nut, and extract the ball stud from the transverse link.
4. Remove the bolts which secure the ball joint to the housing.
5. Extract the ball joint from housing.

To install:

6. Insert the ball joint into housing and tighten to 37 ft. lbs. (50 Nm).
7. Connect the ball joint to transverse link and tighten the castle nut to 30 ft. lbs. (40 Nm).
8. Tighten the castle nut further but within 60° until the hole in ball stud is aligned with a slot in castle nut. Then, insert a new cotter pin and bend it around the castle nut
9. Install the front wheels.

LOWER CONTROL ARM

REMOVAL & INSTALLATION

See Figures 234 and 235.

1. Before servicing the vehicle, refer to the Precautions Section.
2. Disconnect the negative battery cable.
3. Raise and support the vehicle safely.
4. Remove the tire and wheel.
5. Remove the sub frame.
6. Disconnect the stabilizer link from the lower control arm.
7. Remove the bolt securing the ball joint of the lower control arm.
8. Remove the nut (do not remove the bolt) securing the lower control arm to the crossmember.

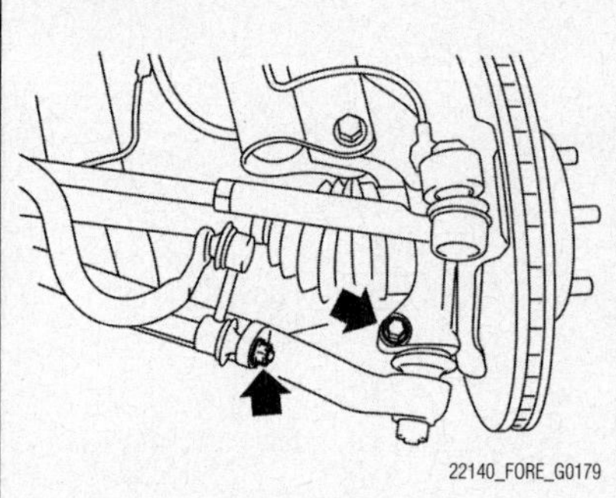

Fig. 234 Stabilizer link nut and bolt securing the ball joint

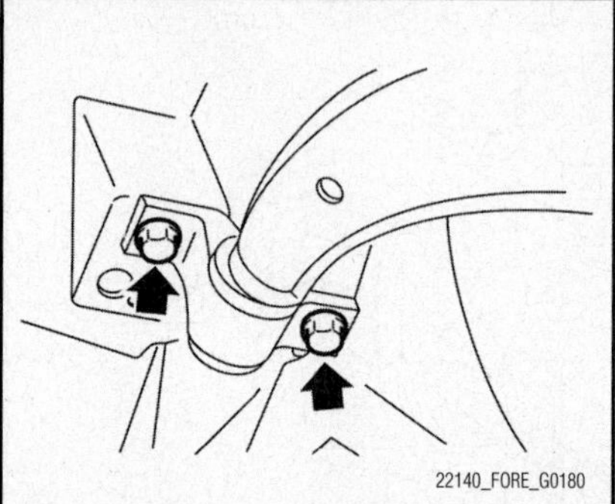

Fig. 235 Stabilizer link nut and bolt securing the ball joint

9. Remove the two bolts securing the bushing bracket of the lower control arm to the vehicle body at the rear bushing.
10. Remove the ball joint from the housing.
11. Remove the bolt securing the lower control arm to the crossmember. Remove the lower control arm from the crossmember.

To install:

12. Install the lower control arm to its mounting.
13. Temporarily tighten the two bolts used to secure the rear bushing of the lower control arm

➡These bolts should be tightened so that they can still move back and forth in the oblong shaped hole in the bracket, which holds the bushing.

14. Continue the installation in the reverse order of the removal procedure.
15. Always fully tighten the rubber bushings when the wheels are in full contact with the ground and the vehicle is at curb height.
16. Tighten stabilizer link nut to 33 ft. lbs. (45 Nm).
17. Tighten lower control arm rear bushing bracket to 92.3 ft. lbs. (125 Nm).
18. Tighten the lower control arm rear bushing to body to 184 ft. lbs. (250 Nm)
19. Check and adjust alignment, as required.

MACPHERSON STRUT

REMOVAL & INSTALLATION

See Figure 236.

1. Before servicing the vehicle, refer to the Precautions Section.
2. Raise and support the vehicle safely.
3. Remove the tire and wheel.
4. Remove the bolt retaining the brake hose to the strut.
5. Make and alignment mark on the camber adjusting bolt which secures the strut to the housing.
6. Remove the bolt retaining the ABS wheel speed sensor harness.
7. Remove the two bolts retaining the strut to the housing.

➡While holding the head of the adjusting bolt, loosen the self locking nut.

8. Remove the three upper strut retaining nuts.
9. Remove the strut from the vehicle.

To install:

10. Tighten the upper retaining nuts to 14.8 ft. lbs. (20 Nm).
11. Position the alignment mark on the camber adjusting bolt with the alignment mark on the lower side of the strut. Install using a new self locking nut.

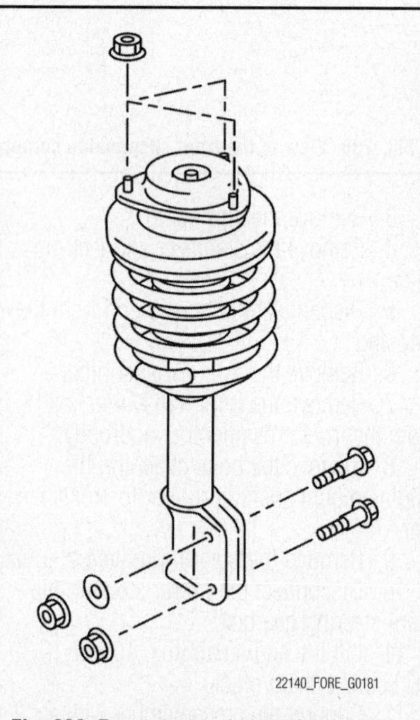

Fig. 236 Front strut and mounting hardware

➡**While holding the head of the adjusting bolt, tighten the self locking nut to 129 ft. lbs. (175 Nm).**

12. Secure the ABS wheel speed sensor harness to the strut and tighten to 24.3 ft. lbs. (33 Nm).
13. Install the bolts which secure the brake hose to the strut. Tighten to 24.3 ft. lbs. (33 Nm).
14. Install the wheel.
15. Check and adjust wheel alignment, as necessary.

OVERHAUL

See Figure 237.

1. Remove the strut from the vehicle.
2. Using a coil spring compressor tool, carefully compress the spring. Remove the self locking nut.
3. Remove the strut mount, upper spring and rubber seat from the strut.
4. Gradually decrease the compression force of the spring compressor tool. Remove the coil spring.
5. Remove the dust cover and helper spring.
6. Check for the presence of air in the damping force generating mechanism.
7. Using the spring compression tool, compress the coil spring.
8. Remove coil spring.

To install:

➡**Be sure to properly install the coil spring.**

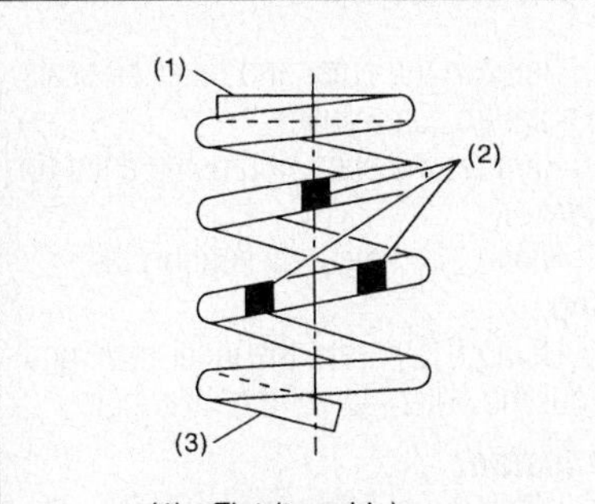

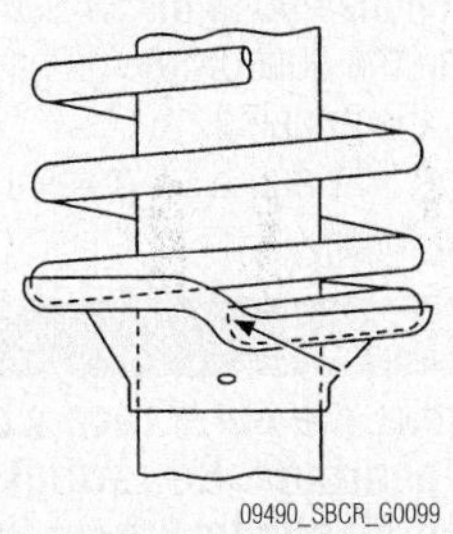

Fig. 237 Strut spring alignment

9. Position the coil spring so that its end face fits good into the spring seat.
10. Install the helper spring and dust cover to the piston rod.
11. Pull the piston rod fully upward, and install the rubber seat and spring seat.
12. Install the strut mount to the piston rod, and then tighten the self locking nut, temporarily. Be sure to use a new self locking nut.
13. Use a hexagon wrench to prevent the strut rod from turning.
14. Tighten the self locking nut to 41 ft. lbs. (55 Nm).
15. Carefully loosen the coil spring.

STEERING KNUCKLE

REMOVAL & INSTALLATION

1. Before servicing the vehicle, refer to the Precautions Section.
2. Raise and safely support the vehicle.
3. Remove the front wheel.
4. While depressing the brake pedal, un-stake and remove axle nut.

⁂ CAUTION

Be sure to loosen and retighten axle nut after removing the wheel from vehicle or damage to the wheel bearings may occur.

5. Remove the stabilizer link.
6. Remove the brake caliper without disconnect the brake housing. Securely hang the caliper from the strut assembly.
7. Remove the brake rotor.
8. Remove the cotter pin and castle nut that secure the tie-rod end to the knuckle.
9. Using a suitable puller, remove the tie rod joint from the knuckle.
10. Remove the ABS sensor assembly, if equipped.
11. Remove the bolt that holds the sensor wiring harness to the strut.
12. Disconnect the lower ball joint from the steering knuckle.
13. Remove the front axle shaft assembly from the hub.
14. If the axle shaft is difficult to remove, use Special Tool 926470000 Puller and 927140000 Plate to assist with the removal.
15. After placing a matchmark on the camber adjusting bolt head, remove the bolts which connect the steering knuckle and strut, and disconnect the wheel knuckle from the strut.
16. Installation is the reverse order of assembly.

STABILIZER BAR

REMOVAL & INSTALLATION

See Figure 238.

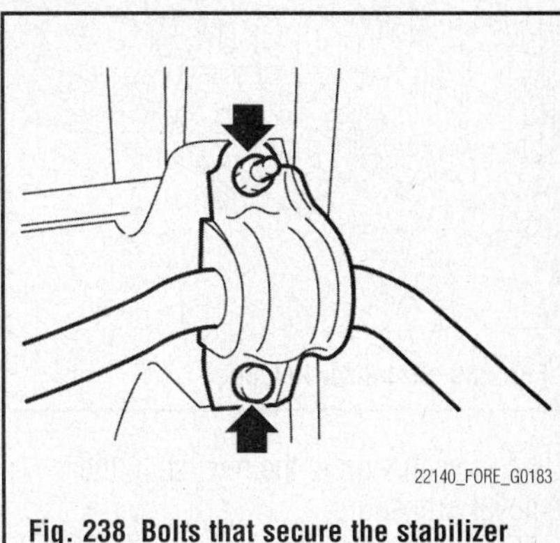

Fig. 238 Bolts that secure the stabilizer bar

1. Before servicing the vehicle, refer to the Precautions Section.
2. Raise and support the vehicle safely.
3. Remove the jack up plate from the lower part of the crossmember.
4. Remove the sub frame.
5. Remove the nut which secures the stabilizer link to the front transverse link.
6. Remove the bolts that secure the stabilizer bar to the crossmember.
7. Remove the stabilizer bar from the vehicle.

To install:

8. Installation is the reverse of the removal procedure.
9. Install the rubber bushing, on the front crossmember side, while aligning it with the paint mark on the stabilizer bar.
10. Be sure that the bushings and the stabilizer have the same identification colors.
11. Always fully tighten the rubber bushings when the wheels are in full contact with the ground and the vehicle is at curb height.
12. Tighten stabilizer link nuts to 33 ft. lbs. (45 Nm).
13. Tighten stabilizer to crossmember to 18 ft. lbs. (25 Nm)

STABILIZER LINKS

REMOVAL & INSTALLATION

See Figure 239.

1. Before servicing the vehicle, refer to the Precautions Section.
2. Remove the retaining nut from the transverse link.
3. Remove the retaining nut from the stabilizer bar.

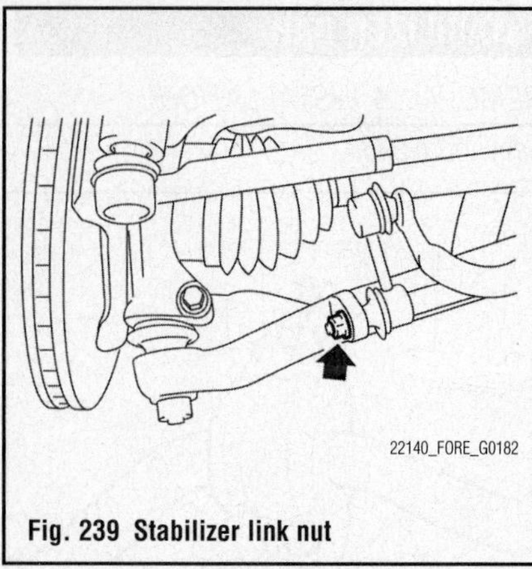

Fig. 239 Stabilizer link nut

4. Installation is the reverse of the removal procedure.
5. Tighten stabilizer link nuts to 33 ft. lbs. (45 Nm).

SUB-FRAME

REMOVAL & INSTALLATION

See Figure 240.

1. Lift up the vehicle.
2. Remove the undercover.
3. Remove the bolt cover.
4. Remove the clip.
5. Loosen bolt (1) and leave a few threads caught, then remove the bolts in the order of (2), (3), (4), (5), and (6).
6. Remove the sub frame.
7. Install in the reverse order of removal.
8. Refer to Graphic for specifications.

WHEEL HUB AND BEARING (SEALED UNIT)

REMOVAL & INSTALLATION

See Figures 241 and 242.

1. Before servicing the vehicle, refer to the Precautions Section.

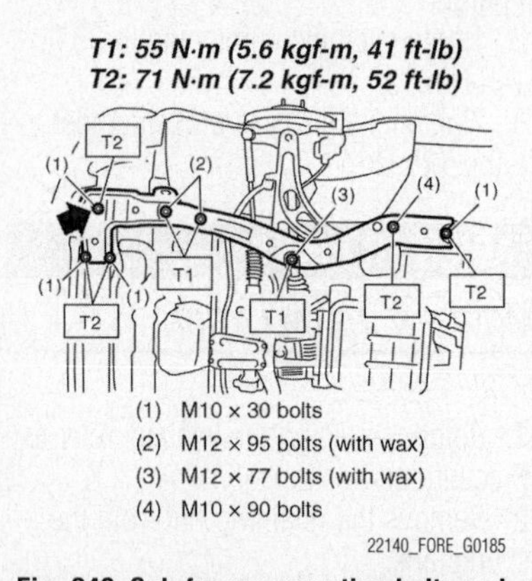

Fig. 240 Sub-frame, mounting bolts and tightening specifications

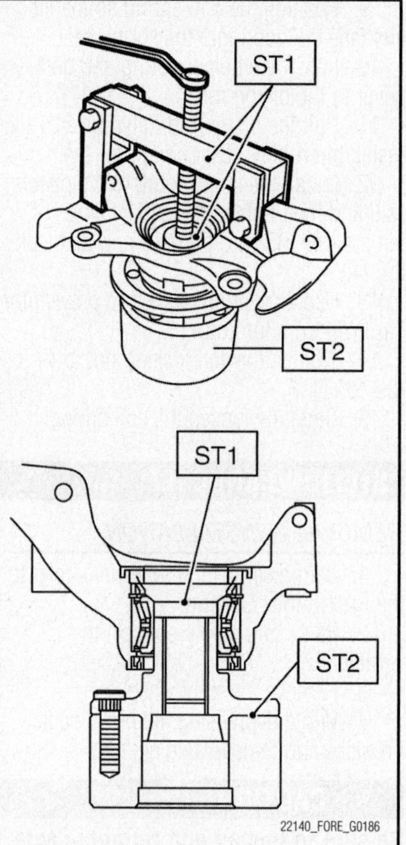

Fig. 241 Special tools ST1 and ST2

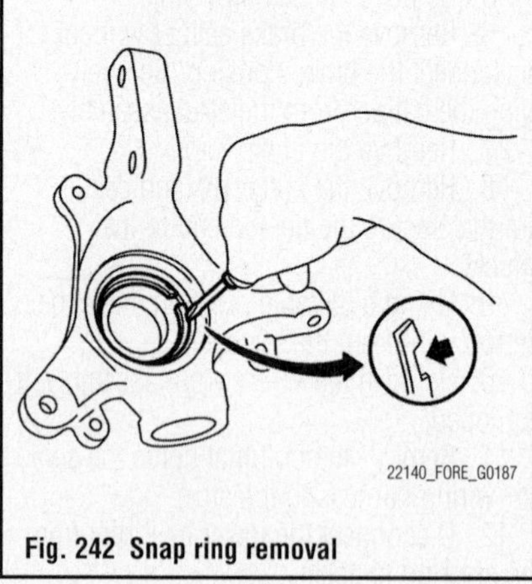

Fig. 242 Snap ring removal

2. Raise and safely support the vehicle.
3. Remove the front wheel.
4. While depressing the brake pedal, un-stake and remove axle nut.

CAUTION

Be sure to loosen and retighten axle nut after removing the wheel from vehicle or damage to the wheel bearings may occur.

5. Remove the stabilizer link.
6. Remove the brake caliper without disconnect the brake housing. Securely hang the caliper from the strut assembly.
7. Remove the brake rotor.
8. Remove the cotter pin and castle nut that secure the tie-rod end to the knuckle.
9. Using a suitable puller, remove the tie rod joint from the knuckle.
10. Remove the ABS sensor assembly, if equipped.
11. Remove the bolt that holds the sensor wiring harness to the strut.
12. Disconnect the lower ball joint from the steering knuckle.
13. Remove the front axle shaft assembly from the hub.

➡If the axle shaft is difficult to remove, use Special Tool 926470000 Puller and 927140000 Plate to assist with the removal.

14. Using ST1, securely support the housing and hub.
15. Attach ST2 to housing and drive hub out.
16. If inner bearing race remains in the hub, remove it with a suitable tool (commercially available tools).
17. Using ST1 and ST2, press a new bearing into place.

CAUTION

Be careful not to scratch the polished area of the hub.

18. Remove the disc cover from housing.
19. Remove the outer and inner oil seals using a flat tip screwdriver.
20. Remove the snap ring using a flat tip screwdriver.
21. Using ST1, securely support the housing.
22. Using ST2, press the inner race, and push out the outer race of the bearing.

To install:

23. When the hub is removed from housing, replace the bearing set and oil seal with new parts.
24. Attach the hub to the ST securely.
25. Clean the dust or foreign particles from inside the housing.
26. Using ST1 and ST2, press a new bearing into place.

CAUTION

Always press the outer race when installing bearings. Be careful not to remove the plastic lock from the inner race when installing the bearings.

27. Using a pliers, securely install the snap ring.
28. Using the ST1 and ST2, press the outer oil seal until it contacts the bottom of housing.
29. Using the ST1 and ST2, press the inner oil seal until it contacts the circlip.
30. Invert the ST and housing (up and down).
31. Apply sufficient grease to the oil seal lip.
32. Install the disc cover to housing with three bolts. Tighten to 13 ft, lbs. (18 Nm).
33. Attach the hub to ST1 securely
34. Clean dust and foreign particles from the polished surface of hub.
35. Place ST2 against the bearing inner race to press the bearing into the hub.
36. The remainder of the installation is the reverse of the removal procedure.

ADJUSTMENT

The wheel bearings are not adjustable.

SUSPENSION

REAR SUSPENSION

See Figure 243.

COIL SPRING

REMOVAL & INSTALLATION

See Figures 244 through 246.

1. Before servicing the vehicle, refer to the Precautions Section.
2. Remove the strut from the vehicle.
3. Using a coil spring compressor tool, carefully compress the spring. Remove the self locking nut.
4. Remove the strut mount, upper spring and rubber seat from the strut.
5. Gradually decrease the compression force of the spring compressor tool.
6. Remove the coil spring.

To install:

7. Before installing the coil spring, strut mount, etc. on the strut, check the condition of air inside the strut damper mechanism to make sure that excessive air is not inhibiting the creation of appropriate damping force.
8. Using a coil spring compressor, compress the coil spring.
9. Set the coil spring correctly so that its end face seats well in the spring seat as shown in the figure.
10. Install the helper and dust cover to the piston rod.

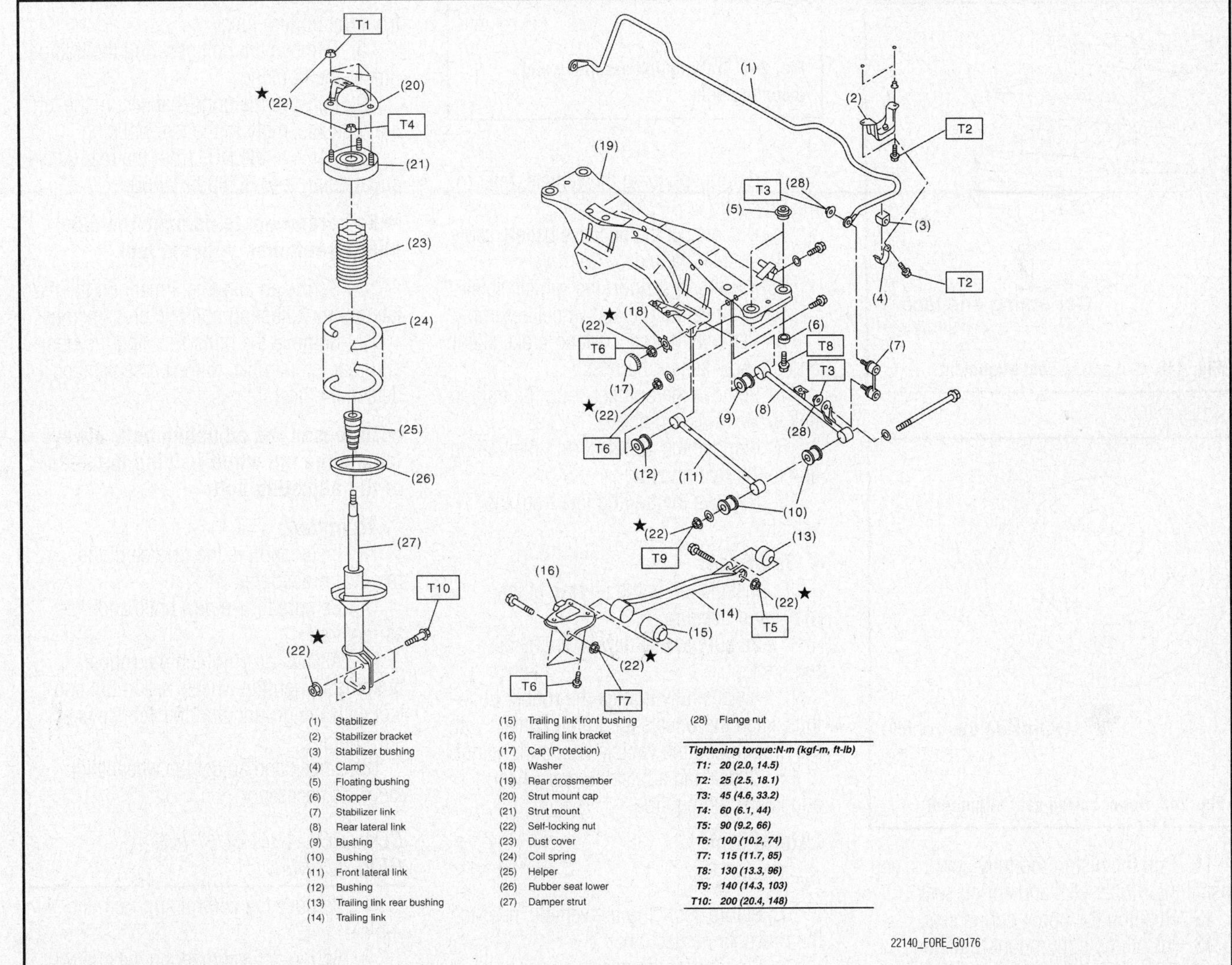

Fig. 243 View of the rear suspension components

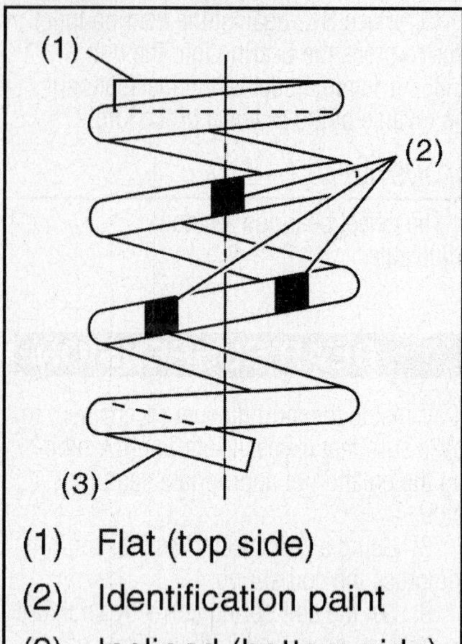

Fig. 244 Coil spring alignment

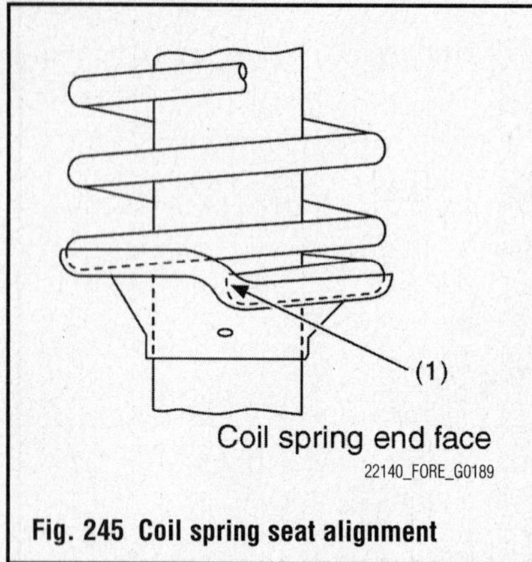

Fig. 245 Coil spring seat alignment

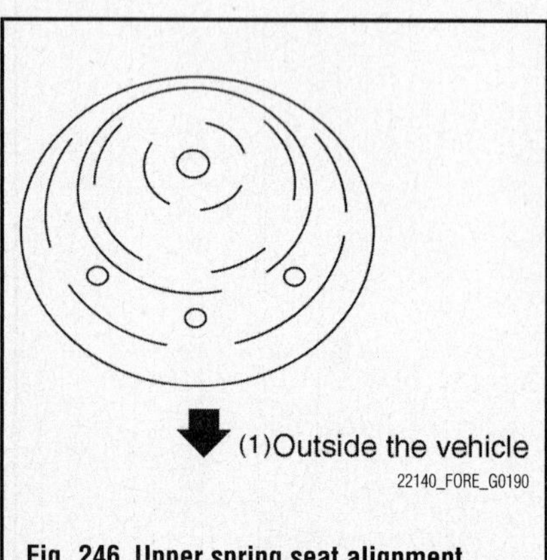

Fig. 246 Upper spring seat alignment

11. Pull the piston rod fully upward, and install the rubber seat and spring seat.
12. Position the upper spring seat.
13. Install the strut mount to piston rod, and temporarily attach and tighten a new self locking nut.
14. Using a hex wrench to prevent strut rod from turning, tighten the self-locking nut with the ST-92776000 strut mount socket. Tighten to 41 ft. lbs. (55 Nm).
15. Carefully loosen the coil spring tension.

CONTROL ARMS/LINKS

REMOVAL & INSTALLATION

Trailing Link

See Figure 247.

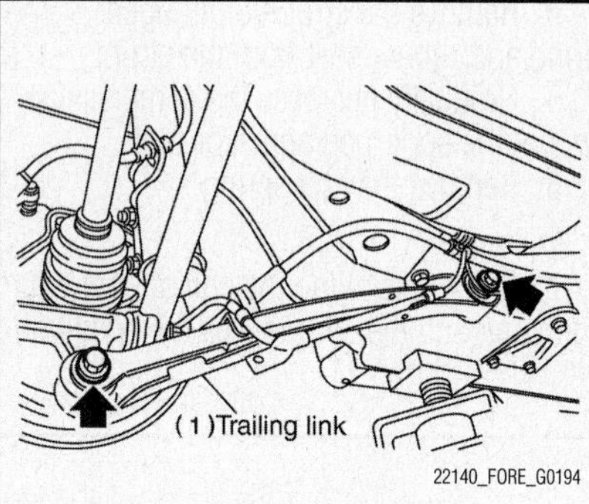

Fig. 247 Trailing link assembly and mounting bolts

1. Before servicing the vehicle, refer to the Precautions Section.
2. Disconnect the negative battery cable.
3. Loosen the rear wheel nuts.
4. Raise and support the vehicle safely.
5. Remove the tire and wheel assembly.
6. Remove the rear parking brake clamp and the ABS sensor harness.
7. Remove the bolt securing the trailing link to the trailing bracket.
8. Remove the bolt securing the trailing link to the rear housing.
9. Remove the trailing link from the vehicle.

To install:

10. Installation is the reverse of the removal procedure.
11. Be sure to use new locknuts, as required.
12. Always fully tighten the rubber bushings when the wheels are in full contact with the ground and the vehicle is at curb height.
13. Check and adjust the rear wheel alignment, as required.

Lateral Link

See Figure 248.

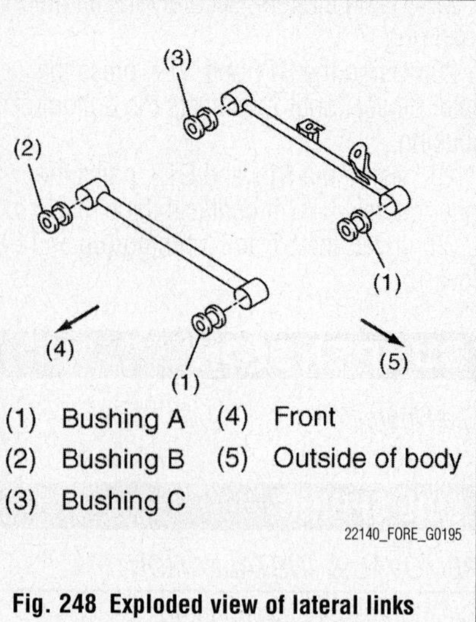

Fig. 248 Exploded view of lateral links

1. Before servicing the vehicle, refer to the Precautions Section.
2. Disconnect the negative battery cable.
3. Loosen the rear wheel nuts.
4. Raise and support the vehicle safely.
5. Remove the tire and wheel assembly.
6. Remove the stabilizer.
7. Remove the ABS sensor harness from the trailing link.
8. Remove the bolt securing the trailing link to the housing.
9. Remove the bolts that secure the lateral link assembly to the rear housing.
10. Remove the DOJ from the rear differential using tool St28099PA100.

➡Be careful not to damage the side bearing retainer, with the tool.

11. Scribe an alignment mark on the rear lateral link adjusting bolt and crossmember.
12. Remove the bolts securing the front and rear lateral links to the crossmember, detach the lateral links.

➡To loosen the adjusting bolt, always loosen the nut while holding the head of the adjusting bolt.

To install:

13. Installation is the reverse of the removal procedure.
14. Be sure to use new bolts and nuts, as required.
15. Always fully tighten the rubber bushings when the wheels are in full contact with the ground and the vehicle is at curb height.
16. Check and adjust the wheel alignment, as necessary.

CONTROL ARM BUSHING REPLACEMENT

1. Remove the control arm from the vehicle.
2. Scribe a matchmark on the control arm and rear bushing.

3. Loosen the nut and remove the rear bushing. Discard the nut.

4. Install the rear bushing to the control arm, making sure to align the marks made during removal.

5. Install a new nut.

MACPHERSON STRUTS

REMOVAL & INSTALLATION

See Figures 249 through 251.

1. Before servicing the vehicle, refer to the Precautions Section.

2. Loosen the rear wheel nuts.

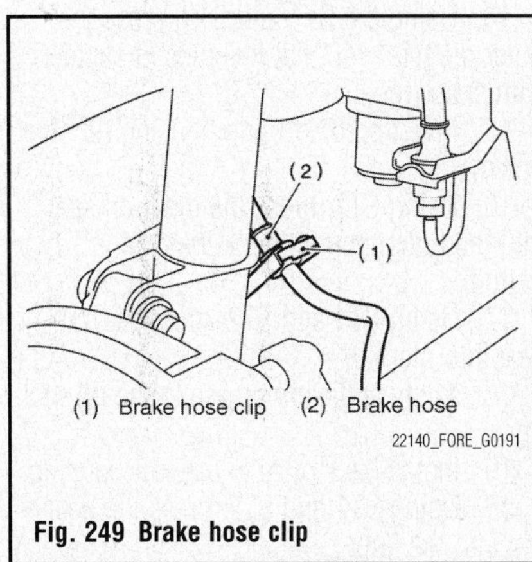

Fig. 249 Brake hose clip

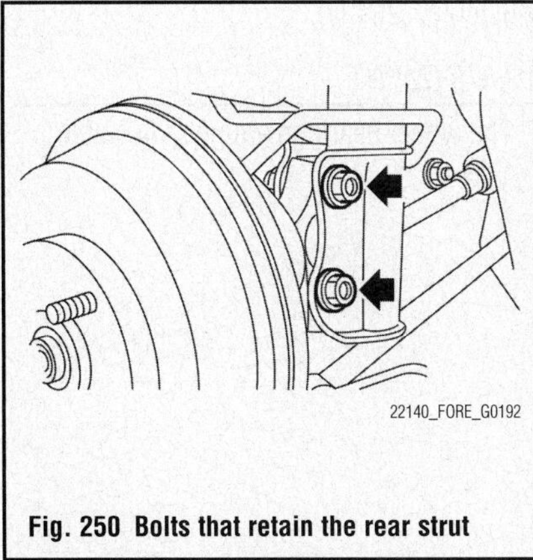

Fig. 250 Bolts that retain the rear strut

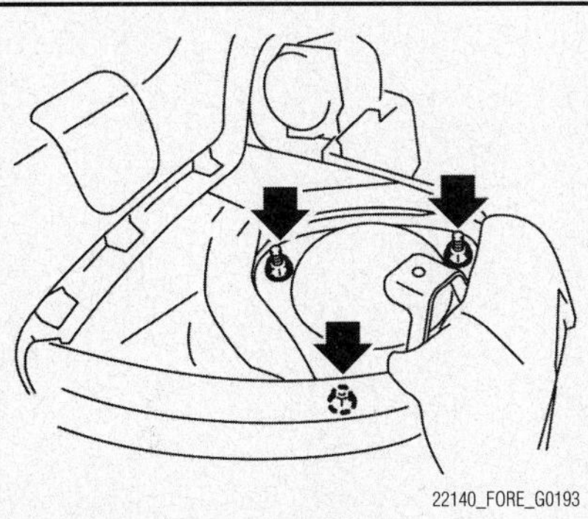

Fig. 251 Nuts which secure the strut mount to the vehicle body

3. Raise and support the vehicle safely.

4. Remove the tire and wheel.

5. Remove the brake hose clip and remove the brake hose from the rear strut.

6. Remove the bolts that retain the rear strut to the housing.

➡Do not apply excessive tension to the brake hose and ABS wheel speed sensor harness.

7. Remove the nuts which secure the strut mount to the vehicle body.

To install:

8. Installation is the reverse of the removal procedure.

9. Be sure to use new self locking locknuts.

10. Tighten the strut mount to body bolts to 14.5 ft. lbs. (20 Nm).

11. Tighten the rear strut to housing bolts to 148 ft. lbs. (200 Nm).

12. Check and adjust the rear wheel alignment, as required.

OVERHAUL

See Figures 252 through 254.

1. Before servicing the vehicle, refer to the Precautions Section.

2. Remove the strut from the vehicle.

3. Using a coil spring compressor tool, carefully compress the spring. Remove the self locking nut.

4. Remove the strut mount, upper spring and rubber seat from the strut.

5. Gradually decrease the compression force of the spring compressor tool.

6. Remove the coil spring.

To install:

7. Before installing the coil spring, strut mount, etc. on the strut, check the condition

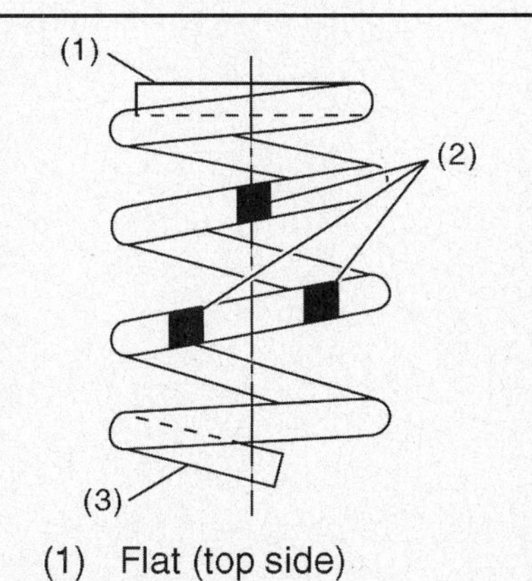

Fig. 252 Coil spring alignment

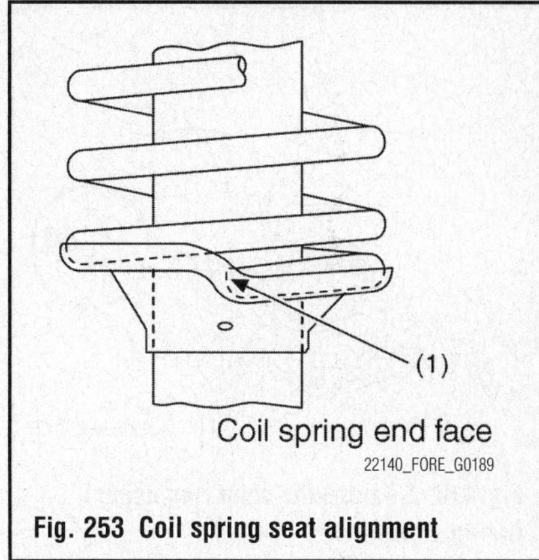

Fig. 253 Coil spring seat alignment

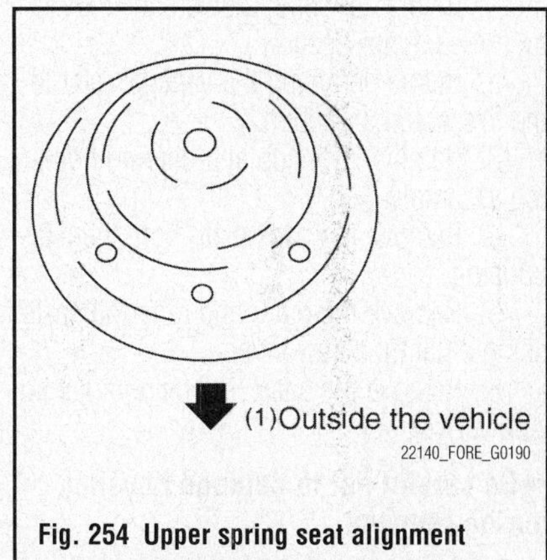

Fig. 254 Upper spring seat alignment

of air inside the strut damper mechanism to make sure that excessive air is not inhibiting the creation of appropriate damping force.

8. Using a coil spring compressor, compress the coil spring.

9. Set the coil spring correctly so that its end face seats well in the spring seat as shown in the figure.

10. Install the helper and dust cover to the piston rod.

11. Pull the piston rod fully upward, and install the rubber seat and spring seat.

12. Position the upper spring seat.

13. Install the strut mount to piston rod, and temporarily attach and tighten a new self locking nut.

14. Using a hex wrench to prevent strut rod from turning, tighten the self-locking nut with the ST-92776000 strut mount socket. Tighten to 41 ft. lbs. (55 Nm).

15. Carefully loosen the coil spring tension.

WHEEL HUB AND BEARING (SEALED UNIT)

REMOVAL & INSTALLATION

See Figures 255 through 257.

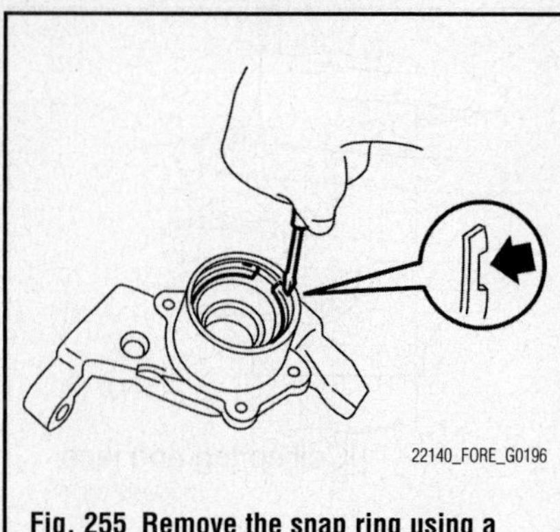

Fig. 255 Remove the snap ring using a flat tip screwdriver

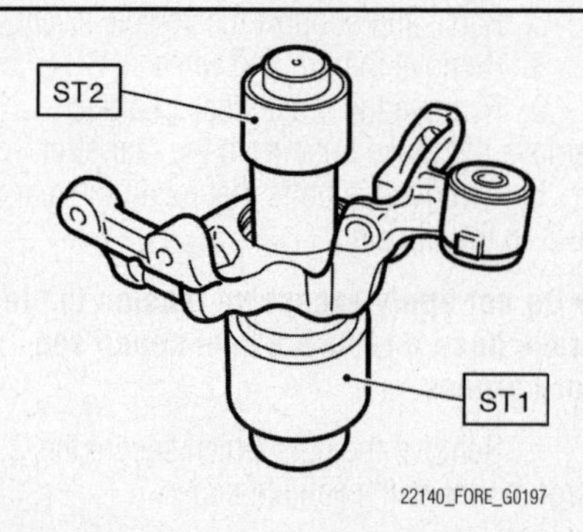

Fig. 256 Remove the bearing by pressing the inner race

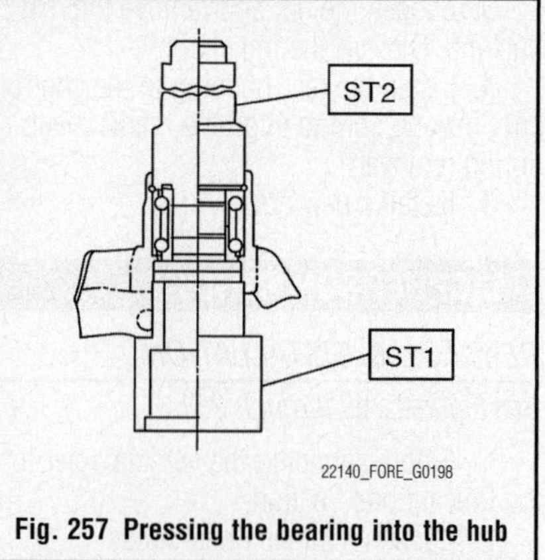

Fig. 257 Pressing the bearing into the hub

1. Before servicing the vehicle, refer to the Precautions Section.
2. Before servicing the vehicle, refer to the Precautions Section.
3. Remove rear hub and bearing housing assembly.
4. Remove the back plate from the rear housing
5. Remove the outer and inner oil seals using a flat tip screwdriver.
6. Remove the snap ring using a flat tip screwdriver.

➡Be careful not to damage housing during removal.

7. Using ST1 and ST2, remove the bearing by pressing the inner race.
8. Remove the tone wheel bolts and remove the tone wheel from the hub.
9. Using ST, press the hub bolt out.

⁂ CAUTION

Be careful not to hammer the hub bolts. Doing so may deform the hub.

To install:

➡When the hub is removed from the housing, replace the bearing set and oil seal.

10. Remove the foreign particles (dust, rust, etc.) from the mating surfaces of hub tone wheel, and install the tone wheel to hub.
11. Clean the housing interior completely. Using ST1 and ST2, press the bearing into the housing.
12. Be careful not to remove the plastic lock from the inner race when installing the bearings.
13. Using snap ring pliers, securely install the snap ring.
14. Using the ST1 and ST2, press the outer oil seal unit until it comes in contact with snap ring.
15. Invert both ST1 and housing (upside down).
16. Using ST2, press the inner oil seal into the housing until it touches the bottom.
17. Using ST1 and ST2, press the sub seal into place.
18. Apply sufficient grease to the oil seal lip.
19. Install back plate to the rear housing.
20. Using ST1 and ST2, press the bearing into the hub.
21. Install the rear hub and bearing housing assembly.

ADJUSTMENT

The wheel bearings are not adjustable.

SUBARU

Diagnostic Trouble Codes

16

DIAGNOSTIC TROUBLE CODES 16-1

DIAGNOSTIC TROUBLE CODES

OBD II VEHICLE APPLICATIONS

SUBARU

Baja

2006
- 2.5L I4 MFI (SOHC) .... VIN 6

2006
- 2.5L I4 MFI (DOHC) .... VIN 6

B9Tribeca

2006-07
- 3.0L V6 MFI .... VIN 8

Tribeca

2008
- 3.6L V6 MFI .... VIN 9

Forester

2006-08
- 2.5L I4 MFI (SOHC) .... VIN 6

2006-08
- 2.5L I4 MFI (DOHC) .... VIN 6

Impreza Models

2006-08
- 2.5L I4 MFI (SOHC) .... VIN 6

2006-08
- 2.5L I4 MFI (DOHC) .... VIN 6

Legacy, Outback

2006-08
- 2.5L I4 MFI (SOHC) .... VIN 6

2006-08
- 2.5L I4 MFI (DOHC) .... VIN 6

2006-08
- 3.0L V6 MFI .... VIN 8

OBD II Trouble Code List (P0XXX Codes)

DTC	Trouble Code Title, Conditions & Possible Causes
DTC: P0011 **2T CCM, MIL: Yes** **Years:** 2006, 2007, 2008 **Engines:** 2.5L DOHC VIN 6, 3.0L VIN 8, 3.6L VIN 9 **Models:** All **Transmissions:** All	**Intake Camshaft Position Timing (over advanced) Bank 1** The battery voltage is between 10.9 volts. Engine coolant temperature is 122°F. (50°C).Target timing is 0°CA and engine speed is 500 RPM. The ECM detects the AVCS system malfunction. Engine stalls. Erroneous idling. **Possible Causes:** • Check any other DTC on display • Using a scan tool, check current data. Check AVCS system. • Oil pipe clog • Clogged oil flow control solenoid valve • Intake camshaft damaged • Timing chain problem
DTC: P0016 **2T CCM, MIL: Yes** **Years:** 2006, 2007, 2008 **Engines:** 2.5L DOHC VIN 6, 3.0L VIN 8, 3.6L VIN 9 **Models:** All **Transmissions:** All	**Crankshaft Position (camshaft position correlation) Bank 1** The battery voltage is between 10.9 volts. Engine coolant temperature is122°F. (50°C).Target timing is 0°CA and engine speed is 500 RPM. The ECM detects the AVCS system malfunction. Engine stalls. Erroneous idling. Engine stalls. Erroneous idling. **Possible Causes:** • Check any other DTC on display • Using a scan tool, check current data. Check AVCS system. • Oil pipe clog • Clogged oil flow control solenoid valve • Intake camshaft damaged • Timing chain problem
DTC: P0018 **2T CCM, MIL: Yes** **Years:** 2006, 2007, 2008 **Engines:** 2.5L DOHC VIN 6, 3.0L VIN 8, 3.6L VIN 9 **Models:** All **Transmissions:** All	**Crankshaft Position (camshaft position correlation) Bank 2** The battery voltage is between 10.9 volts. Engine coolant temperature is122°F. (50°C).Target timing is 0°CA and engine speed is 500 RPM. The ECM detects the AVCS system malfunction. Engine stalls. Erroneous idling. Engine stalls. Erroneous idling. **Possible Causes:** • Check any other DTC on display • Using a scan tool, check current data. Check AVCS system. • Oil pipe clog • Clogged oil flow control solenoid valve • Intake camshaft damaged • Timing chain problem
DTC: P0021 **2T CCM, MIL: Yes** **Years:** 2006, 2007, 2008 **Engines:** 2.5L DOHC VIN 6, 3.0L VIN 8, 3.6L VIN 9 **Models:** All **Transmissions:** All	**Intake Camshaft Position Timing (over advanced) Bank 2** The battery voltage is between 10.9 volts. Engine coolant temperature is 122°F. (50°C).Target timing is 0°CA and engine speed is 500 RPM. The ECM detects the AVCS system malfunction. Engine stalls. Erroneous idling. Engine stalls. Erroneous idling. **Possible Causes:** • Check any other DTC on display • Using a scan tool, check current data. Check AVCS system. • Oil pipe clog • Clogged oil flow control solenoid valve • Dirty engine oil • Intake camshaft damaged • Timing chain problem
DTC: P0026 **2T CCM, MIL: Yes** **Years:** 2006, 2007, 2008 **Engines:** 2.5L SOHC VIN 6, 3.0L VIN 8 **Models:** Baja, B9 Tribeca, Forester, Impreza, Legacy, Outback **Transmissions:** All	**Intake Valve Control Solenoid Circuit Range/Performance Bank 1** The battery voltage is between 10.9 volts. Engine has run for 6 seconds. The engine oil temperature is 0°F. (0°C).Variable lift control is operational. Erroneous idling. **Possible Causes:** • Check any other DTC on display • Check harness between ECM and variable valve lift diagnosis oil pressure switch connector • Check harness between ECM and variable valve lift diagnosis oil pressure switch connector • Check oil switching solenoid valve

DTC	Trouble Code Title, Conditions & Possible Causes
DTC: P0028 **2T CCM, MIL: Yes** **Years:** 2006, 2007, 2008 **Engines:** 2.5L SOHC VIN 6, 3.0L VIN 8 **Models:** Baja, B9 Tribeca, Forester, Impreza, Legacy, Outback **Transmissions:** All	**Intake Valve Control Solenoid Circuit Range/Performance Bank 2** The battery voltage is between 10.9 volts. Engine has run for 6 seconds. The engine oil temperature is 0°F. (0°C).Variable lift control is operational. Erroneous idling. **Possible Causes:** • Check any other DTC on display • Check harness between ECM and variable valve lift diagnosis oil pressure switch connector • Check harness between ECM and variable valve lift diagnosis oil pressure switch connector • Check oil switching solenoid valve
DTC: P0030 **2T CCM, MIL: Yes** **Years:** 2006, 2007, 2008 **Engines:** 2.5L SOHC, DOHC VIN 6, 3.0L VIN 8, 3.6L VIN 9 **Models:** All **Transmissions:** All	**HO2S Heater Control Circuit Bank 1 Sensor 1** The battery voltage is between 10.9 volts. 30000 ms seconds or more have passed since the engine was started. Detect functional errors of the front oxygen (A/F) sensor heater. Poor drive ability. **Possible Causes:** • Check harness between ECM and front oxygen sensor connector • Check harness between main relay and front oxygen sensor connector • Check front oxygen sensor • Check poor contact
DTC: P0031 **2T CCM, MIL: Yes** **Years:** 2006, 2007, 2008 **Engines:** 2.5L SOHC, DOHC VIN 6, 3.0L VIN 8, 3.6L VIN 9 **Models:** All **Transmissions:** All	**HO2S Heater Control Circuit Low Bank 1 Sensor 1** The battery voltage is between 10.9 volts. Detect the open or short circuit of heater. The heater performs duty control, and the output terminal voltage at ON is 0 V and the output terminal voltage at OFF is the battery voltage. Judge NG when the terminal voltage remains Low. **Possible Causes:** • Check power supply to front oxygen sensor • Check ground circuit for ECM • Using a scan tool check current data • Check output signal of ECM • Check front oxygen sensor
DTC: P0032 **2T CCM, MIL: Yes** **Years:** 2006, 2007, 2008 **Engines:** 2.5L SOHC, DOHC VIN 6, 3.0L VIN 8, 3.6L VIN 9 **Models:** All **Transmissions:** All	**HO2S Heater Control Circuit High Bank 1 Sensor 1** The battery voltage is between 10.9 volts. Detect the open or short circuit of heater. The heater performs duty control, and the output terminal voltage at ON is 0 V and the output terminal voltage at OFF is the battery voltage. Judge NG when the terminal voltage remains High. **Possible Causes:** • Check output signal of ECM • Check front oxygen sensor heater current • Faulty ECM
DTC: P0037 **2T CCM, MIL: Yes** **Years:** 2006, 2007, 2008 **Engines:** 2.5L SOHC, DOHC VIN 6, 3.0L VIN 8, 3.6L VIN 9 **Models:** All **Transmissions:** All	**HO2S Heater Control Circuit Low Bank 1 Sensor 2** The battery voltage is between 10.9 volts. Detect the open or short circuit of heater. The heater performs duty control, and the output terminal voltage at ON is 0 V and the output terminal voltage at OFF is the battery voltage. Judge NG when the terminal voltage remains Low. **Possible Causes:** • Check power supply to rear oxygen sensor • Check ground circuit for ECM • Using a scan tool check current data • Check output signal of ECM • Check rear oxygen sensor
DTC: P0038 **Years:** 2006, 2007, 2008 **Engines:** 2.5L SOHC, DOHC VIN 6, 3.0L VIN 8, 3.6L VIN 9 **Models:** All **Transmissions:** All	**HO2S Heater Control Circuit High Bank 1 Sensor 2** The battery voltage is between 10.9 volts. Detect the open or short circuit of heater. The heater performs duty control, and the output terminal voltage at ON is 0 V and the output terminal voltage at OFF is the battery voltage. Judge NG when the terminal voltage remains High. **Possible Causes:** • Check input signal of ECM • Using a scan tool check current data • Check poor contact
DTC: P0050 **2T CCM, MIL: Yes** **Years:** 2006, 2007, 2008 **Engines:** 3.0L VIN 8, 3.6L VIN 9 **Models:** B9 Tribeca, Tribeca, Legacy Outback **Transmissions:** All	**HO2S Heater Control Circuit Bank 2 Sensor 1** The battery voltage is between 10.9 volts. 30000 ms seconds or more have passed since the engine was started. Detect functional errors of the front oxygen (A/F) sensor heater. Poor drive ability. **Possible Causes:** • Check harness between ECM and front oxygen sensor connector • Check harness between main relay and front oxygen sensor connector • Check front oxygen sensor • Check poor contact

DTC	Trouble Code Title, Conditions & Possible Causes
DTC: P0051 **2T CCM, MIL: Yes** **Years:** 2006, 2007, 2008 **Engines:** 3.0L VIN 8, 3.6L VIN 9 **Models:** B9 Tribeca, Tribeca, Legacy Outback **Transmissions:** All	**HO2S Heater Control Circuit Low Bank 2 Sensor 1** The battery voltage is between 10.9 volts. Detect the open or short circuit of heater. The heater performs duty control, and the output terminal voltage at ON is 0 V and the output terminal voltage at OFF is the battery voltage. Judge NG when the terminal voltage remains Low. **Possible Causes:** • Check power supply to front oxygen sensor • Check ground circuit for ECM • Using a scan tool check current data • Check output signal of ECM • Check front oxygen sensor
DTC: P0052 **2T CCM, MIL: Yes** **Years:** 2006, 2007, 2008 **Engines:** 3.0L VIN 8, 3.6L VIN 9 **Models:** B9 Tribeca, Tribeca, Legacy Outback **Transmissions:** All	**HO2S Heater Control Circuit High Bank 2 Sensor 1** The battery voltage is between 10.9 volts. Detect the open or short circuit of heater. The heater performs duty control, and the output terminal voltage at ON is 0 V and the output terminal voltage at OFF is the battery voltage. Judge NG when the terminal voltage remains High. **Possible Causes:** • Check output signal of ECM • Check front oxygen sensor heater current • Check output signal of ECM • Faulty ECM
DTC: P0057 **2T CCM, MIL: Yes** **Years:** 2006, 2007, 2008 **Engines:** 3.0L VIN 8, 3.6L VIN 9 **Models:** B9 Tribeca, Tribeca, Legacy Outback **Transmissions:** All	**HO2S Heater Control Circuit Low Bank 2 Sensor 2** The battery voltage is between 10.9 volts. Detect the open or short circuit of heater. The heater performs duty control, and the output terminal voltage at ON is 0 V and the output terminal voltage at OFF is the battery voltage. Judge NG when the terminal voltage remains Low. Poor drive ability. **Possible Causes:** • Check power supply to rear oxygen sensor • Check ground circuit for ECM • Using a scan tool check current data • Check output signal of ECM • Check rear oxygen sensor
DTC: P0058 **2T CCM, MIL: Yes** **Years:** 2006, 2007, 2008 **Engines:** 3.0L VIN 8, 3.6L VIN 9 **Models:** B9 Tribeca, Tribeca, Legacy Outback **Transmissions:** All	**HO2S Heater Control Circuit High Bank 2 Sensor 2** The battery voltage is between 10.9 volts. Detect the open or short circuit of heater. The heater performs duty control, and the output terminal voltage at ON is 0 V and the output terminal voltage at OFF is the battery voltage. Judge NG when the terminal voltage remains High. Poor drive ability. **Possible Causes:** • Check input signal of ECM • Using a scan tool check current data • Check poor contact
DTC: P0068 **2T CCM, MIL: Yes** **Years:** 2006, 2007, 2008 **Engines:** 2.5L SOHC, DOHC VIN 6, 3.0L VIN 8, 3.6L VIN 9 **Models:** All **Transmissions:** All	**MAP/MAF- Throttle Position Correlation** Engine coolant temperature is 167°F. (75°C). Detect problems in the intake manifold pressure sensor output properties. Determining from the engine condition, if the intake manifold pressure AD value is small even when operating conditions suggest it should be large, or if the intake manifolds pressure AD value is large even though the operating condition suggests it should be small, this is judged as being NG. **Possible Causes:** • Check any other DTC on display • Check condition of the sensor • Check condition of the throttle body
DTC: P0076 **1T CCM, MIL: Yes** **Years:** 2006, 2007, 2008 **Engines:** 2.5L SOHC, DOHC VIN 6, 3.0L VIN 8 **Models:** Baja, B9 Tribeca, Forester, Impreza, Legacy, Outback **Transmissions:** All	**Intake Valve Control Circuit Low Bank 1** The battery voltage is between 10.9 volts. Detect the open circuit of the oil switching solenoid valve. Judge NG when the current is small even though the output duty is large. Judge NG when the current is small even though the output duty is large. Poor idling condition **Possible Causes:** • Check harness between ECM and oil switching solenoid valve • Check oil switching solenoid valve

DTC	Trouble Code Title, Conditions & Possible Causes
DTC: P0077 **1T CCM, MIL: Yes** **Years:** 2006, 2007, 2008 **Engines:** 2.5L SOHC, DOHC VIN 6, 3.0L VIN 8 **Models:** Baja, B9 Tribeca, Forester, Impreza, Legacy, Outback **Transmissions:** All	**Intake Valve Control Circuit High Bank 1** The battery voltage is between 10.9 volts. Detect short circuits of the oil switching solenoid valve. Judge as a short NG when the current is large even though the output duty is small. Poor idling condition. **Possible Causes:** • Check harness between ECM and oil switching solenoid valve • Check oil switching solenoid valve
DTC: P0082 **1T CCM, MIL: Yes** **Years:** 2006, 2007, 2008 **Engines:** 2.5L SOHC, DOHC VIN 6, 3.0L VIN 8 **Models:** Baja, B9 Tribeca, Forester, Impreza, Legacy, Outback **Transmissions:** All	**Intake Valve Control Circuit Low Bank 2** The battery voltage is between 10.9 volts. Detect the open circuit of the oil switching solenoid valve. Judge NG when the current is small even though the output duty is large. Judge NG when the current is small even though the output duty is large. Poor idling condition. **Possible Causes:** • Check harness between ECM and oil switching solenoid valve • Check oil switching solenoid valve
DTC: P0083 **1T CCM, MIL: Yes** **Years:** 2006, 2007, 2008 **Engines:** 2.5L SOHC, DOHC VIN 6, 3.0L VIN 8 **Models:** Baja, B9 Tribeca, Forester, Impreza, Legacy, Outback **Transmissions:** All	**Intake Valve Control Circuit High Bank 2** The battery voltage is between 10.9 volts. Detect short circuits of the oil switching solenoid valve. Judge as a short NG when the current is large even though the output duty is small. Poor idling condition. **Possible Causes:** • Check harness between ECM and oil switching solenoid valve • Check oil switching solenoid valve
DTC P0101 **2T CCM, MIL: Yes** **Years:** 2006, 2007, 2008 **Engines:** 2.5L SOHC, DOHC VIN 6, 3.0L VIN 8, 3.6L VIN 9 **Models:** All **Transmissions:** All	**Mass or Volume Air Flow Circuit Range/Performance** Engine coolant temperature is 167°F. (75°C). Detect for abnormalities in the air flow sensor output properties. Judge as a low side NG when the air flow voltage indicates a small value regardless of running in a state where the air flow voltage increases. Judge as a high side NG when the air flow voltage indicates a large value regardless of running in a state where the air flow voltage decreases. Judge air flow sensor property NG when the Low side or High side becomes NG. Poor idling condition. Engine stalls. Poor driving performance. **Possible Causes:** • Check any other DTC on display • Faulty Mass Air Flow (MAF) sensor
DTC: P0102 **2T CCM, MIL: Yes** **Years:** 2006, 2007, 2008 **Engines:** 2.5L SOHC, DOHC VIN 6, 3.0L VIN 8, 3.6L VIN 9 **Models:** All **Transmissions:** All	**Mass or Volume Air Flow Circuit Low Input** Detects open or short circuits of the air flow sensor. Engine stalls. Poor idling condition. Poor driving performance. **Possible Causes:** • Using a scan tool, check current data • Check input signal of ECM • Check power supply to sensor • Check harness between ECM and sensor connector • Check for poor contacts • Faulty Mass Air Flow (MAF) sensor
DTC: P0103 **1T CCM, MIL: Yes** **Years:** 2006, 2007, 2008 **Engines:** 2.5L SOHC, DOHC VIN 6, 3.0L VIN 8, 3.6L VIN 9 **Models:** All **Transmissions:** All	**Mass or Volume Air Flow Circuit High Input** Detects open or short circuits of the air flow sensor. Engine stalls. Poor idling condition. Poor driving performance. **Possible Causes:** • Using a scan tool, check current data • Check harness between ECM and sensor connector • Check for poor contact • Faulty Mass Air Flow (MAF) sensor
DTC: P0107 **1T CCM, MIL: Yes** **Years:** 2006, 2007, 2008 **Engines:** 2.5L SOHC, DOHC VIN 6, 3.0L VIN 8, 3.6L VIN 9 **Models:** All **Transmissions:** All	**Manifold Absolute Pressure/Barometric Pressure Circuit Low Input** Detects the open or short circuit of intake manifold pressure sensor. **Possible Causes:** • Using a scan tool, check current data • Check for poor contacts • Check output signal of ECM • Check input signal of ECM • Check harness between ECM and sensor connector • Faulty MAP sensor

DTC	Trouble Code Title, Conditions & Possible Causes
DTC: P0108 **1T CCM, MIL: Yes** **Years:** 2006, 2007, 2008 **Engines:** 2.5L SOHC, DOHC VIN 6, 3.0L VIN 8, 3.6L VIN 9 **Models:** All **Transmissions:** All	**Manifold Absolute Pressure/Barometric Pressure Circuit Low Input** Detects the open or short circuit of intake manifold pressure sensor. **Possible Causes:** • Check output signal of ECM • Check input signal of ECM • Check harness between ECM and sensor connector • Check harness between the sensor connector • Check poor contact • Faulty MAP sensor
DTC: P0111 **2T CCM, MIL: Yes** **Years:** 2006, 2007, 2008 **Engines:** 2.5L SOHC, DOHC VIN 6, 3.0L VIN 8, 3.6L VIN 9 **Models:** All **Transmissions:** All	**Intake Air Temperature Circuit Range/Performance** The battery voltage is between 10.9 volts. Engine coolant temperature at start-up is 86°F (30°C). Engine coolant temperature is 203°F (95°C). Detects the malfunction of intake air temperature sensor output property. Judge as NG when the intake air temperature is not varied whereas it seemed to be varied from the viewpoint of engine condition. Poor idling condition. Poor driving performance. **Possible Causes:** • Check any other DTC on display • Check engine coolant temperature
DTC: P0112 **2T CCM, MIL: Yes** **Years:** 2006, 2007, 2008 **Engines:** 2.5L SOHC, DOHC VIN 6, 3.0L VIN 8, 3.6L VIN 9 **Models:** All **Transmissions:** All	**Intake Air Temperature Circuit Low Input** Detect open or short circuit of the intake air temperature sensor. Poor idling condition. Poor driving performance. **Possible Causes:** • Using a scan tool, check current data • Check the harness between the sensor and the ECM connector • Check for poor contacts • Faulty IAT sensor
DTC: P0113 **1T CCM, MIL: Yes** **Years:** 2006, 2007, 2008 **Engines:** 2.5L SOHC, DOHC VIN 6, 3.0L VIN 8, 3.6L VIN 9 **Models:** All **Transmissions:** All	**Intake Air Temperature Circuit High Input** Detect open or short circuit of the intake air temperature sensor. Poor idling condition. Poor driving performance. **Possible Causes:** • Using a scan tool, check current data • Check the harness between the sensor and the ECM connector • Check for poor contacts • Faulty IAT sensor
DTC: P0117 **1T CCM, MIL: Yes** **Years:** 2006, 2007, 2008 **Engines:** 2.5L SOHC, DOHC VIN 6, 3.0L VIN 8, 3.6L VIN 9 **Models:** All **Transmissions:** All	**Engine Coolant Temperature Sensor Circuit Low Input** Detects the open or short circuit of the engine coolant temperature sensor. Hard starting condition. Poor idling condition. Poor driving performance. **Possible Causes:** • Using a scan tool, check current data • Check harness between sensor and ECM connector • Check for poor contacts • Faulty ECT sensor
DTC: P0118 **1T CCM, MIL: Yes** **Years:** 2006, 2007, 2008 **Engines:** 2.5L SOHC, DOHC VIN 6, 3.0L VIN 8, 3.6L VIN 9 **Models:** All **Transmissions:** All	**Engine Coolant Temperature Sensor Circuit High Input** Detects the open or short circuit of the engine coolant temperature sensor. Hard starting condition. Poor idling condition. Poor driving performance. **Possible Causes:** • Using a scan tool, check current data • Check harness between sensor and ECM connector • Check for poor contacts • Faulty ECT sensor
DTC: P0122 **1T CCM, MIL: Yes** **Years:** 2006, 2007, 2008 **Engines:** 2.5L SOHC, DOHC VIN 6, 3.0L VIN 8, 3.6L VIN 9 **Models:** All **Transmissions:** All	**Throttle/Pedal Position Sensor/Switch "A" Circuit Low Input** Ignition switch on and battery voltage at least 6 volts. Detects the open or short circuit of throttle position sensor 1. Poor idling condition. Engine stalls Poor driving performance **Possible Causes:** • Check sensor output • Check for poor contact • Check harness between ECM and electronic throttle control • Faulty electronic throttle control • Faulty ECM

DTC	Trouble Code Title, Conditions & Possible Causes
DTC: P0123 **1T CCM, MIL: Yes** **Years:** 2006, 2007, 2008 **Engines:** 2.5L SOHC, DOHC VIN 6, 3.0L VIN 8, 3.6L VIN 9 **Models:** All **Transmissions:** All	**Throttle/Pedal Position Sensor/Switch "A" Circuit Low Input** Ignition switch on and battery voltage at least 6 volts. Detects the open or short circuit of throttle position sensor 1. Poor idling condition. Engine stalls Poor driving performance. **Possible Causes:** • Check sensor output • Check for poor contact • Check harness between ECM and electronic throttle control • Faulty electronic throttle control • Faulty ECM
DTC: P0125 **2T CCM, MIL: Yes** **Years:** 2006, 2007, 2008 **Engines:** 2.5L SOHC, DOHC VIN 6, 3.0L VIN 8, 3.6L VIN 9 **Models:** All **Transmissions:** All	**Insufficient Coolant Temperature For Closed loop Fuel Control** Battery voltage 10.9 volts. Engine speed 500 RPM. Detects the malfunction of engine coolant temperature output property. Engine does not return to normal idle. **Possible Causes:** • Check any other DTC on display • Check tire size • Check engine coolant • Check thermostat • Faulty ECT sensor
DTC: P0126 **2T CCM, MIL: Yes** **Years:** 2006, 2007, 2008 **Engines:** 2.5L SOHC, DOHC VIN 6, 3.0L VIN 8, 3.6L VIN 9 **Models:** All **Transmissions:** All	**Insufficient Coolant Temperature For Stable Operation** Battery voltage 10.9 volts. Engine coolant temperature at last stop 167°F (75°C). Fuel level at 3.96 gal. (15 L). Detect the malfunction of the engine coolant temperature sensor characteristics. Memorize the engine coolant temperature and fuel temperature at the last engine stop, and use them to judge as NG when the engine coolant temperature does not decrease when it should. **Possible Causes:** • Check any other DTC on display • Check Engine Coolant Temperature (ECT) sensor and ECM connectors • Faulty ECT sensor
DTC: P0128 **2T CCM, MIL: Yes** **Years:** 2006, 2007, 2008 **Engines:** 2.5L SOHC, DOHC VIN 6, 3.0L VIN 8, 3.6L VIN 9 **Models:** All **Transmissions:** All	**Coolant Temperature Thermostat (Coolant Temperature Below Thermostat Regulating Temperature)** Detects malfunctions of the thermostat function. Judge as NG when the engine coolant temperature is lower than the estimated engine coolant temperature and the difference between them is large. Judge as OK when the engine coolant temperature becomes to (75°C) 167°F, and the difference is small, before judging NG. Thermostat remains open **Possible Causes:** • Open or missing thermostat • Check any other DTC on display • Check engine coolant for proper mixture • Check radiator fan (Stuck On)
DTC: P0131 **1T CCM, MIL: Yes** **Years:** 2006, 2007, 2008 **Engines:** 2.5L SOHC, DOHC VIN 6, 3.0L VIN 8, 3.6L VIN 9 **Models:** All **Transmissions:** All	**O2 Sensor Circuit Low Voltage Bank 1 Sensor 1** Engine started, engine running in closed loop at hot idle speed, and the ECM detected an unexpected "low" voltage condition on the A/F sensor signal circuit for over 6 seconds during the CCM test. **Possible Causes:** • Check harness between ECM and sensor • Shorted ground circuit of harness between ECM and front oxygen A/F sensor connector • Check for poor contacts • Faulty front oxygen A/F sensor
DTC: P0132 **2T CCM, MIL: Yes** **Years:** 2006, 2007, 2008 **Engines:** 2.5L SOHC, DOHC VIN 6, 3.0L VIN 8, 3.6L VIN 9 **Models:** All **Transmissions:** All	**O2 Sensor Circuit High Voltage Bank 1 Sensor 1** Engine started, engine running in closed loop at hot idle speed, and the ECM detected an unexpected "high" voltage condition on the A/F sensor signal circuit for over 6 seconds during the CCM test. **Possible Causes:** • Check harness between ECM and sensor • Shorted power circuit of harness between ECM and front oxygen A/F sensor connector • Check for poor contacts • Faulty front oxygen A/F sensor
DTC: P0133 **2T CCM, MIL: Yes** **Years:** 2006, 2007, 2008 **Engines:** 2.5L SOHC, DOHC VIN 6, 3.0L VIN 8, 3.6L VIN 9 **Models:** All **Transmissions:** All	**O2 Sensor Circuit Slow Response Bank 1 Sensor 1** Engine started, engine running in closed loop, ECT sensor signal more than 170°F, and the ECM detected the number of A/F S-11 rich-to-lean or lean-to-rich switches was less than a calibrated amount. **Possible Causes:** • Check any other DTC on display • Check exhaust system • Faulty front oxygen A/F sensor

DTC	Trouble Code Title, Conditions & Possible Causes
DTC: P0134 **1T CCM, MIL: Yes** **Years:** 2006, 2007, 2008 **Engines:** 2.5L SOHC, DOHC VIN 6, 3.0L VIN 8, 3.6L VIN 9 **Models:** All **Transmissions:** All	**O2 Sensor Circuit No Activity Detected Bank 1 Sensor 1** Detects open circuits of the sensor. Judge as NG when the impedance of the element is large. Poor drive ability. **Possible Causes:** • Check harness between ECM and sensor connector • Check poor contact • Faulty front oxygen A/F sensor
DTC: P0137 **2T CCM, MIL: Yes** **Years:** 2006, 2007, 2008 **Engines:** 2.5L SOHC, DOHC VIN 6, 3.0L VIN 8, 3.6L VIN 9 **Models:** All **Transmissions:** All	**O2 Sensor Circuit Low Voltage Bank 1 Sensor 2** Engine started, vehicle driven at a speed of 30-55 MPH for 2 minutes, and the ECM detected an unexpected low voltage condition on the HO2S-12 circuit during the CCM test period. **Possible Causes:** • Check any other DTC on display • Using the a scan tool, check current data • Check harness between ECM and rear oxygen sensor connector • Check exhaust system • Faulty rear oxygen sensor
DTC: P0138 **2T CCM, MIL: Yes** **Years:** 2006, 2007, 2008 **Engines:** 2.5L SOHC, DOHC VIN 6, 3.0L VIN 8, 3.6L VIN 9 **Models:** All **Transmissions:** All.	**O2 Sensor Circuit High Voltage Bank 1 Sensor 2** Engine started, vehicle driven at a speed of 30-55 MPH for 2 minutes, and the ECM detected an unexpected high voltage condition on the HO2S-12 circuit during the CCM test period. **Possible Causes:** • Check any other DTC on display • Using a scan tool, check current data • Check harness between ECM and rear oxygen sensor connector • Check exhaust system • Faulty rear oxygen sensor
DTC: P0139 **2T CCM, MIL: Yes** **Years:** 2006, 2007, 2008 **Engines:** 2.5L SOHC, DOHC VIN 6, 3.0L VIN 8, 3.6L VIN 9 **Models:** All **Transmissions:** All	**O2 Sensor Circuit Slow Response Bank 1 Sensor 2** Engine started, engine running in closed loop, ECT sensor signal more than 170°F, and the ECM detected the number of HO2S-12 rich-to-lean or lean-to-rich switches was less than a calibrated amount. **Possible Causes:** • Check any other DTC on display • Check harness between ECM and rear oxygen sensor connector • Check sensor • Faulty rear oxygen sensor
DTC: P0140 **2T CCM, MIL: Yes** **Years:** 2006, 2007, 2008 **Engines:** 2.5L SOHC, DOHC VIN 6, 3.0L VIN 8, 3.6L VIN 9 **Models:** All **Transmissions:** All	**O2 Sensor Circuit No Activity Detected Bank 1 Sensor 2** Detects open circuits of the sensor. Judge as NG when the impedance of the element is large. Poor drive ability. **Possible Causes:** • Check any other DTC on display • Using a scan tool, check current data • Check Harness between ECM and rear oxygen sensor connector • Check exhaust system • Faulty rear oxygen sensor
DTC: P0151 **1T CCM, MIL: Yes** **Years:** 2006, 2007, 2008 **Engines:** 3.0L VIN 8, 3.6L VIN 9 **Models:** B9 Tribeca, Tribeca **Transmissions:** All	**O2 Sensor Circuit Low Voltage Bank 2 Sensor 1** Engine started, engine running in closed loop at hot idle speed, and the ECM detected an unexpected "low" voltage condition on the front A/F sensor signal circuit for over 6 seconds during the CCM test. **Possible Causes:** • Check any other DTC on display • Using a scan tool, check current data • Check the harness between the ECM and the front sensor connector • Faulty front oxygen A/F sensor
DTC: P0152 **1T CCM, MIL: Yes** **Years:** 2006, 2007, 2008 **Engines:** 3.0L VIN 8, 3.6L VIN 9 **Models:** B9 Tribeca, Tribeca **Transmissions:** All	**O2 Sensor Circuit High Voltage Bank 2 Sensor 1** Engine started, engine running in closed loop at hot idle speed, and the ECM detected an unexpected "high" voltage condition on the front A/F sensor signal circuit for over 6 seconds in the CCM test. **Possible Causes:** • Check any other DTC on display • Using a scan tool, check current data • Check the harness between the ECM and the front sensor connector • Faulty front oxygen A/F sensor

DTC	Trouble Code Title, Conditions & Possible Causes
DTC: P0153 **2T CCM, MIL: Yes** **Years:** 2006, 2007, 2008 **Engines:** 3.0L VIN 8, 3.6L VIN 9 **Models:** B9 Tribeca, Tribeca **Transmissions:** All	**O2 Sensor Circuit Slow Response Bank 2 Sensor 1** Engine started, engine running in closed loop, ECT sensor more than 170°F, and the ECM detected the number of A/F S-21 rich-to-lean or lean-to-rich switches was less than a calibrated amount. **Possible Causes:** • Check any other DTC on display • Check exhaust system • Faulty front oxygen A/F sensor
DTC: P0154 **1T CCM, MIL: Yes** **Years:** 2006, 2007, 2008 **Engines:** 3.0L VIN 8, 3.6L VIN 9 **Models:** B9 Tribeca, Tribeca **Transmissions:** All	**O2 Sensor Circuit No Activity Detected Bank 2 Sensor 1** Detects open circuits of the sensor. Judge as NG when the impedance of the element is large. Poor drive ability. **Possible Causes:** • Check the harness between the ECM and the front sensor • Check poor contact • Faulty front oxygen A/F sensor
DTC: P0157 **2T CCM, MIL: Yes** **Years:** 2006, 2007, 2008 **Engines:** 3.0L VIN 8, 3.6L VIN 9 **Models:** B9 Tribeca, Tribeca **Transmissions:** All	**O2 Sensor Circuit Low Voltage Bank 2 Sensor 2** Engine started, vehicle driven at a speed of 30-55 MPH for 2 minutes, and the ECM detected an unexpected low voltage condition on the HO2S-22 circuit during the CCM test period Poor drive ability. **Possible Causes:** • Check any other DTC on display • Using a scan tool, check data • Check harness between ECM and rear sensor connector • Check harness between rear sensor and ECM connector • Check exhaust system • Faulty rear oxygen sensor
DTC: P0158 **2T CCM, MIL: Yes** **Years:** 2006, 2007, 2008 **Engines:** 3.0L VIN 8, 3.6L VIN 9 **Models:** B9 Tribeca, Tribeca **Transmissions:** All	**O2 Sensor Circuit High Voltage Bank 2 Sensor 2** Engine started, vehicle driven at a speed of 30-55 MPH for 2 minutes, and the ECM detected an unexpected high voltage condition on the HO2S-22 circuit during the CCM test period. **Possible Causes:** • Check any other DTC on display • Using a scan tool, check data • Check harness between ECM and rear sensor connector • Check harness between rear sensor and ECM connector • Check exhaust system • Faulty rear oxygen sensor
DTC: P0159 **2T CCM, MIL: Yes** **Years:** 2006, 2007, 2008 **Engines:** 3.0L VIN 8, 3.6L VIN 9 **Models:** B9 Tribeca, Tribeca **Transmissions:** All	**O2 Sensor Circuit Slow Response Bank 2 Sensor 2** Engine started, engine running in closed loop, ECT sensor signal more than 170°F, and the ECM detected the number of HO2S-22 rich-to-lean or lean-to-rich switches was less than a calibrated amount. Poor drive ability. **Possible Causes:** • Check any other DTC on display • Check harness between ECM and rear sensor connector • Faulty rear oxygen sensor
DTC: P0160 **2T CCM, MIL: Yes** **Years:** 2006, 2007, 2008 **Engines:** 3.0L VIN 8, 3.6L VIN 9 **Models:** B9 Tribeca, Tribeca **Transmissions:** All	**O2 Sensor Circuit No Activity Detected Bank 2 Sensor 2** Detects open circuits of the sensor. Judge as NG when the impedance of the element is large. Poor drive ability. **Possible Causes:** • Check any other DTC on display • Using a scan tool, check data • Check harness between ECM and rear sensor connector • Check harness between rear sensor and ECM connector • Check exhaust system • Faulty rear oxygen sensor

DTC	Trouble Code Title, Conditions & Possible Causes
DTC: P0171 **2T CCM, MIL: Yes** **Years:** 2006, 2007, 2008 **Engines:** 2.5L SOHC, DOHC VIN 6, 3.0L VIN 8, 3.6L VIN 9 **Models:** All **Transmissions:** All	**System Too Lean Bank 1** Engine coolant temperature is 167°F (75°C). The Fuel system is diagnosed by comparing the target air fuel ratio calculated by ECM with the actual air fuel ratio measured by sensor. Improper idling. Engine stalls. Poor driving performance. **Possible Causes:** • Check any other DTC on display • Check exhaust system • Check air intake system • Check fuel pressure • Check fuel injector • Check coolant temperature • Check intake manifold pressure sensor • Check intake air temperature sensor
DTC: P0172 **2T CCM, MIL: Yes** **Years:** 2006, 2007, 2008 **Engines:** 2.5L SOHC, DOHC VIN 6, 3.0L VIN 8, 3.6L VIN 9 **Models:** All **Transmissions:** All	**System Too Rich Bank 1** Engine coolant temperature is 167°F (75°C). The Fuel system is diagnosed by comparing the target air fuel ratio calculated by ECM with the actual air fuel ratio measured by sensor. Improper idling. Engine stalls. Poor driving performance. **Possible Causes:** • Check any other DTC on display • Check exhaust system • Check air intake system • Check fuel pressure • Check fuel injector • Check coolant temperature • Check intake manifold pressure sensor • Check intake air temperature sensor
DTC: P0174 **2T CCM, MIL: Yes** **Years:** 2006, 2007, 2008 **Engines:** 3.0L VIN 8, 3.6L VIN 9 **Models:** B9 Tribeca, Tribeca, Legacy, Outback **Transmissions:** All	**System Too Lean Bank 2** Engine coolant temperature is 167°F (75°C). The Fuel system is diagnosed by comparing the target air fuel ratio calculated by ECM with the actual air fuel ratio measured by sensor. Improper idling. Engine stalls. Poor driving performance. **Possible Causes:** • Check any other DTC on display • Check exhaust system • Check air intake system • Check any other DTC on display • Check exhaust system • Check air intake system • Check fuel pressure • Check fuel injector • Check coolant temperature • Check intake manifold pressure sensor • Check intake air temperature sensor
DTC: P0175 **2T CCM, MIL: Yes** **Years:** 2006, 2007, 2008 **Engines:** 3.0L VIN 8, 3.6L VIN 9 **Models:** B9 Tribeca, Tribeca, Legacy, Outback **Transmissions:** All	**System Too Rich Bank 2** Engine coolant temperature is 167°F (75°C). The Fuel system is diagnosed by comparing the target air fuel ratio calculated by ECM with the actual air fuel ratio measured by sensor. Improper idling. Engine stalls. Poor driving performance. **Possible Causes:** • Check any other DTC on display • Check exhaust system • Check air intake system • Check fuel pressure • Check fuel injector • Check coolant temperature • Check intake manifold pressure sensor • Check intake air temperature sensor
DTC: P0181 **2T CCM, MIL: Yes** **Years:** 2006, 2007, 2008 **Engines:** 2.5L SOHC, DOHC VIN 6, 3.0L VIN 8, 3.6L VIN 9 **Models:** All **Transmissions:** All	**Fuel Temperature Sensor "A" Circuit Range/Performance** Engine started, vehicle driven to a speed of over 3 MPH in closed loop, and the ECM detected the Fuel Temperature Sensor 'A' signal was out-of-range, or the signal was not plausible. **Possible Causes:** • Check any other DTC on display • Faulty fuel temperature sensor

DTC	Trouble Code Title, Conditions & Possible Causes
DTC: P0182 **1T CCM, MIL: Yes** **Years:** 2006, 2007, 2008 **Engines:** 2.5L SOHC, DOHC VIN 6, 3.0L VIN 8, 3.6L VIN 9 **Models:** All **Transmissions:** All	**Fuel Temperature Sensor "A" Circuit Low Input** Key on or engine running; vehicle driven to a speed of over 3 MPH, and the ECM detected the Fuel Temperature Sensor 'A' signal was less than 0.10 volts. **Possible Causes:** • Using a scan tool, check data • Shorted ground circuit of harness between ECM and fuel pump • Faulty fuel temperature sensor
DTC: P0183 **1T CCM, MIL: Yes** **Years:** 2006, 2007, 2008 **Engines:** 2.5L SOHC, DOHC VIN 6, 3.0L VIN 8, 3.6L VIN 9 **Models:** All **Transmissions:** All	**Fuel Temperature Sensor "A" Circuit High Input** Key on or engine running; vehicle driven to a speed of over 3 MPH, and the PCM detected the Fuel Temperature Sensor 'A' signal was more than 4.80 volts. **Possible Causes:** • Using a scan tool, check data • Shorted ground circuit of harness between ECM and fuel pump • Faulty fuel temperature sensor
DTC: P0196 **2T CCM, MIL: Yes** **Years:** 2006, 2007, 2008 **Engines:** 2.5L SOHC, VIN 6, 3.0L VIN 8, 3.6L VIN 9 **Models:** B9 Tribeca, Tribeca, Impreza, Legacy, Outback **Transmissions:** All	**Engine Oil Temperature Sensor Circuit Range/Performance** Detected when two consecutive driving cycles with fault occur. Hard starting condition. Engine stalls. Erroneous idling. Poor driving performance. **Possible Causes:** • Check any other DTC on display • Faulty oil temperature sensor
DTC: P0197 **1T CCM, MIL: Yes** **Years:** 2006, 2007, 2008 **Engines:** 2.5L SOHC, VIN 6, 3.0L VIN 8, 3.6L VIN 9 **Models:** B9 Tribeca, Tribeca, Impreza, Legacy, Outback **Transmissions:** All	**Engine Oil Temperature Sensor Circuit Low** Detect the open or short circuit of the oil temperature sensor Hard starting condition. Engine stalls. Erroneous idling. Poor driving performance. **Possible Causes:** • Check harness between sensor and ECM • Check poor contact • Faulty oil temperature sensor
DTC: P0198 **1T CCM, MIL: Yes** **Engines:** 2.5L SOHC, VIN 6, 3.0L VIN 8, 3.6L VIN 9 **Models:** B9 Tribeca, Tribeca, Impreza, Legacy, Outback **Transmissions:** All	**Engine Oil Temperature Sensor Circuit High** Detects the open or short circuit of the oil temperature sensor Hard starting condition. Engine stalls. Improper idling. Poor driving performance. **Possible Causes:** • Check harness between sensor and ECM • Check poor contact • Faulty oil temperature sensor
DTC: P0222 **1T CCM, MIL: Yes** **Years:** 2006, 2007, 2008 **Engines:** 2.5L SOHC, DOHC VIN 6, 3.0L VIN 8, 3.6L VIN 9 **Models:** All **Transmissions:** All	**Throttle/Pedal Position Sensor/Switch "B" Circuit Low Input** Ignition switch on and battery volts at least 6 volts. Detects the open or short circuit of throttle position sensor 2. Improper idling. Poor driving performance. Engine stalls. **Possible Causes:** • Check sensor output • Check poor contact • Check harness between ECM and electronic throttle control • Check sensor power supply • Check for short circuit in ECM
DTC: P0223 **1T CCM, MIL: Yes** **Years:** 2006, 2007, 2008 **Engines:** 2.5L SOHC, DOHC VIN 6, 3.0L VIN 8, 3.6L VIN 9 **Models:** All **Transmissions:** All	**Throttle/Pedal Position Sensor/Switch "B" Circuit High Input** Ignition switch on and battery volts at least 6 volts. Detects the open or short circuit of throttle position sensor 2. Improper idling. Poor driving performance. Engine stalls. **Possible Causes:** • Check sensor output • Check poor contact • Check harness between ECM and electronic throttle control • Check for short circuit in ECM

DTC	Trouble Code Title, Conditions & Possible Causes
DTC: P0230 **2T CCM, MIL: Yes** **Years:** 2006, 2007, 2008 **Engines:** 2.5L SOHC, DOHC VIN 6, 3.0L VIN 8, 3.6L VIN 9 **Models:** All **Transmissions:** All	**Fuel Pump Primary Circuit** Detect the malfunction of fuel pump control unit. Judge as NG when the NG signal is sent through a diagnostic line coming from the fuel pump control unit. Fuel pump control unit detects the open or short circuit malfunction for each line, and then sends NG signals if one of them is found NG. Poor drive ability. **Possible Causes:** • Check power supply to fuel pump control unit • Check ground circuit of fuel pump control unit • Check harness between the fuel pump control unit and fuel pump connector • Check poor contact • Check if vehicle has previously ran out of fuel • Faulty fuel pump control unit
DTC: P0301 **2T CCM, MIL: Yes** **Years:** 2006, 2007, 2008 **Engines:** 2.5L SOHC, DOHC VIN 6, 3.0L VIN 8, 3.6L VIN 9 **Models:** All **Transmissions:** All	**Cylinder No. 1 Misfire Detected** Engine started, vehicle driven to a speed of over 3 MPH, and the ECM detected a misfire in Cylinder 1 in the 200 (Catalyst) or 1000 RPM (High Emissions) revolution range. Engine stalls. Improper idling. Rough driving condition. **Note: If the misfire is severe, the MIL will flash on/off on the 1st trip!** **Possible Causes:** • Check any other DTC on display • Check output signal of ECM • Check harness between injector and ECM connector • Check fuel injector • Check power supply line • Check power supply to mass air flow sensor • Check harness between ECM and sensor connector • Check installation of camshaft and crankshaft sensors • Check crank plate • Check condition of timing chain • Check fuel level • Check status of MIL light • Check air intake system • Check cause of misfire (engine running) • Check engine condition • Check all cylinders individually, in pairs, in groups and at random
DTC P0302 **2T CCM, MIL: Yes** **Years:** 2006, 2007, 2008 **Engines:** 2.5L SOHC, DOHC VIN 6, 3.0L VIN 8, 3.6L VIN 9 **Models:** All **Transmissions:** All	**Cylinder No. 2 Misfire Detected** Engine started, vehicle driven to a speed of over 3 MPH, and the ECM detected a misfire in Cylinder 1 in the 200 (Catalyst) or 1000 RPM (High Emissions) revolution range. Engine stalls. Improper idling. Rough driving condition. **Note: If the misfire is severe, the MIL will flash on/off on the 1st trip!** **Possible Causes:** • Check any other DTC on display • Check output signal of ECM • Check harness between injector and ECM connector • Check fuel injector • Check power supply line • Check power supply to mass air flow sensor • Check harness between ECM and sensor connector • Check installation of camshaft and crankshaft sensors • Check crank plate • Check condition of timing chain • Check fuel level • Check status of MIL light • Check air intake system • Check cause of misfire (engine running) • Check engine condition • Check all cylinders individually, in pairs, in groups and at random

DTC	Trouble Code Title, Conditions & Possible Causes
DTC: P0303 **2T CCM, MIL: Yes** **Years:** 2006, 2007, 2008 **Engines:** 2.5L SOHC, DOHC VIN 6, 3.0L VIN 8, 3.6L VIN 9 **Models:** All **Transmissions:** All	**Cylinder No. 3 Misfire Detected** Engine started, vehicle driven to a speed of over 3 MPH, and the ECM detected a misfire in Cylinder 1 in the 200 (Catalyst) or 1000 RPM (High Emissions) revolution range. Engine stalls. Improper idling. Rough driving condition. **Note: If the misfire is severe, the MIL will flash on/off on the 1st trip!** **Possible Causes:** • Check any other DTC on display • Check output signal of ECM • Check harness between injector and ECM connector • Check fuel injector • Check power supply line • Check power supply to mass air flow sensor • Check harness between ECM and sensor connector • Check installation of camshaft and crankshaft sensors • Check crank plate • Check condition of timing chain • Check fuel level • Check status of MIL light • Check air intake system • Check cause of misfire (engine running) • Check engine condition • Check all cylinders individually, in pairs, in groups and at random
DTC: P0304 **2T CCM, MIL: Yes** **Years:** 2006, 2007, 2008 **Engines:** 2.5L SOHC, DOHC VIN 6, 3.0L VIN 8, 3.6L VIN 9 **Models:** All **Transmissions:** All	**Cylinder No. 4 Misfire Detected** Engine started, vehicle driven to a speed of over 3 MPH, and the ECM detected a misfire in Cylinder 1 in the 200 (Catalyst) or 1000 RPM (High Emissions) revolution range. Engine stalls. Improper idling. Rough driving condition. **Note: If the misfire is severe, the MIL will flash on/off on the 1st trip!** **Possible Causes:** • Check any other DTC on display • Check output signal of ECM • Check harness between injector and ECM connector • Check fuel injector • Check power supply line • Check power supply to mass air flow sensor • Check harness between ECM and sensor connector • Check installation of camshaft and crankshaft sensors • Check crank plate • Check condition of timing chain • Check fuel level • Check status of MIL light • Check air intake system • Check cause of misfire (engine running) • Check engine condition • Check all cylinders individually, in pairs, in groups and at random
DTC: P0305 **2T CCM, MIL: Yes** **Years:** 2006, 2007, 2008 **Engines:** 3.0L VIN 8, 3.6L VIN 9 **Models:** B9 Tribeca, Tribeca, Legacy, Outback **Transmissions:** All	**Cylinder No. 5 Misfire Detected** Engine started, vehicle driven to a speed of over 3 MPH, and the ECM detected a misfire in Cylinder 1 in the 200 (Catalyst) or 1000 RPM (High Emissions) revolution range. Engine stalls. Improper idling. Rough driving condition. **Note: If the misfire is severe, the MIL will flash on/off on the 1st trip!** **Possible Causes:** • Check any other DTC on display • Check output signal of ECM • Check harness between injector and ECM connector • Check fuel injector • Check power supply line • Check power supply to mass air flow sensor • Check harness between ECM and sensor connector • Check installation of camshaft and crankshaft sensors • Check crank plate • Check condition of timing chain • Check fuel level • Check status of MIL light • Check air intake system • Check cause of misfire (engine running) • Check engine condition • Check all cylinders individually, in pairs, in groups and at random

DTC	Trouble Code Title, Conditions & Possible Causes
DTC: P0306 **2T CCM, MIL: Yes** **Years:** 2006, 2007, 2008 **Engines:** 3.0L VIN 8, 3.6L VIN 9 **Models:** B9 Tribeca, Tribeca, Legacy, Outback **Transmissions:** All	**Cylinder No. 6 Misfire Detected** Engine started, vehicle driven to a speed of over 3 MPH, and the ECM detected a misfire in Cylinder 1 in the 200 (Catalyst) or 1000 RPM (High Emissions) revolution range. Engine stalls. Improper idling. Rough driving condition. **Note: If the misfire is severe, the MIL will flash on/off on the 1st trip!** **Possible Causes:** • Check any other DTC on display • Check output signal of ECM • Check harness between injector and ECM connector • Check fuel injector • Check power supply line • Check power supply to mass air flow sensor • Check harness between ECM and sensor connector • Check installation of camshaft and crankshaft sensors • Check crank plate • Check condition of timing chain • Check fuel level • Check status of MIL light • Check air intake system • Check cause of misfire (engine running) • Check engine condition • Check all cylinders individually, in pairs, in groups and at random
DTC: P0327 **1T CCM, MIL: Yes** **Years:** 2006, 2007, 2008 **Engines:** 2.5L SOHC, DOHC VIN 6, 3.0L VIN 8, 3.6L VIN 9 **Models:** All **Transmissions:** All	**Knock Sensor 1 Circuit Low Input Bank 1 or Single Sensor** Engine started, engine speed over 1200 RPM, system voltage from 11-16 volts, and the ECM detected an unexpected "low" voltage condition on the Knock Sensor (KS) 1 circuit during the CCM test. Poor driving condition. Knocking occurs. **Possible Causes:** • Check harness between sensor and ECM connector • Check sensor installation • Faulty KS
DTC: P0328 **1T CCM, MIL: Yes** **Years:** 2006, 2007, 2008 **Engines:** 2.5L SOHC, DOHC VIN 6, 3.0L VIN 8, 3.6L VIN 9 **Models:** All **Transmissions:** All	**Knock Sensor 1 Circuit High Input Bank 1 or Single Sensor** Engine started, engine speed over 1200 RPM, system voltage from 11-16 volts, and the ECM detected an unexpected "high" voltage condition on the Knock Sensor (KS) 1 circuit during the CCM test. Poor driving condition. Knocking occurs. **Possible Causes:** • Check harness between sensor and ECM connector • Check sensor • Check input signal of ECM • Faulty KS
DTC: P0332 **1T CCM, MIL: Yes** **Years:** 2006, 2007, 2008 **Engines:** 3.0L VIN 8, 3.6L VIN 9 **Models:** B9 Tribeca, Tribeca, Legacy, Outback **Transmissions:** All	**Knock Sensor 2 Circuit Low Input Bank 2** Engine started, engine speed over 1200 RPM, system voltage from 11-16 volts, and the ECM detected an unexpected "low" voltage condition on the Knock Sensor (KS) 2 circuit during the CCM test. Poor driving condition. Knocking occurs. **Possible Causes:** • Check harness between sensor and ECM connector • Check sensor installation • Faulty KS
DTC: P0333 **1T CCM, MIL: Yes** **Years:** 2006, 2007, 2008 **Engines:** 3.0L VIN 8, 3.6L VIN 9 **Models:** B9 Tribeca, Tribeca, Legacy, Outback **Transmissions:** All	**Knock Sensor 2 Circuit High Input Bank 2** Engine started, engine speed over 1200 RPM, system voltage from 11-16 volts, and the ECM detected an unexpected "high" voltage condition on the Knock Sensor (KS) 2 circuit during the CCM test. **Possible Causes:** • Check harness between sensor and ECM connector • Check sensor installation • Faulty KS • Check input signal of ECM
DTC: P0335 **1T CCM, MIL: Yes** **Years:** 2006, 2007, 2008 **Engines:** 2.5L SOHC, DOHC VIN 6, 3.0L VIN 8, 3.6L VIN 9 **Models:** All **Transmissions:** All	**Crankshaft Position Sensor "A" Circuit** Engine cranking; and then the ECM did not detect any Crankshaft Position (CKP) sensor signals, or the CKP sensor signal was interrupted after then engine was running during the CCM test. **Note: The engine will not start without a proper CKP sensor signal.** **Possible Causes:** • Check harness between the sensor and the ECM connector • Check condition of the sensor • Faulty CKP sensor

DTC	Trouble Code Title, Conditions & Possible Causes
DTC: P0336 **2T CCM, MIL: Yes** **Years:** 2006, 2007, 2008 **Engines:** 2.5L SOHC, DOHC VIN 6, 3.0L VIN 8, 3.6L VIN 9 **Models:** All **Transmissions:** All	**Crankshaft Position Sensor "A" Circuit range/Performance** Engine started, and the ECM did not detect any Crankshaft Position Sensor (CKP) signals, or the CKP sensor signal was interrupted after the engine was running during the CCM test. **Note: The engine may not start, or it may stall if it loses the proper CKP sensor signal after is has been running.** **Possible Causes:** • Check any other DTC on display • Check condition of the sensor • Check the crankshaft plate • Check timing chain • Faulty CKP sensor
DTC: P0340 **1T CCM, MIL: Yes** **Years:** 2006, 2007, 2008 **Engines:** 2.5L SOHC, DOHC VIN 6, 3.0L VIN 8, 3.6L VIN 9 **Models:** All **Transmissions:** All	**Camshaft Position Sensor "A" Circuit Bank 1 or Single Sensor** Engine cranking; and then the ECM did not detect any Camshaft Position (CMP) sensor signals, or the CMP sensor signal was interrupted after the engine was running during the CCM test. **Note: The engine may not start without a proper CMP sensor signal.** **Possible Causes:** • Check any other DTC on display • Check power supply • Check harness between sensor connector and ECM • Check sensor installation and condition • Check the sensor • Check for poor contact • Faulty CMP sensor
DTC: P0345 **1T CCM, MIL: Yes** **Years:** 2006, 2007, 2008 **Engines:** 2.5L DOHC VIN 6, 3.0L VIN 8, 3.6L VIN 9 **Models:** All **Transmissions:** All	**Camshaft Position Sensor "A" Circuit Bank 2** Detect the open or short circuit of the camshaft position sensor. Judge as NG when the number of camshaft signals remains abnormal. Engine stalls. Failure of engine to start. **Possible Causes:** • Check any other DTC on display • Check power supply • Check harness between sensor connector and ECM • Check the sensor • Check poor contact • Check the installation and condition of the sensor
DTC: P0365 **1T CCM, MIL: Yes** **Years:** 2008 **Engines:** 3.6L VIN 9 **Models:** Tribeca **Transmissions:** All	**Camshaft Position Sensor "B" Circuit Bank 1** Detect the open or short circuit of the camshaft position sensor. Judge as NG when the number of camshaft signals remains abnormal. Engine stalls. Failure of engine to start. **Possible Causes:** • Check any other DTC on display • Check power supply • Check harness between sensor connector and ECM • Check the sensor • Check poor contact • Check the installation and condition of the sensor
DTC: P0390 **1T CCM, MIL: Yes** **Years:** 2008 **Engines:** 3.6L VIN 9 **Models:** Tribeca **Transmissions:** All	**Camshaft Position Sensor "B" Circuit Bank 2** Detect the open or short circuit of the camshaft position sensor. Judge as NG when the number of camshaft signals remains abnormal. Engine stalls. Failure of engine to start. **Possible Causes:** • Check any other DTC on display • Check power supply • Check harness between sensor connector and ECM • Check the sensor • Check poor contact • Check the installation and condition of the sensor
DTC: P0400 **2T CCM, MIL: Yes** **Years:** 2006, 2007, 2008 **Engines:** 2.5L SOHC, VIN 6, 3.0L, 3.6L VIN 9 **Models:** Baja, Forester Impreza, B9 Tribeca, Tribeca **Transmissions:** All	**Exhaust Gas Recirculation Flow** Movement performance problem when engine is low speed. Erroneous idling. Movement performance problem. **Possible Causes:** • Plugged piping or foreign objects caught in the EGR system • Manifold absolute pressure sensor and throttle body improperly installed • Faulty EGR valve

DTC	Trouble Code Title, Conditions & Possible Causes
DTC: P0410 **2T CCM, MIL: Yes** **Years:** 2006, 2007, 2008 **Engines:** 2.5L DOHC **Models:** Forester, Impreza, Legacy, Outback **Transmissions:** All	**Secondary Air Injection System** Faulty secondary air delivery pipe pressure, pulse of secondary air delivery pipe pressure and secondary air pipe airflow amount. **Possible Causes:** • Blown fuse • Check harness between fuse, pump harness and ground • Damage, clogged or disconnected duct • Faulty secondary air combination valve • Open circuit between air pump relay and pump • Faulty air pump relay • Faulty air pump relay power supply circuit • Faulty air pump
DTC: P0411 **2T CCM, MIL: Yes** **Years:** 2006, 2007, 2008 **Engines:** 2.5L DOHC **Models:** Forester, Impreza, Legacy, Outback **Transmissions:** All	**Secondary Air Injection System Incorrect Flow Detected** Faulty secondary air delivery pipe pressure, pulse of secondary air delivery pipe pressure and secondary air pipe airflow amount. **Possible Causes:** • Blown fuse • Check harness between fuse, pump harness and ground • Damage, clogged or disconnected duct • Faulty secondary air combination valve • Open circuit between air pump relay and pump • Faulty air pump relay • Faulty air pump relay power supply circuit • Faulty air pump
DTC: P0413 **1T CCM, MIL: Yes** **Years:** 2006, 2007, 2008 **Engines:** 2.5L DOHC **Models:** Forester, Impreza, Legacy, Outback **Transmissions:** All	**Secondary Air Injection System Switching Valve "A" Circuit Open** The ECM output level differs from the actual terminal level. **Possible Causes:** • Open circuit of harness between ECM and secondary air combination valve relay • Ground short circuit of harness between ECM and secondary air combination valve relay
DTC: P0414 **1T CCM, MIL: Yes** **Years:** 2006, 2007, 2008 **Engines:** 2.5L DOHC **Models:** Forester, Impreza, Legacy, Outback **Transmissions:** All	**Secondary Air Injection System Switching Valve "A" Circuit Shorted** The ECM output level differs from the actual terminal level. **Possible Causes:** • Short circuit to power in the harness between ECM and secondary air combination valve relay 1 • Poor contact of the ECM connector
DTC: P0416 **1T CCM, MIL: Yes** **Years:** 2006, 2007, 2008 **Engines:** 2.5L DOHC **Models:** Forester, Impreza, Legacy, Outback **Transmissions:** All	**Secondary Air Injection System Switching Valve "B" Circuit Open** The ECM output level differs from the actual terminal level. **Possible Causes:** • Open circuit of harness between ECM and secondary air combination valve relay • Ground short circuit of harness between ECM and secondary air combination valve relay.
DTC: P0417 **1T CCM, MIL: Yes** **Years:** 2006, 2007, 2008 **Engines:** 2.5L DOHC **Models:** Forester, Impreza, Legacy, Outback **Transmissions:** All	**Secondary Air Injection System Switching Valve "B" Circuit Shorted** The ECM output level differs from the actual terminal level. **Possible Causes:** • Short circuit to power in the harness between ECM and secondary air combination valve relay 1 • Poor contact of the ECM connector
DTC: P0418 **1T CCM, MIL: Yes** **Years:** 2006, 2007, 2008 **Engines:** 2.5L DOHC **Models:** Forester, Impreza, Legacy, Outback **Transmissions:** All	**Secondary Air Injection System Control "A" Circuit** The ECM output level differs from the actual terminal level. **Possible Causes:** • Short circuit of harness between ECM and secondary air pump relay • Open circuit of harness between ECM and secondary air pump relay

DTC	Trouble Code Title, Conditions & Possible Causes
DTC: P0420 **2T CCM, MIL: Yes** **Years:** 2006, 2007, 2008 **Engines:** 2.5L SOHC, DOHC VIN 6, 3.0L VIN 8, 3.6L VIN 9 **Models:** All **Transmissions:** All	**Catalyst System Efficiency Below Threshold Bank 1** Engine started, vehicle driven at 45-60 MPH in closed loop for 3-5 minutes, and the ECM detected the amplitudes of the HO2S-11 and HO2S-12 signals were too similar during the test. Engine stalls. Idle mixture is out of specifications. **Possible Causes:** • Check any other DTC on display • Check harness between fuse and rear oxygen sensor connector • Check exhaust system • Faulty catalytic converter
DTC: P0442 **2T CCM, MIL: Yes** **Years:** 2006, 2007, 2008 **Engines:** 2.5L SOHC, DOHC VIN 6, 3.0L VIN 8, 3.6L VIN 9 **Models:** All **Transmissions:** All	**Evaporative Emission Control System Leak Detected (Small Leak)** The ECM detected a vacuum decaying condition existed during the test. There is a hole of more than 0.04 inch (1.0 m m) diameter in evaporation system or fuel tank. Possible fuel odor. **Possible Causes:** • Check any other DTC on display • Check fuel filler cap • Check fuel filler pipe packing • Check the drain valve • Check the purge control solenoid valve • Check the pressure control solenoid valve • Check evaporative emission control system line • Check canister • Check fuel tank • Check any other mechanical components in the evaporative control system
DTC: P0447 **2T CCM, MIL: Yes** **Years:** 2006, 2007, 2008 **Engines:** 2.5L SOHC, DOHC VIN 6, 3.0L VIN 8, 3.6L VIN 9 **Models:** All **Transmissions:** All	**Evaporative Emission Control System Vent Control Circuit Open** Key on or engine running; and the ECM detected an open or short circuit of the drain valve. **Possible Causes:** • Check output signal from ECM • Check poor contact • Check harness between drain valve and ECM connector • Check drain valve • Check power supply to drain valve
DTC: P0448 **2T CCM, MIL: Yes** **Years:** 2006, 2007, 2008 **Engines:** 2.5L SOHC, DOHC VIN 6, 3.0L VIN 8, 3.6L VIN 9 **Models:** All **Transmissions:** All	**Evaporative Emission Control System Vent Control Circuit Shorted** Key on or engine running; and the ECM detected an open or short circuit of the drain valve. **Possible Causes:** • Check input signal from ECM • Check poor contact • Check harness between drain valve and ECM connector • Check drain valve
DTC: P0451 **2T CCM, MIL: Yes** **Years:** 2006, 2007, 2008 **Engines:** 2.5L SOHC, DOHC VIN 6, 3.0L VIN 8, 3.6L VIN 9 **Models:** All **Transmissions:** All	**Evaporative Emission Control System Pressure Sensor Malfunction** Detect the tank pressure sensor output property abnormality. Judge as NG when there is no pressure variation, which should exist in the tank, considering the engine status. **Possible Causes:** • Check any other DTC on display • Check fuel filler cap • Check pressure vacuum line • Faulty fuel tank pressure sensor
DTC: P0452 **2T CCM, MIL: Yes** **Years:** 2006, 2007, 2008 **Engines:** 2.5L SOHC, DOHC VIN 6, 3.0L VIN 8, 3.6L VIN 9 **Models:** All **Transmissions:** All	**Evaporative Emission Control System Pressure Sensor Low Input** Key on or engine running; and the ECM detected Detect an open or short circuit of the fuel tank pressure sensor. **Possible Causes:** • Using a scan tool, check current data • Check power supply to fuel tank pressure sensor • Check input signal to ECM • Check harness between ECM and coupling connector in rear wiring harness • Check fuel tank cord • Check the purge control solenoid valve • Check poor contact • Faulty fuel tank pressure sensor

DTC	Trouble Code Title, Conditions & Possible Causes
DTC P0453 **1T CCM, MIL: Yes** **Years:** 2006, 2007, 2008 **Engines:** 2.5L SOHC, DOHC VIN 6, 3.0L VIN 8, 3.6L VIN 9 **Models:** All **Transmissions:** All	**Evaporative Emission Control System Pressure Sensor High Input** Key on or engine running; and the ECM detected Detect an open or short circuit of the fuel tank pressure sensor. **Possible Causes:** • Using a scan tool, check current data • Check power supply to fuel tank pressure sensor • Check input signal to ECM • Check harness between ECM and coupling connector in rear wiring harness • Check fuel tank cord • Check poor contact • Check harness between ECM and fuel tank pressure sensor connector • Faulty fuel tank pressure sensor
DTC: P0456 **2T CCM, MIL: Yes** **Years:** 2006, 2007, 2008 **Engines:** 2.5L SOHC, DOHC VIN 6, 3.0L VIN 8, 3.6L VIN 9 **Models:** All **Transmissions:** All	**Evaporative Emission Control System Pressure Sensor Very Small Leak Detected** The ECM detected a vacuum decaying condition existed during the test. There is a hole of more than 0.04 inch (1.0 m m) diameter in evaporation system or fuel tank. Possible fuel odor. **Possible Causes:** • Check any other DTC on display • Check fuel filler cap • Check fuel filler pipe packing • Check drain valve • Check purge control solenoid valve • Check pressure control solenoid valve • Check harness between ECM and fuel tank pressure sensor connector • Check evaporative emission control system line • Check canister • Check any other mechanical component in the system
DTC: P0457 **2T CCM, MIL: Yes** **Years:** 2006, 2007, 2008 **Engines:** 2.5L SOHC, DOHC VIN 6, 3.0L VIN 8, 3.6L VIN 9 **Models:** All **Transmissions:** All	**Evaporative Emission Control System Leak Detected (Fuel Cap Loose/Off)** Engine started, and the ECM detected an unexpected low voltage condition on the EVAP Purge Controls solenoid control circuit. **Possible Causes:** • Check any other DTC on display • Check fuel filler cap • Check fuel filler pipe packing • Check drain valve • Check purge control solenoid valve • Check pressure control solenoid valve • Check canister • Check fuel tank • Check any other mechanical component in the system
DTC: P0458 **2T CCM, MIL: Yes** **Years:** 2006, 2007, 2008 **Engines:** 2.5L SOHC, DOHC VIN 6, 3.0L VIN 8, 3.6L VIN 9 **Models:** All **Transmissions:** All	**Evaporative Emission System Purge Control Valve Circuit Low** Battery voltage 10.9 volts. Detect open or short circuit of the purge control solenoid valve. Elapsed time after start up 1 second. Improper idling. **Possible Causes:** • Check output signal of ECM • Check harness between purge control solenoid valve and ECM connector • Check purge control solenoid valve • Check power supply to purge control solenoid valve • Check poor contact
DTC: P0459 **2T CCM, MIL: Yes** **Years:** 2006, 2007, 2008 **Engines:** 2.5L SOHC, DOHC VIN 6, 3.0L VIN 8, 3.6L VIN 9 **Models:** All **Transmissions:** All	**Evaporative Emission System Purge Control Valve Circuit High** Battery voltage 10.9 volts. Detect open or short circuit of the purge control solenoid valve. Elapsed time after start up 1 second. Improper idling. **Possible Causes:** • Check output signal of ECM • Check harness between purge control solenoid valve and ECM connector • Check purge control solenoid valve • Check poor contact • Faulty purge control solenoid valve

DTC	Trouble Code Title, Conditions & Possible Causes
DTC: P0461 **2T CCM, MIL: Yes** **Years:** 2006, 2007, 2008 **Engines:** 2.5L SOHC, DOHC VIN 6, 3.0L VIN 8, 3.6L VIN 9 **Models:** All **Transmissions:** All	**Fuel Level Sensor "A" Circuit Range/Performance** Engine started, then after the vehicle was driven a predetermined amount of miles, the ECM detected the Fuel Level sensor signal changed less than 0.14 volt, or the ECM detected the Fuel Level sensor signal was too low or tool high during the CCM test. **Possible Causes:** • Check any other DTC on display • Faulty fuel level sensor and fuel sub level sensor
DTC: P0462 **2T CCM, MIL: Yes** **Years:** 2006, 2007, 2008 **Engines:** 2.5L SOHC, DOHC VIN 6, 3.0L VIN 8, 3.6L VIN 9 **Models:** All **Transmissions:** All	**Fuel Level Sensor "A" Circuit Low** Key on or engine running; and the ECM detected the Fuel Level sensor signal indicated less than 0.10 volt at any time during the CCM continuous test. **Possible Causes:** • Check speedometer and tachometer operation • Check input signal of ECM • Check input voltage from ECM • Check harness between ECM and combination meter connector • Check fuel tank cord • Check fuel level sensor • Check fuel sub level sensor
DTC: P0463 **2T CCM, MIL: Yes** **Years:** 2006, 2007, 2008 **Engines:** 2.5L SOHC, DOHC VIN 6, 3.0L VIN 8, 3.6L VIN 9 **Models:** All **Transmissions:** All	**Fuel Level Sensor "A" Circuit High** Key on or engine running; and the ECM detected the Fuel Level sensor signal indicated more than 4.60 volts at any time during the CCM continuous test. **Possible Causes:** • Check any other DTC on display • Check harness between ECM and combination meter connector
DTC: P0464 **2T CCM, MIL: Yes** **Years:** 2006, 2007, 2008 **Engines:** 2.5L SOHC, DOHC VIN 6, 3.0L VIN 8, 3.6L VIN 9 **Models:** All **Transmissions:** All	**Fuel level Sensor Circuit Intermittent** Detect the unstable output faults from the fuel level sensor caused by noise. Judge as NG when the Max value and cumulative value of output voltage variation of the fuel level sensor is larger than the threshold value. **Possible Causes:** • Check any other DTC on display • Check harness between ECM and combination meter connector
DTC: P0483 **2T CCM, MIL: Yes** **Years:** 2006, 2007, 2008 **Engines:** 2.5L SOHC, DOHC VIN 6, 3.0L VIN 8 **Models:** Baja, Forester, Impreza, Legacy, Outback **Transmissions:** All	**Fan Rationality Check** Detect the function abnormality of the fan operation. Judge as NG when the engine coolant temperature slowly decreases even when the radiator fan is rotating. **Possible Causes:** • Check any other DTC on display • Check radiator fan and fan motor
DTC: P0500 **1T CCM, MIL: Yes** **Years:** 2006, 2007, 2008 **Engines:** 2.5L SOHC, DOHC VIN 6, 3.0L VIN 8 **Models:** Forester, Impreza, Legacy, Outback **Transmissions:** All	**Vehicle Speed Sensor "A"** Engine started, vehicle driven to a speed of over 30 MPH at medium engine load for 30 seconds, and the ECM did not detect any VSS signals during the CCM Rationality test. **Possible Causes:** • Check DTC of ABS • Check poor contact of the ECM connector.
DTC: P0506 **2T CCM, MIL: Yes** **Engines:** 2.5L SOHC, DOHC VIN 6, 3.0L VIN 8, 3.6L VIN 9 **Models:** All **Transmissions:** All	**Idle Air Control System RPM Lower Than Expected** Engine started. The ECM Detects a malfunction in actual engine speed is not close to target engine speed during idling. The engine may be hard to start, stall or not start when this code is set. Improper idling could also be caused. **Possible Causes:** • Check any other DTC on display • Clogged air cleaner element • Check electronic throttle control

DTC	Trouble Code Title, Conditions & Possible Causes
DTC: P0507 **2T CCM, MIL: Yes** **Engines:** 2.5L SOHC, DOHC VIN 6, 3.0L VIN 8, 3.6L VIN 9 **Models:** All **Transmissions:** All	**Idle Air Control System RPM Higher Than Expected** Engine started The ECM detected the Actual idle speed was 100-200 RPM more than the Target idle speed for over 10 seconds. **Note: Improper idling could also be caused.** **Possible Causes:** • Check any other DTC on display • Check air intake system • Check electronic throttle control
DTC: P0512 **2T CCM, MIL: No** **Engines:** 2.5L SOHC, DOHC VIN 6, 3.0L VIN 8, 3.6L VIN 9 **Models:** All **Transmissions:** All	**Starter Request Circuit** Detect the open or short circuit of starter SW. Judge as ON NG when the starter SW signal remains ON. Engine Fails to start **Possible Causes:** • Check any other DTC on display • Short circuit to power in the harness between the ECM and ignition switch
DTC: P0513 **2T CCM, MIL: No** **Engines:** 2.5L SOHC, DOHC VIN 6, 3.0L VIN 8, 3.6L VIN 9 **Models:** B9 Tribeca, Tribeca, Forester, Impreza, Legacy, Outback **Transmissions:** All	**Incorrect Immobilizer Key** Use of unregistered key in the body integrated unit **Possible Causes:** • Perform teaching operation of ignition key • Check the ignition keys (including transponder) which cannot be registered • Replace the body integrated unit and replace all the ignition keys (including transponder). • Execute the registration procedure next
DTC: P0519 **1T CCM, MIL: Yes** **Years:** 2006, 2007, 2008 **Engines:** 2.5L SOHC, DOHC VIN 6, 3.0L VIN 8 **Models:** Baja, B9, Tribeca Forester, Impreza, Legacy, Outback **Transmissions:** All	**Idle Air Control System Performance** Detect malfunctions in which the engine RPM continues to rise during idling. Engine keeps running at higher speed than specified idle speed. **Possible Causes:** • Check any other DTC on display • Check air intake system • Vacuum leaks • Check electronic throttle control
DTC: P0600 **1T CCM, MIL: Yes** **Years:** 2006, 2007, 2008 **Engines:** 2.5L SOHC, DOHC VIN 6, 3.0L VIN 8, 3.6L VIN 9 **Models:** All **Transmissions:** All	**Serial communication link** Subaru Select Monitor is required for reading DTC, performing diagnosis and reading current data. **Note: Check harness for broken wires or short circuits, shake trouble spot or connector.** **Possible Causes:** • Check the harness between ECM and TCM which might affect body control • Check resistance between ECM connector and chassis • Faulty ECM
DTC: P0602 **1T CCM, MIL: Yes** **Years:** 2006 **Engines:** 2.5L SOHC **Models:** Legacy **Transmissions:** All	**Control Module Programming Error** Engine keeps running at higher speed than specified idle speed. Engine keeps running at lower speed than specified idle speed. Engine stalls. **Possible Causes:** • Check any other DTC on display • Check engine oil level • Check exhaust system • Check all mechanical parts • Check all electrical out put sensors for open or short to ground • Check for poor connections at ECM and sensors • Faulty ECM
DTC: P0604 **1T CCM, MIL: Yes** **Years:** 2006, 2007, 2008 **Engines:** 3.0L VIN 8, 3.6L VIN 9 **Models:** B9 Tribeca, Tribeca **Transmissions:** All	**Internal Control Module Random Access Memory (RAM) Error** Engine stalls. Engine does not start. **Possible Causes:** • Check any other DTC on display • There may be a temporary connector contact failure • Check any other DTC on display • There may be a temporary connector contact failure • Faulty control module

DTC	Trouble Code Title, Conditions & Possible Causes
DTC: P0605 **1T CCM, MIL: Yes** **Years:** 2006, 2007, 2008 **Engines:** 2.5L SOHC, DOHC VIN 6, 3.0L VIN 8, 3.6L VIN 9 **Models:** All **Transmissions:** All	**Internal Control Module Read Only Memory (ROM) Error** SUM value of ROM is outside the standard value. Ignition switch is on. **Possible Causes:** • Check any other DTC on display • There may be a temporary connector contact failure • Faulty control module
DTC: P0607 **2T CCM, MIL: Yes** **Years:** 2006, 2007, 2008 **Engines:** 2.5L SOHC, DOHC VIN 6, 3.0L VIN 8, 3.6L VIN 9 **Models:** All **Transmissions:** All	**Control Module Performance** Judge as NG when any one of the followings is established: When the read value of throttle position sensor 1 signal is mismatched between main CPU and sub CPU. When the read value of accelerator pedal position sensor 1 signal is mismatched between main CPU and sub CPU. When the sub CPU operates abnormally. When the communication between main CPU sub. CPU is abnormal. When the input amplifier circuit of throttle position sensor 1 is abnormal. When the cruise control cannot be canceled correctly. When the signal of brake SW1 and 2 is mismatched. When the directed angle from the main CPU is abnormal. Improper idling. Poor driving conditions. **Possible Causes:** • Open, grounded shorted circuits of power supply circuit • Open, grounded shorted circuits of ground supply circuit • Faulty control module connections • Faulty control module
DTC: P0638 **1T CCM, MIL: Yes** **Years:** 2006, 2007, 2008 **Engines:** 2.5L SOHC, DOHC VIN 6, 3.0L VIN 8, 3.6L VIN 9 **Models:** All **Transmissions:** All	**Throttle Actuator Control range/Performance Bank 1** Judge as NG when the target opening angle and actual opening angle is mismatched or the current to motor is the specified duty or more for specified time continuously. Improper idling. Engine stalls. Poor driving performance. **Possible Causes:** • Check electronic throttle control relay • Check power supply of electronic control relay • Check harness between ECM and electronic throttle control relay • Check sensor output • Check poor contact • Check harness between ECM and electronic throttle control • Check sensor power supply • Check for short in ECM • Check sensor output • Check harness between ECM and electronic throttle control motor • Check electronic throttle control motor harness • Check ground circuit • Check electronic throttle control
DTC: P0691 **2T CCM, MIL: No** **Years:** 2006, 2007, 2008 **Engines:** 2.5L SOHC, DOHC VIN 6, 3.0L VIN 8 **Models:** Baja, B9, Tribeca Forester, Impreza, Legacy, Outback **Transmissions:** All	**Cooling Fan 1 Control Circuit Low** Detects the open or short circuit of radiator fan circuit. Radiator fan does not operate properly. Overheating condition. **Possible Causes:** • Check any other DTC on display • Check radiator fan system • Check for temporary poor contact
DTC: P0692 **2T CCM, MIL: No** **Years:** 2006, 2007, 2008 **Engines:** 2.5L SOHC, DOHC VIN 6, 3.0L VIN 8 **Models:** Baja, B9, Tribeca Forester, Impreza, Legacy, Outback **Transmissions:** All	**Cooling Fan 1 Control Circuit High** Detects the open or short circuit of radiator fan circuit. Radiator fan does not operate properly. Overheating condition. **Possible Causes:** • Check any other DTC on display • Check radiator fan system • Check for temporary poor contact

DTC	Trouble Code Title, Conditions & Possible Causes
DTC: P0700 **1T CCM, MIL: Yes** **Years:** 2006, 2007, 2008 **Engines:** 2.5L SOHC, DOHC VIN 6, 3.0L VIN 8, 3.6L VIN 9 **Models:** All **Transmissions:** All	**Transaxle Control System (MIL Request)** There is CAN communication with the AT and there is a MIL lighting request. Vehicle performance issue. **Possible Causes:** • Road Test • Inhibitor Switch • Time Lag Test • Transfer Clutch Pressure Test • Line pressure test • Stall speed test • Oil leakage • Check engine start failure (does engine run) • Check illumination of MIL light • Check indication of DTC on display • Using the a scan tool, perform diagnosis
DTC: P0705 **2T CCM, MIL: Yes** **Years:** 2006, 2007, 2008 **Engines:** 2.5L SOHC, DOHC VIN 6, 3.0L VIN 8, 3.6L VIN 9 **Models:** All **Transmissions:** A/T	**Transmission Range Sensor Circuit (PRNDL Input)** The inhibitor switch is open or short. Shift characteristics are erroneous. Shift indicator light does not match with select lever. Shift indicator light does not illuminate. N-D, N-R shock occurs. **Possible Causes:** • Short circuit of harness between TCM connector and chassis ground • Open circuit of harness between TCM connector and transmission connector • Short circuit of harness between control valve body connector and transmission connector • Faulty TCM • Control valve body
DTC: P0712 **2T CCM, MIL: Yes** **Years:** 2006, 2007, 2008 **Engines:** 2.5L SOHC, DOHC VIN 6, 3.0L VIN 8, 3.6L VIN 9 **Models:** All **Transmissions:** A/T	**Transmission Fluid Temperature Sensor Circuit Low Input** Input signal circuit to ATF temperature sensor 1 is open. Excessive shift shock **Possible Causes:** • Open circuit of harness between TCM and transmission connector • Open circuit of harness between transmission connector and control valve body connector • Poor contact in ATF temperature sensor 1 circuit • Faulty Sensor • Faulty TCM
DTC: P0713 **2T CCM, MIL: Yes** **Years:** 2006, 2007, 2008 **Engines:** 2.5L SOHC, DOHC VIN 6, 3.0L VIN 8, 3.6L VIN 9 **Models:** All **Transmissions:** A/T	**Transmission Fluid Temperature Sensor Circuit High Input** Input signal circuit to ATF temperature sensor 1 is shorted. Excessive shift shock. **Possible Causes:** • Short circuit of harness between TCM and transmission connector • Short circuit of harness between transmission connector and control valve body connector • Poor contact in ATF temperature sensor 1 circuit • Faulty Sensor • Faulty TCM
DTC: P0715 **2T CCM, MIL: Yes** **Years:** 2006, 2007, 2008 **Engines:** 2.5L SOHC, DOHC VIN 6, 3.0L VIN 8, 3.6L VIN 9 **Models:** All **Transmissions:** A/T	**Input/Turbine Speed Sensor Circuit** Input signal circuit of TCM is open or shorted. Excessive shift shock. Does not shift into 5th. **Possible Causes:** • Open circuit of harness between TCM and transmission connector • Short circuit of harness between TCM and chassis ground • Open or short circuit for power supply and ground • Open circuit of harness between TCM and transmission connector, or poor contact of connector. • Faulty turbine speed sensor 1 • Check for mechanical problems • Faulty TCM
DTC: P0719 **2T CCM, MIL: Yes** **Years:** 2006, 2007, 2008 **Engines:** 2.5L SOHC, DOHC VIN 6, 3.0L VIN 8, 3.6L VIN 9 **Models:** All **Transmissions:** A/T	**Torque Converter/Brake Switch "B" Circuit Low** Brake switch malfunction, open input signal circuit. Brake down control is not operated at SPORT mode. **Possible Causes:** • Poor contact of connector or harness may be the cause • Faulty TCM • Open circuit of harness between body integrated unit and stop light switch • Short circuit of harness between body integrated unit and stop light switch • Faulty Switch

DTC	Trouble Code Title, Conditions & Possible Causes
DTC: P0720 **2T CCM, MIL: Yes** **Years:** 2006, 2007, 2008 **Engines:** 3.0L VIN 8, 3.6L VIN 9 **Models:** B9 Tribeca, Tribeca **Transmissions:** A/T	**Output Speed Sensor Circuit** AT vehicle speed signal is abnormal. The harness connector between TCM and vehicle speed sensor is shorted or open. Deterioration of shifting quality. Driving performance is poor. **Possible Causes:** • Poor contact of connector or harness may be the cause • Open or short circuit for power supply and ground • Open circuit of harness between control valve body connector and transmission connector • Short circuit of harness between transmission connector and transmission ground • Control valve body • Faulty TCM • Vehicle speed sensor
DTC: P0724 **2T CCM, MIL: Yes** **Years:** 2006, 2007, 2008 **Engines:** 2.5L SOHC, DOHC VIN 6, 3.0L VIN 8, 3.6L VIN 9 **Models:** All **Transmissions:** A/T	**Torque Converter/Brake Switch "B" Circuit High** Brake switch malfunction, open input signal circuit. Gear is not shifted down when climbing a hill. **Possible Causes:** • Temporary poor contact of connector or harness may be the cause • Faulty stop light switch • Short circuit of harness between TCM and stop light switch • Faulty TCM
DTC: P0725 **2T CCM, MIL: Yes** **Years:** 2006, 2007, 2008 **Engines:** 2.5L SOHC, DOHC VIN 6, 3.0L VIN 8, 3.6L VIN 9 **Models:** All **Transmissions:** A/T	**Engine Speed Input Circuit** Information of engine speed is not correctly received from ECM. No lock-up occurs. (After engine is warmed-up) **Possible Causes:** • Temporary poor contact of connector or harness may be the cause • Check indication of DTC on display • Faulty TCM
DTC: P0731 **2T CCM, MIL: Yes** **Years:** 2006, 2007, 2008 **Engines:** 2.5L SOHC, DOHC VIN 6, 3.0L VIN 8, 3.6L VIN 9 **Models:** All **Transmissions:** A/T	**Gear 1 Incorrect Ratio** Target gear ratio and actual gear ratio do not match. Shift point is too high or too low. Excessive shift shock. Gear is not changed. The vehicle does not move in D or R range with the engine running at high speed. **Possible Causes:** • Using the a scan tool, perform diagnosis • Temporary poor contact of connector or harness may be the cause • Faulty transmission assembly
DTC: P0732 **2T CCM, MIL: Yes** **Years:** 2006, 2007, 2008 **Engines:** 2.5L SOHC, DOHC VIN 6, 3.0L VIN 8, 3.6L VIN 9 **Models:** All **Transmissions:** A/T	**Gear 2 Incorrect Ratio** Target gear ratio and actual gear ratio do not match. Shift point is too high or too low. Excessive shift shock. Gear is not changed. The vehicle does not move in D or R range with the engine running at high speed. **Possible Causes:** • Using the a scan tool, perform diagnosis • Temporary poor contact of connector or harness may be the cause • Faulty transmission assembly
DTC: P0733 **2T CCM, MIL: Yes** **Years:** 2006, 2007, 2008 **Engines:** 2.5L SOHC, DOHC VIN 6, 3.0L VIN 8, 3.6L VIN 9 **Models:** All **Transmissions:** A/T	**Gear 3 Incorrect Ratio** Target gear ratio and actual gear ratio do not match. Shift point is too high or too low. Excessive shift shock. Gear is not changed. The vehicle does not move in D or R range with the engine running at high speed. **Possible Causes:** • Using the a scan tool, perform diagnosis • Temporary poor contact of connector or harness may be the cause • Faulty transmission assembly
DTC: P0734 **2T CCM, MIL: Yes** **Years:** 2006, 2007, 2008 **Engines:** 2.5L SOHC, DOHC VIN 6, 3.0L VIN 8, 3.6L VIN 9 **Models:** All **Transmissions:** A/T	**Gear 4 Incorrect Ratio** Target gear ratio and actual gear ratio do not match. Shift point is too high or too low. Excessive shift shock. Gear is not changed. The vehicle does not move in D or R range with the engine running at high speed. **Possible Causes:** • Using the a scan tool, perform diagnosis • Temporary poor contact of connector or harness may be the cause • Faulty transmission assembly

DTC	Trouble Code Title, Conditions & Possible Causes
DTC: P0735 **2T CCM, MIL: Yes** **Years:** 2006, 2007, 2008 **Engines:** 2.5L SOHC, DOHC VIN 6, 3.0L VIN 8, 3.6L VIN 9 **Models:** All **Transmissions:** A/T	**Gear 5 Incorrect Ratio** Target gear ratio and actual gear ratio do not match. Shift point is too high or too low. Excessive shift shock. Gear is not changed. The vehicle does not move in D or R range with the engine running at high speed. **Possible Causes:** • Using the a scan tool, perform diagnosis • Temporary poor contact of connector or harness may be the cause • Faulty transmission assembly
DTC: P0736 **2T CCM, MIL: Yes** **Years:** 2006, 2007, 2008 **Engines:** 2.5L SOHC, DOHC VIN 6, 3.0L VIN 8, 3.6L VIN 9 **Models:** All **Transmissions:** A/T	**Reverse Incorrect Ratio** Target gear ratio and actual gear ratio do not match. Shift point is too high or too low. Excessive shift shock. Gear is not changed. The vehicle does not move in D or R range with the engine running at high speed. **Possible Causes:** • Using the a scan tool, perform diagnosis • Temporary poor contact of connector or harness may be the cause • Faulty transmission assembly
DTC: P0741 **2T CCM, MIL: Yes** **Years:** 2006, 2007, 2008 **Engines:** 2.5L SOHC, DOHC VIN 6, 3.0L VIN 8, 3.6L VIN 9 **Models:** All **Transmissions:** A/T	**Torque Converter Clutch Circuit Performance Or Stuck Off** Defective lock-up clutch or torque converter assembly. Defective control valve. Defective turbine speed sensor 1 or 2. No lock-up occurs. (After engine is warmed-up) **Possible Causes:** • Using the a scan tool, perform diagnosis • Temporary poor contact of connector or harness may be the cause • Faulty transmission assembly • Suspect the transmission assembly when DTC P0741 is displayed
DTC: P0743 **2T CCM, MIL: Yes** **Years:** 2006, 2007, 2008 **Engines:** 2.5L SOHC, DOHC VIN 6, 3.0L VIN 8, 3.6L VIN 9 **Models:** All **Transmissions:** A/T	**Torque Converter Clutch Circuit Electrical** The output signal circuit of lock up solenoid is open or shorted. No lock-up occurs. (After engine is warmed-up) **Possible Causes:** • Open circuit of harness between TCM connector and transmission connector • Short circuit of harness between TCM connector and transmission connector • Open circuit of harness between transmission connector and control valve body connector • Short circuit of harness between control valve body connector and transmission ground • Faulty control valve body • Temporary poor contact of connector or harness may be the cause • Faulty TCM
DTC: P0748 **2T CCM, MIL: Yes** **Years:** 2006, 2007, 2008 **Engines:** 2.5L SOHC, DOHC VIN 6, 3.0L VIN 8, 3.6L VIN 9 **Models:** All **Transmissions:** A/T	**Pressure Control Solenoid "A" Electrical** Output signal circuit of line pressure solenoid is open or shorted. Excessive shift shock. **Possible Causes:** • Temporary poor contact of connector or harness may be the cause • Open circuit of harness between TCM connector and transmission connector • Short circuit of harness between TCM connector and transmission connector • Open circuit of harness between transmission connector and control valve body connector • Short circuit of harness between control valve body connector and transmission ground • Faulty control valve body • Faulty TCM
DTC: P0751 **2T CCM, MIL: Yes** **Years:** 2006, 2007, 2008 **Engines:** 3.0L VIN 8, 3.6L VIN 9 **Models:** B9 Tribeca, Tribeca **Transmissions:** All	**Shift Solenoid "A" Performance Or Stuck Off** Output signal of front brake solenoid does not match with oil pressure. Locked to 4th or 5th gear. **Possible Causes:** • Open circuit of harness between TCM and transmission connector • Short circuit of harness between TCM and transmission connector • Faulty control valve body • Transmission harness assembly
DTC: P0753 **2T CCM, MIL: Yes** **Years:** 2006, 2007, 2008 **Engines:** 2.5L SOHC, DOHC VIN 6, 3.0L VIN 8, 3.6L VIN 9 **Models:** All **Transmissions:** A/T	**Shift Solenoid "A" Electrical** Output signal circuit of front brake solenoid is open or shorted. Locked to 4th or 5th gear. **Possible Causes:** • Open circuit of harness between TCM connector and transmission connector • Short circuit of harness between TCM connector and transmission connector • Open circuit of harness between transmission connector and control valve body connector • Short circuit of harness between control valve body and transmission connector • Temporary poor contact of connector or harness may be the cause • Faulty control valve body • Faulty TCM

DTC	Trouble Code Title, Conditions & Possible Causes
DTC: P0756 **2T CCM, MIL: Yes** **Years:** 2006, 2007, 2008 **Engines:** 2.5L SOHC, DOHC VIN 6, 3.0L VIN 8, 3.6L VIN 9 **Models:** All **Transmissions:** A/T	**Shift Solenoid "B" Performance Or Stuck Off** Output signal value of input clutch solenoid and oil pressure does not match. Locked to 4th gear. **Possible Causes:** • Open circuit of harness between TCM and transmission connector • Short circuit of harness between TCM and transmission connector • Temporary poor contact of connector or harness may be the cause • Faulty control valve body • Transmission harness assembly
DTC: P0758 **2T CCM, MIL: Yes** **Years:** 2006, 2007, 2008 **Engines:** 3.0L VIN 8, 3.6L VIN 9 **Models:** B9 Tribeca, Tribeca **Transmissions:** All	**Shift Solenoid "B" Electrical** Output signal circuit of input clutch solenoid is open or shorted. Locked to 4th gear. **Possible Causes:** • Open circuit of harness between TCM and transmission connector • Short circuit of harness between TCM and transmission connector • Temporary poor contact of connector or harness may be the cause • Transmission harness assembly • Faulty TCM
DTC: P0761 **2T CCM, MIL: Yes** **Years:** 2006, 2007, 2008 **Engines:** 2.5L SOHC, DOHC VIN 6, 3.0L VIN 8, 3.6L VIN 9 **Models:** All **Transmissions:** A/T	**Shift Solenoid "C" Performance Or Stuck Off** Output signal value of high & low reverse clutch solenoid and oil pressure does not match. Locked to 4th gear. **Possible Causes:** • Open circuit of harness between TCM and transmission connector • Short circuit of harness between TCM and transmission connector • Faulty control valve body • Transmission harness assembly
DTC: P0763 **2T CCM, MIL: Yes** **Years:** 2006, 2007, 2008 **Engines:** 2.5L SOHC, DOHC VIN 6, 3.0L VIN 8, 3.6L VIN 9 **Models:** All **Transmissions:** A/T	**Shift Solenoid "C" Electrical** Output signal circuit of input clutch solenoid is open or shorted. Locked to 4th gear. **Possible Causes:** • Open circuit of harness between TCM and transmission connector • Short circuit of harness between TCM and transmission connector • Temporary poor contact of connector or harness may be the cause • Transmission harness assembly • Faulty TCM
DTC: P0766 **2T CCM, MIL: Yes** **Years:** 2006, 2007, 2008 **Engines:** 2.5L SOHC, DOHC VIN 6, 3.0L VIN 8, 3.6L VIN 9 **Models:** All **Transmissions:** A/T	**Shift Solenoid "D" Performance Or Stuck Off** Output signal of front brake solenoid does not match with oil pressure. Locked to 4th or 5th gear. **Possible Causes:** • Open circuit of harness between TCM and transmission connector • Short circuit of harness between TCM and transmission connector • Faulty control valve body • Transmission harness assembly
DTC: P0768 **2T CCM, MIL: Yes** **Years:** 2006, 2007, 2008 **Engines:** 2.5L SOHC, DOHC VIN 6, 3.0L VIN 8, 3.6L VIN 9 **Models:** All **Transmissions:** A/T	**Shift Solenoid "D" Electrical** The output signal circuit of direct clutch solenoid is open or shorted. Locked to 4th gear. **Possible Causes:** • Open circuit of harness between TCM and transmission connector • Short circuit of harness between TCM and transmission connector • Temporary poor contact of connector or harness may be the cause • Transmission harness assembly • Faulty TCM
DTC: P0771 **2T CCM, MIL: Yes** **Years:** 2006, 2007, 2008 **Engines:** 2.5L SOHC, DOHC VIN 6, 3.0L VIN 8, 3.6L VIN 9 **Models:** All **Transmissions:** A/T	**Shift Solenoid "E" Performance Or Stuck Off** Output signal value of low coast brake solenoid and oil pressure does not match. Locked to 2nd gear. Engine brake does not function at 1st or 2nd of manual mode. **Possible Causes:** • Open circuit of harness between TCM and transmission connector • Short circuit of harness between TCM and transmission connector • Temporary poor contact of connector or harness may be the cause • Transmission harness assembly • Faulty control valve body

DTC	Trouble Code Title, Conditions & Possible Causes
DTC: P0773 **2T CCM, MIL: Yes** **Years:** 2006, 2007, 2008 **Engines:** 2.5L SOHC, DOHC VIN 6, 3.0L VIN 8, 3.6L VIN 9 **Models:** All **Transmissions:** A/T	**Shift Solenoid "E" Electrical** Output signal circuit of low coast brake solenoid is open or shorted. Locked to 2nd 3rd or 4th gear. Engine brake does not function at 1st or 2nd of manual mode. **Possible Causes:** • Open circuit of harness between TCM connector and transmission connector • Short circuit of harness between TCM connector and transmission connector • Open circuit of harness between transmission connector and control valve body connector • Short circuit of harness between control valve body connector and transmission ground • Faulty control valve body • Faulty TCM
DTC: P0801 **2T CCM, MIL: Yes** **Years:** 2006, 2007, 2008 **Engines:** 2.5L SOHC, DOHC VIN 6, 3.0L VIN 8, 3.6L VIN 9 **Models:** All **Transmissions:** A/T	**Reverse Inhibitor Control Circuit** Shift lock solenoid malfunction, open or short reverse inhibitor control circuit Gear is shifted from "N" range to "R" range during driving at 12MPH or more. Gear can not be shifted from "N" range to "R" range though the vehicle is parked. **Possible Causes:** • Blown fuse number 32 • Short circuit of harness between fuse number 32 and TCM • Open circuit of harness between TCM and shift lock solenoid connector • Short circuit of harness between TCM and shift lock solenoid connector • Open circuit of harness between chassis ground and shift lock solenoid connector • Faulty shift lock solenoid • Poor contact in the reverse inhibitor control circuit • Faulty TCM
DTC: P0817 **2T CCM, MIL: Yes** **Years:** 2006, 2007, 2008 **Engines:** 2.5L SOHC, DOHC VIN 6, 3.0L VIN 8, 3.6L VIN 9 **Models:** All **Transmissions:** A/T	**Starter Disable Circuit** Open or short in P/N signal output circuit. Engine can be started on other than "P" or "N" range. Engine can not be started on "P" or "N" range. **Possible Causes:** • Blown fuse number 32 • Short circuit of harness between fuse number 32 and TCM • Open circuit of harness between TCM and transmission connector, or poor contact of connector • Short circuit of harness between transmission connector and chassis ground • Temporary poor contact of connector or harness may be the cause • Neutral switch of ECM • Faulty TCM
DTC: P0851 **2T CCM, MIL: No** **Years:** 2006, 2007, 2008 **Engines:** 2.5L SOHC, DOHC VIN 6, 3.0L VIN 8, 3.6L VIN 9 **Models:** All **Transmissions:** A/T	**Neutral Switch Input Circuit Low (Automatic Transaxle)** Erroneous idling. **Possible Causes:** • Check any other DTC on display • Check input signal of ECM • Check harness between ECM and transmission harness connector • Check transmission harness connector • Check inhibitor switch • Check selector cable connection
DTC: P0851 **2T CCM, MIL: No** **Years:** 2006, 2007, 2008 **Engines:** 2.5L SOHC, DOHC VIN 6, 3.0L VIN 8, 3.6L VIN 9 **Models:** All **Transmissions:** M/T	**Neutral Switch Input Circuit Low (Manual Transaxle)** Engine started, vehicle driven to 30-40 MPH, and then back to idle, and the PCM detected an unexpected low voltage condition on the Neutral Position switch circuit during the CCM test. **Possible Causes:** • Check input signal of ECM • Check poor contact • Check neutral safety switch • Check harness between ECM and neutral safety switch connector • Check neutral safety switch ground

DTC	Trouble Code Title, Conditions & Possible Causes
DTC: P0852 **2T CCM, MIL: No** **Years:** 2006, 2007, 2008 **Engines:** 2.5L SOHC, DOHC VIN 6, 3.0L VIN 8, 3.6L VIN 9 **Models:** All **Transmissions:** A/T	**Neutral Switch Input Circuit High (Automatic Transaxle)** Engine started, vehicle driven to 30-40 MPH, and then back to idle, and the PCM detected an unexpected low voltage condition on the Neutral Position switch circuit during the CCM test. **Possible Causes:** • Check any other DTC on display • Check input signal of ECM • Check poor contact • Check harness between ECM and inhibitor switch connector • Check inhibitor switch ground • Check inhibitor switch • Check selector cable connection
DTC: P0852 **2T CCM, MIL: No** **Years:** 2006, 2007, 2008 **Engines:** 2.5L SOHC, DOHC VIN 6, 3.0L VIN 8, 3.6L VIN 9 **Models:** All **Transmissions:** M/T	**Neutral Switch Input Circuit High (Manual Transaxle)** Engine started, vehicle driven to 30-40 MPH, and then back to idle, and the PCM detected an unexpected low voltage condition on the Neutral Position switch circuit during the CCM test. **Possible Causes:** • Check input signal of ECM • Check poor contact • Check harness between ECM and transmission harness connector • Check neutral safety switch ground • Check neutral safety switch
DTC: P0864 **2T CCM, MIL: No** **Years:** 2006, 2007, 2008 **Engines:** 2.5L SOHC, DOHC VIN 6, 3.0L VIN 8, 3.6L VIN 9 **Models:** All **Transmissions:** All	**TCM Communication Circuit Range/Performance** Engine started, vehicle driven to 30-40 MPH, and then back to idle, and the PCM detected an unexpected voltage condition on the A/T Diagnosis circuit during the CCM test. **Possible Causes:** • Check driving condition (is AT shift control functioning) • Check accessory
DTC: P0865 **2T CCM, MIL: No** **Years:** 2006, 2007, 2008 **Engines:** 2.5L SOHC, DOHC VIN 6, 3.0L VIN 8, 3.6L VIN 9 **Models:** All **Transmissions:** A/T	**TCM Communication Circuit Low** Engine started, vehicle driven to 30-40 MPH, and then back to idle, and the PCM detected an unexpected low voltage condition on the A/T Diagnosis circuit during the CCM test. **Possible Causes:** • Check harness between ECM and TCM Connector • Check output signal for ECM • Check trouble code for automatic transaxle
DTC: P0866 **2T CCM, MIL: No** **Years:** 2006, 2007, 2008 **Engines:** 2.5L SOHC, DOHC VIN 6, 3.0L VIN 8, 3.6L VIN 9 **Models:** All **Transmissions:** A/T	**TCM Communication Circuit High** Engine started, vehicle driven to 30-40 MPH, and then back to idle, and the PCM detected an unexpected high voltage condition on the A/T Diagnosis circuit during the CCM test. **Possible Causes:** • Check harness between ECM and TCM Connector • Check poor contact
DTC: P0882 **2T CCM, MIL: Yes** **Years:** 2006, 2007, 2008 **Engines:** 2.5L SOHC, DOHC VIN 6, 3.0L VIN 8, 3.6L VIN 9 **Models:** All **Transmissions:** A/T	**TCM Power Input Signal Low** Malfunction of PVIGN power supply relay or open, short circuit of PVIGN power supply circuit. Gear is not changed. **Possible Causes:** • MAIN SBF, SBF 8 or fuse number 12 blown out • Open circuit of harness between fuse (No. 12) and PVIGN relay • Temporary poor contact of connector or harness may be the cause • Faulty PVIGN relay • Faulty TCM
DTC: P0957 **2T CCM, MIL: Yes** **Years:** 2006, 2007, 2008 **Engines:** 2.5L SOHC, DOHC VIN 6, 3.0L VIN 8, 3.6L VIN 9 **Models:** All **Transmissions:** A/T	**Backup Light Relay Circuit Low** Shorted circuits of back-up light relay output circuit. Back-up light does not illuminate in "R" range **Possible Causes:** • Open circuit of harness between TCM and transmission connector, or poor contact of connector • Short circuit of harness between TCM and transmission connector • Faulty back-up light relay • Open or short circuit of harness between fuse number 18 and back-up light relay • Faulty TCM

DTC	Trouble Code Title, Conditions & Possible Causes
DTC: P0958 **2T CCM, MIL: Yes** **Years:** 2006, 2007, 2008 **Engines:** 2.5L SOHC, DOHC VIN 6, 3.0L VIN 8, 3.6L VIN 9 **Models:** All **Transmissions:** A/T	**Backup Light Relay Circuit High** Backup light relay malfunction, or open/short circuit in back-up light relay output circuit. Back-up light does not illuminate in "R" range. • Back-up light always illuminate except in "R" range. **Possible Causes:** • Open circuit of harness between TCM and transmission connector, or poor contact of connector • Short circuit of harness between TCM and transmission connector • Faulty back-up light relay • Open or short circuit of harness between fuse number 18 and back-up light relay • Faulty TCM

OBD II Trouble Code List (P1XXX Codes)

DTC	Trouble Code Title, Conditions & Possible Causes
DTC: P1086 **1T CCM, MIL: Yes** **Years:** 2006, 2007, 2008 **Engines:** 2.5L DOHC **Models:** All **Transmissions:** All	**Tumble Generated Valve Position Sensor 2 Circuit Low** Engine stalls. Erroneous idling. Poor driving performance. **Possible Causes:** • Using a scan tool, check current data • Check input signal for ECM • Check harness between ECM and tumble generator valve position sensor connector • Check poor contact
DTC: P1087 **1T CCM, MIL: Yes** **Years:** 2006, 2007, 2008 **Engines:** 2.5L DOHC **Models:** All **Transmissions:** All	**Tumble Generated Valve Position Sensor 2 Circuit High** Engine stalls. Erroneous idling. Poor driving performance. **Possible Causes:** • Using a scan tool, check current data • Check harness between tumble generator valve position sensor and ECM connector • Check harness between throttle position sensor and ECM connector
DTC: P1088 **1T CCM, MIL: Yes** **Years:** 2006, 2007, 2008 **Engines:** 2.5L DOHC **Models:** All **Transmissions:** All	**Tumble Generated Valve Position Sensor 1 Circuit Low** Engine stalls. Erroneous idling. Poor driving performance. **Possible Causes:** • Using a scan tool, check current data • Check input signal for ECM • Check harness between ECM and tumble generator valve position sensor connector • Check poor contact
DTC: P1089 **1T CCM, MIL: Yes** **2002-04** **Engines:** 2.5L DOHC **Models:** Forester **Transmissions:** All	**Tumble Generated Valve Position Sensor 1 Circuit High** Engine stalls. Erroneous idling. Poor driving performance. **Possible Causes:** • Using a scan tool, check current data • Check harness between tumble generator valve position sensor and ECM connector
DTC: P1090 **1T CCM, MIL: Yes** **Years:** 2006, 2007, 2008 **Engines:** 2.5L DOHC **Models:** All **Transmissions:** All	**Tumble Generated Valve System 1 (Valve Open)** Engine stalls. Erroneous idling. Poor driving performance. **Possible Causes:** • Check any other DTC on display • Check tumble generator valve (RH)
DTC: P1091 **1T CCM, MIL: Yes** **2002-04** **Engines:** 2.5L DOHC **Models:** Forester **Transmissions:** All	**Tumble Generated Valve System 1 (Valve Closed)** Engine stalls. Erroneous idling. Poor driving performance. **Possible Causes:** • Check any other DTC on display • Check tumble generator valve (RH)
DTC: P1092 **1T CCM, MIL: Yes** **Years:** 2006, 2007, 2008 **Engines:** 2.5L DOHC **Models:** All **Transmissions:** All	**Tumble Generated Valve System 2 (Valve Open)** Engine stalls. Erroneous idling. Poor driving performance. **Possible Causes:** • Check any other DTC on display • Check tumble generator valve (RH)

DTC	Trouble Code Title, Conditions & Possible Causes
DTC: P1093 **1T CCM, MIL: Yes** **2002-04** **Engines:** 2.5L DOHC **Models:** Forester **Transmissions:** All	**Tumble Generated Valve System 2 (Valve Closed)** Engine stalls. Erroneous idling. Poor driving performance. **Possible Causes:** • Check any other DTC on display • Check tumble generator valve (RH)
DTC: P1094 **1T CCM, MIL: Yes** **Years:** 2006, 2007, 2008 **Engines:** 2.5L DOHC **Models:** All **Transmissions:** All	**Tumble Generated Valve Signal 1 Circuit Malfunction** Engine stalls. Erroneous idling. Poor driving performance. **Possible Causes:** • Check harness between ECM and tumble generator valve actuator connector • Check poor contact
DTC: P1095 **1T CCM, MIL: Yes** **Years:** 2006, 2007, 2008 **Engines:** 2.5L DOHC **Models:** All **Transmissions:** All	**Tumble Generated Valve Signal 1 Circuit Malfunction (Short)** Engine stalls. Erroneous idling. Poor driving performance. **Possible Causes:** • Check harness between ECM and tumble generator valve actuator connector • Check poor contact
DTC: P1096 **1T CCM, MIL: Yes** **Years:** 2006, 2007, 2008 **Engines:** 2.5L DOHC **Models:** All **Transmissions:** All	**Tumble Generated Valve Signal 2 Circuit Malfunction (Open)** Engine stalls. Erroneous idling. Poor driving performance. **Possible Causes:** • Check harness between ECM and tumble generator valve actuator connector • Check poor contact
DTC: P1097 **1T CCM, MIL: Yes** **Years:** 2006, 2007, 2008 **Engines:** 2.5L DOHC **Models:** All **Transmissions:** All	**Tumble Generated Valve Signal 2 Circuit Malfunction (Short)** Engine stalls. Erroneous idling. Poor driving performance. **Possible Causes:** • Check harness between ECM and tumble generator valve actuator connector • Check poor contact
DTC: P1134 **1T CCM, MIL: Yes** **Years:** 2006, 2007, 2008 **Engines:** 2.5L SOHC **Models:** Baja **Transmissions:** All	**A/F Sensor Micro-Computer Problem** Detect the malfunction of IC communication. Judge NG when the communication to front oxygen (A/F) sensor control IC is unable. **Possible Causes:** • Check any other DTC on display • Faulty ECM
DTC: P1137 **2T CCM, MIL: Yes** **Years:** 2006, 2007, 2008 **Engines:** 2.5L SOHC **Models:** Baja **Transmissions:** All	**02 Sensor Circuit Middle Electric Potential (Bank 1 Sensor 1)** Detect the output property malfunction of from oxygen (A/F) sensor. Judge NG when output voltage does not move to lean side or rich side. **Possible Causes:** • Check any other DTC on display • Open circuit between ECM and front oxygen (A/F) sensor. • Short circuit between ECM and front oxygen (A/F) sensor • Faulty front oxygen (A/F) sensor
DTC: P1152 **2T CCM, MIL: Yes** **Years:** 2006, 2007, 2008 **Engines:** 2.5L SOHC, DOHC VIN 6, 3.0L VIN 8, 3.6L VIN 9 **Models:** All **Transmissions:** All	**02 Sensor Circuit Range/Performance Low (Bank 1 Sensor 1)** Detected when two consecutive driving cycles with fault occur. Poor drive ability. **Possible Causes:** • Open circuit in harness between ECM and front oxygen A/F sensor connector • Poor contact in front oxygen A/F sensor connector • Poor contact in ECM connector • Poor contact of coupling connector • Faulty front oxygen A/F sensor

<table>
<tr><th>DTC</th><th>Trouble Code Title, Conditions & Possible Causes</th></tr>
<tr><td>DTC: P1153
2T CCM, MIL: Yes
Years: 2006, 2007, 2008
Engines: 2.5L SOHC, DOHC VIN 6, 3.0L VIN 8, 3.6L VIN 9
Models: All
Transmissions: All</td><td>02 Sensor circuit Range/Performance High (Bank 1 Sensor 1)
Detected when two consecutive driving cycles with fault occur. Poor drive ability.
Possible Causes:
• Poor contact in front oxygen A/F sensor connector
• Shorted ground circuit of harness between ECM and front oxygen A/F sensor connector
• Short circuit to power in the harness between the ECM and front oxygen A/F sensor connector
• Faulty front oxygen A/F sensor</td></tr>
<tr><td>DTC: P1154
2T CCM, MIL: Yes
Years: 2006, 2007, 2008
Engines: 2.5L SOHC, DOHC VIN 6, 3.0L VIN 8, 3.6L VIN 9
Models: All
Transmissions: All</td><td>02 Sensor circuit Range/Performance Low (Bank 2 Sensor 1)
Detected when two consecutive driving cycles with fault occur. Poor drive ability.
Possible Causes:
• Open circuit in harness between ECM and front oxygen A/F sensor connector
• Poor contact in front oxygen A/F sensor connector
• Poor contact in ECM connector
• Poor contact of coupling connector
• Faulty front oxygen A/F sensor</td></tr>
<tr><td>DTC: P1155
2T CCM, MIL: Yes
Years: 2006, 2007, 2008
Engines: 3.0L VIN 8, 3.6L VIN 9
Models: B9 Tribeca, Tribeca
Transmissions: All</td><td>02 Sensor circuit Range/Performance High (Bank 2 Sensor 1)
Detected when two consecutive driving cycles with fault occur. Poor drive ability.
Possible Causes:
• Poor contact in front oxygen A/F sensor connector
• Shorted ground circuit of harness between ECM and front oxygen A/F sensor connector
• Short circuit to power in the harness between the ECM and front oxygen A/F sensor connector
• Faulty front oxygen A/F sensor</td></tr>
<tr><td>DTC: P1160
1T CCM, MIL: Yes
Years: 2006, 2007, 2008
Engines: 2.5L DOHC 3.0L VIN 8, 3.6L VIN 9
Models: All
Transmissions: All</td><td>Return Spring Failure
Judge as NG when the valve is opened more than the default opening angle, but does not move to the close direction with the motor power stopped. Improper idling. Poor driving performance. Engine stalls.
Possible Causes:
• Check electronic throttle control relay
• Check power supply of electronic control relay
• Check harness between ECM and electronic throttle control relay
• Check sensor output
• Check poor contact
• Check harness between ECM and electronic throttle control
• Check sensor power supply
• Check for short in ECM
• Check sensor output
• Check harness between ECM and electronic throttle control motor
• Check electronic throttle control motor harness
• Check ground circuit</td></tr>
<tr><td>DTC: P1301
1T CCM, MIL: Yes
Years: 2006, 2007, 2008
Engines: 2.5L DOHC
Models: Baja
Transmissions: All</td><td>Misfire Detected (High Temperature Exhaust Gas)
Detect whether the misfire occurred or not. (Exhaust temperature method) Judge NG when the exhaust temperature is high. Poor driving performance. Engine stalls. Poor Idling condition.
Possible Causes:
• Check any other DTC on display
• Check output signal of ECM
• Check harness between injector and ECM connector
• Check fuel injector
• Check power supply line
• Check power supply to mass air flow sensor
• Check harness between ECM and sensor connector
• Check installation of camshaft and crankshaft sensors
• Check crank plate
• Check condition of timing chain
• Check fuel level
• Check status of MIL light
• Check air intake system
• Check cause of misfire (engine running)
• Check engine condition
• Check all cylinders individually, in pairs, in groups and at random</td></tr>
</table>

DTC	Trouble Code Title, Conditions & Possible Causes
DTC: P1312 **1T CCM, MIL: Yes** **Years:** 2006, 2007, 2008 **Engines:** 2.5L DOHC **Models:** Baja **Transmissions:** All	**Exhaust Gas Temperature Sensor Function** Detect the malfunction of exhaust temperature sensor output property. Judge NG when the exhaust temperature remains high or low whereas it seemed to vary from the viewpoint of driving condition. **Possible Causes:** • Check any other DTC on display • Faulty exhaust gas temperature sensor
DTC: P1400 **2T CCM, MIL: Yes** **Years:** 2006, 2007, 2008 **Engines:** 2.5L DOHC 3.0L VIN 8, 3.6L VIN 9 **Models:** Baja, Impreza, Forester, Outback **Transmissions:** All	**Fuel Tank Pressure Control Solenoid Valve Circuit** Detect open or short circuit of pressure control solenoid valve. Judge NG when ECM output level is different from actual terminal level. **Possible Causes:** • Check output signal from ECM • Check for poor contact in ECM connector • Check harness between pressure control solenoid valve and ECM • Faulty pressure control solenoid valve
DTC: P1410 **1T CCM, MIL: Yes** **Years:** 2006, 2007, 2008 **Engines:** 2.5L DOHC **Models:** Baja, Impreza, Forester, Outback **Transmissions:** All	**Secondary Air Injection System Switching Valve Stuck Open** Continually detects for a combination solenoid valve and lead valve stuck open condition. Calculate the integrated value of secondary air supply piping pressure sensor output voltage maximum/minimum values and output voltage deviation for a constant time period after engine start, and if the difference between the maximum/minimum is large and the integrated value is also large, judge as NG. **Possible Causes:** • Check secondary combination valve • Check for poor contact of connector • Faulty secondary air combination valve
DTC: P1418 **1T CCM, MIL: Yes** **Years:** 2006, 2007, 2008 **Engines:** 2.5L DOHC **Models:** Impreza, Forester, Outback **Transmissions:** All	**Secondary Air Injection System Control "A" Circuit Shorted** Judge NG when the ECM output level is different from the actual terminal level. **Possible Causes:** • Check harness between ECM and secondary air pump relay • Short circuit in harness between ECM and secondary air pump relay terminal.
DTC: P1420 **2T CCM, MIL: Yes** **Years:** 2006, 2007, 2008 **Engines:** 2.5L SOHC, DOHC **Models:** Baja, Impreza, Forester, Outback **Transmissions:** All	**Fuel Tank Pressure Control Solenoid Valve Circuit High** Detect open or short circuit of pressure control solenoid valve. Judge NG when ECM output level is different from actual terminal level. **Possible Causes:** • Check input signal for ECM • Check for poor contact in ECM connector • Check harness between pressure control solenoid valve and ECM • Faulty ECM
DTC: P1443 **1T CCM, MIL: Yes** **Years:** 2006, 2007, 2008 **Engines:** 2.5L SOHC, DOHC VIN 6, 3.0L VIN 8, 3.6L VIN 9 **Models:** Baja, B9Tribeca, Tribeca, Forester, Impreza Legacy, Outback **Transmissions:** All	**Vent Control Solenoid Valve Function Problem** Detect the abnormal function (stuck closed) of the drain valve. Judge as NG when fuel tank pressure is low. **Possible Causes:** • Check any other DTC on display • Check vent line hoses • Check drain valve operation
DTC: P1446 **2T CCM, MIL: Yes** **Years:** 2006 **Engines:** 2.5L SOHC, DOHC VIN 6 **Models:** Baja, Impreza **Transmissions:** All	**Fuel Tank Sensor Control Valve Circuit Low** Detect the open or short circuit of tank pressure switching solenoid. Judge NG when the ECM output level is different from actual terminal level. **Possible Causes:** • Check output signal for ECM • Check for poor contact in ECM connector • Check harness between fuel tank sensor control valve and ECM connector • Check fuel tank sensor control valve • Check power supply to sensor control valve • Check for poor contact at sensor control valve connector

DTC	Trouble Code Title, Conditions & Possible Causes
DTC: P1447 **2T CCM, MIL: Yes** **Years:** 2006 **Engines:** 2.5L SOHC, DOHC VIN 6 **Models:** Baja, Impreza **Transmissions:** All	**Fuel Tank Sensor Control Valve Circuit High** Detect the open/short circuit of fuel tank sensor control valve. Judge NG when the ECM output level is different from actual terminal level. **Possible Causes:** • Check any other DTC on display • Check fuel filler cap • Check EVAP hoses and pipes • Faulty fuel tank pressure sensor
DTC: P1448 **2T CCM, MIL: Yes** **Years:** 2006 **Engines:** 2.5L SOHC, DOHC VIN 6 **Models:** Baja, Impreza **Transmissions:** All	**Fuel Tank Sensor Control Valve Range/Performance** Detects the tank pressure switching solenoid function abnormality. The tank pressure sensor is a relative pressure sensor, which normally compares the pressure with the atmospheric pressure. The tank pressure switching solenoid is a solenoid, which shifts the compare space from opening to closed during the EVAP diagnosis. Detect the malfunction that the compare space remains closed. (Not judge NG after enable condition completed but assume NG before enable condition completed.) **Possible Causes:** • Check any other DTC on display • Check fuel filler cap • Check EVAP hoses and pipes • Faulty fuel tank pressure sensor
DTC: P1491 **1T CCM, MIL: Yes** **Years:** 2006, 2007, 2008 **Engines:** 2.5L SOHC, DOHC VIN 6 **Models:** Baja, Forester, Impreza, Legacy, Outback **Transmissions:** All	**Positive Crankcase Ventilation (Blow-By) Function Problem** Detect the blow-by hose release abnormality. Judge NG when the diagnosis terminal voltage is high. Poor idling condition. **Possible Causes:** • Check the blow-by hose for disconnection or cracks • Check harness between PCV diagnosis connector and ECM connector • Check between PCV diagnosis connector and engine ground • Check for poor contact in ECM and PCV diagnosis connector • Faulty PCV diagnosis connector
DTC: P1492 **1T CCM, MIL: Yes** **Years:** 2006, 2007, 2008 **Engines:** 2.5L SOHC, DOHC VIN 6 **Models:** Baja, Forester, Impreza, Legacy, Outback **Transmissions:** All	**EGR Solenoid Valve Signal 1 Circuit Low Input** Detects open or short circuit of EGR. Judge as NG when the ECM output level differs from the actual terminal level. Battery voltage 10.9 volts. EGR target position 0 step. 1 second after startup. Improper idling. Poor driving performance. Engine breathing **Possible Causes:** • Open circuit in harness between EGR valve and main relay connector • Poor contact of coupling connector • Open circuit in harness between ECM and EGR valve connector • Ground short in harness between ECM and EGR valve connector • Poor contact in ECM or EGR valve connector • Faulty EGR valve
DTC: P1493 **1T CCM, MIL: Yes** **Years:** 2006, 2007, 2008 **Engines:** 2.5L SOHC, DOHC VIN 6, 3.0L VIN 8, 3.6L VIN 9 **Models:** Baja, Forester, Impreza, Legacy, Tribeca **Transmissions:** All	**EGR Solenoid Valve Signal 1 Circuit High Input** Detects open or short circuit of EGR. Judge as NG when the ECM output level differs from the actual terminal level. Battery voltage 10.9 volts. EGR target position 0 step. 1 second after startup. Improper idling. Poor driving performance. Engine breathing **Possible Causes:** • Short circuit to power in the harness between the ECM and EGR valve connectors • Poor contact of the ECM connector
DTC: P1494 **1T CCM, MIL: Yes** **Years:** 2006, 2007, 2008 **Engines:** 2.5L SOHC, DOHC VIN 6, 3.0L VIN 8, 3.6L VIN 9 **Models:** Baja, Forester, Impreza, Legacy, Tribeca **Transmissions:** All	**EGR Solenoid Valve Signal 2 Circuit Low Input** Detects open or short circuit of EGR. Judge as NG when the ECM output level differs from the actual terminal level. Battery voltage 10.9 volts. EGR target position 0 step. 1 second after startup. Improper idling. Poor driving performance. Engine breathing **Possible Causes:** • Open circuit in harness between EGR valve and main relay connector • Poor contact of coupling connector • Open circuit in harness between ECM and EGR valve connector • Ground short in harness between ECM and EGR valve connector • Poor contact in ECM or EGR valve connector • Faulty EGR valve

DTC	Trouble Code Title, Conditions & Possible Causes
DTC: P1495 **1T CCM, MIL: Yes** **Years:** 2006, 2007, 2008 **Engines:** 2.5L SOHC, DOHC VIN 6, 3.0L VIN 8, 3.6L VIN 9 **Models:** Baja, Forester, Impreza, Legacy, Tribeca **Transmissions:** All	**EGR Solenoid Valve Signal 2 Circuit High Input** Detects open or short circuit of EGR. Judge as NG when the ECM output level differs from the actual terminal level. Battery voltage 10.9 volts. EGR target position 0 step. 1 second after startup. Improper idling. Poor driving performance. Engine breathing **Possible Causes:** • Short circuit to power in the harness between the ECM and EGR valve connectors • Poor contact of the ECM connector
DTC: P1496 **1T CCM, MIL: Yes** **Years:** 2006, 2007, 2008 **Engines:** 2.5L SOHC, DOHC VIN 6, 3.0L VIN 8, 3.6L VIN 9 **Models:** Baja, Forester, Impreza, Legacy, Tribeca **Transmissions:** All	**EGR Solenoid Valve Signal 3 Circuit Low Input** Detects open or short circuit of EGR. Judge as NG when the ECM output level differs from the actual terminal level. Battery voltage 10.9 volts. EGR target position 0 step. 1 second after startup. Improper idling. Poor driving performance. Engine breathing **Possible Causes:** • Open circuit in harness between EGR valve and main relay connector • Poor contact of coupling connector • Open circuit in harness between ECM and EGR valve connector • Ground short in harness between ECM and EGR valve connector • Poor contact in ECM or EGR valve connector • Faulty EGR valve
DTC: P1497 **1T CCM, MIL: Yes** **Years:** 2006, 2007, 2008 **Engines:** 2.5L SOHC, DOHC VIN 6, 3.0L VIN 8, 3.6L VIN 9 **Models:** Baja, Forester, Impreza, Legacy, Tribeca **Transmissions:** All	**EGR Solenoid Valve Signal 3 Circuit High Input** Detects open or short circuit of EGR. Judge as NG when the ECM output level differs from the actual terminal level. Battery voltage 10.9 volts. EGR target position 0 step. 1 second after startup. Improper idling. Poor driving performance. Engine breathing **Possible Causes:** • Short circuit to power in the harness between the ECM and EGR valve connectors • Poor contact of the ECM connector
DTC: P1498 **1T CCM, MIL: Yes** **Years:** 2006, 2007, 2008 **Engines:** 3.0L VIN 8, 3.6L VIN 9 **Models:** B9 Tribeca, Tribeca **Transmissions:** All	**EGR Solenoid Valve Signal 4 Circuit Low Input** Detects open or short circuit of EGR. Judge as NG when the ECM output level differs from the actual terminal level. Battery voltage 10.9 volts. EGR target position 0 step. 1 second after startup. Improper idling. Poor driving performance. Engine breathing **Possible Causes:** • Open circuit in harness between EGR valve and main relay connector • Poor contact of coupling connector • Open circuit in harness between ECM and EGR valve connector • Ground short in harness between ECM and EGR valve connector • Poor contact in ECM or EGR valve connector • Faulty EGR valve
DTC: P1499 **1T CCM, MIL: Yes** **Years:** 2006, 2007, 2008 **Engines:** 3.0L VIN 8, 3.6L VIN 9 **Models:** B9 Tribeca, Tribeca **Transmissions:** All	**EGR Solenoid Valve Signal 4 Circuit High Input** Detects open or short circuit of EGR. Judge as NG when the ECM output level differs from the actual terminal level. Battery voltage 10.9 volts. EGR target position 0 step. 1 second after startup. Improper idling. Poor driving performance. Engine breathing **Possible Causes:** • Short circuit to power in the harness between the ECM and EGR valve connectors • Poor contact of the ECM connector
DTC: P1510 **1T CCM, MIL: Yes** **Years:** 2006 **Engines:** 2.5L SOHC VIN 6 **Models:** Baja **Transmissions:** All	**ISC Solenoid Valve Signal No. 1 Circuit Malfunction (Low Input)** Key on or engine running; and the PCM detected an unexpected "low" condition on the IAC solenoid Signal 1 circuit during the test. **Possible Causes:** • Check power supply to idle air control solenoid valve • Check harness between ECM and idle control solenoid valve connector • Check poor contact
DTC: P1511 **1T CCM, MIL: Yes** **Years:** 2006 **Engines:** 2.5L SOHC VIN 6 **Models:** Baja **Transmissions:** All	**ISC Solenoid Valve Signal No. 1 Circuit Malfunction (High Input)** Key on or engine running; and the PCM detected an unexpected "low" condition on the IAC solenoid Signal 1 circuit during the test. **Possible Causes:** • Check any other DTC on display • Check ground circuit for ECM • Check harness between ECM and idle control solenoid valve connector

DTC	Trouble Code Title, Conditions & Possible Causes
DTC: P1512 **1T CCM, MIL: Yes** **Years:** 2006 **Engines:** 2.5L SOHC VIN 6 **Models:** Baja **Transmissions:** All	**ISC Solenoid Valve Signal No. 2 Circuit Malfunction (Low Input)** Key on or engine running; and the PCM detected an unexpected "low" condition on the IAC solenoid Signal 2 circuit during the test. **Possible Causes:** • Check power supply to idle air control solenoid valve • Check harness between ECM and idle control solenoid valve connector • Check poor contact
DTC: P1513 **2T CCM, MIL: Yes** **1998-06** **Engines:** 2.5L SOHC VIN 6 **Models:** Baja, Impreza, WRX, Legacy, Outback **Transmissions:** All	**ISC Solenoid Valve Signal No. 2 Circuit Malfunction (High Input)** Key on or engine running; and the PCM detected an unexpected "low" condition on the IAC solenoid Signal 2 circuit during the test. **Possible Causes:** • Check any other DTC on display • Check ground circuit for ECM • Check harness between ECM and idle control solenoid valve connector
DTC: P1514 **2T CCM, MIL: Yes** **Years:** 2006 **Engines:** 2.5L SOHC VIN 6 **Models:** Baja **Transmissions:** All	**ISC Solenoid Valve Signal No. 3 Circuit Malfunction (Low Input)** Key on or engine running; and the PCM detected an unexpected "low" condition on the IAC solenoid Signal 3 circuit during the test. **Possible Causes:** • Check power supply to idle air control solenoid valve • Check harness between ECM and idle control solenoid valve connector • Check poor contact
DTC: P1515 **2T CCM, MIL: Yes** **Years:** 2006 **Engines:** 2.5L SOHC VIN 6 **Models:** Baja **Transmissions:** All	**ISC Solenoid Valve Signal No. 3 Circuit Malfunction (High Input)** Key on or engine running; and the PCM detected an unexpected "low" condition on the IAC solenoid Signal 3 circuit during the test. **Possible Causes:** • Check any other DTC on display • Check ground circuit for ECM • Check harness between ECM and idle control solenoid valve connector
DTC: P1516 **2T CCM, MIL: Yes** **1998-06** **Engines:** 2.5L SOHC VIN 6 **Models:** Baja, Impreza, Legacy, Outback **Transmissions:** All	**ISC Solenoid Valve Signal No. 4 Circuit Malfunction (Low Input)** Key on or engine running; and the PCM detected an unexpected "low" condition on the IAC solenoid Signal 4 circuit during the test. **Possible Causes:** • Check power supply to idle air control solenoid valve • Check harness between ECM and idle control solenoid valve connector • Check poor contact
DTC: P1517 **2T CCM, MIL: Yes** **Years:** 2006 **Engines:** 2.5L SOHC VIN 6 **Models:** Baja **Transmissions:** All	**ISC Solenoid Valve Signal No. 4 Circuit Malfunction (High Input)** Key on or engine running; and the PCM detected an unexpected "low" condition on the IAC solenoid Signal 4 circuit during the test. **Possible Causes:** • Check any other DTC on display • Check ground circuit for ECM • Check harness between ECM and idle control solenoid valve connector
DTC: P1518 **1T CCM, MIL: No** **Years:** 2006, 2007, 2008 **Engines:** 2.5L SOHC, DOHC VIN 6, 3.0L VIN 8, 3.6L VIN 9 **Models:** Baja, Forester, Impreza, Legacy, Tribeca **Transmissions:** All	**Starter Switch Circuit Low Input** Ignition key in crank position, and the PCM detected an unexpected "low" voltage condition on the Starter Switch circuit during the test. **Note: The engine will not start with this condition present.** **Possible Causes:** • Check operation of starter motor circuit • Open or ground short circuit of harness between ECM and starter motor connector
DTC: P1544 **1T CCM, MIL: No** **Years:** 2006 **Engines:** 2.5L SOHC VIN 6 **Models:** Baja **Transmissions:** All	**Exhaust Gas Temperature Too High** Detect the malfunction of high exhaust gas temperature. Judge NG when the exhaust gas becomes too high. Erroneous idling. Poor driving performance. **Possible Causes:** • Check any other DTC on display • Check exhaust system

DTC	Trouble Code Title, Conditions & Possible Causes
DTC: P1560 **1T CCM, MIL: Yes** **Years:** 2006, 2007, 2008 **Engines:** 2.5L SOHC, DOHC VIN 6, 3.0L VIN 8, 3.6L VIN 9 **Models:** All **Transmissions:** A/T	**Back-Up Voltage Circuit Malfunction** Key on, and the ECM detected an unexpected "low" voltage condition on the Battery Backup circuit during the CCM test. **Note: The engine will not start with this condition present.** **Possible Causes:** • Check input signal of ECM • Check harness between ECM and main fuse box connector • Check fuse number 13
DTC: P1570 **1T CCM, MIL: Yes** **Years:** 2006, 2007, 2008 **Engines:** 2.5L SOHC, DOHC VIN 6, 3.0L VIN 8, 3.6L VIN 9 **Models:** B9 Tribeca, Tribeca, Forester, Impreza, Legacy, Outback **Transmissions:** A/T	**Antenna Incorrect Immobilizer Key** When the engine is started. Improper antenna **Possible Causes:** • Perform teaching operation of ignition key • Check the ignition keys (including transponder) which cannot be registered • Replace the body integrated unit and replace all the ignition keys (including transponder). • Execute the registration procedure next
DTC: P1572 **1T CCM, MIL: Yes** **Years:** 2006, 2007, 2008 **Engines:** 2.5L SOHC, DOHC VIN 6, 3.0L VIN 8, 3.6L VIN 9 **Models:** B9 Tribeca, Tribeca, Forester, Impreza, Legacy, Outback **Transmissions:** All	**IMM Circuit Failure (Except Antenna Circuit)** When starting the engine. Communication failure between body integrated unit and ECM. Incorrect immobilizer key (Use of unregistered key in body integrated unit. **Possible Causes:** • Poor connections • Wrong key • Faulty body integrated unit • Faulty ECM
DTC: P1574 **1T CCM, MIL: Yes** **Years:** 2006, 2007, 2008 **Engines:** 2.5L SOHC, DOHC VIN 6, 3.0L VIN 8, 3.6L VIN 9 **Models:** B9 Tribeca, Tribeca, Forester, Impreza, Legacy, Outback **Transmissions:** All	**Key Communication Failure** When starting the engine. Incorrect immobilizer key (Use of unregistered key in body integrated unit). **Possible Causes:** • Poor connections • Wrong key
DTC: P1601 **2T CCM, MIL: Yes** **Years:** 2006, 2007, 2008 **Engines:** 2.5L SOHC, DOHC VIN 6, 3.0L VIN 8, 3.6L VIN 9 **Models:** All **Transmissions:** A/T	**TCM Data Communication Failure** Communication does not complete between control valve memory box. Shifting quality malfunction. **Possible Causes:** • Check loose connection on TCM connector • Check TCM output signal • Check harness between TCM connector and transmission connector • Check circuit of harness between transmission connector and control valve body connector • Faulty transmission assembly
DTC: P1602 **2T CCM, MIL: Yes** **Years:** 2006, 2007, 2008 **Engines:** 2.5L SOHC, DOHC VIN 6, 3.0L VIN 8, 3.6L VIN 9 **Models:** B9 Tribeca, Tribeca, Forester, Impreza, Legacy, Outback **Transmissions:** A/T	**Control module Programming Error** Detect malfunctions of the catalyst advanced idling retard angle control. Judge as NG when ECM is not controlling the angle properly during catalyst advanced idling retard angle control. Engine keeps running at higher speed than specified idle speed. Engine keeps running at a lower speed than the specified idle speed. Engine stalls. **Possible Causes:** • Low engine oil • Exhaust system • Air intake system • Fuel pressure • Engine coolant temperature sensor • Mass air flow and intake air temperature sensor • Ground short circuit of harness between ECM and fuel injector • Faulty harness and connector • Faulty fuel injector • Short circuit to power in the harness between the ECM and fuel injector • Engine mechanical engine problems • Throttle control system

DTC	Trouble Code Title, Conditions & Possible Causes
DTC: P1698 **2T CCM, MIL: Yes** **Years:** 2006, 2007, 2008 **Engines:** 2.5L SOHC, DOHC VIN 6, 3.0L VIN 8, 3.6L VIN 9 **Models:** B9 Tribeca, Tribeca, Forester, Impreza, Legacy, Outback **Transmissions:** A/T	**Engine Torque Control Cut Signal Circuit Malfunction (Low Input)** Engine started, and the ECM detected an unexpected "low" voltage condition on the Engine Torque Control Cut Signal circuit (5-volt) during the CCM test. **Possible Causes:** • Check output signal for ECM • Check harness between ECM and TCM connector
DTC: P1699 **2T CCM, MIL: Yes** **Years:** 2006, 2007, 2008 **Engines:** 2.5L SOHC, DOHC VIN 6, 3.0L VIN 8, 3.6L VIN 9 **Models:** B9 Tribeca, Tribeca, Forester, Impreza, Legacy, Outback **Transmissions:** A/T	**Engine Torque Control Cut Signal Circuit Malfunction (High Input)** Engine started, and the PCM detected an unexpected "high" voltage condition on the Engine Torque Control Cut Signal circuit (5-volt) during the CCM test **Possible Causes:** • Check output signal for ECM • Check harness between ECM and TCM connector
DTC: P1700 **2T CCM, MIL: No** **Years:** 2006 **Engines:** 2.5L SOHC, DOHC VIN 6 **Models:** Baja **Transmissions:** A/T	**Throttle Position Sensor Circuit Malfunction (Automatic Transmission)** Key on or engine running; and the TCM detected an unexpected "low" or "high" voltage condition on the TP Sensor circuit. **Note: The TP sensor signal is shared with the ECM on this circuit.** **Possible Causes:** • Check any other DTC on display • Check throttle position sensor circuit
DTC: P1706 **1T CCM, MIL: Yes** **Years:** 2006, 2007, 2008 **Engines:** 3.0L VIN 8, 3.6L VIN 9 **Models:** B9 Tribeca, Tribeca **Transmissions:** All	**Vehicle Speed Sensor Circuit (Rear Wheels)** Input signal circuit of TCM is open or shorted. Shifting quality malfunction. Tight corner braking phenomenon is occurred. **Possible Causes:** • Open or short circuit for power supply and ground • Open circuit of harness between TCM and transmission connector • Short circuit of harness between TCM and transmission connector • Poor contact of harness in ATF temperature sensor and transmission connector • Faulty transmission harness • Faulty transmission assembly
DTC: P1707 **1T CCM, MIL: Yes** **Years:** 2006, 2007, 2008 **Engines:** 3.0L VIN 8, 3.6L VIN 9 **Models:** All **Transmissions:** All	**AWD Solenoid Valve Circuit** Output signal circuit of transfer solenoid is open or shorted. Tight corner braking phenomenon is occurred. Drivability getting worse. **Possible Causes:** • Open circuit of harness between TCM connector and transmission connector • Short circuit in accelerator pedal position sensor • Open circuit of harness between transmission connector and control valve body connector • Short circuit of harness between control valve body connector and transmission ground • Faulty control valve body • Faulty TCM
DTC: P1708 **1T CCM, MIL: Yes** **Years:** 2006 **Engines:** 2.5L SOHC DOHC VIN 6 **Models:** Baja **Transmissions:** A/T	**Throttle Position Circuit (Low Input)** The input signal circuit of accelerator pedal position sensor is open or shorted. Shift point too high or too low. Excessive shift shock. Excessive tight corner "braking". **Possible Causes:** • Check engine ground connections • Check ECM ground • Check accelerator pedal position sensor connector receptacle's terminals • Check harness between TCM and accelerator pedal position sensor • Check input signal of TCM using a scan tool • Faulty TCM

DTC	Trouble Code Title, Conditions & Possible Causes
DTC: P1709 **1T CCM, MIL: Yes** **Years:** 2006 **Engines:** 2.5L SOHC DOHC VIN 6 **Models:** Baja **Transmissions:** A/T	**Throttle Position Circuit (High Input)** The input signal circuit of accelerator pedal position sensor is shorted. Shift point too high or too low. Excessive shift shock. Excessive tight corner "braking". **Possible Causes:** • Check engine ground connections • Check ECM ground • Check accelerator pedal position sensor connector receptacle's terminals • Check harness between TCM and accelerator pedal position sensor • Check input signal of TCM using a scan tool • Faulty TCM
DTC: P1710 **1T CCM, MIL: Yes** **Years:** 2006, 2007, 2008 **Engines:** 2.5L SOHC, DOHC VIN 6, 3.0L VIN 8, 3.6L VIN 9 **Models:** B9 Tribeca, Tribeca, Forester, Impreza, Legacy, Outback **Transmissions:** A/T	**Torque Converter Turbine 2 Speed Signal Circuit 2** Input signal circuit of TCM is open or shorted. Excessive shift shock. Does not shift to 5th **Possible Causes:** • Open or short circuit for power supply and ground • Open circuit of harness between TCM and transmission connector • Short circuit of harness between TCM and transmission connector • Faulty TCM
DTC: P1817 **1T CCM, MIL: Yes** **Years:** 2006, 2007, 2008 **Engines:** 2.5L SOHC, DOHC VIN 6, 3.0L VIN 8, 3.6L VIN 9 **Models:** B9 Tribeca, Tribeca, Forester, Impreza, Legacy, Outback **Transmissions:** A/T	**Sport Mode Manual Switch Circuit (Manual Switch)** Input signal circuit of manual mode switch is open or shorted. Manual mode can not be set. When shifting to "N" "D", the SPORT shift indicator light illuminates. **Possible Causes:** • Faulty select lever assembly • Short circuit of harness between body integrated unit and manual mode • Temporary poor contact of connector or harness may be the cause • Faulty TCM
DTC: P1840 **1T CCM, MIL: Yes** **Years:** 2006, 2007, 2008 **Engines:** 2.5L SOHC, DOHC VIN 6, 3.0L VIN 8, 3.6L VIN 9 **Models:** B9 Tribeca, Tribeca, Forester, Impreza, Legacy, Outback **Transmissions:** A/T	**Transmission Fluid Pressure Sensor/Switch A Circuit** Output signal of front brake solenoid does not match with oil pressure. Locked to 1st gear. **Possible Causes:** • Open circuit of harness between TCM and transmission connector • Short circuit of harness between TCM and transmission connector • Transmission harness assembly • Faulty control valve body
DTC: P1841 **1T CCM, MIL: Yes** **Years:** 2006, 2007, 2008 **Engines:** 2.5L SOHC, DOHC VIN 6, 3.0L VIN 8, 3.6L VIN 9 **Models:** B9 Tribeca, Tribeca, Forester, Impreza, Legacy, Outback **Transmissions:** A/T	**Transmission Fluid Pressure Sensor/Switch B Circuit** Output signal of forward brake solenoid does not match the oil pressure. Locked to 2nd, 3rd, 4th gear. **Possible Causes:** • Open circuit of harness between TCM and transmission connector • Short circuit of harness between TCM and transmission connector • Transmission harness assembly • Faulty control valve body
DTC: P1842 **1T CCM, MIL: Yes** **Years:** 2006, 2007, 2008 **Engines:** 2.5L SOHC, DOHC VIN 6, 3.0L VIN 8, 3.6L VIN 9 **Models:** B9 Tribeca, Tribeca, Forester, Impreza, Legacy, Outback **Transmissions:** A/T	**Transmission Fluid Pressure Sensor/Switch C Circuit** Output signal value of input clutch solenoid and oil pressure does not match. Locked to 1st or 4th gear. **Possible Causes:** • Open circuit of harness between TCM and transmission connector • Short circuit of harness between TCM and transmission connector • Transmission harness assembly • Faulty control valve body
DTC: P1843 **1T CCM, MIL: Yes** **Years:** 2006, 2007, 2008 **Engines:** 2.5L SOHC, DOHC VIN 6, 3.0L VIN 8, 3.6L VIN 9 **Models:** B9 Tribeca, Tribeca, Forester, Impreza, Legacy, Outback **Transmissions:** A/T	**Transmission Fluid Pressure Sensor/Switch D Circuit** Output signal value of direct clutch solenoid and oil pressure does not match. Locked to 1st or 4th gear. **Possible Causes:** • Open circuit of harness between TCM and transmission connector • Short circuit of harness between TCM and transmission connector • Transmission harness assembly • Faulty control valve body

DTC	Trouble Code Title, Conditions & Possible Causes
DTC: P1844 **1T CCM, MIL: Yes** **Years:** 2006, 2007, 2008 **Engines:** 2.5L SOHC, DOHC VIN 6, 3.0L VIN 8, 3.6L VIN 9 **Models:** B9 Tribeca, Tribeca, Forester, Impreza, Legacy, Outback **Transmissions:** A/T	**Transmission Fluid Pressure Sensor/Switch E Circuit** Output signal value of high & low reverse clutch solenoid and oil pressure does not match. Locked to 1st gear. **Possible Causes:** • Open circuit of harness between TCM and transmission connector • Short circuit of harness between TCM and transmission connector • Transmission harness assembly • Faulty control valve body

OBD II Trouble Code List (P2XXX Codes)

DTC	Trouble Code Title, Conditions & Possible Causes
DTC: P2004 **1T CCM, MIL: Yes** **Years:** 2006, 2007, 2008 **Engines:** 2.5L DOHC VIN 6 **Models:** Baja, Forester, Impreza, Legacy, Outback **Transmissions:** All	**Tumble Generated Valve System 1 (Valve Open)** Vehicle performance issue. **Possible Causes:** • Check any other DTC on display • Check tumble generator valve (RH)
DTC: P2005 **1T CCM, MIL: Yes** **Years:** 2006, 2007, 2008 **Engines:** 2.5L DOHC VIN 6 **Models:** Baja, Forester, Impreza, Legacy, Outback **Transmissions:** All	**Tumble Generated Valve System 2 (Valve Open)** Vehicle performance issue. **Possible Causes:** • Check any other DTC on display • Check tumble generator valve (RH)
DTC: P2006 **1T CCM, MIL: Yes** **Years:** 2006, 2007, 2008 **Engines:** 2.5L DOHC VIN 6 **Models:** Baja, Forester, Impreza, Legacy, Outback **Transmissions:** All	**Tumble Generated Valve System 1 (Valve Closed)** Vehicle performance issue. **Possible Causes:** • Check any other DTC on display • Check tumble generator valve (RH)
DTC: P2007 **1T CCM, MIL: Yes** **Years:** 2006, 2007, 2008 **Engines:** 2.5L DOHC VIN 6 **Models:** Baja, Forester, Impreza, Legacy, Outback **Transmissions:** All	**Tumble Generated Valve System 2 (Valve Closed)** Vehicle performance issue. **Possible Causes:** • Check any other DTC on display • Check tumble generator valve (RH)
DTC: P2008 **1T CCM, MIL: Yes** **Years:** 2006, 2007, 2008 **Engines:** 2.5L DOHC VIN 6 **Models:** Baja, Forester, Impreza, Legacy, Outback **Transmissions:** All	**Tumble Generated Valve System 1 Circuit Malfunction (Open)** Vehicle performance issue. **Possible Causes:** • Check harness between ECM and tumble generator valve actuator connector • Check poor contact
DTC: P2009 **1T CCM, MIL: Yes** **Years:** 2006, 2007, 2008 **Engines:** 2.5L DOHC VIN 6 **Models:** Baja, Forester, Impreza, Legacy, Outback **Transmissions:** All	**Tumble Generated Valve System 1 Circuit Malfunction (Closed)** Vehicle performance issue. **Possible Causes:** • Check harness between ECM and tumble generator valve actuator connector

DTC	Trouble Code Title, Conditions & Possible Causes
DTC: P2011 **1T CCM, MIL: Yes** **Years:** 2006, 2007, 2008 **Engines:** 2.5L DOHC VIN 6 **Models:** Baja, Forester, Impreza, Legacy, Outback **Transmissions:** All	**Tumble Generated Valve System 2 Circuit Malfunction (Dhort)** Vehicle performance issue. **Possible Causes:** • Check harness between ECM and tumble generator valve actuator connector
DTC: P2016 **1T CCM, MIL: Yes** **Years:** 2006, 2007, 2008 **Engines:** 2.5L DOHC VIN 6 **Models:** Baja, Forester, Impreza, Legacy, Outback **Transmissions:** All	**Tumble Generated Valve Position Sensor 1 Circuit Low** Engine stalls. Erroneous idling. Poor driving performance. **Possible Causes:** • Using the a scan tool, check current data • Check input signal for ECM • Check harness between ECM and tumble generator valve position sensor connector • Check harness between ECM and throttle position sensor connector • Check poor contact
DTC: P2017 **1T CCM, MIL: Yes** **Years:** 2006, 2007, 2008 **Engines:** 2.5L DOHC VIN 6 **Models:** Baja, Forester, Impreza, Legacy, Outback **Transmissions:** All	**Tumble Generated Valve Position Sensor 1 Circuit High** Engine stalls. Erroneous idling. Poor driving performance. **Possible Causes:** • Using a scan tool, check current data • Check harness between tumble generator valve position sensor and ECM connector
DTC: P2021 **1T CCM, MIL: Yes** **Years:** 2006, 2007, 2008 **Engines:** 2.5L DOHC VIN 6 **Models:** Baja, Forester, Impreza, Legacy, Outback **Transmissions:** All	**Tumble Generated Valve Position Sensor 2 Circuit Low** Engine stalls. Erroneous idling. Poor driving performance. **Possible Causes:** • Using a scan tool, check current data • Check input signal for ECM • Check harness between ECM and tumble generator valve position sensor connector • Check poor contact
DTC: P2022 **1T CCM, MIL: Yes** **Years:** 2006, 2007, 2008 **Engines:** 2.5L DOHC VIN 6 **Models:** Baja, Forester, Impreza, Legacy, Outback **Transmissions:** All	**Tumble Generated Valve Position Sensor 2 Circuit High** Engine stalls. Erroneous idling. Poor driving performance. **Possible Causes:** • Using a scan tool, check current data • Check harness between tumble generator valve position sensor and ECM connector • Check poor contact
DTC: P2088 **1T CCM, MIL: No** **Years:** 2006, 2007, 2008 **Engines:** 3.0L VIN 8, 3.6L VIN 9 **Models:** B9 Tribeca, Tribeca **Transmissions:** All	**Intake Camshaft Position Actuator Control Circuit Low (Bank 1)** Detect open or short circuit of the oil flow control solenoid valve. Judge as NG when the current is small even though the duty signal is large Improper idling **Possible Causes:** • Open circuit of the harness between the ECM and oil flow control solenoid valve connector • Ground short circuit of harness between ECM and oil flow control solenoid valve connector • Poor contact of the ECM or oil flow control solenoid valve connector • Faulty oil flow control solenoid valve
DTC: P2089 **1T CCM, MIL: No** **Years:** 2006, 2007, 2008 **Engines:** 3.0L VIN 8, 3.6L VIN 9 **Models:** B9 Tribeca, Tribeca **Transmissions:** All	**Intake Camshaft Position Actuator Control Circuit High (Bank 1)** Detect open or short circuit of the oil flow control solenoid valve. Judge as NG when the current is small even though the duty signal is large Improper idling **Possible Causes:** • Short circuit to power in the harness between the ECM and oil flow control solenoid valve connector • Open circuit of the harness between the ECM and oil flow control solenoid valve connector • Poor contact of the ECM or oil flow control solenoid valve connector • Faulty oil flow control solenoid valve

DTC	Trouble Code Title, Conditions & Possible Causes
DTC: P2090 **1T CCM, MIL: No** **Years:** 2006, 2007, 2008 **Engines:** 3.0L VIN 8, 3.6L VIN 9 **Models:** B9 Tribeca, Tribeca **Transmissions:** All	**Exhaust Camshaft Position Actuator Control Circuit Low (Bank 1)** Detect open circuit of the oil flow control solenoid valve. Detect open circuit of the oil flow control solenoid valve. Improper idling **Possible Causes:** • Open circuit of the harness between the ECM and oil flow control solenoid valve connector • Poor contact of coupling connector • Short circuit to ground in harness between ECM and oil flow control solenoid valve connector • Poor contact of ECM or oil flow control solenoid valve connector • Faulty oil flow control solenoid valve
DTC: P2091 **1T CCM, MIL: No** **Years:** 2006, 2007, 2008 **Engines:** 3.0L VIN 8, 3.6L VIN 9 **Models:** B9 Tribeca, Tribeca **Transmissions:** All	**Exhaust Camshaft Position Actuator Control Circuit High (Bank 1)** Detect short circuit of oil flow control solenoid valve. Judge as short NG when the current is large even though the duty signal is small. Improper idling. **Possible Causes:** • Short circuit to power in the harness between the ECM and oil flow control solenoid valve connector • Open circuit of the harness between the ECM and oil flow control solenoid valve connector • Poor contact of the ECM or oil flow control solenoid valve connector • Faulty oil flow control solenoid valve
DTC: P2092 **2T CCM, MIL: No** **Years:** 2006, 2007, 2008 **Engines:** 2.5L DOHC, VIN 6 3.0L Vin 8 3.6L Vin 9 **Models:** Baja B9 Tribeca, Tribeca, Impreza, Legacy, Outback **Transmissions:** All	**OCV Solenoid Valve Signal "A" Circuit Open (Bank 2)** Erroneous idling. **Possible Causes:** • Check harness between ECM and oil flow control solenoid valve • Check oil flow control solenoid valve
DTC: P2093 **2T CCM, MIL: No** **Years:** 2006, 2007, 2008 **Engines:** 2.5L SOHC, DOHC VIN 6 3.0L VIN 8, 3.6L VIN 9 **Models:** Baja, B9 Tribeca, Tribeca, Legacy, Outback **Transmissions:** All	**OCV Solenoid Valve Signal "A" Circuit Short (Bank 2)** Erroneous idling. **Possible Causes:** • Check harness between ECM and oil flow control solenoid valve • Check oil flow control solenoid valve
DTC: P2094 **1T CCM, MIL: No** **Years:** 2006, 2007, 2008 **Engines:** 3.0L VIN 8, 3.6L VIN 9 **Models:** B9 Tribeca, Tribeca **Transmissions:** All	**Exhaust Camshaft Position Actuator Control Circuit Low (Bank 1)** Detect open circuit of the oil flow control solenoid valve. Detect open circuit of the oil flow control solenoid valve. Improper idling **Possible Causes:** • Open circuit of the harness between the ECM and oil flow control solenoid valve connector • Poor contact of coupling connector • Short circuit to ground in harness between ECM and oil flow control solenoid valve connector • Poor contact of ECM or oil flow control solenoid valve connector • Faulty oil flow control solenoid valve
DTC: P2095 **1T CCM, MIL: No** **Years:** 2006, 2007, 2008 **Engines:** 3.0L VIN 8, 3.6L VIN 9 **Models:** B9 Tribeca, Tribeca **Transmissions:** All	**Exhaust Camshaft Position Actuator Control Circuit High (Bank 1)** Detect short circuit of oil flow control solenoid valve. Judge as short NG when the current is large even though the duty signal is small. Improper idling. **Possible Causes:** • Short circuit to power in the harness between the ECM and oil flow control solenoid valve connector • Open circuit of the harness between the ECM and oil flow control solenoid valve connector • Poor contact of the ECM or oil flow control solenoid valve connector • Faulty oil flow control solenoid valve

DTC	Trouble Code Title, Conditions & Possible Causes
DTC: P2096 **2T CCM, MIL: Yes** **Years:** 2006, 2007, 2008 **Engines:** 2.5L SOHC, DOHC VIN 6, 3.0L VIN 8, 3.6L VIN 9 **Models:** Baja, B9 Tribeca, Tribeca, Legacy, Outback **Transmissions:** All	**Post Catalyst Fuel Trim System Too Lean (Bank 1)** Poor Drive ability. **Possible Causes:** • Check any other DTC on display • Using the a scan tool, check front oxygen sensor data • Using the Subaru scan tool, check rear oxygen sensor data • Check exhaust system • Check air intake system • Check fuel pressure • Check engine coolant temperature sensor • Check Mass air flow and intake air temperature • Check harness between ECM and front oxygen sensor connector • Using harness between ECM and rear oxygen sensor connector
DTC: P2097 **2T CCM, MIL: Yes** **Years:** 2006, 2007, 2008 **Engines:** 2.5L SOHC, DOHC VIN 6, 3.0L VIN 8, 3.6L VIN 9 **Models:** Baja B9 Tribeca, Tribeca, Legacy, Outback **Transmissions:** All	**Post Catalyst Fuel Trim System Too Rich (Bank 1)** Poor Drive ability. **Possible Causes:** • Check any other DTC on display • Using a scan tool, check front oxygen sensor data • Using a scan tool, check rear oxygen sensor data • Check exhaust system • Check air intake system • Check fuel pressure • Check engine coolant temperature sensor • Check Mass air flow and intake air temperature • Check harness between ECM and front oxygen sensor connector • Using harness between ECM and rear oxygen sensor connector
DTC: P2098 **2T CCM, MIL: Yes** **Years:** 2006, 2007, 2008 **Engines:** 2.5L SOHC, DOHC VIN 6, 3.0L VIN 8, 3.6L VIN 9 **Models:** B9 Tribeca, Tribeca, Legacy, Outback **Transmissions:** All	**Post Catalyst Fuel Trim System Too Lean (Bank 2)** Poor Drive ability. **Possible Causes:** • Check any other DTC on display • Using a scan tool, check front oxygen sensor data • Using a scan tool, check rear oxygen sensor data • Check exhaust system • Check air intake system • Check fuel pressure • Check engine coolant temperature sensor • Check Mass air flow and intake air temperature • Check harness between ECM and front oxygen sensor connector • Using harness between ECM and rear oxygen sensor connector
DTC: P2099 **2T CCM, MIL: Yes** **Years:** 2006, 2007, 2008 **Engines:** 3.0L VIN 8, 3.6L VIN 9 **Models:** B9 Tribeca, Tribeca, Legacy, Outback **Transmissions:** All	**Post Catalyst Fuel Trim System Too Rich (Bank 2)** Poor Drive ability. **Possible Causes:** • Check any other DTC on display • Using a scan tool, check front oxygen sensor data • Using a scan tool, check rear oxygen sensor data • Check exhaust system • Check air intake system • Check fuel pressure • Check engine coolant temperature sensor • Check Mass air flow and intake air temperature • Check harness between ECM and front oxygen sensor connector • Using harness between ECM and rear oxygen sensor connector

DTC	Trouble Code Title, Conditions & Possible Causes
DTC: P2101 **1T CCM, MIL: Yes** **Years:** 2006, 2007, 2008 **Engines:** 2.5L SOHC, DOHC VIN 6, 3.0L VIN 8, 3.6L VIN 9 **Models:** All **Transmissions:** All	**Throttle Actuator Control Motor Circuit Range/Performance** Judge as NG when the motor current becomes large or drive circuit is heated. Improper idling. Engine stalls. Poor driving performance. **Possible Causes:** • Check electronic throttle control relay • Check power supply of electronic control relay • Check harness between ECM and electronic throttle control relay • Check sensor output • Check poor contact • Check harness between ECM and electronic throttle control • Check sensor power supply • Check for short in ECM • Check sensor output • Check harness between ECM and electronic throttle control motor • Check electronic throttle control motor harness • Check ground circuit • Check electronic throttle control
DTC: P2102 **1T CCM, MIL: Yes** **Years:** 2006, 2007, 2008 **Engines:** 2.5L SOHC, DOHC VIN 6, 3.0L VIN 8, 3.6L VIN 9 **Models:** All **Transmissions:** All	**Throttle Actuator Control Motor Circuit Low** Judge as NG when the electronic throttle control power is not supplied even when ECM sets the electronic throttle control relay to ON Improper idling. Engine stalls. Poor driving performance. **Possible Causes:** • Check electronic throttle control relay • Check power supply of electronic control relay • Check harness between ECM and electronic throttle control relay
DTC: P2103 **1T CCM, MIL: Yes** **Years:** 2006, 2007, 2008 **Engines:** 2.5L SOHC, DOHC VIN 6, 3.0L VIN 8, 3.6L VIN 9 **Models:** All **Transmissions:** All	**Throttle Actuator Control Motor Circuit High** Judge as NG when the electronic throttle control power is supplied even when ECM sets the electronic throttle control relay to OFF. Poor drive ability. **Possible Causes:** • Check electronic throttle control relay • Check for short circuit of the relay and power supply • Check harness between ECM and electronic throttle control relay
DTC: P2109 **1T CCM, MIL: Yes** **Years:** 2006, 2007, 2008 **Engines:** 2.5L SOHC, DOHC VIN 6, 3.0L VIN 8, 3.6L VIN 9 **Models:** All **Transmissions:** All	**Throttle/Pedal Position Sensor "A" Minimum Stop Performance** Judge as NG when full close point learning cannot be conducted or abnormal value is detected Improper idling. Engine stalls. Poor driving performance. **Possible Causes:** • Check electronic throttle control relay • Check power supply of electronic control relay • Check harness between ECM and electronic throttle control relay • Check sensor output • Check poor contact • Check harness between ECM and electronic throttle control • Check sensor power supply • Check for short in ECM • Check sensor output • Check harness between ECM and electronic throttle control motor • Check electronic throttle control motor harness • Check ground circuit • Check electronic throttle control
DTC: P2122 **1T CCM, MIL: Yes** **Years:** 2006, 2007, 2008 **Engines:** 2.5L SOHC, DOHC VIN 6, 3.0L VIN 8, 3.6L VIN 9 **Models:** All **Transmissions:** All	**Throttle/Pedal Position Sensor "D" Circuit Low Input** Detect the open or short circuit of accelerator pedal position sensor 1. Judge as NG if out of specification. Improper idling. Poor driving performance. **Possible Causes:** • Check accelerator pedal position sensor output • Check poor contact • Check harness between ECM and accelerator pedal position sensor • Check power supply of accelerator pedal position sensor • Check sensor

DTC	Trouble Code Title, Conditions & Possible Causes
DTC: P2123 **1T CCM, MIL: Yes** **Years:** 2006, 2007, 2008 **Engines:** 2.5L SOHC, DOHC VIN 6, 3.0L VIN 8, 3.6L VIN 9 **Models:** All **Transmissions:** All	**Throttle/Pedal Position Sensor/Switch "D" Circuit High Input** Detect the open or short circuit of accelerator pedal position sensor 1. Judge as NG if out of specification. Improper idling. Poor driving performance. **Possible Causes:** • Check accelerator pedal position sensor output • Check poor contact • Check harness between ECM and accelerator pedal position sensor
DTC: P2127 **1T CCM, MIL: Yes** **Years:** 2006, 2007, 2008 **Engines:** 2.5L SOHC, DOHC VIN 6, 3.0L VIN 8, 3.6L VIN 9 **Models:** All **Transmissions:** All	**Throttle/Pedal Position Sensor/Switch "E" Circuit Low Input** Detect the open or short circuit of accelerator pedal position sensor 2 Judge as NG if out of specification. Improper idling. Poor driving performance. **Possible Causes:** • Check accelerator pedal position sensor output • Check poor contact • Check harness between ECM and accelerator pedal position sensor • Check sensor
DTC: P2128 **1T CCM, MIL: Yes** **Years:** 2006, 2007, 2008 **Engines:** 3.0L VIN 8, 3.6L VIN 9 **Models:** B9 Tribeca, Tribeca **Transmissions:** All	**Throttle/Pedal Position Sensor/Switch "E" Circuit High Input** Detect the open or short circuit of accelerator pedal position sensor 2. Judge as NG if out of specification. Improper idling. Poor driving performance. **Possible Causes:** • Check accelerator pedal position sensor output • Check poor contact • Check harness between ECM and accelerator pedal position sensor • Check sensor
DTC: P2135 **Years:** 2006, 2007, 2008 **Engines:** 2.5L SOHC, DOHC VIN 6, 3.0L VIN 8, 3.6L VIN 9 **Models:** All **Transmissions:** All	**Throttle/Pedal Position Sensor/Switch "A"/"B" Voltage Correlation** Judge as NG when the signal level of throttle position sensor 1 is different from the throttle position sensor 2. Improper idling. Poor driving performance. **Possible Causes:** • Check sensor output • Check poor contact • Check harness between ECM and electronic throttle control • Check sensor power supply • Check for short circuit in ECM • Check sensor output • Check electronic throttle control harness
DTC: P2138 **1T CCM, MIL: Yes** **Years:** 2006, 2007, 2008 **Engines:** 2.5L SOHC, DOHC VIN 6, 3.0L VIN 8, 3.6L VIN 9 **Models:** All **Transmissions:** All	**Throttle/Pedal Position Sensor/Switch "D"/"E" Voltage Correlation** Judge as NG when the signal level of throttle position sensor 1 is different from the throttle position sensor 2. Improper idling. Poor driving performance. **Possible Causes:** • Check accelerator pedal position sensor output • Check poor contact • Check harness between ECM and sensor • Check sensor • Check sensor output
DTC: P2227 **2T CCM, MIL: Yes** **Years:** 2006, 2007, 2008 **Engines:** 2.5L SOHC, DOHC VIN 6, 3.0L VIN 8, 3.6L VIN 9 **Models:** All **Transmissions:** All	**Barometric Pressure Circuit Range/Performance** Detect the malfunction of barometric pressure sensor output property Judge as NG when the barometric pressure sensor output is largely different from the intake manifold pressure at engine start Poor drive ability. **(Note: The barometric pressure sensor is built into the ECM).** **Possible Causes:** • Check any other DTC on display • Faulty ECM
DTC: P2228 **1T CCM, MIL: Yes** **Years:** 2006, 2007, 2008 **Engines:** 2.5L SOHC, DOHC VIN 6, 3.0L VIN 8, 3.6L VIN 9 **Models:** All **Transmissions:** All	**Barometric Pressure Circuit Low Input** Detect the open/short circuit of the barometric pressure sensor. Judge as NG if out of specification. Poor drive ability. **(Note: The barometric pressure sensor is built into the ECM).** **Possible Causes:** • Check any other DTC on display • Faulty ECM

DTC	Trouble Code Title, Conditions & Possible Causes
DTC: P2229 **1T CCM, MIL: Yes** **Years:** 2006, 2007, 2008 **Engines:** 2.5L SOHC, DOHC VIN 6, 3.0L VIN 8, 3.6L VIN 9 **Models:** All **Transmissions:** All	**Barometric Pressure Circuit High Input** Detect the open/short circuit of the barometric pressure sensor. Judge as NG if out of specification. Poor drive ability. **(Note: The barometric pressure sensor is built into the ECM).** **Possible Causes:** • Check any other DTC on display • Faulty ECM
DTC: P2431 **2T CCM, MIL: Yes** **2006** **Engines:** 2.5L DOHC **Models:** Impreza, WRX **Transmissions:** All	**Secondary Air Injection System Air Flow/Pressure Sensor Circuit Range/Performance** Poor Drive ability. **Possible Causes:** • Check any other DTC on display • Using a scan tool, check front oxygen sensor data
DTC: P2432 **1T CCM, MIL: Yes** **Years:** 2006, 2007, 2008 **Engines:** 2.5L SOHC, DOHC VIN 6, 3.0L VIN 8, 3.6L VIN 9 **Models:** All **Transmissions:** All	**Secondary Air Injection System Air Flow/Pressure Sensor Circuit Low** Poor Drive ability. **Possible Causes:** • Check harness between ECM and valve LH connector
DTC: P2433 **1T CCM, MIL: Yes** **Years:** 2006, 2007, 2008 **Engines:** 2.5L SOHC, DOHC VIN 6, 3.0L VIN 8, 3.6L VIN 9 **Models:** All **Transmissions:** All	**Secondary Air Injection System Air Flow/Pressure Sensor Circuit High** Poor Drive ability. **Possible Causes:** • Check harness between ECM and valve LH connector
DTC: P2440 **2T CCM, MIL: Yes** **Years:** 2006, 2007, 2008 **Engines:** 2.5L SOHC, DOHC VIN 6, 3.0L VIN 8, 3.6L VIN 9 **Models:** All **Transmissions:** All	**Secondary Air Injection System Switching Valve Stock Open (Bank 1)** Poor Drive ability. **Possible Causes:** • Check secondary air combination valve operation • Check duct between secondary air pump and secondary air combination valve • Check pipe between secondary air combination valve and cylinder head • Check power supply to valve • Check harness between secondary air combination valve relay and secondary air combination valve connector terminal • Check valve relay • Check harness between ECM and valve relay connector
DTC: P2441 **2T CCM, MIL: Yes** **Years:** 2006, 2007, 2008 **Engines:** 2.5L SOHC, DOHC VIN 6, 3.0L VIN 8, 3.6L VIN 9 **Models:** All **Transmissions:** All	**Secondary Air Injection System Switching Valve Stock Closed (Bank 1)** Poor Drive ability. **Possible Causes:** • Check secondary air combination valve operation • Check duct between secondary air pump and secondary air combination valve • Check pipe between secondary air combination valve and cylinder head • Check power supply to valve • Check harness between secondary air combination valve relay and secondary air combination valve connector terminal • Check valve relay • Check harness between ECM and valve relay connector
DTC: P2442 **2T CCM, MIL: Yes** **Years:** 2006, 2007, 2008 **Engines:** 2.5L SOHC, DOHC VIN 6, 3.0L VIN 8, 3.6L VIN 9 **Models:** All **Transmissions:** All	**Secondary Air Injection System Switching Valve Stock Open (Bank 2)** Poor Drive ability. **Possible Causes:** • Check secondary air combination valve operation • Check duct between secondary air pump and secondary air combination valve • Check pipe between secondary air combination valve and cylinder head • Check power supply to valve • Check harness between secondary air combination valve relay and secondary air combination valve connector terminal • Check valve relay • Check harness between ECM and valve relay connector

DTC	Trouble Code Title, Conditions & Possible Causes
DTC: P2443 **2T CCM, MIL: Yes** **Years:** 2006, 2007, 2008 **Engines:** 2.5L SOHC, DOHC VIN 6, 3.0L VIN 8, 3.6L VIN 9 **Models:** All **Transmissions:** All	**Secondary Air Injection System Switching Valve Stock Closed (Bank 2)** Poor Drive ability. **Possible Causes:** • Check secondary air combination valve operation • Check duct between secondary air pump and secondary air combination valve • Check pipe between secondary air combination valve and cylinder head • Check power supply to valve • Check harness between secondary air combination valve relay and secondary air combination valve connector terminal • Check valve relay • Check harness between ECM and valve relay connector
DTC: P2444 **1T CCM, MIL: Yes** **Years:** 2006, 2007, 2008 **Engines:** 2.5L SOHC, DOHC VIN 6, 3.0L VIN 8, 3.6L VIN 9 **Models:** All **Transmissions:** All	**Secondary Air Injection System Pump Stuck ON** Poor Drive ability. **Possible Causes:** • Check secondary air piping pressure • Check power supply to valve • Check valve relay
DTC: P2503 **1T CCM, MIL: No** **2005-06** **Engines:** 3.0L **Models:** Outback **Transmissions:** All	**Charging System Voltage Low** Vehicle performance issue. **Possible Causes:** • Check harness between alternator and ECM connector
DTC: P2504 **1T CCM, MIL: No** **Years:** 2006, 2007, 2008 **Engines:** 2.5L SOHC, DOHC VIN 6, 3.0L VIN 8, 3.6L VIN 9 **Models:** All **Transmissions:** All	**Charging System Voltage High** Vehicle performance issue. **Possible Causes:** • Check harness between alternator and ECM connector

OBD II Trouble Code List (U1XXX Codes)

DTC	Trouble Code Title, Conditions & Possible Causes
DTC: U1201 **2T CCM, MIL: Yes** **Years:** 2006, 2007, 2008 **Engines:** 2.5L SOHC, DOHC VIN 6, 3.0L VIN 8, 3.6L VIN 9 **Models:** All **Transmissions:** All	**CAN - HS Counter Abnormal** High speed CAN communication of body integrated unit which monitor the error data and non-received data are faulty. **Possible Causes:** • Connect the Subaru Select Monitor and read all DTCs • Diagnosis according to the DTC of all control modules • Poor contact of suspected module connectors
DTC: U1202 **2T CCM, MIL: Yes** **Years:** 2006, 2007, 2008 **Engines:** 2.5L SOHC, DOHC VIN 6, 3.0L VIN 8, 3.6L VIN 9 **Models:** All **Transmissions:** All	**CAN – HS Bus Off** Find the unit or CAN line in which trouble occurs, and repair and replace it. Not received data and error data may be detected at the same time. Er HC" is displayed in odo/trip meter. **Possible Causes:** • Check the DTC displayed in the body integrated unit • Faulty body integrated unit • Yaw rate sensor • Steering angle sensor • Open circuit in related line of body integrated unit • Open circuit in end resistance of VDCCM • Ground short circuit of the harness • Open circuit in end resistance of ECM

DTC	Trouble Code Title, Conditions & Possible Causes
DTC: U1211 **2T CCM, MIL: Yes** **Years:** 2006, 2007, 2008 **Engines:** 2.5L SOHC, DOHC VIN 6, 3.0L VIN 8, 3.6L VIN 9 **Models:** All **Transmissions:** All	**CAN – HS ECM Data Abnormal** Defective data from ECM. Er HC" or "Er EG" is displayed in odo/trip meter. **Possible Causes:** • Connect the Subaru Select Monitor and read all DTCs • Diagnosis according to the DTC of all control modules • Temporary poor contact • Faulty body integrated unit • Faulty ECM
DTC: U1212 **2T CCM, MIL: Yes** **Years:** 2006, 2007, 2008 **Engines:** 2.5L SOHC, DOHC VIN 6, 3.0L VIN 8, 3.6L VIN 9 **Models:** All **Transmissions:** All	**CAN – HS TCM Data Abnormal** TCM has error, harness between the main harness splice and TCM is open or shorted, connectors are not connected securely, or the terminal has poor crimping. SPORT indicator light blinks. "Er HC" or "Er tC" is displayed in odo/trip meter. **Possible Causes:** • Connect the Subaru Select Monitor and read all DTCs • Diagnosis according to the DTC of all control modules • Temporary poor contact • Faulty body integrated unit • Faulty TCM
DTC: U1213 **2T CCM, MIL: Yes** **Years:** 2006, 2007, 2008 **Engines:** 2.5L SOHC, DOHC VIN 6, 3.0L VIN 8, 3.6L VIN 9 **Models:** All **Transmissions:** All	**CAN – HS VDS/ABS Data Abnormal** VDCCM body has error, the main harness is open or shorted, the connector is not connected properly, or the terminal has poor crimping. ABS warning light and VDC warning light come on. "Er HC" or "Er Ab" is displayed in odo/trip meter. **Possible Causes:** • Connect the Subaru Select Monitor and read all DTCs • Diagnosis according to the DTC of all control modules • Temporary poor contact • Faulty body integrated unit • Faulty VDCCM
DTC: U1221 **2T CCM, MIL: Yes** **Years:** 2006, 2007, 2008 **Engines:** 2.5L SOHC, DOHC VIN 6, 3.0L VIN 8, 3.6L VIN 9 **Models:** All **Transmissions:** All	**CAN – HS ECM No - Receive Data** Defective ECM. (If error is in the main harness, DTC P0600 CAN communication link is input simultaneously.) Malfunction indicator light illuminates. "Er HC" is displayed in odo/trip meter. P1718 TCM and C0057 VDCCM are output. **(Note: When more than two DTC codes are recorded, referring to their combination will make it easy to identify the possible cause).** **Possible Causes:** • Connect the Subaru Select Monitor and read all DTCs • Diagnosis according to the DTC of all control modules • Temporary poor contact • Faulty body integrated unit • Faulty ECM
DTC: U1222 **2T CCM, MIL: Yes** **Years:** 2006, 2007, 2008 **Engines:** 2.5L SOHC, DOHC VIN 6, 3.0L VIN 8, 3.6L VIN 9 **Models:** All **Transmissions:** All	**CAN – HS TCM No - Receive Data** TCM has error, harness between the main harness splice and TCM is open or shorted, connectors are not connected securely, or the terminal has poor crimping. Malfunction indicator light illuminates. "Er HC" is displayed in odo/trip meter. P0600 ECM and C0057 VDCCM are output. **(Note: When more than two DTC codes are recorded, referring to their combination will make it easy to identify the possible cause).** **Possible Causes:** • Connect the Subaru Select Monitor and read all DTCs • Diagnosis according to the DTC of all control modules • Temporary poor contact • Faulty body integrated unit • Faulty TCM
DTC: U1223 **2T CCM, MIL: Yes** **Years:** 2006, 2007, 2008 **Engines:** 2.5L SOHC, DOHC VIN 6, 3.0L VIN 8, 3.6L VIN 9 **Models:** All **Transmissions:** All	**CAN – HS VDS/ABS No - Receive Data** Defective VDCCM. (If error is in the main harness, DTC P0600 High-speed CAN circuit is input at the same time.) ABS warning light and VDC warning light come on. "Er HC" is displayed in odo/trip meter. P0600 ECM and P1718 TCM are output. **(Note: When more than two DTC codes are recorded, referring to their combination will make it easy to identify the possible cause).** **Possible Causes:** • Connect the Subaru Select Monitor and read all DTCs • Diagnosis according to the DTC of all control modules • Temporary poor contact • Faulty body integrated unit • Faulty VDCCM

DTC	Trouble Code Title, Conditions & Possible Causes
DTC: U1300 **2T CCM, MIL: Yes** **Years:** 2006, 2007, 2008 **Engines:** 2.5L SOHC, DOHC VIN 6, 3.0L VIN 8, 3.6L VIN 9 **Models:** All **Transmissions:** All	**CAN – LS Malfunction** Either end of the low speed CAN communication line is open or shorted, the connector is not connected properly, or the terminal has poor crimping. "Er LC" is displayed in odo/trip meter, but communicating function is OK. **Possible Causes:** • Connect the Subaru Select Monitor and read all DTCs • Diagnosis according to the DTC of all control modules • Temporary poor contact • Faulty body integrated unit • Perform the auto A/C self-diagnosis • Faulty MFD • Faulty auto A/C control module • Faulty combination meter
DTC: U1301 **2T CCM, MIL: Yes** **Years:** 2006, 2007, 2008 **Engines:** 2.5L SOHC, DOHC VIN 6, 3.0L VIN 8, 3.6L VIN 9 **Models:** All **Transmissions:** All	**CAN – LS Counter Abnormal** Find the unit in which trouble occurs and open or short CAN line, and repair and replace them. (Free running counter error may be detected at the same time from the unit in which the malfunction occurs.) "Er LC" is displayed in odo/trip meter. **Possible Causes:** • Connect the Subaru Select Monitor and read all DTCs • Diagnosis according to the DTC of all control modules • Temporary poor contact • Faulty body integrated unit • Perform the auto A/C self-diagnosis • Faulty MFD • Faulty auto A/C control module • Faulty combination meter
DTC: U1302 **2T CCM, MIL: Yes** **Years:** 2006, 2007, 2008 **Engines:** 2.5L SOHC, DOHC VIN 6, 3.0L VIN 8, 3.6L VIN 9 **Models:** All **Transmissions:** All	**CAN – LS Bus Off** Because of a lot of error data occurred, some units have been disconnected not to affect other units. Communication failure from the unit in which error is occurred is input at the same time. "Er LC" is displayed in odo/trip meter. **Possible Causes:** • Connect the Subaru Select Monitor and read all DTCs • Diagnosis according to the DTC of all control modules • Temporary poor contact • Faulty body integrated unit • Perform the auto A/C self-diagnosis • Faulty MFD • Faulty auto A/C control module • Faulty combination meter
DTC: U1311 **2T CCM, MIL: Yes** **Years:** 2006, 2007, 2008 **Engines:** 2.5L SOHC, DOHC VIN 6, 3.0L VIN 8, 3.6L VIN 9 **Models:** All **Transmissions:** All	**CAN – LS Meter Unit Data Abnormal** Combination meter has error, the harness between main harness splice and combination meter is open or shorted, the connector is not connected properly, or the terminal has poor crimping, "Er LC" is displayed in odo/trip meter. **Possible Causes:** • Connect the Subaru Select Monitor and read all DTCs • Diagnosis according to the DTC of all control modules • Temporary poor contact • Faulty body integrated unit • Faulty combination meter
DTC: U1313 **2T CCM, MIL: Yes** **Years:** 2006, 2007, 2008 **Engines:** 3.0L VIN 8, 3.6L VIN 9 **Models:** B9 Tribeca, Tribeca **Transmissions:** All	**CAN – LS Monitor Data Abnormal** Center display unit error, or harness between the main harness splice and center display unit is open or shorted, the connector is not connected securely and the terminal has poor crimping. "Er LC" is displayed in odo/trip meter. **Possible Causes:** • Connect the Subaru Select Monitor and read all DTCs • Diagnosis according to the DTC of all control modules • Refer to MFD or navigation display • Refer to navigation body

DTC	Trouble Code Title, Conditions & Possible Causes
DTC: U1321 **2T CCM, MIL: Yes** **Years:** 2006, 2007, 2008 **Engines:** 2.5L SOHC, DOHC VIN 6, 3.0L VIN 8, 3.6L VIN 9 **Models:** All **Transmissions:** All	**CAN – LS Meter No - Receive Data** Combination meter unit error, or harness between the main harness splice and combination meter unit is open or shorted, the connector is not connected properly and the terminal has poor crimping. Fail mode occurs because the data is not received from combination meter unit. **Possible Causes:** • Check communication Line • Check harness • Temporary poor contact • Short circuit of harness • Check combination meter • Faulty combination meter

SUZUKI

Aerio

17

SPECIFICATIONS AND MAINTENANCE CHARTS

VEHICLE AND ENGINE IDENTIFICATION

Engine							Model Year	
Code	Liters (cc)	Cu. in.	Cyl.	Fuel Sys.	Engine Type	Eng. Mfg.	Code	Year
6	2.3 (2290)	140	4	MFI	DOHC	Suzuki	7	2007

DOHC: Dual Overhead Camshaft

MFI: Multiport Fuel Injection

22140_AERI_C0001

GENERAL ENGINE SPECIFICATIONS

Year	Engine ID/VIN	Engine Displacement Liters (cc)	Fuel System Type	Net Horsepower @ rpm	Net Torque @ rpm (ft. lbs.)	Bore x Stroke (in.)	Com-pression Ratio	Oil Pressure @ rpm
2007	6	2.3 (2290)	MFI	155@5400	152@3000	3.54x3.54	9.3:1	55.5@4000

MFI: Multiport Fuel Injection

22140_AERI_C0002

GASOLINE ENGINE TUNE-UP SPECIFICATIONS

Year	Engine Displacement Liters	Engine ID/VIN	Spark Plugs Gap (in.)	Ignition Timing (deg.)		Fuel Pump (psi)	Idle Speed (rpm)		Valve Clearance	
				MT	AT		MT	AT	In.	Ex.
2007	2.3	6	0.040-0.043	5B	5B	31.3-36.9 ①	700-800	700-800	HYD	HYD

Note: The Vehicle Emission Control Information label often reflects specification changes made during production.

The label figures must be used if they differ from those in this chart.

HYD: Hydraulic

B: Before top dead center

① At idle

22140_AERI_C0003

CAPACITIES

Year	Model	Engine Displacement Liters	Engine ID/VIN	Engine Oil with Filter (qts.)	Transmission (pts.) 5-Spd	Transmission (pts.) Auto.	Fuel Tank (gal.)	Cooling System (qts.)
2007	Aerio	2.3	6	5.0	4	15.6 ①	13.2	7.4

Note: All capacities are approximate. Add fluid gradualy and check to be sure a proper fluid level is obtained.

① Specification for automatic transaxle is after complete overhaul. Drain and fill will be less

22140_AERI_C0004

FLUID SPECIFICATIONS

Year	Engine Displacement Liters	Engine ID/VIN	Engine Oil	Auto. Trans.	Manual Trans	Rear Differential (AWD)	Power Steering Fluid	Brake Master Cylinder	Engine Coolant
2007	2.3	6	5W-30	Dexron III	GL-4 75W-90	GL-5 SAE 80W-90	①	DOT 3	Ethylene

DOT: Department Of Transpotation

① An equivalent of DEXRON®-III, DEXRON®-IIE or DEXRON®-II

22140_AERI_C0013

VALVE SPECIFICATIONS

Year	Engine ID/VIN	Engine Displacement Liters	Seat Angle (deg.)	Face Angle (deg.)	Spring Test Pressure (lbs. @ in.)	Spring Installed Height (in.)	Stem-to-Guide Clearance (in.) Intake	Stem-to-Guide Clearance (in.) Exhaust	Stem Diameter (in.) Intake	Stem Diameter (in.) Exhaust
2007	6	2.3	45	45	49-56@1.33	1.744	0.0008-0.0018	0.0018-0.0028	0.2348-0.2354	0.2339-0.2344

22140_AERI_C0005

CAMSHAFT AND BEARING SPECIFICATIONS

All measurements are given in inches.

Year	Engine Displacement Liters	Engine VIN	Journal Diameter	Brg. Oil Clearance	Shaft End-play	Runout	Journal Bore	Lobe Lift Intake	Lobe Lift Exhaust
2007	2.3	6	1.0221 1.0228	0.0008- 0.0029	NA	0.0012	1.0236- 1.0249	1.591- 1.5972	1.5716- 1.5778

NA: Information not available

22140_AERI_C0014

CRANKSHAFT AND CONNECTING ROD SPECIFICATIONS

All measurements are given in inches.

			Crankshaft				Connecting Rod		
Year	Engine ID/VIN	Engine Displacement Liters (cc)	Main Brg. Journal Dia.	Main Brg. Oil Clearance	Shaft End-play	Thrust on No.	Journal Diameter	Oil Clearance	Side Clearance
2007	6	2.3 (2290)	①	0.0013- 0.0019	0.0039- 0.0138	3	②	0.0018- 0.0025	0.0099- 0.0150

① No. 1: 2.2835-2.2837
No. 2: 2.2832-2.2834
No. 3: 2.2830-2.2832

② No. A: 1.9683-1.9685
No. B: 1.9681-1.9682
No. C: 1.9677-1..9680

22140_AERI_C0006

PISTON AND RING SPECIFICATIONS

All measurements are given in inches.

				Ring Gap			Ring Groove Clearance		
Year	Engine ID/VIN	Engine Displacement Liters	Piston Clearance	Top Compression	Bottom Compression	Oil Control	Top Compression	Bottom Compression	Oil Control
2007	6	2.3	0.0008- 0.0015	0.0079- 0.0125	0.0126- 0.0185	0.0079- 0.0196	0.0012- 0.0027	0.0008- 0.0023	NA

NA: Not available

22140_AERI_C0007

TORQUE SPECIFICATIONS

All readings in ft. lbs.

Year	Engine ID/VIN	Engine Displacement Liters	Cylinder Head Bolts	Main Bearing Bolts	Rod Bearing Bolts	Crankshaft Damper Bolts	Flywheel Bolts	Manifold		Spark Plugs	Lug Nut
								Intake	Exhaust		
2007	6	2.3	①	③	④	108.5	51	18	40	18	61.5

① Step 1: 38.5 ft. lbs. (53 Nm)
Step 2: 61 ft. lbs. (84 Nm)
Step 3: loosen bolts 1-10
Step 4: 38.5 ft. lbs. (53 Nm)
Step 5: 76 ft. lbs. (105 Nm)
Step 6 - (M6 bolt) 8 ft. lbs. (11 Nm)

② Step 1 - bolts 1-10: 29 ft. lbs. (40 Nm)
Step 2 - loosen bolts 1-10
Step 3 - bolts 1-10: 29 ft. lbs. (40 Nm)
Step 4 - bolts 1-10: (10mm) 42 ft. lbs. (58 Nm);
bolts 11-22: (8mm) 18 ft. lbs. (25 Nm)

③ Step 1: 11 ft. lbs. (15 Nm)
Step 2: turn head bolts 45 degrees
Step 3: turn head bolts an additional 45 degrees

22140_AERI_C0008

WHEEL ALIGNMENT

Year	Model		Caster		Camber		Toe-in (in.)	Steering Axis Inclination (Deg.)
			Range (+/-Deg.)	Preferred Setting (Deg.)	Range (+/-Deg.)	Preferred Setting (Deg.)		
2007	Aerio	F	+2.00	+2.00	1.00	+0.00	0.039+/-0.079	—
		R	—	—	—	—	—	—

22140_AERI_C0009

TIRE, WHEEL AND BALL JOINT SPECIFICATIONS

Year	Model	OEM Tires		Tire Pressures (psi)		Wheel Size	Ball Joint Inspection
		Standard	Optional	Front	Rear		
2007	Aerio	P185/65R14	None	30	30	5.5JJ	①
	Aerio LX	P195/55R15	None	30	30	5.5JJ	①

OEM: Original Equipment Manufacturer
PSI: Pounds Per Square Inch
STD: Standard
OPT: Optional

22140_AERI_C0010

BRAKE SPECIFICATIONS

All measurements in inches unless noted

		Brake Disc			Brake Drum Diameter			Minimum Lining Thickness		Brake Caliper	
Year	Model	Original Thickness	Minimum Thickness	Maximum Runout	Original Inside Diameter	Max. Wear Limit	Maximum Machine Diameter	Front	Rear	Bracket bolts (ft. lbs.)	Mounting bolts (ft. lbs.)
2007	Aerio	0.790	0.710	0.004	7.87	7.95	7.95	0.080 ①	0.100 ①	62	16

① Measurement is for lining and backing together.

22140_AERI_C0011

SCHEDULED MAINTENANCE INTERVALS

2007 SUZUKI—AERIO

TO BE SERVICED	TYPE OF SERVICE	VEHICLE MILEAGE INTERVAL (x1000)												
		7.5	15	22.5	30	37.5	45	52.5	60	67.5	75	82.5	90	97.5
Engine oil & filter	R	✓	✓	✓	✓	✓	✓	✓	✓	✓	✓	✓	✓	✓
Automatic transmission fluid & filter ①	S/I	✓	✓	✓	✓	✓	✓	✓	✓	✓	✓	✓	✓	✓
Drive axle boots	S/I	✓	✓	✓	✓	✓	✓	✓	✓	✓	✓	✓	✓	✓
Inspect & rotate tires	S/I	✓	✓	✓	✓	✓	✓	✓	✓	✓	✓	✓	✓	✓
Manual transmission oil ②	S/I	✓	✓	✓	✓	✓	✓	✓	✓	✓	✓	✓	✓	✓
Power steering system	S/I	✓	✓	✓	✓	✓	✓	✓	✓	✓	✓	✓	✓	✓
Suspension system	S/I	✓	✓	✓	✓	✓	✓	✓	✓	✓	✓	✓	✓	✓
Brake discs, pads, drums & shoes	S/I	✓		✓		✓		✓		✓		✓		✓
Brake hoses, pipes, brake lever & cable	S/I	✓		✓		✓		✓		✓		✓		✓
Brake fluid ③	S/I		✓		✓		✓		✓		✓		✓	
Brake pedal	S/I		✓		✓		✓		✓		✓		✓	
Cooling system, hoses & connections	S/I		✓		✓		✓		✓		✓		✓	
Fuel tank, cap & lines	S/I		✓		✓		✓		✓		✓		✓	
Transfer case oil	S/I	✓			✓		✓		✓		✓		✓	
Air cleaner filter element	R				✓				✓				✓	
Engine coolant	R				✓				✓				✓	
Spark plugs	R								✓					
Drive belts	S/I				✓				✓				✓	
Exhaust system	S/I				✓				✓				✓	
Automatic transmission fluid hose	R				✓				✓				✓	
Ignition coils	S/I				✓				✓				✓	

R: Replace S/I: Service or Inspect

① Replace every 100,000 miles.

② Replace every 30,000 miles.

③ Replace every 60,000 miles.

FREQUENT OPERATION MAINTENANCE (SEVERE SERVICE)

If a vehicle is operated under any of the following conditions it is considered severe service:

- Extremely dusty areas.

- 50% or more of the vehicle operation is below 32°C (90°F) or higher temperatures, or constant operation in temperatures below 0°C (32°F).

- Prolonged idling (vehicle operation in stop and go traffic).

- Frequent short running periods (engine does not warm to normal operating temperatures).

- Police, taxi, delivery usage or trailer towing usage or other commercial applications

Oil & oil filter: change every 3000 miles.

Brake discs, pads, drums & shoes: service or inspect initially at 3000 miles, 6000 miles, & every 12,000 miles thereafter.

Brake hoses & pipes: service or inspect initially at 3000 miles, 6000 miles & every 12,000 miles thereafter.

Air cleaner filter element: service or inspect ever 3000 miles & replace every 30,000 miles (if not replaced previously).

Automatic transmission fluid & filter: service or inspect every 6000 miles & replace every 15,000 miles (if not replaced previously).

Inspect & rotate tires: service or inspect every 6000 miles.

Manual transmission oil: service or inspect every 6000 miles & replace every 12,000 miles (if not replaced previously).

Power steering system: service or inspect every 6000 miles.

Steering system: service or inspect every 6000 miles.

Suspension system: service or inspect every 6000 miles.

Drive belts: service or inspect every 15,000 miles.

Exhaust system: service or inspect every 15,000 miles.

22140_AERI_C0012

PRECAUTIONS

Before servicing any vehicle, please be sure to read all of the following precautions, which deal with personal safety, prevention of component damage, and important points to take into consideration when servicing a motor vehicle:

- Never open, service or drain the radiator or cooling system when the engine is hot; serious burns can occur from the steam and hot coolant.
- Observe all applicable safety precautions when working around fuel. Whenever servicing the fuel system, always work in a well-ventilated area. Do not allow fuel spray or vapors to come in contact with a spark, open flame, or excessive heat (a hot drop light, for example). Keep a dry chemical fire extinguisher near the work area. Always keep fuel in a container specifically designed for fuel storage; also, always properly seal fuel containers to avoid the possibility of fire or explosion. Refer to the additional fuel system precautions later in this section.
- Fuel injection systems often remain pressurized, even after the engine has been turned **OFF**. The fuel system pressure must be relieved before disconnecting any fuel lines. Failure to do so may result in fire and/or personal injury.
- Brake fluid often contains polyglycol ethers and polyglycols. Avoid contact with the eyes and wash your hands thoroughly after handling brake fluid. If you do get brake fluid in your eyes, flush your eyes with clean, running water for 15 minutes. If eye irritation persists, or if you have taken brake fluid internally, IMMEDIATELY seek medical assistance.
- The EPA warns that prolonged contact with used engine oil may cause a number of skin disorders, including cancer. You should make every effort to minimize your exposure to used engine oil. Protective gloves should be worn when changing oil. Wash your hands and any other exposed skin areas as soon as possible after exposure to used engine oil. Soap and water, or waterless hand cleaner should be used.
- All new vehicles are now equipped with an air bag system, often referred to as a Supplemental Restraint System (SRS) or Supplemental Inflatable Restraint (SIR) system. The system must be disabled before performing service on or around system components, steering column, instrument panel components, wiring and sensors. Failure to follow safety and disabling procedures could result in accidental air bag deployment, possible personal injury and unnecessary system repairs.
- Always wear safety goggles when working with, or around, the air bag system. When carrying a non-deployed air bag, be sure the bag and trim cover are pointed away from your body. When placing a non-deployed air bag on a work surface, always face the bag and trim cover upward, away from the surface. This will reduce the motion of the module if it is accidentally deployed. Refer to the additional air bag system precautions later in this section.
- Clean, high quality brake fluid from a sealed container is essential to the safe and proper operation of the brake system. You should always buy the correct type of brake fluid for your vehicle. If the brake fluid becomes contaminated, completely flush the system with new fluid. Never reuse any brake fluid. Any brake fluid that is removed from the system should be discarded. Also, do not allow any brake fluid to come in contact with a painted surface; it will damage the paint.
- Never operate the engine without the proper amount and type of engine oil; doing so WILL result in severe engine damage.
- Timing belt maintenance is extremely important. Many models utilize an interference-type, non-freewheeling engine. If the timing belt breaks, the valves in the cylinder head may strike the pistons, causing potentially serious (also time-consuming and expensive) engine damage. Refer to the maintenance interval charts for the recommended replacement interval for the timing belt, and to the timing belt section for belt replacement and inspection.
- Disconnecting the negative battery cable on some vehicles may interfere with the functions of the on-board computer system(s) and may require the computer to undergo a relearning process once the negative battery cable is reconnected.
- When servicing drum brakes, only disassemble and assemble one side at a time, leaving the remaining side intact for reference.
- Only an MVAC-trained, EPA-certified automotive technician should service the air conditioning system or its components.

BRAKES — ANTI-LOCK BRAKE SYSTEM (ABS)

GENERAL INFORMATION

The ABS (Antilock Brake System) controls the fluid pressure applied to the wheel cylinder of each brake from the master cylinder so that each wheel is not locked even when hard braking is applied. This ABS has also the following function.

While braking is applied, but before ABS control becomes effective, braking force is distributed between the front and rear so as to prevent the rear wheels from being locked too early for better stability of the vehicle.

The main component parts of this ABS include the following parts in addition to those of the conventional brake system.

- Wheel speed sensor which senses revolution speed of each wheel and outputs its signal.
- ABS warning lamp which lights to inform abnormality when system fails to operate properly.
- ABS hydraulic unit / control module assembly is incorporated ABS control module, ABS hydraulic unit (actuator assembly), fail-safe relay (transistor), pump motor relay (transistor) and G sensor.
- ABS control module which sends operation signal to ABS hydraulic unit to control fluid pressure applied to each wheel cylinder based on signal from each wheel speed sensor so as to prevent wheel from locking.
- ABS hydraulic unit which operates according to signal from ABS control module to control fluid pressure applied to wheel cylinder of each 4 wheels.
- Fail-safe relay (transistor) which supplies power to solenoid valve in ABS hydraulic unit.
- Pump motor relay (transistor) which supplies power to pump motor in ABS hydraulic unit.
- G sensor which detects vehicle deceleration speed in ABS hydraulic unit. (For 4WD model only)

This ABS is equipped with Electronic Brake force Distribution (EBD) system that controls a fluid pressure of rear wheels to best condition, which is the same function as that of proportioning valve, by the signal from wheel sensor independently of change of load due to load capacity and so on. And if the EBD system fails to operate properly, the brake warning lamp lights to inform abnormality.

WHEEL SPEED SENSORS

REMOVAL & INSTALLATION

Front

See Figures 1 and 2.

1. Disconnect negative cable at battery.
2. Disconnect front wheel speed sensor coupler (1).
3. Hoist vehicle and remove wheel.
4. Remove harness clamp bolts (2) and grommet (3).
5. Remove front wheel speed sensor (4) from knuckle.

✲✲ CAUTION

Do not pull wire harness when removing front wheel speed sensor.

✲✲ CAUTION

Do not cause damage to surface of front wheel speed sensor and do not allow dust, etc. to enter its installation hole.

To install:

6. Check that no foreign material is attached to sensor (1) and sensor ring (2).
7. Install it by reversing removal procedure. Tighten the front wheel speed sensor bolt to 17 ft. lbs. (23 Nm).

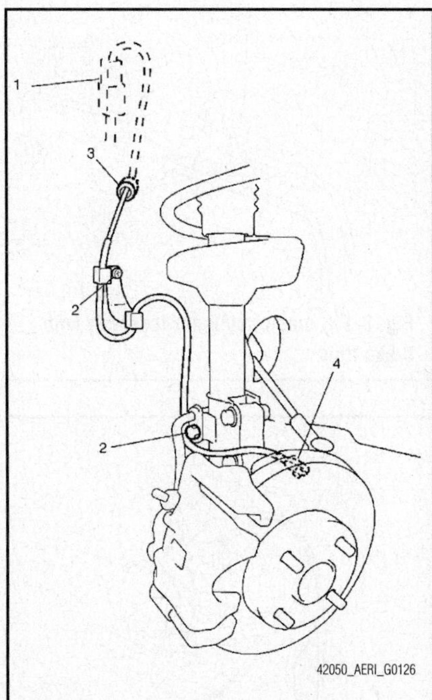

Fig. 1 Disconnect the front wheel speed sensor coupler and remove the harness clamp bolts, grommet and wheel speed sensor

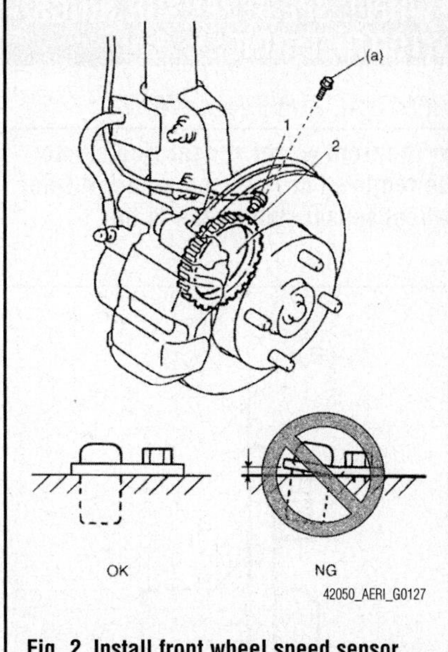

Fig. 2 Install front wheel speed sensor

✲✲ CAUTION

Do not pull or twist wire harness more than necessary when installing front wheel speed sensor.

8. Check that there is no clearance between sensor and knuckle.

Rear

See Figures 3 through 6.

1. Disconnect the negative battery cable.
2. Remove quarter inner trim (1). (A) for wagon model and (B) for sedan model.

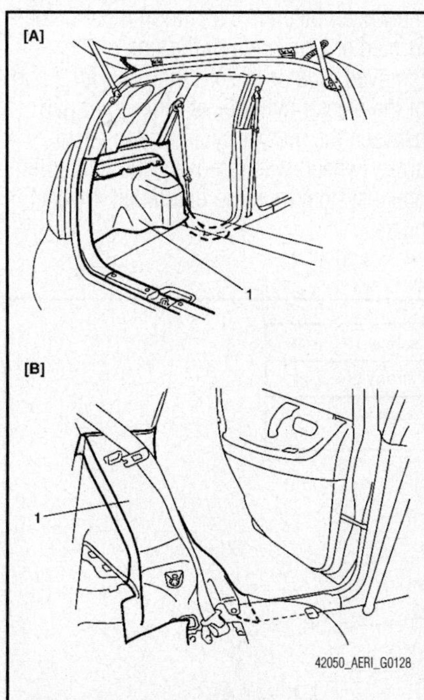

Fig. 3 Remove quarter inner trim

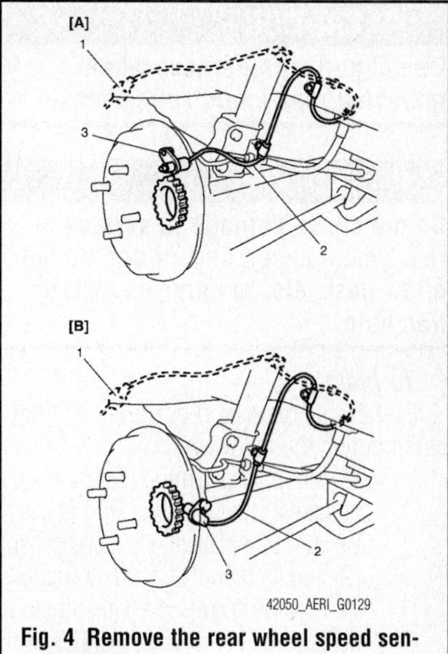

Fig. 4 Remove the rear wheel speed sensor from knuckle

3. Turn over floor carpet.
4. Hoist vehicle.
5. Disconnect rear wheel speed sensor coupler (1).
6. Detach ABS wheel sensor wire harness (2).

➡Do not detach clip of rear wheel speed sensor connector from vehicle body unless replacement is necessary.

7. Remove rear wheel speed sensor (3) from knuckle. (A) for 2WD and (B) for 4WD.

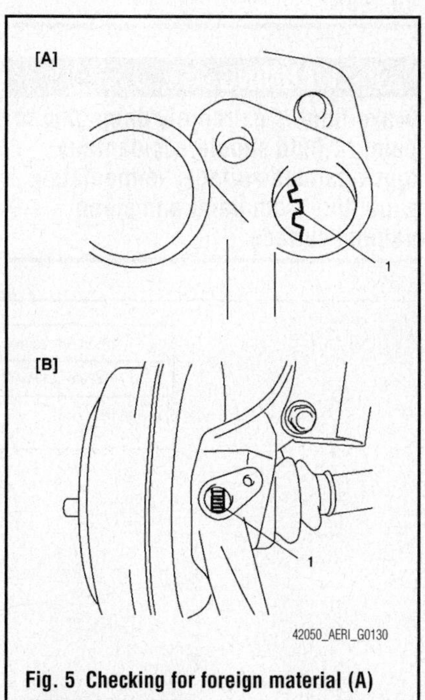

Fig. 5 Checking for foreign material (A) 2WD and (B) 4WD

⁂ CAUTION

Do not pull wire harness when removing rear wheel speed sensor.

⁂ CAUTION

Do not cause damage to surface of rear wheel speed sensor and do not allow dust, etc. to enter its installation hole.

To install:

8. Reverse removal procedure for installation noting the following:
    - Check that no foreign material is attached to sensor and ring (1).
    - Be sure to install wheel speed sensor (2) and its bolt at the correct (upper) position as shown in figure. Tighten sensor bolt (1) to 17 ft. lbs. (23 Nm).

⁂ CAUTION

Do not pull or twist wire harness more than necessary when installing rear wheel speed sensor.

- Check that there is no clearance between sensor and knuckle.

WHEEL SPEED SENSOR RINGS (TOOTHED RINGS)

REMOVAL & INSTALLATION

➡The front wheel sensor ring cannot be removed or replaced alone. If front wheel sensor ring needs to be replaced, replace it as a wheel side joint assembly of driveshaft.

➡The rear wheel sensor ring cannot be removed or replaced alone. If rear wheel sensor ring needs to be replaced, replace it as a wheel hub assembly for 2WD vehicle or wheel side joint assembly of driveshaft for 4WD vehicle

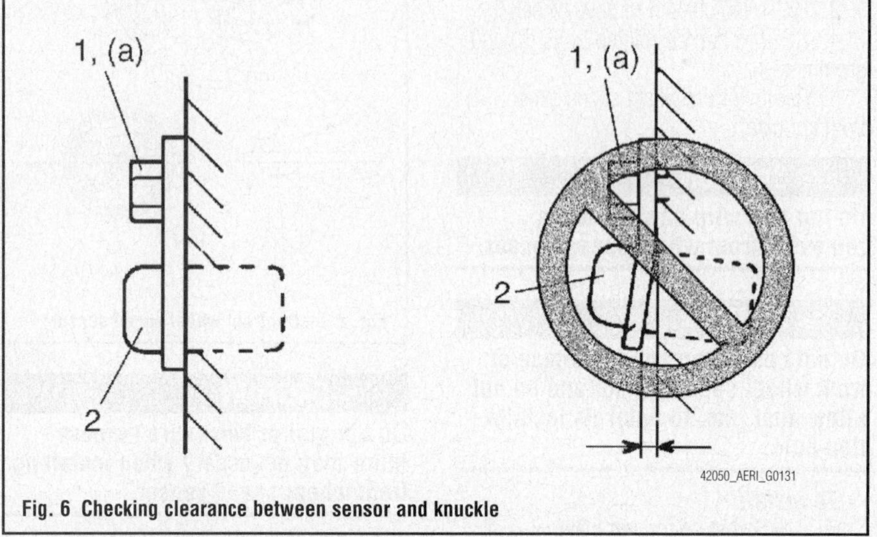

Fig. 6 Checking clearance between sensor and knuckle

BLEEDING THE BRAKE SYSTEM

BLEEDING PROCEDURE

See Figures 7 through 12.

⁂ CAUTION

DOT 3 or SAE J1703 brake fluid must be used.

⁂ WARNING

Brake fluid is extremely damaging to paint. If fluid should accidentally touch painted surface, immediately wipe fluid from paint and clean painted surface.

A bleeding operation is necessary to remove air whenever it entered hydraulic brake system.

Hydraulic lines of brake system are based on the diagonal split system. When a brake pipe or hose was disconnected at the wheel, bleeding operation must be performed at both ends of the line of the removed pipe or hose. When any joint part of the master cylinder of other joint part between the master cylinder and each brake (wheel) was removed, the hydraulic brake system must be bled at all 4 wheel brakes.

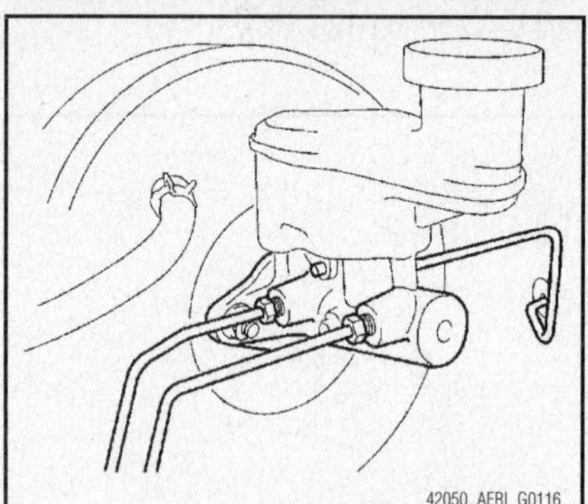

Fig. 8 Fill master cylinder reservoir with brake fluid

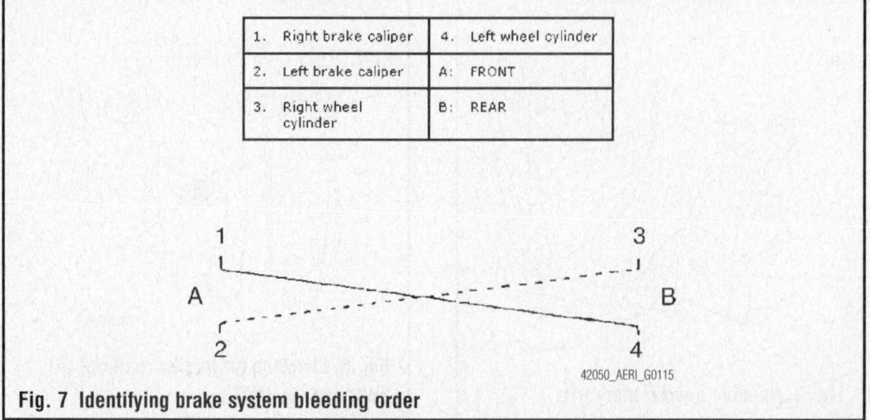

1. Right brake caliper	4. Left wheel cylinder
2. Left brake caliper	A: FRONT
3. Right wheel cylinder	B: REAR

Fig. 7 Identifying brake system bleeding order

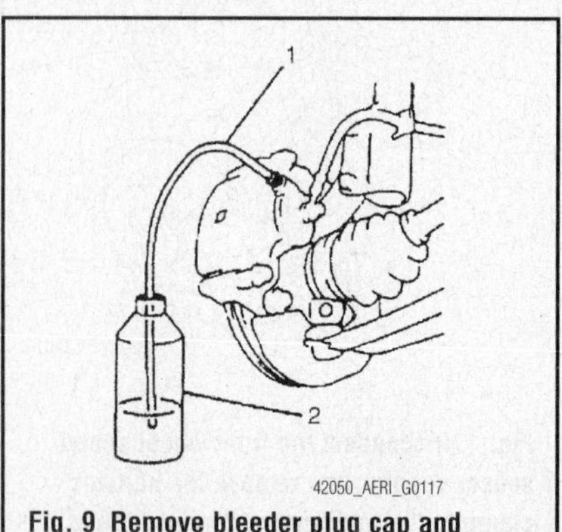

Fig. 9 Remove bleeder plug cap and attach a vinyl tube to bleeder plug

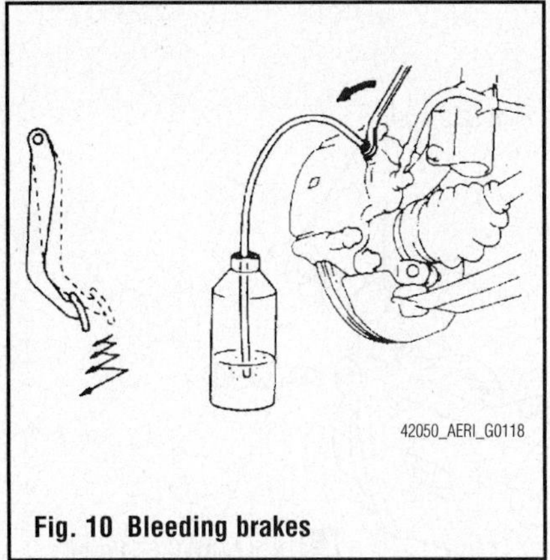

Fig. 10 Bleeding brakes

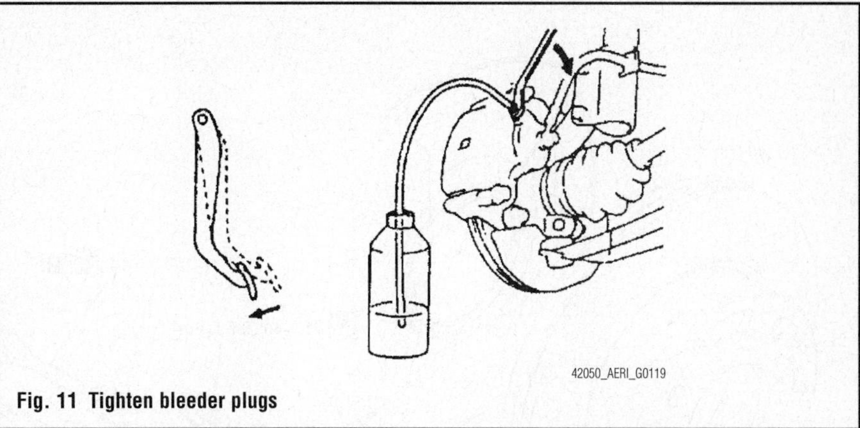

Fig. 11 Tighten bleeder plugs

➡Perform bleeding operation starting with wheel cylinder farthest from master cylinder and then at front caliper of the same brake line. Do the same on the other brake line.

1. Fill master cylinder reservoir with brake fluid (DOT 3) and keep at least one-half full of fluid during bleeding operation.
2. Remove bleeder plug cap. Attach a vinyl tube (1) to bleeder plug, and insert the other end into container (2).
3. Depress brake pedal several times, and then while holding it depressed, loosen bleeder plug about one-third to one-half turn.
4. When fluid pressure in cylinder is almost depleted, retighten bleeder plug.

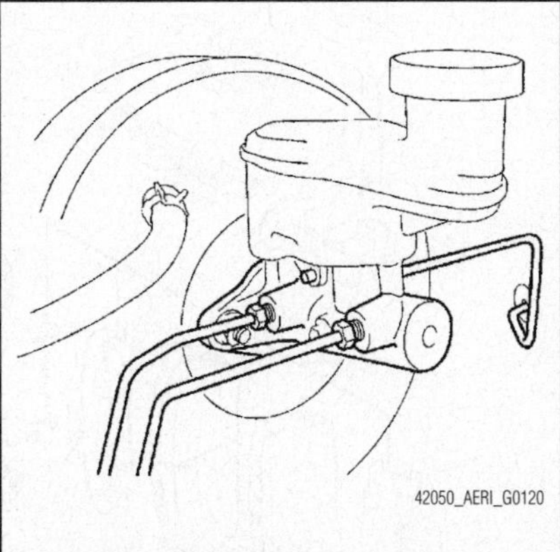

Fig. 12 Apply fluid pressure to pipe line and check for leakage

5. Repeat this operation until there are no more air bubbles in hydraulic line.
6. When bubbles stop, depress and hold brake pedal and tighten the front brake caliper bleeder plug to 8 ft. lbs. (11 Nm) and the rear wheel cylinder bleeder plug to 6 ft. lbs. (8 Nm).
7. Then attach bleeder plug cap.
8. After completing bleeding operation, apply fluid pressure to pipe line and check for leakage.
9. Replenish fluid into reservoir up to specified level.
10. Check brake pedal for sponginess. If found spongy, repeat entire procedure of bleeding.

BRAKES

FRONT DISC BRAKES

⁂ WARNING

All brake fasteners are important attaching parts in that they could affect the performance of vital parts and systems, and/or could result in a major repair expense. Do not use a replacement parts of a lesser quality or substitute design. Torque values must be used as specified during reassembly to assure proper retention of all parts. There is to be no welding as it may result in extensive damage and weakening of the metal.

⁂ CAUTION

Dust and dirt accumulating on brake parts during normal use may contain asbestos fibers from production or aftermarket brake linings. Breathing excessive concentrations of asbestos fibers can cause serious bodily harm. Exercise care when servicing brake parts. Do not sand or grind brake lining unless equipment used is designed to contain the dust residue. Do not clean brake parts with compressed air or by dry brushing. Cleaning should be done by dampening the brake components with a fine mist of water, then wiping the brake components clean with a dampened cloth. Dispose of cloth and all residue containing asbestos fibers in an impermeable container with the appropriate label. Follow practices prescribed by the Occupational Safety and Health Administration (OSHA) and the Environmental Protection Agency (EPA) for the handling, processing, and disposing of dust or debris that may contain asbestos fibers.

BRAKE CALIPER

REMOVAL & INSTALLATION

See Figure 13.

1. Before servicing the vehicle, refer to the Precautions Section.
2. Remove the wheels.
3. Loosen flexible hose joint bolt a little at caliper.

➡Be careful not to twist flexible hose while loosening the bolt.

4. Remove caliper pin bolts.
5. Remove caliper from caliper carrier.
6. Disconnect flexible hose from caliper using care not to twist it. As this will allow brake fluid to flow out of flexible hose, have a container ready beforehand.

To install:

7. Apply grease to slide pin, then install caliper to caliper carrier.
8. Torque caliper pin bolts to 16 ft. lbs. (22 Nm).
9. Connect caliper to flexible hose.
10. Torque flexible hose joint bolt to 17 ft. lbs. (23 Nm).

⁂ WARNING

Make sure that flexible hose is not twisted when tightening joint bolt. If it is twisted, reconnect it using care not to twist it.

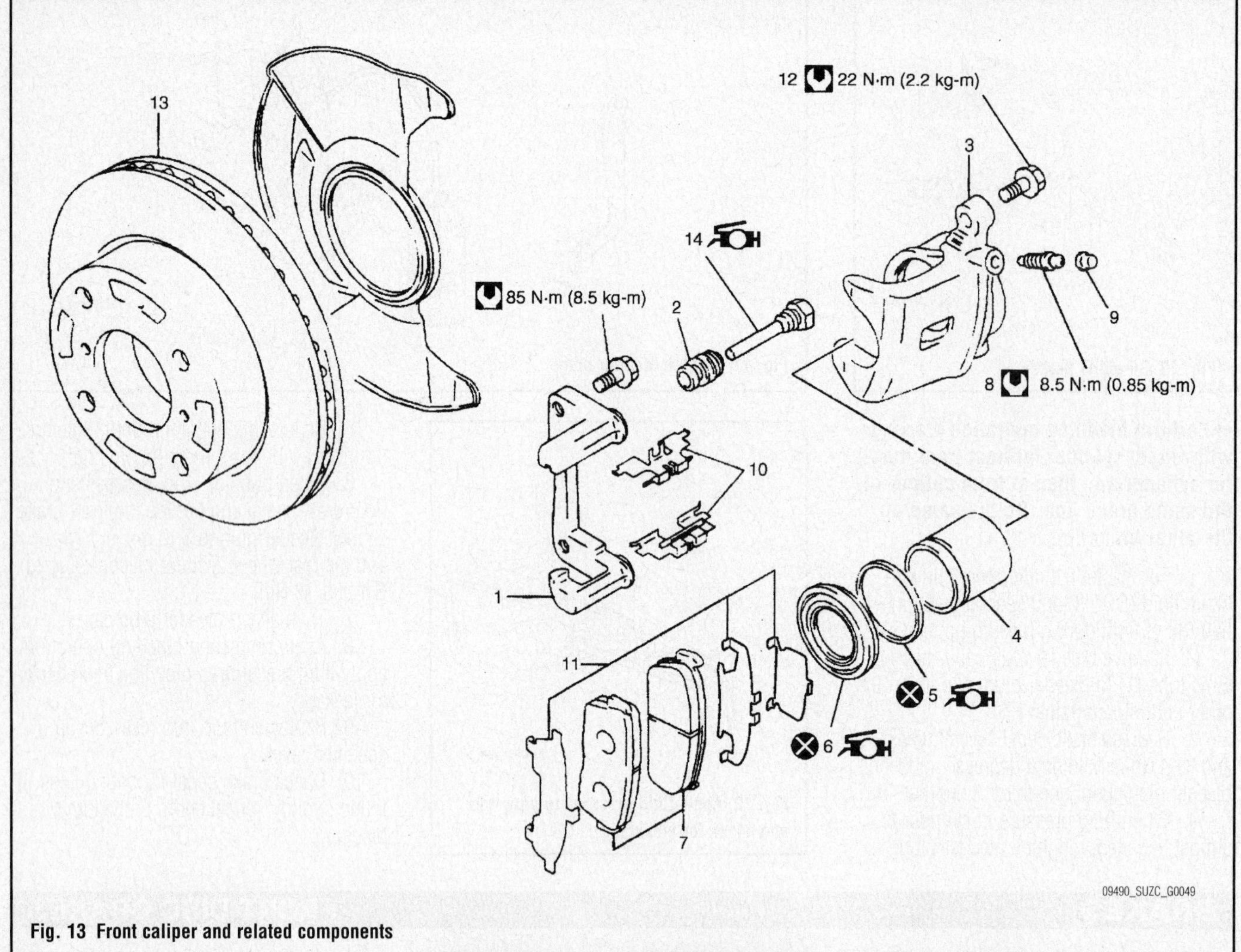

Fig. 13 Front caliper and related components

11. Tighten bleeder plug.

12. Tighten wheel temporarily and lower lift.

13. Tighten wheel nuts to 61.5 ft. lbs. (85 Nm).

14. After completing installation, fill reservoir with brake fluid and bleed air from brake system. Perform brake test and check each installed part for oil leakage.

DISC BRAKE PADS

REMOVAL & INSTALLATION

1. Before servicing the vehicle, refer to the Precautions Section.
2. Remove the wheels.
3. Remove the brake caliper.
4. Remove the brake pads.

To install:

5. Before installing brake pad shim, apply small amount of grease (included in spare parts) to mating surfaces of brake pad and pad shim.
6. Set brake pad springs and shim and install brake pads.
7. Install brake caliper.
8. Tighten wheel temporarily and lower lift.
9. Tighten wheel nuts to 61.5 ft. lbs. (85 Nm).
10. After completion of installation, check for brake effectiveness.

WARNING

All brake fasteners are important attaching parts in that they could affect the performance of vital parts and systems, and/or could result in a major repair expense. Do not use a replacement parts of a lesser quality or substitute design. Torque values must be used as specified during reassembly to assure proper retention of all parts. There is to be no welding as it may result in extensive damage and weakening of the metal.

CAUTION

Dust and dirt accumulating on brake parts during normal use may contain asbestos fibers from production or aftermarket brake linings. Breathing excessive concentrations of asbestos fibers can cause serious bodily harm. Exercise care when servicing brake parts. Do not sand or grind brake lining unless equipment used is designed to contain the dust residue. Do not clean brake parts with compressed air or by dry brushing. Cleaning should be done by dampening the brake components with a fine mist of water, then wiping the brake components clean with a dampened cloth. Dispose of cloth and all residue containing asbestos fibers in an impermeable container with the appropriate label. Follow practices prescribed by the Occupational Safety and Health Administration (OSHA) and the Environmental Protection Agency (EPA) for the handling, processing, and disposing of dust or debris that may contain asbestos fibers.

BRAKE DRUM

REMOVAL & INSTALLATION

See Figures 14 through 18.

1. Before servicing the vehicle, refer to the Precautions Section.
2. Remove the rear wheels.
3. Release parking brake lever.
4. Remove console box and loosen parking brake cable locking nut (1).
5. Remove back plate plug (1) attached to the backside of brake back plate so as to increase clearance between brake shoe and brake drum.

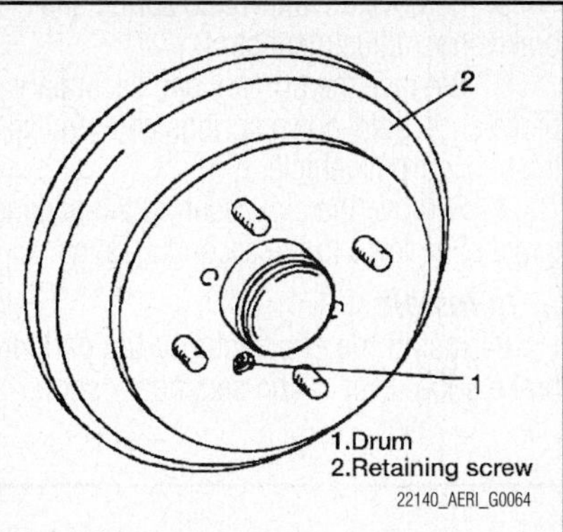

Fig. 14 Rear brake drum

6. Insert a screwdriver into the plug hole until it contacts the shoe hold down spring and push the screwdriver in the arrow direction, with this push, the hold down spring is pushed up and releases the parking shoe lever from the hold down spring, resulting in a larger clearance.
7. Pull brake drum off by using 8 mm bolts (1) (2 pieces).

To install:

8. Before installing brake drum, to maximize brake shoe-to-drum clearance, put screwdriver between rod (1) and ratchet (2) and push down ratchet as shown in figure.

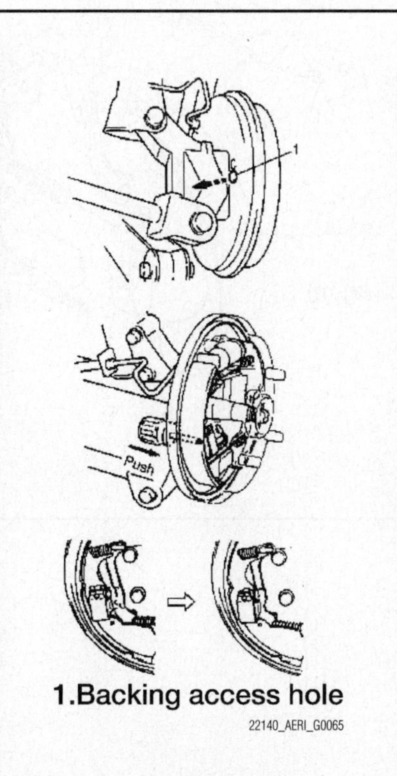

Fig. 15 Rear brake drum removal

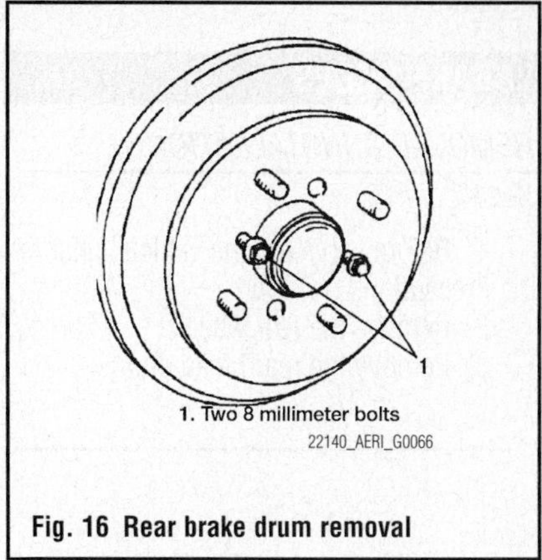

Fig. 16 Rear brake drum removal

9. Put brake shoe hold down spring (2) back to its original position as shown. (Put shoe hold down spring in place by moving shoe lever (3) so that shoe lever comes to the side of shoe hold down spring.)
10. Install the brake drum after making sure that the inside of brake drum and brake shoes are free from dirt and oil.
11. Tighten brake drum screws (1) to 6 ft. lb. (8.2 Nm).

➡Check to ensure that brake drum is free from dragging and proper braking

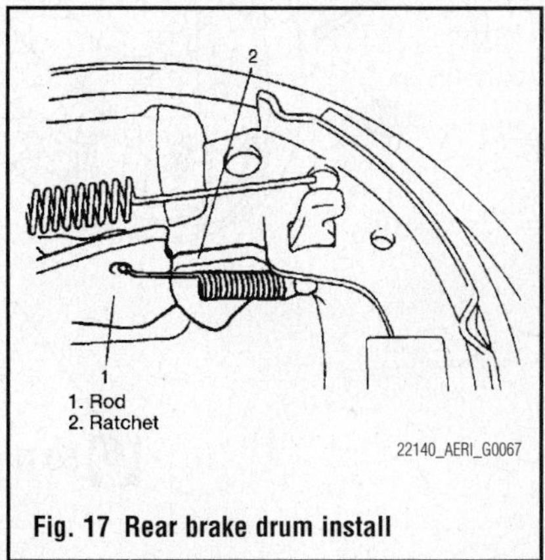

Fig. 17 Rear brake drum install

Fig. 18 Rear brake drum install

is obtained. Then remove vehicle from hoist and test foot brake and parking brake.

BRAKE SHOES

REMOVAL & INSTALLATION

See Figure 19.

1. Before servicing the vehicle, refer to the Precautions Section.
2. Remove the rear wheels.
3. Remove the rear brake drums.
4. Remove the upper and lower springs from the brake shoes.
5. Remove the anti-rattle spring and brake shoe adjustment strut.
6. Remove the primary and secondary brake shoe hold-down springs and remove the shoes from vehicle.
7. Remove the clip securing the parking brake shoe lever to the secondary shoe.

To install:

8. Install the clip securing the parking brake shoe lever to the secondary shoe.
9. Install the primary and secondary brake shoes to the vehicle and secure with the hold-down springs.
10. Install the brake adjustment strut and anti-rattle spring.
11. Install the upper and lower return springs to the primary and secondary brake shoes.
12. Install the brake drum
13. Install the wheels.
14. Press the brake pedal 3–5 times to adjust the brake shoe clearance.
15. Adjust the parking brake cable.

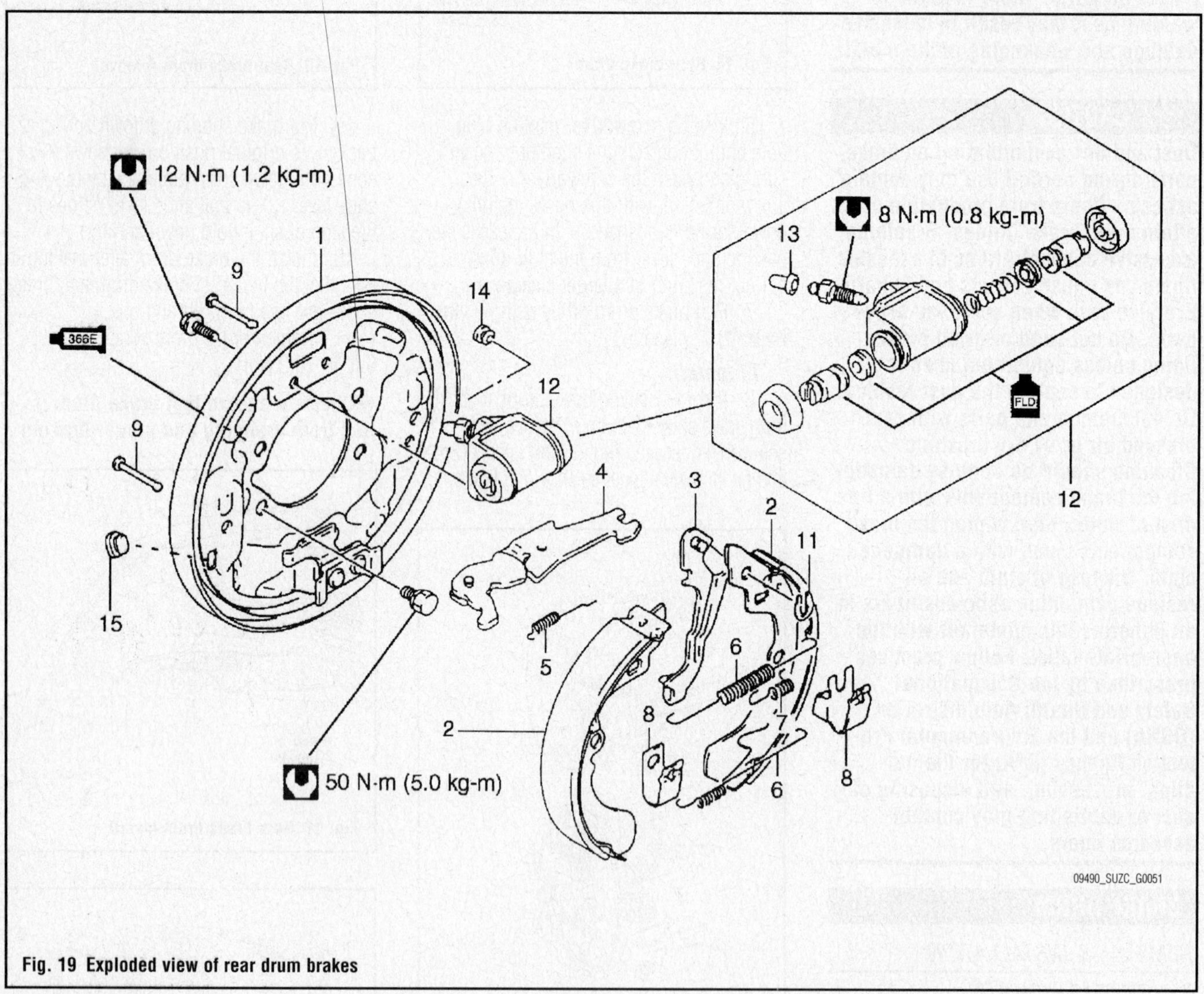

Fig. 19 Exploded view of rear drum brakes

BRAKES

PARKING BRAKE

PARKING BRAKE CABLES

ADJUSTMENT

See Figure 20.

➡**Make sure for the following conditions before cable adjustment:**

- No air is trapped in brake system.
- Brake pedal travel is proper.
- Brake pedal has been depressed a few times with about 66 lbs (300 Nm) load.
- Parking brake lever (1) has been pulled up a few times with about 44 lbs (200 Nm) load.

1. If parking brake cable is replaced with new one, pull up parking brake lever a few times with about 110 lbs (500 Nm) force.
2. Pulling the parking brake lever up, count the ratchet notches. There should be 6 to 8 notches.
3. Check if both right and left rear wheels are locked firmly.

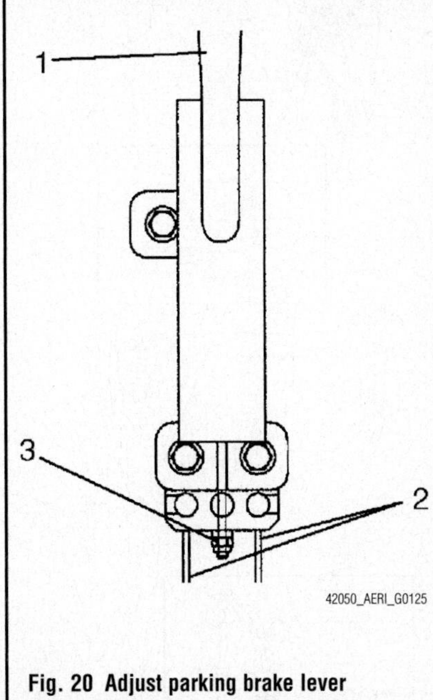

Fig. 20 Adjust parking brake lever

4. To count number of notches easily, listen to click sounds that ratchet makes while pulling parking brake lever without pressing its button. One click sound corresponds to one notch.
5. After confirming that the conditions are all satisfied, adjust parking brake lever stroke by loosening or tightening adjust nut (3).
6. Check brake drum for dragging after adjustment. Parking brake stroke (when lever is pulled up at 44 lbs. (200 Nm).
7. Check tooth tip of each notch for damage or wear.

⁂ WARNING

If any damage or wear is found, replace parking brake lever.

PARKING BRAKE SHOES

The rear drum brake shoes serve as the parking brakes. Refer to the procedures under Rear Drum Brakes.

CHASSIS ELECTRICAL

AIR BAG (SUPPLEMENTAL RESTRAINT SYSTEM)

GENERAL INFORMATION

⁂ CAUTION

Vehicles that are equipped with an air bag system must be disarmed before performing service on, or around, system components. The steering column, instrument panel components, wiring and sensors are some if the components that would require the air bag system to be disarmed. Failure to follow the safety precautions and the disarming procedure could result in accidental air bag deployment, possible injury and unnecessary system repairs.

SERVICE PRECAUTIONS

Disconnect and isolate the battery negative cable before beginning any airbag system component diagnosis, testing, removal, or installation procedures. Allow system capacitor to discharge for two minutes before beginning any component service. This will disable the airbag system. Failure to disable the airbag system may result in accidental airbag deployment, personal injury, or death.

Do not place an intact un-deployed airbag face down on a solid surface. The airbag will propel into the air if accidentally deployed and may result in personal injury or death.

When carrying or handling an un-deployed airbag, the trim side (face) of the airbag should be pointing towards the body to minimize possibility of injury if accidental deployment occurs. Failure to do this may result in personal injury or death.

Replace airbag system components with OEM replacement parts. Substitute parts may appear interchangeable, but internal differences may result in inferior occupant protection. Failure to do so may result in occupant personal injury or death.

Wear safety glasses, rubber gloves, and long sleeved clothing when cleaning powder residue from vehicle after an airbag deployment. Powder residue emitted from a deployed airbag can cause skin irritation. Flush affected area with cool water if irritation is experienced. If nasal or throat irritation is experienced, exit the vehicle for fresh air until the irritation ceases. If irritation continues, see a physician.

Do not use a replacement airbag that is not in the original packaging. This may result in improper deployment, personal injury, or death.

The factory installed fasteners, screws and bolts used to fasten airbag components have a special coating and are specifically designed for the airbag system. Do not use substitute fasteners. Use only original equipment fasteners listed in the parts catalog when fastener replacement is required.

During, and following, any child restraint anchor service, due to impact event or vehicle repair, carefully inspect all mounting hardware, tether straps, and anchors for proper installation, operation, or damage. If a child restraint anchor is found damaged in any way, the anchor must be replaced. Failure to do this may result in personal injury or death.

Deployed and non-deployed airbags may or may not have live pyrotechnic material within the airbag inflator.

Do not dispose of driver/passenger/curtain airbags or seat belt tensioners unless you are sure of complete deployment. Refer to the Hazardous Substance Control System for proper disposal.

Dispose of deployed airbags and tensioners consistent with state, provincial, local, and federal regulations.

After any airbag component testing or service, do not connect the battery negative cable. Personal injury or death may result if the system test is not performed first.

If the vehicle is equipped with the Occupant Classification System (OCS), do not connect the battery negative cable before performing the OCS Verification Test using the scan tool and the appropriate diagnostic information. Personal injury or death may result if the system test is not performed properly.

Never replace both the Occupant Restraint Controller (ORC) and the Occupant Classification Module (OCM) at the same time. If both require replacement, replace one, then perform the Airbag System test before replacing the other.

Both the ORC and the OCM store Occupant Classification System (OCS) calibration data, which they transfer to one another when one of them is replaced. If both are replaced at the same time, an irreversible fault will be set in both modules and the OCS may malfunction and cause personal injury or death.

If equipped with OCS, the Seat Weight Sensor is a sensitive, calibrated unit and must be handled carefully. Do not drop or handle roughly. If dropped or damaged, replace with another sensor. Failure to do so may result in occupant injury or death.

If equipped with OCS, the front passenger seat must be handled carefully as well. When removing the seat, be careful when setting on floor not to drop. If dropped, the sensor may be inoperative, could result in occupant injury, or possibly death.

If equipped with OCS, when the passenger front seat is on the floor, no one should sit in the front passenger seat. This uneven force may damage the sensing ability of the seat weight sensors. If sat on and damaged, the sensor may be inoperative, could result in occupant injury, or possibly death.

DISARMING THE SYSTEM

See Figure 21.

1. Turn the steering wheel so that vehicle's wheels (front tires) and pointing straight ahead.
2. Disconnect the negative battery cable.
3. Turn the ignition switch to "LOCK" position and remove key.
4. Remove "AIR BAG" fuse from fuse box (1).
5. Disconnect yellow connector of contact coil and combination switch assembly as follows.
6. Release locking mechanism of lock slider (2).
7. After unlocked, disconnect connector.
8. In case of passenger air bag (inflator) module, pull out glove box while pushing its stopper from both right and left sides and disconnect yellow connector of passenger air bag (inflator) module as follows.
    a. Release locking of lock slider (2).
    b. After unlocked, disconnect connector.

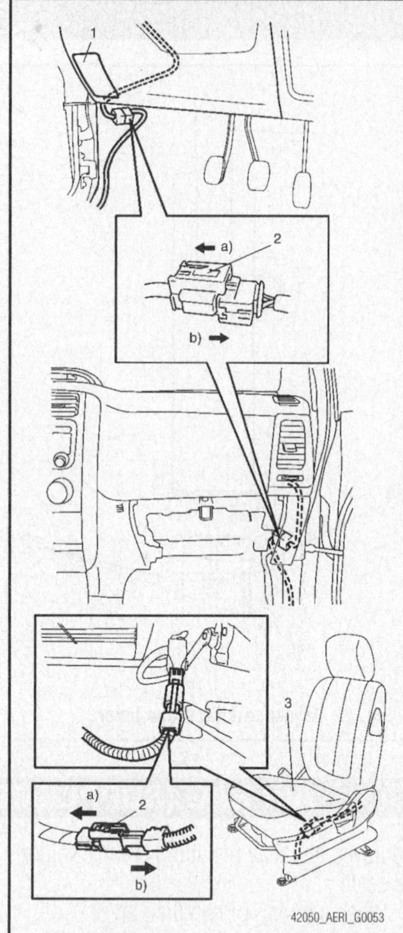

Fig. 21 Disarming air bag supplemental restraint system

9. If equipped with side air bag (inflator) module, disconnect yellow connector of side air bag (inflator) module under front seat cushion (3).
    a. Release locking of lock slider.
    b. After unlocked, disconnect connector.

ARMING THE SYSTEM

See Figures 22 through 24.

1. Confirm that the battery negative (–) cable is disconnected.
2. Turn ignition switch to "LOCK" position and remove key.
3. Connect yellow connector (1) of contact coil and combination switch assembly by pushing connector till clicks heard from it.
4. In case of passenger air bag (inflator) module, connect yellow connector (1) of passenger air bag (inflator) module by pushing connector till click is heard from it.
5. Install glove box.
6. If equipped with side air bag (inflator) module, connect yellow connector (2) of side air bag (inflator) module by pushing connector till click is heard from it.

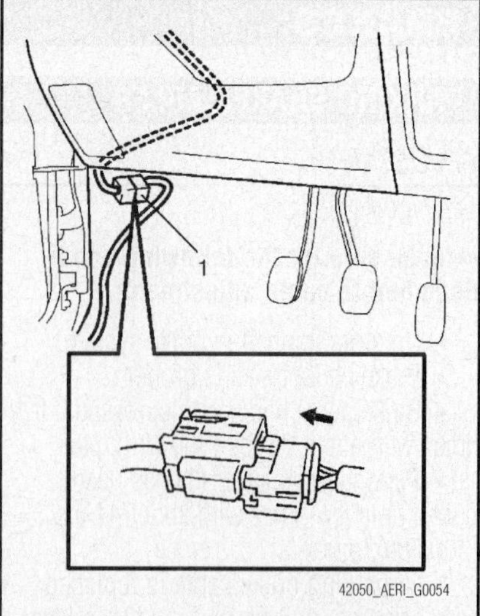

Fig. 22 Attach the yellow connector of contact coil and the combination switch assembly

7. Install "AIR BAG" fuse to fuse box.
8. Connect the negative battery cable.
9. Turn ignition switch to ON position and verify that "AIR BAG" warning lamp flashes 6 times and then turns OFF. If it does not operate as described, perform diagnostic system check.

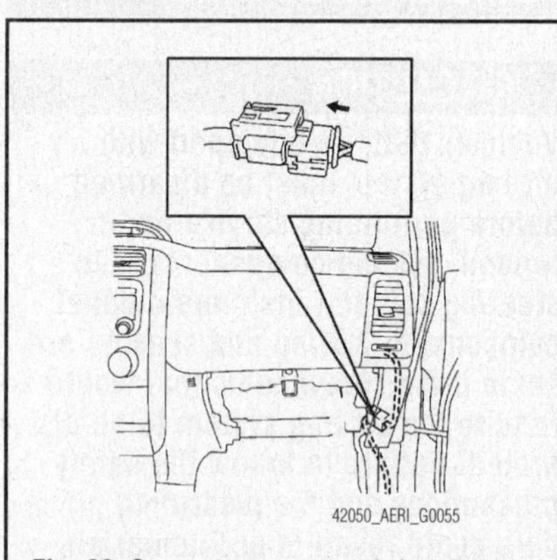

Fig. 23 Connect passenger air bag inflator module

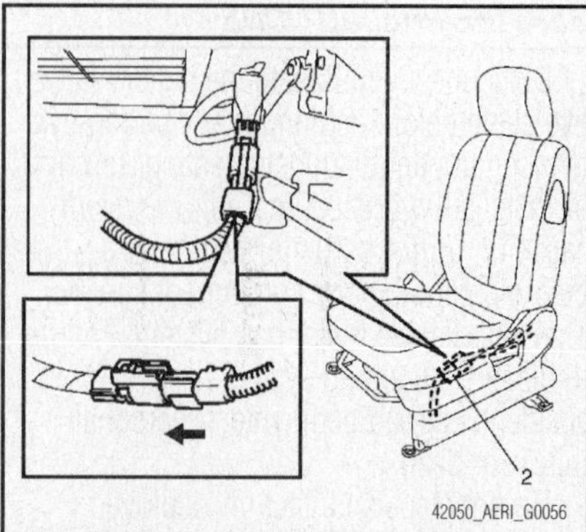

Fig. 24 Connect side air bag inflator module

DRIVETRAIN

AUTOMATIC TRANSAXLE ASSEMBLY

REMOVAL & INSTALLATION

See Figure 25.

1. Before servicing the vehicle, refer to the Precautions Section.
2. Disconnect the negative battery cable.
3. Drain the cooling system, engine oil, transaxle fluid and transfer case fluid (AWD).
4. Remove the engine/transaxle assembly from the vehicle. Refer to the engine procedure earlier in this chapter.
5. Remove or disconnect the following:
   - Engine rear mounting No. 1 bracket and engine rear mounting No. 2 bracket with stiffener (2WD)
   - Transfer case assembly (AWD)
   - Center bearing support bolts and center bearing support with center shaft from differential side gear (2WD)
   - Lower stiffener
   - Torque converter bolts
   - Starter
   - Transaxle with the torque converter from the engine compartment

➡When removing the transaxle from the engine, move it parallel with the crankshaft and use care so not to apply excessive force to the drive plate and torque converter.

⁂ CAUTION

Be sure to keep the transaxle with the torque converter horizontal or facing up throughout the work. Should it be tilted with converter down, the converter may fall off and cause personal injury.

To install:

6. Install or connect the following:
   - Transaxle to the engine assembly
   - Transaxle attaching bolts and nut and tighten to 61.5 ft. lbs. (85 Nm).
   - Torque converter bolts and tighten the bolts to 18 ft. lbs. (25 Nm)
   - Lower stiffener. Tighten the lower stiffener bolts to the transaxle first and the lower stiffener bolt to the engine second. Tighten the bolts to 36.5 ft. lbs. (50 Nm).
   - Starter
   - Center shaft to differential side gear (2WD). Tighten the center bearing support bolts to 40 ft. lbs. (55 Nm).

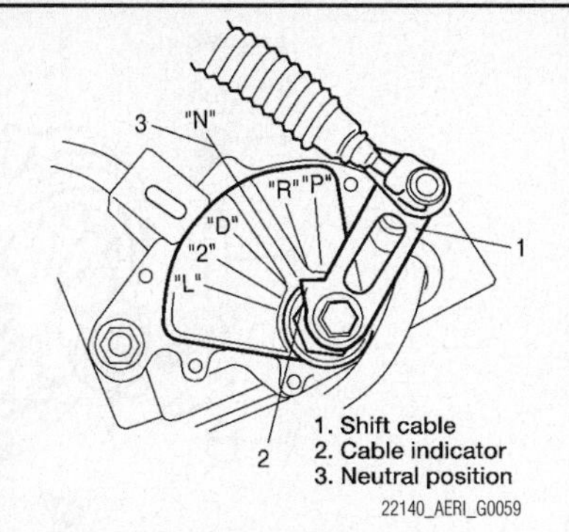

Fig. 25 Automatic transaxle cable adjustment

   - Engine rear mounting No. 1 bracket and engine rear mounting No. 2 bracket with stiffener (2WD)
   - Transfer case assembly (AWD)
7. Install the engine/transaxle assembly into the vehicle. Refer to the engine procedure earlier in this chapter.
8. Fill the cooling system, engine, transaxle and transfer case (AWD).
9. Connect the negative battery cable

MANUAL TRANSAXLE ASSEMBLY

REMOVAL & INSTALLATION

See Figures 26.

1. Before servicing the vehicle, refer to the Precautions Section.
2. Disconnect the negative battery cable.
3. Drain the transaxle oil.
4. Remove or disconnect the following:
   - Clutch operating cylinder with hose still attached
   - Gear control cables
   - All the wiring harness clamps and connectors involved with the transaxle removal, tag if necessary for location to aid during installation
   - Ground cable at the transaxle
   - Starter
   - Exhaust No. 1 pipe bolts
5. Support engine with lifting device.
   - Engine under covers
   - Exhaust No. 1 pipe and exhaust No. 2 pipe
   - Lower stiffener
   - Left and right ball joints from steering knuckles
   - Driveshaft joints
   - Center shaft support and center shaft
   - Dynamic damper
   - Engine rear mounting
   - Engine rear mounting No. 1 bracket with No. 2 bracket
   - Transaxle to engine bolts and nut
6. Support transaxle with a transmission jack.
   - Left engine mounting with bracket

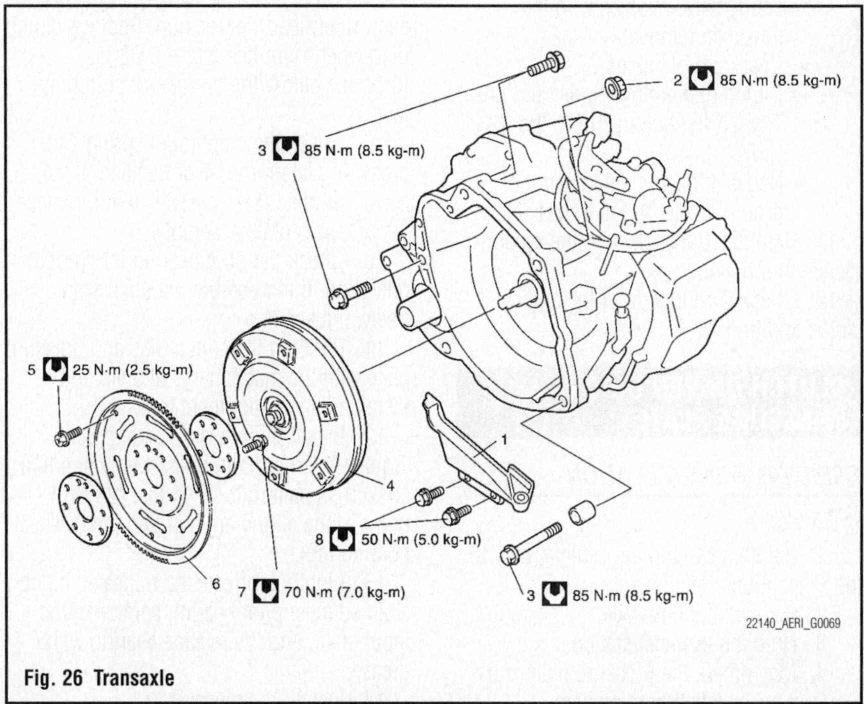

Fig. 26 Transaxle

7. Remove any remaining attached parts from the transaxle.

8. Pull transaxle out so as to disconnect the input shaft from the clutch disc and then remove it.

To install:

9. Install or connect the following:
   - Transaxle assembly. Use care when inserting the input shaft into the clutch assembly. If the spline on the input shaft does not align with the clutch assembly spline, turn the crankshaft slightly to aid in spline alignment.
   - Left engine mounting with bracket
   - Transaxle-to-engine bolts and nut. Tighten the nut and bolts to 44 ft. lbs. (61 Nm).
   - Engine rear mounting No. 1 bracket with No. 2 bracket
   - Engine rear mounting. Tighten the nuts to 32.5 ft. lbs. (45 Nm).
   - Dynamic damper. Tighten the bolts to 18 ft. lbs. (25 Nm).
   - Center shaft support and center shaft
   - Driveshaft joints
   - Left and right ball joints from steering knuckles
   - Lower stiffener. Tighten the bolts to 7 ft. lbs. (10 Nm).
   - Exhaust No. 1 pipe and exhaust No. 2 pipe
   - Engine under covers
10. Remove engine lifting device.
   - Starter
   - All the wiring harness clamps and connectors involved with the transaxle removal
   - Gear control cables
   - Clutch operating cylinder and hose. Tighten the bolts to 16 ft. lbs. (23 Nm).
   - Negative battery cable and the ground cable on the transaxle

11. Refill the transaxle with the recommended lubricant.

12. Check the function of the engine, clutch and transaxle.

CLUTCH DRIVEN DISC & PRESSURE PLATE

REMOVAL & INSTALLATION

See Figure 27.

1. Before servicing the vehicle, refer to the Precautions Section.
2. Remove the transaxle.
3. Hold the flywheel stationary.
4. Matchmark the pressure plate and flywheel for installation reference.

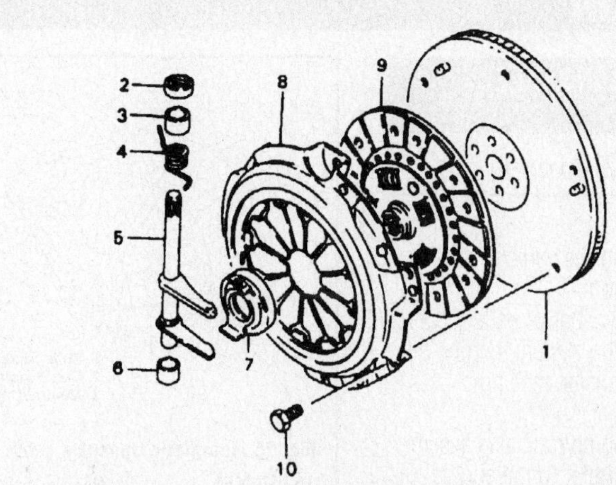

Fig. 27 Clutch component identification

5. Loosen the pressure plate attaching bolts 1 turn at a time (evenly) until the spring pressure is released.
6. Remove the clutch disc and pressure plate.

To install:

7. Clean the flywheel mating surfaces of all oil, grease and metal deposits. Inspect flywheel for cracks, heat checking or other defects and replace or resurface as necessary.
8. Check the wear on the facings of the clutch disc by measuring the depth of each rivet head depression. Replace clutch disc when rivet heads are 0.02 in. (0.5mm) below the surface of clutch surface.
9. Check the diaphragm spring and pressure plate for wear or damage. If the spring or plate is excessively worn, replace the pressure plate assembly.
10. Check the pilot bearing for smooth operation. If the bearing does not spin freely, replace it.
11. Position the clutch disc and pressure plate with the matchmarks aligned and install a clutch alignment tool.
12. Install the pressure plate bolts. Tighten the mounting bolts evenly and in a crisscross pattern to 17 ft. lbs. (23 Nm). Remove the alignment tool and the flywheel holding tool.
13. Lightly lubricate the transaxle input shaft splines, pilot bearing surface of the input shaft, and the release bearing with grease.
14. Install the transaxle.

ADJUSTMENTS

The clutch system is a hydraulic design and requires no manual adjustment.

CLUTCH MASTER CYLINDER

REMOVAL & INSTALLATION

See Figures 28 through 30.

**** CAUTION**

Do not allow fluid to get on painted surfaces.

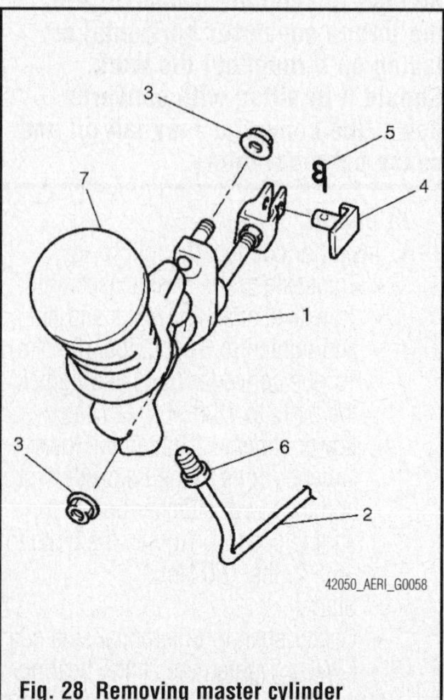

Fig. 28 Removing master cylinder

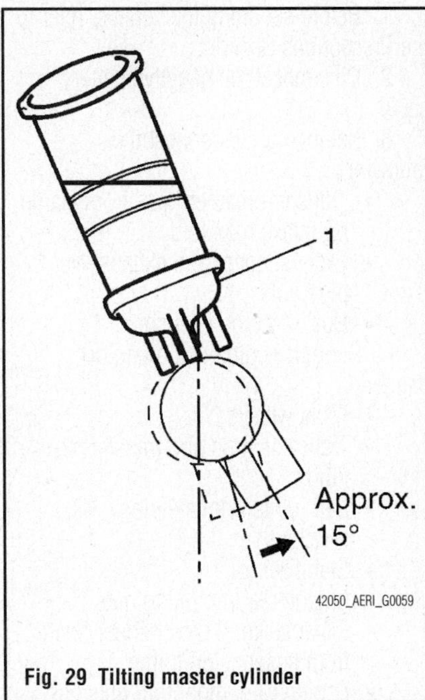

Fig. 29 Tilting master cylinder

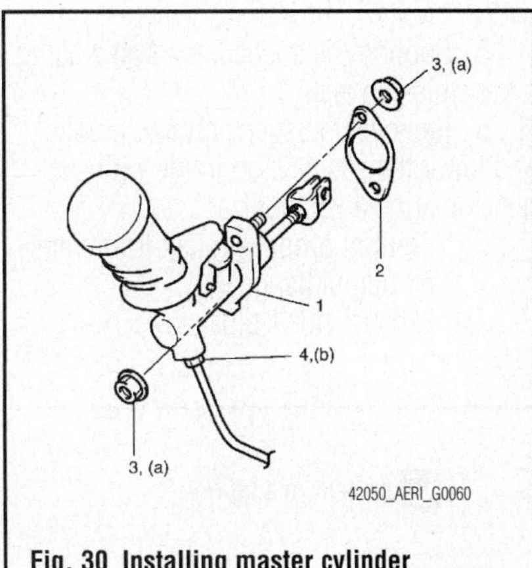

Fig. 30 Installing master cylinder

1. Clean around reservoir cap (7) and take out fluid with syringe or similar removal tool.
2. Remove clevis b-pin (5) and clevis pin (4).
3. Disconnect fluid pipe (2) from master cylinder assembly (1) by loosing flare nut (6).
4. Remove master cylinder attaching nuts (3).
5. Remove master cylinder assembly (1) and gasket.

To install:

6. To bleed air from the master cylinder (1) itself, tilt it as shown and add fluid into it.

➡After bleeding air from the master cylinder, plug pipe hole in it to prevent fluid from spilling out of it until the brake line is connected.

7. Install the master cylinder assembly (1) and new gasket (2) to body, attaching nuts (3).

➡Do not reuse gasket.

8. Tighten the clutch master cylinder attaching nut to 9.5 ft. lbs. (13 Nm).
9. Connect the fluid pipe and tighten flare nut (4) to 11.5 ft. lbs. (16 Nm).
10. Apply grease to clevis pin and install clevis pin.
11. Install clevis b-pin.
12. Fill the master cylinder reservoir with specified brake fluid and check fluid leakage.
13. After installation, bleed air from system.
14. Check clutch pedal free travel and clutch release margin.

CLUTCH SLAVE CYLINDER

REMOVAL & INSTALLATION

See Figures 31 and 32.

✲✲ CAUTION

Do not allow fluid to get on painted surfaces.

1. Clean around the master cylinder reservoir cap and take out fluid with syringe or a similar tool.
2. Disconnect the fluid hose (5) from slave cylinder by loosening union bolt (4).
3. Remove the operating cylinder attaching bolts (2) and slave cylinder assembly (1).

➡For bleeding air from the master cylinder alone, it must be removed from vehicle body. (For procedures of removal and installation of the master cylinder assembly and air bleeding, refer to master cylinder.)

4. Apply small amount of grease to rod tip.

➡Don't allow any grease to be on boot.

5. Install the clutch slave cylinder assembly (1) and tighten attaching bolts (2) to (a) 17 ft. lbs. (23 Nm).
6. Connect the clutch fluid hose (4) and tighten union bolt (3) to (b) 17 ft. lbs (23 Nm).

➡Do not reuse gaskets.

7. Fill the master cylinder reservoir with specified brake fluid and check for fluid leakage.
8. After master cylinder installation, bleed air from system.
9. Check clutch pedal free travel and clutch release.

CLUTCH HYDRAULIC SYSTEM BLEEDING

1. Remove the bleeder plug cap.
2. Connect a vinyl tube to the bleeder plug.
3. Depress the clutch pedal several times, and then loosen the bleeder plug with the pedal depressed.
4. When fluid no longer comes out, tighten the bleeder plug, and then release the clutch pedal.
5. Repeat the previous 2 steps until all the air in the fluid is completely bled.
6. Tighten the bleeder plug.

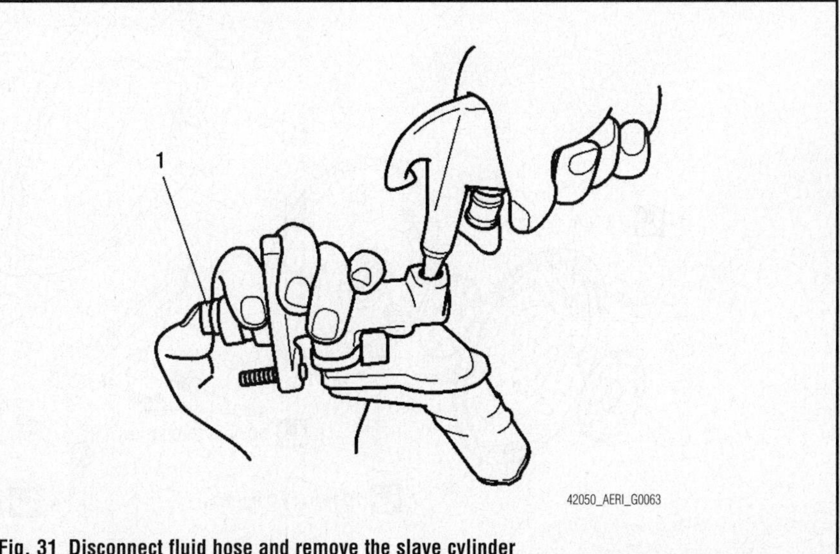

Fig. 31 Disconnect fluid hose and remove the slave cylinder

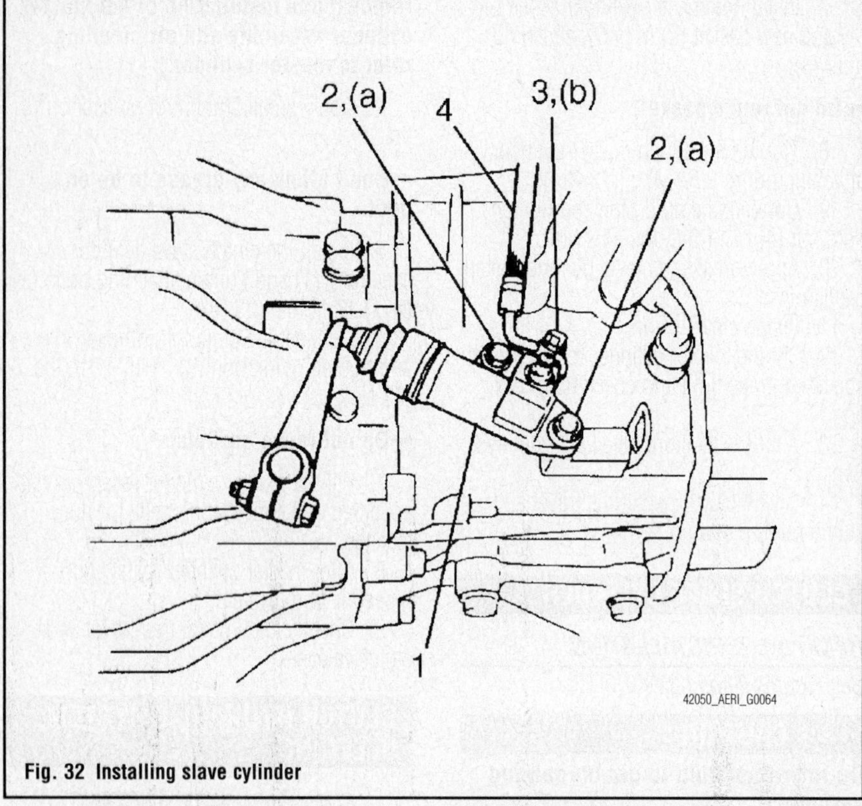

Fig. 32 Installing slave cylinder

7. Install the bleeder plug cap.
8. Check that all of the air has been bled from the clutch line.
9. Check the fluid level and add as necessary.

TRANSFER CASE ASSEMBLY

REMOVAL & INSTALLATION

See Figure 33.

1. Before servicing the vehicle, refer to the Precautions Section.
2. Disconnect the negative battery cable.
3. Remove or disconnect the following:
    - Transfer breather hose from clamp on intake manifold
    - Exhaust cover and exhaust No. 1 pipe nuts
    - Heated oxygen sensor
4. Support engine using engine hanger.
    - Front wheels
    - Transaxle fluid and transfer case fluid
    - Ball joints from steering knuckles
    - Stabilizer joint
    - Engine rear mounting nuts
    - Exhaust No. 2 pipe hanger bush from suspension frame
    - Engine front mounting nuts and mounting member to body bolts
5. Support the suspension frame using a transmission jack.
6. Remove 6 suspension frame bolts and lower the suspension frame with suspension arms and stabilizer.
    - Front of propeller shaft from transfer output flange
    - Exhaust No. 1 pipe stay

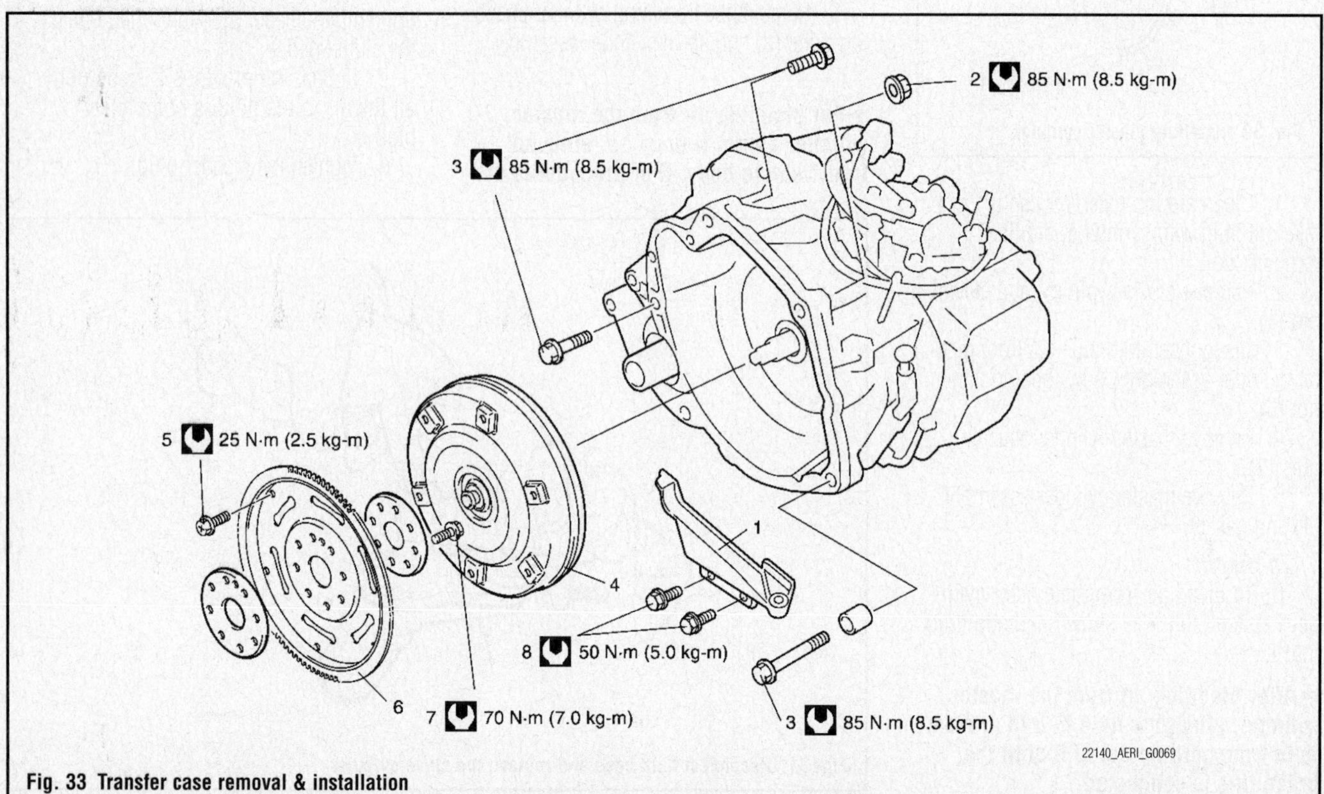

Fig. 33 Transfer case removal & installation

- Exhaust No. 1 pipe
- Exhaust No. 2 pipe
- Right side driveshaft
- Transfer case to transaxle stiffener
- Transfer case to engine stiffener

7. Support the transfer case assembly using a transmission jack.
   - Transfer case mounting bolts
   - Transfer case from transaxle

To install:

8. Remount the transfer case assembly by reversing the removal procedure.
9. Tighten the bolts and nuts to the specified torque as follows:
   - Engine rear mounting bracket nut: 14.5 ft. lbs. (20 Nm)
   - Transfer case mounting bolts: 36.5 ft. lbs. (50 Nm)
   - Transfer case to engine stiffener bolts: 36.5 ft. lbs. (50 Nm)
   - Transfer case to transaxle stiffener bolts: 36.5 ft. lbs. (50 Nm)
   - Propeller shaft No. 1 bolts: 17 ft. lbs. (23 Nm)
   - Exhaust No. 2 pipe to muffler bolts: 31.5 ft. lbs. (43 Nm)
   - Suspension frame bolts: 65 ft. lbs. (90 Nm)
   - Engine rear mounting nuts: 40 ft. lbs. (55 Nm)
   - Ball joint lock nuts: 43.5 ft. lbs. (60 Nm)
   - Exhaust No. 2 pipe to No. 1 pipe bolts: 31.5 ft. lbs. (43 Nm)
   - Exhaust No. 1 pipe stay bolts: 36.5 ft. lbs. (50 Nm)
   - Engine front mounting nuts: 32.5 ft. lbs. (45 Nm)
   - Mounting member to body bolts: 39.5 ft. lbs. (55 Nm)
   - Wheel lug nuts: 61.5 ft. lbs. (85 Nm)
   - Remaining components in the reverse order of removal
10. Fill the transaxle and the transfer case with the correct type and amount of fluid.
11. Connect the negative battery cable.

FRONT AUTOMATIC LOCKING HUBS

REMOVAL & INSTALLATION

1. Before servicing the vehicle, refer to the Precautions Section.
2. Disconnect the negative battery cable.
3. Drain the cooling system, engine oil, transaxle fluid and transfer case fluid (AWD).
4. Remove the engine/transaxle assembly from the vehicle. Refer to the engine procedure earlier in this chapter.
5. Remove or disconnect the following:
   - Engine rear mounting No. 1 bracket and engine rear mounting No. 2 bracket with stiffener (2WD)
   - Transfer case assembly (AWD)
   - Center bearing support bolts and center bearing support with center shaft from differential side gear (2WD)
   - Lower stiffener
   - Torque converter bolts
   - Starter
   - Transaxle with the torque converter from the engine compartment

➡When removing the transaxle from the engine, move it parallel with the crankshaft and use care so not to apply excessive force to the drive plate and torque converter.

⁂ CAUTION

Be sure to keep the transaxle with the torque converter horizontal or facing up throughout the work. Should it be tilted with converter down, the converter may fall off and cause personal injury.

To install:

6. Install or connect the following:
   - Transaxle to the engine assembly
   - Transaxle attaching bolts and nut and tighten to 61.5 ft. lbs. (85 Nm).
   - Torque converter bolts and tighten the bolts to 18 ft. lbs. (25 Nm)
   - Lower stiffener. Tighten the lower stiffener bolts to the transaxle first and the lower stiffener bolt to the engine second. Tighten the bolts to 36.5 ft. lbs. (50 Nm).
   - Starter
   - Center shaft to differential side gear (2WD). Tighten the center bearing support bolts to 40 ft. lbs. (55 Nm).
   - Engine rear mounting No. 1 bracket and engine rear mounting No. 2 bracket with stiffener (2WD)
   - Transfer case assembly (AWD)
7. Install the engine/transaxle assembly into the vehicle. Refer to the engine procedure earlier in this chapter.
8. Fill the cooling system, engine, transaxle and transfer case (AWD).
9. Connect the negative battery cable.

FRONT AXLE SHAFT, BEARING & SEAL

REMOVAL & INSTALLATION

See Figures 34 through 38.

1. Disengage caulk ("A") and remove driveshaft nut (1) and washer.
2. Hoist vehicle.

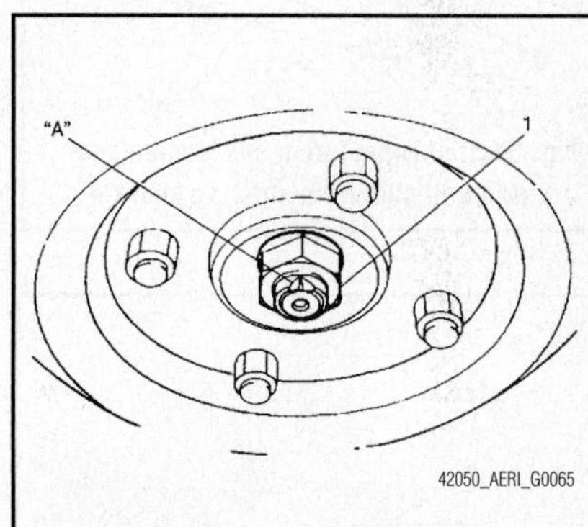

Fig. 34 Disengage caulk and remove driveshaft nut

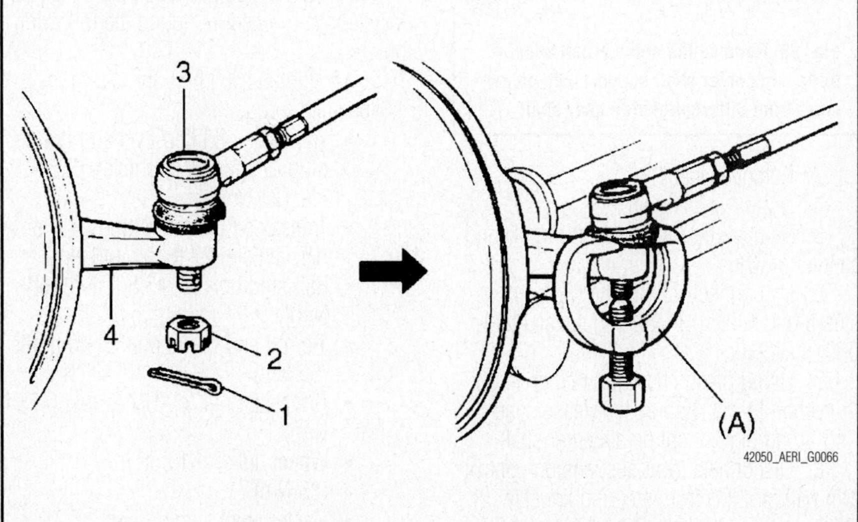

Fig. 35 Remove tie rod end split pin and castle nut then disconnect the tie rod

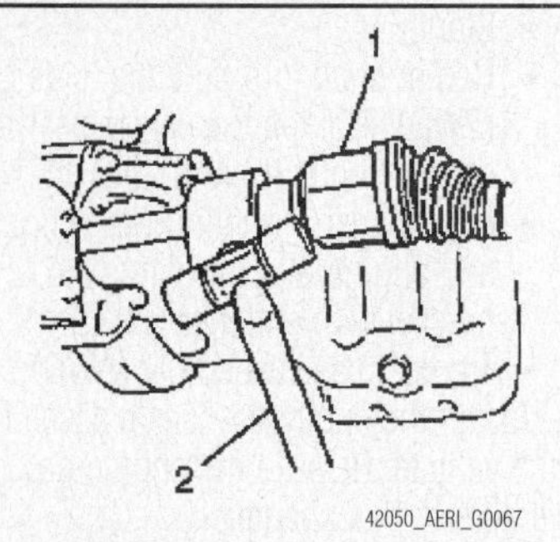

Fig. 36 Use plastic hammer to drive out driveshaft joint to release snap ring fitting of joint spline

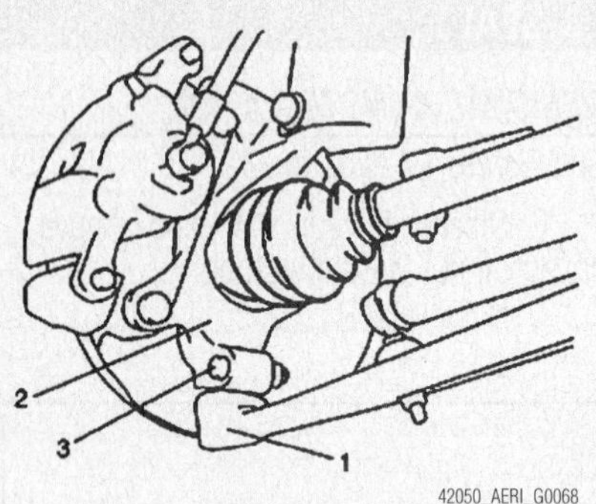

Fig. 37 Disconnect front suspension control arm ball stud from steering knuckle

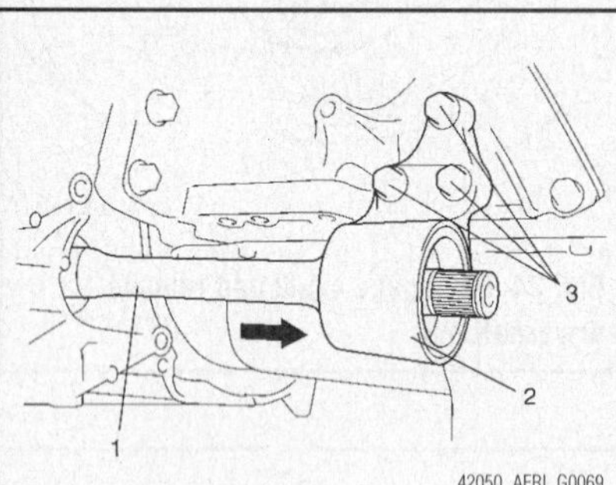

Fig. 38 Remove the center shaft support bolts and center shaft support with center shaft from differential side gear shaft

3. Remove wheel.
4. Drain transaxle oil.
5. Remove tie rod end split pin (1) and castle nut (2).
6. Disconnect tie rod end (3) from steering knuckle (4) by using special tool 09913-65210.
7. Using plastic hammer (2), drive out driveshaft joint (1) so as to release snap ring fitting of joint spline at center shaft.
8. Disconnect front suspension control arm ball stud (1) from steering knuckle (2) after removing ball stud bolt (3) and nut.
9. Remove driveshaft assembly.

**** CAUTION**

To prevent damage of boots, be careful not to contact them with other parts when removing the driveshaft assembly.

10. Remove center shaft support bolts (3) and remove center shaft support (2) with center shaft (1) from differential side gear shaft (if equipped).

To install:

**** CAUTION**

To avoid excessive expansion of boot and consequential disconnection of joint in boot, do not pull differential side joint housing.

**** CAUTION**

Prevent any damage of boots by protecting them from unnecessary contact while installing driveshaft.

**** CAUTION**

Do not hit boot with hammer. Inserting joint only by hand is allowed.

**** CAUTION**

Make sure that driveshaft assembly is inserted fully and its snap ring is installed in its correct position.

**** CAUTION**

Do not damage oil seal when installing and removing center shaft.

11. Install driveshaft assembly by reversing removal procedure noting the following points.
   a. Tighten each bolt and nut to the specified torque.
      - Transaxle oil filler/level and drain plug "a" (for M/T vehicle) 15.5 ft. lbs. (21 Nm)
      - Transaxle fluid drain plug "b" (for A/T vehicle) 29 ft. lbs. (40 Nm)
      - Ball stud bolt "c" 43.5 ft. lbs. (60 Nm)
      - Tie rod end castle nut "d" 40 ft. lbs. (55 Nm)
      - Driveshaft nut "e" 127 ft. lbs. (175 Nm)
      - Wheel nut "f" 61.5 ft. lbs. (85 Nm)
      - Center shaft support bolt "g" 40 ft. lbs. (55 Nm)
   b. Apply sealant "a" 99000-31260 to drain plug (1) and filler / level plug (2) for manual transaxle.
   c. Fill transaxle with oil.

FRONT HALFSHAFTS

REMOVAL & INSTALLATION

See Figure 39.

1. Before servicing the vehicle, refer to the Precautions Section.
2. Disconnect the negative battery cable.
3. Remove driveshaft nut and washer.
4. Raise and safely support the front of the vehicle securely on jackstands.
5. Remove wheel.
6. Drain transaxle oil.
7. Remove tie rod end cotter pin and castle nut.
8. Disconnect tie rod end from steering knuckle by using special tool 09913-65210.
9. Using plastic hammer, drive out driveshaft joint so as to release snap ring fitting of joint spline at center shaft.
10. Disconnect front suspension control arm ball stud from steering knuckle after removing ball stud bolt and nut.
11. Remove driveshaft assembly.

➡To prevent damage of boots, be careful not to contact them with other parts when removing driveshaft assembly.

12. Remove center shaft support bolts and remove center shaft support with center shaft from differential side gear shaft (if equipped).

To install:

13. Install the driveshaft assembly by reversing removal procedure noting the following points:
   a. Tighten each bolt and nut to the specified torque.
      - Transaxle oil filler/level and drain plug (for M/T): 15.5 ft. lbs. (21 Nm)
      - Transaxle fluid drain plug (for A/T): 29 ft. lbs. (40 Nm)
      - Ball stud bolt: 43.5 ft. lbs. (60 Nm)
      - Tie rod end castle nut: 25.5–40 ft. lbs. (35–55 Nm)
      - New driveshaft nut: 127 ft. lbs. (175 Nm)
      - Wheel nut: 61.5 ft. lbs. (85 Nm)
      - Center shaft support bolt: 40 ft. lbs. (55 Nm)
14. Apply sealant to drain plug and filler/level plug for manual transaxle.
15. Fill transaxle with the correct type and amount of oil.
16. Connect the negative battery cable.

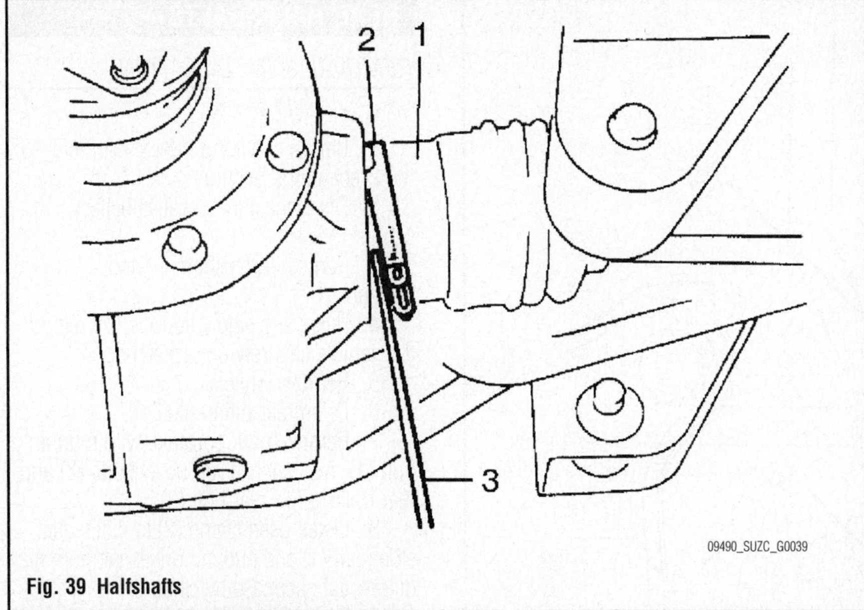

Fig. 39 Halfshafts

REAR AXLE SHAFT, BEARING & SEAL

REMOVAL & INSTALLATION

See Figures 40 through 43.

1. Remove caulking of driveshaft nut (2) and then remove driveshaft nut and washer (1).
2. Remove tire and drain rear differential oil.
3. Hoist vehicle.
4. Remove parking brake wire mounting bolt (1), rear control rod outer bolts (2) and rear trailing rod rear bolt (3).
5. Install the used clamp (2) to differential side joint (1) and pull out driveshaft from rear differential gear case by using tire lever (3).
6. Remove the driveshaft.

To install:

7. Install the driveshaft assembly by reversing removal procedure and noting following points:

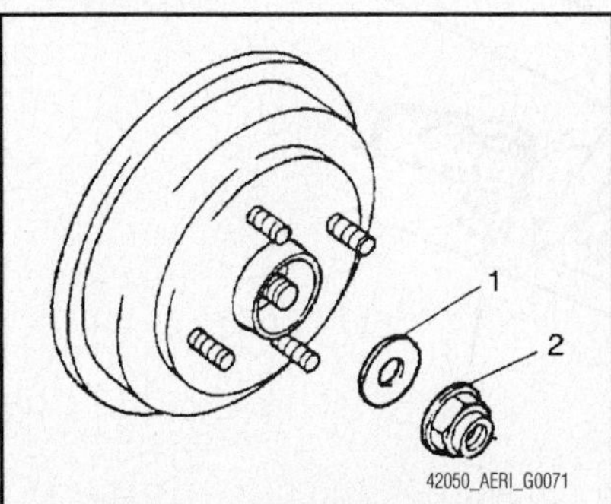

Fig. 40 Remove the caulking of driveshaft nut and then remove the drain shaft nut and washer

a. Be sure to use new driveshaft nut (1).

b. Tighten rear control rod outer bolt (a) to 68.5 ft. lbs. (95 Nm).

c. Tighten rear trailing rod rear bolt (b) to 68.5 ft. lbs. (95 Nm).

8. Tighten driveshaft nut (c) to 126.5 ft. lbs. (175 Nm).
9. Clean rear wheel bearing oil seal and then apply grease (A) 99000-25010. Replace it if required.
10. Apply sealant to drain plug for rear differential gear case.
11. Fill rear differential gear case with oil.

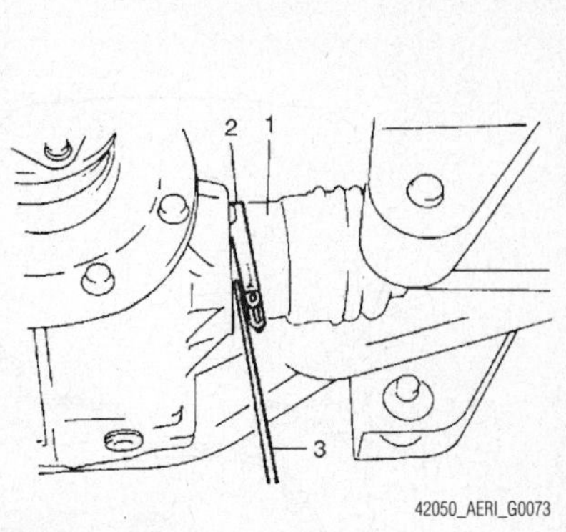

Fig. 42 Pull driveshaft from rear differential gear case

**** CAUTION**

Protect oil seals and boots from any damage, preventing them from unnecessary contact while installing driveshaft.

**** CAUTION**

Do not hit joint boot with hammer. Inserting joint only by hands is allowed.

**** CAUTION**

Make sure that differential side joint is inserted fully and its snap ring is seated as it was.

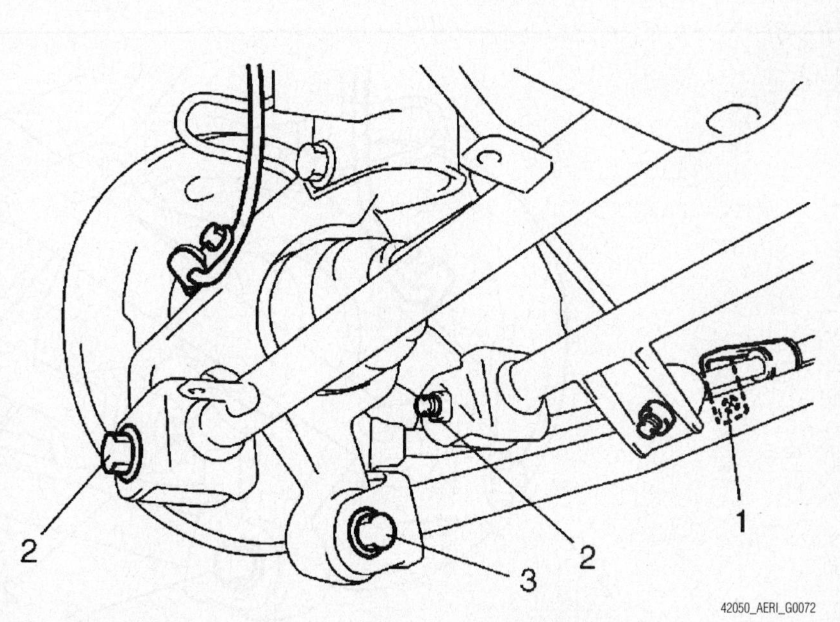

Fig. 41 Remove the parking wire mounting bolt, rear control rod outer bolts and the rear trailing rod rear bolt

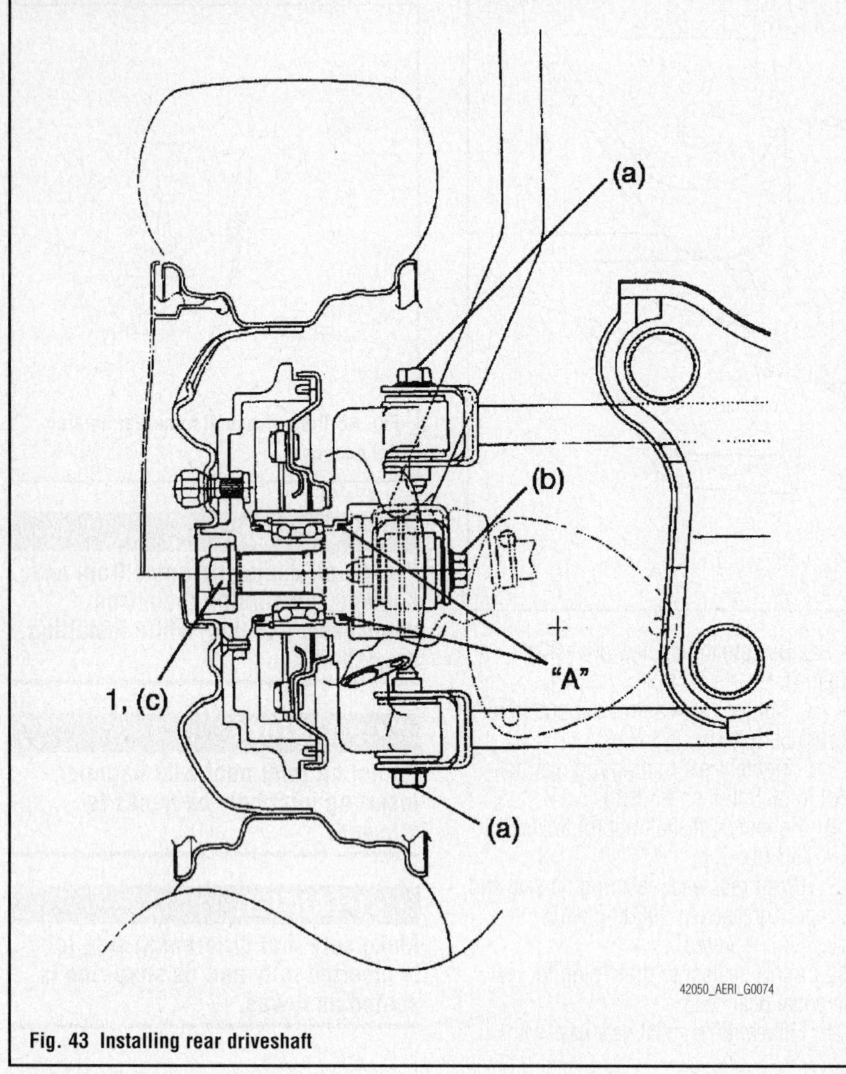

Fig. 43 Installing rear driveshaft

REAR HALFSHAFTS

REMOVAL & INSTALLATION

See Figures 44 and 45.

1. Before servicing the vehicle, refer to the Precautions Section.
2. Disconnect the negative battery cable.
3. Remove driveshaft nut and washer.
4. Raise and safely support the rear of the vehicle securely on jackstands.
5. Remove wheel.
6. Drain rear differential oil.
7. Remove parking brake wire mounting bolt (1), rear control rod outer bolts (2) and rear trailing rod rear bolt (3).
8. Install used clamp (2) to differential side joint (1) and pull out driveshaft from rear differential gear case by using tire lever (3).
9. Remove driveshaft.

To install:

✻✻ WARNING

Protect oil seals and boots from any damage, preventing them from unnecessary contact while installing driveshaft. Do not hit joint boot with hammer. Inserting joint only by hands is allowed. Make sure that differential side joint is inserted fully and its snap ring is seated as it was.

10. Clean rear wheel bearing oil seal and then apply grease. Replace it if required.

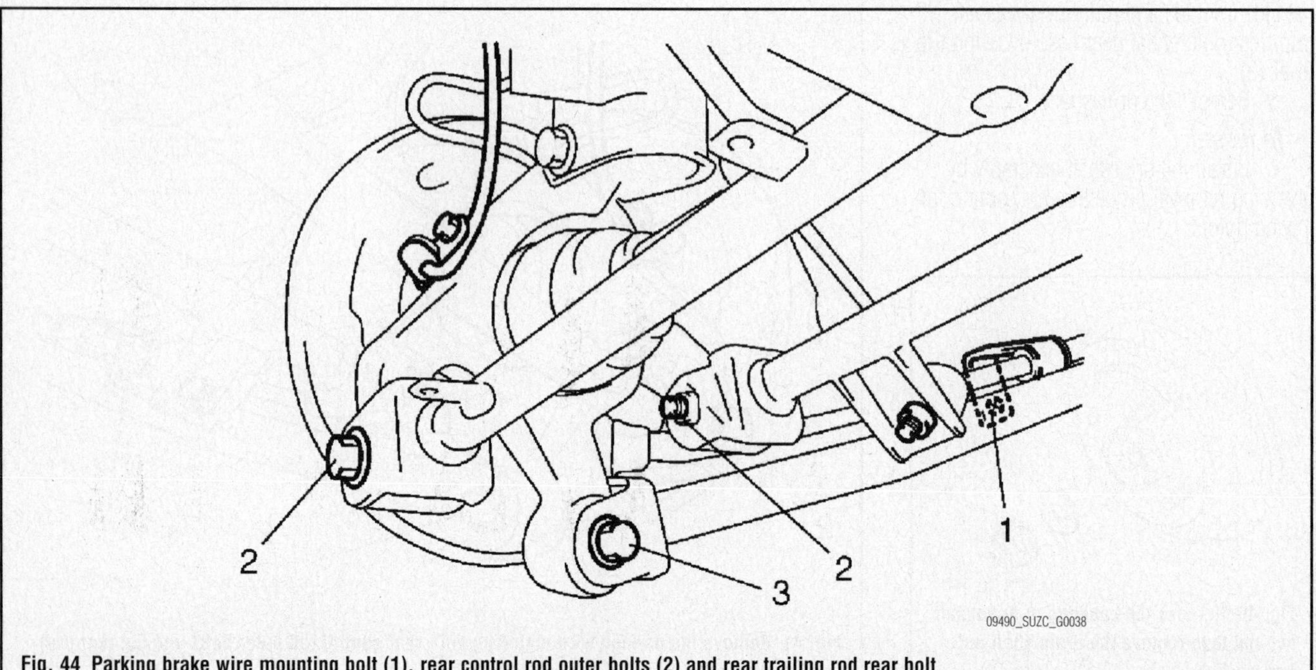

Fig. 44 Parking brake wire mounting bolt (1), rear control rod outer bolts (2) and rear trailing rod rear bolt

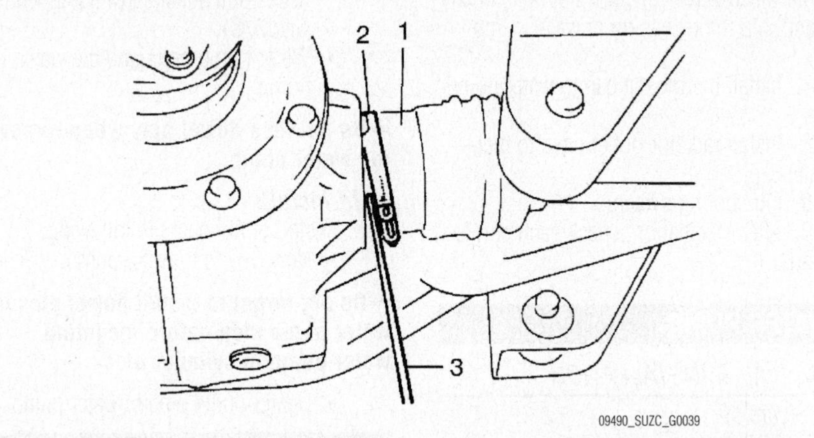

Fig. 45 Pull out drive shaft from rear differential gear case by using tire lever with a used clamp—Aerio with AWD

11. Install driveshaft assembly by reversing removal procedure noting the following points:
    a. Tighten each bolt and nut to the specified torque.
        - Rear control rod outer bolt: 68.5 ft. lbs. (95 Nm)
        - Rear trailing rod rear bolt: 68.5 ft. lbs. (95 Nm)
        - New driveshaft nut: 127 ft. lbs. (175 Nm)
12. Clean rear wheel bearing oil seal and then apply grease. Replace it if required.
13. Apply sealant to drain plug for rear differential gear case.
14. Fill rear differential gear case with the correct type and amount of oil.
15. Connect the negative battery cable.

ENGINE COOLING

ENGINE FAN

REMOVAL & INSTALLATION

See Figure 46.

1. Disconnect the negative battery cable.
2. Remove condenser cooling fan (if equipped with A/C).
3. Disconnect connector of radiator cooling fan motor.
4. Remove radiator cooling fan motor (1) from radiator (2).
5. Installation is the reverse of the removal procedure.

RADIATOR

REMOVAL & INSTALLATION

See Figure 47.

✲✲ CAUTION

To avoid the danger of being burned, do not service the exhaust system while it is hot. Service should be performed only after the system cools down.

1. Before servicing the vehicle, refer to the Precautions Section.

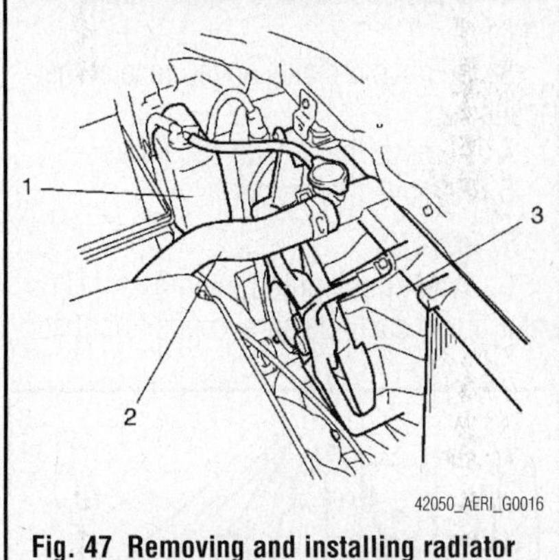

Fig. 47 Removing and installing radiator

✲✲ CAUTION

Be careful not to damage fins of radiator. If radiator fin is bent during installation, straighten it by using a fin comb with proper size spacing.

2. Disconnect the negative battery cable.
3. Drain cooling system by loosening drain plug of radiator.
4. Disconnect connector of cooling fan motor.
5. Remove condenser cooling fan (if equipped with A/C).
6. Remove reservoir (1).
7. Disconnect radiator inlet (2) and outlet hoses from radiator.
8. Disconnect A/T fluid hoses from radiator, if equipped with A/T vehicle.
9. Remove radiator cooling fan assembly (3).
10. Remove radiator.

To install:

11. Reverse removal procedures, noting the following:

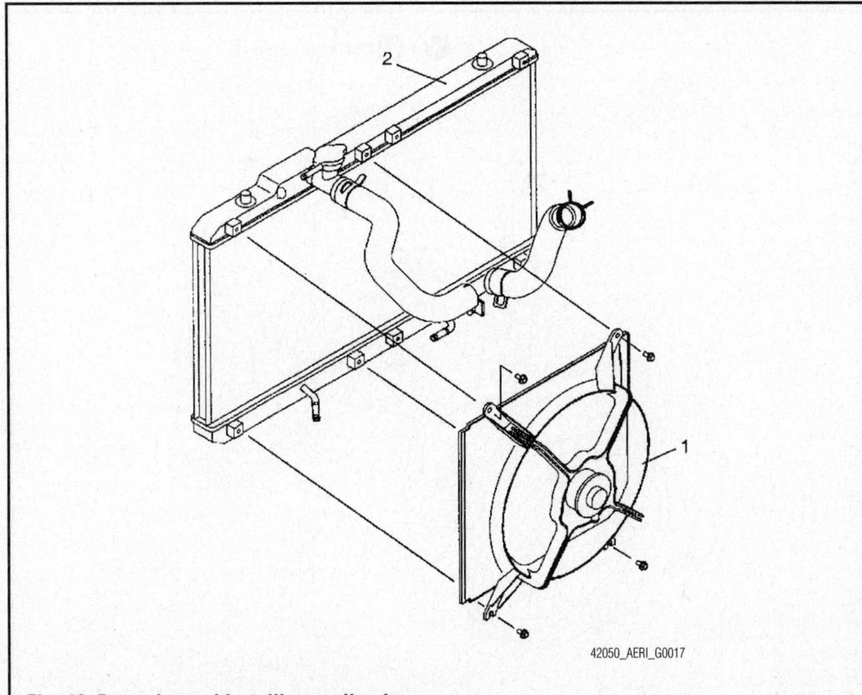

Fig. 46 Removing and installing cooling fan

a. Refill cooling system with proper coolant.

b. After installation, check each joint for leakage.

c. Check automatic fluid level (for A/T vehicle).

THERMOSTAT

REMOVAL & INSTALLATION

See Figures 48 and 49.

CAUTION

To avoid the danger of being burned, do not service the exhaust system while it is hot. Service should be performed only after the system cools down.

1. Before servicing the vehicle, refer to the Precautions Section.
2. Drain coolant and tighten drain plug.
3. Remove radiator outlet hose at thermostat cap.
4. Remove thermostat cap (1).
5. Remove thermostat.

To install:

6. When positioning thermostat (1) on water pump case, be sure to position it so that air bleed valve (2) comes at match mark (3) and into the recession of water pump case.
7. Install thermostat cap to water pump case.
8. Install radiator outlet hose to thermostat cap.
9. Fill cooling system.
10. After installation, check each part for leakage

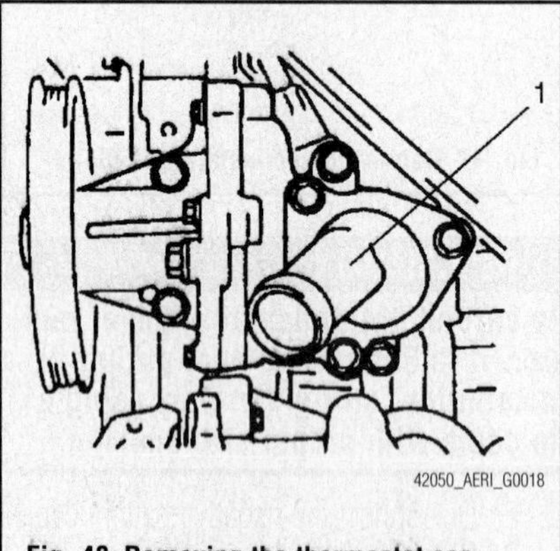

Fig. 48 Removing the thermostat cap

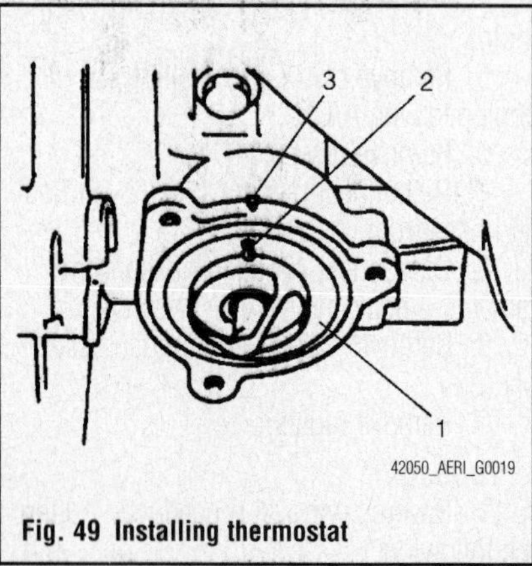

Fig. 49 Installing thermostat

WATER PUMP

REMOVAL & INSTALLATION

See Figure 50.

CAUTION

To avoid the danger of being burned, do not service the exhaust system while it is hot. Service should be performed only after the system cools down.

1. Before servicing the vehicle, refer to the Precautions Section.
2. Disconnect the negative battery cable.
3. Drain the cooling system into a suitable container and tighten the drain plug.
4. Remove or disconnect the following:
   - Condenser cooling fan (if equipped with A/C)
   - Radiator outlet hose from thermostat cap
   - Heater outlet pipe bolt
   - Serpentine belt
   - Compressor from its bracket with hoses still attached (if equipped with A/C)
   - Water pump bolts and the water pump

➡Do not lose dowel pins when removing water pump.

To install:

5. Install or connect the following:
   - New O-ring to water pump.

➡Do not forget to install dowel pins on water pump side before mounting water pump to cylinder block.

   - Heater outlet pipe to water pump

WARNING

Use NEW bolts to install water pump to cylinder block. Failure to do so may result water leakage.

➡Install the water pump with new bolts to cylinder block and tighten to 18 ft. lbs. (25 Nm)

   - Compressor to its bracket (if equipped with A/C)
   - Heater outlet pipe bolt
   - Radiator outlet hose to thermostat cap
   - Serpentine belt
   - Fill the cooling system.
   - Connect the negative battery cable.
   - Start the engine and top off the coolant as necessary.
   - Check the cooling system for leaks.

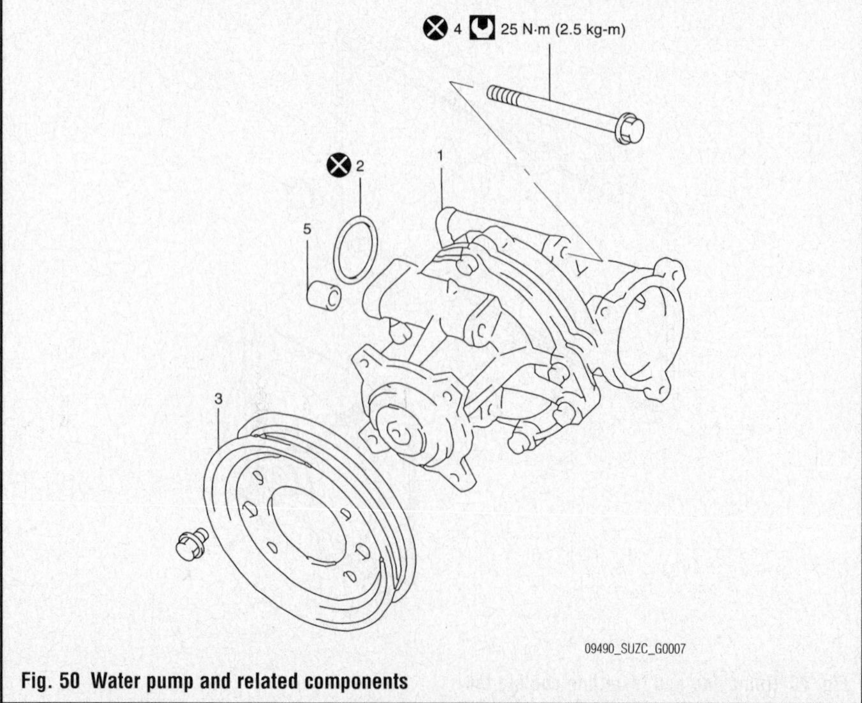

Fig. 50 Water pump and related components

ALTERNATOR

REMOVAL & INSTALLATION

See Figure 51.

1. Disconnect the negative battery cable.
2. Remove air cleaner outlet hose and air cleaner assembly.
3. Remove P/S return pipe clamp bolt.
4. Remove right side driveshaft referring to front driveshaft removal and installation.
5. Disassemble in order shown in figure.
6. Installation is the reverse of the removal procedure.

VOLTAGE REGULATOR

REMOVAL & INSTALLATION

See Figure 52.

The voltage regulator is behind the rear cover and is not replaced separately. Refer to the accompanying illustration for location

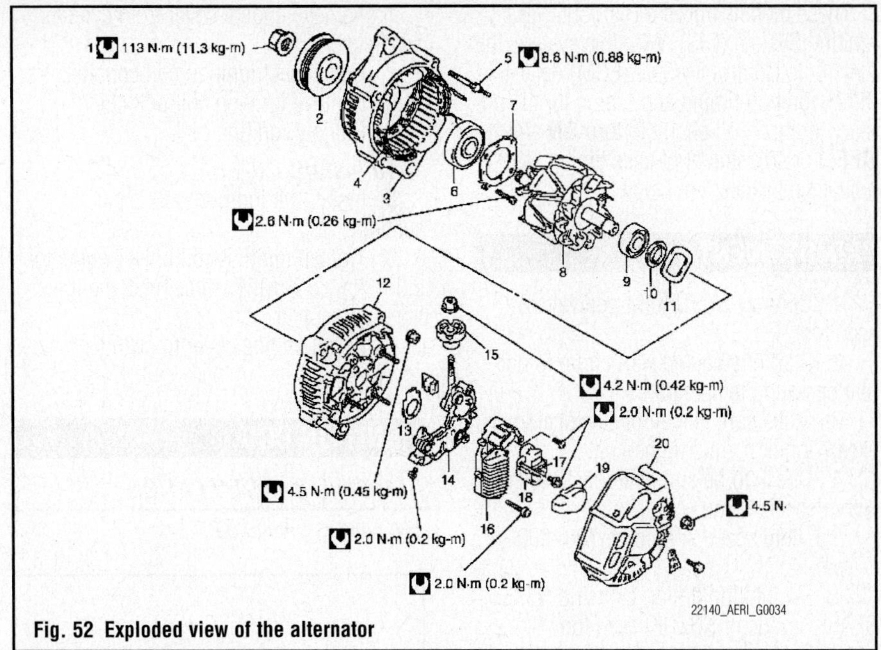

Fig. 52 Exploded view of the alternator

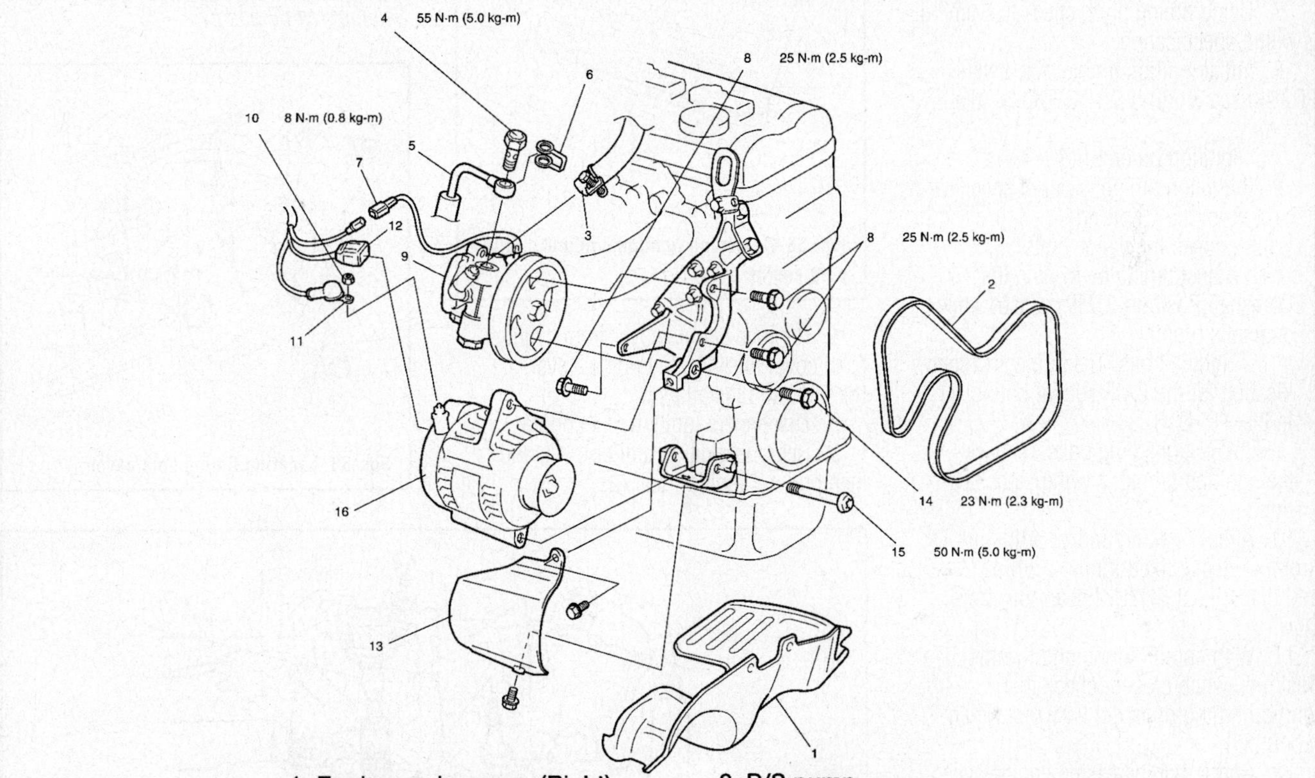

1. Engine under cover (Right)
2. Alternator belt
3. P/S suction hose
4. Union bolt
5. P/S high pressure pipe
6. Gasket
7. P/S pressure switch coupler
8. P/S pump bolt
9. P/S pump
10. "B" terminal nut
11. "B" terminal wire
12. Coupler: do not reuse
13. Alternator cover
14. Alternator bolt
15. Alternator bolt
16. Alternator

22140_AERI_G0033

Fig. 51 Exploded view of the alternator mounting

ENGINE ELECTRICAL

IGNITION SYSTEM

These models utilize a Distributorless Ignition System (DIS). With this system, the Electronic Control Module (ECM) determines ignition timing and timing for the primary ignition coil circuit to turn **ON** and **OFF**. For adjustment of base timing see adjustment procedure below.

ADJUSTMENT

1. Connect SUZUKI (or equivalent) scan tool.
2. Start engine and warm it up to normal operating temperature.
3. Make sure that all of electrical loads except ignition are switched off.
4. Check to be sure that idle speed are within specification:
   a. Idle speed should be 700–800 rpm.
5. Fix ignition timing by using "FIXED SPARK" mode of SUZUKI scan tool.
6. Set timing light to No.1 ignition coil harness.
7. Using timing light, check that timing is within specification.
8. Initial ignition timing (fixed with SUZUKI scan tool) 5 ± 1° BTDC (at idle speed):
   a. Ignition order 1—3—4—2
9. If ignition timing is out of specification, perform the following:
   a. Loosen the flange bolts
   b. Adjust timing by turning the Camshaft Position (CMP) sensor while engine is running.
   c. Tighten bolts. The tightening specification for the CMP sensor bolt is: 11 ft. lbs. (15 Nm).
   d. After tightening bolts, recheck that ignition timing is within specification.
10. After checking and/or adjusting timing, release fixed ignition timing and disconnect SUZUKI scan tool from DLC.
11. With engine idling and throttle position switch closed, check that ignition timing does not vary more then (about 8°).
12. Check that increasing engine speed advances ignition timing.
13. If the check results are not satisfactory, check TP sensor and ECM.

IGNITION COIL

REMOVAL & INSTALLATION

See Figures 53 and 54.

1. Remove engine cover from cylinder head cover.
2. Disconnect ignition coil coupler.
3. Remove ignition coil bolt (1).
4. Pull out ignition coil.

To install:

5. Install the ignition coil (2) securely.
6. Tighten ignition coil bolt. Tighten to 2.5 ft. lbs. (3 Nm). Then connect the ignition coil coupler.
7. Install engine cover to cylinder head cover.

IGNITION COIL

REMOVAL & INSTALLATION

See Figures 53 and 54.

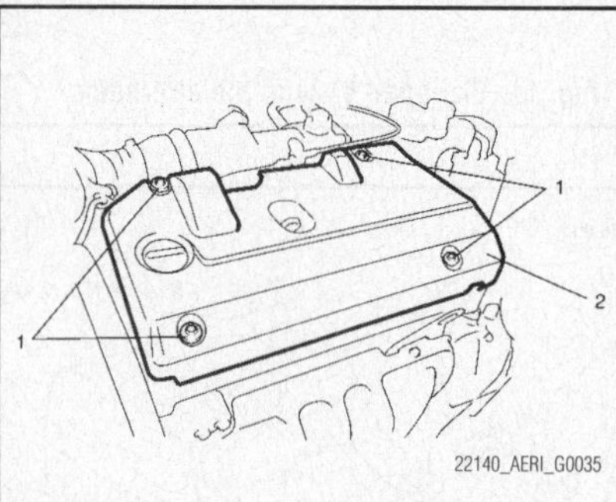

Fig. 53 Removing ignition coil and spark plug engine cover

1. Remove engine cover from cylinder head cover referring to Engine Cover Removal and Installation.
2. Disconnect ignition coil coupler.
3. Remove ignition coil bolt (1), and then pull out ignition coil (2).

FIRING ORDERS

See Figure 55.

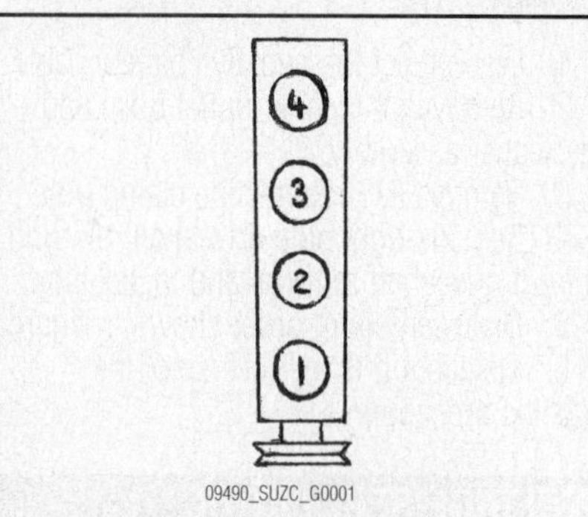

Fig. 55 2.3L engines Firing order: 1–3–4–2 Distributorless ignition system (Coil over each cylinder)

IGNITION TIMING

INSPECTION

See Figures 56 and 57.

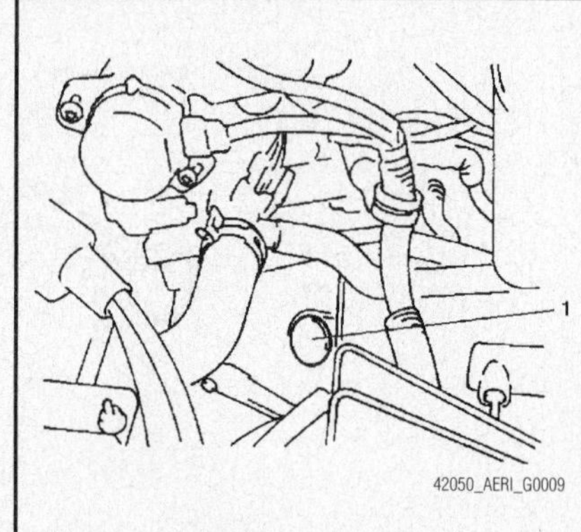

Fig. 56 Locating timing hole cover

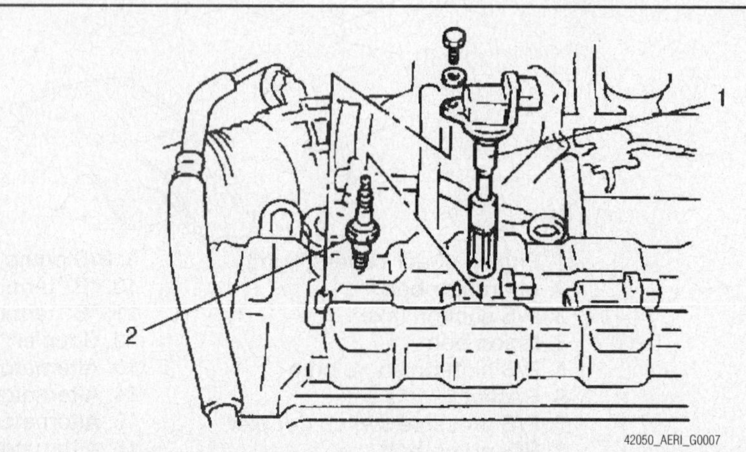

Fig. 54 Removing ignition coil and spark plugs

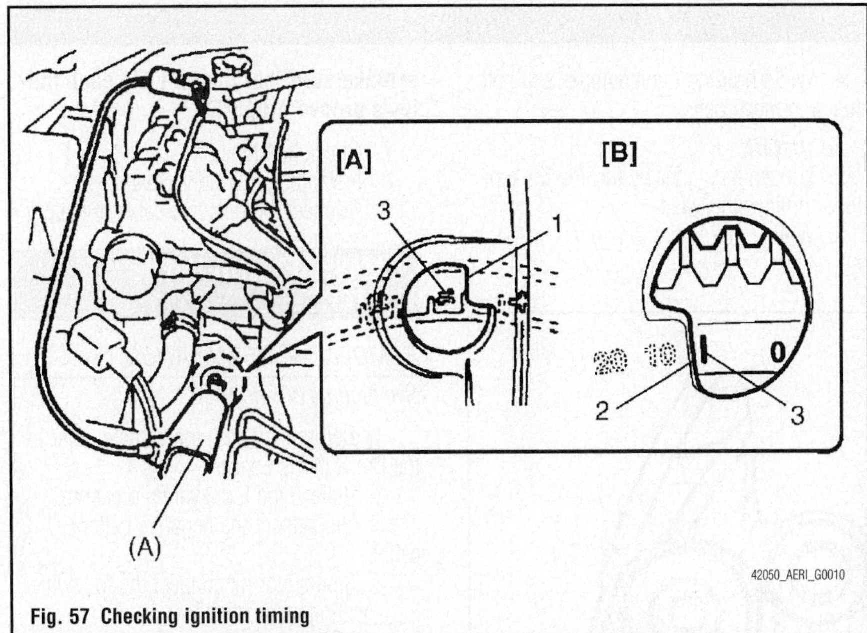

Fig. 57 Checking ignition timing

➡This procedure requires the following special tool, or its equivalent: 09930-76420 (Suzuki scan tool).

➡Before starting engine, place transmission gear shift lever in "Neutral", and set parking brake.

1. Detach timing hole cover (1).
2. Connect SUZUKI scan tool.
3. Start engine and warm it up to normal operating temperature.
4. Make sure that all of electrical loads except ignition are switched off.
5. Check to be sure that idle speed is within specification of 750 +/- 50 RPM.
6. Fix ignition timing by using "FIXED SPARK" mode of SUZUKI scan tool.
7. Set timing light to No.1 ignition coil harness.
8. Using timing light, check that timing is within specification. Initial ignition timing fixed with Suzuki scan tool. 5 +/- 1° BTDC (at idle speed). Ignition order: 1-3-4-2.
9. **Manual transmission:**
   - Timing mark on clutch housing
   - Timing mark (BTDC 5°)
10. **Automatic transmission:**
    - Timing mark on torque converter housing
    - Timing mark (BTDC 5°)

ADJUSTMENT

1. If ignition timing is out of specification, loosen flange bolts (2), adjust timing by turning CMP sensor (1) while engine is running, and then tighten bolts. Tighten CMP sensor bolt to 11 ft. lbs. (15 Nm)
2. After tightening bolts, recheck that ignition timing is within specification.
3. After checking and/or adjusting, release fixed ignition timing and disconnect SUZUKI scan tool from DLC.
4. With engine idling (closed throttle position switch ON and vehicle stopped), check that ignition timing vary more or less of initial ignition timing (about 8°). Also, check that increasing engine speed advances ignition timing. If the check results are not satisfactory, check TP sensor and ECM.

SPARK PLUGS

REMOVAL & INSTALLATION

See Figure 54.

1. Remove engine cover from cylinder head cover.
2. Remove ignition coil by performing the following
   a. Disconnect ignition coil coupler.
   b. Remove ignition coil bolt (1).
   c. Pull out ignition coil.
3. Remove the spark plug (2).

To install:

4. Install spark plug (2). Tighten to 18 ft. lbs. (25 Nm).
5. Install ignition coil (1) by performing the following:
   a. Install the ignition coil (2) securely.
   b. Tighten ignition coil bolt. Then connect the ignition coil coupler.
6. Install engine cover to cylinder head cover.

ENGINE ELECTRICAL

STARTING SYSTEM

STARTER

REMOVAL & INSTALLATION

1. Remove or disconnect the following:
   - Negative battery cable
   - Starter electrical connections
   - 2 starter mounting bolts
   - Starter motor

To install:

2. Install or connect the following:
   - Starter
   - 2 starter mounting bolts and tighten to 96 inch lbs. (10 Nm)
   - Starter electrical connections
   - Negative battery cable
   - Negative Starting motor battery cable nut a: 8 ft. lbs. (11 Nm)
   - Negative Starting motor mount bolt b: 17 ft. lb. (23Nm)

ENGINE MECHANICAL

ACCESSORY DRIVE BELTS

ACCESSORY BELT ROUTING

See Figure 58.

Refer to the accompanying illustration for accessory drive belt routing.

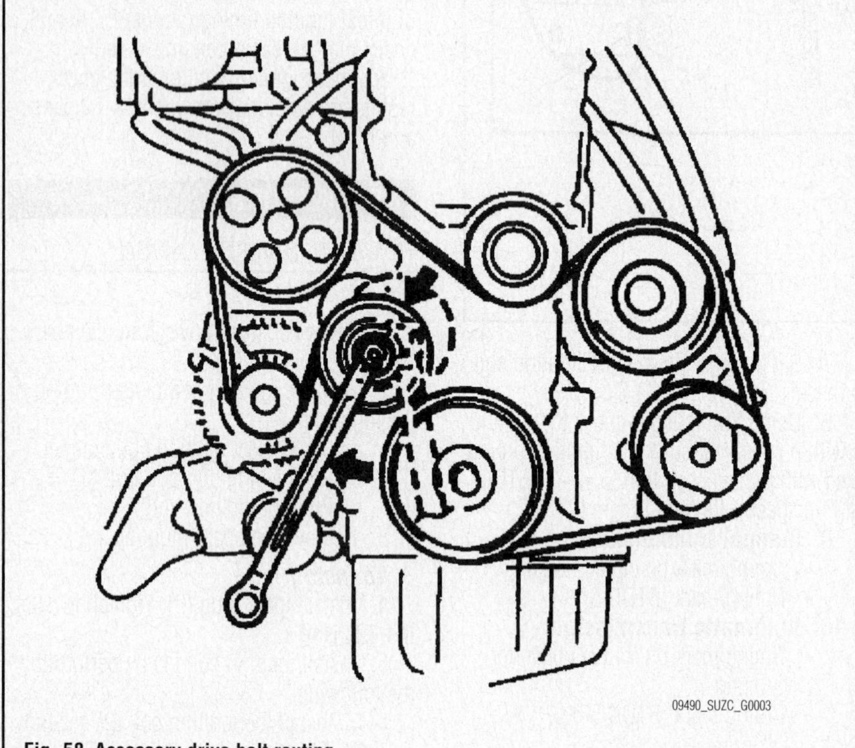

Fig. 58 Accessory drive belt routing

INSPECTION

Inspect the drive belt for signs of glazing or cracking. A glazed belt will be perfectly smooth from slippage, while a good belt will have a slight texture of fabric visible. Cracks will usually start at the inner edge of the belt and run outward. All worn or damaged drive belts should be replaced immediately.

ADJUSTMENT

➡The drive belt is automatically adjusted by a pretention idler pulley. No further adjustment can be made.

REMOVAL & INSTALLATION

See Figure 59.

1. Disconnect the negative battery cable.
2. Remove engine under cover of right side from vehicle body.
3. Loosen tensioner by turning the tensioner pulley clockwise.
4. While holding the tensioner and belt loosen, remove belt.

To install:

5. Loosen tensioner by turning the tensioner pulley clockwise.
6. While holding the tensioner, install belt.

➡Make sure that the belt fits each pulley's groove properly.

7. Check belt tension.
8. Install engine under cover.
9. Connect the negative cable at battery

CAMSHAFT AND VALVE LIFTERS

REMOVAL & INSTALLATION

See Figures 60 and 61.

1. Before servicing the vehicle, refer to the Precautions Section.
2. Relieve the fuel system pressure.
3. Disconnect the negative battery cable.
4. Remove or disconnect the following:
   - Engine assembly
   - Oil pan
   - Timing chains
   - Camshaft position (CMP) sensor
5. To secure work in the following steps, reinstall mounting member, engine front torque bush, rear mounting and rear mounting No. 2 bracket.
6. Set key on crankshaft in position by turning crankshaft. This is to prevent interference between valves and piston.
7. Loosen camshaft housing bolts in such order as indicated in figure and remove them.
   - Camshaft housings
   - Camshafts
   - Valve lash adjusters

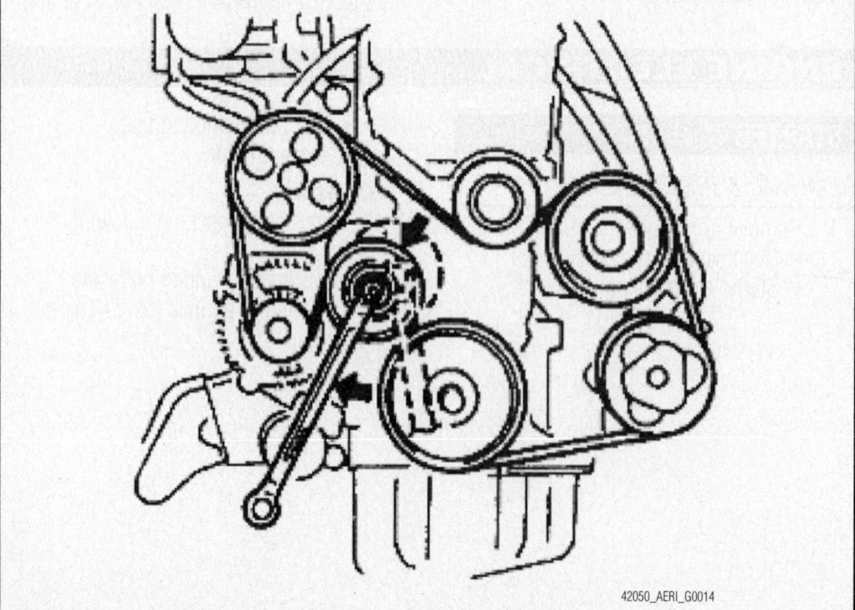

Fig. 59 Removing and installing accessory drive belt

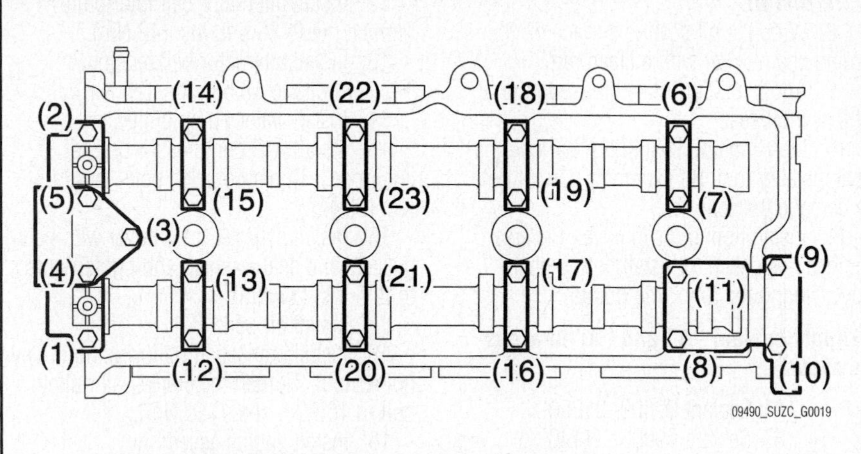

Fig. 60 Loosen the camshaft bearing caps in the sequence shown

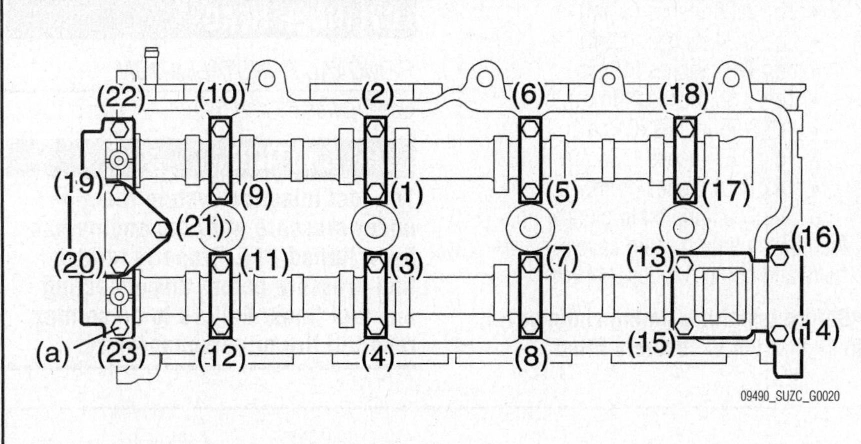

Fig. 61 Camshaft cap bolt tightening sequence

➡Never disassemble hydraulic valve lash adjuster. Don't apply force (1) to body of adjuster; oil in high pressure chamber in adjuster will leak.

8. Immerse removed adjuster in clean engine oil and keep it there until reinstalling it so as to prevent oil leakage. If it is left in air, place it with its bucket body facing down. Don't place on its side or with bucket body facing up.

To install:

9. Before installing valve lash adjuster to cylinder head, fill oil passage of cylinder head with engine oil according to following procedure:
   a. Pour engine oil through oil holes and check that oil comes out from oil holes in sliding part of valve lash adjuster.
   b. Perform this check on both intake and exhaust sides.
10. Install valve lash adjusters to cylinder head. Apply engine oil around valve lash adjuster and then install it to cylinder head.
11. Match matchmark on crank timing sprocket and mating surface of cylinder block and lower crankcase.
12. Lubricate the lobes and journals of the camshaft with clean engine oil.
13. Install or connect the following:
   - Camshafts
14. Apply oil to sliding surface of each camshaft and camshaft journal then install them by aligning match marks on cylinder head and camshafts.

➡Install camshaft in such direction that its end with groove for CMP sensor installation comes to exhaust side.

   - Camshaft housing pins
15. Check position of camshaft housings.
16. Embossed marks are provided on each camshaft housing, indicating position and direction for installation. Install housings as indicated by these marks.
17. Apply sealant to exhaust camshaft end housing sealing surface area.
18. After applying oil to housing bolts, tighten them temporarily first. Then tighten them by following numerical order in figure.
19. Tighten a little at a time and evenly among bolts and repeat tightening sequence two or three times before they are tightened to 8 ft. lbs. (11 Nm).
   - CMP sensor
   - Timing chains
   - Oil pan
   - Engine assembly
   - Negative battery cable
20. Start the engine and check for any water or oil leaks when finished.
21. Check ignition timing as necessary.

➡Don't turn camshafts or start engine (i.e., valves should not be operated) for about half an hour after reinstalling hydraulic valve lash adjusters and camshafts. As it takes time for valves to settle in place, operating engine within half an hour after their installation may cause interference to occur between valves themselves or valves and piston.

CRANKSHAFT DAMPER

REMOVAL & INSTALLATION

1. Remove crankshaft pulley bolt. To lock crankshaft pulley, use special tool 09917-68221 (camshaft pulley-damper assembly holder).

➡Be sure to use the following bolts instead of pins for fixing special tool to crankshaft pulley-damper assembly. Bolt size: M8, P1.25 L = 25 mm (0.98 in.) Strength: 7T

2. Remove crankshaft pulley-damper assembly. To remove crankshaft pulley, use special tools 09944–36011 and 09926–58010 (steering wheel remover and bearing puller attachment).

To install:

3. Install crankshaft pulley-damper assembly using holder tool. Tighten the crankshaft pulley bolt to 108.5 ft. lbs. (150 Nm).

CRANKSHAFT FRONT SEAL

REMOVAL & INSTALLATION

See Figure 62.

1. Before servicing the vehicle, refer to the Precautions Section.
2. Remove engine assembly from vehicle.
3. Remove oil pan.
4. Remove cylinder head cover.

5. Remove timing chain cover as follows:

a. Remove crankshaft pulley bolt. To lock crankshaft pulley, use special tool 09917-68221 (camshaft pulley holder).

➡Be sure to use the following bolts instead of pins for fixing special tool to crankshaft pulley. Bolt size: M8, P1.25 L = 25 mm (0.98 in.) Strength: 7T

b. Remove crankshaft pulley. To remove crankshaft pulley, use special tools 09944–36011 and 09926-58010 (steering wheel remover and bearing puller attachment).

c. Remove A/C compressor bracket.

d. Remove alternator belt idler pulley, water pump pulley and alternator belt tensioner.

e. Remove timing chain cover bolts and nut.

6. Tape the end of a flat-bladed tool to avoid damaging the timing chain cover. Pry out the oil seal using the taped end of the tool.

➡Be careful not to damage the oil seal contact surface when removing or installing the seal.

7. Inspect the oil seal contact surface on the timing chain cover for signs of wear or damage.

To install:

8. Wipe the oil sealing surface on the timing chain cover with a clean rag.

9. Apply multipurpose grease to the lip of a new oil seal.

10. Install the oil seal into place using a seal installer tool. Be extremely careful not to damage the seal.

11. Install timing chain cover. Reverse removal sequence to install timing chain cover noting the following points:

➡Apply sealant "A" and "B" to areas as shown.

- "A": Sealant 99000–31250
- "B": Sealant 99000–31140

a. Apply sealant amount to the following areas.

- "a": 0.12 inches (3mm)
- "b": 0.08 inches (2mm)
- "c": 0.24 inches (6mm)
- "d": 0.63 inches (16mm)
- "e": 0.55 inches (14mm)
- "f": 2.56 inches (65mm)
- "g": 2.87 inches (73mm)
- "h": 0.16 inches (4mm)

b. Apply engine oil to oil seal lip, then install timing chain cover. Tighten bolts and nut to 8 ft. lbs. (11 Nm).

➡Before installing timing chain cover, check that pin is securely fitted.

12. Install alternator belt idler pulley. Tighten nut to 30.5 ft. lbs. (42 Nm).

13. Install alternator belt tensioner. Tighten bolts to 18.5 ft. lbs. (25 Nm).

14. Install water pump pulley.

15. Install A/C compressor bracket (if equipped). Tighten bracket bolts to 40 ft. lbs. (55 Nm).

16. Install cylinder head cover with new gaskets and seal washers and tighten nuts to 8 ft. lbs. (11 Nm).

17. Install oil pan.

18. Install crankshaft pulley using pulley holder tool. Tighten the crankshaft pulley bolt to 108.5 ft. lbs. (150 Nm).

19. Install engine assembly.

20. Install all remaining components in the reverse order of the removal procedure.

CYLINDER HEAD

REMOVAL & INSTALLATION

See Figures 63 and 64.

⁂ CAUTION

The fuel injection system remains under pressure after the engine has been turned OFF. Properly relieve fuel pressure before disconnecting any fuel lines. Failure to do so may result in fire or personal injury.

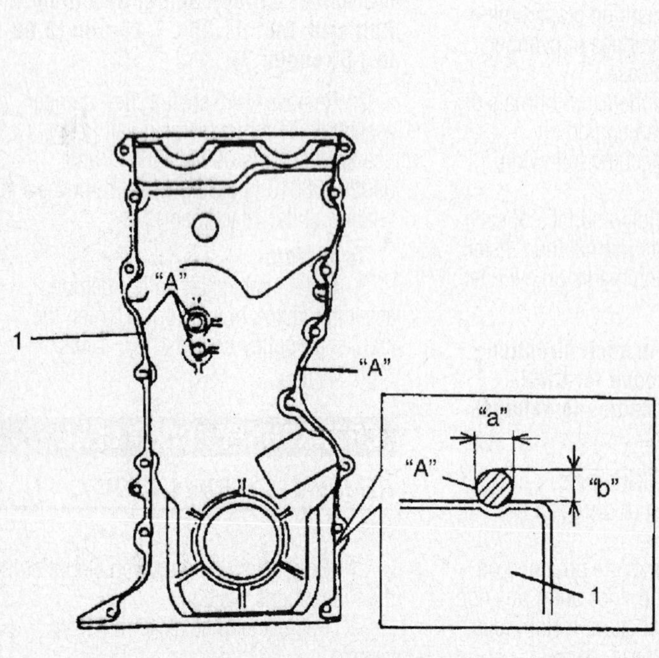

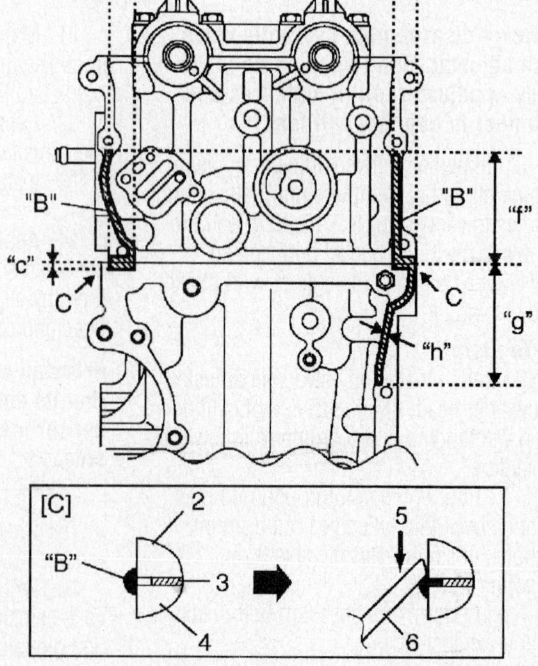

Fig. 62 Sealant application on timing belt cover

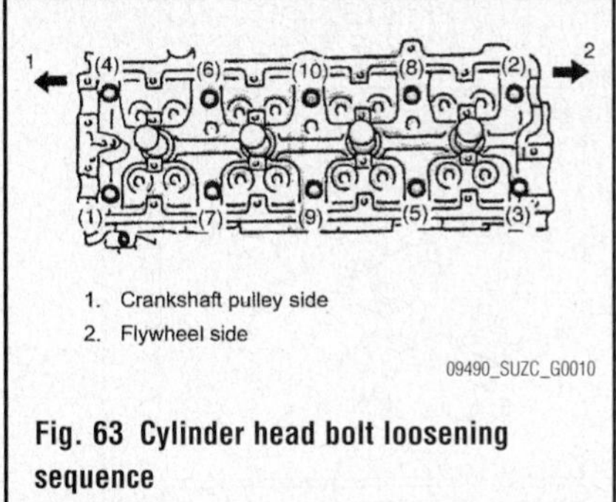

Fig. 63 Cylinder head bolt loosening sequence

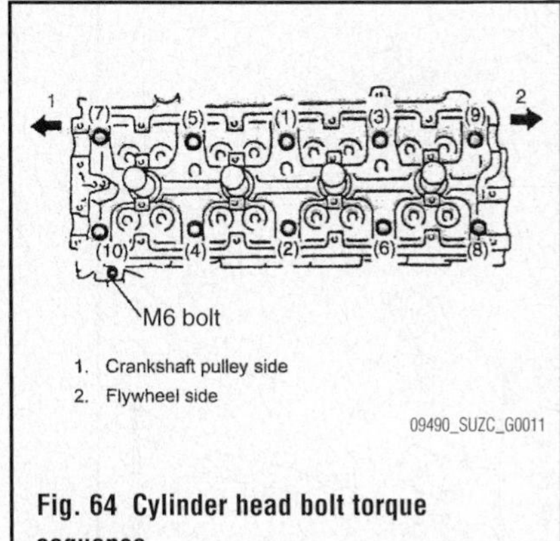

Fig. 64 Cylinder head bolt torque sequence

1. Before servicing the vehicle, refer to the Precautions Section.
2. Disconnect the negative battery cable.
3. Relieve the fuel system pressure.
4. Drain the cooling system.
5. Remove or disconnect the following:
   - Engine assembly
   - Timing chains
   - Camshafts and valve lash adjusters
   - Intake manifold rear stiffener
   - Coolant pipe from intake manifold
   - Right engine mounting bracket
   - Power steering pump bracket
6. Loosen the cylinder head bolts in reverse order of tightening including the M6 bolt, located on the bottom left portion of the cylinder head. Once each bolt is loose, remove the bolts from the cylinder head.
7. Check to be sure all components are removed or disconnected before removing the cylinder head.
8. Remove the cylinder head with the intake manifold, exhaust manifold and water outlet cap as an assembly.
9. Clean the cylinder block and cylinder head mating surfaces of any old gasket material, oil or dust and clean any engine coolant from the cylinders.

To install:

10. Match the matchmark on the crank timing sprocket and the mating surface of the cylinder block and lower crankcase.
11. Install or connect the following:
    - Knock pins to cylinder block
    - New cylinder head gasket with the top mark facing up and toward the crankshaft pulley
    - Cylinder head to the engine block
12. Apply engine oil to the bolt threads and tighten the cylinder head bolts, in sequence, using the following 6 Steps:
    a. Step 1: 38.5 ft. lbs. (53 Nm)
    b. Step 2: 61 ft. lbs. (84 Nm)
    c. Step 3: Loosen all of the cylinder head bolts
    d. Step 4: 38.5 ft. lbs. (53 Nm)
    e. Step 5: 76 ft. lbs. (105 Nm)
    f. Step 6: Bolt M6 (bottom left of cylinder head) - 8 ft. lbs. (11 Nm)
    - Right engine mounting bracket with new mounting bolts. Tighten to 40 ft. lbs. (55 Nm).
    - Power steering pump bracket
    - Coolant pipe from intake manifold
    - Intake manifold rear stiffener
    - Camshafts and valve lash adjusters
    - Timing chains
    - Engine assembly
    - All remaining components in the reverse order of removal
13. Refill the cooling system with coolant.
14. Connect the negative battery cable.
15. Start the engine and check for leaks.

ENGINE ASSEMBLY

REMOVAL & INSTALLATION

See Figure 66.

**** CAUTION**

The fuel injection system remains under pressure after the engine has been turned OFF. Properly relieve fuel pressure before disconnecting any fuel lines. Failure to do so may result in fire or personal injury.

1. Before servicing the vehicle, refer to the Precautions Section.
2. Properly relieve the fuel system pressure.
3. Disconnect the negative battery cable.
4. Remove engine hood after disconnecting windshield washer hose.
5. Drain cooling system.
6. Remove radiator with cooling fan.
7. Remove top engine cover.
8. Remove air cleaner outlet hose.
9. Remove air cleaner case assembly.
10. Disconnect the following cables:
    a. Accelerator cable from throttle body.
    b. Shift and select cable from transmission (M/T).
    c. Gear select cable from transmission (A/T).
11. Remove accelerator cable with its bracket from intake manifold.
12. Remove or disconnect the following:
    - Brake booster hose from intake manifold
    - Heater hose from water outlet cap
    - Fuel feed hose from delivery pipe
    - Vacuum hose from EVAP canister purge valve
    - Fuel return hose from return pipe
    - Heater outlet hose from heater outlet pipe
    - P/S suction hose and high pressure pipe from P/S pump
    - Injector wire
    - Camshaft position (CMP) sensor
    - Ignition coil wire
    - Throttle position (TP) sensor
    - Mass airflow (MAF) sensor
    - Idle air control (IAC) valve
    - Manifold absolute pressure (MAP) sensor
    - Crankshaft position (CKP) sensor
    - Ground terminals from throttle body
    - EVAP canister purge valve
    - EGR valve
    - Heated oxygen sensor
    - Engine coolant temperature (ECT) sensor
    - Alternator
    - Starter
    - Knock sensor
    - Oil pressure switch
    - Cylinder block heater (if equipped)
    - Power steering pressure switch
    - A/C magnetic switch (if equipped)
    - Back-up light switch (M/T)
    - Vehicle speed sensor
    - Input shaft speed sensor (A/T)
    - Output shaft speed sensor (A/T)
    - Battery negative cable from transmission
    - Transmission range switch (A/T)
    - Valve body and trans fluid temperature sensor connector (A/T)
13. Remove clutch operating cylinder from transmission with hose still attached (M/T).

➡Suspend removed clutch operating cylinder at a place free from any possible damage during removal and installation of engine assembly.

14. Remove right and left engine under covers.
15. Remove alternator belt.

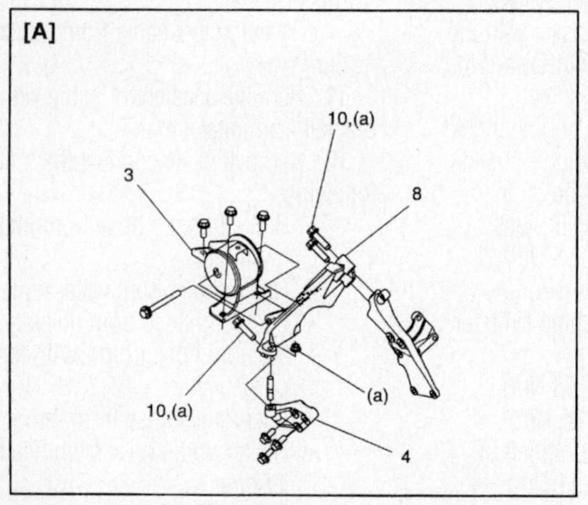

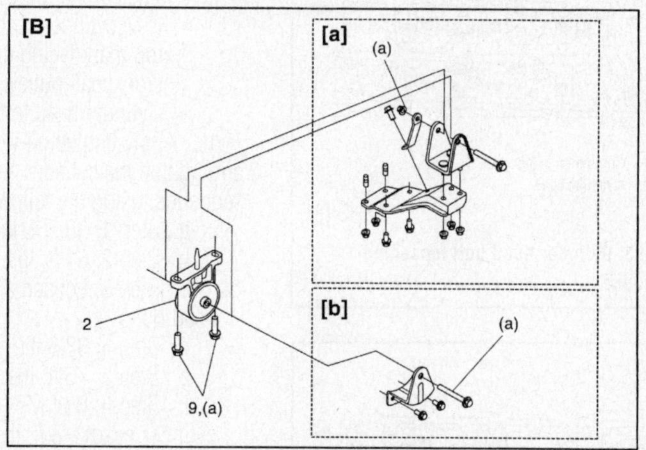

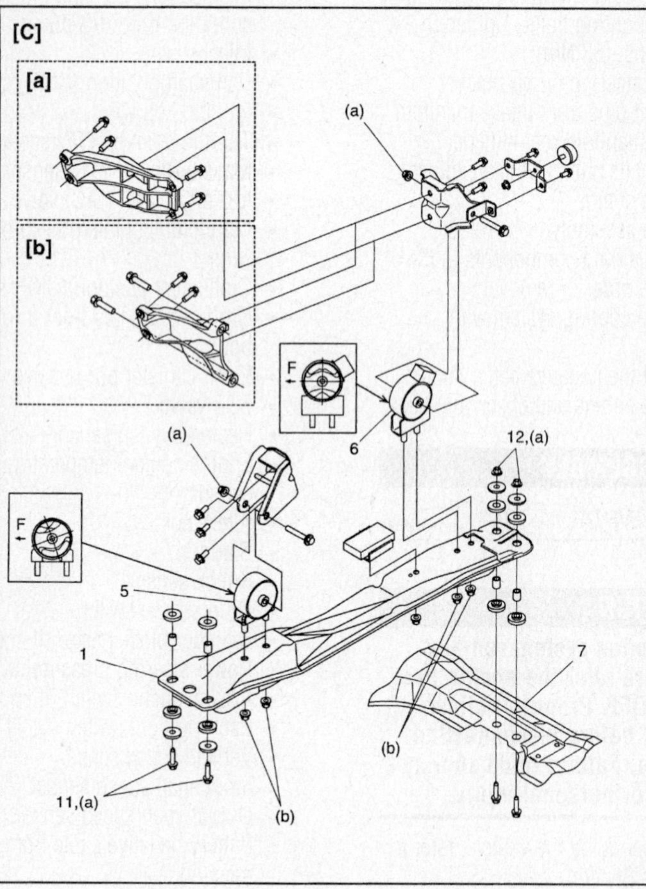

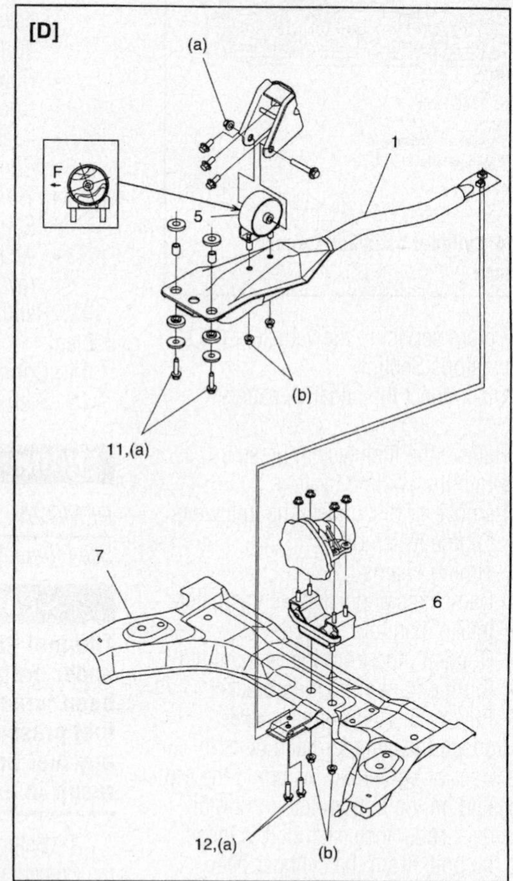

Fig. 65 Engine mounting components

16. Remove exhaust No.1 and No. 2 pipe.
17. Drain engine and transaxle oil.
18. Drain transfer case oil (if equipped).
19. Remove driveshafts.
20. Remove propeller shaft (AWD vehicle).
21. Remove A/C compressor from compressor bracket with hoses still attached.

➡Suspend removed compressor at a place free from any possible damage during removal and installation of engine assembly.

22. Remove intake manifold intake stiffener.
23. Install lifting device.
24. Remove mounting member from front member and suspension frame.
25. Remove suspension frame with stabilizer bar and suspension control arms (AWD).
26. Remove engine left mounting bolts from body.
27. Remove engine right mounting bolts from right mounting bracket.
28. With P/S hose connected, detach P/S pump from its bracket (2.0L).

➡Suspend removed P/S pump at a place free from any possible damage during removal and installation of engine assembly.

Before removing engine with transaxle and transfer case (AWD) from vehicle body, recheck to make sure all hoses, electric wires and cables are disconnected from the full assembly.

29. Lower engine with transaxle and transfer case (if equipped) from vehicle body.

➡Before lowering engine, to avoid damage to A/C compressor, raise it through clearance made on engine crankshaft pulley side. At this time, use care so that no excessive force is applied to hoses.

30. Disconnect the transfer case from the transaxle assembly (AWD vehicles).
31. Disconnect transaxle from engine.
32. Remove clutch cover and clutch disc.

To install:

33. Install or connect the following:
    - Clutch cover and clutch disc. Torque the clutch cover bolts to 17 ft. lbs. (23 Nm).
    - Transaxle to engine assembly
    - Transfer case assembly to transaxle
34. Lift engine with transaxle and transfer case (if equipped) into engine compartment, but do not remove lifting device.
35. Install the engine mounts as follows:
    a. Install engine right mounting No. 1 bracket (8) to right mounting bracket (4), alternator & P/S bracket and engine right mounting (3) with temporal tightening its bolts.
    b. Install engine left mounting bolts (9) with temporal tightening its nut.
    c. Install mounting member (1) to front member and suspension frame (7).
    d. Install suspension frame (7) with stabilizer bar and suspension control arms for AWD model.
    e. Tighten the engine mounting bolts and nuts A to 40 ft. lbs. (55 Nm); and engine front and rear mounting nuts B to 32.5 ft. lbs. (45 Nm).
    f. Be sure to tighten engine right mounting No.1 bracket bolt (10) previously, and tighten engine left mounting bolts (9).
    g. Be sure to tighten mounting member to suspension frame bolts (12) or nuts (12) previously, and tighten mounting member to body bolts (11).
36. Remove lifting device.
37. Push in each driveshaft joint fully so that snap ring engages with differential gear or center bearing support. Use care not to damage oil seal lip when inserting.
38. Clamp electric wire securely.
    - Propeller shaft (AWD)
    - Gear shift control cable
    - Alternator belt
39. Refill the transaxle and transfer case (AWD) with the correct amount and type of fluid.
40. Install the remaining components in the reverse order of removal.
41. Adjust the accelerator cable free-play as follows:
    a. With accelerator pedal depressed fully, check clearance between throttle lever and lever stopper (throttle body).
    b. The clearance between the throttle lever and lever stopper (throttle body) should be within 0.02–0.07 inches (0.5–2.0mm).
    c. If measured value is out of specification, adjust it to specification with cable adjusting nut.
42. Fill the engine with engine oil and the cooling system with coolant.
43. Fill the power steering reservoir and bleed the power steering system.
44. Run the engine and verify that there are no fuel, coolant, transaxle or exhaust leaks.

EXHAUST MANIFOLD

REMOVAL & INSTALLATION

See Figure 66.

✲✲ CAUTION

To avoid the danger of being burned, do not service the exhaust system while it is hot. Service should be performed only after the system cools down.

1. Before servicing the vehicle, refer to the Precautions Section.
2. Remove or disconnect the following:
    - Negative battery cable
    - Heated Oxygen (HO_2S) sensor electrical connector and A/C magnet clutch connector
    - Exhaust manifold cover
    - 2 bolts attaching the exhaust pipe to the exhaust manifold
    - Engine under cover (right side)
    - Exhaust pipe stiffener
    - Exhaust No.1 pipe
    - Air cleaner outlet hose and air cleaner case assembly
    - Wire harness clamp
3. Support engine with engine support jack and remove engine right mounting No.1 bracket.
    - Engine right mounting bracket
    - Exhaust manifold mounting nuts and bolts
    - Exhaust manifold and the gasket

➡Be careful not to damage fins of radiator.

To install:

4. Install or connect the following:
    - New gasket to the cylinder head
    - Exhaust manifold. Tighten manifold bolts and nuts in sequence to 40 ft. lbs. (55 Nm).
    - Engine right mounting bracket and new bracket bolts. Tighten the bracket bolts to 40 ft. lbs. (55 Nm).

➡Use new engine right mounting bracket bolt. Reusing bolt may result exhaust leakage.

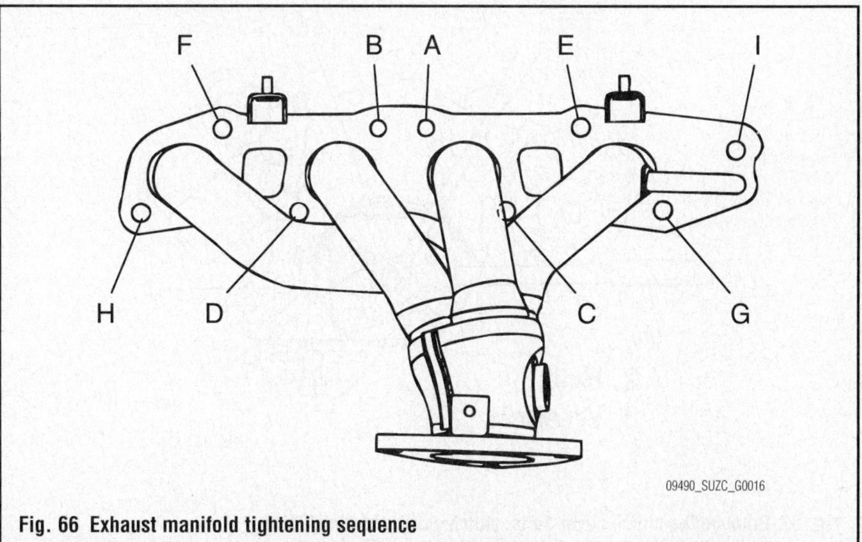

Fig. 66 Exhaust manifold tightening sequence

- Engine right mounting No.1 bracket. Tighten the bolts and nut to 40 ft. lbs. (55 Nm).
- Air cleaner case assembly and air cleaner outlet hose
- New pipe gasket and exhaust No.1 pipe. Tighten the exhaust manifold-to-exhaust pipe nut to 40 ft. lbs. (55 Nm); and the No.1-to-No. 2 exhaust pipe bolt to 31.5 ft. lbs. (43 Nm).
- Exhaust pipe stiffener and tighten exhaust pipe stiffener bolts at the exhaust pipe first, and at the engine next to 40 ft. lbs. (55 Nm)
- Engine under cover (right side)
- Exhaust manifold cover
- Heated Oxygen (HO_2S) sensor electrical connector and A/C magnet clutch connector
- Remaining components in the reverse order of removal

5. Connect the negative battery cable.
6. Run the engine and check for exhaust leaks.

FLYWHEEL/FLEXPLATE

REMOVAL & INSTALLATION

Flywheel

See Figures 67 through 72.

1. Dismount the manual transaxle.
2. Hold flywheel stationary with special tool 09924-17811 and remove clutch cover bolts (2), clutch cover (1) and clutch disc.
3. When pulling out input shaft bearing (1) from flywheel (2), use special tools 09921-20210 and 09930-30104.

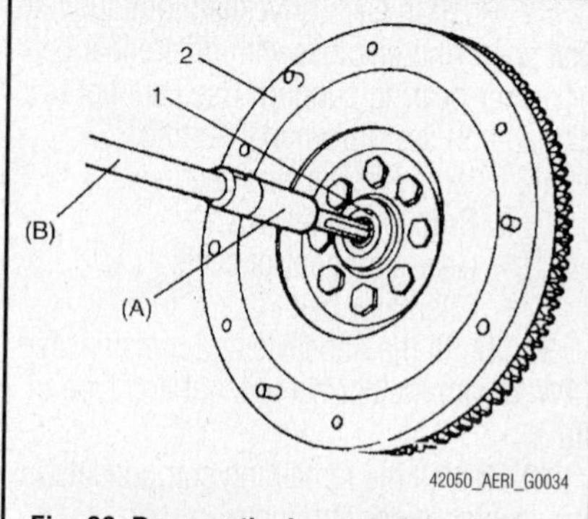

Fig. 68 Remove the input shaft bearing from flywheel with special tools

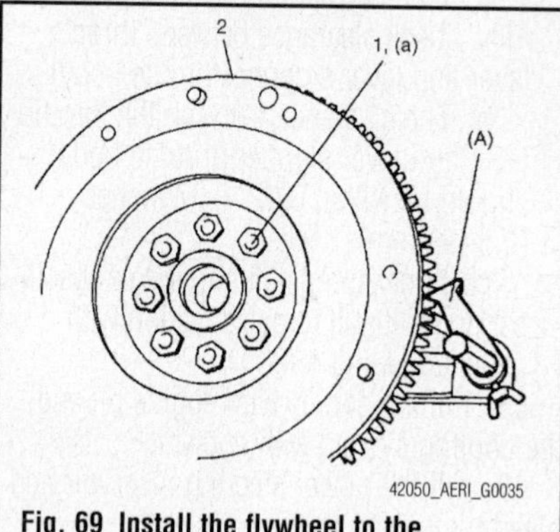

Fig. 69 Install the flywheel to the crankshaft

To install:

➡Before assembling, make sure that flywheel surface and pressure plate surface have been cleaned and dried thoroughly.

4. Install flywheel (2) to crankshaft and tighten bolts (1) to 51 ft. lbs. (70 Nm).

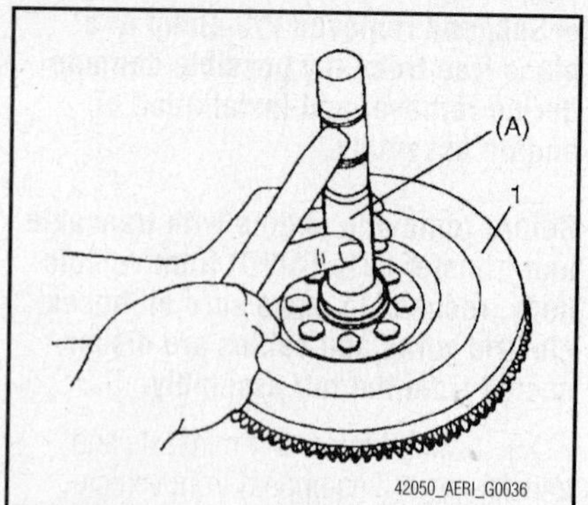

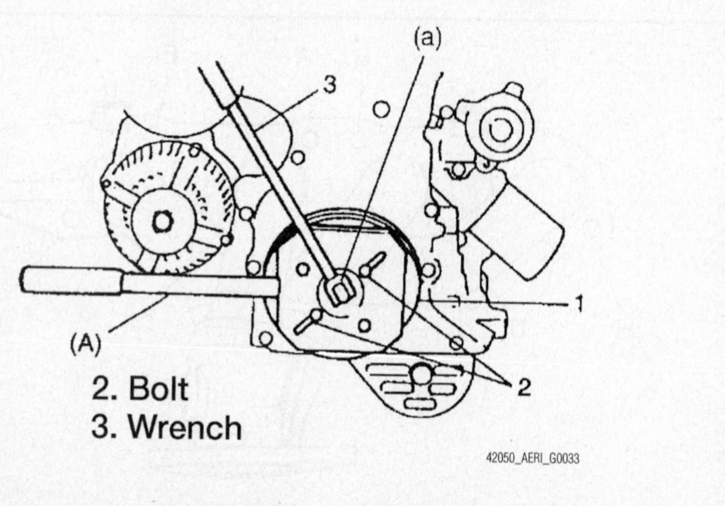

Fig. 67 Remove the clutch cover bolts, clutch cover and clutch disc

Fig. 70 Install the input shaft bearing to the flywheel

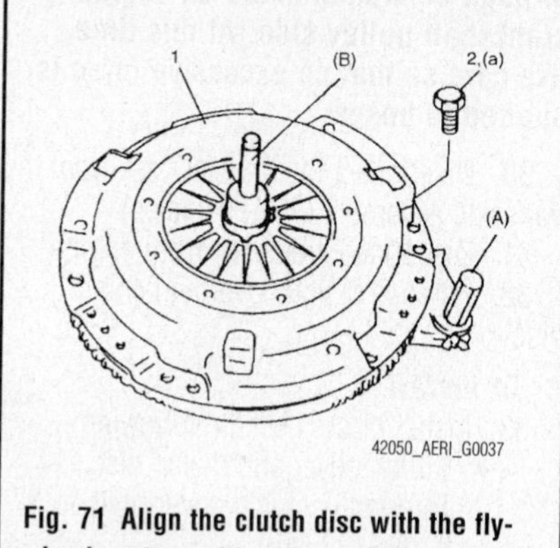

Fig. 71 Align the clutch disc with the flywheel center and install the clutch cover and bolts

5. Using special tool 09913-76010, install input shaft bearing to flywheel (1).
6. Align clutch disc with flywheel center by using special tool, and install clutch cover (1) and bolts (2). Then tighten bolts (2) to 17 ft. lbs. (23 Nm).

➡While tightening clutch cover bolts, compress clutch disc with special tool (clutch center guide) by hand so that the disc is centered.

➡Tighten cover bolts little by little evenly in diagonal order.

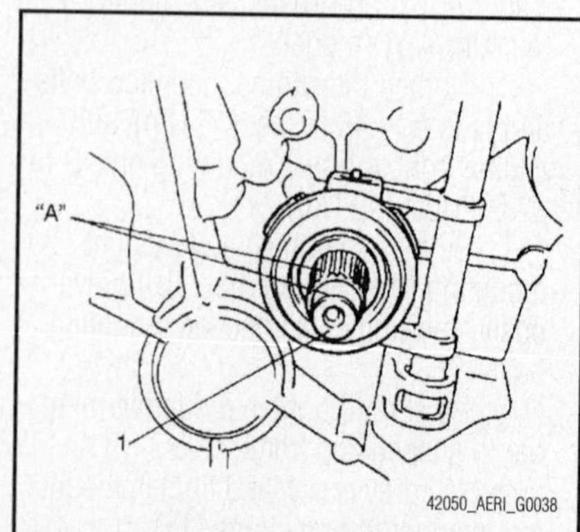

Fig. 72 Apply grease to the input shaft

7. Slightly apply grease (A: Grease 99000–25210) to the input shaft (1), then join the transaxle assembly with to the engine.

➡**When inserting transaxle input shaft to the clutch disc, turn the crankshaft little by little to match the spline mesh.**

INTAKE MANIFOLD

REMOVAL & INSTALLATION

See Figure 73.

✲✲ CAUTION

The fuel system pressure must be relieved before disconnecting any fuel lines. Failure to do so may result in personal injury.

1. Before servicing the vehicle, refer to the Precautions Section.
2. Properly relieve the fuel system pressure.
3. Disconnect the negative battery cable.
4. Drain the coolant from the vehicle.

✲✲ CAUTION

To help avoid the danger of being burned, do not remove the drain plug and the radiator cap while the engine is still hot. Scalding fluid and steam can be blown out under pressure if the plug and cap are taken off too soon.

5. Remove the engine cover.
6. Remove the air cleaner outlet hose.

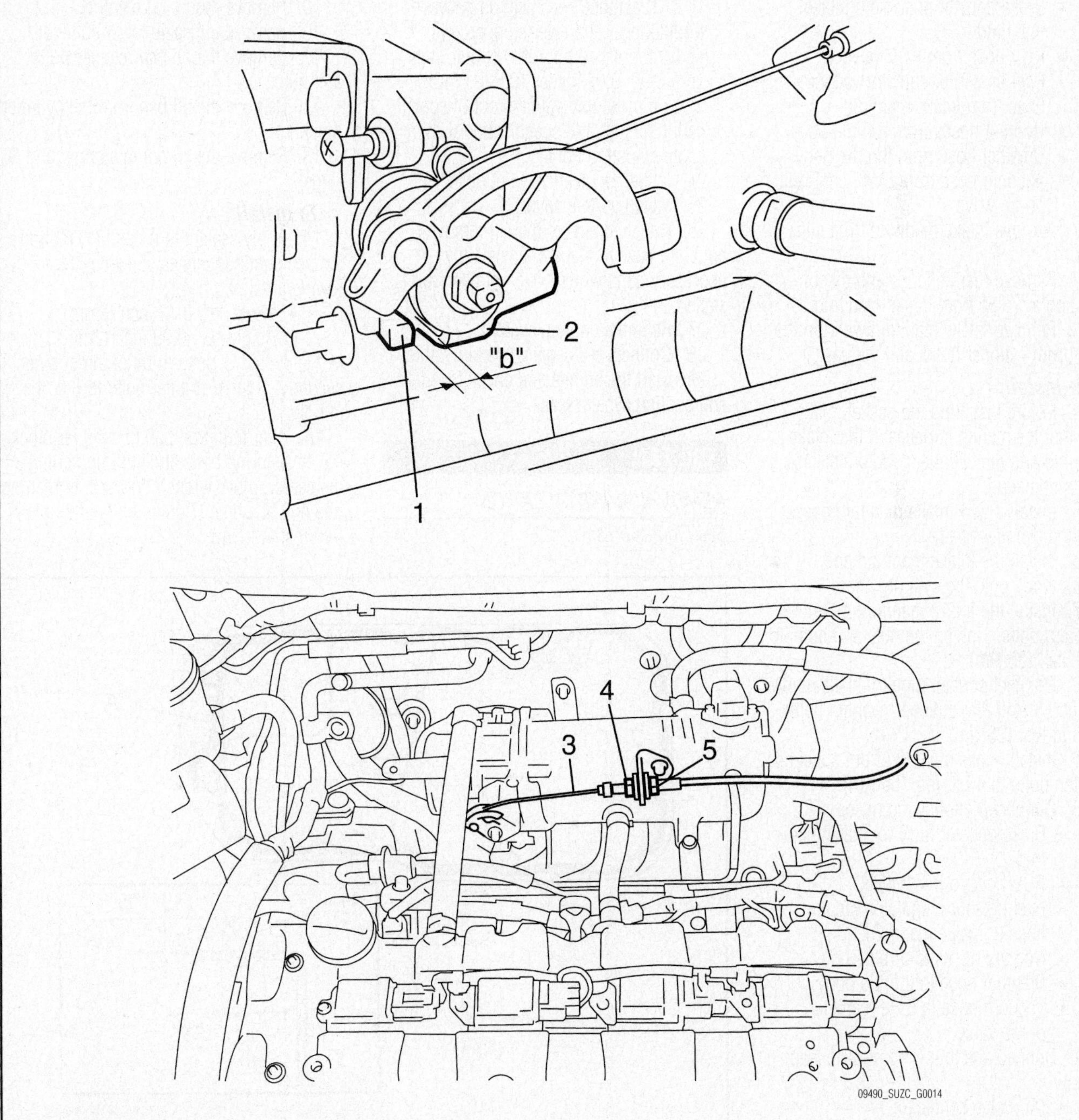

Fig. 73 Proper adjustment of the accelerator cable

7. Disconnect the accelerator cable from throttle valve lever.

8. Remove the accelerator cable, with cable bracket from intake manifold.

9. Disconnect the following electric lead wires:

- IAC valve connector
- TP sensor connector
- EGR valve connector
- EVAP canister purge valve connector
- MAP sensor connector
- Ground terminal from throttle body

10. Disconnect the following hoses:

- Brake booster hose from intake manifold
- PCV hose from PCV valve
- Fuel pressure regulator vacuum hose from intake manifold
- Coolant hoses from throttle body
- Breather hose from throttle body
- Vacuum hose from EVAP canister purge valve

11. Remove intake manifold front stiffener.

12. Remove rear stiffener, and disconnect coolant pipe from intake manifold.

13. Remove intake manifold with throttle body from cylinder head, and the gasket.

To install:

14. Before installing the gasket, make sure that the mating surfaces of the intake manifold and the cylinder head are clean and undamaged.

15. Install a new intake manifold gasket onto the cylinder head.

16. Install the intake manifold and throttle body onto the cylinder head.

17. Install the intake manifold mounting nuts and bolts. Tighten the nuts and bolts to 18 ft. lbs. (25 Nm).

18. Connect coolant pipe to intake manifold and install rear stiffener. Tighten bolts to 18 ft. lbs. (25 Nm).

19. Install intake manifold front stiffener. Tighten bolts to 40 ft. lbs. (55 Nm).

20. Connect the following hoses:

- Brake booster hose to intake manifold
- PCV hose to PCV valve
- Fuel pressure regulator vacuum hose to intake manifold
- Coolant hoses to throttle body
- Breather hose to throttle body
- Vacuum hose to EVAP canister purge valve

21. Connect the following electric lead wires:

- IAC valve connector
- TP sensor connector
- EGR valve connector
- EVAP canister purge valve connector
- MAP sensor connector
- Ground terminal to throttle body

22. Connect accelerator cable to throttle valve lever and install its bracket to the intake manifold.

23. Adjust accelerator cable play as follows:

a. With accelerator pedal depressed fully, check clearance between throttle lever (2) and lever stopper (1) (throttle body) which should be within specification.

b. Clearance specification between throttle lever and lever stopper (with pedal depressed fully) "b" should measure 0.02–0.07 inches (0.5–2.0mm).

c. If measured value is out of specification, adjust it to specification with cable adjusting nut (4).

24. Install air cleaner outlet hose.

25. Install engine cover.

26. Check to ensure that all removed parts are back in place. Reinstall any necessary parts which have not been reinstalled.

27. Refill the cooling system.

28. Connect the negative battery cable.

29. Start the engine and check for fuel and cooling system leaks.

OIL PAN

REMOVAL & INSTALLATION

See Figures 65 and 74.

1. Before servicing the vehicle, refer to the Precautions Section.

2. Disconnect the negative battery cable.

3. Remove the oil level gauge.

4. Raise and safely support the front of the vehicle.

5. Drain engine oil by removing drain plug.

6. Remove engine under cover(s).

7. Remove exhaust No.1 pipe.

8. Support engine and transmission with engine support jack for 2WD vehicle.

9. Remove mounting member.

10. Remove transaxle lower stiffener.

11. Remove the oil pan retainer bolts and nuts.

12. Remove the oil pan from the cylinder block.

13. Remove the oil pump strainer and O-rings.

To install:

14. Apply sealant to oil pan (1) mating surface continuously as shown in the following amount:

- Width "a": 0.12 inch (3mm)
- Height "b": 0.08 inch (2mm)

15. Install O-rings to oil pump strainer securely. Tighten strainer bolts to 8 ft. lbs. (11 Nm).

16. After fitting oil pan to cylinder block, run in securing bolts and start tightening at the center: move wrench outward, tightening one bolt at a time. Tighten bolts and nuts to 8 ft. lbs. (11 Nm).

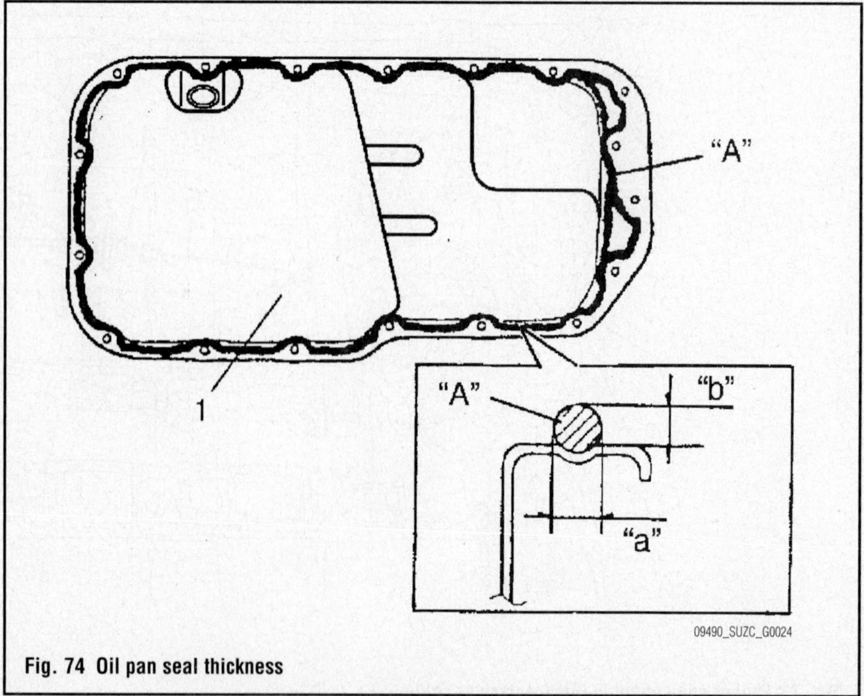

Fig. 74 Oil pan seal thickness

17. Install gasket and drain plug to oil pan after applying engine oil. Tighten drain plug to 36.5 ft. lbs. (50 Nm).
18. Install transmission lower stiffener.
19. Install mounting member for 2WD vehicle. Tighten the engine mounting bolt and nut to 40 ft. lbs. (55 Nm); and tighten the engine front and rear mounting nut to 32.5 ft. lbs. (45 Nm).
20. Remove engine support jack for 2WD vehicle.
21. Install exhaust No.1 pipe.
22. Install engine under cover(s).
23. Install oil level gauge.
24. Refill the engine with oil.
25. Connect the negative battery cable.
26. Start the engine and check for leaks.

OIL PUMP

REMOVAL & INSTALLATION

See Figure 75.

1. Before servicing the vehicle, refer to the Precautions Section.
2. Disconnect the negative battery cable.
3. Drain engine oil.
4. Remove oil pan and oil pump strainer.
5. Remove oil pump sprocket cover.
6. Remove baffle plate from lower crank case.
7. Remove oil pump with sprocket from lower crankcase.

To install:

8. Install oil pump and baffle plate to lower crank case. Tighten the oil pump mounting bolts to 18 ft. lbs. (25 Nm). Tighten the baffle plate bolts to 8 ft. lbs. (11 Nm).

➡When installing oil pump, be careful not to allow pins to fall off.

9. Install oil pump sprocket cover and tighten bolts to 8 ft. lbs. (11 Nm).
10. Install oil pan and oil pump strainer.
11. Refill engine with engine oil.
12. Connect the negative battery cable.
13. Start the engine and check the engine oil pressure.
14. Check that no leaks are present.

INSPECTION

See Figures 76 and 77.

Inspect the following:

- Check outer rotor (3), inner rotor and oil pump cases (1), (2) for excessive wear or damage.

If abnormal condition is found in above checks, replace oil pump assembly.

- Check relief valve (4) for excessive wear or damage. If abnormal condition is found in above checks, replace oil pump relief valve set.
- Measure free length and tension of oil relief spring. If the measured values of length or tension is less than the specification, replace oil pump relief valve set.
- Measure clearance of oil pump rotor and oil pump case.

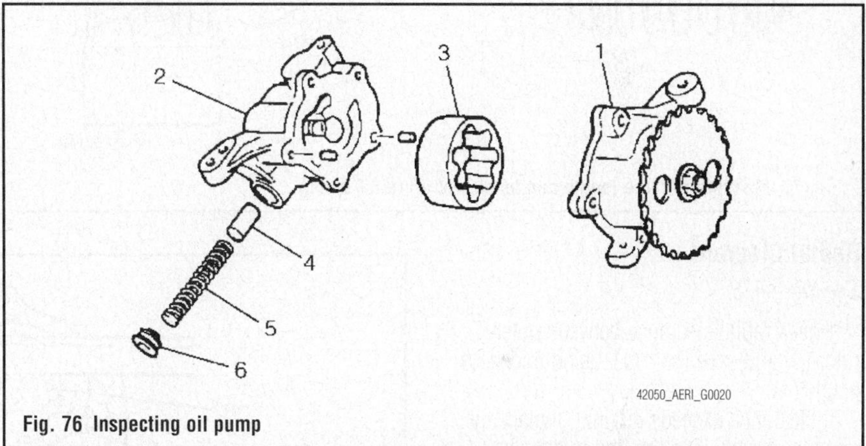

Fig. 76 Inspecting oil pump

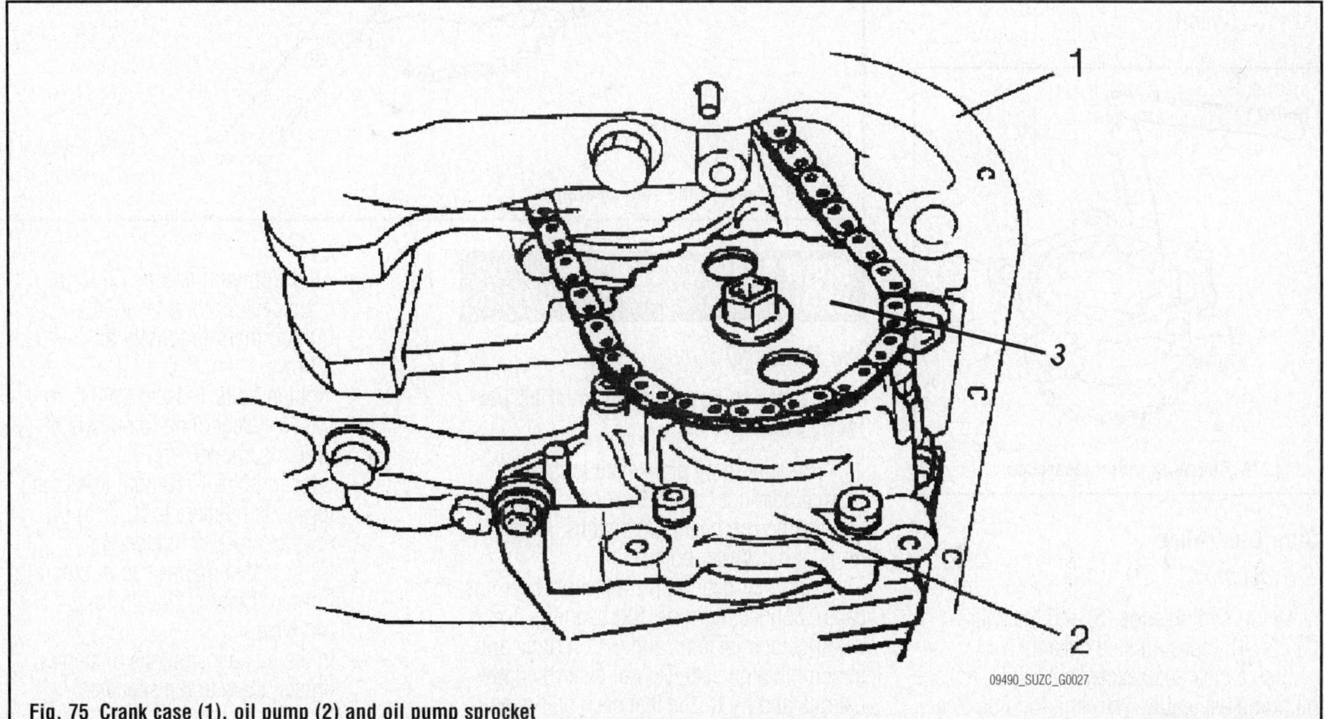

Fig. 75 Crank case (1), oil pump (2) and oil pump sprocket

Item	Standard
Spring free length	63.5 mm (2.5 in.)
Spring preload	85.0 N for 52.0 mm (8.5 kg for 52.0 mm, 19.0 lb/2.05 in.)

42050_AERI_G0021

Fig. 77 Measure the free length and tension of oil relief spring

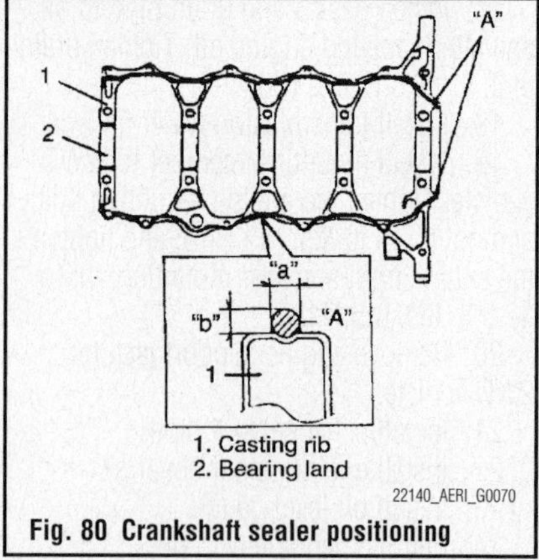

Fig. 80 Crankshaft sealer positioning

- Clean mating surface of cylinder block and lower crankcase, remove oil, old sealant and dust from mating surface.

1. Apply sealant water tight sealant (99000-31250SUZUKI Bond No.1207F) to lower crankcase (1) mating surface area as shown in figure. Sealant amount for lower crankcase:
   - Width "a": 3 mm (0.12 in.)
   - Height "b": 2 mm (0.08 in.)
   - Install lower crankcase (1) to cylinder block.

Radial Clearance

See Figure 78.

Check radial clearance between outer rotor (2) and case No.1 (1), using thickness gauge (4).

If clearance exceeds its limit, replace oil pump assembly. The limit on radial clearance between outer rotor and case is 0.0079 inch (0.20 mm).

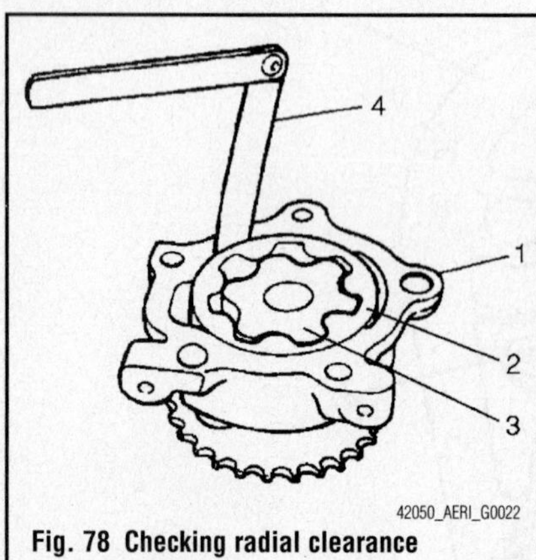

Fig. 78 Checking radial clearance

Side Clearance

See Figure 79.

Using straightedge (5) and thickness gauge (4), measure side clearance.

If side clearance exceeds its limit, replace oil pump assembly. The limit for side clearance is 0.0043 inch (0.11 mm).

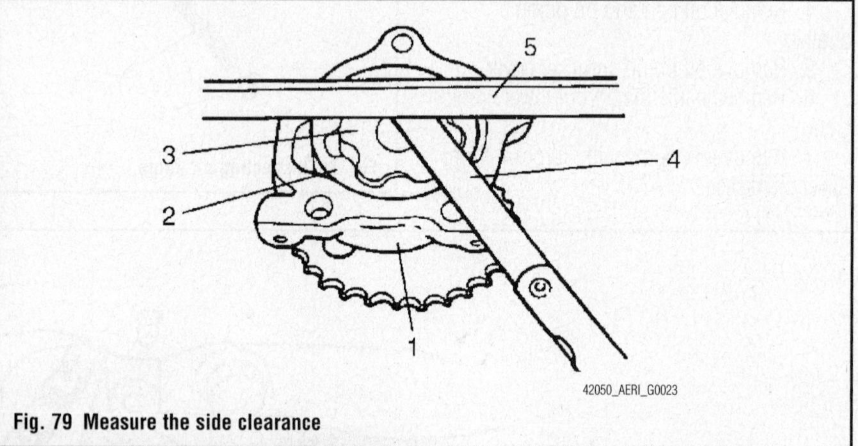

Fig. 79 Measure the side clearance

MAIN BEARING TORQUE SEQUENCE

See Figures 80 and 81.

➡All parts to be installed must be perfectly clean.

- Be sure to oil crankshaft journals, journal bearings, thrust bearings, crankpins, connecting rod bearings, pistons, piston rings and cylinder bores.
- Journal bearings, crankcase (bearings caps), connecting rods, rod bearings, rod bearing caps, pistons and piston rings are in combination sets. Do not disturb combinations and try to see that each part goes back to where it came from, when installing.

- After applying engine oil to all crankcase bolts ((1) –(22)), tighten them gradually as follows.
- Tighten bolts 1–10 to 29.0 ft. lb. (40 Nm).according to numerical order as shown.
- Loosen bolts 1–10 until tightening torque is reduced to 0 (zero) in reverse order of tightening.
- In the same manner as in step a), tighten them to 29.0 ft. lb. (40 Nm).
- In the same manner as in step a), tighten them to the specified torque.

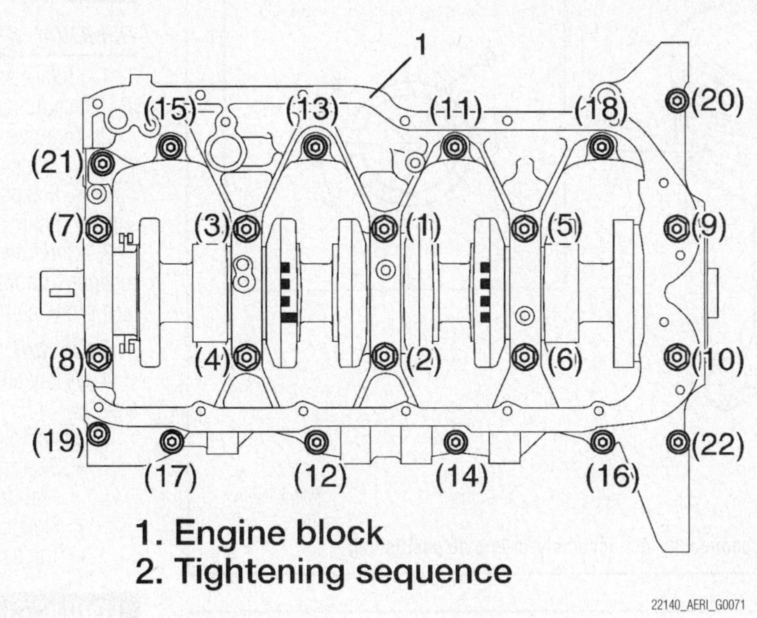

Fig. 81 Crankshaft tightening sequence

- Tighten bolts 11–22 to the specified torque according to numerical order as follows:
- Crankcase bolt with 10 mm thread diameter (1–10): 42 ft. lbs. (58 Nm).
- Crankcase bolt with 8 mm thread diameter (11–22): 18 ft. lbs. (25 Nm).

➡**After tightening crankcase bolts, check to be sure that crankshaft rotates smoothly when turned by hand.**

PISTON AND RING

POSITIONING

See Figures 82 through 86.

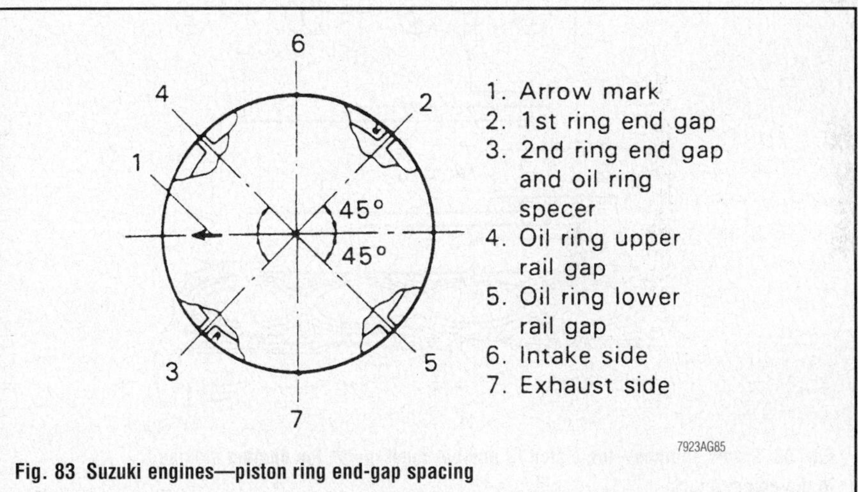

Fig. 83 Suzuki engines—piston ring end-gap spacing

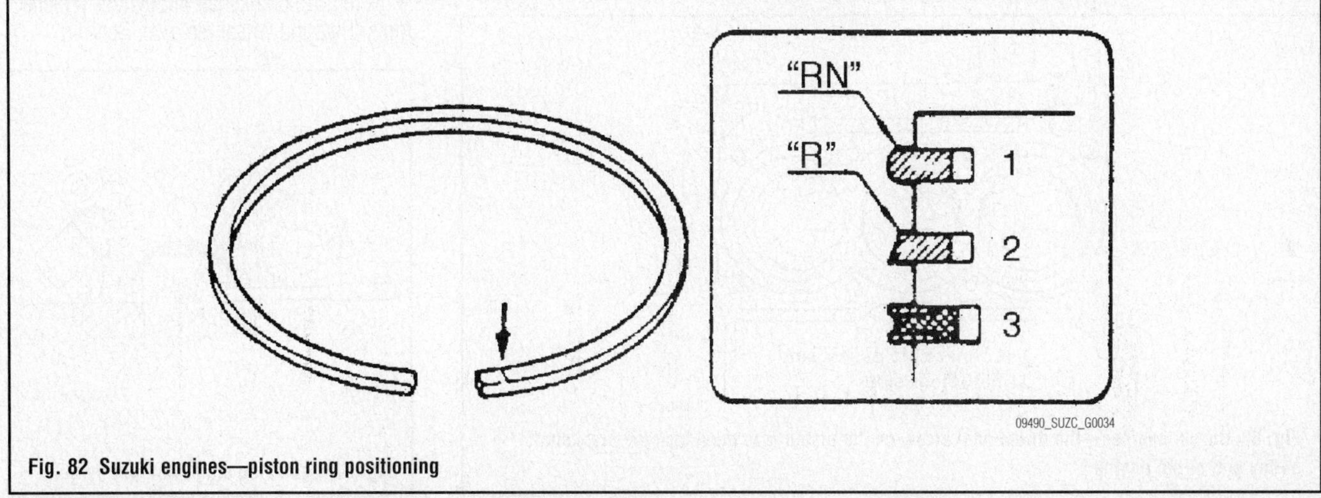

Fig. 82 Suzuki engines—piston ring positioning

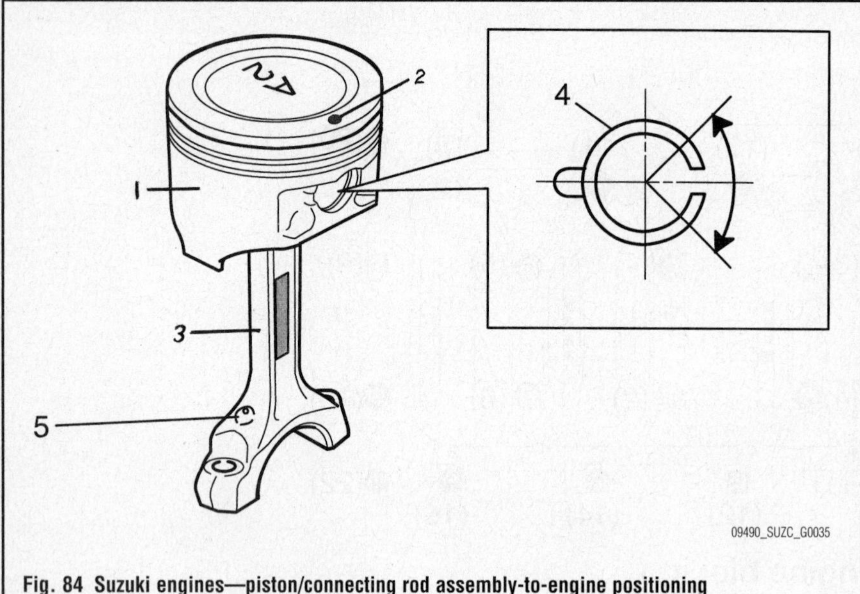

Fig. 84 Suzuki engines—piston/connecting rod assembly-to-engine positioning

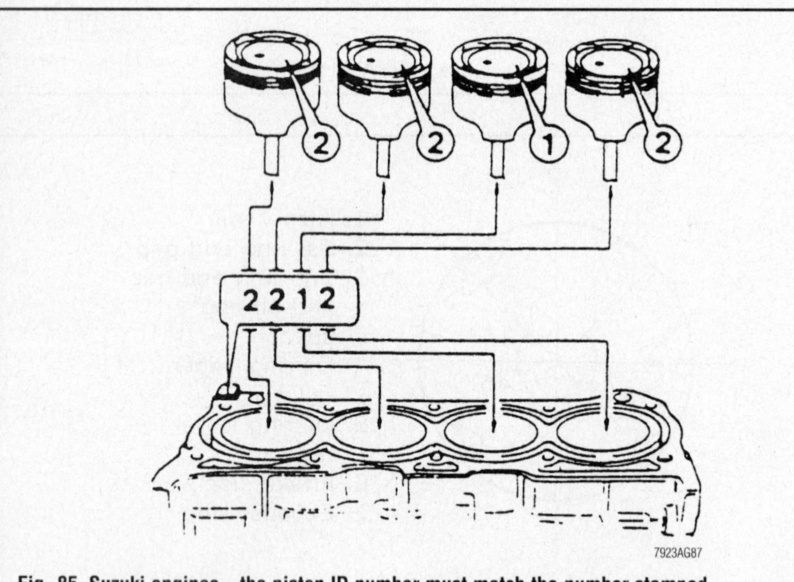

Fig. 85 Suzuki engines—the piston ID number must match the number stamped in the engine block

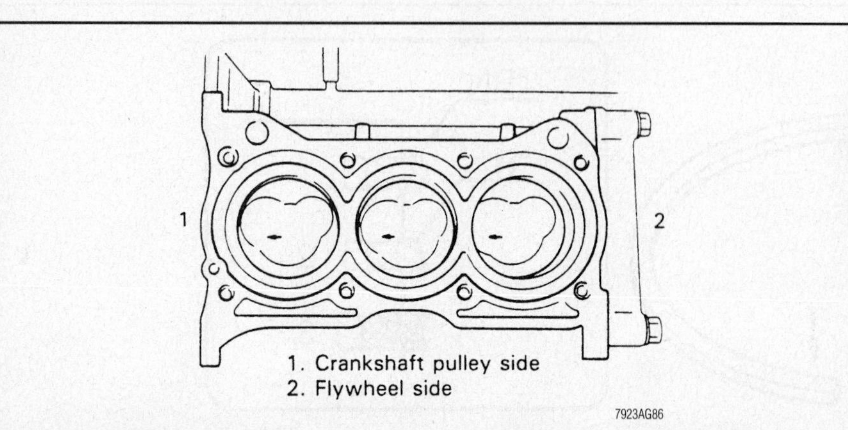

Fig. 86 Suzuki engines—the directional arrow on the piston face must face the crankshaft pulley end of the engine

REAR MAIN SEAL

REMOVAL & INSTALLATION

1. Before servicing the vehicle, refer to the Precautions Section.
2. Remove or disconnect the following:
   - Transaxle assembly
   - Flexplate/flywheel from the crankshaft
3. Carefully pry the oil seal out of the retainer without scratching the sealing surface of the crankshaft.

To install:

4. Apply engine oil the lip of the new seal.
5. Install or connect the following:
   - Seal in the retainer using a suitable seal driver
   - Flexplate/flywheel
   - Transaxle assembly

TIMING CHAIN COVER AND SEAL

REMOVAL & INSTALLATION

See Figures 87 through 94.

1. Remove the engine, as outlined in this section.
2. Remove or disconnect the following:
   - Remove oil pan.
   - Remove cylinder head cover.
   - Remove valve cover
3. Remove crankshaft pulley bolt (2). To lock crankshaft pulley (1), use special tool (camshaft pulley holder) with it as shown in figure.

➡Be sure to use the following bolts instead of pins for fixing special tool to crankshaft pulley. Bolt size: M8, P1.25 L = 25 mm (0.98 in.) Strength: 7T

4. Remove crankshaft pulley (1).

To remove crankshaft pulley, use special tools (Steering wheel remover, Bearing

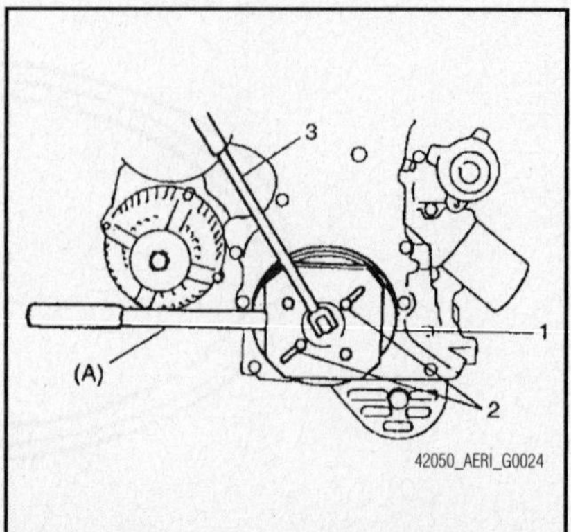

Fig. 87 Removing crankshaft pulley bolt

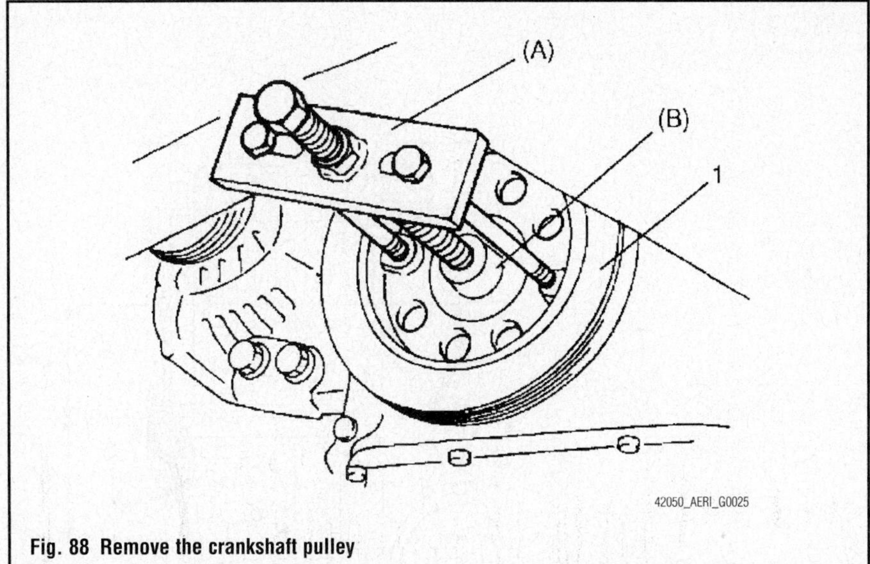

Fig. 88 Remove the crankshaft pulley

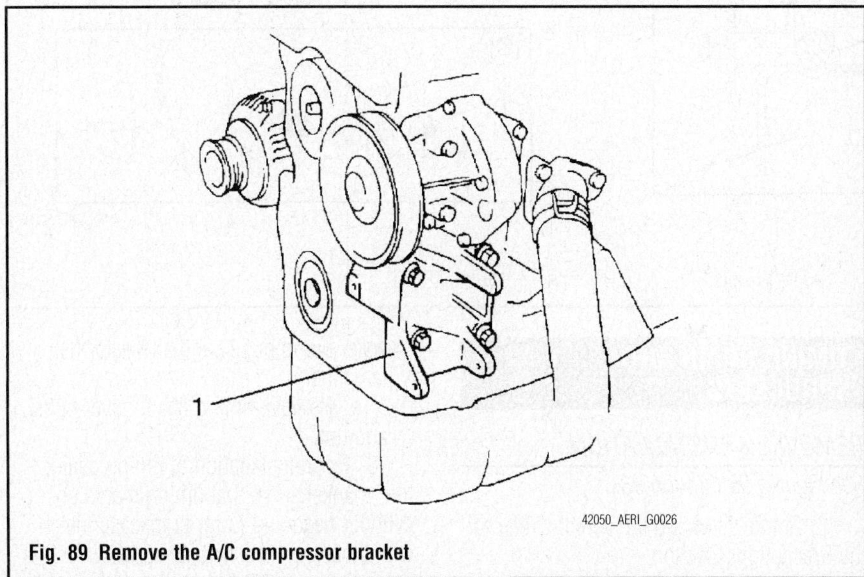

Fig. 89 Remove the A/C compressor bracket

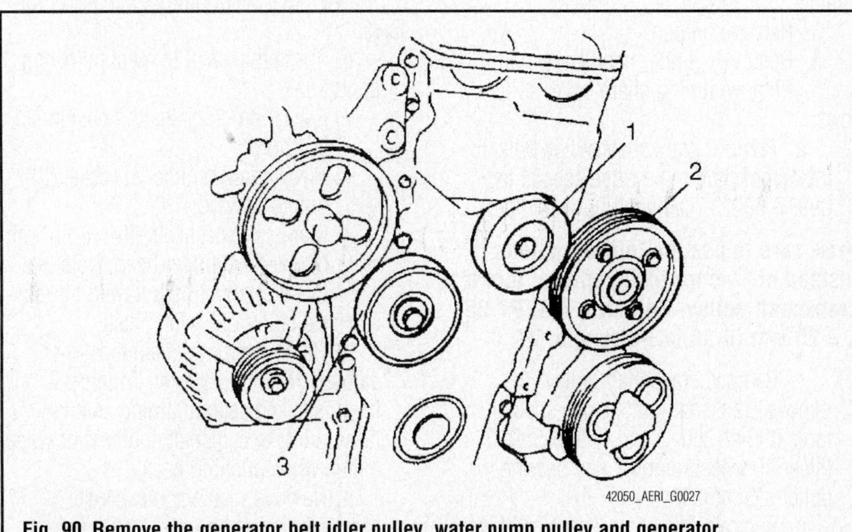

Fig. 90 Remove the generator belt idler pulley, water pump pulley and generator belt tensioner

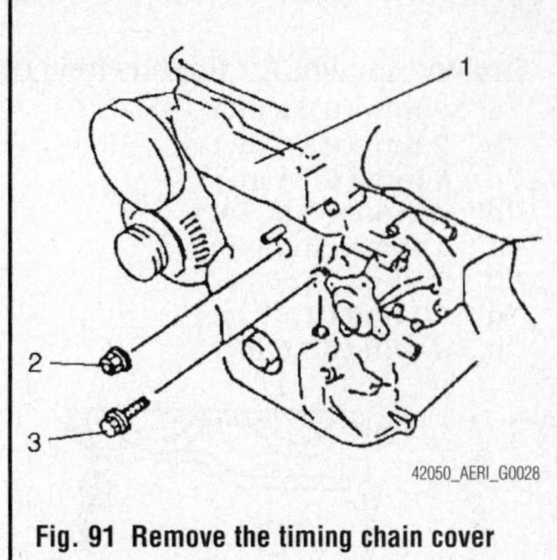

Fig. 91 Remove the timing chain cover

puller attachment) with it as shown in figure.

5. Remove the A/C compressor bracket (1).
6. Remove generator belt idler pulley (1), water pump pulley (2) and generator belt tensioner (3).
7. Remove timing chain cover (1) bolts (3) and nut (2).

To install:

8. Reverse removal sequence to install timing chain cover (1) noting the following steps.
9. Apply sealant "A" (sealant 99000-31250) and "B" (99000-31140) to area as shown in the figure below.
10. Apply engine oil to oil seal lip, then install timing chain cover (1). Tighten bolts (a) and nut (a) to 8 ft. lbs (11 Nm).

➡Before installing timing chain cover, check that pin is securely fitted.

11. Install generator belt idler pulley (1). Tighten nut to specified torque of 30.5 ft. lbs. (42 Nm).
12. Install generator belt tensioner (2). Tighten bolts to specified torque of 18.5 ft. lbs. (25 Nm).
13. Install water pump pulley (3).
14. Install A/C compressor bracket if equipped. Tighten bracket bolts to 40 ft. lbs. (55 Nm).
15. Install cylinder head cover.
16. Install oil pan.
17. Install oil pan and oil pump strainer.
18. Install crankshaft pulley. To lock crankshaft pulley (1), use special tool (camshaft pulley holder) with it as shown in figure. Tighten the crankshaft bolt to 108.5 ft. lbs. (150 Nm).

➡Be sure to use the following bolts instead of pins for fixing special tool to crank pulley. Bolt size: M8, P1.25 L = 25 mm (0.98 in.) Strength: 7T

19. Install the engine assembly.

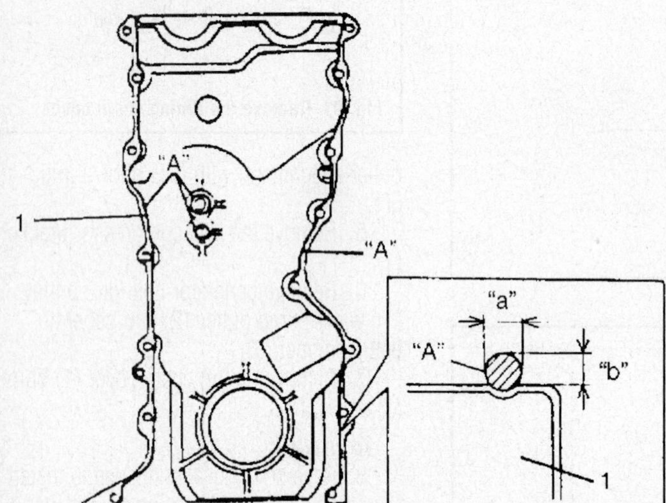

Fig. 92 Applying sealant

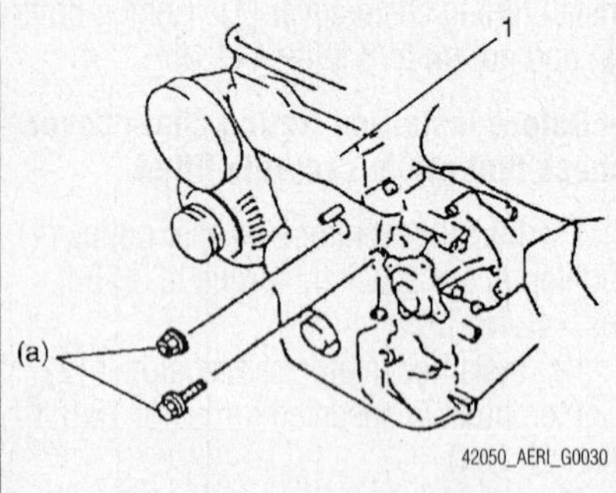

Fig. 93 Apply engine oil to oil seal lip and tighten the bolts and nuts

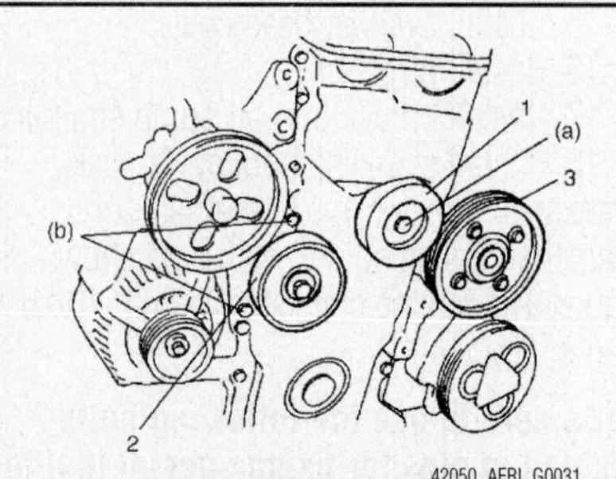

Fig. 94 Install the generator belt idler pulley, generator belt tensioner and water pump pulley

TIMING CHAIN AND SPROCKETS

REMOVAL & INSTALLATION

See Figures 95 through 98.

1. Before servicing the vehicle, refer to the Precautions Section.
2. Remove engine assembly from vehicle.
3. Remove oil pan.
4. Remove cylinder head cover.
5. Remove timing chain cover as follows:
   a. Remove crankshaft pulley bolt. To lock crankshaft pulley, use special tool 09917-68221 (camshaft pulley holder).

➡Be sure to use the following bolts instead of pins for fixing special tool to crankshaft pulley. Bolt size: M8, P1.25 L = 25 mm (0.98 in.) Strength: 7T

   b. Remove crankshaft pulley. To remove crankshaft pulley, use special tools 09944-36011 and 09926-58010 (steering wheel remover and bearing puller attachment).
   c. Remove A/C compressor bracket.
   d. Remove alternator belt idler pulley, water pump pulley and alternator belt tensioner.
   e. Remove timing chain cover bolts and nut.
6. For reinstallation of timing chain, turn crankshaft so that timing marks on cylinder head and lower crankcase match with those on sprockets as shown in figure.
7. Remove second timing chain as follows:
   a. Turn crankshaft to meet following conditions.
      - Key (I) on crankshaft is positioned as shown.
      - Arrow mark on idler sprocket (II) points upward.
      - Marks on sprockets (III) match with marks on cylinder head. Note that this step must be followed for reinstallation of timing chain.
   b. Remove timing chain tensioner adjuster No.2 and gasket. To remove them, slacken second timing chain by turning intake camshaft counterclockwise a little while pushing back pad.
   c. Remove intake and exhaust camshaft timing sprocket bolts. To remove them, fit a spanner to hexagonal

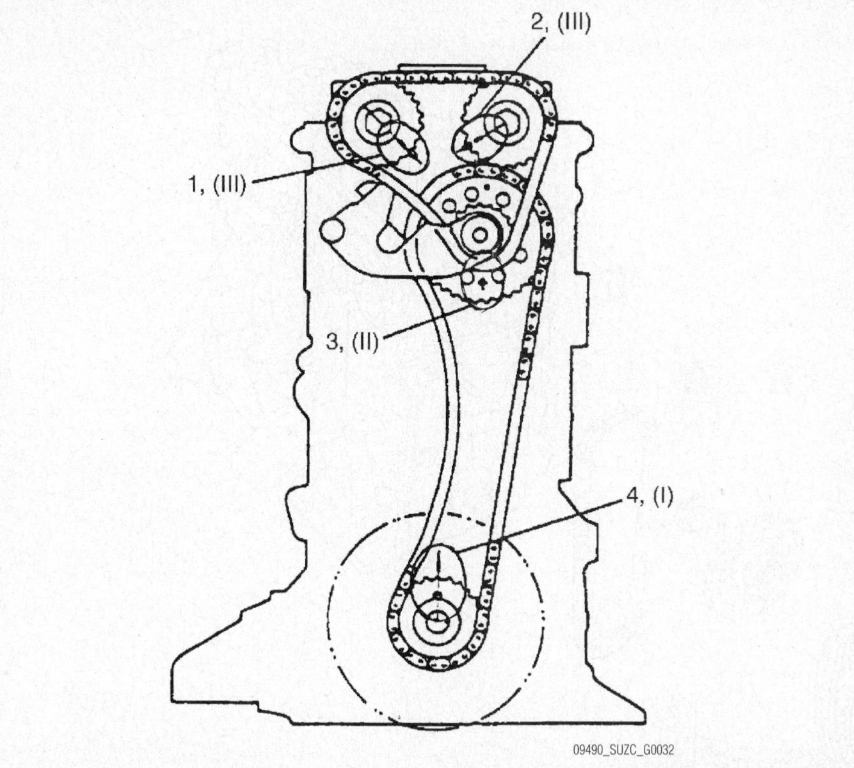

Fig. 95 Turn crankshaft so that timing marks on cylinder head and lower crankcase match with those on sprockets as shown.

part at the center of camshaft to hold it stationary.

d. Remove camshaft timing sprockets and second timing chain.

⁂ WARNING

After second timing chain is removed, never turn intake camshaft, exhaust camshaft and crankshaft independently more than such an extent as shown. If turned, interference may occur between piston and valves and valves themselves, and parts related to piston and valves may be damaged.

8. Remove timing chain guide No.1.
9. Remove timing chain tensioner adjuster No. 1.
10. Remove timing chain tensioner.
11. Remove idler sprocket and first timing chain.
12. Remove crankshaft timing sprocket.

To install:

13. Check that match mark on crankshaft timing sprocket is in match with timing mark on lower crankcase.
14. Install crankshaft timing sprocket as shown in figure.
15. Apply oil to bush of idler sprocket and install idler sprocket and sprocket shaft.
16. Install first timing chain by aligning dark blue plate of first timing chain and match mark on idler sprocket.
17. Bring yellow plate of first timing chain into match with match mark on crankshaft timing sprocket.
18. Apply engine oil to sliding surface of timing chain tensioner and then install it as shown in figure. Tighten tensioner nut to 18 ft. lbs. (25 Nm).
19. With latch of tensioner adjuster No.1 returned and plunger pushed back into body, insert stopper into latch and body.

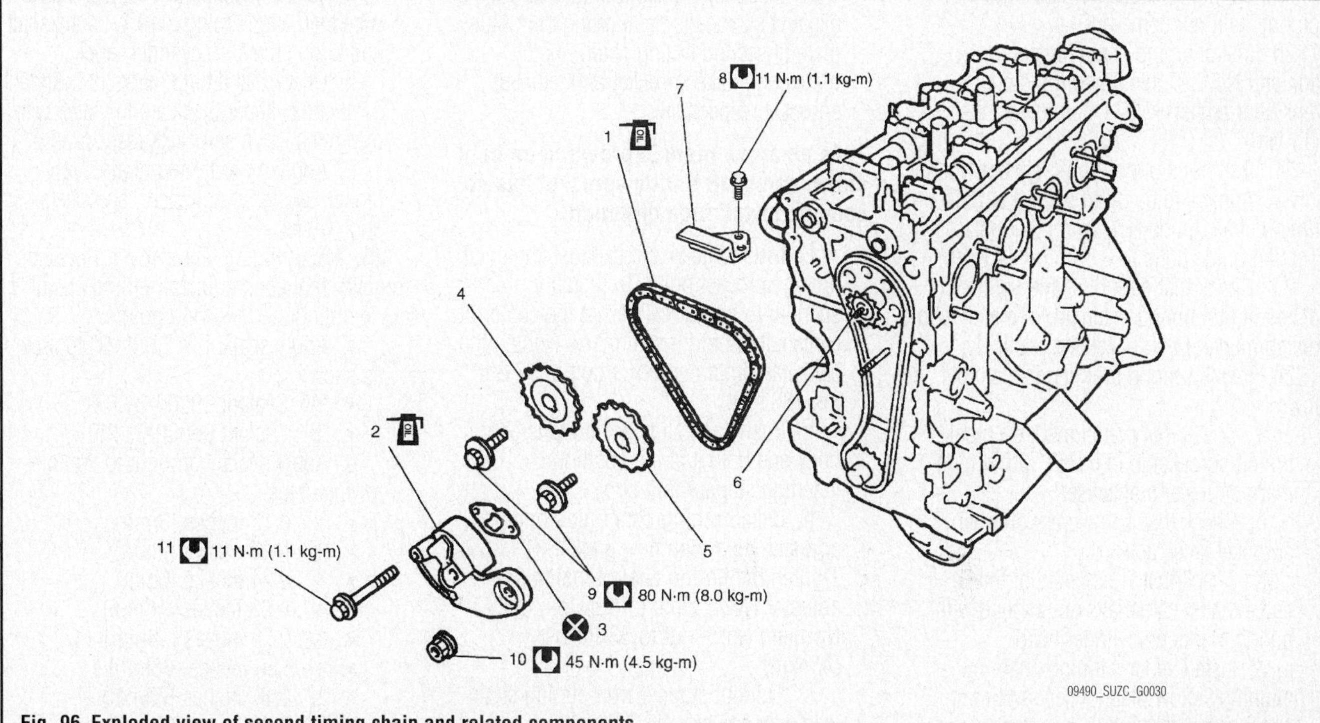

Fig. 96 Exploded view of second timing chain and related components

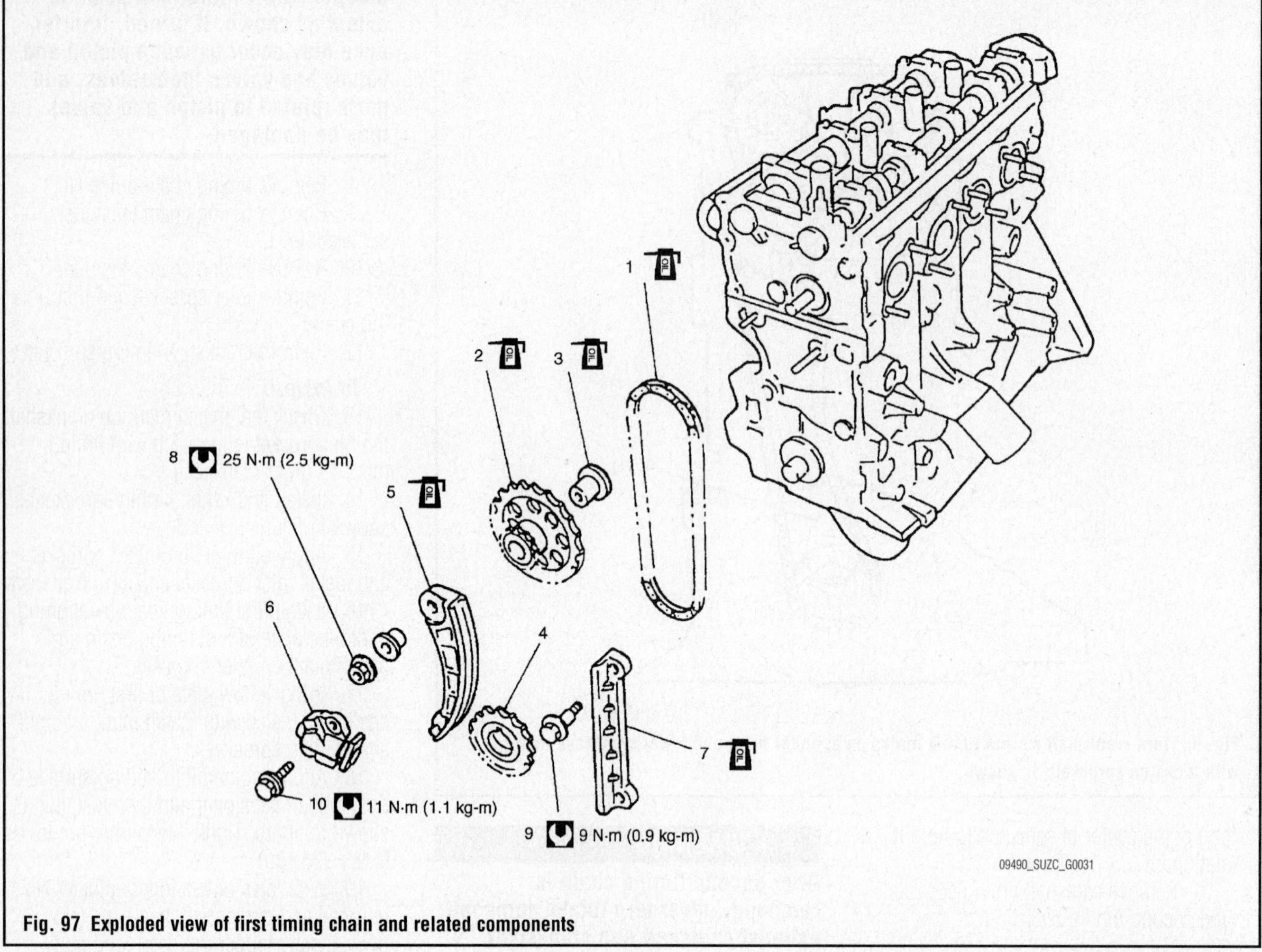

Fig. 97 Exploded view of first timing chain and related components

After inserting it, check to make sure that plunger will not come out.

20. Install timing chain tensioner adjuster No.1. Tighten the timing chain tensioner adjuster No. 1 bolt to 8 ft. lbs. (11 Nm).
21. Pull out stopper from adjuster No.1.
22. Apply engine oil to sliding surface of timing chain guide No.1 and then install it. Tighten guide bolts to 6.5 ft. lbs. (9 Nm).
23. Check that dark blue and yellow plates of first timing chain are in match with match marks on sprockets respectively.
24. Install second timing chain as follows:
    a. Check that match mark on crank timing sprocket is in match with timing mark on lower crankcase.
    b. Check that arrow mark on idler sprocket faces upward.
    c. Check that knock pins of intake and exhaust camshafts are aligned with timing marks on cylinder head.
    d. Install second timing chain by aligning yellow plate of second timing chain and match marks on idler sprocket.
    e. Install sprockets to intake and exhaust camshafts by aligning dark blue plate of second timing chain, match marks on intake sprocket and exhaust sprocket respectively.

➡As an arrow mark is provided on both sides, camshaft timing sprocket has no specific installation direction.

    f. Install intake and exhaust camshaft timing sprocket bolts. To install it, fit a spanner to hexagonal part at the center of camshaft to hold it stationary. Tighten the camshaft timing sprocket bolt to 57.5 ft. lbs. (80 Nm).
    g. Push back plunger into tensioner body and hold it at the position by inserting stopper into body.
    h. Install timing chain tensioner adjuster No.2 with new gasket. Tighten the timing chain tensioner adjuster No. 2 bolts to 8 ft. lbs. (11 Nm) and No. 2 nut to 33 ft. lbs. (45 Nm).
    i. Pull out stopper from timing chain tensioner adjuster No.2.
    j. Turn crankshaft two rotations clockwise then align timing mark on crankshaft and timing mark on cylinder block.
    k. Check that timing marks of cylinder head and cylinder block are in match with match marks on sprockets respectively.
    l. Apply oil to timing chains, tensioner, tensioner adjusters, sprockets, and guides.
25. Install timing chain cover. Reverse removal sequence to install timing chain cover noting the following points:
    a. Apply sealant "A" and "B" to area as shown.
    - "A": Sealant 99000-31250
    - "B": Sealant 99000-31140
    b. Apply sealant amount to the following areas.
    - "a": 0.12 inches (3mm)
    - "b": 0.08 inches (2mm)
    - "c": 0.24 inches (6mm)
    - "d": 0.63 inches (16mm)
    - "e": 0.55 inches (14mm)
    - "f": 2.56 inches (65mm)
    - "g": 2.87 inches (73mm)
    - "h": 0.16 inches (4mm)

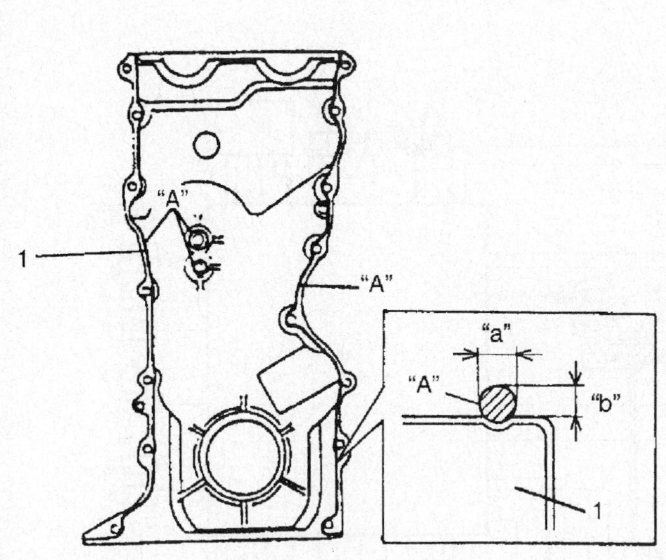

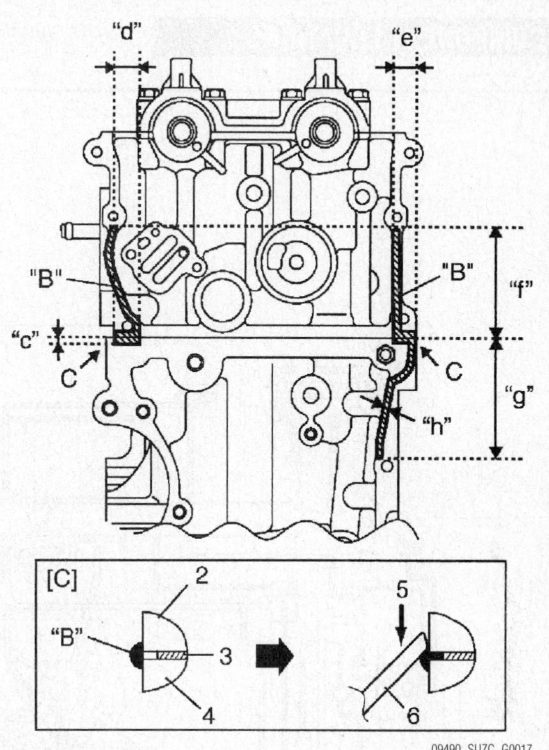

Fig. 98 Sealant application on timing belt cover

c. Apply engine oil to oil seal lip, then install timing chain cover. Tighten bolts and nut to 8 ft. lbs. (11 Nm).

➡Before installing timing chain cover, check that pin is securely fitted.

26. Install alternator belt idler pulley. Tighten nut to 30.5 ft. lbs. (42 Nm).
27. Install alternator belt tensioner. Tighten bolts to 18.5 ft. lbs. (25 Nm).
28. Install water pump pulley.
29. Install A/C compressor bracket (if equipped). Tighten bracket bolts to 40 ft. lbs. (55 Nm).
30. Install cylinder head cover with new gaskets and seal washers and tighten nuts to 8 ft. lbs. (11 Nm).
31. Install oil pan.
32. Install crankshaft pulley using pulley holder tool. Tighten the crankshaft pulley bolt to 108.5 ft. lbs. (150 Nm).
33. Install engine assembly.
34. Install all remaining components in the reverse order of the removal procedure.

VALVE COVERS

REMOVAL & INSTALLATION

1. Disconnect the negative battery cable.
2. Remove engine.
3. Disconnect ignition coil couplers.
4. Remove ignition coils.
5. Remove oil level gauge.
6. Disconnect breather hose and PCV hose from cylinder head cover.
7. Remove cylinder head cover.

To install:

8. Remove oil, old sealant and dust from sealing surfaces on cylinder head and cover. After cleaning, apply sealant "A" to cylinder head sealing surface area as shown in figure.

➡Use: Watertight sealant 99000-31250SUZUKI Bond No.1207F

9. Install new O-rings and new cylinder head cover gasket to cylinder head cover.

➡Be sure to check each of these parts for deterioration or any damage before installation and replace if found defective.

10. Tighten nuts to 8.0 ft. lbs. (11 Nm).
11. Use new seal washers
12. Install oil level gauge.
13. Connect breather hose and PCV hose to cylinder head cover.
14. Install ignition coils and connect ignition coil couplers.
15. Install engine cover.
16. Connect the negative battery cable.

VALVE LASH

ADJUSTMENT

Hydraulic valve lash adjusters are used to adjust the valve clearance to **0** lash automatically at all times. Adjustment is not required.

ENGINE PERFORMANCE & EMISSION CONTROLS

COMPONENT LOCATIONS

See Figures 99 through 101.

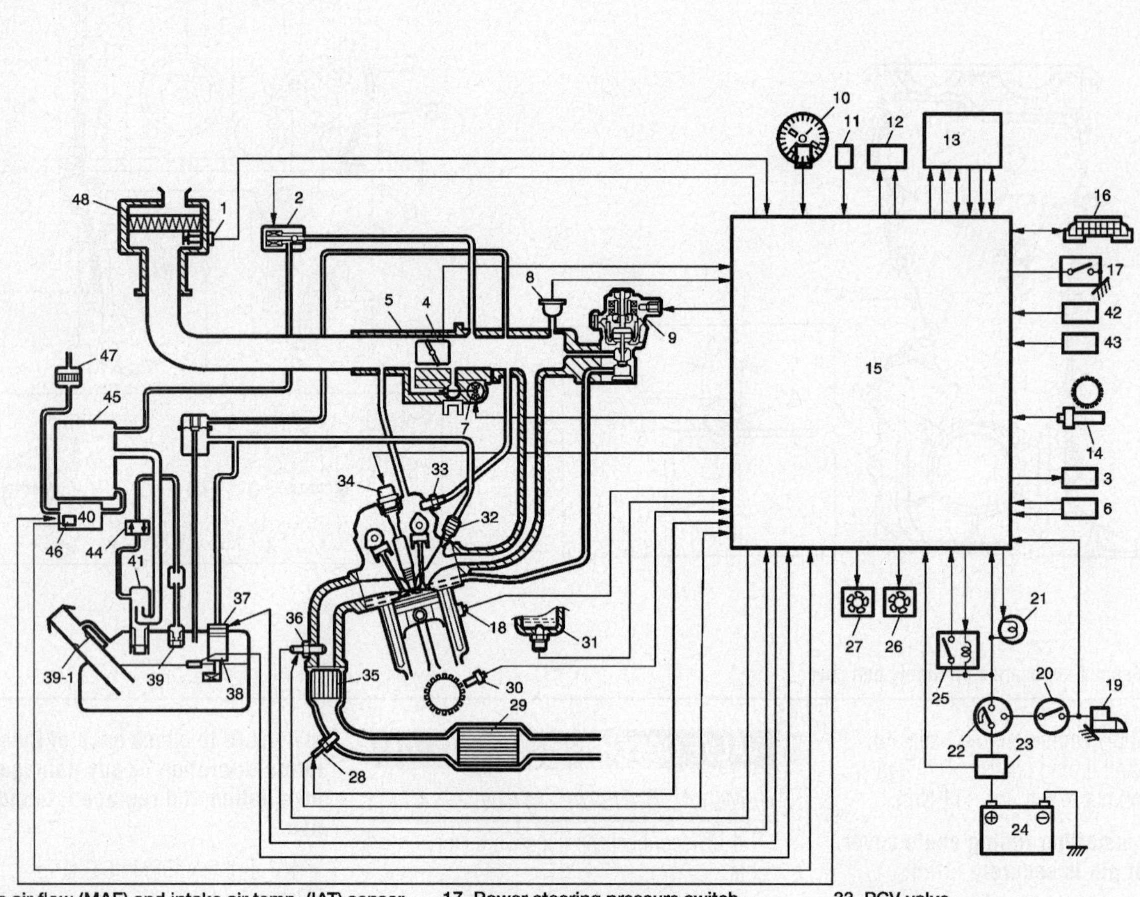

1. Mass air flow (MAF) and intake air temp. (IAT) sensor
2. EVAP canister purge valve
3. A/C compressor
4. Throttle position sensor
5. Throttle body
6. A/C pressure switch
7. Idle air control valve
8. Manifold absolute pressure sensor
9. EGR valve
10. Camshaft position sensor
11. Electric loads
12. Speedometer, engine coolant temp. warning light and tachometer in combination meter
13. Transmission control module
14. Vehicle speed sensor
15. Engine control module
16. Data link connector
17. Power steering pressure switch
18. Knock sensor
19. Starter magnetic switch
20. Clutch pedal position switch (M/T) or park / neutral position switch in transmission range sensor (A/T)
21. Malfunction indicator lamp
22. Main switch (ignition switch)
23. Main fuse box
24. Battery
25. Main relay
26. Radiator fan
27. A/C condenser fan
28. Heated oxygen sensor-2
29. Three way catalytic converter
30. Crankshaft position sensor
31. Engine coolant temp. sensor
32. Fuel injector
33. PCV valve
34. Ignition coil assembly
35. Warm-up three way catalytic converter
36. Heated oxygen sensor-1
37. Fuel pump
38. Fuel level sensor
39. Fuel cut valve
39-1. Fuel tank inlet valve
40. EVAP leak check module (if equipped)
41. Refuel vapor control valve (if equipped)
42. A/C evaporator inlet temp. sensor
43. A/C evaporator outlet temp. sensor -Lighting
44. Tank pressure control valve (if equipped) - Heater blower
45. EVAP canister -Rear defogger
46. EVAP leak check pressure sensor (if equipped)
47. EVAP canister air suction filter
48. Air cleaner

22140_AERI_G0044

Fig. 99 Emission Diagram

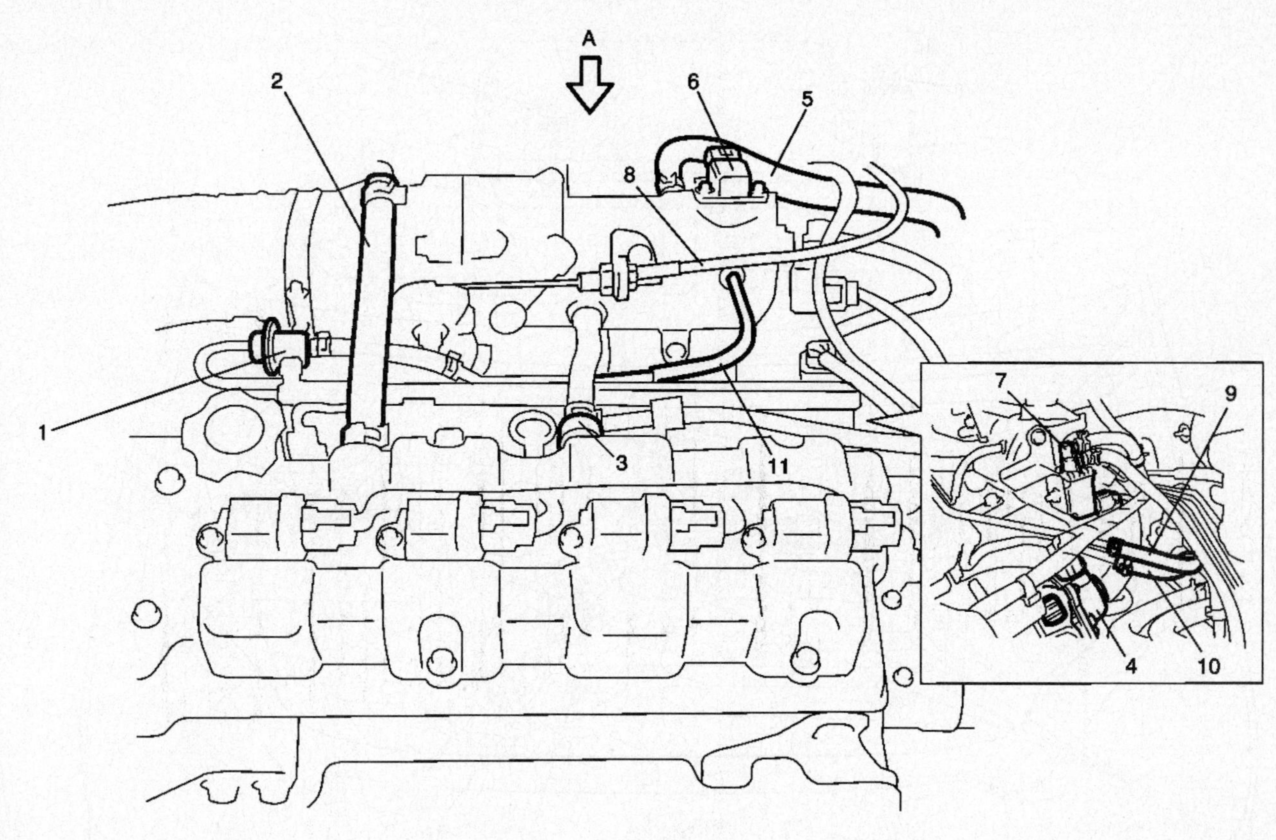

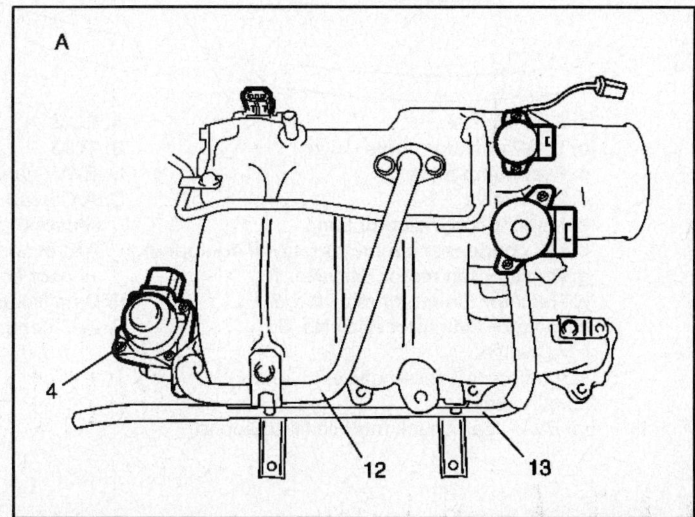

1. Fuel pressure regulator
2. Breather hose
3. PCV valve
4. EGR valve
5. Brake booster hose
6. Manifold absolute pressure sensor
7. EVAP canister purge valve
8. Accelerator cable
9. Fuel return hose
10. Fuel feed hose
11. Fuel pressure regulator hose
12. EGR pipe
13. Coolant pipe
A: View A

22140_AERI_G0073

Fig. 100 Emission diagram piping/components

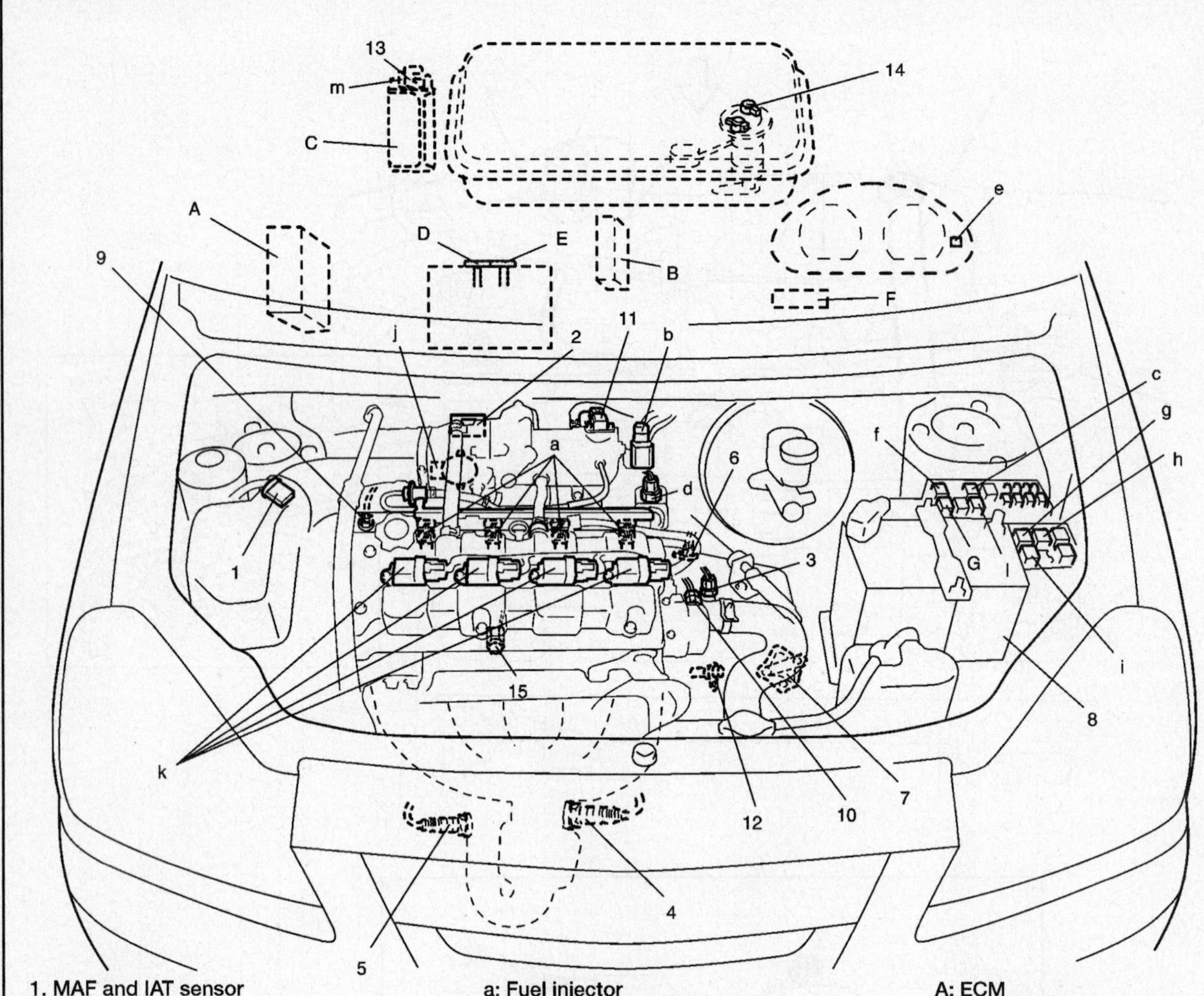

1. MAF and IAT sensor
2. TP sensor
3. ECT sensor
4. Heated oxygen sensor-1
5. Heated oxygen sensor-2
6. VSS
7. Transmission range sensor
8. Battery
9. PSP switch
10. CMP sensor
11. MAP sensor
12. CKP sensor
13. EVAP leak check pressure sensor (if equipped)
14. Fuel level sensor
15. Knock sensor

a: Fuel injector
b: EVAP canister purge valve
c: Fuel pump relay
d: EGR valve
e: Malfunction indicator lamp
f: A/C condenser fan motor relay (if equipped)
g: Radiator fan motor relay No. 1
h: Radiator fan motor relay No. 2
i: Radiator fan motor relay No. 3
j: IAC valve
k: Ignition coil assembly (with ignitor)
l: Main relay
m: EVAP leak check module (if equipped)

A: ECM
B: TCM
C: EVAP canister
D: A/C evaporator inlet air temp. sensor (if equipped)
E: A/C evaporator outlet air temp. sensor (if equipped)
F: Data link connector
G: A/C compressor relay

22140_AERI_G0072

Fig. 101 Electronic Control System Components

CAMSHAFT POSITION (CMP) SENSOR

LOCATION

The Camshaft Position (CMP) sensor is located on the transmission side of cylinder head under motor cover.

OPERATION

The Camshaft Position (CMP) sensor consists of the signal generator (phototransistors) and signal rotor (slits plate). The signal generator generates a reference signal and position signal through two types of slits in the slit plate which turns together with the camshaft. The CMP sensor generates 4 pulses of signals while each has a different waveform length. The camshaft makes one full rotation. Based on these signals the ECM judges which cylinder piston is in the compression stroke. The CMP sensor generates 360° pulses of signals while the camshaft makes one full rotation (i.e., 1 pulse per 1° of movement of the camshaft).

Based on these signals, the ECM judges the waveform length of the Reference signal, the engine speed and piston position.

REMOVAL & INSTALLATION

⁂ CAUTION

Disassembly is prohibited. If anything faulty is found, replace as an assembly unit.

1. Remove engine cover from cylinder head cover.
2. Remove the Camshaft Position Sensor (CMP) by removing the two bolts.

To install:

3. Install a new O-ring with engine oil applied to CMP sensor.
4. Install CMP sensor to camshaft.
5. Fit the dog of CMP sensor coupling into the slots of camshaft, when installing. The dogs of CMP sensor coupling are offset. If the dogs cannot be fitted into the slots, turn the CMP sensor shaft by 180 degrees and try again.
6. Hand tighten the CMP sensor bolts temporarily.
7. Connect the CMP sensor coupler.
8. Check and adjust ignition timing
9. Tighten the CMP sensor bolts to 11 ft. lb. (16.5 Nm)
10. Install engine cover to cylinder head cover

TESTING

1. Check the Camshaft Position Sensor (CMP) sensor power supply voltage, as follows:
    a. Disconnect connector from CMP sensor (1) with ignition switch turned OFF.
    b. Turn the ignition switch **ON**, then measure voltage between "BLK/ORN" and "BLK/RED" wire terminals of CMP sensor connector. The desired voltage is 10–14 Volts. If the specification falls outside of the desired range, the sensor must be replaced.

CRANKSHAFT POSITION (CKP) SENSOR

LOCATION

See Figure 102.

The Crankshaft Position Sensor (CKP) is located behind the flywheel or driveplate attached to the cylinder block.

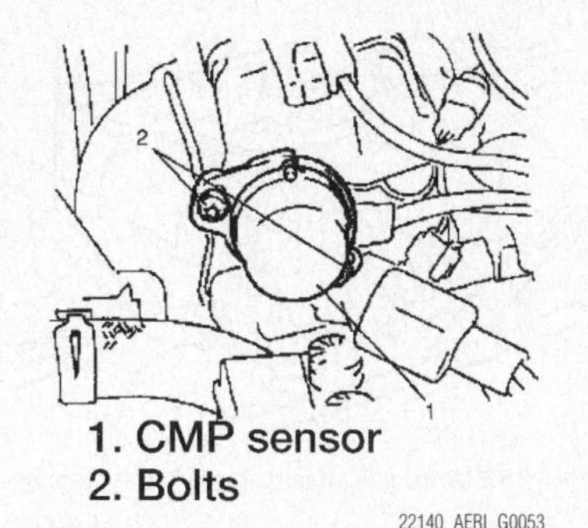

Fig. 102 Crankshaft Position (CKP) sensor location

OPERATION

With the engine running at idle after warming up. The Crankshaft Position Sensor (CKP) will generate 36 pulses per 1 crankshaft revolution. These pulses provide piston location for fuel and ignition timing.

REMOVAL & INSTALLATION

1. Remove engine with transmission from vehicle
2. Remove transmission from engine and then remove flywheel or drive plate from crankshaft.
3. Disconnect connector from Crankshaft Position (CKP) sensor.
4. Remove the CKP sensor from cylinder block.

To install:

5. Reverse removal procedure noting the following.
6. Check to make sure that CKP sensor is free from any metal particles and damage.
7. Apply engine oil to O-ring of sensor.
8. Install CKP to cylinder block with and tighten to 4.5 ft. lb. (5.5 Nm)
9. Attach the sensor connector and fix wire harness with clamp securely.

TESTING

1. With the ignition switch turned **OFF**, connect scan tool.
2. Turn the ignition switch **ON** and clear DTC using scan tool.
3. Start engine and run it for 10 seconds.
4. Check DTCs and pending DTCs.
5. Check Crankshaft Position Sensor (CKP) and starter motor for resistance
6. Disconnect CKP sensor connector with ignition switch turned **OFF**.
7. Then, check for proper connection to CKP sensor connector at "BLK/RED" and "BRN" wire terminals.
8. If OK, measure sensor resistance between terminals.
9. CKP sensor resistance should be 484–656 kohms at 68°F, 20°C.
10. Resistance measure resistance between each terminal and ground.
11. Between CKP sensor terminal and ground resistance should be 1 megaohms or more.
12. If the sensor did not meet specified values, replace sensor.

ELECTRONIC CONTROL MODULE (ECM)

LOCATION

See Figure 103.

The Electronic Control Module (ECM) is located above the right hand kick panel.

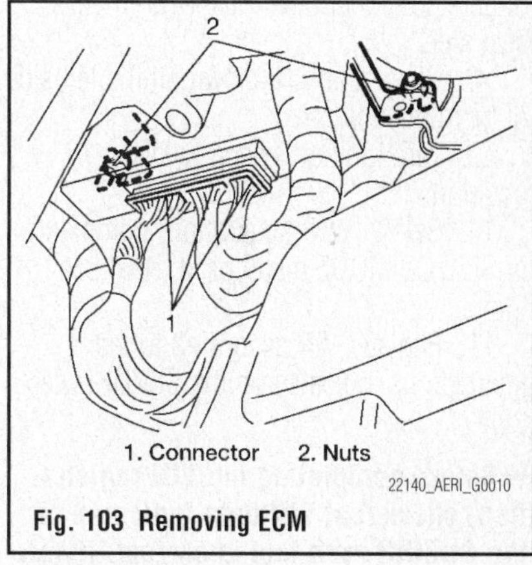

Fig. 103 Removing ECM

OPERATION

Electronic Control Module (ECM) in this vehicle has the following functions:

- When the ignition switch is turned ON with the engine at a stop, the Malfunction Indicator Lamp (MIL) turns ON to check the circuit of the (MIL).
- When the ECM detects a malfunction which gives an adverse effect to vehicle emission while the engine is running, it makes the malfunction indicator lamp in the meter cluster of the instrument panel turn ON or flash (flashing only when detecting a misfire which can cause damage to the catalyst) and stores the malfunction Diagnostic Trouble Code (DTC according to SAE J2012) in its memory.

REMOVAL & INSTALLATION

See Figure 103.

➡If the Electronic Control Module (ECM) is replaced with a new or used ECM, the new VIN data must be registered. Make sure to re-register VIN data correctly by performing VIN registration as outlined in the installation procedure.

1. Disconnect the battery negative cable.

2. Disable air bag system, refer to Air Bag section—Disabling Air Bag System.

3. Remover glove box, see Instrument Panel Removal and Installation.

4. Disconnect connectors from ECM while releasing connectors lock.

5. Unfasten the retaining nuts, then remove the ECM from the body.

To install:

6. Connect connectors to ECM securely until a click is heard. When installing each part, be careful not to catch any cable or wiring harness.

7. Connect SUZUKI scan tool to Data Link Connector (DLC) located on underside of instrument panel at driver's seat side.

8. Check the VIN shown on the left side of instrument panel.

9. Turn ignition switch to ON position.

10. Select "VIN registration" command in SELECT MODE menu of SUZUKI scan tool.

11. Register VIN according to the instructions indicated on the SUZUKI scan tool.

➡Before completing the VIN registration, check that VIN data indicated on the SUZUKI scan tool is correct. If not, re-register VIN data into ECM correctly, otherwise, the MIL remains on even if the engine started or it causes the inconsistency between registered VIN data in ECM and actual VIN.

TESTING

The Data Link Connector (DLC) in compliance with SAE J1962 in its installation position is located 12 inches from the centerline of the steering column. The On Board Diagnostic (OBD)-II serial data line (K) is used for SUZUKI scan tool or OBD-II generic scan tool to communicate with the ECM, module. Testing of the ECM can be done from DLC.

ENGINE COOLANT TEMPERATURE (ECT) SENSOR

LOCATION

See Figure 104.

The Engine Coolant Temperature (ECT) sensor is located in the coolant housing above the upper coolant hose.

OPERATION

As temperature of coolant increases, resistance of Engine Coolant Temperature (ECT) Sensor decreases.

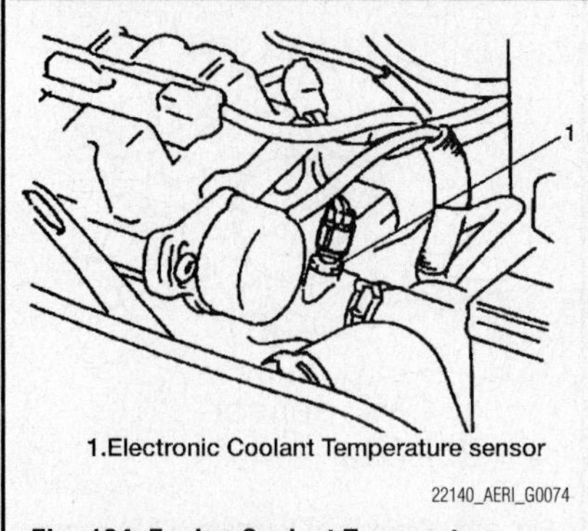

Fig. 104 Engine Coolant Temperature (ECT) sensor (1) location

➡ECT sensor output voltage is 0.16 V or lower for 5 sec. continuously. (High engine coolant temp. / low resistance)

➡ECT sensor output voltage is 5 V or higher for 5 sec. continuously. (Low engine coolant temp. / high resistance)

REMOVAL & INSTALLATION

See Figure 104.

1. Disconnect negative cable from battery.

2. Drain coolant.

3. Disconnect connector from Engine Coolant Temperature (ECT) sensor.

4. Remove ECT sensor from water outlet cap.

To install:

5. Clean mating surfaces of sensor and water outlet cap.

6. Install ECT sensor into water outlet cap.

7. Tighten ECT sensor to: 9.5 ft. lbs. (13.Nm).

8. Attach the connector to sensor securely.

9. Refill cooling system

TESTING

See Figure 104.

Immerse temperature sensing part of the ECT sensor in water and measure resistance between sensor terminals while heating water gradually. If measured resistance doesn't show such characteristic as shown, replace ECT sensor.

FUEL LEVEL SENDING UNIT

LOCATION

See Figure 105.

The fuel level gauge is attached to the pump-housing unit located in the fuel tank. See testing for complete operating instructions.

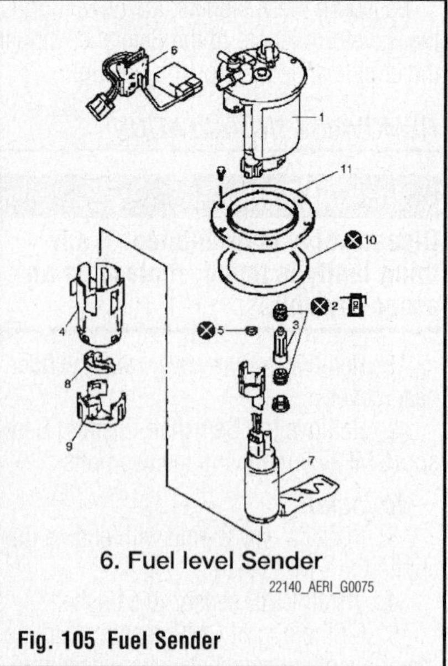

Fig. 105 Fuel Sender

OPERATION

As the fuel decreases the fuel level float drops and resistance increases.

REMOVAL & INSTALLATION

1. Relieve fuel pressure in fuel feed line, as outlined in the Fuel Section.

2. Disconnect the negative battery cable.

3. Remove fuel filler cap.

4. Insert hose of a hand operated pump into fuel filler hose (1) and drain fuel in space "A" in the figure.

CAUTION

Do not force the pump hose into the fuel tank, or pump hose may damage fuel tank inlet valve

- Hoist vehicle.
- Remove muffler.
- Remove propeller shaft.
- Remove EVAP canister from vehicle.
- Disconnect fuel filler hose and breather hose from filler neck.
- Disconnect filter inlet hose from canister air suction filter.

➡Due to absence of fuel tank drain plug, drain fuel tank by pumping fuel out through fuel tank filler. Use hand operated pump device to drain fuel tank.

CAUTION

Do not force pump hose into fuel tank, or pump hose may damage fuel tank inlet valve. Never drain or store fuel in an open container due to possibility of fire or explosion.

5. Disconnect fuel pipe joints from fuel pipes.

➡Remove mud, dust and/or foreign material between pipe and joint by blowing compressed air.

6. Disconnect joint from pipe.

✲✲ WARNING

A small amount of fuel may be released after the fuel hose is disconnected. In order to reduce the chance of personal injury, cover the hose and pipe to be disconnected with a shop cloth. Be sure to put that cloth in an approved container when disconnection is completed.

7. Support fuel tank with jack and appropriate tank fixture, remove its mounting bolts.

8. Lower fuel tank a little to disconnect fuel pump connector and ground wire harness, then remove fuel tank.

To install:

➡If parts have been removed from fuel tank, install them before installing fuel tank to vehicle.

9. Raise fuel tank with jack and connect fuel pump connector and ground wire harness, and then clamp wire harness.

10. Install fuel tank to vehicle.

11. Fuel tank bolt tightening torque: 36.5 ft. lbs. (50 Nm).

12. Connect fuel filler hose breather hose and filter inlet hose and clamp them securely.

13. Fuel filler hose tightening torque for vehicle with ORVR system a: 2.5ft. lbs. (3.5 Nm).

14. Fuel filler hose clamp for vehicle without ORVR system a: 1.0 ft. lbs. (1.5 Nm) .

15. Connect fuel feed pipe, fuel return pipe and purge pipe, clamp them securely.

✲✲ CAUTION

When connecting joints, clean outside surfaces of pipe where joint is to be inserted, push joint into pipe till joint lock clicks and check to ensure that pipes are connected securely, or fuel leaks may occur. Never let the fuel hoses touch the ABS sensor harness.

16. Install EVAP canister to vehicle

17. Install axle shaft

18. Install muffler

19. Connect the negative battery cable. With engine OFF, turn ignition switch to ON position and check for fuel leaks

TESTING

See Figure 106.

1. Check resistance between terminals "a" and "b" at each float position:
   - Fuel level (A) float in the up position, indicating a full tank: 9–11 ohms of resistance.
   - Fuel level (B), float in the down position, indicating an empty tank: 129–131 ohms of resistance.

2. If the measured value is out of specification, replace fuel level sensor.

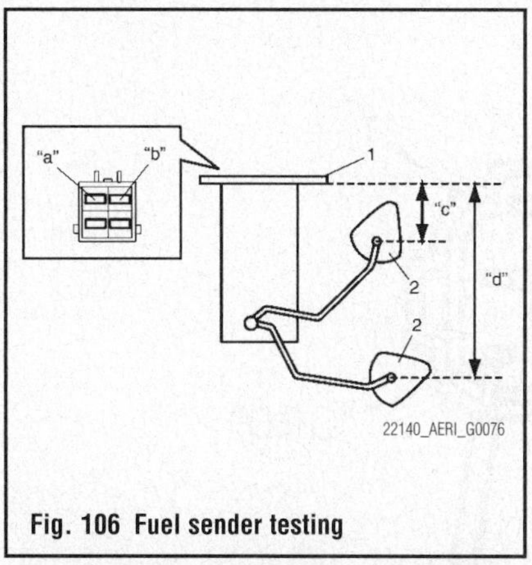

Fig. 106 Fuel sender testing

HEATED OXYGEN (HO2S) SENSOR

LOCATION

See Figure 107.

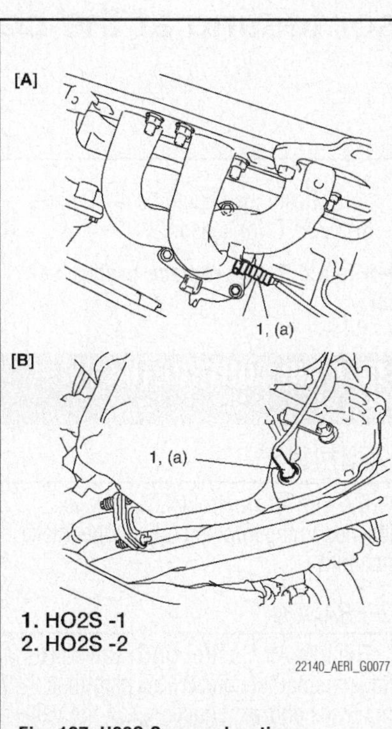

Fig. 107 HO2S Sensors locations

Heated Oxygen Sensor (HO2S) 1 is located in the bottom end of the exhaust manifold. HO2S 2 is located in the rear of the catalytic converter.

OPERATION

The Oxygen (HO2S) sensors compare the oxygen content in the vehicles exhaust to atmospheric oxygen content, this comparison provided the Electronic Control Module (ECM) with information to make adjustments in fuel mixture as demand changes providing the optimum air/fuel ratio.

REMOVAL & INSTALLATION

✲✲ CAUTION

To avoid danger of being burned, do not touch exhaust system when system is hot. Oxygen sensor removal should be performed when system is cool.

1. Disconnect connector of the Heated Oxygen Sensor (HO2S) and release its wire harness from clamps.

2. For HO2S-2, raise and safely support the vehicle.

3. For HO2S-1, remove the exhaust manifold cover.

4. Remove HO2S (1) from exhaust manifold or exhaust No.1 pipe

To install:

5. Install and tighten the HO2S to 32.5 ft. lbs. (35 Nm).

6. Attach the connector to the HO2S and clamp wire harness securely.

7. After installing HO2S, start engine and check that no exhaust gas leakage exists.

TESTING

See Figure 108.

1. Check the HO2S-1 and HO2S-2 signals, as follows:
   a. Connect scan tool to DLC with ignition switch turned OFF.
   b. Warm up engine to normal operating temperature and keep it at 2000 r/min. for 60 sec.
   c. Repeat racing engine (Repeat depressing accelerator pedal 5 to 6 times continuously to enrich A/F mixture and take foot from pedal to lean fuel mixture readings), check HO2S output voltage displayed on scan tool.
   d. Voltage should deflect between below 0.4 V and over 0.6 V repeatedly.

2. Check the Check HO2S-1 and HO2S-2 sensor grounds:
   a. Disconnect connector from Check HO2S-1.

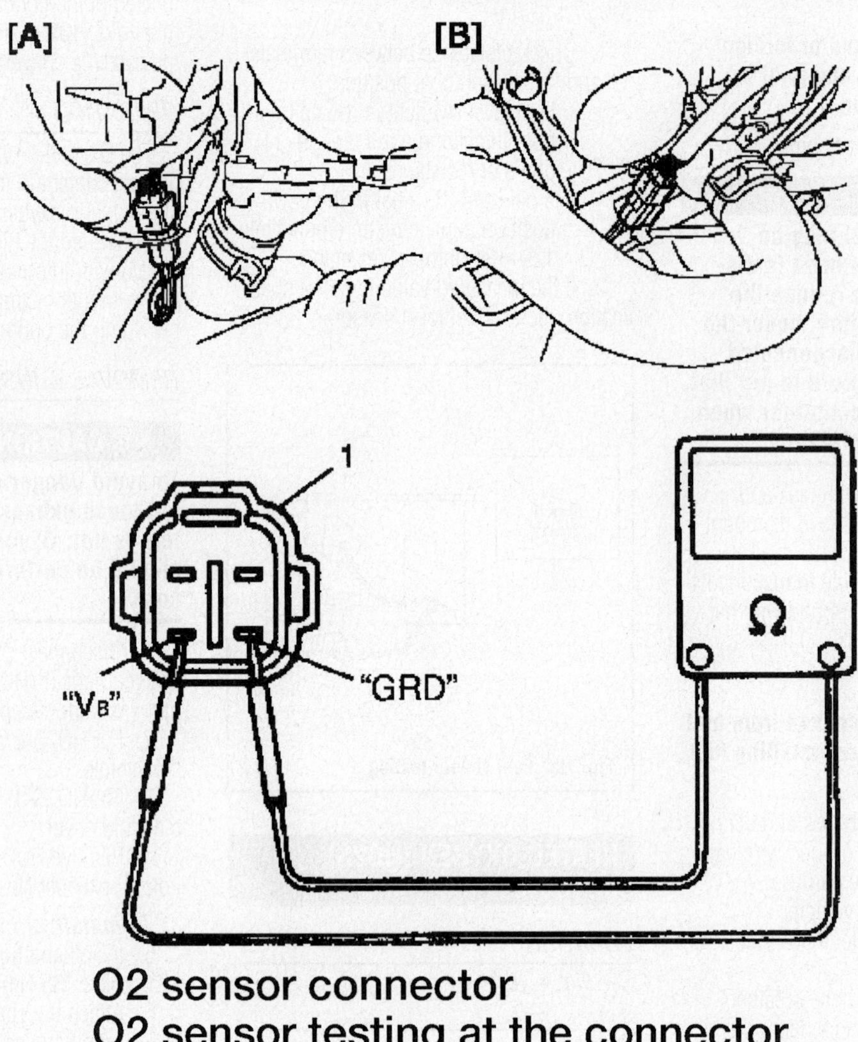

Fig. 108 Testing the HO2S heater

b. With ignition switch turn Off, check for proper connection to HO2S-1 connector at "WHT", "ORN", "PNK/BLK" and "BLK/WHT" wire terminals.

c. If wire and connection are OK, check there is continuity between "ORN" wire terminal and engine ground.

d. With ignition switch turned ON, check voltage between "ORN" wire terminal and engine ground

e. Voltage should be 0.1 volt or less.

3. Check the HO2S heater, as follows:

a. Disconnect sensor connector (1).

b. Using ohmmeter, measure resistance between terminals "VB" and "GND" of sensor connector.

c. Readings should be 2–6 ohms at: 68°F, 20°C for sensor 1.

d. Readings should be 4–10 ohms at: 68°F, 20°C for sensor 2

➡If found faulty, replace oxygen sensor.

IDLE AIR CONTROL (IAC) VALVE

LOCATION

The Idle Air Control (IAC) valve is attached to the throttle body on the intake manifold.

OPERATION

The Idle Air Control (IAC) valve provides a small secondary air path that increases and/or reduces when the valve is opened and/or when it is closed. The IAC adjusts engine idle and maintains idle under high loads, i.e. air-conditioning and power steering with engine at idle while parking.

REMOVAL & INSTALLATION

See Figure 109.

1. Remove the throttle body from the intake manifold. Refer to the procedure in the Fuel Section.

2. Remover the Idle Air Control (IAC) valve from the throttle body.

To install:

3. For installation, reverse removal procedure noting the following:

a. Tighten IAC valve screw to 3 ft. lbs. (4.2 Nm).

b. Install a new gasket (1) to throttle body.

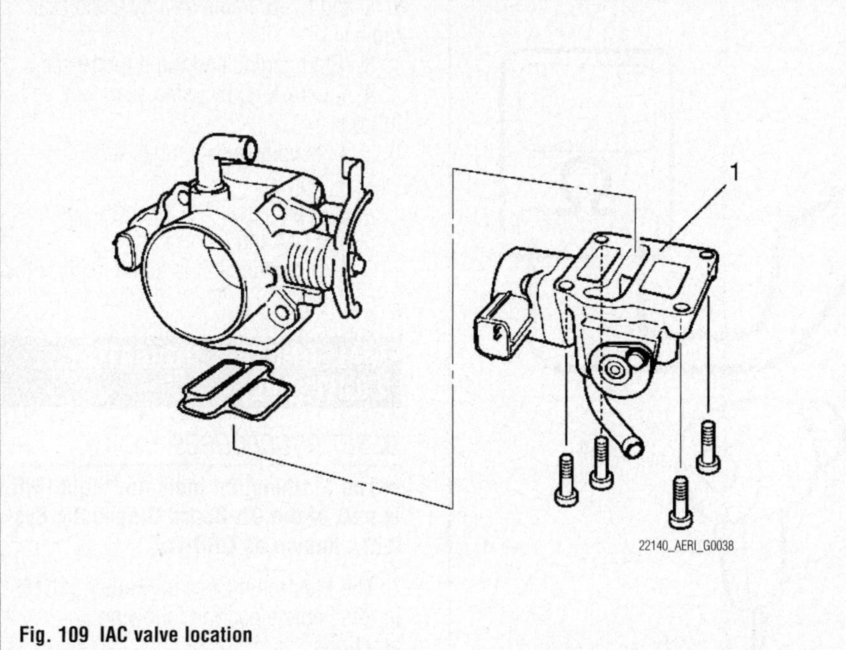

Fig. 109 IAC valve location

TESTING

See Figure 110.

1. If you are using the SUZUKI scan tool, perform the following steps:
   a. Connect SUZUKI scan tool to DLC with ignition switch OFF.
   b. Warm up engine to normal operating temperature.
   c. Clear DTC and select "MISC. TEST" mode on SUZUKI scan tool.
   d. Check that idle speed increases and/or reduces when IAC valve is opened and/or when closed by SUZUKI scan tool.
   e. If idle speed does not change, check IAC valve and wire harness.
2. If you are not Using SUZUKI Scan Tool, perform the following steps:
   a. Warm up engine to normal operating temperature.
   b. Stop engine and then turn ON ignition switch.
   c. Disconnect connector (1) from IAC valve (2).
   d. Start engine.
   e. Check that idle speed increases and/or reduces when connector is connected to IAC valve.

Fig. 110 IAC test socket

   f. If idle speed does not change, check IAC valve and wire harness.
   g. After checking the valve operation, clear DTC and pending DTC.

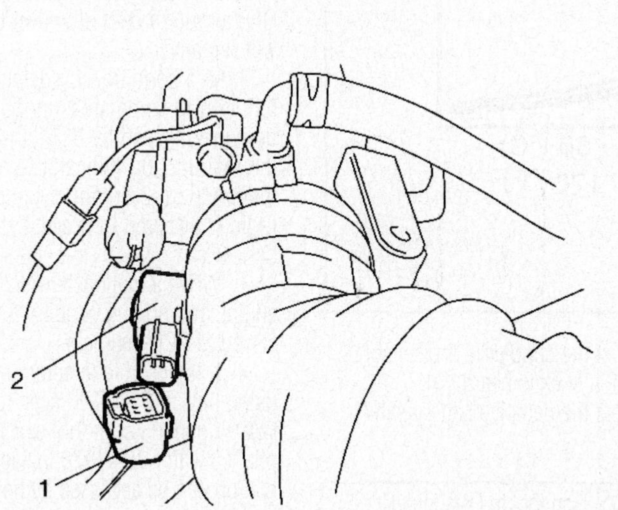

INTAKE AIR TEMPERATURE (IAT) SENSOR

LOCATION

The Intake Air Temperature (IAT) sensor is located in the top of the throttle housing attached to the intake manifold.

OPERATION

The Intake Air Temperature (IAT) sensor monitors ambient air temperature and reduces resistance as temperature rises. The IAT sensor provides air density to the ECM for fuel adjustments.

REMOVAL & INSTALLATION

1. Disconnect the negative battery cable.
2. Disconnect the coupler from Mass Air Flow (MAF) and Intake Air Temperature (IAT) sensor.
3. Remove MAF and IAT sensor from air cleaner assembly.

To install:

4. Connect coupler to MAF and IAT sensor securely until a click is heard.
5. Connect the negative battery cable.

TESTING

See Figure 111.

1. Check the Intake Air Temperature (IAT) sensor O-ring for damage and deterioration. Replace as necessary.
2. Blow hot air to temperature sensing part of Mass Air Flow sensor and IAT sensor using hot air drier
3. Measure resistance between sensor terminals while heating air gradually.
4. If measured resistance is not between 1.80–2.20 kohms at 77°F and 0.50–0.68 kohms at 140°F, replace MAF and IAT sensor.

KNOCK SENSOR (KS)

LOCATION

The Knock Sensor (KS) is located in the engine above the exhaust manifold

OPERATION

If the Knock Sensor (KS) generates voltage of 1.0–3.7 volts or more for 5 seconds, a Diagnostic Trouble Code (DTC) will occur, turning on the Malfunction Indicator Light (MIL) after 2 failures.

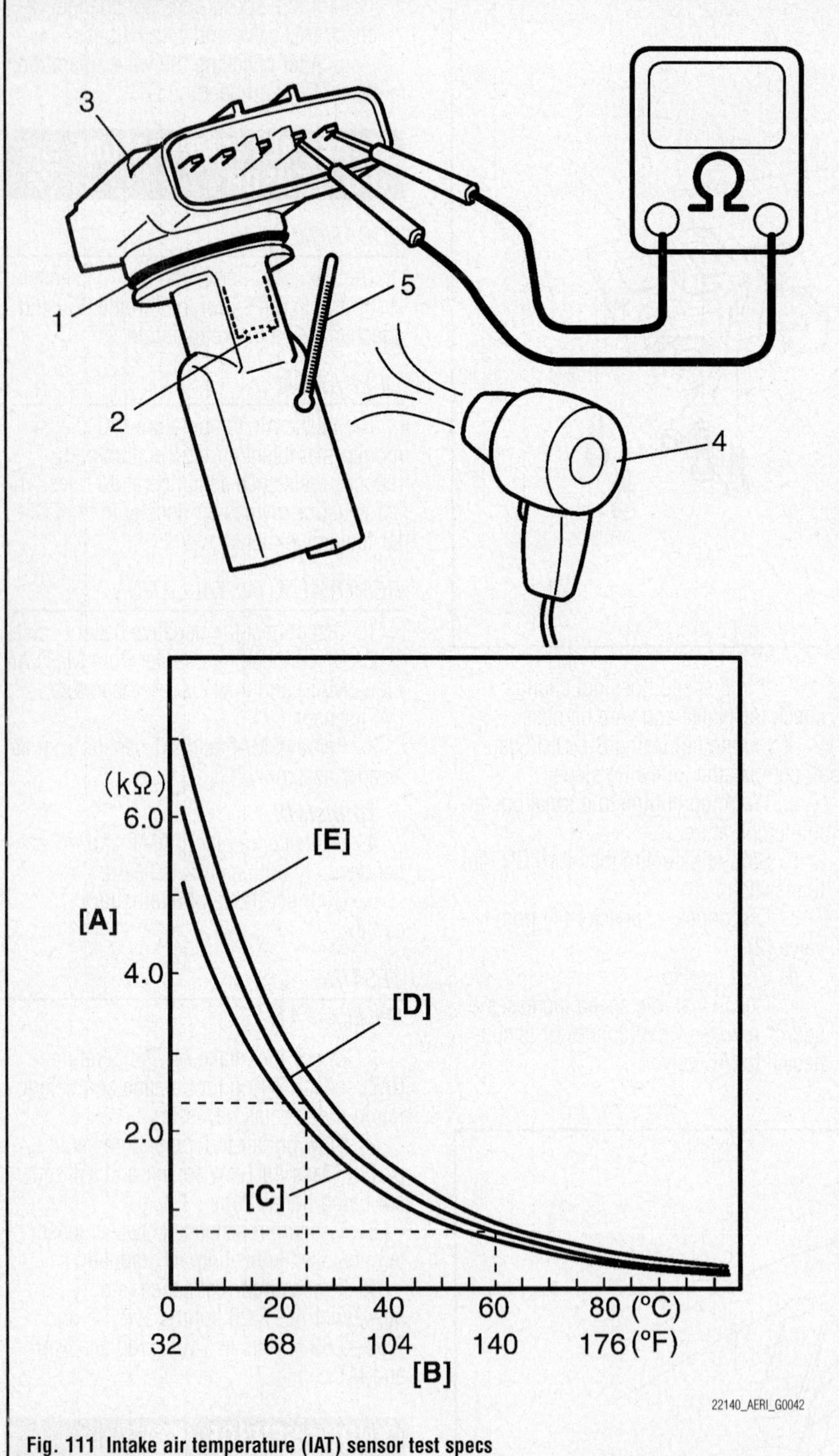

Fig. 111 Intake air temperature (IAT) sensor test specs

REMOVAL & INSTALLATION

1. Disconnect the negative battery cable.
2. Detach the Knock Sensor (KS) electrical connector.
3. Remove the retaining bolt, then remove the KS.

To install:

4. Install the KS and tighten the bolt to 9.5 ft. lbs. (13 Nm).
5. Connect the coupler to the sensor securely until a click is heard
6. Connect the negative battery cable.

TESTING

1. Connect scan tool to Data Link Connector (DLC) with ignition switch turned **OFF**.
2. Turn the ignition switch **ON** and clear Diagnostic Trouble Code (DTC), pending DTC and freeze frame data by using scan tool.
3. Start engine and run it for 10 sec.
4. Check DTC by using scan tool, as follows:
   - Is voltage within 0.5–4.5 volts?
   - If OK substitute a known-good ECM and recheck.
   - If malfunction is found, replace the KS.

MALFUNCTION INDICATOR LIGHT (MIL)

RESET PROCEDURES

➡The Malfunction Indicator Light (MIL) is part of the On-Board Diagnostic System., known as OBD-II.

The Electronic Control Module (ECM) in this vehicle has the following functions:

- When the ignition switch is turned **ON** with the engine at a stop, the Malfunction Indicator Lamp (MIL) turns ON to check the circuit of the malfunction indicator lamp.
- When ECM detects a malfunction which gives an adverse effect to vehicle emission while the engine is running, it makes the malfunction indicator lamp in the meter cluster of the instrument panel turn ON or flash (flashing only when detecting a misfire which can cause damage to the catalyst) and stores the malfunction area (Diagnostic Trouble Code (DTC) according to SAE J2012) in its memory. If it detects that continuously 3 driving cycles are normal after detecting a malfunction, however, it makes MIL (1) turn OFF although DTC stored in its memory will remain.
- As a condition for detecting a malfunction in some areas in the system being monitored by ECM and turning ON the malfunction indicator lamp due to that malfunction, 2 driving cycle detection logic is adopted to prevent erroneous detection.
- When a malfunction is detected, engine and driving conditions then are stored in ECM memory.
- It is possible to communicate by using not only SUZUKI scan tool but also OBD-II generic scan tool which are in compliance with SAE J1978. (Diagnostic information can be accessed by using a scan tool.)

To clear DTC's there are several options:

- Use the SUZUKI scan tool

• Use a compliant OBD-II generic scan tool
• Disconnect the battery cable for 15 minutes
• Remove the ECM fuse or disconnect ECM connectors
• Repair problem and the DTC is not detected again during 40 engine warm-up cycles

MASS AIR FLOW (MAF) SENSOR

LOCATION

See Figure 112.

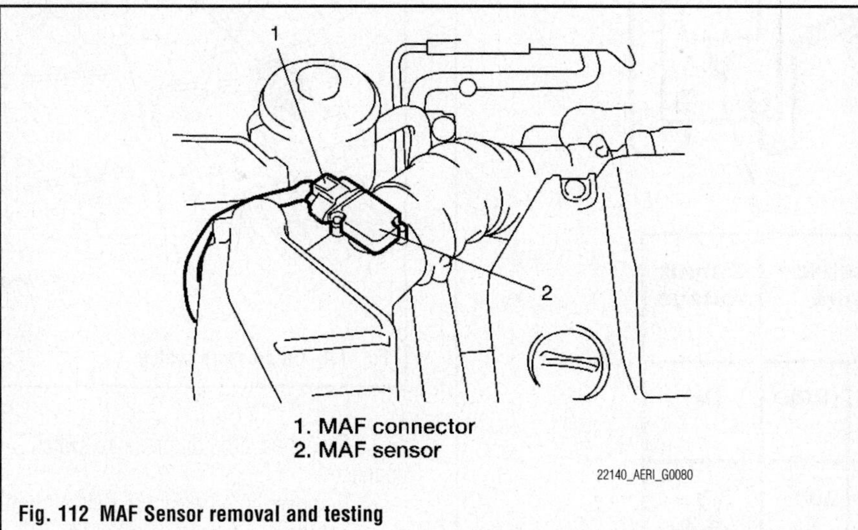

Fig. 112 MAF Sensor removal and testing

The Mass Air Flow (MAF) sensor is located in the top of the throttle housing attached to the intake manifold.

OPERATION

The Mass Air Flow (MAF) sensor measures the volume of air that passes through its body and provides the information to the ECM.

REMOVAL & INSTALLATION

See Figure 113.

1. Disconnect negative battery cable at battery.
2. Disconnect coupler from Mass Air Flow Sensor (MAF) and Intake Air Temperature (IAT) sensor.
3. Remove MAF and IAT sensor from air cleaner assembly.

To install:

4. Install MAF and IAT sensor.
5. Connect coupler to MAF and IAT sensor securely until a click is heard.
6. Connect negative battery cable at battery.

TESTING

See Figure 113.

1. Check Mass Air Flow Sensor (MAF) and Intake Air Temperature (IAT) sensor O-ring for damage and deterioration. Replace as necessary.
2. Blow hot air to temperature sensing part of MAF and IAT sensor using hot air drier and measure resistance between sensor terminals while heating air gradually.
3. If measured resistance does not show such characteristic as shown in specifications below, replace MAF and IAT sensor.
4. Measure MAF/IAT sensor resistance:
   a. At 77° F (25°C): 1.80–2.20 kohms,
   b. At 140°F (60°C): 0.50–0.68 kohms.

MANIFOLD ABSOLUTE PRESSURE (MAP) SENSOR

LOCATION

1. The Manifold Absolute Pressure (MAP) sensor is located on the intake manifold.

OPERATION

The Manifold Absolute Pressure (MAP) sensor measures the barometric pressure and provides an output voltage to the Electronic Control Module (ECM) to assist in fuel metering.

The (MAP) voltage drops as the barometric pressure drops.

REMOVAL & INSTALLATION

1. Disconnect the negative battery cable.
2. Remove Manifold Absolute Pressure (MAP) sensor connector.
3. Remove MAP sensor.

To install:

4. Install MAP sensor securely.
5. Connect MAP sensor connector securely.
6. Connect the negative battery cable.

TESTING

See Figures 114.

1. Disconnect connector from the Manifold Absolute Pressure (MAP) sensor.
2. Remove MAP sensor.
3. Arrange 3 new 1.5 V batteries in series (check that total voltage is 4.5–5.0 V) and connect its positive terminal to "Vin" terminal of sensor and negative terminal to "Ground" terminal. Then check voltage between "Vout" and "Ground".

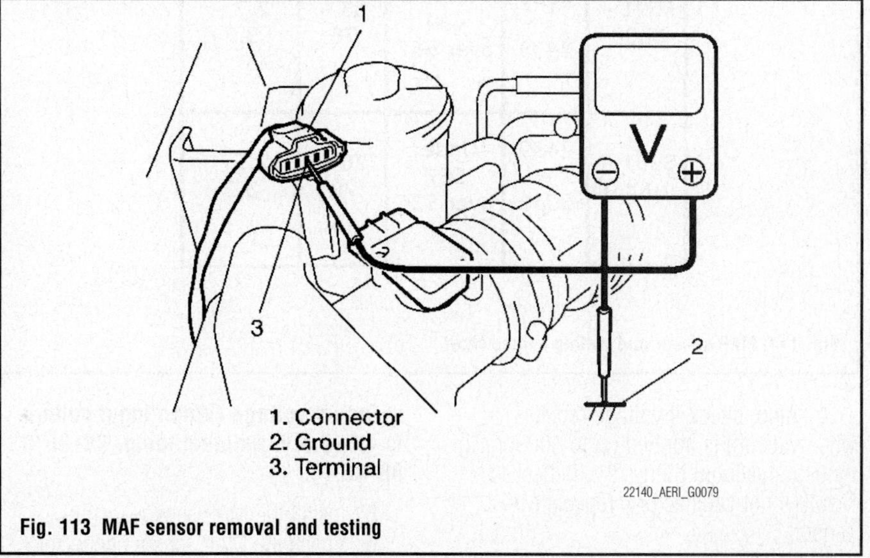

Fig. 113 MAF sensor removal and testing

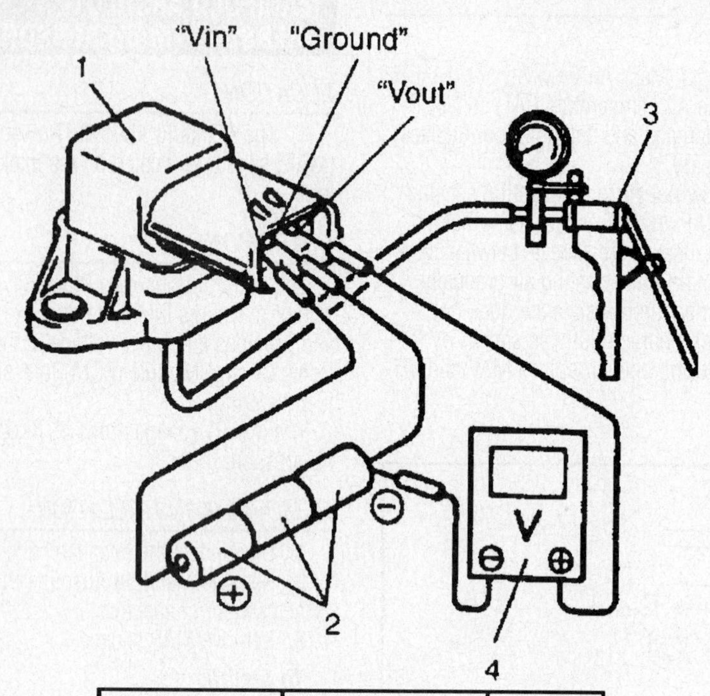

Altitude (Reference)		Barometric pressure		Output voltage
(ft)	(m)	(mmHg)	(kPa)	(V)
0 – 2000	0 – 610	760 – 707	100 – 94	3.3 – 4.3
2001 – 5000	611 – 1524	Under 707 over 634	94 – 85	3.0 – 4.1
5001 – 8000	1525 – 2438	Under 634 over 567	85 – 76	2.7 – 3.7
8001 – 10000	2439 – 3048	Under 567 over 526	76 – 70	2.5 – 3.3

22140_AERI_G0039

Fig. 114 MAP sensor and testing values chart

4. Also, check if voltage reduces when vacuum is applied up to 400 mmHg by using vacuum pump (3). If checked result is not satisfactory, replace MAP sensor.

➡Output voltage (When input voltage is 4.5–5.5 V, ambient temp. 20–30°C, 68–86°F).

5. Install the MAP sensor securely.
6. Attach the MAP sensor connector.

OIL PRESSURE SWITCH

LOCATION

The oil pressure switch is located in the cylinder head under the exhaust port.

REMOVAL & INSTALLATION

See Figure 115.

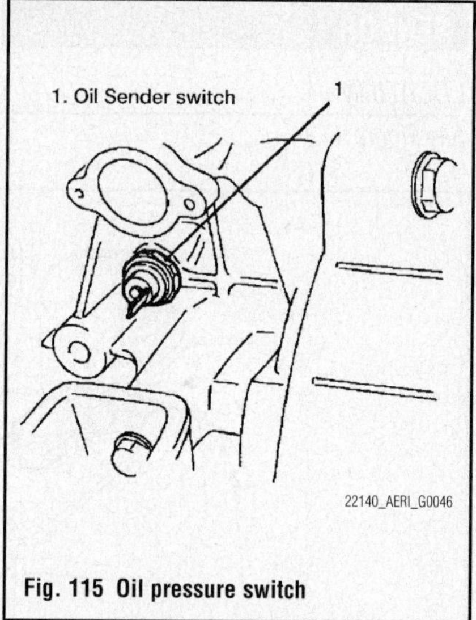

Fig. 115 Oil pressure switch

1. Disconnect the oil pressure switch connector.
2. Remove the oil pressure switch from cylinder block.

To install:

3. Wrap the oil pressure switch screw threads with sealing tape.
4. Install the switch and tighten to 11 ft. lbs. (15 Nm).
5. Attach the oil pressure switch connector.

TESTING

1. Remove the oil pressure switch.
2. Install oil pressure gauge to vacated threaded hole.
3. Start engine and warm it up to normal operating temperature
4. Oil pressure specification must be more than 55.5 psi

THROTTLE POSITION SENSOR (TPS)

LOCATION

See Figure 116.

The Throttle Position Sensor (TPS) is located on the left side of the throttle housing.

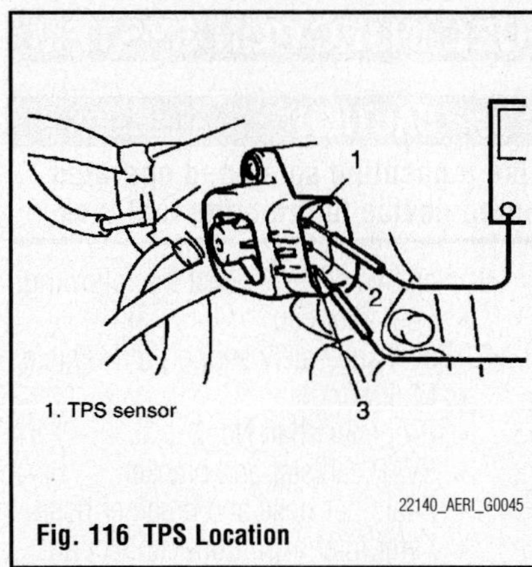

Fig. 116 TPS Location

OPERATION

The Throttle Position (TPS) sensor provides an output voltage to the ECM for proper fuel management.

REMOVAL & INSTALLATION

See Figure 117.

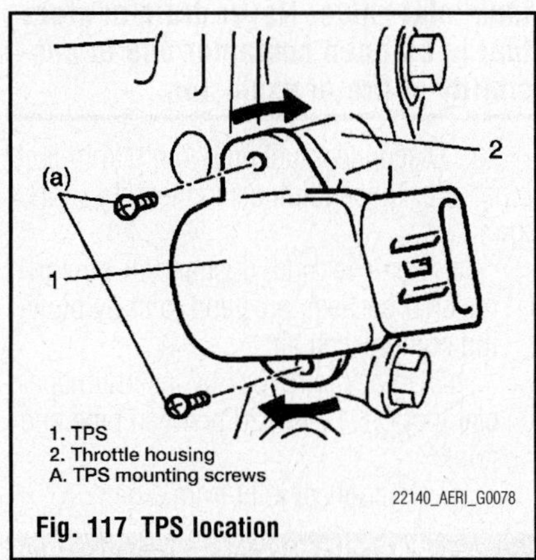

Fig. 117 TPS location

1. Disconnect battery negative battery cable.
2. Disconnect coupler from The Throttle Position Sensor (TPS) sensor.
3. Remove TP sensor from throttle body.

To install:

4. Install TP sensor to throttle body.
5. Fit TPs sensor to throttle body in such way that its holes are a little away from TP sensor screw holes as shown in the figure and turn TP sensor (1) clockwise so that those holes align.
6. Tighten TP sensor screw to specified torque.
7. Tightening torque for the TPS sensor screw is: 1.8 ft. lb. (25 Nm)
8. Connect coupler to TP sensor securely.
9. Connect battery negative cable to battery.

TESTING

See Figure 116.

1. Turn OFF ignition switch and connect SUZUKI scan tool to DLC.
2. Turn ON ignition switch and check TP sensor output voltage when throttle valve is at idle position, voltage should be 0.2–1.0 volts.
3. Turn ON ignition switch and check TP sensor output voltage when throttle valve is at the fully opened position, voltage should be 2.8–4.8 volts.

VEHICLE SPEED SENSOR (VSS)

LOCATION

See Figure 118.

The Vehicle Speed Sensor (VVS) is located in the top of the transmission housing.

OPERATION

Check that VSS signal voltage varies from low to high or from high to low dependent on RPMs of the output shaft of the transmission. Range of voltage is between 0–6 volts.

REMOVAL & INSTALLATION

See Figure 118.

1. Disconnect the negative battery cable.
2. Disconnect the Vehicle Speed Sensor (VSS) coupler.
3. Remove the VSS.

To install:

4. Check O-ring and VSS surface for damage, apply oil to O-ring and then install VSS to transaxle.

➡Tightening torque for VSS bolt is 4 ft. lbs. (5.5 Nm).

TESTING

1. Turn ignition switch to the OFF position, then disconnect the Vehicle Speed Sensor (VSS) connectors.
2. Turn ON ignition switch, without running engine.
3. Measure voltage from "PPL" wire terminal of VSS connector to engine ground.
4. If voltage is between 4–5 volts then check dash display.

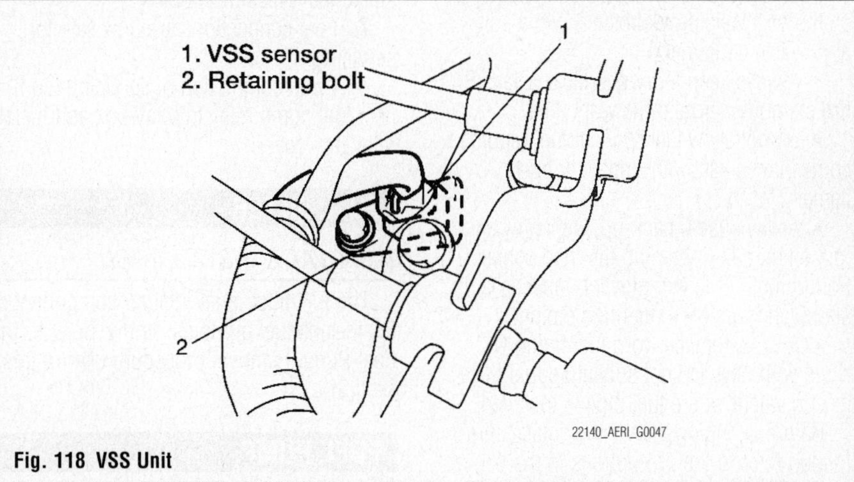

Fig. 118 VSS Unit

FUEL SYSTEM SERVICE PRECAUTIONS

Safety is the most important factor when performing not only fuel system maintenance but any type of maintenance. Failure to conduct maintenance and repairs in a safe manner may result in serious personal injury or death. Maintenance and testing of the vehicle's fuel system components can be accomplished safely and effectively by adhering to the following rules and guidelines.

- To avoid the possibility of fire and personal injury, always disconnect the negative battery cable unless the repair or test procedure requires that battery voltage be applied.
- Always relieve the fuel system pressure prior to disconnecting any fuel system component (injector, fuel rail, pressure regulator, etc.), fitting or fuel line connection. Exercise extreme caution whenever relieving fuel system pressure to avoid exposing skin, face and eyes to fuel spray. Please be advised that fuel under pressure may penetrate the skin or any part of the body that it contacts.
- Always place a shop towel or cloth around the fitting or connection prior to loosening to absorb any excess fuel due to spillage. Ensure that all fuel spillage (should it occur) is quickly removed from engine surfaces. Ensure that all fuel soaked cloths or towels are deposited into a suitable waste container.
- Always keep a dry chemical (Class B) fire extinguisher near the work area.
- Do not allow fuel spray or fuel vapors to come into contact with a spark or open flame.
- Always use a back-up wrench when loosening and tightening fuel line connection fittings. This will prevent unnecessary stress and torsion to fuel line piping.
- Always replace worn fuel fitting O-rings with new. Do not substitute fuel hose or equivalent where fuel pipe is installed.

Before servicing the vehicle, make sure to also refer to the precautions in the beginning of this section as well.

RELIEVING FUEL SYSTEM PRESSURE

See Figure 119.

CAUTION

This work must not be done when engine is hot. If done so, it may cause adverse effect to catalyst.

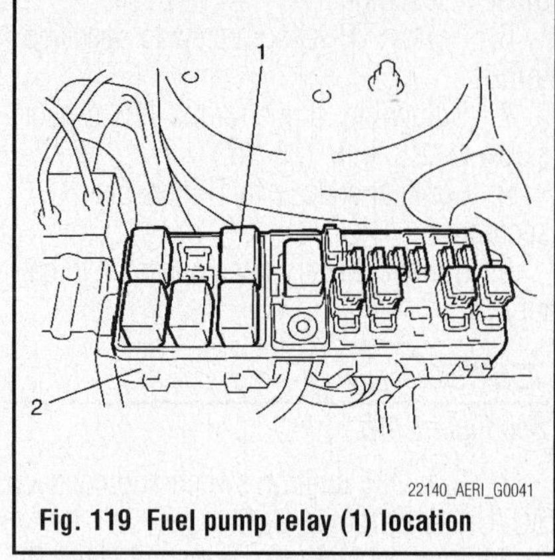

Fig. 119 Fuel pump relay (1) location

1. After making sure that engine is cold, release fuel pressure as follows.
2. Place transmission gearshift lever in "Neutral" (Shift selector lever to "P" range for automatic transmissions model), set parking brake, and block drive wheels.
3. Remove relay box cover.
4. Disconnect fuel pump relay from relay box.
5. Remove fuel filter cap to release fuel vapor pressure in fuel tank and then reinstall it.
6. Start engine and run it till it stops for lack of fuel. Repeat cranking engine 2–3 times for about 3 seconds each time to dissipate fuel pressure in lines.
7. Fuel connections are now safe for servicing.
8. Upon completion of servicing, connect fuel pump relay to relay box and install relay box cover.

FUEL FILTER

REMOVAL & INSTALLATION

The fuel filter is an integral component of the in-tank fuel pump assembly. Refer to the Fuel Pump Removal procedure later in this section.

FUEL PUMP

REMOVAL & INSTALLATION

See Figure 120.

1. Before servicing the vehicle, refer to the Precautions Section.
2. Relieve the pressure from the fuel system.
3. Disconnect the negative battery cable.
4. Drain the fuel from the tank by pumping the fuel out through the fuel tank filler.

CAUTION

Use a gasoline safe hand operated pump device to drain the fuel tank.

5. Remove or disconnect the following:
   - Fuel filler cap
6. Raise and safely support the vehicle.
   - Muffler
   - Propeller shaft No. 2
   - EVAP canister and bracket
   - Fuel filler hose and breather hose
   - Filter inlet hose from canister air suction filter
7. Due to absence of fuel tank drain plug, drain fuel tank by pumping fuel out through fuel tank filler. Use hand operated pump device to drain fuel tank.

CAUTION

Do not force pump hose into fuel tank, or pump hose may damage fuel tank inlet valve. Never drain or store fuel in an open container due to possibility of fire or explosion.

8. Disconnect fuel pipe joints from fuel pipes. For quick joint, disconnect it as follows:
   a. Remove mud, dust and/or foreign material between pipe and joint by blowing compressed air.
   b. Unlock joint lock by inserting special tool 09919-47020 between pipe and joint.
   c. Disconnect joint from pipe.

CAUTION

A small amount of fuel may be released after the fuel hose is disconnected. In order to reduce the chance of personal injury, cover the hose and pipe to be disconnected with a shop cloth. Be sure to put that cloth in an approved container when disconnection is completed.

9. Support the fuel tank using a transmission jack.
   - Two bands that secure the fuel tank
10. Slowly and carefully lower the fuel tank enough to disconnect the fuel pump connector, tank pressure and temperature sensor connector and ground wire harness.
    - Fuel tank from vehicle
    - Fuel cut valve hose, fuel feed line and fuel return pipe from fuel pump assembly
    - Fuel pump assembly from fuel tank

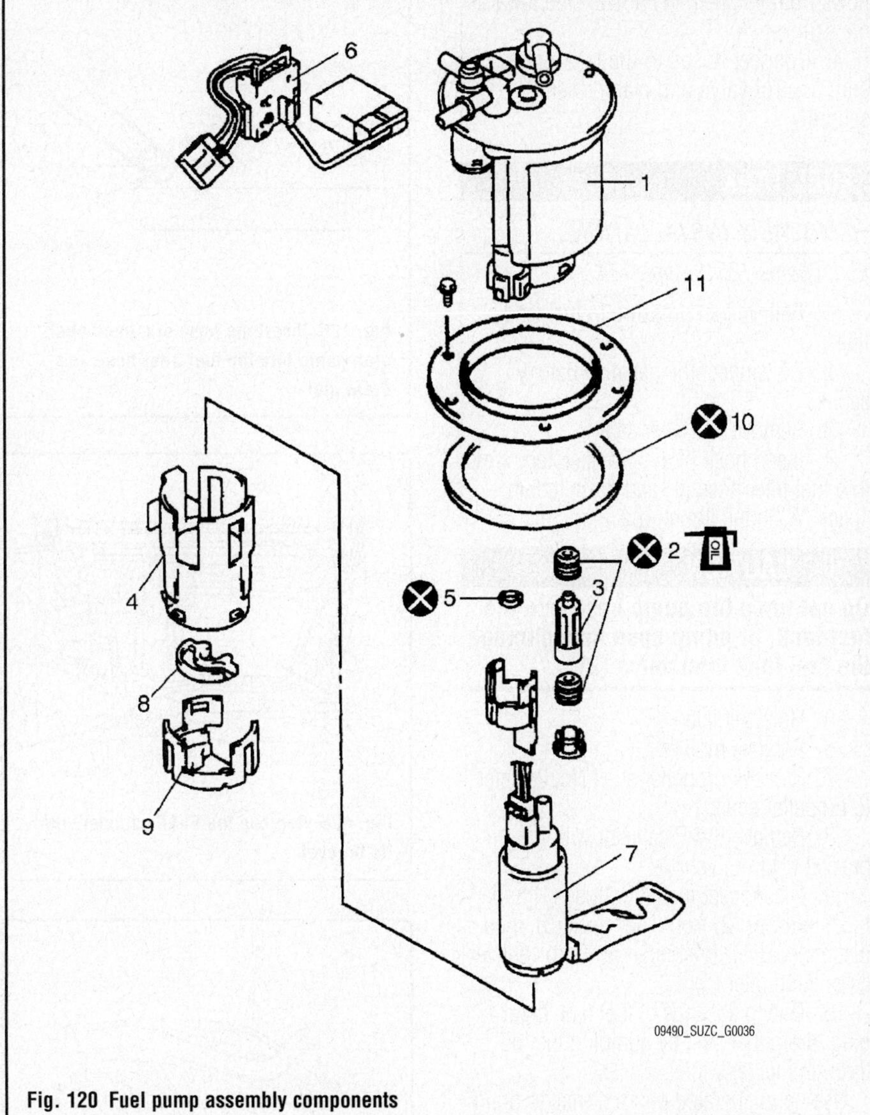

Fig. 120 Fuel pump assembly components

To install:

11. Clean mating surfaces of fuel pump assembly and fuel tank.
12. Put plate on fuel pump assembly by matching the protrusion of fuel pump assembly to plate hole.
13. Install new gasket and fuel pump assembly with plate to fuel tank and tighten the assembly bolts to 7.5 ft. lbs. (10 Nm).
14. Install or connect the following:
    - Fuel cut valve hose, fuel feed line and fuel return pipe to the fuel pump assembly

➡When connecting joint, clean outside surface of pipe where joint is to be inserted, push joint into pipe till joint lock clicks and check to ensure that pipes are connected securely, or fuel leak may occur.

15. Raise fuel tank with jack and connect fuel pump connector, tank pressure and temperature sensor connector and ground wire harness and then clamp wire harness.
    - Fuel tank to the vehicle. Tighten the fuel tank strap bolts to 36.5 ft. lbs. (50 Nm).
    - Fuel filler hose, breather hose and filter inlet hose. Clamp them securely and tighten to 2.5 ft. lbs. (3.5 Nm).
    - Fuel feed pipe, fuel return pipe and purge pipe and clamp securely

➡When connecting joint, clean outside surfaces of pipe where joint is to be inserted, push joint into pipe till joint lock clicks and check to ensure that pipes are connected securely, or fuel leak may occur.

- EVAP canister and bracket
- Propeller shaft No. 2
- Muffler
- Lower the vehicle
- Fuel filler cap
- Negative battery cable

16. Fill the fuel tank enough to check for fuel leaks.
17. Turn the ignition switch to the **ON** position, but leave the engine **OFF** and check for fuel leaks.

FUEL PRESSURE REGULATOR

REMOVAL & INSTALLATION

See Figures 121 through 123.

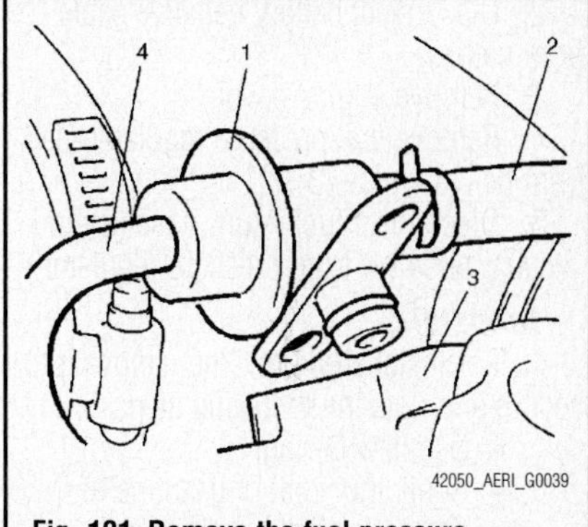

Fig. 121 Remove the fuel pressure regulator

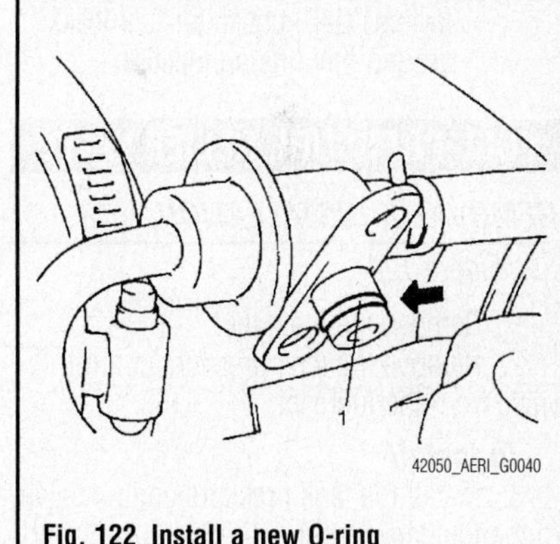

Fig. 122 Install a new O-ring

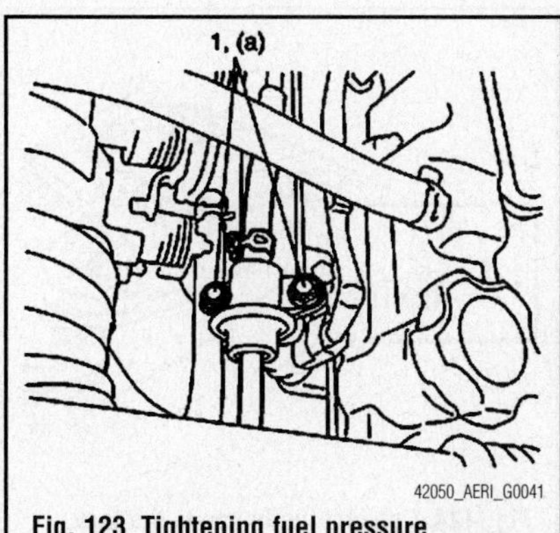

Fig. 123 Tightening fuel pressure regulator bolts

WARNING

A small amount of fuel may be released when it is removed from the delivery pipe. Place a shop cloth under delivery pipe so that released fuel is absorbed in it.

WARNING

A small amount of fuel may be released when hose is disconnected. Cover the hose to be disconnected with a shop cloth.

1. Relieve fuel pressure.
2. Disconnect battery negative cable from battery.
3. Remove engine cover.
4. Remove fuel pressure regulator (1) from delivery pipe (3).
5. Disconnect fuel return hose (2) and vacuum hose (4) from pressure regulator.

To install:

6. For installation, reverse removal procedure and note the following steps:
   - Use new O-ring.
   - Apply thin coat of gasoline to O-ring to facilitate installation.
   - Tighten fuel pressure regulator bolts (1) to 8 ft. lbs. (12 Nm).
   - With engine "OFF" and ignition switch "ON", check for fuel leaks around fuel line connection

FUEL PRESSURE RELIEF VALVE

REMOVAL & INSTALLATION

See Figure 124.

1. Remove the fuel tank.
2. Remove the tank pressure control valve from the fuel tank.

To install:

3. Install the tank pressure control valve to the fuel tank in such direction that vapor flows from its green (1) nozzle side towards the canister (A).
4. Connect hoses to the tank pressure control valve and clamp them securely.

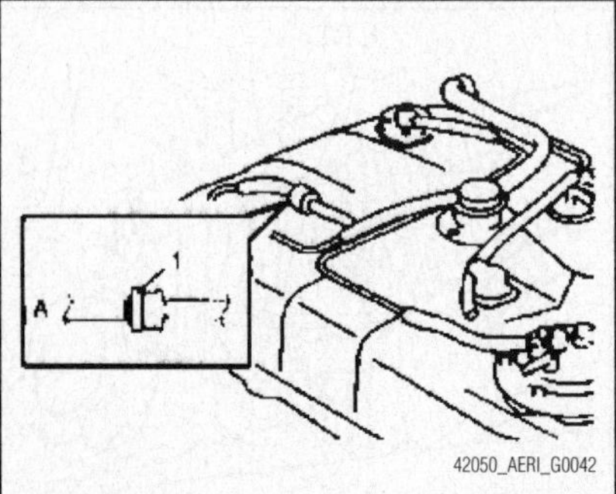

Fig. 124 Connect the hoses to the tank pressure control valve

FUEL TANK

REMOVAL & INSTALLATION

See Figures 125 through 134.

1. Relieve fuel pressure in fuel feed line.
2. Disconnect the negative battery cable.
3. Remove fuel filler cap.
4. Insert hose of a hand operated pump into fuel filler hose (1) and drain fuel in space "A" in the figure.

CAUTION

Do not force the pump hose into the fuel tank, or pump hose may damage the fuel tank inlet valve (2).

5. Hoist vehicle.
6. Remove muffler.
7. Remove propeller shaft No. 2. Refer to propeller shaft.
8. Remove EVAP canister with its bracket (1) from vehicle.
9. Disconnect fuel filler hose (1) and breather hose (2) from filler neck (3), then disconnect filter inlet hose (4) from canister air suction filter (5).
10. Due to absence of fuel tank drain plug, drain fuel tank by pumping fuel out through fuel tank filler.

Use hand operated pump device to drain fuel tank.

CAUTION

Do not force pump hose into the fuel tank, or pump hose may damage fuel tank inlet valve.

CAUTION

Never drain or store fuel in an open container due to possibility of fire or explosion.

11. Disconnect fuel pipe joints from fuel pipes. For quick joint, disconnect it as follows:
    a. Remove mud, dust and/or foreign material between pipe and joint by blowing compressed air.
    b. Unlock joint (1) lock by inserting special tool between pipe (2) and joint (1).
    c. Disconnect joint from pipe.

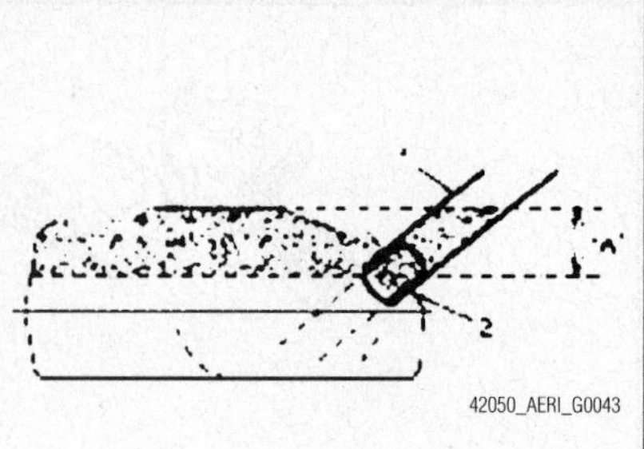

Fig. 125 Insert the hose of a hand operated pump into the fuel filler hose and drain fuel

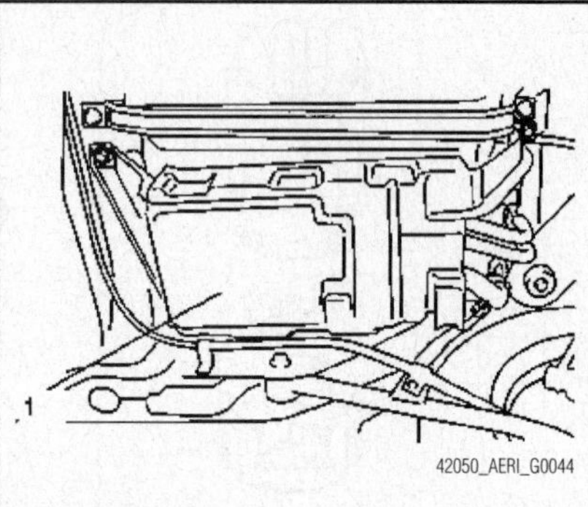

Fig. 126 Remove the EVAP canister with its bracket

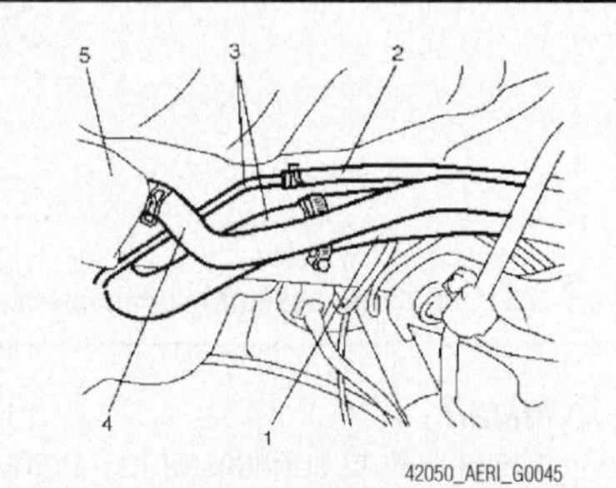

Fig. 127 Disconnect fuel filler hose and breather hose from filler neck, then disconnect filter inlet hose from canister air suction filter

CAUTION

A small amount of fuel may be released after the fuel hose is disconnected. In order to reduce the chance of personal injury, cover the hose and pipe to be disconnected with a shop cloth. Be sure to put that cloth in an approved container when disconnection is completed.

12. Support fuel tank (1) with jack (2) and remove its mounting bolts.
13. Lower fuel tank a little to disconnect fuel pump connector (1), tank pressure and

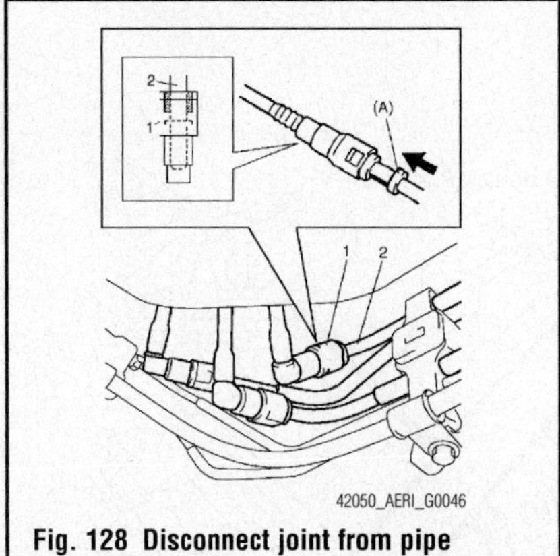

Fig. 128 Disconnect joint from pipe

Fig. 129 Support fuel tank and remove mounting bolts

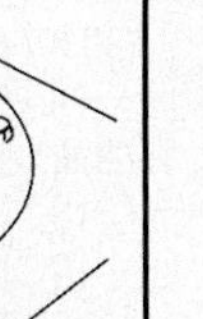

Fig. 130 Disconnect fuel pump connector, tank pressure and temperature sensor connector and ground wire harness

temp. sensor connector (2) and ground wire harness, then remove fuel tank.

To install:

14. If parts have been removed from fuel tank, install them before installing fuel tank to vehicle.
15. Raise fuel tank (1) with jack and connect fuel pump connector, tank pressure and temp. sensor connector and ground wire harness and then clamp wire harness.
16. Install fuel tank to vehicle. Tighten the fuel tank bolt to 36.5 ft. lbs. (50 Nm).
17. Connect fuel filler hose (1), breather hose (2) and filter inlet hose (3) as shown in

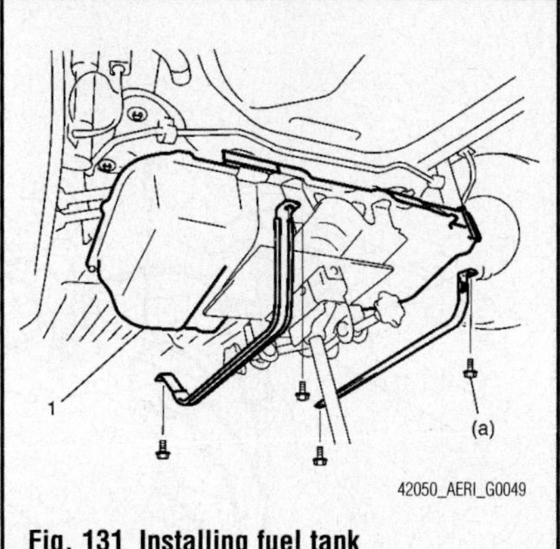

Fig. 131 Installing fuel tank

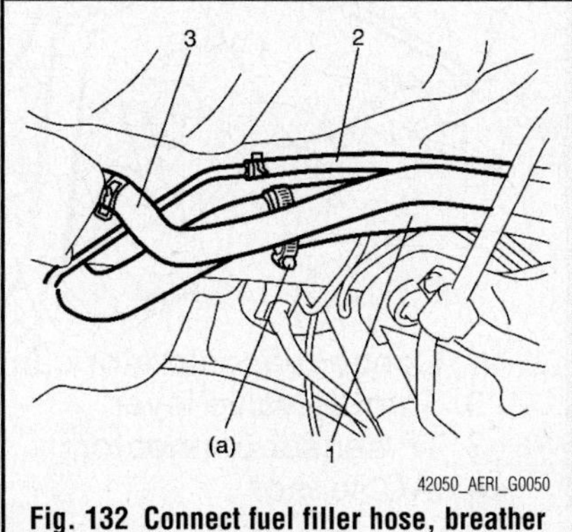

Fig. 132 Connect fuel filler hose, breather hose and filter inlet hose

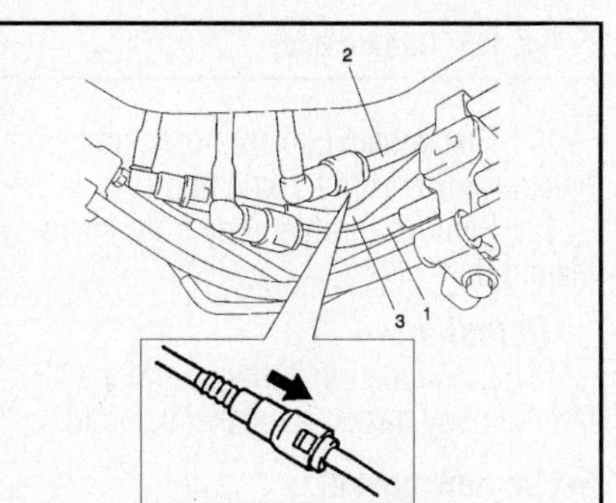

Fig. 133 Connect fuel feed pipe, fuel return pipe and purge pipe

figure and clamp them securely. Tighten fuel filler hose clamp to 2.5 ft. lbs. (3.5 Nm).

18. Connect fuel feed pipe (1), fuel return pipe (3) and purge pipe (2) to each pipe as shown in figure and clamp them securely.

CAUTION

When connecting joint, clean outside surfaces of pipe where joint is to be inserted, push joint into pipe till joint lock clicks and check to ensure that pipes are connected securely, or fuel leak may occur.

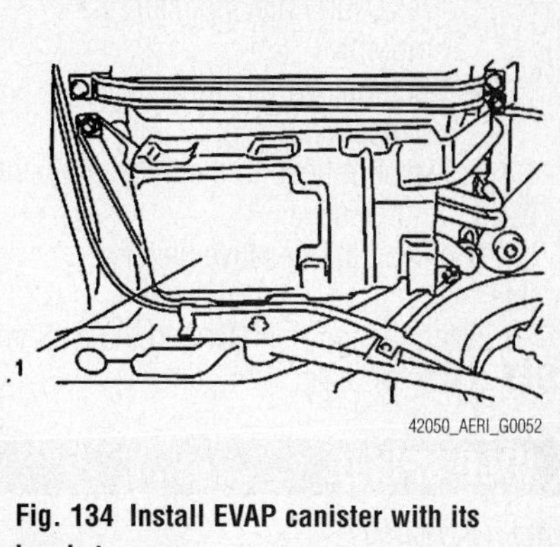

Fig. 134 Install EVAP canister with its bracket

CAUTION

Never let the fuel hoses touch the ABS sensor harness (if equipped).

19. Install EVAP canister with its bracket (1) to vehicle.
20. Install propeller shaft No. 2.
21. Install the muffler.
22. Connect the negative battery cable. With engine OFF, turn ignition switch to ON position and check for fuel leaks.

FUEL RAIL & INJECTORS

REMOVAL & INSTALLATION

1. Before servicing the vehicle, refer to the Precautions Section.
2. Relieve the pressure from the fuel system.
3. Disconnect the negative battery cable.
4. Remove or disconnect the following:
   - Engine cover
   - Vacuum hose from intake manifold
   - Fuel injector electrical connections
   - Fuel rail from cylinder head
   - Fuel injector(s)
   - Injector O-ring and discard

To install:

5. Install or connect the following:
   - Grommet to the injector
   - New injector O-ring
   - Check if fuel rail insulators for damage or scoring and replace if necessary
   - Insulators and cushions to fuel injector and fuel rail
6. Coat the injector O-rings with a thin coat of gasoline.
   - Fuel injector into cylinder head and fuel rail
   - Fuel rail mounting bolts and tighten to 18 ft. lbs. (25 Nm).

Make sure injectors can rotate smoothly.
- Fuel injector electrical connections
- Vacuum hose to the intake manifold
- Engine cover

7. Connect the negative battery cable.
8. With the ignition **ON** and the engine **OFF** check for leaks.

IDLE SPEED

ADJUSTMENT

Idle speed is maintained by the Powertrain Control Module (PCM) and the IAC motor.

THROTTLE BODY

REMOVAL & INSTALLATION

See Figure 135.

WARNING

To help avoid danger of being burned, do not remove coolant drain plug and radiator cap while engine and radiator are still hot. Scalding fluid and steam can be blown out under pressure if plug and cap are taken off too soon.

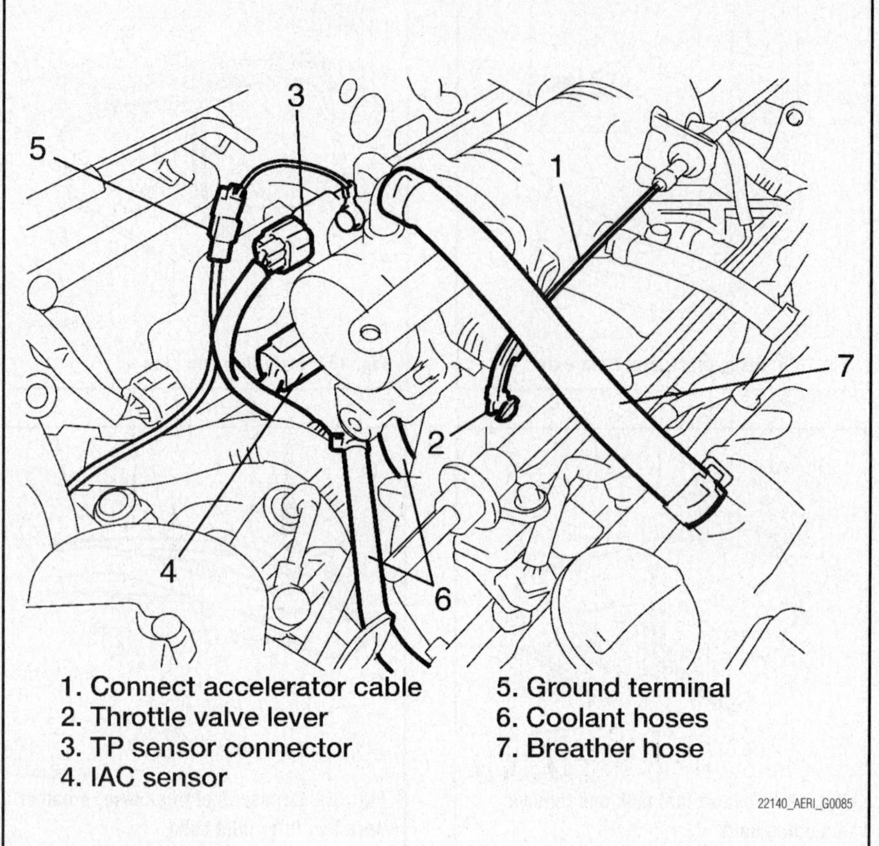

Fig. 135 Throttle Body

1. Before servicing the vehicle, refer to the Precautions Section.
2. Relieve the pressure from the cooling system.
3. Disconnect the negative battery cable.
4. Drain coolant.
5. Remove engine cover.
6. Remove air cleaner outlet hose.
7. Remove generator drive belt.
8. With hose connected, detach P/S pump from its bracket.
9. Disconnect accelerator cable from throttle valve lever.
10. Disconnect couplers of TP sensor, IAC valve and ground connector.
11. Disconnect coolant hoses and breather hose from throttle body.
12. Remove throttle body from intake manifold.

To install:

13. Clean mating surfaces and install throttle body gasket to intake manifold.

➡Use new gasket.

14. Install throttle body to intake manifold and tighten bolts and nuts.
15. Connect coolant hoses and breather hose.
16. Connect couplers of TP sensor, IAC sensor and ground terminal.
17. Connect accelerator cable to throttle valve lever.
18. Install P/S pump to its bracket.
19. Install generator drive belt.
20. Install air cleaner outlet hose.
21. Install engine cover.
22. Refill cooling system.
23. Connect the negative battery cable.
24. Start engine and check for engine coolant leakage.

BLOWER MOTOR

REMOVAL & INSTALLATION

See Figure 136.

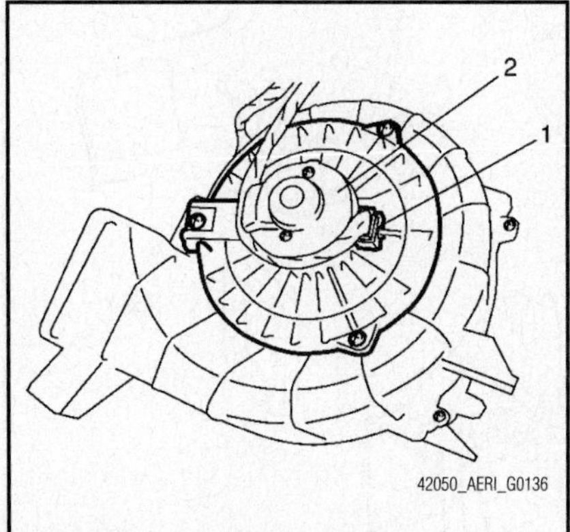

Fig. 136 Remove the blower motor

1. Before servicing the vehicle, refer to the Precautions Section.
2. Remove or disconnect the following:
3. Disconnect negative (-) cable at battery.
4. Remove glove box.
5. Disconnect blower motor connector (1).
6. Remove blower motor (2).

To install:

7. Install blower motor (2).
8. Connect blower motor connector (1).
9. Install the glove box.
10. Connect negative (–) cable at battery.

HEATER CORE

REMOVAL & INSTALLATION

See Figures 137 and 138.

1. Before servicing the vehicle, refer to the Precautions Section.
2. Disconnect the negative battery cable.
3. Recover refrigerant from A/C system.

➡The amount of removed compressor oil must be measured for replenishing compressor oil.

4. Remove suction pipe mounting bolts.
5. Disconnect suction pipe and condenser outlet pipe using special tools 09991—15410 and 09991—15420.

⁂ WARNING

Keep internal parts of air conditioning system free from moisture and dirt. When disconnecting any line from system, install a blind plug or cap to the fitting immediately.

6. Drain engine coolant and disconnect heater hoses from heater unit.
7. Remove heater unit mounting nuts.
8. Remove instrument panel as follows:
   a. Disable air bag system.
   b. Remove steering column hole cover.
   c. Detach steering lower shaft.
   d. Remove glove box and hood latch release lever.
   e. Remove console box.
   f. Remove instrument panel center lower covers.
   g. Remove front pillar trims, front pillar lower garnishes, front side sill scuffs and dash side trims.
   h. Disconnect instrument panel harness connectors and antenna cable which need to be disconnected for removal for instrument panel.

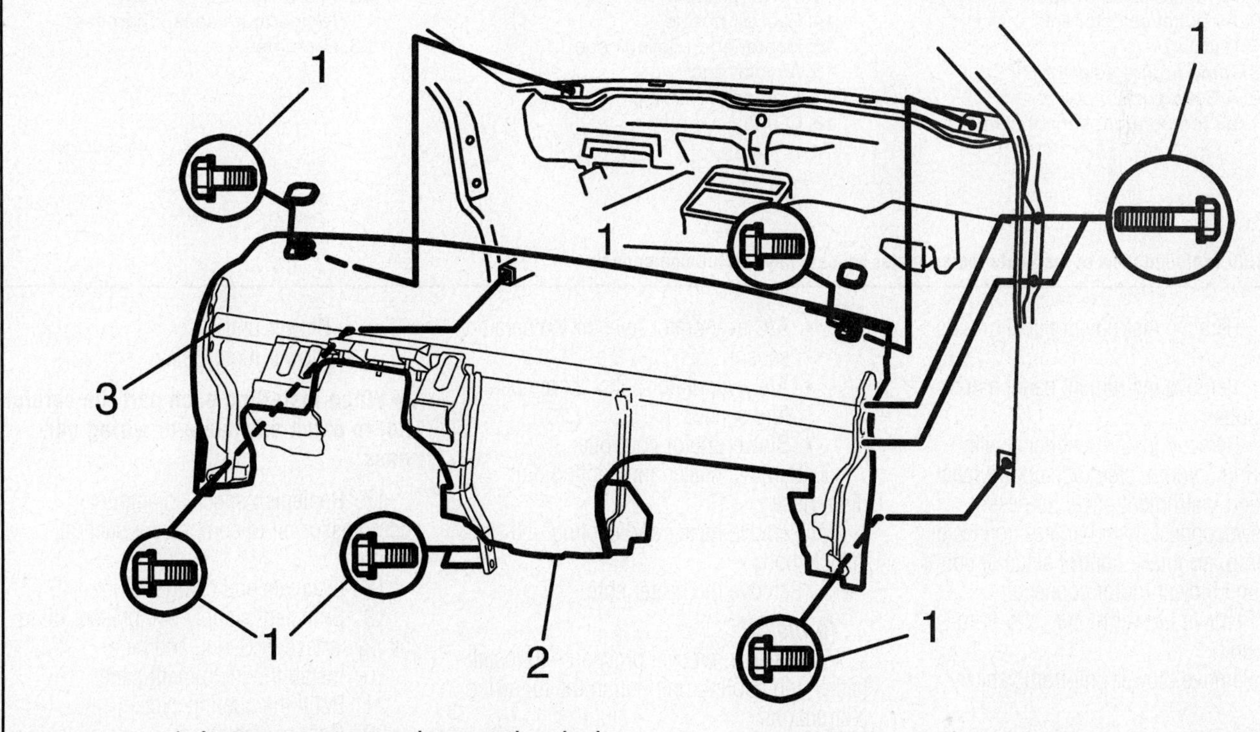

Fig. 137 Instrument panel and mounting bolt locations

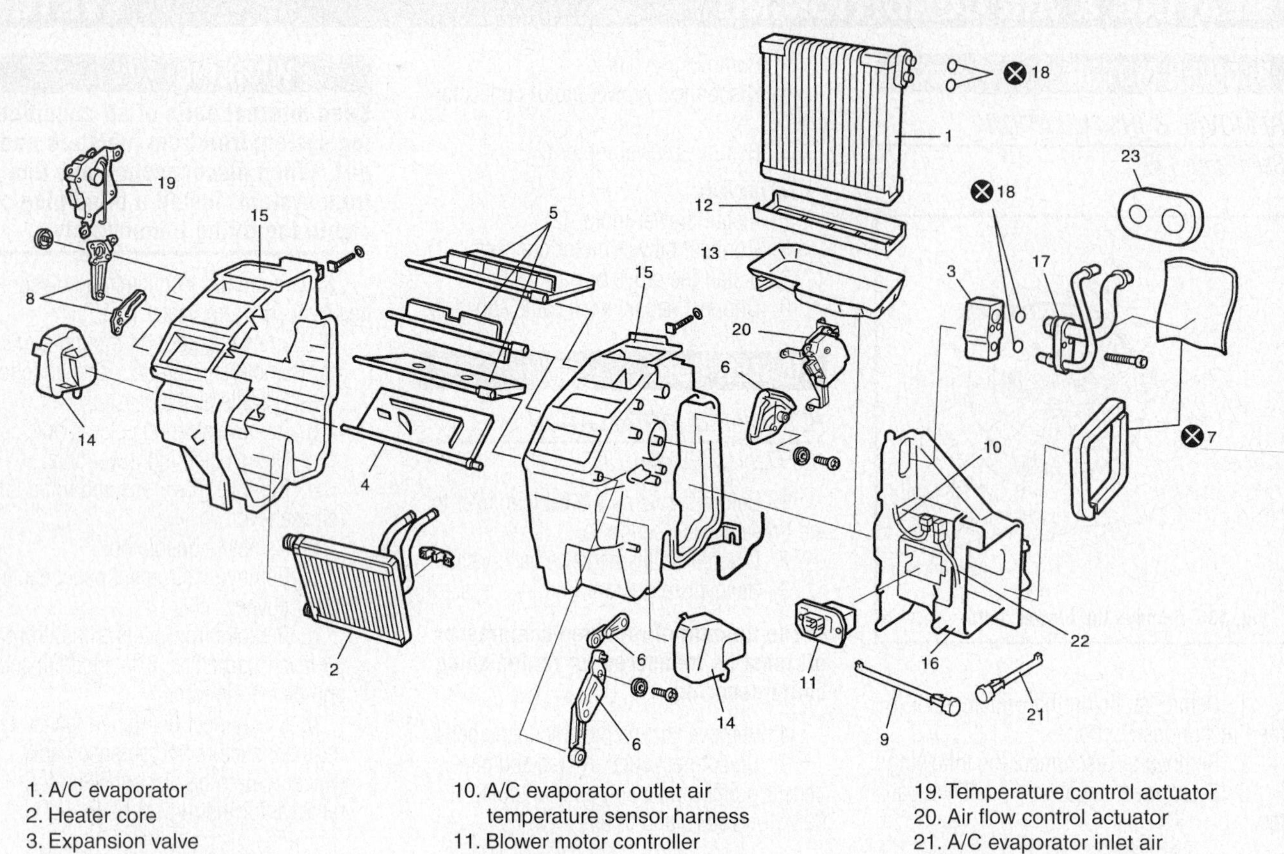

Fig. 138 Exploded view of the heater core, heater housing and related components

i. Remove instrument panel ground wire.

j. Remove instrument panel mounting bolts.

k. Remove instrument panel with steering column, steering support member and instrument panel harness.

9. Disconnect main harness connector and clamp, air intake control actuator connector and blower motor connector.
10. Remove blower motor relay from blower unit.
11. Remove blower unit from vehicle body.
12. Disconnect couplers from the following parts:
    - Temperature control actuator
    - Air flow control actuator
    - A/C evaporator inlet air temperature sensor
    - A/C evaporator outlet air temperature sensor
    - Blower motor controller
13. Remove heater and cooling unit drain hose .
14. Remove heater and cooling unit from vehicle body.
15. Remove the heater core.

To install:

16. Reverse removal procedure to install heater and cooling unit noting the following instructions:
    - Heater core to the heater and cooling unit
    - Heater and cooling unit into the vehicle
    - Blower unit
    - Blower motor relay

➡When installing each part, be careful not to catch any cable or wiring harness.

17. Replenish specified amount of compressor oil to compressor suction side.
18. Evacuate and charge refrigerant.
19. Sufficiently apply compressor oil to fitting surface of O-ring and pipe.
20. Install the instrument panel.
21. Refill the cooling system.
22. Connect the negative battery cable.
23. Operate the engine to normal operating temperatures; then, check the climate control operation and check for leaks.

STEERING

POWER RACK & PINION STEERING GEAR

REMOVAL & INSTALLATION

See Figure 139.

⁂ WARNING

Be sure to set the front wheels straight ahead and remove the ignition key from the cylinder before starting repairs. The contact coil of the air bag system may be damaged if the key is not removed and the wheels are not straight ahead.

1. Before servicing the vehicle, refer to the Precautions Section.
2. Disconnect the negative battery cable.
3. Take out fluid in P/S fluid reservoir with syringe or such.
4. Remove or disconnect the following:
   - Steering column joint cover
   - Steering shaft upper joint bolt, loosen but do not remove
   - Steering shaft lower joint bolt
   - Lower joint from the pinion
   - Front wheels
   - Tie rod ends from the steering knuckles
   - Exhaust pipe No. 2
   - Engine rear mounting together with its bracket from engine and member
   - Engine mounting member
   - Transfer case (AWD)

➡When the lines are disconnected plug the lines or place an oil pan under the vehicle.

   - Cylinder pipes and from steering gear box, using flare nut wrench
   - High pressure pipe and low pressure hose from steering gear box, using flare nut wrench
   - Steering gear box mounting bolts and steering gear box

To install:

⁂ WARNING

Be sure to confirm that steering wheel and front tires (wheels) are in straight position when inserting steering lower joint into steering pinion shaft.

5. Install or connect the following:
   - Steering gear, brackets and mounting bolts. Tighten bolts to 40 ft. lbs. (55 Nm).

➡If a plug was put to disconnected pipe when removing steering gear box, remove that plug before reconnecting pipe.

   - Cylinder lines on the rack and pinion and tighten their fittings to 18 ft. lbs. (25 Nm)
   - High and low pressure lines to the steering gear. Tighten the fittings to 25 ft. lbs. (35 Nm).
   - Transfer case (AWD)
   - Engine mounting member
   - Engine rear mounting together with its bracket to the engine and member
   - Exhaust pipe No. 2
   - Tie rod ends to the steering knuckles. Tighten the castle nuts to 25.5–39.5 ft. lbs. (35–55 Nm).
   - Front wheels
6. Be sure the steering wheel is straight and the front wheels are pointing straight ahead.
   - Steering shaft to the rack and pinion
   - Lower steering shaft-to-rack and pinion clinch bolt and tighten both steering joint bolts (upper and lower) to 18 ft. lbs. (25 Nm)
   - Front wheels
7. Lower the vehicle.
8. Connect the negative battery cable.
9. Fill the power steering system and then bleed the power steering system.
10. Check and adjust the front wheel alignment.

POWER STEERING PUMP

REMOVAL & INSTALLATION

See Figures 140 through 143.

➡Be sure to clean each joint of suction and discharge sides thoroughly before removal.

1. Before servicing the vehicle, refer to the Precautions Section.
2. Set the parking brake.
3. Remove or disconnect the following:

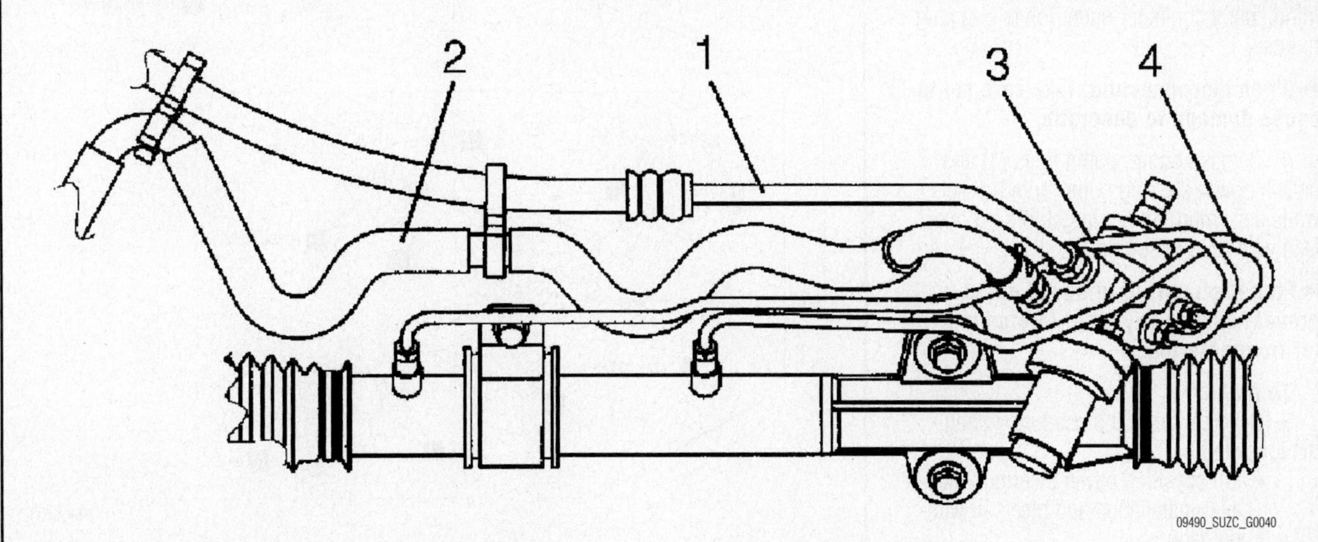

Fig. 139 Fluid pipes and hose to the power steering gear

Fig. 140 Remove air cleaner outlet hose and empty power steering fluid from reservoir

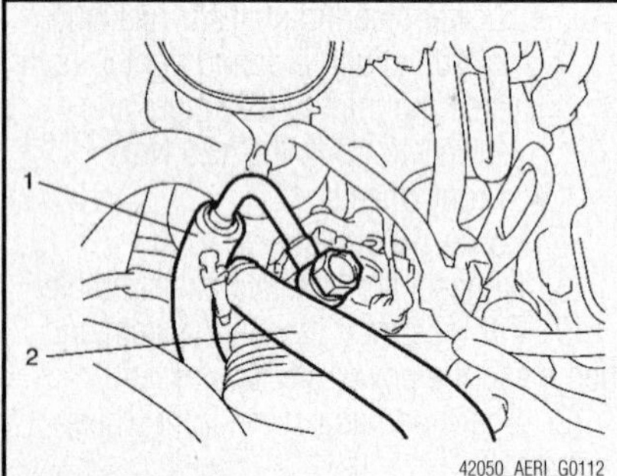

Fig. 141 Disconnect the high pressure hose and suction hose from the power steering pump

4. Remove engine under cover (right side only) and loosen generator belt tensioner pulley then remove generator belt.
5. Remove air cleaner outlet hose (1).
6. Empty power steering fluid from reservoir (2) using syringe or such.
7. Disconnect high pressure hose (1) and suction hose (2) from power steering pump. As fluid flows out of disconnected joints, put a container under joints or a plug to hose.

➡When disconnecting, take care not to cause damage to generator.

8. Remove pump mount bolts (1) and detach power steering pump from bracket, while detaching, do not forget to disconnect lead wire from the pump.

➡Plug each port of removed pump to prevent dust or any other foreign matter from entering.

To install:

9. Reverse removal procedure, noting the following:
   - Fill specified power steering fluid after installation and bleed air without failure.
   - For tightening torques, refer to the accompanying graphic.

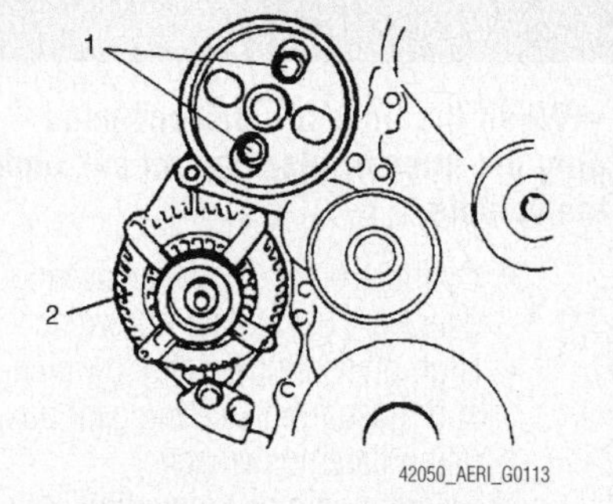

Fig. 142 Remove the pump mount bolts and detach the power steering pump from the bracket

   - Check power steering belt (generator belt) tension.
   - Bleed air from P/S system.

BLEEDING

1. Jack up the front end of vehicle and apply safety stands.
2. Fill P/S fluid reservoir with fluid up to specified level.

➡Before starting engine, place transmission gear shift lever in "Neutral" (shift selector lever to "P" range for A/T model), and set parking brake.

3. After running engine at idling speed for 3 to 5 seconds, stop it and add fluid to satisfy specification.
4. With engine stopped, turn steering wheel to the right and left as far as it stops, repeat it a few times and fill fluid to specified level.
5. With engine running at idling speed, repeat stop-to-stop turn of steering wheel till all foams in P/S fluid reservoir are gone.

➡Make sure to bleed air completely. If air remains in fluid, P/S pump may make humming noise or steering wheel may feel heavy.

6. Finally check to make sure that fluid is filled to specified level.

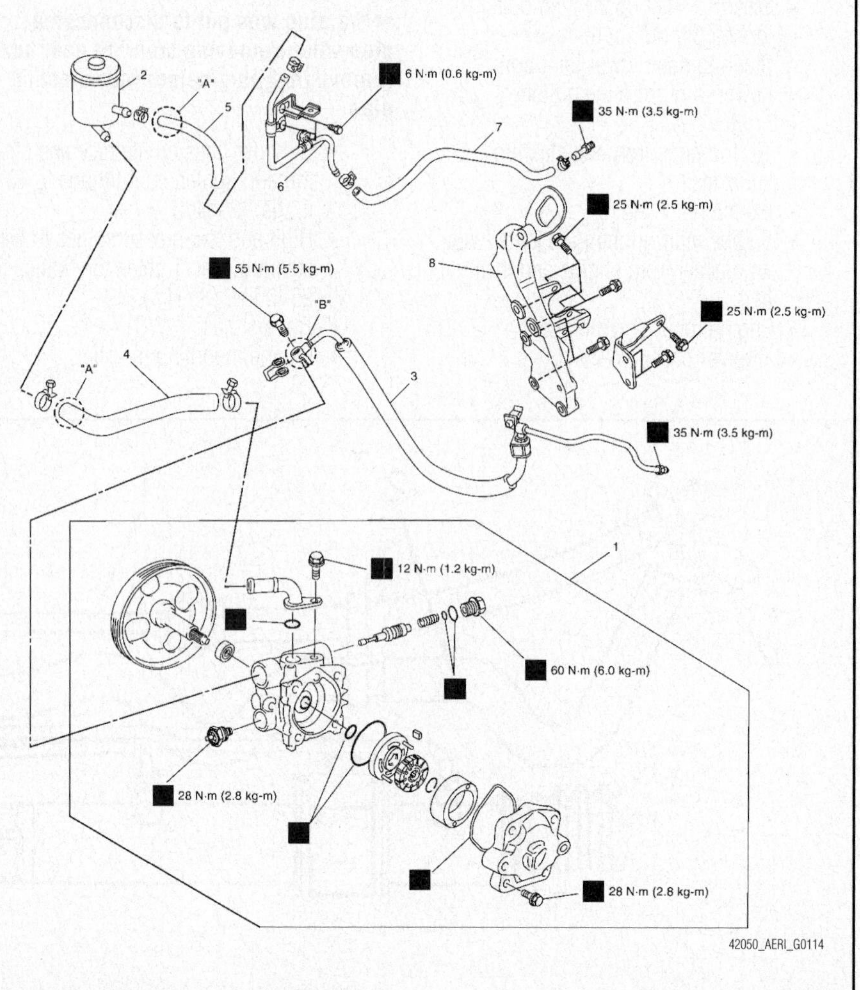

Fig. 143 Install the power steering pump

COIL SPRING

REMOVAL & INSTALLATION

For coil spring service, refer to the strut removal and installation procedure.

LOWER BALL JOINT

REMOVAL & INSTALLATION

The lower ball joint is an integral part of the lower control arm assembly. If the ball joint is found to be defective the whole lower control arm assembly must be replaced.

LOWER CONTROL ARM

REMOVAL & INSTALLATION

See Figures 144 and 145.

The lower control arm and ball joint are a complete unit that will not separate.

1. Before servicing the vehicle, refer to the Precautions Section.
2. Remove or disconnect the following:
   - Front wheels
   - Suspension control arm ball joint bolt and nut
   - Suspension control arm bolts
   - Suspension control arm

To install:

3. Install suspension control arm as shown but tighten suspension control arm bolts only temporarily.
4. Install suspension control arm ball joint to steering knuckle. Align ball stud groove with steering knuckle bolt hole. Then install ball joint bolt from the front direction. Tighten suspension arm ball joint nut to 43.5 ft. lbs. (60 Nm).
5. Lower hoist and vehicle in non-loaded condition, tighten control arm mounting bolts to 65 ft. lbs. (90 Nm).
6. Check front wheel alignment.

CONTROL ARM BUSHING REPLACEMENT

1. Before servicing the vehicle, refer to the precautions in the beginning of this section.
2. Remove the lower control arm.
3. Cut the flange from the front bushing.
4. Use a hydraulic press to remove the front bushing.

To install:

5. Apply a solution of soapy water to the outer diameter of the front bushing, this will aid in installation.

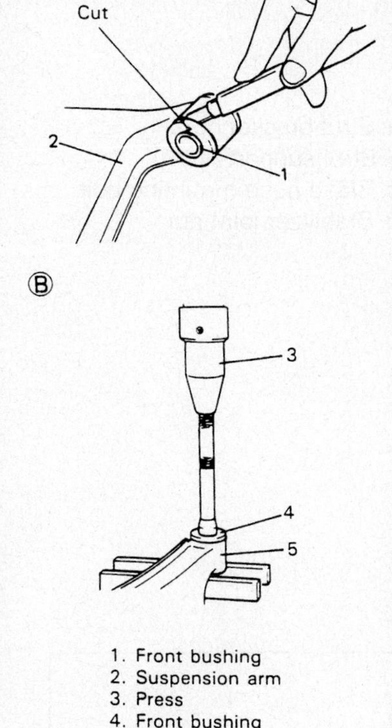

Fig. 144 Cut the flange from the front bushing, then using a suitable hydraulic press; remove the rear lower control arm bushing

6. Press the front bushing into its bore using a hydraulic press until the bushing is equal on the right and left of the arm as shown in the accompanying illustration.

MACPHERSON STRUT

REMOVAL & INSTALLATION

See Figure 146.

1. Before servicing the vehicle, refer to the Precautions Section.
2. Set the parking brake.
3. Remove or disconnect the following:
4. Disconnect the negative battery cable.
5. Raise and safely support the front of the vehicle.
6. Remove or disconnect the following:
   - Wheel
   - Stabilizer joint from strut bracket
   - Brake hose mounting bolt and brake hose
   - ABS wheel speed sensor harness (if equipped) from strut bracket
   - Strut bracket bolts
   - Strut support nuts

➡Hold strut by hand so that it will not fall off.

   - Strut assembly

7. Using a spring compressor tool, turn special tool bolts alternately until spring tension is released. Whether it is released or not can be known by whether strut turns lightly while strut spring is held stationary.

➡Use a commercially available spring compressor and follow the operation

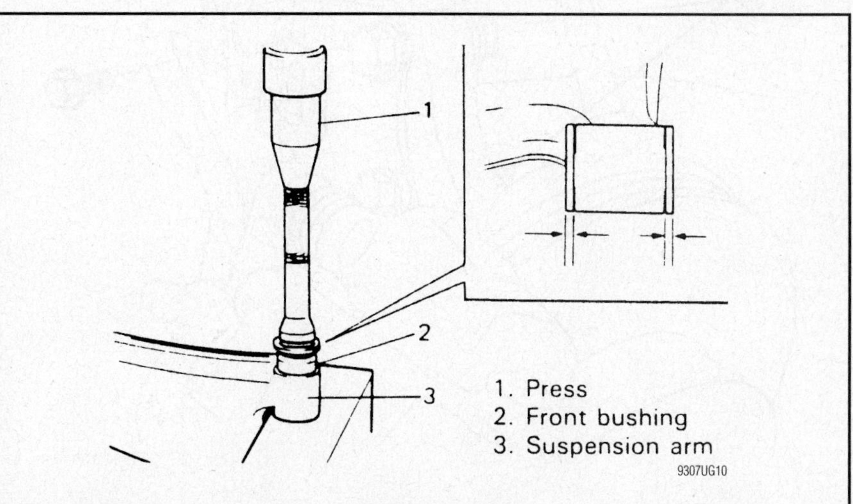

Fig. 145 The front bushing should be positioned equally as shown after being pressed into position

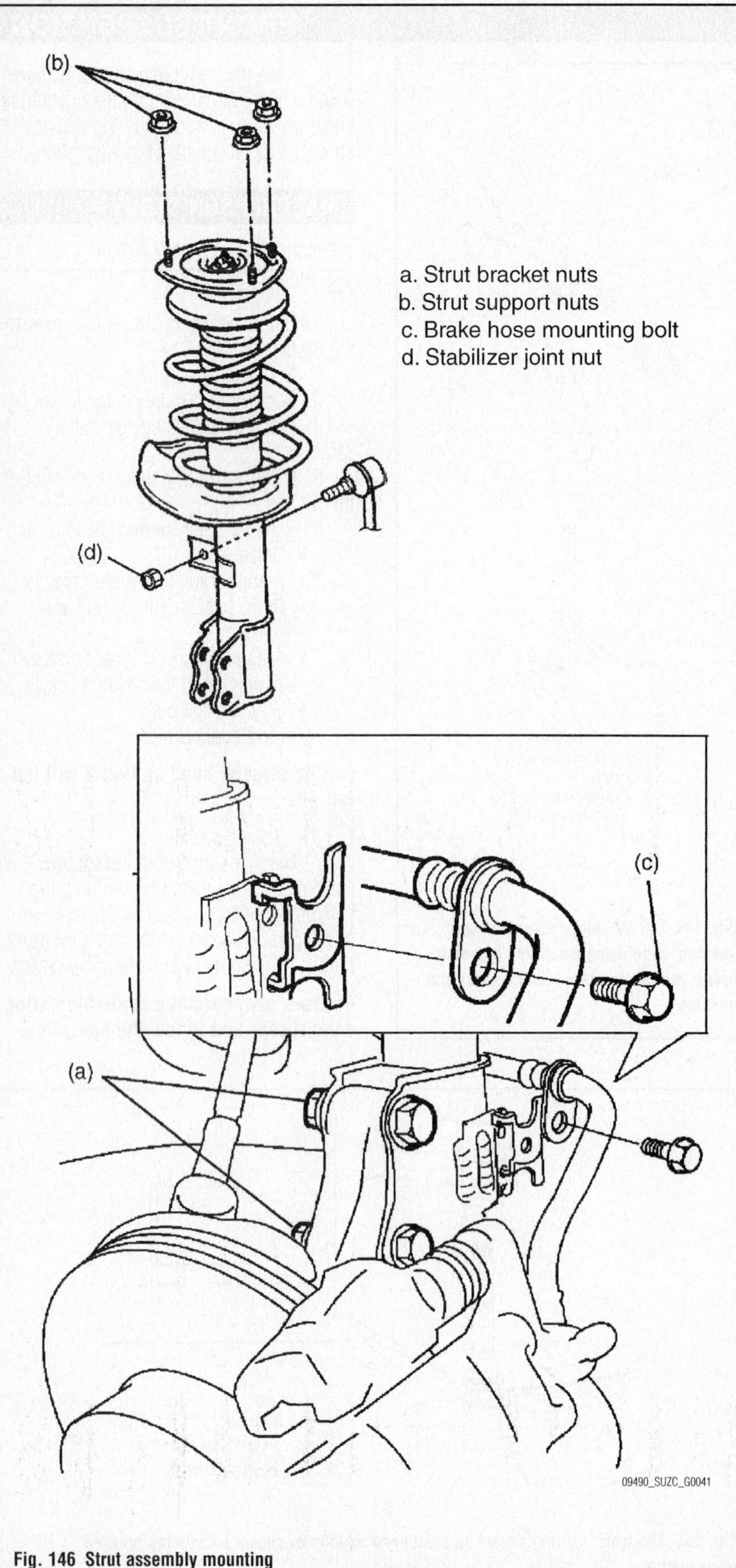

Fig. 146 Strut assembly mounting

procedure described in the Instruction Manual supplied with that spring compressor.

8. While keeping spring compressed, remove strut nut.
9. Disassemble strut assembly.

To install:

10. Assemble the coil spring into the strut assembly as follows:
    a. Compress spring with special tool until total length becomes about 9.8 inches (250mm).
    b. Install coil spring lower seat and compressing coil spring, and mate spring end with stepped part of lower seat.
    c. Install bump stopper and dust cover onto strut rod.
    d. Pull strut rod as far up as possible and use care not to allow it to retract into strut.
    e. Install spring seat on coil spring and then spring upper seat aligning "OUT" mark on spring upper seat and center of strut bracket.
    f. Install bearing, strut support and strut nut in this sequence. Tighten strut nut to 50.5 ft. lbs. (70 Nm).
    g. Install rubber cap.
    - Strut assembly. Tighten the strut bracket nuts to 76 ft. lbs. (105 Nm). Tighten the strut support nuts to 20.5 ft. lbs. (28 Nm).
    - Brake hose and mounting bolt. Tighten the brake hose mounting bolt to 18 ft. lbs. (25 Nm).
    - Stabilizer joint to strut bracket. Tighten the stabilizer joint nut to 36.5 ft. lbs. (50 Nm).
    - ABS wheel speed sensor harness, if equipped
    - Wheels
11. Lower the vehicle.
12. Connect the negative battery cable.
13. After installation, confirm front wheel alignment.

STABILIZER BAR

REMOVAL & INSTALLATION

See Figures 147 and 148.

1. Before servicing the vehicle, refer to the Precautions Section.
2. Set the parking brake.
3. Remove or disconnect the following:
4. Remove stabilizer joints. When loosening joint nut, hold stud with wrench.
5. Remove the stabilizer mount brackets.
6. Remove the stabilizer bar.

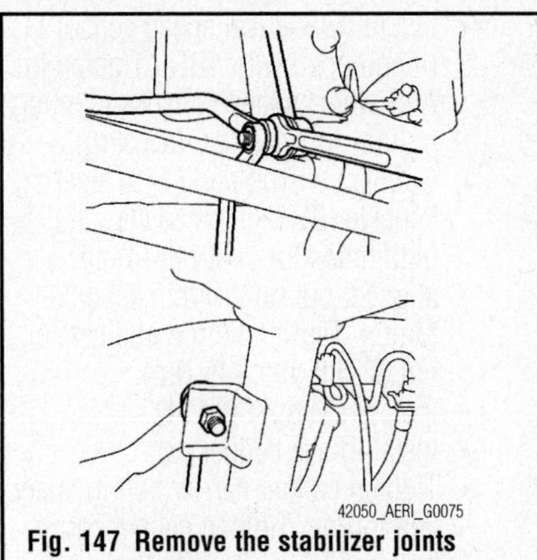
Fig. 147 Remove the stabilizer joints

To install:

7. Loosely assemble all components in reverse order of removal.
8. With painted mark (1) on stabilizer bar and end face of mount bushing (2) aligned, tighten mount bracket bolts to 17 ft. lbs. (23 Nm).

➡If painted mark has gone, align the vehicle center and the center of stabilizer bar.

9. Tighten joint nuts to 36.5 ft. lbs. (50 Nm). When tightening, hold stud with wrench.

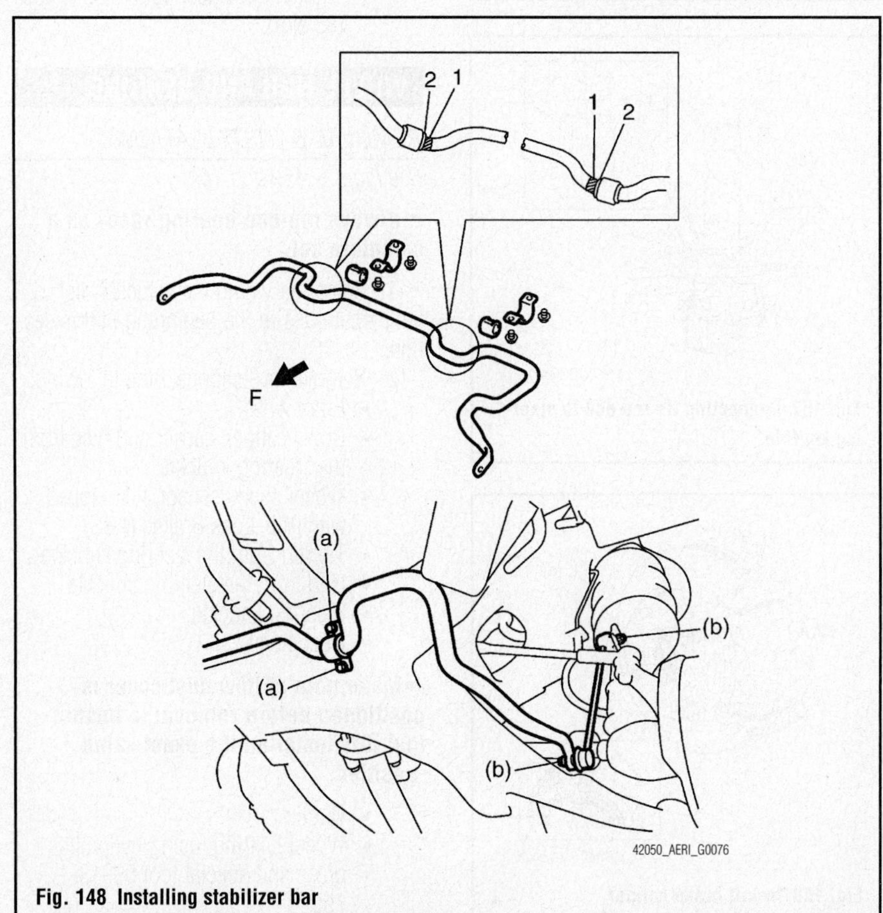

Fig. 148 Installing stabilizer bar

STEERING KNUCKLE

REMOVAL & INSTALLATION

See Figures 149 through 158.

1. Before servicing the vehicle, refer to the Precautions Section.
2. Remove or disconnect the following:
3. Hoist vehicle and remove wheel.
4. Remove driveshaft nut (1).
5. Depress foot brake pedal and hold it there. Remove driveshaft nut (1).
6. Remove caliper carrier bolts.
7. Remove caliper (1) with carrier.

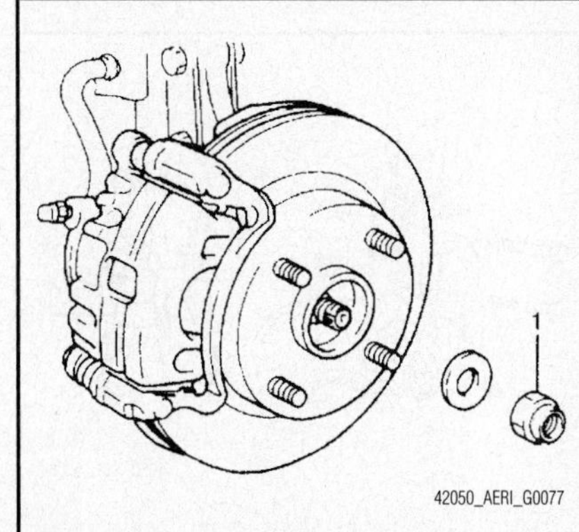

Fig. 149 Identifying driveshaft nut

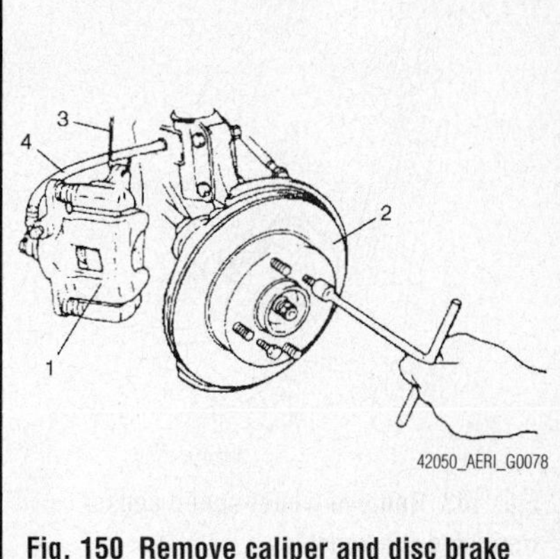

Fig. 150 Remove caliper and disc brake

➡Hang removed caliper with a wire hook or the like (3) so as to prevent brake hose (4) from bending and twisting excessively or being pulled. Don't operate brake pedal with pads removed.

8. Pull brake disc (2) off by using two 8 mm bolts.
9. Pull out wheel hub (1) with special tools (A) 09943-17912 and (B) 09942-15511.

CAUTION

When wheel hub is removed, replace wheel bearing with new one.

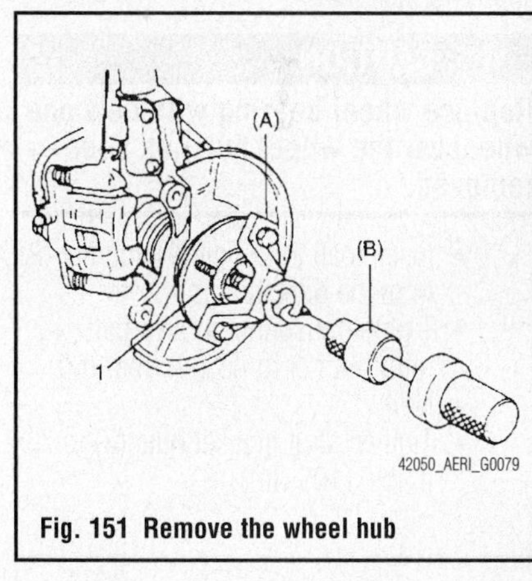

Fig. 151 Remove the wheel hub

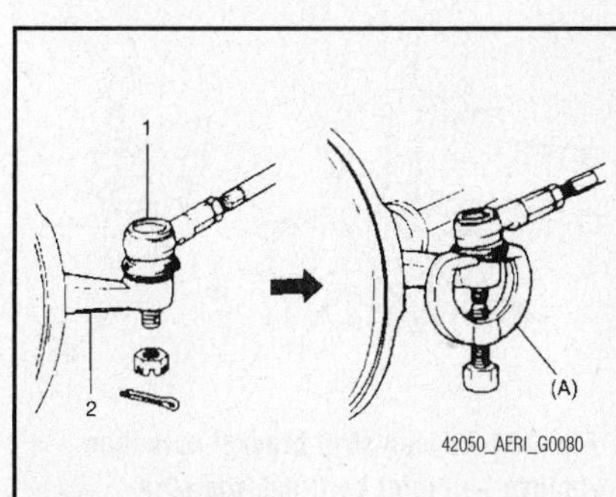

Fig. 152 Disconnect the tie rod end from the steering knuckle

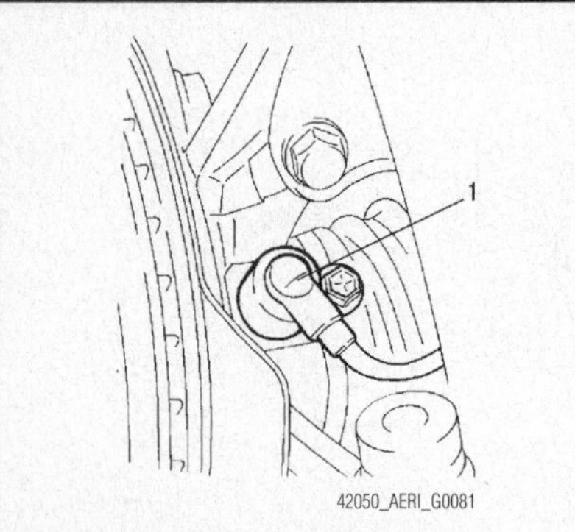

Fig. 153 Remove wheel speed sensor from steering knuckle

10. Disconnect tie-rod end (1) from steering knuckle (2) with special tool (A): 09913-65210.

11. Remove wheel speed sensor (1) from knuckle (if equipped with ABS).

12. Loosen strut bracket nuts (1).

13. Remove ball joint bolt (3) and nut (4).

14. Remove strut bracket bolts from strut bracket and then steering knuckle (2).

To install:

For installation, reverse removal procedure, noting the following instructions.

- Using special tool (A) 09913-75520, drive wheel hub (1) into steering knuckle (2) and then install steering knuckle (2) to strut and lower ball joint.

CAUTION

Replace wheel bearing with new one whenever the wheel hub has been removed.

- Install ball joint bolt (1) and nut (2) from the direction as shown.
- Tighten suspension arm ball joint nut (2) to 43.5 ft. lbs. (60 Nm).
- Tighten strut bracket nuts (3) to 76 ft. lbs. (105 Nm).

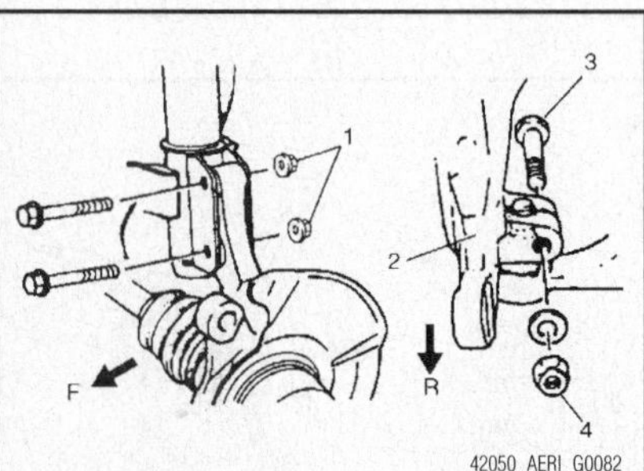

Fig. 154 Loosen strut bracket nuts then remove ball joint bolt, nut and strut bracket bolts from strut bracket then steering knuckle

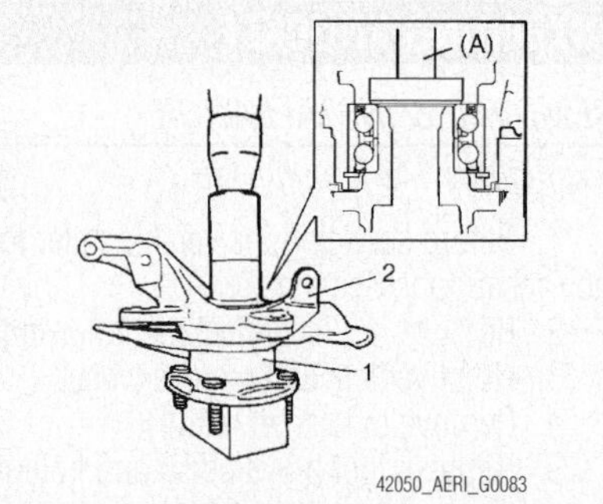

Fig. 155 Drive wheel hub into steering knuckle and steering knuckle to strut and lower ball joint

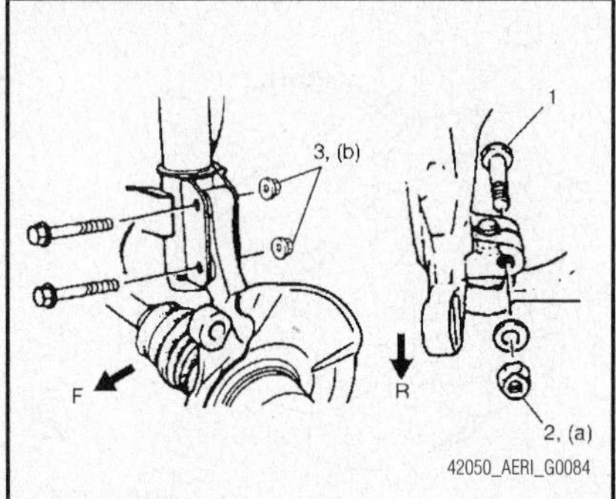

Fig. 156 Install ball joint bolt and nut and strut bracket nuts

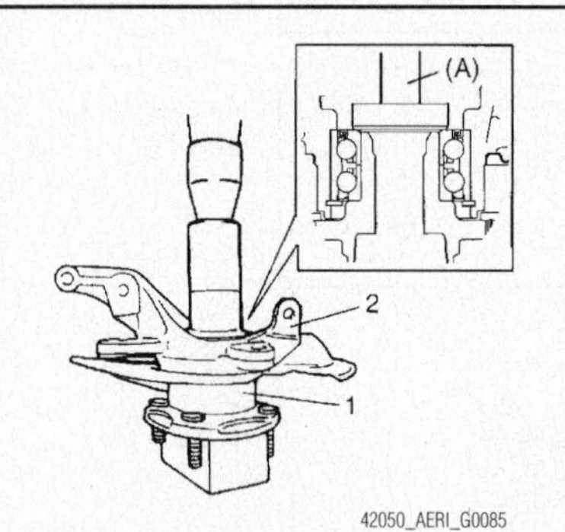

Fig. 157 Connecting tie rod end to steering knuckle

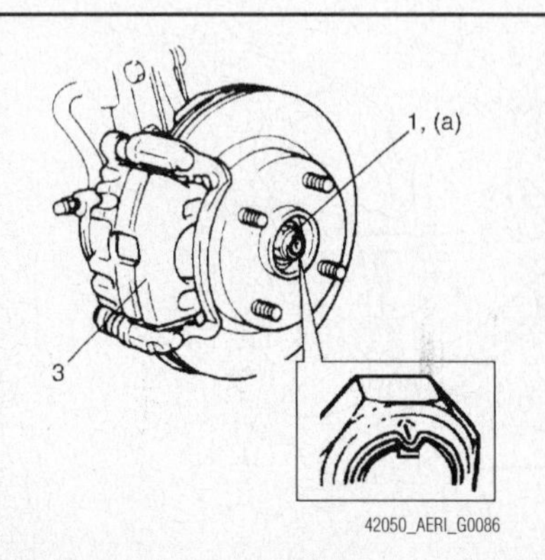

Fig. 158 Install brake caliper

- Install ABS wheel speed sensor (1) (if equipped with ABS). Tighten the ABS wheel speed sensor mounting bolt (a) to 17 ft. lbs. (23 Nm).
- Connect tie-rod end (1) to steering knuckle (2). Tighten castle nut (3) until holes for split pin (4) are aligned, but only within specified torque. Tighten tie rod end castle nut 32.5 ft. lbs. (45 Nm).
- Install new split pin (4).
- Install brake caliper (3).
- Tighten caliper carrier bolt to specified torque. Tighten caliper carrier bolt to 61.5 ft. lbs. (85 Nm)
- Depress foot brake pedal and hold it there. Tighten new driveshaft nut (1) to 127 ft. lbs. (175 Nm).

CAUTION

Never reuse driveshaft nut (1). Caulk driveshaft nut (1) as shown.

CAUTION

Be careful while caulking nut so that no crack will occur in caulked part of nut. Cracked nut must be replaced with new one.

- Tighten the wheel nuts to 62 ft. lbs. (85 Nm).

WHEEL HUB AND BEARING

REMOVAL & INSTALLATION

See Figures 159 and 160.

➡Always replace bearing races as a complete set.

1. Before servicing the vehicle, refer to the precautions in the beginning of this section.

2. Remove or disconnect the following:
   - Front wheel
   - Brake caliper, carrier and disc from the steering knuckle
   - Wheel speed sensor, if equipped with Anti-Lock Brakes (ABS)
   - Tie rod from the steering knuckle
   - Hub from the steering knuckle
   - Steering knuckle
   - Dust cover

➡Make note of how dust cover is positioned before removal to ensure that it is installed the exact same -position.

   - Circlip
   - Wheel bearing, using hydraulic press and special tool 09913-75520

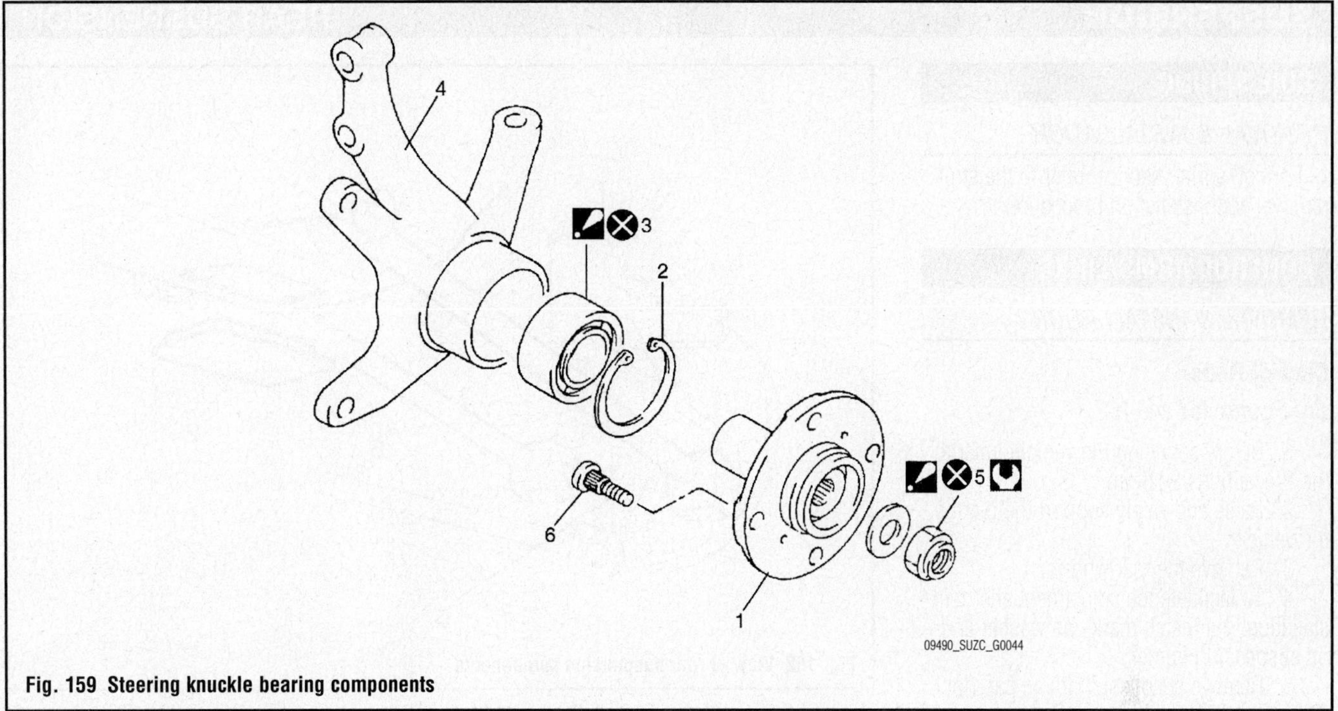

Fig. 159 Steering knuckle bearing components

➡When installing wheel bearing, replace it with new one.

- Wheel bearing outside inner race

To install:

3. Face grooved rubber seal side of new wheel bearing upward as shown in figure and press-fit it into knuckle using special tool 09913-75510.
4. Install or connect the following:
   - Circlip
   - Dust cover. Be sure that it is installed in the same position that it was removed from.

➡When drive in dust cover, be careful not to deform it.

- Caulk with a punch
- Steering knuckle in the vehicle
- Tie rod end
- Brake caliper, carrier and disc on the steering knuckle
- Front wheel

ADJUSTMENT

The front wheel bearings are a cartridge type design and cannot be adjusted.

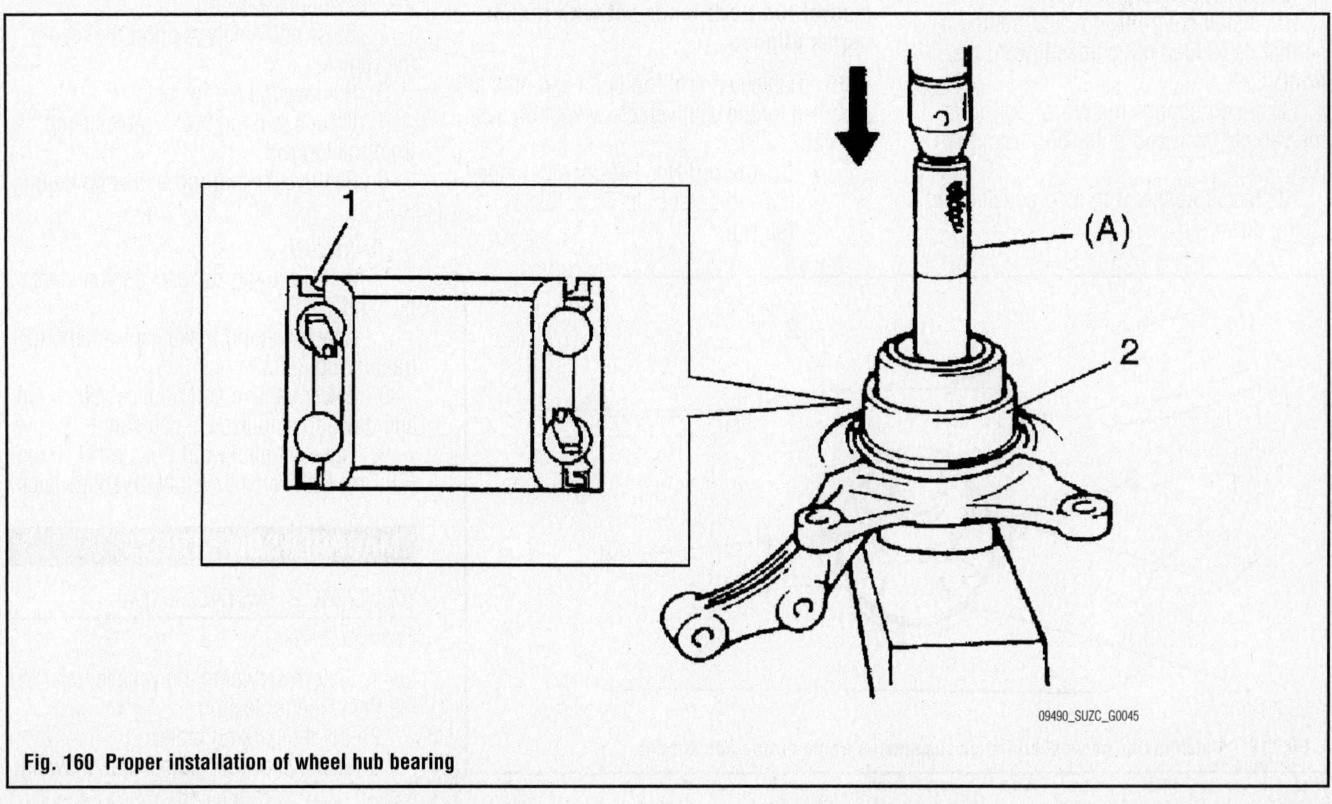

Fig. 160 Proper installation of wheel hub bearing

COIL SPRING

REMOVAL & INSTALLATION

For coil spring service, refer to the strut removal and installation procedure.

CONTROL ARMS/LINKS

REMOVAL & INSTALLATION

Control Rods

See Figures 161 and 162.

1. Before servicing the vehicle, refer to the Precautions Section.
2. Raise and safely support the rear of the vehicle.
3. Remove the rear wheels.
4. To facilitate toe adjustment after reinstallation, put match marks on washer and on suspension frame.
5. Remove suspension frame cap (for 2WD model).
6. Remove control rod No.2 and No.1 from suspension frame and knuckle.

To install:

7. Install control rod No. 1, setting it so that its welded nut comes toward the rear.
8. Insert inner bolt and outer bolt from the vehicle front and tighten them temporarily by hand.
9. Install control rod No. 2.
10. Install control rod No. 2, setting it so that its welded nut comes toward the front.
11. Insert control rod No. 2 bolt from the vehicle front and outer bolt from the rear.
12. Install washer with its graduated part facing down.

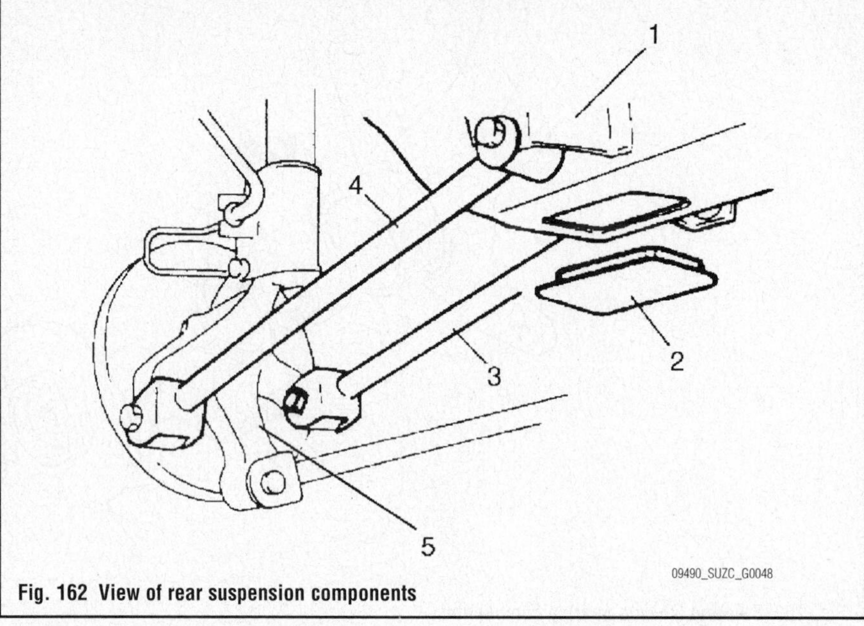

Fig. 162 View of rear suspension components

13. With marks on washer and frame marked before removal aligned to each other, tighten bolts and nut temporarily by hand.
14. Install wheels and tighten wheel nuts to specified torque.
15. Lower hoist and bounce vehicle up and down to stabilize suspension.

➡It is the most desirable to have vehicle off hoist and in non-loaded condition when tightening them. Tighten control rod No.2 inner nut with match marks aligned.

16. Tighten control rod bolts and nuts to specified torque with vehicle weight on suspension:
    - Control rod No. 1 (front side) inner nut and outer bolt: 71 ft. lbs. (98 Nm)
    - Control rod No. 2 (rear side) outer bolt: 71 ft. lbs. (98 Nm)
    - Control rod No. 2 (rear side) inner nut: 65 ft. lbs. (90 Nm)
17. Check rear toe and adjust it as necessary.
18. Install suspension frame cap.

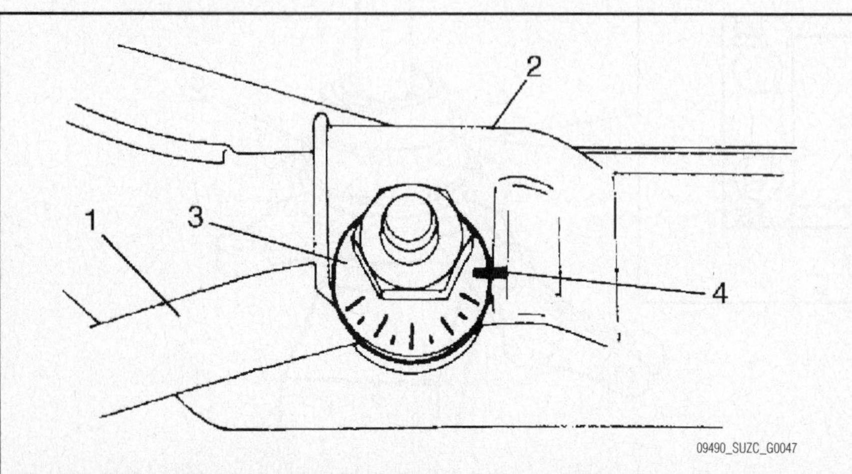

Fig. 161 Match marks on washer and on suspension frame for toe adjustment

Trailing Rod

1. Before servicing the vehicle, refer to the Precautions Section.
2. Raise and safely support the rear of the vehicle.
3. Remove the rear wheels.
4. Detach parking brake cable clamp from trailing rod.
5. Remove trailing rod bolts and trailing rod.

To install:

6. Install trailing rod and tighten bolts by hand.
7. Attach parking brake cable clamp to trailing rod.
8. Lower vehicle and bounce vehicle up and down to stabilize suspension.
9. Tighten trailing rod bolts to 71 ft. lbs. (98 Nm) with vehicle weight on suspension.

MACPHERSON STRUTS

REMOVAL & INSTALLATION

See Figure 163.

1. Before servicing the vehicle, refer to the Precautions Section.
2. Set the parking brake.
3. Remove or disconnect the following:

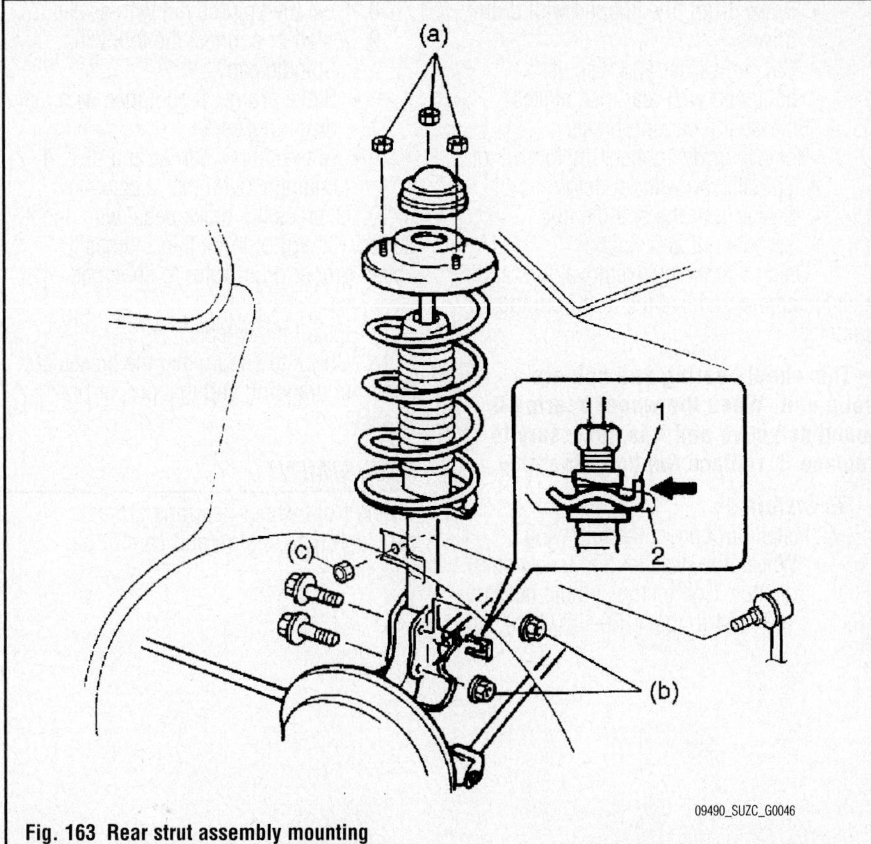

Fig. 163 Rear strut assembly mounting

➡When servicing component parts of strut assembly, loosen strut upper nut a little before removing strut assembly. This will make service work easier. Note, however, nut must not be removed at this point.

4. Raise and safely support the front of the vehicle.
5. Remove or disconnect the following:
   - Rear wheel
   - Stabilizer joint from strut bracket
   - E-ring securing brake hose
   - ABS wheel speed sensor harness mounting bolt (if equipped)
   - Strut bracket bolts and nuts. And then take brake hose off strut bracket using care not to deform brake pipe.
6. Strut support nuts. Hold strut by hand so that it will not fall off.
   - Strut assembly
7. Using a spring compressor tool, turn special tool bolts alternately until spring tension is released. Whether it is released or not can be known by whether strut turns lightly while strut spring is held stationary.

➡Use a commercially available spring compressor and follow the operation procedure described in the Instruction Manual supplied with that spring compressor.

8. While keeping spring compressed, remove strut nut.
9. Disassemble strut assembly.

To install:

10. Assemble the coil spring into the strut assembly as follows:
    a. Compress spring with special tool until total length becomes about 11.42 inches (290mm).
    b. Install coil spring lower seat and compressing coil spring, and mate spring end with stepped part of lower seat.
    c. Install bump stopper and dust cover onto strut rod.
    d. Pull strut rod as far up as possible and use care not to allow it to retract into strut.
    e. Fit coil spring (rubber) seat in coil spring upper seat, making sure that their depth matches all around. No part of rubber seat should stick out higher than upper seat.
    f. With "OUT" mark on spring upper seat and the center of strut bracket aligned, place upper spring seat together with spring (rubber) seat on coil spring. Put strut support on spring upper seat. Tighten strut nut to 50.5 ft. lbs. (70 Nm).
    g. Loosen and remove special tool from compressing coil spring.
    h. While loosening spring compressor tool, recheck that stepped part of spring seat and spring end are in place to each other.
    i. Also, check to make sure that "OUT" mark on upper seat is matched with the center of strut bracket.
11. Install or connect the following:
    - Strut assembly. Install the bracket bolts from the front direction and tighten the strut bracket nuts to 65 ft. lbs. (90 Nm). Tighten the strut support nuts to 20.5 ft. lbs. (28 Nm).
    - E-ring securing brake hose

➡Do not twist brake hose when installing it. Install E-ring as far as it fits to bracket.

   - Stabilizer joint to strut bracket. Tighten the stabilizer joint nut to 36.5 ft. lbs. (50 Nm).
   - ABS wheel speed sensor harness, if equipped
   - Wheels
12. Lower the vehicle.

WHEEL HUB AND BEARING

REMOVAL & INSTALLATION

With Wheel Hubs

1. Before servicing the vehicle, refer to the Precautions Section.
2. Set the parking brake.
3. Remove or disconnect the following:
   - Rear wheels
   - Rear brake drums
4. Use a brass drift and knock the wheel bearings from the drum assembly.

To install:

5. Position the inner wheel bearing on the drum with the sealed side facing out. Using a rear wheel Bearing Installer 09913-76010, install the rear wheel bearing. Install the wheel bearing spacer into the drum.
6. Install or connect the following:
   - Outer wheel bearing with the sealed side facing out, using a wheel bearing installer
7. Fill the space in the brake drum in between the wheel bearings to about 40% capacity with wheel bearing grease.
8. Install or connect the following:
   - Brake drum
   - Spindle washer and a new spindle nut. Tighten the spindle nut to 58–86 ft. lbs. (80–120 Nm).
9. Coat the spindle nut and the spindle dust cap with sealer.

➡When installing the spindle cap, hammer lightly several times on the collar of the cap until the collar comes closely into contact with the brake drum. If the fitting part of the cap is deformed or damaged or if it fits loose, replace the cap with a new one.

10. Depress the brake pedal with about 66 lbs. (30 kg) of force 3 to 5 times to obtain proper drum to shoe clearance.
11. Install the wheels.
12. Check to ensure that the brake drum is free from dragging and proper braking is obtained.

Without Wheel Hubs

1. Before servicing the vehicle, refer to the Precautions Section.
2. Set the parking brake.
3. Remove or disconnect the following:
   - Rear wheels
   - Brake drum, if equipped with drum brakes
   - Caliper, carrier and disc, if equipped with rear disc brakes
4. Release the parking brake.
5. Remove or disconnect the following:
   - Spindle cap without deforming it
   - Sealer from the spindle nut
   - Spindle nut and washer
6. Using a brake hub removal tool and a slide hammer remove the hub from the spindle.

➡The wheel bearing and hub are a solid unit. When the wheel bearing is found defective and it is necessary to replace it, replace the hub assembly.

To install:

7. Install or connect the following:
   - Wheel hub, washer and a new spindle nut. Tighten the spindle nut to 108–144 ft. lbs. (150–200 Nm).
8. Coat the spindle nut with sealer
9. Install or connect the following:
   - Spindle cap
   - Brake drums, if equipped with rear drum brakes
   - Brake caliper carrier and disc, if equipped with rear disc brakes
10. Depress the brake pedal with about 66 lbs. (30 kg) of force 3 to 5 times to obtain proper drum/rotor to shoe/pad clearance.
11. Install the wheels.
12. Check to ensure that the brakes are free from dragging and that proper braking is obtained.

ADJUSTMENT

The front wheel bearings are a cartridge type design and cannot be adjusted.

SUZUKI

Forenza • Reno

18

SPECIFICATIONS AND MAINTENANCE CHARTS

VEHICLE AND ENGINE IDENTIFICATION

Engine Code							Model Year	
Code	Liters (cc)	Cu. in.	Cyl.	Fuel Sys.	Engine Type	Eng. Mfg.	Code	Year
Z	2.0 (1998)	140	4	MFI	DOHC	Suzuki	7	2007
Z	2.0 (1998)	140	4	MFI	DOHC	Suzuki	8	2008

DOHC: Dual Overhead Camshaft

MFI: Multiport Fuel Injection

22140_RENO_C0001

GENERAL ENGINE SPECIFICATIONS

Year	Engine ID/VIN	Engine Displacement Liters (cc)	Fuel System Type	Net Horsepower @ rpm	Net Torque @ rpm (ft. lbs.)	Bore x Stroke (in.)	Com-pression Ratio	Oil Pressure @ rpm
2007	Z	2.0 (1998)	MFI	127@5600	131@4000	3.39x3.39	9.6:1	55.5@4000
2008	Z	2.0 (1998)	MFI	127@5600	131@4000	3.39x3.39	9.6:1	55.5@4000

MFI: Multiport Fuel Injection

22140_RENO_C0002

GASOLINE ENGINE TUNE-UP SPECIFICATIONS

Year	Engine Displacement Liters	Engine ID/VIN	Spark Plugs Gap (in.)	Ignition Timing (deg.)		Fuel Pump (psi) ①	Idle Speed (rpm)		Valve Clearance	
				MT	AT		MT	AT	In.	Ex.
2007	2.0	Z	0.039	8B	8B	31.3-36.9	700-800	700-800	HYD	HYD
2008	2.0	Z	0.039	8B	8B	31.3-36.9	700-800	700-800	HYD	HYD

Note: The Vehicle Emission Control Information label often reflects specification changes made during production.
The label figures must be used if they differ from those in this chart.

HYD: Hydraulic

B Before Top Dead Center (BTDC)

① At idle

22140_RENO_C0003

CAPACITIES

Year	Model	Engine Displacement Liters	Engine ID/VIN	Engine Oil with Filter (qts.)	Transmission (pts.) 5-Spd	Transmission (pts.) Auto. ①	Fuel Tank (gal.)	Cooling System (qts.)
2007	Forenza	2.0	Z	4.3	3.8	14.6	14.5	7.9
	Reno	2.0	Z	4.3	3.8	14.6	14.5	7.9
2008	Forenza	2.0	Z	4.3	3.8	14.6	14.5	7.9
	Reno	2.0	Z	4.3	3.8	14.6	14.5	7.9

Note: All capacities are approximate. Add fluid gradualy and check to be sure a proper fluid level is obtained.

① Specification for automatic transaxle is after complete overhaul. Drain and fill will be less

22140_RENO_C0004

FLUID SPECIFICATIONS

Year	Engine Displacement Liters	Engine ID/VIN	Engine Oil	Auto. Trans.	Manual Trans??	Power Steering Fluid	Brake Master Cylinder	Engine Coolant
2007	2.0	Z	5W-30 SL	①	②	Dexron IID	DOT 3	③
2008	2.0	Z	5W-30 SL	①	②	Dexron IID	DOT 3	③

DOT: Department Of Transpotation

① Esso LT 71141 or total ATF H50235

② SAE 80W / SAE 75W-90 Cold Weather

③ Silicate based coolant Dex-cool

22140_RENO_C0013

VALVE SPECIFICATIONS

Year	Engine ID/VIN	Engine Displacement Liters	Seat Angle (deg.)	Face Angle (deg.)	Spring Test Pressure (lbs. @ in.)	Spring Installed Height (in.)	Stem-to-Guide Clearance (in.) Intake	Stem-to-Guide Clearance (in.) Exhaust	Stem Diameter (in.) Intake	Stem Diameter (in.) Exhaust
2007	Z	2.0	NA	44	NA	NA	0.0019-0.0036	0.0019-0.0036	0.2341-0.2346	0.2341-0.2346
2008	Z	2.0	NA	44	NA	NA	0.0019-0.0036	0.0019-0.0036	0.2341-0.2346	0.2341-0.2346

NA: Not Available

22140_RENO_C0005

CAMSHAFT AND BEARING SPECIFICATIONS

All measurements are given in inches.

Year	Engine Displacement Liters	Engine VIN	Journal Diameter	Brg. Oil Clearance	Shaft End-play	Runout	Journal Bore	Lobe Lift Intake	Lobe Lift Exhaust
2007	2.0	Z	1.0221-1.0228	0.0008-0.0029	NA	0.0012	1.0236-1.0249	1.591-1.5972	1.5716-1.5778
2008	2.0	Z	1.0221-1.0228	0.0008-0.0029	NA	0.0012	1.0236-1.0249	1.591-1.5972	1.5716-1.5778

NA: Not Available

22140_RENO_C0014

CRANKSHAFT AND CONNECTING ROD SPECIFICATIONS

All measurements are given in inches.

Year	Engine ID/VIN	Engine Displacement Liters	Crankshaft Main Brg. Journal Dia.	Crankshaft Main Brg. Oil Clearance	Crankshaft Shaft End-play	Crankshaft Thrust on No.	Connecting Rod Journal Diameter	Connecting Rod Oil Clearance	Connecting Rod Side Clearance
2007	Z	2.0	2.2824- ① 2.2832 ②	0.00059-0.00157	0.0027-0.0118	3	1.9279- ③ 1.9287 ④	NA	NA
2008	Z	2.0	2.2824- ① 2.2832 ②	0.00059-0.00157	0.0027-0.0118	3	1.9279- ③ 1.9287 ④	NA	NA

① All Journals

② Out of Round: 0.0001 in. MAX.

③ All Journals

④ Out of Round: 0.00015 in. MAX.

22140_RENO_C0006

PISTON AND RING SPECIFICATIONS

All measurements are given in inches.

Year	Engine ID/VIN	Engine Displacement Liters	Piston Clearance	Ring Gap Top Compression	Ring Gap Bottom Compression	Ring Gap Oil Control	Ring Groove Clearance Top Compression	Ring Groove Clearance Bottom Compression	Ring Groove Clearance Oil Control
2007	Z	2.0	0.00039-0.0011	0.0080-0.0014	0.011-0.0190	0.0100-0.0300	0.0012-0.0027	0.0008-0.0023	NA
2008	Z	2.0	0.00039-0.0011	0.0080-0.0014	0.011-0.0190	0.0100-0.0300	0.0012-0.0027	0.0008-0.0023	NA

NA: Not available

22140_RENO_C0007

TORQUE SPECIFICATIONS

All readings in ft. lbs.

Year	Engine ID/VIN	Engine Displacement Liters	Cylinder Head Bolts	Main Bearing Bolts	Rod Bearing Bolts	Crankshaft Damper Bolts	Flywheel Bolts	Manifold Intake	Manifold Exhaust	Spark Plugs	Oil Pan Drain Plug
2007	Z	2.0	①	②	③	④	⑤	16	16	15	⑥
2008	Z	2.0	①	②	③	④	⑤	16	16	15	⑥

① Step 1: Tighten head bolts 1 - 10 to 18 ft. lbs. (15 Nm)
Step 2: Turn bolts another 3 turns of 90 degrees using a angular torque gauge
Step 3: Loosen bolts 1-10

② Step 1: Tighten crank shaft bearing cap bolts to to 37 lb-ft. (50 Nm)
Step 2: Using the angular torque gaugue, tigthen the crankshaft bearing cap bolts another 45 degrees
Step 3: Turn bolts another 15 degrees

③ Step 1: Tighten rod bearing bolts to to 26 lb-ft. (35 Nm)
Step 2: Tighten the rod bearing bolts another 45 degrees
Step 3: Tighten the rod bearing bolts an additional 15 degrees

④ Step 1:- Crank Shaft Gear bolt : 107 ft. lbs. (145 Nm)
Step 2: Turn bolts 30 degrees
Step 3: Turn bolts an additional 15 degrees

⑤ Step 1: Flywheel bolts : 48 ft. lbs. (65 Nm)
Step 2: Turn bolts 30 degrees
Step 3: Turn bolts an additional 15 degrees

⑥ 89 inch pounds (10 Nm)

22140_RENO_C0008

WHEEL ALIGNMENT

Year	Model		Caster Range (+/-Deg.)	Caster Preferred Setting (Deg.)	Camber Range (+/-Deg.)	Camber Preferred Setting (Deg.)	Toe-in (in.) ①	Steering Axis Inclination (Deg.)
2007	Forenza	F	4.0° ± 0.75°	—	–0.33°±0.75°	—	0.0° ± 0.17°	—
		R	—	—	–1.0°±0.75°	—	0.2° ± 0.17°	—
	Reno	F	4.0° ± 0.75°	—	–0.33°±0.75°	—	0.0° ± 0.17°	—
		R	—	—	–1.0°±0.75°	—	0.2° ± 0.17°	—
2008	Forenza	F	4.0° ± 0.75°	—	–0.33°±0.75°	—	0.0° ± 0.17°	—
		R	—	—	–1.0°±0.75°	—	0.2° ± 0.17°	—
	Reno	F	4.0° ± 0.75°	—	–0.33°±0.75°	—	0.0° ± 0.17°	—
		R		—	–1.0°±0.75°	—	0.2° ± 0.17°	—

① No person, full tank

22140_RENO_C0009

TIRE, WHEEL AND BALL JOINT SPECIFICATIONS

Year	Model	OEM Tires Standard	OEM Tires Optional	Tire Pressures (psi) Front	Tire Pressures (psi) Rear	Wheel Size	Ball Joint Inspection	Lug Nut (ft. lbs.)
2007	Forenza	P185/55R14	None	30	32	5.5JJ	①	88
	Reno	P195/55R15	None	30	32	5.5JJ	①	88
2008	Forenza	P185/55R14	None	30	32	5.5JJ	①	88
	Reno	P195/55R15	None	30	32	5.5JJ	①	88

OEM: Original Equipment Manufacturer

PSI: Pounds Per Square Inch

Inflation Pressure at Full Load

① Ball joints must be replaced under the following conditions:
- The joint is loose.
- The ball seal is cut.
- The ball stud is disconnected from the knuckle.
- The ball stud is loose at the knuckle.
- The ball stud can be twisted in its socket with finger pressure.

22140_RENO_C0010

BRAKE SPECIFICATIONS

All measurements in inches unless noted

Year	Model		Brake Disc Original Thickness	Brake Disc Minimum Thickness	Brake Disc Maximum Runout ①	Maximum Diameter	Minimum Lining Thickness Front	Brake Caliper Bracket Bolts (ft. lbs.)	Brake Caliper Mounting Bolts (ft. lbs.)
2007	Forenza	F	0.950	0.870	0.002	10.07	0.080	70	20
		R	0.400	0.310	0.001	10.15	0.080	41	20
	Reno	F	0.950	0.870	0.002	10.07	0.080	70	20
		R	0.400	0.310	0.001	10.15	0.080	41	20
2008	Forenza	F	0.950	0.870	0.002	10.07	0.080	70	20
		R	0.400	0.310	0.001	10.15	0.080	41	20
	Reno	F	0.950	0.870	0.002	10.07	0.080	70	20
		R	0.400	0.310	0.001	10.15	0.080	41	20

① Lateral Runout (Installed)

② Measurement is for lining and backing together.

22140_RENO_C0011

SCHEDULED MAINTENANCE INTERVALS

2007-08 SUZUKI—Forenza, Reno

TO BE SERVICED	TYPE OF SERVICE	VEHICLE MILEAGE INTERVAL (x1000)												
		7,5	15	22.5	30	37.5	45	52.5	60	67.5	75	82.5	90	97.5
Engine oil & filter	R	✓	✓	✓	✓	✓	✓	✓	✓	✓	✓	✓	✓	✓
Automatic transmission fluid & filter ①	S/I	✓			✓				✓				✓	
Drive axle boots	S/I	✓	✓	✓	✓	✓	✓	✓	✓	✓	✓	✓	✓	✓
Inspect & rotate tires	S/I	✓	✓	✓	✓	✓	✓	✓	✓	✓	✓	✓	✓	✓
Manual transmission oil ②	S/I	✓			✓				✓				✓	
Power steering system	S/I		✓		✓		✓		✓		✓		✓	
Suspension system	S/I	✓	✓	✓	✓	✓	✓	✓	✓	✓	✓	✓	✓	✓
Brake discs, pads, drums & shoes	S/I	✓	✓	✓	✓	✓	✓	✓	✓	✓	✓	✓	✓	✓
Brake hoses, pipes, brake lever & cable	S/I	✓		✓				✓		✓		✓		✓
Brake fluid ③	R				✓				✓				✓	
Brake pedal	S/I		✓		✓		✓		✓		✓		✓	
Cooling system, hoses & connections	S/I		✓		✓		✓		✓		✓		✓	
Fuel tank, cap & lines	S/I		✓		✓		✓		✓		✓		✓	
Transfer case oil	S/I	✓			✓		✓		✓		✓		✓	
Air cleaner filter element	R				✓				✓				✓	
Engine coolant	R				✓				✓				✓	
Spark plugs	R				✓				✓					
Drive belts	S/I	✓	✓	✓	✓	✓	✓	✓	✓	✓	✓	✓	✓	✓
Exhaust system	S/I		✓		✓				✓				✓	
Automatic transmission fluid hose	R				✓				✓				✓	
Ignition coils	S/I				✓				✓				✓	

R: Replace S/I: Service or Inspect

① Replace every 100,000 miles.

② Replace every 30,000 miles.

③ Replace every 60,000 miles.

FREQUENT OPERATION MAINTENANCE (SEVERE SERVICE)

If a vehicle is operated under any of the following conditions it is considered severe service:

- Extremely dusty areas.

- 50% or more of the vehicle operation is below 32°C (90°F) or higher temps, or constant operation in temps below 0°C (32°F).

- Prolonged idling (vehicle operation in stop and go traffic).

- Frequent short running periods (engine does not warm to normal operating temperatures).

- Police, taxi, delivery usage or trailer towing usage or other commericl applications

Oil & oil filter: change every 3000 miles.

Brake discs, pads, drums & shoes: service or inspect initially at 3000 miles, 6000 miles, & every 12,000 miles thereafter.

Brake hoses & pipes: service or inspect initially at 3000 miles, 6000 miles & every 12,000 miles thereafter.

Air cleaner filter element: service or inspect ever 3000 miles & replace every 30,000 miles (if not replaced previously).

Automatic transmission fluid & filter: service or inspect every 6000 miles & replace every 15,000 miles (if not replaced previously).

Inspect & rotate tires: service or inspect every 6000 miles.

Manual transmission oil: service or inspect every 6000 miles & replace every 12,000 miles (if not replaced previously).

Power steering system: service or inspect every 6000 miles.

Steering system: service or inspect every 6000 miles.

Suspension system: service or inspect every 6000 miles.

Drive belts: service or inspect every 15,000 miles.

Exhaust system: service or inspect every 15,000 miles.

22140_RENO_C0012

PRECAUTIONS

Before servicing any vehicle, please be sure to read all of the following precautions, which deal with personal safety, prevention of component damage, and important points to take into consideration when servicing a motor vehicle:

- Never open, service or drain the radiator or cooling system when the engine is hot; serious burns can occur from the steam and hot coolant.
- Observe all applicable safety precautions when working around fuel. Whenever servicing the fuel system, always work in a well-ventilated area. Do not allow fuel spray or vapors to come in contact with a spark, open flame, or excessive heat (a hot drop light, for example). Keep a dry chemical fire extinguisher near the work area. Always keep fuel in a container specifically designed for fuel storage; also, always properly seal fuel containers to avoid the possibility of fire or explosion. Refer to the additional fuel system precautions later in this section.
- Fuel injection systems often remain pressurized, even after the engine has been turned **OFF**. The fuel system pressure must be relieved before disconnecting any fuel lines. Failure to do so may result in fire and/or personal injury.
- Brake fluid often contains polyglycol ethers and polyglycols. Avoid contact with the eyes and wash your hands thoroughly after handling brake fluid. If you do get brake fluid in your eyes, flush your eyes with clean, running water for 15 minutes. If eye irritation persists, or if you have taken brake fluid internally, IMMEDIATELY seek medical assistance.
- The EPA warns that prolonged contact with used engine oil may cause a number of skin disorders, including cancer. You should make every effort to minimize your exposure to used engine oil. Protective gloves should be worn when changing oil. Wash your hands and any other exposed skin areas as soon as possible after exposure to used engine oil. Soap and water, or waterless hand cleaner should be used.
- All new vehicles are now equipped with an air bag system, often referred to as a Supplemental Restraint System (SRS) or Supplemental Inflatable Restraint (SIR) system. The system must be disabled before performing service on or around system components, steering column, instrument panel components, wiring and sensors. Failure to follow safety and disabling procedures could result in accidental air bag deployment, possible personal injury and unnecessary system repairs.
- Always wear safety goggles when working with, or around, the air bag system. When carrying a non-deployed air bag, be sure the bag and trim cover are pointed away from your body. When placing a non-deployed air bag on a work surface, always face the bag and trim cover upward, away from the surface. This will reduce the motion of the module if it is accidentally deployed. Refer to the additional air bag system precautions later in this section.
- Clean, high quality brake fluid from a sealed container is essential to the safe and proper operation of the brake system. You should always buy the correct type of brake fluid for your vehicle. If the brake fluid becomes contaminated, completely flush the system with new fluid. Never reuse any brake fluid. Any brake fluid that is removed from the system should be discarded. Also, do not allow any brake fluid to come in contact with a painted surface; it will damage the paint.
- Never operate the engine without the proper amount and type of engine oil; doing so WILL result in severe engine damage.
- Timing belt maintenance is extremely important. Many models utilize an interference-type, non-freewheeling engine. If the timing belt breaks, the valves in the cylinder head may strike the pistons, causing potentially serious (also time-consuming and expensive) engine damage. Refer to the maintenance interval charts for the recommended replacement interval for the timing belt, and to the timing belt section for belt replacement and inspection.
- Disconnecting the negative battery cable on some vehicles may interfere with the functions of the on-board computer system(s) and may require the computer to undergo a relearning process once the negative battery cable is reconnected.
- When servicing drum brakes, only disassemble and assemble one side at a time, leaving the remaining side intact for reference.
- Only an MVAC-trained, EPA-certified automotive technician should service the air conditioning system or its components.

BRAKES

ANTI-LOCK BRAKE SYSTEM (ABS)

GENERAL INFORMATION

Before using this section, it is important that you have a basic knowledge of the following items. Without this knowledge, it will be difficult to use the diagnostic procedures contained.

- Basic Electrical Circuits : You should understand the basic theory of electricity and know the meaning of voltage, current (amps), and resistance (ohms). You should understand what happens in a circuit with an open or shorted wire. You should be able to read and understand a wiring diagram.
- Use of Circuit Testing Tools : You should know how to use a test light and how to bypass components to test circuits using fused jumper wires. You should be familiar with a digital multimeter. You should be able to measure voltage, resistance, and current, and be familiar with the controls and how to use them correctly.
- The ABS 5.3 Antilock Braking System (ABS) consists of a conventional hydraulic brake system plus antilock components.
- The conventional brake system includes a vacuum booster, master cylinder, front disc brakes, rear leading/trailing drum brakes, interconnecting hydraulic brake pipes and hoses, brake fluid level sensor and the BRAKE indicator.
- The ABS components include a hydraulic unit, an electronic brake control module (EBCM), two system fuses, four wheel speed sensors (one at each wheel), interconnecting wiring, the ABS indicator, the EBD indicator (which is connected to the parking lamp) and the rear disk brakes.
- The hydraulic unit with the attached EBCM is located between the surge tank and the fire wall on the left side of the vehicle.
- The basic hydraulic unit configuration consists of hydraulic check valves, two solenoid valves for each wheel, a hydraulic pump, two accumulators, and two damper.
- The hydraulic unit controls hydraulic pressure to the front calipers and rear wheel cylinders by modulating hydraulic pressure to prevent wheel lockup.

➡**Nothing in the hydraulic unit or the EBCM is serviceable. In the event of any failure, the entire ABS unit with attached EBCM must be replaced.**

PRECAUTIONS

- Certain components within the ABS system are not intended to be serviced or repaired individually.
- Do not use rubber hoses or other parts not specifically specified for and ABS system. When using repair kits, replace all parts included in the kit. Partial or incorrect repair may lead to functional problems and require the replacement of components.
- Lubricate rubber parts with clean, fresh brake fluid to ease assembly. Do not use shop air to clean parts; damage to rubber components may result.
- Use only DOT 3 brake fluid from an unopened container.
- If any hydraulic component or line is removed or replaced, it may be necessary to bleed the entire system.
- A clean repair area is essential. Always clean the reservoir and cap thoroughly before removing the cap. The slightest amount of dirt in the fluid may plug an orifice and impair the system function. Perform repairs after components have been thoroughly cleaned; use only denatured alcohol to clean components. Do not allow ABS components to come into contact with any substance containing mineral oil; this includes used shop rags.
- The Anti-Lock control unit is a microprocessor similar to other computer units in the vehicle. Ensure that the ignition switch is **OFF** before removing or installing controller harnesses. Avoid static electricity discharge at or near the controller.
- If any arc welding is to be done on the vehicle, the control unit should be unplugged before welding operations begin.

WHEEL SPEED SENSORS

REMOVAL & INSTALLATION

Front Speed Sensor

See Figures 1 and 2.

1. Disconnect the negative battery cable.
2. Disconnect the front wheel speed sensor electrical connector.
3. Raise and suitably support the vehicle.
4. Remove the wheel.
5. Turn the steering wheel to expose the speed sensor. It is located at the rear of the steering knuckle near the tie rod end.
6. Remove the bolt and the front wheel speed sensor from the steering knuckle.
7. Feed the harness through the grommet for the speed sensor harness and the hydraulic pipe from the strut tower. Remove the speed sensor harness from it so that the connector can pass through the hole in the strut tower.
8. Free the sensor harness from the grommet holders and the clamps and pull it through the fender.

To install:

9. Install the front wheel speed sensor to the steering knuckle. Secure it with the bolt.
10. Tighten the front wheel speed sensor bolt to 71 inch lbs. (8 Nm)
11. Feed the sensor harness into the engine compartment, insert it into the grommet, and secure the grommet into the hole in the strut tower.
12. Secure the harness into the grommet holders and the clamps under the fender.
13. Install the wheel.
14. Lower the vehicle.
15. Connect the front wheel speed sensor electrical connector.
16. Connect the negative battery cable.

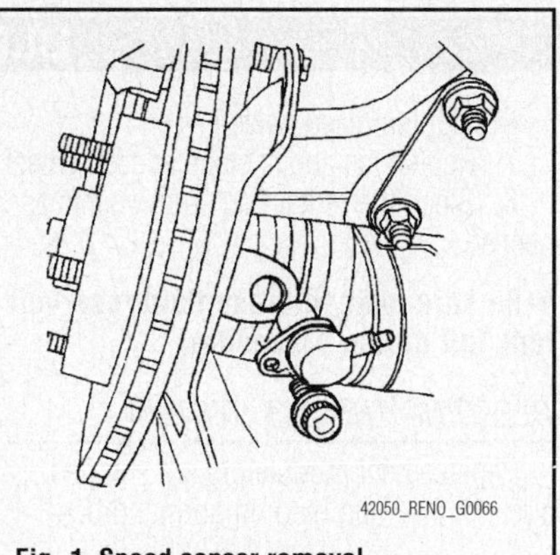

Fig. 1 Speed sensor removal

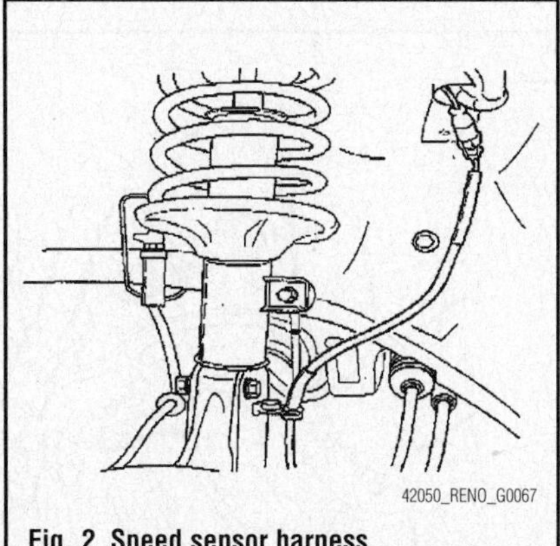

Fig. 2 Speed sensor harness

Rear Speed Sensor

See Figures 3 and 4.

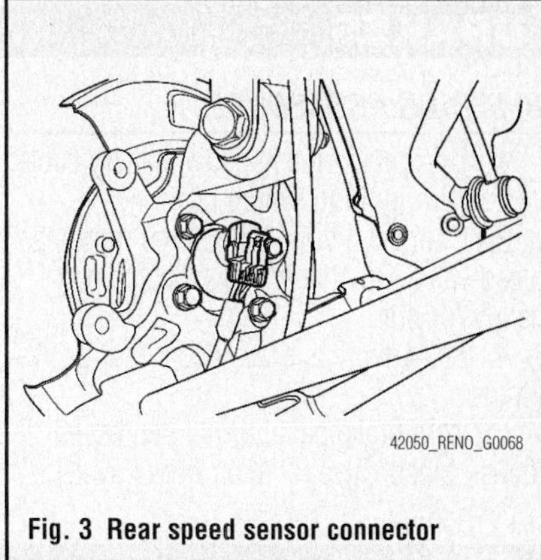

Fig. 3 Rear speed sensor connector

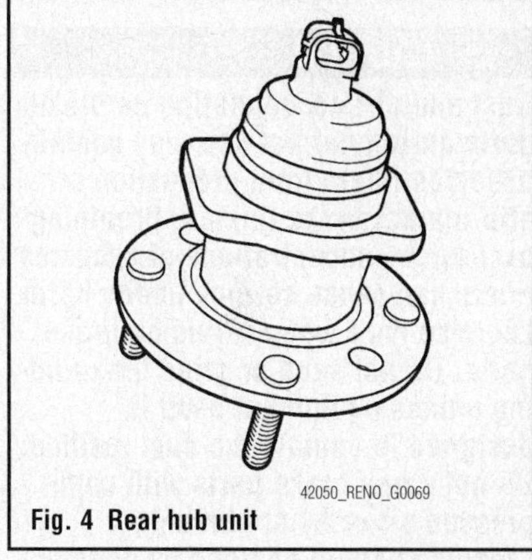

Fig. 4 Rear hub unit

1. Disconnect the negative battery cable.
2. Raise and suitably support the vehicle.
3. Disconnect the rear wheel speed sensor electrical connector.
4. Remove the rear hub unit from the knuckle assembly.

➡The rear wheel speed sensor is not serviceable separately.

To install:

5. Install the rear hub unit to the knuckle assembly.
6. Tighten the rear hub unit to the knuckle assembly to 48 ft. lbs. (65 Nm)
7. Connect the rear wheel speed sensor electrical connector.

WHEEL SPEED SENSOR RINGS (TOOTHED RINGS)

REMOVAL & INSTALLATION

Front tone rings come with front axles. Rear tone rings are not serviceable and are part of rear hub and bearing assembly.

BRAKES — BLEEDING THE BRAKE SYSTEM

BLEEDING PROCEDURE

1. Disconnect the negative battery cable.
2. Turn ignition switch off.
3. Connect a vinyl tube to the right rear bleed valve. Be sure to have a catch pan handy for fluid.
4. Fully depress brake pedal four or five times.
5. With brake pedal firmly depressed loosen bleed valve to let air out of system.
6. Tighten bleed valve.
7. Repeat the above steps at each wheel.
8. Bleed the remaining components as needed.

➡Be sure to keep brake fluid reservoir tank full during procedure.

BLEEDING THE ABS SYSTEM

Replacement modulators are shipped already filled and bled. In normal procedures requiring removal of the modulator, such as to replace the Electronic Brake Control Module (EBCM), air will not enter the modulator, and normal bleeding will be all that is needed. If air enters the hydraulic modulator, or if an unfilled modulator is installed, use the brake bleeding program in the scan tool to bleed the modulator. Manual bleeding of the hydraulic modulator is not possible.

BRAKES — FRONT DISC BRAKES

⁂ CAUTION

Dust and dirt accumulating on brake parts during normal use may contain asbestos fibers from production or aftermarket brake linings. Breathing excessive concentrations of asbestos fibers can cause serious bodily harm. Exercise care when servicing brake parts. Do not sand or grind brake lining unless equipment used is designed to contain the dust residue. Do not clean brake parts with compressed air or by dry brushing. Cleaning should be done by dampening the brake components with a fine mist of water, then wiping the brake components clean with a dampened cloth. Dispose of cloth and all residue containing asbestos fibers in an impermeable container with the appropriate label. Follow practices prescribed by the Occupational Safety and Health Administration (OSHA) and the Environmental Protection Agency (EPA) for the handling, processing, and disposing of dust or debris that may contain asbestos fibers.

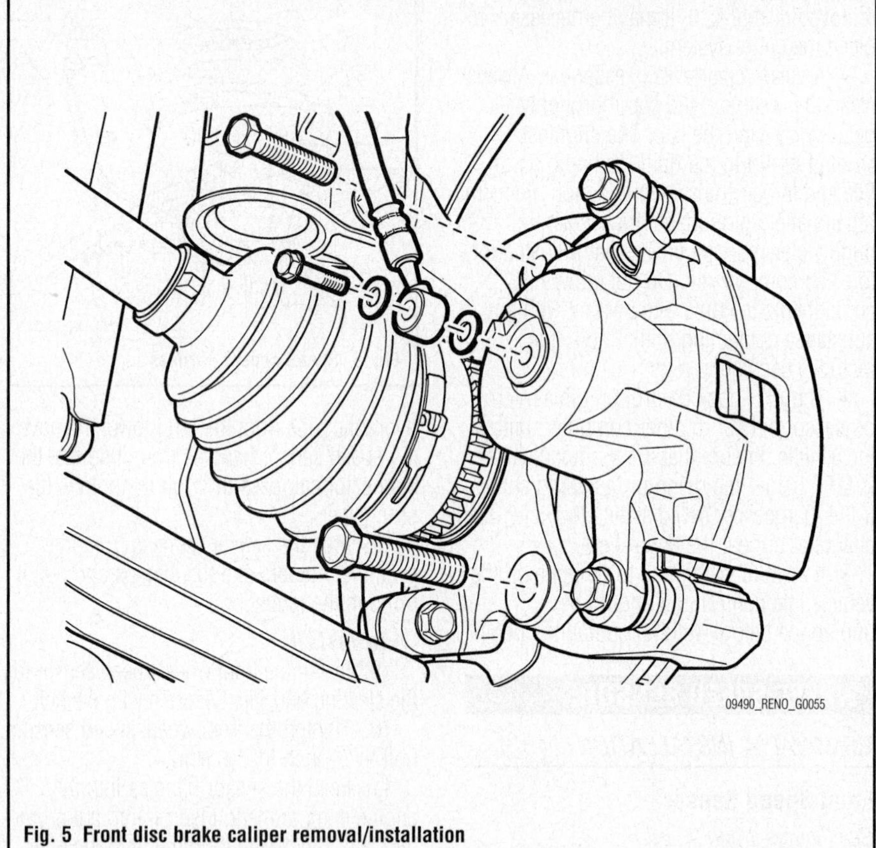

Fig. 5 Front disc brake caliper removal/installation

BRAKE CALIPER

REMOVAL & INSTALLATION

See Figure 5.

1. Before servicing the vehicle, refer to the Precautions Section.
2. Raise and safely support the front of the vehicle.
3. To preserve wheel balance, mark the position of the front wheel relative to the wheel hub.
4. Remove the wheel.
5. Remove the bolt and the washers attaching the brake hose to the caliper.
6. Disconnect the brake hose, and plug the openings in the caliper and the brake hose to prevent fluid loss and contamination.
7. Remove the caliper mounting bolts from the steering knuckle, and remove the caliper assembly.

To install:

8. Install the caliper assembly with the mounting bolts.
9. Tighten the caliper-to-steering knuckle mounting bolts to 70 ft. lbs. (95 Nm).
10. Connect the brake hose to the caliper with bolt and washer. Tighten the caliper brake hose inlet bolt to 30 ft. lbs. (40 Nm).
11. Align the marks that were made when removing the front wheel, and install the wheel.
12. Lower the vehicle.
13. Fill the master cylinder to the proper level with clean brake fluid.
14. Bleed the air out of the brake system.
15. Recheck the fluid level.
16. Repeatedly press the brake pedal to bring the pads in contact with the rotor.

DISC BRAKE PADS

REMOVAL & INSTALLATION

1. Before servicing the vehicle, refer to the Precautions Section.
2. Raise and safely support the front of the vehicle.
3. Remove the wheels.
4. Remove the lower caliper mounting bolt.
5. Pivot the caliper upward.
6. Remove the brake pads.

To install:

7. Measure the minimum brake shoe lining thickness.
8. Install the brake pads.
9. Push the caliper piston inward, if needed.
10. Pivot the caliper downward and install the lower mounting bolt. Tighten the lower caliper mounting bolt to 20 ft. lbs. (27 Nm).
11. Install the wheels.
12. Lower the vehicle.

BRAKES

REAR DISC BRAKES

⁂ CAUTION

Dust and dirt accumulating on brake parts during normal use may contain asbestos fibers from production or aftermarket brake linings. Breathing excessive concentrations of asbestos fibers can cause serious bodily harm. Exercise care when servicing brake parts. Do not sand or grind brake lining unless equipment used is designed to contain the dust residue. Do not clean brake parts with compressed air or by dry brushing. Cleaning should be done by dampening the brake components with a fine mist of water, then wiping the brake components clean with a dampened cloth. Dispose of cloth and all residue containing asbestos fibers in an impermeable container with the appropriate label. Follow practices prescribed by the Occupational Safety and Health Administration (OSHA) and the Environmental Protection Agency (EPA) for the handling, processing, and disposing of dust or debris that may contain asbestos fibers.

BRAKE CALIPER

REMOVAL & INSTALLATION

See Figure 6.

1. Before servicing the vehicle, refer to the Precautions Section.
2. Raise and safely support the rear of the vehicle.
3. To preserve wheel balance, mark the position of the front wheel relative to the wheel hub.
4. Remove the wheels.
5. Remove the bolt and the ring seals that attach the brake hose inlet fitting to the caliper.
6. Disconnect the brake hose. Plug the openings in the caliper and the brake hose to prevent fluid loss or contamination.

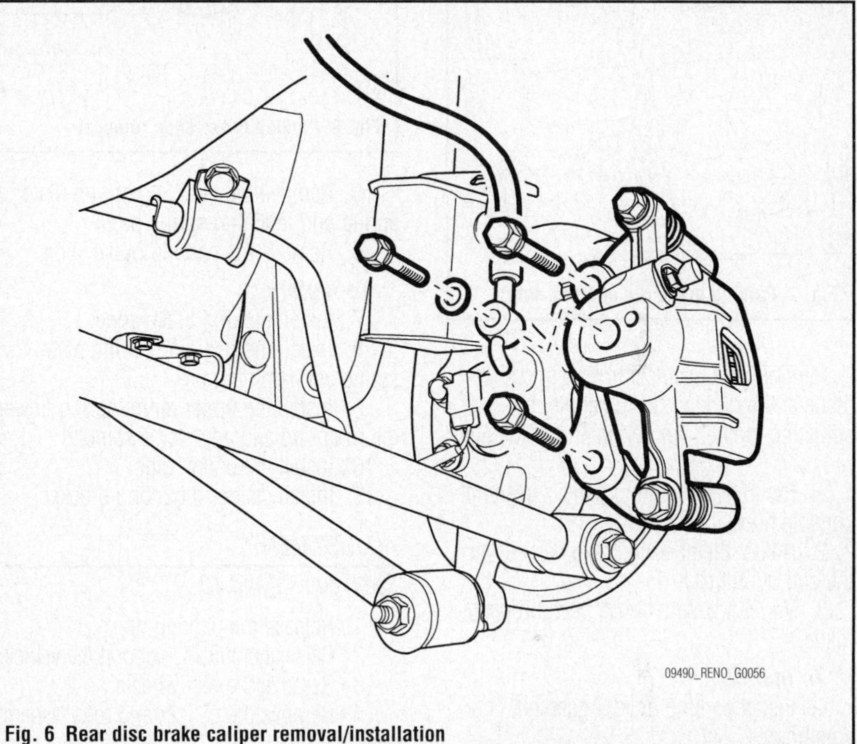

Fig. 6 Rear disc brake caliper removal/installation

7. Remove the caliper mounting bolts from the knuckle.
8. Remove the caliper.

To install:

9. Install the caliper assembly with the mounting bolts.
10. Tighten the caliper mounting bolts to 41 ft. lbs. (56 Nm).
11. Connect the brake hose to the caliper with bolt and ring seals. Tighten the caliper brake hose inlet bolt to 24 ft. lbs. (32 Nm).
12. Align the marks that were made when removing the front wheel, and install the wheel.
13. Lower the vehicle.
14. Fill the master cylinder to the proper level with clean brake fluid.
15. Bleed the air out of the brake system.
16. Recheck the fluid level.
17. Repeatedly press the brake pedal to bring the pads in contact with the rotor.

DISC BRAKE PADS

REMOVAL & INSTALLATION

1. Before servicing the vehicle, refer to the Precautions Section.
2. Raise and safely support the rear of the vehicle.
3. Remove the wheels.
4. Remove the lower caliper guide pin bolt.
5. Pivot the caliper upward.
6. Remove the brake pads.

To install:

7. Measure the minimum brake shoe lining thickness.
8. Install the brake pads into the caliper.
9. Push the piston inward, if needed.
10. Pivot the caliper downward and tighten the bolt to 20 ft. lbs. (27 Nm).
11. Install the rear wheels.
12. Lower the vehicle.

PARKING BRAKE CABLES

ADJUSTMENT

See Figure 7.

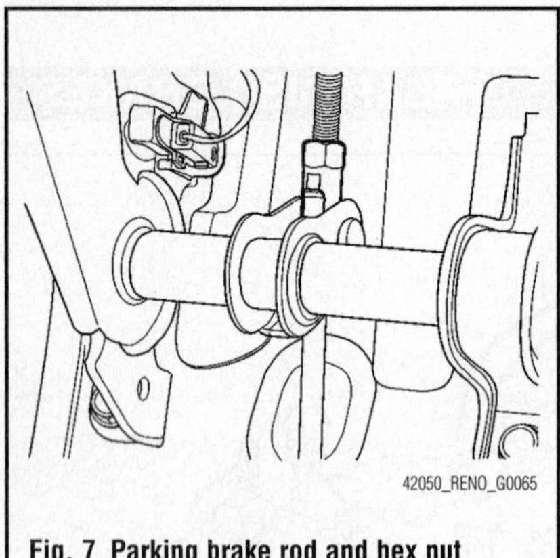

Fig. 7 Parking brake rod and hex nut

Parking brake cable adjustment can be made at the parking brake handle, by adjusting the hex nut on the end of the pull rod.

1. Remove the parking brake / gearshift console hood.
2. Adjust the parking brake hex nut on the end of pull rod.
3. Verify adjustment has been properly made.

To install:

4. Install parking brake / gearshift console hood.

➡Be sure to check parking brake shoe adjustment prior to cable adjustment. This is only for cables that may have stretched over time.

PARKING BRAKE SHOES

REMOVAL & INSTALLATION

See Figure 8.

1. Remove the caliper mounting bracket.
2. Remove the rotor detent screw.
3. Remove the brake rotor.
4. Remove the upper return spring, lower return spring and adjuster assembly.

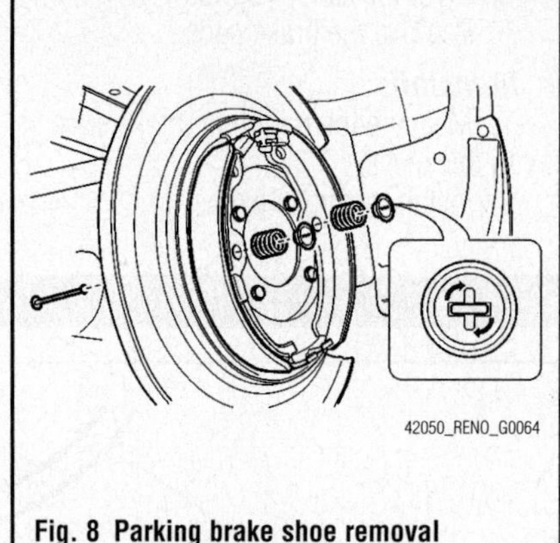

Fig. 8 Parking brake shoe removal

5. Remove the retaining spring plate, spring and retaining spring pin.
6. Remove the parking brake shoe.

To install:

7. Install parking brake shoe.
8. Install the retaining spring plate, spring and retaining pin.
9. Install the upper return spring, lower return spring and adjuster assembly.
10. Install the brake rotor.
11. Install caliper mounting bracket.

ADJUSTMENT

See Figures 9 and 10.

1. Release the parking brake.
2. Raise and suitably support the vehicle.
3. Remove the rear wheels.
4. Remove the caliper and rotor assemblies.

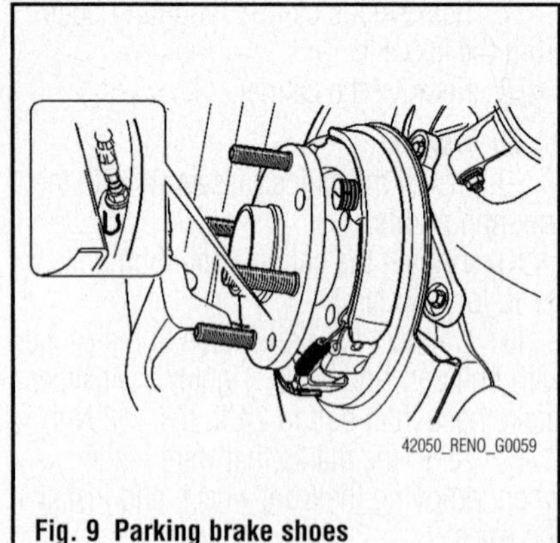

Fig. 9 Parking brake shoes

5. Remove the brake rotors on each side of the vehicle.
6. Disconnect the parking brake cable from the backplate operating lever on each side of the vehicle.

➡Inspect and replace any parts of doubtful strength or quality. This can be shown by discoloration from heat or stress.

7. Using a vernier caliper, adjust the shoe assembly to 6.60–6.61 inches (167.6–167.8 mm) by turning the adjuster nut clockwise to increase the diameter. Measure the shoe assembly diameter as closely as possible to the center of the lining material.
8. Inspect and install the rotors and calipers.

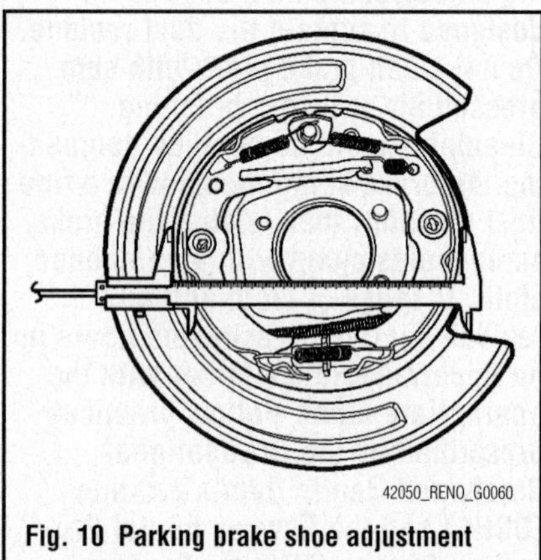

Fig. 10 Parking brake shoe adjustment

9. Install the parking brake cable to the backplate lever on each side of the vehicle.
10. In the vehicle cabin, pull on the parking brake handle. Stop after hearing two clicks.
11. Turn the rear wheel by hand until the wheel begins to drag.
12. Release the parking brake.
13. Turn the rear wheel by hand to check the drag. Readjust the cable, if necessary.
14. Repeat the process for the other rear wheel.
15. Lower the vehicle.

CHASSIS ELECTRICAL AIR BAG (SUPPLEMENTAL RESTRAINT SYSTEM)

GENERAL INFORMATION

See Figure 11.

All vehicles are equipped with an air bag system, often referred to as a Supplemental Restraint System (SRS) or Supplemental Inflatable Restraint (SIR) system. The system must be disabled before performing service on or around system components, steering column, instrument panel components, wiring and sensors. Failure to follow safety and disabling procedures could result in accidental air bag deployment, possible personal injury and unnecessary system repairs.

⁂ WARNING

Always wear safety goggles when working with, or around, the air bag system. When carrying a non-deployed air bag, be sure the bag and trim cover are pointed away from your body. When placing a non-deployed air bag on a work surface, always face the bag and trim cover upward, away from the surface. This will reduce the motion of the module if it is accidentally deployed. Refer to the additional air bag system precautions later in this section.

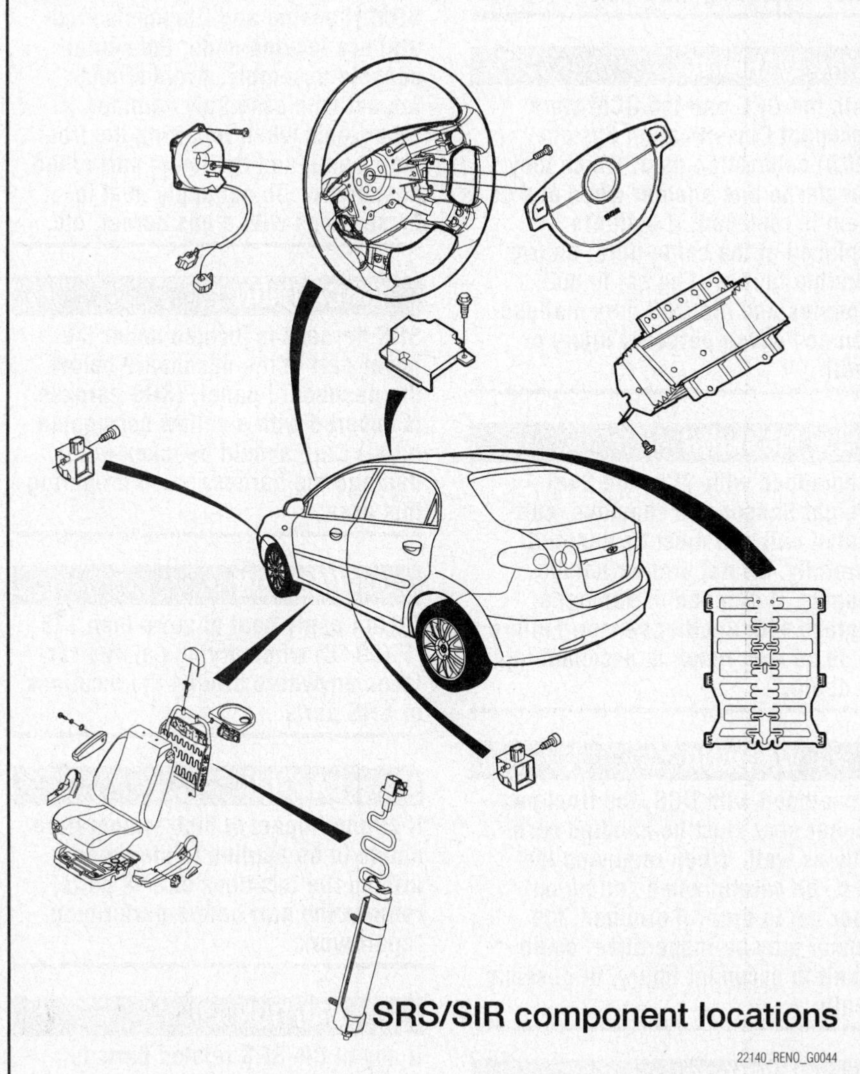

Fig. 11 SRS/SIR component location

SERVICE PRECAUTIONS

⁂ WARNING

Disconnect and isolate the battery negative cable before beginning any airbag system component diagnosis, testing, removal, or installation procedures. Allow system capacitor to discharge for two minutes before beginning any component service. This will disable the airbag system. Failure to disable the airbag system may result in accidental airbag deployment, personal injury, or death.

⁂ WARNING

Do not place an intact undeployed airbag face down on a solid surface. The airbag will propel into the air if accidentally deployed and may result in personal injury or death.

⁂ WARNING

When carrying or handling an undeployed airbag, the trim side (face) of the airbag should be pointing towards the body to minimize possibility of injury if accidental deployment occurs. Failure to do this may result in personal injury or death.

⁂ WARNING

Replace airbag system components with OEM replacement parts. Substitute parts may appear interchangeable, but internal differences may result in inferior occupant protection. Failure to do so may result in occupant personal injury or death.

⁂ CAUTION

Wear safety glasses, rubber gloves, and long sleeved clothing when cleaning powder residue from vehicle after an airbag deployment. Powder residue emitted from a deployed airbag can cause skin irritation. Flush affected area with cool water if irritation is experienced. If nasal or throat irritation is experienced, exit the vehicle for fresh air until the irritation ceases. If irritation continues, see a physician.

⁂ WARNING

Do not use a replacement airbag that is not in the original packaging. This may result in improper deployment, personal injury, or death.

⁂ CAUTION

The factory installed fasteners, screws and bolts used to fasten airbag components have a special coating and are specifically designed for the airbag system. Do not use substitute fasteners. Use only original equipment fasteners listed in the parts catalog when fastener replacement is required.

⁂ WARNING

During, and following, any child restraint anchor service, due to impact event or vehicle repair, carefully inspect all mounting hardware, tether straps, and anchors for proper installation, operation, or damage. If a child restraint anchor is found damaged in any way, the anchor must be replaced. Failure to do this may result in personal injury or death.

⁂ CAUTION

Deployed and non-deployed airbags may or may not have live pyrotechnic material within the airbag inflator.

⁂ CAUTION

Do not dispose of driver/passenger/curtain airbags or seat belt tensioners unless you are sure of complete deployment. Refer to the Hazardous Substance Control System for proper disposal.

⁂ CAUTION

Dispose of deployed airbags and tensioners consistent with state, provincial, local, and federal regulations.

⁂ WARNING

After any airbag component testing or service, do not connect the battery negative cable. Personal injury or death may result if the system test is not performed first.

⁂ WARNING

If the vehicle is equipped with the Occupant Classification System (OCS), do not connect the battery negative cable before performing the OCS Verification Test using the scan tool and the appropriate diagnostic information. Personal injury or death may result if the system test is not performed properly.

⁂ CAUTION

Never replace both the Occupant Restraint Controller (ORC) and the Occupant Classification Module (OCM) at the same time. If both require replacement, replace one, then perform the Airbag System test before replacing the other.

⁂ WARNING

Both the ORC and the OCM store Occupant Classification System (OCS) calibration data, which they transfer to one another when one of them is replaced. If both are replaced at the same time, an irreversible fault will be set in both modules and the OCS may malfunction and cause personal injury or death.

⁂ WARNING

If equipped with OCS, the Seat Weight Sensor is a sensitive, calibrated unit and must be handled carefully. Do not drop or handle roughly. If dropped or damaged, replace with another sensor. Failure to do so may result in occupant injury or death.

⁂ WARNING

If equipped with OCS, the front passenger seat must be handled carefully as well. When removing the seat, be careful when setting on floor not to drop. If dropped, the sensor may be inoperative, could result in occupant injury, or possibly death.

⁂ WARNING

If equipped with OCS, when the passenger front seat is on the floor, no one should sit in the front passenger seat. This uneven force may damage the sensing ability of the seat weight sensors. If sat on and damaged, the sensor may be inoperative, could result in occupant injury, or possibly death.

DISARMING THE SYSTEM

See Figure 12.

⁂ WARNING

When performing service on or around the air bag system components or wiring, disable the air bag system. Failure to follow the procedures could result in possible deployment, personal injury or unneeded system repairs.

⁂ CAUTION

SDM (Sensing and Diagnosis Module) are located under the center console assembly. Avoid strong impact with assembly hammer or other tools when repairing the front side frame and the lower part of the dashboard. Do not apply heat to these areas with a gas burner, etc.

⁂ CAUTION

SRS harness is located under the lower part of the dashboard below the dashboard panel. (SRS harness is covered with a yellow corrugated tube.) Care should be taken not to damage the harness when preparing this area.

⁂ CAUTION

Do not apply heat of more than 176 °F (80 °C) when drying painted surfaces anywhere around the locations or SRS parts.

⁂ CAUTION

If strong impact of high temperature needs to be applied to the areas around the locations of SRS parts, remove the part before performing repair work.

⁂ CAUTION

If any of the SRS related parts is damaged or deformed, be sure to replace it.

1. Before servicing the vehicle, refer to the Precautions Section.

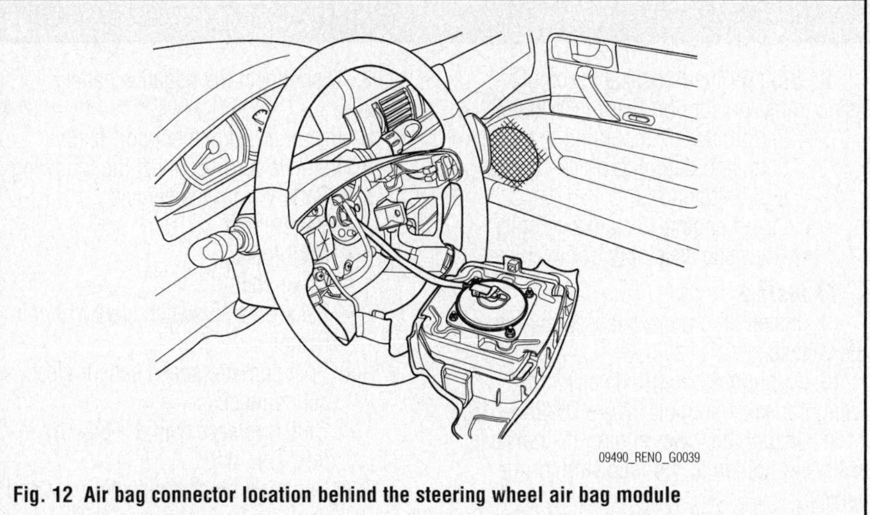

Fig. 12 Air bag connector location behind the steering wheel air bag module

2. Disconnect the negative battery cable.
3. Turn the steering wheel so the wheels are pointing straight ahead.
4. Turn the ignition switch to the **LOCK** position and remove the key.
5. Remove the **AIR BAG-F1** fuse from the instrument panel (I/P) fuse block located at the end of the instrument panel on the driver side.
6. Wait more than 1 minute for SIR capacitor to discharge.
7. Remove the mounting bolts, steering wheel air bag module and disconnect the yellow connector for the driver's side air bag (inflator) module from behind the module.
8. Remove the glove box. Disconnect the yellow connector for the passenger air bag (inflator) module.

ARMING THE SYSTEM

WARNING

When performing service on or around the air bag system components or wiring, disable the air bag system. Failure to follow the procedures could result in possible deployment, personal injury or unneeded system repairs.

1. Before servicing the vehicle, refer to the Precautions Section.
2. Connect the negative battery cable.
3. Turn the ignition switch to the **LOCK** position and remove the key.
4. Connect the yellow connector for the passenger side air bag (inflator) module and the yellow connector for the driver's side air bag (inflator) module.
5. Install the glove box assembly.
6. Install the passenger side air bag module and tighten the mounting bolts to 97 inch lbs. (11 Nm).
7. Install the steering wheel air bag module and tighten the mounting bolts to 11 ft. lbs. (15 Nm).
8. Install the **AIR BAG-F1** fuse into the instrument panel fuse block.
9. Turn the ignition **ON** and verify that the **AIR BAG** indicator lamp flashes 7 times, then turns off. If the system does not operate as described, diagnosis and repairs to the air bag system are necessary.

CLOCKSPRING CENTERING

See Figures 13 and 14.

WARNING

The Sensing and Diagnostic Module (SDM) can maintain sufficient voltage to deploy the airbags and pretensioners for 1 minute after the ignition is OFF and the fuse has been removed. If the airbags and pretensioners are not disconnected, do not begin service until 1 minute has passed after disconnecting power to SDM. If the airbags are disconnected, service can be done immediately without waiting for 1-minute time period to expire. Failure to temporarily disable the SIR during service can result in unexpected deployment, personal injury and unnecessary SIR repairs.

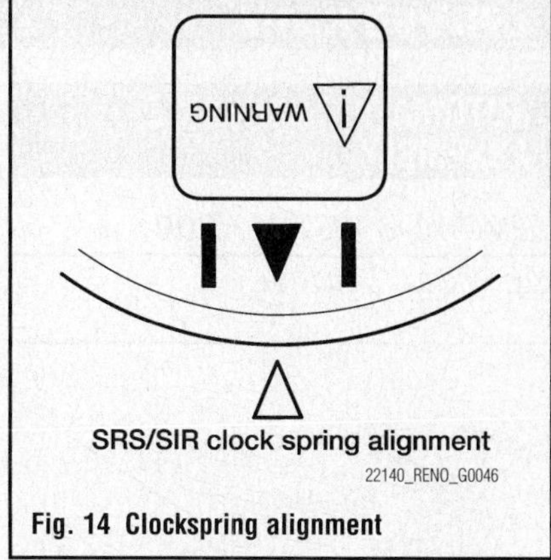

Fig. 14 Clockspring alignment

1. Turn the front wheels straight ahead.
2. Turn the lobe of clockspring clockwise to lock (Do not force).
3. Then turn the lobe of clockspring counterclockwise approximately three turns to the neutral position, with the front of the wheels straight ahead.
4. Properly align the pointed marks on the components of the clockspring.

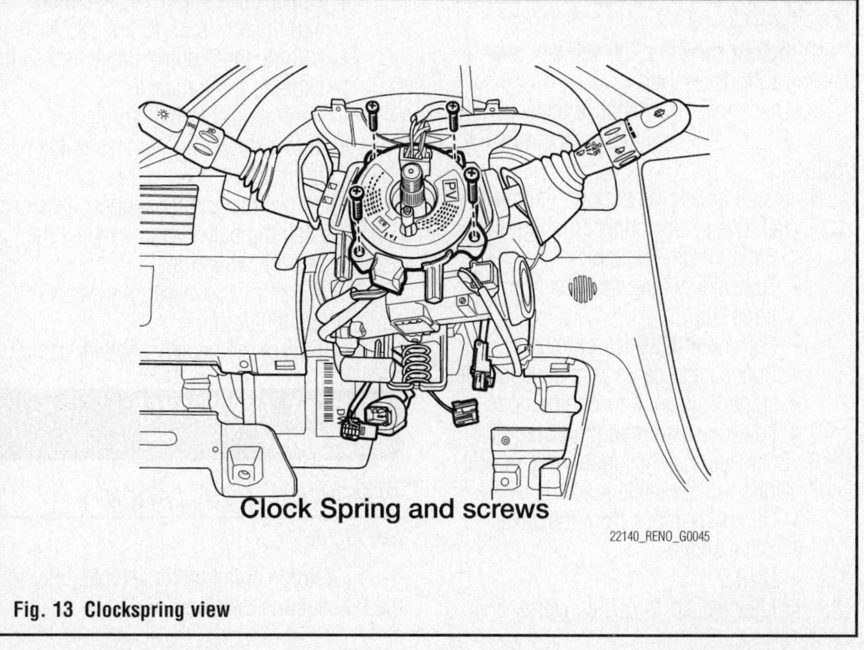

Fig. 13 Clockspring view

DRIVETRAIN

AUTOMATIC TRANSAXLE ASSEMBLY

REMOVAL & INSTALLATION

See Figures 15 and 16.

Fig. 15 Lower engine-to-transaxle bolts—Forenza/Reno

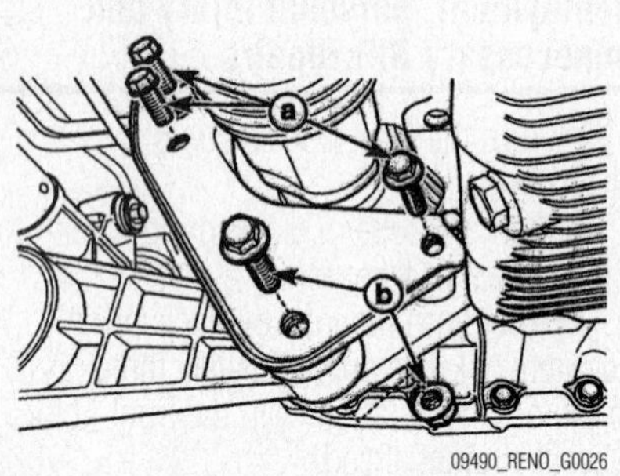

Fig. 16 Rear transaxle mounting bracket bolts (a) and damping block connection bolt and nut (b)—Forenza/Reno

1. Before servicing the vehicle, refer to the Precautions Section.
2. Disconnect the negative battery cable.
3. Position the shift lever in the **PARK** position.
4. Install an engine support fixture.
5. Remove or disconnect the following:
   - Battery from the vehicle
   - Transaxle wiring harness from the transaxle
   - Park/neutral switch connector
   - Shift control cable from the transaxle
   - Upper transaxle-to-engine bolts
   - Left transaxle mounting bracket
6. Raise and safely support the vehicle.
7. Drain the transaxle fluid.
   - Oil cooler pipes from transaxle
   - Drive shafts
   - Starter
   - Flywheel-to-torque converter bolts
8. Support the transaxle assembly using transaxle support fixture DW260-010.
   - Rear transaxle mounting bracket bolts and damping block connection bolt and nut
   - Lower engine-to-transaxle bolts
   - Transaxle assembly from vehicle

To install:

9. Install the transaxle assembly into the vehicle.
10. Support the transaxle assembly using transaxle support fixture DW260-010.
11. Install the lower engine-to-transaxle bolts and tighten to the following torque values:
   - Lower engine-to-transaxle bolts **"a"**: 55 ft. lbs. (75 Nm)
   - Lower engine-to-transaxle bolt **"b"**: 15 ft. lbs. (21 Nm)
   - Lower engine-to-transaxle bolts **"c"**: 23 ft. lbs. (31 Nm)
12. Install or connect the following:
   - Rear transaxle mounting bracket bolts and tighten to 45 ft. lbs. (62 Nm)
   - Damping block connection bolt and nut and tighten to 50 ft. lbs. (68 Nm)
   - Torque converter bolts and tighten the bolts to 33 ft. lbs. (45 Nm)
   - Starter
   - Drive shafts
   - Oil cooler pipes to the transaxle
13. Lower the vehicle.
   - Left transaxle mounting bracket and tighten the bolts to 35 ft. lbs. (48 Nm)
   - Upper transaxle-to-engine bolts and tighten to 55 ft. lbs. (75 Nm)
   - Clip to shift control cable and shift cable to the transaxle
   - Park/neutral switch connector
   - Transaxle wiring harness to the transaxle
14. Remove the engine support fixture.
15. Install the battery and connect the positive battery cable first.
16. Fill the transaxle with the correct amount and type of fluid.
17. Connect the negative battery cable.

MANUAL TRANSAXLE ASSEMBLY

REMOVAL & INSTALLATION

See Figure 17.

1. Before servicing the vehicle, refer to the Precautions Section.
2. Disconnect the negative battery cable.
3. Install an engine support fixture.
4. Remove or disconnect the following:
   - Battery from the vehicle
   - Battery tray
   - Shift linkage
   - Drive shaft
   - Backup lamp switch electrical connector
   - Speedometer speed sensor electrical connector
   - Clutch release cylinder pipe by releasing clip
   - Damping block connection nut and bolt
   - Three rear mounting bracket bolts
   - Rear mounting bracket from the transaxle
   - Two rear damping block retaining bolts
   - Rear damping block from the front cross member
   - Two cage retaining bolts
   - Three transaxle upper mounting bracket bolts
   - Upper mounting bracket and cage
   - Three transaxle upper retaining bolts
5. Support transaxle with a transmission jack.
   - Seven transaxle lower retaining bolts
   - Transaxle assembly from vehicle

To install:

6. Support transaxle with a transmission jack.
7. Install the transaxle assembly. Use care when inserting the input shaft into the clutch assembly. If the spline on the input shaft does not align with the clutch assembly spline, turn the crankshaft slightly to aid in spline alignment.

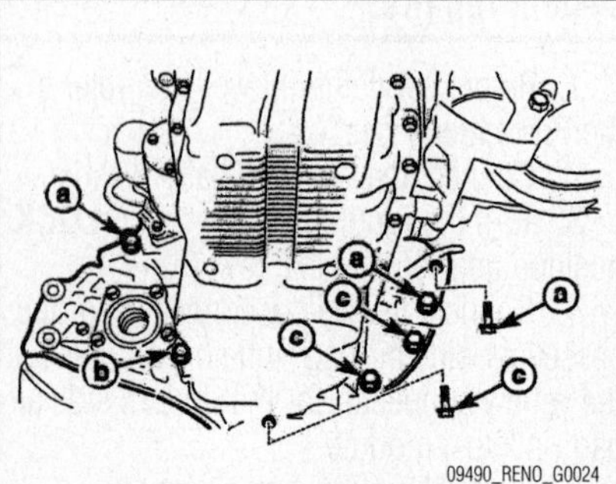

Fig. 17 Manual transaxle lower retaining bolts

8. Install the seven transaxle lower retaining bolts and tighten to the following torque values:

- Lower retaining bolts **"a"**: 54 ft. lbs. (73 Nm)
- Lower retaining bolts **"b"**: 23 ft. lbs. (31 Nm)
- Lower retaining bolts **"c"**: 15 ft. lbs. (21 Nm)

9. Install or connect the following:

- Three transaxle upper retaining bolts and tighten to 54 ft. lbs. (73 Nm)
- Cage retaining bolt and cage
- Three transaxle upper mounting bracket bolts and bracket. Tighten the transaxle upper mounting bracket bolts to 35 ft. lbs. (48 Nm).
- Two rear damping block retaining bolts and tighten to 50 ft. lbs. (68 Nm).
- Rear damping block to the front cross member
- Three rear mounting bracket bolts and the bracket. Tighten the rear mounting bracket bolts to 66 ft. lbs. (90 Nm).
- Damping block connection nut and bolt and tighten to 50 ft. lbs. (68 Nm).
- Clutch release cylinder pipe and secure with clip
- Speedometer speed sensor electrical connector
- Backup lamp switch electrical connector
- Drive shaft
- Shift linkage
- Battery tray
- Battery

10. Remove the engine support fixture.
11. Connect the positive battery cable first.
12. Connect the negative battery cable second.
13. Inspect the fluid level in the transaxle and add, if necessary.
14. Check the function of the engine, clutch and transaxle.

CLUTCH DRIVEN DISC & PRESSURE PLATE

REMOVAL & INSTALLATION

See Figure 18.

1. Before servicing the vehicle, refer to the Precautions Section.
2. Disconnect the negative battery cable.
3. Remove the transaxle.
4. Hold the flywheel stationary.
5. Matchmark the pressure plate and flywheel for installation reference.

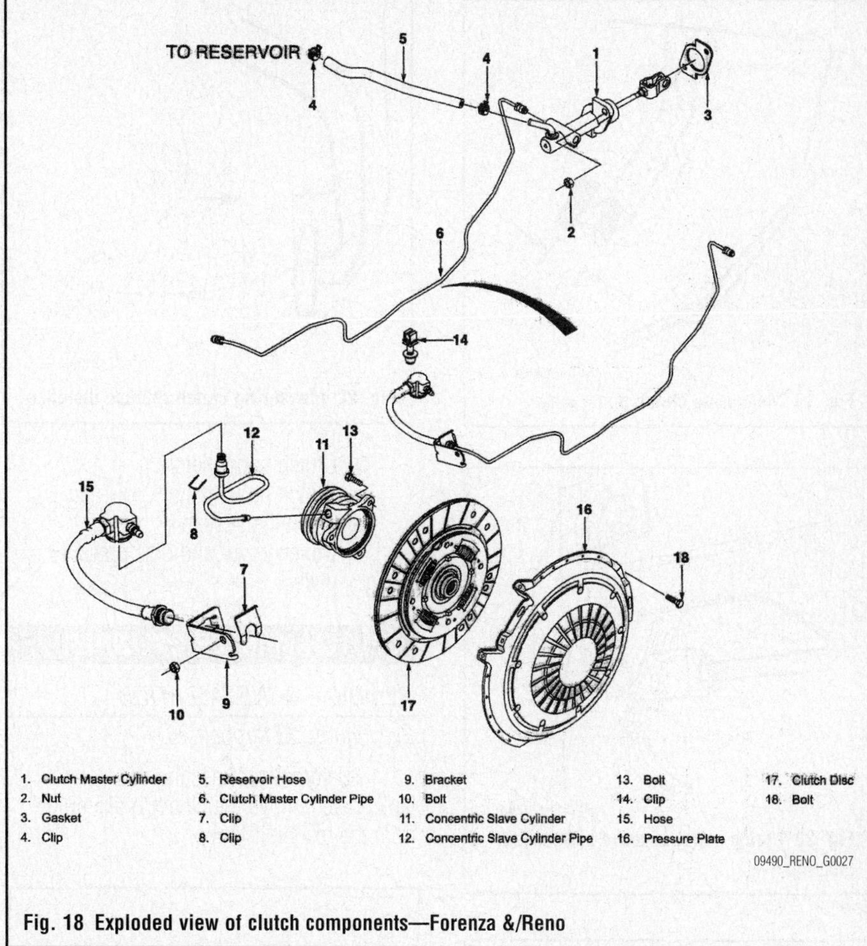

Fig. 18 Exploded view of clutch components—Forenza &/Reno

6. Loosen the pressure plate attaching bolts 1 turn at a time (evenly) until the spring pressure is released.
7. Remove the clutch disc and pressure plate.

To install:

8. Clean the flywheel mating surfaces of all oil, grease and metal deposits. Inspect flywheel for cracks, heat checking or other defects and replace or resurface as necessary.
9. Check the wear on the facings of the clutch disc by measuring the depth of each rivet head depression. Replace clutch disc when rivet heads are 0.012 in. (0.3mm) below the surface of clutch surface.
10. Check the diaphragm spring and pressure plate for wear or damage. If the spring or plate is excessively worn, replace the pressure plate assembly.
11. Check the slave cylinder for smooth operation. If the bearing does not spin freely, replace it.
12. Coat the clutch disc splines with multi-purpose grease.
13. Position the clutch disc and pressure plate with the matchmarks aligned using a clutch alignment tool.
14. Install the pressure plate bolts. Tighten the mounting bolts evenly and in a crisscross pattern to 11 ft. lbs. (15 Nm). Remove the alignment tool.
15. Install the transaxle.
16. Connect the negative battery cable.

ADJUSTMENTS

Clutch Pedal

See Figures 19 through 21.

1. Determine clutch pedal play. Depress the clutch pedal lightly with your hand and measure the distance when you feel resistance
2. Adjust the clutch pedal play. Loosen the locknut and turn the pushrod. Clutch pedal play should measure 0.2–0.5 inch (6–12mm). Tighten the locknut after adjustment.
3. Measure the clutch pedal travel. Press the clutch pedal all the way to the floor. Measure from the starting position to the ending position.
4. Adjust the clutch pedal travel. Loosen the locknut and turn the bolt. Clutch pedal travel should measure 5.1–5.5 inches (130–140mm). Tighten the locknut after adjustment.

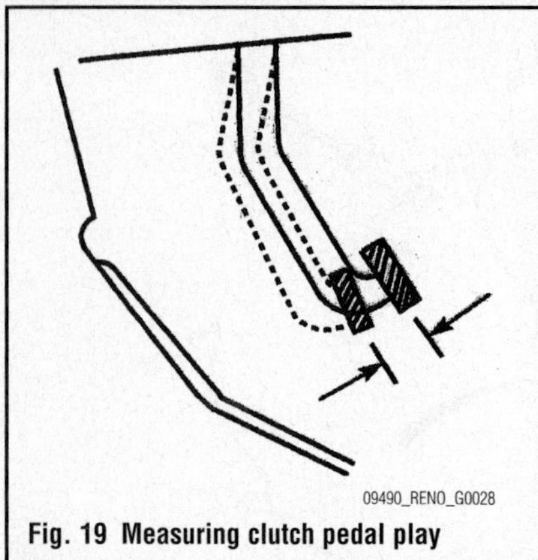

Fig. 19 Measuring clutch pedal play

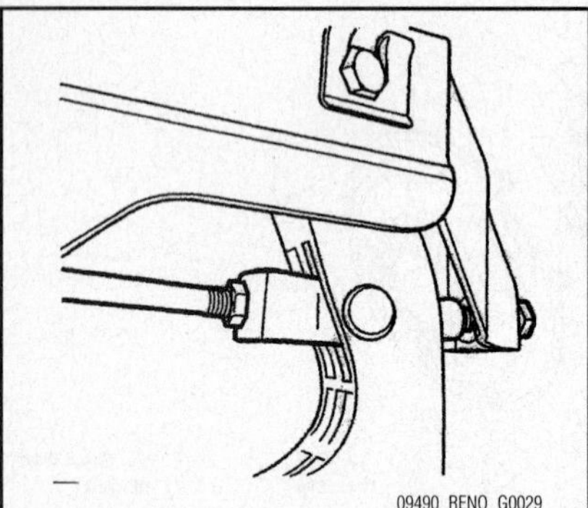

Fig. 20 Clutch pedal pushrod and locknut location

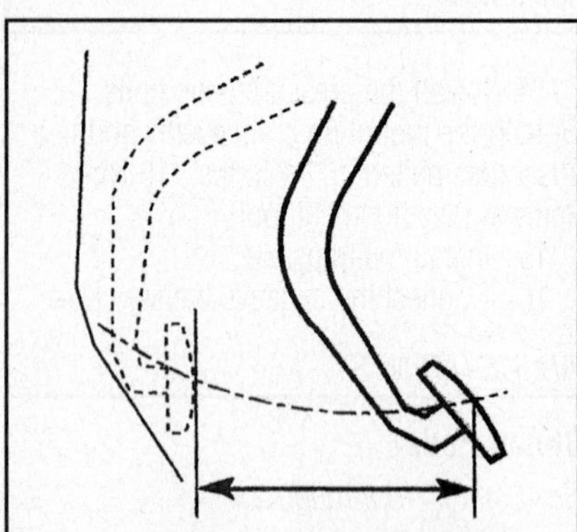

Fig. 21 Measuring clutch pedal travel

Clutch Release Point

See Figure 22.

1. Apply parking brake.
2. Run engine at idle speed.
3. While moving the shifter lever into **REVERSE** position, depress the clutch pedal slowly and measure the distance between the point when gear noise is not heard and the point the clutch pedal is completely depressed. The distance should be 1.2–1.6 inches (30–40mm).
4. If the distance is not within the specified value, check the following:

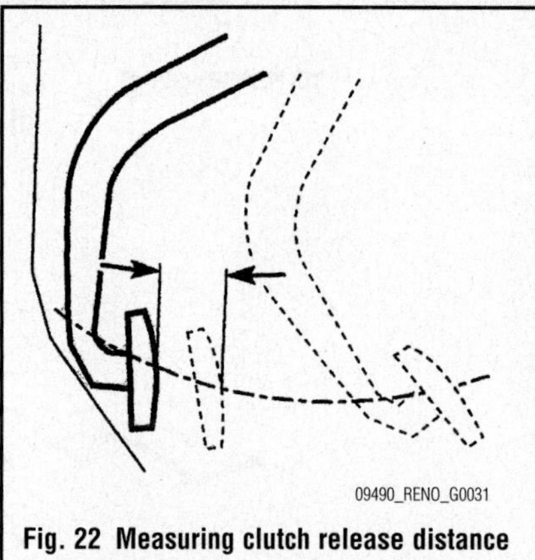

Fig. 22 Measuring clutch release distance

- Clutch pedal height
- Clutch pedal play
- Air in the system
- Clutch cover and disc pressure plate

CLUTCH MASTER CYLINDER

REMOVAL & INSTALLATION

See Figures 23 through 25.

1. Before disconnecting the reservoir tank hose, remove the clutch/brake fluid from the reservoir tank.

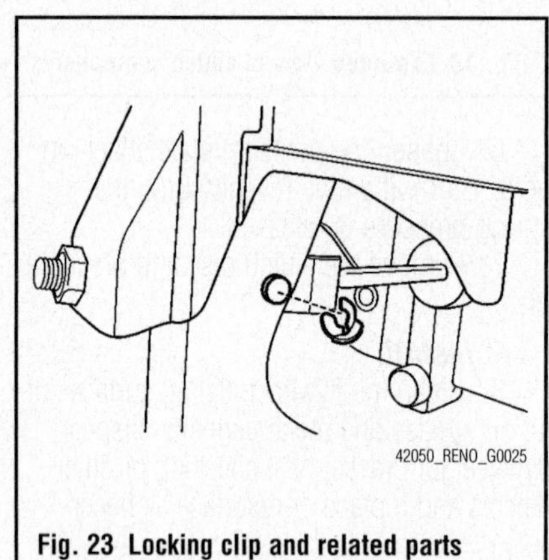

Fig. 23 Locking clip and related parts

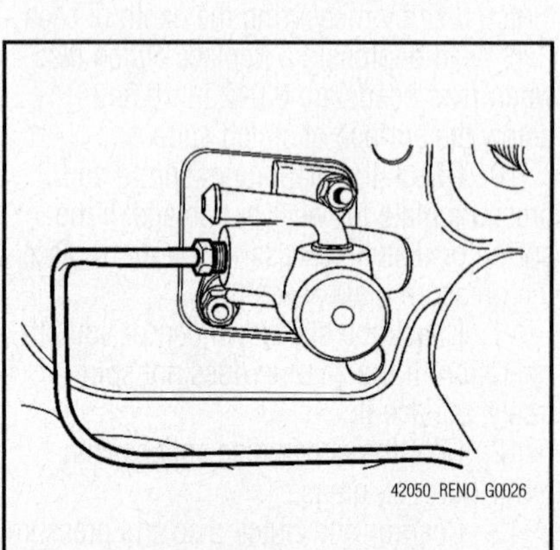
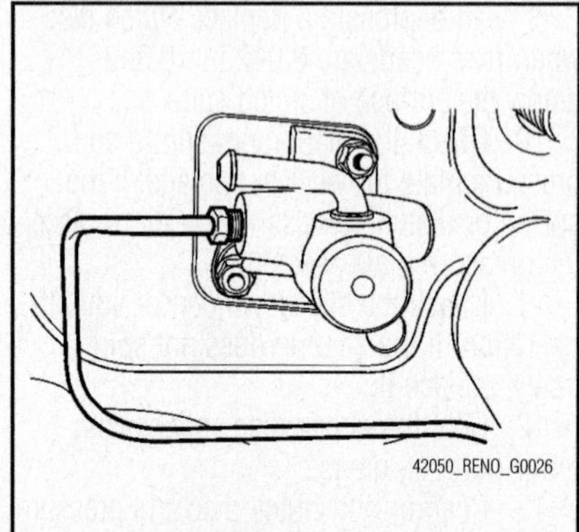

Fig. 24 Master cylinder

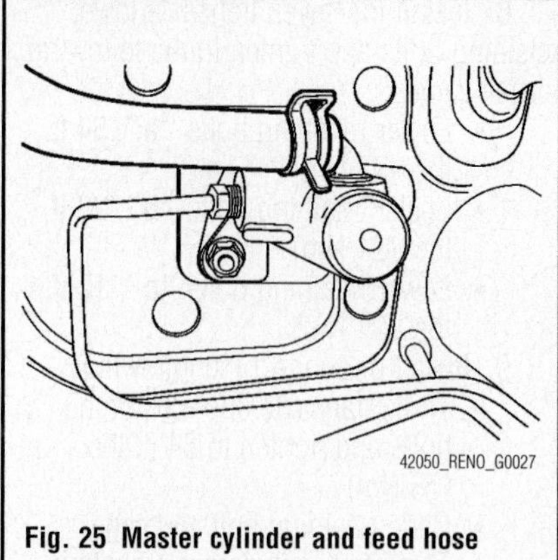

Fig. 25 Master cylinder and feed hose

2. Remove the locking clip.
3. Remove the push rod fixing pin and push rod.
4. Disconnect the hose clamp on the master cylinder.
5. Disconnect the master cylinder hose.
6. Remove the master cylinder pipe.
7. Remove the clutch master cylinder nuts.
8. Remove the clutch master cylinder.

To install:

9. Install the clutch master cylinder and clutch master cylinder nuts.
10. Tighten the clutch master cylinder nuts to 16 ft. lbs. (22 Nm)
11. Install the master cylinder pipe.
12. Connect the master cylinder hose.
13. Connect the hose clamp on the master cylinder.
14. Install the push rod fixing pin and push rod.
15. Install the locking clip.
16. Bleed the air.
17. Adjust the clutch pedal.
18. Fill the reservoir with clutch/brake fluid up to the proper level.

BLEEDING PROCEDURE

See Figure 26.

Bleed the hydraulic system to remove the air which entered when the pipes were disconnected for repairs. The clutch/brake fluid in the clutch/brake reservoir must be maintained at the Minimum level or higher during air bleeding.

1. Step 1: Attach a vinyl hose to the bleeder plug. Place the other end of the vinyl tube in a glass container half-filled with brake fluid.
2. Step 2: Slowly pump the clutch pedal several times.
3. Step 3: While you press the clutch pedal, loosen the bleeder screw until the fluid starts to run out. Close the bleeder screw.

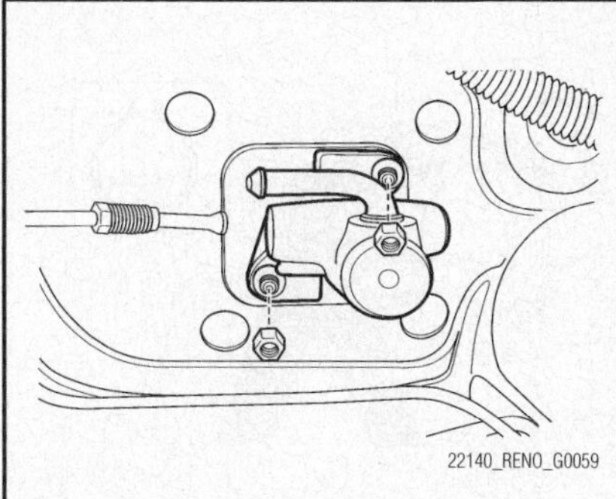

Fig. 26 Clutch master cylinder bleeding procedure

4. Step 4: Repeat Step 3 until there are no air bubbles in the fluid.

5. Step 5: Fill the reservoir with brake fluid up to the proper level.

CLUTCH SLAVE CYLINDER

REMOVAL & INSTALLATION

See Figures 27 through 29.

1. Remove the transaxle from the vehicle.
2. Remove the slave cylinder retaining bolts.
3. Disconnect the slave cylinder pipe from the slave cylinder.
4. Remove the bushing from the slave cylinder pipe in the transaxle housing.
5. Remove the slave cylinder pipe.
6. Remove the O-ring and the slave cylinder.

To install:

7. Install the concentric slave cylinder and the pipe.
8. After installing the pipe screw provisionally, tighten the concentric slave cylinder retaining bolts.
9. Tighten the concentric slave cylinder retaining bolts to 62 inch lbs. (7 Nm)
10. Tighten the pipe screw.

Fig. 27 Slave cylinder and pipe

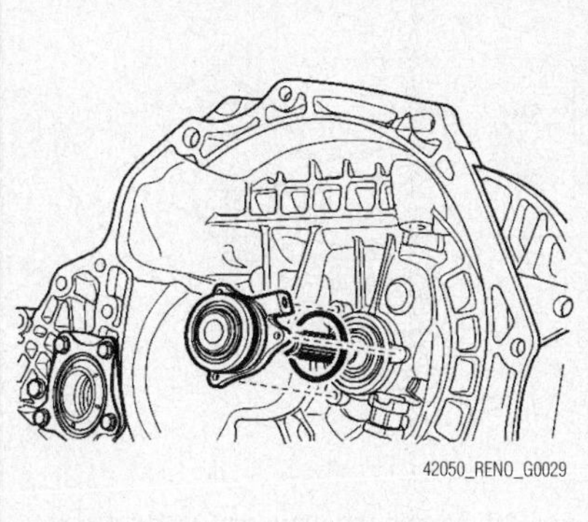

Fig. 28 O-ring and slave cylinder

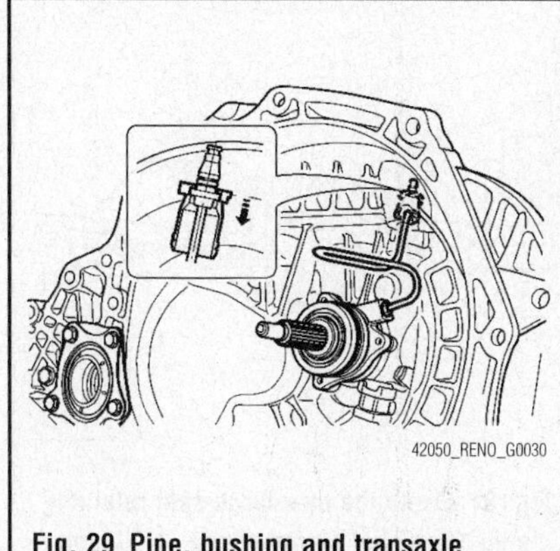

Fig. 29 Pipe, bushing and transaxle

11. Tighten the pipe screw to 11 ft. lbs. (15 Nm)
12. Install the bushing onto the transaxle housing.
13. Connect the pipe with bushing.
14. Install the transaxle into the vehicle.

CLUTCH HYDRAULIC SYSTEM BLEEDING

BLEEDING PROCEDURE

Bleed the hydraulic system to remove the air which entered when the pipes were disconnected for repairs. The clutch/brake fluid in the clutch/brake reservoir must be maintained at the Minimum level or higher during air bleeding.

1. Step 1: Attach a vinyl hose to the bleeder plug. Place the other end of the vinyl tube in a glass container half-filled with brake fluid.
2. Step 2: Slowly pump the clutch pedal several times.
3. Step 3: While you press the clutch pedal, loosen the bleeder screw until the fluid starts to run out.
4. Close the bleeder screw.
5. Step 4: Repeat Step 3 until there are no air bubbles in the fluid.
6. Step 5: Fill the reservoir with brake fluid up to the proper level.

FRONT HALFSHAFTS

REMOVAL & INSTALLATION

1. Before servicing the vehicle, refer to the Precautions Section.
2. Disconnect the negative battery cable.
3. Raise and safely support the vehicle.
4. Remove wheel.
5. Remove drive shaft nut and washer. Discard nut.
6. Disconnect front suspension control arm ball stud from steering knuckle after removing pinch bolt and nut.

➡Use only a tool made specifically for separating the lower ball joint. Failure to use the correct tool may cause damage to the ball joint and the seal.

7. Remove tie rod end nut.
8. Disconnect tie rod end from steering knuckle by using ball joint separator KM-507-B.

➡Use only a tool made specifically for separating the tie rod from the steering knuckle. Failure to use the correct tool may cause damage to the knuckle/strut assembly.

9. If equipped with automatic transaxle, perform the following:
   a. Remove the damping block connection nut and bolt.
   b. Remove the rear mounting bracket bolts and the bracket.
10. Push the drive shaft from the wheel hub.
11. Remove drive shaft assembly from the transaxle using axle shaft removal tool DW340-110 (A/T), or KM-460-A (M/T).

➡Support the unfastened end of the drive axle. Do not allow the drive axle to dangle freely from the transaxle for any length of time after it has been removed from the wheel hub. Place a drain pan below the transaxle to catch the escaping fluid. Cap the transaxle drive opening after the drive axle has been removed to keep the fluid in and any contamination out.

To install:

12. Clean the hub seal and the transaxle seal. Be careful to not damage the seals.
13. Install the drive shaft into the transaxle.
14. Install the wheel hub onto the axle shaft.

15. If equipped with automatic transaxle, perform the following:

16. Install the rear mounting bracket bolts and the bracket and tighten to 45 ft. lbs. (62 Nm).

17. Install the damping block connection nut and bolt and tighten to 50 ft. lbs. (68 Nm).

18. Install the tie rod into the knuckle/strut and install the tie rod nut. Tighten the tie rod nut to 41 ft. lbs. (55 Nm).

19. Install the lower ball joint pinch bolt and nut and tighten to 44 ft. lbs. (60 Nm).

20. Loosely install the washer and a new axle shaft caulking nut. Always use a new nut.

21. Install the wheels. Loosely install the nuts.

22. Lower the vehicle to the floor and tighten the wheel nuts to 74 ft. lbs. (100 Nm).

23. Tighten the new axle shaft caulking nut to 221 ft. lbs. (300 Nm).

24. Peen the new caulking nut with a punch and a hammer until the nut is locked into place on the axle shaft hub.

25. Refill the transaxle fluid to the proper level.

26. Connect the negative battery cable.

CV-JOINTS OVERHAUL

Double Offset Joint (DOJ) Type

See Figures 30 through 33.

The Double Offset Joint (DOJ) type is identified by the outside shape of the differential side joint which has no dent.

1. Before servicing the vehicle, refer to the Precautions Section.
2. Remove the drive shaft.
3. Remove the large seal retaining clamp. Discard the clamp.
4. Remove the small seal retaining clamp. Discard the clamp.
5. Degrease the joint.
6. Spread the snap ring using snap ring pliers and remove the outer joint of the axle shaft.
7. Remove the seal from the joint assembly.
8. Inspect all components for wear and/or damage and replace as necessary.

To install:

9. Clean all components and allow them to completely dry.
10. Install the seal onto the axle shaft.
11. Spread the snap ring using snap ring pliers and install the outer joint the axle shaft.

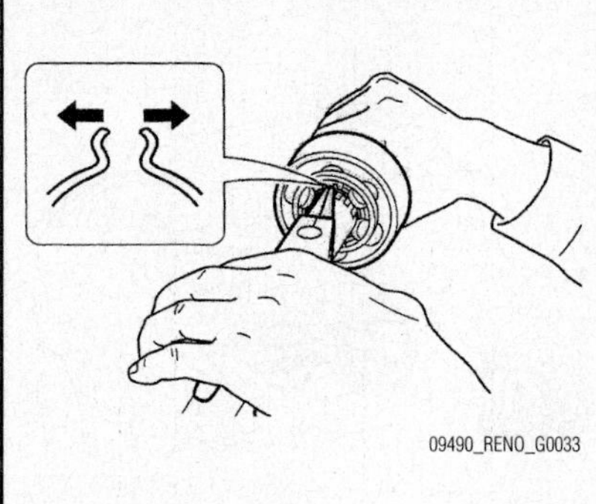

Fig. 30 Spread the snap ring using snap ring pliers—DOJ type

Fig. 32 Spread the snap ring using snap ring pliers—Tripod joint

Fig. 31 Crimp the new large seal retaining clamp using seal clamp pliers—DOJ type

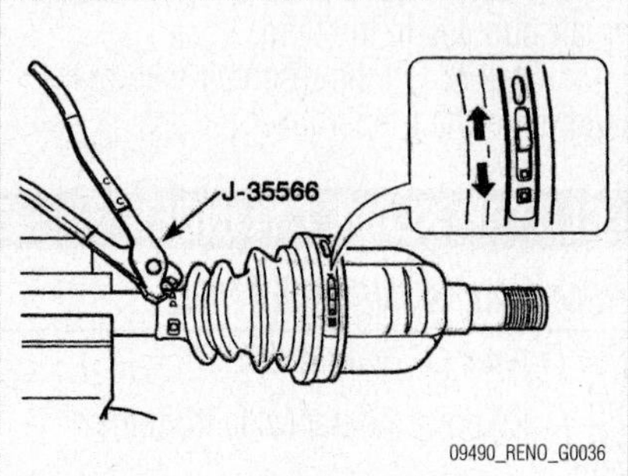

Fig. 33 Crimp the new seal retaining clamps using seal clamp pliers—Tripod joint

12. Fill the joint seal with 4.2–4.9 ounces (120–140g) of the recommended grease.
13. Repack the joint with 4.2–4.9 ounces (120–140g) of the recommended grease.
14. Install a new large seal retaining clamp and a new small seal retaining clamp.
15. Crimp the new small seal retaining clamp and the new large seal retaining clamp using the seal clamp pliers J-35566.
16. Install the drive axle shaft to the vehicle.

Tripod Joint Type

The Tripod joint type can be identified by the 3 dent lines on the outside of the differential side joint.

1. Before servicing the vehicle, refer to the Precautions Section.
2. Remove the drive shaft.
3. Remove the large seal retaining clamp. Discard the clamp.
4. Remove the small seal retaining clamp. Discard the clamp.
5. Separate the joint housing from the boot.
6. Degrease the joint.
7. Remove the shaft retaining ring using the snap ring pliers J-8059.
8. Remove the tripod and the tripod joint retaining ring from the axle shaft.
9. Remove the tripod joint seal from the axle shaft.
10. Inspect all components for wear and/or damage and replace as necessary.

To install:

11. Clean all components and allow them to completely dry.
12. Install a new small seal retaining clamp onto the seal.
13. Install the seal onto the axle shaft.
14. Install the shaft retaining ring onto the axle shaft using the snap ring pliers J-8059.
15. Fill the tripod housing with 6.9–7.6 ounces (195–215g) of the recommended grease.
16. Repack the tripod with 6.9–7.6 ounces (195–215g) of the recommended grease.
17. Install the boot to the joint housing.
18. Install a new large seal retaining clamp. Crimp the large seal retaining clamp using the seal clamp pliers J-35566.
19. Crimp the new small seal retaining clamp using the seal clamp pliers J-35566.
20. Install the drive shaft to the vehicle.

Cross Groove Joint Type

See Figures 34 and 35.

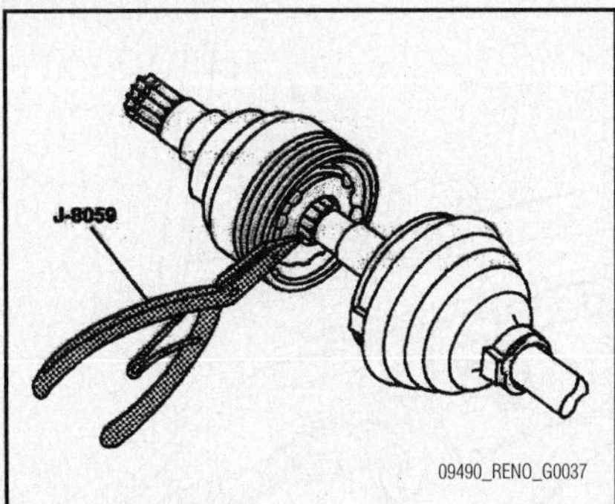

Fig. 34 Spread the snap ring using snap ring pliers—Cross groove joint

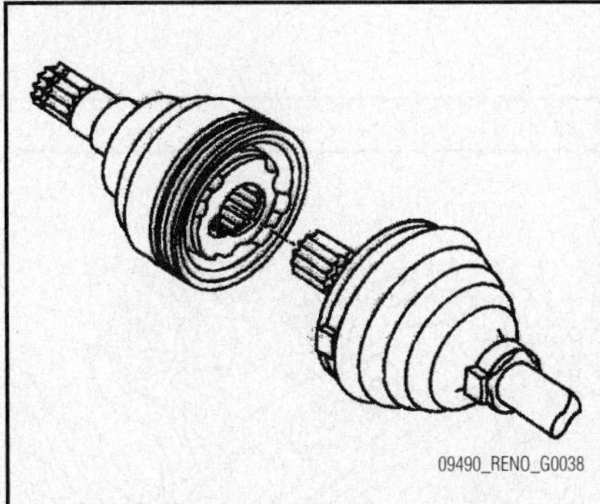

Fig. 35 Remove/install the axle shaft from the joint assembly—Cross groove joint

1. Before servicing the vehicle, refer to the Precautions Section.
2. Remove the drive shaft.
3. Remove the large seal retaining clamp. Discard the clamp.
4. Remove the small seal retaining clamp. Discard the clamp.
5. Degrease the joint.
6. Remove the shaft retaining ring using the snap ring pliers J-8059.
7. Remove the axle shaft from the joint assembly.
8. Remove the seal from the joint assembly.

To install:

9. Clean all components and allow them to completely dry.
10. Install a new small seal retaining clamp onto the seal. Do not crimp.
11. Install the seal onto the axle shaft.
12. Install the joint assembly onto the axle shaft.
13. Install the shaft retaining ring using the snap ring pliers J-8059.
14. Fill the joint assembly with 4.2–4.9 ounces (120–140g) of the recommended grease.
15. Repack the joint with 4.2–4.9 ounces (120–140g) of the recommended grease.
16. Install a new large seal retaining clamp.
17. Crimp the new large seal retaining clamp using the seal clamp pliers J-35566.
18. Crimp the new small retaining clamp using the seal clamp pliers J-35566.

REAR AXLE SHAFT

REMOVAL & INSTALLATION

See Figures 36 and 37.

1. Raise and suitably support the vehicle.
2. Remove the wheels.
3. Remove the axle shaft caulking nut. Discard the nut.
4. Remove the lower ball joint pinch bolt and nut.
5. Separate the steering knuckle from the lower ball joint.
6. Remove the tie rod nut.
7. Separate the tie rod end.
8. Push the drive axle shaft from the wheel hub.
9. Remove the drive axle from the transaxle.

To install:

10. Clean the hub seal and the transaxle seal.
11. Install the drive axle into the transaxle.
12. Install the wheel hub onto the axle shaft.
13. Install the tie rod into the knuckle/strut and install the tie rod nut.
14. Tighten the tie rod nut to 41 ft. lbs. (55 Nm)

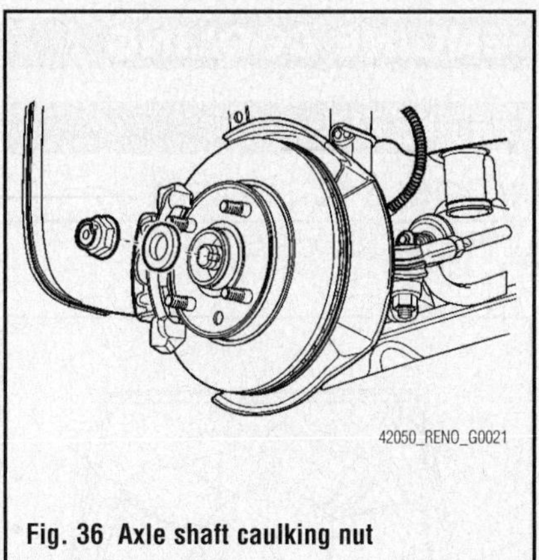

Fig. 36 Axle shaft caulking nut

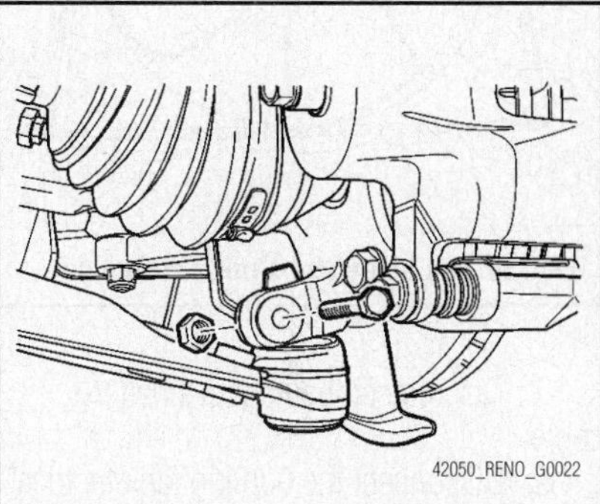

Fig. 37 Lower ball joint pinch nut and related parts

15. Install the lower ball joint pinch bolt and nut.
16. Tighten the lower ball joint pinch bolt and nut to 44 ft. lbs. (60 Nm)
17. Loosely install a new axle shaft caulking nut.

➡Always use a new nut.

18. Install the wheels. Loosely install the nuts.
19. Lower the vehicle to the floor.
20. Tighten the wheel nuts to 74 ft. lbs. (100 Nm)
21. Tighten the axle shaft caulking nut to 221 ft. lbs. (300 Nm)
22. Peen the caulking nut with a punch and a hammer until the nut is locked into place on the axle shaft hub.
23. Refill the transaxle fluid to the proper level.

ENGINE COOLING

ENGINE FAN

REMOVAL & INSTALLATION

See Figures 38 and 39.

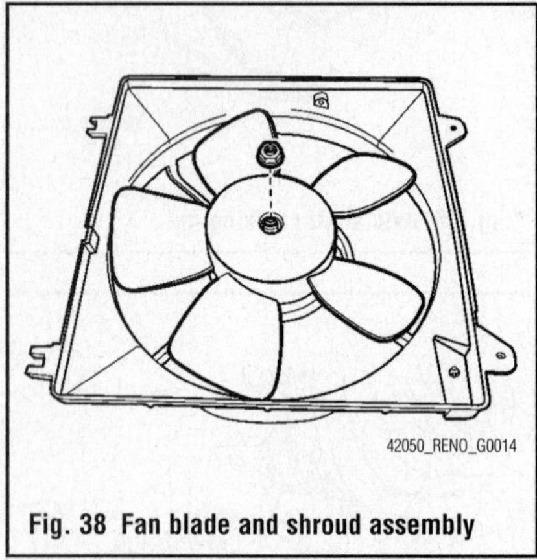

Fig. 38 Fan blade and shroud assembly

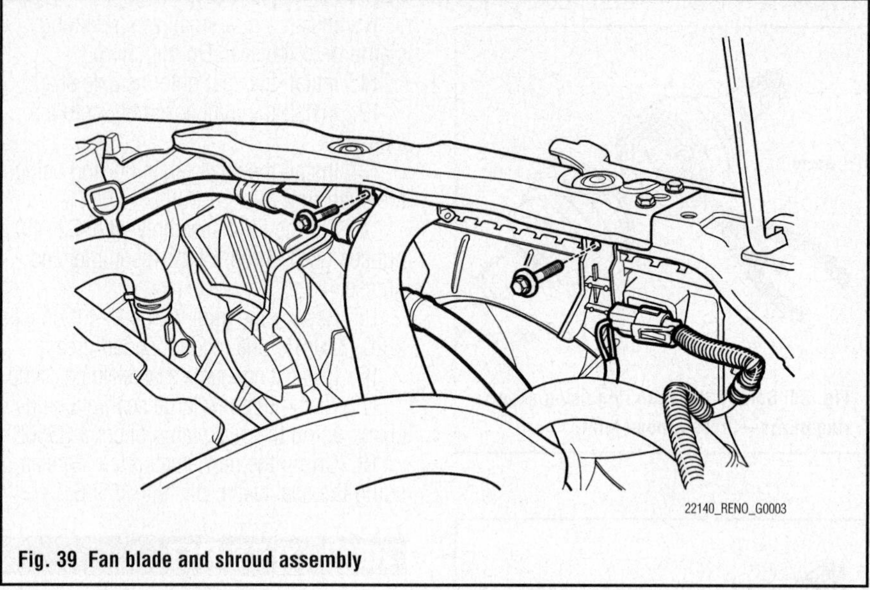

Fig. 39 Fan blade and shroud assembly

1. Disconnect the negative battery cable.
2. Disconnect the cooling fan electrical connector.
3. Remove the fan shroud mounting bolts.
4. Lift the fan shroud assembly upward, and remove the fan shroud assembly from the vehicle.
5. Remove the fan blade from the fan shroud assembly by removing the nut at the center of the fan hub.
6. Turn over the fan shroud assembly.
7. Remove the fan motor retaining screws.
8. Remove the fan motor from the shroud.

To install:

⁂ WARNING

If a fan blade is bent or damaged in any way, no attempt should be made to repair or reuse the damaged part. A bent or damaged fan assembly must be replaced with a new fan assembly. It is essential that fan assemblies remain in proper balance. A fan assembly that is not in proper balance can fail and fly apart during use, creating extreme danger. Proper balance cannot be assured on a fan assembly that has been bent or damaged.

9. Install the fan motor to the shroud.
10. Secure the motor to the shroud with the retaining screws.
11. Tighten the fan motor retaining screws to 35 inch lbs. (4 Nm)
12. Turn over the fan shroud assembly.
13. Install the fan to the fan shroud assembly with the single nut in the center of the fan hub.
14. Tighten the fan motor nut to 28 inch lbs. (3.2 Nm)
15. Install the fan shroud assembly to the radiator.
16. Secure the shroud to the top of the radiator with the mounting bolts.
17. Tighten the fan assembly mounting bolts to 35 inch lbs. (4 Nm)
18. Connect the cooling fan electrical connector.
19. Connect the negative battery cable.

RADIATOR

REMOVAL & INSTALLATION

See Figure 40.

1. Disconnect the negative battery cable.
2. Drain the engine cooling system.
3. Remove the main and the auxiliary cooling fans.
4. Remove the upper radiator retaining bolt.
5. Remove the upper radiator retaining bracket.
6. Remove the upper radiator hose clamp.
7. Disconnect the upper radiator hose from the radiator.
8. Remove the hose clamp from the surge tank hose at the radiator.
9. Disconnect the surge tank hose from the radiator.

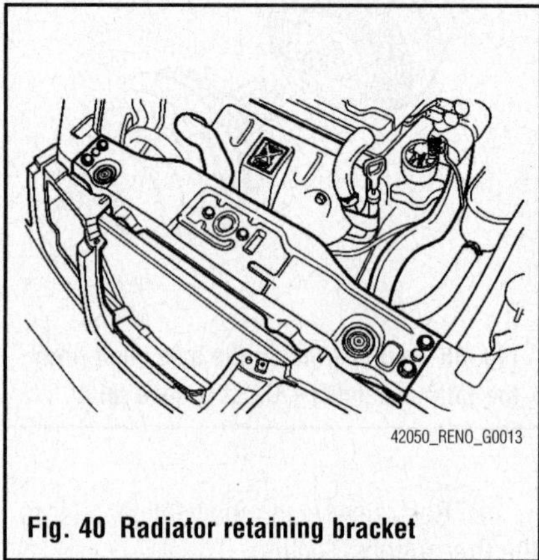

Fig. 40 Radiator retaining bracket

10. Remove the lower radiator hose clamp.
11. Disconnect the lower radiator hose from the radiator.
12. Disconnect the transaxle cooler pipes from the lower radiator tank, if equipped.
13. Remove the bolt and the transaxle pipe support clamp from the radiator.
14. Remove the radiator from the vehicle.

To install:

15. Set the radiator into place in the vehicle with the radiator bottom posts in the rubber shock bumpers.
16. Connect the transaxle cooler pipes to the lower radiator tank, if equipped.
17. Install the transaxle pipe and support clamp to the radiator with a bolt.
18. Connect the surge tank hose to the radiator.
19. Secure the surge tank hose with a hose clamp.

20. Connect the upper radiator hose and the lower radiator hose to the radiator.
21. Secure each hose with a hose clamp.
22. Position the radiator retainers in place.
23. Install the upper radiator retainer bracket.
24. Install the upper radiator retaining bolt.
25. Install the main and the auxiliary cooling fans.
26. Refill the engine cooling system.
27. Connect the negative battery cable.
28. Tighten the radiator retaining bolt to 71 inch lbs. (8 Nm)

THERMOSTAT

REMOVAL & INSTALLATION

See Figures 41 through 43.

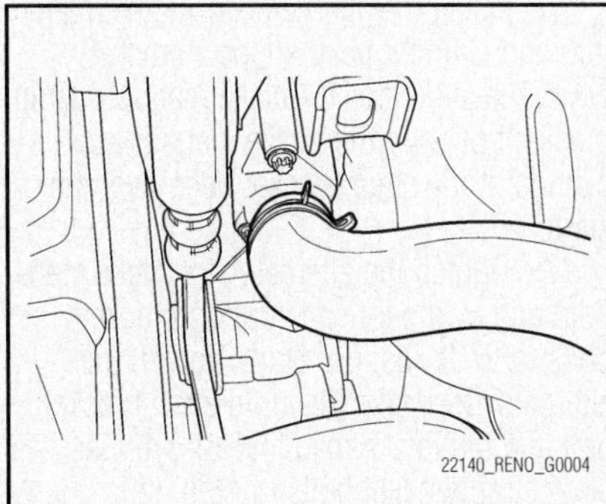

Fig. 41 Upper hose attached to the thermostat housing

1. Drain the coolant.
2. Loosen the hose clamp on the upper radiator hose at the thermostat housing.
3. Disconnect the upper radiator hose from the thermostat housing.
4. Remove the mounting bolts that hold the thermostat housing to the cylinder head.
5. Remove the thermostat housing from the cylinder head.
6. Remove the seal ring from the thermostat housing.
7. Remove the thermostat from the thermostat housing by pressing the thermostat mounting flange downward and then rotating the flange clockwise.
8. Inspect the valve seat for foreign matter that could prevent the valve from sealing properly.
9. Inspect the thermostat for proper operation.
10. Clean the thermostat housing and the cylinder head mating surfaces.

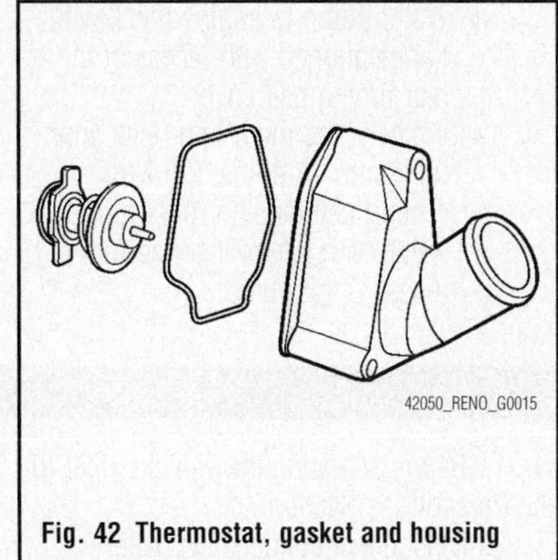

Fig. 42 Thermostat, gasket and housing

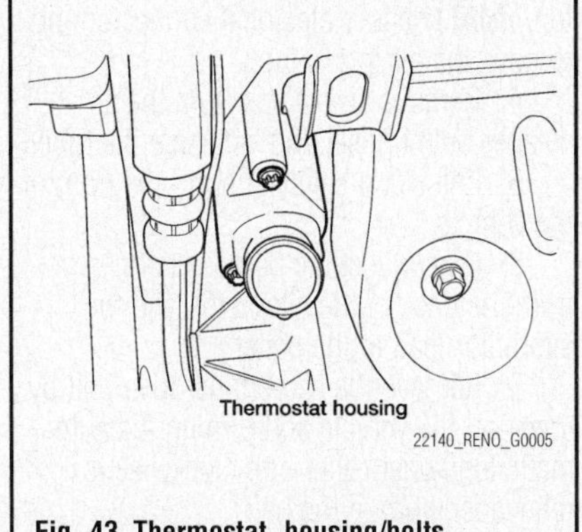

Fig. 43 Thermostat, housing/bolts

To install:

11. Install the thermostat into the thermostat housing by pressing the thermostat mounting flange downward and then rotating the flange counterclockwise.
12. Rotate the thermostat mounting flange until it is seated in the thermostat housing recesses.
13. Coat the sealing surface of a new seal ring with Lubriplate®.
14. Install a new seal ring into the recess in the thermostat housing.

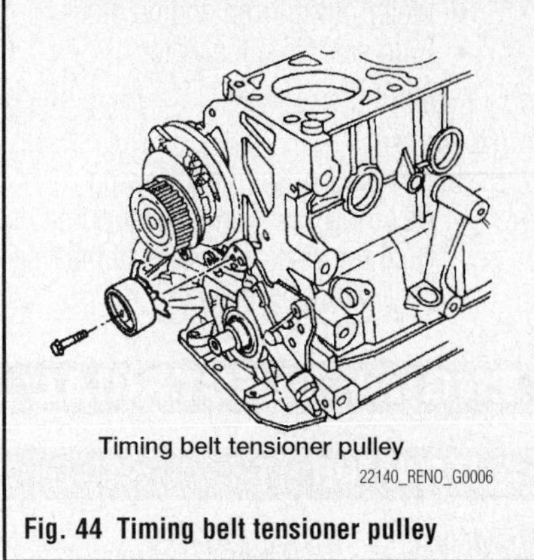

Fig. 44 Timing belt tensioner pulley

15. Install the thermostat housing to the cylinder head.
16. Secure the thermostat housing to the cylinder head with the mounting bolts.
17. Tighten the thermostat housing mounting bolts to 11 ft. lbs. (15 Nm)
18. Connect the upper radiator hose to the thermostat housing.
19. Secure the upper radiator hose to the thermostat housing with a hose clamp.
20. Refill the engine cooling system.

WATER PUMP

REMOVAL & INSTALLATION

See Figures 44 and 45.

1. Before servicing the vehicle, refer to the Precautions Section.
2. Disconnect the negative battery cable.
3. Drain the cooling system into a suitable container and tighten the drain plug.
4. Remove or disconnect the following:
   - Timing belt
   - Timing belt tension roller retaining bolt
   - Timing belt tension roller
   - Water pump mounting bolts

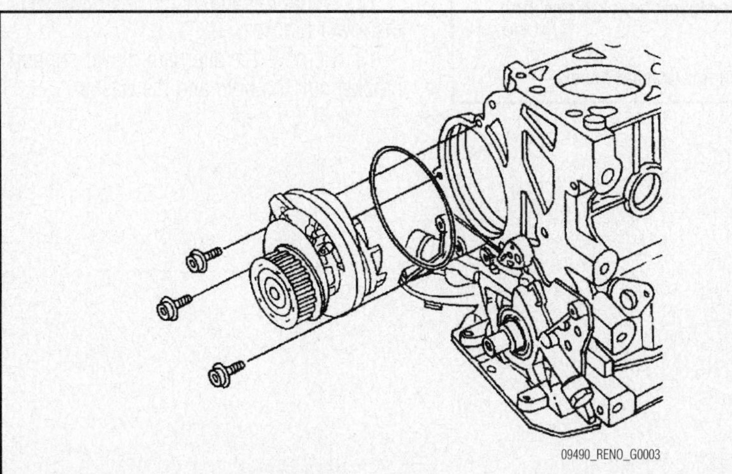

Fig. 45 Water pump, ring seal and mounting bolts

- Water pump from engine block
- Ring seal from the water pump

To install:

5. Install or connect the following:
   - New ring seal to the water pump. Coat the sealing surface of the ring seal with Lubriplate®.
   - Water pump to engine block with flange aligned with recess of the rear timing belt cover
   - Water pump mounting bolts and tighten to 18 ft. lbs. (25 Nm)
   - Timing belt tension roller to oil pump with flange inserted into recess of oil pump
   - Timing belt tension roller bolt. Do not tighten the bolt at this time
   - Timing belt
6. Refill the engine cooling system.
7. Connect the negative battery cable.
8. Start the engine and top off the coolant as necessary.
9. Check the cooling system for leaks.

ENGINE ELECTRICAL

CHARGING SYSTEM

ALTERNATOR

REMOVAL & INSTALLATION

See Figures 46 and 47.

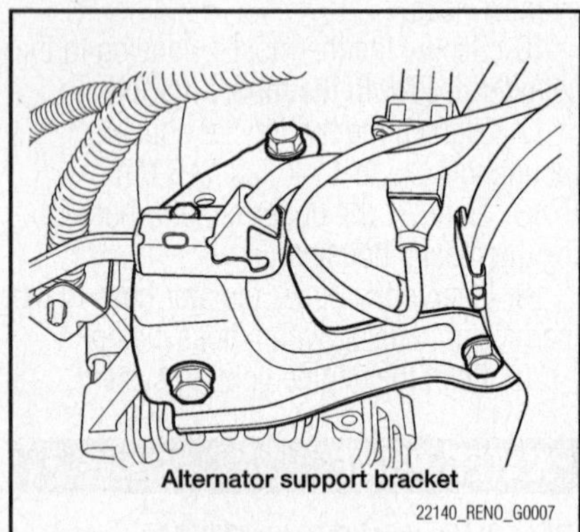

Fig. 46 Alternator bracket upper

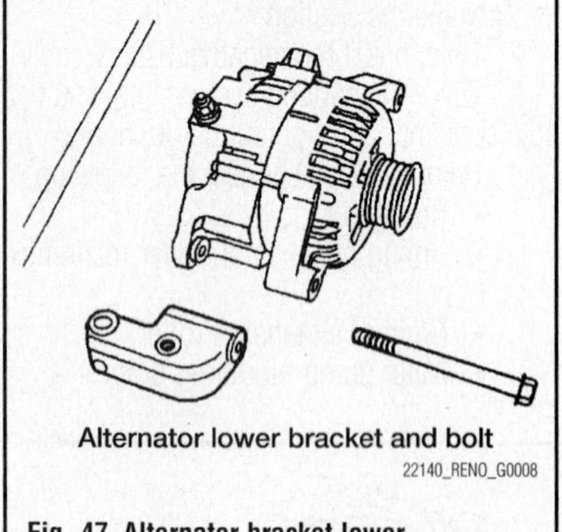

Fig. 47 Alternator bracket lower

1. Before servicing the vehicle, refer to the Precautions Section.
2. Disconnect the negative battery cable.
3. Disconnect the manifold air temperature (MAT) sensor electrical connector and remove the air intake tube.
4. Remove all clamps from the air cleaner outlet hose, and set aside the tube.
5. Raise and safely support the front of the vehicle.
6. Disconnect the harness connector from the back of the alternator, and the alternator lead to the battery.
7. Remove the serpentine drive belt by lowering the vehicle and turning the automatic tensioner roller bolt clockwise to relieve tension on the belt.
8. Push up the power steering reservoir and set it aside.
9. Remove the alternator upper mounting bolts to the intake manifold/cylinder head support bracket, the intake manifold strap bracket, and the intake manifold-to-cylinder body strap bracket.
10. Raise and suitably support the vehicle and remove the nut and washers which hold the alternator lower bracket-to-alternator bolt. Work the bolt loose and remove the alternator.
11. Remove the alternator lower support bracket bolts.
12. Carefully remove the alternator with the lower bracket.
13. Remove the alternator lower support bracket nut, the bolt, and the washer.

To install:

14. Install the alternator to the alternator lower bracket and insert the alternator bolt. Tighten the alternator lower bracket-to-alternator nut to 18 ft. lbs. (25 Nm).
15. Install the alternator and the lower support bracket assembly to the engine block. Tighten the alternator and the lower bracket-to-engine block bolts to 27 ft. lbs. (37 Nm).
16. Install the alternator-to-intake manifold and cylinder head support bracket bolts, the alternator-to-intake manifold strap bracket bolt, and the intake manifold-to-cylinder body strap bracket bolts over the starter.
17. Tighten the alternator-to-intake manifold and cylinder head support bracket bolts to 27 ft. lbs. (37 Nm). Tighten the alternator-to-intake manifold strap bracket bolt and the intake manifold-to-cylinder body strap bracket bolts to 16 ft. lbs. (22 Nm).
18. Route the serpentine accessory drive belt.
19. Relieve tension on the belt by first applying downward pressure on the automatic tension roller bolt and releasing pressure once the belt is in place.
20. Install the power steering reservoir.
21. Install the air cleaner outlet hose and connect the MAT electrical connector.
22. Connect the negative battery cable.

ENGINE ELECTRICAL — IGNITION SYSTEM

FIRING ORDERS

See Figure 48.

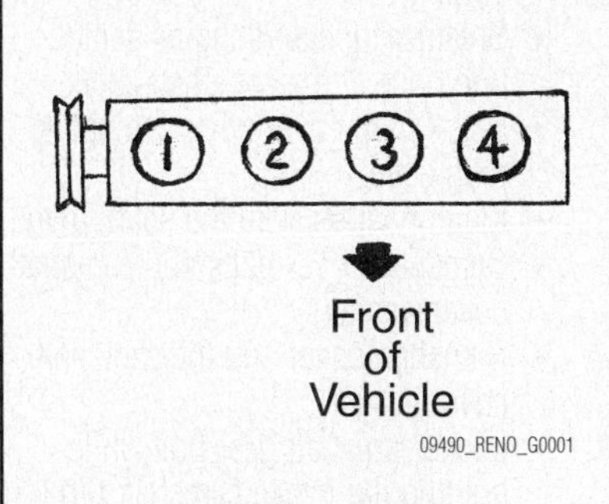

**Fig. 48 Firing order: 1–3–4–2
Distributorless ignition system**

IGNITION COIL

REMOVAL & INSTALLATION

See Figure 49.

1. Disconnect the negative battery cable.
2. Disconnect the Electronic Ignition (EI) system ignition coil connector.
3. Note the ignition wire location and remove the ignition wires.
4. Remove the ignition coil retaining bolts.

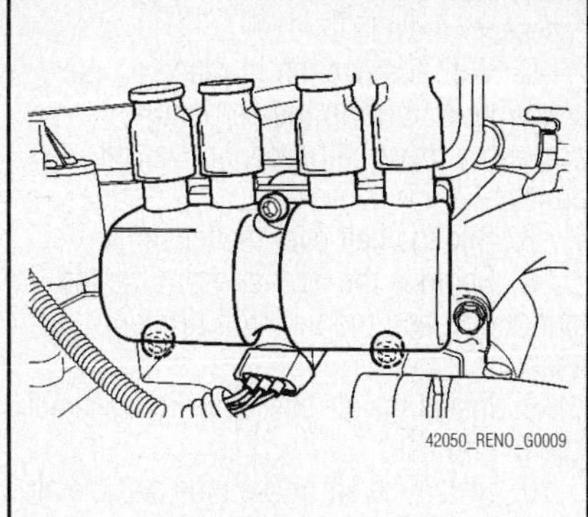

Fig. 49 Ignition coil location

5. Remove the ignition coil.

To install:

6. Install the ignition coil into the mounting location and install the retaining bolts.
7. Tighten ignition coil retaining bolts to 89 inch lbs.
8. Connect the ignition coil connector.
9. Install the ignition wires.
10. Connect the negative battery cable.

IGNITION TIMING

INSPECTION

The Electronic Ignition (EI) system ignition coil provides the spark for two spark plugs simultaneously. The EI system ignition coil is not serviceable and must be replaced as an assembly.

ADJUSTMENT

This ignition system does not use a conventional distributor and coil. It uses a crankshaft position sensor input to the Engine Control Module (ECM). The ECM then determines Electronic Spark Timing (EST) and triggers the direct ignition system ignition coil.

➡Since the CKP sensor is in a fixed position, timing adjustments are not possible or needed.

SPARK PLUGS

REMOVAL & INSTALLATION

1. Disconnect the negative battery cable.
2. Remove spark plug wire.
3. Remove spark plug using a spark plug socket.

To install:

4. Check spark plug for proper gap.
5. Apply a small amount of anti-seize on spark plug threads.
6. Carefully install spark plug and torque to 15 ft. lbs (21 Nm)
7. Install ignition wire.

ENGINE ELECTRICAL — STARTING SYSTEM

STARTER

REMOVAL & INSTALLATION

See Figure 50.

1. Before servicing the vehicle, refer to the Precautions Section.
2. Disconnect the negative battery cable.
3. Remove the nut which secures the starter ground wire to the lower mounting stud and remove the ground wire.
4. Remove the starter-to-engine block mounting bolt and the starter-to-transaxle mounting bolt.
5. Remove the starter solenoid nuts to disconnect the electrical cable.
6. Remove the starter assembly.

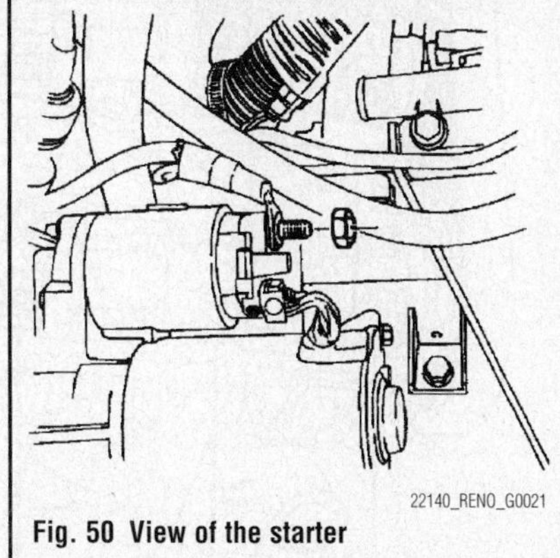

Fig. 50 View of the starter

To install:

7. Place the starter assembly in position using an assistant to prop up the starter to aid in screwing in the upper stud with the weld nut.
8. Install the starter mounting bolts and tighten to 15 ft. lbs. (21 Nm).
9. Position the starter electrical wire on the solenoid terminals and the ground wire on the lower stud.
10. Install the starter solenoid nuts and the ground wire nut. Tighten the starter solenoid terminal-to-battery cable terminal nut to 106 inch lbs. (12 Nm), and the starter solenoid terminal-to-ignition solenoid terminal nut to 53 inch lbs. (6 Nm). Tighten the starter lower mounting stud ground wire nut to 106 inch lbs. (12 Nm).
11. Connect the negative battery cable.

ENGINE MECHANICAL

➡Disconnecting the negative battery cable may interfere with the functions of the on board computer systems and may require the computer to undergo a relearning process, once the negative battery cable is reconnected.

ACCESSORY DRIVE BELTS

ACCESSORY BELT ROUTING

See Figure 51.

Refer to the accompanying illustration for belt routing.

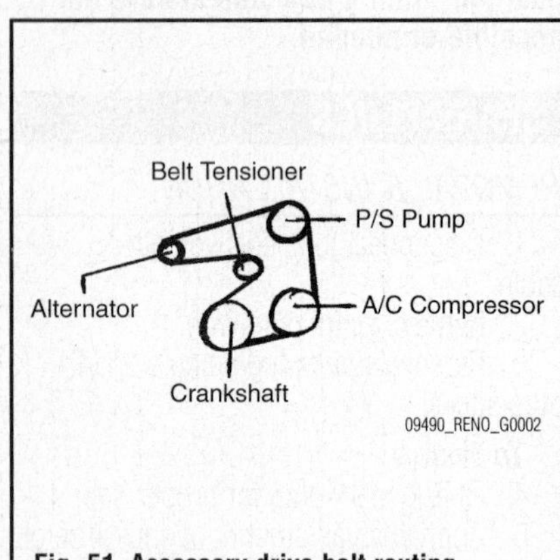

Fig. 51 Accessory drive belt routing

INSPECTION

Inspect the drive belt for signs of glazing or cracking. A glazed belt will be perfectly smooth from slippage, while a good belt will have a slight texture of fabric visible. Cracks will usually start at the inner edge of the belt and run outward. All worn or damaged drive belts should be replaced immediately.

ADJUSTMENT

The belt is automatically adjusted by the auto tensioner.

REMOVAL & INSTALLATION

1. Remove the air filter housing assembly bolts and the air intake tube.
2. Remove the air filter housing assembly from the vehicle.
3. Use a wrench to turn the tensioner bolt clockwise, compressing the tensioner, and releasing the tension on the serpentine accessory drive belt.
4. Remove the serpentine accessory drive belt.

To install:

5. Use a wrench to turn the tensioner bolt clockwise, compressing the tensioner, and releasing the tension on the serpentine accessory drive belt.
6. With the wrench in place on the tensioner bolt, loop the serpentine accessory drive belt loosely over the pulleys.
7. Slip the belt over the tensioner.
8. Remove the wrench from the tensioner bolt and the belt will tighten itself.
9. Install the air filter housing assembly into the vehicle.
10. Install the air intake tube and the air filter housing assembly bolts.
11. Tighten the air filter housing assembly bolts to 53 inch lbs. and nm

CAMSHAFT AND VALVE LIFTERS

REMOVAL & INSTALLATION

See Figure 52.

1. Before servicing the vehicle, refer to the Precautions Section.
2. Relieve the fuel system pressure.
3. Disconnect the negative battery cable.
4. Remove or disconnect the following:
   - Timing belt
   - Breather tube and engine ventilation hose at the camshaft cover
   - Spark plug cover bolts and spark plug cover
   - Ignition wires from the spark plugs
   - Camshaft cover bolts and camshaft cover washers
   - Camshaft cover and the camshaft cover gasket
   - Intake camshaft gear bolt while holding the intake camshaft firmly in place
   - Intake camshaft gear
   - Exhaust camshaft gear bolt while holding the exhaust camshaft firmly in place
   - Exhaust camshaft gear
5. Loosen the camshaft bearing cap bolts in stages of one-half to one turn.
   - Camshaft bearing cap bolts from the cylinder head

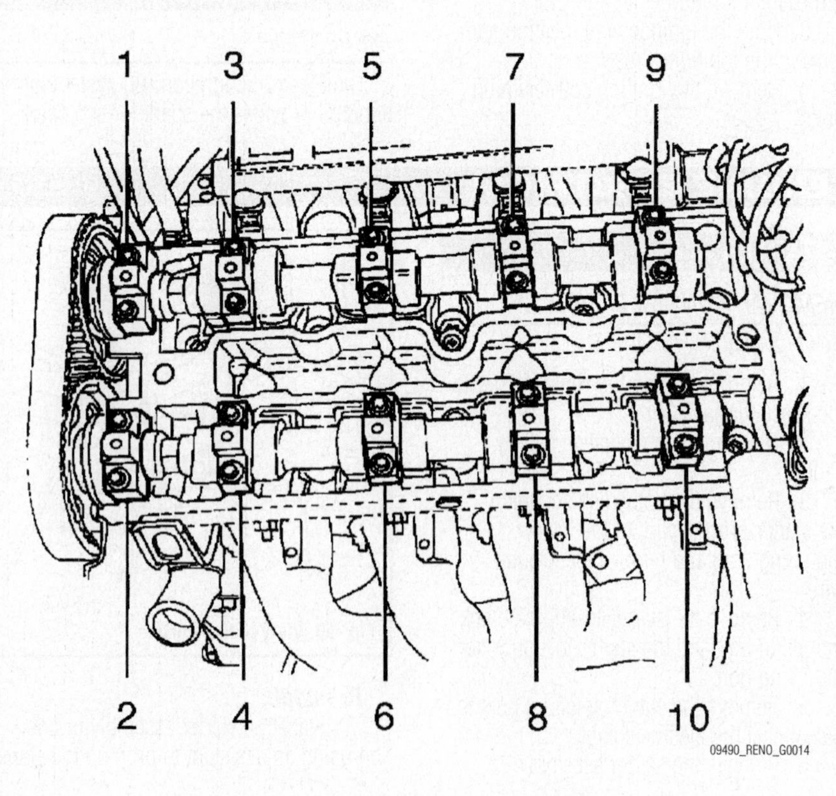

Fig. 52 Camshaft cap bolt tightening sequence

- Camshaft bearing caps. Maintain the correct positions for installation.
- Camshafts
- Seal ring from the camshafts

6. Check the camshaft and bearing seats for wear and replace them if necessary.
7. Remove the valve tappet adjusters. Maintain the correct positions for installation.

To install:

8. Lubricate the valve tappet adjusters with engine oil.
9. Install the valve tappet adjusters.

⁂ WARNING

Take extreme care to prevent any scratches, nicks or damage to the camshafts.

10. Lubricate the camshaft journals and the camshaft caps with engine oil.
11. Install or connect the following:
    - Intake camshaft
    - Intake camshaft caps in their original positions
    - Intake camshaft cap bolts
    - Exhaust camshaft
    - Exhaust camshaft caps in their original positions
    - Exhaust camshaft cap bolts
12. Tighten the camshaft cap bolts gradually and in sequence shown. Tighten the camshaft bearing cap bolts to 71 inch lbs. (8 Nm).
13. Measure the intake camshaft end play and the exhaust camshaft end play. End play should not exceed 0.0015–0.0055 inch (0.040–0.14mm).
14. Install the intake camshaft gear.
15. Tighten the intake camshaft gear bolt while holding the intake camshaft firmly in place. Tighten the intake camshaft gear bolt to 37 ft. lbs. (50 Nm) plus 60 degrees and 15 degrees using the angular torque gauge KM-470-B.
16. Tighten the exhaust camshaft gear.
17. Tighten the exhaust camshaft gear bolt while holding the exhaust camshaft firmly in place. Tighten the exhaust camshaft gear bolt to 37 ft. lbs. (50 Nm) plus 60 degrees and 15 degrees using the angular torque gauge KM-470-B.
18. Install or connect the following:
    - Camshaft cover and gasket
    - Camshaft cover washers and bolts and tighten to 71 inch lbs. (8 Nm)
    - Ignition wires to the spark plugs
    - Spark plug cover and bolts and tighten to 71 inch lbs. (8 Nm)
    - Breather tube and engine ventilation hose to the camshaft cover
    - Timing belt
    - Negative battery cable
19. Start the engine and check for any leaks when finished.

CAMSHAFT BEARING REPLACEMENT

Camshaft caps are utilized as the bearings and the caps must be replaced and machined in the event of a mechanical failure. If damage is too extensive for cap replacement, the cylinder head must be replaced.

CRANKSHAFT DAMPER

REMOVAL & INSTALLATION

1. Disconnect the negative battery cable.
2. Remove the air filter housing assembly bolts and the air intake tube.
3. Remove the air filter housing assembly from the vehicle.
4. Use a wrench to turn the tensioner bolt clockwise, compressing the tensioner, and releasing the tension on the serpentine accessory drive belt.
5. Remove the serpentine accessory drive belts.
6. Remove the crankshaft pulley.

To install:

7. Install the crankshaft pulley.
8. Tighten the crankshaft gear bolt to 107 ft. lbs. (145 Nm) plus 30° plus 15° using the angular torque gauge KM-470-B.
9. Use a wrench to turn the tensioner bolt clockwise, compressing the tensioner, and releasing the tension on the serpentine accessory drive belt.
10. With the wrench in place on the tensioner bolt, loop the serpentine accessory drive belt loosely over the pulleys.
11. Slip the belt over the tensioner.
12. Remove the wrench from the tensioner bolt and the belt will tighten itself.
13. Install the air filter housing assembly into the vehicle.
14. Install the air intake tube and the air filter housing assembly bolts.
15. Tighten the air filter housing assembly bolts to 53inch lbs.(6 Nm).
16. Connect the negative battery cable.

CRANKSHAFT FRONT SEAL

REMOVAL & INSTALLATION

1. Before servicing the vehicle, refer to the Precautions Section.
2. Disconnect the negative battery cable.
3. Drain the engine oil.
4. Remove the timing belt.
5. Remove the oil pump.
6. Remove the front crankshaft seal from the oil pump housing using a suitable prying tool.

⁂ CAUTION

Be careful not to damage the oil pump sealing surface when removing or installing the seal.

To install:

7. Install the oil pump.
8. Install a new oil pump-to-crankshaft seal. Coat the lip of the seal with a thin coat of grease. Use an oil seal guide tool.
9. Install the timing belt.
10. Refill engine with engine oil.
11. Connect the negative battery cable.
12. Start the engine and check the engine oil pressure.
13. Check that no leaks are present.

CYLINDER HEAD

REMOVAL & INSTALLATION

See Figure 53.

⁂ CAUTION

The fuel injection system remains under pressure after the engine has been turned OFF. Properly relieve fuel pressure before disconnecting any fuel lines. Failure to do so may result in fire or personal injury.

1. Before servicing the vehicle, refer to the Precautions Section.
2. Remove the fuel pump fuse.
3. Start the engine. After it stalls, crank the engine for 10 seconds to rid the fuel system of fuel pressure.
4. Disconnect the negative battery cable.
5. Disconnect the Engine Control Module (ECM) ground terminal.
6. Drain the engine coolant.
7. Remove or disconnect the following:
    - Manifold Air Temperature (MAT) sensor connector
    - Breather tube from the camshaft cover
    - Air cleaner outlet hose from the throttle body
    - Direct Ignition System (DIS) coil connector
    - Oxygen (O2) sensor connector
    - Electronic Throttle Control (ETC) connector
    - Engine Coolant Temperature (ECT) sensor connector
    - Coolant Temperature Sensor (CTS) connector

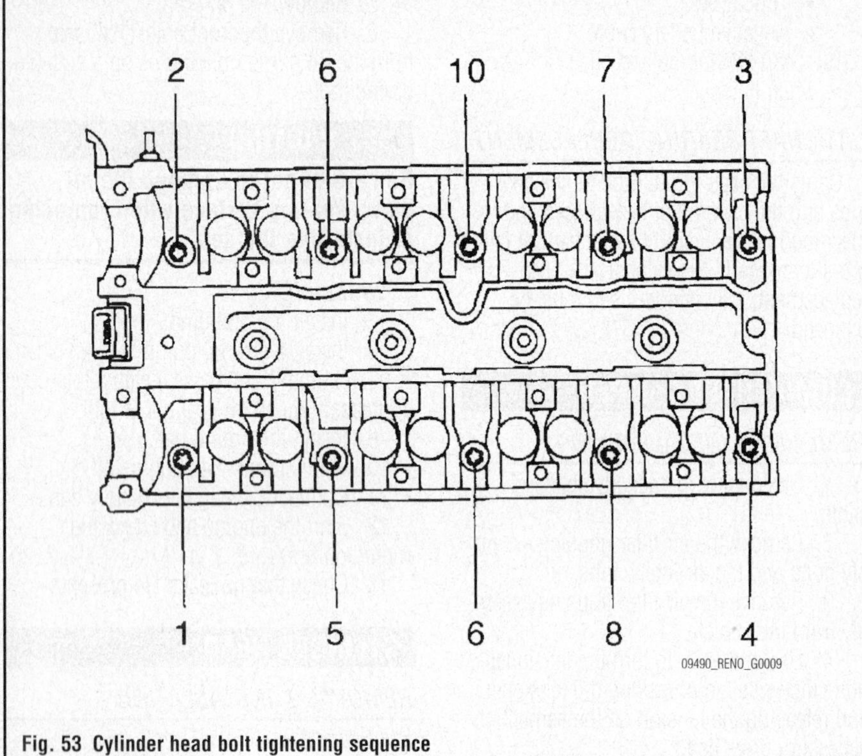

Fig. 53 Cylinder head bolt tightening sequence

- Remove the air cleaner housing bolts and air cleaner housing
- Right front wheel
- Right front wheel well splash shield

8. Install the engine assembly support fixture DW110-060.

- Right engine mount bracket and bolts
- Upper radiator hose at the thermostat housing
- Serpentine drive belt
- Crankshaft pulley bolts and crankshaft pulley
- Front timing belt cover bolts and front timing belt cover
- Timing belt
- Breather tube at the camshaft cover
- Spark plug cover bolts and spark plug cover
- Ignition wires from the spark plugs
- Camshaft cover bolts, camshaft cover and camshaft cover gasket
- Intake camshaft gear bolt, while holding intake camshaft firmly in place
- Intake camshaft gear
- Exhaust camshaft gear bolt, while holding exhaust camshaft firmly in place
- Exhaust camshaft gear
- Timing belt automatic tensioner bolts and timing belt automatic tensioner
- Timing belt idler pulley bolt and nut and timing belt idler pulleys
- Engine mount bolts and engine mount
- Crankshaft gear
- CMP sensor
- Water pump
- Rear timing belt cover bolts and rear timing belt cover
- Exhaust flex pipe retaining nuts at the exhaust manifold studs
- All of the necessary vacuum hoses
- Fuel feed line at the fuel rail
- Alternator adjusting bracket retaining bolt and the bracket
- Coolant hose at the rear cylinder head and ignition coil Exhaust Gas Recirculation (EGR) bracket
- Surge tank coolant hose at the throttle body
- Fuel rail assembly
- Alternator-to-intake manifold support bracket bolts at cylinder head and intake manifold
- Alternator support bracket
- Intake manifold-to-generator strap bracket bolt and loosen bolt on alternator
- Move the strap clear of the intake manifold.
- Charcoal Canister Purge (CCP) and EGR solenoid bracket bolt and move bracket clear
- Throttle cable at throttle body and intake manifold
- Loosen all cylinder head bolts gradually and in reverse order of tightening
- Camshafts
- Cylinder head bolts
- Cylinder head with intake manifold and exhaust manifold attached
- Cylinder head gasket

9. Clean the cylinder block and cylinder head mating surfaces of any old gasket material, oil or dust and clean any engine coolant from the cylinders.

To install:

10. Install or connect the following:

- New cylinder head gasket
- Cylinder head with intake manifold and exhaust manifold attached
- Cylinder head bolts. Tighten the cylinder head bolts gradually and in the sequence shown. Tighten the cylinder head bolts to 18 ft. lbs. (25 Nm) and turn the bolts another 3 turns of 90 degrees using the angular torque gauge KM-470-B.
- Camshafts
- Throttle cable at throttle body and intake manifold
- Alternator-to-intake manifold support bracket
- Alternator-to-intake manifold support bracket bolts and tighten to 27 ft. lbs. (37 Nm)
- Intake manifold support bracket bolt to alternator and tighten to 16 ft. lbs. (22 Nm)
- Surge tank coolant hose at throttle body
- Coolant hose to rear cylinder head and ignition coil EGR bracket
- Fuel feed line at the fuel rail
- Fuel return line at the fuel rail
- All necessary vacuum hoses
- Fuel rail assembly
- Exhaust flex pipe retaining nuts at the exhaust manifold studs and tighten to 26 ft. lbs. (35 Nm)
- Rear timing belt cover. Tighten the rear timing belt cover bolts to 62 inch lbs. (7 Nm).
- Engine mount and retaining bolts. Tighten the engine mount retaining bolts to 33 ft. lbs. (45 Nm).
- CMP sensor
- Crankshaft gear

- Water pump
- Timing belt automatic tensioner. Tighten the timing belt automatic tensioner bolts to 18 ft. lbs. (25 Nm).
- Timing belt idler pulleys, bolt and nut. Tighten the timing belt idler pulley bolt to 18 ft. lbs. (25 Nm).
- Camshaft gears with the timing marks at the front
- Guide pin of the intake camshaft into **"IN"** bore
- Guide pin of the exhaust camshaft into **"EX"** bore
- Camshaft gears by counter holding on the hex of the camshaft with an open-ended wrench
- Intake camshaft gear with a new bolt to the camshaft. Tighten the intake camshaft gear bolt to 37 ft. lbs. (50 Nm) turn the bolt another 60 degrees and 15 degrees using the angular torque gauge.
- Exhaust camshaft gear bolt while holding the exhaust camshaft firmly in place. Tighten the exhaust camshaft gear bolt to 37 ft. lbs. (50 Nm) turn the bolt another 60 degrees and 15 degrees using the angular torque gauge.
- Small amount of gasket sealant to the corners of the front camshaft caps and to the top of the rear camshaft cover-to-cylinder head seal
- Camshaft cover, gasket and washers
- Camshaft cover bolts and tighten to 71 inch lbs. (8 Nm)
- Ignition wires to the spark plugs
- Spark plug cover and tighten the spark plug cover bolts to 71 inch lbs. (8 Nm)
- Breather tube to the camshaft cover
- Timing belt
- Front timing belt cover
- Crankshaft pulley. Tighten the crankshaft pulley bolts to 106 inch lbs. (12 Nm).
- Right engine mount bracket. Tighten the engine mount bracket retaining bolts to 41 ft. lbs. (55 Nm).

11. Remove the engine assembly lift support DW110-060.
    - Serpentine drive belt
    - Upper radiator hose to thermostat housing
    - Right front wheel well splash shield
    - Right front wheel
    - Air cleaner housing
    - Air cleaner outlet hose to the throttle body
    - Breather tube to the camshaft cover
    - MAT sensor connector
    - CTS connector
    - Engine CTS connector
    - ETC connector
    - CCP and EGR solenoid bracket bolt and tighten to 44 inch lbs. (5 Nm)
    - DIS coil connector
    - O2 sensor connector
    - ECM ground terminal
    - Fuel pump fuse
12. Refill the engine cooling system.
13. Connect the negative battery cable.
14. Start the engine and check for leaks.

ENGINE ASSEMBLY

REMOVAL & INSTALLATION

✲✲ CAUTION

The fuel injection system remains under pressure after the engine has been turned OFF. Properly relieve fuel pressure before disconnecting any fuel lines. Failure to do so may result in fire or personal injury.

1. Before servicing the vehicle, refer to the Precautions Section.
2. Properly relieve the fuel system pressure.
3. Disconnect the negative battery cable.
4. Remove the fuel pump fuse.
5. Start the engine. After it stalls, crank the engine for 10 seconds to rid the fuel system of fuel pressure.
6. Drain the engine oil.
7. Discharge the Air Conditioning (A/C) system, if equipped.
8. Drain the engine coolant.
9. Remove or disconnect the following:
    - Hood
    - Manifold Air Temperature (MAT) sensor connector
    - Air cleaner outlet hose from throttle body and air cleaner housing
    - Breather tubes from the camshaft cover
    - Right front wheel.
    - Right front wheel well splash shield
    - Serpentine accessory drive belt.
    - Cooling system radiator and the engine cooling fans
    - Upper radiator hose from the thermostat housing
    - Power steering return hose from the power steering pump
    - Power steering pressure hose from the power steering pump
    - Electrical connector at the direct ignition system (DIS) coil and the electronic control module (ECM) ground terminal and at the starter motor
    - Oxygen (O2) sensor connector, if equipped
    - Idle Air Control (IAC) valve connector
    - Throttle Position Sensor (TPS) connector
    - Engine Coolant Temperature Sensor (CTS) connector
    - Alternator voltage regulator connector and power lead
    - All of the necessary vacuum lines, including the brake booster vacuum hose
    - Fuel return line at the fuel rail
    - Fuel feed line at the fuel rail
    - Fuel rail and injector channel cover as an assembly
    - Throttle cable from the throttle body and the intake manifold bracket
    - Coolant hose at the throttle body
    - Heater outlet hose at the coolant pipe
    - Coolant bypass hose from the cylinder head
    - Surge tank coolant hose from the coolant pipe
    - Lower radiator hose from the coolant pipe
    - Starter solenoid terminal wire and power lead
    - A/C compressor
    - Exhaust flex pipe retaining nuts from the exhaust manifold studs
    - Exhaust flex pipe retaining nuts from the catalytic converter or the connecting pipe
    - Exhaust flex pipe
    - Crankshaft pulley bolts and crankshaft pulley
    - Vacuum lines at the charcoal canister purge solenoid
    - Electrical connector at the charcoal canister purge (CCP) and the exhaust gas recirculation (EGR) solenoid
    - Electrical connector at the oil pressure switch
    - Crankshaft Position Sensor (CPS) connector
    - Knock sensor connector
    - Lower reaction rod bracket bolts
    - Lower reaction rod bracket
    - Lower reaction rod mount bolt
    - Lower reaction rod mount
    - Rubber cover from service hole

- Transaxle torque converter bolts through the service hole (A/T)
- Transaxle bell housing bolts and the oil pan flange bolts

10. Support the transaxle with a floor jack.
11. Install the engine lifting device.
12. Disconnect the right engine mount bracket from the engine mount by removing the retaining bolt.
13. Remove the right engine mount bracket from the engine block and frame mount.
14. Separate the engine block from the transaxle. Remove the engine.

To install:

15. Install the engine into the engine compartment
16. Align the engine alignment pins to the transaxle.
17. Install the transaxle bell housing bolts and tighten to 55 ft. lbs. (75 Nm).
18. Install the oil pan flange-to-transaxle bolts and tighten to 30 ft. lbs. (40 Nm).
19. Install or connect the following:
    - Right engine mount to engine block mount and frame mount
    - Right engine mount bracket retaining bolts and nuts. Tighten the engine mount bracket retaining bolts and nuts to 41 ft. lbs. (55 Nm).
20. Remove the floor jack used for support of the transmission.
21. Remove the engine lifting device.
22. Install or connect the following:
    - Transaxle torque converter bolts (A/T) and tighten to 44 ft. lbs. (60 Nm)
    - Rubber service hole cover
    - Lower reaction rod mount
    - Lower reaction rod mount bolt and tighten to 41 ft. lbs. (55 Nm)
    - Lower reaction rod bracket
    - Lower reaction rod bracket bolts and tighten to 49 ft. lbs. (69 Nm)
    - Vacuum lines at the CCP solenoid
    - Electrical connector to the CCP and the EGR solenoid
    - Oil pressure switch connector
    - Crankshaft pulley and crankshaft pulley bolts. Tighten the crankshaft pulley bolts to 15 ft. lbs. (20 Nm) using a torque wrench.
    - CPS connector
    - Exhaust flex pipe
    - Exhaust flex pipe retaining nuts to the exhaust manifold studs and tighten to 26 ft. lbs. (35 Nm)
    - Exhaust flex pipe retaining nuts to the catalytic converter or the connecting pipe and tighten to 26 ft. lbs. (35 Nm)
    - Power steering pressure hose
    - Power steering return hose
    - A/C compressor, if equipped
    - Serpentine drive belt
    - Right front wheel well splash shield
    - Right front wheel
    - Fuel feed line to the fuel rail
    - Fuel return line to fuel rail
    - Fuel rail and injector channel cover as an assembly
    - All necessary vacuum lines including brake booster vacuum hose
    - O2 sensor connector, if equipped
    - Starter solenoid terminal wire and power lead
    - Alternator voltage regulator connector
    - Engine CTS connector
    - TPS connector
    - IAC valve connector
    - MAP sensor connector
    - Knock sensor, if necessary
    - Electrical connector at the DIS ignition coil and the ECM ground terminal and at the starter motor
    - Air cleaner outlet hose between the throttle body and air cleaner housing
    - Breather tubes to the camshaft cover
    - MAT sensor connector
    - Cooling system radiator and engine cooling fans
    - Lower radiator hose to the coolant pipe
    - Upper radiator hose to the thermostat housing
    - Heater inlet hose to the cylinder head
    - Heater outlet hose to coolant pipe
    - Coolant surge tank hose to coolant pipe
    - Coolant hose to throttle body
    - Throttle cable to throttle body and intake manifold bracket
23. Install the fuel pump fuse.
24. Connect the negative battery cable.
25. Refill the engine crankcase with engine oil.
26. Refill the engine coolant system.
27. Bleed the power steering system.
28. Recharge the A/C refrigerant system, if equipped.
29. Install the hood.
30. Run the engine and verify that there are no fuel, coolant, transaxle or exhaust leaks.

EXHAUST MANIFOLD

REMOVAL & INSTALLATION

See Figures 54 and 55.

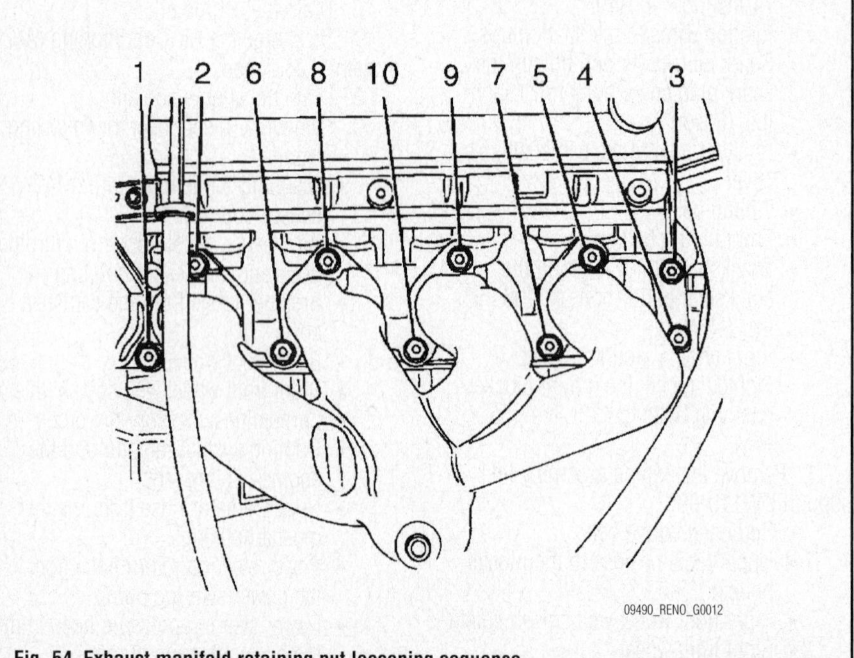

Fig. 54 Exhaust manifold retaining nut loosening sequence

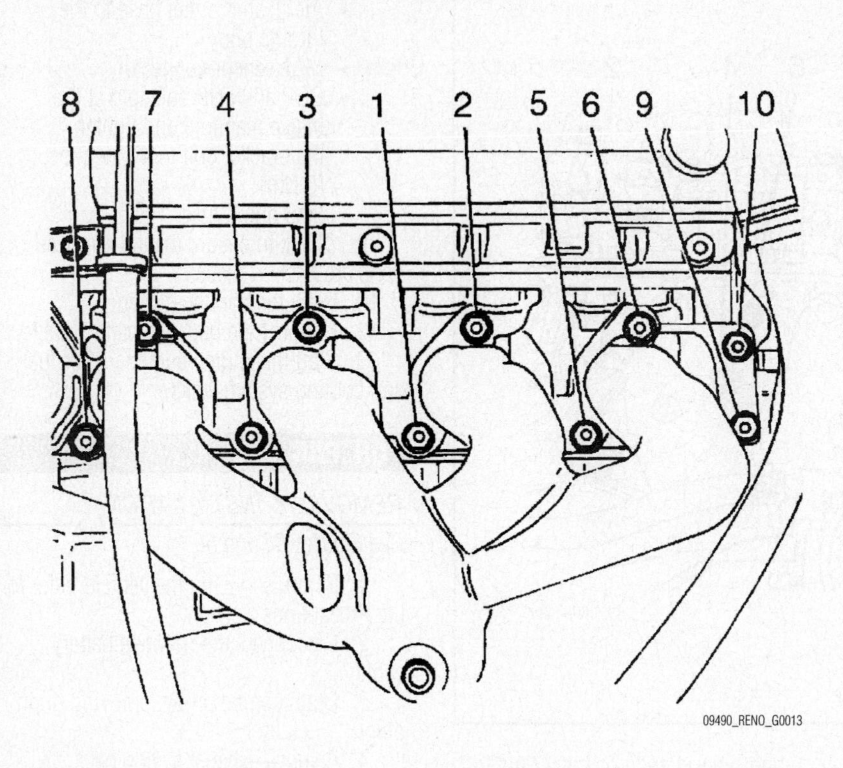

Fig. 55 Exhaust manifold retaining nut tightening sequence

CAUTION

To avoid the danger of being burned, do not service the exhaust system while it is hot. Service should be performed only after the system cools down.

1. Before servicing the vehicle, refer to the Precautions Section.
2. Disconnect the negative battery cable.
3. Remove or disconnect the following:
    - Heated Oxygen (HO2) sensor connector
    - Exhaust manifold heat shield bolts and exhaust manifold heat shield
    - Exhaust flex pipe retaining nuts from exhaust manifold studs
    - Exhaust manifold retaining nuts in the sequence shown
    - Exhaust manifold and gasket
4. Clean the sealing surfaces of the exhaust manifold and the cylinder head.

To install:

5. Install or connect the following:
    - New exhaust manifold gasket
    - Exhaust manifold and retaining nuts. Tighten the exhaust manifold retaining nuts to 16 ft. lbs. (22 Nm) in the sequence shown.
    - Exhaust flex pipe retaining nuts to the exhaust manifold studs and tighten to 26 ft. lbs. (35 Nm)
    - Exhaust manifold heat shield and tighten the bolts to 71 inch lbs. (8 Nm)
    - O2 sensor connector
6. Connect the negative battery cable.
7. Run the engine and check for exhaust leaks.

FLYWHEEL/FLEXPLATE

REMOVAL & INSTALLATION

Flexplate

1. Remove automatic transaxle, as outlined in the Drivetrain Section.
2. Loosen flexplate bolts in a star pattern.
3. Remove flexplate.

To install:

4. Install flexplate.
5. Install flexplate bolts.
6. Tighten flexplate bolts in a star pattern, to 33 ft lbs. (45 Nm)
7. Install automatic transaxle.

Flywheel

1. Remove manual transaxle, as outlined in the Drivetrain Section.
2. Remove pressure plate and clutch disc.
3. Loosen flywheel bolts in a star pattern.
4. Remove flywheel.

To install:

5. Install flywheel.
6. Install flywheel bolts.
7. Tighten flywheel bolts in a star pattern as follows:
    a. Step 1: Tighten to 48 ft lbs. (65 Nm)
    b. Step 2: Plus 30 degrees
    c. Step 3: Plus 15 degrees
8. Install clutch disc and pressure plate.
9. Install manual transaxle.

INTAKE MANIFOLD

REMOVAL & INSTALLATION

See Figures 56 and 57.

CAUTION

The fuel system pressure must be relieved before disconnecting any fuel lines. Failure to do so may result in personal injury.

1. Before servicing the vehicle, refer to the Precautions Section.
2. Remove the fuel pump fuse.
3. Start the engine. After it stalls, crank the engine for 10 seconds to rid the fuel system of fuel pressure.
4. Disconnect the negative battery cable.
5. Drain the engine coolant.
6. Remove or disconnect the following:
    - Charcoal Canister Purge (CCP)
    - Exhaust Gas Recirculation (EGR) solenoid
    - Bracket bolt
    - Manifold Air Temperature (MAT) sensor connector
    - Air cleaner outlet hose from throttle body
    - Idle Air Control (IAC) valve connector
    - Manifold Absolute Pressure (MAP) sensor connector
    - Coolant hoses at the throttle body
    - All necessary vacuum hoses, including the vacuum hose at the fuel pressure regulator and the brake booster vacuum hose at intake manifold
    - Alternator-to-intake manifold strap bracket bolts and strap

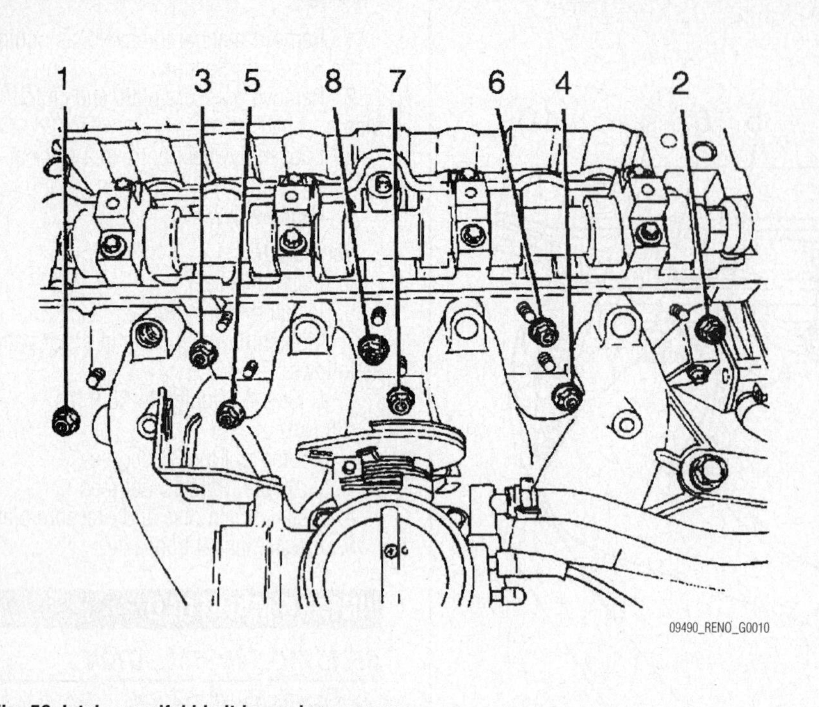

Fig. 56 Intake manifold bolt loosening sequence

- Fuel rail as an assembly
- Alternator-to-intake manifold support bracket bolts
- Alternator-to-intake manifold support bracket
- Intake manifold support bracket bolt at the engine block and the intake manifold
- Intake manifold support bracket
- Intake manifold retaining bolt and nuts in the sequence shown
- Intake manifold and intake manifold gasket

7. Clean the sealing surfaces of the intake manifold and the cylinder head.

To install:

8. Install or connect the following:
   - New intake manifold gasket
   - Intake manifold
   - Intake manifold retaining bolt and nuts and tighten, in sequence, to 16 ft. lbs. (22 Nm).
   - Alternator-to-intake manifold strap bracket and bolts. Tighten the alternator-to-intake manifold strap bracket bolts to 16 ft. lbs. (22 Nm).
   - Intake manifold support bracket
   - Intake manifold support bracket upper bolts to the intake manifold and tighten to 18 ft. lbs. (25 Nm)
   - Intake manifold support bracket lower bolt to the engine block and tighten to 18 ft. lbs. (25 Nm)
   - Alternator-to-intake manifold support bracket and bolts and tighten to 27 ft. lbs. (37 Nm)
   - Fuel rail and injector cover as an assembly
   - All of the necessary vacuum lines that were previously disconnected
   - MAP sensor connector
   - Coolant hoses to the throttle body
   - ETC connector
   - Air cleaner outlet hose to the throttle body
   - MAT sensor connector
   - CCP and EGR solenoid at the intake manifold and tighten the bracket bolt to 44 inch lbs. (5 Nm)
   - Fuel pump fuse
9. Check to ensure that all removed parts are back in place.
10. Refill the cooling system.
11. Connect the negative battery cable.
12. Start the engine and check for fuel and cooling system leaks.

OIL PAN

REMOVAL & INSTALLATION

See Figures 58 and 59.

1. Before servicing the vehicle, refer to the Precautions Section.
2. Disconnect the negative battery cable.
3. Drain engine oil by removing drain plug.
4. Remove exhaust front pipe.
5. Remove the oil pan flange-to-transaxle retaining bolts.
6. Remove the transaxle-to-oil pan flange retaining bolt.
7. Remove transaxle rear mount bracket.
8. Remove the lower reaction rod.

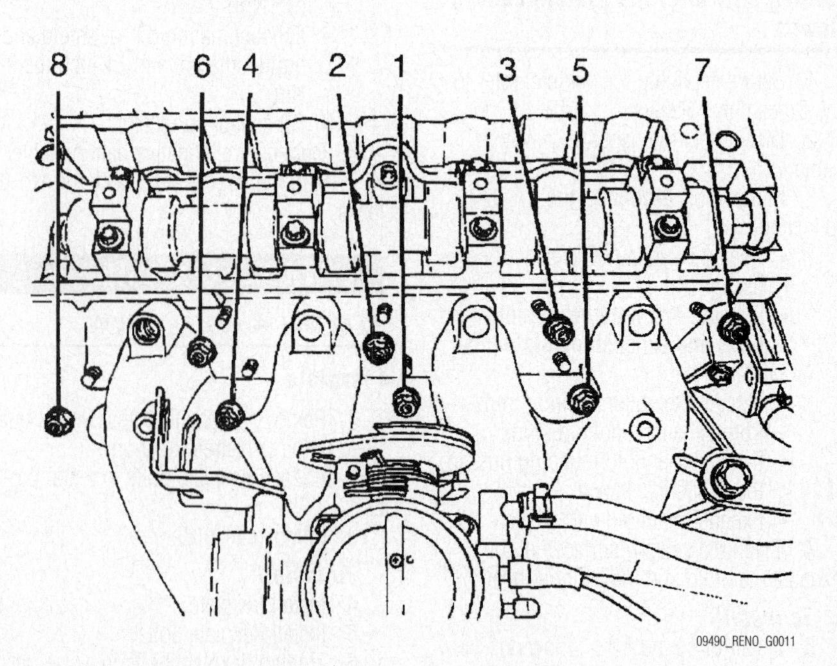

Fig. 57 Intake manifold bolt tightening sequence

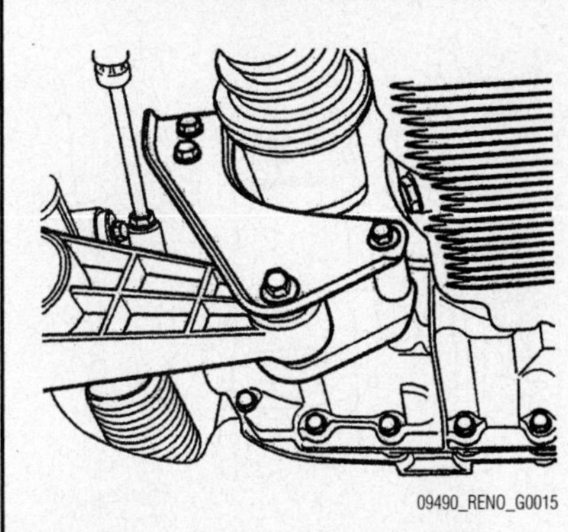

Fig. 58 Transaxle rear mount bracket

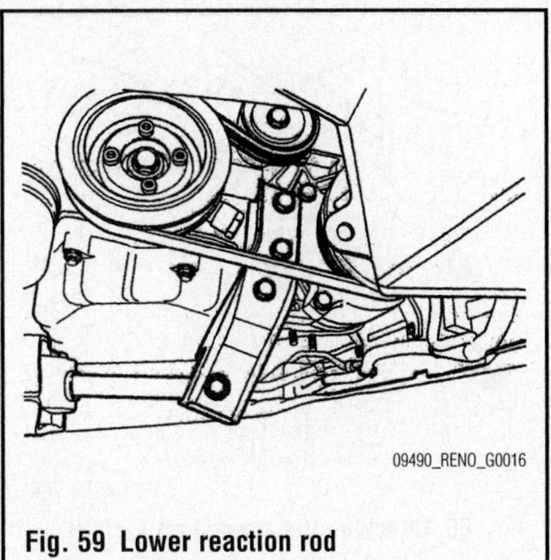

Fig. 59 Lower reaction rod

9. Remove the oil pan retaining bolts.
10. Remove the oil pan from the cylinder block.

To install:

11. Coat the new oil pan gasket with sealant.
12. Install the oil pan within 5 minutes after applying the liquid gasket to the oil pan.
13. Install the oil pan to the cylinder block.
14. Install the oil pan retaining bolts and tighten to 89 inch lbs. (10 Nm).
15. Install the oil pan flange-to-transaxle retaining bolts and tighten to 30 ft. lbs. (40 Nm).
16. Install the transaxle-to-oil pan flange retaining bolt and tighten to 30 ft. lbs. (40 Nm).
17. Install transaxle rear mount bracket.
18. Install the lower reaction rod and bolts. Tighten the lower reaction rod bolts to 49 ft. lbs. (69 Nm).
19. Install exhaust front pipe and tighten the retaining nuts to 26 ft. lbs. (35 Nm).
20. Refill the engine with oil.
21. Connect the negative battery cable.
22. Start the engine and check for leaks.

OIL PUMP

REMOVAL & INSTALLATION

See Figure 60.

1. Before servicing the vehicle, refer to the Precautions Section.
2. Disconnect the negative battery cable.
3. Drain the engine oil.
4. Remove the timing belt.
5. Remove the rear timing belt cover.
6. Disconnect the oil pressure switch connector.
7. Remove the oil pan.
8. Remove the oil suction pipe and support bracket bolts.
9. Remove the oil suction pipe.
10. Remove the oil pump retaining bolts.
11. Carefully separate the oil pump and gasket from the engine block and oil pan.
12. Remove the oil pump.

To install:

13. Apply Loctite® 242 to the oil pump bolts and Room Temperature Vulcanizing (RTV) sealant to the new oil pump gasket.
14. Install the gasket to the oil pump and install the oil pump to the engine block with the bolts and tighten to 89 inch lbs. (10 Nm).
15. Install a new oil pump-to-crankshaft seal. Coat the lip of the seal with a thin coat of grease.
16. Coat the threads of the oil suction pipe and support bracket bolts with Loctite® 242.
17. Install the oil suction pipe and the bolts. Tighten the oil suction pipe bolts to 89 inch lbs. (10 Nm) and support bracket bolts to 53 inch lbs. (6 Nm).
18. Install the oil pan.
19. Connect the oil pressure switch connector.
20. Install the rear timing belt cover.
21. Install the timing belt.
22. Refill engine with engine oil.
23. Connect the negative battery cable.
24. Start the engine and check the engine oil pressure.
25. Check that no leaks are present.

Fig. 60 Oil pump and retaining bolts

INSPECTION

See Figure 61.

1. Clean the oil pump and the engine block gasket mating surface areas.
2. Remove the safety relief valve bolt.
3. Remove the safety relief valve and the spring.
4. Remove the oil pump-to-crankshaft seal.
5. Remove the oil pump rear cover bolts.
6. Remove the rear cover.
7. Clean the oil pump housing and all the oil pump parts.
8. Inspect all the oil pump parts for signs of wear.
9. Replace the worn oil pump parts.
10. Coat all the oil pump parts with clean engine oil. Install the oil pump parts.
11. Apply Loctite® 242 to the oil pump rear cover bolts and install the cover and bolts.
12. Tighten the oil pump rear cover bolts to53inch. lbs.(6 Nm).
13. Install the safety relief valve, spring, washer and bolt.
14. Tighten the safety relief valve bolt to 22ft-lb (30 Nm).

Oil Pump Specifications

Application	Description
Oil Pump:	
Gear Lash	0.10 - 0.20 mm (0.003 - 0.007 in.)
Outer Gear to Body	0.11 - 0.19 mm (0.004 - 0.007 in.)
Outer Gear to Crescent	0.40 - 0.50 mm (0.015 - 0.019 in.)
Inner Gear to Crescent	0.35 - 0.40 mm (0.013 - 0.015 in.)
End Clearance	0.030 - 0.10 mm (0.001 - 0.003 in.)
Sealants and Adhesives:	
Oil Pan Bolts	HN 1256 (Loctite® 242)
Oil Pump Bolts	HN 1256 (Loctite® 242)
Oil Pan Pickup Tube Bolts	HN 1256 (Loctite® 242)
Oil Gallery Plug	HN 1256 (Loctite® 242)

22140_RENO_G0060

Fig. 61 Oil pump specifications

MAIN BEARING TORQUE SEQUENCE

See Figures 62 through 65.

1. Mark the order of the crankshaft bearing caps.
2. Remove the crankshaft bearing cap bolts.
3. Remove the crankshaft bearing caps and the lower crankshaft bearings.

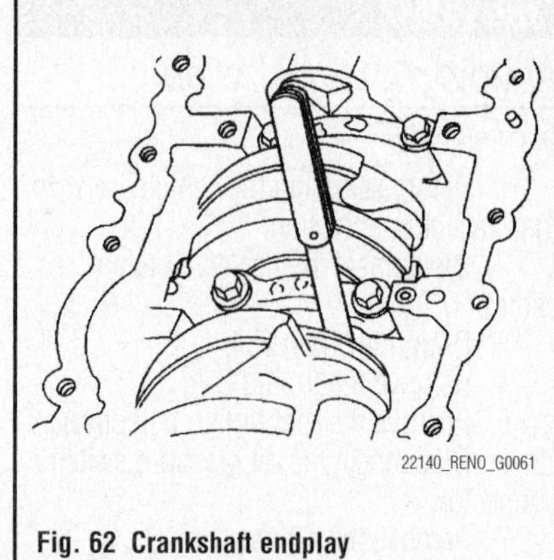

Fig. 62 Crankshaft endplay

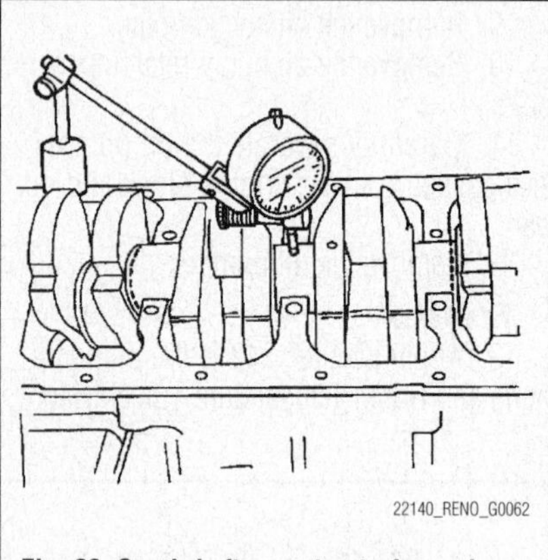

Fig. 63 Crankshaft runout-out-of-round

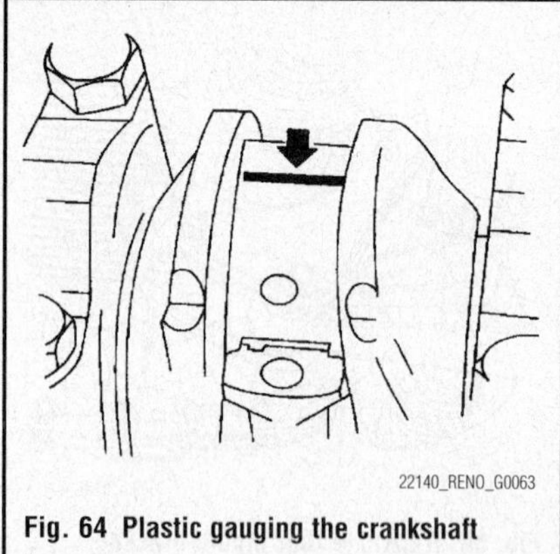

Fig. 64 Plastic gauging the crankshaft journal

Fig. 65 Checking the crankshaft journal clearance with plastic gauge

4. Remove the crankshaft.
5. Clean the parts, as needed.
6. Coat the crankshaft bearings with engine oil.
7. If replacing the crankshaft, transfer the pulse pickup sensor disc to the new crankshaft.
8. Install the crankshaft.
9. Install the lower crankshaft bearings in the bearing caps.
10. Inspect the crankshaft end play with the crankshaft bearings installed.
11. Crankshaft End Play 0.0027 –0.0118 in (0.070–0.302 mm).
12. With the crankshaft mounted on the front and rear crankshaft bearings, check the middle crankshaft journal for permissible out-of-round (runout).

Main Journal Diameter (All) 2.2824–2.2832 in (57.974–57.995 mm.) Limit 0.0001 in (0.003 mm.).

13. Inspect all of the crankshaft bearing clearances using a commercially available plastic gauging (ductile plastic threads).
14. Cut the plastic gauging threads to the length of the bearing width. Lay them axially between the crankshaft journals and the crankshaft bearings.
15. Install the crankshaft bearing caps and the bolts.
16. Tighten the crankshaft bearing cap bolts to 37ft. lbs. (50 Nm) plus 45° and 15°.
17. Remove the crankshaft bearing cap bolts and the caps.
18. Measure the width of the flattened plastic thread of the plastic gauging using a ruler. (Plastic gauging is available for different tolerance ranges.)
19. Inspect the bearing clearance for permissible tolerance ranges.
20. Main Bearing clearances (All) 0.00059–0.00157 in. (0.015–0.040 mm)
21. Apply a bead of adhesive sealing compound to the grooves of the crankshaft bearing caps.
22. Install the crankshaft bearing caps to the engine block.
23. Tighten the crankshaft bearing caps using new bolts.
24. Tighten the crankshaft bearing cap bolts to 37ft. lbs. (50 Nm) using a torque wrench.

➡Use the angular torque gauge KM-470-B to tighten the crankshaft bearings another 45° and 15°.

PISTON AND RING

POSITIONING

See Figures 66 through 70.

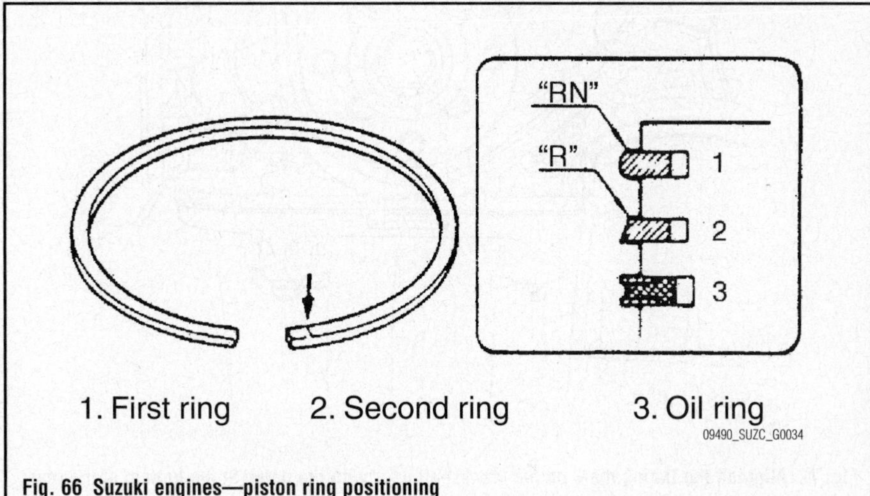

Fig. 66 Suzuki engines—piston ring positioning

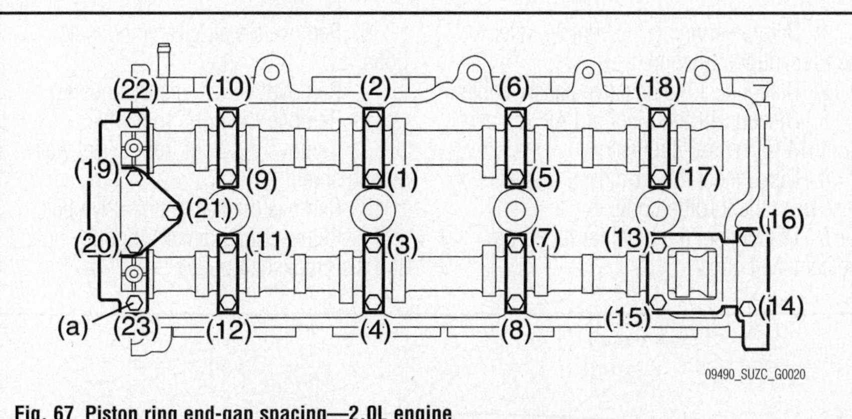

Fig. 67 Piston ring end-gap spacing—2.0L engine

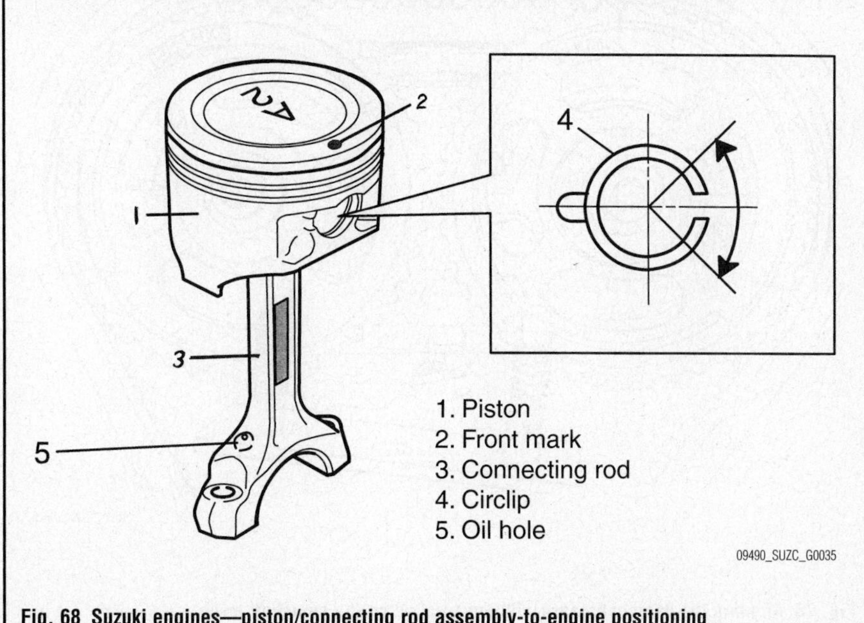

Fig. 68 Suzuki engines—piston/connecting rod assembly-to-engine positioning

REAR MAIN SEAL

REMOVAL & INSTALLATION

See Figure 71.

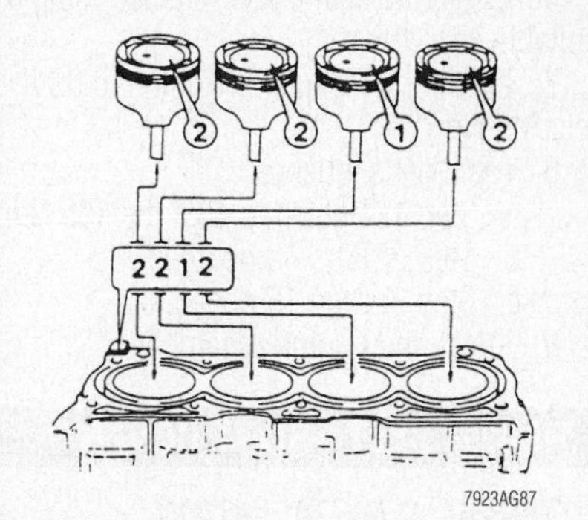

Fig. 69 Suzuki engines—the piston ID number must match the number stamped in the engine block

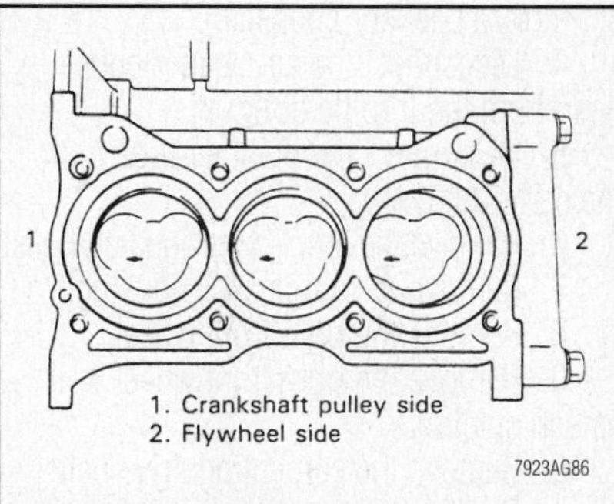

Fig. 70 Suzuki engines—the directional arrow on the piston face must face the crankshaft pulley end of the engine

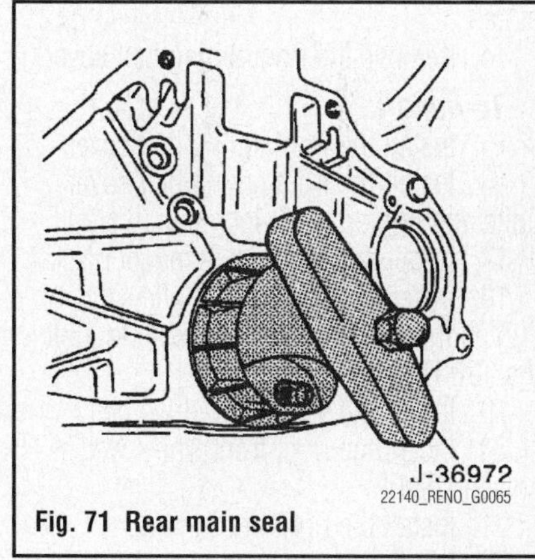

Fig. 71 Rear main seal

1. Before servicing the vehicle, refer to the Precautions Section.
2. Remove or disconnect the following:
   - Engine assembly
   - Flexplate/flywheel from the crankshaft
3. Carefully pry the oil seal out of the retainer without scratching the sealing surface of the crankshaft.

To install:

4. Apply engine oil to the lip of the new seal.
5. Install or connect the following:

6. Install the seal in the retainer using a suitable seal driver
7. Install the Flexplate. Tighten to 33 ft lbs. (45 Nm)
8. Flywheel sequence
   a. Step 1: Tighten to 48 ft lbs. (65 Nm)
   b. Step 2: Plus 30 degrees
   c. Step 3: Plus 15 degrees
9. Install the Engine assembly.

TIMING BELT FRONT COVER

REMOVAL & INSTALLATION

1. Before servicing the vehicle, refer to the Precautions Section.
2. Disconnect the negative battery cable.
3. Disconnect the Manifold Air Temperature (MAT) sensor connector.
4. Disconnect the air cleaner outlet hose from the throttle body.
5. Disconnect the breather tube from the camshaft cover.
6. Remove the air cleaner housing bolts.
7. Remove the air cleaner housing.
8. Remove the right front wheel.
9. Remove the right front wheel well splash shield.
10. Remove the serpentine drive belt.
11. Remove the crankshaft pulley bolts and the crankshaft pulley.
12. Remove the right engine mount bracket.
13. Remove the front timing belt cover bolts.
14. Remove the front timing belt cover.

To install:

15. Install the front timing belt cover.
16. Install the front timing belt cover bolts and tighten to 53 inch lbs. (6 Nm).
17. Install the right engine mount bracket.
18. Install the crankshaft pulley and tighten the crankshaft pulley bolts to 15 ft. lbs. (20 Nm).
19. Install the serpentine drive belt.
20. Install the right front wheel well splash shield.
21. Install the right front wheel.
22. Install the air cleaner housing.
23. Install the air cleaner housing bolts and tighten to 89 inch lbs. (10 Nm).
24. Connect the air cleaner outlet hose to the throttle body.
25. Connect the breather tube to the camshaft cover.
26. Connect the MAT sensor connector.
27. Connect the negative battery cable.

TIMING BELT AND SPROCKETS

REMOVAL & INSTALLATION

See Figures 72 and 73.

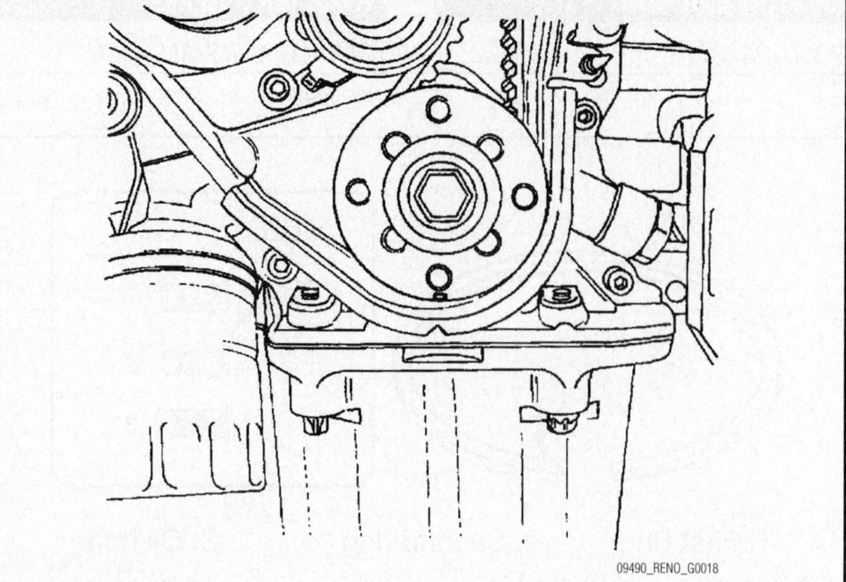

Fig. 72 Aligning the timing mark on the crankshaft gear with the notch at the bottom of the rear timing belt cover.

1. Before servicing the vehicle, refer to the Precautions Section.
2. Disconnect the negative battery cable.
3. Disconnect the Manifold Air Temperature (MAT) sensor connector.
4. Disconnect the air cleaner outlet hose from the throttle body.
5. Disconnect the breather tube from the camshaft cover.
6. Remove the air cleaner housing bolts.
7. Remove the air cleaner housing.
8. Remove the right front wheel.
9. Remove the right front wheel well splash shield.
10. Remove the serpentine drive belt.
11. Remove the crankshaft pulley bolts and the crankshaft pulley.

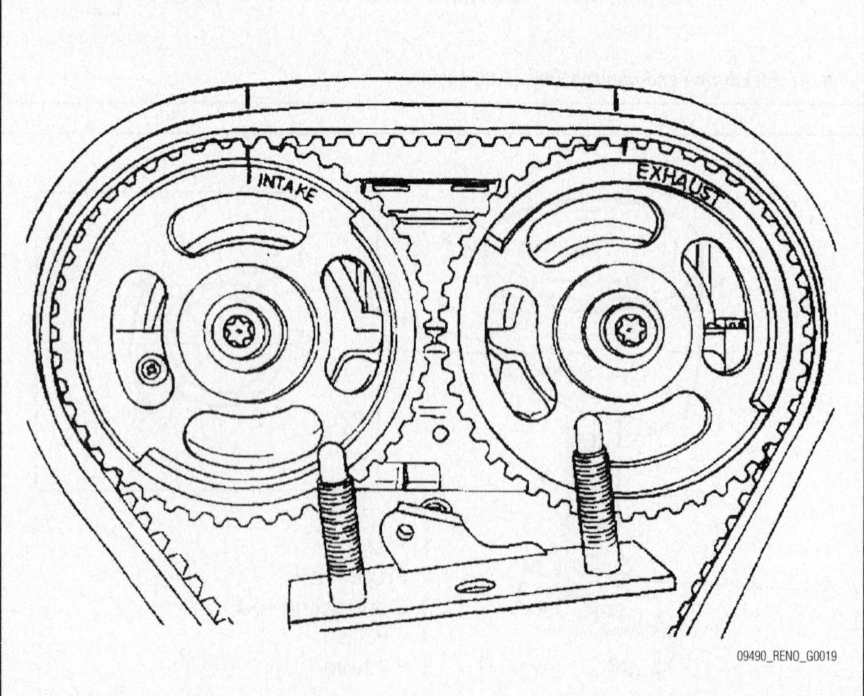

Fig. 73 Aligning the camshaft gears with the notches on the camshaft cover.

12. Remove the right engine mount bracket.
13. Remove the front timing belt cover bolts.
14. Remove the front timing belt cover.
15. Using the crankshaft gear bolt, rotate the crankshaft clockwise until the timing mark on the crankshaft gear is aligned with the notch at the bottom of the rear timing belt cover.
16. Align the camshaft gears with the notch on the camshaft cover.
17. Remove the timing belt.
18. Loosen the automatic tensioner bolt. Turn the hex-key tab to relieve belt tension.

To install:

19. Align the timing mark on the crankshaft gear with the notch on the bottom of the rear timing belt cover.
20. Align the timing marks on the camshaft gears, using the intake gear mark for the intake gear and the exhaust gear mark for the exhaust gear.
21. Install the timing belt.
22. Turn the hex-key tab in a clockwise direction to tension the belt. Turn until the pointer aligns with the notch.
23. Install the automatic tensioner bolt and tighten to 18 ft. lbs. (25 Nm).
24. Rotate the crankshaft two full turns clockwise using the crankshaft pulley bolt.
25. Recheck the automatic tensioner pointer.
26. Install the front timing belt cover.
27. Install the front timing belt cover bolts and tighten to 53 inch lbs. (6 Nm).
28. Install the right engine mount bracket.
29. Install the crankshaft pulley and tighten the crankshaft pulley bolts to 15 ft. lbs. (20 Nm).
30. Install the serpentine drive belt.
31. Install the right front wheel well splash shield.
32. Install the right front wheel.
33. Install the air cleaner housing.
34. Install the air cleaner housing bolts and tighten to 89 inch lbs. (10 Nm).
35. Connect the air cleaner outlet hose to the throttle body.
36. Connect the breather tube to the camshaft cover.
37. Connect the MAT sensor connector.
38. Connect the negative battery cable.

TIMING BELT REAR COVER

REMOVAL & INSTALLATION

See Figures 72 and 73.

➡Tools required. KM-470-B Angular Torque Gauge.

1. Before servicing the vehicle, refer to the Precautions Section.
2. Disconnect the negative battery cable.
3. Disconnect the Manifold Air Temperature (MAT) sensor connector.
4. Disconnect the air cleaner outlet hose from the throttle body.
5. Disconnect the breather tube from the camshaft cover.
6. Remove the air cleaner housing bolts.
7. Remove the air cleaner housing.
8. Remove the right front wheel.
9. Remove the right front wheel well splash shield.
10. Remove the serpentine drive belt.
11. Remove the crankshaft pulley bolts and the crankshaft pulley.
12. Remove the right engine mount bracket.
13. Remove the front timing belt cover bolts.
14. Remove the front timing belt cover.
15. Using the crankshaft gear bolt, rotate the crankshaft clockwise until the timing mark on the crankshaft gear is aligned with the notch at the bottom of the rear timing belt cover.
16. Align the camshaft gears with the notch on the camshaft cover.
17. Remove the timing belt.
18. Loosen the automatic tensioner bolt. Turn the hex-key tab to relieve belt tension.
19. While holding the intake camshaft firmly in place, remove the intake camshaft gear bolt.
20. Remove the intake camshaft gear.
21. Remove the Camshaft Position (CMP) sensor.
22. Remove the water pump.
23. While holding the exhaust camshaft firmly in place, remove the exhaust camshaft gear bolt.
24. Remove the exhaust camshaft gear.
25. Remove the timing belt automatic tensioner bolt.
26. Remove the timing belt automatic tensioner.
27. Remove the timing belt idler pulley bolt and nut.
28. Remove the timing belt idler pulleys.
29. Remove the engine mount retaining bolts.
30. Remove the engine mount.
31. Remove the crankshaft gear bolt.
32. Remove the crankshaft gear.
33. Remove the rear timing belt cover bolts.
34. Remove the rear timing belt cover.

To install:

35. Install the rear timing belt cover.
36. Install the rear timing belt cover bolts.
37. Tighten the rear timing belt cover bolts to 62 inch lbs. (7 Nm)
38. Install the engine mount and retaining bolts.
39. Tighten the engine mount retaining bolts to 33 ft. lbs. (45 Nm)
40. Install the timing belt idler pulleys.
41. Install the timing belt idler pulley bolt and nut.
42. Tighten the timing belt idler pulley bolt to 18 ft. lbs. (25 Nm)
43. Tighten the timing belt idler pulley nut to 18 ft. lbs. (25 Nm)
44. Install the crankshaft timing belt drive gear and bolt.
45. Tighten the crankshaft gear bolt to 107 ft. lbs. (145 Nm) plus 30 degrees and plus 15° using the angular torque gauge KM-470-B.
46. Install the timing belt automatic tensioner and bolt.
47. Install the camshaft gears.
48. Install the timing belt and timing belt cover.
49. Connect the negative battery cable.

VALVE COVERS

REMOVAL & INSTALLATION

See Figures 74 and 75.

➡The valve cover is also referred to as the camshaft cover.

1. Disconnect the negative battery cable.
2. Disconnect the breather tube from the valve cover.
3. Disconnect all of the necessary vacuum lines.
4. Remove the spark plug cover bolts.
5. Remove the spark plug cover.
6. Disconnect the ignition wires from the spark plugs.
7. Disconnect the camshaft position sensor connector.

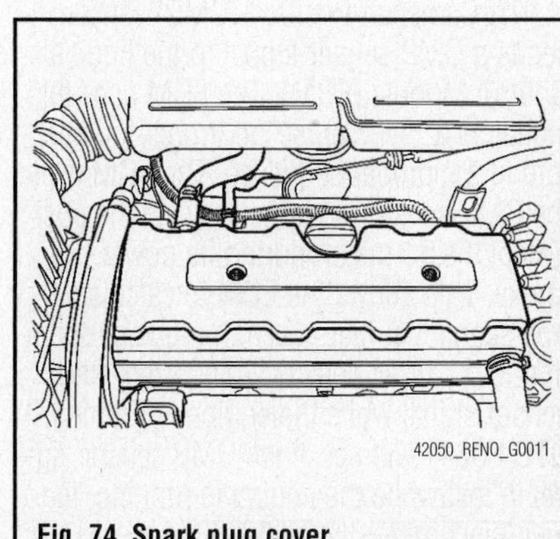

Fig. 74 Spark plug cover

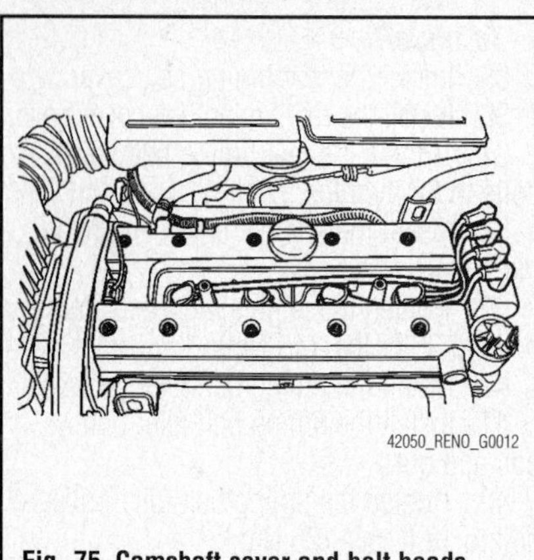

Fig. 75 Camshaft cover and bolt heads

8. Remove the camshaft cover bolts.
9. Remove the camshaft cover.
10. Remove the camshaft cover gasket from the camshaft cover.

To install:

11. Apply a small amount of gasket sealant to the corners of the front camshaft caps and the top of the rear camshaft cover-to-cylinder head seal.
12. Install the new camshaft cover gasket to the camshaft cover.
13. Install the camshaft cover.
14. Install the camshaft cover bolts.
15. Tighten the camshaft cover bolts to 71 inch lbs. (8 Nm)
16. Connect the ignition wires to the spark plugs.
17. Install the spark plug cover.
18. Tighten the spark plug cover bolts to 71 inch lbs. (8 Nm)
19. Install the spark plug cover bolts.
20. Connect the camshaft position sensor connector.
21. Connect all of the necessary vacuum lines.
22. Connect the breather tube to the camshaft cover.
23. Connect the negative battery cable.

VALVE LASH

ADJUSTMENT

Hydraulic valve lash adjusters are used to adjust the valve clearance to **0** lash automatically at all times. Adjustment is not required

ENGINE PERFORMANCE & EMISSION CONTROL

CAMSHAFT POSITION (CMP) SENSOR

LOCATION

See Figure 76.

The Camshaft Position (CMP) sensor is located in the cam cover behind the cam gear.

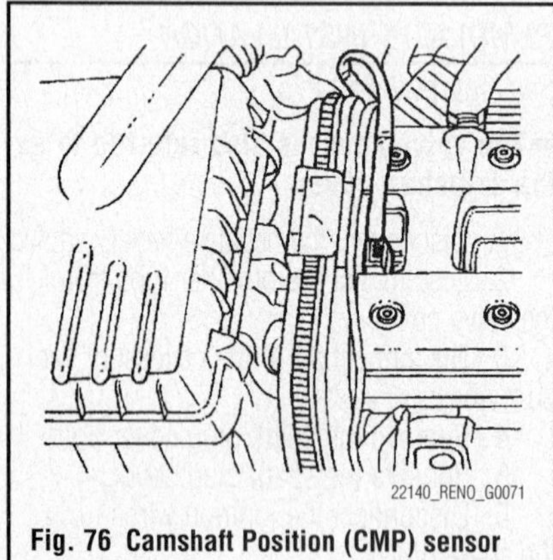

Fig. 76 Camshaft Position (CMP) sensor

OPERATION

The Camshaft Position (CMP) sensor sends a CMP sensor signal to the Engine Control Module (ECM). The ECM uses this signal as a "sync pulse" to trigger the injectors in the proper sequence. The ECM uses the CMP sensor signal to indicate the position of the #1 piston during its power stroke. This allows the ECM to calculate true sequential fuel injection mode of operation. If the ECM detects an incorrect CMP sensor signal while the engine is running, DTC P0341 will set. If the CMP sensor signal is lost while the engine is running, the fuel injection system will shift to a calculated sequential fuel injection mode based on the last fuel injection pulse, and the engine will continue to run. As long as the fault is present, the engine can be restarted. It will run in the calculated sequential mode with a 1-in-6 chance of the injector sequence being correct.

REMOVAL & INSTALLATION

See Figure 76.

1. Disconnect the negative battery cable.
2. Remove the engine cover.
3. Disconnect the sensor electrical connector.
4. Remove the timing belt front cover.
5. Remove the camshaft position sensor bolts.
6. Remove the camshaft position sensor from the top.

To install:

7. Install the camshaft position sensor and bolts.
8. Tighten the camshaft position bolts to71inch lbs. (8 Nm).

TESTING

See Figure 77.

➡An intermittent problem may be caused by a poor connection, rubbed-through wire insulation or a wire that is broken inside the insulation.

Any circuitry that is suspected as causing the complaint should be thoroughly checked for the following conditions:

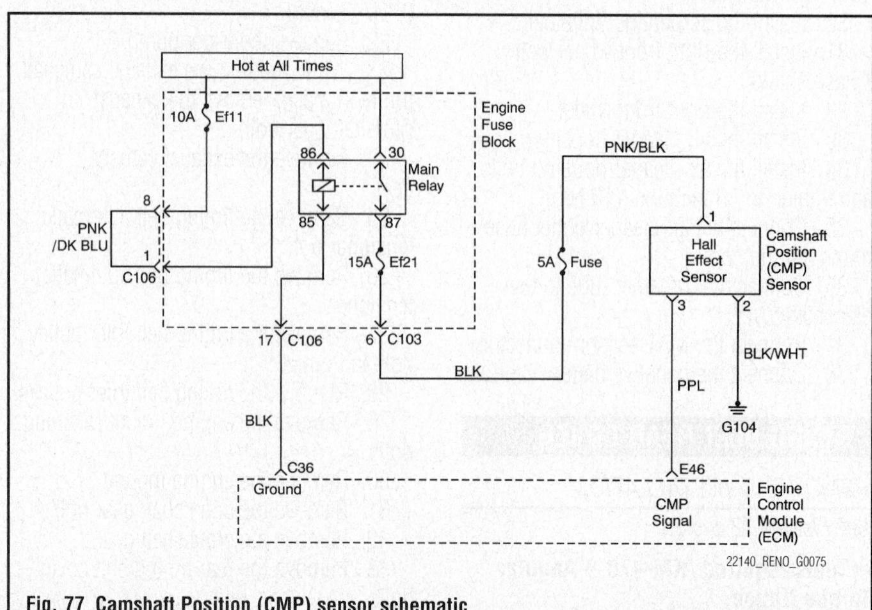

Fig. 77 Camshaft Position (CMP) sensor schematic

- Backed-out terminals
- Improper mating
- Broken locks
- Improperly formed
- Damaged terminals
- Poor terminal-to-wire connection
- Physical damage to the wiring harness

1. To test with a scan tool.
2. Turn the ignition OFF.
3. Install a scan tool to the Data Link Connector (DLC).
4. Idle the engine.
5. The Camshaft Position (CMP) Active Counter should be incrementally counting as the vehicle runs.

CRANKSHAFT POSITION (CKP) SENSOR

LOCATION

See Figure 78.

Refer to the accompanying illustration for Crankshaft Position (CKP) sensor location.

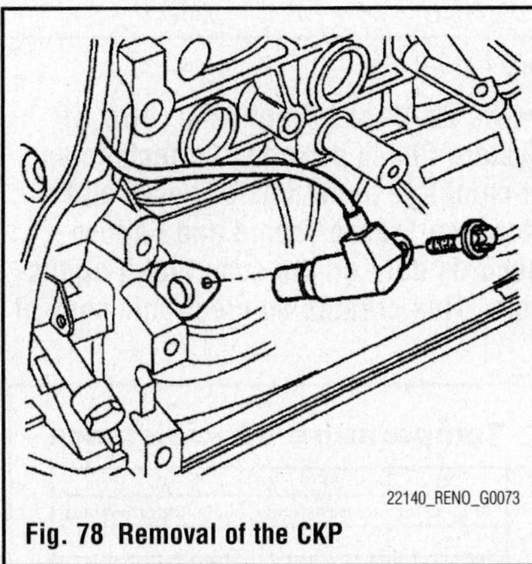

Fig. 78 Removal of the CKP

OPERATION

This direct ignition system uses a magnetic crankshaft position sensor. This sensor protrudes through its mount to within approximately 0.05 inches. (1.3 mm) of the crankshaft reluctor. The reluctor is a special wheel attached to the crankshaft or crankshaft pulley with 58 slots machined into it, 57 of which are equally spaced in 6° intervals. The last slot is wider and serves to generate a "sync pulse." As the crankshaft rotates, the slots in the reluctor change the magnetic field of the sensor, creating an induced voltage pulse. The longer pulse of the 58th slot identifies a specific orientation of the crankshaft and allows the Engine Control Module (ECM) to determine the crankshaft orientation at all times. The ECM uses this information to generate timed ignition and injection pulses that it sends to the ignition coils and to the fuel injectors.

REMOVAL & INSTALLATION

See Figures 78 and 79.

1. Disconnect the negative battery cable.
2. Remove the power steering pump.
3. Remove the A/C compressor.
4. Remove the rear A/C compressor mounting bracket bolts and the rear A/C compressor mounting bracket.

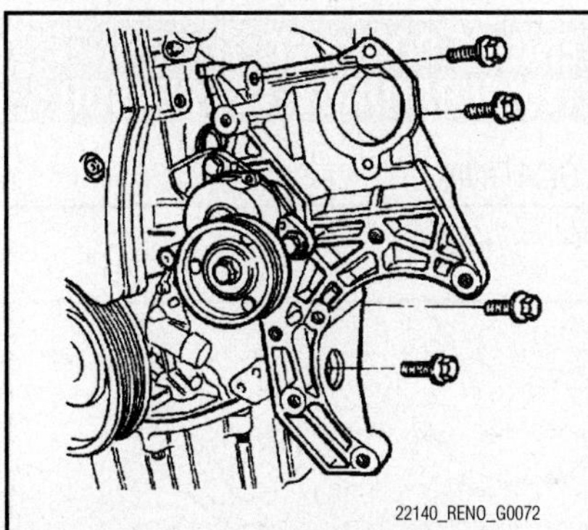

Fig. 79 Removal of the A/C bracket to access the CKP sensor

5. Remove the accessory mounting bracket by removing the bolts.
6. Disconnect the Crankshaft Position (CKP) sensor connector.
7. Remove the CKP sensor retaining bolt.
8. Gently rotate and remove the CKP sensor from the engine block.

To install:

9. Insert the CKP sensor into the engine block.
10. Install the CKP sensor retaining bolt.
11. Tighten the crankshaft position sensor retaining bolt to 71inch lbs. (8 Nm).
12. Connect the CKP sensor connector.
13. Install the accessory mounting bracket with the bolts.
14. Tighten the accessory mounting bracket bolts to 27 ft. lbs. (37 Nm).
15. Install the rear A/C mounting bracket.
16. Tighten the rear A/C mounting bracket bolts to 26 ft. lbs. (35 Nm).
17. Install the A/C compressor.
18. Install the power steering pump.
19. Connect the negative battery cable.

TESTING

1. Using a voltmeter, back probe the ECM connector terminal E36 and E37.
2. Observe the voltage while cranking the engine.
3. Voltage should be between 1.3–1.6volts.
4. If voltage is out of range replace Crankshaft Position (CKP) sensor.

EGR VALVE

REMOVAL & INSTALLATION

See Figure 80.

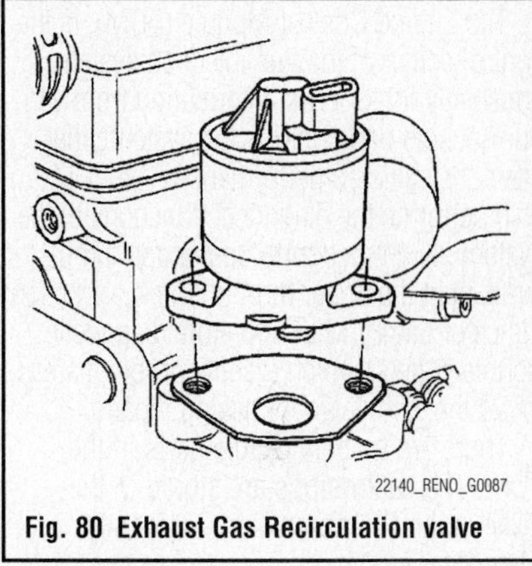

Fig. 80 Exhaust Gas Recirculation valve

1. Disconnect the negative battery cable.
2. Disconnect the vacuum hose from the Exhaust Gas Recirculation (EGR) valve.
3. Remove the EGR valve retaining bolts.
4. Remove the EGR valve from the Electronic Ignition (EI) system ignition coil adapter.

To install:

5. Clean the EI system ignition coil adapter mating surface.
6. Install a new EGR valve gasket.
7. Install the EGR valve with the bolts.
8. Tighten the exhaust gas recirculation valve retaining bolts to 22 ft. lbs. (30 Nm).
9. Connect the vacuum hose to the EGR valve.
10. Connect the negative battery cable.

ELECTRIC FAN SWITCH

OPERATION

The engine cooling fan circuit operates the main cooling fan and the auxiliary cooling fan. The cooling fans are controlled by the Engine Control Module (ECM) based on inputs from the Engine Coolant Temperature (ECT) sensor and the Air Conditioning Pressure (ACP) sensor.

REMOVAL & INSTALLATION

The cooling fans are controlled by the Engine Control Module (ECM)

ELECTRONIC CONTROL MODULE (ECM)

LOCATION

The Engine Control Module (ECM) is located inside the passenger kick-panel.

OPERATION

The Engine Control Module (ECM), is the control center of the fuel injection system. It constantly looks at the information from various sensors and controls the systems that affect the vehicle's performance. The ECM also performs the diagnostic functions of the system. It can recognize operational problems, alert the driver through the Malfunction Indicator Lamp (MIL) and store diagnostic trouble code(s) which identify problem areas to aid the technician in making repairs.

There are no serviceable parts in the ECM. The calibrations are stored in the ECM in the Programmable Read-Only Memory (PROM).

The ECM controls output circuits such as the fuel injectors, the idle air control motor, the A/C clutch relay, etc., by controlling the ground circuit through transistors or a device called a "quad-driver."

REMOVAL & INSTALLATION

➡If the Engine Control Module (ECM) is replaced with new one (VIN data is still not registered) or if an ECM is from a used vehicle (VIN data of other vehicle has been registered previously), make sure to re-register the VIN in the ECM by performing the VIN Registration so that VIN can be read by a scan tool.

1. Disconnect the negative battery cable.
2. Remove the air intake tube and resonator.
3. Disconnect the ECM connectors from the ECM.
4. Remove the engine control module.

To install:

5. Connect the ECM connectors to the ECM.
6. Align the ECM into the mounting base.
7. Snap the ECM into its mounting base.
8. Install the ECM trim locks.
9. Install the air intake tube and resonator.
10. Connect the negative battery cable.
11. Perform a crankshaft position system variation learning procedure.

TESTING

The Engine Control Module (ECM) supplies either 5 or 12 Volts to power the sensors or switches. This is done through resistances in the ECM which are so high in value that a test light will not come on when connected to the circuit. In some cases, even an ordinary shop voltmeter will not give an accurate reading because its resistance is too low. You must use a digital voltmeter with a 10 milliohm input impedance to get accurate voltage readings.

ENGINE COOLANT TEMPERATURE (ECT) SENSOR

LOCATION

See Figure 81.

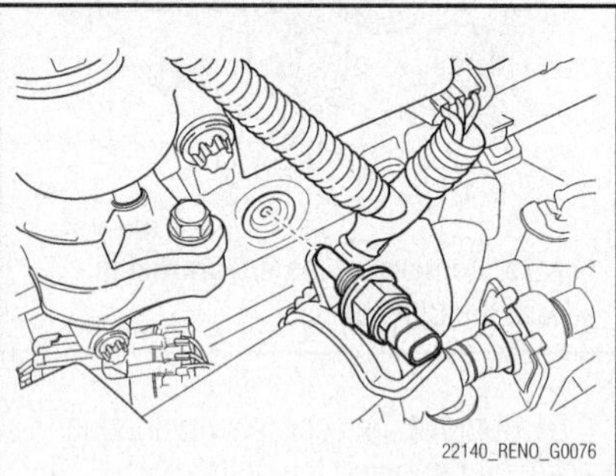

Fig. 81 Engine Coolant Temperature (ECT) sensor

The Engine Coolant Temperature (ECT) sensor is located in the cylinder head in the Electronic Ignition (EI) system ignition coil adapter under the Exhaust Gas Recirculation (EGR) valve.

OPERATION

The Engine Coolant Temperature (ECT) sensor is a thermistor (a resistor which changes value based on temperature) mounted in the engine coolant stream. Low coolant temperature produces a high resistance (100,000 ohms at –40°C [–40°F]) while high temperature causes low resistance (70 ohms at 130°C [266°F]).

The Engine Control Module (ECM) supplies 5 Volts to the ECT sensor through a resistor in the ECM and measures the change in voltage. The voltage will be high when the engine is cold, and low when the engine is hot. By measuring the change in voltage, the ECM can determine the coolant temperature. The engine coolant temperature affects most of the systems that the ECM controls. A failure in the ECT sensor circuit should set a diagnostic trouble code P0117 or P0118. Remember, these diagnostic trouble codes indicate a failure in the ECT sensor circuit, so proper use of the chart will lead either to repairing a wiring problem or to replacing the sensor to repair a problem properly.

REMOVAL & INSTALLATION

See Figure 81.

1. Relieve the coolant system pressure.
2. Disconnect the negative battery cable.
3. Disconnect the Engine Coolant Temperature (ECT) sensor connector.
4. Remove the ECT sensor from the Electronic Ignition (EI) system ignition coil adapter.

To install:

5. Coat the threads of the ECT sensor with sealer.
6. Install the ECT sensor into the EI system ignition coil adapter.
7. Tighten the engine coolant temperature sensor to 15 ft. lbs. (20 Nm).
8. Connect the ECT sensor connector.
9. Fill the coolant system.
10. Connect the negative battery cable.

TESTING

See Figure 82.

➡The On-Board Diagnostic (OBD II) System Check prompts the technician to complete some basic checks and store the Freeze Frame and Failure Records data on the scan tool if applicable. This creates an electronic copy of

Temperature VS Resistance

°C	°F	Ω
Temperature vs Resistance Values (Approximate)		
100	212	177
90	194	241
80	176	332
70	158	467
60	140	667
50	122	973
45	113	1,188
40	104	1,459
35	95	1,802
30	86	2,238
25	77	2,796
20	68	3,520
15	59	4,450
10	50	5,670
5	41	7,280
0	32	9,420
–5	23	12,300
–10	14	16,180
–15	5	21,450
–20	–4	28,680
–30	–22	52,700
–40	–40	100,700

22140_RENO_G0076

Fig. 82 Temperature vs. Resistance chart

the data taken when the malfunction occurred. The information is then stored on the scan tool for later reference.

1. The engine must be allowed to cool fully before the ECT and IAT sensor will read close to the ambient temperature in order to check for a possible skewed sensor.
2. Measure the engine coolant temperature with thermometer to determine the actual value the ECT sensor should be. Take into consideration if the engine has been run and the engine coolant has been warmed without opening the thermostat.
3. Disconnect the ECT sensor electrical connector.
4. Using a Digital Voltmeter (DVM), measure the resistance across the ECT sensor terminals.
5. Check the ECT sensor value to actual coolant temperature using the Temperature vs. Resistance table.
6. The ECT sensor should accurately reflect the actual engine coolant temperature.
7. Replace the sensor if it is out of specification.

FUEL LEVEL SENDING UNIT

LOCATION

The fuel level sender is located in the fuel pump canister in the fuel tank.

REMOVAL & INSTALLATION

See Figures 83 through 89.

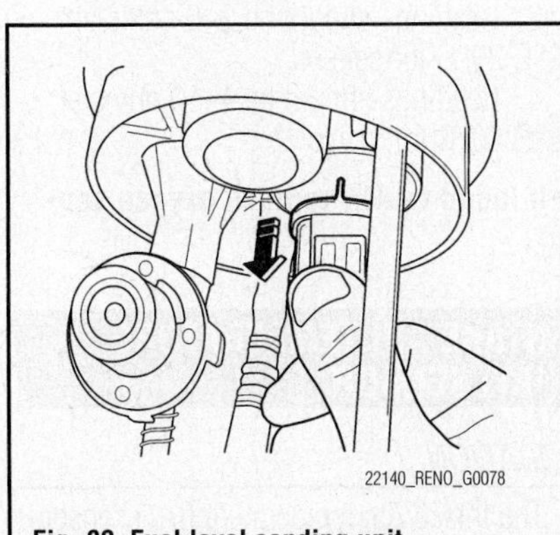

Fig. 83 Fuel level sending unit

1. Disconnect the insulator connector.
2. Push the terminal wedge (1) in the insulator connector.
3. Push the wedge (1) outside and then pull the wires to disconnect from insulator.
4. Remove the fuel-level sensor (1) from the sender-housing (2).
5. Remove the sender-housing (2).
6. Remove fuel sender assembly (3).

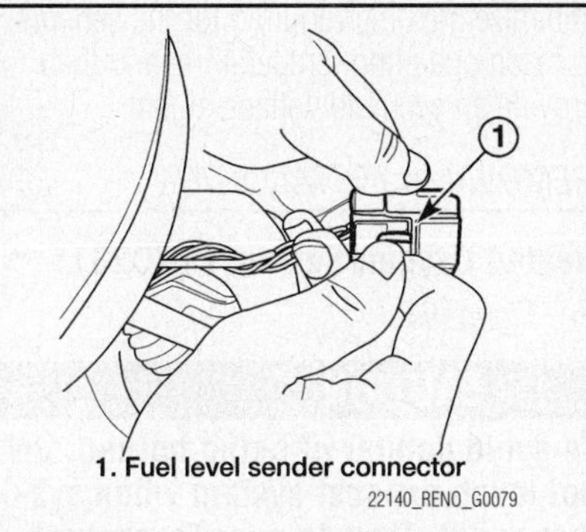

Fig. 84 Fuel sending unit connector assembly

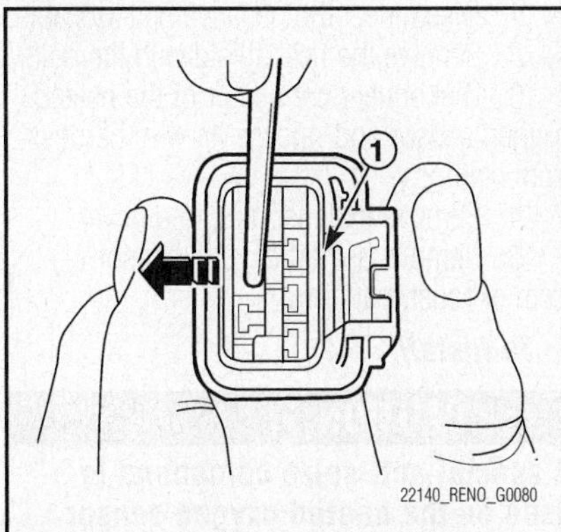

Fig. 85 Fuel sending unit wire removal

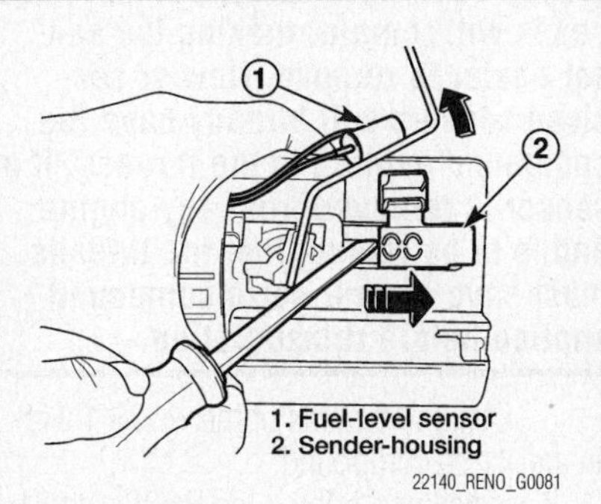

Fig. 86 Fuel sending unit level sensor removal

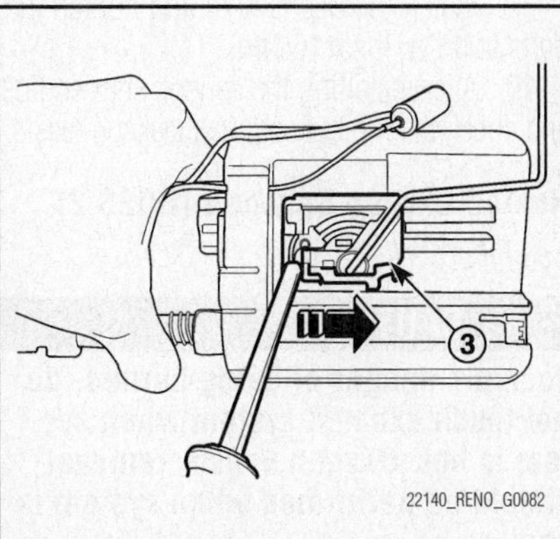

Fig. 87 Fuel sending unit removal

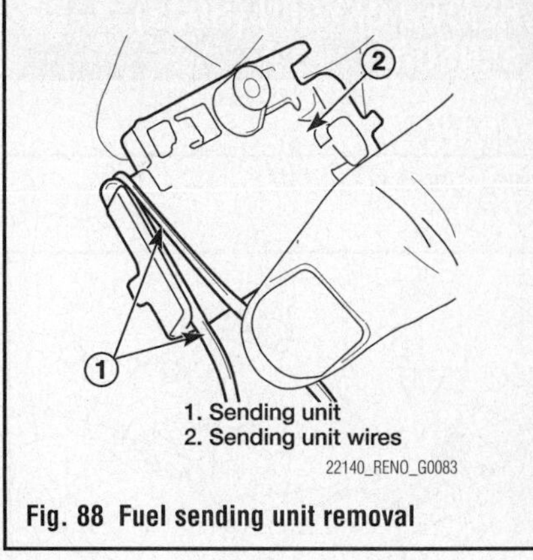

Fig. 88 Fuel sending unit removal

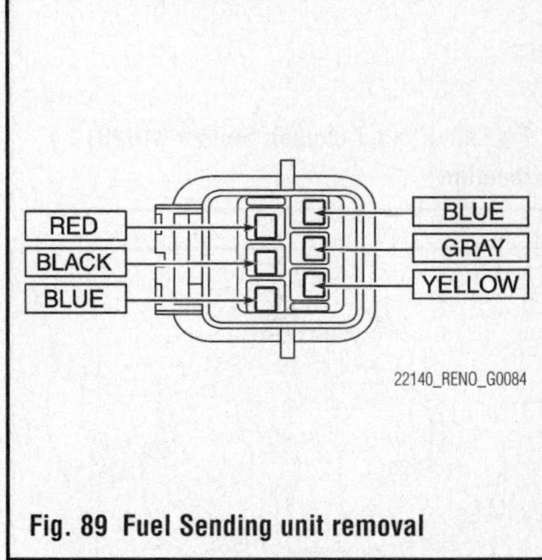

Fig. 89 Fuel Sending unit removal

To install:

7. Wind the wires (1) the sender assembly (2).
8. Install the sender assembly onto the fuel pump assembly (1).
9. Install the fuel-level sensor onto the sender-housing.
10. Connect the wire into the insulator connector
11. Connect the insulator connector.

G SENSOR

OPERATION

The Engine Control Module (ECM) receives rough road information from the G sensor. The ECM uses the rough road information to enable or disable the misfire diagnostic. The misfire diagnostic can be greatly affected by crankshaft speed variations caused by driving on rough road surfaces. The G sensor generates rough road information by producing a signal which is proportional to the movement of a small metal bar inside the sensor. If a fault occurs which causes the ECM to not receive rough road information between 50 and 132 km/h (30 and 80 mph), DTC P1391 will set.

HEATED OXYGEN (HO2S) SENSOR

LOCATION

See Figures 90 and 91.

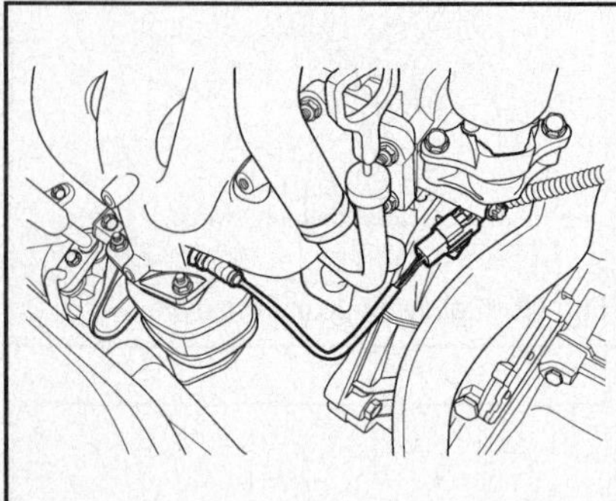
22140_RENO_G0085

Fig. 90 Heated Oxygen Sensor (HO2S) 1 location

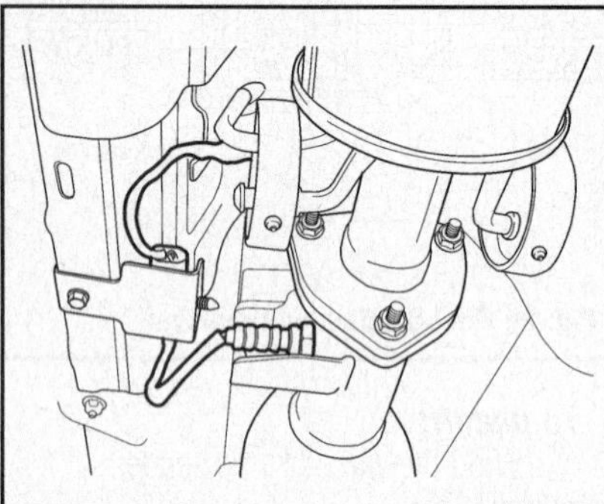
22140_RENO_G0086

Fig. 91 Heated Oxygen Sensor 2 (HO2S 2) location

1. Heated Oxygen Sensor (HO2S) 1 is located in the bottom end of the exhaust manifold. Heated Oxygen Sensor 2 (HO2S 2) is located in the rear of the catalytic converter.

OPERATION

Heated Oxygen Sensors (HO2S) are used for fuel control and post catalyst monitoring. Each HO2S compares the oxygen content of the surrounding air with the oxygen content of the exhaust stream. When the vehicle is first started, the Engine Control Module (ECM) operates in an open loop mode, ignoring the HO2S signal voltage when calculating the air fuel ratio. The ECM supplies the HO2S with a reference voltage of about 0.45 Volts. The HO2S generates a voltage within a range of 0 –1 volts that fluctuates above and below bias voltage once in closed loop. A high HO2S voltage output indicates a rich fuel mixture. A low HO2S voltage output indicates a lean mixture. Heating elements inside the HO2S minimize the time required for the sensors to reach operating temperature, and then provide an accurate voltage signal.

REMOVAL & INSTALLATION

Heated Oxygen Sensor 1 (HO2S1)

See Figure 90.

**** CAUTION**

To avoid danger of being burned, do not touch exhaust system when system is hot. Oxygen sensor removal should be performed when system is cool.

1. Disconnect the negative battery cable.
2. Remove the heat shield with the bolts.
3. Disconnect connector of the heated oxygen sensor and release its wire harness from clamps.
4. Remove exhaust manifold cover.
5. Remove heated oxygen sensor (1) from exhaust manifold.

To install:

**** CAUTION**

A special anti-seize compound is used on the heated oxygen sensor threads. This compound consists of a liquid graphite and glass beads. The graphite will burn away, but the glass beads will remain, making the sensor easier to remove. New or serviced sensors will already have the compound applied to the threads. If a sensor is removed from any engine and is to be reinstalled, the threads must have an anti-seize compound applied before reinstallation.

6. Coat the threads of the HO2S 1 with an anti-seize compound.
7. Install and tighten the HO2S 1 to 31 ft. lbs. (42 Nm).
8. Connect the connector of the HO2S 1and clamp wire harness securely.
9. Install the heat shield and tighten the bolts to 11 ft. lbs. (15 Nm).
10. After installing the sensor, start engine and check that no exhaust gas leakage exists.

Heated Oxygen Sensor 2 (HO2S 2)

See Figure 91.

**** CAUTION**

To avoid danger of being burned, do not touch exhaust system when system is hot. Oxygen sensor removal should be performed when system is cool.

1. Disconnect the negative battery cable.
2. Disconnect connector of the HO2S 2 and release its wire harness from clamps.
3. Remove the HO2S 2 from exhaust pipe.

To install:

4. Install and tighten the HO2S 2 to 31 ft. lbs. (42 Nm).
5. Attach the connector of the HO2S 2 and clamp wire harness securely.
6. Connect the negative battery cable.

TESTING

1. Start the engine.
2. Allow the engine to reach operating temperature.
3. Operate the engine at 1.500 rpm for 30 seconds.
4. While observing the HO2S voltage parameter with a scan tool, quickly cycle the throttle from closed throttle to wide open throttle (WOT), 3 times.
5. The HO2S voltage parameter should change more than 200 mV.
6. If the voltage does not change more that the specified value, the condition is present to set a Diagnostic Trouble Code (DTC).
7. Observe the Freeze Frame/Failure Records for this DTC.

Heater Testing

1. Disconnect the sensor connector.
2. Using ohmmeter, measure resistance between terminals "V_B" and "GND" of sensor connector.
3. Readings should be 2–6 ohms at 68°F, 20°C for sensor.
4. Readings should be 4–10 ohms at 68°F, 20°C for sensor 2.

➡If found faulty, replace oxygen sensor.

INTAKE AIR TEMPERATURE (IAT) SENSOR

LOCATION

The Intake Air Temperature (IAT) sensor is located in the air intake tube on the top of the engine.

OPERATION

The Intake Air Temperature (IAT) sensor is a thermistor, a resistor which changes value based on the temperature of the air entering the engine. Low temperature produces a high resistance (4,500 ohms at –40 °C [–40 °F]), while high temperature causes a low resistance (70 ohms at 130°C [266 °F]).

The Engine Control Module (ECM) provides 5 volts to the IAT sensor through a resistor in the ECM and measures the change in voltage to determine the IAT. The voltage will be high when the manifold air is cold and low when the air is hot. The ECM knows the intake IAT by measuring the voltage.

The IAT sensor is also used to control spark timing when the manifold air is cold.

A failure in the IAT sensor circuit sets a diagnostic trouble code P0112 or P0113.

REMOVAL & INSTALLATION

See Figure 92.

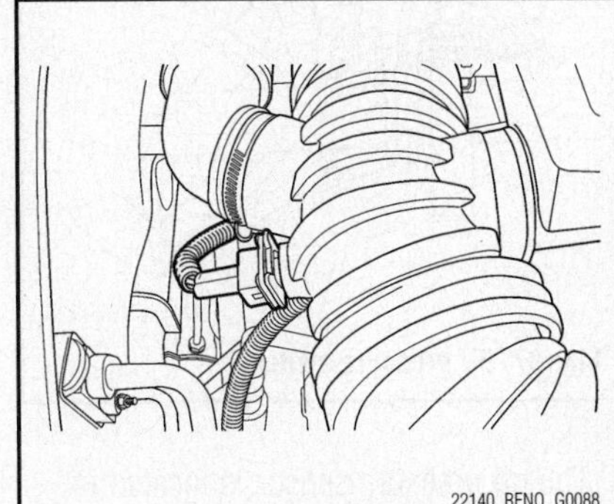

22140_RENO_G0088

Fig. 92 Intake Air Temperature (IAT) sensor

1. Disconnect the negative battery cable.
2. Disconnect the Intake Air Temperature (IAT) sensor connector.
3. Remove the IAT sensor by pulling it out of the air intake tube.
4. Install the IAT sensor into the air intake tube.
5. Connect the IAT connector.
6. Connect the negative battery cable.

TESTING

See Figure 93.

1. Observe the IAT sensor scan tool display continuously while performing the tests below.
2. Disconnect the IAT sensor connector.
3. Remove the IAT sensor from the air filter assembly.
4. Connect the IAT sensor connector to the IAT sensor.
5. Wrap a cool damp shop towel around the IAT sensor in order to lower the temperature while monitoring the IAT sensor parameter on the scan tool
6. A (high resistance value) should be observed, use Temperature vs. Resistance chart to calculate proper values.

Temperature VS Resistance

°C	°F	Ω
Temperature vs Resistance Values (Approximate)		
100	212	177
90	194	241
80	176	332
70	158	467
60	140	667
50	122	973
45	113	1,188
40	104	1,459
35	95	1,802
30	86	2,238
25	77	2,796
20	68	3,520
15	59	4,450
10	50	5,670
5	41	7,280
0	32	9,420
–5	23	12,300
–10	14	16,180
–15	5	21,450
–20	–4	28,680
–30	–22	52,700
–40	–40	100,700

22140_RENO_G0077

Fig. 93 Temperature vs. resistance chart

7. If values do not change replace the IAT sensor.
8. Remove the cool shop towel and place a warm damp shop towel around the IAT sensor while monitoring the IAT sensor parameter on the scan tool (low resistance value).
9. A (low resistance value) should be observed,. use Temperature vs. Resistance chart to calculate proper values.
10. If values do not change replace the IAT sensor.

KNOCK SENSOR (KS)

LOCATION

The Knock Sensor (KS) is mounted in the engine block between the cylinders.

OPERATION

The Knock Sensor (KS) detects abnormal knocking in the engine. The sensor is mounted in the engine block near the cylinders. The sensor produces an AC output voltage which increases with the severity of the knock. This signal is sent to the Engine Control Module (ECM). The ECM then adjusts the ignition timing to reduce the spark knock.

TESTING

➡If the engine has an internal knock or audible noise that causes a knocking type noise on the engine block, the Knock Sensor (KS) may be responding to the noise.

➡The replacement ECM must be reprogrammed. Refer to the latest Techline procedure for ECM reprogramming.

1. The KS produces an AC signal so that under a no-knock condition the signal on the KS circuit measures about 0.007 Volts AC. The KS signal's amplitude and frequency depend upon the amount of knock being experienced.
2. Conditions for Setting the DTC.
3. Vacuum is less than the predetermined value (10–50 kPa, based on RPM).
4. The RPM is less than 1,600.

MALFUNCTION INDICATOR LIGHT (MIL)

RESET PROCEDURES

When the ECM sets a Diagnostic Trouble Code (DTC), the Malfunction Indicator Lamp (MIL) lamp will be turned on only for type A, B and E but a DTC will be stored in the ECM's memory for all types of DTC. If the problem is intermittent, the MIL will go out after 10 sec. if the fault is no longer present. The DTC will stay in the ECM's memory until cleared by scan tool. Removing battery voltage for 10 sec. will clear some stored DTCs.

➡DTCs should be cleared after repairs have been completed. Some diagnostic tables will tell you to clear the codes before using the chart. This allows the ECM to set the DTC while going through the chart, which will help to find the cause of the problem more quickly.

MANIFOLD ABSOLUTE PRESSURE (MAP) SENSOR

LOCATION

See Figure 94.

Refer to the accompanying illustration for Manifold Absolute Pressure (MAP) sensor location.

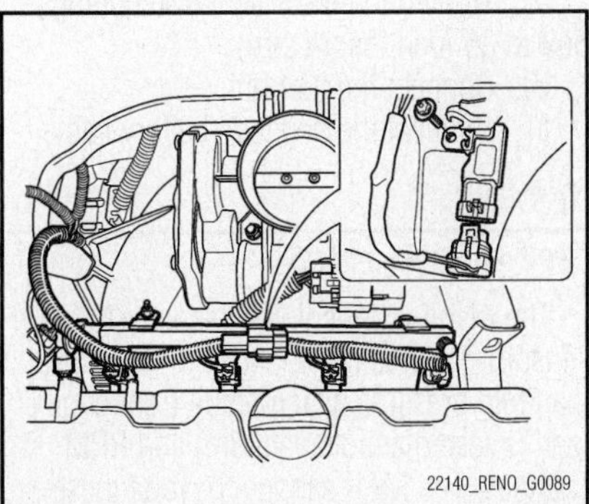

22140_RENO_G0089

Fig. 94 Manifold Absolute Pressure (MAP) location

OPERATION

The Manifold Absolute Pressure (MAP) sensor measures the changes in the intake manifold pressure which result from engine load and speed changes. It converts these to a voltage output.

A closed throttle on engine coast down produces a relatively low MAP output. MAP is the opposite of vacuum. When manifold pressure is high, vacuum is low. The MAP sensor is also used to measure barometric pressure. This is performed as part of MAP sensor calculations. With the ignition ON and the engine not running, the Engine Control Module (ECM) will read the manifold pressure as barometric pressure and adjust the air/fuel ratio accordingly. This compensation for altitude allows the system to maintain driving performance while holding emissions low. The barometric function will update periodically during steady driving or under a wide open throttle condition. In the case of a fault in the barometric portion of the MAP sensor, the ECM will set to the default value.

➡A failure in the MAP sensor circuit sets a diagnostic trouble code P0107 or P0108.

REMOVAL & INSTALLATION

See Figure 94.

1. Disconnect the negative battery cable.
2. Disconnect the fuel rail.
3. Disconnect the Manifold Absolute Pressure (MAP) sensor connector from the MAP sensor.
4. Remove the MAP sensor retaining bolt.
5. Remove the MAP sensor from the intake manifold.

To install:

6. Connect the MAP sensor connector to the MAP sensor.
7. Install the MAP sensor into the intake manifold.
8. Install the MAP sensor retaining bolt.
9. Tighten the MAP sensor retaining bolt to 35 inch lbs. (4 Nm).
10. Connect the fuel rail
11. Connect the negative battery cable.

TESTING

See Figures 95 and 96.

The Manifold Absolute Pressure (MAP) sensor measures the changes in the intake manifold pressure which result from engine load (intake manifold vacuum) and RPM changes. The MAP sensor converts these changes into a voltage output. The Engine Control Module (ECM) sends a 5 volt reference voltage to the MAP sensor. As the intake manifold pressure changes, the output voltage of the MAP sensor also changes. A low voltage (high vacuum) output of 1–2 Volts is present at idle. A high voltage (low vacuum) output of 4.0–4.8 Volts is present at wide open throttle. The MAP sensor is also used under certain conditions to measure barometric pressure. This allows the ECM to make adjustments for altitude changes. The ECM uses the MAP sensor for fuel delivery and ignition timing changes.

1. Using a vacuum gauge check as the intake manifold pressure changes, the output voltage of the MAP sensor also changes.

A low voltage (high vacuum) output of 1–2 Volts is present at idle. A high voltage (low vacuum) output of 4.0–4.8 Volts is present at wide open throttle.

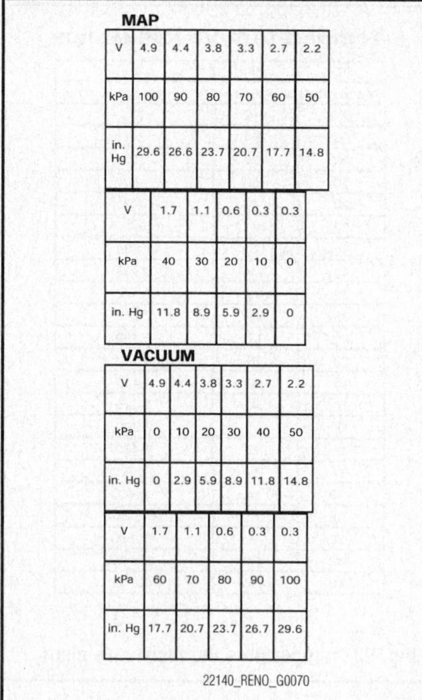

MAP

V	4.9	4.4	3.8	3.3	2.7	2.2
kPa	100	90	80	70	60	50
in. Hg	29.6	26.6	23.7	20.7	17.7	14.8

V	1.7	1.1	0.6	0.3	0.3
kPa	40	30	20	10	0
in. Hg	11.8	8.9	5.9	2.9	0

VACUUM

V	4.9	4.4	3.8	3.3	2.7	2.2
kPa	0	10	20	30	40	50
in. Hg	0	2.9	5.9	8.9	11.8	14.8

V	1.7	1.1	0.6	0.3	0.3
kPa	60	70	80	90	100
in. Hg	17.7	20.7	23.7	26.7	29.6

22140_RENO_G0070

Fig. 95 Manifold Absolute Pressure (MAP) sensor testing

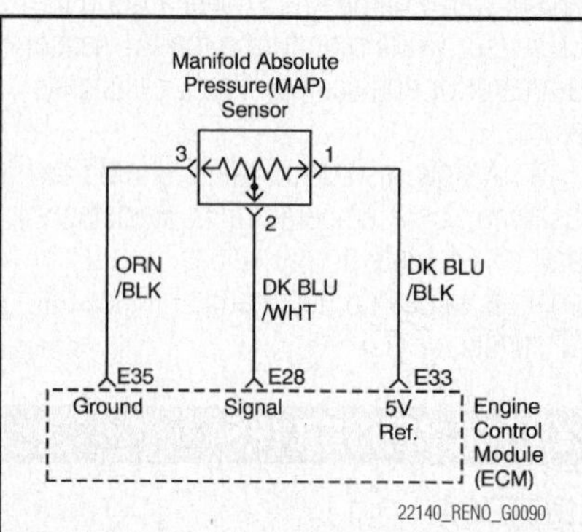

Fig. 96 Manifold Absolute Pressure (MAP) Schematic

2. If sensor is out of range check wiring for an open or short.
3. Check for 5V reference from the ECM.
4. Replace sensor if readings are not within specifications.

OIL PRESSURE SENSOR

LOCATION

See Figure 97.

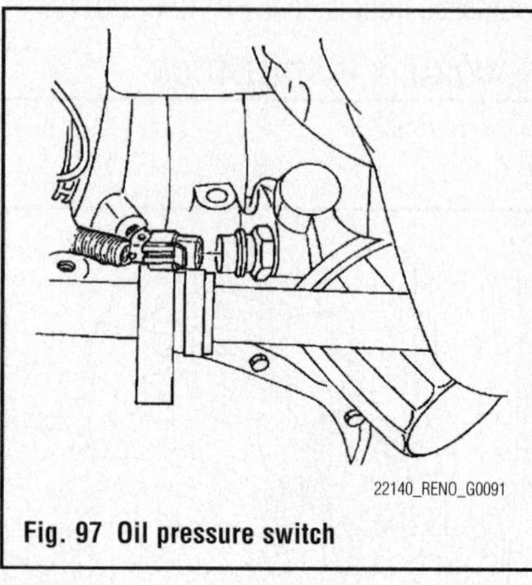

Fig. 97 Oil pressure switch

The oil pressure sensor is located in the front of the engine block behind the right front wheel well oil pan scraper.

REMOVAL & INSTALLATION

See Figure 97.

1. Remove the right front wheel well oil pan scraper.
2. Disconnect the oil pressure switch connector.
3. Remove the oil pressure switch

To install:

4. Install the oil pressure switch.
5. Tighten the oil pressure switch to 30 ft. lbs. (40 Nm).
6. Connect the electrical connector to the oil pressure switch.
7. Install the right front wheel well oil pan scraper.
8. Check the oil level.

TESTING

See Figures 97 and 98.

1. Remove the right front wheel well oil pan scraper.
2. Disconnect the oil pressure switch connector.
3. Remove the oil pressure switch
4. Install the adapter KM-135 in place of the oil pressure switch.
5. Connect the pressure gauge KM-498-B to the adapter.

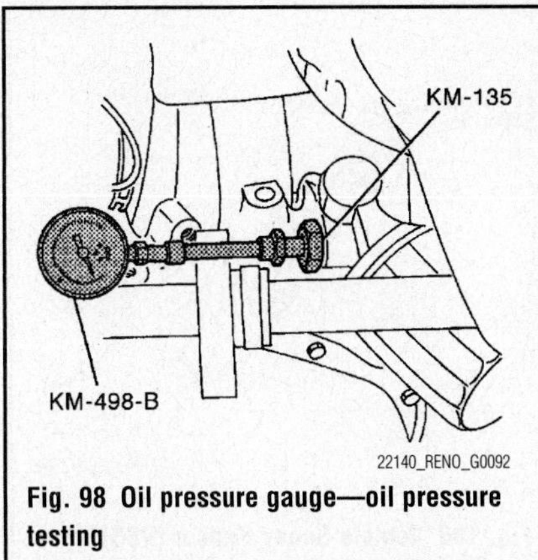

Fig. 98 Oil pressure gauge—oil pressure testing

6. Start the engine and check the oil pressure at idle speed and engine temperature of 176 °F (80°C).

➡The minimum oil pressure should be 4.35 psi.

7. Stop the engine and remove the pressure gauge KM-498-B and the adapter KM-135.
8. Install the oil pressure switch.
9. Tighten the oil pressure switch to 30 ft. lbs. (40 Nm).
10. Connect the electrical connector to the oil pressure switch.
11. Install the right front wheel well oil pan scraper.
12. Check the oil level.

THROTTLE ACTUATOR CONTROL (TAC) SYSTEM

LOCATION

The Throttle Actuator Control (TAC) is located on the throttle housing along with the TP sensor.

OPERATION

The Throttle Actuator Control (TAC) system is used to improve emissions, fuel economy and drivability. The TAC system eliminates the mechanical link between the accelerator pedal and the throttle plate. The TAC system eliminates the need for a cruise control module and idle air control motor. The following is a list of TAC system components:

- The accelerator pedal assembly includes the following components:
- The accelerator pedal
- The Accelerator Pedal Position (APP) sensor
- The APP sensor 2

The throttle body assembly includes the following components

- The Throttle Position (TP) sensor 1
- The TP sensor 2
- The throttle actuator motor
- The throttle plate
- The Engine Control Module (ECM)

The ECM monitors the driver demand for acceleration with 2 APP sensors. The APP sensor 1 signal voltage range is from about 0.7–4.5 Volts as the accelerator pedal is moved from the rest pedal position to the full pedal travel position. The APP sensor 2 range is from about 0.3–2.2 Volts as the accelerator pedal is moved from the rest pedal position to the full pedal travel position. The ECM processes this information along with other sensor inputs to command the throttle plate to a certain position.

The throttle plate is controlled with a direct current motor called a throttle actuator control motor. The ECM can move this motor in the forward or reverse direction by controlling battery voltage and/or ground to 2 internal drivers. The throttle plate is held at a 5.7° TPS rest position using a constant force return spring. This spring holds the throttle plate to the rest position when there is no current flowing to the actuator motor.

The ECM monitors the throttle plate angle with 2 TP sensors. The TP sensor 1 signal voltage range is from about 0.7–4.3 Volts as the throttle plate is moved from 0% to Wide Open Throttle (WOT). The TP sensor 2 voltage range is from about 4.3–0.7 Volts as the throttle plate is moved from 0% to WOT.

The ECM performs diagnostics that monitor the voltage levels of both APP sensors, both TP sensors and the throttle actuator control motor circuit. It also monitors the spring return rate of both return springs that the housed are internal to the throttle body assembly. These diagnostics are performed at different times based on whether the engine is running or not running.

Every ignition cycle, the ECM performs a quick throttle return spring test to make sure the throttle plate can return to the 7% rest position from the 0% position. This is to ensure that the throttle plate can be brought to the rest position in case of an actuator motor circuit failure.

THROTTLE POSITION (TP) SENSOR

LOCATION

The Throttle Position (TP) sensor is located in the intake duct attached to the throttle housing.

OPERATION

The Throttle Position (TP) sensor is a potentiometer connected to the throttle shaft of the throttle body. The TP sensor electrical circuit consists of a 5 Volts supply line and a ground line, both provided by the Engine Control Module (ECM). The ECM calculates the throttle position by monitoring the voltage on this signal line. The TP sensor output changes as the accelerator pedal is moved, changing the throttle valve angle. At a closed throttle position, the output of the TP sensor is low, about 0.5 Volts. As the throttle valve opens, the output increases so that, at Wide Open Throttle (WOT), the output voltage will be about 5 Volts. The ECM can determine fuel delivery based on throttle valve angle (driver demand). A broken or loose TP sensor can cause intermittent bursts of fuel from the injector and an unstable idle, because the ECM thinks the throttle is moving. A problem in any of the TP sensor circuits should set a Diagnostic Trouble Code (DTC) P0121 or P0122. Once the DTC is set, the ECM will substitute a default value for the TP sensor and some vehicle performance will return. A DTC P0121 will cause a high idle speed.

TESTING

See Figure 99.

The Throttle Position (TP) sensor incorporates 2 TP sensors into one housing. TP sensor 1 and TP sensor 2 each have a 5 volt reference circuit supplied by the ECM. The ECM supplies each TP sensor with a low reference circuit. Each TP sensor supplies the ECM with a signal voltage that is proportional to the throttle plate position. The TP signal voltages are opposite from one another. TP sensor 1 is pulled up to reference voltage as the throttle plate is opened. The TP sensor 2 is pulled down to low reference as the throttle plate is opened.

➡The Throttle Position (TP) sensors are part of the Throttle Actuator Control (TAC) system.

1. Scan tool readings for the TP Sensor should read between 0.36– 0.96 Volts at idle to above 4 Volts at WOT.
2. If sensor is out of range check wiring for an open or short.
3. Check for 5Volts reference from the ECM.
4. Replace sensor if readings are not within specifications.

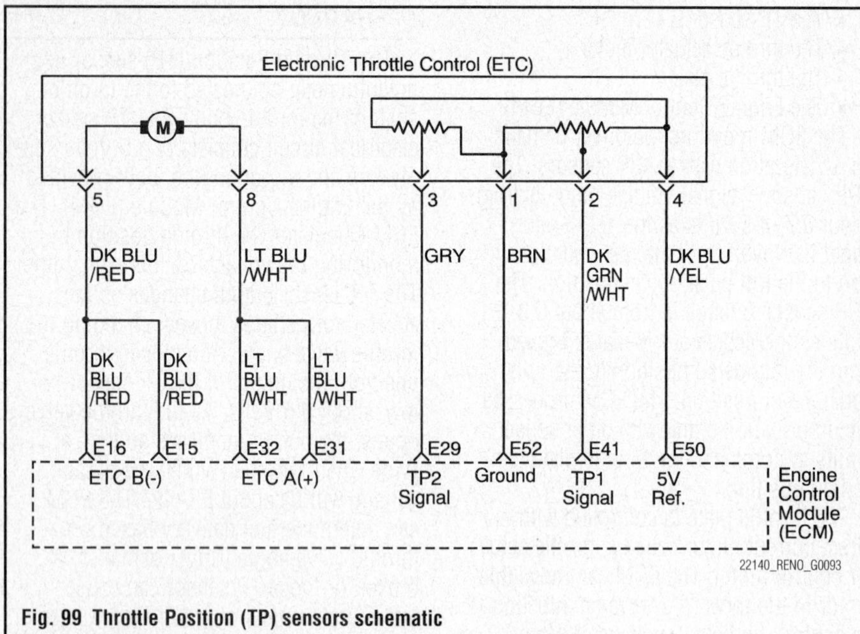

Fig. 99 Throttle Position (TP) sensors schematic

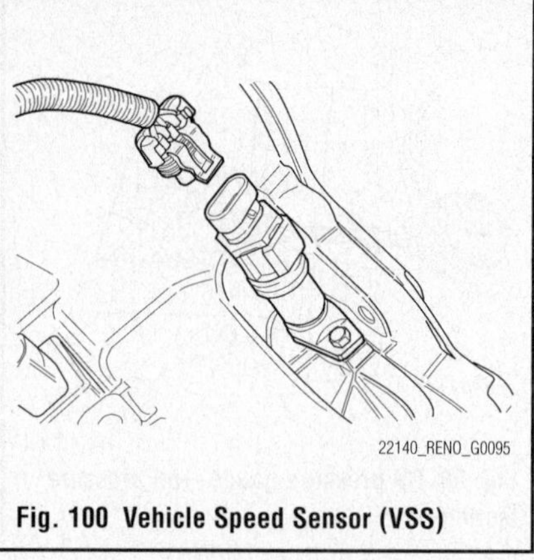

Fig. 100 Vehicle Speed Sensor (VSS)

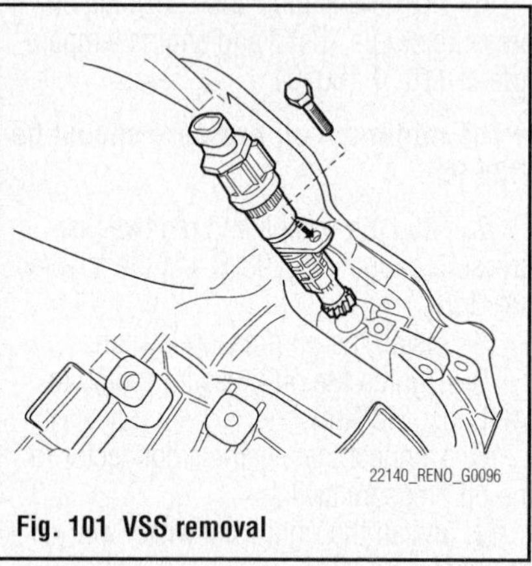

Fig. 101 VSS removal

VEHICLE SPEED SENSOR (VSS)

LOCATION

The Vehicle Speed Sensor (VSS) is a permanent magnet generator that is mounted in the transaxle.

OPERATION

Vehicle speed information is provided to the Engine Control Module (ECM) by the Vehicle Speed Sensor (VSS). The VSS is a permanent magnet generator that is mounted in the transaxle and produces a pulsing voltage whenever vehicle speed is over 3 mph (5 km/h). The Alternating Current (AC) voltage level and the number of pulses increases with vehicle speed. The ECM converts the pulsing voltage into mph (km/h) and then supplies the necessary signal to the instrument panel for speedometer/odometer operation and to the cruise control module and multi-function alarm module operation. This Diagnostic Trouble Code (DTC) will detect if vehicle speed is reasonable according to engine RPM and load. The permanent magnet generator only produces a signal if the drive wheels are turning greater than 5 mph (8 km/h).

REMOVAL & INSTALLATION

See Figures 100 and 101.

1. Disconnect the speedometer speed sensor electrical connector.
2. Remove the speedometer housing retaining bolt.
3. Remove the speedometer-driven gear and the speedometer housing.

To install:

4. Coat the O-ring with petroleum jelly.
5. Install the speedometer-driven gear and the speedometer housing.
6. Install the speedometer housing retaining bolt.
7. Tighten the speedometer housing retaining bolt to 35 inch lbs. (4 Nm).
8. Connect the speedometer speed sensor electrical connector.

TESTING

See Figure 102.

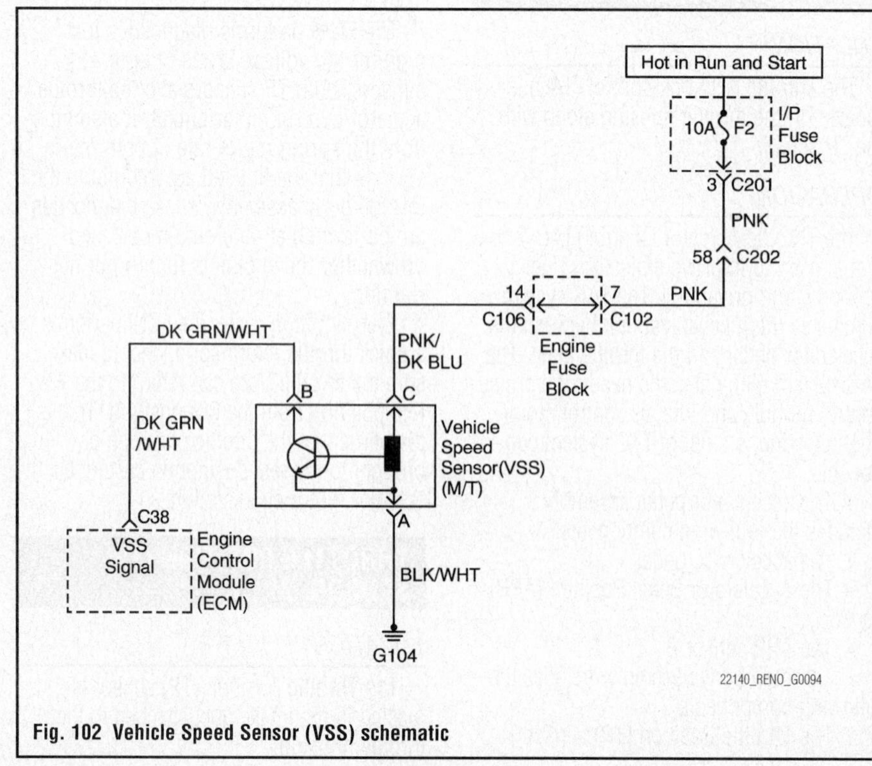

Fig. 102 Vehicle Speed Sensor (VSS) schematic

The permanent magnet generator only produces a signal if the drive wheels are turning greater than 5 mph (8 km/h).

➡**Proper engine loads cannot be achieved in a shop environment to properly run the vehicle within the Freeze Frame Data conditions. It will be necessary to drive the vehicle on the road to obtain the proper engine loads. This step verifies that the ECM is receiving a signal from the vehicle speed sensor.**

1. A resistance reading that is higher than the specified value indicates that the VSS circuitry is open.
2. With the Scan Tool, check that the displayed resistance is less than the 1300 ohms, if so, the VSS high and low circuits are shorted together.
3. This checks the resistance of the VSS if no opens or shorts were found on the VSS high and low circuit.

Power Test with Scan Tool:

- The RPM is between 2,500 and 4,000 rpm
- Throttle Position (TP) sensor is between 25 and 60%
- Manifold Absolute Pressure (MAP) is greater than 8.7 psi (60 kPa)

Deceleration Test with Scan Tool.

- Manifold Absolute Pressure (MAP) is less than 4.4 psi (30 kPa)
- Change in RPM/cycle is greater than 50 rpm/ cycle
- Throttle Position (TP) sensor is less than 0.8%
- The RPM is between 1,800–6,000 rpm

FUEL

GASOLINE FUEL INJECTION SYSTEM

FUEL SYSTEM SERVICE PRECAUTIONS

Safety is the most important factor when performing not only fuel system maintenance but any type of maintenance. Failure to conduct maintenance and repairs in a safe manner may result in serious personal injury or death. Maintenance and testing of the vehicle's fuel system components can be accomplished safely and effectively by adhering to the following rules and guidelines.

**** CAUTION**

To avoid the possibility of fire and personal injury, always disconnect the negative battery cable unless the repair or test procedure requires that battery voltage be applied.

**** WARNING**

Always relieve the fuel system pressure prior to disconnecting any fuel system component (injector, fuel rail, pressure regulator, etc.), fitting or fuel line connection. Exercise extreme caution whenever relieving fuel system pressure to avoid exposing skin, face and eyes to fuel spray. Please be advised that fuel under pressure may penetrate the skin or any part of the body that it contacts.

**** CAUTION**

Always place a shop towel or cloth around the fitting or connection prior to loosening to absorb any excess fuel due to spillage. Ensure that all fuel spillage (should it occur) is quickly removed from engine surfaces. Ensure that all fuel soaked cloths or towels are deposited into a suitable waste container.

**** CAUTION**

Always keep a dry chemical (Class B) fire extinguisher near the work area.

**** WARNING**

Do not allow fuel spray or fuel vapors to come into contact with a spark or open flame.

**** CAUTION**

Always use a back-up wrench when loosening and tightening fuel line connection fittings. This will prevent unnecessary stress and torsion to fuel line piping.

**** CAUTION**

Always replace worn fuel fitting O-rings with new. Do not substitute fuel hose or equivalent where fuel pipe is installed.

**** CAUTION**

Before servicing the vehicle, make sure to also refer to the precautions in the beginning of this section as well.

RELIEVING FUEL SYSTEM PRESSURE

**** CAUTION**

Care should be used when working around the fuel system. DO NOT smoke or expose the fuel system to any open flames. Keep a fire extinguisher handy.

1. Before servicing the vehicle, refer to the Precautions Section.
2. Place the vehicle in **PARK** for automatic transaxle or **NEUTRAL** for manual transaxle.
3. Disconnect the fuel pump fuse from the engine fuse block.
4. Remove the fuel filler cap from the filler neck to release the fuel vapor pressure in the fuel tank.
5. Start the vehicle and allow the engine to run until it stalls.
6. Crank the engine for an additional 10 seconds to eliminate any remaining pressure in the fuel lines.
7. Disconnect the negative battery cable.
8. Connect the fuel pump fuse to the engine fuse block.
9. After servicing the fuel system, connect the negative battery cable.
10. Start the engine and check for leaks in the system.

FUEL FILTER

REMOVAL & INSTALLATION

The fuel filter for Forenza and Reno models is an integral component of the in-tank fuel pump assembly. Refer to the Fuel Pump Removal procedure later in this section.

FUEL INJECTORS

REMOVAL & INSTALLATION

See Figure 103.

1. Before servicing the vehicle, refer to the Precautions Section.
2. Relieve the pressure from the fuel system.
3. Disconnect the negative battery cable.
4. Remove or disconnect the following:
    - Intake Air Temperature (IAT) sensor connector
    - Breather hose from cylinder head cover

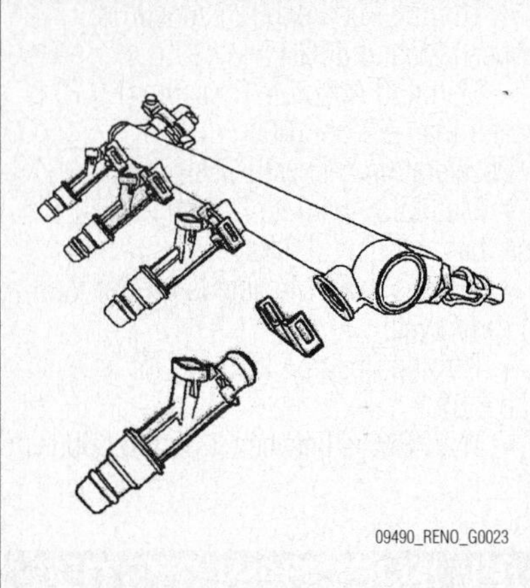

Fig. 103 After removing the retaining clip, pull the injector down and out

- Positive Crankcase Ventilation (PCV) hose
- Throttle cable
- Fuel injector electrical connections
- Fuel feed line at the fuel rail
- Fuel rail retaining bolts
- Fuel rail with injectors attached
- Fuel injector retaining clips
- Fuel injectors from rail by pulling down and out
- Injector O-ring and discard

To install:

➡Different injectors are calibrated for different flow rates. When ordering new fuel injectors, be certain to order the identical part number that is inscribed on the old injector.

5. Lubricate the new injector O-rings with engine oil.
6. Injector to the fuel rail with terminals facing outward, and secure with retaining clip. Be sure that the clips are parallel to the injector harness connector.
7. Install or connect the following:
   - New injector O-rings
   - Fuel injector and fuel rail assembly into the cylinder head
   - Fuel rail retaining bolts and tighten to 18 ft. lbs. (25 Nm)
   - Fuel feed line to the fuel rail
   - Fuel injector electrical connections. Rotate each injector as required.
   - Throttle cable to the throttle body and bracket
   - PCV hose to the cylinder head cover
   - Breather hose to the cylinder head cover
   - IAT sensor connector
8. Connect the negative battery cable.
9. With the ignition **ON** and the engine **OFF** check for leaks.

FUEL PUMP

REMOVAL & INSTALLATION

See Figures 104 and 105.

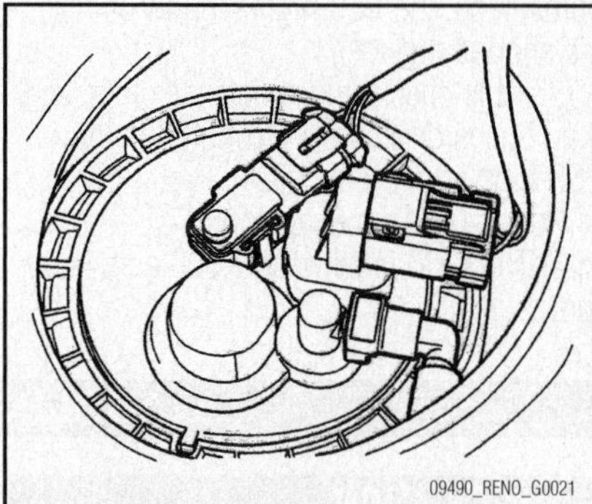

Fig. 104 Fuel pump assembly location under rear seat—Forenza &Reno

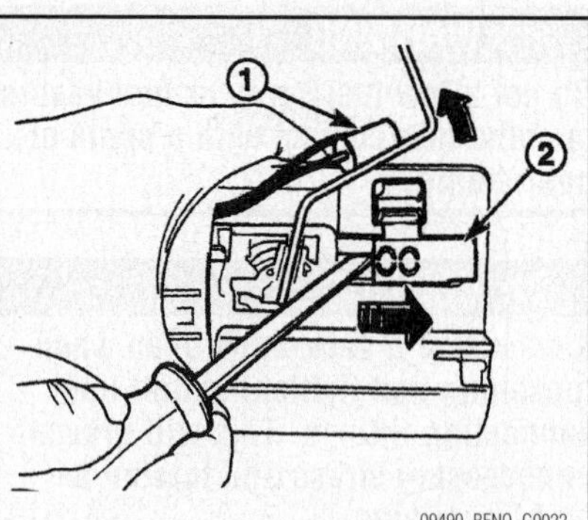

Fig. 105 Fuel level sensor (1) and sender housing (2)—Forenza & Reno

1. Before servicing the vehicle, refer to the Precautions Section.
2. Relieve the pressure from the fuel system.
3. Disconnect the negative battery cable.
4. Remove the rear seat cushion from the floor by lifting it off the retaining brackets and sliding it forward.
5. Remove or disconnect the following:
   - Fuel pump access cover
   - Electrical connector at fuel pump assembly
   - Fuel line
   - Fuel pump assembly clip
   - Fuel pump assembly from fuel tank
6. Remove the fuel sender unit from the fuel pump housing
7. Disconnect the insulator connector.
8. Push the terminal wedge in the insulator connector.
9. Push the wedge outside and then pull the wires to disconnect from insulator.
10. Remove the fuel-level sensor from the sender housing.
11. Remove the sender housing.
12. Remove the fuel sender assembly.

To install:

⁂ WARNING

The components of the fuel sender unit must be installed in the same position as removed or it will perform inaccurately.

13. Install the fuel sender unit onto the fuel pump housing
14. Wires on the sender assembly.
15. Install the sender assembly to the fuel pump assembly.
16. Install the fuel level sensor onto sender housing.
17. Connect the wire into the insulator connector.
18. Connect the insulator connector.
19. Clean the gasket mating surface on the fuel tank.
20. Install or connect the following:
    - New fuel pump mounting gasket
    - Fuel pump assembly into fuel tank
    - Fuel pump assembly clip
    - Electrical connector to fuel pump assembly
    - Fuel line
    - Fuel pump access cover
21. Connect the negative battery cable.
22. Perform operational check of the fuel pump.
23. Install the rear seat cushion by inserting the metal loops into the rear retaining brackets and pressing the front of the seat cushion down.

FUEL PRESSURE REGULATOR

REMOVAL & INSTALLATION

The fuel pressure regulator is built in to the fuel pump.

FUEL PRESSURE RELIEF VALVE

REMOVAL & INSTALLATION

➡There is no specific valve on the fuel system to relieve fuel pressure in the system. The procedure to relieve fuel pressure is outlined below.

⁂ WARNING

The fuel system is under pressure. To avoid fuel spillage and the risk of personal injury or fire, it is necessary to relieve the fuel system pressure before disconnecting the fuel lines

1. Remove the fuel cap.
2. Remove the fuel pump fuse Ef18 from the engine fuse block.
3. Start the engine and allow the engine to stall.
4. Crank the engine for an additional 10 sec.

FUEL TANK

REMOVAL & INSTALLATION

See Figures 106 through 108.

1. Relieve the fuel pressure.
2. Disconnect the negative battery cable.
3. Drain the fuel tank.
4. Remove the front muffler.
5. Remove the fuel tank filler tube clamp at the fuel tank.
6. Disconnect the fuel tank filler tube.
7. Disconnect the fuel line near the right front of the fuel tank.
8. Remove the rear seat.
9. Disconnect the electrical connector at the fuel pump assembly.
10. Disconnect the fuel line.
11. Remove the fuel pump access cover.
12. Remove the parking brake cables support bracket bolt.
13. Remove the parking brake cables from the fuel tank.
14. Disconnect the fuel vapor line from the solenoid valve.
15. Support the fuel tank.
16. Remove the fuel tank strap retaining nuts.
17. Remove the fuel tank strap.
18. Carefully lower the fuel tank.
19. Remove the fuel tank.
20. Transfer any parts as needed.

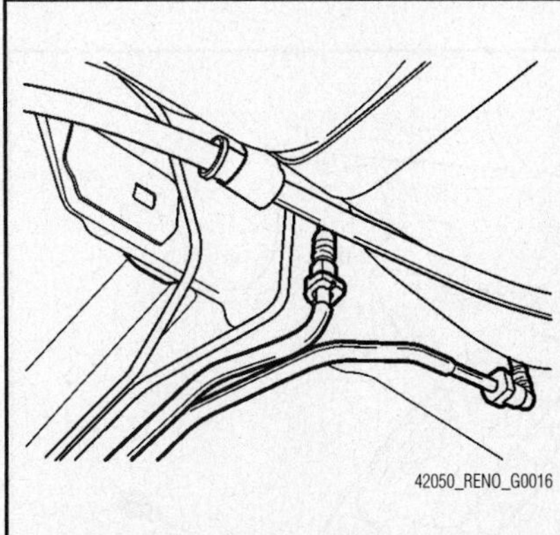

Fig. 106 Fuel line near right front of tank

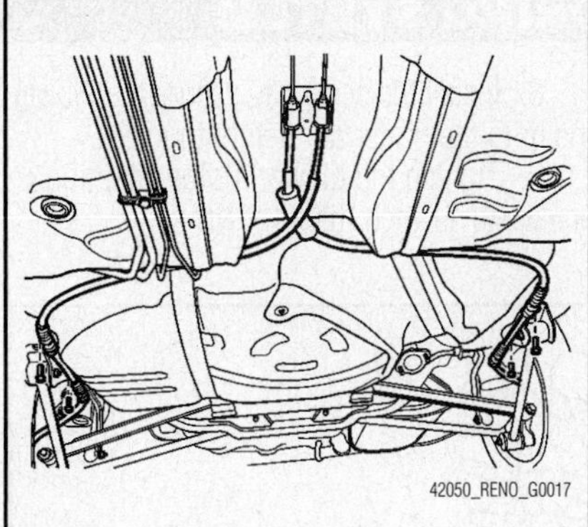

Fig. 107 Parking brake cables

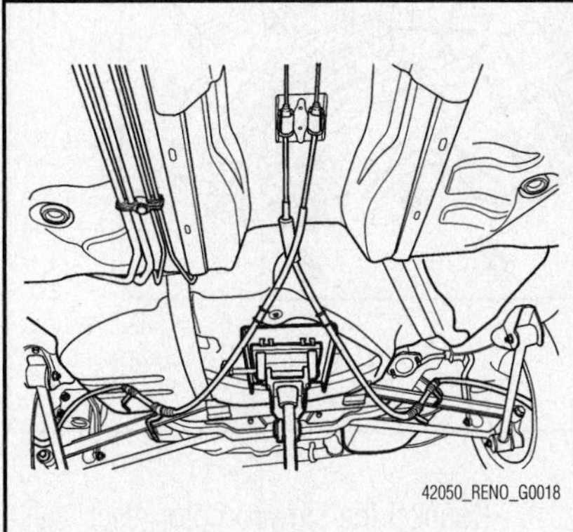

Fig. 108 Fuel tank and support shown

To install:

21. Raise the fuel tank into position.
22. Install the fuel tank straps.
23. Install the fuel tank strap retaining nuts.
24. Tighten the fuel tank strap retaining nuts to 15 ft. lbs. (20 Nm)
25. Connect the fuel vapor line from the solenoid valve.
26. Install the parking brake cables support bracket.
27. Install the parking brake cables from the fuel tank.
28. Connect the fuel line.
29. Connect the electrical connector at the fuel pump assembly.
30. Install the fuel pump access cover.
31. Install the rear seat.
32. Connect the fuel line near the right front of the fuel tank.
33. Connect the fuel tank filler tube.
34. Install the fuel tank filler tube clamp at the fuel tank.
35. Install the front muffler.
36. Connect the negative battery cable.
37. Fill the fuel tank.
38. Perform a leak check of the fuel tank and the fuel line connections.

IDLE SPEED

ADJUSTMENT

Idle speed is maintained by the Powertrain Control Module (PCM). No adjustment is necessary or possible.

THROTTLE BODY

REMOVAL & INSTALLATION

1. Disconnect the negative battery cable.
2. Disconnect the Intake Air Temperature (IAT) sensor connector.
3. Disconnect the breather hose from the valve cover.
4. Remove the air intake tube.
5. Disconnect the throttle cables by opening the throttle and moving the cable through the release slot.
6. Disconnect the vacuum hoses from the throttle body.
7. Disconnect the Throttle Position (TP) sensor and the idle air control valve connectors.
8. Disconnect the coolant hoses from the throttle body.
9. Remove the throttle body and discard the gasket.
10. Remove the TP sensor.
11. Remove the Idle Air Control (IAC) valve.
12. Remove the throttle body retaining nuts.
13. Remove the throttle body.

To install:

14. Install the throttle body.
15. Connect the TP sensor connector and the IAC valve connector.
16. Connect the coolant hoses to the throttle body.
17. Connect the vacuum hoses to the throttle body.
18. Connect the throttle cable.
19. Install the air intake tube.
20. Connect the breather hose to the valve cover.
21. Connect the IAT sensor connector.
22. Connect the negative battery cable.
23. Fill the cooling system.

HEATING & AIR CONDITIONING SYSTEM

BLOWER MOTOR

REMOVAL & INSTALLATION

See Figures 109 and 110.

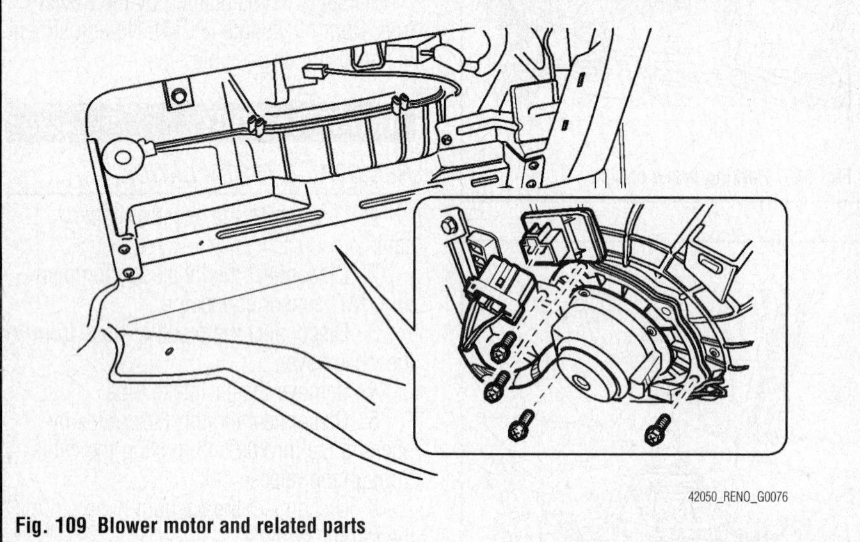

Fig. 109 Blower motor and related parts

1. Disconnect the negative battery cable.
2. Remove the glove box.
3. Put the floor carpet aside for preventing stain.
4. Disconnect the blower motor electrical connector and resistor connector.
5. Remove the screws that secure the motor to the heater/air distributor case.
6. Remove the motor, the seal, and the shock mount pads from the heater/air distributor case by gently pulling the motor straight down and out.

To install:

7. Install the blower motor and the seal with the shock mount pads in the heater/air distributor case. Hold the blower motor in position.

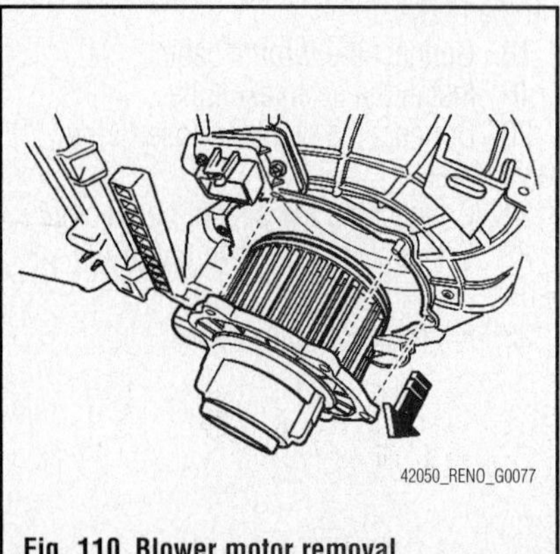

Fig. 110 Blower motor removal

8. Install the screws to secure the blower motor to the heater/air distributor case.
9. Tighten the blower motor retaining screws to 11 inch lbs. (1.2 Nm)
10. Connect the blower motor electrical connector and resistor connector.
11. Connect the negative battery cable.
12. Confirm that the blower motor operates properly.
13. Install the glove box.

HEATER CORE

REMOVAL & INSTALLATION

See Figures 111 through 115.

1. Before servicing the vehicle, refer to the Precautions Section.
2. Disconnect the negative battery cable and let the vehicle sit for 1 minute to deactivate the air bag.
3. Drain the cooling system.
4. Recover refrigerant from A/C system.
5. Tilt the steering wheel to the lowest level.
6. Remove the screws and the instrument cluster trim panel.
7. Remove the A-pillar trim panels.
8. Remove the screw from the rear portion of the floor console.
9. Remove the screws from the front portion of the floor console.
10. Remove the floor console.
11. Disconnect the electrical connectors.
12. Remove the sun sensor.
13. Remove the center molding.
14. Remove the screws and the audio system.
15. Disconnect the audio system electrical connectors.
16. Remove the instrument cluster dimmer switch assembly.
17. Tilt the steering wheel to the lowest level.
18. Remove the screws and the instrument cluster trim panel.
19. Disconnect the electrical connectors.
20. Remove the screws and the instrument cluster.
21. Disconnect the electrical connectors.
22. Remove the screws and separate the hood release cable from the hood latch release handle.
23. Carefully pull the instrument panel under cover until the mounting clips are released and the instrument panel under cover.

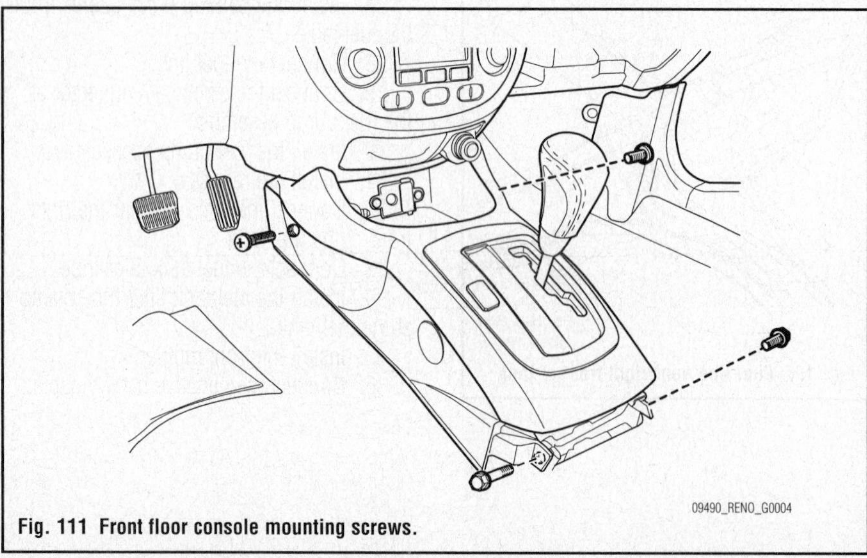

Fig. 111 Front floor console mounting screws.

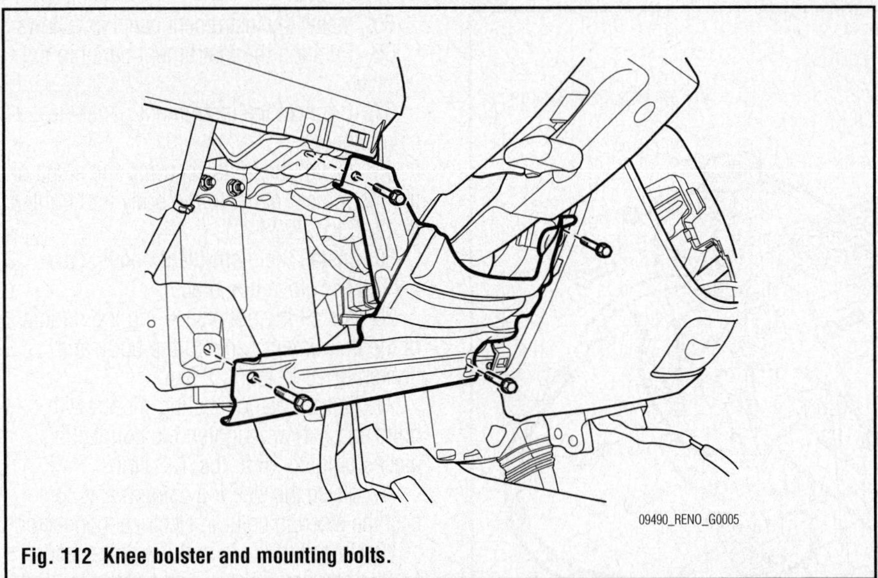

Fig. 112 Knee bolster and mounting bolts.

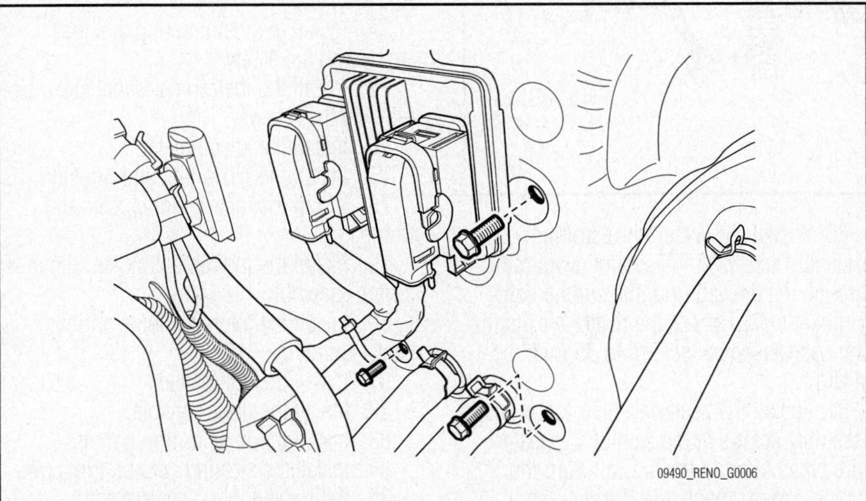

Fig. 113 Remove the screws that secure heater/air distributor case assembly to firewall on firewall side.

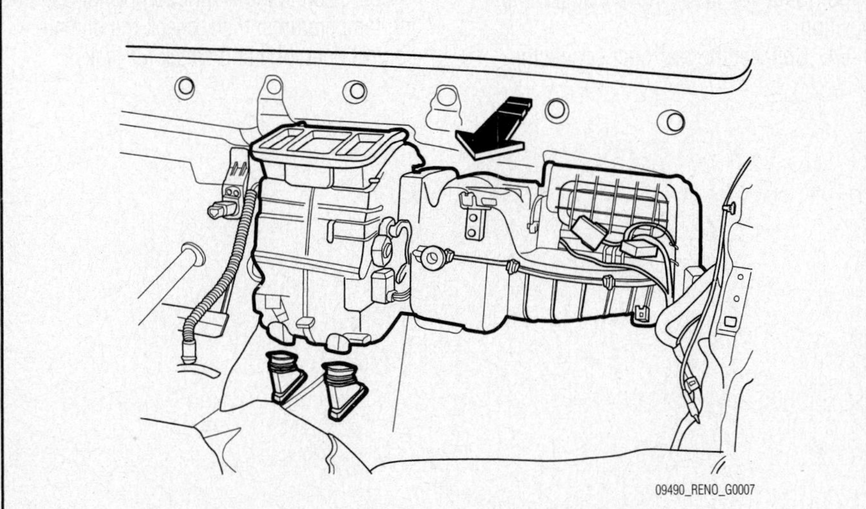

Fig. 114 Remove the heater/air distributor case assembly from the vehicle.

24. Remove the bolts from the driver's side knee bolster.
25. Remove the knee bolster.
26. Remove the screws and the instrument panel side covers.
27. Remove the nuts and the bolts securing the steering column.
28. Disconnect the steering column electrical connector.
29. Lower the steering column.
30. Remove bolts and screws.
31. Remove the connecting pieces.
32. Remove the bolt securing the middle of the instrument panel to the body.
33. Remove the instrument panel screws behind the glove box brace.
34. Remove the bolts securing the sides of the instrument panel to body.
35. Disconnect the instrument panel electrical connectors.
36. Remove the instrument panel.
37. Remove the heater/air distributor case assembly as follows:
38. Compress the heater hose clamps at the firewall and slide the clamps toward the engine.
39. Remove the 2 heater hoses from the core pipes at the firewall.
40. Remove the screws that secure the heater/air distributor case assembly to the firewall on the side of the heater core pipes.
41. Remove the high-pressure and low-pressure pipe from the A/C expansion valve.
42. Remove the evaporator drain hose.

➡Have an assistant support the heater/air distributor case from inside the vehicle.

43. Remove the screws that secure the heater/air distributor case assembly to the firewall on the side of the evaporator.
44. Disconnect the rear duct connector.
45. Remove the heater/air distributor case assembly from the vehicle.
46. Remove the wiring harness and electrical connectors from the heater/air distributor case assembly.
47. Remove the screws that connect the heater core housing and the evaporator housing.
48. Separate the sponge from heater/air distributor case assembly.
49. Remove the heater core cover screws from the heater core housing.
50. Remove the heater core.

To install:

51. Install the heater core into the case.
52. Install the screws on the heater core housing and tighten to 11 inch lbs. (1.2 Nm).
53. Install the sponge onto the heater/air distributor case assembly.

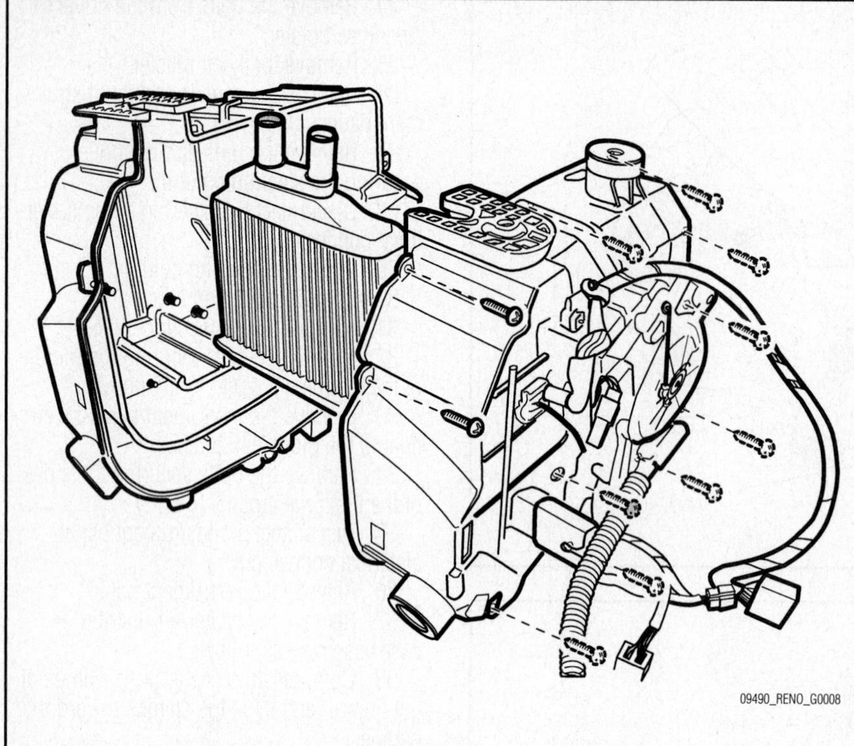

Fig. 115 Remove the heater core from the housing.

54. Install the screws that connect the heater core housing and the evaporator housing. Tighten the heater core housing-to-evaporator housing screws to 11 inch lbs. (1.2 Nm).

55. Connect the wiring harness and electrical connectors.

56. Install the heater/air distributor case assembly as follows:

57. Position the heater/air distributor case assembly in the vehicle.

58. Slowly raise the heater/air distributor case assembly into position against the firewall and hold it in.

59. Install the high-pressure and low-pressure pipe.

60. Install the heater/air distributor case assembly screws at the side of evaporator through the firewall from the engine compartment side. Tighten the heater/air distributor case assembly screws to 35 inch lbs. (4 Nm).

61. Install the heater/air distributor case assembly screws at the side of the heater core pipes through the firewall from the engine compartment side. Tighten the heater/air distributor case assembly screw to 35 inch lbs. (4 Nm).

62. Install the two heater hoses.

63. Slide the heater hose clamps into position.

64. Connect the rear duct connector.

65. Install the instrument panel as follows:

66. Position the instrument panel in the vehicle.

67. Connect the instrument panel electrical connectors.

68. Install the bolts securing the sides of the instrument panel to the body and tighten to 16 ft. lbs. (22 Nm).

69. Install the instrument panel screws behind the glove box brace.

70. Install the bolts securing the middle of the instrument panel to the body and tighten to 16 ft. lbs. (22 Nm).

71. Install the connecting pieces with bolts and screws. Tighten the connecting pieces bolts to 16 ft. lbs. (22 Nm).

72. Raise the steering column and connect the steering column electrical connector.

73. Install the nuts and the bolts securing the steering column and tighten to 16 ft. lbs. (22 Nm).

74. Install the instrument panel fuse block with the screw.

75. Install the instrument panel side covers with the screws.

76. Install the knee bolster.

77. Install the glove box and housing.

78. Install the instrument cluster and trim panel.

79. Install the instrument cluster dimmer switch assembly.

80. Install the audio systems and the center molding.

81. Install the sun sensor.

82. Install the floor console.

83. Install the A-pillar trim panels.

84. Install the steering column trim cover.

85. Refill the cooling system with approved coolant.

86. Recharge the refrigerant.

87. Connect the negative battery cable.

88. Operate the engine to normal operating temperatures; then, check the climate control operation and check for leaks.

STEERING

POWER RACK & PINION STEERING GEAR

REMOVAL & INSTALLATION

See Figures 116 and 117.

1. Before servicing the vehicle, refer to the Precautions Section.
2. Disconnect the negative battery cable.
3. Raise and safely support the vehicle.
4. Remove or disconnect the following:
5. Front wheels.
6. Power steering gear fluid outlet pipe.
7. Place a drain pan under the steering gear to catch the power steering fluid.
8. Power steering gear fluid inlet pipe
9. Position the steering gear straight ahead by turning the steering wheel until the steering wheel spokes are vertical and pointed to the left.
10. Scribe a mark on the stub shaft housing that lines up with a mark on the intermediate shaft lower coupling.
11. Remove or disconnect the following:
    - Intermediate shaft pinch bolt
    - Outer tie rod nuts
    - Tie rod ends from strut assembly using ball joint remover KM-507-B
    - Crossmember assembly
    - Transaxle center bracket (automatic transaxle)
    - Bolts securing transaxle center bracket to transaxle and engine and move center bracket out of the way (manual transaxle)
    - Nuts and bolts from the steering gear mounting bracket
    - Return line from the clip on the crossmember
    - Rack and pinion assembly from crossmember assembly

To install:

12. Install the rack and pinion assembly onto the crossmember. The steering gear must be in a straight-ahead position, and the steering wheel spokes must be vertical and pointing to the left. Align the marks on the shafts to ensure proper positioning. Seat the stub shaft into the intermediate shaft.
13. Install or connect the following:
14. Bolts and nuts on the steering gear mounting bracket. Tighten the steering gear mounting bracket bolts and nuts to 44 ft. lbs. (60 Nm).
15. Return line into clip on the cross-member and tighten to 71 inch lbs. (8 Nm)
16. On vehicles equipped with a manual transaxle, position the transaxle center bracket in place and install the bolts securing the bracket to the engine and the transaxle. Tighten the transaxle center bracket-to-transaxle bolts and the transaxle center bracket-to-engine bolt to 59 ft. lbs. (80 Nm).
17. On vehicles equipped with an automatic transaxle, install the transaxle center bracket.
18. Install the crossmember.
19. Tighten the rear crossmember-to-body bolts to 145 ft. lbs. (196 Nm) and the front crossmember-to-body bolts to 96 ft. lbs. (130 Nm).
20. Install the tie rod ends to strut assembly
21. Install the outer tie rod nuts and tighten to 37 ft. lbs. (50 Nm)
22. Install the lower intermediate shaft pinch bolt and tighten to 18 ft. lbs. (25 Nm)
23. Install the power steering gear fluid inlet and outlet pipes and tighten the pipe fittings to 21 ft. lbs. (28 Nm)
24. Install the front wheels
25. Lower the vehicle.
26. Check and adjust the front wheel alignment.
27. Refill the power steering system and check for leaks. If leaks are found, correct the cause of the leak and bleed the system.
28. Connect the negative battery cable.

09490_RENO_G0040

Fig. 116 Location of intermediate shaft pinch bolt and power steering gear fluid inlet and outlet pipes—Forenza/&Reno

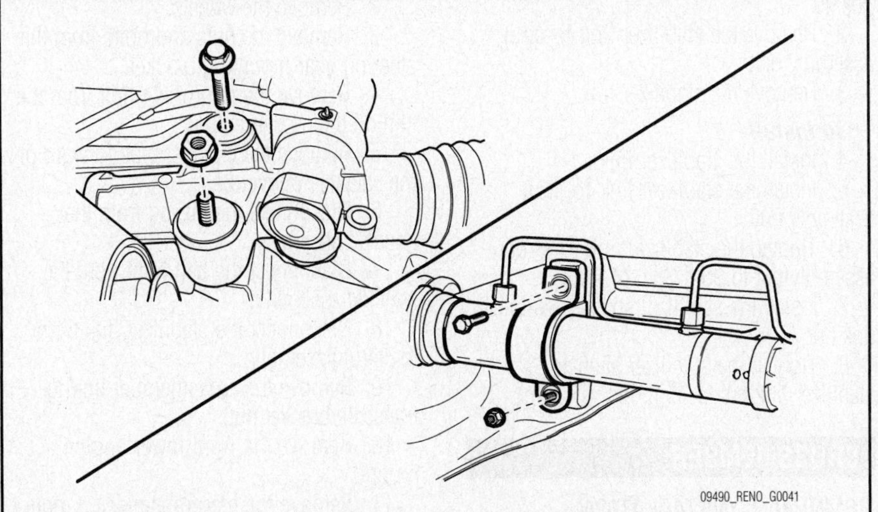

Fig. 117 Power steering gear mounting bracket nuts and bolts—Forenza/Reno

POWER STEERING PUMP

REMOVAL & INSTALLATION

See Figures 118 and 119.

➡The unit repair of the power steering pump in this vehicle is not serviceable. A faulty pump must be replaced with a new one.

1. Remove the serpentine accessory drive belt.
2. Disconnect the electrical connector at the Electronic Variable Orifice solenoid actuator.
3. Position a drain pan to catch the power steering fluid that will drain from the high-pressure hose after it is disconnected from the power steering pump.
4. Disconnect the high-pressure hose fitting from the power steering pump.
5. Position a drain pan to catch the power steering fluid that will drain from the supply hose after it is disconnected from the power steering pump.
6. Disconnect the supply hose from the power steering pump. Remove the bolt from the front right side of the power steering pump.

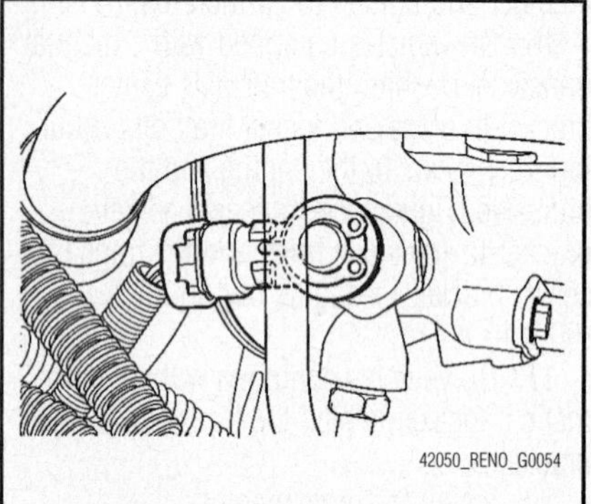

Fig. 118 Electronic variable orifice solenoid actuator

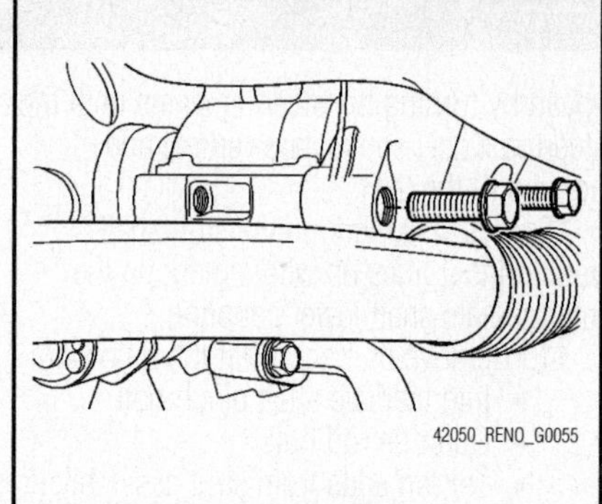

Fig. 119 Power steering pump and bolts

7. Remove the two power steering pump bolts and remove the power steering pump from the vehicle.

To install:

8. Install the power steering pump into the vehicle and install the two power steering pump bolts.
9. Tighten the two power steering pump bolts to 18 ft. lbs. (25 Nm)
10. Swing the power steering pump bracket down and install the front right side bolt.
11. Tighten the both of the right side power steering pump bracket bolts to 26 ft. lbs. (35 Nm)
12. Connect the supply hose to the power steering pump.
13. Connect the high-pressure hose fitting to the power steering pump.
14. Tighten the high-pressure hose fitting to 21 ft. lbs. (28 Nm)
15. Connect the electrical connector at the Electronic Variable Orifice solenoid actuator.
16. Install the serpentine accessory drive belt.
17. Refill the power steering fluid.
18. Bleed the power steering system.

BLEEDING

If the power steering hydraulic system has been serviced, an accurate fluid level reading cannot be obtained until the air is bled from the system. Follow these steps to bleed the air from the system.

1. Turn the wheels all the way to the left and add the power steering fluid to the MIN mark on the fluid level indicator.
2. Start the engine. With the engine running at fast idle, recheck the fluid level. If necessary, add fluid to bring the level up to the MIN mark.
3. Bleed the system by turning the wheels from side to side without reaching the stop at either end. Keep the fluid level at the MIN mark. The air must be eliminated from the fluid before normal steering action can be obtained.
4. Return the wheels to the center position. Continue running the engine for 2 to 3 minutes.
5. Road test the car to be sure the steering functions normally and is free from noise.
6. Recheck the fluid level as described in steps 1 and 2. Make sure the fluid level is at the MAX mark after the system has stabilized at its normal operating temperature. Add fluid as needed.

SUSPENSION

FRONT SUSPENSION

CONTROL LINKS

REMOVAL & INSTALLATION

See Figure 120.

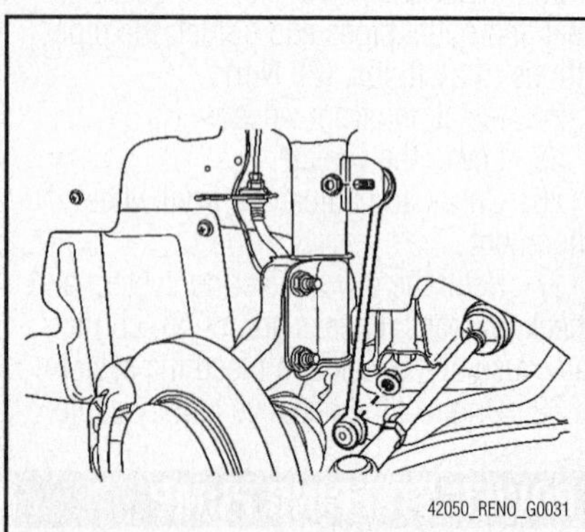

Fig. 120 Stabilizer link and related parts

1. Remove the stabilizer shaft-to-stabilizer link nut.
2. Remove the stabilizer link-to-strut assembly nut.
3. Remove the stabilizer link.

To install:

4. Install the stabilizer link.
5. Install the stabilizer link-to-strut assembly nut.
6. Tighten the stabilizer link-to-strut assembly nut to 35 ft. lbs. (47 Nm)
7. Install the stabilizer shaft-to-stabilizer link nut.
8. Tighten the stabilizer shaft-to-stabilizer link nut to 35 ft. lbs. (47 Nm)

CROSSMEMBER

REMOVAL & INSTALLATION

See Figure 121.

1. Raise and suitably support the vehicle.
2. Remove the wheels.
3. Remove the nuts and bolts from the steering gear mounting bracket.
4. Remove the return line bolt from the clip on the crossmember.
5. Remove the exhaust pipe forward of the catalytic converter.
6. Disconnect the tie rod from the knuckle assembly.
7. Disconnect the ball joint from the knuckle assembly.
8. Disconnect the stabilizer link from the strut assembly.
9. Remove the crossmember link-to-transaxle bracket nut.
10. Remove the right lower engine mount.
11. Remove the rear transmission mount bracket.

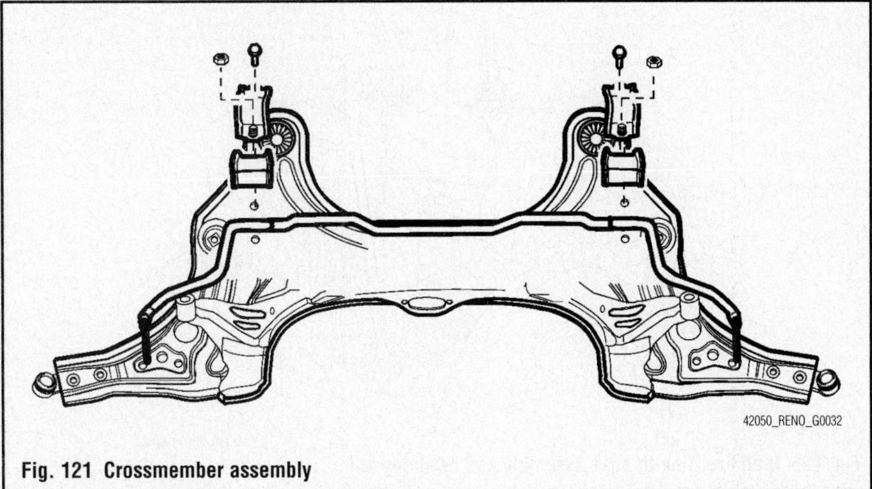

Fig. 121 Crossmember assembly

12. Remove the rear crossmember-to-body bolts.
13. Remove the front crossmember-to-body bolts.
14. Remove the crossmember assembly from the vehicle.

To install:

15. Install the crossmember assembly.
16. Install the rear crossmember-to-body bolts.
17. Tighten the rear crossmember-to-body bolts to 145 ft. lbs. (196 Nm)
18. Install the front crossmember-to-body bolts.
19. Tighten the front crossmember-to-body bolts to 96 ft. lbs. (130 Nm)
20. Install the bolts and nuts on the steering gear mounting brackets.
21. Install the return line bolt into the clip on the crossmember.
22. Install the crossmember link-to-transaxle bracket nut.
23. Tighten the crossmember link-to-transaxle bracket nut to 125 ft. lbs. (169 Nm)
24. Connect the stabilizer link to the strut assembly.
25. Connect the ball joint to the knuckle assembly.
26. Connect the tie rod from the knuckle assembly.
27. Install the exhaust pipe into the vehicle.
28. Install the wheels.
29. Lower the vehicle.

LOWER BALL JOINT

REMOVAL & INSTALLATION

See Figure 122.

1. Before servicing the vehicle, refer to the Precautions Section.
2. Remove the control arm.

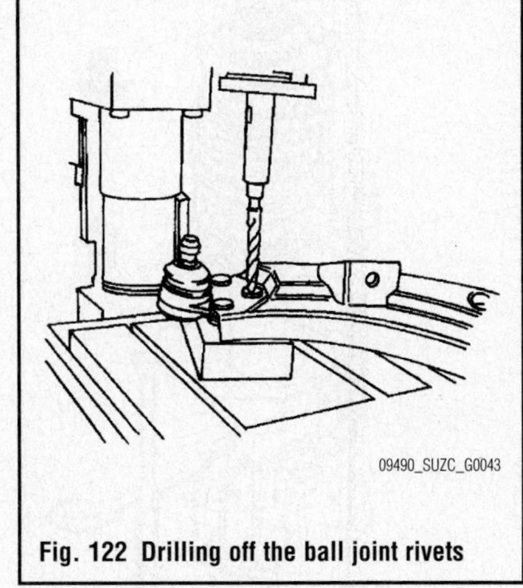

Fig. 122 Drilling off the ball joint rivets

3. Drill off the heads of the three rivets with a 0.47 inch (12mm) drill bit.
4. Punch out the rivets with a drift.

To install:

5. Connect the ball joint to the control arm by inserting 3 ball joint bolts from below the control arm.
6. Tighten the ball joint-to-control arm nuts to 74 ft. lbs. (100 Nm).
7. Install the control arm.

LOWER CONTROL ARM

REMOVAL & INSTALLATION

1. Before servicing the vehicle, refer to the Precautions Section.
2. Disconnect the negative battery cable.
3. Raise and safely support the front of the vehicle. Let the control arms hang free.
4. Remove or disconnect the following:
    - Front wheels
    - Pinch bolt and nut from the ball joint
    - Ball joint from knuckle assembly using ball joint remover KM-507-B
    - Control arm-to-crossmember bolts
    - Control arm from vehicle

To install:

5. Install or connect the following:
    - Control arm
    - Control arm-to-crossmember bolts. Do not tighten the bolts.
    - Ball joint to steering knuckle
    - Ball joint pinch bolt and the nut. Tighten ball joint pinch bolt and nut to 44 ft. lbs. (60 Nm).
    - Control arm-to-crossmember bolts. Tighten the front control arm-to-crossmember bolt to 92 ft. lbs. (125 Nm) and the rear control arm-to-crossmember bolt to 81 ft. lbs. (110 Nm).
6. Install the wheels.
7. Lower the vehicle.
8. Connect the negative battery cable.
9. Check the front wheel alignment.

CONTROL ARM BUSHING REPLACEMENT

See Figure 123.

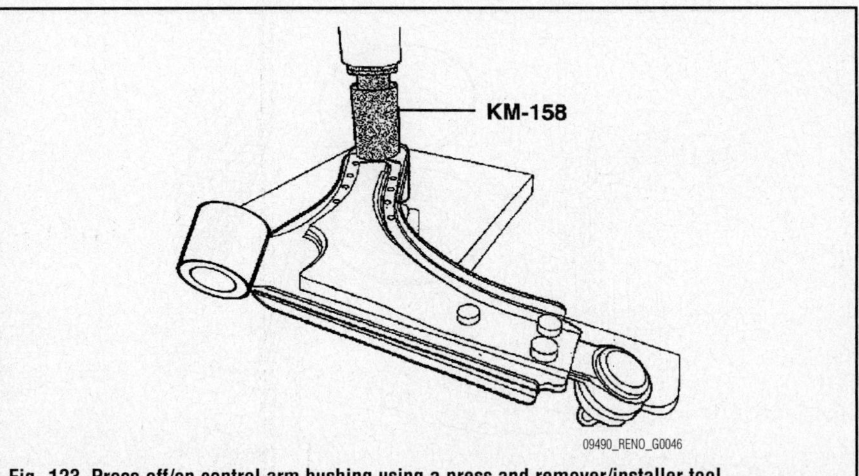

Fig. 123 Press off/on control arm bushing using a press and remover/installer tool

1. Before servicing the vehicle, refer to the Precautions Section.
2. Remove the control arm.
3. Remove the split sleeves from the rear control arm bushing.
4. Press off the control arm rear damping bushing using a press and the remover/installer KM-158.

To install:

5. Press the control arm rear damping bushing into the control arm using a press and the remover/installer KM-158.
6. Install the split sleeves into the rear control arm bushing.
7. Install the control arm.

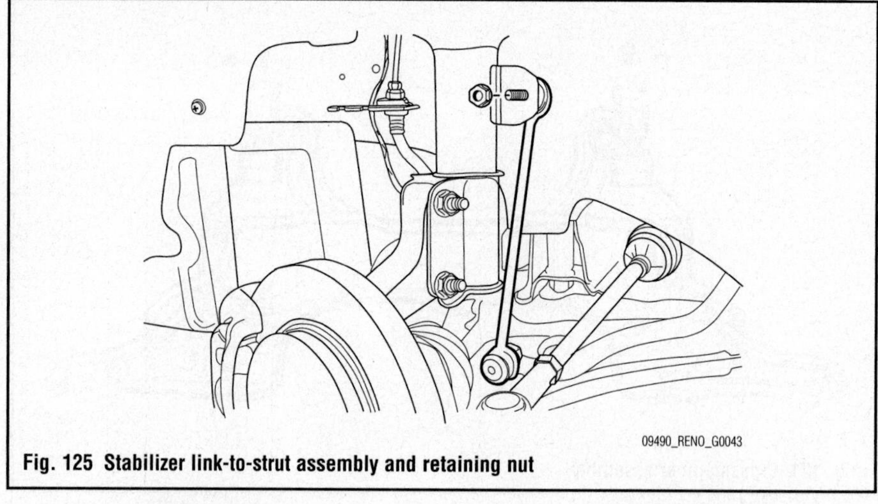

Fig. 125 Stabilizer link-to-strut assembly and retaining nut

MACPHERSON STRUT

REMOVAL & INSTALLATION

See Figures 124 through 127.

1. Before servicing the vehicle, refer to the Precautions Section.
2. Disconnect the negative battery cable.
3. Remove the strut upper cap and nut.
4. Raise and safely support the front of the vehicle.
5. Remove or disconnect the following:
   - Wheels
   - ABS sensor line from strut assembly (if equipped)
   - Brake line from securing bracket on the strut assembly
   - Stabilizer shaft link by removing the stabilizer link-to-strut assembly nut
   - Steering knuckle by removing the steering knuckle-to-strut assembly nuts and bolts
   - Strut assembly
6. Remove the coil spring by disassembling the strut assembly as follows:
   a. Fasten the strut assembly to the spring compressor DW320-010 or KM-329-A. Make sure the hooks are seated on the strut spring properly.
   b. Compress the front spring with the spring compressor.
   c. Use an open end wrench to hold the threaded piston rod while removing the piston rod nut and the washer with a commercially available double ring spanner, sharply offset.
   d. Remove the upper strut mount, the mount bearing, the upper spring seat, the upper spring insulator and the hollow bumper.
   e. Release the spring.
   f. Remove the spring and the lower spring insulator.

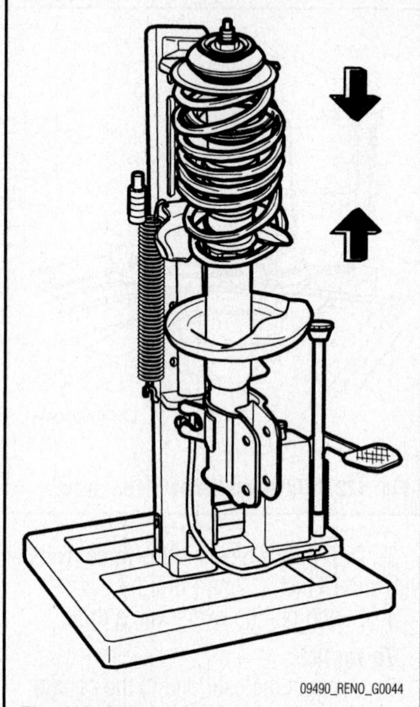

Fig. 126 Strut assembly mounted in spring compressor tool

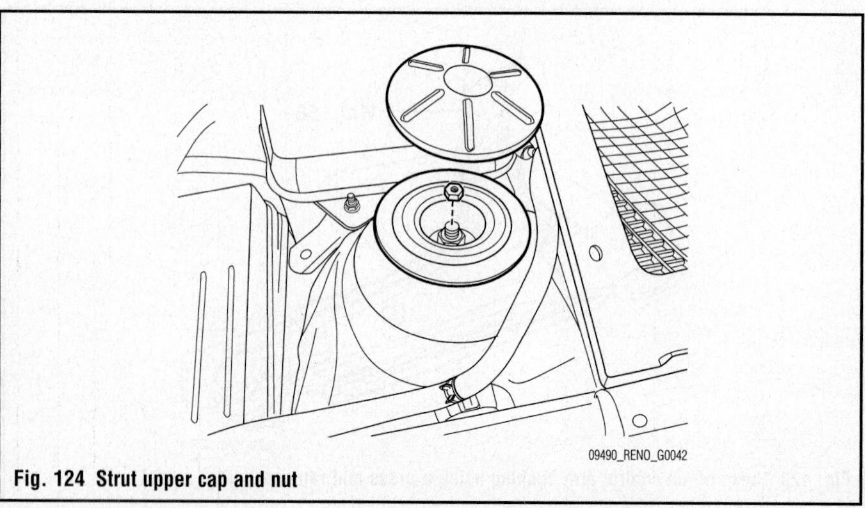

Fig. 124 Strut upper cap and nut

To install:

7. Install the coil spring by assembling the strut assembly as follows:
   a. Install the lower spring insulator and the spring.
   b. Compress the spring using the spring compressor KM-329-A.
   c. Install the hollow bumper, the upper spring insulator, the upper spring seat, the upper strut mount, and the mount bearing. Be sure the upper spring seat is clipped to the front spring locator.
   d. Install the piston rod nut. Tighten the piston rod nut to 55 ft. lbs. (75 Nm).
   e. Release and remove the spring compressor KM-329-A.
8. Install or connect the following:
   - Strut assembly
   - Strut assembly to steering knuckle and tighten the steering knuckle-to-strut assembly nuts and the bolts to 89 ft. lbs. (120 Nm)
   - Stabilizer shaft link to strut assembly and tighten the stabilizer link-to-strut assembly nut to 35 ft. lbs. (47 Nm)
   - Brake line to securing bracket on the strut assembly

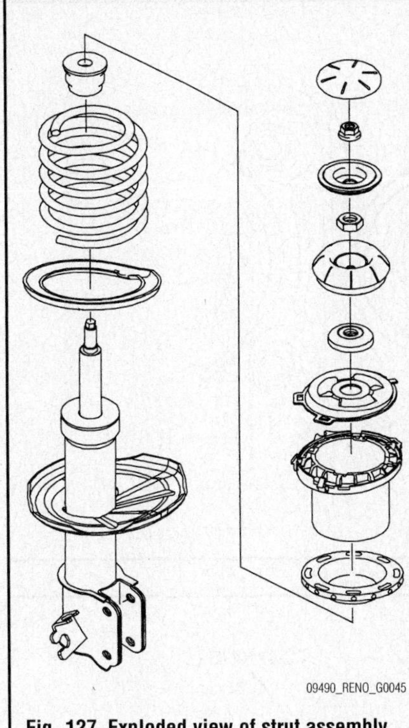
Fig. 127 Exploded view of strut assembly

- ABS sensor line to strut assembly (if equipped)
- Wheels

9. Lower the vehicle.
10. Install the nuts securing the strut assembly to the body of the vehicle and tighten to 48 ft. lbs. (65 Nm).
11. Connect the negative battery cable.

OVERHAUL

Refer to MacPherson Strut removal and installation.

STEERING KNUCKLE

REMOVAL & INSTALLATION

See Figures 128 through 131.

1. Raise and suitably support the vehicle.
2. Remove the wheel.
3. Remove the caulking nut from the axle shaft.
4. Remove the brake caliper from the rotor. Support the caliper so it does not hang from the hydraulic brake hose.
5. Remove the outer tie rod from the knuckle assembly.
6. On vehicles equipped with the antilock braking system, disconnect the speed sensor electrical connection from the knuckle.
7. Remove the ball joint pinch bolt and the nut.
8. Separate the knuckle from the ball joint using the ball joint remover KM-507-B.

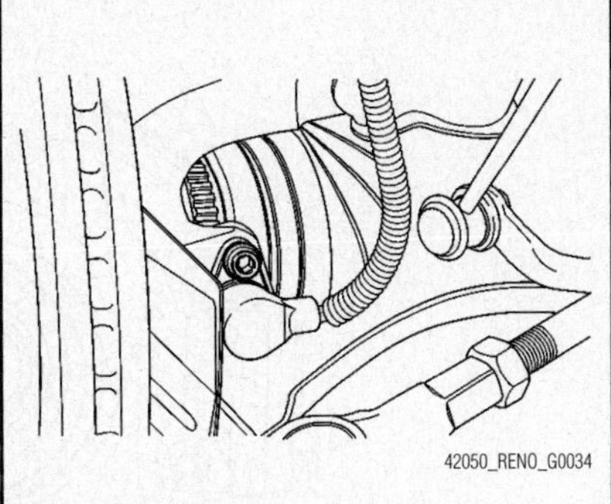
Fig. 128 Antilock speed sensor and related parts

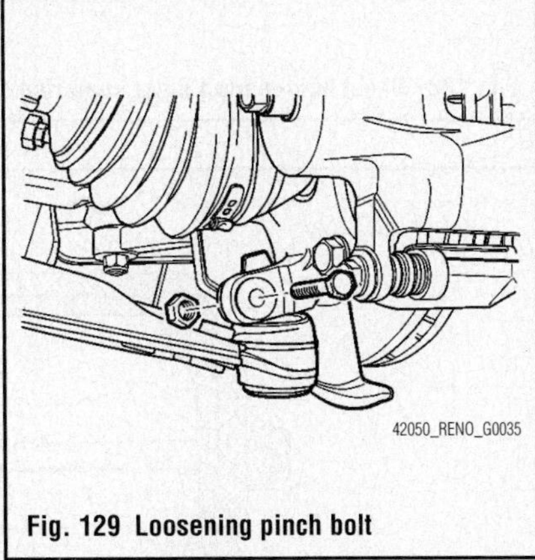
Fig. 129 Loosening pinch bolt

9. Remove the nuts from the bolts that connect the knuckle assembly to the strut assembly.
10. Support the drive axle.
11. Separate the drive axle shaft from the wheel hub.
12. Remove the bolts that connect the knuckle assembly to the strut assembly.
13. Remove the knuckle assembly from the vehicle.

To install:

14. Install the knuckle assembly onto the vehicle.

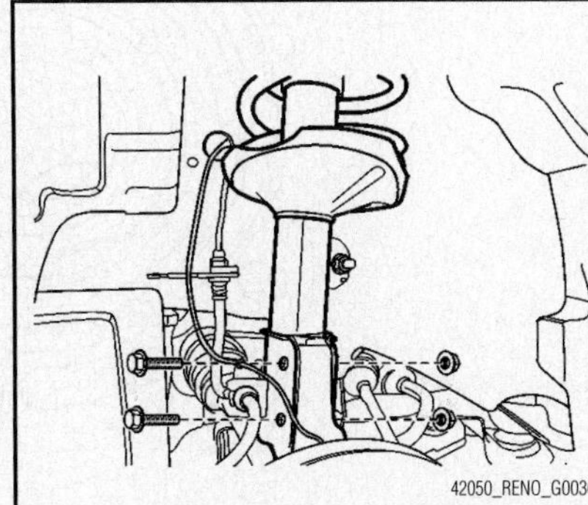
Fig. 130 Removing knuckle-to-strut nuts and bolts

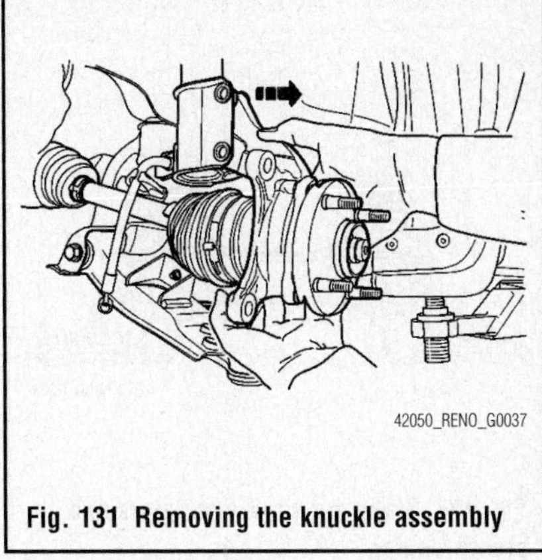
Fig. 131 Removing the knuckle assembly

15. Install the steering knuckle-to-strut assembly nuts.
16. Tighten the steering knuckle-to-strut assembly nuts to 89 ft. lbs. (120 Nm).
17. Connect the drive axle to the front wheel hub.
18. Connect the ball joint to the knuckle assembly.
19. Install the ball joint pinch bolt and the nut.
20. Tighten the ball joint pinch bolt nut to 44 ft. lbs. (60 Nm).
21. Connect the ABS speed sensor electrical connection.
22. Connect the outer tie rod to the knuckle assembly.
23. Install the brake caliper onto the rotor.
24. Install the caulking nut onto the axle shaft.
25. Tighten the drive axle-to-hub caulking nut to 221 ft. lbs. (300 Nm).
26. Install the wheel.
27. Lower the vehicle.

STABILIZER BAR

REMOVAL & INSTALLATION

See Figure 132.

1. Raise and suitably support the vehicle.
2. Remove the stabilizer shaft-to-stabilizer link nut.
3. Remove the crossmember.
4. Remove the stabilizer shaft-to-crossmember clamp bolts.
5. Remove the stabilizer shaft, the stabilizer shaft insulator clamp, and the insulators from the vehicle.

To install:

6. Install the stabilizer shaft.
7. Install the stabilizer shaft insulator clamps, the stabilizer shaft clamp bolt, and the insulators onto the crossmember.
8. Do not tighten the bolt.

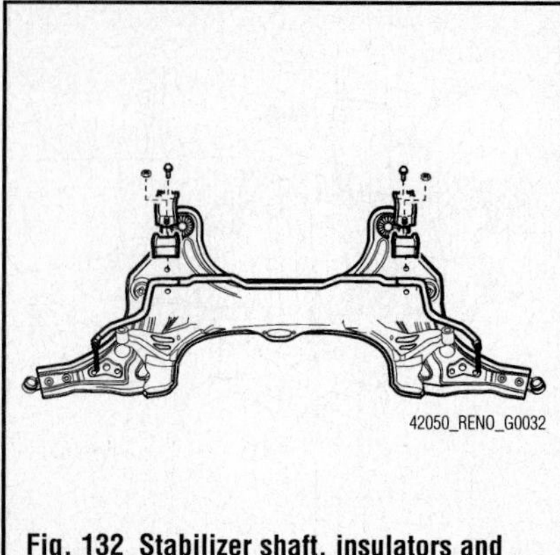

Fig. 132 Stabilizer shaft, insulators and crossmember

9. Install the stabilizer link onto the stabilizer shaft and connect them with the stabilizer shaft-to-stabilizer link nut.
10. Tighten the stabilizer shaft-to-stabilizer link nut to 35 ft. lbs. (47 Nm).
11. Tighten the stabilizer shaft-to-crossmember clamp bolts to 18 ft. lbs. (25 Nm).
12. Install the front wheels.

WHEEL HUB AND BEARING

REMOVAL & INSTALLATION

See Figures 133 through 136.

1. Before servicing the vehicle, refer to the Precautions Section.
2. Remove the drive shaft from the front wheel hub.
3. Remove the inner snap ring.
4. Remove the wheel hub with the support bridge J-37105-B-1, the bearing adapter J-37105-B-2, the hex nut 500-20, and the forcing screw J-36661-2.
5. Remove the brake shield.
6. Remove the outer snap ring.
7. Remove the wheel bearing with the support bridge J-37105-B-1, the bearing adapter J-37105-B-2, the hex nut 500-20, and the forcing screw J-36661-2.
8. Clean the bore of the knuckle.

To install:

9. Install the outer snap ring and push the wheel bearing into place with the support bridge J-37105-B-1, the bearing adapter J-37105-B-2, the hex nut 500-20, and the forcing screw J-36661-2.
10. Install the brake shield.
11. Install the inner snap ring.
12. Push the wheel hub into place with the hub adapter J-37105-B-3, the bearing adapter J-37105-B-2, the hex

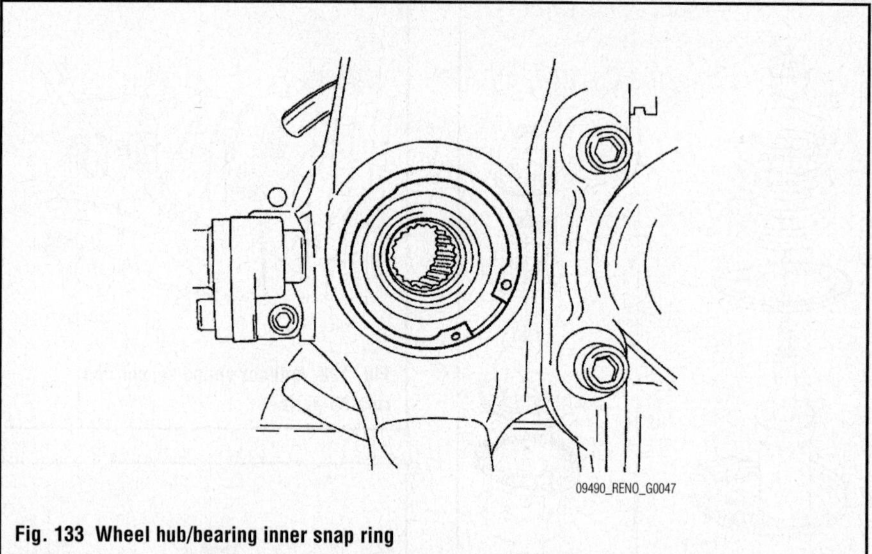

Fig. 133 Wheel hub/bearing inner snap ring

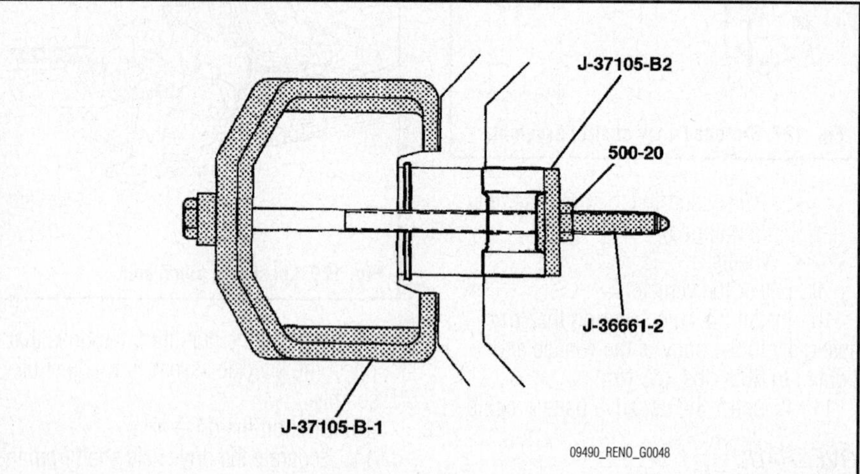

Fig. 134 Support bridge J-37105-B-1, bearing adapter J-37105-B-2, hex nut 500-20, and forcing screw J-6661-2

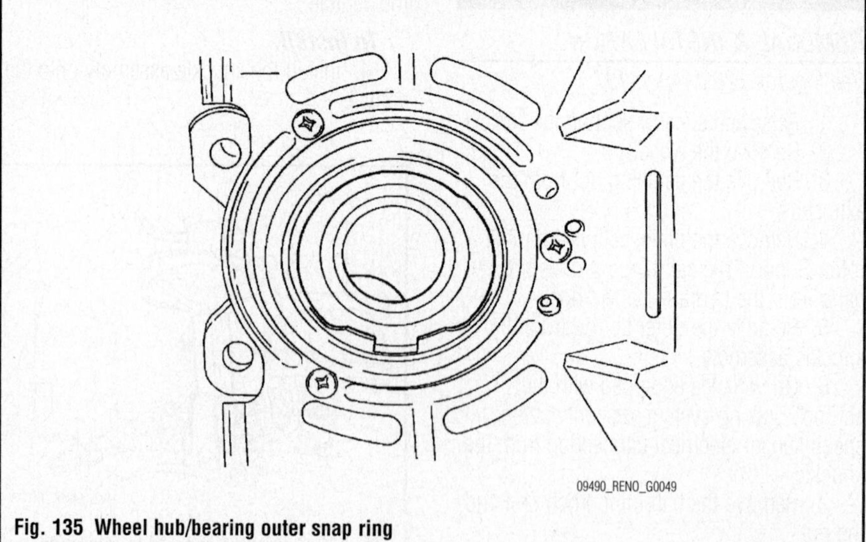

Fig. 135 Wheel hub/bearing outer snap ring

nut 500-20, and the forcing screw J-36661-2.

13. Install the drive axle into the front wheel hub.

ADJUSTMENT

The front wheel bearings are a cartridge type design and cannot be adjusted.

REPACKING

Front wheel bearings are a cartridge type and cannot be repacked.

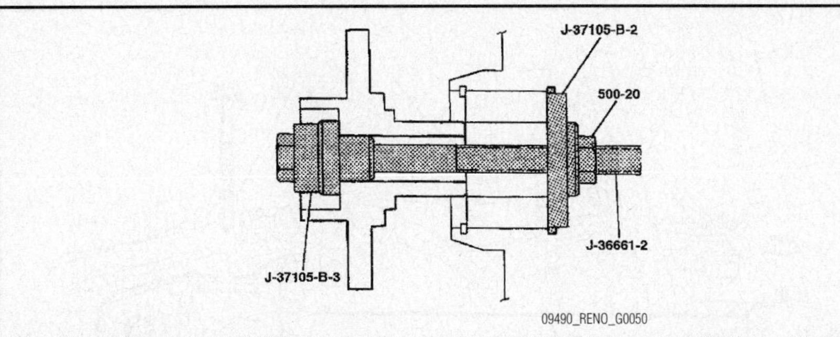

Fig. 136 Hub adapter J-37105-B-3, bearing adapter J-37105-B-2, hex nut 500-20, and forcing screw J-36661-2

SUSPENSION

REAR SUSPENSION

CONTROL ARMS/LINKS

REMOVAL & INSTALLATION

Front Parallel Link

1. Before servicing the vehicle, refer to the Precautions Section.
2. Raise and safely support the rear of the vehicle.
3. Remove the rear wheels.
4. For vehicles equipped with the antilock braking system, remove the ABS sensor from the knuckle and the ABS housing assembly from the front parallel link.
5. Remove the front parallel link bolt from the rear crossmember.
6. Remove the front parallel link bolt from the rear knuckle.
7. Remove the front parallel link.

To install:

8. Install the front parallel link.
9. Install the front parallel link onto the rear knuckle with the bolt.
10. Tighten the front parallel link-to-knuckle bolt to 89 ft. lbs. (120 Nm).
11. Install the front parallel link onto the rear cross-member with the bolt. Do not tighten.
12. For vehicles equipped with the antilock braking system, install the ABS housing assembly onto the front parallel link and the ABS sensor line into the knuckle.
13. Install the rear wheels and lower the vehicle.
14. Perform a rear toe adjustment.

Rear Parallel Link

See Figure 137.

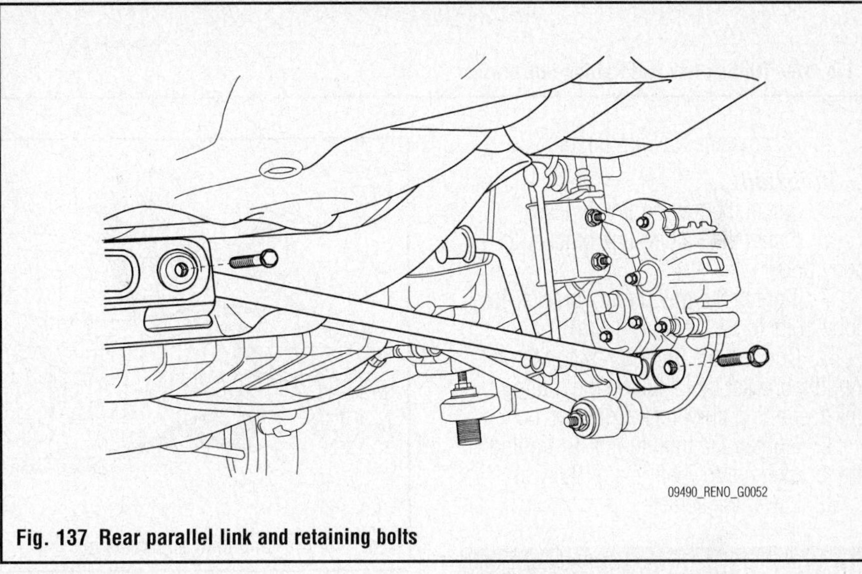

Fig. 137 Rear parallel link and retaining bolts

1. Before servicing the vehicle, refer to the Precautions Section.
2. Raise and safely support the rear of the vehicle.
3. Remove the rear wheel.
4. Remove the rear parallel link bolt from the rear crossmember.
5. Remove the rear parallel link bolt from the rear knuckle.
6. Remove the rear parallel link.

To install:

7. Install the rear parallel link.
8. Install the rear parallel link onto the rear knuckle with the bolt.
9. Tighten the rear parallel link-to-knuckle bolt to 89 ft. lbs. (120 Nm).
10. Install the rear parallel link onto the rear crossmember. Install the rear parallel link-to-crossmember bolt.
11. Tighten the rear parallel link-to-crossmember bolt to 66 ft. lbs. (90 Nm).
12. Install the rear wheel and lower the vehicle.

Trailing Link

See Figure 138.

1. Before servicing the vehicle, refer to the Precautions Section.
2. Raise and safely support the rear of the vehicle.
3. Remove the rear trailing link-to-rear knuckle nut.
4. Remove the rear trailing link-to-trailing link bracket nut and the rear trailing link-to-knuckle bolt.
5. Remove the rear trailing link.

To install:

6. Install the rear trailing link.
7. Install the rear trailing link bracket nut and the bolt. Tighten the rear trailing link-to-trailing link bracket nut to 74 ft. lbs. (100 Nm).
8. Install the trailing link-to-knuckle nut and tighten to 110 ft. lbs. (150 Nm).
9. Lower the vehicle.

Trailing Link Bracket

1. Raise and suitably support the vehicle.
2. Remove the trailing link-to-trailing link bracket nut and the bolt.
3. Remove the trailing link bracket-to-body bolts.

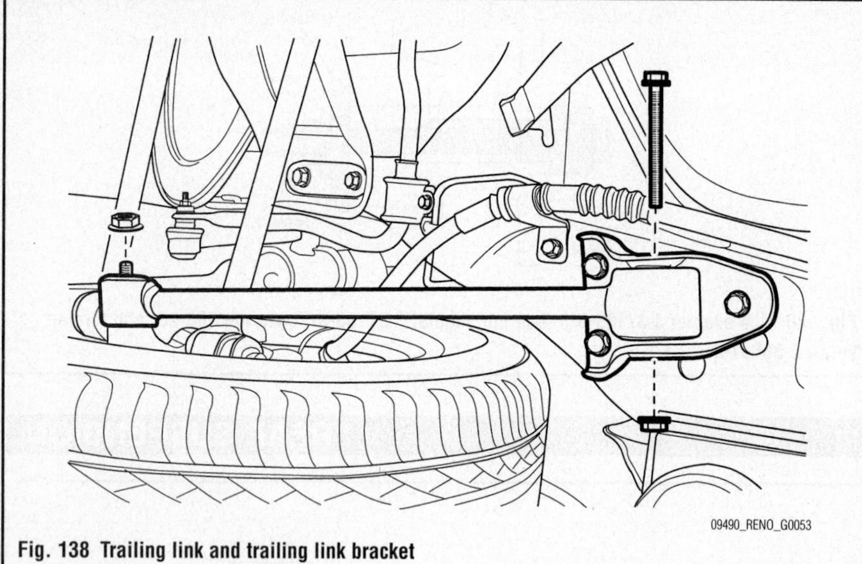

Fig. 138 Trailing link and trailing link bracket

4. Remove the trailing link bracket.

To install:

5. Install the trailing link bracket.
6. Install the trailing link bracket-to-body bolts.
7. Tighten the trailing link bracket-to-body bolts to 52 ft. lbs. (70 Nm)
8. Connect the trailing link to the trailing link bracket by installing the trailing link-to-trailing link bracket nut and bolt.
9. Tighten the trailing link-to-trailing link bracket nut to 74 ft. lbs. (100 Nm)
10. Lower the vehicle.

CROSSMEMBER

REMOVAL & INSTALLATION

See Figures 139 and 140.

1. Raise and suitably support the vehicle.
2. Remove the front parallel link bolt from the crossmember.
3. Remove the rear parallel link bolt from the crossmember.

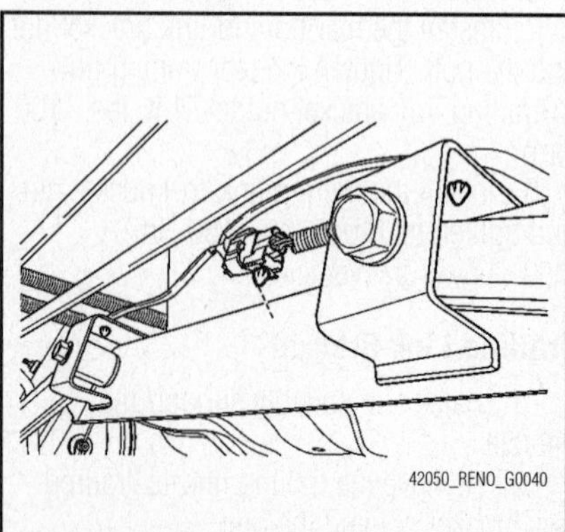

Fig. 139 Speed sensor connector

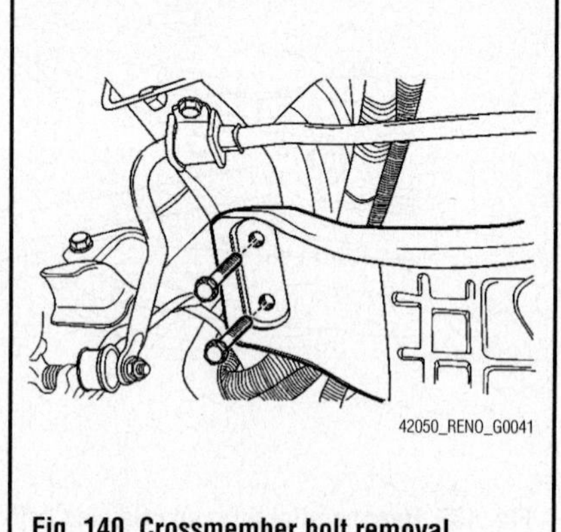

Fig. 140 Crossmember bolt removal

4. Remove the rear wheel speed sensor wiring from the crossmember.
5. Remove the crossmember-to-body bolts.
6. Remove the crossmember.

To install:

7. Install the crossmember.
8. Install the crossmember-to-body bolts.
9. Tighten the crossmember-to-body bolts to 83 ft. lbs. (112 Nm)
10. Route and install the rear wheel speed sensor wiring.
11. Install the rear parallel link bolt onto the crossmember.
12. Install the front parallel link bolt onto the crossmember.
13. Lower the vehicle.

STABILIZER LINKS

REMOVAL & INSTALLATION

See Figure 141.

1. Raise and suitably support the vehicle.

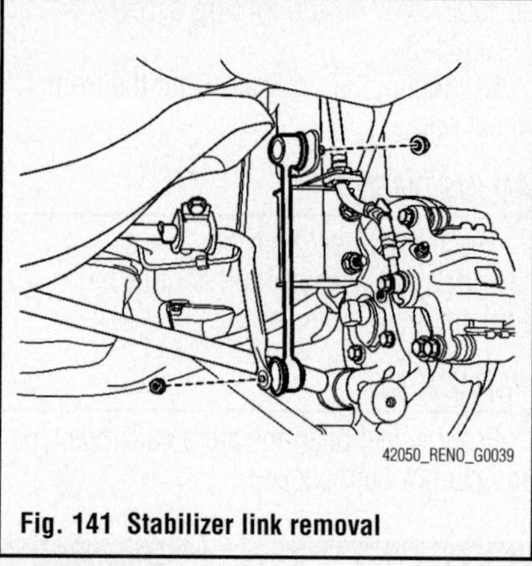

Fig. 141 Stabilizer link removal

2. Remove the wheel.
3. Remove the stabilizer shaft-to-stabilizer link nut.
4. Remove the stabilizer link-to-strut assembly nut.
5. Remove the stabilizer link.

To install:

6. Install the stabilizer link.
7. Install the stabilizer link-to-strut assembly nut.
8. Tighten the stabilizer link-to-strut assembly nut to 35 ft. lbs. (47 Nm)
9. Install the stabilizer shaft-to-stabilizer link nut.
10. Tighten the stabilizer shaft-to-stabilizer link nut to 35 ft. lbs. (47 Nm)
11. Install the wheel.
12. Lower the vehicle.

STABILIZER SHAFT

REMOVAL & INSTALLATION

See Figure 142.

1. Raise and suitably support the vehicle.
2. Remove the stabilizer shaft clamp bolts.
3. Remove the stabilizer shaft-to-stabilizer link nut.

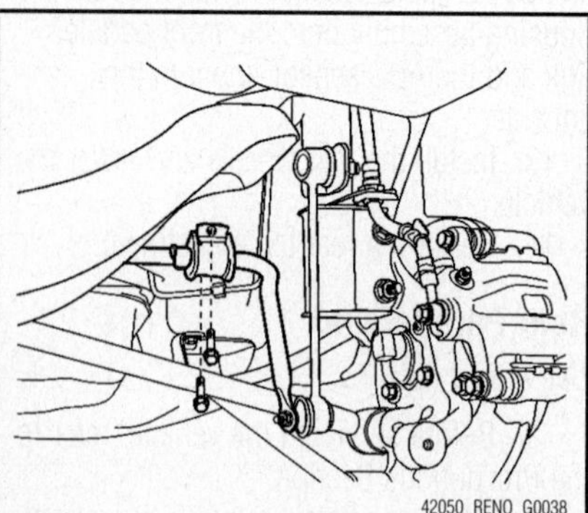

Fig. 142 Stabilizer shaft clamp and bolt removal

4. Remove the stabilizer shaft clamps, the insulators, and the stabilizer shaft from the vehicle.

To install:

5. Install the stabilizer shaft, the insulators, and the stabilizer shaft clamps.
6. Install the stabilizer shaft-to-stabilizer link nut.
7. Tighten the stabilizer shaft-to-stabilizer link nut to 35 ft. lbs. (47 Nm)
8. Install the stabilizer shaft clamp bolts.
9. Tighten the stabilizer shaft clamp bolts to 30 ft. lbs. (40 Nm)
10. Lower the vehicle.

MACPHERSON STRUTS

REMOVAL & INSTALLATION

See Figure 143.

1. Before servicing the vehicle, refer to the Precautions Section.
2. Remove the trunk carpeting that covers the rear strut mounting nuts. For station wagons, remove the panels that cover the luggage compartment wheelhouse trim panel.
3. Remove the rear strut mounting nuts.
4. Raise and suitably support the vehicle.
5. Remove the wheel.
6. Disconnect the parking brake.
7. Remove the clip that holds the brake hose to the strut assembly.
8. Remove the stabilizer link-to-strut assembly nut and disconnect the stabilizer link from the strut assembly.
9. Remove the knuckle-to-strut assembly nuts and the bolts.
10. Remove the rear strut assembly from the vehicle.
11. Remove the coil spring by disassembling the strut assembly as follows:
    a. Mount the rear strut assembly into the spring compressor (KM-329-A or DW320-010). Ensure that the hooks are properly seated.
    b. Compress the spring.
    c. Remove the lock nut from the strut dampener rod by using SST, J-42468.
    d. Remove the rear strut mount.
    e. Remove the rear spring upper seat, the dust cover, and the hollow bumper.
    f. Release the coil spring.
    g. Remove the rear spring and the rear spring lower seat.

To install:

12. Install the coil spring by assembling the strut assembly as follows:
    a. Install the rear spring lower seat and the rear spring.

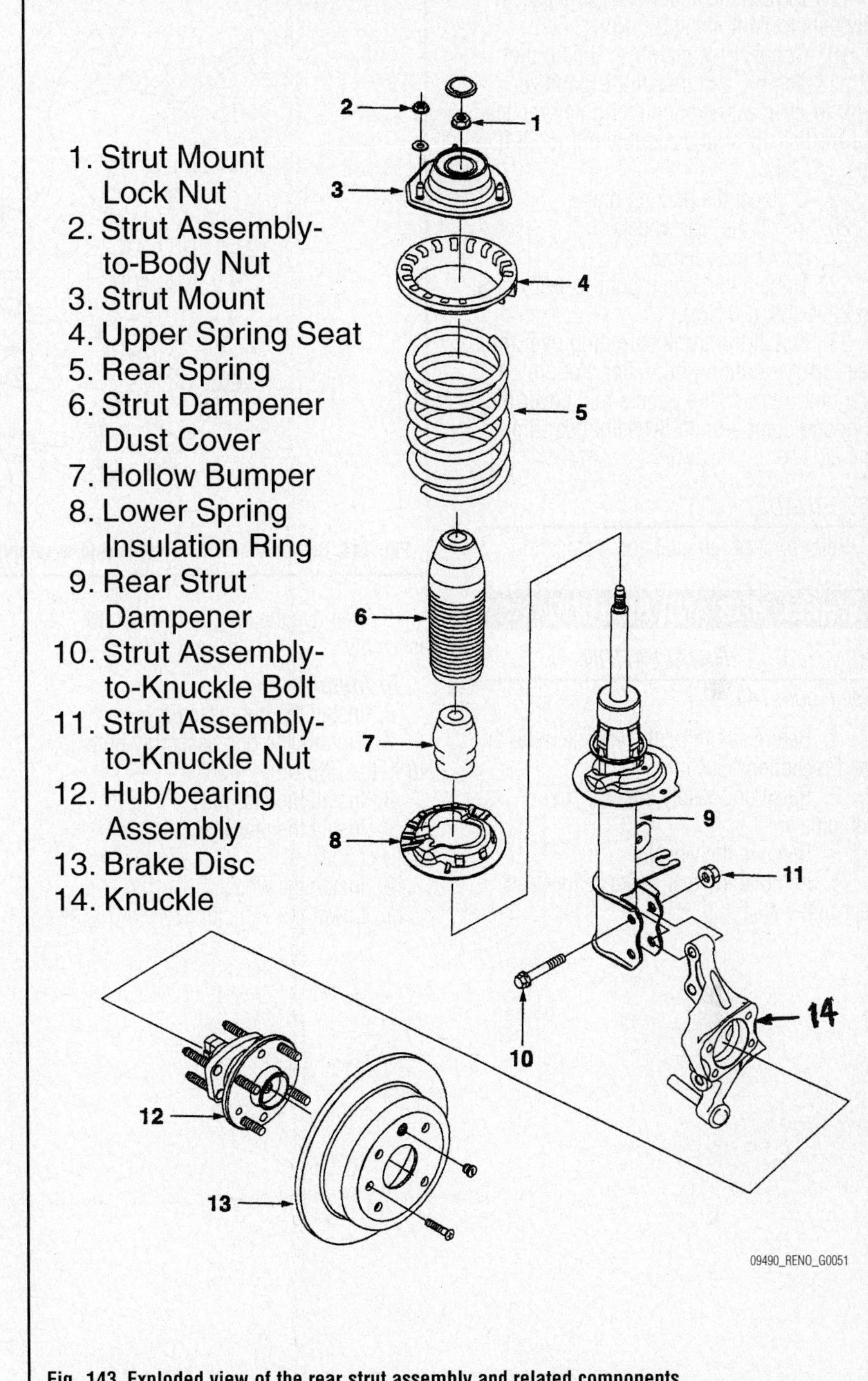

Fig. 143 Exploded view of the rear strut assembly and related components

    b. Compress the spring using the spring compressor tool.
    c. Install the hollow bumper, the dust cover, and the rear spring upper seat.
    d. Install the rear strut mount.
    e. Install the lock nut onto the strut dampener rod by using SST, J-42468.
    f. Tighten the ball joint-to-control arm nuts to 74 ft. lbs. (100 Nm) and the strut dampener-to-strut mount nut to 55 ft. lbs. (75 Nm).
    g. Release the spring.
    h. Remove the strut assembly from the spring compressor.
13. Install the rear strut assembly into the vehicle.
14. Secure the strut assembly by loosely attaching the strut mount-to-body nuts.
15. Install the knuckle-to-strut assembly nuts and bolts. Do not tighten.
16. Install the clip holding the brake hose to the strut assembly.

17. Tighten the knuckle-to-strut assembly nuts to 74 ft. lbs. (100 Nm).
18. Connect the stabilizer link to the strut assembly and install the stabilizer link-to-strut assembly nut. Tighten the stabilizer link-to-strut assembly nut to 35 ft. lbs. (47 Nm).
19. Connect the parking brake.
20. Install the rear wheels.
21. Lower the vehicle.
22. Tighten the strut mount-to-body nuts to 22 ft. lbs. (30 Nm).
23. Install the trunk carpeting over the rear strut mounting nuts. For station wagons, remove the panels that cover the luggage compartment wheelhouse trim panel.

OVERHAUL

Refer to strut removal and installation.

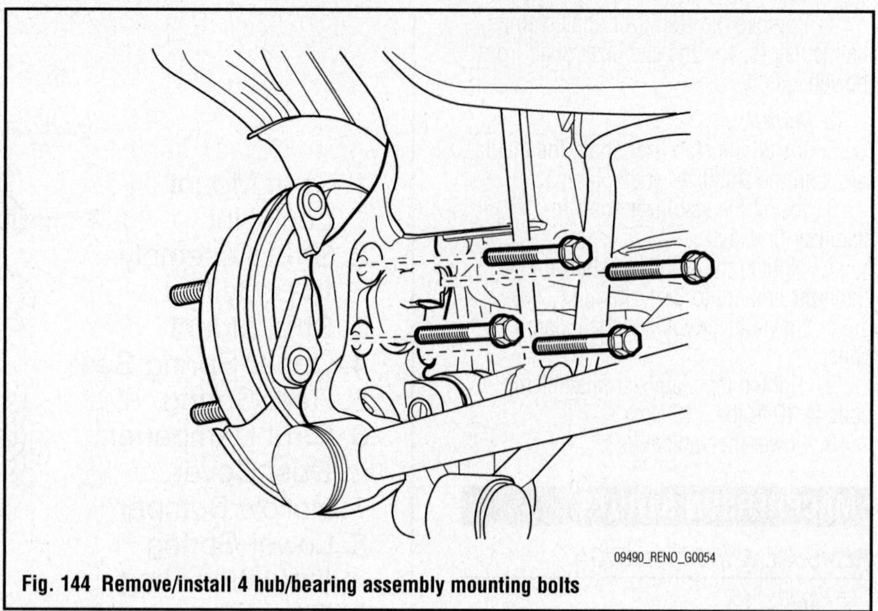

Fig. 144 Remove/install 4 hub/bearing assembly mounting bolts

WHEEL HUB AND BEARING

REMOVAL & INSTALLATION

See Figure 144.

1. Before servicing the vehicle, refer to the Precautions Section.
2. Raise and safely support the vehicle.
3. Remove the wheel.
4. Remove the rear brake caliper and rear brake disc.
5. Remove the hub bolts and hub assembly.

To install:

6. Install the hub assembly.
7. Tighten the hub assembly bolts to 48 ft. lbs. (65 Nm).
8. Install the hub nut.
9. Install the rear brake disc and rear brake caliper.
10. Install the wheel.
11. Lower the vehicle.
12. Check to ensure that the brakes are free from dragging and that proper braking is obtained.

ADJUSTMENT

The front wheel bearings are a cartridge type design and cannot be adjusted.

REPACKING

Front wheel bearings are a cartridge type and cannot be repacked.

SUZUKI

Grand Vitara • XL-7

19

ENGINE PERFORMANCE & EMISSION CONTROLS 19-78

FUEL 19-97

GASOLINE FUEL INJECTION SYSTEM 19-97

HEATING & AIR CONDITIONING SYSTEM....................19-102

PRECAUTIONS....................19-11

SPECIFICATIONS AND MAINTENANCE CHARTS.....19-4

SPECIFICATIONS AND MAINTENANCE CHARTS

ENGINE AND VEHICLE IDENTIFICATION CHART

Engine Code							Model Year	
Code ①	Liters (cc)	Cu. In.	Cyl.	Fuel Sys.	Engine Type	Eng. Mfg.	Code ②	Year
9:H27A	2.7 (2736)	165	6	MFI	DOHC	Suzuki	7	2007
LY7	3.6 (3600)	217	6	SFI	DOHC	Suzuki	8	2008

MFI: Multiport Fuel Injection

DOHC: Dual Overhead Cam

① Stamped on the upper rear of the engine block, just behind a cylinder head.

② 10th digit of the Vehicle Identification Number (VIN)

22140_VITA_C0001

GENERAL ENGINE SPECIFICATIONS

Year	Model	Engine Displacement Liters (cc)	Engine ID	Fuel System Type	Net Horsepower @ rpm	Net Torque @ rpm (ft. lbs.)	Bore x Stroke (in.)	Com-pression Ratio	Oil Pressure @ rpm
2007	Grand Vitara	2.7 (2736)	9:H27A	MFI	185@6000	162@4000	3.31x2.95	10.2:1	36.25@3000
	XL-7	3.6 (3600)	LY7	SFI	252@6400	243@2300	3.70x3.37	10.2:1	55-67@4000
2008	Grand Vitara	2.7 (2736)	9:H27A	MFI	185@6000	184@4500	3.46x2.95	10.2:1	36.25@3000
	XL-7	3.6 (3600)	LY7	SFI	252@6400	243@2300	3.70x3.37	10.2:1	55-67@4000

MFI: Multi-port Fuel Injection

22140_VITA_C0002

ENGINE TUNE-UP SPECIFICATIONS

Year	Engine Displacement Liters	Engine ID	Spark Plugs Gap (in.)	Ignition Timing (deg.)		Fuel Pump (psi)	Idle Speed (rpm)		Valve Clearance	
				MT	AT		MT	AT	In.	Ex.
2007	2.7	9:H27A	0.039-0.043	①	①	38-44	②	②	HYD	HYD
	3.6	LY7	0.0440	①	①	55-60	③	③	HYD	HYD
2008	2.7	9:H27A	0.039-0.043	①	①	38-44	③	③	HYD	HYD
	3.6	LY7	0.0440	①	①	55-60	③	③	HYD	HYD

① The ECM uses engine speed, the MAF sensor signal, CKP and CMP sensors to control dwell, and ignition timing of the spark.

② Engine Idling with throttle closed and vehicle stopped 10°-16°

③ Idle is controlled by the Throttle Actuator Control (TAC)

22140_VITA_C0003

CAPACITIES

Year	Model	Engine Displacement Liters	Engine ID	Engine Oil with Filter (qts.)	Transmission (pts.) 5-Spd	Transmission (pts.) Auto.	Transfer Case (pts.)	Drive Axle Front (pts.)	Drive Axle Rear (pts.)	Fuel Tank (gal.)	Cooling System (qts.)
2007	Grand Vitara	2.7	9:H27A	10.2	4.01	20.9	5.9	2.1	4.6	17.4	8.40
	XL-7	3.6	LY7	11	8.2	16.4	1.7	2.1	5.3	18.5	10.6
2008	Grand Vitara	2.7	9:H27A	10.2	4.01	20.9	5.9	2.1	1.9	17.4	8.65
	XL-7	3.6	LY7	11	8.2	16.4	1.7	2.1	5.3	18.5	10.6

Note: All capacities are approximate. Add fluid gradually and check to be sure a proper fluid level is obtained.

22140_VITA_C0004

FLUID SPECIFICATIONS

Year	Engine Displ. Liters	Engine ID	Engine Oil	Auto. Trans.	Manual Trans.	Transfer Case	Differential Lubricant	Power Steering Fluid	Master Cylinder	Engine Coolant
2007	2.7	9:H27A	10W-30 ①	②	③	④	④	Dexron II	DOT 3	Ethylene glycol
	3.6	LY7	10W-30 ①	⑤	③	④	⑥	Dexron II	DOT 3	Dex-Cool
2008	2.7	9:H27A	10W-30 ①	②	③	④	④	Dexron II	DOT 3	Ethylene glycol
	3.6	LY7	10W-30 ①	⑤	③	④	⑥	Dexron II	DOT 3	Dex-cool

Note: 5-30 American Petroleum Institute (API) must be used if ambient temperatures are below –18 °C (0 °F).

DOT: Department Of Transpotation

① Use only engine oil certified by the American Petroleum Institute (API) for Gasoline Engines with the "starburst" symbol

② SUZUKI ATF 3317 or Mobil ATF 3309

③ API GL-4 75W-90

④ API GL-5 80W-90

⑤ Fluid Type T- IV

⑥ Synthetic Gear Oil GL75W90

22140_VITA_C0015

VALVE SPECIFICATIONS

Year	Engine Displacement Liters	Engine ID	Seat Angle (deg.)	Face Angle (deg.)	Spring Test Pressure (lbs. @ in.)	Spring Installed Height (in.)	Stem-to-Guide Clearance (in.) Intake	Stem-to-Guide Clearance (in.) Exhaust	Stem Diameter (in.) Intake	Stem Diameter (in.) Exhaust
2007	2.7	9:H27A	①	44.25	②	N/A	0.007-0.0028	0.09-0.0035	0.2348-0.2354	0.2348-0.2354
	3.6	LY7	①	44.25	③	1.3779	0.0010-0.0026	0.0014-0.003	0.2344-0.2352	0.2341-0.2348
2008	2.7	9:H27A	①	44.25	②	N/A	0.007-0.0028	0.09-0.0035	0.2348-0.2354	0.2348-0.2354
	3.6	LY7	①	44.25	③	1.3779	0.0010-0.0026	0.0014-0.003	0.2344-0.2352	0.2341-0.2348

① Valve Seat Angle Seating Surface : 45 degre

Valve Seat Angle Relief Surface : 30 degrees

Valve Seat Angle Undercut Surface : 60 degrees

② 46.1- 53.1 lbs @ 1.480 inches

③ 56-61 lbs @ 1.3779 inches

22140_VITA_C0006

CAMSHAFT AND BEARING SPECIFICATIONS

All measurements are given in inches.

Year	Engine Displacement Liters	Engine VIN	Journal Outside Diameter	Brg. Oil Clearance	Shaft End-play	Runout	Journal Bore	Lobe Lift Intake	Lobe Lift Exhaust
2007	2.7	9:H27A	1.0220-1.0228	0.0008-0.0029	0.0018-0.0085	0.002	1.0236-1.0249	1.7719-1.7781	1.7933-1.7996
	3.6	LY7	①	0.0016-0.0033	0.0018-0.0085	0.002	②	1.6687–1.6805	1.6703 1.6821
2008	2.7	9:H27A	1.0220-1.0228	0.0008-0.0029	0.0018-0.0085	0.002	1.0236-1.0249	1.7719-1.7781	1.7933-1.7996
	3.6	LY7	①	0.0016-0.0033	0.0018-0.0085	0.002	②	1.6687–1.6805	1.6703 1.6821

① Journal Bore-Front Number 1: 1.3779-1.3787
Journal Bore-Middle and Rear Number 2-4: 1.0630-1.0638

② Journal Diameter-Front Number 1: 1.3754-1.3764
Journal Diameter-Middle and Rear Number 2-4: 1.0605-1.0614

22140_VITA_C0014

CRANKSHAFT AND CONNECTING ROD SPECIFICATIONS

All measurements are given in inches.

Year	Engine Displacement Liters	Engine ID	Crankshaft Main Brg. Journal Dia.	Crankshaft Main Brg. Oil Clearance	Crankshaft Shaft End-play	Crankshaft Thrust on No.	Connecting Rod Journal Diameter	Connecting Rod Oil Clearance	Connecting Rod Side Clearance
2007	2.7	9:H27A	①	0.0009-0.0015	0.0040-0.0137	3	②	0.0018-0.0024	0.0099-0.0137
	3.6	LY7	2.6768-2.6775	0.0004-0.0024	0.0039-0.0130	3	2.2044-2.2050	0.0004-0.0028	0.0098-0.0177
2008	2.7	9:H27A	①	0.0009-0.0015	0.0040-0.0137	3	②	0.0018-0.0031	0.0099-0.0137
	3.6	LY7	2.6768-2.6775	0.0004-0.0024	0.0039-0.0130	3	2.2044-2.2050	0.0024-0.0028	0.0098-0.0177

NA: Not Available

① Journal (Main Bearing) 1: 2.75590-2.75613 inches
Journal 2: 2.75614-2.75637 inches
Journal 3: 2.75638-2.75660 inches

② Journal (Rod) sizing drepends on 1 of 3 stampings on the rod and its cap.
Journal 1: 2.0867-2.0868 inches
Journal 2: 2.0869-2.0870 inches
Journal 3: 2.0871-2.0873 inches

22140_VITA_C0005

PISTON AND RING SPECIFICATIONS

All measurements are given in inches.

	Engine			Ring Gap			Ring Side Clearance		
	Displacement	Engine	Piston	Top	Bottom	Oil	Top	Bottom	Oil
Year	Liters	ID	Clearance	Compression	Compression	Control	Compression	Compression	Control
2007	2.7	9:H27A	0.0008-0.0015	0.0079-0.0276	0.0138-0.0276	0.0079-0.0709	0.0012-0.0027	0.0008-0.0023	NA
	3.6	LY7	0.0008-0.0015	0.0059-0.0118	0.0110-0.0189	0.0059-0.0236	0.0012-0.0026	0.0006-0.0024	0.0012-0.0067
2008	2.7	9:H27A	0.0008-0.0015	0.0079-0.0276	0.0138-0.0276	0.0079-0.0709	0.0012-0.0027	0.0008-0.0023	NA
	3.6	LY7	0.0008-0.0015	0.0059-0.0118	0.0110-0.0189	0.0059-0.0236	0.0012-0.0026	0.0006-0.0024	0.0012-0.0067

NA: Not Available

22140_VITA_C0007

TORQUE SPECIFICATIONS

All readings in ft. lbs.

	Engine		Cylinder	Main	Rod	Crankshaft					
	Displacement	Engine	Head	Bearing	Bearing	Damper	Flywheel	Manifold		Spark	Lug
Year	Liters	ID	Bolts	Bolts	Bolts	Bolts	Bolts	Intake	Exhaust	Plugs	Nut
2007	2.7	9:H27A	①	②	③	108.5	49.5	16.0	36.5	18	72.5
	3.6	LY7	④	⑤	⑥	⑦	⑧	18.0	15.0	18	100
2008	2.7	9:H27A	①	②	③	108.5	49.5	16.0	36.5	18	72.5
	3.6	LY7	④	⑤	⑥	⑦	⑧	18.0	15.0	18	100

① Step 1: 38.5 ft. lbs.
Step 2: 61 ft. lbs.
Step 3: Loosen in reverse order to 0 ft. lbs.
Step 4: 38 ft. lbs.
Step 5: 76 ft. lbs.

② Inner : First Pass: 15 ft. lbs.
Final Step: 80 degrees.

③ Step 1: 11 ft. lbs.
Step 2: 45 degrees.
Step 3: 45 degrees.

④ First Pass: 33 ft. lbs.
Final Step 2: tighten 120 degrees.

⑤ Step 1: 22 ft. lbs.
Step 2: 30.5 ft. lbs.
Step 3: Tighten bolts 40 degrees.

⑥ First Pass : 22 ft. lbs.
Second Pass: Back off to Zero.
Third Pass: 18 ft-lbs.
Final Step: 110 degrees.

⑦ First Pass:74 ft. lbs.
Final Step: 150 degrees.

⑧ First Pass:22 ft. lbs.
Final Step: 45 degrees.

22140_VITA_C0008

WHEEL ALIGNMENT

Year	Model	Caster Range (+/-Deg.)	Caster Preferred Setting (Deg.)	Camber Range (+/-Deg.)	Camber Preferred Settings Front (Deg.)	Camber Preferred Settings Rear (Deg.)	Toe-in Front (Deg.)	Toe-in Rear (Deg.)
2007	Grand Vitara	0.00	2.00	1.00	0.00	-1.00	0.00	0.00
	XL-7	0.75	3.00	0.75	-0.60	-0.5 ± .75	0.15 ± 0.20	+0.2 ± 0.20
2008	Grand Vitara	0.00	2.00	1.00	0.00	-1.00	0.00	0.00
	XL-7	0.75	3.00	0.75	-0.60	-0.5 ± .75	0.04 ± 0.08	+0.2 ± 0.20

22140_VITA_C0009

TIRE, WHEEL AND BALL JOINT SPECIFICATIONS

Year	Model	OEM Tires Standard	OEM Tires Optional	Tire Pressures (psi) Front	Tire Pressures (psi) Rear	Wheel Size	Ball Joint Inspection
2007	Grand Vitara	P235/60R16	none	26	26	16x6	①
	XL-7	P235/60R16	none	26	26	16x6	①
2008	Grand Vitara	P225/70R16	P225/65R17	26	26	16x6/17x6	①
	XL-7	P235/60R16	none	26	26	16x6	①

OEM: Original Equipment Manufacturer

PSI: Pounds Per Square Inch

① Replace if any measurable movement is found.

22140_VITA_C0010

BRAKE SPECIFICATIONS

All measurements in inches unless noted

Year	Model	Brake Disc Original Thickness	Brake Disc Minimum Thickness	Brake Disc Max. Runout	Brake Drum Diameter Original Inside Diameter	Brake Drum Diameter Max. Wear Limit	Brake Drum Diameter Max. Machine Diameter	Minimum Lining Thickness Front	Minimum Lining Thickness Rear	Brake Caliper Bracket Bolts (ft. lbs.)	Brake Caliper Mounting Bolts (ft. lbs.)
2007	Grand Vitara	0.866	0.080	0.006	8.66	8.74	8.74	0.08	0.04	61.5	①
	XL-7	N/A	0.080	0.002	N/A	N/A	N/A	0.08	0.08	61.5	26
2008	Grand Vitara	0.430	0.080	0.004	N/A	N/A	N/A	0.08	0.04	61.5	26
	XL-7	N/A	0.080	0.002	N/A	N/A	N/A	0.08	0.08	61.5	26

① 10mm: 37 ft. lbs.
12mm: 62 ft. lbs.

22140_VITA_C0011

SCHEDULED MAINTENANCE INTERVALS
SUZUKI—GRAND VITARA & XL-7 2007 & 2008

TO BE SERVICED	TYPE OF SERVICE	VEHICLE MILEAGE INTERVAL (x1000)												
		7.5	15	22.5	30	37.5	45	52.5	60	67.5	75	82.5	90	97.5
Engine oil & filter	R	✓	✓	✓	✓	✓	✓	✓	✓	✓	✓	✓	✓	✓
Automatic transmission fluid ①	S/I	✓	✓	✓	✓	✓	✓	✓	✓	✓	✓	✓	✓	✓
Manual transmission oil ②	S/I	✓	✓	✓	✓	✓	✓	✓	✓	✓	✓	✓	✓	✓
Steering system	S/I	✓	✓	✓	✓	✓	✓	✓	✓	✓	✓	✓	✓	✓
Transfer & differential oil ③	S/I	✓	✓	✓	✓	✓	✓	✓	✓	✓	✓	✓	✓	✓
Wheel discs & free wheeling hubs	S/I	✓	✓	✓	✓	✓	✓	✓	✓	✓	✓	✓	✓	✓
Suspension system	S/I	✓		✓		✓		✓		✓		✓		✓
Brake discs & pads	S/I		✓		✓		✓		✓		✓		✓	
Brake drums & shoes	S/I		✓		✓		✓		✓		✓		✓	
Brake fluid ③	S/I		✓		✓		✓		✓		✓		✓	
Brake hoses & pipes	S/I		✓		✓		✓		✓		✓		✓	
Brake pedal	S/I		✓		✓		✓		✓		✓		✓	
Brake lever & cable	S/I		✓		✓		✓		✓		✓		✓	
Clutch fluid	S/I		✓		✓		✓		✓		✓		✓	
Differential oil and extention case oil (2WD) ④	S/I		✓		✓		✓		✓		✓		✓	
Propeller shafts	S/I		✓		✓		✓		✓		✓		✓	
Valve lash (clearance)	S/I												✓	
Wheel bearings	S/I		✓		✓		✓		✓		✓		✓	
Wheel disc (rotor)	S/I	✓	✓	✓	✓	✓	✓	✓	✓	✓	✓	✓	✓	
Air cleaner filter element	R				✓				✓				✓	
Engine coolant	R				✓				✓				✓	
Fuel filter	R													
Spark plugs	R								✓					
Cooling system hoses	S/I				✓				✓				✓	
Drive belt(s) ③	S/I				✓				✓				✓	
Exhaust pipes & mountings	S/I				✓				✓				✓	
Fuel lines & connections	S/I				✓				✓				✓	
Emission-related hoses & tubes	S/I								✓					
Latches, hinges and locks	S/I	✓	✓	✓	✓	✓	✓	✓	✓	✓	✓	✓	✓	✓
EVAP canister	R	every 150,000 miles												
HVAC air filter	R				✓				✓				✓	
Iginition coil (plug cap)	S/I			✓				✓				✓		
Fuel tank and cap	S/I		✓		✓		✓		✓		✓		✓	
Tires	S/I	✓	✓	✓	✓	✓	✓	✓	✓	✓	✓	✓	✓	✓

R: Replace S/I: Service or Inspect

① Replace at 100,000 miles.

② Replace oil every 30,000 miles.

③ Replace every 60,000 miles.

④ Replace at 7500 miles and Inspect every 15,000 miles

22140_VITA_C0012

SCHEDULED MAINTENANCE INTERVALS

SUZUKI—GRAND VITARA & XL-7 2007 & 2008

Footnotes Continued

FREQUENT OPERATION MAINTENANCE (SEVERE SERVICE)

If a vehicle is operated under any of the following conditions it is considered severe service:

- Extremely dusty areas.

- 50% or more of the vehicle operation is in 32°C (90°F) or higher temperatures, or constant operation in temperatures below 0°C (32°F).

- Prolonged idling (vehicle operation in stop and go traffic).

- Frequent short running periods (engine does not warm to normal operating temperatures).

- Police, taxi, delivery usage or trailer towing usage.

Oil & oil filter: replace every 3000 miles.

Air cleaner filter element: service or inspect every 3000 miles & replace every 15,000 miles.

Steering wheel free play, gear box oil & linkage: service or inspect every 3000 miles.

Brake & nuts on chassis: tighten every 6000 miles.

Brake discs & pads (front): service or inspect every 6000 miles.

Brake drums & shoes (rear): service or inspect every 6000 miles.

Exhaust pipes & mountings: tighten every 6000 miles.

Propeller shafts: service or inspect every 6000 miles.

Automatic transmission fluid & filter: replace every 15,000 miles.

Distributor cap & ignition wires: service or inspect every 15,000 miles.

Drive belt(s): service or inspect every 15,000 miles.

Manual transmission oil: replace every 15,000 miles.

Transfer & differential oil: replace every 15,000 miles.

22140_VITA_C0013

PRECAUTIONS

Before servicing any vehicle, please be sure to read all of the following precautions, which deal with personal safety, prevention of component damage, and important points to take into consideration when servicing a motor vehicle:

- Never open, service or drain the radiator or cooling system when the engine is hot; serious burns can occur from the steam and hot coolant.
- Observe all applicable safety precautions when working around fuel. Whenever servicing the fuel system, always work in a well-ventilated area. Do not allow fuel spray or vapors to come in contact with a spark, open flame, or excessive heat (a hot drop light, for example). Keep a dry chemical fire extinguisher near the work area. Always keep fuel in a container specifically designed for fuel storage; also, always properly seal fuel containers to avoid the possibility of fire or explosion. Refer to the additional fuel system precautions later in this section.
- Fuel injection systems often remain pressurized, even after the engine has been turned **OFF**. The fuel system pressure must be relieved before disconnecting any fuel lines. Failure to do so may result in fire and/or personal injury.
- Brake fluid often contains polyglycol ethers and polyglycols. Avoid contact with the eyes and wash your hands thoroughly after handling brake fluid. If you do get brake fluid in your eyes, flush your eyes with clean, running water for 15 minutes. If eye irritation persists, or if you have taken brake fluid internally, IMMEDIATELY seek medical assistance.
- The EPA warns that prolonged contact with used engine oil may cause a number of skin disorders, including cancer. You should make every effort to minimize your exposure to used engine oil. Protective gloves should be worn when changing oil. Wash your hands and any other exposed skin areas as soon as possible after exposure to used engine oil. Soap and water, or waterless hand cleaner should be used.
- All new vehicles are now equipped with an air bag system, often referred to as a Supplemental Restraint System (SRS) or Supplemental Inflatable Restraint (SIR) system. The system must be disabled before performing service on or around system components, steering column, instrument panel components, wiring and sensors. Failure to follow safety and disabling procedures could result in accidental air bag deployment, possible personal injury and unnecessary system repairs.
- Always wear safety goggles when working with, or around, the air bag system. When carrying a non-deployed air bag, be sure the bag and trim cover are pointed away from your body. When placing a non-deployed air bag on a work surface, always face the bag and trim cover upward, away from the surface. This will reduce the motion of the module if it is accidentally deployed. Refer to the additional air bag system precautions later in this section.
- Clean, high quality brake fluid from a sealed container is essential to the safe and proper operation of the brake system. You should always buy the correct type of brake fluid for your vehicle. If the brake fluid becomes contaminated, completely flush the system with new fluid. Never reuse any brake fluid. Any brake fluid that is removed from the system should be discarded. Also, do not allow any brake fluid to come in contact with a painted surface; it will damage the paint.
- Never operate the engine without the proper amount and type of engine oil; doing so WILL result in severe engine damage.
- Timing belt maintenance is extremely important. Many models utilize an interference-type, non-freewheeling engine. If the timing belt breaks, the valves in the cylinder head may strike the pistons, causing potentially serious (also time-consuming and expensive) engine damage. Refer to the maintenance interval charts for the recommended replacement interval for the timing belt, and to the timing belt section for belt replacement and inspection.
- Disconnecting the negative battery cable on some vehicles may interfere with the functions of the on-board computer system(s) and may require the computer to undergo a relearning process once the negative battery cable is reconnected.
- When servicing drum brakes, only disassemble and assemble one side at a time, leaving the remaining side intact for reference.
- Only an MVAC-trained, EPA-certified automotive technician should service the air conditioning system or its components.

BRAKES

ANTI-LOCK BRAKE SYSTEM (ABS)

GENERAL INFORMATION

ABS is controlled by ABS control module or by ESP® control module (if equipped).

ABS activation and control with ESP® are done by ESP® hydraulic unit/control module assembly.

The ABS (Antilock Brake System) controls the fluid pressure applied to the wheel cylinder of each brake from the master cylinder so that each wheel is not locked even when hard braking is applied.

This ABS has also the following function. While braking is applied, but before ABS control becomes effective, braking force is distributed between the front and rear so as to prevent the rear wheels from being locked too early for better stability of the vehicle.

The main component parts of this ABS include the following parts in addition to those of the conventional brake system.

- Wheel speed sensor which senses revolution speed of each wheel and outputs its signal.
- ABS warning lamp which lights to inform abnormality when system fails to operate properly.
- ABS (ESP®) hydraulic unit/control module assembly is incorporated ABS (ESP®) control module, ABS (ESP®) hydraulic unit (actuator assembly), solenoid valve power supply driver (transistor), solenoid valve driver (transistor), pump motor driver (transistor).
- ABS (ESP®) control module which sends operation signal to ABS (ESP®) hydraulic unit to control fluid pressure applied to each wheel cylinder based on signal from each wheel speed sensor so as to prevent wheel from locking.
- ABS (ESP®) hydraulic unit which operates according to signal from ABS (ESP®) control module to control fluid pressure applied to wheel cylinder of each 4 wheels.
- Solenoid valve power supply driver (transistor) which supplies power to solenoid valve in ABS (ESP®) hydraulic unit.
- Solenoid valve driver (transistor) which controls each solenoid valves in ABS (ESP®) hydraulic unit.
- Pump motor driver (transistor) which supplies power to pump motor in ABS (ESP®) hydraulic unit.
- This ABS is equipped with Electronic Brake force Distribution (EBD) system that controls a fluid pressure of rear wheels to

best condition, which is the same function as that of proportioning valve, by the signal from wheel sensor independently of change of load due to load capacity and so on. And if the EBD system fails to operate properly, the brake warning lamp lights to inform abnormality.

➡It is recommended that entire hydraulic system be thoroughly flushed with clean brake fluid whenever new parts are installed in hydraulic system. Periodical change of brake fluid is also recommended.

WHEEL SPEED SENSORS

REMOVAL & INSTALLATION

Front

2007–08 Grand Vitara

See Figure 1.

1. Disconnect the negative battery cable.
2. Disconnect front wheel speed sensor coupler.
3. Raise vehicle and remove wheel.
4. Remove harness clamp, clamp bolts and grommet.
5. Remove front wheel speed sensor from knuckle.

To install:

6. Check that no foreign material is attached to sensor and mating encoder.
7. Install it by reversing removal procedure.
8. Tighten wheel speed sensor and speed sensor harness clamp bolts to 8 ft. lbs. (11 Nm).
9. Reconnect speed sensor harness coupler.
10. Check that there is no clearance between sensor and knuckle.
11. Connect the negative battery cable.

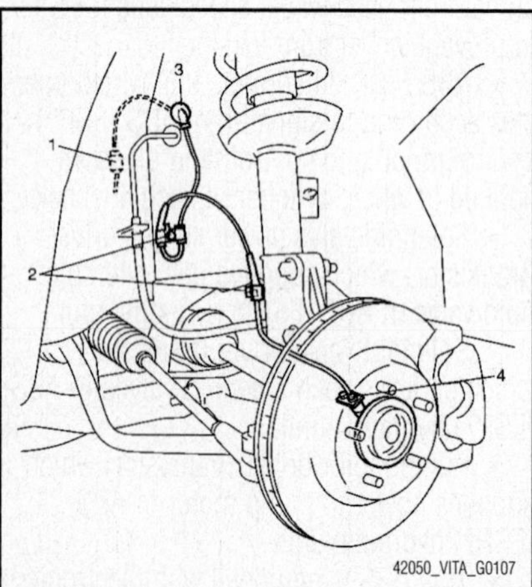

Fig. 1 Front speed sensor and harness

2007–08 XL-7

See Figure 2.

1. Disconnect the negative battery cable.
2. Raise vehicle and remove wheel.
3. Disconnect front wheel speed sensor coupler.
4. Remove sensor bolt.
5. Remove front wheel speed sensor from knuckle.

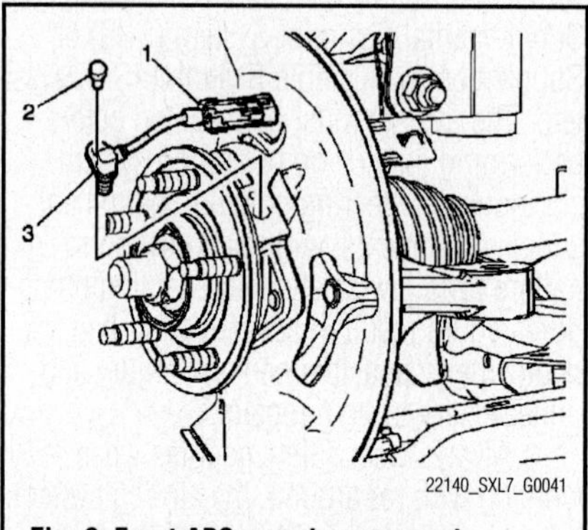

Fig. 2 Front ABS speed sensor and harness

To install:

6. Route the wheel speed sensor electrical harness through the splash shield.
7. Check that no foreign material is attached to sensor and mating encoder.
8. Tighten wheel speed sensor and speed sensor harness clamp bolts to 71 inch lbs. (8 Nm).
9. Reconnect speed sensor harness coupler.
10. Check that there is no clearance between sensor and knuckle.
11. Connect the negative battery cable.

Rear

2007–08 Grand Vitara

See Figures 3 and 4.

1. Disconnect the negative battery cable.
2. Disconnect rear wheel speed sensor coupler.
3. Raise vehicle and remove wheel.
4. Remove harness clamp and clamp bolts.
5. Remove rear wheel speed sensor from knuckle.

To install:

6. Check that no foreign material is attached to sensor and mating encoder.

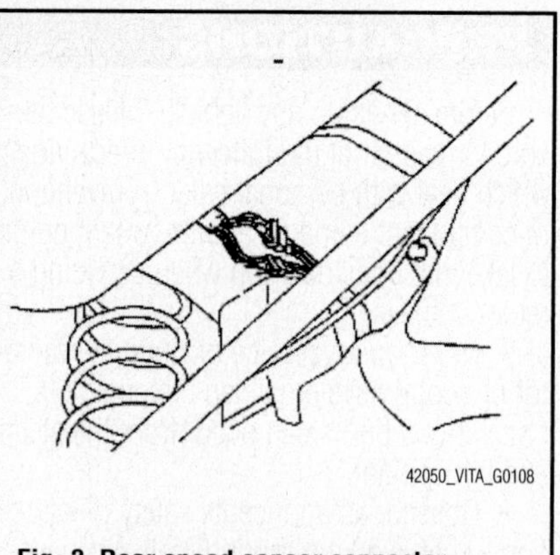

Fig. 3 Rear speed sensor connector

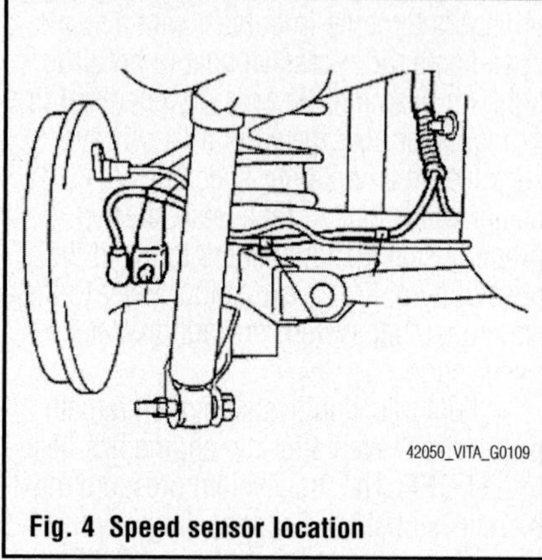

Fig. 4 Speed sensor location

7. Install rear speed sensor and harness, tighten mounting bolts for both to 8 ft. lbs. (11 Nm).
8. Reconnect speed sensor harness coupler.
9. Lower vehicle.
10. Connect the negative battery cable.

2007–08 XL-7

See Figure 5.

1. Disconnect the negative battery cable.
2. Raise vehicle and remove wheel.
3. Remove the park brake shoes assembly.

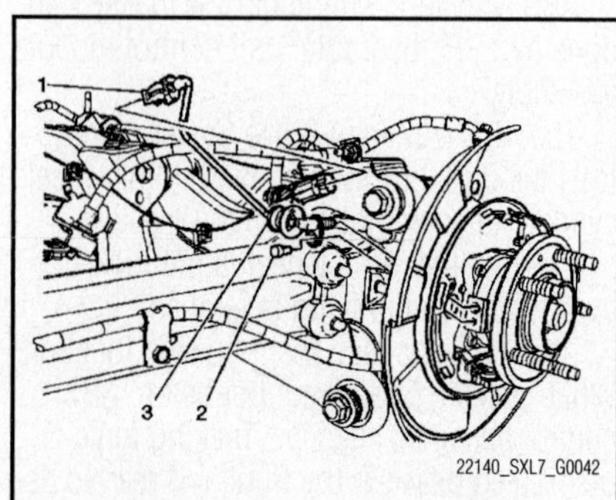

Fig. 5 Rear ABS speed sensor and harness

4. Remove the following:
   - Park Brake Shoe Hold Down Spring (Qty: 2) by compressing the spring and rotating a 1/4 turn to release.
   - Park Brake Shoe Adjuster Spring
   - Park Brake Shoe Adjuster Screw
   - Park Brake Shoe Return Spring
   - Park Brake Shoes
5. Remove the wheel speed sensor electrical connector.
6. Remove the wheel speed sensor bolt.
7. Release the wheel speed sensor electrical harness grommet from the backing plate.
8. Route the wheel speed sensor electrical harness through the backing plate to remove.

To install:

9. Route the wheel speed sensor electrical harness through the backing plate.
10. Install the wheel speed sensor electrical harness grommet into the backing plate.
11. Install the wheel speed sensor bolt and torque to 71 inch lbs. (8 Nm).
12. Connect the Speed Sensor Electrical Connector.
13. Install the parking brake assembly by the following steps:
    - Park brake shoes
    - Park brake shoe return spring.
    - Park brake shoe adjuster screw.
    - Park brake shoe adjuster spring.
    - Park brake shoe hold down spring—rotate ¼ turn to install.
14. Install the wheel and torque to 100 ft. lbs. (140 Nm).
15. Connect the negative battery cable.
16. Test operation of system.

WHEEL SPEED SENSOR RINGS (TOOTHED RINGS)

REMOVAL & INSTALLATION

Tone ring is part of the hub and bearing assembly. Refer to front or rear hub bearing removal and installation, as applicable, in the Suspension Section.

BLEEDING THE BRAKE SYSTEM

BLEEDING PROCEDURE

See Figure 6.

⁂ CAUTION

When adding fluid to the brake master cylinder reservoir, use only GM approved or equivalent DOT-3 brake fluid from a clean, sealed brake fluid container. The use of any type of fluid other than the recommended type of brake fluid may cause contamination which could result in damage to the internal rubber seals and/or rubber linings of hydraulic brake system components.

➡It is recommended that entire hydraulic system be thoroughly flushed with clean brake fluid whenever new parts are installed in hydraulic system. Periodical change of brake fluid is also recommended.

Be sure to bleed air of brake system according to the following procedure when its fluid hydraulic circuit has been disconnected.

Hydraulic lines of brake system consists of two separate lines, one for front wheel brakes and the other for rear wheel brakes.

Air bleeding is necessary at right and left front wheel brakes, left rear wheel brake and LSPV (if equipped without ABS), i.e. 4 places (3 places for vehicle with ABS) in all.

➡For vehicle equipped with ABS, make sure that ignition switch turns OFF.

1. Fill master cylinder reservoir with brake fluid and keep at least one-half full of fluid during bleeding operation.

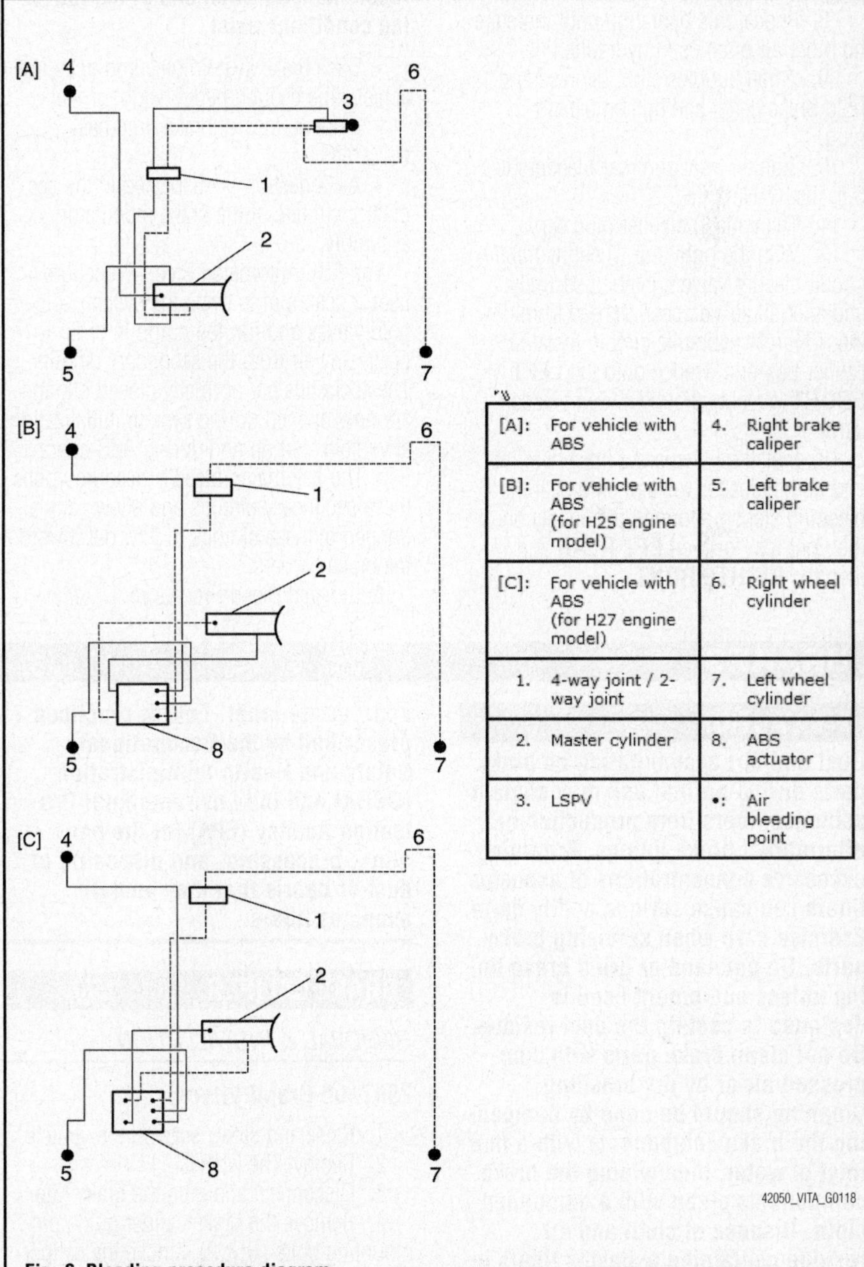

Fig. 6 Bleeding procedure diagram

2. Remove bleeder plug cap.

3. Install a proper box-end wrench onto the **RIGHT REAR** wheel hydraulic circuit bleeder valve.

4. Attach a vinyl tube to bleeder plug of bleeder valve, and insert the other end into container.

5. Have an assistant slowly depress the brake pedal fully and maintain steady pressure on the pedal.

6. Depress brake pedal several times, and then while holding it depressed, loosen bleeder plug about one-third to one half turn.

7. When fluid pressure in the cylinder is almost depleted, retighten bleeder plug.

8. Repeat this operation until there are no more air bubbles in hydraulic line.

9. When bubbles stop, depress and hold brake pedal and tighten bleeder plug.

10. Tighten front and rear bleeders to 5 ft. lbs. (7 Nm)

11. Then attach bleeder plug cap.

12. With the right rear wheel hydraulic circuit bleeder valve tightened securely, and after all air has been purged from the right rear hydraulic circuit, install a proper box-end wrench onto the **LEFT FRONT** wheel hydraulic circuit bleeder valve.

13. Install a transparent hose over the end of the bleeder valve, then repeat the bleeding steps preformed previously on the right rear for the **LEFT REAR** and then the **RIGHT FRONT**

14. After completing bleeding operation, apply fluid pressure to pipe line and check for leakage.

15. Replenish fluid into reservoir up to **MAX** level.

16. Check brake pedal for (sponginess). If found spongy, repeat entire procedure of bleeding.

BLEEDING THE ABS SYSTEM

➡Before performing the antilock brake system (ABS) Automated Bleed Procedure, first perform a manual or pressure bleed of the base brake system. Refer to Hydraulic Brake System Bleeding The automated bleed procedure is recommended when one of the following conditions exist:

- Base brake system bleeding does not achieve the desired pedal height or feel
- Extreme loss of brake fluid has occurred
- Air ingestion is suspected in the secondary circuits of the brake modulator assembly

The ABS Automated Bleed Procedure uses a scan tool to cycle the system solenoid valves and run the pump in order to purge any air from the secondary circuits. These circuits are normally closed off, and are only opened during system initialization at vehicle start up and during ABS operation. The automated bleed procedure opens these secondary circuits and allows any air trapped in these circuits to flow out toward the brake corners.

Automated Bleed Procedure

⁂ CAUTION

The Auto Bleed Procedure may be terminated at any time during the process by pressing the EXIT button. No further Scan Tool prompts pertaining to the Auto Bleed procedure will be given. After exiting the bleed procedure, relieve bleed pressure and disconnect bleed equipment per manufacturer's instructions. Failure to properly relieve pressure may result in spilled brake fluid causing damage to components and painted surfaces.

1. Raise and support the vehicle.

2. Remove all 4 tire and wheel assemblies.

3. Inspect the brake system for leaks and visual damage.

4. Lower the vehicle.

5. Inspect the battery state of charge.

6. Install a scan tool.

7. Turn the ignition ON, with the engine OFF.

8. With the scan tool, establish communications with the ABS system. Select Special Functions.

9. Select Automated Bleed from the Special Functions menu.

10. Raise and support the vehicle.

11. Following the directions given on the scan tool, pressure bleed the base brake system.

12. Follow the scan tool directions until the desired brake pedal height is achieved.

13. Road test the vehicle while inspecting that the pedal remains high and firm.

BRAKES

FRONT DISC BRAKES

⁂ CAUTION

Dust and dirt accumulating on brake parts during normal use may contain asbestos fibers from production or aftermarket brake linings. Breathing excessive concentrations of asbestos fibers can cause serious bodily harm. Exercise care when servicing brake parts. Do not sand or grind brake lining unless equipment used is designed to contain the dust residue. Do not clean brake parts with compressed air or by dry brushing. Cleaning should be done by dampening the brake components with a fine mist of water, then wiping the brake components clean with a dampened cloth. Dispose of cloth and all residue containing asbestos fibers in an impermeable container with the appropriate label. Follow practices prescribed by the Occupational Safety and Health Administration (OSHA) and the Environmental Protection Agency (EPA) for the handling, processing, and disposing of dust or debris that may contain asbestos fibers.

BRAKE CALIPER

REMOVAL & INSTALLATION

2007–08 Grand Vitara

1. Raise and safely support the vehicle.

2. Remove the wheels.

3. Disconnect and plug the brake line.

4. Remove the brake caliper guide pin mounting bolts (Qty: 2) remove the caliper from the vehicle.

To install:

5. Install the caliper on the vehicle. Tighten the mounting bolts as follows:

   a. 10mm bolt: 37 ft. lbs. (50 Nm)
   b. 12mm bolt: 62 ft. lbs. (84 Nm)

6. Connect the hydraulic brake line, using 2 new washers. Torque the union bolt to 17 ft. lbs. (23 Nm).

7. Replace the front wheels.

8. Lower the vehicle.

9. Fill the brake reservoir and bleed the hydraulic brake system.

➡It is recommended that entire hydraulic system be thoroughly flushed with clean brake fluid whenever new parts are installed in hydraulic system. Periodical change of brake fluid is also recommended.

2007–08 XL-7

See Figure 7.

1. Raise and safely support the vehicle.
2. Remove the wheels.
3. Disconnect and plug the brake line and hose to prevent brake fluid loss and contamination.
4. Remove brake hose fitting gasket (Qty: 2)
5. Remove the caliper mounting bolts (guide pins) and remove the caliper from the vehicle.
6. Install an open end wrench to hold the caliper guide pin in line with the brake caliper while removing.

✻✻ WARNING

DO NOT allow the open end wrench to come in contact with the brake caliper. Allowing the open end wrench to come in contact with the brake caliper will cause a pulsation when the brakes are applied.

7. Remove the brake caliper.

To install:

8. Install the caliper on the vehicle.
9. Install an open end wrench to hold the caliper guide pin in line with the brake caliper while installing the caliper guide pin bolt.

✻✻ WARNING

DO NOT allow the open end wrench to come in contact with the brake caliper. Allowing the open end wrench to come in contact with the brake caliper will cause a pulsation when the brakes are applied.

10. Tighten the mounting bolts to 20 ft. lbs. (27 Nm)
11. Connect the hydraulic brake line, using 2 new brake hose fitting gasket (Qty: 2).

✻✻ WARNING

DO NOT reuse the gaskets. Install NEW gaskets.

12. Torque the union bolt to 38 ft. lbs. (52 Nm).
13. Replace the front wheels.
14. Lower the vehicle.
15. Fill the brake reservoir and bleed the hydraulic brake system.

➡It is recommended that entire hydraulic system be thoroughly flushed with clean brake fluid whenever new parts are installed in hydraulic system. Periodical change of brake fluid is also recommended.

DISC BRAKE PADS

REMOVAL & INSTALLATION

2007–08 Grand Vitara

See Figures 8 and 9.

✻✻ CAUTION

Dust and dirt accumulating on brake parts during normal use may contain asbestos fibers from production or aftermarket brake linings. Breathing excessive concentrations of asbestos fibers can cause serious bodily harm. Exercise care when servicing brake parts. Do not sand or grind brake lining unless equipment used is designed to contain the dust residue. Do not clean brake parts with compressed air or by dry brushing. Cleaning should be done by dampening the brake components with a fine mist of water, then wiping the brake components clean with a dampened cloth. Dispose of cloth and all residue containing asbestos fibers in an impermeable container with the appropriate label. Follow practices prescribed by the Occupational Safety and Health Administration (OSHA) and the Environmental Protection Agency (EPA) for the handling, processing, and disposing of dust or debris that may contain asbestos fibers.

✻✻ CAUTION

Support the brake caliper with heavy mechanic wire, or equivalent, whenever it is separated from its mount and the hydraulic flexible brake hose is still connected. Failure to support the caliper in this manner will cause the flexible brake hose to bear the weight of the caliper, which may cause damage to the brake hose and in turn may cause a brake fluid leak.

Inspect the fluid level in the brake master cylinder reservoir. If the brake fluid level is midway between the maximum-full point and the minimum allowable level, no brake fluid needs to be removed from the reservoir before proceeding. If the brake fluid level is higher than midway between the maximum-full point and the minimum allowable level, remove brake fluid to the midway point before proceeding.

1. Siphon about ⅔ of the fluid out of the master cylinder.
2. Raise and safely support the vehicle.
3. Remove the wheels.
4. Remove the brake caliper mounting bolts and remove the caliper from the mounting bracket.
5. Support the caliper with a wire.
6. Using a large pair of pliers or a C-clamp compress the caliper piston back into the bore.
7. Remove the disc brake pads and any shims from the caliper mounting bracket.

To install:

8. Install the brake pads and any shims removed from the caliper mounting bracket.
9. Install the caliper on the mounting bracket and install the mounting bolts.
10. Install the front wheels and lower the vehicle.

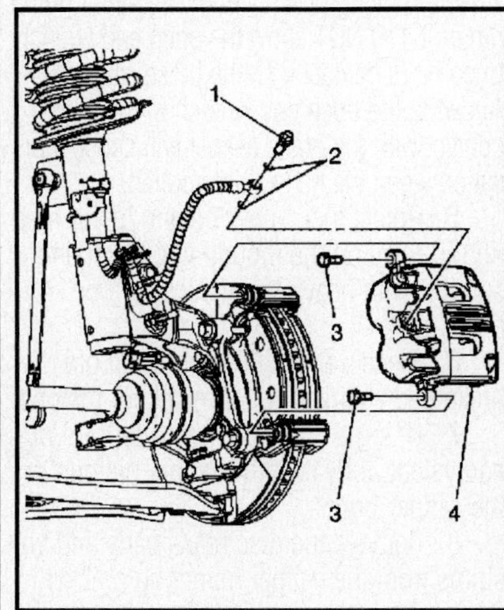

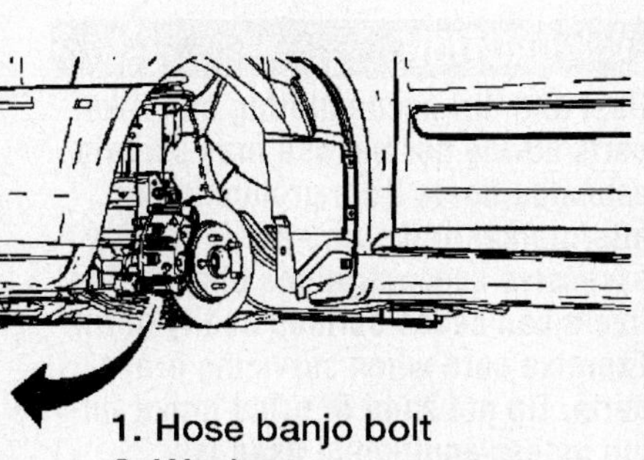

1. Hose banjo bolt
2. Washers
3. Caliper retaining bolts
4. Caliper

22140_SXL7_G0053

Fig. 7 Front brake caliper removal and Installation

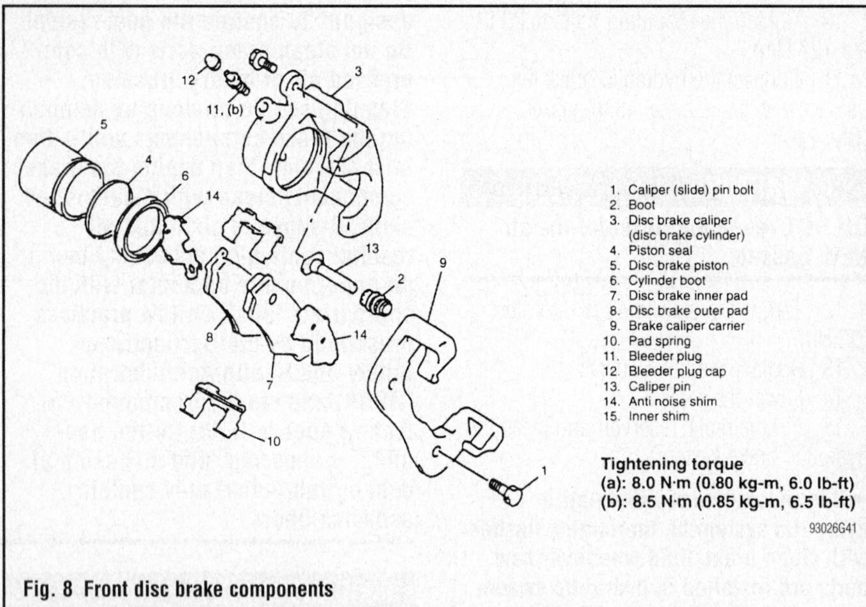

Fig. 8 Front disc brake components

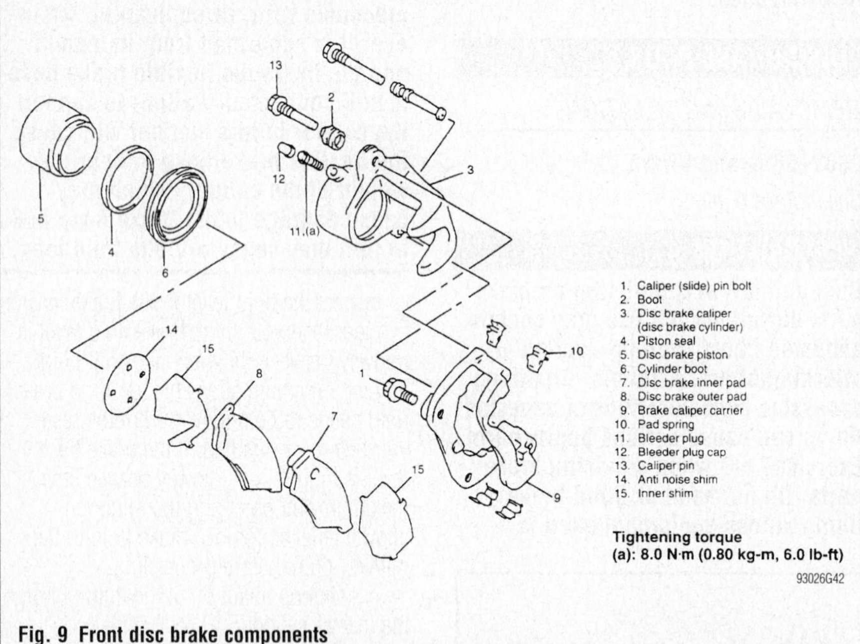

Fig. 9 Front disc brake components

CAUTION

Do not attempt to drive the vehicle until after the following step is performed.

11. Depress the brake pedal repeatedly until a firm pedal is obtained. Do not attempt to drive the vehicle unless a firm pedal is obtained.
12. Check the fluid level in the master cylinder. Add fresh brake fluid, as necessary.
13. Road-test the vehicle.

2007–08 XL-7

See Figure 10.

CAUTION

Dust and dirt accumulating on brake parts during normal use may contain asbestos fibers from production or aftermarket brake linings. Breathing excessive concentrations of asbestos fibers can cause serious bodily harm. Exercise care when servicing brake parts. Do not sand or grind brake lining unless equipment used is designed to contain the dust residue. Do not clean brake parts with compressed air or by dry brushing. Cleaning should be done by dampening the brake components with a fine mist of water, then wiping the brake components clean with a dampened cloth. Dispose of cloth and all residue containing asbestos fibers in an impermeable container with the appropriate label. Follow practices prescribed by the Occupational Safety and Health Administration (OSHA) and the Environmental Protection Agency (EPA) for the handling, processing, and disposing of dust or debris that may contain asbestos fibers.

CAUTION

Support the brake caliper with heavy mechanic wire, or equivalent, whenever it is separated from its mount and the hydraulic flexible brake hose is still connected. Failure to support the caliper in this manner will cause the flexible brake hose to bear the weight of the caliper, which may cause damage to the brake hose and in turn may cause a brake fluid leak.

Inspect the fluid level in the brake master cylinder reservoir. If the brake fluid level is midway between the maximum-full point and the minimum allowable level, no brake fluid needs to be removed from the reservoir before proceeding. If the brake fluid level is higher than midway between the maximum-full point and the minimum allowable level, remove brake fluid to the midway point before proceeding.

1. Raise and safely support the vehicle.
2. Remove the wheel.
3. Remove the brake caliper mounting bolts and remove the caliper from the mounting bracket.
4. Install an open end wrench to hold the caliper guide pin in line with the brake caliper while removing or installing the caliper guide pin bolt. DO NOT allow the open end wrench to come in contact with the brake caliper. Allowing the open end wrench to come in contact with the brake caliper will cause a pulsation when the brakes are applied.
5. Rotate the caliper up and to the rear until it rests on the mounting bracket and support with heavy mechanics wire or equivalent.
6. Place a block of wood or an old brake pad against the brake caliper pistons.
7. Using a brake pad spreader tool or equivalent, fully seat the caliper pistons in the caliper bores.
8. Remove the disc brake pads and any shims from the caliper mounting bracket.

To install:

9. Install the brake pads and **NEW** shims into the caliper mounting bracket.

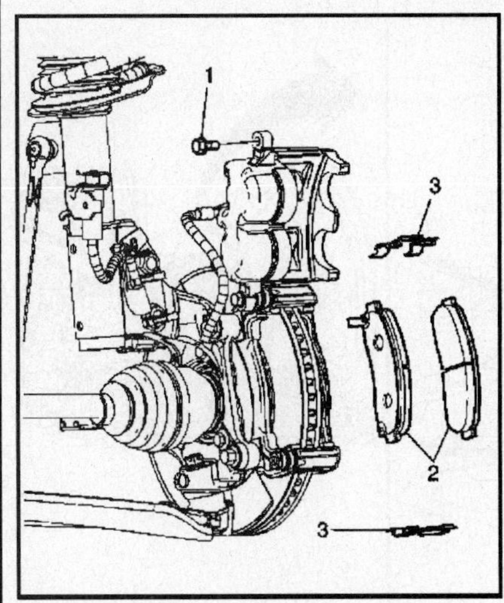

Fig. 10 Front brake caliper and pad removal

10. Install the caliper on the mounting bracket and install the mounting bolts.
11. Torque caliper on the mounting bracket to 20 ft. lbs. (27 Nm).
12. Install the front wheels and lower the vehicle.
13. Torque wheels to 100 ft. lbs. (136 Nm).

⁂ CAUTION

Do not attempt to drive the vehicle until after the following step is performed.

14. Depress the brake pedal repeatedly until a firm pedal is obtained. Do not attempt to drive the vehicle unless a firm pedal is obtained.
15. Check the fluid level in the master cylinder. Add fresh brake fluid, as necessary.
16. Road-test the vehicle.

BRAKES — REAR DISC BRAKES

⁂ CAUTION

Dust and dirt accumulating on brake parts during normal use may contain asbestos fibers from production or aftermarket brake linings. Breathing excessive concentrations of asbestos fibers can cause serious bodily harm. Exercise care when servicing brake parts. Do not sand or grind brake lining unless equipment used is designed to contain the dust residue. Do not clean brake parts with compressed air or by dry brushing. Cleaning should be done by dampening the brake components with a fine mist of water, then wiping the brake components clean with a dampened cloth. Dispose of cloth and all residue containing asbestos fibers in an impermeable container with the appropriate label. Follow practices prescribed by the Occupational Safety and Health Administration (OSHA) and the Environmental Protection Agency (EPA) for the handling, processing, and disposing of dust or debris that may contain asbestos fibers.

BRAKE CALIPER

REMOVAL & INSTALLATION

2007–08 XL-7

See Figure 11.

1. Raise and safely support the vehicle.
2. Remove the wheels.
3. Disconnect and plug the brake line and hose to prevent brake fluid loss and contamination.
4. Remove brake hose fitting gasket (Qty: 2)
5. Remove the caliper mounting bolts (guide pins) and remove the caliper from the vehicle.
6. Install an open end wrench to hold the caliper guide pin in line with the brake caliper while removing.
7. Remove caliber.

➡DO NOT allow the open end wrench to come in contact with the brake caliper. Allowing the open end wrench to come in contact with the brake caliper will cause a pulsation when the brakes are applied.

To install:

8. Install the caliper on the vehicle.
9. Install an open end wrench to hold the caliper guide pin in line with the brake caliper while installing the caliper guide pin bolt.

⁂ WARNING

DO NOT allow the open end wrench to come in contact with the brake caliper. Allowing the open end wrench to come in contact with the brake caliper will cause a pulsation when the brakes are applied.

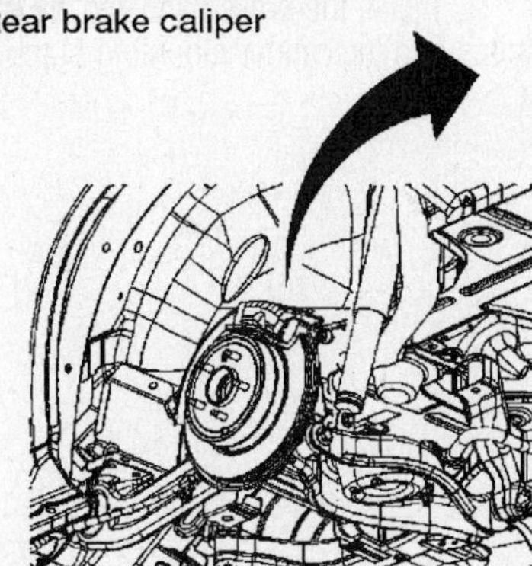

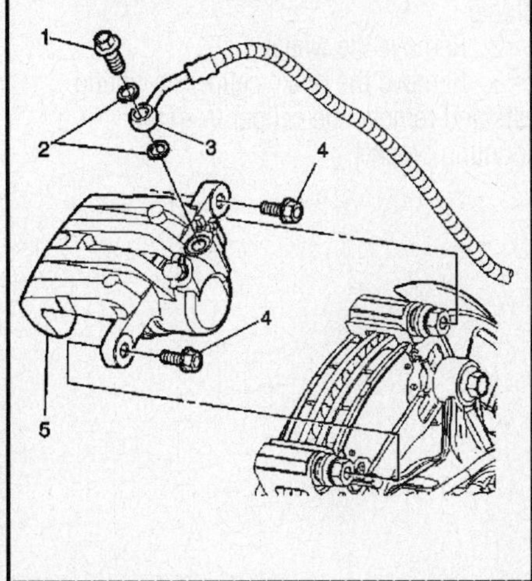

Fig. 11 Rear brake caliper removal

10. Install the caliper.
11. Tighten the mounting bolts to 20 ft. lbs. (27 Nm).
12. Connect the hydraulic brake line, using 2 new brake hose fitting gasket (Qty: 2).

⁂ WARNING

DO NOT reuse the gaskets. Install NEW gaskets.

13. Torque the union bolt to 38 ft. lbs. (52 Nm).
14. Replace the front wheels.
15. Lower the vehicle.
16. Fill the brake reservoir and bleed the hydraulic brake system.

➡It is recommended that entire hydraulic system be thoroughly flushed with clean brake fluid whenever new parts are installed in hydraulic system. Periodical change of brake fluid is also recommended.

DISC BRAKE PADS

REMOVAL & INSTALLATION

2007–08 XL-7

See Figure 12.

Inspect the fluid level in the brake master cylinder reservoir. If the brake fluid level is midway between the maximum-full point and the minimum allowable level, no brake fluid needs to be removed from the reservoir before proceeding. If the brake fluid level is higher than midway between the maximum-full point and the minimum allowable level, remove brake fluid to the midway point before proceeding.

1. Raise and safely support the vehicle.
2. Remove the wheel.
3. Remove the brake caliper mounting bolts and remove the caliper from the mounting bracket.

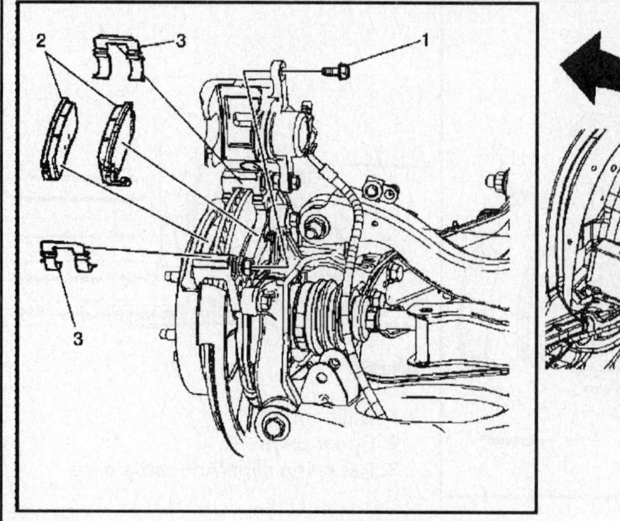

1. Rear caliper retaining bolt
2. Rear brake pads
3. Pad retaining clips /Anti rattle clips

22140_SXL7_G0056

Fig. 12 Rear brake caliper and pad removal

4. Install an open end wrench to hold the caliper guide pin in line with the brake caliper while removing or installing the caliper guide pin bolt. DO NOT allow the open end wrench to come in contact with the brake caliper. Allowing the open end wrench to come in contact with the brake caliper will cause a pulsation when the brakes are applied.
5. Rotate the caliper up and to the rear until it rests on the mounting bracket and support with heavy mechanics wire or equivalent.
6. Place a block of wood or an old brake pad against the brake caliper pistons.
7. Using a brake pad spreader tool or equivalent, fully seat the caliper pistons in the caliper bores.
8. Remove the disc brake pads and any shims from the caliper mounting bracket.

To install:

9. Install the brake pads and **NEW** shims into the caliper mounting bracket.
10. Install the caliper on the mounting bracket and install the mounting bolts.
11. Torque caliper on the mounting bracket to 20 ft. lbs. (27 Nm).
12. Install the rear wheels and lower the vehicle.
13. Torque wheels to 100 ft. lbs. (136 Nm).

⁂ CAUTION

Do not attempt to drive the vehicle until after the following step is performed.

14. Depress the brake pedal repeatedly until a firm pedal is obtained. Do not attempt to drive the vehicle unless a firm pedal is obtained.
15. Check the fluid level in the master cylinder. Add fresh brake fluid, as necessary.
16. Road-test the vehicle.

⁂ CAUTION

Dust and dirt accumulating on brake parts during normal use may contain asbestos fibers from production or aftermarket brake linings. Breathing excessive concentrations of asbestos fibers can cause serious bodily harm. Exercise care when servicing brake parts. Do not sand or grind brake lining unless equipment used is designed to contain the dust residue. Do not clean brake parts with compressed air or by dry brushing. Cleaning should be done by dampening the brake components with a fine mist of water, then wiping the brake components clean with a dampened cloth. Dispose of cloth and all residue containing asbestos fibers in an impermeable container with the appropriate label. Follow practices prescribed by the Occupational Safety and Health Administration (OSHA) and the Environmental Protection Agency (EPA) for the handling, processing, and disposing of dust or debris that may contain asbestos fibers.

BRAKE DRUM

REMOVAL & INSTALLATION

2007–08 Grand Vitara

See Figures 13 through 15.

1. Raise and safely support the vehicle.
2. Remove the rear wheel(s).
3. Release the parking brake.
4. Remove the parking brake lever cover screws and loosen the brake cable locking nut.
5. Install 2, 8mm bolts into the brake drum holes and uniformly tighten each bolt. Tighten each bolt until the brake drum is removed from the vehicle. If there is difficulty in removing the drum, insert a small tool through the hole in the rear of the backing plate, and hold the automatic adjusting lever away from the adjuster. Using another narrow, flat tool at the same time, reduce the brake shoe adjuster by turning the adjusting wheel.

To install:

6. Install the brake drum and pull the parking brake lever all the way up until a clicking sound can no longer be heard.
7. Verify that the rear wheels will not turn. If the rear wheels turn, adjust the parking brake cable as necessary.

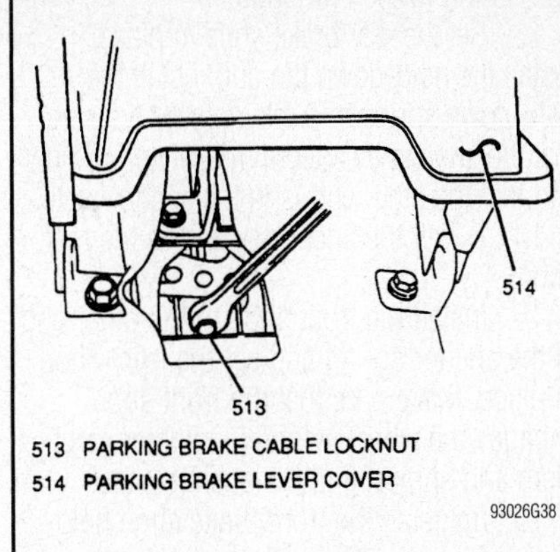

Fig. 13 Reducing the adjuster to remove the brake drum

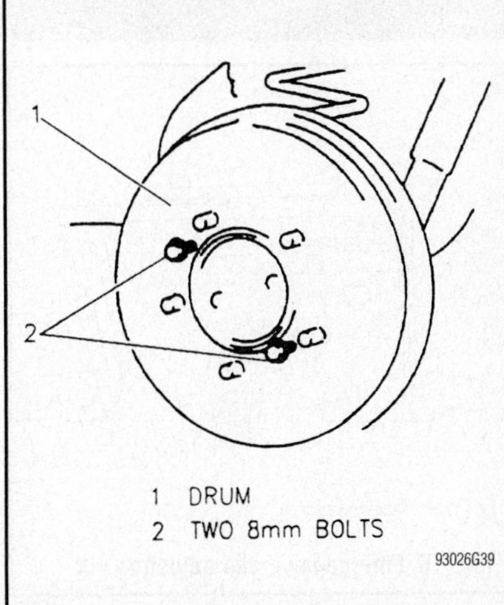

Fig. 14 Removing the brake drum with the two 8mm bolts

8. Release the parking brake and remove the brake drum. Measure the diameter of the brake shoes. Outer diameter should be as follows:
   - For 2 door models: 8.638 +/- 0.0012 inches (219 +/- 0.3mm).
   - For 4 door models: 9.980 +/- 0.0079 inches (253.5 +/- 0.2mm).
9. If the brake shoe clearance is not correct, adjust the brake shoes until the clearance is correct.
10. Reinstall the brake drum, replace the wheel(s), and safely lower the vehicle.
11. Adjust the parking brake and install the cover with the 2 screws.
12. Road-test the vehicle for proper brake operation.

BRAKE SHOES

REMOVAL & INSTALLATION

2007–08 Grand Vitara

1. Raise and safely support the vehicle.
2. Remove the rear wheel(s).
3. Remove the brake drum.
4. Using a suitable tool, remove the brake shoe return spring.
5. Using a brake spring hold-down tool, disengage the hold-down spring and retainers from the front shoe. Remove the hold-down retainer pinch.
6. Disconnect the anchor spring from the front shoe and remove the front shoe.
7. Remove the anchor spring from the rear shoe. Using a brake spring hold-down tool, disengage the hold-down spring and

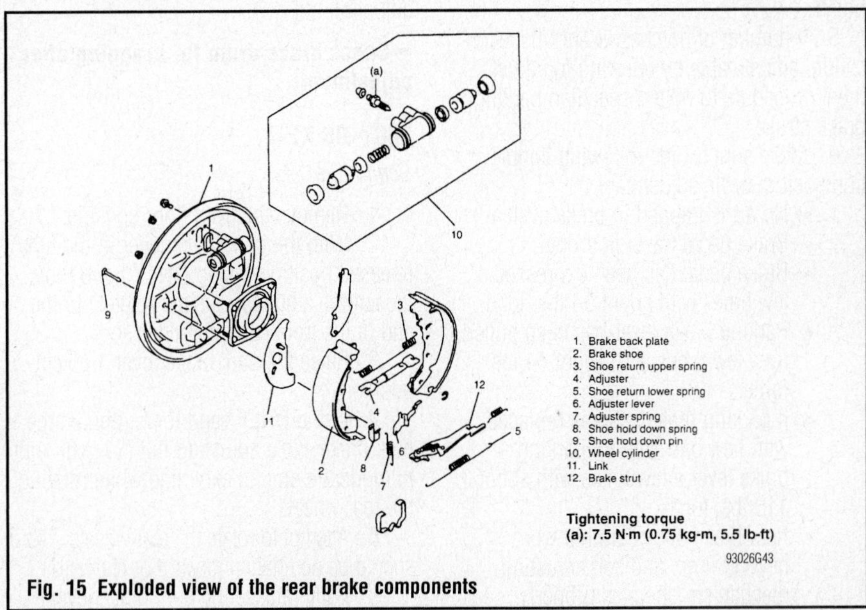

Fig. 15 Exploded view of the rear brake components

retainers from the rear shoe. Remove the hold-down pinch.

8. Disengage the parking brake lever from the parking brake cable and remove the rear shoe.

9. Remove the C-washer, the automatic adjuster lever and spring, the C-washer, and the parking brake lever from the rear shoe.

10. Thoroughly clean the backing plate and brake hardware with brake cleaning solvent. Apply high temperature grease to the backing plate shoe contact points, anchor plate and shoe contact points, adjusting bolt, and adjuster and brake shoe contact points.

To install:

11. Reinstall the automatic adjuster lever and the parking brake lever to the rear shoe using new C-washers.

12. Connect the parking brake lever to the parking brake cable. Set the adjuster and spring to the rear shoe.

13. Set the rear brake shoe in place, install the hold-down pin and install the hold-down spring and retainers. Make sure that the shoe is inserted in the wheel cylinder and that the other end is in the anchor plate.

14. Install the anchor spring to the rear shoe.

15. Install the front shoe to the other end of the anchor spring and set the front shoe in place. Make sure that the front shoe engages the wheel cylinder, adjuster mechanism and spring, and the anchor plate.

16. Reinstall the front brake shoe hold-down pin and secure with the hold-down spring and retainers using a suitable tool.

17. Install the return spring.

18. Install the brake drum and pull the parking brake lever all the way up until a clicking sound can no longer be heard.

19. Verify that the rear wheels will not turn. If the rear wheels turn, adjust the parking brake cable as necessary.

20. Release the parking brake and remove the brake drum. Measure the diameter of the brake shoes. Brake diameter should be as follows:

- 9.980 (+/- 0.0079 inches (253.5 +/- 0.2mm)

21. If the brake shoe clearance is not correct, adjust the brake shoes until the clearance is correct.

22. Reinstall the brake drum, replace the wheel(s), and safely lower the vehicle.

23. Road-test the vehicle for proper brake operation.

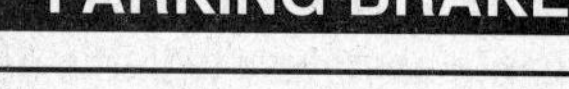

PARKING BRAKE CABLES

ADJUSTMENT

2007–08 Grand Vitara

See Figure 16.

1. Hold center of parking brake lever grip and pull it up with 200 N (20 kg, 44 lbs) force.

2. With parking brake lever pulled up as shown, count ratchet notches. There should be 5 to 7 notches. Also, check if both right and left rear wheels are locked firmly.

3. To count number of notches easily, listen to click sounds that ratchet makes while pulling parking brake lever without pressing its button.

4. One click sound corresponds to one notch.

5. If number of notches is out of specification, adjust cable by referring to adjustment procedure to obtain specified parking brake stroke.

6. Make sure for the following conditions before cable adjustment.

- No air is trapped in brake system.
- Brake pedal travel is proper.
- Brake pedal has been depressed a few times with about 66 lbs. load.
- Parking brake lever has been pulled up a few times with about 44 lbs. force.
- If parking brake cable is replaced with new one, pull up parking brake lever a few times with about 110 lbs. force.
- Rear brake shoes are not worn beyond limit, and self adjusting mechanism operates properly.

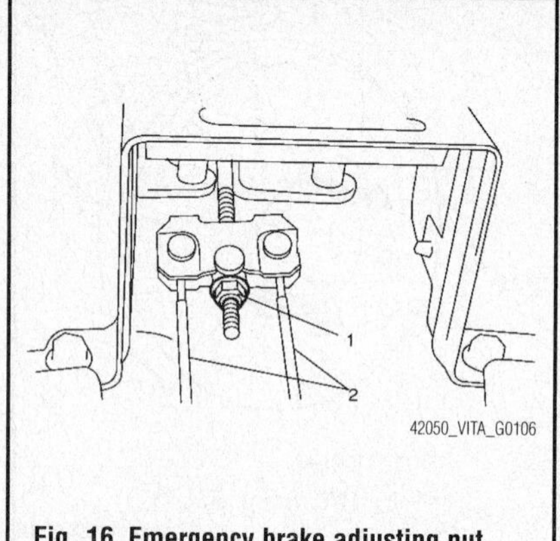

Fig. 16 Emergency brake adjusting nut

7. After confirming that above 5 conditions are all satisfied, adjust parking brake lever stroke by loosening or tightening adjusting nut.

➡Check brake drum for dragging after adjustment.

2007–08 XL-7

See Figure 17.

1. Remove the front floor console.

2. With the park brake lever in the fully released position, using **ONLY** hand tools, loosen the adjusting nut completely to the end of the front cable threaded rod.

3. Raise the park brake lever 1 detent position.

4. Using ONLY hand tools, tighten the park brake cable adjusting nut (1) until light to moderate drag is exhibited while rotating the rear wheels.

5. Attempt to rotate the rear wheels. There should be no rotation forward or rearward.

6. Fully release the park brake lever.

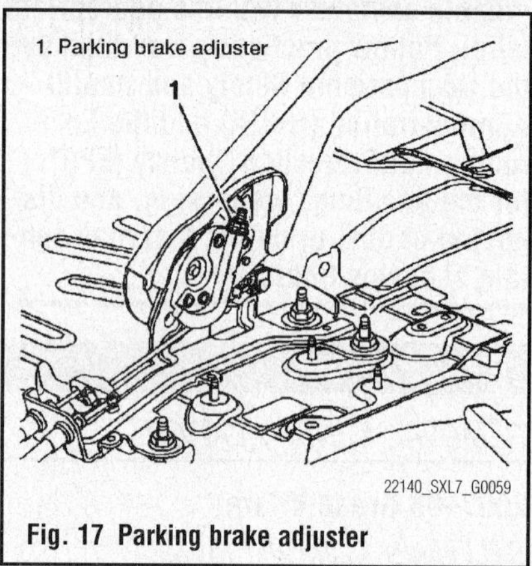

Fig. 17 Parking brake adjuster

7. Verify the park brake is released by rotating the rear wheels. The wheels should rotate freely and exhibit no park brake shoe drag.

8. If the wheels do not rotate freely, repeat the park brake cable adjustment procedure.

9. Raise the park brake lever 3 detent positions and attempt to rotate the rear wheels:

- One of the wheels should not rotate forward or rearward.
- The other wheel should not rotate forward or rearward, or should require substantial effort to rotate.

10. Install the front floor console.

11. Release the park brake lever.

PARKING BRAKE SHOES

REMOVAL & INSTALLATION

2007–08 Grand Vitara

The rear drum brake shoes serve as the parking brakes. Refer to the procedures under Rear Drum Brakes.

2007–08 XL-7

See Figure 18.

1. Remove the park brake shoes assembly.
2. Remove the following:
   - Park Brake Shoe Hold Down Spring (Qty: 2) by compressing the spring and rotating a 1/4 turn to release.
   - Park Brake Shoe Adjuster Spring
   - Park Brake Shoe Adjuster Screw
   - Park Brake Shoe Return Spring
   - Park Brake Shoes
3. Install the parking brake assembly by the following steps:
   - Park Brake Shoes
   - Park Brake Shoe Return Spring
   - Park Brake Shoe Adjuster Screw
   - Park Brake Shoe Adjuster Spring
   - Park Brake Shoe Hold Down Spring—rotate 1/4 turn to install
4. Install the wheel. Torque to 100 ft. lbs. (140 Nm).
5. Connect the negative battery cable.
6. Test operation of system.

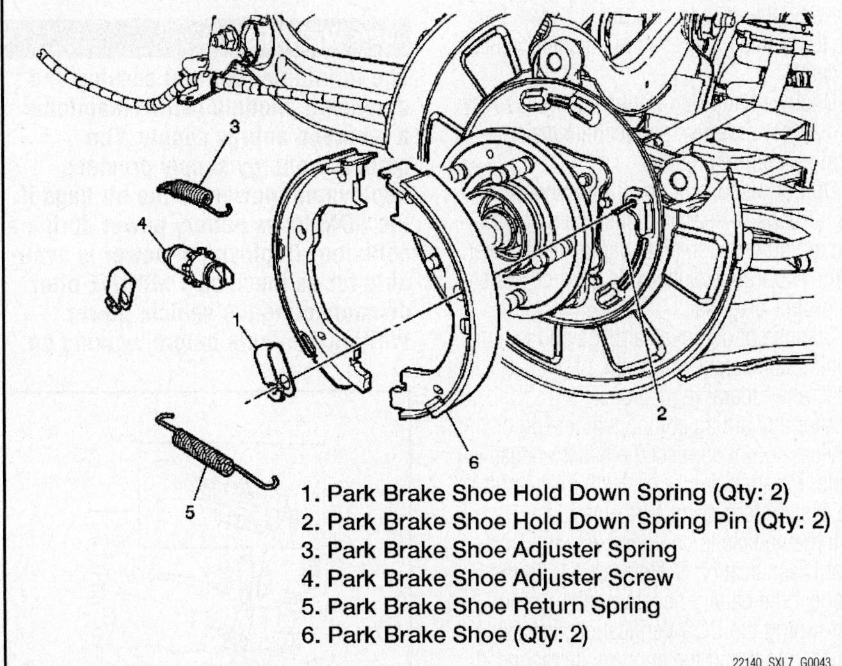

Fig. 18 Rear parking brake shoes—exploded view

ADJUSTMENT

2007–08 XL-7

1. Apply and fully release the park brake lever.
2. Verify that the park brake lever releases completely.
3. Turn **ON** the ignition. Verify that the red **BRAKE** warning indicator lamp is off.
4. Turn **OFF** the ignition.
5. Raise and support the vehicle
6. Remove the rear tire and wheel assemblies.
7. Remove the rear disc brake rotors.
8. Place the inside measurement contacts of the J 21177-A at the widest point of the drum portion of the brake rotor (1).
9. Tighten the set screw on the tool in order to ensure the proper measurement when removing the tool from the drum.
10. Position the outside measurement contacts of the J 21177-A over the park brake shoe (1) at the widest point.

➡If the gap between the adjuster nut and the adjuster screw exceeds 5 mm (0.25 in) during the adjustment procedure, the park brake shoe must be replaced.

11. Adjust the park brake shoe-to-drum clearance by rotating the adjustment nut on the park brake actuator.
12. Adjust to 0.015 inches (0.38mm).
13. Install the rear brake rotors.
14. Install the rear tire and wheel assemblies.
15. Apply the park brake lever. Inspect the rotation of the rear wheels:
    a. The wheels should not rotate forward.
    b. The wheels should drag or not rotate rearward.
16. If the rear tire and wheel assemblies rotate forward or do not exhibit drag rearward, proceed to the park brake cable adjustment.
17. Release the park brake lever. Verify that the wheels rotate freely.

CHASSIS ELECTRICAL

AIR BAG (SUPPLEMENTAL RESTRAINT SYSTEM)

GENERAL INFORMATION

✲✲ CAUTION

Some vehicles are equipped with an air bag system. The system must be disarmed before performing service on, or around, system components, the steering column, instrument panel components, wiring and sensors. Failure to follow the safety precautions and the disarming procedure could result in accidental air bag deployment, possible injury and unnecessary system repairs.

SERVICE PRECAUTIONS

WARNING/CAUTION labels are attached on each part of air bag system components (SDM, air bag (inflator) modules and seat belt pretensioners). Be sure to follow the instructions.

Many of service procedures require disconnection of "A/B" fuse and air bag (inflator) module(s) (driver, passenger, side of both sides and curtain of both sides) from initiator circuit to avoid an accidental deployment.

➡Do not apply power to the air bag system unless all components are connected or a diagnostic flow requests it, as this will set a DTC.

✲✲ WARNING

Never use air bag component parts from another vehicle.

✲✲ WARNING

If the vehicle will be exposed to temperatures over 93°C (200°F) (for example, during a paint baking process), remove the air bag system components beforehand to avoid component damage or unintended system activation.

When handling the air bag (inflator) modules (driver, passenger, side of both sides and curtain of both sides), seat belt pretensioners (driver and passenger), SDM, forward-sensor or side-sensor, be careful not to drop it or apply an impact to it. If an excessive impact was applied (e.g., SDM, forward-sensor and side-sensor are dropped, air bag (inflator) module is dropped from a height of 90 cm (3 ft) or more, seat belt pretensioner (retractor assembly) is dropped from a height of 30 cm (1 ft) or more), never attempt disassembly or repair but replace it with a new one.

When using electric welding, be sure to disconnect air bag (inflator) module connectors (driver, passenger, side of both sides and curtain of both sides) and seat belt pretensioner connectors (driver and passenger) respectively.

When applying paint around the air bag system related parts, use care so that the harness or connector will not be exposed to the paint mist.

Never expose air bag system component parts directly to hot air (drying or baking the vehicle after painting) or flames.

Disconnect and isolate the battery negative cable before beginning any airbag system component diagnosis, testing, removal, or installation procedures. Allow system capacitor to discharge for two minutes before beginning any component service. This will disable the airbag system. Failure to disable the airbag system may result in accidental airbag deployment, personal injury, or death.

Do not place an intact undeployed airbag face down on a solid surface. The airbag will propel into the air if accidentally deployed and may result in personal injury or death.

When carrying or handling an undeployed airbag, the trim side (face) of the airbag should be pointing towards the body to minimize possibility of injury if accidental deployment occurs. Failure to do this may result in personal injury or death.

Replace airbag system components with OEM replacement parts. Substitute parts may appear interchangeable, but internal differences may result in inferior occupant protection. Failure to do so may result in occupant personal injury or death.

Wear safety glasses, rubber gloves, and long sleeved clothing when cleaning powder residue from vehicle after an airbag deployment. Powder residue emitted from a deployed airbag can cause skin irritation. Flush affected area with cool water if irritation is experienced. If nasal or throat irritation is experienced, exit the vehicle for fresh air until the irritation ceases. If irritation continues, see a physician.

Do not use a replacement airbag that is not in the original packaging. This may result in improper deployment, personal injury, or death.

The factory installed fasteners, screws and bolts used to fasten airbag components have a special coating and are specifically designed for the airbag system. Do not use substitute fasteners. Use only original equipment fasteners listed in the parts catalog when fastener replacement is required.

During, and following, any child restraint anchor service, due to impact event or vehicle repair, carefully inspect all mounting hardware, tether straps, and anchors for proper installation, operation, or damage. If a child restraint anchor is found damaged in any way, the anchor must be replaced. Failure to do this may result in personal injury or death.

Deployed and non-deployed airbags may or may not have live pyrotechnic material within the airbag inflator.

Do not dispose of driver/passenger/curtain airbags or seat belt tensioners unless you are sure of complete deployment. Refer to the Hazardous Substance Control System for proper disposal.

Dispose of deployed airbags and tensioners consistent with state, provincial, local, and federal regulations.

After any airbag component testing or service, do not connect the battery negative cable. Personal injury or death may result if the system test is not performed first.

If the vehicle is equipped with the Occupant Classification System (OCS), do not connect the battery negative cable before performing the OCS Verification Test using the scan tool and the appropriate diagnostic information. Personal injury or death may result if the system test is not performed properly.

Never replace both the Occupant Restraint Controller (ORC) and the Occupant Classification Module (OCM) at the same time. If both require replacement, replace one, then perform the Airbag System test before replacing the other.

Both the ORC and the OCM store Occupant Classification System (OCS) calibration data, which they transfer to one another when one of them is replaced. If both are replaced at the same time, an irreversible fault will be set in both modules and the OCS may malfunction and cause personal injury or death.

If equipped with OCS, the Seat Weight Sensor is a sensitive, calibrated unit and must be handled carefully. Do not drop or handle roughly. If dropped or damaged, replace with another sensor. Failure to do so may result in occupant injury or death.

If equipped with OCS, the front passenger seat must be handled carefully as well. When removing the seat, be careful when setting on floor not to drop. If dropped, the sensor may be inoperative, could result in occupant injury, or possibly death.

If equipped with OCS, when the passenger front seat is on the floor, no one should sit in the front passenger seat. This uneven force may damage the sensing ability of the seat weight sensors. If sat on and damaged, the sensor may be inoperative, could result in occupant injury, or possibly death.

DISARMING THE SYSTEM

See Figure 19.

⁂ WARNING

The inflatable restraint sensing and diagnostic module (SDM) maintains a reserved energy supply. The reserved energy supply provides deployment power for the air bags if the SDM loses battery power during a collision. Deployment power is available for as much as 1 MINUTE after disconnecting the vehicle power. Waiting 1 minute before working on

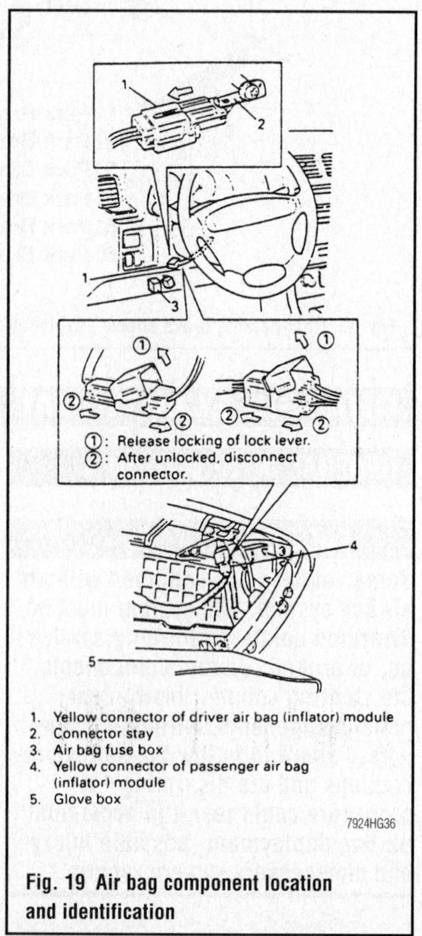

Fig. 19 Air bag component location and identification

the system after disabling the SRS system prevents deployment of the air bags from the reserved energy supply.

1. Before servicing the vehicle, refer to the precautions in the beginning of this section.
2. Wait 1 minute before working on system.
3. Remove or disconnect the following:
   - Negative battery cable
   - **AIR BAG** fuse
   - Driver air bag connector
   - Glove box
   - Passenger air bag connector
   - Side air bag connector (if equipped).
   - Curtain air bag (if equipped).

ARMING THE SYSTEM

1. Place the ignition in the **OFF** position. When repairs are complete, install or connect the following:
   - Install the fuse(s) supplying power to the SDM
   - Passenger air bag connector
   - Glove box
   - Driver air bag connector
   - **AIR BAG** fuse
   - Side air bag connector (if equipped).
   - Curtain air bag (if equipped).
   - Negative battery cable
2. Turn the ignition switch to the **ON** position.

➡The AIR BAG indicator will flash then turn OFF.

3. Perform the Diagnostic System Check.

CLOCKSPRING CENTERING

2007–08 Grand Vitara

See Figure 20.

⁂ WARNING

Removal of the steering wheel allows the contact coil cable assembly to turn freely but do not turn the contact coil cable assembly more than the allowable number of turns (about two and a half turns from the center position clockwise or counterclockwise respectively), or coil will break.

1. Check that vehicle's wheels (front tires) are set at straight-ahead position.
2. Check that ignition switch is at **LOCK** position.
3. Turn contact coil cable assembly counterclockwise slowly with a light force till contact coil cable assembly will not turn any further.

➡Contact coil cable assembly can turn about 5 turns at maximum, that is, if it is at the center position, can turn about two and a half turns both clockwise and counterclockwise.

4. From the position where contact coil cable assembly became unable to turn any further (it stopped), turn it back clockwise about two and a half rotations and align center mark with alignment mark.

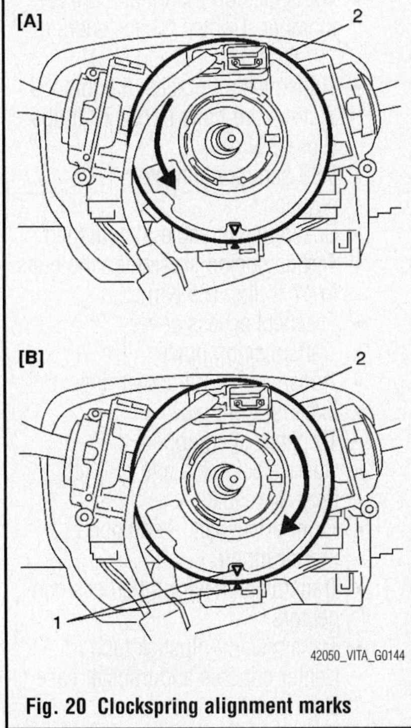

Fig. 20 Clockspring alignment marks

2007–08 XL-7

See Figure 21.

⁂ WARNING

The new SRS coil assembly will be centered. Improper alignment of the SRSSRS coil assembly may damage the unit, causing an inflatable restraint malfunction.

1. Verify the following conditions before centering the Supplemental Restraint System (SRS) steering wheel module coil:
2. The wheels on the vehicle are straight ahead.
3. The block tooth and the centering mark (1) of the steering shaft is in the 12 o'clock position.
4. If available, remove the yellow retaining tab (1) from the SRS steering wheel module coil and save the tab for reassembly.
5. Hold the SRS steering wheel module coil face up by the casing (2).
6. Slowly turn the SRS steering wheel module coil hub (3) clockwise until the coil ribbon stops.
7. Slowly rotate the SRS steering wheel module coil hub (3) counterclockwise 2.5 revolutions until the centering window (4) turns yellow. This indicates the CENTER position.
8. Install the yellow retaining tab (1) to the SRS steering wheel module coil.
9. Slide the centered SRS steering wheel module coil onto the steering shaft.

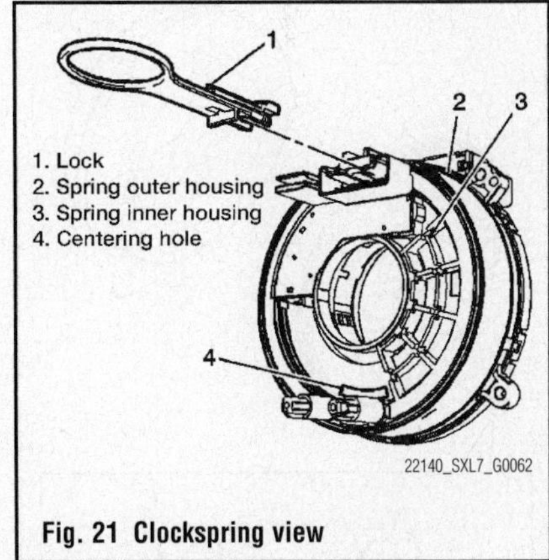

Fig. 21 Clockspring view

DRIVETRAIN

AUTOMATIC TRANSMISSION ASSEMBLY

REMOVAL & INSTALLATION

2.7L Engines

See Figure 22.

1. Before servicing the vehicle, refer to the precautions in the beginning of this section.
2. Drain the transfer case and transmission oil.
3. Remove or disconnect the following:
   - Negative battery cable
   - Center console and transfer case shift lever, if equipped
   - Transmission dipstick tube
   - Transmission wiring harness connectors
   - Exhaust manifold left and right bracket.
   - Cooler hose and cooler pipe No.1.
   - Drive plate cover, and then remove drive plate bolts by holding crankshaft pulley bolt stationary
   - Starter motor
   - Front driveshaft
   - Rear driveshaft
   - Gear select cable and bracket
   - Throttle Valve (TV) cable, if equipped

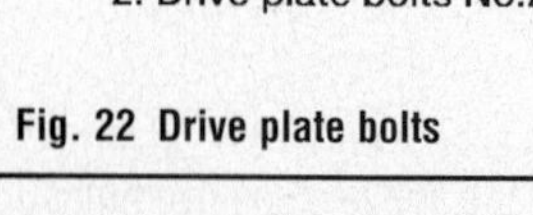

1. Plate bolt No.1
2. Drive plate bolts No.2

22140_VITA_G0079

Fig. 22 Drive plate bolts

   - Exhaust front pipe
   - Transmission oil cooler lines
   - Transmission brace
   - Flywheel access cover
   - Torque converter
   - Speedometer cable, if equipped
   - Vehicle Speed Sensor (VSS) connector, if equipped
   - Transmission flange bolts and nuts
   - Transmission braces, if equipped
4. Support the transmission with a jack and remove the transmission mount and crossmember.
5. Place a wooden block at the rear of the cylinder head as shown to support the engine when the transmission is removed.
6. Lower the transmission away from the vehicle.

To install:

7. Install or connect the following:
   - Transmission. Tighten the flange fasteners to 62–72 ft. lbs. (85–98 Nm).
   - Transmission mount and crossmember. Tighten the fasteners to 29–43 ft. lbs. (40–60 Nm).
   - Transmission braces, if equipped. Tighten the bolts to 62–72 ft. lbs. (85–98 Nm).
   - VSS sensor connector, if equipped
   - Speedometer cable, if equipped
   - Torque converter. Tighten the bolts to 47 ft. lbs. (65 Nm).
   - Flywheel access cover
   - Transmission brace
   - Transmission oil cooler lines
   - Exhaust front pipe
   - TV cable, if equipped
   - Gear select cable and bracket
   - Rear driveshaft
   - Front driveshaft, if equipped
   - Starter motor
   - Transmission wiring harness connectors
   - Transmission dipstick tube
   - Center console and transfer case shift lever
   - Negative battery cable
   - Set each clamp for wiring securely.
8. Fill A/T fluid referring.
9. Connect battery and check function of the engine and transmission.
10. Fill the transfer case.

3.6L Engines

1. Before servicing the vehicle, refer to the precautions in the beginning of this section.
2. Drain the transfer case and transmission oil.
3. Remove the battery tray.
4. Remove the shift control cable bracket.
5. Disconnect the electrical connector from the transaxle control module.
6. Remove the battery negative cable and wire harness ground from the transaxle stud.
7. Remove the nut and fuel line retaining clip from the transaxle.
8. Secure the wire harness, and shift cable up away from the transaxle.
9. Remove the upper 4 transaxle-to-engine bolts.
10. Remove the frame.
11. Raise the vehicle up away from the frame and remove the frame from under the vehicle.
12. Remove or disconnect the following:
    - Drain the transaxle fluid
    - Remove the transaxle oil cooler pipes
    - Remove the engine-to-transaxle brace
    - Remove the starter motor
13. Turn the crankshaft balancer bolt clockwise to gain access to the torque converter-to-flywheel bolts through the starter motor hole.

➡Mark the relation of the flywheel to torque converter for reassembly.

14. Remove or disconnect the following:
    - Remove the torque converter-to-flywheel bolts
    - Remove the front transmission mount from the transmission
    - Remove the left transmission mount bracket from the transmission
    - Remove the rear transmission mount bracket from the transmission
    - Disconnect the right and left drive shafts from the intermediate shaft and transaxle
    - Remove the intermediate shaft, if equipped with FWD

To install:

15. Install or connect the following:
    - Align and install the transaxle to the engine
    - Install the 4 lower transaxle-to-engine bolts
16. Tighten the 4 lower transaxle-to-engine bolts to 55 ft. lbs. (75 Nm).

All Wheel Drive (AWD) Models

1. If vehicle is equipped with All Wheel Drive (AWD) complete the following steps
   - Remove the rear propeller shaft

- Remove the transfer case
- Support the transaxle with a suitable transaxle jack
- Remove the 4 lower transaxle-to-engine bolts
- Slide the transaxle away from the engine until the transaxle torque-converter clears the flywheel

➡**If equipped with a drive axle seal dust cover, discard it and do not replace it.**

2. Lower the transaxle away from the vehicle.

To install:

3. Install or connect the following:
   - Install the transfer case to the transaxle
   - Install the rear propeller shaft
   - If the vehicle is equipped with FWD install the intermediate shaft
   - Install the right and left drive shafts
4. Turn the crankshaft balancer bolt clockwise to gain access to the torque converter-to-flywheel bolts through the starter motor hole.

➡**Align the reference marks on the flywheel and torque converter.**

5. Install or connect the following:
   - Install the torque converter to flywheel bolts to 44 ft. lbs. (60 Nm)
   - Install the starter motor
   - Install the front transmission mount to the transmission
   - Install the left transmission mount bracket to the transmission
   - Install the rear transmission mount bracket to the transmission
   - Install the engine-to-transaxle brace if equipped
   - Install the transaxle oil cooler pipes to the transaxle
   - Install the frame to the vehicle
   - Install the upper 4 transaxle-to-engine bolts torque to 55 ft. lbs. (75 Nm)
   - Install the nut and fuel line retaining clip to the transaxle torque to 18 inch lbs. (25 Nm)
   - Install the battery negative cable and wire harness ground to the transaxle stud
   - Install the nut securing the battery negative cable and wire harness ground to the transaxle stud.
   - Connect the electrical connector to the transaxle control module
   - Install the shift control cable bracket
   - Fill the transaxle with fluid
6. Perform the transmission adaptive learn procedure.

Transmission Adaptive Learn Procedure

See Figure 23.

1. With the vehicle in park "P" and ignition "OFF" set the parking brake and block the tires.
2. Turn the ignition switch to the **ACC** position.
3. Release the shift lock and place the transmission shift lever in the neutral "N" position.
4. Turn the ignition to the **"OFF"** .position.
5. Remove the shift cable from the transaxle range switch lever.
6. Verify the two arrows on the TCM are aligned.
7. Install the shift cable on the transaxle range switch lever.
8. Turn the ignition to **"ON"**,do not start the vehicle.
9. Using a scan tool implement the N-position adaptive learn.
10. If the N-position adaptation is okay skip the next 3 steps.
11. If N-position adaptation does not learn go back to step
12. If N-position adaptation sets a DTC, clear DTC and go back to the shift cable adjustment steps. If the cable cannot be adjusted the cable must be replaced.
13. If still cannot learn N-position after three times replace the TCM.
14. Shift the transmission to park **"P"**.
15. Turn ignition **"OFF"**.
16. Wait 15 seconds.
17. Turn ignition **"ON"**.
18. Verify the shift lever indicator aligns in all gear positions.
19. If they do not align, adjust shift control cable

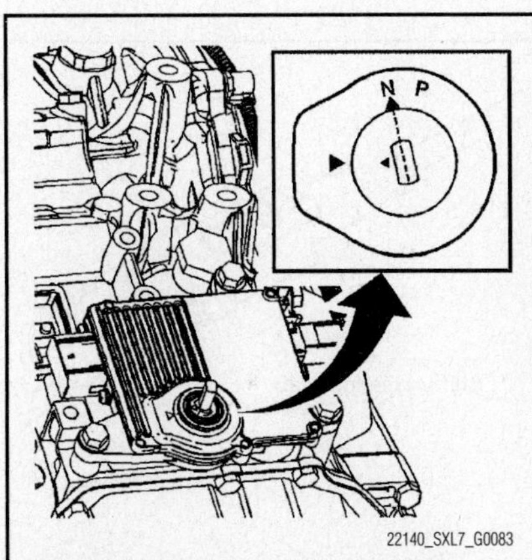

Fig. 23 transmission shift lever in the neutral "N" position.

20. Clear all TCM DTC's.
21. Turn ignition **"OFF"**.
22. Wait 15 seconds.
23. Start the engine.
24. Confirm that all TCM DTC's are clear.
25. Road test vehicle to confirm gear shift positions.

➡**Perform all steps of the adaptive learn procedure. Repeat the steps as many times as indicated. Engine flare during shifting or harsh shifts will occur if not done correctly. When performing the up/down shifting during 2-3 and 3-4 up-shifts, small shift flare may occur the first or second time depending on transaxle internal tolerances.**

Drive the vehicle to warm the transaxle fluid until 65-110°C (150-230°F). The adaptive learn procedure will not work unless transaxle fluid is the correct temperature.

Reset the transaxle adaptive learns using a scan tool. Go to the Transmission Special Functions then Trans. Output Controls menu and select Reset Transmission Adapts.

26. Perform the following steps for the garage shifts adaptive learn:
    - Apply the parking brake and the foot brake.
    - Shift from **NEUTRAL** to **REVERSE** and keep in **REVERSE** for 3 seconds.
    - Shift from **REVERSE** to **NEUTRAL**.
    - Repeat the above two steps five times.
    - Shift from **NEUTRAL** to **DRIVE** and keep in **DRIVE** for 3 seconds.
    - Shift from DRIVE to **NEUTRAL** .
    - Repeat the above two steps five times.
27. Perform the following steps for the Up/Down shifting adaptive learn:
    - Drive the vehicle in **DRIVE** with light (15-20 percent) throttle until above 50 km/h (31 mph) in 4th gear. If vehicle is not connected to a scan tool and throttle percentage cannot be observed, take 30 seconds for reaching the 50 km/h (31 mph).
    - If vehicle is not connected to a scan tool and throttle percentage cannot be observed, take 30 seconds for reaching the 50 km/h (31 mph).
    - Decelerate and apply the brakes until vehicle comes to a stop. Brake the vehicle so that it takes at least 14 seconds.
    - Repeat the above steps five times.

28. Perform the following steps for 2-1 Manual down shift adaptive learn:
    - Drive the vehicle in M or "Tap 2" until over 25 km/h (16 mph) in 2nd with any throttle position.
    - Decelerate, shift from "Tap 2" to "Tap 1" manually and stop the vehicle.
    - Repeat the two steps ten times.

➡If shift quality does not improve, ensure the TCM has the correct transmission calibration.

29. Confirm shift quality.

Transmission Shift Control Cable Adjustment

See Figures 24 through 28.

1. Position the shift lever in **PARK** .
2. Release the shift control cable adjustment lock cover retaining tab (1) on the bottom side.
3. Slide the shift control cable adjustment lock cover (2) forward exposing the adjustment lock.
4. Carefully squeeze the adjustment lock tabs (1) inward to disengage the adjustment lock.

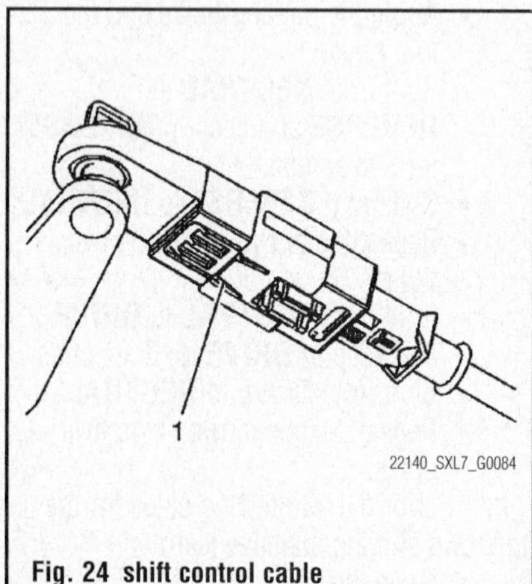

Fig. 24 shift control cable

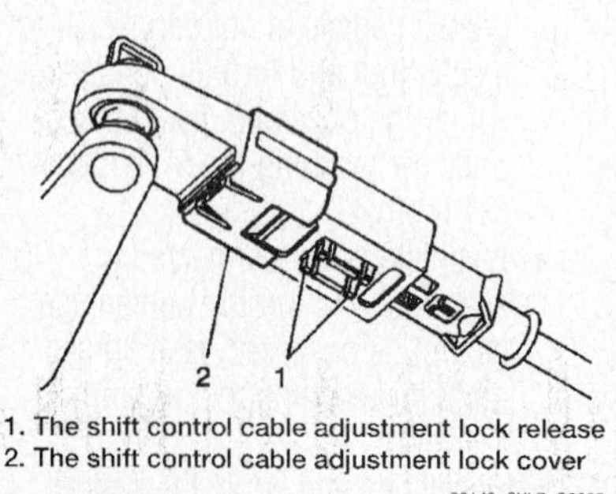

Fig. 25 The shift control cable adjustment lock

5. Lift the shift control cable adjustment lock (3).
6. The shift control cable (1) and cable end (2) are free to slide to proper adjustment. Ensure the shift lever and transmission are both in the **PARK** position.
7. Depress the shift control cable adjustment lock (1) to lock the adjustment.
8. Slide the shift control cable adjustment lock cover (1) over the adjustment lock.

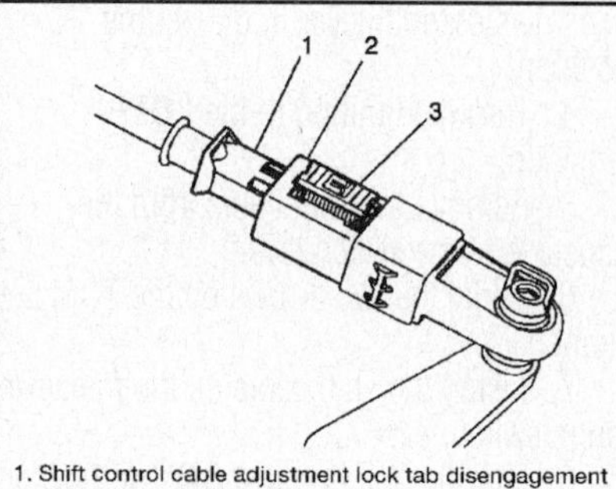

Fig. 26 Shift control cable adjustment lock cover disengagement

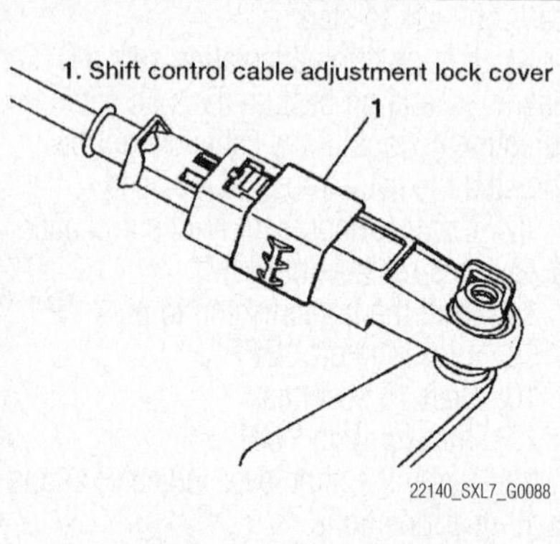

Fig. 27 Control cable adjustment lock (1) to lock the adjustment

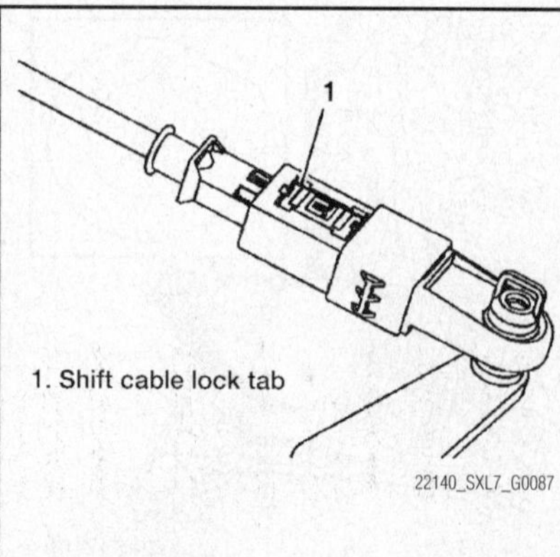

Fig. 28 Adjustment lock cover (1) over the adjustment lock.

9. Inspect the operation of the starting system with the shift lever in each position. The engine should only crank when the lever is in the **PARK** or **NEUTRAL** position.

MANUAL TRANSMISSION ASSEMBLY

REMOVAL & INSTALLATION

2.7L Engines

See Figures 29 through 31.

1. Before servicing the vehicle, refer to the service precautions.
2. Drain the transfer case and transmission oil.
3. Remove transmission and transfer case shift control levers.
4. Remove transfer case assembly from transmission.
5. Remove or disconnect the following:
    - Negative battery cable
    - Center console and transfer case shift lever, if equipped
    - Transmission wiring harness connectors
    - Drive plate cover, and then remove drive plate bolts by holding crankshaft pulley bolt stationary
    - Starter motor
    - Clutch fluid joint from pipe of clutch operating cylinder assembly
6. Hoist vehicle.
7. Drain oil from transmission and transfer or extension case.

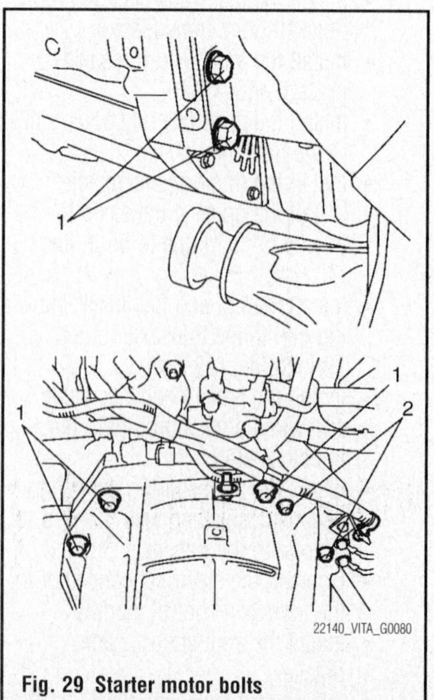

Fig. 29 Starter motor bolts

8. Remove or disconnect the following:

- Remove propeller shafts
- Rear driveshaft
- Gear select cable and bracket
- Exhaust pipe number 2
- Engine under cover.
- Clutch housing lower plates (1).
- Transmission fastening nut (2) and bolts (3).

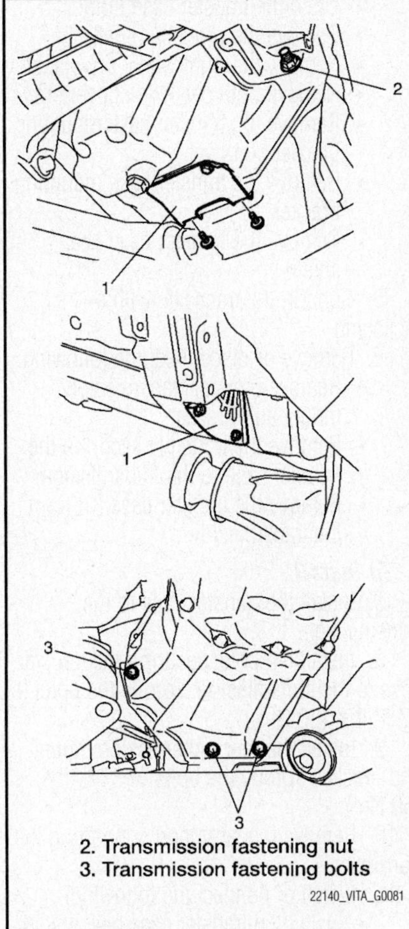

Fig. 30 Transmission bolts and nuts

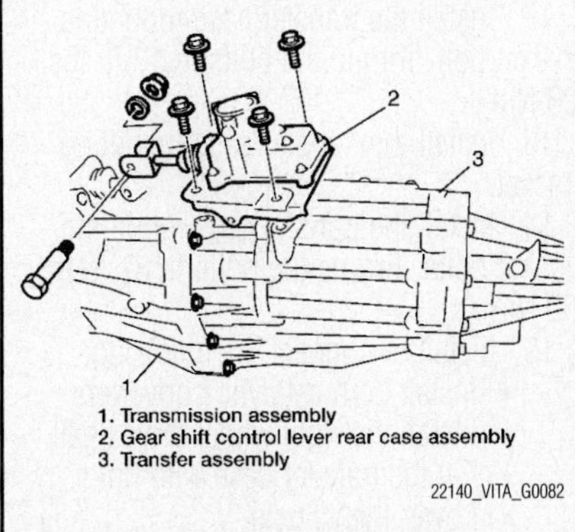

Fig. 31 Rear case assembly, transfer assembly and extension case

- Back up light switch
- Transfer shift actuator (if equipped)
- 4L/N switch (if equipped)
- Center differential lock switch (if equipped)

9. Apply transmission jack (1) and remove engine rear mounting member (2) taking off its bolts.

10. After removing mounting member, move rearward transmission and transfer or extension case assemblies placed on jack and then lower them.

11. Separate gear shift control lever rear case assembly (2) and transfer assembly (3) or extension case from transmission assembly (1).

12. Separate gear shift control lever rear case assembly (2) and transfer assembly (3) or extension case from transmission assembly (1).

To install:

13. Slant the rear of the engine down, using support device (1) and install transmission to engine.

14. Assemble the gear shift control lever rear case assembly (2) and transfer assembly (3) or extension case from transmission assembly (1).

15. Install or connect the following:

- Engine rear mounting member 40 ft. lbs. (55 Nm).
- Transmission mount and crossmember. Tighten the fasteners to 40 ft. lbs. (55 Nm).
- Transmission to engine bolts and nuts 61.5 ft. lbs. (85 Nm).
- Clutch housing lower plate bolts 8 ft. lbs. (11 Nm).
- Clamp for wiring and hose securely.
- Shift control lever
- Clutch fluid joint to pipe of clutch operating cylinder assembly
- Exhaust No.2 pipe
- Transmission oil cooler lines
- Propeller shafts
- Negative battery cable

16. Fill the transmission to the correct level.

17. Fill the transfer case, if equipped.

CLUTCH DRIVEN DISC & PRESSURE PLATE

REMOVAL & INSTALLATION

2.7L Engines

1. Before servicing the vehicle, refer to the precautions in the beginning of this section.

2. Remove the transmission.

3. Loosen the pressure plate mounting bolts in a 2-step crisscross sequence until the spring tension is relieved.

4. Inspect flywheel replace or resurface as recommended.

To install:

5. If flywheel was removed for replacement. Tighten flywheel bolts to 50 ft. lbs. (69 Nm).

6. Using a clutch alignment tool, assemble the clutch disc and pressure plate onto the flywheel.

7. Tighten the pressure plate bolts in multiple passes to 17 ft. lbs. (23 Nm).

8. Install the transmission.

9. Check for proper clutch operation.

ADJUSTMENTS

2.7L Engines

These vehicles are equipped with a hydraulic clutch system. No adjustment is necessary.

CLUTCH MASTER CYLINDER

REMOVAL & INSTALLATION

2.7L Engines

**** WARNING**

Clean, high quality brake fluid is essential to the safe and proper operation of the brake system. You should always buy the highest quality brake fluid that is available. If the brake fluid becomes contaminated, drain and flush the system, then refill the master cylinder with new fluid. Never reuse any brake fluid. Any brake fluid that is removed from the system should be discarded. Also, do not allow any brake fluid to come in contact with a painted surface; it will damage the paint.

1. Clean around reservoir cap and take out fluid with syringe or such.

2. Detach main fuse box.

3. Disconnect fluid pipe from master cylinder assembly.

4. Remove clip and push rod clevis pin.

5. Remove master cylinder attaching nut and bolt.

6. Remove master cylinder assembly and gasket.

7. Bleed air from master cylinder itself.

To install:

8. Install master cylinder assembly and new gasket to body, attaching bolt, nuts and push rod clevis pin.

9. Torque attaching bolt and nut to 17 ft. lbs. (33 Nm).
10. Connect fluid line and torque flare nut to 12 ft. lbs. (16 Nm).
11. Apply grease to clevis pin and install it.
12. Fill reservoir with specified brake fluid and check fluid leakage.
13. After installation, bleed air from system and check clutch pedal free travel.
14. Adjust clutch pedal height to 0.79 inch. (20 mm) by loosening clevis lock nut and turning push rod

BENCH BLEEDING PROCEDURE

See Clutch Hydraulic System Bleeding for bleeding operation.

CLUTCH SLAVE CYLINDER

REMOVAL & INSTALLATION

2.7L Engines

1. Clean around reservoir cap and take out fluid with syringe or such.
2. Dismount transmission—See manual transmission removal.
3. Disconnect fluid hose from slave cylinder.
4. Remove slave cylinder attaching bolts and slave cylinder.

To install:

5. Apply small amount of grease to rod tip of slave cylinder and install.
6. Tighten mounting bolts to 7.5 ft. lbs. (10 Nm).
7. Connect clutch fluid hose and tighten union bolt to 11.5 ft. lbs. (16 Nm).
8. Remount transmission.
9. Fill clutch master reservoir with specified brake fluid.
10. Bleed air from system and check clutch pedal free travel.

CLUTCH HYDRAULIC SYSTEM BLEEDING

BLEEDING PROCEDURE

2.7L Engines

1. Before servicing the vehicle, refer to the precautions in the beginning of this section.
2. Fill the master cylinder reservoir to the MAX line with clean brake fluid and keep it at least half full throughout the bleeding procedure.
3. From beneath the vehicle, remove the bleeder plug cap, then attach a clear vinyl tube to the slave cylinder bleeder plug. Insert the open end of the hose into a container.
4. Have an assistant depress the clutch pedal. Open the bleeder after the pedal is depressed.
5. Close the bleeder before releasing the clutch pedal.
6. Repeat until all air bubbles are gone from the hydraulic fluid.
7. Install the bleeder plug cap.
8. Fill the clutch master cylinder fluid reservoir to the specified full level.

TRANSFER CASE ASSEMBLY

REMOVAL & INSTALLATION

2.7L Engines

1. Before servicing the vehicle, refer to the service precautions.
2. Drain the transfer case oil.
3. Remove or disconnect the following:
   - Negative battery cable
   - Camshaft Position (CMP) sensor, if equipped
   - Center console
   - Transmission shift lever and case, if equipped with a manual transmission
   - Transfer case shift lever
   - Front and rear driveshafts
   - Exhaust center pipe
   - Speedometer cable or Vehicle Speed Sensor (VSS), as equipped
   - Vent hose
   - 4WD switch connector
4. Support the transmission with a jack and remove the transmission mount and crossmember.
5. Place a wooden block at the rear of the cylinder head as shown to support the engine when the transfer case is removed.
6. Lower the transfer case away from the vehicle.

To install:

7. Install or connect the following:
   - Transfer case. Tighten the bolts to 30 ft. lbs. (41 Nm).
   - 4WD switch connector
   - Vent hose
   - Speedometer cable or VSS sensor, as equipped
   - Exhaust center pipe
   - Front and rear driveshafts. Tighten the bolts to 36 ft. lbs. (50 Nm).
   - Transfer case shift lever
   - Transmission shift lever and case, if equipped with a manual transmission
   - Center console
   - Distributor or CMP sensor, if equipped
   - Negative battery cable
8. Fill the transfer case.

3.6L Engines

1. Before servicing the vehicle, refer to the service precautions.
2. Negative battery cable
3. Raise and support the vehicle.
4. Remove or disconnect the following:
   - Drain the transfer case fluid.
   - Remove the propeller shaft
   - Remove the right wheel drive shaft
   - Remove both catalytic converters.
   - Remove the transfer case mounting bracket bolts
   - Remove the transfer case mounting bracket
   - Remove the transfer case heat shield.
5. Support the transaxle with a jackstand
6. Remove or disconnect the following:
   - Remove the rear transmission mount and bracket.
   - Remove the bolts (1) securing the transfer case to the transmission
   - Remove the transfer case (2) from the transmission.

To install:

7. Install the transfer case to the transmission.
8. Install the bolts securing the transfer case to the transmission. Torque the bolts to 37 ft. lbs. (50 Nm).
9. Install the rear transmission mount and bracket. Torque the bolts to 37 ft. lbs. (50 Nm).
10. Remove the jackstand supporting the transaxle.
11. Install or connect the following:
   - Install the transfer case heat shield
   - Install the transfer case mounting bracket. Torque the bolts to 37 ft. lbs. (50 Nm).
12. Install the transfer case mounting bracket bolt. Torque the bolts to 17 ft. lbs. (23 Nm).
13. Install the transfer case mounting bracket.
14. Install the transfer case mounting bracket bolts. Torque the bolts to 37 ft. lbs. (50 Nm).
15. Install or connect the following:
   - Install both catalytic converters
   - Install the right wheel drive shaft
   - Fill the transfer case with fluid
   - Lower the vehicle
   - Connect the negative battery cable
   - Test drive for proper operation

FRONT HALFSHAFTS

REMOVAL & INSTALLATION

2.7L Engines

1. Before servicing the vehicle, refer to the service precautions section.
2. Raise vehicle and remove wheel.
3. Remove halfshaft wheel nut.
4. Mark inboard halfshaft flange, and then remove flange nuts.
5. Remove halfshaft.

To install:

6. Install halfshaft aligning match marks.
7. Tighten inboard halfshaft nuts to 58 ft. lbs. (80 Nm).
8. Tighten halfshaft wheel nut to 145 ft. lbs. (200 Nm).
9. Install wheels and lower vehicle.

3.6L Engines

See Figures 32 and 33.

1. Before servicing the vehicle, refer to the service precautions section.
2. Raise vehicle and remove wheel.
3. Remove halfshaft wheel nut.
4. Remove the outer tie rod end-to-steering knuckle nut. Do not loosen the tie rod end jam nut.

➡Do not use a wedge type tool to separate the tie rod end from the steering knuckle.

5. Separate the tie rod end from the steering knuckle.
6. Mark inboard halfshaft flange, and then remove flange nuts.
7. Remove and discard the cotter pin from the lower ball joint stud.
8. Remove the ball joint stud nut.
9. Separate the lower ball joint stud from the steering knuckle.
10. Using a backup wrench on the stud, remove the nut securing the lower stabilizer bar link and disengage the link.

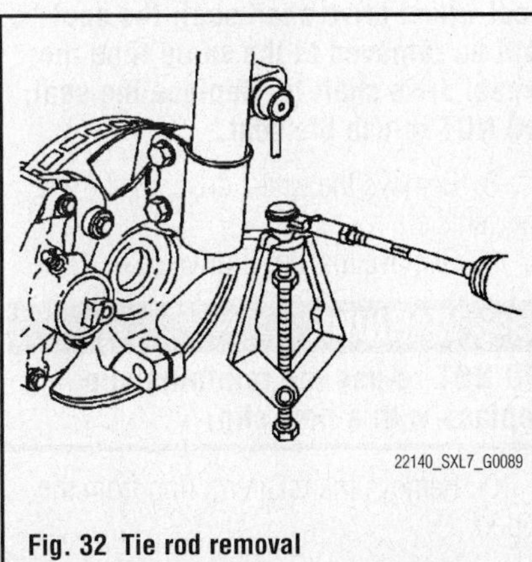

Fig. 32 Tie rod removal

11. Disengage the wheel drive shaft spindle from the wheel hub assembly. If necessary, place a wood block against the end of the wheel drive shaft spindle and tap with a hammer to aid removal.

➡Use care not to damage the joint boot when removing the wheel drive shaft.

➡On vehicles equipped with all-wheel drive (AWD), the stub shaft may disengage from the power takeoff unit (PTU). If necessary, cap the opening in the PTU to prevent fluid loss.

12. Disengage the wheel drive shaft from the transmission or power takeoff unit (PTU), if equipped.
13. Remove the wheel drive shaft from the vehicle.

To install:

14. Install a new wheel drive shaft retaining ring to the output shaft.
15. Install the wheel drive shaft to the output shaft:
    - Guide the wheel drive shaft tripot joint squarely onto the output shaft
    - After the splined end of the wheel drive shaft passes the oil seal, remove the J 44394 from the oil seal
    - Ensure that the tripot joint is fully seated on the output shaft by grasping the tripot joint and attempting to pull free of the output shaft
    - Insert the constant velocity (CV) joint spindle to the wheel hub/bearing assembly of the steering knuckle
    - Hand install a new wheel drive shaft spindle nut
    - Install the lower ball joint stud to the steering knuckle
    - Install the lower ball joint nut to the stud
16. Tighten the ball joint to 30 ft. lbs. (40 Nm).
17. Install the following:
    - Install the cotter pin to the ball joint stud.
    - If necessary, tighten the nut one additional flat at a time until the castle nut aligns with the hole in the ball joint stud.
    - Secure the cotter pin to the ball joint stud by folding one tine over the end of the ball joint stud.
    - Cut off any excess length of the cotter pin tines.
    - Install the lower link to the stabilizer bar.
    - Install a new nut to the stabilizer bar link stud.

➡In order to prevent damaging the stabilizer bar link stud seal, do not allow the stud to rotate while tightening the nut.

18. Tighten the ball joint to 48 ft. lbs. (65 Nm).
19. Install the tie rod end to the steering knuckle.
20. Using the J 44015, pull the stud into steering knuckle.

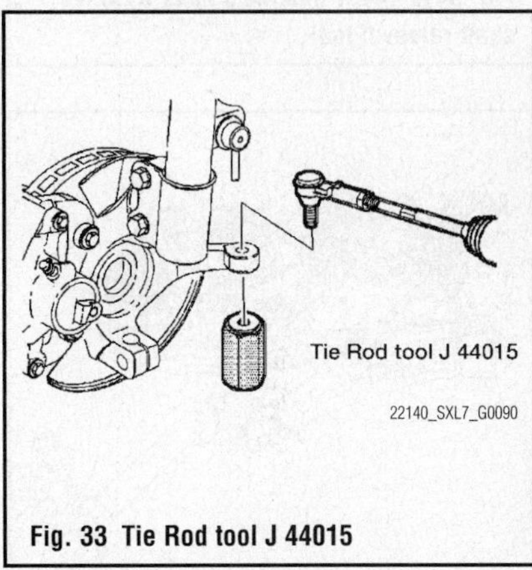

Fig. 33 Tie Rod tool J 44015

21. Tighten the tie rod end to 30 ft. lbs. (40 Nm).
22. Remove the J 44015 from the tie rod end stud.
23. Install a new nut to the tie rod end stud.
24. Tighten the tie rod end to 18 ft. lbs. (25 Nm) plus 90°.
25. Tighten the wheel drive shaft spindle nut.
26. Tighten the wheel drive shaft spindle nut to 151 ft. lbs. (205 Nm)
27. Install the tire and wheel assembly.
28. Lower the vehicle.
29. Inspect the transmission fluid level.

Intermediate Shaft Replacement

See Figures 34 and 35.

1. Before servicing the vehicle, refer to the service precautions section.
2. Raise and support the vehicle.
3. Remove the wheel drive shaft assembly.

➡Refer to front wheel drive shaft replacement.

4. Remove the retaining clip for the wheel drive shaft.
5. Remove the mounting bolts for the intermediate shaft support bracket.
6. Support the wheel drive shaft.

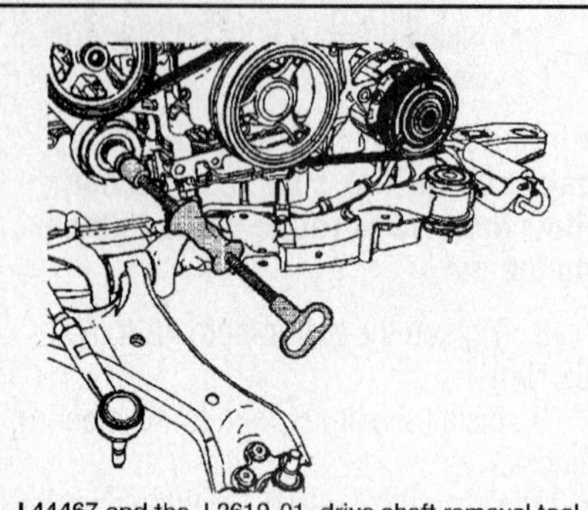

Fig. 34 J 44467 and the J 2619-01,drive shaft removal tool

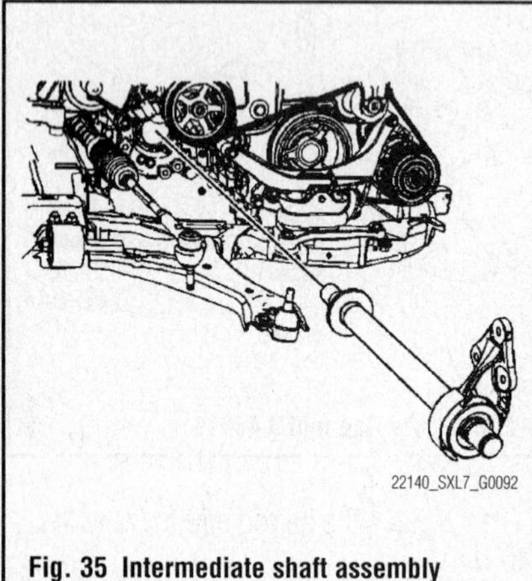

Fig. 35 Intermediate shaft assembly

7. Assemble the J 44467 and the J 2619-01.
8. Install the J 44467 and the J 2619-01 in the retaining ring groove on the wheel drive shaft.
9. Using the J 44467 and the J 2619-01, remove the wheel drive shaft from the transmission.
10. Remove the drive shaft from the vehicle.

To install:

11. Install the intermediate shaft in the transmission.
12. Move the intermediate shaft back and forth to ensure that the intermediate shaft is properly seated.
13. Hand tighten the mounting bolts for the intermediate shaft, then tighten the mounting bolts for the intermediate shaft mounting bracket.
14. Tighten mounting bolts to 26 ft. lbs. (35 Nm).
15. Install the NEW intermediate shaft retaining.
16. Install the wheel drive shaft assembly.
17. Remove the support and lower the vehicle.

FRONT PINION SEAL

REMOVAL & INSTALLATION

See Figure 36.

1. Before servicing the vehicle, refer to the precautions in the beginning of this section.
2. Raise vehicle, remove wheels and brake pads and calipers.
3. Mark propeller shaft and pinion flange for installation.
4. Remove front propeller shaft.
5. Drain front axle, or install catch pan under work area.

➡The brake calipers and pads must be removed so that there is no additional drag when measuring pinion bearing preload.

6. Remove pinion nut and record bearing preload.
7. Make mating marks on drive bevel pinion and companion flange.
8. Remove flange from drive bevel pinion. Use puller tool if it is hard to remove.
9. Remove pinion seal with seal removing tool.
10. Remove pinion bearing.
11. Remove collapsible spacer.

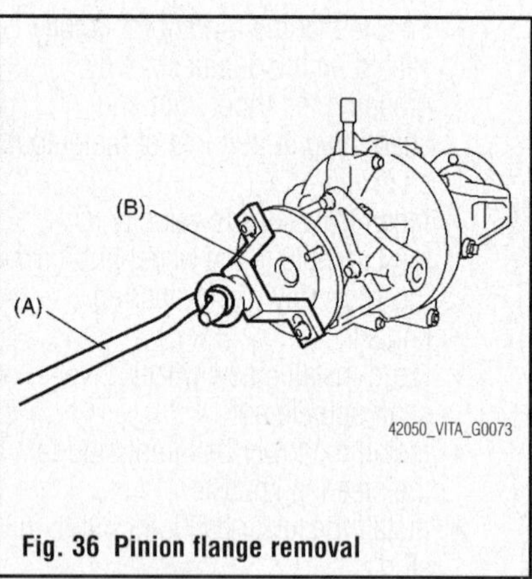

Fig. 36 Pinion flange removal

To install:

12. Install new Collapsible spacer.
13. Install pinion bearing.
14. Lube pinion seal lip and carefully install.
15. Install pinion flange align marks.
16. Check pinion flange nut and install Thread locking liquid.
17. Tighten pinion flange nut gradually then set to recorded preload. Standard preload is 7.8–14.7 inch. lbs. (0.9–1.7 Nm).
18. Install front propeller shaft align mating marks.
19. Tighten propeller shaft mounting bolts tighten to 62 ft. lbs. (85 Nm).
20. Install wheels, brake pads and calipers.
21. Refill front drive axle housing with specified oil.

REAR HALFSHAFT

REMOVAL & INSTALLATION

2.7L Engines

1. Before servicing the vehicle, refer to the precautions in the beginning of this section.
2. Raise vehicle and remove wheel.
3. Remove halfshaft wheel nut.
4. Mark inboard halfshaft flange, and then remove flange nuts.
5. Remove halfshaft.

To install:

6. Install halfshaft aligning match marks.
7. Tighten inboard halfshaft nuts to 58 ft. lbs. (80 Nm).
8. Tighten halfshaft wheel nut to 145 ft. lbs. (200 Nm).
9. Install wheels and lower vehicle.

3.6L Engines

1. Before servicing the vehicle, refer to the precautions in the beginning of this section.
2. Raise vehicle and remove wheel.
3. Remove halfshaft wheel nut.
4. Mark inboard halfshaft flange, and then remove flange nuts.

⁂ WARNING

DO NOT re-use the wheel drive shaft spindle nut. Replace with a NEW nut.

5. Disengage the wheel drive shaft from the wheel hub/bearing.
6. Remove the rear suspension knuckle.
7. Using a suitable tool, carefully release the wheel drive shaft from the rear drive module (RDM).

➡Because of the design of the inner seal wheel drive shaft seal, the seal will be removed at the same time the wheel drive shaft is. Replace the seal, DO NOT re-use the seal.

8. Remove the wheel drive shaft from the vehicle.
9. Remove the wheel drive seal.

⁂ WARNING

DO NOT re-use the retaining clip, replace with a new clip.

10. Remove the retaining ring from the tripod.

To install:

11. Install the new retaining clip on the tripod.

12. Install the new wheel drive shaft seal.

➡When installing the wheel drive shaft, you will notice a slight resistance. This is the wheel drive shaft seal. A snap or click should be heard when the wheel drive shaft is fully seated.

13. Install the wheel drive shaft.

14. Install the rear suspension knuckle.

15. Hand install a new wheel drive shaft spindle nut.

16. Tighten the wheel drive shaft spindle nut.

17. Tighten halfshaft wheel nut to 151 ft. lbs. (205 Nm).

18. Install the tire and wheel assembly.

19. Remove the support and lower the vehicle.

REAR PINION SEAL

REMOVAL & INSTALLATION

2.7L Engines

1. Before servicing the vehicle, refer to the precautions in the beginning of this section.

2. Raise vehicle, remove wheels.

3. Mark propeller shaft and pinion flange for installation.

4. Remove rear propeller shaft.

5. Drain rear axle, or install catch pan under work area.

6. Remove pinion nut and record bearing preload.

➡The brake calipers and pads or brake drum must be removed so that there is no additional drag when measuring pinion bearing preload.

7. Make mating marks on drive bevel pinion and companion flange.

8. Remove flange from drive bevel pinion. Use puller tool if it is hard to remove.

9. Remove pinion seal with seal removing tool.

10. Remove pinion bearing.

11. Remove collapsible spacer.

To install:

12. Install new collapsible spacer.

13. Install pinion bearing.

14. Lube pinion seal lip and carefully install.

15. Install pinion flange align marks.

16. Check pinion flange nut and install Thread locking liquid.

17. Tighten pinion flange nut gradually then set to recorded preload. Standard preload is 8–15 inch. lbs. (0.9–1.7 Nm).

18. Install front propeller shaft align mating marks.

19. Tighten rear propeller shaft mounting bolts tighten to 62 ft. lbs. (85 Nm).

20. Refill rear differential with specified oil.

21. Install wheels, brake pads and calipers.

22. Check pinion seal for leaks.

3.6L Engines

1. Before servicing the vehicle, refer to the service precautions.

2. Raise vehicle, remove wheels.

3. Mark propeller shaft and pinion flange for installation.

4. Remove propeller shaft.

➡The brake calipers and pads or brake drum must be removed so that there is no additional drag when measuring pinion bearing preload.

5. Drain the differential oil.

6. Make mating marks on drive bevel pinion and companion flange.

7. Remove flange from drive bevel pinion. Use puller tool if it is hard to remove.

8. Install a sheet metal screw into Differential Pinion Seal.

9. Attach a pair of pliers or a slider hammer to the screw and remove the seal.

10. Using the slide hammer or pliers, remove the seal.

To install:

11. Using the J 44636-1, install the pinion seal.

12. Install new collapsible spacer.

13. Install pinion bearing.

14. Lube pinion seal lip and carefully install.

15. Install pinion flange align marks.

16. Check pinion flange nut and install Thread locking liquid.

17. Tighten pinion flange nut gradually then set to recorded preload. Standard preload is 8–15 inch. lbs. (0.9–1.7 Nm).

18. Install front propeller shaft align mating marks.

19. Tighten rear propeller shaft mounting bolts tighten to 62 ft. lbs. (85 Nm).

20. Refill rear differential with specified oil to the proper level.

21. Install wheels, brake pads and calipers.

22. Check pinion seal for leaks.

ENGINE COOLING

ENGINE FAN

REMOVAL & INSTALLATION

2.7L Engines

1. Disconnect the negative battery cable.

2. Disconnect connectors of cooling fan motors.

3. With hose connected, detach P/S fluid reservoir with reservoir bracket.

4. Remove air cleaner case and air cleaner suction pipe.

5. Remove reservoir from radiator.

6. Remove cooling fan assembly.

To install:

7. Installation is the reverse of the removal procedures, noting the following steps.

8. Refill and bleed cooling system.

9. After installation, verify there is no coolant leakage at each connection.

3.6L Engines

1. Remove the front bumper fascia.

2. Drain the coolant.

3. Disconnect the electrical connectors from the fan motors.

4. Unclip the wire harness from the fan assembly.

5. Remove the condenser radiator fan module (CRFM) closeout panel retainers from the condenser.

6. Remove the CRFM closeout panel from the condenser.

7. Remove the front impact bar.

8. Remove the CRFM mounting bracket bolts from the radiator support.

9. Remove the CRFM mounting brackets from the radiator support.

10. Remove the radiator inlet hose clamp (1) from the radiator.

11. Remove the radiator inlet hose (2) from the radiator.

12. Disconnect the transmission cooler lines from the radiator.

13. Unclip the transmission cooler lines from the fan assembly.

14. Remove the fan assembly bolts from the radiator.

15. Remove the fan assembly from the radiator.

16. Position the condenser and radiator assembly forward of the right hand front impact bar bracket.

17. Remove the fan assembly from the vehicle.

To install:

18. Install the fan assembly to the vehicle.
19. Return the condenser and radiator assembly to the mounted position.
20. Install the fan assembly to the radiator by guiding the lower tabs into the corresponding hooks on the radiator.
21. Install the fan assembly bolts to the radiator tighten to 80 inch lbs. (9 Nm).
22. Install the radiator inlet hose (2) to the radiator.
23. Install the radiator inlet hose clamp (1) to the radiator.
24. Clip the transmission cooler lines to the fan assembly.
25. Connect the transmission cooler lines to the radiator.
26. Install the CRFM mounting brackets to the radiator support.
27. Install the CRFM mounting bracket bolts to the radiator support tighten the bolts to 15 ft. lbs. (20 Nm).
28. Install the front impact bar.
29. Install the CRFM closeout panel to the condenser.
30. Install the CRFM closeout panel retainers to the condenser.
31. Clip the engine wire harness to fan assembly

⁂ WARNING

Be sure there is proper Clearance between the Compressor Hose and the Exhaust Gas Recirculation (EGR) Pipe.

32. Install the electrical connectors to the fan motors.
33. Install the front bumper fascia.
34. Fill the cooling system.
35. Refill and bleed cooling system.
36. After installation, verify there is no coolant leakage at each connection.

RADIATOR

REMOVAL & INSTALLATION

2.7L Engines

1. Disconnect the negative battery cable.
2. Drain coolant.
3. Remove cooling fan assembly.
4. For A/T, remove A/T fluid cooler inlet and outlet hoses.
5. Remove radiator outlet hose from radiator.
6. Remove condenser bolts from condenser brackets.
7. Remove radiator from vehicle.
8. For A/T, remove A/T fluid cooler inlet and outlet hoses.

To install:

9. Reverse removal procedures, noting the following.
10. Refill A/T fluid (if applicable).
11. Fill with suitable coolant
12. Connect the negative battery cable.
13. Run vehicle and check for proper cooling fan operation and leaks.

3.6L Engines

See Figure 37.

1. Disconnect the negative battery cable.
2. Drain coolant.
3. Raise and suitably support the vehicle as necessary.
4. Remove front fascia support.
5. Remove the front bumper impact bar.
6. Remove radiator closeout panel.
7. Disconnect the compressor hose/pipe from clip at bottom of fan shroud.
8. Remove mounting bolts from condenser using care with the upper left bolt to clear the compressor hose/pipe, reposition and support the condenser.
9. Remove radiator hose inlet.
10. Remove radiator hose outlet.
11. Disconnect transmission oil cooler lines from attaching clips on radiator.
12. Remove fan shroud bolts (Qty: 2) and reposition.

To install:

13. To install the radiator install or connect the following:
    - Fan shroud bolts (Qty: 2)
    - Transmission oil cooler lines
    - Transmission oil cooler line clips to radiator
    - Radiator hose outlet
    - Radiator hose inlet
    - Condenser mounting bolts
    - Compressor hose/pipe and clip at bottom of fan shroud
    - Radiator closeout panel
    - Front bumper impact bar
    - Front fascia support
14. Lower vehicle.
15. Fill with suitable coolant.
16. Connect the negative battery cable.
17. Run vehicle and check for proper cooling fan operation and leaks.

THERMOSTAT

REMOVAL & INSTALLATION

2.7L Engines

See Figure 38.

1. Drain coolant and tighten drain plug.
2. Remove fan shroud with cooling fan after disconnecting radiator inlet hose from radiator.

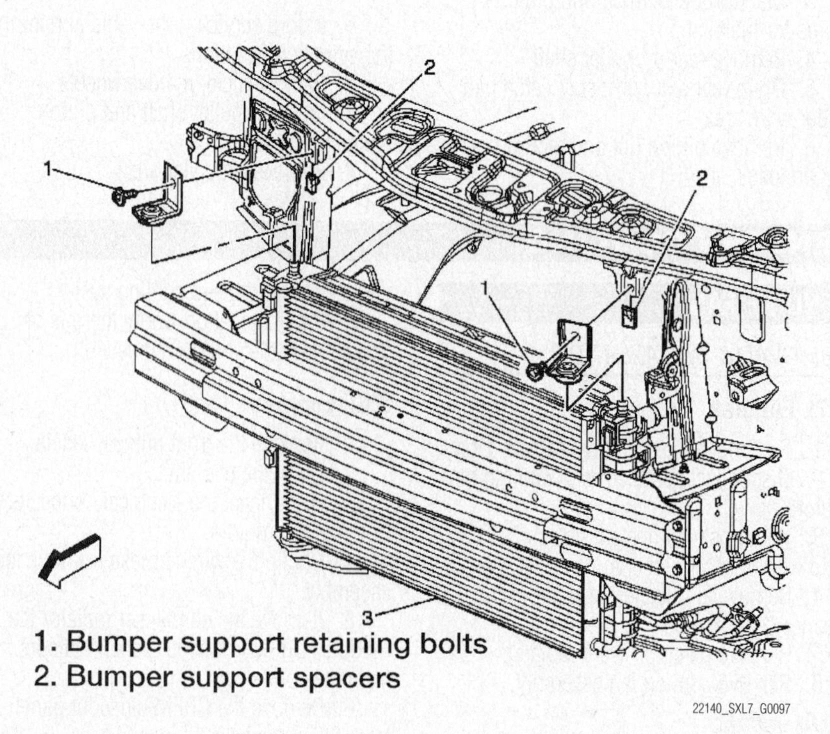

Fig. 37 Bumper support

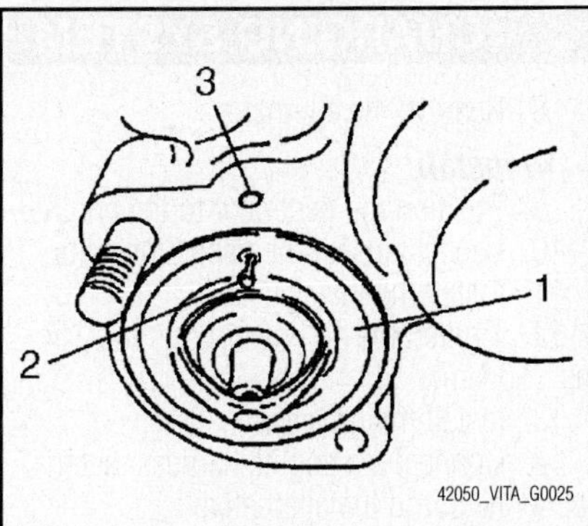

Fig. 38 Thermostat match mark position

3. Remove thermostat housing.
4. When positioning thermostat and O-ring on thermostat housing, be sure to position it so that air bleed valve comes at match mark and into the recession of thermostat case.

To install:

5. Install thermostat into thermostat housing.
6. Install cooling fan and fan shroud, and connect radiator inlet hose to radiator.
7. Fill with suitable coolant
8. Connect the negative battery cable.
9. Run vehicle and check for proper cooling fan operation and leaks.

3.6L Engines

See Figure 39.

1. Drain coolant and tighten drain plug.
2. Remove the heater inlet and outlet pipes.
3. Remove the thermostat housing bolts (1).
4. Remove the housing.

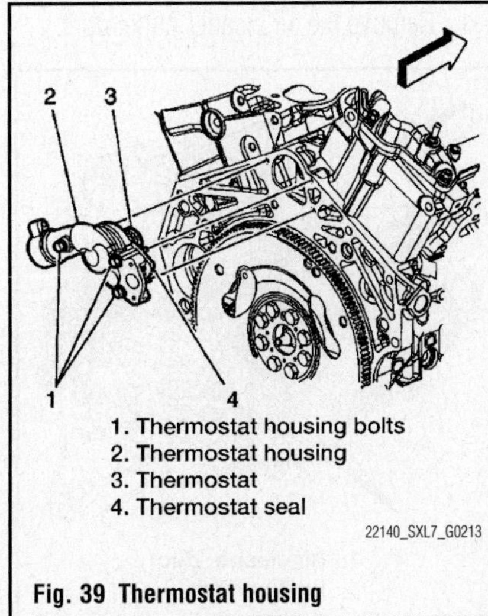

Fig. 39 Thermostat housing

5. Remove the thermostat (3) and discard the thermostat seal (4).

To install:

6. Install the thermostat with a NEW thermostat seal.
7. Install the thermostat housing bolts to 89 inch lbs. (10 Nm).
8. Install the heater inlet and outlet pipes.
9. Fill the cooling system.

WATER PUMP

REMOVAL & INSTALLATION

2.7L Engines

See Figure 40.

1. Before servicing the vehicle, refer to the precautions in the beginning of this section.
2. Drain the cooling system.
3. Remove or disconnect the following:
   - Negative battery cable
   - Accessory drive belts
   - Front timing cover
   - Water pump

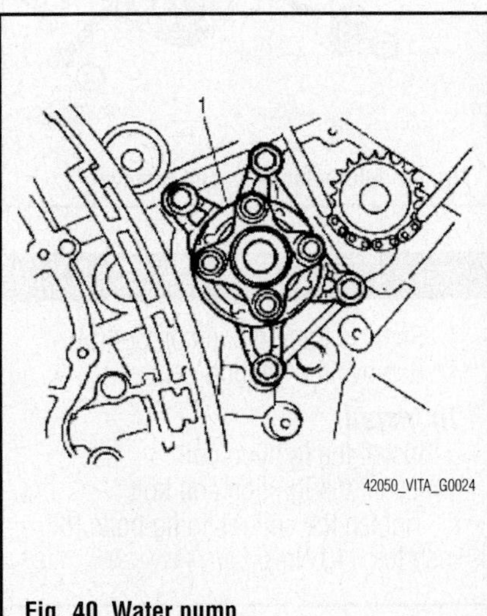

Fig. 40 Water pump

To install:

4. Install or connect the following:
   - Water pump with a new O-ring seal. Tighten the bolts to 19 ft. lbs. (27 Nm).
   - Front timing cover
   - Accessory drive belts
5. Fill with suitable coolant.
6. Connect the negative battery cable.
7. Run vehicle and check for proper cooling fan operation and leaks.

3.6L Engines

See Figure 41.

Fig. 41 EN 46104 water pump pulley tool

1. Before servicing the vehicle, refer to the precautions in the beginning of this section.
2. Drain the cooling system.
3. Remove or disconnect the following:
   - Negative battery cable
   - Accessory drive belts
   - Right engine strut mount
   - Right engine mount strut bracket

➡Use the EN 46104 in order to retain the water pump pulley.

4. Remove or disconnect the following:
   - Water pump pulley bolts
   - Water pump pulley
   - Remove the water pump bolts
   - Remove the water pump
5. Remove and discard the water pump seal.
6. Carefully clean the water pump sealing surfaces.

To install:

7. Install a NEW water pump seal.
8. Install the water pump
9. Install the water pump bolts. Tighten the bolts to 89 inch. lbs. (10 Nm).
10. Install the water pump pulley and the water pump pulley bolts
11. Install the water pump pulley bolts. Tighten the bolts to 89 inch. lbs. (10 Nm)
12. Install the right engine strut mount
13. Install the right engine mount strut bracket
14. Install the drive belt
15. Fill with suitable coolant
16. Connect the negative battery cable.
17. Run vehicle and check for proper cooling fan operation and leaks.

ENGINE ELECTRICAL — CHARGING SYSTEM

ALTERNATOR

REMOVAL & INSTALLATION

2.7L Engines

1. Disconnect negative cable at battery.
2. Remove air cleaner case and hose.
3. Disconnect alternator lead wire and coupler from alternator.
4. Remove water pump and alternator drive belt.
5. Remove alternator bracket bolt and then remove bracket.
6. Remove alternator mounting bolts.
7. Remove alternator.

To install:

8. Install bracket and then tighten alternator bracket bolt to 32 ft. lbs. (45 Nm).
9. Mount alternator on the alternator bracket.
10. Tighten alternator mounting bolts, upper to 18 ft. lbs. (25 Nm), lower to 32 ft. lbs. (45 Nm).
11. Install water pump and alternator drive belt.
12. Connect terminal lead wire and coupler to alternator.
13. Install air cleaner case and hose.
14. Connect the negative battery cable.

3.6L Engines

See Figure 42.

1. Disconnect negative cable at battery.
2. Reposition the positive battery cable boot at the alternator terminal.
3. Remove the positive battery cable nut at the alternator.
4. Remove the positive battery cable terminal from the alternator.
5. Disconnect the engine harness electrical connector from the alternator.
6. Remove the drive.
7. Remove the alternator bolts.

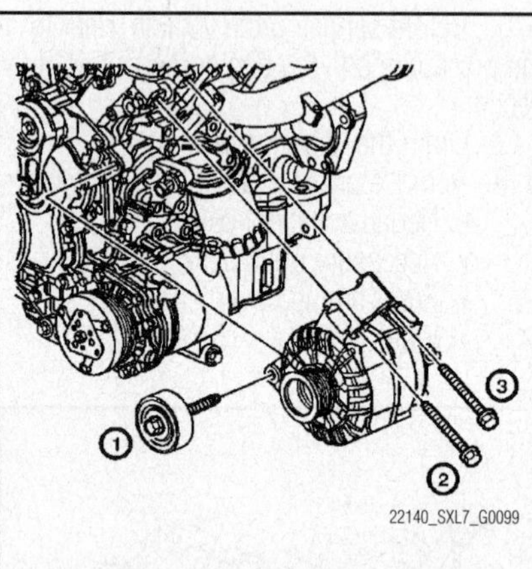

Fig. 42 Alternator bolts and removal

8. Remove the alternator.

To install:

9. Position the generator to the engine.
10. Loosely install the generator bolts.
11. Install the idler pulley.
12. Tighten the alternator bolts to 37 ft. lbs. (50 Nm).
13. Install the drive belt.
14. Connect the engine harness electrical connector to the alternator.
15. Install the positive battery cable terminal to the alternator.
16. Install the positive battery cable nut to the alternator.
17. Tighten the alternator bolts to 15 ft. lbs. (20 Nm).
18. Position the positive battery cable boot at the generator terminal.
19. Connect the negative battery cable.

VOLTAGE REGULATOR

ADJUSTMENT

The voltage regulator is controlled by the electrical power management (EPM) systems computer.

REMOVAL & INSTALLATION

The voltage regulator is integral to the alternator. It is not serviceable.

ENGINE ELECTRICAL — IGNITION SYSTEM

IGNITION COIL

REMOVAL & INSTALLATION

2.7L Engine

1. Remove surge tank cover.
2. Disconnect ignition coil couplers.
3. Remove ignition coil bolts, and then pull out ignition coil assembly.

To install:

4. Install ignition coil assembly.
5. Tighten ignition coil assembly bolts.
6. Connect ignition coil couplers.
7. Install surge tank cover.

3.6L Engines

Bank 1

See Figure 43.

1. Remove the fuel injector sight shield.
2. Disconnect the engine wiring harness electrical connector(s) from the ignition coil(s).
3. Remove the ignition coil bolt(s).
4. Remove the ignition coil

To install:

5. Install the ignition coil.
6. Install the ignition coil bolt.
7. Tighten the coil retaining bolts to 89 inch lbs. (10 Nm).

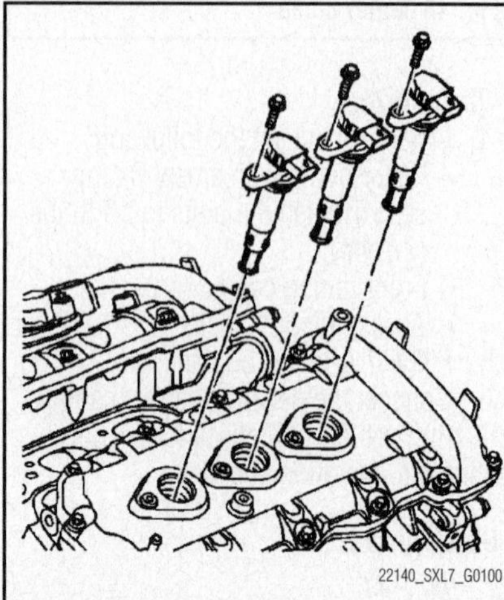

Fig. 43 Ignition coil(s)

8. Connect the engine wiring harness electrical connector(s) to the ignition coil(s).
9. Install the fuel injector sight shield.

Bank 2

See Figure 44.

1. Remove the air cleaner outlet duct.

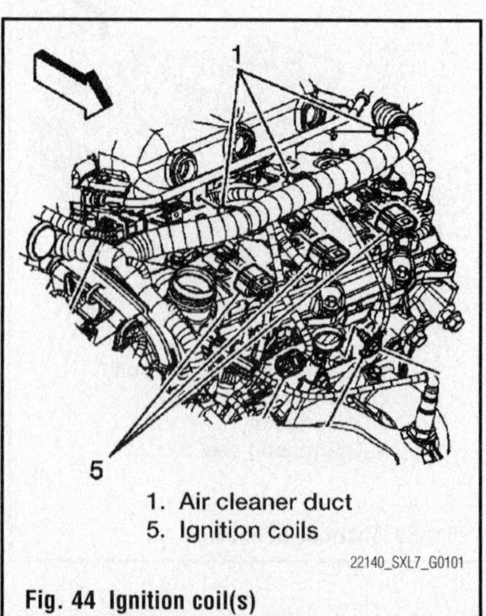

Fig. 44 Ignition coil(s)

2. Disconnect the engine wiring harness electrical connector(s)from the ignition coil(s).
3. Remove the ignition coil bolt(s).
4. Remove the ignition coil

To install:

5. Install the ignition coil.
6. Install the ignition coil bolt.
7. Tighten the coil retaining bolts to 89 inch lbs. (10 Nm).
8. Connect the engine wiring harness electrical connector(s) to the ignition coil(s).
9. Install the air cleaner outlet duct.

FIRING ORDERS

See Figures 45 and 46.

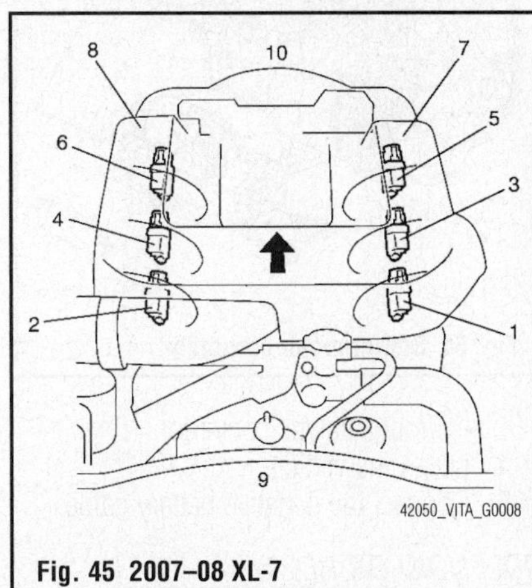

Fig. 45 2007–08 XL-7
1—2—3—4—5—6

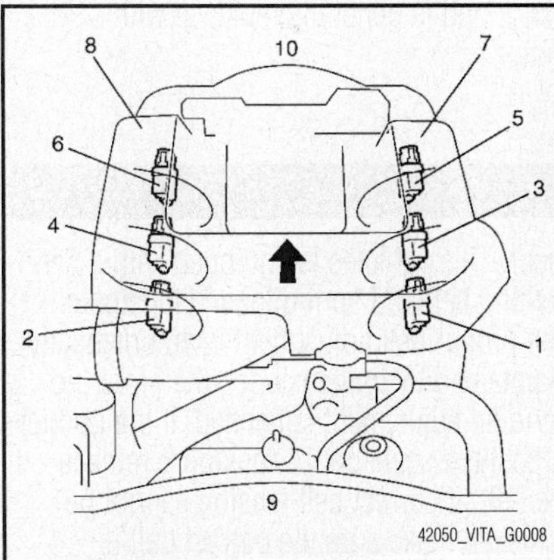

Fig. 46 2007–08 Grand Vitara
Firing order: 1-6-5-4-3-2
Distributorless ignition system
(Coil over each cylinder)

IGNITION TIMING

ADJUSTMENT

2.7L Engines

See Figures 47 and 48.

➡Before starting engine, place transmission gear shift lever in "Neutral" (shift selector lever to "P" range for A/T model), and set parking brake.

1. Start engine and warm it up to normal operating temperature.
2. Make sure that all of electrical loads except ignition are switched **OFF**.
3. Check to be sure that idle speed is within specification.
4. Connect Suzuki scan tool to DLC (1) with ignition switch **OFF**, restart engine and fix ignition timing by using fixed spark mode of Suzuki scan tool.
5. Set timing light to ignition harness for No.1 cylinder.
6. Using timing light, check that timing observed from viewpoint is within specification 5° plus or minus 1°BTDC.
7. If ignition timing is out of specification, check the following:
   - CKP sensor
   - CKP sensor plate

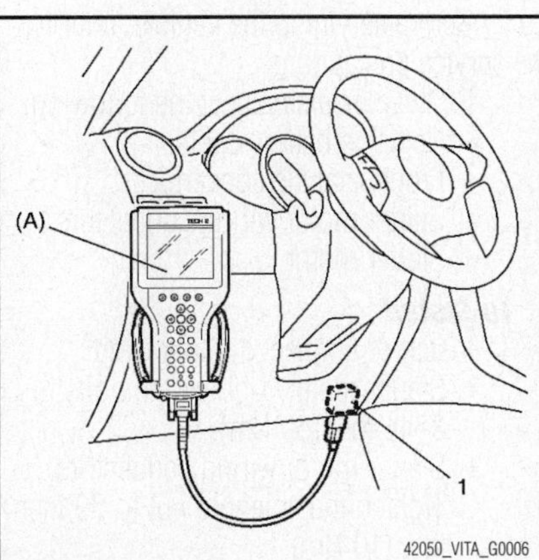

Fig. 47 Diagnostic scan tool and connector

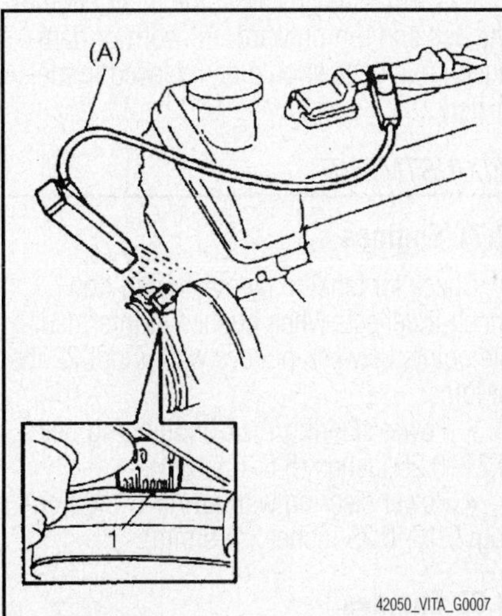

Fig. 48 Timing light and timing mark viewpoint.

   - TP sensor
   - CMP sensor
   - CMP sensor rotor tooth of camshaft
   - Wheel speed sensor
   - Timing chain cover installation

8. After checking initial ignition timing, release ignition timing fixation by using scan tool.
9. With engine idling (throttle opening at closed position and vehicle stopped), check that ignition timing is about 10—16°BTDC. (Constant variation within a few degrees from 10—15° BTDC indicates no abnormality but proves operation of electronic timing control system.) Also, check that increasing engine speed advances ignition timing. If the check results are not satisfactory, check CKP sensor and ECM.

3.6L Engines

The ECM primarily uses engine speed, the MAF sensor signal, and position information from the crankshaft position (CKP) and the camshaft position (CMP) sensors. This controls the sequence, dwell, and timing of the spark. No adjustment is necessary or possible.

SPARK PLUGS

REMOVAL & INSTALLATION

2.7L Engines

See Figure 49.

1. Remove surge tank cover.
2. Disconnect ignition coil couplers.
3. Remove ignition coil bolts, and then pull out ignition coil assembly.
4. Remove spark plugs.

To install:

5. Install spark plugs, and tighten to 18 ft. lbs (25 Nm).
6. Install ignition coil assembly.
7. Tighten ignition coil assembly bolts.

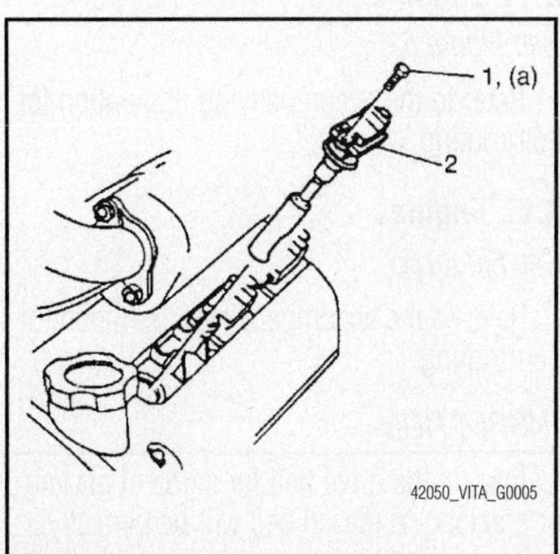

Fig. 49 Ignition coil removal—2.7L engines

8. Connect ignition coil couplers.
9. Install surge tank cover.

3.6L Engines

1. Remove the ignition coils.
2. Use compressed air in order to remove debris from the spark plug cavity.
3. Remove the spark plug.

To install:

4. Install the spark plugs.
5. Set spark gap to .044 inches (1,10mm).
6. Tighten the spark plug to 15 ft. lbs. (20 Nm).
7. Install the ignition coils.

INSPECTION

Inspect the spark plugs for electrode air gap wear, carbon deposits and insulator damage. If any abnormality is found, replace them with new plugs.

ENGINE ELECTRICAL

STARTING SYSTEM

STARTER

REMOVAL & INSTALLATION

2.7L Engines

See Figure 50.

1. Before servicing the vehicle, refer to the precautions in the beginning of this section.
2. Remove or disconnect the following:
   - Negative battery cable

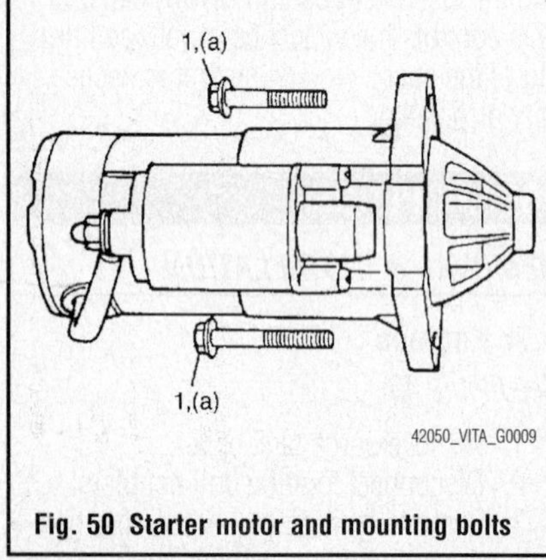

Fig. 50 Starter motor and mounting bolts

   - Starter motor wiring connectors
   - Starter motor

To install:

3. Install or connect the following:
   - Starter motor. Tighten the bolts to 22 ft. lbs. (30 Nm).
   - Starter motor wiring connectors. Tighten the solenoid nut to 11 ft. lbs. (15 Nm).
   - Negative battery cable

3.6L Engines

See Figure 51.

1. Before servicing the vehicle, refer to the service precautions.
2. Remove or disconnect the following:
   - Negative battery cable
   - Front catalytic convertor
   - Starter motor wiring connectors
   - Starter motor

To install:

3. Install or connect the following:
   - Starter motor. Tighten the bolts to 37 ft. lbs. (50 Nm).
   - Starter motor wiring connectors. Tighten the solenoid nut to 89 inch. lbs. (10 Nm)

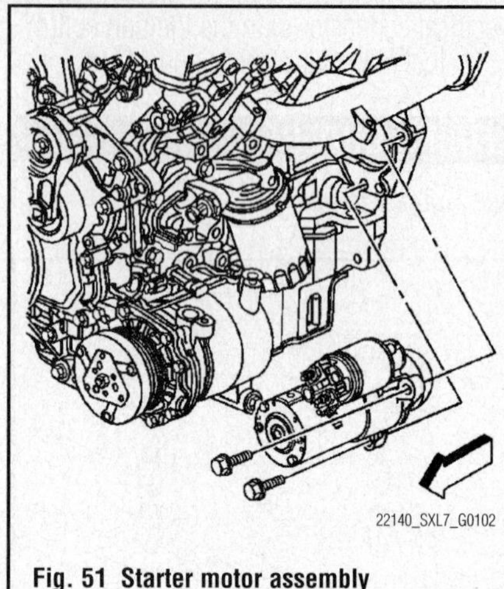

Fig. 51 Starter motor assembly

   - Front catalytic convertor
4. Lower the vehicle.
5. Connect the negative battery cable.

SOLENOID OR RELAY REPLACEMENT

Solenoid is mounted to the starter motor and is generally replaced with motor.

ENGINE MECHANICAL

ACCESSORY DRIVE BELTS

ACCESSORY BELT ROUTING

2.7L Engines

See Figure 52.

Refer to the accompanying illustration for belt routing.

3.6L Engines

See Figure 53.

Refer to the accompanying illustration for belt routing.

INSPECTION

Inspect the drive belt for signs of glazing or cracking. A glazed belt will be perfectly smooth from slippage, while a good belt will have a slight texture of fabric visible. Cracks will usually start at the inner edge of the belt and run outward. All worn or damaged drive belts should be replaced immediately.

ADJUSTMENT

2.7L Engines

Check for tension by measuring how much it deflects when pushed at intermediate points between pulleys with about 22 lbs of force.

- Power steering/Air Conditioning belt: 0.21–0.25 inches (5.5–6.5 mm).
- Power steering without Air Conditioning: 0.16–0.35 inches (4–9 mm).

3.6L Engines

The belt system in the XL-7 is self adjusting and requires no manual adjustment. This vehicles is equipped with a serpentine belt and spring loaded tensioner. The proper belt adjustment is automatically maintained by the tensioner, therefore, no periodic adjustment is needed. If the pointer is past the scale on the tensioner replace the belt. If correct belt tension cannot be achieved make sure the correct belt is installed. If the correct tension is still not achieved and check for proper mounting off all accessory drives.

REMOVAL & INSTALLATION

2.7L Engines

Water Pump and Alternator Belt

1. Disconnect the negative battery cable.
2. Loosen tension pulley bolts and tension pulley adjusting bolt.

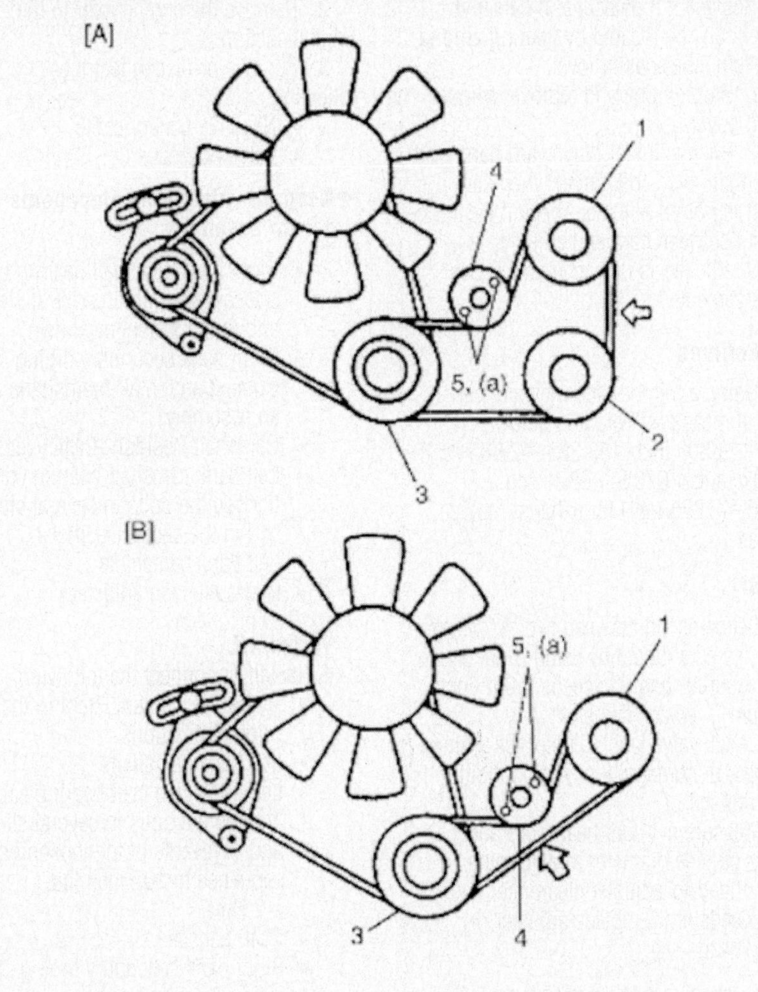

1. P/S pump pulley	5. Tension pulley bolts
2. A/C compressor pulley (if equipped)	[A]: with A/C
3. Crankshaft pulley	[B]: without A/C
4. Tension pulley	

42050_VITA_G0013

Fig. 52 Accessory drive belt routing—2.7L engine with P/S only, with A/C and P/S

3. Remove W/P and alternator drive belt.

To install:

4. Install new drive belt.
5. To adjust W/P and alternator belt tension.
6. Tighten tension pulley bolts to 18 ft. lbs. (25 Nm).
7. Tighten pulley adjusting bolt to 18 ft. lbs. (25 Nm).
8. Connect the negative battery cable.

Power Steering And A/C Compressor (If Equipped) Belt

1. Loosen tension pulley bolts
2. Remove tension pulley and P/S pump drive belt.

To install:

3. Install new drive belt.
4. Tighten tension pulley and P/S pump drive belt.
5. Tighten tension pulley bolts.

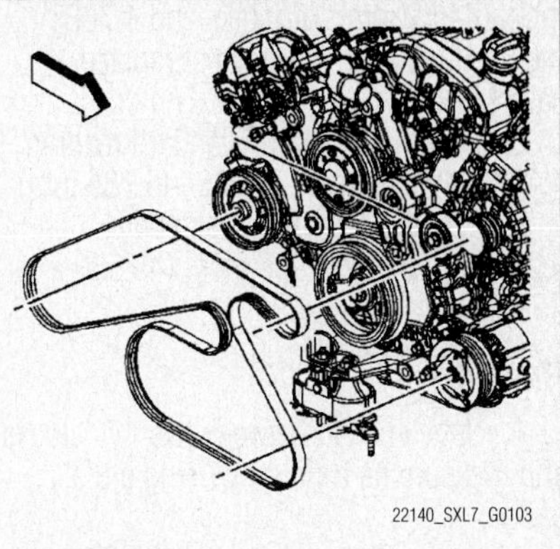

Fig. 53 Drive belt routing—3.6L engines

3.6L Engines

See Figure 53

1. Remove the air cleaner assembly.
2. Install the engine support fixture.
3. Remove the engine mount strut bracket.
4. Raise and support the vehicle.
5. Remove the engine splash shield.
6. Remove the right side engine mount bracket.
7. Lower the vehicle.
8. Rotate the drive belt tensioner clockwise to release the drive belt tension.
9. Slide the drive belt off of the belt idler pulley (1).
10. Slowly release the drive belt tensioner.
11. Remove the drive belt from the accessory drive pulleys.

To install:

12. Install the drive belt to the crankshaft pulley, the tensioner and the generator.
13. Rotate the drive belt tensioner clockwise.
14. Install the drive belt to the idler pulley (1).

➡Ensure the drive belt is properly aligned and seated into the grooves of the accessory drive pulleys.

15. Slowly release the drive belt tensioner.
16. Raise the vehicle.
17. Install the right side engine mount bracket.
18. Lower the vehicle.
19. Install the engine mount strut bracket.
20. Remove the engine support fixture.
21. Install the air cleaner assembly.

CAMSHAFT AND VALVE LIFTERS

REMOVAL & INSTALLATION

2.7L Engines

1. Using a micrometer, measure cam height. If measured height is below 1.7637

inch. (44.89 mm) for intake and 1.7886 inch. (45.43 mm) for exhaust replace camshaft.

2. Intake cam height standard reading 1.7719–1.7781 inch. (4.005–45.165 mm).

3. Exhaust cam height standard reading 1.7933–1.7996 inch. (45.550–45.710 mm).

Runout

1. Set camshaft between two "V" blocks, and measure its runout by using a dial gauge.

If measured runout exceeds 0.002 inch. (0.04 mm), replace camshaft.

2. Check valve lash adjuster for pitting, scratches, or damage. If any malcondition is found, replace.

3. Measure cylinder head bore and adjuster outside diameter to determine cylinder head-to-adjuster clearance. If clearance exceeds limit, replace adjuster or cylinder head.

Valve Lash Adjusters

1. Valve lash adjuster O.D standard reading 1.2778–1.2784 inch. (32.456–32.472mm).

2. Cylinder head bore standard reading 1.2795–1.2808 inch. (32.500–32.525 mm).

3. Cylinder head to adjuster clearance 0.0011–0.0027 inch. (0.028–0.069 mm).

4. Cylinder head adjuster clearance limit 0.0039 inch. (0.10 mm).

5. If case of the following, valve lash adjuster noise may be caused by air trapped into valve lash adjusters:

- Vehicle is left for 24 hours or more.
- Engine oil is changed.
- Hydraulic lash adjuster is replaced or reinstalled.
- Engine is overhauled.
- If noise from valve lash adjusters is suspected, perform the following checks.
- Check engine oil for the followings.
- Oil level in oil pan. If oil level is low, add oil up to full level hole on oil level gauge.
- Oil quality. If oil is discolored, or deteriorated, change it.
- Oil leaks - If leak is found, repair it.
- Oil pressure. If defective pressure is found, repair it.

6. Run engine for about half an hour at about 2000 to 3000 rpm, and then air will be purge and tapping sound will cease.

7. Should tapping sound not cease, it is possible that hydraulic valve lash adjuster is defective.

8. Replace it if defective. If defective adjuster can't be located by hearing among 16 of them. check as follows.

a. Stop engine and remove cylinder head cover.

b. Push adjuster downward hard (with less than 44 ft. lbs. force) when cam crest is not on adjuster check if clearance exists between cam and adjuster. If it does, adjuster is defective and needs replacement.

3.6L Engines

1. Using a micrometer, measure cam height. If measured height is below 1.6687–1.6805 inch. (42.385–42.685 mm) for intake and 1.6703–1.6821inch. (45.425–42.725 mm) for exhaust replace camshaft.

Runout

1. Set camshaft between two "V" blocks, and measure its runout by using a dial gauge.

If measured runout exceeds 0.002 inch. (0.06 mm), replace camshaft.

2. Check valve lash adjuster for pitting, scratches, or damage. If any malcondition is found, replace.

3. Measure cylinder head bore and adjuster outside diameter to determine cylinder head-to-adjuster clearance. If clearance exceeds limit, replace adjuster or cylinder head.

REMOVAL & INSTALLATION

2.7L Engines

See Figures 54 and 55.

1. Before servicing the vehicle, refer to the precautions section.

2. Remove the engine refer to the Engine section.

3. Remove or disconnect the following:

- Negative battery cable
- Valve covers

➡Keep all valvetrain components in order for assembly.

- Right bank camshaft bearing caps. Loosen the bolts in several steps and in the sequence shown.
- Right bank secondary timing chain, exhaust and intake camshafts as an assembly
- Camshaft Position (CMP) sensor
- Left bank camshaft bearing caps. Loosen the bolts in several steps and in the sequence shown.
- Left bank camshafts
- Hydraulic lash adjusters

To install:

4. Install or connect the following:

- Hydraulic lash adjusters in their original positions
- Left bank camshafts
- Left bank camshaft bearing caps. Tighten the bolts in several steps and in reverse of the loosening sequence to 102 inch lbs. (12 Nm).
- CMP sensor
- Right bank secondary timing chain, exhaust and intake camshafts as an assembly
- Right bank camshaft bearing caps. Tighten the bolts in several steps and in reverse of the loosening sequence to 102 inch lbs. (12 Nm).

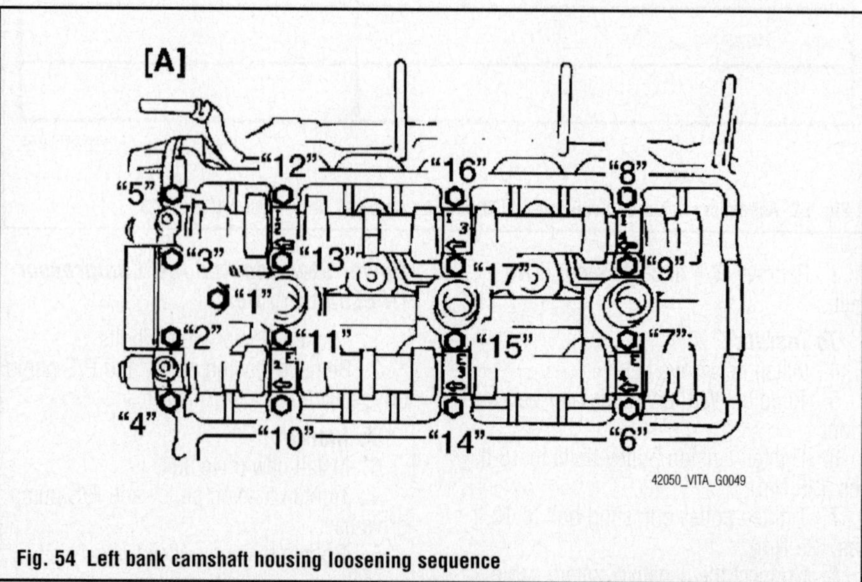

Fig. 54 Left bank camshaft housing loosening sequence

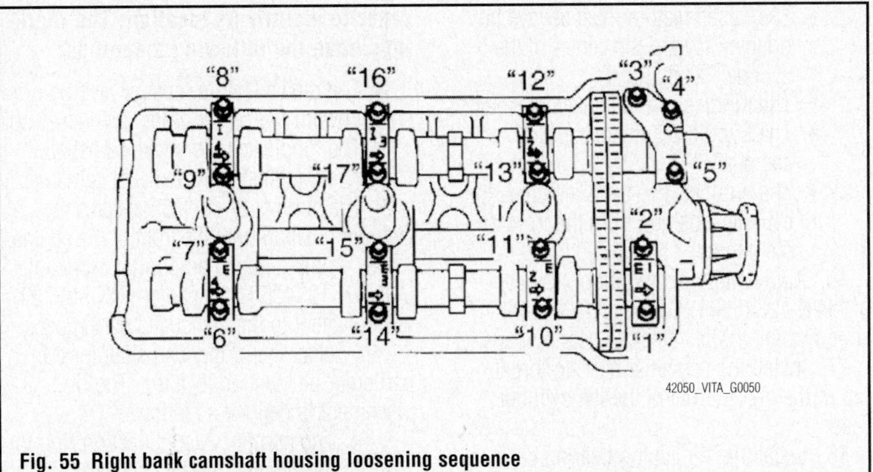

Fig. 55 Right bank camshaft housing loosening sequence

✲✲ WARNING

Wait ½ hour after installing the lash adjusters and camshafts before cranking or starting the engine to allow the lash adjusters to bleed down. Operating the engine before this time period may result in interference between the valves and pistons.

5. Connect the negative battery cable.
6. Fill the crankcase to the correct level.
7. Start the engine and check for leaks.

3.6L Engines

Left Side

See Figures 56 through 62.

1. Before servicing the vehicle, refer to all service precaution.
2. Remove the engine refer to the Engine Assembly section.
3. Remove the timing chain assembly, refer to the Timing Chain section.
4. Remove the camshaft bearing cap bolts
5. Remove the camshaft bearing caps
6. Remove the camshafts

To install:

7. Ensure that the camshaft sealing rings (1) are in place in the camshaft grooves. Camshaft sealing rings must be in place below the surface of the camshaft journal in order to avoid being pinched between the cylinder head and the camshaft caps.
8. Observe the markings on the bearing caps. Each bearing cap is marked in order to identify its location. The markings have the following meanings:

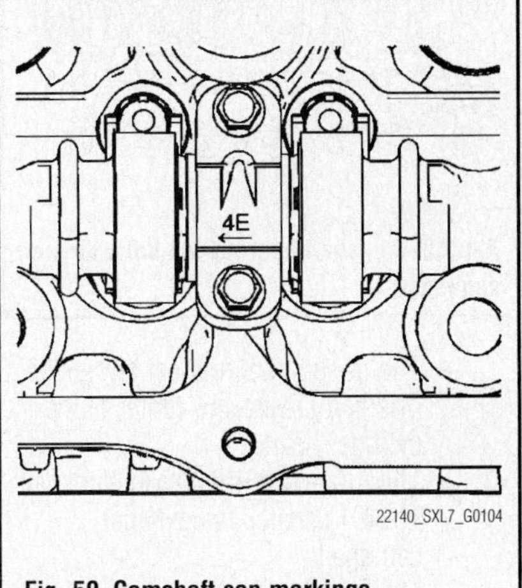

Fig. 59 Camshaft cap markings

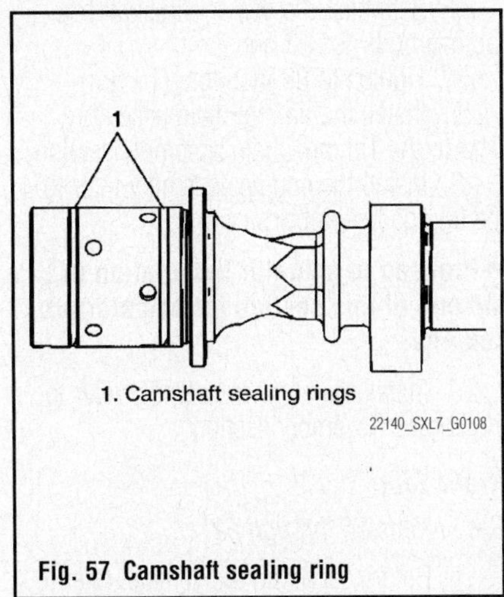

Fig. 57 Camshaft sealing ring

Fig. 60 Camshaft lobes in a neutral position

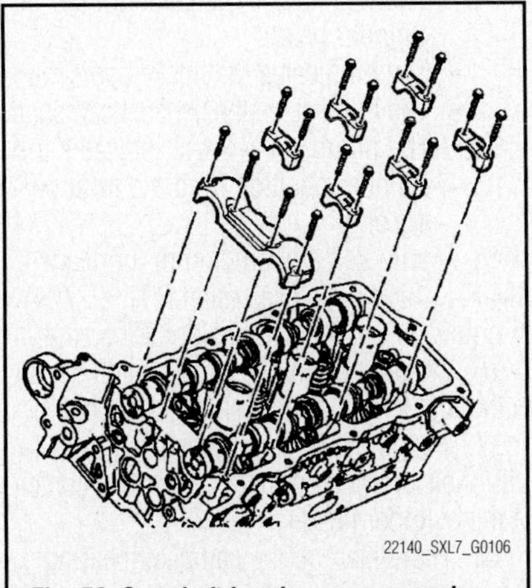

Fig. 56 Camshaft bearing cap removal

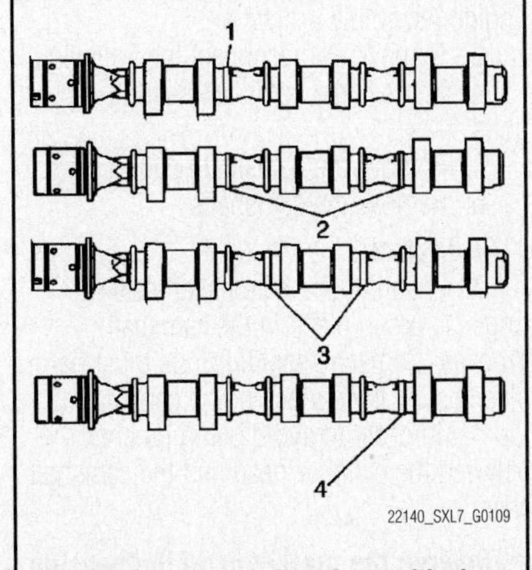

Fig. 58 Camshaft sealing ring positioning

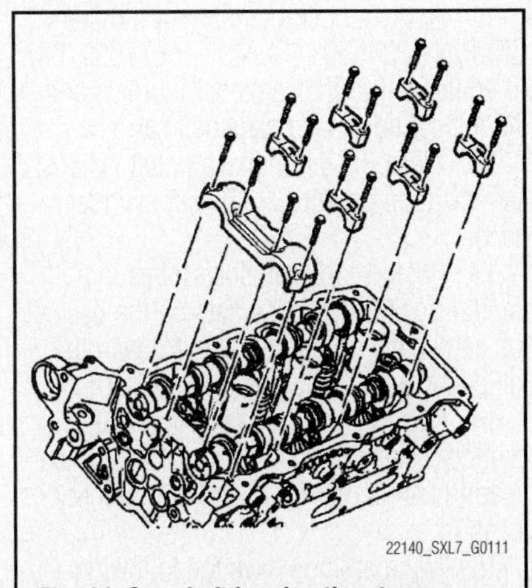

Fig. 61 Camshaft bearing thrust caps

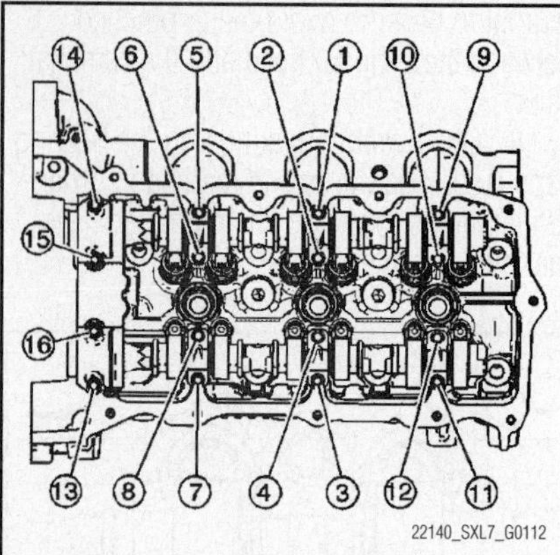

Fig. 62 Camshaft bearing cap bolts torque sequence

- The raised feature must always be oriented toward the center of the cylinder head.
- The I indicates the intake camshaft.
- The E indicates the exhaust camshaft.
- The number 2—4—6 indicates the cylinder position from the front of the engine.

9. Apply a liberal amount of lubricant GM P/N 12345501 (Canadian P/N 992704) or equivalent to the camshaft bearing caps.

10. Select the proper camshaft for the particular installation location. The ring placement is defined as follows:

- The number 2 identification ring for the right exhaust camshaft is machined off (1).
- The number 2 and 5 identification rings for the right intake camshaft is machined off (2).
- The number 3 and 4 identification rings for the left intake camshaft is machined off (3).
- The number 5 identification ring for the left exhaust camshaft is machined off (4).

11. Apply a liberal amount of lubricant GM P/N 12345501 (Canadian P/N 992704) or equivalent to the camshaft journals and the left cylinder head camshaft carriers.

12. Place the left intake and left exhaust camshafts in position in the left cylinder head.

13. Position the camshaft lobes in a neutral position with the flats on the back of the camshafts up and parallel (1) with the left cylinder head camshaft cover rail

14. Observe the markings on the left cylinder head camshaft bearing caps. Each bearing cap is marked in order to identify its location.

15. The markings have the following meanings:

- The raised feature must always be oriented toward the center of the cylinder head.
- The I indicates the intake camshaft.
- The E indicates the exhaust camshaft.
- The number 2—4—6 indicates the cylinder position from the front of the engine.

16. Apply a liberal amount of lubricant GM P/N 12345501 (Canadian P/N 992704) or equivalent to the camshaft bearing caps.

17. Install the camshaft bearing thrust cap in the first journal of the left cylinder head.

18. Install the remaining bearing caps with their orientation mark toward the center of the cylinder head.

19. Hand start all the camshaft bearing cap bolts.

20. Tighten the camshaft bearing cap bolts in the sequence shown.

21. Tighten to 89 inch lbs. (10 Nm).

22. Loosen the center intake camshaft bearing cap bolts 1, 2 and the center exhaust camshaft bearing cap bolts 3 and 4.

23. Retighten the center camshaft bearing cap bolts 1, 2, 3 and 4.

24. Tighten to 89 inch lbs. (10 Nm).

25. Install the timing chain assembly, refer to the Timing Chain assembly section.

26. Install the engine assembly refer to the Engine Assembly section.

➡Proceed to actuator installation in the end of this section for actuator procedures.

27. Install the engine assembly refer to the Engine Assembly section.

Right Side

See Figures 56 through 62

1. Before servicing the vehicle, refer to the precautions section.

2. Remove the engine, refer to the Engine Assembly procedure.

3. Remove or disconnect the following:

4. Remove the camshaft bearing cap bolts

5. Remove the camshaft bearing caps

6. Remove the camshafts

To install:

7. Ensure that the camshaft sealing rings (1) are in place in the camshaft grooves. Camshaft sealing rings must be in place below the surface of the camshaft journal in order to avoid being pinched between the cylinder head and the camshaft caps.

➡Observe the markings on the bearing caps. Each bearing cap is marked in order to identify its location. The markings have the following meanings:

- The raised feature must always be oriented toward the center of the cylinder head.
- The I indicates the intake camshaft.
- The E indicates the exhaust camshaft.
- The number 2—4—6 indicates the cylinder position from the front of the engine.

8. Apply a liberal amount of lubricant GM P/N 12345501 (Canadian P/N 992704) or equivalent to the camshaft bearing caps.

9. Select the proper camshaft for the particular installation location. The ring placement is defined as follows:

- The number 2 identification ring for the right exhaust camshaft is machined off (1).
- The number 2 and 5 identification rings for the right intake camshaft is machined off (2).
- The number 3 and 4 identification rings for the left intake camshaft is machined off (3).
- The number 5 identification ring for the left exhaust camshaft is machined off (4).

10. Apply a liberal amount of lubricant GM P/N 12345501 (Canadian P/N 992704) or equivalent to the camshaft journals and the right cylinder head camshaft carriers.

11. Place the right intake and right exhaust camshafts in position in the right cylinder head.

12. Position the camshaft lobes in a neutral position with the flats on the back of the camshafts up and parallel (1) with the right cylinder head camshaft cover rail

13. Observe the markings on the right cylinder head camshaft bearing caps. Each bearing cap is marked in order to identify its location.

14. The markings have the following meanings:

- The raised feature must always be oriented toward the center of the cylinder head
- The I indicates the intake camshaft
- The E indicates the exhaust camshaft
- The number 1—3—5 indicates the cylinder position from the front of the engine.

15. Apply a liberal amount of lubricant GM P/N 12345501 (Canadian P/N 992704) or equivalent to the camshaft bearing caps.

16. Install the camshaft bearing thrust cap in the first journal of the left cylinder head.

17. Install the remaining bearing caps with their orientation mark toward the center of the cylinder head.

18. Hand start all the camshaft bearing cap bolts.

19. Tighten the camshaft bearing cap bolts in the sequence shown.
20. Tighten to 89 inch lbs. (10 Nm).
21. Loosen the center intake camshaft bearing cap bolts 1, 2 and the center exhaust camshaft bearing cap bolts 3, 4.
22. Retighten the center camshaft bearing cap bolts 1, 2, 3 and 4.
23. Tighten to 89 inch lbs. (10 Nm).

➡Proceed to actuator installation.

Valve Lifter Replacement

See Figure 63.

Following the Camshaft, Camshaft Position Actuator and Rocker Arm removal procedure.

1. Remove the valve lifters.

To install:

➡Do not stroke/cycle the stationary hydraulic lash adjuster plunger without oil in the lower pressure chamber.

➡Do not allow the stationary hydraulic lash adjuster to tip over, plunger down, after the oil fill.

2. Fill the Stationary Hydraulic Lash Adjuster (SHLA) with clean engine oil GM P/N 12345610 (Canadian P/N 993193) or equivalent. Take precautions to prevent scratching the pivot sphere area (1) of the SHLA.
3. Lubricate the SHLA bores in the cylinder head with clean engine oil GM P/N 12345610 (Canadian P/N 993193) or equivalent.
4. Install the SHLAs in the cylinder head.
5. Apply a liberal amount of lubricant GM P/N 12345501 (Canadian P/N 992704) or equivalent to the SHLA pivot spheres

Camshaft Position Actuator—Intake

See Figures 64 through 67.

✱✱ WARNING

Observe the body of the camshaft position actuator for the "EX" or "IN" markings. Ensure the proper timing mark is used.

1. Use an open end wrench on the hex cast into the camshaft in order to prevent engine rotation when loosening the camshaft position actuator bolt.
2. Remove the left intake camshaft position actuator bolt.
3. Remove the left intake camshaft position actuator. When removing the actuator, place the chain on the engine cover side of the actuators.
4. Rotate the actuator in order to align the opening in the actuator reluctor wheel with the cam sensor boss in the front cover, to allow actuator removal.
5. If removing both the exhaust and intake camshaft actuators, the timing chain can be removed once the actuators have been removed.

To install:

✱✱ WARNING

Observe the body of the camshaft position actuator for the "EX" or "IN" markings. Ensure the proper timing mark is used.

✱✱ WARNING

The reluctor wheel on the right intake camshaft position actuator (1) is indexed in a different position compared to the left intake camshaft position actuator (2).

Fig. 64 Camshaft position actuator removal

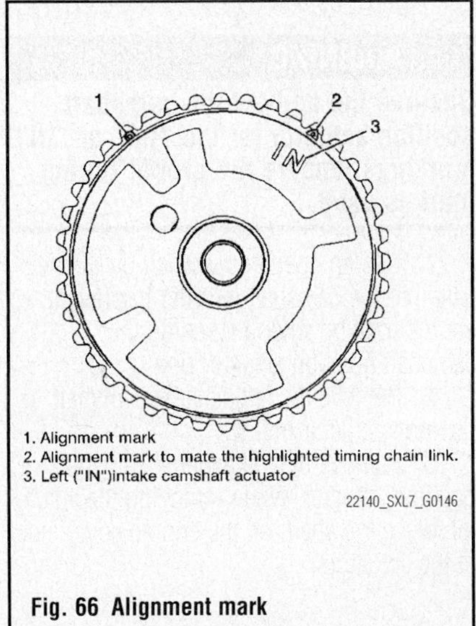

Fig. 66 Alignment mark

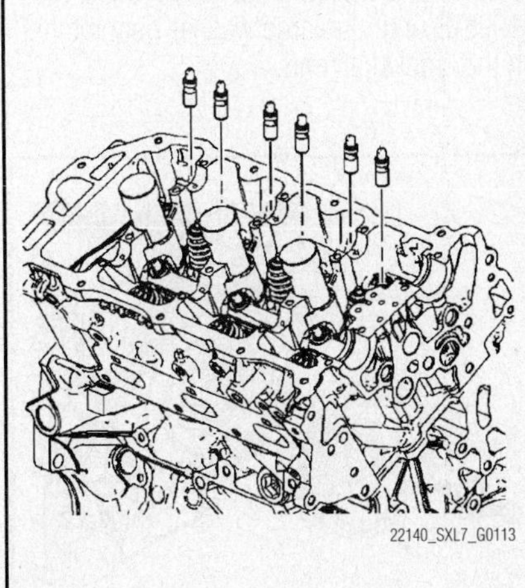

Fig. 63 Valve lifter removal and installation

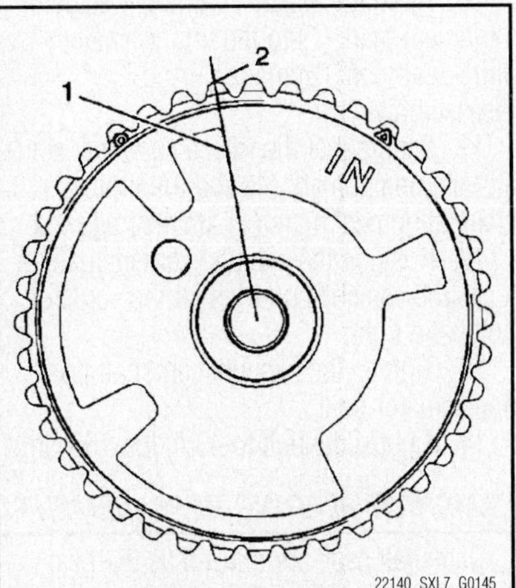

Fig. 65 Camshaft position actuator marking alignment designations

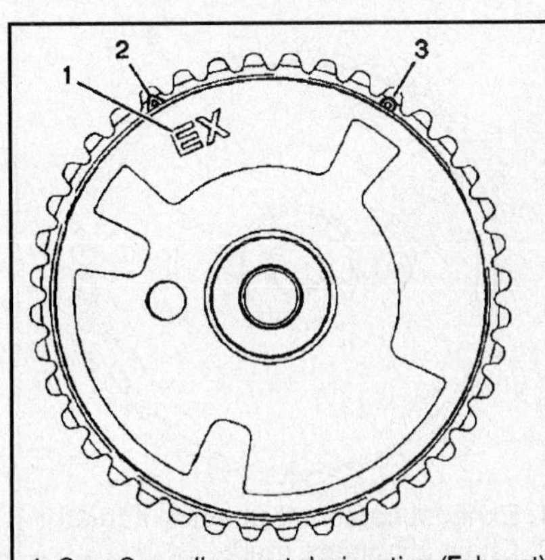

Fig. 67 Camshaft position actuator installation and marking alignment designations

➡The circle marking on the intake gears is for alignment to the highlighted timing chain link.

6. On the left intake camshaft actuator the edge of the reluctor wheel (1) lines up with the peak (2) of the sprocket tooth.

7. Ensure the proper timing mark is used. Observe the outer ring of the camshaft position actuator for the circle marking.

➡Use an open wrench on the hex cast into the camshaft in order to prevent engine rotation when tightening the camshaft position actuator bolt.

8. Install the left intake camshaft position actuator.

9. Install the left intake camshaft position actuator bolt.

10. Tighten to 43 ft. lbs. (58 Nm).

Camshaft Position Actuator—Exhaust

See Figures 68 and 69.

⁂ WARNING

Observe the body of the camshaft position actuator for the "EX" or "IN" markings. Ensure the proper timing mark is used.

1. Use an open end wrench on the hex cast into the camshaft in order to prevent engine rotation when loosening the camshaft position actuator bolt.

2. Remove the left exhaust camshaft position actuator bolt.

3. Remove the left exhaust camshaft position actuator. When removing the actuator, place the chain on the engine cover side of the actuators.

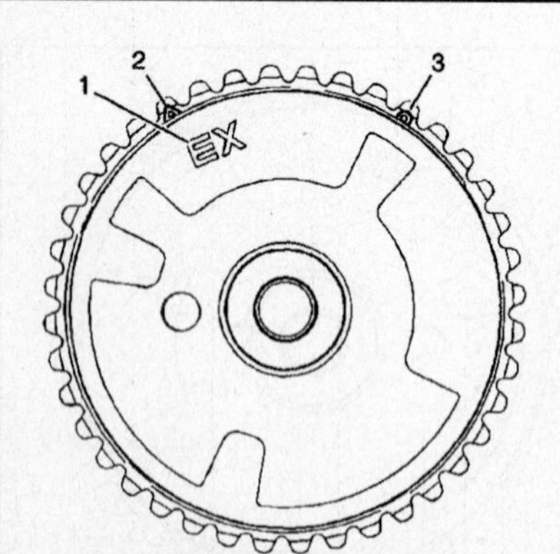

Fig. 68 Camshaft position actuator alignment—Exhaust

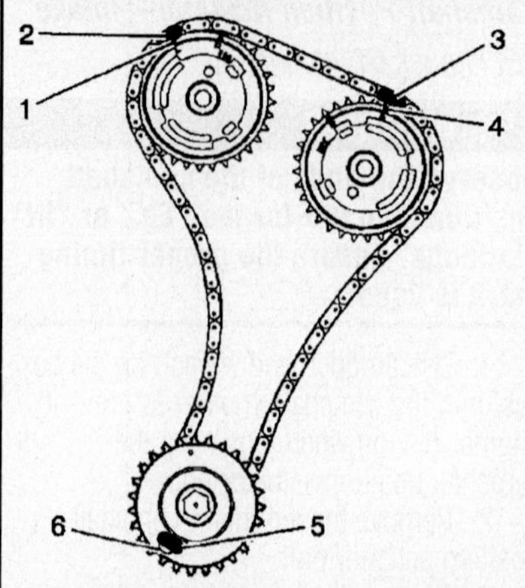

Fig. 69 Camshaft chain alignment—Exhaust

4. Rotate the actuator in order to align the opening in the actuator reluctor wheel with the cam sensor boss in the front cover, to allow actuator removal.

5. If removing both the exhaust and intake camshaft actuators, the timing chain can be removed once the actuators have been removed.

To install:

6. Ensure the proper camshaft position actuator is installed. Observe the body of the camshaft position actuator for the EX marking (1).

7. The marking is for an exhaust camshaft position actuator.

8. Ensure the proper timing mark is used. Observe the outer ring of the camshaft position actuator for the circle marking (3).

9. The marking is for alignment to the highlighted timing chain link on the left side of the engine.

10. Align the exhaust camshaft actuator alignment mark (1) to the timing chain alignment mark (2) made during disassembly.

11. Ensure that the intake camshaft actuator alignment mark (4) and the timing chain alignment mark (3) are also aligned.

12. Position the exhaust camshaft actuator to the camshaft and install the actuator bolt hand tight.

13. Tighten the exhaust camshaft position actuator bolt.

14. Tighten the bolt to 43 ft. lbs. (58 Nm).

CAMSHAFT BEARING REPLACEMENT

Camshaft caps are utilized as the bearings and the caps must be replace and machined in the event of a mechanical failure. If damage is too extensive for cap replacement the cylinder head must be replaced.

CRANKSHAFT DAMPER

REMOVAL & INSTALLATION

2.7L Engines

1. Remove or disconnect the following:
   - Negative battery cable
   - Accessory drive belts
   - Water pump pulley
   - Power steering pump and brackets
   - Crankshaft pulley

To install:

2. Install or connect the following:
   - Crankshaft pulley. Tighten the bolt to 109 ft. lbs. (148 Nm).
   - Power steering pump and brackets
   - Water pump pulley
   - Accessory drive belts
   - Negative battery cable
3. Fill the crankcase to the correct level.
4. Fill the cooling system.
5. Start the engine and check for leaks.

3.6L Engines

See Figure 70.

➡Do not lubricate the crankshaft front oil seal or crankshaft balancer sealing surfaces. The crankshaft balancer is installed into a dry seal. Apply lubricant to the inside of the crankshaft balancer hub bore.

1. Place the crankshaft balancer in position on the crankshaft.

2. Thread the J 41998-B in the crankshaft. Ensure you engage at least 10 threads of the J 41998-B before pressing the crankshaft balancer in place.

3. Push the crankshaft balancer into position by tightening the nut on the J 41998-B until the large washer bottoms out on the crankshaft end.

Fig. 70 Crankshaft damper installation

4. Remove the J 41998-B.
5. Install the crankshaft balancer bolt.
6. Install the crankshaft balancer bolt to 74 ft. lbs. (100 Nm).
7. Install the crankshaft balancer bolt an additional 150°

CRANKSHAFT FRONT SEAL

REMOVAL & INSTALLATION

2.7L Engines

See Figures 71 through 85.

1. Before servicing the vehicle, refer to the precautions in the beginning of this section.
2. Drain the cooling system.
3. Drain the engine oil.
4. Remove engine assembly from vehicle.
5. Remove or disconnect the following:
   - Negative battery cable
   - Intake manifold
   - Ignition coils
   - Valve covers
   - Accessory drive belts
   - Cooling fan and shroud
   - Water pump pulley
   - Radiator
   - Power steering pump and brackets
   - Oil pan and pickup tube
   - Crankshaft pulley
   - Front crankshaft seal
   - Crankshaft Position (CKP) sensor
   - Front cover
6. Remove timing chain tensioner adjuster.
7. To remove it, slacken left (No.1) bank 2nd timing chain by turning intake camshaft counterclockwise a little while pushing back pad.
8. Remove left (No.1) bank intake and exhaust camshaft sprocket bolts.
9. To remove it, fit a spanner to hexagonal part at the center of camshaft to hold it stationary.
10. Remove left (No.1) bank exhaust camshaft sprocket.
11. Remove left (No.1) bank intake camshaft sprocket.
12. Remove left (No.1) bank 2nd timing chain.
13. Remove or disconnect the following.
   - Timing chain guide No.1 and No.2
   - Timing chain tensioner
   - Timing chain tensioner adjuster
   - Idler sprocket No.1 and 1st timing chain
   - Idler sprocket No.2 and sprocket shaft
14. Remove right(No.2) bank 1st timing chain intake camshaft sprocket bolt. To remove it, fit a spanner to hexagonal part at the center of camshaft to hold it stationary.
15. Remove right (No.2) bank 1st timing chain intake camshaft sprocket.
16. Remove 1st timing chain crankshaft sprocket.

To install:

17. Prepare the timing chain tensioners for installation by releasing the latches, compressing the tensioner piston fully into the bore and installing retaining pins.
18. Align the timing chain sprocket matchmarks and colored chain links as shown during assembly.
19. Check timing pulley key of crankshaft is on specified position.
20. Install 1st timing chain crankshaft sprocket to crankshaft.
21. Check timing mark on right (No.2) bank intake camshaft.

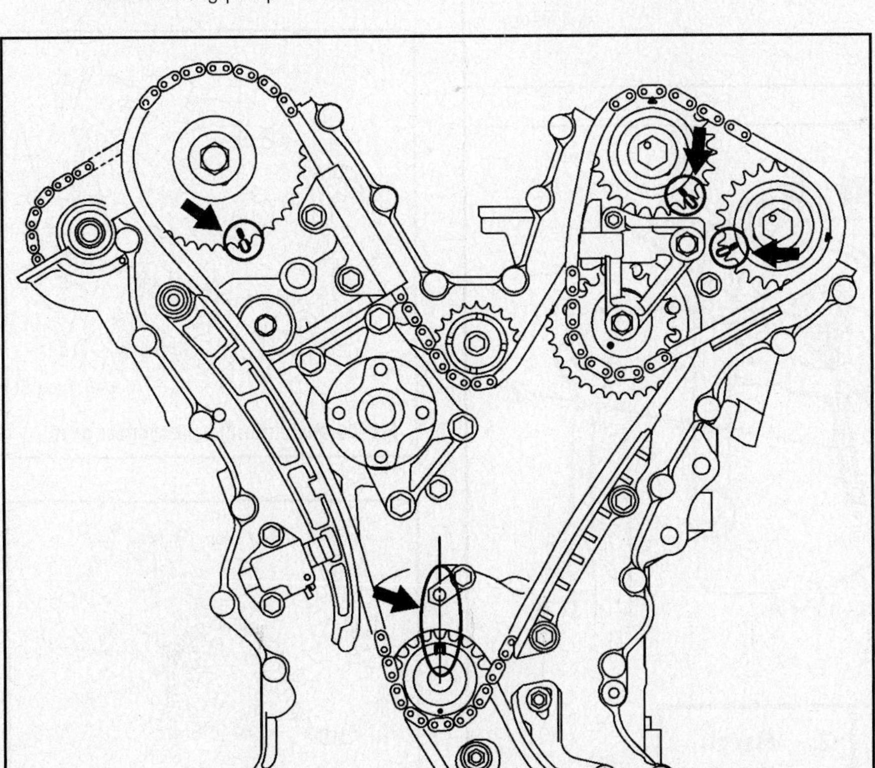

Fig. 71 Timing chain alignment marks

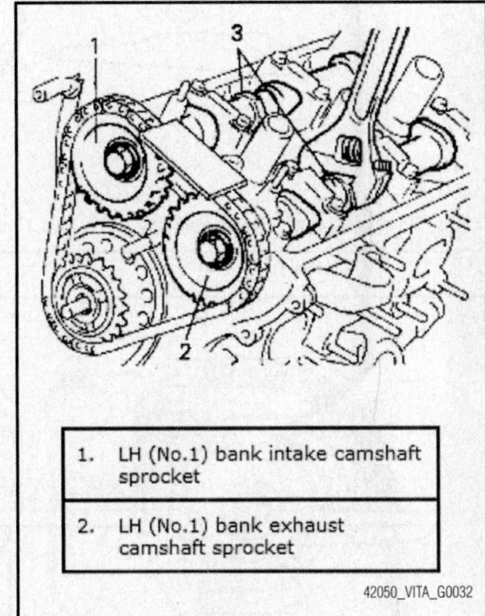

Fig. 72 Camshaft sprocket removal

Fig. 73 Right bank 1st camshaft sprocket

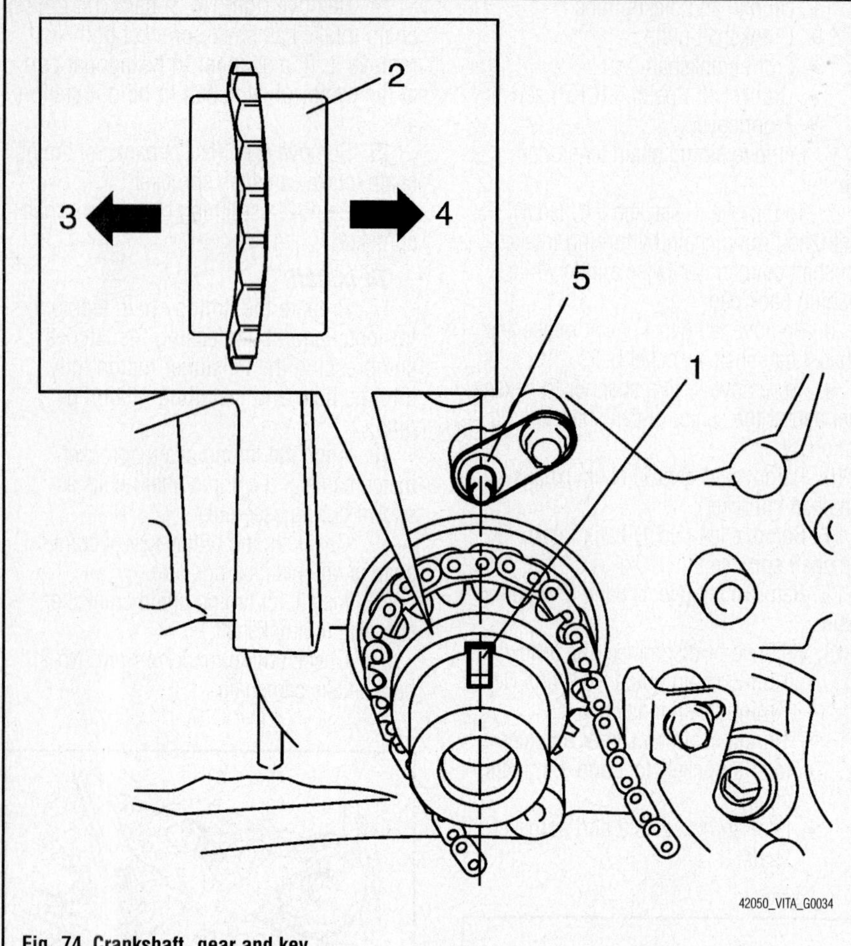

Fig. 74 Crankshaft, gear and key

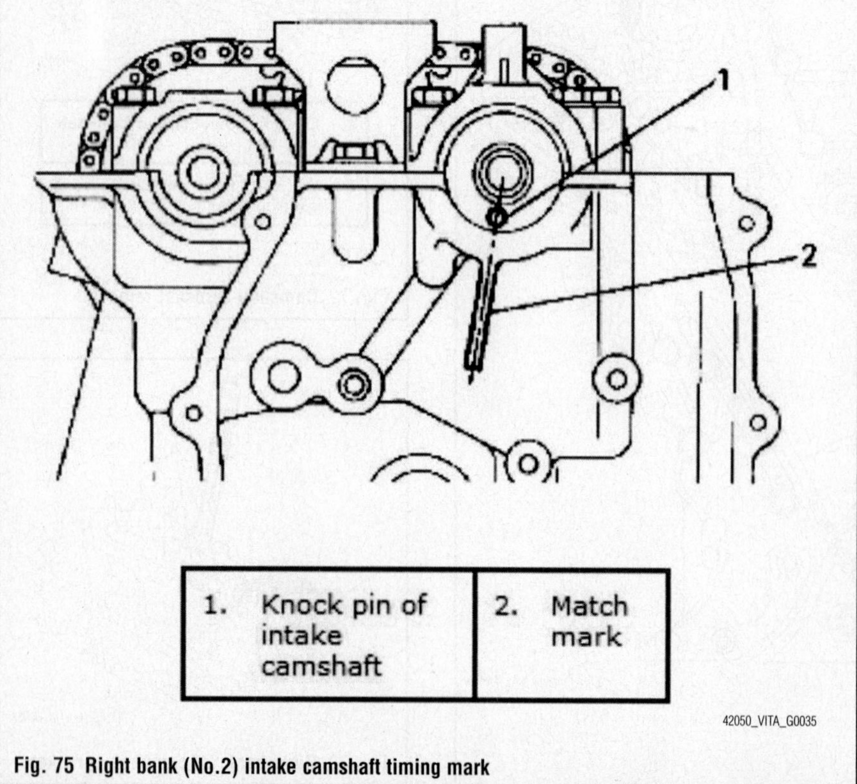

Fig. 75 Right bank (No.2) intake camshaft timing mark

22. Install right (No.2) bank 1st timing chain intake camshaft sprocket noting the following points.
    a. The sprocket should be set in such way that its timing marks can be seen.
    b. Camshaft should be held stationary by using a spanner at its hexagonal parts.
23. Tighten right bank 1st timing chain intake camshaft sprocket bolt to 58 ft. lbs. (80 Nm).
24. Install timing chain tensioner.
25. Tighten timing chain tensioner nut to 18 ft. lbs. (25 Nm).
26. Install 1st timing chain by aligning match marks on right silver plate of 1st timing chain and right (No.2) bank 1st timing chain intake camshaft sprocket.
27. Apply oil to bush of idler sprocket No.2.
28. Install idler sprocket No.2 and sprocket shaft.
29. Install idler sprocket No.2 by aligning match marks on left silver plate of 1st timing chain.
30. Install crankshaft sprocket by aligning match marks on yellow plate of 1st timing chain and crankshaft timing sprocket.
31. To install it, fit a spanner to hexagonal part at the center of right (No.2) bank intake camshaft to turn a little.

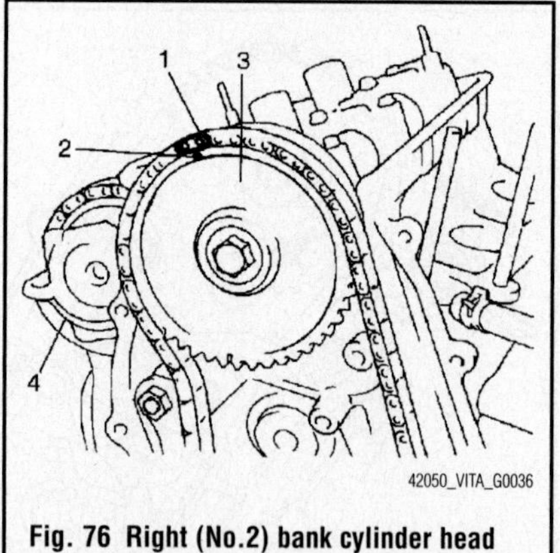

Fig. 76 Right (No.2) bank cylinder head

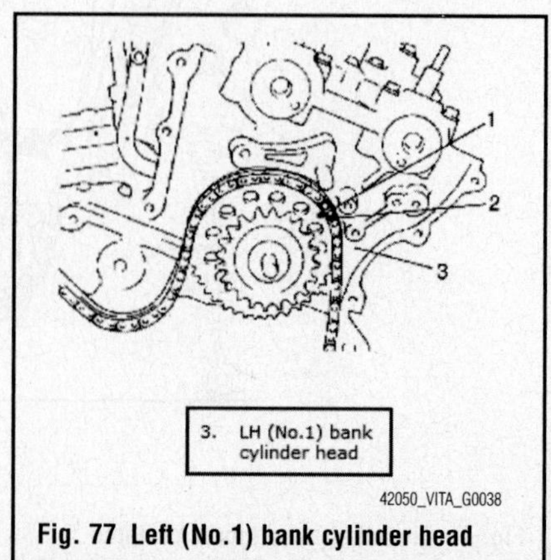

Fig. 77 Left (No.1) bank cylinder head

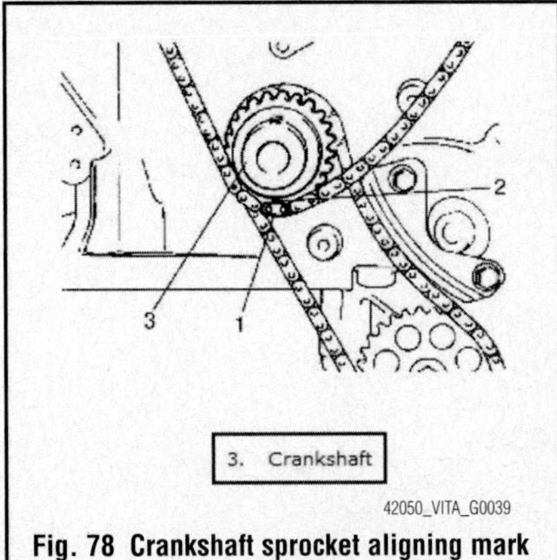

Fig. 78 Crankshaft sprocket aligning mark

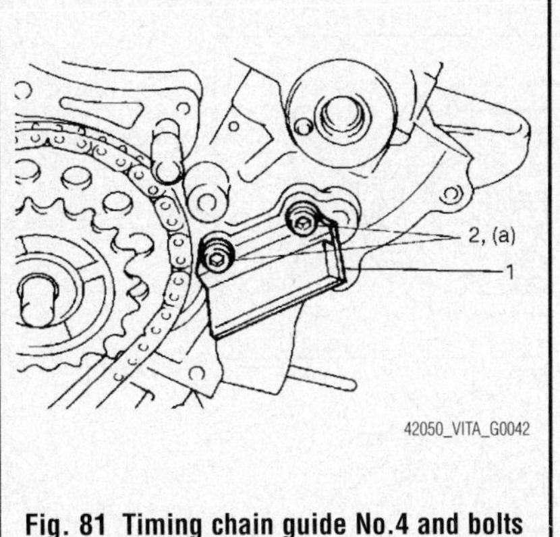

Fig. 81 Timing chain guide No.4 and bolts

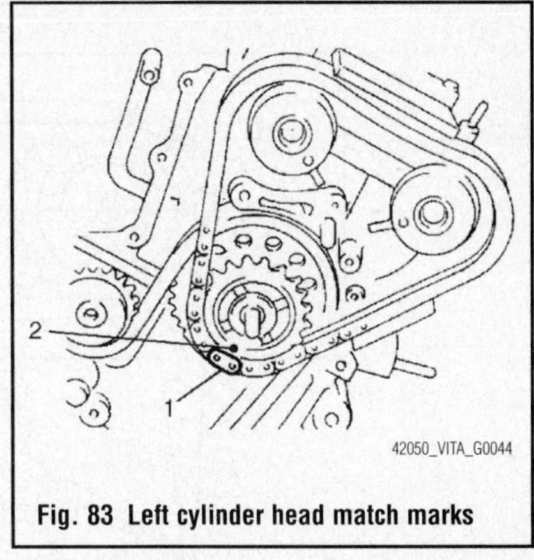

Fig. 83 Left cylinder head match marks

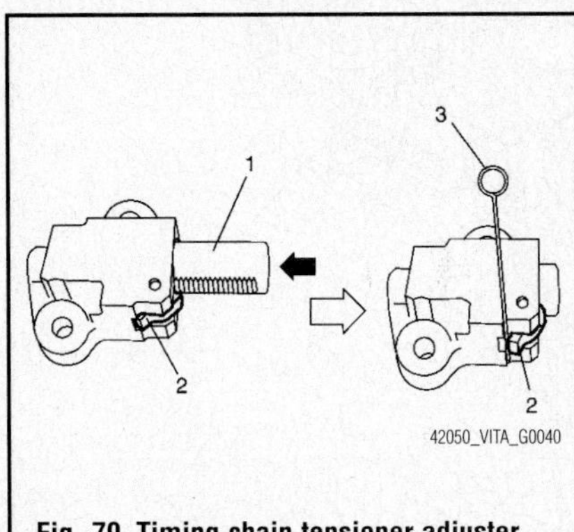

Fig. 79 Timing chain tensioner adjuster No.1

36. Install timing chain tensioner adjuster No.1 and tighten to 8 ft. lbs (11 Nm).
37. Pull out stopper (3) from adjuster No.1 (1).
38. Install timing chain guide No.1 and No.2.
39. Tighten both chain guides to 6 ft. lbs. (9 Nm).
40. Recheck each aligned timing marks.
41. Install timing chain guide No.4 and tighten to 8 ft. lbs (11 Nm).
42. Check that knock-pins of intake and exhaust camshafts are aligned with match marks on cylinder head.

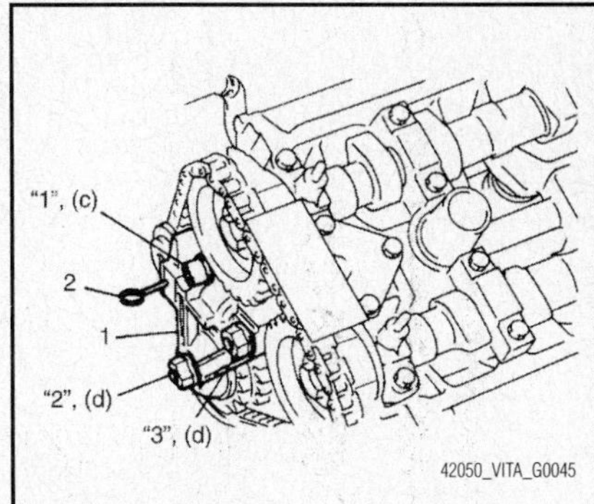

Fig. 84 Timing chain tensioner adjuster No.3 bolt c–d

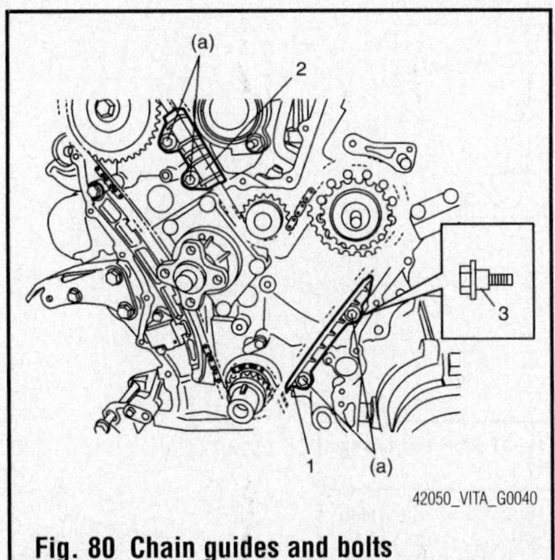

Fig. 80 Chain guides and bolts

WARNING

Do not turn camshaft more than necessary. If turned excessively, valve and piston may get damaged.

32. Apply oil to bearing of idler sprocket No.1.
33. Install idler sprocket No.1 and tighten to 40 ft. lbs. (55 Nm).
34. With latch of tensioner adjuster No.1 returned and plunger pushed back into body, insert stopper into latch and body.
35. After inserting it, check to make sure that plunger will not come out.

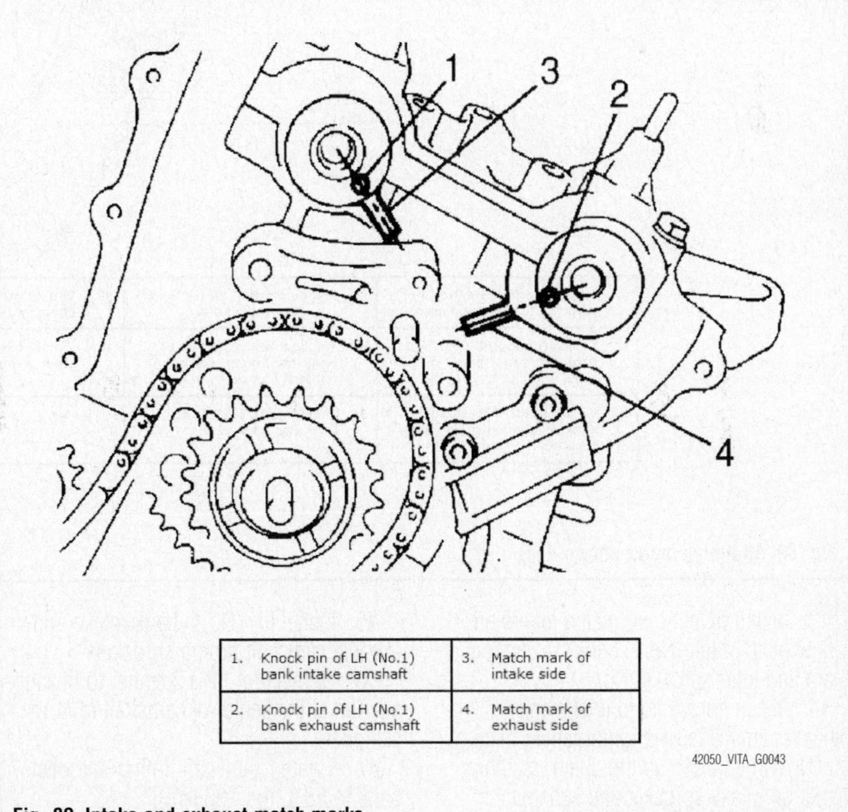

Fig. 82 Intake and exhaust match marks

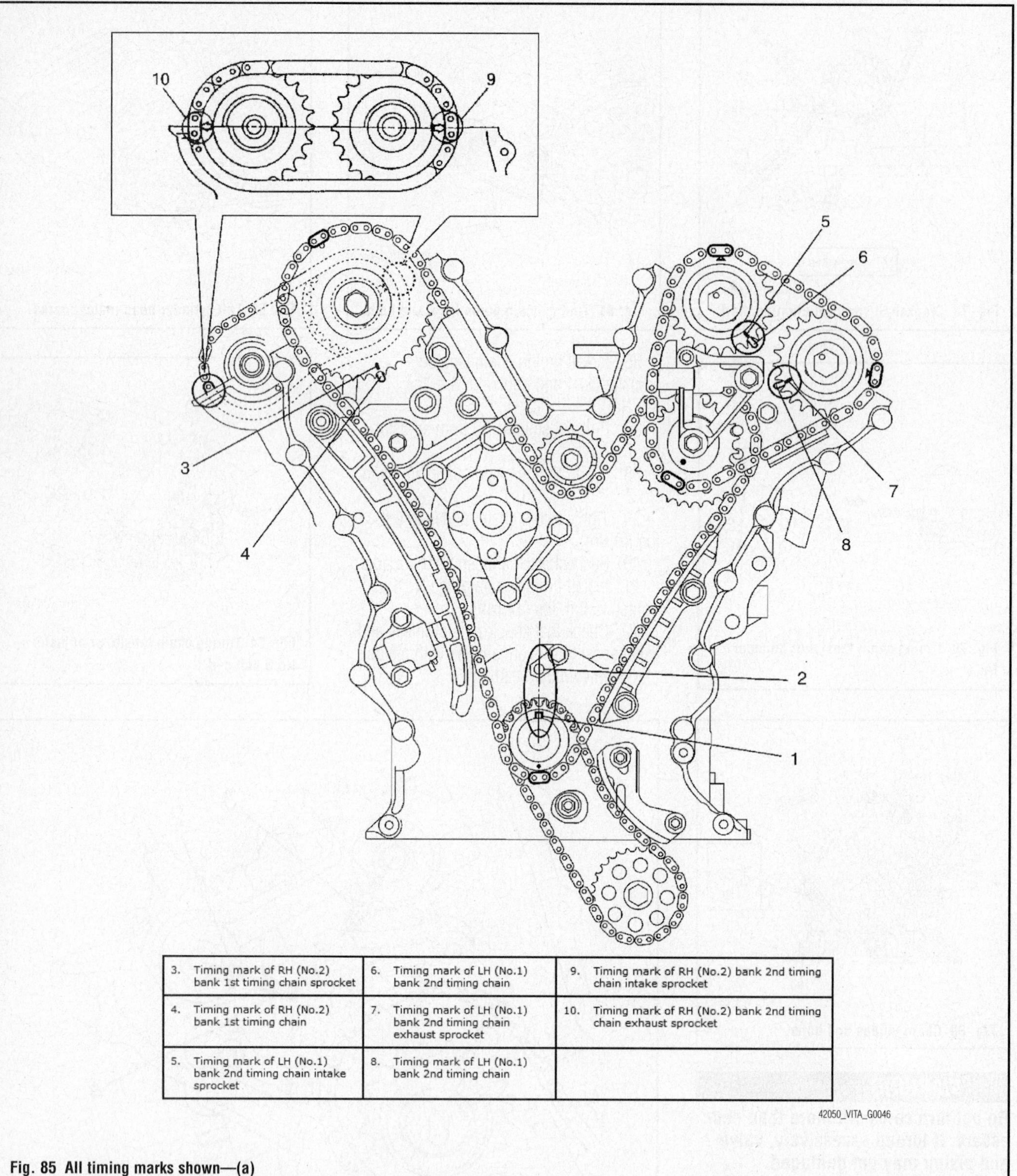

3. Timing mark of RH (No.2) bank 1st timing chain sprocket	6. Timing mark of LH (No.1) bank 2nd timing chain	9. Timing mark of RH (No.2) bank 2nd timing chain intake sprocket
4. Timing mark of RH (No.2) bank 1st timing chain	7. Timing mark of LH (No.1) bank 2nd timing chain exhaust sprocket	10. Timing mark of RH (No.2) bank 2nd timing chain exhaust sprocket
5. Timing mark of LH (No.1) bank 2nd timing chain intake sprocket	8. Timing mark of LH (No.1) bank 2nd timing chain	

Fig. 85 All timing marks shown—(a)

43. Install by aligning match marks on yellow plate of left (No.1) bank 2nd timing chain and idler sprocket No.2.

44. Install sprockets to intake and exhaust camshafts by aligning silver plate of left (No.1) bank 2nd timing chain, match marks on intake sprocket and exhaust sprocket respectively.

45. Install LH (No.1) bank intake and exhaust camshaft timing sprockets.

46. To install it, fit a spanner to hexagonal part at the center of camshaft to hold stationary.

47. Tighten camshaft timing sprocket bolts to 58 ft. lbs. (88 Nm).

48. With latch of tensioner adjuster No.3 returned and plunger pushed back into body, insert stopper (pin) into set hole.

49. After inserting it, check that plunger will not come out.

50. Install timing chain tensioner adjuster No.3.

51. Tighten Timing chain tensioner adjuster No.3 bolt (c) to 7 ft. lbs. (11 Nm).

3. Timing mark of RH (No.2) bank 1st timing chain sprocket	6. Timing mark of LH (No.1) bank 2nd timing chain	9. Timing mark of RH (No.2) bank 2nd timing chain intake sprocket
4. Timing mark of RH (No.2) bank 1st timing chain	7. Timing mark of LH (No.1) bank 2nd timing chain exhaust sprocket	10. Timing mark of RH (No.2) bank 2nd timing chain exhaust sprocket
5. Timing mark of LH (No.1) bank 2nd timing chain intake sprocket	8. Timing mark of LH (No.1) bank 2nd timing chain	

42050_VITA_G0047

Fig. 86 All timing marks key list—(b)

52. Tighten Timing chain tensioner adjuster No.3 nut (d) to 32 ft. lbs. (45 Nm).
53. Tightening order for tensioner adjuster bolt and nuts1–2–3
54. Pull out stopper (pin) from set hole.
55. Turn crankshaft two rotations clockwise then align timing pulley key (1) on crankshaft and oil jet on cylinder block.
56. Recheck all timing marks.
57. Apply oil to timing chains, tensioner adjusters sprockets and guides.
58. Install or connect the following:
    - Front cover. Tighten the bolts to 97 inch lbs. (11 Nm).
    - CKP sensor
    - Front crankshaft seal
    - Crankshaft pulley. Tighten the bolt to 109 ft. lbs. (148 Nm).
    - Oil pan and pickup tube
    - Power steering pump and brackets
    - Radiator
    - Water pump pulley
    - Cooling fan and shroud
    - Accessory drive belts
    - Valve covers
    - Ignition coils
    - Intake manifold
    - Negative battery cable
59. Install engine assembly
60. Fill the crankcase to the correct level.
61. Fill the cooling system.
62. Start the engine and check for leaks.

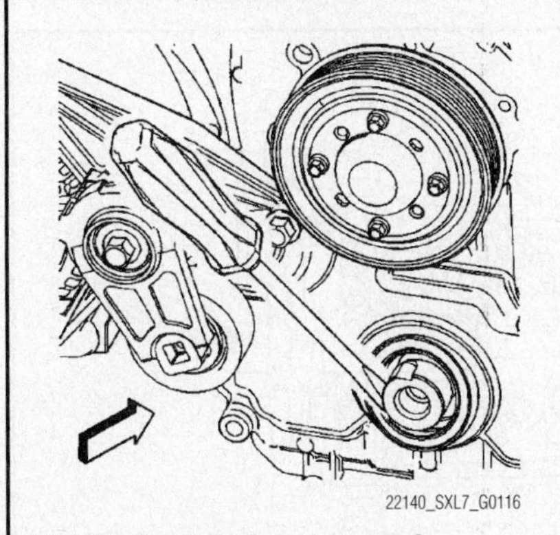

Fig. 87 Crankshaft seal removal

3.6L Engines

See Figures 87 and 88.

1. Remove the crankshaft damper, refer to the Crankshaft Damper procedure.
2. Use a flat-bladed tool in order to remove the crankshaft oil seal. Use care not to damage the engine front cover or the crankshaft.

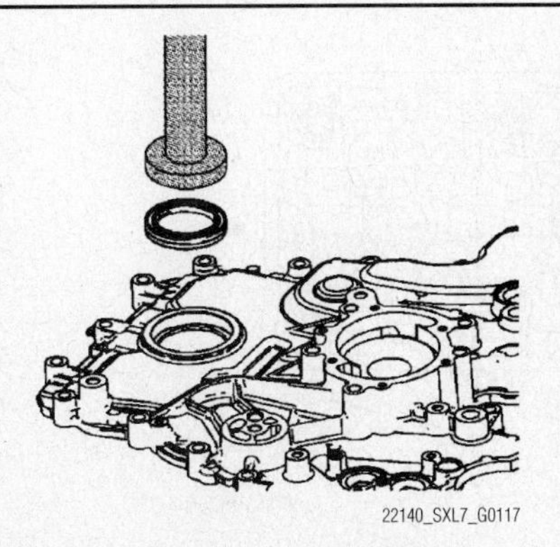

Fig. 88 Crankshaft seal installation

To install:

3. Use the J 29184 or equivalent to install the crankshaft front oil seal.
4. Install the oil seal.
5. Install the Crankshaft balancer, refer to the procedure in this section.

CYLINDER HEAD

REMOVAL & INSTALLATION

2.7L Engines

See Figures 89 through 91.

1. Before servicing the vehicle, refer to all precautions.
2. Disconnect the negative battery cable.
3. Relieve the fuel system pressure.
4. Drain the cooling system and engine oil.
5. Evacuate A/C system.
6. Remove the Engine, refer to the Engine section.
7. Remove or disconnect the following:
    - Camshafts, tappets and shims and RH (No.2) bank 2nd timing chain
    - 1st timing chain
    - LH (No.1) bank 2nd timing chain
    - Timing chain cover
    - Oil pans and oil pump strainer
    - Cylinder head covers
    - Exhaust manifold
    - Radiator outlet cap
    - Intake manifold, intake collector and electric throttle body assembly

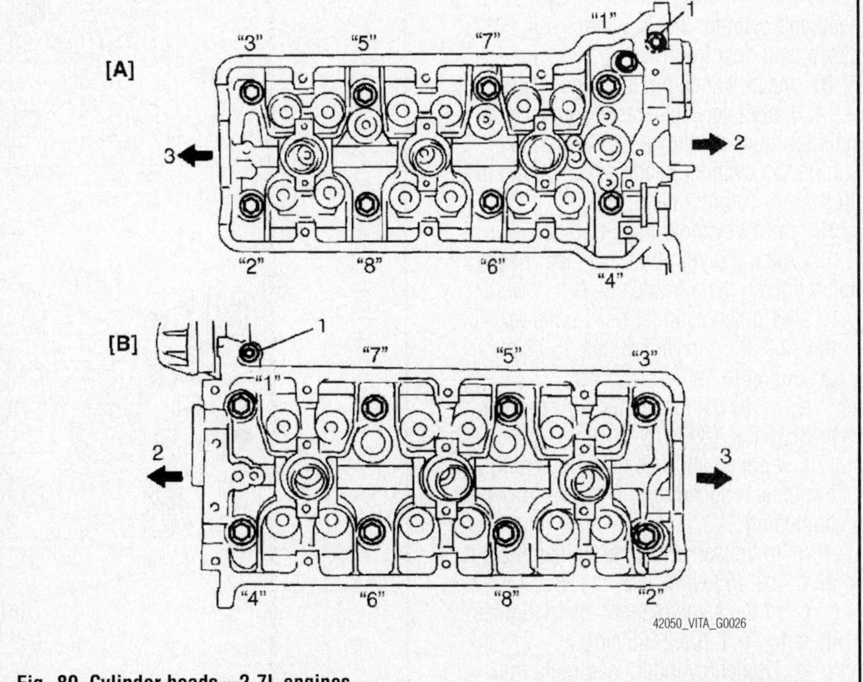

Fig. 89 Cylinder heads—2.7L engines

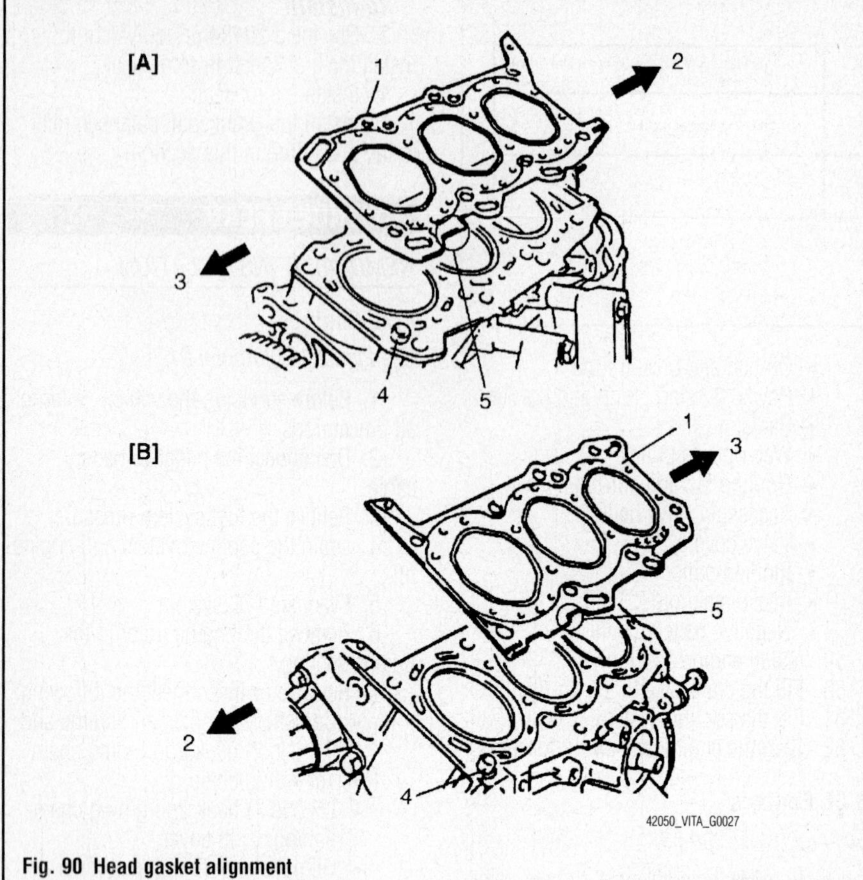

Fig. 90 Head gasket alignment

8. Loosen cylinder head bolts in numerical order (1—8).
   - Cylinder head bolts
   - Cylinder heads

To install:

9. Clean mating surface on cylinder head and cylinder block. Remove oil, old gasket and dust from mating surface.
10. Install knock pin to cylinder block.
11. Install new cylinder head gasket to cylinder block as shown in. Carved lot number on cylinder head gasket should face up (toward cylinder head side).
12. Install cylinder head to block.
13. After applying oil to cylinder head bolts, tighten them gradually as follows:
    a. Tighten cylinder head bolts (except hex hole bolt) to 38.5 ft. lbs. (53 Nm) according to numerical order.
    b. In the same manner, tighten them to 61 ft. lbs. (84 Nm)
    c. Loosen all bolts until tightening torque is reduced to 0 in reverse order of tightening.
    d. In the same manner tighten them to 38 ft. lbs. (53 Nm).
    e. In the same manner again, tighten them to 76 ft. lbs. (105 Nm).
    f. Tighten cylinder head bolts (hex hole bolt) 7 ft. lbs. (11 Nm).
14. Check timing mark on crankshaft.
15. Install or connect the following:
   - Water outlet cap
   - Tappets, shims, camshafts, and RH (No.2) bank 2nd timing chain
   - 1st timing chain
   - LH (No.1) bank 2nd timing chain
   - Timing chain cover
   - Oil pans and oil pump strainer
   - Cylinder head covers
   - Exhaust manifold
   - Radiator outlet pipe
   - Intake manifold, intake collector and electric throttle body assembly
16. Install engine assembly to vehicle.
17. Refill engine oil, cooling system, and any other fluids that may have been lost.
18. Vacuum and recharge A/C system.
19. Connect the negative battery cable.
20. Start the engine and check for leaks.

3.6L Engines

Left Side

See Figures 68, 69 and 92 through 112.

1. Before servicing the vehicle, refer to all precautions.
2. Relieve the fuel system pressure.
3. Drain the cooling system and engine oil.

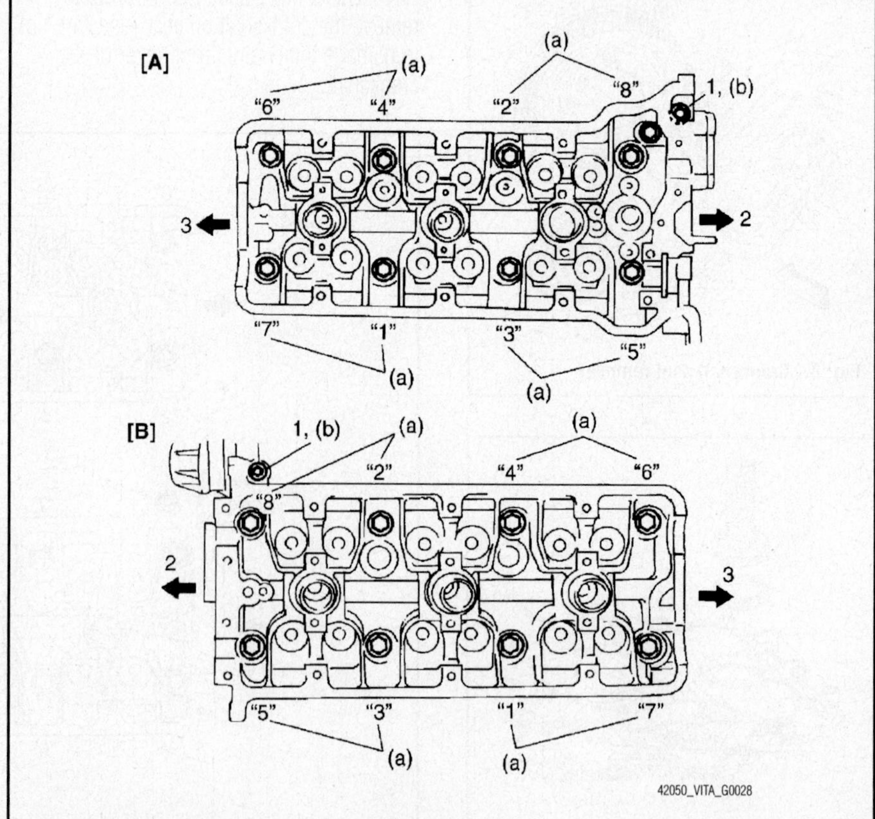

Fig. 91 Cylinder head numerical order–2.7L engines

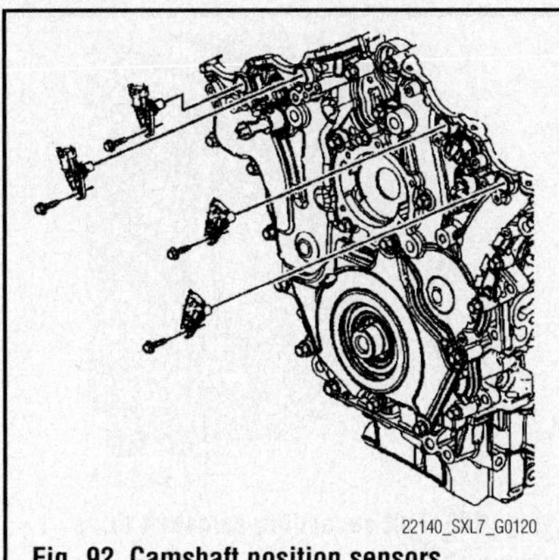
22140_SXL7_G0120

Fig. 92 Camshaft position sensors

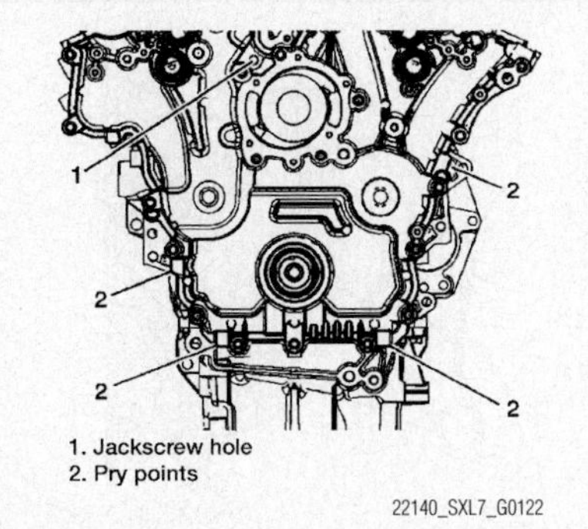

22140_SXL7_G0122

Fig. 93 Engine front cover bolts and pry points

4. Evacuate A/C system.
5. Remove the Engine refer to the Engine Assembly section.
6. Remove or disconnect the following:
   - Engine harness
   - Ignition coil bolts
   - Ignition coils
   - Camshaft cover bolts
   - Camshaft cover
   - Upper intake retaining bolts.
   - Fuel injectors and fuel rail
   - Lower intake manifold bolts
   - Lower intake manifold
   - Water pump pulley bolts
   - Water pump pulley
   - Water pump bolts
   - Water pump from the front cover
   - Camshaft position sensor bolts
   - Camshaft position sensors
   - Camshaft position actuator valve bolts.
   - Camshaft position actuator valves
   - Front cover bolts that hold the engine front cover deadener into position
   - Front cover bolts

✱✱ WARNING

Do not pry between the engine front cover and the camshaft position sensors or the camshaft position actuators in order to separate the RTV. Use the pry points and a bolt in the jackscrew hole in order to remove the engine front cover. Damage to the camshaft position sensors or the camshaft position actuators may occur if the camshaft position sensors or the camshaft position actuators are used to pry against in order to remove the engine front cover.

7. Loosely install a 10 x 1.5 mm bolt in the jackscrew hole (1).
8. Using the pry points (2) located at the edge of the front cover and the jackscrew, separate the room temperature vulcanizing (RTV) sealant
9. Remove the engine front cover.

✱✱ WARNING

Observe the body of the camshaft position actuator for the "EX" or "IN" markings. Ensure the proper timing mark is used prior to disassembly.

10. Use an open end wrench on the hex cast into the camshaft in order to prevent engine rotation when loosening the camshaft position actuator bolt.
11. Remove the left exhaust camshaft position actuator bolt.
12. Remove the left exhaust camshaft position actuator. When removing the actuator, place the chain on the engine cover side of the actuators.
13. Rotate the actuator in order to align the opening in the actuator reluctor wheel

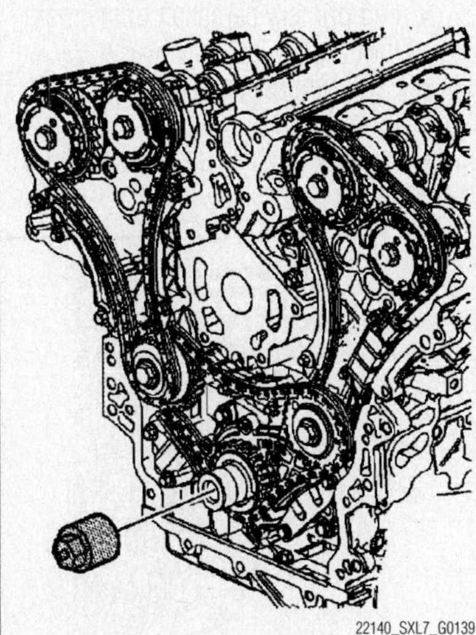
22140_SXL7_G0139

Fig. 94 Crankshaft rotational tool EN 46111

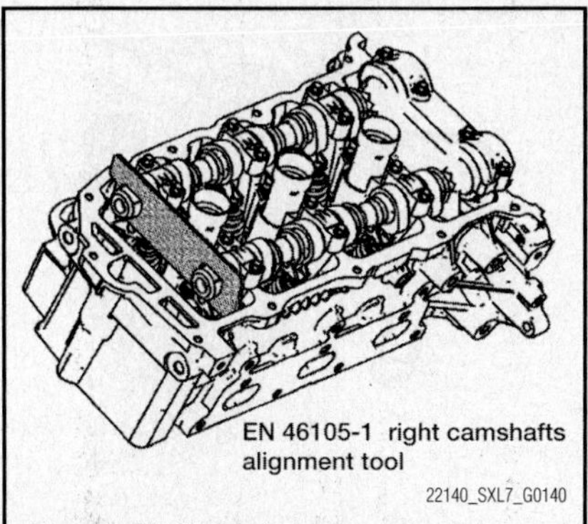

22140_SXL7_G0140

Fig. 95 EN 46105-1 right camshafts alignment tool

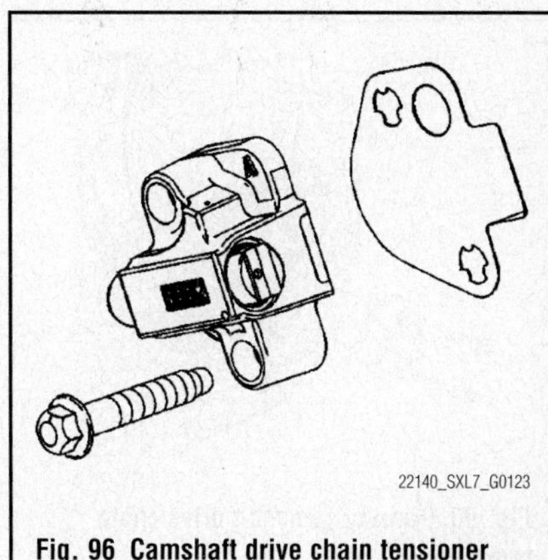
22140_SXL7_G0123

Fig. 96 Camshaft drive chain tensioner

with the cam sensor boss in the front cover, to allow actuator removal.

14. If removing both the exhaust and intake camshaft actuators, the timing chain can be removed once the actuators have been removed.
15. Using the EN 46111, rotate the crankshaft until the right cylinder head camshafts align with the EN 46105-1.
16. Remove the left secondary camshaft drive chain tensioner bolts.
17. Remove the left secondary camshaft drive chain tensioner.
18. Remove and discard the left secondary camshaft drive chain tensioner gasket.
19. Inspect the right secondary camshaft drive chain tensioner mounting surface on the right cylinder head for burrs or any defects that would degrade the sealing of the NEW right secondary camshaft drive chain tensioner gasket.
20. Remove or disconnect the following:
    - Left secondary camshaft drive chain shoe bolts
    - Left secondary camshaft drive chain shoes
    - Left secondary camshaft drive chain guide bolts

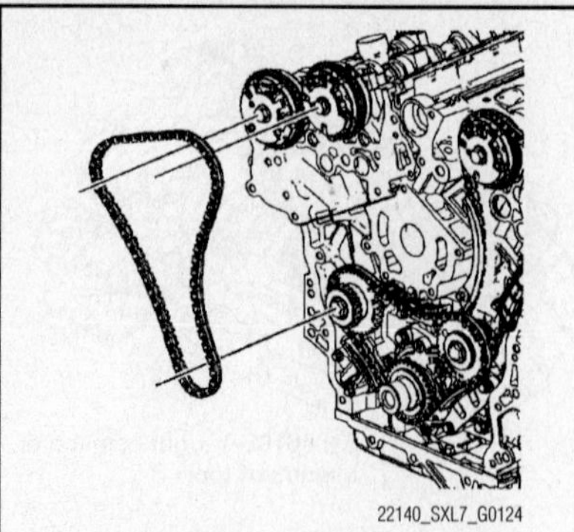

Fig. 97 Camshaft drive chain guide removal

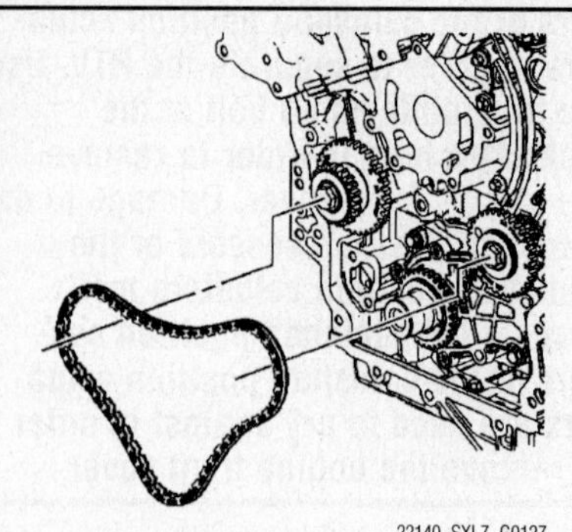

Fig. 100 Left camshaft primary camshaft drive chain

Fig. 103 Left secondary camshaft drive chain shoe

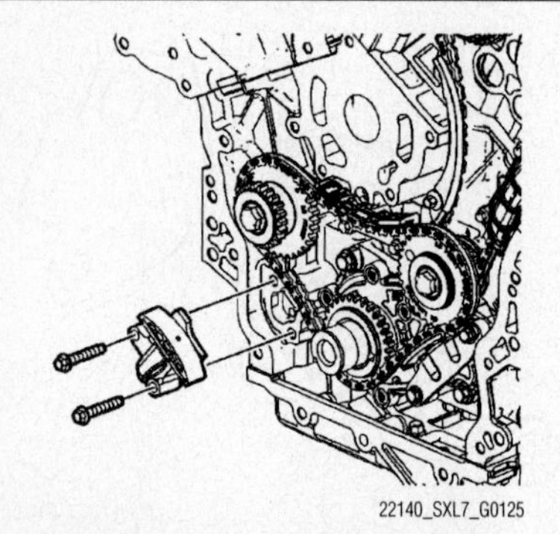

Fig. 98 Primary camshaft drive chain tensioner

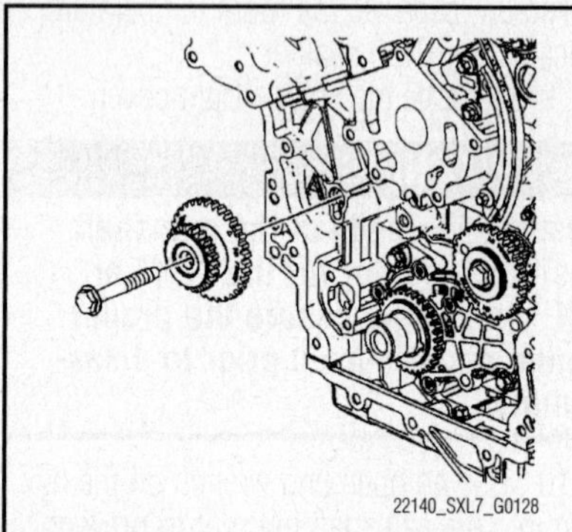

Fig. 101 Camshaft intermediate drive chain idler

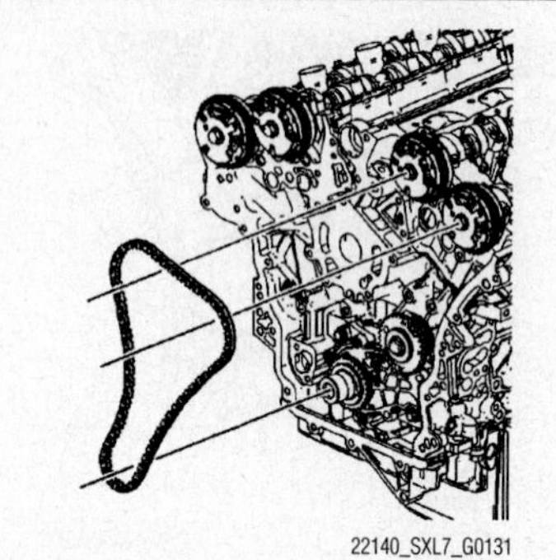

Fig. 104 Left secondary camshaft drive chain

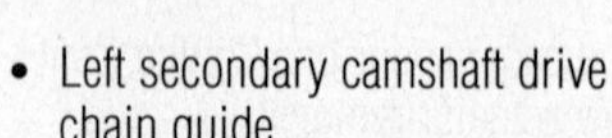

- Left secondary camshaft drive chain guide
- Left camshaft drive chain from the camshaft position actuators
- Left camshaft intermediate drive chain idler sprocket.
- Left primary camshaft drive chain tensioner bolts.
- Left primary camshaft drive chain tensioner.

21. Remove and discard the primary camshaft drive chain tensioner gasket.
22. Inspect the primary camshaft drive chain tensioner mounting surface on the engine block for burrs or any defects that would degrade the sealing of the NEW primary camshaft drive chain tensioner gasket.
23. Remove or disconnect the following:
    - Left primary camshaft drive chain upper guide bolts
    - Left primary camshaft drive chain upper guides
    - Left primary camshaft drive chain lower guide bolts
    - Left primary camshaft drive chain lower guide
    - Left primary camshaft drive chain
    - Left camshaft intermediate drive chain idler bolt
    - Left camshaft intermediate drive chain idler
    - Left secondary camshaft drive chain tensioner bolts
    - Left secondary camshaft drive chain tensioner
24. Discard the left secondary camshaft drive chain tensioner gasket
25. Inspect the left secondary camshaft drive chain tensioner mounting surface on the left cylinder head for burrs or any defects that would degrade the sealing of

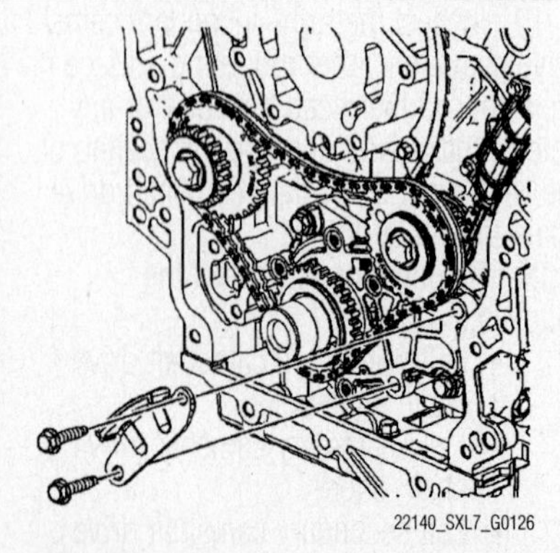

Fig. 99 Camshaft drive chain lower guide.

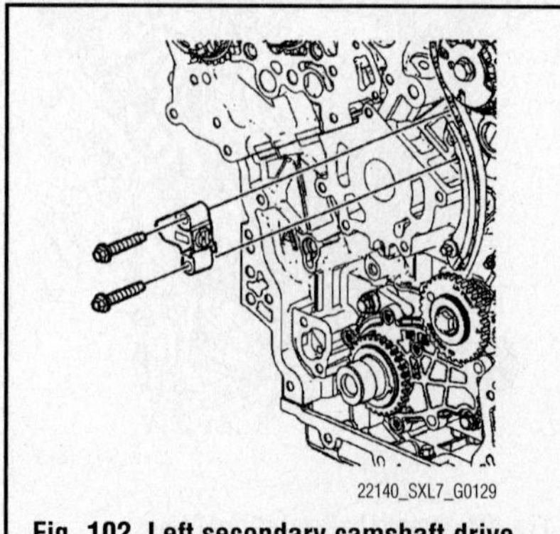

Fig. 102 Left secondary camshaft drive chain tensioner

Fig. 105 Left camshaft intermediate drive chain idler

the NEW left secondary camshaft drive chain tensioner gasket.

26. Remove or disconnect the following:
- Left secondary camshaft drive chain shoe bolt
- Left secondary camshaft drive chain shoe
- Left secondary camshaft drive chain guide bolts
- Left secondary camshaft drive chain guide
- Left secondary camshaft drive chain
- Left camshaft intermediate drive chain idler bolt
- Left camshaft intermediate drive chain idler
- Crankshaft sprocket from the nose of the crankshaft

27. Remove the left intake and left exhaust camshaft position actuator bolts.

➡Use an open wrench on the hex cast into the camshafts in order to prevent engine rotation when loosening the camshaft position actuator bolts.

28. Remove the left intake and left exhaust camshaft position actuators.

29. Remove or disconnect the following:
- Two front M8 left cylinder head bolts
- Left cylinder head bolts

22140_SXL7_G0133

Fig. 106 Crankshaft Sprocket

30. Remove the left cylinder head.

To install:

31. Ensure the cylinder head locating pins are securely mounted in the cylinder block deck face.

32. Install a NEW right cylinder head gasket using the deck face locating pins for retention.

33. Align the right cylinder head with the deck face locating pins.

34. Place the right cylinder head in position on the deck face.

⁂ WARNING

DO NOT allow oil on the cylinder head bolt bosses.

⁂ WARNING

DO NOT reuse the old M11 cylinder head bolts.

35. Install new M11 cylinder head bolts.

36. Tighten the new M11 cylinder head bolts to 22 ft. lbs. (30 Nm).

37. Tighten the new M11 cylinder head bolts in a second pass in the sequence an additional 150°.

38. Install the following:
- Right bank secondary camshaft drive chain. Torque to 17 ft. lbs. (23 Nm).
- Right bank camshaft intermediate drive chain idler

➡Ensure that the right camshaft intermediate drive chain idler (1) is being installed. The recessed hub (4) and the smaller sprocket of the right camshaft intermediate drive chain idler is installed outward. The raised hub and the larger sprocket of the right camshaft intermediate drive chain idler is installed towards the block.

- Intermediate drive chain idler bolt.
- Torque the camshaft intermediate drive chain idler bolt to 43 ft. lbs. (58 Nm)
- Right bank secondary camshaft drive chain guide
- Right bank secondary camshaft drive chain shoe
- Right bank secondary camshaft drive chain tensioner
- Primary camshaft drive chain
- Right bank secondary camshaft drive chain

39. Ensure the proper camshaft position actuator is installed. Observe the body of the camshaft position actuator for the EX marking (1).

40. The marking is for an exhaust camshaft position actuator.

41. Ensure the proper timing mark is used. Observe the outer ring of the camshaft position actuator for the circle marking (3).

42. The marking is for alignment to the highlighted timing chain link on the left side of the engine.

43. Align the exhaust camshaft actuator alignment mark (1) to the timing chain alignment mark (2) made during disassembly.

44. Ensure that the intake camshaft actuator alignment mark (4) and the timing chain alignment mark (3) are also aligned.

45. Position the exhaust camshaft actuator to the camshaft and install the actuator bolt hand tight.

46. Tighten the exhaust camshaft position actuator bolt to 43 ft. lbs. (58 Nm).

➡Use an open wrench on the hex cast into the camshaft in order to prevent engine rotation when tightening the camshaft position actuator bolt.

47. Install the right intake and exhaust camshaft position actuators.

48. Install the right intake and exhaust camshaft position actuator bolts.

49. Install the 0.315 inch (8 mm) guide from the EN 46109 into the cylinder block positions as shown.

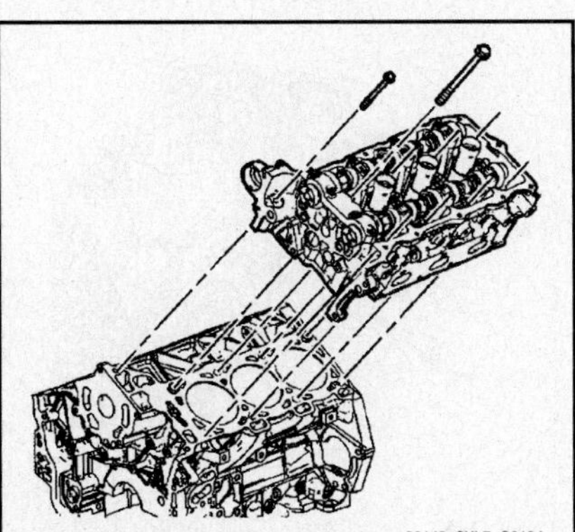

22140_SXL7_G0134

Fig. 107 Cylinder head m8 bolt removal

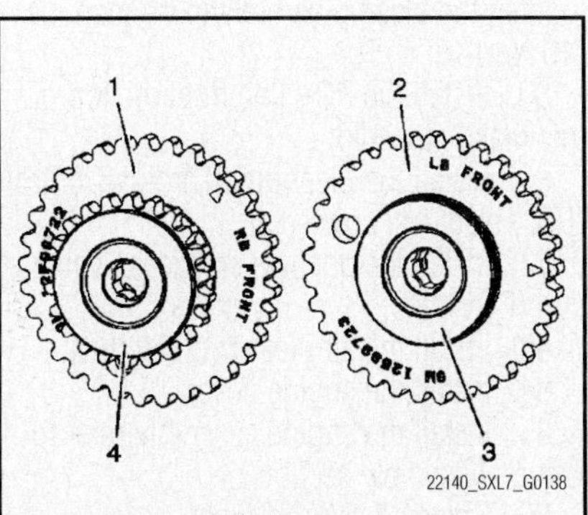

22140_SXL7_G0138

Fig. 108 Right camshaft intermediate drive chain idler orientation

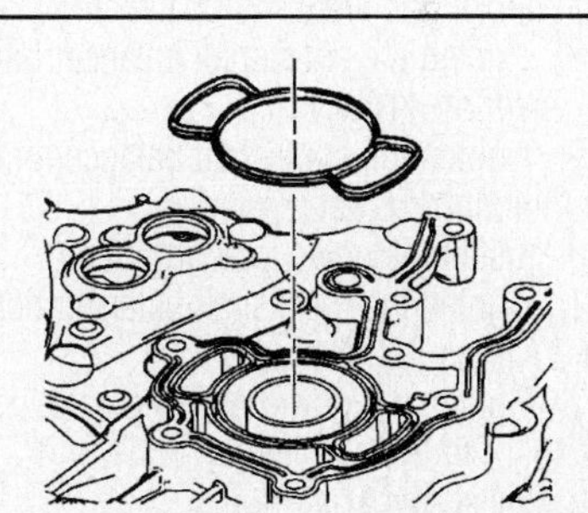

8 mm (0.315 in) front cover guide EN 46109

22140_SXL7_G0141

Fig. 109 0.315 inch (8 mm) front cover installation guide EN 46109

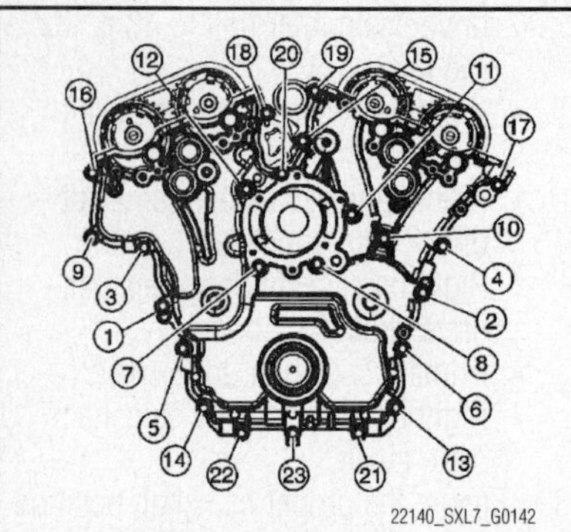

Fig. 110 Engine front cover bolt locations and tightening sequence

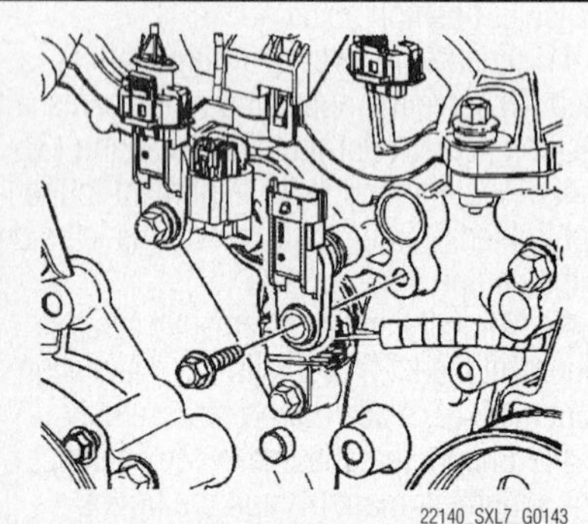

Fig. 111 Camshaft position actuator solenoid valves installation

50. Place a 0.118 inch (3 mm) bead of RTV sealant, GM P/N 12378521 (Canadian P/N 88901148) or equivalent, on the engine.
51. Install the engine front cover and front cover deadener to cylinder block seal.
52. Hand start all the front cover bolts.
53. Tighten the engine front cover bolts in the sequence shown.
54. Tighten the engine front cover bolts in sequence to 17 ft lbs. (23 Nm).
55. Install the camshaft position actuator solenoid valves.
56. Tighten the camshaft position actuators to 89 inch lbs. (10 Nm).
57. Install the crankshaft balancer.
58. Tighten the crankshaft balancer bolt in to 74 ft lbs. (100 Nm).
59. Tighten the crankshaft balancer bolt an additional 150 degrees.
60. Install the power steering pump.
61. Tighten the crankshaft balancer bolt in to 37 ft lbs. (50 Nm).
62. Tighten the power steering gear inlet hose to 18 ft. lbs. (25 Nm).
63. Install the water pump.
64. Tighten the water pump bolts to 89 inch lbs. (10 Nm).
65. Install the camshaft covers.
66. Install new camshaft cover bolt grommets prior to installing the camshaft cover bolts.

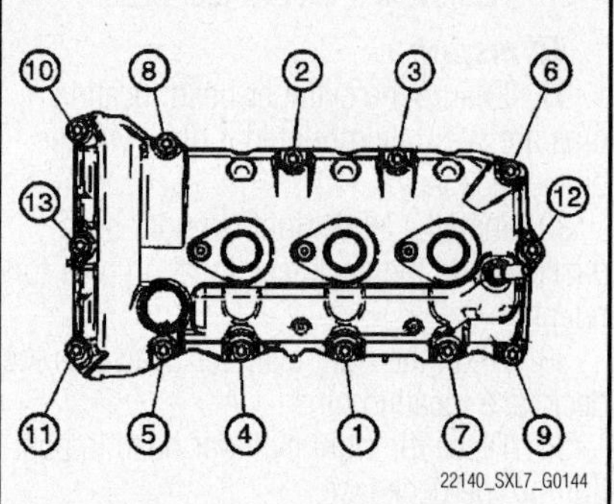

Fig. 112 Camshaft cover bolt tightening sequence

67. Place a bead 0.3150 inch (8 mm) in diameter by0.1575 inch (4 mm) in height of RTV sealant, GM P/N 12378521 (Canadian P/N 88901148) or equivalent, on the engine front cover seams.
68. Tighten the left camshaft cover bolts in the sequence to 89 inch lbs. (10 Nm).
69. Install the lower intake manifold. Tighten the lower intake manifold bolts to 17 ft. lbs. (23 Nm)
70. Install the upper intake manifold gaskets to the lower intake manifold and install the fir tree retainers to retain the upper intake manifold gasket position.
71. Install the upper intake manifold.
72. Apply threadlock to the bolt threads. Tighten the lower intake manifold bolts to 18 ft. lbs. (25 Nm).
73. Install the ignition coils. Tighten the ignition coils to 89 inch lbs. (10 Nm).
74. Install the engine harness.
75. Install the throttle body.
76. Install the EVAP purge solenoid valve.
77. Install the brake booster vacuum hose.
78. Install the evaporative emissions (EVAP) canister purge line.
79. Install the bleed pipe bolts (1). Tighten the bleed pipe bolts to 89 inch lbs. (10 Nm).
80. Install the PCV line from the top of the intake manifold.
81. Install the Electronic Throttle Control (ETC) electrical connector.
82. Install the positive crankcase ventilation (PCV) line.
83. Install the air cleaner outlet duct.
84. Install the engine cover.
85. Install the engine assembly refer to Engine Assembly section.
86. Charge the A/C system.
87. Refill the engine coolant.
88. Run engine and check for leaks

Right Side

See Figures 68, 69 and 92, 93, 96 through 99, 101, 102, 105, 106, 108, 109, 110 through 116

1. Before servicing the vehicle, refer to all precautions.
2. Relieve the fuel system pressure.
3. Drain the cooling system and engine oil.
4. Evacuate A/C system.
5. Remove engine —refer to engine assembly,removal and installation.
6. Remove or disconnect the following:
    - Engine cover
    - Air cleaner outlet duct
    - Positive Crankcase Ventilation (PCV) line
    - Electronic Throttle Control (ETC) electrical connector
    - PCV line from the top of the intake manifold
    - Evaporative emissions (EVAP) canister purge line
    - Bleed pipe bolts (1)
    - Bleed pipe hose clamp
    - Brake booster vacuum hose
    - Engine harness
    - Ignition coil bolts
    - Ignition coils

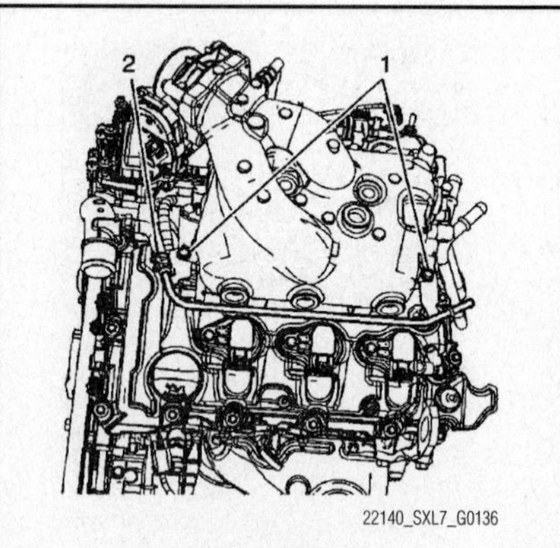

Fig. 113 Bleed pipe bolts

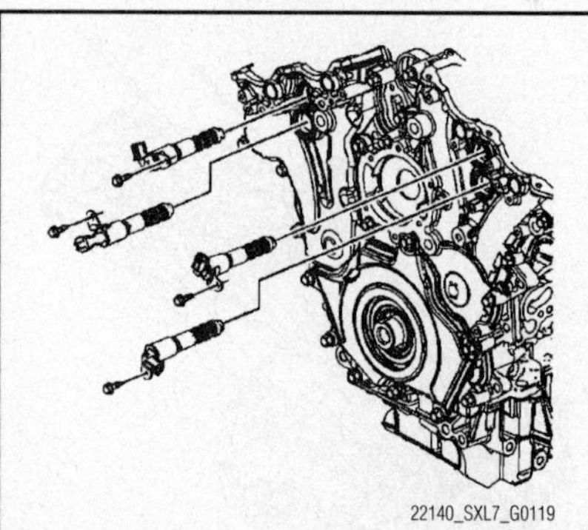

Fig. 114 Camshaft position actuator valves

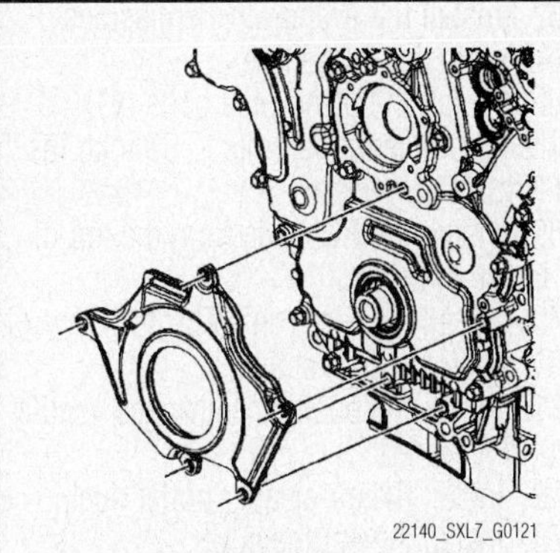
22140_SXL7_G0121

Fig. 115 Engine front cover deadener

- Camshaft cover bolts
- Camshaft cover
- Upper intake retaining bolts.
- Fuel injectors and fuel rail
- Lower intake manifold bolts
- Lower intake manifold
- Water pump pulley bolts
- Water pump pulley
- Water pump bolts
- Water pump from the front cover
- Camshaft position sensor bolts
- Camshaft position sensors
- Camshaft position actuator valve bolts.
- Camshaft position actuator valves
- Front cover bolts that hold the engine front cover deadener into position
- Front cover bolts

⁂ WARNING

Do not pry between the engine front cover and the camshaft position sensors or the camshaft position actuators in order to separate the RTV. Use the pry points and a bolt in the jackscrew hole in order to remove the engine front cover. Damage to the camshaft position sensors or the camshaft position actuators may occur if the camshaft position sensors or the camshaft position actuators are used to pry against in order to remove the engine front cover.

7. Loosely install a 10 x 1.5 mm bolt in the jackscrew hole (1).
8. Using the pry points (2) located at the edge of the front cover and the jackscrew, separate the room temperature vulcanizing (RTV) sealant
9. Remove the engine front cover.
10. Remove the right secondary camshaft drive chain tensioner bolts.
11. Remove the right secondary camshaft drive chain tensioner.
12. Remove and discard the left secondary camshaft drive chain tensioner gasket.
13. Inspect the right secondary camshaft drive chain tensioner mounting surface on the right cylinder head for burrs or any defects that would degrade the sealing of the **NEW** right secondary camshaft drive chain tensioner gasket.
14. Remove or disconnect the following:
    - Right secondary camshaft drive chain shoe bolts
    - Right secondary camshaft drive chain shoes
    - Right secondary camshaft drive chain guide bolts
    - Right secondary camshaft drive chain guide
    - Right camshaft drive chain from the camshaft position actuators
    - Right camshaft intermediate drive chain idler sprocket.
    - Right primary camshaft drive chain tensioner bolts.
    - Right primary camshaft drive chain tensioner.
15. Remove and discard the primary camshaft drive chain tensioner gasket.
16. Inspect the primary camshaft drive chain tensioner mounting surface on the engine block for burrs or any defects that would degrade the sealing of the NEW primary camshaft drive chain tensioner gasket.
17. Remove or disconnect the following:
    - Right primary camshaft drive chain upper guide bolts
    - Right primary camshaft drive chain upper guides
    - Right primary camshaft drive chain lower guide bolts
    - Right primary camshaft drive chain lower guide.
    - Right primary camshaft drive chain
    - Right camshaft intermediate drive chain idler bolt
    - Right camshaft intermediate drive chain idler
    - Right secondary camshaft drive chain tensioner bolts
    - Right secondary camshaft drive chain tensioner
18. Discard the right secondary camshaft drive chain tensioner gasket
19. Inspect the right secondary camshaft drive chain tensioner mounting surface on the left cylinder head for burrs or any defects that would degrade the sealing of the NEW left secondary camshaft drive chain tensioner gasket.
20. Remove or disconnect the following:
    - Right secondary camshaft drive chain shoe bolt

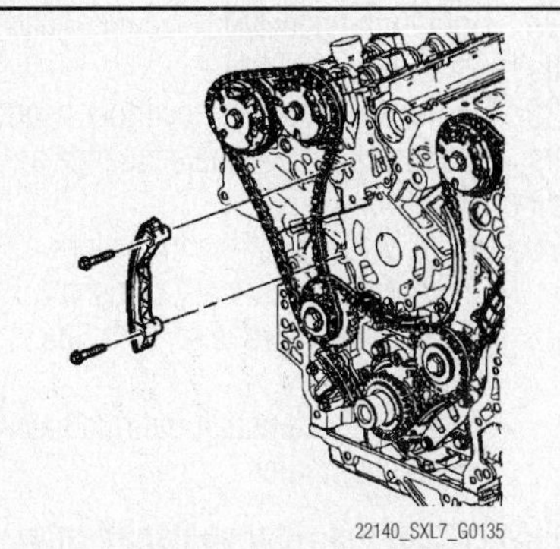
22140_SXL7_G0135

Fig. 116 Right secondary camshaft drive chain shoe

    - Right secondary camshaft drive chain shoe
    - Right secondary camshaft drive chain guide bolts
    - Right secondary camshaft drive chain guide
    - Right secondary camshaft drive chain
    - Right camshaft intermediate drive chain idler bolt
    - Right camshaft intermediate drive chain idler
    - Crankshaft sprocket from the nose of the crankshaft
21. Remove the right intake and right exhaust camshaft position actuator bolts.

➡Use an open wrench on the hex cast into the camshafts in order to prevent engine rotation when loosening the camshaft position actuator bolts.

22. Remove the right intake and right exhaust, camshaft position actuators.

To install:

23. Ensure the cylinder head locating pins are securely mounted in the cylinder block deck face.
24. Install a NEW right cylinder head gasket using the deck face locating pins for retention.
25. Align the right cylinder head with the deck face locating pins.
26. Place the right cylinder head in position on the deck face.

⁂ WARNING

DO NOT allow oil on the cylinder head bolt bosses.

⁂ WARNING

DO NOT reuse the old M11 cylinder head bolts.

27. Install new M11 cylinder head bolts.

28. Tighten the new M11 cylinder head bolts to 22 ft. lbs. (30 Nm).
29. Tighten the new M11 cylinder head bolts in a second pass in the sequence an additional 150°.
30. Install or Connect the following:
- Right bank secondary camshaft drive chain. Torque to 17 ft. lbs. (23 Nm).
- Right bank camshaft intermediate drive chain idler

➡Ensure that the right camshaft intermediate drive chain idler (1) is being installed. The recessed hub (4) and the smaller sprocket of the right camshaft intermediate drive chain idler is installed outward. The raised hub and the larger sprocket of the right camshaft intermediate drive chain idler is installed towards the block.

- Intermediate drive chain idler bolt.
- Torque the camshaft intermediate drive chain idler bolt to 43 ft. lbs. (58 Nm)
- Right bank secondary camshaft drive chain guide
- Right bank secondary camshaft drive chain shoe
- Right bank secondary camshaft drive chain tensioner
- Primary camshaft drive chain
- Right bank secondary camshaft drive chain

➡Use an open wrench on the hex cast into the camshaft in order to prevent engine rotation when tightening the camshaft position actuator bolt.

31. Ensure the proper camshaft position actuator is installed. Observe the body of the camshaft position actuator for the EX marking (1).
32. The marking is for an exhaust camshaft position actuator.
33. Ensure the proper timing mark is used. Observe the outer ring of the camshaft position actuator for the circle marking (3).
34. The marking is for alignment to the highlighted timing chain link on the left side of the engine.
35. Align the exhaust camshaft actuator alignment mark (1) to the timing chain alignment mark (2) made during disassembly.
36. Ensure that the intake camshaft actuator alignment mark (4) and the timing chain alignment mark (3) are also aligned.
37. Position the exhaust camshaft actuator to the camshaft and install the actuator bolt hand tight.
38. Tighten the exhaust camshaft position actuator bolt.
39. Tighten the bolt to 43 ft. lbs. (58 Nm).
40. Install the 0.315 inch (8 mm) guide from the EN 46109 into the cylinder block positions as shown.
41. Place a 0.118 inch (3mm) bead of RTV sealant, GM P/N 12378521 (Canadian P/N 88901148) or equivalent, on the engine.
42. Install the engine front cover and front cover deadener to cylinder block seal.
43. Hand start all the front cover bolts.
44. Tighten the engine front cover bolts in the sequence shown.
45. Tighten the engine front cover bolts in sequence to 17 ft lbs. (23 Nm).
46. Install the camshaft position actuator solenoid valves.
47. Tighten the camshaft position actuators to 89 inch lbs. (10 Nm).
48. Install the crankshaft balancer, and tighten as follows:
    a. Tighten the crankshaft balancer bolt to 74 ft lbs. (100 Nm).
    b. Tighten the crankshaft balancer bolt an additional 150 degrees.
49. Install the power steering pump.
50. Tighten the power steering gear inlet hose to 18 ft. lbs. (25 Nm).
51. Install the water pump.
52. Tighten the water pump bolts to 89 inch lbs. (10 Nm).
53. Install the camshaft covers.
54. Install new camshaft cover bolt grommets prior to installing the camshaft cover bolts.
55. Place a bead 0.3150 inch (8mm) in diameter by0.1575 inch (4mm) in height of RTV sealant, GM P/N 12378521 (Canadian P/N 88901148) or equivalent, on the engine front cover seams.
56. Tighten the left camshaft cover bolts in the sequence to 89 inch lbs. (10 Nm).
57. Install the lower intake manifold. Tighten the lower intake manifold bolts to 17 ft. lbs. (23 Nm)
58. Install the upper intake manifold gaskets to the lower intake manifold and install the fir tree retainers to retain the upper intake manifold gasket position.
59. Install the upper intake manifold.
60. Apply threadlock to the bolt threads.
61. Tighten the lower intake manifold bolts to 18 ft. lbs. (25 Nm).
62. Install the ignition coils. Tighten the ignition coils to 89 inch lbs. (10 Nm).
63. Install the engine harness.
64. Install the throttle body.
65. Install the EVAP purge solenoid valve.
66. Install the brake booster vacuum hose.
67. Install the evaporative emissions (EVAP) canister purge line.
68. Install the bleed pipe bolts (1). Tighten the bleed pipe bolts to 89 inch lbs. (10 Nm).
69. Install the PCV line from the top of the intake manifold.
70. Install the Electronic Throttle Control (ETC) electrical connector.
71. Install the positive crankcase ventilation (PCV) line.
72. Install the air cleaner outlet duct.
73. Install the engine cover.
74. Install the engine assembly refer to Engine Assembly procedure.
75. Charge the A/C system.
76. Refill the engine coolant.
77. Run engine and check for leaks.

ENGINE ASSEMBLY

REMOVAL & INSTALLATION

2.7L Engines

See Figure 117.

1. Before servicing the vehicle, refer to all precautions.
2. Relieve the fuel system pressure.
3. Recover the refrigerant.
4. Drain the cooling system.
5. Drain the engine oil.
6. Remove engine hood.
7. Remove radiator, radiator fan shroud, cooling fan and radiator reservoir.
8. Disconnect accelerator cable from throttle body.
9. Detach fuse/relay box, then remove strut tower bar and surge tank cover.
10. Disconnect IAT sensor coupler and MAF sensor coupler then remove air cleaner upper case, intake air hose, intake air pipe and surge tank pipe as a component.
11. Remove intake control valve.
12. Remove engine oil level gauge guide and A/T fluid level gauge guide (for A/T vehicle).
13. Remove ignition coil covers.
14. Disconnect the following electric lead wires:
- Injector wire coupler
- Camshaft Position Sensor(CMP) sensor coupler
- Ignition coil couplers
- Crankshaft Position Sensor(CKP) sensor coupler
- Manifold Absolute Pressure (MAP) sensor coupler
- Throttle Position (TP) sensor coupler
- Idle Air Control (IAC) valve coupler
- Earth wire from surge tank

- Evaporative Emissions Canister(EVAP) canister purge valve coupler
- Exhaust Gas Recirculation (EGR) valve coupler
- Oxygen sensors
- Coolant temperature sensor coupler
- Knock sensor coupler
- Generator wires
- Starter wires
- Oil pressure wire
- P/S pump wire
- Earth wire from generator bracket
- Engine block heater (if equipped)

15. Disconnect the following hoses:
    - Heater hose from heater water pipe and water outlet cap
    - Tank pressure control solenoid valve hose from intake manifold
    - EVAP canister hose from canister purge pipe
    - Brake booster vacuum hose
16. Remove EVAP canister purge valve.
17. Disconnect the following hoses:
    - Fuel feed hose from fuel feed pipe
    - Fuel return hose from fuel return pipe
18. Remove P/S pump assembly.
19. Remove A/C compressor assembly.
20. Remove steering shaft lower assembly.
21. Raise vehicle.
22. Remove front differential housing with differential from chassis if equipped.
23. Remove front exhaust pipe.
24. Remove exhaust manifold stiffener from transmission.
25. Remove A/T fluid hose clamps from engine mounting bracket (for A/T vehicle).
26. Remove clutch housing lower plate.
27. Remove torque converter bolts (for A/T vehicle).
28. Remove starter motor.
29. Lower vehicle.
30. Support transmission.
31. Remove bolts and nuts fastening cylinder block and transmission.
32. Install lifting device.
33. Disconnect engine side mounting brackets from engine mountings by removing nuts.
34. Before lifting engine, check to ensure all hoses, wires and cables are disconnected from engine.
35. Remove engine assembly from chassis and transmission by sliding toward front, and then, carefully hoist engine assembly.

To install:

36. Reverse removal procedure for installation noting the following points.
37. Tighten engine side mounting bracket nuts to 40 ft. lbs (55 Nm).
38. Tighten transmission case bolts and nut 61.5 ft. lbs. (85 Nm) (for M/T vehicle).
39. Tighten Transmission case bolt and nut 58 ft. lbs. (80 Nm) (for A/T vehicle).
40. Tighten torque converter bolts to 47 ft. lbs. (65 Nm).
41. Tighten bolts and nuts of exhaust pipe to 36.5 ft. lbs. (50 Nm).
42. Install front differential housing with differential to chassis.
43. Install steering shaft lower assembly.
44. Install A/C compressor assembly.
45. Install P/S pump assembly.
46. Adjust accelerator cable play.
47. Refill engine with engine oil.
48. Refill cooling system.
49. Check to ensure that all fasteners and clamps are tightened.
50. Vacuum and recharge A/C system.
51. Upon completion of installation, verify that there is no fuel leakage, coolant leakage, P/S fluid leakage or exhaust gas leakage at each connection.

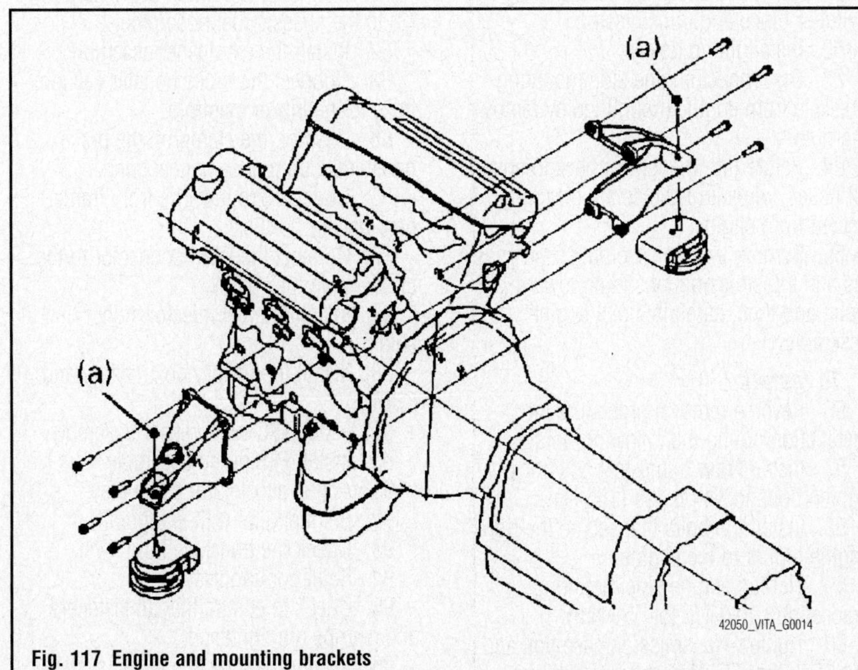

Fig. 117 Engine and mounting brackets

3.6L Engines

See Figure 113

1. Before servicing the vehicle, refer to all precautions.
2. Relieve the fuel system pressure.
3. Drain the cooling system and engine oil.
4. Evacuate A/C system.
5. Drain the engine oil.
6. Remove engine hood.
7. Remove the battery box and carefully set the Engine Control Module (ECM) on top of the engine.
8. Remove radiator, radiator fan shroud, cooling fan and radiator reservoir.
9. Disconnect the ECM connector from the under-hood fuse block.
10. Disconnect ground wire from frame, near battery box.
11. Release the clamp from the brake booster vacuum hose connection.
12. Disconnect the brake booster vacuum hose from the intake manifold
13. Disconnect accelerator cable from throttle body.
14. Detach fuse/relay box, then remove strut tower bar and surge tank cover.
15. Remove or disconnect the following:
    - Engine cover
    - Air cleaner outlet duct
    - Positive crankcase ventilation (PCV) line
    - Electronic Throttle Control (ETC) electrical connector
    - PCV line from the top of the intake manifold
    - Evaporative emissions (EVAP) canister purge line
    - Bleed pipe bolts (1)
    - Bleed pipe hose clamp
    - Brake booster vacuum hose
    - Engine harness
    - Injector wire coupler
    - Camshaft Position Sensor(CMP) sensor coupler
    - Ignition coil couplers

- Crankshaft Position Sensor(CKP) sensor coupler
- Manifold Absolute Pressure (MAP) sensor coupler
- Throttle Position (TP) sensor coupler
- Idle Air Control (IAC) valve coupler
- Earth wire from surge tank
- Evaporative Emissions Canister(EVAP) canister purge valve coupler
- Exhaust Gas Recirculation (EGR) valve coupler
- Oxygen sensors
- Coolant temperature sensor coupler
- Knock sensor coupler
- Generator wires
- Starter wires
- Oil pressure wire
- P/S pump wire
- Earth wire from generator bracket
- Engine block heater (if equipped)

16. Disconnect the following hoses:
    - Heater hose from heater water pipe and water outlet cap
    - Tank pressure control solenoid valve hose from intake manifold
    - EVAP canister hose from canister purge pipe
    - Brake booster vacuum hose
17. Remove EVAP canister purge valve.
18. Disconnect the following hoses.
19. Fuel feed hose from fuel feed pipe
20. Fuel return hose from fuel return pipe
21. Remove P/S pump assembly.
22. Remove A/C compressor assembly. Tie the radiator, A/C condenser, and fan module assembly to the upper radiator support to keep the assembly with the vehicle when the frame and drivetrain is removed.
23. Disconnect the transaxle oil cooler lines from the transaxle and remove the seals.
24. Raise vehicle.
25. Remove front differential housing with differential from chassis if equipped.
26. Remove front catalytic converters and secure the rear half of the exhaust system to the vehicle underbody.
27. Remove the front tires.
28. Remove the right and left engine splash shields.
29. Remove steering shaft lower assembly.
30. Disconnect the steering intermediate shaft from the steering gear.
31. Remove the right and left outer tie rod ends from the steering knuckles.
32. Remove the right and left stabilizer shaft links from the stabilizer shaft.
33. Remove the right and left lower ball joints from the steering knuckles.
34. On front wheel drive (FWD) models, place a drain pan under the transaxle then separate the right and left front wheel drive shafts from the transaxle.
35. On all wheel drive (AWD) models, remove the rear wheel driveshaft.
36. On all models, place a block of wood (1) between the frame and the engine oil pan in order to support the engine once the bolts are removed from the right engine mount
37. Lower the vehicle.
38. Remove the bolts (2) that secure the right engine mount (1) to the engine (3).
39. Place a universal frame support fixture or jackstands under the frame.
40. Lower the vehicle until the frame contacts the frame support fixture or jackstands.
41. Disconnect the wiring harness retaining clips near the right and left shock towers.
42. Remove the drivetrain and front suspension frame reinforcement.
43. Remove the frame-to-body bolts. Discard the bolts.
44. Carefully raise the vehicle body up away from the powertrain.
45. Remove A/T fluid hose clamps from engine mounting bracket (for A/T vehicle).
46. Remove clutch housing lower plate.
47. Remove torque converter bolts (for A/T vehicle).
48. Remove starter motor.
49. Lower vehicle.
50. Support transmission.
51. Remove bolts and nuts fastening cylinder block and transmission.
52. Install lifting device.
53. Disconnect engine side mounting brackets from engine mountings by removing nuts.
54. Before lifting engine, check to ensure all hoses, wires and cables are disconnected from engine.
55. Remove engine assembly from chassis and transmission by sliding toward front, and then, carefully hoist engine assembly.

To install:

56. Reverse removal procedure for installation noting the following points.
57. Install **New** frame-to-body bolts tighten bolts to 144 ft. lbs (155 Nm).
58. Install the bolts that secure the right engine mount to the engine.
59. Tighten engine side mounting bracket nuts to 37 ft. lbs (50 Nm).
60. Tighten Transmission case bolt and nut 55 ft. lbs. (75 Nm) (for A/T vehicle).
61. Tighten torque converter bolts to 44 ft. lbs. (60 Nm).
62. Tighten bolts and nuts of exhaust pipe to 36.5 ft. lbs. (50 Nm).
63. Install steering shaft lower assembly.
64. On all wheel drive (AWD) models, install the rear wheel driveshaft.
65. On front wheel drive models, install the right and left front wheel drive shafts into the transaxle.
66. On all models, install the right and left lower ball joints to the steering knuckles.
67. Install the right and left stabilizer shaft links to the stabilizer shaft.
68. Install the right and left tie rod ends to the steering knuckles.
69. Connect the steering intermediate shaft to the steering gear.
70. Install a NEW pinch bolt to the steering intermediate shaft.
71. Install the right and left engine splash shields.
72. Install the front tires.
73. Install the catalytic converters.
74. Install new seals and connect the transaxle oil cooler lines to the transaxle.
75. Install the radiator outlet hose.
76. Lower the vehicle.
77. Install the radiator inlet hose.
78. Connect the heater hoses to the engine.
79. Untie the radiator, AC condenser, and fan module assembly from the upper radiator support.
80. Connect the transaxle shift control cable to the transaxle.
81. Connect the engine fuel hose/pipe (3) to the chassis fuel hose/pipe.
82. Install the air cleaner assembly.
83. Connect the brake booster vacuum hose to the intake manifold.
84. Position the clamp on the brake booster vacuum hose connection.
85. Connect ground wire from frame, near battery box.
86. Connect the ECM connector to the under-hood fuse block.
87. Install the fuel injector sight shield.
88. Install the battery box, battery and ECM.
89. Install A/C compressor assembly.
90. Install P/S pump assembly.
91. Adjust accelerator cable play.
92. Refill engine with engine oil.
93. Check the transaxle fluid level.
94. Refill cooling system.
95. Check to ensure that all fasteners and clamps are tightened.
96. Vacuum and recharge A/C system.
97. Prime the fuel system.

a. Cycle the ignition ON for 5 seconds then OFF for 10 seconds. Repeat cycling twice.

b. Crank the engine until it starts. The maximum starter motor cranking time is 20 seconds.

c. If the engine does not start, repeat the steps.

98. Upon completion of installation, verify that there is no fuel leakage, coolant leakage, P/S fluid leakage or exhaust gas leakage at each connection.

EXHAUST MANIFOLD

REMOVAL & INSTALLATION

2.7L Engines

1. Disconnect the negative battery cable.
2. Remove air cleaner upper case and intake air hose (for right side exhaust manifold removal).
3. Remove HO2S connectors from bracket, then disconnect them.
4. Disconnect EGR pipe from right (No.2) bank exhaust manifold (for right side exhaust manifold removal).
5. Remove oil level gauge guide (for left side exhaust manifold removal).
6. Detach P/S pump assembly (for left side exhaust manifold removal).
7. Remove exhaust manifold covers from exhaust manifolds.
8. Hoist vehicle.
9. For 4WD vehicle, before disconnecting front propeller shaft, put match mark on joint flange and propeller shaft to facilitate their installation.
10. For 4WD vehicle, disconnect propeller shaft from front differential.
11. Remove exhaust No.1 pipe.
12. Remove exhaust manifold stiffeners from transmission case.
13. Remove stabilizer bar mounting bolt and pull down stabilizer bar (for right side exhaust manifold removal)
14. Remove exhaust manifolds and their gaskets from cylinder heads.

To install:

15. Install new manifold gaskets to cylinder heads and No.1 pipe gasket to exhaust No.1 pipe.
16. Install exhaust manifolds.
17. Always install new bolts with precoated adhesive.
18. Tighten both manifold nuts and bolts to 21.5 ft. lbs. (30 Nm).
19. Install stabilizer bar if it was removed.
20. Install exhaust manifold stiffener.
21. Install exhaust No.1 pipe.
22. Tighten exhaust No.1 pipe bolts and nuts to 36.5 ft. lbs. (50 Nm).
23. Reverse removal procedure to install front propeller shaft if removed.
24. When installing propeller shaft, align match mark.
25. Tighten universal joint flange to 36.5 ft. lbs. (50 Nm).
26. Connect EGR pipe to left (No.1) bank exhaust manifold, if it was removed.
27. Install exhaust manifold covers.
28. Install oil level gauge guide using new O-ring, if it was removed.
29. Install P/S pump assembly if it was removed.

3.6L Engines

1. Remove the exhaust manifold heat shield.
2. Remove the oil level indicator.
3. Remove the catalytic converter to exhaust manifold nuts.
4. Remove the exhaust manifold bolt.
5. Remove the exhaust manifold and gasket.

To install:

6. Install one exhaust manifold bolt to the exhaust manifold.
7. Install the NEW exhaust manifold gasket onto the cylinder head and bolt.
8. Install the exhaust manifold (with gasket) to the catalytic converter and the cylinder head.
9. Install the remaining exhaust manifold bolts.
10. Tighten the exhaust manifold bolts to 37 ft. lbs. (50 Nm).
11. Install the oil level indicator.
12. Tighten the oil level indicator to 89 inch lbs.(10 Nm).
13. Install the exhaust manifold heat shield.
14. Tighten the exhaust manifold heat shield to 89 inch lbs.(10 Nm).

FLEXPLATE

REMOVAL & INSTALLATION

2.7L Engines

1. Disconnect the negative battery cable.
2. Before servicing the vehicle, refer to the precautions in the beginning of this section.
3. Remove torque converter housing lower plate.
4. Hold flywheel stationary with special tool or the like, remove torque converter mounting bolts with wrench.
5. Remove transmission assembly.
6. Unbolt flywheel and remove.

To install:

7. Install flywheel and tighten bolts to 56 ft. lbs. (78 Nm).
8. Install transmission assembly.
9. Tighten torque converter bolts while holding flywheel to 47 ft. lbs. (65 Nm).
10. Install torque converter housing plate.
11. Connect the negative battery cable.

3.6L Engines

1. Remove the transmission refer to the Automatic Transmission section.
2. Remove the engine flexplate bolts and flywheel.
3. Remove the engine flexplate bolts and discard.
4. Remove the engine flexplate from the crankshaft.

To install:

5. Place the engine flywheel in position on the crankshaft.
6. Install 2 NEW bolts in location at the top and bottom of the engine flywheel bolt pattern allowing the engine flywheel to hang in position.
7. Install the remaining NEW engine flywheel bolts.
8. Tighten the NEW engine flywheel bolts to 22 ft. lbs. (30 Nm).
9. Tighten the NEW engine flywheel bolts an additional 45 degrees.

FLYWHEEL

REMOVAL & INSTALLATION

2.7L Engines

1. Disconnect the negative battery cable.
2. Before servicing the vehicle, refer to the precautions in the beginning of this section.
3. Dismount transmission assembly.
4. Hold flywheel stationary with special tool and remove clutch cover bolts, clutch cover and clutch disc.
5. Remove flywheel from crankshaft.

➡Before installation, make sure that flywheel surface and pressure plate surface have been cleaned and dried thoroughly.

To install:

6. Install flywheel to crankshaft and tighten new bolts to 49 ft. lbs. (68.5 Nm).
7. Aligning clutch disc to flywheel center by using tool, install clutch cover and bolts.

8. Then tighten bolts to 17 ft. lbs. (23 Nm).
9. Slightly apply grease to input shaft.
10. Install transmission assembly.
11. Connect the negative battery cable.

INTAKE MANIFOLD

REMOVAL & INSTALLATION

2.7L Engines

See Figures 118 through 120.

1. Release fuel pressure in fuel feed line.
2. Disconnect negative cable at battery.
3. Drain coolant.
4. Detach fuse/relay box and remove strut tower bar.
5. Disconnect coupler from IAT sensor and MAF sensor.
6. Remove surge tank cover.
7. Remove air cleaner upper case, MAF sensor, intake air hose, intake air pipe and surge tank pipe as one component. Do not disassemble them, when removing and reinstalling.
8. Remove intake control valve.
9. Disconnect accelerator cable from throttle body.
10. Disconnect water hoses from throttle body.
11. Disconnect injector wire coupler.
12. Disconnect brake booster hose, fuel pressure regulator vacuum hose and intake control valve hose from intake manifold.
13. Disconnect couplers of TP sensor and IAC valve.
14. Disconnect earth terminal from intake collector.
15. Remove clamp bracket from intake collector.
16. Disconnect couplers from MAP sensor, EVAP canister purge valve, earth terminal and EGR valve.

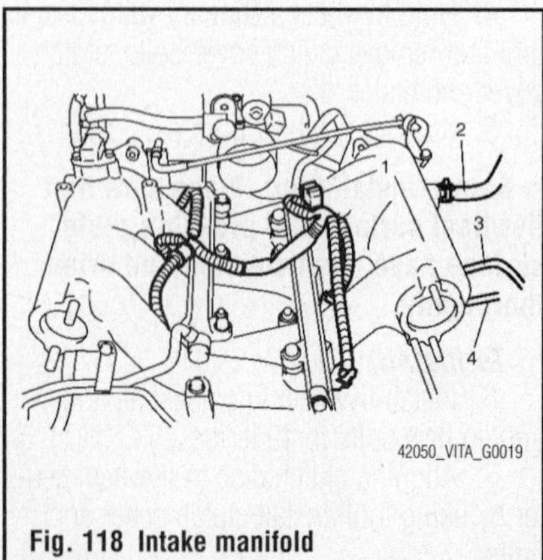

Fig. 118 Intake manifold

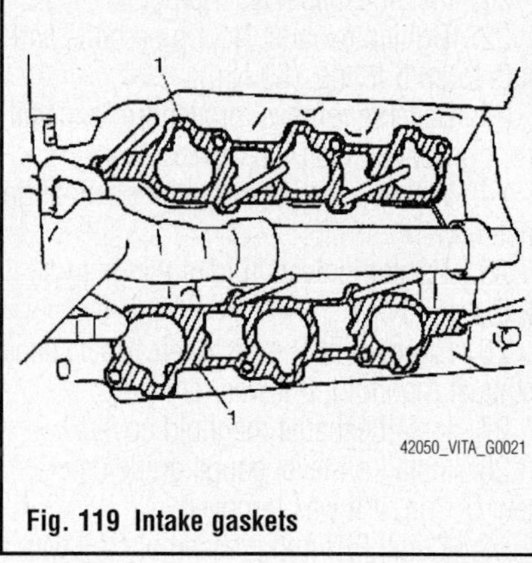

Fig. 119 Intake gaskets

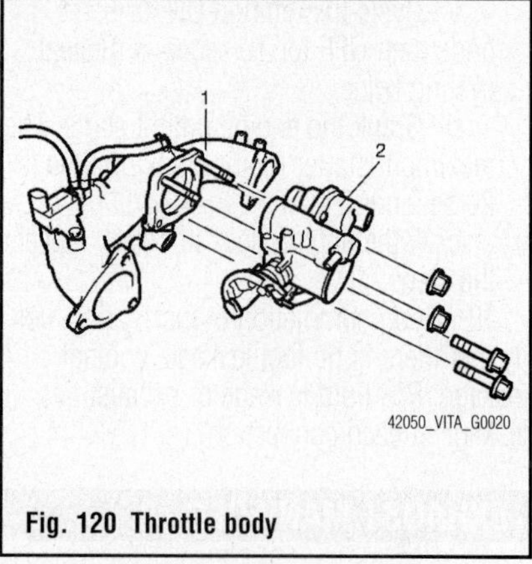

Fig. 120 Throttle body

17. Disconnect PCV hose from cylinder head cover.
18. Disconnect breather hoses from throttle body or cylinder head cover.
19. Disconnect EVAP canister purge valve hose, heater hose and water hose.
20. Remove EGR pipe and wire harness clamps.
21. Disconnect hoses of heater, EVAP canister purge, fuel feed and fuel return.
22. Remove throttle body and intake collector from intake manifold.
23. Disconnect hoses of PCV valve and EVAP canister purge valve from intake collector.
24. Remove throttle body from intake collector.
25. Remove IAC valve from throttle body.
26. Remove EGR valve and EVAP canister purge valve from intake collector.
27. Remove intake manifold bolts and nuts.
28. Remove IAC valve from throttle body.
29. Remove EGR valve and EVAP canister purge valve from intake collector.
30. Remove intake manifold bolts and nuts.
31. Remove intake manifold.

To install:

32. Install new intake manifold gaskets to cylinder heads.
33. Install intake manifold, and tighten bolts and nuts.
34. Install throttle body to intake collector with new throttle body gasket.
35. Install IAC valve, EGR valve, EVAP canister purge valve, MAP sensor and each hoses to intake collector and throttle body if removed.
36. Install throttle body and intake collector assembly to intake manifold with new intake collector gaskets.
37. Connect hoses of heater, EVAP canister, fuel feed and fuel return.
38. Install EGR pipe with new gaskets.
39. Connect hoses of EVAP canister purge valve and heater.
40. Connect hoses of PCV, breather and water.
41. Connect couplers of manifold absolute pressure (MAP) sensor, EVAP canister purge valve, EGR valve and earth terminal.
42. Fix wire harness with clamps.
43. Install clamp bracket to intake collector.
44. Connect earth terminal to intake collector.
45. Connect couplers of TP sensor and IAC valve.
46. Connect brake booster hose, fuel pressure regulator vacuum hose and intake control valve hose to intake manifold.
47. Connect injector wire coupler.
48. Connect water hoses to throttle body.
49. Connect accelerator cable to throttle body
50. Install surge tank pipe and intake control valve to intake manifold with new gaskets and intake air pipe to throttle body.
51. Install surge tank cover
52. Install air cleaner upper case.
53. Connect coupler to intake air temp. sensor and MAF sensor.
54. Install strut tower bar and tighten strut tower bar mounting bolts 36.5 ft. lbs. (50 Nm).
55. Install fuse/relay box.
56. Check to ensure that all removed parts are back in place.
57. Reinstall any necessary parts which have not been reinstalled.
58. Refill cooling system.
59. Connect the negative battery cable.
60. Check to make sure that there are no fuel leakages, performing the following steps.
61. Turn ON ignition switch for 3 seconds (to operate fuel pump), and then turn it **OFF**.
62. Repeat step 3 or 4 times, and apply

fuel pressure to fuel line (till fuel pressure is felt by hand placed on fuel return hose).

63. In this state, check to see that there are no fuel leakages from any part of fuel system.

3.6L Engines

See Figure 121.

1. Release fuel pressure in fuel feed line.
2. Disconnect negative cable at battery.
3. Drain coolant.
4. Remove or disconnect the following:
   - Engine cover
   - Air cleaner outlet duct
   - Positive crankcase ventilation (PCV) line
   - Electronic Throttle Control (ETC) electrical connector
   - PCV line from the top of the intake manifold
   - Evaporative emissions (EVAP) canister purge line
   - Bleed pipe bolts
   - Bleed pipe hose clamp
   - Brake booster vacuum hose
   - Engine harness
   - Upper intake retaining bolts.
   - Upper intake manifold
   - Fuel injectors and fuel rail
   - Lower intake manifold bolts
   - Lower intake manifold

To install:

5. Install or connect the following:
   - Lower intake manifold gasket.
   - Lower intake manifold bolts
   - Torque the lower manifold to 17 ft. lbs. (23 Nm).
6. Install the upper intake manifold gaskets to the lower intake manifold and install the fir tree retainers to retain the upper intake manifold gasket position.
7. Install the upper intake manifold.
8. Apply threadlock to the bolt threads.
9. Torque the upper manifold bolts to 18 ft. lbs. (25 Nm).
10. Install the engine harness retaining clips.
11. Install the brake booster vacuum hose to the intake manifold.
12. Position the bleed pipe.
13. Install the bleed pipe hose clamp (2).
14. Install the bleed pipe bolts (1).
15. Torque the bleed pipe bolts to 89 inch. lbs. (10 Nm).
16. Install or Connect the following:
   - EVAP canister purge line
   - PCV line to the top of the intake manifold.
   - ETC electrical connector
   - Air cleaner outlet duct
   - Fresh air PCV line to the air cleaner inlet duct.
   - Engine cover
   - Negative cable at battery.
17. Fill coolant.
18. Run engine and check for leaks, recheck fill levels.

OIL PAN

REMOVAL & INSTALLATION

2.7L Engines

See Figures 122 and 123.

1. Before servicing the vehicle, refer to the precautions in the beginning of this section.
2. Drain the engine oil.
3. Remove or disconnect the following:
   - Negative battery cable
   - Oil dipstick tube
   - Front wheels
   - Front skid plate, if equipped
   - Steering gear
   - Front differential, if equipped
   - Lower oil pan
   - Oil pickup tube bracket
   - Radiator outlet pipe
   - Upper oil pan and oil pickup tube

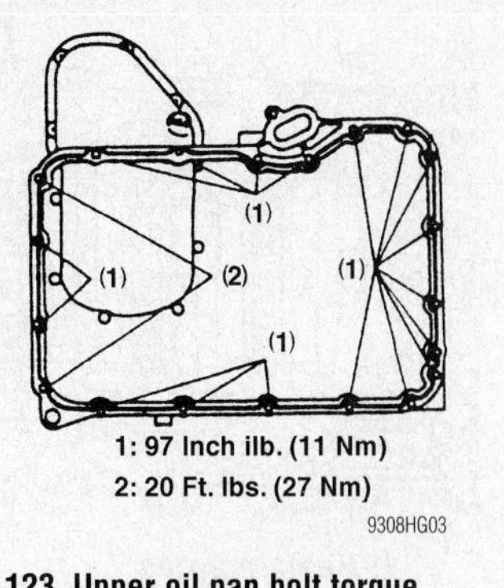

Fig. 123 Upper oil pan bolt torque values—2.7L engine

To install:

4. Install a new O-ring to the lower crankcase.
5. Apply a bead of silicone sealant to the upper oil pan flange.
6. Install or connect the following:
   - New oil pump pickup tube O-ring seals
   - Upper oil pan and oil pump pickup tube and tighten the fasteners as shown
   - Radiator outlet pipe
   - Oil pickup tube bracket
   - Lower oil pan. Tighten the bolts to 97 inch lbs. (11 Nm).
   - Front differential, if equipped
   - Steering gear
   - Front skid plate, if equipped
   - Front wheels
   - Oil dipstick tube
   - Negative battery cable
7. Fill the crankcase to the correct level.
8. Start the engine and check for leaks.

3.6L Engines

See Figures 124 through 126.

1. Disconnect the battery negative cable.
2. Install the engine support fixture.
3. Remove the right side engine mount.

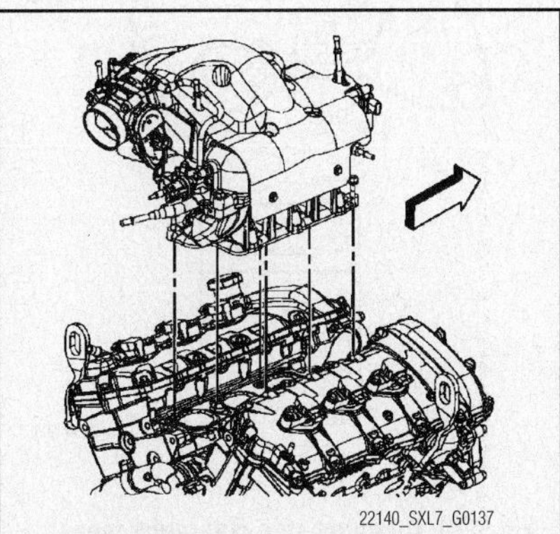

Fig. 121 Upper intake manifold

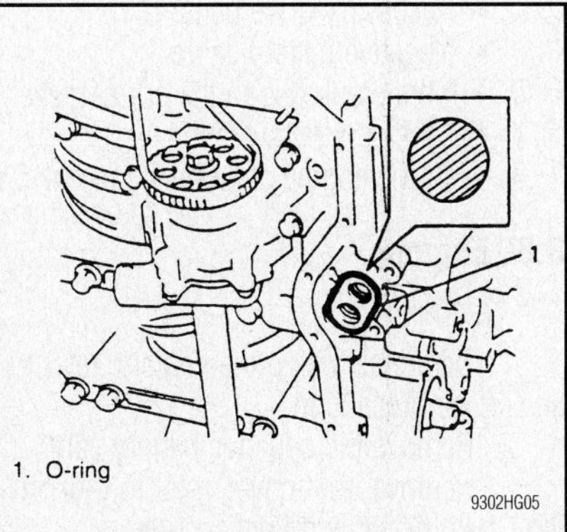

Fig. 122 Lower crankcase O-ring seal

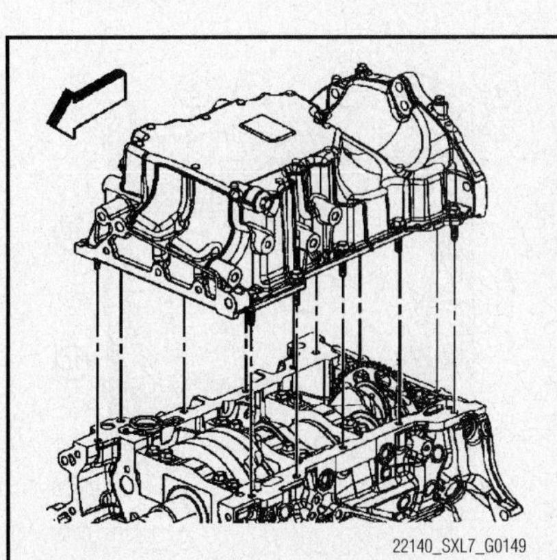

Fig. 124 View of the oil pan

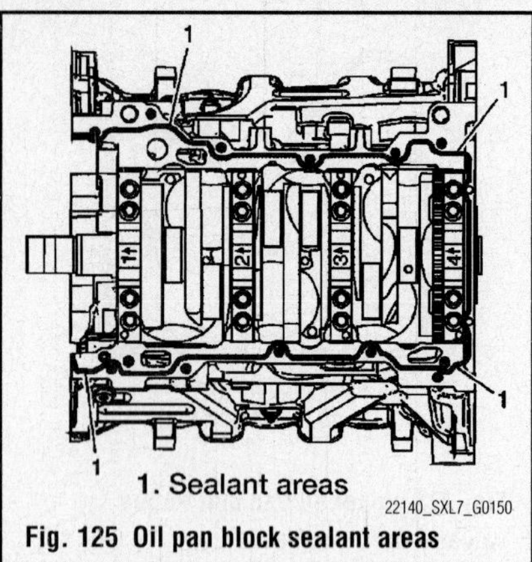

Fig. 125 Oil pan block sealant areas

4. Raise and support the vehicle.
5. Drain the engine oil and remove the oil filter.
6. Remove the catalytic converter.
7. Remove the air conditioning (A/C) compressor.

For removal of the air conditioning compressor refer to the Heating & Air Conditioning section.

8. Remove the oil pan bolts.
9. Remove the oil pan.

To install:

10. Install the 0.315 inch (8 mm) guides from the EN 46109 into the center oil pan rail bolt hole on each side of the engine block.
11. Place a 0.118 in (3 mm) bead (1) of RTV sealant, GM P/N 12378521 (Canadian P/N 88901148) or equivalent, on the block pan rail and the crankshaft rear oil seal housing.
12. Position the oil pan onto the block.
13. Remove the EN 46109 8 mm (0.315 in) guides from the engine block.
14. Loosely install the oil pan bolts.
15. Tighten the oil pan bolts in sequence shown.
16. Tighten the 8mm oil pan bolts 1 through 11 to 17 ft. lbs. (23 Nm).

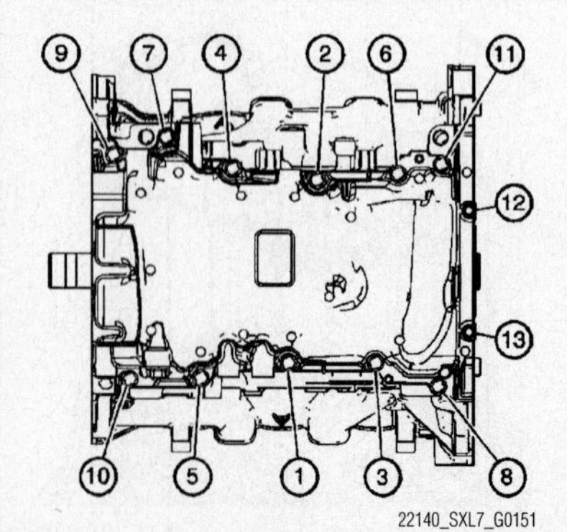

Fig. 126 Oil pan bolt tightening sequence

17. Tighten the 6mm oil pan bolts 12 through 13 to 89 inch lbs. (10 Nm).
18. Install the air conditioning (A/C) compressor.

To charge the air conditioning refer to the Heating & Air Conditioning section.

19. Install the catalytic converter.
20. Lower the vehicle.
21. Refill the engine oil.
22. Install the right side engine mount.
23. Remove the engine support fixture.
24. Connect the battery negative cable.

OIL PUMP

REMOVAL & INSTALLATION

2.7L Engines

1. Before servicing the vehicle, refer to the precautions in the beginning of this section.
2. Drain the cooling system.
3. Drain the engine oil.
4. Remove or disconnect the following:
   - Negative battery cable
   - Accessory drive belts
   - Intake manifold
   - Oil pan and oil pickup tube
   - Front cover
   - Oil pump chain guide
   - Oil pump

⁂ WARNING

Do not remove the sprocket from the oil pump. Damage to the oil pump center shaft and abnormal pump operation may result.

To install:

5. Install or connect the following:
   - Oil pump. Tighten the bolts to 20 ft. lbs. (27 Nm).
   - Oil pump chain guide. Tighten the bolts to 97 inch lbs. (11 Nm).
   - Front cover
   - Oil pan and oil pickup tube
   - Intake manifold
   - Accessory drive belts
   - Negative battery cable
6. Fill the crankcase to the correct level.
7. Fill the cooling system.
8. Start the engine and check for leaks.

3.6L Engines

See Figure 127.

1. Before servicing the vehicle, refer to all service precaution.
2. Remove the negative battery cable.
3. Remove front cover, refer to the Timing Chain Cover and Seal section.
4. Remove the timing chains and actuators, refer to the Timing Chain and Sprockets section.

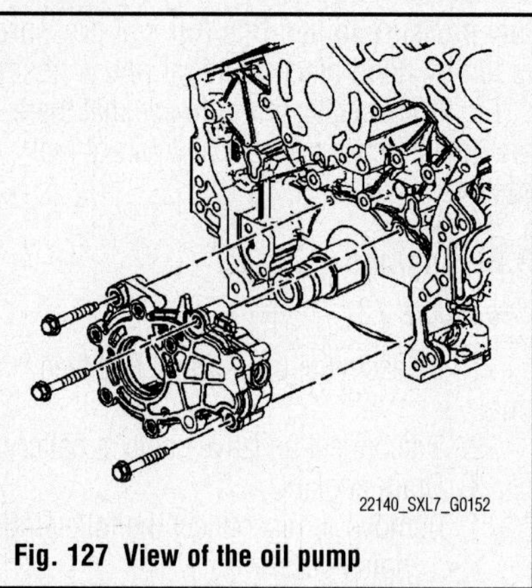

Fig. 127 View of the oil pump

➡Do not remove the left bank idler sprocket.

5. Remove the oil pump bolts
6. Remove the oil pump assembly.

To install:

7. Install the oil pump.
8. Install the timing chain, refer to the Timing Chain and Sprockets procedure.
9. Install the front cover, refer to the Timing Chain Cover and Seal procedure.
10. Install the negative battery cable.
11. Run vehicle and check for leaks and proper operation of engine.

INSPECTION

2.7L Engines

See Figure 128.

1. Check outer rotor, inner rotor and oil pump cases for excessive wear or damage.

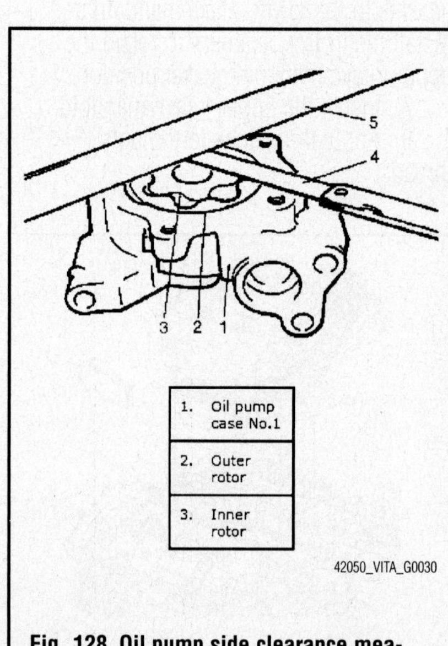

Fig. 128 Oil pump side clearance measurement

2. Check relief valve for excessive wear or damage.

3. Measure free length and tension of oil relief spring.

4. Measure clearance of oil pump rotor and oil pump case using thickness gauge.

5. Check radial clearance between outer rotor and case, using thickness gauge.

6. If clearance exceeds 0.0075 inch. (0.19 mm), replace oil pump assembly.

7. Using straightedge and thickness gauge, measure side clearance. If clearance exceeds 0.0043 inch. (0.11 mm), replace oil pump assembly.

3.6L Engines

See Figures 129 through 134.

There are no serviceable components within the oil pump. Disassemble the pump only to diagnose an oiling concern.

WARNING

A disassembled oil pump must not be reused. A disassembled oil pump must be replaced

1. Inspect the oil pump housing for the following:
   - Damage, scoring, or debris on the housing surface for the driven gear (1)
   - Damage to the oil pump mounting bosses (2)
   - Damage, scoring, or debris on the housing surface for the drive gear (3)
   - Damage, scoring, or debris in the oil pump relief valve port (4)
   - Damage, scoring, or debris in the oil pump intake port (5)
   - Damage, scoring, or debris in the oil pump relief valve bore (6)
   - Damage, scoring, or debris in the oil pump output port (7)
   - Damage to the threads in the oil pump housing for the oil pump cover bolts (8)

2. Inspect the oil pump cover for the following conditions:

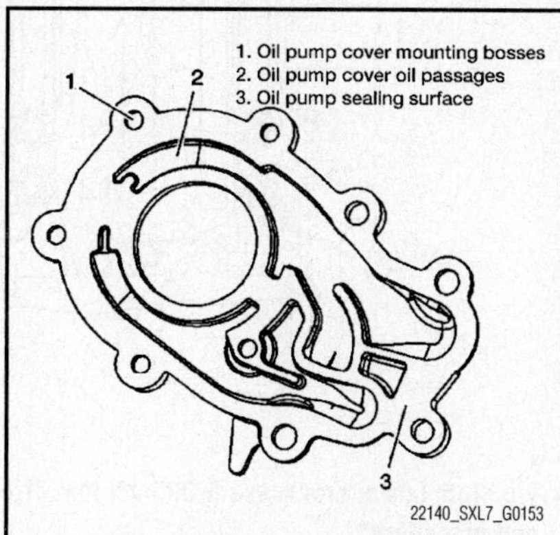

Fig. 130 Oil pump cover inspection areas

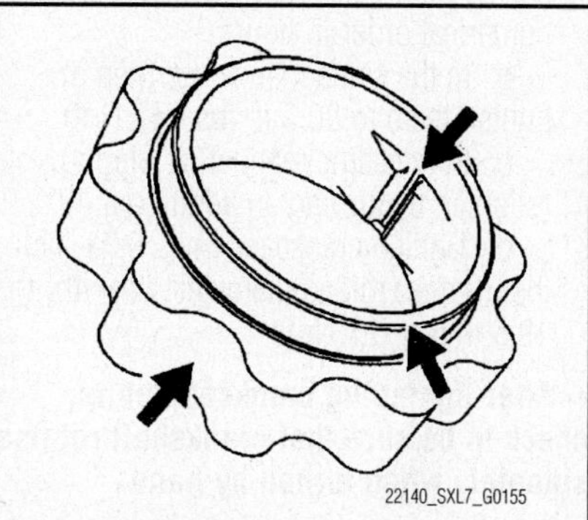

Fig. 131 Oil pump inner drive gear inspection

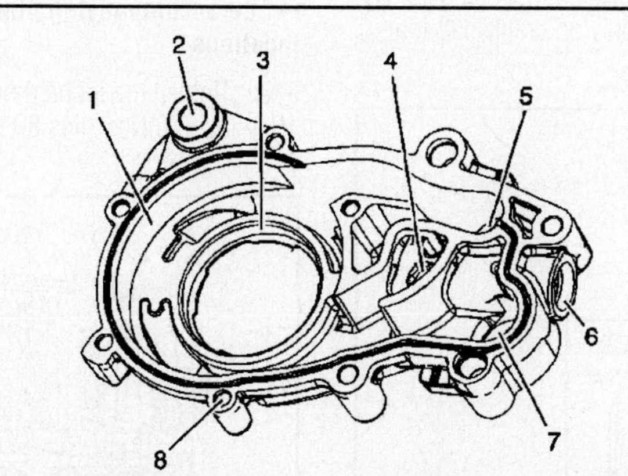

Fig. 129 Oil pump inspection areas

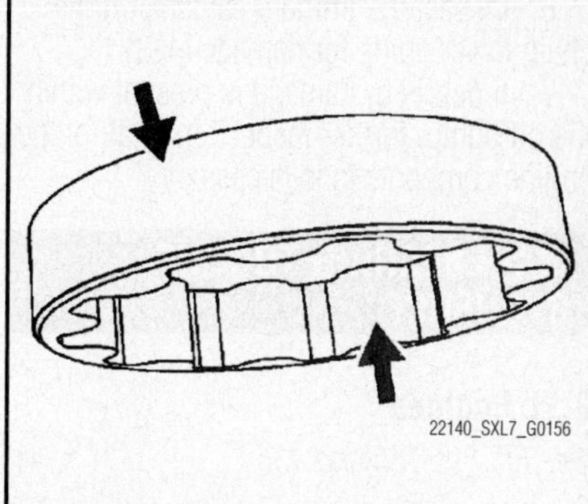

Fig. 132 Oil pump outer drive gear inspection

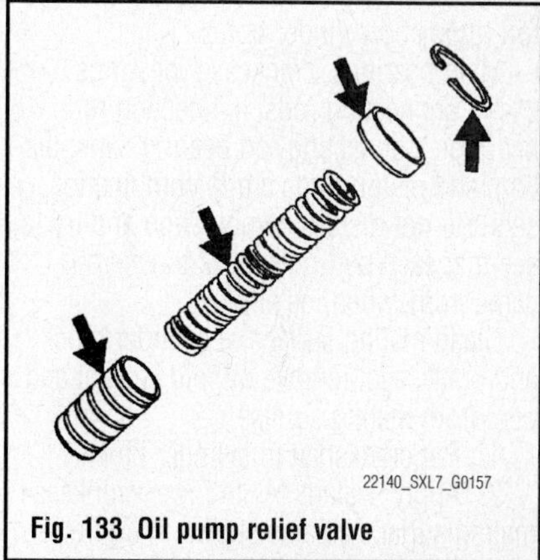

Fig. 133 Oil pump relief valve

- Damage to the oil pump cover mounting bosses (1)
- Damage, scoring, or debris in the oil pump cover oil passages (2)
- Damage to the sealing surface between the oil pump cover and the oil pump housing (3)

3. Inspect the inner drive gear for damage. If inner diameter damage is found, ensure the crankshaft is also inspected

4. Inspect the outer driven gear for damage.

5. Inspect the oil pump relief valve components for debris or damage.

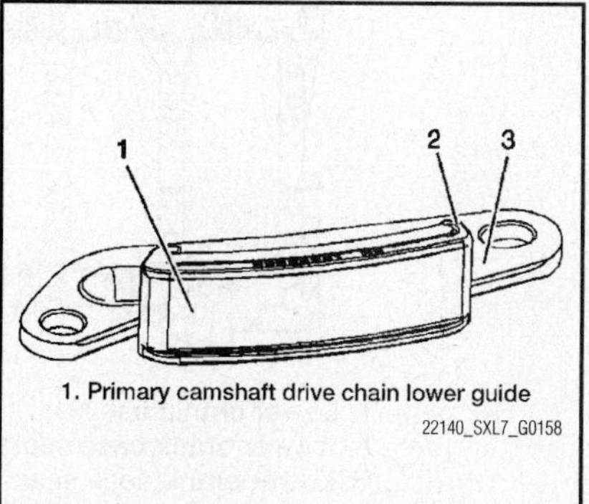

Fig. 134 Primary camshaft drive chain lower guide inspection

6. Inspect the primary camshaft drive chain lower guide for damage (1-3).

7. If debris or damage is present within the oil pump, further inspection of all of the engine components is necessary.

MAIN BEARING TORQUE SEQUENCE

2.7L Engines

See Figures 135 and 136.

All parts to be installed must be perfectly clean. Be sure to apply oil crankshaft journals, main bearings, thrust bearings, crank pins, connecting rod bearings, pistons, piston rings and cylinder bores.

Main bearings, crankcase (bearings caps), connecting rods, connecting rod bearings, connecting rod bearing caps, pistons and piston rings are in combination sets. Do not disturb combination and try to see that each part goes back to where it came from, when installing.

Clean mating surface of cylinder block and crankcase, remove oil, old sealant and dust from mating surface.

1. Put crankshaft to cylinder block.
2. Apply sealant "A" to lower crankcase mating surface area as shown in figure.
3. Install lower crankcase (1) to cylinder block. Apply oil to crankcase bolts before installing them.

➡Tighten 0.39 inch lbs. (10 mm.) thread diameter bolts first (the following the order shown in figure), then tighten 8 mm (0.31 in.) thread diameter bolts.

4. Tighten crankcase bolts gradually as follows:

a. Tighten crankcase bolts (1–16) 0.39 inch lbs. (10 mm.) thread diameter) (2a) to 22 ft. lbs. (30 Nm) according to numerical order in figure.

b. In the same manner as Step a), tighten them to 30.5 ft. lbs. (42 Nm).

c. In the same manner as Step a), retighten by turning an additional 40°.

d. Tighten crankcase bolts 0.31 inch lbs. (8 mm) thread diameter) (3b–4b) to 19.5 ft. lbs. (27 Nm).

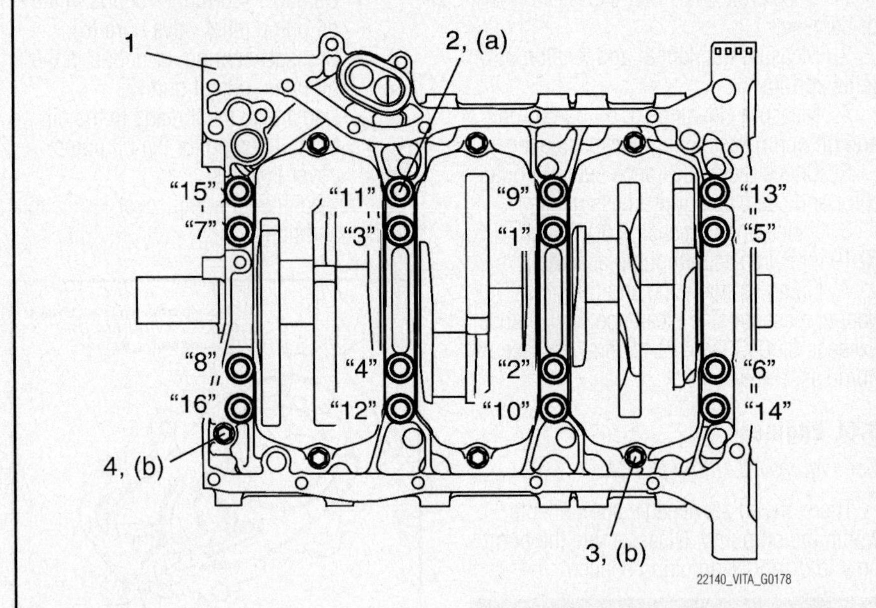

Fig. 136 Lower crankcase 0.39 inch lbs. (10 mm.) thread diameter) (2a torque specifications and procedure

➡After tightening crankcase bolts, check to be sure that crankshaft rotates smoothly when turned by hand.

3.6L Engines

See Figures 137 through 139.

All parts to be installed must be perfectly clean. Be sure to apply oil crankshaft journals, main bearings, thrust bearings, crank pins, connecting rod bearings, pistons, piston rings and cylinder bores.

Main bearings, crankcase (bearings caps), connecting rods, connecting rod bearings, connecting rod bearing caps, pistons and piston rings are in combination sets. Do not disturb combination and try to see that each part goes back to where it came from, when installing.

Clean mating surface of cylinder block and crankcase, remove oil, old sealant and dust from mating surface.

1. Tighten the inboard bolts (1—8)\ to 15 ft. lbs.
(20 Nm) plus 80 degrees.

➡See accompanying diagram for bolt locations.

2. Tighten the inboard bolts (9—16) to 15 ft. lbs. (20 Nm) plus 80 degrees.

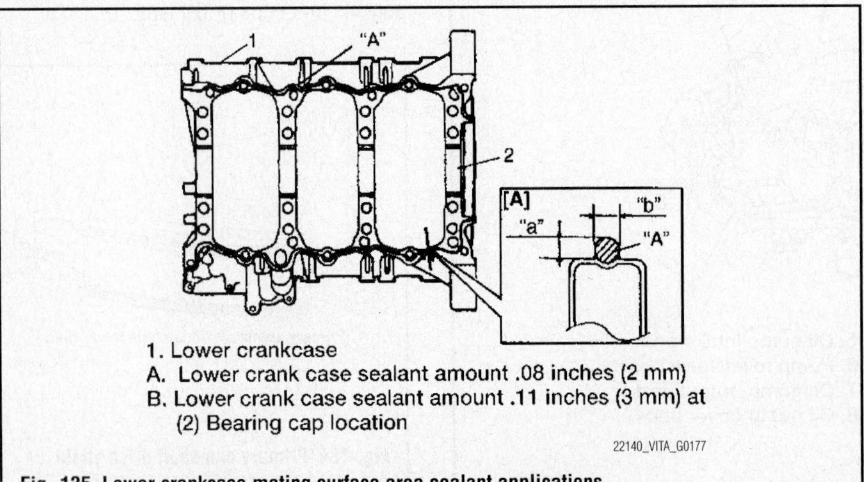

Fig. 135 Lower crankcase mating surface area sealant applications

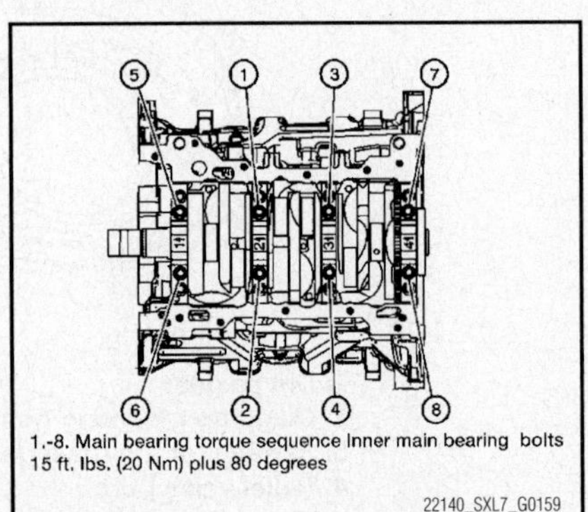

Fig. 137 Main bearing bolt torque-inboard bolts

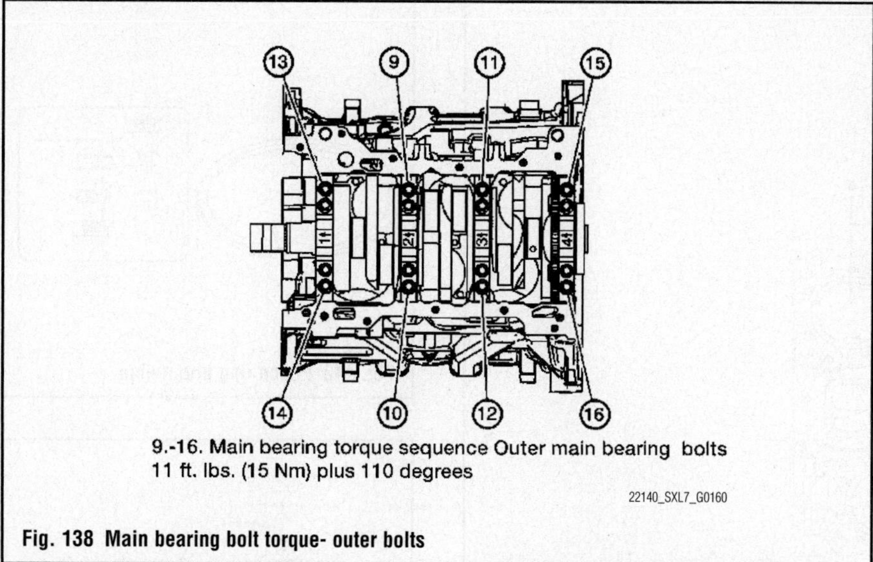

Fig. 138 Main bearing bolt torque- outer bolts

➡See accompanying diagram for bolt locations.

3. Tighten the original short/inner side main cap bolts (17—20) to 22 ft. lbs. (30 Nm) plus 60 degrees.

4. Tighten the original short/inner side main cap bolts (21—24) to 22 ft. lbs. (30 Nm) plus 60 degrees.

➡Ensure that the crankshaft turns without binding or noise.

PISTON AND RING

POSITIONING

2.7L Engines

See Figures 140 through 143.

Two sizes of piston are available as standard size spare part so as to ensure proper piston-to-cylinder clearance. When installing a standard size piston, make sure to match piston with cylinder as follows.

1. Each piston (8) has stamped number (9) on its piston head. It represents outer diameter of piston.
2. There are also stamped numbers (10) or painted color (10) on cylinder block as shown in figure.
3. The stamped number on piston and painted color (stamped number) on cylinder block must correspond. That is, install number "2" stamped piston to cylinder which is identified with blue painted (or "2" stamped) and a number "1" piston to cylinder with red
4. Match pistons with "2" indicators to blue cylinder paint marks

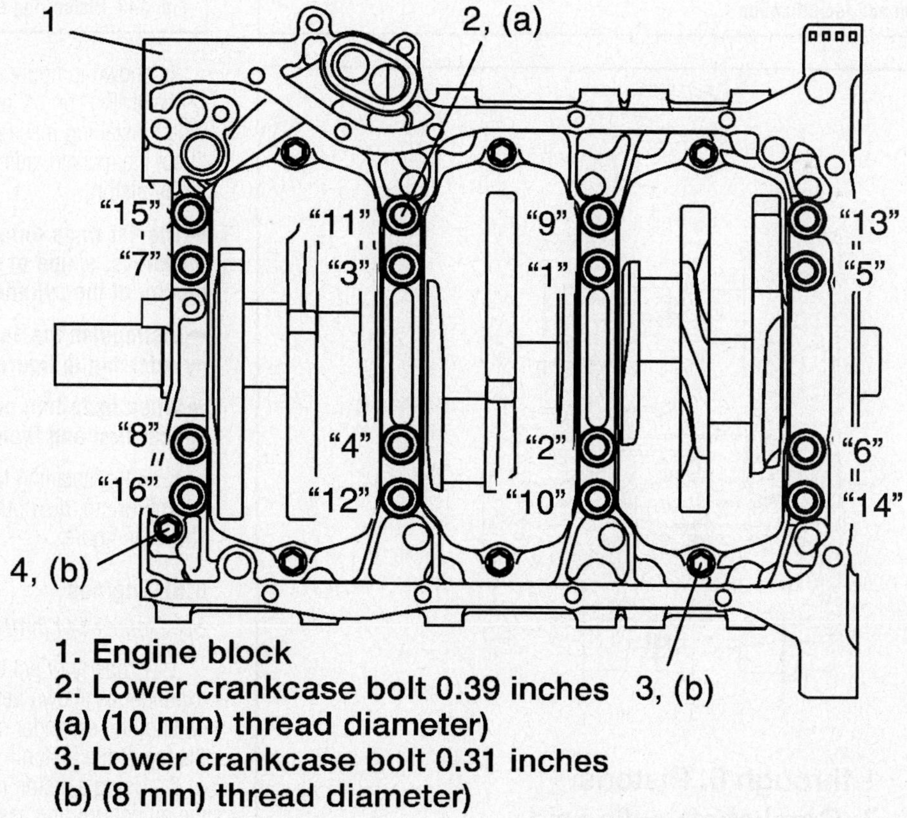

1. Engine block
2. Lower crankcase bolt 0.39 inches (a) (10 mm) thread diameter)
3. Lower crankcase bolt 0.31 inches (b) (8 mm) thread diameter)

22140_SXL7_G0162

Fig. 139 Side main bearing bolt torque sequence— short bolts and long bolts

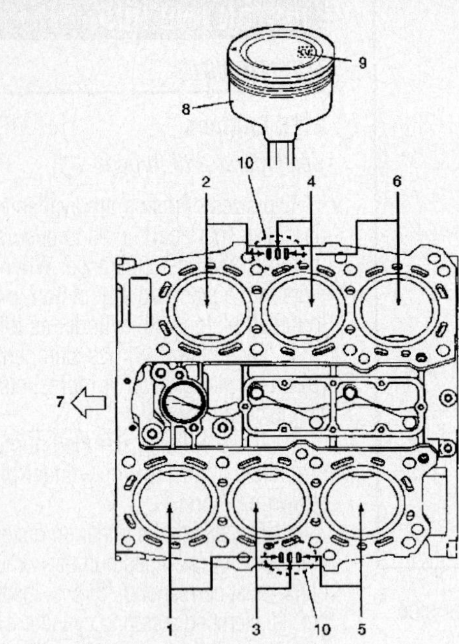

Fig. 140 Piston location and identification

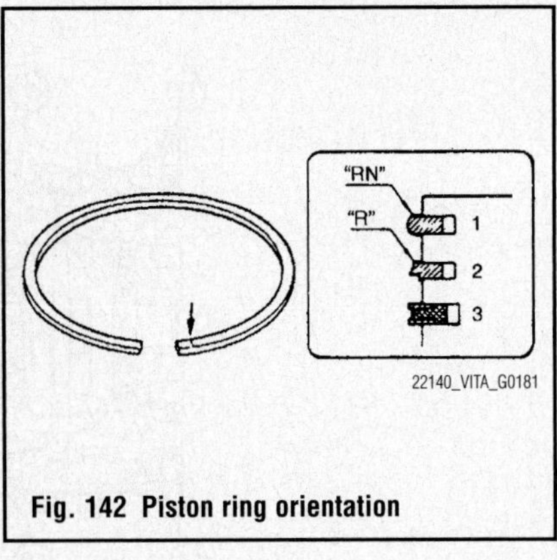

Fig. 142 Piston ring orientation

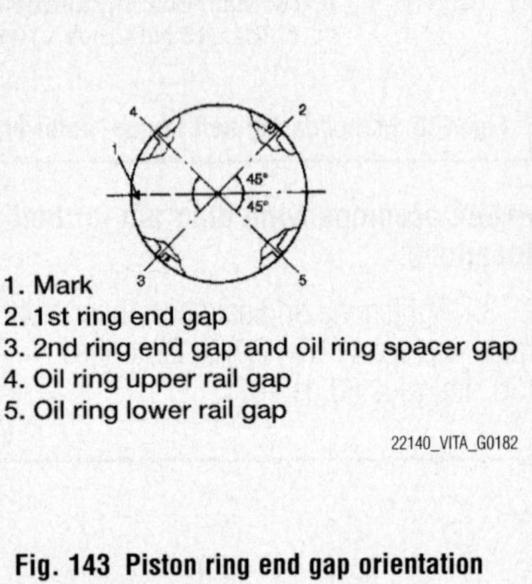

Fig. 143 Piston ring end gap orientation

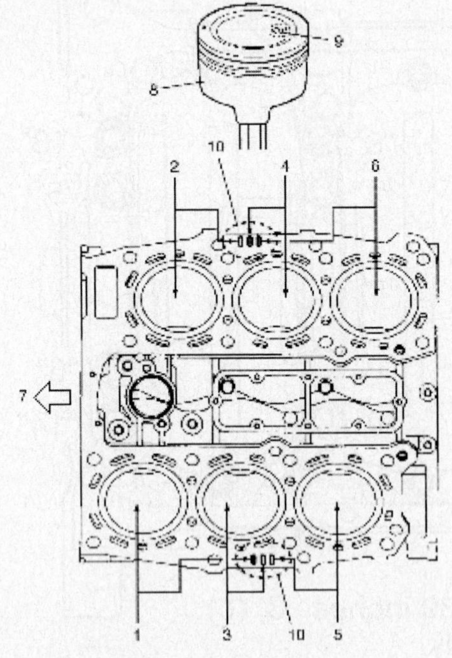

Fig. 141 Piston location and identification with chart for locations of pistons into the cylinder block

As shown in figure, 1st (1) and 2nd rings (2) have "RN" or "R" mark respectively. When installing these piston rings to piston, direct the marked side of each ring toward top of piston.

➡The 1st rings differs from 2nd ring in thickness, shape of surface and contacting of the cylinder wall.

➡Distinguish the 1st ring from 2nd ring by referring to figure.

➡When installing oil ring (3), install spacer first and then 2 rails.

5. After installing three rings (1st, 2nd and oil rings), distribute their end gaps as shown in figure.

3.6L Engines

See Figures 144 through 146.

1. Properly orient the oil control ring expander as shown before installation. The ends of the expander must be facing toward the top of the piston.
2. Using a piston ring expander, install the oil control ring assembly using the following procedure:
   - Install the expander ring (3).
   - Install the 2 oil scraper rings (4).

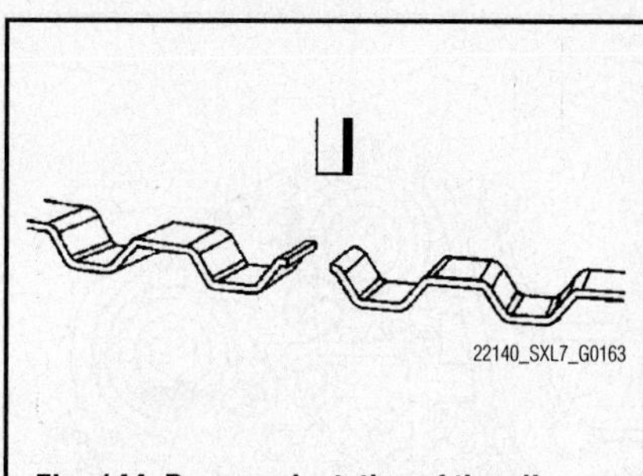

Fig. 144 Proper orientation of the oil expander ring

3. Expand the rings only enough to clear the piston diameter.

WARNING

Over expanding the piston rings will distort or crack the rings.

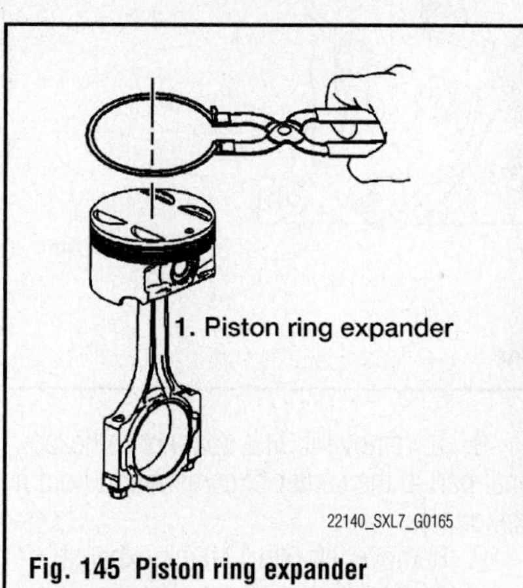

Fig. 145 Piston ring expander

4. Install the second and top piston rings using the ring expander.
5. Once the rings are installed, set the ring gaps for the oil control, second and top ring. Use the piston location arrow in the diagram for reference.
   - Lower oil control ring - position 1
   - Upper oil control ring - position 2
   - Top Ring - position 3
   - Oil control ring expander - position 4
   - Second ring - position 5

REAR MAIN SEAL

REMOVAL & INSTALLATION

2.7L Engines

1. Before servicing the vehicle, refer to all precaution.
2. Disconnect the negative battery cable.
3. Dismount transmission assembly, refer to the Transmissions section.
4. Hold flywheel stationary with special tool and remove clutch cover bolts, clutch cover and clutch disc.
5. Using special tool (flywheel holder—09924—17811) remove flywheel (M/T vehicle) or drive plate (A/T vehicle).
6. Carefully pry oil seal from rear housing

To install:

7. Using special tools (Oil seal installer 09911-97710 and oil seal guide 09911-97811), install rear oil seal.
8. Install flywheel (M/T vehicle) or drive plate (A/T vehicle). Using special tool, lock flywheel or drive. plate, and tighten new flywheel bolts or new drive plate bolts to 51 ft. lbs. (70 Nm).
9. Install the transmission assembly, refer to the Transmission section.

3.6L Engines

See Figures 147 through 152.

1. Remove the negative battery cable.
2. Remove the engine flywheel.
3. For removal of the flywheel refer to the Flywheel section.
4. Remove the oil pan, refer to the Oil Pan section.
5. Remove the crankshaft rear oil seal housing bolts.
6. Use the pry points located at the edge of the crankshaft rear oil seal housing to separate the RTV sealant.
7. Remove the crankshaft rear oil seal and housing.

To install:

8. Install the 6 mm (0.236 in) guides from the EN 46109 into the 2 crankshaft rear oil seal housing corner bolt holes of the engine block
9. Install the EN-47839 with the J 42183 (1, 2) onto the rear of the crankshaft flange.
10. Place a 0.118 inch (3 mm) bead of RTV sealant, GM P/N 12378521 (Canadian

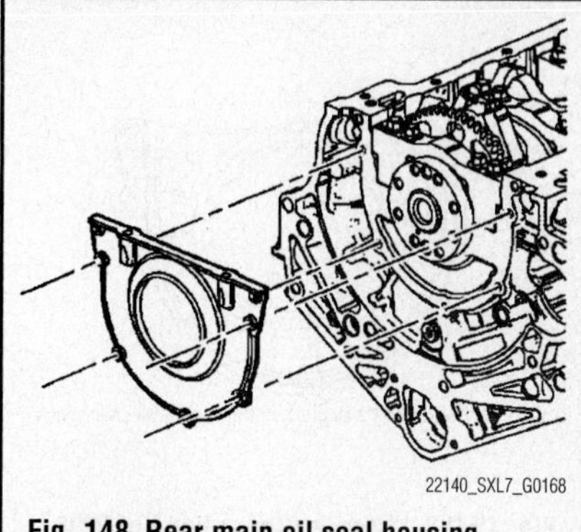

Fig. 148 Rear main oil seal housing removed

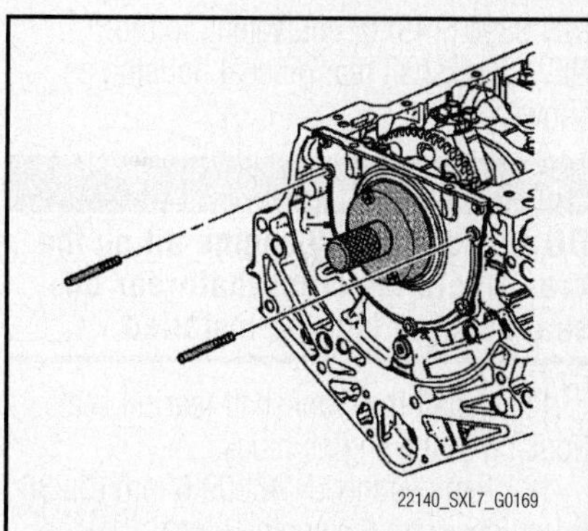

Fig. 149 0.236 in (6 mm) rear main seal installer guides

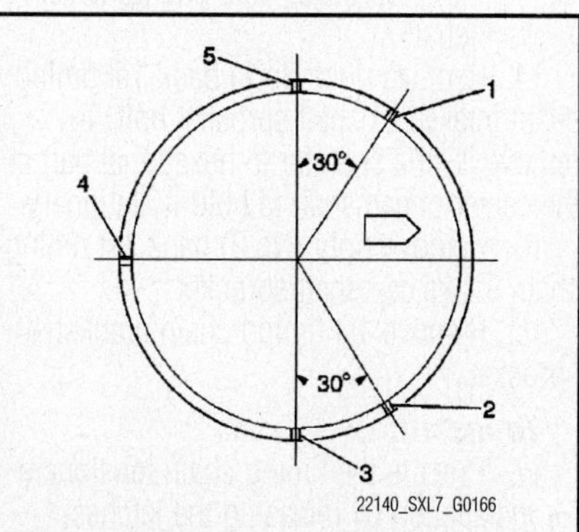

Fig. 146 Ring end gap positioning in reference to each other

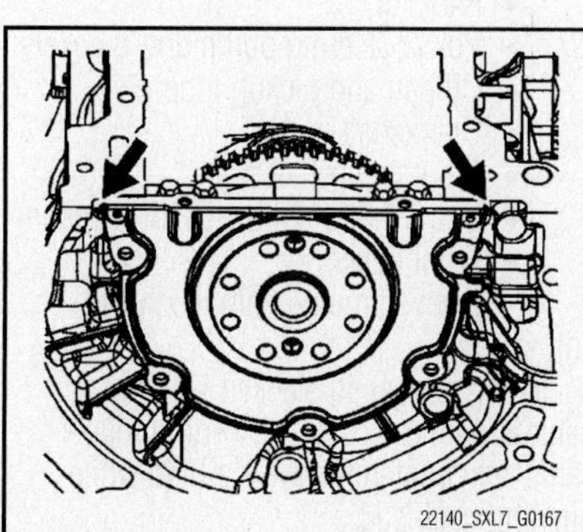

Fig. 147 Rear main oil seal housing pry points

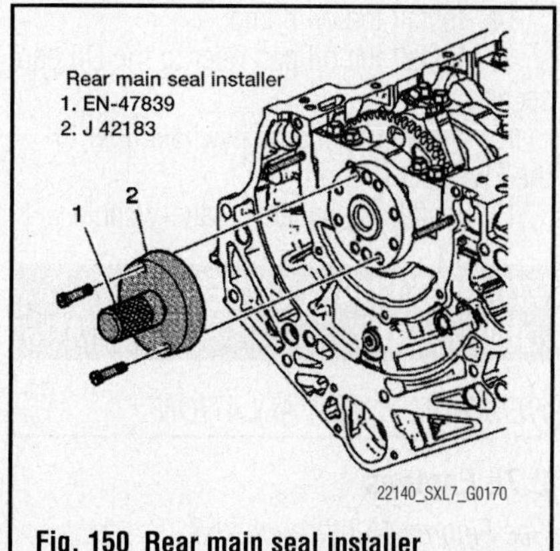

Fig. 150 Rear main seal installer

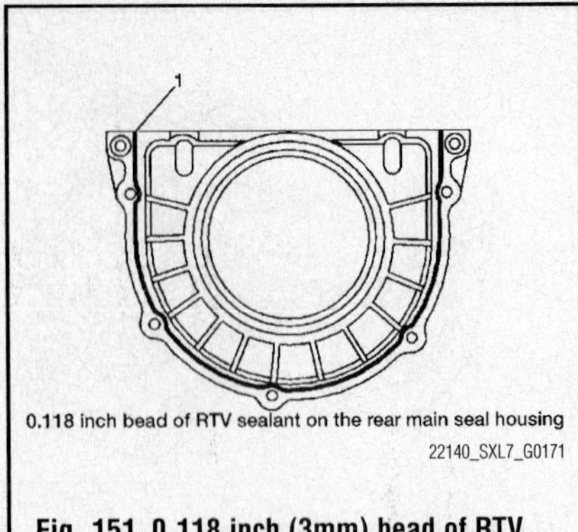

Fig. 151 0.118 inch (3mm) bead of RTV sealant on the rear main seal housing

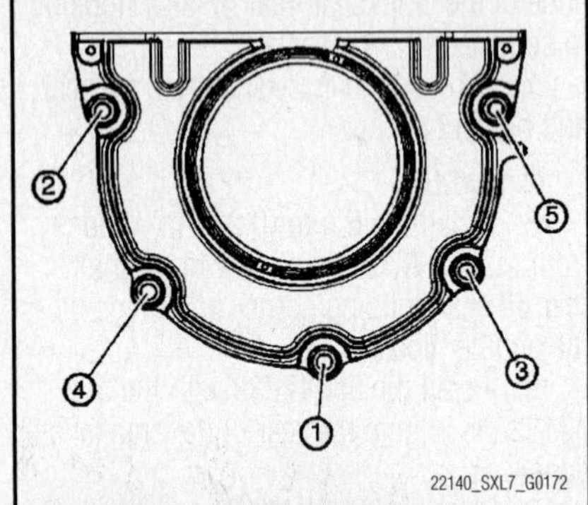

Fig. 152 Crankshaft rear main oil seal housing bolts torque sequence

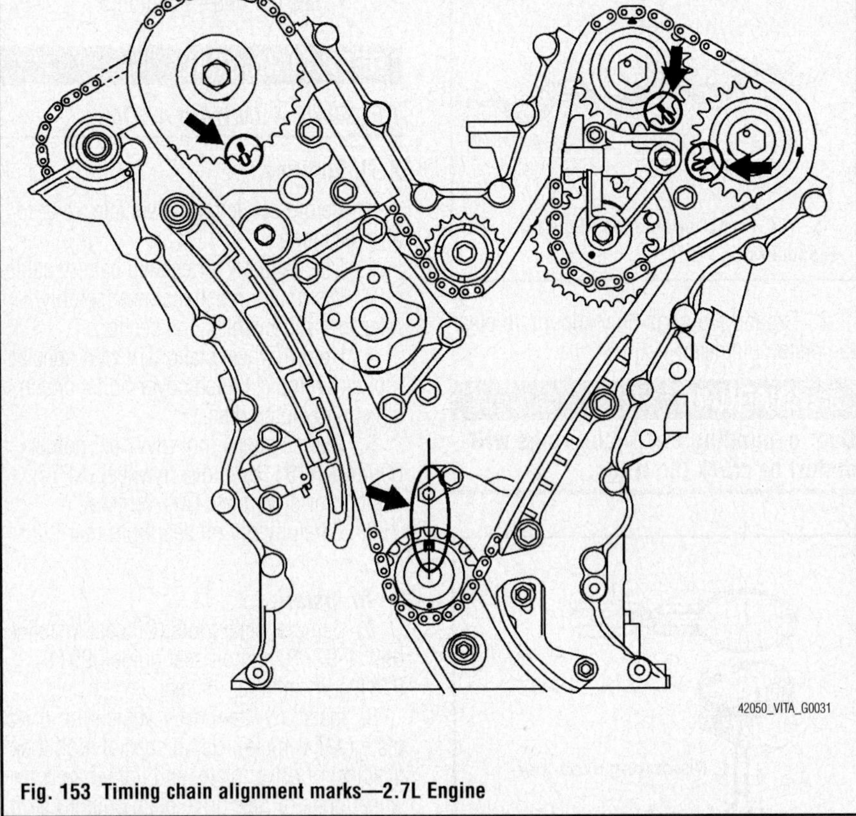

Fig. 153 Timing chain alignment marks—2.7L Engine

P/N 88901148) or equivalent, to the NEW crankshaft rear oil seal housing as shown (1).

WARNING

DO NOT allow any engine oil on the area where the crankshaft rear oil seal housing is to be installed.

11. Install the crankshaft rear oil seal housing to the engine block.
12. Remove the EN 46109 6 mm (0.236 in) guides from the engine block.
13. Tighten the crankshaft rear oil seal housing bolts to 89 inch lbs. (10 Nm) in the sequence shown.
14. Install the oil pump.
15. Install the oil pan refer to the Oil pan section.
16. Install the engine flywheel refer to the Flywheel section.
17. Install the negative battery cable.

TIMING CHAIN COVER AND SEAL

REMOVAL & INSTALLATION

2.7L Engines

See Figures 153 through 168.

1. Before servicing the vehicle, refer to the precautions in the beginning of this section.
2. Drain the cooling system.
3. Drain the engine oil.
4. Remove engine assembly from vehicle.
5. Remove or disconnect the following:
    - Negative battery cable
    - Intake manifold
    - Ignition coils
    - Valve covers
    - Accessory drive belts
    - Cooling fan and shroud
    - Water pump pulley
    - Radiator
    - Power steering pump and brackets
    - Oil pan and pickup tube
    - Crankshaft pulley
    - Front crankshaft seal
    - Crankshaft Position (CKP) sensor
    - Front cover
6. Remove timing chain tensioner adjuster.
7. To remove it, slacken left (No.1) bank 2nd timing chain by turning intake camshaft counterclockwise a little while pushing back pad.
8. Remove left (No.1) bank intake and exhaust camshaft sprocket bolts.
9. To remove it, fit a spanner to hexagonal part at the center of camshaft to hold it stationary.
10. Remove left (No.1) bank exhaust camshaft sprocket.
11. Remove left (No.1) bank intake camshaft sprocket.
12. Remove left (No.1) bank 2nd timing chain.
13. Remove or disconnect the following.
    - Timing chain guide No.1 and No.2
    - Timing chain tensioner
    - Timing chain tensioner adjuster
    - Idler sprocket No.1 and 1st timing chain
    - Idler sprocket No.2 and sprocket shaft
14. Remove right(No.2) bank 1st timing chain intake camshaft sprocket bolt. To remove it, fit a spanner to hexagonal part at the center of camshaft to hold it stationary.
15. Remove right (No.2) bank 1st timing chain intake camshaft sprocket.
16. Remove 1st timing chain crankshaft sprocket.

To install:

17. Prepare the timing chain tensioners for installation by releasing the latches, compressing the tensioner piston fully into the bore and installing retaining pins.

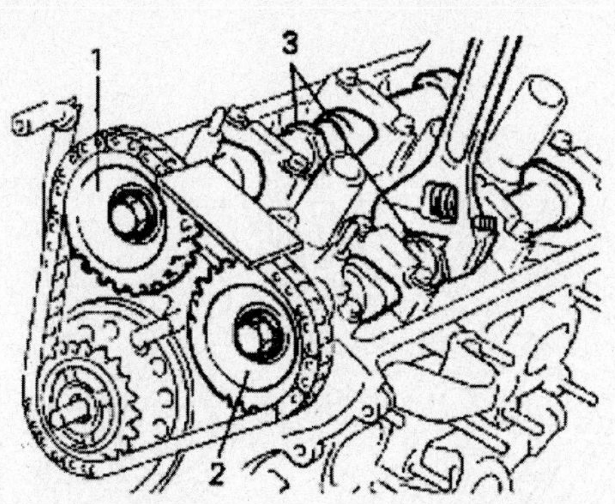

1.	LH (No.1) bank intake camshaft sprocket
2.	LH (No.1) bank exhaust camshaft sprocket

42050_VITA_G0032

Fig. 154 Camshaft sprocket removal—2.7L Engine

18. Align the timing chain sprocket matchmarks and colored chain links as shown during assembly.
19. Check timing pulley key of crankshaft is on specified position.
20. Install 1st timing chain crankshaft sprocket to crankshaft.
21. Check timing mark on right (No.2) bank intake camshaft.
22. Install right (No.2) bank 1st timing chain intake camshaft sprocket noting the following points.
    a. The sprocket should be set in such way that its timing marks can be seen.
    b. Camshaft should be held stationary by using a spanner at its hexagonal parts.
23. Tighten right bank 1st timing chain intake camshaft sprocket bolt to 58 ft. lbs. (80 Nm).
24. Install timing chain tensioner.
25. Tighten timing chain tensioner nut to 18 ft. lbs. (25 Nm).
26. Install 1st timing chain by aligning match marks on right silver plate of 1st timing chain and right (No.2) bank 1st timing chain intake camshaft sprocket.
27. Apply oil to bush of idler sprocket No.2.
28. Install idler sprocket No.2 and sprocket shaft.

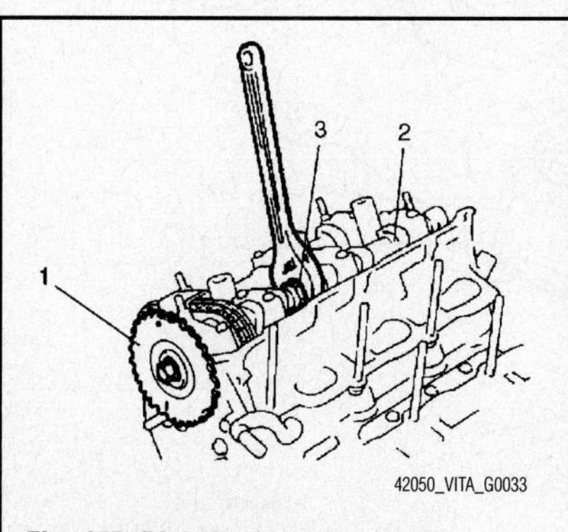

Fig. 155 Right bank 1st camshaft sprocket—2.7L Engine

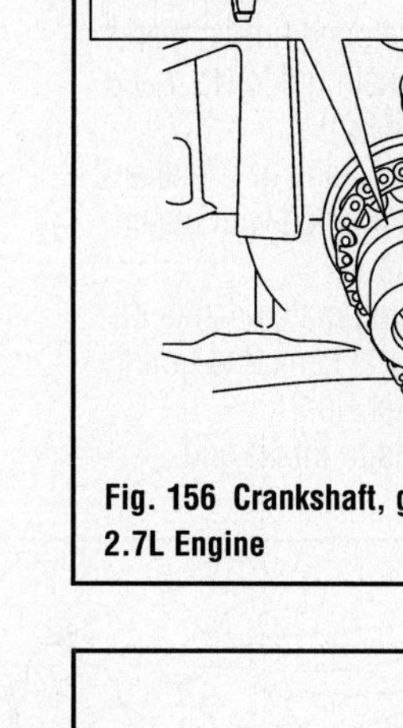

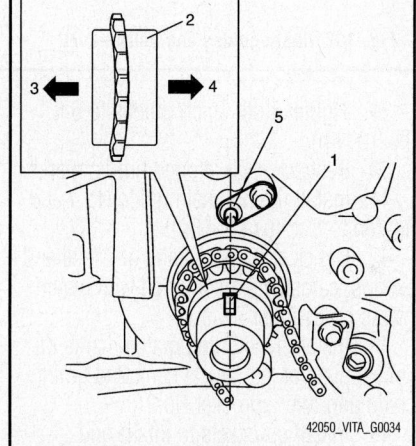

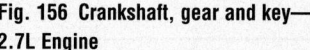

Fig. 156 Crankshaft, gear and key—2.7L Engine

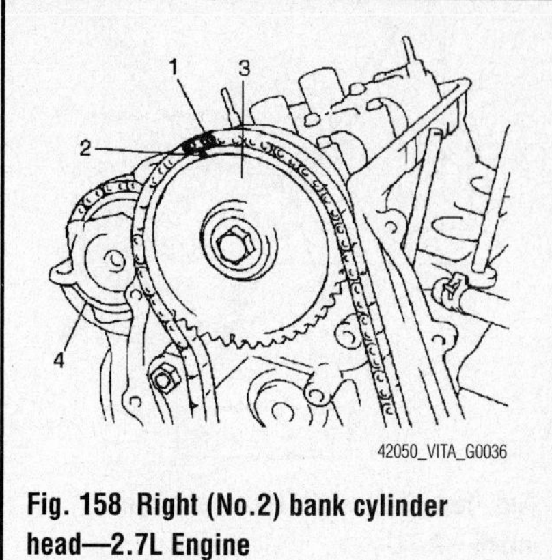

Fig. 158 Right (No.2) bank cylinder head—2.7L Engine

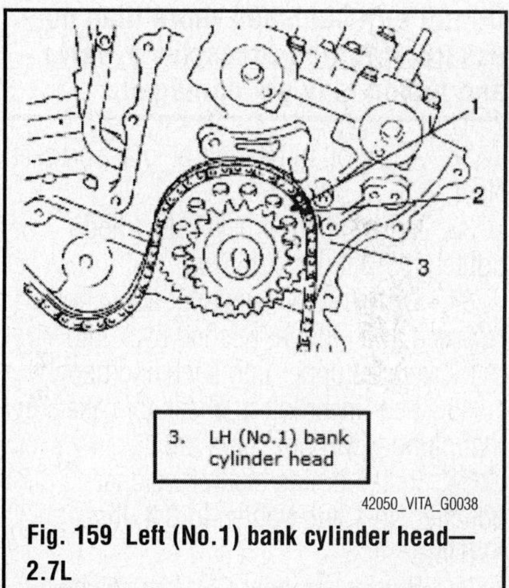

Fig. 159 Left (No.1) bank cylinder head—2.7L

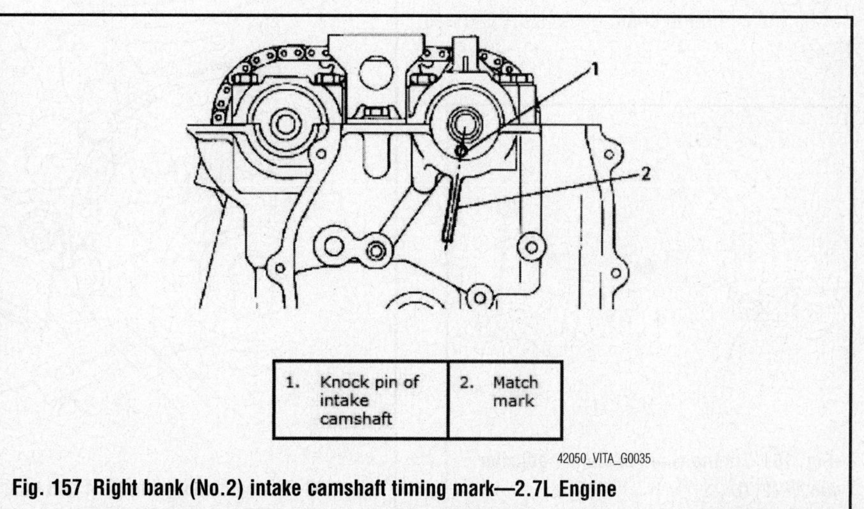

Fig. 157 Right bank (No.2) intake camshaft timing mark—2.7L Engine

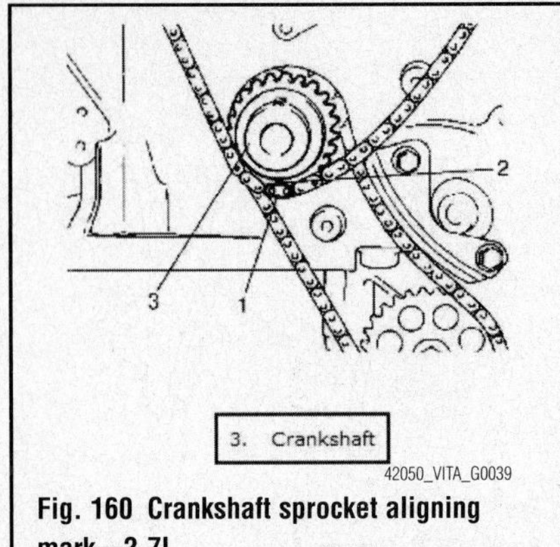

Fig. 160 Crankshaft sprocket aligning mark—2.7L

29. Install idler sprocket No.2 by aligning match marks on left silver plate of 1st timing chain.

30. Install crankshaft sprocket by aligning match marks on yellow plate of 1st timing chain and crankshaft timing sprocket.

31. To install it, fit a spanner to hexagonal part at the center of right (No.2) bank intake camshaft to turn a little.

**** WARNING**

Do not turn camshaft more than necessary. If turned excessively, valve and piston may get damaged.

32. Apply oil to bearing of idler sprocket No.1.

33. Install idler sprocket No.1 and tighten to 40 ft. lbs. (55 Nm).

34. With latch of tensioner adjuster No.1 returned and plunger pushed back into body, insert stopper into latch and body.

35. After inserting it, check to make sure that plunger will not come out.

36. Install timing chain tensioner adjuster No.1 and tighten to 8 ft. lbs (11Nm).

37. Pull out stopper (3) from adjuster No.1 (1).

38. Install timing chain guide No.1 and No.2.

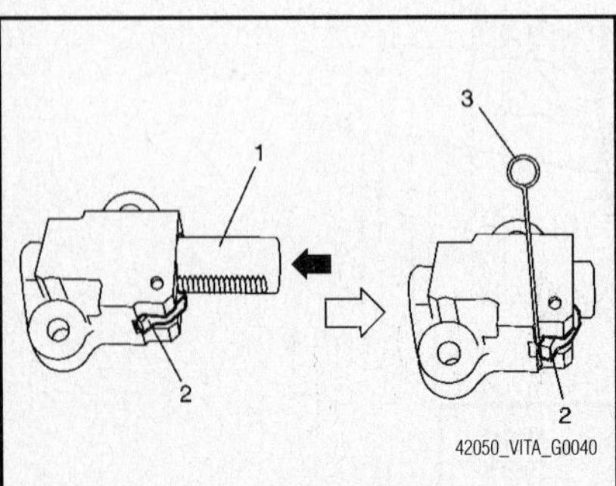

Fig. 161 Timing chain tensioner adjuster No.1—2.7L

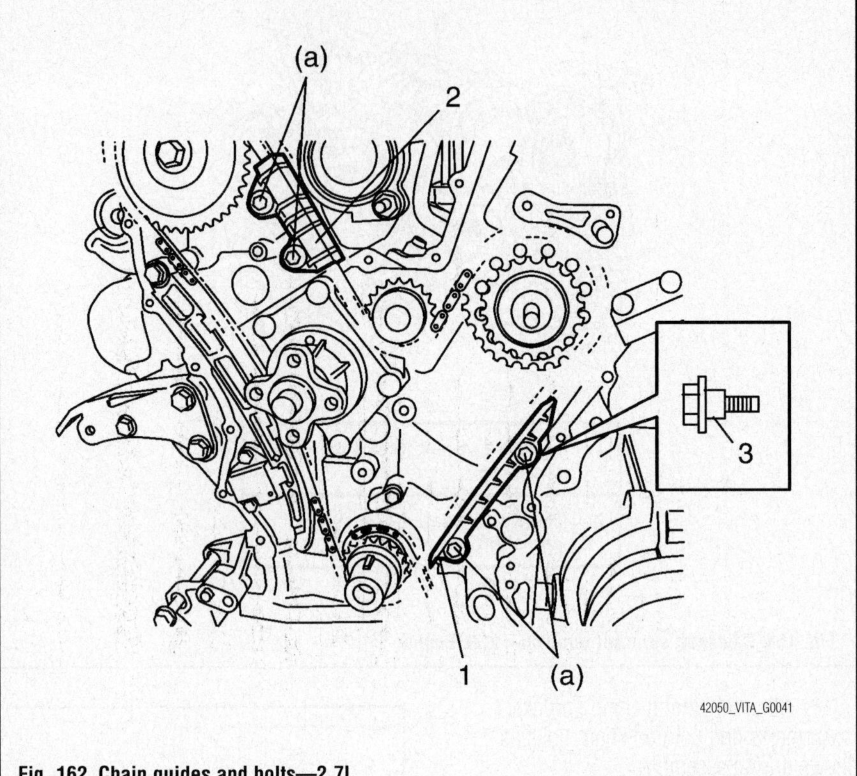

Fig. 162 Chain guides and bolts—2.7L

39. Tighten both chain guides to 6 ft. lbs. (9 Nm).

40. Recheck each aligned timing marks.

41. Install timing chain guide No.4 and tighten to 8 ft. lbs (11 Nm).

42. Check that knock-pins of intake and exhaust camshafts are aligned with match marks on cylinder head.

43. Install by aligning match marks on yellow plate of left (No.1) bank 2nd timing chain and idler sprocket No.2.

44. Install sprockets to intake and exhaust camshafts by aligning silver plate of left (No.1) bank 2nd timing chain, match marks on intake sprocket and exhaust sprocket respectively.

45. Install LH (No.1) bank intake and exhaust camshaft timing sprockets.

46. To install it, fit a spanner to hexagonal part at the center of camshaft to hold stationary.

47. Tighten camshaft timing sprocket bolts to 58 ft. lbs. (88 Nm).

48. With latch of tensioner adjuster No.3

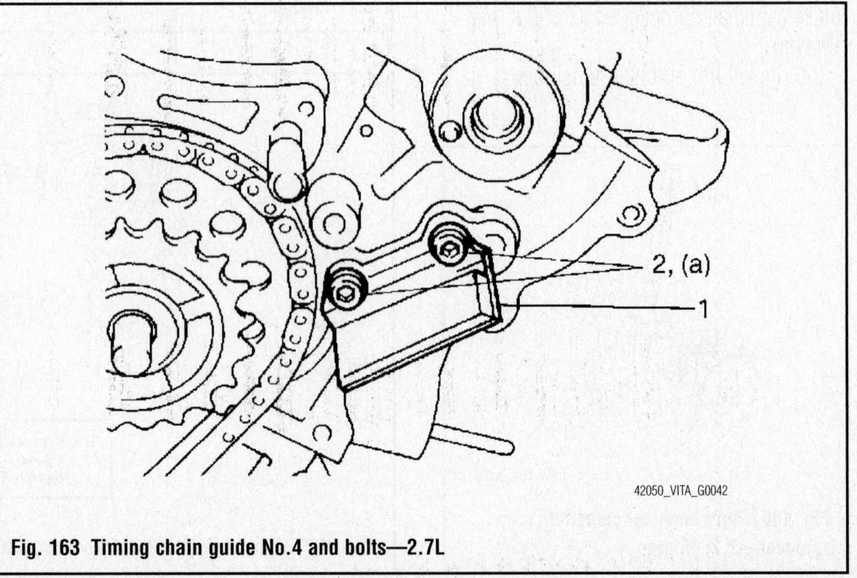

Fig. 163 Timing chain guide No.4 and bolts—2.7L

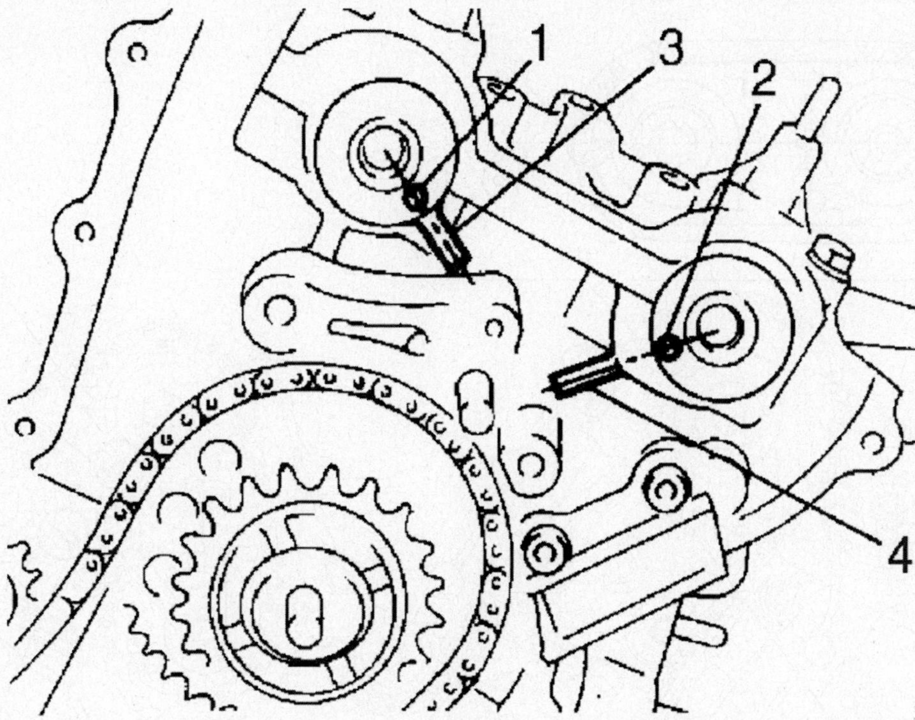

1. Knock pin of LH (No.1) bank intake camshaft	3. Match mark of intake side
2. Knock pin of LH (No.1) bank exhaust camshaft	4. Match mark of exhaust side

42050_VITA_G0043

Fig. 164 Intake and exhaust match marks—2.7L

returned and plunger pushed back into body, insert stopper (pin) into set hole.

49. After inserting it, check that plunger will not come out.

42050_VITA_G0044

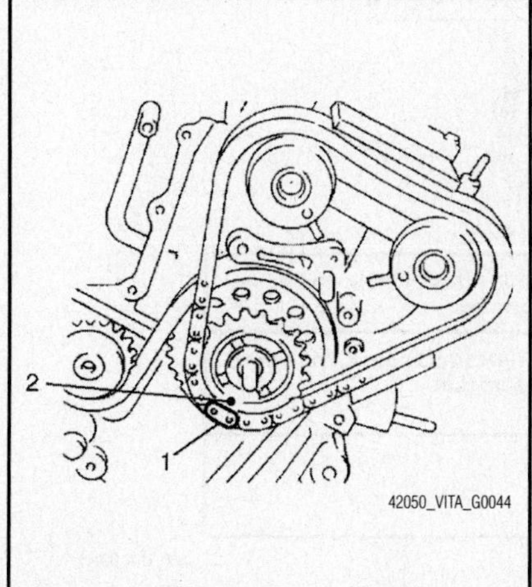

Fig. 165 Left cylinder head match marks—2.7L

50. Install timing chain tensioner adjuster No.3.

51. Tighten Timing chain tensioner adjuster No.3 bolt (c) to 7 ft. lbs. (11 Nm).

52. Tighten Timing chain tensioner adjuster No.3 nut (d) to 32 ft. lbs. (45 Nm).

53. Tightening order for tensioner adjuster bolt and nuts1–2–3

54. Pull out stopper (pin) from set hole.

55. Turn crankshaft two rotations clockwise then align timing pulley key (1) on crankshaft and oil jet on cylinder block.

56. Recheck all timing marks.

57. Apply oil to timing chains, tensioner adjusters sprockets and guides.

58. Install or connect the following:

- Front cover. Tighten the bolts to 97 inch lbs. (11 Nm).
- CKP sensor
- Front crankshaft seal
- Crankshaft pulley. Tighten the bolt to 109 ft. lbs. (148 Nm).
- Oil pan and pickup tube
- Power steering pump and brackets
- Radiator
- Water pump pulley
- Cooling fan and shroud
- Accessory drive belts
- Valve covers
- Ignition coils
- Intake manifold
- Negative battery cable

42050_VITA_G0045

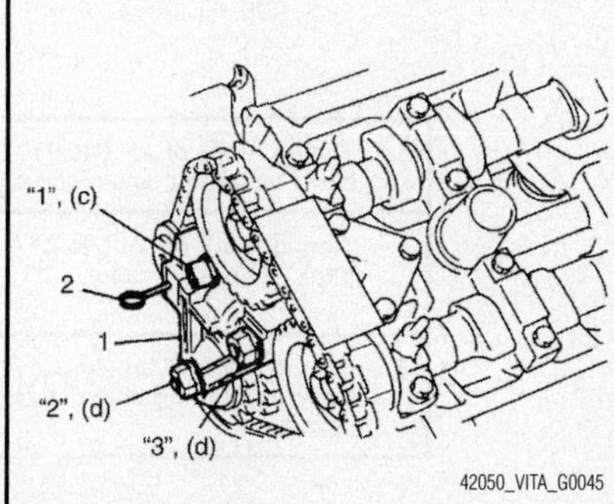

Fig. 166 Timing chain tensioner adjuster No.3 bolt c–d—2.7L

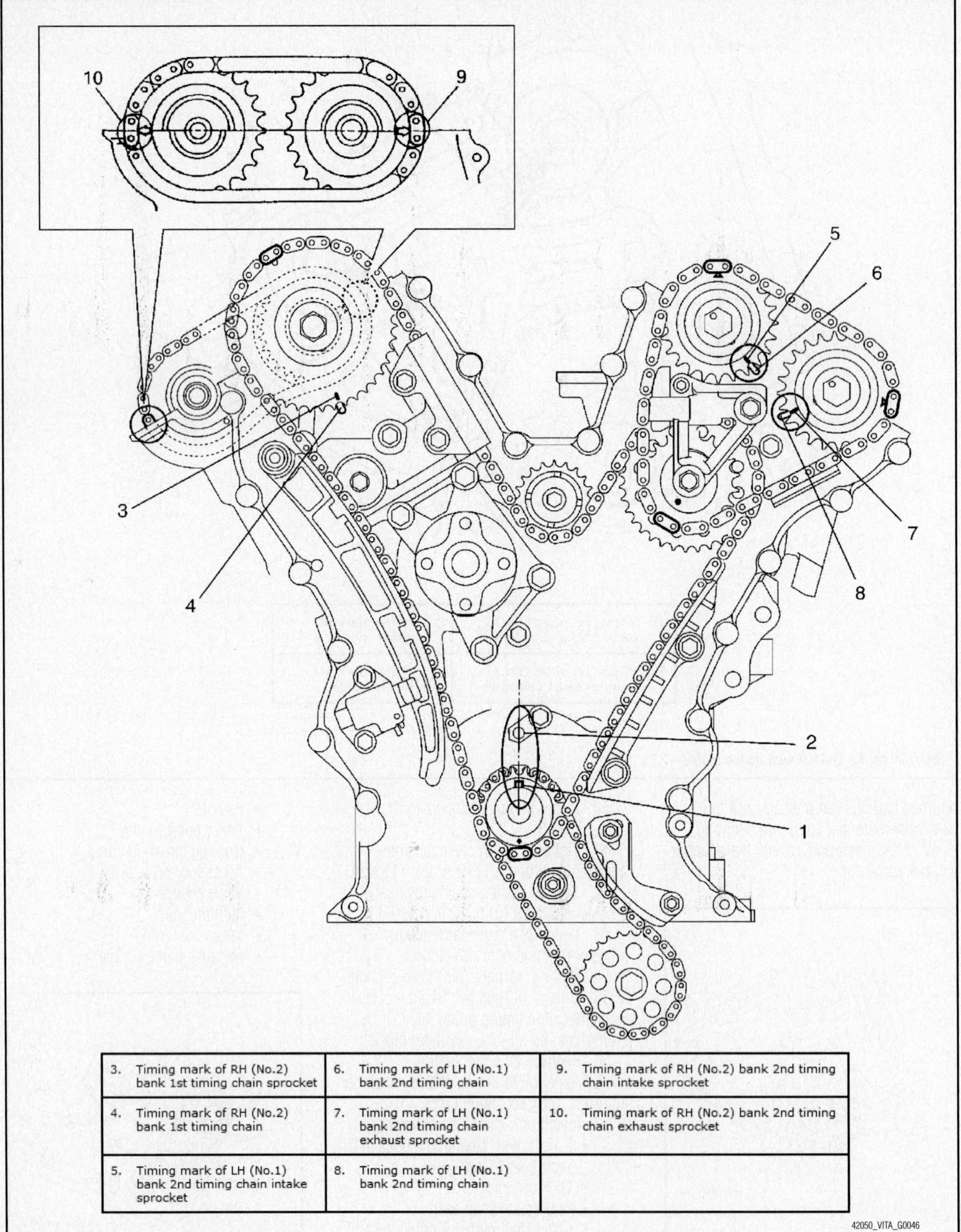

3. Timing mark of RH (No.2) bank 1st timing chain sprocket	6. Timing mark of LH (No.1) bank 2nd timing chain	9. Timing mark of RH (No.2) bank 2nd timing chain intake sprocket
4. Timing mark of RH (No.2) bank 1st timing chain	7. Timing mark of LH (No.1) bank 2nd timing chain exhaust sprocket	10. Timing mark of RH (No.2) bank 2nd timing chain exhaust sprocket
5. Timing mark of LH (No.1) bank 2nd timing chain intake sprocket	8. Timing mark of LH (No.1) bank 2nd timing chain	

Fig. 167 All timing marks shown—2.7L(a)

3. Timing mark of RH (No.2) bank 1st timing chain sprocket	6. Timing mark of LH (No.1) bank 2nd timing chain	9. Timing mark of RH (No.2) bank 2nd timing chain intake sprocket
4. Timing mark of RH (No.2) bank 1st timing chain	7. Timing mark of LH (No.1) bank 2nd timing chain exhaust sprocket	10. Timing mark of RH (No.2) bank 2nd timing chain exhaust sprocket
5. Timing mark of LH (No.1) bank 2nd timing chain intake sprocket	8. Timing mark of LH (No.1) bank 2nd timing chain	

42050_VITA_G0047

Fig. 168 All timing marks key list—2.7L(b)

59. Install engine assembly
60. Fill the crankcase to the correct level.
61. Fill the cooling system.
62. Start the engine and check for leaks.

3.6L Engines

See Figures 169 through 173.

1. Before servicing the vehicle, refer to all service precautions.
2. Relieve the fuel system pressure.
3. Drain the cooling system and engine oil.
4. Evacuate A/C system.
5. Remove engine refer to the engine assembly section.
6. Remove or disconnect the following:
   - Engine cover
   - Air cleaner outlet duct
   - Positive crankcase ventilation (PCV) line
   - Electronic Throttle Control (ETC) electrical connector
   - PCV line from the top of the intake manifold
   - Evaporative emissions (EVAP) canister purge line
   - Bleed pipe bolts
   - Bleed pipe hose clamp
   - Brake booster vacuum hose
   - Engine harness
   - Ignition coil bolts
   - Ignition coils
   - Camshaft cover bolts
   - Camshaft cover
   - Upper intake retaining bolts.
   - Fuel injectors and fuel rail
   - Lower intake manifold bolts
   - Lower intake manifold
   - Water pump pulley bolts
   - Water pump pulley
   - Water pump bolts
   - Water pump from the front cover
   - Camshaft position sensor bolts
   - Camshaft position sensors
   - Camshaft position actuator valve bolts.
   - Camshaft position actuator valves
   - Front cover bolts that hold the engine front cover deadener into position
   - Front cover bolts

⁂ WARNING

Do not pry between the engine front cover and the camshaft position sensors or the camshaft position actuators in order to separate the RTV. Use the pry points and a bolt in the jackscrew hole in order to remove the engine front cover. Damage to the camshaft position sensors or the camshaft position actuators may occur if the camshaft position sensors or the camshaft position actuators are used to pry against in order to remove the engine front cover.

7. Loosely install a 10 x 1.5 mm bolt in the jackscrew hole (1).
8. Using the pry points (2) located at the edge of the front cover and the jackscrew, separate the room temperature vulcanizing (RTV) sealant
9. Remove the engine front cover.

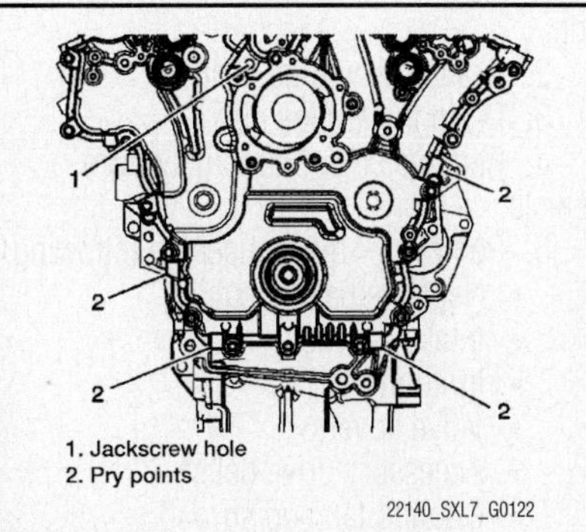

Fig. 169 Engine front cover bolts and pry points

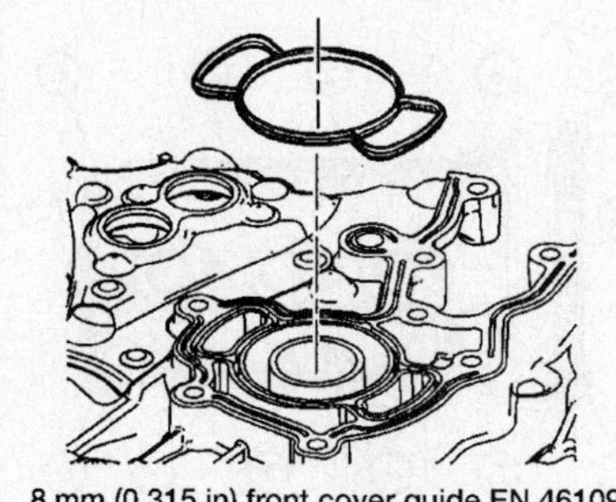

Fig. 170 0.315 inch (8 mm) front cover installation guide EN 46109

To install:

10. Install the 0.315 inch (8 mm) guide from the EN 46109 into the cylinder block positions as shown.
11. Place a 0.118 inch (3 mm) bead of RTV sealant, GM P/N 12378521 (Canadian P/N 88901148) or equivalent, on the engine.
12. Install the engine front cover and front cover deadener to cylinder block seal.
13. Hand start all the front cover bolts.
14. Tighten the engine front cover bolts in the sequence shown.
15. Tighten the engine front cover bolts in sequence to 17 ft lbs. (23 Nm).

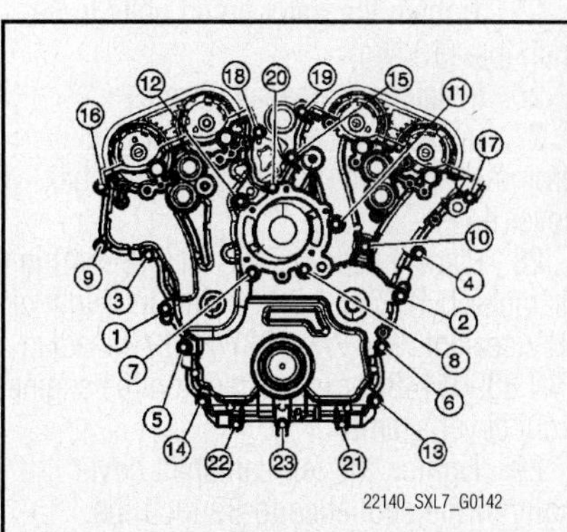

Fig. 171 Engine front cover bolt locations and tightening sequence

Fig. 172 Camshaft position actuator solenoid valves installation

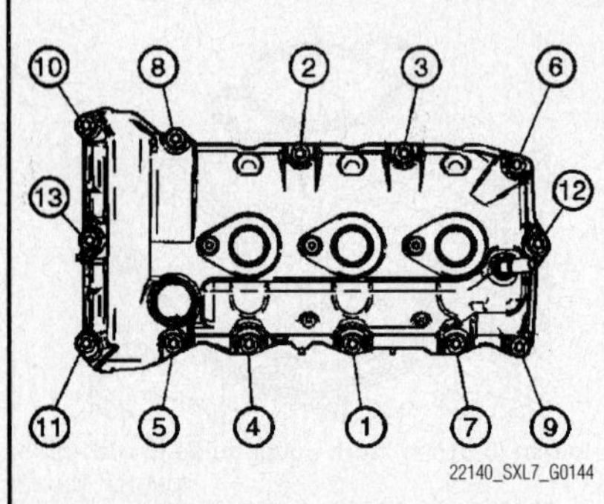

Fig. 173 Camshaft cover bolt tightening sequence

16. Install the camshaft position actuator solenoid valves.
17. Tighten the camshaft position actuators to 89 inch lbs. (10 Nm).
18. Install the crankshaft balancer.
19. Tighten the crankshaft balancer bolt in to 74 ft lbs. (100 Nm).
20. Tighten the crankshaft balancer bolt an additional 150 degrees.
21. Install the power steering pump.
22. Tighten the crankshaft balancer bolt in to 37 ft lbs. (50 Nm).
23. Tighten the power steering gear inlet hose to 18 ft. lbs. (25 Nm).
24. Install the water pump.
25. Tighten the water pump bolts to 89 inch lbs. (10 Nm).
26. Install the camshaft covers.
27. Install new camshaft cover bolt grommets prior to installing the camshaft cover bolts.
28. Place a bead 0.3150 inch (8 mm) in diameter by0.1575 inch (4 mm) in height of RTV sealant, GM P/N 12378521 (Canadian P/N 88901148) or equivalent, on the engine front cover seams.
29. Tighten the left camshaft cover bolts in the sequence to 89 inch lbs. (10 Nm).
30. Install the lower intake manifold.
31. Tighten the lower intake manifold bolts to 17 ft. lbs. (23 Nm)
32. Install the upper intake manifold gaskets to the lower intake manifold and install the fir tree retainers to retain the upper intake manifold gasket position.
33. Install the upper intake manifold.
34. Apply threadlock to the bolt threads.
35. Tighten the lower intake manifold bolts to 18 ft. lbs. (25 Nm).
36. Install the ignition coils.
37. Tighten the ignition coils to 89 inch lbs. (10 Nm).
38. Install the engine harness.
39. Install the throttle body.
40. Install the EVAP purge solenoid valve.
41. Install the brake booster vacuum hose.
42. Install the evaporative emissions (EVAP) canister purge line.
43. Install the bleed pipe bolts (1).
44. Tighten the bleed pipe bolts to 89 inch lbs. (10 Nm).
45. Install the PCV line from the top of the intake manifold.
46. Install the Electronic Throttle Control (ETC) electrical connector.
47. Install the positive crankcase ventilation (PCV) line.
48. Install the air cleaner outlet duct.
49. Install the engine cover.
50. Install the engine assembly refer to the Engine Assembly section.
51. Charge the A/C system.
52. Refill the engine coolant.
53. Run engine and check for leaks

TIMING CHAIN AND SPROCKETS

REMOVAL & INSTALLATION

2.7L Engines

See Figures 71 through 79, 81 and 85 through 162.

1. Before servicing the vehicle, refer to the precautions in the beginning of this section.
2. Drain the cooling system.
3. Drain the engine oil.
4. Remove engine assembly from vehicle.
5. Remove or disconnect the following:
   - Negative battery cable
   - Intake manifold
   - Ignition coils
   - Valve covers
   - Accessory drive belts
   - Cooling fan and shroud
   - Water pump pulley
   - Radiator
   - Power steering pump and brackets
   - Oil pan and pickup tube
   - Crankshaft pulley
   - Front crankshaft seal
   - Crankshaft Position (CKP) sensor
   - Front cover
6. Remove timing chain tensioner adjuster.
7. To remove it, slacken left (No.1) bank 2nd timing chain by turning intake camshaft counterclockwise a little while pushing back pad.
8. Remove left (No.1) bank intake and exhaust camshaft sprocket bolts.
9. To remove it, fit a spanner to hexagonal part at the center of camshaft to hold it stationary.
10. Remove left (No.1) bank exhaust camshaft sprocket.
11. Remove left (No.1) bank intake camshaft sprocket.
12. Remove left (No.1) bank 2nd timing chain.
13. Remove or disconnect the following.
   - Timing chain guide No.1 and No.2
   - Timing chain tensioner
   - Timing chain tensioner adjuster
   - Idler sprocket No.1 and 1st timing chain
   - Idler sprocket No.2 and sprocket shaft
14. Remove right(No.2) bank 1st timing chain intake camshaft sprocket bolt. To remove it, fit a spanner to hexagonal part at the center of camshaft to hold it stationary.
15. Remove right (No.2) bank 1st timing chain intake camshaft sprocket.
16. Remove 1st timing chain crankshaft sprocket.

To install:

17. Prepare the timing chain tensioners for installation by releasing the latches, compressing the tensioner piston fully into the bore and installing retaining pins.
18. Align the timing chain sprocket matchmarks and colored chain links as shown during assembly.
19. Check timing pulley key of crankshaft is on specified position.
20. Install 1st timing chain crankshaft sprocket to crankshaft.
21. Check timing mark on right (No.2) bank intake camshaft.
22. Install right (No.2) bank 1st timing chain intake camshaft sprocket noting the following points.
   a. The sprocket should be set in such way that its timing marks can be seen.
   b. Camshaft should be held stationary by using a spanner at its hexagonal parts.

23. Tighten right bank 1st timing chain intake camshaft sprocket bolt to 58 ft. lbs. (80 Nm).

24. Install timing chain tensioner.

25. Tighten timing chain tensioner nut to 18 ft. lbs. (25 Nm).

26. Install 1st timing chain by aligning match marks on right silver plate of 1st timing chain and right (No.2) bank 1st timing chain intake camshaft sprocket.

27. Apply oil to bush of idler sprocket No.2.

28. Install idler sprocket No.2 and sprocket shaft.

29. Install idler sprocket No.2 by aligning match marks on left silver plate of 1st timing chain.

30. Install crankshaft sprocket by aligning match marks on yellow plate of 1st timing chain and crankshaft timing sprocket.

31. To install it, fit a spanner to hexagonal part at the center of right (No.2) bank intake camshaft to turn a little.

⁂ WARNING

Do not turn camshaft more than necessary. If turned excessively, valve and piston may get damaged.

32. Apply oil to bearing of idler sprocket No.1.

33. Install idler sprocket No.1 and tighten to 40 ft. lbs. (55 Nm).

34. With latch of tensioner adjuster No.1 returned and plunger pushed back into body, insert stopper into latch and body.

35. After inserting it, check to make sure that plunger will not come out.

36. Install timing chain tensioner adjuster No.1 and tighten to 8 ft. lbs (11 Nm).

37. Pull out stopper (3) from adjuster No.1 (1).

38. Install timing chain guide No.1 and No.2.

39. Tighten both chain guides to 6 ft. lbs. (9 Nm).

40. Recheck each aligned timing marks.

41. Install timing chain guide No.4 and tighten to 8 ft. lbs (11 Nm).

42. Check that knock-pins of intake and exhaust camshafts are aligned with match marks on cylinder head.

43. Install by aligning match marks on yellow plate of left (No.1) bank 2nd timing chain and idler sprocket No.2.

44. Install sprockets to intake and exhaust camshafts by aligning silver plate of left (No.1) bank 2nd timing chain, match marks on intake sprocket and exhaust sprocket respectively.

45. Install LH (No.1) bank intake and exhaust camshaft timing sprockets.

46. To install it, fit a spanner to hexagonal part at the center of camshaft to hold stationary.

47. Tighten camshaft timing sprocket bolts to 58 ft. lbs. (88 Nm).

48. With latch of tensioner adjuster No.3 returned and plunger pushed back into body, insert stopper (pin) into set hole.

49. After inserting it, check that plunger will not come out.

50. Install timing chain tensioner adjuster No.3.

51. Tighten Timing chain tensioner adjuster No.3 bolt (c) to 7 ft. lbs. (11 Nm).

52. Tighten Timing chain tensioner adjuster No.3 nut (d) to 32 ft. lbs. (45 Nm).

53. Tightening order for tensioner adjuster bolt and nuts1–2–3

54. Pull out stopper (pin) from set hole.

55. Turn crankshaft two rotations clockwise then align timing pulley key (1) on crankshaft and oil jet on cylinder block.

56. Recheck all timing marks.

57. Apply oil to timing chains, tensioner adjusters sprockets and guides.

58. Install or connect the following:
- Front cover. Tighten the bolts to 97 inch lbs. (11 Nm).
- CKP sensor
- Front crankshaft seal
- Crankshaft pulley. Tighten the bolt to 109 ft. lbs. (148 Nm).
- Oil pan and pickup tube
- Power steering pump and brackets
- Radiator
- Water pump pulley
- Cooling fan and shroud
- Accessory drive belts
- Valve covers
- Ignition coils
- Intake manifold
- Negative battery cable

59. Install engine assembly

60. Fill the crankcase to the correct level.

61. Fill the cooling system.

62. Start the engine and check for leaks.

3.6L Engines

See Figures 174 through 186.

1. Remove the right secondary camshaft drive chain tensioner bolts.

2. Remove the right secondary camshaft drive chain tensioner.

3. Remove and discard the left secondary camshaft drive chain tensioner gasket.

4. Inspect the right secondary camshaft drive chain tensioner mounting surface on the right cylinder head for burrs or any defects that would degrade the sealing of the NEW right secondary camshaft drive chain tensioner gasket.

5. Remove or disconnect the following:
- Right secondary camshaft drive chain shoe bolts
- Right secondary camshaft drive chain shoes
- Right secondary camshaft drive chain guide bolts
- Right secondary camshaft drive chain guide
- Right camshaft drive chain from the camshaft position actuators

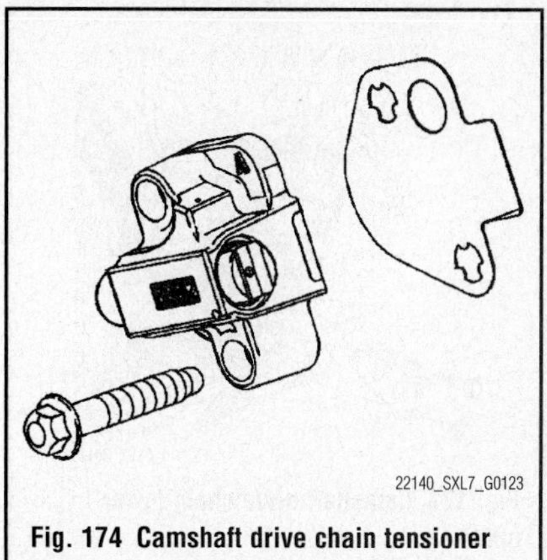

Fig. 174 Camshaft drive chain tensioner

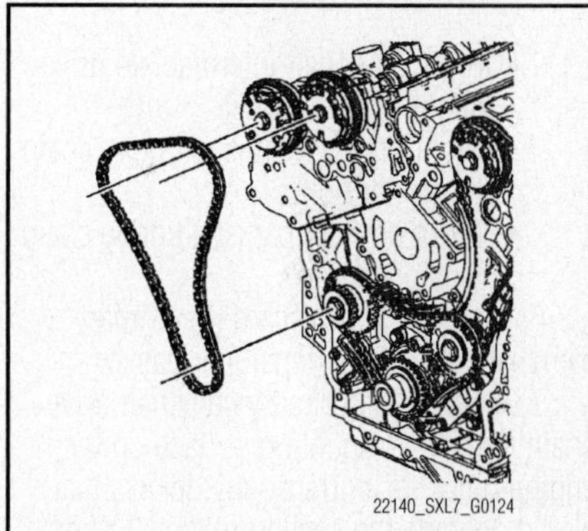

Fig. 175 Camshaft drive chain guide removal

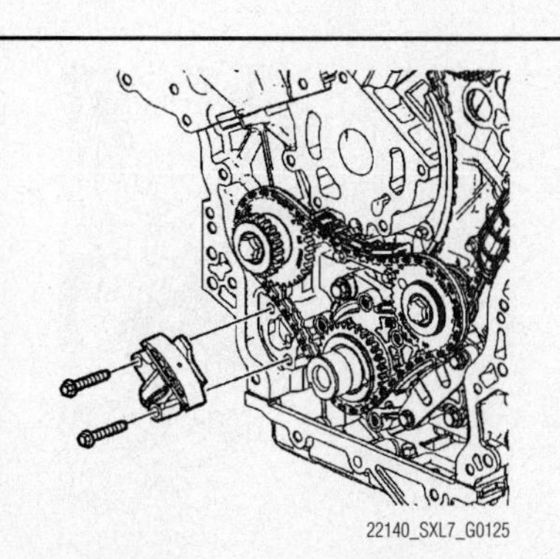

Fig. 176 Primary camshaft drive chain tensioner

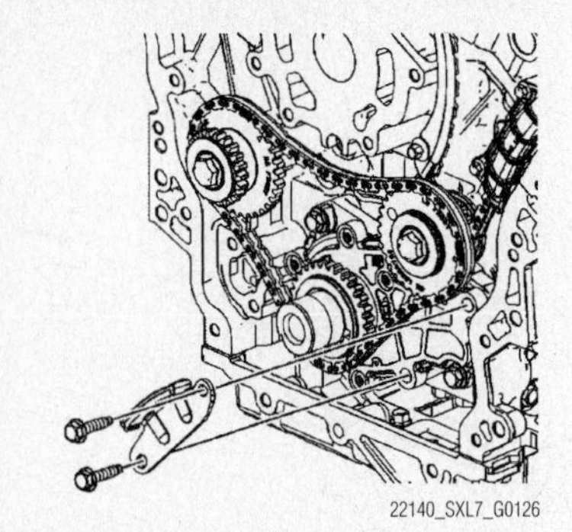

Fig. 177 Camshaft drive chain lower guide

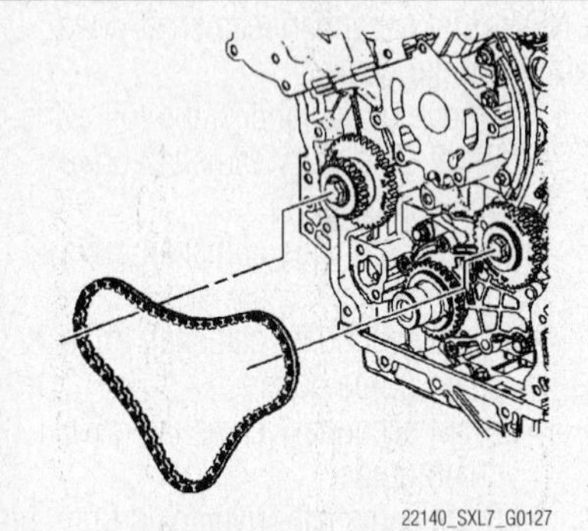

Fig. 178 Camshaft primary camshaft drive chain

- Right camshaft intermediate drive chain idler sprocket.
- Right primary camshaft drive chain tensioner bolts.
- Right primary camshaft drive chain tensioner.

6. Remove and discard the primary camshaft drive chain tensioner gasket.
7. Inspect the primary camshaft drive chain tensioner mounting surface on the engine block for burrs or any defects that would degrade the sealing of the NEW primary camshaft drive chain tensioner gasket.
8. Remove or disconnect the following:
   - Right primary camshaft drive chain upper guide bolts
   - Right primary camshaft drive chain upper guides
   - Right primary camshaft drive chain lower guide bolts
   - Right primary camshaft drive chain lower guide.
   - Right primary camshaft drive chain
   - Right camshaft intermediate drive chain idler bolt
   - Right camshaft intermediate drive chain idler
   - Right secondary camshaft drive chain tensioner bolts
   - Right secondary camshaft drive chain tensioner
9. Discard the right secondary camshaft drive chain tensioner gasket
10. Inspect the right secondary camshaft drive chain tensioner mounting surface on the left cylinder head for burrs or any defects that would degrade the sealing of the NEW left secondary camshaft drive chain tensioner gasket.
11. Remove or disconnect the following:
    - Right secondary camshaft drive chain shoe bolt
    - Right secondary camshaft drive chain shoe
    - Right secondary camshaft drive chain guide bolts
    - Right secondary camshaft drive chain guide
    - Right secondary camshaft drive chain
    - Right camshaft intermediate drive chain idler bolt
    - Right camshaft intermediate drive chain idler
    - Crankshaft sprocket from the nose of the crankshaft
12. Remove the right intake and right exhaust camshaft position actuator bolts.

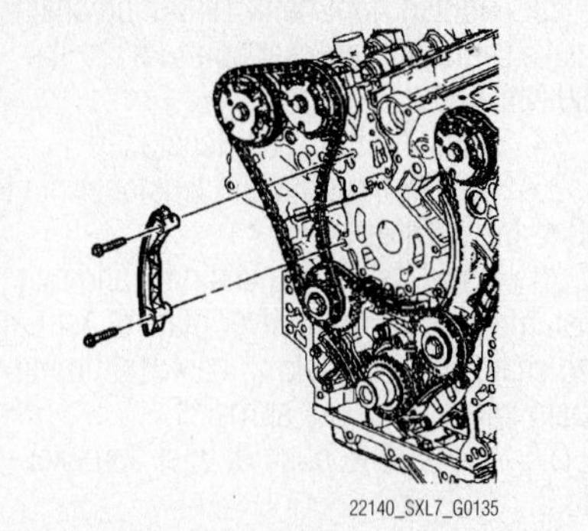

Fig. 181 Right secondary camshaft drive chain shoe

Fig. 182 Left camshaft intermediate drive chain idler

➡Use an open wrench on the hex cast into the camshafts in order to prevent engine rotation when loosening the camshaft position actuator bolts.

13. Remove the right intake and right exhaust, camshaft position actuators.

To install:

14. Install or connect the following:
    - Right bank secondary camshaft drive chain. Torque to 17 ft. lbs. (23 Nm).
    - Right bank camshaft intermediate drive chain idler

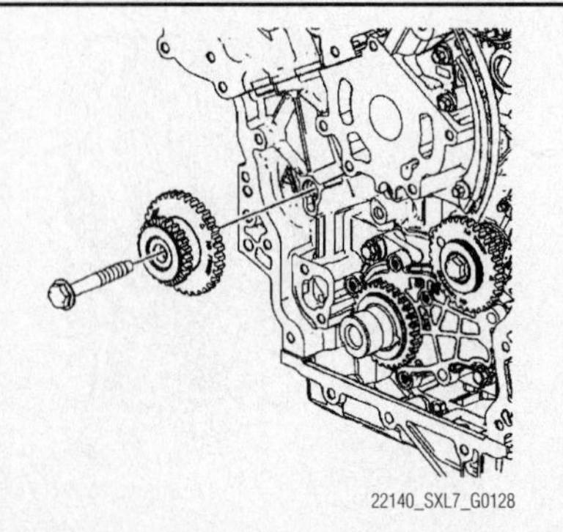

Fig. 179 Camshaft intermediate drive chain idler

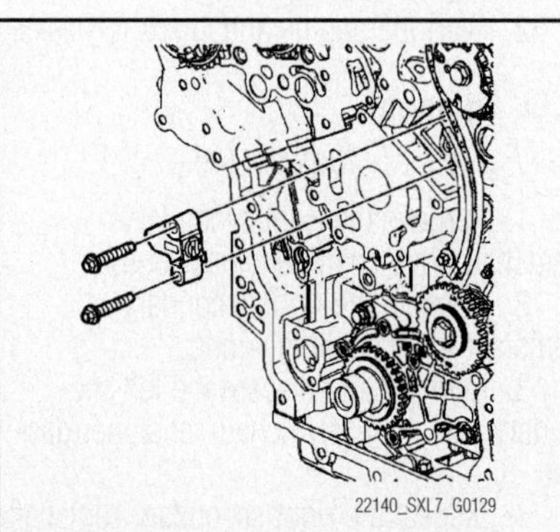

Fig. 180 Left secondary camshaft drive chain tensioner

Fig. 183 Crankshaft sprocket

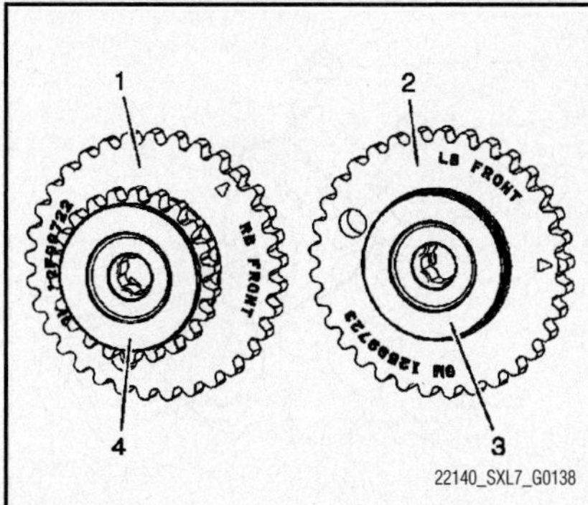

Fig. 184 Right camshaft intermediate drive chain idler orientation

➡Ensure that the right camshaft intermediate drive chain idler (1) is being installed. The recessed hub (4) and the smaller sprocket of the right camshaft intermediate drive chain idler is installed outward. The raised hub and the larger sprocket of the right camshaft intermediate drive chain idler is installed towards the block.

- Intermediate drive chain idler bolt.
- Torque the camshaft intermediate drive chain idler bolt to 43 ft. lbs. (58 Nm)
- Right bank secondary camshaft drive chain guide
- Right bank secondary camshaft drive chain shoe
- Right bank secondary camshaft drive chain tensioner
- Primary camshaft drive chain
- Right bank secondary camshaft drive chain

15. Ensure the proper camshaft position actuator is installed. Observe the body of the camshaft position actuator for the EX marking (1).
16. The marking is for an exhaust camshaft position actuator.

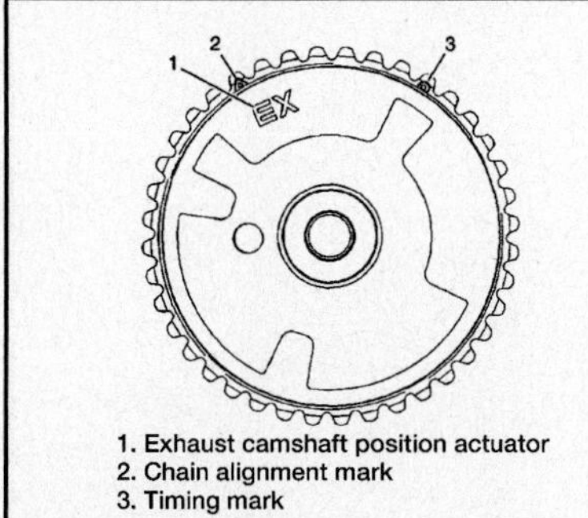

Fig. 185 Camshaft position actuator alignment—Exhaust

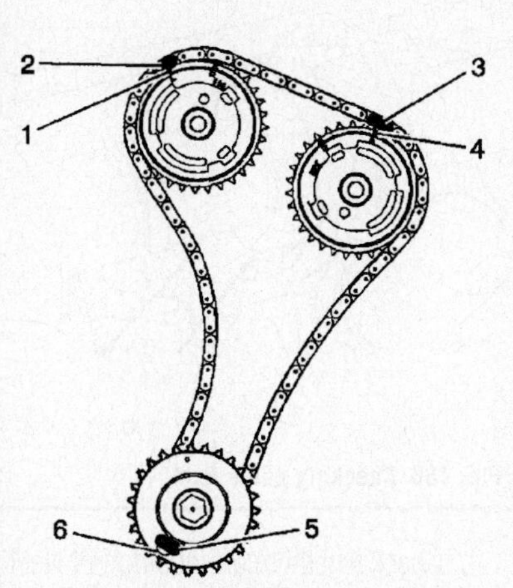

Fig. 186 Camshaft chain alignment—Exhaust

17. Ensure the proper timing mark is used. Observe the outer ring of the camshaft position actuator for the circle marking (3).
18. The marking is for alignment to the highlighted timing chain link on the left side of the engine.
19. Align the exhaust camshaft actuator alignment mark (1) to the timing chain alignment mark (2) made during disassembly.
20. Ensure that the intake camshaft actuator alignment mark (4) and the timing chain alignment mark (3) are also aligned.
21. Position the exhaust camshaft actuator to the camshaft and install the actuator bolt hand tight.
22. Tighten the exhaust camshaft position actuator bolt.
23. Tighten the bolt to 43 ft. lbs. (58 Nm).

➡Use an open wrench on the hex cast into the camshaft in order to prevent engine rotation when tightening the camshaft position actuator bolt.

24. Install the right intake and exhaust camshaft position actuators.
25. Install the right intake and exhaust camshaft position actuator bolts.

VALVE COVERS

REMOVAL & INSTALLATION

2.7L Engines

See Figure 187.

➡The valve covers are also referred to as camshaft covers.

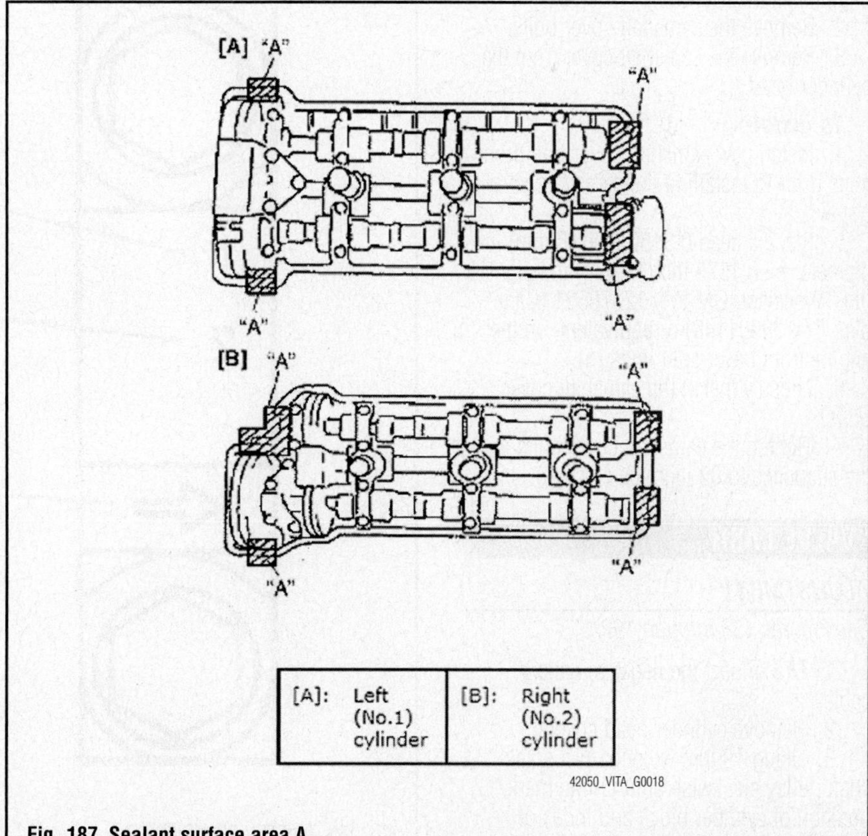

Fig. 187 Sealant surface area A

1. Disconnect the negative battery cable.
2. Remove throttle body and intake manifold.
3. Remove ignition coil covers.
4. Remove camshaft covers.
5. Clean sealing surfaces on cylinder heads and covers.
6. Remove oil, old sealant, and dust from sealing surfaces. After cleaning, apply sealant A to cylinder heads sealing surface area as shown in figure.

To install:

7. Install new camshaft cover gaskets to camshaft covers.
8. Install camshaft covers to cylinder heads.
9. Using new seal washers, tighten nuts to 7.5 ft. lbs. (10.5 Nm).
10. Install ignition coils and connect ignition coil couplers.
11. Install ignition coil covers.
12. Install intake manifold with throttle body.
13. Connect the negative battery cable.

3.6L Engines

➡The valve covers are also referred to as camshaft covers.

1. Remove the ignition coil bolts and coils. For more information, refer to the Engine Electrical Section.
2. Remove the camshaft cover bolts.
3. Remove the camshaft cover from the cylinder head.

To install:

4. Install new camshaft cover bolt grommets prior to installing the camshaft cover bolts.
5. Place a bead 0.3150 inch (8 mm) in diameter by 0.1575 inch (4 mm) in height of RTV sealant, GM P/N 12378521 (Canadian P/N 88901148) or equivalent, on the engine front cover split lines (1).
6. Loosely install the camshaft cover bolts.
7. Tighten the camshaft cover bolts in the sequence to 89 inch lbs (10 Nm).

VALVE LASH

ADJUSTMENT

See Figures 188 through 192.

1. Disconnect the negative battery cable.
2. Remove cylinder head covers.
3. Using 19 mm wrench, turn crankshaft pulley clockwise until timing mark "O" position of cylinder block and index of crankshaft pulley are aligned.

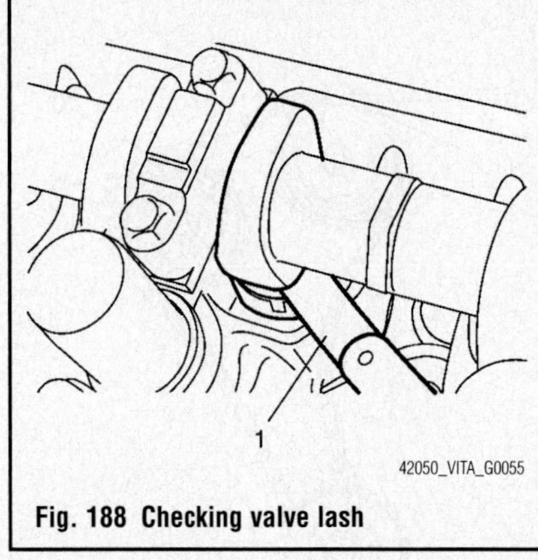

Fig. 188 Checking valve lash

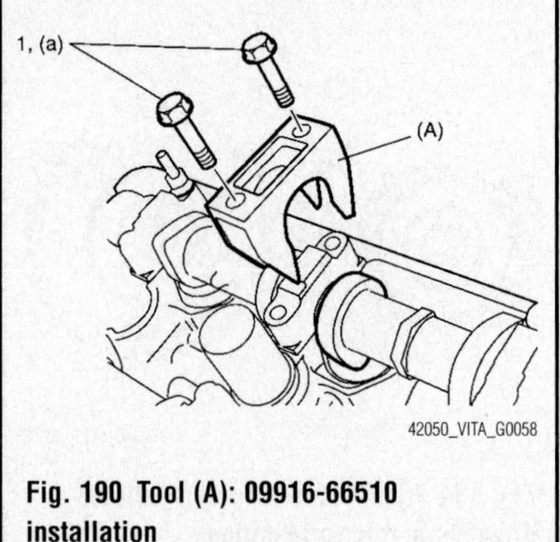

Fig. 190 Tool (A): 09916-66510 installation

4. Check whether cam position of No.1 cylinder is at the specified position [A] as shown in figure.
5. If cam position is [B], locate cam position to [A] by turning crankshaft one rotation.
6. Check valve lashes with thickness gauge according to the following procedure.
   a. Check valve lashes of cylinder No.1.
   b. Turn crankshaft pulley by 120° clockwise.
   c. Check valve lashes of cylinder No.6.
   d. In the same manner as (b)—(c), check valve lashes of cylinder No.5, cylinder No.4, cylinder No.3 then cylinder No.2.
7. If valve lash is out of specification, record valve lash and adjust it to specification by replacing shim.
8. Cold engine intake valve clearance specification 0.007—0.008 inch. (0.18—0.22 mm).
9. Hot engine intake valve clearance specification 0.008—0.011 inch. (0.21—0.27 mm).
10. Cold engine exhaust valve clearance specification 0.011—0.012 inch. (0.28—0.32 mm).

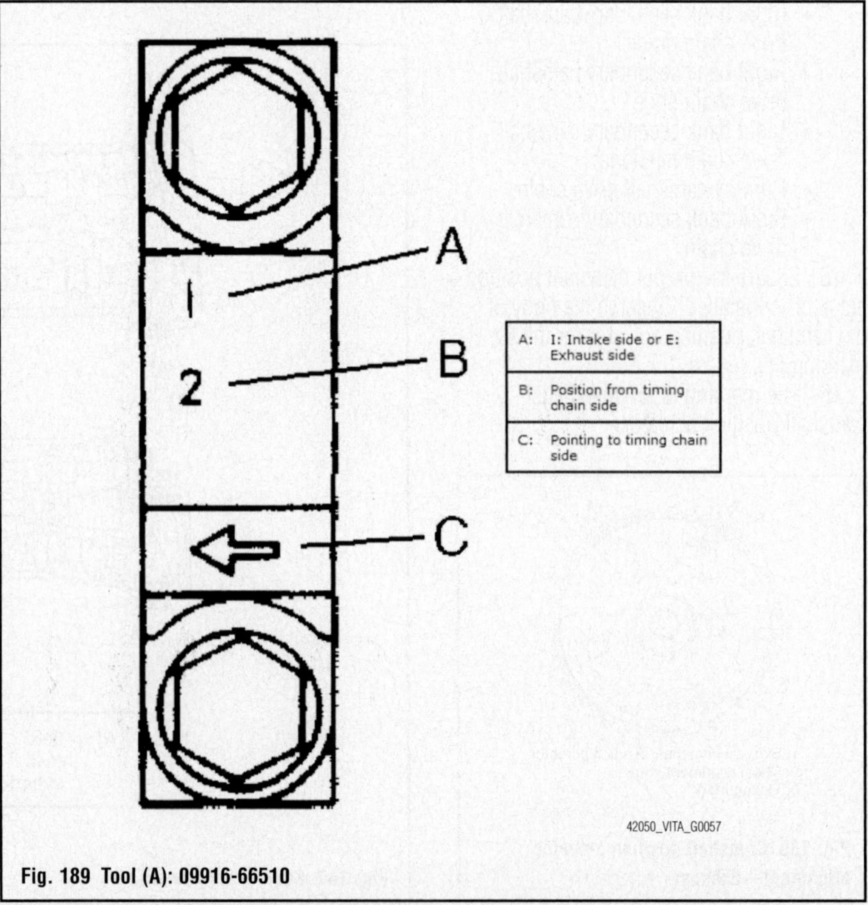

Fig. 189 Tool (A): 09916-66510

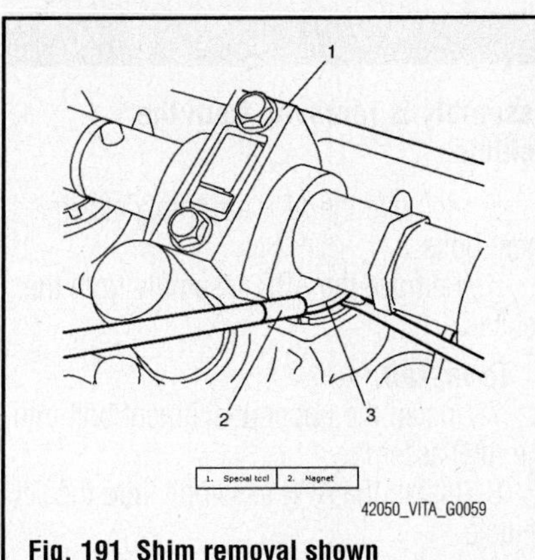

Fig. 191 Shim removal shown

11. Hot engine exhaust valve clearance specification 0.012—0.014 inch. (0.30—0.36 mm).

12. Close the valve whose shim is to be replaced by turning crankshaft, then turn tappet till its cut section faces inside.

13. Lift down the valve by turning crankshaft to 360°.

14. Hold tappet at that position using special tool (A): 09916-66510 as follows.

   a. Remove its housing bolts.

   b. Check housing No. and select special tool corresponding to housing No.

   c. Numbers on camshaft housing are marked for intake and exhaust. Select corresponding tool for intake or exhaust.

15. Hold down the tappet so as not to contact the shim by installing special tool on camshaft housing with housing bolt tighten housing bolts to 7 ft. lbs. (11 Nm).

16. Turn camshaft by approximately 90° clockwise and remove shim.

17. Using a micrometer, measure the thickness of the removed shim, and determine replacement shim by calculating the thickness of new shim with the following formula and table.

- Intake side:
- A = B + C - 0.008 inch. (0.200 mm)
- Exhaust side:
- A = B + C - 0.012 inch. (0.300 mm)
- A: Thickness of new shim
- B: Thickness of removed shim
- C: Measured valve clearance

➡For example of intake side:

- When thickness of removed shim is 0.094 inch. (2.400 mm), and measured valve clearance is 0.018 inch. (0.450 mm).
- A = 0.094 inch. (2.400 mm) + 0.018 inch. (0.450 mm)—0.008 inch. (0.200 mm) = 0.104 inch. (2.650 mm)
- Calculated thickness of new shim = 0.104 inch. (2.650 mm)

18. Select new shim No. (1) with a thickness as close as possible to calculated value.

19. Install new shim facing shim No. side with tappet.

20. Lift valve by turning crankshaft counterclockwise (in opposite direction against above Step and remove special tool.

21. Install camshaft housing and tighten bolts to 8 ft. lbs. (11 Nm).

22. Turn crankshaft pulley more than 4 rotations.

23. Check valve clearance again after adjusting it.

24. After checking and adjusting all valves.

25. Install cylinder head covers.

26. Connect the negative battery cable.

Thickness mm (in.)	Shim No.	Thickness mm (in.)	Shim No.
2.175 (0.0856)	218	2.600 (0.1024)	260
2.200 (0.0866)	220	2.625 (0.1033)	263
2.225 (0.0876)	223	2.650 (0.1043)	265
2.250 (0.0886)	225	2.675 (0.1053)	268
2.275 (0.0896)	228	2.700 (0.1063)	270
2.300 (0.0906)	230	2.725 (0.1073)	273
2.325 (0.0915)	233	2.750 (0.1083)	275
2.350 (0.0925)	235	2.775 (0.1093)	278
2.375 (0.0935)	238	2.800 (0.1102)	280
2.400 (0.0945)	240	2.825 (0.1112)	283
2.425 (0.0955)	243	2.850 (0.1122)	285
2.450 (0.0965)	245	2.875 (0.1132)	288
2.475 (0.0974)	248	2.900 (0.1142)	290
2.500 (0.0984)	250	2.925 (0.1152)	293
2.525 (0.0994)	253	2.950 (0.1161)	295
2.550 (0.1004)	255	2.975 (0.1171)	298
2.575 (0.1014)	258	3.000 (0.1181)	300

42050_VITA_G0060

Fig. 192 Valve lash shim chart

ENGINE PERFORMANCE & EMISSION CONTROLS

ACCELERATOR PEDAL POSITION (APP) SENSOR

LOCATION

The Accelerator Pedal Position (APP) sensor 1 and the APP sensor 2 are located on the accelerator pedal assembly.

OPERATION

2.7L Engines

The ECM monitors the driver demand for acceleration with 2 Accelerator Pedal Position (APP) sensors. The APP sensor 1 signal voltage range is from about 1–4.16 Volts as the accelerator pedal is moved from the rest pedal position to the full pedal travel position. The APP sensor 2 range is from about 0.49–2.08 Volts as the accelerator pedal is moved from the rest pedal position to the full pedal travel position. The ECM processes this information along with other sensor inputs to command the throttle plate to a certain position.

3.6L Engines

The Accelerator Pedal Position (APP) sensor provides driver demand for acceleration to the ECM by 2 APP sensors. The APP sensor 1 signal voltage range is from about 0.98-4.16 volts as the accelerator pedal is moved from the rest pedal position to the full pedal travel position. The APP sensor 2 range is from about 0.49-2.08 volts as the accelerator pedal is moved from the rest pedal position to the full pedal travel position. The ECM processes this information along with other sensor inputs to command the throttle plate to a certain position.

REMOVAL & INSTALLATION

2.7L Engines

See Figure 193.

1. Disconnect negative cable at battery.
2. Disconnect connector from Accelerator Pedal Position (APP) sensor assembly.
3. Remove Accelerator Pedal Position (APP) sensor assembly from its bracket

To install:

4. Install the Accelerator Pedal Position (APP) sensor assembly to its bracket.
5. Tighten Accelerator Pedal Position (APP) sensor assembly upper nut (1) first and then lower nuts (2) to 4.5 ft. lbs. (6 Nm).
6. Attach the connector to Accelerator Pedal Position (APP) sensor assembly securely.

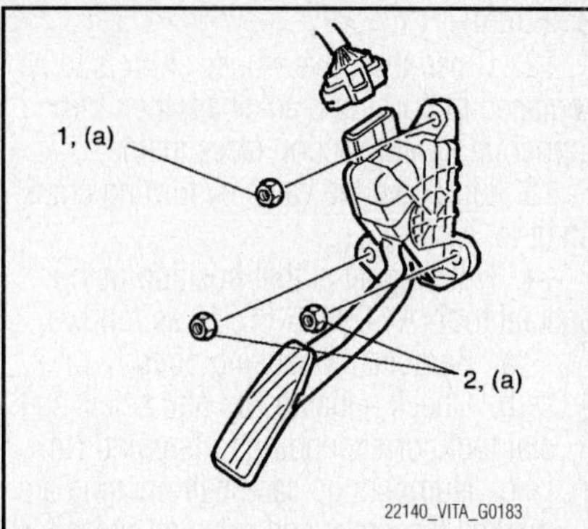

Fig. 193 Accelerator Pedal Position (APP) sensor

7. Connect the negative battery cable.

3.6L Engines

See Figure 194.

1. Remove the instrument panel left lower closeout panel.
2. Remove the driver knee bolster.
3. Remove the Connector Position Assurance (CPA) from the Accelerator Pedal Position (APP) sensor connector.
4. Disconnect the APP sensor harness connector.

➡Due to clearance issues, the upper attachment bolt cannot be removed from the accelerator pedal assembly. Loose then bolt completely and leave the bolt in the component until the assembly is removed from the vehicle.

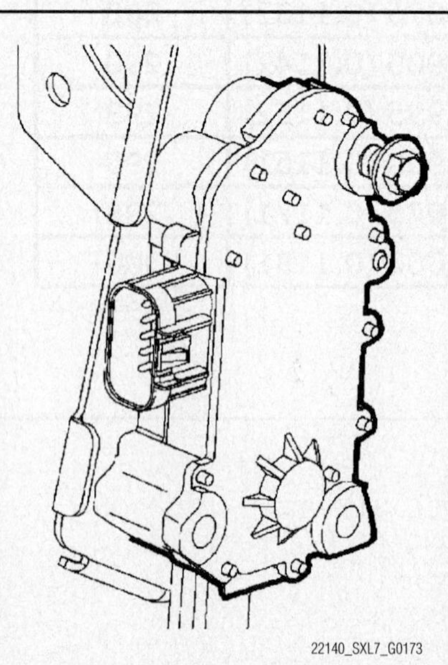

Fig. 194 Accelerator Pedal Position (APP) sensor

5. Remove the APP assembly attachment bolts.
6. Remove the APP assembly from the vehicle.

To install:

7. Install the upper attachment bolt into the APP assembly.
8. Install the APP assembly into the vehicle.
9. Install the remaining attachment bolts into the APP assembly
10. Tighten bolts to 18 ft. lbs. (25 Nm).
11. Connect the APP sensor harness connector. Push the connector in until the lock position is felt, then pull back to confirm engagement.
12. Install the CPA to the APP sensor harness connector.
13. Install the driver knee bolster.
14. Install the instrument panel left lower closeout panel

TESTING

2.7L Engines

See Figures 195 and 196.

Check Accelerator Pedal Position (APP) sensor (main and sub) output voltage as following steps.

1. For Accelerator Pedal Position (APP) sensor (main), arrange 3 new 1.5 V batteries (1) in series (check that total voltage is 4.7–5.0 V) and connect its positive terminal to "Vin 1" terminal (2) and negative terminal to "Ground 1" terminal (3) of sensor. Then using voltmeter, connect positive terminal to "Vout 1" terminal (4) of sensor and negative terminal to battery.
2. For Accelerator Pedal Position (APP) sensor (sub), arrange 3 new 1.5 V batteries (1) in series (check that total voltage is 4.7–5.0 V) and connect its positive terminal to "Vin 2" terminal (2) and negative terminal to "Ground 2" terminal (3) of sensor. Then using voltmeter, connect positive terminal to "Vout 2" terminal (4) of sensor and negative terminal to battery.
3. Measure output voltage variation while accelerator pedal is no depressed and fully depressed as following specification.
4. If sensor voltage is out of specified value or does not vary linearly as the following graph, replace Accelerator Pedal Position (APP) sensor assembly.

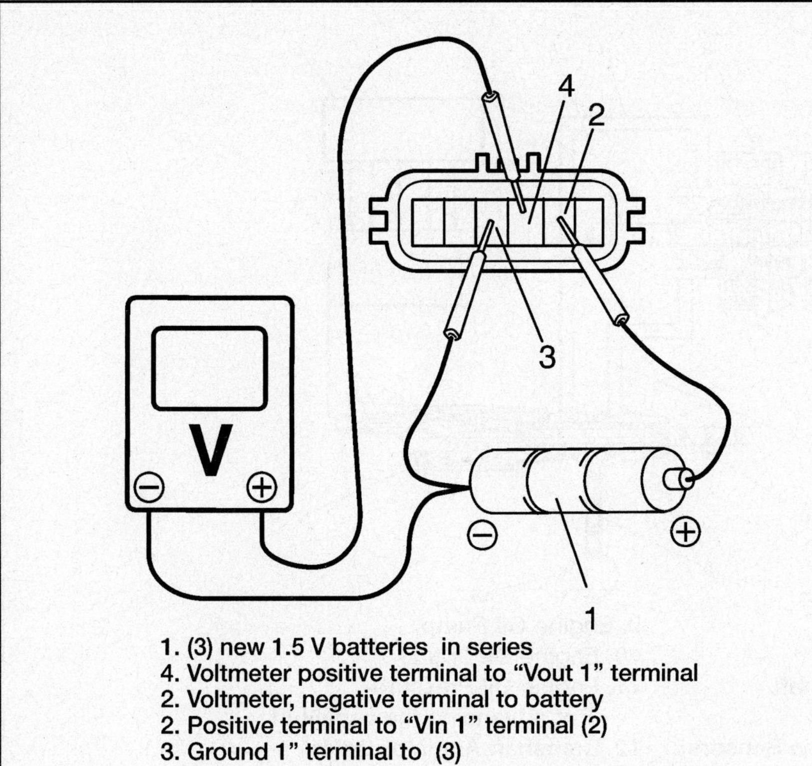

Fig. 195 Accelerator Pedal Position (APP) sensor testing

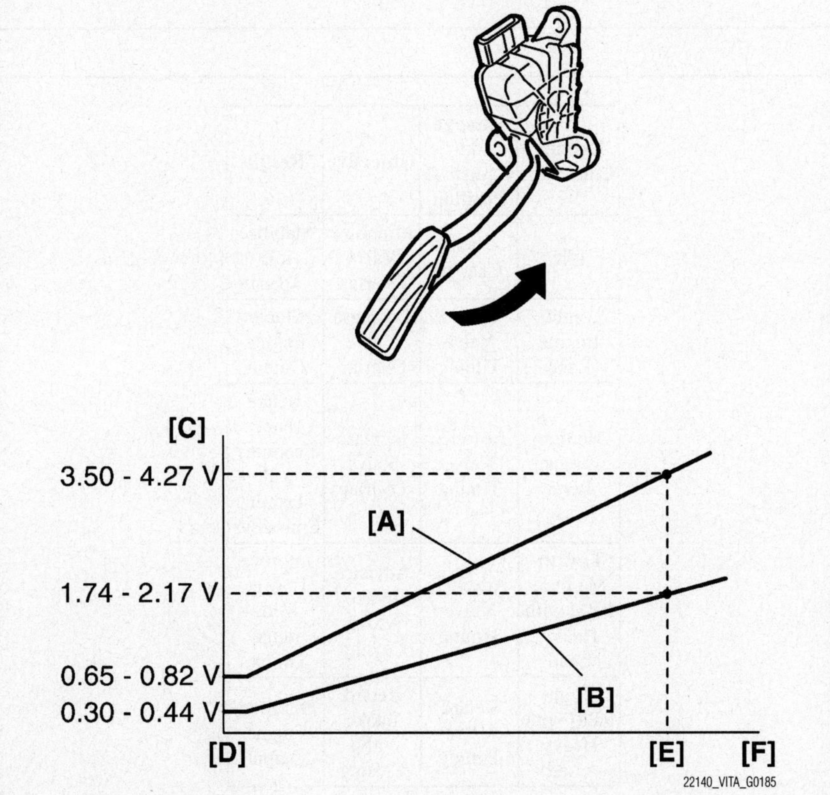

[C]
3.50 - 4.27 V
[A]
1.74 - 2.17 V
0.65 - 0.82 V
0.30 - 0.44 V
[B]
[D] [E] [F]
22140_VITA_G0185

Fig. 196 Accelerator Pedal Position (APP) sensor output voltage variation graph

5. Accelerator Pedal Position (APP) sensor (main) output voltage variation [A]: 0.82–3.50 V at minimum variation, varying according to depressed extent of accelerator pedal

6. Accelerator Pedal Position (APP) sensor (sub) output voltage variation [B]: 0.44–1.74 V at minimum variation, varying according to depressed extent of accelerator pedal

CAMSHAFT POSITION (CMP) ACTUATORS

LOCATION

3.6L Engines

The Camshaft Position (CMP) actuator assembly is driven by an engine timing chain. Inside the assembly is a rotor with fixed vanes that is attached to the camshaft.

OPERATION

3.6L Engines

See Figure 197.

The Camshaft Position (CMP) actuator assembly has an outer housing that is driven by an engine timing chain. Inside the assembly is a rotor with fixed vanes that is attached to the camshaft. Oil pressure that is applied to the fixed vanes will rotate a specific camshaft in relationship to the crankshaft. The movement of the intake camshafts will advance the intake valve timing up to a maximum of 50 crankshaft degrees. The movement of the exhaust camshafts will retard the exhaust valve timing up to a maximum of 50 crankshaft degrees. When oil pressure is applied to the return side of the vanes, the camshafts will return to 0 crankshaft degrees, or Top Dead Center (TDC). The CMP actuator solenoid valve directs the oil flow that controls the camshaft movement. The ECM commands the CMP solenoid to move the solenoid plunger and spool valve until oil flows from the advance passage (11). Oil flowing thru the CMP actuator assembly from the CMP solenoid advance passage applies pressure to the advance side of the vanes in the CMP actuator assembly. When the camshaft position is retarded, the CMP actuator solenoid valve directs oil to flow into the CMP actuator assembly from the retard passage (3). The ECM can also command the CMP actuator solenoid valve to stop oil flow from both passages in order to hold the current camshaft position.

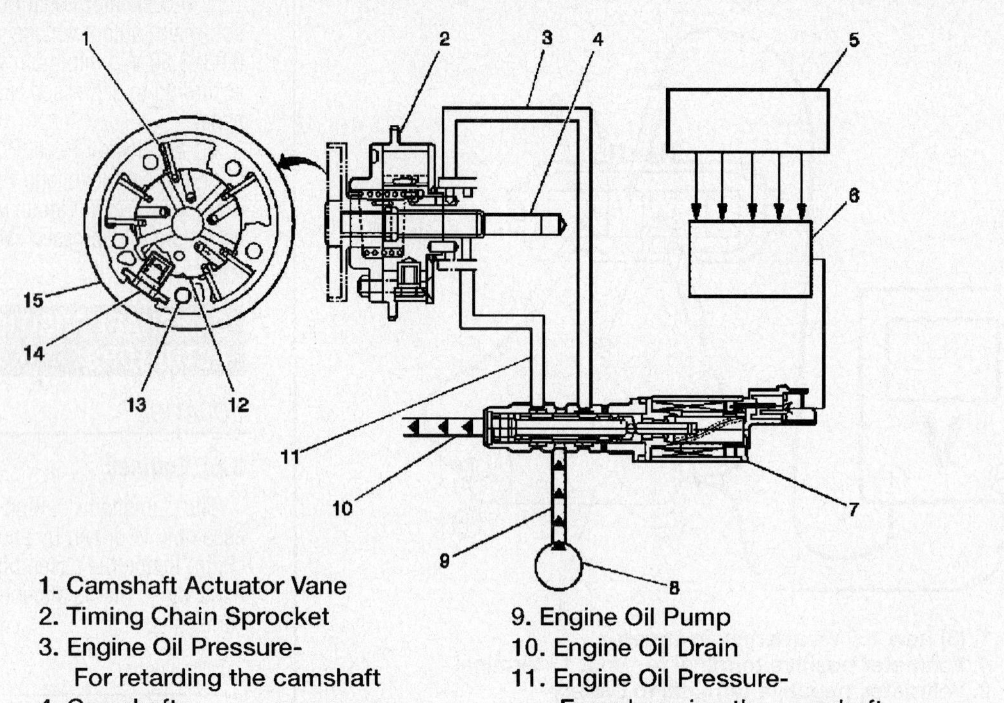

1. Camshaft Actuator Vane
2. Timing Chain Sprocket
3. Engine Oil Pressure- For retarding the camshaft
4. Camshaft
5. Input Signals from Engine Sensors
6. Engine Control Module (ECM)
7. Camshaft Actuator Solenoid
8. Engine Oil Pump
9. Engine Oil Pump
10. Engine Oil Drain
11. Engine Oil Pressure- For advancing the camshaft
12. Camshaft Actuator Rotor
13. Camshaft Position Sensor Reluctor
14. Camshaft Actuator Lock Pin
15. Camshaft Actuator Housing

22140_SXL7_G0174

Fig. 197 Camshaft actuator system overview

The ECM operates the CMP actuator solenoid valve by pulse width modulation (PWM) of the solenoid coil. The higher the PWM duty cycle, the larger the change in camshaft timing. The CMP actuator assembly also contains a lock pin (14) that prevents movement between the outer housing and the rotor vane assembly. The lock pin is released by oil pressure before any movement in the CMP actuator assembly takes place. The ECM is continuously comparing CMP sensor inputs with CKP sensor input in order to monitor camshaft position and detect any system malfunctions. If a condition exists in either the intake or exhaust camshaft actuator system, the opposite bank, intake or exhaust, camshaft actuator will default to 0 crankshaft degrees.

REMOVAL & INSTALLATION

3.6L Engines

See Figure 198.

For Camshaft actuator removal and installation, refer to the Cylinder Head Removal & Installation procedure in the Engine Section.

Driving Condition	Change in Camshaft Position	Objective	Result
Idle	No Change	Minimize Valve Overlap	Stabilize Idle Speed
Light Engine Load	Retard Valve Timing	Decrease Valve Overlap	Stable Engine Output
Medium Engine Load	Advance Valve Timing	Increase Valve Overlap	Better Fuel Economy with Lower Emissions
Low to Medium RPM with Heavy Load	Advance Valve Timing	Advance Intake Valve Closing	Improve Low to Mid-range Torque
High RPM with Heavy Load	Retard Valve Timing	Retard Intake Valve Closing	Improve Engine Output

22140_SXL7_G0175

Fig. 198 Camshaft actuator connectors

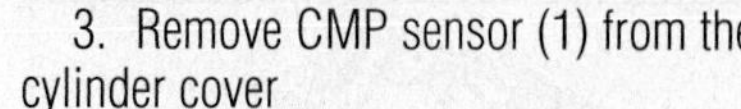

CAMSHAFT POSITION (CMP) SENSOR

LOCATION

2.7L Engines

See Figure 199.

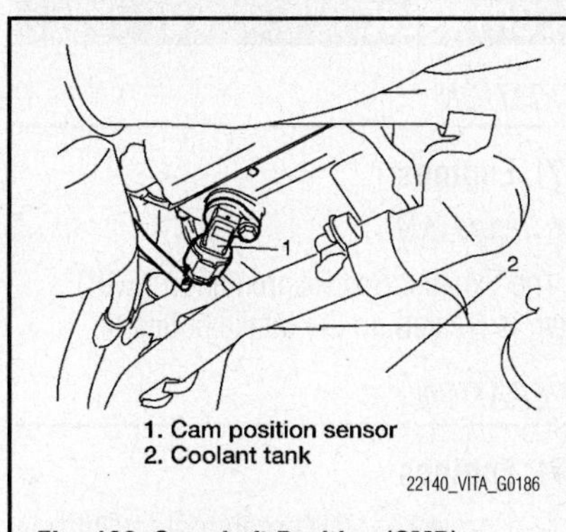

Fig. 199 Camshaft Position (CMP) sensor—2.7L engine

The Camshaft Position (CMP) sensor is located in the side of the cylinder head

3.6L Engines

See Figure 200.

The Camshaft Position (CMP) sensors are located in the front engine cover.

OPERATION

The Camshaft Position (CMP) sensor sends a CMP sensor signal to the Engine Control Module (ECM). The ECM uses this signal as a "sync pulse" to trigger the injectors in the proper sequence. The ECM uses the CMP sensor signal to indicate the position of the #1 piston during its power stroke. This allows the ECM to calculate true sequential fuel injection mode of operation. If the ECM detects an incorrect CMP sensor signal while the engine is running, a DTC will set. If the CMP sensor signal is lost while the engine is running, the fuel injection system will shift to a calculated sequential fuel injection mode based on the last fuel injection pulse, and the engine will continue to run. As long as the fault is present, the engine can be restarted. It will run in the calculated sequential mode with a 1-in-6 chance of the injector sequence being correct.

REMOVAL & INSTALLATION

2.7L Engines

See Figure 199.

1. Disconnect negative cable at battery.
2. Disconnect connector from CMP sensor.
3. Remove CMP sensor (1) from the cylinder cover

To install:

4. Reverse removal procedure noting the following.
5. Tighten CMP sensor bolt to 8 ft. lbs. (11 Nm).
6. Connect CMP sensor connector securely.

3.6L Engines

See Figure 200.

1. Remove the air cleaner assembly.
2. Disconnect the engine wiring harness electrical connector from the bank 1 exhaust Camshaft Position (CMP) sensor.
3. Remove the CMP sensor bolt.
4. Remove the CMP sensor.

To install:

5. Install the CMP sensor.
6. Install the CMP sensor bolt and torque to 89 inch lbs. (10 Nm).
7. Connect the engine wiring harness electrical connector (1) to the bank 1 exhaust CMP sensor.
8. Install the air cleaner assembly.

Fig. 200 Camshaft Position (CMP) sensor connectors and locations—3.6L engines

CRANKSHAFT POSITION (CKP) SENSOR

LOCATION

2.7L Engines

See Figure 201.

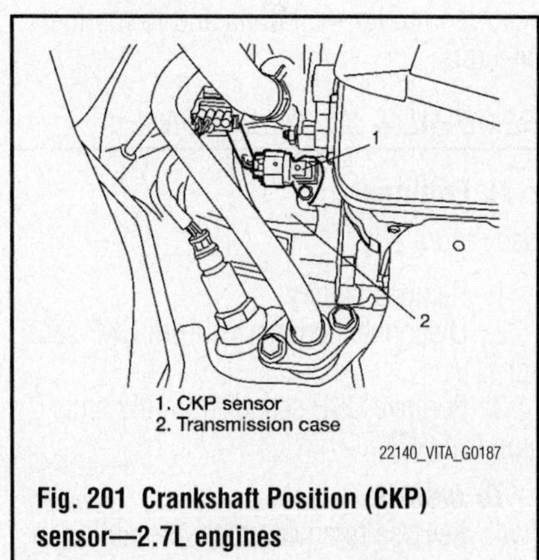

Fig. 201 Crankshaft Position (CKP) sensor—2.7L engines

The Crankshaft Position (CKP) sensor is located in the transmission housing.

3.6L Engines

See Figure 202.

The Crankshaft Position (CKP) sensor is located in the right side of the engine block.

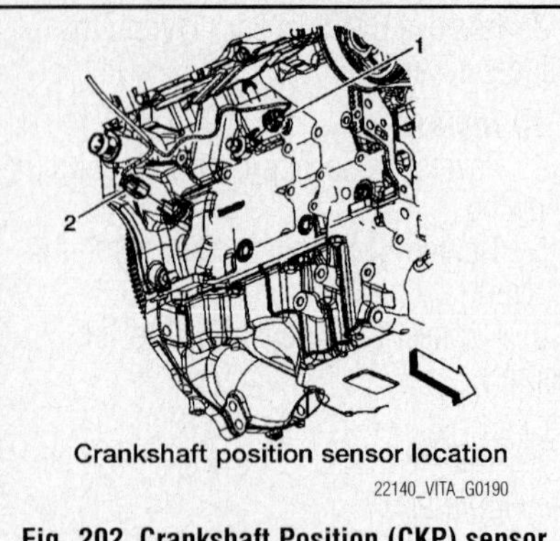

Fig. 202 Crankshaft Position (CKP) sensor location—3.6L engines

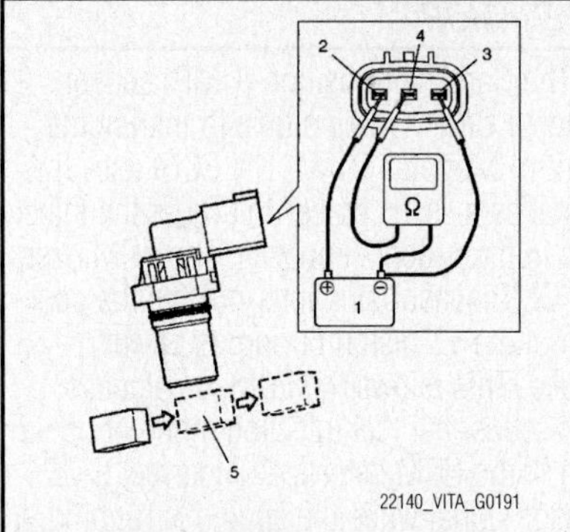

Fig. 203 Crankshaft Position (CKP) sensor—3.6L engine

OPERATION

These vehicles have a direct ignition system which uses a magnetic Crankshaft Position (CKP) sensor. This sensor protrudes through its mount to within approximately 0.05 inches. (1.3 mm) of the crankshaft reluctor. The reluctor is a special wheel attached to the crankshaft or crankshaft pulley with slots machined into it, which are equally spaced in 6° intervals. The last slot is wider and serves to generate a "sync pulse". As the crankshaft rotates, the slots in the reluctor change the magnetic field of the sensor, creating an induced voltage pulse. The longer pulse of the last slot identifies a specific orientation of the crankshaft and allows the Engine Control Module (ECM) to determine the crankshaft orientation at all times. The ECM uses this information to generate timed ignition and injection pulses that it sends to the ignition coils and to the fuel injectors.

REMOVAL & INSTALLATION

2.7L Engines

See Figure 201.

1. Remove battery.
2. Disconnect connector from CKP sensor (1).
3. Remove CKP sensor from transmission case (2).

To install:

4. Reverse removal procedure noting the following:
5. Tighten CKP sensor bolt to 8 ft. lbs. (11 Nm).
6. Attach the connector to the CKP sensor securely.

3.6L Engines

See Figure 203.

1. Raise and support the vehicle

➡If equipped with all wheel drive (AWD), remove the transfer case.

2. Disconnect the engine wiring harness electrical connector (2) from the crankshaft position (CKP) sensor.
3. Remove the crankshaft sensor bolt.
4. Remove the crankshaft sensor.

To install:

5. Install the crankshaft position sensor.
6. Install the crankshaft position sensor bolt.
7. Tighten the crankshaft position sensor bolt to 89 inch lbs. (10 Nm).
8. Connect the engine wiring harness electrical connector to the CKP sensor.

➡If equipped with all AWD, install the transfer case.

TESTING

1. Using a voltmeter, back probe the Crankshaft Position (CKP) sensor connector.
2. Observe the voltage while cranking the engine.
3. Voltage should be between 1.3–1.6 volts.
4. If voltage is out of range replace CKP sensor.

EXHAUST GAS RECIRCULATION (EGR) VALVE

LOCATION

2.7L Engines

See Figure 204.

The Exhaust Gas Recirculation (EGR) valve is located on the intake collector.

OPERATION

2.7L Engines

1. Connect SUZUKI scan tool to DLC (1) with ignition switch turned **OFF**.
2. Start engine and warm it up to normal operating temperature.
3. With engine idling (without depressing accelerator pedal), open EGR valve by using "MISC. TEST" mode.
4. In this state, the EGR valve opening increases as engine idle speed drops. If not, a possible cause is a clogged EGR gas passage or a stuck or faulty EGR valve.

REMOVAL & INSTALLATION

2.7L Engines

See Figure 204.

1. Remove intake collector from intake manifold.

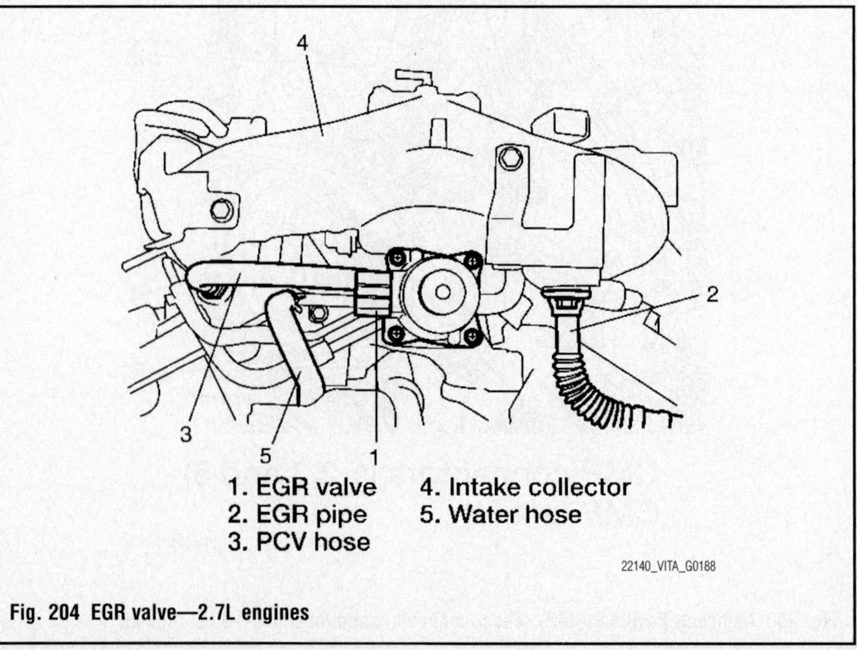

Fig. 204 EGR valve—2.7L engines

2. Remove EGR valve (1) and gasket from intake collector (2).

To install:

3. Install the EGR valve (1) and gasket to the intake collector (2).
4. Torque to 89 inch lbs. (10 Nm).
5. Install the intake collector.

TESTING

2.7L Engines

See Figure 205.

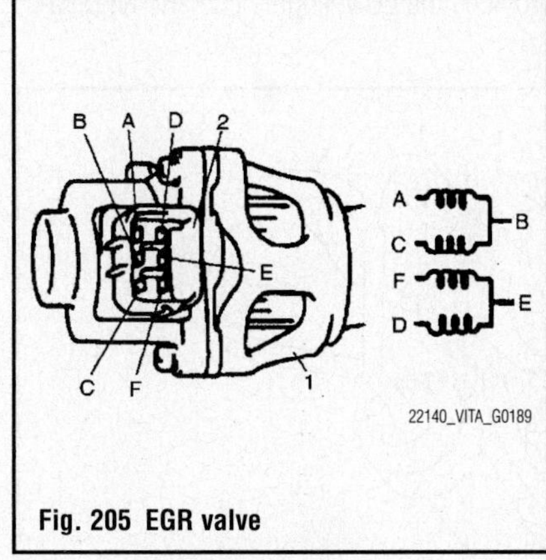

Fig. 205 EGR valve

1. Check resistance between the following terminals of the EGR valve (1) in each pair.
2. EGR valve resistance is as follows:
   - Terminal "A" and terminal "B": 20–24 ohms at 20°C (68°F)
   - Terminal "C" and terminal "B": 20–24 ohms at 20°C (68°F)
   - Terminal "F" and terminal "E": 20 - 24 ohms at 20°C (68°F)
   - Terminal "D" and terminal "E": 20 - 24 ohms at 20°C (68°F)
   - Terminal "B" and valve body: infinity
   - Terminal "E" and valve body: infinity
3. If found faulty, replace the EGR valve assembly.

ELECTRIC FAN SWITCH

The engine cooling fan is operated by a relay that is controlled through the Engine Control Module No switch is used, the Engine Coolant Temperature (ECT) sensor provides the coolant temperature to the ECM and the ECM does the switching of fan operation and speed.

ELECTRONIC CONTROL MODULE (ECM)

If the Engine Control Module (ECM) is replaced with new one (VIN data is still not registered) or with used one (VIN data of other vehicle has been registered previously), make sure to re-register VIN data correctly by performing a VIN Registration. Otherwise, the MIL light remains on even if the engine started, it will cause an inconsistency between registered VIN data in ECM and the actual VIN.

➡For vehicle equipped with immobilizer control systems. If the ECM is replaced with new one or with another one, make sure to register immobilizer transponder code to the ECM.

LOCATION

2.7L Engines

See Figure 206.

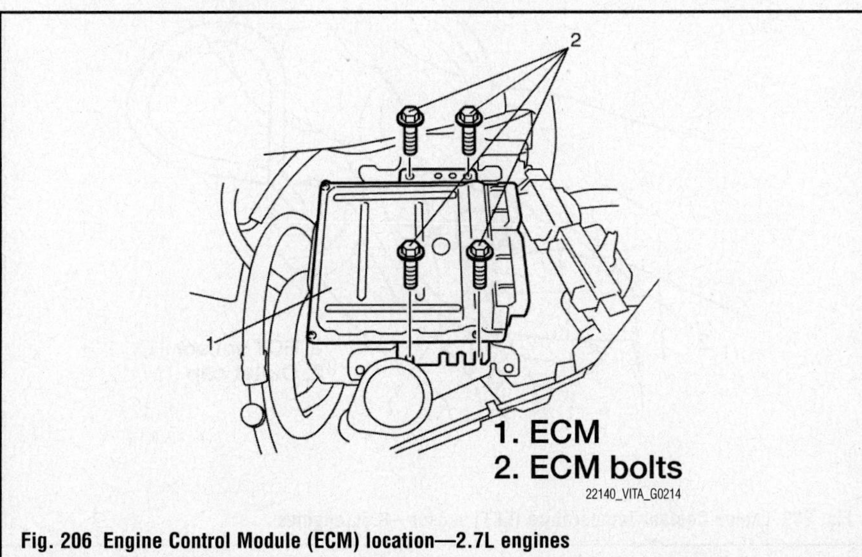

Fig. 206 Engine Control Module (ECM) location—2.7L engines

Refer to the accompanying illustration for Engine Control Module (ECM) location.

3.6L Engines

See Figure 207.

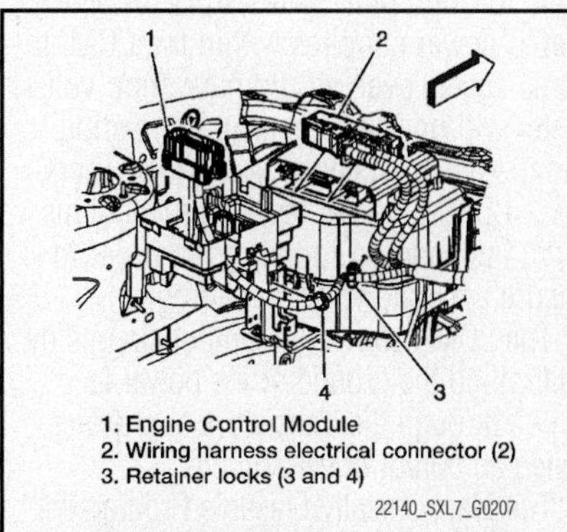

Fig. 207 Engine Control Module (ECM) location—3.6L engines

Refer to the accompanying illustration for Engine Control Module (ECM) location.

OPERATION

The Engine Control Module (ECM) interacts with many more emission related components and systems, and monitors emission related components and systems for deterioration. OBD II diagnostics monitor the system performance and a Diagnostic Trouble Code (DTC) sets if the system performance degrades.

The Malfunction Indicator Lamp (MIL) operation and the DTC storage are dictated by the DTC type. A DTC is ranked as a Type A or Type B if the DTC is emissions related. Type C is a non-emissions related DTC.

The ECM is in the engine compartment. The ECM is the control center of the engine controls system. The ECM controls the following components:

- The fuel injection system
- The ignition system
- The emission control systems
- The on-board diagnostics
- The A/C and fan systems
- The Throttle Actuation Control (TAC) system

The ECM constantly monitors the information from various sensors and other inputs, and controls the systems that affect the vehicle performance and the emissions. The ECM also performs diagnostic tests on various parts of the system. The ECM can recognize operational problems and alert the driver via the MIL.

When the ECM detects a malfunction, the ECM stores a DTC. The condition area is identified by the particular DTC that is set. This aids the technician in making repairs.

The ECM can supply 5 volts or 12 volts to the various sensors or switches. This is done through pull-up resistors to the regulated power supplies within the ECM. In some cases, even an ordinary shop voltmeter will not give an accurate reading because the resistance is too low. Therefore, a DMM with at least 10 megaohms input impedance is required in order to ensure accurate voltage readings.

The ECM controls the output circuits by controlling the ground or the power feed circuit through the transistors or a device called an output driver module.

The Electronically Erasable Programmable Read Only Memory (EEPROM) is a permanent memory that is physically part of the ECM. The EEPROM contains program and calibration information that the ECM needs in order to control the powertrain operation.

Special equipment, as well as the correct program and calibration for the vehicle, are required in order to reprogram the ECM.

REMOVAL & INSTALLATION

1. Disconnect negative cable from battery.
2. Remove ECM cover from bracket.
3. Disconnect connectors from ECM.
4. Remove ECM (1) from its bracket by removing its mounting bolts (2).

To install:

5. Attach the connectors to ECM as follows:
   - Make sure that lock lever (1) of ECM connector is at unlock position.
   - In this state, insert connectors to ECM securely.
   - Lock ECM connectors securely by turning its lock lever to left until it stops.

TESTING

2.7L Engines

For vehicle equipped with immobilizer control systems. If the ECM is replaced with new one or with another one, make sure to register immobilizer transponder code to the ECM.

After the ECM is replaced with new one or used one, the transponder code in the transponder that is built in the ignition key has to be registered with the ECM. To register transponder code in the ignition key with ECM, perform “Replace New ECM” mode of SUZUKI scan tool referring to “SUZUKI Tech 2 Operator’s Manual”.

ENGINE COOLANT TEMPERATURE (ECT) SENSOR

LOCATION

2.7L Engines

See Figure 208.

The Engine Coolant Temperature (ECT) sensor is located in the water outlet cap (2).

3.6L Engines

See Figure 209.

Refer to the accompanying illustration for Engine Coolant Temperature (ECT) sensor location.

OPERATION

2.7L Engines

The Engine Coolant Temperature (ECT) provides the ECM engine coolant tempera-

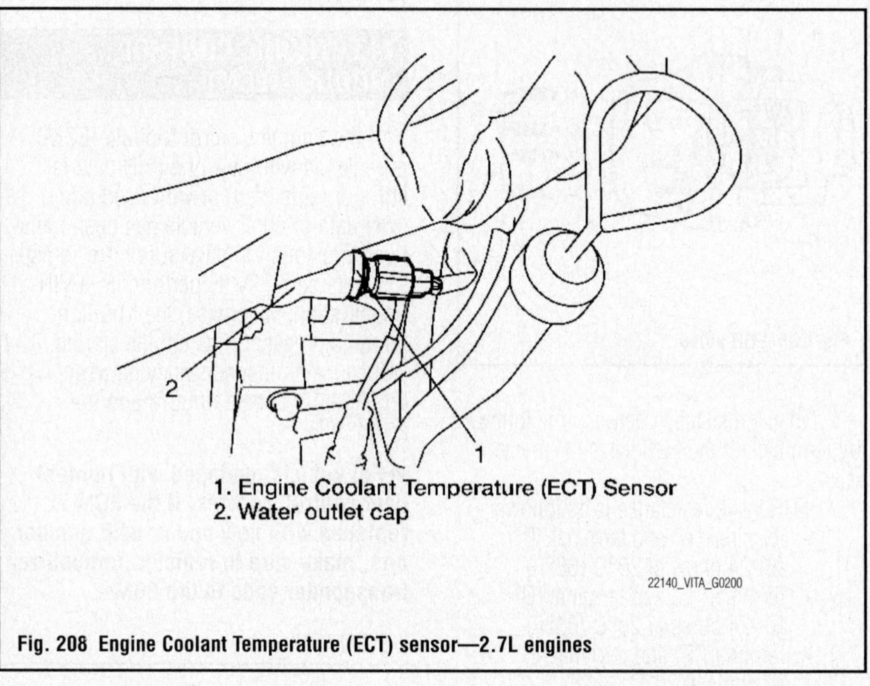

Fig. 208 Engine Coolant Temperature (ECT) sensor—2.7L engines

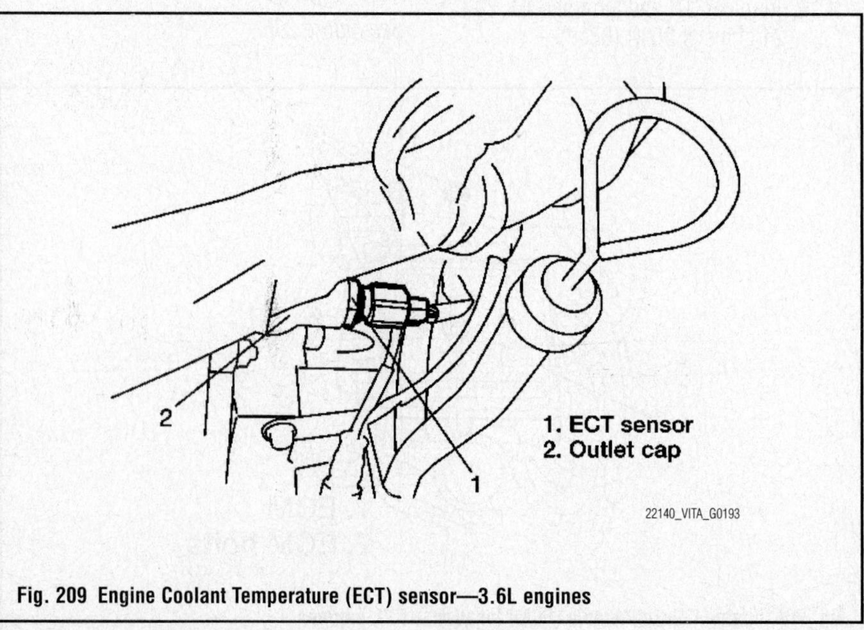

Fig. 209 Engine Coolant Temperature (ECT) sensor—3.6L engines

ture the ECM controls the cooling fans dependent on input from the ECT.

3.6L Engines

The Engine Coolant Temperature (ECT) provides the ECM engine coolant temperature the ECM controls the cooling fans dependent on input from the ECT.

REMOVAL & INSTALLATION

2.7L Engines

See Figure 208.

1. Disconnect negative cable from battery.
2. Drain cooling system.
3. Disconnect coupler from ECT sensor (1).
4. Remove ECT sensor (1) from water outlet cap (2).

To install:

5. Clean mating surfaces of sensor and water outlet cap (1).
6. Use new O-ring.
7. Tighten ECT sensor to 8.5 ft. lbs. (12 Nm).
8. Connect coupler to sensor securely.
9. Refill cooling system.

3.6L Engines

1. Disconnect the engine wiring harness electrical connector (4) from the Engine Coolant Temperature (ECT) sensor.
2. Remove the ECT sensor.

To install:

3. Install the ECT sensor.
4. Tighten the sensor to 16 ft. lbs. (22 Nm).
5. Connect the engine wiring harness electrical connector (4) to the ECT sensor.

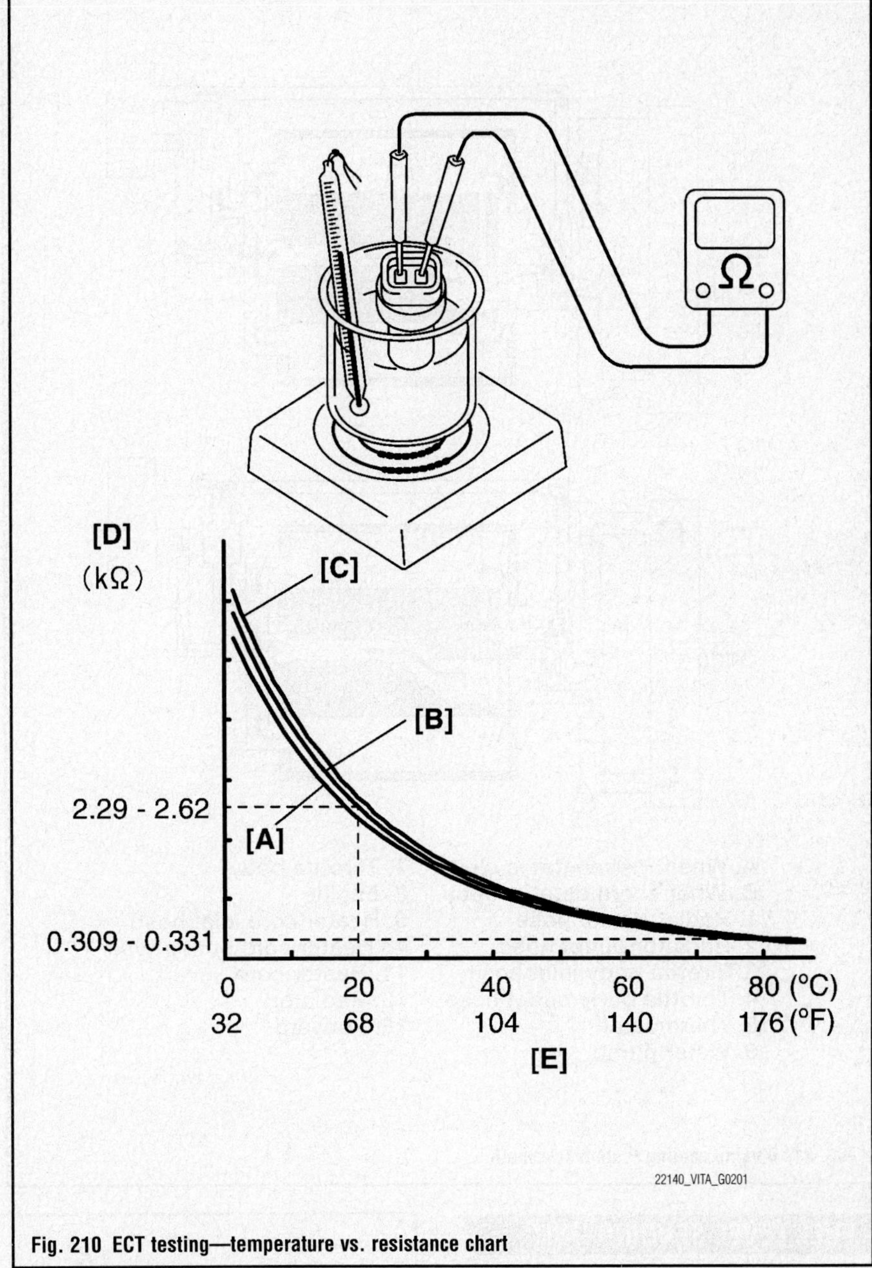

Fig. 210 ECT testing—temperature vs. resistance chart

TESTING

2.7L Engines

See Figure 210.

➡The On-Board Diagnostic (OBD II) System Check prompts the technician to complete some basic checks and store the Freeze Frame and Failure Records data on the scan tool if applicable. This creates an electronic copy of the data taken when the malfunction occurred. The information is then stored on the scan tool for later reference.

1. The engine must be allowed to cool fully before the ECT sensor will read close to the ambient temperature in order to check for a possible skewed sensor.
2. Measure the engine coolant temperature with thermometer to determine the actual value the ECT sensor should be. Take into consideration if the engine has been run and the engine coolant has been warmed without opening the thermostat.
3. Disconnect the ECT sensor electrical connector.
4. Using a Digital Voltmeter (DVM), measure the resistance across the ECT sensor terminals.
5. Check the ECT sensor value to actual coolant temperature using the Temperature vs. Resistance table.
6. The ECT sensor should accurately reflect the actual engine coolant temperature.
7. Replace the sensor if it is out of specification.

3.6L Engines

See Figure 211.

Refer to the accompanying schematic.

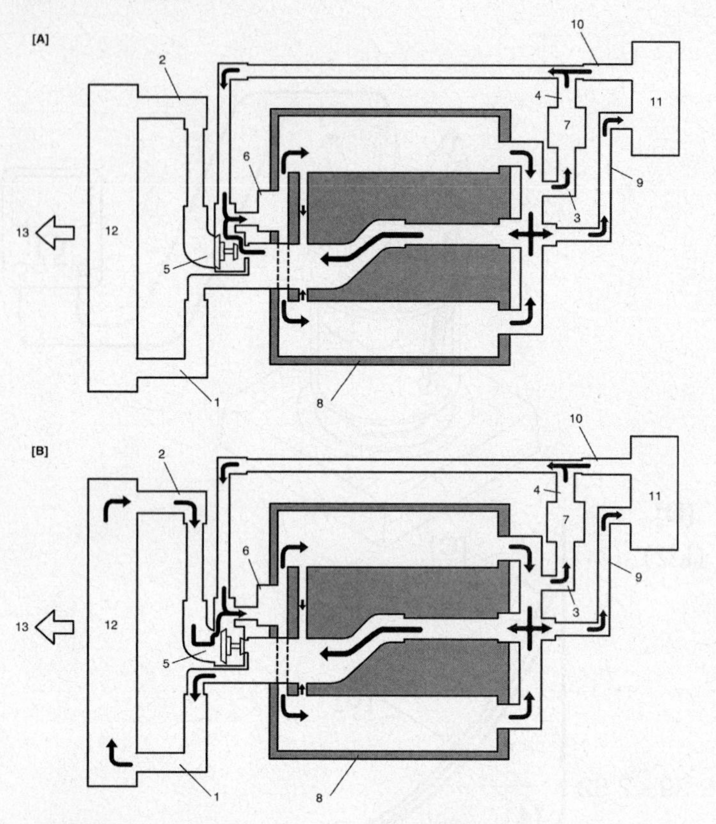

Fig. 211 Engine cooling system schematic

FUEL LEVEL SENDING UNIT

LOCATION

2.7L Engines

See Figure 212.

All fuel senders are located in the fuel tank.

3.6L Engines

See Figure 213.

The fuel sender is located in the fuel tank. There are 2 fuel level sender unit and float assemblies in the fuel tank. There is 1 located on each fuel pump module. The fuel level sender unit and float is NOT the same for each of the fuel pump modules.

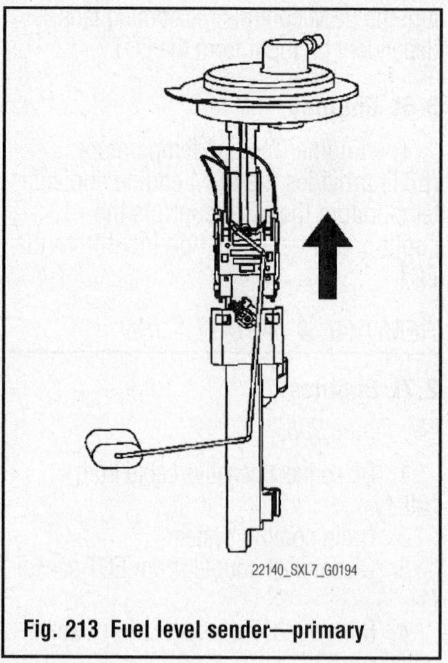

Fig. 213 Fuel level sender—primary

OPERATION

Clean all of the following areas before performing any disconnections in order to avoid possible contamination in the system:

1. The fuel pipe connections
2. The hose connections
3. The areas surrounding the connections

REMOVAL & INSTALLATION

⁂ WARNING

Clean all fuel pipe and hose connections and surrounding areas before disassembling to avoid possible contamination of the fuel system. Spray the fuel pump module cam-lock ring tang with penetrating oil prior to attempting removal.

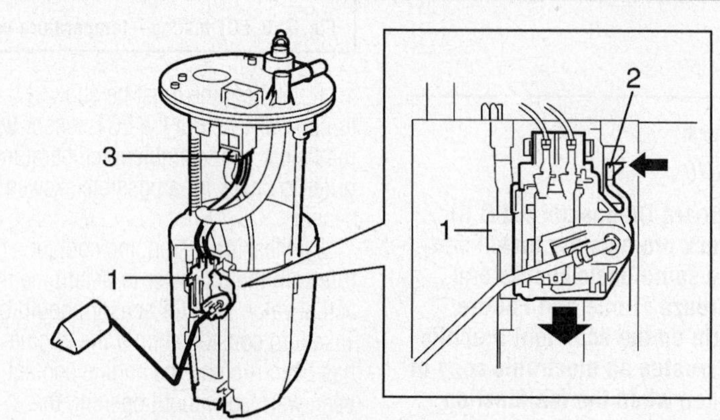

Fig. 212 Fuel tank and level sender

CAUTION

Whenever fuel line fittings are loosened or removed, wrap a shop cloth around the fitting and have an approved container available to collect any fuel.

2.7L Engines

Main Fuel Sending Unit

See Figure 213.

1. Remove fuel pump assembly from fuel tank.
2. Disconnect main fuel level gauge connector
3. With pressing snap-fit part (2), remove main fuel level gauge (1) by sliding it in the arrow direction as shown in figure.

To install:

4. With pressing snap-fit part (2), install main fuel level gauge (1) by sliding it up onto the main fuel pump assembly.

Sub-Level Fuel Sending Unit

See Figure 214.

1. Remove fuel tank from vehicle
2. Disconnect sub fuel level gauge connector (1).
3. Remove sub fuel level gauge (2).

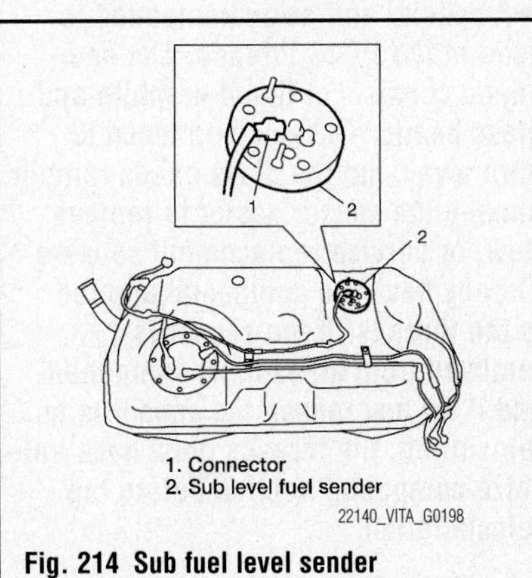

Fig. 214 Sub fuel level sender

To install:

4. Replace O-ring with new one using care not to damage it.
5. Apply thin coat of fuel to O-ring, and then install sub fuel level gauge
6. Install sub fuel level gauge (2).
7. Connect sub fuel level gauge connector (1).
8. Install the fuel tank assembly.

3.6L Engines

Primary

See Figure 213.

1. Remove the fuel pump module
2. Disconnect the fuel level sender unit and the float electrical connector.
3. Release the retaining tabs and remove the level sensor by sliding up.

To install:

4. Install the fuel level sender unit and float onto the fuel pump module. Make sure that the sender cap snaps into place.
5. Connect the fuel level sender unit and the float electrical connector.
6. Install the fuel pump module into the fuel tank.

Secondary—AWD Models

See Figure 215.

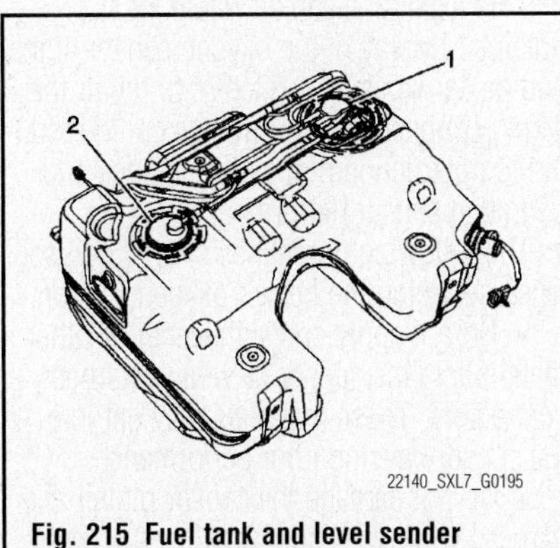

Fig. 215 Fuel tank and level sender (secondary)—AWD models

1. Remove the fuel tank, refer to Fuel Tank Replacement (AWD).
2. Disconnect the secondary fuel level sender unit and the float electrical connector.
3. Release the retaining tabs and remove the level sensor by sliding up.
4. Disconnect the suction port attaching tube by pressing down on the tab.

WARNING

To prevent bending of the sending unit float arm during removal, lift the pump module up slightly to disengage the orientation tabs in the tank and rotate the module 45 degrees.

5. Remove the secondary fuel pump module.
6. Install the fuel level sender unit and float onto the fuel pump module. Make sure that the sender cap snaps into place.
7. Connect the fuel level sender unit and the float electrical connector.
8. Install the fuel pump module into the fuel tank.
9. Connect the EVAP line quick connect.

TESTING

2.7L Engines

Main Fuel Sending Unit

See Figure 216.

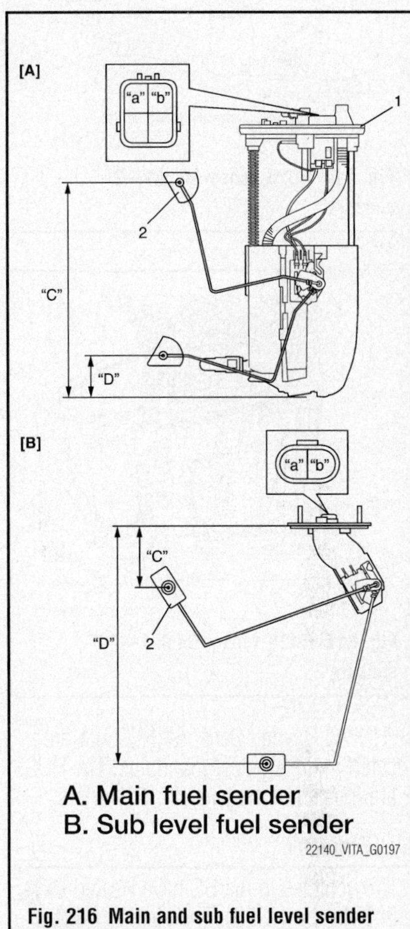

Fig. 216 Main and sub fuel level sender

1. Check resistance between terminals "a" and "b" at each float position.
2. Check that resistance between terminals "a" and "b" of the main fuel level sensor changes with change of the float position. 19.0 –21.0 Ohms at "C" 8.11 inch (206 mm) and 129.8–133.2 Ohms at "D". 1.50 inch (38 mm)

Sub-Level Fuel Sending Unit

See Figure 216.

1. Sub fuel level sensor [B] specifications —19.0 –21.0 Ohms at "C" 2.20 inch (56 mm) and 146.8–150.2 Ohms at "D". 9.45 inch (240 mm)

HEATED OXYGEN (HO2S) SENSOR

LOCATION

See Figures 217 and 218.

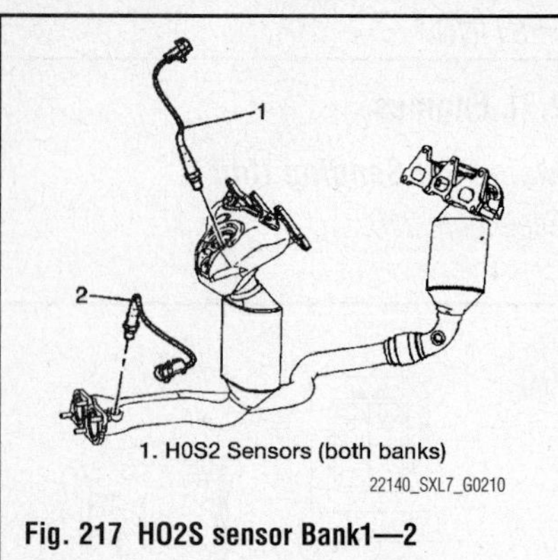

Fig. 217 HO2S sensor Bank1—2 sensors 1

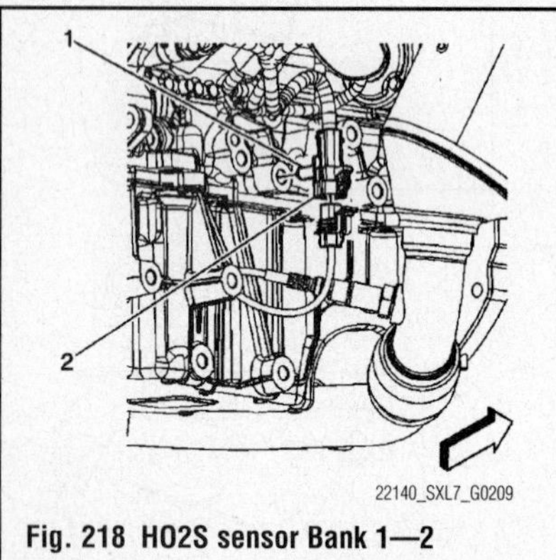

Fig. 218 HO2S sensor Bank 1—2 sensors 2

Heated Oxygen Sensors (HO2S) 1 are located in the exhaust manifolds. The HO2S 2 are located in the catalytic converters.

OPERATION

All vehicles in the US have Heated Oxygen Sensors (HO2S). The need to bring a vehicle into closed loop requires heating the 02 sensors to speed the ability to function and provide feedback to the ECM.

HO2S are used for fuel control and post catalyst monitoring. Each HO2S compares the oxygen content of the surrounding air with the oxygen content of the exhaust stream. When the vehicle is first started, the Engine Control Module (ECM) operates in an open loop mode, ignoring the HO2S signal voltage when calculating the air fuel ratio. The ECM supplies the HO2S with a reference voltage of about 0.45 Volts. The HO2S generates a voltage within a range of 0 –1 volts that fluctuates above and below bias voltage once in closed loop. A high HO2S voltage output indicates a rich fuel mixture. A low HO2S voltage output indicates a lean mixture. Heating elements inside the HO2S minimize the time required for the sensors to reach operating temperature, and then provide an accurate voltage signal.

REMOVAL & INSTALLATION

Do not remove the pigtail from the Heated Oxygen Sensor (HO2S). Removing the pigtail or the connector will affect sensor operation.

➡Handle the oxygen sensor carefully. Do not drop the HO2S. Keep the in-line electrical connector and the louvered end free of grease, dirt, or other contaminants. Do not use cleaning solvents of any type.

➡Do not repair the wiring, connector or terminals. Replace the oxygen sensor if the pigtail wiring, connector, or terminal is damaged.

This external clean air reference is obtained by way of the oxygen sensor signal and heater wires. Any attempt to repair the wires, connectors, or terminals could result in the obstruction of the air reference and degraded sensor performance.

The following guidelines should be used when servicing the heated oxygen sensor:

- Do not apply contact cleaner or other materials to the sensor or vehicle harness connectors. These materials may get into the sensor causing poor performance.
- Do not damage the sensor pigtail and harness wires in such a way that the wires inside are exposed. This could provide a path for foreign materials to enter the sensor and cause performance problems.
- Ensure the sensor or vehicle lead wires are not bent sharply or kinked. Sharp bends or kinks could block the reference air path through the lead wire.
- Do not remove or defeat the oxygen sensor ground wire, where applicable. Vehicles that utilize the ground wired sensor may rely on this ground as the only ground contact to the sensor. Removal of the ground wire will cause poor engine performance.
- Ensure that the peripheral seal remains intact on the vehicle harness connector in order to prevent damage due to water intrusion. The engine harness may be repaired using Packard's Crimp and Splice Seals Terminal Repair Kit. Under no circumstances should repairs be soldered since this could result in the air reference being obstructed.

⁂ CAUTION

To avoid danger of being burned, do not touch exhaust system when system is hot. Oxygen sensor removal should be performed when system is cool.

2.7L Engines

1. Disconnect the negative battery cable.
2. Disconnect connector of the heated oxygen sensor and release its wire harness from clamps.
3. For sensor-1, remove exhaust manifold cover.
4. Remove heated oxygen sensor (1) from exhaust manifold.

To install:

5. Install the heated oxygen sensor (1) from exhaust manifold.
6. Tighten oxygen sensor to 32 ft. lbs. (45 Nm).

3.6L Engines

Bank 1—2 Sensor 1

See Figure 217.

1. Disconnect the negative battery cable.
2. Remove the fuel injector sight shield, if necessary.
3. Disconnect connector of the heated oxygen sensor and release its wire harness from clamps.
4. Remove heated oxygen sensor from the exhaust manifold.

To install:

➡A special anti-seize compound is used in the HO2S threads. The compound consists of liquid graphite and glass beads. The graphite tends to burn away, but the glass beads remain, making the sensor easier to remove. New, or service replacement sensors already have the compound applied to the threads. If the sensor is removed from an exhaust component and if for any reason the sensor is to reinstalled, the threads must have anti-seize compound applied before the reinstallation.

5. If reinstalling the old sensor, coat the threads with anti-seize compound GM P/N 12377953, or equivalent.
6. Install the heated oxygen sensor into the exhaust manifold.
7. Tighten oxygen sensor to 31 ft. lbs. (42 Nm).

Bank 1—2 Sensor 2

See Figure 218.

1. Disconnect the negative battery cable.
2. Raise and support the vehicle.
3. Remove the fuel injector sight shield, if necessary.

4. Disconnect connector of the heated oxygen sensor and release its wire harness from clamps.

5. Remove heated oxygen sensor from the exhaust manifold.

To install:

➡A special anti-seize compound is used in the HO2S threads. The compound consists of liquid graphite and glass beads. The graphite tends to burn away, but the glass beads remain, making the sensor easier to remove. New, or service replacement sensors already have the compound applied to the threads. If the sensor is removed from an exhaust component and if for any reason the sensor is to reinstalled, the threads must have anti-seize compound applied before the reinstallation.

6. If reinstalling the old sensor, coat the threads with anti-seize compound GM P/N 12377953, or equivalent.

7. Install the heated oxygen sensor into the exhaust manifold.

8. Tighten oxygen sensor to 31 ft. lbs. (42 Nm).

9. Connect the HO2S electrical connector (2) to the engine wiring harness electrical connector.

10. Lower the vehicle.

11. Connect the negative battery cable.

TESTING

2.7L Engines

Sensor Testing

See Figure 219.

1. Start the engine.

2. Allow the engine to reach operating temperature.

3. Operate the engine at 1.500 rpm for 30 seconds.

4. While observing the HO2S voltage parameter with a scan tool, quickly cycle the throttle from closed throttle to wide open throttle (WOT), 3 times.

5. The HO2S voltage parameter should change more than 200 mV.

6. If the voltage does not change more that the specified value, the condition is present to set a Diagnostic Trouble Code (DTC).

7. Observe the Freeze Frame/Failure Records for this DTC.

Heater Testing

See Figure 220.

1. Disconnect sensor connector (1).

2. Using ohmmeter, measure resistance between terminals "VB" and "GND" of sensor connector.

➡Temperature of sensor affects resistance value largely. Make sure that sensor heater is at correct temperature

3. Readings should be 5–6.4 ohms at 68°F, 20°C for sensors 1 and 2.

➡If found faulty, replace oxygen sensor.

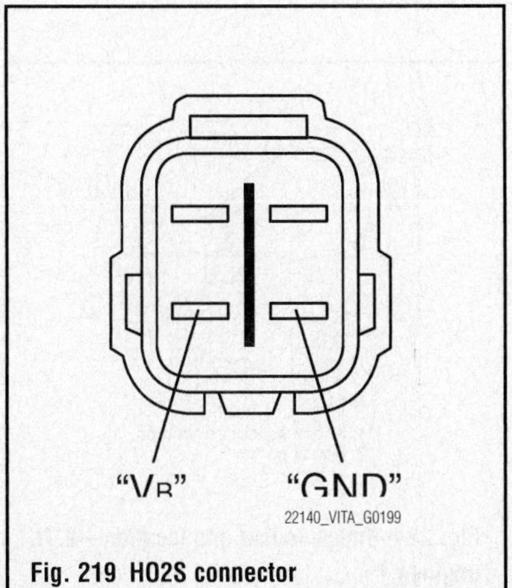

Fig. 219 HO2S connector

INTAKE AIR TEMPERATURE (IAT) SENSOR

LOCATION

See Figure 220

The Intake Air Temperature (IAT) sensor is located in the Mass Air Flow (MAF) sensor in the air cleaner outlet duct.

OPERATION

➡The Mass Air Flow (MAF) sensor also houses an integrated Intake Air Temperature (IAT) sensor.

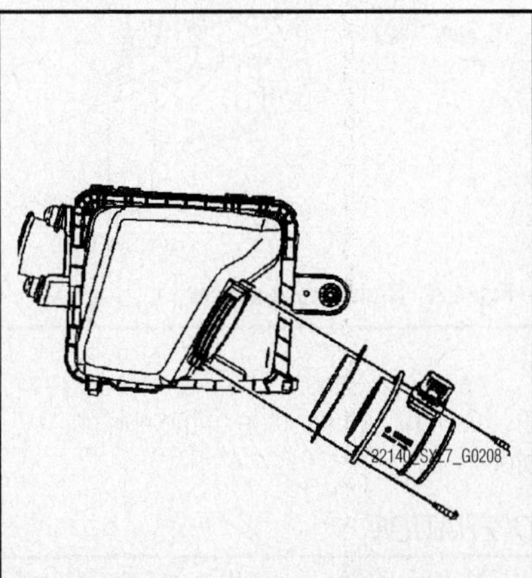

Fig. 220 Mass Air Flow (MAF)/Intake Air Temperature (IAT) sensor

The MAF sensor measures the amount of air coming into the engine. This direct airflow measurement is more accurate than the calculated airflow information obtained from the other sensor inputs. The MAF sensor also houses an integrated Intake Air Temperature (IAT) sensor. The MAF sensor uses the following circuits:

- An ignition voltage circuit
- A signal circuit
- A ground circuit
- An IAT signal circuit
- An IAT low reference circuit

The MAF sensor that is used on this vehicle is a hot film type and is used in order to measure the air flow rate. The air flow through the sensor passes over a temperature sensor, is then heated, and then passes over another temperature sensor. The difference in air temperature before and after the heater is measured. The air temperature difference is proportional to the amount of air flow. This air temperature difference also allow for determination of whether the air is flowing in the forward or reverse directions. As the air flow increases the delta temperature between the two sensors increases.

The MAF sensor converts the temperature difference into a frequency signal that the ECM monitors. The ECM calculates the air flow based on this signal.

The ECM monitors the MAF sensor signal frequency and can determine if the sensor signal is too low or too high. The ECM can also detect airflow that is inappropriate for a given operating condition based on the signal frequency.

REMOVAL & INSTALLATION

Refer to the Mass Air Flow (MAF) Sensor procedure in this section.

TESTING

2.7L Engines

See Figure 221

1. Check sensor O-ring for damage and deterioration. Replace as necessary.

2. Blow hot air to temperature sensing part (1) of MAF and IAT sensor (2) using hot air drier (3) and measure resistance between IAT sensor terminals (5) and (6) while heating air gradually.

3. If measured resistance does not show such characteristic as shown, replace MAF and IAT sensor.

4. Intake air temperature sensor resistance:
- -4°F (-20°C): 12.5–16.9 k Ohms
- 77°F (25°C) : 1.8–2.2 k Ohms
- 140°F (60°C): 0.5– 0.68 k Ohms

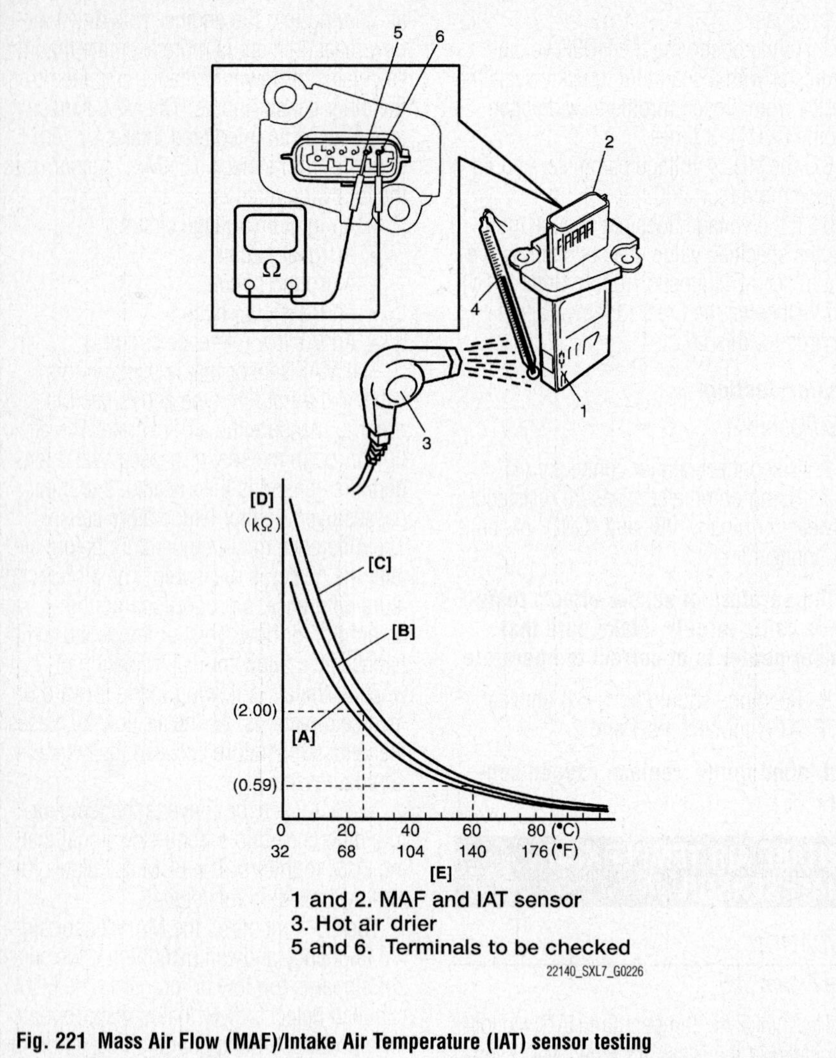

Fig. 221 Mass Air Flow (MAF)/Intake Air Temperature (IAT) sensor testing

3.6L Engines

The scan tool displays the MAF sensor value in grams per second (g/s) and hertz (Hz). Values should change rather quickly on acceleration, but should remain fairly stable at any given engine speed. If the ECM detects a condition with the MAF sensor circuits, the following DTCs set:

- P0100 Mass Air Flow (MAF) Sensor Circuit
- P0101 Mass Air Flow (MAF) Sensor Performance
- P0102 Mass Air Flow (MAF) Sensor Circuit Low Voltage
- P0103 Mass Air Flow (MAF) Sensor Circuit High Voltage

KNOCK SENSOR (KS)

LOCATION

3.6L Engines

See Figures 222 and 223.

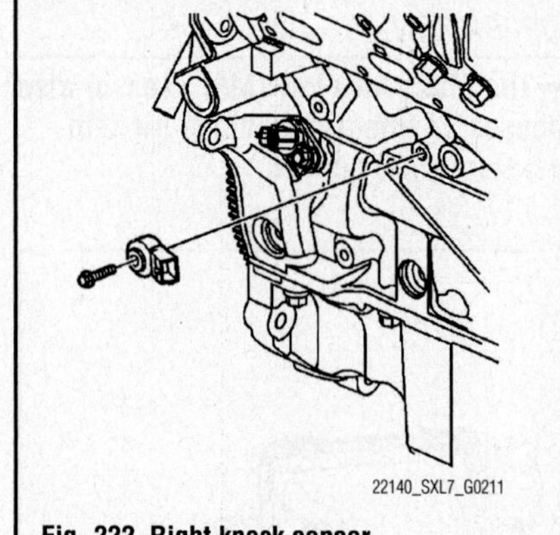

Fig. 222 Right knock sensor

The Knock Sensors (KS) are located on the right and left side of the engine block.

OPERATION

The Knock Sensor (KS) system detects engine knocking or pinging. The ECM will retard the spark timing based on the signals

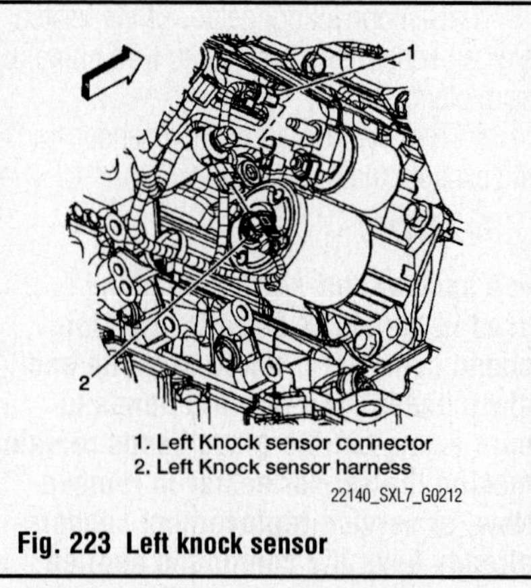

Fig. 223 Left knock sensor

from the KS system. The KS produce an AC voltage that is sent to the Engine Control Module (ECM). The amount of the AC voltage produced is proportional to the amount of knock.

The ECM monitors the voltage of the sensors after each cylinder has fired. If knock occurs in any of the cylinders, the ignition will be retarded for that particular cylinder. If the knocking then stops, the ignition will be restored to what it was before in steps. Should knocking continue in the same cylinder in spite of the ignition being retarded, the ECM will retard the ignition an additional steps, and so on, up to a maximum of 12 degrees of retard. The ignition will also be retarded at high ambient temperatures in order to counteract knocking tendencies provoked by high intake air temperatures. Should either bank 1 or bank 2 sensor fail to work, or should an internal circuit problem occur, the ignition timing will then use a default strategy. The default strategy will retard the ignition the maximum allowed amount to protect the engine from possible damage.

REMOVAL & INSTALLATION

2.7L Engines

See Figure 224.

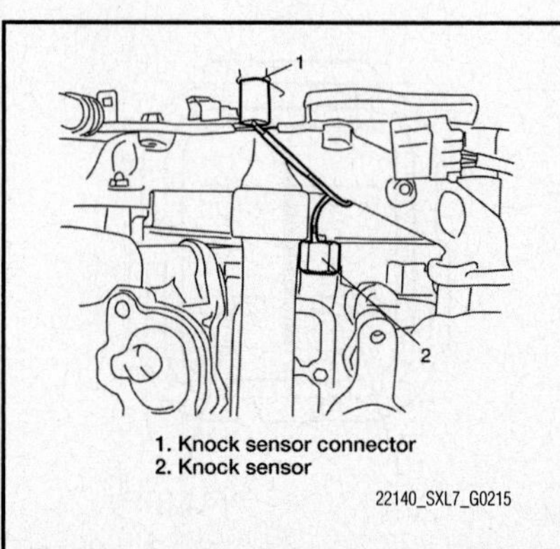

Fig. 224 Knock sensor and location—2.7L engines

1. Disconnect negative cable at battery.
2. Remove the intake manifold from cylinder head.
3. Disconnect knock sensor connector (1).
4. Remove knock sensor (2) from cylinder block.

To install:

5. Install the knock sensor (2) from cylinder block.
6. Torque the knock sensor to 16 ft. lbs. (22 Nm).
7. Install the intake manifold to the cylinder head.
8. Connect negative cable at the battery.

3.6L Engines

Bank 1

See Figure 222.

1. Raise and support the vehicle.
2. Remove the rear exhaust manifold heat shield bolts and shield.
3. Remove the right catalytic converter.
4. Disconnect the engine wiring harness electrical connector (2) from the bank 1 knock sensor.
5. Loosen the knock sensor bolt and remove the knock sensor.

To install:

6. Position the knock sensor and tighten the knock sensor bolt to 17 ft. lbs. (23 Nm).
7. Connect the engine wiring harness electrical connector (2) to the bank 1 knock sensor.
8. Position the rear exhaust manifold heat shield to the engine and install the bolts and tighten to 37 ft. lbs. (50 Nm).
9. Install the right catalytic converter.
10. Lower the vehicle.

Bank 2

See Figure 223.

1. Raise and support the vehicle.
2. Disconnect the engine wiring harness electrical connector (2) from the bank 2 knock sensor.
3. Loosen the knock sensor bolt and remove the knock sensor.

To install:

4. Position the knock sensor and tighten the knock sensor bolt to 17 ft. lbs. (23 Nm).
5. Connect the engine wiring harness electrical connector (2) to the bank 2 knock sensor.
6. Lower the vehicle.

MALFUNCTION INDICATOR LIGHT (MIL)

RESET PROCEDURES

The Malfunction Indicator Light (MIL) can be reset by the steps listed below.

- The control module turns **OFF** the MIL after 4 consecutive ignition cycles that the diagnostic runs and does not fail.
- A current Diagnostic Trouble Code (DTC), Last Test Failed, clears when the diagnostic runs and passes.
- A history DTC clears after 40 consecutive warm-up cycles, if no failures are reported by this or any other emission related diagnostic.
- Clear the MIL and the DTC with a scan tool.

MASS AIR FLOW (MAF) SENSOR

LOCATION

The Mass Air Flow (MAF) sensor is located on the air cleaner attached to the intake duct leading to the throttle housing.

OPERATION

The Mass Air Flow (MAF) sensor measures the amount of air coming into the engine. This direct airflow measurement is more accurate than the calculated airflow information obtained from the other sensor inputs. The MAF sensor also houses an integrated Intake Air Temperature (IAT) sensor. The MAF sensor uses the following circuits:

- An ignition voltage circuit
- A signal circuit
- A ground circuit
- An IAT signal circuit
- An IAT low reference circuit

The MAF sensor that is used on this vehicle is a hot film type and is used in order to measure the air flow rate. The air flow through the sensor passes over a temperature sensor, is then heated, and then passes over another temperature sensor. The difference in air temperature before and after the heater is measured. The air temperature difference is proportional to the amount of air flow. This air temperature difference also allow for determination of whether the air is flowing in the forward or reverse directions. As the air flow increases the delta temperature between the two sensors increases. The MAF sensor converts the temperature difference into a frequency signal that the ECM monitors. The ECM calculates the air flow based on this signal. The ECM monitors the MAF sensor signal frequency and can determine if the sensor signal is too low or too high. The ECM can also detect airflow that is inappropriate for a given operating condition based on the signal frequency.

REMOVAL & INSTALLATION

2.7L Engines

See Figure 225.

1. Disconnect the engine wiring harness electrical connector from the mass air flow (MAF)/Intake Air Temperature (IAT) sensor.
2. Remove the MAF/IAT sensor screws.
3. Remove the MAF/IAT sensor from the air cleaner assembly.

To install:

4. Install the MAF/IAT sensor from the air cleaner assembly.
5. Install the MAF/IAT sensor screws.
6. Tighten MAF and IAT sensor screws to 1.1 ft. lbs. (1.5 Nm).
7. Connect MAF and IAT sensor connector securely.

3.6L Engines

See Figure 225.

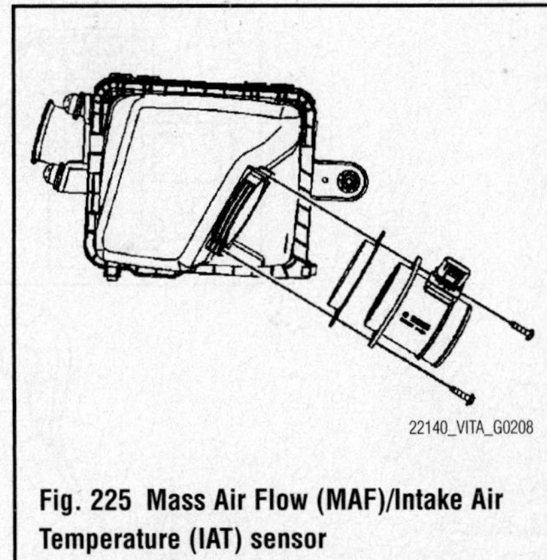

Fig. 225 Mass Air Flow (MAF)/Intake Air Temperature (IAT) sensor

1. Disconnect negative cable at battery.
2. Remove the air cleaner outlet duct.
3. Disconnect the engine wiring harness electrical connector from the Mass Air Flow (MAF)/Intake Air Temperature (IAT) sensor.
4. Remove the MAF/IAT sensor screws.
5. Remove the MAF/IAT sensor from the air cleaner assembly.
6. Remove the MAF/IAT sensor seal.

To install:

7. Install the MAF/IAT sensor seal to the MAF/IAT sensor.
8. Install the MAF/IAT sensor to the air cleaner assembly.
9. Install the MAF/IAT sensor screws.

10. Tighten the screws to 35 inch lbs. (4 Nm).

11. Connect the engine wiring harness electrical connector to the MAF/IAT sensor.

12. Install the air cleaner outlet duct.

TESTING

2.7L Engines

See Figure 226

1. Remove ECM cover from bracket.
2. With ignition switch **OFF**, disconnect MAF and IAT sensor connector (1).
3. Connect voltmeter between C13-2 terminal (2) and C13-3 terminal (3) of MAF and IAT sensor connector (1) disconnected.
4. Turn ignition switch **ON** and check that sensor power supply is battery voltage. If not, check if wire harness is open or connection is poor.
5. Turn ignition switch **OFF** and connect connector to MAF and IAT sensor.
6. Check MAF sensor output voltage 1.2–3.6 Volts, Engine running at specified idle speed after warmed up engine.
7. If MAF sensor output voltage is not as specified, cause may lie in wire harness, coupler connection, MAF and IAT sensor or ECM.

3.6L Engines

The scan tool displays the MAF sensor value in grams per second (g/s) and hertz (Hz). Values should change rather quickly on acceleration, but should remain fairly stable at any given engine speed. If the ECM detects a condition with the MAF sensor circuits, the following DTCs set:

- P0100 Mass Air Flow (MAF) Sensor Circuit
- P0101 Mass Air Flow (MAF) Sensor Performance
- P0102 Mass Air Flow (MAF) Sensor Circuit Low Voltage
- P0103 Mass Air Flow (MAF) Sensor Circuit High Voltage

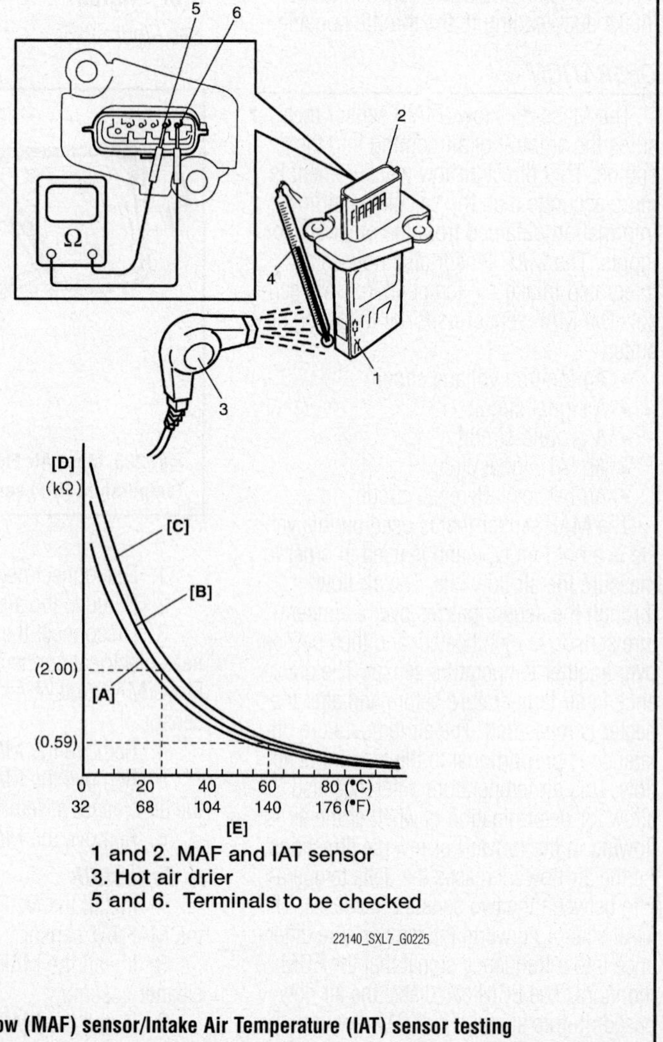

Fig. 226 Mass Air Flow (MAF) sensor/Intake Air Temperature (IAT) sensor testing

MANIFOLD ABSOLUTE PRESSURE (MAP) SENSOR

LOCATION

2.7L Engines

See Figure 227.

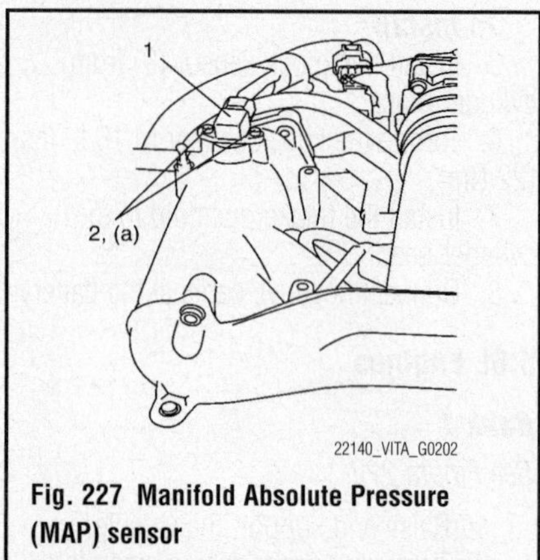

Fig. 227 Manifold Absolute Pressure (MAP) sensor

Refer to the accompanying illustration for Manifold Absolute Pressure (MAP) sensor location.

OPERATION

The Manifold Absolute Pressure (MAP) sensor provides a varying voltage to the ECM for fuel and spark management dependent on manifold pressure voltage reduces when vacuum is applied up to 400 mmHg.

REMOVAL & INSTALLATION

2.7L Engines

See Figure 227 and 228

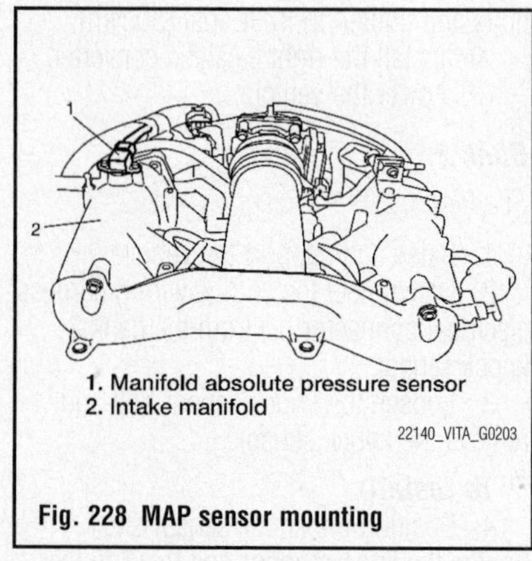

Fig. 228 MAP sensor mounting

1. Disconnect negative cable at battery.
2. Disconnect connector from manifold absolute pressure sensor (1).

3. Remove manifold absolute pressure sensor (1) from intake manifold (2).

To install:

4. Confirm that vacuum passage on intake manifold is free from clog.
5. Apply engine oil to O-ring of manifold absolute pressure sensor.
6. Install manifold absolute pressure sensor (1) onto intake manifold.
7. Tighten MAP sensor bolt (2) to 3.5 ft. lbs. (5 Nm).
8. Attach the connector to manifold absolute pressure sensor securely.

TESTING

2.7L Engines

See Figures 229 through 231.

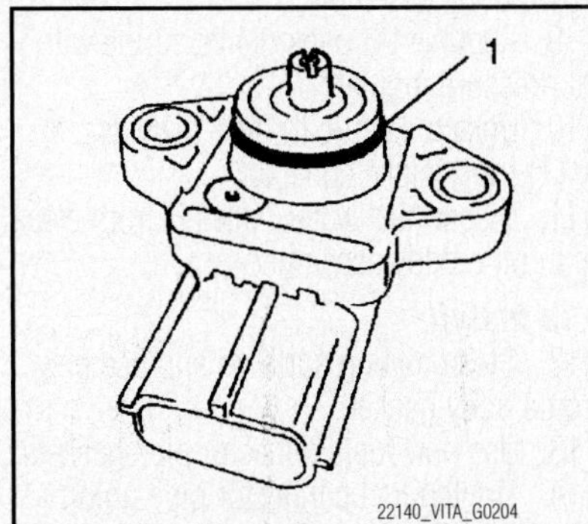

Fig. 229 Manifold Absolute Pressure (MAP) sensor O-ring inspection

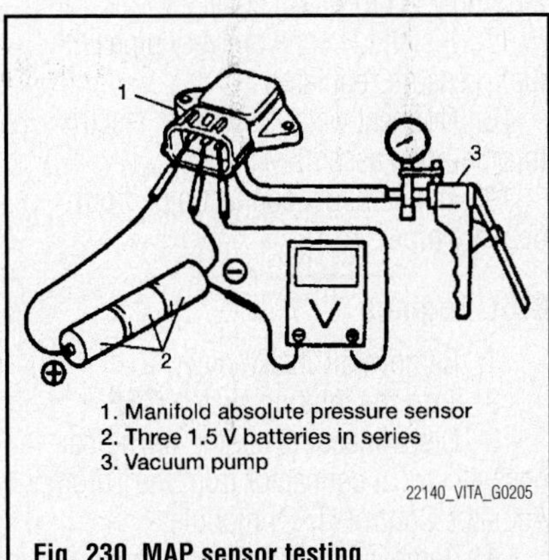

Fig. 230 MAP sensor testing

Check the Manifold Absolute Pressure (MAP) sensor O-ring (1) for damage and deterioration. Replace as necessary.

Arrange new three 1.5 V batteries (2) in series and connect its positive terminal to "Vin" terminal and negative terminal to "Ground" terminal. Then, check voltage between "Vout" and "Ground". Also, check if voltage reduces when vacuum is slowly applied up to 400 mmHg by using vacuum pump (3). If check result is not satisfactory, replace manifold absolute pressure sensor (1).

Altitude (Reference)	Barometric pressure		Output voltage
(ft (m))	(mmHg)	(kPa)	(V)
0 – 2000 (0 – 610)	760 – 707	100 – 94	3.3 – 4.3
2001 – 5000 (611 – 1524)	Under 707, Over 634	94 – 85	3.0 – 4.1
5001 – 8000 (1525 – 2438)	Under 634, Over 567	85 – 76	2.7 – 3.7
8001 – 10000 (2439 – 3048)	Under 567, Over 526	76 – 70	2.5 – 3.3

22140_VITA_G0206

Fig. 231 MAP testing chart

OIL PRESSURE SWITCH

LOCATION

See Figure 232.

Refer to the accompanying illustration for oil pressure switch location.

OPERATION

The Engine Oil Pressure (EOP) switch is a normally closed switch that opens with the proper oil pressure. With the ignition switch turned **ON** and the engine not running, the Engine Control Module (ECM) should detect a low signal voltage input. With the engine running, the pressure switch opens, and the ECM should detect a high signal voltage input.

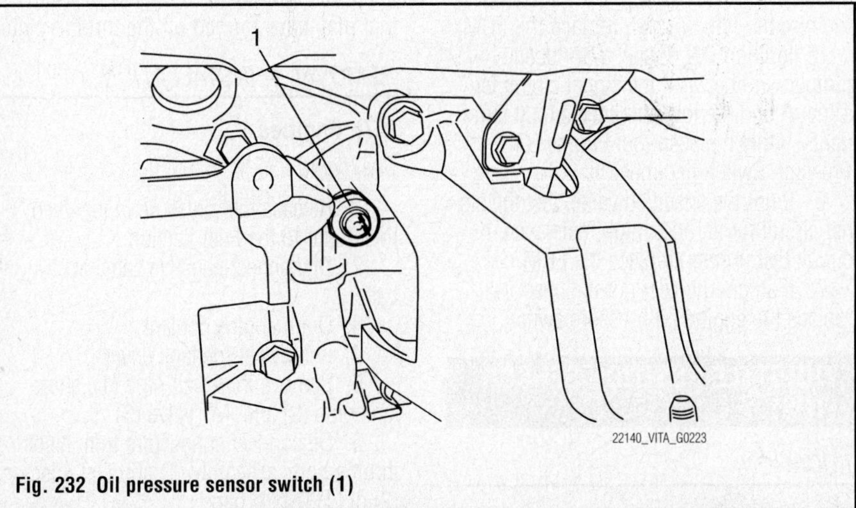

Fig. 232 Oil pressure sensor switch (1)

REMOVAL & INSTALLATION

See Figure 233

1. Turn the ignition **OFF** .
2. Disconnect the oil pressure sensor electrical connector.
3. Remove the oil pressure sensor (1).

To install:

4. Install the oil pressure sensor (1).
5. Tighten the oil pressure sensor to 15 ft. lbs. (20 Nm).
6. Connect the oil pressure sensor electrical connector.

TESTING

2.7L Engines

See Figure 233.

1. Disconnect oil pressure switch (1) lead wire.

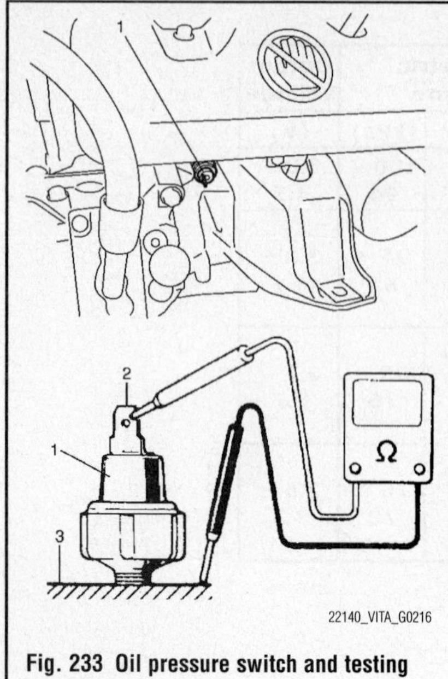

Fig. 233 Oil pressure switch and testing

2. Check for continuity between oil pressure switch terminal (2) and cylinder block (3) as shown.

3. If check result is not as specified, replace oil pressure switch (1).

3.6L Engines

1. With the ignition **OFF**, disconnect the harness connector at the Engine Oil Pressure (EOP) switch.

2. With the ignition **OFF**, test for less than 1 ohm of resistance between the low reference circuit terminal **B** and ground.

3. If greater than the specified range, test the low reference circuit for an open/high resistance. If the circuit tests normal, replace the ECM.

4. With the ignition **ON**, verify the scan tool Engine Oil Pressure Switch parameter is **OK** .

If not the specified value, test the signal circuit terminal A for a short to ground. If the circuit tests normal, replace the ECM.

5. Ignition **ON**, install a 3A fused jumper wire between the signal circuit terminal **A** and the low reference circuit terminal B. Verify the scan tool Engine Oil Pressure Switch parameter is Low.

6. If not the specified value, test the signal circuit for an open/high resistance. If circuit test normal, replace the ECM.

7. If all circuits test normal, test or replace the engine oil pressure switch.

THROTTLE POSITION SENSOR (TPS)

LOCATION

See Figure 234.

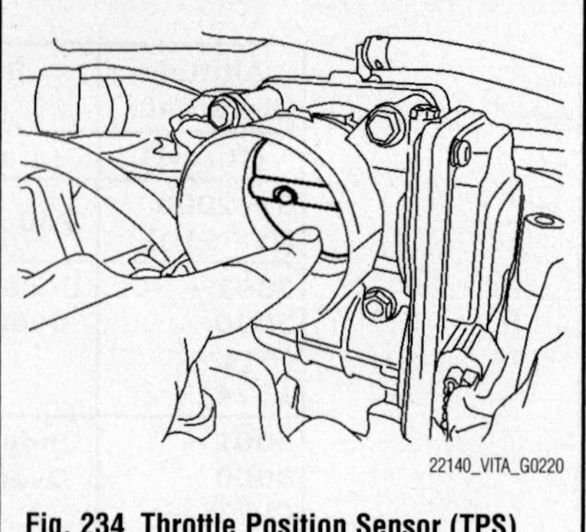

Fig. 234 Throttle Position Sensor (TPS) location

The Throttle Position Sensor (TPS) is integrated with the throttle actuator (motor) on the throttle housing.

OPERATION

The ECM monitors the throttle plate angle with 2 Throttle Position Sensors (TPS). The ECM performs diagnostics that monitor the voltage levels of both APP sensors, both TPS, and the throttle actuator control motor circuit. It also monitors the spring return rate of both return springs that are housed internal to the throttle body assembly. These diagnostics are performed at different times based on whether the engine is running, not running, or whether the ECM is currently in a throttle body relearn procedure.

With every ignition cycle, the ECM performs a quick throttle return spring test to make sure the throttle plate can return to the 7 percent rest position from the 0 percent position. This is to ensure that the throttle plate can be brought to the rest position in case of an actuator motor circuit failure. Observe, under cold conditions, the ECM commands the throttle plate to 7 percent with the ignition **ON** and the engine **OFF** to release any ice that may have formed on the throttle plate.

REMOVAL & INSTALLATION

2.7L Engines

See Figure 235.

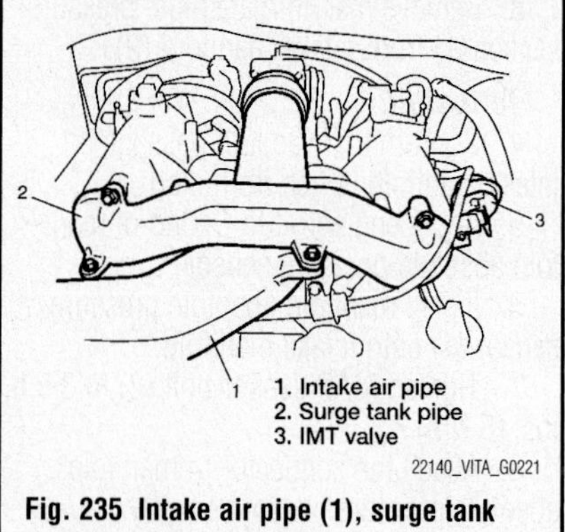

Fig. 235 Intake air pipe (1), surge tank cover and IMT valve (3)

1. Release fuel pressure in fuel feed line. Refer to the Fuel Section.
2. Disconnect negative cable at battery.
3. Drain engine coolant.
4. Remove surge tank cover.
5. Remove intake air pipe (1), surge tank pipe (2) and IMT valve (3).
6. Disconnect connectors from electric throttle body assembly (1) and fuel injectors (2) of LH (No.1) bank.
7. Disconnect fuel connect pipe from delivery pipe.
8. Disconnect water hose from electric throttle body assembly.
9. Remove LH bank delivery pipe with fuel pressure regulator.
10. Detach electric throttle body assembly (1) from intake collector.
11. Disconnect water hose (2) from electric throttle body assembly.

To install:

12. Clean mating surfaces and use new throttle body gasket.
13. Use new fuel connector pipe gaskets.
14. Tighten fuel connector pipe union bolts and delivery pipe bolts.
15. Connect water hose (2) from electric throttle body assembly.
16. Connect electric throttle body assembly (1) to intake collector.
17. Install LH bank delivery pipe with fuel pressure regulator.
18. Connect water hose from electric throttle body assembly.
19. Connect fuel connect pipe from delivery pipe.

3.6L Engines

1. Remove air intake pipe.
2. Turn the ignition switch **OFF**.
3. Disconnect the engine wiring harness electrical connector from the Throttle Actuator Control (TAC) module
4. Remove the throttle body bolts.
5. Remove the throttle body and gasket. Discard the gasket.

To install:

6. Clean mating surfaces and use new throttle body gasket.
7. Position the throttle body to the upper intake manifold.
8. Install the throttle body bolts.
9. Tighten the bolts to 89 inch lbs. (10 Nm).
10. Connect the engine wiring harness electrical connector to the TAC module.
11. Install the air cleaner outlet duct.

TESTING

See Figures 236 through 238.

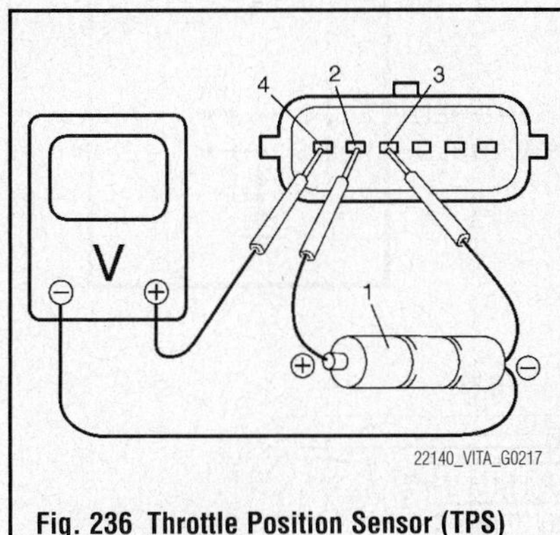

Fig. 236 Throttle Position Sensor (TPS) performance check

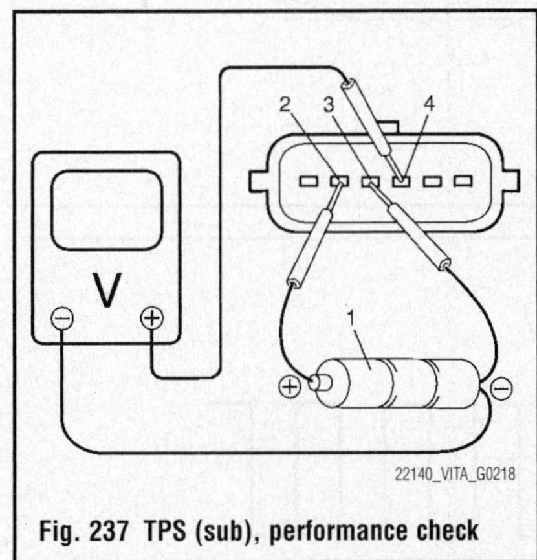

Fig. 237 TPS (sub), performance check

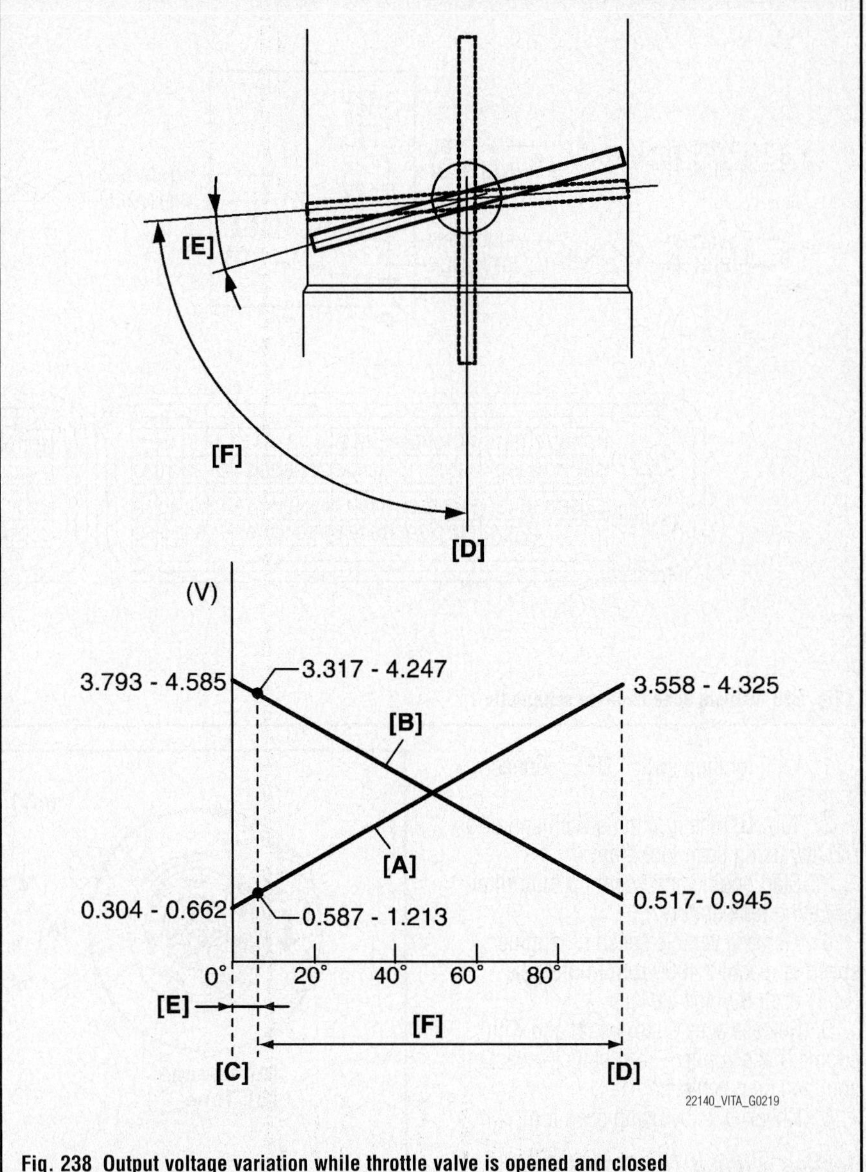

Fig. 238 Output voltage variation while throttle valve is opened and closed

The TP sensor 1 signal voltage range is from about 4.86–0.86 volts as the throttle plate is moved from 0 percent to wide open throttle (WOT). The TP sensor 2 voltage range is from about 0.82–4.14 volts as the throttle plate is moved from 0 percent to WOT.

1. Check throttle position sensor (main and sub) output voltage using the following steps.
2. For throttle position sensor (main), arrange 3 new 1.5 Volts batteries (1) in series (check that total voltage is 4.7–5.0 Volts. and connect its positive terminal to "Vin" terminal (2) and negative terminal to "Ground" terminal (3) of sensor. Then using voltmeter, connect positive terminal to "Vout 1" terminal (4) of sensor and negative terminal to battery. Sweep the TP from 0 percent to wide open throttle (WOT) and look for range 0.82–4.14 volts on the volt meter.
3. For throttle position sensor (sub), arrange, 3 new 1.5 V batteries (1) in series (check that total voltage is 4.7–5.0 Volts and connect its positive terminal to "Vin" terminal (2) and negative terminal to "Ground" terminal (3) of sensor. Then using voltmeter, connect positive terminal to "Vout 2" terminal (4) of sensor and negative terminal to battery.
   a. Measure output voltage variation while throttle valve is opened and closed as following specification.
4. If sensor voltage is out of specified value and linear variation is as shown in the following graph, replace electric throttle body assembly.
5. Throttle position sensor output voltage
6. Throttle position sensor (main) variation [A]: 0.304–4.325 Volts at minimum variation, varying according to throttle valve opening by finger (Voltage should vary by about 0.04 Volts for each 1°valve opening).
7. Throttle position sensor (sub) variation [B]: 4.585–0.517 Volts at minimum variation, varying according to throttle valve opening by finger (Voltage should vary by about 0.04 V for each 1° valve opening)
   a. Throttle valve is completely closed position.
   b. Throttle valve is fully open position.
   c. Position where throttle valve is open by 10° from completely closed position (default position).
   d. Angle obtained when accelerator pedal is depressed fully (84°).

VEHICLE SPEED SENSOR (VSS)

TESTING

2007–08 Grand Vitara

Testing With The Suzuki Scan Tool

See Figure 239.

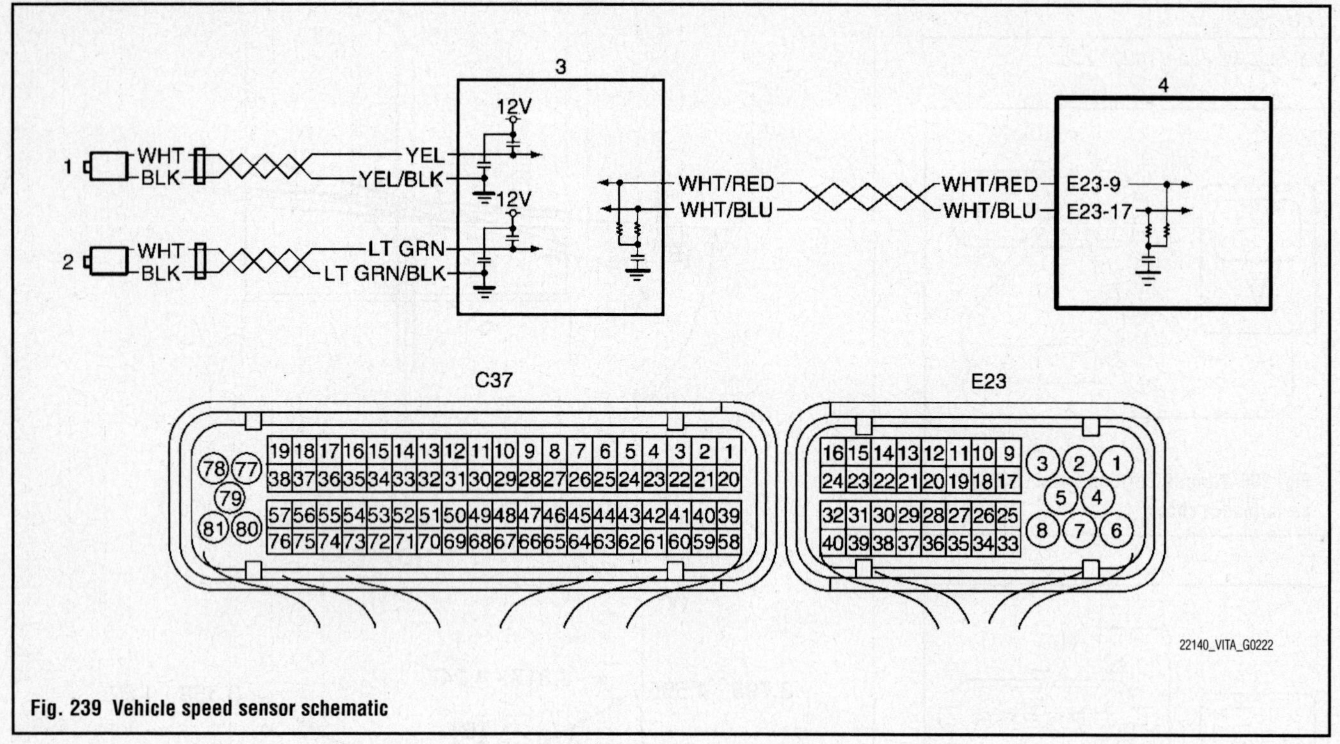

Fig. 239 Vehicle speed sensor schematic

1. With ignition switch **OFF**, connect scan tool.
2. Turn **ON** the ignition switch and clear DTC by using scan tool if any.
3. Start engine and warm up to normal operating temperature.
4. Increase vehicle speed till engine speed is reached 4000 rpm in 3rd gear (M/T) or 2nd range (A/T).
5. Release accelerator pedal and with engine brake applied, keep vehicle coasting and then stop vehicle.
6. Check DTC by using scan tool.

Testing Without The Suzuki Scan Tool

See Figure 240.

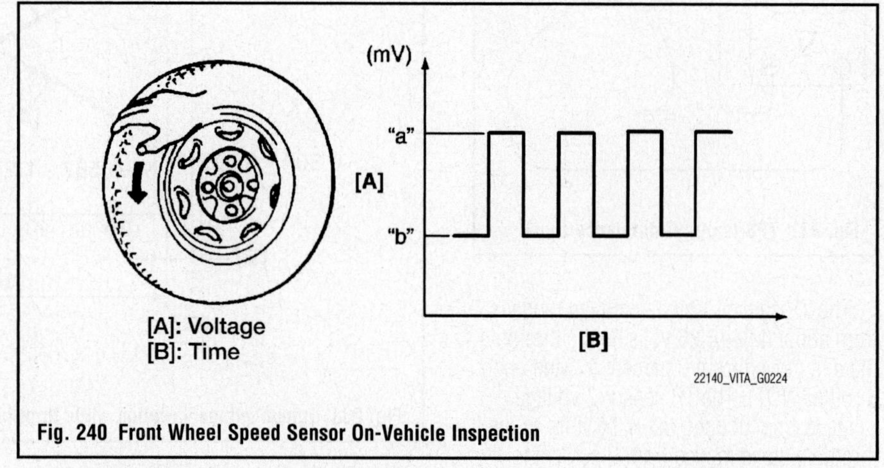

Fig. 240 Front Wheel Speed Sensor On-Vehicle Inspection

1. Measure voltage at resistance with wheel rotation and confirm voltage alternately changes between high and low voltages.
2. If voltage does not change with wheel rotation, check sensor, mating encoder and their installation conditions.
3. Voltage at the resistance (115 ohms) with wheel rotation
4. High voltage "a": 1360 to 1930 mV
5. Low voltage "b": 680 to 960 mV

FUEL SYSTEM SERVICE PRECAUTIONS

Safety is the most important factor when performing not only fuel system maintenance but any type of maintenance. Failure to conduct maintenance and repairs in a safe manner may result in serious personal injury or death. Maintenance and testing of the vehicle's fuel system components can be accomplished safely and effectively by adhering to the following rules and guidelines.

- To avoid the possibility of fire and personal injury, always disconnect the negative battery cable unless the repair or test procedure requires that battery voltage be applied.
- Always relieve the fuel system pressure prior to disconnecting any fuel system component (injector, fuel rail, pressure regulator, etc.), fitting or fuel line connection. Exercise extreme caution whenever relieving fuel system pressure to avoid exposing skin, face and eyes to fuel spray. Please be advised that fuel under pressure may penetrate the skin or any part of the body that it contacts.
- Always place a shop towel or cloth around the fitting or connection prior to loosening to absorb any excess fuel due to spillage. Ensure that all fuel spillage (should it occur) is quickly removed from engine surfaces. Ensure that all fuel soaked cloths or towels are deposited into a suitable waste container.
- Always keep a dry chemical (Class B) fire extinguisher near the work area.
- Do not allow fuel spray or fuel vapors to come into contact with a spark or open flame.
- Always use a back-up wrench when loosening and tightening fuel line connection fittings. This will prevent unnecessary stress and torsion to fuel line piping.
- Always replace worn fuel fitting O-rings with new. Do not substitute fuel hose or equivalent where fuel pipe is installed.

Before servicing the vehicle, make sure to also refer to the precautions in the beginning of this section as well.

**** WARNING**

For any operation requiring removal of the fuel tank, there should be no more than 11.4 L (3 gal) of fuel remaining. This minimizes the weight of the fuel tank and eases handling. The fuel level can be determined by reading the fuel level gage. A reading below 1/4 full indicates that no more than 11.4 L (3 gal) are remaining.

RELIEVING FUEL SYSTEM PRESSURE

1. Before servicing the vehicle, refer to the precautions in the beginning of this section.
2. Detach the wiring harness connector from the fuel pump relay, located under the left-hand side of the instrument panel near the ECM.
3. Start the engine and run it until it stops from lack of fuel. Crank the engine 2–3 times for a 3 second period. The fuel lines should now be depressurized.
4. After servicing, reattach the wiring harness connector to the fuel pump relay.

FUEL FILTER

REMOVAL & INSTALLATION

The fuel filter is located in the primary fuel tank module. The paper filter element traps particles in the fuel that may damage the fuel injection system. The filter housing is made to withstand maximum fuel system pressure, exposure to fuel additives, and changes in temperature. Refer to the Fuel Pump Removal & Installation procedure in this section.

FUEL PUMP

REMOVAL & INSTALLATION

**** WARNING**

Clean all fuel pipe and hose connections and surrounding areas before disassembling to avoid possible contamination of the fuel system. Spray the fuel pump module cam-lock ring tang with penetrating oil prior to attempting removal.

**** CAUTION**

Whenever fuel line fittings are loosened or removed, wrap a shop cloth around the fitting and have an approved container available to collect any fuel.

2007–08 Grand Vitara

See Figure 241.

1. Before servicing the vehicle, refer to tall service precautions.
2. Disconnect the negative battery cable.
3. Relieve fuel pressure in fuel feed line.
4. Disconnect the negative battery cable.
5. Hoist vehicle.
6. Remove exhaust center pipe.
7. Remove rear propeller shaft.
8. With cable connected, detach parking brake cable clamp from fuel tank cover.
9. Disconnect fuel filler hose and breather hose from filler neck.
10. Disconnect fuel pump connector.
11. Due to absence of fuel tank drain plug, drain fuel tank by pumping fuel out through fuel tank filler.
12. Use hand operated pump device to drain fuel tank.
13. Disconnect fuel hoses and EVAP canister hose from each pipe.
14. Support fuel tank with jack and remove fuel tank.
15. Remove fuel pump assembly from fuel tank.
16. Disconnect main fuel level gauge connector
17. With pressing snap-fit part (2), remove main fuel level gauge (1) by sliding it in the arrow direction as shown in figure.

To install:

18. With pressing snap-fit part (2), install main fuel level gauge (1) by sliding it up onto the main fuel pump assembly.
19. Install assembly.
20. Raise fuel tank with jack, and connect connectors of fuel pump and sub fuel level gauge and clamp wire harness.
21. Install fuel tank to vehicle.
22. Tighten fuel tank mounting bolts to 36 ft. lbs (50 Nm).
23. Connect fuel filler hose and breather hose to filler neck, and clamp them securely.
24. Connect fuel pump connector.
25. Connect fuel hoses and EVAP canister hose to each pipe, and clamp them securely.
26. Install parking brake cable clamp to fuel tank cover.
27. Install rear propeller shaft.
Install exhaust center pipe
28. Connect the negative battery cable.
29. With engine OFF, turn ignition switch to ON position and check for fuel leaks.

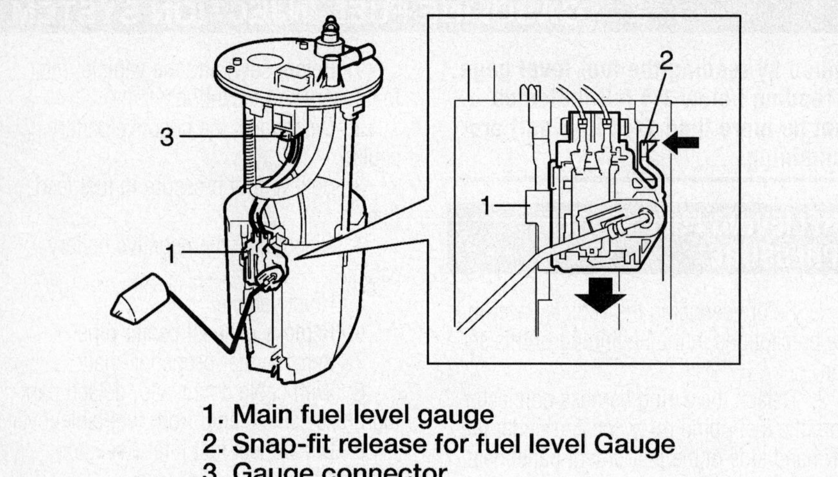

Fig. 241 Fuel pump and level sender

2007–08 XL-7

See Figure 242.

1. Before servicing the vehicle, refer to all service precautions.
2. Disconnect the negative battery cable.
3. Relieve fuel pressure in fuel feed line.
4. Disconnect the negative battery cable.
5. Disconnect fuel filler hose and breather hose from filler neck.
6. Disconnect fuel pump connector.
7. Disconnect fuel hoses and EVAP canister hose from each pipe.
8. Support fuel tank with jack and remove fuel tank.
9. Remove fuel pump assembly from fuel tank.
10. Remove the fuel pump module
11. Disconnect the fuel level sender unit and the float electrical connector.
12. Release the retaining tabs and remove the level sensor by sliding up.
13. Support the fuel tank.
14. Remove the fuel tank strap bolts and fuel tank straps.
15. Lower the fuel tank from the underbody of the vehicle.

To install:

16. Install the fuel level sender unit and float onto the fuel pump module. Make sure that the sender cap snaps into place.
17. Connect the fuel level sender unit and the float electrical connector.
18. Install the fuel pump module into the fuel tank.
19. Raise fuel tank with jack, and connect connectors of fuel pump and sub fuel level gauge and clamp wire harness.
20. Install fuel tank to vehicle.
21. Tighten fuel tank mounting bolts to 18 ft. lbs (25 Nm).
22. Connect fuel filler hose and breather hose to filler neck, and clamp them securely.
23. Tighten fuel filler hose and tighten to 4 inch lbs. (5 Nm).
24. Connect fuel pump connector.
25. Connect fuel hoses and EVAP canister hose to each pipe, and clamp them securely.

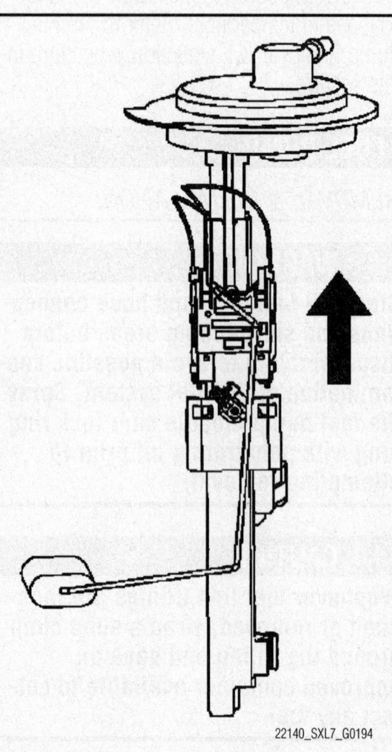

Fig. 242 Fuel pump and level sender (primary)—XL-7

26. Lower the vehicle.
27. Fill the fuel tank with gasoline.
28. Connect the negative battery cable.
29. Prime the fuel system:
    a. Cycle the ignition ON for 5 seconds and then OFF for 10 seconds.
    b. Repeat the previous step twice.
    c. Crank the engine until it starts. The maximum starter motor cranking time is 20 seconds.
    d. If the engine does not start, repeat steps 15.1–15.3.
30. Install the fuel injector sight shield
31. With engine **OFF**, turn ignition switch to **ON** position and check for fuel leaks.

FUEL PRESSURE REGULATOR

REMOVAL & INSTALLATION

2007–08 Grand Vitara

See Figure 243.

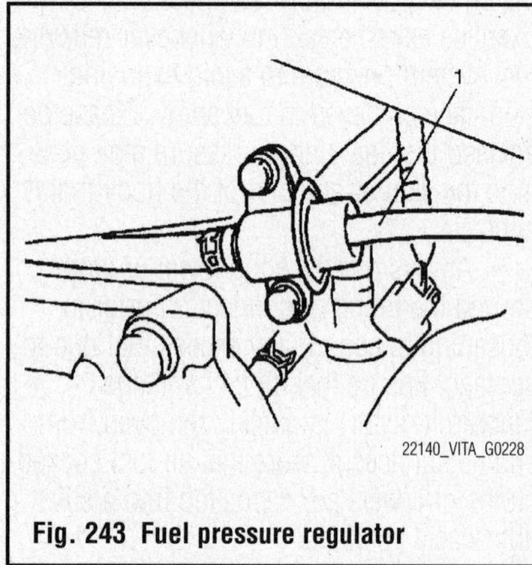

Fig. 243 Fuel pressure regulator

1. Before servicing the vehicle, refer to all safety precautions.
2. Relieve the fuel system pressure.
3. Disconnect the negative battery cable.
4. Remove surge tank cover (if applicable).
5. Disconnect vacuum hose from fuel pressure regulator.
6. Remove fuel pressure regulator from fuel rail.

**** CAUTION**

Fuel may be released when removing fuel pressure regulator. Keep shop towel under regulator and fuel rail to catch fuel. Remove shop towel when fuel release stops.

7. Disconnect fuel return hose from fuel pressure regulator.

To install:

8. Reconnect fuel return hose to fuel pressure regulator.

9. Install fuel pressure regulator to fuel rail and tighten mounting bolts to 8 ft. lbs. (11 Nm).

10. Reconnect vacuum hose to fuel pressure regulator.

11. Install surge tank cover if removed.

12. Connect the negative battery cable.

13. After installation, with engine **OFF** and ignition switch **ON**, check for fuel leaks around fuel line connection.

2007–08 XL-7

The fuel pressure regulator is located in the primary fuel tank pump assembly. See the Fuel Pump Removal & Installation procedure in this section.

FUEL PRESSURE RELIEF VALVE

REMOVAL & INSTALLATION

2007–08 Grand Vitara

See Figure 244.

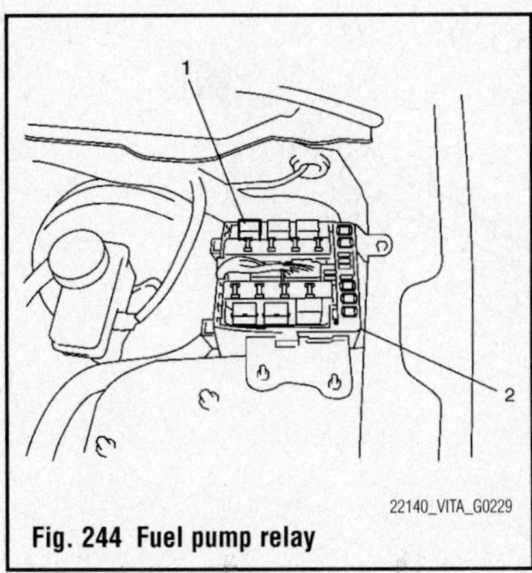

Fig. 244 Fuel pump relay

1. Make sure that engine is cold.

2. Shift transaxle gear shift lever in "Neutral" (shift select lever in **P** range for **A/T** model), set parking brake and block drive wheels.

3. Remove fuse box No.2 cover.

4. Disconnect fuel pump relay (1) from fuse box No.2 (2).

5. Remove fuel filter cap in order to release fuel vapor pressure in fuel tank, and then reinstall it.

6. Start engine and run it until engine stops for lack of fuel. Repeat cranking engine 2 – 3 times for about 3 seconds each time in order to dissipate fuel pressure in lines. Fuel connections are now safe for servicing.

7. After servicing, connect fuel pump relay (1) to fuse box No.2 and install fuse box No.2 cover

2007–08 XL-7

With J 34730-1A

1. Remove the engine cover, if required.

2. Loosen the fuel fill cap in order to relieve the fuel tank vapor pressure.

3. Remove the fuel rail service port cap.

4. Wrap a shop towel around the fuel rail service port.

5. Connect the J 34730-1A to the fuel rail service port.

6. Place the hose on the J 34730-1A into an approved gasoline container.

7. Open the valve on the J 34730-1A in order to bleed any fuel from the fuel rail.

8. Close the valve on the J 34730-1A.

9. Remove the hose on the J 34730-1A from the approved gasoline container.

10. Disconnect the J 34730-1A from the fuel rail service port.

11. Remove the shop towel from around the fuel rail service port, and place in an approved gasoline container.

12. Install the fuel rail service port cap.

13. Install the engine cover, if required.

14. Tighten the fuel fill cap.

Without J 34730-1A

1. Loosen the fuel fill cap in order to relieve the fuel tank vapor pressure.

2. Remove the engine cover, if required.

3. Remove the fuel rail service port cap.

4. Wrap a shop towel around the fuel rail service port and using a small flat bladed tool, depress (open) the fuel rail test port valve.

5. Remove the shop towel from around the fuel rail service port, and place in an approved gasoline container.

6. Install the fuel rail service port cap.

7. Install the engine cover, if required.

8. Tighten the fuel fill cap.

FUEL TANK

REMOVAL & INSTALLATION

2007–08 Grand Vitara

See Figure 245.

1. Before servicing the vehicle, refer to all safety precautions.

2. Relieve the fuel system pressure.

3. Raise vehicle.

4. Remove exhaust center pipe.

5. Remove rear propeller shaft.

6. With cable connected, detach parking brake cable clamp from fuel tank cover

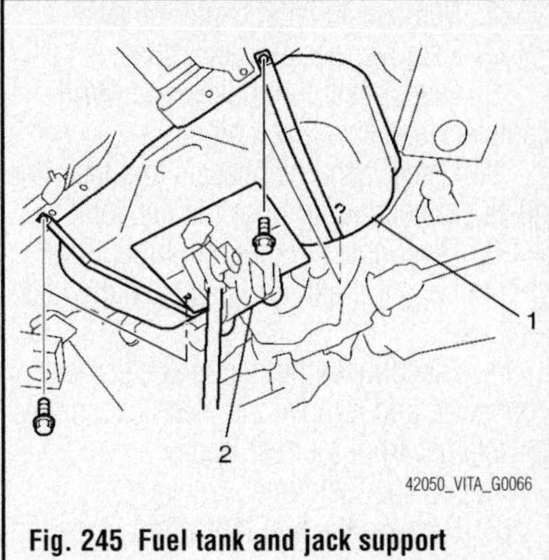

Fig. 245 Fuel tank and jack support

7. Disconnect fuel filler hose and breather hose from filler neck.

8. Disconnect fuel pump connector.

9. Due to absence of fuel tank drain plug, drain fuel tank by pumping fuel out through fuel tank filler.

10. Use hand operated pump device to drain fuel tank.

11. Disconnect fuel hoses and EVAP canister hose from each pipe.

12. Support fuel tank with jack and remove fuel tank.

To install:

13. If parts have been removed from fuel tank, install them before installing fuel tank to vehicle.

14. Raise fuel tank with jack, and connect connectors of fuel pump and sub fuel level gauge and clamp wire harness.

15. Install fuel tank to vehicle.

16. Tighten fuel tank mounting bolts to 37 ft. lbs. (50 Nm).

17. Connect fuel filler hose and breather hose to filler neck as shown in figure, and clamp them securely.

18. Connect fuel pump connector.

19. Connect fuel hoses and EVAP canister hose to each pipe as shown in figure, and clamp them securely.

20. Install parking brake cable clamp to fuel tank cover.

21. Install rear propeller shaft.

22. Install exhaust center pipe.

23. Connect the negative battery cable.

24. After installation, with engine OFF and ignition switch ON, check for fuel leaks around fuel line connections.

2007–08 XL-7

1. Before servicing the vehicle, refer to all safety precautions.

2. Relieve the fuel system pressure.

3. Raise vehicle.

4. Remove exhaust center pipe.

5. Disconnect the negative battery cable.

6. Remove exhaust center pipe.
7. Remove rear propeller shaft.
8. Disconnect fuel hoses and EVAP canister hose from each pipe.
9. Disconnect the chassis fuel feed line quick connect fitting from the fuel tank.
10. Disconnect the fuel fill tube, EVAP vent hose, and fresh air hose from the fuel tank.
11. Disconnect the fuel tank electrical connector and remove the electrical connector retainer from the rear frame.

Support the fuel tank.

12. Remove the fuel tank strap bolts and fuel tank straps.
13. Lower the fuel tank from the underbody of the vehicle.

To install:

14. If parts have been removed from fuel tank, install them before installing fuel tank to vehicle.
15. Raise fuel tank with jack, and connect connectors of fuel pump and sub fuel level gauge and clamp wire harness.
16. Install fuel tank to vehicle.
17. Install the fuel tank straps and the fuel tank strap-to-body bolts.
18. Tighten fuel tank mounting bolts to 18 ft. lbs. (25 Nm).
19. Connect the fuel tank electrical connector and install the electrical connector retainer to the rear frame.
20. Connect the EVAP vent, fresh air hoses and fuel fill to the fuel tank.
21. Connect the fuel fill tube to the fuel tank.
22. Connect the chassis fuel feed line quick connect fitting to the fuel tank.
23. Fill the fuel tank with gasoline.
24. Connect the negative battery cable.
25. Tighten the fuel fill cap.
26. Prime the fuel system:
    a. Cycle the ignition ON for 5 seconds and then OFF for 10 seconds.
    b. Repeat the previous step twice.
    c. Crank the engine until it starts. The maximum starter motor cranking time is 20 seconds.
    d. If the engine does not start, repeat steps 15.1–15.3.
27. Install the fuel injector sight shield.

FUEL RAIL & INJECTORS

REMOVAL & INSTALLATION

2.7L Engines

See Figure 246.

1. Before servicing the vehicle, refer to the precautions in the beginning of this section.
2. Relieve the fuel system pressure.
3. Disconnect the negative battery cable.
4. Remove surge tank cover.
5. Remove intake air pipe and surge tank pipe.
6. Disconnect fuel injector couplers and release fuel injector harness from clamps.
7. Disconnect fuel return hose from pressure regulator.
8. Disconnect vacuum hose from pressure regulator.
9. Remove fuel rail bolt.
10. Remove fuel union bolt of fuel rail (right side).
11. Remove fuel rails (right and left) from intake manifold.

To install:

12. Reverse removal procedure for installation noting the following.
13. Replace injector O-ring and insulator with new one using care not to damage it.
14. Check if cushion is scored or damaged. If it is, replace with new one.
15. Apply O-ring oil to O-rings, and then install injectors into fuel rail and intake manifold.
16. Make sure that injectors rotate smoothly. If not, probable cause is incorrect installation of O-ring.
17. Replace O-ring with new one.
18. Tighten fuel rail bolts to 18 ft. lbs. (25 Nm) and fuel union bolts to 22 ft. lbs. (30 Nm) and make sure that injectors rotate smoothly.
19. Connect the negative battery cable.
20. After installation, with engine **OFF** and ignition switch **ON**, check for fuel leaks around fuel line connection.

3.6L Engines

1. Disconnect the fuel feed pipe quick connect fitting (3) from the fuel rail.

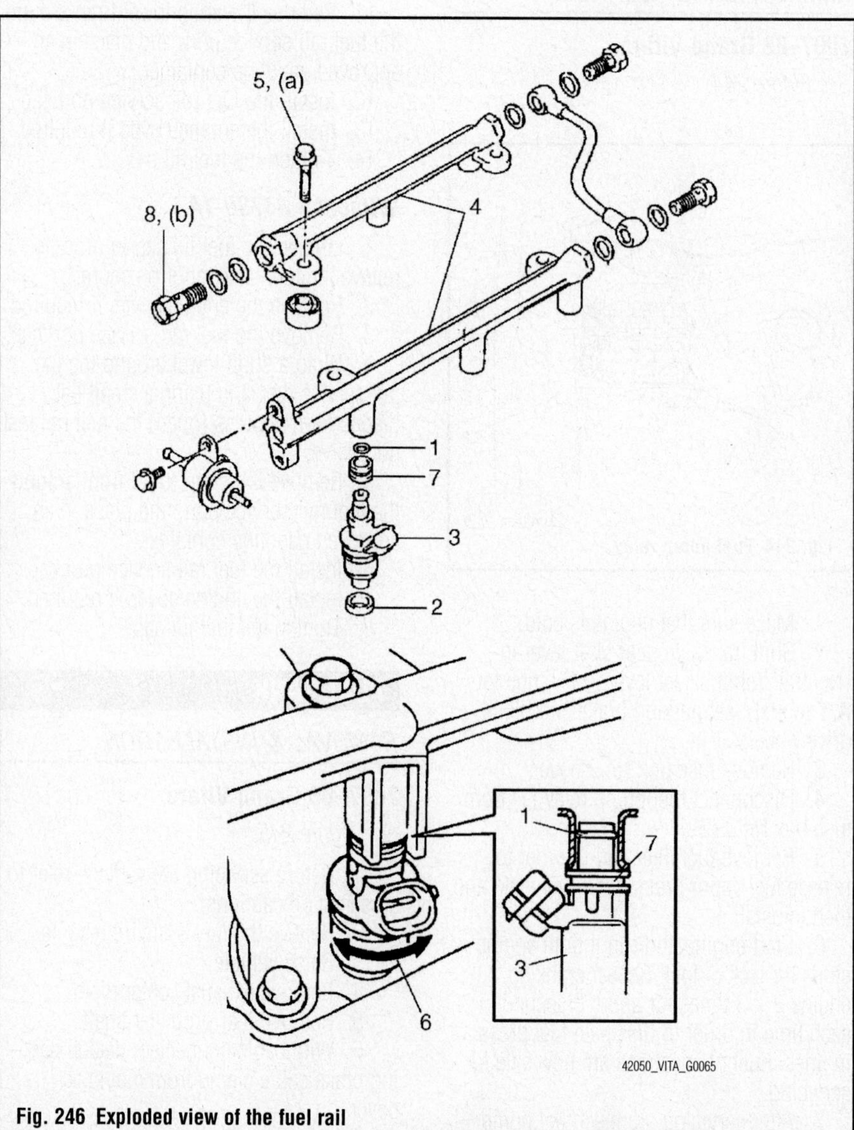

Fig. 246 Exploded view of the fuel rail

2. Use compressed air in order to remove any debris from the around the area where the fuel injectors enter the lower intake manifold.

3. Remove the fuel rail bolts.

4. Remove the fuel rail with fuel injectors from the lower intake manifold.

5. Disengage the fuel injector electrical connector lock.

6. Disconnect the fuel injector electrical connector.

7. Remove the fuel injector retainer clip.

8. Remove the fuel injector.

9. Remove and discard the fuel injector seals.

To install:

10. Install **NEW** fuel injector seals.

11. Install the fuel injector

12. Install the fuel injector retainer clip.

13. Install the fuel injector electrical connector.

14. Engage the fuel injector electrical connector lock.

15. Install the fuel rail with fuel injectors to the lower intake manifold.

16. Tighten the bolts to 89 inch lbs. (10 Nm).

17. Connect the fuel feed pipe quick connect fitting (3) to the fuel rail.

IDLE SPEED

ADJUSTMENT

Idle speed is maintained by the Powertrain Control Module (PCM). No adjustment is necessary or possible.

THROTTLE BODY

REMOVAL & INSTALLATION

2.7L Engines

See Figures 247 and 248.

1. Disconnect the negative battery cable.

2. Relieve the fuel system pressure.

3. Drain cooling system.

4. Remove surge tank cover.

5. Remove intake air pipe, surge tank pipe and IMT valve.

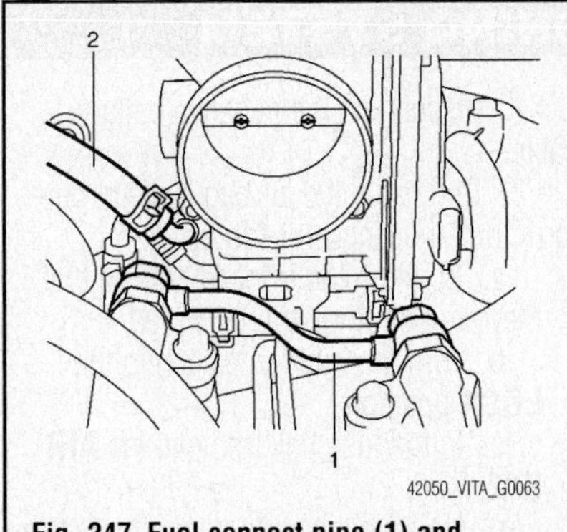

Fig. 247 Fuel connect pipe (1) and coolant hose (2) shown

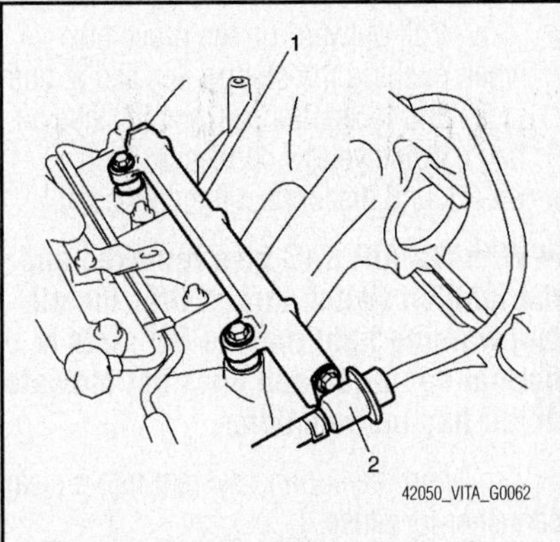

Fig. 248 Left bank fuel rail and regulator

6. Disconnect connectors from electric throttle body assembly and fuel injectors of left (No.1) bank.

7. Disconnect fuel rail connect pipe from left fuel rail (No.1) bank.

8. Disconnect coolant hose from electric throttle body assembly.

9. Remove left (No.1) bank fuel rail and fuel pressure regulator.

10. Remove electric throttle body assembly from intake collector.

To install:

11. Clean mating surfaces and use new throttle body gasket.

12. Install electric throttle body assembly.

13. Install left (No.1) bank fuel rail and fuel pressure regulator.

14. Tighten fuel rail mounting bolts to 18 ft. lbs. (25 Nm).

15. Reconnect coolant hose at throttle body.

16. Install fuel rail connect pipe to left fuel rail (No.1) bank.

17. Tighten fuel union bolts to 22 ft. lbs. (30 Nm).

18. Reconnect connectors to electric throttle body assembly and fuel injectors of left (No.1) bank.

19. Install intake air pipe, surge tank pipe and IMT valve.

20. Install surge tank cover.

21. Check to ensure that all removed parts are back in place.

22. Refill cooling system.

23. Connect the negative battery cable.

24. Verify that there is no fuel leakage at each connection.

3.6L Engines

1. Disconnect the negative battery cable.

2. Relieve the fuel system pressure.

3. Drain cooling system.

4. Disconnect the engine wiring harness electrical connector (1) from the throttle actuator control (TAC) module.

5. Remove the throttle body bolts.

6. Remove the throttle body and gasket. Discard the gasket.

To install:

7. Clean mating surfaces and use new throttle body gasket.

8. Install electric throttle body assembly.

9. Position a NEW throttle body gasket to the upper intake manifold.

10. Position the throttle body to the upper intake manifold.

11. Install the throttle body bolts.

12. Tighten the bolts to 89 inch lbs. (10 Nm).

13. 4.Connect the engine wiring harness electrical connector (1) to the TAC module.

14. Install the air cleaner outlet duct.

HEATING & AIR CONDITIONING SYSTEM

BLOWER MOTOR

REMOVAL & INSTALLATION

2007–08 Grand Vitara

1. Disconnect the negative battery cable.
2. Disable air bag system.
3. Remove ECM/PCM with bracket from blower unit, if applicable.
4. Disconnect blower motor lead wire at coupler.
5. Remove blower motor from blower unit.

To install:

6. Install blower motor.
7. Reconnect blower motor lead.
8. Reconnect ECM/PCM with bracket to blower unit, if removed.
9. Enable air bag system.
10. Connect the negative battery cable.

2007—2008 XL-7

See Figure 249.

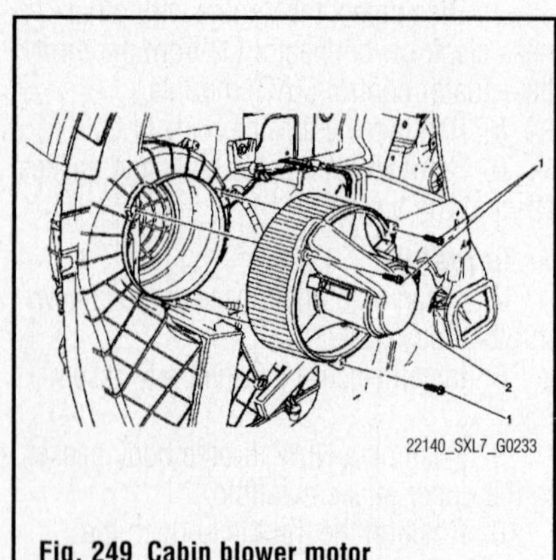

Fig. 249 Cabin blower motor

1. Remove the left rear trim panel.
2. Remove the auxiliary blower motor screws (Qty 3).
3. Disconnect electrical connection.

To install:

4. Connect electrical connection.
5. Install the auxiliary blower motor screws (Qty 3).
6. Tighten the screws to 13 inch lbs. (1.5 Nm).
7. Install the left rear trim panel.

HEATER CORE

REMOVAL & INSTALLATION

2007–08 Grand Vitara

See Figures 250 through 252.

1. Disconnect the negative battery cable.
2. To disable the air bag system, perform the following procedure:
   a. Position the front wheels so that they are pointing straight ahead.
   b. Turn the ignition switch to the **LOCK** position.
   c. In the fuse box, remove the **AIR BAG** fuse.
   d. Under the steering column, locate the contact coil/combination switch assembly's yellow connector; then, unlock and disconnect the connector.
   e. Pull outward on the glove box while pushing the stopper located at both sides and locate the passenger's side air bag module yellow connector; then, unlock and disconnect the connector.

➡With the AIR BAG fuse removed and the ignition switch turned ON ; the air bag warning light may be ON ; this is normal operation and does not indicate an air bag malfunction.

3. Drain the cooling system into a clean container for reuse.
4. Disconnect the heater hoses from the heater core.
5. Remove the instrument panel by performing the following procedure:
   a. Remove the console.
   b. Remove the glove box and the column hole cover.
   c. Disconnect the electrical connector and cables from the heater housing and blower motor assembly.
   d. Remove the steering column.
   e. Disconnect the speedometer connector and the speedometer assembly.
   f. Remove the hood opener.
   g. Disconnect the instrument panel electrical connectors.
   h. Remove the instrument panel-to-chassis screws and bolts.
   i. Using an assistant, remove the instrument panel.
6. If equipped with air conditioning, perform the following procedure:
   a. Discharge and recover the air conditioning refrigerant.
   b. If equipped with a G16 or a J20 engine, disconnect the suction pipe and liquid pipe from the air conditioning housing. Plug the openings to prevent contamination.
   c. If equipped with an H25 engine, disconnect the compressor suction pipe and receiver/drier outlet pipe from the air conditioning housing. Plug the openings to prevent contamination.
   d. Remove the blower motor assembly.
   e. Disconnect the thermistor wire coupler.
   f. Remove the air conditioning housing.
7. Disconnect the rear duct from the heater housing.
8. Disconnect the mode actuator electrical connectors.
9. If equipped, remove the air conditioning controller.
10. If equipped, disconnect and remove the Sensing and Diagnostic Module (SDM) or air bag controller module.
11. Remove the heater housing.
12. Remove the heater core pipe clamps and grommet.
13. Remove the heater core from the heater housing.

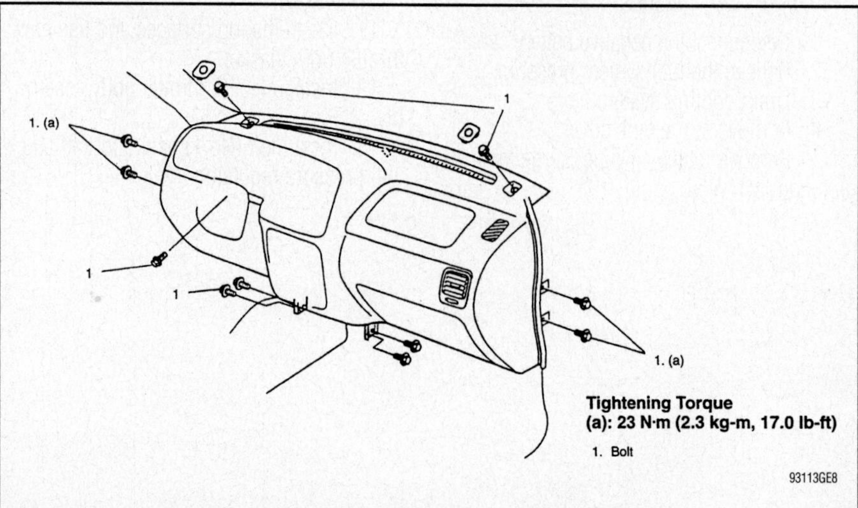

Fig. 250 View of the instrument panel and fasteners—Suzuki Grand Vitara

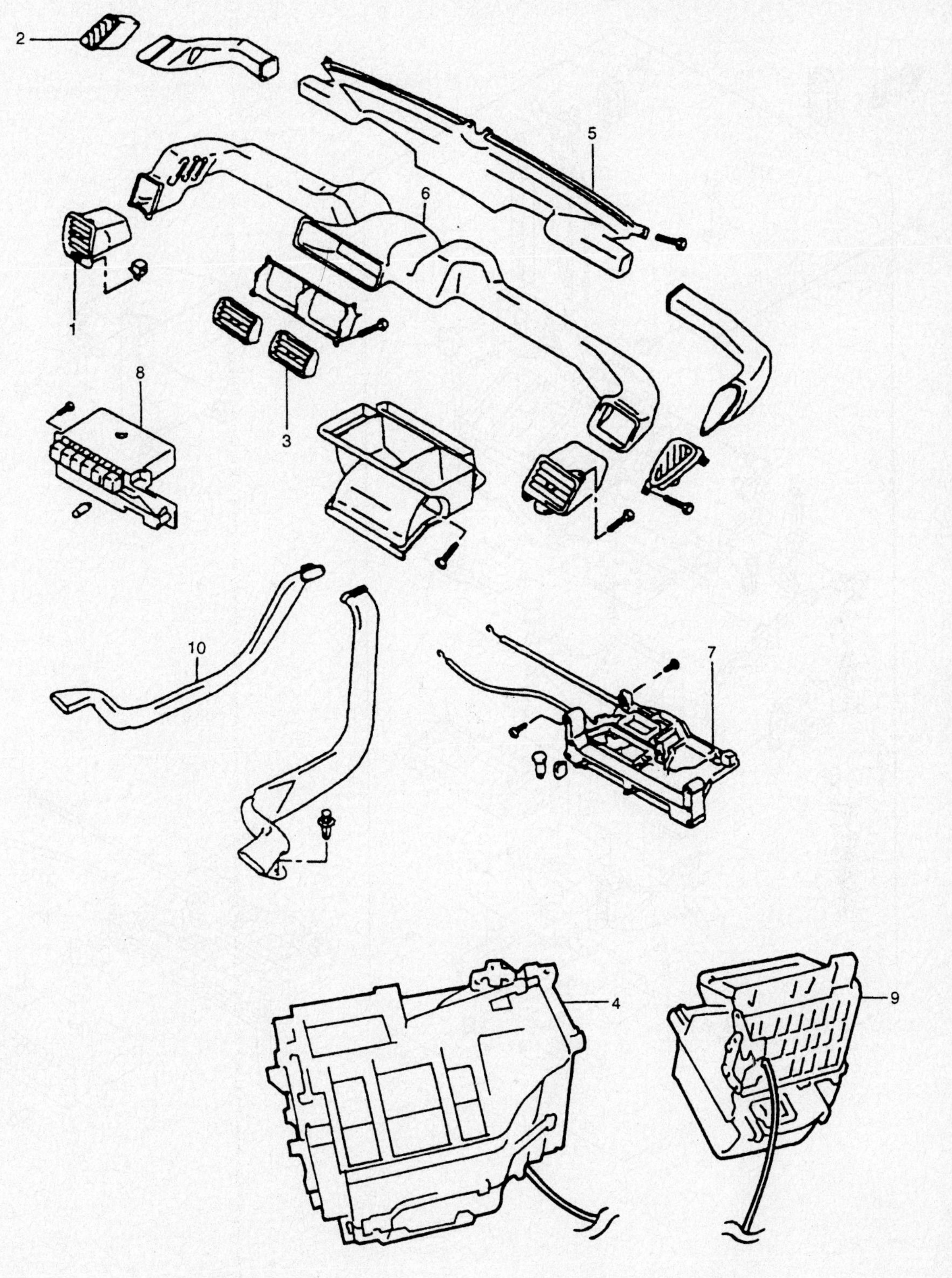

1. Side ventilator outlet
2. Side defroster outlet
3. Center ventilatior outlet
4. Heater unit
5. Defroster duct
6. Ventilator duct
7. Control lever
8. Mode control switch
9. Blower unit
10. Rear duct

93113GE9

Fig. 251 Exploded view of the heater housing and ventilation ducts—Suzuki Grand Vitara

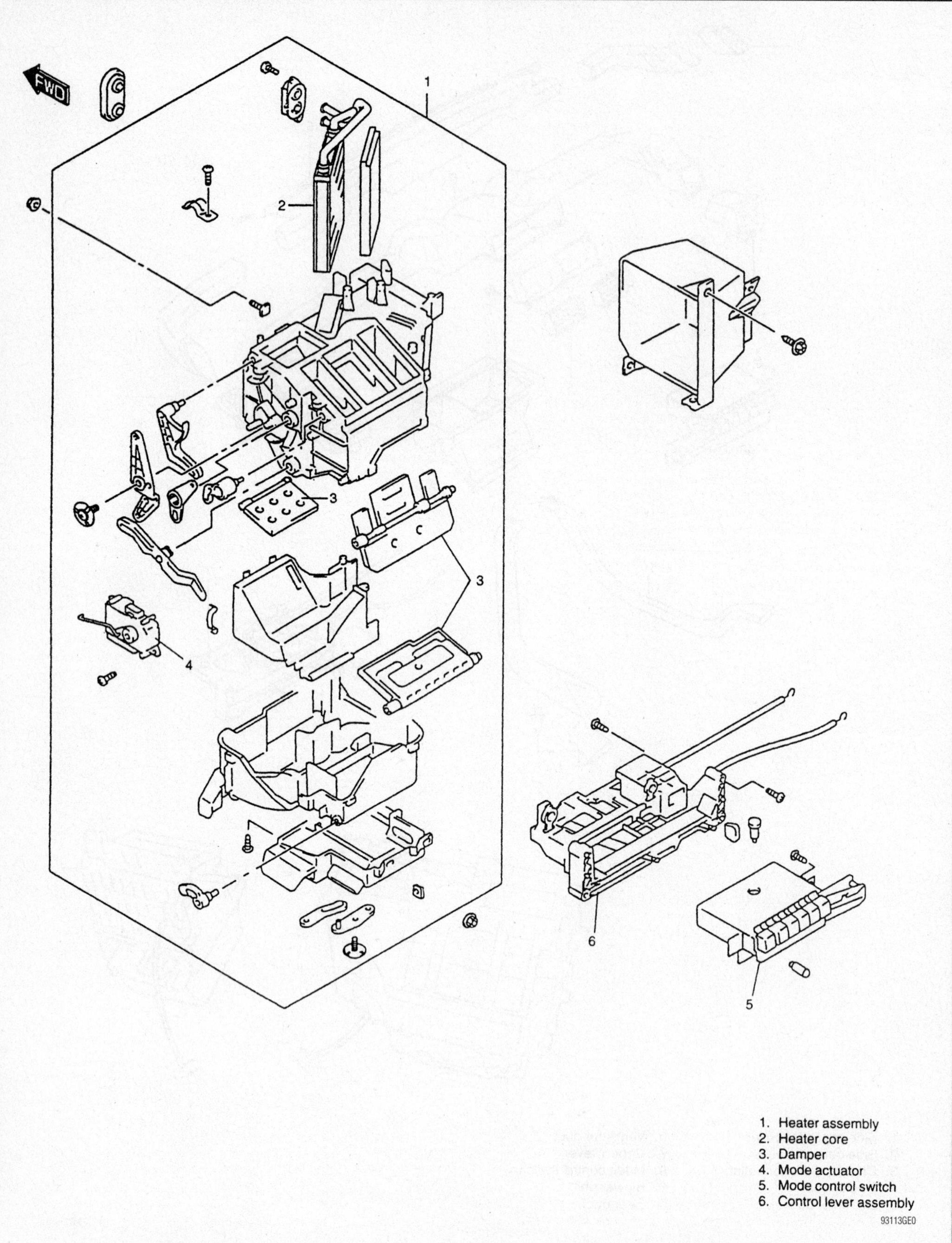

Fig. 252 Exploded view of the heater core, heater housing and related components—Suzuki Grand Vitara

To install:

14. Install the heater core to the heater housing.
15. Install the heater core pipe clamps and grommet.
16. Install the heater housing.
17. If equipped, connect and install the Sensing and Diagnostic Module (SDM) or air bag controller module.
18. If equipped, install the air conditioning controller.
19. Connect the mode actuator electrical connectors.
20. Connect the rear duct from the heater housing.
21. If equipped with air conditioning, perform the following procedure:
    a. Install the air conditioning housing.
    b. Connect the thermistor wire coupler.
    c. Install the blower motor assembly.
    d. If equipped with an H25 engine, connect the compressor suction pipe and receiver/drier outlet pipe to the air conditioning housing.
    e. If equipped with a G16 or a J20 engine, connect the suction pipe and liquid pipe to the air conditioning housing.
    f. Evacuate and charge the air conditioning system.
22. Install the instrument panel by performing the following procedure:
    a. Using an assistant, install the instrument panel.
    b. Install the instrument panel-to-chassis screws and bolts.
    c. Connect the instrument panel electrical connectors.
    d. Install the hood opener.
    e. Connect the speedometer connector and the speedometer assembly.
    f. Install the steering column.
    g. Connect the electrical connector and cables to the heater housing and blower motor assembly.
    h. Install the glove box and the column hole cover.
    i. Install the console.
23. Connect the heater hoses to the heater core.
24. Refill the cooling system.
25. To enable the air bag system, perform the following procedure:
    a. Push inward on the glove box while pushing the stopper located at both sides and connect the passenger's side air bag module yellow connector and lock it.
    b. Under the steering column, connect the contact coil/combination switch assembly's yellow connector.
    c. In the fuse box, install the **AIR BAG** fuse.
    d. Turn the ignition switch ON and verify that the **AIR BAG** warning light flashes 7 times and turns **OFF** ; if the system does not operate as described, perform the Air Bad Diagnostic System Check.
26. Connect the negative battery cable.
27. Run the engine to normal operating temperatures; then, check the climate control operation and check for leaks.

2007—2008 XL-7

See Figures 253 through 258.

1. Disconnect the negative battery cable.
2. To disable the air bag system, perform the following procedure:
    a. Position the front wheels so that they are pointing straight ahead.
    b. Turn the ignition switch to the LOCK position.
    c. In the fuse box, remove the AIR BAG fuse.
    d. Under the steering column, locate the contact coil/combination switch assembly's yellow connector; then, unlock and disconnect the connector.
3. Drain the cooling system into a clean container for reuse.
4. Disconnect the heater hoses from the heater core.
5. Remove the evaporator outlet hose and liquid line nut from the thermal expansion valve (TXV).
6. Remove the evaporator outlet hose from the TXV.
7. Remove and discard the sealing washer from the evaporator outlet hose.
8. Remove the evaporator outlet hose and liquid line from the TXV.
9. Remove and discard the sealing washer from the liquid line.
10. Install a protective caps to the evaporator outlet hose and the liquid line to prevent contamination and desiccant saturation
11. Reposition the heater outlet hose clamp at the heater core.
12. Remove the heater outlet hose from the heater core.
13. Reposition the heater inlet hose clamp at the heater core.
14. Remove the heater inlet hose at the heater core.
15. Plug the heater core and the evaporator core with clean towels to prevent spillage when the HVAC module is removed.
16. Remove the HVAC module seal nuts from the front of dash.
17. Remove the Instrument Panel (I/P) retainer.
18. Remove the shift control bracket.
19. Remove the center floor air outlet duct by sliding the duct forward then up at the rear.
20. Remove the center I/P air outlet duct retainers from the instrument panel tie bar.
21. Remove the center I/P air outlet duct from the instrument panel tie bar.
22. Remove the instrument panel tie bar.
23. Disconnect the blower motor electrical connector from the I/P wire harness.

22140_SXL7_G0247

Fig. 254 HVAC module seal nuts

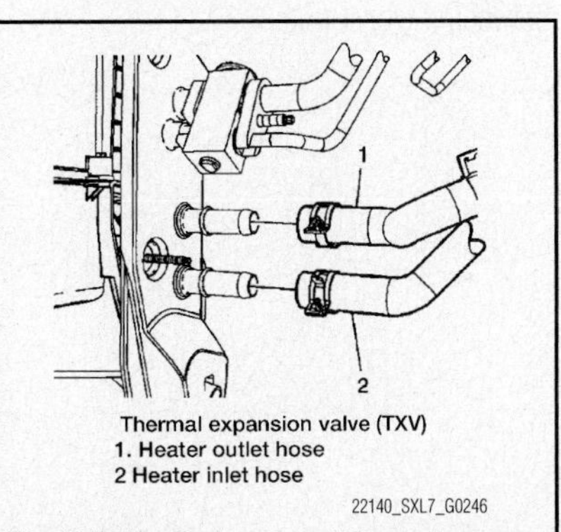

Fig. 253 Heater hoses and thermal expansion valve (TXV)

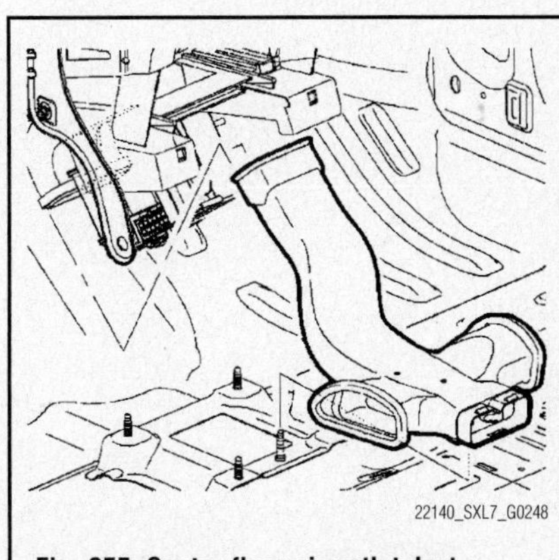

Fig. 255 Center floor air outlet duct

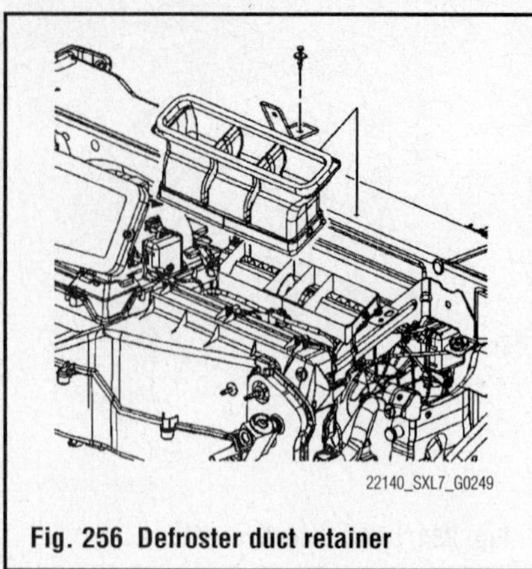

Fig. 256 Defroster duct retainer

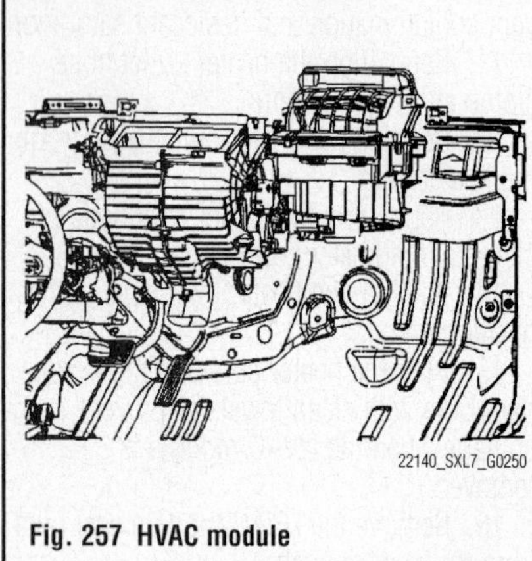

Fig. 257 HVAC module

24. Disconnect the blower motor control module electrical connector from the I/P wire harness.
25. Disconnect the HVAC module electrical connector from the I/P wire harness.
26. Remove the defroster duct retainer from the instrument panel tie bar.
27. Remove the defroster duct from the HVAC module.
28. Disconnect the I/P wire harness clips from the HVAC module.
29. Remove the HVAC module from the vehicle.
30. Remove the heater core cover screws from the HVAC module.

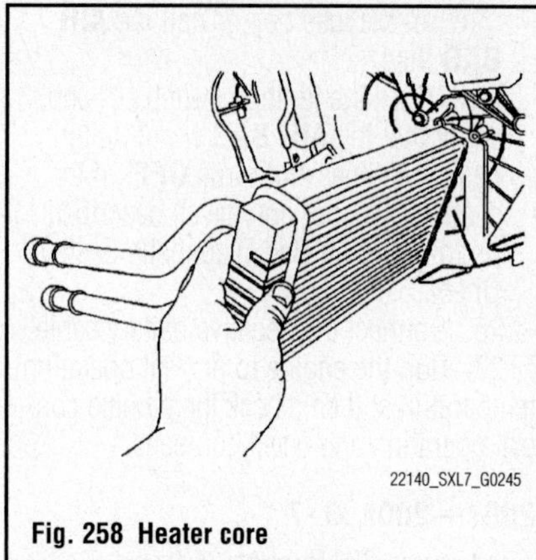

Fig. 258 Heater core

31. Remove the heater core cover from the HVAC module.

To install:

32. Install the heater core to the HVAC module.
33. Install the heater core cover to the HVAC module.
34. Install the heater core cover screws to the HVAC module.
35. Tighten the screw to 13 inch lbs. (1.5 Nm).

➡Make sure the HVAC module seals are flush and even as they meet their mating surfaces. This will reduce the chance of leaks and ensure proper fit.

36. Inspect the front of dash seal for proper alignment.
37. Inspect the seal mating surfaces to ensure there are no obstructions.
38. Position the HVAC module in the vehicle.
39. Install, but do not tighten, the seal nuts to the front of dash.
40. Install the instrument panel tie bar.

➡New front of dash seal nuts must be used to prevent leaks.

41. Tighten the HVAC module seal nuts to the front of dash. Draw the HVAC module to the front of dash evenly by alternating between the seal nuts.
42. Tighten the nuts to 80 inch lbs. (9 Nm).
43. Install the defroster duct to the HVAC module.
44. Install the defroster duct retainer to the instrument panel tie bar.
45. Connect the I/P wire harness clips to the HVAC module.
46. Connect the blower motor electrical connector to the I/P wire harness.
47. Connect the blower motor control module electrical connector to the I/P wire harness.
48. Connect the HVAC module electrical connector to the I/P wire harness.
49. Install the center I/P duct to the instrument panel tie bar.
50. Install the center I/P duct retainers to the instrument panel tie bar.
51. Install the center floor air outlet by sliding forward onto the front floor air outlet then down and rearward over the rear floor air outlet.
52. Install the shift control bracket
53. Install the I/P retainer.
54. Install the heater inlet hose to the heater core.
55. Install the heater inlet hose clamp to the heater core.
56. Install the heater outlet hose to heater core outlet.
57. Install the heater outlet hose clamp to the heater core.

➡Ensure the mating surfaces are clean and free of debris, and install new seal washers to the evaporator outlet hose and the liquid line.

58. Install the evaporator outlet hose and the liquid line to the TXV.
59. Install the evaporator outlet hose and liquid line nut to the TXV.
60. Enable the Supplemental Inflatable Restraints (SRS) system.
61. Fill the coolant.
62. Evacuate and charge the A/C system.
63. Test the affected A/C joints for leaks.

AUXILIARY HEATING & AIR CONDITIONING SYSTEM

BLOWER MOTOR

REMOVAL & INSTALLATION

2007–08 XL-7

See Figure 259.

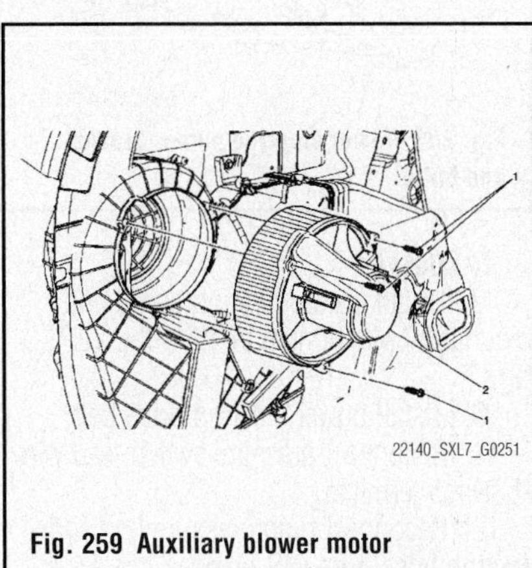

Fig. 259 Auxiliary blower motor

1. Remove the left rear trim panel.
2. Remove the auxiliary blower motor screws (Qty 3).
3. Disconnect electrical connection.
4. Remove the auxiliary blower motor.

To install:

5. Install the auxiliary blower motor.
6. Connect electrical connection.
7. Install the auxiliary blower motor screws.
8. Tighten the auxiliary blower motor screws to 13 inch lbs. (1.5 Nm).

HEATER CORE

REMOVAL & INSTALLATION

2007–08 XL-7

See Figure 260.

1. Drain the cooling system.
2. Recover the refrigerant.
3. Remove the left rear upper garnish molding.
4. Remove the air ducts from the HVAC module.
5. Disconnect the wiring connections.
6. Raise and suitably support the vehicle.

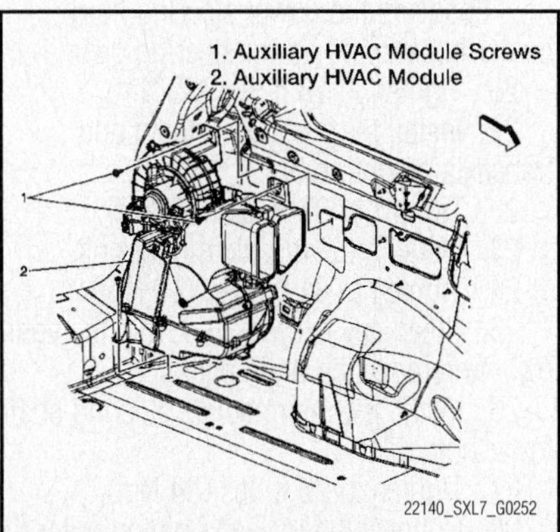

Fig. 260 Auxiliary HVAC Module

7. Remove the tire and wheel.
8. Remove the left inner wheelhouse liner.
9. Remove the rubber exhaust hangers to gain access to the A/C and the 3 heater pipes.
10. Disconnect the A/C and the heater pipes from the underside of the HVAC module.
11. Lower the vehicle.
12. Remove the HVAC module.

To install:

13. Install the HVAC module.
14. Tighten the HVAC module screws to 13 inch lbs. (1.5 Nm).
15. Raise and support the vehicle.
16. Connect the A/C and the heater pipes from the underside of the HVAC module.
17. Install the rubber exhaust hangers that were removed to gain access to the A/C and the 3 heater pipes.
18. Install the left inner wheelhouse liner.
19. Install the tire and wheel.
20. Connect the wiring connections
21. Install the air ducts to the HVAC module.
22. Install the left rear upper garnish molding.
23. Fill the cooling system
24. Evacuate and charge the A/C system.
25. Test the affected A/C joints for leaks

STEERING

POWER RACK & PINION STEERING GEAR

REMOVAL & INSTALLATION

2007–08 Grand Vitara

See Figures 261 and 262.

⁂ WARNING

Never turn steering wheel while steering column is removed or shafts are removed. Turning steering wheel more than about two and a half turns will break contact coil.

1. Disconnect the negative battery cable.
2. Before servicing the vehicle, refer to the precautions in the beginning of this section.
3. Take out P/S fluid in reservoir with syringe or such.
4. Raise vehicle and remove both right and left wheels.
5. Disconnect both right and left tie-rod ends from steering knuckles.
6. Disconnect steering lower shaft assembly from pinion shaft of P/S gear case.
7. Disconnect high pressure hose from P/S gear case assembly.
8. Disconnect low pressure hose from P/S gear case assembly.
9. Remove P/S gear case cylinder hoses from P/S gear case assembly.
10. Remove stabilizer bar mount bolt and stabilizer link.
11. Remove stabilizer bar mount bracket from left side of front suspension frame.
12. Remove bolts and then take off P/S gear case assembly from left side of vehicle.

➡P/S gear case assembly cannot be removed from the right side of vehicle.

To install:

13. Install power steering gear assembly and install and tighten bolts to 76 ft. lbs. (105 Nm).
14. Install stabilizer bar mount bracket and bolt, tighten bolt to 43 ft. lbs. (60 Nm).

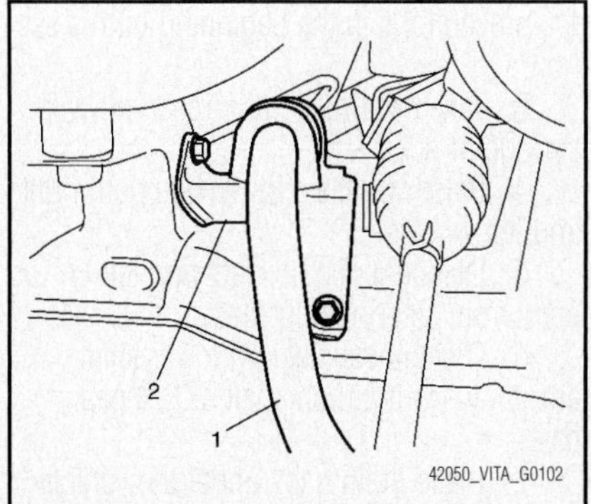

Fig. 261 Stabilizer bar and mounting bracket

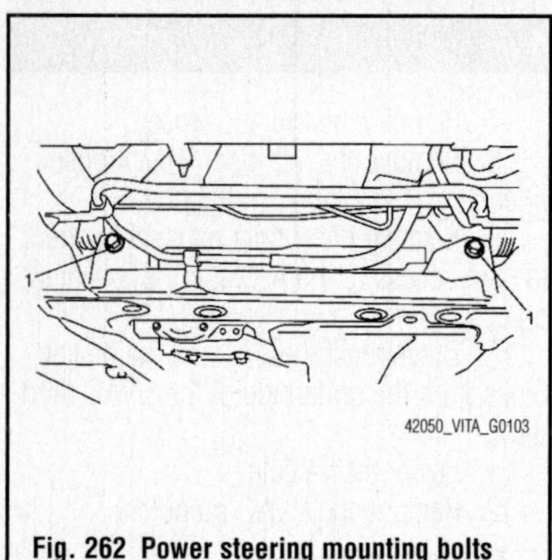

Fig. 262 Power steering mounting bolts

15. Install stabilizer control link and tighten to 44 ft. lbs. (60 Nm).
16. Reconnect power steering hoses and tighten to 26 ft. lbs. (35 Nm).
17. Reconnect steering lower shaft assembly from pinion shaft of P/S gear case.
18. Tighten pinch bolt to 19 ft. lbs. (25 Nm).
19. Reconnect both right and left tie-rod ends from steering knuckles.
20. Tighten tie rod end nuts to 31 ft. lbs. (43 Nm).
21. Lower vehicle and install both right and left wheels.
22. Refill power steering fluid reservoir with specified oil and bleed system.
23. Check for connections for any leaks.
24. Connect the negative battery cable.

2007–08 XL-7

⁂ WARNING

Never turn steering wheel while steering column is removed or shafts are removed. Turning steering wheel more than about two and a half turns will break contact coil.

1. Disconnect the negative battery cable.
2. Before servicing the vehicle, refer to the precautions in the beginning of this section.
3. Take out P/S fluid in reservoir with syringe or such.
4. Raise vehicle and remove both right and left wheels.
5. Disconnect both right and left tie-rod ends from steering knuckles.
6. Disconnect steering lower shaft assembly from pinion shaft of P/S gear case.
7. Place drain pans under the vehicle as needed.
8. Remove the transaxle mount bolt.
9. Remove the transaxle mount bolts and position the transaxle mount aside.
10. Disconnect the power steering gear inlet hose and the power steering cooler hose from the power steering gear
11. Remove the power steering gear hoses bracket bolt.
12. Remove the power steering gear bolts.
13. Remove the power steering gear through the left wheel house area.

To install:

14. Position the power steering gear into the vehicle through the left wheel house area.
15. Install the power steering gear bolts.
16. Tighten to 81 ft. lbs.(110 Nm).
17. Install the power steering gear hoses bracket bolt.
18. Tighten to 80 inch. lbs.(9 Nm).
19. Connect the power steering gear inlet hose and the power steering fluid cooler hose to the power steering gear.
20. Tighten to 18 ft. lbs.(25 Nm).
21. Install the transaxle mount and transaxle mount bolts.
22. Tighten to 37 ft. lbs.(50 Nm).
23. Install the transaxle mount bolt.
24. Tighten to 81 ft. lbs.(110 Nm).
25. Clean any excess fluid from the vehicle and remove the drain pans.
26. Install the intermediate steering shaft lower bolt.
27. Tighten to 25 ft. lbs.(34 Nm).
28. Connect the rack and pinion outer tie rods to the steering knuckles.
29. Tighten to 18 ft. lbs.(25 Nm) plus 90 degrees.
30. Fill and bleed the power steering system.
31. Adjust the front toe.

POWER STEERING PUMP

REMOVAL & INSTALLATION

2007–08 Grand Vitara

See Figure 263.

1. Before servicing the vehicle, refer to all service precaution.
2. Disconnect the negative battery cable.
3. Take out P/S fluid in reservoir with syringe or such.
4. Disconnect high pressure line and suction hose from P/S pump.
5. Disconnect pressure switch lead wire at switch terminal.
6. Remove P/S drive belt.
7. Remove P/S pump mounting bolts.
8. Remove P/S pump.

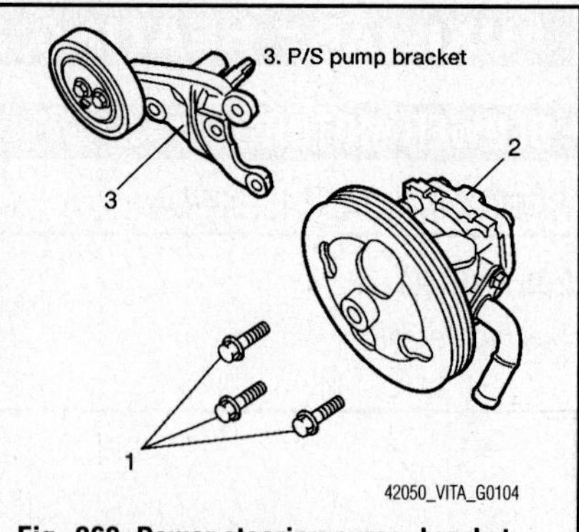

Fig. 263 Power steering pump, bracket and bolts

To install:

9. Install power steering pump and mounting bolts tighten to 19 ft. lbs. (25 Nm).
10. Install power steering drive belt.
11. Reconnect pressure switch lead wire at switch terminal.
12. Reconnect high pressure line and suction hose from P/S pump.
13. Tighten suction hose. Then high pressure line union bolt to 44 ft. lbs. (60 Nm).
14. Tighten power steering drive belt.
15. Refill power steering reservoir with specified oil and bleed system.
16. Connect the negative battery cable.

2007–08 XL-7

1. Before servicing the vehicle, refer to all service precaution.
2. Disconnect the negative battery cable.
3. Take out P/S fluid in reservoir with syringe or such.

Remove the right side engine splash shield

Place drain pans under the vehicle as needed.

4. Disconnect high pressure line and suction hose from P/S pump.
5. Disconnect pressure switch lead wire at switch terminal.
6. Remove P/S drive belt.
7. Remove P/S pump mounting bolts.
8. Remove P/S pump.

To install:

9. Install power steering pump and mounting bolts tighten to 7 ft. lbs. (50 Nm).
10. Install power steering drive belt.
11. Reconnect pressure switch lead wire at switch terminal.
12. Reconnect high pressure line and suction hose from P/S pump.
13. Tighten suction hose. Then high pressure line union bolt to 18 ft. lbs. (25 Nm).

14. Tighten power steering drive belt.
15. Refill power steering reservoir with specified oil and bleed system.
16. Connect the negative battery cable.

BLEEDING

1. Raise the front end of vehicle and apply safety stands.
2. Fill P/S fluid reservoir with fluid up to specified level.

➡Before starting engine, place transmission gear shift lever in "Neutral" (shift selector lever to "P" range for A/T model), and set parking brake.

3. After running engine at idling speed for 3 to 5 seconds, stop it and add fluid to satisfy specification.
4. With engine stopped, turn steering wheel to the right and left as far as it stops, repeat it a few times and fill fluid to specified level.
5. With engine running at idling speed, repeat stop-to-stop turn of steering wheel till all foams in P/S fluid reservoir are gone.

➡Make sure to bleed air completely. If air remains in fluid, P/S pump may make humming noise or steering wheel may feel heavy.

6. Finally check to make sure that fluid is filled to specified level.

STEERING LINKAGE

REMOVAL & INSTALLATION

2007–08 Grand Vitara

Upper Shaft

See Figures 264 through 267.

⁂ WARNING

Never turn steering wheel while steering column is removed or shafts are removed. Turning steering wheel more than about two and a half turns will break contact coil.

1. Before servicing the vehicle, refer to the precautions in the beginning of this section.
2. Turn steering wheel so that vehicle's front tires are at straight-ahead position.
3. Turn ignition switch to **LOCK** position and remove key.
4. Make alignment marks on shaft joint and shaft (upper shaft assembly side) for a guide during reinstallation.
5. After removing bolt on upper shaft assembly side of shaft joint and loosening

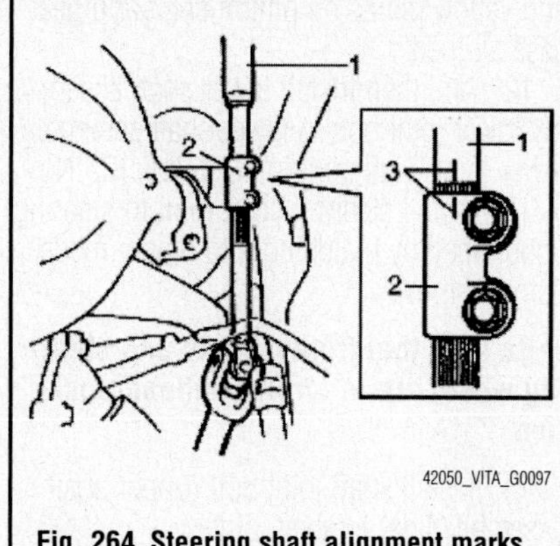

Fig. 264 Steering shaft alignment marks

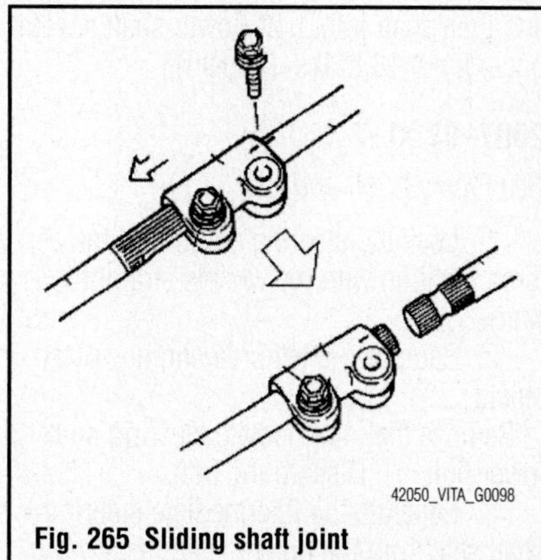

Fig. 265 Sliding shaft joint

bolt on its lower shaft assembly side, move shaft joint to lower shaft assembly side.
6. Remove steering upper shaft upper joint bolt and nut and disconnect steering upper shaft upper joint.
7. Remove steering upper shaft mounting bolts.
8. Remove steering upper shaft assembly from vehicle.

To install:

9. Be sure that front tires and steering wheel are in straight-ahead position.
10. Install steering upper shaft assembly to dash panel.

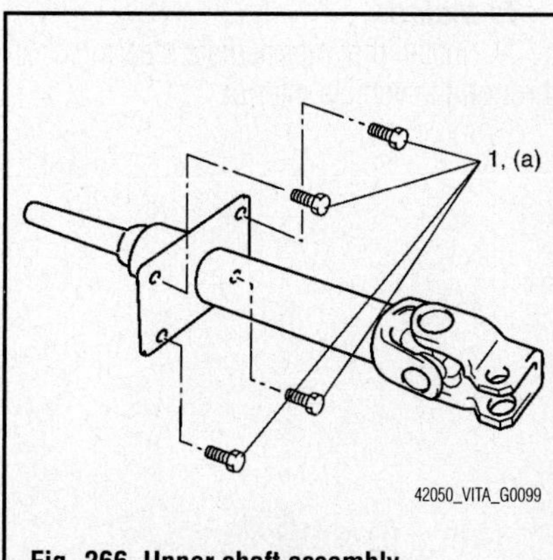

Fig. 266 Upper shaft assembly

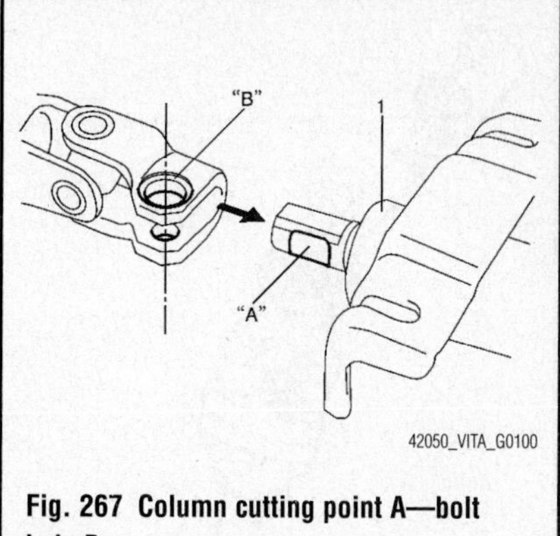

Fig. 267 Column cutting point A—bolt hole B

11. Tighten steering upper shaft mounting bolts to 17 ft. lbs. (23 Nm).
12. Align cutting point A of steering column assembly with bolt hole A of steering upper shaft upper joint as shown in the figure. Then connect steering upper shaft upper joint.
13. Install new steering upper shaft upper joint bolt and nut. Tighten steering upper shaft upper joint nut to 17 ft. lbs. (23 Nm).

➡Do not reuse steering upper shaft upper joint bolt and nut. Be sure to use new bolt and nut when installing.

14. Install steering shaft joint to steering upper shaft by matching it to marks made before removal.
15. Install shaft joint bolt (upper shaft assembly side) to steering shaft joint.
16. Tighten shaft joint bolt (upper shaft assembly side) 17 ft. lbs. (23 Nm), first and then shaft joint bolt (lower shaft assembly side) to 18 ft. lbs. (25 Nm).

Lower Shaft

See Figure 268.

⁂ WARNING

Never turn steering wheel while steering column is removed or shafts are removed. Turning steering wheel more than about two and a half turns will break contact coil.

1. Before servicing the vehicle, refer to the precautions in the beginning of this section.
2. Turn steering wheel so that vehicle's front tires are at straight-ahead position.
3. Turn ignition switch to LOCK position and remove key.
4. Make alignment marks on shaft joint and shaft (upper shaft assembly side) for a guide during reinstallation.
5. After removing bolt on upper shaft assembly side of shaft joint and loosening

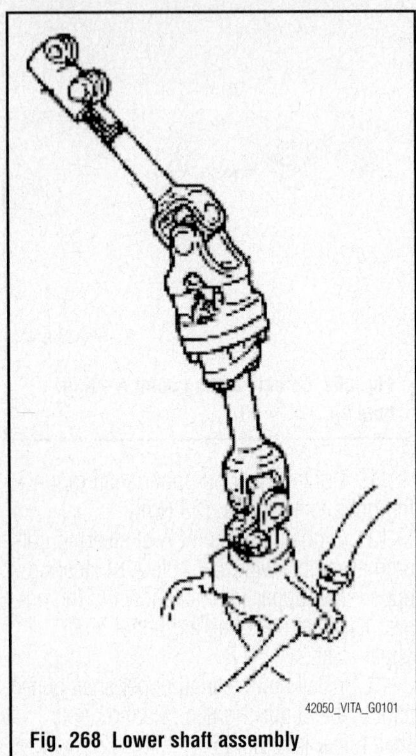
Fig. 268 Lower shaft assembly

bolt on its lower shaft assembly side, move shaft joint to lower shaft assembly side.

6. Remove lower shaft assembly lower joint bolt and then remove lower shaft assembly.

7. Remove shaft joint bolt (lower shaft assembly side) from shaft joint and then remove shaft joint from lower shaft assembly.

To install:

8. Be sure that front wheels and steering wheel are in straight-ahead position.

9. Align flat part of lower shaft assembly with bolt hole of shaft joint as shown in the figure. Then insert shaft joint into lower shaft assembly.

10. Install shaft joint bolt (lower shaft assembly side) to shaft joint. Then tighten it by hand.

11. Insert pinion shaft into lower shaft assembly lower joint with slit of lower joint and yellow marks on pinion shaft and gear case aligned.

12. And then install lower shaft assembly lower joint bolt to lower shaft assembly lower joint. Tighten it to 18 ft. lbs. (25 Nm).

13. Install steering shaft joint to steering upper shaft by matching it to marks made before removal.

➡Be sure that front wheels and steering wheel are in straight-ahead position.

14. Install shaft joint bolt (upper shaft assembly side) to shaft joint.

15. Tighten shaft joint bolt (upper shaft assembly side) to 18 ft. lbs. (25 Nm), first and then shaft joint bolt (lower shaft assembly side) to 18 ft. lbs. (25 Nm).

2007–08 XL-7

See Figures 269 and 270.

1. Lock the steering column in the centered position with the wheels straight forward.

2. Remove the left side engine splash shield.

Remove the intermediate steering shaft lower bolt (1). Discard the bolt.

3. Separate the intermediate steering shaft seal from the body panel pass through.

4. Separate the intermediate steering shaft from the power steering gear and collapse it into its shortest position.

5. Remove the driver knee bolster reinforcement.

6. Remove the intermediate steering shaft bolt (1) and nut (2).

➡Place scribe marks on the intermediate steering shaft to the steering column connection prior to removal.

7. Separate the intermediate steering shaft from the steering column.

8. Remove the intermediate steering shaft through the vehicle interior.

To install:

9. Install the intermediate steering shaft through the vehicle interior.

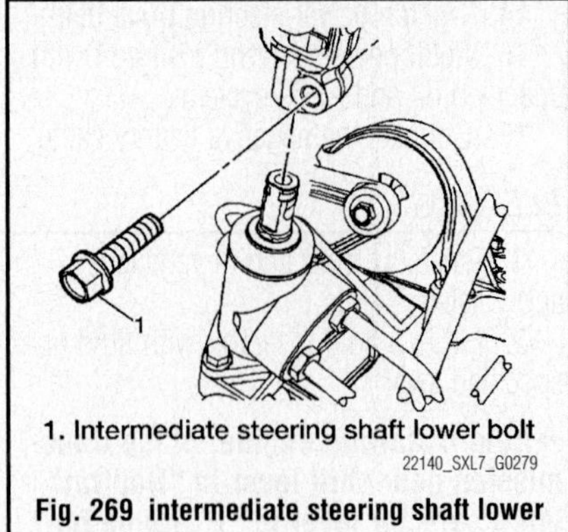

Fig. 269 Intermediate steering shaft lower bolt

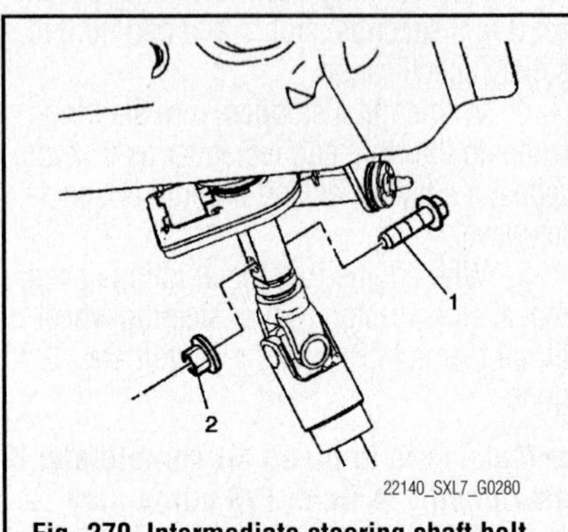

Fig. 270 Intermediate steering shaft bolt and nut

10. Connect the intermediate steering shaft to the steering column.

11. Install the intermediate steering shaft bolt (1) and nut (2).

12. Tighten the bolt and nut to 25 ft. lbs. (34 Nm).

13. Install the driver knee bolster reinforcement.

14. Connect the intermediate steering shaft to the power steering gear.

15. Install a new intermediate steering shaft lower bolt (1).

16. Tighten the bolt to 25 ft lbs. (34 Nm).

17. Seat the intermediate steering shaft seal into the body panel pass through.

18. Install the left side engine splash shield.

COIL SPRING

REMOVAL & INSTALLATION

1. Before servicing the vehicle, refer to the Precautions Section.
2. Remove the strut from the vehicle and install a spring compressor.
3. Remove the front cap.
4. Compress the coil spring so that the end of the spring comes away from the spring seat.
5. Remove or disconnect the following:
    - Upper strut mounting nut
    - Insulator
    - Upper spring seat
    - Compressed spring from the strut
    - Spring from the spring compressor

CAUTION

Do not use an impact gun.

To install:

6. Compress the spring and install it on the strut.
7. Install or connect the following:
    - Upper spring seat
    - Insulator
    - Upper strut mount and torque the nut to 40 ft. lbs. (55 Nm).
    - Strut to the vehicle
8. Check and/or adjust the wheel alignment.

LOWER BALL JOINT

REMOVAL & INSTALLATION

2007–08 Grand Vitara

The lower ball joint is serviced with the lower control arm as an assembly. If there is any damage to either part, the control arm assembly must be replaced as a complete unit.

2007–08 XL-7

See Figures 271 and 272.

1. Remove the lower control arm.
2. Place the control arm in a vise or suitable holding device.
3. Remove the ball joint rivets using the following procedure.
    a. Drill through the rivets using a 5/16 in (8 mm) drill bit.
    b. Enlarge the hole using a 31/64 in (12 mm) drill bit.
    c. Remove any remaining burs from the control arm.
    d. Remove the ball joint from the control arm. Note the position of the ball joint for reassembly.

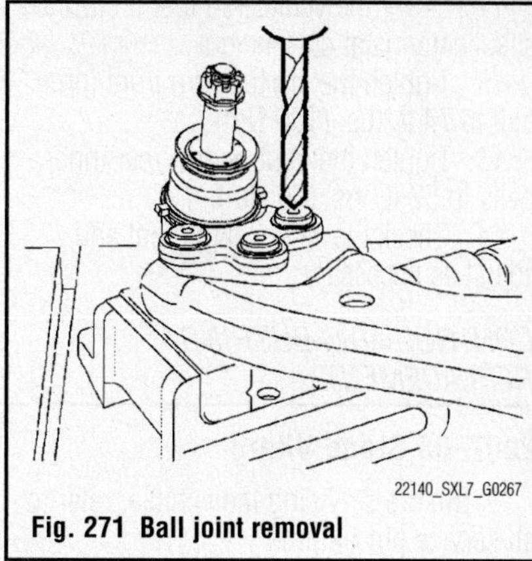

Fig. 271 Ball joint removal

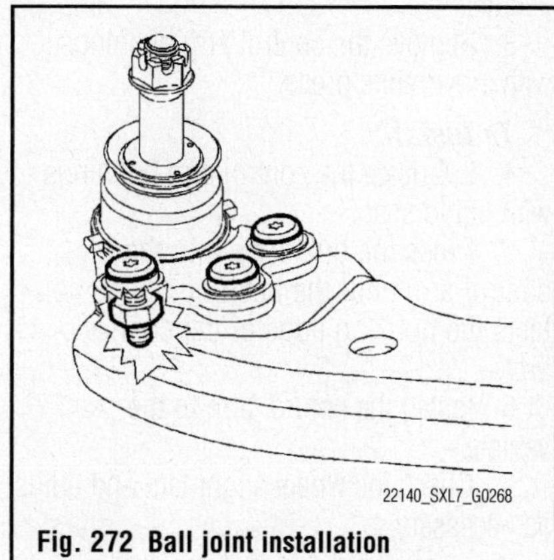

Fig. 272 Ball joint installation

To install:

➡**The control arm must be clean and free of debris.**

4. Install the ball joint to the control arm.

➡**Only use hardware provided with the new ball joint.**

➡**The bolts must be installed with the bolt head on top of the ball joint.**

5. Install the ball joint to control arm bolts.
6. Tighten the bolts and nuts to 50 ft. lbs. (68 Nm).
7. Install the lower control arm.

LOWER CONTROL ARM

REMOVAL & INSTALLATION

2007–08 Grand Vitara

1. Before servicing the vehicle, refer all service precautions.
2. Support the vehicle at the frame with a hoist or jackstand.
3. Support the control arm with a floor jack.
4. Remove or disconnect the following:
    - Front wheel
    - Brake caliper and rotor
    - Axle shaft snap ring and thrust washer, if equipped
    - Wheel speed sensor
    - Stabilizer bar link
    - Tie-rod end from steering knuckle
    - Lower ball joint
    - Strut bracket bolts
5. Remove the control arm-to-frame front bolt and nut. Discard the bolt and nut.
6. Lower the floor jack and remove the coil spring.
7. Remove the inner control arm bolts and remove the control arm.

To install:

8. Install the inner control arm bolts.
9. Install the coil spring onto the control arm and raise the floor jack.
10. Install or connect the following:
    - Strut bracket bolts. Tighten them to 70 ft. lbs. (95 Nm).
    - Lower ball joint. Tighten the nut to 40 ft. lbs. (55 Nm).
    - Stabilizer bar link. Tighten the nut to 21 ft. lbs. (29 Nm).
    - Wheel speed sensor, if equipped
    - Connect tie-rod end to steering knuckle. Tighten the nut to 21 ft. lbs. (43 Nm).
    - Brake caliper and rotor
    - Depress foot brake pedal and hold it there. Tighten new drive shaft nut to 159.5 ft. lbs. (220 Nm).
    - Front wheel
11. Lower the vehicle so that the front suspension is at curb height.
12. Tighten the front inner bolt to 62 ft. lbs. (85 Nm) and the rear inner bolt to 92 ft. lbs. (127 Nm).
13. Check the wheel alignment and adjust as necessary.

2007–08 XL-7

See Figures 273 and 274.

1. Before servicing the vehicle, refer all service precautions.
2. Support the vehicle at the frame with a hoist or jackstand.
3. Support the control arm with a floor jack.

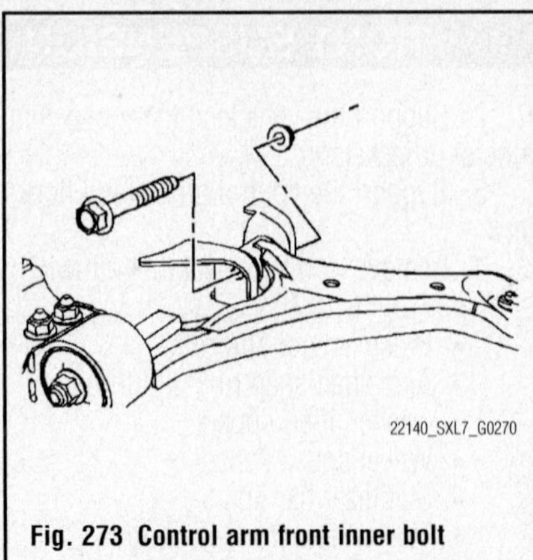

Fig. 273 Control arm front inner bolt

4. Remove or disconnect the following:
   - Front wheel
   - Brake caliper and rotor
   - Axle shaft nut
   - Wheel speed sensor, if equipped
   - Stabilizer bar link
   - Lower ball joint
   - Strut bracket bolts
5. Remove the control arm-to-frame front bolt and nut. Discard the bolt and nut.
6. Lower the floor jack and remove the coil spring.
7. Remove the inner control arm bolts and remove the control arm.

To install:

8. Install the inner control arm bolts.
9. Install the control arm and raise the floor jack.
10. Install or connect the following:
    - Strut bracket bolts. Tighten them to 70 ft. lbs. (95 Nm).
    - Lower ball joint. Tighten the nut to 30 ft. lbs. (40 Nm).
    - Stabilizer bar link. Tighten the nut to 21 ft. lbs. (29 Nm).
    - Wheel speed sensor, if equipped
    - Axle shaft snap ring and thrust washer, if equipped
    - Locking hub or drive flange, if equipped
    - Brake caliper and rotor
    - Front wheel
11. Lower the vehicle so that the front suspension is at curb height.
12. Tighten the control arm front inner bolt to 74 ft. lbs. (100 Nm).
13. Tighten the control arm rear inner bolts to 52 ft. lbs. (70 Nm).
14. Check the wheel alignment and adjust as necessary.

CONTROL ARM BUSHING REPLACEMENT

2007–08 Grand Vitara

1. Before servicing the vehicle, refer to all service precautions.
2. Remove the control arm from the vehicle.
3. Remove the control arm bushings with a hydraulic press.

To install:

4. Lubricate the control arm bushings with liquid soap.
5. Press the bushings into the control arm until the bushing flange contacts the housing edge of the control arm.
6. Install the control arm to the vehicle.
7. Check the wheel alignment and adjust as necessary.

2007–08 XL-7

See Figure 275.

1. Before servicing the vehicle, refer to all service precautions.
2. Remove the control arm from the vehicle.
3. Remove the control arm rear bushing nut.
4. Remove the rear bushing.

To install:

5. Lubricate the control arm bushings with liquid soap.
6. Press the bushings into the control arm until the bushing flange contacts the housing edge of the control arm.
7. Install the control arm to the vehicle.
8. Tighten the nut to 110 ft. lbs. (150 Nm).
9. Check the wheel alignment and adjust as necessary.

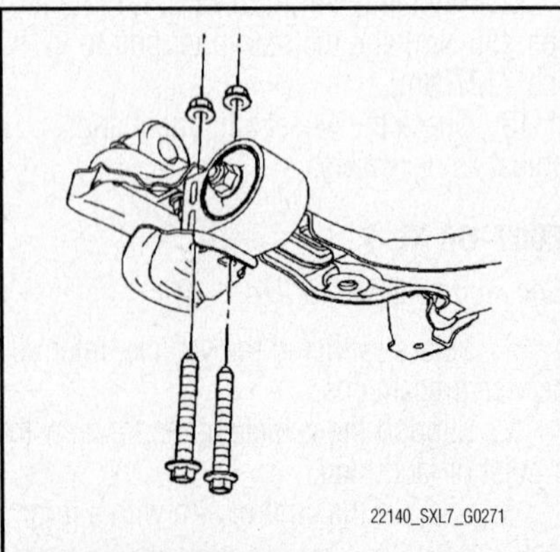

Fig. 274 Control arm rear inner bolts

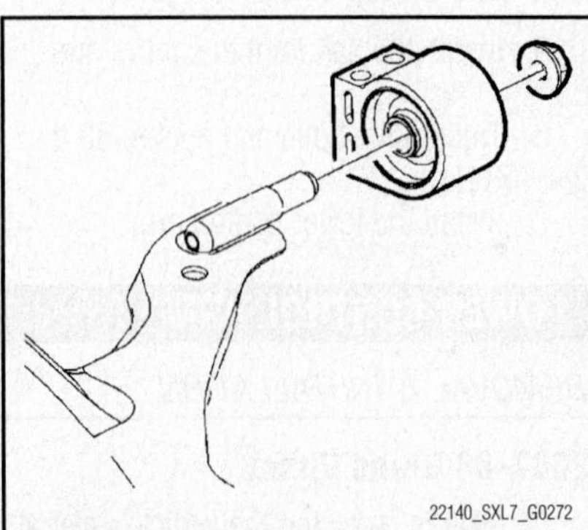

Fig. 275 Lower control arm rear bushing replacement

MACPHERSON STRUT

REMOVAL & INSTALLATION

2007–08 Grand Vitara

1. Before servicing the vehicle, refer to all service precautions.
2. Support the control arm with a stand or floor jack.
3. Remove or disconnect the following:
   - Front wheel
   - Brake hose bracket
   - Wheel speed sensor harness clamp bolt
   - Speed sensor harness
   - Stabilizer joint from strut bracket
   - Strut bracket bolts
   - Upper strut mount nuts
   - Strut

To install:

4. Install or connect the following:
   - Strut. Tighten the upper mount nuts to 40 ft. lbs. (55 Nm) and the bracket bolts to 98 ft. lbs. (135 Nm).
   - Brake hose bracket. Tighten brake hose mounting bolt to 18 ft. lbs. (25 Nm)
   - Speed sensor harness. Tighten speed sensor harness clamp mounting bolt to 8 ft. lbs. (10 Nm)
   - Stabilizer joint to strut bracket. Tighten stabilizer joint to strut bracket bolt to 43 ft. lbs. (60 Nm).
   - Front wheel. Tighten wheel nuts to 72.5 ft. lbs. (100 Nm)
5. Check the wheel alignment and adjust as necessary.

2007–08 XL-7

1. Before servicing the vehicle, refer to the precautions
2. Support the control arm with a stand or floor jack.
3. Remove or disconnect the following:
   - Front wheel
   - Brake hose bracket
   - Strut bracket bolts
   - Upper strut mount nuts
   - Strut

To install:

4. Install or connect the following:
   - Strut. Tighten the upper mount nuts to 40 ft. lbs. (55 Nm) and the bracket bolts to 70 ft. lbs. (95 Nm).
   - Brake hose bracket
   - Front wheel
5. Check the wheel alignment and adjust as necessary.

OVERHAUL

See Figures 276 through 281.

1. Attach special tool (A) to coil spring as shown.

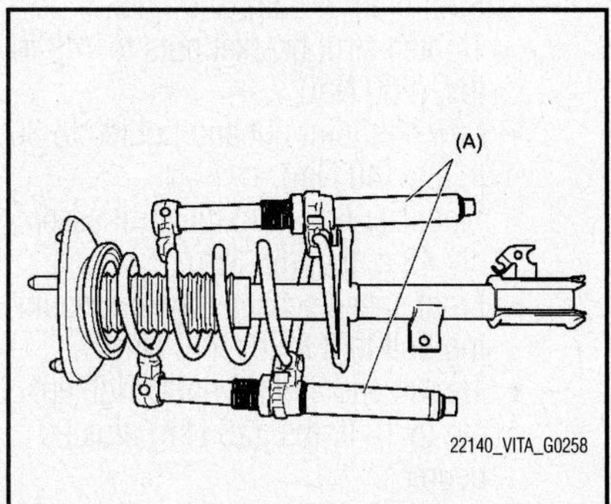

Fig. 276 Strut removal with spring compressed

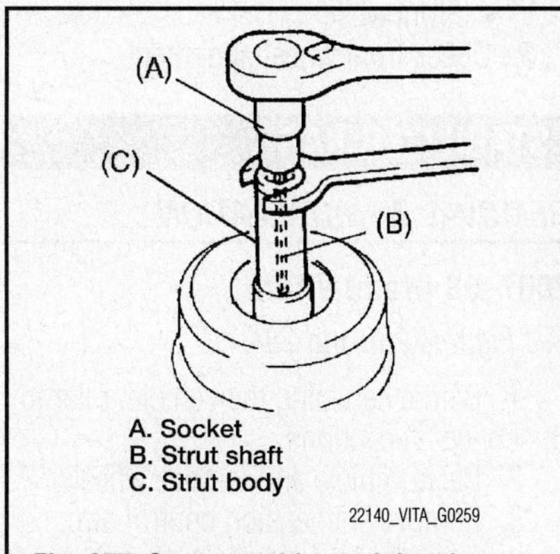

Fig. 277 Strut nut with special tools

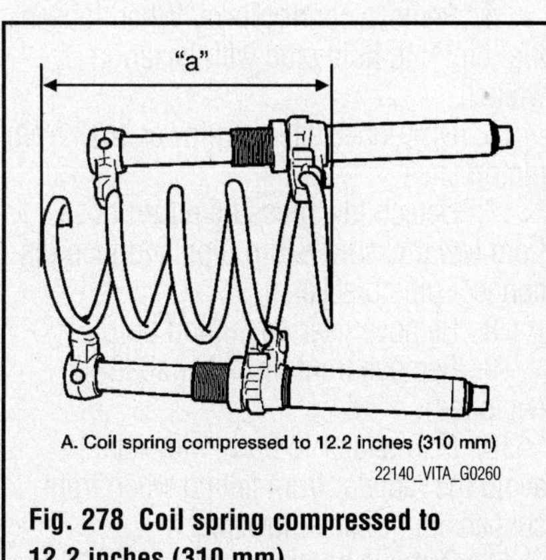

Fig. 278 Coil spring compressed to 12.2 inches (310 mm)

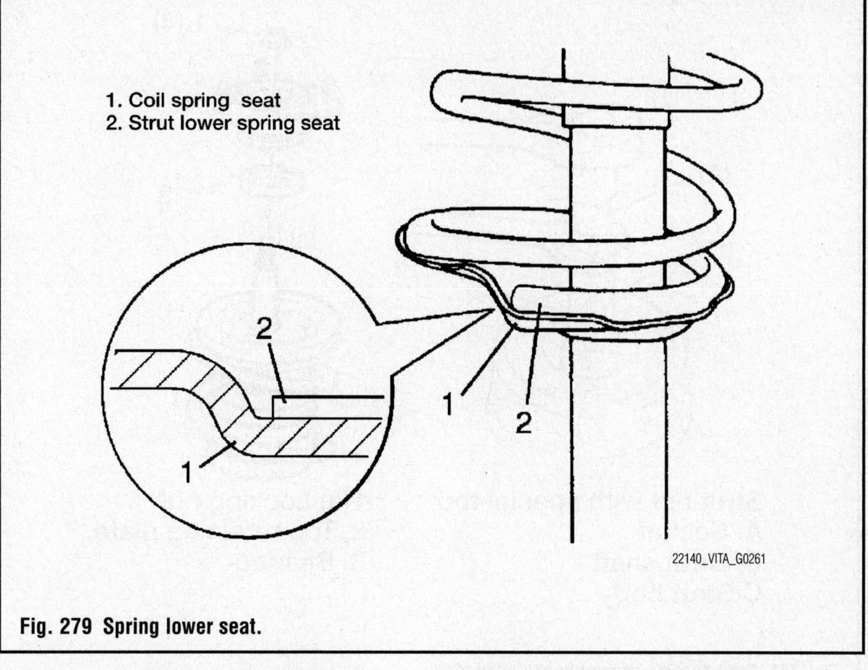

Fig. 279 Spring lower seat.

2. Turn special tool bolts alternately until coil spring tension is released.
3. Rotate the strut around its axis to confirm that the coil spring is released or not.
4. While keeping coil spring compressed, remove strut nut with special tools as shown.
5. Disassemble strut assembly.

To install:

6. Compress coil spring with special tool (A) until total length becomes about 310 mm (12.2 in.) as shown.
7. Install bump stopper onto strut rod.
8. Install compressed coil spring to strut, and place coil spring end (2) onto spring lower seat (1) as shown.

➡End of coil spring must not interfere with step of spring lower seat.

9. Pull strut rod as far up as possible and use care not to allow it to retract into strut.
10. Attach coil spring seat to coil spring upper seat and then install strut dust cover firmly.
11. Install coil spring upper seat with strut dust cover on coil spring and then spring upper seat (1) aligning **OUT** mark (3) on spring upper seat and center of strut bracket (2).
12. Install bearing (3), strut support (2) and strut nut (1) in this sequence.
13. Tighten strut nut (1) holding stud with special tools.

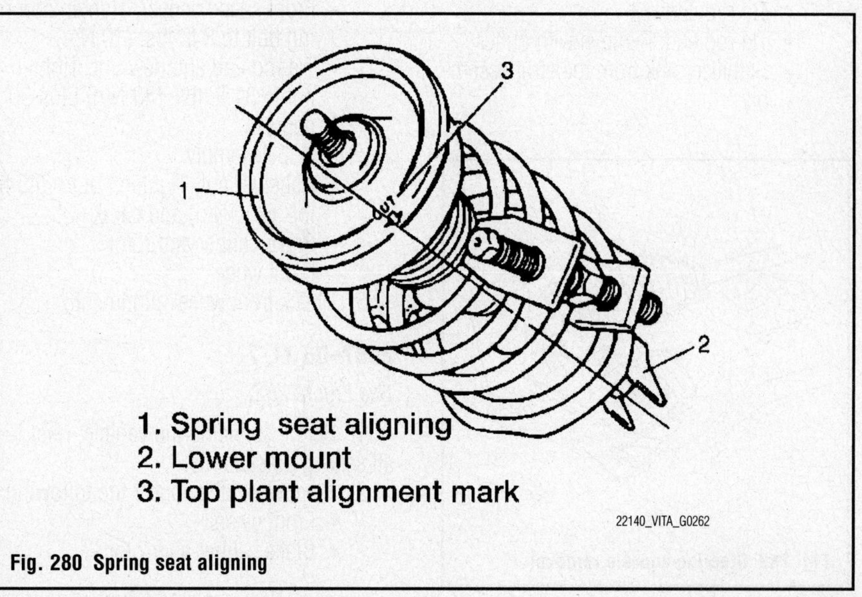

Fig. 280 Spring seat aligning

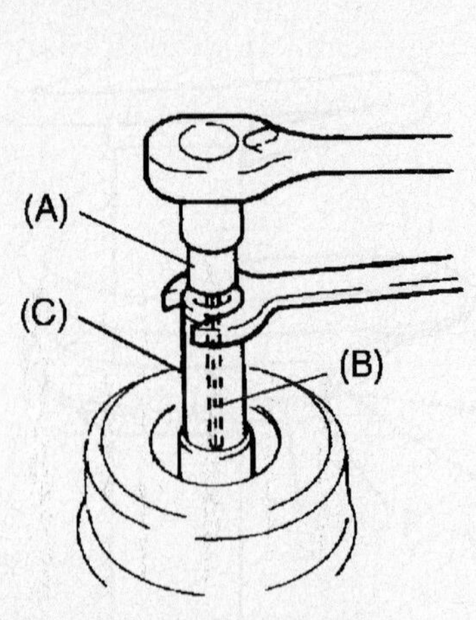

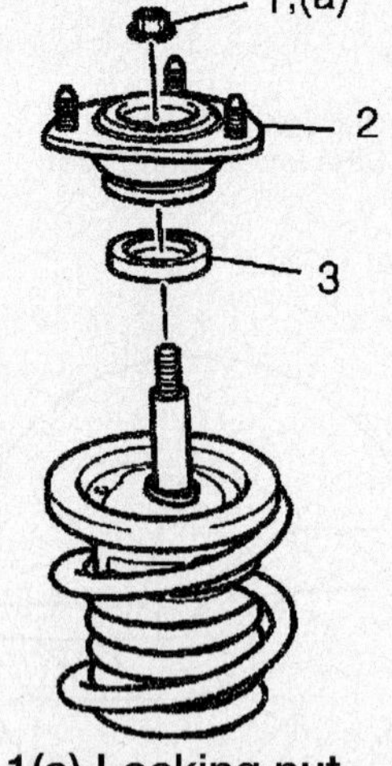

Strut nut with special tools
A. Socket
B. Strut shaft
C Strut body

1(a) Locking nut
2. Top retaining plate
3. Bearing

22140_VITA_G0263

Fig. 281 Strut bearing assembly exploded view

14. Tighten strut nut to 51 ft. lbs. (70 Nm).

STEERING KNUCKLE

REMOVAL & INSTALLATION

2007–08 Grand Vitara

See Figure 282.

1. Before servicing the vehicle, refer to all service precautions.
2. Remove or disconnect the following:
   - Front wheel
   - Brake caliper and rotor
   - Halfshaft nut
   - Wheel bearing—Hub assembly
   - Tie rod end nut
   - Tie rod end, Remove with puller
   - Stabilizer link from the strut assembly
   - Front wheel speed sensor
   - Ball joint nut
   - Strut bracket bolts and nuts
3. Disconnect ball joint from steering knuckle with puller and then remove steering knuckle.

To install:

4. Install or connect the following:
   - Steering knuckle to suspension arm
   - Strut bracket bolts and nuts, Tighten strut bracket nuts to 98 ft. lbs. (135 Nm).
   - New ball joint nut and tighten to 40 ft. lbs. (55 Nm).
   - Stabilizer link from the strut assembly 48 ft. lbs. (65 Nm).
   - Front speed sensor, Tighten mounting bolt to 8 ft. lbs. (10 Nm).
   - Tie rod end and new nut, Tighten nut to 31 ft. lbs. (43 Nm) plus 90 degrees.
   - Hub assembly
   - Halfshaft nut, Tighten nut to 245 ft. lbs. (200 Nm) and caulk nut.
   - Brake caliper and rotor
   - Front wheel
5. Check front wheel alignment.

42050_VITA_G0078

Fig. 282 Steering knuckle removal

2007–08 XL-7

See Figure 282.

1. Before servicing the vehicle, refer to all service precautions.
2. Remove or disconnect the following:
   - Front wheel
   - Brake caliper and rotor
   - Halfshaft nut
   - Wheel bearing—Hub assembly
   - Tie rod end nut
   - Tie rod end, Remove with puller
   - Stabilizer link from the strut assembly
   - Front wheel speed sensor
   - Ball joint nut
   - Strut bracket bolts and nuts
3. Disconnect ball joint from steering knuckle with puller and then remove steering knuckle.

To install:

4. Install or connect the following:
   - Steering knuckle to suspension arm
   - Strut bracket bolts and nuts, Tighten strut bracket nuts to 148 ft. lbs. (200 Nm).
   - New ball joint nut and tighten to 30 ft. lbs. (40 Nm).
   - Stabilizer link from the strut assembly 48 ft. lbs. (65 Nm).
   - Front speed sensor, Tighten mounting bolt to 8 ft. lbs. (10 Nm).
   - Tie rod end and new nut, Tighten nut to 18 ft. lbs. (25 Nm) plus 90 degrees.
   - Hub assembly
   - Halfshaft nut, Tighten nut to 245 ft. lbs. (200 Nm) and caulk nut.
   - Brake caliper and rotor
   - Front wheel
5. Check front wheel alignment.

STABILIZER BAR

REMOVAL & INSTALLATION

2007–08 Grand Vitara

See Figures 283 and 284.

1. Before servicing the vehicle, refer to all service precautions.
2. Raise vehicle and remove wheels.
3. Remove suspension control arm.
4. Remove right side and left side front halfshafts.
5. Remove control links. When loosening joint nut, hold stud with hexagon wrench.
6. Disconnect steering lower shaft from pinion shaft.
7. Detach low pressure return hose from low pressure return pipe and then disconnect pipe bracket.
8. Remove gear box union bolt.
9. Remove front propeller shaft (if equipped).
10. Fix radiator to body with rope to avoid the radiator from falling when front suspension frame is lowered.
11. Remove hood.

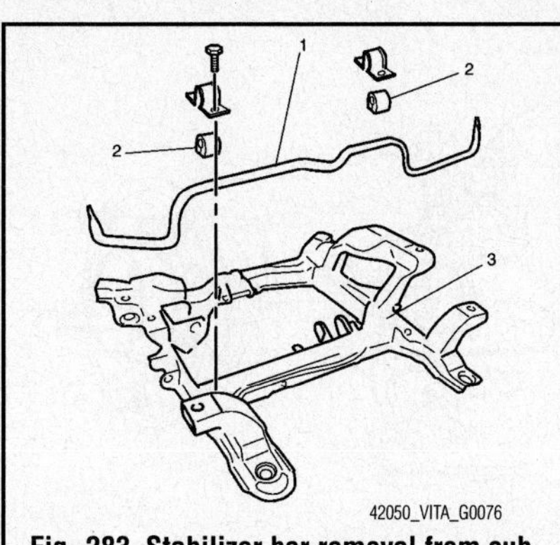

Fig. 283 Stabilizer bar removal from subframe

12. Support engine assemble by using chain hoist or engine support system.
13. Support suspension frame at the specified positions.
14. Remove engine front body side mounting nuts.
15. Remove suspension frame mounting bolts, and then lower suspension frame with stabilizer bar, P/S gear box assembly and front differential assembly (if equipped).
16. Remove P/S gear box assembly and front differential assembly (if equipped).
17. Remove stabilizer bar and bushing from suspension frame.
18. When installing stabilizer, loosely assemble all components while insuring that stabilizer is centered, side-to-side.
19. Install stabilizer bar, stabilizer bushing and stabilizer mounting bracket to suspension frame.

To install:

➡Install the stabilizer bar whose mark (5) is to front.

20. Tighten stabilizer bar mounting bracket bolts to 37 ft. lbs. (50 Nm).
21. Install P/S gear box assembly and front differential assembly (if equipped).
22. Install suspension frame.
23. Tighten suspension frame mounting bolts to 98 ft. lbs. (135 Nm). and engine front body side mounting nuts to 40 ft. lbs. (55 Nm).
24. Remove chain hoist or engine support system from engine.
25. Install hood.
26. Install front propeller shaft (if equipped).
27. Tighten pipe bracket bolts to 8 ft. lbs. (11 Nm) and then insert low pressure return hose to low pressure return pipe.
28. Tighten union gear box bolt to 26 ft. lbs. (35 Nm).
29. Connect steering lower shaft from pinion shaft.
30. Install control link and tighten nuts to 44 ft. lbs. (60 Nm).
31. Install right and left front halfshaft assembly (if equipped).
32. Install suspension control arm
33. Install engine under cover.
34. Install wheels and lower vehicle.
35. After installation, be sure to fill specified power steering fluid and bleed system.

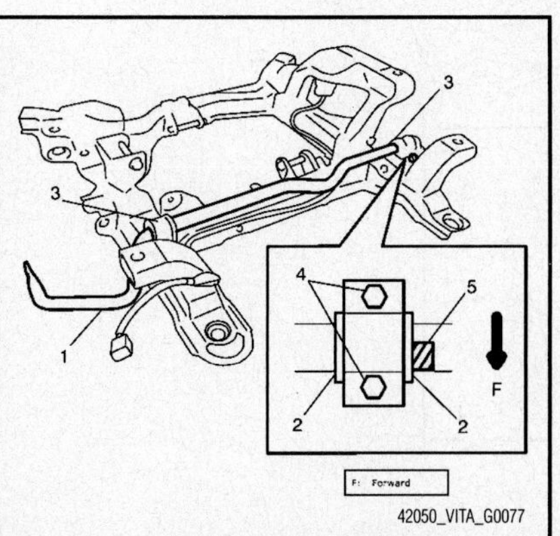

Fig. 284 Stabilizer bar positioning

2007–08 XL-7

See Figures 285 through 287.

1. Before servicing the vehicle, refer to all service precautions.

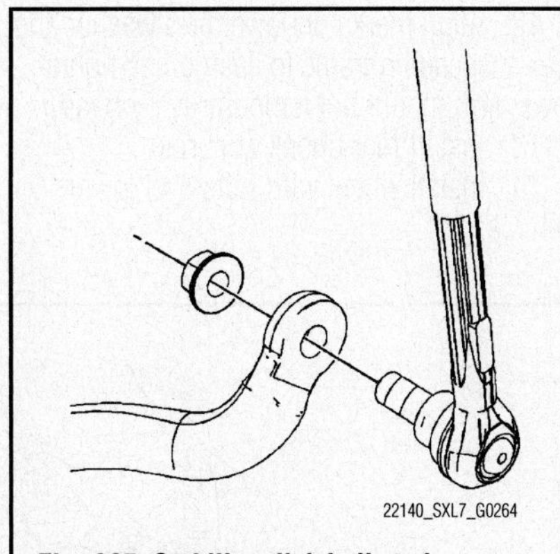

Fig. 285 Stabilizer link ball stud

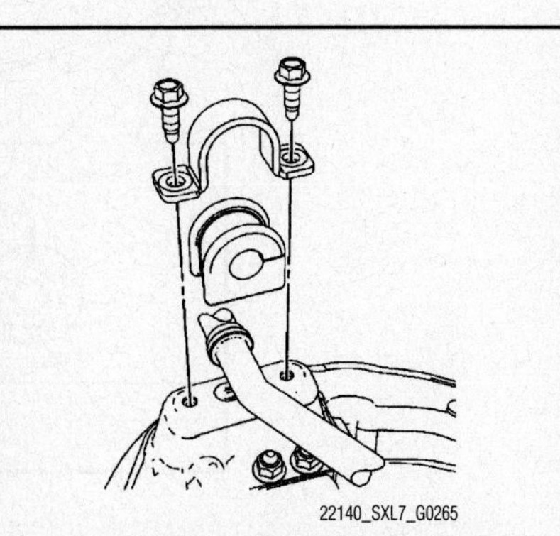

Fig. 286 Stabilizer shaft clamp and bushings

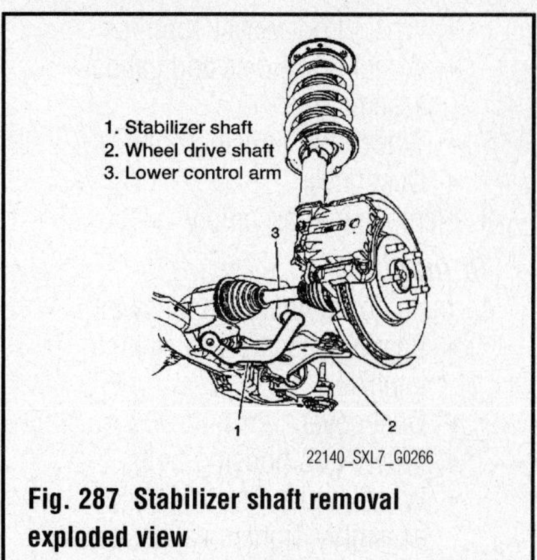

Fig. 287 Stabilizer shaft removal exploded view

2. Raise vehicle and remove wheels.
3. Remove stabilizer shaft links.
4. Use the proper size Allen wrench to keep the stabilizer link ball stud from rotate while removing or installing the nut.
5. Remove the outer tie rod ends from the knuckle.
6. Rotate the left and right knuckle all the way forward.
7. Remove the stabilizer shaft clamp bolts.
8. Remove the stabilizer shaft clamp and bushings.
9. Maneuver the stabilizer shaft (1) so that it is positioned between the wheel drive shaft (3) and the front lower control arm (2).
10. Remove the stabilizer shaft from the vehicle through the left wheel opening.

To install:

11. Position the stabilizer shaft (1) so that it is positioned between the wheel drive shaft (3) and the front lower control arm (2).
12. Maneuver the stabilizer shaft until it is properly position on the frame.
13. Install the stabilizer shaft bushing and clamps.
14. Tighten the stabilizer shaft clamp bolts to 37 ft. lbs. (50 Nm).
15. Install the stabilizer shaft links.
16. Install the outer tie rod ends.
17. Install the tires and wheels.
18. Remove the support.
19. Lower the vehicle.

WHEEL HUB AND BEARING

REMOVAL & INSTALLATION

2007–08 Grand Vitara

1. Before servicing the vehicle, refer all service precautions.
2. Remove or disconnect the following:
   - Front wheel
   - Brake caliper and rotor
   - Wheel speed sensor

- Wheel front wheel spindle
- Wheel brake disc and caliber assembly
- Wheel hub housing bolts
- Dust cover

3. Remove hub assembly

To install:

4. Install or connect the following:
   - Bearing assembly to steering knuckle
   - Dust cover
   - Wheel hub housing bolts
   - Wheel brake disc and caliber assembly Tighten caliper carrier bolt to 62 ft. lbs. (85 Nm).
   - Front wheel spindle
   - Drive shaft. Tighten the drive shaft nut to 160 ft. lbs. (220 Nm).
   - Wheel speed sensor. Tighten speed sensor bracket to 8 ft. lbs. (10 Nm).
   - Front wheel. Tighten the wheel nuts to 73 ft. lbs. (100 Nm).

2007–08 XL-7

See Figure 288.

1. Before servicing the vehicle, refer all service precautions.
2. Remove or disconnect the following:
   - Front wheel
   - Brake caliper and rotor
   - Wheel speed sensor
   - Front wheel drive shaft spindle nut.
   - Wheel hub and bearing assembly

To install:

3. Install or connect the following:
   - Snap ring
   - Wheel hub and bearing assembly
   - Wheel bearing locknut.
   - Tighten the nut to 151 ft. lbs. (205 Nm).
   - Wheel speed sensor. Tighten speed sensor bracket to 8 ft. lbs. (10 Nm).
   - Brake caliper and rotor

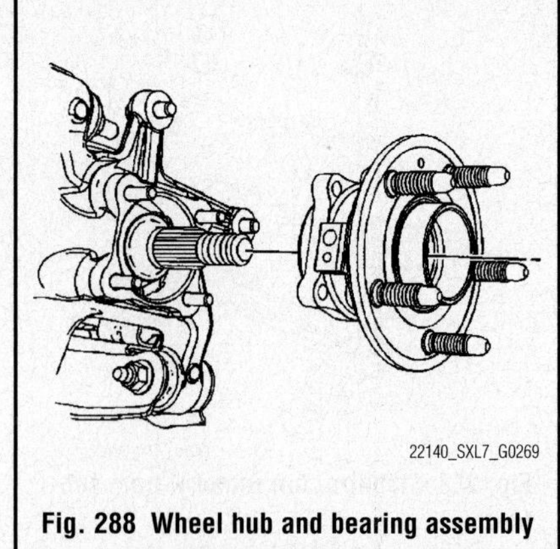

Fig. 288 Wheel hub and bearing assembly

- Front wheel. Tighten the wheel nuts to 73 ft. lbs. (100 Nm).

ADJUSTMENT

Wheel bearings are a sealed unit and do not require any adjustments. If play is detected in the bearing assembly the unit must be replaced.

SUSPENSION

REAR SUSPENSION

COIL SPRING

REMOVAL & INSTALLATION

2007–08 Grand Vitara

See Figures 289 and 290.

1. Before servicing the vehicle, refer to all service precautions.
2. Raise vehicle, allowing rear suspension to hang free.
3. Remove rear wheels.
4. Remove rear shock absorber
5. Put match marks on lower arm washer and on suspension frame to install the bolts correctly in position.
6. Loosen lower arm mount nut.
7. Remove lower arm outer bolt.
8. Lower jack and then remove rear coil spring and coil spring rubber seat.

To install:

9. Installing coil spring on lower arm and place coil spring end onto lower arm.

⁂ WARNING

Flat end of coil spring is upward. Upper end of coil spring has to be firmly mated to coil spring rubber seat. Lower end of coil spring has to be match to marking (3) in lower figure. End of the coil spring must not interfere with step of spring lower seat.

10. Support lower arm with jack.
11. Hoist jack and then install lower arm outer bolt and tighten bolt temporarily by hand.
12. Align marks on lower arm washer and rear suspension frame to each other, tighten lower arm mount nut temporarily by hand.
13. Install rear shock absorber
14. Install wheel with nuts and lower vehicle.
15. Tighten wheel nuts to 72 ft. lbs. (100 Nm).
16. Tighten lower arm outer bolt and lower arm mount nut to 98 ft. lbs. (135 Nm).

➡Tighten lower arm nut and bolt in non loaded condition.

17. Tighten shock absorber upper bolt to 44 ft. lbs. (60 Nm).

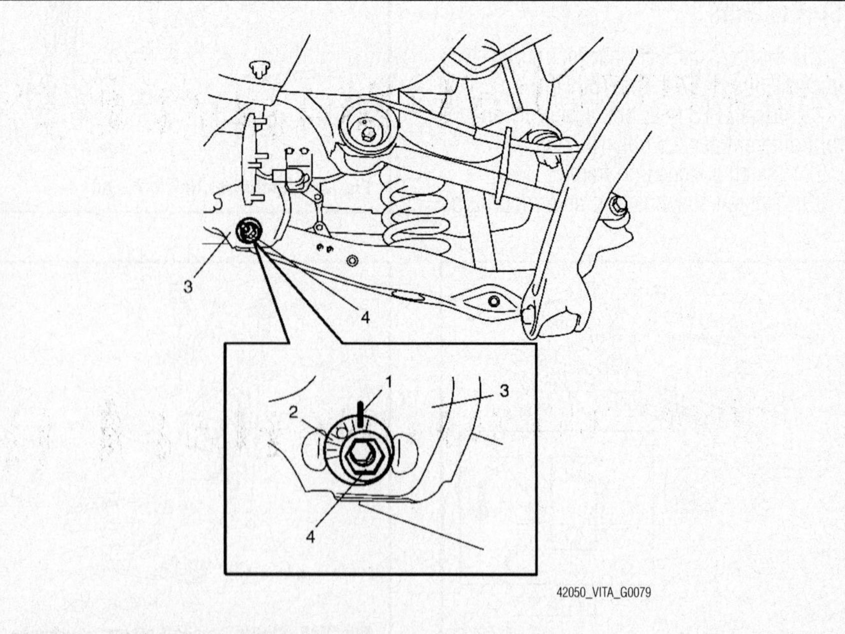

Fig. 289 Lower control arm alignment mark

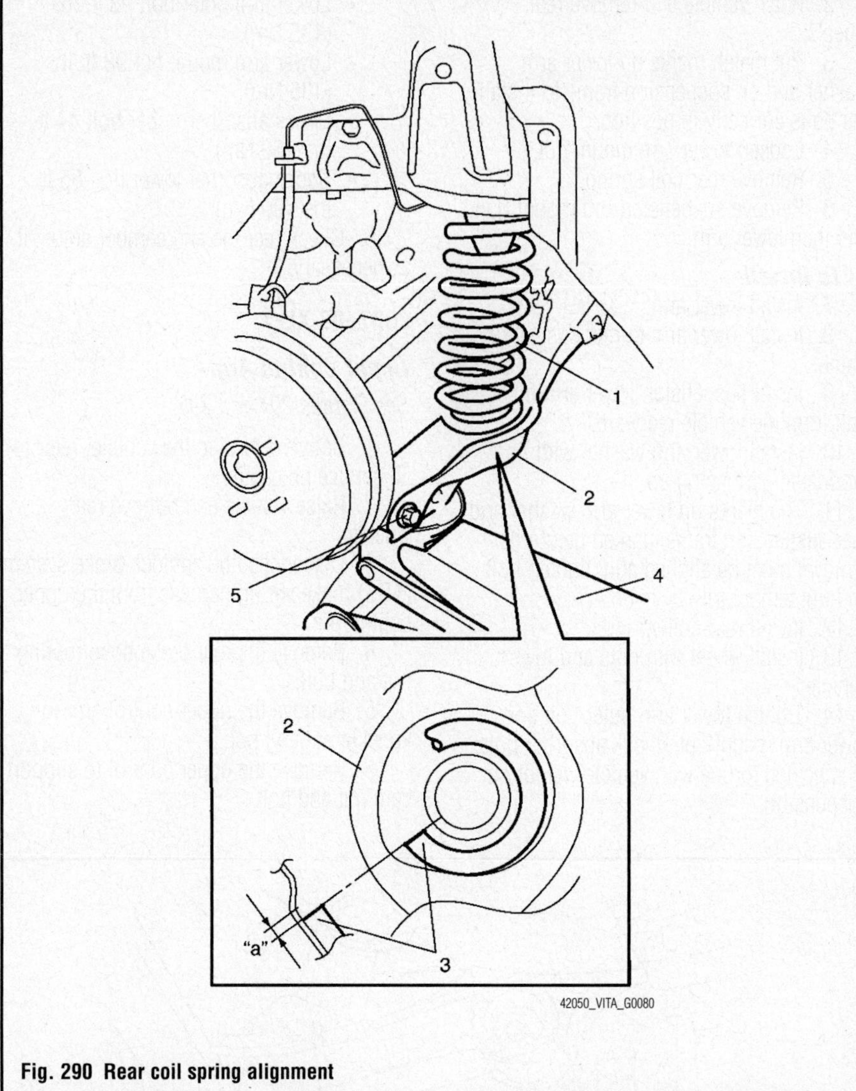

Fig. 290 Rear coil spring alignment

18. Tighten lower shock absorber bolt to 65 ft. lbs. (90 Nm). with vehicle weight on suspension.

19. Check rear toe and camber adjust it as necessary

2007–08 XL-7

See Figures 289 and 290.

1. Before servicing the vehicle, refer to all service precautions.
2. Raise vehicle, allowing rear suspension to hang free.
3. Remove rear wheels.
4. Remove the stabilizer shaft link.
5. Position a jackstand underneath the lower control arm.
6. Raise the jackstand slightly to compress the coil spring.
7. Remove rear shock absorber
8. Put match marks on lower arm washer and on suspension frame to install the bolts correctly in position.
9. Loosen lower arm mount nut.
10. Remove lower arm outer bolt.
11. Slowly lower the jack and then remove rear coil spring and coil spring rubber seat.

To install:

12. Installing coil spring on lower arm and place coil spring end onto lower arm.

⁂ WARNING

The flat end of coil spring is upward. Upper end of coil spring has to be firmly mated to coil spring rubber seat. Lower end of coil spring has to be match to marking (3) in lower figure. End of the coil spring must not interfere with step of spring lower seat.

13. Support lower arm with jack.
14. Hoist jack and then install lower arm outer bolt and tighten bolt temporarily by hand.
15. Align marks on lower arm washer and rear suspension frame to each other, tighten lower arm mount nut temporarily by hand.
16. Install rear shock absorber
17. Install wheel with nuts and lower vehicle.
18. Tighten wheel nuts to 71 ft. lbs. (100 Nm).
19. Tighten lower arm to knuckle bolt and nut to 118 ft. lbs. (160 Nm).

➡Tighten lower arm to support nut and bolt to 81 ft. lbs. (110 Nm).

20. Install the stabilizer link to lower control arm nut.
21. Tighten the nut to 11 ft. lbs. (15 Nm).
22. Tighten shock absorber upper bolt to 44 ft. lbs. (60 Nm).
23. Tighten lower shock absorber bolt to 65 ft. lbs. (90 Nm). with vehicle weight on suspension.
24. Check rear toe and camber adjust it as necessary

CONTROL ARMS/LINKS

REMOVAL & INSTALLATION

2007–08 Grand Vitara

Upper Control Arm

See Figure 291.

1. Before servicing the vehicle, refer to all service precautions.
2. Raise vehicle and remove rear wheels.
3. Remove control rod.
4. Remove trailing rod.
5. Remove lower arm.
6. Remove rear suspension knuckle.
7. Remove wheel sensor bolts from upper arm.
8. Remove upper arm bolts and then upper arm.

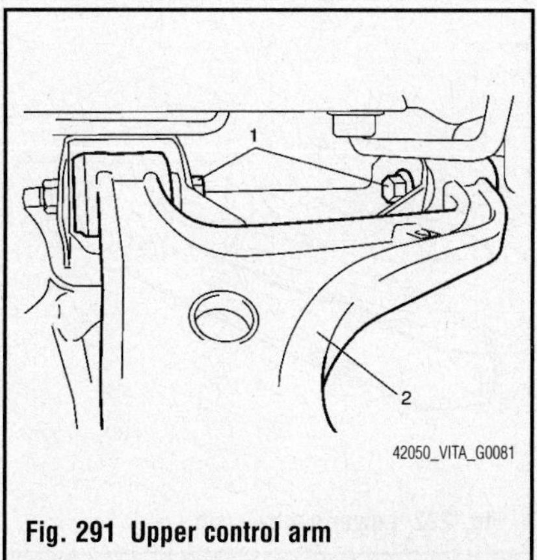

Fig. 291 Upper control arm

To install:

9. Install upper arm to rear suspension frame.
10. Insert upper arm bolt from the upper arm inside.
11. Tighten upper arm mount nuts temporarily by hand.
12. Install rear suspension knuckle
13. Install trailing rod.
14. Install control rod.
15. Install lower arm
16. Install rear wheels and lower vehicle.
17. Tighten each bolts and nuts to specified torque with vehicle weight on suspension.
    - Wheel sensor bolt: 8 ft. lbs. (11 Nm)
    - Upper arm mount nut: 98 ft. lbs. (135 Nm)
    - Shock absorber upper bolt: 44 ft. lbs. (60 Nm)
    - Shock absorber lower bolt: 65 ft. lbs. (90 Nm)
    - Lower arm outer bolt: 98 ft. lbs. (135 Nm)
    - Lower arm mount nut 98 ft. lbs. (135 Nm).
    - Control rod outer bolt 98 ft. lbs. (135 Nm).
    - Control rod mount nut 98 ft. lbs. (135 Nm).
    - Trailing rod rear bolt 98 ft. lbs. (135 Nm).
    - Trailing rod mount nut 98 ft. lbs. (135 Nm).
    - Rear wheels and lower hoist
    - Tighten wheel nuts to 3 ft. lbs. (100 Nm).
18. Check rear toe and camber adjust it as necessary.

Lower Control Arm

See Figures 292 and 293.

1. Before servicing the vehicle, refer to all service precautions.
2. Raise vehicle and remove rear wheels.
3. Put match marks on lower arm washer and on suspension frame to install the bolts correctly in position.
4. Loosen lower arm mount nut.
5. Remove rear coil spring.
6. Remove suspension rod mount bolt and then lower arm.

To install:

7. Install lower arm.
8. Install lower arm to rear suspension frame.
9. Insert suspension lower arm inner bolt from the vehicle rearward.
10. Install lower arm washer with its graduated part facing up.
11. The marks on lower arm washer and rear suspension frame marked before its removal must be aligned and, tighten bolt and nut temporarily by hand.
12. Install rear coil spring.
13. Install wheel with nuts and lower vehicle.
14. Tighten lower arm outer bolt and lower arm mount nut, shock absorber bolts to specified torque with vehicle weight on suspension.
    - Lower arm outer bolt 98 ft. lbs. (135 Nm).
    - Lower arm mount nut 98 ft. lbs. (135 Nm).
    - Shock absorber upper bolt 44 ft. lbs. (60 Nm).
    - Shock absorber lower bolt 65 ft. lbs. (90 Nm).
15. Check rear toe and camber adjust it as necessary.

2007–08 XL-7

Upper Control Arm

See Figures 294 and 295.

1. Before servicing the vehicle, refer to all service precautions.
2. Raise vehicle and remove rear wheels.
3. Disconnect the antilock brake system (ABS) brake wiring harness from the upper control arm.
4. Remove the rear brake hose routing nut and bolt.
5. Remove the upper control arm to knuckle nut and bolt.
6. Remove the upper control to support cam nut and bolt.

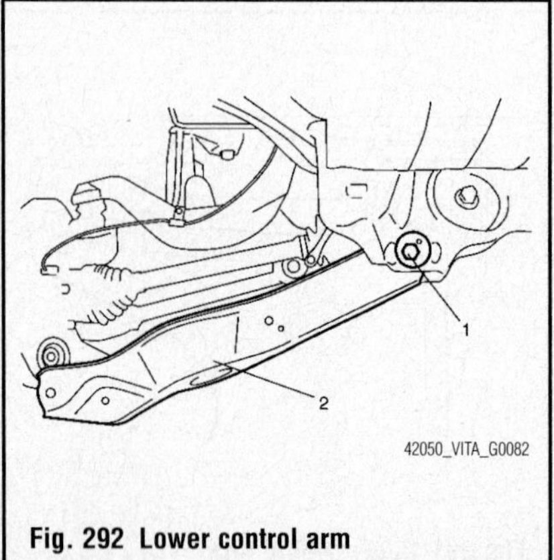

Fig. 292 Lower control arm

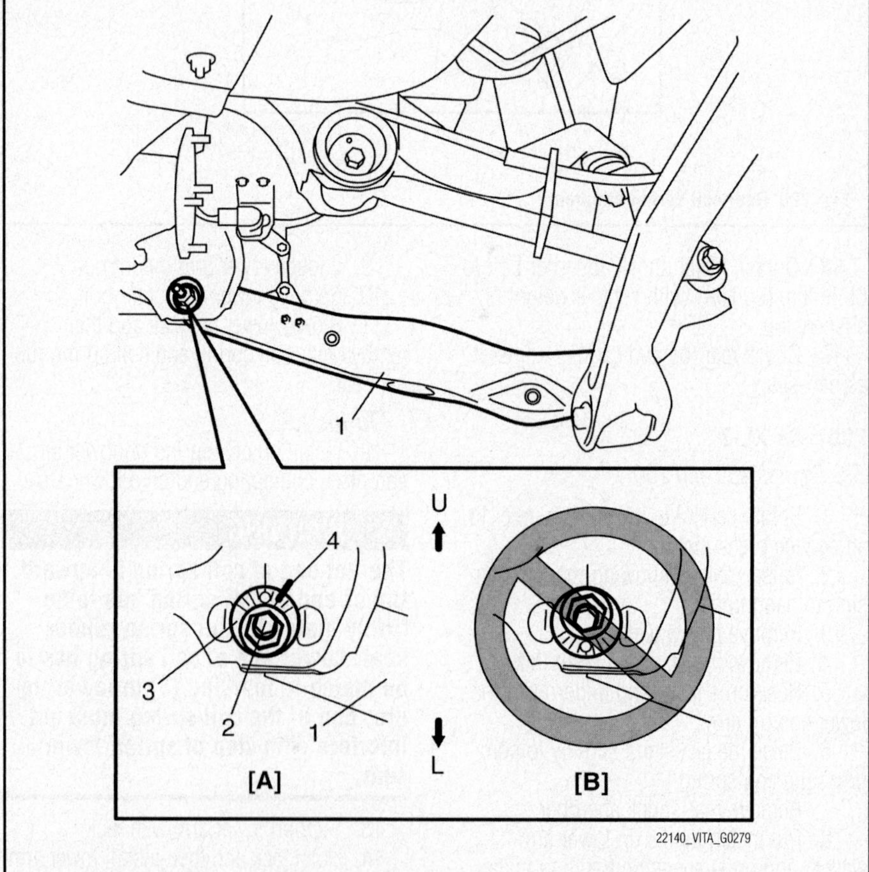

Fig. 293 Lower control arm washer with its graduated part facing up

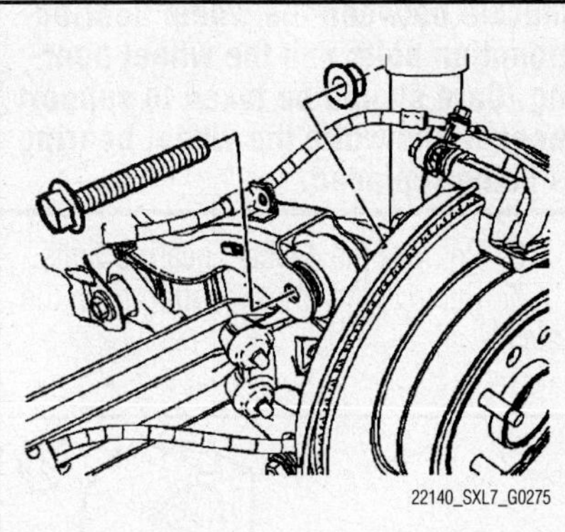
22140_SXL7_G0275

Fig. 294 Upper control arm to knuckle nut and bolt

7. Remove the upper control arm

To install:

8. Install the upper control arm to the knuckle.
9. Loosely install the upper control arm to knuckle nut and bolt.
10. Install the upper control to support bolt and cam nut.

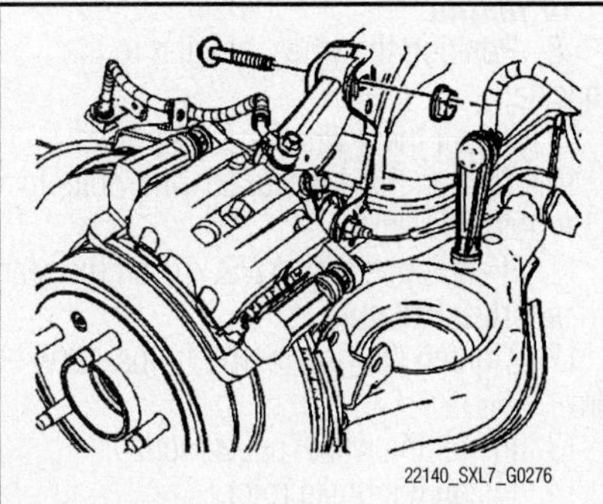
22140_SXL7_G0276

Fig. 295 Upper control to support cam nut and bolt

11. Tighten the upper control arm to knuckle nut and bolt to 118 ft. lbs. (160 Nm).
12. Tighten the upper control arm to support bolt to 121 ft. lbs. (164 Nm).
13. Install the rear brake hose routing nut and bolt to 106 inch. lbs. (12 Nm).
14. Connect the ABS brake wiring harness to the upper control arm.
15. Install the rear tire and wheel assembly.
16. Lower the vehicle.
17. Check the rear alignment.

Lower Control Arm

See Figures 296 and 297.

1. Before servicing the vehicle, refer to all service precautions.
2. Raise vehicle and remove rear wheels.
3. Remove the stabilizer shaft link.
4. Position a jackstand underneath the lower control arm.
5. Raise the jackstand slightly to compress the coil spring.
6. Remove the lower shock bolt and nut.
7. Loosen the lower control arm to support frame nut and bolt.
8. Remove the lower control arm to knuckle nut and bolt.
9. Slowly lower the control arm in order to unload the coil spring.
10. Remove the coil spring.
11. Remove the jackstand.
12. Remove the lower control arm to support frame nut and bolt.
13. Remove the lower control arm.

To install:

14. Inspect the coil spring upper and lower insulators, if damage exists replace the insulators.
15. Position the lower control arm to the support frame and loosely install the nut and bolt.
16. Position the jackstand under the lower control arm.
17. Position the spring with the rubber insulators into the vehicle.
18. Raise the jackstand to compress the spring.
19. Position the lower control arm to the knuckle and install the nut and bolt.
20. Tighten the nut and bolt to 118 ft. lbs. (160 Nm).
21. Tighten the lower control arm to support nut and bolt to 81 ft. lbs. (110 Nm).
22. Install the shock to the lower control arm nut and bolt. Tighten to 81 ft. lbs. (110 Nm).
23. Remove the jackstand from under the vehicle.
24. Install the stabilizer shaft link.
25. Install the rear tire and wheel assembly.
26. Lower the vehicle.
27. Check the rear alignment.

SHOCK ABSORBER

REMOVAL & INSTALLATION

2007–08 Grand Vitara

1. Before servicing the vehicle, refer to all service precautions.
2. Raise vehicle, allowing rear suspension to hang free.
3. Remove wheel.
4. Support lower arm with jack and remove shock absorber bolts.
5. Compress the shock absorber enough to remove it from body.

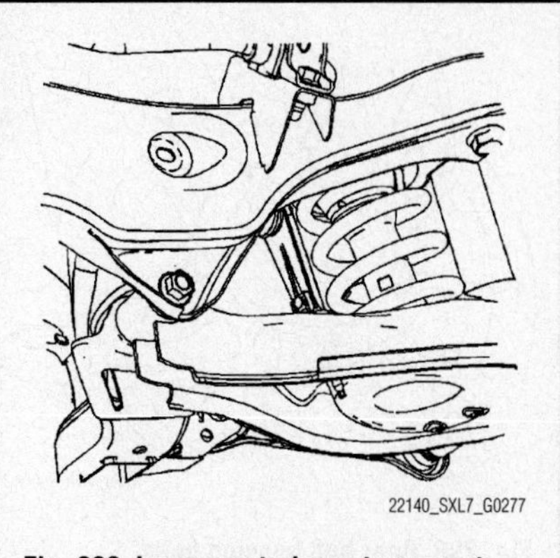
22140_SXL7_G0277

Fig. 296 Lower control arm to support frame nut and bolt

To install:

6. Install shock absorber and mounting bolts.
7. Tighten shock absorber upper bolt to 44 ft. lbs. (60 Nm).
8. Tighten lower shock absorber bolt to 65 ft. lbs. (90 Nm) with vehicle weight on suspension.
9. Remove lower arm support jack.
10. Install rear wheel.
11. Tighten rear wheel to73 ft. lbs. (100 Nm).

2007–08 XL-7

1. Before servicing the vehicle, refer to all service precautions.
2. Raise vehicle and remove rear wheels.
3. Remove the rear wheelhouse from the vehicle.
4. Remove the lower shock bolt and nut.
5. Remove the upper shock bolt.
6. Remove the shock from the vehicle.

To install:

7. Position the shock to the vehicle.
8. Install the upper shock bolt.
9. Tighten the upper shock bolt to 81 ft. lbs. (110 Nm).

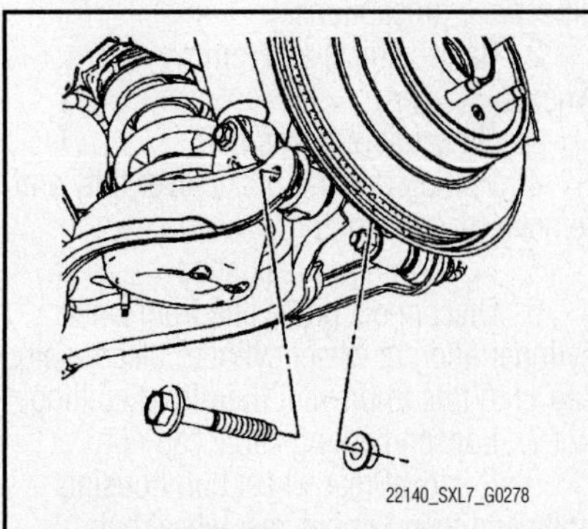
22140_SXL7_G0278

Fig. 297 Lower control arm to knuckle nut and bolt

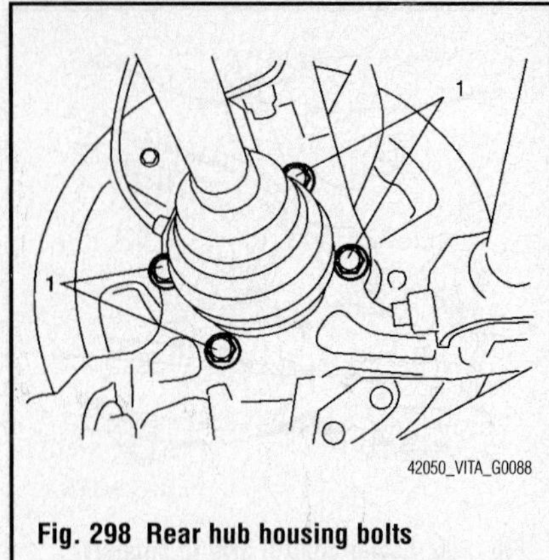

Fig. 298 Rear hub housing bolts

10. Install the lower shock bolt and nut.
11. Tighten the lower shock bolt to 81 ft. lbs. (110 Nm).
12. Install the rear wheelhouse to the vehicle.
13. Install rear wheel assembly and back plate to rear suspension knuckle and tighten rear wheel hub housing bolts to 37 ft. lbs. (50 Nm).

TESTING

1. Test drive vehicle.
2. Inspect each shock absorber for external fluid leakage.
3. Inspect for deformation or damage.
4. Inspect bushings for wear or damage.
5. Use your hands in order to lift up and push down each corner of the vehicle 3 times.
6. Remove your hands from the vehicle.
7. Replace any shock that exceeds more than two bounces.

WHEEL HUB AND BEARING

REMOVAL & INSTALLATION

2007–08 Grand Vitara

See Figures 299 and 300.

1. Before servicing the vehicle, refer to all service precautions.
2. Raise vehicle and remove rear wheels.
3. Un caulk rear axle nut.
4. Pull up parking brake lever fully and remove rear axle nut.
5. Remove rear break shoe
6. Disconnect brake line from wheel cylinder and put wheel cylinder bleeder plug cap onto line to prevent fluid from spilling.
7. Loosen parking cable cap nut.
8. Removal rear wheel hub housing bolts and then remove rear wheel hub assembly and back plate.

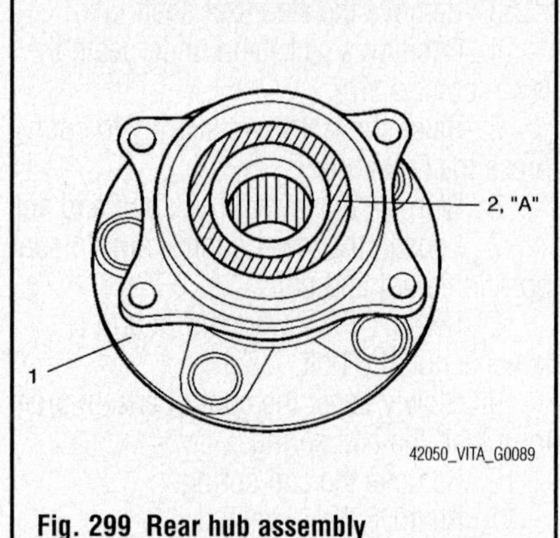

Fig. 299 Rear hub assembly

To install:

9. Install rear wheel assembly and back plate to rear suspension knuckle and tighten rear wheel hub housing bolts to 37 ft. lbs. (50 Nm).
10. Tighten parking cable cap nut to 8 ft. lbs. (11 Nm).
11. Connect brake line to wheel cylinder and tighten brake line flare nut to 12 ft. lbs. (16 Nm). and install bleeder plug cap taken off line back to bleeder plug.
12. Install rear break shoe.
13. Pull up parking brake lever fully and tighten new rear axle nut to 145 ft. lbs. (200 Nm).
14. Caulk rear axle nut.
15. Fill reservoir with brake fluid and bleed brake system.
16. Upon completion of all jobs, depress brake pedal with about 66 ft. lbs.(30 kg). load three to five times so as to obtain proper drum-to-shoe clearance. Adjust parking brake cable.
17. Install rear wheels.
18. Check to ensure that brake drum is free from dragging and proper braking is obtained.
19. Check brake line and bleeder for any fluid leaks.

2007–08 XL-7

See Figures 300 and 301.

1. Before servicing the vehicle, refer to all service precautions.
2. Raise vehicle and remove rear wheels.
3. Remove the brake rotor.
4. Remove the wheel speed sensor.
5. If equipped with all wheel drive (AWD), remove the rear wheel driveshaft nut.

WARNING

The splash shield and park brake assembly are supported to the knuckle between the wheel bearing mounting bolts and the wheel bearing. Care should be taken to support these items while the wheel bearing is being replaced.

6. Remove the 4 wheel bearing bolts.
7. Remove the wheel bearing from the knuckle.

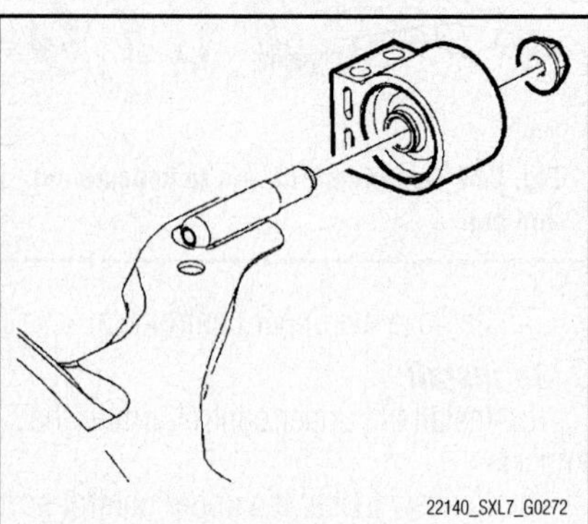

Fig. 300 4 wheel bearing bolts and spindle/ knuckle view

To install:

8. Position the wheel bearing to the knuckle.
9. Install the 4 wheel bearing bolts.
10. Tighten the 4 wheel bearing bolts to 55 ft. lbs. (75 Nm).
11. If equipped with AWD, install the rear wheel driveshaft nut.
12. Tighten the bolt to 151 ft. lbs. (205 Nm).
13. Install the wheel speed sensor.
14. Install the brake rotor.
15. Install the rear tire and wheel assembly.
16. Lower the vehicle.

ADJUSTMENT

The ear bearings are sealed units and are not adjustable.

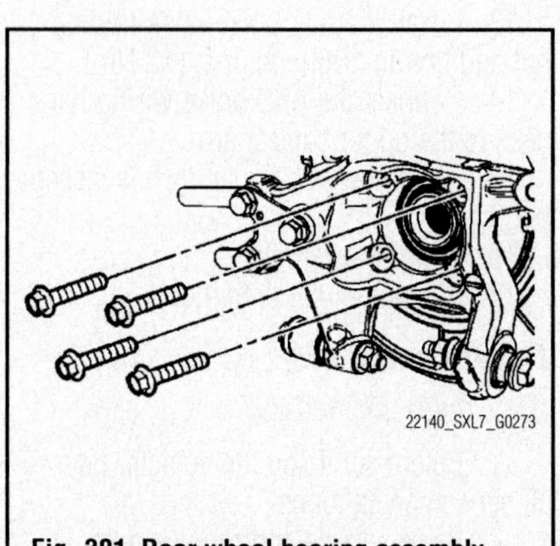

Fig. 301 Rear wheel bearing assembly

SUZUKI

Diagnostic Trouble Codes

20

DIAGNOSTIC TROUBLE CODES

DIAGNOSTIC TROUBLE CODES

OBD II VEHICLE APPLICATIONS

SUZUKI

Aerio
2007
- 2.3L I4 .... VIN Z

Forenza
2007-2008
- 2.0L I4 .... VIN Z

Grand Vitara
2007-2008
- 2.7L V6. .... VIN 9

Reno
2007-2008
- 2.0L I4 .... VIN Z

XL-7
2007-2008
- 3.6L V6. .... VIN 7

Gas Engine OBD II Trouble Code List (P0xxx Codes)

DTC	Trouble Code Title, Conditions & Possible Causes
DTC: P0016 **1T CCM; MIL: Yes** **Years:** 2007, 2008 **Models:** Reno, Forenza **Engines:** 2.0L VIN Z **Transmissions:** All	**Crankshaft Position (CKP) – Camshaft Position (CMP) Correlation** The CMP Sensor reference pulse is not detected at the correct interval every 4 cylinders. Engine is running. **Possible Causes:** • Backed-out terminals • Improper mating • Damaged terminals • Physical damage to the wiring harness
DTC: P0030 **2T OBD/HO2S: Yes** **Years:** 2007, 2008 **Models:** Reno, Forenza **Engines:** 2.0L VIN Z **Transmissions:** All	**Upstream HO2S (Bank 1) Heater Circuit Low Input** Engine started, vehicle driven to a speed over 45 MPH for 6 minutes at an engine speed from 2500-3000 RPM, followed by a deceleration period of 3 seconds, fuel level from 15-85 percent, BARO sensor more than 75 kPa, IAT sensor more than 6.8°F, and the PCM detected the maximum HO2S voltage was less than 300 mv during the test. **Possible Causes:** • HO2S signal ground circuit is open • HO2S is contaminated or has failed • Vehicle driven until it ran out of fuel, or while very low on fuel • PCM has failed
DTC: P0031 **2T CCM; MIL: Yes** **Years:** 2007, 2008 **Models:** Aerio, Forenza, Reno, Vitara, XL-7, **Engines:** 2.3 VIN 6, 2.0L VIN Z, 2.7L VIN 9, 3.6L VIN 7, **Transmissions:** All	**HO2S-11 (Bank 1 Sensor 1) Heater Circuit Low Input** Engine started, vehicle driven to a speed over 45 MPH for 6 minutes at an engine speed from 2500-3000 RPM, followed by a deceleration period of 3 seconds, fuel level from 15-85 percent, BARO sensor more than 75 kPa, IAT sensor more than 6.8°F, and the PCM detected the maximum HO2S-11 voltage was less than 300 mv during the test. **Possible Causes:** • HO2S signal ground circuit is open • HO2S is contaminated or has failed • Vehicle driven until it ran out of fuel, or while very low on fuel • PCM has failed
DTC: P0032 **2T CCM; MIL: Yes** **Years:** 2007, 2008 **Models:** Forenza, Reno, Vitara, XL-7, **Engines:** 2.0L VIN Z, 2.7L VIN 9, 3.6L VIN 7, **Transmissions:** All	**HO2S-11 (Bank 1 Sensor 1) Heater Circuit High Input** Engine started, vehicle driven to a speed over 45 MPH for 6 minutes at an engine speed from 2500-3000 RPM, followed by a deceleration period of 3 seconds, fuel level from 15-85 percent, BARO sensor more than 75 kPa, IAT sensor more than 6.8°F, and the PCM detected the HO2S-11 signal did not go below 600 mv during the CCM test. **Possible Causes:** • HO2S signal circuit is open or shorted to system power • Fuel supply system is too rich (fuel injector or regulator leaking) • HO2S is contaminated, damaged or has failed • PCM has failed • TSB TS015 (12/95) contains information related to this code
DTC: P0037 **2T CCM; MIL: Yes** **Years:** 2007, 2008 **Models:** Aerio, Forenza, Reno, Vitara, XL-7, **Engines:** 2.3 VIN 6, 2.0L VIN Z, 2.7L VIN 9, 3.6L VIN 7, **Transmissions:** All	**HO2S-12 (Bank 1 Sensor 2) Heater Circuit Low Input** Engine started, vehicle driven to a speed over 45 MPH for 6 minutes at an engine speed from 2500-3000 RPM, followed by a deceleration period of 3 seconds, fuel level from 15-85 percent, BARO sensor more than 75 kPa, IAT sensor more than 6.8°F, and the PCM detected the maximum HO2S-12 voltage was less than 300 mv during the test. **Possible Causes:** • HO2S signal ground circuit is open • HO2S is contaminated or has failed • Vehicle driven until it ran out of fuel, or while very low on fuel • PCM has failed
DTC: P0038 **2T CCM; MIL: Yes** **Years:** 2007, 2008 **Models:** Forenza, Reno, Vitara, XL-7, **Engines:** 2.0L VIN Z, 2.7L VIN 9, 3.6L VIN 7, **Transmissions:** All	**HO2S-12 (Bank 1 Sensor 2) Heater Circuit High Input** Engine started, vehicle driven to a speed over 45 MPH for 6 minutes at an engine speed from 2500-3000 RPM, followed by a deceleration period of 3 seconds, fuel level from 15-85 percent, BARO sensor more than 75 kPa, IAT sensor more than 6.8°F, and the PCM detected the HO2S-12 signal did not go below 600 mv during the CCM test. **Possible Causes:** • HO2S signal circuit is open or shorted to system power • Fuel supply system is too rich (fuel injector or regulator leaking) • HO2S is contaminated, damaged or has failed • PCM has failed • TSB TS015 (12/95) contains information related to this code

DTC	Trouble Code Title, Conditions & Possible Causes
DTC: P0051 **2T CCM; MIL: Yes** **Years:** 2007, 2008 **Models:** Vitara, XL-7 **Engines:** 2.7L VIN 9, 3.6L VIN 7 **Transmissions:** A/T, M/T	**HO2S-12 (Bank 2 Sensor 1) Heater Circuit Low Input** Engine started, vehicle driven to a speed over 45 MPH for 6 minutes at an engine speed from 2500-3000 RPM, followed by a deceleration period of 3 seconds, fuel level from 15-85 percent, BARO sensor more than 75 kPa, IAT sensor more than 6.8°F, and the PCM detected the maximum HO2S-21 voltage was less than 300 mv during the test. **Possible Causes:** • HO2S signal ground circuit is open • HO2S is contaminated or has failed • Vehicle driven until it ran out of fuel, or while very low on fuel • PCM has failed
DTC: P0052 **2T CCM; MIL: Yes** **Years:** 2007, 2008 **Models:** Vitara, XL-7 **Engines:** 2.7L VIN 9, 3.6L VIN 7 **Transmissions:** A/T, M/T	**HO2S-12 (Bank 2 Sensor 1) Heater Circuit High Input** Engine started, vehicle driven to a speed over 45 MPH for 6 minutes at an engine speed from 2500-3000 RPM, followed by a deceleration period of 3 seconds, fuel level from 15-85 percent, BARO sensor more than 75 kPa, IAT sensor more than 6.8°F, and the PCM detected the HO2S-21 signal did not go below 600 mv during the CCM test. **Possible Causes:** • HO2S signal circuit is open or shorted to system power • Fuel supply system is too rich (fuel injector or regulator leaking) • HO2S is contaminated, damaged or has failed • PCM has failed • TSB TS015 (12/95) contains information related to this code
DTC: P0057 **2T CCM; MIL: Yes** **Years:** 2007, 2008 **Models:** Vitara, XL-7 **Engines:** 2.7L VIN 9, 3.6L VIN 7 **Transmissions:** A/T, M/T	**HO2S-12 (Bank 2 Sensor 2) Heater Circuit Low Input** Engine started, vehicle driven to a speed over 45 MPH for 6 minutes at an engine speed from 2500-3000 RPM, followed by a deceleration period of 3 seconds, fuel level from 15-85 percent, BARO sensor more than 75 kPa, IAT sensor more than 6.8°F, and the PCM detected the maximum HO2S-22 voltage was less than 300 mv during the test. **Possible Causes:** • HO2S signal ground circuit is open • HO2S is contaminated or has failed • Vehicle driven until it ran out of fuel, or while very low on fuel • PCM has failed
DTC: P0058 **2T CCM; MIL: Yes** **Years:** 2007, 2008 **Models:** Vitara, XL-7 **Engines:** 2.7L VIN 9, 3.6L VIN 7 **Transmissions:** A/T, M/T	**HO2S-12 (Bank 2 Sensor 2) Heater Circuit High Input** Engine started, vehicle driven to a speed over 45 MPH for 6 minutes at an engine speed from 2500-3000 RPM, followed by a deceleration period of 3 seconds, fuel level from 15-85 percent, BARO sensor more than 75 kPa, IAT sensor more than 6.8°F, and the PCM detected the HO2S-22 signal did not go below 600 mv during the CCM test. **Possible Causes:** • HO2S signal circuit is open or shorted to system power • Fuel supply system is too rich (fuel injector or regulator leaking) • HO2S is contaminated, damaged or has failed • PCM has failed • TSB TS015 (12/95) contains information related to this code
DTC: P0068 **2T CCM, MIL: Yes** **Years:** 2007, 2008 **Models:** Reno, Forenza **Engines:** 2.0L VIN Z **Transmissions:** All	**Throttle Body Airflow Performance** Engine is running. • No MAP or IAT fault exists. • Difference between measured airflow and estimated airflow is greater than 9 g/s. **Possible Causes:** • Vacuum hoses for splits, kinks and improper connections • Air leaks at the throttle body • Intake manifold sealing surfaces • Restrictions in the air intake system, including the filter
DTC: P0101 **2T CCM, MIL: Yes** **Years:** 2007, 2008 **Models:** Vitara, XL-7 **Engines:** 2.7L VIN 9, 3.6L VIN 7 **Transmissions:** A/T, M/T	**Mass Airflow Sensor Range/Performance** Engine started, vehicle driven at a speed of over 35 MPH in closed loop, then back to idle speed, altitude less than 8000 ft, ECT sensor from 18-230°F, IAT sensor more than 6.8°F, and the PCM detected the MAF signal was less than 0.82-1.08 lbs/min at 2500 RPM, or that it was less than 0.22-0.44 lbs/min at idle speed during the CCM test. **Possible Causes:** • Intake air system is restricted, or the air filter is clogged • MAF sensor signal or ground circuit has high resistance • MAF sensor is contaminated, damaged or has failed • PCM has failed

DTC	Trouble Code Title, Conditions & Possible Causes
DTC: P0102 **1T CCM, MIL: Yes** **Years:** 2007, 2008 **Models:** Vitara, XL-7 **Engines:** 2.7L VIN 9, 3.6L VIN 7 **Transmissions:** A/T, M/T	**Mass Airflow Sensor Circuit Low Input** Engine started, engine running in closed loop, altitude less than 8000 ft, ECT signal from 18-230°F, IAT signal from 18-122°F, and the PCM detected the MAF signal indicated less than (Scan Tool reads close to '0' lbs/minute) for 1 second during the CCM test. **Possible Causes:** • MAF sensor signal circuit is open or shorted to ground • MAF sensor power circuit is open • MAF sensor is damaged or has failed • PCM has failed
DTC: P0103 **1T CCM, MIL: Yes** **Years:** 2007, 2008 **Models:** Vitara, XL-7 **Engines:** 2.7L VIN 9, 3.6L VIN 7 **Transmissions:** A/T, M/T	**Mass Airflow Sensor Circuit High Input** Engine started engine running at idle speed in closed loop, and the PCM detected the MAF sensor indicated more than 4.90v (Scan Tool reads 28 lbs/minute) for 1 second during the CCM test. **Possible Causes:** • MAF sensor ground circuit is open • MAF sensor signal is shorted to VREF or system power (B+) • MAF sensor is damaged or has failed • PCM has failed
DTC: P0106 **2T CCM; MIL: Yes** **Years:** 2007, 2008 **Models:** Aerio, Forenza, Reno, Vitara, XL-7, **Engines:** 2.3 VIN 6, 2.0L VIN Z, 2.7L VIN 9, 3.6L VIN 7, **Transmissions:** All	**Manifold Air Pressure Sensor Range/Performance** Engine cranking, altitude less than 8000 ft, ambient temperature over 14°F, IAT signal less than 158°F, ECT sensor from 158-230°F, Closed Throttle switch "on", and the PCM detected the MAP signal differed from the startup MAP signal by less than 1.39 kPa, or with the vehicle driven to over 2000 RPM in closed loop, and then back to idle with the Closed Throttle switch indicating "on", it detected the BARO sensor signal differed from MAP sensor signal by less than 33 kPa during the CCM test. **Possible Causes:** • Air leak present between vacuum passage and the MAP sensor • MAP sensor vacuum line clogged between sensor and manifold • MAP sensor is damaged, out-of-calibration or has failed • PCM has failed
DTC: P0107 **1T CCM; MIL: Yes** **Years:** 2007, 2008 **Models:** Aerio, Forenza, Reno, Vitara, XL-7, **Engines:** 2.3 VIN 6, 2.0L VIN Z, 2.7L VIN 9, 3.6L VIN 7, **Transmissions:** All	**MAP Sensor Circuit Low Input (High Vacuum)** Engine started, engine running at idle speed for 1 minute, and the PCM detected the MAP sensor was less than 5 kPa during the test. **Note: If DTC P0122 and P0450 are set with this code, the sensor power (VREF) circuit may be open between the sensor and PCM.** **Possible Causes:** • MAP sensor 5-volt power circuit is open or shorted to ground • MAP Sensor signal circuit is shorted to ground • MAP Sensor is damaged or has failed • PCM has failed
DTC: P0108 **1T CCM; MIL: Yes** **Years:** 2007, 2008 **Models:** Aerio, Forenza, Reno, Vitara, XL-7, **Engines:** 2.3 VIN 6, 2.0L VIN Z, 2.7L VIN 9, 3.6L VIN 7, **Transmissions:** All	**MAP Sensor Circuit High Input (Low Vacuum)** Engine started, engine running at idle speed for 1 minute, and the PCM detected the MAP sensor was more than 129.3 kPa in the test. **Note: If DTC P0113, P0118 and P0450 are also set, the sensor ground circuit may be open between the sensor and the PCM.** **Possible Causes:** • MAP sensor signal circuit is open between sensor and the PCM • MAP Sensor ground circuit is open between sensor and PCM • MAP sensor signal circuit is shorted to VREF or system power • MAP Sensor is damaged or has failed • PCM has failed
DTC: P0110 **1T CCM; MIL: Yes** **Years:** 2007, 2008 **Models:** Reno, Forenza **Engines:** 2.0L VIN Z **Transmissions:** All	**Intake Air Temperature (IAT) Sensor Circuit** (Driving Condition) The throttle position is greater than 0.2 percent, vehicle speed is 40 km/h or greater, airflow is greater than 15 g /sec and stuck drive test counter is 300 seconds or greater. (Idling Condition) The throttle position is 0.2 percent or less, vehicle speed is 2 km/h or less and stuck idle test counter is 60 seconds or greater. Change in intake air temperature (IAT) is less than or equal to 3°C (5.4°F) with no disabling faults present, engine is running, soak time is greater than or equal to 480 minutes and Intake air temperature (IAT) is stored on previous trip. **Possible Causes:** • IAT sensor signal circuit has high resistance • IAT sensor ground circuit has high resistance • IAT sensor is contaminated, damaged or has failed • PCM has failed

DTC	Trouble Code Title, Conditions & Possible Causes
DTC: P0111 **2T CCM; MIL: Yes** **Years:** 2007, 2008 **Models:** Aerio, Forenza, Reno, Vitara, XL-7, **Engines:** 2.3 VIN 6, 2.0L VIN Z, 2.7L VIN 9, 3.6L VIN 7, **Transmissions:** All	**Intake Air Temperature Sensor Range/Performance** Engine started with the ECT sensor less than 86°F, engine runtime over 5 minutes, and the PCM detected too small an amount of change in the IAT sensor signal under these conditions in the test. **Possible Causes:** • IAT sensor signal circuit has high resistance • IAT sensor ground circuit has high resistance • IAT sensor is contaminated, damaged or has failed • PCM has failed
DTC: P0112 **1T CCM; MIL: Yes** **Years:** 2007, 2008 **Models:** Aerio, Forenza, Reno, Vitara, XL-7, **Engines:** 2.3 VIN 6, 2.0L VIN Z, 2.7L VIN 9, 3.6L VIN 7, **Transmissions:** All	**Intake Air Temperature Sensor Circuit Low Input** Key on or engine running, and the PCM detected an unexpected "low" voltage condition on the IAT sensor circuit (Scan Tool reads 246°F or higher) during the CCM test. **Possible Causes:** • IAT sensor signal circuit is shorted to sensor ground • IAT sensor signal circuit is shorted to chassis ground • IAT sensor is damaged or has failed (it may be shorted) • PCM has failed
DTC: P0113 **1T CCM; MIL: Yes** **Years:** 2007, 2008 **Models:** Aerio, Forenza, Reno, Vitara, XL-7, **Engines:** 2.3 VIN 6, 2.0L VIN Z, 2.7L VIN 9, 3.6L VIN 7, **Transmissions:** All	**Intake Air Temperature Sensor Circuit High Input** Key on or engine running, and the PCM detected the IAT sensor was more than 4.90v (Scan Tool reads −40°F or lower) in the test. **Possible Causes:** • IAT sensor signal circuit or ground circuit is open • IAT sensor signal circuit is shorted to VREF or system power • IAT sensor is damaged or has failed • PCM has failed
DTC: P0116 **2T CCM; MIL: Yes** **Years:** 2007, 2008 **Models:** Aerio, Forenza, Reno, Vitara, XL-7, **Engines:** 2.3 VIN 6, 2.0L VIN Z, 2.7L VIN 9, 3.6L VIN 7, **Transmissions:** All	**Engine Coolant Temperature Sensor Range/Performance** Engine started, engine runtime over 5 minutes, and the PCM detected the ECT sensor indicated too small an amount of change when compared to a calibrated amount in memory during the test. **Possible Causes:** • Check the operation of the thermostat (it may be stuck open) • Inspect for low coolant level or an incorrect coolant mixture • ECT sensor signal circuit has high resistance • ECT sensor has failed
DTC: P0117 **1T CCM; MIL: Yes** **Years:** 2007, 2008 **Models:** Aerio, Forenza, Reno, Vitara, XL-7, **Engines:** 2.3 VIN 6, 2.0L VIN Z, 2.7L VIN 9, 3.6L VIN 7, **Transmissions:** All	**Engine Coolant Temperature Sensor Circuit Low Input** Key on and engine running, and the PCM detected the ECT signal indicated more than 0.20v (Scan Tool reads 246°F) during the test. **Possible Causes:** • ECT sensor signal circuit is shorted to sensor ground • ECT sensor signal circuit is shorted to chassis ground • ECT sensor is damaged or has failed (it may be shorted) • PCM has failed
DTC: P0118 **1T CCM; MIL: Yes** **Years:** 2007, 2008 **Models:** Aerio, Forenza, Reno, Vitara, XL-7, **Engines:** 2.3 VIN 6, 2.0L VIN Z, 2.7L VIN 9, 3.6L VIN 7, **Transmissions:** All	**Engine Coolant Temperature Sensor Circuit High Input** Key on or engine running, and the PCM detected the ECT sensor was more than 4.90v (Scan Tool reads −40°F or lower) in the test. **Possible Causes:** • ECT sensor signal circuit is open between sensor and the PCM • ECT sensor signal circuit is shorted to VREF or system power • ECT sensor ground circuit is open between sensor and PCM • ECT sensor is damaged or has failed • PCM has failed
DTC: P0121 **2T CCM; MIL: Yes** **Years:** 2007, 2008 **Models:** Vitara, XL-7 **Engines:** 2.7L VIN 9, 3.6L VIN 7 **Transmissions:** A/T, M/T	**Throttle Position Sensor Range/Performance** Engine started, vehicle driven to a speed of 30-40 MPH, ambient temperature more than 14°F, IAT sensor less than 122°F, and the PCM detected the difference between the Actual TP sensor signal and the Calculated target signal (from the MAP and TP inputs) was more than a calibrated amount stored in memory during the test. **Possible Causes:** • MAF sensor or TP sensor ground circuit has high resistance • MAF sensor or TP sensor is out-of-calibration • TP sensor is damaged or has failed • PCM has failed

DTC	Trouble Code Title, Conditions & Possible Causes
DTC: P0122 **1T CCM; MIL: Yes** **Years:** 2007, 2008 **Models:** Aerio, Forenza, Reno, Vitara, XL-7, **Engines:** 2.3 VIN 6, 2.0L VIN Z, 2.7L VIN 9, 3.6L VIN 7, **Transmissions:** All	**Throttle Position Sensor Circuit Malfunction** Engine started, engine running at idle for 1 minute, and the PCM detected the TP sensor indicated less than 2 percent during the CCM test. **Possible Causes:** • TP sensor signal circuit is shorted to ground (intermittent fault) • TP sensor power circuit is open between sensor and the PCM • TP sensor is damaged or has failed (perform a "sweep" test) • PCM has failed
DTC: P0123 **1T CCM; MIL: Yes** **Years:** 2007, 2008 **Models:** Aerio, Forenza, Reno, Vitara, XL-7, **Engines:** 2.3 VIN 6, 2.0L VIN Z, 2.7L VIN 9, 3.6L VIN 7, **Transmissions:** All	**Throttle Position Sensor Circuit High Input** Engine started, engine running at idle for 1 minute, and the PCM detected the TP sensor indicated more than 96 percent during the test. **Possible Causes:** • TP sensor signal circuit is open between sensor and the PCM • TP sensor signal circuit is shorted to VREF (intermittent fault) • TP sensor ground circuit is open between sensor and the PCM • TP sensor is damaged or has failed (perform a "sweep" test) • PCM has failed
DTC: P0125 **2T OBD/ECT; MIL: Yes** **Years:** 2007, 2008 **Models:** Aerio, Forenza, Reno, Vitara, XL-7, **Engines:** 2.3 VIN 6, 2.0L VIN Z, 2.7L VIN 9, 3.6L VIN 7, **Transmissions:** All	**Insufficient Coolant Temperature For Closed Loop** Engine started, vehicle driven at over 35 MPH for 15 minutes if the ECT sensor indicated less than 5°F at startup, or vehicle driven for over 5 minutes if the ECT sensor indicated over 5°F at startup, and the PCM detected the ECT sensor did not reach a closed loop value. **Possible Causes:** • Inspect for low coolant level or an incorrect coolant mixture • Check the operation of the thermostat (it may be stuck open) • ECT sensor signal circuit has high resistance • ECT sensor has failed
DTC: P0125 **2T OBD/ECT; MIL: Yes** **Years:** 2007, 2008 **Models:** Vitara, XL-7 **Engines:** 2.7L VIN 9, 3.6L VIN 7 **Transmissions:** A/T, M/T	**Insufficient Coolant Temperature For Closed Loop** Engine started, ECT sensor from 18-230°F, IAT sensor over 18°F, BARO sensor more than 75 kPa, vehicle driven at over 35 MPH for 5 minutes under normal stop and go or cruise conditions, then back to idle speed for 5 minutes, and the PCM detected the engine coolant temperature did not reach a value that allows closed loop operation. **Possible Causes:** • Inspect for low coolant level or an incorrect coolant mixture • Check the operation of the thermostat (it may be stuck open) • ECT sensor signal circuit has high resistance • ECT sensor has failed
DTC: P0128 **2T OBD/ECT; MIL: Yes** **Years:** 2007, 2008 **Models:** Aerio, Forenza, Reno, Vitara, XL-7, **Engines:** 2.3 VIN 6, 2.0L VIN Z, 2.7L VIN 9, 3.6L VIN 7, **Transmissions:** All	**Coolant Temperature Below Thermostat Regulating Temperature** Engine started, ECT sensor from 18-230°F, IAT sensor over 18°F, BARO sensor more than 75 kPa, vehicle driven at over 35 MPH for 5 minutes under normal stop and go or cruise conditions, then back to idle speed for 5 minutes, and the PCM detected the engine coolant temperature did not reach a value over that of the thermostat regulating temperature. **Possible Causes:** • Inspect for low coolant level or an incorrect coolant mixture • Check the operation of the thermostat (it may be stuck open) • ECT sensor signal circuit has high resistance • ECT sensor has failed
DTC: P0131 **2T CCM; MIL: Yes** **Years:** 2007, 2008 **Models:** Aerio, Forenza, Reno, Vitara, XL-7, **Engines:** 2.3 VIN 6, 2.0L VIN Z, 2.7L VIN 9, 3.6L VIN 7, **Transmissions:** All	**HO2S-11 (Bank 1 Sensor 1) Circuit Low Input** Engine started, vehicle driven to a speed over 45 MPH for 6 minutes at an engine speed from 2500-3000 RPM, followed by a deceleration period of 3 seconds, fuel level from 15-85 percent, BARO sensor more than 75 kPa, IAT sensor more than 6.8°F, and the PCM detected the maximum HO2S-11 voltage was less than 300 mv during the test. **Possible Causes:** • HO2S signal ground circuit is open • HO2S is contaminated or has failed • Vehicle driven until it ran out of fuel, or while very low on fuel • PCM has failed

DTC	Trouble Code Title, Conditions & Possible Causes
DTC: P0132 **2T CCM; MIL: Yes** **Years:** 2007, 2008 **Models:** Forenza, Reno, Vitara, XL-7, **Engines:** 2.0L VIN Z, 2.7L VIN 9, 3.6L VIN 7, **Transmissions:** All	**HO2S-11 (Bank 1 Sensor 1) Circuit High Input** Engine started, vehicle driven to a speed over 45 MPH for 6 minutes at an engine speed from 2500-3000 RPM, followed by a deceleration period of 3 seconds, fuel level from 15-85 percent, BARO sensor more than 75 kPa, IAT sensor more than 6.8°F, and the PCM detected the HO2S-11 signal did not go below 600 mv during the CCM test. **Possible Causes:** • HO2S signal circuit is open or shorted to system power • Fuel supply system is too rich (fuel injector or regulator leaking) • HO2S is contaminated, damaged or has failed • PCM has failed • TSB TS015 (12/95) contains information related to this code
DTC: P0133 **2T OBD/O2S; MIL: Yes** **Years:** 2007, 2008 **Models:** Aerio, Forenza, Reno, Vitara, XL-7, **Engines:** 2.3 VIN 6, 2.0L VIN Z, 2.7L VIN 9, 3.6L VIN 7, **Transmissions:** All	**HO2S-11 (Bank 1 Sensor 1) Slow Response** DTC P0131 and P0132 not set, vehicle driven at a speed of 35 MPH for 2 minutes in closed loop, then back to idle speed for 2 minutes, ambient air temperature over 14°F, fuel level from 15-85 percent, BARO sensor more than 75 kPa, IAT sensor more than 14°F, and the PCM detected the HO2S-11 signal response time to change from rich-to-lean or lean-to-rich was more than 1 second during the CCM test. **Possible Causes:** • Air leaks in the intake manifold or exhaust manifold or pipes • IAT sensor or MAF sensor has deteriorated (out of calibration) • HO2S signal circuit is open or shorted to ground (intermittent) • HO2S contaminated with wrong fuel, has deteriorated or failed • PCM has failed • TSB TS015 (12/95) contains information related to this code
DTC: P0134 **2T OBD/O2S; MIL: Yes** **Years:** 2007, 2008 **Models:** Aerio, Forenza, Reno, Vitara, XL-7, **Engines:** 2.3 VIN 6, 2.0L VIN Z, 2.7L VIN 9, 3.6L VIN 7, **Transmissions:** All	**HO2S-11 (Bank 1 Sensor 1) No Activity Detected** DTC P0131 and P0132 not set, vehicle driven at a speed of 35 MPH for 2 minutes in closed loop, then back to idle speed for 2 minutes, ambient air temperature over 14°F, fuel level from 15-85 percent, BARO sensor more than 75 kPa, IAT sensor more than 14°F, and the PCM detected the HO2S-11 signal remained "high" or "low" in the test. **Possible Causes:** • Air leaks in the intake manifold or exhaust manifold or pipes • HO2S is contaminated (wrong fuel), has deteriorated or failed • HO2S heater element is damaged or has failed • PCM has failed • TSB TS015 (12/95) contains information related to this code
DTC: P0135 **2T OBD/O2S; MIL: Yes** **Years:** 2007, 2008 **Models:** Aerio, Forenza, Reno, Vitara, XL-7, **Engines:** 2.3 VIN 6, 2.0L VIN Z, 2.7L VIN 9, 3.6L VIN 7, **Transmissions:** All	**HO2S-11 (Bank 1 Sensor 1) Heater Circuit Malfunction** Engine started, engine runtime 1 minute, and the PCM detected the HO2S-11 heater signal was "low" with the heater commanded "off", or that it was "high: with the heater commanded "on" during the test. **Possible Causes:** • HO2S heater control circuit is open or shorted to ground • HO2S heater power circuit is open (check the IG fuse) • HO2S heater control circuit is shorted to power • HO2S heater is damaged or has failed • PCM has failed
DTC: P0135 **2T OBD/O2S; MIL: Yes** **Years:** 2007, 2008 **Models:** Aerio, Forenza, Reno, Vitara, XL-7, **Engines:** 2.3 VIN 6, 2.0L VIN Z, 2.7L VIN 9, 3.6L VIN 7, **Transmissions:** All	**HO2S-11 (Bank 1 Sensor 1) Heater Circuit Malfunction** Engine started, engine runtime over 1 minute, IAT sensor more than 14°F, ECT sensor from 18-230°F, and the PCM detected the HO2S-11 heater current level was more than 5.3A or less than 0.09A. **Possible Causes:** • HO2S heater control circuit is open or shorted to ground • HO2S heater power circuit is open (check the IG fuse) • HO2S heater control circuit is shorted to power • HO2S heater is damaged or has failed • PCM has failed

DTC	Trouble Code Title, Conditions & Possible Causes
DTC: P0136 **2T CCM; MIL: Yes** **Years:** 2007, 2008 **Models:** Vitara, XL-7 **Engines:** 2.7L VIN 9, 3.6L VIN 7 **Transmissions:** A/T, M/T	**HO2S-12 (Bank 1 Sensor 2) Circuit Low Input** Engine started, vehicle driven to a speed over 45 MPH for 6 minutes at an engine speed from 2500-3000 RPM, followed by a deceleration period of 3 seconds, fuel level from 15-85 percent, BARO sensor more than 75 kPa, IAT sensor more than 6.8°F, and the PCM detected the maximum HO2S-12 voltage was less than 600 mv, or the minimum voltage was more than 400 mv, or the maximum HO2S-12 voltage was more than 4.5v during the CCM test. **Possible Causes:** • HO2S signal circuit open (sensor signal reads over 4.5v) • HO2S is shorted to ground • HO2S signal ground circuit is open • HO2S is contaminated or has failed • Vehicle driven until it ran out of fuel, or while very low on fuel • PCM has failed
DTC: P0137 **2T CCM; MIL: Yes** **Years:** 2007, 2008 **Models:** Vitara, XL-7 **Engines:** 2.7L VIN 9, 3.6L VIN 7 **Transmissions:** A/T, M/T	**HO2S-12 (Bank 1 Sensor 2) Circuit Low Input** Engine started, vehicle driven to a speed over 45 MPH for 6 minutes at an engine speed from 2500-3000 RPM, followed by a deceleration period of 3 seconds, fuel level from 15-85 percent, BARO sensor more than 75 kPa, IAT sensor more than 6.8°F, and the PCM detected the maximum HO2S-12 voltage was less than 600 mv, or the minimum voltage was more than 400 mv, or the maximum HO2S-12 voltage was more than 4.5v during the CCM test. **Possible Causes:** • HO2S signal circuit open (sensor signal reads over 4.5v) • HO2S is shorted to ground • HO2S signal ground circuit is open • HO2S is contaminated or has failed • Vehicle driven until it ran out of fuel, or while very low on fuel • PCM has failed
DTC: P0138 **2T CCM; MIL: Yes** **Years:** 2007, 2008 **Models:** Aerio, Forenza, Reno, Vitara, XL-7, **Engines:** 2.3 VIN 6, 2.0L VIN Z, 2.7L VIN 9, 3.6L VIN 7, **Transmissions:** All	**HO2S-12 (Bank 1 Sensor 2) Circuit High Input** Engine started, vehicle driven to a speed over 45 MPH for 6 minutes at an engine speed from 2500-3000 RPM, followed by a deceleration period of 3 seconds, fuel level from 15-85 percent, BARO sensor more than 75 kPa, IAT sensor more than 6.8°F, and the PCM detected the maximum HO2S-12 voltage was less than 600 mv, or the minimum voltage was more than 400 mv, or the maximum HO2S-12 voltage was more than 4.5v during the CCM test. **Possible Causes:** • HO2S signal circuit open (sensor signal reads over 4.5v) • HO2S is shorted to ground • HO2S signal ground circuit is open • HO2S is contaminated or has failed • Vehicle driven until it ran out of fuel, or while very low on fuel • PCM has failed
DTC: P0140 **2T CCM; MIL: Yes** **Years:** 2007, 2008 **Models:**Forenza, Reno, Vitara, XL-7, **Engines:** 2.0L VIN Z, 2.7L VIN 9, 3.6L VIN 7, **Transmissions:** All	**HO2S-12 (Bank 1 Sensor 2) Circuit, No Activity Detected** DTC P0131 and P0132 not set, vehicle driven at a speed of 35 MPH for 2 minutes in closed loop, then back to idle speed for 2 minutes, ambient air temperature over 14°F, fuel level from 15-85 percent, BARO sensor more than 75 kPa, IAT sensor more than 14°F, and the PCM detected the HO2S-11 signal remained "high" or "low" in the test. **Possible Causes:** • Air leaks in the intake manifold or exhaust manifold or pipes • HO2S is contaminated (wrong fuel), has deteriorated or failed • HO2S heater element is damaged or has failed • PCM has failed • TSB TS015 (12/95) contains information related to this code
DTC: P0141 **2T OBD/O2S; MIL: Yes** **Years:** 2007, 2008 **Models:** Aerio, Forenza, Reno, Vitara, XL-7, **Engines:** 2.3 VIN 6, 2.0L VIN Z, 2.7L VIN 9, 3.6L VIN 7, **Transmissions:** All	**HO2S-12 (Bank 1 Sensor2) Heater Circuit Malfunction** Engine started, engine runtime over 1 minute, IAT sensor more than 14°F, ECT sensor from 18-230°F, and the PCM detected the HO2S-12 heater current level was more than 5.3A or less than 0.09A. **Possible Causes:** • HO2S heater control circuit is open or shorted to ground • HO2S heater power circuit is open (check the IG fuse) • HO2S heater control circuit is shorted to power • HO2S heater is damaged or has failed • PCM has failed

DTC	Trouble Code Title, Conditions & Possible Causes
DTC: P0151 **2T CCM; MIL: Yes** **Years:** 2007, 2008 **Models:** Vitara, XL-7 **Engines:** 2.7L VIN 9, 3.6L VIN 7 **Transmissions:** A/T, M/T	**HO2S-21 (Bank 2 Sensor 1) Circuit Low Input** Engine started, vehicle driven at a speed of over 35 MPH for 2 minutes, then back to idle speed for 1 minute, BARO sensor more than 75 kPa, ECT sensor from 18-230°F, IAT sensor more than 18°F, fuel level from 15-85 percent, and the PCM detected the maximum HO2S-12 voltage was less than 300 mv during the CCM test. **Possible Causes:** • HO2S is shorted to ground • HO2S is contaminated, damaged or it has failed • Vehicle driven until it ran out of fuel, or while very low on fuel • PCM has failed
DTC: P0152 **2T CCM; MIL: Yes** **Years:** 2007, 2008 **Models:** Vitara, XL-7 **Engines:** 2.7L VIN 9, 3.6L VIN 7 **Transmissions:** A/T, M/T	**HO2S-21 (Bank 2 Sensor 1) Circuit High Input** Engine started, vehicle driven at a speed of over 35 MPH for 2 minutes, then back to idle speed for 1 minute, BARO sensor more than 75 kPa, ECT sensor from 18-230°F, IAT sensor more than 18°F, fuel level from 15-85 percent, and the PCM detected the maximum HO2S-12 voltage was less than 600 mv during the CCM test. **Possible Causes:** • HO2S signal circuit is open or shorted to system power • Fuel supply system is too rich (fuel injector or regulator leaking) • HO2S is contaminated, damaged or has failed • PCM has failed
DTC: P0153 **2T CCM; MIL: Yes** **Years:** 2007, 2008 **Models:** Vitara, XL-7 **Engines:** 2.7L VIN 9, 3.6L VIN 7 **Transmissions:** A/T, M/T	**HO2S-21 (Bank 2 Sensor 1) Slow Response** Engine started, vehicle driven at a constant speed of over 35 MPH for 2 minutes in closed loop, then back to idle speed for 1 minute, ambient air temperature over 14°F, fuel level from 15-85 percent, BARO sensor more than 75 kPa, IAT sensor from 18-230°F, and the PCM detected the response rate of the HO2S-21 signal was too long during the CCM test. **Possible Causes:** • Air leaks in the intake manifold or exhaust manifold or pipes • IAT sensor or MAF sensor has deteriorated (out of calibration) • HO2S signal circuit is open or shorted to ground (intermittent) • HO2S contaminated with wrong fuel, has deteriorated or failed • PCM has failed
DTC: P0154 **2T CCM; MIL: Yes** **Years:** 2007, 2008 **Models:** Vitara, XL-7 **Engines:** 2.7L VIN 9, 3.6L VIN 7 **Transmissions:** A/T, M/T	**HO2S-21 (Bank 2 Sensor 1) No Activity Detected** Engine started, vehicle driven at a constant speed of over 35 MPH for 2 minutes in closed loop, then back to idle speed for 1 minute, ambient air temperature over 14°F, fuel level from 15-85 percent, BARO sensor more than 75 kPa, IAT sensor more than 18°F, and the PCM detected the HO2S-21 signal did not exceed 0.45v during the test. **Possible Causes:** • Air leaks in the intake manifold or exhaust manifold or pipes • HO2S is contaminated (wrong fuel), has deteriorated or failed • HO2S heater element is damaged or has failed • PCM has failed
DTC: P0155 **2T OBD/O2S; MIL: Yes** **Years:** 2007, 2008 **Models:** Vitara, XL-7 **Engines:** 2.7L VIN 9, 3.6L VIN 7 **Transmissions:** A/T, M/T	**HO2S-21 (Bank 2 Sensor 1) Heater Circuit Malfunction** Engine started, engine runtime over 1 minute, IAT sensor more than 14°F, ECT sensor from 18-230°F, and the PCM detected the HO2S-21 heater current level was more than 5.3A or less than 0.09A. **Possible Causes:** • HO2S heater control circuit is open or shorted to ground • HO2S heater power circuit is open (check the IG fuse) • HO2S heater control circuit is shorted to power • HO2S heater is damaged or has failed • PCM has failed
DTC: P0156 **2T CCM; MIL: Yes** **Years:** 2007, 2008 **Models:** Vitara, XL-7 **Engines:** 2.7L VIN 9, 3.6L VIN 7 **Transmissions:** A/T, M/T	**HO2S-22 (Bank 2 Sensor 2) Range/Performance** Engine started, vehicle driven to a constant speed of over 45 MPH for 5 minutes, then back to idle speed for 1 minute, IAT sensor more than 18°F, ECT sensor from 14-230°F, BARO sensor more than 75 kPa, fuel level from 15-85 percent, and the PCM detected the HO2S-22 signal was more than 4.5v, or the maximum HO2S-22 signal voltage was too low, or that the minimum HO2S-22 signal voltage was more than 300 mv during the CCM test. **Possible Causes:** • HO2S signal circuit open (sensor signal reads over 4.5v) • HO2S is shorted to ground • HO2S signal ground circuit is open • HO2S is contaminated or has failed • Vehicle driven until it ran out of fuel, or while very low on fuel • PCM has failed

DTC	Trouble Code Title, Conditions & Possible Causes
DTC: P0157 **2T CCM; MIL: Yes** **Years:** 2007, 2008 **Models:** Vitara, XL-7 **Engines:** 2.7L VIN 9, 3.6L VIN 7 **Transmissions:** A/T, M/T	**HO2S-22 (Bank 2 Sensor 2) Circuit Low Input** Engine started, vehicle driven at a speed of over 35 MPH for 2 minutes, then back to idle speed for 1 minute, BARO sensor more than 75 kPa, ECT sensor from 18-230°F, IAT sensor more than 18°F, fuel level from 15-85 percent, and the PCM detected the maximum HO2S-12 voltage was less than 600 mv during the CCM test. **Possible Causes:** • HO2S signal circuit is open or shorted to system power • Fuel supply system is too rich (fuel injector or regulator leaking) • HO2S is contaminated, damaged or has failed • PCM has failed
DTC: P0158 **2T CCM; MIL: Yes** **Years:** 2007, 2008 **Models:** Vitara, XL-7 **Engines:** 2.7L VIN 9, 3.6L VIN 7 **Transmissions:** A/T, M/T M/T	**HO2S-21 (Bank 2 Sensor 2) Circuit High Input** Engine started, vehicle driven at a speed of over 35 MPH for 2 minutes, then back to idle speed for 1 minute, BARO sensor more than 75 kPa, ECT sensor from 18-230°F, IAT sensor more than 18°F, fuel level from 15-85 percent, and the PCM detected the maximum HO2S-12 voltage was less than 600 mv during the CCM test. **Possible Causes:** • HO2S signal circuit is open or shorted to system power • Fuel supply system is too rich (fuel injector or regulator leaking) • HO2S is contaminated, damaged or has failed • PCM has failed
DTC: P0160 **2T CCM; MIL: Yes** **Years:** 2007, 2008 **Models:** Vitara, XL-7 **Engines:** 2.7L VIN 9, 3.6L VIN 7 **Transmissions:** A/T, M/T	**HO2S-12 (Bank 2 Sensor 2) Circuit, No Activity Detected** DTC P0131 and P0132 not set, vehicle driven at a speed of 35 MPH for 2 minutes in closed loop, then back to idle speed for 2 minutes, ambient air temperature over 14°F, fuel level from 15-85 percent, BARO sensor more than 75 kPa, IAT sensor more than 14°F, and the PCM detected the HO2S-11 signal remained "high" or "low" in the test. **Possible Causes:** • Air leaks in the intake manifold or exhaust manifold or pipes • HO2S is contaminated (wrong fuel), has deteriorated or failed • HO2S heater element is damaged or has failed • PCM has failed • TSB TS015 (12/95) contains information related to this code
DTC: P0161 **2T OBD/O2S; MIL: Yes** **Years:** 2007, 2008 **Models:** Vitara, XL-7 **Engines:** 2.7L VIN 9, 3.6L VIN 7 **Transmissions:** A/T, M/T	**HO2S-22 (Bank 2 Sensor 2) Heater Circuit Malfunction** Engine started, engine runtime over 1 minute, IAT sensor more than 14°F, ECT sensor from 18-230°F, and the PCM detected the HO2S-22 heater current level was more than 11 mA or less than 0.32 mA. **Possible Causes:** • HO2S heater control circuit is open or shorted to ground • HO2S heater power circuit is open (check the IG fuse) • HO2S heater control circuit is shorted to power • HO2S heater is damaged or has failed • PCM has failed
DTC: P0171 **2T MIL: Yes** **Years:** 2007, 2008 **Models:** Aerio, Forenza, Reno, Vitara, XL-7, **Engines:** 2.3 VIN 6, 2.0L VIN Z, 2.7L VIN 9, 3.6L VIN 7, **Transmissions:** All	**Fuel System Too Lean (Bank 1)** DTC P0101, P0102, P0103, P0106, P0107, P0108, P0112, P0113, P0117, P0118, P0121, P0131, P0132, P0133, P0135, P0136, P0141, P0301-306, P0400, P0400 and P0601 not set, engine started, vehicle driven at a speed of 30-40 MPH for 5 minutes at an engine speed from 2500-3000 RPM, BARO sensor more than 75 kPa, IAT sensor from 14-158°F, ambient temperature over 14°F, fuel level from 15-85 percent, and the PCM detected the Short Term fuel trim was over +20 percent, or the total of the Short Term and Long Term fuel trims values (added together) exceeded +33 percent during the test. **Possible Causes:** • Air leaks after the MAF sensor, or in the EGR or PCV system • Base engine "mechanical" fault affecting one or more cylinders • Exhaust leaks before or near where the front HO2S is mounted • Fuel control sensor is out of calibration (i.e., ECT, IAT or MAP) • Fuel delivery system supplying too little fuel during cruise or idle periods (e.g., faulty fuel pump or dirty, restricted fuel filter) • Fuel injector (one or more) dirty or pressure regulator has failed • HO2S is contaminated, deteriorated or it has failed • Vehicle driven low on fuel or until it ran out of fuel

DTC	Trouble Code Title, Conditions & Possible Causes
DTC: P0171 **2T MIL: Yes** **Models:** Aerio, Forenza, Reno, Vitara, XL-7, **Engines:** 2.3 VIN 6, 2.0L VIN Z, 2.7L VIN 9, 3.6L VIN 7, **Transmissions:** All	**Fuel System Too Lean (Bank 1)** DTC P0101, P0102, P0103, P0106, P0107, P0108, P0112, P0113, P0117, P0118, P0121, P0131, P0132, P0133, P0135, P0136, P0141, P0301-306, P0400, P0400 and P0601 not set, engine started, vehicle driven at a speed of 30-40 MPH for 5 minutes at an engine speed from 2500-3000 RPM, BARO sensor more than 75 kPa, IAT sensor from 14-158°F, ambient temperature over 14°F, fuel level from 15-85 percent, and the PCM detected the total fuel trim (Short and Long Term fuel trim total) exceed +31 percent in the Fuel System test. **Possible Causes:** • Air leaks after the MAF sensor, or in the EGR or PCV system • Base engine "mechanical" fault affecting one or more cylinders • Exhaust leaks before or near where the front HO2S is mounted • Fuel control sensor is out of calibration (i.e., ECT, IAT or MAP) • Fuel delivery system supplying too little fuel during cruise or idle periods (e.g., faulty fuel pump or dirty, restricted fuel filter) • Fuel injector (one or more) dirty or pressure regulator has failed • HO2S is contaminated, deteriorated or it has failed • Vehicle driven low on fuel or until it ran out of fuel
DTC: P0172 **2T MIL: Yes** **Years:** 2007, 2008 **Models:** Forenza, Reno, Vitara, XL-7, **Engines:** 2.0L VIN Z, 2.7L VIN 9, 3.6L VIN 7 **Transmissions:** All	**Fuel System Too Rich (Bank 1)** DTC P0101, P0102, P0103, P0106, P0107, P0108, P0112, P0113, P0117, P0118, P0121, P0131, P0132, P0133, P0135, P0136, P0141, P0301-306, P0400, P0400 and P0601 not set, engine started, vehicle driven at a speed of 30-40 MPH for 5 minutes at an engine speed from 2500-3000 RPM, BARO sensor more than 75 kPa, IAT sensor from 14-1588°F, ambient temperature over 14°F, fuel level from 15-85 percent, and the PCM detected the Short Term fuel trim was over −20 percent, or the total of the Short Term and Long Term fuel trims values (added together) exceeded −33 percent during the test. **Possible Causes:** • Base engine "mechanical" fault affecting one or more cylinders • EVAP system component has failed or canister fuel saturated • Fuel control sensor is out of calibration (i.e., ECT, IAT or MAP) • Fuel delivery system supplying too much fuel during cruise or idle periods (e.g., faulty fuel pump, or faulty pressure regulator) • Fuel injector(s) is leaking or stuck partially open (one or more) • HO2S is contaminated, deteriorated or it has failed
DTC: P0174 **2T MIL: Yes** **Years:** 2007, 2008 **Models:** Vitara, XL-7 **Engines:** 2.7L VIN 9, 3.6L VIN 7 **Transmissions:** A/T, M/T	**Fuel System Too Lean (Bank 2)** DTC P0101, P0102, P0103, P0106, P0107, P0108, P0112, P0113, P0117, P0118, P0121, P0131, P0132, P0133, P0135, P0136, P0141, P0301-306, P0400, P0400 and P0601 not set, engine started, vehicle driven at a speed of 30-40 MPH for 5 minutes at an engine speed from 2500-3000 RPM, BARO sensor more than 75 kPa, IAT sensor from 14-158°F, ambient temperature over 14°F, fuel level from 15-85 percent, and the PCM detected the total fuel trim (Short and Long Term fuel trim total) exceed +31 percent in the Fuel System test. **Possible Causes:** • Air leaks after the MAF sensor, or in the EGR or PCV system • Base engine "mechanical" fault affecting one or more cylinders • Exhaust leaks before or near where the front HO2S is mounted • Fuel control sensor is out of calibration (i.e., ECT, IAT or MAP) • Fuel delivery system supplying too little fuel during cruise or idle periods (e.g., faulty fuel pump or dirty, restricted fuel filter) • Fuel injector (one or more) dirty or pressure regulator has failed • HO2S is contaminated, deteriorated or it has failed • Vehicle driven low on fuel or until it ran out of fuel
DTC: P0175 **2T MIL: Yes** **Years:** 2007, 2008 **Models:** Vitara, XL-7 **Engines:** 2.7L VIN 9, 3.6L VIN 7 **Transmissions:** A/T, M/T	**Fuel System Too Rich (Bank 2)** DTC P0101, P0102, P0103, P0106, P0107, P0108, P0112, P0113, P0117, P0118, P0121, P0131, P0132, P0133, P0135, P0136, P0141, P0301-306, P0400, P0400 and P0601 not set, engine started, vehicle driven at a speed of 30-40 MPH for 5 minutes at an engine speed from 2500-3000 RPM, BARO sensor more than 75 kPa, IAT sensor from 14-158°F, ambient temperature over 14°F, fuel level from 15-85 percent, and the PCM detected the total fuel trim (Short and Long Term fuel trim total) exceed −31 percent in the Fuel System test. **Possible Causes:** • Base engine "mechanical" fault affecting one or more cylinders • EVAP system component has failed or canister fuel saturated • Fuel control sensor is out of calibration (i.e., ECT, IAT or MAP) • Fuel delivery system supplying too much fuel during cruise or idle periods (e.g., faulty fuel pump, or faulty pressure regulator) • Fuel injector(s) is leaking or stuck partially open (one or more) • HO2S is contaminated, deteriorated or it has failed

DTC	Trouble Code Title, Conditions & Possible Causes
DTC: P0217 **2T CCM; MIL: Yes** **Years:** 2007, 2008 **Models:** Reno, Forenza **Engines:** 2.0L VIN Z **Transmissions:** All	**Engine Coolant Over Temperature** Engine is running, engine coolant temperature is higher than 50°C (122°F), Intake Air Temperature (IAT) is higher than 35°C (95°F), engine soak time is higher than 360 minutes, or start-up coolant temperature is less than 45°C (113°F) and Engine Coolant Temperature (ECT) sensor reading is greater than 107°C (224.6°F). **Possible Causes:** • ECT sensor electrical connector • Engine Control Module (ECM) electrical terminals E25 and E51 • ECM • ECT sensor
DTC: P0223 **2T CCM; MIL: Yes** **Years:** 2007, 2008 **Models:** Reno, Forenza **Engines:** 2.0L VIN Z **Transmissions:** All	**Throttle Position (TP) Sensor 2 Circuit High Voltage** Ignition ON, no TP sensor fault exists and MTIA voltage is higher than 4.79 V **Possible Causes:** • Backed-out terminals • Improper mating • Broken locks • Damaged terminals • Poor terminals to wire connection • Physical damage to the wiring harness
DTC: P0261 **1T MIL: Yes** **Years:** 2007, 2008 **Models:** Reno, Forenza **Engines:** 2.5L VIN 6 **Transmissions:** All	**Fuel Injector 1 Low Input** Engine started, engine speed less than 1000 RPM, TP sensor less than 0.7v, Actuator Tests all "off", and the PCM detected an unexpected voltage on the injector coil. **Possible Causes:** • Injector control circuit is open or shorted to ground • Injector power circuit is open (check power from the MFI relay) • Fuel injector is damaged or has failed
DTC: P0262 **1T MIL: Yes** **Years:** 2007, 2008 **Models:** Reno, Forenza **Engines:** 2.5L VIN 6 **Transmissions:** All	**Fuel Injector 1 High Input** Engine started, engine speed less than 1000 RPM, TP sensor less than 0.7v, Actuator Tests all "off", and the PCM detected an unexpected voltage on the injector coil. **Possible Causes:** • Injector control circuit is open or shorted to ground • Injector power circuit is open (check power from the MFI relay) • Fuel injector is damaged or has failed
DTC: P0264 **1T MIL: Yes** **Years:** 2007, 2008 **Models:** Reno, Forenza **Engines:** 2.5L VIN 6 **Transmissions:** All	**Fuel Injector 2 Low Input** Engine started, engine speed less than 1000 RPM, TP sensor less than 0.7v, Actuator Tests all "off", and the PCM detected an unexpected voltage on the injector coil. **Possible Causes:** • Injector control circuit is open or shorted to ground • Injector power circuit is open (check power from the MFI relay) • Fuel injector is damaged or has failed
DTC: P0265 **1T MIL: Yes** **Years:** 2007, 2008 **Models:** Reno, Forenza **Engines:** 2.5L VIN 6 **Transmissions:** All	**Fuel Injector 2 High Input** Engine started, engine speed less than 1000 RPM, TP sensor less than 0.7v, Actuator Tests all "off", and the PCM detected an unexpected voltage on the injector coil. **Possible Causes:** • Injector control circuit is open or shorted to ground • Injector power circuit is open (check power from the MFI relay) • Fuel injector is damaged or has failed
DTC: P0267 **1T MIL: Yes** **Years:** 2007, 2008 **Models:** Reno, Forenza **Engines:** 2.5L VIN 6 **Transmissions:** All	**Fuel Injector 3 Low Input** Engine started, engine speed less than 1000 RPM, TP sensor less than 0.7v, Actuator Tests all "off", and the PCM detected an unexpected voltage on the injector coil. **Possible Causes:** • Injector control circuit is open or shorted to ground • Injector power circuit is open (check power from the MFI relay) • Fuel injector is damaged or has failed

DTC	Trouble Code Title, Conditions & Possible Causes
DTC: P0268 **1T MIL: Yes** **Years:** 2007, 2008 **Models:** Reno, Forenza **Engines:** 2.5L VIN 6 **Transmissions:** All	**Fuel Injector 3 High Input** Engine started, engine speed less than 1000 RPM, TP sensor less than 0.7v, Actuator Tests all "off", and the PCM detected an unexpected voltage on the injector coil. **Possible Causes:** • Injector control circuit is open or shorted to ground • Injector power circuit is open (check power from the MFI relay) • Fuel injector is damaged or has failed
DTC: P0270 **1T MIL: Yes** **Years:** 2007, 2008 **Models:** Reno, Forenza **Engines:** 2.5L VIN 6 **Transmissions:** All	**Fuel Injector 4 Low Input** Engine started, engine speed less than 1000 RPM, TP sensor less than 0.7v, Actuator Tests all "off", and the PCM detected an unexpected voltage on the injector coil. **Possible Causes:** • Injector control circuit is open or shorted to ground • Injector power circuit is open (check power from the MFI relay) • Fuel injector is damaged or has failed
DTC: P0271 **1T MIL: Yes** **Years:** 2007, 2008 **Models:** Reno, Forenza **Engines:** 2.5L VIN 6 **Transmissions:** All	**Fuel Injector 4 High Input** Engine started, engine speed less than 1000 RPM, TP sensor less than 0.7v, Actuator Tests all "off", and the PCM detected an unexpected voltage on the injector coil. **Possible Causes:** • Injector control circuit is open or shorted to ground • Injector power circuit is open (check power from the MFI relay) • Fuel injector is damaged or has failed
DTC: P0300 **2T CCM; MIL: Yes** **Years:** 2007, 2008 **Models:** Aerio, Forenza, Reno, Vitara, XL-7, **Engines:** 2.3 VIN 6, 2.0L VIN Z, 2.7L VIN 9, 3.6L VIN 7 **Transmissions:** All	**Multiple Cylinder Misfire Detected** DTC P0335 and P0340 not set, altitude less than 9150 feet, fuel level from 15-85 percent, IAT signal from 6.8-158°F, ambient temperature over 14°F, ECT sensor from 14-230°F, engine runtime over 1 minute, and the PCM detected a misfire in more than 1 cylinders in the 200 (Catalyst) or 1000-RPM (High Emissions) revolution range. **Note: If the misfire is severe, the MIL will flash on/off on the 1st trip!** **Possible Causes:** • Air leak in the intake manifold, or in the EGR or PCM system • Base engine mechanical fault that affects one or more cylinders • Fuel delivery component fault that affects one or more cylinders (i.e., one or more contaminated, dirty or sticking fuel injectors) • Ignition system problem (coil or plug) in one or more cylinders
DTC: P0301 **2T CCM; MIL: Yes** **Years:** 2007, 2008 **Models:** Forenza, Reno, Vitara, XL-7, **Engines:** 2.0L VIN Z, 2.7L VIN 9, 3.6L VIN 7 **Transmissions:** All	**Cylinder 1 Misfire Detected** DTC P0106, P0107, P0108, P0117, P0118, P0121, P0122, P0123, P0335, P0340 and P0500 not set, BARO sensor over 75 kPa, fuel level from 15-85 percent, IAT signal from 6.8-158°F, ambient temperature over 14°F, engine running, and the PCM detected a misfire in one cylinder in the 200 (Catalyst) or 1000-RPM (High Emissions) range. **Note: If the misfire is severe, the MIL will flash on/off on the 1st trip!** **Possible Causes:** • Air leak in the intake manifold, or in the EGR or PCM system • Base engine mechanical fault that affects only one cylinder • Fuel delivery component fault that affects only one cylinder (i.e., a contaminated, dirty or sticking fuel injector) • Ignition system problem (coil or plug) that affects one cylinder
DTC: P0301 **2T CCM; MIL: Yes** **Years:** 2007, 2008 **Models:** Forenza, Reno, Vitara, XL-7, **Engines:** 2.0L VIN Z, 2.7L VIN 9, 3.6L VIN 7 **Transmissions:** All	**Cylinder 1 Misfire Detected** DTC P0335 and P0340 not set, BARO sensor more than 75 kPa, fuel level from 15-85 percent, IAT signal more than 18°F, ECT sensor from 18-230°F, engine running, and the PCM detected a misfire in one cylinder in the 200 (Catalyst) or 1000-RPM (High Emissions) range. **Note: If the misfire is severe, the MIL will flash on/off on the 1st trip!** **Possible Causes:** • Air leak in the intake manifold, or in the EGR or PCM system • Base engine mechanical fault that affects only one cylinder • Fuel delivery component fault that affects only one cylinder (i.e., a contaminated, dirty or sticking fuel injector) • Ignition system problem (coil or plug) that affects one cylinder

DTC	Trouble Code Title, Conditions & Possible Causes
DTC: P0302 **2T CCM; MIL: Yes** **Years:** 2007, 2008 **Models:** Forenza, Reno, Vitara, XL-7, **Engines:** 2.0L VIN Z, 2.7L VIN 9, 3.6L VIN 7 **Transmissions:** All	**Cylinder 2 Misfire Detected** DTC P0106, P0107, P0108, P0117, P0118, P0121, P0122, P0123, P0335, P0340 and P0500 not set, BARO sensor over 75 kPa, fuel level from 15-85 percent, IAT signal from 6.8-158°F, ambient temperature over 14°F, engine running, and the PCM detected a misfire in one cylinder in the 200 (Catalyst) or 1000-RPM (High Emissions) range. **Note: If the misfire is severe, the MIL will flash on/off on the 1st trip!** **Possible Causes:** • Air leak in the intake manifold, or in the EGR or PCM system • Base engine mechanical fault that affects only one cylinder • Fuel delivery component fault that affects only one cylinder (i.e., a contaminated, dirty or sticking fuel injector) • Ignition system problem (coil or plug) that affects one cylinder
DTC: P0302 **2T CCM; MIL: Yes** **Years:** 2007, 2008 **Models:** Forenza, Reno, Vitara, XL-7, **Engines:** 2.0L VIN Z, 2.7L VIN 9, 3.6L VIN 7 **Transmissions:** All	**Cylinder 2 Misfire Detected** DTC P0335 and P0340 not set, BARO sensor more than 75 kPa, fuel level from 15-85 percent, IAT signal more than 18°F, ECT sensor from 18-230°F, engine running, and the PCM detected a misfire in one cylinder in the 200 (Catalyst) or 1000-RPM (High Emissions) range. **Note: If the misfire is severe, the MIL will flash on/off on the 1st trip!** **Possible Causes:** • Air leak in the intake manifold, or in the EGR or PCM system • Base engine mechanical fault that affects only one cylinder • Fuel delivery component fault that affects only one cylinder (i.e., a contaminated, dirty or sticking fuel injector) • Ignition system problem (coil or plug) that affects one cylinder
DTC: P0303 **2T CCM; MIL: Yes** **Years:** 2007, 2008 **Models:** Forenza, Reno, Vitara, XL-7, **Engines:** 2.0L VIN Z, 2.7L VIN 9, 3.6L VIN 7 **Transmissions:** All	**Cylinder 3 Misfire Detected** DTC P0335 and P0340 not set, BARO sensor more than 75 kPa, fuel level from 15-85 percent, IAT signal more than 18°F, ECT sensor from 18-230°F, engine running, and the PCM detected a misfire in one cylinder in the 200 (Catalyst) or 1000-RPM (High Emissions) range. **Note: If the misfire is severe, the MIL will flash on/off on the 1st trip!** **Possible Causes:** • Air leak in the intake manifold, or in the EGR or PCM system • Base engine mechanical fault that affects only one cylinder • Fuel delivery component fault that affects only one cylinder (i.e., a contaminated, dirty or sticking fuel injector) • Ignition system problem (coil or plug) that affects one cylinder
DTC: P0303 **2T CCM; MIL: Yes** **Years:** 2007, 2008 **Models:** Forenza, Reno, Vitara, XL-7, **Engines:** 2.0L VIN Z, 2.7L VIN 9, 3.6L VIN 7 **Transmissions:** All	**Cylinder 3 Misfire Detected** DTC P0106, P0107, P0108, P0117, P0118, P0121, P0122, P0123, P0335, P0340 and P0500 not set, BARO sensor over 75 kPa, fuel level from 15-85 percent, IAT signal from 6.8-158°F, ambient temperature over 14°F, engine running, and the PCM detected a misfire in one cylinder in the 200 (Catalyst) or 1000-RPM (High Emissions) range. **Note: If the misfire is severe, the MIL will flash on/off on the 1st trip!** **Possible Causes:** • Air leak in the intake manifold, or in the EGR or PCM system • Base engine mechanical fault that affects only one cylinder • Fuel delivery component fault that affects only one cylinder (i.e., a contaminated, dirty or sticking fuel injector) • Ignition system problem (coil or plug) that affects one cylinder
DTC: P0304 **2T CCM; MIL: Yes** **Years:** 2007, 2008 **Models:** Forenza, Reno, Vitara, XL-7, **Engines:** 2.0L VIN Z, 2.7L VIN 9, 3.6L VIN 7 **Transmissions:** All	**Cylinder 4 Misfire Detected** DTC P0106, P0107, P0108, P0117, P0118, P0121, P0122, P0123, P0335, P0340 and P0500 not set, BARO sensor over 75 kPa, fuel level from 15-85 percent, IAT signal from 6.8-158°F, ambient temperature over 14°F, engine running, and the PCM detected a misfire in one cylinder in the 200 (Catalyst) or 1000-RPM (High Emissions) range. **Note: If the misfire is severe, the MIL will flash on/off on the 1st trip!** **Possible Causes:** • Air leak in the intake manifold, or in the EGR or PCM system • Base engine mechanical fault that affects only one cylinder • Fuel delivery component fault that affects only one cylinder (i.e., a contaminated, dirty or sticking fuel injector) • Ignition system problem (coil or plug) that affects one cylinder
DTC: P0304 **2T CCM; MIL: Yes** **Years:** 2007, 2008 **Models:** Forenza, Reno, Vitara, XL-7, **Engines:** 2.0L VIN Z, 2.7L VIN 9, 3.6L VIN 7 **Transmissions:** All	**Cylinder 4 Misfire Detected** DTC P0335 and P0340 not set, BARO sensor more than 75 kPa, fuel level from 15-85 percent, IAT signal more than 18°F, ECT sensor from 18-230°F, engine running, and the PCM detected a misfire in one cylinder in the 200 (Catalyst) or 1000-RPM (High Emissions) range. **Note: If the misfire is severe, the MIL will flash on/off on the 1st trip!** **Possible Causes:** • Air leak in the intake manifold, or in the EGR or PCM system • Base engine mechanical fault that affects only one cylinder • Fuel delivery component fault that affects only one cylinder (i.e., a contaminated, dirty or sticking fuel injector) • Ignition system problem (coil or plug) that affects one cylinder

DTC	Trouble Code Title, Conditions & Possible Causes
DTC: P0305 **2T CCM; MIL: Yes** **Years:** 2007, 2008 **Models:** Vitara, XL-7 **Engines:** 2.7L VIN 9, 3.6L VIN 7 **Transmissions:** A/T, M/T	**Cylinder 5 Misfire Detected** DTC P0335 and P0340 not set, BARO sensor more than 75 kPa, fuel level from 15-85 percent, IAT signal more than 18°F, ECT sensor from 18-230°F, engine running, and the PCM detected a misfire in one cylinder in the 200 (Catalyst) or 1000-RPM (High Emissions) range. **Note: If the misfire is severe, the MIL will flash on/off on the 1st trip!** **Possible Causes:** • Air leak in the intake manifold, or in the EGR or PCM system • Base engine mechanical fault that affects only one cylinder • Fuel delivery component fault that affects only one cylinder (i.e., a contaminated, dirty or sticking fuel injector) • Ignition system problem (coil or plug) that affects one cylinder
DTC: P0314 **2T CCM; MIL: Yes** **Years:** 2007, 2008 **Models:** Reno, Forenza **Engines:** 2.0L VIN Z **Transmissions:** All	**Misfire Detected with Camshaft Position (CMP) Sensor Error** The engine is running, and the ECM detects that the misfire is present with the CMP sensor error. **Possible Causes:** • CMP signal circuit open between the sensor and ECM • CMP sensor ground circuit is open • CMP sensor is damaged or has failed • ECM has failed
DTC: P0315 **2T CCM; MIL: Yes** **Years:** 2007, 2008 **Models:** Reno, Forenza **Engines:** 2.0L VIN Z **Transmissions:** All	**Crankshaft Position (CKP) System Variation Not Learned** Key on or engine running, and the PCM detects tooth error sample counter is less than 30. **Possible Causes:** • CKP sensor signal circuit open between the sensor and PCM • CKP sensor signal circuit is shorted to VREF or system power • CKP sensor ground circuit is open • CKP sensor is damaged or has failed • PCM
DTC: P0317 **2T CCM; MIL: Yes** **Years:** 2007, 2008 **Models:** Reno, Forenza **Engines:** 2.0L VIN Z **Transmissions:** All	**Rough Road Sensor Sensing System Input Not Present** The ECM can not detect any rough road source, the engine run time is greater than 10 seconds and the vehicle speed is greater than 10 km/h (6 MPH). **Possible Causes:** • Open signal circuit of G sensor • Open serial data line between the ECM and the EBCM • PCM
DTC: P0324 **2T CCM; MIL: Yes** **Years:** 2007, 2008 **Models:** Reno, Forenza **Engines:** 2.0L VIN Z **Transmissions:** All	**Knock Sensor (KS) Module Performance** Vacuum is less than the predetermined value (10 – 50 kPa, based on RPM), RPMs are less than 1,600 RPM or knock filtered value is less than 25 or greater than 80. **Possible Causes:** • Backed-out terminals • Improper mating • Broken locks • Improperly formed • Damaged terminals • Poor terminal-to-wire connection • Physical damage to the wiring harness
DTC: P0325 **2T CCM; MIL: Yes** **Years:** 2007, 2008 **Models:** Forenza, Reno, Vitara, XL-7, **Engines:** 2.0L VIN Z, 2.7L VIN 9, 3.6L VIN 7 **Transmissions:** All	**Knock Sensor Circuit** Vacuum is less than the predetermined value (10 – 50 kPa, based on RPM), the RPM is less than 1,600 RPM or the filter coefficient is less than 1.0 percent. **Possible Causes:** • Backed-out terminals • Improper mating • Broken locks • Improperly formed • Damaged terminals • Poor terminal-to-wire connection • Physical damage to the wiring harness
DTC: P0327 **1T CCM; MIL: Yes** **Years:** 2007, 2008 **Models:** Aerio, Vitara, XL-7 **Engines:** 2.3 VIN 6, 2.7L VIN 9, 3.6L VIN 7 **Transmissions:** A/T, M/T	**Knock Sensor Circuit Low Input** Key on or engine running, and the PCM detected an unexpected "high" voltage condition (0.90v or less) on the Knock Sensor circuit. **Possible Causes:** • Knock sensor signal circuit is shorted to sensor ground • Knock sensor signal circuit is shorted to chassis ground • Knock sensor is damaged or has failed • PCM has failed

DTC	Trouble Code Title, Conditions & Possible Causes
DTC: P0328 **1T CCM; MIL: Yes** **Years:** 2007, 2008 **Models:** Vitara, XL-7 **Engines:** 2.7L VIN 9, 3.6L VIN 7 **Transmissions:** A/T, M/T	**Knock Sensor Circuit High Input** Key on or engine running, and the PCM detected an unexpected "low" voltage condition (3.98v or less) on the Knock Sensor circuit. **Possible Causes:** • Knock sensor signal circuit open between the sensor and PCM • Knock sensor signal circuit is shorted to VREF or system power • Knock sensor ground circuit is open • Knock sensor is damaged or has failed • PCM has failed
DTC: P0335 **1T CCM; MIL: Yes** **Years:** 2007, 2008 **Models:** Aerio, Forenza, Reno, Vitara, XL-7, **Engines:** 2.3 VIN 6, 2.0L VIN Z, 2.7L VIN 9, 3.6L VIN 7 **Transmissions:** All	**Crankshaft Position Sensor Circuit Malfunction** Engine started, and the PCM did not detect any CKP signals for 3 seconds or after 100 CMP pulses were received in the CCM test. **Possible Causes:** • CKP sensor positive (+) circuit is open or shorted to ground • CKP sensor negative (−) circuit is open or shorted to ground • CKP sensor is damaged or has failed • PCM has failed
DTC: P0336 **1T CCM; MIL: Yes** **Years:** 2007, 2008 **Models:** Reno, Forenza, **Engines:** 2.0L VIN Z **Transmissions:** All	**Crankshaft Position Sensor Malfunction (No Plausible Signal)** Engine started, and the PCM did not detect any CKP signals for 3 seconds or after 100 CMP pulses were received in the CCM test. **Possible Causes:** • CKP sensor positive (+) circuit is open or shorted to ground • CKP sensor negative (−) circuit is open or shorted to ground • CKP sensor is damaged or has failed • PCM has failed
DTC: P0340 **1T CCM; MIL: Yes** **Years:** 2007, 2008 **Models:** Aerio, Forenza, Reno, Vitara, XL-7, **Engines:** 2.3 VIN 6, 2.0L VIN Z, 2.7L VIN 9, 3.6L VIN 7 **Transmissions:** All	**Camshaft Position Sensor Circuit Malfunction** Engine cranking, engine start signal received (from starter circuit), and the PCM did not detect any CMP sensor signals for 5 seconds. **Possible Causes:** • CMP sensor signal circuit is open between sensor and PCM • CMP sensor signal circuit is shorted to ground • CMP sensor power circuit is open (check power from the relay) • CMP sensor is damaged or has failed • PCM has failed
DTC: P0351 **1T CCM; MIL: Yes** **Years:** 2007, 2008 **Models:** Reno, Forenza **Engines:** 2.0L VIN Z **Transmissions:** All	**Ignition Coil 1 and 4 Control Circuit** Ignition ON monitor fault feedback signal must receive more than 40 failures within 80 test cycles. Ignition voltage is between 11 and 16 V. **Possible Causes:** • Backed-out terminals • Improper mating • Broken locks • Improperly formed terminals • Damaged terminals • Poor terminal-to-wire connection • Physical damage to the wiring harness
DTC: P0352 **1T CCM; MIL: Yes** **Years:** 2007, 2008 **Models:** Reno, Forenza **Engines:** 2.0L VIN Z **Transmissions:** All	**Camshaft Position Sensor Circuit No Signal** Engine cranking, engine start signal received (from starter circuit), and the PCM did not detect any CMP sensor signals for 5 seconds. **Possible Causes:** • CMP sensor signal circuit is open between sensor and PCM • CMP sensor signal circuit is shorted to ground • CMP sensor power circuit is open (check power from the relay) • CMP sensor is damaged or has failed • PCM has failed
DTC: P0400 **2T CCM; MIL: Yes** **Years:** 2007, 2008 **Models:** Vitara, XL-7 **Engines:** 2.7L VIN 9, 3.6L VIN 7 **Transmissions:** A/T, M/T	**EGR System Fault (Too Much or Too Little Flow)** Engine started, BARO sensor over 75 kPa, IAT signal more than 18°F, ECT sensor from 18-230°F, vehicle driven at a speed of 35-40 MPH with the throttle steady for 3 minutes, followed by a deceleration period with Fuel Cut enabled, and the PCM detected the change in the MAP sensor signal was too small after the EGR valve was opened momentarily, and then closed during the EGR flow test. **Possible Causes:** • EGR stepper motor control circuit(s) open or shorted to ground • EGR stepper motor assembly is damaged or has failed • Exhaust connection to EGR valve is clogged or restricted • Catalytic converter or exhaust system is clogged or restricted • PCM has failed

DTC	Trouble Code Title, Conditions & Possible Causes
DTC: P0401 **2T CCM; MIL: Yes** **Years:** 2007, 2008 **Models:** Aerio, Forenza, Reno, Vitara, XL-7, **Engines:** 2.3 VIN 6, 2.0L VIN Z, 2.7L VIN 9, 3.6L VIN 7 **Transmissions:** All	**EGR System Fault (Too Much or Too Little Flow)** Engine started, BARO sensor over 75 kPa, IAT signal more than 18°F, ECT sensor from 18-230°F, vehicle driven at a speed of 35-40 MPH with the throttle steady for 3 minutes, followed by a deceleration period with Fuel Cut enabled, and the PCM detected the change in the MAP sensor signal was too small after the EGR valve was opened momentarily, and then closed during the EGR flow test. **Possible Causes:** • EGR stepper motor control circuit(s) open or shorted to ground • EGR stepper motor assembly is damaged or has failed • Exhaust connection to EGR valve is clogged or restricted • Catalytic converter or exhaust system is clogged or restricted • PCM has failed
DTC: P0402 **2T CCM; MIL: Yes** **Years:** 2007, 2008 **Models:** Forenza, Reno, Vitara, XL-7, **Engines:** 2.0L VIN Z, 2.7L VIN 9, 3.6L VIN 7 **Transmissions:** All	**EGR System Fault (Too Much or Too Little Flow)** Engine started, BARO sensor over 75 kPa, IAT signal more than 18°F, ECT sensor from 18-230°F, vehicle driven at a speed of 35-40 MPH with the throttle steady for 3 minutes, followed by a deceleration period with Fuel Cut enabled, and the PCM detected the change in the MAP sensor signal was too small after the EGR valve was opened momentarily, and then closed during the EGR flow test. **Possible Causes:** • EGR stepper motor control circuit(s) open or shorted to ground • EGR stepper motor assembly is damaged or has failed • Exhaust connection to EGR valve is clogged or restricted • Catalytic converter or exhaust system is clogged or restricted • PCM has failed
DTC: P0403 **1T CCM; MIL: Yes** **Years:** 2007, 2008 **Models:** Aerio, Forenza, Reno, Vitara, XL-7, **Engines:** 2.3 VIN 6, 2.0L VIN Z, 2.7L VIN 9, 3.6L VIN 7 **Transmissions:** All	**EGR Solenoid Control Circuit Malfunction** Engine started, IAT sensor over 18°F, ECT sensor from 18-230°F, BARO sensor over 75 kPa, and the PCM detected an unexpected voltage condition on the EGR solenoid control circuit during the test. **Possible Causes:** • One or more EGR stepper motor circuits is open • One or more EGR stepper motor circuits is shorted to ground • EGR stepper motor power circuit is open (check the J/B fuse) • EGR stepper motor is damaged or has failed • PCM has failed
DTC: P0404 **2T CCM; MIL: Yes** **Years:** 2007, 2008 **Models:** Forenza, Reno, Vitara, XL-7, **Engines:** 2.0L VIN Z, 2.7L VIN 9, 3.6L VIN 7 **Transmissions:** All	**Exhaust Gas Recirculation Open Valve Position Error** Engine started, BARO sensor over 75 kPa, IAT signal more than 18°F, ECT sensor from 18-230°F, vehicle driven at a speed of 35-40 MPH with the throttle steady for 3 minutes, followed by a deceleration period with Fuel Cut enabled, and the PCM detected the EGR valve failed. **Possible Causes:** • EGR stepper motor control circuit(s) open or shorted to ground • EGR stepper motor assembly is damaged or has failed • Exhaust connection to EGR valve is clogged or restricted • Catalytic converter or exhaust system is clogged or restricted • PCM has failed
DTC: P0405 **2T CCM; MIL: Yes** **Years:** 2007, 2008 **Models:** Reno, Forenza **Engines:** 2.0L VIN Z **Transmissions:** All	**Exhaust Gas Recirculation Pintle Position Low Voltage** Engine started, BARO sensor over 75 kPa, IAT signal more than 18°F, ECT sensor from 18-230°F, vehicle driven at a speed of 35-40 MPH with the throttle steady for 3 minutes, followed by a deceleration period with Fuel Cut enabled, and the PCM detected an unexpected voltage. **Possible Causes:** • EGR stepper motor control circuit(s) open or shorted to ground • EGR stepper motor assembly is damaged or has failed • Exhaust connection to EGR valve is clogged or restricted • Catalytic converter or exhaust system is clogged or restricted • PCM has failed
DTC: P0406 **2T CCM; MIL: Yes** **Years:** 2007, 2008 **Models:** Reno, Forenza **Engines:** 2.0L VIN Z **Transmissions:** All	**Exhaust Gas Recirculation Pintle Position High Voltage** Engine started, BARO sensor over 75 kPa, IAT signal more than 18°F, ECT sensor from 18-230°F, vehicle driven at a speed of 35-40 MPH with the throttle steady for 3 minutes, followed by a deceleration period with Fuel Cut enabled, and the PCM detected an unexpected voltage. **Possible Causes:** • EGR stepper motor control circuit(s) open or shorted to ground • EGR stepper motor assembly is damaged or has failed • Exhaust connection to EGR valve is clogged or restricted • Catalytic converter or exhaust system is clogged or restricted • PCM has failed

DTC	Trouble Code Title, Conditions & Possible Causes
DTC: P0420 **2T CCM; MIL: Yes** **Years:** 2007, 2008 **Models:** Aerio, Forenza, Reno, Vitara, XL-7, **Engines:** 2.3 VIN 6, 2.0L VIN Z, 2.7L VIN 9, 3.6L VIN 7 **Transmissions:** All	**Catalyst Efficiency Below Normal (Bank 1)** DTC P0101, P0102, P0103, P0111-P0113, P0116-P0118, P0131, P0132, P0133, P0134, P0135, P0136, P0141, P0151, P0152, P0153, P0154, P0155, P0156, P0161, P0335, P0461, P0463, P0500, P1450 and P1451 not set, IAT signal more than 18°F, ECT sensor from 1858-230°F, BARO sensor over 75 kPa, vehicle driven at over 35 MPH, then at 55-60 MPH for 5 minutes; and the PCM detected the HO2S-12 and HO2S-11 signals were too similar. **Possible Causes:** • Air leaks at the exhaust manifold or in the exhaust pipes • Catalytic converter is damaged, contaminated or has failed • Front HO2S or rear HO2S is contaminated with fuel or moisture • Front HO2S and/or the rear HO2S is loose in the mounting hole • Front HO2S is older (aged) than the rear HO2S (HO2S is lazy)
DTC: P0430 **2T CCM; MIL: Yes** **Years:** 2007, 2008 **Models:** Forenza, Reno, Vitara, XL-7, **Engines:** 2.0L VIN Z, 2.7L VIN 9, 3.6L VIN 7 **Transmissions:** All	**Catalyst Efficiency Below Normal (Bank 2)** DTC P0101, P0102, P0103, P0111-P0113, P0116-P0118, P0131, P0132, P0133, P0134, P0135, P0136, P0141, P0151, P0152, P0153, P0154, P0155, P0156, P0161, P0335, P0461, P0463, P0500, P1450 and P1451 not set, IAT signal more than 18°F, ECT sensor from 158°F-230°F, BARO sensor over 75 kPa, vehicle driven at over 35 MPH, then at 55-60 MPH for 5 minutes; and the PCM detected the HO2S-22 and HO2S-21 signals were too similar. **Possible Causes:** • Air leaks at the exhaust manifold or in the exhaust pipes • Catalytic converter is damaged, contaminated or has failed • Front HO2S or rear HO2S is contaminated with fuel or moisture • Front HO2S and/or the rear HO2S is loose in the mounting hole • Front HO2S is older (aged) than the rear HO2S (HO2S is lazy)
DTC: P0440 **2T CCM; MIL: Yes** **Years:** 2007, 2008 **Models:** Vitara, XL-7 **Engines:** 2.7L VIN 9, 3.6L VIN 7 **Transmissions:** A/T, M/T	**EVAP System Malfunction** Engine started, vehicle driven to a speed of 35-40 for 20 minutes under conditions that do not include high-load, high engine speed, rapid acceleration or deceleration events, then at a constant speed of 30-40 MPH for 3 minutes, fuel level from 25-75 percent, BARO sensor over 75 kPa, ambient temperature over 14°F, IAT sensor less than 158°F, ECT sensor from 158°F-230°F, then after the pressure in the EVAP system reached a predetermined value, and with the Purge valve command closed, the PCM detected a leak in the system. **Possible Causes:** • Canister air valve (solenoid) is damaged or has failed • Charcoal canister is clogged, loaded with fuel or moisture • Fuel filler cap loose, cross-threaded, incorrect part or damaged • Fuel tank pressure sensor is damaged or has failed • Fuel tank or fuel tank sender assembly 'O' ring is leaking • Fuel tank vapor line(s) blocked, damaged or disconnected • Tank pressure control solenoid is contaminated or damaged • Purge solenoid control circuit is open or shorted to ground • Purge solenoid power circuit is open (test FI fuse in relay box) • Purge solenoid is contaminated, damaged or has failed • PCM has failed
DTC: P0441 **2T CCM; MIL: Yes** **Years:** 2007, 2008 **Models:** Aerio, Vitara, XL-7 **Engines:** 2.3 VIN 6, 2.7L VIN 9, 3.6L VIN 7 **Transmissions:** A/T, M/T	**EVAP System Leak Detected (very smAll leak)** Engine started, vehicle driven to a speed of 35-40 for 20 minutes under conditions that do not include high-load, high engine speed, rapid acceleration or deceleration events, then at a constant speed of 30-40 MPH for 3 minutes, fuel level from 25-75 percent, BARO sensor over 75 kPa, ambient temperature over 14°F, IAT sensor less than 158°F, ECT sensor from 158°F-230°F, then after the pressure in the EVAP system reached a predetermined value, and with the Purge valve command closed, the PCM detected a leak in the system. **Possible Causes:** • Canister air valve (solenoid) is damaged or has failed • Charcoal canister is clogged, loaded with fuel or moisture • Fuel filler cap loose, cross-threaded, incorrect part or damaged • Fuel tank pressure sensor is damaged or has failed • Fuel tank or fuel tank sender assembly 'O' ring is leaking • Fuel tank vapor line(s) blocked, damaged or disconnected • Tank pressure control solenoid is contaminated or damaged • Purge solenoid control circuit is open or shorted to ground • Purge solenoid power circuit is open (test FI fuse in relay box) • Purge solenoid is contaminated, damaged or has failed • PCM has failed

DTC	Trouble Code Title, Conditions & Possible Causes
DTC: P0442 **2T CCM; MIL: Yes** **Years:** 2007, 2008 **Models:** Forenza, Reno, Vitara, XL-7, **Engines:** 2.0L VIN Z, 2.7L VIN 9, 3.6L VIN 7 **Transmissions:** All	**EVAP System Leak Detected (Small Leak)** Engine started, vehicle driven to a speed of 35-40 for 20 minutes under conditions that do not include high-load, high engine speed, rapid acceleration or deceleration events, then at a constant speed of 30-40 MPH for 3 minutes, fuel level from 25-75 percent, BARO sensor over 75 kPa, ambient temperature over 14°F, IAT sensor less than 158°F, ECT sensor from 158°F-230°F, then after the pressure in the EVAP system reached a predetermined value, and with the Purge valve command closed, the PCM detected a leak in the system. **Possible Causes:** • Canister air valve (solenoid) is damaged or has failed • Charcoal canister is clogged, loaded with fuel or moisture • Fuel filler cap loose, cross-threaded, incorrect part or damaged • Fuel tank pressure sensor is damaged or has failed • Fuel tank or fuel tank sender assembly 'O' ring is leaking • Fuel tank vapor line(s) blocked, damaged or disconnected • Tank pressure control solenoid is contaminated or damaged • Purge solenoid control circuit is open or shorted to ground • Purge solenoid power circuit is open (test FI fuse in relay box) • Purge solenoid is contaminated, damaged or has failed • PCM has failed
DTC: P0443 **2T CCM; MIL: Yes** **Years:** 2007, 2008 **Models:** Vitara, XL-7 **Engines:** 2.7L VIN 9, 3.6L VIN 7 **Transmissions:** A/T, M/T	**EVAP Canister Purge Solenoid Circuit Malfunction** Engine started, IAT signal less than 122°F, ambient temperature over 14°F, engine running at a stable idle speed, fuel level from 25-75 percent, all accessory loads off, and the PCM detected an unexpected voltage condition on the EVAP Canister Purge solenoid circuit. **Possible Causes:** • Purge solenoid control circuit is open or shorted to ground • Purge solenoid power circuit is open (check IG fuse in the J/B) • Purge solenoid is damaged or has failed • PCM has failed
DTC: P0444 **2T CCM; MIL: Yes** **Years:** 2007, 2008 **Models:** Aerio, Vitara, XL-7 **Engines:** 2.3 VIN 6, 2.7L VIN 9, 3.6L VIN 7 **Transmissions:** A/T, M/T	**EVAP Canister Purge Control Valve Circuit Open** Engine started, IAT signal less than 122°F, ambient temperature over 14°F, engine running at a stable idle speed, fuel level from 25-75 percent, all accessory loads off, and the PCM detected an unexpected voltage condition on the EVAP Canister Purge solenoid circuit. **Possible Causes:** • Purge solenoid control circuit is open or shorted to ground • Purge solenoid power circuit is open (check IG fuse in the J/B) • Purge solenoid is damaged or has failed • PCM has failed
DTC: P0445 **2T CCM; MIL: Yes** **Years:** 2007, 2008 **Models:** Aerio, Vitara, XL-7 **Engines:** 2.3 VIN 6, 2.7L VIN 9, 3.6L VIN 7 **Transmissions:** A/T, M/T	**EVAP Canister Purge Control Valve Circuit Shorted** Engine started, IAT signal less than 122°F, ambient temperature over 14°F, engine running at a stable idle speed, fuel level from 25-75 percent, all accessory loads off, and the PCM detected an unexpected voltage condition on the EVAP Canister Purge solenoid circuit. **Possible Causes:** • Purge solenoid control circuit is open or shorted to ground • Purge solenoid power circuit is open (check IG fuse in the J/B) • Purge solenoid is damaged or has failed • PCM has failed
DTC: P0446 **2T CCM; MIL: Yes** **Years:** 2007, 2008 **Models:** Forenza, Reno, Vitara, XL-7, **Engines:** 2.0L VIN Z, 2.7L VIN 9, 3.6L VIN 7 **Transmissions:** All	**EVAP Canister Control System Vent Control Circuit Malfunction** Engine started, IAT signal less than 122°F, ambient temperature over 14°F, engine running at a stable idle speed, fuel level from 25-75 percent, all accessory loads off, and the PCM detected an unexpected voltage condition on the EVAP Canister Purge solenoid circuit. **Possible Causes:** • Purge solenoid control circuit is open or shorted to ground • Purge solenoid power circuit is open (check IG fuse in the J/B) • Purge solenoid is damaged or has failed • PCM has failed

DTC	Trouble Code Title, Conditions & Possible Causes
DTC: P0447 **2T CCM; MIL: Yes** **Years:** 2007, 2008 **Models:** Vitara, XL-7 **Engines:** 2.7L VIN 9, 3.6L VIN 7 **Transmissions:** A/T, M/T	**EVAP Canister Control System Vent Control Valve Circuit Open** Engine started, IAT signal less than 122°F, ambient temperature over 14°F, engine running at a stable idle speed, fuel level from 25-75 percent, all accessory loads off, and the PCM detected an unexpected voltage condition on the EVAP Canister Purge solenoid circuit. **Possible Causes:** • Purge solenoid control circuit is open or shorted to ground • Purge solenoid power circuit is open (check IG fuse in the J/B) • Purge solenoid is damaged or has failed • PCM has failed
DTC: P0448 **2T CCM; MIL: Yes** **Years:** 2007, 2008 **Models:** Vitara, XL-7 **Engines:** 2.7L VIN 9, 3.6L VIN 7 **Transmissions:** A/T, M/T	**EVAP Canister Control System Vent Control Valve Circuit Shorted** Engine started, IAT signal less than 122°F, ambient temperature over 14°F, engine running at a stable idle speed, fuel level from 25-75 percent, all accessory loads off, and the PCM detected an unexpected voltage condition on the EVAP Canister Purge solenoid circuit. **Possible Causes:** • Purge solenoid control circuit is open or shorted to ground • Purge solenoid power circuit is open (check IG fuse in the J/B) • Purge solenoid is damaged or has failed • PCM has failed
DTC: P0449 **2T CCM; MIL: Yes** **Years:** 2007, 2008 **Models:** Aerio, Vitara, XL-7 **Engines:** 2.3 VIN 6, 2.7L VIN 9, 3.6L VIN 7 **Transmissions:** A/T, M/T	**EVAP Canister Control System Vent Control Valve/Solenoid Circuit Malfunction** Engine started, IAT signal less than 122°F, ambient temperature over 14°F, engine running at a stable idle speed, fuel level from 25-75 percent, all accessory loads off, and the PCM detected an unexpected voltage condition on the EVAP Canister Purge solenoid circuit. **Possible Causes:** • Purge solenoid control circuit is open or shorted to ground • Purge solenoid power circuit is open (check IG fuse in the J/B) • Purge solenoid is damaged or has failed • PCM has failed
DTC: P0450 **2T CCM; MIL: Yes** **Years:** 2007, 2008 **Models:** Vitara, XL-7 **Engines:** 2.7L VIN 9, 3.6L VIN 7 **Transmissions:** A/T, M/T	**EVAP Pressure Sensor Circuit Malfunction** Engine started, then driven in stop and go conditions for 5 minutes, fuel level from 25-75 percent, ambient temperature over 14°F, IAT signal less than 122°F, and the PCM detected an unexpected voltage condition on the Pressure sensor circuit during the CCM test. **Possible Causes:** • Pressure sensor signal circuit is open or shorted to ground • Pressure sensor signal is shorted to VREF or system power • Pressure sensor is damaged or has failed • PCM has failed
DTC: P0451 **2T CCM; MIL: Yes** **Years:** 2007, 2008 **Models:** Forenza, Reno, Vitara, XL-7, **Engines:** 2.0L VIN Z, 2.7L VIN 9, 3.6L VIN 7 **Transmissions:** All	**EVAP Pressure Sensor Performance** Engine started, vehicle driven to a speed over 45 MPH for 6 minutes at an engine speed from 2500-3000 RPM, followed by a deceleration period of 3 seconds, fuel level from 15-85 percent, BARO sensor more than 75 kPa, IAT sensor more than 158°F, ambient temperature over 14°F, ECT sensor from 158-230°F, and the PCM detected too small of a Pressure sensor change under these test conditions. **Possible Causes:** • Pressure sensor signal circuit is open or shorted to ground • Pressure sensor signal is shorted to VREF or system power • Pressure sensor is damaged or has failed • PCM has failed
DTC: P0452 **2T CCM; MIL: Yes** **Years:** 2007, 2008 **Models:** Forenza, Reno, Vitara, XL-7, **Engines:** 2.0L VIN Z, 2.7L VIN 9, 3.6L VIN 7 **Transmissions:** All	**EVAP Pressure Sensor Low Input** Engine started, vehicle driven to a speed over 45 MPH for 6 minutes at an engine speed from 2500-3000 RPM, followed by a deceleration period of 3 seconds, fuel level from 15-85 percent, BARO sensor more than 75 kPa, IAT sensor more than 158°F, ambient temperature over 14°F, ECT sensor from 158-230°F, and the PCM detected an unexpected voltage change under these test conditions. **Possible Causes:** • Pressure sensor signal circuit is open or shorted to ground • Pressure sensor signal is shorted to VREF or system power • Pressure sensor is damaged or has failed • PCM has failed

DTC	Trouble Code Title, Conditions & Possible Causes
DTC: P0453 **2T CCM; MIL: Yes** **Years:** 2007, 2008 **Models:** Forenza, Reno, Vitara, XL-7, **Engines:** 2.0L VIN Z, 2.7L VIN 9, 3.6L VIN 7 **Transmissions:** All	**EVAP Pressure Sensor High Input** Engine started, vehicle driven to a speed over 45 MPH for 6 minutes at an engine speed from 2500-3000 RPM, followed by a deceleration period of 3 seconds, fuel level from 15-85 percent, BARO sensor more than 75 kPa, IAT sensor more than 158°F, ambient temperature over 14°F, ECT sensor from 158-230°F, and the PCM detected an unexpected voltage change under these test conditions. **Possible Causes:** • Pressure sensor signal circuit is open or shorted to ground • Pressure sensor signal is shorted to VREF or system power • Pressure sensor is damaged or has failed • PCM has failed
DTC: P0454 **2T CCM; MIL: Yes** **Years:** 2007, 2008 **Models:** Reno, Forenza **Engines:** 2.0L VIN Z **Transmissions:** All	**Fuel Tank Pressure (FTP) Sensor Circuit Intermittent** Engine is in idle condition, Intake Air Temperature (IAT) is greater than 0°C (32°F) and no relevant DTCs are set. **Possible Causes:** • Sensor signal circuit is open or shorted to ground • Sensor is shorted to power • Sensor is damaged or has failed • PCM
DTC: P0455 **2T CCM; MIL: Yes** **Years:** 2007, 2008 **Models:** Forenza, Reno, Vitara, XL-7, **Engines:** 2.0L VIN Z, 2.7L VIN 9, 3.6L VIN 7 **Transmissions:** All	**EVAP System Large Leak (0.080') Detected** Cold engine startup, vehicle driven at 35-55 MPH for 2 minutes, then returned to idle speed, ambient temperature over 14°F, ECT signal from 18-230°F, and the PCM detected a large change in the FTP sensor indicating a large leak (0.080') present during the leak test. **Possible Causes:** • Canister vent (CV) solenoid is stuck open • EVAP canister tube, EVAP canister purge outlet tube or EVAP return tube disconnected or cracked, or canister is damaged • EVAP canister purge valve stuck closed, or canister damaged • Fuel filler cap missing, loose (not tightened) or the wrong part • Fuel vapor hoses/tubes blocked or restricted, or fuel vapor control valve tube or fuel vapor vent valve assembly blocked • Fuel tank pressure (FTP) sensor has failed (mechanical fault) • Fuel tank control valve is contaminated, damaged or has failed
DTC: P0456 **2T CCM; MIL: Yes** **Years:** 2007, 2008 **Models:** Aerio, Forenza, Reno, Vitara, XL-7, **Engines:** 2.3 VIN 6, 2.0L VIN Z, 2.7L VIN 9, 3.6L VIN 7 **Transmissions:** All	**EVAP System Small Leak (0.080') Detected** Cold engine startup, vehicle driven at 35-55 MPH for 2 minutes, then returned to idle speed, ambient temperature over 14°F, ECT signal from 18-230°F, and the PCM detected a large change in the FTP sensor indicating a small leak (0.040') present during the leak test. **Possible Causes:** • Canister vent (CV) solenoid is stuck open • EVAP canister tube, EVAP canister purge outlet tube or EVAP return tube disconnected or cracked, or canister is damaged • EVAP canister purge valve stuck closed, or canister damaged • Fuel filler cap missing, loose (not tightened) or the wrong part • Fuel vapor hoses/tubes blocked or restricted, or fuel vapor control valve tube or fuel vapor vent valve assembly blocked • Fuel tank pressure (FTP) sensor has failed (mechanical fault) • Fuel tank control valve is contaminated, damaged or has failed
DTC: P0458 **2T CCM; MIL: Yes** **Years:** 2007, 2008 **Models:** Reno, Forenza **Engines:** 2.0L VIN Z **Transmissions:** All	**Evaporative Emission (EVAP) Purge Solenoid Control Circuit Low Voltage** Engine is running, Ignition voltage is between 11 and 16 V, enable time delay is greater than 0.5 seconds. **Possible Causes:** • Ignition 1 voltage circuit of the EVAP canister purge valve • control circuit of the EVAP canister purge valve. • ECM

DTC	Trouble Code Title, Conditions & Possible Causes
DTC: P0459 **2T CCM; MIL: Yes** **Years:** 2007, 2008 **Models:** Reno, Forenza **Engines:** 2.0L VIN Z **Transmissions:** All	**EVAP System SmAll Leak (0.080') Detected** Cold engine startup, vehicle driven at 35-55 MPH for 2 minutes, then returned to idle speed, ambient temperature over 14°F, ECT signal from 18-230°F, and the PCM detected a large change in the FTP sensor indicating a small leak (0.040') present during the leak test. **Possible Causes:** • Canister vent (CV) solenoid is stuck open • EVAP canister tube, EVAP canister purge outlet tube or EVAP return tube disconnected or cracked, or canister is damaged • EVAP canister purge valve stuck closed, or canister damaged • Fuel filler cap missing, loose (not tightened) or the wrong part • Fuel vapor hoses/tubes blocked or restricted, or fuel vapor control valve tube or fuel vapor vent valve assembly blocked • Fuel tank pressure (FTP) sensor has failed (mechanical fault) • Fuel tank control valve is contaminated, damaged or has failed
DTC: P0461 **2T CCM; MIL: Yes** **Years:** 2007, 2008 **Models:** Aerio, Forenza, Reno, Vitara, XL-7, **Engines:** 2.3 VIN 6, 2.0L VIN Z, 2.7L VIN 9, 3.6L VIN 7 **Transmissions:** All	**Fuel Level Sensor Performance** DTC P0461 not set, engine started, then after the vehicle was driven a predetermined amount of miles, the PCM detected the fuel level signal value changed less than 0.14v, or the PCM detected a signal from the Fuel Level sensor that was too "low" or too "high" during the CCM Rationality test. **Possible Causes:** • Fuel tank empty or overfull (fuel sender is stuck mechanically) • Wrong fuel gauge is installed, or instrument panel is damaged • Fuel gauge sender unit is damaged or has failed • PCM has failed
DTC: P0462 **2T CCM; MIL: Yes** **Years:** 2007, 2008 **Models:** Aerio, Reno, Forenza **Engines:** 2.3 VIN 6, 2.0L VIN Z **Transmissions:** All	**Fuel Level Sensor Circuit Low Input** Key on or engine running, and the PCM detected the Fuel Level Sensor signal indicated an unexpected voltage during the CCM test. **Possible Causes:** • Fuel level signal circuit is open between the splice and the I/P • Wrong fuel gauge is installed, or instrument panel is damaged • Fuel gauge sender unit is damaged or has failed • PCM has failed
DTC: P0463 **2T CCM; MIL: Yes** **Years:** 2007, 2008 **Models:** Aerio, Forenza, Reno, Vitara, XL-7, **Engines:** 2.3 VIN 6, 2.0L VIN Z, 2.7L VIN 9, 3.6L VIN 7 **Transmissions:** All	**Fuel Level Sensor Circuit High Input** Key on or engine running, and the PCM detected the Fuel Level Sensor signal indicated more than 4.60v during the CCM test. **Possible Causes:** • Fuel level signal circuit is open between the splice and the I/P • Wrong fuel gauge is installed, or instrument panel is damaged • Fuel gauge sender unit is damaged or has failed • PCM has failed
DTC: P0464 **2T CCM; MIL: Yes** **Years:** 2007, 2008 **Models:** Aerio, Forenza, Reno, Vitara, XL-7, **Engines:** 2.3 VIN 6, 2.0L VIN Z, 2.7L VIN 9, 3.6L VIN 7 **Transmissions:** All	**Fuel Level Sensor Circuit Intermittent** Key on or engine running, and the PCM detected the Fuel Level Sensor signal indicated an intermittent voltage during the CCM test. **Possible Causes:** • Fuel level signal circuit is open between the splice and the I/P • Wrong fuel gauge is installed, or instrument panel is damaged • Fuel gauge sender unit is damaged or has failed • PCM has failed
DTC: P0480 **2 CCM; MIL: Yes** **Years:** 2007, 2008 **Models:** Aerio, Reno, Forenza **Engines:** 2.3 VIN 6, 2.0L VIN Z **Transmissions:** All	**Radiator Fan Control System Performance** Engine started, engine running with the ECT sensor indicated a temperature of less than 200°F, and the PCM detected the Radiator Fan Control circuit indicated a "low" voltage condition during the test. **Possible Causes:** • Radiator fan relay control circuit is open • Radiator fan relay control circuit is shorted to ground • Radiator fan relay power circuit is open (check the relay fuse) • Radiator fan relay is damaged or has failed • PCM has failed

DTC	Trouble Code Title, Conditions & Possible Causes
DTC: P0481 **2T CCM; MIL: Yes** **Years:** 2007, 2008 **Models:** Aerio, Reno, Forenza **Engines:** 2.3 VIN 6, 2.0L VIN Z **Transmissions:** All	**A/C Cooling Relay Control Circuit Malfunction** Engine started, engine running, A/C switch in "off" position, ECT sensor signal less than 230°F, and the PCM detected the Radiator Fan Control circuit indicated a "low" voltage condition during the test. **Possible Causes:** • Condenser relay control circuit is open • Condenser fan relay control circuit is shorted to ground • Condenser fan relay power circuit is open (check relay fuse) • Radiator fan relay is damaged or has failed • PCM has failed
DTC: P0496 **2T CCM; MIL: Yes** **Years:** 2007, 2008 **Models:** Reno, Forenza **Engines:** 2.0L VIN Z **Transmissions:** All	**Evaporative Emission (EVAP) System Flow During Non-Purge** DTC(s) P0106, P0107, P0108, P0112, P0113, P0117, P0118, P0122, P0123, P0131, P0132, P0133, P0134, P0135, P0137, P0138, P0140, P0141, P0201, P0202, P0203, P0204, P0300, P0402, P0404, P0405, P0406, P0443, P0452, P0453, P0488, P0502, P0462, P0463, P0506, P0507, P2195 and P2196 will not set. (Common EVAP Enable Criteria) System voltage is between 10 and 16 V, Barometric Pressure (BARO) is greater than 72 kPa (10.4 psi), engine soak time is greater than 720 minutes. Or, at startup IAT – ECT is less than 12°C (53.6°F) at start up, Engine Coolant Temperature (ECT) and Intake Air Temperature (IAT) are between 0°C (32°F) and 40°C (104°F). Startup IAT – IAT is less than 3°C (5.4°F) purge enable time is less than pre-determined value based on startup ECT, fuel level is between 6 percent and 93 percent and engine run time is between 1 second and 300 seconds plus purge enable time. (Continuous Purge Flow Enable Criteria) Engine run time is between 1 second and 100 seconds. **Possible Causes:** • Backed-out terminals • Improper mating • Broken locks • Improperly formed • Damaged terminals • Poor terminal-to-wire connection
DTC: P0498 **2T CCM; MIL: Yes** **Years:** 2007, 2008 **Models:** Reno, Forenza **Engines:** 2.0L VIN Z **Transmissions:** All	**Evaporative Emission (EVAP) Vent Solenoid Control Circuit Low Voltage** Engine is running, battery voltage is between 11 and 16 V. An open in the EVAP vent solenoid circuit exists. **Possible Causes:** • Move the related harnesses and connectors, with the engine operating, while monitoring the circuit status for the component with a scan tool
DTC: P0499 **2T CCM; MIL: Yes** **Years:** 2007, 2008 **Models:** Reno, Forenza **Engines:** 2.0L VIN Z **Transmissions:** All	**Evaporative Emission (EVAP) Vent Solenoid Control Circuit High Voltage** Engine is running, battery voltage is between 11 and 16 V. An open in the EVAP vent solenoid circuit exists. **Possible Causes:** • Move the related harnesses and connectors, with the engine operating, while monitoring the circuit status for the component with a scan tool
DTC: P0499 **2T CCM; MIL: Yes** **Years:** 2007, 2008 **Models:** Reno, Forenza **Engines:** 2.0L VIN Z **Transmissions:** All	**Evaporative Emission (EVAP) Vent Solenoid Control Circuit High Voltage** Engine is running, battery voltage is between 11 and 16 V. An open in the EVAP vent solenoid circuit exists. **Possible Causes:** • Move the related harnesses and connectors, with the engine operating, while monitoring the circuit status for the component with a scan tool
DTC: P0500 **1T CCM; MIL: Yes** **Years:** 2007, 2008 **Models:** Aerio, Vitara, XL-7 **Engines:** 2.3 VIN 6, 2.7L VIN 9, 3.6L VIN 7 **Transmissions:** A/T, M/T	**Vehicle Speed Sensor Circuit Malfunction** Engine started, vehicle driven in 'D' range for 1 minute, or under "fuel cut" conditions, and the PCM did not detect any VSS signals. **Possible Causes:** • VSS signal circuit is open or shorted to ground • VSS ground circuit is open • VSS signal circuit from Combo Meter open or shorted to ground • VSS is damaged or has failed (drive gear may be damaged) • PCM has failed • TSB TS4-23 (6/96) contains information related to this code

DTC	Trouble Code Title, Conditions & Possible Causes
DTC: P0502 **2T CCM; MIL: Yes** **Years:** 2007, 2008 **Models:** Reno, Forenza **Engines:** 2.0L VIN Z **Transmissions:** All	**Engine Vehicle Speed Sensor (VSS) Circuit Low Voltage** Engine is running, engine Coolant Temperature (ECT) is greater than 60°C (140°F). Ignition voltage is between 11 and 16 V. Power Test- the RPM is between 2,500 and 4,000 RPM, throttle Position (TP) sensor is between 25and 60 percent, manifold Absolute Pressure (MAP) is greater than 60 kPa (8.7 psi). Deceleration Test-manifold Absolute Pressure (MAP) is less than 30 kPa (4.4 psi),change in RPM/cycle is greater than 50 RPM/cycle, throttle Position (TP) sensor is less than 0.8 percent, the RPM is between 1,800 and 6,000 RPM. No relevant DTCs are set. **Possible Causes:** • Backed-out terminals • Improper mating • Broken locks • Improperly formed • Damaged terminals • Poor terminal-to-wire connection • VSS improper torque to the transmission housing
DTC: P0504 **2T CCM; MIL: Yes** **Years:** 2007, 2008 **Models:** Reno, Forenza **Engines:** 2.0L VIN Z **Transmissions:** All	**Brake Switch Circuit 1 – 2 Correlation** Engine is running, if time from state change of one brake input to when the other brake input changes state (making switch states equal again) is greater than 0.5 seconds, fail counts increase by 3. When the fail counts reach 39, this DTC sets. **Possible Causes:** • Inspect for proper adjustment of the stop lamp switch. • Check for internments and poor connections.
DTC: P0505 **2T CCM; MIL: Yes** **Years:** 2007, 2008 **Models:** Aerio, Vitara, XL-7 **Engines:** 2.3 VIN 6, 2.7L VIN 9, 3.6L VIN 7 **Transmissions:** A/T, M/T	**Idle Speed Control System** Engine started, ECT sensor signal more than 140°F, engine running at idle speed with the throttle closed, and the PCM detected the Actual idle speed was more than or less than the Target idle speed by too great an amount during the test. **Possible Causes:** • IAC valve control circuit is open or shorted to ground • IAC valve power circuit is open (check power from the relay) • IAC valve ground circuit is open • IAC valve is damaged or has failed • PCM has failed
DTC: P0506 **2T CCM; MIL: Yes** **Years:** 2007, 2008 **Models:** Aerio, Forenza, Reno, Vitara, XL-7, **Engines:** 2.3 VIN 6, 2.0L VIN Z, 2.7L VIN 9, 3.6L VIN 7 **Transmissions:** All	**Idle Speed Control System Lower Than Expected** Engine started, engine running at idle speed, throttle switch is "on", IAT signal less than 158°F, ambient temperature over 14°F, ECT sensor from 158-230°F, and the PCM detected the Actual idle speed was over 100 RPM less than the Desired idle speed in the test. **Possible Causes:** • IAC valve control circuit is open or shorted to ground • IAC valve power circuit is open (intermittent fault) • IAC valve ground circuit is open (intermittent fault) • Throttle plate is carbon fouled (it may need to be cleaned) • PCM has failed
DTC: P0507 **2T CCM; MIL: Yes** **Years:** 2007, 2008 **Models:** Aerio, Forenza, Reno, Vitara, XL-7, **Engines:** 2.3 VIN 6, 2.0L VIN Z, 2.7L VIN 9, 3.6L VIN 7 **Transmissions:** All	**Idle Speed Control System Higher Than Expected** Engine started, engine running at idle speed, throttle switch is "on", IAT signal less than 122°F, ambient temperature over 14°F, ECT sensor from 158-230°F, and the PCM detected the Actual idle speed was over 100 RPM more than the Desired idle speed in the test. **Possible Causes:** • IAC valve control circuit is open (intermittent fault) • IAC valve power circuit is open (intermittent fault) • IAC valve ground circuit is open (intermittent fault) • Throttle plate is carbon fouled (it may need to be cleaned) • PCM has failed
DTC: P0508 **2T CCM; MIL: Yes** **Years:** 2007, 2008 **Models:** Reno, Forenza **Engines:** 2.0L VIN Z **Transmissions:** All	**A/C Pressure Sensor High Voltage** Engine started, vehicle driven, and the PCM detected an unexpected voltage on the A/C pressure sensor circuit **Possible Causes:** • Sensor signal circuit is open or shorted to ground • Sensor ground circuit is open • Sensor is damaged or has failed (drive gear may be damaged) • PCM has failed

DTC	Trouble Code Title, Conditions & Possible Causes
DTC: P0532 **2T CCM; MIL: Yes** **Years:** 2007, 2008 **Models:** Reno, Forenza **Engines:** 2.0L VIN Z **Transmissions:** All	**System Voltage Too Low (Engine Side)** Engine started, vehicle driven, and the PCM detected an unexpected voltage on the system. **Possible Causes:** • There is a signal circuit open or shorted to ground • There is a ground circuit is open • Check the battery voltage • PCM has failed
DTC: P0533 **2T CCM; MIL: Yes** **Years:** 2007, 2008 **Models:** Reno, Forenza **Engines:** 2.0L VIN Z **Transmissions:** All	**System Voltage Too High (Engine Side)** Engine started, vehicle driven, and the PCM detected an unexpected voltage on the system. **Possible Causes:** • There is a signal circuit open or shorted to ground • There is a ground circuit is open • Check the battery voltage • PCM has failed
DTC: P0562 **2T CCM; MIL: Yes** **Years:** 2007, 2008 **Models:** Reno, Forenza **Engines:** 2.0L VIN Z **Transmissions:** All	**PCM Memory Check Sum Error** Key on, and the PCM detected a Memory Check Sum Error in the initial check during the CCM test. **Note: The engine may not start when this trouble code is set.** **Possible Causes:** • Clear the trouble code, and recheck for the same code. • If this trouble code resets, the PCM must be replaced
DTC: P0563 **2T CCM; MIL: Yes** **Years:** 2007, 2008 **Models:** Reno, Forenza **Engines:** 2.0L VIN Z **Transmissions:** All	**System Voltage High** System voltage is greater than 16 V, ignition switch is turned to ON. **Note: The engine may not start when this trouble code is set.** **Possible Causes:** • Clear the trouble code, and recheck for the same code. • If this trouble code resets, the PCM must be replaced • Backed-out terminals • Improper mating • Broken locks • Improperly formed • Damaged terminals • Poor terminal-to-wire connection
DTC: P0565 **2T; MIL: Yes** **Years:** 2007, 2008 **Models**: Reno, Forenza **Engines:** 2.0L VIN Z **Transmissions:** All	**Cruise Control Switch Circuit** Engine is running, ignition voltage is greater than 7.9956 V. The ECM detects the switch is switching too quickly for too long. **Possible Causes:** • Cruise control switch • ON/OFF switch signal circuit • ECM • Improperly formed • Damaged terminals • Poor terminal-to-wire connection
DTC: P0567 **2T; MIL: Yes** **Years:** 2007, 2008 **Models**: Reno, Forenza **Engines:** 2.0L VIN Z **Transmissions:** All	**Cruise Control Resume Switch Circuit** Engine is running, ignition voltage is greater than 7.9956 V. The ECM detects the switch is switching too quickly for too long. **Possible Causes:** • Cruise control resume/accel. switch • ON/OFF switch signal circuit • ECM • Internally open or shorted contact coil • Damaged terminals • Poor terminal-to-wire connection
DTC: P0568 **2T; MIL: Yes** **Years:** 2007, 2008 **Models**: Reno, Forenza **Engines:** 2.0L VIN Z **Transmissions:** All	**Cruise Control Switch Circuit** Engine is running, ignition voltage is greater than 7.9956 V. The ECM detects the switch is switching too quickly for too long. **Possible Causes:** • Cruise control switch • ON/OFF switch signal circuit • ECM • Improperly formed • Damaged terminals • Poor terminal-to-wire connection

DTC	Trouble Code Title, Conditions & Possible Causes
DTC: P0571 **2T; MIL: Yes** **Years:** 2007, 2008 **Models**: Reno, Forenza **Engines:** 2.0L VIN Z **Transmissions:** All	**Brake Switch Circuit** Engine is running **Possible Causes:** • Operation of the stop lamps • Backed-out terminals • Improper mating • Broken locks • Improperly formed • Damaged terminals • Poor terminal-to-wire connection
DTC: P0602 **1T; MIL: Yes** **Years:** 2007, 2008 **Models**: Aerio, Reno, Forenza **Engines:** 2.3 VIN 6, 2.0L VIN Z **Transmissions:** All	**Control Module Not Programmed** Calibration ID is not equal to the value in the software level. **Possible Causes:** • Engine Control Module (ECM).
DTC: P0604 **1T; MIL: Yes** **Years:** 2007, 2008 **Models**: Reno, Forenza **Engines:** 2.0L VIN Z **Transmissions:** All	**ECM RAM Memory Error** Key on, and the ECM detected a RAM Memory Error. **Possible Causes:** • The contents of the ECM have changed • Clear the trouble code, and recheck for the same code. • If this trouble code resets, the ECM must be replaced
DTC: P0606 **1T; MIL: Yes** **Years:** 2007, 2008 **Models**: Reno, Forenza **Engines:** 2.0L VIN Z **Transmissions:** All	**Control Module Internal Performance** Ignition switch is turned to ON, battery voltage is between 7 and 20 V. **Possible Causes:** • ECM • Clear the trouble code, and recheck for the same code. • If this trouble code resets, the ECM must be replaced
DTC: P0616 **1T; MIL: Yes** **Years:** 2007, 2008 **Models:** Aerio, Vitara, XL-7 **Engines:** 2.3 VIN 6, 2.7L VIN 9, 3.6L VIN 7 **Transmissions:** A/T, M/T	**Starter Relay Circuit Low** Key on, and the PCM detected an unexpected voltage on the starter relay circuit. **Possible Causes:** • There is a short or an open circuit • The starter relay and/or motor is faulty • The PCM is faulty
DTC: P0617 **1T; MIL: Yes** **Years:** 2007, 2008 **Models:** Aerio, Vitara, XL-7 **Engines:** 2.3 VIN 6, 2.7L VIN 9, 3.6L VIN 7 **Transmissions:** A/T, M/T	**Starter Relay Circuit Low** Key on, and the PCM detected an unexpected voltage on the starter relay circuit. **Possible Causes:** • There is a short or an open circuit • The starter relay and/or motor is faulty • The PCM is faulty
DTC: P0630 **1T; MIL: Yes** **Years:** 2007, 2008 **Models:** Aerio, Vitara, XL-7 **Engines:** 2.3 VIN 6, 2.7L VIN 9, 3.6L VIN 7 **Transmissions:** A/T, M/T	**VIN Not Programmed or Incompatible-ECM/PCM** Key on, and the PCM detected incompatibility error in regards to the VIN. **Possible Causes:** • The contents of the ECM have changed • Check the validity of the VIN • PCM is faulty
DTC: P0641 **1T; MIL: Yes** **Years:** 2007, 2008 **Models**: Reno, Forenza **Engines:** 2.0L VIN Z **Transmissions:** All	**5 Volt Reference 1 Circuit** Ignition switch is turned to ON, The 5 volt reference 1 voltage is greater than 5.5 V or less than 4.5 V for a certain period of time. **Possible Causes:** • ECM • Clear the trouble code, and recheck for the same code. • If this trouble code resets, the ECM must be replaced

DTC	Trouble Code Title, Conditions & Possible Causes
DTC: P0646 **1T; MIL: Yes** **Years:** 2007, 2008 **Models**: Reno, Forenza **Engines:** 2.0L VIN Z **Transmissions:** All	**Control Module Internal Performance** Engine is running, A/C clutch is engaged, ignition voltage is between 11 and 16 V. An open or a short to ground condition in the A/C compressor relay circuit exists. **Possible Causes:** • Operation of the stop lamps • Backed-out terminals • Improper mating • Broken locks • Improperly formed • Damaged terminals • Poor terminal-to-wire connection
DTC: P0647 **1T; MIL: Yes** **Years:** 2007, 2008 **Models**: Reno, Forenza **Engines:** 2.0L VIN Z **Transmissions:** All	**Air Conditioning (A/C) Clutch Relay Control Circuit High Voltage** Ignition switch is turned to ON, engine running, battery voltage is between 11 and 16 V. An open or a short to ground condition in the A/C compressor relay circuit exists. **Possible Causes:** • Operation of the stop lamps • Backed-out terminals • Improper mating • Broken locks • Improperly formed • Damaged terminals • Poor terminal-to-wire connection
DTC: P0650 **1T; MIL: Yes** **Years:** 2007, 2008 **Models**: Reno, Forenza **Engines:** 2.0L VIN Z **Transmissions:** All	**Malfunction Indicator Lamp (MIL) Control Circuit** Engine is running, ignition voltage is between 11 and 16 V. The ECM detects an open, short to ground or short to battery in the MIL control circuit. **Possible Causes:** • Operation of the stop lamps • Backed-out terminals • Improper mating • Broken locks • Improperly formed • Damaged terminals • Poor terminal-to-wire connection
DTC: P0651 **1T; MIL: Yes** **Years:** 2007, 2008 **Models**: Reno, Forenza **Engines:** 2.0L VIN Z **Transmissions:** All	**5 Volt Reference 2 Circuit** Ignition switch is turned to ON, The 5 volt reference 2 voltage is greater than 5.5 V or less than 4.5 V for a certain period of time. **Possible Causes:** • ECM • Operation of the stop lamps • Backed-out terminals • Improper mating • Broken locks • Improperly formed • Damaged terminals • Poor terminal-to-wire connection
DTC: P0660 **2T; MIL: Yes** **Years:** 2007, 2008 **Models:** Vitara, XL-7 **Engines:** 2.7L VIN 9, 3.6L VIN 7 **Transmissions:** A/T, M/T	**Intake Manifold Tuning Valve Control Circuit Open** The PCM detected an unexpected voltage across the intake manifold tuning valve control circuit **Possible Causes:** • Faulty valve • There is a short or an open circuit • PCM is faulty
DTC: P0685 **1T; MIL: Yes** **Years:** 2007, 2008 **Models**: Reno, Forenza **Engines:** 2.0L VIN Z **Transmissions:** All	**Engine Controls Ignition Relay Control Circuit** Ignition switch is turned to ON, ignition voltage is between 11 and 16 V. The ECM detects an open, a short to ground, or a short to battery condition in the main relay control circuit. **Possible Causes:** • ECM • The DTC P0685 diagnostic table assumes that the vehicle battery is fully charged.

DTC	Trouble Code Title, Conditions & Possible Causes
DTC: P0705 **2T CCM; MIL: Yes** **Years:** 2007, 2008 **Models:** Vitara, XL-7 **Engines:** 2.7L VIN 9, 3.6L VIN 7 **Transmissions:** A/T, M/T	**Transmission Range Sensor Circuit Malfunction** Engine started, then driven to a speed of over 8 MPH, and the PCM did not detect 'R', 'N', 'D' or 2nd gear TR switch inputs for 5 seconds, or it detected multiple TR switch signals during the CCM test. **Note: This is a 1-trip code when multiple TR switch inputs are detected.** **Possible Causes:** • TR range switch signal is open • TR range switch signal shorted to another switch position signal • TR range switch is damaged or has failed • PCM has failed
DTC: P0707 **2T CCM; MIL: Yes** **Years:** 2007, 2008 **Models:** Vitara, XL-7 **Engines:** 2.7L VIN 9, 3.6L VIN 7 **Transmissions:** A/T, M/T	**Transmission Range Sensor Circuit Low Input** Engine started, then driven to a speed of over 8 MPH, and the PCM did not detect 'R', 'N', 'D' or 2nd gear TR switch inputs for 5 seconds, or it detected multiple TR switch signals during the CCM test. **Note: This is a 1-trip code when multiple TR switch inputs are detected.** **Possible Causes:** • TR range switch signal is open • TR range switch signal shorted to another switch position signal • TR range switch is damaged or has failed • PCM has failed
DTC: P0715 **2T CCM; MIL: Yes** **Years:** 2007, 2008 **Models:** Vitara, XL-7 **Engines:** 2.7L VIN 9, 3.6L VIN 7 **Transmissions:** A/T, M/T	**TCM Input Speed Sensor Circuit Malfunction** Engine started, vehicle driven at of speed of 30 MPH in 'D', 2nd or 'L', and then at over 30 MPH in Drive for 30 seconds, and the TCM did not detect any ISS (A/T VSS) signals during the test. **Note: The Output Speed sensor resistance is 387-475 ohms at 68°F.** **Possible Causes:** • ISS signal (+) circuit is open or shorted to ground • ISS signal (−) circuit is open or shorted to ground • ISS is damaged or has failed • TCM has failed
DTC: P0717 **2T CCM; MIL: Yes** **Years:** 2007, 2008 **Models:** Vitara, XL-7 **Engines:** 2.7L VIN 9, 3.6L VIN 7 **Transmissions:** A/T, M/T	**TCM Input/Turbine Speed Sensor Circuit No Signal** Engine started, vehicle driven at of speed of 30 MPH in 'D', 2nd or 'L', and then at over 30 MPH in Drive for 30 seconds, and the TCM did not detect any ISS (A/T VSS) signals during the test. **Note: The Output Speed sensor resistance is 387-475 ohms at 68°F.** **Possible Causes:** • ISS signal (+) circuit is open or shorted to ground • ISS signal (−) circuit is open or shorted to ground • ISS is damaged or has failed • TCM has failed
DTC: P0720 **2T CCM; MIL: Yes** **Years:** 2007, 2008 **Models:** Vitara, XL-7 **Engines:** 2.7L VIN 9, 3.6L VIN 7 **Transmissions:** A/T, M/T	**TCM Output Speed Sensor Circuit Malfunction** Engine started, vehicle driven at of speed of 30 MPH in 'D', 2nd or 'L', and then at over 30 MPH in Drive for 30 seconds, and the TCM did not detect any OSS signals during the CCM the test. **Note: The Output Speed sensor resistance is 648-792 ohms at 68°F.** **Possible Causes:** • OSS signal (+) circuit is open or shorted to ground • OSS signal (−) circuit is open or shorted to ground • OSS is damaged or has failed • TCM has failed
DTC: P0722 **2T CCM; MIL: Yes** **Years:** 2007, 2008 **Models:** Vitara, XL-7 **Engines:** 2.7L VIN 9, 3.6L VIN 7 **Transmissions:** A/T, M/T	**TCM Output Speed Sensor Circuit No Signal** Engine started, vehicle driven at of speed of 30 MPH in 'D', 2nd or 'L', and then at over 30 MPH in Drive for 30 seconds, and the TCM did not detect any OSS signals during the CCM the test. **Note: The Output Speed sensor resistance is 648-792 ohms at 68°F.** **Possible Causes:** • OSS signal (+) circuit is open or shorted to ground • OSS signal (−) circuit is open or shorted to ground • OSS is damaged or has failed • TCM has failed

DTC	Trouble Code Title, Conditions & Possible Causes
DTC: P0741 **2T CCM; MIL: Yes** **Years:** 2007, 2008 **Models:** Vitara, XL-7 **Engines:** 2.7L VIN 9, 3.6L VIN 7 **Transmissions:** A/T, M/T	**TCC Solenoid Performance (Stuck Off)** Engine started, vehicle driven in 'D' range (4th gear), and the PCM detected the difference in RPM between the engine and A/T input speeds was too large with the TCC commanded "on", or with the engine running in 'D' range (3rd gear), the difference between the engine and A/T input speeds was too small during the test. **Possible Causes:** • TCC solenoid has a mechanical failure • TCC solenoid has a hydraulic failure • PCM has failed
DTC: P0742 **2T CCM; MIL: Yes** **Years:** 2007, 2008 **Models:** Vitara, XL-7 **Engines:** 2.7L VIN 9, 3.6L VIN 7 **Transmissions:** A/T, M/T	**TCC Solenoid Performance (Stuck On)** Engine started, vehicle driven in 'D' range (4th gear), and the PCM detected the difference in RPM between the engine and A/T input speeds was too large with the TCC commanded "on", or with the engine running in 'D' range (3rd gear), the difference between the engine and A/T input speeds was too small during the test. **Possible Causes:** • TCC solenoid has a mechanical failure • TCC solenoid has a hydraulic failure • PCM has failed
DTC: P0743 **1T CCM; MIL: Yes** **Years:** 2007, 2008 **Models:** Vitara, XL-7 **Engines:** 2.7L VIN 9, 3.6L VIN 7 **Transmissions:** A/T, M/T	**Torque Converter Clutch Solenoid Circuit Malfunction** Engine started, engine running in closed loop, O/D Cut Switch "on" (O/D Lamp ON), gearshift in Drive at least 20 seconds, vehicle driven in 4th gear in 'D' range at less than a 40 percent throttle opening at a vehicle speed of over 45 MPH or more for 10 seconds. This step needs to be repeated at least three times for this code to set. **Note: The TCC solenoid resistance is 11-15 ohms at 68°F.** **Possible Causes:** • TCC solenoid control circuit is open or shorted to ground • TCC solenoid control circuit is shorted to system power (B+) • TCC solenoid power circuit is open (check main relay & fuse) • TCC solenoid is damaged or has failed • PCM has failed
DTC: P0751 **2T CCM; MIL: Yes** **Years:** 2007, 2008 **Models:** Vitara, XL-7 **Engines:** 2.7L VIN 9, 3.6L VIN 7 **Transmissions:** A/T, M/T	**A/T Shift Solenoid 'A' Performance (Stuck Off)** Engine started, vehicle driven to a speed of over 30 MPH, and the PCM detected a transmission gear ratio equal to 3rd gear with 2nd gear commanded "on" or a transmission gear ratio equal to 2nd gear with 3rd gear commanded "on" during the CCM Rationality test. **Possible Causes:** • SSA is stuck in "off" position (mechanical problem) • SSA is damaged is damaged or has failed (mechanical fault) • SSA has a hydraulic problem • PCM has failed
DTC: P0752 **2T CCM; MIL: Yes** **Years:** 2007, 2008 **Models:** Vitara, XL-7 **Engines:** 2.7L VIN 9, 3.6L VIN 7 **Transmissions:** A/T, M/T	**A/T Shift Solenoid 'A' Performance (Stuck On)** Engine started, vehicle driven to a speed of over 30 MPH, and the PCM detected a transmission gear ratio equal to 3rd gear with 2nd gear commanded "on" or a transmission gear ratio equal to 2nd gear with 3rd gear commanded "on" during the CCM Rationality test. **Possible Causes:** • SSA is stuck in "off" position (mechanical problem) • SSA is damaged is damaged or has failed (mechanical fault) • SSA has a hydraulic problem • PCM has failed
DTC: P0753 **1T CCM; MIL: Yes** **Years:** 2007, 2008 **Models:** Vitara, XL-7 **Engines:** 2.7L VIN 9, 3.6L VIN 7 **Transmissions:** A/T, M/T	**A/T Shift Solenoid 'A' Circuit Malfunction** Engine started, vehicle driven to over 40 MPH for 10 seconds, and the PCM detected a "high" voltage condition on the SSA circuit with the solenoid commanded "off", or the SSA CCM detected a "low" voltage condition with the SSA commanded "on" in the CCM test. **Note: The Shift Solenoid 'A' resistance is 11-15 ohms at 68°F.** **Possible Causes:** • SSA control circuit is open or shorted to ground • SSA ground circuit is open, or the SSA power circuit is open • SSA is damaged or has failed • PCM has failed

DTC	Trouble Code Title, Conditions & Possible Causes
DTC: P0756 **2T CCM; MIL: Yes** **Years:** 2007, 2008 **Models:** Vitara, XL-7 **Engines:** 2.7L VIN 9, 3.6L VIN 7 **Transmissions:** A/T, M/T	**A/T Shift Solenoid 'B' Performance (Stuck Off)** Engine started, gear selector in 'D' position for 20 seconds, then after the vehicle was driven through a 1st, 2nd, 3rd and 4th gear shift change, and the PCM detected a transmission gear ratio equal to 4th gear with 3rd gear commanded "on" (SSA is "off", SSB is "on"), or a transmission gear ratio equal to 3rd gear with 4th gear commanded "on" (SSA is "off", SSB is "off") during the CCM test. **Possible Causes:** • SSB is stuck in "off" position (mechanical problem) • SSB is damaged is damaged or has failed (mechanical fault) • SSB has a hydraulic problem • PCM has failed
DTC: P0757 **1T CCM; MIL: Yes** **Years:** 2007, 2008 **Models:** Vitara, XL-7 **Engines:** 2.7L VIN 9, 3.6L VIN 7 **Transmissions:** A/T, M/T	**A/T Shift Solenoid 'B' Circuit Malfunction** Engine started, vehicle driven to over 40 MPH fro 10 seconds, and the PCM detected a "high" voltage condition on the SSB circuit with the solenoid commanded "off", or the CCM detected a "low" voltage condition with the SSB commanded "on" during the CCM test. **Note: The Shift Solenoid 'B' resistance is 11-15 ohms at 68°F.** **Possible Causes:** • SSB control circuit is open or shorted to ground • SSB ground circuit is open, or the SSA power circuit is open • SSB is damaged or has failed • PCM has failed
DTC: P0758 **1T CCM; MIL: Yes** **Years:** 2007, 2008 **Models:** Vitara, XL-7 **Engines:** 2.7L VIN 9, 3.6L VIN 7 **Transmissions:** A/T, M/T	**A/T Shift Solenoid 'B' Circuit Malfunction** Engine started, vehicle driven to over 40 MPH fro 10 seconds, and the PCM detected a "high" voltage condition on the SSB circuit with the solenoid commanded "off", or the CCM detected a "low" voltage condition with the SSB commanded "on" during the CCM test. **Note: The Shift Solenoid 'B' resistance is 11-15 ohms at 68°F.** **Possible Causes:** • SSB control circuit is open or shorted to ground • SSB ground circuit is open, or the SSA power circuit is open • SSB is damaged or has failed • PCM has failed
DTC: P0771 **2T CCM; MIL: Yes** **Years:** 2007, 2008 **Models:** Vitara, XL-7 **Engines:** 2.7L VIN 9, 3.6L VIN 7 **Transmissions:** A/T, M/T	**A/T Shift Solenoid 'E' Performance (Stuck Off)** Engine started, gear selector in 'D' position for 20 seconds, then after the vehicle was driven through a 1st, 2nd, 3rd and 4th gear shift change, and the PCM detected a transmission gear ratio equal to 4th gear with 3rd gear commanded "on" (SSA is "off", SSB is "on"), or a transmission gear ratio equal to 3rd gear with 4th gear commanded "on" (SSA is "off", SSB is "off") during the CCM test. **Possible Causes:** • SSB is stuck in "off" position (mechanical problem) • SSB is damaged is damaged or has failed (mechanical fault) • SSB has a hydraulic problem • PCM has failed
DTC: P0772 **1T CCM; MIL: Yes** **Years:** 2007, 2008 **Models:** Vitara, XL-7 **Engines:** 2.7L VIN 9, 3.6L VIN 7 **Transmissions:** A/T, M/T	**A/T Shift Solenoid 'E' Performance (Stuck On)** Engine started, vehicle driven to over 40 MPH fro 10 seconds, and the PCM detected a "high" voltage condition on the SSE circuit with the solenoid commanded "on". **Note: The Shift Solenoid 'E' resistance is 11-15 ohms at 68°F.** **Possible Causes:** • SSB control circuit is open or shorted to ground • SSB ground circuit is open, or the SSA power circuit is open • SSB is damaged or has failed • PCM has failed
DTC: P834 **1T; MIL: Yes** **Years:** 2007, 2008 **Models:** Reno, Forenza **Engines:** 2.0L VIN Z **Transmissions:** M/T	**Clutch Pedal Switch Circuit Low Voltage** Ignition switch is turned to ON, battery voltage is between 11 and 16 V, vehicle speed is between 3 and 52 km/h, the ECM detects that the clutch pedal is depressed and the ECM detects no clutch pedal switch transition. **Possible Causes:** • ECM • Operation of the stop lamps • Backed-out terminals • Improper mating • Broken locks • Improperly formed • Damaged terminals • Poor terminal-to-wire connection

DTC	Trouble Code Title, Conditions & Possible Causes
DTC: P0835 **1T; MIL: Yes** **Years:** 2007, 2008 **Models**: Reno, Forenza **Engines:** 2.0L VIN Z **Transmissions:** M/T	**Control Module Internal Performance** Ignition voltage is between 11 and 16 V, vehicle speed is between 3 and 52 km/h, the ECM detects that the clutch pedal is released, the ECM detects no clutch pedal switch transition. **Possible Causes:** • ECM • Operation of the stop lamps • Backed-out terminals • Improper mating • Broken locks • Improperly formed • Damaged terminals • Poor terminal-to-wire connection
DTC: P0973 **1T CCM; MIL: Yes** **Years:** 2007, 2008 **Models:** Vitara, XL-7 **Engines:** 2.7L VIN 9, 3.6L VIN 7 **Transmissions:** A/T, M/T	**A/T Shift Solenoid 'A' Control Circuit Low** Engine started, vehicle driven to over 40 MPH fro 10 seconds, and the PCM detected a "high" voltage condition on the SSA circuit with the solenoid commanded "off", or the CCM detected a "low" voltage condition with the SSA commanded "on" during the CCM test. **Note: The Shift Solenoid 'B' resistance is 11-15 ohms at 68°F.** **Possible Causes:** • SSA control circuit is open or shorted to ground • SSA ground circuit is open, or the SSA power circuit is open • SSA is damaged or has failed • PCM has failed
DTC: P0974 **1T CCM; MIL: Yes** **Years:** 2007, 2008 **Models:** Vitara, XL-7 **Engines:** 2.7L VIN 9, 3.6L VIN 7 **Transmissions:** A/T, M/T	**A/T Shift Solenoid 'A' Control Circuit High** Engine started, vehicle driven to over 40 MPH fro 10 seconds, and the PCM detected a "high" voltage condition on the SSA circuit with the solenoid commanded "off", or the CCM detected a "high" voltage condition with the SSA commanded "on" during the CCM test. **Note: The Shift Solenoid 'B' resistance is 11-15 ohms at 68°F.** **Possible Causes:** • SSA control circuit is open or shorted to ground • SSA ground circuit is open, or the SSA power circuit is open • SSA is damaged or has failed • PCM has failed
DTC: P0976 **1T CCM; MIL: Yes** **Years:** 2007, 2008 **Models:** Vitara, XL-7 **Engines:** 2.7L VIN 9, 3.6L VIN 7 **Transmissions:** A/T, M/T	**A/T Shift Solenoid 'B' Control Circuit Low** Engine started, vehicle driven to over 40 MPH fro 10 seconds, and the PCM detected a "high" voltage condition on the SSB circuit with the solenoid commanded "off", or the CCM detected a "low" voltage condition with the SSB commanded "on" during the CCM test. **Note: The Shift Solenoid 'B' resistance is 11-15 ohms at 68°F.** **Possible Causes:** • SSB control circuit is open or shorted to ground • SSB ground circuit is open, or the SSA power circuit is open • SSB is damaged or has failed • PCM has failed
DTC: P0977 **1T CCM; MIL: Yes** **Years:** 2007, 2008 **Models:** Vitara, XL-7 **Engines:** 2.7L VIN 9, 3.6L VIN 7 **Transmissions:** A/T, M/T	**A/T Shift Solenoid 'B' Control Circuit High** Engine started, vehicle driven to over 40 MPH fro 10 seconds, and the PCM detected a "high" voltage condition on the SSB circuit with the solenoid commanded "off", or the CCM detected a "high" voltage condition with the SSB commanded "on" during the CCM test. **Note: The Shift Solenoid 'B' resistance is 11-15 ohms at 68°F.** **Possible Causes:** • SSB control circuit is open or shorted to ground • SSB ground circuit is open, or the SSA power circuit is open • SSB is damaged or has failed • PCM has failed

Gas Engine OBD II Trouble Code List (P1xxx Codes)

DTC	Trouble Code Title, Conditions & Possible Causes
DTC: P1011 **2T CCM; MIL: Yes** **Years:** 2007, 2008 **Models:** XL-7 **Engines:** 3.6L VIN 7 **Transmissions:** A/T	**Intake Camshaft Position (CMP) Actuator Park Position Bank 1** The ECM detects that a CMP actuator is not in the parked position on an engine start-up. **Possible Causes:** • Correct oil level • Correct oil pressure • (CMP) actuator • ECM

DTC	Trouble Code Title, Conditions & Possible Causes
DTC: P1012 **2T CCM; MIL: Yes** **Years:** 2007, 2008 **Models:** XL-7 **Engines:** 3.6L VIN 7 **Transmissions:** A/T	**Exhaust Camshaft Position (CMP) Actuator Park Position Bank 1** The ECM detects that a CMP actuator is not in the parked position on an engine start-up. **Possible Causes:** • Correct oil level • Correct oil pressure • (CMP) actuator • ECM
DTC: P1013 **2T CCM; MIL: Yes** **Years:** 2007, 2008 **Models:** XL-7 **Engines:** 3.6L VIN 7 **Transmissions:** A/T	**Intake Camshaft Position (CMP) Actuator Park Position Bank 2** The ECM detects that a CMP actuator is not in the parked position on an engine start-up. **Possible Causes:** • Correct oil level • Correct oil pressure • (CMP) actuator • ECM
DTC: P1014 **2T CCM; MIL: Yes** **Years:** 2007, 2008 **Models:** XL-7 **Engines:** 3.6L VIN 7 **Transmissions:** A/T	**Exhaust Camshaft Position (CMP) Actuator Park Position Bank 2** The ECM detects that a CMP actuator is not in the parked position on an engine start-up. **Possible Causes:** • Correct oil level • Correct oil pressure • (CMP) actuator • ECM
DTC: P1107 **2T CCM; MIL: Yes** **Years:** 2007, 2008 **Models:** Aerio, Grand Vitara **Engines:** 2.3 VIN 6, 2.7L VIN 9 **Transmissions:** A/T, M/T	**MAP Sensor Circuit Performance Problem Low** Intake air temperature: 14° (–10°C) or higher, engine coolant temperature: higher than 113°F (45°C) The following DTC(s) are not detected: MAF sensor, ECT sensor and MAP sensor (other than P1107 and P1108). Input voltage of MAP sensor circuit is lower than the 1 V even though vehicle runs under engine load factor of 43 percent or more. **Possible Causes:** • MAP sensor 5-volt power circuit is open or shorted to ground • MAP Sensor signal circuit is shorted to ground • MAP Sensor is damaged or has failed • ECM has failed
DTC: P1108 **2T CCM; MIL: Yes** **Years:** 2007, 2008 **Models:** Aerio, Grand Vitara **Engines:** 2.3 VIN 6, 2.7L VIN 9 **Transmissions:** A/T, M/T	**MAP Sensor Circuit Performance Problem High** Intake air temperature: 14°F (–10°C) or higher, engine coolant temperature: higher than 113°F (45°C) The following DTC(s) are not detected: MAF sensor, ECT sensor and MAP sensor (other than P1107 and P1108). Input voltage of MAP sensor circuit is more than the specified 2.7 V even though vehicle runs under engine load factor of 27 percent or less. **Possible Causes:** • MAP sensor 5-volt power circuit is open or shorted to ground • MAP Sensor signal circuit is shorted to ground • MAP Sensor is damaged or has failed • ECM has failed
DTC: P1116 **2T CCM; MIL: Yes** **Years:** 2007, 2008 **Models:** Grand Vitara **Engines:** 2.7L VIN 9 **Transmissions:** A/T, M/T	**Engine Coolant Temperature Circuit Performance at Engine Soaking** The difference between ECT and IAT is more than 95°F (35°C) or less than 4°F (–20°C) after 5 hours from engine stop. **Possible Causes:** • ECT sensor and its circuit • Thermostat • IAT sensor and its circuit • ECM
DTC: P1137 **1T; OBD:H02S Yes** **Years:** 2007, 2008 **Models**: Reno, Forenza **Engines:** 2.0L VIN Z **Transmissions:** All	**HO2S Circuit Low Voltage During Power Enrichment Sensor 2** Mass air flow is greater than 6 g/s, ignition voltage is greater than 10 V, the engine is running for greater than 10 seconds, power enrichment conditions are met, coolant temperature is greater than 60°C, HO2S filtered voltage is less than 0.3501 V, air fuel ratio is less than 13.5 : 1. **Possible Causes:** • ECM • Clear the trouble code, and recheck for the same code. • If this trouble code resets, the ECM must be **replaced**

DTC	Trouble Code Title, Conditions & Possible Causes
DTC: P01138 **1T; OBD:H02S Yes** **Years:** 2007, 2008 **Models**: Reno, Forenza **Engines:** 2.0L VIN Z **Transmissions:** All	**HO2S Circuit High Voltage During Decel Fuel Cut-Off (DFCO) Sensor 2** Mass air flow is greater than 6 g /s, ignition voltage is greater than 10 V, the engine is running for greater than 10 seconds, power enrichment conditions are met, coolant temperature is greater than 60°C, HO2S filtered voltage is greater than 0.125 V. **Possible Causes:** • ECM • Clear the trouble code, and recheck for the same code. • If this trouble code resets, the ECM must be replaced
DTC: P01166 **1T; OBD:H02S: Yes** **Years:** 2007, 2008 **Models**: Reno, Forenza **Engines:** 2.0L VIN Z **Transmissions:** All	**HO2S Circuit Low Voltage During Power Enrichment Sensor 1** HO2S1 voltage is less than 0.35 V in Power Enrichment (PE) mode, Engine is operating in Closed Loop and in PE mode, Engine Coolant Temperature (ECT) is greater than 60°C (140°F), system voltage is greater than 10 V, Air/Fuel ration is less than or equal to 13.5 : 1, airflow is greater than 6 g /s, 2 second delay after in PE mode. **Possible Causes:** • Faulty or plugged injector(s). • Water or alcohol fuel • Inadequate amount of fuel in the Tank
DTC: P01391 **1T; MIL: Yes** **Years:** 2007, 2008 **Models**: Reno, Forenza **Engines:** 2.0L VIN Z **Transmissions:** All	**Rough Road Sensor Performance** Engine is running more than or equal to 10 seconds and idle, vehicle speed is less than or equal to 5 km/h (3.1 MPH), G sensor output at idle indicates below 1.1 V or above 3.7 V. OR Engine is running more than 10 seconds and vehicle speed is between 50 km/h (30 MPH) and 112 km/h (70 MPH), G sensor signal changes less than 0.0002 V while driving. **Possible Causes:** • Backed-out terminals • Improper mating • Broken locks • Improperly formed • Damaged terminals • Poor terminal-to-wire connection • Damaged A/C Pressure Sensor harness • Damaged A/C Pressure Sensor • G sensor
DTC: P01392 **1T; MIL: Yes** **Years:** 2007, 2008 **Models**: Reno, Forenza **Engines:** 2.0L VIN Z **Transmissions:** All	**Rough Road Sensor Circuit Low Voltage** G sensor output indicates below 0.1 V, engine is running more than or equal to 10 seconds. **Possible Causes:** • Backed-out terminals • Improper mating • Broken locks • Improperly formed • Damaged terminals • Poor terminal-to-wire connection • G sensor must be level and mounted securely to its bracket • Damaged A/C Pressure Sensor harness • Damaged A/C Pressure Sensor • G sensor
DTC: P01393 **1T; MIL: Yes** **Years:** 2007, 2008 **Models**: Reno, Forenza **Engines:** 2.0L VIN Z **Transmissions:** All	**Rough Road Sensor Circuit High Voltage** G sensor output indicates above 4.9 V, engine is running more than or equal to 10 seconds. **Possible Causes:** • Backed-out terminals • Improper mating • Broken locks • Improperly formed • Damaged terminals • Poor terminal-to-wire connection • G sensor must be level and mounted securely to its bracket • Damaged A/C Pressure Sensor harness • Damaged A/C Pressure Sensor • G sensor

DTC	Trouble Code Title, Conditions & Possible Causes
DTC: P01396 **1T; MIL: Yes** **Years:** 2007, 2008 **Models**: Reno, Forenza **Engines:** 2.0L VIN Z **Transmissions:** All	**Rough Road System Fault** The wheel speed variation test detects at least one missing edge from the wheel speed sensor signal, No relevant DTCs are set, coolant temperature is greater than or equal to 60°C (140°F), vehicle voltage is between 11 and 16 V, engine run time is greater than 10 seconds. **Possible Causes:** • Open signal circuit of open PWM serial data line between the ECM and the EBCM • Improper mating • Broken locks • Improperly formed • Damaged terminals • Poor terminal-to-wire connection
DTC: P01397 **1T; MIL: Yes** **Years:** 2007, 2008 **Models**: Reno, Forenza **Engines:** 2.0L VIN Z **Transmissions:** All	**WSSD Rough Road – Serial Data Fault** The wheel speed variation test detects failed wheel speed sensor under power conditions, No relevant DTCs are set, coolant temperature is greater than or equal to 60°C (140°F), vehicle voltage is between 11 and 16 V, engine run time is greater than 10 seconds. **Possible Causes:** • Open signal circuit of open PWM serial data line between the ECM and the EBCM • Improper mating • Broken locks • Improperly formed • Damaged terminals • Poor terminal-to-wire connection
DTC: P1420 **2T CCM; MIL: Yes** **Years:** 2007, 2008 **Models:** Aerio, Grand Vitara **Engines:** 2.3 VIN 6, 2.7L VIN 9 **Transmissions:** A/T, M/T	**Evaporative Emission System Vent Valve Stuck Open** Even if the EVAP vent valve is changed off (open) to on (close) with vacuum pump on, variation of pressure sensor is less than 0.2 kPa and pressure difference between minimum pressure of EVAP system and reference pressure is higher than the threshold. **Possible Causes:** • EVAP leak check module and its circuit • ECM
DTC: P1421 **2T CCM; MIL: Yes** **Years:** 2007, 2008 **Models:** Grand Vitara **Engines:** 2.7L VIN 9 **Transmissions:** A/T, M/T	**Evaporative Emission System Vent Valve Stuck Closed** Even if EVAP vent valve is changed off (open) to on (close) with vacuum pump on, variation of pressure sensor is less than 0.2 kPa and pressure difference between minimum pressure of EVAP system and reference pressure is lower than the threshold. **Possible Causes:** • EVAP leak check module and its circuit • ECM
DTC: P1422 **2T CCM; MIL: Yes** **Years:** 2007, 2008 **Models:** Aerio, Grand Vitara **Engines:** 2.3 VIN 6, 2.7L VIN 9 **Transmissions:** A/T, M/T	**Evaporative Emission System Leak Detect Pump Stuck OFF** Even if vacuum pump is operated, variation of EVAP leak pressure sensor is less than 0.2 kPa. (2 driving cycle detection logic, monitoring once per driving cycle) **Possible Causes:** • EVAP leak check module and its circuit • ECM
DTC: P1424 **2T CCM; MIL: Yes** **Years:** 2007, 2008 **Models:** Aerio, Grand Vitara **Engines:** 2.3 VIN 6, 2.7L VIN 9 **Transmissions:** A/T, M/T	**Evaporative Emission System Leak Detect Pump Flow High** Difference pressure between initial pressure and reference pressure in EVAP system is higher than the threshold. (2 driving cycle detection logic, monitoring once per driving cycle) **Possible Causes:** • EVAP leak check module and its circuit • ECM
DTC: P1425 **2T CCM; MIL: Yes** **Years:** 2007, 2008 **Models:** Grand Vitara **Engines:** 2.7L VIN 9 **Transmissions:** A/T, M/T	**Evaporative Emission System Leak Detect Pump Flow Low** Difference pressure between initial pressure and reference pressure in EVAP system is lower than the 1 kPa and pressure difference between maximum pressure of EVAP system and reference pressure is higher than the 0.2 kPa when EVAP canister vent valve is changed off (open) to on (close). (2 driving cycle detection logic, monitoring once per driving cycle) **Possible Causes:** • EVAP leak check module and its circuit • ECM

DTC	Trouble Code Title, Conditions & Possible Causes
DTC: P1510 **1T CCM; MIL: Yes** **Years:** 2007, 2008 **Models:** Aerio, Grand Vitara **Engines:** 2.3 VIN 6, 2.7L VIN 9 **Transmissions:** A/T, M/T	**PCM Backup Power Supply Circuit Malfunction** The ECM detects the TP sensor angle is less than 1.8 percent or greater than 13 percent when the throttle actuator control motor is deactivated. The condition exists for greater than 1 second or a cumulative of 10 seconds. **Possible Causes:** • Throttle valve • ECM
DTC: P01516 **1T; MIL: Yes** **Years:** 2007, 2008 **Models**: Reno, Forenza **Engines:** 2.0L VIN Z **Transmissions:** All	**Throttle Actuator Control (TAC) Module Throttle Actuator Position Performance** Engine is running, the ECM detects that the desired and actual throttle positions are not within a calibrated range of each other. **Possible Causes:** • Signs of water intrusion. • System voltage • ECM
DTC: P1551 **1T CCM; MIL: Yes** **Years:** 2007, 2008 **Models:** XL-7 **Engines:** 3.6L **Transmissions:** A/T	**Throttle Valve Rest Position Not Reached During Learn** Start engine and run it for 10 sec. Backup power voltage of internal circuit is lower than 3 V or more than 7 V. **Possible Causes:** • Backup power supply circuit • ECM
DTC: P1600 **1T CCM; MIL: Yes** **Years:** 2007 **Models:** Aerio, **Engines:** 2.3 VIN 6 **Transmissions:** All	**Serial Communication Problem Between ECM and TCM** Start engine and run it for 10 sec. Backup power voltage of internal circuit is lower than 3 V or more than 7 V. **Possible Causes:** • Improper mating • Broken locks • Improperly formed • Damaged terminals • Poor terminal-to-wire connection • ECM /TCM
DTC: P1603 **1T PCM; MIL: Yes** **Years:** 2007, 2008 **Models:** Grand Vitara **Engines:** 2.7L VIN 9 **Transmissions:** A/T	**TCM Trouble Code Detected** When ECM receives a trouble code from TMC, which indicates that some problem occurred in sensor circuits and its calculated values used for operations such as Idle speed control, engine power control, and so on by TCM, ECM sets DTC P1603. (TCM outputs the trouble code to ECM when TCM can not compute the engine control signal due to malfunctions of sensor circuits used for gear shift control.) **Possible Causes:** • TCM sensor circuits • TCM has failed • PCM has failed
DTC: P01618 **1T; MIL: Yes** **Years:** 2007, 2008 **Models**: Reno, Forenza **Engines:** 2.0L VIN Z **Transmissions:** All	**Control Module Torque Performance** Test 1 (Torque Control)The engine is operating, the ECM detects that the delivered torque does not match the desired torque. Test 2 (Steady State Torque Control) The engine is operating, engine speed > 800 RPM, desired Flywheel torque is within 20 N·m band for 1 sec, the ECM detects that the delivered torque does not match the desired torque. Test 3 (Redundant Torque Return) Pedal Position < 0.8 percent, engine Speed – Desired Engine Speed > 0, driver input is the desired torque source, injectors are enabled, the ECM detects that the fuel flow exceeds requirements. **Possible Causes:** • ECM • Clear the trouble code, and recheck for the same code. • Actuator Control (TAC) system.
DTC: P1717 **2T CCM; MIL: Yes** **Years:** 2007 **Models:** Aerio, **Engines:** 2.3L VIN 6 **Transmissions:** All	**A/T Driver Range Signal Circuit Malfunction** D" range signal not received by (Park / Neutral position signal inputted) the ECM while vehicle running. **Possible Causes:** • Transmission range sensor (shift switch) • "GRY/RED" circuit open • "R", "D", "2" or "L" range signal circuit open • TCM power or ground circuit open • ECM

Gas Engine OBD II Trouble Code List (P2xxx Codes)

DTC	Trouble Code Title, Conditions & Possible Causes
DTC: P2096 **2T CCM; MIL: Yes** **Years:** 2007, 2008 **Models:** Forenza, Reno, Vitara, XL-7, **Engines:** 2.0L VIN Z, 2.7L VIN 9, 3.6L VIN 7 **Transmissions:** All	**Post Catalyst Fuel Trim System Too Lean (Bank 1)** With engine at normal operating temperature Drive vehicle at 45 MPH (70 km/h) or higher. (engine speed: 1500 – 3000 r/min.) Keep above vehicle speed for 4 min. or more. (Throttle valve opening is kept constant in this step.) Check DTCs and pending DTCs. Correction index of A/F (bank-1) feedback estimated based on HO2S (bank-1) signal is higher than correction upper limit for more than 10 sec. **Possible Causes:** • Vacuum leakage • Exhaust gas leakage • Fuel injector malfunction • HO2S and its circuit • A/F sensor and its circuit • Intake & exhaust valve clearance • Intake & exhaust valve • Valve timing • ECM
DTC: P2097 **2T CCM; MIL: Yes** **Years:** 2007, 2008 **Models:** Forenza, Reno, Vitara, XL-7, **Engines:** 2.0L VIN Z, 2.7L VIN 9, 3.6L VIN 7 **Transmissions:** All	**Post Catalyst Fuel Trim System Too Rich (Bank 1)** With the engine at normal operating temperature Drive vehicle at 45 MPH (70 km/h) or higher. (engine speed: 1500 – 3000 RPM.) Keep above vehicle speed for 4 min. or more. (Throttle valve opening is kept constant in this step.) Check DTCs and pending DTCs. Correction index of A/F (bank-1) feedback estimated based on HO2S (bank-1) signal is lower than correction lower limit for more than 10 sec. **Possible Causes:** • Vacuum leakage • Exhaust gas leakage • Fuel injector malfunction • HO2S and its circuit • A/F sensor and its circuit • Intake & exhaust valve clearance • Intake & exhaust valve • Valve timing • ECM
DTC: P2099 **2T CCM; MIL: Yes** **Years:** 2007, 2008 **Models:** Grand Vitara, XL-7 **Engines:** 2.7L VIN 9, 3.6L VIN 7 **Transmissions:** A/T, M/T	**Post Catalyst Fuel Trim System Too Rich (Bank 2)** With the engine at normal operating temperature Drive vehicle at 45 MPH (70 km/h) or higher. (engine speed: 1500–3000 RPM) Keep above vehicle speed for 4 min. or more. (Throttle valve opening is kept constant in this step.) Check DTCs and pending DTCs. Correction index of A/F (bank-2) feedback estimated based on HO2S (bank-2) signal is lower than correction lower limit for more than 10 sec. **Possible Causes:** • Vacuum leakage • Exhaust gas leakage • Fuel injector malfunction • HO2S and its circuit • A/F sensor and its circuit • Intake & exhaust valve clearance • Intake & exhaust valve • Valve timing • ECM
DTC: P2101 **1T CCM; MIL: Yes** **Years:** 2007, 2008 **Models:** Forenza, Reno, Vitara, XL-7, **Engines:** 2.0L VIN Z, 2.7L VIN 9, 3.6L VIN 7 **Transmissions:** All	**Throttle Actuator Control Motor Circuit Range / Performance** Turn ON ignition switch and keep the accelerator pedal at idle position for 2 seconds. Keep the accelerator pedal at fully depressed position for 2 seconds. Repeat 3 times. The difference between actual throttle valve position and target throttle valve position is more than 2° for more than 2 seconds. **Possible Causes:** • Throttle actuator circuit • Electric throttle body assembly • ECM

DTC	Trouble Code Title, Conditions & Possible Causes
DTC: P2102 **1T CCM; MIL: Yes** **Years:** 2007, 2008 **Models:** Grand Vitara **Engines:** 2.7L VIN 9 **Transmissions:** A/T, M/T	**Throttle Actuator Control Motor Circuit Low** Keep ignition switch is at ON position for 5 seconds or longer check DTC. Power supply voltage of throttle actuator control circuit is lower than 6 V for more than 0.5 sec. even if throttle actuator control relay turned on. **Possible Causes:** • Throttle actuator control relay and its circuit • ECM
DTC: P2103 **1T CCM; MIL: Yes** **Years:** 2007, 2008 **Models:** Grand Vitara **Engines:** 2.7L VIN 9 **Transmissions:** A/T, M/T	**Throttle Actuator Control Motor Circuit High** Keep ignition switch is at ON position for 5 seconds or longer check DTC. Power supply current of throttle actuator control circuit is higher than specification for more than 0.5 sec. **Possible Causes:** • Throttle actuator control relay and its circuit • Electric throttle body assembly • ECM
DTC: P2104 **1T; MIL: Yes** **Years:** 2007, 2008 **Models:** Forenza, Reno, **Engines:** 2.0L VIN Z **Transmissions:** All	**Throttle Actuator Control (TAC) System – Forced Idle** Ignition ON, the ECM detects the system is in Forced Idle Mode. **Possible Causes:** • ECM • P0502, P0504, P0571, P2122, P2123, P2127, P2128, P2138 • If this trouble code resets, the ECM must be replaced
DTC: P2105 **1T CCM; MIL: Yes** **Years:** 2007, 2008 **Models:** Forenza, Reno, Vitara, XL-7, **Engines:** 2.0L VIN Z, 2.7L VIN 9, 3.6L VIN 7 **Transmissions:** All	**Throttle Actuator Control (TAC) System – Forced Engine Shutdown** The ECM detects an incorrect voltage level at the ignition voltage supply circuits for greater than 1 second or The ECM detects an internal communication error. **Possible Causes:** • Throttle Valve • TAC unit • ECM
DTC: P02106 **1T; MIL: Yes** **Years:** 2007, 2008 **Models**: Reno, Forenza **Engines:** 2.0L VIN Z **Transmissions:** All	**Throttle Actuator Control (TAC) System Performance – Throttle Limitation Active** The ECM detects the system is in Limit Performance Mode. **Possible Causes:** • Throttle Position (TP) sensor • Cracked or bent throttle shaft • ECM
DTC: P2108 **1T CCM, MIL: Yes** **Years:** 2007, 2008 **Models:** Grand Vitara, **Engines:** 2.7L VIN 9 **Transmissions:** A/T, M/T	**Throttle Actuator Control Module Performance** Keep ignition switch is at ON position for 5 seconds or longer check DTC. Throttle actuator power supply voltage is 3 V or less even though its control relay is turned OFF, or throttle actuator power supply voltage is 8.4 V or more even though the relay is turned ON. **Possible Causes:** • Throttle actuator control relay and its circuit • ECM
DTC: P2119 **1T CCM, MIL: Yes** **Years:** 2007, 2008 **Models:** Grand Vitara, **Engines:** 2.7L VIN 9 **Transmissions:** A/T, M/T	**Throttle Actuator Control Throttle Body Range / Performance** Keep ignition switch is at ON position for 5 seconds or longer check DTC. Even though the throttle motor control is interrupted (throttle valve default position) at the time of detecting failure in the electric throttle body, TP is 12 degrees or more, or MAF is 1400 g /min or more. Even though the throttle motor control is interrupted (throttle valve default position) at the time of IG OFF control, TP is 5 degrees or less, or 15 degrees or more. **Possible Causes:** • Electric throttle body assembly • ECM
DTC: P2122 **1T CCM, MIL: Yes** **Years:** 2007, 2008 **Models:** Grand Vitara, XL-7 **Engines:** 2.7L VIN 9, 3.6L VIN 7 **Transmissions:** A/T, M/T	**Accelerator Pedal Position (APP) Sensor 1 Circuit Low Voltage** Turn ON ignition switch and keep the accelerator pedal at idle position for 2 seconds. Keep the accelerator pedal at fully depressed position for 2 seconds. Repeat 3 times. Output voltage of pedal position sensor (main) is less than 0.45 V. **Possible Causes:** • APP sensor and its circuit • ECM

DTC	Trouble Code Title, Conditions & Possible Causes
DTC: P2123 **1T CCM, MIL: Yes** **Years:** 2007, 2008 **Models:** Grand Vitara, XL-7 **Engines:** 2.7L VIN 9, 3.6L VIN 7 **Transmissions:** A/T, M/T	**Pedal Position Sensor (Main) Circuit High** Turn ON ignition switch and keep the accelerator pedal at idle position for 2 seconds. Keep the accelerator pedal at fully depressed position for 2 seconds. Repeat 3 times. Output voltage of pedal position sensor (main) is more than 4.8 V. **Possible Causes:** • APP sensor and its circuit • ECM
DTC: P2127 **1T CCM, MIL: Yes** **Years:** 2007, 2008 **Models:** Grand Vitara, XL-7 **Engines:** 2.7L VIN 9, 3.6L VIN 7 **Transmissions:** A/T, M/T	**Pedal Position Sensor (Sub) Circuit Low** Turn ON ignition switch and keep the accelerator pedal at idle position for 2 seconds. Keep the accelerator pedal at fully depressed position for 2 seconds. Repeat 3 times. Output voltage of pedal position sensor (sub) is lower than 0.23 V. **Possible Causes:** • APP sensor and its circuit • ECM
DTC: P2128 **1T CCM, MIL: Yes** **Years:** 2007, 2008 **Models:** Grand Vitara, XL-7 **Engines:** 2.7L VIN 9, 3.6L VIN 7 **Transmissions:** A/T, M/T	**Pedal Position Sensor (Sub) Circuit High** Turn ON ignition switch and keep the accelerator pedal at idle position for 2 seconds. Keep the accelerator pedal at fully depressed position for 2 seconds. Repeat 3 times. Output voltage of pedal position sensor (sub) is more than 2.4 V. **Possible Causes:** • APP sensor and its circuit • ECM
DTC: P2135 **1T CCM, MIL: Yes** **Years:** 2007, 2008 **Models:** Grand Vitara, **Engines:** 2.7L VIN 99 **Transmissions:** A/T, M/T	**Throttle Position Sensor (Main / Sub) Voltage Correlation** Turn ON ignition switch and keep the accelerator pedal at idle position for 2 seconds. Keep the accelerator pedal at fully depressed position for 2 seconds. Repeat 3 times. Index of correlation between main throttle position sensor and sub throttle position sensor (main throttle position sensor voltage minus sub throttle position sensor voltage) is more than 0.2 V. **Possible Causes:** • TP sensor (main and sub) circuit • Electric throttle body assembly • ECM
DTC: P2138 **1T CCM; MIL: Yes** **Years:** 2007, 2008 **Models:** Grand Vitara, XL-7 **Engines:** 2.7L VIN 9, 3.6L VIN 7 **Transmissions:** A/T, M/T	**Pedal Position Sensor (Main / Sub) Voltage Correlation** Turn ON ignition switch and keep the accelerator pedal at idle position for 2 seconds. Keep the accelerator pedal at fully depressed position for 2 seconds. Repeat 3 times. Index of correlation between the main accelerator pedal position sensor and sub accelerator pedal position sensor (1/2 main accelerator pedal position sensor voltage minus sub accelerator pedal position sensor voltage) is more than 0.35 V or less than –0.35 V. **Possible Causes:** • APP sensor and its circuit • ECM failure
DTC: P2176 **1T CCM; MIL: Yes** **Years:** 2007, 2008 **Models:**, XL-7 **Engines:** 3.6L VIN 7 **Transmissions:** A/T	**Minimum Throttle Position Not Learned** The ECM detects that the TP sensor 1 voltage is not between 0.2–0.86 volts during the throttle learn procedure. The ECM detects that the TP sensor 2 voltage is not between 4.1–4.8 volts during the throttle learn procedure. The minimum throttle position is not learned after an ECM replacement. The ECM detects that any of the above conditions exist for greater than 1 second or a cumulative of 10 seconds. **Possible Causes:** • APP sensor and its circuit • ECM failure
DTC: P2177 **1T CCM; MIL: Yes** **Years:** 2007, 2008 **Models:**, XL-7 **Engines:** 3.6L VIN 7 **Transmissions:** A/T	**Fuel Trim System Lean at Cruise or Accel Bank 1** The Total Fuel Trim Avg. is less than −22 percent. The condition exists for 4 seconds or for a cumulative of 30 seconds. **Possible Causes:** • APP sensor and its circuit • ECM failure
DTC: P2178 **2T CCM; MIL: Yes** **Years:** 2007, 2008 **Models:**, XL-7 **Engines:** 3.6L VIN 7 **Transmissions:** A/	**Fuel Trim System Rich at Cruise or Accel Bank 1** The Total Fuel Trim Avg. is greater than 23 percent for 4 seconds or for a cumulative time of 30 seconds. **Possible Causes:** • APP sensor and its circuit • ECM failure

DTC	Trouble Code Title, Conditions & Possible Causes
DTC: P2179 **1T CCM; MIL: Yes** **Years:** 2007, 2008 **Models:**, XL-7 **Engines:** 3.6L VIN 7 **Transmissions:** A/T	**Fuel Trim System Lean at Cruise or Accel Bank 2** The Total Fuel Trim Avg. is greater than 23 percent for 4 seconds or for a cumulative time of 30 seconds. **Possible Causes:** • APP sensor and its circuit • ECM failure
DTC: P2180 **2T CCM; MIL: Yes** **Years:** 2007, 2008 **Models:**, XL-7 **Engines:** 3.6L VIN 7 **Transmissions:** A/T	**Fuel Trim System Rich at Cruise or Accel Bank 2** The Total Fuel Trim Avg. is less than −22 percent. The condition exists for 4 seconds or for a cumulative of 30 seconds. **Possible Causes:** • APP sensor and its circuit • ECM failure
DTC: P2187 **1T CCM; MIL: Yes** **Years:** 2007, 2008 **Models:**, XL-7 **Engines:** 3.6L VIN 7 **Transmissions:** A/T	**Fuel Trim System Lean at Idle Bank 1** The Total Fuel Trim Avg. is greater than 40 percent, or the LT FT Idle/Decel is greater than 7 percent. Either condition exists for 4 seconds or for a cumulative of 30 seconds. **Possible Causes:** • APP sensor and its circuit • ECM failure
DTC: P2188 **2T CCM; MIL: Yes** **Years:** 2007, 2008 **Models:**, XL-7 **Engines:** 3.6L VIN 7 **Transmissions:** A/T	**Fuel Trim System Rich at Idle Bank 1** The Total Fuel Trim Avg. is less than −40 percent. The LT FT Idle/Decel is less than −7 percent. The condition exists for 4 seconds or for a cumulative of 40 seconds. **Possible Causes:** • APP sensor and its circuit • ECM failure
DTC: P2189 **1T CCM; MIL: Yes** **Years:** 2007, 2008 **Models:**, XL-7 **Engines:** 3.6L VIN 7 **Transmissions:** A/T	**Fuel Trim System Lean at Idle Bank 2** The Total Fuel Trim Avg. is greater than 40 percent, or the LT FT Idle/Decel is greater than 7 percent. Either condition exists for 4 seconds or for a cumulative of 30 seconds. **Possible Causes:** • APP sensor and its circuit • ECM failure
DTC: P2190 **2T CCM; MIL: Yes** **Years:** 2007, 2008 **Models:**, XL-7 **Engines:** 3.6L VIN 7 **Transmissions:** A/T	**Fuel Trim System Rich at Idle Bank 2** The Total Fuel Trim Avg. is less than −40 percent. The LT FT Idle/Decel is less than −7 percent. The condition exists for 4 seconds or for a cumulative of 40 seconds. **Possible Causes:** • APP sensor and its circuit • ECM failure
DTC: P2195 **2T OBD/O2S; MIL: Yes** **Years:** 2007, 2008 **Models:** Grand Vitara **Engines:** 2.7L VIN 9 **Transmissions:** A/T, M/T	**O2 Sensor Signal Stuck Lean (Bank 1 Sensor 1)** Drive vehicle at 40 MPH (60 km/h) or higher. (engine speed: 2500 – 3000 RPM.) Keep the vehicle speed for 4 min. or more. (Throttle valve opening is kept constant in this step.) Short term fuel trim is higher than 130 percent for 30 seconds even though HO2S (bank-1) signal voltage is higher than 0.5 V. The Intake air temperature must be between 14°F (−10°C) to 176 °F (80 °C) Engine coolant temperature: 68°F (20°C) or more, atmospheric pressure: higher than 75 kPa (560 mmHg) (Altitude: lower than 2790 m (9150 ft) **Possible Causes:** • A/F sensor circuit • A/F sensor • ECM
DTC: P2196 **2T OBD/O2S; MIL: Yes** **Years:** 2007, 2008 **Models:** Grand Vitara **Engines:** 2.7L VIN 9 **Transmissions:** A/T, M/T	**O2 Sensor Signal Stuck Rich (Sensor-1, Bank-1)** Drive vehicle at 40 MPH (60 km/h) or higher. (engine speed: 2500 – 3000 RPM.) Keep the vehicle speed for 4 min. or more. (Throttle valve opening is kept constant in this step.) Short-term fuel trim is lower than 70 percent for 30 seconds even though HO2S (bank-1) signal voltage is lower than 0.5 V. The Intake air temperature must be between 14°F (−10°C) to 176 °F (80 °C)\ Engine coolant temperature: 68°F (20°C) or more Atmospheric pressure: higher than 75 kPa (560 mmHg) (Altitude: lower than 2790 m (9150 ft) **Possible Causes:** • A/F sensor circuit • A/F sensor • ECM

DTC	Trouble Code Title, Conditions & Possible Causes
DTC: P2197 **2T OBD/O2S; MIL: Yes** **Years:** 2007, 2008 **Models:** Grand Vitara **Engines:** 2.7L VIN 9 **Transmissions:** A/T, M/T	**O2 Sensor Signal Stuck Lean (Bank 2 Sensor 1)** Drive vehicle at 40 MPH (60 km/h) or higher. (engine speed: 2500 – 3000 RPM.) Keep the vehicle speed for 4 min. or more. (Throttle valve opening is kept constant in this step.) Short term fuel trim is higher than 130 percent for 30 seconds even though HO2S (bank-2) signal voltage is higher than 0.5 V. The Intake air temperature must be between 14°F (−10°C) to 176 °F (80 °C) Engine coolant temperature: 68°F (20°C) or more Atmospheric pressure: higher than 75 kPa (560 mmHg) (Altitude: lower than 2790 m (9150 ft) **Possible Causes:** • A/F sensor circuit • A/F sensor • ECM
DTC: P2198 **2T OBD/O2S; MIL: Yes** **Years:** 2007, 2008 **Models:** Grand Vitara **Engines:** 2.7L VIN 9 **Transmissions:** A/T, M/T	**O2 Sensor Signal Stuck Rich (Sensor-2, Bank-1)** Drive vehicle at 40 MPH (60 km/h) or higher. (engine speed: 2500 – 3000 RPM.) Keep the vehicle speed for 4 min. or more. (Throttle valve opening is kept constant in this step.) Short-term fuel trim is lower than 70 percent for 30 seconds even though HO2S (bank-2) signal voltage is lower than 0.5 V. The Intake air temperature must be between 14°F(−10°C) to 176 °F (80 °C) Engine coolant temperature: 68°F (20°C) or more Atmospheric pressure: higher than 75 kPa (560 mmHg) (Altitude: lower than 2790 m (9150 ft) **Possible Causes:** • A/F sensor circuit • A/F sensor • ECM
DTC: P2227 **2T CCM; MIL: Yes** **Years:** 2007, 2008 **Models:** Arieo, Grand Vitara, XL-7 **Engines:** 2.3L VIN 6, 2.7L VIN 9, 3.6L VIN 7 **Transmissions:** A/T, M/T	**Barometric Pressure Circuit Range / Performance** Difference between measured barometric pressure and estimated barometric pressure based on MAP is more than 25.3 kPa at engine start. Intake air temperature must be between 14°F and 113°F (–10°C and 45°C) Engine coolant temperature must be between –10°C and 110°C (14°F and 230°F) **Possible Causes:** • EVAP leak check module • MAP Sensor
DTC: P2228 **2T CCM; MIL: Yes** **Years:** 2007, 2008 **Models:** Grand Vitara, XL-7 **Engines:** 2.7L VIN 9, 3.6L VIN 7 **Transmissions:** A/T, M/T	**Barometric Pressure Circuit Low** Barometric pressure sensor voltage is less than 0.1 V for more than 19 sec. Intake air temperature must be between: 14°F and 113°F (–10°C and 45°C) Engine coolant temperature must be between: –10°C and 110°C (14°F and 230°F) **Possible Causes:** • EVAP leak check module • MAP Sensor
DTC: P2229 **2T CCM; MIL: Yes** **Years:** 2007, 2008 **Models:** Grand Vitara, XL-7 **Engines:** 2.7L VIN 9, 3.6L VIN 7 **Transmissions:** A/T, M/T	**Barometric Pressure Circuit High** Barometric pressure sensor voltage is more than 5.1 V for more than 19 sec. Intake air temperature must be between: 14°F and 113°F (–10°C and 45°C) Engine coolant temperature must be between: –10°C and 110°C (14°-230°F) **Possible Causes:** • EVAP leak check module • MAP Sensor
DTC: P2232 **2T CCM; MIL: Yes** **Years:** 2007, 2008 **Models:** XL-7 **Engines:** 3.6L VIN 7 **Transmissions:** A/T	**HO2S Signal Circuit Shorted to Heater Circuit Bank 1 Sensor 2** The ECM detects that the HO2S signal voltage increases greater than 2 volts within 0.04 second, in 4 out of 6 HO2S heater switch OFF samples. The condition exists for greater than 1 second, or for a cumulative of 10 seconds. **Possible Causes:** • HO2S Harness Connector • HO2S Sensor • ECM

DTC	Trouble Code Title, Conditions & Possible Causes
DTC: P2235 **2T CCM; MIL: Yes** **Years:** 2007, 2008 **Models:** XL-7 **Engines:** 3.6L VIN 7 **Transmissions:** A/T	**HO2S Signal Circuit Shorted to Heater Circuit Bank 2 Sensor 2** The ECM detects that the HO2S signal voltage increases greater than 2 volts within 0.04 second, in 4 out of 6 HO2S heater switch OFF samples. The condition exists for greater than 1 second, or for a cumulative of 10 seconds. **Possible Causes:** • HO2S Harness Connector • HO2S Sensor • ECM
DTC: P2237 **1T OBD/O2S; MIL: Yes** **Years:** 2007, 2008 **Models:** Grand Vitara **Engines:** 2.7L VIN 9 **Transmissions:** A/T, M/T	**O2 Sensor Positive Current Control Circuit / Open (Bank 1 Sensor 1)** Start engine and warm up to normal operating temperature, run engine at idle speed for 1 min. or more. Impedance of A/F sensor element is higher than 250 Ω for more than 5 sec even though A/F sensor heater is turned ON with engine running. **Possible Causes:** • A/F sensor circuit • A/F sensor • ECM
DTC: P2238 **1T OBD/O2S; MIL: Yes** **Years:** 2007, 2008 **Models:** Grand Vitara **Engines:** 2.7L VIN 9 **Transmissions:** A/T, M/T	**O2 Sensor Positive Current Control Circuit Low (Bank 1 Sensor 1)** Start engine and warm up to normal operating temperature, run engine at idle speed for 1 min. or more. Circuit voltage of A/F sensor (+) terminal is lower than 0.4 V or A/F sensor circuit is shorted between (+) terminal wire and (−) terminal wire with engine running. **Possible Causes:** • A/F sensor circuit • A/F sensor • ECM
DTC: P2239 **1T OBD/O2S; MIL: Yes** **Years:** 2007, 2008 **Models:** Grand Vitara **Engines:** 2.7L VIN 9 **Transmissions:** A/T, M/T	**O2 Sensor Positive Current Control Circuit High (Bank 1 Sensor 1)** Start engine and warm up to normal operating temperature, run engine at idle speed for 1 min. or more. Circuit voltage of A/F sensor (+) terminal is higher than 4.4 V with engine running **Possible Causes:** • A/F sensor circuit • A/F sensor • ECM
DTC: P2240 **1T OBD/O2S; MIL: Yes** **Years:** 2007, 2008 **Models:** Grand Vitara, **Engines:** 2.7L VIN 9 **Transmissions:** A/T, M/T	**O2 Sensor Positive Current Control Circuit / Open(Bank 2 Sensor 1)** Start engine and warm up to normal operating temperature, run engine at idle speed for 1 min. or more. Impedance of A/F sensor element is higher than 250 Ω for more than 5 sec. even though A/F sensor heater is turned ON with engine running. **Possible Causes:** • A/F sensor circuit • A/F sensor • ECM
DTC: P2241 **1T OBD/O2S; MIL: Yes** **Years:** 2007, 2008 **Models:** Grand Vitara **Engines:** 2.7L VIN 9 **Transmissions:** A/T, M/T	**O2 Sensor Positive Current Control Circuit Low (Bank 2 Sensor 1)** Start engine and warm up to normal operating temperature, run engine at idle speed for 1 min. or more. Circuit voltage of A/F sensor (+) terminal is lower than 0.4 V or A/F sensor circuit is shorted between (+) terminal wire and (−) terminal wire with engine running. **Possible Causes:** • A/F sensor circuit • A/F sensor • ECM
DTC: P2242 **1T OBD/O2S; MIL: Yes** **Years:** 2007, 2008 **Models:** Grand Vitara, **Engines:** 2.7L VIN 9 **Transmissions:** A/T, M/T	**O2 Sensor Positive Current Control Circuit High (Bank 2 Sensor 1)** Start engine and warm up to normal operating temperature, run engine at idle speed for 1 min. or more. Circuit voltage of A/F sensor (+) terminal is higher than 4.4 V with engine running **Possible Causes:** • A/F sensor circuit • A/F sensor • ECM

DTC	Trouble Code Title, Conditions & Possible Causes
DTC: P2252 **1T OBD/O2S; MIL: Yes** **Years:** 2007, 2008 **Models:** Grand Vitara, **Engines:** 2.7L VIN 9 **Transmissions:** A/T, M/T	**O2 Sensor Negative Current Control Circuit Low (Bank 1 Sensor 1)** Start engine and warm up to normal operating temperature, run engine at idle speed for 1 min. or more. Circuit voltage of A/F sensor (−) terminal is lower than 0.4 V with engine running. **Possible Causes:** • A/F sensor circuit • A/F sensor • ECM
DTC: P2253 **1T OBD/O2S; MIL: Yes** **Years:** 2007, 2008 **Models:** Grand Vitara, **Engines:** 2.7L VIN 9 **Transmissions:** A/T, M/T	**O2 Sensor Negative Current Control Circuit High (Bank 1 Sensor 1)** Start engine and warm up to normal operating temperature, run engine at idle speed for 1 min. or more. Circuit voltage of A/F sensor (-) terminal is higher than 4.4 V with engine running **Possible Causes:** • A/F sensor circuit • A/F sensor • ECM
DTC: P2255 **1T OBD/O2S; MIL: Yes** **Years:** 2007, 2008 **Models:** Grand Vitara **Engines:** 2.7L VIN 9 **Transmissions:** A/T, M/T	**O2 Sensor Negative Current Control Circuit Low (Bank 2 Sensor 1)** Start engine and warm up to normal operating temperature, run engine at idle speed for 1 min. or more. Circuit voltage of A/F sensor (-) terminal is higher than 0.4 V with engine running **Possible Causes:** • A/F sensor circuit • A/F sensor • ECM
DTC: P2256 **1T OBD/O2S; MIL: Yes** **Years:** 2007, 2008 **Models:** Grand Vitara **Engines:** 2.7L VIN 9 **Transmissions:** A/T, M/T	**O2 Sensor Negative Current Control Circuit High (Bank 2 Sensor 1)** Start engine and warm up to normal operating temperature, run engine at idle speed for 1 min. or more. Circuit voltage of A/F sensor (-) terminal is higher than 4.4 V with engine running **Possible Causes:** • A/F sensor circuit • A/F sensor • ECM
DTC: P2270 **2T OBD/O2S; MIL: Yes** **Years:** 2007, 2008 **Models:** XL-7 **Engines:** 3.6L VIN 7 **Transmissions:** A/T	**O2 Sensor Negative Current Control Circuit High (Bank 2 Sensor 1)** The ECM detects that the rear HO2S voltage is less than 630 mV for greater than 100 seconds, then an intrusive test is performed. The ECM will enrich the fuel mixture up to 20 percent and then wait for 10 seconds at 30 percent. The ECM detects that the HO2S voltage is less than 630 mV during the intrusive test for greater than 4 second or for a cumulative of 30 seconds. **Possible Causes:** • Contaminated HO2S – Silicon • Water intrusion in the HO2S harness connector • Low fuel system pressure • Fuel that is contaminated • Engine mechanical condition • Vacuum hoses for splits, kinks, and proper connection • The air intake system after the MAF sensor for vacuum leaks • Exhaust system leaks • ECM
DTC: P2271 **2T OBD/O2S; MIL: Yes** **Years:** 2007, 2008 **Models:** XL-7 **Engines:** 3.6L VIN 7 **Transmissions:** A/T	**HO2S Signal Stuck Rich Bank 1 Sensor 2** The ECM detects that the rear HO2S voltage is greater than 630 mV for greater than 100 seconds, then an intrusive test is performed. The ECM will lean the fuel mixture up to −7 percent and then wait for 10 seconds at −7 percent. If the ECM detects that the HO2S voltage is still greater than 630 mV, the ECM then tests the HO2S at the next decel fuel cut-off. If the ECM detects the HO2S voltage is greater than 200 mV after 4 seconds or for a cumulative of 30 seconds. **Possible Causes:** • Contaminated HO2S – Silicon • Water intrusion in the HO2S harness connector • Low fuel system pressure • Fuel that is contaminated • Engine mechanical condition • Vacuum hoses for splits, kinks, and proper connection • The air intake system after the MAF sensor for vacuum leaks • Exhaust system leaks • ECM

DTC	Trouble Code Title, Conditions & Possible Causes
DTC: P2272 **2T OBD/O2S; MIL: Yes** **Years:** 2007, 2008 **Models:** XL-7 **Engines:** 3.6L VIN 7 **Transmissions:** A/T	**O2 Sensor Negative Current Control Circuit High (Bank 2 Sensor 2)** The ECM detects that the rear HO2S voltage is less than 630 mV for greater than 100 seconds, then an intrusive test is performed. The ECM will enrich the fuel mixture up to 20 percent and then wait for 10 seconds at 30 percent. The ECM detects that the HO2S voltage is less than 630 mV during the intrusive test for greater than 4 second or for a cumulative of 30 seconds. **Possible Causes:** • Contaminated HO2S – Silicon • Water intrusion in the HO2S harness connector • Low fuel system pressure • Fuel that is contaminated • Engine mechanical condition • Vacuum hoses for splits, kinks, and proper connection • The air intake system after the MAF sensor for vacuum leaks • Exhaust system leaks • ECM
DTC: P2273 **2T OBD/O2S; MIL: Yes** **Years:** 2007, 2008 **Models:** XL-7 **Engines:** 3.6L VIN 7 **Transmissions:** A/T	**HO2S Signal Stuck Rich Bank 2 Sensor 2** The ECM detects that the rear HO2S voltage is greater than 630 mV for greater than 100 seconds, then an intrusive test is performed. The ECM will lean the fuel mixture up to −7 percent and then wait for 10 seconds at −7 percent. If the ECM detects that the HO2S voltage is still greater than 630 mV, the ECM then tests the HO2S at the next decel fuel cut-off. If the ECM detects the HO2S voltage is greater than 200 mV after 4 seconds or for a cumulative of 30 seconds. **Possible Causes:** • Fuel in the crankcase • Collapsed air intake duct • Water intrusion in the HO2S harness connector • Rich fuel injectors • Fuel that is contaminated • Engine mechanical condition • High fuel system pressure • Restricted air filter element • Exhaust system restrictions • ECM
DTC: P2282 **2T CCM; MIL: Yes** **Years:** 2007, 2008 **Models:** Grand Vitara, XL-7 **Engines:** 2.7L VIN 9, 3.6L VIN 7 **Transmissions:** A/T, M/T	**Air Leak Between Throttle Body and Intake Valves** Start engine and warm up to normal operating temperature, run engine at idle speed for 1 min. or more. Pressure difference between MAP and estimated MAP is larger than the specification for more than 20 sec. when short term fuel trim is 110 percent with engine speed variation by 25 RPM/ 0.05 sec. or more. **Possible Causes:** • PVC system • Throttle Body • Intake Manifold • EGR system and /or canister purge system • MAF Sensor • ECM
DTC: P02297 **1T; MIL: Yes** **Years:** 2007, 2008 **Models**: Reno, Forenza **Engines:** 2.0L VIN Z **Transmissions:** All	**Performance During Decel Fuel Cut-Off (DFCO) Sensor 1** HO2S1 voltage is greater than 0.55 V in Decel Fuel Cutoff (DFCO) mode,system voltage is greater than 10 V, engine running time is greater than 10 seconds, engine Coolant Temperature (ECT) is greater than 60°C (140°F), no relevant DTCs are set, 2 second delay after in DFCO mode. **Possible Causes:** • Leaking injector • A leaking or malfunctioning injector • ECM
DTC: P2400 **1T CCM; MIL: Yes** **Years:** 2007, 2008 **Models:** Aerio, Grand Vitara, XL-7 **Engines:** 2.3L VIN 6 2.7L VIN 9, 3.6L VIN 7 **Transmissions:** A/T, M/T	**Evaporative Emission System Leak Detection Pump Control Circuit/Open** Start engine and warm up to normal operating temperature, run engine at idle speed for 1 min. or more. EVAP leak detecting pump circuit voltage is higher than 4 V while detection pump is turned OFF, or lower than 2 V while detection pump is turned ON. **Possible Causes:** • EVAP Leak Check module and its circuit • ECM

DTC	Trouble Code Title, Conditions & Possible Causes
DTC: P02610 **1T; MIL: Yes** **Years:** 2007, 2008 **Models**: Reno, Forenza **Engines:** 2.0L VIN Z **Transmissions:** All	**Control Module Ignition Off Timer Performance** Ignition switch is turned to ON, Ignition voltage is greater than 11 V, engine is running more than 10 seconds. **Possible Causes:** • ECM • Clear the trouble code, and recheck for the same code. • If this trouble code resets, the ECM must be replaced
DTC: P2636 **2T CCM; MIL: Yes** **Years:** 2007, 2008 **Models:** XL-7 **Engines:** 3.6L VIN 7 **Transmissions:** A/T	**Fuel Transfer Pump Flow Insufficient** The ECM detects that the right side fuel level is less than 11 percent and the left side fuel level is more than 23 percent. The above condition is present for greater than 4 seconds or for a cumulative of 50 seconds. **Possible Causes:** • EVAP Leak Check module and its circuit • ECM
DTC: P2A00 **2T OBD/O2S; MIL: Yes** **Years:** 2007, 2008 **Models:** Aerio, Grand Vitara **Engines:** 2.3L VIN 6, 2.7L VIN 9 **Transmissions:** A/T, M/T	**O2 Sensor Circuit Range/Performance (Bank 1 Sensor 1)** Start engine and warm up to normal operating temperature, run engine at idle speed for 1 min. or more. Amplitude index of A/F sensor signal between rich and lean is more than 1.3 or less than 0.7 while A/F (fuel trim) is shifted from rich to lean and lean to rich with specified diagnosis frequency under specified running. **Possible Causes:** • A/F sensor circuit • A/F sensor • ECM
DTC: P2A01 **2T OBD/O2S; MIL: Yes** **Years:** 2007, 2008 **Models:** Aerio, Grand Vitara **Engines:** 2.3L VIN 6, 2.7L VIN 9 **Transmissions:** A/T, M/T	**O2 Sensor Circuit Range/Performance (Bank 1 Sensor 2)** Start engine and warm up to normal operating temperature, drive vehicle at 40 MPH (60 km/h) or higher. (engine speed: 2500 – 3000 RPM.) Keep above vehicle speed for 6 min. or more. (Throttle valve opening is kept constant in this step.) Release accelerator pedal and with engine brake applied, keep vehicle coasting (with fuel cut for 3 sec. or more) and then stop vehicle. Circuit voltage of HO2S (bank-1) signal is less than 0.35 V while A/F (fuel trim) is shifted from rich to lean and lean to rich with specified diagnosis frequency under specified running, or Circuit voltage of HO2S (bank-1) signal is more than 0.1 V for more than 6 sec even though vehicle is running with fuel cut mode below 4000 RPM. **Possible Causes:** • H02S circuit • H02S • ECM
DTC: P2A03 **2T OBD/O2S; MIL: Yes** **Years:** 2007, 2008 **Models:** Grand Vitara **Engines:** 2.7L VIN 9 **Transmissions:** A/T, M/T	**O2 Sensor Circuit Range/Performance (Bank 2 Sensor 1)** Start engine and warm up to normal operating temperature, drive vehicle at 40 MPH (60 km/h) or higher. (engine speed: 2500 – 3000 r/min.) Keep above vehicle speed for 6 min. or more. (Throttle valve opening is kept constant in this step.) Release accelerator pedal and with engine brake applied, keep vehicle coasting (with fuel cut for 3 sec. or more) and then stop vehicle. Amplitude index of A/F sensor signal between rich and lean is more than 1.3 or less than 0.7 while A/F (fuel trim) is shifted from rich to lean and lean to rich with specified diagnosis frequency under specified running. **Possible Causes:** • A/F sensor circuit • A/F sensor • ECM
DTC: P2A04 **2T OBD/O2S; MIL: Yes** **Years:** 2007, 2008 **Models:** Grand Vitara **Engines:** 2.7L VIN 9 **Transmissions:** A/T, M/T	**O2 Sensor Circuit Range/Performance (Bank 2 Sensor 2)** Start engine and warm up to normal operating temperature, drive vehicle at 40 MPH (60 km/h) or higher. (engine speed: 2500 – 3000 RPM) Keep above vehicle speed for 6 min. or more. (Throttle valve opening is kept constant in this step.) Release accelerator pedal and with engine brake applied, keep vehicle coasting (with fuel cut for 3 sec. or more) and then stop vehicle.

Commonly Used Abbreviations

2

2WD	Two Wheel Drive

4

4WD	Four Wheel Drive

A

A/C	Air Conditioning
ABDC	After Bottom Dead Center
ABS	Anti-lock Brakes
AC	Alternating Current
ACL	Air cleaner
ACT	Air Charge Temperature
AIR	Secondary Air Injection
ALCL	Assembly Line Communications Link
ALDL	Assembly Line Diagnostic Link
AT	Automatic Transaxle/Transmission
ATDC	After Top Dead Center
ATF	Automatic Transmission Fluid
ATS	Air Temperature Sensor
AWD	All Wheel Drive

B

BAP	Barometric Absolute Pressure
BARO	Barometric Pressure
BBDC	Before Bottom Dead Center
BCM	Body Control Module
BDC	Bottom Dead Center
BPT	Backpressure Transducer
BTDC	Before Top Dead Center
BVSV	Bimetallic Vacuum Switching Valve

C

CAC	Charge Air Cooler
CARB	California Air Resources Board
CAT	Catalytic Converter
CCC	Computer Command Control
CCCC	Computer Controlled Catalytic Converter
CCCI	Computer Controlled Coil Ignition
CCD	Computer Controlled Dwell
CDI	Capacitor Discharge Ignition
CEC	Computerized Engine Control
CFI	Continuous Fuel Injection
CIS	Continuous Injection System
CIS-E	Continuous Injection System - Electronic
CKP	Crankshaft Position
CL	Closed Loop
CMP	Camshaft Position
CPP	Clutch Pedal Position
CTOX	Continuous Trap Oxidizer System
CTP	Closed Throttle Position
CVC	Constant Vacuum Control
CYL	Cylinder

D

DBC	Dual Bed Catalyst
DC	Direct Current
DFI	Direct Fuel Injection
DIS	Distributorless Ignition System
DLC	Data Link Connector
DMM	Digital Multimeter
DOHC	Double Overhead Camshaft
DRB	Diagnostic Readout Box
DTC	Diagnostic Trouble Code
DTM	Diagnostic Test Mode
DVOM	Digital Volt/Ohmmeter

E

EBCM	Electronic Brake Control Module
ECM	Engine Control Module
ECT	Engine Coolant Temperature
ECU	Engine Control Unit or Electronic Control Unit
EDIS	Electronic Distributorless Ignition System
EEC	Electronic Engine Control
EEPROM	Electrically Erasable Programmable Read Only Memory
EFE	Early Fuel Evaporation
EGR	Exhaust Gas Recirculation
EGRT	Exhaust Gas Recirculation Temperature
EGRVC	EGR Valve Control
EPROM	Erasable Programmable Read Only Memory
EVAP	Evaporative Emissions
EVP	EGR Valve Position

F

FBC	Feedback Carburetor
FEEPROM	Flash Electrically Erasable Programmable Read Only Memory
FF	Flexible Fuel
FI	Fuel Injection
FT	Fuel Trim
FWD	Front Wheel Drive

G

GND	Ground

H

HAC	High Altitude Compensation
HEGO	Heated Exhaust Gas Oxygen sensor
HEI	High Energy Ignition
HO2 Sensor	Heated Oxygen Sensor

I

IAC	Idle Air Control
IAT	Intake Air Temperature
ICM	Ignition Control Module
IFI	Indirect Fuel Injection
IFS	Inertia Fuel Shutoff
ISC	Idle Speed Control
IVSV	Idle Vacuum Switching Valve

Commonly Used Abbreviations

K

KOEO	Key On, Engine Off
KOER	Key ON, Engine Running
KS	Knock Sensor

M

MAF	Mass Air Flow
MAP	Manifold Absolute Pressure
MAT	Manifold Air Temperature
MC	Mixture Control
MDP	Manifold Differential Pressure
MFI	Multiport Fuel Injection
MIL	Malfunction Indicator Lamp or Maintenance
MST	Manifold Surface Temperature
MVZ	Manifold Vacuum Zone

N

NVRAM	Nonvolatile Random Access Memory

O

O2 Sensor	Oxygen Sensor
OBD	On-Board Diagnostic
OC	Oxidation Catalyst
OHC	Overhead Camshaft
OL	Open Loop

P

P/S	Power Steering
PAIR	Pulsed Secondary Air Injection
PCM	Powertrain Control Module
PCS	Purge Control Solenoid
PCV	Positive Crankcase Ventilation
PIP	Profile Ignition Pick-up
PNP	Park/Neutral Position
PROM	Programmable Read Only Memory
PSP	Power Steering Pressure
PTO	Power Take-Off
PTOX	Periodic Trap Oxidizer System

R

RABS	Rear Anti-lock Brake System
RAM	Random Access Memory
ROM	Read Only Memory
RPM	Revolutions Per Minute
RWAL	Rear Wheel Anti-lock Brakes
RWD	Rear Wheel Drive

S

SBC	Single Bed Converter
SBEC	Single Board Engine Controller
SC	Supercharger
SCB	Supercharger Bypass
SFI	Sequential Multiport Fuel Injection
SIR	Supplemental Inflatible Restraint
SOHC	Single Overhead Camshaft
SPL	Smoke Puff Limiter
SPOUT	Spark Output
SRI	Service Reminder Indicator
SRS	Supplemental Restraint System
SRT	System Readiness Test
SSI	Solid State Ignition
ST	Scan Tool
STO	Self-Test Output

T

TAC	Thermostatic Air Clearner
TBI	Throttle Body Fuel Injection
TC	Turbocharger
TCC	Torque Converter Clutch
TCM	Transmission Control Module
TDC	Top Dead Center
TFI	Thick Film Ignition
TP	Throttle Position
TR Sensor	Transaxle/Transmission Range Sensor
TVV	Thermal Vacuum Valve
TWC	Three-way Catalytic Converter

V

VAF	Volume Air Flow, or Vane Air Flow
VAPS	Variable Assist Power Steering
VRV	Vacuum Regulator Valve
VSS	Vehicle Speed Sensor
VSV	Vacuum Switching Valve

W

WOT	Wide Open Throttle
WU-TWC	Warm Up Three-way Catalytic Converter

ENGLISH TO METRIC CONVERSION: TORQUE

To convert foot-pounds (ft. lbs.) to Newton-meters (Nm), multiply the number of ft. lbs. by 1.36
To convert Newton-meters (Nm) to foot-pounds (ft. lbs.), multiply the number of Nm by 0.7376

ft. lbs.	Nm	ft. lbs.	Nm	ft. lbs.	Nm	ft. lbs.	Nm
0.1	0.1	34	46.2	76	103.4	118	160.5
0.2	0.3	35	47.6	77	104.7	119	161.8
0.3	0.4	36	49.0	78	106.1	120	163.2
0.4	0.5	37	50.3	79	107.4	121	164.6
0.5	0.7	38	51.7	80	108.8	122	165.9
0.6	0.8	39	53.0	81	110.2	123	167.3
0.7	1.0	40	54.4	82	111.5	124	168.6
0.8	1.1	41	55.8	83	112.9	125	170.0
0.9	1.2	42	57.1	84	114.2	126	171.4
1	1.4	43	58.5	85	115.6	127	172.7
2	2.7	44	59.8	86	117.0	128	174.1
3	4.1	45	61.2	87	118.3	129	175.4
4	5.4	46	62.6	88	119.7	130	176.8
5	6.8	47	63.9	89	121.0	131	178.2
6	8.2	48	65.3	90	122.4	132	179.5
7	9.5	49	66.6	91	123.8	133	180.9
8	10.9	50	68.0	92	125.1	134	182.2
9	12.2	51	69.4	93	126.5	135	183.6
10	13.6	52	70.7	94	127.8	136	185.0
11	15.0	53	72.1	95	129.2	137	186.3
12	16.3	54	73.4	96	130.6	138	187.7
13	17.7	55	74.8	97	131.9	139	189.0
14	19.0	56	76.2	98	133.3	140	190.4
15	20.4	57	77.5	99	134.6	141	191.8
16	21.8	58	78.9	100	136.0	142	193.1
17	23.1	59	80.2	101	137.4	143	194.5
18	24.5	60	81.6	102	138.7	144	195.8
19	25.8	61	83.0	103	140.1	145	197.2
20	27.2	62	84.3	104	141.4	146	198.6
21	28.6	63	85.7	105	142.8	147	199.9
22	29.9	64	87.0	106	144.2	148	201.3
23	31.3	65	88.4	107	145.5	149	202.6
24	32.6	66	89.8	108	146.9	150	204.0
25	34.0	67	91.1	109	148.2	151	205.4
26	35.4	68	92.5	110	149.6	152	206.7
27	36.7	69	93.8	111	151.0	153	208.1
28	38.1	70	95.2	112	152.3	154	209.4
29	39.4	71	96.6	113	153.7	155	210.8
30	40.8	72	97.9	114	155.0	156	212.2
31	42.2	73	99.3	115	156.4	157	213.5
32	43.5	74	100.6	116	157.8	158	214.9
33	44.9	75	102.0	117	159.1	159	216.2

METRIC TO ENGLISH CONVERSION: TORQUE

To convert foot-pounds (ft. lbs.) to Newton-meters (Nm), multiply the number of ft. lbs. by 1.36
To convert Newton-meters (Nm) to foot-pounds (ft. lbs.), multiply the number of Nm by 0.7376

Nm	ft. lbs.	Nm	ft. lbs.	Nm	ft. lbs.	Nm	ft. lbs.	Nm	ft. lbs.
0.1	0.1	34	25.0	76	55.9	118	86.8	160	117.6
0.2	0.1	35	25.7	77	56.6	119	87.5	161	118.4
0.3	0.2	36	26.5	78	57.4	120	88.2	162	119.1
0.4	0.3	37	27.2	79	58.1	121	89.0	163	119.9
0.5	0.4	38	27.9	80	58.8	122	89.7	164	120.6
0.6	0.4	39	28.7	81	59.6	123	90.4	165	121.3
0.7	0.5	40	29.4	82	60.3	124	91.2	166	122.1
0.8	0.6	41	30.1	83	61.0	125	91.9	167	122.8
0.9	0.7	42	30.9	84	61.8	126	92.6	168	123.5
1	0.7	43	31.6	85	62.5	127	93.4	169	124.3
2	1.5	44	32.4	86	63.2	128	94.1	170	125.0
3	2.2	45	33.1	87	64.0	129	94.9	171	125.7
4	2.9	46	33.8	88	64.7	130	95.6	172	126.5
5	3.7	47	34.6	89	65.4	131	96.3	173	127.2
6	4.4	48	35.3	90	66.2	132	97.1	174	127.9
7	5.1	49	36.0	91	66.9	133	97.8	175	128.7
8	5.9	50	36.8	92	67.6	134	98.5	176	129.4
9	6.6	51	37.5	93	68.4	135	99.3	177	130.1
10	7.4	52	38.2	94	69.1	136	100.0	178	130.9
11	8.1	53	39.0	95	69.9	137	100.7	179	131.6
12	8.8	54	39.7	96	70.6	138	101.5	180	132.4
13	9.6	55	40.4	97	71.3	139	102.2	181	133.1
14	10.3	56	41.2	98	72.1	140	102.9	182	133.8
15	11.0	57	41.9	99	72.8	141	103.7	183	134.6
16	11.8	58	42.6	100	73.5	142	104.4	184	135.3
17	12.5	59	43.4	101	74.3	143	105.1	185	136.0
18	13.2	60	44.1	102	75.0	144	105.9	186	136.8
19	14.0	61	44.9	103	75.7	145	106.6	187	137.5
20	14.7	62	45.6	104	76.5	146	107.4	188	138.2
21	15.4	63	46.3	105	77.2	147	108.1	189	139.0
22	16.2	64	47.1	106	77.9	148	108.8	190	139.7
23	16.9	65	47.8	107	78.7	149	109.6	191	140.4
24	17.6	66	48.5	108	79.4	150	110.3	192	141.2
25	18.4	67	49.3	109	80.1	151	111.0	193	141.9
26	19.1	68	50.0	110	80.9	152	111.8	194	142.6
27	19.9	69	50.7	111	81.6	153	112.5	195	143.4
28	20.6	70	51.5	112	82.4	154	113.2	196	144.1
29	21.3	71	52.2	113	83.1	155	114.0	197	144.9
30	22.1	72	52.9	114	83.8	156	114.7	198	145.6
31	22.8	73	53.7	115	84.6	157	115.4	199	146.3
32	23.5	74	54.4	116	85.3	158	116.2	200	147.1
33	24.3	75	55.1	117	86.0	159	116.9	201	147.8

ENGLISH/METRIC CONVERSION: TEMPERATURE

To convert Fahrenheit (F°) to Celsius (C°), take F° temperature and subtract 32, multiply the result by 5 and divide the result by 9

To convert Celsius (C°) to Fahrenheit (F°), take C° temperature and multiply it by 9, divide the result by 5 and add 32

F°	C°	F°	C°	C°	F°	C°	F°
-40	-40.0	150	65.6	-38	-36.4	46	114.8
-35	-37.2	155	68.3	-36	-32.8	48	118.4
-30	-34.4	160	71.1	-34	-29.2	50	122
-25	-31.7	165	73.9	-32	-25.6	52	125.6
-20	-28.9	170	76.7	-30	-22	54	129.2
-15	-26.1	175	79.4	-28	-18.4	56	132.8
-10	-23.3	180	82.2	-26	-14.8	58	136.4
-5	-20.6	185	85.0	-24	-11.2	60	140
0	-17.8	190	87.8	-22	-7.6	62	143.6
1	-17.2	195	90.6	-20	-4	64	147.2
2	-16.7	200	93.3	-18	-0.4	66	150.8
3	-16.1	205	96.1	-16	3.2	68	154.4
4	-15.6	210	98.9	-14	6.8	70	158
5	-15.0	212	100.0	-12	10.4	72	161.6
10	-12.2	215	101.7	-10	14	74	165.2
15	-9.4	220	104.4	-8	17.6	76	168.8
20	-6.7	225	107.2	-6	21.2	78	172.4
25	-3.9	230	110.0	-4	24.8	80	176
30	-1.1	235	112.8	-2	28.4	82	179.6
35	1.7	240	115.6	0	32	84	183.2
40	4.4	245	118.3	2	35.6	86	186.8
45	7.2	250	121.1	4	39.2	88	190.4
50	10.0	255	123.9	6	42.8	90	194
55	12.8	260	126.7	8	46.4	92	197.6
60	15.6	265	129.4	10	50	94	201.2
65	18.3	270	132.2	12	53.6	96	204.8
70	21.1	275	135.0	14	57.2	98	208.4
75	23.9	280	137.8	16	60.8	100	212
80	26.7	285	140.6	18	64.4	102	215.6
85	29.4	290	143.3	20	68	104	219.2
90	32.2	295	146.1	22	71.6	106	222.8
95	35.0	300	148.9	24	75.2	108	226.4
100	37.8	305	151.7	26	78.8	110	230
105	40.6	310	154.4	28	82.4	112	233.6
110	43.3	315	157.2	30	86	114	237.2
115	46.1	320	160.0	32	89.6	116	240.8
120	48.9	325	162.8	34	93.2	118	244.4
125	51.7	330	165.6	36	96.8	120	248
130	54.4	335	168.3	38	100.4	122	251.6
135	57.2	340	171.1	40	104	124	255.2
140	60.0	345	173.9	42	107.6	126	258.8
145	62.8	350	176.7	44	111.2	128	262.4

LENGTH CONVERSION

To convert inches (in.) to millimeters (mm), multiply the number of inches by 25.4
To convert millimeters (mm) to inches (in.), multiply the number of millimeters by 0.04

Inches	Millimeters	Inches	Millimeters	Inches	Millimeters	Inches	Millimeters
0.0001	0.00254	0.005	0.1270	0.09	2.286	4	101.6
0.0002	0.00508	0.006	0.1524	0.1	2.54	5	127.0
0.0003	0.00762	0.007	0.1778	0.2	5.08	6	152.4
0.0004	0.01016	0.008	0.2032	0.3	7.62	7	177.8
0.0005	0.01270	0.009	0.2286	0.4	10.16	8	203.2
0.0006	0.01524	0.01	0.254	0.5	12.70	9	228.6
0.0007	0.01778	0.02	0.508	0.6	15.24	10	254.0
0.0008	0.02032	0.03	0.762	0.7	17.78	11	279.4
0.0009	0.02286	0.04	1.016	0.8	20.32	12	304.8
0.001	0.0254	0.05	1.270	0.9	22.86	13	330.2
0.002	0.0508	0.06	1.524	1	25.4	14	355.6
0.003	0.0762	0.07	1.778	2	50.8	15	381.0
0.004	0.1016	0.08	2.032	3	76.2	16	406.4

ENGLISH/METRIC CONVERSION: LENGTH

To convert inches (in.) to millimeters (mm), multiply the number of inches by 25.4

To convert millimeters (mm) to inches (in.), multiply the number of millimeters by 0.04

Inches		Millimeters	Inches		Millimeters	Inches		Millimeters
Fraction	Decimal	Decimal	Fraction	Decimal	Decimal	Fraction	Decimal	Decimal
1/64	0.016	0.397	11/32	0.344	8.731	11/16	0.688	17.463
1/32	0.031	0.794	23/64	0.359	9.128	45/64	0.703	17.859
3/64	0.047	1.191	3/8	0.375	9.525	23/32	0.719	18.256
1/16	0.063	1.588	25/64	0.391	9.922	47/64	0.734	18.653
5/64	0.078	1.984	13/32	0.406	10.319	3/4	0.750	19.050
3/32	0.094	2.381	27/64	0.422	10.716	49/64	0.766	19.447
7/64	0.109	2.778	7/16	0.438	11.113	25/32	0.781	19.844
1/8	0.125	3.175	29/64	0.453	11.509	51/64	0.797	20.241
9/64	0.141	3.572	15/32	0.469	11.906	13/16	0.813	20.638
5/32	0.156	3.969	31/64	0.484	12.303	53/64	0.828	21.034
11/64	0.172	4.366	1/2	0.500	12.700	27/32	0.844	21.431
3/16	0.188	4.763	33/64	0.516	13.097	55/64	0.859	21.828
13/64	0.203	5.159	17/32	0.531	13.494	7/8	0.875	22.225
7/32	0.219	5.556	35/64	0.547	13.891	57/64	0.891	22.622
15/64	0.234	5.953	9/16	0.563	14.288	29/32	0.906	23.019
1/4	0.250	6.350	37/64	0.578	14.684	59/64	0.922	23.416
17/64	0.266	6.747	19/32	0.594	15.081	15/16	0.938	23.813
9/32	0.281	7.144	39/64	0.609	15.478	61/64	0.953	24.209
19/64	0.297	7.541	5/8	0.625	15.875	31/32	0.969	24.606
5/16	0.313	7.938	41/64	0.641	16.272	63/64	0.984	25.003
21/64	0.328	8.334	21/32	0.656	16.669	1/1	1.000	25.400
			43/64	0.672	17.066			

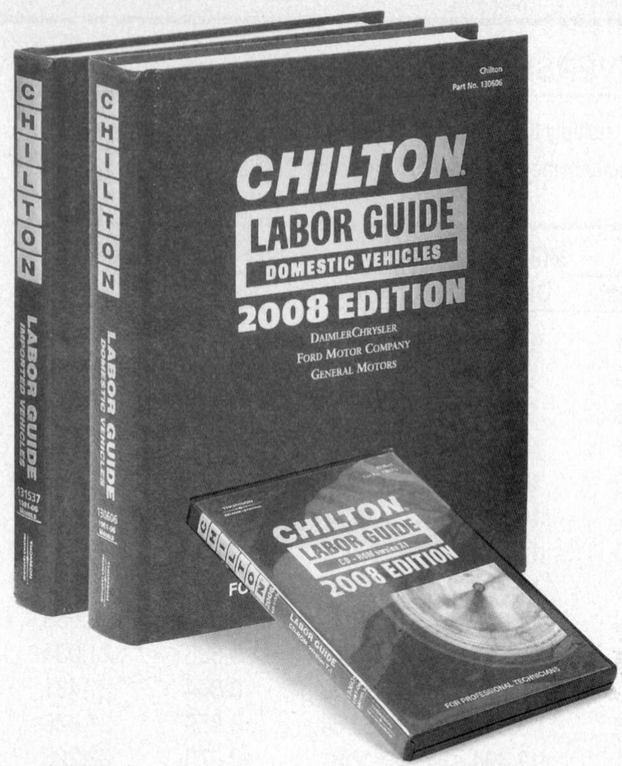

Chilton labor times have been a trusted standard in the industry for over 60 years. That legacy continues in the Chilton 2008 Labor Guide Manuals and CD-ROM. Chilton labor times take into account the real world environment in which technicians work. Three labor times are available for many models: Chilton's standard and severe service times, plus OEM warranty time. Chilton labor times are accepted by most insurance and extended warranty companies. Vehicle makes and models conform to current Automotive Aftermarket Industry Association (AAIA) standards. Get your hands on the newest version of a classic.

Hardcover Manuals are 8 1/2" x 11", ©2007

Chilton 2008 Labor Guide Manuals
978-1-4283-2035-2 (1-4283-2035-0) Part No.142035

Chilton 2008 Labor Guide CD-ROM
978-1-4283-2041-3 (1-4283-2041-5) Part No.142041

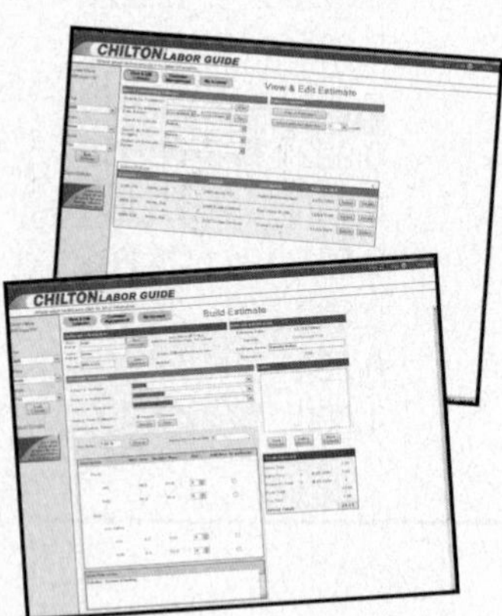

Labor Guide CD-ROM Benefits:

- save time with automatically calculated labor charges, taxes, & parts as total job estimated
- create professional estimates for your customers and worksheets for your technicians, printing them whenever needed
- keep track of customers, prior estimates, and your own parts or package jobs with less paper
- estimate all your vehicles with one tool- labor times from 1981 through current domestic and imported vehicles

Labor Guide CD-ROM Benefits:

- enjoy quicker referencing than ever by using separate, easier-to-handle, domestic and imported vehicle manuals filling more than 2500 pages
- find the times fast by using:
  - tabs that display contents by manufacturer and model
  - two indexes--labor operations and systems--in each model group
  - manufacturers arranged alphabetically, and page numbering that includes manufacturer code so you know where you are in the book
- manufacturers are arranged alphabetically within each volume
- estimate all your vehicles with one tool- labor times from 1981 through current domestic and imported vehicles

CHILTON 2008 SERVICE MANUALS

Chilton 2008® Service Manuals include 11 manuals covering DaimlerChrysler, Ford, General Motors, Asian, and European vehicles. Users will be expertly provided with the most currently available information to assist in daily activities. These new, reliable, and comprehensive manuals provide essential information, allowing users to accurately and efficiently diagnose and repair. Step-by-step procedures and helpful illustrations provide easy references for jobs. These new service manuals cover 2006 and 2007 models, plus any available 2008 models.

Service Manual Benefits:

- multi-volume manual set, organized by vehicle manufacturer, provides more than 2000 pages of expertly written content
- access new year, make, and model information without repeating previous edition's content
- comprehensive, technically detailed content—including exploded view illustrations, diagnostics and specification charts—arranged alphabetically by model group for quick, easy access

Chilton 2008 DaimlerChrysler Service Manuals (2 volume set)—ISBN 978-1-4283-2204-2 (1-4283-2204-3) Part No. 142204
Chilton 2008 Ford Service Manuals (2 volume set)—ISBN 978-1-4283-2208-0 (1-4283-2208-6) Part No. 142208
Chilton 2008 General Motors Service Manuals (2 volume set)—ISBN 978-1-4283-2211-0 (1-4283-2211-6) Part No. 142211
Chilton 2008 Asian Service Manuals (4 volume set)—ISBN 978-1-4283-2214-1 (1-4283-2214-0) Part No. 142214
Chilton 2008 European Service Manual—ISBN 978-1-4283-2220-2 (1-4283-2220-5) Part No. 142220

Manuals are 8 1/2" x 11"

CHILTON SERVICE MANUALS —PERENNIAL EDITIONS

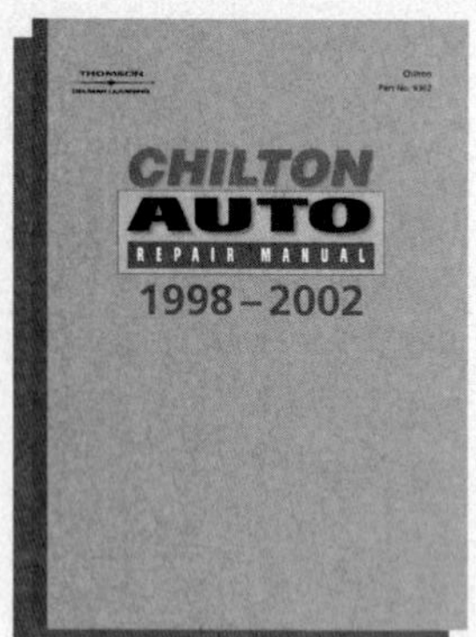

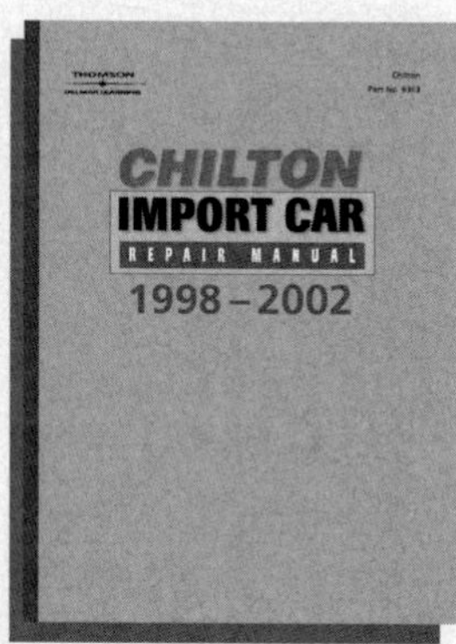

The *Chilton® Perennial Editions* contain repair and maintenance information for popular mechanical systems that may not be available elsewhere. They offer a wide range of repair information on cars, trucks, vans, and SUVs dating back to the early 1960s, and as current as 2002. Information for 1993 and later model years includes scheduled maintenance interval charts.

Benefits:

- covers the most common vehicle models found in the repair aftermarket today
- allows quick understanding of systems using exploded-view illustrations, diagrams, and charts
- simplify tough jobs with easy-to-follow removal and installation instructions for heater core and other components
- gives complete coverage of repair procedures from drive train to chassis and associated components

Auto Repair Manual, 1998-2002, 1,426 pages
ISBN 978-0-8019-9362-6 (0-8019-9362-8) Part No. 9362
Auto Repair Manual, 1993-1997, 2,064 pages
ISBN 978-0-8019-7919-4 (0-8019-7919-6) Part No. 7919
Auto Repair Manual, 1980-1987, 1,344 pages
ISBN 978-0-8019-7670-4 (0-8019-7670-7) Part No. 7670
Import Car Repair Manual, 1998-2002, 1,792 pps
ISBN 978-0-8019-9363-3 ISBN 0-8019-9363-6/Part No. 9363
Import Car Repair Manual, 1993-1997, 2,080 pps
ISBN 978-0-8019-7920-0 (0-8019-7920-X) Part No. 7920
Import Car Repair Manual, 1988-1992, 1,632 pages
ISBN 978-0-8019-7907-1 (0-8019-7907-2) Part No. 7907
Import Car Repair Manual, 1980-1987, 1,488 pages
ISBN 978-0-8019-7672-8 (0-8019-7672-3) Part No. 7672

Truck & Van Repair Manual, 1998-2002, 1,408 pages
ISBN 978-0-8019-9364-0 (0-8019-9364-4) Part No. 9364
Truck & Van Repair Manual, 1993-1997, 2,096 pages
ISBN 978-0-8019-7921-7 (0-8019-7921-8) Part No. 7921
Truck & Van Repair Manual, 1991-1995, 1,664 pages
ISBN 978-0-8019-7911-8 (0-8019-7911-0) Part No. 7911
Truck & Van Repair Manual, 1986-1990, 1,536 pages
ISBN 978-0-8019-7902-6 (0-8019-7902-1) Part No. 7902
Truck & Van Repair Manual, 1979-1986, 1,440 pages
ISBN 978-0-8019-7655-1 (0-8019-7655-3) Part No. 7655

SUV Repair Manual, 1998-2002, 1,292 pages
ISBN 978-0-8019-9365-7 (0-8019-9365-2)Part No. 9365

Hardcover manuals are 8 1/2" x 11".

Chilton Collector's Editions—*Reference Manuals for Vintage Vehicles*
Auto Repair Manual, 1964-1971, ISBN 978-0-8019-5974-5 (0-8019-5974-8) Part No. 5974,
Truck & Van Repair Manual, 1971-1978, ISBN 978-0-8019-7012-2 (0-8019-7012-1) Part No. 7012

Chilton Timing Belts, 1985-2005

Chilton
ISBN 978-1-4018-9880-9 (1-4018-9880-7) Part No. 129880

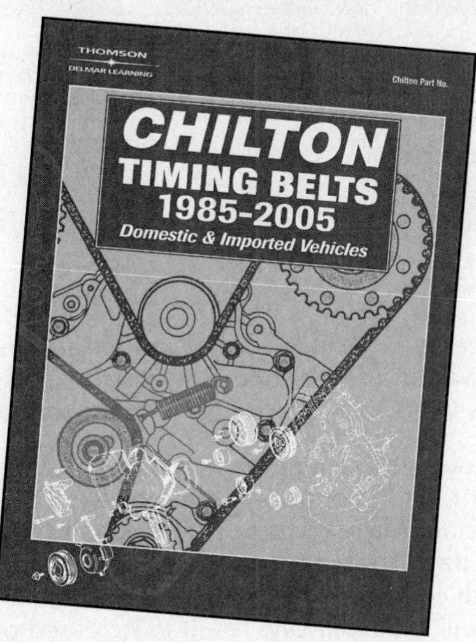

Timing belt procedures can represent increased profits for automotive repair shops and service stations, and this manual contains all the information automotive technicians need to properly service timing belts on domestic and imported cars, vans, and light trucks through 2005 models. Clear, straightforward procedures, illustrations, and specifications help to communicate 20 years of vehicle applications for fast, accurate inspection, replacement, and tensioning of timing belts. Readers will learn step-by-step how to perform key procedures both quickly and safely, while learning the correct labor time to charge for the service. OEM-recommended replacement intervals for proper maintenance of customer's vehicles are also featured.

Benefits:
- detailed illustrations clearly demonstrate important concepts, such as how to correctly align camshaft and crankshaft timing marks, and how to simplify serpentine belt installation
- readers are made aware of potential hazards and time-wasting practices that can impede safe and profitable service procedures
- special tools are identified so that completing the service is as easy and quick as possible

544 pp, 8 1/2" x 11", softcover, ©2006

ALSO AVAILABLE:

Quick-Reference Manuals

Chilton

The Chilton Professional Series offers Quick-Reference Manuals for the automotive professional, providing complete coverage on repair and maintenance, adjustments, and diagnostic procedures for specific systems and components.

Benefits:
- step-by-step procedures
- easy-to-use manufacturer and model indexing
- detailed illustrations and exploded views
- handy specifications or data charts

Heater Core Service 1990-2000,
13-Digit ISBN 978-0-8019-9311-4 Part No. 9311
(10-Digit ISBN 0-8019-9311-3) 560 pp

Brake Specifications and Service 1990-2000
13-Digit ISBN 978-0-8019-9312-1 Part No. 9312
(10-Digit ISBN 0-8019-9312-1) 520 pp

Electric Cooling Fans, Accessory Drive Belts & Water Pumps, 1995-1999,
13-Digit ISBN 978-0-8019-9126-4/Part No. 9126
(10-Digit ISBN 0-8019-9126-9) 312 pp

Powertrain Codes & Oxygen Sensors, 1990-1999,
13-Digit ISBN 978-0-8019-9127-1/Part No. 9127
(10-Digit ISBN 0-8019-9127-7) 400 pp

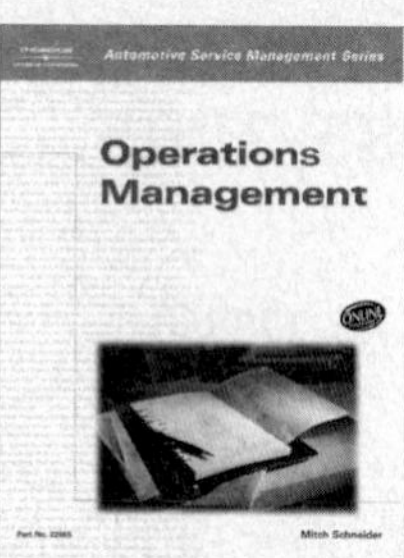

This pioneering eight-book series offers automotive service shop owners and those wanting to be shop owners the necessary business and customer service skills to run a successful automotive service facility.

The series covers three main topical areas: personnel management, business management, and sales and marketing. Each book provides a framework to help technicians make consistent, high-quality, and productive service a part of every day shop operations. According to the author, "Great performance coupled with increased customer loyalty, trust, and operational excellence will almost always result in increased profits."

Automotive Service Management Series Benefits:

- real-world approach reflects author's experience as a fourth generation technician, a repair & service company owner, and an automotive industry trainer
- all-inclusive coverage spans from designing an automotive repair facility floor plan through financial management techniques, customer/staff relations, and more
- length of each book makes it easy to incorporate this series into workshops, seminars, and training/education courses
- information is available "as is" or for customization

ABOUT THE AUTHOR

Mitch Schneider is a fourth generation mechanic/technician and is a frequent speaker at major conventions and meetings of automotive industry trade organizations. Schneider is also an award-winning journalist and is a regular contributor and senior contributing editor for *Motor Age* magazine. He provides commentary on the evolving relationship between service dealers, jobbers, warehouse directors and manufacturers.

Schneider has also appeared on the TNN cable show "Truckin' USA" where he hosted the "Tech Tips" segment. In addition to operating the award-winning Schneider's Automotive for 22 years in Simi Valley, CA, he is also the president and founder of Schneider's Future-Tech, a service company specializing in conducting management seminars for automotive service dealers, jobbers, warehouse distribution companies, and manufacturers.

Total Customer Relationship Management
ISBN 978-1-4018-2657-4 (1-4018-2657-1) Part No. 22657

From Intent to Implementation
ISBN 978-1-4018-2658-1 (1-4018-2658-X) Part No. 22658

Operational Excellence
ISBN 978-1-4018-2659-8 (1-4018-2659-8) Part No. 22659

Building a Team
ISBN 978-4018-2660-4 (1-4018-2660-1) Part No. 22660

The High Performance Shop
ISBN 978-1-4018-2661-1 (1-4018-2661-X) Part No. 22661

Safety Communications
ISBN 978-1-4018-2662-8 (1-4018-2662-8) Part No. 22662

Managing Dollars with Sense
ISBN 978-1-4018-2663-5 (1-4018-2663-6) Part No. 22663

Operations Management
ISBN 978-1-4018-2665-9 (1-4018-2665-2) Part No. 22665

Entire Set of 8 Books
ISBN 978-1-4018-2499-0 (1-4018-2499-4) Part No. 2499

Softcover manuals are 8 1/2" x 11", ©2003

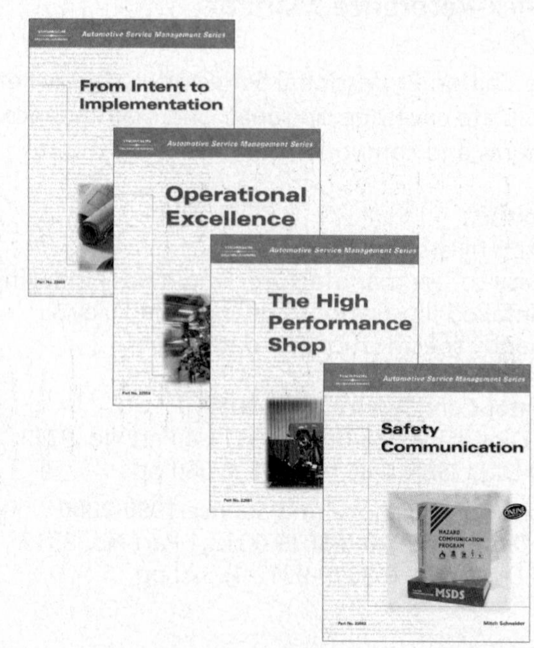

ASE CERTIFICATION TEST PREPARATION

You Deserve The Best When You Are Putting Your Skills To The Test!

ASE Test Preparation

Title	ISBN
(A1) Engine Repair, 4E	1-4180-3878-4
(A2) Transmissions and Transaxles, 4E	1-4180-3879-2
(A3) Manual Drive Train and Axles, 4E	1-4180-3880-6
(A4) Suspension and Steering, 4E	1-4180-3881-4
(A5) Brakes, 4E	1-4180-3882-2
(A6) Electrical-Electronic Systems, 4E	1-4180-3883-0
(A7) Heating and Air Conditioning, 4E	1-4180-3884-9
(A8) Engine Performance, 4E	1-4180-3885-7
(L1) Advanced Engine Performance, 4E	1-4180-3888-1
(X1) Exhaust Systems, 4E	1-4180-3886-5
(P2) Parts Specialist, 4E	1-4180-3887-3
(C1) Service Consultant, 2E	1-4180-3889-X
(B2) Painting and Refinishing, 3E	1-4018-3664-X
(B3) Non Structural Analysis and Damage Repair, 3E	1-4018-3665-8
(B4) Structural Analysis and Damage Repair, 3E	1-4018-3666-6
(B5) Mechanical and Electrical Components, 3E	1-4018-3667-4
(B6) Damage Analysis and Estimation, 3E	1-4018-3668-2
(M1) Cylinder Head Specialist	0-7668-6280-1
(M2) Cylinder Block Specialist	0-7668-6281-X
(M3) Assembly Specialist	0-7668-6282-8
(T1) Gasoline Engines, 4E	1-4180-4828-3
(T2) Diesel Engines, 4E	1-4180-4829-1
(T3) Drive Train, 4E	1-4180-4830-5
(T4) Brakes, 4E	1-4180-4831-3
(T5) Suspension and Steering, 4E	1-4180-4832-1
(T6) Electrical and Electronic Systems, 4E	1-4180-4834-8
(T7) Heating, Ventilation, and Air Conditioning, 4E	1-4180-4835-6
(T8) Preventative Maintenance, 4E	1-4180-4836-4
(S2) Diesel Engines	1-4018-1822-6
(S4) Brakes	1-4018-1824-2
(S5) Suspension and Steering	1-4018-1825-0
(H2) Diesel Engines	1-4180-4998-0
(H3) Drive Train	1-4354-5376-X
(H4) Brakes	1-4180-4998-0
(H5) Suspension & Steering	1-4283-4011-4
(H6) Electrical-Electronic Systems	1-4180-4999-9
(H7) Electrical/Electronic	1-4180-4999-7

ASE Test Preparation in Spanish

Title	ISBN
Spanish (A1) Engine Repair	1-4018-1014-4
Spanish (A2) Transmissions and Transaxle	1-4018-1015-2
Spanish (A3) Manual Drive Train and Axles	1-4018-1016-0
Spanish (A4) Suspension and Steering	1-4018-1017-9
Spanish (A5) Brakes	1-4018-1018-7
Spanish (A6) Electrical-Electronic Systems	1-4018-1019-5
Spanish (A7) Heating and Air Conditioning	1-4018-1020-9
Spanish (A8) Engine Performance	1-4018-1021-7
Spanish (L1) Advanced Engine Performance	1-4018-1022-5
Spanish (X1) Exhaust Systems	1-4018-1024-1
Spanish (P2) Parts Specialist	1-4018-1023-3
Spanish (B2) Painting and Refinishing	1-4018-9255-8
Spanish (B3) Non-Structural Analysis and Damage Repair	1-4018-2544-3
Spanish (B4) Structural Analysis and Damage Repair	1-4018-9131-4
Spanish (B5) Mechanical and Electrical Components	1-4018-7759-1
Spanish (B6) Damage Analysis and Estimation	1-4018-6573-9

*Switch between English & Spanish at the click of a button!

Online ASE Test Preparation

Place your order online at www.techniciantestprep.com

Title	ISBN
*Online (A1) Automotive Engine Repair	1-4180-1305-6
*Online (A2) Automatic Transmissions & Transaxles	1-4180-1306-4
*Online (A3) Automotive Manual Drive Trains & Axles	1-4180-1307-2
*Online (A4) Automotive Suspension & Steering	1-4180-1308-0
*Online (A5) Automotive Brakes	1-4180-1309-9
*Online (A6) Automotive Electrical/Electronic Systems	1-4180-1310-2
*Online (A7) Automotive Heating & Air Conditioning	1-4180-1311-0
*Online (A8) Automotive Engine Performance	1-4180-1312-9
*Online (X1) Exhaust Systems	1-4180-1313-7
*Online (P2) Automobile Parts Specialist	1-4180-1314-5
*Online (L1) Automotive Advance Engine Performance	1-4180-1315-3
*Online (C1) Service Consultant	1-4180-1316-1
Online (T1) Gasoline Engines	1-4018-7897-0
Online (T2) Diesel Engines	1-4018-7898-9
Online (T3) Drive Train	1-4018-7900-4
Online (T4) Brakes	1-4018-7901-2
Online (T5) Suspension & Steering	1-4018-7903-9
Online (T6) Electrical & Electronics	1-4180-1879-1
Online (T7) Heating, Ventilation, & Air Conditioning	1-4180-1880-5
Online (T8) Preventative Maintenance	1-4018-7906-3

Complete Series

	IBSN
ASE Test Preparation Manuals for Automotive (A1-A8, X1, P2, L1, C1)	1-4180-3954-3
ASE Test Preparation Manuals for Automotive (A1-A8 & L1)	1-4180-6139-5
ASE Test Preparation Manuals for Automotive (A1-A8, L1, & P2)	1-4180-4197-1
ASE Test Preparation Manuals for Automotive (A1-A8)	1-4180-6237-5
ASE Test Preparation Manuals for Automotive (A1-A8, L1, P2 & X1)	1-4180-6335-5
Online ASE Test Preparation for Automotive (A1-A8, X1, P2, L1, C1)	1-4180-1344-7
Place your order online at www.techniciantestprep.com	
ASE Test Preparation Manuals for Medium/Heavy Duty Truck (T1-T8)	1-4180-4934-4
Online ASE Test Preparation for Medium/Heavy Duty Truck Package (T1-T8)	1-4180-0611-4
Place your order online at www.techniciantestprep.com	
ASE Test Preparation Manuals for Collision (B2-B6)	1-4018-5120-7
ASE Test Preparation Manuals for Collision in Spanish (B2-B6))	1-4018-4155-4
ASE Test Preparation Manuals for Engine Machinist (M1-M3)	0-7668-6283-6

TAKE YOUR TECHNICIAN TRAINING TO THE NEXT LEVEL

Comprehensive Skill Assessment Tool (CSAT) — Automotive
Delmar

The online *Comprehensive Skill Assessment Tool-Automotive Series* helps instructors and trainers implement the necessary training programs for individual areas needing improvement over various key automotive topics. As a true skill gap analysis tool, within each key topic, strategic learning areas are measured for knowledge of theory, hands-on application, and diagnostic skill. Areas of strength and areas needing improvement are identified. The combined phases of education and training, and post-assessment allow instructors to track skill level growth and target specific areas needing development.

Benefits

- available tests include Engine Repair, Transmissions and Transaxles, Manual Drive Train and Axles, Suspension and Steering, Brakes, Electrical/Electronic Systems, Heating and Air Conditioning, Engine Performance, Advanced Engine Performance, and Exhaust Systems
- can be utilized by companies to measure the technical skill level of individuals against an "ideal" to identify areas of strength and creates a skill gap analysis to help users address areas needing improvement
- questions are written and reviewed by experts in the industry and offer users the opportunity to receive instant feedback
- account set-up that enables instructors and trainers to assess and track the results of individual students
- acts as a true return on investment (ROI) tool for companies to ensure they invest their training dollars in the most appropriate areas

Call Your Delmar Sales Rep for Part Numbers & Pricing

Visit **www.skillanalysis.com** for a free demo!

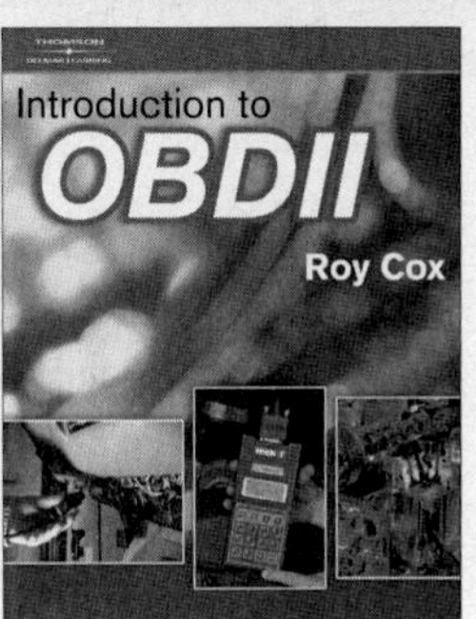

Introduction to OBDII
Roy Cox
ISBN 978-1-4180-1220-5 (1-4180-1220-3) Part No. 131220

Here's an easy-to-understand, logical guide to the diagnosis and repair of today's complex and sophisticated automotive control systems! *Introduction to On-Board Diagnostics (OBD II)* readers will learn the fundamentals of how to perform diagnostic procedures, and be provided with valuable reference material for diagnosing and troubleshooting components and circuits. This book provides a simple, logical approach to explain the operation of the OBD II process and will teach the reader how to quickly spot problems and identify components that are not functioning correctly. In addition, the interrelationships between the fuel delivery, emission control, ignition, and accessory systems are clearly addressed and explained. This truly unique introduction to OBDII systems, and the troubleshooting of components and circuits, features an accompanying interactive diagnostic CD-ROM that leads readers through realistic trouble-code scenarios to reinforce its diagnostic and repair.

Benefits

- "quick hit" troubleshooting tricks teach readers how to diagnose problems when there is no stored OBDII trouble code, as well as how to handle situations where the trouble code is actually set by a basic mechanical problem rather than a failure of the indicated component
- information is useful for those who wish to expand their capabilities from more basic, mechanical repairs to complex electronics and drivability diagnosis and repair
- a substantial portion of the content focuses on logical troubleshooting that can be done without expensive, complicated test equipment and special tools
- also useful as a preparation manual for the ASE Certification exams

CONTENTS
Chapter 1- Introduction, Chapter 2- Evolution of OBD, Chapter 3- OBDII Terminology, Chapter 4- System Operating Protocols, Chapter 5- System Monitors, Chapter 6- Drive Cycles, Chapter 7- Diagnostic Trouble Codes (DTC's), Chapter 8- Diagnostic Routines

256 pp, 8 1/2" x 11", softcover, ©2006

SUPPLEMENTS
Diagnostic Tool CD-ROM 978-1-4180-1221-2 (1-4180-1221-1) Part No. 131221
Instructor's Guide 978-1-4180-1222-9 (1-4180-1222-X)

▶ PROFESSIONAL AUTOMOTIVE TECHNICIAN TRAINING SERIES

Delmar Learning, the leader in providing first-rate educational materials for automotive technicians, now offers this exciting self-paced learning series. Choose the delivery method that best suits your needs– CD-ROM or Web-based product – and receive more than 8.5 hours worth of quality instruction. Combining theory, diagnosis, and repair information into one easy-to-use training tool, this highly interactive product helps technicians receive the most applicable delivery method for their needs, regardless of technical infrastructure.

KEY FEATURES

- attention-grabbing animations and learner interactions keep users interested and engaged throughout the course of the program
- bookmarking technology enables users to track their progress from beginning to end
- periodic progress checks and end-of-section reviews are integrated throughout to ensure the highest level of retention
- a certificate of completion can be printed by users achieving a score of 80% or higher on the final review of the course
- all material is completely AICC and SCORM compliant
- all material follows the latest ASE and NATEF standards

Basic Automotive Service and Maintenance Web Based Training
13-Digit ISBN 978-1-4180-4101-4
(10 Digit ISBN 1-4180-4101-7

Basic Automotive Service and Maintenance Computer Based Training
13-Digit ISBN 978-1-4180-4100-7
(10 Digit ISBN 1-4180-4100-9)

Electricity and Electronics Web Based Training
13-Digit ISBN 978-1-4180-4242-4
(10 Digit ISBN 1-4180-4242-0)

Electricity and Electronics Computer Based Training
13-Digit ISBN 978-1-4180-4241-7
(10 Digit ISBN 1-4180-4241-2)

Brakes Web Based Training
13-Digit ISBN 978-1-4180-4236-3
(10 Digit ISBN 1-4180-4236-6)

Brakes Computer Based Training
13-Digit ISBN 978-1-4180-4235-6
(10 Digit ISBN 1-4180-4235-8)

Engine Performance Web Based Training
13-Digit ISBN 978-1-4180-4240-0
(10 Digit ISBN 1-4180-4240-4)

Engine Performance Computer Based Training
(10 Digit ISBN 1-4180-4239-0)
13-Digit ISBN 978-1-4180-4239-4

Suspension and Steering Web Based Training
13-Digit ISBN 978-1-4180-4238-7
(10 Digit ISBN 1-4180-4238-2)

Suspension and Steering Computer Based Training
13-Digit ISBN 978-1-4180-4237-0
(10 Digit ISBN 1-4180-4237-4)

Automatic Transmissions Web Based Training
13-Digit ISBN 978-1-4180-4244-8
(10 Digit ISBN 1-4180-4244-7)

Automatic Transmissions Computer Based Training
13-Digit ISBN 978-1-4180-4243-1
(10 Digit ISBN 1-4180-4243-9)

Service Consultant Web Based Training
13-Digit ISBN 978-1-4180-4249-3
(10 Digit ISBN 1-4180-4249-8)

Service Consultant Computer Based Training
13-Digit ISBN 978-1-4180-4247-9
(10 Digit ISBN 1-4180-4247-1)

Engine Repair Web Based Training
13-Digit ISBN 978-1-4180-4254-7
(10 Digit ISBN 1-4180-4254-4)

Engine Repair Computer Based Training
13-Digit ISBN 978-1-4180-4253-0
(10 Digit ISBN 1-4180-4253-6)

Parts Specialist Web Based Training
13-Digit ISBN 978-1-4180-4252-3
(10 Digit ISBN 1-4180-4252-8)

Parts Specialist Computer Based Training
(10 Digit ISBN 1-4180-4250-1)
13-Digit ISBN 978-1-4180-4250-9

Heating and Air Conditioning Web Based Training
13-Digit ISBN 978-1-4180-4246-2
(10 Digit ISBN 1-4180-4246-3)

Heating and Air Conditioning Computer Based Training
13-Digit ISBN 978-1-4180-4245-5
(10 Digit ISBN 1-4180-4245-5)

Manual Transmissions Web Based Training
13-Digit ISBN 978-1-4180-4256-1
(10 Digit ISBN 1-4180-4246-3)

Manual Transmissions Computer Based Training
13-Digit ISBN 978-1-4180-4255-4
(10 Digit ISBN 1-4180-4255-2)

Advanced Engine Performance Web Based Training
13-Digit ISBN 978-1-4283-2098-7
(10 Digit ISBN 1-4283-2098-9)

Advanced Engine Performance Computer Based Traini
13-Digit ISBN 978-1-4283-2097-0
(10 Digit ISBN 1-4283-2097-0)